DATE DUE

DEMCO 38-296

M·A·N·U·F·A·C·T·U·R·I·N·G
USA

Industry Analyses,
Statistics, and Leading Companies

ISSN 1044-7024

A *W*ard's Business Directory™

M·A·N·U·F·A·C·T·U·R·I·N·G
USA

Industry Analyses,
Statistics, and Leading Companies

Fifth Edition

- A comprehensive guide to economic activity in 458 manufacturing industries

- Provides unique analysis and synthesis of federal statistics

- Includes more than 21,000 top manufacturing corporations taken from *Ward's Business Directory of U.S. Private and Public Companies*

Volume 2
SIC 3312 - 3999

Arsen J. Darnay, Editor

GALE

DETROIT · NEW YORK · TORONTO · LONDON

Arsen J. Darnay, *Editor*

Editorial Code and Data, Inc. Staff

Kenneth Muth, *Data Processing*
Nancy Ratliff, *Data Entry*
Marlita A. Reddy, *Research*

Gale Research Inc. Staff

Kristin Hart, *Editorial Coordinator*
Lynn Osborn, *Assistant Editorial Coordinator*

Mary Beth Trimper, *Production Director*
Shanna Heilveil, *Production Assistant*
Tracey Rowens, *Art Director*
Sherrell Hobbs, *Desktop Publisher*

Manufacturing USA: Industry Analyses, Statistics, and Leading Companies is published by Gale Research Inc. under license from Information Access Company. *Ward's Business Directory* is a trademark of Information Access Company.

Ward's Business Directory™ utilizes an intensive research approach. Information on companies listed in the directory was gathered from annual reports, questionnaires, banks, trade commissions, newsletters, government documents, and telephone interviews. When sales data are unavailable from private companies, *Ward's* offers an estimate based on several considerations. Estimates are so noted with an asterisk (*).

While an extensive verification and proofing process preceded the printing of this directory, Information Access Company makes no warranties or representation regarding its accuracy or completeness, and each subscriber or user of the directory understands that Information Access Company disclaims any liability for any damages (even if Information Access Company has been advised of such damages) in connection with its use.

ISBN 0-8103-6453-0 (Vol. 1)
ISBN 0-8103-6454-9 (Vol. 2)
ISSN 1044-7024

Printed in the United States of America
Published in the United States by Gale Research

Contents

Introduction

Manufacturing USA: Industry Analyses, Statistics, and Leading Companies (MUSA) presents statistics on 458 U.S. manufacturing industries drawn from a variety of federal sources and combined with information on leading public and private corporations obtained from *Ward's Business Directory of U.S. Private and Public Companies*.

MUSA represents a unique synthesis of relevant data from the *Census of Manufactures*, the *Annual Survey of Manufactures*, the *County Business Patterns* data series, the *U.S. Industrial Outlook*, the *Benchmark Input-Output Accounts for the U.S. Economy, 1982*, and the *Industry-Occupation Matrix* produced by the U.S. Department of Labor. Data on leading private and public corporations are drawn from *Ward's*, as mentioned above. Together, these materials, in preanalyzed presentation, provide a one-stop and well-indexed access to the most recently available data on manufacturing in the United States.

Features

In its first edition, *MUSA* was cited as an *Outstanding Reference Source* by the American Library Association's Reference and Adult Services Division (RASD). The fifth edition of *MUSA* retains the award-winning features of the first edition—detailed industry profiles, illustrated with graphics and maps, combining the most recent governmental data with information on corporations, and arranged so that significance can be instantly recognized. Projections are now also provided. New features include—

- Updated Census data from 1982 through 1994, including all new detail data from the 1992 Economic Census.

- Projections for the 1995-1998 period. Data for these years are projected by the editors for most of the industry series.

- 17-year continuous series. With updates and projections, *MUSA* now presents a continuous 17-year series from 1982 through 1998.

- Updated industry ratios. Industry ratios are now based on 1994 (the last Census survey data available).

- Updated occupational data for 1994 with projections to the year 2005. In the last edition, data for 1992 were shown projected to 2005.

'The Most Current Data Available'

MUSA reports the most current data available at the time of the book's preparation. The objective is to present *hard* information—data based on actual survey by authoritative bodies—for all manufacturing industries on a *comparable* basis. A few industries may collect more recent information through their industry associations or other bodies. Similarly, estimates are published on this or that industry based on the analyses and guesses of knowledgeable individuals. These data are rarely in the same format as the Federal data and are not available for a large cross section of industry. Therefore, the data in *MUSA* are, indeed, the most current at this level of detail and spanning the entirety of manufacturing activity. It is meant to serve as the foundation on which others can base their own projections.

In addition to presenting current survey data, the editors also provide projected data for most categories from 1995 through 1998—continuing the popular feature introduced in the 4th edition. The methodology used to make these projections is discussed in the next chapter.

Scope and Coverage

MUSA presents statistical data on 458 manufacturing industries nationally and in all 50 states (when the industry is present). Data are presented, in addition, on more than 2,000 products/materials, approximately 400 occupational groupings employed in manufacturing, and more than 21,000 public and private companies. The industry count is down by one in this edition because data for *SIC 2067 - Chewing gum* is no longer being collected by the Bureau of the Census.

Data are shown for the years 1982 to 1994 from the Economic Census surveys (1982, 1987, 1992) and from the *Annual Survey of Manufacturing* (*ASM*) for other years. Occupational data are presented for 1994 (updated from 1992) and projected (by the Department of Labor) to the year 2005 (the same year as in the last edition). Input-Output data for 1982 are taken from the latest I-O study (released in 1991). Corporate data are drawn from the 1996 edition of the *Ward's Business Directory*.

Where possible, data are projected through 1998. A special discussion of the methods used to obtain projections is included in the next chapter.

MUSA follows the 1987 classification conventions published by the Office of Management and Budget (*Standard Industrial Classification Manual: 1987*).

The SIC convention divides economic activity hierarchically into major industry groups (2-digit code), industry groups (3-digit), and industries (4-digit). Most data presented in *MUSA* are shown at the 4-digit industry level. Exceptions are occupational data (presented in 3-digit aggregation or in groups of 3-digit industries) and more detailed product information drawn from 5-digit product class codes and 7-digit product codes.

Organization and Content

MUSA is now divided into two volumes, as follows:

- **Volume I** — covers *SICs 2011 - Meatpacking plants* through *SIC 3299 - Nonmetallic mineral products*. The *User's Guide* is placed in Volume I.

- **Volume II** — covers *SICs 3312 - Blast furnaces and steel mills* through *SIC 3999 - Miscellaneous manufacturing industries, nec*. Indexes are placed in Volume II.

Within each volume, *MUSA* is organized by industry. Within each industry, data are presented in nine tables and two graphics as follows:

1	Trends Graphic	Graphs showing shipments and employment.
2	General Statistics	National statistics.
3	Indices of Change	National data in index format.
4	Selected Ratios	Twenty ratios for the industry.
5	Leading Companies	Up to 75 companies in this industry.
6	Materials Consumed	Purchases of materials and products by quantity and cost.
7	Product Share Details	Product categories within industry in percent of total.
8	Inputs and Outputs	Economic sectors that sell to and buy from industry.
9	Occupations Employed	Occupations employed by 3-digit industry group.
10	Maps	States and regions where industry is active.
11	Industry Data by State	State level statistics.

Each industry begins on a new page. The order of graphics and tables is invariable. In a few instances, tables are split between pages.

The four indexes (found in Volume II) are:

- Standard Industrial Classification (SIC) Code Index
- Product Index
- Company Index
- Occupation Index

The SIC Index is in two parts. The first part is arranged in SIC code sequence followed by the name of the industry and the page number on which it begins; the second part is arranged alphabetically by industry name.

For detailed information on *MUSA*'s industry profiles and indexes, please consult the *Overview of Content and Sources*—found in both volumes.

Comments and Suggestions Are Welcome

Comments on or suggestions for improvement of the usefulness, format, and coverage of *MUSA* are always welcome. Although every effort is made to maintain accuracy, errors may occasionally occur; the editor will be grateful if these are called to his attention. Please contact the editor below with comments and suggestions or, to have technical questions answered, call the editor directly at ECDI at (313) 961-2926.

Editor
Manufacturing USA: Industry Analyses, Statistics, and Leading Companies
Gale Research
835 Penobscot Building
Detroit, MI 48226-4094
Phone:(313) 961-2242
 (800) 347-GALE
Fax: (313) 961-6815

Overview of Content and Sources

The 1987 SIC Structure

Data in *MUSA* are ordered in conformity with the 1987 Standard Industrial Classification (SIC) system. This version of the SIC structure is now well entrenched. In relation to the older 1982 SIC system, the user should note five cases:

Case 1. The industry has not changed in any way in the move from the 1982 to the 1987 system; neither SIC number nor the components have changed. In these cases, *MUSA* presents data from 1982 forward to the most currently available year.

Case 2. The industry components have not changed but the SIC number has. Treatment of the industry is the same as in Case 1, but the industry designation is the 1987 SIC.

Case 3. The industry components have changed. However, the Department of Commerce (DOC) has restated the data for earlier years so that a complete series is available. *MUSA* presents these data as reported.

Case 4. The industry SIC *number* has not changed, the industry components *have* changed, and DOC has *not* restated the numbers for earlier years. In these cases, *MUSA* presents historical data, from 1982 to 1986, based on the 1977 components of the SIC. From 1987 forward, the new components are assumed. These industries show an asterisk (*) next to the year 1987, and the source note explains the situation. In these cases, also, the graphic illustrating shipments and employment will have a broken line to show discontinuity of data. This treatment was adopted to show as much historical data as possible, indicative of general trends, even if detailed comparison between early and later years is ill advised.

Case 5. The industry was new at the time of the 1987 SIC structure's unveiling; DOC does not report data earlier than 1987. In this case, the time series begins in 1987.

Recent Change - SIC 2067. A recent change, reflected for the first time in this, the 5th edition, is the elimination of *SIC 2067 - Chewing gum*. In the 1992 Economic Census, the Bureau of the Census did not report data for this industry. Product-level data were collected but were reported as part of *SIC 2064 - Candy & other confectionery products*. This change will undoubtedly be formalized in the next edition of the SIC Manual.

As a consequence of this change, *SIC 2067* has been removed, and *MUSA* now features 458 industries.

Industry Profiles

Each industry profile contains the tables and graphics listed in the *Introduction*. A detailed discussion of each graphic display and table follows; the meaning of each data element is explained, and the sources from which the data were obtained are cited.

1. Trends Graphics: Shipments and Employment

At the beginning of each industry profile, two graphs are presented showing industry shipments and employment plotted for the years 1982 to 1998 (or an earlier date) on logarithmic scale. The curves are provided primarily to give the user an at-a-glance assessment of important trends in the industry. The logarithmic scale ensures that the shipment trends and employment trends can be compared visually despite different magnitudes and denominations of the data (millions of dollars for shipments and thousands of employees for employment); in this mode of presentation, if two curves have the same slope, the values are growing or declining at the same rate. If the values fall within a single cycle (1 to 9, 10 to 90, etc.), a single cycle is shown; if the values bridge two cycles, both are shown.

The data graphed are derived from the first table, *General Statistics*. As many years are plotted as are available. If data gaps appear in the series, missing points are calculated using a least-squares curve fitting algorithm.

In the case of a few industries, data discontinuities are present in the general statistics; data for the 1982-1986 period are not strictly comparable to the data for 1987 and later. In such cases, the line of the graph is interrupted between 1986 and 1987 to show this discontinuity.

Those portions of curves based on projections by the editors are shown in a dotted-line format.

2. General Statistics

This table shows national statistics for the industry for the years 1982-1998 under five groupings: Companies, Establishments, Employment, Compensation, and Production. The last four groupings are further subdivided, as described below.

Data for 1982, 1987, and 1992 are from Economic Census held in each of those years. Data for other years, through and including 1994, are from the *Annual Survey of*

Manufactures (*ASM*). Establishment counts in the *ASM* years are from the *County Business Patterns* for those years; exceptions are the years 1983 and 1984; establishment data for these years are extrapolations of data from the 1982 and 1985 values. New industries created in the 1987 SIC reclassification will not show data earlier than 1987. In the case of some industries, an asterisk appears next to the year 1987. This indicates that data for earlier years are not directly comparable with data for 1987 and later years due to the reclassification of the subcomponents of the industry.

Please note that data from the *U.S. Industrial Outlook* are no longer part of *MUSA*. In its ill-advised downsizing, the Federal government has seen fit to replace this useful publication with another that lacks the necessary numbers in the necessary detail.

Data for the period 1995-1998 are projected by the editors. A discussion of the methods of projection is presented below. Projected data are followed by the letter P.

Company counts are available only from the full *Census of Manufactures* conducted every five years. Data for the 1992 Economic Census add the third data point on company counts for the first time in this edition.

Establishment data are provided for 1982 through 1993; projections are shown thereafter. Establishments counts in the Census years (1982, 1987, 1992) are from the Economic Census. In other years, values are from the *County Business Patterns*. Establishment counts are typically higher than company counts because many companies operate from more than one facility. Total establishments are shown together with establishments that employ 20 or more people. Comparing the number of large establishments with total establishments will tell the user whether the industry is populated by relatively small operations or is dominated by large facilities. Values shown are absolute numbers of establishments.

The **Employment** grouping is subdivided into total employment, shown in 1,000 employees (thus a value of 134.9 means that the industry employs 134,900 people), production workers (in thousands), and production hours worked (in millions of hours). Dividing hours worked by production workers produces hours worked by a production worker in the year. This value is precalculated for the user in the table of Selected Ratios. A value of around 1,940 hours indicates full-time employment—on average; obviously such aggregate data hide the finer details of day-to-day industrial operation: the presence of part-time workers, overtime clocked, etc.

The **Compensation** grouping shows the industry's total payroll (in millions of dollars) and wages (in dollars per hour). The payroll value includes all forms of compensation subject to federal taxes, including wages, salaries, commissions, bonuses, etc. The *Census of Manufactures* provides payroll and wage data as aggregates. The wages per hour were calculated by dividing the Census wage aggregate by the total hours worked in production.

The interested user can reverse this calculation ($/hour times hours will produce wages-in-the-aggregate). Additional calculations can be used to determine the salaries of those employees who are not production workers. The procedure is to calculate aggregate wages and to deduct the result from payroll to obtain salaries paid; next, salaried employees can be calculated by deducting production workers from total employment; finally, salaries paid divided by salaried employment will produce the average annual salary of the administrative/technical work force in the industry.

The **Production** grouping shows cost of materials, value added in manufacturing, value of shipments, and capital investments, all in millions of dollars; thus a value of 0.9 means that the actual value is $900,000.

Cost of materials includes cost of raw materials, fuels, freight, and contract work associated with production and excludes costs of services (e.g., advertising, insurance), overhead, depreciation, rents, royalties, and capital expenditures.

Value Added by Manufacture represents Value of Shipments less cost of materials, supplies, containers, fuel, purchased electrical power, and contract work plus income for services rendered. The result is adjusted by adding the difference between the cost and sales price of the merchandise by merchandising operations plus net change in finished goods and work-in-process inventories between the beginning and the end of the year. Value Added is a good measure of *net* value of production because it avoids the duplications inherent in the Value of Shipments measure (below).

Value of Shipments is the net selling value of products leaving production plants in an industry. In industries where two or more production stages for a product are included under the same SIC, the Value of Shipments measure will tend to overstate the economic importance of the industry. Value of Shipments, however, corresponds to the sales volume of the industry.

The Capital Investments column shows capital expenditures for equipment and structures made by the industry provided that these expenditures are depreciated rather than expensed in the year of acquisition.

3. Indices of Change

The data presented in the *General Statistics* table are restated as indices in the *Indices of Change* table. The purpose of the table is to show the user rapidly how different categories of the industry have changed since 1992.

The year 1992 is used as the base and is therefore shown as 100 in every category. The values in the years 1982-1991 and 1993-1998 are then expressed in relation to the 1992 value. For example, Shipments in 1992 were $50,434.4 million for *SIC 2011, Meat packing plants*. This value is taken as the base for the Shipments column of the *Indices of Change* table. Shipments in 1988 in that industry were $47,333.2 million. That value, divided by $50,434.4 and multiplied by 100 is 94, which is the index for Shipments in 1988.

Thus, for the years 1982-1991 and 1993-1998, values of 100 indicate no change in relation to the 1992 base year; values above 100 mean better and values below 100 indicate worse performance—all relative to the 1992 base. Note, however, that these are *indices* rather than compounded annual rates of growth or decline. Note, also, that the base used in the last edition was 1987. The most recent full Economic Census year is always used as the index reference.

Indexes based on projections by the editors are followed by a P.

4. Selected Ratios

To understand an industry, analysts calculate ratios of various kinds so that the absolute numbers can be placed in a more global perspective. Twenty important industrial ratios are precalculated for the user in the *Selected Ratios* table. Additionally, the same ratios are also provided for the average of all manufacturing industries; an index, comparing the two categories, is also provided.

The ratios are calculated for the most recent complete year available; that year is usually 1994. In calculations using establishments, projected 1994 values are used because 1994 establishment counts were unavailable.

The categories—"Employees per Establishment," "Hours per Production Worker," "Shipments per Production Worker," etc.—represent a division between the first and the second element (Total Employees divided by Total Establishments); the exception is "Wages per Hour," which is reproduced without calculation from the *General Statistics* table.

The first column of values represents the **Average of All Manufacturing**. These ratios are calculated by (1) adding all categories for manufacturing and (2) following the method for ratio calculation described above.

The second column of values shows the ratios for the **Analyzed Industry**, i.e., the industry currently under consideration.

The third column is an **Index** comparing the Analyzed Industry to the Average of All Manufacturing Industries. The index is useful for determining quickly and consistently how the Analyzed Industry stands in relation to all manufacturing. Index values of 100 mean that the Analyzed Industry, within a given ratio, is identical to the average of all 458 manufacturing industries. An index value of 500 means that the Analyzed Industry is five times the average—for instance, that it has five times as many employees per establishment or pays five times as much. An index value of 50 would indicate that the Analyzed Industry is half of the average of all industries (50%). Similarly, an index of 105 means 5% above average and 95 indicates 5% below. The best use of these index values is discussed further in the ''User's Guide'' (Volume I).

5. Leading Companies

The table of *Leading Companies* shows up to 75 companies that participate in the industry. The listings are sorted in descending order of sales and show the company name, address, name of the Chief Executive Officer, telephone, company type, sales (in millions of dollars) and employment (in thousands of employees). The number of companies shown, their total sales, and total employment are summed at the top of the table for the user's convenience.

The data are from the *Ward's Business Directory of U.S. Private and Public Companies* for 1996, Volumes 1, 2, and 3. Public and private corporations, divisions, subsidiaries, joint ventures, and corporate groups are shown. Thus a listing for an industry may show the parent company as well as important divisions and subsidiaries of the same company (usually in a different location).

While this method of presentation has the disadvantage of duplication (the sales of a parent corporation include the sales of any divisions listed separately), it has the advantage of providing the user with information on major components of an enterprise at different locations. In any event, the user should *not* assume that the sum of the sales (or employment) shown in the *Leading Companies* table represents the total sales (or employment) of an industry. The Shipments column of the *General Statistics* table is a better guide to industry sales.

The company's type (private, public, division, etc.) is shown on the table under the column headed ''Co Type,'' thus providing the user with a means of roughly determining the total ''net'' sales (or employment) represented in the table; this can be accomplished by adding the values and then deducting values corresponding to divisions and subsidiaries of parent organizations also shown in the table. The code used is as follows:

P	Public corporation
R	Private corporation
S	Subsidiary
D	Division
J	Joint venture
G	Corporate group

An asterisk (*) placed behind the sales volume indicates an estimate; the absence of an asterisk indicates that the sales value has been obtained from annual reports, other formal submissions to regulatory bodies, or from the corporation. The symbol " < " appears in front of some employment values to indicate that the actual value is "less than" the value shown. Thus the value of " <0.1" means that the company employs fewer than 100 people.

6. Materials Consumed

The *Materials Consumed* table is a completely updated table drawn from the 1992 Economic Census; it reports the quantities of materials and products (e.g., containers, packaging) used by the industry. The delivered cost of the materials, in millions of dollars, is also shown. Data are not available for all industries. Where data are missing, the table header is reproduced with the notation that data are not available.

A number of symbols are used to indicate why data are omitted or their basis. (D) means that data are withheld to avoid disclosure of competitive data; "na" is used when data are "not available." (S) means that data are withheld because statistical norms were not met; (X) stands for "not applicable;" (Z) means that less than half of the unit quantity is consumed; "nec" means "not elsewhere classified," and "nsk" abbreviates "not specified by kind." A single asterisk (*) shows instances where 10-19 percent of the data were estimated; two asterisks (**) show a 20-29 percent estimate.

7. Product Share Details

The table of *Product Share Details* shows what products the industry makes and what percentile of total shipments these products represent. The data are new, reporting updated 1992 results for the first time. The source is the 1992 Economic Census.

The products shown come primarily from the industry under consideration but include the same products if manufactured under another SIC classification; shipment data based on *products* is not always identical to shipment data based on *industry*—because no one industry makes only products that are classified under its SIC. The volume of such inter-industry transfers is relatively small; thus applying product share percentages to Shipment data (to determine, say, the dollar volume of a particular product shipped) will yield reasonably good (but not precise) results.

The table combines data based on products and product classes (an aggregation of products). This method avoids the need for presenting very extensive tables while allowing details to be shown when appropriate. In many product categories, the ''product-level'' is very finely subdivided—into canned beans by size of container, for instance—so that reproducing all products would make *Manufacturing USA* unmanageably bulky. In other instances, the ''product class'' does not provide sufficient detail for a reasonable overview of the industry's product mix. Decisions on which level to use were made item by item with the aim of presenting a good overall picture of the industry's product array. Once again, in this 5th edition, detail has been expanded. *Product tables are the most detailed ever published in MUSA.*

In some tables, the first item represents the overall category and is shown as 100%. The sum of the following product/product class entries will be 100; sums slightly above or below 100 are due to rounding. Note, however, that the values of *indented* subcategories are summed in the main heading above them.

In some instances, the symbol (D) will appear instead of a value; the symbol appears when data are withheld to prevent disclosure of competitive information. The abbreviation ''nsk'' stands for ''not specified by kind.''

8. Inputs and Outputs

One of the more useful tools for tracing economic activity from industry to industry and from one economic sector to another is an economic data structure known as the input-output table. For this reason, *MUSA* provides an extract from the *Benchmark Study of the U.S. Economy, 1982*, published in the fall of 1991 by the Department of Commerce. Input-Output studies are expensive, complex enterprises; consequently, such data are published infrequently and at a substantial lag in time; nevertheless, these data show the fundamental structural relationships between economic activities and are a useful if imprecise adjunct to market, developmental, locational, and other analyses.

The table of *Inputs and Outputs* is an *extraction* because it records only those transactions between manufacturing industries and other sectors of the economy that represent at least one tenth of one percent of total inputs to an industry or outputs from an industry. All lesser transactions are suppressed in order to conserve space.

The table is in two parts. The first part (column 1) shows economic sectors or industries that supply an industry with goods and services; the second part (column 2) shows economic sectors or industries that purchase the outputs of the industry. (Since some of the economic units shown are not ''industries''—for instance State and Local Government—the term ''sector'' is used to denote such economic entities.) Within each part, the sector/industry is shown first followed by a percentile of input or output; the final entry categorizes the sector/industry by type (Manufacturing, Wholesale or Retail Trade, Services, Utilities, etc.).

The Input-Output accounts are based on the 1977 SIC structure which was in force in 1982. For presentation in *MUSA*, data have been associated with the 1987 SIC structure as closely as possible; the user, however, must keep in mind that two different schemes of classification have been coordinated; for this reason, some distortions are likely.

The sectors/industries in each part of the table are sorted in descending order of importance: the largest supplying and purchasing sectors/industries will be placed at the top of the table. If the percentile is the same, the data are shown in sector order following the SIC classification (from agriculture down to government); if the sector is also the same, sorting is alphabetically by name of activity.

The user should note the following when using this table:

- Most industries buy from and sell to themselves. These transactions may at times be the largest inputs and one of the larger outputs of an industry. The reason for this phenomenon is simply that a 4-digit industry is very often *many* specialized industries that supply each other in a cascade of components.

- Not all inputs to an industry are shown—only those that account for 0.1 percent or more of inputs. The user may know an industry that uses large amounts of electricity—yet the table does not show ''Electric service utilities'' as an input. The reason, most likely, is that electric power is too small a proportion of purchases in relation to other raw material obtained.

- The magnitude of certain inputs may be surprising and difficult to explain without detailed knowledge of an industry. For instance, ''Real estate'' is a major input into industries that produce tobacco products—because acquisition of appropriate land for growing tobacco is an important aspect of that industry's activity.

- Some of the output categories used in the Input-Output study hide the ultimate user. Examples are ''Exports'' and ''Gross private fixed investment''; it is not always possible to know which element of the economy made the investment. This limitation derives from the manner in which the Input-Output accounts are organized by the Department of Commerce.

- The input and output ratios are not likely to be accurate for industries where there has been dramatic change (technological or other) since the early 1980s.

9. Occupations Employed by Industry Group

Manufacturing USA presents data on 200 occupation categories employed by manufacturing industries; since most of these categories combine two or three occupations, nearly 460 occupations are covered. The information presented is an extract from the *Industry-Occupation Matrix* produced by the Bureau of Labor Statistics (BLS), Department of Labor.

The table on *Occupations Employed* presents an extract; showing the entire matrix would have required too much space. Thus only those occupations are included that represent 1% or more of total employment in an industry. The advantage of this method is that the data are kept manageable while *most* of an industry's employment is defined by occupation. The disadvantage is that certain occupations, although employed by an industry, do not make the "cutoff" of 1% of total employment.

The data are shown for 1994 (updated from 1992) in percent of Total Employment for an industry group (3-digit industry level or groupings of 3-digit industries). Also shown is the Bureau of Labor Statistics' projection of the anticipated growth or decline of the occupation to the year 2005. This value is reported as a percent change to 2005; a value of 5.5, for instance, means that overall employment, in the industry group, will increase 5.5% between 1992 and 2005; a negative value indicates a corresponding decline. Note that these are *not* rates of annual change. The last update by BLS of this database moved the starting point from 1992 to 1994 but did not move the target year. In the last edition of *MUSA* projections to 2005 were reported but from a 1992 base.

BLS does not provide occupational data at the 4-digit SIC level. Consequently, the same table of *Occupations Employed* is reproduced for each industry which is in the same 3-digit grouping. This approach has been adopted so that the user will find the occupations associated with a 4-digit industry with other data on that industry.

The user should note the following:

- As already stated, the occupations shown are a subset of total occupations employed: those that account for 1% or more of employment in the industry group.

- Since the data are for *groups*, some occupations listed will appear out of place in a particular 4-digit industry; that is because those occupations are employed by a related 4-digit industry in the same group.

- Growth or decline indicated for an occupation within an industry group does not mean that the occupation is growing or declining overall. Also, changes introduced by BLS between editions of this series can be quite drastic. An occupation that grew by several percentage points in an industry two years ago is shown suddenly declining to the year 2005 now.

10. Map Graphics

The geographical presentation of data begins with two maps titled *Location by State* and *Regional Concentration*. In the first map, all states in which the industry is present are shaded. In the second, the industry's concentration is shown by Census region. The two maps, together, tell the user at a glance where the industry is active and which regions rank first, second, and third in value of shipments or in number of establishments; establishment counts are used for ranking in those industries where shipment data are withheld (the (D) symbol) for the majority of states. In the case of some industries, only one or two regions are shaded because the industry is concentrated in a few states. The data for ranking are taken from the table on *Industry Data by State* which immediately follows the maps.

The regional boundaries are those of the Census Regions and are named, from left to right and top to bottom as follows:

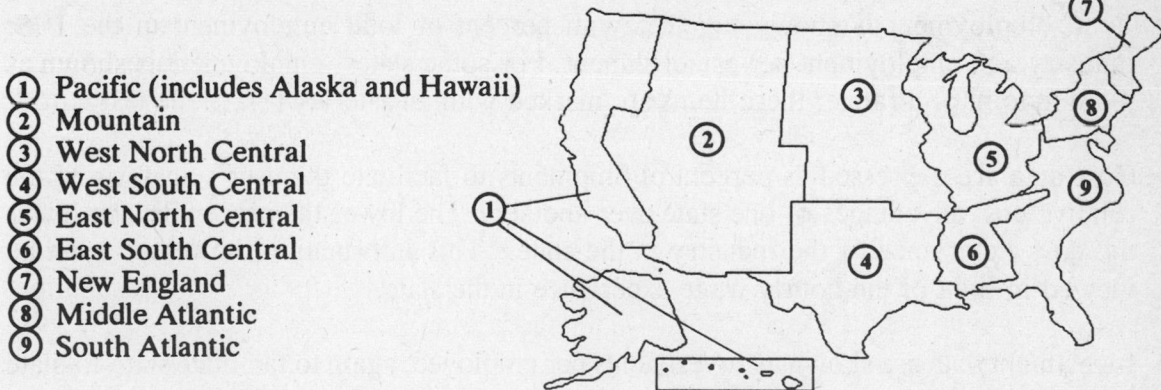

(1) Pacific (includes Alaska and Hawaii)
(2) Mountain
(3) West North Central
(4) West South Central
(5) East North Central
(6) East South Central
(7) New England
(8) Middle Atlantic
(9) South Atlantic

In the case of the Pacific region, all parts of the region are shaded (including Alaska and Hawaii), even if the basis for the ranking is the industry's predominance in California (the usual case).

Although regional data are only graphed and not reported in a separate table, the table of *Industry Data by State* provides all the necessary information for constructing a regional table.

11. Industry Data by State

The table on *Industry Data by State* provides ten data elements for each state in which the industry is active. The data are updated in this edition. They come from the 1992 Economic Census, the most recently available *complete* data set on states. Even in this series, certain data elements are suppressed by the Bureau of the Census to prevent disclosure of competitive information. This may come about in instances where only a few operations are present in the state or they are operated by a small number of companies. The states are shown in descending order of shipments. The categories of Establishments, Shipments, Total Employment, and Wages are identical to those in the table of *General Statistics*. In addition, six elements of information are provided so that the user can more easily compare the size, performance, and characteristics of the industry from one state to the next:

- Shipments are expressed in millions of dollars and as a percent of the total U.S. shipments for the industry. This is useful for determining the relative importance of the state in the industry as a whole. Shipments per Establishment are also provided; this measure gives an insight into the relative size of the factories in the state.

- Total employment is shown together with percent of total employment in the U.S. industry and employment per establishment. For some states, employment is shown as the midpoint of a range; these items are marked with an asterisk (*).

- Cost data are expressed as percent of Shipments to facilitate the user's analysis of the relative cost advantages of one state over another. The lower the percentile, the lower the cost experience of the industry in the state. This information, however, must be viewed in light of the hourly wage experience in the state.

- Investment data are shown as Investments per employee, again to facilitate state-to-state comparisons.

The symbol (D) is used when data are withheld to prevent disclosure. Dashes are used to indicate that the corresponding data element cannot be calculated because some part of the ratio is missing or withheld.

Projected Data Series

Beginning with the 4th edition of *MUSA*, sufficient data points became available to permit the projection of data to the current year and beyond. For this reason, and as a service to the busy user of the book, the editors have introduced a new feature—trend projections of available data.

How Projections Were Made

Projections are based on a curve-fitting algorithm using the least-squares method. In essence, the algorithm calculates a trend line for the data using existing data points. Extensions of the trend line are used to predict future years of data. The method is illustrated in the chart below. It shows actual data values plotted for 1982 through 1994 and the least-squares trend line drawn to overlay the points. The trend line is extended through 1998. The values for 1995 through 1998 represent the "trend" of the earlier years.

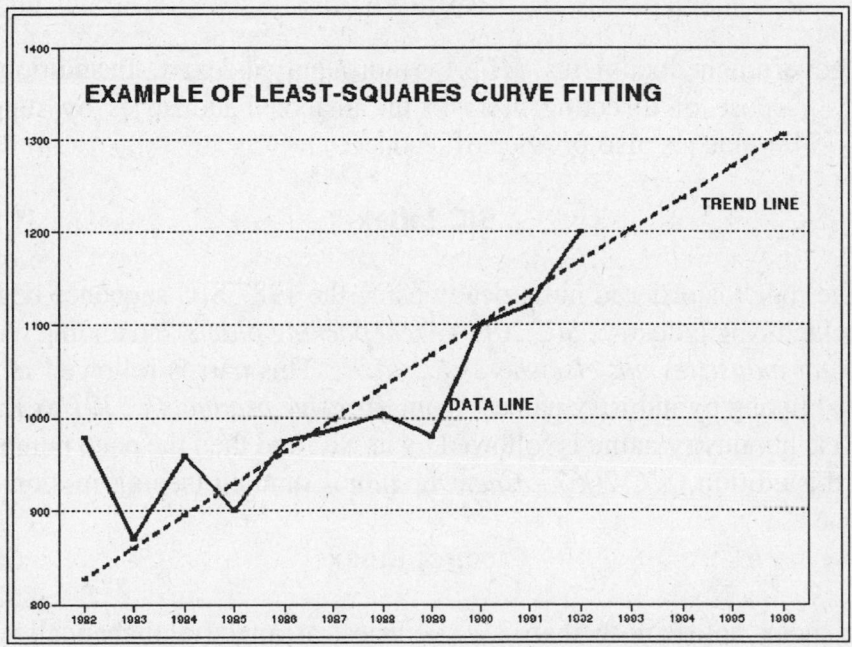

What Values Were Projected

In this edition, every category (column) reported under General Statistics has been subject to projection. In those cases where a coherent series exists from 1982 to the present, the entire series was used. In those cases where the industry definition underwent a change in 1987, trends are calculated from 1987 forward.

Cost of Materials and Value Added by Manufacturer were calculated using the 1994 ratio of costs or value added to shipments in 1994 and then applying that ratio to other years using the projected shipments values for those years. Costs and value added were treated in this manner because averaging these data for a long period (1982-1992, for example) would not properly reflect cost savings and productivity changes achieved most recently. Therefore the use of a ratio, based on the most recent survey year, seemed more appropriate.

Limitations of Projections

Projections are simply means of detecting trends—that may or may not hold in the future. The projections in *MUSA*, therefore, are not as reliable as actual survey data. Most analysts trying to project the future routinely turn to trend projection. In this edition of *MUSA*, the work of doing the projections has been done for the user in advance.

Indexes

Unlike most government documents, *MUSA* is thoroughly indexed. In addition to fulfilling their primary purpose of directing users to the analyzed industries by supplying page numbers, *MUSA*'s indexes also provide SIC codes.

SIC Index

Part one of the index is ordered numerically using the 1987 SIC sequence beginning with the first manufacturing industry, *SIC 2011 - Meat packing plants* and ending with *SIC 3999 - Manufacturing industries not elsewhere classified*. This part is followed immeditely by an alphabetical listing by industry name—from *Abrasive products - 3291* to *Yarn spinning mills - 2281*; each industry name is followed by its SIC and then the page number on which it begins. In this edition, *SIC 2067 - Chewing gum* is omitted (see discussion above).

Product Index

The Product Index holds more than 2,200 entries, arranged alphabetically, identifying products, materials, and chemicals manufactured in the United States. The names were largely, but not exclusively, obtained from the actual product names as published by the *Census of Manufactures*. Additional keywords have been added when the Federal terminology was too technical or obscure. Each product name is followed by one or more SIC codes indicating in which industry or industries the product is manufactured. The references are arranged in SIC order and do not necessarily indicate, by their arrangement, which industry is the predominant supplier of the product.

Company Index

This index shows more than 21,000 company names arranged in alphabetical order; company names that begin with a numeral (3M, etc.) precede company names that begin with the letter A. Company names are followed by page references and a listing of SICs (within brackets).

Occupation Index

The Occupation index shows nearly 460 occupations under 200 occupational groupings. This index does not attempt to refer the user to *every* industry in which an occupation occurs; that approach would render the index unwieldy. "Inspectors, testers, and graders," for instance, are employed in 121 3-digit industries, "Bookkeeping clerks" in 75, etc. Rather than to make the index unmanageable, the total number of 3-digit industries employing the occupation is shown, in parentheses, following the name of the occupation; thereafter, the top ten (or fewer) industry groups are shown in their order of importance; the most important group (that which employs the largest number) is shown first.

The user should note, again, that—

- Occupations are reported by 3-digit industry group; a reference to industry 372, for instance, means that the user can find the occupation under *SIC 3721 - Aircraft*; *3724 - Aircraft and engine parts*; and *3728 - Aircraft equipment, not elsewhere classified*.

- Only those occupations are included which represent at least 1% of employment in a 3-digit industry group. As an example, "Librarians, professional" are employed by manufacturing companies and at a growing rate; but as highly specialized professionals among others, their numbers do not reach the "reporting threshold" used in *Manufacturing USA*.

Abbreviations, Codes, and Symbols

U.S. State Postal Codes

AK	Alaska	MT	Montana
AL	Alabama	NC	North Carolina
AR	Arkansas	ND	North Dakota
AZ	Arizona	NE	Nebraska
CA	California	NH	New Hampshire
CO	Colorado	NJ	New Jersey
CT	Connecticut	NM	New Mexico
CZ	Canal Zone	NV	Nevada
DC	District of Columbia	NY	New York
DE	Delaware	OH	Ohio
FL	Florida	OK	Oklahoma
GA	Georgia	OR	Oregon
GU	Guam	PA	Pennsylvania
HI	Hawaii	PR	Puerto Rico
IA	Iowa	RI	Rhode Island
ID	Idaho	SC	South Carolina
IL	Illinois	SD	South Dakota
IN	Indiana	TN	Tennessee
KS	Kansas	TX	Texas
KY	Kentucky	UT	Utah
LA	Louisiana	VA	Virginia
MA	Massachusetts	VI	Virgin Islands
MD	Maryland	VT	Vermont
ME	Maine	WA	Washington
MI	Michigan	WI	Wisconsin
MN	Minnesota	WV	West Virginia
MO	Missouri	WY	Wyoming
MS	Mississippi		

M·A·N·U·F·A·C·T·U·R·I·N·G
USA

Industry Analyses,
Statistics, and Leading Companies

3312 - BLAST FURNACES & STEEL MILLS

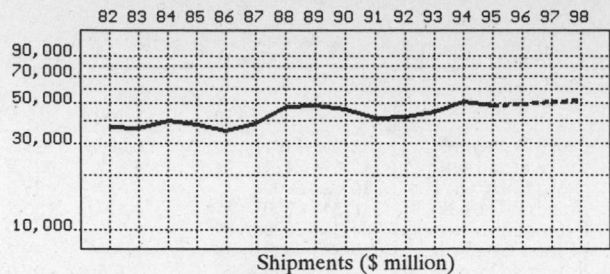

Shipments ($ million)

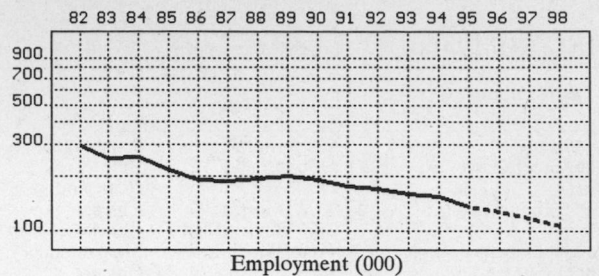

Employment (000)

GENERAL STATISTICS

Year	Companies	Establishments		Employment			Compensation		Production ($ million)			
		Total	with 20 or more employees	Total (000)	Production Workers (000)	Hours (Mil)	Payroll ($ mil)	Wages ($/hr)	Cost of Materials	Value Added by Manufacture	Value of Shipments	Capital Invest.
1982	219	301	225	295.8	215.2	375.7	8,677.9	16.45	23,568.7	11,763.3	36,824.4	2,170.2
1983		340	233	249.2	186.1	349.3	7,353.3	15.15	23,639.9	12,380.5	36,449.7	1,986.9
1984		379	241	257.0	197.3	382.2	7,724.7	14.84	25,807.3	14,563.2	40,335.2	1,691.5
1985		419	250	217.9	167.7	330.6	6,984.6	15.68	24,719.8	12,763.5	38,080.4	2,044.3
1986		427	250	192.2	147.7	294.5	6,281.5	15.96	21,677.5	12,453.0	34,924.9	924.9
1987	270	342	222	188.9	147.6	307.1	6,450.8	15.83	22,949.8	15,820.3	38,663.0	1,220.3
1988		451	243	195.2	151.2	324.3	7,033.5	16.46	28,826.9	19,555.7	47,589.6	1,893.0
1989		442	236	199.7	148.8	347.7	7,160.8	15.41	28,958.6	19,337.8	47,853.4	2,498.4
1990		493	248	191.6	146.0	314.7	7,328.4	17.37	27,959.6	18,283.0	45,950.4	2,554.1
1991		514	260	177.3	135.3	278.7	6,939.5	18.26	25,770.7	14,567.5	40,775.3	2,986.3
1992	136	217	197	170.6	131.4	277.0	7,035.7	18.99	25,392.2	16,569.1	42,215.2	2,210.5
1993		343	215	159.1	123.2	267.5	7,000.9	19.77	26,899.3	17,908.3	44,777.1	1,778.4
1994		408P	227P	156.1	121.7	265.5	7,272.1	20.72	30,232.8	20,692.8	50,633.8	2,614.8
1995		411P	226P	136.7P	107.7P	256.3P	6,777.0P	19.89P	28,969.9P	19,828.4P	48,518.6P	2,469.6P
1996		413P	225P	127.1P	100.9P	247.7P	6,720.5P	20.31P	29,532.0P	20,213.1P	49,460.1P	2,530.4P
1997		416P	224P	117.5P	94.1P	239.1P	6,664.0P	20.72P	30,094.1P	20,597.9P	50,401.5P	2,591.2P
1998		419P	223P	107.9P	87.3P	230.5P	6,607.5P	21.14P	30,656.2P	20,982.6P	51,342.9P	2,652.0P

Sources: 1982, 1987, 1992 *Economic Census*; *Annual Survey of Manufactures*, 83-86, 88-91, 93-94. Establishment counts for non-Census years are from *County Business Patterns*; establishment values for 83-84 are extrapolations. 'P's show projections by the editors. Industries reclassified in 87 will not have data for prior years.

INDICES OF CHANGE

Year	Companies	Establishments		Employment			Compensation		Production ($ million)			
		Total	with 20 or more employees	Total (000)	Production Workers (000)	Hours (Mil)	Payroll ($ mil)	Wages ($/hr)	Cost of Materials	Value Added by Manufacture	Value of Shipments	Capital Invest.
1982	161	139	114	173	164	136	123	87	93	71	87	98
1983		157	118	146	142	126	105	80	93	75	86	90
1984		175	122	151	150	138	110	78	102	88	96	77
1985		193	127	128	128	119	99	83	97	77	90	92
1986		197	127	113	112	106	89	84	85	75	83	42
1987	199	158	113	111	112	111	92	83	90	95	92	55
1988		208	123	114	115	117	100	87	114	118	113	86
1989		204	120	117	113	126	102	81	114	117	113	113
1990		227	126	112	111	114	104	91	110	110	109	116
1991		237	132	104	103	101	99	96	101	88	97	135
1992	100	100	100	100	100	100	100	100	100	100	100	100
1993		158	109	93	94	97	100	104	106	108	106	80
1994		188P	115P	92	93	96	103	109	119	125	120	118
1995		189P	115P	80P	82P	93P	96P	105P	114P	120P	115P	112P
1996		191P	114P	75P	77P	89P	96P	107P	116P	122P	117P	114P
1997		192P	114P	69P	72P	86P	95P	109P	119P	124P	119P	117P
1998		193P	113P	63P	66P	83P	94P	111P	121P	127P	122P	120P

Sources: Same as General Statistics. Values reflect change from the base year, 1992. Values above 100 mean greater than 92, values below 100 mean less than 92, and a value of 100 in the 82-91 or 93-98 period means same as 92. 'P's mark projections by the editors.

SELECTED RATIOS

For 1994	Avg. of All Manufact.	Analyzed Industry	Index	For 1994	Avg. of All Manufact.	Analyzed Industry	Index
Employees per Establishment	49	383	781	Value Added per Production Worker	134,084	170,031	127
Payroll per Establishment	1,500,273	17,835,697	1,189	Cost per Establishment	5,045,178	74,149,565	1,470
Payroll per Employee	30,620	46,586	152	Cost per Employee	102,970	193,676	188
Production Workers per Establishment	34	298	870	Cost per Production Worker	146,988	248,421	169
Wages per Establishment	853,319	13,492,254	1,581	Shipments per Establishment	9,576,895	124,185,463	1,297
Wages per Production Worker	24,861	45,203	182	Shipments per Employee	195,460	324,368	166
Hours per Production Worker	2,056	2,182	106	Shipments per Production Worker	279,017	416,054	149
Wages per Hour	12.09	20.72	171	Investment per Establishment	321,011	6,413,110	1,998
Value Added per Establishment	4,602,255	50,751,572	1,103	Investment per Employee	6,552	16,751	256
Value Added per Employee	93,930	132,561	141	Investment per Production Worker	9,352	21,486	230

Sources: Same as General Statistics. The 'Average of All Manufacturing' column represents the average of all manufacturing industries reported for the most recent complete year available. The Index shows the relationship between the Average and the Analyzed Industry. For example, 100 means that they are equal; 500 that the Analyzed Industry is five times the average; 50 means that the Analyzed Industry is half the national average. The abbreviation 'na' is used to show that data are 'not available'.

LEADING COMPANIES   Number shown: 75   Total sales ($ mil): 53,765   Total employment (000): 209.9

Company Name	Address				CEO Name	Phone	Co. Type	Sales ($ mil)	Empl. (000)
US Steel Group	600 Grant St	Pittsburgh	PA	15219	Paul J Wilhelm	412-433-1121	S	6,066	21.3
Bethlehem Steel Corp	1170 8th Av	Bethlehem	PA	18016	Curtis H Barnette	215-694-2424	P	4,819	19.9
LTV Corp	PO Box 6778	Cleveland	OH	44101	David H Hoag	216-622-5000	P	4,529	17.0
Inland Steel Industries Inc	30 W Monroe St	Chicago	IL	60603	Robert J Darnall	312-346-0300	P	4,497	15.5
Nucor Corp	2100 Rexford Rd	Charlotte	NC	28211	F Kenneth Iverson	704-366-7000	P	2,976	5.9
National Steel Corp	4100 Edison Lks	Mishawaka	IN	46545	W John Goodwin	219-273-7000	P	2,700	9.7
Inland Steel Co	30 W Monroe St	Chicago	IL	60603	Robert J Darnall	312-346-0300	P	2,175	12.0
AK Steel Corp	703 Curtis St	Middletown	OH	45043	Thomas C Graham	513-425-5000	S	2,017	6.0
AK Steel Holding Corp	703 Curtis St	Middletown	OH	45043	Thomas C Graham	513-425-5000	P	2,017	6.0
Rouge Steel Co	PO Box 1699	Dearborn	MI	48121	Carl L Valdiserri	313-390-6877	P	1,236	3.2
Weirton Steel Corp	400 3 Springs Dr	Weirton	WV	26062	Herbert Elish	304-797-2000	P	1,201	6.0
Allegheny Ludlum Corp	100 Six PPG Pl	Pittsburgh	PA	15222	Robert P Bozzone	412-394-2800	P	1,077	6.0
Lukens Inc	50 S 1st Av	Coatesville	PA	19320	R William Van Sant	610-383-2000	P	947	4.1
Oregon Steel Mills Inc	1000 SW Broadway	Portland	OR	97205	Thomas B Boklund	503-223-9228	P	838	3.0
Armco Inc	1 Oxford Centre	Pittsburgh	PA	15219	Robert C Purdum	412-255-9800	P	788	4.6
WCI Steel Inc	1040 Pine Av SE	Warren	OH	44483	John R Scheessele	216-841-8311	P	709	2.3
Texas Industries Inc	7610 Stemmons Fwy	Dallas	TX	75247	Robert D Rogers	214-647-6700	P	707	2.7
Birmingham Steel Corp	1000 Urban Ctr	Birmingham	AL	35242	James A Todd Jr	205-970-1200	P	703	2.0
USS/KOBE Steel Co	1807 E 28th St	Lorain	OH	44055	Ralph E Fifield	216-277-2000	J	700	2.6
Northwestern Steel and Wire Co	121 Wallace St	Sterling	IL	61081	Robert N Gurnitz	815-625-2500	P	604	2.5
Georgetown Industries Inc	1901 Roxborough	Charlotte	NC	28211	R R Regelbrugge	704-366-6901	R	600*	3.5
National Steel Corp	20th & State St	Granite City	IL	62040	Kenneth J Leonard	618-451-3456	D	600	3.3
North Star Steel Co	PO Box 9300	Minneapolis	MN	55440	Robert Garvey	612-742-5471	S	570*	3.2
FLS Holdings Inc	PO Box 31328	Tampa	FL	33631	Koichi Takashima	813-251-8811	S	550	2.4
Florida Steel Corp	PO Box 31328	Tampa	FL	33631	Koichi Takashima	813-251-8811	S	547	2.4
Acme Metals Inc	13500 S Perry Av	Riverdale	IL	60627	Brian WH Marsden	708-849-2500	P	523	2.8
Gulf States Steel Incorporated	174 S 26th St	Gadsden	AL	35904	John D Lefler	205-543-6100	S	500	1.9
Geneva Steel	10 S Geneva Rd	Vineyard	UT	84058	Joseph A Cannon	801-227-9000	P	465	2.6
Chaparral Steel Co	300 Ward Rd	Midlothian	TX	76065	Gordon E Forward	214-775-8241	P	462	1.0
Lukens Steel Co	ARC Bldg	Coatesville	PA	19320	Dennis M Oates	215-383-2000	S	418	2.2
Carpenter Technology Corp	PO Box 14662	Reading	PA	19612	Robert W Cardy	215-208-2000	D	410*	2.6
Laclede Steel Co	1 Metropolitan Sq	St Louis	MO	63102	John B McKinney	314-425-1400	P	341	1.9
Pennsylvania Steel Technologies	215 S Front St	Steelton	PA	17113	Andrew W Futchko	717-986-2000	S	275	1.5
Schnitzer Steel Industries Inc	PO Box 10047	Portland	OR	97210	Leonard Schnitzer	503-224-9900	P	262	0.9
Nooter Corp	PO Box 451	St Louis	MO	63166	Gene R Smith Hays	314-621-6000	R	260	1.2
Georgetown Steel Corp	PO Box 619	Georgetown	SC	29442	Don B Daily	803-546-2525	S	256	0.7
Inland Steel Co	3300 Dickey Rd	East Chicago	IN	46312	Dale Wiersbe	219-399-7100	D	250	1.2
MacSteel	One Jackson Sq	Jackson	MI	49201	Robert V Kelly Jr	517-782-0415	D	245	0.6
American Steel and Wire Corp	4300 E 49th St	Cuyahoga H	OH	44125	Thomas N Tyrrell	216-883-3800	S	235	0.6
Roanoke Electric Steel Corp	PO Box 13948	Roanoke	VA	24038	Donald G Smith	703-342-1831	P	216	0.8
CSC Industries Inc	4000 Mahoning NW	Warren	OH	44483	Donald J Caiazza	216-841-6011	P	209	1.2
Copperweld Steel Co	4000 Mahoning Av	Warren	OH	44483	Donald J Caiazza	216-841-6011	S	209	1.2
Atlantic Steel Industries Inc	PO Box 1714	Atlanta	GA	30301	Jesse J Webb	404-897-4500	S	200	0.9
Lee Capital Holdings	1 International Pl	Boston	MA	02110	Jonathan O Lee	617-345-0477	R	190*	1.1
ABC Rail Products Corp	200 S Michigan Av	Chicago	IL	60604	Donald W Grinter	312-322-0360	P	187	1.4
AlTech Specialty Steel Corp	PO Box 152	Dunkirk	NY	14048	Tom Parker	716-366-1000	S	170	1.0
Armco Coshocton Operations	PO Box 190	Coshocton	OH	43812	CN Parr	614-829-2341	D	170*	0.5
Scot Industries Inc	1532 W Galena St	Milwaukee	WI	53205	Steven Wilmeth	414-342-4310	R	170*	1.0
Freedom Forge Corp	500 N Walnut St	Burnham	PA	17009	Herbert C Graves	717-248-4911	R	165	1.0
Bayou Steel Corp	PO Box 5000	La Place	LA	70069	Howard M Meyers	504-652-4900	P	161	0.5
Newport Steel Corp	PO Box 1670	Newport	KY	41072	Ron Noel	606-292-6000	~S	150	0.7
Huntco Inc	14323 S Outer 40	Chesterfield	MO	63017	BD Hunter	314-878-0155	P	146	0.3
Acme Packaging Corp	13500 S Perry Av	Riverdale	IL	60627	Robert W Dyke	708-849-2500	S	140	0.4
Pilot Industries Inc	PO Box 476	Dexter	MI	48130	Robert A Davis	313-426-4376	R	140*	0.8
Huntco Steel Inc	14323 S Outer 40	Chesterfield	MO	63017	Terry J Heinz	314-878-0155	S	137	0.3
Charter Steel	1658 Cold Spring	Saukville	WI	53080	Bill Gano	414-268-2500	D	130	0.4
Koppel Steel Corp	PO Box 750	Beaver Falls	PA	15010	Paul C Borland Jr	412-847-4059	S	130	0.7
Latrobe Steel Co	2626 S Ligonier St	Latrobe	PA	15650	C Philip Weigel	412-537-7711	S	130*	0.8
Cable Design Technologies Inc	661 Andersen Dr	Pittsburg	PA	15220	Paul Olson	412-937-2300	R	127	0.7
Cascade Steel Rolling Mills Inc	PO Box 687	McMinnville	OR	97128	Kurt Zetzsche	503-472-4181	S	124	0.5
Citisteel USA Inc	4001 Phila Pk	Claymont	DE	19703	Zengxin Mi	302-792-5400	R	115	0.3
Charter Manufacturing Company	PO Box 217	Mequon	WI	53092	John A Mellowes	414-243-4700	R	110	0.6
Synalloy Corp	PO Box 5627	Spartanburg	SC	29304	James G Lane Jr	803-585-3605	P	103	0.5
Kentucky Electric Steel Inc	PO Box 3500	Ashland	KY	41105	Charles C Hanebuth	606-928-6441	P	103	0.5
Crucible Specialty Metals	PO Box 977	Syracuse	NY	13201	William K Grant	315-487-4111	D	100	0.6
Marion Steel Co	912 Cheney Av	Marion	OH	43302	James Conway	614-383-4011	R	100	0.4
NexTech	300 Braddock Av	Turtle Creek	PA	15145	Wilson Farmerie	412-825-4300	R	100	<0.1
Pinole Point Steel Co	PO Box 4050	Richmond	CA	94804	Peter E Wais	510-223-8883	S	100	0.2
Alpha/Beta Tube Corp	PO Box 352799	Toledo	OH	43635	Steve Jansto	419-865-8031	S	85	0.1
Connecticut Steel Corp	PO Box 928	Wallingford	CT	06492	W Fergus Porter	203-265-0615	R	80	0.2
SMI Steel Inc	PO Box 321188	Birmingham	AL	35232	D Morrison	205-592-8981	S	71	0.4
Affiliated Metals Co	PO Box 1306	Granite City	IL	62040	Pat Notestine	618-451-4700	R	62	<0.1
Sloss Industries Corp	PO Box 5327	Birmingham	AL	35207	Lee C Houlditeh	205-808-7914	S	55	0.5
CHS Acquisition Corp	211 E Main St	Chicago Hts	IL	60411	Frank L Corral	708-756-5619	R	53*	0.3
Schuylkill Holdings Inc	PO Box 74040	Baton Rouge	LA	70874	Kent Hudson	504-775-3040	R	53*	0.4

Source: Ward's Business Directory of U.S. Private and Public Companies, Volumes 1 and 2, 1996. The company type code used is as follows: P - Public, R - Private, S - Subsidiary, D - Division, J - Joint Venture, A - Affiliate, G - Group. Sales are in millions of dollars, employees are in thousands. An asterisk (*) indicates an estimated sales volume. The symbol < stands for 'less than'. Company names and addresses are truncated, in some cases, to fit into the available space.

MATERIALS CONSUMED

Material	Quantity	Delivered Cost ($ million)
Materials, ingredients, containers, and supplies	(X)	20,513.3
Steel rod for wiredrawing	1,000 s tons	(D)	(D)
Steel wire for redrawing	1,000 s tons	(D)	(D)
All other steel shapes and forms (except castings, forgings, and fabricated metal products)	1,000 s tons	(S)	3,055.3
Pig iron shapes and forms	1,000 s tons	(D)	(D)
Ferromanganese, silicomanganese, and manganese shapes and forms	1,000 s tons	714.7	403.6
Ferrochromium shapes and forms	1,000 s tons	663.8	221.2
Ferrosilicon (more than 8 percent silicon) shapes and forms	1,000 s tons	235.5	126.8
Other ferroalloy shapes and forms	1,000 s tons	286.9	342.1
All other ferrous shapes and forms	1,000 s tons	41.5	80.8
Aluminum and aluminum-base alloy shapes and forms	1,000 s tons	136.8*	171.4
Nickel (except ferronickel) shapes and forms	1,000 s tons	43.3	281.6
Tin shapes and forms	1,000 s tons	8.1*	44.1
Zinc and zinc-base alloy shapes and forms	1,000 s tons	168.4	199.6
All other nonferrous shapes and forms	1,000 s tons	60.4**	89.6
Gray iron ingot molds and stools	1,000 s tons	503.4**	83.3
All other iron and steel castings	(X)	226.1
Clay refractories	(X)	235.6
Nonclay refractories	(X)	216.0
Limestone fluxes	1,000 s tons	(S)	71.5
Lime fluxes, including quicklime	1,000 s tons	3,307.8	257.2
Dead-burned dolomite fluxes	1,000 s tons	(S)	105.1
Fluorspar fluxes	(X)	11.3
Other fluxes	(X)	134.0
All other stone, clay, glass, and concrete products	(X)	36.6
Industrial chemicals (except sulfuric acid and oxygen)	(X)	93.9
Oxygen (including high and low purity) (liquid oxygen should be converted to its gaseous equivalent)	mil cu ft	(S)	377.3
Sulfuric acid (new and spent) (100 percent H2SO4)	1,000 s tons	(S)	13.0
All other chemicals and allied products	(X)	32.4
Mining crude iron ore and iron ore concentrates, including pelletized and manganiferous (gross weight)	1,000 s tons	25,153.0*	1,216.9
Mining iron ore agglomerates, including pelletized and manganiferous (gross weight)	1,000 s tons	44,092.6*	1,520.9
Mining manganese ore (including 10 percent manganese) and ferroalloy ore	(X)	9.6
All other metal mining	(X)	(D)
Coal used in the production of coke	1,000 s tons	35,164.3*	1,756.5
Iron and steel scrap, excluding home scrap	1,000 s tons	44,894.0	4,099.9
Lubricating oils and greases and other petroleum products	(X)	142.6
Industrial dies, molds, jigs, and fixtures	(X)	60.4
Carbon and graphite electrodes	(X)	246.7
All other materials and components, parts, containers, and supplies	(X)	3,853.6
Materials, ingredients, containers, and supplies, nsk	(X)	219.4

Source: 1992 *Economic Census*. Explanation of symbols used: (D): Withheld to avoid disclosure of competitive data; na: Not available; (S): Withheld because statistical norms were not met; (X): Not applicable; (Z): Less than half the unit shown; nec: Not elsewhere classified; nsk: Not specified by kind; - : zero; * : 10-19 percent estimated; ** : 20-29 percent estimated.

PRODUCT SHARE DETAILS

Product or Product Class	% Share	Product or Product Class	% Share
Blast furnaces and steel mills	100.00	made in steel mills producing wire rods or hot-rolled bars	0.88
Coke oven and blast furnace products, including ferroalloys	4.12	Steel pipe and tubes, made in steel mills producing semifinished shapes or plate	4.03
Coke oven coke, except screenings and breeze	67.12	Cold-rolled steel sheet and strip, made in steel mills producing hot-rolled sheet or strip	14.57
Coke oven screenings and breeze	4.67		
Coke oven gas	(D)	Cold-finished steel bars and bar shapes, made in steel mills producing hot-rolled bars and bar shapes	1.56
Coke oven crude tar	4.89		
Coke oven crude light oil	2.05	Seamless rolled ring forgings, ferrous (made in steel mills)	0.18
Other coke oven products (including tar derivatives, ammonia, and light oil derivations)	0.88	Seamless carbon steel and alloy steel rolled ring forgings (excluding stainless and hi-temp), ferrous (made in steel mills)	60.67
Blast furnace pig iron, except ferroalloy (including pig iron with silicon content up to and including 6 percent silicon)	(D)		
Blast furnace slag, except ferroalloy	0.31	Seamless stainless and hi-temp (iron, nickel, or cobalt-base alloy) rolled ring forgings, ferrous (made in steel mills)	39.33
Blast furnace sinter from ore, flue dust, and other materials, except ferroalloy	(D)	Open die or smith forgings (hammer or press), ferrous (made in steel mills)	0.27
Blast furnace gas, except ferroalloy	(D)	Carbon and alloy steel (excluding stainless and hi-temp) open die or smith forgings (hammer or press), ferrous (made in steel mills)	94.47
Other blast furnace products, except ferroalloy	0.63		
Coke oven and blast furnace products, including ferroalloys, nsk	5.56		
Steel ingot and semifinished shapes and forms	8.68	Stainless and hi-temp (iron, nickel, or cobalt-base alloy) open die or smith forgings (hammer or press), ferrous (made in steel mills)	5.61
Hot-rolled sheet and strip, including tin mill products, tinplate, blackplate, terneplate, and tin-free steel	41.62	Other steel mill products, including steel rails, except wire products	0.91
Hot-rolled bars and bar shapes, plates, structural shapes, and piling (including concrete reinforcing and tool steel bars)	23.18	Blast furnaces and steel mills, nsk	0.01
Steel wire, including galvanized and other coated wire,			

Source: 1992 *Economic Census*. The values shown are percent of total shipments in an industry. Values of indented subcategories are summed in the main heading. The symbol (D) appears when data are withheld to prevent disclosure of competitive information. The abbreviation nsk stands for 'not specified by kind' and nec for 'not elsewhere classified'.

INPUTS AND OUTPUTS FOR BLAST FURNACES & STEEL MILLS

Economic Sector or Industry Providing Inputs	%	Sector	Economic Sector or Industry Buying Outputs	%	Sector
Imports	29.2	Foreign	Petroleum & natural gas well drilling	6.8	Constr.
Blast furnaces & steel mills	10.0	Manufg.	Blast furnaces & steel mills	6.5	Manufg.
Coal	5.6	Mining	Automotive stampings	5.1	Manufg.
Wholesale trade	5.6	Trade	Motor vehicle parts & accessories	4.5	Manufg.
Electric services (utilities)	5.5	Util.	Metal cans	4.1	Manufg.
Gas production & distribution (utilities)	4.6	Util.	Fabricated structural metal	4.0	Manufg.
Iron & ferroalloy ores	4.1	Mining	Steel pipe & tubes	3.3	Manufg.
Scrap	3.3	Scrap	Cold finishing of steel shapes	3.2	Manufg.
Advertising	3.1	Services	Fabricated plate work (boiler shops)	2.9	Manufg.
Railroads & related services	3.0	Util.	Exports	2.7	Foreign
Maintenance of nonfarm buildings nec	2.7	Constr.	Sheet metal work	2.6	Manufg.
Electrometallurgical products	2.4	Manufg.	Pipe, valves, & pipe fittings	2.3	Manufg.
Cyclic crudes and organics	2.2	Manufg.	Metal stampings, nec	2.1	Manufg.
Petroleum refining	1.2	Manufg.	Screw machine and related products	1.8	Manufg.
Motor freight transportation & warehousing	1.1	Util.	Steel wire & related products	1.6	Manufg.
Water transportation	1.1	Util.	Iron & steel forgings	1.5	Manufg.
Miscellaneous repair shops	1.1	Services	Miscellaneous fabricated wire products	1.5	Manufg.
Primary nonferrous metals, nec	0.9	Manufg.	Miscellaneous metal work	1.5	Manufg.
Carbon & graphite products	0.7	Manufg.	Oil field machinery	1.3	Manufg.
Iron & steel foundries	0.7	Manufg.	Fabricated metal products, nec	1.2	Manufg.
Water supply & sewage systems	0.7	Util.	Farm machinery & equipment	1.2	Manufg.
Industrial gases	0.5	Manufg.	Industrial buildings	1.1	Constr.
Metal heat treating	0.5	Manufg.	Construction machinery & equipment	1.1	Manufg.
Primary zinc	0.5	Manufg.	Prefabricated metal buildings	1.0	Manufg.
Lime	0.4	Manufg.	Ship building & repairing	1.0	Manufg.
Lubricating oils & greases	0.4	Manufg.	Special dies & tools & machine tool accessories	1.0	Manufg.
Banking	0.4	Fin/R.E.	Hardware, nec	0.9	Manufg.
Equipment rental & leasing services	0.4	Services	Machinery, except electrical, nec	0.9	Manufg.
Ball & roller bearings	0.3	Manufg.	Refrigeration & heating equipment	0.9	Manufg.
Industrial inorganic chemicals, nec	0.3	Manufg.	Office buildings	0.8	Constr.
Machinery, except electrical, nec	0.3	Manufg.	Motors & generators	0.8	Manufg.
Metal coating & allied services	0.3	Manufg.	Electric utility facility construction	0.7	Constr.
Pumps & compressors	0.3	Manufg.	Metal partitions & fixtures	0.7	Manufg.
Communications, except radio & TV	0.3	Util.	Pumps & compressors	0.7	Manufg.
Chemical & fertilizer mineral	0.2	Mining	Ball & roller bearings	0.6	Manufg.
Dimension, crushed & broken stone	0.2	Mining	Hand & edge tools, nec	0.6	Manufg.
Blowers & fans	0.2	Manufg.	Metal barrels, drums, & pails	0.6	Manufg.
Fabricated metal products, nec	0.2	Manufg.	Motor vehicles & car bodies	0.6	Manufg.
Industrial controls	0.2	Manufg.	Wiring devices	0.6	Manufg.
Metal stampings, nec	0.2	Manufg.	Architectural metal work	0.5	Manufg.
Nonferrous wire drawing & insulating	0.2	Manufg.	Metal office furniture	0.5	Manufg.
Power transmission equipment	0.2	Manufg.	Power transmission equipment	0.5	Manufg.
Primary aluminum	0.2	Manufg.	Special industry machinery, nec	0.5	Manufg.
Screw machine and related products	0.2	Manufg.	Transformers	0.5	Manufg.
Special dies & tools & machine tool accessories	0.2	Manufg.	Maintenance of nonfarm buildings nec	0.4	Constr.
Sanitary services, steam supply, irrigation	0.2	Util.	Maintenance of railroads	0.4	Constr.
Eating & drinking places	0.2	Trade	Aircraft & missile engines & engine parts	0.4	Manufg.
Computer & data processing services	0.2	Services	Blowers & fans	0.4	Manufg.
U.S. Postal Service	0.2	Gov't	General industrial machinery, nec	0.4	Manufg.
Abrasive products	0.1	Manufg.	Heating equipment, except electric	0.4	Manufg.
Chemical preparations, nec	0.1	Manufg.	Household cooking equipment	0.4	Manufg.
Minerals, ground or treated	0.1	Manufg.	Household laundry equipment	0.4	Manufg.
Pipe, valves, & pipe fittings	0.1	Manufg.	Lighting fixtures & equipment	0.4	Manufg.
Small arms ammunition	0.1	Manufg.	Metal coating & allied services	0.4	Manufg.
Insurance carriers	0.1	Fin/R.E.	Metal doors, sash, & trim	0.4	Manufg.
Engineering, architectural, & surveying services	0.1	Services	Railroad equipment	0.4	Manufg.
Noncomparable imports	0.1	Foreign	Highway & street construction	0.3	Constr.
			Maintenance of electric utility facilities	0.3	Constr.
			Conveyors & conveying equipment	0.3	Manufg.
			Food products machinery	0.3	Manufg.
			Household appliances, nec	0.3	Manufg.
			Household refrigerators & freezers	0.3	Manufg.
			Internal combustion engines, nec	0.3	Manufg.
			Iron & steel foundries	0.3	Manufg.
			Metal household furniture	0.3	Manufg.
			Printing trades machinery	0.3	Manufg.
			Switchgear & switchboard apparatus	0.3	Manufg.
			Tanks & tank components	0.3	Manufg.
			Truck & bus bodies	0.3	Manufg.
			Truck trailers	0.3	Manufg.
			Turbines & turbine generator sets	0.3	Manufg.
			Welding apparatus, electric	0.3	Manufg.
			Construction of educational buildings	0.2	Constr.
			Construction of hospitals	0.2	Constr.
			Construction of stores & restaurants	0.2	Constr.
			Farm service facilities	0.2	Constr.
			Residential 1-unit structures, nonfarm	0.2	Constr.
			Sewer system facility construction	0.2	Constr.

Continued on next page.

INPUTS AND OUTPUTS FOR BLAST FURNACES & STEEL MILLS - Continued

Economic Sector or Industry Providing Inputs	%	Sector	Economic Sector or Industry Buying Outputs	%	Sector
			Warehouses	0.2	Constr.
			Ammunition, except for small arms, nec	0.2	Manufg.
			Cyclic crudes and organics	0.2	Manufg.
			Drapery hardware & blinds & shades	0.2	Manufg.
			Electric housewares & fans	0.2	Manufg.
			Electronic components nec	0.2	Manufg.
			Engine electrical equipment	0.2	Manufg.
			Furniture & fixtures, nec	0.2	Manufg.
			Hoists, cranes, & monorails	0.2	Manufg.
			Industrial furnaces & ovens	0.2	Manufg.
			Industrial trucks & tractors	0.2	Manufg.
			Lawn & garden equipment	0.2	Manufg.
			Machine tools, metal cutting types	0.2	Manufg.
			Manufacturing industries, nec	0.2	Manufg.
			Mechanical measuring devices	0.2	Manufg.
			Mining machinery, except oil field	0.2	Manufg.
			Motorcycles, bicycles, & parts	0.2	Manufg.
			Paper industries machinery	0.2	Manufg.
			Radio & TV communication equipment	0.2	Manufg.
			Service industry machines, nec	0.2	Manufg.
			Sporting & athletic goods, nec	0.2	Manufg.
			Steel springs, except wire	0.2	Manufg.
			Maintenance of water supply facilities	0.1	Constr.
			Residential additions/alterations, nonfarm	0.1	Constr.
			Residential garden apartments	0.1	Constr.
			Aircraft & missile equipment, nec	0.1	Manufg.
			Burial caskets & vaults	0.1	Manufg.
			Carburetors, pistons, rings, & valves	0.1	Manufg.
			Concrete products, nec	0.1	Manufg.
			Electronic computing equipment	0.1	Manufg.
			Hand saws & saw blades	0.1	Manufg.
			Industrial inorganic chemicals, nec	0.1	Manufg.
			Machine tools, metal forming types	0.1	Manufg.
			Metal sanitary ware	0.1	Manufg.
			Metalworking machinery, nec	0.1	Manufg.
			Power driven hand tools	0.1	Manufg.
			Public building furniture	0.1	Manufg.
			Textile machinery	0.1	Manufg.
			Transportation equipment, nec	0.1	Manufg.
			Typewriters & office machines, nec	0.1	Manufg.
			Woodworking machinery	0.1	Manufg.

Source: Benchmark Input-Output Accounts for the U.S. Economy, 1982, U.S. Department of Commerce, Washington, D.C., July 1991. Data, as reported in the source, are organized by the 1977 SIC structure in use in 1982 but have been matched, as closely as is possible, to the 1987 SIC structure used in this book.

OCCUPATIONS EMPLOYED BY SIC 331 - BLAST FURNACES AND BASIC STEEL PRODUCTS

Occupation	% of Total 1994	Change to 2005	Occupation	% of Total 1994	Change to 2005
Helpers, laborers, & material movers nec	7.2	-37.8	Machinists	1.8	-37.8
Machine tool cutting & forming etc. nec	6.2	-6.7	Machine feeders & offbearers	1.8	-44.0
Blue collar worker supervisors	5.8	-40.7	Assemblers, fabricators, & hand workers nec	1.8	-37.8
Industrial machinery mechanics	4.7	-0.4	Welders & cutters	1.4	-50.2
Crane & tower operators	3.5	-37.8	Sales & related workers nec	1.4	-37.8
Inspectors, testers, & graders, precision	3.2	-37.8	Freight, stock, & material movers, hand	1.4	-50.2
Metal & plastic machine workers nec	3.1	-33.8	Truck drivers light & heavy	1.3	-35.8
Electricians	2.6	-41.6	Machine tool cutting operators, metal & plastic	1.2	-48.2
Millwrights	2.4	-31.5	Production, planning, & expediting clerks	1.2	-37.8
Machine forming operators, metal & plastic	2.2	-31.6	Material moving equipment operators nec	1.1	-37.7
Maintenance repairers, general utility	1.9	-44.0	Machine operators nec	1.1	-45.1
Industrial truck & tractor operators	1.9	-37.8	General managers & top executives	1.1	-40.9
Furnace operators	1.9	-31.6			

Source: Industry-Occupation Matrix, Bureau of Labor Statistics. These data relate to one or more 3-digit SIC industry groups rather than to a single 4-digit SIC. The change reported for each occupation to the year 2005 is a percent of growth or decline as estimated by the Bureau of Labor Statistics. The abbreviation nec stands for 'not elsewhere classified'.

LOCATION BY STATE AND REGIONAL CONCENTRATION

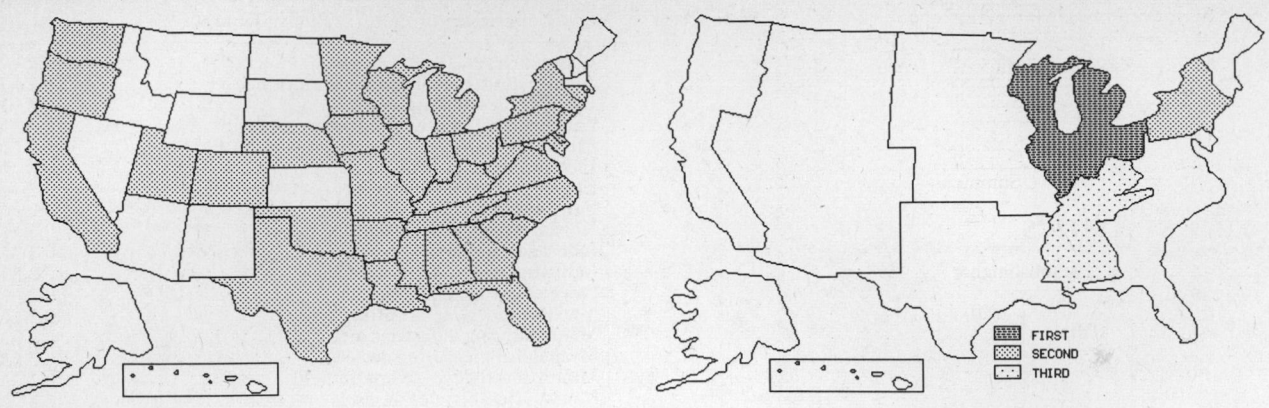

FIRST
SECOND
THIRD

INDUSTRY DATA BY STATE

| State | Establish-ments | Shipments | | | Employment | | | | Cost as % of Shipments | Investment per Employee ($) |
		Total ($ mil)	% of U.S.	Per Establ.	Total Number	% of U.S.	Per Establ.	Wages ($/hour)		
Indiana	12	8,445.2	20.0	703.8	32,500	19.1	2,708	20.99	53.1	16,772
Ohio	25	7,660.2	18.1	306.4	28,100	16.5	1,124	19.17	55.1	15,192
Pennsylvania	47	7,313.8	17.3	155.6	34,900	20.5	743	18.99	65.2	10,350
Michigan	10	2,754.9	6.5	275.5	9,700	5.7	970	19.64	67.3	-
Illinois	17	2,734.5	6.5	160.9	12,800	7.5	753	17.35	63.5	6,109
Alabama	11	1,798.6	4.3	163.5	6,300	3.7	573	16.78	61.4	7,127
Texas	12	1,464.8	3.5	122.1	5,000	2.9	417	16.17	65.1	10,960
Kentucky	8	1,023.8	2.4	128.0	5,000	2.9	625	18.25	49.0	-
New York	9	906.0	2.1	100.7	3,500	2.1	389	19.06	49.0	6,371
California	7	604.9	1.4	86.4	1,500	0.9	214	15.03	86.7	21,600
Arkansas	5	551.1	1.3	110.2	1,400	0.8	280	17.72	55.4	8,857
South Carolina	6	514.6	1.2	85.8	1,700	1.0	283	18.28	79.7	31,000
Maryland	5	(D)	-	-	7,500 *	4.4	1,500	-	-	-
New Jersey	4	(D)	-	-	750 *	0.4	188	-	-	-
Tennessee	4	(D)	-	-	750 *	0.4	188	-	-	-
West Virginia	4	(D)	-	-	7,500 *	4.4	1,875	-	-	-
Virginia	3	(D)	-	-	750 *	0.4	250	-	-	-
Florida	2	(D)	-	-	375 *	0.2	188	-	-	-
Georgia	2	(D)	-	-	750 *	0.4	375	-	-	-
Minnesota	2	(D)	-	-	375 *	0.2	188	-	-	-
Mississippi	2	(D)	-	-	175 *	0.1	88	-	-	-
Missouri	2	(D)	-	-	750 *	0.4	375	-	-	-
Oklahoma	2	(D)	-	-	375 *	0.2	188	-	-	-
Oregon	2	(D)	-	-	1,750 *	1.0	875	-	-	-
Utah	2	(D)	-	-	3,750 *	2.2	1,875	-	-	-
Washington	2	(D)	-	-	375 *	0.2	188	-	-	-
Colorado	1	(D)	-	-	1,750 *	1.0	1,750	-	-	-
Connecticut	1	(D)	-	-	175 *	0.1	175	-	-	-
Delaware	1	(D)	-	-	375 *	0.2	375	-	-	-
Iowa	1	(D)	-	-	375 *	0.2	375	-	-	-
Louisiana	1	(D)	-	-	375 *	0.2	375	-	-	-
Nebraska	1	(D)	-	-	375 *	0.2	375	-	-	-
North Carolina	1	(D)	-	-	175 *	0.1	175	-	-	-
Rhode Island	1	(D)	-	-	175 *	0.1	175	-	-	-
Wisconsin	1	(D)	-	-	175 *	0.1	175	-	-	-

Source: 1992 *Economic Census*. The states are in descending order of shipments or establishments (if shipment data are missing for the majority). The symbol (D) appears when data are withheld to prevent disclosure of competitive information. States marked with (D) are sorted by number of establishments. A dash (-) indicates that the data element cannot be calculated; * indicates the midpoint of a range.

3313 - ELECTROMETALLURGICAL PRODUCTS

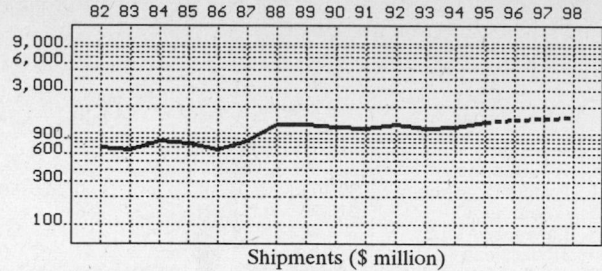

Shipments ($ million)

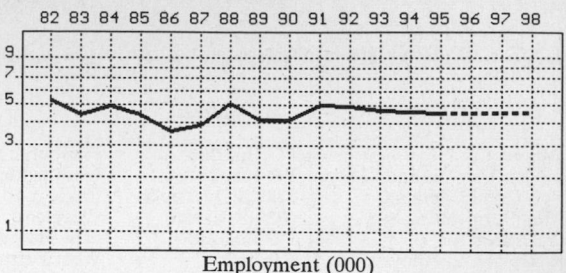

Employment (000)

GENERAL STATISTICS

| Year | Com-panies | Establishments | | Employment | | | Compensation | | Production ($ million) | | | |
		Total	with 20 or more employees	Total (000)	Production Workers (000)	Hours (Mil)	Payroll ($ mil)	Wages ($/hr)	Cost of Materials	Value Added by Manufacture	Value of Shipments	Capital Invest.
1982	31	41	27	5.3	3.9	7.0	123.5	12.21	494.2	180.3	707.5	23.8
1983		37	26	4.4	3.3	6.8	100.9	10.41	428.8	192.5	661.5	11.8
1984		33	25	4.9	3.6	7.6	126.5	11.80	525.7	331.5	835.8	18.6
1985		30	24	4.4	3.3	6.7	117.3	12.03	481.9	313.9	783.1	32.4
1986		30	25	3.6	2.6	5.5	99.4	12.13	433.1	207.1	667.9	33.0
1987	25	30	24	3.9	2.8	5.9	112.6	12.66	511.1	295.6	843.8	24.7
1988		27	24	5.1	3.6	7.8	154.9	13.33	769.1	508.4	1,275.8	41.0
1989		27	23	4.1	3.8	8.6	170.7	13.24	782.6	519.7	1,269.4	40.5
1990		30	24	4.1	3.6	7.8	168.6	13.82	734.4	431.2	1,180.4	37.0
1991		29	24	5.0	3.4	7.4	165.2	14.39	743.7	373.9	1,145.6	26.4
1992	31	37	28	4.9	3.5	7.2	171.2	15.60	816.9	414.2	1,263.8	36.9
1993		36	29	4.7	3.3	7.3	167.9	15.18	764.4	382.0	1,165.2	23.9
1994		30P	26P	4.6	3.3	7.5	170.1	15.17	803.8	400.9	1,197.8	25.3
1995		30P	26P	4.5P	3.4P	7.7P	185.8P	15.84P	921.6P	459.6P	1,373.3P	34.6P
1996		29P	26P	4.6P	3.4P	7.7P	192.0P	16.21P	957.4P	477.5P	1,426.6P	35.4P
1997		29P	26P	4.6P	3.3P	7.8P	198.3P	16.58P	993.2P	495.4P	1,480.0P	36.2P
1998		29P	26P	4.6P	3.3P	7.9P	204.5P	16.96P	1,029.0P	513.2P	1,533.4P	37.0P

Sources: 1982, 1987, 1992 *Economic Census; Annual Survey of Manufactures*, 83-86, 88-91, 93-94. Establishment counts for non-Census years are from *County Business Patterns*; establishment values for 83-84 are extrapolations. 'P's show projections by the editors. Industries reclassified in 87 will not have data for prior years.

INDICES OF CHANGE

| Year | Com-panies | Establishments | | Employment | | | Compensation | | Production ($ million) | | | |
		Total	with 20 or more employees	Total (000)	Production Workers (000)	Hours (Mil)	Payroll ($ mil)	Wages ($/hr)	Cost of Materials	Value Added by Manufacture	Value of Shipments	Capital Invest.
1982	100	111	96	108	111	97	72	78	60	44	56	64
1983		100	93	90	94	94	59	67	52	46	52	32
1984		89	89	100	103	106	74	76	64	80	66	50
1985		81	86	90	94	93	69	77	59	76	62	88
1986		81	89	73	74	76	58	78	53	50	53	89
1987	81	81	86	80	80	82	66	81	63	71	67	67
1988		73	86	104	103	108	90	85	94	123	101	111
1989		73	82	84	109	119	100	85	96	125	100	110
1990		81	86	84	103	108	98	89	90	104	93	100
1991		78	86	102	97	103	96	92	91	90	91	72
1992	100	100	100	100	100	100	100	100	100	100	100	100
1993		97	104	96	94	101	98	97	94	92	92	65
1994		81P	92P	94	94	104	99	97	98	97	95	69
1995		80P	93P	93P	96P	107P	109P	102P	113P	111P	109P	94P
1996		80P	93P	93P	96P	108P	112P	104P	117P	115P	113P	96P
1997		79P	93P	93P	96P	109P	116P	106P	122P	120P	117P	98P
1998		78P	94P	93P	96P	110P	119P	109P	126P	124P	121P	100P

Sources: Same as General Statistics. Values reflect change from the base year, 1992. Values above 100 mean greater than 92, values below 100 mean less than 92, and a value of 100 in the 82-91 or 93-98 period means same as 92. 'P's mark projections by the editors.

SELECTED RATIOS

For 1994	Avg. of All Manufact.	Analyzed Industry	Index	For 1994	Avg. of All Manufact.	Analyzed Industry	Index
Employees per Establishment	49	153	312	Value Added per Production Worker	134,084	121,485	91
Payroll per Establishment	1,500,273	5,652,870	377	Cost per Establishment	5,045,178	26,712,387	529
Payroll per Employee	30,620	36,978	121	Cost per Employee	102,970	174,739	170
Production Workers per Establishment	34	110	320	Cost per Production Worker	146,988	243,576	166
Wages per Establishment	853,319	3,781,042	443	Shipments per Establishment	9,576,895	39,806,042	416
Wages per Production Worker	24,861	34,477	139	Shipments per Employee	195,460	260,391	133
Hours per Production Worker	2,056	2,273	111	Shipments per Production Worker	279,017	362,970	130
Wages per Hour	12.09	15.17	125	Investment per Establishment	321,011	840,785	262
Value Added per Establishment	4,602,255	13,322,961	289	Investment per Employee	6,552	5,500	84
Value Added per Employee	93,930	87,152	93	Investment per Production Worker	9,352	7,667	82

Sources: Same as General Statistics. The 'Average of All Manufacturing' column represents the average of all manufacturing industries reported for the most recent complete year available. The Index shows the relationship between the Average and the Analyzed Industry. For example, 100 means that they are equal; 500 that the Analyzed Industry is five times the average; 50 means that the Analyzed Industry is half the national average. The abbreviation 'na' is used to show that data are 'not available'.

LEADING COMPANIES   Number shown: 12   Total sales ($ mil): 651   Total employment (000): 4.0

Company Name	Address				CEO Name	Phone	Co. Type	Sales ($ mil)	Empl. (000)
Elkem Metals Co	PO Box 266	Pittsburgh	PA	15230	Jerry Hurt	412-778-3600	S	190	2.0
Steel of West Virginia Inc	PO Box 2547	Huntington	WV	25726	R L Bunting Jr	304-696-8200	P	124	0.6
Thompson Creek Metals Co	5241 S Quebec St	Englewood	CO	80111	F Stephen Mooney	303-740-9022	R	100	0.3
Globe Metallurgical Inc	6450 Rockside	Cleveland	OH	44131	Arden Sims	216-328-0145	R	56*	0.3
American Alloys Inc	PO Box 218	New Haven	WV	25265	Lawrence McKee	304-882-2161	R	50	0.2
Reactive Metal and Alloys Corp	Rte 168	West Pittsburg	PA	16160	JR Jackman	412-535-4357	R	45	0.2
Macalloy Corp	PO Box 130	Charleston	SC	29402	Norris McFarlane	803-722-8355	R	37*	0.2
Reading Alloys Inc	PO Box 53	Robesonia	PA	19551	Marjorie Perfect	215-693-5822	R	24	<0.1
Delta Centrifugal Corp	PO Box 1043	Temple	TX	76501	Mark Anderson	817-773-9055	R	12	0.1
North Metal and Chemical Co	PO Box 1904	York	PA	17405	Fred C Fay II	717-845-8646	R	6	<0.1
Buffalo Tungsten Inc	PO Box 397	Depew	NY	14043	Ralph Showalter	716-683-9170	R	4	<0.1
Applied Indust Materials	Pk Ridge	Pittsburgh	PA	15275	Charles W Kopec	412-788-1860	S	3*	<0.1

Source: Ward's Business Directory of U.S. Private and Public Companies, Volumes 1 and 2, 1996. The company type code used is as follows: P - Public, R - Private, S - Subsidiary, D - Division, J - Joint Venture, A - Affiliate, G - Group. Sales are in millions of dollars, employees are in thousands. An asterisk (*) indicates an estimated sales volume. The symbol < stands for 'less than'. Company names and addresses are truncated, in some cases, to fit into the available space.

MATERIALS CONSUMED

Material		Quantity	Delivered Cost ($ million)
Materials, ingredients, containers, and supplies		(X)	596.5
All other steel shapes and forms	1,000 s tons	(D)	(D)
Pig iron shapes and forms	1,000 s tons	(D)	(D)
Ferromanganese, silicomanganese, and manganese shapes and forms	1,000 s tons	(D)	(D)
Ferrochromium shapes and forms	1,000 s tons	5.9	3.7
Ferrosilicon (more than 8 percent silicon) shapes and forms	1,000 s tons	(D)	(D)
Other ferroalloy shapes and forms	1,000 s tons	(D)	(D)
All other ferrous shapes and forms	1,000 s tons	(D)	(D)
Aluminum and aluminum-base alloy shapes and forms	1,000 s tons	(S)	18.3
Nickel (except ferronickel) shapes and forms	1,000 s tons	3.0	21.4
All other nonferrous shapes and forms	1,000 s tons	142.6	41.2
Gray iron ingot molds and stools	1,000 s tons	(D)	(D)
All other iron and steel castings		(X)	0.6
Clay refractories		(X)	2.7
Nonclay refractories		(X)	6.3
Limestone fluxes	1,000 s tons	(D)	(D)
Lime fluxes, including quicklime	1,000 s tons	(D)	(D)
Dead-burned dolomite fluxes	1,000 s tons	(D)	(D)
Fluorspar fluxes		(X)	(D)
Other fluxes		(X)	1.2
All other stone, clay, glass, and concrete products		(X)	(D)
Industrial chemicals (except sulfuric acid and oxygen)		(X)	10.5
Oxygen (including high and low purity) (liquid oxygen should be converted to its gaseous equivalent)	mil cu ft	(D)	(D)
Sulfuric acid (new and spent) (100 percent H2SO4)	1,000 s tons	(S)	2.8
All other chemicals and allied products		(X)	4.0
Mining crude iron ore and iron ore concentrates, including pelletized and manganiferous (gross weight)	1,000 s tons	(D)	(D)
Mining manganese ore (including 10 percent manganese) and ferroalloy ore		(X)	(D)
Coal used in the production of coke	1,000 s tons	(S)	12.8
Iron and steel scrap, excluding home scrap	1,000 s tons	335.0	34.8
Lubricating oils and greases and other petroleum products		(X)	0.9
Industrial dies, molds, jigs, and fixtures		(X)	0.7
Carbon and graphite electrodes		(X)	23.7
All other materials and components, parts, containers, and supplies		(X)	237.3
Materials, ingredients, containers, and supplies, nsk		(X)	18.7

Source: 1992 Economic Census. Explanation of symbols used: (D): Withheld to avoid disclosure of competitive data; na: Not available; (S): Withheld because statistical norms were not met; (X): Not applicable; (Z): Less than half the unit shown; nec: Not elsewhere classified; nsk: Not specified by kind; - : zero; * : 10-19 percent estimated; ** : 20-29 percent estimated.

PRODUCT SHARE DETAILS

Product or Product Class	% Share	Product or Product Class	% Share
Electrometallurgical products	100.00	Other ferroalloys	(D)
Ferrochromium (including briquets), ferrochromium silicon, exothermic chromium additives, and other chromium alloys	(D)	Ferromanganese (including briquets), manganese metal, and silicomanganese	(D)
		Superalloys	78.59
Ferrochromium (including briquets), ferrochromium silicon, exothermic chromium additives, and other chromium alloys	(D)	Other ferroalloys (including silvery iron and spiegeleisen)	35.38
		Other products made in electric and other furnaces	51.32
		Other ferroalloys, nsk	5.85
Ferrosilicon (including briquets) and other silicon alloys	29.55	Electrometallurgical products, nsk	0.83

Source: 1992 Economic Census. The values shown are percent of total shipments in an industry. Values of indented subcategories are summed in the main heading. The symbol (D) appears when data are withheld to prevent disclosure of competitive information. The abbreviation nsk stands for 'not specified by kind' and nec for 'not elsewhere classified'.

INPUTS AND OUTPUTS FOR ELECTROMETALLURGICAL PRODUCTS

Economic Sector or Industry Providing Inputs	%	Sector	Economic Sector or Industry Buying Outputs	%	Sector
Imports	43.6	Foreign	Blast furnaces & steel mills	70.3	Manufg.
Electric services (utilities)	15.4	Util.	Iron & steel foundries	17.8	Manufg.
Electrometallurgical products	7.3	Manufg.	Electrometallurgical products	5.3	Manufg.
Iron & ferroalloy ores	4.5	Mining	Exports	3.9	Foreign
Noncomparable imports	4.0	Foreign	Nonferrous castings, nec	2.1	Manufg.
Nonferrous metal ores, except copper	3.3	Mining	Aluminum castings	0.5	Manufg.
Gas production & distribution (utilities)	1.9	Util.			
Maintenance of nonfarm buildings nec	1.8	Constr.			
Carbon & graphite products	1.7	Manufg.			
Wholesale trade	1.6	Trade			
Motor freight transportation & warehousing	1.4	Util.			
Scrap	1.4	Scrap			
Primary nonferrous metals, nec	1.3	Manufg.			
Railroads & related services	1.1	Util.			
Water transportation	0.9	Util.			
Nonmetallic mineral products, nec	0.8	Manufg.			
Petroleum refining	0.8	Manufg.			
Metal heat treating	0.6	Manufg.			
Chemical & fertilizer mineral	0.5	Mining			
Coal	0.4	Mining			
Machinery, except electrical, nec	0.4	Manufg.			
Mechanical measuring devices	0.4	Manufg.			
Miscellaneous repair shops	0.4	Services			
Wood pallets & skids	0.3	Manufg.			
Communications, except radio & TV	0.3	Util.			
Eating & drinking places	0.3	Trade			
Banking	0.3	Fin/R.E.			
Equipment rental & leasing services	0.3	Services			
Abrasive products	0.2	Manufg.			
Blast furnaces & steel mills	0.2	Manufg.			
Special dies & tools & machine tool accessories	0.2	Manufg.			
Sanitary services, steam supply, irrigation	0.2	Util.			
Advertising	0.2	Services			
U.S. Postal Service	0.2	Gov't			
Sand & gravel	0.1	Mining			
Royalties	0.1	Fin/R.E.			
Engineering, architectural, & surveying services	0.1	Services			
Legal services	0.1	Services			

Source: Benchmark Input-Output Accounts for the U.S. Economy, 1982, U.S. Department of Commerce, Washington, D.C., July 1991. Data, as reported in the source, are organized by the 1977 SIC structure in use in 1982 but have been matched, as closely as is possible, to the 1987 SIC structure used in this book.

OCCUPATIONS EMPLOYED BY SIC 331 - BLAST FURNACES AND BASIC STEEL PRODUCTS

Occupation	% of Total 1994	Change to 2005	Occupation	% of Total 1994	Change to 2005
Helpers, laborers, & material movers nec	7.2	-37.8	Machinists	1.8	-37.8
Machine tool cutting & forming etc. nec	6.2	-6.7	Machine feeders & offbearers	1.8	-44.0
Blue collar worker supervisors	5.8	-40.7	Assemblers, fabricators, & hand workers nec	1.8	-37.8
Industrial machinery mechanics	4.7	-0.4	Welders & cutters	1.4	-50.2
Crane & tower operators	3.5	-37.8	Sales & related workers nec	1.4	-37.8
Inspectors, testers, & graders, precision	3.2	-37.8	Freight, stock, & material movers, hand	1.4	-50.2
Metal & plastic machine workers nec	3.1	-33.8	Truck drivers light & heavy	1.3	-35.8
Electricians	2.6	-41.6	Machine tool cutting operators, metal & plastic	1.2	-48.2
Millwrights	2.4	-31.5	Production, planning, & expediting clerks	1.2	-37.8
Machine forming operators, metal & plastic	2.2	-31.6	Material moving equipment operators nec	1.1	-37.7
Maintenance repairers, general utility	1.9	-44.0	Machine operators nec	1.1	-45.1
Industrial truck & tractor operators	1.9	-37.8	General managers & top executives	1.1	-40.9
Furnace operators	1.9	-31.6			

Source: Industry-Occupation Matrix, Bureau of Labor Statistics. These data relate to one or more 3-digit SIC industry groups rather than to a single 4-digit SIC. The change reported for each occupation to the year 2005 is a percent of growth or decline as estimated by the Bureau of Labor Statistics. The abbreviation nec stands for 'not elsewhere classified'.

LOCATION BY STATE AND REGIONAL CONCENTRATION

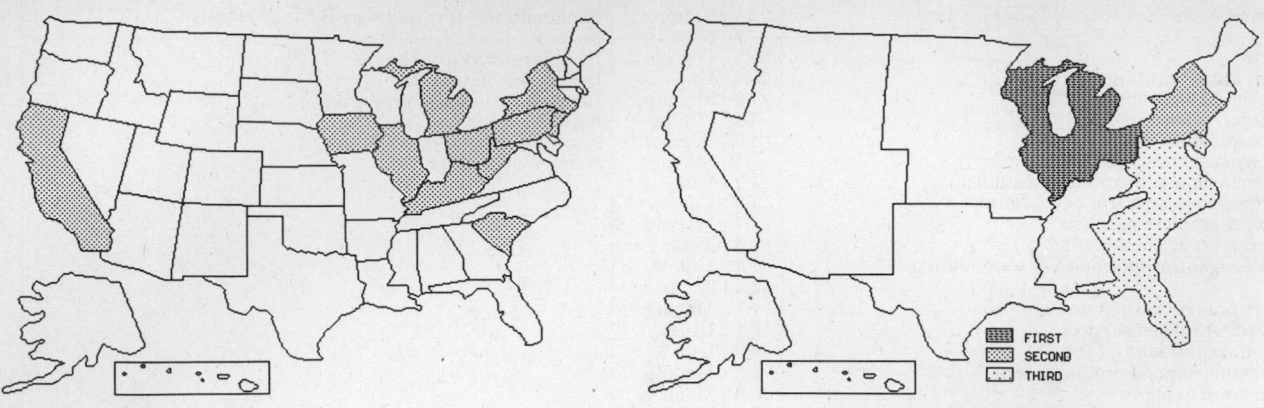

INDUSTRY DATA BY STATE

| State | Establish-ments | Shipments | | | Employment | | | | Cost as % of Shipments | Investment per Employee ($) |
		Total ($ mil)	% of U.S.	Per Establ.	Total Number	% of U.S.	Per Establ.	Wages ($/hour)		
Ohio	9	483.8	38.3	53.8	2,000	40.8	222	16.48	64.1	6,500
Pennsylvania	4	46.7	3.7	11.7	200	4.1	50	12.25	29.1	-
California	3	(D)	-	-	175 *	3.6	58	-	-	-
Michigan	3	(D)	-	-	175 *	3.6	58	-	-	-
New Jersey	3	(D)	-	-	750 *	15.3	250	-	-	-
New York	3	(D)	-	-	375 *	7.7	125	-	-	-
West Virginia	3	(D)	-	-	750 *	15.3	250	-	-	-
Illinois	2	(D)	-	-	175 *	3.6	88	-	-	-
Iowa	1	(D)	-	-	175 *	3.6	175	-	-	-
Kentucky	1	(D)	-	-	175 *	3.6	175	-	-	-
South Carolina	1	(D)	-	-	175 *	3.6	175	-	-	-

Source: 1992 *Economic Census*. The states are in descending order of shipments or establishments (if shipment data are missing for the majority). The symbol (D) appears when data are withheld to prevent disclosure of competitive information. States marked with (D) are sorted by number of establishments. A dash (-) indicates that the data element cannot be calculated; * indicates the midpoint of a range.

3315 - STEEL WIREDRAWING & STEEL NAILS

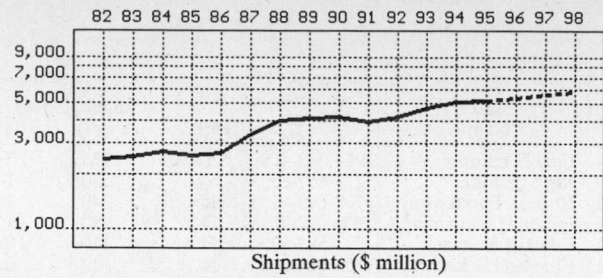

Shipments ($ million)

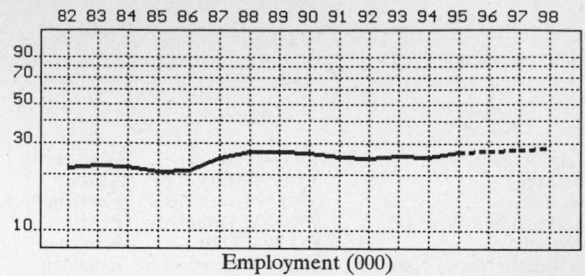

Employment (000)

GENERAL STATISTICS

Year	Com-panies	Establishments		Employment			Compensation		Production ($ million)			
		Total	with 20 or more employees	Total (000)	Production Workers (000)	Hours (Mil)	Payroll ($ mil)	Wages ($/hr)	Cost of Materials	Value Added by Manufacture	Value of Shipments	Capital Invest.
1982	251	311	185	22.0	16.3	31.0	442.9	9.57	1,434.3	938.3	2,415.4	61.1
1983		312	187	22.5	17.1	32.6	461.5	9.74	1,494.8	981.6	2,490.9	74.6
1984		313	189	22.3	17.2	34.0	491.9	10.08	1,578.0	1,131.0	2,689.6	95.2
1985		315	192	20.9	15.9	31.6	470.8	10.30	1,542.1	998.5	2,549.8	80.5
1986		317	195	21.1	15.8	32.5	488.7	10.32	1,612.6	1,007.7	2,624.6	63.6
1987	274	343	206	24.7	18.7	40.1	592.0	10.16	1,901.5	1,422.9	3,330.3	73.2
1988		346	205	27.0	20.7	44.0	678.7	10.56	2,357.0	1,696.7	4,019.9	98.5
1989		344	214	27.1	21.3	44.4	680.6	10.84	2,497.8	1,651.1	4,123.3	207.9
1990		342	212	26.6	21.4	44.5	679.6	11.07	2,462.1	1,723.4	4,179.7	208.3
1991		348	210	25.2	19.9	40.8	655.0	11.52	2,331.5	1,597.5	3,944.3	111.2
1992	269	349	208	24.7	19.0	39.9	691.6	12.21	2,455.1	1,727.1	4,206.4	124.5
1993		362	221	25.5	19.9	41.5	735.7	12.64	2,717.1	2,042.7	4,745.4	119.9
1994		364P	222P	25.3	19.5	42.6	749.3	12.33	2,733.1	2,367.7	5,064.0	170.1
1995		369P	225P	26.9P	21.2P	45.9P	794.9P	12.62P	2,783.4P	2,411.2P	5,157.1P	172.7P
1996		374P	228P	27.3P	21.6P	46.9P	822.6P	12.87P	2,905.9P	2,517.4P	5,384.1P	181.1P
1997		379P	231P	27.7P	21.9P	48.0P	850.2P	13.12P	3,028.4P	2,623.5P	5,611.2P	189.4P
1998		383P	235P	28.1P	22.3P	49.1P	877.8P	13.37P	3,150.9P	2,729.7P	5,838.2P	197.7P

Sources: 1982, 1987, 1992 *Economic Census*; *Annual Survey of Manufactures*, 83-86, 88-91, 93-94. Establishment counts for non-Census years are from *County Business Patterns*; establishment values for 83-84 are extrapolations. 'P's show projections by the editors. Industries reclassified in 87 will not have data for prior years.

INDICES OF CHANGE

Year	Com-panies	Establishments		Employment			Compensation		Production ($ million)			
		Total	with 20 or more employees	Total (000)	Production Workers (000)	Hours (Mil)	Payroll ($ mil)	Wages ($/hr)	Cost of Materials	Value Added by Manufacture	Value of Shipments	Capital Invest.
1982	93	89	89	89	86	78	64	78	58	54	57	49
1983		89	90	91	90	82	67	80	61	57	59	60
1984		90	91	90	91	85	71	83	64	65	64	76
1985		90	92	85	84	79	68	84	63	58	61	65
1986		91	94	85	83	81	71	85	66	58	62	51
1987	102	98	99	100	98	101	86	83	77	82	79	59
1988		99	99	109	109	110	98	86	96	98	96	79
1989		99	103	110	112	111	98	89	102	96	98	167
1990		98	102	108	113	112	98	91	100	100	99	167
1991		100	101	102	105	102	95	94	95	92	94	89
1992	100	100	100	100	100	100	100	100	100	100	100	100
1993		104	106	103	105	104	106	104	111	118	113	96
1994		104P	107P	102	103	107	108	101	111	137	120	137
1995		106P	108P	109P	112P	115P	115P	103P	113P	140P	123P	139P
1996		107P	110P	111P	114P	118P	119P	105P	118P	146P	128P	145P
1997		108P	111P	112P	115P	120P	123P	107P	123P	152P	133P	152P
1998		110P	113P	114P	117P	123P	127P	109P	128P	158P	139P	159P

Sources: Same as General Statistics. Values reflect change from the base year, 1992. Values above 100 mean greater than 92, values below 100 mean less than 92, and a value of 100 in the 82-91 or 93-98 period means same as 92. 'P's mark projections by the editors.

SELECTED RATIOS

For 1994	Avg. of All Manufact.	Analyzed Industry	Index	For 1994	Avg. of All Manufact.	Analyzed Industry	Index
Employees per Establishment	49	69	142	Value Added per Production Worker	134,084	121,421	91
Payroll per Establishment	1,500,273	2,056,462	137	Cost per Establishment	5,045,178	7,501,023	149
Payroll per Employee	30,620	29,617	97	Cost per Employee	102,970	108,028	105
Production Workers per Establishment	34	54	156	Cost per Production Worker	146,988	140,159	95
Wages per Establishment	853,319	1,441,576	169	Shipments per Establishment	9,576,895	13,898,204	145
Wages per Production Worker	24,861	26,936	108	Shipments per Employee	195,460	200,158	102
Hours per Production Worker	2,056	2,185	106	Shipments per Production Worker	279,017	259,692	93
Wages per Hour	12.09	12.33	102	Investment per Establishment	321,011	466,841	145
Value Added per Establishment	4,602,255	6,498,179	141	Investment per Employee	6,552	6,723	103
Value Added per Employee	93,930	93,585	100	Investment per Production Worker	9,352	8,723	93

Sources: Same as General Statistics. The 'Average of All Manufacturing' column represents the average of all manufacturing industries reported for the most recent complete year available. The Index shows the relationship between the Average and the Analyzed Industry. For example, 100 means that they are equal; 500 that the Analyzed Industry is five times the average; 50 means that the Analyzed Industry is half the national average. The abbreviation 'na' is used to show that data are 'not available'.

LEADING COMPANIES   Number shown: **65**   Total sales ($ mil): **3,121**   Total employment (000): **15.8**

Company Name	Address				CEO Name	Phone	Co. Type	Sales ($ mil)	Empl. (000)
BICC Cables Corp	1 Crosfield Av	West Nyack	NY	10994	Carl E Painter	914-353-4000	S	735	3.1
Bekaert Corp	3200 W Market St	Akron	OH	44333	Jan Smolders	216-867-3325	S	350*	1.4
Duo-Fast Corp	3702 N River Rd	Franklin Park	IL	60131	Robert Torstenson	708-678-0100	R	240	1.5
Keystone Steel and Wire Co	7000 SW Adams St	Peoria	IL	61641	James H Kauffman	309-697-7020	D	190	1.6
Sandvik Steel Co	PO Box 1220	Scranton	PA	18501	Edward Nuzzaci	717-587-5191	D	125	0.5
Dickson Weatherproof Nail Co	1900 Greenwood St	Evanston	IL	60201	Charles Dickson	708-864-2060	R	110	0.1
Florida Wire and Cable Inc	PO Box 6835	Jacksonville	FL	32236	John E Burnsworth	904-781-9224	S	110	0.4
American Spring Wire Corp	PO Box 46510	Bedford Hts	OH	44146	Joseph L Kwasny	216-292-4620	R	100	0.5
Brand-Rex	1600 W Main St	Willimantic	CT	06226	John Macchia	203-456-8000	D	90	0.8
Industrial Wire Products Corp	PO Box 1710	Pomona	CA	91769	Bruce Yost	310-945-3556	R	80	0.1
Techalloy Company Inc	370 Franklin Tpk	Mahwah	NJ	07430	P L Maitrepierre	201-529-0900	S	80	0.3
Taubensee Steel and Wire Co	600 Diens Dr	Wheeling	IL	60090	Dale T Taubensee	708-459-5100	R	60	0.2
MCM Enterprises Inc	PO Box 392	Crawfordsville	IN	47933	EM Marberg	317-362-2200	R	55	0.5
Indiana Steel and Wire Co	PO Box 2647	Muncie	IN	47307	A Randy Nahvi	317-288-3601	R	50	0.3
Walker Wire and Steel Corp	660 E 10 Mile Rd	Ferndale	MI	48220	JD King	313-399-4800	S	50	0.2
Industrial Alloys Inc	PO Box 1839	Pomona	CA	91769	Bruce Yost	909-594-7511	R	49	0.2
Maryland Specialty Wire Inc	100 Cockeysville Rd	Cockeysville	MD	21030	R Nash Jr	410-785-2500	S	42	0.3
Midstates Wire	PO Box 392	Crawfordsville	IN	47933	EM Marberg	317-362-2200	S	40	0.4
O and K American Corp	4630 W 55th St	Chicago	IL	60632	Kazuta Oku	312-767-2500	S	34*	0.2
Sherman Wire	PO Box 729	Sherman	TX	75091	Billy J Johnson	903-893-0191	D	33	0.2
Etherington Industries	2 Clark St	Old Saybrook	CT	06475	G Etherington	203-388-5629	R	30	0.3
Grayline Housewares	455 Kehoe Blv	Carol Stream	IL	60188	Fred Rosen	708-682-3330	R	30*	0.2
Mid-South Wire Company Inc	1040 Visco Dr	Nashville	TN	37210	John T Johnson Sr	615-244-5258	R	30	<0.1
BICC Cables Corp	PO Box 391	Yonkers	NY	10702	Steve Rigby	914-963-8200	D	28*	0.1
Atlantic Wire Co	1 Church St	Branford	CT	06405	Robert J Lawlor	203-488-8331	S	25	0.1
Felsted	PO Box 68	Holmesville	OH	44633	Richard Trowman	216-279-3711	D	22*	0.1
Kanthal Corp	PO Box 281	Bethel	CT	06801	E Roger Clark	203-744-1440	S	20	0.1
Mount Joy Wire Corp	1000 E Main St	Mount Joy	PA	17552	Fred B Krieger	717-653-1461	R	20	0.1
Superior Rope and Sling Inc	PO Box 93702	Atlanta	GA	30377	Milton Mitchell	404-351-6141	D	17	<0.1
Semiconductor Packaging Mat	431 Sayette Av	Mamaroneck	NY	10543	Ken Hoth	914-698-5353	P	16	<0.1
Loos and Company Inc	Rte 101 & Cable Rd	Pomfret	CT	06258	August W Loos	203-928-7981	R	16	0.2
Sherman Wire of Caldwell Inc	PO Box 879	Caldwell	TX	77836	Nick Gilley	409-567-7916	S	16	<0.1
A-1 Wire Tech Co	840 39th Av	Rockford	IL	61109	William Pigott	815-226-0477	R	15	<0.1
Consolidated Products Corp	PO Box 67	Idyllwild	CA	92549	Gordon W Brown	909-659-2183	R	15	<0.1
Delta Wire Corp	110 Industrial Dr	Clarksdale	MS	38614	GF Walker	601-627-7853	R	15	0.1
City Wire Cloth Inc	PO Box 1410	Paramount	CA	90723	Bob Milmoe	310-630-8050	S	14	<0.1
Branford Wire & Mfg Co	PO Box 677	Mountain H	NC	28758	Richard Harcke	704-692-5791	R	12	0.1
Florida Steel Corp	Rte 4	Lancaster	SC	29720	Richard Danhoff	803-285-8444	D	12*	<0.1
Lynn Electronics Corp	915 Pennsylvania	Feasterville	PA	19053	Michael Rosen	215-355-8200	R	11	0.2
Midway Wire Inc	4630 W 55th St	Chicago	IL	60632	Kazuta Oku	312-767-2500	S	11	<0.1
Wire Products Co	PO Box 130	Hortonville	WI	54944	James J Monroe	414-779-4544	D	10	<0.1
WH Maze Co	PO Box 449	Peru	IL	61354	Peter G Loveland	815-223-8290	D	10	0.1
WW Cross	PO Box 365	Jaffrey	NH	03452	John Morton	603-532-8332	D	10*	0.1
Wilson Steel and Wire Co	4840 S Western Av	Chicago	IL	60609	EM Marberg	312-523-1221	D	9	<0.1
Molecu Wire Corp	PO Box 495	Farmingdale	NJ	07727	TH Horsman	908-938-9473	R	8	<0.1
Chicago Steel and Wire	10257 S Torrence	Chicago	IL	60617	E Marberg	312-768-2140	D	7*	<0.1
Phoenix Wire Cloth Inc	PO Box 610	Troy	MI	48099	John D Holmes	810-585-6350	R	7*	<0.1
Northeast Wire & Machine Prod	PO Box 9007	Forestville	CT	06011	A Weaver	203-589-2700	R	7	<0.1
Cavert/Ace Baling Wire Co	PO Box 487	Levittown	PA	19058	Ted Granson	215-949-2299	D	6	<0.1
WH Maze Co	PO Box 449	Peru	IL	61354	Peter G Loveland	815-223-8290	R	6	0.1
Ace Fence Co	15135 Salt Lake Av	City of Industry	CA	91746	America Tang	818-333-0727	R	5	<0.1
ARI Industries Inc	381 ARI Ct	Addison	IL	60101	DJ Mackenzie	708-953-9100	S	5	<0.1
WJ Young	181 Elliott St	Beverly	MA	01915	PV Kenney	508-922-5552	R	5	<0.1
Anchor Post Products of Texas	803 E Whitney St	Houston	TX	77022	Edward Brooks	713-699-0723	R	4*	<0.1
Great Northern Mfg Corp	100 Eastern Av	Chelsea	MA	02150	A M Gesamondo	617-284-4444	R	4	<0.1
Wachusett Wire Company Inc	641 Cambridge St	Worcester	MA	01610	Richard Heeps	508-757-6076	R	4	<0.1
Hamden Metal Service Co	2 Broadway	Hamden	CT	06518	JM Hirsch	203-281-1522	R	3	<0.1
TWP Inc	2831 10th St	Berkeley	CA	94710	Colleen Heffley	510-548-4434	R	3*	<0.1
Tremont Nail	PO Box 111	Wareham	MA	02571	William Driscoll	508-295-0038	D	3	<0.1
ES Products Inc	PO Box 810	Bristol	RI	02809	John Barker	401-253-8600	R	2	<0.1
Ottawa Steel and Wire Inc	PO Box 821	Ottawa	IL	61350	Kazuta Oku	815-434-5581	S	2*	<0.1
Micro Wire Products Inc	156 Porter St	East Boston	MA	02128	Arnold Jacobson	617-569-2000	R	2	<0.1
Radcliff Wire Inc	PO Box 603	Bristol	CT	06011	JE Radcliff	203-583-1305	R	1	<0.1
TKM Specialty Fasteners Inc	6219 Hwy 62	Trinity	NC	27370	Terry G Dills	910-431-1191	R	1	<0.1
Cavaler Wire Products Inc	5914 Rosa Pks Blvd	Detroit	MI	48208	R Cavaler	313-871-8180	R	0	<0.1

Source: *Ward's Business Directory of U.S. Private and Public Companies*, Volumes 1 and 2, 1996. The company type code used is as follows: P - Public, R - Private, S - Subsidiary, D - Division, J - Joint Venture, A - Affiliate, G - Group. Sales are in millions of dollars, employees are in thousands. An asterisk (*) indicates an estimated sales volume. The symbol < stands for 'less than'. Company names and addresses are truncated, in some cases, to fit into the available space.

MATERIALS CONSUMED

Material	Quantity		Delivered Cost ($ million)
Materials, ingredients, containers, and supplies		(X)	2,201.7
Steel rod for wiredrawing	1,000 s tons	3,788.5	1,263.8
Steel wire for redrawing	1,000 s tons	(S)	284.5
All other steel shapes and forms	1,000 s tons	(S)	169.2
Ferrochromium shapes and forms	1,000 s tons	(D)	(D)
All other ferrous shapes and forms	1,000 s tons	(D)	(D)
Aluminum and aluminum-base alloy shapes and forms	1,000 s tons	(D)	(D)
Nickel (except ferronickel) shapes and forms	1,000 s tons	0.4*	3.7
Tin shapes and forms	1,000 s tons	(D)	(D)
Zinc and zinc-base alloy shapes and forms	1,000 s tons	20.4*	24.6
All other nonferrous shapes and forms	1,000 s tons	(S)	22.8
All other iron and steel castings		(X)	(D)
Clay refractories		(X)	(D)
Lime fluxes, including quicklime	1,000 s tons	(S)	0.1
Other fluxes		(X)	(D)
Industrial chemicals (except sulfuric acid and oxygen)		(X)	9.3
Oxygen (including high and low purity) (liquid oxygen should be converted to its gaseous equivalent)	mil cu ft	(D)	(D)
Sulfuric acid (new and spent) (100 percent H2SO4)	1,000 s tons	6.1	0.5
All other chemicals and allied products		(X)	10.5
Iron and steel scrap, excluding home scrap	1,000 s tons	(D)	(D)
Lubricating oils and greases and other petroleum products		(X)	5.8
Industrial dies, molds, jigs, and fixtures		(X)	9.0
All other materials and components, parts, containers, and supplies		(X)	304.5
Materials, ingredients, containers, and supplies, nsk		(X)	75.1

Source: 1992 *Economic Census.* Explanation of symbols used: (D): Withheld to avoid disclosure of competitive data; na: Not available; (S): Withheld because statistical norms were not met; (X): Not applicable; (Z): Less than half the unit shown; nec: Not elsewhere classified; nsk: Not specified by kind; - : zero; * : 10-19 percent estimated; ** : 20-29 percent estimated.

PRODUCT SHARE DETAILS

Product or Product Class	% Share	Product or Product Class	% Share
Steel wire and related products	100.00	Steel tacks (wire and cut)	1.16
Noninsulated ferrous wire rope, cable, and strand (made in wiredrawing plants)	22.10	Steel spikes and brads (including track spikes)	2.19
		Steel nails, staples, tacks, spikes, and brads, nsk	1.70
Noninsulated ferrous wire rope and cable made from steel wire (excluding fabricated wire rope assemblies) (made in wiredrawing plants)	46.40	Steel wire, including galvanized and other coated wire made in steel mills not producing wire rods or hot-rolled bars	37.21
		Steel fencing and fence gates (made in wiredrawing plants)	8.08
Noninsulated fabricated ferrous wire rope assemblies (including lifting slings) (made in wiredrawing plants)	4.73	Steel chain link fencing, excluding posts, gates, and fittings (including galvanized and plastics coated) (made in wiredrawing plants)	25.40
Noninsulated ferrous wire forms (excluding rope assemblies) and composite rope and cable (made in wiredrawing plants)	5.40	Woven and welded steel fencing, excluding posts, gates, and fittings (including galvanized and plastics coated) (made in wiredrawing plants)	58.30
Noninsulated ferrous wire strand (including strand for prestressed concrete, composite wire strand except ACSR, and guard rail cable) (made in wiredrawing plants)	43.44	Steel fence gates, posts, and fittings (made in wiredrawing plants)	14.96
Noninsulated ferrous wire rope, cable, and strand (made in wiredrawing plants), nsk	0.03	Ornamental steel lawn fence (excluding posts, gates, and fittings) (made in wiredrawing plants)	(D)
Steel nails, staples, tacks, spikes, and brads	14.60	Fencing and fence gates (made in wiredrawing plants), nsk	(D)
Round steel wire nails, collated, prepackaged	27.93	Ferrous wire cloth and other ferrous woven wire products (made in wiredrawing plants)	2.38
Galvanized round steel wire nails, not collated, smooth shank, coated, plated, or painted	9.36	Other fabricated ferrous wire products (except springs) (made in wiredrawing plants)	13.33
Vinyl, resin, or cement coated round steel wire nails, not collated, smooth shank, coated, plated, or painted	7.26	Ferrous wire chain (including tire chain, stud-link, and welded-link) (made in wiredrawing plants)	(D)
Other coated, plated, or painted round steel wire nails, not collated, smooth shank	2.91	Barbed and twisted ferrous wire (made in wiredrawing plants)	24.77
Round steel wire nails, not coated, plated, or painted, not collated, smooth shank	4.89	Ferrous wire bale ties (made in wiredrawing plants)	7.42
Galvanized round steel wire nails, not collated, other than smooth shank, coated, plated, or painted	(D)	Welded steel wire fabrics (including concrete reinforcing mesh) (made in wiredrawing plants)	18.99
Other coated, plated, or painted round steel wire nails (including vinyl, resin, or cement coated), not collated, other than smooth shank	(D)	Ferrous wire garment hangers (made in wiredrawing plants)	20.48
Round steel wire nails, not coated, plated, or painted, not collated, other than smooth shank	1.43	Ferrous wire carts (including household, grocery, and industrial) (made in wiredrawing plants)	(D)
Steel cut nails (including horseshoe nails)	12.21	Other fabricated ferrous wire products (guards, baskets, florists' designs, paperclips, kitchenware, wire shelving, wire rack, etc.), except springs (made in wiredrawing plants)	18.53
Steel wire staples	25.24	Steel wire and related products, nsk	2.29

Source: 1992 *Economic Census.* The values shown are percent of total shipments in an industry. Values of indented subcategories are summed in the main heading. The symbol (D) appears when data are withheld to prevent disclosure of competitive information. The abbreviation nsk stands for 'not specified by kind' and nec for 'not elsewhere classified'.

INPUTS AND OUTPUTS FOR STEEL WIRE & RELATED PRODUCTS

Economic Sector or Industry Providing Inputs	%	Sector	Economic Sector or Industry Buying Outputs	%	Sector
Blast furnaces & steel mills	41.2	Manufg.	Residential 1-unit structures, nonfarm	21.7	Constr.
Wholesale trade	9.8	Trade	Nonfarm residential structure maintenance	12.5	Constr.
Imports	9.2	Foreign	Maintenance of nonfarm buildings nec	12.2	Constr.
Special dies & tools & machine tool accessories	7.9	Manufg.	Residential garden apartments	7.1	Constr.
Chemical preparations, nec	3.3	Manufg.	Residential additions/alterations, nonfarm	5.6	Constr.
Electric services (utilities)	2.7	Util.	Office buildings	3.0	Constr.
Cyclic crudes and organics	2.4	Manufg.	Electric utility facility construction	2.9	Constr.
Motor freight transportation & warehousing	1.8	Util.	Residential 2-4 unit structures, nonfarm	2.6	Constr.
Gas production & distribution (utilities)	1.6	Util.	Exports	2.6	Foreign
Ball & roller bearings	1.4	Manufg.	Construction of stores & restaurants	2.3	Constr.
Pumps & compressors	1.3	Manufg.	Hotels & motels	2.0	Constr.
Maintenance of nonfarm buildings nec	1.2	Constr.	Construction of educational buildings	1.8	Constr.
Wood products, nec	1.1	Manufg.	Residential high-rise apartments	1.7	Constr.
Power transmission equipment	1.0	Manufg.	Steel wire & related products	1.4	Manufg.
Railroads & related services	1.0	Util.	Wholesale trade	1.1	Trade
Miscellaneous repair shops	1.0	Services	S/L Govt. purch., correction	1.1	S/L Govt
Screw machine and related products	0.9	Manufg.	Construction of hospitals	1.0	Constr.
Petroleum refining	0.8	Manufg.	Industrial buildings	1.0	Constr.
Primary zinc	0.8	Manufg.	Wood household furniture	1.0	Manufg.
Coal	0.7	Mining	Dormitories & other group housing	0.9	Constr.
Nonferrous wire drawing & insulating	0.6	Manufg.	Warehouses	0.9	Constr.
Fabricated metal products, nec	0.5	Manufg.	Upholstered household furniture	0.9	Manufg.
Metal heat treating	0.5	Manufg.	Personal consumption expenditures	0.8	
Steel wire & related products	0.5	Manufg.	Construction of religious buildings	0.8	Constr.
Primary metal products, nec	0.4	Manufg.	Maintenance of farm residential buildings	0.7	Constr.
Communications, except radio & TV	0.4	Util.	Maintenance of farm service facilities	0.7	Constr.
Banking	0.4	Fin/R.E.	Wood products, nec	0.7	Manufg.
Equipment rental & leasing services	0.4	Services	Farm service facilities	0.6	Constr.
Industrial controls	0.3	Manufg.	Garage & service station construction	0.6	Constr.
Machinery, except electrical, nec	0.3	Manufg.	Maintenance of railroads	0.6	Constr.
Paperboard containers & boxes	0.3	Manufg.	Wood pallets & skids	0.6	Manufg.
Primary copper	0.3	Manufg.	Farm housing units & additions & alterations	0.4	Constr.
Real estate	0.3	Fin/R.E.	Resid. & other health facility construction	0.4	Constr.
Business services nec	0.3	Services	S/L Govt. purch., higher education	0.4	S/L Govt
Water transportation	0.2	Util.	Amusement & recreation building construction	0.3	Constr.
Eating & drinking places	0.2	Trade	Construction of nonfarm buildings nec	0.3	Constr.
Advertising	0.2	Services	Highway & street construction	0.3	Constr.
Computer & data processing services	0.2	Services	Gas utility facility construction	0.2	Constr.
Engineering, architectural, & surveying services	0.2	Services	Maintenance of military facilities	0.2	Constr.
Nonmetallic mineral services	0.1	Mining	Maintenance of nonbuilding facilities nec	0.2	Constr.
Abrasive products	0.1	Manufg.	Furniture & fixtures, nec	0.2	Manufg.
Sawmills & planning mills, general	0.1	Manufg.	Hand & edge tools, nec	0.2	Manufg.
Sanitary services, steam supply, irrigation	0.1	Util.	Public building furniture	0.2	Manufg.
Insurance carriers	0.1	Fin/R.E.	Wood containers	0.2	Manufg.
U.S. Postal Service	0.1	Gov't	Wood office furniture	0.2	Manufg.
			Federal Government purchases, nondefense	0.2	Fed Govt
			Maintenance of sewer facilities	0.1	Constr.
			Commercial printing	0.1	Manufg.
			Mobile homes	0.1	Manufg.
			Prefabricated wood buildings	0.1	Manufg.
			Motor freight transportation & warehousing	0.1	Util.
			S/L Govt. purch., other education & libraries	0.1	S/L Govt

Source: Benchmark Input-Output Accounts for the U.S. Economy, 1982, U.S. Department of Commerce, Washington, D.C., July 1991. Data, as reported in the source, are organized by the 1977 SIC structure in use in 1982 but have been matched, as closely as is possible, to the 1987 SIC structure used in this book.

OCCUPATIONS EMPLOYED BY SIC 331 - BLAST FURNACES AND BASIC STEEL PRODUCTS

Occupation	% of Total 1994	Change to 2005	Occupation	% of Total 1994	Change to 2005
Helpers, laborers, & material movers nec	7.2	-37.8	Machinists	1.8	-37.8
Machine tool cutting & forming etc. nec	6.2	-6.7	Machine feeders & offbearers	1.8	-44.0
Blue collar worker supervisors	5.8	-40.7	Assemblers, fabricators, & hand workers nec	1.8	-37.8
Industrial machinery mechanics	4.7	-0.4	Welders & cutters	1.4	-50.2
Crane & tower operators	3.5	-37.8	Sales & related workers nec	1.4	-37.8
Inspectors, testers, & graders, precision	3.2	-37.8	Freight, stock, & material movers, hand	1.4	-50.2
Metal & plastic machine workers nec	3.1	-33.8	Truck drivers light & heavy	1.3	-35.8
Electricians	2.6	-41.6	Machine tool cutting operators, metal & plastic	1.2	-48.2
Millwrights	2.4	-31.5	Production, planning, & expediting clerks	1.2	-37.8
Machine forming operators, metal & plastic	2.2	-31.6	Material moving equipment operators nec	1.1	-37.7
Maintenance repairers, general utility	1.9	-44.0	Machine operators nec	1.1	-45.1
Industrial truck & tractor operators	1.9	-37.8	General managers & top executives	1.1	-40.9
Furnace operators	1.9	-31.6			

Source: Industry-Occupation Matrix, Bureau of Labor Statistics. These data relate to one or more 3-digit SIC industry groups rather than to a single 4-digit SIC. The change reported for each occupation to the year 2005 is a percent of growth or decline as estimated by the Bureau of Labor Statistics. The abbreviation nec stands for 'not elsewhere classified'.

LOCATION BY STATE AND REGIONAL CONCENTRATION

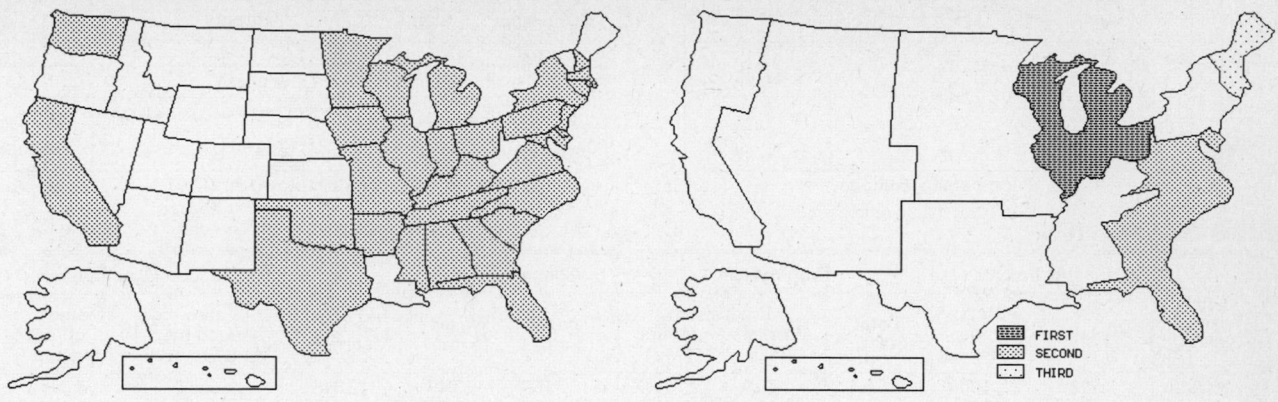

FIRST
SECOND
THIRD

INDUSTRY DATA BY STATE

State	Establish-ments	Shipments			Employment				Cost as % of Shipments	Investment per Employee ($)
		Total ($ mil)	% of U.S.	Per Establ.	Total Number	% of U.S.	Per Establ.	Wages ($/hour)		
Ohio	15	326.3	7.8	21.8	1,600	6.5	107	15.83	50.8	1,688
Pennsylvania	22	313.1	7.4	14.2	2,200	8.9	100	11.03	53.1	3,136
Missouri	9	264.4	6.3	29.4	1,300	5.3	144	11.64	60.7	-
California	32	237.4	5.6	7.4	1,200	4.9	38	12.42	65.8	4,167
Tennessee	8	224.0	5.3	28.0	700	2.8	88	12.00	63.4	11,571
Texas	20	222.1	5.3	11.1	1,300	5.3	65	10.61	70.1	3,385
Georgia	7	212.4	5.0	30.3	1,600	6.5	229	11.52	47.6	-
Florida	15	193.5	4.6	12.9	900	3.6	60	11.20	59.9	3,000
Indiana	17	185.9	4.4	10.9	1,100	4.5	65	13.29	60.6	-
Kentucky	5	172.3	4.1	34.5	1,100	4.5	220	12.84	50.6	3,909
Arkansas	8	163.1	3.9	20.4	700	2.8	88	15.27	51.4	-
North Carolina	14	99.3	2.4	7.1	700	2.8	50	10.00	67.0	1,714
South Carolina	8	94.8	2.3	11.9	600	2.4	75	11.89	69.4	7,833
Maryland	5	81.1	1.9	16.2	500	2.0	100	12.11	61.4	-
Massachusetts	19	77.1	1.8	4.1	500	2.0	26	11.86	65.4	-
New Jersey	10	71.0	1.7	7.1	500	2.0	50	10.78	57.9	2,400
Alabama	5	69.7	1.7	13.9	500	2.0	100	11.67	61.4	-
New York	13	54.8	1.3	4.2	400	1.6	31	11.60	65.1	-
Illinois	31	(D)	-	-	1,750 *	7.1	56	-	-	-
Connecticut	17	(D)	-	-	750 *	3.0	44	-	-	-
Michigan	13	(D)	-	-	750 *	3.0	58	-	-	3,467
Iowa	6	(D)	-	-	375 *	1.5	63	-	-	1,600
Oklahoma	6	(D)	-	-	750 *	3.0	125	-	-	-
Rhode Island	5	(D)	-	-	1,750 *	7.1	350	-	-	-
Washington	5	(D)	-	-	175 *	0.7	35	-	-	-
New Hampshire	4	(D)	-	-	175 *	0.7	44	-	-	-
Wisconsin	4	(D)	-	-	750 *	3.0	188	-	-	-
Minnesota	3	(D)	-	-	375 *	1.5	125	-	-	-
Mississippi	3	(D)	-	-	375 *	1.5	125	-	-	-
Virginia	3	(D)	-	-	175 *	0.7	58	-	-	-
Delaware	2	(D)	-	-	175 *	0.7	88	-	-	-

Source: 1992 Economic Census. The states are in descending order of shipments or establishments (if shipment data are missing for the majority). The symbol (D) appears when data are withheld to prevent disclosure of competitive information. States marked with (D) are sorted by number of establishments. A dash (-) indicates that the data element cannot be calculated; * indicates the midpoint of a range.

3316 - COLD FINISHING OF STEEL SHAPES

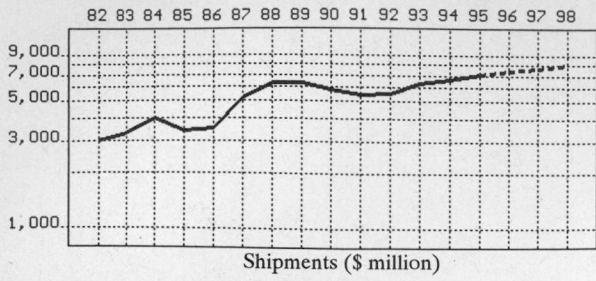

Shipments ($ million)

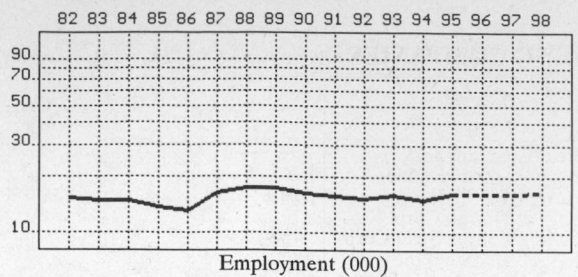

Employment (000)

GENERAL STATISTICS

Year	Com-panies	Establishments		Employment			Compensation		Production ($ million)			
		Total	with 20 or more employees	Total (000)	Production Workers (000)	Hours (Mil)	Payroll ($ mil)	Wages ($/hr)	Cost of Materials	Value Added by Manufacture	Value of Shipments	Capital Invest.
1982	145	192	106	15.4	10.9	19.7	367.9	12.47	2,348.2	623.3	3,005.1	45.2
1983		188	107	14.8	10.8	20.9	396.9	13.06	2,267.3	996.4	3,310.9	41.8
1984		184	108	14.7	10.8	21.1	404.1	13.53	2,677.6	1,346.1	3,992.1	60.7
1985		179	108	13.8	9.8	19.4	383.2	13.50	2,341.3	1,030.7	3,432.3	57.1
1986		185	111	13.1	9.4	18.7	387.0	14.50	2,377.6	1,108.5	3,514.5	51.2
1987	154	191	101	16.4	11.7	24.6	524.1	15.03	3,605.6	1,617.3	5,216.3	290.4
1988		191	105	17.3	12.3	25.5	566.8	15.47	4,430.6	2,051.4	6,323.4	163.3
1989		203	118	17.3	11.9	24.9	549.7	15.18	4,522.3	1,798.7	6,311.9	117.0
1990		195	117	16.4	11.4	24.0	554.6	15.78	4,173.6	1,620.8	5,842.2	148.8
1991		201	112	15.8	11.0	22.8	538.7	15.91	3,786.6	1,599.2	5,451.5	94.9
1992	161	190	86	15.3	11.0	22.8	551.2	16.80	3,779.9	1,725.1	5,510.7	115.6
1993		219	112	15.9	11.4	24.6	610.3	17.40	4,187.3	2,017.1	6,223.1	154.6
1994		206P	107P	15.1	11.1	24.0	631.4	18.11	4,509.8	2,138.9	6,538.4	198.1
1995		208P	107P	16.2P	11.5P	25.3P	653.7P	18.16P	4,854.4P	2,302.3P	7,038.0P	189.0P
1996		210P	107P	16.3P	11.6P	25.7P	676.0P	18.59P	5,057.7P	2,398.8P	7,332.8P	199.1P
1997		212P	107P	16.5P	11.7P	26.1P	698.3P	19.02P	5,261.0P	2,495.2P	7,627.5P	209.2P
1998		214P	107P	16.6P	11.7P	26.5P	720.7P	19.45P	5,464.3P	2,591.6P	7,922.3P	219.3P

Sources: 1982, 1987, 1992 *Economic Census*; *Annual Survey of Manufactures*, 83-86, 88-91, 93-94. Establishment counts for non-Census years are from *County Business Patterns*; establishment values for 83-84 are extrapolations. 'P's show projections by the editors. Industries reclassified in 87 will not have data for prior years.

INDICES OF CHANGE

Year	Com-panies	Establishments		Employment			Compensation		Production ($ million)			
		Total	with 20 or more employees	Total (000)	Production Workers (000)	Hours (Mil)	Payroll ($ mil)	Wages ($/hr)	Cost of Materials	Value Added by Manufacture	Value of Shipments	Capital Invest.
1982	90	101	123	101	99	86	67	74	62	36	55	39
1983		99	124	97	98	92	72	78	60	58	60	36
1984		97	126	96	98	93	73	81	71	78	72	53
1985		94	126	90	89	85	70	80	62	60	62	49
1986		97	129	86	85	82	70	86	63	64	64	44
1987	96	101	117	107	106	108	95	89	95	94	95	251
1988		101	122	113	112	112	103	92	117	119	115	141
1989		107	137	113	108	109	100	90	120	104	115	101
1990		103	136	107	104	105	101	94	110	94	106	129
1991		106	130	103	100	100	98	95	100	93	99	82
1992	100	100	100	100	100	100	100	100	100	100	100	100
1993		115	130	104	104	108	111	104	111	117	113	134
1994		108P	124P	99	101	105	115	108	119	124	119	171
1995		110P	124P	106P	105P	111P	119P	108P	128P	133P	128P	163P
1996		111P	124P	107P	105P	113P	123P	111P	134P	139P	133P	172P
1997		112P	124P	108P	106P	115P	127P	113P	139P	145P	138P	181P
1998		113P	124P	108P	107P	116P	131P	116P	145P	150P	144P	190P

Sources: Same as General Statistics. Values reflect change from the base year, 1992. Values above 100 mean greater than 92, values below 100 mean less than 92, and a value of 100 in the 82-91 or 93-98 period means same as 92. 'P's mark projections by the editors.

SELECTED RATIOS

For 1994	Avg. of All Manufact.	Analyzed Industry	Index	For 1994	Avg. of All Manufact.	Analyzed Industry	Index
Employees per Establishment	49	73	150	Value Added per Production Worker	134,084	192,694	144
Payroll per Establishment	1,500,273	3,063,922	204	Cost per Establishment	5,045,178	21,884,185	434
Payroll per Employee	30,620	41,815	137	Cost per Employee	102,970	298,662	290
Production Workers per Establishment	34	54	157	Cost per Production Worker	146,988	406,288	276
Wages per Establishment	853,319	2,109,127	247	Shipments per Establishment	9,576,895	31,728,138	331
Wages per Production Worker	24,861	39,157	158	Shipments per Employee	195,460	433,007	222
Hours per Production Worker	2,056	2,162	105	Shipments per Production Worker	279,017	589,045	211
Wages per Hour	12.09	18.11	150	Investment per Establishment	321,011	961,297	299
Value Added per Establishment	4,602,255	10,379,193	226	Investment per Employee	6,552	13,119	200
Value Added per Employee	93,930	141,649	151	Investment per Production Worker	9,352	17,847	191

Sources: Same as General Statistics. The 'Average of All Manufacturing' column represents the average of all manufacturing industries reported for the most recent complete year available. The Index shows the relationship between the Average and the Analyzed Industry. For example, 100 means that they are equal; 500 that the Analyzed Industry is five times the average; 50 means that the Analyzed Industry is half the national average. The abbreviation 'na' is used to show that data are 'not available'.

LEADING COMPANIES   Number shown: **63**   Total sales ($ mil): **11,692**   Total employment (000): **52.5**

Company Name	Address				CEO Name	Phone	Co. Type	Sales ($ mil)	Empl. (000)
LTV Steel Company Inc	25 W Prospect Av	Cleveland	OH	44115	David Hoag	216-622-5000	S	3,565	16.5
Worthington Industries Inc	1205 Dearborn Dr	Columbus	OH	43085	John P McConnell	614-438-3210	P	1,285	7.0
WHX Corp	110 E 59th St	New York	NY	10022	James L Wareham	212-355-5200	P	1,194	5.5
Wheeling-Pittsburgh Corp	110 E 59th St	New York	NY	10022	James L Wareham	212-355-5200	S	1,047	5.4
Wheeling-Pittsburgh Steel Corp	1134 Market St	Wheeling	WV	26003	James L Wareham	304-234-2400	S	930	5.0
J and L Specialty Steel Inc	PO Box 3373	Pittsburgh	PA	15230	Claude F Kronk	412-338-1600	P	648	1.3
Heidtman Steel Products Inc	2401 Front St	Toledo	OH	43605	John Bates	419-691-4646	R	350	0.6
NS Group Inc	PO Box 1670	Newport	KY	41072	Clifford R Borland	606-292-6809	P	303	1.6
Cold Metal Products Inc	8526 South Av	Youngstown	OH	44514	James R Harpster	216-758-1194	P	221	0.7
Contech Construction Products	1001 Grove St	Middletown	OH	45044	George R Gage	513-425-5896	R	210	1.0
Greer Industries Inc	PO Box 1900	Morgantown	WV	26507	John R Raese	304-296-1751	R	150*	0.7
Tuscaloosa Steel Corp	1700 Holt Rd	Tuscaloosa	AL	35404	Dave Tarasevich	205-556-1310	S	150	0.3
Thomas Steel Strip Corp	Delaware NW	Warren	OH	44485	Mike T Schmader	216-841-6111	S	140*	0.6
Bliss and Laughlin Steel Co	281 E 155th St	Harvey	IL	60426	Gregory H Parker	708-333-1220	S	125	0.3
Gerrard and Co	400 E Touhy Av	Des Plaines	IL	60018	A Tako	708-299-8000	R	120	0.5
Interstate Steel Co	401 E Touhy Av	Des Plaines	IL	60017	David S Soble	708-827-5151	S	116	0.2
American Strip Steel Inc	55 Passaic Av	Kearny	NJ	07032	James A Daniell	201-991-1500	R	100*	0.1
Border Steel Mills Inc	PO Box 12843	El Paso	TX	79912	AW Lupia	915-886-2000	R	91	0.5
Thompson Steel Company Inc	120 Royal St	Canton	MA	02021	G Ryan	617-828-8800	R	90	0.5
Samuel-Whittar Inc	20001 Sherwood	Detroit	MI	48234	Jacques Langlois	313-893-5000	R	67*	0.3
Feralloy Corp	2500 Nameoki Dr	Granite City	IL	62040	Raymond Bello	618-452-2500	D	60	<0.1
Corey Steel Co	PO Box 5137	Chicago	IL	60680	PJ Darling II	708-863-8000	R	50	0.2
Niagara Cold Drawn Corp	PO Box 399	Buffalo	NY	14240	Frank Archer	716-827-7010	S	45	0.1
Hannibal Industries	3851 S Santa Fe Av	Los Angeles	CA	90058	Robert Tennant	213-588-4261	R	42	0.1
Somers Thin Strip/Olin Brass	215 Piedmont St	Waterbury	CT	06706	John Moritz	203-597-5000	D	40	0.2
Shape Corp	1900 Hayes St	Grand Haven	MI	49417	Peter Sturrus	616-846-8700	R	39*	0.5
Dale Industries Inc	6455 Kingsley Av	Dearborn	MI	48126	Ronald Garris	313-846-9400	R	35	0.1
Precision Industries Inc	PO Box 711	Washington	PA	15301	D Jack Milhollan	412-222-2100	R	31*	0.1
Southwest Steel Company Inc	PO Drawer 1890	Catoosa	OK	74015	Joe Handrahan	918-266-3850	R	31	0.1
Rome Strip Steel Company Inc	530 Henry St	Rome	NY	13440	Kirk Hinman	315-336-5500	R	30	0.2
Saegertown Manufacturing Corp	1 Crawford St	Saegertown	PA	16433	Chalmer C Jorden	814-763-2655	R	30	0.2
California Steel and Tube	16049 Stephens St	City of Industry	CA	91745	JE Byers	818-968-5511	S	27	0.1
Charter Wire	114 N Jackson St	Milwaukee	WI	53202	Bob Melstrand	414-276-0672	D	27*	0.2
SIVACO New York Inc	PO Box 646	Tonawanda	NY	14151	Ian Cudmore	716-874-5681	S	26*	<0.1
Pacific Tube Co	PO Box 91-1222	Los Angeles	CA	90091	Robert J Porthan	213-728-2611	R	25	0.3
St Louis Cold Drawn Inc	1060 Pershall Rd	St Louis	MO	63137	William L McNair	314-867-4301	R	25*	<0.1
Caldwell Culvert Co	PO Box 1337	Greenville	MS	38702	Bill Lovelace	601-332-2625	R	20*	0.1
Washington Spec Metals Corp	1400 E Lake Cook	Buffalo Grove	IL	60089	James Norton	708-459-6060	S	20*	0.1
Chicago Steel	700 Chase St	Gary	IN	46404	Bruce Mannakee	219-949-1111	R	18*	<0.1
Moltrup Steel Products Company	PO Box 331	Beaver Falls	PA	15010	Mike Pitterich	412-846-3100	R	17*	<0.1
Exactacut Steel Company Inc	PO Box 3000	Niles	OH	44446	JR Swager	216-652-2640	R	15	0.1
Hynes Industries Inc	3760 Oakwood Av	Youngstown	OH	44509	WW Bresnahan	216-799-3221	D	15	0.1
Marwas Steel Co	1 Mount Pleasant	Scottdale	PA	15683	Marvin Waspe	412-887-8090	R	12*	<0.1
Elliott Brothers Steel Co	PO Box 551	New Castle	PA	16103	TC Elliott	412-658-5561	R	10*	<0.1
Rathbone Precision Metals Inc	241 Park St	Palmer	MA	01069	Frank Carter	413-283-8961	R	10*	0.1
Mansfield Structural	429 Park Av E	Mansfield	OH	44905	Richard Gash	419-522-5911	R	9	<0.1
Fort Howard Steel Inc	PO Box 11934	Green Bay	WI	54307	Vance VanLaanen	414-339-8835	R	9	<0.1
Precision-Kidd Steel Co	Drawer D	Aliquippa	PA	15001	Thomas Milhollan	412-378-0084	R	9	<0.1
Circlemaster Inc	8747 Magnolia Av	Santee	CA	92071	NT Henkel Jr	619-449-9087	R	8*	<0.1
Technical Metals Co	PO Box 3210	Melvindale	MI	48122	Sherwood Merril	313-388-1880	R	8*	<0.1
Hartman-Fabco Inc	1415 Lake Lansing	Lansing	MI	48912	John Lima	517-485-9493	R	7	<0.1
Shane Steel Processing Inc	17495 Malyn	Fraser	MI	48026	John Hartley	313-296-1990	R	7*	<0.1
Pittsburgh Tool Steel Inc	1535 Beaver Av	Monaca	PA	15061	Larry M Megan	412-773-7000	R	6	<0.1
Rigidized Metals Corp	658 Ohio St	Buffalo	NY	14203	RS Smith Jr	716-849-4760	R	6*	<0.1
Exmet Corp	PO Box 1266	Naugatuck	CT	06770	Roger Cozens	203-723-1514	R	4	<0.1
Rolled Wire Products Co	3762 Oakwood Av	Youngstown	OH	44509	WW Bresnahan	216-799-3221	D	3*	<0.1
Strip Steel Inc	11525 Shoemaker	Santa Fe Sprgs	CA	90670	Harold Allen	310-944-3077	R	3	<0.1
Unitron Products Inc	905 Brush Av	Bronx	NY	10465	DI Ilich	718-863-7000	R	3	<0.1
Ramco Metal Forming Inc	3165 E Slauson Av	Vernon	CA	90058	R Lytle	213-584-7040	D	3	<0.1
Marshall Steel Inc	1555 Harbor Av	Memphis	TN	38113	Elton North	901-946-1124	S	2*	<0.1
St Paul Corrugating Co	700 39th Av NE	Columbia H	MN	55421	IN Burhans	612-788-9271	R	2	<0.1
David Witherspoon Inc	901 Maryville	Knoxville	TN	37920	David Witherspoon	615-577-1613	R	1	<0.1
Lanco Metal	1503 Adelia Av	S El Monte	CA	91733	S P Saurenman	818-243-1393	S	0*	<0.1

Source: Ward's Business Directory of U.S. Private and Public Companies, Volumes 1 and 2, 1996. The company type code used is as follows: P - Public, R - Private, S - Subsidiary, D - Division, J - Joint Venture, A - Affiliate, G - Group. Sales are in millions of dollars, employees are in thousands. An asterisk (*) indicates an estimated sales volume. The symbol < stands for 'less than'. Company names and addresses are truncated, in some cases, to fit into the available space.

MATERIALS CONSUMED

Material	Quantity	Delivered Cost ($ million)
Materials, ingredients, containers, and supplies	(X)	3,518.4
Steel rod for wiredrawing ..... 1,000 s tons	164.1*	76.8
Steel wire for redrawing ..... 1,000 s tons	(D)	(D)
All other steel shapes and forms ..... 1,000 s tons	(S)	2,411.9
Other ferroalloy shapes and forms ..... 1,000 s tons	(D)	(D)
All other ferrous shapes and forms ..... 1,000 s tons	(D)	(D)
Aluminum and aluminum-base alloy shapes and forms ..... 1,000 s tons	(D)	(D)
Nickel (except ferronickel) shapes and forms ..... 1,000 s tons	(D)	(D)
Tin shapes and forms ..... 1,000 s tons	(D)	(D)
Zinc and zinc-base alloy shapes and forms ..... 1,000 s tons	(D)	(D)
All other nonferrous shapes and forms ..... 1,000 s tons	(D)	(D)
All other iron and steel castings	(X)	3.5
Clay refractories	(X)	(D)
Lime fluxes, including quicklime ..... 1,000 s tons	(D)	(D)
Industrial chemicals (except sulfuric acid and oxygen)	(X)	10.5
Oxygen (including high and low purity) (liquid oxygen should be converted to its gaseous equivalent) . mil cu ft	(D)	(D)
Sulfuric acid (new and spent) (100 percent H2SO4) ..... 1,000 s tons	2.5*	0.2
All other chemicals and allied products	(X)	1.7
Iron and steel scrap, excluding home scrap ..... 1,000 s tons	(D)	(D)
Lubricating oils and greases and other petroleum products	(X)	11.8
Industrial dies, molds, jigs, and fixtures	(X)	1.2
All other materials and components, parts, containers, and supplies	(X)	759.6
Materials, ingredients, containers, and supplies, nsk	(X)	94.2

Source: 1992 *Economic Census*. Explanation of symbols used: (D): Withheld to avoid disclosure of competitive data; na: Not available; (S): Withheld because statistical norms were not met; (X): Not applicable; (Z): Less than half the unit shown; nec: Not elsewhere classified; -: zero; * : 10-19 percent estimated; ** : 20-29 percent estimated.

PRODUCT SHARE DETAILS

Product or Product Class	% Share	Product or Product Class	% Share
Cold-finishing of steel shapes	100.00	Cold-finished steel bars and bar shapes, made in steel mills not producing hot-rolled bars and bar shapes	25.57
Cold-rolled steel sheet and strip, made in steel mills not producing hot-rolled sheet or strip	72.45	Cold-finishing of steel shapes, nsk	1.98

Source: 1992 *Economic Census*. The values shown are percent of total shipments in an industry. Values of indented subcategories are summed in the main heading. The symbol (D) appears when data are withheld to prevent disclosure of competitive information. The abbreviation nsk stands for 'not specified by kind' and nec for 'not elsewhere classified'.

INPUTS AND OUTPUTS FOR COLD FINISHING OF STEEL SHAPES

Economic Sector or Industry Providing Inputs	%	Sector	Economic Sector or Industry Buying Outputs	%	Sector
Blast furnaces & steel mills	60.0	Manufg.	Maintenance of nonfarm buildings nec	16.1	Constr.
Wholesale trade	11.0	Trade	Residential 1-unit structures, nonfarm	11.5	Constr.
Cyclic crudes and organics	5.0	Manufg.	Nonfarm residential structure maintenance	7.7	Constr.
Screw machine and related products	2.0	Manufg.	Office buildings	7.6	Constr.
Electric services (utilities)	2.0	Util.	Exports	5.8	Foreign
Blowers & fans	1.7	Manufg.	Residential additions/alterations, nonfarm	5.1	Constr.
Motor freight transportation & warehousing	1.7	Util.	Industrial buildings	4.7	Constr.
Gas production & distribution (utilities)	1.6	Util.	Machinery, except electrical, nec	4.1	Manufg.
Pumps & compressors	1.3	Manufg.	Mobile homes	4.0	Manufg.
Railroads & related services	1.1	Util.	Construction of hospitals	3.8	Constr.
Power transmission equipment	1.0	Manufg.	Construction of stores & restaurants	3.3	Constr.
Metal coating & allied services	0.8	Manufg.	Construction of educational buildings	3.0	Constr.
Ball & roller bearings	0.7	Manufg.	Residential garden apartments	2.5	Constr.
Chemical preparations, nec	0.7	Manufg.	Ship building & repairing	2.0	Manufg.
Maintenance of nonfarm buildings nec	0.6	Constr.	Mineral wool	1.9	Manufg.
Fabricated metal products, nec	0.6	Manufg.	Hotels & motels	1.3	Constr.
Industrial controls	0.6	Manufg.	Residential 2-4 unit structures, nonfarm	1.1	Constr.
Industrial inorganic chemicals, nec	0.6	Manufg.	Amusement & recreation building construction	1.0	Constr.
Miscellaneous fabricated wire products	0.6	Manufg.	Electric utility facility construction	1.0	Constr.
Sanitary services, steam supply, irrigation	0.5	Util.	Miscellaneous plastics products	1.0	Manufg.
Banking	0.5	Fin/R.E.	Warehouses	0.9	Constr.
Petroleum refining	0.4	Manufg.	Electric housewares & fans	0.9	Manufg.
Fabricated rubber products, nec	0.3	Manufg.	Refrigeration & heating equipment	0.9	Manufg.
Metal heat treating	0.3	Manufg.	Resid. & other health facility construction	0.8	Constr.
Primary nonferrous metals, nec	0.3	Manufg.	Household cooking equipment	0.8	Manufg.
Sawmills & planning mills, general	0.3	Manufg.	Prefabricated wood buildings	0.8	Manufg.
Communications, except radio & TV	0.3	Util.	Construction of nonfarm buildings nec	0.7	Constr.
Business services nec	0.3	Services	Household refrigerators & freezers	0.6	Manufg.
Engineering, architectural, & surveying services	0.3	Services	Change in business inventories	0.6	In House
Equipment rental & leasing services	0.3	Services	Farm housing units & additions & alterations	0.4	Constr.
Miscellaneous repair shops	0.3	Services	Highway & street construction	0.4	Constr.
Machinery, except electrical, nec	0.2	Manufg.	Household laundry equipment	0.4	Manufg.
Nonferrous wire drawing & insulating	0.2	Manufg.	Residential high-rise apartments	0.3	Constr.
Transformers	0.2	Manufg.	Motor vehicles & car bodies	0.3	Manufg.
Water transportation	0.2	Util.	Motors & generators	0.3	Manufg.
Paperboard containers & boxes	0.1	Manufg.	Truck trailers	0.3	Manufg.
Eating & drinking places	0.1	Trade	Construction of religious buildings	0.2	Constr.
Real estate	0.1	Fin/R.E.	Telephone & telegraph facility construction	0.2	Constr.

Source: Benchmark Input-Output Accounts for the U.S. Economy, 1982, U.S. Department of Commerce, Washington, D.C., July 1991. Data, as reported in the source, are organized by the 1977 SIC structure in use in 1982 but have been matched, as closely as is possible, to the 1987 SIC structure used in this book.

OCCUPATIONS EMPLOYED BY SIC 331 - BLAST FURNACES AND BASIC STEEL PRODUCTS

Occupation	% of Total 1994	Change to 2005	Occupation	% of Total 1994	Change to 2005
Helpers, laborers, & material movers nec	7.2	-37.8	Machinists	1.8	-37.8
Machine tool cutting & forming etc. nec	6.2	-6.7	Machine feeders & offbearers	1.8	-44.0
Blue collar worker supervisors	5.8	-40.7	Assemblers, fabricators, & hand workers nec	1.8	-37.8
Industrial machinery mechanics	4.7	-0.4	Welders & cutters	1.4	-50.2
Crane & tower operators	3.5	-37.8	Sales & related workers nec	1.4	-37.8
Inspectors, testers, & graders, precision	3.2	-37.8	Freight, stock, & material movers, hand	1.4	-50.2
Metal & plastic machine workers nec	3.1	-33.8	Truck drivers light & heavy	1.3	-35.8
Electricians	2.6	-41.6	Machine tool cutting operators, metal & plastic	1.2	-48.2
Millwrights	2.4	-31.5	Production, planning, & expediting clerks	1.2	-37.8
Machine forming operators, metal & plastic	2.2	-31.6	Material moving equipment operators nec	1.1	-37.7
Maintenance repairers, general utility	1.9	-44.0	Machine operators nec	1.1	-45.1
Industrial truck & tractor operators	1.9	-37.8	General managers & top executives	1.1	-40.9
Furnace operators	1.9	-31.6			

Source: Industry-Occupation Matrix, Bureau of Labor Statistics. These data relate to one or more 3-digit SIC industry groups rather than to a single 4-digit SIC. The change reported for each occupation to the year 2005 is a percent of growth or decline as estimated by the Bureau of Labor Statistics. The abbreviation nec stands for 'not elsewhere classified'.

LOCATION BY STATE AND REGIONAL CONCENTRATION

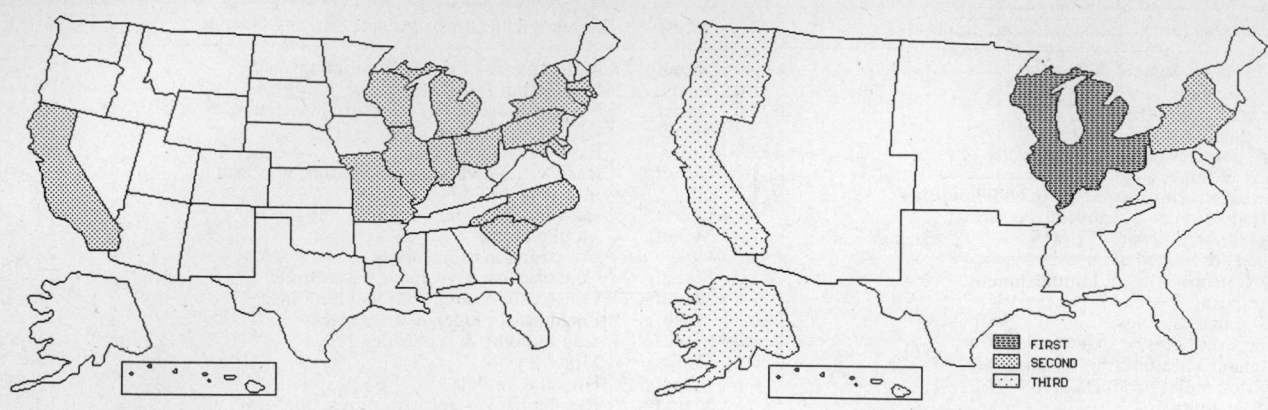

FIRST
SECOND
THIRD

INDUSTRY DATA BY STATE

State	Establish-ments	Shipments			Employment				Cost as % of Shipments	Investment per Employee ($)
		Total ($ mil)	% of U.S.	Per Establ.	Total Number	% of U.S.	Per Establ.	Wages ($/hour)		
Ohio	35	1,157.5	21.0	33.1	3,800	24.8	109	16.71	66.6	13,158
Pennsylvania	15	1,033.7	18.8	68.9	2,600	17.0	173	18.36	50.1	6,500
Illinois	17	804.1	14.6	47.3	1,700	11.1	100	15.92	68.4	3,941
Michigan	17	514.6	9.3	30.3	900	5.9	53	18.47	88.9	5,222
Indiana	6	284.9	5.2	47.5	1,100	7.2	183	17.93	67.8	-
Connecticut	10	278.2	5.0	27.8	1,000	6.5	100	17.82	100.4	2,900
New York	6	224.5	4.1	37.4	800	5.2	133	15.73	66.6	7,000
Wisconsin	6	100.5	1.8	16.7	300	2.0	50	11.80	74.0	-
California	16	(D)	-	-	1,750 *	11.4	109	-	-	-
South Carolina	6	(D)	-	-	175 *	1.1	29	-	-	-
Maryland	4	(D)	-	-	175 *	1.1	44	-	-	-
Massachusetts	4	(D)	-	-	375 *	2.5	94	-	-	-
Missouri	4	(D)	-	-	175 *	1.1	44	-	-	571
North Carolina	2	(D)	-	-	175 *	1.1	88	-	-	-

Source: 1992 *Economic Census*. The states are in descending order of shipments or establishments (if shipment data are missing for the majority). The symbol (D) appears when data are withheld to prevent disclosure of competitive information. States marked with (D) are sorted by number of establishments. A dash (-) indicates that the data element cannot be calculated; * indicates the midpoint of a range.

3317 - STEEL PIPE & TUBES

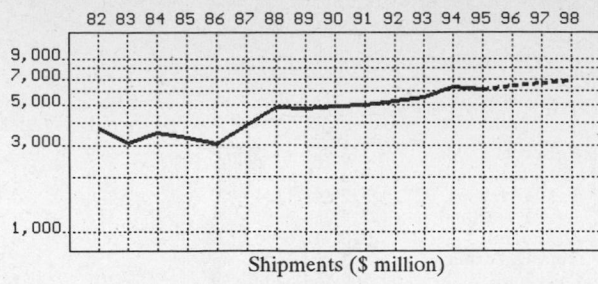

82 83 84 85 86 87 88 89 90 91 92 93 94 95 96 97 98

Shipments ($ million)

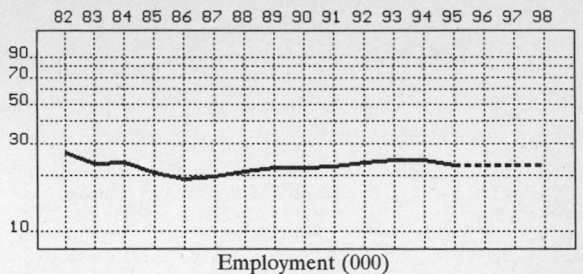

82 83 84 85 86 87 88 89 90 91 92 93 94 95 96 97 98

Employment (000)

GENERAL STATISTICS

Year	Companies	Establishments		Employment			Compensation		Production ($ million)			
		Total	with 20 or more employees	Total (000)	Production Workers (000)	Hours (Mil)	Payroll ($ mil)	Wages ($/hr)	Cost of Materials	Value Added by Manufacture	Value of Shipments	Capital Invest.
1982	169	223	167	27.0	20.3	37.6	648.1	12.25	2,451.7	1,212.7	3,762.2	124.5
1983		223	162	23.1	17.5	32.0	537.1	11.56	2,047.3	959.9	3,057.0	134.4
1984		223	157	23.7	18.3	36.0	537.1	10.36	2,313.7	1,178.3	3,522.2	77.8
1985		223	153	21.0	16.0	30.8	510.6	11.59	2,116.7	1,092.8	3,291.9	60.3
1986		229	163	19.1	14.6	28.7	469.5	11.39	1,953.9	1,085.3	3,055.1	72.2
1987	153	221	161	19.6	15.4	31.2	515.4	11.81	2,522.7	1,345.0	3,856.3	59.0
1988		224	161	21.2	16.7	33.5	581.4	12.51	3,051.4	1,914.5	4,850.2	67.5
1989		220	164	22.1	16.8	34.5	589.3	11.93	2,991.6	1,761.3	4,765.5	137.4
1990		229	173	22.3	17.1	36.7	654.3	12.28	3,266.7	1,707.7	4,968.5	99.6
1991		241	171	22.7	17.3	37.3	677.8	12.58	3,450.3	1,547.0	5,010.6	144.8
1992	166	233	181	23.7	17.9	38.3	739.5	13.26	3,512.1	1,714.4	5,250.2	114.3
1993		252	179	24.2	18.7	40.6	787.7	13.66	3,747.6	1,833.1	5,555.0	113.3
1994		241P	177P	24.6	19.1	41.6	820.1	13.87	4,257.1	2,064.7	6,294.6	148.3
1995		243P	179P	22.8P	17.7P	39.7P	776.0P	13.65P	4,135.0P	2,005.5P	6,114.1P	126.0P
1996		244P	181P	22.9P	17.7P	40.3P	798.2P	13.85P	4,300.4P	2,085.7P	6,358.6P	129.2P
1997		246P	183P	22.9P	17.7P	41.0P	820.4P	14.05P	4,465.7P	2,165.9P	6,603.0P	132.3P
1998		248P	185P	22.9P	17.8P	41.6P	842.6P	14.25P	4,631.0P	2,246.0P	6,847.4P	135.4P

Sources: 1982, 1987, 1992 *Economic Census*; *Annual Survey of Manufactures*, 83-86, 88-91, 93-94. Establishment counts for non-Census years are from *County Business Patterns*; establishment values for 83-84 are extrapolations. 'P's show projections by the editors. Industries reclassified in 87 will not have data for prior years.

INDICES OF CHANGE

Year	Companies	Establishments		Employment			Compensation		Production ($ million)			
		Total	with 20 or more employees	Total (000)	Production Workers (000)	Hours (Mil)	Payroll ($ mil)	Wages ($/hr)	Cost of Materials	Value Added by Manufacture	Value of Shipments	Capital Invest.
1982	102	96	92	114	113	98	88	92	70	71	72	109
1983		96	90	97	98	84	73	87	58	56	58	118
1984		96	87	100	102	94	73	78	66	69	67	68
1985		96	85	89	89	80	69	87	60	64	63	53
1986		98	90	81	82	75	63	86	56	63	58	63
1987	92	95	89	83	86	81	70	89	72	78	73	52
1988		96	89	89	93	87	79	94	87	112	92	59
1989		94	91	93	94	90	80	90	85	103	91	120
1990		98	96	94	96	96	88	93	93	100	95	87
1991		103	94	96	97	97	92	95	98	90	95	127
1992	100	100	100	100	100	100	100	100	100	100	100	100
1993		108	99	102	104	106	107	103	107	107	106	99
1994		103P	98P	104	107	109	111	105	121	120	120	130
1995		104P	99P	96P	99P	104P	105P	103P	118P	117P	116P	110P
1996		105P	100P	96P	99P	105P	108P	104P	122P	122P	121P	113P
1997		106P	101P	97P	99P	107P	111P	106P	127P	126P	126P	116P
1998		107P	102P	97P	99P	109P	114P	107P	132P	131P	130P	118P

Sources: Same as General Statistics. Values reflect change from the base year, 1992. Values above 100 mean greater than 92, values below 100 mean less than 92, and a value of 100 in the 82-91 or 93-98 period means same as 92. 'P's mark projections by the editors.

SELECTED RATIOS

For 1994	Avg. of All Manufact.	Analyzed Industry	Index	For 1994	Avg. of All Manufact.	Analyzed Industry	Index
Employees per Establishment	49	102	209	Value Added per Production Worker	134,084	108,099	81
Payroll per Establishment	1,500,273	3,406,974	227	Cost per Establishment	5,045,178	17,685,441	351
Payroll per Employee	30,620	33,337	109	Cost per Employee	102,970	173,053	168
Production Workers per Establishment	34	79	231	Cost per Production Worker	146,988	222,885	152
Wages per Establishment	853,319	2,397,021	281	Shipments per Establishment	9,576,895	26,149,909	273
Wages per Production Worker	24,861	30,209	122	Shipments per Employee	195,460	255,878	131
Hours per Production Worker	2,056	2,178	106	Shipments per Production Worker	279,017	329,560	118
Wages per Hour	12.09	13.87	115	Investment per Establishment	321,011	616,089	192
Value Added per Establishment	4,602,255	8,577,466	186	Investment per Employee	6,552	6,028	92
Value Added per Employee	93,930	83,931	89	Investment per Production Worker	9,352	7,764	83

Sources: Same as General Statistics. The 'Average of All Manufacturing' column represents the average of all manufacturing industries reported for the most recent complete year available. The Index shows the relationship between the Average and the Analyzed Industry. For example, 100 means that they are equal; 500 that the Analyzed Industry is five times the average; 50 means that the Analyzed Industry is half the national average. The abbreviation 'na' is used to show that data are 'not available'.

LEADING COMPANIES   Number shown: 75   Total sales ($ mil): 5,549   Total employment (000): 27.0

Company Name	Address				CEO Name	Phone	Co. Type	Sales ($ mil)	Empl. (000)
Quanex Corp	1900 W Loop S	Houston	TX	77027	Robert C Snyder	713-961-4600	P	699	2.6
Bundy Corp	12345 E 9 Mile Rd	Warren	MI	48090	Bill Iaule	313-758-4511	S	500	2.8
UNR Industries Inc	332 S Michigan Av	Chicago	IL	60604	T A Gildehaus	312-341-1234	P	363	2.1
Lone Star Steel Co	PO Box 803546	Dallas	TX	75380	John P Harbin	214-386-3981	S	357	1.6
LTV Steel Tubular Products Co	PO Box 1000	Youngstown	OH	44501	DL Carroll	216-742-6000	S	300	0.6
Copperweld Corp	4 Gateway Ctr	Pittsburgh	PA	15222	John D Turner	412-263-3200	S	250	1.3
John Maneely Co	900 Haddon Av	Collingswood	NJ	08108	C Richard Wahl	609-854-5400	R	200	1.0
Wheatland Tube Co	900 Haddon Av	Collingswood	NJ	08108	C Richard Wahl	609-854-5400	S	190	1.0
Sawhill Tubular	PO Box 11	Sharon	PA	16146	David A Higbee	412-347-7771	D	170	0.8
Copperweld Shelby	132 W Main St	Shelby	OH	44875	Dave Coppock	419-342-1200	D	130	0.6
Maverick Tube Corp	400 Chesterfield Ctr	Chesterfield	MO	63017	Gregg M Eisenberg	314-537-1314	P	125	0.6
American Steel Pipe	PO Box 2727	Birmingham	AL	35202	Van L Ritchey	205-325-7742	D	100	<0.1
Palmer Tube Mills Inc	1855 E 122nd St	Chicago	IL	60633	Craig Lammarino	312-646-4500	S	100	0.3
Welded Tube of America	1855 E 122nd St	Chicago	IL	60633	Grahame Milton	312-646-4500	S	100	0.3
Christiana Companies Inc	777 E Wisconsin	Milwaukee	WI	53202	Sheldon B Lubar	414-291-9000	P	90	0.8
Trent Tube	PO Box 77	East Troy	WI	53120	Brent Ward	414-642-7321	D	90	0.5
Pittsburgh Tube Co	2060 Penn Av	Monaca	PA	15061	Richard C Huemme	412-774-6830	R	85	0.5
Metal-Matic Inc	629 2nd St SE	Minneapolis	MN	55414	GJ Bliss	612-378-0411	R	79	0.4
Webco Industries Inc	PO Box 100	Sand Springs	OK	74063	F William Weber	918-241-1000	P	70	0.6
Gulf States Tube	PO Box 952	Rosenberg	TX	77471	Les Whitver	713-342-5401	D	67*	0.3
Bristol Metals Inc	PO Box 1589	Bristol	TN	37621	Joseph N Avento	615-968-2151	S	66	0.4
Northwest Pipe and Casing Co	PO Box 83149	Portland	OR	97283	William R Tagmyer	503-285-1400	R	65	0.3
UNR-Leavitt	1717 W 115th St	Chicago	IL	60643	Roy A Herman	312-239-7700	D	57*	0.4
Salem Tube Inc	PO Box 144	Greenville	PA	16125	Robert F Green	412-646-4303	R	52*	0.3
Damascus-Bishop Tube Co	795 Reynolds	Greenville	PA	16125	George F Werner	412-646-1500	S	50	0.2
Jackson Tube Service Inc	PO Box 1650	Piqua	OH	45356	Gordon T Ralph	513-773-8550	R	50	0.3
Tex-Tube	PO Box 7705	Houston	TX	77270	Carl Farnsworth	713-686-4351	D	50	0.1
United Industries Inc	PO Box 118	Beloit	WI	53511	Roger West	608-365-8891	R	48*	0.3
UTI Corp	200 W 7th Av	Collegeville	PA	19426	GB Hattersley	215-489-0300	D	48	0.4
Romac Metals Inc	PO Box 836	Troutman	NC	28166	Harold Todd	704-528-4531	S	45	0.2
Bock Industries Inc	2700 S Nappanne St	Elkhart	IN	46517		219-295-8070	R	42*	0.1
Saw Pipes USA Inc	PO Box 2349	Baytown	TX	77522	Dilip Bhargava	713-383-3300	S	40	0.1
Pittsburgh International	PO Box 9	Fairbury	IL	61739	RC Huemme	815-692-2311	D	38	0.1
Southwestern Pipe Inc	PO Box 2002	Houston	TX	77252	Philip C Lewis	713-863-4300	S	38	0.2
Spax Inc	PO Box 2002	Houston	TX	77252	Philip C Lewis	713-863-4300	R	38	0.2
AP Parts Manufacturing Co	401 E 5th St	Pinconning	MI	48650	ME Brooks	517-879-2611	D	35	0.3
Parthenon Metal Works	PO Box 307	La Vergne	TN	37086	George T Johnston	615-793-6801	D	31*	0.2
Indiana Tube Corp	PO Box 3005	Evansville	IN	47730	Jerry Stohler	812-424-9028	S	30	0.2
Swepco Tube Corp	PO Box 1899	Clifton	NJ	07011	Alfred Ridella	201-778-3000	R	30	0.2
Tectron Tube Corp	PO Box 360	De Pere	WI	54115	Don Clancy Jr	414-336-3318	R	30	<0.1
Valex Corp	6080 Leland St	Ventura	CA	93003	Daniel A Mangan	805-658-0944	S	29	0.2
Blissfield Manufacturing Co	626 Depot St	Blissfield	MI	49228	OH Farver	517-486-2121	R	28	0.3
Bent Tube Inc	PO Box 709	Fowlerville	MI	48836	WR Doak	517-521-4330	R	26*	0.2
Oakley Industries Inc	3211 W Bear Creek	Englewood	CO	80110	Gary A Oakley	303-761-1835	R	26*	0.2
Quality Tubing Inc	PO Box 9819	Houston	TX	77213	David Daniel	713-456-0751	R	26	<0.1
National Metalwares Inc	900 N Russell Av	Aurora	IL	60506	Gary Hill	708-892-9000	S	25	0.3
Standard Fittings Co	PO Box 1268	Opelousas	LA	70571	Malcolm Marcus	318-948-3671	R	24	0.2
West Coast Tube and Pipe	28445 Driver Av	Agoura Hills	CA	91301	Wayne Moore	818-991-6000	R	23	<0.1
Le Fiell Manufacturing Co	13700 Firestone	Santa Fe Sprgs	CA	90670	George Ray	310-921-3411	R	22	0.2
Stupp	PO Box 3558	Baton Rouge	LA	70821	JP Stupp Jr	504-775-8800	D	21*	0.1
Middletown Tube Works Inc	PO Box 529	Middletown	OH	45044	Ralph Phillips	513-425-3725	R	20*	<0.1
Public Tube Mills Inc	1855 E 122nd St	Chicago	IL	60633	Graham Milton	312-646-4500	R	20*	0.1
Georgia Tubular Products	109 Dent Dr	Cartersville	GA	30120	Charles Coughlan	404-386-2553	S	19	<0.1
M and R Industries Inc	80 Field Point Rd	Greenwich	CT	06830	Maurice J Cunniffe	203-622-6600	R	19*	0.1
Roscoe Moss Co	PO Box 31064	Los Angeles	CA	90063	Bob Van Valer	213-261-4185	R	19	0.1
D-Velco Manufacturing	401 S 36th St	Phoenix	AZ	85034	John Marris	602-275-4406	S	17*	0.1
Naylor Pipe Co	1262 E 92nd St	Chicago	IL	60619	William B Skeates	312-721-9400	R	17	0.1
Davis Pipe and Metal Fabricators	PO Box 565	Blountville	TN	37617	George McGee	615-323-2171	R	16	0.1
General Tube Co	PO Box 27036	Chicago	IL	60627	Ben Boyd	708-389-1925	D	16	<0.1
Albach Company Inc	PO Box 1159	Chalmette	LA	70044	B S Brupbacher Jr	504-271-1113	R	15	0.1
Copperweld Corp	7401 S Linder Av	Chicago	IL	60638		708-458-4820	D	15*	0.1
Independence Tube Corp	6226 W 74th St	Chicago	IL	60638	DF Grohne	708-496-0380	R	15	<0.1
Central Nebraska Tubing	14631 US Hwy 6	Waverly	NE	68462	Scott Brown	402-786-5005	D	14	<0.1
Acme Tube Inc	1 Howard Av	Somerset	NJ	08873	James MacMahon	908-560-8111	R	13*	<0.1
Irby Steel	PO Box 2275	Gulfport	MS	39505	JD Tate	601-863-7733	R	13*	<0.1
KMI Inc	PO Box 225	Reed City	MI	49677	John H Kinnally	616-832-5562	R	12	0.2
Tubetech Inc	PO Box 470	East Palestine	OH	44413	Stephen D Oliphant	216-426-9476	R	12	<0.1
Valley Metals	PO Box 1111	El Cajon	CA	92022	J Patrick Hodgetts	619-442-5514	S	12*	<0.1
Acme Roll Forming Company	PO Box 706	Sebewaing	MI	48759	J Wineman	517-883-2050	R	10	<0.1
Kraftube Inc	PO Box 225	Reed City	MI	49677	John H Kinnally	616-832-5562	S	10	0.2
Metal Culverts Inc	PO Box 330	Jefferson City	MO	65102	M Mozelle Bielski	314-636-7312	R	10	<0.1
Munroe Inc	1820 N Franklin St	Pittsburgh	PA	15233	Phil F Muck	412-231-0600	R	10	0.2
Intern Specialty Tube Corp	260 S Crawford	Detroit	MI	48209	Clayton J Wallace	313-841-6900	S	9	<0.1
Triple A Tube Inc	PO Box 5	Jonesville	MI	49250	David Palmer	517-849-9945	R	9	0.1
Burtco Inc	PO Box 40	Westminster St	VT	05159	Roland Scott	802-722-3358	R	8	<0.1

Source: Ward's Business Directory of U.S. Private and Public Companies, Volumes 1 and 2, 1996. The company type code used is as follows: P - Public, R - Private, S - Subsidiary, D - Division, J - Joint Venture, A - Affiliate, G - Group. Sales are in millions of dollars, employees are in thousands. An asterisk (*) indicates an estimated sales volume. The symbol < stands for 'less than'. Company names and addresses are truncated, in some cases, to fit into the available space.

MATERIALS CONSUMED

Material	Quantity	Delivered Cost ($ million)
Materials, ingredients, containers, and supplies	(X)	3,174.5
All other steel shapes and forms . . . 1,000 s tons	(S)	2,147.0
Pig iron shapes and forms . . . 1,000 s tons	(D)	(D)
Ferromanganese, silicomanganese, and manganese shapes and forms . . . 1,000 s tons	(D)	(D)
Ferrochromium shapes and forms . . . 1,000 s tons	(D)	(D)
Ferrosilicon (more than 8 percent silicon) shapes and forms . . . 1,000 s tons	(D)	(D)
All other ferrous shapes and forms . . . 1,000 s tons	(S)	41.9
Aluminum and aluminum-base alloy shapes and forms . . . 1,000 s tons	(D)	(D)
Nickel (except ferronickel) shapes and forms . . . 1,000 s tons	(D)	(D)
Tin shapes and forms . . . 1,000 s tons	(D)	(D)
Zinc and zinc-base alloy shapes and forms . . . 1,000 s tons	(D)	(D)
All other nonferrous shapes and forms . . . 1,000 s tons	57.8*	32.2
Gray iron ingot molds and stools . . . 1,000 s tons	(D)	(D)
All other iron and steel castings	(X)	(D)
Clay refractories	(X)	(D)
Nonclay refractories	(X)	(D)
Lime fluxes, including quicklime . . . 1,000 s tons	(D)	(D)
Other fluxes	(X)	(Z)
All other stone, clay, glass, and concrete products	(X)	(D)
Industrial chemicals (except sulfuric acid and oxygen)	(X)	5.6
Oxygen (including high and low purity) (liquid oxygen should be converted to its gaseous equivalent) . mil cu ft	0.6*	0.1
Sulfuric acid (new and spent) (100 percent H2SO4) . . . 1,000 s tons	(S)	0.9
All other chemicals and allied products	(X)	(D)
Mining crude iron ore and iron ore concentrates, including pelletized and manganiferous (gross weight) 1,000 s tons	(D)	(D)
Iron and steel scrap, excluding home scrap . . . 1,000 s tons	(D)	(D)
Lubricating oils and greases and other petroleum products	(X)	5.1
Industrial dies, molds, jigs, and fixtures	(X)	6.1
Carbon and graphite electrodes	(X)	(D)
All other materials and components, parts, containers, and supplies	(X)	409.8
Materials, ingredients, containers, and supplies, nsk	(X)	490.6

Source: 1992 *Economic Census*. Explanation of symbols used: (D): Withheld to avoid disclosure of competitive data; na: Not available; (S): Withheld because statistical norms were not met; (X): Not applicable; (Z): Less than half the unit shown; nec: Not elsewhere classified; nsk: Not specified by kind; - : zero; * : 10-19 percent estimated; ** : 20-29 percent estimated.

PRODUCT SHARE DETAILS

Product or Product Class	% Share	Product or Product Class	% Share
Steel pipe and tubes	100.00		

Source: 1992 *Economic Census*. The values shown are percent of total shipments in an industry. Values of indented subcategories are summed in the main heading. The symbol (D) appears when data are withheld to prevent disclosure of competitive information. The abbreviation nsk stands for 'not specified by kind' and nec for 'not elsewhere classified'.

INPUTS AND OUTPUTS FOR STEEL PIPE & TUBES

Economic Sector or Industry Providing Inputs	%	Sector	Economic Sector or Industry Buying Outputs	%	Sector
Blast furnaces & steel mills	51.6	Manufg.	Maintenance of nonfarm buildings nec	16.1	Constr.
Cyclic crudes and organics	15.1	Manufg.	Residential 1-unit structures, nonfarm	11.5	Constr.
Wholesale trade	9.4	Trade	Nonfarm residential structure maintenance	7.7	Constr.
Communications, except radio & TV	2.8	Util.	Office buildings	7.6	Constr.
Electric services (utilities)	1.7	Util.	Exports	5.8	Foreign
Motor freight transportation & warehousing	1.6	Util.	Residential additions/alterations, nonfarm	5.1	Constr.
Gas production & distribution (utilities)	1.5	Util.	Industrial buildings	4.7	Constr.
Railroads & related services	1.3	Util.	Machinery, except electrical, nec	4.1	Manufg.
Maintenance of nonfarm buildings nec	1.2	Constr.	Mobile homes	4.0	Manufg.
Power transmission equipment	1.1	Manufg.	Construction of hospitals	3.8	Constr.
Miscellaneous repair shops	1.1	Services	Construction of stores & restaurants	3.3	Constr.
Screw machine and related products	1.0	Manufg.	Construction of educational buildings	3.0	Constr.
Chemical preparations, nec	0.8	Manufg.	Residential garden apartments	2.5	Constr.
Primary copper	0.7	Manufg.	Ship building & repairing	2.0	Manufg.
Coal	0.4	Mining	Mineral wool	1.9	Manufg.
Ball & roller bearings	0.4	Manufg.	Hotels & motels	1.3	Constr.
Metal coating & allied services	0.4	Manufg.	Residential 2-4 unit structures, nonfarm	1.1	Constr.
Metal heat treating	0.4	Manufg.	Amusement & recreation building construction	1.0	Constr.
Water transportation	0.4	Util.	Electric utility facility construction	1.0	Constr.
Banking	0.4	Fin/R.E.	Miscellaneous plastics products	1.0	Manufg.
Business services nec	0.4	Services	Warehouses	0.9	Constr.
Alkalies & chlorine	0.3	Manufg.	Electric housewares & fans	0.9	Manufg.
Carbon & graphite products	0.3	Manufg.	Refrigeration & heating equipment	0.9	Manufg.
Copper rolling & drawing	0.3	Manufg.	Resid. & other health facility construction	0.8	Constr.
Fabricated metal products, nec	0.3	Manufg.	Household cooking equipment	0.8	Manufg.
Industrial controls	0.3	Manufg.	Prefabricated wood buildings	0.8	Manufg.
Machinery, except electrical, nec	0.3	Manufg.	Construction of nonfarm buildings nec	0.7	Constr.

Continued on next page.

INPUTS AND OUTPUTS FOR STEEL PIPE & TUBES - Continued

Economic Sector or Industry Providing Inputs	%	Sector	Economic Sector or Industry Buying Outputs	%	Sector
Petroleum refining	0.3	Manufg.	Household refrigerators & freezers	0.6	Manufg.
Primary zinc	0.3	Manufg.	Change in business inventories	0.6	In House
Pumps & compressors	0.3	Manufg.	Farm housing units & additions & alterations	0.4	Constr.
Special dies & tools & machine tool accessories	0.3	Manufg.	Highway & street construction	0.4	Constr.
Equipment rental & leasing services	0.3	Services	Household laundry equipment	0.4	Manufg.
Lubricating oils & greases	0.2	Manufg.	Residential high-rise apartments	0.3	Constr.
Primary lead	0.2	Manufg.	Motor vehicles & car bodies	0.3	Manufg.
Primary metal products, nec	0.2	Manufg.	Motors & generators	0.3	Manufg.
Primary nonferrous metals, nec	0.2	Manufg.	Truck trailers	0.3	Manufg.
Eating & drinking places	0.2	Trade	Construction of religious buildings	0.2	Constr.
Computer & data processing services	0.2	Services	Telephone & telegraph facility construction	0.2	Constr.
Abrasive products	0.1	Manufg.	Household appliances, nec	0.2	Manufg.
Real estate	0.1	Fin/R.E.	Gaskets, packing & sealing devices	0.1	Manufg.
Engineering, architectural, & surveying services	0.1	Services	Metal stampings, nec	0.1	Manufg.

Source: Benchmark Input-Output Accounts for the U.S. Economy, 1982, U.S. Department of Commerce, Washington, D.C., July 1991. Data, as reported in the source, are organized by the 1977 SIC structure in use in 1982 but have been matched, as closely as is possible, to the 1987 SIC structure used in this book.

OCCUPATIONS EMPLOYED BY SIC 331 - BLAST FURNACES AND BASIC STEEL PRODUCTS

Occupation	% of Total 1994	Change to 2005	Occupation	% of Total 1994	Change to 2005
Helpers, laborers, & material movers nec	7.2	-37.8	Machinists	1.8	-37.8
Machine tool cutting & forming etc. nec	6.2	-6.7	Machine feeders & offbearers	1.8	-44.0
Blue collar worker supervisors	5.8	-40.7	Assemblers, fabricators, & hand workers nec	1.8	-37.8
Industrial machinery mechanics	4.7	-0.4	Welders & cutters	1.4	-50.2
Crane & tower operators	3.5	-37.8	Sales & related workers nec	1.4	-37.8
Inspectors, testers, & graders, precision	3.2	-37.8	Freight, stock, & material movers, hand	1.4	-50.2
Metal & plastic machine workers nec	3.1	-33.8	Truck drivers light & heavy	1.3	-35.8
Electricians	2.6	-41.6	Machine tool cutting operators, metal & plastic	1.2	-48.2
Millwrights	2.4	-31.5	Production, planning, & expediting clerks	1.2	-37.8
Machine forming operators, metal & plastic	2.2	-31.6	Material moving equipment operators nec	1.1	-37.7
Maintenance repairers, general utility	1.9	-44.0	Machine operators nec	1.1	-45.1
Industrial truck & tractor operators	1.9	-37.8	General managers & top executives	1.1	-40.9
Furnace operators	1.9	-31.6			

Source: Industry-Occupation Matrix, Bureau of Labor Statistics. These data relate to one or more 3-digit SIC industry groups rather than to a single 4-digit SIC. The change reported for each occupation to the year 2005 is a percent of growth or decline as estimated by the Bureau of Labor Statistics. The abbreviation nec stands for 'not elsewhere classified'.

LOCATION BY STATE AND REGIONAL CONCENTRATION

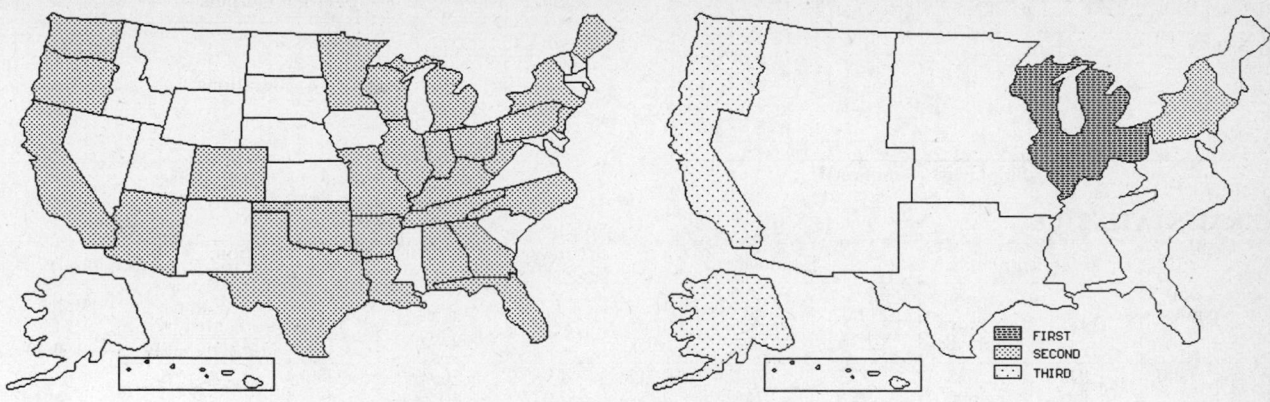

FIRST
SECOND
THIRD

INDUSTRY DATA BY STATE

| State | Establish-ments | Shipments | | | Employment | | | | Cost as % of Shipments | Investment per Employee ($) |
		Total ($ mil)	% of U.S.	Per Establ.	Total Number	% of U.S.	Per Establ.	Wages ($/hour)		
Pennsylvania	20	831.4	15.8	41.6	3,500	14.8	175	14.95	64.2	-
Ohio	25	808.5	15.4	32.3	3,900	16.5	156	14.20	62.7	3,692
Illinois	19	683.6	13.0	36.0	2,200	9.3	116	15.58	72.2	4,727
California	24	566.6	10.8	23.6	1,900	8.0	79	13.30	81.9	6,895
Indiana	15	246.4	4.7	16.4	1,200	5.1	80	12.37	59.0	4,083
Michigan	18	234.9	4.5	13.1	1,600	6.8	89	12.79	56.0	6,188
Texas	10	201.1	3.8	20.1	1,000	4.2	100	12.06	69.1	8,300
Wisconsin	6	173.4	3.3	28.9	900	3.8	150	13.83	58.7	4,556
Oklahoma	6	141.9	2.7	23.6	700	3.0	117	10.33	69.0	7,429
New Jersey	10	91.4	1.7	9.1	500	2.1	50	13.71	65.2	5,000
West Virginia	4	78.1	1.5	19.5	300	1.3	75	11.25	78.0	-
Alabama	7	71.6	1.4	10.2	400	1.7	57	14.83	72.6	-
Georgia	7	59.6	1.1	8.5	200	0.8	29	9.75	64.3	2,000
Tennessee	8	(D)	-	-	1,750 *	7.4	219	-	-	-
Colorado	6	(D)	-	-	375 *	1.6	63	-	-	-
Florida	6	(D)	-	-	375 *	1.6	63	-	-	-
Arkansas	4	(D)	-	-	750 *	3.2	188	-	-	-
Kentucky	4	(D)	-	-	375 *	1.6	94	-	-	-
Louisiana	4	(D)	-	-	375 *	1.6	94	-	-	-
New York	4	(D)	-	-	375 *	1.6	94	-	-	3,467
North Carolina	3	(D)	-	-	375 *	1.6	125	-	-	-
Arizona	2	(D)	-	-	175 *	0.7	88	-	-	571
Minnesota	2	(D)	-	-	375 *	1.6	188	-	-	-
Missouri	2	(D)	-	-	375 *	1.6	188	-	-	-
Oregon	2	(D)	-	-	175 *	0.7	88	-	-	-
Maine	1	(D)	-	-	175 *	0.7	175	-	-	-
Washington	1	(D)	-	-	175 *	0.7	175	-	-	-

Source: 1992 *Economic Census*. The states are in descending order of shipments or establishments (if shipment data are missing for the majority). The symbol (D) appears when data are withheld to prevent disclosure of competitive information. States marked with (D) are sorted by number of establishments. A dash (-) indicates that the data element cannot be calculated; * indicates the midpoint of a range.

3321 - GRAY IRON FOUNDRIES

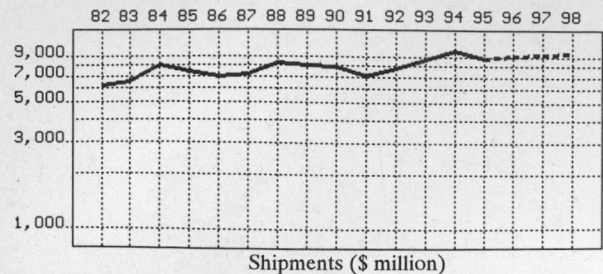

Shipments ($ million)

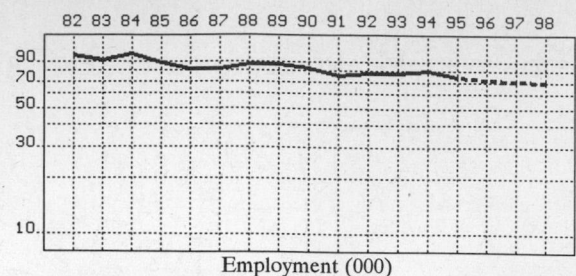

Employment (000)

GENERAL STATISTICS

Year	Companies	Establishments		Employment			Compensation		Production ($ million)			
		Total	with 20 or more employees	Total (000)	Production Workers (000)	Hours (Mil)	Payroll ($ mil)	Wages ($/hr)	Cost of Materials	Value Added by Manufacture	Value of Shipments	Capital Invest.
1982	800	925	635	97.3	78.2	139.8	1,965.0	10.46	2,840.4	3,310.5	6,202.2	348.0
1983		881	605	91.8	74.8	143.5	2,072.2	10.97	3,006.9	3,465.9	6,498.3	223.1
1984		837	575	98.4	81.1	162.4	2,429.0	11.69	3,522.6	4,432.2	7,959.6	273.9
1985		792	546	88.6	72.5	146.1	2,312.9	12.25	3,285.8	3,989.7	7,295.7	329.8
1986		747	516	81.1	65.9	132.2	2,180.4	12.73	3,119.7	3,769.0	6,919.1	393.2
1987	693	774	493	82.4	67.9	140.2	2,289.9	12.67	3,198.9	4,044.4	7,213.1	357.0
1988		722	482	87.3	72.8	154.2	2,529.9	13.08	3,861.0	4,473.9	8,277.1	359.5
1989		706	483	87.2	69.8	144.5	2,472.3	13.33	3,823.8	4,219.1	8,027.4	406.5
1990		690	476	82.9	66.4	136.2	2,428.2	13.74	3,697.8	4,111.2	7,825.3	347.0
1991		698	459	75.7	61.4	123.7	2,262.3	13.87	3,388.4	3,674.6	7,078.8	316.2
1992	641	713	427	77.1	63.2	130.4	2,461.4	14.50	3,483.6	4,331.6	7,789.2	339.3
1993		685	430	77.8	64.2	139.0	2,631.3	14.68	3,654.6	4,958.8	8,602.8	395.6
1994		632P	394P	79.6	66.0	147.6	2,897.3	15.22	4,245.0	5,458.3	9,685.5	425.2
1995		612P	376P	73.9P	60.8P	135.4P	2,727.3P	15.55P	3,901.8P	5,017.0P	8,902.5P	405.1P
1996		591P	358P	72.3P	59.5P	134.6P	2,777.0P	15.91P	3,980.6P	5,118.3P	9,082.3P	413.3P
1997		571P	340P	70.7P	58.3P	133.7P	2,826.7P	16.27P	4,059.4P	5,219.6P	9,262.0P	421.6P
1998		551P	322P	69.1P	57.0P	132.8P	2,876.4P	16.63P	4,138.2P	5,321.0P	9,441.8P	429.9P

Sources: 1982, 1987, 1992 *Economic Census*; *Annual Survey of Manufactures*, 83-86, 88-91, 93-94. Establishment counts for non-Census years are from *County Business Patterns*; establishment values for 83-84 are extrapolations. 'P's show projections by the editors. Industries reclassified in 87 will not have data for prior years.

INDICES OF CHANGE

Year	Companies	Establishments		Employment			Compensation		Production ($ million)			
		Total	with 20 or more employees	Total (000)	Production Workers (000)	Hours (Mil)	Payroll ($ mil)	Wages ($/hr)	Cost of Materials	Value Added by Manufacture	Value of Shipments	Capital Invest.
1982	125	130	149	126	124	107	80	72	82	76	80	103
1983		124	142	119	118	110	84	76	86	80	83	66
1984		117	135	128	128	125	99	81	101	102	102	81
1985		111	128	115	115	112	94	84	94	92	94	97
1986		105	121	105	104	101	89	88	90	87	89	116
1987	108	109	115	107	107	108	93	87	92	93	93	105
1988		101	113	113	115	118	103	90	111	103	106	106
1989		99	113	113	110	111	100	92	110	97	103	120
1990		97	111	108	105	104	99	95	106	95	100	102
1991		98	107	98	97	95	92	96	97	85	91	93
1992	100	100	100	100	100	100	100	100	100	100	100	100
1993		96	101	101	102	107	107	101	105	114	110	117
1994		89P	92P	103	104	113	118	105	122	126	124	125
1995		86P	88P	96P	96P	104P	111P	107P	112P	116P	114P	119P
1996		83P	84P	94P	94P	103P	113P	110P	114P	118P	117P	122P
1997		80P	80P	92P	92P	103P	115P	112P	117P	121P	119P	124P
1998		77P	75P	90P	90P	102P	117P	115P	119P	123P	121P	127P

Sources: Same as General Statistics. Values reflect change from the base year, 1992. Values above 100 mean greater than 92, values below 100 mean less than 92, and a value of 100 in the 82-91 or 93-98 period means same as 92. 'P's mark projections by the editors.

SELECTED RATIOS

For 1994	Avg. of All Manufact.	Analyzed Industry	Index	For 1994	Avg. of All Manufact.	Analyzed Industry	Index
Employees per Establishment	49	126	257	Value Added per Production Worker	134,084	82,702	62
Payroll per Establishment	1,500,273	4,583,456	306	Cost per Establishment	5,045,178	6,715,484	133
Payroll per Employee	30,620	36,398	119	Cost per Employee	102,970	53,329	52
Production Workers per Establishment	34	104	304	Cost per Production Worker	146,988	64,318	44
Wages per Establishment	853,319	3,553,863	416	Shipments per Establishment	9,576,895	15,322,220	160
Wages per Production Worker	24,861	34,037	137	Shipments per Employee	195,460	121,677	62
Hours per Production Worker	2,056	2,236	109	Shipments per Production Worker	279,017	146,750	53
Wages per Hour	12.09	15.22	126	Investment per Establishment	321,011	672,656	210
Value Added per Establishment	4,602,255	8,634,895	188	Investment per Employee	6,552	5,342	82
Value Added per Employee	93,930	68,572	73	Investment per Production Worker	9,352	6,442	69

Sources: Same as General Statistics. The 'Average of All Manufacturing' column represents the average of all manufacturing industries reported for the most recent complete year available. The Index shows the relationship between the Average and the Analyzed Industry. For example, 100 means that they are equal; 500 that the Analyzed Industry is five times the average; 50 means that the Analyzed Industry is half the national average. The abbreviation 'na' is used to show that data are 'not available'.

LEADING COMPANIES   Number shown: 75   Total sales ($ mil): 5,910   Total employment (000): 55.3

Company Name	Address				CEO Name	Phone	Co. Type	Sales ($ mil)	Empl. (000)
AMSTED Industries Inc	205 Michigan	Chicago	IL	60601	Gordon Lohman	312-645-1700	R	827	7.9
Intermet Corp	2859 Paces Ferry Rd	Atlanta	GA	30339	John Doddridge	404-431-6000	P	501	4.4
Tyler Corp	3200 S Jacinto	Dallas	TX	75201	Joseph F McKinney	214-754-7800	P	358	4.0
Intermet Foundries Inc	PO Box 6200	Lynchburg	VA	24505	G W Mathews Jr	804-528-8200	D	322	3.3
United States Pipe & Foundry	PO Box 10406	Birmingham	AL	35202	WN Temple	205-254-7000	S	300	2.7
Waupaca Foundry Inc	PO Box 249	Waupaca	WI	54981	Don G Brunner	715-258-6611	S	280	1.7
Sudbury Inc	30100 Chagrin Blv	Cleveland	OH	44124	Jacques R Sardas	216-464-7026	P	273	2.2
American Cast Iron Pipe Co	PO Box 2727	Birmingham	AL	35202	Van L Richey	205-325-7701	R	250*	2.6
Grede Foundries Inc	PO Box 26499	Milwaukee	WI	53226	Bruce Jacobs	414-257-3600	R	220	3.0
Griffin Pipe Products Co	1400 Opus Pl	Downers Grove	IL	60515	T C Fitzgerald	708-719-6500	D	175	1.3
Citation Corp	2 Office Park Cir	Birmingham	AL	35223	T Morris Hackney	205-871-5731	P	150	2.0
Neenah Foundry Co	PO Box 729	Neenah	WI	54957	Bill Aylward Jr	414-725-7000	R	120	1.2
Brillion Iron Works Inc	200 Park Av	Brillion	WI	54110	Richard Larson	414-756-2121	S	109	1.0
Castwell Products	7800 N Austin Av	Skokie	IL	60077	Marshall K Wells	708-966-5050	D	100*	0.5
Wells Manufacturing Co	7800 N Austin Av	Skokie	IL	60077	Marshall K Wells	708-966-5050	R	100	0.7
Waupaca Foundry Inc	805 Ogden St	Marinette	WI	54143	Gary Gigante	715-735-4999	D	85	0.5
Kurdziel Industries Inc	801 W Norton Av	Muskegon	MI	49441	Don Huizenga	616-739-4349	R	70	0.5
Grede Foundry Inc	711 W Alexander St	Greenwood	SC	29648	Tim Flanagan	803-227-1713	S	67*	0.4
Dalton Foundries Inc	PO Box 1388	Warsaw	IN	46580	K Davidson	219-267-8111	R	60	0.6
Talladega Foundry and Machine	PO Box 579	Talladega	AL	35160	Ray Robbins	205-362-0056	R	60	<0.1
New Haven Foundry	PO Box 480310	New Haven	MI	48048	Delbert W Mullens	810-749-5111	R	56	0.4
Donsco Inc	PO Box 2001	Wrightsville	PA	17368	Arthur K Mann	717-252-1561	R	50	0.7
Charlotte Pipe and Foundry Co	PO Box 35430	Charlotte	NC	28235	Edward Hardison	704-372-5030	R	48*	0.5
Teledyne Casting Service	300 Philadelphia St	La Porte	IN	46350	WF Lange Jr	219-362-6267	D	46	0.5
Baker Manufacturing Co	133 Enterprise St	Evansville	WI	53536	Peter Sears	608-882-5100	R	45	0.3
Alabama Ductile Casting Co	PO Box 1649	Brewton	AL	36427	Allan Adams	205-867-5481	D	40	0.4
EMI Co	603 W 12th St	Erie	PA	16501	Michael J Steiner	814-452-6431	R	40	0.4
General Casting Co	PO Box 220	Delaware	OH	43015	C Frank DeMeo	614-363-1941	R	38	0.4
Cast-Fab Technologies Inc	3040 Forrer St	Cincinnati	OH	45209	James Bushman	513-758-1000	R	37*	0.4
Advanced Cast Products Inc	PO Box 417	Meadville	PA	16335	J Hildebrand	814-724-2600	R	35	0.3
Berlin Foundry Corp	242 S Pearl St	Berlin	WI	54923	T C Butterbroudt	414-361-2220	R	35	0.4
CWC Castings	1085 W Sherman	Muskegon	MI	49441	John L Kelly	616-733-1331	D	35	0.3
Dock Foundry Co	428 4th St	Three Rivers	MI	49093	Fitz Coghlin	616-278-1765	R	35	0.1
Urick Foundry Co	15th St & Cherry St	Erie	PA	16501	W Kearns	814-454-2461	S	35	0.1
Osco Industries Inc	PO Box 1388	Portsmouth	OH	45662	Jeffery Burke	614-354-3183	R	33	0.4
Electron Corp	PO Box 318	Littleton	CO	80160	Michael Norwood	303-794-4392	R	31	0.3
AB and I	7825 San Leandro	Oakland	CA	94621	A Boscacci	510-632-3467	R	30	0.2
Southern Ductile Casting Co	2217 Carolina Av	Bessemer	AL	35020	John Ballinger	205-424-4030	D	30	0.3
East Jordan Iron Works Inc	PO Box 439	East Jordan	MI	49727	Fred F Malpass	616-536-2261	R	29*	0.3
Golden Casting Corp	PO Box 364	Columbus	IN	47202	Joel R Parson	812-372-3701	S	29	0.5
Buck Company Inc	897 Lancaster Pike	Quarryville	PA	17566	Richard McMinn	717-284-4114	S	29	0.4
Elyria Foundry	120 Filbert St	Elyria	OH	44036	Gregg Foster	216-322-4657	R	28	0.3
Anaheim Foundry Co	800 E Orangethorpe	Anaheim	CA	92801	Cecil Sills	714-870-9000	R	26	0.3
Burnham Corp	PO Box 2398	Zanesville	OH	43702	Robert B Balfantz	614-452-9371	D	26	0.3
Aarrow General Inc	2900 Richmond St	Shawano	WI	54166	Michael Jensen	715-526-3600	R	25	0.3
Aarrowcast Inc	2900 Richmond St	Shawano	WI	54166	Michael Jensen	715-526-3600	S	25	0.3
Great Lakes Castings Corp	PO Box 520	Ludington	MI	49431	Gerald Robertson	616-843-2501	R	25	0.2
Gregg Industries Inc	10460 Hickson St	El Monte	CA	91734	Robert C Gregg	818-575-7664	R	25	0.3
Vulcan Foundry Inc	PO Box 905	Denham Sp	LA	70727	James H Jenkins Jr	504-665-6151	R	25	0.2
Western Enterprises	875 Basset Rd	Westlake	OH	44145	Byron Crompton	216-933-2171	D	25	0.4
Acme Foundry Inc	PO Box 908	Coffeyville	KS	67337	Dick Tatman	316-251-6800	R	24	0.4
US Foundry & Mfg Corp	8351 NW 93rd St	Miami	FL	33166	Alex Debogory Jr	305-885-0301	R	24*	0.3
Foundry Service Co	PO Box 748	Biscoe	NC	27209	Randy Wertz	919-428-2111	S	23	0.2
Grede Perm Cast Inc	PO Box 220	Cynthiana	KY	41031	Brian Schlump	606-234-1704	D	22*	0.3
Quaker City Castings Inc	PO Box 37	Salem	OH	44460	Pat Clarke	216-332-1566	R	22	0.2
Grede-Vassar Foundry	700 E Huron Av	Vassar	MI	48768	Bill Riner	517-823-8411	D	20	0.3
Motor Castings Co	1323 S 65th St	Milwaukee	WI	53214	JJ O'Sullivan	414-476-1434	R	20	0.2
Versa Cos	867 Forest St	St Paul	MN	55106	John Moffat	612-778-3300	R	20	0.2
Durametal Corp	PO Box 606	Tualatin	OR	97062	Donald Dauterman	503-692-0850	R	19	0.2
Frazier and Frazier Inc	PO Box 279	Coolidge	TX	76635	C W Frazier Sr	817-786-2293	R	19*	0.2
Prospect Foundry Inc	1225 Winter St NE	Minneapolis	MN	55413	Richard Sitarz	612-331-9282	R	19	0.2
Sioux City Foundry Co	PO Box 3067	Sioux City	IA	51102	AM Galinsky	712-252-4181	R	19*	0.2
Universal Cast Iron Mfg Co	5404 Tweedy Pl	South Gate	CA	90280	Raul Gonzalez	213-569-8151	R	19*	0.2
Blackhawk Foundry and Machine	323 S Clark St	Davenport	IA	52803	JR Grafton Jr	319-323-3621	R	19	0.3
CENTEC	RR 5	Bethlehem	PA	18015	Ronald E Saukko	610-694-7655	J	18	0.1
Denver Thomas	PO Box 96	Birmingham	AL	35201	George D Sheehy	205-599-6600	S	18	0.2
Hamilton Foundry & Machine	200 Industrial Dr	Harrison	OH	45030	CE Rentschler	513-367-6900	R	18*	0.3
Hodge Foundry Division	PO Box 550	Greenville	PA	16125	Scott Hodge	412-588-4100	D	18	0.2
Cyprus Rod Chicago	2324 S Kenneth Av	Chicago	IL	60623	Dick Barstow	312-522-5036	D	16	0.1
Lodge Manufacturing Co	PO Box 380	South Pittsburg	TN	37380	Bob Kellermann	615-837-7181	R	16	0.3
B and B Foundry Inc	Keyst-Benrs	Philadelphia	PA	19135	Eugene Scarpa	215-333-7100	R	15	<0.1
Castalloy Corp	PO Box 827	Waukesha	WI	53187	RE Janke	414-547-0070	R	15	0.1
Decatur Casting	PO Box 450	Decatur	IN	46733	Dennis M Porda	219-724-3191	D	15	0.2
Goldens Foundry & Machine	PO Box 96	Columbus	GA	31993	George G Boyd	706-323-0471	R	15	0.2
Harper Foundry&Machine Co	PO Box 992	Jackson	MS	39205	NS Harper	601-353-8391	R	15	0.1

Source: Ward's Business Directory of U.S. Private and Public Companies, Volumes 1 and 2, 1996. The company type code used is as follows: P - Public, R - Private, S - Subsidiary, D - Division, J - Joint Venture, A - Affiliate, G - Group. Sales are in millions of dollars, employees are in thousands. An asterisk (*) indicates an estimated sales volume. The symbol < stands for 'less than'. Company names and addresses are truncated, in some cases, to fit into the available space.

MATERIALS CONSUMED

Material	Quantity	Delivered Cost ($ million)
Materials, ingredients, containers, and supplies	(X)	2,637.4
Pig iron shapes and forms	1,000 s tons (S)	82.1
Ferrochromium shapes and forms	(X)	(D)
Ferromanganese, silicomanganese, and manganese shapes and forms	(X)	30.9
Ferrosilicon (more than 8 percent silicon) shapes and forms	(X)	87.0
All other ferrous shapes and forms	(X)	45.3
Nickel and nickel-base alloy shapes and forms	(X)	17.1
Cobalt-base alloy shapes and forms	(X)	(D)
All other nonferrous shapes and forms	(X)	52.3
Iron and steel scrap, excluding home scrap	1,000 s tons 7,063.9*	762.0
All other purchased scrap, excluding home scrap	1,000 s tons 252.0*	23.1
Clay refractories	(X)	31.9
Grinding wheels and other abrasive products, except industrial diamonds	(X)	38.6
Nonclay refractories	(X)	9.5
All other stone, clay, glass, and concrete products	(X)	6.7
Industrial patterns	(X)	47.5
Industrial dies, molds, jigs, and fixtures	(X)	6.0
All other industrial and commercial machinery and computer equipment	(X)	22.7
Sand	(X)	110.2
All other materials and components, parts, containers, and supplies	(X)	1,109.1
Materials, ingredients, containers, and supplies, nsk	(X)	149.4

Source: 1992 Economic Census. Explanation of symbols used: (D): Withheld to avoid disclosure of competitive data; na: Not available; (S): Withheld because statistical norms were not met; (X): Not applicable; (Z): Less than half the unit shown; nec: Not elsewhere classified; nsk: Not specified by kind; - : zero; * : 10-19 percent estimated; ** : 20-29 percent estimated.

PRODUCT SHARE DETAILS

Product or Product Class	% Share	Product or Product Class	% Share
Gray and ductile iron foundries	100.00	diameter).	20.09
Ductile iron pressure pipe and fittings	15.05	Cast iron pressure pipe and fittings, nsk	4.70
Ductile iron pressure pipe, less than 14 in. (inside diameter)	78.01	Cast iron soil pipe and fittings (including special fittings)	2.53
Ductile iron pressure pipe, 14 in. or more (inside diameter).	8.65	Cast iron soil pipe, 3 in. or less (inside diameter)	19.67
Ductile iron fittings, less than 14 in. (inside diameter).	9.49	Cast iron soil pipe, more than 3 in. up to but not including 5 in. (inside diameter)	25.44
Ductile iron fittings, 14 in. or more (inside diameter)	3.85	Cast iron soil pipe, 5 in. or more (inside diameter).	18.88
Other ductile iron castings	23.10	Cast iron soil pipe fittings (including special fittings), 3 in. or less (inside diameter).	11.58
Other ductile iron castings for automotive uses	59.57		
Other ductile iron castings for construction and utility uses	4.13	Cast iron soil pipe fittings (including special fittings), more than 3 in. up to but not including 5 in. (inside diameter)	13.91
Other ductile iron castings for machinery uses	16.41		
Other ductile iron castings for valve uses	3.34	Cast iron soil pipe fittings (including special fittings), 5 in. or more (inside diameter)	9.31
Other ductile iron castings for all other uses, including electric and electronic equipment uses, heat-resistant parts, and coke oven door parts	16.03	Cast iron soil pipe and fittings (including special fittings), nsk	1.22
Other ductile iron castings, nsk	0.52	Other gray iron castings	52.99
Molds and stools for heavy steel ingots	1.50	Gray iron rolls for rolling mills	3.01
Cast iron pressure pipe and fittings	0.63	Other gray iron castings for automotive uses.	47.50
Cast iron pressure pipe, less than 14 in. (inside diameter)	12.18	Other gray iron castings for construction and utility uses.	12.37
Cast iron pressure pipe, 14 in. or more (inside diameter).	4.06	Other gray iron castings for all other uses.	36.19
Cast iron pressure pipe fittings, less than 14 in. (inside diameter).	58.97	Other gray iron castings, nsk	0.93
Cast iron pressure pipe fittings, 14 in. or more (inside		Gray and ductile iron foundries, nsk	4.20

Source: 1992 Economic Census. The values shown are percent of total shipments in an industry. Values of indented subcategories are summed in the main heading. The symbol (D) appears when data are withheld to prevent disclosure of competitive information. The abbreviation nsk stands for 'not specified by kind' and nec for 'not elsewhere classified'.

INPUTS AND OUTPUTS FOR IRON & STEEL FOUNDRIES

Economic Sector or Industry Providing Inputs	%	Sector	Economic Sector or Industry Buying Outputs	%	Sector
Electric services (utilities)	10.7	Util.	Motor vehicle parts & accessories	21.9	Manufg.
Miscellaneous repair shops	10.0	Services	Internal combustion engines, nec	7.6	Manufg.
Wholesale trade	7.2	Trade	Construction machinery & equipment	5.0	Manufg.
Scrap	7.1	Scrap	Pipe, valves, & pipe fittings	4.9	Manufg.
Electrometallurgical products	4.4	Manufg.	Pumps & compressors	4.2	Manufg.
Gas production & distribution (utilities)	4.2	Util.	Motor vehicles & car bodies	4.1	Manufg.
Advertising	4.2	Services	Oil field machinery	3.9	Manufg.
Maintenance of nonfarm buildings nec	4.1	Constr.	Water supply facility construction	3.1	Constr.
Imports	3.1	Foreign	Blast furnaces & steel mills	2.5	Manufg.
Primary nonferrous metals, nec	3.0	Manufg.	Farm machinery & equipment	2.5	Manufg.
Communications, except radio & TV	3.0	Util.	Sewer system facility construction	2.1	Constr.
Nonferrous metal ores, except copper	2.8	Mining	Exports	1.8	Foreign
Blast furnaces & steel mills	2.8	Manufg.	Railroad equipment	1.5	Manufg.
Railroads & related services	2.0	Util.	Machine tools, metal cutting types	1.4	Manufg.
Motor freight transportation & warehousing	1.8	Util.	Residential 1-unit structures, nonfarm	1.3	Constr.

Continued on next page.

INPUTS AND OUTPUTS FOR IRON & STEEL FOUNDRIES - Continued

Economic Sector or Industry Providing Inputs	%	Sector	Economic Sector or Industry Buying Outputs	%	Sector
Sand & gravel	1.7	Mining	Power transmission equipment	1.2	Manufg.
Power transmission equipment	1.6	Manufg.	Railroads & related services	1.2	Util.
Abrasive products	1.5	Manufg.	Aircraft & missile engines & engine parts	1.1	Manufg.
Metal heat treating	1.3	Manufg.	Refrigeration & heating equipment	1.1	Manufg.
Computer & data processing services	1.3	Services	Machinery, except electrical, nec	1.0	Manufg.
Petroleum refining	1.1	Manufg.	Office buildings	0.8	Constr.
Special dies & tools & machine tool accessories	1.1	Manufg.	Motors & generators	0.8	Manufg.
Carbon & graphite products	0.9	Manufg.	Printing trades machinery	0.8	Manufg.
Industrial patterns	0.8	Manufg.	Turbines & turbine generator sets	0.8	Manufg.
Machinery, except electrical, nec	0.8	Manufg.	Maintenance of water supply facilities	0.7	Constr.
Primary metal products, nec	0.7	Manufg.	Residential garden apartments	0.7	Constr.
Eating & drinking places	0.7	Trade	Electronic computing equipment	0.7	Manufg.
Banking	0.7	Fin/R.E.	Heating equipment, except electric	0.7	Manufg.
Equipment rental & leasing services	0.7	Services	Mining machinery, except oil field	0.7	Manufg.
Metal coating & allied services	0.6	Manufg.	General industrial machinery, nec	0.6	Manufg.
Miscellaneous plastics products	0.6	Manufg.	Industrial trucks & tractors	0.6	Manufg.
Sanitary services, steam supply, irrigation	0.6	Util.	Special industry machinery, nec	0.6	Manufg.
Fabricated metal products, nec	0.5	Manufg.	Copper ore	0.5	Mining
Internal combustion engines, nec	0.5	Manufg.	Industrial buildings	0.5	Constr.
Iron & steel foundries	0.5	Manufg.	Food products machinery	0.5	Manufg.
Sawmills & planning mills, general	0.5	Manufg.	Hardware, nec	0.5	Manufg.
Industrial controls	0.4	Manufg.	Paper industries machinery	0.5	Manufg.
Metal stampings, nec	0.4	Manufg.	Special dies & tools & machine tool accessories	0.5	Manufg.
Primary copper	0.4	Manufg.	Tanks & tank components	0.5	Manufg.
Water transportation	0.4	Util.	Highway & street construction	0.4	Constr.
Engineering, architectural, & surveying services	0.4	Services	Maintenance of nonbuilding facilities nec	0.4	Constr.
U.S. Postal Service	0.4	Gov't	Ball & roller bearings	0.4	Manufg.
Clay, ceramic, & refractory minerals	0.3	Mining	Carburetors, pistons, rings, & valves	0.4	Manufg.
Chemical preparations, nec	0.3	Manufg.	Hand & edge tools, nec	0.4	Manufg.
Cyclic crudes and organics	0.3	Manufg.	Machine tools, metal forming types	0.4	Manufg.
Lubricating oils & greases	0.3	Manufg.	Mechanical measuring devices	0.4	Manufg.
Screw machine and related products	0.3	Manufg.	Iron & ferroalloy ores	0.3	Mining
Welding apparatus, electric	0.3	Manufg.	Construction of educational buildings	0.3	Constr.
Real estate	0.3	Fin/R.E.	Construction of hospitals	0.3	Constr.
Electrical repair shops	0.3	Services	Nonfarm residential structure maintenance	0.3	Constr.
Legal services	0.3	Services	Blowers & fans	0.3	Manufg.
Management & consulting services & labs	0.3	Services	Conveyors & conveying equipment	0.3	Manufg.
Chemical & fertilizer mineral	0.2	Mining	Fabricated metal products, nec	0.3	Manufg.
Ball & roller bearings	0.2	Manufg.	Fabricated plate work (boiler shops)	0.3	Manufg.
Nonferrous wire drawing & insulating	0.2	Manufg.	Fabricated rubber products, nec	0.3	Manufg.
Primary aluminum	0.2	Manufg.	Iron & steel foundries	0.3	Manufg.
Pumps & compressors	0.2	Manufg.	Motorcycles, bicycles, & parts	0.3	Manufg.
Air transportation	0.2	Util.	Small arms	0.3	Manufg.
Insurance carriers	0.2	Fin/R.E.	Textile machinery	0.3	Manufg.
Accounting, auditing & bookkeeping	0.2	Services	Federal Government purchases, national defense	0.3	Fed Govt
Automotive repair shops & services	0.2	Services	Electric utility facility construction	0.2	Constr.
State & local government enterprises, nec	0.2	Gov't	Maintenance of nonfarm buildings nec	0.2	Constr.
Coal	0.1	Mining	Residential high-rise apartments	0.2	Constr.
Blowers & fans	0.1	Manufg.	Telephone & telegraph facility construction	0.2	Constr.
Industrial gases	0.1	Manufg.	Elevators & moving stairways	0.2	Manufg.
Manifold business forms	0.1	Manufg.	Hoists, cranes, & monorails	0.2	Manufg.
Nonclay refractories	0.1	Manufg.	Household laundry equipment	0.2	Manufg.
Primary lead	0.1	Manufg.	Lighting fixtures & equipment	0.2	Manufg.
Rubber & plastics hose & belting	0.1	Manufg.	Miscellaneous metal work	0.2	Manufg.
Wood pallets & skids	0.1	Manufg.	Mobile homes	0.2	Manufg.
Water supply & sewage systems	0.1	Util.	Radio & TV communication equipment	0.2	Manufg.
Royalties	0.1	Fin/R.E.	Service industry machines, nec	0.2	Manufg.
Automotive rental & leasing, without drivers	0.1	Services	Ship building & repairing	0.2	Manufg.
Laundry, dry cleaning, shoe repair	0.1	Services	Surgical appliances & supplies	0.2	Manufg.
			Welding apparatus, electric	0.2	Manufg.
			Wiring devices	0.2	Manufg.
			Woodworking machinery	0.2	Manufg.
			Construction of stores & restaurants	0.1	Constr.
			Hotels & motels	0.1	Constr.
			Residential 2-4 unit structures, nonfarm	0.1	Constr.
			Aircraft & missile equipment, nec	0.1	Manufg.
			Fabricated structural metal	0.1	Manufg.
			Lawn & garden equipment	0.1	Manufg.
			Measuring & dispensing pumps	0.1	Manufg.
			Metal stampings, nec	0.1	Manufg.
			Metalworking machinery, nec	0.1	Manufg.
			Plumbing fixture fittings & trim	0.1	Manufg.
			Scales & balances	0.1	Manufg.
			Sheet metal work	0.1	Manufg.
			Steel springs, except wire	0.1	Manufg.

Continued on next page.

INPUTS AND OUTPUTS FOR IRON & STEEL FOUNDRIES - Continued

Economic Sector or Industry Providing Inputs	%	Sector	Economic Sector or Industry Buying Outputs	%	Sector
			Surgical & medical instruments	0.1	Manufg.
			Typewriters & office machines, nec	0.1	Manufg.

Source: Benchmark Input-Output Accounts for the U.S. Economy, 1982, U.S. Department of Commerce, Washington, D.C., July 1991. Data, as reported in the source, are organized by the 1977 SIC structure in use in 1982 but have been matched, as closely as is possible, to the 1987 SIC structure used in this book.

OCCUPATIONS EMPLOYED BY SIC 332 - IRON AND STEEL FOUNDRIES

Occupation	% of Total 1994	Change to 2005	Occupation	% of Total 1994	Change to 2005
Assemblers, fabricators, & hand workers nec	6.4	-28.8	Metal & plastic machine workers nec	2.2	-37.0
Grinders & polishers, hand	6.4	24.5	Precision metal workers nec	2.0	-28.9
Metal molding machine workers	6.2	-11.1	Grinding machine operators, metal & plastic	2.0	-53.8
Helpers, laborers, & material movers nec	5.9	-28.8	Maintenance repairers, general utility	1.7	-36.0
Precision workers nec	5.4	-28.9	Electricians	1.7	-33.2
Blue collar worker supervisors	5.2	-31.8	Machinists	1.5	-28.9
Foundry mold assembly & shakeout workers	5.0	-28.9	General managers & top executives	1.5	-32.5
Inspectors, testers, & graders, precision	4.2	-28.9	Crane & tower operators	1.3	-28.9
Industrial machinery mechanics	3.3	-21.8	Sales & related workers nec	1.1	-28.8
Welders & cutters	2.9	-28.9	Machine tool cutting operators, metal & plastic	1.1	-40.8
Furnace operators	2.5	-14.6	Machine operators nec	1.1	-37.3
Industrial truck & tractor operators	2.4	-28.8	Janitors & cleaners, incl maids	1.0	-43.0

Source: Industry-Occupation Matrix, Bureau of Labor Statistics. These data relate to one or more 3-digit SIC industry groups rather than to a single 4-digit SIC. The change reported for each occupation to the year 2005 is a percent of growth or decline as estimated by the Bureau of Labor Statistics. The abbreviation nec stands for 'not elsewhere classified'.

LOCATION BY STATE AND REGIONAL CONCENTRATION

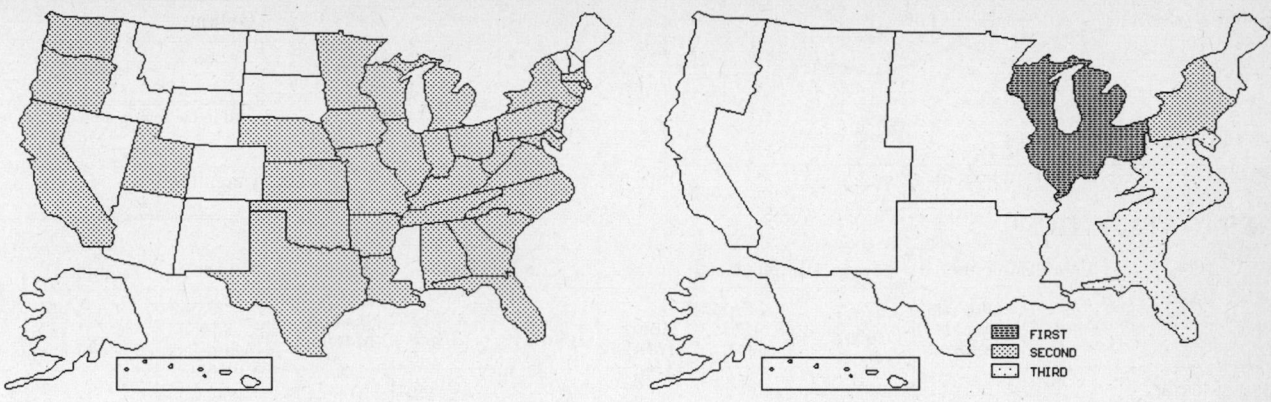

FIRST
SECOND
THIRD

INDUSTRY DATA BY STATE

State	Establish-ments	Shipments			Employment				Cost as % of Shipments	Investment per Employee ($)
		Total ($ mil)	% of U.S.	Per Establ.	Total Number	% of U.S.	Per Establ.	Wages ($/hour)		
Ohio	84	1,317.4	16.9	15.7	11,900	15.4	142	19.19	41.9	6,748
Wisconsin	42	834.0	10.7	19.9	8,000	10.4	190	13.53	45.1	4,063
Alabama	37	833.0	10.7	22.5	6,800	8.8	184	12.38	60.1	4,250
Michigan	54	757.7	9.7	14.0	8,300	10.8	154	17.89	39.6	4,711
Indiana	43	731.0	9.4	17.0	6,900	8.9	160	15.07	41.0	2,609
Illinois	36	469.4	6.0	13.0	4,600	6.0	128	18.20	41.8	3,000
Pennsylvania	65	416.6	5.3	6.4	4,500	5.8	69	12.21	40.1	4,022
Texas	30	353.1	4.5	11.8	4,700	6.1	157	11.80	42.6	2,489
Tennessee	23	273.4	3.5	11.9	3,100	4.0	135	11.63	50.4	3,774
Iowa	12	203.7	2.6	17.0	1,800	2.3	150	18.17	49.2	-
New Jersey	10	182.1	2.3	18.2	900	1.2	90	12.67	49.0	5,222
California	39	167.5	2.2	4.3	1,600	2.1	41	11.87	47.1	7,000
Missouri	14	104.6	1.3	7.5	1,000	1.3	71	11.39	33.2	-
Minnesota	17	73.3	0.9	4.3	800	1.0	47	11.79	36.4	4,250
Oklahoma	11	70.1	0.9	6.4	900	1.2	82	11.79	52.4	2,000
Massachusetts	18	41.4	0.5	2.3	600	0.8	33	11.78	42.0	1,333
New York	17	35.4	0.5	2.1	500	0.6	29	11.29	38.4	1,400
Florida	11	30.9	0.4	2.8	300	0.4	27	11.80	43.4	1,667
Arkansas	6	21.2	0.3	3.5	300	0.4	50	8.50	49.1	-
North Carolina	19	(D)	-	-	1,750 *	2.3	92	-	-	-
Georgia	16	(D)	-	-	1,750 *	2.3	109	-	-	-
Virginia	16	(D)	-	-	3,750 *	4.9	234	-	-	-
Kansas	11	(D)	-	-	750 *	1.0	68	-	-	-
South Carolina	9	(D)	-	-	750 *	1.0	83	-	-	-
Washington	7	(D)	-	-	175 *	0.2	25	-	-	3,429
West Virginia	7	(D)	-	-	175 *	0.2	25	-	-	-
Connecticut	6	(D)	-	-	175 *	0.2	29	-	-	-
Louisiana	6	(D)	-	-	175 *	0.2	29	-	-	1,143
Oregon	6	(D)	-	-	175 *	0.2	29	-	-	571
Utah	6	(D)	-	-	375 *	0.5	63	-	-	-
Nebraska	5	(D)	-	-	375 *	0.5	75	-	-	-
Kentucky	4	(D)	-	-	375 *	0.5	94	-	-	-
Maryland	4	(D)	-	-	175 *	0.2	44	-	-	-

Source: 1992 *Economic Census*. The states are in descending order of shipments or establishments (if shipment data are missing for the majority). The symbol (D) appears when data are withheld to prevent disclosure of competitive information. States marked with (D) are sorted by number of establishments. A dash (-) indicates that the data element cannot be calculated; * indicates the midpoint of a range.

3322 - MALLEABLE IRON FOUNDRIES

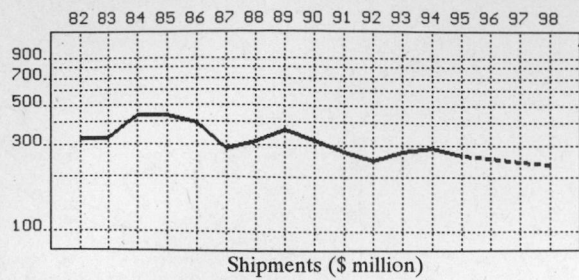

Shipments ($ million)

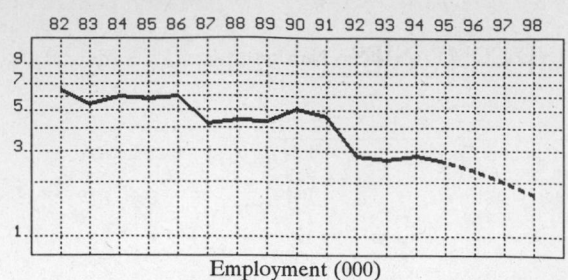

Employment (000)

GENERAL STATISTICS

Year	Companies	Establishments Total	Establishments with 20 or more employees	Employment Total (000)	Employment Production Workers (000)	Employment Hours (Mil)	Compensation Payroll ($ mil)	Compensation Wages ($/hr)	Production Cost of Materials	Production Value Added by Manufacture	Production Value of Shipments	Capital Invest.
1982	46	50	31	6.5	4.8	8.2	135.7	10.89	112.9	206.3	323.2	11.9
1983		46	30	5.4	4.0	7.1	125.2	11.07	111.9	213.3	324.0	6.6
1984		42	29	6.0	4.7	9.2	157.5	11.27	165.4	270.1	431.9	22.1
1985		39	29	5.8	4.5	8.5	159.8	12.09	168.2	264.6	432.9	16.9
1986		38	27	6.0	4.2	8.0	156.6	12.25	172.2	221.3	396.7	15.6
1987	27	28	19	4.2	3.0	5.9	121.4	13.12	102.6	181.6	283.4	3.6
1988		27	19	4.4	3.3	6.5	111.3	13.20	111.1	202.4	314.3	7.1
1989		28	18	4.3	3.8	7.3	145.1	13.37	130.5	233.3	363.9	4.5
1990		28	20	5.1	3.5	6.8	160.3	14.96	124.8	196.8	320.6	16.2
1991		30	19	4.6	3.3	6.2	144.0	14.56	121.1	156.0	276.5	10.6
1992	24	24	11	2.8	2.4	4.8	100.9	17.67	102.3	144.3	247.7	3.8
1993		22	10	2.7	2.2	5.0	110.5	18.54	120.3	156.4	275.3	4.1
1994		18P	9P	2.8	2.4	5.0	112.3	18.90	112.4	180.3	292.1	4.8
1995		16P	8P	2.6P	2.1P	4.7P	116.3P	18.76P	101.5P	162.8P	263.7P	4.3P
1996		13P	6P	2.3P	1.9P	4.4P	113.8P	19.45P	97.9P	157.0P	254.4P	3.5P
1997		11P	4P	2.0P	1.7P	4.1P	111.3P	20.13P	94.3P	151.2P	245.0P	2.7P
1998		9P	2P	1.7P	1.5P	3.8P	108.8P	20.81P	90.7P	145.4P	235.6P	1.9P

Sources: 1982, 1987, 1992 *Economic Census*; *Annual Survey of Manufactures*, 83-86, 88-91, 93-94. Establishment counts for non-Census years are from *County Business Patterns*; establishment values for 83-84 are extrapolations. 'P's show projections by the editors. Industries reclassified in 87 will not have data for prior years.

INDICES OF CHANGE

Year	Companies	Establishments Total	Establishments with 20 or more employees	Employment Total (000)	Employment Production Workers (000)	Employment Hours (Mil)	Compensation Payroll ($ mil)	Compensation Wages ($/hr)	Production Cost of Materials	Production Value Added by Manufacture	Production Value of Shipments	Capital Invest.
1982	192	208	282	232	200	171	134	62	110	143	130	313
1983		192	273	193	167	148	124	63	109	148	131	174
1984		175	264	214	196	192	156	64	162	187	174	582
1985		163	264	207	187	177	158	68	164	183	175	445
1986		158	245	214	175	167	155	69	168	153	160	411
1987	113	117	173	150	125	123	120	74	100	126	114	95
1988		113	173	157	137	135	110	75	109	140	127	187
1989		117	164	154	158	152	144	76	128	162	147	118
1990		117	182	182	146	142	159	85	122	136	129	426
1991		125	173	164	137	129	143	82	118	108	112	279
1992	100	100	100	100	100	100	100	100	100	100	100	100
1993		92	91	96	92	104	110	105	118	108	111	108
1994		76P	86P	100	100	104	111	107	110	125	118	126
1995		66P	69P	93P	89P	98P	115P	106P	99P	113P	106P	112P
1996		56P	52P	82P	80P	91P	113P	110P	96P	109P	103P	91P
1997		46P	34P	71P	72P	85P	110P	114P	92P	105P	99P	70P
1998		36P	17P	61P	64P	79P	108P	118P	89P	101P	95P	49P

Sources: Same as General Statistics. Values reflect change from the base year, 1992. Values above 100 mean greater than 92, values below 100 mean less than 92, and a value of 100 in the 82-91 or 93-98 period means same as 92. 'P's mark projections by the editors.

SELECTED RATIOS

For 1994	Avg. of All Manufact.	Analyzed Industry	Index	For 1994	Avg. of All Manufact.	Analyzed Industry	Index
Employees per Establishment	49	154	315	Value Added per Production Worker	134,084	75,125	56
Payroll per Establishment	1,500,273	6,191,980	413	Cost per Establishment	5,045,178	6,197,494	123
Payroll per Employee	30,620	40,107	131	Cost per Employee	102,970	40,143	39
Production Workers per Establishment	34	132	386	Cost per Production Worker	146,988	46,833	32
Wages per Establishment	853,319	5,210,526	611	Shipments per Establishment	9,576,895	16,105,764	168
Wages per Production Worker	24,861	39,375	158	Shipments per Employee	195,460	104,321	53
Hours per Production Worker	2,056	2,083	101	Shipments per Production Worker	279,017	121,708	44
Wages per Hour	12.09	18.90	156	Investment per Establishment	321,011	264,662	82
Value Added per Establishment	4,602,255	9,941,353	216	Investment per Employee	6,552	1,714	26
Value Added per Employee	93,930	64,393	69	Investment per Production Worker	9,352	2,000	21

Sources: Same as General Statistics. The 'Average of All Manufacturing' column represents the average of all manufacturing industries reported for the most recent complete year available. The Index shows the relationship between the Average and the Analyzed Industry. For example, 100 means that they are equal; 500 that the Analyzed Industry is five times the average; 50 means that the Analyzed Industry is half the national average. The abbreviation 'na' is used to show that data are 'not available'.

LEADING COMPANIES   Number shown: 8   Total sales ($ mil): 377   Total employment (000): 4.2

Company Name	Address				CEO Name	Phone	Co. Type	Sales ($ mil)	Empl. (000)
Wagner Castings Co	PO Box 1319	Decatur	IL	62525	James F Mason	217-428-7791	S	114	1.1
Duriron Company Inc	450 N Findlay St	Dayton	OH	45404	Mark E Armstrong	513-226-4000	D	100*	1.0
Texas Foundries	PO Box 3718	Lufkin	TX	75903	James Milstead	409-632-4451	D	70	0.9
CMI Cast Parts Inc	1500 4th Av	Cadillac	MI	49601	Ray Witt	616-779-9600	S	52	0.6
Lancaster Malleable Castings	1170 Lititz Av	Lancaster	PA	17601	J Robert Hess	717-295-8200	R	20	0.3
Belcher	558 Foundry St	Easton	MA	02334	Richard Porter	508-238-2072	D	10*	0.1
Milwaukee Malleable	PO Box 2039	Milwaukee	WI	53201	RL Gutenkunst	414-645-0200	S	7	<0.1
Staver Foundry Company Inc	100 S 10th St	Virginia	MN	55792	Leland Russell	218-741-4122	R	5	<0.1

Source: Ward's Business Directory of U.S. Private and Public Companies, Volumes 1 and 2, 1996. The company type code used is as follows: P - Public, R - Private, S - Subsidiary, D - Division, J - Joint Venture, A - Affiliate, G - Group. Sales are in millions of dollars, employees are in thousands. An asterisk (*) indicates an estimated sales volume. The symbol < stands for 'less than'. Company names and addresses are truncated, in some cases, to fit into the available space.

MATERIALS CONSUMED

Material	Quantity		Delivered Cost ($ million)
Materials, ingredients, containers, and supplies		(X)	72.6
Pig iron shapes and forms	1,000 s tons	(S)	1.9
Ferrochromium shapes and forms		(X)	(D)
Ferrosilicon (more than 8 percent silicon) shapes and forms		(X)	(D)
All other ferrous shapes and forms		(X)	0.3
Nickel and nickel-base alloy shapes and forms		(X)	(D)
All other nonferrous shapes and forms		(X)	(D)
Iron and steel scrap, excluding home scrap	1,000 s tons	43.7	13.9
All other purchased scrap, excluding home scrap	1,000 s tons	(D)	(D)
Clay refractories		(X)	(D)
Grinding wheels and other abrasive products, except industrial diamonds		(X)	0.3
All other stone, clay, glass, and concrete products		(X)	(D)
Industrial patterns		(X)	(D)
Industrial dies, molds, jigs, and fixtures		(X)	(D)
All other industrial and commercial machinery and computer equipment		(X)	(D)
Sand		(X)	0.5
All other materials and components, parts, containers, and supplies		(X)	50.5
Materials, ingredients, containers, and supplies, nsk		(X)	2.6

Source: 1992 Economic Census. Explanation of symbols used: (D): Withheld to avoid disclosure of competitive data; na: Not available; (S): Withheld because statistical norms were not met; (X): Not applicable; (Z): Less than half the unit shown; nec: Not elsewhere classified; nsk: Not specified by kind; - : zero; * : 10-19 percent estimated; ** : 20-29 percent estimated.

PRODUCT SHARE DETAILS

Product or Product Class	% Share	Product or Product Class	% Share
Malleable iron foundries	100.00	Pearlitic malleable iron castings	34.95
Standard malleable iron castings	62.50	Malleable iron foundries, nsk	2.55

Source: 1992 Economic Census. The values shown are percent of total shipments in an industry. Values of indented subcategories are summed in the main heading. The symbol (D) appears when data are withheld to prevent disclosure of competitive information. The abbreviation nsk stands for 'not specified by kind' and nec for 'not elsewhere classified'.

INPUTS AND OUTPUTS FOR IRON & STEEL FOUNDRIES

Economic Sector or Industry Providing Inputs	%	Sector	Economic Sector or Industry Buying Outputs	%	Sector
Electric services (utilities)	10.7	Util.	Motor vehicle parts & accessories	21.9	Manufg.
Miscellaneous repair shops	10.0	Services	Internal combustion engines, nec	7.6	Manufg.
Wholesale trade	7.2	Trade	Construction machinery & equipment	5.0	Manufg.
Scrap	7.1	Scrap	Pipe, valves, & pipe fittings	4.9	Manufg.
Electrometallurgical products	4.4	Manufg.	Pumps & compressors	4.2	Manufg.
Gas production & distribution (utilities)	4.2	Util.	Motor vehicles & car bodies	4.1	Manufg.
Advertising	4.2	Services	Oil field machinery	3.9	Manufg.
Maintenance of nonfarm buildings nec	4.1	Constr.	Water supply facility construction	3.1	Constr.
Imports	3.1	Foreign	Blast furnaces & steel mills	2.5	Manufg.
Primary nonferrous metals, nec	3.0	Manufg.	Farm machinery & equipment	2.5	Manufg.
Communications, except radio & TV	3.0	Util.	Sewer system facility construction	2.1	Constr.
Nonferrous metal ores, except copper	2.8	Mining	Exports	1.8	Foreign
Blast furnaces & steel mills	2.8	Manufg.	Railroad equipment	1.5	Manufg.
Railroads & related services	2.0	Util.	Machine tools, metal cutting types	1.4	Manufg.
Motor freight transportation & warehousing	1.8	Util.	Residential 1-unit structures, nonfarm	1.3	Constr.
Sand & gravel	1.7	Mining	Power transmission equipment	1.2	Manufg.

Continued on next page.

INPUTS AND OUTPUTS FOR IRON & STEEL FOUNDRIES - Continued

Economic Sector or Industry Providing Inputs	%	Sector	Economic Sector or Industry Buying Outputs	%	Sector
Power transmission equipment	1.6	Manufg.	Railroads & related services	1.2	Util.
Abrasive products	1.5	Manufg.	Aircraft & missile engines & engine parts	1.1	Manufg.
Metal heat treating	1.3	Manufg.	Refrigeration & heating equipment	1.1	Manufg.
Computer & data processing services	1.3	Services	Machinery, except electrical, nec	1.0	Manufg.
Petroleum refining	1.1	Manufg.	Office buildings	0.8	Constr.
Special dies & tools & machine tool accessories	1.1	Manufg.	Motors & generators	0.8	Manufg.
Carbon & graphite products	0.9	Manufg.	Printing trades machinery	0.8	Manufg.
Industrial patterns	0.8	Manufg.	Turbines & turbine generator sets	0.8	Manufg.
Machinery, except electrical, nec	0.8	Manufg.	Maintenance of water supply facilities	0.7	Constr.
Primary metal products, nec	0.7	Manufg.	Residential garden apartments	0.7	Constr.
Eating & drinking places	0.7	Trade	Electronic computing equipment	0.7	Manufg.
Banking	0.7	Fin/R.E.	Heating equipment, except electric	0.7	Manufg.
Equipment rental & leasing services	0.7	Services	Mining machinery, except oil field	0.7	Manufg.
Metal coating & allied services	0.6	Manufg.	General industrial machinery, nec	0.6	Manufg.
Miscellaneous plastics products	0.6	Manufg.	Industrial trucks & tractors	0.6	Manufg.
Sanitary services, steam supply, irrigation	0.6	Util.	Special industry machinery, nec	0.6	Manufg.
Fabricated metal products, nec	0.5	Manufg.	Copper ore	0.5	Mining
Internal combustion engines, nec	0.5	Manufg.	Industrial buildings	0.5	Constr.
Iron & steel foundries	0.5	Manufg.	Food products machinery	0.5	Manufg.
Sawmills & planning mills, general	0.5	Manufg.	Hardware, nec	0.5	Manufg.
Industrial controls	0.4	Manufg.	Paper industries machinery	0.5	Manufg.
Metal stampings, nec	0.4	Manufg.	Special dies & tools & machine tool accessories	0.5	Manufg.
Primary copper	0.4	Manufg.	Tanks & tank components	0.5	Manufg.
Water transportation	0.4	Util.	Highway & street construction	0.4	Constr.
Engineering, architectural, & surveying services	0.4	Services	Maintenance of nonbuilding facilities nec	0.4	Constr.
U.S. Postal Service	0.4	Gov't	Ball & roller bearings	0.4	Manufg.
Clay, ceramic, & refractory minerals	0.3	Mining	Carburetors, pistons, rings, & valves	0.4	Manufg.
Chemical preparations, nec	0.3	Manufg.	Hand & edge tools, nec	0.4	Manufg.
Cyclic crudes and organics	0.3	Manufg.	Machine tools, metal forming types	0.4	Manufg.
Lubricating oils & greases	0.3	Manufg.	Mechanical measuring devices	0.4	Manufg.
Screw machine and related products	0.3	Manufg.	Iron & ferroalloy ores	0.3	Mining
Welding apparatus, electric	0.3	Manufg.	Construction of educational buildings	0.3	Constr.
Real estate	0.3	Fin/R.E.	Construction of hospitals	0.3	Constr.
Electrical repair shops	0.3	Services	Nonfarm residential structure maintenance	0.3	Constr.
Legal services	0.3	Services	Blowers & fans	0.3	Manufg.
Management & consulting services & labs	0.3	Services	Conveyors & conveying equipment	0.3	Manufg.
Chemical & fertilizer mineral	0.2	Mining	Fabricated metal products, nec	0.3	Manufg.
Ball & roller bearings	0.2	Manufg.	Fabricated plate work (boiler shops)	0.3	Manufg.
Nonferrous wire drawing & insulating	0.2	Manufg.	Fabricated rubber products, nec	0.3	Manufg.
Primary aluminum	0.2	Manufg.	Iron & steel foundries	0.3	Manufg.
Pumps & compressors	0.2	Manufg.	Motorcycles, bicycles, & parts	0.3	Manufg.
Air transportation	0.2	Util.	Small arms	0.3	Manufg.
Insurance carriers	0.2	Fin/R.E.	Textile machinery	0.3	Manufg.
Accounting, auditing & bookkeeping	0.2	Services	Federal Government purchases, national defense	0.3	Fed Govt
Automotive repair shops & services	0.2	Services	Electric utility facility construction	0.2	Constr.
State & local government enterprises, nec	0.2	Gov't	Maintenance of nonfarm buildings nec	0.2	Constr.
Coal	0.1	Mining	Residential high-rise apartments	0.2	Constr.
Blowers & fans	0.1	Manufg.	Telephone & telegraph facility construction	0.2	Constr.
Industrial gases	0.1	Manufg.	Elevators & moving stairways	0.2	Manufg.
Manifold business forms	0.1	Manufg.	Hoists, cranes, & monorails	0.2	Manufg.
Nonclay refractories	0.1	Manufg.	Household laundry equipment	0.2	Manufg.
Primary lead	0.1	Manufg.	Lighting fixtures & equipment	0.2	Manufg.
Rubber & plastics hose & belting	0.1	Manufg.	Miscellaneous metal work	0.2	Manufg.
Wood pallets & skids	0.1	Manufg.	Mobile homes	0.2	Manufg.
Water supply & sewage systems	0.1	Util.	Radio & TV communication equipment	0.2	Manufg.
Royalties	0.1	Fin/R.E.	Service industry machines, nec	0.2	Manufg.
Automotive rental & leasing, without drivers	0.1	Services	Ship building & repairing	0.2	Manufg.
Laundry, dry cleaning, shoe repair	0.1	Services	Surgical appliances & supplies	0.2	Manufg.
			Welding apparatus, electric	0.2	Manufg.
			Wiring devices	0.2	Manufg.
			Woodworking machinery	0.2	Manufg.
			Construction of stores & restaurants	0.1	Constr.
			Hotels & motels	0.1	Constr.
			Residential 2-4 unit structures, nonfarm	0.1	Constr.
			Aircraft & missile equipment, nec	0.1	Manufg.
			Fabricated structural metal	0.1	Manufg.
			Lawn & garden equipment	0.1	Manufg.
			Measuring & dispensing pumps	0.1	Manufg.
			Metal stampings, nec	0.1	Manufg.
			Metalworking machinery, nec	0.1	Manufg.
			Plumbing fixture fittings & trim	0.1	Manufg.
			Scales & balances	0.1	Manufg.
			Sheet metal work	0.1	Manufg.
			Steel springs, except wire	0.1	Manufg.

Continued on next page.

INPUTS AND OUTPUTS FOR IRON & STEEL FOUNDRIES - Continued

Economic Sector or Industry Providing Inputs	%	Sector	Economic Sector or Industry Buying Outputs	%	Sector
			Surgical & medical instruments	0.1	Manufg.
			Typewriters & office machines, nec	0.1	Manufg.

Source: Benchmark Input-Output Accounts for the U.S. Economy, 1982, U.S. Department of Commerce, Washington, D.C., July 1991. Data, as reported in the source, are organized by the 1977 SIC structure in use in 1982 but have been matched, as closely as is possible, to the 1987 SIC structure used in this book.

OCCUPATIONS EMPLOYED BY SIC 332 - IRON AND STEEL FOUNDRIES

Occupation	% of Total 1994	Change to 2005	Occupation	% of Total 1994	Change to 2005
Assemblers, fabricators, & hand workers nec	6.4	-28.8	Metal & plastic machine workers nec	2.2	-37.0
Grinders & polishers, hand	6.4	24.5	Precision metal workers nec	2.0	-28.9
Metal molding machine workers	6.2	-11.1	Grinding machine operators, metal & plastic	2.0	-53.8
Helpers, laborers, & material movers nec	5.9	-28.8	Maintenance repairers, general utility	1.7	-36.0
Precision workers nec	5.4	-28.9	Electricians	1.7	-33.2
Blue collar worker supervisors	5.2	-31.8	Machinists	1.5	-28.9
Foundry mold assembly & shakeout workers	5.0	-28.9	General managers & top executives	1.5	-32.5
Inspectors, testers, & graders, precision	4.2	-28.9	Crane & tower operators	1.3	-28.9
Industrial machinery mechanics	3.3	-21.8	Sales & related workers nec	1.1	-28.8
Welders & cutters	2.9	-28.9	Machine tool cutting operators, metal & plastic	1.1	-40.8
Furnace operators	2.5	-14.6	Machine operators nec	1.1	-37.3
Industrial truck & tractor operators	2.4	-28.8	Janitors & cleaners, incl maids	1.0	-43.0

Source: Industry-Occupation Matrix, Bureau of Labor Statistics. These data relate to one or more 3-digit SIC industry groups rather than to a single 4-digit SIC. The change reported for each occupation to the year 2005 is a percent of growth or decline as estimated by the Bureau of Labor Statistics. The abbreviation nec stands for 'not elsewhere classified'.

LOCATION BY STATE AND REGIONAL CONCENTRATION

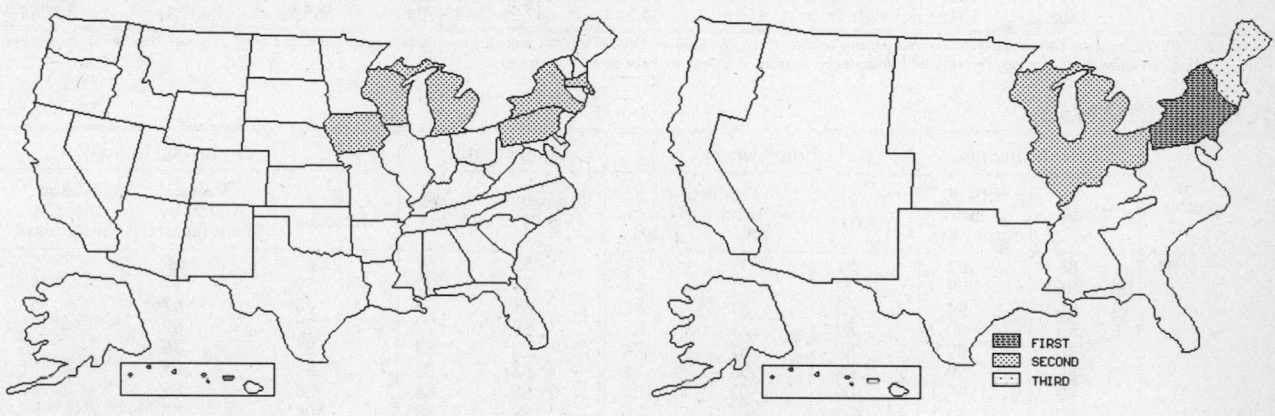

FIRST
SECOND
THIRD

INDUSTRY DATA BY STATE

State	Establish-ments	Shipments Total ($ mil)	Shipments % of U.S.	Shipments Per Establ.	Employment Total Number	Employment % of U.S.	Employment Per Establ.	Wages ($/hour)	Cost as % of Shipments	Investment per Employee ($)
Massachusetts	3	(D)	-	-	175 *	6.3	58	-	-	-
Pennsylvania	3	(D)	-	-	750 *	26.8	250	-	-	-
New York	2	(D)	-	-	175 *	6.3	88	-	-	-
Wisconsin	2	(D)	-	-	175 *	6.3	88	-	-	-
Iowa	1	(D)	-	-	175 *	6.3	175	-	-	-
Michigan	1	(D)	-	-	1,750 *	62.5	1,750	-	-	-

*Source: 1992 Economic Census. The states are in descending order of shipments or establishments (if shipment data are missing for the majority). The symbol (D) appears when data are withheld to prevent disclosure of competitive information. States marked with (D) are sorted by number of establishments. A dash (-) indicates that the data element cannot be calculated; * indicates the midpoint of a range.*

3324 - STEEL INVESTMENT FOUNDRIES

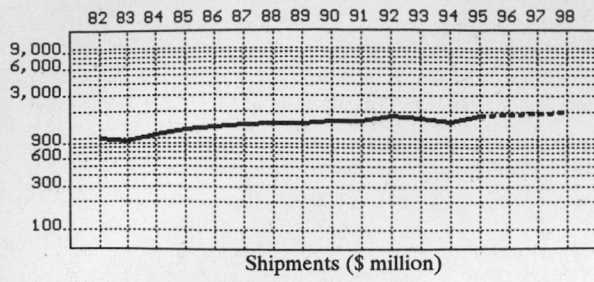

Shipments ($ million)

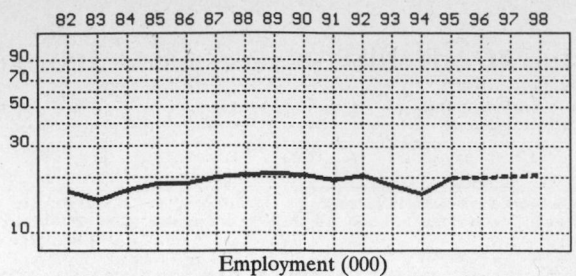

Employment (000)

GENERAL STATISTICS

| Year | Companies | Establishments | | Employment | | | Compensation | | Production ($ million) | | | |
		Total	with 20 or more employees	Total (000)	Production Workers (000)	Hours (Mil)	Payroll ($ mil)	Wages ($/hr)	Cost of Materials	Value Added by Manufacture	Value of Shipments	Capital Invest.
1982	117	132	108	16.8	12.6	24.4	299.4	8.08	363.3	635.7	1,024.6	32.9
1983		132	104	15.2	11.3	22.2	291.9	8.60	354.9	589.0	939.6	21.5
1984		132	100	17.2	13.4	26.9	352.8	8.75	421.2	727.7	1,124.5	25.0
1985		132	95	18.5	14.1	29.4	394.9	9.22	477.4	798.5	1,262.9	51.4
1986		131	94	18.6	13.9	27.7	418.5	9.97	489.3	884.6	1,359.3	45.2
1987	120	135	95	20.3	15.6	31.1	470.6	10.31	495.7	988.8	1,450.8	36.5
1988		137	99	20.8	16.4	33.0	488.0	10.09	532.6	941.2	1,469.4	58.4
1989		138	98	21.2	16.1	32.5	507.7	10.56	656.7	952.6	1,515.8	51.5
1990		132	99	20.7	15.1	30.7	503.7	11.37	563.5	1,033.5	1,592.1	51.2
1991		134	101	19.5	14.8	30.9	509.3	11.44	520.5	1,052.5	1,585.0	47.0
1992	139	153	106	20.8	16.0	33.1	574.7	12.02	605.8	1,164.6	1,783.1	52.8
1993		146	104	18.2	14.1	29.2	515.9	12.14	526.5	1,112.3	1,686.9	46.1
1994		145P	101P	16.4	13.0	27.7	468.5	11.59	430.9	1,096.5	1,530.4	56.4
1995		146P	101P	20.1P	15.6P	32.7P	583.3P	12.69P	514.9P	1,310.4P	1,828.9P	59.3P
1996		147P	101P	20.3P	15.7P	33.2P	602.9P	13.02P	531.8P	1,353.3P	1,888.8P	61.4P
1997		149P	101P	20.5P	15.9P	33.7P	622.5P	13.36P	548.7P	1,396.2P	1,948.7P	63.5P
1998		150P	101P	20.7P	16.1P	34.2P	642.2P	13.70P	565.5P	1,439.1P	2,008.6P	65.7P

Sources: 1982, 1987, 1992 *Economic Census*; *Annual Survey of Manufactures*, 83-86, 88-91, 93-94. Establishment counts for non-Census years are from *County Business Patterns*; establishment values for 83-84 are extrapolations. 'P's show projections by the editors. Industries reclassified in 87 will not have data for prior years.

INDICES OF CHANGE

| Year | Companies | Establishments | | Employment | | | Compensation | | Production ($ million) | | | |
		Total	with 20 or more employees	Total (000)	Production Workers (000)	Hours (Mil)	Payroll ($ mil)	Wages ($/hr)	Cost of Materials	Value Added by Manufacture	Value of Shipments	Capital Invest.
1982	84	86	102	81	79	74	52	67	60	55	57	62
1983		86	98	73	71	67	51	72	59	51	53	41
1984		86	94	83	84	81	61	73	70	62	63	47
1985		86	90	89	88	89	69	77	79	69	71	97
1986		86	89	89	87	84	73	83	81	76	76	86
1987	86	88	90	98	98	94	82	86	82	85	81	69
1988		90	93	100	102	100	85	84	88	81	82	111
1989		90	92	102	101	98	88	88	108	82	85	98
1990		86	93	100	94	93	88	95	93	89	89	97
1991		88	95	94	93	93	89	95	86	90	89	89
1992	100	100	100	100	100	100	100	100	100	100	100	100
1993		95	98	88	88	88	90	101	87	96	95	87
1994		95P	95P	79	81	84	82	96	71	94	86	107
1995		96P	95P	97P	97P	99P	101P	106P	85P	113P	103P	112P
1996		96P	95P	98P	98P	100P	105P	108P	88P	116P	106P	116P
1997		97P	95P	99P	99P	102P	108P	111P	91P	120P	109P	120P
1998		98P	95P	100P	101P	103P	112P	114P	93P	124P	113P	124P

Sources: Same as General Statistics. Values reflect change from the base year, 1992. Values above 100 mean greater than 92, values below 100 mean less than 92, and a value of 100 in the 82-91 or 93-98 period means same as 92. 'P's mark projections by the editors.

SELECTED RATIOS

For 1994	Avg. of All Manufact.	Analyzed Industry	Index	For 1994	Avg. of All Manufact.	Analyzed Industry	Index
Employees per Establishment	49	113	231	Value Added per Production Worker	134,084	84,346	63
Payroll per Establishment	1,500,273	3,235,430	216	Cost per Establishment	5,045,178	2,975,766	59
Payroll per Employee	30,620	28,567	93	Cost per Employee	102,970	26,274	26
Production Workers per Establishment	34	90	262	Cost per Production Worker	146,988	33,146	23
Wages per Establishment	853,319	2,217,101	260	Shipments per Establishment	9,576,895	10,568,840	110
Wages per Production Worker	24,861	24,696	99	Shipments per Employee	195,460	93,317	48
Hours per Production Worker	2,056	2,131	104	Shipments per Production Worker	279,017	117,723	42
Wages per Hour	12.09	11.59	96	Investment per Establishment	321,011	389,495	121
Value Added per Establishment	4,602,255	7,572,355	165	Investment per Employee	6,552	3,439	52
Value Added per Employee	93,930	66,860	71	Investment per Production Worker	9,352	4,338	46

Sources: Same as General Statistics. The 'Average of All Manufacturing' column represents the average of all manufacturing industries reported for the most recent complete year available. The Index shows the relationship between the Average and the Analyzed Industry. For example, 100 means that they are equal; 500 that the Analyzed Industry is five times the average; 50 means that the Analyzed Industry is half the national average. The abbreviation 'na' is used to show that data are 'not available'.

LEADING COMPANIES   Number shown: **26**   Total sales ($ mil): **3,591**   Total employment (000): **36.7**

Company Name	Address				CEO Name	Phone	Co. Type	Sales ($ mil)	Empl. (000)
Pechiney Corp	475 Steamboat Rd	Greenwich	CT	06830	Michel Simonnard	203-661-4600	S	2,780	28.0
Precision Castparts Corp	4600 SE Harney Dr	Portland	OR	97206	W C McCormick	503-777-3881	P	420	4.3
Howmet Whitehall Casting	1 Misco Dr	Whitehall	MI	49461	BD Albrechtsen	616-894-4066	D	110	0.9
Dolphin Inc	PO Box 6514	Phoenix	AZ	85005	Louis Chacopulos	602-272-6747	S	39	0.5
Signicast Corp	9000 N 55th St	Milwaukee	WI	53223	W Lutz	414-354-1212	R	36	0.3
Gray Syracuse Inc	901 E Genesee St	Chittenango	NY	13037	Robert J Barbero	315-687-0014	R	27	0.4
Southern Tool Inc	PO Box 2248	Anniston	AL	36202	HM Burt	205-831-2811	S	25	0.4
Stainless Foundry & Engineering	5150 N 35th St	Milwaukee	WI	53209	J K McBroom Jr	414-462-7400	R	25	0.4
Aero Metals Inc	402 Darlington St	La Porte	IN	46350	JW Fleming	219-326-1976	R	20*	0.3
Amcast Precision	11000 Jersey Blv	R Cucamonga	CA	91730	Randy Caraway	909-987-4621	D	15*	0.2
Wilson Sporting Goods Co	810 Lawrence Dr	Newbury Park	CA	91320	Salvador Aguayo	805-498-2168	D	15	<0.1
PED Manufacturing Ltd	PO Box 5299	Oregon City	OR	97045	Richard A Day	503-656-9653	R	14	0.1
Consolidated Casting Corp	1501 S I-45	Hutchins	TX	75141	Mal Engleby	214-225-7305	S	12	0.1
Pennsylvania Precision Cast Parts	PO Box 1429	Lebanon	PA	17042	Richard Miller Jr	717-273-3338	R	10*	0.1
Waltek Inc	14310 NW Sunfish	Ramsey	MN	55303	Eugene A Walters	612-427-3181	R	10*	<0.1
Bescast Inc	4600 E 355th St	Willoughby	OH	44094	John Gallagher	216-946-5300	R	8*	<0.1
Bimac Corp	3034 Dryden Rd	Dayton	OH	45439	Roger Reedy	513-299-7333	R	5	<0.1
Independent Steel Castings	US 12 and Grand	New Buffalo	MI	49117	Ronald Gregory	616-469-2100	R	4	<0.1
Becker Metal Works Inc	800 Fred W Moore	St Clair	MI	48079	Rudolph Becker	810-329-9310	R	3	<0.1
Barroncast Inc	PO Box 138	Oxford	MI	48371	Paul E Barron	810-628-4300	R	3*	<0.1
Casting Technology Inc	5323 N Lakeland	Minneapolis	MN	55429	Mike Knight	612-533-2201	R	3	<0.1
Dal-Air Investments Castings Inc	PO Box 330	Point	TX	75472	Oren Northcutt	903-598-2226	R	2*	<0.1
Delvest Inc	PO Box 747	West Chester	PA	19381	Anthony Micola	610-436-6380	R	2	<0.1
Acra Cast Inc	1837 First St	Bay City	MI	48708	ML Smith Jr	517-893-3961	R	1	<0.1
Nova Precision Casting Corp	RD 1	Auburn	PA	17922	Richard W Boyd	717-366-2679	R	1	<0.1
Tulsa Fittings Inc	2300 Falling Springs	Sauget	IL	62206	Earl J Bewig	618-337-2255	S	1*	<0.1

Source: *Ward's Business Directory of U.S. Private and Public Companies*, Volumes 1 and 2, 1996. The company type code used is as follows: P - Public, R - Private, S - Subsidiary, D - Division, J - Joint Venture, A - Affiliate, G - Group. Sales are in millions of dollars, employees are in thousands. An asterisk (*) indicates an estimated sales volume. The symbol < stands for 'less than'. Company names and addresses are truncated, in some cases, to fit into the available space.

MATERIALS CONSUMED

Material	Quantity		Delivered Cost ($ million)
Materials, ingredients, containers, and supplies		(X)	457.0
Pig iron shapes and forms	1,000 s tons	(S)	3.8
Ferrochromium shapes and forms		(X)	2.4
Ferromanganese, silicomanganese, and manganese shapes and forms		(X)	0.8
Ferrosilicon (more than 8 percent silicon) shapes and forms		(X)	1.3
All other ferrous shapes and forms		(X)	19.4
Nickel and nickel-base alloy shapes and forms		(X)	88.0
Cobalt-base alloy shapes and forms		(X)	24.4
All other nonferrous shapes and forms		(X)	25.3
Iron and steel scrap, excluding home scrap	1,000 s tons	94.3	16.8
All other purchased scrap, excluding home scrap	1,000 s tons	(S)	1.9
Clay refractories		(X)	2.4
Grinding wheels and other abrasive products, except industrial diamonds		(X)	15.6
Nonclay refractories		(X)	12.3
All other stone, clay, glass, and concrete products		(X)	1.5
Industrial patterns		(X)	6.9
Industrial dies, molds, jigs, and fixtures		(X)	10.0
All other industrial and commercial machinery and computer equipment		(X)	0.5
Sand		(X)	11.9
All other materials and components, parts, containers, and supplies		(X)	174.6
Materials, ingredients, containers, and supplies, nsk		(X)	37.1

Source: 1992 *Economic Census*. Explanation of symbols used: (D): Withheld to avoid disclosure of competitive data; na: Not available; (S): Withheld because statistical norms were not met; (X): Not applicable; (Z): Less than half the unit shown; nec: Not elsewhere classified; nsk: Not specified by kind; - : zero; * : 10-19 percent estimated; ** : 20-29 percent estimated.

PRODUCT SHARE DETAILS

Product or Product Class	% Share	Product or Product Class	% Share
Steel investment foundries	100.00	Stainless steel investment castings	21.62
Carbon (including low alloy) steel investment castings	15.75	Hi-temp metal investment castings (iron, nickel, or cobalt-	
Alloy (excluding stainless) steel investment castings	9.22	base alloys)	47.41

Source: 1992 *Economic Census*. The values shown are percent of total shipments in an industry. Values of indented subcategories are summed in the main heading. The symbol (D) appears when data are withheld to prevent disclosure of competitive information. The abbreviation nsk stands for 'not specified by kind' and nec for 'not elsewhere classified'.

INPUTS AND OUTPUTS FOR IRON & STEEL FOUNDRIES

Economic Sector or Industry Providing Inputs	%	Sector	Economic Sector or Industry Buying Outputs	%	Sector
Electric services (utilities)	10.7	Util.	Motor vehicle parts & accessories	21.9	Manufg.
Miscellaneous repair shops	10.0	Services	Internal combustion engines, nec	7.6	Manufg.
Wholesale trade	7.2	Trade	Construction machinery & equipment	5.0	Manufg.
Scrap	7.1	Scrap	Pipe, valves, & pipe fittings	4.9	Manufg.
Electrometallurgical products	4.4	Manufg.	Pumps & compressors	4.2	Manufg.
Gas production & distribution (utilities)	4.2	Util.	Motor vehicles & car bodies	4.1	Manufg.
Advertising	4.2	Services	Oil field machinery	3.9	Manufg.
Maintenance of nonfarm buildings nec	4.1	Constr.	Water supply facility construction	3.1	Constr.
Imports	3.1	Foreign	Blast furnaces & steel mills	2.5	Manufg.
Primary nonferrous metals, nec	3.0	Manufg.	Farm machinery & equipment	2.5	Manufg.
Communications, except radio & TV	3.0	Util.	Sewer system facility construction	2.1	Constr.
Nonferrous metal ores, except copper	2.8	Mining	Exports	1.8	Foreign
Blast furnaces & steel mills	2.8	Manufg.	Railroad equipment	1.5	Manufg.
Railroads & related services	2.0	Util.	Machine tools, metal cutting types	1.4	Manufg.
Motor freight transportation & warehousing	1.8	Util.	Residential 1-unit structures, nonfarm	1.3	Constr.
Sand & gravel	1.7	Mining	Power transmission equipment	1.2	Manufg.
Power transmission equipment	1.6	Manufg.	Railroads & related services	1.2	Util.
Abrasive products	1.5	Manufg.	Aircraft & missile engines & engine parts	1.1	Manufg.
Metal heat treating	1.3	Manufg.	Refrigeration & heating equipment	1.1	Manufg.
Computer & data processing services	1.3	Services	Machinery, except electrical, nec	1.0	Manufg.
Petroleum refining	1.1	Manufg.	Office buildings	0.8	Constr.
Special dies & tools & machine tool accessories	1.1	Manufg.	Motors & generators	0.8	Manufg.
Carbon & graphite products	0.9	Manufg.	Printing trades machinery	0.8	Manufg.
Industrial patterns	0.8	Manufg.	Turbines & turbine generator sets	0.8	Manufg.
Machinery, except electrical, nec	0.8	Manufg.	Maintenance of water supply facilities	0.7	Constr.
Primary metal products, nec	0.7	Manufg.	Residential garden apartments	0.7	Constr.
Eating & drinking places	0.7	Trade	Electronic computing equipment	0.7	Manufg.
Banking	0.7	Fin/R.E.	Heating equipment, except electric	0.7	Manufg.
Equipment rental & leasing services	0.7	Services	Mining machinery, except oil field	0.7	Manufg.
Metal coating & allied services	0.6	Manufg.	General industrial machinery, nec	0.6	Manufg.
Miscellaneous plastics products	0.6	Manufg.	Industrial trucks & tractors	0.6	Manufg.
Sanitary services, steam supply, irrigation	0.6	Util.	Special industry machinery, nec	0.6	Manufg.
Fabricated metal products, nec	0.5	Manufg.	Copper ore	0.5	Mining
Internal combustion engines, nec	0.5	Manufg.	Industrial buildings	0.5	Constr.
Iron & steel foundries	0.5	Manufg.	Food products machinery	0.5	Manufg.
Sawmills & planning mills, general	0.5	Manufg.	Hardware, nec	0.5	Manufg.
Industrial controls	0.4	Manufg.	Paper industries machinery	0.5	Manufg.
Metal stampings, nec	0.4	Manufg.	Special dies & tools & machine tool accessories	0.5	Manufg.
Primary copper	0.4	Manufg.	Tanks & tank components	0.5	Manufg.
Water transportation	0.4	Util.	Highway & street construction	0.4	Constr.
Engineering, architectural, & surveying services	0.4	Services	Maintenance of nonbuilding facilities nec	0.4	Constr.
U.S. Postal Service	0.4	Gov't	Ball & roller bearings	0.4	Manufg.
Clay, ceramic, & refractory minerals	0.3	Mining	Carburetors, pistons, rings, & valves	0.4	Manufg.
Chemical preparations, nec	0.3	Manufg.	Hand & edge tools, nec	0.4	Manufg.
Cyclic crudes and organics	0.3	Manufg.	Machine tools, metal forming types	0.4	Manufg.
Lubricating oils & greases	0.3	Manufg.	Mechanical measuring devices	0.4	Manufg.
Screw machine and related products	0.3	Manufg.	Iron & ferroalloy ores	0.3	Mining
Welding apparatus, electric	0.3	Manufg.	Construction of educational buildings	0.3	Constr.
Real estate	0.3	Fin/R.E.	Construction of hospitals	0.3	Constr.
Electrical repair shops	0.3	Services	Nonfarm residential structure maintenance	0.3	Constr.
Legal services	0.3	Services	Blowers & fans	0.3	Manufg.
Management & consulting services & labs	0.3	Services	Conveyors & conveying equipment	0.3	Manufg.
Chemical & fertilizer mineral	0.2	Mining	Fabricated metal products, nec	0.3	Manufg.
Ball & roller bearings	0.2	Manufg.	Fabricated plate work (boiler shops)	0.3	Manufg.
Nonferrous wire drawing & insulating	0.2	Manufg.	Fabricated rubber products, nec	0.3	Manufg.
Primary aluminum	0.2	Manufg.	Iron & steel foundries	0.3	Manufg.
Pumps & compressors	0.2	Manufg.	Motorcycles, bicycles, & parts	0.3	Manufg.
Air transportation	0.2	Util.	Small arms	0.3	Manufg.
Insurance carriers	0.2	Fin/R.E.	Textile machinery	0.3	Manufg.
Accounting, auditing & bookkeeping	0.2	Services	Federal Government purchases, national defense	0.3	Fed Govt
Automotive repair shops & services	0.2	Services	Electric utility facility construction	0.2	Constr.
State & local government enterprises, nec	0.2	Gov't	Maintenance of nonfarm buildings nec	0.2	Constr.
Coal	0.1	Mining	Residential high-rise apartments	0.2	Constr.
Blowers & fans	0.1	Manufg.	Telephone & telegraph facility construction	0.2	Constr.
Industrial gases	0.1	Manufg.	Elevators & moving stairways	0.2	Manufg.
Manifold business forms	0.1	Manufg.	Hoists, cranes, & monorails	0.2	Manufg.
Nonclay refractories	0.1	Manufg.	Household laundry equipment	0.2	Manufg.
Primary lead	0.1	Manufg.	Lighting fixtures & equipment	0.2	Manufg.
Rubber & plastics hose & belting	0.1	Manufg.	Miscellaneous metal work	0.2	Manufg.
Wood pallets & skids	0.1	Manufg.	Mobile homes	0.2	Manufg.
Water supply & sewage systems	0.1	Util.	Radio & TV communication equipment	0.2	Manufg.
Royalties	0.1	Fin/R.E.	Service industry machines, nec	0.2	Manufg.
Automotive rental & leasing, without drivers	0.1	Services	Ship building & repairing	0.2	Manufg.
Laundry, dry cleaning, shoe repair	0.1	Services	Surgical appliances & supplies	0.2	Manufg.
			Welding apparatus, electric	0.2	Manufg.
			Wiring devices	0.2	Manufg.
			Woodworking machinery	0.2	Manufg.
			Construction of stores & restaurants	0.1	Constr.

Continued on next page.

INPUTS AND OUTPUTS FOR IRON & STEEL FOUNDRIES - Continued

Economic Sector or Industry Providing Inputs	%	Sector	Economic Sector or Industry Buying Outputs	%	Sector
			Hotels & motels	0.1	Constr.
			Residential 2-4 unit structures, nonfarm	0.1	Constr.
			Aircraft & missile equipment, nec	0.1	Manufg.
			Fabricated structural metal	0.1	Manufg.
			Lawn & garden equipment	0.1	Manufg.
			Measuring & dispensing pumps	0.1	Manufg.
			Metal stampings, nec	0.1	Manufg.
			Metalworking machinery, nec	0.1	Manufg.
			Plumbing fixture fittings & trim	0.1	Manufg.
			Scales & balances	0.1	Manufg.
			Sheet metal work	0.1	Manufg.
			Steel springs, except wire	0.1	Manufg.
			Surgical & medical instruments	0.1	Manufg.
			Typewriters & office machines, nec	0.1	Manufg.

Source: Benchmark Input-Output Accounts for the U.S. Economy, 1982, U.S. Department of Commerce, Washington, D.C., July 1991. Data, as reported in the source, are organized by the 1977 SIC structure in use in 1982 but have been matched, as closely as is possible, to the 1987 SIC structure used in this book.

OCCUPATIONS EMPLOYED BY SIC 332 - IRON AND STEEL FOUNDRIES

Occupation	% of Total 1994	Change to 2005	Occupation	% of Total 1994	Change to 2005
Assemblers, fabricators, & hand workers nec	6.4	-28.8	Metal & plastic machine workers nec	2.2	-37.0
Grinders & polishers, hand	6.4	24.5	Precision metal workers nec	2.0	-28.9
Metal molding machine workers	6.2	-11.1	Grinding machine operators, metal & plastic	2.0	-53.8
Helpers, laborers, & material movers nec	5.9	-28.8	Maintenance repairers, general utility	1.7	-36.0
Precision workers nec	5.4	-28.9	Electricians	1.7	-33.2
Blue collar worker supervisors	5.2	-31.8	Machinists	1.5	-28.9
Foundry mold assembly & shakeout workers	5.0	-28.9	General managers & top executives	1.5	-32.5
Inspectors, testers, & graders, precision	4.2	-28.9	Crane & tower operators	1.3	-28.9
Industrial machinery mechanics	3.3	-21.8	Sales & related workers nec	1.1	-28.8
Welders & cutters	2.9	-28.9	Machine tool cutting operators, metal & plastic	1.1	-40.8
Furnace operators	2.5	-14.6	Machine operators nec	1.1	-37.3
Industrial truck & tractor operators	2.4	-28.8	Janitors & cleaners, incl maids	1.0	-43.0

Source: Industry-Occupation Matrix, Bureau of Labor Statistics. These data relate to one or more 3-digit SIC industry groups rather than to a single 4-digit SIC. The change reported for each occupation to the year 2005 is a percent of growth or decline as estimated by the Bureau of Labor Statistics. The abbreviation nec stands for 'not elsewhere classified'.

LOCATION BY STATE AND REGIONAL CONCENTRATION

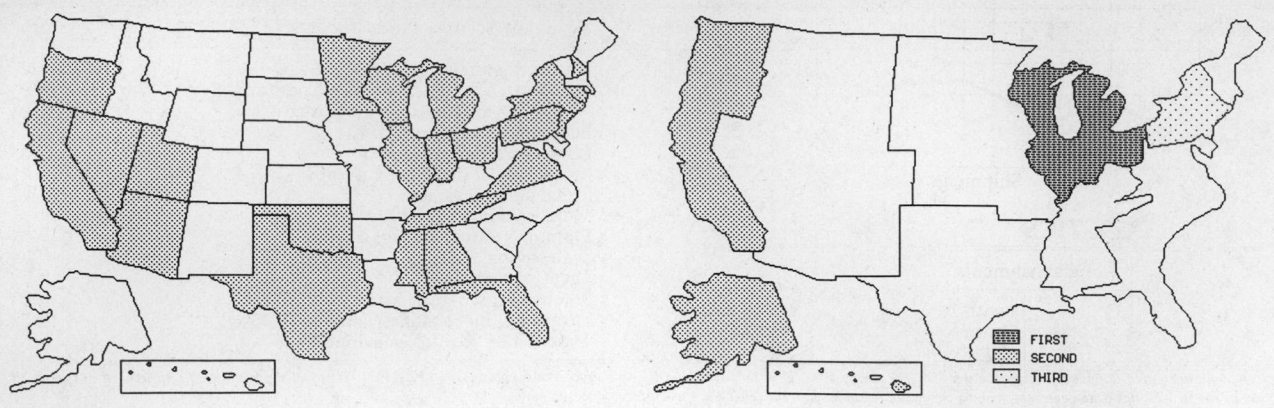

FIRST
SECOND
THIRD

INDUSTRY DATA BY STATE

| State | Establish-ments | Shipments | | | Employment | | | | Cost as % of Shipments | Investment per Employee ($) |
		Total ($ mil)	% of U.S.	Per Establ.	Total Number	% of U.S.	Per Establ.	Wages ($/hour)		
California	25	188.2	10.6	7.5	2,500	12.0	100	9.51	36.2	4,960
Michigan	14	170.6	9.6	12.2	1,800	8.7	129	14.57	34.8	-
Texas	11	129.5	7.3	11.8	1,500	7.2	136	9.74	37.3	-
Oklahoma	4	10.6	0.6	2.7	100	0.5	25	7.67	36.8	2,000
Ohio	17	(D)	-	-	1,750 *	8.4	103	-	-	-
Pennsylvania	9	(D)	-	-	750 *	3.6	83	-	-	-
Wisconsin	9	(D)	-	-	1,750 *	8.4	194	-	-	-
Tennessee	7	(D)	-	-	175 *	0.8	25	-	-	-
Indiana	5	(D)	-	-	1,750 *	8.4	350	-	-	-
Connecticut	4	(D)	-	-	750 *	3.6	188	-	-	-
Florida	4	(D)	-	-	175 *	0.8	44	-	-	-
New Jersey	4	(D)	-	-	1,750 *	8.4	438	-	-	-
New York	4	(D)	-	-	375 *	1.8	94	-	-	-
Oregon	4	(D)	-	-	1,750 *	8.4	438	-	-	-
Arizona	3	(D)	-	-	375 *	1.8	125	-	-	-
Minnesota	3	(D)	-	-	175 *	0.8	58	-	-	-
New Hampshire	3	(D)	-	-	1,750 *	8.4	583	-	-	-
Alabama	2	(D)	-	-	750 *	3.6	375	-	-	-
Illinois	2	(D)	-	-	175 *	0.8	88	-	-	-
Mississippi	1	(D)	-	-	375 *	1.8	375	-	-	-
Nevada	1	(D)	-	-	175 *	0.8	175	-	-	-
Utah	1	(D)	-	-	175 *	0.8	175	-	-	-
Virginia	1	(D)	-	-	1,750 *	8.4	1,750	-	-	-

Source: 1992 *Economic Census*. The states are in descending order of shipments or establishments (if shipment data are missing for the majority). The symbol (D) appears when data are withheld to prevent disclosure of competitive information. States marked with (D) are sorted by number of establishments. A dash (-) indicates that the data element cannot be calculated; * indicates the midpoint of a range.

3325 - STEEL FOUNDRIES, NEC

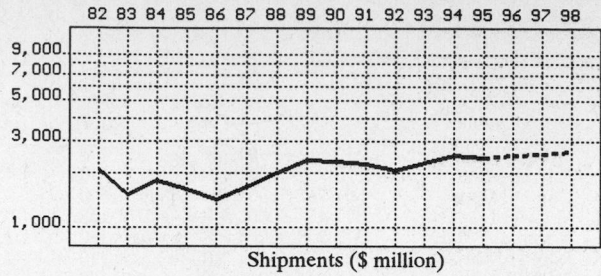

Shipments ($ million)

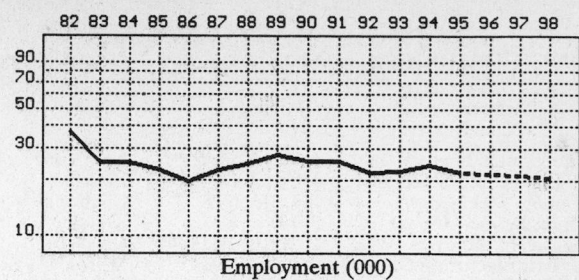

Employment (000)

GENERAL STATISTICS

Year	Companies	Establishments		Employment			Compensation		Production ($ million)			
		Total	with 20 or more employees	Total (000)	Production Workers (000)	Hours (Mil)	Payroll ($ mil)	Wages ($/hr)	Cost of Materials	Value Added by Manufacture	Value of Shipments	Capital Invest.
1982	291	331	229	36.9	28.4	49.6	713.0	10.06	826.3	1,209.0	2,091.4	99.6
1983		318	218	25.4	19.0	35.1	540.8	10.68	676.1	815.5	1,518.4	45.3
1984		305	207	25.0	19.7	38.9	575.5	10.71	748.6	1,067.2	1,799.4	61.1
1985		291	196	22.8	17.8	34.6	507.9	10.55	680.8	920.3	1,618.2	48.0
1986		272	183	19.7	15.3	29.4	457.3	10.73	629.9	774.8	1,424.5	43.4
1987	270	294	176	22.9	17.9	35.5	544.1	10.63	685.7	1,004.9	1,680.4	37.8
1988		297	184	24.8	20.0	41.1	607.3	10.75	795.3	1,229.8	2,005.6	37.1
1989		283	181	27.9	21.9	44.3	673.5	10.72	1,008.4	1,374.4	2,372.7	54.9
1990		284	182	25.7	21.5	43.3	688.5	11.21	959.6	1,350.0	2,326.5	69.2
1991		296	185	25.6	20.3	41.0	664.0	11.68	918.1	1,310.8	2,231.4	71.1
1992	271	288	166	22.2	17.6	36.3	610.2	11.76	817.4	1,253.4	2,059.3	71.7
1993		294	170	22.7	18.2	37.5	643.3	12.14	928.6	1,360.3	2,281.4	70.6
1994		279P	159P	24.2	19.3	42.2	697.8	11.77	1,026.2	1,506.7	2,523.8	60.8
1995		277P	155P	22.2P	18.1P	39.6P	671.8P	12.04P	995.4P	1,461.5P	2,448.1P	62.1P
1996		274P	150P	21.7P	17.9P	39.7P	680.7P	12.18P	1,021.8P	1,500.2P	2,512.9P	62.5P
1997		272P	145P	21.3P	17.6P	39.8P	689.6P	12.33P	1,048.1P	1,538.8P	2,577.6P	62.9P
1998		269P	141P	20.9P	17.4P	39.9P	698.5P	12.47P	1,074.4P	1,577.5P	2,642.4P	63.3P

Sources: 1982, 1987, 1992 *Economic Census*; *Annual Survey of Manufactures*, 83-86, 88-91, 93-94. Establishment counts for non-Census years are from *County Business Patterns*; establishment values for 83-84 are extrapolations. 'P's show projections by the editors. Industries reclassified in 87 will not have data for prior years.

INDICES OF CHANGE

Year	Companies	Establishments		Employment			Compensation		Production ($ million)			
		Total	with 20 or more employees	Total (000)	Production Workers (000)	Hours (Mil)	Payroll ($ mil)	Wages ($/hr)	Cost of Materials	Value Added by Manufacture	Value of Shipments	Capital Invest.
1982	107	115	138	166	161	137	117	86	101	96	102	139
1983		110	131	114	108	97	89	91	83	65	74	63
1984		106	125	113	112	107	94	91	92	85	87	85
1985		101	118	103	101	95	83	90	83	73	79	67
1986		94	110	89	87	81	75	91	77	62	69	61
1987	100	102	106	103	102	98	89	90	84	80	82	53
1988		103	111	112	114	113	100	91	97	98	97	52
1989		98	109	126	124	122	110	91	123	110	115	77
1990		99	110	116	122	119	113	95	117	108	113	97
1991		103	111	115	115	113	109	99	112	105	108	99
1992	100	100	100	100	100	100	100	100	100	100	100	100
1993		102	102	102	103	103	105	103	114	109	111	98
1994		97P	96P	109	110	116	114	100	126	120	123	85
1995		96P	93P	100P	103P	109P	110P	102P	122P	117P	119P	87P
1996		95P	90P	98P	102P	109P	112P	104P	125P	120P	122P	87P
1997		94P	88P	96P	100P	110P	113P	105P	128P	123P	125P	88P
1998		93P	85P	94P	99P	110P	114P	106P	131P	126P	128P	88P

Sources: Same as General Statistics. Values reflect change from the base year, 1992. Values above 100 mean greater than 92, values below 100 mean less than 92, and a value of 100 in the 82-91 or 93-98 period means same as 92. 'P's mark projections by the editors.

SELECTED RATIOS

For 1994	Avg. of All Manufact.	Analyzed Industry	Index	For 1994	Avg. of All Manufact.	Analyzed Industry	Index
Employees per Establishment	49	87	177	Value Added per Production Worker	134,084	78,067	58
Payroll per Establishment	1,500,273	2,498,497	167	Cost per Establishment	5,045,178	3,674,345	73
Payroll per Employee	30,620	28,835	94	Cost per Employee	102,970	42,405	41
Production Workers per Establishment	34	69	201	Cost per Production Worker	146,988	53,171	36
Wages per Establishment	853,319	1,778,430	208	Shipments per Establishment	9,576,895	9,036,554	94
Wages per Production Worker	24,861	25,735	104	Shipments per Employee	195,460	104,289	53
Hours per Production Worker	2,056	2,187	106	Shipments per Production Worker	279,017	130,767	47
Wages per Hour	12.09	11.77	97	Investment per Establishment	321,011	217,697	68
Value Added per Establishment	4,602,255	5,394,792	117	Investment per Employee	6,552	2,512	38
Value Added per Employee	93,930	62,260	66	Investment per Production Worker	9,352	3,150	34

Sources: Same as General Statistics. The 'Average of All Manufacturing' column represents the average of all manufacturing industries reported for the most recent complete year available. The Index shows the relationship between the Average and the Analyzed Industry. For example, 100 means that they are equal; 500 that the Analyzed Industry is five times the average; 50 means that the Analyzed Industry is half the national average. The abbreviation 'na' is used to show that data are 'not available'.

LEADING COMPANIES Number shown: 75 Total sales ($ mil): 3,908 Total employment (000): 26.1

Company Name	Address				CEO Name	Phone	Co. Type	Sales ($ mil)	Empl. (000)
Inco United States Inc	1 New York Plz	New York	NY	10004	Dyer S Wadsworth	212-612-5500	S	776	2.3
Republic Engineered Steels Inc	410 Oberlin Rd SW	Massillon	OH	44647	Russell Maier	216-837-6000	R	670*	4.7
Nucor-Yamato Steel Co	PO Box 1228	Blytheville	AR	72316	Dan DiMicco	501-762-5500	S	600	0.7
Haynes International Inc	PO Box 9013	Kokomo	IN	46904	Michael D Austin	317-456-6000	R	163	0.9
American Steel Foundries	10 S Riverside Plz	Chicago	IL	60606	Norman A Berg	312-258-8000	D	150	1.5
Hitchiner Manufacturing Co	PO Box 2001	Milford	NH	03055	John H Morison III	603-673-1100	R	110	2.2
Buckeye Steel Castings Co	2211 Parsons Av	Columbus	OH	43207	Joe W Harden	614-444-2121	S	90	1.1
Atchison Casting Corp	PO Box 188	Atchison	KS	66002	Hugh H Aiken	913-367-2121	P	83	0.7
GH Hensley Industries Inc	PO Box 29779	Dallas	TX	75229	William R Smith	214-241-2321	S	63*	0.6
Ervin Industries Inc	PO Box 1168	Ann Arbor	MI	48106	JE Pearson	313-769-4600	R	60	0.2
BR Holdings Ltd	1442 N Mem Dr	Racine	WI	53404	Jeffrey Renzoni	414-634-3341	R	56*	0.5
Amalloy Corp	PO Box 90	Goshen	NY	10924	Arthur Borin	914-294-1600	R	50	0.7
Harrison Steel Casting Co	PO Box 60	Attica	IN	47918	W Harrison II	317-762-2481	R	50	0.6
Rafferty-Brown Steel Co	PO Box 18927	Greensboro	NC	27419	Robert B Wood	910-855-6300	R	50	0.1
SMS Concast Inc	100 Sandusky St	Pittsburgh	PA	15212	Herbert Fastert	412-231-6800	R	50	<0.1
Pyramid Northern	7406 Fullerton St	Jacksonville	FL	32256	Otto Byett	904-363-9988	D	45	0.5
Atlas Foundry and Machine Co	3021 S Wilkeson St	Tacoma	WA	98409	James E Reder	206-475-4600	S	40	0.3
Commercial Steel Corp	PO Box U	Glassport	PA	15045	Somi Alhuwalia	412-664-6330	R	40	<0.1
Johnstown Corp	545 Central Av	Johnstown	PA	15902	Robert M Funk	814-535-9000	R	39	0.4
Maynard Steel Casting Co	2856 S 27th St	Milwaukee	WI	53215	ED Wabiszewski	414-645-0440	R	38	0.5
Duraloy Technologies Inc	PO Box 81	Scottdale	PA	15683	Vince Schiavoni	412-887-5100	S	35	0.3
Pacific Steel Casting Co	1333 2nd St	Berkeley	CA	94710	Robert Delsol	510-525-9200	R	35	0.3
Jersey Shore Steel Inc	PO Box 5055	Jersey Shore	PA	17740	John C Schultz	717-398-0220	R	34*	0.4
Texas Steel Co	PO Box 2976	Fort Worth	TX	76113	T K Armstrong Jr	817-923-4611	R	30*	0.4
Huron Casting Inc	PO Box 679	Pigeon	MI	48755	L Wurst	517-453-3933	R	28	0.3
Kutztown Foundry	230 Railroad St	Kutztown	PA	19530	Buddy Bell	215-683-7351	D	25	0.3
Quality Electric Steel Castings	252 McCarty Dr	Houston	TX	77029	JD Sturkie	713-672-6625	R	25	0.2
Sivyer Steel Corp	225 S 33rd St	Bettendorf	IA	52722	Claude D Robinson	319-355-1811	R	25	0.3
Wall Colmonoy Corp	30261ephenson Hwy	Madison H	MI	48071	WP Clark	810-585-6400	R	23	0.2
Davy-Clicem	912 Fort Duquesne	Pittsburgh	PA	15222	John Koontz	412-566-4500	D	22*	0.2
Post Precision Castings Inc	21 Walnut St	Strausstown	PA	19559	John R Post	610-488-1011	R	22	0.3
Waukesha Foundry Inc	1300 Lincoln Av	Waukesha	WI	53186	B Kerwin	414-542-0741	R	20	0.2
CMI Quaker Alloy Inc	720 S Cherry St	Myerstown	PA	17067	Dinny Kinloch	717-866-6511	S	18	0.2
Pelton Casteel Inc	2929 S Chase Av	Milwaukee	WI	53207	Larry S Krueger	414-481-3400	R	17	0.2
Pennsylvania Steel	PO Box 128	Hamburg	PA	19526	DM Goodyear	215-562-7533	R	16*	0.3
Eagle Foundry Company Inc	PO Box 250	Eagle Creek	OR	97022	R Pursel	503-637-3048	R	15	0.1
General Plug & Mfg Co	PO Box 26	Grafton	OH	44044	Kevin J Flanigan	216-926-2411	R	15	0.1
Spokane Industries Inc	Spokane Industrial	Spokane	WA	99216	Greg Tenold	509-928-0750	R	15	0.2
Crown Castings Inc	108 Greenwood Av	Midland Park	NJ	07432	Paul Suhadolc	201-444-8228	R	14	0.2
Wollaston Alloys Inc	205 Wood Rd	Braintree	MA	02184	David P Bernardi	617-848-3333	S	13	0.1
Empire Steel Company Inc	PO Box 139	Reading	PA	19603	Edward J Crowley	215-921-8101	R	13	0.3
St Louis Steel Casting Inc	2300 Falling Springs	Sauget	IL	62206	Earl J Bewig	618-337-2255	P	13	0.1
Durbin-Durco Inc	PO Box 16966	St Louis	MO	63105	Marty Durbin	314-725-8900	R	12*	0.1
National Eastern Corp	75 Neal Ct	Plainville	CT	06062	Warren Kart	203-747-3700	S	12	0.1
Omaha Steel Castings Co	PO Box 6276	Omaha	NE	68106	Ronald L Howlett	402-558-6000	R	12*	0.2
Aelco Foundries Inc	1930 S 4th St	Milwaukee	WI	53204	John Wyatt	414-645-3771	D	10	<0.1
Braeburn Alloy Steel	101 Braeburn Rd	Lower Burrell	PA	15068	Thomas R Wiseman	412-224-6900	D	10	0.1
Spokane Steel Foundry	PO Box 3305	Spokane	WA	99220	Robert Tenold	509-924-0440	D	10	0.1
Columbiana Foundry Co	PO Box 98	Columbiana	OH	44408	FJ Boston	216-482-3336	R	8	<0.1
Tube-Alloy Corp	3106 Grand Caillou	Houma	LA	70363	Gerald Beard	504-876-2886	R	8*	<0.1
Westlectric Castings Inc	2040 Camfield Av	City of Com	CA	90040	John Heine	213-722-8000	R	8	<0.1
Arneson Foundry Inc	3303 66th St	Kenosha	WI	53142	Richard Arneson	414-657-6108	R	7	<0.1
Brighton Electric	PO Box 206	Beaver Falls	PA	15010	Robert Lewis	412-846-7377	S	7	<0.1
Electric Steel Castings Co	PO Box 24524	Indianapolis	IN	46224	Vernon H King	317-248-2551	R	7	0.1
Cast Masters	1145 Fairview Av	Bowling Green	OH	43402	Richard C Lyle	419-352-4631	D	6	<0.1
Galt Alloys Inc	122 Central Plz N	Canton	OH	44702	Scott Jackson	216-456-9929	R	6*	<0.1
May Foundry and Machine	454 W 600 N	Salt Lake City	UT	84103	Jack May	801-531-8931	R	6	<0.1
Quali-Cast Foundry Inc	PO Box 976	Chehalis	WA	98532	Bruce Roberts	206-748-6645	R	6	<0.1
Ski-Way Machine Products Co	24460 Lakeland	Euclid	OH	44132	Ralph Fross	216-732-9000	R	6	<0.1
Star Brass Foundry	PO Box 25597	Salt Lake City	UT	84125	ES McGrath	801-972-5881	R	6	<0.1
Acme Industrial Co	441 Maple Av	Carpentersville	IL	60110	John Evans	708-428-3911	S	5	0.1
Composite Technical Alloys Inc	1 Mill St	Attleboro	MA	02703	Robert Jones	508-226-0420	R	5	<0.1
General Bearing Corp	30156 W 8 Mile Rd	Farmington Hls	MI	48336	John Hanaway	810-478-1746	R	5	0.1
Jaxson Roll Forming Inc	145 Dixon Av	Amityville	NY	11701	Alex Trink	516-777-7775	R	5*	<0.1
Joy Technologies Inc	Grissom Ln	Claremont	NH	03743	Dave McDowell	603-543-1228	D	5	<0.1
Meltec	3444 13th Av SW	Seattle	WA	98134	R Lindberg	206-623-3274	D	5	<0.1
Texas Metal Works Inc	PO Box 3607	Beaumont	TX	77704	G Perry Culp	409-833-5601	R	5	<0.1
Talladega Castings	228 N Court St	Talladega	AL	35160	J W Heacock Jr	205-362-5550	D	5	<0.1
Bahr Brothers Manufacturing	PO Box 411	Marion	IN	46952	WM Jackson	317-664-6235	R	4	<0.1
Damascus Steel Casting Co	PO Box 257	New Brighton	PA	15066	Joseph A Presto	412-846-2770	R	4	<0.1
Midwest Alloys Foundry Inc	2300 Falling Springs	Sauget	IL	62206	Earl J Bewig	618-337-2255	S	4*	<0.1
Philbrick Booth Spencer Inc	PO Box 320520	Hartford	CT	06132	Edgar B Spencer III	203-522-9271	R	4	<0.1
Smith Steel Casting Co	PO Box 969	Marshall	TX	75671	Gerald Smith	903-935-5266	R	4*	0.1
Southern Coil Processing Inc	PO Box 232	Fairfield	AL	35064	Ed Puchy	205-785-0066	R	4*	<0.1
Sterling Steel Foundry Inc	2300 Falling Spring	Sauget	IL	62206	R Lussow	618-337-6123	S	4	<0.1

Source: Ward's Business Directory of U.S. Private and Public Companies, Volumes 1 and 2, 1996. The company type code used is as follows: P - Public, R - Private, S - Subsidiary, D - Division, J - Joint Venture, A - Affiliate, G - Group. Sales are in millions of dollars, employees are in thousands. An asterisk (*) indicates an estimated sales volume. The symbol < stands for 'less than'. Company names and addresses are truncated, in some cases, to fit into the available space.

MATERIALS CONSUMED

Material	Quantity	Delivered Cost ($ million)
Materials, ingredients, containers, and supplies	(X)	605.4
Pig iron shapes and forms 1,000 s tons	12.3**	2.4
Ferrochromium shapes and forms	(X)	23.6
Ferromanganese, silicomanganese, and manganese shapes and forms	(X)	11.3
Ferrosilicon (more than 8 percent silicon) shapes and forms	(X)	6.7
All other ferrous shapes and forms	(X)	11.4
Nickel and nickel-base alloy shapes and forms	(X)	28.7
Cobalt-base alloy shapes and forms	(X)	1.3
All other nonferrous shapes and forms	(X)	10.1
Iron and steel scrap, excluding home scrap 1,000 s tons	1,024.1**	104.4
All other purchased scrap, excluding home scrap 1,000 s tons	(S)	13.9
Clay refractories	(X)	9.5
Grinding wheels and other abrasive products, except industrial diamonds	(X)	12.4
Nonclay refractories	(X)	12.6
All other stone, clay, glass, and concrete products	(X)	4.7
Industrial patterns	(X)	11.6
Industrial dies, molds, jigs, and fixtures	(X)	3.3
All other industrial and commercial machinery and computer equipment	(X)	5.1
Sand	(X)	22.4
All other materials and components, parts, containers, and supplies	(X)	245.9
Materials, ingredients, containers, and supplies, nsk	(X)	64.0

Source: 1992 Economic Census. Explanation of symbols used: (D): Withheld to avoid disclosure of competitive data; na: Not available; (S): Withheld because statistical norms were not met; (X): Not applicable; (Z): Less than half the unit shown; nec: Not elsewhere classified; nsk: Not specified by kind; - : zero; * : 10-19 percent estimated; ** : 20-29 percent estimated.

PRODUCT SHARE DETAILS

Product or Product Class	% Share	Product or Product Class	% Share
Steel foundries, nec	100.00	Other high alloy steel castings, except investment	76.23
Other carbon steel castings, except investment	44.63	Other alloy steel castings, except investment	22.60
Cast carbon steel railroad car wheels and railway specialties	42.32	Other alloy steel railway specialties castings, except	
Carbon steel rolls for rolling mills	(D)	investment	0.27
Other carbon steel castings, except investment	55.25	Other alloy steel rolls for rolling mills, except investment	8.89
Other carbon steel castings, except investment, nsk	(D)	All other alloy steel castings, except investment	90.53
High alloy steel castings, except investment	22.01	Other alloy steel castings, except investment, nsk	0.31
High manganese alloy steel castings, except investment	23.77	Steel foundries, nec, nsk	10.75

Source: 1992 Economic Census. The values shown are percent of total shipments in an industry. Values of indented subcategories are summed in the main heading. The symbol (D) appears when data are withheld to prevent disclosure of competitive information. The abbreviation nsk stands for 'not specified by kind' and nec for 'not elsewhere classified'.

INPUTS AND OUTPUTS FOR IRON & STEEL FOUNDRIES

Economic Sector or Industry Providing Inputs	%	Sector	Economic Sector or Industry Buying Outputs	%	Sector
Electric services (utilities)	10.7	Util.	Motor vehicle parts & accessories	21.9	Manufg.
Miscellaneous repair shops	10.0	Services	Internal combustion engines, nec	7.6	Manufg.
Wholesale trade	7.2	Trade	Construction machinery & equipment	5.0	Manufg.
Scrap	7.1	Scrap	Pipe, valves, & pipe fittings	4.9	Manufg.
Electrometallurgical products	4.4	Manufg.	Pumps & compressors	4.2	Manufg.
Gas production & distribution (utilities)	4.2	Util.	Motor vehicles & car bodies	4.1	Manufg.
Advertising	4.2	Services	Oil field machinery	3.9	Manufg.
Maintenance of nonfarm buildings nec	4.1	Constr.	Water supply facility construction	3.1	Constr.
Imports	3.1	Foreign	Blast furnaces & steel mills	2.5	Manufg.
Primary nonferrous metals, nec	3.0	Manufg.	Farm machinery & equipment	2.5	Manufg.
Communications, except radio & TV	3.0	Util.	Sewer system facility construction	2.1	Constr.
Nonferrous metal ores, except copper	2.8	Mining	Exports	1.8	Foreign
Blast furnaces & steel mills	2.8	Manufg.	Railroad equipment	1.5	Manufg.
Railroads & related services	2.0	Util.	Machine tools, metal cutting types	1.4	Manufg.
Motor freight transportation & warehousing	1.8	Util.	Residential 1-unit structures, nonfarm	1.3	Constr.
Sand & gravel	1.7	Mining	Power transmission equipment	1.2	Manufg.
Power transmission equipment	1.6	Manufg.	Railroads & related services	1.2	Util.
Abrasive products	1.5	Manufg.	Aircraft & missile engines & engine parts	1.1	Manufg.
Metal heat treating	1.3	Manufg.	Refrigeration & heating equipment	1.1	Manufg.
Computer & data processing services	1.3	Services	Machinery, except electrical, nec	1.0	Manufg.
Petroleum refining	1.1	Manufg.	Office buildings	0.8	Constr.
Special dies & tools & machine tool accessories	1.1	Manufg.	Motors & generators	0.8	Manufg.
Carbon & graphite products	0.9	Manufg.	Printing trades machinery	0.8	Manufg.
Industrial patterns	0.8	Manufg.	Turbines & turbine generator sets	0.8	Manufg.
Machinery, except electrical, nec	0.8	Manufg.	Maintenance of water supply facilities	0.7	Constr.
Primary metal products, nec	0.7	Manufg.	Residential garden apartments	0.7	Constr.
Eating & drinking places	0.7	Trade	Electronic computing equipment	0.7	Manufg.
Banking	0.7	Fin/R.E.	Heating equipment, except electric	0.7	Manufg.
Equipment rental & leasing services	0.7	Services	Mining machinery, except oil field	0.7	Manufg.

Continued on next page.

INPUTS AND OUTPUTS FOR IRON & STEEL FOUNDRIES - Continued

Economic Sector or Industry Providing Inputs	%	Sector	Economic Sector or Industry Buying Outputs	%	Sector
Metal coating & allied services	0.6	Manufg.	General industrial machinery, nec	0.6	Manufg.
Miscellaneous plastics products	0.6	Manufg.	Industrial trucks & tractors	0.6	Manufg.
Sanitary services, steam supply, irrigation	0.6	Util.	Special industry machinery, nec	0.6	Manufg.
Fabricated metal products, nec	0.5	Manufg.	Copper ore	0.5	Mining
Internal combustion engines, nec	0.5	Manufg.	Industrial buildings	0.5	Constr.
Iron & steel foundries	0.5	Manufg.	Food products machinery	0.5	Manufg.
Sawmills & planning mills, general	0.5	Manufg.	Hardware, nec	0.5	Manufg.
Industrial controls	0.4	Manufg.	Paper industries machinery	0.5	Manufg.
Metal stampings, nec	0.4	Manufg.	Special dies & tools & machine tool accessories	0.5	Manufg.
Primary copper	0.4	Manufg.	Tanks & tank components	0.5	Manufg.
Water transportation	0.4	Util.	Highway & street construction	0.4	Constr.
Engineering, architectural, & surveying services	0.4	Services	Maintenance of nonbuilding facilities nec	0.4	Constr.
U.S. Postal Service	0.4	Gov't	Ball & roller bearings	0.4	Manufg.
Clay, ceramic, & refractory minerals	0.3	Mining	Carburetors, pistons, rings, & valves	0.4	Manufg.
Chemical preparations, nec	0.3	Manufg.	Hand & edge tools, nec	0.4	Manufg.
Cyclic crudes and organics	0.3	Manufg.	Machine tools, metal forming types	0.4	Manufg.
Lubricating oils & greases	0.3	Manufg.	Mechanical measuring devices	0.4	Manufg.
Screw machine and related products	0.3	Manufg.	Iron & ferroalloy ores	0.3	Mining
Welding apparatus, electric	0.3	Manufg.	Construction of educational buildings	0.3	Constr.
Real estate	0.3	Fin/R.E.	Construction of hospitals	0.3	Constr.
Electrical repair shops	0.3	Services	Nonfarm residential structure maintenance	0.3	Constr.
Legal services	0.3	Services	Blowers & fans	0.3	Manufg.
Management & consulting services & labs	0.3	Services	Conveyors & conveying equipment	0.3	Manufg.
Chemical & fertilizer mineral	0.2	Mining	Fabricated metal products, nec	0.3	Manufg.
Ball & roller bearings	0.2	Manufg.	Fabricated plate work (boiler shops)	0.3	Manufg.
Nonferrous wire drawing & insulating	0.2	Manufg.	Fabricated rubber products, nec	0.3	Manufg.
Primary aluminum	0.2	Manufg.	Iron & steel foundries	0.3	Manufg.
Pumps & compressors	0.2	Manufg.	Motorcycles, bicycles, & parts	0.3	Manufg.
Air transportation	0.2	Util.	Small arms	0.3	Manufg.
Insurance carriers	0.2	Fin/R.E.	Textile machinery	0.3	Manufg.
Accounting, auditing & bookkeeping	0.2	Services	Federal Government purchases, national defense	0.3	Fed Govt
Automotive repair shops & services	0.2	Services	Electric utility facility construction	0.2	Constr.
State & local government enterprises, nec	0.2	Gov't	Maintenance of nonfarm buildings nec	0.2	Constr.
Coal	0.1	Mining	Residential high-rise apartments	0.2	Constr.
Blowers & fans	0.1	Manufg.	Telephone & telegraph facility construction	0.2	Constr.
Industrial gases	0.1	Manufg.	Elevators & moving stairways	0.2	Manufg.
Manifold business forms	0.1	Manufg.	Hoists, cranes, & monorails	0.2	Manufg.
Nonclay refractories	0.1	Manufg.	Household laundry equipment	0.2	Manufg.
Primary lead	0.1	Manufg.	Lighting fixtures & equipment	0.2	Manufg.
Rubber & plastics hose & belting	0.1	Manufg.	Miscellaneous metal work	0.2	Manufg.
Wood pallets & skids	0.1	Manufg.	Mobile homes	0.2	Manufg.
Water supply & sewage systems	0.1	Util.	Radio & TV communication equipment	0.2	Manufg.
Royalties	0.1	Fin/R.E.	Service industry machines, nec	0.2	Manufg.
Automotive rental & leasing, without drivers	0.1	Services	Ship building & repairing	0.2	Manufg.
Laundry, dry cleaning, shoe repair	0.1	Services	Surgical appliances & supplies	0.2	Manufg.
			Welding apparatus, electric	0.2	Manufg.
			Wiring devices	0.2	Manufg.
			Woodworking machinery	0.2	Manufg.
			Construction of stores & restaurants	0.1	Constr.
			Hotels & motels	0.1	Constr.
			Residential 2-4 unit structures, nonfarm	0.1	Constr.
			Aircraft & missile equipment, nec	0.1	Manufg.
			Fabricated structural metal	0.1	Manufg.
			Lawn & garden equipment	0.1	Manufg.
			Measuring & dispensing pumps	0.1	Manufg.
			Metal stampings, nec	0.1	Manufg.
			Metalworking machinery, nec	0.1	Manufg.
			Plumbing fixture fittings & trim	0.1	Manufg.
			Scales & balances	0.1	Manufg.
			Sheet metal work	0.1	Manufg.
			Steel springs, except wire	0.1	Manufg.
			Surgical & medical instruments	0.1	Manufg.
			Typewriters & office machines, nec	0.1	Manufg.

Source: Benchmark Input-Output Accounts for the U.S. Economy, 1982, U.S. Department of Commerce, Washington, D.C., July 1991. Data, as reported in the source, are organized by the 1977 SIC structure in use in 1982 but have been matched, as closely as is possible, to the 1987 SIC structure used in this book.

OCCUPATIONS EMPLOYED BY SIC 332 - IRON AND STEEL FOUNDRIES

Occupation	% of Total 1994	Change to 2005	Occupation	% of Total 1994	Change to 2005
Assemblers, fabricators, & hand workers nec	6.4	-28.8	Metal & plastic machine workers nec	2.2	-37.0
Grinders & polishers, hand	6.4	24.5	Precision metal workers nec	2.0	-28.9
Metal molding machine workers	6.2	-11.1	Grinding machine operators, metal & plastic	2.0	-53.8
Helpers, laborers, & material movers nec	5.9	-28.8	Maintenance repairers, general utility	1.7	-36.0
Precision workers nec	5.4	-28.9	Electricians	1.7	-33.2
Blue collar worker supervisors	5.2	-31.8	Machinists	1.5	-28.9
Foundry mold assembly & shakeout workers	5.0	-28.9	General managers & top executives	1.5	-32.5
Inspectors, testers, & graders, precision	4.2	-28.9	Crane & tower operators	1.3	-28.9
Industrial machinery mechanics	3.3	-21.8	Sales & related workers nec	1.1	-28.8
Welders & cutters	2.9	-28.9	Machine tool cutting operators, metal & plastic	1.1	-40.8
Furnace operators	2.5	-14.6	Machine operators nec	1.1	-37.3
Industrial truck & tractor operators	2.4	-28.8	Janitors & cleaners, incl maids	1.0	-43.0

Source: Industry-Occupation Matrix, Bureau of Labor Statistics. These data relate to one or more 3-digit SIC industry groups rather than to a single 4-digit SIC. The change reported for each occupation to the year 2005 is a percent of growth or decline as estimated by the Bureau of Labor Statistics. The abbreviation nec stands for 'not elsewhere classified'.

LOCATION BY STATE AND REGIONAL CONCENTRATION

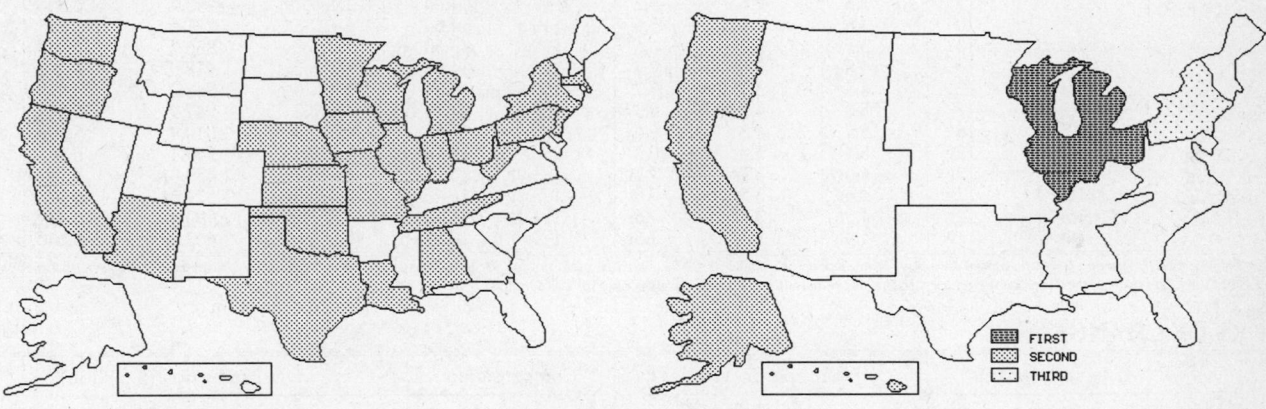

FIRST
SECOND
THIRD

INDUSTRY DATA BY STATE

State	Establish-ments	Shipments Total ($ mil)	Shipments % of U.S.	Shipments Per Establ.	Employment Total Number	Employment % of U.S.	Employment Per Establ.	Wages ($/hour)	Cost as % of Shipments	Investment per Employee ($)
Ohio	24	285.1	13.8	11.9	2,900	13.1	121	13.86	38.5	2,793
Wisconsin	18	233.4	11.3	13.0	2,800	12.6	156	11.42	39.5	2,464
Pennsylvania	32	190.8	9.3	6.0	2,400	10.8	75	10.41	41.6	1,875
Texas	18	143.6	7.0	8.0	1,900	8.6	106	10.19	38.5	-
Oregon	8	124.3	6.0	15.5	1,500	6.8	188	13.95	39.0	-
Illinois	11	105.1	5.1	9.6	700	3.2	64	11.38	46.1	-
Washington	16	97.9	4.8	6.1	900	4.1	56	12.50	38.2	13,444
Indiana	9	67.5	3.3	7.5	1,100	5.0	122	10.81	33.8	-
California	25	58.8	2.9	2.4	700	3.2	28	11.83	28.6	1,857
Missouri	10	43.6	2.1	4.4	500	2.3	50	9.33	37.2	-
Oklahoma	5	24.5	1.2	4.9	400	1.8	80	8.00	40.0	2,000
New York	6	11.3	0.5	1.9	100	0.5	17	12.00	39.8	3,000
Michigan	23	(D)	-	-	750 *	3.4	33	-	-	3,600
Alabama	12	(D)	-	-	1,750 *	7.9	146	-	-	-
Iowa	6	(D)	-	-	750 *	3.4	125	-	-	1,333
Minnesota	6	(D)	-	-	375 *	1.7	63	-	-	-
Kansas	5	(D)	-	-	750 *	3.4	150	-	-	-
Louisiana	5	(D)	-	-	375 *	1.7	75	-	-	1,067
New Jersey	5	(D)	-	-	175 *	0.8	35	-	-	-
Arizona	4	(D)	-	-	375 *	1.7	94	-	-	-
Massachusetts	4	(D)	-	-	175 *	0.8	44	-	-	-
Tennessee	4	(D)	-	-	375 *	1.7	94	-	-	-
Nebraska	2	(D)	-	-	175 *	0.8	88	-	-	-
West Virginia	2	(D)	-	-	375 *	1.7	188	-	-	-

Source: 1992 Economic Census. The states are in descending order of shipments or establishments (if shipment data are missing for the majority). The symbol (D) appears when data are withheld to prevent disclosure of competitive information. States marked with (D) are sorted by number of establishments. A dash (-) indicates that the data element cannot be calculated; * indicates the midpoint of a range.

3331 - PRIMARY COPPER

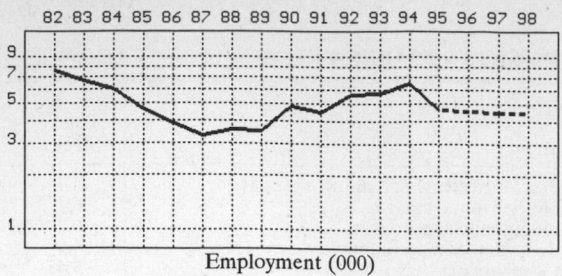

82 83 84 85 86 87 88 89 90 91 92 93 94 95 96 97 98
Shipments ($ million)

82 83 84 85 86 87 88 89 90 91 92 93 94 95 96 97 98
Employment (000)

GENERAL STATISTICS

Year	Com-panies	Establishments		Employment			Compensation		Production ($ million)			
		Total	with 20 or more employees	Total (000)	Production Workers (000)	Hours (Mil)	Payroll ($ mil)	Wages ($/hr)	Cost of Materials	Value Added by Manufacture	Value of Shipments	Capital Invest.
1982	7	22	20	7.6	5.9	12.0	216.9	14.08	2,630.9	440.4	3,077.5	112.8
1983		19	18	6.7	5.3	10.8	203.7	14.22	2,763.1	656.9	3,467.0	272.5
1984		16	16	6.0	4.6	9.5	181.1	14.34	2,532.0	177.9	2,753.3	187.7
1985		14	13	4.6	3.4	6.9	149.7	16.13	1,795.8	327.2	2,239.1	138.9
1986		12	10	3.9	3.1	6.0	116.7	14.87	1,847.0	194.9	2,065.0	13.8
1987	8	13	12	3.3	2.6	5.2	97.8	14.04	2,177.1	443.4	2,556.9	33.6
1988		13	10	3.6	2.8	5.6	110.5	14.64	3,122.5	721.6	3,825.4	
1989		15	11	3.5	3.0	6.6	119.0	13.50	3,315.8	836.2	4,146.8	44.5
1990		17	14	4.8	3.6	7.7	145.5	14.38	3,216.2	918.0	4,201.2	95.5
1991		15	12	4.5	3.5	7.7	152.5	14.83	2,987.0	942.2	3,898.1	110.3
1992	12	20	17	5.6	4.2	9.3	188.6	14.99	4,598.7	947.9	5,578.2	195.5
1993		20	16	5.7	4.3	9.0	199.6	16.58	4,527.3	1,010.7	5,596.0	312.8
1994		16P	12P	6.4	5.0	10.3	238.9	18.14	4,719.4	1,549.1	6,185.1	702.7
1995		16P	12P	4.6P	3.5P	7.7P	171.9P	16.26P	4,419.9P	1,450.8P	5,792.6P	
1996		16P	12P	4.6P	3.5P	7.6P	173.2P	16.45P	4,635.6P	1,521.6P	6,075.2P	
1997		16P	12P	4.5P	3.4P	7.6P	174.4P	16.63P	4,851.2P	1,592.4P	6,357.8P	
1998		16P	11P	4.4P	3.4P	7.5P	175.7P	16.81P	5,066.8P	1,663.1P	6,640.4P	

Sources: 1982, 1987, 1992 *Economic Census*; *Annual Survey of Manufactures*, 83-86, 88-91, 93-94. Establishment counts for non-Census years are from *County Business Patterns*; establishment values for 83-84 are extrapolations. 'P's show projections by the editors. Industries reclassified in 87 will not have data for prior years.

INDICES OF CHANGE

Year	Com-panies	Establishments		Employment			Compensation		Production ($ million)			
		Total	with 20 or more employees	Total (000)	Production Workers (000)	Hours (Mil)	Payroll ($ mil)	Wages ($/hr)	Cost of Materials	Value Added by Manufacture	Value of Shipments	Capital Invest.
1982	58	110	118	136	140	129	115	94	57	46	55	58
1983		95	106	120	126	116	108	95	60	69	62	139
1984		80	94	107	110	102	96	96	55	19	49	96
1985		70	76	82	81	74	79	108	39	35	40	71
1986		60	59	70	74	65	62	99	40	21	37	7
1987	67	65	71	59	62	56	52	94	47	47	46	17
1988		65	59	64	67	60	59	98	68	76	69	
1989		75	65	63	71	71	63	90	72	88	74	23
1990		85	82	86	86	83	77	96	70	97	75	49
1991		75	71	80	83	83	81	99	65	99	70	56
1992	100	100	100	100	100	100	100	100	100	100	100	100
1993		100	94	102	102	97	106	111	98	107	100	160
1994		82P	73P	114	119	111	127	121	103	163	111	359
1995		82P	71P	83P	85P	83P	91P	108P	96P	153P	104P	
1996		82P	70P	81P	83P	82P	92P	110P	101P	161P	109P	
1997		82P	68P	80P	82P	81P	92P	111P	105P	168P	114P	
1998		82P	67P	79P	80P	81P	93P	112P	110P	175P	119P	

Sources: Same as General Statistics. Values reflect change from the base year, 1992. Values above 100 mean greater than 92, values below 100 mean less than 92, and a value of 100 in the 82-91 or 93-98 period means same as 92. 'P's mark projections by the editors.

SELECTED RATIOS

For 1994	Avg. of All Manufact.	Analyzed Industry	Index	For 1994	Avg. of All Manufact.	Analyzed Industry	Index
Employees per Establishment	49	390	795	Value Added per Production Worker	134,084	309,820	231
Payroll per Establishment	1,500,273	14,545,572	970	Cost per Establishment	5,045,178	287,343,542	5,695
Payroll per Employee	30,620	37,328	122	Cost per Employee	102,970	737,406	716
Production Workers per Establishment	34	304	887	Cost per Production Worker	146,988	943,880	642
Wages per Establishment	853,319	11,375,989	1,333	Shipments per Establishment	9,576,895	376,583,579	3,932
Wages per Production Worker	24,861	37,368	150	Shipments per Employee	195,460	966,422	494
Hours per Production Worker	2,056	2,060	100	Shipments per Production Worker	279,017	1,237,020	443
Wages per Hour	12.09	18.14	150	Investment per Establishment	321,011	42,784,317	13,328
Value Added per Establishment	4,602,255	94,317,897	2,049	Investment per Employee	6,552	109,797	1,676
Value Added per Employee	93,930	242,047	258	Investment per Production Worker	9,352	140,540	1,503

Sources: Same as General Statistics. The 'Average of All Manufacturing' column represents the average of all manufacturing industries reported for the most recent complete year available. The Index shows the relationship between the Average and the Analyzed Industry. For example, 100 means that they are equal; 500 that the Analyzed Industry is five times the average; 50 means that the Analyzed Industry is half the national average. The abbreviation 'na' is used to show that data are 'not available'.

LEADING COMPANIES   Number shown: 4   Total sales ($ mil): 177   Total employment (000): 0.8

Company Name	Address				CEO Name	Phone	Co. Type	Sales ($ mil)	Empl. (000)
Phelps Dodge Refining Corp	PO Box 20001	El Paso	TX	79998	HW Konerko	915-778-9881	S	91	0.5
Colonial Metals Co	PO Box 311	Columbia	PA	17512	Phillip Serls	717-684-2311	R	70	0.2
GA Avril Co	4445 Kings Run Dr	Cincinnati	OH	45232	Thomas B Avril	513-641-0566	R	10	<0.1
Cox Creek Refining Co	1000 Kembo Rd	Baltimore	MD	21226	Yuzo Yamamoto	410-360-3035	S	6*	<0.1

Source: Ward's Business Directory of U.S. Private and Public Companies, Volumes 1 and 2, 1996. The company type code used is as follows: P - Public, R - Private, S - Subsidiary, D - Division, J - Joint Venture, A - Affiliate, G - Group. Sales are in millions of dollars, employees are in thousands. An asterisk (*) indicates an estimated sales volume. The symbol < stands for 'less than'. Company names and addresses are truncated, in some cases, to fit into the available space.

MATERIALS CONSUMED

Material		Quantity	Delivered Cost ($ million)
Materials, ingredients, containers, and supplies		(X)	4,291.2
All other aluminum and aluminum-base alloy shapes and forms		(X)	(D)
Refined unalloyed copper (cathodes, ingots, cakes, slabs, etc.) and blister or anode copper	1,000 s tons	2,337.1	1,775.8
All other copper and copper-base alloy shapes and forms		(X)	(D)
Refined unalloyed lead shapes and forms		(X)	(D)
All other zinc and zinc-base alloy shapes and forms		(X)	(D)
Refined unalloyed tin shapes and forms		(X)	(D)
Refined unalloyed precious metal shapes and forms		(X)	(D)
All other precious metal and precious metal alloy shapes and forms		(X)	(D)
All other magnesium and magnesium-base alloy shapes and forms		(X)	(D)
All other nonferrous shapes and forms		(X)	(D)
Mining copper ores, concentrates and precipitates (gross weight)	1,000 s tons	4,228.7	2,148.8
Mining precious metal ores and concentrates		(X)	(D)
Copper and copper-base alloy scrap (except home scrap)	1,000 s tons	(D)	(D)
All other materials and components, parts, containers, and supplies		(X)	128.2
Materials, ingredients, containers, and supplies, nsk		(X)	(D)

Source: 1992 Economic Census. Explanation of symbols used: (D): Withheld to avoid disclosure of competitive data; na: Not available; (S): Withheld because statistical norms were not met; (X): Not applicable; (Z): Less than half the unit shown; nec: Not elsewhere classified; nsk: Not specified by kind; - : zero; * : 10-19 percent estimated; ** : 20-29 percent estimated.

PRODUCT SHARE DETAILS

Product or Product Class	% Share	Product or Product Class	% Share
Primary copper	100.00	copper, etc.	39.53
Primary copper smelter products, not of commercial grade, produced for further refining, including blister or anode		Refined primary copper and copper-base alloy.	60.42
		Primary copper, nsk	0.05

Source: 1992 Economic Census. The values shown are percent of total shipments in an industry. Values of indented subcategories are summed in the main heading. The symbol (D) appears when data are withheld to prevent disclosure of competitive information. The abbreviation nsk stands for 'not specified by kind' and nec for 'not elsewhere classified'.

INPUTS AND OUTPUTS FOR PRIMARY COPPER

Economic Sector or Industry Providing Inputs	%	Sector	Economic Sector or Industry Buying Outputs	%	Sector
Primary copper	35.7	Manufg.	Primary copper	25.2	Manufg.
Copper ore	29.9	Mining	Copper rolling & drawing	16.7	Manufg.
Imports	18.2	Foreign	Primary nonferrous metals, nec	8.5	Manufg.
Gas production & distribution (utilities)	3.1	Util.	Surgical & medical instruments	6.2	Manufg.
Railroads & related services	1.7	Util.	Nonferrous wire drawing & insulating	6.1	Manufg.
Wholesale trade	1.7	Trade	Switchgear & switchboard apparatus	6.0	Manufg.
Electric services (utilities)	1.5	Util.	Secondary nonferrous metals	4.1	Manufg.
Petroleum refining	1.1	Manufg.	Brass, bronze, & copper castings	3.5	Manufg.
Scrap	1.0	Scrap	Change in business inventories	3.3	In House
Business services nec	0.6	Services	Motor vehicle parts & accessories	3.0	Manufg.
Primary lead	0.5	Manufg.	Plating & polishing	2.0	Manufg.
Motor freight transportation & warehousing	0.4	Util.	Carbon & graphite products	1.8	Manufg.
Miscellaneous repair shops	0.4	Services	Pipe, valves, & pipe fittings	1.6	Manufg.
Coal	0.3	Mining	Exports	1.6	Foreign
Miscellaneous plastics products	0.3	Manufg.	Primary aluminum	1.5	Manufg.
Pipe, valves, & pipe fittings	0.3	Manufg.	Miscellaneous plastics products	1.2	Manufg.
Sanitary services, steam supply, irrigation	0.3	Util.	Electrical equipment & supplies, nec	0.8	Manufg.
Banking	0.3	Fin/R.E.	Hardware, nec	0.8	Manufg.
Chemical preparations, nec	0.2	Manufg.	Metal heat treating	0.7	Manufg.
Cyclic crudes and organics	0.2	Manufg.	Fabricated metal products, nec	0.6	Manufg.
Water transportation	0.2	Util.	Aluminum castings	0.5	Manufg.

Continued on next page.

INPUTS AND OUTPUTS FOR PRIMARY COPPER - Continued

Economic Sector or Industry Providing Inputs	%	Sector	Economic Sector or Industry Buying Outputs	%	Sector
Advertising	0.2	Services	Industrial controls	0.5	Manufg.
Maintenance of nonfarm buildings nec	0.1	Constr.	Iron & steel foundries	0.5	Manufg.
Automotive rental & leasing, without drivers	0.1	Services	Nonferrous rolling & drawing, nec	0.5	Manufg.
			Steel pipe & tubes	0.5	Manufg.
			Plumbing fixture fittings & trim	0.3	Manufg.
			Power transmission equipment	0.3	Manufg.
			Primary metal products, nec	0.3	Manufg.
			Aluminum rolling & drawing	0.2	Manufg.
			Metal sanitary ware	0.2	Manufg.
			Pumps & compressors	0.2	Manufg.
			Federal Government purchases, nondefense	0.2	Fed Govt
			Blast furnaces & steel mills	0.1	Manufg.
			Steel wire & related products	0.1	Manufg.

Source: Benchmark Input-Output Accounts for the U.S. Economy, 1982, U.S. Department of Commerce, Washington, D.C., July 1991. Data, as reported in the source, are organized by the 1977 SIC structure in use in 1982 but have been matched, as closely as is possible, to the 1987 SIC structure used in this book.

OCCUPATIONS EMPLOYED BY SIC 333 - PRIMARY METALS NEC

Occupation	% of Total 1994	Change to 2005	Occupation	% of Total 1994	Change to 2005
Blue collar worker supervisors	7.8	-15.9	Crushing & mixing machine operators	1.8	-9.1
Furnace operators	6.7	22.2	Sales & related workers nec	1.6	-8.6
Heat treating machine operators, metal & plastic	6.2	-8.1	Secretaries, ex legal & medical	1.6	-17.5
Assemblers, fabricators, & hand workers nec	4.5	-10.4	Machine operators nec	1.5	-22.2
Industrial machinery mechanics	4.4	-1.5	Industrial production managers	1.5	-9.1
Metal & plastic machine workers nec	3.6	-19.5	Freight, stock, & material movers, hand	1.4	-27.9
Maintenance repairers, general utility	3.6	-19.1	Metal molding machine workers	1.4	-1.2
Inspectors, testers, & graders, precision	2.7	-9.0	Machinists	1.3	-10.2
General managers & top executives	2.5	-13.2	Traffic, shipping, & receiving clerks	1.2	-11.9
Industrial truck & tractor operators	2.3	-9.9	Science & mathematics technicians	1.2	-10.0
Electricians	2.2	-16.7	Separating & still machine operators	1.1	-11.4
Truck drivers light & heavy	2.0	-5.5	Bookkeeping, accounting, & auditing clerks	1.1	-31.5
Machine tool cutting & forming etc. nec	1.9	-9.7	Metallurgists & ceramic & materials engineers	1.0	2.2
Crane & tower operators	1.8	-10.8			

Source: Industry-Occupation Matrix, Bureau of Labor Statistics. These data relate to one or more 3-digit SIC industry groups rather than to a single 4-digit SIC. The change reported for each occupation to the year 2005 is a percent of growth or decline as estimated by the Bureau of Labor Statistics. The abbreviation nec stands for 'not elsewhere classified'.

LOCATION BY STATE AND REGIONAL CONCENTRATION

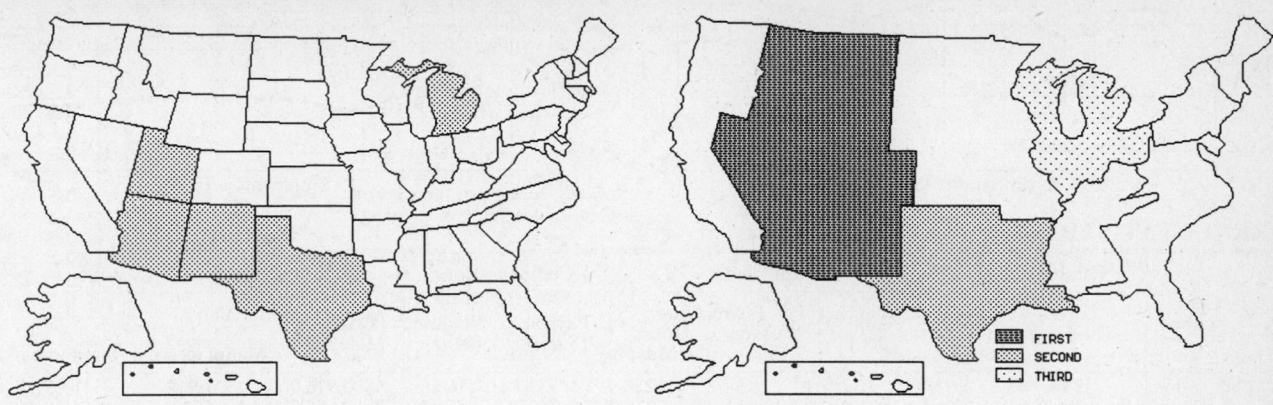

INDUSTRY DATA BY STATE

| State | Establish-ments | Shipments | | | Employment | | | | Cost as % of Shipments | Investment per Employee ($) |
		Total ($ mil)	% of U.S.	Per Establ.	Total Number	% of U.S.	Per Establ.	Wages ($/hour)		
Arizona	6	(D)	-	-	1,750 *	31.3	292	-	-	-
Texas	3	(D)	-	-	1,750 *	31.3	583	-	-	-
Michigan	2	(D)	-	-	175 *	3.1	88	-	-	-
New Mexico	2	(D)	-	-	750 *	13.4	375	-	-	-
Utah	2	(D)	-	-	750 *	13.4	375	-	-	-

Source: 1992 *Economic Census*. The states are in descending order of shipments or establishments (if shipment data are missing for the majority). The symbol (D) appears when data are withheld to prevent disclosure of competitive information. States marked with (D) are sorted by number of establishments. A dash (-) indicates that the data element cannot be calculated; * indicates the midpoint of a range.

3334 - PRIMARY ALUMINUM

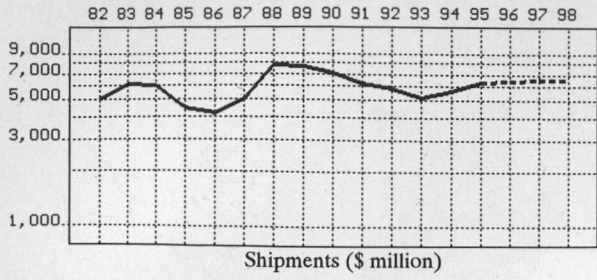

Shipments ($ million)

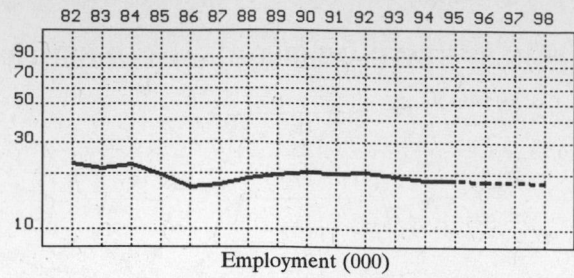

Employment (000)

GENERAL STATISTICS

| Year | Com-panies | Establishments | | Employment | | | Compensation | | Production ($ million) | | | |
		Total	with 20 or more employees	Total (000)	Production Workers (000)	Hours (Mil)	Payroll ($ mil)	Wages ($/hr)	Cost of Materials	Value Added by Manufacture	Value of Shipments	Capital Invest.
1982	15	34	29	22.9	16.9	32.3	733.1	16.26	3,916.0	1,133.9	5,037.1	181.2
1983		38	30	21.4	16.5	32.0	712.9	16.53	4,158.7	1,566.8	6,089.9	390.7
1984		42	31	22.5	17.8	35.9	767.2	16.20	4,516.4	1,722.5	6,011.2	154.4
1985		45	31	19.7	15.3	30.0	664.8	16.41	3,423.2	877.1	4,488.4	122.8
1986		52	35	16.9	13.1	23.9	570.4	16.97	3,014.5	1,003.9	4,241.8	71.1
1987	34	49	27	17.3	13.8	27.7	568.5	15.18	3,049.3	1,900.1	5,005.7	173.5
1988		49	30	18.9	15.1	32.0	668.8	15.78	4,171.2	3,663.6	7,806.8	146.8
1989		52	29	19.7	14.9	30.3	717.0	17.69	4,732.4	2,931.5	7,641.6	181.9
1990		54	30	20.5	15.4	32.1	750.1	17.50	4,844.3	2,205.6	7,033.9	187.8
1991		57	32	20.0	15.8	32.1	776.4	18.02	4,542.6	1,649.6	6,199.2	207.6
1992	30	41	28	20.4	16.2	32.3	805.9	18.81	4,226.9	1,609.8	5,848.9	215.0
1993		44	27	19.4	14.8	29.0	727.9	18.52	4,015.4	1,147.4	5,170.5	167.1
1994		53P	29P	18.5	13.9	26.9	706.1	19.20	3,797.6	1,794.3	5,576.6	120.7
1995		54P	29P	18.5P	14.4P	29.1P	740.3P	18.94P	4,286.2P	2,025.2P	6,294.1P	150.0P
1996		55P	28P	18.3P	14.2P	28.9P	745.3P	19.20P	4,328.6P	2,045.2P	6,356.4P	145.9P
1997		56P	28P	18.2P	14.1P	28.7P	750.3P	19.46P	4,371.1P	2,065.3P	6,418.7P	141.8P
1998		57P	28P	18.0P	13.9P	28.5P	755.3P	19.71P	4,413.5P	2,085.3P	6,481.1P	137.7P

Sources: 1982, 1987, 1992 *Economic Census*; *Annual Survey of Manufactures*, 83-86, 88-91, 93-94. Establishment counts for non-Census years are from *County Business Patterns*; establishment values for 83-84 are extrapolations. 'P's show projections by the editors. Industries reclassified in 87 will not have data for prior years.

INDICES OF CHANGE

| Year | Com-panies | Establishments | | Employment | | | Compensation | | Production ($ million) | | | |
		Total	with 20 or more employees	Total (000)	Production Workers (000)	Hours (Mil)	Payroll ($ mil)	Wages ($/hr)	Cost of Materials	Value Added by Manufacture	Value of Shipments	Capital Invest.
1982	50	83	104	112	104	100	91	86	93	70	86	84
1983		93	107	105	102	99	88	88	98	97	104	182
1984		102	111	110	110	111	95	86	107	107	103	72
1985		110	111	97	94	93	82	87	81	54	77	57
1986		127	125	83	81	74	71	90	71	62	73	33
1987	113	120	96	85	85	86	71	81	72	118	86	81
1988		120	107	93	93	99	83	84	99	228	133	68
1989		127	104	97	92	94	89	94	112	182	131	85
1990		132	107	100	95	99	93	93	115	137	120	87
1991		139	114	98	98	99	96	96	107	102	106	97
1992	100	100	100	100	100	100	100	100	100	100	100	100
1993		107	96	95	91	90	90	98	95	71	88	78
1994		129P	103P	91	86	83	88	102	90	111	95	56
1995		132P	102P	91P	89P	90P	92P	101P	101P	126P	108P	70P
1996		134P	101P	90P	88P	89P	92P	102P	102P	127P	109P	68P
1997		136P	101P	89P	87P	89P	93P	103P	103P	128P	110P	66P
1998		139P	100P	88P	86P	88P	94P	105P	104P	130P	111P	64P

Sources: Same as General Statistics. Values reflect change from the base year, 1992. Values above 100 mean greater than 92, values below 100 mean less than 92, and a value of 100 in the 82-91 or 93-98 period means same as 92. 'P's mark projections by the editors.

SELECTED RATIOS

For 1994	Avg. of All Manufact.	Analyzed Industry	Index	For 1994	Avg. of All Manufact.	Analyzed Industry	Index
Employees per Establishment	49	349	713	Value Added per Production Worker	134,084	129,086	96
Payroll per Establishment	1,500,273	13,337,894	889	Cost per Establishment	5,045,178	71,734,860	1,422
Payroll per Employee	30,620	38,168	125	Cost per Employee	102,970	205,276	199
Production Workers per Establishment	34	263	765	Cost per Production Worker	146,988	273,209	186
Wages per Establishment	853,319	9,756,062	1,143	Shipments per Establishment	9,576,895	105,339,325	1,100
Wages per Production Worker	24,861	37,157	149	Shipments per Employee	195,460	301,438	154
Hours per Production Worker	2,056	1,935	94	Shipments per Production Worker	279,017	401,194	144
Wages per Hour	12.09	19.20	159	Investment per Establishment	321,011	2,279,966	710
Value Added per Establishment	4,602,255	33,893,475	736	Investment per Employee	6,552	6,524	100
Value Added per Employee	93,930	96,989	103	Investment per Production Worker	9,352	8,683	93

Sources: Same as General Statistics. The 'Average of All Manufacturing' column represents the average of all manufacturing industries reported for the most recent complete year available. The Index shows the relationship between the Average and the Analyzed Industry. For example, 100 means that they are equal; 500 that the Analyzed Industry is five times the average; 50 means that the Analyzed Industry is half the national average. The abbreviation 'na' is used to show that data are 'not available'.

LEADING COMPANIES   Number shown: 10   Total sales ($ mil): 20,157   Total employment (000): 111.3

Company Name	Address				CEO Name	Phone	Co. Type	Sales ($ mil)	Empl. (000)
Aluminum Company of America	425 Sixth Av	Pittsburgh	PA	15219	Paul H O'Neill	412-553-4545	P	9,904	61.7
Reynolds Metals Co	PO Box 27003	Richmond	VA	23261	Richard G Holder	804-281-2000	P	5,879	29.0
Alumax Inc	5565 Peachtree	Norcross	GA	30092	Allen Born	404-246-6600	P	2,754	14.1
Noranda Aluminum Inc	750 Old Hickory	Brentwood	TN	37027	Elzie Z Borders	615-377-4300	S	1,500	5.6
Girard Extrusion	PO Box 60	Girard	OH	44420	Michael Haggerty	216-545-4311	D	43	0.4
TST Inc	11601 Etiwanda Av	Fontana	CA	92337	Robert Caldwell	909-685-2155	R	41*	0.2
Midamerica Extrusions	4925 Aluminum Dr	Indianapolis	IN	46218	T D Strickland	317-545-1221	D	15*	<0.1
M Kimerling and Sons Inc	PO Box 10184	Birmingham	AL	35202	Joseph Kimerling	205-841-6706	R	15*	<0.1
Dix Metals and Plastics Inc	1361 Bell Av	Tustin	CA	92680	RJ Dix	714-259-1611	R	5	<0.1
Specified Components Inc	6102 Milwee St	Houston	TX	77092	Kelly E Johnson	713-957-0391	R	0*	<0.1

Source: Ward's Business Directory of U.S. Private and Public Companies, Volumes 1 and 2, 1996. The company type code used is as follows: P - Public, R - Private, S - Subsidiary, D - Division, J - Joint Venture, A - Affiliate, G - Group. Sales are in millions of dollars, employees are in thousands. An asterisk (*) indicates an estimated sales volume. The symbol < stands for 'less than'. Company names and addresses are truncated, in some cases, to fit into the available space.

MATERIALS CONSUMED

Material	Quantity		Delivered Cost ($ million)
Materials, ingredients, containers, and supplies		(X)	2,501.5
Aluminum and aluminum-base alloy ingot	1,000 s tons	98.5	42.8
All other aluminum and aluminum-base alloy shapes and forms		(X)	40.3
Refined unalloyed copper (cathodes, ingots, cakes, slabs, etc.) and blister or anode copper	1,000 s tons	(D)	(D)
All other copper and copper-base alloy shapes and forms		(X)	0.8
All other zinc and zinc-base alloy shapes and forms		(X)	(D)
All other tin and tin-base alloy shapes and forms		(X)	(D)
Magnesium and magnesium-base alloy ingot		(X)	19.9
All other magnesium and magnesium-base alloy shapes and forms		(X)	2.2
All other nonferrous shapes and forms		(X)	25.1
Mining all other nonferrous metal ores and concentrates		(X)	(D)
Aluminum and aluminum-base alloy scrap from other establishments	1,000 s tons	(D)	(D)
Aluminum and aluminum-base alloy scrap from all other sources	1,000 s tons	(D)	(D)
Copper and copper-base alloy scrap (except home scrap)	1,000 s tons	(D)	(D)
Lead and lead-base alloy scrap		(X)	(D)
Zinc and zinc-base alloy scrap (including drosses and skimmings)		(X)	(D)
All other nonferrous metal and metal-base alloy scrap		(X)	(D)
Alumina (gross weight)	1,000 s tons	11,048.3	1,401.5
All other materials and components, parts, containers, and supplies		(X)	936.4
Materials, ingredients, containers, and supplies, nsk		(X)	1.6

Source: 1992 Economic Census. Explanation of symbols used: (D): Withheld to avoid disclosure of competitive data; na: Not available; (S): Withheld because statistical norms were not met; (X): Not applicable; (Z): Less than half the unit shown; nec: Not elsewhere classified; nsk: Not specified by kind; - : zero; * : 10-19 percent estimated; ** : 20-29 percent estimated.

PRODUCT SHARE DETAILS

Product or Product Class	% Share	Product or Product Class	% Share
Primary aluminum.	100.00	Primary aluminum extrusion ingot (billet), produced in primary aluminum reduction plants.	19.63
Primary aluminum ingot, produced in primary aluminum reduction plants (including pigs, sows, and molten metal, excluding billet).	80.31	Primary aluminum, nsk	0.05

Source: 1992 Economic Census. The values shown are percent of total shipments in an industry. Values of indented subcategories are summed in the main heading. The symbol (D) appears when data are withheld to prevent disclosure of competitive information. The abbreviation nsk stands for 'not specified by kind' and nec for 'not elsewhere classified'.

INPUTS AND OUTPUTS FOR PRIMARY ALUMINUM

Economic Sector or Industry Providing Inputs	%	Sector	Economic Sector or Industry Buying Outputs	%	Sector
Imports	24.1	Foreign	Aluminum rolling & drawing	47.4	Manufg.
Primary aluminum	20.6	Manufg.	Primary aluminum	16.0	Manufg.
Electric services (utilities)	19.0	Util.	Metal cans	6.7	Manufg.
Cyclic crudes and organics	7.3	Manufg.	Aluminum castings	6.5	Manufg.
Products of petroleum & coal, nec	4.4	Manufg.	Exports	6.1	Foreign
Nonferrous metal ores, except copper	3.0	Mining	Nonferrous wire drawing & insulating	5.3	Manufg.
Motor freight transportation & warehousing	3.0	Util.	Metal doors, sash, & trim	1.8	Manufg.
Railroads & related services	2.7	Util.	Abrasive products	1.1	Manufg.
Wholesale trade	2.2	Trade	Nonferrous forgings	1.1	Manufg.
Gas production & distribution (utilities)	1.6	Util.	Internal combustion engines, nec	1.0	Manufg.

Continued on next page.

INPUTS AND OUTPUTS FOR PRIMARY ALUMINUM - Continued

Economic Sector or Industry Providing Inputs	%	Sector	Economic Sector or Industry Buying Outputs	%	Sector
Banking	1.5	Fin/R.E.	Secondary nonferrous metals	0.9	Manufg.
Carbon & graphite products	1.0	Manufg.	Blast furnaces & steel mills	0.8	Manufg.
Industrial inorganic chemicals, nec	0.9	Manufg.	Fabricated plate work (boiler shops)	0.8	Manufg.
Petroleum refining	0.9	Manufg.	Fabricated metal products, nec	0.5	Manufg.
Primary copper	0.9	Manufg.	Motor vehicle parts & accessories	0.5	Manufg.
Miscellaneous plastics products	0.5	Manufg.	Motors & generators	0.4	Manufg.
Alkalies & chlorine	0.4	Manufg.	Industrial controls	0.3	Manufg.
Pumps & compressors	0.4	Manufg.	Miscellaneous metal work	0.3	Manufg.
Water transportation	0.4	Util.	Architectural metal work	0.2	Manufg.
Business services nec	0.4	Services	Copper rolling & drawing	0.2	Manufg.
Maintenance of nonfarm buildings nec	0.3	Constr.	Engine electrical equipment	0.2	Manufg.
Lime	0.3	Manufg.	Nonferrous castings, nec	0.2	Manufg.
Miscellaneous repair shops	0.3	Services	Sheet metal work	0.2	Manufg.
Pipe, valves, & pipe fittings	0.2	Manufg.	Brass, bronze, & copper castings	0.1	Manufg.
Primary nonferrous metals, nec	0.2	Manufg.	Heating equipment, except electric	0.1	Manufg.
Air transportation	0.2	Util.	Motor vehicles & car bodies	0.1	Manufg.
Communications, except radio & TV	0.2	Util.	Primary metal products, nec	0.1	Manufg.
Eating & drinking places	0.2	Trade	Switchgear & switchboard apparatus	0.1	Manufg.
Automotive rental & leasing, without drivers	0.2	Services			
Engineering, architectural, & surveying services	0.2	Services			
Scrap	0.2	Scrap			
Chemical preparations, nec	0.1	Manufg.			
Machinery, except electrical, nec	0.1	Manufg.			
Real estate	0.1	Fin/R.E.			
Automotive repair shops & services	0.1	Services			
Equipment rental & leasing services	0.1	Services			

Source: Benchmark Input-Output Accounts for the U.S. Economy, 1982, U.S. Department of Commerce, Washington, D.C., July 1991. Data, as reported in the source, are organized by the 1977 SIC structure in use in 1982 but have been matched, as closely as is possible, to the 1987 SIC structure used in this book.

OCCUPATIONS EMPLOYED BY SIC 333 - PRIMARY METALS NEC

Occupation	% of Total 1994	Change to 2005	Occupation	% of Total 1994	Change to 2005
Blue collar worker supervisors	7.8	-15.9	Crushing & mixing machine operators	1.8	-9.1
Furnace operators	6.7	22.2	Sales & related workers nec	1.6	-8.6
Heat treating machine operators, metal & plastic	6.2	-8.1	Secretaries, ex legal & medical	1.6	-17.5
Assemblers, fabricators, & hand workers nec	4.5	-10.4	Machine operators nec	1.5	-22.2
Industrial machinery mechanics	4.4	-1.5	Industrial production managers	1.5	-9.1
Metal & plastic machine workers nec	3.6	-19.5	Freight, stock, & material movers, hand	1.4	-27.9
Maintenance repairers, general utility	3.6	-19.1	Metal molding machine workers	1.4	-1.2
Inspectors, testers, & graders, precision	2.7	-9.0	Machinists	1.3	-10.2
General managers & top executives	2.5	-13.2	Traffic, shipping, & receiving clerks	1.2	-11.9
Industrial truck & tractor operators	2.3	-9.9	Science & mathematics technicians	1.2	-10.0
Electricians	2.2	-16.7	Separating & still machine operators	1.1	-11.4
Truck drivers light & heavy	2.0	-5.5	Bookkeeping, accounting, & auditing clerks	1.1	-31.5
Machine tool cutting & forming etc. nec	1.9	-9.7	Metallurgists & ceramic & materials engineers	1.0	2.2
Crane & tower operators	1.8	-10.8			

Source: Industry-Occupation Matrix, Bureau of Labor Statistics. These data relate to one or more 3-digit SIC industry groups rather than to a single 4-digit SIC. The change reported for each occupation to the year 2005 is a percent of growth or decline as estimated by the Bureau of Labor Statistics. The abbreviation nec stands for 'not elsewhere classified'.

LOCATION BY STATE AND REGIONAL CONCENTRATION

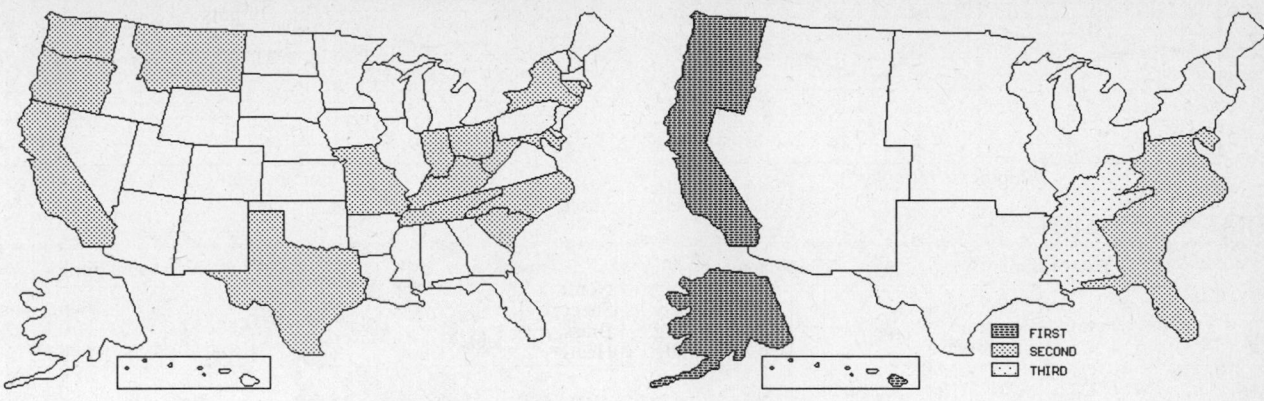

INDUSTRY DATA BY STATE

| State | Establish-ments | Shipments | | | Employment | | | | Cost as % of Shipments | Investment per Employee ($) |
		Total ($ mil)	% of U.S.	Per Establ.	Total Number	% of U.S.	Per Establ.	Wages ($/hour)		
Washington	9	1,563.3	26.7	173.7	5,700	27.9	633	16.95	65.8	-
California	3	(D)	-	-	750 *	3.7	250	-	-	-
Kentucky	3	(D)	-	-	1,750 *	8.6	583	-	-	-
New York	3	(D)	-	-	1,750 *	8.6	583	-	-	-
Maryland	2	(D)	-	-	750 *	3.7	375	-	-	-
North Carolina	2	(D)	-	-	750 *	3.7	375	-	-	-
Ohio	2	(D)	-	-	1,750 *	8.6	875	-	-	-
Oregon	2	(D)	-	-	750 *	3.7	375	-	-	-
Tennessee	2	(D)	-	-	750 *	3.7	375	-	-	-
Texas	2	(D)	-	-	1,750 *	8.6	875	-	-	-
Connecticut	1	(D)	-	-	750 *	3.7	750	-	-	-
Indiana	1	(D)	-	-	1,750 *	8.6	1,750	-	-	-
Missouri	1	(D)	-	-	1,750 *	8.6	1,750	-	-	-
Montana	1	(D)	-	-	750 *	3.7	750	-	-	-
South Carolina	1	(D)	-	-	750 *	3.7	750	-	-	-
West Virginia	1	(D)	-	-	750 *	3.7	750	-	-	-

Source: 1992 *Economic Census*. The states are in descending order of shipments or establishments (if shipment data are missing for the majority). The symbol (D) appears when data are withheld to prevent disclosure of competitive information. States marked with (D) are sorted by number of establishments. A dash (-) indicates that the data element cannot be calculated; * indicates the midpoint of a range.

3339 - PRIMARY NONFERROUS METALS, NEC

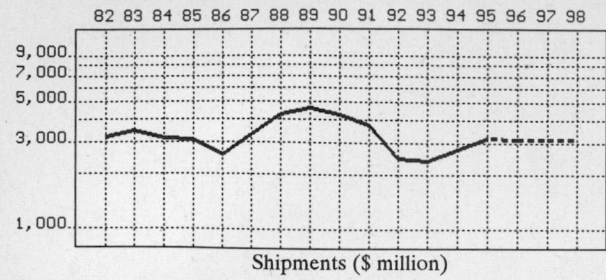

Shipments ($ million)

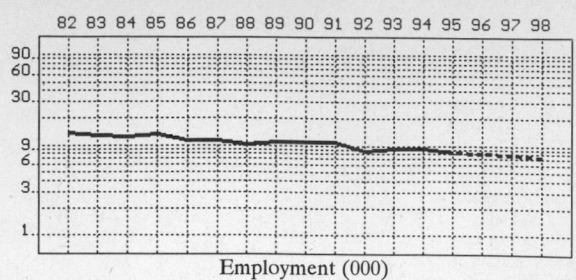

Employment (000)

GENERAL STATISTICS

Year	Com-panies	Establishments		Employment			Compensation		Production ($ million)			
		Total	with 20 or more employees	Total (000)	Production Workers (000)	Hours (Mil)	Payroll ($ mil)	Wages ($/hr)	Cost of Materials	Value Added by Manufacture	Value of Shipments	Capital Invest.
1982		105	45	13.4	9.2	17.8	356.6	13.16	2,510.5	736.7	3,206.2	118.4
1983		97	41	12.7	8.5	15.9	338.9	13.37	2,542.2	889.2	3,465.7	69.8
1984		89	37	12.2	8.8	16.8	351.5	13.86	2,516.7	674.9	3,179.5	111.7
1985		82	34	13.2	9.3	17.9	380.5	13.90	2,485.2	426.4	3,060.0	277.2
1986		77	38	11.2	7.9	15.1	323.9	13.70	1,913.5	546.5	2,569.5	113.8
1987	96	108	49	11.0	7.6	15.4	329.2	13.22	2,621.6	654.5	3,306.5	154.3
1988		111	49	10.1	7.5	15.0	325.3	14.66	3,110.6	1,196.9	4,286.2	
1989		104	49	10.7	7.7	15.9	358.2	15.11	3,382.8	1,306.9	4,650.7	116.1
1990		103	48	10.9	7.8	16.0	385.2	16.02	3,123.7	1,142.2	4,272.0	146.2
1991		104	51	10.7	7.6	16.4	381.2	14.98	2,776.3	1,001.4	3,740.2	108.3
1992	101	113	53	8.6	6.2	13.6	337.9	16.31	1,621.6	828.4	2,468.6	146.3
1993		117	49	9.1	6.6	13.6	329.0	15.49	1,590.9	788.5	2,391.8	111.7
1994		114P	53P	9.3	6.6	14.3	329.8	14.25	1,913.9	854.7	2,784.2	107.0
1995		116P	55P	8.5P	6.2P	13.8P	344.0P	15.88P	2,189.3P	977.7P	3,184.9P	
1996		118P	56P	8.1P	6.0P	13.6P	343.4P	16.08P	2,174.4P	971.0P	3,163.1P	
1997		120P	57P	7.8P	5.8P	13.3P	342.8P	16.28P	2,159.5P	964.4P	3,141.4P	
1998		122P	58P	7.4P	5.6P	13.1P	342.1P	16.48P	2,144.5P	957.7P	3,119.7P	

Sources: 1982, 1987, 1992 *Economic Census*; *Annual Survey of Manufactures*, 83-86, 88-91, 93-94. Establishment counts for non-Census years are from *County Business Patterns*; establishment values for 83-84 are extrapolations. 'P's show projections by the editors. Industries reclassified in 87 will not have data for prior years.

INDICES OF CHANGE

Year	Com-panies	Establishments		Employment			Compensation		Production ($ million)			
		Total	with 20 or more employees	Total (000)	Production Workers (000)	Hours (Mil)	Payroll ($ mil)	Wages ($/hr)	Cost of Materials	Value Added by Manufacture	Value of Shipments	Capital Invest.
1982		93	85	156	148	131	106	81	155	89	130	81
1983		86	77	148	137	117	100	82	157	107	140	48
1984		79	70	142	142	124	104	85	155	81	129	76
1985		73	64	153	150	132	113	85	153	51	124	189
1986		68	72	130	127	111	96	84	118	66	104	78
1987	95	96	92	128	123	113	97	81	162	79	134	105
1988		98	92	117	121	110	96	90	192	144	174	
1989		92	92	124	124	117	106	93	209	158	188	79
1990		91	91	127	126	118	114	98	193	138	173	100
1991		92	96	124	123	121	113	92	171	121	152	74
1992	100	100	100	100	100	100	100	100	100	100	100	100
1993		104	92	106	106	100	97	95	98	95	97	76
1994		101P	101P	108	106	105	98	87	118	103	113	73
1995		102P	103P	99P	100P	102P	102P	97P	135P	118P	129P	
1996		104P	105P	95P	97P	100P	102P	99P	134P	117P	128P	
1997		106P	108P	90P	93P	98P	101P	100P	133P	116P	127P	
1998		108P	110P	86P	90P	96P	101P	101P	132P	116P	126P	

Sources: Same as General Statistics. Values reflect change from the base year, 1992. Values above 100 mean greater than 92, values below 100 mean less than 92, and a value of 100 in the 82-91 or 93-98 period means same as 92. 'P's mark projections by the editors.

SELECTED RATIOS

For 1994	Avg. of All Manufact.	Analyzed Industry	Index	For 1994	Avg. of All Manufact.	Analyzed Industry	Index
Employees per Establishment	49	82	167	Value Added per Production Worker	134,084	129,500	97
Payroll per Establishment	1,500,273	2,898,375	193	Cost per Establishment	5,045,178	16,819,893	333
Payroll per Employee	30,620	35,462	116	Cost per Employee	102,970	205,796	200
Production Workers per Establishment	34	58	169	Cost per Production Worker	146,988	289,985	197
Wages per Establishment	853,319	1,790,832	210	Shipments per Establishment	9,576,895	24,468,336	255
Wages per Production Worker	24,861	30,875	124	Shipments per Employee	195,460	299,376	153
Hours per Production Worker	2,056	2,167	105	Shipments per Production Worker	279,017	421,848	151
Wages per Hour	12.09	14.25	118	Investment per Establishment	321,011	940,346	293
Value Added per Establishment	4,602,255	7,511,345	163	Investment per Employee	6,552	11,505	176
Value Added per Employee	93,930	91,903	98	Investment per Production Worker	9,352	16,212	173

Sources: Same as General Statistics. The 'Average of All Manufacturing' column represents the average of all manufacturing industries reported for the most recent complete year available. The Index shows the relationship between the Average and the Analyzed Industry. For example, 100 means that they are equal; 500 that the Analyzed Industry is five times the average; 50 means that the Analyzed Industry is half the national average. The abbreviation 'na' is used to show that data are 'not available'.

LEADING COMPANIES   Number shown: 36   Total sales ($ mil): 3,555   Total employment (000): 21.8

Company Name	Address				CEO Name	Phone	Co. Type	Sales ($ mil)	Empl. (000)
Handy and Harman	555 T Fremd	Rye	NY	10580	Richard N Daniel	212-661-2400	P	781	4.8
OSi Specialties Inc	39 Old Ridgebury	Danbury	CT	06810	DI Barton	203-794-4300	S	400	1.2
Cookson America Inc	1 Cookson Pl	Providence	RI	02903	Donald L Carcieri	401-521-1000	S	370	6.0
Stern Metals Inc	PO Box 2018	Attleboro	MA	02703	Fredric J Hammerle	508-222-7400	S	370	1.0
Brush Wellman Inc	17876 St Clair Av	Cleveland	OH	44110	Gordon D Harnett	216-486-4200	P	346	1.8
Ravenswood Aluminum Corp	PO Box 98	Ravenswood	WV	26164	Gerald A Meyers	304-273-6564	R	300	2.0
Teledyne Wah Chang Albany	PO Box 460	Albany	OR	97321	Al E Riesen	503-926-4211	D	150	1.1
Special Metals Corp	4317 Middle	New Hartford	NY	13413	Donald R Muzyka	315-798-2900	S	94	0.6
I Schumann and Co	PO Box 46271	Bedford	OH	44146	M A Schumann	216-439-2300	R	80	0.2
Oregon Metallurgical Corp	PO Box 580	Albany	OR	97321	Carlos E Aguirre	503-926-4281	P	71	0.5
Sipi Metals Corp	1720 N Elston Av	Chicago	IL	60622	Leslie Pinsof	312-276-0070	R	65	0.1
Sanders Lead Company Inc	PO Box 707	Troy	AL	36081	Wiley C Sanders Jr	205-566-1563	R	61	0.4
HC Starck Inc	45 Industrial Pl	Newton	MA	02164	Wolf Albrecht	617-630-5800	S	52	0.3
Glines and Rhodes Inc	PO Box 2285	Attleboro	MA	02703	RE Crowell	508-226-2000	R	49	<0.1
Big River Zinc Co	Rte 3-Monsan	Sauget	IL	62201	Lonny Ludwig	618-274-5000	D	43*	0.4
Axel Johnson Metals Inc	215 Welsh Pool Rd	Exton	PA	19341	William C Acton	610-363-0330	S	40*	0.2
Victory White Metal Co	6100 Roland Av	Cleveland	OH	44127	Joseph B Sturman	216-271-1400	R	40	0.1
Pease and Curren Inc	75 Pennsylvania Av	Warwick	RI	02888	Robert Pease	401-739-6350	R	28	<0.1
Miller Co	PO Box 1010	Meriden	CT	06450	B G Tremaine III	203-235-4474	R	27*	0.2
Guardian Metal Sales Inc	6116-32 W Oakton	Morton Grove	IL	60053	William Bohnen	708-967-7400	R	25	<0.1
Metropolitan Alloys Corp	17385 Ryan Rd	Detroit	MI	48212	Murray Spilman	313-366-4443	R	20	<0.1
Semco Enterprises Inc	475 S Wilson Way	City of Industry	CA	91744	James Seminoff	818-333-2237	R	20	<0.1
United Refining&Smelting Co	3700 N Runge Av	Franklin Park	IL	60131	R Glavin	708-455-8800	S	19	<0.1
Belmont Metals Inc	330 Belmont Av	Brooklyn	NY	11207	RV Henning	718-342-4900	R	16	0.1
NuSil Technology	1040 Cindy Ln	Carpinteria	CA	93013	Richard Compton	805-684-8780	R	15*	0.1
Attleboro Refining	PO Box 1390	Attleboro	MA	02703	Brian W Gates	508-226-1000	D	12	<0.1
Sierra Alloys Company Inc	5467 Ayon Av	Irwindale	CA	91706	JP Augustyn	818-969-6711	R	12	<0.1
L-S Plate and Wire Corp	70-17 51st Av	Woodside	NY	11377	C Strominger	718-458-9312	R	10	<0.1
Mandel Metals Inc	333 N California	Chicago	IL	60612	Richard Mandel	312-722-4900	R	9	<0.1
Taracorp Evans Inc	740 Lambert Dr	Atlanta	GA	30324	Mark Taylor	404-875-5636	D	6*	<0.1
Interamerican Zinc Inc	401 Gulf St	Adrian	MI	49221	Larry L Parkinson	517-263-8984	S	5	<0.1
Magnesium Alloy Products Co	PO Box 4668	Compton	CA	90224	JW Long	213-774-1590	R	5	<0.1
Metallic Resources Inc	PO Box 368	Twinsburg	OH	44087	Stanley Rothschild	216-425-3155	R	5	<0.1
North American Oxide Inc	480 Arcata Blv	Clarksville	TN	37040	Charles Kimbel	615-552-8080	R	4	<0.1
WE Mowery Company Inc	1435 University Av	St Paul	MN	55104	Walter F Hirschey	612-646-1895	R	4*	<0.1
Carolmet Inc	PO Box 1329	Laurinburg	NC	28353	Luke Gellens	919-844-5614	S	1*	<0.1

Source: *Ward's Business Directory of U.S. Private and Public Companies*, Volumes 1 and 2, 1996. The company type code used is as follows: P - Public, R - Private, S - Subsidiary, D - Division, J - Joint Venture, A - Affiliate, G - Group. Sales are in millions of dollars, employees are in thousands. An asterisk (*) indicates an estimated sales volume. The symbol < stands for 'less than'. Company names and addresses are truncated, in some cases, to fit into the available space.

MATERIALS CONSUMED

Material	Quantity		Delivered Cost ($ million)
Materials, ingredients, containers, and supplies		(X)	1,402.0
Aluminum and aluminum-base alloy ingot	1,000 s tons	(D)	(D)
All other aluminum and aluminum-base alloy shapes and forms		(X)	(D)
Refined unalloyed copper (cathodes, ingots, cakes, slabs, etc.) and blister or anode copper	1,000 s tons	(D)	(D)
All other copper and copper-base alloy shapes and forms		(X)	(D)
Refined unalloyed lead shapes and forms		(X)	(D)
All other lead and lead-base alloy shapes and forms		(X)	(D)
Refined unalloyed zinc shapes and forms	1,000 s tons	24.7	21.7
All other zinc and zinc-base alloy shapes and forms		(X)	(D)
Refined unalloyed tin shapes and forms		(X)	20.1
All other tin and tin-base alloy shapes and forms		(X)	(D)
Refined unalloyed precious metal shapes and forms		(X)	(D)
All other precious metal and precious metal alloy shapes and forms		(X)	(D)
All other magnesium and magnesium-base alloy shapes and forms		(X)	(D)
All other nonferrous shapes and forms		(X)	(D)
Mining copper ores, concentrates and precipitates (gross weight)	1,000 s tons	(D)	(D)
Mining lead ores and concentrates		(X)	(D)
Mining zinc ores and concentrates		(X)	(D)
Mining precious metal ores and concentrates		(X)	(D)
Mining all other nonferrous metal ores and concentrates		(X)	27.6
Aluminum and aluminum-base alloy scrap from other establishments	1,000 s tons	(D)	(D)
Aluminum and aluminum-base alloy scrap from all other sources	1,000 s tons	2.1*	1.9
Copper and copper-base alloy scrap (except home scrap)	1,000 s tons	(D)	(D)
Lead and lead-base alloy scrap		(X)	(D)
Zinc and zinc-base alloy scrap (including drosses and skimmings)		(X)	(D)
Tinplate scrap (including shredded steel can scrap)		(X)	(D)
Precious metal and precious metal alloy scrap		(X)	(D)
All other nonferrous metal and metal-base alloy scrap		(X)	(D)
All other materials and components, parts, containers, and supplies		(X)	361.2
Materials, ingredients, containers, and supplies, nsk		(X)	108.1

Source: 1992 *Economic Census*. Explanation of symbols used: (D): Withheld to avoid disclosure of competitive data; na: Not available; (S): Withheld because statistical norms were not met; (X): Not applicable; (Z): Less than half the unit shown; nec: Not elsewhere classified; nsk: Not specified by kind; - : zero; * : 10-19 percent estimated; ** : 20-29 percent estimated.

PRODUCT SHARE DETAILS

Product or Product Class	% Share	Product or Product Class	% Share
Primary nonferrous metals, nec	100.00	Primary platinum and platinum alloys (including platinum-group metals)	2.43
Primary zinc residues and other zinc smelter products not of commercial grade, produced for further refining, including base bullion, matte, speiss, etc.	(D)	Other primary nonferrous metals, nec	54.15
Primary zinc residues and other zinc smelter products not of commercial grade, produced for further refining, including base bullion, matte, speiss, etc.	(D)	Primary lead smelter products not of commercial grade, produced for further refining including base bullion, matte, speiss, etc.	(D)
Refined primary zinc	(D)	Primary lead and lead-base alloys	16.59
Refined primary unalloyed zinc slab, excluding remelt zinc (including all ASTM specification zinc)	14.71	Primary magnesium and magnesium-base alloys	24.00
Refined primary unalloyed remelt zinc slab (including all ASTM specification zinc)	(D)	Primary nickel and nickel-base alloys	(D)
		Primary unalloyed tin	(D)
Refined primary unalloyed zinc dust (including all ASTM specification zinc)	(D)	Primary unalloyed silicon	9.28
		Other primary unrefined nonferrous metals (including metal bearing furnace residues and other metal products).	(D)
Refined primary zinc-base alloys.	2.14	Other primary refined nonferrous metals and their alloys (including cadmium, antimony, cobalt, molybdenum, titanium sponge, etc.).	24.55
Primary precious metals and precious metal alloys	24.31		
Primary gold and gold alloys	53.12	Other primary nonferrous metals, nec, nsk	3.79
Primary silver and silver alloys	44.43	Primary nonferrous metals, nec, nsk	4.22

Source: 1992 *Economic Census*. The values shown are percent of total shipments in an industry. Values of indented subcategories are summed in the main heading. The symbol (D) appears when data are withheld to prevent disclosure of competitive information. The abbreviation nsk stands for 'not specified by kind' and nec for 'not elsewhere classified'.

INPUTS AND OUTPUTS FOR PRIMARY NONFERROUS METALS, NEC

Economic Sector or Industry Providing Inputs	%	Sector	Economic Sector or Industry Buying Outputs	%	Sector
Imports	45.4	Foreign	Nonferrous rolling & drawing, nec	27.8	Manufg.
Noncomparable imports	18.1	Foreign	Jewelry, precious metal	16.1	Manufg.
Nonferrous metal ores, except copper	17.0	Mining	Exports	10.3	Foreign
Primary copper	7.7	Manufg.	Electronic components nec	5.7	Manufg.
Motor freight transportation & warehousing	2.6	Util.	Blast furnaces & steel mills	5.3	Manufg.
Primary nonferrous metals, nec	2.5	Manufg.	Federal Government purchases, national defense	5.0	Fed Govt
Electric services (utilities)	1.9	Util.	Telephone & telegraph apparatus	2.8	Manufg.
Wholesale trade	1.6	Trade	Dental equipment & supplies	2.7	Manufg.
Gas production & distribution (utilities)	1.1	Util.	Semiconductors & related devices	2.6	Manufg.
Advertising	0.4	Services	Iron & steel foundries	2.5	Manufg.
Railroads & related services	0.2	Util.	Aluminum rolling & drawing	2.4	Manufg.
Business services nec	0.2	Services	Primary nonferrous metals, nec	2.1	Manufg.
Maintenance of nonfarm buildings nec	0.1	Constr.	Copper rolling & drawing	1.6	Manufg.
Petroleum refining	0.1	Manufg.	Chemical preparations, nec	1.3	Manufg.
Equipment rental & leasing services	0.1	Services	Silverware & plated ware	1.2	Manufg.
			Secondary nonferrous metals	1.1	Manufg.
			Wiring devices	1.1	Manufg.
			Costume jewelry	0.8	Manufg.
			Pipe, valves, & pipe fittings	0.6	Manufg.
			Primary metal products, nec	0.6	Manufg.
			Surgical & medical instruments	0.6	Manufg.
			Photographic equipment & supplies	0.5	Manufg.
			Radio & TV communication equipment	0.5	Manufg.
			Metal doors, sash, & trim	0.4	Manufg.
			Nonferrous castings, nec	0.4	Manufg.
			Nonferrous wire drawing & insulating	0.4	Manufg.
			Fabricated metal products, nec	0.3	Manufg.
			Fabricated plate work (boiler shops)	0.3	Manufg.
			Paints & allied products	0.3	Manufg.
			Primary aluminum	0.3	Manufg.
			Cold finishing of steel shapes	0.2	Manufg.
			Electrometallurgical products	0.2	Manufg.
			Instruments to measure electricity	0.2	Manufg.
			Machinery, except electrical, nec	0.2	Manufg.
			Motors & generators	0.2	Manufg.
			Plating & polishing	0.2	Manufg.
			Watches, clocks, & parts	0.2	Manufg.
			Aluminum castings	0.1	Manufg.
			Electron tubes	0.1	Manufg.
			Welding apparatus, electric	0.1	Manufg.

Source: Benchmark Input-Output Accounts for the U.S. Economy, 1982, U.S. Department of Commerce, Washington, D.C., July 1991. Data, as reported in the source, are organized by the 1977 SIC structure in use in 1982 but have been matched, as closely as is possible, to the 1987 SIC structure used in this book.

OCCUPATIONS EMPLOYED BY SIC 333 - PRIMARY METALS NEC

Occupation	% of Total 1994	Change to 2005	Occupation	% of Total 1994	Change to 2005
Blue collar worker supervisors	7.8	-15.9	Crushing & mixing machine operators	1.8	-9.1
Furnace operators	6.7	22.2	Sales & related workers nec	1.6	-8.6
Heat treating machine operators, metal & plastic	6.2	-8.1	Secretaries, ex legal & medical	1.6	-17.5
Assemblers, fabricators, & hand workers nec	4.5	-10.4	Machine operators nec	1.5	-22.2
Industrial machinery mechanics	4.4	-1.5	Industrial production managers	1.5	-9.1
Metal & plastic machine workers nec	3.6	-19.5	Freight, stock, & material movers, hand	1.4	-27.9
Maintenance repairers, general utility	3.6	-19.1	Metal molding machine workers	1.4	-1.2
Inspectors, testers, & graders, precision	2.7	-9.0	Machinists	1.3	-10.2
General managers & top executives	2.5	-13.2	Traffic, shipping, & receiving clerks	1.2	-11.9
Industrial truck & tractor operators	2.3	-9.9	Science & mathematics technicians	1.2	-10.0
Electricians	2.2	-16.7	Separating & still machine operators	1.1	-11.4
Truck drivers light & heavy	2.0	-5.5	Bookkeeping, accounting, & auditing clerks	1.1	-31.5
Machine tool cutting & forming etc. nec	1.9	-9.7	Metallurgists & ceramic & materials engineers	1.0	2.2
Crane & tower operators	1.8	-10.8			

Source: Industry-Occupation Matrix, Bureau of Labor Statistics. These data relate to one or more 3-digit SIC industry groups rather than to a single 4-digit SIC. The change reported for each occupation to the year 2005 is a percent of growth or decline as estimated by the Bureau of Labor Statistics. The abbreviation nec stands for 'not elsewhere classified'.

LOCATION BY STATE AND REGIONAL CONCENTRATION

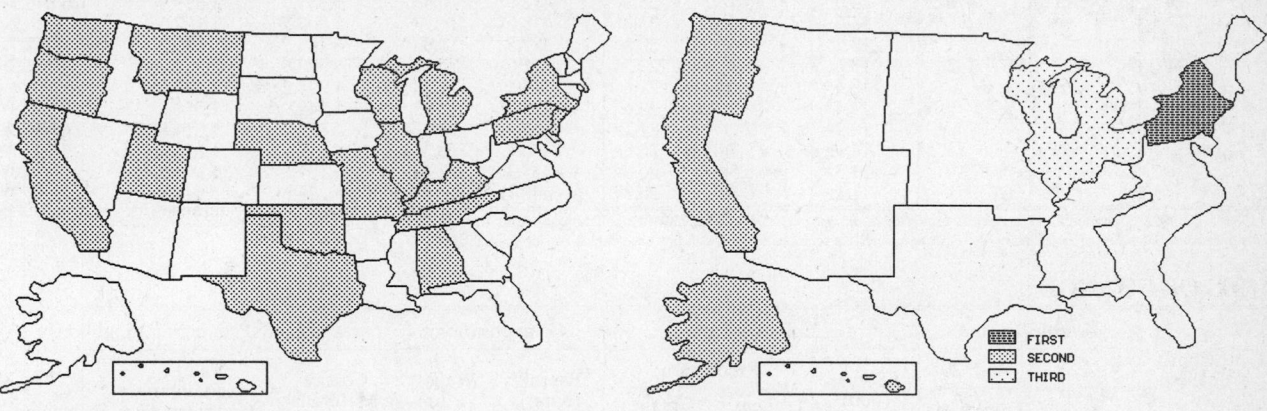

FIRST
SECOND
THIRD

INDUSTRY DATA BY STATE

State	Establish- ments	Shipments			Employment				Cost as % of Shipments	Investment per Employee ($)
		Total ($ mil)	% of U.S.	Per Establ.	Total Number	% of U.S.	Per Establ.	Wages ($/hour)		
Pennsylvania	8	342.4	13.9	42.8	1,100	12.8	138	14.35	54.4	-
Washington	5	148.4	6.0	29.7	800	9.3	160	17.33	52.1	-
Wisconsin	5	23.9	1.0	4.8	200	2.3	40	16.00	55.6	2,000
New York	17	(D)	-	-	375 *	4.4	22	-	-	-
California	10	(D)	-	-	175 *	2.0	18	-	-	-
New Jersey	8	(D)	-	-	375 *	4.4	47	-	-	-
Illinois	5	(D)	-	-	375 *	4.4	75	-	-	-
Michigan	4	(D)	-	-	375 *	4.4	94	-	-	-
Missouri	4	(D)	-	-	750 *	8.7	188	-	-	-
Texas	4	(D)	-	-	1,750 *	20.3	438	-	-	-
Alabama	3	(D)	-	-	375 *	4.4	125	-	-	-
Tennessee	3	(D)	-	-	375 *	4.4	125	-	-	-
Utah	3	(D)	-	-	750 *	8.7	250	-	-	-
Nebraska	2	(D)	-	-	175 *	2.0	88	-	-	-
Oklahoma	2	(D)	-	-	375 *	4.4	188	-	-	-
Oregon	2	(D)	-	-	375 *	4.4	188	-	-	-
Kentucky	1	(D)	-	-	175 *	2.0	175	-	-	-
Montana	1	(D)	-	-	375 *	4.4	375	-	-	-

*Source: 1992 Economic Census. The states are in descending order of shipments or establishments (if shipment data are missing for the majority). The symbol (D) appears when data are withheld to prevent disclosure of competitive information. States marked with (D) are sorted by number of establishments. A dash (-) indicates that the data element cannot be calculated; * indicates the midpoint of a range.*

3341 - SECONDARY SMELTING NONFERROUS METALS

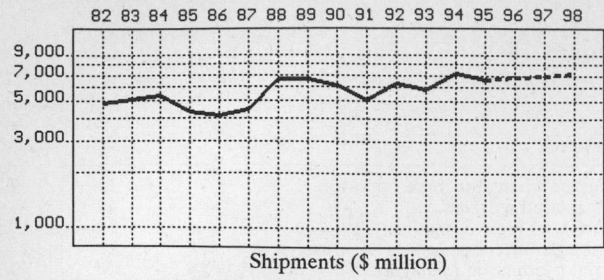

82 83 84 85 86 87 88 89 90 91 92 93 94 95 96 97 98

9,000.
7,000.
5,000.

3,000.

1,000.

Shipments ($ million)

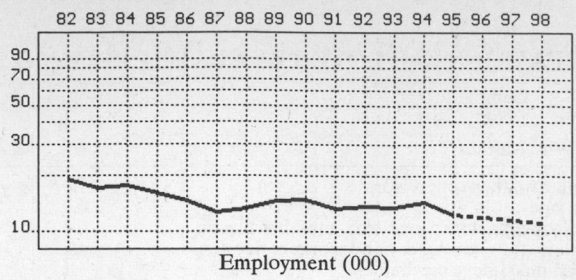

82 83 84 85 86 87 88 89 90 91 92 93 94 95 96 97 98

90.
70.
50.

30.

10.

Employment (000)

GENERAL STATISTICS

| Year | Companies | Establishments | | Employment | | | Compensation | | Production ($ million) | | | |
		Total	with 20 or more employees	Total (000)	Production Workers (000)	Hours (Mil)	Payroll ($ mil)	Wages ($/hr)	Cost of Materials	Value Added by Manufacture	Value of Shipments	Capital Invest.
1982	396	458	212	19.2	13.5	26.3	402.2	9.37	4,134.7	619.8	4,851.9	146.4
1983				17.2	12.3	24.2	376.3	9.51	4,181.9	833.1	5,135.9	54.4
1984				17.7	12.9	26.0	404.6	9.78	4,387.7	940.5	5,308.7	116.7
1985				16.0	11.6	23.5	371.4	9.61	3,486.8	782.6	4,315.1	103.0
1986				14.6	10.5	21.0	351.1	10.00	3,323.4	833.3	4,143.0	58.2
1987	366	398	162	12.5	9.1	19.3	312.0	9.64	3,450.7	947.3	4,431.4	62.6
1988				13.2	9.4	20.5	362.2	10.34	5,528.7	1,180.3	6,663.4	67.1
1989		383	176	14.6	10.4	23.2	412.6	10.63	5,335.7	1,353.7	6,646.3	108.3
1990				14.7	10.8	23.0	418.2	10.73	5,003.7	1,107.5	6,130.2	103.9
1991				13.2	9.6	20.8	387.5	11.00	4,167.7	815.0	5,040.7	72.2
1992	347	387	162	13.6	9.9	22.5	411.9	11.22	4,970.6	1,272.7	6,223.7	154.5
1993				13.4	9.6	21.6	414.5	11.78	4,394.2	1,420.0	5,770.4	109.8
1994				14.4	10.6	23.6	465.7	12.23	5,249.9	1,954.9	7,151.8	103.2
1995				12.2P	9.0P	21.1P	425.6P	12.02P	4,873.7P	1,814.8P	6,639.3P	105.2P
1996				11.9P	8.7P	20.8P	430.4P	12.25P	4,990.7P	1,858.4P	6,798.7P	106.3P
1997				11.5P	8.5P	20.6P	435.3P	12.47P	5,107.6P	1,901.9P	6,958.0P	107.5P
1998				11.1P	8.2P	20.3P	440.1P	12.70P	5,224.6P	1,945.5P	7,117.3P	108.7P

Sources: 1982, 1987, 1992 *Economic Census*; *Annual Survey of Manufactures*, 83-86, 88-91, 93-94. Establishment counts for non-Census years are from *County Business Patterns*; establishment values for 83-84 are extrapolations. 'P's show projections by the editors. Industries reclassified in 87 will not have data for prior years.

INDICES OF CHANGE

| Year | Companies | Establishments | | Employment | | | Compensation | | Production ($ million) | | | |
		Total	with 20 or more employees	Total (000)	Production Workers (000)	Hours (Mil)	Payroll ($ mil)	Wages ($/hr)	Cost of Materials	Value Added by Manufacture	Value of Shipments	Capital Invest.
1982	114	118	131	141	136	117	98	84	83	49	78	95
1983				126	124	108	91	85	84	65	83	35
1984				130	130	116	98	87	88	74	85	76
1985				118	117	104	90	86	70	61	69	67
1986				107	106	93	85	89	67	65	67	38
1987	105	103	100	92	92	86	76	86	69	74	71	41
1988				97	95	91	88	92	111	93	107	43
1989		99	109	107	105	103	100	95	107	106	107	70
1990				108	109	102	102	96	101	87	98	67
1991				97	97	92	94	98	84	64	81	47
1992	100	100	100	100	100	100	100	100	100	100	100	100
1993				99	97	96	101	105	88	112	93	71
1994				106	107	105	113	109	106	154	115	67
1995				90P	91P	94P	103P	107P	98P	143P	107P	68P
1996				87P	88P	93P	104P	109P	100P	146P	109P	69P
1997				84P	85P	91P	106P	111P	103P	149P	112P	70P
1998				81P	83P	90P	107P	113P	105P	153P	114P	70P

Sources: Same as General Statistics. Values reflect change from the base year, 1992. Values above 100 mean greater than 92, values below 100 mean less than 92, and a value of 100 in the 82-91 or 93-98 period means same as 92. 'P's mark projections by the editors.

SELECTED RATIOS

For 1992	Avg. of All Manufact.	Analyzed Industry	Index	For 1992	Avg. of All Manufact.	Analyzed Industry	Index
Employees per Establishment	46	35	77	Value Added per Production Worker	122,353	128,556	105
Payroll per Establishment	1,332,320	1,064,341	80	Cost per Establishment	4,239,462	12,843,928	303
Payroll per Employee	29,181	30,287	104	Cost per Employee	92,853	365,485	394
Production Workers per Establishment	31	26	81	Cost per Production Worker	135,003	502,081	372
Wages per Establishment	734,496	652,326	89	Shipments per Establishment	8,100,800	16,081,912	199
Wages per Production Worker	23,390	25,500	109	Shipments per Employee	177,425	457,625	258
Hours per Production Worker	2,025	2,273	112	Shipments per Production Worker	257,966	628,657	244
Wages per Hour	11.55	11.22	97	Investment per Establishment	278,244	399,225	143
Value Added per Establishment	3,842,210	3,288,630	86	Investment per Employee	6,094	11,360	186
Value Added per Employee	84,153	93,581	111	Investment per Production Worker	8,861	15,606	176

Sources: Same as General Statistics. The 'Average of All Manufacturing' column represents the average of all manufacturing industries reported for the most recent complete year available. The Index shows the relationship between the Average and the Analyzed Industry. For example, 100 means that they are equal; 500 that the Analyzed Industry is five times the average; 50 means that the Analyzed Industry is half the national average. The abbreviation 'na' is used to show that data are 'not available'.

LEADING COMPANIES   Number shown: 64   Total sales ($ mil): 4,085   Total employment (000): 11.9

Company Name	Address				CEO Name	Phone	Co. Type	Sales ($ mil)	Empl. (000)
Commercial Metals Co	PO Box 1046	Dallas	TX	75221	Stanley A Rabin	214-689-4300	P	1,658	4.3
US Reduction Co	9200 Calumet Av	Munster	IN	46321	Richard S Neufield	219-836-0555	R	300	0.5
Tredegar Aluminum	PO Box 428	Newnan	GA	30264	Doug Monk	404-253-2020	D	250	1.0
RSR Corp	2777 Stemmons Fwy	Dallas	TX	75207	Albert P Lospinoso	214-631-6070	R	160	0.8
Gulf Met Holdings Corp	PO Box 611	Houston	TX	77001	Russ Robinson	713-926-1705	R	120	0.3
US Zinc Corp	PO Box 611	Houston	TX	77001	Russ Robinson	713-926-1705	S	120	0.2
AMG Resources Corp	4100 Grand Av	Pittsburgh	PA	15225	Allan M Goldstein	412-331-0770	R	100	0.2
Futura Corp	PO Box 7968	Boise	ID	83707	Brent S Lloyd	208-336-0150	R	100	0.4
Sadoff Iron and Metal Co	PO Box 1138	Fond du Lac	WI	54935	Sheldon J Lasky	414-921-2070	D	100	0.1
Ansam Metals Corp	PO Box 3408	Baltimore	MD	21225	Samuel S Kahan	410-355-8220	R	80	<0.1
L & P Aluminum Smelting	PO Box 455	Steele	AL	35987	Robert H Pine	205-538-7817	S	80	0.1
Miller Compressing Co	PO Box 369	Milwaukee	WI	53201	Robert Miller	414-671-5980	R	62*	0.2
Fry Metals	6th Av and 41st	Altoona	PA	16602	Harry Kurec	814-946-1611	D	60	0.2
H Kramer and Co	1343 W 21st St	Chicago	IL	60608	HK Chapman Jr	312-226-6600	R	50	0.1
Schuylkill Metals Corp	PO Box 74040	Baton Rouge	LA	70874	EK Hudson	504-775-3040	S	50	0.3
Cousins Inc	PO Box 787	Mansfield	OH	44901	Steven Senser	419-525-0011	R	49*	<0.1
Indium Corporation of America	1676 Lincoln Av	Utica	NY	13502	Gregory P Evans	315-853-4900	R	49	0.2
Cannon-Muskegon Corp	PO Box 506	Muskegon	MI	49443	Joseph Snowden	616-755-1681	S	46	0.1
Vac Air Alloys	PO Box 650	Frewsburg	NY	14738	Roger Payne	716-569-0700	D	46*	0.2
Southwire Co	PO Box 1000	Carrollton	GA	30119	Roy Long	404-832-5130	D	45	0.4
National Nickel Alloy Corp	660 4th St	Greenville	PA	16125	S Greenberger	412-646-4561	S	40	<0.1
Roth Brothers Smelting Corp	PO Box 639	East Syracuse	NY	13057	J Roth	315-463-9500	R	40*	0.1
Manitoba Corp	PO Box 385	Lancaster	NY	14086	Joseph Baker	716-685-7000	R	35	<0.1
Charleston Steel and Metal Co	PO Box 814	Charleston	SC	29402	S Steinberg	803-722-7278	R	31*	0.1
Midland Industries Inc	1424 N Halsted St	Chicago	IL	60622	Laurence S Spector	312-664-7300	R	30	<0.1
River Recycling Industries	4195 Bradley Rd	Cleveland	OH	44109	WA Grodin	216-459-2100	R	25	<0.1
Allied Metal Co	2059 S Canal St	Chicago	IL	60616	Marvin Fink	312-225-2800	R	24	0.1
Aerochem Inc	1885 N Batavia St	Orange	CA	92665	Bob Hahn	714-637-4401	S	22	0.3
Federal Metal Co	7250 Division St	Bedford	OH	44146	David R Nagusky	216-232-8700	R	20	<0.1
Bermco Aluminum	616 33rd Pl N	Birmingham	AL	35222	Steven Weinstein	205-320-2419	D	18*	<0.1
Gulf Chemical	PO Box 2290	Freeport	TX	77541	William G Deering	409-233-7882	S	18	<0.1
Plasma Processing Corp	109 Westpark Dr	Brentwood	TN	37027	Terry Moore	615-221-2460	S	18	<0.1
Custom Alloy Scrap Sales Inc	2730 Peralta St	Oakland	CA	94607	Chal Sulprizio	510-893-6476	R	16*	<0.1
Gettysburg Foundry Spec Co	2664 Emmitsburg	Gettysburg	PA	17325	Creed White	717-334-7661	R	16	<0.1
Halaco Engineering Co	6200 Perkins Rd	Oxnard	CA	93033	CW Haack	805-488-3684	R	15	<0.1
Schilberg Integrated Metals	47 Milk St	Willimantic	CT	06226	Ben Schilberg	203-423-2562	R	15*	<0.1
WJ Bullock Inc	PO Box 539	Fairfield	AL	35064	W E Bullock Jr	205-788-6586	R	15	<0.1
MC Canfield Sons	1000 Brighton St	Union	NJ	07083	Robert McIntire	908-688-5050	R	14	0.1
Warrenton Resources Inc	1710 Daniel Boone	Warrenton	MO	63383	Bob Hartman	314-456-3488	D	13*	<0.1
S-G Metals Industries Inc	PO Box 2039	Kansas City	KS	66110	L G Galambra Jr	913-621-4100	R	12	<0.1
Micro Metallics Corp	1695 Monterey Hwy	San Jose	CA	95112		408-998-4930	S	11*	<0.1
Arco Alloys Corp	1891 Trombly St	Detroit	MI	48211	HB Aronow	313-871-2680	R	10*	<0.1
ARK Foundry & Mfg Co	1321 S Walker Av	Oklahoma City	OK	73109	Bea Ramos	405-235-5505	R	10*	<0.1
Greenville Metals Inc	RD 2	Transfer	PA	16154	Eugene A March	412-646-0654	R	10.	<0.1
Norfab Inc	14830 27th Av N	Minneapolis	MN	55447	Elmer Chumer	612-559-2510	R	10*	<0.1
Amerway Inc	3701 Beale Av	Altoona	PA	16601	Gerry Buck	814-944-0200	R	9*	<0.1
General Smelting and Refining	PO Box 37	College Grove	TN	37046	William G Cole Jr	615-368-7125	R	8	<0.1
NEY Smelting and Refining Inc	269 Freeman St	Brooklyn	NY	11222	Henry Meyer	718-389-4900	R	7	<0.1
Thames River Recycling Inc	330 Middle St	Middletown	CT	06457	Frank Clare	203-346-3217	R	7	<0.1
Vulcan Lead Products Company	1400 W Pierce St	Milwaukee	WI	53204	Charles Yanke	414-645-2040	R	7	<0.1
Alloy Metals Company Inc	1000 E 60th St	Los Angeles	CA	90001	HS Lim	213-234-9235	R	5	<0.1
Grand Rapids Alloys Inc	392 54th St SW	Wyoming	MI	49548	John Barth	616-532-2301	R	5*	<0.1
Ultramet	12173 Montague St	Pacoima	CA	91331	Richard B Kaplan	818-899-0236	R	5	<0.1
American Nickel Alloy Mfg Corp	30 Vesey St	New York	NY	10007	RD Grunebaum	212-267-6420	R	5	<0.1
Hallmark Precious Metals Inc	PO Box 1180	Mercer Island	WA	98040	ME Senff	206-232-0472	R	3	<0.1
Proler International Corp	6000 W Marginal	Seattle	WA	98106	Jack Force	206-767-4337	D	3	<0.1
Mercury Refining Co	1218 Central Av	Albany	NY	12205	David Cohen	518-459-0820	R	3	<0.1
Merlan Inc	615 Section St	Danville	IL	61832	Lou Mervis	217-431-3201	R	2	<0.1
Indiana Rolling Mill Baling Corp	PO Box 11451	Fort Wayne	IN	46858	M Krel	219-432-2549	R	2	<0.1
Champion Agate Company Inc	PO Box 516	Pennsboro	WV	26415	Helen Michels	304-659-2861	R	1	<0.1
Industrial Lead & Plastics Constr	10332 E Rush St	S El Monte	CA	91733	AW Sheller	818-448-6063	R	1	<0.1
Martin Metals Inc	1321 Wilson St	Los Angeles	CA	90021	H G Martin Jr	213-627-7755	R	1	<0.1
Transmet Corp	4290 Perimeter Dr	Columbus	OH	43228	Doug Shull	614-276-5522	R	1	<0.1
Quemetco Inc	720 S 7th Av	La Puente	CA	91746	Bob Finn	818-330-2294	S	0*	0.2

Source: Ward's Business Directory of U.S. Private and Public Companies, Volumes 1 and 2, 1996. The company type code used is as follows: P - Public, R - Private, S - Subsidiary, D - Division, J - Joint Venture, A - Affiliate, G - Group. Sales are in millions of dollars, employees are in thousands. An asterisk (*) indicates an estimated sales volume. The symbol < stands for 'less than'. Company names and addresses are truncated, in some cases, to fit into the available space.

MATERIALS CONSUMED

Material		Quantity	Delivered Cost ($ million)
Materials, ingredients, containers, and supplies		(X)	4,607.5
Aluminum and aluminum-base alloy ingot	1,000 s tons	116.8**	104.5
All other aluminum and aluminum-base alloy shapes and forms		(X)	(D)
Refined unalloyed copper (cathodes, ingots, cakes, slabs, etc.) and blister or anode copper	1,000 s tons	1.3**	2.7
All other copper and copper-base alloy shapes and forms		(X)	5.3
Refined unalloyed lead shapes and forms		(X)	(D)
All other lead and lead-base alloy shapes and forms		(X)	0.1
Refined unalloyed zinc shapes and forms		(D)	(D)
All other zinc and zinc-base alloy shapes and forms		(X)	(D)
Refined unalloyed tin shapes and forms		(X)	22.2
All other tin and tin-base alloy shapes and forms		(X)	(D)
Refined unalloyed precious metal shapes and forms		(X)	(D)
All other precious metal and precious metal alloy shapes and forms		(X)	(D)
All other nonferrous shapes and forms		(X)	6.6
Mining precious metal ores and concentrates		(X)	(D)
Mining all other nonferrous metal ores and concentrates		(X)	(D)
Aluminum and aluminum-base alloy scrap from other establishments	1,000 s tons	(D)	(D)
Aluminum and aluminum-base alloy scrap from all other sources	1,000 s tons	1,655.6*	1,273.6
Copper and copper-base alloy scrap (except home scrap)	1,000 s tons	452.7**	652.3
Lead and lead-base alloy scrap		(X)	280.9
Zinc and zinc-base alloy scrap (including drosses and skimmings)		(X)	77.4
Tinplate scrap (including shredded steel can scrap)		(X)	(D)
Precious metal and precious metal alloy scrap		(X)	585.8
All other nonferrous metal and metal-base alloy scrap		(X)	76.7
Alumina (gross weight)	1,000 s tons	(D)	(D)
All other materials and components, parts, containers, and supplies		(X)	256.5
Materials, ingredients, containers, and supplies, nsk		(X)	578.1

Source: 1992 *Economic Census.* Explanation of symbols used: (D): Withheld to avoid disclosure of competitive data; na: Not available; (S): Withheld because statistical norms were not met; (X): Not applicable; (Z): Less than half the unit shown; nec: Not elsewhere classified; - : zero; * : 10-19 percent estimated; ** : 20-29 percent estimated.

PRODUCT SHARE DETAILS

Product or Product Class	% Share	Product or Product Class	% Share
Secondary nonferrous metals	100.00	Secondary zinc dust (including all ASTM specification zinc).	(D)
Secondary copper (pig, ingot, shot, etc.)	14.24	Secondary zinc-base alloys	40.36
Secondary unalloyed copper cathode, wire bar, ingot, and ingot bar	16.13	Zinc, nsk	0.35
		Secondary precious metals and precious metal alloys	22.42
Other secondary unalloyed copper (including cakes, slabs, shot, etc.)	41.64	Secondary gold and gold alloys	74.68
Secondary copper-base alloys	41.29	Secondary silver and silver alloys	7.19
Copper, nsk	0.94	Secondary platinum and platinum alloys (including platinum-group metals)	15.84
Secondary lead	9.75	Precious metals and precious metal alloys, nsk	2.29
Secondary unalloyed lead (pig, ingots, shot, etc.)	61.68	Other secondary nonferrous metals	1.81
Secondary antimonial lead- and tin-base alloys	18.06	Secondary magnesium and magnesium-base alloys	12.28
Secondary lead- and tin-base alloy babbitt metal	0.50	Secondary nickel and nickel-base alloys	29.71
Secondary lead- and tin-base alloy solder	16.04	Other secondary nonferrous metals and their alloys (cadmium, antimony, cobalt, titanium sponge, etc.)	39.83
Other secondary lead- and tin-base alloys (including type metal)	2.39	Other secondary nonferrous metals, nsk	18.18
Lead, nsk	1.34	Aluminum ingot (including pigs, sows, and molten metal, excluding billet), produced by secondary smelters	33.60
Secondary zinc	6.29	Aluminum extrusion ingot (billet), produced by secondary smelters	2.44
Secondary unalloyed zinc slab, excluding remelt zinc (including all ASTM specification zinc)	(D)	Secondary nonferrous metals, nsk	9.44
Secondary unalloyed remelt zinc slab (including all ASTM specification zinc)	(D)		

Source: 1992 *Economic Census.* The values shown are percent of total shipments in an industry. Values of indented subcategories are summed in the main heading. The symbol (D) appears when data are withheld to prevent disclosure of competitive information. The abbreviation nsk stands for 'not specified by kind' and nec for 'not elsewhere classified'.

INPUTS AND OUTPUTS FOR SECONDARY NONFERROUS METALS

Economic Sector or Industry Providing Inputs	%	Sector	Economic Sector or Industry Buying Outputs	%	Sector
Scrap	37.1	Scrap	Storage batteries	30.5	Manufg.
Wholesale trade	13.2	Trade	Primary lead	12.7	Manufg.
Motor freight transportation & warehousing	10.8	Util.	Typesetting	6.2	Manufg.
Primary copper	6.4	Manufg.	Lithographic platemaking & services	5.1	Manufg.
Gas production & distribution (utilities)	3.4	Util.	Small arms ammunition	4.4	Manufg.
Primary metal products, nec	3.1	Manufg.	Adhesives & sealants	3.8	Manufg.
Primary aluminum	2.7	Manufg.	Exports	3.5	Foreign
Railroads & related services	2.6	Util.	Sheet metal work	3.2	Manufg.
Carbon & graphite products	2.5	Manufg.	Paints & allied products	3.0	Manufg.
Primary nonferrous metals, nec	2.3	Manufg.	Commercial printing	2.8	Manufg.

Continued on next page.

INPUTS AND OUTPUTS FOR SECONDARY NONFERROUS METALS - Continued

Economic Sector or Industry Providing Inputs	%	Sector	Economic Sector or Industry Buying Outputs	%	Sector
Primary zinc	2.3	Manufg.	Nonferrous rolling & drawing, nec	2.0	Manufg.
Electric services (utilities)	2.2	Util.	Nonferrous wire drawing & insulating	1.9	Manufg.
Power transmission equipment	1.9	Manufg.	Printing trades machinery	1.9	Manufg.
Miscellaneous plastics products	1.4	Manufg.	Fabricated plate work (boiler shops)	1.6	Manufg.
Petroleum refining	1.0	Manufg.	Metal cans	1.5	Manufg.
Miscellaneous repair shops	1.0	Services	Primary batteries, dry & wet	1.4	Manufg.
Chemical preparations, nec	0.6	Manufg.	Primary copper	1.3	Manufg.
Maintenance of nonfarm buildings nec	0.5	Constr.	Metal foil & leaf	1.2	Manufg.
Primary lead	0.5	Manufg.	Blast furnaces & steel mills	1.0	Manufg.
Banking	0.4	Fin/R.E.	Secondary nonferrous metals	1.0	Manufg.
Equipment rental & leasing services	0.4	Services	Special dies & tools & machine tool accessories	1.0	Manufg.
Coal	0.3	Mining	Nonferrous castings, nec	0.9	Manufg.
Gum & wood chemicals	0.3	Manufg.	Aluminum rolling & drawing	0.8	Manufg.
Communications, except radio & TV	0.3	Util.	Motor vehicle parts & accessories	0.8	Manufg.
Hand saws & saw blades	0.2	Manufg.	Primary metal products, nec	0.8	Manufg.
Machinery, except electrical, nec	0.2	Manufg.	Truck & bus bodies	0.8	Manufg.
Water transportation	0.2	Util.	Book printing	0.7	Manufg.
Eating & drinking places	0.2	Trade	Fabricated metal products, nec	0.6	Manufg.
Blast furnaces & steel mills	0.1	Manufg.	Iron & steel foundries	0.5	Manufg.
Metal heat treating	0.1	Manufg.	Steel pipe & tubes	0.5	Manufg.
Special dies & tools & machine tool accessories	0.1	Manufg.	Federal Government purchases, national defense	0.5	Fed Govt
Sanitary services, steam supply, irrigation	0.1	Util.	Metal heat treating	0.4	Manufg.
Insurance carriers	0.1	Fin/R.E.	Copper rolling & drawing	0.2	Manufg.
Real estate	0.1	Fin/R.E.	Photoengraving, electrotyping & stereotyping	0.2	Manufg.
Computer & data processing services	0.1	Services	Industrial buildings	0.1	Constr.
U.S. Postal Service	0.1	Gov't	Office buildings	0.1	Constr.

Source: Benchmark Input-Output Accounts for the U.S. Economy, 1982, U.S. Department of Commerce, Washington, D.C., July 1991. Data, as reported in the source, are organized by the 1977 SIC structure in use in 1982 but have been matched, as closely as is possible, to the 1987 SIC structure used in this book.

OCCUPATIONS EMPLOYED BY SIC 334 - PRIMARY METALS NEC

Occupation	% of Total 1994	Change to 2005	Occupation	% of Total 1994	Change to 2005
Blue collar worker supervisors	7.8	-15.9	Crushing & mixing machine operators	1.8	-9.1
Furnace operators	6.7	22.2	Sales & related workers nec	1.6	-8.6
Heat treating machine operators, metal & plastic	6.2	-8.1	Secretaries, ex legal & medical	1.6	-17.5
Assemblers, fabricators, & hand workers nec	4.5	-10.4	Machine operators nec	1.5	-22.2
Industrial machinery mechanics	4.4	-1.5	Industrial production managers	1.5	-9.1
Metal & plastic machine workers nec	3.6	-19.5	Freight, stock, & material movers, hand	1.4	-27.9
Maintenance repairers, general utility	3.6	-19.1	Metal molding machine workers	1.4	-1.2
Inspectors, testers, & graders, precision	2.7	-9.0	Machinists	1.3	-10.2
General managers & top executives	2.5	-13.2	Traffic, shipping, & receiving clerks	1.2	-11.9
Industrial truck & tractor operators	2.3	-9.9	Science & mathematics technicians	1.2	-10.0
Electricians	2.2	-16.7	Separating & still machine operators	1.1	-11.4
Truck drivers light & heavy	2.0	-5.5	Bookkeeping, accounting, & auditing clerks	1.1	-31.5
Machine tool cutting & forming etc. nec	1.9	-9.7	Metallurgists & ceramic & materials engineers	1.0	2.2
Crane & tower operators	1.8	-10.8			

Source: Industry-Occupation Matrix, Bureau of Labor Statistics. These data relate to one or more 3-digit SIC industry groups rather than to a single 4-digit SIC. The change reported for each occupation to the year 2005 is a percent of growth or decline as estimated by the Bureau of Labor Statistics. The abbreviation nec stands for 'not elsewhere classified'.

LOCATION BY STATE AND REGIONAL CONCENTRATION

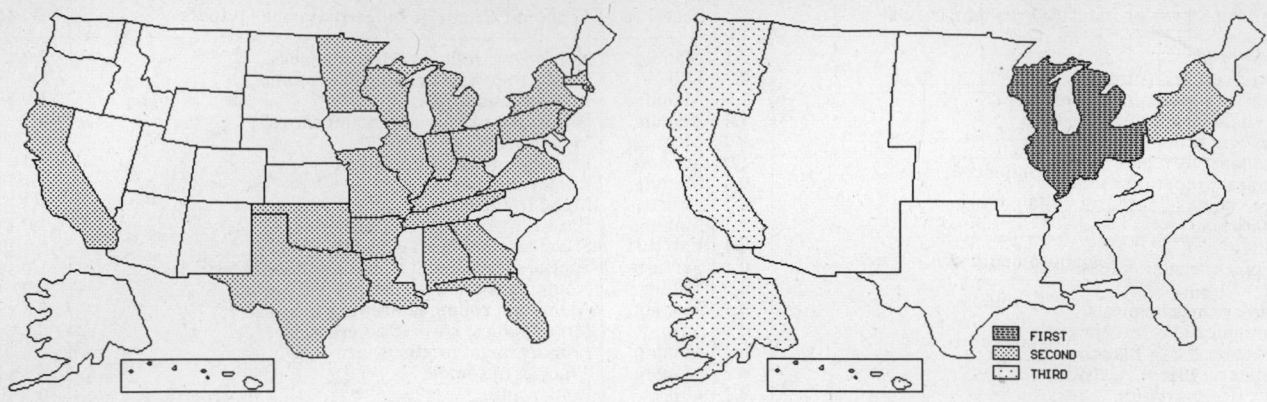

INDUSTRY DATA BY STATE

| State | Establish-ments | Shipments | | | Employment | | | | Cost as % of Shipments | Investment per Employee ($) |
		Total ($ mil)	% of U.S.	Per Establ.	Total Number	% of U.S.	Per Establ.	Wages ($/hour)		
Illinois	28	758.5	12.2	27.1	1,200	8.8	43	11.26	86.2	9,417
Pennsylvania	34	632.5	10.2	18.6	1,600	11.8	47	12.42	76.2	9,313
Alabama	10	566.2	9.1	56.6	1,000	7.4	100	11.50	85.4	4,700
Ohio	33	441.6	7.1	13.4	1,000	7.4	30	12.50	79.8	-
California	37	405.4	6.5	11.0	1,300	9.6	35	11.12	73.8	11,462
Indiana	21	397.0	6.4	18.9	1,300	9.6	62	11.43	76.3	8,385
New York	18	380.4	6.1	21.1	1,100	8.1	61	11.06	80.5	10,636
New Jersey	13	335.4	5.4	25.8	400	2.9	31	13.50	45.1	3,500
Michigan	17	261.5	4.2	15.4	600	4.4	35	8.92	81.3	-
Tennessee	16	174.0	2.8	10.9	600	4.4	38	9.78	73.2	12,167
Georgia	7	153.2	2.5	21.9	200	1.5	29	8.25	89.8	-
Missouri	9	118.2	1.9	13.1	300	2.2	33	10.17	84.2	-
Wisconsin	9	105.6	1.7	11.7	200	1.5	22	12.67	86.8	3,000
Minnesota	8	72.4	1.2	9.1	200	1.5	25	10.67	82.7	-
Texas	23	61.3	1.0	2.7	200	1.5	9	8.25	75.9	9,000
Oklahoma	7	37.1	0.6	5.3	300	2.2	43	10.60	54.7	4,000
Florida	13	20.4	0.3	1.6	100	0.7	8	11.00	49.0	-
Massachusetts	9	(D)	-	-	375 *	2.8	42	-	-	4,533
Arkansas	5	(D)	-	-	175 *	1.3	35	-	-	-
Rhode Island	5	(D)	-	-	175 *	1.3	35	-	-	571
Kentucky	2	(D)	-	-	175 *	1.3	88	-	-	-
Louisiana	2	(D)	-	-	175 *	1.3	88	-	-	-
Virginia	1	(D)	-	-	175 *	1.3	175	-	-	-

Source: 1992 Economic Census. The states are in descending order of shipments or establishments (if shipment data are missing for the majority). The symbol (D) appears when data are withheld to prevent disclosure of competitive information. States marked with (D) are sorted by number of establishments. A dash (-) indicates that the data element cannot be calculated; * indicates the midpoint of a range.

3351 - COPPER ROLLING DRAWING & EXTRUDING

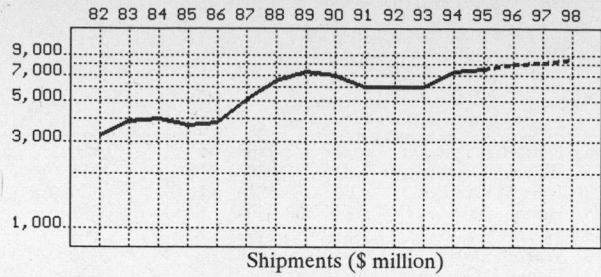

Shipments ($ million)

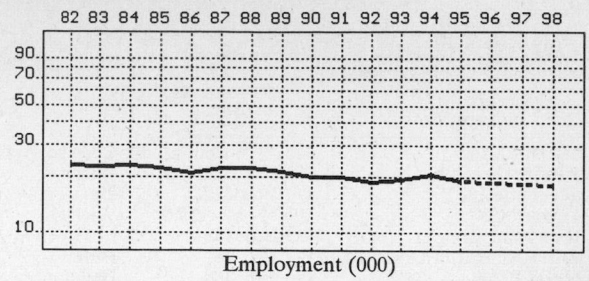

Employment (000)

GENERAL STATISTICS

Year	Companies	Establishments Total	Establishments with 20 or more employees	Employment Total (000)	Employment Production Workers (000)	Employment Hours (Mil)	Compensation Payroll ($ mil)	Compensation Wages ($/hr)	Production Cost of Materials	Production Value Added by Manufacture	Production Value of Shipments	Production Capital Invest.
1982	101	137	95	23.3	17.2	32.6	468.2	9.90	2,267.0	957.7	3,270.0	123.9
1983		133	97	22.8	17.1	33.9	511.2	10.55	2,663.5	1,189.2	3,855.1	94.0
1984		129	99	23.4	18.0	37.8	554.3	10.56	2,817.5	1,285.7	4,020.2	110.0
1985		126	100	22.4	17.1	34.5	517.8	10.64	2,611.2	1,061.2	3,671.8	132.0
1986		120	94	21.2	16.5	34.2	525.1	11.06	2,629.7	1,182.1	3,820.6	85.1
1987	94	121	91	22.6	17.6	36.8	589.1	11.36	3,617.9	1,610.2	5,134.0	90.4
1988		120	92	22.5	17.4	37.4	622.9	11.88	4,890.8	1,807.8	6,521.0	128.6
1989		115	92	21.4	17.4	37.3	655.0	12.31	5,505.0	1,667.5	7,259.5	179.5
1990		117	90	20.1	16.3	35.6	662.8	12.94	5,223.2	1,679.3	6,880.2	126.1
1991		120	90	20.1	15.3	33.4	640.0	13.42	4,551.2	1,362.4	6,005.1	178.6
1992	88	115	86	18.9	14.5	32.9	606.1	12.91	4,487.1	1,453.3	6,000.1	126.8
1993		112	84	19.4	14.9	32.6	638.7	14.07	4,267.9	1,612.5	5,998.2	127.4
1994		109P	85P	21.0	16.5	35.8	721.9	14.56	5,009.4	2,362.0	7,266.8	192.2
1995		107P	84P	19.2P	15.2P	34.7P	711.6P	14.63P	5,230.2P	2,466.1P	7,587.1P	167.1P
1996		105P	82P	18.9P	15.1P	34.7P	728.5P	15.00P	5,449.4P	2,569.5P	7,905.1P	172.3P
1997		103P	81P	18.5P	14.9P	34.6P	745.4P	15.38P	5,668.5P	2,672.8P	8,223.0P	177.6P
1998		101P	80P	18.2P	14.7P	34.6P	762.3P	15.75P	5,887.7P	2,776.1P	8,540.9P	182.8P

Sources: 1982, 1987, 1992 *Economic Census*; *Annual Survey of Manufactures*, 83-86, 88-91, 93-94. Establishment counts for non-Census years are from *County Business Patterns*; establishment values for 83-84 are extrapolations. 'P's show projections by the editors. Industries reclassified in 87 will not have data for prior years.

INDICES OF CHANGE

Year	Companies	Establishments Total	Establishments with 20 or more employees	Employment Total (000)	Employment Production Workers (000)	Employment Hours (Mil)	Compensation Payroll ($ mil)	Compensation Wages ($/hr)	Production Cost of Materials	Production Value Added by Manufacture	Production Value of Shipments	Production Capital Invest.
1982	115	119	110	123	119	99	77	77	51	66	54	98
1983		116	113	121	118	103	84	82	59	82	64	74
1984		112	115	124	124	115	91	82	63	88	67	87
1985		110	116	119	118	105	85	82	58	73	61	104
1986		104	109	112	114	104	87	86	59	81	64	67
1987	107	105	106	120	121	112	97	88	81	111	86	71
1988		104	107	119	120	114	103	92	109	124	109	101
1989		100	107	113	120	113	108	95	123	115	121	142
1990		102	105	106	112	108	109	100	116	116	115	99
1991		104	105	106	106	102	106	104	101	94	100	141
1992	100	100	100	100	100	100	100	100	100	100	100	100
1993		97	98	103	103	99	105	109	95	111	100	100
1994		95P	99P	111	114	109	119	113	112	163	121	152
1995		93P	97P	102P	105P	106P	117P	113P	117P	170P	126P	132P
1996		92P	96P	100P	104P	105P	120P	116P	121P	177P	132P	136P
1997		90P	95P	98P	102P	105P	123P	119P	126P	184P	137P	140P
1998		88P	93P	96P	101P	105P	126P	122P	131P	191P	142P	144P

Sources: Same as General Statistics. Values reflect change from the base year, 1992. Values above 100 mean greater than 92, values below 100 mean less than 92, and a value of 100 in the 82-91 or 93-98 period means same as 92. 'P's mark projections by the editors.

SELECTED RATIOS

For 1994	Avg. of All Manufact.	Analyzed Industry	Index	For 1994	Avg. of All Manufact.	Analyzed Industry	Index
Employees per Establishment	49	192	392	Value Added per Production Worker	134,084	143,152	107
Payroll per Establishment	1,500,273	6,602,744	440	Cost per Establishment	5,045,178	45,817,683	908
Payroll per Employee	30,620	34,376	112	Cost per Employee	102,970	238,543	232
Production Workers per Establishment	34	151	440	Cost per Production Worker	146,988	303,600	207
Wages per Establishment	853,319	4,767,512	559	Shipments per Establishment	9,576,895	66,464,634	694
Wages per Production Worker	24,861	31,591	127	Shipments per Employee	195,460	346,038	177
Hours per Production Worker	2,056	2,170	106	Shipments per Production Worker	279,017	440,412	158
Wages per Hour	12.09	14.56	120	Investment per Establishment	321,011	1,757,927	548
Value Added per Establishment	4,602,255	21,603,659	469	Investment per Employee	6,552	9,152	140
Value Added per Employee	93,930	112,476	120	Investment per Production Worker	9,352	11,648	125

Sources: Same as General Statistics. The 'Average of All Manufacturing' column represents the average of all manufacturing industries reported for the most recent complete year available. The Index shows the relationship between the Average and the Analyzed Industry. For example, 100 means that they are equal; 500 that the Analyzed Industry is five times the average; 50 means that the Analyzed Industry is half the national average. The abbreviation 'na' is used to show that data are 'not available'.

LEADING COMPANIES   Number shown: **41**   Total sales ($ mil): **8,315**   Total employment (000): **51.7**

Company Name	Address				CEO Name	Phone	Co. Type	Sales ($ mil)	Empl. (000)
Marmon Group Inc	225 W Washington	Chicago	IL	60606	Robert A Pritzker	312-372-9500	R	3,900	27.0
Mueller Industries Inc	2959 North Rock	Wichita	KS	67226	William D O'Hagen	316-636-6300	P	550	2.3
Wolverine Tube Inc	PO Box 2202	Decatur	AL	35609	John M Quarles	205-353-1310	P	526	2.3
Mueller Brass Co	2959 N Rock Rd	Wichita	KS	67226	Harvey Carp	316-636-6300	S	502	1.9
Chase Brass and Copper	PO Box 152	Montpelier	OH	43543	Martin V Alonzo	419-485-3193	S	300	0.3
Outokumpu American Brass Co	PO Box 981	Buffalo	NY	14240	Warren Bartel	716-879-6700	S	300	0.9
Cerro Metal Products Co	PO Box 388	Bellefonte	PA	16823	David A Gardiner	814-355-6217	S	250	1.2
Gaston Copper Recycling Corp	PO Box 318	Gaston	SC	29053	John Avery	803-796-4720	R	224	0.4
Revere Copper Products Inc	PO Box 300	Rome	NY	13442	MB O'Shaughnessy	315-338-2022	R	220	0.7
Chase Brass Industries Inc	State Rte 15	Montpelier	OH	43543	Martin V Alonzo	419-485-3193	P	212	0.3
GTI Corp	9171 Towne Ctr Dr	San Diego	CA	92122	Gary L Luick	619-546-0531	P	141	8.6
Olin Brass, Indianapolis	1800 S Holt Rd	Indianapolis	IN	46241	Larry Stewart	317-244-1111	S	130	0.9
Encore Wire Corp	1410 Millwood Rd	McKinney	TX	75069	Donald M Spurgin	214-562-9473	P	123	0.2
Cerro Copper Products Co	Hwy 3 & Mississippi	Sauget	IL	62201	Henry L Schwich	618-337-6000	S	110*	0.8
Hussey Copper Ltd	100 Washington St	Leetsdale	PA	15056	RD Allen	412-251-4200	R	100	0.4
Copperweld Fayetteville	PO Box 70	Fayetteville	TN	37334	AL Smith	615-433-7177	D	80	0.1
Small Tube Products Company	PO Box 1674	Altoona	PA	16603	JA Woomer Jr	814-695-4491	R	70	0.3
Stuarts Draft	RR 1	Stuarts Draft	VA	24477	Karl Rose	703-337-1213	D	65*	0.3
Ansonia Copper and Brass Inc	75 Liberty St	Ansonia	CT	06401	George R Wilson	203-732-6600	R	50	0.5
Reading Tube Corp	PO Box 14026	Reading	PA	19612	James H Burton	215-926-4141	R	47	0.5
Valleycast Inc	PO Box 1714	Appleton	WI	54913	RA Biersteker	414-749-3838	S	40	0.2
Western Reserve Mfg Co	5311 W River N	Lorain	OH	44055	WD Nielsen	216-277-1226	R	40	0.1
Southwire Specialty Products	PO Box 643	Osceola	AR	72370	Joe O Williams	501-563-5207	D	35	0.2
SPD Magnet Wire Co	909 Industrial Dr	Edmonton	KY	42129	Jim Tsukagoshi	502-432-2233	J	33*	<0.1
Howell Metal Co	PO Box 218	New Market	VA	22844	AL Howell	703-740-3111	S	31*	0.2
Owl Wire and Cable Inc	PO Box 187	Canastota	NY	13032	Philip Kemper	315-697-2011	R	30*	0.2
Technical Materials Inc	5 Wellington Rd	Lincoln	RI	02865	Al Lubrano	401-333-1700	S	30	0.2
Algonquin Industries Inc	PO Box 441	Guilford	CT	06437	WT Gorman	203-453-4348	S	25	<0.1
Guardian Products Company Inc	PO Box 554	Watkinsville	GA	30677	Ed Skukalek	706-769-5611	S	23*	0.1
H & H Tube & Mfg Co	PO Box 455	Vanderbilt	MI	49795	Larry B Higgins	517-983-2800	R	20	0.2
Precision Tube Company Inc	287 Wissahickon	North Wales	PA	19454	HE Passmore	215-699-5801	R	20	0.1
Brand-Rex Nonotuck	75 Canal St	South Hadley	MA	01075	John Guzik	413-534-0271	D	15*	<0.1
Phelps Dodge Corp	PO Box 648	Elizabeth	NJ	07207	William Spellman	908-351-3200	D	15*	0.1
Spinco Metal Products Inc	PO Box 231	Newark	NY	14513	CR Straubing	315-331-6285	R	15	0.2
Waterbury Rolling Mills Inc	PO Box 550	Waterbury	CT	06720	Malcolm D Mogul	203-754-0151	S	15*	0.1
MWS Wire Industries	31200 Cedar Val Dr	Westlake Vil	CA	91362	AH Friedman	818-991-8553	R	8	<0.1
Drawn Metal Tube Co	PO Box 370	Thomaston	CT	06787	EW Hartley Jr	203-283-4345	R	5	<0.1
Hi-Mill Manufacturing Co	1704 E Highland Rd	Highland	MI	48356	Robert Beard	810-887-4191	R	5	<0.1
R and F Alloy Wires Inc	PO Box 10003	Fairfield	NJ	07004	L Michael Halleran	201-227-6222	R	5	<0.1
Trojan Tube Company Inc	PO Box 496	Farmingdale	NJ	07727	Emil A Schroth Jr	908-938-5687	R	5	<0.1
Mueller Streamline	PO Box 789761	Wichita	KS	67278	Fernando Alava	316-682-6300	D	1*	<0.1

Source: Ward's Business Directory of U.S. Private and Public Companies, Volumes 1 and 2, 1996. The company type code used is as follows: P - Public, R - Private, S - Subsidiary, D - Division, J - Joint Venture, A - Affiliate, G - Group. Sales are in millions of dollars, employees are in thousands. An asterisk (*) indicates an estimated sales volume. The symbol < stands for 'less than'. Company names and addresses are truncated, in some cases, to fit into the available space.

MATERIALS CONSUMED

Material	Quantity	Delivered Cost ($ million)
Materials, ingredients, containers, and supplies	(X)	4,252.7
Unalloyed aluminum and aluminum-base alloy ingot, pig, and shot .... 1,000 s tons	(D)	(D)
Alloyed aluminum and aluminum-base alloy ingot, pig, and shot .... 1,000 s tons	(D)	(D)
Aluminum and aluminum-base alloy sheet, plate, foil, and welded tubing .... mil lb	(D)	(D)
Aluminum and aluminum-base alloy extruded shapes, including extruded rod, bar, pipe, tube, etc. ... mil lb	1.3	1.7
All other aluminum and aluminum-base alloy shapes and forms	(X)	(D)
Copper and copper-base alloy cathodes .... 1,000 s tons	1,167.1	957.1
Copper and copper-base alloy wire bar .... 1,000 s tons	13.5	47.6
Copper and copper-base alloy ingot and ingot bar .... 1,000 s tons	11.4	32.4
All other copper and copper-base alloy shapes and forms .... 1,000 s tons	577.8	969.8
Magnesium and magnesium-base alloy shapes and forms .... 1,000 s tons	(D)	(D)
Nickel and nickel-base alloy shapes and forms .... 1,000 s tons	1.6	12.9
Zinc and zinc-base alloy shapes and forms .... 1,000 s tons	51.9	70.5
Tin and tin-base alloy shapes and forms .... mil lb	1.5*	4.4
Titanium and titanium-base alloy shapes and forms .... 1,000 s tons	(D)	(D)
Precious metals and precious metal alloy shapes and forms .... 1,000 troy ounces	(D)	(D)
Brass shapes and forms .... 1,000 s tons	131.6	340.5
All other nonferrous shapes and forms	(X)	37.5
Aluminum and aluminum-base alloy scrap from other establishments .... 1,000 s tons	(D)	(D)
Aluminum and aluminum-base alloy scrap from all other sources .... 1,000 s tons	(D)	(D)
Copper and copper-base alloy scrap (except home scrap) .... 1,000 s tons	679.9	1,372.1
Other nonferrous metal scrap (except home scrap) .... 1,000 s tons	17.1**	20.1
All other materials and components, parts, containers, and supplies	(X)	224.4
Materials, ingredients, containers, and supplies, nsk	(X)	132.0

Source: 1992 *Economic Census.* Explanation of symbols used: (D): Withheld to avoid disclosure of competitive data; na: Not available; (S): Withheld because statistical norms were not met; (X): Not applicable; (Z): Less than half the unit shown; nec: Not elsewhere classified; nsk: Not specified by kind; - : zero; * : 10-19 percent estimated; ** : 20-29 percent estimated.

PRODUCT SHARE DETAILS

Product or Product Class	% Share	Product or Product Class	% Share
Copper rolling and drawing.	100.00	plate	23.40
Copper wire, bare and tinned (nonelectrical)	4.14	Alloyed copper and copper-base alloy sheet, strip, and plate (including military cups and discs)	76.40
Unalloyed copper wire, bare and tinned (nonelectrical) . .	46.12	Copper and copper-base alloy sheet, strip, and plate, nsk	0.20
Alloyed copper wire, bare and tinned (nonelectrical) . . .	50.94	Copper and copper-base alloy pipe and tube	27.90
Copper wire, bare and tinned (nonelectrical), nsk	2.98	Unalloyed copper and copper-base alloy plumbing pipe and tube	50.19
Copper and copper-base alloy rod, bar, and shapes . . .	42.03	Other unalloyed copper and copper-base alloy pipe and tube	21.93
Unalloyed copper and copper-base alloy rod, bar, and shapes (except electric rod)	35.89	Alloyed copper and copper-base alloy plumbing pipe and tube	27.68
Alloyed copper and copper-base alloy rod, bar, and shapes (except electric rod)	64.10	Copper and copper-base alloy pipe and tube, nsk	0.20
Copper and copper-base alloy rod, bar, and shapes, nsk . .	0.01	Copper rolling and drawing, nsk	1.36
Copper and copper-base alloy sheet, strip, and plate . . .	24.58		
Unalloyed copper and copper-base alloy sheet, strip, and			

Source: 1992 *Economic Census*. The values shown are percent of total shipments in an industry. Values of indented subcategories are summed in the main heading. The symbol (D) appears when data are withheld to prevent disclosure of competitive information. The abbreviation nsk stands for 'not specified by kind' and nec for 'not elsewhere classified'.

INPUTS AND OUTPUTS FOR COPPER ROLLING & DRAWING

Economic Sector or Industry Providing Inputs	%	Sector	Economic Sector or Industry Buying Outputs	%	Sector
Primary copper	25.5	Manufg.	Nonferrous wire drawing & insulating	18.0	Manufg.
Scrap	16.6	Scrap	Refrigeration & heating equipment	6.0	Manufg.
Imports	10.6	Foreign	Pipe, valves, & pipe fittings	5.2	Manufg.
Wholesale trade	7.8	Trade	Motor vehicle parts & accessories	5.0	Manufg.
Motor freight transportation & warehousing	4.9	Util.	Exports	4.5	Foreign
Copper rolling & drawing	3.5	Manufg.	Plumbing fixture fittings & trim	3.9	Manufg.
Primary nonferrous metals, nec	3.1	Manufg.	Electronic components nec	3.7	Manufg.
Electric services (utilities)	3.0	Util.	Copper rolling & drawing	2.9	Manufg.
Copper ore	2.9	Mining	Switchgear & switchboard apparatus	2.9	Manufg.
Cyclic crudes and organics	2.6	Manufg.	Screw machine and related products	2.6	Manufg.
Railroads & related services	1.6	Util.	Metal stampings, nec	2.5	Manufg.
Gas production & distribution (utilities)	1.4	Util.	Wiring devices	2.5	Manufg.
Primary zinc	1.3	Manufg.	Hardware, nec	2.0	Manufg.
Power transmission equipment	1.2	Manufg.	Pumps & compressors	1.8	Manufg.
Miscellaneous plastics products	0.8	Manufg.	Turbines & turbine generator sets	1.8	Manufg.
Petroleum refining	0.8	Manufg.	Fabricated plate work (boiler shops)	1.3	Manufg.
Advertising	0.8	Services	Industrial controls	1.2	Manufg.
Miscellaneous repair shops	0.7	Services	Special industry machinery, nec	0.9	Manufg.
Aluminum rolling & drawing	0.6	Manufg.	Maintenance of nonfarm buildings nec	0.8	Constr.
Industrial controls	0.6	Manufg.	Residential 1-unit structures, nonfarm	0.8	Constr.
Primary aluminum	0.6	Manufg.	Blowers & fans	0.8	Manufg.
Maintenance of nonfarm buildings nec	0.4	Constr.	Machinery, except electrical, nec	0.8	Manufg.
Fabricated metal products, nec	0.4	Manufg.	Nonferrous forgings	0.8	Manufg.
Metal coating & allied services	0.4	Manufg.	Paper industries machinery	0.8	Manufg.
Communications, except radio & TV	0.4	Util.	Ship building & repairing	0.8	Manufg.
Industrial inorganic chemicals, nec	0.3	Manufg.	Transformers	0.8	Manufg.
Machinery, except electrical, nec	0.3	Manufg.	Nonfarm residential structure maintenance	0.7	Constr.
Metal stampings, nec	0.3	Manufg.	Office buildings	0.7	Constr.
Nonferrous wire drawing & insulating	0.3	Manufg.	Engine electrical equipment	0.7	Manufg.
Paperboard containers & boxes	0.3	Manufg.	Industrial furnaces & ovens	0.7	Manufg.
Wood products, nec	0.3	Manufg.	Needles, pins, & fasteners	0.7	Manufg.
Eating & drinking places	0.3	Trade	Power transmission equipment	0.7	Manufg.
Banking	0.3	Fin/R.E.	Printing trades machinery	0.7	Manufg.
Equipment rental & leasing services	0.3	Services	Small arms ammunition	0.7	Manufg.
U.S. Postal Service	0.3	Gov't	General industrial machinery, nec	0.6	Manufg.
Ball & roller bearings	0.2	Manufg.	Miscellaneous fabricated wire products	0.6	Manufg.
Metal heat treating	0.2	Manufg.	Silverware & plated ware	0.6	Manufg.
Paints & allied products	0.2	Manufg.	Federal Government purchases, nondefense	0.6	Fed Govt
Primary metal products, nec	0.2	Manufg.	Electronic computing equipment	0.5	Manufg.
Rubber & plastics hose & belting	0.2	Manufg.	Farm machinery & equipment	0.5	Manufg.
Sawmills & planning mills, general	0.2	Manufg.	Lighting fixtures & equipment	0.5	Manufg.
Screw machine and related products	0.2	Manufg.	Machine tools, metal forming types	0.5	Manufg.
Special dies & tools & machine tool accessories	0.2	Manufg.	Mechanical measuring devices	0.5	Manufg.
Air transportation	0.2	Util.	Motors & generators	0.5	Manufg.
Water transportation	0.2	Util.	Surgical & medical instruments	0.5	Manufg.
Colleges, universities, & professional schools	0.2	Services	Residential garden apartments	0.4	Constr.
Pumps & compressors	0.1	Manufg.	Environmental controls	0.4	Manufg.
Sanitary services, steam supply, irrigation	0.1	Util.	Fabricated metal products, nec	0.4	Manufg.
Insurance carriers	0.1	Fin/R.E.	Heating equipment, except electric	0.4	Manufg.
Royalties	0.1	Fin/R.E.	Metal sanitary ware	0.4	Manufg.
Computer & data processing services	0.1	Services	Metalworking machinery, nec	0.4	Manufg.
Legal services	0.1	Services	Semiconductors & related devices	0.4	Manufg.
Management & consulting services & labs	0.1	Services	Telephone & telegraph apparatus	0.4	Manufg.
			Welding apparatus, electric	0.4	Manufg.
			Residential additions/alterations, nonfarm	0.3	Constr.
			Automotive stampings	0.3	Manufg.
			Ball & roller bearings	0.3	Manufg.

Continued on next page.

INPUTS AND OUTPUTS FOR COPPER ROLLING & DRAWING - Continued

Economic Sector or Industry Providing Inputs	%	Sector	Economic Sector or Industry Buying Outputs	%	Sector
			Electron tubes	0.3	Manufg.
			Household refrigerators & freezers	0.3	Manufg.
			Internal combustion engines, nec	0.3	Manufg.
			Radio & TV communication equipment	0.3	Manufg.
			Railroad equipment	0.3	Manufg.
			Sheet metal work	0.3	Manufg.
			Steel pipe & tubes	0.3	Manufg.
			Construction of stores & restaurants	0.2	Constr.
			Industrial buildings	0.2	Constr.
			Carburetors, pistons, rings, & valves	0.2	Manufg.
			Cutlery	0.2	Manufg.
			Electrical equipment & supplies, nec	0.2	Manufg.
			Electrical industrial apparatus, nec	0.2	Manufg.
			Food products machinery	0.2	Manufg.
			Instruments to measure electricity	0.2	Manufg.
			Manufacturing industries, nec	0.2	Manufg.
			Photographic equipment & supplies	0.2	Manufg.
			Special dies & tools & machine tool accessories	0.2	Manufg.
			Surgical appliances & supplies	0.2	Manufg.
			Construction of educational buildings	0.1	Constr.
			Construction of nonfarm buildings nec	0.1	Constr.
			Residential high-rise apartments	0.1	Constr.
			Construction machinery & equipment	0.1	Manufg.
			Conveyors & conveying equipment	0.1	Manufg.
			Elevators & moving stairways	0.1	Manufg.
			Explosives	0.1	Manufg.
			Fabricated structural metal	0.1	Manufg.
			Hand & edge tools, nec	0.1	Manufg.
			Household appliances, nec	0.1	Manufg.
			Miscellaneous metal work	0.1	Manufg.
			Motor vehicles & car bodies	0.1	Manufg.
			Oil field machinery	0.1	Manufg.
			Public building furniture	0.1	Manufg.
			Sporting & athletic goods, nec	0.1	Manufg.

Source: Benchmark Input-Output Accounts for the U.S. Economy, 1982, U.S. Department of Commerce, Washington, D.C., July 1991. Data, as reported in the source, are organized by the 1977 SIC structure in use in 1982 but have been matched, as closely as is possible, to the 1987 SIC structure used in this book.

OCCUPATIONS EMPLOYED BY SIC 335 - NONFERROUS ROLLING AND DRAWING

Occupation	% of Total 1994	Change to 2005	Occupation	% of Total 1994	Change to 2005
Machine tool cutting & forming etc. nec	11.0	-12.4	Machinists	1.6	-20.3
Assemblers, fabricators, & hand workers nec	5.9	-20.4	Machine tool cutting operators, metal & plastic	1.4	-33.7
Blue collar worker supervisors	5.0	-22.9	Extruding & forming machine workers	1.3	-12.4
Machine forming operators, metal & plastic	5.0	7.5	Coating, painting, & spraying machine workers	1.3	-20.4
Metal & plastic machine workers nec	3.9	-15.3	General managers & top executives	1.2	-24.4
Inspectors, testers, & graders, precision	3.7	-20.3	Combination machine tool operators	1.2	-12.4
Helpers, laborers, & material movers nec	3.6	-20.4	Traffic, shipping, & receiving clerks	1.1	-23.3
Machine operators nec	3.1	-29.8	Industrial production managers	1.1	-20.4
Industrial machinery mechanics	2.9	-12.4	Packaging & filling machine operators	1.1	-20.4
Industrial truck & tractor operators	2.8	-20.3	Secretaries, ex legal & medical	1.1	-27.5
Maintenance repairers, general utility	2.1	-28.3	Machine feeders & offbearers	1.1	-28.3
Sales & related workers nec	1.9	-20.3	Production, planning, & expediting clerks	1.0	-20.3
Hand packers & packagers	1.7	-31.7	Metal molding machine workers	1.0	19.5
Electricians	1.7	-25.3			

Source: Industry-Occupation Matrix, Bureau of Labor Statistics. These data relate to one or more 3-digit SIC industry groups rather than to a single 4-digit SIC. The change reported for each occupation to the year 2005 is a percent of growth or decline as estimated by the Bureau of Labor Statistics. The abbreviation nec stands for 'not elsewhere classified'.

LOCATION BY STATE AND REGIONAL CONCENTRATION

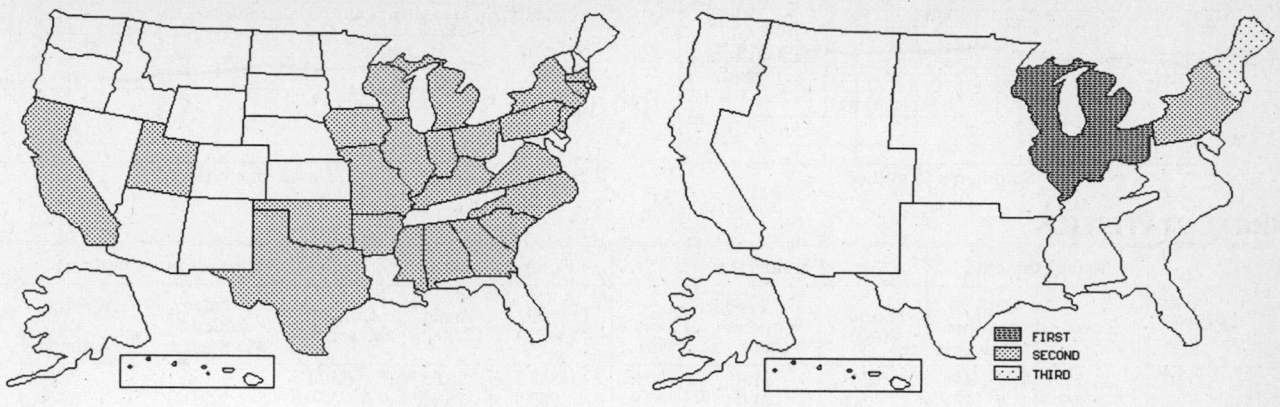

FIRST
SECOND
THIRD

INDUSTRY DATA BY STATE

State	Establish-ments	Shipments			Employment				Cost as % of Shipments	Investment per Employee ($)
		Total ($ mil)	% of U.S.	Per Establ.	Total Number	% of U.S.	Per Establ.	Wages ($/hour)		
Pennsylvania	12	574.9	9.6	47.9	2,700	14.3	225	13.82	71.2	3,593
Ohio	5	324.7	5.4	64.9	800	4.2	160	14.00	74.8	8,500
Connecticut	12	274.6	4.6	22.9	1,200	6.3	100	14.63	62.4	4,583
Wisconsin	5	216.1	3.6	43.2	1,200	6.3	240	13.90	59.2	4,000
New Jersey	6	80.1	1.3	13.4	600	3.2	100	5.55	64.8	1,333
Massachusetts	7	50.2	0.8	7.2	300	1.6	43	13.00	54.2	4,333
Illinois	8	(D)	-	-	3,750 *	19.8	469	-	-	-
New York	7	(D)	-	-	1,750 *	9.3	250	-	-	-
California	6	(D)	-	-	175 *	0.9	29	-	-	2,286
Michigan	5	(D)	-	-	750 *	4.0	150	-	-	-
North Carolina	5	(D)	-	-	750 *	4.0	150	-	-	-
Alabama	4	(D)	-	-	1,750 *	9.3	438	-	-	-
Mississippi	4	(D)	-	-	750 *	4.0	188	-	-	-
Texas	4	(D)	-	-	375 *	2.0	94	-	-	-
Indiana	3	(D)	-	-	750 *	4.0	250	-	-	-
Kentucky	3	(D)	-	-	175 *	0.9	58	-	-	-
Missouri	2	(D)	-	-	175 *	0.9	88	-	-	-
Oklahoma	2	(D)	-	-	375 *	2.0	188	-	-	-
Virginia	2	(D)	-	-	175 *	0.9	88	-	-	-
Arkansas	1	(D)	-	-	750 *	4.0	750	-	-	-
Georgia	1	(D)	-	-	375 *	2.0	375	-	-	-
Iowa	1	(D)	-	-	175 *	0.9	175	-	-	-
South Carolina	1	(D)	-	-	375 *	2.0	375	-	-	-
Utah	1	(D)	-	-	175 *	0.9	175	-	-	-

Source: 1992 *Economic Census*. The states are in descending order of shipments or establishments (if shipment data are missing for the majority). The symbol (D) appears when data are withheld to prevent disclosure of competitive information. States marked with (D) are sorted by number of establishments. A dash (-) indicates that the data element cannot be calculated; * indicates the midpoint of a range.

3353 - ALUMINUM SHEET, PLATE & FOIL

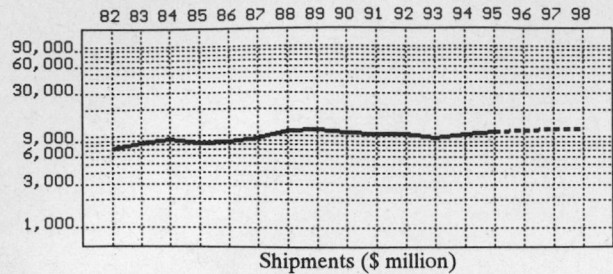

Shipments ($ million)

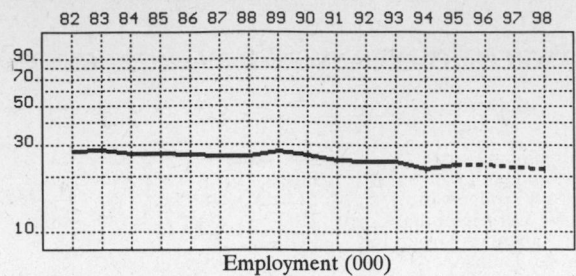

Employment (000)

GENERAL STATISTICS

Year	Companies	Establishments		Employment			Compensation		Production ($ million)			
		Total	with 20 or more employees	Total (000)	Production Workers (000)	Hours (Mil)	Payroll ($ mil)	Wages ($/hr)	Cost of Materials	Value Added by Manufacture	Value of Shipments	Capital Invest.
1982	40	57	44	27.8	21.0	40.0	863.7	16.03	5,911.3	1,156.2	7,228.7	260.4
1983		56	43	28.1	21.9	42.9	907.4	16.27	6,859.9	1,503.9	8,453.5	296.9
1984		55	42	26.9	21.0	41.6	914.5	16.73	7,222.7	1,720.9	9,118.8	359.3
1985		53	42	26.9	20.8	42.0	926.6	16.40	6,382.5	1,797.7	8,424.8	348.0
1986		50	41	26.7	20.4	39.7	910.7	16.88	6,523.6	2,240.1	8,699.4	439.4
1987	39	56	44	26.1	19.8	40.2	912.3	16.39	7,841.4	1,790.0	9,497.2	439.2
1988		67	52	26.1	19.5	41.2	963.4	16.72	9,925.2	1,959.7	11,647.1	524.0
1989		62	53	27.7	19.3	40.8	993.0	17.17	9,667.4	1,864.7	11,820.6	551.9
1990		63	51	26.3	18.7	39.7	1,021.0	18.13	8,682.0	2,508.5	11,121.5	681.3
1991		72	54	24.9	18.4	39.3	999.7	17.84	7,823.7	2,579.8	10,773.0	567.2
1992	45	63	44	24.4	18.3	39.9	1,014.3	17.98	7,273.1	3,229.2	10,648.7	418.3
1993		63	45	24.2	18.2	38.6	1,023.6	18.59	6,890.3	2,691.5	9,672.4	241.4
1994		68P	51P	22.2	16.7	36.5	981.1	19.03	8,212.3	2,701.6	10,577.2	272.9
1995		69P	51P	23.4P	17.0P	38.0P	1,041.1P	18.87P	9,014.9P	2,965.6P	11,610.9P	464.9P
1996		70P	52P	23.0P	16.6P	37.7P	1,053.2P	19.10P	9,213.3P	3,030.9P	11,866.5P	472.0P
1997		71P	53P	22.6P	16.3P	37.4P	1,065.3P	19.33P	9,411.8P	3,096.2P	12,122.1P	479.1P
1998		72P	53P	22.3P	15.9P	37.1P	1,077.4P	19.56P	9,610.2P	3,161.5P	12,377.7P	486.1P

Sources: 1982, 1987, 1992 *Economic Census*; *Annual Survey of Manufactures*, 83-86, 88-91, 93-94. Establishment counts for non-Census years are from *County Business Patterns*; establishment values for 83-84 are extrapolations. 'P's show projections by the editors. Industries reclassified in 87 will not have data for prior years.

INDICES OF CHANGE

Year	Companies	Establishments		Employment			Compensation		Production ($ million)			
		Total	with 20 or more employees	Total (000)	Production Workers (000)	Hours (Mil)	Payroll ($ mil)	Wages ($/hr)	Cost of Materials	Value Added by Manufacture	Value of Shipments	Capital Invest.
1982	89	90	100	114	115	100	85	89	81	36	68	62
1983		89	98	115	120	108	89	90	94	47	79	71
1984		87	95	110	115	104	90	93	99	53	86	86
1985		84	95	110	114	105	91	91	88	56	79	83
1986		79	93	109	111	99	90	94	90	69	82	105
1987	87	89	100	107	108	101	90	91	108	55	89	105
1988		106	118	107	107	103	95	93	136	61	109	125
1989		98	120	114	105	102	98	95	133	58	111	132
1990		100	116	108	102	99	101	101	119	78	104	163
1991		114	123	102	101	98	99	99	108	80	101	136
1992	100	100	100	100	100	100	100	100	100	100	100	100
1993		100	102	99	99	97	101	103	95	83	91	58
1994		107P	115P	91	91	91	97	106	113	84	99	65
1995		109P	117P	96P	93P	95P	103P	105P	124P	92P	109P	111P
1996		111P	118P	94P	91P	94P	104P	106P	127P	94P	111P	113P
1997		113P	120P	93P	89P	94P	105P	108P	129P	96P	114P	115P
1998		115P	121P	91P	87P	93P	106P	109P	132P	98P	116P	116P

Sources: Same as General Statistics. Values reflect change from the base year, 1992. Values above 100 mean greater than 92, values below 100 mean less than 92, and a value of 100 in the 82-91 or 93-98 period means same as 92. 'P's mark projections by the editors.

SELECTED RATIOS

For 1994	Avg. of All Manufact.	Analyzed Industry	Index	For 1994	Avg. of All Manufact.	Analyzed Industry	Index
Employees per Establishment	49	328	670	Value Added per Production Worker	134,084	161,772	121
Payroll per Establishment	1,500,273	14,515,266	968	Cost per Establishment	5,045,178	121,500,067	2,408
Payroll per Employee	30,620	44,194	144	Cost per Employee	102,970	369,923	359
Production Workers per Establishment	34	247	720	Cost per Production Worker	146,988	491,754	335
Wages per Establishment	853,319	10,276,456	1,204	Shipments per Establishment	9,576,895	156,488,500	1,634
Wages per Production Worker	24,861	41,593	167	Shipments per Employee	195,460	476,450	244
Hours per Production Worker	2,056	2,186	106	Shipments per Production Worker	279,017	633,365	227
Wages per Hour	12.09	19.03	157	Investment per Establishment	321,011	4,037,525	1,258
Value Added per Establishment	4,602,255	39,969,872	868	Investment per Employee	6,552	12,293	188
Value Added per Employee	93,930	121,694	130	Investment per Production Worker	9,352	16,341	175

Sources: Same as General Statistics. The 'Average of All Manufacturing' column represents the average of all manufacturing industries reported for the most recent complete year available. The Index shows the relationship between the Average and the Analyzed Industry. For example, 100 means that they are equal; 500 that the Analyzed Industry is five times the average; 50 means that the Analyzed Industry is half the national average. The abbreviation 'na' is used to show that data are 'not available'.

LEADING COMPANIES   Number shown: 17   Total sales ($ mil): 7,720   Total employment (000): 22.3

Company Name	Address				CEO Name	Phone	Co. Type	Sales ($ mil)	Empl. (000)
Alcan Aluminum Corp	100 Erieview Plz	Cleveland	OH	44114	Robert L Ball	216-523-6800	S	3,700	6.0
Kaiser Aluminum Corp	PO Box 572887	Houston	TX	77257	GT Haymaker	713-267-3777	S	1,781	9.5
Alumax Mill Products Inc	12 Salt Creek Ln	Hinsdale	IL	60521	George P Stoe	708-654-0500	S	800	1.7
Commonwealth Aluminum Corp	1200 Meidinger Twr	Louisville	KY	40202	Mark V Kaminski	502-589-8100	P	496	1.1
CasTech Aluminum Group Inc	PO Box 3740	Akron	OH	44314	Norman E Wells Jr	216-848-5555	P	263	0.9
Homeshield Fabricated	901 N Batavia Av	Batavia	IL	60510	TF Olt	708-406-5410	D	201	0.9
Golden Aluminum Co	1600 Jackson St	Golden	CO	80401	Joseph Coors Jr	303-277-7500	S	129	0.6
Hydro Aluminum-Adrian	PO Box 5083	Southfield	MI	48037	Joseph J Galambos	313-355-8293	S	100	0.6
Quanex Corp	1725 Rockingham	Davenport	IA	52802	MF Friemel	319-324-2121	D	80	0.2
Insilco Corp	PO Box 3253	Montgomery	AL	36109	John H Marshall	205-277-1810	S	60	0.3
Petersen Aluminum Corp	1005 Tonne Rd	Elk Grove Vill	IL	60007	Maury R Petersen	708-228-7150	R	47	0.2
Alumax Inc Building Products	206 Kesco St	Bristol	IN	46507	Steve Salzer	219-848-7431	D	15*	<0.1
United Aluminum Corp	PO Box 215	North Haven	CT	06473	JS Lapides	203-239-5881	R	15	<0.1
AJ Oster Foils Inc	2081 McCrea St	Alliance	OH	44601	Wiliam Ingellis	216-823-1700	D	14	<0.1
All Metals Service	848 Damar Rd NE	Marietta	GA	30062	John Dill	404-421-6680	R	8	0.1
JL Clark Tube	2300 Wisconsin Av	Downers Grove	IL	60515	John P Benton	708-969-6100	D	7*	0.1
Alufoil Products Company Inc	135 Oser Av	Hauppauge	NY	11788	A Simon	516-231-4141	R	4	<0.1

Source: Ward's Business Directory of U.S. Private and Public Companies, Volumes 1 and 2, 1996. The company type code used is as follows: P - Public, R - Private, S - Subsidiary, D - Division, J - Joint Venture, A - Affiliate, G - Group. Sales are in millions of dollars, employees are in thousands. An asterisk (*) indicates an estimated sales volume. The symbol < stands for 'less than'. Company names and addresses are truncated, in some cases, to fit into the available space.

MATERIALS CONSUMED

Material		Quantity	Delivered Cost ($ million)
Materials, ingredients, containers, and supplies		(X)	6,590.2
Unalloyed aluminum and aluminum-base alloy ingot, pig, and shot	1,000 s tons	1,424.9	1,634.6
Alloyed aluminum and aluminum-base alloy ingot, pig, and shot	1,000 s tons	999.8	1,743.1
Aluminum and aluminum-base alloy sheet, plate, foil, and welded tubing	mil lb	1,004.5	812.4
Aluminum and aluminum-base alloy extruded shapes, including extruded rod, bar, pipe, tube, etc.	mil lb	(D)	(D)
All other aluminum and aluminum-base alloy shapes and forms		(X)	255.1
Copper and copper-base alloy wire bar	1,000 s tons	(D)	(D)
Copper and copper-base alloy ingot and ingot bar	1,000 s tons	(D)	(D)
All other copper and copper-base alloy shapes and forms	1,000 s tons	(S)	0.5
Magnesium and magnesium-base alloy shapes and forms	1,000 s tons	46.6	111.0
Zinc and zinc-base alloy shapes and forms	1,000 s tons	2.9	3.5
Titanium and titanium-base alloy shapes and forms	1,000 s tons	0.9	2.9
Precious metals and precious metal alloy shapes and forms	1,000 troy ounces	(D)	(D)
Brass shapes and forms	1,000 s tons	(D)	(D)
All other nonferrous shapes and forms		(X)	25.5
Aluminum and aluminum-base alloy scrap from other establishments	1,000 s tons	270.2	305.9
Aluminum and aluminum-base alloy scrap from all other sources	1,000 s tons	1,168.6	1,168.3
Copper and copper-base alloy scrap (except home scrap)	1,000 s tons	(D)	(D)
All other materials and components, parts, containers, and supplies		(X)	495.4

Source: 1992 Economic Census. Explanation of symbols used: (D): Withheld to avoid disclosure of competitive data; na: Not available; (S): Withheld because statistical norms were not met; (X): Not applicable; (Z): Less than half the unit shown; nec: Not elsewhere classified; nsk: Not specified by kind; - : zero; * : 10-19 percent estimated; ** : 20-29 percent estimated.

PRODUCT SHARE DETAILS

Product or Product Class	% Share	Product or Product Class	% Share
Aluminum sheet, plate, and foil	100.00	and strip (including continuous cast)	3.81
Aluminum plate (thickness of 0.25 in. or more) (including continuous cast)	6.11	Coiled, heat-treatable aluminum sheet and strip (including continuous cast)	16.03
Heat-treatable aluminum plate (thickness of 0.25 in. or more) (including continuous cast)	72.68	Coiled, nonheat-treatable, bare aluminum sheet and strip (including continuous cast)	55.15
Nonheat-treatable aluminum plate (thickness of 0.25 in. or more) (including continuous cast)	27.32	Coiled, nonheat-treatable, precoated (including only permanent finishes such as enameling and vinyl coatings) aluminum sheet and strip (including continuous cast)	22.68
Aluminum sheet and strip (including continuous cast)	86.58	Plain aluminum foil (less than .006 in. thick)	(D)
Flat, heat-treatable aluminum sheet and strip (including continuous cast)	2.33	Plain aluminum foil (less than .006 in. thick)	8.44
Flat, nonheat-treatable, bare and precoated aluminum sheet		Aluminum welded tube	(D)

Source: 1992 Economic Census. The values shown are percent of total shipments in an industry. Values of indented subcategories are summed in the main heading. The symbol (D) appears when data are withheld to prevent disclosure of competitive information. The abbreviation nsk stands for 'not specified by kind' and nec for 'not elsewhere classified'.

INPUTS AND OUTPUTS FOR ALUMINUM ROLLING & DRAWING

Economic Sector or Industry Providing Inputs	%	Sector	Economic Sector or Industry Buying Outputs	%	Sector
Primary aluminum	45.4	Manufg.	Metal cans	17.9	Manufg.
Scrap	8.8	Scrap	Metal doors, sash, & trim	7.7	Manufg.
Wholesale trade	6.6	Trade	Aluminum rolling & drawing	6.1	Manufg.
Aluminum rolling & drawing	6.1	Manufg.	Sheet metal work	5.0	Manufg.
Imports	4.7	Foreign	Exports	5.0	Foreign
Motor freight transportation & warehousing	3.2	Util.	Metal foil & leaf	4.4	Manufg.
Gas production & distribution (utilities)	2.6	Util.	Nonferrous wire drawing & insulating	3.5	Manufg.
Special dies & tools & machine tool accessories	2.2	Manufg.	Refrigeration & heating equipment	2.9	Manufg.
Electric services (utilities)	2.2	Util.	Metal stampings, nec	2.1	Manufg.
Petroleum refining	1.5	Manufg.	Aircraft	2.0	Manufg.
Primary nonferrous metals, nec	1.4	Manufg.	Tanks & tank components	1.9	Manufg.
Cyclic crudes and organics	1.3	Manufg.	Truck trailers	1.5	Manufg.
Railroads & related services	1.3	Util.	Electronic components nec	1.4	Manufg.
Power transmission equipment	1.2	Manufg.	Aircraft & missile equipment, nec	1.3	Manufg.
Miscellaneous plastics products	1.0	Manufg.	Fabricated metal products, nec	1.3	Manufg.
Nonferrous wire drawing & insulating	0.7	Manufg.	Lighting fixtures & equipment	1.3	Manufg.
Miscellaneous repair shops	0.7	Services	Miscellaneous metal work	1.3	Manufg.
Metal coating & allied services	0.5	Manufg.	Automotive stampings	1.2	Manufg.
Screw machine and related products	0.5	Manufg.	Prefabricated metal buildings	1.2	Manufg.
Maintenance of nonfarm buildings nec	0.4	Constr.	Radio & TV communication equipment	1.1	Manufg.
Ball & roller bearings	0.4	Manufg.	Crowns & closures	1.0	Manufg.
Fabricated metal products, nec	0.4	Manufg.	Architectural metal work	0.9	Manufg.
Advertising	0.4	Services	Drapery hardware & blinds & shades	0.9	Manufg.
Industrial controls	0.3	Manufg.	Machinery, except electrical, nec	0.9	Manufg.
Metal stampings, nec	0.3	Manufg.	Hardware, nec	0.8	Manufg.
Air transportation	0.3	Util.	Metal household furniture	0.8	Manufg.
Automotive rental & leasing, without drivers	0.3	Services	Motor vehicle parts & accessories	0.8	Manufg.
Chemical preparations, nec	0.2	Manufg.	Motor vehicles & car bodies	0.8	Manufg.
Industrial inorganic chemicals, nec	0.2	Manufg.	Truck & bus bodies	0.8	Manufg.
Paints & allied products	0.2	Manufg.	Electronic computing equipment	0.6	Manufg.
Primary metal products, nec	0.2	Manufg.	Fabricated plate work (boiler shops)	0.6	Manufg.
Rubber & plastics hose & belting	0.2	Manufg.	General industrial machinery, nec	0.6	Manufg.
Communications, except radio & TV	0.2	Util.	Photographic equipment & supplies	0.6	Manufg.
Eating & drinking places	0.2	Trade	Pipe, valves, & pipe fittings	0.6	Manufg.
Banking	0.2	Fin/R.E.	Screw machine and related products	0.6	Manufg.
Automotive repair shops & services	0.2	Services	Signs & advertising displays	0.6	Manufg.
Equipment rental & leasing services	0.2	Services	Sporting & athletic goods, nec	0.6	Manufg.
Machinery, except electrical, nec	0.1	Manufg.	Travel trailers & campers	0.6	Manufg.
Metal heat treating	0.1	Manufg.	Manufacturing industries, nec	0.5	Manufg.
Paperboard containers & boxes	0.1	Manufg.	Metal barrels, drums, & pails	0.5	Manufg.
Primary lead	0.1	Manufg.	Transformers	0.5	Manufg.
Sawmills & planning mills, general	0.1	Manufg.	Aircraft & missile engines & engine parts	0.4	Manufg.
Wood products, nec	0.1	Manufg.	Blowers & fans	0.4	Manufg.
Water transportation	0.1	Util.	Boat building & repairing	0.4	Manufg.
Insurance carriers	0.1	Fin/R.E.	Electric housewares & fans	0.4	Manufg.
Computer & data processing services	0.1	Services	Fabricated structural metal	0.4	Manufg.
Hotels & lodging places	0.1	Services	Mechanical measuring devices	0.4	Manufg.
			Paperboard containers & boxes	0.4	Manufg.
			Printing trades machinery	0.4	Manufg.
			Special industry machinery, nec	0.4	Manufg.
			Switchgear & switchboard apparatus	0.4	Manufg.
			Wiring devices	0.4	Manufg.
			Ammunition, except for small arms, nec	0.3	Manufg.
			Carburetors, pistons, rings, & valves	0.3	Manufg.
			Food products machinery	0.3	Manufg.
			Household refrigerators & freezers	0.3	Manufg.
			Motors & generators	0.3	Manufg.
			Public building furniture	0.3	Manufg.
			Pumps & compressors	0.3	Manufg.
			Ship building & repairing	0.3	Manufg.
			Special dies & tools & machine tool accessories	0.3	Manufg.
			Copper rolling & drawing	0.2	Manufg.
			Electrical equipment & supplies, nec	0.2	Manufg.
			Furniture & fixtures, nec	0.2	Manufg.
			Games, toys, & children's vehicles	0.2	Manufg.
			Heating equipment, except electric	0.2	Manufg.
			Household cooking equipment	0.2	Manufg.
			Industrial controls	0.2	Manufg.
			Instruments to measure electricity	0.2	Manufg.
			Metal office furniture	0.2	Manufg.
			Metal partitions & fixtures	0.2	Manufg.
			Mobile homes	0.2	Manufg.
			Needles, pins, & fasteners	0.2	Manufg.
			Nonferrous forgings	0.2	Manufg.
			Paper coating & glazing	0.2	Manufg.
			Service industry machines, nec	0.2	Manufg.
			Small arms	0.2	Manufg.
			Surgical appliances & supplies	0.2	Manufg.

Continued on next page.

INPUTS AND OUTPUTS FOR ALUMINUM ROLLING & DRAWING - Continued

Economic Sector or Industry Providing Inputs	%	Sector	Economic Sector or Industry Buying Outputs	%	Sector
			Textile machinery	0.2	Manufg.
			Transportation equipment, nec	0.2	Manufg.
			Typewriters & office machines, nec	0.2	Manufg.
			Federal Government purchases, national defense	0.2	Fed Govt
			Maintenance of nonfarm buildings nec	0.1	Constr.
			Construction machinery & equipment	0.1	Manufg.
			Conveyors & conveying equipment	0.1	Manufg.
			Elevators & moving stairways	0.1	Manufg.
			Engineering & scientific instruments	0.1	Manufg.
			Household appliances, nec	0.1	Manufg.
			Industrial furnaces & ovens	0.1	Manufg.
			Miscellaneous fabricated wire products	0.1	Manufg.
			Telephone & telegraph apparatus	0.1	Manufg.

Source: Benchmark Input-Output Accounts for the U.S. Economy, 1982, U.S. Department of Commerce, Washington, D.C., July 1991. Data, as reported in the source, are organized by the 1977 SIC structure in use in 1982 but have been matched, as closely as is possible, to the 1987 SIC structure used in this book.

OCCUPATIONS EMPLOYED BY SIC 335 - NONFERROUS ROLLING AND DRAWING

Occupation	% of Total 1994	Change to 2005	Occupation	% of Total 1994	Change to 2005
Machine tool cutting & forming etc. nec	11.0	-12.4	Machinists	1.6	-20.3
Assemblers, fabricators, & hand workers nec	5.9	-20.4	Machine tool cutting operators, metal & plastic	1.4	-33.7
Blue collar worker supervisors	5.0	-22.9	Extruding & forming machine workers	1.3	-12.4
Machine forming operators, metal & plastic	5.0	7.5	Coating, painting, & spraying machine workers	1.3	-20.4
Metal & plastic machine workers nec	3.9	-15.3	General managers & top executives	1.2	-24.4
Inspectors, testers, & graders, precision	3.7	-20.3	Combination machine tool operators	1.2	-12.4
Helpers, laborers, & material movers nec	3.6	-20.4	Traffic, shipping, & receiving clerks	1.1	-23.3
Machine operators nec	3.1	-29.8	Industrial production managers	1.1	-20.4
Industrial machinery mechanics	2.9	-12.4	Packaging & filling machine operators	1.1	-20.4
Industrial truck & tractor operators	2.8	-20.3	Secretaries, ex legal & medical	1.1	-27.5
Maintenance repairers, general utility	2.1	-28.3	Machine feeders & offbearers	1.1	-28.3
Sales & related workers nec	1.9	-20.3	Production, planning, & expediting clerks	1.0	-20.3
Hand packers & packagers	1.7	-31.7	Metal molding machine workers	1.0	19.5
Electricians	1.7	-25.3			

Source: Industry-Occupation Matrix, Bureau of Labor Statistics. These data relate to one or more 3-digit SIC industry groups rather than to a single 4-digit SIC. The change reported for each occupation to the year 2005 is a percent of growth or decline as estimated by the Bureau of Labor Statistics. The abbreviation nec stands for 'not elsewhere classified'.

LOCATION BY STATE AND REGIONAL CONCENTRATION

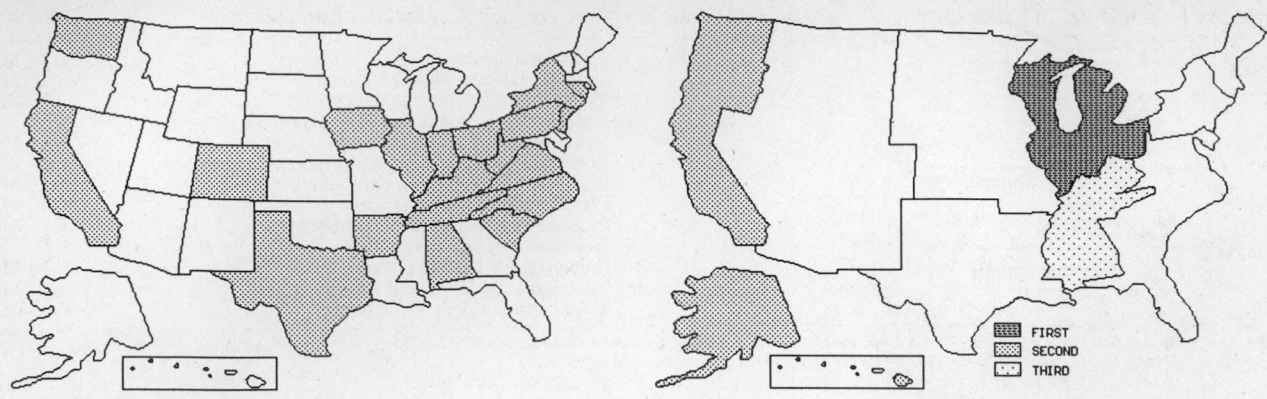

FIRST
SECOND
THIRD

INDUSTRY DATA BY STATE

State	Establish-ments	Shipments			Employment				Cost as % of Shipments	Investment per Employee ($)
		Total ($ mil)	% of U.S.	Per Establ.	Total Number	% of U.S.	Per Establ.	Wages ($/hour)		
Tennessee	4	1,215.4	11.4	303.9	2,100	8.6	525	19.05	59.0	-
Ohio	5	607.7	5.7	121.5	1,400	5.7	280	17.95	77.6	-
California	9	183.6	1.7	20.4	600	2.5	67	13.11	68.8	6,333
Alabama	3	(D)	-	-	3,750 *	15.4	1,250	-	-	-
Illinois	3	(D)	-	-	1,750 *	7.2	583	-	-	-
Indiana	3	(D)	-	-	3,750 *	15.4	1,250	-	-	-
Kentucky	3	(D)	-	-	1,750 *	7.2	583	-	-	-
New York	3	(D)	-	-	750 *	3.1	250	-	-	-
Pennsylvania	3	(D)	-	-	1,750 *	7.2	583	-	-	-
Arkansas	2	(D)	-	-	375 *	1.5	188	-	-	-
Connecticut	2	(D)	-	-	175 *	0.7	88	-	-	-
Iowa	2	(D)	-	-	3,750 *	15.4	1,875	-	-	-
North Carolina	2	(D)	-	-	175 *	0.7	88	-	-	-
Texas	2	(D)	-	-	750 *	3.1	375	-	-	-
Virginia	2	(D)	-	-	175 *	0.7	88	-	-	-
West Virginia	2	(D)	-	-	1,750 *	7.2	875	-	-	-
Colorado	1	(D)	-	-	375 *	1.5	375	-	-	-
New Jersey	1	(D)	-	-	375 *	1.5	375	-	-	-
South Carolina	1	(D)	-	-	375 *	1.5	375	-	-	-
Washington	1	(D)	-	-	1,750 *	7.2	1,750	-	-	-

Source: 1992 *Economic Census.* The states are in descending order of shipments or establishments (if shipment data are missing for the majority). The symbol (D) appears when data are withheld to prevent disclosure of competitive information. States marked with (D) are sorted by number of establishments. A dash (-) indicates that the data element cannot be calculated; * indicates the midpoint of a range.

3354 - ALUMINUM EXTRUDED PRODUCTS

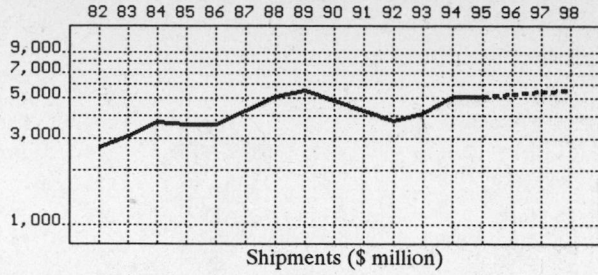

Shipments ($ million)

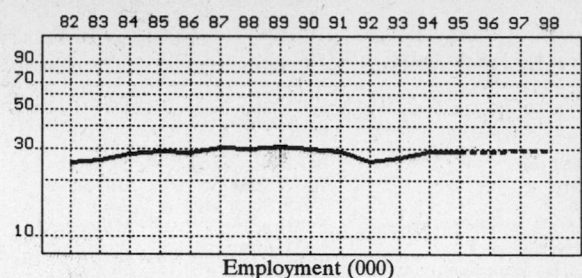

Employment (000)

GENERAL STATISTICS

Year	Companies	Establishments		Employment			Compensation		Production ($ million)			
		Total	with 20 or more employees	Total (000)	Production Workers (000)	Hours (Mil)	Payroll ($ mil)	Wages ($/hr)	Cost of Materials	Value Added by Manufacture	Value of Shipments	Capital Invest.
1982	134	193	151	25.4	19.3	38.0	499.0	9.06	1,778.5	859.3	2,673.1	114.7
1983		192	151	26.0	20.1	40.9	538.4	9.29	2,193.7	915.6	3,057.3	96.8
1984		191	151	28.2	22.1	45.6	607.8	9.55	2,566.1	1,194.1	3,718.7	125.0
1985		189	152	29.1	22.7	47.3	639.0	9.52	2,384.1	1,173.8	3,575.7	120.6
1986		186	151	28.4	21.8	45.6	658.4	9.89	2,461.2	1,102.9	3,573.6	109.0
1987	132	204	161	30.7	23.5	50.0	737.6	10.05	2,908.8	1,404.7	4,292.8	116.9
1988		190	156	30.0	22.9	49.4	740.5	10.24	3,599.5	1,516.2	5,077.6	
1989		192	162	30.8	24.3	50.7	802.6	10.57	3,974.7	1,514.4	5,509.4	
1990		194	158	30.3	23.8	51.3	793.0	10.36	3,335.2	1,466.9	4,850.3	128.2
1991		195	149	29.2	22.4	48.6	767.1	10.45	2,781.8	1,422.1	4,260.1	113.3
1992	134	193	153	25.6	19.7	42.8	687.2	10.66	2,358.7	1,374.0	3,735.6	150.0
1993		196	153	27.0	20.9	45.2	750.6	11.16	2,517.2	1,664.0	4,158.5	108.4
1994		195P	156P	29.2	23.4	51.7	840.3	11.18	3,137.2	2,051.2	5,130.5	146.7
1995		196P	156P	29.3P	23.0P	50.9P	856.5P	11.34P	3,136.6P	2,050.8P	5,129.5P	
1996		196P	156P	29.4P	23.1P	51.5P	879.2P	11.50P	3,224.4P	2,108.2P	5,273.2P	
1997		196P	157P	29.5P	23.2P	52.1P	902.0P	11.67P	3,312.3P	2,165.7P	5,416.8P	
1998		197P	157P	29.6P	23.3P	52.7P	924.8P	11.84P	3,400.1P	2,223.1P	5,560.4P	

Sources: 1982, 1987, 1992 *Economic Census*; *Annual Survey of Manufactures*, 83-86, 88-91, 93-94. Establishment counts for non-Census years are from *County Business Patterns*; establishment values for 83-84 are extrapolations. 'P's show projections by the editors. Industries reclassified in 87 will not have data for prior years.

INDICES OF CHANGE

Year	Companies	Establishments		Employment			Compensation		Production ($ million)			
		Total	with 20 or more employees	Total (000)	Production Workers (000)	Hours (Mil)	Payroll ($ mil)	Wages ($/hr)	Cost of Materials	Value Added by Manufacture	Value of Shipments	Capital Invest.
1982	100	100	99	99	98	89	73	85	75	63	72	76
1983		99	99	102	102	96	78	87	93	67	82	65
1984		99	99	110	112	107	88	90	109	87	100	83
1985		98	99	114	115	111	93	89	101	85	96	80
1986		96	99	111	111	107	96	93	104	80	96	73
1987	99	106	105	120	119	117	107	94	123	102	115	78
1988		98	102	117	116	115	108	96	153	110	136	
1989		99	106	120	123	118	117	99	169	110	147	
1990		101	103	118	121	120	115	97	141	107	130	85
1991		101	97	114	114	114	112	98	118	104	114	76
1992	100	100	100	100	100	100	100	100	100	100	100	100
1993		102	100	105	106	106	109	105	107	121	111	72
1994		101P	102P	114	119	121	122	105	133	149	137	98
1995		101P	102P	114P	116P	119P	125P	106P	133P	149P	137P	
1996		101P	102P	115P	117P	120P	128P	108P	137P	153P	141P	
1997		102P	102P	115P	118P	122P	131P	110P	140P	158P	145P	
1998		102P	103P	116P	118P	123P	135P	111P	144P	162P	149P	

Sources: Same as General Statistics. Values reflect change from the base year, 1992. Values above 100 mean greater than 92, values below 100 mean less than 92, and a value of 100 in the 82-91 or 93-98 period means same as 92. 'P's mark projections by the editors.

SELECTED RATIOS

For 1994	Avg. of All Manufact.	Analyzed Industry	Index	For 1994	Avg. of All Manufact.	Analyzed Industry	Index
Employees per Establishment	49	150	305	Value Added per Production Worker	134,084	87,658	65
Payroll per Establishment	1,500,273	4,305,551	287	Cost per Establishment	5,045,178	16,074,466	319
Payroll per Employee	30,620	28,777	94	Cost per Employee	102,970	107,438	104
Production Workers per Establishment	34	120	349	Cost per Production Worker	146,988	134,068	91
Wages per Establishment	853,211	2,961,602	347	Shipments per Establishment	9,576,895	26,287,788	274
Wages per Production Worker	24,861	24,701	99	Shipments per Employee	195,460	175,702	90
Hours per Production Worker	2,056	2,209	107	Shipments per Production Worker	279,017	219,252	79
Wages per Hour	12.09	11.18	92	Investment per Establishment	321,011	751,665	234
Value Added per Establishment	4,602,255	10,509,991	228	Investment per Employee	6,552	5,024	77
Value Added per Employee	93,930	70,247	75	Investment per Production Worker	9,352	6,269	67

Sources: Same as General Statistics. The 'Average of All Manufacturing' column represents the average of all manufacturing industries reported for the most recent complete year available. The Index shows the relationship between the Average and the Analyzed Industry. For example, 100 means that they are equal; 500 that the Analyzed Industry is five times the average; 50 means that the Analyzed Industry is half the national average. The abbreviation 'na' is used to show that data are 'not available'.

3354 - Aluminum Extruded Products

Manufacturing USA, 5th Edition

LEADING COMPANIES

Number shown: 65   **Total sales ($ mil): 5,403**   **Total employment (000): 34.8**

Company Name	Address				CEO Name	Phone	Co. Type	Sales ($ mil)	Empl. (000)
MAXXAM Inc	PO Box 572887	Houston	TX	77257	Charles E Hurwitz	713-975-7600	P	2,116	12.6
Easco Corp	706 S State St	Girard	OH	44420	Michael M Hagerty	216-545-4311	S	317	2.0
Easco Inc	706 S State St	Girard	OH	44420	Michael M Hagerty	216-545-4311	P	317	2.0
Aluminum Company of America	PO Box 150	Massena	NY	13662	TS Mock	315-764-4732	D	240	1.5
Alumax Extrusions Inc	2700 International	West Chicago	IL	60185	Richard L Brown	708-584-1000	S	200	1.4
William L Bonnell Company Inc	PO Box 428	Newnan	GA	30264	Doug Monk	404-253-2020	S	200	0.8
Wells Aluminum Corp	809 Glen Eagles Ct	Baltimore	MD	21286	Edward H Heiser	410-494-4500	R	180*	1.0
Cressona Aluminum Co	53 Pottsville St	Cressona	PA	17929	James M Stine	717-385-5000	R	150*	0.9
VAW of America Inc	PO Box 667	Ellenville	NY	12428	Albert F Styring	914-647-7510	S	150	1.4
Macklanburg-Duncan Co	PO Box 25188	Oklahoma City	OK	73125	Mike Samis	405-528-4411	R	140	1.2
Barmet Aluminum Corp	PO Box 3740	Akron	OH	44314	Norman Wells	216-753-7701	S	130	0.5
Construction Specialties Inc	49 Meeker Av	Cranford	NJ	07016	Ronald F Dadd	908-272-5200	R	120*	0.8
Hoogovens Aluminium Corp	PO Box 2127	Secaucus	NJ	07096	R Johnson	201-866-7776	S	80	<0.1
Anodizing Inc	7933 NE 21st Av	Portland	OR	97211	dennis Merkel	503-285-0404	R	70	0.7
International Extrusion Corp	1000 Meridian Av	Alhambra	CA	91803	Robert Taylor	818-576-2424	S	46	0.3
Penn Aluminum International	PO Box 490	Murphysboro	IL	62966	Charles Ohl	618-684-2146	S	45	0.4
Alcan Pipe USA	PO Box 309	York	NE	68467	Kenneth Nordlund	402-362-6651	S	42	0.1
AFCO Industries Inc	PO Box 5085	Alexandria	LA	71307	Ray Bordelon	318-448-1651	R	40	0.4
Easton Aluminum Inc	7800 Haskell Av	Van Nuys	CA	91406	Erik Watts	818-782-6445	R	40	0.4
Patrick Metals	PO Box 1330	Mishawaka	IN	46544	David Lung	219-255-9692	D	37	0.3
Universal Molding Co	PO Box 292	Lynwood	CA	90262	S Sossin	310-886-1750	R	35	0.3
Mideast Aluminum Indust	PO Box 98	Mountain Top	PA	18707	D Smith	717-474-5935	D	34	0.3
Wells Aluminum Corp	5575 N Riverview	Kalamazoo	MI	49004	Bob Cecil	616-349-6626	D	34*	0.3
Light Metals Corp	PO Box 902	Wyoming	MI	49509	George T Boylan	616-538-3030	R	33*	0.2
Wellstream Corp	1700 Ceder Av	Panama City	FL	32401	Robert Miller	904-769-9471	R	33*	0.2
Capitol Products Corp	PO Box 106	Kentland	IN	47951	Ken Wampler	219-474-5136	S	31	0.3
Cardinal Aluminum Co	PO Box 19987	Louisville	KY	40259	WE Edwards III	502-969-9302	R	30	0.3
Columbia Pacific	PO Box 1587	City of Industry	CA	91749	Eddie Gaut	818-964-3411	S	30	0.2
Extrusion Painting Inc	5800 Venoy Rd	Garden City	MI	48135	Nicholas Noecker	313-427-8700	R	27	0.1
Anaheim Extrusion Company	PO Box 6380	Anaheim	CA	92816	Cliff Lotzenhiser	714-630-3111	S	25	<0.1
General Extrusions Inc	PO Box 2669	Youngstown	OH	44507	HF Schuler	216-783-0270	R	25	0.3
Moultrie	PO Box 1470	Moultrie	GA	31768	Ed Heiser	912-985-9889	D	25	0.2
Wells Aluminum Corp	PO Box 519	North Liberty	IN	46554	Ron Berton	219-656-8111	D	23	0.2
Halethorpe Extrusions Inc	2000 Halethorpe	Baltimore	MD	21227	Vernon N Ford	410-242-2800	S	22*	0.1
Reliable Products	PO Box 580	Geneva	AL	36340	BR Cross	205-684-3621	D	22	0.3
Cliff Impact	33800 Lakeland	Eastlake	OH	44095	Gary Karlson	216-946-9092	D	20	0.1
Florida Extruders International	2540 Jewett Ln	Sanford	FL	32771	JG Lehman	407-323-3300	R	20	0.2
Mid-States Aluminum Corp	PO Box 548	Fond du Lac	WI	54936	Joseph P Colwin	414-922-7207	R	20	0.2
Futura Industries Corp	PO Box 1506	Clearfield	UT	84016	Susan Johnson	801-773-6282	S	19	0.2
Tower Extrusions Ltd	PO Box 218	Olney	TX	76374	W L McClelland	817-564-5681	R	18	<0.1
United Electric Co	501 Galveston St	Wichita Falls	TX	76301	Maurice McCall	817-767-8333	R	18	0.2
Temroc Metals Inc	4375 Willow Dr	Hamel	MN	55340	P Cienciwa	612-478-6360	R	17	0.2
Central Aluminum Co	2045 Broehm Rd	Columbus	OH	43207	Dale Rashon	614-491-5700	R	15	0.1
Keller Extrusions of Texas	PO Box 709	Woodville	TX	75979	Donald Williams	409-283-3761	D	15*	<0.1
Taber Metals LP	PO Box 1418	Russellville	AR	72811	Robert E Rains	501-968-1021	S	15	<0.1
Precision Extrusions Inc	720 E Green Av	Bensenville	IL	60106	R Ziehm	708-766-0340	R	12	0.1
Whitehall Industries	PO Box 368	Ludington	MI	49431	K Rocco	616-845-5101	R	12	0.2
American Modern Metals	25 Belgrove Dr	Kearny	NJ	07032	IS Rosalsky	201-991-2100	R	11*	0.2
Aham Tor Inc	PO Box 9019	Temecula	CA	92589	William Goodman	909-676-4151	S	9*	<0.1
Metal Impact Corp	5500 Milton Ct	Rosemont	IL	60018	TM Ostrom	708-671-6680	R	9	<0.1
Dant Clayton Corp	PO Box 740008	Louisville	KY	40201	Bruce C Merrick	502-634-3626	R	8*	<0.1
Mapes Industries Inc	2929 Cornhusker	Lincoln	NE	68504	Bill Cintani	402-466-1985	R	8	<0.1
Easco Aluminum Corp	PO Box 177	Winton	NC	27986	Robert Keathley	919-358-5811	D	7	<0.1
Industrial Louvers Inc	511 S 7th St	Delano	MN	55328	James W Sterriker	612-972-2981	R	7	<0.1
Michigan Extruded Aluminum	205 Watts Rd	Jackson	MI	49203	J Jacobs	517-764-5400	R	7*	<0.1
Briteline Extrusion Inc	575 Beech Hill Rd	Summerville	SC	29485	George O Hayes Jr	803-873-4410	R	6*	<0.1
Extrusions Inc	2401 Main St	Fort Scott	KS	66701	JA Ida	316-223-1111	R	6	0.1
Minalex Corp	PO Box 247	Wh House Stat	NJ	08889	James Casey	908-534-4044	R	6	<0.1
Extrusion Technology Inc	80 Trim Way	Randolph	MA	02368	James M Sharpe	617-963-7200	R	5	<0.1
Judson A Smith Co	PO Box 563	Boyertown	PA	19512	Duane Ottolini	215-367-2021	S	5	<0.1
National Extrusion & Mfg Co	PO Box 460	Bellefontaine	OH	43311	Daniel Juhl	513-592-9010	S	5*	<0.1
United States Aluminum	8000 Farrow Rd	Columbia	SC	29203	WJ Arvay	803-754-3100	R	5	<0.1
Aluma Trim Inc	239 Richmond St	Brooklyn	NY	11208	D Yarmeisch	718-647-5700	R	4	<0.1
Bristol Aluminum Co	5514 Emilie Rd	Levittown	PA	19057	Paul Mathias Jr	215-673-0282	R	4*	<0.1
UCO Inc	9225 151st Av NE	Redmond	WA	98052	GL Draper	206-883-6600	R	2	<0.1

Source: Ward's Business Directory of U.S. Private and Public Companies, Volumes 1 and 2, 1996. The company type code used is as follows: P - Public, R - Private, S - Subsidiary, D - Division, J - Joint Venture, A - Affiliate, G - Group. Sales are in millions of dollars, employees are in thousands. An asterisk (*) indicates an estimated sales volume. The symbol < stands for 'less than'. Company names and addresses are truncated, in some cases, to fit into the available space.

MATERIALS CONSUMED

Material	Quantity		Delivered Cost ($ million)
Materials, ingredients, containers, and supplies		(X)	2,135.4
Unalloyed aluminum and aluminum-base alloy ingot, pig, and shot	1,000 s tons	313.0	389.4
Alloyed aluminum and aluminum-base alloy ingot, pig, and shot	1,000 s tons	503.8*	471.6
Aluminum and aluminum-base alloy sheet, plate, foil, and welded tubing	mil lb	0.8	6.0
Aluminum and aluminum-base alloy extruded shapes, including extruded rod, bar, pipe, tube, etc. . .	mil lb	393.0	271.1
All other aluminum and aluminum-base alloy shapes and forms		(X)	307.2
Copper and copper-base alloy wire bar	1,000 s tons	(D)	(D)
Copper and copper-base alloy ingot and ingot bar	1,000 s tons	(D)	(D)
All other copper and copper-base alloy shapes and forms	1,000 s tons	0.5	0.9
Magnesium and magnesium-base alloy shapes and forms	1,000 s tons	3.8	8.8
Nickel and nickel-base alloy shapes and forms	1,000 s tons	(D)	(D)
Zinc and zinc-base alloy shapes and forms	1,000 s tons	(D)	(D)
Tin and tin-base alloy shapes and forms	mil lb	(D)	(D)
Titanium and titanium-base alloy shapes and forms	1,000 s tons	0.7	2.3
All other nonferrous shapes and forms		(X)	5.0
Aluminum and aluminum-base alloy scrap from other establishments	1,000 s tons	57.0	66.9
Aluminum and aluminum-base alloy scrap from all other sources	1,000 s tons	157.4*	165.9
Copper and copper-base alloy scrap (except home scrap)	1,000 s tons	(D)	(D)
Other nonferrous metal scrap (except home scrap)	1,000 s tons	(D)	(D)
All other materials and components, parts, containers, and supplies		(X)	290.3
Materials, ingredients, containers, and supplies, nsk		(X)	135.2

Source: 1992 Economic Census. Explanation of symbols used: (D): Withheld to avoid disclosure of competitive data; na: Not available; (S): Withheld because statistical norms were not met; (X): Not applicable; (Z): Less than half the unit shown; nec: Not elsewhere classified; nsk: Not specified by kind; - : zero; * : 10-19 percent estimated; ** : 20-29 percent estimated.

PRODUCT SHARE DETAILS

Product or Product Class	% Share	Product or Product Class	% Share
Aluminum extruded products	100.00	Extruded aluminum rod, bar, and other extruded shapes, nsk	2.24
Extruded aluminum rod, bar, and other extruded shapes . .	80.13	Extruded and drawn aluminum tube	14.53
Extruded aluminum rod and bar, alloys other than 2000 and 7000 series	16.02	Extruded and drawn aluminum tube, alloys other than 2000 and 7000 series	75.97
Extruded aluminum rod and bar, alloys in 2000 and 7000 series	6.32	Extruded and drawn aluminum tube, alloys in 2000 and 7000 series	23.87
Other extruded aluminum shapes (except tube), alloys other than 2000 and 7000 series	68.22	Extruded and drawn aluminum tube, nsk	0.19
Other extruded aluminum shapes (except tube), alloys in 2000 and 7000 series	7.19	Aluminum extruded products, nsk	5.34

Source: 1992 Economic Census. The values shown are percent of total shipments in an industry. Values of indented subcategories are summed in the main heading. The symbol (D) appears when data are withheld to prevent disclosure of competitive information. The abbreviation nsk stands for 'not specified by kind' and nec for 'not elsewhere classified'.

INPUTS AND OUTPUTS FOR ALUMINUM ROLLING & DRAWING

Economic Sector or Industry Providing Inputs	%	Sector	Economic Sector or Industry Buying Outputs	%	Sector
Primary aluminum	45.4	Manufg.	Metal cans	17.9	Manufg.
Scrap	8.8	Scrap	Metal doors, sash, & trim	7.7	Manufg.
Wholesale trade	6.6	Trade	Aluminum rolling & drawing	6.1	Manufg.
Aluminum rolling & drawing	6.1	Manufg.	Sheet metal work	5.0	Manufg.
Imports	4.7	Foreign	Exports	5.0	Foreign
Motor freight transportation & warehousing	3.2	Util.	Metal foil & leaf	4.4	Manufg.
Gas production & distribution (utilities)	2.6	Util.	Nonferrous wire drawing & insulating	3.5	Manufg.
Special dies & tools & machine tool accessories	2.2	Manufg.	Refrigeration & heating equipment	2.9	Manufg.
Electric services (utilities)	2.2	Util.	Metal stampings, nec	2.1	Manufg.
Petroleum refining	1.5	Manufg.	Aircraft	2.0	Manufg.
Primary nonferrous metals, nec	1.4	Manufg.	Tanks & tank components	1.9	Manufg.
Cyclic crudes and organics	1.3	Manufg.	Truck trailers	1.5	Manufg.
Railroads & related services	1.3	Util.	Electronic components nec	1.4	Manufg.
Power transmission equipment	1.2	Manufg.	Aircraft & missile equipment, nec	1.3	Manufg.
Miscellaneous plastics products	1.0	Manufg.	Fabricated metal products, nec	1.3	Manufg.
Nonferrous wire drawing & insulating	0.7	Manufg.	Lighting fixtures & equipment	1.3	Manufg.
Miscellaneous repair shops	0.7	Services	Miscellaneous metal work	1.3	Manufg.
Metal coating & allied services	0.5	Manufg.	Automotive stampings	1.2	Manufg.
Screw machine and related products	0.5	Manufg.	Prefabricated metal buildings	1.2	Manufg.
Maintenance of nonfarm buildings nec	0.4	Constr.	Radio & TV communication equipment	1.1	Manufg.
Ball & roller bearings	0.4	Manufg.	Crowns & closures	1.0	Manufg.
Fabricated metal products, nec	0.4	Manufg.	Architectural metal work	0.9	Manufg.
Advertising	0.4	Services	Drapery hardware & blinds & shades	0.9	Manufg.
Industrial controls	0.3	Manufg.	Machinery, except electrical, nec	0.9	Manufg.
Metal stampings, nec	0.3	Manufg.	Hardware, nec	0.8	Manufg.
Air transportation	0.3	Util.	Metal household furniture	0.8	Manufg.
Automotive rental & leasing, without drivers	0.3	Services	Motor vehicle parts & accessories	0.8	Manufg.

Continued on next page.

INPUTS AND OUTPUTS FOR ALUMINUM ROLLING & DRAWING - Continued

Economic Sector or Industry Providing Inputs	%	Sector	Economic Sector or Industry Buying Outputs	%	Sector
Chemical preparations, nec	0.2	Manufg.	Motor vehicles & car bodies	0.8	Manufg.
Industrial inorganic chemicals, nec	0.2	Manufg.	Truck & bus bodies	0.8	Manufg.
Paints & allied products	0.2	Manufg.	Electronic computing equipment	0.6	Manufg.
Primary metal products, nec	0.2	Manufg.	Fabricated plate work (boiler shops)	0.6	Manufg.
Rubber & plastics hose & belting	0.2	Manufg.	General industrial machinery, nec	0.6	Manufg.
Communications, except radio & TV	0.2	Util.	Photographic equipment & supplies	0.6	Manufg.
Eating & drinking places	0.2	Trade	Pipe, valves, & pipe fittings	0.6	Manufg.
Banking	0.2	Fin/R.E.	Screw machine and related products	0.6	Manufg.
Automotive repair shops & services	0.2	Services	Signs & advertising displays	0.6	Manufg.
Equipment rental & leasing services	0.2	Services	Sporting & athletic goods, nec	0.6	Manufg.
Machinery, except electrical, nec	0.1	Manufg.	Travel trailers & campers	0.6	Manufg.
Metal heat treating	0.1	Manufg.	Manufacturing industries, nec	0.5	Manufg.
Paperboard containers & boxes	0.1	Manufg.	Metal barrels, drums, & pails	0.5	Manufg.
Primary lead	0.1	Manufg.	Transformers	0.5	Manufg.
Sawmills & planning mills, general	0.1	Manufg.	Aircraft & missile engines & engine parts	0.4	Manufg.
Wood products, nec	0.1	Manufg.	Blowers & fans	0.4	Manufg.
Water transportation	0.1	Util.	Boat building & repairing	0.4	Manufg.
Insurance carriers	0.1	Fin/R.E.	Electric housewares & fans	0.4	Manufg.
Computer & data processing services	0.1	Services	Fabricated structural metal	0.4	Manufg.
Hotels & lodging places	0.1	Services	Mechanical measuring devices	0.4	Manufg.
			Paperboard containers & boxes	0.4	Manufg.
			Printing trades machinery	0.4	Manufg.
			Special industry machinery, nec	0.4	Manufg.
			Switchgear & switchboard apparatus	0.4	Manufg.
			Wiring devices	0.4	Manufg.
			Ammunition, except for small arms, nec	0.3	Manufg.
			Carburetors, pistons, rings, & valves	0.3	Manufg.
			Food products machinery	0.3	Manufg.
			Household refrigerators & freezers	0.3	Manufg.
			Motors & generators	0.3	Manufg.
			Public building furniture	0.3	Manufg.
			Pumps & compressors	0.3	Manufg.
			Ship building & repairing	0.3	Manufg.
			Special dies & tools & machine tool accessories	0.3	Manufg.
			Copper rolling & drawing	0.2	Manufg.
			Electrical equipment & supplies, nec	0.2	Manufg.
			Furniture & fixtures, nec	0.2	Manufg.
			Games, toys, & children's vehicles	0.2	Manufg.
			Heating equipment, except electric	0.2	Manufg.
			Household cooking equipment	0.2	Manufg.
			Industrial controls	0.2	Manufg.
			Instruments to measure electricity	0.2	Manufg.
			Metal office furniture	0.2	Manufg.
			Metal partitions & fixtures	0.2	Manufg.
			Mobile homes	0.2	Manufg.
			Needles, pins, & fasteners	0.2	Manufg.
			Nonferrous forgings	0.2	Manufg.
			Paper coating & glazing	0.2	Manufg.
			Service industry machines, nec	0.2	Manufg.
			Small arms	0.2	Manufg.
			Surgical appliances & supplies	0.2	Manufg.
			Textile machinery	0.2	Manufg.
			Transportation equipment, nec	0.2	Manufg.
			Typewriters & office machines, nec	0.2	Manufg.
			Federal Government purchases, national defense	0.2	Fed Govt
			Maintenance of nonfarm buildings nec	0.1	Constr.
			Construction machinery & equipment	0.1	Manufg.
			Conveyors & conveying equipment	0.1	Manufg.
			Elevators & moving stairways	0.1	Manufg.
			Engineering & scientific instruments	0.1	Manufg.
			Household appliances, nec	0.1	Manufg.
			Industrial furnaces & ovens	0.1	Manufg.
			Miscellaneous fabricated wire products	0.1	Manufg.
			Telephone & telegraph apparatus	0.1	Manufg.

Source: Benchmark Input-Output Accounts for the U.S. Economy, 1982, U.S. Department of Commerce, Washington, D.C., July 1991. Data, as reported in the source, are organized by the 1977 SIC structure in use in 1982 but have been matched, as closely as is possible, to the 1987 SIC structure used in this book.

OCCUPATIONS EMPLOYED BY SIC 335 - NONFERROUS ROLLING AND DRAWING

Occupation	% of Total 1994	Change to 2005	Occupation	% of Total 1994	Change to 2005
Machine tool cutting & forming etc. nec	11.0	-12.4	Machinists	1.6	-20.3
Assemblers, fabricators, & hand workers nec	5.9	-20.4	Machine tool cutting operators, metal & plastic	1.4	-33.7
Blue collar worker supervisors	5.0	-22.9	Extruding & forming machine workers	1.3	-12.4
Machine forming operators, metal & plastic	5.0	7.5	Coating, painting, & spraying machine workers	1.3	-20.4
Metal & plastic machine workers nec	3.9	-15.3	General managers & top executives	1.2	-24.4
Inspectors, testers, & graders, precision	3.7	-20.3	Combination machine tool operators	1.2	-12.4
Helpers, laborers, & material movers nec	3.6	-20.4	Traffic, shipping, & receiving clerks	1.1	-23.3
Machine operators nec	3.1	-29.8	Industrial production managers	1.1	-20.4
Industrial machinery mechanics	2.9	-12.4	Packaging & filling machine operators	1.1	-20.4
Industrial truck & tractor operators	2.8	-20.3	Secretaries, ex legal & medical	1.1	-27.5
Maintenance repairers, general utility	2.1	-28.3	Machine feeders & offbearers	1.1	-28.3
Sales & related workers nec	1.9	-20.3	Production, planning, & expediting clerks	1.0	-20.3
Hand packers & packagers	1.7	-31.7	Metal molding machine workers	1.0	19.5
Electricians	1.7	-25.3			

Source: Industry-Occupation Matrix, Bureau of Labor Statistics. These data relate to one or more 3-digit SIC industry groups rather than to a single 4-digit SIC. The change reported for each occupation to the year 2005 is a percent of growth or decline as estimated by the Bureau of Labor Statistics. The abbreviation nec stands for 'not elsewhere classified'.

LOCATION BY STATE AND REGIONAL CONCENTRATION

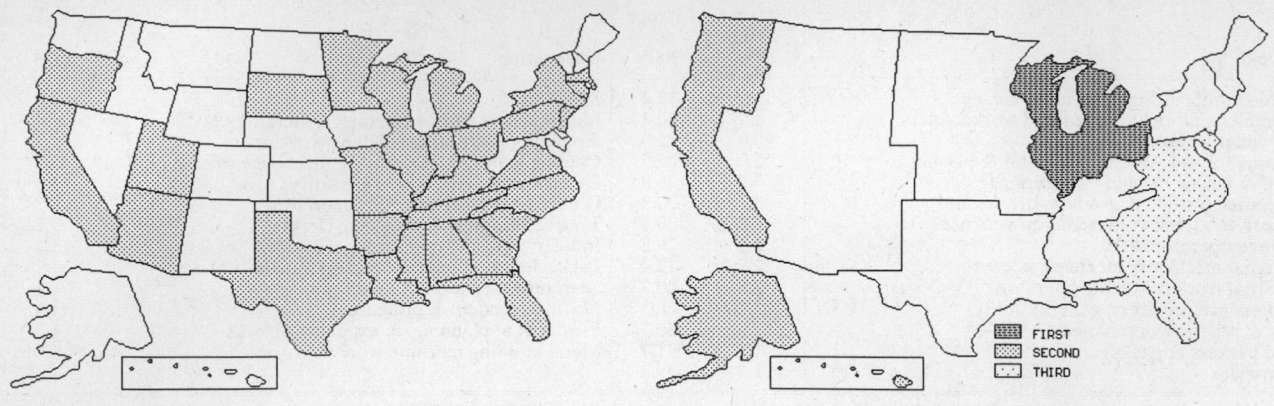

FIRST
SECOND
THIRD

INDUSTRY DATA BY STATE

State	Establish-ments	Shipments			Employment				Cost as % of Shipments	Investment per Employee ($)
		Total ($ mil)	% of U.S.	Per Establ.	Total Number	% of U.S.	Per Establ.	Wages ($/hour)		
Indiana	15	409.1	11.0	27.3	2,400	9.4	160	12.97	56.5	2,375
California	30	372.1	10.0	12.4	2,300	9.0	77	11.97	67.3	2,870
Ohio	21	298.0	8.0	14.2	2,000	7.8	95	9.61	61.2	3,300
New York	6	258.3	6.9	43.1	1,500	5.9	250	13.12	60.3	-
Pennsylvania	7	254.1	6.8	36.3	1,400	5.5	200	12.65	63.2	-
Michigan	16	198.6	5.3	12.4	1,600	6.3	100	12.75	53.3	5,563
Texas	11	188.5	5.0	17.1	1,100	4.3	100	10.25	62.5	1,364
Tennessee	5	179.5	4.8	35.9	1,000	3.9	200	10.06	64.3	2,100
Florida	6	161.1	4.3	26.9	1,100	4.3	183	8.55	70.1	11,273
North Carolina	7	145.7	3.9	20.8	900	3.5	129	7.53	93.1	4,111
Illinois	6	114.8	3.1	19.1	1,000	3.9	167	10.44	62.2	5,200
Kentucky	2	71.5	1.9	35.8	500	2.0	250	9.00	60.4	2,400
Alabama	3	51.8	1.4	17.3	400	1.6	133	10.00	63.5	-
Massachusetts	3	15.6	0.4	5.2	100	0.4	33	13.00	71.8	-
Georgia	9	(D)	-	-	1,750 *	6.8	194	-	-	-
Arizona	6	(D)	-	-	750 *	2.9	125	-	-	-
New Jersey	5	(D)	-	-	175 *	0.7	35	-	-	1,714
Mississippi	4	(D)	-	-	750 *	2.9	188	-	-	3,467
Missouri	4	(D)	-	-	750 *	2.9	188	-	-	-
Arkansas	3	(D)	-	-	175 *	0.7	58	-	-	-
Minnesota	3	(D)	-	-	375 *	1.5	125	-	-	-
Oregon	3	(D)	-	-	375 *	1.5	125	-	-	-
South Carolina	3	(D)	-	-	375 *	1.5	125	-	-	-
Virginia	3	(D)	-	-	750 *	2.9	250	-	-	-
Wisconsin	2	(D)	-	-	375 *	1.5	188	-	-	-
Connecticut	1	(D)	-	-	175 *	0.7	175	-	-	-
Louisiana	1	(D)	-	-	375 *	1.5	375	-	-	-
South Dakota	1	(D)	-	-	375 *	1.5	375	-	-	-
Utah	1	(D)	-	-	175 *	0.7	175	-	-	-

Source: 1992 *Economic Census*. The states are in descending order of shipments or establishments (if shipment data are missing for the majority). The symbol (D) appears when data are withheld to prevent disclosure of competitive information. States marked with (D) are sorted by number of establishments. A dash (-) indicates that the data element cannot be calculated; * indicates the midpoint of a range.

3355 - ALUMINUM ROLLING & DRAWING, NEC

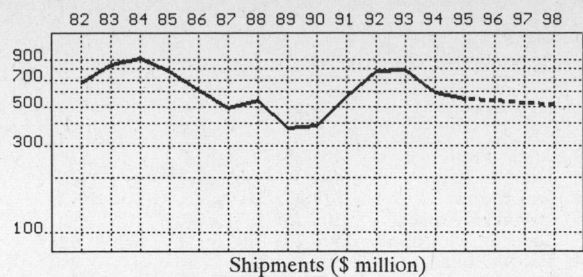

Shipments ($ million)

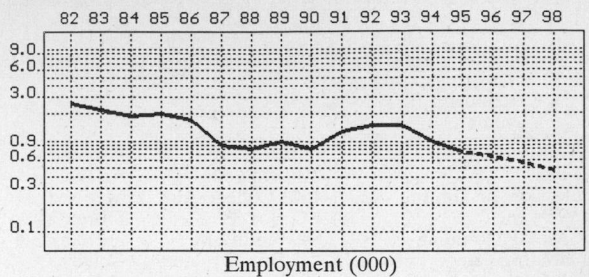

Employment (000)

GENERAL STATISTICS

| Year | Companies | Establishments | | Employment | | | Compensation | | Production ($ million) | | | |
		Total	with 20 or more employees	Total (000)	Production Workers (000)	Hours (Mil)	Payroll ($ mil)	Wages ($/hr)	Cost of Materials	Value Added by Manufacture	Value of Shipments	Capital Invest.
1982	24	27	15	2.6	1.9	3.5	65.0	12.63	535.3	30.7	670.8	5.5
1983		28	16	2.2	1.6	3.2	57.9	12.72	725.9	139.3	844.3	9.0
1984		29	17	1.9	1.5	2.9	56.0	13.45	702.0	166.7	905.2	13.8
1985		29	18	2.0	1.5	3.1	55.0	12.26	599.3	168.8	783.4	13.2
1986		29	16	1.7	1.3	2.7	51.5	13.11	481.6	122.7	613.7	7.5
1987	29	29	14	0.9	0.6	1.4	24.2	12.14	392.6	109.3	482.3	2.8
1988		33	16	0.8	0.6	1.4	23.5	12.14	381.8	162.7	541.1	
1989		29	13	1.0	0.6	1.3	22.9	12.00	326.5	50.4	374.7	
1990		29	12	0.8	0.6	1.2	22.6	12.17	333.1	54.6	388.3	1.9
1991		33	14	1.3	0.9	2.2	40.4	9.86	479.7	90.1	569.5	3.4
1992	27	29	11	1.5	1.1	2.7	51.0	11.30	651.1	128.5	778.0	14.7
1993		34	15	1.5	1.1	2.8	50.1	10.64	597.0	191.4	791.4	8.2
1994		33P	12P	1.0	0.8	1.5	31.6	15.67	417.5	178.7	590.7	6.4
1995		33P	12P	0.8P	0.6P	1.5P	28.5P	11.93P	386.4P	165.4P	546.6P	
1996		33P	12P	0.7P	0.5P	1.4P	26.5P	11.88P	376.8P	161.3P	533.2P	
1997		34P	11P	0.6P	0.4P	1.3P	24.5P	11.82P	367.3P	157.2P	519.7P	
1998		34P	11P	0.5P	0.3P	1.2P	22.5P	11.77P	357.8P	153.1P	506.2P	

Sources: 1982, 1987, 1992 *Economic Census*; *Annual Survey of Manufactures*, 83-86, 88-91, 93-94. Establishment counts for non-Census years are from *County Business Patterns*; establishment values for 83-84 are extrapolations. 'P's show projections by the editors. Industries reclassified in 87 will not have data for prior years.

INDICES OF CHANGE

| Year | Companies | Establishments | | Employment | | | Compensation | | Production ($ million) | | | |
		Total	with 20 or more employees	Total (000)	Production Workers (000)	Hours (Mil)	Payroll ($ mil)	Wages ($/hr)	Cost of Materials	Value Added by Manufacture	Value of Shipments	Capital Invest.
1982	89	93	136	173	173	130	127	112	82	24	86	37
1983		97	145	147	145	119	114	113	111	108	109	61
1984		100	155	127	136	107	110	119	108	130	116	94
1985		100	164	133	136	115	108	108	92	131	101	90
1986		100	145	113	118	100	101	116	74	95	79	51
1987	107	100	127	60	55	52	47	107	60	85	62	19
1988		114	145	53	55	52	46	107	59	127	70	
1989		100	118	67	55	48	45	106	50	39	48	
1990		100	109	53	55	44	44	108	51	42	50	13
1991		114	127	87	82	81	79	87	74	70	73	23
1992	100	100	100	100	100	100	100	100	100	100	100	100
1993		117	136	100	100	104	98	94	92	149	102	56
1994		112P	113P	67	73	56	62	139	64	139	76	44
1995		114P	110P	51P	50P	56P	56P	106P	59P	129P	70P	
1996		115P	106P	44P	43P	52P	52P	105P	58P	126P	69P	
1997		116P	103P	37P	36P	47P	48P	105P	56P	122P	67P	
1998		118P	100P	31P	29P	43P	44P	104P	55P	119P	65P	

Sources: Same as General Statistics. Values reflect change from the base year, 1992. Values above 100 mean greater than 92, values below 100 mean less than 92, and a value of 100 in the 82-91 or 93-98 period means same as 92. 'P's mark projections by the editors.

SELECTED RATIOS

For 1994	Avg. of All Manufact.	Analyzed Industry	Index	For 1994	Avg. of All Manufact.	Analyzed Industry	Index
Employees per Establishment	49	31	63	Value Added per Production Worker	134,084	223,375	167
Payroll per Establishment	1,500,273	971,855	65	Cost per Establishment	5,045,178	12,840,168	255
Payroll per Employee	30,620	31,600	103	Cost per Employee	102,970	417,500	405
Production Workers per Establishment	34	25	72	Cost per Production Worker	146,988	521,875	355
Wages per Establishment	853,319	722,894	85	Shipments per Establishment	9,576,895	18,166,915	190
Wages per Production Worker	24,861	29,381	118	Shipments per Employee	195,460	590,700	302
Hours per Production Worker	2,056	1,875	91	Shipments per Production Worker	279,017	738,375	265
Wages per Hour	12.09	15.67	130	Investment per Establishment	321,011	196,831	61
Value Added per Establishment	4,602,255	5,495,899	119	Investment per Employee	6,552	6,400	98
Value Added per Employee	93,930	178,700	190	Investment per Production Worker	9,352	8,000	86

Sources: Same as General Statistics. The 'Average of All Manufacturing' column represents the average of all manufacturing industries reported for the most recent complete year available. The Index shows the relationship between the Average and the Analyzed Industry. For example, 100 means that they are equal; 500 that the Analyzed Industry is five times the average; 50 means that the Analyzed Industry is half the national average. The abbreviation 'na' is used to show that data are 'not available'.

LEADING COMPANIES   Number shown: **13**   Total sales ($ mil): **567**   Total employment (000): **3.1**

Company Name	Address				CEO Name	Phone	Co. Type	Sales ($ mil)	Empl. (000)
Intalco Aluminum Corp	PO Box 937	Ferndale	WA	98248	James Frederick	206-384-7061	S	210*	1.1
Alcan Cable	3 Ravinia Dr	Atlanta	GA	30346	Douglas Bland	404-394-9886	D	120	0.8
Spectrulite Consortium Inc	1001 College Av	Madison	IL	62060	WA Barnes Sr	618-452-5190	R	72	0.4
Nichols Wire Inc	PO Box 38	Florence	AL	35631	Donald K Larsen	205-764-4271	R	30	0.2
Hygrade Metal Moulding	540 Smith St	Farmingdale	NY	11735	Vincent A Pagano	516-293-8797	R	25	<0.1
Colmac Holding Co	PO Box 72	Colville	WA	99114	RK McMillan	509-684-4506	R	24	0.1
Lakeside Metals Inc	7000 Adams St	Willowbrook	IL	60521	Ira Nadler	708-850-3800	R	20	<0.1
Roll Forming Corp	PO Box 369	Shelbyville	KY	40065	BW Brooks Jr	502-633-4435	R	14	0.1
Shaped Wire Inc	900 Douglas Rd	Batavia	IL	60510	RB Troendly	708-406-0800	R	14	<0.1
United States Alumoweld	115 USAC Dr	Duncan	SC	29334	JJ Underwood	803-848-1901	S	12*	<0.1
Erickson Metals Corp	25 Knotter Dr	Cheshire	CT	06410	RH Erickson	203-272-2918	R	10	<0.1
Beneke Wire Co	5559 National Tpk	Louisville	KY	40214	J David Beneke	502-367-6434	R	9*	<0.1
Airflo Industries Inc	130 W Victoria St	Gardena	CA	90248	Sanford Steinberg	310-217-9900	R	7	<0.1

Source: *Ward's Business Directory of U.S. Private and Public Companies*, Volumes 1 and 2, 1996. The company type code used is as follows: P - Public, R - Private, S - Subsidiary, D - Division, J - Joint Venture, A - Affiliate, G - Group. Sales are in millions of dollars, employees are in thousands. An asterisk (*) indicates an estimated sales volume. The symbol < stands for 'less than'. Company names and addresses are truncated, in some cases, to fit into the available space.

MATERIALS CONSUMED

Material	Quantity	Delivered Cost ($ million)
Materials, ingredients, containers, and supplies	(X)	628.0
Unalloyed aluminum and aluminum-base alloy ingot, pig, and shot	1,000 s tons	(D)
Alloyed aluminum and aluminum-base alloy ingot, pig, and shot	1,000 s tons 7.5	9.4
Aluminum and aluminum-base alloy sheet, plate, foil, and welded tubing	mil lb	(D)
Aluminum and aluminum-base alloy extruded shapes, including extruded rod, bar, pipe, tube, etc. . .	mil lb	(D)
All other aluminum and aluminum-base alloy shapes and forms	(X)	(D)
Copper and copper-base alloy cathodes	1,000 s tons	(D)
All other copper and copper-base alloy shapes and forms	1,000 s tons	(D)
Magnesium and magnesium-base alloy shapes and forms	1,000 s tons 1.2	3.2
Nickel and nickel-base alloy shapes and forms	1,000 s tons	(D)
Zinc and zinc-base alloy shapes and forms	1,000 s tons	(D)
Titanium and titanium-base alloy shapes and forms	1,000 s tons	(D)
All other nonferrous shapes and forms	(X)	7.7
Aluminum and aluminum-base alloy scrap from other establishments	1,000 s tons 23.0	23.1
Aluminum and aluminum-base alloy scrap from all other sources	1,000 s tons	(D)
Copper and copper-base alloy scrap (except home scrap)	1,000 s tons	(D)
Other nonferrous metal scrap (except home scrap)	1,000 s tons	(D)
All other materials and components, parts, containers, and supplies	(X)	(D)
Materials, ingredients, containers, and supplies, nsk	(X)	66.0

Source: 1992 *Economic Census*. Explanation of symbols used: (D): Withheld to avoid disclosure of competitive data; na: Not available; (S): Withheld because statistical norms were not met; (X): Not applicable; (Z): Less than half the unit shown; nec: Not elsewhere classified; nsk: Not specified by kind; - : zero; * : 10-19 percent estimated; ** : 20-29 percent estimated.

PRODUCT SHARE DETAILS

Product or Product Class	% Share	Product or Product Class	% Share
Aluminum rolling and drawing, nec	100.00	Rolled aluminum rod and bar, except continuous cast . . .	12.09
Aluminum and aluminum-base alloy wire and cable, except covered or insulated (including ASCR), produced in aluminum rolling mills	(D)	Rolled continuous cast aluminum rod and bar . . . . . Aluminum ingot (excluding billet), produced in aluminum rolling mills.	(D) 38.33
Aluminum and aluminum-base alloy wire and cable, except covered or insulated (including ACSR), produced in aluminum rolling mills	(D)	Aluminum extrusion ingot (billet), produced in aluminum rolling mills.	14.14
Rolled aluminum rod, bar (including continuous cast) . . .	(D)	Aluminum rolling and drawing, nec, nsk	4.93

Source: 1992 *Economic Census*. The values shown are percent of total shipments in an industry. Values of indented subcategories are summed in the main heading. The symbol (D) appears when data are withheld to prevent disclosure of competitive information. The abbreviation nsk stands for 'not specified by kind' and nec for 'not elsewhere classified'.

INPUTS AND OUTPUTS FOR ALUMINUM ROLLING & DRAWING

Economic Sector or Industry Providing Inputs	%	Sector	Economic Sector or Industry Buying Outputs	%	Sector
Primary aluminum	45.4	Manufg.	Metal cans	17.9	Manufg.
Scrap	8.8	Scrap	Metal doors, sash, & trim	7.7	Manufg.
Wholesale trade	6.6	Trade	Aluminum rolling & drawing	6.1	Manufg.
Aluminum rolling & drawing	6.1	Manufg.	Sheet metal work	5.0	Manufg.
Imports	4.7	Foreign	Exports	5.0	Foreign
Motor freight transportation & warehousing	3.2	Util.	Metal foil & leaf	4.4	Manufg.
Gas production & distribution (utilities)	2.6	Util.	Nonferrous wire drawing & insulating	3.5	Manufg.
Special dies & tools & machine tool accessories	2.2	Manufg.	Refrigeration & heating equipment	2.9	Manufg.
Electric services (utilities)	2.2	Util.	Metal stampings, nec	2.1	Manufg.
Petroleum refining	1.5	Manufg.	Aircraft	2.0	Manufg.
Primary nonferrous metals, nec	1.4	Manufg.	Tanks & tank components	1.9	Manufg.
Cyclic crudes and organics	1.3	Manufg.	Truck trailers	1.5	Manufg.
Railroads & related services	1.3	Util.	Electronic components nec	1.4	Manufg.
Power transmission equipment	1.2	Manufg.	Aircraft & missile equipment, nec	1.3	Manufg.
Miscellaneous plastics products	1.0	Manufg.	Fabricated metal products, nec	1.3	Manufg.
Nonferrous wire drawing & insulating	0.7	Manufg.	Lighting fixtures & equipment	1.3	Manufg.
Miscellaneous repair shops	0.7	Services	Miscellaneous metal work	1.3	Manufg.
Metal coating & allied services	0.5	Manufg.	Automotive stampings	1.2	Manufg.
Screw machine and related products	0.5	Manufg.	Prefabricated metal buildings	1.2	Manufg.
Maintenance of nonfarm buildings nec	0.4	Constr.	Radio & TV communication equipment	1.1	Manufg.
Ball & roller bearings	0.4	Manufg.	Crowns & closures	1.0	Manufg.
Fabricated metal products, nec	0.4	Manufg.	Architectural metal work	0.9	Manufg.
Advertising	0.4	Services	Drapery hardware & blinds & shades	0.9	Manufg.
Industrial controls	0.3	Manufg.	Machinery, except electrical, nec	0.9	Manufg.
Metal stampings, nec	0.3	Manufg.	Hardware, nec	0.8	Manufg.
Air transportation	0.3	Util.	Metal household furniture	0.8	Manufg.
Automotive rental & leasing, without drivers	0.3	Services	Motor vehicle parts & accessories	0.8	Manufg.
Chemical preparations, nec	0.2	Manufg.	Motor vehicles & car bodies	0.8	Manufg.
Industrial inorganic chemicals, nec	0.2	Manufg.	Truck & bus bodies	0.8	Manufg.
Paints & allied products	0.2	Manufg.	Electronic computing equipment	0.6	Manufg.
Primary metal products, nec	0.2	Manufg.	Fabricated plate work (boiler shops)	0.6	Manufg.
Rubber & plastics hose & belting	0.2	Manufg.	General industrial machinery, nec	0.6	Manufg.
Communications, except radio & TV	0.2	Util.	Photographic equipment & supplies	0.6	Manufg.
Eating & drinking places	0.2	Trade	Pipe, valves, & pipe fittings	0.6	Manufg.
Banking	0.2	Fin/R.E.	Screw machine and related products	0.6	Manufg.
Automotive repair shops & services	0.2	Services	Signs & advertising displays	0.6	Manufg.
Equipment rental & leasing services	0.2	Services	Sporting & athletic goods, nec	0.6	Manufg.
Machinery, except electrical, nec	0.1	Manufg.	Travel trailers & campers	0.6	Manufg.
Metal heat treating	0.1	Manufg.	Manufacturing industries, nec	0.5	Manufg.
Paperboard containers & boxes	0.1	Manufg.	Metal barrels, drums, & pails	0.5	Manufg.
Primary lead	0.1	Manufg.	Transformers	0.5	Manufg.
Sawmills & planning mills, general	0.1	Manufg.	Aircraft & missile engines & engine parts	0.4	Manufg.
Wood products, nec	0.1	Manufg.	Blowers & fans	0.4	Manufg.
Water transportation	0.1	Util.	Boat building & repairing	0.4	Manufg.
Insurance carriers	0.1	Fin/R.E.	Electric housewares & fans	0.4	Manufg.
Computer & data processing services	0.1	Services	Fabricated structural metal	0.4	Manufg.
Hotels & lodging places	0.1	Services	Mechanical measuring devices	0.4	Manufg.
			Paperboard containers & boxes	0.4	Manufg.
			Printing trades machinery	0.4	Manufg.
			Special industry machinery, nec	0.4	Manufg.
			Switchgear & switchboard apparatus	0.4	Manufg.
			Wiring devices	0.4	Manufg.
			Ammunition, except for small arms, nec	0.3	Manufg.
			Carburetors, pistons, rings, & valves	0.3	Manufg.
			Food products machinery	0.3	Manufg.
			Household refrigerators & freezers	0.3	Manufg.
			Motors & generators	0.3	Manufg.
			Public building furniture	0.3	Manufg.
			Pumps & compressors	0.3	Manufg.
			Ship building & repairing	0.3	Manufg.
			Special dies & tools & machine tool accessories	0.3	Manufg.
			Copper rolling & drawing	0.2	Manufg.
			Electrical equipment & supplies, nec	0.2	Manufg.
			Furniture & fixtures, nec	0.2	Manufg.
			Games, toys, & children's vehicles	0.2	Manufg.
			Heating equipment, except electric	0.2	Manufg.
			Household cooking equipment	0.2	Manufg.
			Industrial controls	0.2	Manufg.
			Instruments to measure electricity	0.2	Manufg.
			Metal office furniture	0.2	Manufg.
			Metal partitions & fixtures	0.2	Manufg.
			Mobile homes	0.2	Manufg.
			Needles, pins, & fasteners	0.2	Manufg.
			Nonferrous forgings	0.2	Manufg.
			Paper coating & glazing	0.2	Manufg.
			Service industry machines, nec	0.2	Manufg.
			Small arms	0.2	Manufg.
			Surgical appliances & supplies	0.2	Manufg.

Continued on next page.

INPUTS AND OUTPUTS FOR ALUMINUM ROLLING & DRAWING - Continued

Economic Sector or Industry Providing Inputs	%	Sector	Economic Sector or Industry Buying Outputs	%	Sector
			Textile machinery	0.2	Manufg.
			Transportation equipment, nec	0.2	Manufg.
			Typewriters & office machines, nec	0.2	Manufg.
			Federal Government purchases, national defense	0.2	Fed Govt
			Maintenance of nonfarm buildings nec	0.1	Constr.
			Construction machinery & equipment	0.1	Manufg.
			Conveyors & conveying equipment	0.1	Manufg.
			Elevators & moving stairways	0.1	Manufg.
			Engineering & scientific instruments	0.1	Manufg.
			Household appliances, nec	0.1	Manufg.
			Industrial furnaces & ovens	0.1	Manufg.
			Miscellaneous fabricated wire products	0.1	Manufg.
			Telephone & telegraph apparatus	0.1	Manufg.

Source: *Benchmark Input-Output Accounts for the U.S. Economy, 1982*, U.S. Department of Commerce, Washington, D.C., July 1991. Data, as reported in the source, are organized by the 1977 SIC structure in use in 1982 but have been matched, as closely as is possible, to the 1987 SIC structure used in this book.

OCCUPATIONS EMPLOYED BY SIC 335 - NONFERROUS ROLLING AND DRAWING

Occupation	% of Total 1994	Change to 2005	Occupation	% of Total 1994	Change to 2005
Machine tool cutting & forming etc. nec	11.0	-12.4	Machinists	1.6	-20.3
Assemblers, fabricators, & hand workers nec	5.9	-20.4	Machine tool cutting operators, metal & plastic	1.4	-33.7
Blue collar worker supervisors	5.0	-22.9	Extruding & forming machine workers	1.3	-12.4
Machine forming operators, metal & plastic	5.0	7.5	Coating, painting, & spraying machine workers	1.3	-20.4
Metal & plastic machine workers nec	3.9	-15.3	General managers & top executives	1.2	-24.4
Inspectors, testers, & graders, precision	3.7	-20.3	Combination machine tool operators	1.2	-12.4
Helpers, laborers, & material movers nec	3.6	-20.4	Traffic, shipping, & receiving clerks	1.1	-23.3
Machine operators nec	3.1	-29.8	Industrial production managers	1.1	-20.4
Industrial machinery mechanics	2.9	-12.4	Packaging & filling machine operators	1.1	-20.4
Industrial truck & tractor operators	2.8	-20.3	Secretaries, ex legal & medical	1.1	-27.5
Maintenance repairers, general utility	2.1	-28.3	Machine feeders & offbearers	1.1	-28.3
Sales & related workers nec	1.9	-20.3	Production, planning, & expediting clerks	1.0	-20.3
Hand packers & packagers	1.7	-31.7	Metal molding machine workers	1.0	19.5
Electricians	1.7	-25.3			

Source: *Industry-Occupation Matrix*, Bureau of Labor Statistics. These data relate to one or more 3-digit SIC industry groups rather than to a single 4-digit SIC. The change reported for each occupation to the year 2005 is a percent of growth or decline as estimated by the Bureau of Labor Statistics. The abbreviation nec stands for 'not elsewhere classified'.

LOCATION BY STATE AND REGIONAL CONCENTRATION

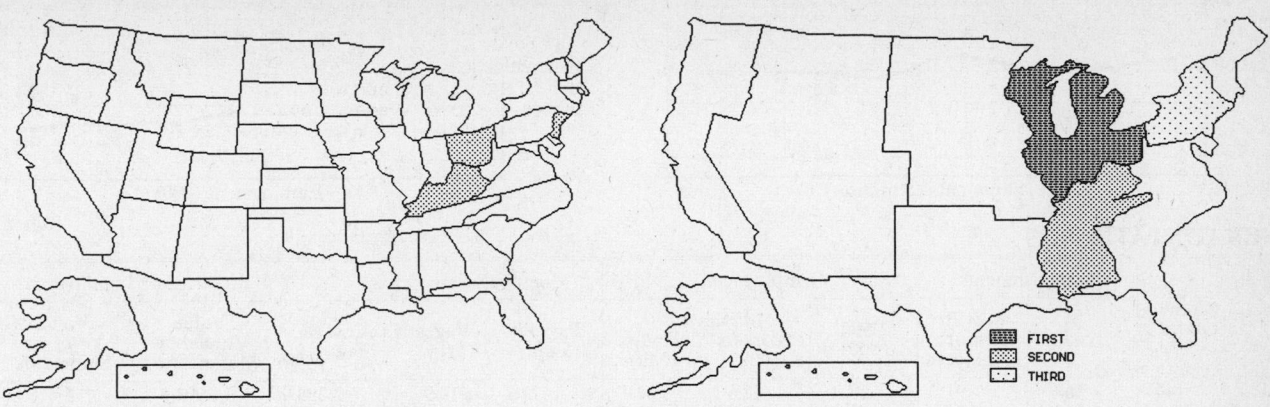

FIRST
SECOND
THIRD

INDUSTRY DATA BY STATE

| State | Establish-ments | Shipments | | | Employment | | | | Cost as % of Shipments | Investment per Employee ($) |
		Total ($ mil)	% of U.S.	Per Establ.	Total Number	% of U.S.	Per Establ.	Wages ($/hour)		
Kentucky	2	(D)	-	-	175 *	11.7	88	-	-	-
Ohio	2	(D)	-	-	375 *	25.0	188	-	-	-
New Jersey	1	(D)	-	-	750 *	50.0	750	-	-	-

Source: 1992 *Economic Census*. The states are in descending order of shipments or establishments (if shipment data are missing for the majority). The symbol (D) appears when data are withheld to prevent disclosure of competitive information. States marked with (D) are sorted by number of establishments. A dash (-) indicates that the data element cannot be calculated; * indicates the midpoint of a range.

3356 - NONFERROUS ROLLING & DRAWING, NEC

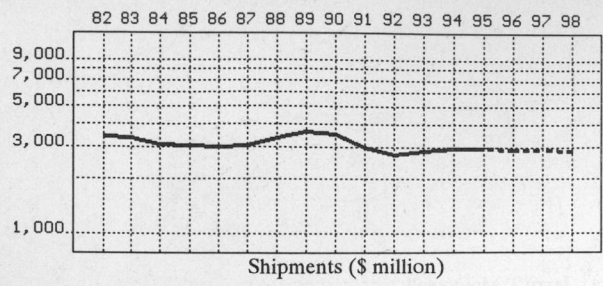

Shipments ($ million)

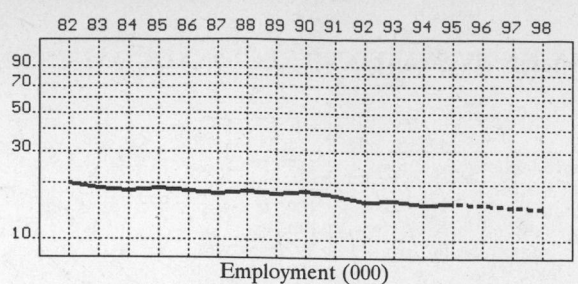

Employment (000)

GENERAL STATISTICS

Year	Com- panies	Establishments		Employment			Compensation		Production ($ million)			
		Total	with 20 or more employees	Total (000)	Production Workers (000)	Hours (Mil)	Payroll ($ mil)	Wages ($/hr)	Cost of Materials	Value Added by Manufacture	Value of Shipments	Capital Invest.
1982	144	169	91	20.0	12.6	24.0	472.2	10.99	2,293.9	993.3	3,418.3	134.4
1983		164	91	18.9	12.0	22.7	466.3	11.46	2,299.1	1,049.8	3,323.8	99.3
1984		159	91	18.2	11.4	23.2	474.4	11.62	2,061.0	1,009.1	3,103.7	91.1
1985		154	92	18.9	11.8	23.5	508.2	12.13	1,920.9	1,086.4	3,034.4	91.7
1986		149	86	18.3	11.9	24.0	523.0	12.45	1,901.5	1,137.8	3,008.2	74.7
1987	140	172	93	17.9	11.8	25.2	527.7	12.24	1,876.7	1,187.9	3,049.4	96.2
1988		170	94	18.4	12.2	25.6	559.0	12.93	2,029.6	1,378.3	3,328.5	104.4
1989		157	93	17.8	12.6	26.4	589.2	13.73	2,138.5	1,541.9	3,635.0	100.5
1990		163	95	18.1	12.4	25.6	612.1	14.57	1,967.8	1,502.5	3,481.2	106.0
1991		162	94	17.4	11.8	23.8	579.7	14.73	1,709.9	1,188.1	3,023.6	82.4
1992	159	182	92	16.0	10.7	22.2	545.3	14.96	1,466.4	1,193.8	2,713.3	91.8
1993		179	100	16.4	11.0	22.7	571.1	15.62	1,499.4	1,340.6	2,842.4	91.5
1994		173P	96P	15.5	10.3	21.8	549.0	15.48	1,522.9	1,449.4	2,941.9	62.5
1995		174P	97P	15.8P	11.0P	23.4P	603.0P	16.17P	1,522.7P	1,449.2P	2,941.5P	77.9P
1996		176P	98P	15.5P	10.9P	23.4P	612.4P	16.58P	1,507.6P	1,434.8P	2,912.2P	75.5P
1997		177P	98P	15.2P	10.8P	23.3P	621.9P	16.99P	1,492.4P	1,420.4P	2,883.0P	73.2P
1998		178P	99P	14.9P	10.6P	23.2P	631.4P	17.40P	1,477.2P	1,405.9P	2,853.7P	70.8P

Sources: 1982, 1987, 1992 *Economic Census*; *Annual Survey of Manufactures*, 83-86, 88-91, 93-94. Establishment counts for non-Census years are from *County Business Patterns*; establishment values for 83-84 are extrapolations. 'P's show projections by the editors. Industries reclassified in 87 will not have data for prior years.

INDICES OF CHANGE

Year	Com- panies	Establishments		Employment			Compensation		Production ($ million)			
		Total	with 20 or more employees	Total (000)	Production Workers (000)	Hours (Mil)	Payroll ($ mil)	Wages ($/hr)	Cost of Materials	Value Added by Manufacture	Value of Shipments	Capital Invest.
1982	91	93	99	125	118	108	87	73	156	83	126	146
1983		90	99	118	112	102	86	77	157	88	123	108
1984		87	99	114	107	105	87	78	141	85	114	99
1985		85	100	118	110	106	93	81	131	91	112	100
1986		82	93	114	111	108	96	83	130	95	111	81
1987	88	95	101	112	110	114	97	82	128	100	112	105
1988		93	102	115	114	115	103	86	138	115	123	114
1989		86	101	111	118	119	108	92	146	129	134	109
1990		90	103	113	116	115	112	97	134	126	128	115
1991		89	102	109	110	107	106	98	117	100	111	90
1992	100	100	100	100	100	100	100	100	100	100	100	100
1993		98	109	102	103	102	105	104	102	112	105	100
1994		95P	105P	97	96	98	101	103	104	121	108	68
1995		96P	105P	99P	103P	106P	111P	108P	104P	121P	108P	85P
1996		97P	106P	97P	101P	105P	112P	111P	103P	120P	107P	82P
1997		97P	107P	95P	100P	105P	114P	114P	102P	119P	106P	80P
1998		98P	107P	93P	99P	105P	116P	116P	101P	118P	105P	77P

Sources: Same as General Statistics. Values reflect change from the base year, 1992. Values above 100 mean greater than 92, values below 100 mean less than 92, and a value of 100 in the 82-91 or 93-98 period means same as 92. 'P's mark projections by the editors.

SELECTED RATIOS

For 1994	Avg. of All Manufact.	Analyzed Industry	Index	For 1994	Avg. of All Manufact.	Analyzed Industry	Index
Employees per Establishment	49	90	183	Value Added per Production Worker	134,084	140,718	105
Payroll per Establishment	1,500,273	3,170,079	211	Cost per Establishment	5,045,178	8,793,648	174
Payroll per Employee	30,620	35,419	116	Cost per Employee	102,970	98,252	95
Production Workers per Establishment	34	59	173	Cost per Production Worker	146,988	147,854	101
Wages per Establishment	853,319	1,948,611	228	Shipments per Establishment	9,576,895	16,987,349	177
Wages per Production Worker	24,861	32,763	132	Shipments per Employee	195,460	189,800	97
Hours per Production Worker	2,056	2,117	103	Shipments per Production Worker	279,017	285,621	102
Wages per Hour	12.09	15.48	128	Investment per Establishment	321,011	360,892	112
Value Added per Establishment	4,602,255	8,369,239	182	Investment per Employee	6,552	4,032	62
Value Added per Employee	93,930	93,510	100	Investment per Production Worker	9,352	6,068	65

Sources: Same as General Statistics. The 'Average of All Manufacturing' column represents the average of all manufacturing industries reported for the most recent complete year available. The Index shows the relationship between the Average and the Analyzed Industry. For example, 100 means that they are equal; 500 that the Analyzed Industry is five times the average; 50 means that the Analyzed Industry is half the national average. The abbreviation 'na' is used to show that data are 'not available'.

LEADING COMPANIES   Number shown: **34**   Total sales ($ mil): **1,157**   Total employment (000): **7.9**

Company Name	Address				CEO Name	Phone	Co. Type	Sales ($ mil)	Empl. (000)
Titanium Metals Corp	1999 Broadway	Denver	CO	80202	Kirby Adams	303-296-5600	S	151	1.1
Tremont Corp	1999 Broadway	Denver	CO	80202	J Landis Martin	303-296-5652	P	146	1.1
RMI Titanium Co	1000 Warren Av	Niles	OH	44446	L Frederick Geig Jr	216-652-9951	P	143	0.8
Teledyne Allvac/Vasco	PO Box 5030	Monroe	NC	28111	JV Andrews	704-289-4511	S	120	1.0
Unistrut International Corp	3971 Research Pk	Ann Arbor	MI	48108	C J Malfese	313-930-0030	R	75	0.6
Driver-Harris Co	308 Middlesex St	Harrison	NJ	07029	Frank L Driver III	201-483-4802	P	61	0.1
Zinc Products Co	PO Box 1890	Greeneville	TN	37744	Jerry T McDowell	615-639-8111	D	61	0.3
Alpha Metals Inc	600 Rte 440	Jersey City	NJ	07304	Ray Sharp	201-434-6778	S	60*	0.3
CCX Inc	1901 Roxborough	Charlotte	NC	28211	Richard A Rinaldi	704-365-0560	R	57	0.6
Sandvik Special Metals Corp	PO Box 6027	Kennewick	WA	99336	Kirk P Galbraith	509-586-4131	S	35	0.3
Reade Manufacturing	100 Ridgeway Blv	Lakehurst	NJ	08733	Mickey Markovich	908-657-6451	D	25	<0.1
Gardiner Metal Co	4820 S Campbell	Chicago	IL	60632	Scott Gardiner	312-847-0100	R	23	<0.1
Vermont American Corp	715 E Gray St	Louisville	KY	40202	John Lundy	502-589-3781	D	22	0.1
Jackson Wheeler Metals	270 Georgia Av	Brooklyn	NY	11207	Cliff Jackson	718-342-5000	R	16	<0.1
Sigmund Cohn Corp	121 S Columbus	Mount Vernon	NY	10553	Richard Cohn	914-664-5300	R	15	<0.1
LaSalle Rolling Mills Inc	1375 9th St	La Salle	IL	61301	Fred Carus	815-224-2174	R	14*	0.1
Platt Brothers and Co	PO Box 1030	Waterbury	CT	06706	James Behuniak	203-753-4194	R	14	0.1
Hamilton Precision Metals Inc	1780 Rohrerstown	Lancaster	PA	17601	Juergen Schlenker	717-569-7061	S	12	<0.1
Hood and Company Inc	PO Box 485	Hamburg	PA	19526	Stanley Golemo Jr	215-562-3841	R	12*	<0.1
Sunshine Precious Metals Inc	877 Main St	Boise	ID	83702	John Simko	208-345-0660	S	11*	<0.1
Pittsburgh Flatroll Co	31st St & AVRR	Pittsburgh	PA	15201		412-765-3322	R	10	<0.1
Seafab Metal Corp	2700 16th Av SW	Seattle	WA	98134	DJ Del Dotto	206-447-2718	R	10	<0.1
Bow Solder Products Co	900 Hobson St	Union	NJ	07083	Joe Noonan	908-686-6040	R	9	<0.1
Braxton Manufacturing Company	PO Box 429	Watertown	CT	06795	Braxton T Nelson	203-274-6781	R	9	0.1
Basic Carbide Corp	900 Blythedale Rd	Buena Vista	PA	15018	John Goodrum	412-754-0060	R	8*	<0.1
Lawrence Aviation Industries	Sheep Pasture Rd	P Jef Stat	NY	11776	G Cohen	516-473-1800	R	8	0.2
Hyper Alloys Inc	PO Box 247	Roseville	MI	48066	Pamela E Neumann	810-772-0571	R	6	<0.1
Division Lead LP	7742 W 61st Pl	Summit Argo	IL	60501	Philip Weiss	708-458-4528	R	5	<0.1
II Cross Co	363 Park Av	Weehawken	NJ	07087	George McClary	201-863-1134	R	5	<0.1
Sandvik Rhenium Alloys Inc	PO Box 245	Elyria	OH	44036	Jan C Carlen	216-365-7388	S	4	<0.1
L and LK Inc	2413-27 Federal St	Philadelphia	PA	19146	Sam Katz	215-732-2614	R	3*	<0.1
Metallurgical Processing Inc	PO Box 2320	New Britain	CT	06050	Elena T Ritoli	203-224-2648	R	3*	<0.1
Ray-Bar Engineering Corp	PO Box 415	Azusa	CA	91702	Ron A Wohler	818-969-1818	R	3	<0.1
APPCO Process Equipment	PO Box 23449	Charlotte	NC	28227	Tony Bell	704-545-7777	R	2	<0.1

Source: Ward's Business Directory of U.S. Private and Public Companies, Volumes 1 and 2, 1996. The company type code used is as follows: P - Public, R - Private, S - Subsidiary, D - Division, J - Joint Venture, A - Affiliate, G - Group. Sales are in millions of dollars, employees are in thousands. An asterisk (*) indicates an estimated sales volume. The symbol < stands for 'less than'. Company names and addresses are truncated, in some cases, to fit into the available space.

MATERIALS CONSUMED

Material	Quantity	Delivered Cost ($ million)
Materials, ingredients, containers, and supplies	(X)	1,280.3
Unalloyed aluminum and aluminum-base alloy ingot, pig, and shot	1,000 s tons (D)	(D)
Alloyed aluminum and aluminum-base alloy ingot, pig, and shot	1,000 s tons (D)	(D)
Aluminum and aluminum-base alloy sheet, plate, foil, and welded tubing	mil lb (S)	0.2
Aluminum and aluminum-base alloy extruded shapes, including extruded rod, bar, pipe, tube, etc.	mil lb (D)	(D)
All other aluminum and aluminum-base alloy shapes and forms	(X)	4.4
Copper and copper-base alloy cathodes	1,000 s tons (D)	(D)
All other copper and copper-base alloy shapes and forms	1,000 s tons 4.0	11.6
Magnesium and magnesium-base alloy shapes and forms	1,000 s tons 21.2	57.8
Nickel and nickel-base alloy shapes and forms	1,000 s tons (S)	145.3
Zinc and zinc-base alloy shapes and forms	1,000 s tons 33.2	38.1
Molybdenum and molybdenum-base alloy shapes and forms	mil lb 2.8	15.8
Tin and tin-base alloy shapes and forms	mil lb (S)	23.1
Titanium and titanium-base alloy shapes and forms	1,000 s tons 18.8*	101.5
Precious metals and precious metal alloy shapes and forms	1,000 troy ounces 838.4	248.8
Brass shapes and forms	1,000 s tons (D)	(D)
All other nonferrous shapes and forms	(X)	167.1
Copper and copper-base alloy scrap (except home scrap)	1,000 s tons (D)	(D)
Other nonferrous metal scrap (except home scrap)	1,000 s tons 9.9	16.4
All other materials and components, parts, containers, and supplies	(X)	256.0
Materials, ingredients, containers, and supplies, nsk	(X)	155.5

Source: 1992 Economic Census. Explanation of symbols used: (D): Withheld to avoid disclosure of competitive data; na: Not available; (S): Withheld because statistical norms were not met; (X): Not applicable; (Z): Less than half the unit shown; nec: Not elsewhere classified; nsk: Not specified by kind; - : zero; * : 10-19 percent estimated; ** : 20-29 percent estimated.

PRODUCT SHARE DETAILS

Product or Product Class	% Share	Product or Product Class	% Share
Nonferrous rolling and drawing, nec	100.00	Other precious metal mill shapes, including platinum-group metals (excluding wire)	14.28
Nickel and nickel-base alloy mill shapes (including nickel-copper alloys)	26.32	Precious metal wire	15.26
Nickel and nickel-base alloy plate, sheet, and strip (excluding nickel-copper alloys)	35.59	Precious metal mill shapes, nsk	24.09
Other nickel and nickel-base alloy mill shapes (excluding nickel-copper alloys and wire)	55.64	All other nonferrous metal mill shapes	28.37
Nickel-copper alloy mill shapes and forms (except wire)	3.18	Magnesium and magnesium-base alloy mill shapes (excluding wire)	9.96
Nickel and nickel alloy wire	5.58	Lead and lead-base alloy plate, sheet, and strip	1.43
Titanium and titanium-base alloy mill shapes (excluding wire)	19.35	Other rolled, drawn, or extruded lead and lead-base alloy mill shapes, including pipe, tubing, traps, and bends (excluding wire)	9.00
Titanium and titanium-base alloy ingot	24.90	Tungsten and tungsten-base alloy mill shapes (excluding wire)	0.95
Forging and extrusion titanium and titanium-alloy ingot (billet)	29.05	Molybdenum and molybdenum-base alloy mill shapes (excluding wire)	3.26
Other titanium and titanium-base alloy mill shapes, including sheet, plate, tubing, bar, etc. (excluding wire)	44.08	Other nonferrous metal rolled, drawn, and extruded shapes, including zinc (excluding wire)	43.86
Titanium and titanium-base alloy mill shapes (excluding wire), nsk	1.97	Other nonferrous wire (except copper, aluminum, nickel, and precious metals)	23.07
Precious metal mill shapes	20.53	All other nonferrous metal mill shapes, nsk	8.47
Gold mill shapes (excluding wire)	23.49	Nonferrous rolling and drawing, nec, nsk	5.43
Silver mill shapes (excluding wire)	22.89		

Source: 1992 *Economic Census.* The values shown are percent of total shipments in an industry. Values of indented subcategories are summed in the main heading. The symbol (D) appears when data are withheld to prevent disclosure of competitive information. The abbreviation nsk stands for 'not specified by kind' and nec for 'not elsewhere classified'.

INPUTS AND OUTPUTS FOR NONFERROUS ROLLING & DRAWING, NEC

Economic Sector or Industry Providing Inputs	%	Sector	Economic Sector or Industry Buying Outputs	%	Sector
Primary nonferrous metals, nec	59.1	Manufg.	Semiconductors & related devices	21.1	Manufg.
Wholesale trade	6.4	Trade	Exports	11.4	Foreign
Imports	3.6	Foreign	Photographic equipment & supplies	9.0	Manufg.
Cyclic crudes and organics	3.5	Manufg.	Telephone & telegraph apparatus	7.6	Manufg.
Special dies & tools & machine tool accessories	2.7	Manufg.	Nonferrous forgings	4.7	Manufg.
Electric services (utilities)	2.0	Util.	Chemical preparations, nec	3.8	Manufg.
Nonferrous rolling & drawing, nec	1.2	Manufg.	Jewelers' materials & lapidary work	3.4	Manufg.
Communications, except radio & TV	1.2	Util.	Aircraft & missile engines & engine parts	2.9	Manufg.
Advertising	1.2	Services	Iron & steel forgings	2.6	Manufg.
Miscellaneous plastics products	1.1	Manufg.	Aircraft & missile equipment, nec	2.2	Manufg.
Gas production & distribution (utilities)	1.1	Util.	Dental equipment & supplies	1.9	Manufg.
Ball & roller bearings	1.0	Manufg.	Nonferrous wire drawing & insulating	1.6	Manufg.
Nonferrous wire drawing & insulating	1.0	Manufg.	Electronic components nec	1.5	Manufg.
Primary lead	0.9	Manufg.	Aircraft	1.3	Manufg.
Primary copper	0.8	Manufg.	Refrigeration & heating equipment	1.0	Manufg.
Petroleum refining	0.7	Manufg.	Mechanical measuring devices	0.9	Manufg.
Power transmission equipment	0.7	Manufg.	Nonferrous rolling & drawing, nec	0.9	Manufg.
Maintenance of nonfarm buildings nec	0.5	Constr.	Electric housewares & fans	0.8	Manufg.
Chemical preparations, nec	0.5	Manufg.	Electronic computing equipment	0.8	Manufg.
Fabricated metal products, nec	0.5	Manufg.	Fabricated metal products, nec	0.8	Manufg.
Metal coating & allied services	0.5	Manufg.	Gaskets, packing & sealing devices	0.8	Manufg.
Primary zinc	0.5	Manufg.	Power driven hand tools	0.8	Manufg.
Pumps & compressors	0.5	Manufg.	Primary batteries, dry & wet	0.8	Manufg.
Motor freight transportation & warehousing	0.5	Util.	Storage batteries	0.8	Manufg.
Banking	0.5	Fin/R.E.	Guided missiles & space vehicles	0.7	Manufg.
Miscellaneous repair shops	0.5	Services	Federal Government purchases, nondefense	0.7	Fed Govt
Industrial controls	0.4	Manufg.	Ophthalmic goods	0.6	Manufg.
Industrial inorganic chemicals, nec	0.4	Manufg.	Power transmission equipment	0.6	Manufg.
Metal stampings, nec	0.4	Manufg.	Watches, clocks, & parts	0.6	Manufg.
Paints & allied products	0.3	Manufg.	X-ray apparatus & tubes	0.6	Manufg.
Primary aluminum	0.3	Manufg.	Metalworking machinery, nec	0.5	Manufg.
Primary metal products, nec	0.3	Manufg.	Mining machinery, except oil field	0.5	Manufg.
Rubber & plastics hose & belting	0.3	Manufg.	Special industry machinery, nec	0.5	Manufg.
Screw machine and related products	0.3	Manufg.	Costume jewelry	0.4	Manufg.
Air transportation	0.3	Util.	Electron tubes	0.4	Manufg.
Equipment rental & leasing services	0.3	Services	General industrial machinery, nec	0.4	Manufg.
Machinery, except electrical, nec	0.2	Manufg.	Hardware, nec	0.4	Manufg.
Metal heat treating	0.2	Manufg.	Machinery, except electrical, nec	0.4	Manufg.
Paperboard containers & boxes	0.2	Manufg.	Miscellaneous plastics products	0.4	Manufg.
Sawmills & planning mills, general	0.2	Manufg.	Motor vehicle parts & accessories	0.4	Manufg.
Railroads & related services	0.2	Util.	Pumps & compressors	0.4	Manufg.
Water transportation	0.2	Util.	Screw machine and related products	0.4	Manufg.
Eating & drinking places	0.2	Trade	Service industry machines, nec	0.4	Manufg.
Royalties	0.2	Fin/R.E.	Typewriters & office machines, nec	0.4	Manufg.
Business services nec	0.2	Services	Fabricated plate work (boiler shops)	0.3	Manufg.
Sanitary services, steam supply, irrigation	0.1	Util.	Games, toys, & children's vehicles	0.3	Manufg.
Insurance carriers	0.1	Fin/R.E.	Internal combustion engines, nec	0.3	Manufg.
Automotive rental & leasing, without drivers	0.1	Services	Ship building & repairing	0.3	Manufg.
Hotels & lodging places	0.1	Services	Industrial buildings	0.2	Constr.

Continued on next page.

INPUTS AND OUTPUTS FOR NONFERROUS ROLLING & DRAWING, NEC - Continued

Economic Sector or Industry Providing Inputs	%	Sector	Economic Sector or Industry Buying Outputs	%	Sector
U.S. Postal Service	0.1	Gov't	Boat building & repairing	0.2	Manufg.
Noncomparable imports	0.1	Foreign	Drugs	0.2	Manufg.
Scrap	0.1	Scrap	Electrical equipment & supplies, nec	0.2	Manufg.
			Engine electrical equipment	0.2	Manufg.
			Industrial furnaces & ovens	0.2	Manufg.
			Motor homes (made on purchased chassis)	0.2	Manufg.
			Radio & TV communication equipment	0.2	Manufg.
			Railroad equipment	0.2	Manufg.
			Sheet metal work	0.2	Manufg.
			Surgical appliances & supplies	0.2	Manufg.
			Tanks & tank components	0.2	Manufg.
			Textile machinery	0.2	Manufg.
			Wiring devices	0.2	Manufg.
			Household appliances, nec	0.1	Manufg.
			Instruments to measure electricity	0.1	Manufg.
			Optical instruments & lenses	0.1	Manufg.
			Printing trades machinery	0.1	Manufg.
			Special dies & tools & machine tool accessories	0.1	Manufg.
			Sporting & athletic goods, nec	0.1	Manufg.
			Surgical & medical instruments	0.1	Manufg.
			Turbines & turbine generator sets	0.1	Manufg.

Source: Benchmark Input-Output Accounts for the U.S. Economy, 1982, U.S. Department of Commerce, Washington, D.C., July 1991. Data, as reported in the source, are organized by the 1977 SIC structure in use in 1982 but have been matched, as closely as is possible, to the 1987 SIC structure used in this book.

OCCUPATIONS EMPLOYED BY SIC 335 - NONFERROUS ROLLING AND DRAWING

Occupation	% of Total 1994	Change to 2005	Occupation	% of Total 1994	Change to 2005
Machine tool cutting & forming etc. nec	11.0	-12.4	Machinists	1.6	-20.3
Assemblers, fabricators, & hand workers nec	5.9	-20.4	Machine tool cutting operators, metal & plastic	1.4	-33.7
Blue collar worker supervisors	5.0	-22.9	Extruding & forming machine workers	1.3	-12.4
Machine forming operators, metal & plastic	5.0	7.5	Coating, painting, & spraying machine workers	1.3	-20.4
Metal & plastic machine workers nec	3.9	-15.3	General managers & top executives	1.2	-24.4
Inspectors, testers, & graders, precision	3.7	-20.3	Combination machine tool operators	1.2	-12.4
Helpers, laborers, & material movers nec	3.6	-20.4	Traffic, shipping, & receiving clerks	1.1	-23.3
Machine operators nec	3.1	-29.8	Industrial production managers	1.1	-20.4
Industrial machinery mechanics	2.9	-12.4	Packaging & filling machine operators	1.1	-20.4
Industrial truck & tractor operators	2.8	-20.3	Secretaries, ex legal & medical	1.1	-27.5
Maintenance repairers, general utility	2.1	-28.3	Machine feeders & offbearers	1.1	-28.3
Sales & related workers nec	1.9	-20.3	Production, planning, & expediting clerks	1.0	-20.3
Hand packers & packagers	1.7	-31.7	Metal molding machine workers	1.0	19.5
Electricians	1.7	-25.3			

Source: Industry-Occupation Matrix, Bureau of Labor Statistics. These data relate to one or more 3-digit SIC industry groups rather than to a single 4-digit SIC. The change reported for each occupation to the year 2005 is a percent of growth or decline as estimated by the Bureau of Labor Statistics. The abbreviation nec stands for 'not elsewhere classified'.

LOCATION BY STATE AND REGIONAL CONCENTRATION

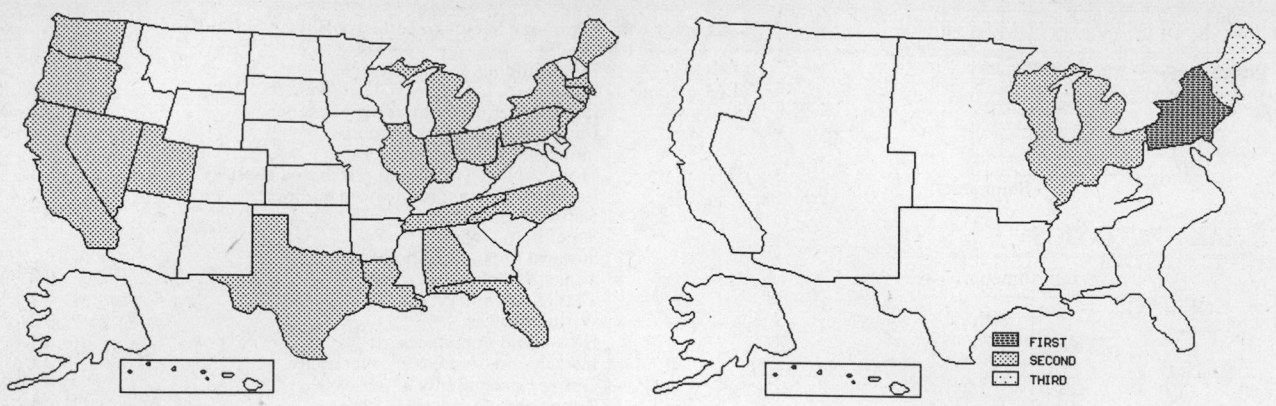

FIRST
SECOND
THIRD

INDUSTRY DATA BY STATE

| State | Establish- ments | Shipments | | | Employment | | | | Cost as % of Shipments | Investment per Employee ($) |
		Total ($ mil)	% of U.S.	Per Establ.	Total Number	% of U.S.	Per Establ.	Wages ($/hour)		
New York	13	152.6	5.6	11.7	800	5.0	62	17.10	66.5	-
New Jersey	14	144.0	5.3	10.3	800	5.0	57	15.70	53.5	-
Washington	5	121.7	4.5	24.3	600	3.8	120	13.44	52.6	-
Illinois	9	115.8	4.3	12.9	800	5.0	89	13.55	56.0	2,125
Connecticut	9	107.6	4.0	12.0	700	4.4	78	16.63	50.4	2,429
Michigan	13	79.5	2.9	6.1	400	2.5	31	11.14	67.2	5,500
Tennessee	4	70.1	2.6	17.5	300	1.9	75	11.60	58.6	-
California	10	33.9	1.2	3.4	200	1.3	20	12.00	63.7	3,000
Florida	9	24.6	0.9	2.7	100	0.6	11	8.50	50.8	8,000
Pennsylvania	22	(D)	-	-	1,750 *	10.9	80	-	-	3,314
Ohio	12	(D)	-	-	1,750 *	10.9	146	-	-	3,314
Rhode Island	9	(D)	-	-	375 *	2.3	42	-	-	-
Texas	8	(D)	-	-	175 *	1.1	22	-	-	3,429
Massachusetts	7	(D)	-	-	750 *	4.7	107	-	-	3,733
Indiana	4	(D)	-	-	750 *	4.7	188	-	-	-
Nevada	3	(D)	-	-	750 *	4.7	250	-	-	-
North Carolina	3	(D)	-	-	750 *	4.7	250	-	-	-
Oregon	2	(D)	-	-	1,750 *	10.9	875	-	-	-
Alabama	1	(D)	-	-	175 *	1.1	175	-	-	-
Louisiana	1	(D)	-	-	175 *	1.1	175	-	-	-
Maine	1	(D)	-	-	375 *	2.3	375	-	-	-
Utah	1	(D)	-	-	375 *	2.3	375	-	-	-
West Virginia	1	(D)	-	-	1,750 *	10.9	1,750	-	-	-

Source: 1992 *Economic Census*. The states are in descending order of shipments or establishments (if shipment data are missing for the majority). The symbol (D) appears when data are withheld to prevent disclosure of competitive information. States marked with (D) are sorted by number of establishments. A dash (-) indicates that the data element cannot be calculated; * indicates the midpoint of a range.

3357 - DRAWING & INSULATING NONFERROUS WIRE

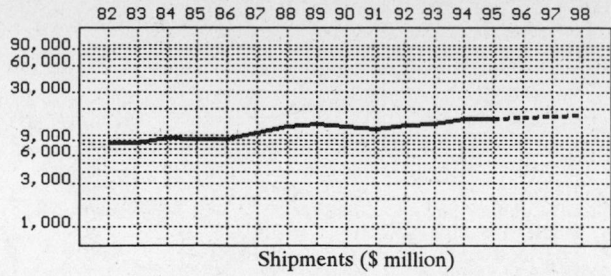

Shipments ($ million)

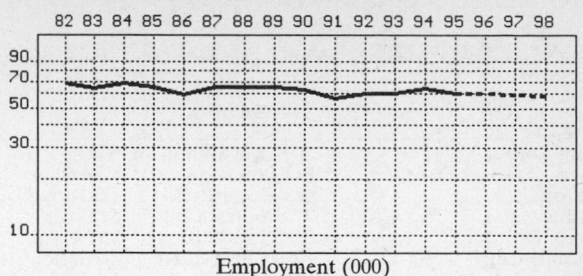

Employment (000)

GENERAL STATISTICS

Year	Companies	Establishments		Employment			Compensation		Production ($ million)			
		Total	with 20 or more employees	Total (000)	Production Workers (000)	Hours (Mil)	Payroll ($ mil)	Wages ($/hr)	Cost of Materials	Value Added by Manufacture	Value of Shipments	Capital Invest.
1982	271	441	341	67.7	50.0	95.7	1,271.6	8.75	5,266.4	2,831.1	8,224.5	294.7
1983		437	341	63.9	47.1	93.6	1,291.9	9.09	5,522.7	2,762.5	8,320.8	219.9
1984		433	341	68.5	51.7	102.8	1,460.2	9.55	5,954.3	3,461.7	9,323.1	249.9
1985		428	342	64.4	47.9	96.1	1,411.4	9.72	5,706.4	3,427.0	9,217.4	308.4
1986		414	325	59.0	44.3	89.9	1,352.5	10.00	5,743.8	3,309.8	9,055.8	219.6
1987	340	487	355	64.9	48.4	99.2	1,504.3	10.18	6,722.3	4,229.5	10,826.5	243.9
1988		475	338	64.9	48.6	101.7	1,588.6	10.48	7,968.0	4,684.7	12,485.4	248.2
1989		471	348	65.0	48.5	98.7	1,636.8	10.95	8,589.2	4,838.1	13,459.9	282.1
1990		472	344	62.5	44.6	90.1	1,600.8	11.51	7,995.6	4,620.7	12,609.4	300.3
1991		495	348	57.1	41.9	86.3	1,575.2	11.90	7,307.4	4,313.1	11,686.4	277.9
1992	379	528	354	60.6	44.8	94.1	1,724.4	11.91	8,229.1	4,867.7	13,106.7	364.2
1993		529	357	60.1	44.9	95.8	1,799.0	12.23	8,208.3	5,432.7	13,618.8	329.8
1994		527P	354P	64.3	48.2	103.9	1,952.3	12.36	9,553.6	6,018.8	15,346.6	482.9
1995		536P	355P	60.0P	44.4P	95.8P	1,889.8P	12.86P	9,483.1P	5,974.4P	15,233.4P	380.3P
1996		545P	357P	59.5P	44.1P	95.8P	1,938.1P	13.18P	9,830.3P	6,193.1P	15,791.1P	392.6P
1997		554P	358P	59.1P	43.7P	95.8P	1,986.5P	13.49P	10,177.5P	6,411.8P	16,348.8P	405.0P
1998		563P	359P	58.6P	43.3P	95.8P	2,034.8P	13.80P	10,524.7P	6,630.6P	16,906.5P	417.3P

Sources: 1982, 1987, 1992 *Economic Census*; *Annual Survey of Manufactures*, 83-86, 88-91, 93-94. Establishment counts for non-Census years are from *County Business Patterns*; establishment values for 83-84 are extrapolations. 'P's show projections by the editors. Industries reclassified in 87 will not have data for prior years.

INDICES OF CHANGE

Year	Companies	Establishments		Employment			Compensation		Production ($ million)			
		Total	with 20 or more employees	Total (000)	Production Workers (000)	Hours (Mil)	Payroll ($ mil)	Wages ($/hr)	Cost of Materials	Value Added by Manufacture	Value of Shipments	Capital Invest.
1982	72	84	96	112	112	102	74	73	64	58	63	81
1983		83	96	105	105	99	75	76	67	57	63	60
1984		82	96	113	115	109	85	80	72	71	71	69
1985		81	97	106	107	102	82	82	69	70	70	85
1986		78	92	97	99	96	78	84	70	68	69	60
1987	90	92	100	107	108	105	87	85	82	87	83	67
1988		90	95	107	108	108	92	88	97	96	95	68
1989		89	98	107	108	105	95	92	104	99	103	77
1990		89	97	103	100	96	93	97	97	95	96	82
1991		94	98	94	94	92	91	100	89	89	89	76
1992	100	100	100	100	100	100	100	100	100	100	100	100
1993		100	101	99	100	102	104	103	100	112	104	91
1994		100P	100P	106	108	110	113	104	116	124	117	133
1995		101P	100P	99P	99P	102P	110P	108P	115P	123P	116P	104P
1996		103P	101P	98P	98P	102P	112P	111P	119P	127P	120P	108P
1997		105P	101P	97P	98P	102P	115P	113P	124P	132P	125P	111P
1998		107P	102P	97P	97P	102P	118P	116P	128P	136P	129P	115P

Sources: Same as General Statistics. Values reflect change from the base year, 1992. Values above 100 mean greater than 92, values below 100 mean less than 92, and a value of 100 in the 82-91 or 93-98 period means same as 92. 'P's mark projections by the editors.

SELECTED RATIOS

For 1994	Avg. of All Manufact.	Analyzed Industry	Index	For 1994	Avg. of All Manufact.	Analyzed Industry	Index
Employees per Establishment	49	122	249	Value Added per Production Worker	134,084	124,871	93
Payroll per Establishment	1,500,273	3,707,432	247	Cost per Establishment	5,045,178	18,142,356	360
Payroll per Employee	30,620	30,362	99	Cost per Employee	102,970	148,579	144
Production Workers per Establishment	34	92	267	Cost per Production Worker	146,988	198,207	135
Wages per Establishment	853,319	2,438,713	286	Shipments per Establishment	9,576,895	29,143,306	304
Wages per Production Worker	24,861	26,643	107	Shipments per Employee	195,460	238,672	122
Hours per Production Worker	2,056	2,156	105	Shipments per Production Worker	279,017	318,394	114
Wages per Hour	12.09	12.36	102	Investment per Establishment	321,011	917,031	286
Value Added per Establishment	4,602,255	11,429,745	248	Investment per Employee	6,552	7,510	115
Value Added per Employee	93,930	93,605	100	Investment per Production Worker	9,352	10,019	107

Sources: Same as General Statistics. The 'Average of All Manufacturing' column represents the average of all manufacturing industries reported for the most recent complete year available. The Index shows the relationship between the Average and the Analyzed Industry. For example, 100 means that they are equal; 500 that the Analyzed Industry is five times the average; 50 means that the Analyzed Industry is half the national average. The abbreviation 'na' is used to show that data are 'not available'.

LEADING COMPANIES   Number shown: **75**   Total sales ($ mil): **8,219**   Total employment (000): **51.2**

Company Name	Address				CEO Name	Phone	Co. Type	Sales ($ mil)	Empl. (000)
Raychem Corp	300 Constitution Dr	Menlo Park	CA	94025	Robert J Saldich	415-361-3333	P	1,462	10.8
Southwire Co	PO Box 1000	Carrollton	GA	30119	Roy Richards Jr	404-832-4242	R	1,330	5.0
Siecor Corp	489 Siecor Park	Hickory	NC	28601	JD Hicks	704-327-5000	S	470*	3.0
Belden Inc	7701 Forsyth Blv	St Louis	MO	63105	CB Cunningham	314-854-8000	P	440	2.9
Reliance COMM-TEC	5875 Landerbrook	Cleveland	OH	44124	Dudley P Sheffler	216-266-5300	S	439	3.9
Pirelli Cable Corp	705 Industrial Dr	Lexington	SC	29072	Eugenio Razelli	803-951-4800	S	400	1.0
CommScope Inc	PO Box 1729	Hickory	NC	28602	Frank M Drendel	704-324-2200	S	347	1.6
Rea Wire Industries Inc	PO Box 6128	Fort Wayne	IN	46896	Tim Zumbaugh	219-424-4252	R	200	0.8
Times Fiber Communications	PO Box 384	Wallingford	CT	06492	Timothy F Cohane	203-265-8500	S	195	0.7
Phelps Dodge Magnet Wire Co	2131 S Collesium	Fort Wayne	IN	46803	Kirk C Kemmish	219-421-5400	S	190*	1.2
Tandy Electronics	200 Taylor St	Fort Worth	TX	76102	Robert M McClure	817-390-3690	D	190	4.3
Triangle Wire and Cable Inc	95 Grand Av	Pawtucket	RI	02861	Ray Pearman	401-729-5400	S	170*	1.1
Camden Wire Company Inc	12 Masonic Av	Camden	NY	13316	Terry M French	315-245-3800	S	151	1.0
Okonite Company Inc	PO Box 340	Ramsey	NJ	07446	Alfred Coppola	201-825-0300	R	150	1.0
Simplex Technologies Inc	PO Box 479	Portsmouth	NH	03802	Peter L Bergeron	603-436-6100	S	135	0.7
Rome Cable Corp	421 Ridge St	Rome	NY	13440	DE Harvey	315-337-3000	R	130	0.5
Superior Cable	150 Interstate N	Atlanta	GA	30339	Justin F Deedy Jr	404-953-8338	D	120	0.5
AFC Cable Systems Inc	55 Samuel Barnet	New Bedford	MA	02745	Harry M Crump	508-998-1131	P	114	0.6
Cerro Wire and Cable Company	PO Box 1049	Hartselle	AL	35640	Michael Hartley	205-773-2522	S	100	0.5
Coleman Cable Systems Inc	2500 Commonw	North Chicago	IL	60064	James Coleman	708-689-9090	S	91*	0.4
Trilogy Communications Inc	PO Box 5918	Pearl	MS	39208	S Shinn Lee	601-932-4461	R	75	0.3
Amoco Technology Co	55 Shuman Blv	Naperville	IL	60563	RC Carr	708-717-2489	S	70	0.7
Westinghouse Electric Corp	18259 Westinghouse	Abingdon	VA	24210	Roy A Thomas	703-676-9100	D	63*	0.4
Tandy Wire and Cable	3500 McCart Av	Fort Worth	TX	76110	Al Esquivel	817-924-5789	D	62	0.7
Kalas Manufacturing Inc	PO Box 328	Denver	PA	17517	RP Witwer	717-336-5575	R	60	0.3
Kerite Co	PO Box 452	Seymour	CT	06483	TH Pluff	203-888-2591	S	55	0.4
Tamaqua Cable Products Corp	PO Box 347	Schuylkill H	PA	17972	WH Combs III	717-385-4381	R	50	0.2
Communication Cable Inc	PO Box 729	Siler City	NC	27344	James R Fore	919-663-2629	P	48	0.4
Telco Systems Inc	63 Nahatan St	Norwood	MA	02062	Bruce W Young	617-551-0300	D	46	0.2
Champlain Cable Corp	12 Hercules Dr	Colchester	VT	05446	David Binch	802-655-2121	S	43	0.3
Judd Wire Inc	Turnpike Rd	Turners Falls	MA	01376	Hiro Miyauchi	413-863-4357	S	43	0.2
Aetna Insulated Wire Co	PO Box 638	Bellmawr	NJ	08099	Edward Sproat Jr	609-933-1000	R	39	0.3
Chromatic Technologies Inc	9 Forge Park	Franklin	MA	02038	Garo Artinian	508-520-1200	R	33	0.1
C and M Corp	PO Box 348	Wauregan	CT	06387	MG Andrews	203-774-4812	R	31*	0.5
Tri-Tec Engineering Inc	14500 S Broadway	Gardena	CA	90248	Robert Lavigne	310-327-3960	S	31*	0.2
Tensolite Co	100 Tensolite Dr	St Augustine	FL	32092	John Berlin	904-829-5600	S	30	0.2
Wyre-Wynd	77 Anthony St	Jewett City	CT	06351	T Randolph	203-376-2516	D	30	0.2
Philadelphia Insulated Wire Co	333 New Albany Rd	Moorestown	NJ	08057	Rich Rale	609-235-6700	S	28	0.1
Amercable	350 Bailey Rd	El Dorado	AR	71730	Bob Hogan	501-862-4919	D	27	0.2
Consolidated Electronic	11044 King St	Franklin Park	IL	60131	TA Mann	708-455-8830	R	26	0.1
Optical Cable Corp	PO Box 11967	Roanoke	VA	24022	Robert Kopstein	703-265-0690	R	26	<0.1
Vector Cable Co	555 Industrial Blv	Sugar Land	TX	77478	E Collet	713-275-7790	D	25	0.2
Precision Cable Mfg Corp	PO Box 1448	Rockwall	TX	75087	Dale B Fleming	214-771-1233	R	24	0.2
Prestolite Wire Corp	PO Box 259	Sidney	NE	69162	Tim Ryder	308-254-5310	D	23*	0.2
New England Electric Wire Corp	PO Box 264	Lisbon	NH	03585	W W Jesseman	603-838-6628	S	22	0.3
LF Gaubert	PO Box 50500	New Orleans	LA	70150	KR Charboneau	504-254-0754	D	20	<0.1
Remee Products Corp	186 N Main St	Florida	NY	10921	Al Muhlrad	914-651-4431	R	18	0.2
Remfo	186 N Main St	Florida	NY	10921	A Muhlrad	914-651-4431	D	18	0.2
CE Shepherd Company Inc	PO Box 9445	Houston	TX	77261	Charles Shepherd	713-928-3763	R	16*	0.2
Dynatronic Cable Engineering	136 San Fernando	Los Angeles	CA	90031	AJ Evangelista	213-225-5611	D	16	<0.1
National Wire and Cable Corp	136 San Fernando	Los Angeles	CA	90031	AJ Evangelista	213-225-5611	R	16*	0.2
Cove Fourslide & Stamping	195 E Merrick Rd	Freeport	NY	11520	Barry Jaffee	516-379-4232	R	15	0.2
Harbor Electronics Inc	650 Danbury Rd	Ridgefield	CT	06877	R Marsilio	203-438-9625	R	15*	0.2
Lawrence Technology Inc	PO Box 945	Lawrence	KS	66044	William Russell	913-841-5610	S	15*	0.1
Nagle Industries Inc	PO Box 658	White House	TN	37188	Terry Nagle	615-672-4943	R	15*	0.1
Radix Wire Co	26260 Lakeland	Cleveland	OH	44132	C Vermerris	216-731-9191	R	15	<0.1
South Bay Cable	PO Box 67	Idyllwild	CA	92549	Gordon W Brown	909-659-2183	D	15	<0.1
Berkshire Electric Cable	PO Box 306	Leeds	MA	01053	GR Fields	413-584-3853	R	14	0.1
Horning Wire Corp	66 N Buesching Rd	Lake Zurich	IL	60047	Carol A Horning	708-438-8844	R	14	<0.1
Electro Fiberoptics Corp	45 Bartlett St	Marlborough	MA	01752	K Lisi	508-229-8312	R	12*	<0.1
Rapco International Inc	3581 Larch Ln	Jackson	MO	63755	Pete Marsac	314-243-1433	R	11	0.1
Montgomery Wire Corp	PO Box 227	Littleton	NH	03561	Doron Tamir	603-444-0550	R	11	0.1
Specialty Cable Corp	PO Box 50	Wallingford	CT	06492	James Leahy	203-265-7126	R	11*	<0.1
Apex Wire and Cable Corp	PO Box 11127	Hauppauge	NY	11788	Harvey Cooper	516-273-3322	R	10	<0.1
Atlas Wire Corp	9525 River St	Schiller Park	IL	60176	JW Herley	708-678-1210	R	10	<0.1
Connecting Devices Inc	PO Box 92619	Long Beach	CA	90809	John G Dunbabin	310-498-0901	R	10	<0.1
Cooner Wire Co	9265 Owensmouth	Chatsworth	CA	91311	Patrick G Weir	818-882-8311	R	10	<0.1
General Wire Products Inc	425 Shrewsbury St	Worcester	MA	01604	T P Andrews Sr	508-752-8260	R	10	<0.1
Oswego Wire Inc	1 Wire Dr	Oswego	NY	13126	Thomas Sasso	315-343-0524	R	10	0.1
Philatron International	15315 Cornet Av	Santa Fe Sprgs	CA	90670	Phillip M Ramos Jr	310-802-2570	R	10	0.1
Raynet Corp	155 Constitution Dr	Menlo Park	CA	94025	Robert Kelsch	415-324-6400	S	10	0.8
Storm Products Company Inc	PO Box 1668	Inglewood	CA	90308	John A Storm	310-649-6141	R	10*	<0.1
United Wire and Cable Corp	425 Shrewsbury St	Worcester	MA	01604	Thomas P Andrews	508-757-3872	R	10	<0.1
Olympic Wire and Cable	7 Madison Rd	Fairfield	NJ	07004	Robert H Burke	201-227-7996	R	9	<0.1
American Electric Cable	181 Appleton St	Holyoke	MA	01040	George Lowell	413-539-9893	R	9	<0.1

Source: Ward's *Business Directory of U.S. Private and Public Companies*, Volumes 1 and 2, 1996. The company type code used is as follows: P - Public, R - Private, S - Subsidiary, D - Division, J - Joint Venture, A - Affiliate, G - Group. Sales are in millions of dollars, employees are in thousands. An asterisk (*) indicates an estimated sales volume. The symbol < stands for 'less than'. Company names and addresses are truncated, in some cases, to fit into the available space.

MATERIALS CONSUMED

Material	Quantity		Delivered Cost ($ million)
Materials, ingredients, containers, and supplies		(X)	7,747.2
Bare steel wire	1,000 s tons	70.9**	103.9
All other steel shapes and forms	1,000 s tons	36.7	40.2
Unalloyed copper and copper-base alloy rods	mil lb	2,127.1	2,194.8
Alloyed copper and copper-base alloy rods	mil lb	357.6	307.9
Copper and copper-base alloy wire for redrawing	mil lb	249.8	272.5
Bare copper and copper-base alloy wire, electrical (except wire for redrawing)	mil lb	(D)	
Insulated copper wire and cable		(X)	125.2
Copper and copper-base alloy cathodes	1,000 s tons	151.1	159.0
All other copper and copper-base alloy shapes and forms	mil lb	(S)	164.4
Aluminum and aluminum-base alloy rods	mil lb	521.4	344.7
Aluminum and aluminum-base alloy wire for redrawing	mil lb	22.5	21.4
Bare aluminum and aluminum-base alloy wire, except for redrawing	mil lb	41.8	52.7
All other aluminum and aluminum-base alloy shapes and forms	mil lb	2,958.1	145.2
Refined unalloyed tin shapes and forms		(X)	6.7
All other nonferrous shapes and forms	mil lb	185.3	55.6
Plastics resins consumed in the form of granules, pellets, powders, liquids, etc.	mil lb	1,287.3	891.9
Synthetic rubber	mil lb	635.0	93.4
All other chemicals and allied products	mil lb	62.4*	76.3
Optical fiber, data and nondata transmission		(X)	246.3
Fiberglass insulating materials		(X)	11.6
All other stone, clay, glass, and concrete products		(X)	5.5
Natural rubber	mil lb	11.2	10.7
Plastics products consumed in the form of sheets, rods, tubes, film, and other shapes		(X)	64.7
Cotton yarns	mil lb	(S)	8.3
Connectors		(X)	38.4
All other materials and components, parts, containers, and supplies		(X)	1,031.2
Materials, ingredients, containers, and supplies, nsk		(X)	824.8

Source: 1992 *Economic Census*. Explanation of symbols used: (D): Withheld to avoid disclosure of competitive data; na: Not available; (S): Withheld because statistical norms were not met; (X): Not applicable; (Z): Less than half the unit shown; nec: Not elsewhere classified; nsk: Not specified by kind; - : zero; * : 10-19 percent estimated; ** : 20-29 percent estimated.

PRODUCT SHARE DETAILS

Product or Product Class	% Share	Product or Product Class	% Share
Nonferrous wiredrawing and insulating	100.00	Apparatus wire and cord and flexible cord sets, except wiring harnesses	6.65
Aluminum and aluminum-base alloy wire and cable (except covered or insulated, including ACSR), produced in nonferrous wiredrawing plants	3.12	Magnet wire	8.52
		Power wire and cable	10.05
Copper and copper-base alloy wire, strand, and cable for electrical transmission	4.83	Fiber optic cable	8.69
Bare unalloyed copper wire for electrical transmission	37.61	Fiber optic cable for communication (telephone, telegraph, and electronic)	89.53
Bare alloyed copper wire for electrical transmission	8.27	Fiber optic cable for all other uses	10.32
Copper and copper-base alloy bare strand and cable for electrical transmission	43.39	Fiber optic cable, nsk	0.15
		Electronic wire and cable	14.82
Other copper and copper-base alloy wire, strand, and cable for electrical transmission (including electrical wire rod)	4.60	Telephone and telegraph wire and cable	12.53
		Control and signal wire and cable	3.14
Copper and copper-base alloy wire, strand, and cable, (for electrical transmission), nsk	6.11	Building wire and cable	16.49
		Other insulated wire and cable, nec, including automotive	5.63
Other bare nonferrous metal wire made in nonferrous wiredrawing plants	1.09	Nonferrous wiredrawing and insulating, nsk, total	(D)
Nonferrous wire cloth and other woven wire products made in nonferrous wiredrawing plants	(D)	Nonferrous wiredrawing and insulating, nsk, for non-administrative record establishments	(D)
Nonferrous wire cloth and other woven wire products made in nonferrous wiredrawing plants	(D)	Nonferrous wiredrawing and insulating, nsk, for administrative record establishments	(D)

Source: 1992 *Economic Census*. The values shown are percent of total shipments in an industry. Values of indented subcategories are summed in the main heading. The symbol (D) appears when data are withheld to prevent disclosure of competitive information. The abbreviation nsk stands for 'not specified by kind' and nec for 'not elsewhere classified'.

INPUTS AND OUTPUTS FOR NONFERROUS WIRE DRAWING & INSULATING

Economic Sector or Industry Providing Inputs	%	Sector	Economic Sector or Industry Buying Outputs	%	Sector
Nonferrous wire drawing & insulating	15.5	Manufg.	Telephone & telegraph facility construction	14.9	Constr.
Copper rolling & drawing	10.5	Manufg.	Nonferrous wire drawing & insulating	10.5	Manufg.
Plastics materials & resins	9.6	Manufg.	Exports	5.4	Foreign
Wholesale trade	9.0	Trade	Electric utility facility construction	4.7	Constr.
Primary aluminum	8.1	Manufg.	Semiconductors & related devices	4.0	Manufg.
Aluminum rolling & drawing	5.7	Manufg.	Office buildings	3.3	Constr.
Imports	5.7	Foreign	Motors & generators	3.1	Manufg.
Primary copper	4.5	Manufg.	Maintenance of nonfarm buildings nec	2.9	Constr.
Cyclic crudes and organics	2.5	Manufg.	Residential additions/alterations, nonfarm	2.5	Constr.
Electric services (utilities)	2.3	Util.	Maintenance of electric utility facilities	1.9	Constr.

Continued on next page.

INPUTS AND OUTPUTS FOR NONFERROUS WIRE DRAWING & INSULATING - Continued

Economic Sector or Industry Providing Inputs	%	Sector	Economic Sector or Industry Buying Outputs	%	Sector
Motor freight transportation & warehousing	1.9	Util.	Residential 1-unit structures, nonfarm	1.8	Constr.
Miscellaneous plastics products	1.8	Manufg.	Industrial buildings	1.6	Constr.
Advertising	1.6	Services	Nonfarm residential structure maintenance	1.6	Constr.
Power transmission equipment	1.1	Manufg.	Transformers	1.6	Manufg.
Railroads & related services	1.1	Util.	Electronic components nec	1.5	Manufg.
Synthetic rubber	1.0	Manufg.	Construction of hospitals	1.4	Constr.
Metal coating & allied services	0.9	Manufg.	Signs & advertising displays	1.2	Manufg.
Nonferrous rolling & drawing, nec	0.9	Manufg.	Federal Government purchases, national defense	1.2	Fed Govt
Nonmetallic mineral products, nec	0.9	Manufg.	Maintenance of telephone & telegraph facilities	1.0	Constr.
Primary zinc	0.8	Manufg.	General industrial machinery, nec	1.0	Manufg.
Banking	0.8	Fin/R.E.	Radio & TV communication equipment	1.0	Manufg.
Ball & roller bearings	0.7	Manufg.	Blast furnaces & steel mills	0.9	Manufg.
Maintenance of nonfarm buildings nec	0.6	Constr.	Blowers & fans	0.9	Manufg.
Gas production & distribution (utilities)	0.6	Util.	Construction of educational buildings	0.8	Constr.
Fabricated metal products, nec	0.5	Manufg.	Aluminum rolling & drawing	0.8	Manufg.
Petroleum refining	0.5	Manufg.	Electric lamps	0.8	Manufg.
Communications, except radio & TV	0.5	Util.	Engine electrical equipment	0.8	Manufg.
Miscellaneous repair shops	0.5	Services	Primary metal products, nec	0.8	Manufg.
Glass & glass products, except containers	0.4	Manufg.	Telephone & telegraph apparatus	0.8	Manufg.
Machinery, except electrical, nec	0.4	Manufg.	Construction of stores & restaurants	0.7	Constr.
Primary lead	0.4	Manufg.	Aircraft	0.7	Manufg.
Primary nonferrous metals, nec	0.4	Manufg.	Electronic computing equipment	0.7	Manufg.
Screw machine and related products	0.4	Manufg.	Fabricated plate work (boiler shops)	0.7	Manufg.
Eating & drinking places	0.4	Trade	Pipe, valves, & pipe fittings	0.7	Manufg.
Equipment rental & leasing services	0.4	Services	Water transportation	0.7	Util.
Chemical preparations, nec	0.3	Manufg.	Highway & street construction	0.6	Constr.
Fabricated rubber products, nec	0.3	Manufg.	Hotels & motels	0.6	Constr.
Industrial controls	0.3	Manufg.	Residential garden apartments	0.6	Constr.
Industrial inorganic chemicals, nec	0.3	Manufg.	Electric housewares & fans	0.6	Manufg.
Metal heat treating	0.3	Manufg.	Electrical equipment & supplies, nec	0.6	Manufg.
Pumps & compressors	0.3	Manufg.	Fabricated metal products, nec	0.6	Manufg.
Royalties	0.3	Fin/R.E.	Electric services (utilities)	0.6	Util.
Noncomparable imports	0.3	Foreign	Federal Government purchases, nondefense	0.6	Fed Govt
Paints & allied products	0.2	Manufg.	Farm service facilities	0.5	Constr.
Paperboard containers & boxes	0.2	Manufg.	Sewer system facility construction	0.5	Constr.
Special dies & tools & machine tool accessories	0.2	Manufg.	Motor vehicle parts & accessories	0.5	Manufg.
Water transportation	0.2	Util.	Pumps & compressors	0.5	Manufg.
Real estate	0.2	Fin/R.E.	Tanks & tank components	0.5	Manufg.
U.S. Postal Service	0.2	Gov't	Gross private fixed investment	0.5	Cap Inv
Scrap	0.2	Scrap	Personal consumption expenditures	0.4	
Asbestos products	0.1	Manufg.	Warehouses	0.4	Constr.
Carbon & graphite products	0.1	Manufg.	Lighting fixtures & equipment	0.4	Manufg.
Lubricating oils & greases	0.1	Manufg.	Ship building & repairing	0.4	Manufg.
Narrow fabric mills	0.1	Manufg.	Special industry machinery, nec	0.4	Manufg.
Paper mills, except building paper	0.1	Manufg.	Wiring devices	0.4	Manufg.
Sawmills & planning mills, general	0.1	Manufg.	Communications, except radio & TV	0.4	Util.
Yarn mills & finishing of textiles, nec	0.1	Manufg.	Coal	0.3	Mining
Air transportation	0.1	Util.	Maintenance of highways & streets	0.3	Constr.
Accounting, auditing & bookkeeping	0.1	Services	Resid. & other health facility construction	0.3	Constr.
Legal services	0.1	Services	Electrical industrial apparatus, nec	0.3	Manufg.
Management & consulting services & labs	0.1	Services	Industrial controls	0.3	Manufg.
			Nonferrous rolling & drawing, nec	0.3	Manufg.
			Switchgear & switchboard apparatus	0.3	Manufg.
			Amusement & recreation building construction	0.2	Constr.
			Construction of nonfarm buildings nec	0.2	Constr.
			Maintenance of farm service facilities	0.2	Constr.
			Maintenance of military facilities	0.2	Constr.
			Maintenance of railroads	0.2	Constr.
			Residential 2-4 unit structures, nonfarm	0.2	Constr.
			Residential high-rise apartments	0.2	Constr.
			Food products machinery	0.2	Manufg.
			Instruments to measure electricity	0.2	Manufg.
			Logging camps & logging contractors	0.2	Manufg.
			Mechanical measuring devices	0.2	Manufg.
			Mobile homes	0.2	Manufg.
			Power driven hand tools	0.2	Manufg.
			Refrigeration & heating equipment	0.2	Manufg.
			Turbines & turbine generator sets	0.2	Manufg.
			Welding apparatus, electric	0.2	Manufg.
			X-ray apparatus & tubes	0.2	Manufg.
			Local transit facility construction	0.1	Constr.
			Maintenance of local transit facilities	0.1	Constr.
			Maintenance of nonbuilding facilities nec	0.1	Constr.
			Maintenance of petroleum & natural gas wells	0.1	Constr.
			Aircraft & missile equipment, nec	0.1	Manufg.

Continued on next page.

INPUTS AND OUTPUTS FOR NONFERROUS WIRE DRAWING & INSULATING - Continued

Economic Sector or Industry Providing Inputs	%	Sector	Economic Sector or Industry Buying Outputs	%	Sector
			Automatic merchandising machines	0.1	Manufg.
			Copper rolling & drawing	0.1	Manufg.
			Drapery hardware & blinds & shades	0.1	Manufg.
			Household cooking equipment	0.1	Manufg.
			Household vacuum cleaners	0.1	Manufg.
			Iron & steel foundries	0.1	Manufg.
			Miscellaneous fabricated wire products	0.1	Manufg.
			Motor vehicles & car bodies	0.1	Manufg.
			Radio & TV receiving sets	0.1	Manufg.
			Railroad equipment	0.1	Manufg.
			Steel wire & related products	0.1	Manufg.
			Surgical appliances & supplies	0.1	Manufg.

Source: Benchmark Input-Output Accounts for the U.S. Economy, 1982, U.S. Department of Commerce, Washington, D.C., July 1991. Data, as reported in the source, are organized by the 1977 SIC structure in use in 1982 but have been matched, as closely as is possible, to the 1987 SIC structure used in this book.

OCCUPATIONS EMPLOYED BY SIC 335 - NONFERROUS ROLLING AND DRAWING

Occupation	% of Total 1994	Change to 2005	Occupation	% of Total 1994	Change to 2005
Machine tool cutting & forming etc. nec	11.0	-12.4	Machinists	1.6	-20.3
Assemblers, fabricators, & hand workers nec	5.9	-20.4	Machine tool cutting operators, metal & plastic	1.4	-33.7
Blue collar worker supervisors	5.0	-22.9	Extruding & forming machine workers	1.3	-12.4
Machine forming operators, metal & plastic	5.0	7.5	Coating, painting, & spraying machine workers	1.3	-20.4
Metal & plastic machine workers nec	3.9	-15.3	General managers & top executives	1.2	-24.4
Inspectors, testers, & graders, precision	3.7	-20.3	Combination machine tool operators	1.2	-12.4
Helpers, laborers, & material movers nec	3.6	-20.4	Traffic, shipping, & receiving clerks	1.1	-23.3
Machine operators nec	3.1	-29.8	Industrial production managers	1.1	-20.4
Industrial machinery mechanics	2.9	-12.4	Packaging & filling machine operators	1.1	-20.4
Industrial truck & tractor operators	2.8	-20.3	Secretaries, ex legal & medical	1.1	-27.5
Maintenance repairers, general utility	2.1	-28.3	Machine feeders & offbearers	1.1	-28.3
Sales & related workers nec	1.9	-20.3	Production, planning, & expediting clerks	1.0	-20.3
Hand packers & packagers	1.7	-31.7	Metal molding machine workers	1.0	19.5
Electricians	1.7	-25.3			

Source: Industry-Occupation Matrix, Bureau of Labor Statistics. These data relate to one or more 3-digit SIC industry groups rather than to a single 4-digit SIC. The change reported for each occupation to the year 2005 is a percent of growth or decline as estimated by the Bureau of Labor Statistics. The abbreviation nec stands for 'not elsewhere classified'.

LOCATION BY STATE AND REGIONAL CONCENTRATION

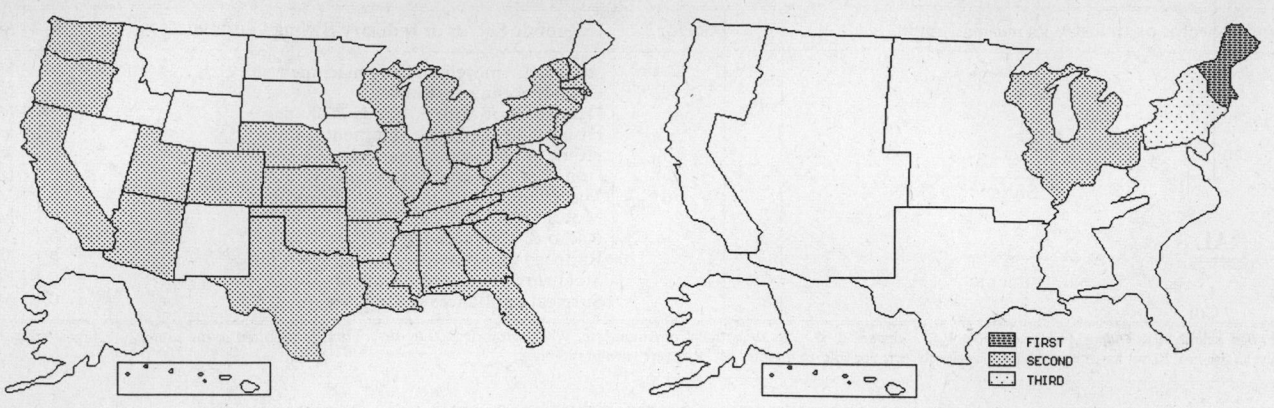

FIRST
SECOND
THIRD

INDUSTRY DATA BY STATE

State	Establish-ments	Shipments Total ($ mil)	Shipments % of U.S.	Shipments Per Establ.	Employment Total Number	Employment % of U.S.	Employment Per Establ.	Wages ($/hour)	Cost as % of Shipments	Investment per Employee ($)
Indiana	27	1,302.8	9.9	48.3	4,600	7.6	170	12.38	66.3	4,696
Georgia	14	1,260.8	9.6	90.1	4,100	6.8	293	16.78	54.2	-
North Carolina	21	975.7	7.4	46.5	4,300	7.1	205	10.04	61.3	8,512
Illinois	34	677.2	5.2	19.9	3,600	5.9	106	11.48	61.5	4,222
Texas	33	625.6	4.8	19.0	3,500	5.8	106	9.85	64.9	3,429
New York	40	622.8	4.8	15.6	3,100	5.1	78	10.65	70.1	4,032
Massachusetts	46	606.1	4.6	13.2	3,500	5.8	76	12.72	59.0	2,857
Kentucky	11	565.7	4.3	51.4	1,900	3.1	173	12.12	62.4	7,579
Connecticut	34	542.5	4.1	16.0	3,200	5.3	94	12.23	62.5	4,844
Arkansas	12	527.2	4.0	43.9	2,100	3.5	175	9.15	69.1	3,524
Pennsylvania	19	518.0	4.0	27.3	2,700	4.5	142	9.77	64.9	6,926
Rhode Island	8	497.2	3.8	62.2	2,700	4.5	338	11.48	55.4	1,593
California	60	439.7	3.4	7.3	2,900	4.8	48	10.83	68.8	3,724
Arizona	9	429.0	3.3	47.7	2,100	3.5	233	17.87	67.9	-
New Jersey	23	327.7	2.5	14.2	2,000	3.3	87	12.26	55.6	3,600
South Carolina	6	304.3	2.3	50.7	1,700	2.8	283	10.07	64.5	11,118
Ohio	21	272.7	2.1	13.0	1,200	2.0	57	12.10	64.7	5,500
Missouri	5	244.6	1.9	48.9	600	1.0	120	11.88	77.4	6,167
Mississippi	6	241.0	1.8	40.2	900	1.5	150	10.71	58.0	-
Virginia	7	209.6	1.6	29.9	1,200	2.0	171	11.94	57.5	4,750
Alabama	6	183.8	1.4	30.6	600	1.0	100	10.33	80.3	-
Vermont	7	103.1	0.8	14.7	700	1.2	100	8.00	50.6	-
Michigan	10	87.5	0.7	8.8	700	1.2	70	11.09	51.1	-
Florida	8	46.1	0.4	5.8	400	0.7	50	9.40	69.6	-
Wisconsin	3	27.7	0.2	9.2	200	0.3	67	12.50	65.3	3,000
New Hampshire	13	(D)	-	-	1,750 *	2.9	135	-	-	-
Washington	7	(D)	-	-	375 *	0.6	54	-	-	-
Kansas	5	(D)	-	-	750 *	1.2	150	-	-	-
Tennessee	5	(D)	-	-	750 *	1.2	150	-	-	-
Oklahoma	4	(D)	-	-	175 *	0.3	44	-	-	-
Oregon	4	(D)	-	-	375 *	0.6	94	-	-	-
Colorado	3	(D)	-	-	175 *	0.3	58	-	-	-
Nebraska	3	(D)	-	-	750 *	1.2	250	-	-	-
Utah	3	(D)	-	-	175 *	0.3	58	-	-	-
Delaware	2	(D)	-	-	375 *	0.6	188	-	-	-
Louisiana	2	(D)	-	-	175 *	0.3	88	-	-	-
West Virginia	1	(D)	-	-	175 *	0.3	175	-	-	-

Source: 1992 *Economic Census*. The states are in descending order of shipments or establishments (if shipment data are missing for the majority). The symbol (D) appears when data are withheld to prevent disclosure of competitive information. States marked with (D) are sorted by number of establishments. A dash (-) indicates that the data element cannot be calculated; * indicates the midpoint of a range.

3363 - ALUMINUM DIE-CASTINGS

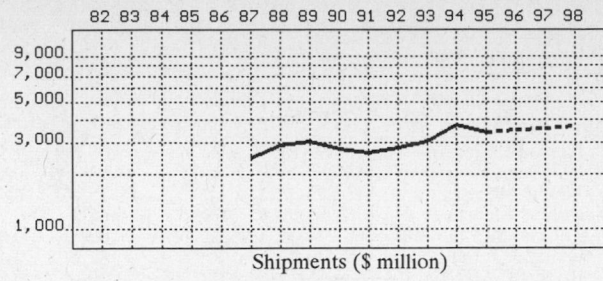

Shipments ($ million)

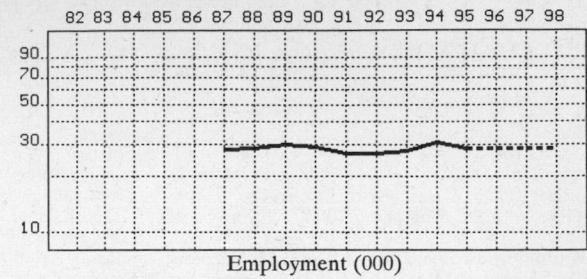

Employment (000)

GENERAL STATISTICS

| Year | Com- panies | Establishments | | Employment | | | Compensation | | Production ($ million) | | | |
		Total	with 20 or more employees	Total (000)	Production Workers (000)	Hours (Mil)	Payroll ($ mil)	Wages ($/hr)	Cost of Materials	Value Added by Manufacture	Value of Shipments	Capital Invest.
1982												
1983												
1984												
1985												
1986												
1987	397	412	231	28.1	22.7	46.4	696.8	11.12	1,216.8	1,267.8	2,468.8	94.8
1988		391	243	28.9	23.4	47.9	736.6	11.49	1,534.1	1,373.9	2,888.9	94.8
1989		371	240	30.1	24.5	49.8	756.5	11.42	1,625.4	1,416.1	3,044.0	152.2
1990		356	236	29.3	23.6	47.6	749.1	11.74	1,448.4	1,326.6	2,779.5	96.1
1991		355	239	27.0	22.0	44.0	714.1	12.09	1,302.9	1,306.6	2,628.0	100.5
1992	301	333	217	27.1	22.3	45.9	744.5	11.91	1,290.9	1,536.1	2,817.4	122.5
1993		323	208	27.6	22.9	48.7	797.2	12.19	1,387.9	1,668.4	3,035.9	131.7
1994		306P	213P	31.0	25.8	56.5	900.3	11.79	1,736.8	2,049.1	3,753.5	210.4
1995		292P	209P	28.8P	24.0P	51.5P	850.6P	12.26P	1,574.9P	1,858.1P	3,403.6P	174.1P
1996		278P	204P	28.8P	24.1P	52.2P	870.3P	12.37P	1,623.9P	1,915.9P	3,509.5P	184.9P
1997		263P	200P	28.8P	24.3P	52.9P	890.1P	12.49P	1,672.9P	1,973.7P	3,615.4P	195.7P
1998		249P	196P	28.9P	24.4P	53.7P	909.8P	12.61P	1,721.9P	2,031.5P	3,721.3P	206.5P

Sources: 1982, 1987, 1992 *Economic Census*; *Annual Survey of Manufactures*, 83-86, 88-91, 93-94. Establishment counts for non-Census years are from *County Business Patterns*; establishment values for 83-84 are extrapolations. 'P's show projections by the editors. Industries reclassified in 87 will not have data for prior years.

INDICES OF CHANGE

| Year | Com- panies | Establishments | | Employment | | | Compensation | | Production ($ million) | | | |
		Total	with 20 or more employees	Total (000)	Production Workers (000)	Hours (Mil)	Payroll ($ mil)	Wages ($/hr)	Cost of Materials	Value Added by Manufacture	Value of Shipments	Capital Invest.
1982												
1983												
1984												
1985												
1986												
1987	132	124	106	104	102	101	94	93	94	83	88	77
1988		117	112	107	105	104	99	96	119	89	103	77
1989		111	111	111	110	108	102	96	126	92	108	124
1990		107	109	108	106	104	101	99	112	86	99	78
1991		107	110	100	99	96	96	102	101	85	93	82
1992	100	100	100	100	100	100	100	100	100	100	100	100
1993		97	96	102	103	106	107	102	108	109	108	108
1994		92P	98P	114	116	123	121	99	135	133	133	172
1995		88P	96P	106P	108P	112P	114P	103P	122P	121P	121P	142P
1996		83P	94P	106P	108P	114P	117P	104P	126P	125P	125P	151P
1997		79P	92P	106P	109P	115P	120P	105P	130P	128P	128P	160P
1998		75P	90P	106P	109P	117P	122P	106P	133P	132P	132P	169P

Sources: Same as General Statistics. Values reflect change from the base year, 1992. Values above 100 mean greater than 92, values below 100 mean less than 92, and a value of 100 in the 82-91 or 93-98 period means same as 92. 'P's mark projections by the editors.

SELECTED RATIOS

For 1994	Avg. of All Manufact.	Analyzed Industry	Index	For 1994	Avg. of All Manufact.	Analyzed Industry	Index
Employees per Establishment	49	101	207	Value Added per Production Worker	134,084	79,422	59
Payroll per Establishment	1,500,273	2,942,157	196	Cost per Establishment	5,045,178	5,675,817	112
Payroll per Employee	30,620	29,042	95	Cost per Employee	102,970	56,026	54
Production Workers per Establishment	34	84	246	Cost per Production Worker	146,988	67,318	46
Wages per Establishment	853,319	2,176,912	255	Shipments per Establishment	9,576,895	12,266,340	128
Wages per Production Worker	24,861	25,819	104	Shipments per Employee	195,460	121,081	62
Hours per Production Worker	2,056	2,190	107	Shipments per Production Worker	279,017	145,484	52
Wages per Hour	12.09	11.79	98	Investment per Establishment	321,011	687,582	214
Value Added per Establishment	4,602,255	6,696,405	146	Investment per Employee	6,552	6,787	104
Value Added per Employee	93,930	66,100	70	Investment per Production Worker	9,352	8,155	87

Sources: Same as General Statistics. The 'Average of All Manufacturing' column represents the average of all manufacturing industries reported for the most recent complete year available. The Index shows the relationship between the Average and the Analyzed Industry. For example, 100 means that they are equal; 500 that the Analyzed Industry is five times the average; 50 means that the Analyzed Industry is half the national average. The abbreviation 'na' is used to show that data are 'not available'.

LEADING COMPANIES   Number shown: **75**   Total sales ($ mil): **1,764**   Total employment (000): **16.2**

Company Name	Address				CEO Name	Phone	Co. Type	Sales ($ mil)	Empl. (000)
Doehler-Jarvis Inc	PO Box 902	Toledo	OH	43697	Ronald L Stewart	419-470-8177	R	200	1.5
ITT Automotive	3000 University Dr	Auburn Hills	MI	48326	RD Davis	810-340-3000	S	90*	0.8
Stahl Specialty Co	PO Box 6	Kingsville	MO	64061	Jack R Moore	816-597-3322	R	75	0.9
Cascade Die Casting Group Inc	3040 Charlevoix S	Grand Rapids	MI	49546	Robert Karban Jr	616-956-0966	S	65	0.7
Madison-Kipp Corp	PO Box 3037	Madison	WI	53704	Thomas W Caldwell	608-244-3511	R	60	0.6
Spartan Aluminum Products Inc	510 E McClurken	Sparta	IL	62286	Henry A Jubel	618-443-4346	R	52	0.4
Aallied Die Casting & Mfg	3021 Cullerton Dr	Franklin Park	IL	60131	Robert C Marconi	708-455-1950	R	50	0.3
Small Assemblies Inc	4140 W Belmont	Chicago	IL	60641	Vincent Blakeley	312-202-9000	R	50	0.1
PHB Die Castings Inc	7900 W Ridge Rd	Fairview	PA	16415	W M Hilbert Sr	814-474-5511	S	43	0.5
Le Sueur Inc	PO Box 149	Le Sueur	MN	56058	Mark Mueller	612-665-6204	R	42	0.5
Tool Products	5100 Boone Av N	Minneapolis	MN	55428	Robert W Carlson	612-536-5500	S	40	0.4
Du-Wel Products Inc	901 Spruce St	Dowagiac	MI	49047	Dave Harrington	616-782-2108	R	35	0.3
Inverness Casting Group Inc	25700 Science Pk Dr	Beachwood	OH	44122	William R Seelbach	216-464-2900	R	35*	0.3
Magnesium Casting Co	98 Business St	Hyde Park	MA	02136	Harvey B Berman	617-361-1710	R	35	0.2
Tecumseh Products Co	PO Box 0909	Sheboygan Fls	WI	53085	Ken Miller	414-467-6161	D	33*	0.3
Cambridge Tool & Mfg Co	67 Faulkner St	North Billerica	MA	01862	Dieter Morlock	508-667-8400	R	30	0.3
Liston Aluminum Brick	PO Box 1869	Corona	CA	91718	Walter Hall	909-277-4221	R	30	<0.1
Progress Casting Group Inc	2600 Niagara Ln	Plymouth	MN	55447	Edwin Miguelucci	612-557-1000	R	30	0.3
Diecast Corp	522 Hupp Av	Jackson	MI	49203	RB Honer	517-788-6100	R	26	0.3
Empire Die Casting Company	19800 Miles Av	Cleveland	OH	44128	Richard P Rogel	216-662-8700	R	26	0.3
Impact Industries Inc	1212 E 6th St	Sandwich	IL	60548	Gary Miller	815-786-9810	S	25	0.3
Top Die Casting Co	PO Box 188	South Beloit	IL	61080	LaVerne McCurdy	815-389-2599	R	25	0.2
Heick Die Casting Corp	6550 W Diversey	Chicago	IL	60635	Fred Heick	312-637-1100	R	24	0.2
Pemco Die Casting Corp	PO Box 367	Bridgman	MI	49106	T Allen	616-465-6000	R	22	0.2
Acme Die Casting Corp	5626 21st St	Racine	WI	53406	RJ Dovorany	414-554-8887	R	20	0.3
Dee Inc	PO Box 627	Crookston	MN	56716	James Ellinger	218-281-5811	R	20*	0.2
Dycast Inc	320 E Main St	Lake Zurich	IL	60047	Stephen L Nyers III	708-438-8214	R	20	0.2
Premier Tool and Die Cast Corp	PO Box 210	Berrien Springs	MI	49103	Ken Nitz	616-471-7715	R	20*	0.2
Quad City Die Casting Co	3800 River Dr	Moline	IL	61265	D Debrey	309-762-7346	R	20	0.2
Twin City Die Castings Co	1070 33rd Av SE	Minneapolis	MN	55414	Steven J Harmon	612-645-3611	R	20*	0.2
Southern Die Casting & Eng	PO Box 8045	High Point	NC	27264	L Mark Kuchinic	910-882-0186	D	19	0.2
Alpha Technology Corp	PO Box 168	Howell	MI	48844	Stephen Sweda	517-546-9700	S	18	0.1
Central Die Casting & Mfg	2935 W 47th St	Chicago	IL	60632	John Olson	312-523-6515	R	18	0.2
Hollander Industries	PO Box 135	Dayton	OH	45404	Larry Hollander	513-223-8817	R	18	0.2
C and H Die Casting Inc	PO Box 1170	Temple	TX	76501	CE Hinkle	817-938-2541	R	16	0.3
Hartzell Manufacturing Inc	Industrial Park	Turtle Lake	WI	54889	Dwain Kasel	715-986-4407	D	16	0.2
Hartzell Manufacturing Inc	PO Box 64529	St Paul	MN	55164	Dwain Kasel	612-646-9456	R	16	0.2
Basic Aluminum Castings Co	1325 E 168th St	Cleveland	OH	44110	Thomas M Byrne	216-481-5606	R	15	<0.1
Imperial Die Casting	2249 Old Liberty	Liberty	SC	29657	Robert Marconi	803-859-0202	D	15	0.2
Port City Diecast Inc	1985 E Laketon Av	Muskegon	MI	49442	Bruce J Essex	616-777-3941	R	15	<0.1
Ajax Die-Casting	PO Box 1	Broken Arrow	OK	74013	Anthony Nondorf	918-251-5366	D	14	0.2
Cast-Matic Corp	PO Box 251	Stevensville	MI	49127	D E Shembarger	616-429-1554	S	14	0.2
Blue Ridge Pressure Casting Inc	PO Box 208	Lehighton	PA	18235	A Donald Behler	610-377-2510	R	13*	0.1
Alloy Die Casting Company Inc	6550 Caballero Blv	Buena Park	CA	90620	WE Holmes	714-521-9800	R	12*	0.1
Buchanan Industries Inc	21411 Civic Ctr Dr	Southfield	MI	48076	Donald L Munoz	313-352-6030	R	12	0.1
Cast-Rite Corp	515 E Airline Way	Gardena	CA	90248	Gardner L Bickford	310-532-2080	R	12	0.2
Dilesco Corp	1806 Beidler St	Muskegon	MI	49441	Tom Larsen	616-726-4002	R	12	0.1
Production Aluminum Co	9983 Sparta Av	Sparta	MI	49345	B Karban	616-887-1771	D	12	0.1
Bardane Manufacturing Co	PO Box 70	Jermyn	PA	18433	Peter Horvick	717-876-4844	R	11	<0.1
Kiowa Corp	PO Box 657	Marshalltown	IA	50158	Greg Brown	515-753-5566	R	11*	0.2
ADCO Die Cast Corp	PO Box 365	Bridgman	MI	49106	Ron Gelesko	616-465-5000	R	10	0.2
Advance Die Casting Co	3760 N Holton St	Milwaukee	WI	53212	Richard H Omann	414-964-0284	R	10	<0.1
Advance Pressure Castings	276 Rte 53	Denville	NJ	07834	Kenneth Scripta	201-627-6600	D	10	0.1
Albany-Chicago Co	8200 100 St	Kenosha	WI	53142	Jeff Richardson	414-947-7600	R	10	0.1
Angola Die Casting Inc	410 Weatherhead St	Angola	IN	46703	Gary J Malloy	219-665-9481	R	10*	<0.1
Die Cast Products Inc	621 W Rosecrans	Gardena	CA	90248	Jerry Plantz	213-321-1561	R	10	0.1
Dynacast Manufacturing Co	PO Box 311355	New Braunfels	TX	78131	Dan Ogle	210-625-7357	S	10	0.1
Hamilton Die Cast Inc	999 East Av	Hamilton	OH	45011	Joseph A Woltering	513-856-9500	R	10	0.1
Kennedy Die Castings Inc	15 Coppage Dr	Worcester	MA	01603	PS Kennedy	508-791-5594	R	10*	0.1
Krone Casting Corp	925 ML King	North Chicago	IL	60064	Ray Cross Sr	708-473-4040	R	10	0.1
Taylor-Pohlman Inc	PO Box 648	Orchard Park	NY	14127	Peter Taylor	716-662-3113	R	10	0.1
West Irving Die Co	970 Pauly Dr	Elk Grove Vill	IL	60007	J Klement	708-678-5500	R	10	0.1
Dundee Casting Co	500 Ypsilanti St	Dundee	MI	48131	Ed T Crawley	313-529-2455	R	9*	<0.1
Prima Die Casting Inc	5300 115th Av N	Clearwater	FL	34620	Ray Eichhof	813-572-7040	R	9*	<0.1
Los Angeles Die Casting	333 Central Av	Los Angeles	CA	90013	M Hand	213-624-2322	D	8*	<0.1
Tool Die Engineering Co	28850 Aurora Rd	Solon	OH	44139	JJ Schmader III	216-248-8020	R	8	<0.1
American Tool and Mold Inc	4133 S M-139 Hwy	St Joseph	MI	49085	Les Carroll	616-428-2442	R	7*	<0.1
Arrow Acme Co	PO Box 218	Webster City	IA	50595	Donald Peyser	515-832-3120	D	7	<0.1
C and D Die Casting Company	9649 Owensmouth	Chatsworth	CA	91311	MH Clifford	818-341-9822	R	7*	<0.1
Deckerville Die-Form Co	PO Box 95	Deckerville	MI	48427	Henry Sobell	810-376-2245	S	7*	<0.1
SKS Die Casting and Machining	1849 Oak St	Alameda	CA	94501	JW Keating	510-523-2541	R	7*	<0.1
US Die Casting Inc	3315 N Knox Av	Chicago	IL	60641	John Gegenheimer	312-685-2600	S	7	<0.1
A and B Die Casting Co	1417 4th St	Berkeley	CA	94710	BE Dathe	510-525-8991	R	6	<0.1
AL Johnson Co	4671 Calle Carga	Camarillo	CA	93012	Richard R Carlson	818-358-0196	S	6	<0.1
Louisiana Manufacturing Co	1404 N Carolina	Louisiana	MO	63353	Gary Hart	314-754-4191	D	6	<0.1

Source: *Ward's Business Directory of U.S. Private and Public Companies*, Volumes 1 and 2, 1996. The company type code used is as follows: P - Public, R - Private, S - Subsidiary, D - Division, J - Joint Venture, A - Affiliate, G - Group. Sales are in millions of dollars, employees are in thousands. An asterisk (*) indicates an estimated sales volume. The symbol < stands for 'less than'. Company names and addresses are truncated, in some cases, to fit into the available space.

MATERIALS CONSUMED

Material	Quantity	Delivered Cost ($ million)
Materials, ingredients, containers, and supplies	(X)	1,033.7
Copper and copper-base alloy shapes and forms	1,000 s tons (S)	37.2
Aluminum and aluminum-base alloy shapes and forms	1,000 s tons (S)	486.7
Zinc and zinc-base alloy shapes and forms	(X)	32.3
Magnesium and magnesium-base alloy shapes and forms	(X)	(D)
All other nonferrous shapes and forms	(X)	(D)
Copper and copper-base alloy scrap (except home scrap)	(X)	(D)
Aluminum and aluminum-base alloy scrap (except home scrap)	(X)	134.0
Other nonferrous metal scrap (except home scrap)	(X)	(D)
Industrial patterns	(X)	2.0
Industrial dies, molds, jigs, and fixtures	(X)	26.0
All other industrial and commercial machinery and computer equipment	(X)	4.5
Sand	(X)	0.7
Grinding wheels and other abrasive products, except industrial diamonds	(X)	1.4
All other materials and components, parts, containers, and supplies	(X)	192.2
Materials, ingredients, containers, and supplies, nsk	(X)	100.4

Source: 1992 *Economic Census.* Explanation of symbols used: (D): Withheld to avoid disclosure of competitive data; na: Not available; (S): Withheld because statistical norms were not met; (X): Not applicable; (Z): Less than half the unit shown; nec: Not elsewhere classified; nsk: Not specified by kind; - : zero; * : 10-19 percent estimated; ** : 20-29 percent estimated.

PRODUCT SHARE DETAILS

Product or Product Class	% Share	Product or Product Class	% Share
Aluminum die-castings	100.00		

Source: 1992 *Economic Census.* The values shown are percent of total shipments in an industry. Values of indented subcategories are summed in the main heading. The symbol (D) appears when data are withheld to prevent disclosure of competitive information. The abbreviation nsk stands for 'not specified by kind' and nec for 'not elsewhere classified'.

INPUTS AND OUTPUTS FOR ALUMINUM CASTINGS

Economic Sector or Industry Providing Inputs	%	Sector	Economic Sector or Industry Buying Outputs	%	Sector
Primary aluminum	38.9	Manufg.	Motor vehicle parts & accessories	23.1	Manufg.
Special dies & tools & machine tool accessories	10.2	Manufg.	Internal combustion engines, nec	8.7	Manufg.
Communications, except radio & TV	6.6	Util.	Machinery, except electrical, nec	4.2	Manufg.
Electric services (utilities)	5.2	Util.	Electronic computing equipment	3.5	Manufg.
Gas production & distribution (utilities)	5.2	Util.	Aircraft & missile engines & engine parts	3.2	Manufg.
Scrap	3.3	Scrap	Pipe, valves, & pipe fittings	3.2	Manufg.
Wholesale trade	3.0	Trade	Refrigeration & heating equipment	2.7	Manufg.
Maintenance of nonfarm buildings nec	1.9	Constr.	Pumps & compressors	2.5	Manufg.
Advertising	1.8	Services	Radio & TV communication equipment	2.5	Manufg.
Petroleum refining	1.7	Manufg.	Federal Government purchases, national defense	2.4	Fed Govt
Primary copper	1.6	Manufg.	Motor vehicles & car bodies	2.3	Manufg.
Motor freight transportation & warehousing	1.6	Util.	Exports	1.8	Foreign
Primary zinc	1.5	Manufg.	Lighting fixtures & equipment	1.7	Manufg.
Miscellaneous repair shops	1.3	Services	Motors & generators	1.7	Manufg.
Banking	1.1	Fin/R.E.	Photographic equipment & supplies	1.5	Manufg.
Electrical repair shops	1.1	Services	Household cooking equipment	1.4	Manufg.
Railroads & related services	1.0	Util.	Mechanical measuring devices	1.4	Manufg.
Eating & drinking places	0.8	Trade	Aircraft	1.3	Manufg.
Real estate	0.7	Fin/R.E.	Power driven hand tools	1.3	Manufg.
Royalties	0.7	Fin/R.E.	Carburetors, pistons, rings, & valves	1.2	Manufg.
Equipment rental & leasing services	0.7	Services	Lawn & garden equipment	1.1	Manufg.
Machinery, except electrical, nec	0.6	Manufg.	Wiring devices	1.1	Manufg.
Abrasive products	0.5	Manufg.	Aircraft & missile equipment, nec	0.9	Manufg.
Metal heat treating	0.5	Manufg.	Electronic components nec	0.9	Manufg.
Primary nonferrous metals, nec	0.5	Manufg.	Sporting & athletic goods, nec	0.9	Manufg.
Sand & gravel	0.4	Mining	Construction machinery & equipment	0.8	Manufg.
Electrometallurgical products	0.4	Manufg.	General industrial machinery, nec	0.8	Manufg.
Industrial patterns	0.4	Manufg.	Hardware, nec	0.8	Manufg.
Paperboard containers & boxes	0.4	Manufg.	Special industry machinery, nec	0.8	Manufg.
Aluminum castings	0.3	Manufg.	Blowers & fans	0.7	Manufg.
Lubricating oils & greases	0.3	Manufg.	Power transmission equipment	0.7	Manufg.
Air transportation	0.3	Util.	Printing trades machinery	0.7	Manufg.
Automotive rental & leasing, without drivers	0.3	Services	Federal Government purchases, nondefense	0.7	Fed Govt
Automotive repair shops & services	0.3	Services	Farm machinery & equipment	0.6	Manufg.
Legal services	0.3	Services	Food products machinery	0.6	Manufg.
Management & consulting services & labs	0.3	Services	Household laundry equipment	0.6	Manufg.
U.S. Postal Service	0.3	Gov't	Switchgear & switchboard apparatus	0.6	Manufg.
Manifold business forms	0.2	Manufg.	Woodworking machinery	0.6	Manufg.
Wood products, nec	0.2	Manufg.	S/L Govt. purch., health & hospitals	0.6	S/L Govt
Sanitary services, steam supply, irrigation	0.2	Util.	Electric housewares & fans	0.5	Manufg.
Credit agencies other than banks	0.2	Fin/R.E.	Engine electrical equipment	0.5	Manufg.

Continued on next page.

INPUTS AND OUTPUTS FOR ALUMINUM CASTINGS - Continued

Economic Sector or Industry Providing Inputs	%	Sector	Economic Sector or Industry Buying Outputs	%	Sector
Insurance carriers	0.2	Fin/R.E.	Hand & edge tools, nec	0.5	Manufg.
Accounting, auditing & bookkeeping	0.2	Services	Metal stampings, nec	0.5	Manufg.
Business/professional associations	0.2	Services	Tanks & tank components	0.5	Manufg.
Computer & data processing services	0.2	Services	Transportation equipment, nec	0.5	Manufg.
Laundry, dry cleaning, shoe repair	0.2	Services	Personal consumption expenditures	0.4	
Blast furnaces & steel mills	0.1	Manufg.	Fabricated plate work (boiler shops)	0.4	Manufg.
Machine tools, metal cutting types	0.1	Manufg.	Industrial controls	0.4	Manufg.
Engineering, architectural, & surveying services	0.1	Services	Service industry machines, nec	0.4	Manufg.
State & local government enterprises, nec	0.1	Gov't	Textile machinery	0.4	Manufg.
Imports	0.1	Foreign	Typewriters & office machines, nec	0.4	Manufg.
			Calculating & accounting machines	0.3	Manufg.
			Engineering & scientific instruments	0.3	Manufg.
			Environmental controls	0.3	Manufg.
			Fabricated metal products, nec	0.3	Manufg.
			Hoists, cranes, & monorails	0.3	Manufg.
			Household refrigerators & freezers	0.3	Manufg.
			Instruments to measure electricity	0.3	Manufg.
			Measuring & dispensing pumps	0.3	Manufg.
			Special dies & tools & machine tool accessories	0.3	Manufg.
			S/L Govt. purch., other general government	0.3	S/L Govt
			Guided missiles & space vehicles	0.2	Manufg.
			Household vacuum cleaners	0.2	Manufg.
			Metal doors, sash, & trim	0.2	Manufg.
			Metalworking machinery, nec	0.2	Manufg.
			Motorcycles, bicycles, & parts	0.2	Manufg.
			Oil field machinery	0.2	Manufg.
			Radio & TV receiving sets	0.2	Manufg.
			Scales & balances	0.2	Manufg.
			Sheet metal work	0.2	Manufg.
			Small arms	0.2	Manufg.
			Telephone & telegraph apparatus	0.2	Manufg.
			Water transportation	0.2	Util.
			Eating & drinking places	0.2	Trade
			Aluminum castings	0.1	Manufg.
			Boat building & repairing	0.1	Manufg.
			Conveyors & conveying equipment	0.1	Manufg.
			Heating equipment, except electric	0.1	Manufg.
			Industrial patterns	0.1	Manufg.
			Industrial trucks & tractors	0.1	Manufg.
			Machine tools, metal cutting types	0.1	Manufg.
			Machine tools, metal forming types	0.1	Manufg.
			Railroad equipment	0.1	Manufg.
			Screw machine and related products	0.1	Manufg.
			S/L Govt. purch., correction	0.1	S/L Govt

Source: Benchmark Input-Output Accounts for the U.S. Economy, 1982, U.S. Department of Commerce, Washington, D.C., July 1991. Data, as reported in the source, are organized by the 1977 SIC structure in use in 1982 but have been matched, as closely as is possible, to the 1987 SIC structure used in this book.

OCCUPATIONS EMPLOYED BY SIC 336 - NONFERROUS FOUNDRIES (CASTINGS)

Occupation	% of Total 1994	Change to 2005	Occupation	% of Total 1994	Change to 2005
Metal molding machine workers	12.2	-2.1	Machine tool cutting & forming etc. nec	1.7	-10.9
Assemblers, fabricators, & hand workers nec	7.2	-11.0	Combination machine tool operators	1.6	-2.1
Grinders & polishers, hand	5.6	-11.0	Precision metal workers nec	1.6	-11.0
Blue collar worker supervisors	5.2	-15.7	Machine forming operators, metal & plastic	1.5	-10.9
Inspectors, testers, & graders, precision	5.1	-11.0	Traffic, shipping, & receiving clerks	1.4	-14.3
Precision workers nec	4.6	-19.9	Industrial production managers	1.4	-10.9
Tool & die makers	3.3	-28.1	Metal & plastic machine workers nec	1.3	-21.0
Foundry mold assembly & shakeout workers	3.1	-19.9	Hand packers & packagers	1.3	-23.7
General managers & top executives	2.7	-15.5	Janitors & cleaners, incl maids	1.1	-28.8
Machine tool cutting operators, metal & plastic	2.7	-25.8	Industrial truck & tractor operators	1.1	-11.0
Grinding machine operators, metal & plastic	2.5	20.2	Sales & related workers nec	1.1	-10.9
Furnace operators	2.5	-19.9	Bookkeeping, accounting, & auditing clerks	1.1	-33.3
Machinists	2.5	-11.0	Punching machine operators, metal & plastic	1.0	-28.8
Industrial machinery mechanics	2.1	-2.1	General office clerks	1.0	-24.1
Maintenance repairers, general utility	1.8	-19.9			

Source: Industry-Occupation Matrix, Bureau of Labor Statistics. These data relate to one or more 3-digit SIC industry groups rather than to a single 4-digit SIC. The change reported for each occupation to the year 2005 is a percent of growth or decline as estimated by the Bureau of Labor Statistics. The abbreviation nec stands for 'not elsewhere classified'.

LOCATION BY STATE AND REGIONAL CONCENTRATION

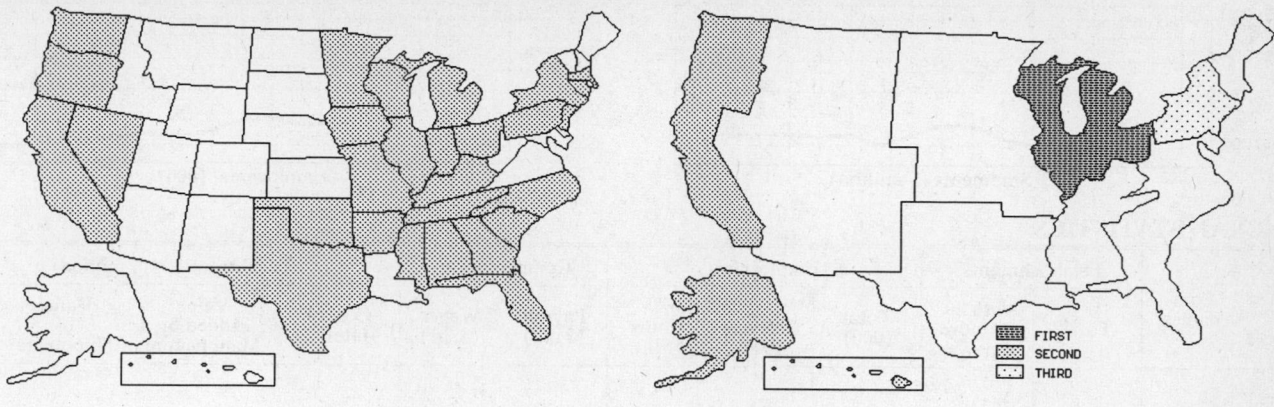

FIRST
SECOND
THIRD

INDUSTRY DATA BY STATE

| State | Establish-ments | Shipments | | | Employment | | | | Cost as % of Shipments | Investment per Employee ($) |
		Total ($ mil)	% of U.S.	Per Establ.	Total Number	% of U.S.	Per Establ.	Wages ($/hour)		
Wisconsin	22	411.0	14.6	18.7	3,400	12.5	155	12.76	52.4	5,235
Michigan	35	357.7	12.7	10.2	3,000	11.1	86	12.65	45.5	3,167
Ohio	31	327.2	11.6	10.6	2,700	10.0	87	13.07	40.6	9,000
Illinois	32	258.4	9.2	8.1	2,900	10.7	91	9.68	41.5	2,862
Indiana	12	248.5	8.8	20.7	2,000	7.4	167	22.59	61.9	-
Pennsylvania	15	149.4	5.3	10.0	1,500	5.5	100	12.88	40.3	1,133
California	50	123.4	4.4	2.5	1,700	6.3	34	11.48	48.1	1,588
Minnesota	8	85.5	3.0	10.7	900	3.3	113	12.93	37.5	-
Tennessee	7	75.5	2.7	10.8	800	3.0	114	9.93	47.2	4,750
Missouri	10	73.2	2.6	7.3	1,100	4.1	110	8.25	37.8	-
Texas	13	45.1	1.6	3.5	700	2.6	54	8.23	47.2	3,857
Massachusetts	8	44.2	1.6	5.5	600	2.2	75	8.10	36.4	1,167
South Carolina	4	42.5	1.5	10.6	500	1.8	125	8.89	37.4	3,000
New York	8	35.6	1.3	4.4	600	2.2	75	17.56	57.6	-
Alabama	7	27.2	1.0	3.9	200	0.7	29	10.00	49.6	2,500
Connecticut	5	23.7	0.8	4.7	300	1.1	60	12.80	38.4	-
North Carolina	4	20.3	0.7	5.1	200	0.7	50	10.00	35.5	-
New Jersey	8	14.8	0.5	1.9	200	0.7	25	16.00	44.6	1,000
Iowa	5	11.4	0.4	2.3	200	0.7	40	12.67	32.5	-
Arkansas	7	(D)	-	-	1,750 *	6.5	250	-	-	-
Georgia	6	(D)	-	-	375 *	1.4	63	-	-	-
Kentucky	5	(D)	-	-	1,750 *	6.5	350	-	-	-
Oregon	5	(D)	-	-	175 *	0.6	35	-	-	-
Florida	4	(D)	-	-	175 *	0.6	44	-	-	-
Nevada	2	(D)	-	-	175 *	0.6	88	-	-	-
Oklahoma	2	(D)	-	-	175 *	0.6	88	-	-	-
Washington	2	(D)	-	-	175 *	0.6	88	-	-	-
Mississippi	1	(D)	-	-	175 *	0.6	175	-	-	-

Source: 1992 *Economic Census.* The states are in descending order of shipments or establishments (if shipment data are missing for the majority). The symbol (D) appears when data are withheld to prevent disclosure of competitive information. States marked with (D) are sorted by number of establishments. A dash (-) indicates that the data element cannot be calculated; * indicates the midpoint of a range.

3364 - NONFERROUS DIE-CASTINGS EX ALUMINUM

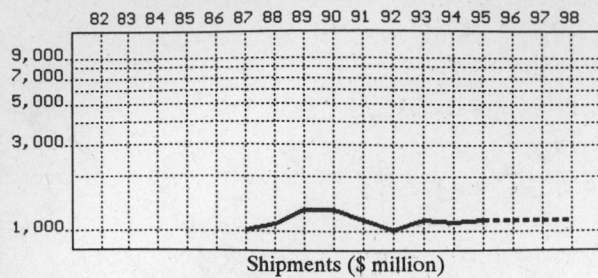

Shipments ($ million)

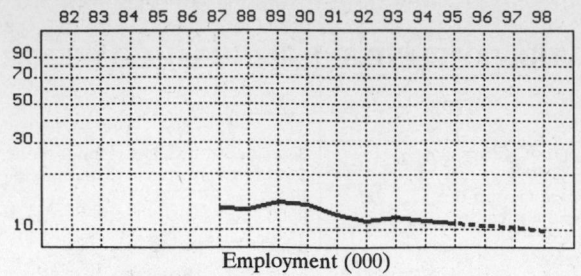

Employment (000)

GENERAL STATISTICS

| Year | Com-panies | Establishments | | Employment | | | Compensation | | Production ($ million) | | | |
		Total	with 20 or more employees	Total (000)	Production Workers (000)	Hours (Mil)	Payroll ($ mil)	Wages ($/hr)	Cost of Materials	Value Added by Manufacture	Value of Shipments	Capital Invest.
1982												
1983												
1984												
1985												
1986												
1987	292	304	156	12.9	10.4	20.7	260.3	8.52	482.9	534.0	1,009.6	26.0
1988		291	158	12.8	10.1	20.4	266.7	8.67	534.2	575.5	1,109.3	32.5
1989		285	166	14.2	10.7	21.8	296.6	8.95	694.0	638.0	1,320.7	34.2
1990		276	158	13.7	10.5	20.8	302.3	9.46	715.5	598.2	1,317.9	26.7
1991		285	154	12.1	9.7	19.7	287.5	9.86	603.0	559.1	1,165.2	26.0
1992	253	263	120	11.1	8.8	17.9	272.6	10.17	481.2	541.7	1,020.8	33.5
1993		258	123	11.6	9.4	17.8	288.4	11.05	550.0	599.1	1,153.7	44.5
1994		253P	121P	11.1	9.5	18.2	311.3	12.79	528.5	604.5	1,121.4	45.4
1995		246P	114P	10.9P	9.0P	17.3P	306.0P	12.39P	541.9P	619.8P	1,149.8P	43.9P
1996		239P	108P	10.5P	8.8P	16.8P	310.5P	12.94P	541.6P	619.5P	1,149.2P	46.2P
1997		232P	101P	10.2P	8.6P	16.3P	315.0P	13.48P	541.3P	619.2P	1,148.6P	48.5P
1998		225P	94P	9.8P	8.4P	15.8P	319.5P	14.03P	541.1P	618.9P	1,148.1P	50.8P

Sources: 1982, 1987, 1992 *Economic Census*; *Annual Survey of Manufactures*, 83-86, 88-91, 93-94. Establishment counts for non-Census years are from *County Business Patterns*; establishment values for 83-84 are extrapolations. 'P's show projections by the editors. Industries reclassified in 87 will not have data for prior years.

INDICES OF CHANGE

| Year | Com-panies | Establishments | | Employment | | | Compensation | | Production ($ million) | | | |
		Total	with 20 or more employees	Total (000)	Production Workers (000)	Hours (Mil)	Payroll ($ mil)	Wages ($/hr)	Cost of Materials	Value Added by Manufacture	Value of Shipments	Capital Invest.
1982												
1983												
1984												
1985												
1986												
1987	115	116	130	116	118	116	95	84	100	99	99	78
1988		111	132	115	115	114	98	85	111	106	109	97
1989		108	138	128	122	122	109	88	144	118	129	102
1990		105	132	123	119	116	111	93	149	110	129	80
1991		108	128	109	110	110	105	97	125	103	114	78
1992	100	100	100	100	100	100	100	100	100	100	100	100
1993		98	102	105	107	99	106	109	114	111	113	133
1994		96P	101P	100	108	102	114	126	110	112	110	136
1995		93P	95P	98P	102P	97P	112P	122P	113P	114P	113P	131P
1996		91P	90P	95P	100P	94P	114P	127P	113P	114P	113P	138P
1997		88P	84P	91P	98P	91P	116P	133P	112P	114P	113P	145P
1998		85P	79P	88P	96P	88P	117P	138P	112P	114P	112P	152P

Sources: Same as General Statistics. Values reflect change from the base year, 1992. Values above 100 mean greater than 92, values below 100 mean less than 92, and a value of 100 in the 82-91 or 93-98 period means same as 92. 'P's mark projections by the editors.

SELECTED RATIOS

For 1994	Avg. of All Manufact.	Analyzed Industry	Index	For 1994	Avg. of All Manufact.	Analyzed Industry	Index
Employees per Establishment	49	44	90	Value Added per Production Worker	134,084	63,632	47
Payroll per Establishment	1,500,273	1,232,523	82	Cost per Establishment	5,045,178	2,092,477	41
Payroll per Employee	30,620	28,045	92	Cost per Employee	102,970	47,613	46
Production Workers per Establishment	34	38	110	Cost per Production Worker	146,988	55,632	38
Wages per Establishment	853,319	921,632	108	Shipments per Establishment	9,576,895	4,439,932	46
Wages per Production Worker	24,861	24,503	99	Shipments per Employee	195,460	101,027	52
Hours per Production Worker	2,056	1,916	93	Shipments per Production Worker	279,017	118,042	42
Wages per Hour	12.09	12.79	106	Investment per Establishment	321,011	179,751	56
Value Added per Establishment	4,602,255	2,393,382	52	Investment per Employee	6,552	4,090	62
Value Added per Employee	93,930	54,459	58	Investment per Production Worker	9,352	4,779	51

Sources: Same as General Statistics. The 'Average of All Manufacturing' column represents the average of all manufacturing industries reported for the most recent complete year available. The Index shows the relationship between the Average and the Analyzed Industry. For example, 100 means that they are equal; 500 that the Analyzed Industry is five times the average; 50 means that the Analyzed Industry is half the national average. The abbreviation 'na' is used to show that data are 'not available'.

LEADING COMPANIES Number shown: **45** Total sales ($ mil): **821** Total employment (000): **7.2**

Company Name	Address				CEO Name	Phone	Co. Type	Sales ($ mil)	Empl. (000)
Gibbs Die Casting Corp	369 Community Dr	Henderson	KY	42420	Nick Gibbs	502-827-7854	R	135	1.0
Contech	5340 Holiday	Kalamazoo	MI	49002	Gary Walker	616-375-1013	D	100	0.5
Fall River Group Inc	PO Box 40	Mequon	WI	53092	Charles F Wright	414-241-8020	R	75	0.5
TCH Industries Inc	3040 Charlevoix S	Grand Rapids	MI	49546	Robert Karban Jr	616-942-0505	R	65	0.7
Halex Co	23901 Aurora Rd	Bedford Hts	OH	44146	Tim Medford	216-439-1616	D	42	0.2
Fishercast	PO Box 837	Watertown	NY	13601	William F Fisher	315-788-8100	D	31*	0.4
Brillcast Inc	3400 Wentw SW	Grand Rapids	MI	49509	RJ Brill	616-534-4977	R	28	0.3
Chicago White Metal Casting Inc	Rte 83	Bensenville	IL	60106	WG Treiber	708-595-4424	R	23	0.2
Deco Products Co	506 Sanford St	Decorah	IA	52101	LO Storlie	319-382-4264	R	22	0.3
Lunt Manufacturing Co	601-605 Lunt Av	Schaumburg	IL	60193	Helmunt Brandt	708-529-5542	R	22	0.2
Arlington Industries Inc	Stauffer Indust'l Pk	Scranton	PA	18517	Thomas Stark	717-562-0270	R	20	0.2
Auto Cast Inc	4565 Spartan	Grandville	MI	49418	Wesley Wiggers	616-534-4941	R	20	0.1
Stroh Die Casting Company Inc	11123 W Burleigh St	Milwaukee	WI	53222	Michael E Stroh	414-771-7100	R	18*	0.2
Gougler Industries Inc	705 Lake St	Kent	OH	44240	A Tsuneishi	216-673-5821	R	16	0.2
Blaser Die Casting Co	PO Box 80286	Seattle	WA	98108	PT Foley	206-767-7800	R	15	0.3
Newton-New Haven Co	222 McDermott Rd	North Haven	CT	06473	WG Newton III	203-772-1100	R	15	0.2
Falcon Foundry Co	PO Box 301	Lowellville	OH	44436	Gary S Slaven	216-536-6221	R	13	<0.1
Del Mar Die Casting Company	12901 S Western	Gardena	CA	90249	Doug R Taylor	213-321-0600	R	12	0.2
Wolverine Cast LP	22550 Nagel Av	Warren	MI	48089	William H Selecman	313-757-1900	R	12	0.1
Aurora Industries Inc	1995 Greenfield Av	Montgomery	IL	60538	James D Pearson	708-844-4900	S	11	0.1
Magnesium Aluminum Corp	3425 Service Rd	Cleveland	OH	44111	Chet Scholtz	216-476-0710	R	10	0.1
Sall-Eclipse Inc	2320 Kishwaukee St	Rockford	IL	61104	Bud Berner	815-964-5687	S	10*	<0.1
Lyons Diecasting Co	PO Box 620	Buckner	MO	64016	HE Jackson Jr	816-650-3146	R	9	0.1
Batesville Products Inc	PO Box 29	Lawrenceburg	IN	47025	Rick Weber	812-537-2275	R	8	<0.1
Micro Industries Inc	200 W 2nd St	Rock Falls	IL	61071	CT Robinson	815-625-8000	R	8	13.0
General Die Casting Co	10750 Capital Av	Oak Park	MI	48237	Richard Spicka	313-548-5555	R	8	<0.1
Serv-All Die and Tool Co	PO Box 159	Crystal Lake	IL	60014	John Groden	815-459-2900	R	7*	<0.1
Tru Die Cast Corp	PO Box 366	New Troy	MI	49119	Bruno Lehmann	616-426-3361	R	7*	<0.1
Able Casting Inc	4551 Furman Av	Bronx	NY	10470	Arthur S Puhn	718-325-5650	S	6	<0.1
Century Brass Works Inc	PO Box 302	Belleville	IL	62222	Jeff R Lutz	618-233-0182	R	6*	<0.1
Mid-State Die Casting Co	7750 S Division	Grand Rapids	MI	49548	Harold Stacy	616-455-4010	S	6	<0.1
Gerity Schultz Corp	1810 Clinton St	Toledo	OH	43607	JJ Murtagh	419-244-4971	R	5	<0.1
Joy-Mark Inc	2121 E Norse Av	Cudahy	WI	53110	Ozzie Clemmons	414-769-8155	R	5*	<0.1
Ray Lewis and Son Inc	PO Box 399	Marysville	OH	43040	James E Brown	513-644-4015	R	5	<0.1
Petoskey Manufacturing Co	PO Box G	Petoskey	MI	49770	Mike Olson	616-347-2538	R	4	<0.1
Cast Specialties Inc	26711 Miles Av	Cleveland	OH	44128	Gerald C Hershey	216-292-7393	R	3	<0.1
Craft Die Casting Corp	1831 N Lorel St	Chicago	IL	60639	Hewitt N Sanabria	312-237-9710	R	3*	<0.1
Dieco Die Casting Corp	338 Moffat St	Brooklyn	NY	11237	G Egor	718-821-7700	R	3	<0.1
Engineers Tool & Mfg Co	265 Benton St	Stratford	CT	06497	LW Gordon	203-375-4465	R	3	<0.1
Industrial Castings	635 E Highland Rd	Macedonia	OH	44056	Robert Rogel	216-467-0750	D	3	0.1
Manger Die Casting Company	PO Box 322	Derby	CT	06418	PK Manger	203-735-7881	R	3	<0.1
Bearium Metals Corp	1170 Chili Av	Rochester	NY	14624	JW Adams	716-235-5360	R	2	0.1
Advance Aluminum & Brass	1001 E Slauson Av	Los Angeles	CA	90011	MJ Welther	213-231-9301	R	2	<0.1
Amcast Inc	350 Meyer Rd	Bensenville	IL	60106	John C Kopp	708-766-7450	R	1	<0.1
Fall River Tool and Die	PO Box 4070	Fall River	MA	02723	OA Fontaine	508-674-4621	R	1	<0.1

Source: Ward's Business Directory of U.S. Private and Public Companies, Volumes 1 and 2, 1996. The company type code used is as follows: P - Public, R - Private, S - Subsidiary, D - Division, J - Joint Venture, A - Affiliate, G - Group. Sales are in millions of dollars, employees are in thousands. An asterisk (*) indicates an estimated sales volume. The symbol < stands for 'less than'. Company names and addresses are truncated, in some cases, to fit into the available space.

MATERIALS CONSUMED

Material	Quantity	Delivered Cost ($ million)
Materials, ingredients, containers, and supplies	(X)	365.2
Copper and copper-base alloy shapes and forms	(S)	24.4
Aluminum and aluminum-base alloy shapes and forms	1,000 s tons	13.0
Zinc and zinc-base alloy shapes and forms	10.4* 1,000 s tons	104.0
Magnesium and magnesium-base alloy shapes and forms	(X)	24.8
Lead-base alloy shapes and forms	(X)	15.9
All other nonferrous shapes and forms	(X)	2.3
Copper and copper-base alloy scrap (except home scrap)	(X)	(D)
Aluminum and aluminum-base alloy scrap (except home scrap)	(X)	(D)
Other nonferrous metal scrap (except home scrap)	(X)	(D)
Industrial patterns	(X)	(D)
Industrial dies, molds, jigs, and fixtures	(X)	3.4
All other industrial and commercial machinery and computer equipment	(X)	2.7
Sand	(X)	(D)
Grinding wheels and other abrasive products, except industrial diamonds	(X)	0.7
All other materials and components, parts, containers, and supplies	(X)	110.0
Materials, ingredients, containers, and supplies, nsk	(X)	61.4

Source: 1992 *Economic Census.* Explanation of symbols used: (D): Withheld to avoid disclosure of competitive data; na: Not available; (S): Withheld because statistical norms were not met; (X): Not applicable; (Z): Less than half the unit shown; nec: Not elsewhere classified; nsk: Not specified by kind; - : zero; * : 10-19 percent estimated; ** : 20-29 percent estimated.

PRODUCT SHARE DETAILS

Product or Product Class	% Share	Product or Product Class	% Share
Nonferrous die-castings, except aluminum	100.00	Magnesium and magnesium-base alloy die-castings	9.53
Copper and copper-base alloy die-castings (including		Lead and lead-base alloy die-castings	4.70
bearings and bushings)	7.33	Other nonferrous metal die-castings, except aluminum	2.60
Zinc and zinc-base alloy die-castings	63.08		

Source: 1992 *Economic Census*. The values shown are percent of total shipments in an industry. Values of indented subcategories are summed in the main heading. The symbol (D) appears when data are withheld to prevent disclosure of competitive information. The abbreviation nsk stands for 'not specified by kind' and nec for 'not elsewhere classified'.

INPUTS AND OUTPUTS FOR NONFERROUS CASTINGS, NEC

Economic Sector or Industry Providing Inputs	%	Sector	Economic Sector or Industry Buying Outputs	%	Sector
Primary zinc	23.1	Manufg.	Federal Government purchases, national defense	24.3	Fed Govt
Special dies & tools & machine tool accessories	8.6	Manufg.	Aircraft & missile engines & engine parts	13.3	Manufg.
Nonmetallic mineral products, nec	7.0	Manufg.	Aircraft & missile equipment	5.0	Manufg.
Wholesale trade	4.9	Trade	Machinery, except electrical, nec	4.4	Manufg.
Electrometallurgical products	4.7	Manufg.	Pipe, valves, & pipe fittings	3.3	Manufg.
Primary aluminum	4.1	Manufg.	Electric housewares & fans	3.2	Manufg.
Electric services (utilities)	3.8	Util.	Pumps & compressors	2.5	Manufg.
Primary nonferrous metals, nec	3.7	Manufg.	Power transmission equipment	2.4	Manufg.
Petroleum refining	3.0	Manufg.	Carburetors, pistons, rings, & valves	2.2	Manufg.
Machine tools, metal forming types	2.8	Manufg.	Motor vehicle parts & accessories	2.2	Manufg.
Gas production & distribution (utilities)	2.8	Util.	Oil field machinery	1.9	Manufg.
Advertising	2.8	Services	Wiring devices	1.8	Manufg.
Primary lead	2.0	Manufg.	Lighting fixtures & equipment	1.6	Manufg.
Cyclic crudes and organics	1.8	Manufg.	Plumbing fixture fittings & trim	1.5	Manufg.
Banking	1.7	Fin/R.E.	Special industry machinery, nec	1.3	Manufg.
Metal coating & allied services	1.4	Manufg.	Fabricated metal products, nec	1.2	Manufg.
Maintenance of nonfarm buildings nec	1.3	Constr.	Internal combustion engines, nec	1.2	Manufg.
Real estate	1.3	Fin/R.E.	Metal stampings, nec	1.2	Manufg.
Fabricated metal products, nec	1.2	Manufg.	Miscellaneous metal work	1.1	Manufg.
Air transportation	1.2	Util.	Ball & roller bearings	0.9	Manufg.
Royalties	1.2	Fin/R.E.	Electronic computing equipment	0.9	Manufg.
Paperboard containers & boxes	1.0	Manufg.	General industrial machinery, nec	0.9	Manufg.
Motor freight transportation & warehousing	1.0	Util.	Motors & generators	0.9	Manufg.
Machinery, except electrical, nec	0.9	Manufg.	Power driven hand tools	0.9	Manufg.
Eating & drinking places	0.9	Trade	Primary batteries, dry & wet	0.9	Manufg.
Metal heat treating	0.8	Manufg.	Radio & TV communication equipment	0.9	Manufg.
Communications, except radio & TV	0.8	Util.	Woodworking machinery	0.9	Manufg.
Miscellaneous repair shops	0.8	Services	Automotive stampings	0.8	Manufg.
Railroads & related services	0.7	Util.	Blowers & fans	0.8	Manufg.
Hotels & lodging places	0.7	Services	Boat building & repairing	0.8	Manufg.
Chemical preparations, nec	0.6	Manufg.	Food products machinery	0.8	Manufg.
Automotive rental & leasing, without drivers	0.6	Services	Household laundry equipment	0.7	Manufg.
Equipment rental & leasing services	0.6	Services	Engine electrical equipment	0.6	Manufg.
Security & commodity brokers	0.4	Fin/R.E.	Farm machinery & equipment	0.6	Manufg.
Automotive repair shops & services	0.4	Services	Household appliances, nec	0.6	Manufg.
Legal services	0.4	Services	Industrial controls	0.6	Manufg.
Management & consulting services & labs	0.4	Services	Lawn & garden equipment	0.6	Manufg.
Abrasive products	0.3	Manufg.	Textile machinery	0.6	Manufg.
Lubricating oils & greases	0.3	Manufg.	Electronic components nec	0.5	Manufg.
Primary copper	0.3	Manufg.	Household refrigerators & freezers	0.5	Manufg.
U.S. Postal Service	0.3	Gov't	Mechanical measuring devices	0.5	Manufg.
Industrial inorganic chemicals, nec	0.2	Manufg.	Screw machine and related products	0.5	Manufg.
Manifold business forms	0.2	Manufg.	Special dies & tools & machine tool accessories	0.5	Manufg.
Nonferrous castings, nec	0.2	Manufg.	Switchgear & switchboard apparatus	0.5	Manufg.
Insurance carriers	0.2	Fin/R.E.	Telephone & telegraph apparatus	0.5	Manufg.
Accounting, auditing & bookkeeping	0.2	Services	Turbines & turbine generator sets	0.5	Manufg.
Business/professional associations	0.2	Services	Household cooking equipment	0.4	Manufg.
Computer & data processing services	0.2	Services	Motor vehicles & car bodies	0.4	Manufg.
Laundry, dry cleaning, shoe repair	0.2	Services	Storage batteries	0.4	Manufg.
Motor vehicle parts & accessories	0.1	Manufg.	Architectural metal work	0.3	Manufg.
Tires & inner tubes	0.1	Manufg.	Commercial laundry equipment	0.3	Manufg.
Sanitary services, steam supply, irrigation	0.1	Util.	Construction machinery & equipment	0.3	Manufg.
Retail trade, except eating & drinking	0.1	Trade	Engineering & scientific instruments	0.3	Manufg.
Credit agencies other than banks	0.1	Fin/R.E.	Environmental controls	0.3	Manufg.
Engineering, architectural, & surveying services	0.1	Services	Fabricated plate work (boiler shops)	0.3	Manufg.
			Mining machinery, except oil field	0.3	Manufg.
			Automatic merchandising machines	0.2	Manufg.
			Guided missiles & space vehicles	0.2	Manufg.
			Industrial patterns	0.2	Manufg.
			Refrigeration & heating equipment	0.2	Manufg.
			Exports	0.2	Foreign
			Measuring & dispensing pumps	0.1	Manufg.
			Nonferrous castings, nec	0.1	Manufg.
			Typewriters & office machines, nec	0.1	Manufg.

Source: Benchmark Input-Output Accounts for the U.S. Economy, 1982, U.S. Department of Commerce, Washington, D.C., July 1991. Data, as reported in the source, are organized by the 1977 SIC structure in use in 1982 but have been matched, as closely as is possible, to the 1987 SIC structure used in this book.

OCCUPATIONS EMPLOYED BY SIC 336 - NONFERROUS FOUNDRIES (CASTINGS)

Occupation	% of Total 1994	Change to 2005	Occupation	% of Total 1994	Change to 2005
Metal molding machine workers	12.2	-2.1	Machine tool cutting & forming etc. nec	1.7	-10.9
Assemblers, fabricators, & hand workers nec	7.2	-11.0	Combination machine tool operators	1.6	-2.1
Grinders & polishers, hand	5.6	-11.0	Precision metal workers nec	1.6	-11.0
Blue collar worker supervisors	5.2	-15.7	Machine forming operators, metal & plastic	1.5	-10.9
Inspectors, testers, & graders, precision	5.1	-11.0	Traffic, shipping, & receiving clerks	1.4	-14.3
Precision workers nec	4.6	-19.9	Industrial production managers	1.4	-10.9
Tool & die makers	3.3	-28.1	Metal & plastic machine workers nec	1.3	-21.0
Foundry mold assembly & shakeout workers	3.1	-19.9	Hand packers & packagers	1.3	-23.7
General managers & top executives	2.7	-15.5	Janitors & cleaners, incl maids	1.1	-28.8
Machine tool cutting operators, metal & plastic	2.7	-25.8	Industrial truck & tractor operators	1.1	-11.0
Grinding machine operators, metal & plastic	2.5	20.2	Sales & related workers nec	1.1	-10.9
Furnace operators	2.5	-19.9	Bookkeeping, accounting, & auditing clerks	1.1	-33.3
Machinists	2.5	-11.0	Punching machine operators, metal & plastic	1.0	-28.8
Industrial machinery mechanics	2.1	-2.1	General office clerks	1.0	-24.1
Maintenance repairers, general utility	1.8	-19.9			

Source: Industry-Occupation Matrix, Bureau of Labor Statistics. These data relate to one or more 3-digit SIC industry groups rather than to a single 4-digit SIC. The change reported for each occupation to the year 2005 is a percent of growth or decline as estimated by the Bureau of Labor Statistics. The abbreviation nec stands for 'not elsewhere classified'.

LOCATION BY STATE AND REGIONAL CONCENTRATION

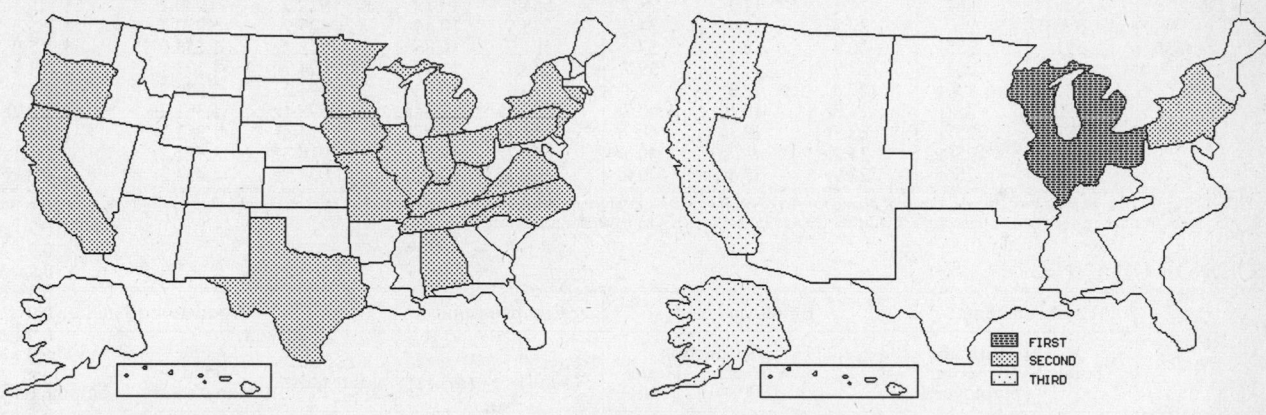

FIRST
SECOND
THIRD

INDUSTRY DATA BY STATE

State	Establish-ments	Shipments Total ($ mil)	Shipments % of U.S.	Shipments Per Establ.	Employment Total Number	Employment % of U.S.	Employment Per Establ.	Wages ($/hour)	Cost as % of Shipments	Investment per Employee ($)
Michigan	27	166.4	16.3	6.2	1,700	15.3	63	9.64	58.4	1,412
Illinois	24	127.0	12.4	5.3	1,500	13.5	63	9.43	42.5	4,600
California	34	117.8	11.5	3.5	1,200	10.8	35	10.29	43.1	1,583
Ohio	20	103.5	10.1	5.2	1,000	9.0	50	10.50	53.5	4,200
New York	23	64.4	6.3	2.8	600	5.4	26	11.60	44.4	1,167
Pennsylvania	12	42.6	4.2	3.5	700	6.3	58	10.56	34.0	4,143
Tennessee	6	25.7	2.5	4.3	300	2.7	50	10.80	60.3	-
North Carolina	6	17.3	1.7	2.9	100	0.9	17	7.50	53.2	-
Connecticut	8	17.0	1.7	2.1	200	1.8	25	9.00	40.0	1,500
Alabama	6	16.8	1.6	2.8	200	1.8	33	6.00	47.0	-
New Jersey	8	13.7	1.3	1.7	200	1.8	25	9.33	38.7	500
Texas	9	(D)	-	-	175 *	1.6	19	-	-	-
Indiana	6	(D)	-	-	375 *	3.4	63	-	-	-
Missouri	6	(D)	-	-	750 *	6.8	125	-	-	-
Oregon	5	(D)	-	-	175 *	1.6	35	-	-	-
Iowa	4	(D)	-	-	375 *	3.4	94	-	-	-
Kentucky	4	(D)	-	-	375 *	3.4	94	-	-	-
Minnesota	3	(D)	-	-	375 *	3.4	125	-	-	-
Virginia	2	(D)	-	-	175 *	1.6	88	-	-	-

*Source: 1992 Economic Census. The states are in descending order of shipments or establishments (if shipment data are missing for the majority). The symbol (D) appears when data are withheld to prevent disclosure of competitive information. States marked with (D) are sorted by number of establishments. A dash (-) indicates that the data element cannot be calculated; * indicates the midpoint of a range.*

3365 - ALUMINUM FOUNDRIES

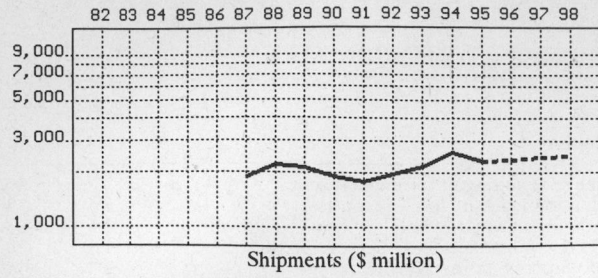

Shipments ($ million)

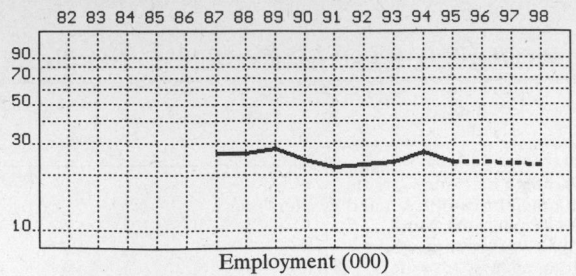

Employment (000)

GENERAL STATISTICS

| Year | Com- panies | Establishments | | Employment | | | Compensation | | Production ($ million) | | | |
		Total	with 20 or more employees	Total (000)	Production Workers (000)	Hours (Mil)	Payroll ($ mil)	Wages ($/hr)	Cost of Materials	Value Added by Manufacture	Value of Shipments	Capital Invest.
1982												
1983												
1984												
1985												
1986												
1987	551	583	255	26.3	21.7	43.2	568.0	9.68	840.9	1,035.9	1,871.7	52.1
1988		572	267	26.4	21.7	43.5	578.7	9.69	1,108.1	1,133.2	2,218.7	76.6
1989		566	271	28.1	20.8	42.1	586.9	9.76	1,022.8	1,108.5	2,130.4	60.7
1990		557	264	24.5	19.0	38.7	564.8	10.19	929.0	980.7	1,919.1	65.0
1991		570	257	22.3	18.0	34.9	538.6	10.44	825.9	950.7	1,783.3	52.5
1992	566	591	255	22.9	18.1	37.1	581.7	10.85	812.2	1,153.0	1,965.0	65.6
1993		594	261	23.9	19.0	39.7	619.7	11.17	994.7	1,191.1	2,181.5	89.3
1994		587P	259P	27.0	22.3	48.1	699.6	10.83	1,094.2	1,530.2	2,600.3	146.1
1995		590P	258P	23.8P	19.1P	40.7P	650.3P	11.34P	973.4P	1,361.2P	2,313.2P	114.8P
1996		592P	257P	23.5P	18.9P	40.7P	663.3P	11.57P	994.8P	1,391.2P	2,364.1P	123.4P
1997		595P	256P	23.2P	18.7P	40.6P	676.2P	11.79P	1,016.3P	1,421.2P	2,415.1P	132.0P
1998		598P	256P	22.9P	18.4P	40.6P	689.1P	12.02P	1,037.7P	1,451.2P	2,466.1P	140.6P

Sources: 1982, 1987, 1992 *Economic Census*; *Annual Survey of Manufactures*, 83-86, 88-91, 93-94. Establishment counts for non-Census years are from *County Business Patterns*; establishment values for 83-84 are extrapolations. 'P's show projections by the editors. Industries reclassified in 87 will not have data for prior years.

INDICES OF CHANGE

| Year | Com- panies | Establishments | | Employment | | | Compensation | | Production ($ million) | | | |
		Total	with 20 or more employees	Total (000)	Production Workers (000)	Hours (Mil)	Payroll ($ mil)	Wages ($/hr)	Cost of Materials	Value Added by Manufacture	Value of Shipments	Capital Invest.
1982												
1983												
1984												
1985												
1986												
1987	97	99	100	115	120	116	98	89	104	90	95	79
1988		97	105	115	120	117	99	89	136	98	113	117
1989		96	106	123	115	113	101	90	126	96	108	93
1990		94	104	107	105	104	97	94	114	85	98	99
1991		96	101	97	99	94	93	96	102	82	91	80
1992	100	100	100	100	100	100	100	100	100	100	100	100
1993		101	102	104	105	107	107	103	122	103	111	136
1994		99P	101P	118	123	130	120	100	135	133	132	223
1995		100P	101P	104P	105P	110P	112P	105P	120P	118P	118P	175P
1996		100P	101P	103P	104P	110P	114P	107P	122P	121P	120P	188P
1997		101P	101P	101P	103P	110P	116P	109P	125P	123P	123P	201P
1998		101P	100P	100P	102P	109P	118P	111P	128P	126P	126P	214P

Sources: Same as General Statistics. Values reflect change from the base year, 1992. Values above 100 mean greater than 92, values below 100 mean less than 92, and a value of 100 in the 82-91 or 93-98 period means same as 92. 'P's mark projections by the editors.

SELECTED RATIOS

For 1994	Avg. of All Manufact.	Analyzed Industry	Index	For 1994	Avg. of All Manufact.	Analyzed Industry	Index
Employees per Establishment	49	46	94	Value Added per Production Worker	134,084	68,619	51
Payroll per Establishment	1,500,273	1,192,113	79	Cost per Establishment	5,045,178	1,864,508	37
Payroll per Employee	30,620	25,911	85	Cost per Employee	102,970	40,526	39
Production Workers per Establishment	34	38	111	Cost per Production Worker	146,988	49,067	33
Wages per Establishment	853,319	887,649	104	Shipments per Establishment	9,576,895	4,430,891	46
Wages per Production Worker	24,861	23,360	94	Shipments per Employee	195,460	96,307	49
Hours per Production Worker	2,056	2,157	105	Shipments per Production Worker	279,017	116,605	42
Wages per Hour	12.09	10.83	90	Investment per Establishment	321,011	248,953	78
Value Added per Establishment	4,602,255	2,607,449	57	Investment per Employee	6,552	5,411	83
Value Added per Employee	93,930	56,674	60	Investment per Production Worker	9,352	6,552	70

Sources: Same as General Statistics. The 'Average of All Manufacturing' column represents the average of all manufacturing industries reported for the most recent complete year available. The Index shows the relationship between the Average and the Analyzed Industry. For example, 100 means that they are equal; 500 that the Analyzed Industry is five times the average; 50 means that the Analyzed Industry is half the national average. The abbreviation 'na' is used to show that data are 'not available'.

LEADING COMPANIES Number shown: 75 Total sales ($ mil): 2,394 Total employment (000): 19.0

Company Name	Address				CEO Name	Phone	Co. Type	Sales ($ mil)	Empl. (000)
CMI International Inc	30333 Southfield Rd	Southfield	MI	48076	Ray H Witt	313-642-9450	R	440	3.7
Wabash Alloys	PO Box 466	Wabash	IN	46992	J Viland	219-563-7461	D	300	0.7
Amcast Industrial Corp	PO Box 98	Dayton	OH	45401	Leo W Ladehoff	513-298-5251	P	272	2.1
Columbia Aluminum Corp	1220 Main St	Vancouver	WA	98660	Kenneth D Peterson	206-693-1336	R	130•	1.0
General Housewares Corp	PO Box 4066	Terre Haute	IN	47804	Paul A Saxton	812-232-1000	P	97	0.7
Consolidated Metco Inc	PO Box 83201	Portland	OR	97283	EJ Oeltjen	503-286-5741	S	95	0.5
Gichner Systems Group Inc	10946 Golden W Dr	Hunt Valley	MD	21031	Charles Atwood	410-771-9679	R	80	0.9
Teksid Aluminum Foundry Inc	1635 Old Columbia	Dickson	TN	37055	Franco Polastro	615-446-8110	S	76•	0.6
PHB Inc	7900 W Ridge Rd	Fairview	PA	16415	W M Hilbert Sr	814-474-1591	R	75	0.9
Northland Aluminum Products	Hwy 7 & Hwy 100	Minneapolis	MN	55416	H David Dalquist	612-920-2888	R	69	<0.1
Kelch Corp	1025 W Glen Oaks	Mequon	WI	53092	Dennis Nourse	414-375-5600	R	60	0.6
Southern Aluminum Castings	PO Box 1209	Bay Minette	AL	36507	J Hunt	205-937-6756	R	38•	0.3
Fansteel Wellman Dynamics	PO Box 147	Creston	IA	50801	R McGinnis	515-782-8521	S	37	0.3
Aluminum Casting & Eng	2039 S Lenox St	Milwaukee	WI	53207	EG Grohmann	414-744-3902	R	35•	0.4
Winters Industries	4125 Mahoning NE	Canton	OH	44705	Daniel Rhodes	216-456-4321	D	34•	0.3
Progress Castings	2600 Niagara Ln N	Plymouth	MN	55447	Edwin Mignelucci	612-557-1000	D	30	0.3
Ross Aluminum Foundries	PO Box 609	Sidney	OH	45365	E Coberly	513-492-4134	D	27	0.3
Bodine Aluminum Inc	2100 Walton Rd	St Louis	MO	63114	Robert Lloyd	314-423-8200	S	25	0.3
General Housewares Corp	440 Fair Rd	Sidney	OH	45365	T G Boerger	513-492-6155	D	20	0.2
Ohio Aluminum Industries Inc	4840 Warner Rd	Cleveland	OH	44125	JR Hubman	216-641-8865	D	20	0.2
Adalet-PLM	4801 W 150th St	Cleveland	OH	44135	D Scott Reeder	216-267-9000	D	19	0.2
Consolidated Metco Die Casting	10448 Hwy 212	Clackamas	OR	97015	Ed Oeltjen	503-657-4183	D	19	0.1
Slyman Industries Inc	820 W Liberty	Medina	OH	44256	Peter Slyman	216-723-3251	R	18	0.3
Arrow Aluminum Castings Co	33659 Walker Rd	Avon Lake	OH	44012	Peter Petto Jr	216-933-2020	R	17	0.2
Oberdorfer Industries Inc	PO Box 4811	Syracuse	NY	13221	Kenneth L Evans	315-437-7588	R	16	0.3
Kelch Corp	622 Madison	Cedarburg	WI	53012	Dennis Nourse	414-375-5600	D	15	0.2
Maco Corp	PO Box 709	Huntington	IN	46750	CB Huesing	219-356-3900	R	14	0.1
Watry Industries Inc	PO Box 131	Sheboygan	WI	53082	Michael F Hammett	414-457-4886	R	14	0.2
Permold Corp	PO Box P	Medina	OH	44258	Dennis R Bergeron	216-723-3251	S	14	0.2
Akron Foundry Co	PO Box 27028	Akron	OH	44301	G Ostich	216-745-3101	R	12•	0.2
Tec-Cast Inc	440 Meadow Ln	Carlstadt	NJ	07072	Edgar Gotthold	201-935-3885	R	12•	<0.1
CMI Noren Inc	PO Box 309	Fruitport	MI	49415	Ray H Witt	616-865-3192	S	11•	0.1
Quality Aluminum Casting Co	PO Box 1622	Waukesha	WI	53187	Jeffrey E Pauly	414-542-0731	R	11	0.1
Springfield Aluminum Co	PO Box 588	Nixa	MO	65714	Ron Mawhiney	417-725-2667	R	11	0.1
Jerebar Corp	20304 S Alameda St	R Dominguez	CA	90220	GH Bork	213-979-4011	R	10•	<0.1
Frontier Foundries Inc	221 S Perry St	Titusville	PA	16354	W E Gephardt Jr	814-827-1800	R	9	0.2
Littlestown Hardware & Foundry	PO Box 69	Littlestown	PA	17340	LR Snyder	717-359-4141	R	9	0.1
Dura-Ware of America Inc	81 Spring St	New York	NY	10012	Harry Bolger	212-226-4065	R	9	<0.1
Hater Industries Inc	240 Stille Dr	Cincinnati	OH	45233	HJ Hater Jr	513-941-6600	R	8	0.2
Precision Enterprises Inc	29W 130 Butterfield	Warrenville	IL	60555	Jim Schrader	708-393-3050	R	8•	<0.1
Wisconsin Aluminum Foundry	PO Box 246	Manitowoc	WI	54221	I Schwartz	414-682-8286	R	8•	0.1
Bremer Manufacturing Company	W2002 Hwy Q	Elkhart Lake	WI	53020	Jere Bremer	414-894-2944	R	8	<0.1
Boose Aluminum Foundry	77 Reamstown Rd	Reamstown	PA	17567	Roger Boose	717-336-5581	R	7•	0.1
Buddy Bar Casting Corp	10801 S Sessler St	South Gate	CA	90280	Bill Fell	213-773-4838	S	7	<0.1
Portage Casting and Mold	PO Box 53	Portage	WI	53901	J Griep	608-742-7137	R	7	<0.1
Sigma	925 S Charlie Rd	City of Industry	CA	91748	Hermann Pawelka	818-965-2457	D	7	0.1
Techni-Cast Corp	11220 S Garfield	South Gate	CA	90280	Bryn J Van Hiel	310-923-4585	R	7	<0.1
American Precision Castings Inc	2400 S Laflin St	Chicago	IL	60608	William W Lamb Jr	312-243-8000	R	6	<0.1
Ball Brass & Aluminum Foundry	PO Box 110	Auburn	IN	46706	Eric B Forst	219-925-3515	S	6	<0.1
Cast Aluminum and Brass Corp	667 Whitney St	San Leandro	CA	94577	Earl Wettstein	510-562-5711	R	6	<0.1
Hollywood Alloy Casting Co	PO Box 2067	Montebello	CA	90640	James A Reid	310-537-7030	R	6	<0.1
Premier Aluminum Inc	PO Box 316	Allenton	WI	53002	Michael Allik	414-629-5503	R	6	<0.1
Thompson Casting Co	4850 Chaincraft Rd	Garfield Hts	OH	44125	RA Thomas	216-581-9200	R	6	<0.1
Carley Foundry Inc	8301 Coral Sea NE	Blaine	MN	55449	Michael F Carley	612-780-5123	R	5•	<0.1
Foundry Inc	440 Ledyard St	Hartford	CT	06114	James J Hlavecek	203-296-0695	R	5	<0.1
KP Iron Foundry Inc	PO Box 2926	Fresno	CA	93745	Gary Kearney	209-233-2591	S	5•	<0.1
M & W Electric Mfg Co	PO Box 350	East Palestine	OH	44413	H Madden	216-426-9456	R	5	<0.1
MA Harrison Manufacturing	PO Box 38	Birmingham	OH	44816	JA Harrison	216-965-4306	R	5	<0.1
Pressure Castings Inc	21500 St Clair Av	Cleveland	OH	44117	Loren F Weiss	216-481-5400	R	5	<0.1
Texas Metal Casting Company	PO Box 3259	Lufkin	TX	75903	DE Smith	409-639-1131	R	5	<0.1
Empire Castings Inc	3195 Regonial Pkwy	Santa Rosa	CA	95403	Ken Clark	707-523-9815	R	5•	<0.1
Meloon Foundries Inc	PO Box 1182	Syracuse	NY	13201	L Iannettoni	315-454-3231	R	5	<0.1
Aluminum Housewares Company	PO Box 1599	Maryland H	MO	63043	James Seidel	314-872-8855	R	4	<0.1
Alum-Alloy Company Inc	603 S Hope Av	Ontario	CA	91761	David W Howell	909-986-0410	R	4	<0.1
AL Johnson North	4671 Calle Carga	Camarillo	CA	93012	Bud Jackson	805-389-4631	D	4	<0.1
Eagleware Manufacturing	2835 E Ana St	R Dominguez	CA	90221	JR Gross	310-604-0404	R	4•	<0.1
Foley Pattern Company Inc	PO Box 150	Auburn	IN	46706	Ellen E Stahly	219-925-4113	R	4	<0.1
Marlborough Foundry Inc	555 Maple Av	Marlboro	MA	01752	HB Nye	508-485-2848	R	4•	<0.1
Multi-Cast Corp	PO Box 111	Wauseon	OH	43567	EJ Metzger, Jr	419-335-0010	R	4	<0.1
Phillips Foundry Inc	PO Box 187	Binghamton	NY	13904	Gordon Phillips	607-723-6483	R	4	<0.1
Sunset Foundry Company Inc	PO Box 927	Kent	WA	98035	Leo Powers	206-872-5110	R	4	<0.1
Woodland Alloy Casting Inc	PO Box 085866	Racine	WI	53408	Hank Adams	414-637-6754	R	4•	<0.1
National Aluminum & Brass	PO Box 290	Independence	MO	64050	Robert Fenton	816-833-4500	R	4	<0.1
Olympia Foundry&Fabrication	913 Broadway	Monett	MO	65708	J Hudson Jr	417-235-7828	R	4	<0.1
AMI Inc	PO Box 292	Lyons	IL	60534	Edwin E Kaminski	708-442-7920	R	3	<0.1

Source: Ward's Business Directory of U.S. Private and Public Companies, Volumes 1 and 2, 1996. The company type code used is as follows: P - Public, R - Private, S - Subsidiary, D - Division, J - Joint Venture, A - Affiliate, G - Group. Sales are in millions of dollars, employees are in thousands. An asterisk (•) indicates an estimated sales volume. The symbol < stands for 'less than'. Company names and addresses are truncated, in some cases, to fit into the available space.

MATERIALS CONSUMED

Material	Quantity	Delivered Cost ($ million)
Materials, ingredients, containers, and supplies	(X)	631.4
Copper and copper-base alloy shapes and forms . . . . 1,000 s tons	(S)	40.8
Aluminum and aluminum-base alloy shapes and forms . . . . 1,000 s tons	233.9	280.2
Zinc and zinc-base alloy shapes and forms	(X)	1.5
Magnesium and magnesium-base alloy shapes and forms	(X)	4.1
Lead-base alloy shapes and forms	(X)	(D)
All other nonferrous shapes and forms	(X)	0.9
Copper and copper-base alloy scrap (except home scrap)	(X)	(D)
Aluminum and aluminum-base alloy scrap (except home scrap)	(X)	34.1
Other nonferrous metal scrap (except home scrap)	(X)	1.2
Industrial patterns	(X)	15.1
Industrial dies, molds, jigs, and fixtures	(X)	7.2
All other industrial and commercial machinery and computer equipment	(X)	11.3
Sand	(X)	15.0
Grinding wheels and other abrasive products, except industrial diamonds	(X)	5.4
All other materials and components, parts, containers, and supplies	(X)	130.5
Materials, ingredients, containers, and supplies, nsk	(X)	83.1

Source: 1992 Economic Census. Explanation of symbols used: (D): Withheld to avoid disclosure of competitive data; na: Not available; (S): Withheld because statistical norms were not met; (X): Not applicable; (Z): Less than half the unit shown; nec: Not elsewhere classified; nsk: Not specified by kind; - : zero; * : 10-19 percent estimated; ** : 20-29 percent estimated.

PRODUCT SHARE DETAILS

Product or Product Class	% Share	Product or Product Class	% Share
Aluminum foundries	100.00	(except cast aluminum cooking utensils)	5.48
Aluminum and aluminum-base alloy sand castings (except cast aluminum cooking utensils)	42.97	Other aluminum and aluminum-base alloy castings, excluding die-castings (except cast aluminum cooking utensils)	4.14
Aluminum and aluminum-base alloy permanent and semipermanent mold castings (except cast aluminum cooking utensils)	34.05	Nonelectric cast aluminum pressure cookers (household-type) and nonelectric cast aluminum cooking utensils . . .	1.96
Aluminum and aluminum-base alloy investment castings			

Source: 1992 Economic Census. The values shown are percent of total shipments in an industry. Values of indented subcategories are summed in the main heading. The symbol (D) appears when data are withheld to prevent disclosure of competitive information. The abbreviation nsk stands for 'not specified by kind' and nec for 'not elsewhere classified'.

INPUTS AND OUTPUTS FOR ALUMINUM CASTINGS

Economic Sector or Industry Providing Inputs	%	Sector	Economic Sector or Industry Buying Outputs	%	Sector
Primary aluminum	38.9	Manufg.	Motor vehicle parts & accessories	23.1	Manufg.
Special dies & tools & machine tool accessories	10.2	Manufg.	Internal combustion engines, nec	8.7	Manufg.
Communications, except radio & TV	6.6	Util.	Machinery, except electrical, nec	4.2	Manufg.
Electric services (utilities)	5.2	Util.	Electronic computing equipment	3.5	Manufg.
Gas production & distribution (utilities)	5.2	Util.	Aircraft & missile engines & engine parts	3.2	Manufg.
Scrap	3.3	Scrap	Pipe, valves, & pipe fittings	3.2	Manufg.
Wholesale trade	3.0	Trade	Refrigeration & heating equipment	2.7	Manufg.
Maintenance of nonfarm buildings nec	1.9	Constr.	Pumps & compressors	2.5	Manufg.
Advertising	1.8	Services	Radio & TV communication equipment	2.5	Manufg.
Petroleum refining	1.7	Manufg.	Federal Government purchases, national defense	2.4	Fed Govt
Primary copper	1.6	Manufg.	Motor vehicles & car bodies	2.3	Manufg.
Motor freight transportation & warehousing	1.6	Util.	Exports	1.8	Foreign
Primary zinc	1.5	Manufg.	Lighting fixtures & equipment	1.7	Manufg.
Miscellaneous repair shops	1.3	Services	Motors & generators	1.7	Manufg.
Banking	1.1	Fin/R.E.	Photographic equipment & supplies	1.5	Manufg.
Electrical repair shops	1.1	Services	Household cooking equipment	1.4	Manufg.
Railroads & related services	1.0	Util.	Mechanical measuring devices	1.4	Manufg.
Eating & drinking places	0.8	Trade	Aircraft	1.3	Manufg.
Real estate	0.7	Fin/R.E.	Power driven hand tools	1.3	Manufg.
Royalties	0.7	Fin/R.E.	Carburetors, pistons, rings, & valves	1.2	Manufg.
Equipment rental & leasing services	0.7	Services	Lawn & garden equipment	1.1	Manufg.
Machinery, except electrical, nec	0.6	Manufg.	Wiring devices	1.1	Manufg.
Abrasive products	0.5	Manufg.	Aircraft & missile equipment, nec	0.9	Manufg.
Metal heat treating	0.5	Manufg.	Electronic components nec	0.9	Manufg.
Primary nonferrous metals, nec	0.5	Manufg.	Sporting & athletic goods, nec	0.9	Manufg.
Sand & gravel	0.4	Mining	Construction machinery & equipment	0.8	Manufg.
Electrometallurgical products	0.4	Manufg.	General industrial machinery, nec	0.8	Manufg.
Industrial patterns	0.4	Manufg.	Hardware, nec	0.8	Manufg.
Paperboard containers & boxes	0.4	Manufg.	Special industry machinery, nec	0.8	Manufg.
Aluminum castings	0.3	Manufg.	Blowers & fans	0.7	Manufg.
Lubricating oils & greases	0.3	Manufg.	Power transmission equipment	0.7	Manufg.
Air transportation	0.3	Util.	Printing trades machinery	0.7	Manufg.
Automotive rental & leasing, without drivers	0.3	Services	Federal Government purchases, nondefense	0.7	Fed Govt
Automotive repair shops & services	0.3	Services	Farm machinery & equipment	0.6	Manufg.

Continued on next page.

INPUTS AND OUTPUTS FOR ALUMINUM CASTINGS - Continued

Economic Sector or Industry Providing Inputs	%	Sector	Economic Sector or Industry Buying Outputs	%	Sector
Legal services	0.3	Services	Food products machinery	0.6	Manufg.
Management & consulting services & labs	0.3	Services	Household laundry equipment	0.6	Manufg.
U.S. Postal Service	0.3	Gov't	Switchgear & switchboard apparatus	0.6	Manufg.
Manifold business forms	0.2	Manufg.	Woodworking machinery	0.6	Manufg.
Wood products, nec	0.2	Manufg.	S/L Govt. purch., health & hospitals	0.6	S/L Govt
Sanitary services, steam supply, irrigation	0.2	Util.	Electric housewares & fans	0.5	Manufg.
Credit agencies other than banks	0.2	Fin/R.E.	Engine electrical equipment	0.5	Manufg.
Insurance carriers	0.2	Fin/R.E.	Hand & edge tools, nec	0.5	Manufg.
Accounting, auditing & bookkeeping	0.2	Services	Metal stampings, nec	0.5	Manufg.
Business/professional associations	0.2	Services	Tanks & tank components	0.5	Manufg.
Computer & data processing services	0.2	Services	Transportation equipment, nec	0.5	Manufg.
Laundry, dry cleaning, shoe repair	0.2	Services	Personal consumption expenditures	0.4	
Blast furnaces & steel mills	0.1	Manufg.	Fabricated plate work (boiler shops)	0.4	Manufg.
Machine tools, metal cutting types	0.1	Manufg.	Industrial controls	0.4	Manufg.
Engineering, architectural, & surveying services	0.1	Services	Service industry machines, nec	0.4	Manufg.
State & local government enterprises, nec	0.1	Gov't	Textile machinery	0.4	Manufg.
Imports	0.1	Foreign	Typewriters & office machines, nec	0.4	Manufg.
			Calculating & accounting machines	0.3	Manufg.
			Engineering & scientific instruments	0.3	Manufg.
			Environmental controls	0.3	Manufg.
			Fabricated metal products, nec	0.3	Manufg.
			Hoists, cranes, & monorails	0.3	Manufg.
			Household refrigerators & freezers	0.3	Manufg.
			Instruments to measure electricity	0.3	Manufg.
			Measuring & dispensing pumps	0.3	Manufg.
			Special dies & tools & machine tool accessories	0.3	Manufg.
			S/L Govt. purch., other general government	0.3	S/L Govt
			Guided missiles & space vehicles	0.2	Manufg.
			Household vacuum cleaners	0.2	Manufg.
			Metal doors, sash, & trim	0.2	Manufg.
			Metalworking machinery, nec	0.2	Manufg.
			Motorcycles, bicycles, & parts	0.2	Manufg.
			Oil field machinery	0.2	Manufg.
			Radio & TV receiving sets	0.2	Manufg.
			Scales & balances	0.2	Manufg.
			Sheet metal work	0.2	Manufg.
			Small arms	0.2	Manufg.
			Telephone & telegraph apparatus	0.2	Manufg.
			Water transportation	0.2	Util.
			Eating & drinking places	0.2	Trade
			Aluminum castings	0.1	Manufg.
			Boat building & repairing	0.1	Manufg.
			Conveyors & conveying equipment	0.1	Manufg.
			Heating equipment, except electric	0.1	Manufg.
			Industrial patterns	0.1	Manufg.
			Industrial trucks & tractors	0.1	Foundry.
			Machine tools, metal cutting types	0.1	Manufg.
			Machine tools, metal forming types	0.1	Manufg.
			Railroad equipment	0.1	Manufg.
			Screw machine and related products	0.1	Manufg.
			S/L Govt. purch., correction	0.1	S/L Govt

Source: Benchmark Input-Output Accounts for the U.S. Economy, 1982, U.S. Department of Commerce, Washington, D.C., July 1991. Data, as reported in the source, are organized by the 1977 SIC structure in use in 1982 but have been matched, as closely as is possible, to the 1987 SIC structure used in this book.

OCCUPATIONS EMPLOYED BY SIC 336 - NONFERROUS FOUNDRIES (CASTINGS)

Occupation	% of Total 1994	Change to 2005	Occupation	% of Total 1994	Change to 2005
Metal molding machine workers	12.2	-2.1	Machine tool cutting & forming etc. nec	1.7	-10.9
Assemblers, fabricators, & hand workers nec	7.2	-11.0	Combination machine tool operators	1.6	-2.1
Grinders & polishers, hand	5.6	-11.0	Precision metal workers nec	1.6	-11.0
Blue collar worker supervisors	5.2	-15.7	Machine forming operators, metal & plastic	1.5	-10.9
Inspectors, testers, & graders, precision	5.1	-11.0	Traffic, shipping, & receiving clerks	1.4	-14.3
Precision workers nec	4.6	-19.9	Industrial production managers	1.4	-10.9
Tool & die makers	3.3	-28.1	Metal & plastic machine workers nec	1.3	-21.0
Foundry mold assembly & shakeout workers	3.1	-19.9	Hand packers & packagers	1.3	-23.7
General managers & top executives	2.7	-15.5	Janitors & cleaners, incl maids	1.1	-28.8
Machine tool cutting operators, metal & plastic	2.7	-25.8	Industrial truck & tractor operators	1.1	-11.0
Grinding machine operators, metal & plastic	2.5	20.2	Sales & related workers nec	1.1	-10.9
Furnace operators	2.5	-19.9	Bookkeeping, accounting, & auditing clerks	1.1	-33.3
Machinists	2.5	-11.0	Punching machine operators, metal & plastic	1.0	-28.8
Industrial machinery mechanics	2.1	-2.1	General office clerks	1.0	-24.1
Maintenance repairers, general utility	1.8	-19.9			

Source: Industry-Occupation Matrix, Bureau of Labor Statistics. These data relate to one or more 3-digit SIC industry groups rather than to a single 4-digit SIC. The change reported for each occupation to the year 2005 is a percent of growth or decline as estimated by the Bureau of Labor Statistics. The abbreviation nec stands for 'not elsewhere classified'.

LOCATION BY STATE AND REGIONAL CONCENTRATION

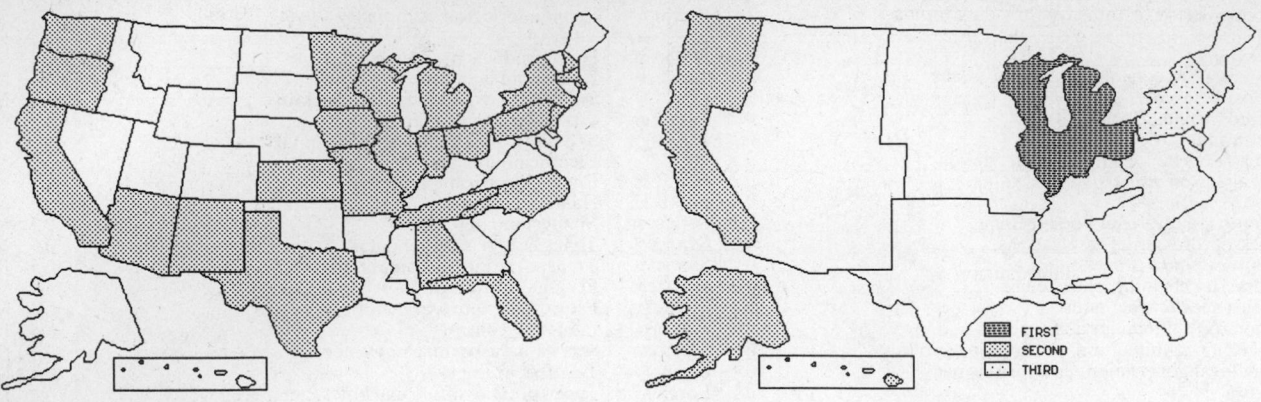

FIRST
SECOND
THIRD

INDUSTRY DATA BY STATE

| State | Establish-ments | Shipments | | | Employment | | | | Cost as % of Shipments | Investment per Employee ($) |
		Total ($ mil)	% of U.S.	Per Establ.	Total Number	% of U.S.	Per Establ.	Wages ($/hour)		
Indiana	35	260.4	13.3	7.4	2,000	8.7	57	12.35	55.8	5,450
Ohio	70	257.9	13.1	3.7	3,300	14.4	47	10.96	39.7	1,697
Michigan	31	187.3	9.5	6.0	1,700	7.4	55	12.48	44.4	4,588
Wisconsin	32	181.4	9.2	5.7	2,000	8.7	63	11.29	45.2	3,800
California	82	157.3	8.0	1.9	2,200	9.6	27	10.00	28.6	1,136
Minnesota	17	139.4	7.1	8.2	1,700	7.4	100	12.12	37.4	2,353
Tennessee	10	135.5	6.9	13.5	1,100	4.8	110	12.06	49.4	-
Pennsylvania	46	124.3	6.3	2.7	1,600	7.0	35	9.63	32.0	2,250
Missouri	17	80.0	4.1	4.7	1,000	4.4	59	10.05	38.6	5,100
New York	29	57.0	2.9	2.0	800	3.5	28	13.36	29.6	1,250
Texas	24	49.8	2.5	2.1	700	3.1	29	8.91	39.4	1,857
Illinois	32	44.8	2.3	1.4	700	3.1	22	9.80	35.5	1,429
Washington	8	40.8	2.1	5.1	400	1.7	50	13.67	44.4	-
Alabama	12	33.1	1.7	2.8	400	1.7	33	7.43	47.7	2,000
New Hampshire	10	27.1	1.4	2.7	400	1.7	40	10.33	24.4	750
New Jersey	13	25.5	1.3	2.0	400	1.7	31	11.50	28.6	250
Iowa	14	24.5	1.2	1.8	300	1.3	21	9.17	40.4	667
Arizona	7	20.3	1.0	2.9	300	1.3	43	9.00	28.6	1,333
Massachusetts	15	13.4	0.7	0.9	200	0.9	13	12.00	34.3	1,000
Florida	13	11.4	0.6	0.9	200	0.9	15	8.67	44.7	500
Kansas	7	7.6	0.4	1.1	200	0.9	29	8.50	48.7	-
North Carolina	7	(D)	-	-	175 *	0.8	25	-	-	-
Oregon	7	(D)	-	-	175 *	0.8	25	-	-	-
New Mexico	1	(D)	-	-	175 *	0.8	175	-	-	-

Source: 1992 Economic Census. The states are in descending order of shipments or establishments (if shipment data are missing for the majority). The symbol (D) appears when data are withheld to prevent disclosure of competitive information. States marked with (D) are sorted by number of establishments. A dash (-) indicates that the data element cannot be calculated; * indicates the midpoint of a range.

3366 - COPPER FOUNDRIES

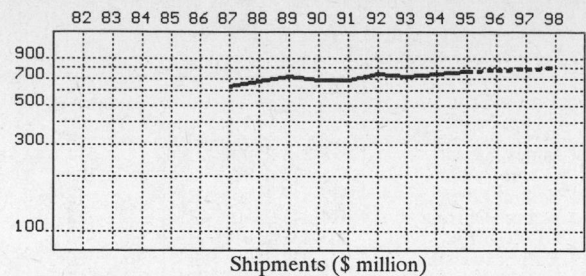

Shipments ($ million)

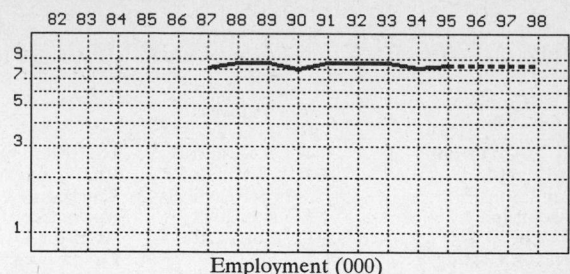

Employment (000)

GENERAL STATISTICS

Year	Companies	Establishments Total	Establishments with 20 or more employees	Employment Total (000)	Employment Production Workers (000)	Employment Hours (Mil)	Compensation Payroll ($ mil)	Compensation Wages ($/hr)	Production Cost of Materials	Production Value Added by Manufacture	Production Value of Shipments	Capital Invest.
1982												
1983												
1984												
1985												
1986												
1987	330	334	122	8.2	6.5	12.8	175.3	9.72	285.3	341.9	625.2	13.2
1988		322	120	8.7	6.9	13.6	184.9	9.65	312.0	366.6	673.5	
1989		305	122	8.6	7.5	14.2	199.9	10.30	321.0	378.5	709.9	
1990		300	123	7.9	7.3	14.7	187.7	9.03	325.5	353.1	677.8	27.3
1991		308	117	8.6	6.5	13.4	193.8	9.53	324.8	362.3	685.6	20.1
1992	324	329	127	8.7	6.8	13.3	205.1	10.38	332.6	413.3	744.3	19.8
1993		323	128	8.6	6.9	13.5	207.3	10.33	289.4	436.0	723.3	15.8
1994		315P	127P	8.1	6.5	13.9	195.5	9.67	311.8	404.9	744.6	3.1
1995		314P	128P	8.4P	6.7P	13.8P	208.4P	10.03P	319.1P	414.4P	762.1P	
1996		314P	128P	8.4P	6.7P	13.9P	211.7P	10.07P	325.1P	422.1P	776.3P	
1997		313P	129P	8.4P	6.6P	13.9P	215.0P	10.12P	331.0P	429.9P	790.6P	
1998		313P	130P	8.4P	6.6P	14.0P	218.2P	10.16P	337.0P	437.6P	804.8P	

Sources: 1982, 1987, 1992 *Economic Census*; *Annual Survey of Manufactures*, 83-86, 88-91, 93-94. Establishment counts for non-Census years are from *County Business Patterns*; establishment values for 83-84 are extrapolations. 'P's show projections by the editors. Industries reclassified in 87 will not have data for prior years.

INDICES OF CHANGE

Year	Companies	Establishments Total	Establishments with 20 or more employees	Employment Total (000)	Employment Production Workers (000)	Employment Hours (Mil)	Compensation Payroll ($ mil)	Compensation Wages ($/hr)	Production Cost of Materials	Production Value Added by Manufacture	Production Value of Shipments	Capital Invest.
1982												
1983												
1984												
1985												
1986												
1987	102	102	96	94	96	96	85	94	86	83	84	67
1988		98	94	100	101	102	90	93	94	89	90	
1989		93	96	99	110	107	97	99	97	92	95	
1990		91	97	91	107	111	92	87	98	85	91	138
1991		94	92	99	96	101	94	92	98	88	92	102
1992	100	100	100	100	100	100	100	100	100	100	100	100
1993		98	101	99	101	102	101	100	87	105	97	80
1994		96P	100P	93	96P	105	95	93	94	98	100	16
1995		96P	100P	97P	99P	104P	102P	97P	96P	100P	102P	
1996		95P	101P	97P	98P	104P	103P	97P	98P	102P	104P	
1997		95P	102P	97P	98P	105P	105P	97P	100P	104P	106P	
1998		95P	103P	97P	97P	105P	106P	98P	101P	106P	108P	

Sources: Same as General Statistics. Values reflect change from the base year, 1992. Values above 100 mean greater than 92, values below 100 mean less than 92, and a value of 100 in the 82-91 or 93-98 period means same as 92. 'P's mark projections by the editors.

SELECTED RATIOS

For 1994	Avg. of All Manufact.	Analyzed Industry	Index	For 1994	Avg. of All Manufact.	Analyzed Industry	Index
Employees per Establishment	49	26	52	Value Added per Production Worker	134,084	62,292	46
Payroll per Establishment	1,500,273	620,635	41	Cost per Establishment	5,045,178	989,841	20
Payroll per Employee	30,620	24,136	79	Cost per Employee	102,970	38,494	37
Production Workers per Establishment	34	21	60	Cost per Production Worker	146,988	47,969	33
Wages per Establishment	853,319	426,708	50	Shipments per Establishment	9,576,895	2,363,810	25
Wages per Production Worker	24,861	20,679	83	Shipments per Employee	195,460	91,926	47
Hours per Production Worker	2,056	2,138	104	Shipments per Production Worker	279,017	114,554	41
Wages per Hour	12.09	9.67	80	Investment per Establishment	321,011	9,841	3
Value Added per Establishment	4,602,255	1,285,397	28	Investment per Employee	6,552	383	6
Value Added per Employee	93,930	49,988	53	Investment per Production Worker	9,352	477	5

Sources: Same as General Statistics. The 'Average of All Manufacturing' column represents the average of all manufacturing industries reported for the most recent complete year available. The Index shows the relationship between the Average and the Analyzed Industry. For example, 100 means that they are equal; 500 that the Analyzed Industry is five times the average; 50 means that the Analyzed Industry is half the national average. The abbreviation 'na' is used to show that data are 'not available'.

LEADING COMPANIES Number shown: **41**   Total sales ($ mil): **1,073**   Total employment (000): **7.2**

Company Name	Address				CEO Name	Phone	Co. Type	Sales ($ mil)	Empl. (000)
Olin Corp Brass Group	427 N Shamrock St	East Alton	IL	62024	James Hascall	618-258-2000	D	654	3.3
SW Centrifugal Inc	4106 S Creek Rd	Chattanooga	TN	37406	Carter Paden	615-622-4131	R	100*	0.9
R Lavin and Sons Inc	3426 S Kedzie Av	Chicago	IL	60623	Bennet Lavin	312-847-1800	R	42*	0.4
Ampco Metal Inc	PO Box 2004	Milwaukee	WI	53201	Robert Darling	414-645-3750	R	36	0.3
Lee Brass Co	PO Box 1229	Anniston	AL	36202	DD Renfroe	205-831-2501	R	36	0.4
Piad Precision Casting	RD 12	Greensburg	PA	15601	Karl Schweisthal	412-838-5500	R	20*	0.1
Grand Haven Brass Foundry	230 N Hopkins	Grand Haven	MI	49417	Bob Mersereau	616-842-4100	D	17	0.2
Jennison Enterprises Inc	PO Box 1500	Chino	CA	91710	James M Jennison	909-591-4832	R	15	0.2
Western Brass Industries	1309 W Sepulveda	Torrance	CA	90501	Walter Storm	310-534-5232	R	14*	0.1
Western Brass Works	1440 N Spring St	Los Angeles	CA	90012	David Snell	213-223-3101	D	14*	0.1
Brass Foundry Co	713 SW Adams St	Peoria	IL	61602	W Mehlenbeck	309-676-2157	R	13*	0.1
Magnacast Corp	1117 E Algonquin	Arlington H	IL	60005	Michael M Darling	708-437-6000	S	10	<0.1
Lawran Foundry Company Inc	PO Box 341398	Milwaukee	WI	53234	Thomas E Woehlke	414-645-4070	R	10	<0.1
Martin Brass Foundry Inc	2341 Jefferson St	Torrance	CA	90501	John Martin	213-775-3803	R	10*	<0.1
Erie Bronze and Aluminum Co	PO Box 8099	Erie	PA	16505	Kevin O'Connell	814-838-8602	S	8*	<0.1
Excal Inc	PO Box 3030	Mills	WY	82644	Dennis H Harper	307-237-0920	R	7	<0.1
Webster Foundry Corp	Rte 3	West Franklin	NH	03235	John Noble	603-934-3860	S	7	<0.1
Brontel/Bearing Bronze Co	9314 Elizabeth Av	Cleveland	OH	44105	Tammy S Flaherty	216-641-6520	R	6	<0.1
MA Bell Co	217 Lombard St	St Louis	MO	63102	Daniel F Aubuchon	314-421-2414	R	6	<0.1
Federal Bronze Casting Indust	9 Backus St	Newark	NJ	07105	John W Burk	201-589-7575	R	5	<0.1
St Paul Brass Foundry Co	954 W Minnehaha	St Paul	MN	55104	Peter Ryan	612-488-5567	R	5	<0.1
Richmond Industries Inc	PO Box 4398	Metuchen	NJ	08840	E Hennessy	908-548-4582	R	4	<0.1
Tampa Brass & Aluminum Corp	8511 Florida Mining	Tampa	FL	33634	Sam S Leto Jr	813-885-6064	R	4	<0.1
Midland Manufacturing Co	4800 Esco Dr	Fort Worth	TX	76140	George O Westhoff	817-478-4848	R	4	<0.1
Hackett Brass Foundry	1200 Lilibridge St	Detroit	MI	48214	Allen Wright	313-822-1214	R	3	<0.1
Rochester Bronze and Aluminum	57 Sherer St	Rochester	NY	14611		716-328-5500	R	2	<0.1
Ballard Pattern & Brass Foundry	PO Box 70561	Seattle	WA	98107	VW Rowe	206-784-0855	R	2	<0.1
Crosbie Foundry Company Inc	1600 Mishawaka St	Elkhart	IN	46514	Daniel J Crosbie	219-262-1502	R	2	<0.1
El Monte Alloys Inc	2023 N Chico Av	S El Monte	CA	91733	MJ Welther	818-448-3164	R	2	<0.1
Faunt Foundry Co	8524 S Vincennes	Chicago	IL	60620	TJ Doyle	312-488-2900	R	2*	<0.1
National Brass Works Inc	2140 E 25th St	Los Angeles	CA	90058	Marvin S Dunn	213-587-9123	R	2	<0.1
'A' Brass Foundry Inc	2052 E Vernon Av	Los Angeles	CA	90058	FM Lee	213-231-1101	R	2*	<0.1
Calumet Brass Foundry Inc	PO Box 158	Dolton	IL	60419	Catherine Lencki	708-849-3040	R	2	<0.1
Quad Industries Inc	PO Box 8	Bradford	TN	38316	A Kerns	901-742-3903	R	1	<0.1
Down River Casting Co	32840 Cleveland St	Rockwood	MI	48173	Robert Schuster	313-379-9666	R	1	<0.1
Duluth Brass and Aluminum Co	6900 Polk St	Duluth	MN	55807	JA Holt	218-628-1065	R	1	<0.1
Halves/Coppersource	3427 Enterprise Av	Hayward	CA	94545	Ken Cecil	510-786-4200	R	1*	<0.1
Lansing Foundry Inc	16150 Suntone Dr	South Holland	IL	60473	Michael P Gucciard	708-339-0082	R	1	<0.1
McGraw Manufacturing Inc	322 N Orange Av	Brea	CA	92621	EK Venugopal	714-529-2022	R	1*	<0.1
Richards Corp	PO Box 368	Boston	MA	02101	Kenneth A Wernick	617-322-2231	R	1	<0.1
Builders Brass Works Corp	3528 Emery St	Los Angeles	CA	90023	Ira Simon	213-269-8111	S	0*	0.1

Source: Ward's Business Directory of U.S. Private and Public Companies, Volumes 1 and 2, 1996. The company type code used is as follows: P - Public, R - Private, S - Subsidiary, D - Division, J - Joint Venture, A - Affiliate, G - Group. Sales are in millions of dollars, employees are in thousands. An asterisk (*) indicates an estimated sales volume. The symbol < stands for 'less than'. Company names and addresses are truncated, in some cases, to fit into the available space.

MATERIALS CONSUMED

Material	Quantity		Delivered Cost ($ million)
Materials, ingredients, containers, and supplies		(X)	271.3
Copper and copper-base alloy shapes and forms	1,000 s tons	(S)	92.0
Aluminum and aluminum-base alloy shapes and forms	1,000 s tons	5.3*	8.8
Zinc and zinc-base alloy shapes and forms		(X)	0.2
Magnesium and magnesium-base alloy shapes and forms		(X)	(D)
Lead-base alloy shapes and forms		(X)	(D)
All other nonferrous shapes and forms		(X)	3.5
Copper and copper-base alloy scrap (except home scrap)		(X)	28.5
Aluminum and aluminum-base alloy scrap (except home scrap)		(X)	0.1
Other nonferrous metal scrap (except home scrap)		(X)	0.7
Industrial patterns		(X)	2.0
Industrial dies, molds, jigs, and fixtures		(X)	4.4
All other industrial and commercial machinery and computer equipment		(X)	0.6
Sand		(X)	2.9
Grinding wheels and other abrasive products, except industrial diamonds		(X)	1.9
All other materials and components, parts, containers, and supplies		(X)	42.6
Materials, ingredients, containers, and supplies, nsk		(X)	82.9

Source: 1992 *Economic Census*. Explanation of symbols used: (D): Withheld to avoid disclosure of competitive data; na: Not available; (S): Withheld because statistical norms were not met; (X): Not applicable; (Z): Less than half the unit shown; nec: Not elsewhere classified; nsk: Not specified by kind; - : zero; * : 10-19 percent estimated; ** : 20-29 percent estimated.

PRODUCT SHARE DETAILS

Product or Product Class	% Share	Product or Product Class	% Share
Copper foundries	100.00	leaded yellow brasses, nickel tin bronzes, nickel silvers, lead bronzes, and special alloys (except bearings and bushings)	6.97
Copper-base alloy sand castings (except bearings and bushings)	16.35	Copper and copper-base alloy permanent and semipermanent mold castings (except bearings and bushings)	3.98
Leaded red and semi-red brass sand castings (except bearings and bushings)	13.27	Copper and copper-base alloy centrifugal castings (except bearings and bushings)	4.48
Tin bronze sand castings, including leaded and high-leaded (except bearings and bushings)	5.60	Copper and copper-base alloy investment castings (except bearings and bushings)	3.11
Copper and high-copper alloy sand castings (except bearings and bushings)	6.97	Other copper and copper-base alloy castings, excluding die-castings (except bearings and bushings)	3.70
Engineered copper alloy sand castings, including manganese bronzes, silicon bronzes and brasses, aluminum bronzes, and copper nickels (except bearings and bushings) . . .	12.27	Copper-base alloy bearings and bushings, nonmachined . .	4.54
Other copper alloy sand castings, including yellow and			

Source: 1992 *Economic Census*. The values shown are percent of total shipments in an industry. Values of indented subcategories are summed in the main heading. The symbol (D) appears when data are withheld to prevent disclosure of competitive information. The abbreviation nsk stands for 'not specified by kind' and nec for 'not elsewhere classified'.

INPUTS AND OUTPUTS FOR BRASS, BRONZE, & COPPER CASTINGS

Economic Sector or Industry Providing Inputs	%	Sector	Economic Sector or Industry Buying Outputs	%	Sector
Primary copper	40.0	Manufg.	Pipe, valves, & pipe fittings	12.9	Manufg.
Special dies & tools & machine tool accessories	8.3	Manufg.	Plumbing fixture fittings & trim	10.9	Manufg.
Electric services (utilities)	5.0	Util.	Pumps & compressors	5.2	Manufg.
Wholesale trade	4.1	Trade	Ship building & repairing	4.5	Manufg.
Advertising	3.6	Services	Exports	4.4	Foreign
Primary aluminum	3.2	Manufg.	Power transmission equipment	3.8	Manufg.
Scrap	3.2	Scrap	Machinery, except electrical, nec	3.2	Manufg.
Brass, bronze, & copper castings	2.6	Manufg.	General industrial machinery, nec	3.1	Manufg.
Gas production & distribution (utilities)	2.4	Util.	Fabricated structural metal	3.0	Manufg.
Petroleum refining	1.9	Manufg.	Mechanical measuring devices	3.0	Manufg.
Maintenance of nonfarm buildings nec	1.7	Constr.	X-ray apparatus & tubes	2.9	Manufg.
Motor freight transportation & warehousing	1.5	Util.	Federal Government purchases, national defense	2.9	Fed Govt
Royalties	1.5	Fin/R.E.	Hardware, nec	2.6	Manufg.
Chemical preparations, nec	1.2	Manufg.	Ball & roller bearings	2.3	Manufg.
Machinery, except electrical, nec	1.2	Manufg.	Musical instruments	2.2	Manufg.
Eating & drinking places	1.2	Trade	Special industry machinery, nec	2.1	Manufg.
Communications, except radio & TV	1.1	Util.	Refrigeration & heating equipment	1.7	Manufg.
Metal heat treating	1.0	Manufg.	Food products machinery	1.5	Manufg.
Railroads & related services	1.0	Util.	Boat building & repairing	1.4	Manufg.
Paperboard containers & boxes	0.9	Manufg.	Brass, bronze, & copper castings	1.4	Manufg.
Miscellaneous repair shops	0.9	Services	Blowers & fans	1.3	Manufg.
Abrasive products	0.8	Manufg.	Miscellaneous metal work	1.3	Manufg.
Air transportation	0.7	Util.	Switchgear & switchboard apparatus	1.2	Manufg.
Insurance carriers	0.7	Fin/R.E.	Special dies & tools & machine tool accessories	1.1	Manufg.
Real estate	0.6	Fin/R.E.	Aircraft & missile engines & engine parts	1.0	Manufg.
Sand & gravel	0.5	Mining	Metal stampings, nec	1.0	Manufg.
Primary zinc	0.5	Manufg.	Farm machinery & equipment	0.9	Manufg.
Equipment rental & leasing services	0.5	Services	Wiring devices	0.9	Manufg.
Legal services	0.5	Services	Miscellaneous plastics products	0.8	Manufg.
Management & consulting services & labs	0.5	Services	Motors & generators	0.8	Manufg.
U.S. Postal Service	0.5	Gov't	Welding apparatus, electric	0.8	Manufg.
Industrial patterns	0.4	Manufg.	Internal combustion engines, nec	0.7	Manufg.
Lubricating oils & greases	0.4	Manufg.	Lighting fixtures & equipment	0.7	Manufg.
Banking	0.4	Fin/R.E.	Service industry machines, nec	0.7	Manufg.
Hotels & lodging places	0.4	Services	Construction machinery & equipment	0.6	Manufg.
Blast furnaces & steel mills	0.3	Manufg.	Oil field machinery	0.6	Manufg.
Electrometallurgical products	0.3	Manufg.	Power driven hand tools	0.6	Manufg.
Machine tools, metal cutting types	0.3	Manufg.	Aircraft & missile equipment, nec	0.5	Manufg.
Accounting, auditing & bookkeeping	0.3	Services	Carburetors, pistons, rings, & valves	0.5	Manufg.
Automotive rental & leasing, without drivers	0.3	Services	Engine electrical equipment	0.5	Manufg.
Automotive repair shops & services	0.3	Services	Fabricated metal products, nec	0.5	Manufg.
Business/professional associations	0.3	Services	Machine tools, metal forming types	0.5	Manufg.
Manifold business forms	0.2	Manufg.	Motor vehicle parts & accessories	0.5	Manufg.
Sanitary services, steam supply, irrigation	0.2	Util.	Paper industries machinery	0.5	Manufg.
Security & commodity brokers	0.2	Fin/R.E.	Architectural metal work	0.4	Manufg.
Computer & data processing services	0.2	Services	Automotive stampings	0.4	Manufg.
Engineering, architectural, & surveying services	0.2	Services	Electronic components nec	0.4	Manufg.
Laundry, dry cleaning, shoe repair	0.2	Services	Fabricated plate work (boiler shops)	0.4	Manufg.
Gaskets, packing & sealing devices	0.1	Manufg.	Manufacturing industries, nec	0.4	Manufg.
Primary nonferrous metals, nec	0.1	Manufg.	Industrial patterns	0.3	Manufg.
Water supply & sewage systems	0.1	Util.	Metalworking machinery, nec	0.3	Manufg.
Detective & protective services	0.1	Services	Mining machinery, except oil field	0.3	Manufg.
Personnel supply services	0.1	Services	Motor vehicles & car bodies	0.3	Manufg.
Services to dwellings & other buildings	0.1	Services	Turbines & turbine generator sets	0.3	Manufg.
			Railroads & related services	0.3	Util.
			Electronic computing equipment	0.2	Manufg.
			Environmental controls	0.2	Manufg.
			Lawn & garden equipment	0.2	Manufg.
			Machine tools, metal cutting types	0.2	Manufg.
			Metal doors, sash, & trim	0.2	Manufg.
			Radio & TV communication equipment	0.2	Manufg.
			Surgical appliances & supplies	0.2	Manufg.
			Heating equipment, except electric	0.1	Manufg.
			Hoists, cranes, & monorails	0.1	Manufg.
			Rolling mill machinery	0.1	Manufg.

Source: Benchmark Input-Output Accounts for the U.S. Economy, 1982, U.S. Department of Commerce, Washington, D.C., July 1991. Data, as reported in the source, are organized by the 1977 SIC structure in use in 1982 but have been matched, as closely as is possible, to the 1987 SIC structure used in this book.

OCCUPATIONS EMPLOYED BY SIC 336 - NONFERROUS FOUNDRIES (CASTINGS)

Occupation	% of Total 1994	Change to 2005	Occupation	% of Total 1994	Change to 2005
Metal molding machine workers	12.2	-2.1	Machine tool cutting & forming etc. nec	1.7	-10.9
Assemblers, fabricators, & hand workers nec	7.2	-11.0	Combination machine tool operators	1.6	-2.1
Grinders & polishers, hand	5.6	-11.0	Precision metal workers nec	1.6	-11.0
Blue collar worker supervisors	5.2	-15.7	Machine forming operators, metal & plastic	1.5	-10.9
Inspectors, testers, & graders, precision	5.1	-11.0	Traffic, shipping, & receiving clerks	1.4	-14.3
Precision workers nec	4.6	-19.9	Industrial production managers	1.4	-10.9
Tool & die makers	3.3	-28.1	Metal & plastic machine workers nec	1.3	-21.0
Foundry mold assembly & shakeout workers	3.1	-19.9	Hand packers & packagers	1.3	-23.7
General managers & top executives	2.7	-15.5	Janitors & cleaners, incl maids	1.1	-28.8
Machine tool cutting operators, metal & plastic	2.7	-25.8	Industrial truck & tractor operators	1.1	-11.0
Grinding machine operators, metal & plastic	2.5	20.2	Sales & related workers nec	1.1	-10.9
Furnace operators	2.5	-19.9	Bookkeeping, accounting, & auditing clerks	1.1	-33.3
Machinists	2.5	-11.0	Punching machine operators, metal & plastic	1.0	-28.8
Industrial machinery mechanics	2.1	-2.1	General office clerks	1.0	-24.1
Maintenance repairers, general utility	1.8	-19.9			

Source: Industry-Occupation Matrix, Bureau of Labor Statistics. These data relate to one or more 3-digit SIC industry groups rather than to a single 4-digit SIC. The change reported for each occupation to the year 2005 is a percent of growth or decline as estimated by the Bureau of Labor Statistics. The abbreviation nec stands for 'not elsewhere classified'.

LOCATION BY STATE AND REGIONAL CONCENTRATION

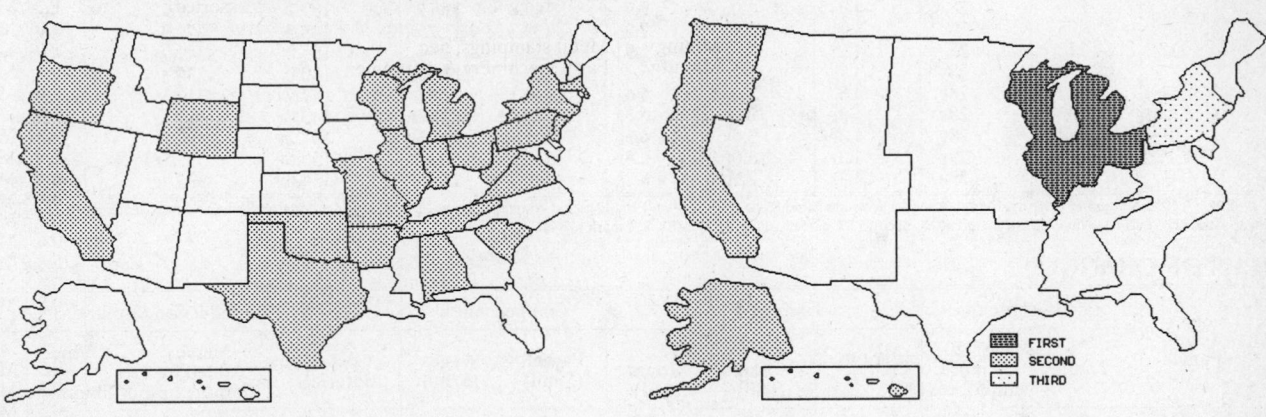

FIRST
SECOND
THIRD

INDUSTRY DATA BY STATE

State	Establish-ments	Shipments Total ($ mil)	Shipments % of U.S.	Shipments Per Establ.	Employment Total Number	Employment % of U.S.	Employment Per Establ.	Wages ($/hour)	Cost as % of Shipments	Investment per Employee ($)
Illinois	30	87.5	11.8	2.9	1,100	12.6	37	9.94	47.4	2,182
Pennsylvania	24	79.9	10.7	3.3	700	8.0	29	11.40	50.3	4,286
California	41	76.1	10.2	1.9	1,000	11.5	24	10.00	41.9	2,300
Wisconsin	13	62.6	8.4	4.8	600	6.9	46	11.63	48.9	-
Michigan	17	55.6	7.5	3.3	600	6.9	35	9.90	41.4	1,667
Indiana	11	51.3	6.9	4.7	600	6.9	55	10.22	42.3	3,167
New York	15	16.8	2.3	1.1	300	3.4	20	9.50	34.5	-
Texas	18	16.8	2.3	0.9	200	2.3	11	11.33	44.6	1,000
New Jersey	7	15.8	2.1	2.3	200	2.3	29	11.67	49.4	2,000
Oklahoma	6	13.1	1.8	2.2	200	2.3	33	8.67	44.3	-
Massachusetts	12	10.6	1.4	0.9	100	1.1	8	11.50	33.0	2,000
Oregon	9	8.7	1.2	1.0	100	1.1	11	11.00	31.0	3,000
Alabama	4	8.6	1.2	2.2	100	1.1	25	9.00	46.5	-
Ohio	23	(D)	-	-	750 *	8.6	33	-	-	-
Missouri	8	(D)	-	-	175 *	2.0	22	-	-	571
Virginia	5	(D)	-	-	175 *	2.0	35	-	-	571
West Virginia	5	(D)	-	-	175 *	2.0	35	-	-	-
South Carolina	4	(D)	-	-	175 *	2.0	44	-	-	-
Arkansas	2	(D)	-	-	175 *	2.0	88	-	-	-
Tennessee	2	(D)	-	-	175 *	2.0	88	-	-	-
Wyoming	2	(D)	-	-	175 *	2.0	88	-	-	-

Source: 1992 Economic Census. The states are in descending order of shipments or establishments (if shipment data are missing for the majority). The symbol (D) appears when data are withheld to prevent disclosure of competitive information. States marked with (D) are sorted by number of establishments. A dash (-) indicates that the data element cannot be calculated; * indicates the midpoint of a range.

3369 - NONFERROUS FOUNDRIES, NEC

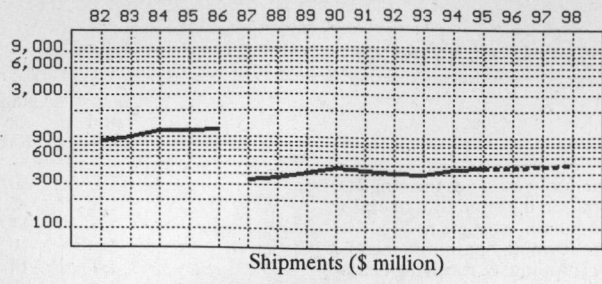

Shipments ($ million)

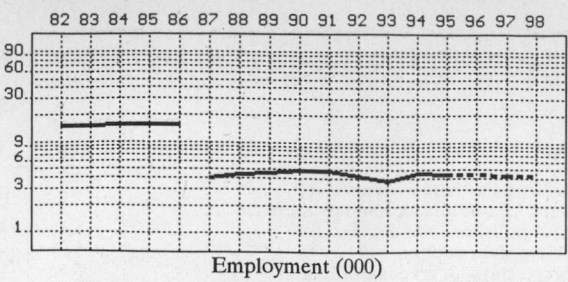

Employment (000)

GENERAL STATISTICS

| Year | Com-panies | Establishments | | Employment | | | Compensation | | Production ($ million) | | | |
		Total	with 20 or more employees	Total (000)	Production Workers (000)	Hours (Mil)	Payroll ($ mil)	Wages ($/hr)	Cost of Materials	Value Added by Manufacture	Value of Shipments	Capital Invest.
1982	351	358	148	14.9	11.8	22.3	264.9	8.11	427.8	478.5	916.1	27.5
1983		348	151	15.1	12.2	23.5	283.2	8.20	472.5	567.0	1,027.2	38.0
1984		338	154	15.4	12.1	23.8	289.7	8.53	542.4	637.1	1,177.2	33.7
1985		328	158	15.3	11.9	23.1	303.6	9.13	551.5	651.2	1,203.1	40.1
1986		318	155	15.5	12.0	23.9	338.3	9.71	586.2	658.6	1,229.0	30.5
1987*	50	56	26	4.0	2.9	5.6	101.3	11.36	127.8	214.9	339.9	9.0
1988		56	27	4.4	3.3	6.7	115.5	10.88	159.7	217.1	366.8	
1989		58	25	4.5	3.6	6.8	127.9	12.13	184.0	234.8	412.5	
1990		70	28	4.9	4.0	7.9	149.8	11.77	187.2	289.5	465.0	13.0
1991		75	30	4.9	3.5	7.3	139.3	11.25	148.3	275.0	430.6	11.6
1992	113	118	25	4.2	3.0	5.9	121.2	12.64	155.4	249.7	412.6	10.8
1993		112	28	3.7	2.7	5.3	108.2	12.36	140.0	236.3	398.7	2.4
1994		122P	28P	4.5	3.1	6.6	144.1	12.67	141.7	310.1	451.8	13.6
1995		133P	28P	4.3P	3.1P	6.3P	138.4P	12.82P	143.8P	314.6P	458.4P	
1996		144P	28P	4.3P	3.0P	6.3P	141.1P	13.03P	147.2P	322.1P	469.2P	
1997		155P	29P	4.3P	3.0P	6.3P	143.9P	13.24P	150.6P	329.5P	480.1P	
1998		166P	29P	4.3P	2.9P	6.2P	146.7P	13.45P	154.0P	336.9P	490.9P	

Sources: 1982, 1987, 1992 *Economic Census*; *Annual Survey of Manufactures*, 83-86, 88-91, 93-94. Establishment counts are from *County Business Patterns* for non-Census years; establishment counts for 83-84 are extrapolations. * indicates that industry content changed in 87; earlier years use 77 SICs. 'P's mark projections.

INDICES OF CHANGE

| Year | Com-panies | Establishments | | Employment | | | Compensation | | Production ($ million) | | | |
		Total	with 20 or more employees	Total (000)	Production Workers (000)	Hours (Mil)	Payroll ($ mil)	Wages ($/hr)	Cost of Materials	Value Added by Manufacture	Value of Shipments	Capital Invest.
1982	311	303	592	355	393	378	219	64	275	192	222	255
1983		295	604	360	407	398	234	65	304	227	249	352
1984		286	616	367	403	403	239	67	349	255	285	312
1985		278	632	364	397	392	250	72	355	261	292	371
1986		269	620	369	400	405	279	77	377	264	298	282
1987*	44	47	104	95	97	95	84	90	82	86	82	83
1988		47	108	105	110	114	95	86	103	87	89	
1989		49	100	107	120	115	106	96	118	94	100	
1990		59	112	117	133	134	124	93	120	116	113	120
1991		64	120	117	117	124	115	89	95	110	104	107
1992	100	100	100	100	100	100	100	100	100	100	100	100
1993		95	112	88	90	90	89	98	90	95	97	22
1994		103P	112P	107	103	112	119	100	91	124	110	126
1995		113P	113P	103P	102P	107P	114P	101P	93P	126P	111P	
1996		122P	114P	103P	100P	107P	116P	103P	95P	129P	114P	
1997		131P	115P	103P	99P	106P	119P	105P	97P	132P	116P	
1998		141P	116P	103P	97P	105P	121P	106P	99P	135P	119P	

Sources: Same as General Statistics. Values reflect change from the base year, 1992. Values above 100 mean greater than 92, values below 100 mean less than 92, and a value of 100 in the 82-91 or 93-98 period means same as 92. * indicates that industry content changed in 87. Data for earlier years are in 77 SIC format.

SELECTED RATIOS

For 1994	Avg. of All Manufact.	Analyzed Industry	Index	For 1994	Avg. of All Manufact.	Analyzed Industry	Index
Employees per Establishment	49	37	75	Value Added per Production Worker	134,084	100,032	75
Payroll per Establishment	1,500,273	1,181,148	79	Cost per Establishment	5,045,178	1,161,475	23
Payroll per Employee	30,620	32,022	105	Cost per Employee	102,970	31,489	31
Production Workers per Establishment	34	25	74	Cost per Production Worker	146,988	45,710	31
Wages per Establishment	853,319	685,426	80	Shipments per Establishment	9,576,895	3,703,279	39
Wages per Production Worker	24,861	26,975	109	Shipments per Employee	195,460	100,400	51
Hours per Production Worker	2,056	2,129	104	Shipments per Production Worker	279,017	145,742	52
Wages per Hour	12.09	12.67	105	Investment per Establishment	321,011	111,475	35
Value Added per Establishment	4,602,255	2,541,803	55	Investment per Employee	6,552	3,022	46
Value Added per Employee	93,930	68,911	73	Investment per Production Worker	9,352	4,387	47

Sources: Same as General Statistics. The 'Average of All Manufacturing' column represents the average of all manufacturing industries reported for the most recent complete year available. The Index shows the relationship between the Average and the Analyzed Industry. For example, 100 means that they are equal; 500 that the Analyzed Industry is five times the average; 50 means that the Analyzed Industry is half the national average. The abbreviation 'na' is used to show that data are 'not available'.

LEADING COMPANIES Number shown: 38 Total sales ($ mil): 504 Total employment (000): 4.7

Company Name	Address				CEO Name	Phone	Co. Type	Sales ($ mil)	Empl. (000)
Ohio Decorative Products Inc	220 S Elizabeth Av	Spencerville	OH	45887	Charles L Moeller	419-647-4191	R	150*	1.5
NGK Metals Corp	PO Box 13367	Reading	PA	19612	EA Lilley	610-921-5000	S	50	0.2
Bunting Bearings Corp	PO Box 729	Holland	OH	43528	John C Gyorgyi	419-866-7000	R	40	0.4
Machining Enterprises Inc	11400 Toepfer Rd	Warren	MI	48089	C H Proctor Jr	810-755-3180	R	35	0.2
Fall River Foundry Co	670 S Main St	Fall River	WI	53932	RR Robbins	414-484-3311	D	25	0.1
Grand Rapids Die Casting	PO Box Q	Grand Rapids	MI	49501	Michael Slack	616-784-6100	S	24	0.3
Johnson Brass & Machine	PO Box 219	Saukville	WI	53080	Leland Johnson	414-377-9440	R	15*	0.1
Chromalloy Casting Miami Corp	4701 NW 77th Av	Miami	FL	33166	Thomas Wendt	305-592-6300	S	12	0.1
Newell Industries Inc	530 Steves Av	San Antonio	TX	78210	Scott Newell	210-227-9090	R	12*	0.2
TCI Aluminum	PO Box 2069	Gardena	CA	90247	BE Belzer	310-323-5613	R	12	<0.1
Ceramet	2175 Av	Bethlehem	PA	18017	Brian Gimson	215-266-0270	D	11	0.2
Radial Casting Corp	70 Pennsylvania Av	Kearny	NJ	07032	Ronald H Landau	201-344-0333	R	11*	0.1
Bronze-Craft Corp	PO Box 788	Nashua	NH	03061	A Jack Atkinson	603-883-7747	R	10	0.2
Custom Metal Crafters Inc	815 N Mountain Rd	Newington	CT	06111	John Bourget	203-953-4210	R	10*	0.1
TiLine Inc	PO Box 908	Albany	OR	97321	Jack Byrne	503-926-7711	S	10*	0.1
Presto Casting Co	PO Box 1059	Glendale	AZ	85311	Harvey Starr	602-939-9441	R	9*	0.1
Gulf Reduction	PO Box 611	Houston	TX	77001	Howard Robinson	713-926-1705	D	7	<0.1
Magparts	1545 Roosevelt St	Azusa	CA	91702	RH Emerson	818-334-7897	R	6	<0.1
Crescent Brass Mfg Corp	132 Angelica St	Reading	PA	19611	IL Jeffery	610-372-7834	R	6	<0.1
Lite Metals Company Inc	PO Box 829	Ravenna	OH	44266	DE McCoy	216-296-6110	R	5*	<0.1
ABM Manufacturing Inc	415 N Marshall St	Sedalia	MO	65301	Kevin Loeffelmen	816-827-3434	S	4	<0.1
Eutectic Engineering Co	6350 E Davison	Detroit	MI	48212	Charles E Baer Jr	313-892-2248	R	4*	<0.1
Kief Industries Inc	100 June Av	Blandon	PA	19510	R Bokovoy	610-926-2128	R	4	<0.1
Lincoln Foundry Inc	PO Box 8156	Erie	PA	16505	Charles E Grappy	814-833-0404	R	4*	<0.1
Prime Alloy Castings Inc	717 Industrial Av	Port Hueneme	CA	93041	Carol Fehr	805-488-6451	R	4	<0.1
El Monte Non-Ferrous Foundry	245 Turnbull	City of Industry	CA	91745	Dana Roukoz	818-330-5061	R	3	<0.1
Non-Ferrous Metals Inc	2905 13th Av SW	Seattle	WA	98134	JW McGee	206-624-8414	R	3	<0.1
Ryder-Heil Bronze Inc	PO Box 647	Bucyrus	OH	44820	J M Quaintance	419-562-2841	R	3	<0.1
US Bronze Foundry & Machine	PO Box 458	Meadville	PA	16335	Dan Higham	814-337-4234	R	3	<0.1
Chicago Magnesium Casting Co	PO Box 237	Blue Island	IL	60406	Ron Larson	708-597-1300	R	3	<0.1
Brass and Bronze Casting Co	387-B Sandy Hill	Irwin	PA	15642	Stu Zarembo	412-864-6830	R	2*	<0.1
Dee Foundries Inc	PO Box 8727	Houston	TX	77249	Melvin Myers	713-222-0236	R	2	<0.1
Precision Cast Products Inc	PO Box 23884	Oakland	CA	94623	D Torkington	510-891-9078	R	2	<0.1
Joseff-Hollywood	129 E Providencia	Burbank	CA	91502	Joan C Joseff	818-846-0157	R	1	<0.1
Phillips Pattern and Castings	311 N Morrison Rd	Muncie	IN	47304	Greg Phillips	317-289-2816	R	1	<0.1
Atherton Foundry Products	13000 S Halsted St	Riverdale	IL	60627	Owen F Smith	708-849-4615	R	1	<0.1
Soligen Inc	19408 Londelius St	Northridge	CA	91324	Yehoram Uziel	818-718-1221	S	1	<0.1
Controlled Castings Corp	31 Commercial Ct	Plainview	NY	11803	P Fratello	516-349-1717	R	0	<0.1

Source: Ward's Business Directory of U.S. Private and Public Companies, Volumes 1 and 2, 1996. The company type code used is as follows: P - Public, R - Private, S - Subsidiary, D - Division, J - Joint Venture, A - Affiliate, G - Group. Sales are in millions of dollars, employees are in thousands. An asterisk (*) indicates an estimated sales volume. The symbol < stands for 'less than'. Company names and addresses are truncated, in some cases, to fit into the available space.

MATERIALS CONSUMED

Material	Quantity	Delivered Cost ($ million)
Materials, ingredients, containers, and supplies	(X)	131.2
Copper and copper-base alloy shapes and forms	1,000 s tons (D)	(D)
Aluminum and aluminum-base alloy shapes and forms	1,000 s tons (S)	1.1
Zinc and zinc-base alloy shapes and forms	(X)	4.3
Magnesium and magnesium-base alloy shapes and forms	(X)	4.4
All other nonferrous shapes and forms	(X)	29.9
Copper and copper-base alloy scrap (except home scrap)	(X)	(D)
Other nonferrous metal scrap (except home scrap)	(X)	(D)
Industrial patterns	(X)	(D)
Industrial dies, molds, jigs, and fixtures	(X)	(D)
All other industrial and commercial machinery and computer equipment	(X)	(D)
Sand	(X)	4.2
Grinding wheels and other abrasive products, except industrial diamonds	(X)	2.0
All other materials and components, parts, containers, and supplies	(X)	57.3
Materials, ingredients, containers, and supplies, nsk	(X)	15.2

Source: 1992 *Economic Census.* Explanation of symbols used: (D): Withheld to avoid disclosure of competitive data; na: Not available; (S): Withheld because statistical norms were not met; (X): Not applicable; (Z): Less than half the unit shown; nec: Not elsewhere classified; nsk: Not specified by kind; - : zero; * : 10-19 percent estimated; ** : 20-29 percent estimated.

PRODUCT SHARE DETAILS

Product or Product Class	% Share	Product or Product Class	% Share
Nonferrous foundries, nec	100.00	Nickel and nickel-base alloy castings (excluding die-castings)	21.80
Zinc and zinc-base alloy castings (excluding die-castings)	7.87	Titanium and titanium-base alloy castings (excluding die-castings)	(D)
Magnesium and magnesium-base alloy sand castings (excluding die-castings)	7.11	Other nonferrous metal castings (excluding die-castings)	32.28
Other magnesium and magnesium-base alloy castings, including permanent and semipermanent mold and cast anodes (excluding die-castings)	2.98	Nonferrous foundries, nec, nsk, for non-administrative record establishments	(D)

Source: 1992 *Economic Census*. The values shown are percent of total shipments in an industry. Values of indented subcategories are summed in the main heading. The symbol (D) appears when data are withheld to prevent disclosure of competitive information. The abbreviation nsk stands for 'not specified by kind' and nec for 'not elsewhere classified'.

INPUTS AND OUTPUTS FOR NONFERROUS CASTINGS, NEC

Economic Sector or Industry Providing Inputs	%	Sector	Economic Sector or Industry Buying Outputs	%	Sector
Primary zinc	23.1	Manufg.	Federal Government purchases, national defense	24.3	Fed Govt
Special dies & tools & machine tool accessories	8.6	Manufg.	Aircraft & missile engines & engine parts	13.3	Manufg.
Nonmetallic mineral products, nec	7.0	Manufg.	Aircraft & missile equipment, nec	5.0	Manufg.
Wholesale trade	4.9	Trade	Machinery, except electrical, nec	4.4	Manufg.
Electrometallurgical products	4.7	Manufg.	Pipe, valves, & pipe fittings	3.3	Manufg.
Primary aluminum	4.1	Manufg.	Electric housewares & fans	3.2	Manufg.
Electric services (utilities)	3.8	Util.	Pumps & compressors	2.5	Manufg.
Primary nonferrous metals, nec	3.7	Manufg.	Power transmission equipment	2.4	Manufg.
Petroleum refining	3.0	Manufg.	Carburetors, pistons, rings, & valves	2.2	Manufg.
Machine tools, metal forming types	2.8	Manufg.	Motor vehicle parts & accessories	2.2	Manufg.
Gas production & distribution (utilities)	2.8	Util.	Oil field machinery	1.9	Manufg.
Advertising	2.8	Services	Wiring devices	1.8	Manufg.
Primary lead	2.0	Manufg.	Lighting fixtures & equipment	1.6	Manufg.
Cyclic crudes and organics	1.8	Manufg.	Plumbing fixture fittings & trim	1.5	Manufg.
Banking	1.7	Fin/R.E.	Special industry machinery, nec	1.3	Manufg.
Metal coating & allied services	1.4	Manufg.	Fabricated metal products, nec	1.2	Manufg.
Maintenance of nonfarm buildings nec	1.3	Constr.	Internal combustion engines, nec	1.2	Manufg.
Real estate	1.3	Fin/R.E.	Metal stampings, nec	1.2	Manufg.
Fabricated metal products, nec	1.2	Manufg.	Miscellaneous metal work	1.1	Manufg.
Air transportation	1.2	Util.	Ball & roller bearings	0.9	Manufg.
Royalties	1.2	Fin/R.E.	Electronic computing equipment	0.9	Manufg.
Paperboard containers & boxes	1.0	Manufg.	General industrial machinery, nec	0.9	Manufg.
Motor freight transportation & warehousing	1.0	Util.	Motors & generators	0.9	Manufg.
Machinery, except electrical, nec	0.9	Manufg.	Power driven hand tools	0.9	Manufg.
Eating & drinking places	0.9	Trade	Primary batteries, dry & wet	0.9	Manufg.
Metal heat treating	0.8	Manufg.	Radio & TV communication equipment	0.9	Manufg.
Communications, except radio & TV	0.8	Util.	Woodworking machinery	0.9	Manufg.
Miscellaneous repair shops	0.8	Services	Automotive stampings	0.8	Manufg.
Railroads & related services	0.7	Util.	Blowers & fans	0.8	Manufg.
Hotels & lodging places	0.7	Services	Boat building & repairing	0.8	Manufg.
Chemical preparations, nec	0.6	Manufg.	Food products machinery	0.8	Manufg.
Automotive rental & leasing, without drivers	0.6	Services	Household laundry equipment	0.7	Manufg.
Equipment rental & leasing services	0.6	Services	Engine electrical equipment	0.6	Manufg.
Security & commodity brokers	0.4	Fin/R.E.	Farm machinery & equipment	0.6	Manufg.
Automotive repair shops & services	0.4	Services	Household appliances, nec	0.6	Manufg.
Legal services	0.4	Services	Industrial controls	0.6	Manufg.
Management & consulting services & labs	0.4	Services	Lawn & garden equipment	0.6	Manufg.
Abrasive products	0.3	Manufg.	Textile machinery	0.6	Manufg.
Lubricating oils & greases	0.3	Manufg.	Electronic components nec	0.5	Manufg.
Primary copper	0.3	Manufg.	Household refrigerators & freezers	0.5	Manufg.
U.S. Postal Service	0.3	Gov't	Mechanical measuring devices	0.5	Manufg.
Industrial inorganic chemicals, nec	0.2	Manufg.	Screw machine and related products	0.5	Manufg.
Manifold business forms	0.2	Manufg.	Special dies & tools & machine tool accessories	0.5	Manufg.
Nonferrous castings, nec	0.2	Manufg.	Switchgear & switchboard apparatus	0.5	Manufg.
Insurance carriers	0.2	Fin/R.E.	Telephone & telegraph apparatus	0.5	Manufg.
Accounting, auditing & bookkeeping	0.2	Services	Turbines & turbine generator sets	0.5	Manufg.
Business/professional associations	0.2	Services	Household cooking equipment	0.4	Manufg.
Computer & data processing services	0.2	Services	Motor vehicles & car bodies	0.4	Manufg.
Laundry, dry cleaning, shoe repair	0.2	Services	Storage batteries	0.4	Manufg.
Motor vehicle parts & accessories	0.1	Manufg.	Architectural metal work	0.3	Manufg.
Tires & inner tubes	0.1	Manufg.	Commercial laundry equipment	0.3	Manufg.
Sanitary services, steam supply, irrigation	0.1	Util.	Construction machinery & equipment	0.3	Manufg.
Retail trade, except eating & drinking	0.1	Trade	Engineering & scientific instruments	0.3	Manufg.
Credit agencies other than banks	0.1	Fin/R.E.	Environmental controls	0.3	Manufg.
Engineering, architectural, & surveying services	0.1	Services	Fabricated plate work (boiler shops)	0.3	Manufg.
			Mining machinery, except oil field	0.3	Manufg.
			Automatic merchandising machines	0.2	Manufg.
			Guided missiles & space vehicles	0.2	Manufg.
			Industrial patterns	0.2	Manufg.
			Refrigeration & heating equipment	0.2	Manufg.
			Exports	0.2	Foreign
			Measuring & dispensing pumps	0.1	Manufg.
			Nonferrous castings, nec	0.1	Manufg.
			Typewriters & office machines, nec	0.1	Manufg.

Source: Benchmark Input-Output Accounts for the U.S. Economy, 1982, U.S. Department of Commerce, Washington, D.C., July 1991. Data, as reported in the source, are organized by the 1977 SIC structure in use in 1982 but have been matched, as closely as is possible, to the 1987 SIC structure used in this book.

OCCUPATIONS EMPLOYED BY SIC 336 - NONFERROUS FOUNDRIES (CASTINGS)

Occupation	% of Total 1994	Change to 2005	Occupation	% of Total 1994	Change to 2005
Metal molding machine workers	12.2	-2.1	Machine tool cutting & forming etc. nec	1.7	-10.9
Assemblers, fabricators, & hand workers nec	7.2	-11.0	Combination machine tool operators	1.6	-2.1
Grinders & polishers, hand	5.6	-11.0	Precision metal workers nec	1.6	-11.0
Blue collar worker supervisors	5.2	-15.7	Machine forming operators, metal & plastic	1.5	-10.9
Inspectors, testers, & graders, precision	5.1	-11.0	Traffic, shipping, & receiving clerks	1.4	-14.3
Precision workers nec	4.6	-19.9	Industrial production managers	1.4	-10.9
Tool & die makers	3.3	-28.1	Metal & plastic machine workers nec	1.3	-21.0
Foundry mold assembly & shakeout workers	3.1	-19.9	Hand packers & packagers	1.3	-23.7
General managers & top executives	2.7	-15.5	Janitors & cleaners, incl maids	1.1	-28.8
Machine tool cutting operators, metal & plastic	2.7	-25.8	Industrial truck & tractor operators	1.1	-11.0
Grinding machine operators, metal & plastic	2.5	20.2	Sales & related workers nec	1.1	-10.9
Furnace operators	2.5	-19.9	Bookkeeping, accounting, & auditing clerks	1.1	-33.3
Machinists	2.5	-11.0	Punching machine operators, metal & plastic	1.0	-28.8
Industrial machinery mechanics	2.1	-2.1	General office clerks	1.0	-24.1
Maintenance repairers, general utility	1.8	-19.9			

Source: Industry-Occupation Matrix, Bureau of Labor Statistics. These data relate to one or more 3-digit SIC industry groups rather than to a single 4-digit SIC. The change reported for each occupation to the year 2005 is a percent of growth or decline as estimated by the Bureau of Labor Statistics. The abbreviation nec stands for 'not elsewhere classified'.

LOCATION BY STATE AND REGIONAL CONCENTRATION

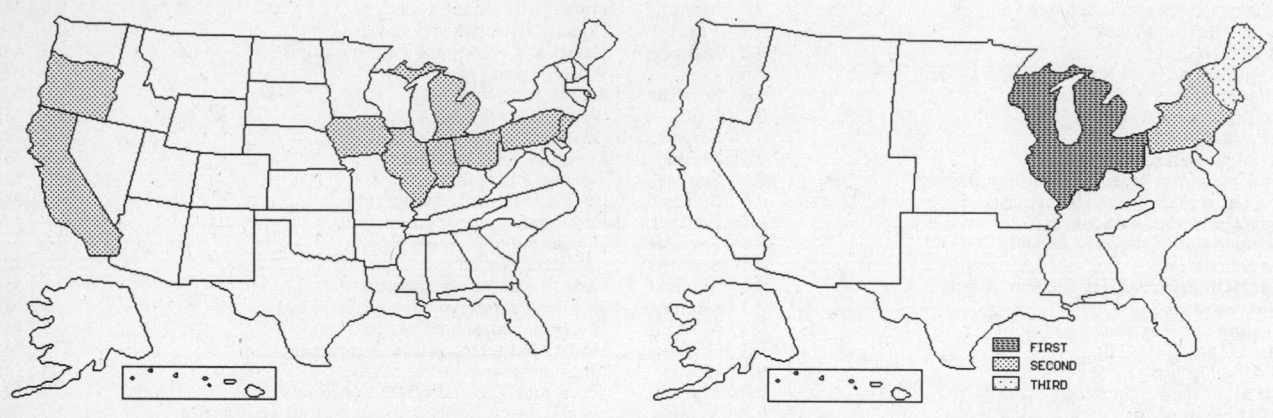

FIRST
SECOND
THIRD

INDUSTRY DATA BY STATE

State	Establish-ments	Shipments			Employment				Cost as % of Shipments	Investment per Employee ($)
		Total ($ mil)	% of U.S.	Per Establ.	Total Number	% of U.S.	Per Establ.	Wages ($/hour)		
Pennsylvania	11	23.5	5.7	2.1	200	4.8	18	13.33	55.3	-
Michigan	11	20.2	4.9	1.8	200	4.8	18	12.25	35.1	2,000
Illinois	11	19.9	4.8	1.8	200	4.8	18	11.33	46.2	3,500
California	9	14.3	3.5	1.6	200	4.8	22	11.33	34.3	4,000
Rhode Island	17	8.8	2.1	0.5	100	2.4	6	8.50	44.3	2,000
Ohio	9	(D)	-	-	1,750 *	41.7	194	-	-	-
New Jersey	7	(D)	-	-	175 *	4.2	25	-	-	-
Oregon	4	(D)	-	-	750 *	17.9	188	-	-	-
Indiana	3	(D)	-	-	175 *	4.2	58	-	-	-
Iowa	1	(D)	-	-	375 *	8.9	375	-	-	-

Source: 1992 Economic Census. The states are in descending order of shipments or establishments (if shipment data are missing for the majority). The symbol (D) appears when data are withheld to prevent disclosure of competitive information. States marked with (D) are sorted by number of establishments. A dash (-) indicates that the data element cannot be calculated; * indicates the midpoint of a range.

3398 - METAL HEAT TREATING

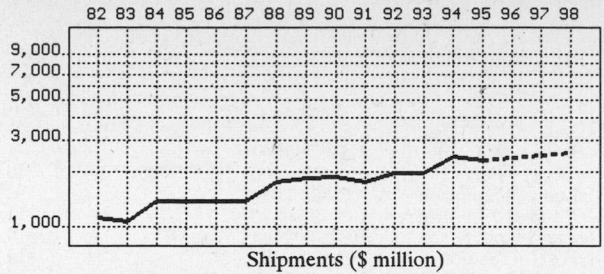

Shipments ($ million)

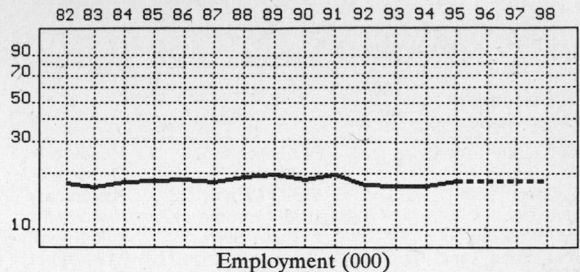

Employment (000)

GENERAL STATISTICS

| Year | Com-panies | Establishments | | Employment | | | Compensation | | Production ($ million) | | | |
		Total	with 20 or more employees	Total (000)	Production Workers (000)	Hours (Mil)	Payroll ($ mil)	Wages ($/hr)	Cost of Materials	Value Added by Manufacture	Value of Shipments	Capital Invest.
1982	668	758	289	17.7	13.5	26.9	324.2	8.05	416.0	684.5	1,128.2	42.5
1983		752	296	16.9	12.5	24.6	314.9	8.56	377.9	691.7	1,064.1	26.8
1984		746	303	18.0	13.7	28.0	373.7	9.03	520.1	918.4	1,397.0	71.3
1985		739	309	18.4	14.2	29.2	387.1	8.88	445.9	971.2	1,379.3	79.9
1986		722	310	18.6	14.3	29.9	411.9	9.14	408.0	973.5	1,381.7	84.6
1987	631	725	294	18.0	13.4	27.9	421.6	9.47	383.3	1,015.4	1,397.3	81.6
1988		703	316	19.3	14.1	31.7	487.5	9.52	630.0	1,135.2	1,757.7	49.8
1989		686	316	19.6	15.7	33.1	558.1	10.46	649.1	1,205.7	1,847.0	81.8
1990		681	297	18.5	15.4	32.8	566.1	10.82	603.9	1,274.0	1,871.7	84.1
1991		688	299	19.9	14.9	32.2	548.4	11.02	568.1	1,194.4	1,767.4	63.7
1992	634	731	299	17.4	13.0	27.7	513.9	11.65	772.4	1,194.1	1,952.4	66.9
1993		707	300	17.2	13.1	29.4	538.7	12.11	675.2	1,285.7	1,954.6	75.2
1994		684P	305P	17.2	12.7	29.3	585.8	12.48	833.8	1,565.4	2,404.8	117.2
1995		679P	305P	18.3P	14.0P	31.6P	624.7P	12.61P	793.9P	1,490.5P	2,289.8P	95.2P
1996		673P	305P	18.3P	14.0P	32.0P	647.6P	12.97P	826.2P	1,551.1P	2,382.8P	98.6P
1997		667P	306P	18.3P	14.0P	32.3P	670.6P	13.33P	858.4P	1,611.6P	2,475.8P	102.0P
1998		662P	306P	18.3P	14.0P	32.6P	693.6P	13.69P	890.7P	1,672.2P	2,568.8P	105.4P

Sources: 1982, 1987, 1992 *Economic Census*; *Annual Survey of Manufactures*, 83-86, 88-91, 93-94. Establishment counts for non-Census years are from *County Business Patterns*; establishment values for 83-84 are extrapolations. 'P's show projections by the editors. Industries reclassified in 87 will not have data for prior years.

INDICES OF CHANGE

| Year | Com-panies | Establishments | | Employment | | | Compensation | | Production ($ million) | | | |
		Total	with 20 or more employees	Total (000)	Production Workers (000)	Hours (Mil)	Payroll ($ mil)	Wages ($/hr)	Cost of Materials	Value Added by Manufacture	Value of Shipments	Capital Invest.
1982	105	104	97	102	104	97	63	69	54	57	58	64
1983		103	99	97	96	89	61	73	49	58	55	40
1984		102	101	103	105	101	73	78	67	77	72	107
1985		101	103	106	109	105	75	76	58	81	71	119
1986		99	104	107	110	108	80	78	53	82	71	126
1987	100	99	98	103	103	101	82	81	50	85	72	122
1988		96	106	111	108	114	95	82	82	95	90	74
1989		94	106	113	121	119	109	90	84	101	95	122
1990		93	99	106	118	118	110	93	78	107	96	126
1991		94	100	114	115	116	107	95	74	100	91	95
1992	100	100	100	100	100	100	100	100	100	100	100	100
1993		97	100	99	101	106	105	104	87	108	100	112
1994		94P	102P	99	98	106	114	107	108	131	123	175
1995		93P	102P	105P	107P	114P	122P	108P	103P	125P	117P	142P
1996		92P	102P	105P	107P	115P	126P	111P	107P	130P	122P	147P
1997		91P	102P	105P	108P	116P	130P	114P	111P	135P	127P	152P
1998		91P	102P	105P	108P	118P	135P	118P	115P	140P	132P	158P

Sources: Same as General Statistics. Values reflect change from the base year, 1992. Values above 100 mean greater than 92, values below 100 mean less than 92, and a value of 100 in the 82-91 or 93-98 period means same as 92. 'P's mark projections by the editors.

SELECTED RATIOS

For 1994	Avg. of All Manufact.	Analyzed Industry	Index	For 1994	Avg. of All Manufact.	Analyzed Industry	Index
Employees per Establishment	49	25	51	Value Added per Production Worker	134,084	123,260	92
Payroll per Establishment	1,500,273	856,414	57	Cost per Establishment	5,045,178	1,218,979	24
Payroll per Employee	30,620	34,058	111	Cost per Employee	102,970	48,477	47
Production Workers per Establishment	34	19	54	Cost per Production Worker	146,988	65,654	45
Wages per Establishment	853,319	534,585	63	Shipments per Establishment	9,576,895	3,515,712	37
Wages per Production Worker	24,861	28,792	116	Shipments per Employee	195,460	139,814	72
Hours per Production Worker	2,056	2,307	112	Shipments per Production Worker	279,017	189,354	68
Wages per Hour	12.09	12.48	103	Investment per Establishment	321,011	171,341	53
Value Added per Establishment	4,602,255	2,288,546	50	Investment per Employee	6,552	6,814	104
Value Added per Employee	93,930	91,012	97	Investment per Production Worker	9,352	9,228	99

Sources: Same as General Statistics. The 'Average of All Manufacturing' column represents the average of all manufacturing industries reported for the most recent complete year available. The Index shows the relationship between the Average and the Analyzed Industry. For example, 100 means that they are equal; 500 that the Analyzed Industry is five times the average; 50 means that the Analyzed Industry is half the national average. The abbreviation 'na' is used to show that data are 'not available'.

LEADING COMPANIES Number shown: **72** Total sales ($ mil): **1,236** Total employment (000): **9.9**

Company Name	Address				CEO Name	Phone	Co. Type	Sales ($ mil)	Empl. (000)
Carpenter Technology Corp	PO Box 14662	Reading	PA	19612	Robert W Cardy	215-208-2000	P	629	3.7
TC Industries Inc	PO Box 477	Crystal Lake	IL	60039	George Barry IV	815-459-2400	S	60	0.5
Lindberg Corp	6133 N River Rd	Rosemont	IL	60018	Leo G Thompson	708-823-2021	P	56	0.5
Cooperheat Inc	1021 Centennial Av	Piscataway	NJ	08854	Emmett J Lescroart	908-981-0800	S	52	0.6
FPM Heat Treating	PO Box 896	Elk Grove Vill	IL	60009	Earl Felpner	708-228-2525	D	34•	0.4
Commercial Steel Treating Corp	31440ephenson Hwy	Madison H	MI	48071	RJ Hoensheid	313-588-3300	R	29	0.1
Paulo Products Co	5711 W Park Av	St Louis	MO	63110	Ben F Rassieur III	314-647-7500	R	29•	0.3
Hinderliter Heat Treating Inc	3890 W Northwest	Dallas	TX	75202	J Hubbard	214-352-8422	R	25	0.4
Carolina Coml Heat Treating	PO Box 240867	Charlotte	NC	28224	H J Hendershot	704-527-2876	R	21•	0.2
Super Steel Treating Co	6227 Rinke	Warren	MI	48091	Terence D Farrar	810-755-9140	R	21•	0.1
Commonwealth Industries	5900 Commonw	Detroit	MI	48208	G Chantres	313-872-7900	D	15	0.1
Lindberg Heat Treating Co	2900 S Sunol Dr	Los Angeles	CA	90023	Paul McCarren	213-264-0111	D	15•	<0.1
Kentube Engineered Prod LLC	4150 S Elwood Av	Tulsa	OK	74107	Bob Hanson	918-446-4561	D	12	<0.1
Industrial Steel Treating Co	613 Carroll St	Jackson	MI	49202	Brenard Levy	517-787-6312	R	11	0.1
General Plasma Inc	12 Thompson Rd	East Windsor	CT	06088	Daniel W Parker	203-623-9901	S	10•	0.2
Erlanger Tubular Corp	5610 Bird Creek	Catoosa	OK	74015	Richard L Carter	918-266-3970	R	9•	0.1
Industrial Steel Treating Co	3370 Benedict Way	Huntington Pk	CA	90255	Gary L Colbert	213-583-1231	R	8	0.1
JW Rex Co	8th St & Val Forge	Lansdale	PA	19446	John W Rex	215-855-1131	R	8	<0.1
Ipsen Commercial Heat Treating	PO Box 6225	Rockford	IL	61125	Bill Denning	815-332-4961	D	8	<0.1
Fountain Inc	PO Drawer 1368	Fountain Inn	SC	29644	J H Hendershot	803-862-3516	D	7•	<0.1
Heat Treat Corporation	1120 W 119th St	Chicago	IL	60643	Marke Garfien	312-264-1234	S	7	<0.1
Metallurgical Service Inc	2681 E River Rd	Dayton	OH	45439	WR Miller	513-294-3212	R	7•	<0.1
Advanced Material Process Corp	3850 Howe Rd	Wayne	MI	48184	Roger Simpson	313-729-4500	R	6	<0.1
Master Appliance Corp	2420 18th St	Racine	WI	53403	S Radwill	414-633-7791	R	6	<0.1
Queen City Steel Treating Co	2980 Spring Grove	Cincinnati	OH	45225	Ed C Stenger	513-541-6300	R	6•	<0.1
Bennett Heat Treating & Brazing	690 Ferry St	Newark	NJ	07105	D Quaglia	201-589-0590	R	6	<0.1
Flame Metals Processing Corp	7317 W Lake St	Minneapolis	MN	55426	L McGuirk	612-929-7815	S	5	<0.1
Milastar Corp	9 Via Parigi	Palm Beach	FL	33480	L Michael McGurk	407-655-9590	P	5	<0.1
Aerospace/Alumatherm	7474 Garden Grove	Westminster	CA	92683	David Janes	714-893-6561	R	5•	<0.1
Aluminum Dip Braze Co	2537 N Ontario St	Burbank	CA	91504	Barry P Beckmann	818-845-6964	R	5	<0.1
Alum-A-Therm Heat Treating	7474 Garden Grove	Westminster	CA	92683	Jerry Cline	714-893-6561	R	5	<0.1
Atmosphere Processing Inc	100 N Fairbanks	Holland	MI	49423	Gail Hering	616-392-7017	R	5	<0.1
Contour Hardening Inc	7898 Zionsville Rd	Indianapolis	IN	46268	John M Storm	317-876-1630	R	5	<0.1
Lindberg Heat Treating Co	675 Christian Ln	Berlin	CT	06037	Roger Fabian	203-225-7691	D	5•	<0.1
Metal Treating Inc	710 Burns St	Cincinnati	OH	45204	O Williams Jr	513-921-2300	R	5•	<0.1
Progressive Steel Treating Inc	922 Lawn Dr	Loves Park	IL	61111	Rick Simonovich	815-877-2571	R	5	<0.1
Sun Steel Treating Inc	PO Box U	South Lyon	MI	48178	Robert K Wright	810-471-0844	R	5	<0.1
Thurner Heat Treating Corp	1501 N Mayfair Rd	Milwaukee	WI	53226	Scott P Thurner	414-771-1600	R	5	<0.1
Med-Tek Inc	2639 2nd St NE	Minneapolis	MN	55418	MP Paulson	612-789-3527	R	4	<0.1
Controlled Atmosphere	15550 Idaho St	Detroit	MI	48238	William M Keough	313-865-6500	R	4	<0.1
Industrial Heat Treating	PO Box 98	Quincy	MA	02171	George O'Brien	617-328-1010	R	4	<0.1
Lee Controls Inc	727 South Av	Piscataway	NJ	08854	RP Luce Jr	908-752-5200	R	4	<0.1
Nor-Cote Inc	11425 Timken Av	Warren	MI	48089	Stanley C Grouse	313-756-1200	R	4	<0.1
Robert Wooler Company Inc	PO Box 300	Dresher	PA	19025	Phillip C Keidel	215-542-7600	R	4•	<0.1
Specialty Steel Treating Inc	PO Box K	East Granby	CT	06026	Harold Cox	203-653-0061	R	4	<0.1
Precision Heat Treatment	PO Box 70	Southampton	PA	18966	Charles J Schaefer	215-355-0100	R	4	<0.1
Aerocraft Heat Treating Inc	15701 Minnesota	Paramount	CA	90723	David Dickson	310-634-3311	R	3•	<0.1
Diamond Heat Treating Co	5660 Jefferson	Detroit	MI	48209	AR Blaine	313-843-6570	R	3	<0.1
General Metal Heat Treating Inc	941 Addison Rd	Cleveland	OH	44103	A Torok	216-391-0886	R	3•	<0.1
Specialty Steel Treating Inc	34501 Com Rd	Fraser	MI	48026	Donald G Cox	810-293-5355	R	3•	<0.1
Steel Treaters Inc	100 Furnace St	Oriskany	NY	13424	Kenneth R Hinckley	315-736-3081	S	3	<0.1
Techni-Braze Inc	PO Box 2886	Santa Fe Sprgs	CA	90670	K Morrison	310-693-7733	R	3	<0.1
Massachusetts Steel Treating	112 Harding St	Worcester	MA	01604	Donald E Weldon	508-792-6400	D	3	<0.1
Burbank Steel Treating Company	415 S Varney St	Burbank	CA	91502	Mildred Bennet	213-849-7480	R	2	<0.1
Heat Treating of Minnesota	10150 Crosstown	Eden Prairie	MN	55344	Bob Manhatton	612-944-5500	D	2•	<0.1
Metal Improvement Company	PO Box 487	Windsor	CT	06095	G Hyndman	203-688-6201	S	2	<0.1
Newton Heat Treating Co	19235 E Walnut Dr	City of Industry	CA	91748	Ted R Newton	818-964-3491	R	2	<0.1
State Heat Treat Inc	520 32nd St SE	Grand Rapids	MI	49548	Robert E Orchard	616-243-0178	R	2	<0.1
Supreme Steel Treating Inc	2466 Seaman Av	S El Monte	CA	91733	Neal Begerow	818-350-5865	R	2•	<0.1
Thermo Techniques LP	42 Oakwood Av	Danville	IL	61832	Wade Candler	217-446-1407	S	2•	<0.1
Hydro Honing Laboratories Inc	8 Eastern Park Rd	East Hartford	CT	06128	Thomas Beach	203-289-4328	R	2	<0.1
Kenney Steel Treating Co	100 Quincy Pl	Kearny	NJ	07032	Jack P Dunphy	201-998-4420	R	2	<0.1
Modern Steel Treating Co	1010 W 122nd St	Chicago	IL	60643	George B Glickley	312-928-9200	R	2	<0.1
O and W Heat Treat Inc	1 Bidwell Rd	South Windsor	CT	06074	Harold Ohlheiser Jr	203-528-9239	R	1	<0.1
Al-Mag Heat Treat Inc	1902 Penn Mar Av	S El Monte	CA	91733	Beverly Bonano	818-442-8570	R	1	<0.1
Edwards Heat Treating Inc	642 McCormick St	San Leandro	CA	94577	Don W Edwards	510-638-4140	R	1	<0.1
Eklund Metal Treating Inc	721 Beacon St	Loves Park	IL	61111	A Eklund Jr	815-877-7436	R	1•	<0.1
Flame Treating & Eng Co	702 Oakwood Av	West Hartford	CT	06110	Thomas Benoit	203-953-3519	R	1•	<0.1
Therm Alliance Co	701 S Post Av	Detroit	MI	48209	GH Willett	313-843-1545	R	1•	<0.1
B and W Heat Treating Company	2780 Kenmore Av	Tonawanda	NY	14150	Clifford Calvello	716-876-8184	R	1	<0.1
Owego Heat Treat Inc	1646 Marshland Rd	Apalachin	NY	13732	E L Engelhard	607-785-8061	R	1•	<0.1
Oakland Metal Treating	450 Derby Av	Oakland	CA	94601	Richard B Nelson	510-261-9675	R	1	<0.1

Source: Ward's Business Directory of U.S. Private and Public Companies, Volumes 1 and 2, 1996. The company type code used is as follows: P - Public, R - Private, S - Subsidiary, D - Division, J - Joint Venture, A - Affiliate, G - Group. Sales are in millions of dollars, employees are in thousands. An asterisk (•) indicates an estimated sales volume. The symbol < stands for 'less than'. Company names and addresses are truncated, in some cases, to fit into the available space.

MATERIALS CONSUMED

Material	Quantity	Delivered Cost ($ million)
No Materials Consumed data available for this industry.		

Source: 1992 *Economic Census*. Explanation of symbols used: (D): Withheld to avoid disclosure of competitive data; na: Not available; (S): Withheld because statistical norms were not met; (X): Not applicable; (Z): Less than half the unit shown; nec: Not elsewhere classified; nsk: Not specified by kind; - : zero; * : 10-19 percent estimated; ** : 20-29 percent estimated.

PRODUCT SHARE DETAILS

Product or Product Class	% Share	Product or Product Class	% Share
Metal heat treating	100.00		

Source: 1992 *Economic Census*. The values shown are percent of total shipments in an industry. Values of indented subcategories are summed in the main heading. The symbol (D) appears when data are withheld to prevent disclosure of competitive information. The abbreviation nsk stands for 'not specified by kind' and nec for 'not elsewhere classified'.

INPUTS AND OUTPUTS FOR METAL HEAT TREATING

Economic Sector or Industry Providing Inputs	%	Sector	Economic Sector or Industry Buying Outputs	%	Sector
Chemical preparations, nec	12.2	Manufg.	Blast furnaces & steel mills	15.5	Manufg.
Electric services (utilities)	11.5	Util.	Iron & steel foundries	5.8	Manufg.
Wholesale trade	8.9	Trade	Motor vehicle parts & accessories	3.4	Manufg.
Petroleum refining	8.7	Manufg.	Motor vehicles & car bodies	3.4	Manufg.
Gas production & distribution (utilities)	8.3	Util.	Aircraft	2.8	Manufg.
Primary copper	8.2	Manufg.	Metal stampings, nec	2.7	Manufg.
Primary metal products, nec	5.2	Manufg.	Machinery, except electrical, nec	2.6	Manufg.
Paints & allied products	2.8	Manufg.	Special dies & tools & machine tool accessories	2.6	Manufg.
Industrial gases	2.5	Manufg.	Screw machine and related products	2.2	Manufg.
Motor freight transportation & warehousing	2.4	Util.	Automotive stampings	2.1	Manufg.
Cyclic crudes and organics	2.2	Manufg.	Hardware, nec	1.9	Manufg.
Miscellaneous repair shops	2.1	Services	Nonferrous wire drawing & insulating	1.7	Manufg.
Metal heat treating	1.9	Manufg.	Oil field machinery	1.6	Manufg.
Maintenance of nonfarm buildings nec	1.8	Constr.	Construction machinery & equipment	1.5	Manufg.
Coal	1.6	Mining	Aircraft & missile engines & engine parts	1.4	Manufg.
Communications, except radio & TV	1.6	Util.	Farm machinery & equipment	1.4	Manufg.
Primary lead	1.3	Manufg.	Federal Government purchases, national defense	1.4	Fed Govt
Real estate	1.2	Fin/R.E.	Steel pipe & tubes	1.3	Manufg.
Equipment rental & leasing services	1.2	Services	Aircraft & missile equipment, nec	1.2	Manufg.
Banking	1.1	Fin/R.E.	Refrigeration & heating equipment	1.2	Manufg.
Detective & protective services	1.1	Services	Hand & edge tools, nec	1.0	Manufg.
Primary nonferrous metals, nec	1.0	Manufg.	Iron & steel forgings	1.0	Manufg.
Railroads & related services	1.0	Util.	Machine tools, metal cutting types	1.0	Manufg.
Advertising	1.0	Services	Aluminum rolling & drawing	0.9	Manufg.
Machinery, except electrical, nec	0.9	Manufg.	Pipe, valves, & pipe fittings	0.9	Manufg.
Eating & drinking places	0.9	Trade	Pumps & compressors	0.9	Manufg.
Special dies & tools & machine tool accessories	0.5	Manufg.	Steel wire & related products	0.9	Manufg.
U.S. Postal Service	0.5	Gov't	Internal combustion engines, nec	0.8	Manufg.
Computer & data processing services	0.4	Services	Aluminum castings	0.7	Manufg.
Legal services	0.4	Services	Cold finishing of steel shapes	0.7	Manufg.
Management & consulting services & labs	0.4	Services	Electronic components nec	0.7	Manufg.
Abrasive products	0.3	Manufg.	Metal heat treating	0.7	Manufg
Industrial inorganic chemicals, nec	0.3	Manufg.	Wholesale trade	0.7	Trade
Power transmission equipment	0.3	Manufg.	Copper rolling & drawing	0.6	Manufg.
Royalties	0.3	Fin/R.E.	Fabricated metal products, nec	0.6	Manufg.
Lubricating oils & greases	0.2	Manufg.	Guided missiles & space vehicles	0.6	Manufg.
Manifold business forms	0.2	Manufg.	Radio & TV communication equipment	0.6	Manufg.
Metal coating & allied services	0.2	Manufg.	Special industry machinery, nec	0.6	Manufg.
Miscellaneous fabricated wire products	0.2	Manufg.	S/L Govt. purch., other general government	0.6	S/L Govt
Paperboard containers & boxes	0.2	Manufg.	Ball & roller bearings	0.5	Manufg.
Air transportation	0.2	Util.	Electrometallurgical products	0.5	Manufg.
Water transportation	0.2	Util.	General industrial machinery, nec	0.5	Manufg.
Accounting, auditing & bookkeeping	0.2	Services	Miscellaneous fabricated wire products	0.5	Manufg.
Automotive repair shops & services	0.2	Services	Plating & polishing	0.5	Manufg.
Business/professional associations	0.2	Services	Power transmission equipment	0.5	Manufg.
Laundry, dry cleaning, shoe repair	0.2	Services	Semiconductors & related devices	0.5	Manufg.
Pipelines, except natural gas	0.1	Util.	Blowers & fans	0.4	Manufg.
Sanitary services, steam supply, irrigation	0.1	Util.	Brass, bronze, & copper castings	0.4	Manufg.
Automotive rental & leasing, without drivers	0.1	Services	Conveyors & conveying equipment	0.4	Manufg.
Engineering, architectural, & surveying services	0.1	Services	Food products machinery	0.4	Manufg.
Services to dwellings & other buildings	0.1	Services	Industrial trucks & tractors	0.4	Manufg.
			Metal cans	0.4	Manufg.
			Mining machinery, except oil field	0.4	Manufg.
			Nonferrous castings, nec	0.4	Manufg.
			Nonferrous rolling & drawing, nec	0.4	Manufg.
			Power driven hand tools	0.4	Manufg.
			Primary aluminum	0.4	Manufg.

Continued on next page.

INPUTS AND OUTPUTS FOR METAL HEAT TREATING - Continued

Economic Sector or Industry Providing Inputs	%	Sector	Economic Sector or Industry Buying Outputs	%	Sector
			Printing trades machinery	0.4	Manufg.
			Railroad equipment	0.4	Manufg.
			Turbines & turbine generator sets	0.4	Manufg.
			Retail trade, except eating & drinking	0.4	Trade
			Carburetors, pistons, rings, & valves	0.3	Manufg.
			Cutlery	0.3	Manufg.
			Fabricated plate work (boiler shops)	0.3	Manufg.
			Glass & glass products, except containers	0.3	Manufg.
			Machine tools, metal forming types	0.3	Manufg.
			Motors & generators	0.3	Manufg.
			Secondary nonferrous metals	0.3	Manufg.
			Service industry machines, nec	0.3	Manufg.
			Truck & bus bodies	0.3	Manufg.
			Miscellaneous repair shops	0.3	Services
			Ammunition, except for small arms, nec	0.2	Manufg.
			Crowns & closures	0.2	Manufg.
			Electric housewares & fans	0.2	Manufg.
			Electrical equipment & supplies, nec	0.2	Manufg.
			Elevators & moving stairways	0.2	Manufg.
			Fabricated structural metal	0.2	Manufg.
			Glass containers	0.2	Manufg.
			Hand saws & saw blades	0.2	Manufg.
			Heating equipment, except electric	0.2	Manufg.
			Hoists, cranes, & monorails	0.2	Manufg.
			Industrial controls	0.2	Manufg.
			Industrial furnaces & ovens	0.2	Manufg.
			Lawn & garden equipment	0.2	Manufg.
			Manufacturing industries, nec	0.2	Manufg.
			Metal coating & allied services	0.2	Manufg.
			Metal doors, sash, & trim	0.2	Manufg.
			Metalworking machinery, nec	0.2	Manufg.
			Nonferrous forgings	0.2	Manufg.
			Paper industries machinery	0.2	Manufg.
			Plumbing fixture fittings & trim	0.2	Manufg.
			Primary copper	0.2	Manufg.
			Primary metal products, nec	0.2	Manufg.
			Primary nonferrous metals, nec	0.2	Manufg.
			Sheet metal work	0.2	Manufg.
			Telephone & telegraph apparatus	0.2	Manufg.
			Textile machinery	0.2	Manufg.
			Truck trailers	0.2	Manufg.
			Typewriters & office machines, nec	0.2	Manufg.
			Wiring devices	0.2	Manufg.
			Federal Government purchases, nondefense	0.2	Fed Govt
			Industrial buildings	0.1	Constr.
			Calculating & accounting machines	0.1	Manufg.
			Commercial printing	0.1	Manufg.
			Cyclic crudes and organics	0.1	Manufg.
			Engine electrical equipment	0.1	Manufg.
			Household cooking equipment	0.1	Manufg.
			Household laundry equipment	0.1	Manufg.
			Household refrigerators & freezers	0.1	Manufg.
			Instruments to measure electricity	0.1	Manufg.
			Lighting fixtures & equipment	0.1	Manufg.
			Mechanical measuring devices	0.1	Manufg.
			Miscellaneous plastics products	0.1	Manufg.
			Rolling mill machinery	0.1	Manufg.
			Ship building & repairing	0.1	Manufg.
			Small arms	0.1	Manufg.
			Switchgear & switchboard apparatus	0.1	Manufg.
			Tanks & tank components	0.1	Manufg.
			Transformers	0.1	Manufg.
			Travel trailers & campers	0.1	Manufg.
			Woodworking machinery	0.1	Manufg.
			Electric services (utilities)	0.1	Util.

Source: Benchmark Input-Output Accounts for the U.S. Economy, 1982, U.S. Department of Commerce, Washington, D.C., July 1991. Data, as reported in the source, are organized by the 1977 SIC structure in use in 1982 but have been matched, as closely as is possible, to the 1987 SIC structure used in this book.

OCCUPATIONS EMPLOYED BY SIC 339 - PRIMARY METALS NEC

Occupation	% of Total 1994	Change to 2005	Occupation	% of Total 1994	Change to 2005
Blue collar worker supervisors	7.8	-15.9	Crushing & mixing machine operators	1.8	-9.1
Furnace operators	6.7	22.2	Sales & related workers nec	1.6	-8.6
Heat treating machine operators, metal & plastic	6.2	-8.1	Secretaries, ex legal & medical	1.6	-17.5
Assemblers, fabricators, & hand workers nec	4.5	-10.4	Machine operators nec	1.5	-22.2
Industrial machinery mechanics	4.4	-1.5	Industrial production managers	1.5	-9.1
Metal & plastic machine workers nec	3.6	-19.5	Freight, stock, & material movers, hand	1.4	-27.9
Maintenance repairers, general utility	3.6	-19.1	Metal molding machine workers	1.4	-1.2
Inspectors, testers, & graders, precision	2.7	-9.0	Machinists	1.3	-10.2
General managers & top executives	2.5	-13.2	Traffic, shipping, & receiving clerks	1.2	-11.9
Industrial truck & tractor operators	2.3	-9.9	Science & mathematics technicians	1.2	-10.0
Electricians	2.2	-16.7	Separating & still machine operators	1.1	-11.4
Truck drivers light & heavy	2.0	-5.5	Bookkeeping, accounting, & auditing clerks	1.1	-31.5
Machine tool cutting & forming etc. nec	1.9	-9.7	Metallurgists & ceramic & materials engineers	1.0	2.2
Crane & tower operators	1.8	-10.8			

Source: Industry-Occupation Matrix, Bureau of Labor Statistics. These data relate to one or more 3-digit SIC industry groups rather than to a single 4-digit SIC. The change reported for each occupation to the year 2005 is a percent of growth or decline as estimated by the Bureau of Labor Statistics. The abbreviation nec stands for 'not elsewhere classified'.

LOCATION BY STATE AND REGIONAL CONCENTRATION

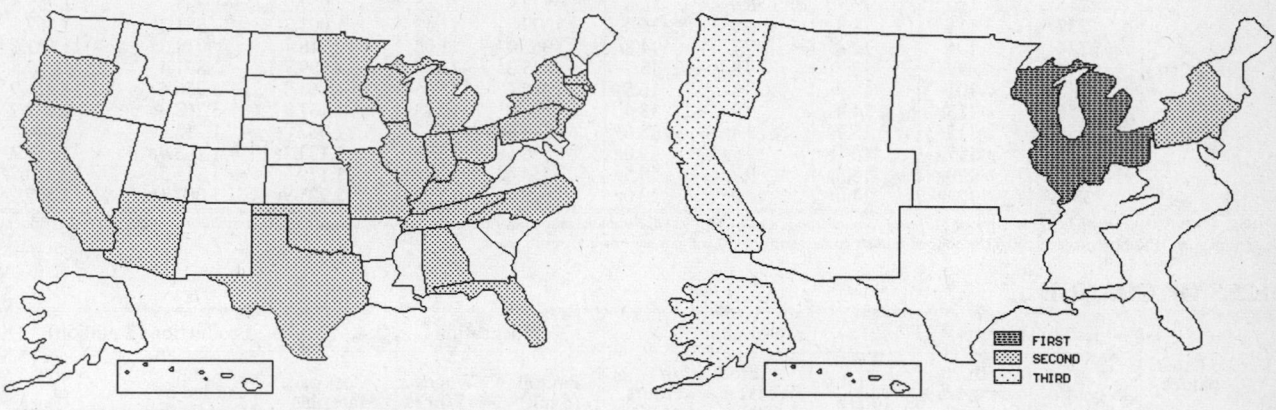

FIRST
SECOND
THIRD

INDUSTRY DATA BY STATE

State	Establish-ments	Shipments			Employment				Cost as % of Shipments	Investment per Employee ($)
		Total ($ mil)	% of U.S.	Per Establ.	Total Number	% of U.S.	Per Establ.	Wages ($/hour)		
Ohio	87	312.5	16.0	3.6	2,500	14.4	29	11.55	67.5	4,040
Indiana	39	280.0	14.3	7.2	1,400	8.0	36	13.39	41.8	4,714
Michigan	105	271.7	13.9	2.6	2,900	16.7	28	12.28	23.3	3,517
Pennsylvania	46	188.6	9.7	4.1	1,100	6.3	24	13.81	57.2	3,455
California	79	134.4	6.9	1.7	1,700	9.8	22	11.29	23.8	3,176
Illinois	47	129.2	6.6	2.7	1,300	7.5	28	10.67	38.6	4,154
Texas	52	108.0	5.5	2.1	1,200	6.9	23	10.72	28.8	1,667
Connecticut	33	82.3	4.2	2.5	700	4.0	21	12.27	40.6	-
Wisconsin	28	58.3	3.0	2.1	700	4.0	25	11.73	21.1	-
New York	28	41.0	2.1	1.5	400	2.3	14	10.86	19.5	5,750
New Jersey	20	39.3	2.0	2.0	400	2.3	20	10.86	24.9	2,750
Alabama	6	33.1	1.7	5.5	200	1.1	33	13.00	65.3	-
North Carolina	16	28.4	1.5	1.8	300	1.7	19	8.80	31.0	3,667
Minnesota	12	23.1	1.2	1.9	300	1.7	25	12.00	23.4	-
Tennessee	12	20.8	1.1	1.7	200	1.1	17	9.00	18.8	-
Oregon	7	11.2	0.6	1.6	100	0.6	14	10.50	20.5	3,000
Florida	7	10.9	0.6	1.6	100	0.6	14	9.67	23.9	-
Massachusetts	21	(D)	-	-	375 *	2.2	18	-	-	-
Arizona	10	(D)	-	-	175 *	1.0	18	-	-	-
Missouri	8	(D)	-	-	175 *	1.0	22	-	-	-
Oklahoma	7	(D)	-	-	175 *	1.0	25	-	-	2,857
Kentucky	5	(D)	-	-	175 *	1.0	35	-	-	-

Source: 1992 Economic Census. The states are in descending order of shipments or establishments (if shipment data are missing for the majority). The symbol (D) appears when data are withheld to prevent disclosure of competitive information. States marked with (D) are sorted by number of establishments. A dash (-) indicates that the data element cannot be calculated; * indicates the midpoint of a range.

3399 - PRIMARY METAL PRODUCTS, NEC

82 83 84 85 86 87 88 89 90 91 92 93 94 95 96 97 98

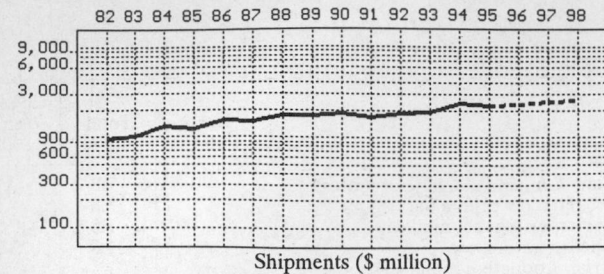

Shipments ($ million)

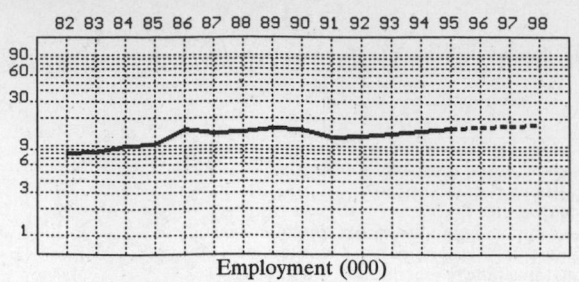

Employment (000)

GENERAL STATISTICS

Year	Companies	Establishments		Employment			Compensation		Production ($ million)			
		Total	with 20 or more employees	Total (000)	Production Workers (000)	Hours (Mil)	Payroll ($ mil)	Wages ($/hr)	Cost of Materials	Value Added by Manufacture	Value of Shipments	Capital Invest.
1982	239	249	80	8.2	5.6	10.6	173.4	9.57	595.2	313.5	938.1	51.7
1983		233	83	8.4	5.9	11.2	200.0	10.78	592.0	427.6	1,025.0	26.9
1984		217	86	9.9	6.7	12.7	227.0	10.86	670.1	662.4	1,301.3	52.8
1985		202	89	10.1	7.0	13.2	235.5	10.91	628.5	613.0	1,234.0	42.3
1986		187	93	14.8	9.6	17.6	366.4	10.48	791.3	741.6	1,538.0	53.2
1987	238	252	99	13.8	8.9	16.8	353.2	10.77	802.1	728.2	1,510.1	58.2
1988		258	112	14.2	9.5	18.1	378.9	10.22	925.4	839.5	1,757.2	101.8
1989		246	110	15.7	8.8	16.8	373.4	10.85	957.9	805.7	1,755.4	84.8
1990		237	111	14.8	8.8	16.3	379.3	11.42	1,001.3	881.3	1,867.1	67.8
1991		224	104	12.4	7.7	14.8	363.0	11.03	918.4	714.3	1,681.6	74.6
1992	260	273	99	12.7	7.8	15.3	415.3	12.35	999.7	871.4	1,869.5	78.9
1993		258	101	13.4	8.3	16.5	440.4	12.06	964.0	956.4	1,914.9	70.8
1994		256P	113P	14.4	9.0	18.4	502.0	12.73	1,137.8	1,268.9	2,382.7	92.8
1995		259P	115P	15.7P	9.4P	18.6P	506.6P	12.37P	1,087.0P	1,212.3P	2,276.4P	93.7P
1996		262P	117P	16.1P	9.6P	19.0P	530.5P	12.56P	1,133.3P	1,263.9P	2,373.3P	97.7P
1997		265P	120P	16.6P	9.8P	19.5P	554.4P	12.74P	1,179.6P	1,315.5P	2,470.2P	101.7P
1998		268P	122P	17.0P	10.0P	20.0P	578.4P	12.93P	1,225.9P	1,367.1P	2,567.1P	105.6P

Sources: 1982, 1987, 1992 *Economic Census*; *Annual Survey of Manufactures*, 83-86, 88-91, 93-94. Establishment counts for non-Census years are from *County Business Patterns*; establishment values for 83-84 are extrapolations. 'P's show projections by the editors. Industries reclassified in 87 will not have data for prior years.

INDICES OF CHANGE

Year	Companies	Establishments		Employment			Compensation		Production ($ million)			
		Total	with 20 or more employees	Total (000)	Production Workers (000)	Hours (Mil)	Payroll ($ mil)	Wages ($/hr)	Cost of Materials	Value Added by Manufacture	Value of Shipments	Capital Invest.
1982	92	91	81	65	72	69	42	77	60	36	50	66
1983		85	84	66	76	73	48	87	59	49	55	34
1984		79	87	78	86	83	55	88	67	76	70	67
1985		74	90	80	90	86	57	88	63	70	66	54
1986		68	94	117	123	115	88	85	79	85	82	67
1987	92	92	100	109	114	110	85	87	80	84	81	74
1988		95	113	112	122	118	91	83	93	96	94	129
1989		90	111	124	113	110	90	88	96	92	94	107
1990		87	112	117	113	107	91	92	100	101	100	86
1991		82	105	98	99	97	87	89	92	82	90	95
1992	100	100	100	100	100	100	100	100	100	100	100	100
1993		95	102	106	106	108	106	98	96	110	102	90
1994		94P	114P	113	115	120	121	103	114	146	127	118
1995		95P	116P	124P	121P	121P	122P	100P	109P	139P	122P	119P
1996		96P	118P	127P	123P	124P	128P	102P	113P	145P	127P	124P
1997		97P	121P	131P	126P	127P	134P	103P	118P	151P	132P	129P
1998		98P	123P	134P	128P	131P	139P	105P	123P	157P	137P	134P

Sources: Same as General Statistics. Values reflect change from the base year, 1992. Values above 100 mean greater than 92, values below 100 mean less than 92, and a value of 100 in the 82-91 or 93-98 period means same as 92. 'P's mark projections by the editors.

SELECTED RATIOS

For 1994	Avg. of All Manufact.	Analyzed Industry	Index	For 1994	Avg. of All Manufact.	Analyzed Industry	Index
Employees per Establishment	49	56	115	Value Added per Production Worker	134,084	140,989	105
Payroll per Establishment	1,500,273	1,960,821	131	Cost per Establishment	5,045,178	4,444,268	88
Payroll per Employee	30,620	34,861	114	Cost per Employee	102,970	79,014	77
Production Workers per Establishment	34	35	102	Cost per Production Worker	146,988	126,422	86
Wages per Establishment	853,319	914,915	107	Shipments per Establishment	9,576,895	9,306,871	97
Wages per Production Worker	24,861	26,026	105	Shipments per Employee	195,460	165,465	85
Hours per Production Worker	2,056	2,044	99	Shipments per Production Worker	279,017	264,744	95
Wages per Hour	12.09	12.73	105	Investment per Establishment	321,011	362,479	113
Value Added per Establishment	4,602,255	4,956,347	108	Investment per Employee	6,552	6,444	98
Value Added per Employee	93,930	88,118	94	Investment per Production Worker	9,352	10,311	110

Sources: Same as General Statistics. The 'Average of All Manufacturing' column represents the average of all manufacturing industries reported for the most recent complete year available. The Index shows the relationship between the Average and the Analyzed Industry. For example, 100 means that they are equal; 500 that the Analyzed Industry is five times the average; 50 means that the Analyzed Industry is half the national average. The abbreviation 'na' is used to show that data are 'not available'.

LEADING COMPANIES   Number shown: 59   Total sales ($ mil): 1,565   Total employment (000): 11.9

Company Name	Address				CEO Name	Phone	Co. Type	Sales ($ mil)	Empl. (000)
Myers Industries Inc	1293 S Main St	Akron	OH	44301	Stephen E Myers	216-253-5592	P	274	1.6
Keystone Carbon Co	1935 State St	St Marys	PA	15857	Richard J Reuscher	814-781-1591	R	115	1.5
Hoeganaes Corp	River & Taylors Ln	Riverton	NJ	08077	Robert J Fulton	609-829-2220	S	110	0.5
SSI Technologies Inc	PO Box 5002	Janesville	WI	53547	David S Baum	608-755-1900	R	110	1.0
Pennsylvania Pressed Metals Inc	RR 2	Emporium	PA	15834	Donald L LeVault	814-486-3314	S	80	0.8
Lamination Specialties Corp	235 N Artesian Av	Chicago	IL	60612	A Delighter	312-243-2181	R	62*	0.3
NN Ball and Roller Co	800 Tennessee Rd	Erwin	TN	37650	Richard D Ennen	615-743-9151	P	61	0.3
National Lamination Co	555 Santa Rosa Dr	Des Plaines	IL	60018	R Del Boccio	708-298-7676	D	50	0.2
AlliedSignal Inc	500 Westinghouse	Pendleton	SC	29670	Dennis Ford	803-646-8311	D	46*	0.4
Industrial Powder Coatings Inc	202 Republic St	Norwalk	OH	44857	R M Warner Sr	419-668-4436	S	41	0.6
Alpha Sintered Metals	PO Box 43D	Ridgway	PA	15853	RM Hasselman	814-773-3191	R	30	0.2
Presmet Corp	112 Harding St	Worcester	MA	01604	SH Clinch	508-792-6400	R	30*	0.3
Vacumet Corp	20 Edison Dr	Wayne	NJ	07470	RT Korowicki	201-628-0400	S	30	0.1
Lanxide Corp	PO Box 6077	Newark	DE	19714	Marc S Newkirk	302-456-6200	R	25	0.4
US Bronze Powders Inc	PO Box 31	Flemington	NJ	08822	KC Ramsey	908-782-5454	R	25	0.1
Pyron Corp	PO Box 310	Niagara Falls	NY	14304	G Russell Lewis	716-285-3451	S	25	0.1
National Die and Button Mould	239 W 39th St	New York	NY	10018	George Eisen	212-398-0263	R	24*	0.2
Abbott Ball Company Inc	PO Box 330100	West Hartford	CT	06133	Roger Bond	203-236-5901	R	20	0.1
Alcan Powder and Pigments	901 Lehigh Av	Union	NJ	07083	Edul Daver	908-851-4500	D	20	<0.1
Alcan-Toyo America Inc	1717 N Naper Blv	Naperville	IL	60563	Peter Ortleb	708-505-2160	S	20	0.1
Clarion Sintered Metals Inc	Montmorenci Rd	Ridgway	PA	15853	Howard Peterson	814-773-3124	R	20	0.3
Friction Products Co	920 Lake Rd	Medina	OH	44256	Doug Wilson	216-725-4941	S	20	<0.1
Homogeneous Metals Inc	Main St	Clayville	NY	13322	PS Sbarra	315-839-5421	S	20	0.1
Nuclear Metals Inc	2229 Main St	Concord	MA	01742	Robert E Quinn	508-369-5410	P	19	0.2
Brico Metals Inc	5800 Wolf Creek Pk	Dayton	OH	45426	Ben James	513-837-2671	S	18	0.2
Obron Atlantic Corp	72 Corwin Dr	Painesville	OH	44077	Jon Fisher	216-354-0400	S	18*	<0.1
Advanced Forming Technology	2150 Miller Dr	Longmont	CO	80502	Robert P Reed	303-651-6557	D	17	0.2
Helsel Inc	PO Box 68	Campbellsburg	IN	47108	JF Helsel	812-755-4501	R	17	0.1
Berger Holdings Ltd	805 Pennsylvania	Feasterville	PA	19053	T A Schwartz	215-355-1200	P	17	<0.1
Aluma-Form Inc	3625 Old Getwell	Memphis	TN	38118	Wayne Mitchell	901-362-0100	R	16*	0.2
John Hassall Inc	Cantiague Rock Rd	Westbury	NY	11590	TB Smith Jr	516-334-6200	R	15	0.1
Fairview Sintered Metals Inc	PO Box D	St Marys	PA	15857	GA Wehler	814-781-6500	R	14	0.2
Inland Sintered Metals	Box 214E Rte 1ate	Hazen	AR	72064	Jim Grant	501-255-3531	D	14*	0.1
Hoosier Magnetics Group	6545 W Central Av	Toledo	OH	43617	T Shirk	419-841-7173	R	12	0.1
Atlantic Sintered Metals Inc	10 Cushing Dr	Wrentham	MA	02093	Mark Paullin	508-384-3100	R	12*	0.1
Clendenin Brothers Inc	4309 Erdman Av	Baltimore	MD	21213	John C Corckran Jr	410-327-4500	R	12*	0.1
Norwalk Powdered Metals Inc	PO Box 271	Norwalk	CT	06851	Thomas Blumenthel	203-847-4537	R	10*	0.1
Conservatek Industries Inc	498 Loop 336 E	Conroe	TX	77301	R John Stanton	713-353-2434	R	9*	<0.1
Electroply Inc	139 Illinois St	El Segundo	CA	90245	Charles L Proctor	310-322-5647	R	9*	<0.1
MD-Both Industries	PO Box 306	Ashland	MA	01721	CT Skeels	508-881-4100	R	9	<0.1
Technetics Corp	1600 Industrial Dr	DeLand	FL	32724	JS Singleton	904-736-7373	R	8*	<0.1
Danco Precision Inc	PO Box 448	Phoenixville	PA	19460	Nicholas B Fagan	610-933-8981	R	7	<0.1
Allied Sinterings Inc	29 Briar Ridge Rd	Danbury	CT	06810	Gifford H Foster	203-743-7502	R	6	<0.1
Hart Metals Inc	PO Box 428	Tamaqua	PA	18252	RJ Hart	717-668-0001	R	6*	<0.1
Klean Metals and Steel	PO Box 5462	Jackson	MS	39208	Teja Jouhal	601-939-1347	R	6	<0.1
Valimet Inc	PO Box 6186	Stockton	CA	95206	Kurt Leopold	209-982-4870	R	6	<0.1
Permacor Inc	9540 Tulley Av	Oak Lawn	IL	60453	P Tsoutsas	708-422-3353	R	5	<0.1
Rush Metals	PO Box 398	Billings	OK	74630	Jim Lee	405-725-3295	D	5	<0.1
Colorado Sintered Metals	PO Box 15468	Co Springs	CO	80935	JA Gerzina	719-596-0110	R	5	0.1
Perry Tool and Research Inc	3415 Enterprise Av	Hayward	CA	94545	Perry Fusselman	510-782-9226	R	4	<0.1
Arthur Court Designs Inc	PO Box 459	Brisbane	CA	94005	Arthur Court	415-468-8599	R	3*	<0.1
Coates Steel Products Company	1937 W Franklin St	Greenville	IL	62246	Timothy F Duft	618-664-1000	R	3	<0.1
TecSyn PMP Inc	4950 Gilmer Dr	Huntsville	AL	35805	Sidney O Nicholls	205-722-9092	S	3	<0.1
Castite Systems Inc	5007 Superior Av	Cleveland	OH	44103	Joan E Lamson	216-432-1111	R	1	<0.1
Connelly-GPM Inc	3154 S California	Chicago	IL	60608	Miles M Klein	312-247-7231	R	1	<0.1
Structural Laminates Co	510 Constitution	N Kensington	PA	15068	Joseph W Evancho	412-339-6977	J	1	<0.1
TyKel Inc	150 Hamilton	Hamilton	IL	62341	James Ruffcorn	217-847-6631	R	1	<0.1
Gold Leaf and Metallic Powders	74 Trinity Pl	New York	NY	10006	Peter Olley	212-267-4900	R	1*	<0.1
Exotherm Corp	1035 Line St	Camden	NJ	08103	Alex Laschiver	609-541-1949	R	0*	<0.1

Source: Ward's Business Directory of U.S. Private and Public Companies, Volumes 1 and 2, 1996. The company type code used is as follows: P - Public, R - Private, S - Subsidiary, D - Division, J - Joint Venture, A - Affiliate, G - Group. Sales are in millions of dollars, employees are in thousands. An asterisk (*) indicates an estimated sales volume. The symbol < stands for 'less than'. Company names and addresses are truncated, in some cases, to fit into the available space.

MATERIALS CONSUMED

Material	Quantity	Delivered Cost ($ million)
No Materials Consumed data available for this industry.		

Source: 1992 Economic Census. Explanation of symbols used: (D): Withheld to avoid disclosure of competitive data; na: Not available; (S): Withheld because statistical norms were not met; (X): Not applicable; (Z): Less than half the unit shown; nec: Not elsewhere classified; nsk: Not specified by kind; - : zero; * : 10-19 percent estimated; ** : 20-29 percent estimated.

PRODUCT SHARE DETAILS

Product or Product Class	% Share	Product or Product Class	% Share
Primary metal products, nec	100.00	Primary molybdenum powders, paste, and flakes	1.59
Primary metal powders, paste, and flakes	67.81	Primary titanium powders, paste, and flakes	2.83
Primary aluminum and aluminum-base alloy powders, paste, and flakes	12.88	Primary precious metal and precious-metal-base alloy powders, paste, and flakes (gold, silver, platinum, etc.)	12.33
Primary copper and copper-base alloy powders, paste, and flakes	10.18	Other primary nonferrous powders, paste, and flakes	15.26
Primary iron and steel powders, paste, and flakes	22.57	Metal powders, paste, and flakes, nsk	3.62
Primary nickel-cobalt-base superalloy material powders, paste, and flakes	6.38	Primary metal products, nec	22.90
		Primary nonferrous nails, brads, tacks, and staples	19.15
Primary tungsten and tungsten-base alloy powders, paste, and flakes	12.35	Other primary metal products, nec	79.73
		Primary metal products, nec, nsk	1.12
		Primary metal products, nec, nsk	9.29

Source: 1992 *Economic Census*. The values shown are percent of total shipments in an industry. Values of indented subcategories are summed in the main heading. The symbol (D) appears when data are withheld to prevent disclosure of competitive information. The abbreviation nsk stands for 'not specified by kind' and nec for 'not elsewhere classified'.

INPUTS AND OUTPUTS FOR PRIMARY METAL PRODUCTS, NEC

Economic Sector or Industry Providing Inputs	%	Sector	Economic Sector or Industry Buying Outputs	%	Sector
Imports	17.2	Foreign	Special dies & tools & machine tool accessories	13.8	Manufg.
Nonferrous wire drawing & insulating	11.9	Manufg.	Secondary nonferrous metals	7.9	Manufg.
Iron & ferroalloy ores	10.6	Mining	Exports	6.8	Foreign
Cyclic crudes and organics	9.9	Manufg.	Motor vehicle parts & accessories	5.7	Manufg.
Wood products, nec	5.5	Manufg.	Miscellaneous plastics products	5.3	Manufg.
Electric services (utilities)	5.5	Util.	Federal Government purchases, national defense	4.6	Fed Govt
Primary nonferrous metals, nec	5.3	Manufg.	Oil field machinery	4.5	Manufg.
Wholesale trade	5.2	Trade	Electronic components nec	3.3	Manufg.
Gas production & distribution (utilities)	3.0	Util.	Iron & steel foundries	2.9	Manufg.
Primary copper	2.3	Manufg.	Fabricated metal products, nec	2.8	Manufg.
Railroads & related services	1.9	Util.	Mining machinery, except oil field	2.6	Manufg.
Scrap	1.8	Scrap	Paints & allied products	2.3	Manufg.
Primary aluminum	1.7	Manufg.	Blast furnaces & steel mills	2.0	Manufg.
Metal coating & allied services	1.6	Manufg.	Fabricated plate work (boiler shops)	2.0	Manufg.
Primary lead	1.6	Manufg.	Metal heat treating	1.9	Manufg.
Communications, except radio & TV	1.4	Util.	Metal stampings, nec	1.8	Manufg.
Industrial inorganic chemicals, nec	1.2	Manufg.	Power driven hand tools	1.8	Manufg.
Petroleum refining	1.0	Manufg.	Aluminum rolling & drawing	1.6	Manufg.
Miscellaneous repair shops	0.9	Services	Pipe, valves, & pipe fittings	1.5	Manufg.
Maintenance of nonfarm buildings nec	0.8	Constr.	Gross private fixed investment	1.5	Cap Inv
Paperboard containers & boxes	0.8	Manufg.	Coal	1.2	Mining
Power transmission equipment	0.8	Manufg.	Iron & steel forgings	1.2	Manufg.
Motor freight transportation & warehousing	0.8	Util.	Pumps & compressors	1.1	Manufg.
Water transportation	0.7	Util.	Aircraft & missile equipment, nec	0.9	Manufg.
Machinery, except electrical, nec	0.5	Manufg.	Radio & TV receiving sets	0.9	Manufg.
Eating & drinking places	0.5	Trade	Welding apparatus, electric	0.9	Manufg.
Banking	0.5	Fin/R.E.	Lawn & garden equipment	0.8	Manufg.
Equipment rental & leasing services	0.5	Services	Radio & TV communication equipment	0.8	Manufg.
Metal heat treating	0.4	Manufg.	Power transmission equipment	0.7	Manufg.
U.S. Postal Service	0.4	Gov't	Refrigeration & heating equipment	0.7	Manufg.
Special dies & tools & machine tool accessories	0.3	Manufg.	Steel wire & related products	0.7	Manufg.
Alkalies & chlorine	0.2	Manufg.	Asbestos products	0.6	Manufg.
Industrial furnaces & ovens	0.2	Manufg.	Electric housewares & fans	0.6	Manufg.
Advertising	0.2	Services	Hardware, nec	0.6	Manufg.
Legal services	0.2	Services	Nonferrous rolling & drawing, nec	0.6	Manufg.
Management & consulting services & labs	0.2	Services	Steel pipe & tubes	0.6	Manufg.
Abrasive products	0.1	Manufg.	Surgical & medical instruments	0.6	Manufg.
Lubricating oils & greases	0.1	Manufg.	Surgical appliances & supplies	0.6	Manufg.
Air transportation	0.1	Util.	Construction machinery & equipment	0.5	Manufg.
Sanitary services, steam supply, irrigation	0.1	Util.	Copper rolling & drawing	0.5	Manufg.
Accounting, auditing & bookkeeping	0.1	Services	Household laundry equipment	0.5	Manufg.
Computer & data processing services	0.1	Services	Railroad equipment	0.5	Manufg.
			Aircraft & missile engines & engine parts	0.4	Manufg.
			Miscellaneous fabricated wire products	0.4	Manufg.
			Small arms ammunition	0.4	Manufg.
			Steel springs, except wire	0.4	Manufg.
			Industrial buildings	0.3	Constr.
			Office buildings	0.3	Constr.
			Aircraft	0.3	Manufg.
			Ammunition, except for small arms, nec	0.3	Manufg.
			Carbon & graphite products	0.3	Manufg.
			Motor vehicles & car bodies	0.3	Manufg.
			Special industry machinery, nec	0.3	Manufg.
			Ball & roller bearings	0.2	Manufg.
			Guided missiles & space vehicles	0.2	Manufg.
			Industrial trucks & tractors	0.2	Manufg.
			Machine tools, metal forming types	0.2	Manufg.
			Semiconductors & related devices	0.2	Manufg.
			Retail trade, except eating & drinking	0.2	Trade
			Wholesale trade	0.2	Trade
			Maintenance of nonfarm buildings nec	0.1	Constr.
			Hoists, cranes, & monorails	0.1	Manufg.
			Household cooking equipment	0.1	Manufg.
			Metalworking machinery, nec	0.1	Manufg.

Source: Benchmark Input-Output Accounts for the U.S. Economy, 1982, U.S. Department of Commerce, Washington, D.C., July 1991. Data, as reported in the source, are organized by the 1977 SIC structure in use in 1982 but have been matched, as closely as is possible, to the 1987 SIC structure used in this book.

OCCUPATIONS EMPLOYED BY SIC 339 - PRIMARY METALS NEC

Occupation	% of Total 1994	Change to 2005	Occupation	% of Total 1994	Change to 2005
Blue collar worker supervisors	7.8	-15.9	Crushing & mixing machine operators	1.8	-9.1
Furnace operators	6.7	22.2	Sales & related workers nec	1.6	-8.6
Heat treating machine operators, metal & plastic	6.2	-8.1	Secretaries, ex legal & medical	1.6	-17.5
Assemblers, fabricators, & hand workers nec	4.5	-10.4	Machine operators nec	1.5	-22.2
Industrial machinery mechanics	4.4	-1.5	Industrial production managers	1.5	-9.1
Metal & plastic machine workers nec	3.6	-19.5	Freight, stock, & material movers, hand	1.4	-27.9
Maintenance repairers, general utility	3.6	-19.1	Metal molding machine workers	1.4	-1.2
Inspectors, testers, & graders, precision	2.7	-9.0	Machinists	1.3	-10.2
General managers & top executives	2.5	-13.2	Traffic, shipping, & receiving clerks	1.2	-11.9
Industrial truck & tractor operators	2.3	-9.9	Science & mathematics technicians	1.2	-10.0
Electricians	2.2	-16.7	Separating & still machine operators	1.1	-11.4
Truck drivers light & heavy	2.0	-5.5	Bookkeeping, accounting, & auditing clerks	1.1	-31.5
Machine tool cutting & forming etc. nec	1.9	-9.7	Metallurgists & ceramic & materials engineers	1.0	2.2
Crane & tower operators	1.8	-10.8			

Source: Industry-Occupation Matrix, Bureau of Labor Statistics. These data relate to one or more 3-digit SIC industry groups rather than to a single 4-digit SIC. The change reported for each occupation to the year 2005 is a percent of growth or decline as estimated by the Bureau of Labor Statistics. The abbreviation nec stands for 'not elsewhere classified'.

LOCATION BY STATE AND REGIONAL CONCENTRATION

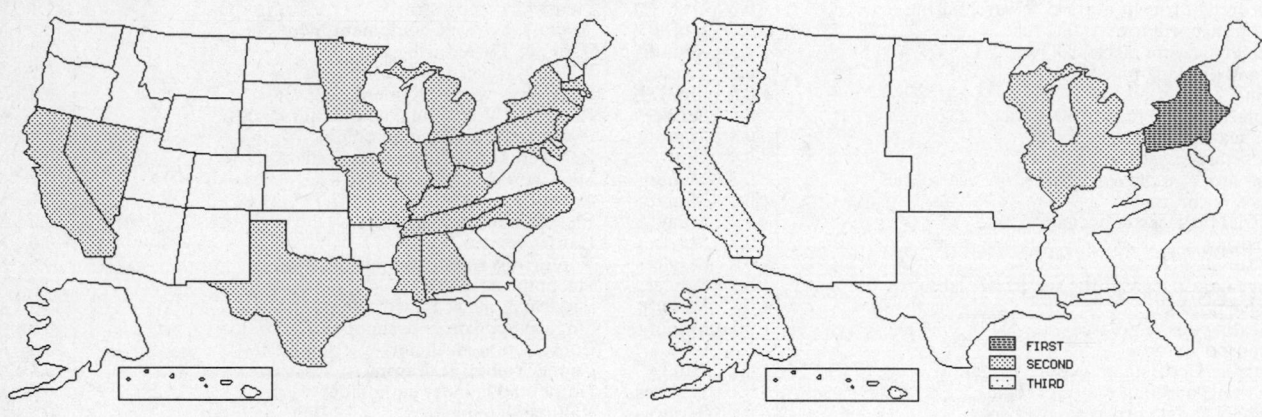

FIRST
SECOND
THIRD

INDUSTRY DATA BY STATE

State	Establish- ments	Shipments			Employment				Cost as % of Shipments	Investment per Employee ($)
		Total ($ mil)	% of U.S.	Per Establ.	Total Number	% of U.S.	Per Establ.	Wages ($/hour)		
Pennsylvania	35	343.4	18.4	9.8	2,200	17.3	63	12.97	59.4	6,091
New York	16	185.4	9.9	11.6	500	3.9	31	14.00	67.1	16,200
New Jersey	17	171.1	9.2	10.1	800	6.3	47	15.67	59.0	16,375
North Carolina	6	84.4	4.5	14.1	400	3.1	67	10.20	56.4	1,750
Texas	12	67.3	3.6	5.6	200	1.6	17	14.00	40.0	2,500
California	24	55.6	3.0	2.3	400	3.1	17	9.20	49.6	4,250
Michigan	21	34.9	1.9	1.7	400	3.1	19	11.00	59.6	2,000
Tennessee	7	27.3	1.5	3.9	200	1.6	29	8.67	75.1	-
Connecticut	9	17.3	0.9	1.9	200	1.6	22	12.67	37.0	-
Minnesota	6	14.3	0.8	2.4	100	0.8	17	9.50	76.9	-
Ohio	18	(D)	-	-	750 *	5.9	42	-	-	5,867
Indiana	12	(D)	-	-	375 *	3.0	31	-	-	-
Massachusetts	10	(D)	-	-	3,750 *	29.5	375	-	-	-
Illinois	6	(D)	-	-	375 *	3.0	63	-	-	-
Alabama	5	(D)	-	-	375 *	3.0	75	-	-	-
Kentucky	4	(D)	-	-	175 *	1.4	44	-	-	-
Maryland	4	(D)	-	-	175 *	1.4	44	-	-	-
Missouri	4	(D)	-	-	175 *	1.4	44	-	-	-
Mississippi	2	(D)	-	-	175 *	1.4	88	-	-	-
Nevada	2	(D)	-	-	175 *	1.4	88	-	-	-

Source: 1992 Economic Census. The states are in descending order of shipments or establishments (if shipment data are missing for the majority). The symbol (D) appears when data are withheld to prevent disclosure of competitive information. States marked with (D) are sorted by number of establishments. A dash (-) indicates that the data element cannot be calculated; * indicates the midpoint of a range.

3411 - METAL CANS

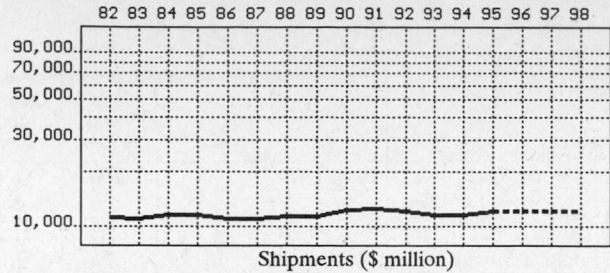

Shipments ($ million)

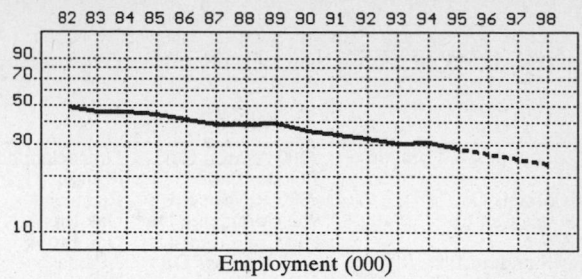

Employment (000)

GENERAL STATISTICS

Year	Companies	Establishments		Employment			Compensation		Production ($ million)			
		Total	with 20 or more employees	Total (000)	Production Workers (000)	Hours (Mil)	Payroll ($ mil)	Wages ($/hr)	Cost of Materials	Value Added by Manufacture	Value of Shipments	Capital Invest.
1982	168	397	294	49.0	40.8	81.5	1,334.5	13.08	7,046.8	4,071.7	11,132.8	247.0
1983		387	291	46.2	38.5	78.3	1,344.3	13.57	7,276.2	3,757.0	10,961.5	178.1
1984		377	288	45.7	38.2	78.9	1,364.9	13.77	7,778.8	3,883.2	11,623.2	204.2
1985		368	286	44.2	37.1	76.5	1,399.1	14.57	7,886.8	3,736.3	11,643.9	263.8
1986		368	280	41.8	35.0	72.3	1,370.0	15.24	7,351.2	3,773.9	11,068.7	362.7
1987	161	369	259	39.4	32.7	69.5	1,325.4	15.22	7,194.6	3,816.0	11,013.6	334.2
1988		358	268	39.0	32.9	71.8	1,361.5	15.07	7,492.2	3,920.3	11,407.1	303.0
1989		345	262	39.6	31.1	68.7	1,342.1	15.75	7,985.9	3,418.5	11,389.3	359.5
1990		337	256	36.0	30.5	67.6	1,319.4	15.93	8,676.4	3,668.4	12,342.4	285.9
1991		330	255	34.6	29.3	65.5	1,315.0	16.01	8,977.9	3,557.3	12,449.6	233.3
1992	132	324	240	32.3	27.2	58.3	1,261.9	17.71	8,797.9	3,290.1	12,112.2	350.6
1993		304	232	30.5	25.9	55.5	1,211.8	18.05	8,359.8	3,012.3	11,497.8	455.2
1994		306P	232P	30.9	26.4	56.0	1,255.6	18.56	8,305.7	3,248.9	11,610.1	411.1
1995		299P	226P	28.4P	24.0P	54.2P	1,251.0P	18.55P	8,621.7P	3,372.5P	12,051.8P	412.0P
1996		291P	221P	26.8P	22.7P	52.0P	1,240.7P	18.98P	8,672.1P	3,392.2P	12,122.3P	427.1P
1997		284P	216P	25.3P	21.5P	49.8P	1,230.3P	19.40P	8,722.6P	3,412.0P	12,192.9P	442.1P
1998		276P	210P	23.7P	20.2P	47.7P	1,219.9P	19.82P	8,773.1P	3,431.7P	12,263.4P	457.1P

Sources: 1982, 1987, 1992 *Economic Census*; *Annual Survey of Manufactures*, 83-86, 88-91, 93-94. Establishment counts for non-Census years are from *County Business Patterns*; establishment values for 83-84 are extrapolations. 'P's show projections by the editors. Industries reclassified in 87 will not have data for prior years.

INDICES OF CHANGE

Year	Companies	Establishments		Employment			Compensation		Production ($ million)			
		Total	with 20 or more employees	Total (000)	Production Workers (000)	Hours (Mil)	Payroll ($ mil)	Wages ($/hr)	Cost of Materials	Value Added by Manufacture	Value of Shipments	Capital Invest.
1982	127	123	123	152	150	140	106	74	80	124	92	70
1983		119	121	143	142	134	107	77	83	114	90	51
1984		116	120	141	140	135	108	78	88	118	96	58
1985		114	119	137	136	131	111	82	90	114	96	75
1986		114	117	129	129	124	109	86	84	115	91	103
1987	122	114	108	122	120	119	105	86	82	116	91	95
1988		110	112	121	121	123	108	85	85	119	94	86
1989		106	109	123	114	118	106	89	91	104	94	103
1990		104	107	111	112	116	105	90	99	111	102	82
1991		102	106	107	108	112	104	90	102	108	103	67
1992	100	100	100	100	100	100	100	100	100	100	100	100
1993		94	97	94	95	95	96	102	95	92	95	130
1994		95P	97P	96	97	96	100	105	94	99	96	117
1995		92P	94P	88P	88P	93P	99P	105P	98P	103P	100P	118P
1996		90P	92P	83P	84P	89P	98P	107P	99P	103P	100P	122P
1997		88P	90P	78P	79P	86P	97P	110P	99P	104P	101P	126P
1998		85P	88P	73P	74P	82P	97P	112P	100P	104P	101P	130P

Sources: Same as General Statistics. Values reflect change from the base year, 1992. Values above 100 mean greater than 92, values below 100 mean less than 92, and a value of 100 in the 82-91 or 93-98 period means same as 92. 'P's mark projections by the editors.

SELECTED RATIOS

For 1994	Avg. of All Manufact.	Analyzed Industry	Index	For 1994	Avg. of All Manufact.	Analyzed Industry	Index
Employees per Establishment	49	101	206	Value Added per Production Worker	134,084	123,064	92
Payroll per Establishment	1,500,273	4,098,195	273	Cost per Establishment	5,045,178	27,109,253	537
Payroll per Employee	30,620	40,634	133	Cost per Employee	102,970	268,793	261
Production Workers per Establishment	34	86	251	Cost per Production Worker	146,988	314,610	214
Wages per Establishment	853,319	3,392,402	398	Shipments per Establishment	9,576,895	37,894,595	396
Wages per Production Worker	24,861	39,370	158	Shipments per Employee	195,460	375,731	192
Hours per Production Worker	2,056	2,121	103	Shipments per Production Worker	279,017	439,777	158
Wages per Hour	12.09	18.56	153	Investment per Establishment	321,011	1,341,803	418
Value Added per Establishment	4,602,255	10,604,194	230	Investment per Employee	6,552	13,304	203
Value Added per Employee	93,930	105,142	112	Investment per Production Worker	9,352	15,572	167

Sources: Same as General Statistics. The 'Average of All Manufacturing' column represents the average of all manufacturing industries reported for the most recent complete year available. The Index shows the relationship between the Average and the Analyzed Industry. For example, 100 means that they are equal; 500 that the Analyzed Industry is five times the average; 50 means that the Analyzed Industry is half the national average. The abbreviation 'na' is used to show that data are 'not available'.

LEADING COMPANIES   Number shown: 31   Total sales ($ mil): 10,192   Total employment (000): 48.9

Company Name	Address				CEO Name	Phone	Co. Type	Sales ($ mil)	Empl. (000)
Crown Cork and Seal Company	9300 Ashton Rd	Philadelphia	PA	19136	William J Avery	215-698-5100	P	4,452	22.4
Ball Corp	PO Box 2407	Muncie	IN	47307	George Sissel	317-747-6100	P	2,595	12.9
Reynolds Metals Co	7900 Reycan Rd	Richmond	VA	23237	WE Leahey Jr	804-743-5371	D	1,200	3.1
United States Can Co	900 Commerce Dr	Oak Brook	IL	60521	William J Smith	708-571-2500	S	553	2.7
US Can Corp	900 Commerce Dr	Oak Brook	IL	60521	William J Smith	708-571-2500	P	553	3.4
Brockway Standard Inc	8607 Roberts Dr	Atlanta	GA	30350	John T Stirrup	404-587-0888	R	255	1.1
Ellisco Inc	3 Bala Plz	Bala Cynwyd	PA	19004	Charles E Foster	215-664-7400	S	89	0.4
Milton Can Company Inc	PO Box 1100	Elizabeth	NJ	07207	JW Milton	908-289-8100	R	57	0.4
Finger Lakes Packaging	PO Box 637	Lyons	NY	14489	Ronald Fithen	315-946-4826	D	47	0.2
Penny Plate Inc	PO Box 3003	Haddonfield	NJ	08033	George J Buff	609-429-7583	R	40	0.2
Strauss Industries Inc	PO Box 6543	Wheeling	WV	26003	Carter Strauss	304-232-8770	R	35	<0.1
Taylor-Wharton Cryogenics	PO Box 568	Theodore	AL	36590	Ron Kaplan	205-443-8680	D	30	0.3
Prospect Industries Inc	1202 Airport Rd	N Brunswick	NJ	08902	Ken Sokoloff	908-247-6700	R	27	0.2
Northern Can Systems Inc	2121 Warner	Canton	OH	44707	Maury Corp	216-430-4900	S	26	<0.1
Blitz USA Inc	404 26th Av NW	Miami	OK	74354	John R Elmburg	918-540-1515	R	25*	0.2
Container Supply Company Inc	PO Box 5367	Garden Grove	CA	92645	RS Hurtt Jr	714-892-8321	R	25*	0.2
Eagle Manufacturing Company	25th & Charles Sts	Wellsburg	WV	26070	James Paull	304-737-3171	R	25*	0.2
North America Packaging	PO Box 2146	Merced	CA	95344	Michael Watt	209-383-4396	S	24	0.1
Independent Can Co	1300 Brass Mill Rd	Belcamp	MD	21017	D Huether	410-272-0090	R	20	0.1
Protectoseal Co	225 Foster Av	Bensenville	IL	60106	W Smith	708-595-0800	R	20	0.2
South Texas Can Company Inc	501 S Pleasantview	Weslaco	TX	78596	Bruce G Keen	210-968-7568	R	20	<0.1
Eagle Can Co	PO Box 753	Peabody	MA	01960	James J Milton	508-532-0400	D	16	0.1
Olive Can Co	1111 Bowes Rd	Elgin	IL	60123	Albert Armato	708-468-7474	R	15	0.2
New Can Company Inc	PO Box 421	Holbrook	MA	02343	Richard Marson	617-767-1650	S	13	0.1
Syme Inc	300 Lake Rd	Medina	OH	44256	Robert Syme	216-723-6000	R	10	<0.1
Crawford Container Co	1500 Chamberlain	Conneaut	OH	44030	Edward Crawford	216-593-5300	R	6	<0.1
Hunter Container Corp	PO Box 706	Vernalis	CA	95385	NW Hunter	209-835-6955	R	6*	<0.1
Powell Systems Inc	PO Box 345	Fowler	IN	47944	T McIntee	317-884-0613	D	4*	<0.1
EH Kneen Co	PO Box 205	Ansonia	CT	06401	M H Saffran	203-735-9305	R	3	<0.1
Mason Can Co	43 Dexter Rd	E Providence	RI	02914	RM Kosten	401-434-2810	R	1*	<0.1
JusTins Ltd	613 W Cheltenham	Melrose Park	PA	19027	Kenneth Olsho	215-635-7170	R	0*	<0.1

Source: Ward's Business Directory of U.S. Private and Public Companies, Volumes 1 and 2, 1996. The company type code used is as follows: P - Public, R - Private, S - Subsidiary, D - Division, J - Joint Venture, A - Affiliate, G - Group. Sales are in millions of dollars, employees are in thousands. An asterisk (*) indicates an estimated sales volume. The symbol < stands for 'less than'. Company names and addresses are truncated, in some cases, to fit into the available space.

MATERIALS CONSUMED

Material	Quantity	Delivered Cost ($ million)
Materials, ingredients, containers, and supplies	(X)	8,390.3
Lids, ends, and parts for metal cans (except castings and forgings)	(X)	631.9
All other fabricated metal products (except castings and forgings)	(X)	(D)
Castings (rough and semifinished)	(X)	(D)
Steel sheet, strip, and tin mill products	(X)	2,025.2
Steel wire and wire products	(X)	0.2
All other steel shapes and forms	(X)	(D)
Aluminum and aluminum-base alloy sheet, plate, foil, and welded tubing	(X)	2,567.8
Other aluminum and aluminum-base alloy shapes and forms	(X)	2,019.4
Copper mechanical wire (including extruded and/or drawn shapes)	(X)	24.6
All other copper and copper-base alloy shapes and forms	(X)	0.9
Other nonferrous shapes and forms	(X)	1.0
Paints, varnishes, lacquers, stains, shellacs, japans, enamels, and allied products	(X)	301.4
Adhesives and sealants	(X)	31.3
Printing ink	(X)	43.0
All other chemicals and allied products	(X)	35.3
Paperboard containers, boxes, and corrugated paperboard	(X)	24.2
All other materials and components, parts, containers, and supplies	(X)	261.6
Materials, ingredients, containers, and supplies, nsk	(X)	303.6

Source: 1992 Economic Census. Explanation of symbols used: (D): Withheld to avoid disclosure of competitive data; na: Not available; (S): Withheld because statistical norms were not met; (X): Not applicable; (Z): Less than half the unit shown; nec: Not elsewhere classified; nsk: Not specified by kind; - : zero; * : 10-19 percent estimated; ** : 20-29 percent estimated.

PRODUCT SHARE DETAILS

Product or Product Class	% Share	Product or Product Class	% Share
Metal cans	100.00	Steel cans and tinware products, nsk	1.02
Steel cans and tinware products	40.93	Aluminum cans, including lids, ends, and parts shipped separately	57.84
Steel cans, including lids, ends, and parts shipped separately	81.83	Metal cans, nsk	1.23
Tinware end products, including ice cream cans, but excluding cooking and kitchen utensils	17.16		

Source: 1992 Economic Census. The values shown are percent of total shipments in an industry. Values of indented subcategories are summed in the main heading. The symbol (D) appears when data are withheld to prevent disclosure of competitive information. The abbreviation nsk stands for 'not specified by kind' and nec for 'not elsewhere classified'.

INPUTS AND OUTPUTS FOR METAL CANS

Economic Sector or Industry Providing Inputs	%	Sector	Economic Sector or Industry Buying Outputs	%	Sector
Blast furnaces & steel mills	28.9	Manufg.	Malt beverages	24.2	Manufg.
Aluminum rolling & drawing	23.5	Manufg.	Bottled & canned soft drinks	21.1	Manufg.
Metal cans	9.5	Manufg.	Canned fruits & vegetables	11.9	Manufg.
Primary aluminum	8.2	Manufg.	Metal cans	6.9	Manufg.
Wholesale trade	7.6	Trade	Canned specialties	5.8	Manufg.
Commercial printing	3.1	Manufg.	Paints & allied products	2.9	Manufg.
Paints & allied products	2.6	Manufg.	Roasted coffee	2.9	Manufg.
Advertising	2.6	Services	Dog, cat, & other pet food	2.8	Manufg.
Electric services (utilities)	1.6	Util.	Wholesale trade	2.1	Trade
Motor freight transportation & warehousing	1.4	Util.	Condensed & evaporated milk	1.4	Manufg.
Metal coating & allied services	1.2	Manufg.	Frozen fruits, fruit juices & vegetables	1.3	Manufg.
Gas production & distribution (utilities)	1.0	Util.	Sausages & other prepared meats	1.3	Manufg.
Air transportation	0.6	Util.	Canned & cured seafoods	1.2	Manufg.
Railroads & related services	0.6	Util.	Drugs	1.2	Manufg.
Petroleum refining	0.5	Manufg.	Toilet preparations	1.2	Manufg.
Banking	0.5	Fin/R.E.	Chocolate & cocoa products	1.1	Manufg.
Paperboard containers & boxes	0.4	Manufg.	Meat packing plants	1.1	Manufg.
Printing ink	0.4	Manufg.	Food preparations, nec	0.9	Manufg.
Maintenance of nonfarm buildings nec	0.3	Constr.	Chemical preparations, nec	0.8	Manufg.
Primary lead	0.3	Manufg.	Polishes & sanitation goods	0.8	Manufg.
Sanitary services, steam supply, irrigation	0.3	Util.	Adhesives & sealants	0.6	Manufg.
Real estate	0.3	Fin/R.E.	Flavoring extracts & syrups, nec	0.6	Manufg.
Machinery, except electrical, nec	0.2	Manufg.	Business services nec	0.6	Services
Plastics materials & resins	0.2	Manufg.	Exports	0.6	Foreign
Communications, except radio & TV	0.2	Util.	Agricultural chemicals, nec	0.5	Manufg.
Eating & drinking places	0.2	Trade	Confectionery products	0.4	Manufg.
Royalties	0.2	Fin/R.E.	Lubricating oils & greases	0.4	Manufg.
Business services nec	0.2	Services	Shortening & cooking oils	0.4	Manufg.
Computer & data processing services	0.2	Services	Fluid milk	0.3	Manufg.
Engineering, architectural, & surveying services	0.2	Services	Petroleum refining	0.3	Manufg.
Equipment rental & leasing services	0.2	Services	Poultry & egg processing	0.3	Manufg.
Abrasive products	0.1	Manufg.	Soap & other detergents	0.3	Manufg.
Cyclic crudes and organics	0.1	Manufg.	Cyclic crudes and organics	0.2	Manufg.
Fabricated rubber products, nec	0.1	Manufg.	Metal barrels, drums, & pails	0.2	Manufg.
Soap & other detergents	0.1	Manufg.	Printing ink	0.2	Manufg.
Special dies & tools & machine tool accessories	0.1	Manufg.	Federal Government purchases, national defense	0.2	Fed Govt
Wood pallets & skids	0.1	Manufg.	Cereal breakfast foods	0.1	Manufg.
Credit agencies other than banks	0.1	Fin/R.E.	Cheese, natural & processed	0.1	Manufg.
Insurance carriers	0.1	Fin/R.E.	Surface active agents	0.1	Manufg.
Hotels & lodging places	0.1	Services	Surgical appliances & supplies	0.1	Manufg.

Source: Benchmark Input-Output Accounts for the U.S. Economy, 1982, U.S. Department of Commerce, Washington, D.C., July 1991. Data, as reported in the source, are organized by the 1977 SIC structure in use in 1982 but have been matched, as closely as is possible, to the 1987 SIC structure used in this book.

OCCUPATIONS EMPLOYED BY SIC 341 - METAL CANS AND SHIPPING CONTAINERS

Occupation	% of Total 1994	Change to 2005	Occupation	% of Total 1994	Change to 2005
Industrial machinery mechanics	11.8	-27.2	Assemblers, fabricators, & hand workers nec	2.4	-33.8
Machine forming operators, metal & plastic	6.7	-33.9	Welding machine setters, operators	1.9	-40.4
Industrial truck & tractor operators	5.9	-33.9	Printing press machine setters, operators	1.8	-33.9
Machine feeders & offbearers	5.6	-40.4	Electricians	1.8	-37.9
Helpers, laborers, & material movers nec	4.9	-33.8	Industrial production managers	1.7	-33.9
Inspectors, testers, & graders, precision	4.8	-33.8	Freight, stock, & material movers, hand	1.7	-47.1
Machine tool cutting & forming etc. nec	4.2	-33.9	Punching machine operators, metal & plastic	1.6	-47.0
Packaging & filling machine operators	2.9	-33.8	General managers & top executives	1.5	-37.3
Coating, painting, & spraying machine workers	2.8	-33.8	Material moving equipment operators nec	1.4	-33.8
Machine operators nec	2.7	-41.6	Managers & administrators nec	1.2	-33.8
Machinists	2.5	-33.9	Secretaries, ex legal & medical	1.1	-39.8
Hand packers & packagers	2.5	-43.3	Sales & related workers nec	1.1	-34.0
Maintenance repairers, general utility	2.5	-40.5	Truck drivers light & heavy	1.1	-31.8
Metal & plastic machine workers nec	2.4	-41.3			

Source: Industry-Occupation Matrix, Bureau of Labor Statistics. These data relate to one or more 3-digit SIC industry groups rather than to a single 4-digit SIC. The change reported for each occupation to the year 2005 is a percent of growth or decline as estimated by the Bureau of Labor Statistics. The abbreviation nec stands for 'not elsewhere classified'.

LOCATION BY STATE AND REGIONAL CONCENTRATION

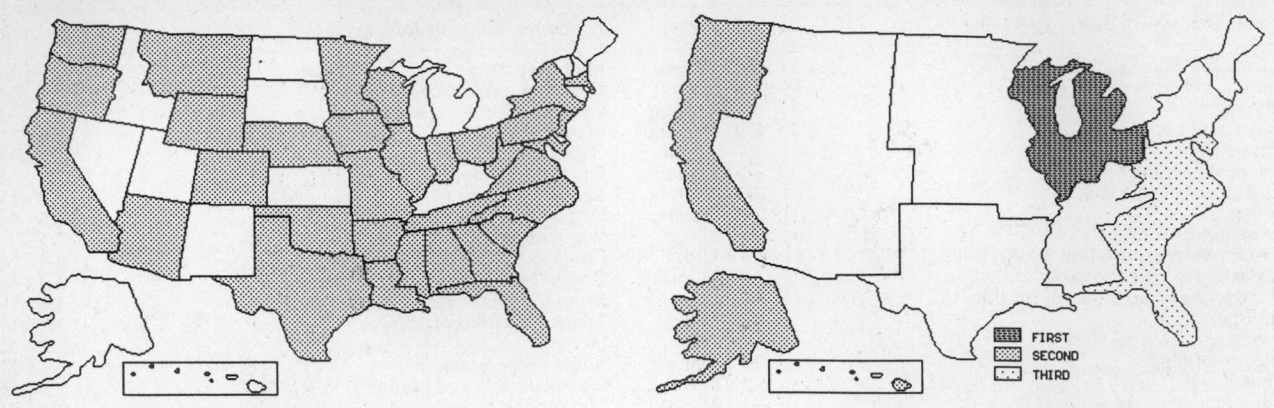

FIRST
SECOND
THIRD

INDUSTRY DATA BY STATE

| State | Establish-ments | Shipments | | | Employment | | | | Cost as % of Shipments | Investment per Employee ($) |
		Total ($ mil)	% of U.S.	Per Establ.	Total Number	% of U.S.	Per Establ.	Wages ($/hour)		
California	55	1,790.2	14.8	32.5	5,100	15.8	93	18.04	74.4	8,706
Ohio	19	1,213.9	10.0	63.9	3,400	10.5	179	16.98	69.0	12,235
Texas	21	799.0	6.6	38.0	1,900	5.9	90	15.94	68.8	9,579
Illinois	22	577.3	4.8	26.2	2,500	7.7	114	17.52	70.5	3,200
Pennsylvania	14	501.2	4.1	35.8	1,700	5.3	121	14.73	66.4	4,941
New York	12	471.0	3.9	39.2	1,000	3.1	83	17.63	73.9	-
New Jersey	21	461.0	3.8	22.0	1,500	4.6	71	16.59	68.0	3,867
North Carolina	6	367.8	3.0	61.3	900	2.8	150	17.82	72.9	-
Georgia	10	326.3	2.7	32.6	900	2.8	90	17.44	68.7	-
Tennessee	5	74.5	0.6	14.9	200	0.6	40	10.00	65.6	-
Wisconsin	19	(D)	-	-	1,750 *	5.4	92	-	-	-
Florida	12	(D)	-	-	1,750 *	5.4	146	-	-	-
Maryland	10	(D)	-	-	750 *	2.3	75	-	-	-
Indiana	8	(D)	-	-	750 *	2.3	94	-	-	10,133
Missouri	8	(D)	-	-	1,750 *	5.4	219	-	-	-
Washington	7	(D)	-	-	750 *	2.3	107	-	-	-
Minnesota	6	(D)	-	-	750 *	2.3	125	-	-	-
South Carolina	6	(D)	-	-	375 *	1.2	63	-	-	-
Alabama	5	(D)	-	-	375 *	1.2	75	-	-	-
Mississippi	5	(D)	-	-	750 *	2.3	150	-	-	-
Oklahoma	4	(D)	-	-	375 *	1.2	94	-	-	-
Virginia	4	(D)	-	-	750 *	2.3	188	-	-	15,867
Colorado	3	(D)	-	-	750 *	2.3	250	-	-	-
Iowa	3	(D)	-	-	175 *	0.5	58	-	-	-
Massachusetts	3	(D)	-	-	375 *	1.2	125	-	-	-
Oregon	3	(D)	-	-	175 *	0.5	58	-	-	-
Arkansas	2	(D)	-	-	375 *	1.2	188	-	-	-
Montana	2	(D)	-	-	175 *	0.5	88	-	-	-
West Virginia	2	(D)	-	-	175 *	0.5	88	-	-	-
Arizona	1	(D)	-	-	175 *	0.5	175	-	-	-
Louisiana	1	(D)	-	-	175 *	0.5	175	-	-	-
Nebraska	1	(D)	-	-	375 *	1.2	375	-	-	-
Wyoming	1	(D)	-	-	175 *	0.5	175	-	-	-

Source: 1992 *Economic Census*. The states are in descending order of shipments or establishments (if shipment data are missing for the majority). The symbol (D) appears when data are withheld to prevent disclosure of competitive information. States marked with (D) are sorted by number of establishments. A dash (-) indicates that the data element cannot be calculated; * indicates the midpoint of a range.

3412 - METAL BARRELS, DRUMS & PAILS

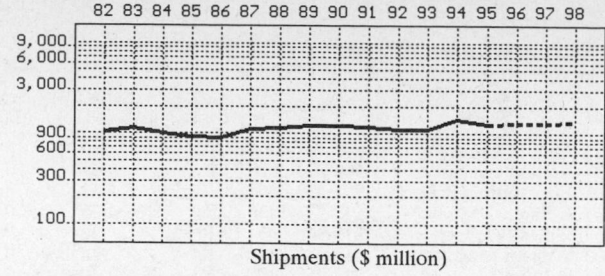

Shipments ($ million)

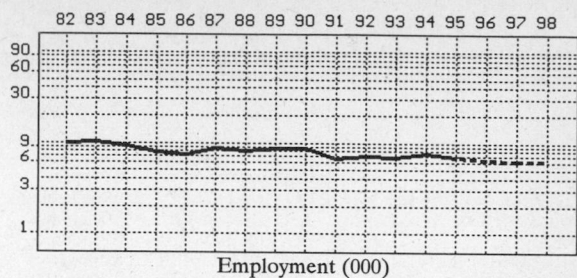

Employment (000)

GENERAL STATISTICS

Year	Companies	Establishments		Employment			Compensation		Production ($ million)			
		Total	with 20 or more employees	Total (000)	Production Workers (000)	Hours (Mil)	Payroll ($ mil)	Wages ($/hr)	Cost of Materials	Value Added by Manufacture	Value of Shipments	Capital Invest.
1982	115	169	109	9.9	7.6	14.8	198.6	9.21	663.0	371.7	1,038.7	22.7
1983		169	107	10.2	8.1	15.9	200.1	9.04	753.7	439.6	1,152.7	23.6
1984		169	105	9.2	7.2	15.1	194.8	8.70	652.6	369.6	1,020.4	19.0
1985		170	102	7.8	6.1	13.1	177.1	9.13	577.8	341.1	923.1	17.0
1986		152	100	7.3	5.6	12.1	169.7	9.09	547.6	339.5	888.2	15.3
1987	120	168	99	8.7	6.5	14.0	202.0	9.79	702.0	397.4	1,100.0	14.0
1988		161	98	8.2	6.3	14.0	198.4	9.74	740.3	408.4	1,148.5	13.3
1989		155	100	8.8	6.0	13.2	195.2	9.36	802.3	434.5	1,238.6	28.5
1990		154	101	8.7	5.4	12.0	184.3	9.82	782.2	422.1	1,213.3	23.1
1991		154	101	6.7	5.0	10.8	174.9	10.41	756.4	426.7	1,192.0	23.4
1992	116	155	91	7.2	5.5	12.0	201.0	10.97	693.6	437.9	1,133.8	22.0
1993		151	85	6.9	5.3	11.6	193.8	11.02	682.7	447.4	1,128.6	32.5
1994		149P	90P	7.7	5.9	13.1	237.3	11.86	897.2	521.4	1,415.2	33.7
1995		147P	88P	6.8P	4.8P	11.2P	203.7P	11.39P	813.8P	472.9P	1,283.6P	28.8P
1996		146P	87P	6.6P	4.6P	10.9P	205.0P	11.60P	828.4P	481.4P	1,306.6P	29.7P
1997		144P	85P	6.4P	4.5P	10.6P	206.3P	11.82P	842.9P	489.9P	1,329.6P	30.7P
1998		142P	84P	6.2P	4.3P	10.4P	207.7P	12.04P	857.5P	498.3P	1,352.6P	31.6P

Sources: 1982, 1987, 1992 *Economic Census*; *Annual Survey of Manufactures*, 83-86, 88-91, 93-94. Establishment counts for non-Census years are from *County Business Patterns*; establishment values for 83-84 are extrapolations. 'P's show projections by the editors. Industries reclassified in 87 will not have data for prior years.

INDICES OF CHANGE

Year	Companies	Establishments		Employment			Compensation		Production ($ million)			
		Total	with 20 or more employees	Total (000)	Production Workers (000)	Hours (Mil)	Payroll ($ mil)	Wages ($/hr)	Cost of Materials	Value Added by Manufacture	Value of Shipments	Capital Invest.
1982	99	109	120	138	138	123	99	84	96	85	92	103
1983		109	118	142	147	133	100	82	109	100	102	107
1984		109	115	128	131	126	97	79	94	84	90	86
1985		110	112	108	111	109	88	83	83	78	81	77
1986		98	110	101	102	101	84	83	79	78	78	70
1987	103	108	109	121	118	117	100	89	101	91	97	64
1988		104	108	114	115	117	99	89	107	93	101	60
1989		100	110	122	109	110	97	85	116	99	109	130
1990		99	111	121	98	100	92	90	113	96	107	105
1991		99	111	93	91	90	87	95	109	97	105	106
1992	100	100	100	100	100	100	100	100	100	100	100	100
1993		97	93	96	96	97	96	100	98	102	100	148
1994		96P	99P	107	107	109	118	108	129	119	125	153
1995		95P	97P	94P	88P	93P	101P	104P	117P	108P	113P	131P
1996		94P	95P	91P	84P	91P	102P	106P	119P	110P	115P	135P
1997		93P	94P	88P	81P	89P	103P	108P	122P	112P	117P	139P
1998		92P	92P	86P	77P	86P	103P	110P	124P	114P	119P	144P

Sources: Same as General Statistics. Values reflect change from the base year, 1992. Values above 100 mean greater than 92, values below 100 mean less than 92, and a value of 100 in the 82-91 or 93-98 period means same as 92. 'P's mark projections by the editors.

SELECTED RATIOS

For 1994	Avg. of All Manufact.	Analyzed Industry	Index	For 1994	Avg. of All Manufact.	Analyzed Industry	Index
Employees per Establishment	49	52	105	Value Added per Production Worker	134,084	88,373	66
Payroll per Establishment	1,500,273	1,591,970	106	Cost per Establishment	5,045,178	6,019,028	119
Payroll per Employee	30,620	30,818	101	Cost per Employee	102,970	116,519	113
Production Workers per Establishment	34	40	115	Cost per Production Worker	146,988	152,068	103
Wages per Establishment	853,319	1,042,301	122	Shipments per Establishment	9,576,895	9,494,125	99
Wages per Production Worker	24,861	26,333	106	Shipments per Employee	195,460	183,792	94
Hours per Production Worker	2,056	2,220	108	Shipments per Production Worker	279,017	239,864	86
Wages per Hour	12.09	11.86	98	Investment per Establishment	321,011	226,083	70
Value Added per Establishment	4,602,255	3,497,906	76	Investment per Employee	6,552	4,377	67
Value Added per Employee	93,930	67,714	72	Investment per Production Worker	9,352	5,712	61

Sources: Same as General Statistics. The 'Average of All Manufacturing' column represents the average of all manufacturing industries reported for the most recent complete year available. The Index shows the relationship between the Average and the Analyzed Industry. For example, 100 means that they are equal; 500 that the Analyzed Industry is five times the average; 50 means that the Analyzed Industry is half the national average. The abbreviation 'na' is used to show that data are 'not available'.

LEADING COMPANIES   Number shown: 28   Total sales ($ mil): 527   Total employment (000): 3.5

Company Name	Address				CEO Name	Phone	Co. Type	Sales ($ mil)	Empl. (000)
Russell-Stanley Corp	230 Half Mile Rd	Red Bank	NJ	07701	John Priesing	908-741-6366	R	110*	0.5
Evans Industries Inc	PO Drawer 68	Harvey	LA	70059	Ronald Evans	504-374-6000	R	67	0.5
Florida Drum Company Inc	PO Box 1951	Pensacola	FL	32589	Christopher R Long	904-433-7601	R	51	0.2
Astro Container Co	2795 Sharon Rd	Cincinnati	OH	45241	A Paul	513-771-1230	S	35	<0.1
Galbreath Inc	PO Box 220	Winamac	IN	46996	James R Herrman	219-946-6631	R	32	0.3
Meyer Steel Drum Inc	3201 S Millard St	Chicago	IL	60623	William Meyer	312-376-8376	R	28*	0.2
Nesco Container Corp	2391 Cassens Dr	Fenton	MO	63026	D Cann	314-452-1190	R	20*	<0.1
Republic Container Co	PO Box 37	Nitro	WV	25143	Jim McKnight	304-755-4325	R	16*	<0.1
Acme Barrel Co	2300 W 13th St	Chicago	IL	60608	Philip A Pearlman	312-829-3838	R	15*	0.2
Florida Drum Delta Co	3300 N Hutchinson	Pine Bluff	AR	71602	Chris Long	501-247-2800	D	15	<0.1
Sirco Systems Inc	PO Box 367	Birmingham	AL	35201	JD Brown Jr	205-731-7846	S	15*	0.1
Queen City Barrel Co	1937 South St	Cincinnati	OH	45204	Edward Paul	513-921-8811	R	12	0.2
American Steel Container Co	4445 W 5th Av	Chicago	IL	60624	MB Spitz	312-826-8080	R	11	<0.1
Calig Steel Drum Co	1400 Fleming Av	McKees Rocks	PA	15136	Lee H Calig	412-771-6440	R	10*	<0.1
Chicago Steel Container Corp	1846 S Kilbourn	Chicago	IL	60623	Louis Pileggi	312-277-2244	R	10	<0.1
Allied-Hastings Barrell&Drum	915 W 37th St	Chicago	IL	60609	RC Krause	312-847-3435	R	9*	<0.1
Container Research Corp	PO Box 159	Glen Riddle	PA	19037	Stanley E Rines	610-459-2160	R	9*	0.2
Mid-America Steel Drum Co	8570 S Chicago Rd	Oak Creek	WI	53154	E Simonis	414-762-1114	R	9	<0.1
CDF Corp	77 Industrial Pk Rd	Plymouth	MA	02360	Joseph Sullivan	508-747-5858	R	7	<0.1
Skolnik Industries Inc	4900 S Kilbourn	Chicago	IL	60632	Howard B Skolnik	312-735-0700	R	7	<0.1
A and J Manufacturing Company	PO Box 90596	Los Angeles	CA	90009	AL Young	213-678-3053	R	6	<0.1
Erie Engineered Products Inc	908 Niagara Falls	N Tonawanda	NY	14120	Lorne Weil	716-694-2020	R	6	<0.1
Pemco Engineers	2398 Railroad St	Corona	CA	91720	Matt Gold	909-273-9696	D	6*	0.1
Mirax Chemical Products Corp	4999 Fyler Av	St Louis	MO	63139	OC Clerc Sr	314-752-1500	R	5	<0.1
Northwest Cooperage Company	7152 1st Av S	Seattle	WA	98108	J Trotsky	206-763-2345	R	5	<0.1
Inductametals Corp	54 W Hubbard St	Chicago	IL	60610	Theodore Krengel	312-467-1031	R	4	<0.1
National Drum and Barrel Corp	35-A Beadle St	Brooklyn	NY	11222	Irwin Tartus	718-388-5810	R	4	<0.1
Greif Bros Corp	346 Goddell St	Wyandotte	MI	48192	Hadley Saathoff	313-284-7022	D	3*	<0.1

Source: Ward's Business Directory of U.S. Private and Public Companies, Volumes 1 and 2, 1996. The company type code used is as follows: P - Public, R - Private, S - Subsidiary, D - Division, J - Joint Venture, A - Affiliate, G - Group. Sales are in millions of dollars, employees are in thousands. An asterisk (*) indicates an estimated sales volume. The symbol < stands for 'less than'. Company names and addresses are truncated, in some cases, to fit into the available space.

MATERIALS CONSUMED

Material	Quantity	Delivered Cost ($ million)
Materials, ingredients, containers, and supplies	(X)	654.6
Lids, ends, and parts for metal cans (except castings and forgings)	(X)	30.8
All other fabricated metal products (except castings and forgings)	(X)	21.7
Steel sheet, strip, and tin mill products	(X)	342.9
Steel wire and wire products	(X)	1.0
All other steel shapes and forms	(X)	32.8
Other aluminum and aluminum-base alloy shapes and forms	(X)	(D)
Other nonferrous shapes and forms	(X)	(D)
Paints, varnishes, lacquers, stains, shellacs, japans, enamels, and allied products	(X)	26.8
Adhesives and sealants	(X)	0.9
Printing ink	(X)	0.2
All other chemicals and allied products	(X)	5.0
Paperboard containers, boxes, and corrugated paperboard	(X)	5.3
All other materials and components, parts, containers, and supplies	(X)	63.9
Materials, ingredients, containers, and supplies, nsk	(X)	122.0

Source: 1992 *Economic Census*. Explanation of symbols used: (D): Withheld to avoid disclosure of competitive data; na: Not available; (S): Withheld because statistical norms were not met; (X): Not applicable; (Z): Less than half the unit shown; nec: Not elsewhere classified; nsk: Not specified by kind; - : zero; * : 10-19 percent estimated; ** : 20-29 percent estimated.

PRODUCT SHARE DETAILS

Product or Product Class	% Share	Product or Product Class	% Share
Metal barrels, drums, and pails	100.00	(more than 12 gal capacity)	67.53
Steel pails (1 to 12 gal capacity)	23.05	All other metal barrels	4.94
Steel shipping barrels and drums, excluding beer barrels		Metal barrels, drums, and pails, nsk	4.48

Source: 1992 *Economic Census*. The values shown are percent of total shipments in an industry. Values of indented subcategories are summed in the main heading. The symbol (D) appears when data is withheld to prevent disclosure of competitive information. The abbreviation nsk stands for 'not specified by kind' and nec for 'not elsewhere classified'.

INPUTS AND OUTPUTS FOR METAL BARRELS, DRUMS, & PAILS

Economic Sector or Industry Providing Inputs	%	Sector	Economic Sector or Industry Buying Outputs	%	Sector
Blast furnaces & steel mills	39.5	Manufg.	Paints & allied products	15.4	Manufg.
Wholesale trade	9.2	Trade	Industrial gases	10.7	Manufg.
Imports	8.4	Foreign	Wholesale trade	9.8	Trade
Aluminum rolling & drawing	6.8	Manufg.	Cyclic crudes and organics	9.2	Manufg.
Metal coating & allied services	5.5	Manufg.	Fluid milk	8.7	Manufg.
Metal cans	3.5	Manufg.	Exports	5.5	Foreign
Paints & allied products	3.4	Manufg.	Soap & other detergents	4.8	Manufg.
Electric services (utilities)	1.8	Util.	Federal Government purchases, national defense	4.8	Fed Govt
Gas production & distribution (utilities)	1.7	Util.	Drugs	4.3	Manufg.
Fabricated metal products, nec	1.6	Manufg.	Lubricating oils & greases	3.5	Manufg.
Plastics materials & resins	1.6	Manufg.	Polishes & sanitation goods	3.3	Manufg.
Motor freight transportation & warehousing	1.6	Util.	Petroleum refining	2.9	Manufg.
Communications, except radio & TV	1.5	Util.	Gross private fixed investment	2.9	Cap Inv
Petroleum refining	1.2	Manufg.	Industrial inorganic chemicals, nec	2.1	Manufg.
Advertising	1.1	Services	Agricultural chemicals, nec	2.0	Manufg.
Miscellaneous fabricated wire products	0.8	Manufg.	Plastics materials & resins	1.8	Manufg.
Metal barrels, drums, & pails	0.7	Manufg.	Adhesives & sealants	1.4	Manufg.
Air transportation	0.7	Util.	Surface active agents	1.2	Manufg.
Railroads & related services	0.7	Util.	Shortening & cooking oils	1.0	Manufg.
Machinery, except electrical, nec	0.6	Manufg.	Chemical preparations, nec	0.7	Manufg.
Metal stampings, nec	0.6	Manufg.	Fresh or frozen packaged fish	0.6	Manufg.
Maintenance of nonfarm buildings nec	0.5	Constr.	Alkalies & chlorine	0.5	Manufg.
Eating & drinking places	0.5	Trade	Metal barrels, drums, & pails	0.5	Manufg.
Paperboard containers & boxes	0.4	Manufg.	Printing ink	0.4	Manufg.
Banking	0.4	Fin/R.E.	Inorganic pigments	0.3	Manufg.
Abrasive products	0.3	Manufg.	Fertilizers, mixing only	0.2	Manufg.
Special dies & tools & machine tool accessories	0.3	Manufg.	Metal stampings, nec	0.2	Manufg.
Real estate	0.3	Fin/R.E.	Paving mixtures & blocks	0.2	Manufg.
Equipment rental & leasing services	0.3	Services	Soybean oil mills	0.2	Manufg.
Noncomparable imports	0.3	Foreign	Synthetic rubber	0.2	Manufg.
Machine tools, metal forming types	0.2	Manufg.	Explosives	0.1	Manufg.
Primary lead	0.2	Manufg.	Food preparations, nec	0.1	Manufg.
Insurance carriers	0.2	Fin/R.E.			
Automotive rental & leasing, without drivers	0.2	Services			
Automotive repair shops & services	0.2	Services			
Computer & data processing services	0.2	Services			
Engineering, architectural, & surveying services	0.2	Services			
Hotels & lodging places	0.2	Services			
Legal services	0.2	Services			
Management & consulting services & labs	0.2	Services			
Adhesives & sealants	0.1	Manufg.			
Lubricating oils & greases	0.1	Manufg.			
Metal heat treating	0.1	Manufg.			
Water supply & sewage systems	0.1	Util.			
Accounting, auditing & bookkeeping	0.1	Services			
U.S. Postal Service	0.1	Gov't			

Source: Benchmark Input-Output Accounts for the U.S. Economy, 1982, U.S. Department of Commerce, Washington, D.C., July 1991. Data, as reported in the source, are organized by the 1977 SIC structure in use in 1982 but have been matched, as closely as is possible, to the 1987 SIC structure used in this book.

OCCUPATIONS EMPLOYED BY SIC 341 - METAL CANS AND SHIPPING CONTAINERS

Occupation	% of Total 1994	Change to 2005	Occupation	% of Total 1994	Change to 2005
Industrial machinery mechanics	11.8	-27.2	Assemblers, fabricators, & hand workers nec	2.4	-33.8
Machine forming operators, metal & plastic	6.7	-33.9	Welding machine setters, operators	1.9	-40.4
Industrial truck & tractor operators	5.9	-33.9	Printing press machine setters, operators	1.8	-33.9
Machine feeders & offbearers	5.6	-40.4	Electricians	1.8	-37.9
Helpers, laborers, & material movers nec	4.9	-33.8	Industrial production managers	1.7	-33.9
Inspectors, testers, & graders, precision	4.8	-33.8	Freight, stock, & material movers, hand	1.7	-47.1
Machine tool cutting & forming etc. nec	4.2	-33.9	Punching machine operators, metal & plastic	1.6	-47.0
Packaging & filling machine operators	2.9	-33.8	General managers & top executives	1.5	-37.3
Coating, painting, & spraying machine workers	2.8	-33.8	Material moving equipment operators nec	1.4	-33.8
Machine operators nec	2.7	-41.6	Managers & administrators nec	1.2	-33.8
Machinists	2.5	-33.9	Secretaries, ex legal & medical	1.1	-39.8
Hand packers & packagers	2.5	-43.3	Sales & related workers nec	1.1	-34.0
Maintenance repairers, general utility	2.5	-40.5	Truck drivers light & heavy	1.1	-31.8
Metal & plastic machine workers nec	2.4	-41.3			

Source: Industry-Occupation Matrix, Bureau of Labor Statistics. These data relate to one or more 3-digit SIC industry groups rather than to a single 4-digit SIC. The change reported for each occupation to the year 2005 is a percent of growth or decline as estimated by the Bureau of Labor Statistics. The abbreviation nec stands for 'not elsewhere classified'.

LOCATION BY STATE AND REGIONAL CONCENTRATION

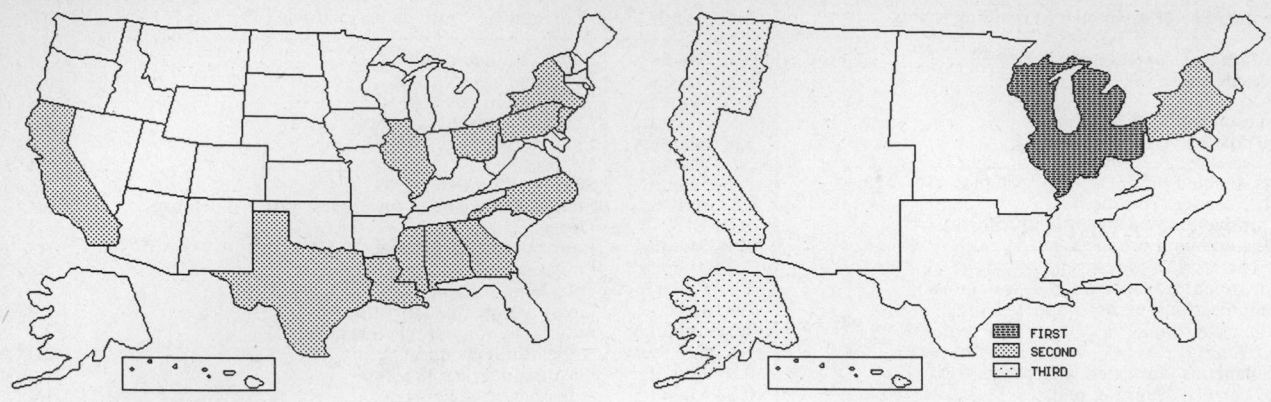

INDUSTRY DATA BY STATE

| State | Establish-ments | Shipments | | | Employment | | | | Cost as % of Shipments | Investment per Employee ($) |
		Total ($ mil)	% of U.S.	Per Establ.	Total Number	% of U.S.	Per Establ.	Wages ($/hour)		
Illinois	19	180.5	15.9	9.5	1,300	18.1	68	13.05	58.9	4,000
Ohio	20	148.9	13.1	7.4	1,000	13.9	50	10.38	60.7	2,600
Texas	11	145.2	12.8	13.2	1,000	13.9	91	8.82	61.0	2,500
California	17	106.4	9.4	6.3	600	8.3	35	13.00	63.0	2,000
Pennsylvania	11	101.4	8.9	9.2	500	6.9	45	13.00	57.1	2,000
Georgia	5	85.1	7.5	17.0	600	8.3	120	9.33	66.7	-
New Jersey	8	78.0	6.9	9.8	400	5.6	50	11.43	57.3	3,500
Louisiana	4	55.0	4.9	13.8	200	2.8	50	12.33	71.8	-
North Carolina	3	26.4	2.3	8.8	200	2.8	67	10.00	64.8	-
New York	5	13.0	1.1	2.6	100	1.4	20	9.50	59.2	-
Alabama	6	(D)	-	-	175 *	2.4	29	-	-	-
Mississippi	1	(D)	-	-	175 *	2.4	175	-	-	-

Source: 1992 *Economic Census*. The states are in descending order of shipments or establishments (if shipment data are missing for the majority). The symbol (D) appears when data are withheld to prevent disclosure of competitive information. States marked with (D) are sorted by number of establishments. A dash (-) indicates that the data element cannot be calculated; * indicates the midpoint of a range.

3421 - CUTLERY

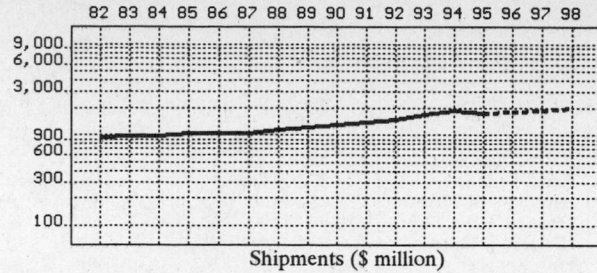

Shipments ($ million)

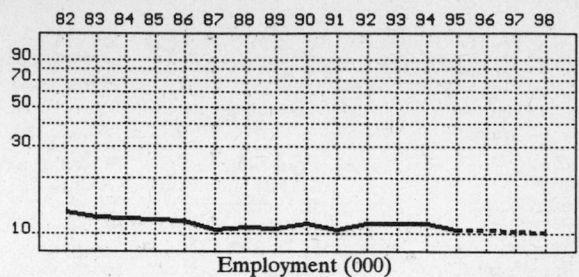

Employment (000)

GENERAL STATISTICS

Year	Com-panies	Establishments		Employment			Compensation		Production ($ million)			
		Total	with 20 or more employees	Total (000)	Production Workers (000)	Hours (Mil)	Payroll ($ mil)	Wages ($/hr)	Cost of Materials	Value Added by Manufacture	Value of Shipments	Capital Invest.
1982	123	132	69	13.0	10.5	20.1	223.7	8.02	256.3	683.6	938.0	44.0
1983		128	68	12.2	9.7	18.3	222.1	8.51	238.1	723.5	972.3	34.2
1984		124	67	12.1	9.6	18.1	231.8	8.93	289.9	717.9	992.4	35.2
1985		119	65	11.8	9.4	17.3	239.9	9.51	303.9	731.7	1,043.4	43.9
1986		117	66	11.7	9.1	17.2	251.2	9.90	269.5	772.9	1,045.1	47.3
1987	131	141	62	10.5	7.9	15.1	241.0	10.46	250.7	803.1	1,054.0	57.9
1988		126	65	10.8	8.4	16.5	270.2	10.46	278.8	884.1	1,142.2	60.6
1989		118	67	10.6	8.3	16.2	263.4	11.28	306.3	940.9	1,235.2	82.2
1990		129	61	11.3	8.6	18.0	292.3	11.42	360.6	977.8	1,320.5	63.9
1991		132	61	10.4	8.1	16.2	274.9	11.90	351.6	1,027.3	1,391.2	38.2
1992	127	133	54	11.2	8.2	15.9	292.1	12.22	430.5	1,077.0	1,509.9	56.2
1993		136	56	11.4	8.3	15.9	313.3	13.30	435.3	1,251.6	1,692.1	53.3
1994		132P	56P	11.2	8.0	16.0	310.9	12.98	522.6	1,374.9	1,853.9	52.9
1995		133P	55P	10.5P	7.5P	15.2P	318.6P	13.68P	491.4P	1,292.7P	1,743.1P	62.0P
1996		133P	54P	10.4P	7.4P	15.0P	326.5P	14.11P	511.4P	1,345.4P	1,814.2P	63.5P
1997		134P	53P	10.2P	7.2P	14.7P	334.3P	14.54P	531.4P	1,398.2P	1,885.3P	65.0P
1998		135P	51P	10.1P	7.0P	14.5P	342.2P	14.96P	551.5P	1,450.9P	1,956.4P	66.5P

Sources: 1982, 1987, 1992 *Economic Census*; *Annual Survey of Manufactures*, 83-86, 88-91, 93-94. Establishment counts for non-Census years are from *County Business Patterns*; establishment values for 83-84 are extrapolations. 'P's show projections by the editors. Industries reclassified in 87 will not have data for prior years.

INDICES OF CHANGE

Year	Com-panies	Establishments		Employment			Compensation		Production ($ million)			
		Total	with 20 or more employees	Total (000)	Production Workers (000)	Hours (Mil)	Payroll ($ mil)	Wages ($/hr)	Cost of Materials	Value Added by Manufacture	Value of Shipments	Capital Invest.
1982	97	99	128	116	128	126	77	66	60	63	62	78
1983		96	126	109	118	115	76	70	55	67	64	61
1984		93	124	108	117	114	79	73	67	67	66	63
1985		89	120	105	115	109	82	78	71	68	69	78
1986		88	122	104	111	108	86	81	63	72	69	84
1987	103	106	115	94	96	95	83	86	58	75	70	103
1988		95	120	96	102	104	93	86	65	82	76	108
1989		89	124	95	101	102	90	92	71	87	82	146
1990		97	113	101	105	113	100	93	84	91	87	114
1991		99	113	93	99	102	94	97	82	95	92	68
1992	100	100	100	100	100	100	100	100	100	100	100	100
1993		102	104	102	101	100	107	109	101	116	112	95
1994		99P	104P	100	98	101	106	106	121	128	123	94
1995		100P	102P	94P	92P	96P	109P	112P	114P	120P	115P	110P
1996		100P	100P	93P	90P	94P	112P	115P	119P	125P	120P	113P
1997		101P	97P	91P	88P	93P	114P	119P	123P	130P	125P	116P
1998		101P	95P	90P	86P	91P	117P	122P	128P	135P	130P	118P

Sources: Same as General Statistics. Values reflect change from the base year, 1992. Values above 100 mean greater than 92, values below 100 mean less than 92, and a value of 100 in the 82-91 or 93-98 period means same as 92. 'P's mark projections by the editors.

SELECTED RATIOS

For 1994	Avg. of All Manufact.	Analyzed Industry	Index	For 1994	Avg. of All Manufact.	Analyzed Industry	Index
Employees per Establishment	49	85	173	Value Added per Production Worker	134,084	171,862	128
Payroll per Establishment	1,500,273	2,353,952	157	Cost per Establishment	5,045,178	3,956,820	78
Payroll per Employee	30,620	27,759	91	Cost per Employee	102,970	46,661	45
Production Workers per Establishment	34	61	176	Cost per Production Worker	146,988	65,325	44
Wages per Establishment	853,319	1,572,431	184	Shipments per Establishment	9,576,895	14,036,641	147
Wages per Production Worker	24,861	25,960	104	Shipments per Employee	195,460	165,527	85
Hours per Production Worker	2,056	2,000	97	Shipments per Production Worker	279,017	231,737	83
Wages per Hour	12.09	12.98	107	Investment per Establishment	321,011	400,528	125
Value Added per Establishment	4,602,255	10,409,935	226	Investment per Employee	6,552	4,723	72
Value Added per Employee	93,930	122,759	131	Investment per Production Worker	9,352	6,613	71

Sources: Same as General Statistics. The 'Average of All Manufacturing' column represents the average of all manufacturing industries reported for the most recent complete year available. The Index shows the relationship between the Average and the Analyzed Industry. For example, 100 means that they are equal; 500 that the Analyzed Industry is five times the average; 50 means that the Analyzed Industry is half the national average. The abbreviation 'na' is used to show that data are 'not available'.

LEADING COMPANIES   Number shown: 27   Total sales ($ mil): **6,533**   Total employment (000): **36.8**

Company Name	Address				CEO Name	Phone	Co. Type	Sales ($ mil)	Empl. (000)
Gillette Co	Prud Bldg	Boston	MA	02199	Alfred Zeien	617-421-7000	P	6,070	32.8
Alcas Corp	1116 E State St	Olean	NY	14760	Erick J Laine	716-372-3111	R	75	0.4
Schick Safety Razor	10 Webster Rd	Milford	CT	06460	Saunders M Morton	203-882-2100	D	53	0.6
Buck Knives Inc	PO Box 1267	El Cajon	CA	92022	Charles T Buck	619-449-1100	R	46	0.4
Hyde Manufacturing Co	54 Eastford Rd	Southbridge	MA	01550	Richard B Hardy	508-764-4344	R	43	0.3
Revlon Implement Corp	196 & 199 Coit St	Irvington	NJ	07111	Paul Spindel	201-373-5803	S	40	0.3
Zenith Cutter Co	PO Box 2252	Rockford	IL	61131	CW Blazer	815-282-5200	R	28*	0.2
Regent-Sheffield Ltd	70 Schmitt Blv	Farmingdale	NY	11735	Jay M Cotler	516-293-8200	S	22	0.1
WR Case and Sons Cutlery Co	Owens Way	Bradford	PA	16701	George T Brinkley	814-523-6350	S	20	0.2
Ardell Industries	555 Lehigh Av	Union	NJ	07083		908-687-5900	D	19	0.1
Russell Harrington Cutlery Inc	44 River St	Southbridge	MA	01550	Edmond Neal	508-765-0201	S	18	0.3
WE Bassett Co	100 Trap Falls Rd	Shelton	CT	06484	William C Bassett	203-929-8483	R	17	0.2
Camillus Cutlery Co	54 Main St	Camillus	NY	13031	James W Furgal	315-672-8111	R	12	0.2
Cook Bates Company Inc	PO Box 4703	Sarasota	FL	34230	Michael Thomas	813-365-1600	S	11*	<0.1
Clauss Cutlery Co	223 N Prospect St	Fremont	OH	43420	R N Olmstead	419-332-7344	D	10	<0.1
Quikut	PO Box 29	Walnut Ridge	AR	72476	C Nathan Howard	501-886-6774	D	10	<0.1
Pacific Handy Cutter Inc	PO Box 10869	Costa Mesa	CA	92627	G Gerry Schmidt	714-662-1033	R	8	<0.1
Queen Cutlery Co	PO Box 500	Franklinville	NY	14737	John Wyllie	716-676-5527	S	7	<0.1
Midwest Tool and Cutlery Co	PO Box 160	Sturgis	MI	49091	DG Schmick	616-651-2476	R	6	<0.1
Lamson & Goodnow Mfg Co	45 Conway St	Shelburne Falls	MA	01370	JR Anderson Jr	413-625-6331	S	5	<0.1
Aero Metalcraft Inc	PO Box 380	Nashville	AR	71852	J Richard Musgrave	501-845-4075	R	3*	0.1
ML Rongo Inc	4817 W Lake St	Melrose Park	IL	60160	Michael L Rongo	708-343-8820	R	3	<0.1
Pumpkin Ltd	PO Box 61456	Denver	CO	80206	Kea Bardeen	303-722-4442	R	3	<0.1
Knives and Blades Inc	11411 W Theodore	Milwaukee	WI	53214	John Hoffman	414-258-6100	R	2*	<0.1
Crestwood Cutter Inc	1050 Spice Islands	Sparks	NV	89431	Kenneth M Ferjo	702-331-1930	R	1	<0.1
R Murphy Co	13 Groton Harvard	Ayer	MA	01432	DA Bethke	508-772-3481	R	1	<0.1
L Hardy Co	17 Mill St	Worcester	MA	01603	NL Monks	508-756-1511	R	1	0.1

Source: Ward's Business Directory of U.S. Private and Public Companies, Volumes 1 and 2, 1996. The company type code used is as follows: P - Public, R - Private, S - Subsidiary, D - Division, J - Joint Venture, A - Affiliate, G - Group. Sales are in millions of dollars, employees are in thousands. An asterisk (*) indicates an estimated sales volume. The symbol < stands for 'less than'. Company names and addresses are truncated, in some cases, to fit into the available space.

MATERIALS CONSUMED

Material	Quantity	Delivered Cost ($ million)
Materials, ingredients, containers, and supplies	(X)	327.2
Metal bolts, nuts, screws, washers, rivets, and other screw machine products	(X)	8.5
Other fabricated metal products (except castings and forgings)	(X)	13.3
Iron and steel castings (rough and semifinished)	(X)	22.4
Aluminum and aluminum-base alloy castings (rough and semifinished)	(X)	3.5
Other nonferrous castings (rough and semifinished)	(X)	(D)
Iron and steel forgings	(X)	13.0
Steel bars, bar shapes, and plates	(X)	2.6
Steel sheet, strip, and tin mill products	(X)	22.7
Steel wire and wire products	(X)	0.7
All other steel shapes and forms	(X)	16.2
Copper and copper-base alloy shapes and forms	(X)	0.1
Aluminum and aluminum-base alloy shapes and forms	(X)	1.4
Other nonferrous shapes and forms	(X)	(D)
Wood parts, including handles	(X)	7.1
Plastics resins consumed in the form of granules, pellets, powders, liquids, etc.	(X)	28.1
Plastics products (film, sheet, rod, tube, and fabricated shapes, including parts, handles, grips, etc.)	(X)	29.6
Paper and paperboard containers, including shipping sacks and other paper packaging supplies	(X)	43.5
All other materials and components, parts, containers, and supplies	(X)	66.8
Materials, ingredients, containers, and supplies, nsk	(X)	44.3

Source: 1992 Economic Census. Explanation of symbols used: (D): Withheld to avoid disclosure of competitive data; na: Not available; (S): Withheld because statistical norms were not met; (X): Not applicable; (Z): Less than half the unit shown; nec: Not elsewhere classified; nsk: Not specified by kind; - : zero; * : 10-19 percent estimated; ** : 20-29 percent estimated.

PRODUCT SHARE DETAILS

Product or Product Class	% Share	Product or Product Class	% Share
Cutlery	100.00	BX and wire filament cutters)	4.34
Cutlery, scissors, shears, trimmers, and snips	42.88	All other scissors and shears (including hedge and grass	
Table cutlery (knives, forks, spoons, etc.) for food serving		shears and pruners)	17.15
and eating, with handles of materials other than metal	4.87	Other knives (including pocket, pen, and replacement blade	
Kitchen cutlery (including knives, forks, cleavers, butchers,		knives)	29.75
and meat packing cutlery), excluding carving sets	12.09	Cutlery, scissors, shears, trimmers, and snips, nsk	0.46
Other cutlery (including knife blades sold separately)	4.54	Razor blades and razors, except electric	54.55
Household scissors and barber shears, pinking shears, and		Razors, except electric	(D)
tailoring shears	12.36	Razor blades, single and double edge for shaving	(D)
Manicure and pedicure scissors and implements (including		Razor blades for all other uses	3.58
tweezers)	14.46	Razor blades and razors, except electric, nsk	0.01
Metal cutting shears (including aviation and tinners' snips,		Cutlery, nsk	2.57

Source: 1992 Economic Census. The values shown are percent of total shipments in an industry. Values of indented subcategories are summed in the main heading. The symbol (D) appears when data are withheld to prevent disclosure of competitive information. The abbreviation nsk stands for 'not specified by kind' and nec for 'not elsewhere classified'.

INPUTS AND OUTPUTS FOR CUTLERY

Economic Sector or Industry Providing Inputs	%	Sector	Economic Sector or Industry Buying Outputs	%	Sector
Imports	35.4	Foreign	Personal consumption expenditures	69.4	
Blast furnaces & steel mills	10.1	Manufg.	Exports	5.7	Foreign
Advertising	7.0	Services	Retail trade, except eating & drinking	3.8	Trade
Paperboard containers & boxes	4.6	Manufg.	Federal Government purchases, national defense	3.2	Fed Govt
Wholesale trade	4.1	Trade	Beauty & barber shops	2.9	Services
Plastics materials & resins	3.6	Manufg.	S/L Govt. purch., other education & libraries	2.3	S/L Govt
Plating & polishing	3.5	Manufg.	S/L Govt. purch., health & hospitals	2.2	S/L Govt
Electric services (utilities)	3.0	Util.	S/L Govt. purch., natural resource & recreation.	1.1	S/L Govt
Metal stampings, nec	1.7	Manufg.	Meat packing plants	0.6	Manufg.
Copper rolling & drawing	1.6	Manufg.	Eating & drinking places	0.6	Trade
Screw machine and related products	1.4	Manufg.	Wholesale trade	0.6	Trade
Wood products, nec	1.3	Manufg.	Change in business inventories	0.6	In House
Gas production & distribution (utilities)	1.3	Util.	Religious organizations	0.5	Services
Fabricated metal products, nec	1.1	Manufg.	Maintenance of nonfarm buildings nec	0.4	Constr.
Iron & steel forgings	1.1	Manufg.	Nonfarm residential structure maintenance	0.4	Constr.
Abrasive products	1.0	Manufg.	Poultry dressing plants	0.4	Manufg.
Aluminum rolling & drawing	1.0	Manufg.	Business/professional associations	0.4	Services
Machinery, except electrical, nec	1.0	Manufg.	Residential 1-unit structures, nonfarm	0.3	Constr.
Maintenance of nonfarm buildings nec	0.9	Constr.	Sausages & other prepared meats	0.3	Manufg.
Fabricated rubber products, nec	0.9	Manufg.	Federal Government purchases, nondefense	0.3	Fed Govt
Industrial patterns	0.9	Manufg.	S/L Govt. purch., elem. & secondary education	0.3	S/L Govt
Miscellaneous plastics products	0.9	Manufg.	S/L Govt. purch., higher education	0.3	S/L Govt
Petroleum refining	0.8	Manufg.	Cutlery	0.2	Manufg.
Communications, except radio & TV	0.8	Util.	Water transportation	0.2	Util.
Motor freight transportation & warehousing	0.8	Util.	Real estate	0.2	Fin/R.E.
Eating & drinking places	0.8	Trade	Amusement & recreation services nec	0.2	Services
Job training & related services	0.8	Services	S/L Govt. purch., other general government	0.2	S/L Govt
Metal coating & allied services	0.7	Manufg.	Apparel made from purchased materials	0.1	Manufg.
Metal heat treating	0.7	Manufg.	Hotels & lodging places	0.1	Services
Banking	0.6	Fin/R.E.	Job training & related services	0.1	Services
Cutlery	0.4	Manufg.	Membership sports & recreation clubs	0.1	Services
Signs & advertising displays	0.4	Manufg.	Portrait, photographic studios	0.1	Services
U.S. Postal Service	0.4	Gov't	Social services, nec	0.1	Services
Real estate	0.3	Fin/R.E.	S/L Govt. purch., fire	0.1	S/L Govt
Royalties	0.3	Fin/R.E.			
Engineering, architectural, & surveying services	0.3	Services			
Equipment rental & leasing services	0.3	Services			
Legal services	0.3	Services			
Management & consulting services & labs	0.3	Services			
Railroads & related services	0.2	Util.			
Sanitary services, steam supply, irrigation	0.2	Util.			
Accounting, auditing & bookkeeping	0.2	Services			
Computer & data processing services	0.2	Services			
Lubricating oils & greases	0.1	Manufg.			
Machine tools, metal forming types	0.1	Manufg.			
Manifold business forms	0.1	Manufg.			
Nonferrous rolling & drawing, nec	0.1	Manufg.			
Special dies & tools & machine tool accessories	0.1	Manufg.			
Veneer & plywood	0.1	Manufg.			
Air transportation	0.1	Util.			
State & local government enterprises, nec	0.1	Gov't			

Source: Benchmark Input-Output Accounts for the U.S. Economy, 1982, U.S. Department of Commerce, Washington, D.C., July 1991. Data, as reported in the source, are organized by the 1977 SIC structure in use in 1982 but have been matched, as closely as is possible, to the 1987 SIC structure used in this book.

OCCUPATIONS EMPLOYED BY SIC 342 - CUTLERY, HANDTOOLS, AND HARDWARE

Occupation	% of Total 1994	Change to 2005	Occupation	% of Total 1994	Change to 2005
Assemblers, fabricators, & hand workers nec	16.9	-25.4	Industrial machinery mechanics	1.7	-17.9
Machine forming operators, metal & plastic	4.1	-62.7	Grinding machine operators, metal & plastic	1.4	-40.4
Machine tool cutting & forming etc. nec	3.7	-25.4	Machine operators nec	1.3	-34.3
Blue collar worker supervisors	3.5	-30.4	Secretaries, ex legal & medical	1.3	-32.1
Machine tool cutting operators, metal & plastic	3.2	-37.9	Packaging & filling machine operators	1.2	-25.5
Machinists	2.8	-25.4	Punching machine operators, metal & plastic	1.2	-62.7
Hand packers & packagers	2.7	-36.1	Lathe & turning machine tool operators	1.1	-40.4
Combination machine tool operators	2.5	-18.0	Freight, stock, & material movers, hand	1.1	-40.3
Sales & related workers nec	2.4	-25.4	Welding machine setters, operators	1.1	-32.9
Tool & die makers	2.3	-39.8	Bookkeeping, accounting, & auditing clerks	1.1	-44.1
Metal & plastic machine workers nec	2.2	-40.5	Metal molding machine workers	1.1	-18.0
General managers & top executives	2.1	-29.3	Production, planning, & expediting clerks	1.1	-25.4
Inspectors, testers, & graders, precision	2.1	-25.4	General office clerks	1.1	-36.4
Grinders & polishers, hand	2.0	-32.9	Stock clerks	1.0	-39.4
Traffic, shipping, & receiving clerks	1.9	-28.2	Industrial truck & tractor operators	1.0	-25.4
Precision metal workers nec	1.7	-25.4	Industrial production managers	1.0	-25.5

Source: *Industry-Occupation Matrix*, Bureau of Labor Statistics. These data relate to one or more 3-digit SIC industry groups rather than to a single 4-digit SIC. The change reported for each occupation to the year 2005 is a percent of growth or decline as estimated by the Bureau of Labor Statistics. The abbreviation nec stands for 'not elsewhere classified'.

LOCATION BY STATE AND REGIONAL CONCENTRATION

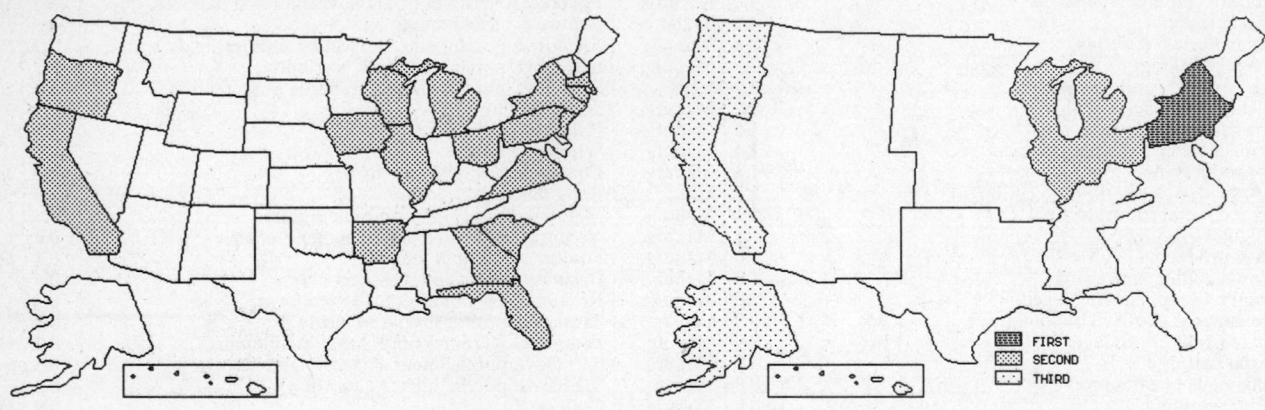

FIRST
SECOND
THIRD

INDUSTRY DATA BY STATE

State	Establish-ments	Shipments			Employment				Cost as % of Shipments	Investment per Employee ($)
		Total ($ mil)	% of U.S.	Per Establ.	Total Number	% of U.S.	Per Establ.	Wages ($/hour)		
New York	16	160.4	10.6	10.0	1,600	14.3	100	9.00	41.8	2,000
California	9	77.3	5.1	8.6	800	7.1	89	12.56	30.5	2,625
Florida	5	39.9	2.6	8.0	300	2.7	60	5.86	44.9	-
Pennsylvania	8	32.9	2.2	4.1	500	4.5	63	9.29	28.3	1,600
Ohio	11	16.8	1.1	1.5	200	1.8	18	6.50	32.7	3,500
Arkansas	5	13.8	0.9	2.8	200	1.8	40	11.00	29.0	-
Michigan	3	7.2	0.5	2.4	100	0.9	33	9.50	36.1	-
New Jersey	9	(D)	-	-	375 *	3.3	42	-	-	-
Illinois	7	(D)	-	-	175 *	1.6	25	-	-	-
Connecticut	6	(D)	-	-	750 *	6.7	125	-	-	-
Oregon	5	(D)	-	-	375 *	3.3	75	-	-	-
Georgia	4	(D)	-	-	750 *	6.7	188	-	-	-
Massachusetts	4	(D)	-	-	3,750 *	33.5	938	-	-	-
South Carolina	3	(D)	-	-	375 *	3.3	125	-	-	-
Wisconsin	3	(D)	-	-	750 *	6.7	250	-	-	-
Iowa	2	(D)	-	-	175 *	1.6	88	-	-	-
Virginia	2	(D)	-	-	750 *	6.7	375	-	-	-

Source: 1992 *Economic Census*. The states are in descending order of shipments or establishments (if shipment data are missing for the majority). The symbol (D) appears when data are withheld to prevent disclosure of competitive information. States marked with (D) are sorted by number of establishments. A dash (-) indicates that the data element cannot be calculated; * indicates the midpoint of a range.

3423 - HAND & EDGE TOOLS

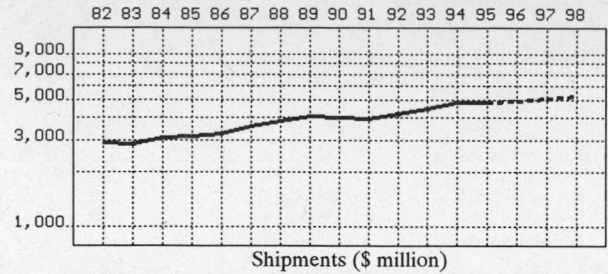

Shipments ($ million)

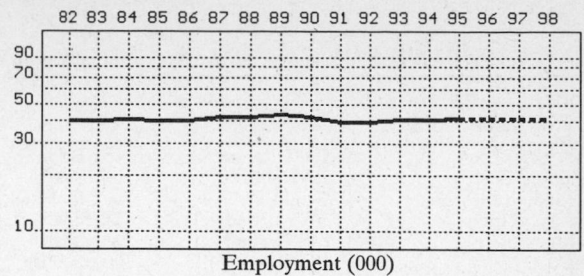

Employment (000)

GENERAL STATISTICS

Year	Companies	Establishments		Employment			Compensation		Production ($ million)			
		Total	with 20 or more employees	Total (000)	Production Workers (000)	Hours (Mil)	Payroll ($ mil)	Wages ($/hr)	Cost of Materials	Value Added by Manufacture	Value of Shipments	Capital Invest.
1982	703	786	322	40.4	29.8	55.4	706.5	8.10	1,103.5	1,785.2	2,915.5	93.2
1983		756	316	40.0	29.6	56.0	735.0	8.25	1,079.2	1,751.9	2,869.6	79.7
1984		726	310	40.8	30.5	59.7	806.8	8.60	1,211.9	1,865.5	3,060.1	94.0
1985		697	303	40.6	30.7	58.2	810.7	9.16	1,183.2	1,948.2	3,136.6	96.3
1986		679	301	40.3	30.2	59.8	839.7	9.20	1,278.1	1,968.6	3,242.3	105.4
1987	731	810	318	41.9	31.3	62.8	922.0	9.47	1,381.7	2,233.2	3,605.6	117.0
1988		773	313	42.0	31.4	63.2	975.2	9.79	1,469.4	2,336.3	3,815.5	118.7
1989		752	323	43.4	30.6	62.5	994.7	10.28	1,517.7	2,519.9	4,058.4	106.0
1990		750	328	42.3	29.9	60.6	991.7	10.26	1,549.1	2,392.8	3,966.7	129.9
1991		765	315	39.6	29.0	58.9	1,006.3	10.62	1,578.6	2,400.8	3,949.5	102.4
1992	828	915	340	39.7	29.3	59.5	1,058.3	10.98	1,608.0	2,584.5	4,208.8	123.4
1993		902	345	40.7	31.2	64.4	1,102.6	11.19	1,703.5	2,761.1	4,458.8	106.6
1994		854P	335P	41.0	31.6	64.5	1,147.3	11.39	1,978.5	2,997.8	4,879.1	131.7
1995		866P	338P	41.2P	30.7P	64.2P	1,178.7P	11.76P	1,949.5P	2,953.8P	4,807.5P	128.8P
1996		878P	340P	41.2P	30.7P	64.8P	1,214.1P	12.05P	2,013.3P	3,050.6P	4,965.0P	131.7P
1997		890P	342P	41.2P	30.8P	65.3P	1,249.6P	12.33P	2,077.2P	3,147.4P	5,122.5P	134.7P
1998		902P	345P	41.3P	30.8P	65.9P	1,285.0P	12.61P	2,141.1P	3,244.1P	5,280.0P	137.7P

Sources: 1982, 1987, 1992 *Economic Census*; *Annual Survey of Manufactures*, 83-86, 88-91, 93-94. Establishment counts for non-Census years are from *County Business Patterns*; establishment values for 83-84 are extrapolations. 'P's show projections by the editors. Industries reclassified in 87 will not have data for prior years.

INDICES OF CHANGE

Year	Companies	Establishments		Employment			Compensation		Production ($ million)			
		Total	with 20 or more employees	Total (000)	Production Workers (000)	Hours (Mil)	Payroll ($ mil)	Wages ($/hr)	Cost of Materials	Value Added by Manufacture	Value of Shipments	Capital Invest.
1982	85	86	95	102	102	93	67	74	69	69	69	76
1983		83	93	101	101	94	69	75	67	68	68	65
1984		79	91	103	104	100	76	78	75	72	73	76
1985		76	89	102	105	98	77	83	74	75	75	78
1986		74	89	102	103	101	79	84	79	76	77	85
1987	88	89	94	106	107	106	87	86	86	86	86	95
1988		84	92	106	107	106	92	89	91	90	91	96
1989		82	95	109	104	105	94	94	94	98	96	86
1990		82	96	107	102	102	94	93	96	93	94	105
1991		84	93	100	99	99	95	97	98	93	94	83
1992	100	100	100	100	100	100	100	100	100	100	100	100
1993		99	101	103	106	108	104	102	106	107	106	86
1994		93P	99P	103	108	108	108	104	123	116	116	107
1995		95P	99P	104P	105P	108P	111P	107P	121P	114P	114P	104P
1996		96P	100P	104P	105P	109P	115P	110P	125P	118P	118P	107P
1997		97P	101P	104P	105P	110P	118P	112P	129P	122P	122P	109P
1998		99P	101P	104P	105P	111P	121P	115P	133P	126P	125P	112P

Sources: Same as General Statistics. Values reflect change from the base year, 1992. Values above 100 mean greater than 92, values below 100 mean less than 92, and a value of 100 in the 82-91 or 93-98 period means same as 92. 'P's mark projections by the editors.

SELECTED RATIOS

For 1994	Avg. of All Manufact.	Analyzed Industry	Index	For 1994	Avg. of All Manufact.	Analyzed Industry	Index
Employees per Establishment	49	48	98	Value Added per Production Worker	134,084	94,867	71
Payroll per Establishment	1,500,273	1,343,753	90	Cost per Establishment	5,045,178	2,317,279	46
Payroll per Employee	30,620	27,983	91	Cost per Employee	102,970	48,256	47
Production Workers per Establishment	34	37	108	Cost per Production Worker	146,988	62,611	43
Wages per Establishment	853,319	860,450	101	Shipments per Establishment	9,576,895	5,714,550	60
Wages per Production Worker	24,861	23,249	94	Shipments per Employee	195,460	119,002	61
Hours per Production Worker	2,056	2,041	99	Shipments per Production Worker	279,017	154,402	55
Wages per Hour	12.09	11.39	94	Investment per Establishment	321,011	154,251	48
Value Added per Establishment	4,602,255	3,511,114	76	Investment per Employee	6,552	3,212	49
Value Added per Employee	93,930	73,117	78	Investment per Production Worker	9,352	4,168	45

Sources: Same as General Statistics. The 'Average of All Manufacturing' column represents the average of all manufacturing industries reported for the most recent complete year available. The Index shows the relationship between the Average and the Analyzed Industry. For example, 100 means that they are equal; 500 that the Analyzed Industry is five times the average; 50 means that the Analyzed Industry is half the national average. The abbreviation 'na' is used to show that data are 'not available'.

1311

LEADING COMPANIES   Number shown: **75**   Total sales ($ mil): **6,201**   Total employment (000): **51.6**

Company Name	Address				CEO Name	Phone	Co. Type	Sales ($ mil)	Empl. (000)
Stanley Works	1000 Stanley Dr	New Britain	CT	06053	Richard H Ayers	203-225-5111	P	2,511	20.0
Snap-on Tools Corp	2801 80th St	Kenosha	WI	53141	Robert A Cornog	414-656-5200	P	1,194	9.0
Stanley-Bostitch Inc	Briggs Dr	East Greenwich	RI	02818	Bruce Behnke	401-884-2500	S	400	2.5
Senco Products Inc	8485 Broadwell Rd	Cincinnati	OH	45244	William Hess	513-388-2000	S	180	1.9
Sencorp Inc	8485 Broadwell Rd	Cincinnati	OH	45244	William Hess	513-388-2000	R	180	1.9
O Ames Co	PO Box 1774	Parkersburg	WV	26101	Rick Keup	304-424-3000	S	150	1.1
Klein Tools Inc	PO Box 599033	Chicago	IL	60659	Michael S Klein	708-677-9500	R	100	1.2
Ridge Tool Co	400 Clark St	Elyria	OH	44036	Patrick J Sly	216-323-5581	S	100	1.0
Stanley-Proto Industrial Tools	14117 Industrial Pk	Covington	GA	30209	J Nichol	404-787-3800	S	75*	0.3
Mac Tools Inc	PO Box 370	Wash Ct House	OH	43160	Charles Blossom	614-335-4112	S	67*	0.6
Wilton Corp	300 S Hicks Rd	Palatine	IL	60067	Alexander J Vogl	708-934-6000	R	65	0.5
Arrow Fastener Company Inc	271 Mayhill St	Saddle Brook	NJ	07663	Allan Abrams	201-843-6900	R	60*	0.5
American Tool Companies Inc	301 S 13th St	Lincoln	NE	68508	AD Petersen	402-435-3300	R	50	0.5
Melnor Industries Inc	1 Carol Pl	Moonachie	NJ	07074	Phillip S Griffin	201-641-5000	R	45*	0.3
Red Devil Inc	2400 Vauxhall Rd	Union	NJ	07083	D Mac Pherson	908-688-6900	R	44	0.4
Republic Tool & Mfg Co	6212 Abeto	Carlsbad	CA	92009	Guy W McRoskey	619-438-7835	R	43*	0.4
Universal Tool and Stamping	PO Box 100	Butler	IN	46721	Larry Kipp	219-868-2147	S	42	0.4
Fletcher Terry Co	65 Spring Ln	Farmington	CT	06032	Terry B Fletcher	203-677-7331	R	40	0.1
SK Hand Tool Corp	3535 W 47th St	Chicago	IL	60632	IR Forte	312-523-1300	R	40	0.3
Utica Tool Co	PO Box 1807	Orangeburg	SC	29116	Dave Dolak	803-534-7010	S	40*	0.6
General Tools Manufacturing	80 White St	New York	NY	10013	Peter Charnley	212-431-6100	R	38	<0.1
UnionTools Inc	PO Box 1930	Columbus	OH	43216	Gabe Mihaly	614-222-4400	R	32	0.6
Grobet File Company of America	750 Washington Av	Carlstadt	NJ	07072	John Canzoneri	201-939-6700	R	30	0.2
Channellock Inc	1306 S Main St	Meadville	PA	16335	WS Dearment	814-724-8700	R	27	0.5
George W McGuire	3300 PJA Blv	P Bch Gardens	FL	33410	Paul R McGuire	407-626-0600	S	27*	0.3
Lund International Corp	571 Main St	Hudson	MA	01749	William P Dulmaine	508-562-4444	R	25	0.3
Marshalltown Trowel Co	PO Box 738	Marshalltown	IA	50158	LJ McComber	515-753-5999	R	25	0.2
Hannaco Knives and Saws	PO Box 100535	Florence	SC	29501	H Brautigam	803-662-6345	D	23*	0.2
Leatherman Tool Group Inc	PO Box 20595	Portland	OR	97220	Tim Leatherman	503-253-7826	R	23*	0.2
Mann Edge Tool Co	PO Box 351	Lewistown	PA	17044	JA Waddell	717-248-9628	R	22	0.2
Cooper Power Tools	PO Box 952	Dayton	OH	45401		513-222-7871	D	21*	0.2
Moeller Manufacturing Company	43938ymouth Oaks	Plymouth	MI	48170	David Moellering	313-416-0000	R	21	0.2
Armstrong Bros Tool Co	5200 W Armstrong	Chicago	IL	60646	Paul Armstrong	312-763-3333	R	20	0.4
Rigid Products	PO Box 150	Orange	VA	22960	Keith L Smith	703-672-5150	D	20*	0.2
Hutchinson/Mayrath	PO Box 629	Clay Center	KS	67432	Larry Tripp	913-632-3133	D	19	0.2
A and E Manufacturing Co	PO Box 1616	Racine	WI	53401	Dan R Peterson	414-554-2300	R	18	0.2
Turner-Cooper Hand Tools	821 Park Av	Sycamore	IL	60178	Daniel Sheasby	815-895-4545	D	18	0.1
Woodings-Verona Tool Works	PO Box 126	Verona	PA	15147	Gerald Potts	412-828-7902	R	18	0.2
Milbar Corp	530 E Washington	Chagrin Falls	OH	44022	Jack A Bares	216-247-4600	R	17	0.2
Magna Industrial Tools	1001 W Park Rd	Elizabethtown	KY	42701	Tim Miller	502-737-3311	D	16	0.2
Ungar	5620 Knott Av	Buena Park	CA	90621	Robert Stephenson	714-994-2510	D	16*	0.2
General Machine Products	3111 Old Lincoln	Trevose	PA	19053	G Nelson Pfundt	215-357-5500	R	15	0.1
Rugg Manufacturing Company	PO Box 507	Greenfield	MA	01302	Michael A Fritz	413-773-5471	R	15	<0.1
Cal-Van Tool	1500 Walter Av	Fremont	OH	43420	James R Labenne	419-334-2692	D	14	<0.1
Seymour Manufacturing	PO Box 248	Seymour	IN	47274	Berl Grant	812-522-2900	R	14	0.1
Bondhus Corp	PO Box 660	Monticello	MN	55362	John Bondhus	612-295-2162	R	13	0.1
Arrow Pneumatics Inc	500 N Oakwood Rd	Lake Zurich	IL	60047	JR Brown	708-438-9100	R	12	0.1
Products Engineering Corp	PO Box 45009	Los Angeles	CA	90045	Martin Luboviski	213-776-1885	R	12	0.2
Edlund Company Inc	PO Box 929	Burlington	VT	05402	WS Foster	802-862-9661	R	11*	0.1
Estwing Manufacturing Co	2467 8th St	Rockford	IL	61109	Robert Mayer	815-397-9521	R	11*	0.2
Malco Products Inc	55 & 136	Annandale	MN	55302	G Keymer	612-274-8246	R	11	0.1
OK Industries Inc	4 Executive Plz	Yonkers	NY	10701	M Gouldsmith	914-969-6800	R	11*	<0.1
Asko Inc	501 W 7th Av	Homestead	PA	15120	William H Rackoff	412-461-4110	R	10*	0.4
ATI Tools Inc	2425 W Vineyard	Escondido	CA	92029		619-746-8301	S	10	0.1
Nupla Corp	PO Box 1165	Sun Valley	CA	91352	Alfred Carmien	818-768-6800	R	9*	0.1
Quaker Manufacturing Corp	187 Georgetown Rd	Salem	OH	44460	Christopher Smith	216-332-4631	R	9*	<0.1
Consolidated Devices Inc	19220 San Jose Av	City of Industry	CA	91748	Bosko Grabovac	818-965-0668	R	8*	0.1
Council Tool Company Inc	PO Box 165	Lk Waccamaw	NC	28450	John M Council Jr	910-646-3011	R	8*	<0.1
K-Line Industries Inc	315 Garden Av	Holland	MI	49424	DJ Kammeraad	616-396-3564	R	8*	0.2
Swing-A-Way Manufacturing	4100 Beck Av	St Louis	MO	63116	Albert Packer	314-773-1487	R	8*	<0.1
Harrington Tools Inc	4316 Alger St	Los Angeles	CA	90039	EM Harrington	213-245-2142	R	7	0.1
Kedman Co	762 S Redwood Rd	Salt Lake City	UT	84104	G Michael Edwards	801-973-9112	R	7*	0.1
Mayes Brothers Tool Mfg	PO Box 1018	Johnson City	TN	37605	John H Stevens	615-926-6171	S	7	<0.1
Milwaukee Tool & Equip Co	PO Box 2039	Milwaukee	WI	53201	RL Gutenkunst	414-645-0200	R	7	0.1
Skyo Industries Inc	171 Brook Av	Deer Park	NY	11729	A Eugene Anderson	516-586-4702	R	7*	<0.1
United Gasket Corp	1633 S 55th Av	Cicero	IL	60650	Sy Peale	708-656-3700	R	7	<0.1
Kemper Enterprises Inc	PO Box 696	Chino	CA	91710	Herbert H Stampfl	909-627-6191	R	6*	<0.1
Dasco Pro Inc	2215 Kishwaukee St	Rockford	IL	61104	DS Marth	815-962-3727	R	6	<0.1
Janel Inc	1320 Valley Rd	Stirling	NJ	07980	Len Alpert	908-647-5600	R	6	<0.1
H and H Tooling	430 S Navajo St	Denver	CO	80223	John Morgan	303-744-6304	D	5*	<0.1
John J Adams Die Corp	10 Nebraska St	Worcester	MA	01613	R Adams	508-757-3894	R	5*	<0.1
Ken Specialties Inc	PO Box 8017	Wood Dale	IL	60191	Dale Kengott	708-766-6226	R	5	<0.1
Lancaster Knives Inc	165 Court St	Lancaster	NY	14086	John F Cant	716-683-5050	R	5	<0.1
ME Heuck Company Inc	PO Box 23036	Cincinnati	OH	45223	Roger Heuck	513-681-1774	R	5*	0.1
Reed Manufacturing Co	PO Box 1321	Erie	PA	16512	Ralph Wright	814-452-3691	R	5*	0.1

Source: Ward's Business Directory of U.S. Private and Public Companies, Volumes 1 and 2, 1996. The company type code used is as follows: P - Public, R - Private, S - Subsidiary, D - Division, J - Joint Venture, A - Affiliate, G - Group. Sales are in millions of dollars, employees are in thousands. An asterisk (*) indicates an estimated sales volume. The symbol < stands for 'less than'. Company names and addresses are truncated, in some cases, to fit into the available space.

MATERIALS CONSUMED

Material	Quantity	Delivered Cost ($ million)
Materials, ingredients, containers, and supplies	(X)	1,180.5
Metal bolts, nuts, screws, washers, rivets, and other screw machine products	(X)	42.3
Other fabricated metal products (except castings and forgings)	(X)	74.9
Iron and steel castings (rough and semifinished)	(X)	65.1
Aluminum and aluminum-base alloy castings (rough and semifinished)	(X)	11.5
Other nonferrous castings (rough and semifinished)	(X)	5.6
Iron and steel forgings	(X)	30.4
Steel bars, bar shapes, and plates	(X)	136.0
Steel sheet, strip, and tin mill products	(X)	104.7
Steel wire and wire products	(X)	21.4
All other steel shapes and forms	(X)	34.4
Copper and copper-base alloy shapes and forms	(X)	11.9
Aluminum and aluminum-base alloy shapes and forms	(X)	15.4
Other nonferrous shapes and forms	(X)	10.2
Wood parts, including handles	(X)	38.1
Plastics resins consumed in the form of granules, pellets, powders, liquids, etc.	(X)	24.0
Plastics products (film, sheet, rod, tube, and fabricated shapes, including parts, handles, grips, etc.)	(X)	35.3
Paper and paperboard containers, including shipping sacks and other paper packaging supplies	(X)	56.6
All other materials and components, parts, containers, and supplies	(X)	346.2
Materials, ingredients, containers, and supplies, nsk	(X)	116.5

Source: 1992 *Economic Census*. Explanation of symbols used: (D): Withheld to avoid disclosure of competitive data; na: Not available; (S): Withheld because statistical norms were not met; (X): Not applicable; (Z): Less than half the unit shown; nec: Not elsewhere classified; nsk: Not specified by kind; - : zero; * : 10-19 percent estimated; ** : 20-29 percent estimated.

PRODUCT SHARE DETAILS

Product or Product Class	% Share	Product or Product Class	% Share
Hand and edge tools, nec	100.00	Edge tools, hand-operated, nsk	1.11
Mechanics' hand service tools	43.94	Dies and interchangeable cutting tools, for machines and power-driven handtools.	15.67
Mechanics' slip joint pliers	3.13	Steel rule dies (except metal cutting), for machines and power-driven handtools	30.00
Mechanics' solid joint pliers	5.41	Other cutting dies, for use in cutting cloth, paper, leathers, etc. (excluding dies for cutting metal), for machines and power-driven handtools	28.81
Mechanics' ball peen hammers	0.64		
Mechanics' socket wrenches, including sockets, drives (ratchet and other), extensions, etc., for hand-operated socket wrenches	22.23	Veneer knives and chipper knives (except metal cutting), for machines and power-driven handtools	2.74
Mechanics' open-end and box wrenches	2.78	All other machines knives	17.29
Mechanics' torque wrenches	1.95	Countersink, drill, and router bits for woodcutting	14.35
Mechanics' adjustable wrenches, including pipe wrenches	4.59	All other woodcutting machine tools (including milling cutters)	6.63
Mechanics' combination open-end and box wrenches	4.60		
All other mechanics' wrenches	3.45	Dies and interchangeable cutting tools, for machines and power-driven handtools, nsk	0.17
Mechanics' screwdrivers	9.18	Other handtools, nec	25.54
Automobile jacks, mechanical (excluding hydraulic and pneumatic)	5.39	Shovels, spades, scoops, telegraph spoons, and scrapers	9.90
Mechanics' tools for automotive use (excluding jacks, but including wheel or gear pullers, valve tools, body or fender tools, etc.)	6.38	Light forged hammers, less than 4 lb (excluding ball peen hammers)	7.91
Other mechanics' hand service tools (including blow torches)	29.37	Heavy forged handtools, sledges (4 lb or more), picks, pick mattocks, and mauls	2.18
Mechanics' hand service tools, nsk	0.91	Steel handtool goods (forks, hoes, rakes, weeders, etc.)	11.10
Edge tools, hand-operated	8.38	Soldering irons (electric)	3.92
Axes, adzes, hatchets, and chisels (hand-operated)	15.56	Clamps and vises (excluding machine tool accessories)	5.95
Professional and craft edge handtools (palette knives, paperhanger knives, putty knives, scrapers, trimmers, etc.)	64.50	Other handtools (including woodworking and metalworking files and rasps, including precision files, except edge tools)	58.64
Kitchen hand-operated edge tools (including nonelectric can openers, peelers, slicers, dicers, etc.)	2.25	Other handtools, nec, nsk	0.41
Other hand-operated edge tools (including agricultural and forestry edge handtools)	16.57	Hand and edge tools, nec, nsk	6.48

Source: 1992 *Economic Census*. The values shown are percent of total shipments in an industry. Values of indented subcategories are summed in the main heading. The symbol (D) appears when data are withheld to prevent disclosure of competitive information. The abbreviation nsk stands for 'not specified by kind' and nec for 'not elsewhere classified'.

INPUTS AND OUTPUTS FOR HAND & EDGE TOOLS, NEC

Economic Sector or Industry Providing Inputs	%	Sector	Economic Sector or Industry Buying Outputs	%	Sector
Imports	26.0	Foreign	Personal consumption expenditures	23.4	
Blast furnaces & steel mills	19.4	Manufg.	Automotive repair shops & services	13.0	Services
Wholesale trade	10.3	Trade	Retail trade, except eating & drinking	10.9	Trade
Advertising	4.8	Services	Exports	6.8	Foreign
Communications, except radio & TV	3.4	Util.	Wholesale trade	2.9	Trade
Plating & polishing	2.8	Manufg.	Meat animals	2.1	Agric.
Electric services (utilities)	2.4	Util.	Federal Government purchases, national defense	1.8	Fed Govt
Iron & steel foundries	2.3	Manufg.	Feed grains	1.7	Agric.
Paperboard containers & boxes	2.3	Manufg.	Motor freight transportation & warehousing	1.5	Util.
Fabricated metal products, nec	2.1	Manufg.	S/L Govt. purch., natural resource & recreation.	1.4	S/L Govt
Iron & steel forgings	2.0	Manufg.	Air transportation	1.2	Util.
Wood products, nec	1.5	Manufg.	S/L Govt. purch., other general government	1.0	S/L Govt
Motor freight transportation & warehousing	1.1	Util.	Railroads & related services	0.9	Util.
Machinery, except electrical, nec	1.0	Manufg.	Dairy farm products	0.8	Agric.
Miscellaneous plastics products	1.0	Manufg.	Miscellaneous repair shops	0.8	Services
Nonferrous forgings	1.0	Manufg.	Federal Government purchases, nondefense	0.8	Fed Govt
Gas production & distribution (utilities)	1.0	Util.	Aircraft & missile engines & engine parts	0.7	Manufg.
Abrasive products	0.9	Manufg.	Oil bearing crops	0.6	Agric.
Aluminum castings	0.8	Manufg.	Veneer & plywood	0.6	Manufg.
Plastics materials & resins	0.8	Manufg.	Wood household furniture	0.6	Manufg.
Eating & drinking places	0.8	Trade	S/L Govt. purch., elem. & secondary education	0.6	S/L Govt
Hand & edge tools, nec	0.7	Manufg.	S/L Govt. purch., highways	0.6	S/L Govt
Metal heat treating	0.7	Manufg.	Food grains	0.5	Agric.
Metal stampings, nec	0.7	Manufg.	Industrial buildings	0.5	Constr.
Banking	0.7	Fin/R.E.	Logging camps & logging contractors	0.5	Manufg.
Equipment rental & leasing services	0.7	Services	Millwork	0.5	Manufg.
Maintenance of nonfarm buildings nec	0.6	Constr.	Motor vehicle parts & accessories	0.5	Manufg.
Petroleum refining	0.6	Manufg.	Motor vehicles & car bodies	0.5	Manufg.
Metal coating & allied services	0.5	Manufg.	Telephone & telegraph apparatus	0.5	Manufg.
Aluminum rolling & drawing	0.4	Manufg.	Upholstered household furniture	0.5	Manufg.
Machine tools, metal cutting types	0.4	Manufg.	Electrical repair shops	0.5	Services
Special dies & tools & machine tool accessories	0.4	Manufg.	Poultry & eggs	0.4	Agric.
Railroads & related services	0.4	Util.	Vegetables	0.4	Agric.
Real estate	0.4	Fin/R.E.	Crude petroleum & natural gas	0.4	Mining
U.S. Postal Service	0.4	Gov't	Electric utility facility construction	0.4	Constr.
Engineering, architectural, & surveying services	0.3	Services	Office buildings	0.4	Constr.
Legal services	0.3	Services	Apparel made from purchased materials	0.4	Manufg.
Management & consulting services & labs	0.3	Services	Hand & edge tools, nec	0.4	Manufg.
Copper rolling & drawing	0.2	Manufg.	Transit & bus transportation	0.4	Util.
Screw machine and related products	0.2	Manufg.	Real estate	0.4	Fin/R.E.
Royalties	0.2	Fin/R.E.	Automotive rental & leasing, without drivers	0.4	Services
Accounting, auditing & bookkeeping	0.2	Services	Local government passenger transit	0.4	Gov't
Computer & data processing services	0.2	Services	Fruits	0.3	Agric.
Hotels & lodging places	0.2	Services	Coal	0.3	Mining
Job training & related services	0.2	Services	Sewer system facility construction	0.3	Constr.
Lubricating oils & greases	0.1	Manufg.	Aircraft	0.3	Manufg.
Machine tools, metal forming types	0.1	Manufg.	Cutstone & stone products	0.3	Manufg.
Manifold business forms	0.1	Manufg.	Internal combustion engines, nec	0.3	Manufg.
Air transportation	0.1	Util.	Lawn & garden equipment	0.3	Manufg.
Insurance carriers	0.1	Fin/R.E.	Wood kitchen cabinets	0.3	Manufg.
			Electric services (utilities)	0.3	Util.
			Amusement & recreation services nec	0.3	Services
			Automobile parking & car washes	0.3	Services
			S/L Govt. purch., higher education	0.3	S/L Govt
			S/L Govt. purch., other education & libraries	0.3	S/L Govt
			Cotton	0.2	Agric.
			Greenhouse & nursery products	0.2	Agric.
			Tobacco	0.2	Agric.
			Maintenance of nonfarm buildings nec	0.2	Constr.
			Nonfarm residential structure maintenance	0.2	Constr.
			Petroleum & natural gas well drilling	0.2	Constr.
			Motors & generators	0.2	Manufg.
			Paperboard containers & boxes	0.2	Manufg.
			Special dies & tools & machine tool accessories	0.2	Manufg.
			Wiring devices	0.2	Manufg.
			Wood partitions & fixtures	0.2	Manufg.
			Freight forwarders	0.2	Util.
			Hotels & lodging places	0.2	Services
			Job training & related services	0.2	Services
			Membership sports & recreation clubs	0.2	Services
			Portrait, photographic studios	0.2	Services
			State & local government enterprises, nec	0.2	Gov't
			S/L Govt. purch., sanitation	0.2	S/L Govt
			Farm service facilities	0.1	Constr.
			Maintenance of electric utility facilities	0.1	Constr.
			Residential 1-unit structures, nonfarm	0.1	Constr.
			Aircraft & missile equipment, nec	0.1	Manufg.
			Aluminum rolling & drawing	0.1	Manufg.

Continued on next page.

INPUTS AND OUTPUTS FOR HAND & EDGE TOOLS, NEC - Continued

Economic Sector or Industry Providing Inputs	%	Sector	Economic Sector or Industry Buying Outputs	%	Sector
			Blast furnaces & steel mills	0.1	Manufg.
			Fabricated plate work (boiler shops)	0.1	Manufg.
			Structural wood members, nec	0.1	Manufg.
			Communications, except radio & TV	0.1	Util.
			Eating & drinking places	0.1	Trade
			Business services nec	0.1	Services
			Colleges, universities, & professional schools	0.1	Services
			Watch, clock, jewelry, & furniture repair	0.1	Services
			U.S. Postal Service	0.1	Gov't
			S/L Govt. purch., health & hospitals	0.1	S/L Govt
			S/L Govt. purch., police	0.1	S/L Govt

Source: Benchmark Input-Output Accounts for the U.S. Economy, 1982, U.S. Department of Commerce, Washington, D.C., July 1991. Data, as reported in the source, are organized by the 1977 SIC structure in use in 1982 but have been matched, as closely as is possible, to the 1987 SIC structure used in this book.

OCCUPATIONS EMPLOYED BY SIC 342 - CUTLERY, HANDTOOLS, AND HARDWARE

Occupation	% of Total 1994	Change to 2005	Occupation	% of Total 1994	Change to 2005
Assemblers, fabricators, & hand workers nec	16.9	-25.4	Industrial machinery mechanics	1.7	-17.9
Machine forming operators, metal & plastic	4.1	-62.7	Grinding machine operators, metal & plastic	1.4	-40.4
Machine tool cutting & forming etc. nec	3.7	-25.4	Machine operators nec	1.3	-34.3
Blue collar worker supervisors	3.5	-30.4	Secretaries, ex legal & medical	1.3	-32.1
Machine tool cutting operators, metal & plastic	3.2	-37.9	Packaging & filling machine operators	1.2	-25.5
Machinists	2.8	-25.4	Punching machine operators, metal & plastic	1.2	-62.7
Hand packers & packagers	2.7	-36.1	Lathe & turning machine tool operators	1.1	-40.4
Combination machine tool operators	2.5	-18.0	Freight, stock, & material movers, hand	1.1	-40.3
Sales & related workers nec	2.4	-25.4	Welding machine setters, operators	1.1	-32.9
Tool & die makers	2.3	-39.8	Bookkeeping, accounting, & auditing clerks	1.1	-44.1
Metal & plastic machine workers nec	2.2	-40.5	Metal molding machine workers	1.1	-18.0
General managers & top executives	2.1	-29.3	Production, planning, & expediting clerks	1.1	-25.4
Inspectors, testers, & graders, precision	2.1	-25.4	General office clerks	1.1	-36.4
Grinders & polishers, hand	2.0	-32.9	Stock clerks	1.0	-39.4
Traffic, shipping, & receiving clerks	1.9	-28.2	Industrial truck & tractor operators	1.0	-25.4
Precision metal workers nec	1.7	-25.4	Industrial production managers	1.0	-25.5

Source: Industry-Occupation Matrix, Bureau of Labor Statistics. These data relate to one or more 3-digit SIC industry groups rather than to a single 4-digit SIC. The change reported for each occupation to the year 2005 is a percent of growth or decline as estimated by the Bureau of Labor Statistics. The abbreviation nec stands for 'not elsewhere classified'.

LOCATION BY STATE AND REGIONAL CONCENTRATION

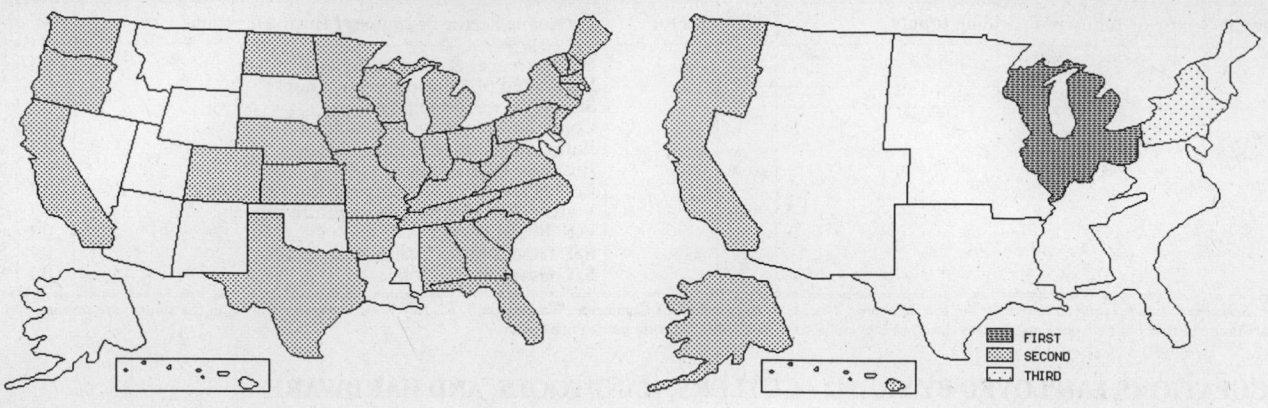

FIRST
SECOND
THIRD

INDUSTRY DATA BY STATE

| State | Establish-ments | Shipments | | | Employment | | | | Cost as % of Shipments | Investment per Employee ($) |
		Total ($ mil)	% of U.S.	Per Establ.	Total Number	% of U.S.	Per Establ.	Wages ($/hour)		
Ohio	77	565.0	13.4	7.3	4,400	11.1	57	12.61	36.3	3,227
Illinois	64	360.7	8.6	5.6	3,900	9.8	61	12.07	32.3	3,436
Michigan	57	262.7	6.2	4.6	1,700	4.3	30	9.64	56.0	2,353
Minnesota	33	241.6	5.7	7.3	2,000	5.0	61	11.56	50.5	2,450
North Carolina	24	229.2	5.4	9.6	1,800	4.5	75	10.76	48.5	4,000
California	122	208.6	5.0	1.7	2,400	6.0	20	9.85	38.8	1,667
Connecticut	30	199.0	4.7	6.6	1,300	3.3	43	13.69	30.7	-
South Carolina	10	183.4	4.4	18.3	1,900	4.8	190	10.27	31.2	2,895
New York	49	180.3	4.3	3.7	1,600	4.0	33	8.62	46.0	2,313
Massachusetts	44	143.9	3.4	3.3	1,600	4.0	36	11.60	32.5	3,875
New Jersey	33	136.6	3.2	4.1	1,800	4.5	55	7.37	41.7	1,500
Pennsylvania	54	135.3	3.2	2.5	1,800	4.5	33	11.08	32.7	1,722
Indiana	26	120.1	2.9	4.6	1,100	2.8	42	9.79	43.1	5,545
Texas	26	109.4	2.6	4.2	1,000	2.5	38	8.58	44.7	1,500
Arkansas	6	57.4	1.4	9.6	800	2.0	133	10.06	42.5	-
Kansas	11	56.4	1.3	5.1	500	1.3	45	10.40	46.6	800
Georgia	11	49.8	1.2	4.5	500	1.3	45	9.83	33.1	2,600
Oregon	8	27.6	0.7	3.5	200	0.5	25	10.67	26.4	-
Florida	21	23.1	0.5	1.1	300	0.8	14	8.75	23.8	667
North Dakota	3	12.1	0.3	4.0	100	0.3	33	7.50	33.9	-
Maine	6	9.7	0.2	1.6	100	0.3	17	11.00	38.1	1,000
Wisconsin	32	(D)	-	-	1,750 *	4.4	55	-	-	-
Missouri	28	(D)	-	-	750 *	1.9	27	-	-	3,067
Washington	18	(D)	-	-	175 *	0.4	10	-	-	4,000
Tennessee	14	(D)	-	-	1,750 *	4.4	125	-	-	-
Kentucky	13	(D)	-	-	375 *	0.9	29	-	-	-
Colorado	11	(D)	-	-	750 *	1.9	68	-	-	-
Iowa	10	(D)	-	-	375 *	0.9	38	-	-	-
Virginia	10	(D)	-	-	175 *	0.4	18	-	-	7,429
New Hampshire	7	(D)	-	-	175 *	0.4	25	-	-	-
Nebraska	6	(D)	-	-	750 *	1.9	125	-	-	-
West Virginia	4	(D)	-	-	750 *	1.9	188	-	-	-
Vermont	3	(D)	-	-	175 *	0.4	58	-	-	-
Alabama	2	(D)	-	-	750 *	1.9	375	-	-	-

Source: 1992 *Economic Census*. The states are in descending order of shipments or establishments (if shipment data are missing for the majority). The symbol (D) appears when data are withheld to prevent disclosure of competitive information. States marked with (D) are sorted by number of establishments. A dash (-) indicates that the data element cannot be calculated; * indicates the midpoint of a range.

3425 - HAND SAWS & SAW BLADES

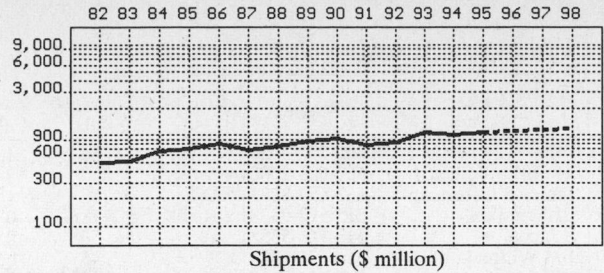

Shipments ($ million)

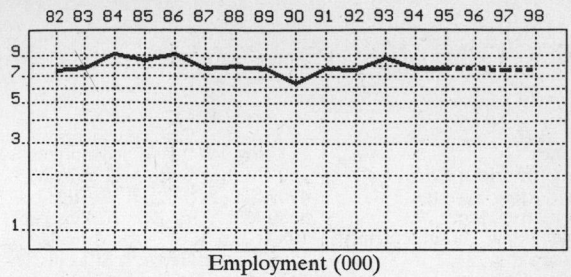

Employment (000)

GENERAL STATISTICS

Year	Com-panies	Establishments		Employment			Compensation		Production ($ million)			
		Total	with 20 or more employees	Total (000)	Production Workers (000)	Hours (Mil)	Payroll ($ mil)	Wages ($/hr)	Cost of Materials	Value Added by Manufacture	Value of Shipments	Capital Invest.
1982	120	136	63	7.5	5.3	10.6	135.5	7.95	227.8	255.0	498.2	24.2
1983		133	60	7.8	5.6	11.0	145.1	8.24	211.5	300.6	509.3	13.3
1984		130	57	9.3	6.8	13.5	191.5	9.05	270.7	394.2	657.0	17.8
1985		127	55	8.5	6.3	13.0	183.0	8.81	270.1	417.6	701.3	28.1
1986		128	59	9.3	7.0	14.1	215.6	9.80	323.3	497.8	805.6	14.2
1987	128	139	61	7.8	5.7	11.2	179.3	10.34	270.6	417.7	682.5	22.9
1988		137	57	7.9	5.8	11.7	197.8	10.54	283.9	459.8	748.2	19.3
1989		136	62	7.8	6.3	12.6	210.4	10.78	355.1	535.5	874.2	40.2
1990		127	49	6.4	6.1	11.6	224.3	11.79	379.5	540.3	916.8	38.4
1991		132	53	7.8	5.5	11.2	200.9	10.71	368.3	449.4	789.6	55.5
1992	128	139	57	7.6	5.3	11.1	218.4	11.65	341.1	507.2	847.2	34.2
1993		137	58	8.8	5.9	12.1	251.8	12.20	445.1	632.0	1,075.8	33.2
1994		136P	55P	7.8	5.6	12.2	237.2	11.70	419.9	601.7	1,026.0	38.1
1995		136P	54P	7.7P	5.7P	11.9P	251.4P	12.69P	436.1P	624.9P	1,065.5P	44.4P
1996		136P	54P	7.7P	5.7P	11.8P	258.8P	13.04P	452.8P	648.8P	1,106.4P	46.6P
1997		137P	53P	7.6P	5.6P	11.8P	266.2P	13.38P	469.5P	672.8P	1,147.2P	48.8P
1998		137P	53P	7.6P	5.6P	11.8P	273.7P	13.73P	486.2P	696.8P	1,188.1P	51.0P

Sources: 1982, 1987, 1992 *Economic Census*; *Annual Survey of Manufactures*, 83-86, 88-91, 93-94. Establishment counts for non-Census years are from *County Business Patterns*; establishment values for 83-84 are extrapolations. 'P's show projections by the editors. Industries reclassified in 87 will not have data for prior years.

INDICES OF CHANGE

Year	Com-panies	Establishments		Employment			Compensation		Production ($ million)			
		Total	with 20 or more employees	Total (000)	Production Workers (000)	Hours (Mil)	Payroll ($ mil)	Wages ($/hr)	Cost of Materials	Value Added by Manufacture	Value of Shipments	Capital Invest.
1982	94	98	111	99	100	95	62	68	67	50	59	71
1983		96	105	103	106	99	66	71	62	59	60	39
1984		94	100	122	128	122	88	78	79	78	78	52
1985		91	96	112	119	117	84	76	79	82	83	82
1986		92	104	122	132	127	99	84	95	98	95	42
1987	100	100	107	103	108	101	82	89	79	82	81	67
1988		99	100	104	109	105	91	90	83	91	88	56
1989		98	109	103	119	114	96	93	104	106	103	118
1990		91	86	84	115	105	103	101	111	107	108	112
1991		95	93	103	104	101	92	92	108	89	93	162
1992	100	100	100	100	100	100	100	100	100	100	100	100
1993		99	102	116	111	109	115	105	130	125	127	97
1994		98P	96P	103	106	110	109	100	123	119	121	111
1995		98P	95P	102P	107P	107P	115P	109P	128P	123P	126P	130P
1996		98P	94P	101P	107P	107P	118P	112P	133P	128P	131P	136P
1997		98P	93P	100P	106P	106P	122P	115P	138P	133P	135P	143P
1998		99P	92P	100P	106P	106P	125P	118P	143P	137P	140P	149P

Sources: Same as General Statistics. Values reflect change from the base year, 1992. Values above 100 mean greater than 92, values below 100 mean less than 92, and a value of 100 in the 82-91 or 93-98 period means same as 92. 'P's mark projections by the editors.

SELECTED RATIOS

For 1994	Avg. of All Manufact.	Analyzed Industry	Index	For 1994	Avg. of All Manufact.	Analyzed Industry	Index
Employees per Establishment	49	57	117	Value Added per Production Worker	134,084	107,446	80
Payroll per Establishment	1,500,273	1,747,817	116	Cost per Establishment	5,045,178	3,094,049	61
Payroll per Employee	30,620	30,410	99	Cost per Employee	102,970	53,833	52
Production Workers per Establishment	34	41	120	Cost per Production Worker	146,988	74,982	51
Wages per Establishment	853,319	1,051,785	123	Shipments per Establishment	9,576,895	7,560,121	79
Wages per Production Worker	24,861	25,489	103	Shipments per Employee	195,460	131,538	67
Hours per Production Worker	2,056	2,179	106	Shipments per Production Worker	279,017	183,214	66
Wages per Hour	12.09	11.70	97	Investment per Establishment	321,011	280,741	87
Value Added per Establishment	4,602,255	4,433,650	96	Investment per Employee	6,552	4,885	75
Value Added per Employee	93,930	77,141	82	Investment per Production Worker	9,352	6,804	73

Sources: Same as General Statistics. The 'Average of All Manufacturing' column represents the average of all manufacturing industries reported for the most recent complete year available. The Index shows the relationship between the Average and the Analyzed Industry. For example, 100 means that they are equal; 500 that the Analyzed Industry is five times the average; 50 means that the Analyzed Industry is half the national average. The abbreviation 'na' is used to show that data are 'not available'.

LEADING COMPANIES   Number shown: **25**   Total sales ($ mil): **1,597**   Total employment (000): **12.7**

Company Name	Address				CEO Name	Phone	Co. Type	Sales ($ mil)	Empl. (000)
Blount Inc	PO Box 949	Montgomery	AL	36101	John M Panettiere	205-272-8020	P	588	4.6
Vermont American Corp	100 E Liberty St	Louisville	KY	40202	Timothy T Shea	502-625-2000	S	450*	3.8
Oregon Cutting Systems	4909 SE	Portland	OR	97222	James S Osterman	503-653-8881	D	180	1.0
RAF Industries Inc	8380 Old York Rd	Elkins Park	PA	19027	Robert Fox	215-572-0738	R	78*	0.7
Great Neck Saw Manufacturing	165 E 2nd St	Mineola	NY	11501	S Jacoff	516-746-5352	R	50*	0.5
Kasco Corp	1569 Tower Grove	St Louis	MO	63110	Jim Whiteaker	314-771-1550	S	45	0.3
MK Morse Co	PO Box 8677	Canton	OH	44711	MK Morse	216-453-8187	R	32	0.3
Sandvik Milford Corp	PO Box 817	Branford	CT	06405	Peter Renwick	203-481-4281	R	30	0.3
Contour Saws Inc	1217 Thacker St	Des Plaines	IL	60016	ML Wilkie	708-824-1146	R	25	0.2
Southern Saw Service Inc	PO Box 11000	Atlanta	GA	30310	Joe H Hall	404-752-6000	R	21	0.2
Armstrong-Blum Mfg Co	1441 Business Ctr	Mt Prospect	IL	60056	Stanley A Woleben	708-803-4000	R	20	0.2
Anderson Products Inc	1040 Southbridge St	Worcester	MA	01610	Alex Vogl	508-755-6100	S	15	0.1
Allway Tools Inc	1255 Seabury Av	Bronx	NY	10462	Donald Gringer	718-792-3636	R	10	0.1
Carlton Co	PO Box 68309	Milwaukie	OR	97268	Ray R Carlton	503-659-8911	R	10	0.1
D and D Saw Works Inc	6162 Mission Gorge	San Diego	CA	92120	Georgia Dutro	619-280-9320	R	7	<0.1
Spartan Saw Works	PO Box 527	Agawam	MA	01001	Bob Whitfield	413-786-9665	D	6	<0.1
Cardinal Industries Inc	PO Box 450	Conshohocken	PA	19428	EA Zuzelo	215-825-2200	R	5	<0.1
Relton Corp	PO Box 779	Arcadia	CA	91066	WC Kinard	818-446-8201	R	5	0.1
California Saw and Knife Works	721 Brannan St	San Francisco	CA	94103	Warren M Bird	415-861-0644	R	4*	<0.1
Disston Precision Inc	6795 State Rd	Philadelphia	PA	19135	Joe Lukes	215-338-1203	S	4*	<0.1
Primark Tool Group	1350 S 15th St	Louisville	KY	40210	Ralph Cox	502-635-8100	S	4*	<0.1
Drake Corp	2723 Ivanhoe Av	St Louis	MO	63139	Jim Braunecker	314-645-3539	R	3*	<0.1
Unicut Corp	1770 W Berteau St	Chicago	IL	60613	Gerhard G Kolb	312-525-4210	R	3	<0.1
Green Mountain Products Inc	Muller Park	Norwalk	CT	06351	Robert Johnson	203-846-9505	R	1	<0.1
ET Lippert Saw Co	608 Lincoln Av	Pittsburgh	PA	15209	RD Pfischner	412-821-6400	R	1*	<0.1

Source: *Ward's Business Directory of U.S. Private and Public Companies*, Volumes 1 and 2, 1996. The company type code used is as follows: P - Public, R - Private, S - Subsidiary, D - Division, J - Joint Venture, A - Affiliate, G - Group. Sales are in millions of dollars, employees are in thousands. An asterisk (*) indicates an estimated sales volume. The symbol < stands for 'less than'. Company names and addresses are truncated, in some cases, to fit into the available space.

MATERIALS CONSUMED

Material	Quantity	Delivered Cost ($ million)
Materials, ingredients, containers, and supplies	(X)	259.5
Metal bolts, nuts, screws, washers, rivets, and other screw machine products	(X)	1.5
Other fabricated metal products (except castings and forgings)	(X)	9.5
Iron and steel castings (rough and semifinished)	(X)	3.8
Aluminum and aluminum-base alloy castings (rough and semifinished)	(X)	(D)
Other nonferrous castings (rough and semifinished)	(X)	7.4
Steel bars, bar shapes, and plates	(X)	10.3
Steel sheet, strip, and tin mill products	(X)	74.6
Steel wire and wire products	(X)	5.3
All other steel shapes and forms	(X)	10.1
Copper and copper-base alloy shapes and forms	(X)	(D)
Aluminum and aluminum-base alloy shapes and forms	(X)	(D)
Other nonferrous shapes and forms	(X)	(D)
Wood parts, including handles	(X)	2.2
Plastics resins consumed in the form of granules, pellets, powders, liquids, etc.	(X)	1.9
Plastics products (film, sheet, rod, tube, and fabricated shapes, including parts, handles, grips, etc.)	(X)	5.3
Paper and paperboard containers, including shipping sacks and other paper packaging supplies	(X)	16.4
All other materials and components, parts, containers, and supplies	(X)	82.2
Materials, ingredients, containers, and supplies, nsk	(X)	18.6

Source: 1992 *Economic Census*. Explanation of symbols used: (D): Withheld to avoid disclosure of competitive data; na: Not available; (S): Withheld because statistical norms were not met; (X): Not applicable; (Z): Less than half the unit shown; nec: Not elsewhere classified; nsk: Not specified by kind; - : zero; * : 10-19 percent estimated; ** : 20-29 percent estimated.

PRODUCT SHARE DETAILS

Product or Product Class	% Share	Product or Product Class	% Share
Saw blades and handsaws	100.00	spring temper metal cutting, and high-speed metal cutting)	20.03
Power circular saw blades for woodworking, solid tooth	5.95	Other metalworking power saw blades (saber, reciprocating, etc.)	7.07
Power circular saw blades for woodworking, inserted tooth	10.84		
Power band saw blades for woodworking	5.35	All other power saw blades	11.72
Teeth for inserted power woodworking saws, sold separately	2.82	Hand-operated hacksaws	2.57
All other woodworking power saw blades (scroll, jig, etc.)	14.78	Hand-operated carpenter crosscut saws and ripsaws	3.16
Power circular saw blades for metalworking (including metal teeth and cutting segments sold separately)	6.97	Other handsaws (heavy handsaws, crosscut, buck, miter, coping, pruning, compass, etc., including frames, and blades)	4.25
Power hack saw blades for metalworking	0.92		
Power band saw blades for metalworking (flexible back,			

Source: 1992 *Economic Census*. The values shown are percent of total shipments in an industry. Values of indented subcategories are summed in the main heading. The symbol (D) appears when data are withheld to prevent disclosure of competitive information. The abbreviation nsk stands for 'not specified by kind' and nec for 'not elsewhere classified'.

INPUTS AND OUTPUTS FOR HAND SAWS & SAW BLADES

Economic Sector or Industry Providing Inputs	%	Sector	Economic Sector or Industry Buying Outputs	%	Sector
Blast furnaces & steel mills	29.1	Manufg.	Gross private fixed investment	19.7	Cap Inv
Imports	19.3	Foreign	Logging camps & logging contractors	6.7	Manufg.
Wholesale trade	13.0	Trade	Personal consumption expenditures	5.3	
Miscellaneous plastics products	6.9	Manufg.	Exports	5.3	Foreign
Plating & polishing	4.1	Manufg.	Maintenance of nonfarm buildings nec	3.8	Constr.
Electric services (utilities)	3.9	Util.	Residential 1-unit structures, nonfarm	3.3	Constr.
Advertising	2.1	Services	Nonfarm residential structure maintenance	3.2	Constr.
Paperboard containers & boxes	1.7	Manufg.	Job training & related services	2.1	Services
Communications, except radio & TV	1.5	Util.	Residential additions/alterations, nonfarm	1.9	Constr.
Abrasive products	1.0	Manufg.	Sewer system facility construction	1.9	Constr.
Machinery, except electrical, nec	1.0	Manufg.	S/L Govt. purch., highways	1.8	S/L Govt
Gas production & distribution (utilities)	1.0	Util.	S/L Govt. purch., other education & libraries	1.8	S/L Govt
Real estate	1.0	Fin/R.E.	Farm service facilities	1.6	Constr.
Maintenance of nonfarm buildings nec	0.9	Constr.	Federal Government purchases, national defense	1.6	Fed Govt
Hand saws & saw blades	0.9	Manufg.	Maintenance of farm residential buildings	1.5	Constr.
Motor freight transportation & warehousing	0.9	Util.	Residential garden apartments	1.5	Constr.
Eating & drinking places	0.9	Trade	Construction of stores & restaurants	1.3	Constr.
Metal coating & allied services	0.8	Manufg.	Ship building & repairing	1.3	Manufg.
Banking	0.8	Fin/R.E.	S/L Govt. purch., elem. & secondary education	1.3	S/L Govt
Metal heat treating	0.7	Manufg.	Power driven hand tools	1.1	Manufg.
Equipment rental & leasing services	0.7	Services	Secondary nonferrous metals	1.1	Manufg.
Iron & steel foundries	0.6	Manufg.	S/L Govt. purch., correction	1.1	S/L Govt
Water supply & sewage systems	0.6	Util.	Paperboard mills	1.0	Manufg.
Special dies & tools & machine tool accessories	0.5	Manufg.	Retail trade, except eating & drinking	1.0	Trade
Wood products, nec	0.5	Manufg.	Screw machine and related products	0.9	Manufg.
Computer & data processing services	0.5	Services	Industrial buildings	0.8	Constr.
Railroads & related services	0.4	Util.	Hardwood dimension & flooring mills	0.8	Manufg.
Sanitary services, steam supply, irrigation	0.4	Util.	Office buildings	0.7	Constr.
U.S. Postal Service	0.4	Gov't	Miscellaneous fabricated wire products	0.7	Manufg.
Iron & steel forgings	0.3	Manufg.	Ordnance & accessories nec	0.7	Manufg.
Engineering, architectural, & surveying services	0.3	Services	Maintenance of sewer facilities	0.6	Constr.
Legal services	0.3	Services	Water supply facility construction	0.6	Constr.
Management & consulting services & labs	0.3	Services	Aircraft	0.6	Manufg.
Aluminum castings	0.2	Manufg.	Fabricated plate work (boiler shops)	0.6	Manufg.
Fabricated rubber products, nec	0.2	Manufg.	Wood containers	0.6	Manufg.
Machine tools, metal cutting types	0.2	Manufg.	Hand saws & saw blades	0.5	Manufg.
Manifold business forms	0.2	Manufg.	Pulp mills	0.5	Manufg.
Petroleum refining	0.2	Manufg.	Sheet metal work	0.5	Manufg.
Accounting, auditing & bookkeeping	0.2	Services	Wholesale trade	0.5	Trade
Gaskets, packing & sealing devices	0.1	Manufg.	Construction of educational buildings	0.4	Constr.
Air transportation	0.1	Util.	Maintenance of farm service facilities	0.4	Constr.
Insurance carriers	0.1	Fin/R.E.	Residential 2-4 unit structures, nonfarm	0.4	Constr.
Royalties	0.1	Fin/R.E.	Fabricated structural metal	0.4	Manufg.
Electrical repair shops	0.1	Services	Metal doors, sash, & trim	0.4	Manufg.
Laundry, dry cleaning, shoe repair	0.1	Services	Federal Government purchases, nondefense	0.4	Fed Govt
			S/L Govt. purch., health & hospitals	0.4	S/L Govt
			S/L Govt. purch., natural resource & recreation.	0.4	S/L Govt
			S/L Govt. purch., other general government	0.4	S/L Govt
			S/L Govt. purch., sanitation	0.4	S/L Govt
			Meat animals	0.3	Agric.
			Construction of hospitals	0.3	Constr.
			Hotels & motels	0.3	Constr.
			Residential high-rise apartments	0.3	Constr.
			Aircraft & missile engines & engine parts	0.3	Manufg.
			Aircraft & missile equipment, nec	0.3	Manufg.
			Blast furnaces & steel mills	0.3	Manufg.
			Boat building & repairing	0.3	Manufg.
			Motor vehicle parts & accessories	0.3	Manufg.
			Motor vehicles & car bodies	0.3	Manufg.
			Structural wood members, nec	0.3	Manufg.
			Wood preserving	0.3	Manufg.
			Wood TV & radio cabinets	0.3	Manufg.
			Railroads & related services	0.3	Util.
			Miscellaneous repair shops	0.3	Services
			S/L Govt. purch., higher education	0.3	S/L Govt
			Feed grains	0.2	Agric.
			Crude petroleum & natural gas	0.2	Mining
			Maintenance of telephone & telegraph facilities	0.2	Constr.
			Maintenance of water supply facilities	0.2	Constr.
			Warehouses	0.2	Constr.
			Aluminum rolling & drawing	0.2	Manufg.
			Household furniture, nec	0.2	Manufg.
			Iron & steel foundries	0.2	Manufg.
			Machinery, except electrical, nec	0.2	Manufg.
			Manufacturing industries, nec	0.2	Manufg.
			Radio & TV communication equipment	0.2	Manufg.
			Special product sawmills, nec	0.2	Manufg.
			Wood household furniture	0.2	Manufg.

Continued on next page.

INPUTS AND OUTPUTS FOR HAND SAWS & SAW BLADES - Continued

Economic Sector or Industry Providing Inputs	%	Sector	Economic Sector or Industry Buying Outputs	%	Sector
			Advertising	0.2	Services
			Dairy farm products	0.1	Agric.
			Construction of dams & reservoirs	0.1	Constr.
			Construction of nonfarm buildings nec	0.1	Constr.
			Architectural metal work	0.1	Manufg.
			Upholstered household furniture	0.1	Manufg.
			Air transportation	0.1	Util.
			Electric services (utilities)	0.1	Util.
			Real estate	0.1	Fin/R.E.
			S/L Govt. purch., fire	0.1	S/L Govt

Source: Benchmark Input-Output Accounts for the U.S. Economy, 1982, U.S. Department of Commerce, Washington, D.C., July 1991. Data, as reported in the source, are organized by the 1977 SIC structure in use in 1982 but have been matched, as closely as is possible, to the 1987 SIC structure used in this book.

OCCUPATIONS EMPLOYED BY SIC 342 - CUTLERY, HANDTOOLS, AND HARDWARE

Occupation	% of Total 1994	Change to 2005	Occupation	% of Total 1994	Change to 2005
Assemblers, fabricators, & hand workers nec	16.9	-25.4	Industrial machinery mechanics	1.7	-17.9
Machine forming operators, metal & plastic	4.1	-62.7	Grinding machine operators, metal & plastic	1.4	-40.4
Machine tool cutting & forming etc. nec	3.7	-25.4	Machine operators nec	1.3	-34.3
Blue collar worker supervisors	3.5	-30.4	Secretaries, ex legal & medical	1.3	-32.1
Machine tool cutting operators, metal & plastic	3.2	-37.9	Packaging & filling machine operators	1.2	-25.5
Machinists	2.8	-25.4	Punching machine operators, metal & plastic	1.2	-62.7
Hand packers & packagers	2.7	-36.1	Lathe & turning machine tool operators	1.1	-40.4
Combination machine tool operators	2.5	-18.0	Freight, stock, & material movers, hand	1.1	-40.3
Sales & related workers nec	2.4	-25.4	Welding machine setters, operators	1.1	-32.9
Tool & die makers	2.3	-39.8	Bookkeeping, accounting, & auditing clerks	1.1	-44.1
Metal & plastic machine workers nec	2.2	-40.5	Metal molding machine workers	1.1	-18.0
General managers & top executives	2.1	-29.3	Production, planning, & expediting clerks	1.1	-25.4
Inspectors, testers, & graders, precision	2.1	-25.4	General office clerks	1.1	-36.4
Grinders & polishers, hand	2.0	-32.9	Stock clerks	1.0	-39.4
Traffic, shipping, & receiving clerks	1.9	-28.2	Industrial truck & tractor operators	1.0	-25.4
Precision metal workers nec	1.7	-25.4	Industrial production managers	1.0	-25.5

Source: Industry-Occupation Matrix, Bureau of Labor Statistics. These data relate to one or more 3-digit SIC industry groups rather than to a single 4-digit SIC. The change reported for each occupation to the year 2005 is a percent of growth or decline as estimated by the Bureau of Labor Statistics. The abbreviation nec stands for 'not elsewhere classified'.

LOCATION BY STATE AND REGIONAL CONCENTRATION

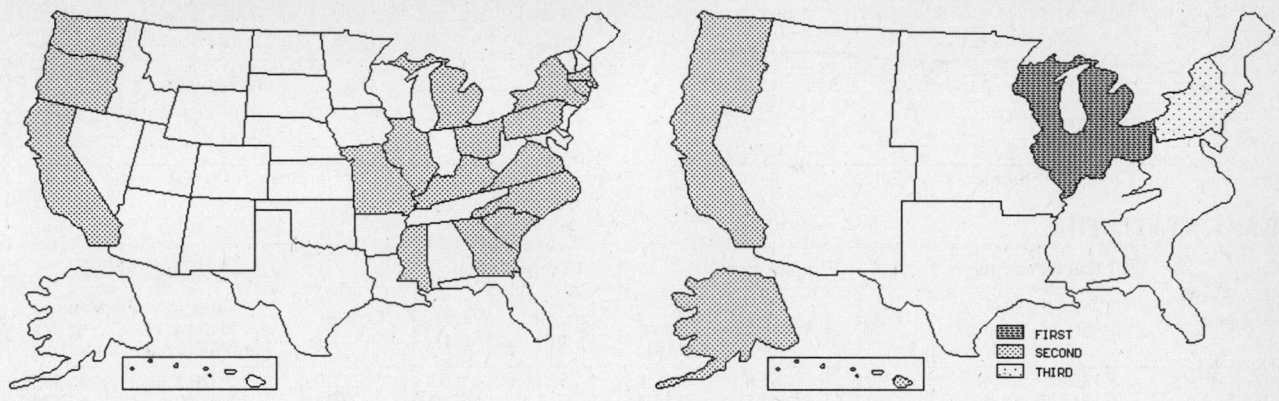

FIRST
SECOND
THIRD

INDUSTRY DATA BY STATE

| State | Establish- ments | Shipments | | | Employment | | | | Cost as % of Shipments | Investment per Employee ($) |
		Total ($ mil)	% of U.S.	Per Establ.	Total Number	% of U.S.	Per Establ.	Wages ($/hour)		
Ohio	14	75.8	8.9	5.4	600	7.9	43	10.75	45.3	4,000
California	19	51.3	6.1	2.7	500	6.6	26	12.20	49.5	2,800
South Carolina	3	43.6	5.1	14.5	400	5.3	133	10.17	57.3	-
Pennsylvania	8	37.0	4.4	4.6	300	3.9	38	10.00	31.1	2,333
New York	7	25.4	3.0	3.6	200	2.6	29	9.25	42.9	2,000
Michigan	8	15.6	1.8	2.0	200	2.6	25	12.50	38.5	-
Massachusetts	8	(D)	-	-	1,750 *	23.0	219	-	-	-
Illinois	7	(D)	-	-	175 *	2.3	25	-	-	-
Oregon	7	(D)	-	-	1,750 *	23.0	250	-	-	-
North Carolina	6	(D)	-	-	375 *	4.9	63	-	-	-
Mississippi	4	(D)	-	-	175 *	2.3	44	-	-	-
Georgia	3	(D)	-	-	375 *	4.9	125	-	-	-
Kentucky	3	(D)	-	-	750 *	9.9	250	-	-	-
Missouri	3	(D)	-	-	375 *	4.9	125	-	-	-
Connecticut	2	(D)	-	-	175 *	2.3	88	-	-	-
Virginia	2	(D)	-	-	175 *	2.3	88	-	-	-
Washington	2	(D)	-	-	175 *	2.3	88	-	-	-

Source: 1992 *Economic Census*. The states are in descending order of shipments or establishments (if shipment data are missing for the majority). The symbol (D) appears when data are withheld to prevent disclosure of competitive information. States marked with (D) are sorted by number of establishments. A dash (-) indicates that the data element cannot be calculated; * indicates the midpoint of a range.

3429 - HARDWARE, NEC

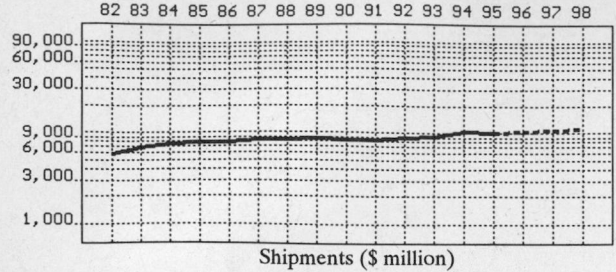

Shipments ($ million)

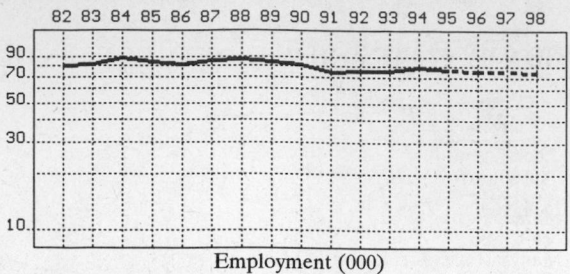

Employment (000)

GENERAL STATISTICS

| Year | Com-panies | Establishments | | Employment | | | Compensation | | Production ($ million) | | | |
		Total	with 20 or more employees	Total (000)	Production Workers (000)	Hours (Mil)	Payroll ($ mil)	Wages ($/hr)	Cost of Materials	Value Added by Manufacture	Value of Shipments	Capital Invest.
1982	1,085	1,185	487	80.3	61.0	114.1	1,520.9	9.03	2,370.6	3,320.1	5,740.9	174.3
1983		1,151	491	82.3	63.5	125.1	1,717.4	9.59	2,843.0	3,947.9	6,752.8	162.3
1984		1,117	495	87.9	67.8	134.0	1,977.0	10.26	3,189.6	4,415.0	7,505.9	215.3
1985		1,084	499	84.7	65.0	128.8	2,032.4	11.01	3,448.4	4,411.9	7,860.9	233.7
1986		1,052	505	82.3	61.9	122.7	1,961.4	10.90	3,393.6	4,311.4	7,688.9	236.5
1987	1,128	1,240	520	85.6	66.1	129.2	2,056.4	10.95	3,575.0	4,602.7	8,175.0	252.1
1988		1,246	547	86.8	66.3	128.7	2,124.0	11.34	3,862.2	4,564.6	8,403.5	182.0
1989		1,196	519	84.4	66.3	130.6	2,042.2	10.72	4,041.6	4,707.0	8,706.1	226.7
1990		1,181	518	81.6	60.6	119.7	2,016.2	11.14	3,865.6	4,593.3	8,462.3	217.9
1991		1,168	501	73.8	56.2	111.5	1,937.5	11.28	3,887.1	4,278.2	8,223.0	187.5
1992	1,172	1,278	485	75.4	56.2	113.3	2,063.1	11.64	3,951.1	4,860.9	8,842.8	253.3
1993		1,221	478	76.1	56.9	113.9	2,131.2	11.96	4,162.0	5,096.4	9,230.0	278.2
1994		1,241P	505P	79.1	60.2	121.8	2,259.3	11.97	4,947.5	5,585.2	10,517.7	313.4
1995		1,251P	505P	76.8P	57.8P	116.8P	2,243.6P	12.29P	4,736.1P	5,346.6P	10,068.4P	278.1P
1996		1,260P	505P	76.1P	57.2P	116.0P	2,280.1P	12.49P	4,864.2P	5,491.2P	10,340.7P	285.6P
1997		1,270P	506P	75.5P	56.6P	115.2P	2,316.7P	12.69P	4,992.3P	5,635.8P	10,613.0P	293.1P
1998		1,280P	506P	74.8P	55.9P	114.4P	2,353.2P	12.89P	5,120.4P	5,780.4P	10,885.3P	300.6P

Sources: 1982, 1987, 1992 *Economic Census*; *Annual Survey of Manufactures*, 83-86, 88-91, 93-94. Establishment counts for non-Census years are from *County Business Patterns*; establishment values for 83-84 are extrapolations. 'P's show projections by the editors. Industries reclassified in 87 will not have data for prior years.

INDICES OF CHANGE

| Year | Com-panies | Establishments | | Employment | | | Compensation | | Production ($ million) | | | |
		Total	with 20 or more employees	Total (000)	Production Workers (000)	Hours (Mil)	Payroll ($ mil)	Wages ($/hr)	Cost of Materials	Value Added by Manufacture	Value of Shipments	Capital Invest.
1982	93	93	100	106	109	101	74	78	60	68	65	69
1983		90	101	109	113	110	83	82	72	81	76	64
1984		87	102	117	121	118	96	88	81	91	85	85
1985		85	103	112	116	114	99	95	87	91	89	92
1986		82	104	109	110	108	95	94	86	89	87	93
1987	96	97	107	114	118	114	100	94	90	95	92	100
1988		97	113	115	118	114	103	97	98	94	95	72
1989		94	107	112	118	115	99	92	102	97	98	89
1990		92	107	108	108	106	98	96	98	94	96	86
1991		91	103	98	100	98	94	97	98	88	93	74
1992	100	100	100	100	100	100	100	100	100	100	100	100
1993		96	99	101	101	101	103	103	105	105	104	110
1994		97P	104P	105	107	108	110	103	125	115	119	124
1995		98P	104P	102P	103P	103P	109P	106P	120P	110P	114P	110P
1996		99P	104P	101P	102P	102P	111P	107P	123P	113P	117P	113P
1997		99P	104P	100P	101P	102P	112P	109P	126P	116P	120P	116P
1998		100P	104P	99P	100P	101P	114P	111P	130P	119P	123P	119P

Sources: Same as General Statistics. Values reflect change from the base year, 1992. Values above 100 mean greater than 92, values below 100 mean less than 92, and a value of 100 in the 82-91 or 93-98 period means same as 92. 'P's mark projections by the editors.

SELECTED RATIOS

For 1994	Avg. of All Manufact.	Analyzed Industry	Index	For 1994	Avg. of All Manufact.	Analyzed Industry	Index
Employees per Establishment	49	64	130	Value Added per Production Worker	134,084	92,777	69
Payroll per Establishment	1,500,273	1,821,059	121	Cost per Establishment	5,045,178	3,987,824	79
Payroll per Employee	30,620	28,563	93	Cost per Employee	102,970	62,547	61
Production Workers per Establishment	34	49	141	Cost per Production Worker	146,988	82,184	56
Wages per Establishment	853,319	1,175,145	138	Shipments per Establishment	9,576,895	8,477,562	89
Wages per Production Worker	24,861	24,218	97	Shipments per Employee	195,460	132,967	68
Hours per Production Worker	2,056	2,023	98	Shipments per Production Worker	279,017	174,713	63
Wages per Hour	12.09	11.97	99	Investment per Establishment	321,011	252,609	79
Value Added per Establishment	4,602,255	4,501,828	98	Investment per Employee	6,552	3,962	60
Value Added per Employee	93,930	70,609	75	Investment per Production Worker	9,352	5,206	56

Sources: Same as General Statistics. The 'Average of All Manufacturing' column represents the average of all manufacturing industries reported for the most recent complete year available. The Index shows the relationship between the Average and the Analyzed Industry. For example, 100 means that they are equal; 500 that the Analyzed Industry is five times the average; 50 means that the Analyzed Industry is half the national average. The abbreviation 'na' is used to show that data are 'not available'.

LEADING COMPANIES Number shown: 75 Total sales ($ mil): 4,590 Total employment (000): 37.8

Company Name	Address				CEO Name	Phone	Co. Type	Sales ($ mil)	Empl. (000)
Securitas Lock Group Inc	103-00 Foster Av	Brooklyn	NY	11236	C Svanberg	718-257-4700	R	1,000	5.0
Michigan Automotive	PO Box 69	Parma	MI	49269	Mike Makino	517-531-5500	J	165	0.4
Knape & Vogt Mfg Co	2700 Oak Indrial Dr	Grand Rapids	MI	49505	Raymond E Knape	616-459-3311	P	160	1.3
Simpson Manufacturing	4637 Chabot Dr	Pleasanton	CA	94588	Thomas J Fitzmyers	510-460-9912	P	151	1.0
Master Lock Co	2600 N 32nd St	Milwaukee	WI	53210	James H Beardsley	414-444-2800	S	150	1.6
National Manufacturing Co	PO Box 577	Sterling	IL	61081	Keith W Benson III	815-625-1320	R	125	0.8
Amerock Corp	PO Box 7018	Rockford	IL	61125	Jack Teela	815-963-9631	S	120	1.5
Stanley Hardware	480 Myrtle St	New Britain	CT	06050	Tom Mahoney	203-225-5111	D	110	2.0
American Consumer Products	31100 Solon Rd	Solon	OH	44139	Stephan W Cole	216-248-7000	P	103	1.0
Sargent Manufacturing Co	100 Sargent Dr	New Haven	CT	06511	John Figurelli	203-562-2151	S	100	0.8
Von Duprin Inc	2720 Tobey Dr	Indianapolis	IN	46219	Brian Jellison	317-897-9944	S	100	0.6
Simpson Strong-Tie Company	4637 Chabot Dr	Pleasanton	CA	94588	Thomas J Fitzmyers	510-460-9912	S	81	0.4
Belwith International Ltd	4300 Gerald R Ford	Grandville	MI	49418	Mark D Pelley	616-531-4300	S	80	0.2
Eaton Corp	PO Box 6688	Cleveland	OH	44101	Larry Iwan	216-220-5361	D	77	0.4
Truth Hardware Corp	700 W Bridge St	Owatonna	MN	55060	Jarrett Dawald	507-451-5620	D	77*	0.8
Baldwin Hardware Corp	PO Box 15048	Reading	PA	19612	Tony Sturrus	215-777-7811	S	70	0.7
International Marine Holdings	PO Box 308	Guilford	CT	06437	Ernest Wong	203-453-4374	S	70*	0.6
Valley Forge Corp	100 Smith Ranch	San Rafael	CA	94903	David R Brining	415-492-1500	P	66	0.5
National Cabinet Lock Inc	PO Box 200	Mauldin	SC	29662	David A Bowers	803-297-6655	S	64	0.6
Super Sagless Inc	PO Box 197	Tupelo	MS	38802	Jay Creamer	601-842-5704	S	63	0.8
HB Ives Co	PO Box 5035	Wallingford	CT	06492	Robert Hoagland	203-294-4837	S	60	0.5
Weiser Lock	6700 Weiser Lock	Tucson	AZ	85746	Vern Schroeder	602-741-6200	D	60	0.5
Hi-Shear Industries Inc	3333 New Hyde Pk	North Hills	NY	11042	David A Wingate	516-627-8600	P	57	0.6
Attwood Corp	1016 N Monroe	Lowell	MI	49331	Don Rocheleau	616-897-9241	S	50	0.4
C Hager & Sons Hinge Mfg	139 Victor St	St Louis	MO	63104	AW Hager III	314-772-4400	R	50	0.9
Faultless Caster Co	1421 N Garvin St	Evansville	IN	47711	RL Dame	812-425-1011	D	50	0.5
Hettich America LP	1607 Anaconda Rd	Harrisonville	MO	64701	Ben Burks	816-380-3456	S	50	0.3
Hurd Lock & Manufacturing Co	PO Box 1450	Greeneville	TN	37744	Chris Edwards	615-639-4101	S	50	1.0
Keeler Hardware Group	955 Godfrey SW	Grand Rapids	MI	49503	John H Watson	616-247-4000	S	50	0.5
Perko Inc	16490 NW 13th Av	Miami	FL	33169	Marvin Perkins	305-621-7525	R	50	0.5
United Fixtures Co	601 N 8th St	Niles	MI	49120	Richard Wakeknight	616-683-9585	R	50	0.2
Accuride International	12311 Shoemaker	Santa Fe Sprgs	CA	90670	Jerry Barr	310-903-0200	R	48	0.5
Hoover Group Inc	2001 Westside Pkwy	Alpharetta	GA	30201	Larry D Clay	404-664-4047	R	48*	0.5
Melard Manufacturing Corp	153 Linden St	Passaic	NJ	07055	Michael Duggin	201-472-8888	R	46*	0.4
Avibank Manufacturing Inc	PO Box 391	Burbank	CA	91503	Milton I Berman	818-843-4330	R	40	0.5
Arrow Lock Manufacturing Co	103-00 Foster Av	Brooklyn	NY	11236	Clas Thelin	718-257-4700	S	35	0.4
Master Lock Co	300 Webster Rd	Auburn	AL	36830	Jim Beardsley	205-826-3300	D	35	0.3
Colson Caster Corp	3700 Airport Rd	Jonesboro	AR	72401	EF Gill Jr	501-932-4501	S	33	0.4
Band-It-Idex Inc	PO Box 16307	Denver	CO	80216	Peter Merkel	303-320-4555	D	30	0.2
Handy Button Machine Co	1750 N 25th Av	Melrose Park	IL	60160	Lenard Baritz	708-450-9000	R	30*	0.4
KYB Corporation of America	901 Oak Creek Dr	Lombard	IL	60148	Takashi Sanada	708-620-5555	S	30	<0.1
Royal Development Co	325 Kettering Rd	High Point	NC	27263	Joseph T Alef	910-889-2569	D	30	0.3
Credo Co	2765 National Way	Woodburn	OR	97071	J W Woodhouse	503-982-0100	R	29*	0.1
Folger Adam Co	16300 W 103rd St	Lemont	IL	60439	Roger E Greene	708-739-3900	R	29	0.3
Lawrence Brothers Inc	2 1st Av	Sterling	IL	61081	Jay Lawrence	815-625-0360	R	28	0.4
American Tack & Hardware Co	25 Robert Pitt Dr	Monsey	NY	10952	EH Weinberg	914-352-2400	R	27	0.2
Medeco Security Locks Inc	PO Box 3075	Salem	VA	24153	Tim Layton	703-380-5000	S	26	0.5
Wiggins Connectors	5000 Trigg St	Los Angeles	CA	90022	Ralph White	213-269-9181	D	26*	0.1
Global Import Co	PO Box 400	El Toro	CA	92630	Howard Furst	714-951-0051	R	25	0.2
Jonathan Manufacturing Corp	PO Box 3-J	Fullerton	CA	92634	Ronald Burch	714-526-4651	R	25	0.3
Sargent and Greenleaf Inc	1 Security Dr	Nicholasville	KY	40356	Jerry Morgan	606-885-9411	S	25	0.2
Adams Rite Manufacturing Co	PO Box 1301	City of Industry	CA	91744	Peter D Adams	310-699-0511	R	24	0.2
Shepherd Products US Inc	203 Kerth St	St Joseph	MI	49085	Byron Hill	616-983-7351	S	24	0.3
Bommer Industries Inc	PO Box 187	Landrum	SC	29356	Peter Frohlich	803-457-3301	R	22*	0.2
Standard-Keil Hardware Mfg Co	Rte 34	Allenwood	NJ	08720	Dennis Magnotta	908-449-3700	D	21*	0.2
Tri/Mark Corp	Industrial Park	New Hampton	IA	50659	D Nelson	515-394-3188	R	21	0.2
American Lock Co	3400 W Exchange	Crete	IL	60417	William F Noone	708-534-2000	R	20	0.4
Dorma Door Controls Inc	Dorma Dr	Reamstown	PA	17567	Don Bixby	717-336-3881	R	20*	0.2
Jarvis East	127 S Main St	Palmer	MA	01069	Ken Carlson	413-283-7601	D	20	0.2
RG Ray Corp	900 Busch Pkwy	Buffalo Grove	IL	60089	Robert J Fabsik	708-459-5900	R	20	0.2
Simplex Access Controls	PO Box 4114	Winston-Salem	NC	27115	Aaron M Fish	910-725-1331	D	20	<0.1
Rotor Clamp Company Inc	187 Davidson Av	Somerset	NJ	08875	R Slass	908-469-7333	S	19	0.2
Blackhawk Automotive Inc	PO Box 1606	Waukesha	WI	53225	JV Russell	414-542-6611	S	18	0.2
Capstan Industries Inc	14000 Avalon Blv	Los Angeles	CA	90061	Mark Paullin	213-321-4595	R	18*	0.2
Chicago Lock Co	4311 W Belmont	Chicago	IL	60641	LB Shinn Jr	312-282-7177	R	16	0.3
De-Sta-Co	165 Kirts Blv	Troy	MI	48007	William Rogerson	810-589-2008	D	16*	0.4
LaBarge Electronics	1505 Maiden Ln	Joplin	MO	64802	V Anderson	417-781-3200	D	16*	0.2
Murray Corp	260 Schilling Cir	Cockeysville	MD	21030	HT Gould	410-771-0380	R	16	0.1
Arrow Tru-Line Inc	PO Box 218	Archbold	OH	43502	David W Shaffer	419-446-2785	R	15	0.2
ESP Lock Products Inc	375 Harvard St	Leominster	MA	01453	A Boucher	508-537-6121	R	15	0.3
Fort Lock Corp	3000 N River Rd	River Grove	IL	60171	L Falk	708-456-1100	R	15	0.3
Harloc Inc	501 Railroad St	Taylorsville	KY	40071	Jeffrey Kates	502-477-8822	S	15	<0.1
Harris Hardware Sales Corp	4 Harbor Park Dr	Pt Washington	NY	11050	K Henin	516-484-4440	R	15	<0.1
Hudson Lock Inc	81 Apsley St	Hudson	MA	01749	Charles Megan	508-562-3481	S	15	0.2
Ilco Unican Corp	2941 Indiana Av	Winston-Salem	NC	27105	William Schmidt	910-725-1331	D	15	0.1

Source: Ward's Business Directory of U.S. Private and Public Companies, Volumes 1 and 2, 1996. The company type code used is as follows: P - Public, R - Private, S - Subsidiary, D - Division, J - Joint Venture, A - Affiliate, G - Group. Sales are in millions of dollars, employees are in thousands. An asterisk (*) indicates an estimated sales volume. The symbol < stands for 'less than'. Company names and addresses are truncated, in some cases, to fit into the available space.

MATERIALS CONSUMED

Material	Quantity	Delivered Cost ($ million)
Materials, ingredients, containers, and supplies	(X)	3,398.8
Metal hardware, including hinges, handles, locks, casters, etc. (except castings and forgings)	(X)	231.7
Metal bolts, nuts, screws, washers, rivets, and other screw machine products	(X)	156.3
Metal stampings	(X)	149.3
Fabricated metal wire products (including wire rope, cable, springs, etc.)	(X)	39.0
All other fabricated metal products (except forgings)	(X)	224.2
Iron and steel castings (rough and semifinished)	(X)	67.7
Aluminum and aluminum-base alloy castings (rough and semifinished)	(X)	20.7
Copper and copper-base alloy castings (rough and semifinished)	(X)	21.8
Zinc and zinc-base alloy castings (rough and semifinished)	(X)	81.9
All other nonferrous castings (rough and semifinished)	(X)	13.2
Forgings	(X)	35.2
Steel bars, bar shapes, and plates	(X)	108.6
Steel sheet, strip, and tin mill products	(X)	408.6
Steel wire and wire products	(X)	55.6
All other steel shapes and forms	(X)	52.3
Copper and copper-base alloy rod, bar, and mechanical wire, including extruded and/or drawn shapes	(X)	30.9
Copper and copper-base alloy sheet, strip, and plate	(X)	69.7
All other copper and copper-base alloy shapes and forms	(X)	21.9
Aluminum and aluminum-base alloy sheet, plate, foil, and welded tubing	(X)	47.6
Aluminum and aluminum-base alloy extruded shapes, including extruded rod, bar, pipe, tube, etc.	(X)	35.8
Other aluminum and aluminum-base alloy shapes and forms	(X)	8.1
Zinc and zinc-base alloy shapes and forms	(X)	86.0
All other nonferrous shapes and forms	(X)	19.4
Metal powders	(X)	19.6
Fabricated rubber products, except tires, tubes, hose, belting, and gaskets	(X)	31.1
Plastics products (film, sheet, rod, tube, and fabricated shapes, including parts, handles, grips, etc.)	(X)	136.0
All other rubber and miscellaneous plastics products	(X)	33.7
Paperboard containers, boxes, and corrugated paperboard	(X)	85.5
Plastics resins consumed in the form of granules, pellets, powders, liquids, etc.	(X)	84.4
Glass and glass products	(X)	18.3
Electric motors, generators, and parts	(X)	188.9
All other materials and components, parts, containers, and supplies	(X)	437.5
Materials, ingredients, containers, and supplies, nsk	(X)	378.3

Source: 1992 *Economic Census*. Explanation of symbols used: (D): Withheld to avoid disclosure of competitive data; na: Not available; (S): Withheld because statistical norms were not met; (X): Not applicable; (Z): Less than half the unit shown; nec: Not elsewhere classified; nsk: Not specified by kind; - : zero; * : 10-19 percent estimated; ** : 20-29 percent estimated.

PRODUCT SHARE DETAILS

Product or Product Class	% Share	Product or Product Class	% Share
Hardware, nec	100.00	hinges), 3 1/2 in. x 3 1/2 in. or less	1.42
Furniture hardware (excluding cabinet hardware).	8.31	Butt hinges (excluding cabinet hinges, but including spring hinges), more than 3 1/2 in. x 3 1/2 in., either dimension .	3.50
Furniture sleeper mechanisms	(D)		
Rotating and tilting furniture fixtures and bases . .	9.84	Other hinges, excluding cabinet hinges, but including spring hinges	3.37
Furniture hardware, including drawer pulls and handles, etc., (excluding furniture and drawer slides)	34.55	Cabinet hinges.	1.48
Furniture and drawer slides	18.52	Cabinet locks	1.54
Furniture casters	12.65	Cabinet knobs, pulls, and catches	2.56
Other floor protective furniture hardware devices (including slides, glides, furniture rests, and desk leg cups)	(D)	Other cabinet hardware (including drawer slides, etc.) . .	2.68
Furniture hardware (excluding cabinet hardware), nsk . .	3.29	Hangers, tracks, and related builders' hardware items (except sliding and folding door hardware), residential and commercial	2.17
Vacuum and insulated bottles, jugs, and chests (except those made principally of foam plastics)	(D)	Sliding and folding door hardware (residential and commercial)	2.74
Vacuum and insulated bottles, jugs, and chests (except those made principally of foam plastics)	(D)	Door holders and stops (overhead, surface, and concealed; floor and wall mounted)	0.42
Builders' hardware	40.51	Rim locks and other locking devices, nec	0.54
Pin tumbler padlocks	4.11	Other builders' hardware	9.63
Nonpin tumbler padlocks	(D)	Builders' hardware, nsk.	0.88
Combination padlocks	(D)	Motor vehicle hardware (lock units, door and window handles, window regulators, hinges, license plate brackets, etc.)	30.29
Bored doorlocks, locksets, and lock trim, cylindrical and tubular (except deadlocks)	16.63		
Mortised doorlocks, locksets, and lock trim, except mortise deadlocks.	3.85	Other transportation equipment hardware (except motor vehicle hardware)	(D)
Tubular and mortised deadlocks	5.48	Marine hardware (including shackles, rope sockets, tackle blocks, wire rope clips, clamps, and joiners' hardware) .	7.99
Electronically or electrically operated doorlocks, locksets, and lock trim	1.53	Aircraft hardware	5.45
All other doorlocks, locksets, and lock trim types . . .	4.70	Other transportation equipment hardware (including railroad car hardware)	(D)
Architectural trim (sold separately) (including protection plates, push plates, pulls, push-pull bars, lock trim, etc.) .	2.05	Other transportation equipment hardware (except motor vehicle hardware), nsk	0.24
Key blanks	2.64	Other hardware, nec	9.60
Exit devices (including fire exit hardware)	4.75	Casket and casket shell hardware	4.67
Screen and storm door hardware (including pneumatic and hydraulic closers)	1.88	Casters and wheels, for dollies and industrial handtrucks. .	22.15
Window hardware (including window locks)	8.48	Fireplace fixtures and equipment, andirons, screens, tongs, and other fire tools	15.50
Miscellaneous closet hardware (including shelving other than wire and decorative shelving)	1.38	Hose fittings and couplings, excluding fittings and couplings used in fluid power systems	15.36
Surface applied door controls, closers, and checking devices	4.40	Refrigerator and stove hardware.	4.99
Concealed (overhead, in the door, or on the floor) door controls, closers, and checking devices	1.45	Other hardware (including saddlery and harness hardware, but excluding drapery hardware)	32.25
Electromechanical-pneumatic door controls, closers, and checking devices (with hold-open mechanism released by integral or remote smoke detector)	1.21	Other hardware, nec, nsk	5.07
Butt hinges (excluding cabinet hinges, but including spring		Hardware, nec, nsk	5.19

Source: 1992 *Economic Census*. The values shown are percent of total shipments in an industry. Values of indented subcategories are summed in the main heading. The symbol (D) appears when data are withheld to prevent disclosure of competitive information. The abbreviation nsk stands for 'not specified by kind' and nec for 'not elsewhere classified'.

INPUTS AND OUTPUTS FOR HARDWARE, NEC

Economic Sector or Industry Providing Inputs	%	Sector	Economic Sector or Industry Buying Outputs	%	Sector
Blast furnaces & steel mills	14.1	Manufg.	Automotive repair shops & services	17.0	Services
Communications, except radio & TV	10.6	Util.	Motor vehicles & car bodies	14.2	Manufg.
Wholesale trade	8.0	Trade	Personal consumption expenditures	7.5	
Imports	7.8	Foreign	Exports	4.7	Foreign
Maintenance of nonfarm buildings nec	7.1	Constr.	Residential 1-unit structures, nonfarm	4.2	Constr.
Plating & polishing	6.3	Manufg.	Office buildings	4.0	Constr.
Plastics materials & resins	2.6	Manufg.	Maintenance of nonfarm buildings nec	3.7	Constr.
Hardware, nec	2.5	Manufg.	Nonfarm residential structure maintenance	2.1	Constr.
Aluminum rolling & drawing	2.4	Manufg.	Millwork	1.9	Manufg.
Copper rolling & drawing	2.2	Manufg.	Wood household furniture	1.9	Manufg.
Electric services (utilities)	2.2	Util.	Federal Government purchases, national defense	1.9	Fed Govt
Motors & generators	2.1	Manufg.	Industrial buildings	1.7	Constr.
Screw machine and related products	1.6	Manufg.	Construction of hospitals	1.6	Constr.
Paperboard containers & boxes	1.5	Manufg.	Residential additions/alterations, nonfarm	1.5	Constr.
Computer & data processing services	1.5	Services	Residential garden apartments	1.4	Constr.
Iron & steel foundries	1.4	Manufg.	Hardware, nec	1.4	Manufg.
Advertising	1.4	Services	Metal doors, sash, & trim	1.2	Manufg.
Metal coating & allied services	1.2	Manufg.	Construction of educational buildings	1.1	Constr.
Glass containers	1.1	Manufg.	Household cooking equipment	1.1	Manufg.
Primary zinc	1.1	Manufg.	Construction of stores & restaurants	0.9	Constr.
Primary copper	1.0	Manufg.	Boat building & repairing	0.9	Manufg.
Motor freight transportation & warehousing	1.0	Util.	Metal office furniture	0.9	Manufg.
Abrasive products	0.9	Manufg.	Wood kitchen cabinets	0.9	Manufg.
Machinery, except electrical, nec	0.9	Manufg.	Aircraft	0.8	Manufg.
Metal stampings, nec	0.9	Manufg.	Burial caskets & vaults	0.8	Manufg.
Wood products, nec	0.8	Manufg.	Maintenance of electric utility facilities	0.7	Constr.
Gas production & distribution (utilities)	0.8	Util.	Telephone & telegraph apparatus	0.7	Manufg.

Continued on next page.

INPUTS AND OUTPUTS FOR HARDWARE, NEC - Continued

Economic Sector or Industry Providing Inputs	%	Sector	Economic Sector or Industry Buying Outputs	%	Sector
Eating & drinking places	0.8	Trade	Fabricated metal products, nec	0.6	Manufg.
Miscellaneous plastics products	0.7	Manufg.	Radio & TV receiving sets	0.6	Manufg.
Special dies & tools & machine tool accessories	0.7	Manufg.	Signs & advertising displays	0.6	Manufg.
Banking	0.7	Fin/R.E.	Wood office furniture	0.6	Manufg.
Aluminum castings	0.6	Manufg.	Wood partitions & fixtures	0.6	Manufg.
Brass, bronze, & copper castings	0.6	Manufg.	Water transportation	0.6	Util.
Metal heat treating	0.6	Manufg.	Electric utility facility construction	0.5	Constr.
Petroleum refining	0.6	Manufg.	Hotels & motels	0.5	Constr.
Engineering, architectural, & surveying services	0.6	Services	Electric housewares & fans	0.5	Manufg.
Equipment rental & leasing services	0.5	Services	Industrial trucks & tractors	0.5	Manufg.
U.S. Postal Service	0.5	Gov't	Luggage	0.5	Manufg.
Chemical preparations, nec	0.4	Manufg.	Metal household furniture	0.5	Manufg.
Fabricated metal products, nec	0.4	Manufg.	Resid. & other health facility construction	0.4	Constr.
Glass & glass products, except containers	0.4	Manufg.	Residential 2-4 unit structures, nonfarm	0.4	Constr.
Nonferrous rolling & drawing, nec	0.4	Manufg.	Household appliances, nec	0.4	Manufg.
Railroads & related services	0.4	Util.	Mobile homes	0.4	Manufg.
Real estate	0.4	Fin/R.E.	Amusement & recreation building construction	0.3	Constr.
Detective & protective services	0.4	Services	Residential high-rise apartments	0.3	Constr.
Electrical repair shops	0.4	Services	Aircraft & missile equipment, nec	0.3	Manufg.
Fabricated rubber products, nec	0.3	Manufg.	Conveyors & conveying equipment	0.3	Manufg.
Miscellaneous fabricated wire products	0.3	Manufg.	Furniture & fixtures, nec	0.3	Manufg.
Paints & allied products	0.3	Manufg.	Metal partitions & fixtures	0.3	Manufg.
Management & consulting services & labs	0.3	Services	Miscellaneous plastics products	0.3	Manufg.
Machine tools, metal cutting types	0.2	Manufg.	Motor homes (made on purchased chassis)	0.3	Manufg.
Primary metal products, nec	0.2	Manufg.	Pipe, valves, & pipe fittings	0.3	Manufg.
Wood pallets & skids	0.2	Manufg.	Public building furniture	0.3	Manufg.
Air transportation	0.2	Util.	Refrigeration & heating equipment	0.3	Manufg.
Sanitary services, steam supply, irrigation	0.2	Util.	Truck & bus bodies	0.3	Manufg.
Credit agencies other than banks	0.2	Fin/R.E.	Upholstered household furniture	0.3	Manufg.
Accounting, auditing & bookkeeping	0.2	Services	Construction of nonfarm buildings nec	0.2	Constr.
Legal services	0.2	Services	Highway & street construction	0.2	Constr.
Lubricating oils & greases	0.1	Manufg.	Sewer system facility construction	0.2	Constr.
Manifold business forms	0.1	Manufg.	Aluminum rolling & drawing	0.2	Manufg.
Insurance carriers	0.1	Fin/R.E.	Automotive stampings	0.2	Manufg.
			Fabricated plate work (boiler shops)	0.2	Manufg.
			Household refrigerators & freezers	0.2	Manufg.
			Leather goods, nec	0.2	Manufg.
			Miscellaneous fabricated wire products	0.2	Manufg.
			Motor vehicle parts & accessories	0.2	Manufg.
			Prefabricated wood buildings	0.2	Manufg.
			Railroad equipment	0.2	Manufg.
			Rubber & plastics hose & belting	0.2	Manufg.
			Ship building & repairing	0.2	Manufg.
			Wood products, nec	0.2	Manufg.
			Communications, except radio & TV	0.2	Util.
			Railroads & related services	0.2	Util.
			Federal Government purchases, nondefense	0.2	Fed Govt
			S/L Govt. purch., other education & libraries	0.2	S/L Govt
			Crude petroleum & natural gas	0.1	Mining
			Construction of religious buildings	0.1	Constr.
			Nonbuilding facilities nec	0.1	Constr.
			Warehouses	0.1	Constr.
			Metal stampings, nec	0.1	Manufg.
			Service industry machines, nec	0.1	Manufg.
			Structural wood members, nec	0.1	Manufg.
			Surgical & medical instruments	0.1	Manufg.
			Surgical appliances & supplies	0.1	Manufg.
			Travel trailers & campers	0.1	Manufg.
			Wood TV & radio cabinets	0.1	Manufg.
			Job training & related services	0.1	Services

Source: Benchmark Input-Output Accounts for the U.S. Economy, 1982, U.S. Department of Commerce, Washington, D.C., July 1991. Data, as reported in the source, are organized by the 1977 SIC structure in use in 1982 but have been matched, as closely as is possible, to the 1987 SIC structure used in this book.

OCCUPATIONS EMPLOYED BY SIC 342 - CUTLERY, HANDTOOLS, AND HARDWARE

Occupation	% of Total 1994	Change to 2005	Occupation	% of Total 1994	Change to 2005
Assemblers, fabricators, & hand workers nec	16.9	-25.4	Industrial machinery mechanics	1.7	-17.9
Machine forming operators, metal & plastic	4.1	-62.7	Grinding machine operators, metal & plastic	1.4	-40.4
Machine tool cutting & forming etc. nec	3.7	-25.4	Machine operators nec	1.3	-34.3
Blue collar worker supervisors	3.5	-30.4	Secretaries, ex legal & medical	1.3	-32.1
Machine tool cutting operators, metal & plastic	3.2	-37.9	Packaging & filling machine operators	1.2	-25.5
Machinists	2.8	-25.4	Punching machine operators, metal & plastic	1.2	-62.7
Hand packers & packagers	2.7	-36.1	Lathe & turning machine tool operators	1.1	-40.4
Combination machine tool operators	2.5	-18.0	Freight, stock, & material movers, hand	1.1	-40.3
Sales & related workers nec	2.4	-25.4	Welding machine setters, operators	1.1	-32.9
Tool & die makers	2.3	-39.8	Bookkeeping, accounting, & auditing clerks	1.1	-44.1
Metal & plastic machine workers nec	2.2	-40.5	Metal molding machine workers	1.1	-18.0
General managers & top executives	2.1	-29.3	Production, planning, & expediting clerks	1.1	-25.4
Inspectors, testers, & graders, precision	2.1	-25.4	General office clerks	1.1	-36.4
Grinders & polishers, hand	2.0	-32.9	Stock clerks	1.0	-39.4
Traffic, shipping, & receiving clerks	1.9	-28.2	Industrial truck & tractor operators	1.0	-25.4
Precision metal workers nec	1.7	-25.4	Industrial production managers	1.0	-25.5

Source: *Industry-Occupation Matrix*, Bureau of Labor Statistics. These data relate to one or more 3-digit SIC industry groups rather than to a single 4-digit SIC. The change reported for each occupation to the year 2005 is a percent of growth or decline as estimated by the Bureau of Labor Statistics. The abbreviation nec stands for 'not elsewhere classified'.

LOCATION BY STATE AND REGIONAL CONCENTRATION

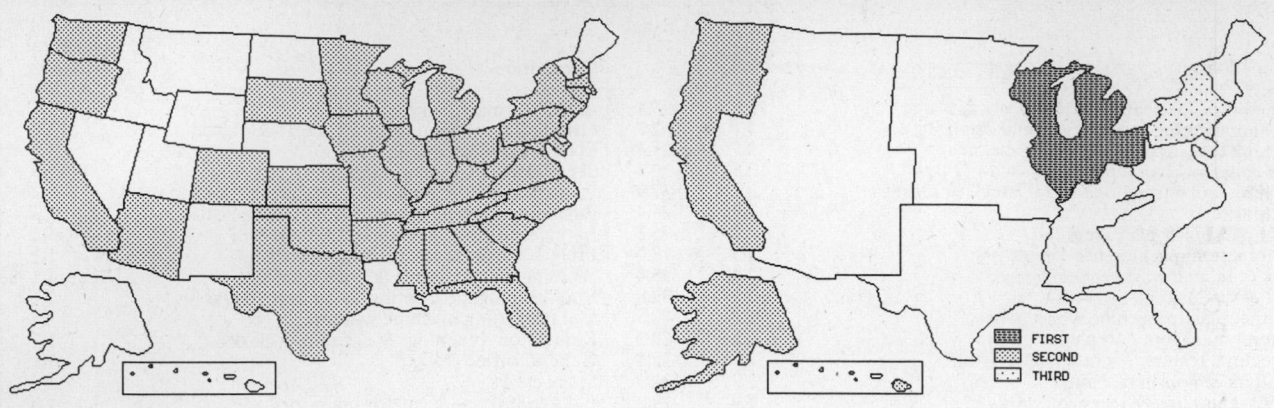

FIRST
SECOND
THIRD

INDUSTRY DATA BY STATE

State	Establish-ments	Shipments			Employment				Cost as % of Shipments	Investment per Employee ($)
		Total ($ mil)	% of U.S.	Per Establ.	Total Number	% of U.S.	Per Establ.	Wages ($/hour)		
Michigan	107	1,216.9	13.8	11.4	7,500	9.9	70	14.13	56.9	3,853
Illinois	100	920.0	10.4	9.2	8,300	11.0	83	9.87	41.6	5,410
California	198	897.8	10.2	4.5	8,200	10.9	41	11.53	39.0	2,524
Tennessee	28	587.2	6.6	21.0	4,400	5.8	157	9.42	56.3	3,477
Ohio	63	575.4	6.5	9.1	4,800	6.4	76	14.84	45.6	2,875
Indiana	47	440.9	5.0	9.4	3,600	4.8	77	11.02	44.6	6,139
Connecticut	49	415.9	4.7	8.5	4,100	5.4	84	12.57	36.1	1,634
New York	75	409.9	4.6	5.5	3,700	4.9	49	15.50	41.0	1,459
Wisconsin	38	403.1	4.6	10.6	3,700	4.9	97	14.09	33.6	4,324
New Jersey	31	377.2	4.3	12.2	2,600	3.4	84	16.82	53.7	500
North Carolina	33	364.3	4.1	11.0	3,400	4.5	103	9.08	48.2	2,324
Pennsylvania	46	309.5	3.5	6.7	2,800	3.7	61	10.67	33.5	2,143
Kentucky	15	199.9	2.3	13.3	1,600	2.1	107	11.17	46.8	2,188
Alabama	16	169.1	1.9	10.6	1,700	2.3	106	9.28	49.3	-
Texas	73	143.5	1.6	2.0	1,300	1.7	18	10.19	59.7	4,769
Missouri	25	125.7	1.4	5.0	1,300	1.7	52	10.05	43.8	-
Minnesota	29	108.4	1.2	3.7	1,000	1.3	34	10.92	42.1	-
Massachusetts	34	107.5	1.2	3.2	1,300	1.7	38	9.22	40.2	-
Florida	53	100.2	1.1	1.9	1,400	1.9	26	8.38	44.5	1,786
Mississippi	9	85.1	1.0	9.5	500	0.7	56	11.12	55.0	1,600
Arkansas	8	83.0	0.9	10.4	1,000	1.3	125	9.67	39.2	1,500
Virginia	10	76.4	0.9	7.6	800	1.1	80	12.30	17.4	-
Washington	25	56.2	0.6	2.2	600	0.8	24	13.73	21.4	-
Oregon	17	54.9	0.6	3.2	500	0.7	29	8.00	43.5	2,400
Iowa	14	54.8	0.6	3.9	600	0.8	43	8.30	48.2	2,500
Georgia	16	49.6	0.6	3.1	600	0.8	38	8.13	43.5	500
Maryland	9	38.0	0.4	4.2	200	0.3	22	8.00	51.6	3,000
West Virginia	4	23.6	0.3	5.9	300	0.4	75	7.80	39.8	-
South Dakota	5	22.2	0.3	4.4	300	0.4	60	8.75	56.8	1,000
New Hampshire	6	13.0	0.1	2.2	200	0.3	33	18.50	32.3	-
Arizona	10	(D)	-	-	750 *	1.0	75	-	-	-
South Carolina	9	(D)	-	-	375 *	0.5	42	-	-	-
Colorado	7	(D)	-	-	750 *	1.0	107	-	-	-
Kansas	7	(D)	-	-	175 *	0.2	25	-	-	-
Oklahoma	7	(D)	-	-	750 *	1.0	107	-	-	-

Source: 1992 *Economic Census*. The states are in descending order of shipments or establishments (if shipment data are missing for the majority). The symbol (D) appears when data are withheld to prevent disclosure of competitive information. States marked with (D) are sorted by number of establishments. A dash (-) indicates that the data element cannot be calculated; * indicates the midpoint of a range.

3431 - ENAMELED IRON & METAL SANITARY WARE

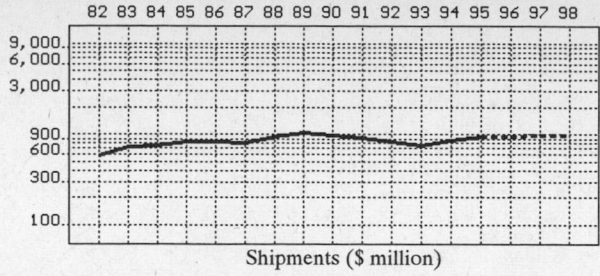

Shipments ($ million)

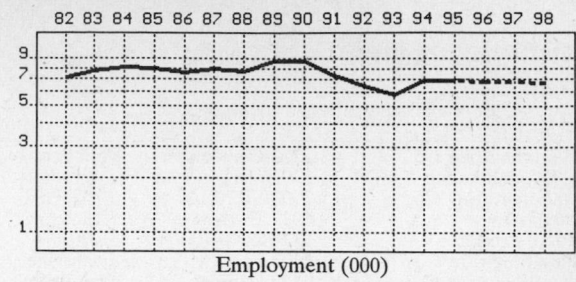

Employment (000)

GENERAL STATISTICS

| Year | Companies | Establishments | | Employment | | | Compensation | | Production ($ million) | | | |
		Total	with 20 or more employees	Total (000)	Production Workers (000)	Hours (Mil)	Payroll ($ mil)	Wages ($/hr)	Cost of Materials	Value Added by Manufacture	Value of Shipments	Capital Invest.
1982	78	85	34	7.2	5.6	10.6	133.1	9.05	281.2	293.7	583.6	15.3
1983		88	36	7.7	6.0	12.6	159.2	9.21	356.6	392.8	737.2	11.1
1984		91	38	8.1	6.4	13.2	172.1	9.49	351.0	409.3	755.2	16.8
1985		95	41	8.0	6.2	12.8	178.0	10.05	359.4	463.3	827.4	26.1
1986	90	100	41	7.6	5.8	12.5	188.9	11.06	367.3	482.5	845.8	23.6
1987	90	97	42	8.0	6.0	12.3	193.3	10.15	359.1	444.1	803.4	20.4
1988		94	47	7.7	6.0	12.3	223.0	11.21	431.3	529.3	945.6	21.3
1989		99	51	8.8	6.0	13.0	228.7	10.98	548.9	505.6	1,051.6	36.0
1990		95	51	8.8	5.9	12.3	233.6	10.93	456.0	524.8	980.0	33.9
1991		105	46	7.3	5.5	12.0	211.4	11.48	410.6	499.3	903.7	28.0
1992	75	81	27	6.5	5.0	10.6	200.8	13.00	321.6	499.4	829.4	18.3
1993		81	27	5.8	4.3	10.1	194.6	12.59	258.4	480.7	749.6	18.1
1994		92P	40P	6.9	5.4	9.8	219.2	14.81	317.9	559.1	866.8	18.8
1995		92P	40P	6.9P	5.0P	10.7P	234.8P	13.78P	346.5P	609.3P	944.7P	26.1P
1996		92P	40P	6.8P	4.9P	10.5P	240.5P	14.17P	352.1P	619.3P	960.1P	26.7P
1997		92P	40P	6.7P	4.8P	10.4P	246.2P	14.56P	357.8P	629.2P	975.5P	27.3P
1998		92P	39P	6.7P	4.8P	10.2P	251.8P	14.94P	363.4P	639.2P	990.9P	27.8P

Sources: 1982, 1987, 1992 *Economic Census*; *Annual Survey of Manufactures*, 83-86, 88-91, 93-94. Establishment counts for non-Census years are from *County Business Patterns*; establishment values for 83-84 are extrapolations. 'P's show projections by the editors. Industries reclassified in 87 will not have data for prior years.

INDICES OF CHANGE

| Year | Companies | Establishments | | Employment | | | Compensation | | Production ($ million) | | | |
		Total	with 20 or more employees	Total (000)	Production Workers (000)	Hours (Mil)	Payroll ($ mil)	Wages ($/hr)	Cost of Materials	Value Added by Manufacture	Value of Shipments	Capital Invest.
1982	104	105	126	111	112	100	66	70	87	59	70	84
1983		109	133	118	120	119	79	71	111	79	89	61
1984		112	141	125	128	125	86	73	109	82	91	92
1985		117	152	123	124	121	89	77	112	93	100	143
1986		123	152	117	116	118	94	85	114	97	102	129
1987	120	120	156	123	120	116	96	78	112	89	97	111
1988		116	174	118	120	116	111	86	134	106	114	116
1989		122	189	135	120	123	114	84	171	101	127	197
1990		117	189	135	118	116	116	84	142	105	118	185
1991		130	170	112	110	113	105	88	128	100	109	153
1992	100	100	100	100	100	100	100	100	100	100	100	100
1993		100	100	89	86	95	97	97	80	96	90	99
1994		114P	147P	106	108	92	109	114	99	112	105	103
1995		114P	147P	107P	101P	101P	117P	106P	108P	122P	114P	143P
1996		114P	147P	105P	99P	99P	120P	109P	109P	124P	116P	146P
1997		114P	146P	104P	97P	98P	123P	112P	111P	126P	118P	149P
1998		114P	146P	102P	95P	96P	125P	115P	113P	128P	119P	152P

Sources: Same as General Statistics. Values reflect change from the base year, 1992. Values above 100 mean greater than 92, values below 100 mean less than 92, and a value of 100 in the 82-91 or 93-98 period means same as 92. 'P's mark projections by the editors.

SELECTED RATIOS

For 1994	Avg. of All Manufact.	Analyzed Industry	Index	For 1994	Avg. of All Manufact.	Analyzed Industry	Index
Employees per Establishment	49	75	153	Value Added per Production Worker	134,084	103,537	77
Payroll per Establishment	1,500,273	2,376,347	158	Cost per Establishment	5,045,178	3,446,353	68
Payroll per Employee	30,620	31,768	104	Cost per Employee	102,970	46,072	45
Production Workers per Establishment	34	59	171	Cost per Production Worker	146,988	58,870	40
Wages per Establishment	853,319	1,573,441	184	Shipments per Establishment	9,576,895	9,396,978	98
Wages per Production Worker	24,861	26,877	108	Shipments per Employee	195,460	125,623	64
Hours per Production Worker	2,056	1,815	88	Shipments per Production Worker	279,017	160,519	58
Wages per Hour	12.09	14.81	122	Investment per Establishment	321,011	203,811	63
Value Added per Establishment	4,602,255	6,061,202	132	Investment per Employee	6,552	2,725	42
Value Added per Employee	93,930	81,029	86	Investment per Production Worker	9,352	3,481	37

Sources: Same as General Statistics. The 'Average of All Manufacturing' column represents the average of all manufacturing industries reported for the most recent complete year available. The Index shows the relationship between the Average and the Analyzed Industry. For example, 100 means that they are equal; 500 that the Analyzed Industry is five times the average; 50 means that the Analyzed Industry is half the national average. The abbreviation 'na' is used to show that data are 'not available'.

LEADING COMPANIES   Number shown: 15   Total sales ($ mil): 360   Total employment (000): 3.4

Company Name	Address				CEO Name	Phone	Co. Type	Sales ($ mil)	Empl. (000)
Elkay Manufacturing Co	2222 Camden Ct	Oak Brook	IL	60521	Ronald C Katz	708-574-8484	R	130	1.5
Creative Specialties Inc	12930 Bradley Av	Sylmar	CA	91342	C Sweetman	818-367-2131	R	50	<0.1
Elkay Manufacturing Co	2700 S 17th Av	Broadview	IL	60153	Ronald Gatehouse	708-681-1880	D	42*	0.5
Fiat Products Inc	1235 Hartrey	Evanston	IL	60202	Kevin Phelan	708-864-7600	D	30*	0.4
UNR Home Products	PO Box 1010	Ruston	LA	71273	JM Buske	318-255-5600	D	22	0.2
Bootz Plumbingware Co	1400 Park St	Evansville	IN	47719	William R Frank	812-423-5401	R	20	0.1
Elkay Manufacturing Co	2222 Camden Ct	Oak Brook	IL	60521		708-573-8870	D	20	0.2
Microphor Inc	PO Box 1460	Willits	CA	95490	John M Mayfield Jr	707-459-5563	R	12	0.1
Accurate Partitions Corp	PO Box 287	Lyons	IL	60534	OA Testa	708-442-6801	R	7*	<0.1
Amer Sanitary Partition Corp	PO Drawer 99	Ocoee	FL	34761	Gerald Birkenmaier	407-656-0611	R	7*	<0.1
Plumbing Products Company Inc	7230 Oxford Way	City of Com	CA	90040	G Yavitz	213-722-3845	R	6*	<0.1
Vance Industries Inc	7401 W Wilson Av	Chicago	IL	60656	WM Rapp	708-867-6000	R	6	0.1
CWECO	1156 W 135th St	Gardena	CA	90247	David Gursky	310-538-9440	R	3	<0.1
Sunset Plastics Inc	6270 Parrellel Rd	Anderson	CA	96007	Charles M Ehn	916-365-5494	R	3	<0.1
Crane Plumbing	17025 E Gale Av	City of Industry	CA	91745	Reed Beidler	818-964-3475	D	2*	<0.1

Source: Ward's Business Directory of U.S. Private and Public Companies, Volumes 1 and 2, 1996. The company type code used is as follows: P - Public, R - Private, S - Subsidiary, D - Division, J - Joint Venture, A - Affiliate, G - Group. Sales are in millions of dollars, employees are in thousands. An asterisk (*) indicates an estimated sales volume. The symbol < stands for 'less than'. Company names and addresses are truncated, in some cases, to fit into the available space.

MATERIALS CONSUMED

Material	Quantity	Delivered Cost ($ million)
Materials, ingredients, containers, and supplies	(X)	258.3
Metal stampings	(X)	12.8
All other fabricated metal products (except forgings)	(X)	9.4
Iron and steel castings (rough and semifinished)	(X)	(D)
Copper and copper-base alloy castings (rough and semifinished)	(X)	(D)
Other nonferrous castings (rough and semifinished)	(X)	(D)
Forgings	(X)	(D)
Steel shapes and forms	(X)	54.8
Copper and copper-base alloy refinery shapes	(X)	(D)
Copper and copper-base alloy rod, bar, and mechanical wire, including extruded and/or drawn shapes	(X)	(D)
Copper and copper-base alloy pipe and tube	(X)	0.8
All other copper and copper-base alloy shapes and forms	(X)	(D)
All other nonferrous shapes and forms	(X)	(D)
Plastics resins consumed in the form of granules, pellets, powders, liquids, etc.	(X)	1.1
Fabricated plastics products (except gaskets, hoses, and belting)	(X)	6.5
Paperboard containers, boxes, and corrugated paperboard	(X)	19.1
All other materials and components, parts, containers, and supplies	(X)	81.0
Materials, ingredients, containers, and supplies, nsk	(X)	29.8

Source: 1992 Economic Census. Explanation of symbols used: (D): Withheld to avoid disclosure of competitive data; na: Not available; (S): Withheld because statistical norms were not met; (X): Not applicable; (Z): Less than half the unit shown; nec: Not elsewhere classified; nsk: Not specified by kind; - : zero; * : 10-19 percent estimated; ** : 20-29 percent estimated.

PRODUCT SHARE DETAILS

Product or Product Class	% Share	Product or Product Class	% Share
Metal sanitary ware	100.00	Other enameled iron and metal plumbing fixtures, including portable chemical toilets, water closet tanks, etc.	12.25
Enameled iron and metal plumbing fixtures	83.60		

Source: 1992 Economic Census. The values shown are percent of total shipments in an industry. Values of indented subcategories are summed in the main heading. The symbol (D) appears when data are withheld to prevent disclosure of competitive information. The abbreviation nsk stands for 'not specified by kind' and nec for 'not elsewhere classified'.

INPUTS AND OUTPUTS FOR METAL SANITARY WARE

Economic Sector or Industry Providing Inputs	%	Sector	Economic Sector or Industry Buying Outputs	%	Sector
Blast furnaces & steel mills	23.3	Manufg.	Maintenance of nonfarm buildings nec	17.7	Constr.
Wholesale trade	9.2	Trade	Nonfarm residential structure maintenance	14.6	Constr.
Metal stampings, nec	5.6	Manufg.	Residential 1-unit structures, nonfarm	13.4	Constr.
Copper rolling & drawing	5.5	Manufg.	Residential garden apartments	8.0	Constr.
Gas production & distribution (utilities)	4.8	Util.	Exports	7.0	Foreign
Electric services (utilities)	4.7	Util.	Office buildings	6.9	Constr.
Paperboard containers & boxes	4.1	Manufg.	Residential additions/alterations, nonfarm	6.6	Constr.
Primary zinc	3.7	Manufg.	Construction of hospitals	4.4	Constr.
Chemical preparations, nec	3.1	Manufg.	Residential high-rise apartments	2.5	Constr.
Primary copper	3.0	Manufg.	Construction of educational buildings	2.4	Constr.
Screw machine and related products	2.6	Manufg.	Residential 2-4 unit structures, nonfarm	2.3	Constr.
Scrap	2.6	Scrap	Construction of stores & restaurants	2.2	Constr.
Plating & polishing	2.1	Manufg.	Industrial buildings	2.2	Constr.
Motor freight transportation & warehousing	1.5	Util.	Hotels & motels	1.9	Constr.
Maintenance of nonfarm buildings nec	1.4	Constr.	Construction of religious buildings	0.9	Constr.
Banking	1.3	Fin/R.E.	Electric utility facility construction	0.8	Constr.
Machinery, except electrical, nec	1.2	Manufg.	Maintenance of highways & streets	0.8	Constr.
Pipe, valves, & pipe fittings	1.2	Manufg.	Resid. & other health facility construction	0.8	Constr.
Eating & drinking places	1.1	Trade	Maintenance of sewer facilities	0.6	Constr.
Advertising	1.1	Services	Maintenance of nonbuilding facilities nec	0.5	Constr.
Paints & allied products	0.9	Manufg.	Metal sanitary ware	0.5	Manufg.
Railroads & related services	0.9	Util.	Construction of nonfarm buildings nec	0.4	Constr.
Metal sanitary ware	0.8	Manufg.	Telephone & telegraph facility construction	0.4	Constr.
Miscellaneous plastics products	0.8	Manufg.	Dormitories & other group housing	0.3	Constr.
Communications, except radio & TV	0.8	Util.	Maintenance of water supply facilities	0.3	Constr.
Real estate	0.8	Fin/R.E.	Amusement & recreation building construction	0.2	Constr.
Equipment rental & leasing services	0.7	Services	Farm housing units & additions & alterations	0.2	Constr.
U.S. Postal Service	0.7	Gov't	Maintenance, conservation & development facilities	0.2	Constr.
Abrasive products	0.6	Manufg.	Highway & street construction	0.1	Constr.
Special dies & tools & machine tool accessories	0.6	Manufg.	Maintenance of farm service facilities	0.1	Constr.
Welding apparatus, electric	0.5	Manufg.	Maintenance of military facilities	0.1	Constr.
Aluminum rolling & drawing	0.4	Manufg.	Sewer system facility construction	0.1	Constr.
Metal coating & allied services	0.4	Manufg.	Warehouses	0.1	Constr.
Sanitary services, steam supply, irrigation	0.4	Util.	Water supply facility construction	0.1	Constr.
Legal services	0.4	Services			
Management & consulting services & labs	0.4	Services			
Aluminum castings	0.3	Manufg.			
Lubricating oils & greases	0.3	Manufg.			
Machine tools, metal forming types	0.3	Manufg.			
Metal heat treating	0.3	Manufg.			
Petroleum refining	0.3	Manufg.			
Accounting, auditing & bookkeeping	0.3	Services			
Computer & data processing services	0.3	Services			
Iron & steel foundries	0.2	Manufg.			
Machine tools, metal cutting types	0.2	Manufg.			
Manifold business forms	0.2	Manufg.			
Air transportation	0.2	Util.			
Insurance carriers	0.2	Fin/R.E.			
Royalties	0.2	Fin/R.E.			
Laundry, dry cleaning, shoe repair	0.2	Services			
Miscellaneous repair shops	0.2	Services			
State & local government enterprises, nec	0.2	Gov't			
Coal	0.1	Mining			
Sand & gravel	0.1	Mining			
Brass, bronze, & copper castings	0.1	Manufg.			
Fabricated rubber products, nec	0.1	Manufg.			
Industrial gases	0.1	Manufg.			
Miscellaneous fabricated wire products	0.1	Manufg.			
Plumbing fixture fittings & trim	0.1	Manufg.			
Soap & other detergents	0.1	Manufg.			
Security & commodity brokers	0.1	Fin/R.E.			
Automotive rental & leasing, without drivers	0.1	Services			
Automotive repair shops & services	0.1	Services			
Personnel supply services	0.1	Services			

Source: Benchmark Input-Output Accounts for the U.S. Economy, 1982, U.S. Department of Commerce, Washington, D.C., July 1991. Data, as reported in the source, are organized by the 1977 SIC structure in use in 1982 but have been matched, as closely as is possible, to the 1987 SIC structure used in this book.

OCCUPATIONS EMPLOYED BY SIC 343 - PLUMBING AND HEATING, EX ELECTRIC

Occupation	% of Total 1994	Change to 2005	Occupation	% of Total 1994	Change to 2005
Assemblers, fabricators, & hand workers nec	18.4	-15.9	Industrial truck & tractor operators	1.8	-15.9
Helpers, laborers, & material movers nec	4.3	-15.9	Industrial machinery mechanics	1.7	-7.5
Inspectors, testers, & graders, precision	3.3	-15.9	Secretaries, ex legal & medical	1.6	-23.4
Hand packers & packagers	2.8	-27.9	Combination machine tool operators	1.6	-7.5
Sales & related workers nec	2.7	-15.9	Metal molding machine workers	1.5	-7.4
Machine tool cutting & forming etc. nec	2.5	-15.9	Welding machine setters, operators	1.4	-24.4
Grinders & polishers, hand	2.4	-16.0	Machinists	1.3	-15.9
Traffic, shipping, & receiving clerks	2.4	-19.1	Maintenance repairers, general utility	1.3	-24.2
Metal & plastic machine workers nec	2.4	-25.5	General office clerks	1.3	-28.3
Machine forming operators, metal & plastic	2.3	-15.9	Bookkeeping, accounting, & auditing clerks	1.3	-36.9
Freight, stock, & material movers, hand	2.3	-32.7	Production, planning, & expediting clerks	1.2	-15.9
Welders & cutters	2.3	-15.9	Lathe & turning machine tool operators	1.1	-32.7
Sheet metal workers & duct installers	2.1	-15.9	Drafters	1.1	-34.4
General managers & top executives	2.0	-20.2	Stock clerks	1.1	-31.7
Machine tool cutting operators, metal & plastic	2.0	-30.0	Tool & die makers	1.1	-32.2

Source: Industry-Occupation Matrix, Bureau of Labor Statistics. These data relate to one or more 3-digit SIC industry groups rather than to a single 4-digit SIC. The change reported for each occupation to the year 2005 is a percent of growth or decline as estimated by the Bureau of Labor Statistics. The abbreviation nec stands for 'not elsewhere classified'.

LOCATION BY STATE AND REGIONAL CONCENTRATION

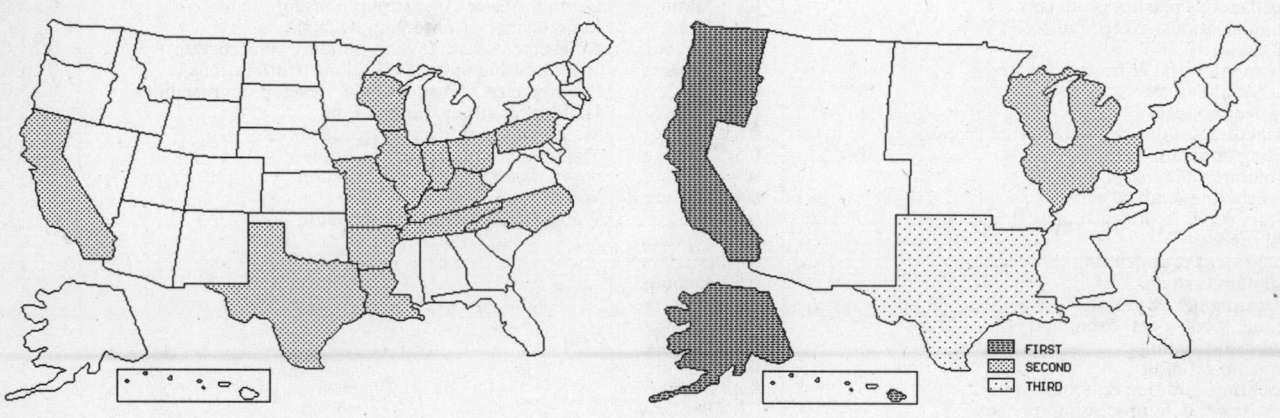

FIRST
SECOND
THIRD

INDUSTRY DATA BY STATE

State	Establish-ments	Shipments			Employment				Cost as % of Shipments	Investment per Employee ($)
		Total ($ mil)	% of U.S.	Per Establ.	Total Number	% of U.S.	Per Establ.	Wages ($/hour)		
California	19	85.6	10.3	4.5	700	10.8	37	12.80	51.2	1,571
Texas	6	(D)	-	-	175 *	2.7	29	-	-	-
Illinois	4	(D)	-	-	750 *	11.5	188	-	-	-
Indiana	4	(D)	-	-	175 *	2.7	44	-	-	-
Ohio	4	(D)	-	-	750 *	11.5	188	-	-	-
Tennessee	4	(D)	-	-	375 *	5.8	94	-	-	-
Pennsylvania	3	(D)	-	-	175 *	2.7	58	-	-	-
Arkansas	2	(D)	-	-	175 *	2.7	88	-	-	-
Missouri	2	(D)	-	-	175 *	2.7	88	-	-	-
North Carolina	2	(D)	-	-	175 *	2.7	88	-	-	-
Wisconsin	2	(D)	-	-	1,750 *	26.9	875	-	-	-
Kentucky	1	(D)	-	-	375 *	5.8	375	-	-	-
Louisiana	1	(D)	-	-	175 *	2.7	175	-	-	-

Source: 1992 Economic Census. The states are in descending order of shipments or establishments (if shipment data are missing for the majority). The symbol (D) appears when data are withheld to prevent disclosure of competitive information. States marked with (D) are sorted by number of establishments. A dash (-) indicates that the data element cannot be calculated; * indicates the midpoint of a range.

3432 - PLUMBING FITTINGS & BRASS GOODS

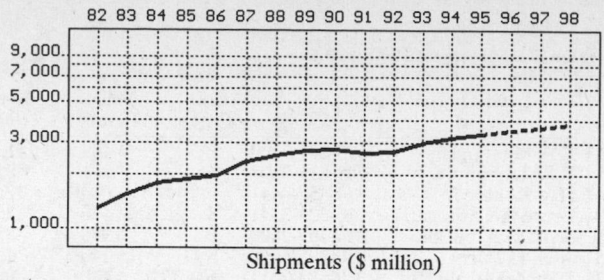

Shipments ($ million)

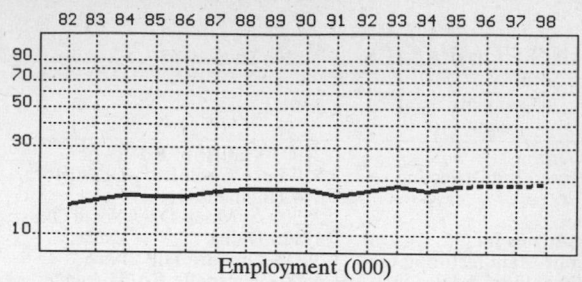

Employment (000)

GENERAL STATISTICS

Year	Companies	Establishments		Employment			Compensation		Production ($ million)			
		Total	with 20 or more employees	Total (000)	Production Workers (000)	Hours (Mil)	Payroll ($ mil)	Wages ($/hr)	Cost of Materials	Value Added by Manufacture	Value of Shipments	Capital Invest.
1982	161	186	103	14.4	11.1	21.2	237.9	7.40	613.8	672.1	1,300.5	37.2
1983		182	104	15.4	12.0	23.4	265.6	7.73	759.3	817.8	1,557.4	28.3
1984		178	105	16.3	12.8	25.0	296.8	8.12	884.9	942.6	1,788.4	44.8
1985		175	105	16.0	12.4	24.9	308.3	8.46	903.8	966.4	1,863.0	45.1
1986		163	97	15.9	12.2	25.3	320.2	8.57	909.6	1,082.5	1,982.3	24.7
1987	161	180	98	17.1	13.1	27.5	355.6	8.56	1,127.4	1,265.9	2,355.4	61.9
1988		177	100	17.9	13.9	29.1	393.8	8.97	1,240.9	1,346.1	2,563.1	67.0
1989		172	98	17.9	13.7	27.3	401.9	9.51	1,340.5	1,406.6	2,733.2	65.5
1990		174	98	17.6	12.9	24.8	405.8	10.25	1,342.7	1,400.1	2,749.9	42.5
1991		168	91	16.4	12.1	23.5	414.7	10.83	1,307.5	1,323.4	2,655.0	40.5
1992	163	185	103	17.4	13.3	26.7	444.5	10.67	1,317.6	1,358.5	2,700.7	74.2
1993		188	101	18.6	13.9	28.4	472.5	10.86	1,398.8	1,650.3	3,002.7	75.1
1994		177P	97P	17.4	12.7	27.8	504.4	10.68	1,379.0	1,890.7	3,202.4	66.0
1995		177P	96P	18.5P	13.6P	28.3P	515.6P	11.47P	1,448.9P	1,986.6P	3,364.8P	72.9P
1996		177P	96P	18.7P	13.8P	28.7P	536.2P	11.78P	1,511.8P	2,072.8P	3,510.8P	75.9P
1997		177P	95P	19.0P	13.9P	29.0P	556.9P	12.09P	1,574.7P	2,159.0P	3,656.8P	78.9P
1998		177P	94P	19.2P	14.0P	29.4P	577.6P	12.41P	1,637.6P	2,245.2P	3,802.8P	82.0P

Sources: 1982, 1987, 1992 *Economic Census*; *Annual Survey of Manufactures*, 83-86, 88-91, 93-94. Establishment counts for non-Census years are from *County Business Patterns*; establishment values for 83-84 are extrapolations. 'P's show projections by the editors. Industries reclassified in 87 will not have data for prior years.

INDICES OF CHANGE

Year	Companies	Establishments		Employment			Compensation		Production ($ million)			
		Total	with 20 or more employees	Total (000)	Production Workers (000)	Hours (Mil)	Payroll ($ mil)	Wages ($/hr)	Cost of Materials	Value Added by Manufacture	Value of Shipments	Capital Invest.
1982	99	101	100	83	83	79	54	69	47	49	48	50
1983		98	101	89	90	88	60	72	58	60	58	38
1984		96	102	94	96	94	67	76	67	69	66	60
1985		95	102	92	93	93	69	79	69	71	69	61
1986		88	94	91	92	95	72	80	69	80	73	33
1987	99	97	95	98	98	103	80	80	86	93	87	83
1988		96	97	103	105	109	89	84	94	99	95	90
1989		93	95	103	103	102	90	89	102	104	101	88
1990		94	95	101	97	93	91	96	102	103	102	57
1991		91	88	94	91	88	93	101	99	97	98	55
1992	100	100	100	100	100	100	100	100	100	100	100	100
1993		102	98	107	105	106	106	102	106	121	111	101
1994		96P	94P	100	95	104	113	100	105	139	119	89
1995		96P	93P	106P	102P	106P	116P	107P	110P	146P	125P	98P
1996		96P	93P	108P	103P	107P	121P	110P	115P	153P	130P	102P
1997		96P	92P	109P	104P	109P	125P	113P	120P	159P	135P	106P
1998		96P	92P	110P	105P	110P	130P	116P	124P	165P	141P	110P

Sources: Same as General Statistics. Values reflect change from the base year, 1992. Values above 100 mean greater than 92, values below 100 mean less than 92, and a value of 100 in the 82-91 or 93-98 period means same as 92. 'P's mark projections by the editors.

SELECTED RATIOS

For 1994	Avg. of All Manufact.	Analyzed Industry	Index	For 1994	Avg. of All Manufact.	Analyzed Industry	Index
Employees per Establishment	49	98	200	Value Added per Production Worker	134,084	148,874	111
Payroll per Establishment	1,500,273	2,845,090	190	Cost per Establishment	5,045,178	7,778,310	154
Payroll per Employee	30,620	28,989	95	Cost per Employee	102,970	79,253	77
Production Workers per Establishment	34	72	209	Cost per Production Worker	146,988	108,583	74
Wages per Establishment	853,319	1,674,700	196	Shipments per Establishment	9,576,895	18,063,277	189
Wages per Production Worker	24,861	23,378	94	Shipments per Employee	195,460	184,046	94
Hours per Production Worker	2,056	2,189	106	Shipments per Production Worker	279,017	252,157	90
Wages per Hour	12.09	10.68	88	Investment per Establishment	321,011	372,276	116
Value Added per Establishment	4,602,255	10,664,576	232	Investment per Employee	6,552	3,793	58
Value Added per Employee	93,930	108,661	116	Investment per Production Worker	9,352	5,197	56

Sources: Same as General Statistics. The 'Average of All Manufacturing' column represents the average of all manufacturing industries reported for the most recent complete year available. The Index shows the relationship between the Average and the Analyzed Industry. For example, 100 means that they are equal; 500 that the Analyzed Industry is five times the average; 50 means that the Analyzed Industry is half the national average. The abbreviation 'na' is used to show that data are 'not available'.

LEADING COMPANIES   Number shown: **63**   Total sales ($ mil): **9,056**   Total employment (000): **90.6**

Company Name	Address				CEO Name	Phone	Co. Type	Sales ($ mil)	Empl. (000)
Masco Corp	21001 Van Born Rd	Taylor	MI	48180	R A Manoogian	313-274-7400	P	4,468	51.3
MasterBrand Industries Inc	510 Lake Cook Rd	Deerfield	IL	60015	R W Larrimore	708-831-0808	S	1,271	8.5
Kohler Co	444 Highland Dr	Kohler	WI	53044	H V Kohler Jr	414-457-4441	R	1,080•	12.5
Moen Inc	25300 Al Moen Dr	North Olmsted	OH	44070	Bruce A Carbonari	216-323-3341	S	440•	3.0
Eljer Industries Inc	17120 Dallas	Dallas	TX	75248	Scott G Arbuckle	214-407-2600	P	388	3.9
Brass-Craft Manufacturing Co	39600 Orchard Hill	Novi	MI	48375	Alan H Barry	810-305-6000	S	140•	1.2
AY McDonald Industries Inc	4800 Chavenelle Rd	Dubuque	IA	52002	J M McDonald III	319-583-7311	R	130•	0.7
Sterling Plumbing Group Inc	2900 Golf Rd	Rolling Mdws	IL	60008	John E Mawdsley	708-734-1777	S	130	1.7
Acorn Engineering Co	PO Box 3527	City of Industry	CA	91744	Donald E Morris	818-336-4561	R	100•	0.5
US Brass Corp	PO Box 879001	Dallas	TX	75287	Scott Arbuckle	214-407-2600	S	100•	0.8
Zurn Industries Inc	1801 Pittsburgh Av	Erie	PA	16514	DF Fessler	814-455-0921	D	75	0.2
Fisher Manufacturing Co	PO Box 60	Tulare	CA	93275	Ray Fisher Sr	209-685-5200	R	60	<0.1
Chicago Faucet Co	2100 S Clearwater	Des Plaines	IL	60018	Alan Lougee	708-803-5000	R	46	0.4
Franklin Brass Mfg Co	5353 Grosvenor	Los Angeles	CA	90066	Norton Sharpe	310-306-5944	R	30	0.3
T and S Brass and Bronze Works	PO Box 1088	Travelers Rest	SC	29690	CI Theisen	803-834-4102	R	30	0.2
AY McDonald Mfg Co	4800 Chavenelle Rd	Dubuque	IA	52002	RD McDonald	319-583-7311	S	28•	0.3
Smith Industries Inc	PO Box 3237	Montgomery	AL	36109	Jay L Smith	334-277-8520	R	26	0.3
American Granby Inc	1111 Vine St	Liverpool	NY	13088	Sam Daniels	315-451-1100	S	25	<0.1
Harden Industries Inc	13915 S Main St	Los Angeles	CA	90061	Barbara Rodstien	310-532-7850	R	25	0.2
Symmons Industries Inc	31 Brooks Dr	Braintree	MA	02184	William O'Keeffe	617-848-2250	R	25	0.2
Alsons Corp	42 Union St	Hillsdale	MI	49242	Donald J Milroy	517-439-1411	S	22•	0.2
Fort Recovery Industries Inc	2440 State Rte 49	Fort Recovery	OH	45846	Wesley Jetter	419-375-4121	R	20	0.3
Josam Co	PO Box T	Michigan City	IN	46360	B Hodgekins	219-872-5531	R	20•	<0.1
URC Faucets	4250 N 124th St	Milwaukee	WI	53222	Roger Silknitter	414-461-8700	D	20	0.1
William Steinen Mfg Co	29 E Halsey Rd	Parsippany	NJ	07054	William Steinen	201-887-6400	R	18	0.3
Anderson Fittings Inc	255 E Industry Av	Frankfort	IL	60423	Gary Kinsella	815-469-2211	S	17•	0.1
Central Brass Manufacturing Co	2950 E 55th St	Cleveland	OH	44127	Richard A Chandler	216-883-0220	R	17	0.2
Norca Corp	185 Great Neck Rd	Great Neck	NY	11022	Russel Stern	516-466-9500	R	17•	0.2
Speakman Co	PO Box 191	Wilmington	DE	19899	W Speakman III	302-764-9100	R	17	0.3
Connecticut Stamping & Bending	206 Newington Av	New Britain	CT	06051	B Barlow	203-225-4637	S	16	0.2
Broadway Industries Inc	PO Box 1210	Olathe	KS	66051	Charles D Miller	913-782-6244	R	16•	0.2
Broadway Collection	PO Box 1210	Olathe	KS	66051	Charles D Miller	913-782-6244	D	15	0.2
Modern Faucet Mfg Co	1700 E 58th Pl	Los Angeles	CA	90001	Claude Smith	213-582-6286	R	15•	0.2
Sanitary-Dash Manufacturing	929 Riverside Dr	N Grosvenordl	CT	06255	M E Schumann	203-923-9533	R	15	0.1
Creed Co	PO Box 700	Concordville	PA	19331	Michael Lavin	215-459-8600	D	14	0.1
Green Garden Inc	PO Box 275-D	Somerset	PA	15501	Jack Ketcham	814-443-3611	R	12•	0.1
Kraft Hardware Inc	306 E 61st St	New York	NY	10021	Carol Saperstein	212-838-2214	R	12	<0.1
Telsco Industries	3301 W Kingsley Rd	Garland	TX	75041	LO Snoddy	214-278-6131	R	11	0.1
Weather-Matic	PO Box 180205	Dallas	TX	75218	LO Snoddy	214-278-6131	D	11•	0.1
Elias Industries Inc	605 Epsilon Dr	Pittsburgh	PA	15238	Norman Elias	412-782-4300	R	10	<0.1
Frost Co	6523 14th Av	Kenosha	WI	53143	JJ Frost Jr	414-658-4301	R	10	0.1
Just Manufacturing Co	9233 King St	Franklin Park	IL	60131	John J Collins	708-678-5150	R	10•	0.1
Tapco Plumbing	605 Epsilon Dr	Pittsburgh	PA	15230	Norman Elias	412-782-4300	D	10	<0.1
Arrowhead Brass Products Inc	5147 Alhambra Av	Los Angeles	CA	90032	Frank Enterante	213-221-9137	R	9	<0.1
Crest/Good Manufacturing	325 Underhill Blv	Syosset	NY	11791	RB Goerler	516-921-7260	R	8	<0.1
Hago Manufacturing Company	1120 Globe Av	Mountainside	NJ	07092	Elinor Smith	908-232-8687	R	8	<0.1
Newperl Inc	407 Brookside Rd	Waterbury	CT	06720	Frederick Luedke	203-756-8895	R	8	<0.1
Northeastern Culvert	PO Box 40	Westminster St	VT	05159	Roland Scott	802-722-3358	D	8	<0.1
Woodford Manufacturing Co	2121 Waynoka Rd	Co Springs	CO	80915	Joseph Woodford	719-574-0600	R	8	0.1
Champion Irrigation Products	1460 Naud St	Los Angeles	CA	90012	A Pejsa	213-221-2108	R	6•	<0.1
Pacific Home Products Inc	PO Box 4348	Burlingame	CA	94011	Jim Semitekol	415-692-3062	R	6•	<0.1
US Tap Inc	PO Box 369	Frankfort	IN	46041	Svend Demant	317-659-3341	S	6	<0.1
Water Saver Faucet Co	701 W Erie St	Chicago	IL	60610	Samuel Kersten Jr	312-666-5500	R	6	0.1
Aqua-Trol Corp	11 Ralph Av	Copiague	NY	11726	Ilan Weiss	516-842-8833	R	5	<0.1
Fillpro Products Inc	55 E 111th St	Indianapolis	IN	46280	RA Manoogian	317-848-1812	S	4	<0.1
Chatham Brass Company Inc	5 Olsen Av	Edison	NJ	08820	Allen Leighton	908-494-7107	R	3	<0.1
Empire Brass Co	5000 Superior Av	Cleveland	OH	44103	R C McConville	216-431-6567	R	3	<0.1
All American Manufacturing Co	2201 E 51st St	Los Angeles	CA	90058	JF Norton	213-581-6293	R	2	<0.1
Manville Manufacturing Corp	342 Rockwell Av	Pontiac	MI	48341	Robert E Guldi	810-334-4583	R	2	<0.1
Bathroom Jewelry	16030 Arthur St	Cerritos	CA	90701	GG Michel	310-407-2707	D	1•	<0.1
Drainage Products Inc	383 S Main St	Windsor Locks	CT	06096	Paul L Tarko	203-668-5108	R	1•	<0.1
Frugal Flush Inc	1209 E Washington	Phoenix	AZ	85034	R L Schnakenberg	602-253-6275	R	1	<0.1
Savoy Brass	335 Underhill Blv	Syosset	NY	11791	RB Goerler	516-496-4600	S	1•	<0.1

Source: Ward's Business Directory of U.S. Private and Public Companies, Volumes 1 and 2, 1996. The company type code used is as follows: P - Public, R - Private, S - Subsidiary, D - Division, J - Joint Venture, A - Affiliate, G - Group. Sales are in millions of dollars, employees are in thousands. An asterisk (•) indicates an estimated sales volume. The symbol < stands for 'less than'. Company names and addresses are truncated, in some cases, to fit into the available space.

MATERIALS CONSUMED

Material	Quantity	Delivered Cost ($ million)
Materials, ingredients, containers, and supplies	(X)	1,181.8
Metal stampings	(X)	93.1
All other fabricated metal products (except forgings)	(X)	113.2
Iron and steel castings (rough and semifinished)	(X)	15.0
Copper and copper-base alloy castings (rough and semifinished)	(X)	67.5
Other nonferrous castings (rough and semifinished)	(X)	23.8
Forgings	(X)	9.3
Steel shapes and forms	(X)	17.4
Copper and copper-base alloy refinery shapes	(X)	3.7
Copper and copper-base alloy rod, bar, and mechanical wire, including extruded and/or drawn shapes	(X)	142.9
Copper and copper-base alloy pipe and tube	(X)	38.7
All other copper and copper-base alloy shapes and forms	(X)	31.3
Plastics resins consumed in the form of granules, pellets, powders, liquids, etc.	(X)	36.1
Fabricated plastics products (except gaskets, hoses, and belting)	(X)	71.9
Paperboard containers, boxes, and corrugated paperboard	(X)	52.8
All other materials and components, parts, containers, and supplies	(X)	304.0
Materials, ingredients, containers, and supplies, nsk	(X)	161.2

Source: 1992 *Economic Census*. Explanation of symbols used: (D): Withheld to avoid disclosure of competitive data; na: Not available; (S): Withheld because statistical norms were not met; (X): Not applicable; (Z): Less than half the unit shown; nec: Not elsewhere classified; nsk: Not specified by kind; - : zero; * : 10-19 percent estimated; ** : 20-29 percent estimated.

PRODUCT SHARE DETAILS

Product or Product Class	% Share	Product or Product Class	% Share
Plumbing fixture fittings and trim	100.00	greater than 4 in., with pop-up drain (metallic and nonmetallic)	12.28
Single lever plumbing fixture controls, two or three handle bath or shower fittings, and anti-scald bath or shower valves (brass goods)	30.99	Lavatory fittings (except basin cocks and single control), greater than 4 in., without pop-up drain (metallic and nonmetallic)	1.71
Lavatory single lever controls (metallic and nonmetallic)	19.07	Lavatory and sink basin cocks (one supply line only)	0.85
Kitchen with spray single lever controls (metallic and nonmetallic)	17.78	Wallmount combination sink faucet fittings	4.18
Kitchen without spray single lever controls (metallic and nonmetallic)	8.76	3 in. to 4 in. deck faucet sink fittings, exposed type (rough or plated, with or without hose end), excluding double laundry-tray faucets	5.82
Shower combination single lever controls, mechanical	6.99	6 in. to 8 in. deck faucet sink fittings, exposed type, with spray (metallic and nonmetallic)	7.86
Shower-tub combination single lever controls, mechanical	9.58	6 in. to 8 in. deck faucet sink fittings, exposed type, without spray (metallic and nonmetallic)	5.44
Tub filler single lever controls, mechanical	(D)	Deck faucet sink fittings, concealed type, with spray	6.19
Shower only single lever controls	(D)	Deck faucet sink fittings, concealed type, without spray	3.89
Two and three handle bathtub fillers	4.36	Single sink faucet fittings (solid flanged female and adjustable male flange)	1.14
Two and three handle shower fittings with shower heads	4.59	Other sink fittings (including sink strainers sold separately)	1.92
Two and three handle shower heads sold separately	4.76	Drains and overflows (metallic and nonmetallic), including pop-up drains for bath and shower, lavatory, and sink	7.95
Two and three handle bathtub and shower diverter spout	3.97	Lavatory and sink fittings (except single control), including drains and overflows (brass goods), nsk	2.92
Two and three handle bathtub and shower three valve diverter	5.98	Miscellaneous plumbing fixtures, fittings, and trim (brass goods)	38.98
Two and three handle personal showers (handheld)	3.06	IPS mechanical connecting plumbing fittings	1.74
Two and three handle tub and shower control only	0.85	Plumbing compression stops, including those with drains	(D)
Other two and three handle bath and shower fittings, including stall and gang	1.57	Solder connecting plumbing fittings	(D)
Thermostatic controlled anti-scald bath and shower fittings	(D)	Sediment, hydrant, lawn, hose bibb, and sill cock faucets	3.41
Thermo/pressure controlled anti-scald bath and shower fittings	0.10	Metallic plumbing P-traps	2.33
Pressure balanced controlled anti-scald bath and shower fittings	6.08	Nonmetallic plumbing P-traps	0.30
Single lever controls, two or three handle bath or shower fittings, and anti-scald bath or shower valves, nsk	0.87	Metallic plumbing S-traps	1.65
Lavatory and sink fittings (except single control), including drains and overflows (brass goods)	22.67	Nonmetallic plumbing S-traps	0.02
Metallic lavatory fittings (except basin cocks and single control), 4 in. center-set with pop-up drains	26.64	Lawn hose nozzles and lawn sprinklers	26.34
Nonmetallic lavatory fittings (except basin cocks and single control), 4 in. center-set with pop-up drains	0.48	Flushometer valves and flush valves (for gravity-type flush tanks)	7.32
Metallic lavatory fittings (except basin cocks and single control), 4 in. center-set without pop-up drains	9.99	Other miscellaneous plumbing items and accessories (water closet tank flushing controls, double laundry-tray faucets, etc.)	53.04
Nonmetallic lavatory fittings (except basin cocks and single control), 4 in. center-set without pop-up drains	0.76	Plumbing fixture fittings and trim, nsk	7.35
Lavatory fittings (except basin cocks and single control),			

Source: 1992 *Economic Census*. The values shown are percent of total shipments in an industry. Values of indented subcategories are summed in the main heading. The symbol (D) appears when data are withheld to prevent disclosure of competitive information. The abbreviation nsk stands for 'not specified by kind' and nec for 'not elsewhere classified'.

INPUTS AND OUTPUTS FOR PLUMBING FIXTURE FITTINGS & TRIM

Economic Sector or Industry Providing Inputs	%	Sector	Economic Sector or Industry Buying Outputs	%	Sector
Copper rolling & drawing	20.2	Manufg.	Residential 1-unit structures, nonfarm	24.9	Constr.
Brass, bronze, & copper castings	11.4	Manufg.	Nonfarm residential structure maintenance	17.6	Constr.
Screw machine and related products	10.4	Manufg.	Maintenance of nonfarm buildings nec	12.4	Constr.
Plating & polishing	7.7	Manufg.	Residential additions/alterations, nonfarm	6.6	Constr.
Wholesale trade	7.5	Trade	Office buildings	6.3	Constr.
Miscellaneous plastics products	5.6	Manufg.	Owner-occupied dwellings	4.0	Fin/R.E.
Blast furnaces & steel mills	2.7	Manufg.	Residential garden apartments	3.0	Constr.
Nonferrous castings, nec	2.1	Manufg.	Mobile homes	3.0	Manufg.
Electric services (utilities)	2.0	Util.	Construction of hospitals	2.6	Constr.
Primary copper	1.9	Manufg.	Construction of stores & restaurants	2.4	Constr.
Iron & steel foundries	1.6	Manufg.	Real estate	2.4	Fin/R.E.
Metal coating & allied services	1.5	Manufg.	Residential 2-4 unit structures, nonfarm	1.9	Constr.
Advertising	1.5	Services	Exports	1.9	Foreign
Maintenance of nonfarm buildings nec	1.2	Constr.	Electric utility facility construction	1.3	Constr.
Machinery, except electrical, nec	1.2	Manufg.	Wholesale trade	1.3	Trade
Paperboard containers & boxes	1.2	Manufg.	Construction of educational buildings	1.1	Constr.
Metal stampings, nec	1.1	Manufg.	Hotels & motels	1.1	Constr.
Eating & drinking places	1.1	Trade	Industrial buildings	1.1	Constr.
Banking	1.1	Fin/R.E.	Residential high-rise apartments	0.7	Constr.
Fabricated rubber products, nec	1.0	Manufg.	Travel trailers & campers	0.5	Manufg.
Gaskets, packing & sealing devices	1.0	Manufg.	Construction of religious buildings	0.4	Constr.
Motor freight transportation & warehousing	0.9	Util.	Farm housing units & additions & alterations	0.4	Constr.
Imports	0.9	Foreign	Maintenance of electric utility facilities	0.4	Constr.
Chemical preparations, nec	0.8	Manufg.	Resid. & other health facility construction	0.4	Constr.
Petroleum refining	0.7	Manufg.	Elementary & secondary schools	0.3	Services
Gas production & distribution (utilities)	0.7	Util.	Construction of nonfarm buildings nec	0.2	Constr.
Scrap	0.7	Scrap	Dormitories & other group housing	0.2	Constr.
Abrasive products	0.6	Manufg.	Maintenance of farm residential buildings	0.2	Constr.
Paints & allied products	0.6	Manufg.	Telephone & telegraph facility construction	0.2	Constr.
Special dies & tools & machine tool accessories	0.6	Manufg.	Warehouses	0.1	Constr.
Communications, except radio & TV	0.6	Util.	Motor homes (made on purchased chassis)	0.1	Manufg.
Aluminum rolling & drawing	0.5	Manufg.			
Welding apparatus, electric	0.5	Manufg.			
Equipment rental & leasing services	0.5	Services			
Engineering, architectural, & surveying services	0.4	Services			
Legal services	0.4	Services			
Management & consulting services & labs	0.4	Services			
Machine tools, metal cutting types	0.3	Manufg.			
Metal heat treating	0.3	Manufg.			
Real estate	0.3	Fin/R.E.			
Royalties	0.3	Fin/R.E.			
Accounting, auditing & bookkeeping	0.3	Services			
U.S. Postal Service	0.3	Gov't			
Aluminum castings	0.2	Manufg.			
Lubricating oils & greases	0.2	Manufg.			
Manifold business forms	0.2	Manufg.			
Miscellaneous fabricated wire products	0.2	Manufg.			
Air transportation	0.2	Util.			
Railroads & related services	0.2	Util.			
Insurance carriers	0.2	Fin/R.E.			
Computer & data processing services	0.2	Services			
Primary aluminum	0.1	Manufg.			
Sanitary services, steam supply, irrigation	0.1	Util.			
Security & commodity brokers	0.1	Fin/R.E.			
Automotive rental & leasing, without drivers	0.1	Services			
Automotive repair shops & services	0.1	Services			

Source: Benchmark Input-Output Accounts for the U.S. Economy, 1982, U.S. Department of Commerce, Washington, D.C., July 1991. Data, as reported in the source, are organized by the 1977 SIC structure in use in 1982 but have been matched, as closely as is possible, to the 1987 SIC structure used in this book.

OCCUPATIONS EMPLOYED BY SIC 343 - PLUMBING AND HEATING, EX ELECTRIC

Occupation	% of Total 1994	Change to 2005	Occupation	% of Total 1994	Change to 2005
Assemblers, fabricators, & hand workers nec	18.4	-15.9	Industrial truck & tractor operators	1.8	-15.9
Helpers, laborers, & material movers nec	4.3	-15.9	Industrial machinery mechanics	1.7	-7.5
Inspectors, testers, & graders, precision	3.3	-15.9	Secretaries, ex legal & medical	1.6	-23.4
Hand packers & packagers	2.8	-27.9	Combination machine tool operators	1.6	-7.5
Sales & related workers nec	2.7	-15.9	Metal molding machine workers	1.5	-7.4
Machine tool cutting & forming etc. nec	2.5	-15.9	Welding machine setters, operators	1.4	-24.4
Grinders & polishers, hand	2.4	-16.0	Machinists	1.3	-15.9
Traffic, shipping, & receiving clerks	2.4	-19.1	Maintenance repairers, general utility	1.3	-24.2
Metal & plastic machine workers nec	2.4	-25.5	General office clerks	1.3	-28.3
Machine forming operators, metal & plastic	2.3	-15.9	Bookkeeping, accounting, & auditing clerks	1.3	-36.9
Freight, stock, & material movers, hand	2.3	-32.7	Production, planning, & expediting clerks	1.2	-15.9
Welders & cutters	2.3	-15.9	Lathe & turning machine tool operators	1.1	-32.7
Sheet metal workers & duct installers	2.1	-15.9	Drafters	1.1	-34.4
General managers & top executives	2.0	-20.2	Stock clerks	1.1	-31.7
Machine tool cutting operators, metal & plastic	2.0	-30.0	Tool & die makers	1.1	-32.2

Source: Industry-Occupation Matrix, Bureau of Labor Statistics. These data relate to one or more 3-digit SIC industry groups rather than to a single 4-digit SIC. The change reported for each occupation to the year 2005 is a percent of growth or decline as estimated by the Bureau of Labor Statistics. The abbreviation nec stands for 'not elsewhere classified'.

LOCATION BY STATE AND REGIONAL CONCENTRATION

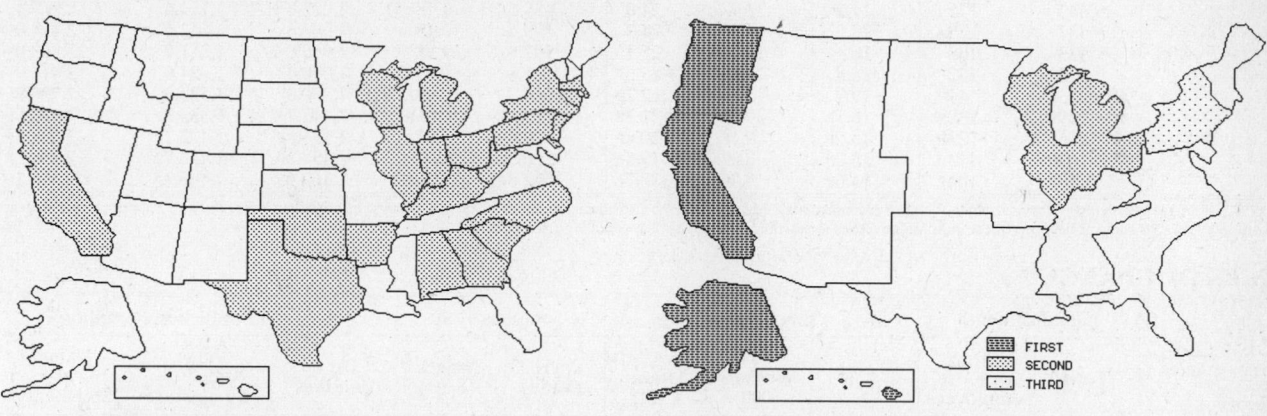

FIRST
SECOND
THIRD

INDUSTRY DATA BY STATE

State	Establish-ments	Shipments			Employment				Cost as % of Shipments	Investment per Employee ($)
		Total ($ mil)	% of U.S.	Per Establ.	Total Number	% of U.S.	Per Establ.	Wages ($/hour)		
California	47	633.2	23.4	13.5	4,700	27.0	100	10.64	43.3	4,489
Illinois	13	236.8	8.8	18.2	1,200	6.9	92	13.18	49.3	2,833
Texas	11	157.9	5.8	14.4	1,100	6.3	100	9.56	49.9	-
Michigan	6	91.2	3.4	15.2	500	2.9	83	8.50	49.7	3,000
New York	12	71.1	2.6	5.9	500	2.9	42	8.56	66.7	-
Connecticut	12	61.4	2.3	5.1	600	3.4	50	9.22	50.8	1,333
Ohio	15	(D)	-	-	750 *	4.3	50	-	-	10,800
Pennsylvania	8	(D)	-	-	750 *	4.3	94	-	-	-
Indiana	7	(D)	-	-	1,750 *	10.1	250	-	-	-
Massachusetts	5	(D)	-	-	375 *	2.2	75	-	-	-
Wisconsin	5	(D)	-	-	175 *	1.0	35	-	-	-
Georgia	4	(D)	-	-	175 *	1.0	44	-	-	-
North Carolina	4	(D)	-	-	1,750 *	10.1	438	-	-	-
Alabama	2	(D)	-	-	375 *	2.2	188	-	-	-
Arkansas	2	(D)	-	-	375 *	2.2	188	-	-	-
New Jersey	2	(D)	-	-	750 *	4.3	375	-	-	-
Oklahoma	2	(D)	-	-	375 *	2.2	188	-	-	-
South Carolina	2	(D)	-	-	375 *	2.2	188	-	-	-
Delaware	1	(D)	-	-	175 *	1.0	175	-	-	-
Kentucky	1	(D)	-	-	375 *	2.2	375	-	-	-
West Virginia	1	(D)	-	-	375 *	2.2	375	-	-	-

Source: 1992 Economic Census. The states are in descending order of shipments or establishments (if shipment data are missing for the majority). The symbol (D) appears when data are withheld to prevent disclosure of competitive information. States marked with (D) are sorted by number of establishments. A dash (-) indicates that the data element cannot be calculated; * indicates the midpoint of a range.

3433 - HEATING EQUIPMENT, EXCEPT ELECTRIC

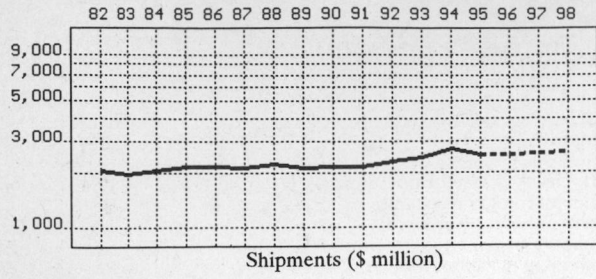

Shipments ($ million)

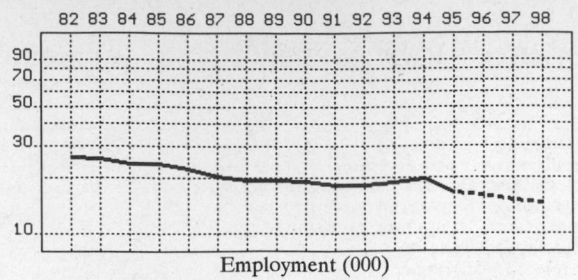

Employment (000)

GENERAL STATISTICS

Year	Com-panies	Establishments		Employment			Compensation		Production ($ million)			
		Total	with 20 or more employees	Total (000)	Production Workers (000)	Hours (Mil)	Payroll ($ mil)	Wages ($/hr)	Cost of Materials	Value Added by Manufacture	Value of Shipments	Capital Invest.
1982	875	905	260	26.0	17.1	31.8	443.4	7.71	1,023.5	1,042.4	2,082.5	44.0
1983		820	249	25.9	17.3	33.4	447.4	7.51	965.4	997.4	1,964.5	32.1
1984		735	238	24.0	16.0	31.2	464.8	8.35	1,028.2	1,061.3	2,084.4	47.0
1985		651	227	24.0	16.3	32.5	490.4	8.51	1,081.4	1,097.6	2,183.7	58.6
1986		608	216	22.4	15.2	30.1	476.4	9.00	1,078.3	1,086.2	2,178.8	45.7
1987	544	556	182	20.4	13.2	26.7	455.7	9.01	980.4	1,143.8	2,123.9	43.5
1988		503	182	19.5	12.5	25.9	467.1	9.32	1,024.7	1,212.3	2,221.2	31.0
1989		464	179	19.5	12.2	24.4	439.6	9.50	955.5	1,149.6	2,127.7	56.5
1990		443	175	19.2	11.9	24.0	455.6	9.78	1,022.8	1,154.6	2,167.4	48.7
1991		437	173	18.3	11.5	24.3	479.4	10.18	1,042.8	1,142.4	2,174.0	45.1
1992	405	423	168	18.1	11.0	25.1	502.6	9.71	1,050.8	1,251.9	2,303.0	37.3
1993		412	162	18.8	11.5	25.6	525.8	9.65	1,070.4	1,324.6	2,416.9	48.2
1994		293P	141P	19.7	12.5	22.2P	564.7	10.43P	1,230.0	1,522.3	2,736.9	42.1
1995		249P	132P	16.5P	9.9P	21.3P	523.3P	10.65P	1,115.7P	1,380.8P	2,482.5P	44.9P
1996		205P	123P	15.9P	9.4P	20.4P	529.8P	10.87P	1,133.0P	1,402.2P	2,521.0P	45.0P
1997		161P	113P	15.2P	8.8P	19.6P	536.3P	11.08P	1,150.3P	1,423.7P	2,559.6P	45.0P
1998		117P	104P	14.6P	8.3P	18.7P	542.8P	11.30P	1,167.6P	1,445.1P	2,598.1P	45.1P

Sources: 1982, 1987, 1992 *Economic Census*; *Annual Survey of Manufactures*, 83-86, 88-91, 93-94. Establishment counts for non-Census years are from *County Business Patterns*; establishment values for 83-84 are extrapolations. 'P's show projections by the editors. Industries reclassified in 87 will not have data for prior years.

INDICES OF CHANGE

Year	Com-panies	Establishments		Employment			Compensation		Production ($ million)			
		Total	with 20 or more employees	Total (000)	Production Workers (000)	Hours (Mil)	Payroll ($ mil)	Wages ($/hr)	Cost of Materials	Value Added by Manufacture	Value of Shipments	Capital Invest.
1982	216	214	155	144	155	127	88	79	97	83	90	118
1983		194	148	143	157	133	89	77	92	80	85	86
1984		174	142	133	145	124	92	86	98	85	91	126
1985		154	135	133	148	129	98	88	103	88	95	157
1986		144	129	124	138	120	95	93	103	87	95	123
1987	134	131	108	113	120	106	91	93	93	91	92	117
1988		119	108	108	114	103	93	96	98	97	96	83
1989		110	107	108	111	97	87	98	91	92	92	151
1990		105	104	106	108	96	91	101	97	92	94	131
1991		103	103	101	105	97	95	105	99	91	94	121
1992	100	100	100	100	100	100	100	100	100	100	100	100
1993		97	96	104	105	102	105	99	102	106	105	129
1994		69P	84P	109	114	88P	112	107P	117	122	119	113
1995		59P	78P	91P	90P	85P	104P	110P	106P	110P	108P	120P
1996		48P	73P	88P	85P	81P	105P	112P	108P	112P	109P	121P
1997		38P	68P	84P	80P	78P	107P	114P	109P	114P	111P	121P
1998		28P	62P	80P	75P	74P	108P	116P	111P	115P	113P	121P

Sources: Same as General Statistics. Values reflect change from the base year, 1992. Values above 100 mean greater than 92, values below 100 mean less than 92, and a value of 100 in the 82-91 or 93-98 period means same as 92. 'P's mark projections by the editors.

SELECTED RATIOS

For 1994	Avg. of All Manufact.	Analyzed Industry	Index	For 1994	Avg. of All Manufact.	Analyzed Industry	Index
Employees per Establishment	49	67	137	Value Added per Production Worker	134,084	121,784	91
Payroll per Establishment	1,500,273	1,925,810	128	Cost per Establishment	5,045,178	4,194,698	83
Payroll per Employee	30,620	28,665	94	Cost per Employee	102,970	62,437	61
Production Workers per Establishment	34	43	124	Cost per Production Worker	146,988	98,400	67
Wages per Establishment	853,319	789,650	93	Shipments per Establishment	9,576,895	9,333,716	97
Wages per Production Worker	24,861	18,524	75	Shipments per Employee	195,460	138,929	71
Hours per Production Worker	2,056	1,776	86	Shipments per Production Worker	279,017	218,952	78
Wages per Hour	12.09	10.43	86	Investment per Establishment	321,011	143,575	45
Value Added per Establishment	4,602,255	5,191,536	113	Investment per Employee	6,552	2,137	33
Value Added per Employee	93,930	77,274	82	Investment per Production Worker	9,352	3,368	36

Sources: Same as General Statistics. The 'Average of All Manufacturing' column represents the average of all manufacturing industries reported for the most recent complete year available. The Index shows the relationship between the Average and the Analyzed Industry. For example, 100 means that they are equal; 500 that the Analyzed Industry is five times the average; 50 means that the Analyzed Industry is half the national average. The abbreviation 'na' is used to show that data are 'not available'.

LEADING COMPANIES   Number shown: 75   Total sales ($ mil): 2,259   Total employment (000): 18.6

Company Name	Address				CEO Name	Phone	Co. Type	Sales ($ mil)	Empl. (000)
Falcon Building Products Inc	2 N Riverside Plz	Chicago	IL	60606	William K Hall	312-906-9700	P	372	3.2
Sterling Radiator	260 N Elm St	Westfield	MA	01085	John E Reed	413-568-9571	D	112	1.4
Martin Industries Inc	PO Box 128	Florence	AL	35631	Jim Wilson	205-767-0330	R	100	0.7
Tampella Power Corp	PO Box 3308	Williamsport	PA	17701	William Pollock	717-326-3361	S	94	0.5
Gensco Inc	4402 20th St E	Tacoma	WA	98424	Charles E Walters	206-922-3003	R	85	0.4
Weil-McLain Co	500 Blaine St	Michigan City	IN	46360	WL Jackson	219-879-6561	D	73*	0.7
Eclipse Inc	1665 Elmwood Rd	Rockford	IL	61103	Douglas C Perks	815-877-3031	R	70	0.5
Majco Building Specialties LP	1000 E Market St	Huntington	IN	46750	Raymond E Deasy	219-356-8000	R	70	0.6
Wayne Home Equipment	801 Glasgow Av	Fort Wayne	IN	46803	Philip Jones	219-425-9200	D	64	<0.1
Climate Master Inc	PO Box 25788	Oklahoma City	OK	73125	Barry Golsen	405-745-6000	S	50	0.3
Maxon Corp	201 E 18th St	Muncie	IN	47302	RM Smitson	317-284-3304	R	50	0.3
North American Mfg Co	4455 E 71st St	Cleveland	OH	44105	Robert J Neville	216-271-6000	R	50	0.5
RW Beckett Corp	PO Box 1289	Elyria	OH	44036	John Beckett	216-327-1060	R	50	0.2
Vermont Castings Inc	PO Box 501	Bethel	VT	05032	Dennis Dillen	802-234-3000	S	45	0.3
Ducane Corp	118 W Main St	Blackville	SC	29817	John Ducate Jr	803-284-3322	R	44*	0.5
Suburban Manufacturing Co	PO Box 399	Dayton	TN	37321	James H Peden	615-775-2131	S	44	0.5
WFI Industries Ltd	9000 Conserv	Fort Wayne	IN	46809	Daniel L Ellis	219-478-5667	P	40	0.2
Harsco Corp	PO Box 458	E Stroudsburg	PA	18301	Timothy M Tauer	717-421-7500	D	39	0.2
Temtex Fireplace Products Inc	301 S Perimeter Pk	Nashville	TN	37211	ER Buford	615-728-4001	S	35	0.3
WaterFurnace International Inc	9000 Conserv	Fort Wayne	IN	46809	Daniel L Ellis	219-478-5667	S	32	0.2
Raypak Inc	31111 Agoura Rd	Westlake Vil	CA	91361	T Ots	818-889-1500	S	27	0.3
Consolidated Industries Corp	PO Box 7800	Lafayette	IN	47903	Robert McAfee	317-447-9500	S	25	0.1
Dunkirk Radiator Corp	85 Middle Rd	Dunkirk	NY	14048	TE Reed	716-366-5500	R	25*	0.2
Industrial Eng & Equipment Co	425 Hanley Indrial	St Louis	MO	63144	F Epstein	314-644-4300	R	25*	0.4
Peerless Industries Inc	PO Box 388	Boyertown	PA	19512	Richard M Smith	215-367-2153	R	23	0.2
Pyro Industries Inc	695 Pease Rd	Burlington	WA	98233	Jerry Whitfield	360-757-9728	R	23	0.2
Eclipse Combustion	1665 Elmwood Rd	Rockford	IL	61103	Richard Russell	815-877-3031	D	22	0.3
Peabody Engineering Corp	39 Maple Tree Av	Stamford	CT	06906	Peter Miller	203-327-7000	S	22	0.2
Aerco International Inc	PO Box 128	Northvale	NJ	07647	Basem Hishmeh	201-768-2400	R	20	<0.1
Aladdin Steel Products Inc	401 N Wynne St	Colville	WA	99114	Alan Trusler	509-684-3745	R	20*	0.1
Burner Systems International Inc	2806 E 50th St	Chattanooga	TN	37407	Michael L Frost	615-867-5787	R	20	0.3
Empire Comfort Systems Inc	PO Box 529	Belleville	IL	62222	Brian M Bauer	618-233-7420	R	20	0.2
Power Flame Inc	2001 S 21st St	Parsons	KS	67357	William Wiener	316-421-0480	R	20	0.2
Schutte and Koerting	2233 State Rd	Bensalem	PA	19020	Dominick V Scott	215-639-0900	D	20	0.2
Williams Furnace Co	225 Acacia St	Colton	CA	92324	Brett Austin	909-825-0993	S	20	0.2
Rapid Engineering Inc	1100 7 Mile Rd NW	Comstock Park	MI	49321	James V Dirkes	616-784-0500	R	18	0.1
Roberts-Gordon Inc	PO Box 44	Buffalo	NY	14240	Paul A Dines	716-852-4400	R	18	0.1
Smith Cast Iron Boilers	260 N Elm St	Westfield	MA	01085	John Reed	413-562-9631	R	16*	0.1
Hauck Manufacturing Co	PO Box 90	Lebanon	PA	17042	Louis Etschmaier	717-272-3051	R	16	0.2
Fuchs Systems Inc	PO Box 379	Salisbury	NC	28145	Manfred Haissig	704-633-2141	S	15*	0.2
Hurst Boiler and Welding	PO Drawer 529	Coolidge	GA	31738	Clifton E Hurst	912-346-3545	R	15	0.1
Vulcan Radiator Corp	260 N Elm St	Westfield	MA	01085	John Kaddanas	413-568-9571	S	14*	0.1
Johnston Boiler Co	PO Box 300	Ferrysburg	MI	49409	Stephen Parker	616-842-5050	R	13	0.1
Perfection-Schwank Inc	PO Box 749	Waynesboro	GA	30830	Bernd Schwank	706-554-2101	R	13	0.2
Haydon Corp	2 Jasper St	Paterson	NJ	07522	Richard Denison	201-904-0800	R	12	0.1
Louisville Tin and Stove	737 S 13th St	Louisville	KY	40201	N Jenkins	502-589-5380	R	12	0.1
PVI Industries Inc	PO Box 7124	Fort Worth	TX	76111	Thomas G McCoy	817-335-9531	R	12	0.2
Ebner Furnaces Inc	224 Quadral Dr	Wadsworth	OH	44281	Robert Ebner	216-335-2311	S	11	0.1
Vestal Manufacturing Co	PO Box 420	Sweetwater	TN	37874	David Vestal	615-337-6125	S	11	0.2
FAFCO Inc	2690 Middlefield Rd	Redwood City	CA	94063	Freeman A Ford	415-363-2690	P	10	<0.1
Sid E Parker Boiler Mfg	5930 Bandini Blv	Los Angeles	CA	90040	Sid E Danenhauer	213-727-9800	R	10	<0.1
Airtherm Products Inc	PO Box 7039	St Louis	MO	63132	ST Kohlbry	314-993-3400	S	10	0.1
Ajax Boiler Inc	14059 Stage Rd	Santa Fe Sprgs	CA	90670	Ed Cancilla	310-926-5161	R	10	0.1
Aqua-Chem Inc	351 21st St	Monroe	WI	53566	T Hanna	608-325-3141	D	10	0.1
ARGO Industries Inc	554 Wilbur Cross	Berlin	CT	06037	Arthur Godbout	203-828-6334	R	10	<0.1
Johnson Gas Appliance Co	520 E Av NW	Cedar Rapids	IA	52405	SB O'Donnel	319-365-5267	R	10	<0.1
LB White Company Inc	W 6636 LB White	Onalaska	WI	54650	Tony Wilson	608-783-5691	R	10	0.1
Preferred Utilities Mfg Corp	11 South St	Danbury	CT	06810	Robert G Bohn	203-743-6741	R	10	<0.1
Webster Eng & Manufacturing	PO Drawer 748	Winfield	KS	67156	Robert D Brown	316-221-7464	R	10	<0.1
Country Flame Inc	PO Box 151	Mount Vernon	MO	65712	Jean Holladay	417-466-7161	R	9	0.1
Heat-N-Glo Fireplaces	6665 W Hwy 13	Savage	MN	55370	D Shimek	612-890-8367	R	9	0.2
Midco International Inc	4140 W Victoria St	Chicago	IL	60646	SZ Beyer	312-604-8700	R	8	0.1
New Buck Corp	PO Box 69	Spruce Pine	NC	28777	Robert Bailey	704-765-6144	R	8	<0.1
Sid Harvey Industries Inc	605 Locust St	Garden City	NY	11530	Ronald Stahl	516-745-9200	R	8*	<0.1
United Solar Systems Corp	1100 W Maple Rd	Troy	MI	48084	S Nagashirma	810-362-4170	R	8*	<0.1
Mountain Safety Research Inc	PO Box 24547	Seattle	WA	98124	David Bartholomew	206-624-7048	S	7*	<0.1
Scheu Manufacturing Co	PO Box 250	Upland	CA	91786	A Scheu	909-982-8933	R	7*	0.2
Watlow/AOV Inc	4545 E La Palma	Anaheim	CA	92807	Joseph L Riley	714-779-2252	S	7	<0.1
Embassy Industries Inc	300 Smith St	Farmingdale	NY	11735	Richard A Horowitz	516-694-1800	S	6	<0.1
Bock Water Heaters Inc	110 S Dickinson St	Madison	WI	53703	John C Bock	608-257-2225	R	6	<0.1
Fulton Thermal Corp	PO Box 257	Pulaski	NY	13142	Ronald B Palm	315-298-6597	R	6	<0.1
General Machine Corp	302 S Fourth St	Emmaus	PA	18049	Jeffrey Perelman	215-965-9041	S	6	<0.1
Electro-Flex Heat Inc	PO Box 88	Bloomfield	CT	06002	Harold Davis Jr	203-242-6287	R	6	0.1
AGF Inc	PO Box 496	Elizabeth	NJ	07207	Michael A Dorio	908-352-2120	R	5	<0.1
Dixstar Inc	PO Box 4756	Chattanooga	TN	37405	Terry Kocher	615-265-3441	R	5	<0.1

Source: *Ward's Business Directory of U.S. Private and Public Companies*, Volumes 1 and 2, 1996. The company type code used is as follows: P - Public, R - Private, S - Subsidiary, D - Division, J - Joint Venture, A - Affiliate, G - Group. Sales are in millions of dollars, employees are in thousands. An asterisk (*) indicates an estimated sales volume. The symbol < stands for 'less than'. Company names and addresses are truncated, in some cases, to fit into the available space.

MATERIALS CONSUMED

Material	Quantity	Delivered Cost ($ million)
Materials, ingredients, containers, and supplies	(X)	950.4
Metal bolts, nuts, screws, washers, rivets, and other screw machine products	(X)	14.8
Fabricated metal pipe (except castings and forgings)	(X)	14.5
Fabricated metal valves and pipe fittings (except castings and forgings)	(X)	20.7
Metal parts and attachments specially designed for heating equipment, except electric	(X)	35.2
Other fabricated metal products (except forgings)	(X)	44.9
Iron and steel castings (rough and semifinished)	(X)	56.5
Nonferrous (aluminum, copper, etc.) castings (rough and semifinished)	(X)	7.5
Steel bars, bar shapes, and plates	(X)	21.6
Steel sheet and strip, including tin plate	(X)	91.7
Steel structural shapes	(X)	1.7
All other steel shapes and forms	(X)	16.2
Copper and copper-base alloy shapes and forms	(X)	6.9
Aluminum and aluminum-base alloy shapes and forms	(X)	7.4
Other nonferrous shapes and forms	(X)	2.5
Electric motors and generators (all types)	(X)	28.2
Automatic temperature controls (thermostats, regulators, etc.)	(X)	94.8
Paperboard containers, boxes, and corrugated paperboard	(X)	21.1
All other materials and components, parts, containers, and supplies	(X)	262.9
Materials, ingredients, containers, and supplies, nsk	(X)	201.3

Source: 1992 *Economic Census.* Explanation of symbols used: (D): Withheld to avoid disclosure of competitive data; na: Not available; (S): Withheld because statistical norms were not met; (X): Not applicable; (Z): Less than half the unit shown; nec: Not elsewhere classified; nsk: Not specified by kind; - : zero; * : 10-19 percent estimated; ** : 20-29 percent estimated.

PRODUCT SHARE DETAILS

Product or Product Class	% Share	Product or Product Class	% Share
Heating equipment, except electric	100.00	electric)	6.58
Cast iron heating boilers, radiators, and convectors, except parts	14.58	Gas-fired infrared heaters	14.97
Gas-fired cast iron boilers, excluding parts	43.81	Mechanical stokers (except electric)	2.06
Other cast iron boilers, excluding electric and parts	39.91	Parts for floor and wall furnaces, unit heaters, gas-fired infrared heaters, and mechanical stokers	9.00
Cast iron and steel radiators and convectors, excluding parts	3.69	Floor and wall furnaces, unit heaters, infrared heaters, and mechanical stokers, nsk	4.54
Aluminum and other nonferrous metal radiators and convectors, including baseboard and finned tube type, residential, industrial, and special type (excluding electric and parts)	12.60	Other heating equipment, except electric (including parts for nonelectric heating equipment and oil burners)	35.19
Domestic heating stoves, except electric (excluding parts)	15.20	Oil burners (sold separately, except range-type), residential-type (oil consumption rate less than 6 gal/hr)	(D)
Gas (all types) domestic heating stoves, including vented and unvented circulators, radiants and laundry stoves (excluding parts)	25.56	Oil burners (sold seperately, except range-type), commercial-and industrial-type (oil consumption rate 6 gal/hr or more)	1.33
Coal, kerosene, gasoline, and fuel oil domestic heating stoves (excluding parts)	10.32	Oil burners (shipped with furnaces or boilers, except range-type), residential-type (oil consumption rate less than 6 gal/hr)	(D)
Freestanding wood fireplaces and domestic heating stoves (all types), including circulating and radiants (excluding parts)	31.82	Oil burners (shipped with furnaces or boilers, except range-type), commercial-and industrial-type (oil consumption rate 6 gal/hr or more)	0.80
Wood pellet fuel-burning domestic heaters (excluding parts)	10.95	Range-type oil burners (sleeve- and pot-type, for use in water heaters, stoves, ranges, etc.)	(D)
Heat exchangers and zero clearance (factory built) domestic wood fireplaces	12.98	Parts and attachments for oil burners	5.43
Domestic wood fireplace inserts	3.77	Gas burners and gas conversion burners (including parts), 400,000 Btu/hr or less	8.51
Domestic heating stoves, except electric (excluding parts), nsk	4.61	Gas burners and gas conversion burners (including parts), 400,001 Btu/hr or more	15.02
Steel heating boilers (15 p.s.i. or less), and all hot water heating boilers (except electric), excluding parts	5.56	Parts and attachments for gas burners (sold separately)	3.25
Steel heating boilers (15 p.s.i. or less), and all hot water heating boilers (except electric), excluding parts, 400,000 Btu/hr or less	56.97	Heat transfer coils (blast coils)	0.61
		Range boilers, expansion tanks (including basement tanks) and hot water storage tanks (except electric)	(D)
Steel heating boilers (15 p.s.i. or less), and all hot water heating boilers (except electric), excluding parts, 400,001 Btu/hr or more, including scotch type and horizontal fire box	37.88	Heating unit ventilators (except electric)	8.97
Steel heating boilers (15 p.s.i. or less), and all hot water heating boilers, excluding parts, nsk	5.07	Tanks for direct-fired water heaters, except electric (sold separately)	(D)
Floor and wall furnaces, unit heaters, infrared heaters, and mechanical stokers (except electric)	16.00	Nonelectric prefabricated metal fireplaces for commercial and industrial use	6.51
Floor and wall furnaces (gas and oil)	37.12	Solar energy collectors (water or air)	4.42
Gas-fired unit heaters, 400,000 Btu/hr or less	25.73	Other heating equipment (dual-fired gas and oil burners, etc.) and other parts for heating equipment, except electric	27.73
Steam or hot water heating element, centrifugal fan-type (blower) and propeller fan-type unit heaters (except		Other heating equipment, except electric, (including parts for nonelectric heating equipment and oil burners), nsk	7.03
		Heating equipment, except electric, nsk	13.47

Source: 1992 *Economic Census.* The values shown are percent of total shipments in an industry. Values of indented subcategories are summed in the main heading. The symbol (D) appears when data are withheld to prevent disclosure of competitive information. The abbreviation nsk stands for 'not specified by kind' and nec for 'not elsewhere classified'.

INPUTS AND OUTPUTS FOR HEATING EQUIPMENT, EXCEPT ELECTRIC

Economic Sector or Industry Providing Inputs	%	Sector	Economic Sector or Industry Buying Outputs	%	Sector
Imports	35.1	Foreign	Personal consumption expenditures	32.3	
Blast furnaces & steel mills	15.5	Manufg.	Office buildings	9.7	Constr.
Wholesale trade	6.3	Trade	Exports	7.0	Foreign
Iron & steel foundries	4.9	Manufg.	Residential additions/alterations, nonfarm	5.9	Constr.
Heating equipment, except electric	3.6	Manufg.	Nonfarm residential structure maintenance	5.8	Constr.
Refrigeration & heating equipment	3.6	Manufg.	Maintenance of nonfarm buildings nec	5.6	Constr.
Blowers & fans	3.0	Manufg.	Industrial buildings	5.1	Constr.
Environmental controls	2.3	Manufg.	Residential 1-unit structures, nonfarm	4.5	Constr.
Motors & generators	1.9	Manufg.	Construction of hospitals	4.1	Constr.
Aluminum rolling & drawing	1.3	Manufg.	Construction of stores & restaurants	3.0	Constr.
Fabricated metal products, nec	1.3	Manufg.	Residential garden apartments	2.4	Constr.
Electric services (utilities)	1.1	Util.	Mobile homes	2.3	Manufg.
Copper rolling & drawing	0.9	Manufg.	Heating equipment, except electric	2.2	Manufg.
Motor freight transportation & warehousing	0.9	Util.	Federal Government purchases, national defense	1.4	Fed Govt
Advertising	0.9	Services	Construction of educational buildings	1.2	Constr.
Metal stampings, nec	0.8	Manufg.	Hotels & motels	0.8	Constr.
Paperboard containers & boxes	0.8	Manufg.	Residential 2-4 unit structures, nonfarm	0.8	Constr.
Primary aluminum	0.8	Manufg.	Travel trailers & campers	0.7	Manufg.
Banking	0.8	Fin/R.E.	Construction of nonfarm buildings nec	0.6	Constr.
Maintenance of nonfarm buildings nec	0.7	Constr.	Warehouses	0.6	Constr.
Machinery, except electrical, nec	0.7	Manufg.	Amusement & recreation building construction	0.5	Constr.
Petroleum refining	0.7	Manufg.	Resid. & other health facility construction	0.5	Constr.
Pipe, valves, & pipe fittings	0.7	Manufg.	Residential high-rise apartments	0.5	Constr.
Real estate	0.7	Fin/R.E.	Refrigeration & heating equipment	0.4	Manufg.
Industrial controls	0.6	Manufg.	Electric utility facility construction	0.3	Constr.
Communications, except radio & TV	0.6	Util.	Construction of religious buildings	0.2	Constr.
Gas production & distribution (utilities)	0.6	Util.	Farm housing units & additions & alterations	0.2	Constr.
Eating & drinking places	0.6	Trade	Maintenance of farm residential buildings	0.2	Constr.
Screw machine and related products	0.5	Manufg.	Maintenance of military facilities	0.2	Constr.
Equipment rental & leasing services	0.5	Services	Dormitories & other group housing	0.1	Constr.
Gaskets, packing & sealing devices	0.4	Manufg.	Telephone & telegraph facility construction	0.1	Constr.
Special dies & tools & machine tool accessories	0.4	Manufg.			
Welding apparatus, electric	0.4	Manufg.			
Railroads & related services	0.4	Util.			
Abrasive products	0.3	Manufg.			
Paints & allied products	0.3	Manufg.			
Signs & advertising displays	0.3	Manufg.			
Engineering, architectural, & surveying services	0.3	Services			
U.S. Postal Service	0.3	Gov't			
Scrap	0.3	Scrap			
Aluminum castings	0.2	Manufg.			
Lubricating oils & greases	0.2	Manufg.			
Metal heat treating	0.2	Manufg.			
Pumps & compressors	0.2	Manufg.			
Wood pallets & skids	0.2	Manufg.			
Royalties	0.2	Fin/R.E.			
Legal services	0.2	Services			
Management & consulting services & labs	0.2	Services			
Internal combustion engines, nec	0.1	Manufg.			
Machine tools, metal forming types	0.1	Manufg.			
Air transportation	0.1	Util.			
Insurance carriers	0.1	Fin/R.E.			
Accounting, auditing & bookkeeping	0.1	Services			

Source: Benchmark Input-Output Accounts for the U.S. Economy, 1982, U.S. Department of Commerce, Washington, D.C., July 1991. Data, as reported in the source, are organized by the 1977 SIC structure in use in 1982 but have been matched, as closely as is possible, to the 1987 SIC structure used in this book.

OCCUPATIONS EMPLOYED BY SIC 343 - PLUMBING AND HEATING, EX ELECTRIC

Occupation	% of Total 1994	Change to 2005	Occupation	% of Total 1994	Change to 2005
Assemblers, fabricators, & hand workers nec	18.4	-15.9	Industrial truck & tractor operators	1.8	-15.9
Helpers, laborers, & material movers nec	4.3	-15.9	Industrial machinery mechanics	1.7	-7.5
Inspectors, testers, & graders, precision	3.3	-15.9	Secretaries, ex legal & medical	1.6	-23.4
Hand packers & packagers	2.8	-27.9	Combination machine tool operators	1.6	-7.5
Sales & related workers nec	2.7	-15.9	Metal molding machine workers	1.5	-7.4
Machine tool cutting & forming etc. nec	2.5	-15.9	Welding machine setters, operators	1.4	-24.4
Grinders & polishers, hand	2.4	-16.0	Machinists	1.3	-15.9
Traffic, shipping, & receiving clerks	2.4	-19.1	Maintenance repairers, general utility	1.3	-24.2
Metal & plastic machine workers nec	2.4	-25.5	General office clerks	1.3	-28.3
Machine forming operators, metal & plastic	2.3	-15.9	Bookkeeping, accounting, & auditing clerks	1.3	-36.9
Freight, stock, & material movers, hand	2.3	-32.7	Production, planning, & expediting clerks	1.2	-15.9
Welders & cutters	2.3	-15.9	Lathe & turning machine tool operators	1.1	-32.7
Sheet metal workers & duct installers	2.1	-15.9	Drafters	1.1	-34.4
General managers & top executives	2.0	-20.2	Stock clerks	1.1	-31.7
Machine tool cutting operators, metal & plastic	2.0	-30.0	Tool & die makers	1.1	-32.2

Source: Industry-Occupation Matrix, Bureau of Labor Statistics. These data relate to one or more 3-digit SIC industry groups rather than to a single 4-digit SIC. The change reported for each occupation to the year 2005 is a percent of growth or decline as estimated by the Bureau of Labor Statistics. The abbreviation nec stands for 'not elsewhere classified'.

LOCATION BY STATE AND REGIONAL CONCENTRATION

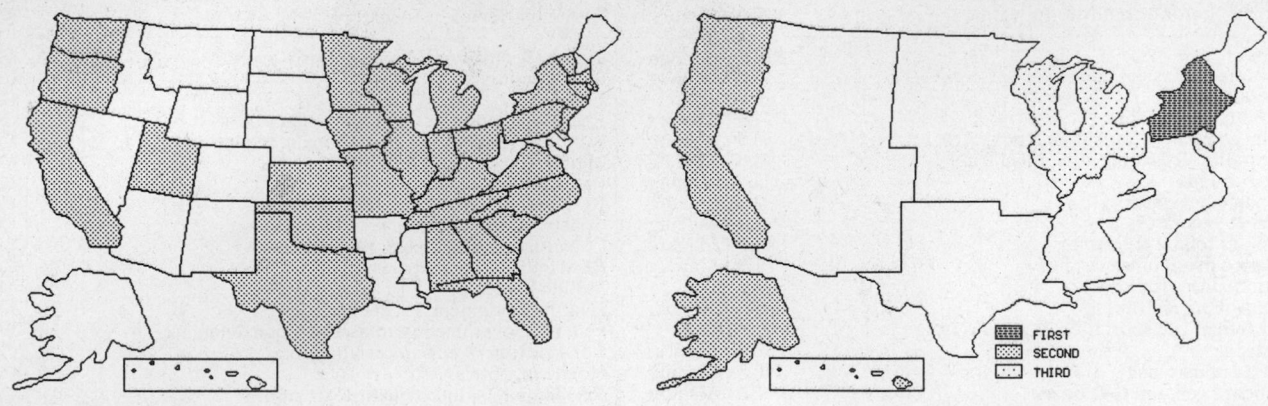

INDUSTRY DATA BY STATE

State	Establish-ments	Shipments			Employment				Cost as % of Shipments	Investment per Employee ($)
		Total ($ mil)	% of U.S.	Per Establ.	Total Number	% of U.S.	Per Establ.	Wages ($/hour)		
Pennsylvania	40	322.0	14.0	8.1	2,200	12.2	55	12.42	43.9	2,682
Indiana	11	229.7	10.0	20.9	1,400	7.7	127	15.00	36.4	2,643
California	44	170.9	7.4	3.9	1,600	8.8	36	4.39	46.1	1,563
New York	29	157.9	6.9	5.4	1,100	6.1	38	11.43	53.1	1,636
Illinois	17	127.1	5.5	7.5	1,000	5.5	59	11.46	41.1	-
Ohio	12	123.8	5.4	10.3	1,000	5.5	83	11.45	47.3	-
Minnesota	22	87.6	3.8	4.0	800	4.4	36	12.89	48.4	1,750
Washington	17	71.5	3.1	4.2	500	2.8	29	10.57	37.1	2,200
New Jersey	21	67.4	2.9	3.2	500	2.8	24	13.29	43.3	-
Wisconsin	9	45.2	2.0	5.0	300	1.7	33	8.75	50.0	3,000
North Carolina	12	38.2	1.7	3.2	400	2.2	33	6.88	49.7	1,500
Massachusetts	6	36.1	1.6	6.0	300	1.7	50	11.00	36.6	1,667
Connecticut	6	33.1	1.4	5.5	300	1.7	50	13.00	50.5	1,667
Florida	9	12.6	0.5	1.4	100	0.6	11	14.00	44.4	-
Michigan	24	(D)	-	-	750 *	4.1	31	-	-	533
Texas	14	(D)	-	-	375 *	2.1	27	-	-	1,333
Virginia	14	(D)	-	-	375 *	2.1	27	-	-	-
Oregon	13	(D)	-	-	175 *	1.0	13	-	-	-
Oklahoma	12	(D)	-	-	375 *	2.1	31	-	-	1,867
Missouri	9	(D)	-	-	175 *	1.0	19	-	-	-
Tennessee	8	(D)	-	-	750 *	4.1	94	-	-	-
Iowa	7	(D)	-	-	375 *	2.1	54	-	-	-
Vermont	6	(D)	-	-	375 *	2.1	63	-	-	-
Alabama	5	(D)	-	-	750 *	4.1	150	-	-	-
South Carolina	5	(D)	-	-	175 *	1.0	35	-	-	-
Georgia	4	(D)	-	-	375 *	2.1	94	-	-	-
Utah	4	(D)	-	-	175 *	1.0	44	-	-	-
Kansas	3	(D)	-	-	375 *	2.1	125	-	-	-
Kentucky	3	(D)	-	-	750 *	4.1	250	-	-	-

Source: 1992 Economic Census. The states are in descending order of shipments or establishments (if shipment data are missing for the majority). The symbol (D) appears when data are withheld to prevent disclosure of competitive information. States marked with (D) are sorted by number of establishments. A dash (-) indicates that the data element cannot be calculated; * indicates the midpoint of a range.

3441 - FABRICATED STRUCTURAL METAL

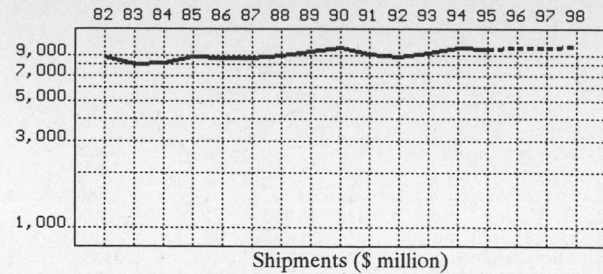

Shipments ($ million)

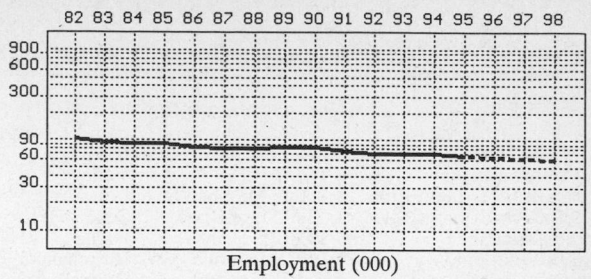

Employment (000)

GENERAL STATISTICS

Year	Com-panies	Establishments		Employment			Compensation		Production ($ million)			
		Total	with 20 or more employees	Total (000)	Production Workers (000)	Hours (Mil)	Payroll ($ mil)	Wages ($/hr)	Cost of Materials	Value Added by Manufacture	Value of Shipments	Capital Invest.
1982	2,590	2,739	1,160	103.5	75.4	151.4	1,988.8	8.56	5,010.2	3,631.9	8,840.2	187.9
1983		2,615	1,127	93.8	67.2	134.5	1,842.4	8.70	4,759.6	3,219.1	7,959.4	133.6
1984		2,491	1,094	92.1	67.4	133.6	1,848.1	8.97	4,981.8	3,340.8	8,198.3	166.3
1985		2,367	1,060	90.6	66.9	130.3	1,949.6	9.72	5,389.0	3,626.3	8,910.4	154.7
1986		2,318	1,050	84.6	61.3	122.5	1,904.6	9.93	5,236.0	3,378.5	8,731.5	161.8
1987	2,334	2,454	1,014	80.9	57.8	117.3	1,880.5	9.91	5,071.5	3,616.4	8,678.0	130.6
1988		2,403	999	80.1	57.9	118.9	1,944.2	10.20	5,317.3	3,811.7	8,999.3	174.0
1989		2,377	1,013	83.7	58.1	120.4	2,036.5	10.39	5,437.0	3,934.9	9,377.1	158.8
1990		2,440	1,021	82.7	59.1	121.6	2,151.7	10.76	5,725.8	4,035.4	9,788.1	159.8
1991		2,564	986	77.0	54.2	113.7	2,064.4	11.05	5,206.2	3,899.9	9,078.6	118.7
1992	2,437	2,539	912	72.0	50.8	107.8	1,958.9	10.98	4,872.8	3,934.4	8,898.4	132.2
1993		2,527	886	70.7	50.8	107.6	1,963.9	11.16	5,247.5	3,926.1	9,244.7	129.9
1994		2,440P	890P	71.3	51.5	109.1	2,008.5	11.25	5,606.7	4,242.0	9,831.7	148.7
1995		2,433P	869P	66.7P	47.0P	101.4P	2,048.0P	11.76P	5,517.7P	4,174.6P	9,675.5P	132.3P
1996		2,426P	848P	64.4P	45.2P	98.5P	2,059.9P	11.99P	5,575.6P	4,218.5P	9,777.2P	129.7P
1997		2,419P	827P	62.0P	43.4P	95.5P	2,071.8P	12.23P	5,633.5P	4,262.3P	9,878.8P	127.1P
1998		2,412P	806P	59.6P	41.5P	92.5P	2,083.7P	12.46P	5,691.5P	4,306.1P	9,980.4P	124.5P

Sources: 1982, 1987, 1992 *Economic Census*; *Annual Survey of Manufactures*, 83-86, 88-91, 93-94. Establishment counts for non-Census years are from *County Business Patterns*; establishment values for 83-84 are extrapolations. 'P's show projections by the editors. Industries reclassified in 87 will not have data for prior years.

INDICES OF CHANGE

Year	Com-panies	Establishments		Employment			Compensation		Production ($ million)			
		Total	with 20 or more employees	Total (000)	Production Workers (000)	Hours (Mil)	Payroll ($ mil)	Wages ($/hr)	Cost of Materials	Value Added by Manufacture	Value of Shipments	Capital Invest.
1982	106	108	127	144	148	140	102	78	103	92	99	142
1983		103	124	130	132	125	94	79	98	82	89	101
1984		98	120	128	133	124	94	82	102	85	92	126
1985		93	116	126	132	121	100	89	111	92	100	117
1986		91	115	117	121	114	97	90	107	86	98	122
1987	96	97	111	112	114	109	96	90	104	92	98	99
1988		95	110	111	114	110	99	93	109	97	101	132
1989		94	111	116	114	112	104	95	112	100	105	120
1990		96	112	115	116	113	110	98	118	103	110	121
1991		101	108	107	107	105	105	101	107	99	102	90
1992	100	100	100	100	100	100	100	100	100	100	100	100
1993		100	97	98	100	100	100	102	108	100	104	98
1994		96P	98P	99	101	101	103	102	115	108	110	112
1995		96P	95P	93P	93P	94P	105P	107P	113P	106P	109P	100P
1996		96P	93P	89P	89P	91P	105P	109P	114P	107P	110P	98P
1997		95P	91P	86P	85P	89P	106P	111P	116P	108P	111P	96P
1998		95P	88P	83P	82P	86P	106P	114P	117P	109P	112P	94P

Sources: Same as General Statistics. Values reflect change from the base year, 1992. Values above 100 mean greater than 92, values below 100 mean less than 92, and a value of 100 in the 82-91 or 93-98 period means same as 92. 'P's mark projections by the editors.

SELECTED RATIOS

For 1994	Avg. of All Manufact.	Analyzed Industry	Index	For 1994	Avg. of All Manufact.	Analyzed Industry	Index
Employees per Establishment	49	29	60	Value Added per Production Worker	134,084	82,369	61
Payroll per Establishment	1,500,273	823,023	55	Cost per Establishment	5,045,178	2,297,457	46
Payroll per Employee	30,620	28,170	92	Cost per Employee	102,970	78,635	76
Production Workers per Establishment	34	21	61	Cost per Production Worker	146,988	108,868	74
Wages per Establishment	853,319	502,941	59	Shipments per Establishment	9,576,895	4,028,735	42
Wages per Production Worker	24,861	23,833	96	Shipments per Employee	195,460	137,892	71
Hours per Production Worker	2,056	2,118	103	Shipments per Production Worker	279,017	190,907	68
Wages per Hour	12.09	11.25	93	Investment per Establishment	321,011	60,933	19
Value Added per Establishment	4,602,255	1,738,244	38	Investment per Employee	6,552	2,086	32
Value Added per Employee	93,930	59,495	63	Investment per Production Worker	9,352	2,887	31

Sources: Same as General Statistics. The 'Average of All Manufacturing' column represents the average of all manufacturing industries reported for the most recent complete year available. The Index shows the relationship between the Average and the Analyzed Industry. For example, 100 means that they are equal; 500 that the Analyzed Industry is five times the average; 50 means that the Analyzed Industry is half the national average. The abbreviation 'na' is used to show that data are 'not available'.

LEADING COMPANIES   Number shown: **75**   Total sales ($ mil): **4,424**   Total employment (000): **38.9**

Company Name	Address				CEO Name	Phone	Co. Type	Sales ($ mil)	Empl. (000)
FKI Industries Inc	425 Post Rd	Fairfield	CT	06430	Jeffrey Courier	203-255-7100	S	790	7.8
Valmont Industries Inc	PO Box 358	Valley	NE	68064	Mogens C Bay	402-359-2201	P	305	4.0
Everett Smith Investment Ltd	800 N Marshall St	Milwaukee	WI	53202	Everett G Smith	414-273-3421	R	300*	4.1
Interlock Industries Inc	7800 State Road 60	Sellersburg	IN	47172	JL Mackin	812-246-1935	R	240	0.8
Trinity Industries Inc	PO Box 568887	Dallas	TX	75356	Don H Johnson	214-631-4420	D	150	1.5
Sherman International	2131 Magnolia Av S	Birmingham	AL	35205	Thomas Holton	205-252-6900	R	125*	1.0
Cives Corp	411 Rouse Ln	Roswell	GA	30076	Ronald W Shaw	404-993-4424	R	110	0.9
Fabwel Inc	1838 Middlebury St	Elkhart	IN	46516	EP Welter	219-522-8473	R	110	0.4
Havens Steel Co	7219 E 17th	Kansas City	MO	64126	Dennis Haven	816-231-5724	R	90*	0.4
John S Frey Enterprises	1900 E 64th St	Los Angeles	CA	90001	JS Frey	213-583-4061	R	88	0.6
Troxel Co	Hwy 57 W	Moscow	TN	38057	Bobby N Rowlett	901-877-6875	R	85	0.6
Modern Welding Company Inc	2880 New Hartford	Owensboro	KY	42301	John W Jones	502-685-4400	R	82	0.7
Braden Manufacturing	PO Box 1229	Tulsa	OK	74101	Gene Schockemoehl	918-272-5371	D	73	0.5
IKG Industries	270 Terminal Av	Clark	NJ	07066	Fred S Beatty	908-815-9500	D	70	0.7
Hirschfeld Steel Company Inc	PO Box 3768	San Angelo	TX	76902	Dave L Hirschfeld	915-653-3211	R	66	0.6
Worthington Steel and Wave Co	PO Box 3050	Malvern	PA	19355	Ralph Roberts	215-644-6700	S	66*	0.7
Stupp Bros Bridge and Iron Co	PO Box 6600	St Louis	MO	63125	Robert P Stupp	314-638-5000	R	61*	0.6
Chicago Steel Construction	1600 N 25th Av	Melrose Park	IL	60160	Larry L Curtis	708-681-5181	D	50	0.3
PDM Bridge	PO Box 1545	Eau Claire	WI	54702	John Grzybowski	715-835-2250	D	50	0.3
Aker Gulf Marine	PO Box C	Ingleside	TX	78362	Myron J Rodrigue	512-776-7551	D	45*	0.6
J Allan Steel Co	829 Beaver Av	Pittsburgh	PA	15233	Joseph Gnazzo	412-321-3111	R	45	0.1
Union Metal Corp	PO Box 9920	Canton	OH	44711	Richard P Barker	216-456-7653	S	45	0.3
Temtex Industries Inc	3010 LBJ Fwy	Dallas	TX	75234	ER Buford	214-484-1845	P	44	0.6
Barker Steel Company Inc	200 Dexter Av	Watertown	MA	02172	Robert B Brack	617-926-0105	R	41	0.3
Engineered Products Inc	1844 Ardmore Blv	Pittsburgh	PA	15221	OW Abraham	412-242-6900	R	40	0.1
Queensboro Steel Corp	PO Box 1769	Wilmington	NC	28402	SL Alper	919-763-6237	R	37*	0.2
Riblet Products Corp	11555 Harter Dr	Middlebury	IN	46540	Nachum Stein	219-825-9466	R	37*	0.3
Berlin Steel Construction Co	PO Box 428	Berlin	CT	06037	Carl A Johnson	203-828-3531	R	36	0.2
Addison Steel Inc	PO Box 3629	Albany	GA	31708	Glen Davis	912-883-4506	D	35	0.2
Addison Steel Inc	PO Box 3629	Albany	GA	31708	Glen Davis	912-883-4506	R	35	0.2
Hausman Corp	11590 N Meridian	Carmel	IN	46032	Richard Raymond	317-844-6044	R	35	0.4
LeJeune Steel Co	PO Box 19070	Minneapolis	MN	55419	Michael LeJeune	612-861-3321	S	35	0.2
Shelby Steel Fabricators Inc	PO Box G	Vincent	AL	35178	Harold Ridgeway	205-672-2249	R	35*	0.3
Economy Forms Corp	1800 NE Broadway	Des Moines	IA	50313	Albert L Jennings	515-266-1141	R	33	0.3
Structural Steel Services Inc	PO Box 2929	Meridian	MS	39301	TE Dulaney	601-483-5381	R	33*	0.3
Graham Steel Corp	PO Box 658	Kirkland	WA	98033	James Graham	206-823-5656	R	31	0.3
Patz Sales Inc	PO Box 7	Pound	WI	54161	Clifford Patz	414-897-2251	D	31*	0.3
Washington Ornamental Iron	17926 S Broadway	Gardena	CA	90248	Dan Welsh	213-321-8373	R	31*	0.3
Cheler Corp	PO Box 1750	Seattle	WA	98111	Chad E Ernst	206-624-9699	R	30	1.5
Gayle Manufacturing Co	PO Box 1365	Woodland	CA	95695	James Deblasio	916-662-0284	R	30	0.1
New Columbia Joist Co	PO Box 31	New Columbia	PA	17856	Nicholas J Bouras	717-568-6761	R	30	0.2
Pathway Bellows Inc	115 Franklin Rd	Oak Ridge	TN	37830	James K Cole	615-483-7444	S	30	0.3
South Carolina Steel Corp	PO Box 71	Greenville	SC	29602	EE Garvin	803-244-2860	R	30	0.1
Southern Ohio Fabricators Inc	10333 Wayne Av	Cincinnati	OH	45215	Jack J Kling	513-771-1600	R	30*	0.3
Watson Bowman Acme Corp	95 Pineview Dr	Amherst	NY	14228	Arthur C Wotiz	716-691-7566	S	30	0.1
Globe Iron Construction	PO Box 2354	Norfolk	VA	23501	Arthur Peregoff	804-625-2542	R	29	0.2
Angeles Metal Trim Co	4817 E Sheila St	Los Angeles	CA	90040	Alan P MacQuoid	213-268-1777	R	25	0.2
Fabwel Inc	1802 Cheyenne	Nappanee	IN	46550	Roger Gowdy	219-773-7981	D	25	<0.1
Rodney Hunt Co	46 Mill St	Orange	MA	01364	David F Waskiewicz	508-544-2511	R	25	0.2
Brocker Rebar Company Inc	1700 7th Av	York	PA	17403	WD Brocker	717-846-7865	R	24*	0.2
Fontana Steel Co	PO Box 2219	R Cucamonga	CA	91729	Don Ware	909-899-9993	R	24*	0.2
Iroquois Investors Inc	500 Skokie Blv	Northbrook	IL	60062	Dale Pinkert	708-272-0009	R	24*	0.2
PSP Industries Inc	300 Montague	Milpitas	CA	95035	Roger Schwab	408-942-1155	R	24*	0.2
W and W Steel Company Inc	PO Box 25369	Oklahoma City	OK	73125	Bert Cooper	405-235-3621	R	22	0.2
Brenner Companies Inc	3415 Glenn Av	Winston-Salem	NC	27105	Abe Brenner	919-725-8333	R	20	0.1
Cives Corp	PO Box 2778	Winchester	VA	22604	Donn Besselievre	703-667-3480	D	20	0.2
Drake-Williams Steel Inc	2301 Hickory St	Omaha	NE	68108	John A Williams	402-342-1043	R	20	0.1
Egger Steel Co	PO Box E	Sioux Falls	SD	57101	Steve E Egger	605-336-2490	R	20	0.2
Gateway Construction Company	3233 W Grand Av	Chicago	IL	60651	George Weiland	312-533-1100	R	20	0.2
Ohio Steel Industries Inc	2575 Ferris Rd	Columbus	OH	43224	WA Hays	614-471-4800	R	20	0.2
Steelfab Inc	PO Box 19289	Charlotte	NC	28219	Ron Sherrill	704-394-5376	R	20	0.1
Temcor	PO Box 6256	Carson	CA	90749	CE Miller	310-549-4311	R	20	0.1
Herr and Sacco Inc	PO Box 99	Landisville	PA	17538	James A Miller	717-898-0111	R	19	0.1
Arrow Inc	H-12 Freeport Ctr	Clearfield	UT	84016	Peter Damisel	801-825-1611	R	18	0.1
Liphart Steel Company Inc	PO Box 6326	Richmond	VA	23230	E C Jennings Jr	804-355-7481	R	18	<0.1
Albany Steel Inc	PO Box 4006	Albany	NY	12204	Peter Hess	518-436-4851	R	17	0.1
Grand Isle Shipyard Inc	PO Box 1000	Grand Isle	LA	70358	Robert Pregeant	504-787-2801	R	17	0.3
Kewaunee Engineering Corp	N Main St	Kewaunee	WI	54216	Gerry Lamer	414-388-2000	S	17	0.3
Safety Steel Services Inc	PO Box 2298	Victoria	TX	77902	C Maxwell Jr	512-575-4561	S	17	0.3
Unarco Industries Inc	PO Box 547	Springfield	TN	37172	Bruce Wise	615-384-3531	D	17	0.2
Mid America Steel Inc	PO Box 2807	Fargo	ND	58108	Donald L Clark	701-232-8831	R	16	0.1
Steel Inc	405 N Clarendon	Scottdale	GA	30079	Gene T Holloway	404-292-7373	R	16*	<0.1
Cives Steel Co	PO Box 850	Augusta	ME	04332	LJ Morgan	207-622-6141	D	15	0.2
Epic Metals Corp	11 Talbot Av	Rankin	PA	15104	DH Landis	412-351-3913	R	15	<0.1
Geiger and Peters Inc	PO Box 33807	Indianapolis	IN	46203	J Caldwell	317-359-9521	R	15	<0.1

Source: *Ward's Business Directory of U.S. Private and Public Companies*, Volumes 1 and 2, 1996. The company type code used is as follows: P - Public, R - Private, S - Subsidiary, D - Division, J - Joint Venture, A - Affiliate, G - Group. Sales are in millions of dollars, employees are in thousands. An asterisk (*) indicates an estimated sales volume. The symbol < stands for 'less than'. Company names and addresses are truncated, in some cases, to fit into the available space.

MATERIALS CONSUMED

Material	Quantity	Delivered Cost ($ million)
Materials, ingredients, containers, and supplies	(X)	3,562.9
Fabricated metal pipe (except castings and forgings)	(X)	52.9
Fabricated metal valves and pipe fittings (except castings and forgings)	(X)	2.1
Fabricated metal parts specially designed for steel power boilers, nec (except castings and forgings)	(X)	(D)
All other fabricated metal products (except castings and forgings)	(X)	214.0
Iron and steel castings (rough and semifinished)	(X)	30.3
Aluminum and aluminum-base alloy castings (rough and semifinished)	(X)	3.0
Other nonferrous castings (rough and semifinished)	(X)	3.7
Forgings	(X)	5.3
Steel bars and bar shapes	(X)	116.2
Steel concrete reinforcing bars	(X)	77.1
Steel sheet and strip, including tin plate	(X)	258.6
Steel plate	(X)	297.3
Wide flange steel structural beams	(X)	574.0
All other steel structural shapes (except sheet pilings, castings, forgings, and fabricated steel products)	(X)	272.8
All other steel shapes and forms	(X)	163.2
Nonferrous refinery shapes	(X)	(D)
Copper and copper-base alloy pipe and tube	(X)	0.6
All other copper and copper-base alloy shapes and forms	(X)	2.4
Aluminum and aluminum-base alloy sheet, plate, foil, and welded tubing	(X)	32.1
All other aluminum and aluminum-base alloy shapes and forms	(X)	17.4
All other nonferrous shapes and forms	(X)	7.7
Paints, varnishes, lacquers, stains, shellacs, japans, enamels, and allied products	(X)	38.1
Welding electrodes	(X)	24.0
All other materials and components, parts, containers, and supplies	(X)	424.1
Materials, ingredients, containers, and supplies, nsk	(X)	944.7

Source: 1992 *Economic Census*. Explanation of symbols used: (D): Withheld to avoid disclosure of competitive data; na: Not available; (S): Withheld because statistical norms were not met; (X): Not applicable; (Z): Less than half the unit shown; nec: Not elsewhere classified; nsk: Not specified by kind; - : zero; * : 10-19 percent estimated; ** : 20-29 percent estimated.

PRODUCT SHARE DETAILS

Product or Product Class	% Share	Product or Product Class	% Share
Fabricated structural metal	100.00	Fabricated structural iron and steel for transmission towers, substations, radio antenna towers, and supporting structures	10.95
Fabricated structural metal for buildings	52.86		
Fabricated structural iron and steel for industrial buildings	36.90	Fabricated structural iron and steel for offshore oil and gas platforms	14.65
Fabricated structural iron and steel for commercial buildings	34.86		
Fabricated structural iron and steel for residential buildings	3.52	Fabricated structural iron and steel for tunneling and subway work	1.62
Fabricated structural iron and steel for institutional, medical, and religious buildings	6.86	Fabricated structural iron and steel for aerospace and defense	8.46
Fabricated structural iron and steel for public and educational buildings	7.94	Other fabricated structural iron and steel	27.86
Fabricated structural iron and steel for public utilities	1.92	Fabricated structural aluminum for ships, boats, barges, transmission towers, and other structures	2.58
Fabricated structural aluminum for buildings (all types)	1.20		
Fabricated structural metal for buildings, nsk	6.79	Fabricated structural metal other than iron, steel, or aluminum	0.61
Fabricated structural metal for bridges	6.24	Other fabricated structural metal, nsk	6.79
Other fabricated structural metal	25.06	Fabricated structural metal, nsk	15.84
Fabricated structural iron and steel for ships, boats, and barges	26.47		

Source: 1992 *Economic Census*. The values shown are percent of total shipments in an industry. Values of indented subcategories are summed in the main heading. The symbol (D) appears when data are withheld to prevent disclosure of competitive information. The abbreviation nsk stands for 'not specified by kind' and nec for 'not elsewhere classified'.

INPUTS AND OUTPUTS FOR FABRICATED STRUCTURAL METAL

Economic Sector or Industry Providing Inputs	%	Sector	Economic Sector or Industry Buying Outputs	%	Sector
Blast furnaces & steel mills	37.6	Manufg.	Office buildings	11.2	Constr.
Metal stampings, nec	11.8	Manufg.	Industrial buildings	10.9	Constr.
Wholesale trade	8.6	Trade	Electric utility facility construction	8.0	Constr.
Sheet metal work	4.9	Manufg.	Petroleum & natural gas well drilling	7.2	Constr.
Screw machine and related products	4.4	Manufg.	Crude petroleum & natural gas	5.8	Mining
Communications, except radio & TV	4.1	Util.	Highway & street construction	5.2	Constr.
Imports	3.2	Foreign	Maintenance of nonfarm buildings nec	3.6	Constr.
Fabricated structural metal	2.8	Manufg.	Federal Government purchases, national defense	3.5	Fed Govt
Metal coating & allied services	2.1	Manufg.	Exports	3.2	Foreign
Maintenance of nonfarm buildings nec	1.6	Constr.	Construction of educational buildings	2.7	Constr.
Electric services (utilities)	1.6	Util.	Construction of stores & restaurants	2.5	Constr.
Petroleum refining	1.4	Manufg.	Construction machinery & equipment	2.1	Manufg.
Miscellaneous plastics products	1.3	Manufg.	Construction of hospitals	1.9	Constr.
Motor freight transportation & warehousing	1.3	Util.	Warehouses	1.9	Constr.
Hotels & lodging places	0.8	Services	Fabricated structural metal	1.9	Manufg.

Continued on next page.

INPUTS AND OUTPUTS FOR FABRICATED STRUCTURAL METAL - Continued

Economic Sector or Industry Providing Inputs	%	Sector	Economic Sector or Industry Buying Outputs	%	Sector
Aluminum rolling & drawing	0.7	Manufg.	Sheet metal work	1.9	Manufg.
Railroads & related services	0.7	Util.	Ship building & repairing	1.7	Manufg.
Banking	0.7	Fin/R.E.	Boat building & repairing	1.6	Manufg.
Real estate	0.6	Fin/R.E.	Maintenance of electric utility facilities	1.5	Constr.
Air transportation	0.5	Util.	Maintenance of highways & streets	1.5	Constr.
Eating & drinking places	0.5	Trade	Sewer system facility construction	1.3	Constr.
Adhesives & sealants	0.4	Manufg.	Coal	1.2	Mining
Ball & roller bearings	0.4	Manufg.	Pipe, valves, & pipe fittings	1.2	Manufg.
Brass, bronze, & copper castings	0.4	Manufg.	Signs & advertising displays	1.0	Manufg.
Paints & allied products	0.4	Manufg.	Chemical & fertilizer mineral	0.8	Mining
Welding apparatus, electric	0.4	Manufg.	Local transit facility construction	0.7	Constr.
Gas production & distribution (utilities)	0.4	Util.	Residential 1-unit structures, nonfarm	0.7	Constr.
Royalties	0.4	Fin/R.E.	Telephone & telegraph facility construction	0.7	Constr.
Equipment rental & leasing services	0.4	Services	Conveyors & conveying equipment	0.7	Manufg.
Machinery, except electrical, nec	0.3	Manufg.	Miscellaneous metal work	0.7	Manufg.
Business services nec	0.3	Services	Amusement & recreation building construction	0.6	Constr.
Industrial patterns	0.2	Manufg.	Maintenance of railroads	0.6	Constr.
Iron & steel foundries	0.2	Manufg.	Residential high-rise apartments	0.6	Constr.
Pipe, valves, & pipe fittings	0.2	Manufg.	Water supply facility construction	0.6	Constr.
Special dies & tools & machine tool accessories	0.2	Manufg.	Prefabricated wood buildings	0.5	Manufg.
Water transportation	0.2	Util.	Copper ore	0.4	Mining
Insurance carriers	0.2	Fin/R.E.	Hotels & motels	0.4	Constr.
Automotive rental & leasing, without drivers	0.2	Services	Nonbuilding facilities nec	0.4	Constr.
Automotive repair shops & services	0.2	Services	Aircraft	0.4	Manufg.
Legal services	0.2	Services	Hoists, cranes, & monorails	0.4	Manufg.
Management & consulting services & labs	0.2	Services	Internal combustion engines, nec	0.4	Manufg.
U.S. Postal Service	0.2	Gov't	Turbines & turbine generator sets	0.4	Manufg.
Abrasive products	0.1	Manufg.	Nonferrous metal ores, except copper	0.3	Mining
Fabricated metal products, nec	0.1	Manufg.	Construction of conservation facilities	0.3	Constr.
Industrial gases	0.1	Manufg.	Construction of dams & reservoirs	0.3	Constr.
Power transmission equipment	0.1	Manufg.	Mining machinery, except oil field	0.3	Manufg.
Primary aluminum	0.1	Manufg.	Radio & TV communication equipment	0.3	Manufg.
Soap & other detergents	0.1	Manufg.	Structural wood members, nec	0.3	Manufg.
Accounting, auditing & bookkeeping	0.1	Services	Federal Government purchases, nondefense	0.3	Fed Govt
Advertising	0.1	Services	Dimension, crushed & broken stone	0.2	Mining
Business/professional associations	0.1	Services	Construction of nonfarm buildings nec	0.2	Constr.
Computer & data processing services	0.1	Services	Gas utility facility construction	0.2	Constr.
			Maintenance of telephone & telegraph facilities	0.2	Constr.
			Petroleum, gas, & solid mineral exploration	0.2	Constr.
			Aircraft & missile equipment, nec	0.2	Manufg.
			Iron & ferroalloy ores	0.1	Mining
			Sand & gravel	0.1	Mining
			Dormitories & other group housing	0.1	Constr.
			Garage & service station construction	0.1	Constr.
			Maintenance of nonbuilding facilities nec	0.1	Constr.
			Maintenance of petroleum & natural gas wells	0.1	Constr.
			Nonfarm residential structure maintenance	0.1	Constr.
			Resid. & other health facility construction	0.1	Constr.
			Residential additions/alterations, nonfarm	0.1	Constr.
			Industrial trucks & tractors	0.1	Manufg.
			Miscellaneous plastics products	0.1	Manufg.
			Oil field machinery	0.1	Manufg.

Source: Benchmark Input-Output Accounts for the U.S. Economy, 1982, U.S. Department of Commerce, Washington, D.C., July 1991. Data, as reported in the source, are organized by the 1977 SIC structure in use in 1982 but have been matched, as closely as is possible, to the 1987 SIC structure used in this book.

OCCUPATIONS EMPLOYED BY SIC 344 - FABRICATED STRUCTURAL METAL PRODUCTS

Occupation	% of Total 1994	Change to 2005	Occupation	% of Total 1994	Change to 2005
Assemblers, fabricators, & hand workers nec	8.8	-17.6	Precision metal workers nec	1.8	-9.4
Welders & cutters	7.0	-17.6	Secretaries, ex legal & medical	1.8	-25.0
Sheet metal workers & duct installers	6.8	-58.8	Machinists	1.6	-17.6
Metal fabricators, structural metal products	5.2	-17.6	General office clerks	1.5	-29.7
Blue collar worker supervisors	4.4	-24.6	Bookkeeping, accounting, & auditing clerks	1.5	-38.2
Machine tool cutting & forming etc. nec	4.2	-1.1	Cost estimators	1.4	-9.4
Welding machine setters, operators	3.7	-17.6	Coating, painting, & spraying machine workers	1.4	-1.1
General managers & top executives	3.4	-21.8	Traffic, shipping, & receiving clerks	1.4	-20.7
Sales & related workers nec	2.8	-17.6	Inspectors, testers, & graders, precision	1.1	-17.6
Helpers, laborers, & material movers nec	2.5	-17.6	Industrial production managers	1.1	-17.6
Drafters	2.3	-35.8	Freight, stock, & material movers, hand	1.1	-34.1
Fitters, structural metal, precision	2.1	-50.6	Boilermakers	1.1	11.2
Machine forming operators, metal & plastic	1.9	-46.4	Combination machine tool operators	1.1	-17.6
Truck drivers light & heavy	1.9	-15.0	Financial managers	1.0	-17.6

Source: Industry-Occupation Matrix, Bureau of Labor Statistics. These data relate to one or more 3-digit SIC industry groups rather than to a single 4-digit SIC. The change reported for each occupation to the year 2005 is a percent of growth or decline as estimated by the Bureau of Labor Statistics. The abbreviation nec stands for 'not elsewhere classified'.

LOCATION BY STATE AND REGIONAL CONCENTRATION

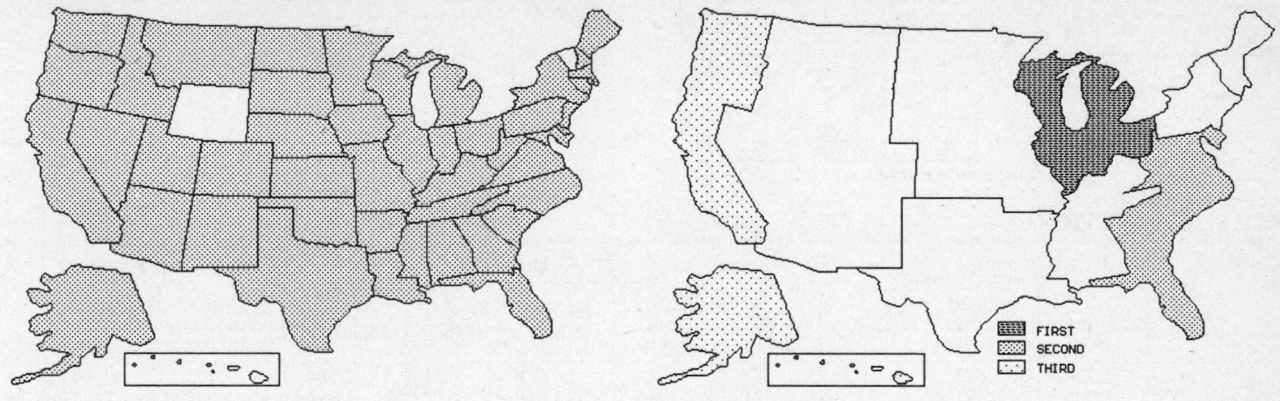

FIRST
SECOND
THIRD

INDUSTRY DATA BY STATE

| State | Establish-ments | Shipments | | | Employment | | | | Cost as % of Shipments | Investment per Employee ($) |
		Total ($ mil)	% of U.S.	Per Establ.	Total Number	% of U.S.	Per Establ.	Wages ($/hour)		
Texas	194	712.5	8.0	3.7	6,300	8.8	32	9.08	55.3	1,238
California	240	710.0	8.0	3.0	5,500	7.6	23	12.65	49.9	1,291
Pennsylvania	140	549.2	6.2	3.9	4,500	6.3	32	11.47	52.4	1,178
Ohio	143	452.9	5.1	3.2	3,200	4.4	22	11.76	57.3	1,875
Illinois	115	446.8	5.0	3.9	3,100	4.3	27	11.58	52.2	1,968
Louisiana	46	398.3	4.5	8.7	3,600	5.0	78	10.47	47.7	2,167
Alabama	91	320.1	3.6	3.5	3,100	4.3	34	9.40	49.1	2,323
North Carolina	75	296.9	3.3	4.0	2,100	2.9	28	10.47	67.1	1,476
Missouri	64	283.3	3.2	4.4	1,700	2.4	27	11.04	60.7	1,765
New York	127	265.1	3.0	2.1	2,200	3.1	17	12.90	54.8	2,364
Florida	88	255.4	2.9	2.9	2,100	2.9	24	9.90	63.8	1,238
Virginia	65	234.0	2.6	3.6	2,100	2.9	32	9.91	60.4	714
South Carolina	51	232.9	2.6	4.6	1,700	2.4	33	9.50	67.1	706
Indiana	66	229.1	2.6	3.5	1,500	2.1	23	11.45	60.7	2,067
Wisconsin	60	228.7	2.6	3.8	1,700	2.4	28	10.76	58.1	5,706
Michigan	96	222.2	2.5	2.3	1,700	2.4	18	10.88	53.2	1,412
Georgia	57	193.4	2.2	3.4	1,600	2.2	28	9.17	63.9	1,125
Tennessee	71	188.1	2.1	2.6	1,800	2.5	25	10.03	50.0	1,944
New Jersey	70	176.1	2.0	2.5	1,300	1.8	19	13.11	49.7	923
Mississippi	28	162.4	1.8	5.8	1,500	2.1	54	11.30	56.0	4,533
Utah	50	153.4	1.7	3.1	1,200	1.7	24	10.78	59.1	2,583
Oklahoma	50	152.0	1.7	3.0	1,400	1.9	28	9.05	53.6	2,214
Oregon	49	150.6	1.7	3.1	1,100	1.5	22	12.19	59.2	1,364
Minnesota	41	146.7	1.6	3.6	800	1.1	20	12.45	59.2	2,250
Iowa	28	132.8	1.5	4.7	1,000	1.4	36	9.36	59.2	4,700
Washington	65	132.1	1.5	2.0	1,000	1.4	15	12.31	53.2	1,000
Arizona	24	106.4	1.2	4.4	900	1.3	38	9.50	62.4	1,556
Massachusetts	41	97.8	1.1	2.4	800	1.1	20	12.92	43.4	1,625
Arkansas	21	93.7	1.1	4.5	700	1.0	33	9.55	47.8	2,857
Kansas	23	91.3	1.0	4.0	700	1.0	30	9.82	51.5	1,714
Connecticut	27	76.3	0.9	2.8	500	0.7	19	14.17	37.5	1,400
Colorado	33	73.2	0.8	2.2	600	0.8	18	10.00	52.9	4,667
West Virginia	20	67.1	0.8	3.4	700	1.0	35	10.89	57.8	1,857
Maryland	33	64.1	0.7	1.9	600	0.8	18	10.22	53.8	833
Kentucky	28	57.4	0.6	2.0	700	1.0	25	9.75	63.4	714
Maine	14	53.9	0.6	3.8	400	0.6	29	11.67	42.5	1,750
Nebraska	15	50.4	0.6	3.4	500	0.7	33	10.50	54.6	2,600
New Hampshire	13	47.1	0.5	3.6	300	0.4	23	11.50	61.1	2,667
Delaware	7	24.6	0.3	3.5	200	0.3	29	8.50	60.2	1,000
Idaho	11	24.2	0.3	2.2	200	0.3	18	11.33	64.9	1,000
New Mexico	14	20.7	0.2	1.5	200	0.3	14	8.50	59.4	3,500
North Dakota	7	18.7	0.2	2.7	200	0.3	29	8.67	56.7	2,000
Montana	10	(D)	-	-	375 *	0.5	38	-	-	533
Rhode Island	9	(D)	-	-	3,750 *	5.2	417	-	-	-
Nevada	7	(D)	-	-	175 *	0.2	25	-	-	-
South Dakota	4	(D)	-	-	175 *	0.2	44	-	-	1,143
Alaska	3	(D)	-	-	175 *	0.2	58	-	-	-

Source: 1992 *Economic Census.* The states are in descending order of shipments or establishments (if shipment data are missing for the majority). The symbol (D) appears when data are withheld to prevent disclosure of competitive information. States marked with (D) are sorted by number of establishments. A dash (-) indicates that the data element cannot be calculated; * indicates the midpoint of a range.

3442 - METAL DOORS, SASH & TRIM

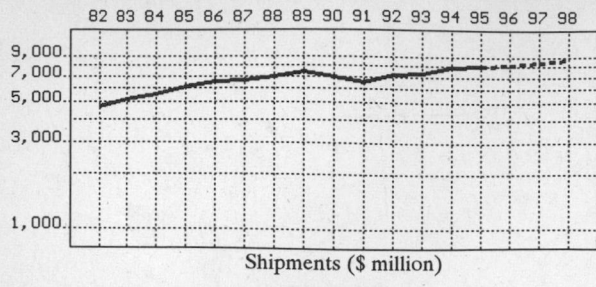

Shipments ($ million)

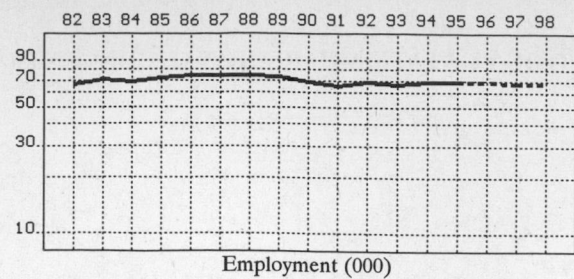

Employment (000)

GENERAL STATISTICS

Year	Companies	Establishments Total	Establishments with 20 or more employees	Employment Total (000)	Employment Production Workers (000)	Employment Hours (Mil)	Compensation Payroll ($ mil)	Compensation Wages ($/hr)	Production Cost of Materials	Production Value Added by Manufacture	Production Value of Shipments	Production Capital Invest.
1982	1,564	1,738	673	66.3	47.6	92.0	1,015.8	6.61	2,514.1	2,175.1	4,685.3	98.6
1983		1,703	691	70.5	52.0	98.5	1,110.9	6.82	2,857.4	2,385.6	5,202.7	95.0
1984		1,668	709	68.4	50.0	97.0	1,146.1	7.20	3,108.6	2,456.4	5,561.7	129.5
1985		1,632	726	71.2	51.8	101.5	1,268.0	7.63	3,321.8	2,800.5	6,137.2	115.0
1986		1,578	720	73.4	52.9	107.4	1,377.6	7.81	3,560.3	2,894.9	6,453.7	139.2
1987	1,431	1,592	695	74.6	54.5	110.6	1,403.8	7.71	3,600.1	2,993.8	6,591.0	131.3
1988		1,531	680	74.2	54.3	110.2	1,447.2	7.92	3,837.4	3,136.0	6,952.8	116.0
1989		1,465	672	73.2	54.2	109.6	1,489.7	8.10	4,134.8	3,245.5	7,347.6	129.8
1990		1,435	629	68.6	51.3	105.6	1,486.8	8.36	3,881.7	3,053.5	6,981.5	92.6
1991		1,428	601	64.4	46.0	95.9	1,378.0	8.45	3,554.5	2,877.5	6,438.3	89.8
1992	1,254	1,416	579	67.8	48.8	99.2	1,528.2	9.15	3,905.0	3,221.2	7,123.3	108.7
1993		1,393	561	66.3	48.5	98.4	1,529.6	9.19	4,015.4	3,261.1	7,284.7	105.0
1994		1,334P	579P	68.1	48.9	103.9	1,605.4	9.18	4,336.9	3,580.4	7,868.9	115.1
1995		1,301P	566P	68.1P	49.5P	104.5P	1,668.0P	9.51P	4,403.5P	3,635.3P	7,989.7P	108.7P
1996		1,268P	553P	67.8P	49.3P	104.9P	1,710.8P	9.73P	4,520.0P	3,731.5P	8,201.0P	108.1P
1997		1,235P	540P	67.6P	49.1P	105.2P	1,753.6P	9.94P	4,636.5P	3,827.7P	8,412.4P	107.6P
1998		1,202P	528P	67.4P	48.9P	105.5P	1,796.5P	10.15P	4,753.0P	3,923.9P	8,623.8P	107.0P

Sources: 1982, 1987, 1992 *Economic Census*; *Annual Survey of Manufactures*, 83-86, 88-91, 93-94. Establishment counts for non-Census years are from *County Business Patterns*; establishment values for 83-84 are extrapolations. 'P's show projections by the editors. Industries reclassified in 87 will not have data for prior years.

INDICES OF CHANGE

Year	Companies	Establishments Total	Establishments with 20 or more employees	Employment Total (000)	Employment Production Workers (000)	Employment Hours (Mil)	Compensation Payroll ($ mil)	Compensation Wages ($/hr)	Production Cost of Materials	Production Value Added by Manufacture	Production Value of Shipments	Production Capital Invest.
1982	125	123	116	98	98	93	66	72	64	68	66	91
1983		120	119	104	107	99	73	75	73	74	73	87
1984		118	122	101	102	98	75	79	80	76	78	119
1985		115	125	105	106	102	83	83	85	87	86	106
1986		111	124	108	108	108	90	85	91	90	91	128
1987	114	112	120	110	112	111	92	84	92	93	93	121
1988		108	117	109	111	111	95	87	98	97	98	107
1989		103	116	108	111	110	97	89	106	101	103	119
1990		101	109	101	105	106	97	91	99	95	98	85
1991		101	104	95	94	97	90	92	91	89	90	83
1992	100	100	100	100	100	100	100	100	100	100	100	100
1993		98	97	98	99	99	100	100	103	101	102	97
1994		94P	100P	100	100	105	105	100	111	111	110	106
1995		92P	98P	100P	101P	105P	109P	104P	113P	113P	112P	100P
1996		90P	96P	100P	101P	106P	112P	106P	116P	116P	115P	99P
1997		87P	93P	100P	101P	106P	115P	109P	119P	119P	118P	99P
1998		85P	91P	99P	100P	106P	118P	111P	122P	122P	121P	98P

Sources: Same as General Statistics. Values reflect change from the base year, 1992. Values above 100 mean greater than 92, values below 100 mean less than 92, and a value of 100 in the 82-91 or 93-98 period means same as 92. 'P's mark projections by the editors.

SELECTED RATIOS

For 1994	Avg. of All Manufact.	Analyzed Industry	Index	For 1994	Avg. of All Manufact.	Analyzed Industry	Index
Employees per Establishment	49	51	104	Value Added per Production Worker	134,084	73,219	55
Payroll per Establishment	1,500,273	1,203,776	80	Cost per Establishment	5,045,178	3,251,936	64
Payroll per Employee	30,620	23,574	77	Cost per Employee	102,970	63,684	62
Production Workers per Establishment	34	37	107	Cost per Production Worker	146,988	88,689	60
Wages per Establishment	853,319	715,189	84	Shipments per Establishment	9,576,895	5,900,334	62
Wages per Production Worker	24,861	19,505	78	Shipments per Employee	195,460	115,549	59
Hours per Production Worker	2,056	2,125	103	Shipments per Production Worker	279,017	160,918	58
Wages per Hour	12.09	9.18	76	Investment per Establishment	321,011	86,305	27
Value Added per Establishment	4,602,255	2,684,690	58	Investment per Employee	6,552	1,690	26
Value Added per Employee	93,930	52,576	56	Investment per Production Worker	9,352	2,354	25

Sources: Same as General Statistics. The 'Average of All Manufacturing' column represents the average of all manufacturing industries reported for the most recent complete year available. The Index shows the relationship between the Average and the Analyzed Industry. For example, 100 means that they are equal; 500 that the Analyzed Industry is five times the average; 50 means that the Analyzed Industry is half the national average. The abbreviation 'na' is used to show that data are 'not available'.

LEADING COMPANIES   Number shown: 75   Total sales ($ mil): 5,791   Total employment (000): 59.9

Company Name	Address				CEO Name	Phone	Co. Type	Sales ($ mil)	Empl. (000)
Tomkins Industries Inc	PO Box 943	Dayton	OH	45401	Anthony J Reading	513-253-7171	S	880*	9.5
Esstar Holdings Inc	555 Long Wharf Dr	New Haven	CT	06511	R A Haversat Sr	203-777-2274	R	540	5.8
Esstar Inc	555 Long Wharf Dr	New Haven	CT	06511	R A Haversat Sr	203-777-2274	S	540	5.8
Robertson-Ceco Corp	222 Berkeley St	Boston	MA	02116	Andrew GC Sage II	617-424-5500	P	309	1.6
AmeriMark Inc	3101 Poplarwood Ct	Raleigh	NC	27604	Roger Scott	919-876-9333	R	200	0.8
Keller Industries Inc	3499 NW 53rd St	Ft Lauderdale	FL	33309	Wayne Doss	305-777-2060	S	170*	1.0
International Aluminum Corp	767 Monterey	Monterey Park	CA	91754	C C Vanderstar	213-264-1670	P	152	1.7
Croft Metals Inc	PO Box 826	McComb	MS	39648	Joseph C Bancroft	601-684-6121	R	140*	1.5
Caradon Better-Bilt Inc	PO Box 277	Smyrna	TN	37167	Roy K Anderson	615-459-4161	S	130	1.4
Keller Holding Inc	3499 NW 53rd St	Ft Lauderdale	FL	33309	Wayne Doss	305-777-2060	R	120*	1.0
Redman Building Products Inc	2550 Walnut Hill Ln	Dallas	TX	75229	Jack Morris	214-353-3600	R	112	1.4
Stanley Door Systems Inc	1225 E Maple Rd	Troy	MI	48084	Dick Dandurand	810-528-1400	S	110	0.8
CGF Industries Inc	Bank IV Twr	Topeka	KS	66603	Jack H Hamilton	913-233-0541	R	100	1.3
Fojtasek Companies Inc	PO Box 226957	Dallas	TX	75222	Joe Fojtasek	214-438-4787	R	100	0.9
Philips Products	3221 Magnum Dr	Elkhart	IN	46516	Thomas Parrish	219-296-0000	D	100	2.0
EFCO Corp	PO Box 609	Monett	MO	65708	WT Fuldner	417-235-3193	R	90	1.1
Silver Line Building Products	207 Pond Av	Middlesex	NJ	08846	Kenneth Silverman	908-752-8600	R	85	1.1
Raynor Garage Door Co	PO Box 448	Dixon	IL	61021	RH Neisewander	815-288-1431	R	84	0.7
Drew Industries Inc	200 Mamaroneck	White Plains	NY	10601	Leigh J Abrams	914-428-9098	P	83	1.0
Steelcraft Manufacturing Co	9017 Blue Ash Rd	Cincinnati	OH	45242	William Moore	513-745-6400	S	80	1.0
Windsor Door	PO Box 8915	Little Rock	AR	72219	RH Burns	501-562-1872	D	70	0.5
Kinro Inc	4381 Green Oaks W	Arlington	TX	76016	David L Webster	817-483-7791	S	67	0.9
Loxcreen Company Inc	PO Box 4004	West Columbia	SC	29171	John W Parrish Jr	803-822-8200	R	65	0.7
Viking Aluminum Products Inc	33-39 John St	New Britain	CT	06051	Morris Trachten	203-225-6478	R	60	0.5
Carefree Aluminum Products	1023 Reynolds Rd	Charlotte	MI	48813	Eric Bacon	517-543-0430	R	52	0.6
Southeastern Metals Mfg Co	PO Box 26347	Jacksonville	FL	32226	Nadine Gramling	904-757-4200	R	51*	0.6
Kinco Ltd	PO Box 6429	Jacksonville	FL	32236	Fred King	904-355-1476	R	50	0.4
Three Rivers Aluminum Co	71 Progress Av	Cranberry	PA	16066	Robert Randall	412-776-7000	R	50	0.8
Wausau Metals Corp	PO Box 1746	Wausau	WI	54401	Larry J Niederhofer	715-845-2161	S	45	0.5
Fenestra Corp	PO Box 8189	Erie	PA	16505	Jerome D Hughes	814-838-2001	S	43*	0.4
Lausell Aluminum Jalousies Inc	PO Box 938	Bayamon	PR	00960	Ernest Reyes	809-798-7610	R	43	0.7
Eagle Window and Door Inc	PO Box 1072	Dubuque	IA	52004	John Straub	319-556-2270	S	42*	0.5
General Products Company Inc	PO Box 7387	Fredericksburg	VA	22404	H Smith McKann	703-898-5700	R	40*	0.4
New York Wire	152 N Main St	Mount Wolf	PA	17347	Todd Root	717-266-5626	D	40	0.4
Tomkins Industries Inc	PO Box 1240	Clarksville	TX	75426	William Boggs Jr	903-427-2256	D	40	0.4
Aluminum Products Company	PO Box 1869	Welcome	NC	27374	Doug Cross	704-731-6877	R	37*	0.4
McKee Door Inc	PO Box 1930	Aurora	IL	60507	Skip Smidt	708-897-9600	R	37*	0.4
Horton Automatics	4242 Baldwin Blv	Corpus Christi	TX	78405	Lew Hewitt	512-888-5591	D	35	0.3
Peerless Products Inc	PO Box 2469	Shawnee Msn	KS	66201	Dennis Shrewsbury	913-432-2232	R	33	0.4
Skotty Aluminum Products	PO Box 226957	Dallas	TX	75222	Randall Fojtasek	214-438-4787	D	33*	0.3
Acme Steel Partition Co	513 Porter Av	Brooklyn	NY	11222	Bill Moore	718-384-7800	S	32	0.4
Bilt-Rite Steel Buck Corp	95 Hopper St	Westbury	NY	11590	Steven Glaser	516-333-4333	R	30*	0.2
General American Door Co	5050 Baseline Rd	Montgomery	IL	60538	Joseph Kee	708-859-3000	R	30	0.3
ChamberDoor Industries Inc	PO Box H	Hot Springs	AR	71901	PJ Staun	501-262-1550	S	28*	0.3
Pease Industries Inc	7100 Dixie Hwy	Fairfield	OH	45014	LW Cavens	513-870-3600	R	28*	0.3
Republic Aluminum Inc	1725 W Diversy	Chicago	IL	60614	Ronald Spielman	312-525-6000	R	28*	0.3
Jordan Cos	4661 Burbank Rd	Memphis	TN	38118	Robert E Hawkins	901-363-2121	D	27	0.3
Columbia Manufacturing Corp	14400 S San Pedro	Gardena	CA	90248	Curt Shoup	310-327-9300	R	25	0.2
Door Systems Inc	751 Expressway Dr	Itasca	IL	60143	JH Weeks	708-437-0800	R	25	0.3
Graham Architectural	1551 Mount Rose	York	PA	17403	Georges Thiret	717-849-8100	S	25*	0.2
Cookson Co	2417 S 50th Av	Phoenix	AZ	85043	RC Cookson	602-272-4244	R	24	0.3
United States Aluminum Corp	3663 Bandini Blv	Vernon	CA	90023	John Kinas	213-268-4230	S	24*	0.1
Gallatin Aluminum Products Co	PO Box 1987	Gallatin	TN	37066	Jack Morris	615-452-4550	D	23	0.3
Northwest Windows	PO Box 886	Kent	WA	98035	Steve Jones	206-854-3970	S	23*	0.3
Binnings-Pan Amer Aluminum	PO Box 630038	Miami	FL	33163	Sam Bammon	305-931-2350	R	22	0.5
International Window Corp	5625 E Firestone	South Gate	CA	90280	Stewart Bender	310-928-6411	S	22	0.2
Pemko Manufacturing Co	4226 Transport St	Ventura	CA	93003	Paul Kops	805-642-2600	R	22	0.2
Curries Co	PO Box 1648	Mason City	IA	50401	Jerry Currie	515-423-1334	S	20	0.6
State Wide Aluminum Inc	PO Box 987	Elkhart	IN	46515	Mark F Fessenden	219-262-2594	R	20	0.2
Security Aluminum	5100 NW 72nd Av	Miami	FL	33166	Selig Golen	305-591-8990	R	19	0.3
RC Aluminum Inc	2805 NW 75th Av	Miami	FL	33122	Raul Casares	305-592-1515	R	18	0.1
Rollyson Aluminum Products	719 County Rd 1	South Point	OH	45680	Warren French	614-377-4351	S	18*	0.2
Emco Specialties Inc	PO Box 853	Des Moines	IA	50304	Daniel Ogden	515-265-6101	S	17	0.1
Jamison Door Co	PO Box 70	Hagerstown	MD	21741	Roland F Smith	301-733-3100	R	17*	0.2
Keller Aluminum Products	PO Box 709	Woodville	TX	75979	John Craine	409-283-2545	D	17	0.5
Elixir Industries Inc	PO Box 150	Douglas	GA	31533	Archie Brown	912-384-2078	D	16*	0.2
SDS Industries Inc	10241 Norris Av	Pacoima	CA	91331	Robert L Day	818-896-3094	R	16*	0.2
Crawford Doors	3395 Addison Dr	Pensacola	FL	32514	Dwayne Davidson	904-474-9890	D	15	0.2
Nu-Air Manufacturing Co	PO Box 15436	Tampa	FL	33684	Connie Horner	813-885-1654	R	15*	0.2
Peelle Co	50 Inez Dr	Bay Shore	NY	11706	HE Peelle III	516-231-6000	R	15	0.2
Windowmaster Products	PO Box 609	El Cajon	CA	92022	David L Long	619-588-1144	R	15	0.2
Fimbel Door Corp	Box 96	Whitehouse	NJ	08888	Edward Fimbel	908-534-4151	R	14*	0.1
Hopes Architectural Products	PO Box 580	Jamestown	NY	14702	Frank A Farrell Jr	716-665-5124	R	14	0.2
Koolvent Aluminum Products	9001 Rico Rd	Monroeville	PA	15146	Gary Iskra	412-858-3500	R	14	0.3
Northwest Aluminum Products	E 5414 Broadway St	Spokane	WA	99212	Bernard Timmer	509-535-3015	D	14*	0.2

Source: Ward's Business Directory of U.S. Private and Public Companies, Volumes 1 and 2, 1996. The company type code used is as follows: P - Public, R - Private, S - Subsidiary, D - Division, J - Joint Venture, A - Affiliate, G - Group. Sales are in millions of dollars, employees are in thousands. An asterisk (*) indicates an estimated sales volume. The symbol < stands for 'less than'. Company names and addresses are truncated, in some cases, to fit into the available space.

MATERIALS CONSUMED

Material	Quantity	Delivered Cost ($ million)
Materials, ingredients, containers, and supplies	(X)	3,564.7
Builders' hardware (including door locks, locksets, lock trim, screen hardware, etc.)	(X)	207.5
Metal bolts, nuts, screws, washers, rivets, and other screw machine products	(X)	41.7
Other fabricated metal products (except castings and forgings)	(X)	117.9
Iron and steel castings (rough and semifinished)	(X)	9.6
Nonferrous (aluminum, copper, etc.) castings (rough and semifinished)	(X)	21.7
Steel bars, bar shapes, and plates	(X)	15.1
Steel sheet and strip, including tin plate	(X)	319.5
Steel structural shapes	(X)	55.0
Steel wire and wire products	(X)	13.7
All other steel shapes and forms	(X)	112.6
Copper and copper-base alloy shapes and forms	(X)	0.6
Aluminum and aluminum-base alloy sheet, plate, foil, and welded tubing	(X)	168.9
Aluminum and aluminum-base alloy extruded shapes, including extruded rod, bar, pipe, tube, etc.	(X)	600.2
All other aluminum and aluminum-base alloy shapes and forms, including refinery shapes	(X)	186.3
Other nonferrous shapes and forms	(X)	41.1
Aluminum and aluminum-base alloy scrap (excluding home scrap)	(X)	88.2
Plastics products consumed in the form of sheets, rods, tubes, film, and other shapes	(X)	109.3
Flat glass (plate, float, and sheet)	(X)	291.1
All other materials and components, parts, containers, and supplies	(X)	745.4
Materials, ingredients, containers, and supplies, nsk	(X)	419.0

Source: 1992 *Economic Census*. Explanation of symbols used: (D): Withheld to avoid disclosure of competitive data; na: Not available; (S): Withheld because statistical norms were not met; (X): Not applicable; (Z): Less than half the unit shown; nec: Not elsewhere classified; nsk: Not specified by kind; - : zero; * : 10-19 percent estimated; ** : 20-29 percent estimated.

PRODUCT SHARE DETAILS

Product or Product Class	% Share	Product or Product Class	% Share
Metal doors, sash, and trim.	100.00	Steel door frames, lighter than 16 gauge (including trim sold as an integral part of the door frame, except storm door frames)	1.29
Metal doors and frames (except storm doors)	47.54	Shower doors and tub enclosures (all metals)	5.67
Overhead industrial aluminum doors	1.06	Metal doors and frames (except storm doors), nsk . . .	0.33
Sliding industrial aluminum doors	1.13	Metal window sash and frames (except storm sash) . . .	24.16
All other industrial aluminum doors	0.79	Residential steel window sash and frames (except storm sash)	2.66
Swinging residential aluminum doors (excluding shower doors, tub enclosures, and storm doors)	3.32	Residential aluminum single and double hung window sash and frames (except storm sash)	39.96
Sliding residential aluminum doors (glass, patio-type) (excluding shower doors, tub enclosures, and storm doors).	6.31	Residential aluminum awning window sash and frames (except storm sash)	3.02
All other residential aluminum doors (including garage and closet doors, excluding shower doors, tub enclosures, and storm doors)	2.11	Residential aluminum horizontal sliding window sash and frames (except storm sash)	15.24
Overhead and sliding commercial and institutional aluminum doors (excluding shower doors, tub enclosures, and storm doors)	2.02	All other residential aluminum window sash and frames (including jalousie, excluding storm sash)	9.28
Swinging commercial and institutional aluminum doors (excluding shower doors, tub enclosures, and storm doors)	2.08	Other steel window sash and frames (including commercial, industrial, etc.) (except storm sash)	2.90
All other commercial and institutional aluminum doors (excluding shower doors, tub enclosures, and storm doors)	1.65	Other aluminum single and double hung sash and frames (including commercial, industrial, etc.) (except storm sash)	7.13
Overhead industrial iron and steel doors	7.34	Other aluminum awning window sash and frames (including commercial, industrial, etc.) (except storm sash) . . .	2.24
Swinging industrial iron and steel doors	2.32	Other aluminum projected window sash and frames (including commercial, industrial, etc.) (except storm sash)	4.07
All other industrial iron and steel doors (including sliding) .	2.51		
Residential iron and steel garage doors	19.80	Other aluminum window sash and frames (including commercial, industrial, etc.) (except storm sash) . . .	10.66
Residential steel composite doors (steel clad with foam wood components) (excluding shower doors, tub enclosures, and storm doors)	(D)	Metal window sash and frames, other than steel or aluminum (except storm sash)	2.45
Residential insulated steel entrance doors (except storm doors).	14.09	Metal window sash and frames (except storm sash), nsk . .	0.39
All other residential iron and steel doors (including slide, swing, and closet doors, excluding shower doors, tub enclosures, and storm doors)	(D)	Metal molding and trim and store fronts	4.86
		Steel molding and trim	21.77
		Aluminum molding and trim	60.63
Overhead and sliding commercial and institutional iron and steel doors (excluding shower doors, tub enclosures, and storm doors)	2.48	Metal store fronts, sold complete at factory	14.84
		Metal molding and trim and store fronts, nsk	2.79
Swinging commercial and institutional iron and steel doors (excluding shower doors, tub enclosures, and storm doors)	9.12	Metal combination screen, storm sash, and storm doors . .	6.09
		Metal storm sash (except combination)	4.17
All other commercial and institutional iron and steel doors (excluding shower doors, tub enclosures, and storm doors)	2.00	Metal combination screen and storm sash.	30.13
		Metal storm doors	65.62
Metal doors other than steel or aluminum (excluding shower doors, tub enclosures, and storm doors)	0.26	Metal combination screen, storm sash, and storm doors, nsk	0.10
Aluminum door frames (including trim sold as an integral part of the door frame, except storm door frames) . . .	2.68	Metal window and door screens (except combination), and metal weather strip	3.97
		Metal door screens	25.51
Steel door frames, 16 gauge and heavier (including trim sold as an integral part of the door frame, except storm door frames)	4.98	Metal window screens, with metal frames (including tension and roll types)	49.45
		Metal weather strip	24.29
		Metal window and door screens, (except combination) and metal weather strip, nsk	0.75
		Metal doors, sash, and trim, nsk	13.39

Source: 1992 *Economic Census*. The values shown are percent of total shipments in an industry. Values of indented subcategories are summed in the main heading. The symbol (D) appears when data are withheld to prevent disclosure of competitive information. The abbreviation nsk stands for 'not specified by kind' and nec for 'not elsewhere classified'.

INPUTS AND OUTPUTS FOR METAL DOORS, SASH, & TRIM

Economic Sector or Industry Providing Inputs	%	Sector	Economic Sector or Industry Buying Outputs	%	Sector
Aluminum rolling & drawing	30.2	Manufg.	Maintenance of nonfarm buildings nec	16.7	Constr.
Blast furnaces & steel mills	8.9	Manufg.	Residential additions/alterations, nonfarm	12.4	Constr.
Glass & glass products, except containers	8.0	Manufg.	Nonfarm residential structure maintenance	12.1	Constr.
Wholesale trade	7.9	Trade	Residential 1-unit structures, nonfarm	11.6	Constr.
Primary aluminum	6.7	Manufg.	Office buildings	8.6	Constr.
Maintenance of nonfarm buildings nec	2.9	Constr.	Industrial buildings	5.5	Constr.
Advertising	2.9	Services	Construction of educational buildings	4.0	Constr.
Hardware, nec	2.7	Manufg.	Residential garden apartments	4.0	Constr.
Miscellaneous fabricated wire products	1.8	Manufg.	Warehouses	2.4	Constr.
Petroleum refining	1.7	Manufg.	Construction of stores & restaurants	2.3	Constr.
Fabricated metal products, nec	1.6	Manufg.	Construction of hospitals	2.2	Constr.
Electric services (utilities)	1.6	Util.	Mobile homes	2.2	Manufg.
Motor freight transportation & warehousing	1.5	Util.	Exports	2.1	Foreign
Metal stampings, nec	1.1	Manufg.	Hotels & motels	1.7	Constr.
Miscellaneous plastics products	1.0	Manufg.	Resid. & other health facility construction	1.4	Constr.
Communications, except radio & TV	1.0	Util.	Residential 2-4 unit structures, nonfarm	1.2	Constr.
Eating & drinking places	1.0	Trade	Amusement & recreation building construction	1.0	Constr.
Hotels & lodging places	1.0	Services	Farm service facilities	0.9	Constr.
Real estate	0.9	Fin/R.E.	Prefabricated wood buildings	0.9	Manufg.
Primary nonferrous metals, nec	0.8	Manufg.	Construction of nonfarm buildings nec	0.8	Constr.

Continued on next page.

INPUTS AND OUTPUTS FOR METAL DOORS, SASH, & TRIM - Continued

Economic Sector or Industry Providing Inputs	%	Sector	Economic Sector or Industry Buying Outputs	%	Sector
Gas production & distribution (utilities)	0.8	Util.	Dormitories & other group housing	0.8	Constr.
Banking	0.8	Fin/R.E.	Farm housing units & additions & alterations	0.8	Constr.
Paperboard containers & boxes	0.7	Manufg.	Residential high-rise apartments	0.8	Constr.
Equipment rental & leasing services	0.7	Services	Travel trailers & campers	0.7	Manufg.
Machinery, except electrical, nec	0.6	Manufg.	Construction of religious buildings	0.4	Constr.
Metal doors, sash, & trim	0.6	Manufg.	Electric utility facility construction	0.4	Constr.
Plating & polishing	0.6	Manufg.	Garage & service station construction	0.3	Constr.
Screw machine and related products	0.6	Manufg.	Metal doors, sash, & trim	0.3	Manufg.
Air transportation	0.6	Util.	Local transit facility construction	0.2	Constr.
Welding apparatus, electric	0.5	Manufg.	Federal Government purchases, national defense	0.2	Fed Govt
Railroads & related services	0.5	Util.	Telephone & telegraph facility construction	0.1	Constr.
Management & consulting services & labs	0.4	Services	Motor homes (made on purchased chassis)	0.1	Manufg.
U.S. Postal Service	0.4	Gov't			
Abrasive products	0.3	Manufg.			
Aluminum castings	0.3	Manufg.			
Ball & roller bearings	0.3	Manufg.			
Industrial gases	0.3	Manufg.			
Metal coating & allied services	0.3	Manufg.			
Paper mills, except building paper	0.3	Manufg.			
Special dies & tools & machine tool accessories	0.3	Manufg.			
Royalties	0.3	Fin/R.E.			
Automotive rental & leasing, without drivers	0.3	Services			
Legal services	0.3	Services			
Paints & allied products	0.2	Manufg.			
Insurance carriers	0.2	Fin/R.E.			
Accounting, auditing & bookkeeping	0.2	Services			
Automotive repair shops & services	0.2	Services			
Scrap	0.2	Scrap			
Cyclic crudes and organics	0.1	Manufg.			
Lubricating oils & greases	0.1	Manufg.			
Machine tools, metal forming types	0.1	Manufg.			
Manifold business forms	0.1	Manufg.			
Soap & other detergents	0.1	Manufg.			
Sanitary services, steam supply, irrigation	0.1	Util.			
Retail trade, except eating & drinking	0.1	Trade			
Computer & data processing services	0.1	Services			
Engineering, architectural, & surveying services	0.1	Services			
Imports	0.1	Foreign			

Source: Benchmark Input-Output Accounts for the U.S. Economy, 1982, U.S. Department of Commerce, Washington, D.C., July 1991. Data, as reported in the source, are organized by the 1977 SIC structure in use in 1982 but have been matched, as closely as is possible, to the 1987 SIC structure used in this book.

OCCUPATIONS EMPLOYED BY SIC 344 - FABRICATED STRUCTURAL METAL PRODUCTS

Occupation	% of Total 1994	Change to 2005	Occupation	% of Total 1994	Change to 2005
Assemblers, fabricators, & hand workers nec	8.8	-17.6	Precision metal workers nec	1.8	-9.4
Welders & cutters	7.0	-17.6	Secretaries, ex legal & medical	1.8	-25.0
Sheet metal workers & duct installers	6.8	-58.8	Machinists	1.6	-17.6
Metal fabricators, structural metal products	5.2	-17.6	General office clerks	1.5	-29.7
Blue collar worker supervisors	4.4	-24.6	Bookkeeping, accounting, & auditing clerks	1.5	-38.2
Machine tool cutting & forming etc. nec	4.2	-1.1	Cost estimators	1.4	-9.4
Welding machine setters, operators	3.7	-17.6	Coating, painting, & spraying machine workers	1.4	-1.1
General managers & top executives	3.4	-21.8	Traffic, shipping, & receiving clerks	1.4	-20.7
Sales & related workers nec	2.8	-17.6	Inspectors, testers, & graders, precision	1.1	-17.6
Helpers, laborers, & material movers nec	2.5	-17.6	Industrial production managers	1.1	-17.6
Drafters	2.3	-35.8	Freight, stock, & material movers, hand	1.1	-34.1
Fitters, structural metal, precision	2.1	-50.6	Boilermakers	1.1	11.2
Machine forming operators, metal & plastic	1.9	-46.4	Combination machine tool operators	1.1	-17.6
Truck drivers light & heavy	1.9	-15.0	Financial managers	1.0	-17.6

Source: Industry-Occupation Matrix, Bureau of Labor Statistics. These data relate to one or more 3-digit SIC industry groups rather than to a single 4-digit SIC. The change reported for each occupation to the year 2005 is a percent of growth or decline as estimated by the Bureau of Labor Statistics. The abbreviation nec stands for 'not elsewhere classified'.

LOCATION BY STATE AND REGIONAL CONCENTRATION

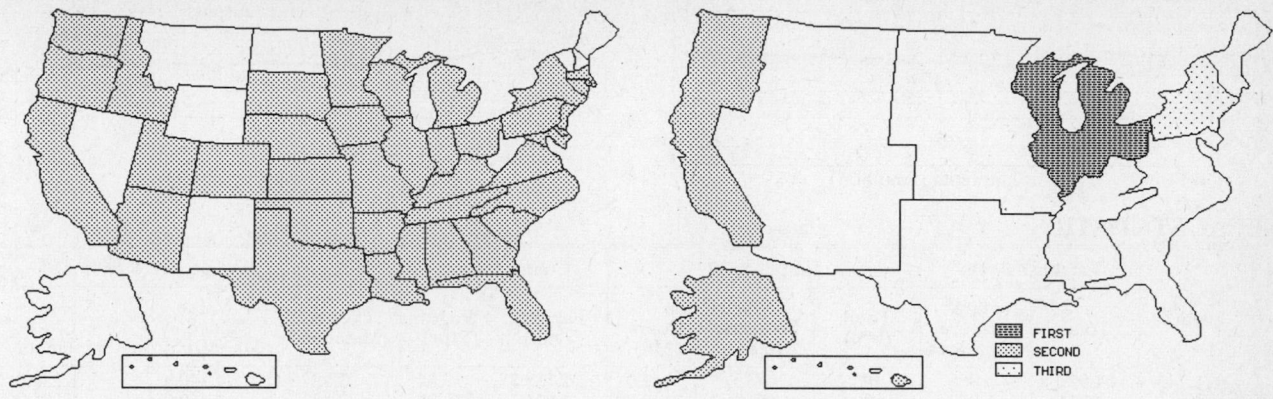

FIRST
SECOND
THIRD

INDUSTRY DATA BY STATE

State	Establish-ments	Shipments			Employment				Cost as % of Shipments	Investment per Employee ($)
		Total ($ mil)	% of U.S.	Per Establ.	Total Number	% of U.S.	Per Establ.	Wages ($/hour)		
California	192	756.8	10.6	3.9	7,700	11.4	40	9.21	53.7	1,039
Ohio	62	601.2	8.4	9.7	4,200	6.2	68	10.67	54.9	2,405
Texas	105	545.5	7.7	5.2	5,600	8.3	53	7.51	58.6	750
Pennsylvania	67	460.5	6.5	6.9	4,600	6.8	69	10.08	59.2	1,370
Tennessee	42	403.4	5.7	9.6	3,600	5.3	86	9.56	48.8	1,667
Florida	116	386.2	5.4	3.3	4,200	6.2	36	7.80	54.8	1,500
Indiana	40	382.0	5.4	9.6	2,900	4.3	73	9.41	56.4	1,828
Georgia	42	296.8	4.2	7.1	2,700	4.0	64	8.55	57.0	1,889
Illinois	64	276.7	3.9	4.3	2,400	3.5	38	9.86	51.9	2,083
New York	102	260.7	3.7	2.6	2,900	4.3	28	10.19	45.3	724
Michigan	55	245.6	3.4	4.5	2,200	3.2	40	10.84	66.7	3,500
Wisconsin	27	232.6	3.3	8.6	2,200	3.2	81	9.87	55.5	1,591
Arkansas	19	198.0	2.8	10.4	1,600	2.4	84	9.52	61.8	1,500
Missouri	32	184.4	2.6	5.8	1,900	2.8	59	7.67	61.3	1,421
New Jersey	55	180.3	2.5	3.3	2,000	2.9	36	8.61	49.1	2,100
North Carolina	27	164.2	2.3	6.1	1,900	2.8	70	8.65	49.3	1,105
Oklahoma	11	154.3	2.2	14.0	1,300	1.9	118	12.20	45.0	2,077
Oregon	19	135.2	1.9	7.1	1,100	1.6	58	8.72	53.1	2,273
Arizona	31	133.9	1.9	4.3	1,500	2.2	48	8.90	50.9	1,533
Iowa	15	119.5	1.7	8.0	1,000	1.5	67	9.71	52.6	3,200
Washington	30	95.8	1.3	3.2	900	1.3	30	10.38	53.9	1,111
Kansas	17	92.3	1.3	5.4	900	1.3	53	7.19	59.4	-
Mississippi	16	87.1	1.2	5.4	1,100	1.6	69	6.52	42.9	1,091
Virginia	13	83.1	1.2	6.4	700	1.0	54	9.44	54.6	-
Connecticut	12	81.0	1.1	6.8	900	1.3	75	10.67	53.1	1,111
Utah	9	68.0	1.0	7.6	500	0.7	56	11.71	50.6	5,000
Alabama	21	64.2	0.9	3.1	700	1.0	33	8.80	55.0	1,286
South Dakota	6	55.6	0.8	9.3	500	0.7	83	8.43	69.4	-
Nebraska	8	54.8	0.8	6.8	300	0.4	38	10.50	73.0	-
Minnesota	22	52.1	0.7	2.4	500	0.7	23	8.25	52.6	1,400
South Carolina	10	43.8	0.6	4.4	400	0.6	40	7.33	58.7	-
Massachusetts	27	43.1	0.6	1.6	500	0.7	19	10.83	46.9	1,000
Louisiana	17	34.4	0.5	2.0	400	0.6	24	7.00	50.6	1,250
Colorado	16	24.8	0.3	1.5	300	0.4	19	8.25	55.6	667
Maryland	14	24.7	0.3	1.8	300	0.4	21	11.75	39.7	1,000
Kentucky	11	23.9	0.3	2.2	300	0.4	27	7.00	51.0	1,000
Rhode Island	9	13.5	0.2	1.5	200	0.3	22	8.00	54.8	-
Delaware	2	(D)	-	-	175 *	0.3	88	-	-	-
Idaho	1	(D)	-	-	175 *	0.3	175	-	-	-

Source: 1992 *Economic Census*. The states are in descending order of shipments or establishments (if shipment data are missing for the majority). The symbol (D) appears when data are withheld to prevent disclosure of competitive information. States marked with (D) are sorted by number of establishments. A dash (-) indicates that the data element cannot be calculated; * indicates the midpoint of a range.

3443 - FABRICATED PLATE WORK, BOILER SHOPS

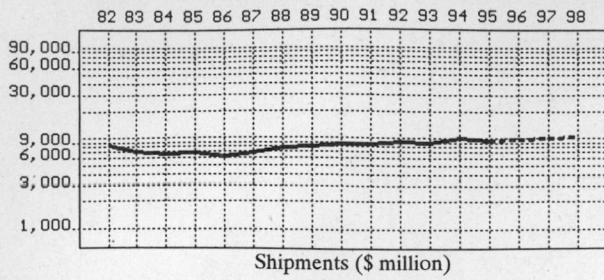

Shipments ($ million)

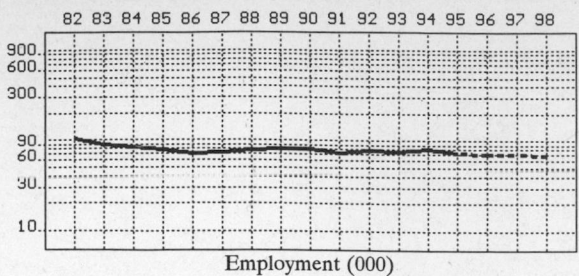

Employment (000)

GENERAL STATISTICS

| Year | Companies | Establishments | | Employment | | | Compensation | | Production ($ million) | | | |
		Total	with 20 or more employees	Total (000)	Production Workers (000)	Hours (Mil)	Payroll ($ mil)	Wages ($/hr)	Cost of Materials	Value Added by Manufacture	Value of Shipments	Capital Invest.
1982	1,743	1,928	966	103.2	73.5	142.0	2,188.3	9.84	3,927.7	4,140.9	8,230.0	200.7
1983		1,844	923	88.5	61.6	119.3	1,921.1	10.07	3,433.7	3,459.9	7,076.7	135.8
1984		1,760	880	82.2	57.5	114.3	1,848.0	10.06	3,281.1	3,427.7	6,649.1	145.8
1985		1,675	836	79.0	55.3	109.7	1,854.8	10.43	3,285.2	3,482.7	6,801.2	150.2
1986		1,587	796	71.2	48.2	97.5	1,723.1	10.60	3,038.1	3,023.3	6,153.9	110.1
1987	1,584	1,740	802	74.6	51.4	101.6	1,841.8	10.90	3,295.4	3,467.4	6,795.4	174.5
1988		1,652	823	78.5	55.0	107.7	1,997.1	11.30	3,965.1	4,032.4	7,814.3	156.8
1989		1,561	840	80.9	51.5	104.8	1,969.7	11.57	4,150.3	4,013.7	8,118.7	145.9
1990		1,595	834	81.5	53.1	107.2	2,113.2	11.93	4,448.2	4,198.5	8,653.7	220.6
1991		1,694	847	73.5	51.2	105.5	2,126.5	12.33	4,207.4	4,396.3	8,683.6	205.8
1992	1,800	1,942	880	79.3	56.1	115.1	2,345.6	12.25	4,259.2	4,877.6	9,117.6	207.8
1993		1,922	869	76.2	54.3	111.2	2,319.3	12.64	4,269.4	4,645.8	8,940.9	146.5
1994		1,737P	823P	80.4	57.8	122.5	2,426.3	12.37	4,816.3	5,262.6	10,079.4	283.1
1995		1,736P	818P	73.0P	50.5P	106.6P	2,326.2P	13.02P	4,566.8P	4,990.0P	9,557.3P	220.1P
1996		1,735P	812P	71.9P	49.8P	105.8P	2,365.4P	13.27P	4,677.8P	5,111.3P	9,789.5P	226.4P
1997		1,735P	807P	70.8P	49.0P	105.0P	2,404.6P	13.52P	4,788.7P	5,232.5P	10,021.7P	232.8P
1998		1,734P	801P	69.7P	48.3P	104.3P	2,443.8P	13.77P	4,899.7P	5,353.7P	10,253.9P	239.1P

Sources: 1982, 1987, 1992 *Economic Census*; *Annual Survey of Manufactures*, 83-86, 88-91, 93-94. Establishment counts for non-Census years are from *County Business Patterns*; establishment values for 83-84 are extrapolations. 'P's show projections by the editors. Industries reclassified in 87 will not have data for prior years.

INDICES OF CHANGE

| Year | Companies | Establishments | | Employment | | | Compensation | | Production ($ million) | | | |
		Total	with 20 or more employees	Total (000)	Production Workers (000)	Hours (Mil)	Payroll ($ mil)	Wages ($/hr)	Cost of Materials	Value Added by Manufacture	Value of Shipments	Capital Invest.
1982	97	99	110	130	131	123	93	80	92	85	90	97
1983		95	105	112	110	104	82	82	81	71	78	65
1984		91	100	104	102	99	79	82	77	70	73	70
1985		86	95	100	99	95	79	85	77	71	75	72
1986		82	90	90	86	85	73	87	71	62	67	53
1987	88	90	91	94	92	88	79	89	77	71	75	84
1988		85	94	99	98	94	85	92	93	83	86	75
1989		80	95	102	92	91	84	94	97	82	89	70
1990		82	95	103	95	93	90	97	104	86	95	106
1991		87	96	93	91	92	91	101	99	90	95	99
1992	100	100	100	100	100	100	100	100	100	100	100	100
1993		99	99	96	97	97	99	103	100	95	98	71
1994		89P	94P	101	103	106	103	101	113	108	111	136
1995		89P	93P	92P	90P	93P	99P	106P	107P	102P	105P	106P
1996		89P	92P	91P	89P	92P	101P	108P	110P	105P	107P	109P
1997		89P	92P	89P	87P	91P	103P	110P	112P	107P	110P	112P
1998		89P	91P	88P	86P	91P	104P	112P	115P	110P	112P	115P

Sources: Same as General Statistics. Values reflect change from the base year, 1992. Values above 100 mean greater than 92, values below 100 mean less than 92, and a value of 100 in the 82-91 or 93-98 period means same as 92. 'P's mark projections by the editors.

SELECTED RATIOS

For 1994	Avg. of All Manufact.	Analyzed Industry	Index	For 1994	Avg. of All Manufact.	Analyzed Industry	Index
Employees per Establishment	49	46	94	Value Added per Production Worker	134,084	91,048	68
Payroll per Establishment	1,500,273	1,396,955	93	Cost per Establishment	5,045,178	2,773,011	55
Payroll per Employee	30,620	30,178	99	Cost per Employee	102,970	59,904	58
Production Workers per Establishment	34	33	97	Cost per Production Worker	146,988	83,327	57
Wages per Establishment	853,319	872,457	102	Shipments per Establishment	9,576,895	5,803,270	61
Wages per Production Worker	24,861	26,217	105	Shipments per Employee	195,460	125,366	64
Hours per Production Worker	2,056	2,119	103	Shipments per Production Worker	279,017	174,384	62
Wages per Hour	12.09	12.37	102	Investment per Establishment	321,011	162,996	51
Value Added per Establishment	4,602,255	3,029,971	66	Investment per Employee	6,552	3,521	54
Value Added per Employee	93,930	65,455	70	Investment per Production Worker	9,352	4,898	52

Sources: Same as General Statistics. The 'Average of All Manufacturing' column represents the average of all manufacturing industries reported for the most recent complete year available. The Index shows the relationship between the Average and the Analyzed Industry. For example, 100 means that they are equal; 500 that the Analyzed Industry is five times the average; 50 means that the Analyzed Industry is half the national average. The abbreviation 'na' is used to show that data are 'not available'.

LEADING COMPANIES   Number shown: **75**   Total sales ($ mil): **6,757**   Total employment (000): **48.3**

Company Name	Address				CEO Name	Phone	Co. Type	Sales ($ mil)	Empl. (000)
CBI Industries Inc	800 Jorie Blv	Oak Brook	IL	60522	John E Jones	708-572-7000	P	1,672	13.9
Chicago Bridge and Iron Co	800 Jorie Blv	Oak Brook	IL	60522	Lewis Akin	708-572-7000	S	760	6.8
Zurn Industries Inc	1422 East Av	Erie	PA	16503	Ducan Cox	814-452-6421	D	500	0.7
Alfa Laval Inc	6133 N River Rd	Rosemont	IL	60018	Quint Jackson	708-318-7340	S	350	1.6
Aqua-Chem Inc	PO Box 421	Milwaukee	WI	53201	Robert W Agnew	414-359-0600	S	200	1.4
Riley Consolidated Inc	5 Neponset St	Worcester	MA	01615	Mike Bray	508-852-7100	S	185	0.8
Asea Brown Boveri	911 W Main St	Chattanooga	TN	37402	John Albright	615-752-2444	D	150	1.0
Heatcraft Inc	PO Box 948	Grenada	MS	38901	Dennis Custance	601-226-3421	D	150*	1.5
Worthington Cylinder Corp	PO Box 391	Columbus	OH	43085	John P McConnell	614-438-3013	S	150*	0.2
Cleaver Brooks	PO Box 421	Milwaukee	WI	53201	John M Plant	414-359-0600	D	140	1.0
Minnesota Valley Engineering	8011 34th Av	Bloomington	MN	55425	RE Cieslukowski	612-853-9600	S	120	0.8
Knolls Atomic Power Laboratory	PO Box 1072	Schenectady	NY	12301	John Freeh	518-395-4000	D	110*	1.0
Senior Engineering Company Inc	PO Box 54940	Los Angeles	CA	90054	Jon Gagg	213-726-0641	S	105	0.5
Alpine Engineered Products Inc	PO Box 2225	Pompano Bch	FL	33061	Charles W Harnden	305-781-3333	R	100	0.7
Asea Brown Boveri Inc	PO Box 372	Wellsville	NY	14895	Edward Bysiek	716-593-2700	D	90	0.7
Ohmstede Inc	PO Box 2431	Beaumont	TX	77704	Will L Ohmstede	409-833-6375	R	85	0.6
Industrial Services Technologies	370 17th St	Denver	CO	80202	Gary Schmitt	303-572-5000	R	81	1.0
Astrotech International Corp	960 Penn Av	Pittsburgh	PA	15222	S Kent Rockwell	412-391-1896	P	69	0.7
Serra Corp	4174 Technology Dr	Fremont	CA	94538	Terry Mahuron	510-651-7333	R	58*	0.5
Astro Metallurgical Inc	PO Drawer 520	Wooster	OH	44691	Dominic M Strollo	216-264-8639	R	50	0.3
Beaird Industries Inc	601 Benton Kelly St	Shreveport	LA	71106	Harry W Hinkle	318-865-6351	S	50	0.7
Cleveland Steel Container Corp	12818 Coit Rd	Cleveland	OH	44108	Chris Page	216-541-1700	R	50	0.3
Williams Enterprise of Georgia	PO Box 756	Smyrna	GA	30081	John Murphy	404-436-1596	R	50	0.6
Clawson Tank Co	PO Box 350	Clarkston	MI	48347	RT Harding Jr	810-625-8700	R	48	0.3
Graver Tank and Manufacturing	PO Box 1764	Houston	TX	77251	Tim McDavid	713-474-5121	R	45	0.3
Brenner Tank Inc	PO Box 670	Fond du Lac	WI	54935	Tim Grahl	414-922-5020	S	41*	0.4
Brown-Minneapolis	PO Box 64670	St Paul	MN	55164	John C Wirt	612-454-6750	S	40	0.3
GEA Rainey Corp	5202 W Channel Rd	Catoosa	OK	74015	CE Curtis	918-266-3060	S	40	0.2
Peabody TecTank Inc	PO Box 996	Parsons	KS	67357	John R Farris	316-421-0200	S	40	0.4
Walker Stainless Equipment Co	625 State St	New Lisbon	WI	53950	Lynn Walker	608-562-3151	R	40	0.4
AO Smith Harvestore Products	345 Harvestore Dr	De Kalb	IL	60115	James Schaap	815-756-1551	S	38	0.2
Universal Fabricators Inc	PO Box 11308	New Iberia	LA	70562	Dailey J Berard	318-367-8291	R	38	0.3
Fabsco Inc	8100 New Sapulpa	Sapulpa	OK	74066	Leon James	918-224-7550	R	36	0.2
Flame Refractories Inc	PO Box 649	Oakboro	NC	28129	Edward D Hunter	704-485-3371	R	35	0.1
Highland Tank & Mfg Co	1 Highland Rd	Stoystown	PA	15563	Robert E Jacob	814-893-5701	R	35	0.3
Ahlstrom Pyropower Inc	8925 Rehco Rd	San Diego	CA	92121	Eric Oakes	619-458-3092	S	34	0.3
Chemi-Trol Chemical Co	721 Graham Dr	Fremont	OH	43420	John P Simcox	419-334-2664	D	33	0.1
Pioneer Astro Industries Inc	3410 N Prospect	Co Springs	CO	80907	Daniel League Jr	719-473-4186	R	31	0.2
Wayne Metal Products Co	400 E Logan St	Markle	IN	46770	Dan Bloom	219-758-3121	R	31	0.3
CP Industries Inc	2214 Walnut St	McKeesport	PA	15132	JT Croushore	412-664-6604	D	30	0.1
Gardner Cryogenics	2136 City Line Rd	Bethlehem	PA	18017	CC Cornell	610-264-4523	D	30	0.2
Holman Boiler Works Inc	1956 Singleton Blv	Dallas	TX	75212	Jerry M Lang	214-637-0020	S	30	0.1
Ketema Inc	2300 W Marshall Dr	Grand Prairie	TX	75051	Jim Evans	214-647-2626	D	30	0.1
LeBlanc Communications	12801 N Central	Dallas	TX	75243	John Miller	214-934-1894	R	30	0.3
Steeltech Manufacturing Inc	2700 W North Av	Milwaukee	WI	53208	Charles L Wallace	414-263-6310	R	30	0.2
Bark River Culvert & Equip Co	PO Box 10947	Green Bay	WI	54307	Fred H Lindner Jr	414-435-6676	R	29	0.1
Plant Maintenance Service Corp	PO Box 280883	Memphis	TN	38168	R Baker	901-353-9880	R	29	0.4
Tranter Inc	1054 Claussen Rd	Augusta	GA	30907	Ken L Kaltz	706-738-7900	S	29	0.3
National Dynamics Corp	PO Box 80404	Lincoln	NE	68501	Dan Scully	402-434-2000	R	28	0.2
CVI Inc	4200 Lyman Ct	Columbus	OH	43026	James B Peeples	614-876-7381	S	28	0.2
Caldwell Tanks Inc	4000 Tower Rd	Louisville	KY	40219	Bernard S Fineman	502-964-3361	R	27	0.2
Eastern Technologies Ltd	10 Industrial Way	Amesbury	MA	01913	Harley W Waite Jr	508-388-5662	R	26	0.1
Mitternight Boiler Works Inc	PO Box 489	Satsuma	AL	36572	Walter F Merae	205-675-2550	R	26	0.2
Basco	2777 Walden Av	Buffalo	NY	14225	C Van Tine	716-684-6700	D	25	0.2
Dethmers Manufacturing Co	PO Box 189	Boyden	IA	51234	JE Koerselman	712-725-2311	R	25	0.3
International Engineers	3501 W 11th St	Houston	TX	77008	Harold Sides	713-803-4700	S	25	0.2
Nebraska Boiler Co	PO Box 82287	Lincoln	NE	68501	Harry W Kumpula	402-434-2000	D	25	0.2
Struthers Industries Inc	1500 34th St	Gulfport	MS	39501	JB Ethridge	601-864-5410	R	25*	0.2
Eaton Metal Products Co	PO Box 16405	Denver	CO	80216	TJ Travis	303-296-4800	R	23*	0.2
Peerless Heater Co	PO Box 388	Boyertown	PA	19512	Richard Smith	215-367-2153	S	23	0.2
Taylor Forge Engineered Syst	1st & Iron St S	Paola	KS	66071	RG Kilkenny	913-294-5331	R	22	0.3
Trusco Tank Inc	PO Box 11925	Fresno	CA	93775	Leslie Scott	209-264-4741	R	22	0.2
Goodhart Sons Inc	2515 Horseshoe Rd	Lancaster	PA	17601	Gary W Goodhart	717-656-2404	R	21*	0.2
Ace Tank and Equipment Co	PO Box 9039	Seattle	WA	98109	R Allan Reese	206-281-5000	R	20*	0.2
Bigbee Steel and Tank Co	99 W Elizabethtown	Manheim	PA	17545	RE Jacob	717-664-0600	S	20	0.2
Burgess-Manning Inc	227 Thorne Av	Orchard Park	NY	14127	Warner G Merton	716-662-6540	S	20	0.2
Capital Industries Inc	5801 3rd Av S	Seattle	WA	98108	RS Taylor	206-762-8585	R	20	0.1
Fisher Tank Co	3131 W 4th St	Chester	PA	19013	Robert M Borst	610-494-7200	R	20	0.1
Hackney	PO Box 568887	Dallas	TX	75356	John Sanford	214-589-8177	D	20	0.3
Industrial Boiler Company Inc	221 Law St	Thomasville	GA	31792	Bob Burtnett	912-226-3024	S	20	0.1
Mohawk Metal Products	2175 Beechgrove Pl	Utica	NY	13501	John B Millet Jr	315-793-3000	R	20	0.2
Phoenix Fabricators and Erectors	182 S County 900	Indianapolis	IN	46234	Jeffery A Short	317-271-7002	R	20	0.1
Pressed Steel Tank Company Inc	PO Box 2180	Milwaukee	WI	53201	GG Finch	414-476-0500	R	20	0.2
Superior Fabricators Inc	PO Box 539	Baldwin	LA	70514	Pierre Larroque	318-923-7271	R	20	0.3
Thermal Transfer Products Ltd	5215 21st St	Racine	WI	53406	A Royse Myers	414-554-8330	R	20	0.2

Source: Ward's Business Directory of U.S. Private and Public Companies, Volumes 1 and 2, 1996. The company type code used is as follows: P - Public, R - Private, S - Subsidiary, D - Division, J - Joint Venture, A - Affiliate, G - Group. Sales are in millions of dollars, employees are in thousands. An asterisk (*) indicates an estimated sales volume. The symbol < stands for 'less than'. Company names and addresses are truncated, in some cases, to fit into the available space.

MATERIALS CONSUMED

Material	Quantity	Delivered Cost ($ million)
Materials, ingredients, containers, and supplies	(X)	3,667.3
Fabricated metal pipe (except castings and forgings)	(X)	97.0
Fabricated metal valves and pipe fittings (except castings and forgings)	(X)	106.4
Fabricated metal parts specially designed for steel power boilers, nec (except castings and forgings)	(X)	61.9
All other fabricated metal products (except castings and forgings)	(X)	307.2
Iron and steel castings (rough and semifinished)	(X)	93.2
Aluminum and aluminum-base alloy castings (rough and semifinished)	(X)	17.3
Other nonferrous castings (rough and semifinished)	(X)	10.9
Forgings	(X)	38.8
Steel bars and bar shapes	(X)	86.4
Steel sheet and strip, including tin plate	(X)	350.5
Steel plate	(X)	455.7
Wide flange steel structural beams	(X)	30.4
All other steel structural shapes (except sheet pilings, castings, forgings, and fabricated steel products)	(X)	41.8
All other steel shapes and forms	(X)	127.6
Nonferrous refinery shapes	(X)	8.0
Copper and copper-base alloy pipe and tube	(X)	30.4
All other copper and copper-base alloy shapes and forms	(X)	16.6
Aluminum and aluminum-base alloy sheet, plate, foil, and welded tubing	(X)	59.8
All other aluminum and aluminum-base alloy shapes and forms	(X)	48.0
All other nonferrous shapes and forms	(X)	41.0
Paints, varnishes, lacquers, stains, shellacs, japans, enamels, and allied products	(X)	43.0
Welding electrodes	(X)	49.2
All other materials and components, parts, containers, and supplies	(X)	695.1
Materials, ingredients, containers, and supplies, nsk	(X)	851.3

Source: 1992 *Economic Census*. Explanation of symbols used: (D): Withheld to avoid disclosure of competitive data; na: Not available; (S): Withheld because statistical norms were not met; (X): Not applicable; (Z): Less than half the unit shown; nec: Not elsewhere classified; nsk: Not specified by kind; - : zero; * : 10-19 percent estimated; ** : 20-29 percent estimated.

PRODUCT SHARE DETAILS

Product or Product Class	% Share	Product or Product Class	% Share
Fabricated plate work (boiler shops).	100.00	Water tube steel power boilers (stationary and marine), more than 15 p.s.i. steam working pressure, 10,001 lb/hr to 100,000 lb/hr, super heated (except for nuclear applications).	5.27
Fabricated heat exchangers and steam condensers (except for nuclear applications)	19.16	Water tube steel power boilers (stationary and marine), more than 15 p.s.i. steam working pressure, 100,001 lb/hr to 250,000 lb/hr, saturated (except for nuclear applications).	(D)
Fabricated bare tube industrial heat exchangers, closed types (except for nuclear applications)	45.37		
Fabricated fin tube industrial heat exchangers, closed types (except for nuclear applications)	41.51	Water tube steel power boilers (stationary and marine), more than 15 p.s.i. steam working pressure, 100,001 lb/hr to 250,000 lb/hr, super heated (except for nuclear applications).	13.96
Fabricated steam condensers (except for nuclear applications).	8.02		
Heat exchangers and steam condensers (except for nuclear applications), nsk	5.11	Water tube steam power boilers (stationary and marine), more than 15 p.s.i. steam working pressure, 250,001 lb/hr or more, saturated (except for nuclear applications)	1.42
Fabricated steel plate (stacks and weldments)	18.53	Water tube steam power boilers (stationary and marine), more than 15 p.s.i. steam working pressure, 250,001 lb/hr or more, super heated (except for nuclear applications)	14.12
Fabricated steel plate shielding for use in nuclear reactor buildings	1.11	Other water tube steam power boilers (stationary and marine), including 10,001 lb/hr or more with 15 p.s.i. steam working pressure or less (except for nuclear applications).	(D)
Fabricated steel plate pipe, penstocks, tunnel lining, stacks, and breeching	7.28		
Fabricated steel plate containers (trash and other), less than 13 gal	1.30	Fire tube steam power boilers (stationary and marine), horizontal return tubular, 15 p.s.i. steam working pressure or less (except for nuclear applications).	(D)
Fabricated steel plate containers (trash and other), more than 79 gal	6.70	Fire tube steam power boilers (stationary and marine), horizontal return tubular, more than 15 p.s.i. steam working pressure (except for nuclear applications)	(D)
Fabricated steel plate sound control equipment for jet engine test facilites (including hush houses, demountable run-up silencers, demountable test cells, etc.)	2.78	Fire tube steam power boilers (stationary and marine), firebox, 15 p.s.i. steam working pressure or less (except for nuclear applications).	(D)
Fabricated steel plate sound control equipment for gas turbine sound systems (enclosed) (including natural gas compression, electric generation, marine propulsion, etc.).	4.99	Fire tube steam power boilers (stationary and marine), firebox, more than 15 p.s.i. steam working pressure (except for nuclear applications).	(D)
Other fabricated steel plate sound control equipment (including sound panels, one piece enclosures, industrial silencers, and air duct silencers)	2.94	Fire tube steam power boilers (stationary and marine), scotch type pressure, 15 p.s.i. steam working pressure or less (except for nuclear applications).	(D)
Weldments and fabricated steel plate for other purposes.	69.14	Fire tube steam power boilers (stationary and marine), scotch type pressure, more than 15 p.s.i. steam working pressure (except for nuclear applications)	4.32
Fabricated steel plate (stacks and weldments), nsk	3.76	Vertical and other fire tube type steam power boilers (stationary and marine) (except for nuclear applications).	(D)
Steel power boilers (stationary and marine), parts and attachments (except for nuclear applications)	9.34	Other steel power boilers (stationary and marine) (except for nuclear applications).	2.24
Water tube steel power boilers (stationary and marine), 10,000 lb/hr or less 15 p.s.i. steam working pressure or less (except for nuclear applications)	0.11	Parts and attachments for steel power boilers (sold separately) (except for nuclear applications)	31.96
Water tube steel power boilers (stationary and marine), more than 15 p.s.i. steam working pressure, 10,000 lb/hr or less, saturated (except for nuclear applications)	(D)		
Water tube steel power boilers (stationary and marine), more than 15 p.s.i. steam working pressure, 10,000 lb/hr or less, super heated (except for nuclear applications).	(D)		
Water tube steel power boilers (stationary and marine), more than 15 p.s.i. steam working pressure, 10,001 lb/hr to 100,000 lb/hr, saturated (except for nuclear applications).	6.38		

Continued on next page.

PRODUCT SHARE DETAILS - Continued

Product or Product Class	% Share	Product or Product Class	% Share
Steel power boilers, (stationary and marine) parts and attachments (except nuclear applications), nsk	2.03	industries	19.00
Gas cylinders	4.19	Nonferrous metal process pressure vessels, tanks, and kettles for refineries, chemical plants, paper mills (more than 24 in. o.d. and not less than 5 cu ft cap.), custom fabricated at the factory	7.06
Seamless ferrous and nonferrous gas cylinders	53.87		
Welded ferrous and nonferrous gas cylinders	45.71		
Gas cylinders, nsk	0.42	Liquefied petroleum gas tanks, ferrous and nonferrous, custom fabricated at the factory	4.24
Metal tanks, complete at factory (standard line pressure)	5.76		
Liquefied petroleum gas tanks (all types) ferrous and nonferrous metal, complete at factory (standard line pressure)	21.91	All other ferrous metal tanks and vessels, custom fabricated at the factory	26.14
Air receivers (tanks), ferrous and nonferrous metal, complete at factory (standard line pressure)	16.36	All other nonferrous metal tanks and vessels, custom fabricated at the factory	8.72
Other pressure tanks (including anhydrous ammonia tanks) ferrous and nonferrous metal, complete at factory (standard line pressure)	52.73	Metal tanks and vessels, custom fabricated at the factory, nsk	3.86
		Metal tanks and vessels, custom fabricated and field erected	8.92
Metal tanks, complete at factory (standard line pressure), nsk	8.98	Ferrous metal bulk storage tanks, custom fabricated and field erected, elevated type, for dry materials	3.17
Nuclear reactor steam supply systems, heat exchangers and condensers, pressurizers, components, and auxiliary equipment	2.82	Ferrous metal bulk storage tanks, custom fabricated and field erected, elevated type, for water	18.40
Metal storage tanks, complete at factory (standard line nonpressure)	5.49	Ferrous metal bulk storage tanks, custom fabricated and field erected, elevated type, for other liquids	6.63
Ferrous metal storage tanks, complete at factory (standard line nonpressure), 4,000 gal capacity or less	30.40	Nonferrous metal bulk storage tanks, custom fabricated and field erected, elevated type	0.03
Nonferrous metal storage tanks, complete at factory (standard line nonpressure), 4,000 gal capacity or less	6.39	Ferrous metal bulk storage tanks, custom fabricated and field erected, ground storage type, for dry materials	2.02
Ferrous metal storage tanks, complete at factory (standard line nonpressure), more than 4,000 gal capacity	22.30	Ferrous metal bulk storage tanks, custom fabricated and field erected, ground storage type, for petroleum products	15.57
Nonferrous metal storage tanks, complete at factory (standard line nonpressure), more than 4,000 gal capacity	4.70	Ferrous metal bulk storage tanks, custom fabricated and field erected, ground storage type, for water	9.15
Other ferrous metal nonpressure storage tanks, complete at factory (including tanks for trailers, metal septic tanks, etc.)	25.62	Ferrous metal bulk storage tanks, custom fabricated and field erected, ground storage type, for other materials	5.74
Other nonferrous metal nonpressure storage tanks, complete at factory (including tanks for trailers, metal septic tanks, etc.)	2.55	Nonferrous metal bulk storage tanks, custom fabricated and field erected, ground storage type	0.17
Metal tanks, complete at factory (standard line nonpressure), nsk	8.04	Ferrous metal pressure vessels and tanks (including gas holders and process vessels, etc.), custom fabricated and field erected, for refineries, chemical plants, and paper mills	19.45
Metal tanks and vessels, custom fabricated at the factory	12.71	Ferrous metal pressure vessels and tanks (including gas holders and process vessels, etc.), custom fabricated and field erected, for other processing industries	12.66
Ferrous metal pressure tanks and vessels (more than 24 in. outside diameter and not less than 5 cu ft capacity), custom fabricated at the factory, for refineries, chemical plants, and paper mills	30.97	Nonferrous metal pressure vessels and tanks (including gas holders and process vessels, etc.), custom fabricated and field erected	5.03
Ferrous metal pressure tanks and vessels (more than 24 in. outside diameter and not less than 5 cu ft capacity), custom fabricated at the factory, for other processing		Metal tanks and vessels, custom fabricated and field erected, nsk	1.99
		Fabricated plate work (boiler shops), nsk	13.08

Source: 1992 *Economic Census*. The values shown are percent of total shipments in an industry. Values of indented subcategories are summed in the main heading. The symbol (D) appears when data are withheld to prevent disclosure of competitive information. The abbreviation nsk stands for 'not specified by kind' and nec for 'not elsewhere classified'.

INPUTS AND OUTPUTS FOR FABRICATED PLATE WORK (BOILER SHOPS)

Economic Sector or Industry Providing Inputs	%	Sector	Economic Sector or Industry Buying Outputs	%	Sector
Blast furnaces & steel mills	30.1	Manufg.	Gross private fixed investment	39.3	Cap Inv
Wholesale trade	12.0	Trade	Electric utility facility construction	9.3	Constr.
Fabricated plate work (boiler shops)	10.9	Manufg.	Exports	8.5	Foreign
Maintenance of nonfarm buildings nec	2.6	Constr.	Fabricated plate work (boiler shops)	6.9	Manufg.
Pipe, valves, & pipe fittings	2.3	Manufg.	Industrial buildings	4.2	Constr.
Imports	2.0	Foreign	Maintenance of nonfarm buildings nec	3.3	Constr.
Fabricated metal products, nec	1.9	Manufg.	Construction machinery & equipment	3.1	Manufg.
Petroleum refining	1.9	Manufg.	Maintenance of electric utility facilities	1.9	Constr.
Electric services (utilities)	1.8	Util.	Service industry machines, nec	1.4	Manufg.
Primary aluminum	1.4	Manufg.	Federal Government purchases, nondefense	1.4	Fed Govt
Motor freight transportation & warehousing	1.4	Util.	Office buildings	1.3	Constr.
Business services nec	1.4	Services	Pumps & compressors	1.3	Manufg.
Nonferrous forgings	1.3	Manufg.	Federal Government purchases, national defense	1.3	Fed Govt
Aluminum rolling & drawing	1.2	Manufg.	Nonfarm residential structure maintenance	1.2	Constr.
Blowers & fans	1.2	Manufg.	Ship building & repairing	1.1	Manufg.
Nonferrous wire drawing & insulating	1.2	Manufg.	Sewer system facility construction	0.8	Constr.
Hotels & lodging places	1.2	Services	Water supply facility construction	0.8	Constr.
Miscellaneous plastics products	1.1	Manufg.	Machine tools, metal cutting types	0.8	Manufg.
Copper rolling & drawing	0.9	Manufg.	Mining machinery, except oil field	0.8	Manufg.
Metal coating & allied services	0.9	Manufg.	Power driven hand tools	0.7	Manufg.
Gas production & distribution (utilities)	0.9	Util.	Special industry machinery, nec	0.7	Manufg.
Eating & drinking places	0.9	Trade	Construction of stores & restaurants	0.6	Constr.
Screw machine and related products	0.8	Manufg.	Residential 1-unit structures, nonfarm	0.6	Constr.
Communications, except radio & TV	0.8	Util.	Internal combustion engines, nec	0.6	Manufg.

Continued on next page.

INPUTS AND OUTPUTS FOR FABRICATED PLATE WORK (BOILER SHOPS) - Continued

Economic Sector or Industry Providing Inputs	%	Sector	Economic Sector or Industry Buying Outputs	%	Sector
Metalworking machinery, nec	0.7	Manufg.	Warehouses	0.5	Constr.
Welding apparatus, electric	0.7	Manufg.	General industrial machinery, nec	0.5	Manufg.
Air transportation	0.7	Util.	Printing trades machinery	0.5	Manufg.
Banking	0.7	Fin/R.E.	Crude petroleum & natural gas	0.4	Mining
Nonferrous metal ores, except copper	0.6	Mining	Construction of educational buildings	0.4	Constr.
Railroads & related services	0.6	Util.	Construction of hospitals	0.4	Constr.
Iron & steel foundries	0.5	Manufg.	Metalworking machinery, nec	0.4	Manufg.
Machinery, except electrical, nec	0.5	Manufg.	Transformers	0.4	Manufg.
Miscellaneous fabricated wire products	0.5	Manufg.	Hotels & motels	0.3	Constr.
Primary metal products, nec	0.5	Manufg.	Industrial furnaces & ovens	0.3	Manufg.
Equipment rental & leasing services	0.5	Services	Gas utility facility construction	0.2	Constr.
Ball & roller bearings	0.4	Manufg.	Maintenance of petroleum & natural gas wells	0.2	Constr.
Machine tools, metal forming types	0.4	Manufg.	Drugs	0.2	Manufg.
Paints & allied products	0.4	Manufg.	Food products machinery	0.2	Manufg.
Power transmission equipment	0.4	Manufg.	Hoists, cranes, & monorails	0.2	Manufg.
Primary lead	0.4	Manufg.	Machine tools, metal forming types	0.2	Manufg.
Carbon & graphite products	0.3	Manufg.	Turbines & turbine generator sets	0.2	Manufg.
Glass & glass products, except containers	0.3	Manufg.	Woodworking machinery	0.2	Manufg.
Primary nonferrous metals, nec	0.3	Manufg.	Meat animals	0.1	Agric.
Pumps & compressors	0.3	Manufg.	Construction of conservation facilities	0.1	Constr.
Real estate	0.3	Fin/R.E.	Farm service facilities	0.1	Constr.
Advertising	0.3	Services	Maintenance of farm service facilities	0.1	Constr.
Automotive rental & leasing, without drivers	0.3	Services	Maintenance of nonbuilding facilities nec	0.1	Constr.
Automotive repair shops & services	0.3	Services	Maintenance of railroads	0.1	Constr.
Legal services	0.3	Services	Residential additions/alterations, nonfarm	0.1	Constr.
Management & consulting services & labs	0.3	Services	Residential garden apartments	0.1	Constr.
U.S. Postal Service	0.3	Gov't	Surface active agents	0.1	Manufg.
Abrasive products	0.2	Manufg.			
Aluminum castings	0.2	Manufg.			
Chemical preparations, nec	0.2	Manufg.			
Fabricated structural metal	0.2	Manufg.			
General industrial machinery, nec	0.2	Manufg.			
Hardware, nec	0.2	Manufg.			
Industrial gases	0.2	Manufg.			
Motors & generators	0.2	Manufg.			
Nonferrous rolling & drawing, nec	0.2	Manufg.			
Special dies & tools & machine tool accessories	0.2	Manufg.			
Insurance carriers	0.2	Fin/R.E.			
Accounting, auditing & bookkeeping	0.2	Services			
Cyclic crudes and organics	0.1	Manufg.			
Industrial patterns	0.1	Manufg.			
Machine tools, metal cutting types	0.1	Manufg.			
Manifold business forms	0.1	Manufg.			
Metal stampings, nec	0.1	Manufg.			
Paperboard containers & boxes	0.1	Manufg.			
Refrigeration & heating equipment	0.1	Manufg.			
Radio & TV broadcasting	0.1	Util.			
Water transportation	0.1	Util.			
Retail trade, except eating & drinking	0.1	Trade			

Source: Benchmark Input-Output Accounts for the U.S. Economy, 1982, U.S. Department of Commerce, Washington, D.C., July 1991. Data, as reported in the source, are organized by the 1977 SIC structure in use in 1982 but have been matched, as closely as is possible, to the 1987 SIC structure used in this book.

OCCUPATIONS EMPLOYED BY SIC 344 - FABRICATED STRUCTURAL METAL PRODUCTS

Occupation	% of Total 1994	Change to 2005	Occupation	% of Total 1994	Change to 2005
Assemblers, fabricators, & hand workers nec	8.8	-17.6	Precision metal workers nec	1.8	-9.4
Welders & cutters	7.0	-17.6	Secretaries, ex legal & medical	1.8	-25.0
Sheet metal workers & duct installers	6.8	-58.8	Machinists	1.6	-17.6
Metal fabricators, structural metal products	5.2	-17.6	General office clerks	1.5	-29.7
Blue collar worker supervisors	4.4	-24.6	Bookkeeping, accounting, & auditing clerks	1.5	-38.2
Machine tool cutting & forming etc. nec	4.2	-1.1	Cost estimators	1.4	-9.4
Welding machine setters, operators	3.7	-17.6	Coating, painting, & spraying machine workers	1.4	-1.1
General managers & top executives	3.4	-21.8	Traffic, shipping, & receiving clerks	1.4	-20.7
Sales & related workers nec	2.8	-17.6	Inspectors, testers, & graders, precision	1.1	-17.6
Helpers, laborers, & material movers nec	2.5	-17.6	Industrial production managers	1.1	-17.6
Drafters	2.3	-35.8	Freight, stock, & material movers, hand	1.1	-34.1
Fitters, structural metal, precision	2.1	-50.6	Boilermakers	1.1	11.2
Machine forming operators, metal & plastic	1.9	-46.4	Combination machine tool operators	1.1	-17.6
Truck drivers light & heavy	1.9	-15.0	Financial managers	1.0	-17.6

Source: Industry-Occupation Matrix, Bureau of Labor Statistics. These data relate to one or more 3-digit SIC industry groups rather than to a single 4-digit SIC. The change reported for each occupation to the year 2005 is a percent of growth or decline as estimated by the Bureau of Labor Statistics. The abbreviation nec stands for 'not elsewhere classified'.

LOCATION BY STATE AND REGIONAL CONCENTRATION

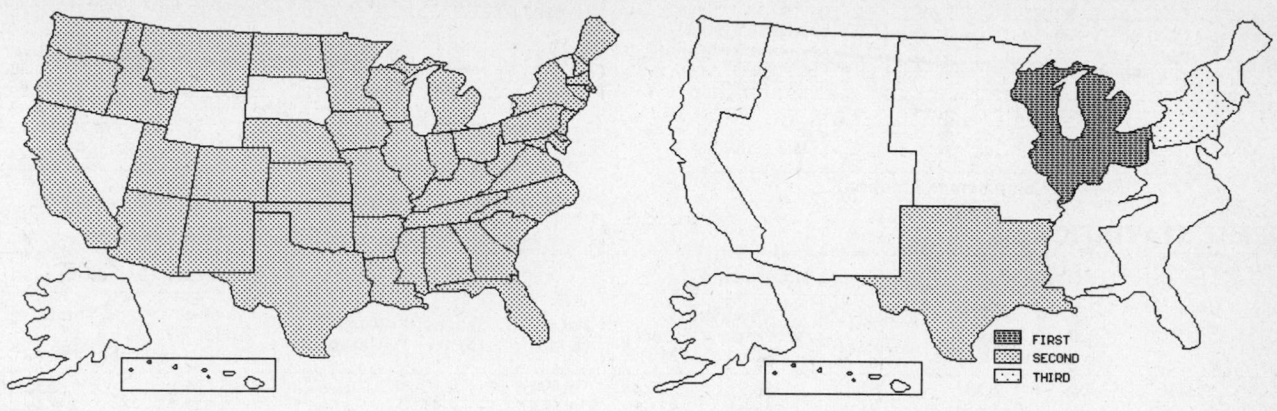

FIRST
SECOND
THIRD

INDUSTRY DATA BY STATE

State	Establish- ments	Shipments			Employment				Cost as % of Shipments	Investment per Employee ($)
		Total ($ mil)	% of U.S.	Per Establ.	Total Number	% of U.S.	Per Establ.	Wages ($/hour)		
Texas	210	987.5	10.8	4.7	8,700	11.0	41	11.30	48.9	2,011
Pennsylvania	127	864.3	9.5	6.8	6,700	8.4	53	13.65	50.1	4,194
California	181	576.7	6.3	3.2	4,700	5.9	26	13.14	40.8	2,915
Ohio	111	575.5	6.3	5.2	4,900	6.2	44	13.31	45.2	1,898
Oklahoma	83	549.0	6.0	6.6	4,500	5.7	54	11.57	50.3	1,933
New York	91	530.4	5.8	5.8	4,900	6.2	54	15.05	40.1	2,041
Illinois	90	422.0	4.6	4.7	3,000	3.8	33	11.87	45.1	3,767
Wisconsin	87	383.5	4.2	4.4	4,000	5.0	46	11.24	44.0	2,000
Missouri	43	325.8	3.6	7.6	3,100	3.9	72	13.55	51.0	1,516
Tennessee	37	303.2	3.3	8.2	2,400	3.0	65	11.71	43.9	4,208
Kentucky	34	293.9	3.2	8.6	2,400	3.0	71	12.62	51.1	4,083
Minnesota	37	276.4	3.0	7.5	2,000	2.5	54	12.44	51.6	2,200
Michigan	67	240.6	2.6	3.6	2,300	2.9	34	12.75	42.7	3,652
Indiana	52	231.7	2.5	4.5	2,400	3.0	46	10.87	41.1	2,792
Alabama	41	217.8	2.4	5.3	1,900	2.4	46	11.67	49.2	2,474
New Jersey	52	193.3	2.1	3.7	1,700	2.1	33	13.81	44.9	2,412
Georgia	34	193.2	2.1	5.7	1,600	2.0	47	11.48	54.5	3,750
Louisiana	39	180.5	2.0	4.6	2,100	2.6	54	11.18	51.4	1,238
Kansas	27	175.5	1.9	6.5	1,600	2.0	59	11.91	46.6	1,188
Mississippi	17	172.7	1.9	10.2	1,700	2.1	100	11.85	33.0	-
Massachusetts	42	135.5	1.5	3.2	1,300	1.6	31	12.17	44.8	2,385
Utah	21	123.7	1.4	5.9	1,000	1.3	48	11.07	50.8	-
North Carolina	43	121.5	1.3	2.8	1,100	1.4	26	10.80	50.9	4,000
Florida	51	120.7	1.3	2.4	1,100	1.4	22	10.33	51.6	3,091
Iowa	21	115.0	1.3	5.5	800	1.0	38	11.08	38.8	1,125
South Carolina	18	92.4	1.0	5.1	800	1.0	44	10.14	43.4	1,250
Virginia	31	90.0	1.0	2.9	800	1.0	26	11.50	52.0	4,625
Washington	32	81.9	0.9	2.6	800	1.0	25	14.09	48.5	1,375
Nebraska	8	68.5	0.8	8.6	600	0.8	75	13.86	58.2	
Arkansas	15	62.9	0.7	4.2	600	0.8	40	10.33	58.5	833
Oregon	30	54.5	0.6	1.8	500	0.6	17	11.83	45.3	600
Colorado	18	50.0	0.5	2.8	500	0.6	28	11.14	39.6	2,600
New Hampshire	11	36.2	0.4	3.3	300	0.4	27	12.80	37.8	1,333
Connecticut	24	31.8	0.3	1.3	300	0.4	13	11.20	39.6	2,667
West Virginia	11	29.7	0.3	2.7	400	0.5	36	13.20	44.1	2,250
Maryland	16	25.6	0.3	1.6	200	0.3	13	14.00	48.4	1,000
Maine	9	17.4	0.2	1.9	100	0.1	11	11.00	46.6	3,000
Arizona	16	16.9	0.2	1.1	200	0.3	13	12.50	36.7	1,000
Idaho	15	13.9	0.2	0.9	200	0.3	13	10.00	45.3	2,000
Montana	7	13.5	0.1	1.9	200	0.3	29	9.33	54.1	-
Delaware	6	11.9	0.1	2.0	100	0.1	17	19.00	36.1	3,000
New Mexico	12	10.0	0.1	0.8	100	0.1	8	12.00	49.0	3,000
Rhode Island	6	(D)	-	-	375 *	0.5	63	-	-	-
North Dakota	2	(D)	-	-	175 *	0.2	88	-	-	-

Source: 1992 *Economic Census*. The states are in descending order of shipments or establishments (if shipment data are missing for the majority). The symbol (D) appears when data are withheld to prevent disclosure of competitive information. States marked with (D) are sorted by number of establishments. A dash (-) indicates that the data element cannot be calculated; * indicates the midpoint of a range.

3444 - SHEET METAL WORK

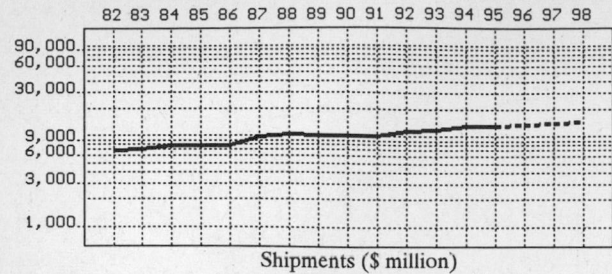

Shipments ($ million)

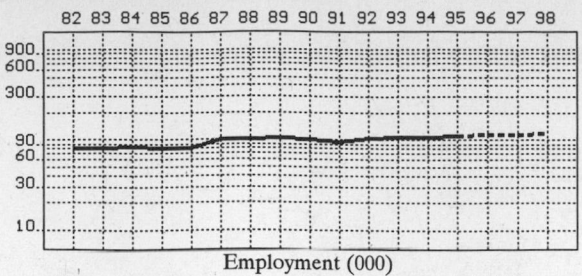

Employment (000)

GENERAL STATISTICS

Year	Com-panies	Establishments		Employment			Compensation		Production ($ million)			
		Total	with 20 or more employees	Total (000)	Production Workers (000)	Hours (Mil)	Payroll ($ mil)	Wages ($/hr)	Cost of Materials	Value Added by Manufacture	Value of Shipments	Capital Invest.
1982	3,579	3,795	1,131	81.1	58.9	114.4	1,494.6	8.30	3,767.8	3,024.1	6,853.9	145.6
1983		3,724	1,154	81.8	58.9	114.8	1,584.7	8.54	4,006.2	3,282.7	7,166.1	135.0
1984		3,653	1,177	84.0	61.4	122.1	1,689.1	8.79	4,373.3	3,426.7	7,795.3	177.4
1985		3,582	1,199	81.3	60.0	117.7	1,703.4	9.21	4,296.2	3,399.2	7,686.1	159.9
1986		3,518	1,159	79.9	58.4	117.4	1,728.0	9.31	4,117.8	3,658.6	7,724.8	126.2
1987	4,073	4,297	1,388	100.3	74.9	149.4	2,237.1	9.62	4,962.4	4,814.2	9,700.3	209.6
1988		4,094	1,393	102.9	77.3	154.9	2,403.3	10.05	5,489.7	4,912.4	10,343.9	136.0
1989		3,995	1,411	103.8	73.7	149.8	2,373.4	10.29	5,403.3	4,885.9	10,236.8	244.1
1990		4,061	1,419	102.2	73.5	150.5	2,445.3	10.40	5,425.8	4,867.4	10,249.1	208.9
1991		4,152	1,345	94.9	70.7	144.0	2,410.9	10.78	5,188.3	4,752.9	9,960.8	210.1
1992	4,452	4,702	1,474	104.3	75.6	157.5	2,850.9	11.12	5,791.8	5,743.8	11,482.2	224.5
1993		4,607	1,469	106.4	77.6	162.1	2,910.3	11.08	5,990.2	5,983.6	11,949.7	250.5
1994		4,580P	1,529P	108.0	79.9	165.7	3,049.4	11.49	6,549.9	6,760.1	13,163.2	279.2
1995		4,667P	1,563P	112.2P	82.3P	172.0P	3,156.1P	11.80P	6,460.2P	6,667.5P	12,982.9P	266.6P
1996		4,754P	1,596P	114.7P	84.1P	176.6P	3,289.6P	12.06P	6,703.3P	6,918.4P	13,471.5P	277.2P
1997		4,840P	1,630P	117.2P	86.0P	181.1P	3,423.1P	12.33P	6,946.5P	7,169.4P	13,960.1P	287.7P
1998		4,927P	1,664P	119.7P	87.8P	185.7P	3,556.6P	12.60P	7,189.6P	7,420.3P	14,448.8P	298.2P

Sources: 1982, 1987, 1992 *Economic Census*; *Annual Survey of Manufactures*, 83-86, 88-91, 93-94. Establishment counts for non-Census years are from *County Business Patterns*; establishment values for 83-84 are extrapolations. 'P's show projections by the editors. Industries reclassified in 87 will not have data for prior years.

INDICES OF CHANGE

Year	Com-panies	Establishments		Employment			Compensation		Production ($ million)			
		Total	with 20 or more employees	Total (000)	Production Workers (000)	Hours (Mil)	Payroll ($ mil)	Wages ($/hr)	Cost of Materials	Value Added by Manufacture	Value of Shipments	Capital Invest.
1982	80	81	77	78	78	73	52	75	65	53	60	65
1983		79	78	78	78	73	56	77	69	57	62	60
1984		78	80	81	81	78	59	79	76	60	68	79
1985		76	81	78	79	75	60	83	74	59	67	71
1986		75	79	77	77	75	61	84	71	64	67	56
1987	91	91	94	96	99	95	78	87	86	84	84	93
1988		87	95	99	102	98	84	90	95	86	90	61
1989		85	96	100	97	95	83	93	93	85	89	109
1990		86	96	98	97	96	86	94	94	85	89	93
1991		88	91	91	94	91	85	97	90	83	87	94
1992	100	100	100	100	100	100	100	100	100	100	100	100
1993		98	100	102	103	103	102	100	103	104	104	112
1994		97P	104P	104	106	105	107	103	113	118	115	124
1995		99P	106P	108P	109P	109P	111P	106P	112P	116P	113P	119P
1996		101P	108P	110P	111P	112P	115P	108P	116P	120P	117P	123P
1997		103P	111P	112P	114P	115P	120P	111P	120P	125P	122P	128P
1998		105P	113P	115P	116P	118P	125P	113P	124P	129P	126P	133P

Sources: Same as General Statistics. Values reflect change from the base year, 1992. Values above 100 mean greater than 92, values below 100 mean less than 92, and a value of 100 in the 82-91 or 93-98 period means same as 92. 'P's mark projections by the editors.

SELECTED RATIOS

For 1994	Avg. of All Manufact.	Analyzed Industry	Index	For 1994	Avg. of All Manufact.	Analyzed Industry	Index
Employees per Establishment	49	24	48	Value Added per Production Worker	134,084	84,607	63
Payroll per Establishment	1,500,273	665,841	44	Cost per Establishment	5,045,178	1,430,180	28
Payroll per Employee	30,620	28,235	92	Cost per Employee	102,970	60,647	59
Production Workers per Establishment	34	17	51	Cost per Production Worker	146,988	81,976	56
Wages per Establishment	853,319	415,718	49	Shipments per Establishment	9,576,895	2,874,204	30
Wages per Production Worker	24,861	23,828	96	Shipments per Employee	195,460	121,881	62
Hours per Production Worker	2,056	2,074	101	Shipments per Production Worker	279,017	164,746	59
Wages per Hour	12.09	11.49	95	Investment per Establishment	321,011	60,964	19
Value Added per Establishment	4,602,255	1,476,078	32	Investment per Employee	6,552	2,585	39
Value Added per Employee	93,930	62,594	67	Investment per Production Worker	9,352	3,494	37

Sources: Same as General Statistics. The 'Average of All Manufacturing' column represents the average of all manufacturing industries reported for the most recent complete year available. The Index shows the relationship between the Average and the Analyzed Industry. For example, 100 means that they are equal; 500 that the Analyzed Industry is five times the average; 50 means that the Analyzed Industry is half the national average. The abbreviation 'na' is used to show that data are 'not available'.

LEADING COMPANIES   Number shown: 75   Total sales ($ mil): 2,859   Total employment (000): 22.4

Company Name	Address				CEO Name	Phone	Co. Type	Sales ($ mil)	Empl. (000)
Consolidated Systems Inc	PO Box 1756	Columbia	SC	29202	Steve Holtschlag	803-771-7920	R	160	0.4
Alcan Building Products	PO Box 511	Warren	OH	44482	Donald Sperry	216-393-1192	D	120*	1.2
Bouras Industries Inc	PO Box 662	Summit	NJ	07901	Nicholas J Bouras	908-277-1617	R	110	0.5
Harrow Corp	2627 E Beltline Av	Grand Rapids	MI	49546	Larry R Adams	616-942-1440	S	110	1.2
Hart and Cooley Inc	500 E 8th St	Holland	MI	49423	Larry Lee	616-392-7855	S	100	1.2
Syro Steel Co	1170 N State St	Girard	OH	44420	Harry A Syak	216-545-4373	S	100	0.4
Symons Corp	251 S Lake Av	Pasadena	CA	91101	Merrill L Nash	818-577-0564	R	84	0.9
ASC Pacific Inc	2110 Enterprise	W Sacramento	CA	95691	Warren Birks	916-372-6851	S	80	0.3
Coastline Distribution Inc	601 Codisco Way	Sanford	FL	32771	R G Brocklemann	407-323-8500	R	80	0.4
AWH Corp	119 Brkstwn	Winston-Salem	NC	27101	J Paul Sticht	919-722-5195	R	75	0.9
Elano Corp	2455 Dayton-Xenia	Dayton	OH	45434	Robert Hessell	513-426-0621	S	70	0.6
United McGill Corp	1 Mission Park	Groveport	OH	43125	James McGill	614-836-9981	R	65	0.6
Rollex Corp	2001 Lunt Av	Elk Grove Vill	IL	60007	Bruce Stevens	708-437-3000	R	60*	0.4
Ruskin Manufacturing Co	3900 Dr Greaves Rd	Grandview	MO	64030	Terry O'Halloran	816-761-7476	S	60	0.4
Hawthorne Metal Products Co	4336 Coolidge Av	Royal Oak	MI	48073	Joseph M Bione	810-549-3800	D	58	0.3
Carnes Company Inc	PO Box 930040	Verona	WI	53593	Gregory Cichon	608-845-6411	R	51*	0.5
Carlisle Engineered Metals	PO Box 968	Stafford	TX	77497	Bob Pim	713-499-5611	R	50	0.3
Childers Products Co	35555 Curtis Blv	Eastlake	OH	44095	James Atterholt	216-953-5200	R	50	0.2
Lomanco Inc	PO Box 519	Jacksonville	AR	72078	Lynn Cooper	501-982-6511	R	50	0.3
NJ Bouras Inc	PO Box 662	Summit	NJ	07901	Nicholas J Bouras	908-277-1617	S	50	0.2
Integrated Metal Technology Inc	17155 Van Wagoner	Spring Lake	MI	49456	Ken Goodson	616-842-2600	S	48	0.5
Morton Metalcraft Co	PO Box 429	Morton	IL	61550	Bill Morton	309-266-7176	R	48	0.6
Suckle Corp	733 Davis St	Scranton	PA	18505	Charles Hagan	717-346-3871	R	44	0.3
Atlantic Metal Products Inc	21 Fadem Rd	Springfield	NJ	07081	Gary Moskovciak	201-379-6200	R	41	0.4
Acme Manufacturing Co	7500 State Rd	Philadelphia	PA	19136	Eugene Feiner	215-338-2850	R	40	0.5
Dayton T Brown Inc	555 Church St	Bohemia	NY	11716	Dayton T Brown Jr	516-589-6300	R	39*	0.4
White Aluminum Products Inc	1107 N Thomas Rd	Leesburg	FL	34748	Richard W Gerber	904-787-7766	R	38	0.3
Watertown Metal Products	1141 10th St	Watertown	WI	53094	Ron Morris	414-261-0660	D	37*	0.4
Tri-Mark Metal Corp	10106 Grinnell	Detroit	MI	48213	RA Krueger	313-925-4900	R	35*	0.1
Excelsior Mfg & Supply	1465 E Indrial Dr	Itasca	IL	60143	John Brady	708-773-5500	R	33	0.2
Noll Manufacturing Co	1900 7th St	Richmond	CA	94804	Barry Miller	510-235-1014	R	32*	0.3
Climatemp Inc	315 N May St	Chicago	IL	60607	John W Comforte	312-829-3131	R	30*	0.3
Cowden Metal Specialties Inc	PO Box 336	Chino	CA	91708	Earl E Payton	909-597-7861	R	30	0.4
Napco Inc	PO Box 208	Valencia	PA	16059	James H Wolf	412-898-1511	R	30	0.3
Snappy Air Distribution Products	1011 11th Av	Detroit Lakes	MN	56501	Paul J Schornack	218-847-9258	D	30	0.2
SOCAR Inc	PO Box 671	Florence	SC	29503	Frank Key Jr	803-669-5183	S	30	0.2
Accra-Fab Inc	PO Box 11895	Spokane	WA	99211	Don Hemmer	509-534-1717	R	25	0.2
Klauer Manufacturing Co	PO Box 59	Dubuque	IA	52001	WR Klauer	319-582-7201	R	25	0.2
Walker Systems Inc	PO Box 1828	Parkersburg	WV	26101	Marc Hofer	304-485-1611	S	25	0.3
Ball and Schafer Metal Products	1320 Egbert Av	San Francisco	CA	94124	Pauline Ball	415-822-8800	R	23	0.1
Bright Sheet Metal Co	2749 Tobey Dr	Indianapolis	IN	46219	Gary S Aletto	317-895-3939	R	22	0.2
Leigh Products	411 64th Av	Coopersville	MI	49404	Richard Liden	616-837-8141	D	22	0.2
Balco/Metalines Inc	5551 NW 5th St	Oklahoma City	OK	73127	Roger Rumsey	405-946-9721	R	21*	0.2
Appleton Supply Company Inc	1050 S Grider St	Appleton	WI	54912	JE Fowler	414-733-1373	R	21	0.1
Hercules Sheet Metal Inc	PO Box 439	Chalmette	LA	70044	Gene Le Bouef	504-277-7541	R	21*	0.2
Kleco Corp	6161 Halle Dr	Cleveland	OH	44125	Walter S Klevay Sr	216-524-7776	R	21*	0.2
Bud Industries Inc	4605 E 355th St	Willoughby	OH	44094	Blair Haas	216-946-3200	R	20	0.2
Du-Mont Co	1122 W Pioneer	Peoria	IL	61615	Philip L Graves	309-692-7240	R	20	0.1
Flameco	PO Box 1534	Ogden	UT	84402	C Stuart Kale	801-621-8960	D	20	0.1
Frank M Booth Inc	PO Box 5	Marysville	CA	95901	F Martin Booth III	916-742-7134	R	20	<0.1
United Dominion	5100 E Grand Av	Dallas	TX	75223	WP Farrer	214-827-1740	D	20	<0.1
Diamond Perforated Metals Inc	PO Box 5003	Visalia	CA	93278	Ron J Edwards	209-651-1889	S	18	0.1
Dynamic Metal Products Co	967 Parker St	Manchester	CT	06040	Howard Miller	203-646-4048	R	18*	0.2
Knecht Inc	PO Box 1346	Camden	NJ	08105	Marion T Knecht	609-966-3636	R	18*	0.2
Semco Inc	1800 E Pointe Dr	Columbia	MO	65201	Bill Thurman	314-443-1481	R	18	0.3
Air Systems Inc	7400 S 28th St	Fort Smith	AR	72906	Chris Witt	501-646-8386	R	17*	0.2
LE Schwartz & Son Inc	PO Box 4223	Macon	GA	31208	Melvin Kruger	912-745-6563	R	17	0.2
Sheet Metal Manufacturing	1080 Wyckoff Av	Ridgewood	NY	11385	Frederick Rippili	718-366-2000	R	17	<0.1
Berger Building Products Corp	805 Pennslyvania	Feasterville	PA	19053	Joseph Weiderman	215-355-1200	S	17	<0.1
AB Myr Industries Inc	39635 Detroit	Belleville	MI	48111	Richard P Marshke	313-941-2200	R	16*	0.1
Banner Metals Inc	PO Box 431	Stroudsburg	PA	18360	Michael J Stirr	717-421-4110	R	15	0.2
Electro Space Fabricator Inc	PO Box 67	Topton	PA	19562	William Straccia	610-682-7181	R	15	0.2
Highway Safety Corp	239 Commerce St	Glastonbury	CT	06033	WP Gregory Jr	203-633-9445	R	15	<0.1
John W McDougall Company	PO Box 90447	Nashville	TN	37209	JW McDougall Jr	615-321-3900	R	15	0.2
MSM Industries Inc	60 Concord St	North Reading	MA	01864	Leonard Sebell	617-944-7292	R	15*	<0.1
Streimer Sheet Metal Works Inc	PO Box 12125	Portland	OR	97212	Mike Streimer	503-288-9393	R	15*	0.2
UNC All Fab Inc	Paine Field	Everett	WA	98204	Ron Savage	206-353-8080	R	15	0.2
Venderbush Industrial Corp	39200 Groesbeck	Clinton	MI	48036	David L Babe	313-468-7800	R	15	0.1
Gary Steel Products Corp	2700 E 5th Av	Gary	IN	46402	Theodore Primich	219-885-3232	R	14	0.1
Crown Products Company Inc	6390 Phillips Hwy	Jacksonville	FL	32216	WP Tuggle Jr	904-737-7144	R	13	0.2
Sheetmetal Inc	PO Box 4067	Albany	NY	12204	William Tougher	518-465-3426	D	13	0.1
SKI Industries Inc	14665 23 Mile Rd	Shelby	MI	48315	Dennis A Haller	810-247-7100	R	13*	<0.1
Logic Design Metals Inc	PO Box 472944	Garland	TX	75047	Walter K Wilemon	214-239-1361	S	12	0.2
Dahlstrom Manufacturing	PO Box 640	Jamestown	NY	14702	Craig Colburn	716-487-0111	R	12	0.2
Fashion Inc	PO Box 1050	Ottawa	KS	66067	Lonnie L King	913-242-8111	R	12	0.1

Source: Ward's Business Directory of U.S. Private and Public Companies, Volumes 1 and 2, 1996. The company type code used is as follows: P - Public, R - Private, S - Subsidiary, D - Division, J - Joint Venture, A - Affiliate, G - Group. Sales are in millions of dollars, employees are in thousands. An asterisk (*) indicates an estimated sales volume. The symbol < stands for 'less than'. Company names and addresses are truncated, in some cases, to fit into the available space.

MATERIALS CONSUMED

Material	Quantity	Delivered Cost ($ million)
Materials, ingredients, containers, and supplies	(X)	4,930.3
Metal bolts, nuts, screws, washers, rivets, and other screw machine products	(X)	86.5
Other fabricated metal products (except castings and forgings)	(X)	346.0
Iron and steel castings (rough and semifinished)	(X)	30.2
Nonferrous (aluminum, copper, etc.) castings (rough and semifinished)	(X)	12.3
Forgings	(X)	1.2
Steel bars, bar shapes, and plates	(X)	63.4
Steel sheet and strip, including tin plate	(X)	1,327.6
Steel structural shapes	(X)	72.5
All other steel shapes and forms	(X)	208.5
Copper and copper-base alloy shapes and forms	(X)	7.8
Aluminum and aluminum-base alloy sheet, plate, foil, and welded tubing	(X)	657.6
Aluminum and aluminum-base alloy extruded shapes, including extruded rod, bar, pipe, tube, etc.	(X)	77.1
All other aluminum and aluminum-base alloy shapes and forms, including refinery shapes	(X)	152.0
Other nonferrous shapes and forms	(X)	20.7
Scrap, including iron, steel, aluminum and aluminum-base alloy (excluding home scrap)	(X)	19.5
Flat glass (plate, float, and sheet)	(X)	17.4
Paperboard containers, boxes, and corrugated paperboard	(X)	52.8
Paints, varnishes, lacquers, stains, shellacs, japans, enamels, and allied products	(X)	78.6
All other materials and components, parts, containers, and supplies	(X)	554.2
Materials, ingredients, containers, and supplies, nsk	(X)	1,144.5

Source: 1992 *Economic Census*. Explanation of symbols used: (D): Withheld to avoid disclosure of competitive data; na: Not available; (S): Withheld because statistical norms were not met; (X): Not applicable; (Z): Less than half the unit shown; nec: Not elsewhere classified; nsk: Not specified by kind; - : zero; * : 10-19 percent estimated; ** : 20-29 percent estimated.

PRODUCT SHARE DETAILS

Product or Product Class	% Share	Product or Product Class	% Share
Sheet metal work	100.00	homes)	32.29
Sheet metal air-conditioning ducts and stove pipe	9.62	Other aluminum sheet metal siding (commercial, industrial, farm buildings, etc.)	7.99
Steel sheet metal air-conditioning ducts (including dust collection ducts)	58.96	Other sheet metal siding	0.49
		Metal flooring and siding, nsk	3.75
Aluminum sheet metal air-conditioning ducts (including dust collection ducts)	5.94	Sheet metal awnings, canopies, cornices, and soffits	4.63
Steel sheet metal stove pipe, furnace smoke pipe, and elbows	26.83	Steel sheet metal awnings, canopies, carports, and patios	24.86
Aluminum sheet metal stove pipe, furnace smoke pipe, and elbows	5.51	Aluminum sheet metal awnings, canopies, carports, and patios	16.34
Air-conditioning ducts and stove pipe, nsk	2.77	Sheet metal cornices, skylights, domes, and copings (steel and aluminum)	20.88
Sheet metal culverts, flumes, irrigation pipes, etc.	3.08	Sheet metal soffits, fascia, and shutters (steel and aluminum)	36.09
Steel sheet metal culverts, flumes, irrigation pipes, etc.	81.46	Metal awnings, canopies, cornices, and soffits, nsk	1.85
Aluminum sheet metal culverts, flumes, irrigation pipes, etc.	13.69	Sheet metal electronic enclosures	15.60
Other sheet metal culverts, flumes, irrigation pipes, etc.	2.13	Steel sheet metal computer and peripheral equipment enclosures	40.31
Culverts, flumes, irrigation pipes, etc., nsk	2.71	Aluminum sheet metal computer and peripheral equipment enclosures	17.94
Sheet metal bins and vats	1.97	Other sheet metal electronic enclosures (including machine and motor housings, panels, and guards) steel and aluminum	38.45
Sheet metal grain bins and vats, excluding drying floors, fans, and heaters (steel and aluminum)	49.31	Electronic enclosures, nsk	3.30
Other sheet metal bins and vats, including feed storage bins and sheet metal vats (steel and aluminum)	49.12	Other sheet metal work	26.71
Bins and vats, nsk	1.57	Sheet metal roof ventilators	5.52
Sheet metal roofing and roof drainage equipment	10.65	Sheet metal louvers and dampers for heating, ventilation, and air-conditioning (steel and aluminum)	8.12
Steel sheet metal roofing, all types	59.04	Steel restaurant and hotel kitchen sheet metal equipment	9.16
Aluminum and other sheet metal roofing, all types	9.71	Aluminum restaurant and hotel kitchen sheet metal equipment	0.65
Steel sheet metal roof drainage equipment (including eave troughs, etc.)	10.16	Other steel sheet metal work	51.16
Aluminum sheet metal roof drainage equipment (including eave troughs, etc.)	16.25	Other aluminum sheet metal work	15.02
All other sheet metal roof drainage equipment (including eave troughs, etc.)	2.57	Other sheet metal work (metals other than steel or aluminum)	8.40
Metal roofing and roof drainage equipment, nsk	2.26	Other sheet metal work, nsk	1.95
Sheet metal flooring and siding	7.89	Sheet metal work, nsk	19.85
Fabricated sheet metal flooring	21.42		
Steel sheet metal siding	34.05		
Residential aluminum sheet metal siding (including mobile			

Source: 1992 *Economic Census*. The values shown are percent of total shipments in an industry. Values of indented subcategories are summed in the main heading. The symbol (D) appears when data are withheld to prevent disclosure of competitive information. The abbreviation nsk stands for 'not specified by kind' and nec for 'not elsewhere classified'.

INPUTS AND OUTPUTS FOR SHEET METAL WORK

Economic Sector or Industry Providing Inputs	%	Sector	Economic Sector or Industry Buying Outputs	%	Sector
Blast furnaces & steel mills	35.8	Manufg.	Nonfarm residential structure maintenance	10.6	Constr.
Aluminum rolling & drawing	12.8	Manufg.	Maintenance of nonfarm buildings nec	8.5	Constr.
Wholesale trade	9.9	Trade	Gross private fixed investment	6.4	Cap Inv
Fabricated structural metal	4.1	Manufg.	Residential additions/alterations, nonfarm	4.9	Constr.
Metal coating & allied services	3.9	Manufg.	Residential 1-unit structures, nonfarm	4.6	Constr.
Maintenance of nonfarm buildings nec	2.6	Constr.	Office buildings	4.3	Constr.
Motor freight transportation & warehousing	1.7	Util.	Fabricated structural metal	4.3	Manufg.
Communications, except radio & TV	1.4	Util.	Electronic computing equipment	3.9	Manufg.
Electric services (utilities)	1.4	Util.	Highway & street construction	3.8	Constr.
Miscellaneous plastics products	1.3	Manufg.	Electric utility facility construction	3.6	Constr.
Paints & allied products	1.3	Manufg.	Industrial buildings	3.5	Constr.
Petroleum refining	1.3	Manufg.	Paperboard containers & boxes	2.6	Manufg.
Screw machine and related products	1.2	Manufg.	Maintenance of highways & streets	2.3	Constr.
Sheet metal work	1.1	Manufg.	Radio & TV communication equipment	2.3	Manufg.
Primary lead	1.0	Manufg.	Construction of hospitals	2.1	Constr.
Hotels & lodging places	1.0	Services	Mobile homes	2.1	Manufg.
Metal stampings, nec	0.9	Manufg.	Motor vehicle parts & accessories	2.1	Manufg.
Paperboard containers & boxes	0.9	Manufg.	Federal Government purchases, national defense	2.1	Fed Govt
Gas production & distribution (utilities)	0.9	Util.	Maintenance of electric utility facilities	2.0	Constr.
Eating & drinking places	0.9	Trade	Exports	1.5	Foreign
Sawmills & planning mills, general	0.8	Manufg.	Construction of stores & restaurants	1.3	Constr.
Railroads & related services	0.8	Util.	Machinery, except electrical, nec	1.2	Manufg.
Banking	0.8	Fin/R.E.	Water supply facility construction	1.1	Constr.
Real estate	0.7	Fin/R.E.	Sewer system facility construction	1.0	Constr.
Equipment rental & leasing services	0.6	Services	Conveyors & conveying equipment	0.9	Manufg.
Building paper & board mills	0.5	Manufg.	Electronic components nec	0.9	Manufg.
Machinery, except electrical, nec	0.5	Manufg.	Maintenance of nonbuilding facilities nec	0.8	Constr.
Primary aluminum	0.5	Manufg.	Residential garden apartments	0.8	Constr.
Welding apparatus, electric	0.5	Manufg.	Special industry machinery, nec	0.8	Manufg.
Air transportation	0.5	Util.	Construction of nonfarm buildings nec	0.7	Constr.
Motors & generators	0.4	Manufg.	Hotels & motels	0.7	Constr.
Pumps & compressors	0.4	Manufg.	Nonbuilding facilities nec	0.7	Constr.
Management & consulting services & labs	0.4	Services	Sheet metal work	0.7	Manufg.
Copper rolling & drawing	0.3	Manufg.	Construction of educational buildings	0.6	Constr.
Iron & steel foundries	0.3	Manufg.	Instruments to measure electricity	0.6	Manufg.
Special dies & tools & machine tool accessories	0.3	Manufg.	Mechanical measuring devices	0.6	Manufg.
Advertising	0.3	Services	Amusement & recreation building construction	0.5	Constr.
Legal services	0.3	Services	Industrial controls	0.5	Manufg.
Abrasive products	0.2	Manufg.	Service industry machines, nec	0.5	Manufg.
Aluminum castings	0.2	Manufg.	Dormitories & other group housing	0.4	Constr.
Glass & glass products, except containers	0.2	Manufg.	Maintenance of water supply facilities	0.4	Constr.
Industrial gases	0.2	Manufg.	Resid. & other health facility construction	0.4	Constr.
Motor vehicles & car bodies	0.2	Manufg.	Warehouses	0.4	Constr.
Refrigeration & heating equipment	0.2	Manufg.	Automatic merchandising machines	0.4	Manufg.
Insurance carriers	0.2	Fin/R.E.	Elevators & moving stairways	0.4	Manufg.
Accounting, auditing & bookkeeping	0.2	Services	Typewriters & office machines, nec	0.4	Manufg.
Automotive rental & leasing, without drivers	0.2	Services	Construction of religious buildings	0.3	Constr.
Automotive repair shops & services	0.2	Services	Farm service facilities	0.3	Constr.
Computer & data processing services	0.2	Services	Maintenance of sewer facilities	0.3	Constr.
U.S. Postal Service	0.2	Gov't	Residential 2-4 unit structures, nonfarm	0.3	Constr.
Scrap	0.2	Scrap	Residential high-rise apartments	0.3	Constr.
Hard surface floor coverings	0.1	Manufg.	Telephone & telegraph facility construction	0.3	Constr.
Manifold business forms	0.1	Manufg.	Switchgear & switchboard apparatus	0.3	Manufg.
Nonferrous rolling & drawing, nec	0.1	Manufg.	Gas utility facility construction	0.2	Constr.
Pipe, valves, & pipe fittings	0.1	Manufg.	Maintenance of farm service facilities	0.2	Constr.
Plating & polishing	0.1	Manufg.	Engineering & scientific instruments	0.2	Manufg.
Primary zinc	0.1	Manufg.	Hoists, cranes, & monorails	0.2	Manufg.
Retail trade, except eating & drinking	0.1	Trade	Prefabricated wood buildings	0.2	Manufg.
Business/professional associations	0.1	Services	Semiconductors & related devices	0.2	Manufg.
			Travel trailers & campers	0.2	Manufg.
			Farm housing units & additions & alterations	0.1	Constr.
			Local transit facility construction	0.1	Constr.
			Maintenance of farm residential buildings	0.1	Constr.
			Maintenance of railroads	0.1	Constr.
			Wholesale trade	0.1	Trade

Source: Benchmark Input-Output Accounts for the U.S. Economy, 1982, U.S. Department of Commerce, Washington, D.C., July 1991. Data, as reported in the source, are organized by the 1977 SIC structure in use in 1982 but have been matched, as closely as is possible, to the 1987 SIC structure used in this book.

OCCUPATIONS EMPLOYED BY SIC 344 - FABRICATED STRUCTURAL METAL PRODUCTS

Occupation	% of Total 1994	Change to 2005	Occupation	% of Total 1994	Change to 2005
Assemblers, fabricators, & hand workers nec	8.8	-17.6	Precision metal workers nec	1.8	-9.4
Welders & cutters	7.0	-17.6	Secretaries, ex legal & medical	1.8	-25.0
Sheet metal workers & duct installers	6.8	-58.8	Machinists	1.6	-17.6
Metal fabricators, structural metal products	5.2	-17.6	General office clerks	1.5	-29.7
Blue collar worker supervisors	4.4	-24.6	Bookkeeping, accounting, & auditing clerks	1.5	-38.2
Machine tool cutting & forming etc. nec	4.2	-1.1	Cost estimators	1.4	-9.4
Welding machine setters, operators	3.7	-17.6	Coating, painting, & spraying machine workers	1.4	-1.1
General managers & top executives	3.4	-21.8	Traffic, shipping, & receiving clerks	1.4	-20.7
Sales & related workers nec	2.8	-17.6	Inspectors, testers, & graders, precision	1.1	-17.6
Helpers, laborers, & material movers nec	2.5	-17.6	Industrial production managers	1.1	-17.6
Drafters	2.3	-35.8	Freight, stock, & material movers, hand	1.1	-34.1
Fitters, structural metal, precision	2.1	-50.6	Boilermakers	1.1	11.2
Machine forming operators, metal & plastic	1.9	-46.4	Combination machine tool operators	1.1	-17.6
Truck drivers light & heavy	1.9	-15.0	Financial managers	1.0	-17.6

Source: Industry-Occupation Matrix, Bureau of Labor Statistics. These data relate to one or more 3-digit SIC industry groups rather than to a single 4-digit SIC. The change reported for each occupation to the year 2005 is a percent of growth or decline as estimated by the Bureau of Labor Statistics. The abbreviation nec stands for 'not elsewhere classified'.

LOCATION BY STATE AND REGIONAL CONCENTRATION

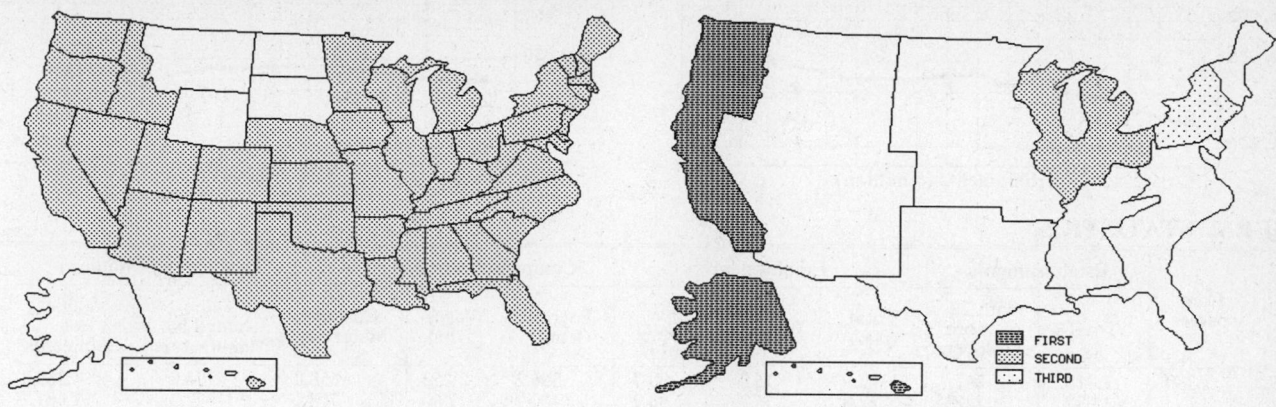

FIRST
SECOND
THIRD

INDUSTRY DATA BY STATE

| State | Establish-ments | Shipments | | | Employment | | | | Cost as % of Shipments | Investment per Employee ($) |
		Total ($ mil)	% of U.S.	Per Establ.	Total Number	% of U.S.	Per Establ.	Wages ($/hour)		
California	731	1,395.9	12.2	1.9	14,400	13.8	20	11.30	44.4	2,278
Ohio	239	1,127.3	9.8	4.7	8,400	8.1	35	11.78	51.7	2,048
Illinois	198	798.8	7.0	4.0	6,100	5.8	31	11.57	54.9	2,541
Texas	360	754.5	6.6	2.1	7,100	6.8	20	10.43	52.3	1,958
Pennsylvania	215	718.2	6.3	3.3	5,900	5.7	27	11.40	53.0	2,034
New York	270	553.3	4.8	2.0	5,800	5.6	21	12.02	42.6	1,810
New Jersey	183	439.1	3.8	2.4	3,400	3.3	19	11.95	50.5	2,118
Florida	234	436.0	3.8	1.9	4,700	4.5	20	9.58	51.7	1,468
Indiana	143	433.6	3.8	3.0	3,700	3.5	26	11.57	52.5	2,216
Massachusetts	179	351.5	3.1	2.0	3,700	3.5	21	13.48	45.6	2,054
Georgia	107	323.6	2.8	3.0	2,100	2.0	20	9.85	59.6	1,381
Michigan	175	303.7	2.6	1.7	3,100	3.0	18	11.21	46.4	2,484
Wisconsin	102	298.0	2.6	2.9	3,100	3.0	30	11.15	46.3	1,935
Minnesota	111	293.5	2.6	2.6	2,900	2.8	26	12.02	46.6	3,552
North Carolina	106	280.6	2.4	2.6	2,600	2.5	25	9.71	53.3	2,500
Washington	104	243.1	2.1	2.3	1,900	1.8	18	11.82	57.3	2,895
Missouri	96	210.5	1.8	2.2	1,800	1.7	19	10.93	51.4	1,833
Alabama	78	188.5	1.6	2.4	1,600	1.5	21	8.79	53.2	2,000
Tennessee	62	159.1	1.4	2.6	1,500	1.4	24	9.27	54.6	2,400
Kansas	41	152.4	1.3	3.7	1,200	1.2	29	10.05	51.7	2,333
Oregon	84	151.4	1.3	1.8	1,600	1.5	19	11.09	47.2	1,750
Arizona	82	151.0	1.3	1.8	1,400	1.3	17	9.65	48.3	2,643
Connecticut	88	134.7	1.2	1.5	1,100	1.1	13	12.12	51.7	1,727
Colorado	65	131.9	1.1	2.0	1,400	1.3	22	10.43	47.1	2,214
Maryland	53	129.9	1.1	2.5	1,300	1.2	25	12.81	49.2	2,462
Kentucky	52	116.7	1.0	2.2	900	0.9	17	10.85	55.8	2,111
Virginia	61	111.9	1.0	1.8	1,400	1.3	23	11.39	44.1	1,714
Mississippi	30	108.3	0.9	3.6	1,100	1.1	37	7.88	50.0	2,091
Arkansas	37	105.0	0.9	2.8	800	0.8	22	8.75	41.4	1,625
Iowa	30	82.4	0.7	2.7	600	0.6	20	10.33	52.1	4,167
South Carolina	49	78.4	0.7	1.6	1,000	1.0	20	10.00	46.2	1,800
Oklahoma	57	73.5	0.6	1.3	700	0.7	12	10.70	49.7	1,571
Utah	33	70.7	0.6	2.1	800	0.8	24	10.23	46.4	2,875
Louisiana	32	61.2	0.5	1.9	800	0.8	25	8.45	49.5	1,000
Idaho	17	54.2	0.5	3.2	500	0.5	29	8.70	41.3	1,600
New Hampshire	39	51.2	0.4	1.3	600	0.6	15	12.33	37.7	2,500
Delaware	12	46.0	0.4	3.8	500	0.5	42	13.50	47.0	1,200
Nebraska	11	43.7	0.4	4.0	300	0.3	27	9.67	70.3	-
Nevada	21	43.0	0.4	2.0	400	0.4	19	12.20	73.5	4,250
Maine	17	42.4	0.4	2.5	400	0.4	24	12.20	55.7	1,000
Vermont	13	22.0	0.2	1.7	300	0.3	23	11.00	45.9	667
Hawaii	6	17.7	0.2	3.0	100	0.1	17	13.50	53.7	3,000
Rhode Island	20	16.6	0.1	0.8	200	0.2	10	9.67	48.2	2,500
New Mexico	16	13.8	0.1	0.9	200	0.2	13	9.50	52.9	500
West Virginia	13	(D)	-	-	750 *	0.7	58	-	-	-

Source: 1992 Economic Census. The states are in descending order of shipments or establishments (if shipment data are missing for the majority). The symbol (D) appears when data are withheld to prevent disclosure of competitive information. States marked with (D) are sorted by number of establishments. A dash (-) indicates that the data element cannot be calculated; * indicates the midpoint of a range.

3446 - ARCHITECTURAL METAL WORK

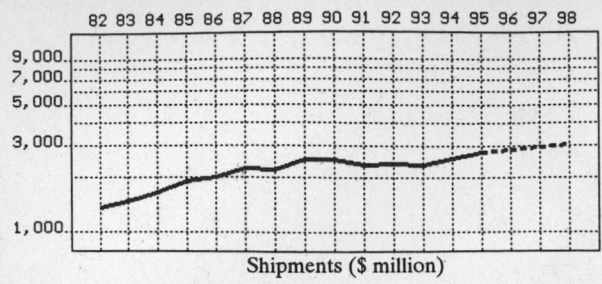

Shipments ($ million)

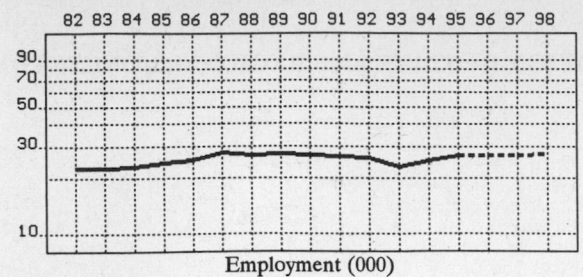

Employment (000)

GENERAL STATISTICS

Year	Com-panies	Establishments		Employment			Compensation		Production ($ million)			
		Total	with 20 or more employees	Total (000)	Production Workers (000)	Hours (Mil)	Payroll ($ mil)	Wages ($/hr)	Cost of Materials	Value Added by Manufacture	Value of Shipments	Capital Invest.
1982	1,380	1,418	241	22.9	16.0	30.7	366.2	7.52	658.8	706.0	1,369.4	26.9
1983		1,357	242	22.8	16.7	33.9	400.9	7.60	708.1	772.2	1,481.7	21.6
1984		1,296	243	23.3	16.7	33.7	426.2	7.93	820.5	842.5	1,648.0	41.3
1985		1,234	243	24.4	17.8	35.5	473.1	8.45	954.7	966.8	1,921.1	48.1
1986		1,199	264	25.3	17.9	33.7	505.6	9.17	979.1	1,022.2	2,017.2	44.4
1987	1,301	1,344	334	27.9	19.7	39.0	571.3	8.72	1,072.0	1,188.9	2,259.3	43.6
1988		1,259	317	26.9	19.2	38.1	580.3	9.03	1,095.7	1,117.1	2,199.6	38.2
1989		1,216	338	27.5	20.5	41.2	689.0	9.76	1,217.3	1,302.4	2,511.1	58.8
1990		1,282	322	27.2	21.6	40.5	695.0	10.14	1,155.4	1,350.8	2,492.9	39.5
1991		1,304	295	26.5	19.0	36.7	636.6	9.91	1,085.2	1,277.5	2,356.1	38.9
1992	1,429	1,476	314	26.1	17.9	35.9	644.8	10.13	1,098.2	1,278.6	2,383.4	34.0
1993		1,450	294	23.0	16.2	33.4	593.6	10.25	1,103.2	1,252.9	2,337.7	31.3
1994		1,358P	337P	25.4	18.1	35.6	656.4	10.58	1,159.7	1,394.5	2,553.9	36.7
1995		1,364P	345P	26.7P	19.3P	38.1P	732.5P	11.01P	1,255.7P	1,510.0P	2,765.4P	40.9P
1996		1,370P	352P	26.9P	19.4P	38.4P	757.6P	11.27P	1,297.7P	1,560.5P	2,857.9P	41.2P
1997		1,376P	360P	27.1P	19.6P	38.7P	782.6P	11.53P	1,339.7P	1,611.0P	2,950.4P	41.5P
1998		1,381P	368P	27.4P	19.7P	39.0P	807.7P	11.79P	1,381.7P	1,661.5P	3,042.9P	41.8P

Sources: 1982, 1987, 1992 *Economic Census*; *Annual Survey of Manufactures*, 83-86, 88-91, 93-94. Establishment counts for non-Census years are from *County Business Patterns*; establishment values for 83-84 are extrapolations. 'P's show projections by the editors. Industries reclassified in 87 will not have data for prior years.

INDICES OF CHANGE

Year	Com-panies	Establishments		Employment			Compensation		Production ($ million)			
		Total	with 20 or more employees	Total (000)	Production Workers (000)	Hours (Mil)	Payroll ($ mil)	Wages ($/hr)	Cost of Materials	Value Added by Manufacture	Value of Shipments	Capital Invest.
1982	97	96	77	88	89	86	57	74	60	55	57	79
1983		92	77	87	93	94	62	75	64	60	62	64
1984		88	77	89	93	94	66	78	75	66	69	121
1985		84	77	93	99	99	73	83	87	76	81	141
1986		81	84	97	100	94	78	91	89	80	85	131
1987	91	91	106	107	110	109	89	86	98	93	95	128
1988		85	101	103	107	106	90	89	100	87	92	112
1989		82	108	105	115	115	107	96	111	102	105	173
1990		87	103	104	121	113	108	100	105	106	105	116
1991		88	94	102	106	102	99	98	99	100	99	114
1992	100	100	100	100	100	100	100	100	100	100	100	100
1993		98	94	88	91	93	92	101	100	98	98	92
1994		92P	107P	97	101	99	102	104	106	109	107	108
1995		92P	110P	102P	108P	106P	114P	109P	114P	118P	116P	120P
1996		93P	112P	103P	109P	107P	117P	111P	118P	122P	120P	121P
1997		93P	115P	104P	109P	108P	121P	114P	122P	126P	124P	122P
1998		94P	117P	105P	110P	109P	125P	116P	126P	130P	128P	123P

Sources: Same as General Statistics. Values reflect change from the base year, 1992. Values above 100 mean greater than 92, values below 100 mean less than 92, and a value of 100 in the 82-91 or 93-98 period means same as 92. 'P's mark projections by the editors.

SELECTED RATIOS

For 1994	Avg. of All Manufact.	Analyzed Industry	Index	For 1994	Avg. of All Manufact.	Analyzed Industry	Index
Employees per Establishment	49	19	38	Value Added per Production Worker	134,084	77,044	57
Payroll per Establishment	1,500,273	483,401	32	Cost per Establishment	5,045,178	854,053	17
Payroll per Employee	30,620	25,843	84	Cost per Employee	102,970	45,657	44
Production Workers per Establishment	34	13	39	Cost per Production Worker	146,988	64,072	44
Wages per Establishment	853,319	277,380	33	Shipments per Establishment	9,576,895	1,880,801	20
Wages per Production Worker	24,861	20,809	84	Shipments per Employee	195,460	100,547	51
Hours per Production Worker	2,056	1,967	96	Shipments per Production Worker	279,017	141,099	51
Wages per Hour	12.09	10.58	87	Investment per Establishment	321,011	27,027	8
Value Added per Establishment	4,602,255	1,026,969	22	Investment per Employee	6,552	1,445	22
Value Added per Employee	93,930	54,902	58	Investment per Production Worker	9,352	2,028	22

Sources: Same as General Statistics. The 'Average of All Manufacturing' column represents the average of all manufacturing industries reported for the most recent complete year available. The Index shows the relationship between the Average and the Analyzed Industry. For example, 100 means that they are equal; 500 that the Analyzed Industry is five times the average; 50 means that the Analyzed Industry is half the national average. The abbreviation 'na' is used to show that data are 'not available'.

LEADING COMPANIES   Number shown: 75   Total sales ($ mil): 2,827   Total employment (000): 22.6

Company Name	Address				CEO Name	Phone	Co. Type	Sales ($ mil)	Empl. (000)
Harsco Corp	PO Box 8888	Camp Hill	PA	17001	Derek C Hathaway	717-763-7064	P	1,358	13.0
Aluma Systems USA Inc	6721 Port West Dr	Houston	TX	77024	Trevor Gosney	713-796-9125	S	625*	2.8
Patent Construction Systems	1 Mack Centre Dr	Paramus	NJ	07652	James P Mitchell	201-261-5600	D	107	1.0
Leslie Building Products Inc	4501 Circle 75 Pkwy	Atlanta	GA	30339	Leigh J Abrams	404-953-6366	P	71	0.3
Leslie-Locke Inc	4501 Circle 75 Pkwy	Atlanta	GA	30339	Ralph C Pepper	404-953-6366	S	64	0.3
Safway Steel Products	PO Box 1991	Milwaukee	WI	53201	MJ Wilson	414-523-6500	D	55*	0.5
WACO Scaffolding & Equip Co	4545 Spring Rd	Cleveland	OH	44131	George Mally	216-749-8900	R	50	0.4
Standard Steel Specialty Co	260 Parkway E	Duncan	SC	29334	TG Armstrong	803-486-9500	R	40	0.3
GS Metals Corp	RR 4	Pinckneyville	IL	62274	KW Coco	618-357-5353	R	35	0.2
Klemp Corp	1132 W Blackhawk	Chicago	IL	60622	William McClure	312-440-3855	R	28	0.2
Milcor LP	1150 N Cable Rd	Lima	OH	45805	Richard Liden	419-228-1411	R	22	0.2
Spider Staging Corp	23500 64th Av S	Kent	WA	98032	Ronald W Tarrant	206-850-3500	S	19	0.2
Bil-Jax Inc	595 E Lugbill Rd	Archbold	OH	43502	SJ Wyse	419-445-8915	R	18	<0.1
Carnes Corp	200 Williams St	Sanford	NC	27330	Donald E Sechler	919-776-4211	S	14	0.2
MM Systems Corp	4520 Elmdale Dr	Tucker	GA	30084	JJ Attaway	404-938-7570	M	13*	<0.1
Wooster Products Inc	PO Box 6005	Wooster	OH	44691	GK Arora	216-264-2844	R	12	<0.1
All American Metal Corp	200 Buffalo Av	Freeport	NY	11520	B Pechter	516-623-0222	R	10	<0.1
ATAS Aluminum Corp	6612 Snowdrift Rd	Allentown	PA	18106	TA Bus	610-395-8445	R	10*	<0.1
Newman Brothers Inc	5609 Ctr Hill Av	Cincinnati	OH	45216	DA Newman	513-242-0011	R	10	<0.1
Superior Aluminum Products Inc	105 Francis St	Russia	OH	45363	Edward Borghers	513-526-4065	R	10	<0.1
Washington Aluminum Co	1330 Knecht Av	Baltimore	MD	21229	RL Pickens	410-242-1000	R	10	<0.1
Brandt-Airflex Corp	937 Conklin St	E Farmingdale	NY	11735	Fred Fogelman	516-752-1234	R	9	0.1
Chemgrate Corp	19240 144th NE	Woodinville	WA	98072	PK Schoening	206-483-9797	R	9*	<0.1
Gulf Aluminum Products Inc	12255 62nd St N	Largo	FL	34643	William Silverstein	813-531-0411	R	9*	<0.1
Lawrence Metal Products	PO Box 400-M	Bay Shore	NY	11706	Stephen Lawrence	516-666-0300	R	9	<0.1
Morton Manufacturing Co	PO Box 640	Libertyville	IL	60048	William C Morton	708-362-5400	R	8*	0.1
Standard Iron Inc	101 Spring Rd	Chattanooga	TN	37405	Larry J Hopper	615-756-0940	S	8	<0.1
Crowley Company Inc	10630 Nassau St NE	Minneapolis	MN	55449	Michael Allen	612-784-1120	R	7	<0.1
Livers Bronze Co	PO Box 266490	Kansas City	MO	64126	Richard Livers Jr	816-833-2828	R	7	0.1
Metal Vent Manufacturing	1025 Firestone St	Memphis	TN	38107	Harold Francis	901-577-9700	R	7	<0.1
Tru-Link Fence Co	5404 N Kedzie Av	Chicago	IL	60625	Michael Levin	312-463-7010	R	7	0.1
McGregor Architectural Iron	46 Line St	Dunmore	PA	18512	Robert McGregor	717-343-2436	R	7	<0.1
Tesko Enterprises	7350 W Montrose	Norridge	IL	60634	Robert Skonieczny	708-452-0045	R	7	<0.1
Consolidated Steel	316 N 12th St	Kenilworth	NJ	07033	Paul Cacicedo Jr	908-272-6262	R	6	<0.1
Custom Enclosures Inc	11631 W Grand Av	Northlake	IL	60164	Daniel Schultz	708-409-1111	R	6	<0.1
Dura-Bilt Products Inc	PO Box 188	Wellsburg	NY	14894	RA Chalk	716-596-2000	R	6	<0.1
James River Iron Co	900 E 4th St	Richmond	VA	23224	H G Kennamer	804-233-7708	R	6*	<0.1
Perry Manufacturing Company	1102 W 16th St	Indianapolis	IN	46208	JE Meadors	317-231-9037	R	6	<0.1
Able Builders Equipment	7451 NW 63th St	Miami	FL	33166	Stanford Freedman	305-592-5940	S	5	<0.1
Artisan House Inc	1755 Glendale Blv	Los Angeles	CA	90026	Henry Goldman	213-664-1111	R	5	<0.1
David Architectural Metals Inc	3100 S Kilbourn	Chicago	IL	60623	Alan Schneider	312-376-3200	R	5	<0.1
Duvinage Corp	PO Box 828	Hagerstown	MD	21741	CR Pedersen	301-733-8255	R	5	0.1
Gadsden Scaffold Company Inc	PO Box 1188	Gadsden	AL	35902	GW Jones	205-547-6918	R	5	<0.1
Moultrie Manufacturing Co	PO Box 1179	Moultrie	GA	31776	William E Smith Jr	912-985-1312	R	5	<0.1
Qual Craft Industries	PO Box 559	Stoughton	MA	02072	Norman Katz	617-344-1000	R	5	<0.1
Rippel Architectural Metals Inc	1525 N Kilpatrick	Chicago	IL	60651	Marvin E Gollob	312-772-0600	R	5*	<0.1
Royal Metal Products Inc	West Rd	Surprise	NY	12176	David Johannesen	518-966-4442	R	5*	<0.1
Steel Ceilings Inc	500 N 3rd St	Coshocton	OH	43812	George Irving	614-622-4655	S	5	<0.1
Universal Manufacturing Corp	PO Box 220	Zelienople	PA	16063	R L Carbeau Jr	412-452-8300	R	5	<0.1
Wesanco Inc	14870 Desman Rd	La Mirada	CA	90638	Andrew D Shyer	714-739-4989	R	5	<0.1
Western Architectural Iron Co	3455 Elston Av	Chicago	IL	60618	Jon F Rundgren	312-463-1500	R	5*	<0.1
Blakeway Metal Works Inc	101 Cargo Way	San Francisco	CA	94124	John Uth	415-647-4494	R	5	<0.1
Fenpro Inc	2601 NW Market St	Seattle	WA	98107	Ed Robinson	206-789-2800	R	5	<0.1
American Architectural Iron Co	80 Liverpool St	East Boston	MA	02128	Bruce J Kapsten	617-567-0011	R	4	<0.1
American Porcelain Enamel Co	3506 Singleton Blv	Dallas	TX	75212	John P Hampton	214-637-4775	R	4	<0.1
Bustin Industrial Products Inc	401 Oak St	E Stroudsburg	PA	18301	Mark Lichty	717-424-6500	R	4	<0.1
Gilpin Inc	PO Box 471	Decatur	IN	46733	Max Gilpin	219-724-9155	R	4	<0.1
Church and Chapel Metal Arts	2616 W Grand Av	Chicago	IL	60612	Joe Taddeo	312-489-3700	R	3	<0.1
Post Road Iron Works Inc	345 W Putnam Av	Greenwich	CT	06830	William Gasparrini	203-869-6322	R	3	<0.1
Anderson Iron Works Inc	5455 State Hwy 169	Plymouth	MN	55442	Richard Grover	612-559-4533	R	3	<0.1
Northwest Grating Products Inc	9230 4th Av S	Seattle	WA	98108	R Breiwick	206-767-3000	R	2	<0.1
Blumcraft of Pittsburgh	460 Melwood St	Pittsburgh	PA	15213	Max Blum	412-681-2400	R	2*	<0.1
Industrial Structures Inc	PO Box 905	Ventura	CA	93002	JP Noga	805-641-1200	R	2*	<0.1
JD Wilkins Co	1130 W Lee St	Greensboro	NC	27403	TC Doss	919-275-4551	R	2	<0.1
Lapeyre Stair Inc	PO Box 50699	New Orleans	LA	70150	Warren Hudson	504-733-6009	S	2	<0.1
Seidelhuber Metal Products Inc	23679 Bernhardt St	Hayward	CA	94545	M J Seidelhuber	510-293-0733	R	2	<0.1
Valley Stairway Inc	PO Box 245	Clovis	CA	93613	A De George Sr	209-299-0151	R	2	<0.1
WS Molnar Co	2545 Beaufait St	Detroit	MI	48207	WS Molnar	313-923-0400	R	2*	<0.1
Barnett-Bates Corp	500 Mills Rd	Joliet	IL	60433	RH Barnett	815-726-5223	R	2	<0.1
Rollform Inc	PO Box 1065	Ann Arbor	MI	48106	GC Adams	313-971-1700	R	2	<0.1
PIW Corp	15765 Annico Dr	Lockport	IL	60441	Gloria Shepherd	708-301-5100	R	2	<0.1
Criterion Gate & Manufacturing	4614 E Washington	Los Angeles	CA	90040	Allen C Ward	213-261-6141	R	1	<0.1
AG Stafford Company Inc	PO Box 8877	Canton	OH	44711	Russell B Caldwell	216-453-8431	R	1	<0.1
Baer Industries Inc	2600 N 2nd St	Philadelphia	PA	19133	Randy Rapoport	215-426-5727	R	1*	<0.1
Grossman Steel	375 Western Hwy	Tappan	NY	10983	Marc E Schreiber	914-359-4300	R	1*	<0.1

Source: Ward's Business Directory of U.S. Private and Public Companies, Volumes 1 and 2, 1996. The company type code used is as follows: P - Public, R - Private, S - Subsidiary, D - Division, J - Joint Venture, A - Affiliate, G - Group. Sales are in millions of dollars, employees are in thousands. An asterisk (*) indicates an estimated sales volume. The symbol < stands for 'less than'. Company names and addresses are truncated, in some cases, to fit into the available space.

MATERIALS CONSUMED

Material	Quantity	Delivered Cost ($ million)
Materials, ingredients, containers, and supplies	(X)	944.1
Metal bolts, nuts, screws, washers, rivets, and other screw machine products	(X)	14.0
Other fabricated metal products (except castings and forgings)	(X)	27.8
Castings (rough and semifinished)	(X)	21.2
Forgings	(X)	1.7
Steel bars, bar shapes, and plates	(X)	49.8
Steel concrete reinforcing bars	(X)	4.3
Steel sheet and strip, including tin plate	(X)	207.2
Steel structural shapes	(X)	38.2
All other steel shapes and forms	(X)	44.5
Copper and copper-base alloy shapes and forms	(X)	4.2
Aluminum and aluminum-base alloy sheet, plate, foil, and welded tubing	(X)	27.1
Aluminum and aluminum-base alloy extruded shapes, including extruded rod, bar, pipe, tube, etc.	(X)	54.9
All other aluminum and aluminum-base alloy shapes and forms, including refinery shapes	(X)	9.7
Other nonferrous shapes and forms	(X)	5.0
Iron and steel scrap, excluding home scrap	(X)	0.6
Paints, varnishes, lacquers, stains, shellacs, japans, enamels, and allied products	(X)	17.8
All other materials and components, parts, containers, and supplies	(X)	133.5
Materials, ingredients, containers, and supplies, nsk	(X)	282.9

Source: 1992 Economic Census. Explanation of symbols used: (D): Withheld to avoid disclosure of competitive data; na: Not available; (S): Withheld because statistical norms were not met; (X): Not applicable; (Z): Less than half the unit shown; nec: Not elsewhere classified; nsk: Not specified by kind; - : zero; * : 10-19 percent estimated; ** : 20-29 percent estimated.

PRODUCT SHARE DETAILS

Product or Product Class	% Share	Product or Product Class	% Share
Architectural metal work	100.00	Nonload-bearing studs (iron, steel, and aluminum)	15.64
Metal grilles, registers, and air diffusers	13.33	Load-bearing studs (iron, steel, and aluminum)	6.81
Iron and steel warm air or air-conditioning grills, registers, and air diffusers	65.54	Metal scaffolding, and shoring and forming for concrete work	9.82
Aluminum warm air or air-conditioning grills, registers, and air diffusers	24.67	Suspended scaffolding (including midpoint, two-point, multilevel, boatswain chairs, etc.) (iron, steel, and aluminum)	20.77
Other iron and steel grills (including open mesh partitions)	3.76	Access scaffolding, including tube and coupler system, prefabricated mobil scaffolds, etc. (iron, steel, and aluminum)	20.60
Other aluminum grills (including open mesh partitions)	4.57		
Grilles, registers, and air diffusers, nsk	1.49	Shoring (including flying forms, postshores, ellis clamps, reshores, etc.) (iron, steel, and aluminum)	7.72
Metal stairs, railings, fences, and gates (other than wire)	23.10		
Iron and steel stairs, staircases, and fire escapes	47.25	Forming (including modular, prefabricated, etc.) (iron, steel, aluminum, and all other material-metal combinations)	50.95
Aluminum stairs, staircases, and fire escapes	7.72		
Iron and steel fences and gates (other than wire)	22.24	Other metal architectural and ornamental work	26.96
Aluminum fences and gates (other than wire)	3.03	Other iron and steel architectural and ornamental work	65.00
Metal railings and window guards (iron, steel, and aluminum, other than wire)	19.01	Other aluminum architectural and ornamental work	20.81
Stairs, railings, fences, and gates (other than wire), nsk	0.75	Other metal architectural and ornamental work (other than iron, steel, or aluminum)	7.69
Open metal flooring, grating, and studs	13.21	Other architectural and ornamental work, nsk	6.52
Open iron and steel flooring and grating for building construction	69.04	Architectural metal work, nsk	13.58
Open aluminum flooring and grating for building construction	8.54		

Source: 1992 Economic Census. The values shown are percent of total shipments in an industry. Values of indented subcategories are summed in the main heading. The symbol (D) appears when data are withheld to prevent disclosure of competitive information. The abbreviation nsk stands for 'not specified by kind' and nec for 'not elsewhere classified'.

INPUTS AND OUTPUTS FOR ARCHITECTURAL METAL WORK

Economic Sector or Industry Providing Inputs	%	Sector	Economic Sector or Industry Buying Outputs	%	Sector
Blast furnaces & steel mills	25.9	Manufg.	Office buildings	15.4	Constr.
Industrial patterns	12.7	Manufg.	Maintenance of nonfarm buildings nec	12.9	Constr.
Aluminum rolling & drawing	9.3	Manufg.	Industrial buildings	9.0	Constr.
Wholesale trade	8.1	Trade	Construction of hospitals	7.1	Constr.
Miscellaneous plastics products	6.0	Manufg.	Highway & street construction	5.2	Constr.
Metal stampings, nec	4.0	Manufg.	Nonfarm residential structure maintenance	5.2	Constr.
Screw machine and related products	3.2	Manufg.	Construction of stores & restaurants	4.8	Constr.
Paperboard containers & boxes	2.9	Manufg.	Construction of educational buildings	3.8	Constr.
Maintenance of nonfarm buildings nec	2.8	Constr.	Electric utility facility construction	3.3	Constr.
Primary aluminum	2.2	Manufg.	Residential 1-unit structures, nonfarm	3.3	Constr.
Motor freight transportation & warehousing	1.8	Util.	Residential additions/alterations, nonfarm	3.1	Constr.
Electric services (utilities)	1.5	Util.	Hotels & motels	2.7	Constr.
Architectural metal work	1.4	Manufg.	Residential garden apartments	2.5	Constr.
Petroleum refining	1.2	Manufg.	Federal Government purchases, national defense	1.9	Fed Govt
Eating & drinking places	1.0	Trade	Warehouses	1.8	Constr.
Paints & allied products	0.8	Manufg.	Exports	1.6	Foreign
Hotels & lodging places	0.8	Services	Construction of nonfarm buildings nec	1.4	Constr.

Continued on next page.

INPUTS AND OUTPUTS FOR ARCHITECTURAL METAL WORK - Continued

Economic Sector or Industry Providing Inputs	%	Sector	Economic Sector or Industry Buying Outputs	%	Sector
Communications, except radio & TV	0.7	Util.	Residential high-rise apartments	1.2	Constr.
Banking	0.7	Fin/R.E.	Resid. & other health facility construction	1.1	Constr.
Machinery, except electrical, nec	0.6	Manufg.	Residential 2-4 unit structures, nonfarm	1.0	Constr.
Gas production & distribution (utilities)	0.6	Util.	Architectural metal work	0.9	Manufg.
Real estate	0.6	Fin/R.E.	Dormitories & other group housing	0.8	Constr.
Fabricated metal products, nec	0.5	Manufg.	Maintenance of water supply facilities	0.8	Constr.
Sawmills & planning mills, general	0.5	Manufg.	Nonbuilding facilities nec	0.8	Constr.
Welding apparatus, electric	0.5	Manufg.	Sewer system facility construction	0.8	Constr.
Air transportation	0.5	Util.	Federal Government purchases, nondefense	0.8	Fed Govt
Railroads & related services	0.5	Util.	Construction of conservation facilities	0.7	Constr.
Iron & steel foundries	0.4	Manufg.	Maintenance of farm service facilities	0.7	Constr.
Miscellaneous fabricated wire products	0.4	Manufg.	Maintenance of highways & streets	0.7	Constr.
Equipment rental & leasing services	0.4	Services	Telephone & telegraph facility construction	0.6	Constr.
Legal services	0.4	Services	Construction of religious buildings	0.5	Constr.
Management & consulting services & labs	0.4	Services	Amusement & recreation building construction	0.4	Constr.
Abrasive products	0.3	Manufg.	Gas utility facility construction	0.4	Constr.
Brass, bronze, & copper castings	0.3	Manufg.	Local transit facility construction	0.4	Constr.
Industrial gases	0.3	Manufg.	Maintenance of local transit facilities	0.3	Constr.
Nonferrous castings, nec	0.3	Manufg.	Maintenance of railroads	0.3	Constr.
Special dies & tools & machine tool accessories	0.3	Manufg.	Maintenance, conservation & development facilities	0.3	Constr.
Advertising	0.3	Services	Farm housing units & additions & alterations	0.2	Constr.
Chemical preparations, nec	0.2	Manufg.	Maintenance of electric utility facilities	0.2	Constr.
Machine tools, metal forming types	0.2	Manufg.	Maintenance of gas utility facilities	0.2	Constr.
Metal coating & allied services	0.2	Manufg.	Maintenance of nonbuilding facilities nec	0.2	Constr.
Paper coating & glazing	0.2	Manufg.	Water supply facility construction	0.2	Constr.
Water supply & sewage systems	0.2	Util.	Maintenance of sewer facilities	0.1	Constr.
Water transportation	0.2	Util.	Mobile homes	0.1	Manufg.
Accounting, auditing & bookkeeping	0.2	Services			
Automotive rental & leasing, without drivers	0.2	Services			
Automotive repair shops & services	0.2	Services			
Business/professional associations	0.2	Services			
U.S. Postal Service	0.2	Gov't			
Manifold business forms	0.1	Manufg.			
Soap & other detergents	0.1	Manufg.			
Retail trade, except eating & drinking	0.1	Trade			
Insurance carriers	0.1	Fin/R.E.			
Computer & data processing services	0.1	Services			
Engineering, architectural, & surveying services	0.1	Services			
Laundry, dry cleaning, shoe repair	0.1	Services			

Source: Benchmark Input-Output Accounts for the U.S. Economy, 1982, U.S. Department of Commerce, Washington, D.C., July 1991. Data, as reported in the source, are organized by the 1977 SIC structure in use in 1982 but have been matched, as closely as is possible, to the 1987 SIC structure used in this book.

OCCUPATIONS EMPLOYED BY SIC 344 - FABRICATED STRUCTURAL METAL PRODUCTS

Occupation	% of Total 1994	Change to 2005	Occupation	% of Total 1994	Change to 2005
Assemblers, fabricators, & hand workers nec	8.8	-17.6	Precision metal workers nec	1.8	-9.4
Welders & cutters	7.0	-17.6	Secretaries, ex legal & medical	1.8	-25.0
Sheet metal workers & duct installers	6.8	-58.8	Machinists	1.6	-17.6
Metal fabricators, structural metal products	5.2	-17.6	General office clerks	1.5	-29.7
Blue collar worker supervisors	4.4	-24.6	Bookkeeping, accounting, & auditing clerks	1.5	-38.2
Machine tool cutting & forming etc. nec	4.2	-1.1	Cost estimators	1.4	-9.4
Welding machine setters, operators	3.7	-17.6	Coating, painting, & spraying machine workers	1.4	-1.1
General managers & top executives	3.4	-21.8	Traffic, shipping, & receiving clerks	1.4	-20.7
Sales & related workers nec	2.8	-17.6	Inspectors, testers, & graders, precision	1.1	-17.6
Helpers, laborers, & material movers nec	2.5	-17.6	Industrial production managers	1.1	-17.6
Drafters	2.3	-35.8	Freight, stock, & material movers, hand	1.1	-34.1
Fitters, structural metal, precision	2.1	-50.6	Boilermakers	1.1	11.2
Machine forming operators, metal & plastic	1.9	-46.4	Combination machine tool operators	1.1	-17.6
Truck drivers light & heavy	1.9	-15.0	Financial managers	1.0	-17.6

Source: Industry-Occupation Matrix, Bureau of Labor Statistics. These data relate to one or more 3-digit SIC industry groups rather than to a single 4-digit SIC. The change reported for each occupation to the year 2005 is a percent of growth or decline as estimated by the Bureau of Labor Statistics. The abbreviation nec stands for 'not elsewhere classified'.

LOCATION BY STATE AND REGIONAL CONCENTRATION

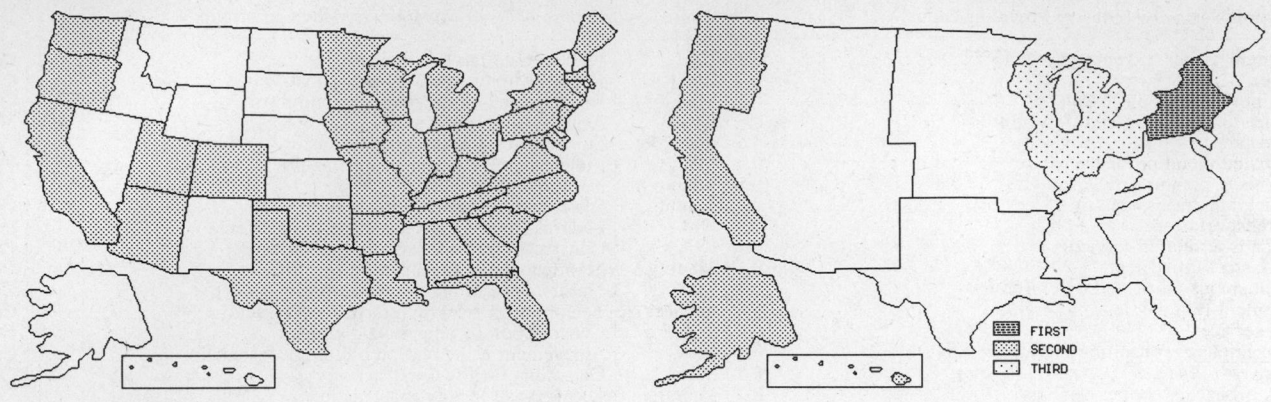

FIRST
SECOND
THIRD

INDUSTRY DATA BY STATE

State	Establish-ments	Shipments			Employment				Cost as % of Shipments	Investment per Employee ($)
		Total ($ mil)	% of U.S.	Per Establ.	Total Number	% of U.S.	Per Establ.	Wages ($/hour)		
California	197	211.5	8.9	1.1	2,100	8.0	11	11.10	46.7	1,381
Texas	100	208.2	8.7	2.1	2,300	8.8	23	8.45	41.5	1,261
Illinois	63	199.7	8.4	3.2	1,500	5.7	24	12.14	52.7	1,133
Ohio	78	195.5	8.2	2.5	1,800	6.9	23	10.80	47.4	1,500
Pennsylvania	95	169.9	7.1	1.8	2,700	10.3	28	10.37	48.1	1,148
Alabama	40	158.4	6.6	4.0	1,700	6.5	43	9.83	46.1	882
New York	115	129.5	5.4	1.1	1,400	5.4	12	11.83	46.4	1,071
Florida	104	109.7	4.6	1.1	1,400	5.4	13	7.36	40.3	714
Michigan	35	103.1	4.3	2.9	1,100	4.2	31	9.50	32.4	4,000
Maryland	27	71.1	3.0	2.6	600	2.3	22	13.13	55.6	833
Connecticut	20	65.9	2.8	3.3	600	2.3	30	13.43	47.5	667
Arizona	40	59.8	2.5	1.5	700	2.7	18	8.23	44.0	1,143
Georgia	33	52.2	2.2	1.6	600	2.3	18	9.25	49.8	2,333
Wisconsin	25	50.3	2.1	2.0	700	2.7	28	9.89	46.5	429
New Jersey	46	47.0	2.0	1.0	400	1.5	9	11.50	55.3	1,750
Oklahoma	15	44.5	1.9	3.0	500	1.9	33	10.22	29.9	-
Massachusetts	38	42.0	1.8	1.1	400	1.5	11	10.83	41.2	750
Missouri	29	40.2	1.7	1.4	500	1.9	17	11.83	49.3	600
Louisiana	17	30.4	1.3	1.8	300	1.1	18	11.33	60.2	1,333
Washington	30	29.0	1.2	1.0	300	1.1	10	10.20	46.9	1,667
Minnesota	21	28.8	1.2	1.4	300	1.1	14	11.25	49.7	667
Utah	14	24.8	1.0	1.8	300	1.1	21	9.50	51.2	-
Kentucky	19	24.8	1.0	1.3	400	1.5	21	6.86	45.6	500
North Carolina	29	23.9	1.0	0.8	300	1.1	10	10.00	43.5	-
Colorado	27	19.4	0.8	0.7	200	0.8	7	11.00	52.1	500
Arkansas	12	7.7	0.3	0.6	100 *	0.4	8	7.00	49.4	-
Indiana	30	(D)	-	-	750 *	2.9	25	-	-	400
Oregon	25	(D)	-	-	175 *	0.7	7	-	-	1,143
Tennessee	25	(D)	-	-	375 *	1.4	15	-	-	-
South Carolina	17	(D)	-	-	375 *	1.4	22	-	-	533
Iowa	13	(D)	-	-	375 *	1.4	29	-	-	-
Virginia	13	(D)	-	-	175 *	0.7	13	-	-	571
Maine	4	(D)	-	-	175 *	0.7	44	-	-	-

Source: 1992 *Economic Census*. The states are in descending order of shipments or establishments (if shipment data are missing for the majority). The symbol (D) appears when data are withheld to prevent disclosure of competitive information. States marked with (D) are sorted by number of establishments. A dash (-) indicates that the data element cannot be calculated; * indicates the midpoint of a range.

3448 - PREFABRICATED METAL BUILDINGS

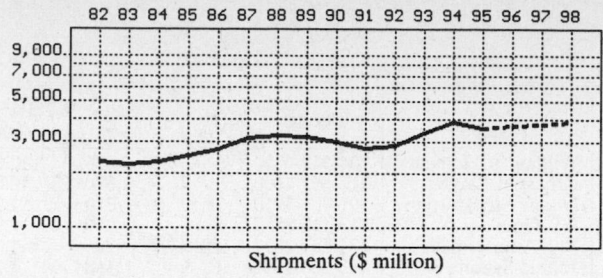

Shipments ($ million)

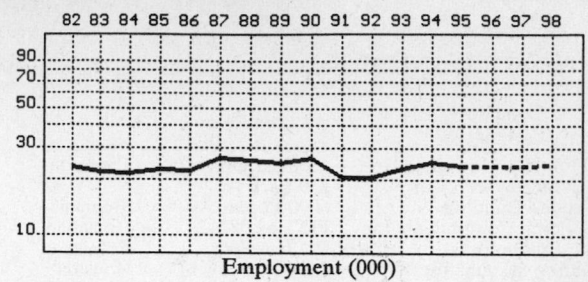

Employment (000)

GENERAL STATISTICS

Year	Com-panies	Establishments		Employment			Compensation		Production ($ million)			
		Total	with 20 or more employees	Total (000)	Production Workers (000)	Hours (Mil)	Payroll ($ mil)	Wages ($/hr)	Cost of Materials	Value Added by Manufacture	Value of Shipments	Capital Invest.
1982	518	569	238	23.5	15.0	29.6	433.0	8.13	1,406.4	873.2	2,315.6	46.1
1983		553	239	22.3	14.6	28.3	424.3	8.13	1,300.1	970.2	2,249.7	49.3
1984		537	240	22.0	14.6	29.4	431.7	8.26	1,384.8	978.8	2,329.0	51.7
1985		521	240	22.8	15.7	30.7	459.6	8.58	1,465.3	1,049.8	2,499.6	50.4
1986		493	231	22.7	15.7	30.6	484.8	8.86	1,589.8	1,143.4	2,736.0	47.4
1987	494	561	246	26.4	17.6	34.3	559.7	9.06	1,802.3	1,340.6	3,146.5	50.8
1988		545	250	25.7	17.5	34.5	565.9	9.17	1,931.5	1,336.9	3,269.1	58.4
1989		499	225	24.9	15.3	30.8	518.6	9.47	1,905.6	1,297.6	3,182.4	31.3
1990		511	230	26.2	14.8	28.7	518.4	9.89	1,781.7	1,183.0	2,984.1	33.2
1991		514	213	21.0	13.5	26.3	470.0	9.80	1,642.4	1,086.2	2,745.7	31.0
1992	487	536	211	20.8	13.4	28.1	527.2	10.08	1,673.0	1,181.8	2,845.9	28.8
1993		555	222	23.2	15.6	28.7	595.9	10.69	1,912.5	1,403.4	3,313.9	34.3
1994		521P	217P	25.3	17.1	34.0	626.6	10.51	2,193.5	1,668.0	3,856.5	50.5
1995		519P	214P	24.0P	15.5P	30.4P	603.5P	10.84P	2,031.3P	1,544.6P	3,571.2P	33.9P
1996		517P	212P	24.1P	15.5P	30.4P	617.0P	11.06P	2,087.2P	1,587.2P	3,669.6P	32.5P
1997		516P	209P	24.1P	15.5P	30.4P	630.5P	11.28P	2,143.2P	1,629.7P	3,768.0P	31.1P
1998		514P	207P	24.2P	15.5P	30.5P	644.0P	11.50P	2,199.1P	1,672.3P	3,866.4P	29.8P

Sources: 1982, 1987, 1992 *Economic Census*; *Annual Survey of Manufactures*, 83-86, 88-91, 93-94. Establishment counts for non-Census years are from *County Business Patterns*; establishment values for 83-84 are extrapolations. 'P's show projections by the editors. Industries reclassified in 87 will not have data for prior years.

INDICES OF CHANGE

Year	Com-panies	Establishments		Employment			Compensation		Production ($ million)			
		Total	with 20 or more employees	Total (000)	Production Workers (000)	Hours (Mil)	Payroll ($ mil)	Wages ($/hr)	Cost of Materials	Value Added by Manufacture	Value of Shipments	Capital Invest.
1982	106	106	113	113	112	105	82	81	84	74	81	160
1983		103	113	107	109	101	80	81	78	82	79	171
1984		100	114	106	109	105	82	82	83	83	82	180
1985		97	114	110	117	109	87	85	88	89	88	175
1986		92	109	109	117	109	92	88	95	97	96	165
1987	101	105	117	127	131	122	106	90	108	113	111	176
1988		102	118	124	131	123	107	91	115	113	115	203
1989		93	107	120	114	110	98	94	114	110	112	109
1990		95	109	126	110	102	98	98	106	100	105	115
1991		96	101	101	101	94	89	97	98	92	96	108
1992	100	100	100	100	100	100	100	100	100	100	100	100
1993		104	105	112	116	102	113	106	114	119	116	119
1994		97P	103P	122	128	121	119	104	131	141	136	175
1995		97P	102P	115P	116P	108P	114P	108P	121P	131P	125P	118P
1996		97P	100P	116P	116P	108P	117P	110P	125P	134P	129P	113P
1997		96P	99P	116P	116P	108P	120P	112P	128P	138P	132P	108P
1998		96P	98P	116P	116P	108P	122P	114P	131P	142P	136P	103P

Sources: Same as General Statistics. Values reflect change from the base year, 1992. Values above 100 mean greater than 92, values below 100 mean less than 92, and a value of 100 in the 82-91 or 93-98 period means same as 92. 'P's mark projections by the editors.

SELECTED RATIOS

For 1994	Avg. of All Manufact.	Analyzed Industry	Index	For 1994	Avg. of All Manufact.	Analyzed Industry	Index
Employees per Establishment	49	49	99	Value Added per Production Worker	134,084	97,544	73
Payroll per Establishment	1,500,273	1,202,442	80	Cost per Establishment	5,045,178	4,209,316	83
Payroll per Employee	30,620	24,767	81	Cost per Employee	102,970	86,700	84
Production Workers per Establishment	34	33	96	Cost per Production Worker	146,988	128,275	87
Wages per Establishment	853,319	685,734	80	Shipments per Establishment	9,576,895	7,400,605	77
Wages per Production Worker	24,861	20,897	84	Shipments per Employee	195,460	152,431	78
Hours per Production Worker	2,056	1,988	97	Shipments per Production Worker	279,017	225,526	81
Wages per Hour	12.09	10.51	87	Investment per Establishment	321,011	96,909	30
Value Added per Establishment	4,602,255	3,200,884	70	Investment per Employee	6,552	1,996	30
Value Added per Employee	93,930	65,929	70	Investment per Production Worker	9,352	2,953	32

Sources: Same as General Statistics. The 'Average of All Manufacturing' column represents the average of all manufacturing industries reported for the most recent complete year available. The Index shows the relationship between the Average and the Analyzed Industry. For example, 100 means that they are equal; 500 that the Analyzed Industry is five times the average; 50 means that the Analyzed Industry is half the national average. The abbreviation 'na' is used to show that data are 'not available'.

LEADING COMPANIES   Number shown: **72**   Total sales ($ mil): **5,825**   Total employment (000): **37.0**

Company Name	Address				CEO Name	Phone	Co. Type	Sales ($ mil)	Empl. (000)
United Dominion Industries Inc	2300 1st Union Ctr	Charlotte	NC	28202	William R Holland	704-347-6800	P	2,036	12.0
Dyson-Kissner-Moran Corp	230 Park Av	New York	NY	10169	Ernest H Larch	212-661-4600	R	1,370*	10.0
Varco-Pruden Buildings	6000 Poplar Av	Memphis	TN	38119	Duane Stockburger	901-767-5910	D	340	1.7
Metal Building Components Inc	PO Box 38217	Houston	TX	77238	AR Ginn	713-445-8555	S	250	0.9
American Buildings Co	PO Box 800	Eufaula	AL	36072	R T Ammerman	334-687-2032	P	205	1.3
NCI Building Systems Inc	7301 Fairview St	Houston	TX	77041	Johnie Schulte Jr	713-466-7788	P	168	1.2
Scotsman Group Inc	8211 Town Ctr Dr	Baltimore	MD	21236	Barry P Gossett	410-931-6000	R	125	0.4
Metal Sales Manufacturing Corp	7800 State Rd 60	Sellersburg	IN	47172	VT Morris	812-246-0819	S	110	0.3
Star Building Systems	PO Box 94910	Oklahoma City	OK	73143	Jack Taylor	405-636-2010	D	100	0.6
Arrow Group Industries Inc	1680 Rte 23 N	Wayne	NJ	07474	George J Smith	201-696-6900	S	95	0.4
Chief Industries Inc	PO Box 2078	Grand Island	NE	68802	Robert G Eihusen	308-382-8820	R	83	1.2
Kawada Industries USA Inc	3000 Airway Av	Costa Mesa	CA	92626	Tommy Naritomi	714-540-3420	S	68*	0.5
Steelox Systems Inc	PO Box 8181	Mason	OH	45040	Frank B Chapman	513-573-5200	S	68	0.5
Metallic Building Co	PO Box 40338	Houston	TX	77240	Leonard George	713-466-7788	D	60	0.3
Gulf States Manufacturers Inc	PO Box 1128	Starkville	MS	39759	Clayton Richardson	601-323-8026	S	50	0.4
Whirlwind Steel Building Inc	PO Box 75280	Houston	TX	77234	Jack Sturdivant	713-946-7140	R	50	0.3
Kirby Building Systems Inc	Kirby Dr	Portland	TN	37148	George F King	615-325-4165	R	41*	0.3
Chief Industries Inc	PO Box 2078	Grand Island	NE	68802	Donald Anderson	308-382-8820	D	40	0.3
Miller Building Systems Inc	PO Box 1283	Elkhart	IN	46515	Edward C Craig	219-295-1214	P	39	0.4
Wedgcor Inc	1515 Bsn Lp E	Jamestown	ND	58401	Denton Wirth	701-252-7380	R	32*	0.3
Inland Buildings	2141 2nd Av SW	Cullman	AL	35055	Dean Grant	205-739-6827	R	30	0.1
Southern Structures Inc	PO Box 52005	Lafayette	LA	70505	A Leonpacher	318-856-5981	R	26	0.2
Aluma Shield Industries Inc	405 Fentress Blv	Daytona Beach	FL	32114	MS Sastri	904-255-5391	S	25*	0.2
Henges Associates Inc	12100 Prichard	Maryland H	MO	63043	Ron Henges	314-739-2600	R	24	0.1
Ruffin Building System Inc	6914 Hwy 2	Oak Grove	LA	71263	Shelton Ruffin	318-428-2305	R	24	0.1
American Steel Building	PO Box 14244	Houston	TX	77221	MB Tankersley	713-433-5661	R	21	0.2
Pascoe Building Systems Inc	PO Box 7186	Columbus	GA	31908	John B Grot	706-324-3562	R	20	0.2
Delta Scientific Corp	24901 W Stanford	Valencia	CA	91355	Harry D Dickinson	805-257-1800	R	19	<0.1
Stuppy Inc	120 E 12th Av	N Kansas City	MO	64116	Jim Stuppy	816-842-3071	R	19*	0.1
Porta-Kamp Manufacturing	PO Box 7064	Houston	TX	77248	William Bigelow	713-674-3163	R	17	0.2
Midco Manufacturing Inc	PO Box 19248	Spokane	WA	99219	CW Savite	509-244-5611	R	14	0.1
Trachte Building Systems Inc	314 Wilburn Rd	Sun Prairie	WI	53590	Steve Pagelow	608-837-7899	R	14*	<0.1
Kaplan Building Systems Inc	Rte 443	Pine Grove	PA	17963	Morris Kaplan	717-345-4635	D	14	<0.1
Ludwig Building Systems Inc	PO Box 23134	Harahan	LA	70183	EB Ludwig Jr	504-733-6260	R	13	0.1
Exhibit Crafts Inc	690 W Manville	Compton	CA	90220	Vito Di Giorgio	213-774-6295	R	11*	<0.1
Stuppy Greenhouse Mfg	120 E 12th Av	N Kansas City	MO	64116	Jim Stuppy	816-842-3071	S	11*	<0.1
A and M Building Systems Inc	PO Box 1450	Clovis	NM	88101	Nels Anderson	505-769-2611	R	10	0.1
Dean Steel Buildings Inc	2929 Industrial Av	Fort Myers	FL	33901	Charles Dean	813-334-1051	R	10	0.1
Dowcraft Corp	65 S Dow St	Falconer	NY	14733	H B Nicholson Jr	716-665-6210	R	10*	0.1
Truss-T Structures Inc	2100 N Pacific Hwy	Woodburn	OR	97071	Floyd Lenhardt Jr	503-981-9581	R	10	<0.1
Tyler Building Systems Co	PO Box 130819	Tyler	TX	75707	Robert C Curtis	903-561-3000	R	10	<0.1
American Panel Corp	5800 SE 78th St	Ocala	FL	34472	Danny E Duncan	904-245-7055	R	9	0.1
Porta-Fab Corp	PO Box 1084	Chesterfield	MO	63006	WR McGee	314-537-5555	R	9	<0.1
Stainless Inc	210 S 3rd St	North Wales	PA	19454	Dave Donelson	215-699-4871	S	9	0.1
Parkline Inc	PO Box 65	Winfield	WV	25213	FE Flanigan	304-586-2113	R	8	0.1
Poly-Tex Inc	PO Box 458	Castle Rock	MN	55010	Terrence Crombie	507-663-0362	R	8	<0.1
Jewell Building Systems Inc	PO Box 397	Dallas	NC	28034	Everett G Jewell	704-922-8652	R	7*	<0.1
Mark Correctional Systems	87 Rte 17 N	Maywood	NJ	07607	Carl Coppola	201-368-8118	D	7	<0.1
Red Dot Corp	PO Box 1240	Athens	TX	75751	LE Bush	903-675-9181	R	7	<0.1
Winandy Greenhouse Company	2211 Peacock Rd	Richmond	IN	47374	Henry R Doherty	317-935-2111	R	7	<0.1
Alabama Metal	PO Box 2529	Fontana	CA	92334	Ed Mooney	909-350-9280	D	6*	<0.1
Beitel Displays and Exhibits Inc	180 Canal Rd	Fairless Hills	PA	19030	William J Beitel	215-736-2100	R	6	<0.1
Desco Steel Buildings Inc	PO Box 38114-102	Houston	TX	77238	Edward de Santiago	713-449-4908	R	6*	<0.1
Haz-Stor Co	2454 Dempster St	Des Plaines	IL	60016	Garry McGovern	708-294-1000	D	6*	<0.1
Enviro-Industries Inc	6501 Ardman Av	Baltimore	MD	21205	Wayne Long	410-483-5600	R	5	<0.1
Lark Builders Inc	PO Box 409	Lyons	GA	30436	Robert L Moore Jr	912-526-3371	R	5	<0.1
Manufactured Structures Corp	PO Box 159	Bristol	IN	46507	Steve A Sabo	219-825-9518	R	5	<0.1
XS Smith Inc	Old Deal Rd	Eatontown	NJ	07724	Richard W Smith Jr	908-222-4600	R	5	<0.1
Miracle Steel Corp	PO Box 1266	Watertown	SD	57201	D Dekker	605-886-7885	R	4	<0.1
Package Industries Inc	15 Harback Rd	Sutton	MA	01590	Daniel E Moroney	508-865-5871	R	4	<0.1
WH Porter Inc	PO Box 1138	Holland	MI	49422	W Porter	616-399-1963	R	4	<0.1
Essex Structural Steel Company	115 Port Watson St	Cortland	NY	13045	Anthony Barbetta	607-753-9384	R	3	<0.1
Arch Technology Corp	PO Box 70006	Plato Center	IL	60170	Victor T Lee	708-464-5656	R	3	<0.1
National Greenhouse Co	PO Box 500	Pana	IL	62557	Cheryl Longthin	217-562-9333	S	2	<0.1
Shenango Steel Buildings Inc	PO Box 268	West Middlesex	PA	16159	JM Campbell	412-528-9925	R	2	<0.1
Super Secure Manufacturing Co	PO Box 3527	La Puente	CA	91749	Donald E Morris	818-333-2543	S	2	<0.1
Modular Systems Inc	12300 Twinbrook	Rockville	MD	20852	Steven Berg	301-984-3334	R	1	<0.1
Craftsmen Steel Buildings Co	1845 Island Av	San Diego	CA	92102	Cliff Houston	619-233-4000	R	1	<0.1
George's Steel Inc	PO Box 23760	Phoenix	AZ	85063	Charles E George	602-278-3501	R	1	<0.1
Omega Sunspaces Inc	3852 Hawkins NE	Albuquerque	NM	87109	Bruno Sommariva	505-344-0333	R	1*	<0.1
Besteel Industries Inc	320 Recold Rd	Walterboro	SC	29488	Robert A Foster	803-538-2351	D	1*	<0.1
Modular Panel Company Inc	63 David St	New Bedford	MA	02744	James Chadwick	508-993-9955	R	1	<0.1

Source: Ward's Business Directory of U.S. Private and Public Companies, Volumes 1 and 2, 1996. The company type code used is as follows: P - Public, R - Private, S - Subsidiary, D - Division, J - Joint Venture, A - Affiliate, G - Group. Sales are in millions of dollars, employees are in thousands. An asterisk (*) indicates an estimated sales volume. The symbol < stands for 'less than'. Company names and addresses are truncated, in some cases, to fit into the available space.

MATERIALS CONSUMED

Material	Quantity	Delivered Cost ($ million)
Materials, ingredients, containers, and supplies	(X)	1,510.5
Metal bolts, nuts, screws, washers, rivets, and other screw machine products	(X)	45.8
Other fabricated metal products (except castings and forgings)	(X)	58.5
Castings (rough and semifinished)	(X)	1.0
Forgings	(X)	(D)
Steel bars, bar shapes, and plates	(X)	135.4
Steel concrete reinforcing bars	(X)	(D)
Steel sheet and strip, including tin plate	(X)	559.6
Steel structural shapes	(X)	51.7
All other steel shapes and forms	(X)	166.9
Copper and copper-base alloy shapes and forms	(X)	(D)
Aluminum and aluminum-base alloy sheet, plate, foil, and welded tubing	(X)	60.8
Aluminum and aluminum-base alloy extruded shapes, including extruded rod, bar, pipe, tube, etc. .	(X)	24.3
All other aluminum and aluminum-base alloy shapes and forms, including refinery shapes	(X)	26.3
Other nonferrous shapes and forms	(X)	2.2
Iron and steel scrap, excluding home scrap	(X)	(D)
Paints, varnishes, lacquers, stains, shellacs, japans, enamels, and allied products	(X)	18.1
All other materials and components, parts, containers, and supplies	(X)	135.9
Materials, ingredients, containers, and supplies, nsk	(X)	152.8

Source: 1992 Economic Census. Explanation of symbols used: (D): Withheld to avoid disclosure of competitive data; na: Not available; (S): Withheld because statistical norms were not met; (X): Not applicable; (Z): Less than half the unit shown; nec: Not elsewhere classified; nsk: Not specified by kind; - : zero; * : 10-19 percent estimated; ** : 20-29 percent estimated.

PRODUCT SHARE DETAILS

Product or Product Class	% Share	Product or Product Class	% Share
Prefabricated metal buildings	100.00	aluminum	3.30
Prefabricated metal building systems (excluding farm service buildings, residential buildings, and parts)	64.25	Other prefabricated and portable farm service buildings (livestock shelters, machinery storage, etc.), steel and aluminum	6.85
Industrial and commercial prefabricated metal building systems (excluding farm service buildings and parts) . .	86.64	Prefabricated and portable dwellings, steel and aluminum (including vacation homes and camps) . . .	4.15
Institutional, medical, and religious prefabricated metal building systems (excluding farm service buildings and parts)	4.70	Prefabricated and portable small steel utility buildings (including toolsheds, cabanas, storage houses, etc.) . . .	10.17
Public and educational prefabricated metal building systems (excluding farm service buildings and parts)	7.46	Prefabricated and portable small aluminum utility buildings (including toolsheds, cabanas, storage houses, etc.) . . .	8.18
Prefabricated metal building systems (excluding farm service buildings, residential buildings, and parts), nsk . .	1.20	Other prefabricated and portable steel buildings	10.48
Other prefabricated and portable metal buildings and parts .	27.21	Other prefabricated and portable aluminum buildings . .	15.03
Prefabricated and portable greenhouses, steel and aluminum	11.51	Panels, parts, or sections for prefabricated buildings, not sold as a complete unit, steel and aluminum	28.88
Prefabricated and portable grain storage buildings (including farm and commercial types), steel and		Other prefabricated and portable metal buildings and parts, nsk	1.42
		Prefabricated metal buildings, nsk	8.54

Source: 1992 Economic Census. The values shown are percent of total shipments in an industry. Values of indented subcategories are summed in the main heading. The symbol (D) appears when data is withheld to prevent disclosure of competitive information. The abbreviation nsk stands for 'not specified by kind' and nec for 'not elsewhere classified'.

INPUTS AND OUTPUTS FOR PREFABRICATED METAL BUILDINGS

Economic Sector or Industry Providing Inputs	%	Sector	Economic Sector or Industry Buying Outputs	%	Sector
Blast furnaces & steel mills	33.0	Manufg.	Industrial buildings	19.9	Constr.
Metal coating & allied services	14.8	Manufg.	Office buildings	19.3	Constr.
Wholesale trade	8.2	Trade	Warehouses	12.7	Constr.
Aluminum rolling & drawing	7.1	Manufg.	Exports	10.4	Foreign
Veneer & plywood	5.5	Manufg.	Farm service facilities	8.0	Constr.
Metal stampings, nec	3.7	Manufg.	Construction of stores & restaurants	5.7	Constr.
Screw machine and related products	3.0	Manufg.	Personal consumption expenditures	4.2	
Fabricated metal products, nec	2.3	Manufg.	Construction of educational buildings	2.8	Constr.
Prefabricated metal buildings	2.1	Manufg.	Residential high-rise apartments	2.6	Constr.
Motor freight transportation & warehousing	1.6	Util.	Residential 1-unit structures, nonfarm	1.8	Constr.
Maintenance of nonfarm buildings nec	1.4	Constr.	Prefabricated metal buildings	1.6	Manufg.
Libraries, vocation education	1.2	Services	Construction of religious buildings	1.4	Constr.
Industrial patterns	1.1	Manufg.	Nonfarm residential structure maintenance	1.4	Constr.
Electric services (utilities)	1.0	Util.	Nonbuilding facilities nec	1.3	Constr.
Miscellaneous fabricated wire products	0.9	Manufg.	Resid. & other health facility construction	1.2	Constr.
Petroleum refining	0.9	Manufg.	Garage & service station construction	1.0	Constr.
Railroads & related services	0.9	Util.	Residential additions/alterations, nonfarm	1.0	Constr.
Hotels & lodging places	0.9	Services	Residential garden apartments	0.9	Constr.
Paints & allied products	0.7	Manufg.	Farm housing units & additions & alterations	0.5	Constr.
Communications, except radio & TV	0.7	Util.	Maintenance of farm service facilities	0.5	Constr.
Equipment rental & leasing services	0.7	Services	Sewer system facility construction	0.5	Constr.
Banking	0.6	Fin/R.E.	Federal Government purchases, national defense	0.5	Fed Govt

Continued on next page.

INPUTS AND OUTPUTS FOR PREFABRICATED METAL BUILDINGS - Continued

Economic Sector or Industry Providing Inputs	%	Sector	Economic Sector or Industry Buying Outputs	%	Sector
Iron & steel foundries	0.5	Manufg.	Maintenance of nonbuilding facilities nec	0.3	Constr.
Air transportation	0.5	Util.	Maintenance of nonfarm buildings nec	0.3	Constr.
Gas production & distribution (utilities)	0.5	Util.	Residential 2-4 unit structures, nonfarm	0.3	Constr.
Eating & drinking places	0.5	Trade			
Gaskets, packing & sealing devices	0.4	Manufg.			
Welding apparatus, electric	0.4	Manufg.			
Glass & glass products, except containers	0.3	Manufg.			
Machinery, except electrical, nec	0.3	Manufg.			
Real estate	0.3	Fin/R.E.			
Copper rolling & drawing	0.2	Manufg.			
Machine tools, metal forming types	0.2	Manufg.			
Water transportation	0.2	Util.			
Credit agencies other than banks	0.2	Fin/R.E.			
Advertising	0.2	Services			
Automotive rental & leasing, without drivers	0.2	Services			
Legal services	0.2	Services			
Management & consulting services & labs	0.2	Services			
Abrasive products	0.1	Manufg.			
Industrial gases	0.1	Manufg.			
Soap & other detergents	0.1	Manufg.			
Special dies & tools & machine tool accessories	0.1	Manufg.			
Insurance carriers	0.1	Fin/R.E.			
Accounting, auditing & bookkeeping	0.1	Services			
Automotive repair shops & services	0.1	Services			
Computer & data processing services	0.1	Services			
U.S. Postal Service	0.1	Gov't			
Scrap	0.1	Scrap			

Source: Benchmark Input-Output Accounts for the U.S. Economy, 1982, U.S. Department of Commerce, Washington, D.C., July 1991. Data, as reported in the source, are organized by the 1977 SIC structure in use in 1982 but have been matched, as closely as is possible, to the 1987 SIC structure used in this book.

OCCUPATIONS EMPLOYED BY SIC 344 - FABRICATED STRUCTURAL METAL PRODUCTS

Occupation	% of Total 1994	Change to 2005	Occupation	% of Total 1994	Change to 2005
Assemblers, fabricators, & hand workers nec	8.8	-17.6	Precision metal workers nec	1.8	-9.4
Welders & cutters	7.0	-17.6	Secretaries, ex legal & medical	1.8	-25.0
Sheet metal workers & duct installers	6.8	-58.8	Machinists	1.6	-17.6
Metal fabricators, structural metal products	5.2	-17.6	General office clerks	1.5	-29.7
Blue collar worker supervisors	4.4	-24.6	Bookkeeping, accounting, & auditing clerks	1.5	-38.2
Machine tool cutting & forming etc. nec	4.2	-1.1	Cost estimators	1.4	-9.4
Welding machine setters, operators	3.7	-17.6	Coating, painting, & spraying machine workers	1.4	-1.1
General managers & top executives	3.4	-21.8	Traffic, shipping, & receiving clerks	1.4	-20.7
Sales & related workers nec	2.8	-17.6	Inspectors, testers, & graders, precision	1.1	-17.6
Helpers, laborers, & material movers nec	2.5	-17.6	Industrial production managers	1.1	-17.6
Drafters	2.3	-35.8	Freight, stock, & material movers, hand	1.1	-34.1
Fitters, structural metal, precision	2.1	-50.6	Boilermakers	1.1	11.2
Machine forming operators, metal & plastic	1.9	-46.4	Combination machine tool operators	1.1	-17.6
Truck drivers light & heavy	1.9	-15.0	Financial managers	1.0	-17.6

Source: Industry-Occupation Matrix, Bureau of Labor Statistics. These data relate to one or more 3-digit SIC industry groups rather than to a single 4-digit SIC. The change reported for each occupation to the year 2005 is a percent of growth or decline as estimated by the Bureau of Labor Statistics. The abbreviation nec stands for 'not elsewhere classified'.

LOCATION BY STATE AND REGIONAL CONCENTRATION

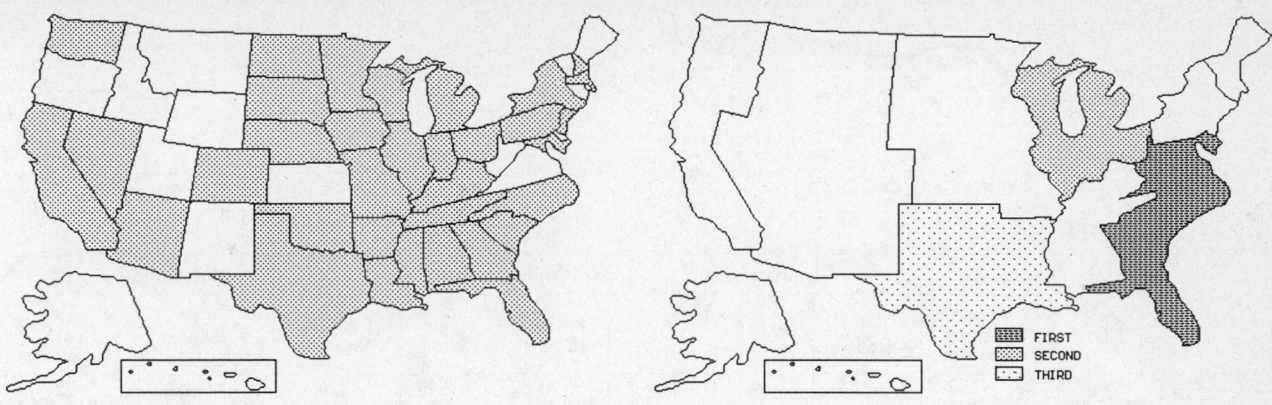

INDUSTRY DATA BY STATE

| State | Establish-ments | Shipments | | | Employment | | | | Cost as % of Shipments | Investment per Employee ($) |
		Total ($ mil)	% of U.S.	Per Establ.	Total Number	% of U.S.	Per Establ.	Wages ($/hour)		
Illinois	18	261.3	9.2	14.5	1,300	6.3	72	12.29	74.1	692
Texas	46	245.6	8.6	5.3	1,900	9.1	41	7.96	59.9	895
California	57	213.0	7.5	3.7	1,700	8.2	30	9.67	55.0	1,412
North Carolina	18	166.6	5.9	9.3	1,000	4.8	56	11.63	58.6	1,500
Iowa	13	153.3	5.4	11.8	1,000	4.8	77	9.80	52.4	-
Pennsylvania	22	139.8	4.9	6.4	1,100	5.3	50	11.56	54.4	2,818
Wisconsin	12	135.6	4.8	11.3	900	4.3	75	11.73	65.5	1,556
Florida	42	119.0	4.2	2.8	900	4.3	21	8.36	56.0	1,111
Alabama	17	111.8	3.9	6.6	900	4.3	53	8.70	52.9	444
Ohio	24	99.6	3.5	4.2	800	3.8	33	10.87	59.9	1,750
New York	26	97.7	3.4	3.8	1,000	4.8	38	10.15	47.6	2,100
Nebraska	4	95.3	3.3	23.8	1,000	4.8	250	8.31	56.7	-
Georgia	25	92.3	3.2	3.7	800	3.8	32	8.64	60.2	500
Mississippi	5	85.8	3.0	17.2	500	2.4	100	8.00	54.4	1,600
New Jersey	12	68.4	2.4	5.7	500	2.4	42	12.17	57.2	1,400
Missouri	13	63.3	2.2	4.9	400	1.9	31	10.20	52.0	750
Oklahoma	11	56.6	2.0	5.1	500	2.4	45	9.14	47.2	1,000
Arkansas	10	49.0	1.7	4.9	300	1.4	30	13.00	61.4	4,667
Michigan	26	45.7	1.6	1.8	500	2.4	19	9.71	51.9	1,400
Louisiana	9	40.2	1.4	4.5	400	1.9	44	7.00	68.2	750
Maryland	8	33.8	1.2	4.2	200	1.0	25	10.50	53.6	1,500
Washington	9	21.3	0.7	2.4	200	1.0	22	11.50	63.8	500
Minnesota	17	19.9	0.7	1.2	200	1.0	12	10.00	51.3	500
Kentucky	11	19.2	0.7	1.7	100	0.5	9	8.50	58.3	2,000
South Dakota	4	17.0	0.6	4.3	200	1.0	50	7.33	51.8	-
Colorado	11	14.4	0.5	1.3	100	0.5	9	10.00	68.1	1,000
Arizona	9	10.9	0.4	1.2	200	1.0	22	8.00	67.9	-
Indiana	8	(D)	-	-	750 *	3.6	94	-	-	400
Tennessee	7	(D)	-	-	375 *	1.8	54	-	-	-
South Carolina	6	(D)	-	-	175 *	0.8	29	-	-	-
Massachusetts	4	(D)	-	-	175 *	0.8	44	-	-	-
Nevada	2	(D)	-	-	175 *	0.8	88	-	-	-
New Hampshire	1	(D)	-	-	375 *	1.8	375	-	-	-
North Dakota	1	(D)	-	-	175 *	0.8	175	-	-	-

Source: 1992 *Economic Census.* The states are in descending order of shipments or establishments (if shipment data are missing for the majority). The symbol (D) appears when data are withheld to prevent disclosure of competitive information. States marked with (D) are sorted by number of establishments. A dash (-) indicates that the data element cannot be calculated; * indicates the midpoint of a range.

3449 - MISCELLANEOUS METAL WORK

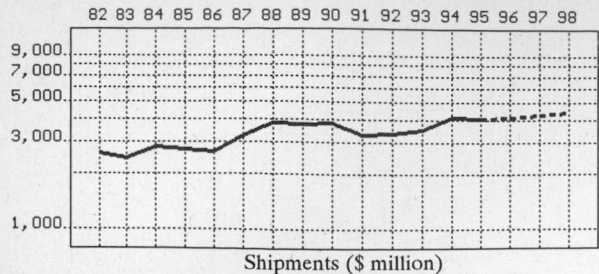

Shipments ($ million)

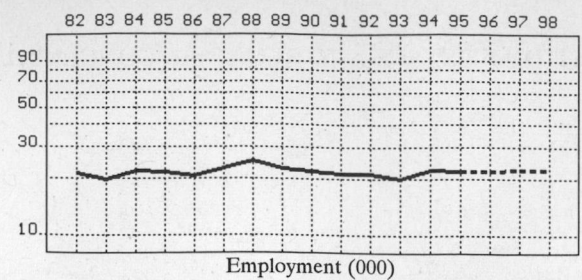

Employment (000)

GENERAL STATISTICS

Year	Companies	Establishments Total	Establishments with 20 or more employees	Employment Total (000)	Employment Production Workers (000)	Employment Hours (Mil)	Compensation Payroll ($ mil)	Compensation Wages ($/hr)	Production Cost of Materials	Production Value Added by Manufacture	Production Value of Shipments	Production Capital Invest.
1982	401	492	257	21.5	14.1	26.7	414.7	8.79	1,653.9	909.9	2,594.6	44.9
1983		477	254	19.9	13.0	25.4	404.8	8.76	1,548.3	914.4	2,441.8	33.2
1984		462	251	22.1	15.0	29.5	459.4	9.13	1,859.1	995.9	2,815.1	58.6
1985		448	248	21.8	13.8	27.8	474.2	9.74	1,731.2	996.9	2,727.8	56.6
1986		433	229	20.7	13.3	26.3	455.6	9.95	1,652.5	931.5	2,616.8	68.2
1987	494	597	278	22.9	15.6	32.3	537.7	9.81	1,999.3	1,250.6	3,231.4	62.0
1988		609	297	25.1	17.2	35.4	611.4	10.38	2,486.1	1,334.1	3,810.6	49.9
1989		570	284	22.8	16.4	32.6	570.3	10.88	2,401.2	1,342.0	3,733.6	52.0
1990		560	298	22.1	15.6	31.7	580.2	11.24	2,537.6	1,245.9	3,786.8	62.7
1991		568	284	21.6	14.6	28.7	556.3	11.97	2,233.0	1,022.8	3,285.3	52.9
1992	564	658	292	21.6	14.4	29.1	589.8	12.04	2,222.1	1,139.1	3,366.4	60.0
1993		648	294	20.3	14.0	28.9	589.5	12.12	2,297.7	1,251.2	3,514.9	67.0
1994		659P	304P	22.5	15.7	32.2	695.9	12.74	2,731.1	1,406.7	4,098.6	39.5
1995		676P	309P	22.2P	15.6P	32.2P	674.6P	12.98P	2,695.0P	1,388.1P	4,044.5P	58.7P
1996		694P	314P	22.3P	15.7P	32.5P	694.7P	13.33P	2,772.3P	1,427.9P	4,160.5P	59.3P
1997		712P	319P	22.3P	15.8P	32.8P	714.8P	13.67P	2,849.6P	1,467.7P	4,276.5P	59.9P
1998		730P	324P	22.4P	15.9P	33.2P	735.0P	14.01P	2,926.9P	1,507.6P	4,392.4P	60.5P

Sources: 1982, 1987, 1992 *Economic Census*; *Annual Survey of Manufactures*, 83-86, 88-91, 93-94. Establishment counts for non-Census years are from *County Business Patterns*; establishment values for 83-84 are extrapolations. 'P's show projections by the editors. Industries reclassified in 87 will not have data for prior years.

INDICES OF CHANGE

Year	Companies	Establishments Total	Establishments with 20 or more employees	Employment Total (000)	Employment Production Workers (000)	Employment Hours (Mil)	Compensation Payroll ($ mil)	Compensation Wages ($/hr)	Production Cost of Materials	Production Value Added by Manufacture	Production Value of Shipments	Production Capital Invest.
1982	71	75	88	100	98	92	70	73	74	80	77	75
1983		72	87	92	90	87	69	73	70	80	73	55
1984		70	86	102	104	101	78	76	84	87	84	98
1985		68	85	101	96	96	80	81	78	88	81	94
1986		66	78	96	92	90	77	83	74	82	78	114
1987	88	91	95	106	108	111	91	81	90	110	96	103
1988		93	102	116	119	122	104	86	112	117	113	83
1989		87	97	106	114	112	97	90	108	118	111	87
1990		85	102	102	108	109	98	93	114	109	112	105
1991		86	97	100	101	99	94	99	100	90	98	88
1992	100	100	100	100	100	100	100	100	100	100	100	100
1993		98	101	94	97	99	100	101	103	110	104	112
1994		100P	104P	104	109	111	118	106	123	123	122	66
1995		103P	106P	103P	108P	110P	114P	108P	121P	122P	120P	98P
1996		105P	108P	103P	109P	112P	118P	111P	125P	125P	124P	99P
1997		108P	109P	103P	110P	113P	121P	114P	128P	129P	127P	100P
1998		111P	111P	104P	111P	114P	125P	116P	132P	132P	130P	101P

Sources: Same as General Statistics. Values reflect change from the base year, 1992. Values above 100 mean greater than 92, values below 100 mean less than 92, and a value of 100 in the 82-91 or 93-98 period means same as 92. 'P's mark projections by the editors.

SELECTED RATIOS

For 1994	Avg. of All Manufact.	Analyzed Industry	Index	For 1994	Avg. of All Manufact.	Analyzed Industry	Index
Employees per Establishment	49	34	70	Value Added per Production Worker	134,084	89,599	67
Payroll per Establishment	1,500,273	1,056,431	70	Cost per Establishment	5,045,178	4,146,025	82
Payroll per Employee	30,620	30,929	101	Cost per Employee	102,970	121,382	118
Production Workers per Establishment	34	24	69	Cost per Production Worker	146,988	173,955	118
Wages per Establishment	853,319	622,758	73	Shipments per Establishment	9,576,895	6,221,998	65
Wages per Production Worker	24,861	26,129	105	Shipments per Employee	195,460	182,160	93
Hours per Production Worker	2,056	2,051	100	Shipments per Production Worker	279,017	261,057	94
Wages per Hour	12.09	12.74	105	Investment per Establishment	321,011	59,964	19
Value Added per Establishment	4,602,255	2,135,482	46	Investment per Employee	6,552	1,756	27
Value Added per Employee	93,930	62,520	67	Investment per Production Worker	9,352	2,516	27

Sources: Same as General Statistics. The 'Average of All Manufacturing' column represents the average of all manufacturing industries reported for the most recent complete year available. The Index shows the relationship between the Average and the Analyzed Industry. For example, 100 means that they are equal; 500 that the Analyzed Industry is five times the average; 50 means that the Analyzed Industry is half the national average. The abbreviation 'na' is used to show that data are 'not available'.

LEADING COMPANIES Number shown: **41**   Total sales ($ mil): **510**   Total employment (000): **4.6**

Company Name	Address				CEO Name	Phone	Co. Type	Sales ($ mil)	Empl. (000)
Superior Metal Products Inc	1005 W Grand Av	Lima	OH	45801	Ed Kasody	419-228-1145	R	59	0.6
Smarte Carte Inc	4455 White Bear	St Paul	MN	55110	Grant McLennen	612-429-3614	R	56*	0.6
Unimast Inc	9595 Grand Av	Franklin Park	IL	60131	Garen W Smith	708-451-1410	S	40	0.4
Western Metal Lath	6510 General Dr	Riverside	CA	92509	Donald R Moody	714-360-3500	S	35	<0.1
Upper Midwest Industries Inc	730 30th Av SE	Minneapolis	MN	55414	Chuck Carlsen	612-378-1071	R	28*	0.3
Saginaw Control and Engineering	95 Midland Rd	Saginaw	MI	48603	H C Baldauf Sr	517-799-6871	R	25	0.4
Morris Bean and Co	PO Box 108	Yellow Springs	OH	45387	Edward Myers	513-767-7301	R	18*	0.2
Ohio Moulding Corp	30396 Lakeland	Wickliffe	OH	44092	Edward F Gleason	216-944-2100	R	18	0.1
RWM Casters Co	PO Box 668	Gastonia	NC	28053	Bob Hillebrand	704-866-8533	R	16	0.2
National Meter Parts Inc	PO Box 908	Lancaster	OH	43130	James E Fosnaugh	614-653-8828	R	15*	0.2
PSB Co	PO Box 1089	Columbus	OH	43216	Robert Johns	614-228-5781	D	14	0.1
Bratton Corp	2801 E 85th St	Kansas City	MO	64132	RD Long	816-363-1015	R	13*	0.1
Picut Manufacturing Company	140 Mt Betchel Rd	Warren	NJ	07060	FR Picut	908-754-1333	R	12	<0.1
North Star Company Inc	14912 S Broadway	Gardena	CA	90248	WE Sornborger	310-515-2200	R	11	<0.1
Ryan Iron Works Inc	PO Box 159	Raynham	MA	02767	HF Shea	508-822-8001	R	11	<0.1
Akron Rebar Co	PO Box 3708	Akron	OH	44314	Dennis Stump	216-745-7100	R	10	<0.1
Consolidated Metal Products Inc	1028 Depot St	Cincinnati	OH	45204	Hugh M Gallagher	513-251-2624	R	10*	0.1
Klemp Corp	PO Box 1720	Orem	UT	84059	Doug Clark	801-225-9350	S	10	<0.1
Spantek	1520 S 5th St	Hopkins	MN	55343	Craig A Swanson	612-935-8431	D	10	<0.1
Phillips Construction Co	4601 S 76th Cir	Omaha	NE	68127	George J Kubat	402-339-3800	R	9*	0.1
Southwest Rebar	1285 N McQueen	Gilbert	AZ	85233	MJ Toone	602-892-0375	R	9	<0.1
Hoertig Iron Works Inc	PO Box 2189	City of Industry	CA	91746	Richard K Hoertig	818-968-1574	R	8	<0.1
Crescent Iron Works Co	4901 Grays Av	Philadelphia	PA	19143	Joseph W Milani	215-729-1204	R	7*	<0.1
CL Rieckhoff Company Inc	26265 Northline Dr	Taylor	MI	48180	CL Rieckhoff	313-946-8220	R	7	<0.1
Ornamental Iron Work Co	2900 Newpark Dr	Norton	OH	44203	William D Boesche	216-745-6071	R	6	<0.1
WSF Industries Inc	PO Box 400	Buffalo	NY	14217	J Hettrick	716-692-4930	R	6	<0.1
Alvarado Manufacturing	12660 Colony St	Chino	CA	91710	JP Armatas	909-591-8431	R	6	<0.1
Cowelco	1634 W 14th St	Long Beach	CA	90813	Nancy E McCrabb	310-432-5766	R	5	<0.1
Hydro-Craft Inc	1821 Rochester	Rochester Hills	MI	48309	William R Walker	313-652-8100	R	5	<0.1
Rollform of Jamestown Inc	181 Blackstone Av	Jamestown	NY	14701	Edward Ruttenberg	716-665-5310	S	5	<0.1
Anchor Fabricators Inc	386 Talmadge Rd	Clayton	OH	45315	WA Semmelman	513-836-5117	R	4	<0.1
Tarpenning-LaFollette Company	404 W Gimber St	Indianapolis	IN	46225	KR Martin	317-780-1500	R	4	<0.1
Tubular Specialties Mfg	13011 S Spring St	Los Angeles	CA	90061	Marcia Huntley	310-515-4801	R	4	<0.1
Johnson Bros Metal Forming	5500 McDermott Dr	Berkeley	IL	60163	EO Johnson	708-449-7050	R	4	<0.1
Johnstone Company Inc	222 Sackett Pt Rd	North Haven	CT	06473	David R Johnstone	203-239-5834	R	3	<0.1
Ornamental Metal Works Co	PO Box 977	Decatur	IL	62525	MF Wallace Jr	217-428-3446	R	3	<0.1
ATR Technologies Inc	PO Box 118	San Gabriel	CA	91778	Don Terry	818-570-1740	R	2*	<0.1
Formetal Inc	230 Houghton St	Oak Harbor	OH	43449	William L Briggs	419-898-2211	R	2	<0.1
GTE Metal Erectors Inc	PO Box 877	Canby	OR	97013	Leo E Thatcher	503-266-6433	R	1*	<0.1
O'Brien Consolidated Industries	PO Box 139	Lewiston	ME	04243	SD Lagueux	207-783-8543	R	1	<0.1
Pioneer Steel Co	10608 S Santa Fe	South Gate	CA	90280	Ralph Peet	213-564-3303	R	1*	<0.1

Source: Ward's Business Directory of U.S. Private and Public Companies, Volumes 1 and 2, 1996. The company type code used is as follows: P - Public, R - Private, S - Subsidiary, D - Division, J - Joint Venture, A - Affiliate, G - Group. Sales are in millions of dollars, employees are in thousands. An asterisk (*) indicates an estimated sales volume. The symbol < stands for 'less than'. Company names and addresses are truncated, in some cases, to fit into the available space.

MATERIALS CONSUMED

Material	Quantity	Delivered Cost ($ million)
Materials, ingredients, containers, and supplies	(X)	2,004.8
Metal bolts, nuts, screws, washers, rivets, and other screw machine products	(X)	12.7
Other fabricated metal products (except castings and forgings)	(X)	43.1
Castings (rough and semifinished)	(X)	16.6
Forgings	(X)	(D)
Steel bars, bar shapes, and plates	(X)	326.0
Steel concrete reinforcing bars	(X)	263.7
Steel sheet and strip, including tin plate	(X)	475.8
Steel structural shapes	(X)	68.4
All other steel shapes and forms	(X)	135.6
Copper and copper-base alloy shapes and forms	(X)	8.2
Aluminum and aluminum-base alloy sheet, plate, foil, and welded tubing	(X)	72.6
Aluminum and aluminum-base alloy extruded shapes, including extruded rod, bar, pipe, tube, etc.	(X)	19.2
All other aluminum and aluminum-base alloy shapes and forms, including refinery shapes	(X)	59.2
Other nonferrous shapes and forms	(X)	(D)
Iron and steel scrap, excluding home scrap	(X)	30.8
Paints, varnishes, lacquers, stains, shellacs, japans, enamels, and allied products	(X)	22.5
All other materials and components, parts, containers, and supplies	(X)	168.3
Materials, ingredients, containers, and supplies, nsk	(X)	251.3

Source: 1992 *Economic Census*. Explanation of symbols used: (D): Withheld to avoid disclosure of competitive data; na: Not available; (S): Withheld because statistical norms were not met; (X): Not applicable; (Z): Less than half the unit shown; nec: Not elsewhere classified; nsk: Not specified by kind; - : zero; * : 10-19 percent estimated; ** : 20-29 percent estimated.

PRODUCT SHARE DETAILS

Product or Product Class	% Share	Product or Product Class	% Share
Miscellaneous metal work	100.00	Aluminum curtain wall	61.97
Fabricated metal bar joists and concrete reinforcing bars	31.18	All other metal curtain wall (including combination of metals)	0.40
Fabricated metal bar joists, long span	19.19	Curtain wall, nsk	5.74
Fabricated metal bar joists, short span (open web)	19.63	Custom metal roll form products	49.02
Fabricated metal concrete reinforcing bars	58.43	Custom carbon steel roll form products	61.05
Fabricated bar joists and concrete reinforcing bars, nsk	2.76	Custom stainless steel roll form products	10.77
Metal plaster bases	2.13	Custom aluminum roll form products	16.61
Expanded metal plaster lath	33.28	Other custom metal roll form products	10.39
Metal plaster base accessories (including corner beads, screeds, grounds, etc.)	66.72	Custom roll form products, nsk	1.17
Metal curtain wall	7.08	Miscellaneous metal work, nsk	10.59
Steel curtain wall (including stainless)	31.89		

Source: 1992 *Economic Census*. The values shown are percent of total shipments in an industry. Values of indented subcategories are summed in the main heading. The symbol (D) appears when data are withheld to prevent disclosure of competitive information. The abbreviation nsk stands for 'not specified by kind' and nec for 'not elsewhere classified'.

INPUTS AND OUTPUTS FOR MISCELLANEOUS METAL WORK

Economic Sector or Industry Providing Inputs	%	Sector	Economic Sector or Industry Buying Outputs	%	Sector
Blast furnaces & steel mills	50.9	Manufg.	Office buildings	30.9	Constr.
Wholesale trade	10.9	Trade	Electric utility facility construction	9.2	Constr.
Aluminum rolling & drawing	8.2	Manufg.	Highway & street construction	7.2	Constr.
Fabricated structural metal	3.9	Manufg.	Construction of stores & restaurants	5.8	Constr.
Electric services (utilities)	2.0	Util.	Maintenance of nonfarm buildings nec	5.3	Constr.
Primary aluminum	1.9	Manufg.	Construction of hospitals	3.9	Constr.
Motor freight transportation & warehousing	1.7	Util.	Industrial buildings	3.6	Constr.
Scrap	1.4	Scrap	Residential garden apartments	3.1	Constr.
Metal stampings, nec	1.2	Manufg.	Construction of educational buildings	2.4	Constr.
Iron & steel foundries	1.1	Manufg.	Residential additions/alterations, nonfarm	2.4	Constr.
Railroads & related services	1.0	Util.	Federal Government purchases, national defense	2.1	Fed Govt
Miscellaneous metal work	0.9	Manufg.	Hotels & motels	2.0	Constr.
Fabricated metal products, nec	0.8	Manufg.	Residential 1-unit structures, nonfarm	2.0	Constr.
Petroleum refining	0.8	Manufg.	Maintenance of highways & streets	1.7	Constr.
Banking	0.8	Fin/R.E.	Amusement & recreation building construction	1.6	Constr.
Metal coating & allied services	0.7	Manufg.	Warehouses	1.6	Constr.
Nonferrous castings, nec	0.7	Manufg.	Residential high-rise apartments	1.4	Constr.
Hotels & lodging places	0.7	Services	Telephone & telegraph facility construction	1.4	Constr.
Maintenance of nonfarm buildings nec	0.6	Constr.	Sewer system facility construction	1.3	Constr.
Brass, bronze, & copper castings	0.6	Manufg.	Construction of nonfarm buildings nec	0.9	Constr.
Paints & allied products	0.6	Manufg.	Local transit facility construction	0.9	Constr.
Communications, except radio & TV	0.6	Util.	Resid. & other health facility construction	0.8	Constr.
Gas production & distribution (utilities)	0.6	Util.	Construction of conservation facilities	0.7	Constr.
Welding apparatus, electric	0.5	Manufg.	Water supply facility construction	0.7	Constr.
Air transportation	0.4	Util.	Radio & TV communication equipment	0.7	Manufg.
Real estate	0.4	Fin/R.E.	Nonfarm residential structure maintenance	0.6	Constr.
Equipment rental & leasing services	0.4	Services	Construction of religious buildings	0.5	Constr.
Copper rolling & drawing	0.3	Manufg.	Garage & service station construction	0.5	Constr.
Glass & glass products, except containers	0.3	Manufg.	Maintenance of farm service facilities	0.5	Constr.
Machine tools, metal cutting types	0.3	Manufg.	Miscellaneous metal work	0.5	Manufg.
Paperboard containers & boxes	0.3	Manufg.	Farm service facilities	0.4	Constr.
Advertising	0.3	Services	Exports	0.4	Foreign
Personnel supply services	0.3	Services	Maintenance of nonbuilding facilities nec	0.3	Constr.
U.S. Postal Service	0.3	Gov't	Nonbuilding facilities nec	0.3	Constr.
Aluminum castings	0.2	Manufg.	Dormitories & other group housing	0.2	Constr.
Chemical preparations, nec	0.2	Manufg.	Gas utility facility construction	0.2	Constr.
Eating & drinking places	0.2	Trade	Maintenance of electric utility facilities	0.2	Constr.
Detective & protective services	0.2	Services	Maintenance of water supply facilities	0.2	Constr.
Engineering, architectural, & surveying services	0.2	Services	Railroad construction	0.2	Constr.
Machinery, except electrical, nec	0.1	Manufg.	Residential 2-4 unit structures, nonfarm	0.2	Constr.
Soap & other detergents	0.1	Manufg.	Federal Government purchases, nondefense	0.2	Fed Govt
Wood pallets & skids	0.1	Manufg.	Construction of dams & reservoirs	0.1	Constr.
Water transportation	0.1	Util.	Maintenance of gas utility facilities	0.1	Constr.
Insurance carriers	0.1	Fin/R.E.	Maintenance of local transit facilities	0.1	Constr.
Automotive rental & leasing, without drivers	0.1	Services	Maintenance of telephone & telegraph facilities	0.1	Constr.
Automotive repair shops & services	0.1	Services			
Business services nec	0.1	Services			
Computer & data processing services	0.1	Services			

Source: *Benchmark Input-Output Accounts for the U.S. Economy, 1982*, U.S. Department of Commerce, Washington, D.C., July 1991. Data, as reported in the source, are organized by the 1977 SIC structure in use in 1982 but have been matched, as closely as is possible, to the 1987 SIC structure used in this book.

OCCUPATIONS EMPLOYED BY SIC 344 - FABRICATED STRUCTURAL METAL PRODUCTS

Occupation	% of Total 1994	Change to 2005	Occupation	% of Total 1994	Change to 2005
Assemblers, fabricators, & hand workers nec	8.8	-17.6	Precision metal workers nec	1.8	-9.4
Welders & cutters	7.0	-17.6	Secretaries, ex legal & medical	1.8	-25.0
Sheet metal workers & duct installers	6.8	-58.8	Machinists	1.6	-17.6
Metal fabricators, structural metal products	5.2	-17.6	General office clerks	1.5	-29.7
Blue collar worker supervisors	4.4	-24.6	Bookkeeping, accounting, & auditing clerks	1.5	-38.2
Machine tool cutting & forming etc. nec	4.2	-1.1	Cost estimators	1.4	-9.4
Welding machine setters, operators	3.7	-17.6	Coating, painting, & spraying machine workers	1.4	-1.1
General managers & top executives	3.4	-21.8	Traffic, shipping, & receiving clerks	1.4	-20.7
Sales & related workers nec	2.8	-17.6	Inspectors, testers, & graders, precision	1.1	-17.6
Helpers, laborers, & material movers nec	2.5	-17.6	Industrial production managers	1.1	-17.6
Drafters	2.3	-35.8	Freight, stock, & material movers, hand	1.1	-34.1
Fitters, structural metal, precision	2.1	-50.6	Boilermakers	1.1	11.2
Machine forming operators, metal & plastic	1.9	-46.4	Combination machine tool operators	1.1	-17.6
Truck drivers light & heavy	1.9	-15.0	Financial managers	1.0	-17.6

Source: Industry-Occupation Matrix, Bureau of Labor Statistics. These data relate to one or more 3-digit SIC industry groups rather than to a single 4-digit SIC. The change reported for each occupation to the year 2005 is a percent of growth or decline as estimated by the Bureau of Labor Statistics. The abbreviation nec stands for 'not elsewhere classified'.

LOCATION BY STATE AND REGIONAL CONCENTRATION

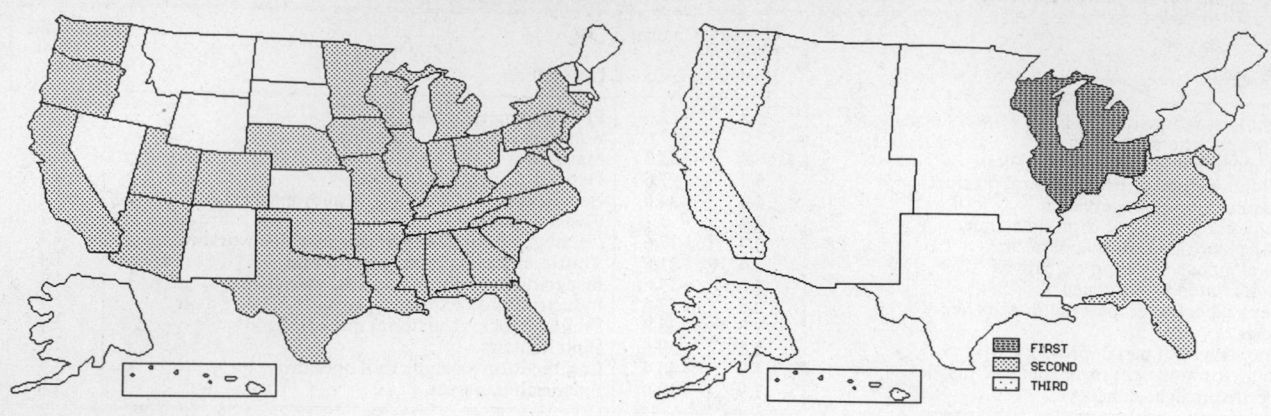

FIRST
SECOND
THIRD

INDUSTRY DATA BY STATE

State	Establish-ments	Shipments			Employment				Cost as % of Shipments	Investment per Employee ($)
		Total ($ mil)	% of U.S.	Per Establ.	Total Number	% of U.S.	Per Establ.	Wages ($/hour)		
Texas	57	381.3	11.3	6.7	2,200	10.2	39	11.31	71.4	1,727
Ohio	44	296.7	8.8	6.7	1,900	8.8	43	12.97	61.7	5,316
California	61	250.6	7.4	4.1	1,400	6.5	23	11.06	64.2	3,286
Pennsylvania	34	224.0	6.7	6.6	1,600	7.4	47	12.42	65.6	1,688
Indiana	22	194.5	5.8	8.8	1,300	6.0	59	13.18	81.3	2,615
Florida	41	187.0	5.6	4.6	900	4.2	22	12.45	69.6	5,667
Illinois	36	152.8	4.5	4.2	1,200	5.6	33	11.05	58.1	2,917
Nebraska	5	129.0	3.8	25.8	500	2.3	100	12.71	69.1	-
Georgia	22	124.5	3.7	5.7	700	3.2	32	10.78	68.6	1,857
Alabama	22	113.1	3.4	5.1	1,000	4.6	45	13.07	61.2	1,000
South Carolina	15	103.9	3.1	6.9	800	3.7	53	15.73	66.0	875
Virginia	15	92.8	2.8	6.2	800	3.7	53	11.73	68.6	500
New York	26	92.4	2.7	3.6	500	2.3	19	12.14	60.5	3,800
Missouri	10	76.7	2.3	7.7	500	2.3	50	11.83	63.6	1,000
New Jersey	16	71.2	2.1	4.4	400	1.9	25	12.83	55.1	3,250
Arkansas	6	66.9	2.0	11.1	700	3.2	117	8.30	59.9	-
Kentucky	8	60.9	1.8	7.6	300	1.4	38	11.20	64.9	3,667
Washington	15	60.3	1.8	4.0	300	1.4	20	11.50	67.5	1,667
Tennessee	12	52.8	1.6	4.4	500	2.3	42	8.50	65.0	800
North Carolina	16	52.3	1.6	3.3	300	1.4	19	11.75	64.6	-
Maryland	11	51.6	1.5	4.7	400	1.9	36	11.20	60.7	500
Wisconsin	14	48.6	1.4	3.5	400	1.9	29	11.00	58.4	2,000
Michigan	28	38.0	1.1	1.4	300	1.4	11	11.50	47.9	1,667
Arizona	11	36.0	1.1	3.3	300	1.4	27	8.75	71.9	1,333
Oregon	9	25.5	0.8	2.8	200	0.9	22	10.33	61.2	-
Minnesota	10	25.3	0.8	2.5	300	1.4	30	14.75	54.9	667
Colorado	9	23.6	0.7	2.6	200	0.9	22	9.00	70.3	1,500
Louisiana	9	22.4	0.7	2.5	200	0.9	22	11.00	67.9	500
Oklahoma	5	16.1	0.5	3.2	200	0.9	40	7.67	66.5	-
Connecticut	12	(D)	-	-	175 *	0.8	15	-	-	-
Utah	7	(D)	-	-	375 *	1.7	54	-	-	-
Iowa	6	(D)	-	-	175 *	0.8	29	-	-	-
Mississippi	3	(D)	-	-	175 *	0.8	58	-	-	1,143

Source: 1992 *Economic Census*. The states are in descending order of shipments or establishments (if shipment data are missing for the majority). The symbol (D) appears when data are withheld to prevent disclosure of competitive information. States marked with (D) are sorted by number of establishments. A dash (-) indicates that the data element cannot be calculated; * indicates the midpoint of a range.

3451 - SCREW MACHINE PRODUCTS

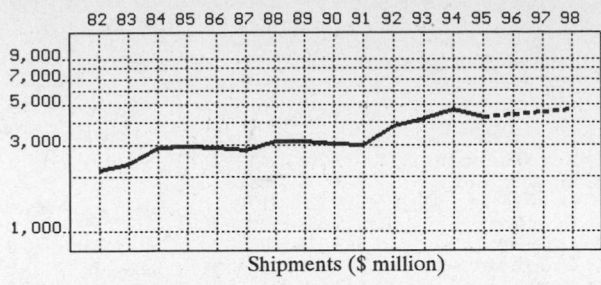

Shipments ($ million)

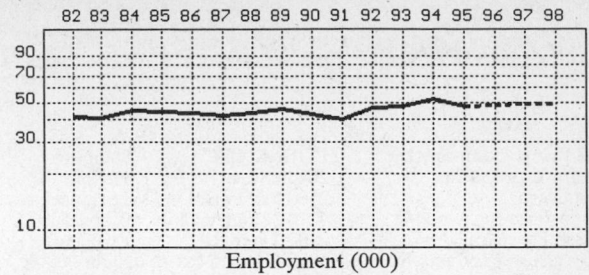

Employment (000)

GENERAL STATISTICS

Year	Companies	Establishments		Employment			Compensation		Production ($ million)			
		Total	with 20 or more employees	Total (000)	Production Workers (000)	Hours (Mil)	Payroll ($ mil)	Wages ($/hr)	Cost of Materials	Value Added by Manufacture	Value of Shipments	Capital Invest.
1982	1,744	1,787	644	41.8	33.7	65.2	718.9	7.72	831.6	1,328.0	2,173.1	77.2
1983		1,755	655	40.9	33.0	66.6	737.6	7.78	898.9	1,448.6	2,337.8	59.9
1984		1,723	666	45.2	36.9	73.3	882.1	8.75	1,167.1	1,778.0	2,903.8	91.9
1985		1,691	676	44.4	35.8	71.4	892.8	8.96	1,179.5	1,870.2	2,947.8	109.4
1986		1,655	658	43.5	34.6	69.2	908.2	9.52	1,158.7	177.7	2,928.3	98.5
1987	1,608	1,635	635	42.7	34.1	72.9	945.3	9.14	1,048.9	1,770.5	2,806.2	99.7
1988		1,584	630	43.7	34.7	74.5	1,019.7	9.43	1,159.6	2,015.8	3,158.8	81.8
1989		1,550	651	45.9	35.8	75.9	1,049.9	9.77	1,132.3	2,024.6	3,124.3	129.0
1990		1,514	607	42.9	34.3	73.4	1,035.6	10.13	1,100.5	1,956.0	3,034.4	125.4
1991		1,492	577	40.6	32.3	70.5	1,017.0	10.23	1,110.3	1,888.1	2,974.7	101.4
1992	1,671	1,706	646	46.4	36.5	77.6	1,270.8	11.29	1,443.5	2,393.5	3,830.1	135.1
1993		1,667	655	47.1	38.5	81.5	1,334.0	11.36	1,524.2	2,655.8	4,169.6	152.2
1994		1,541P	620P	51.8	41.9	91.8	1,461.3	11.25	1,716.6	2,974.0	4,669.1	225.3
1995		1,525P	617P	47.7P	38.1P	84.1P	1,395.0P	11.75P	1,563.5P	2,708.7P	4,252.6P	175.2P
1996		1,509P	613P	48.2P	38.4P	85.6P	1,448.4P	12.05P	1,620.9P	2,808.3P	4,408.9P	183.9P
1997		1,493P	610P	48.6P	38.8P	87.0P	1,501.8P	12.36P	1,678.4P	2,907.8P	4,565.2P	192.6P
1998		1,477P	607P	49.1P	39.2P	88.4P	1,555.3P	12.66P	1,735.9P	3,007.4P	4,721.6P	201.3P

Sources: 1982, 1987, 1992 *Economic Census*; *Annual Survey of Manufactures*, 83-86, 88-91, 93-94. Establishment counts for non-Census years are from *County Business Patterns*; establishment values for 83-84 are extrapolations. 'P's show projections by the editors. Industries reclassified in 87 will not have data for prior years.

INDICES OF CHANGE

Year	Companies	Establishments		Employment			Compensation		Production ($ million)			
		Total	with 20 or more employees	Total (000)	Production Workers (000)	Hours (Mil)	Payroll ($ mil)	Wages ($/hr)	Cost of Materials	Value Added by Manufacture	Value of Shipments	Capital Invest.
1982	104	105	100	90	92	84	57	68	58	55	57	57
1983		103	101	88	90	86	58	69	62	61	61	44
1984		101	103	97	101	94	69	78	81	74	76	68
1985		99	105	96	98	92	70	79	82	78	77	81
1986		97	102	94	95	89	71	84	80	7	76	73
1987	96	96	98	92	93	94	74	81	73	74	73	74
1988		93	98	94	95	96	80	84	80	84	82	61
1989		91	101	99	98	98	83	87	78	85	82	95
1990		89	94	92	94	95	81	90	76	82	79	93
1991		87	89	87	88	91	80	91	77	79	78	75
1992	100	100	100	100	100	100	100	100	100	100	100	100
1993		98	101	102	105	105	105	101	106	111	109	113
1994		90P	96P	112	115	118	115	100	119	124	122	167
1995		89P	95P	103P	104P	108P	110P	104P	108P	113P	111P	130P
1996		88P	95P	104P	105P	110P	114P	107P	112P	117P	115P	136P
1997		88P	94P	105P	106P	112P	118P	109P	116P	121P	119P	143P
1998		87P	94P	106P	107P	114P	122P	112P	120P	126P	123P	149P

Sources: Same as General Statistics. Values reflect change from the base year, 1992. Values above 100 mean greater than 92, values below 100 mean less than 92, and a value of 100 in the 82-91 or 93-98 period means same as 92. 'P's mark projections by the editors.

SELECTED RATIOS

For 1994	Avg. of All Manufact.	Analyzed Industry	Index	For 1994	Avg. of All Manufact.	Analyzed Industry	Index
Employees per Establishment	49	34	69	Value Added per Production Worker	134,084	70,979	53
Payroll per Establishment	1,500,273	948,047	63	Cost per Establishment	5,045,178	1,113,678	22
Payroll per Employee	30,620	28,210	92	Cost per Employee	102,970	33,139	32
Production Workers per Establishment	34	27	79	Cost per Production Worker	146,988	40,969	28
Wages per Establishment	853,319	670,017	79	Shipments per Establishment	9,576,895	3,029,171	32
Wages per Production Worker	24,861	24,648	99	Shipments per Employee	195,460	90,137	46
Hours per Production Worker	2,056	2,191	107	Shipments per Production Worker	279,017	111,434	40
Wages per Hour	12.09	11.25	93	Investment per Establishment	321,011	146,168	46
Value Added per Establishment	4,602,255	1,929,441	42	Investment per Employee	6,552	4,349	66
Value Added per Employee	93,930	57,413	61	Investment per Production Worker	9,352	5,377	57

Sources: Same as General Statistics. The 'Average of All Manufacturing' column represents the average of all manufacturing industries reported for the most recent complete year available. The Index shows the relationship between the Average and the Analyzed Industry. For example, 100 means that they are equal; 500 that the Analyzed Industry is five times the average; 50 means that the Analyzed Industry is half the national average. The abbreviation 'na' is used to show that data are 'not available'.

LEADING COMPANIES   Number shown: **75**   Total sales ($ mil): **1,251**   Total employment (000): **12.9**

Company Name	Address				CEO Name	Phone	Co. Type	Sales ($ mil)	Empl. (000)
Standard Insert Company Inc	POBox 1000	Danboro	PA	18916	Ken Swanstrom	215-766-0960	P	101	1.0
Horizon Enterprises Inc	20400 Superior Rd	Taylor	MI	48180	Gregory Bird	313-374-9200	R	66	0.5
Hi-Shear Corp	2600 Sky Park Dr	Torrance	CA	90509	Leo Reagan	310-326-8110	S	57	0.6
Kelco Industries Inc	9210 Country Club	Woodstock	IL	60098	Kevin Kelly	815-338-5521	R	55*	0.7
MascoTech Precision	2800 Tyler Rd	Ypsilanti	MI	48198	G Thanopoulos	313-274-5800	S	35*	0.2
Amtec Precision Products Inc	1875 Holmes Rd	Elgin	IL	60123	George Dressel	708-695-8030	R	32	0.4
Len Industries Inc	815 Rice St	Leslie	MI	49251	Leonard Len	517-589-8241	R	32	0.3
B and G Manufacturing	3067 Unionville Pk	Hatfield	PA	19440	R F Edmonds Sr	215-822-1925	R	30	0.3
C Thorrez Industries Inc	4909 W Michigan	Jackson	MI	49201	Camiel Thorrez	517-750-3160	R	30	0.2
Huron Inc	6554 Lakeshore Rd	Lexington	MI	48450	Robert Bales	810-359-5344	S	30	0.4
IW Industries Inc	35 Melville Park Rd	Melville	NY	11747	J Warshawsky	516-293-9494	R	30	0.3
Mitchel and Scott Machine	1841 Ludlow Av	Indianapolis	IN	46201	Tom L Mitchel	317-639-5331	R	30	0.4
Metal Seal and Products Inc	4323 Hamann Pkwy	Willoughby	OH	44094	Ed Diemer	216-946-8500	R	25	0.3
RW Screw Products Inc	PO Box 704	Massillon	OH	44648	James Wooley	216-837-9211	R	25*	0.3
Greystone Inc	1 Greystone Dr	N Providence	RI	02911	Everett H Fernald	401-231-5770	R	22	0.4
Camcraft Inc	1080 Muirfield Dr	Hanover Park	IL	60103	B Bertsche	708-582-6000	R	20	0.2
Federal Screw Works	425 Congdon St	Chelsea	MI	48118	Jeff Harness	313-963-2477	D	20	0.1
Frisby PMC Inc	1500 Chase Av	Elk Grove Vill	IL	60007	David A Shotts	708-439-1150	S	20	0.2
Metric Machining	1622 S Magnolia	Monrovia	CA	91016	RR Parker	818-357-2345	R	20	0.2
General Automation Inc	3300 Oakton St	Skokie	IL	60076	MJ Starr	708-676-4004	R	18	0.1
Ness Precision Products	677 Buffalo Rd	Rochester	NY	14611	Robert A Dewey	716-235-0880	D	18	0.2
Black River Manufacturing Inc	2625 20th St	Port Huron	MI	48060	Issac Lang Jr	810-982-9812	R	18	0.1
Dawlen Corp	PO Box 884	Jackson	MI	49204	FF Small	517-787-2200	R	18	<0.1
Quality Control Corp	7315 W Wilson Av	Harwood H	IL	60656	Gregory M Willard	708-867-5400	R	17*	0.2
Horizon Technology Group	27991 Northline Rd	Romulus	MI	48174	Dick Kaspers	313-941-4600	S	16*	0.1
Ashley F Ward Inc	7490 Easy St	Mason	OH	45040	William Ward	513-771-7990	R	15*	0.2
AT and G Co	30790 W 8 Mile Rd	Farmington Hls	MI	48336	BV Bonner	810-474-6330	R	15	0.2
Creed Monarch Inc	PO Box 550	New Britain	CT	06050	Richard Creed	203-225-7884	R	15*	0.2
Rima Manufacturing Co	PO Box 29	Hudson	MI	49247	EJ Engle Jr	517-448-8921	R	15*	0.2
Kerr Lakeside Inc	PO Box 32220	Euclid	OH	44132	RW Kerr	216-261-2100	R	14	0.1
Dexter Automatic Products Co	2500 Bishop Cir E	Dexter	MI	48130	WE Tupper	313-426-8900	R	14	0.2
Alger Manufacturing Company	724 S Bon View Av	Ontario	CA	91761	Richard A Krause	909-986-4591	S	13	0.1
Athanor Group Inc	3452 E Foothill	Pasadena	CA	91107	Robert W Miller	818-440-1602	P	13	0.1
Cape Industries Inc	24055 Mound Rd	Warren	MI	48091	Robert Hayes	313-754-0898	R	13	<0.1
DuPage Machine Products Inc	99 International	Glendale H	IL	60139	David R Knuepfer	708-690-5400	R	12	<0.1
K and K Screw Products Inc	730 Baker Dr	Itasca	IL	60143	Jack Emerick	708-773-9011	R	12*	0.2
MKM Machine Tool Company	PO Box 2307	Clarksville	IN	47129	Robert Moore	812-282-6627	R	12*	0.1
RB Royal Industries Inc	442 Arlington Av	Fond du Lac	WI	54935	JW Neumann Jr	414-921-1550	R	12	0.1
Allstar Fasteners Inc	1550 Arthur Av	Elk Grove Vill	IL	60007	Allan Vodicka	708-640-7827	R	11	<0.1
Dabko Industries Inc	61 E Main St	Forestville	CT	06010	R G Dabkowski	203-589-0756	R	11	0.1
Elyria Manufacturing Corp	PO Box 479	Elyria	OH	44036	J B Ohlemacher	216-365-4171	R	11	<0.1
Northwest Automatic Products	501 Royalston Av	Minneapolis	MN	55405	John E Allen	612-339-7521	S	11	0.2
Allan Tool and Machine Co	1822 E Maple Rd	Troy	MI	48083	Jeffery M Scott	313-585-2910	R	10	<0.1
Dirksen Screw Products Co	14490 23 Mile Rd	Shelby	MI	48315	CS Dirksen	810-247-5400	R	10	<0.1
Enoch Manufacturing Co	PO Box 98	Clackamas	OR	97015	Richard A Dawes	503-659-2660	R	10	0.1
Keystone Threaded Products	PO Box 31059	Independence	OH	44131	JW Krejci	216-524-9626	R	10	<0.1
Northwest Swiss-Matic Inc	7600 32nd Av N	Minneapolis	MN	55427	GL Martin	612-544-4222	R	10	0.1
Safety Socket Screw Corp	6501 N Avondale	Chicago	IL	60631	R Payne	312-763-2020	R	9	0.1
Goe Engineering Inc	1425 S Vineyard	Ontario	CA	91761	Richard Parker	909-947-9222	R	9*	0.1
Gates Albert Inc	3434 Union St	North Chili	NY	14514	Robert Brinkman	716-594-9401	R	8	<0.1
Cass Screw Machine Prod Co	4748 France Av N	Minneapolis	MN	55429	M E Greenwald	612-535-0501	R	8*	<0.1
Duffin Manufacturing Co	PO Box 4036	Elyria	OH	44036	MB Duffin	216-323-4681	S	8	<0.1
H and H Products Inc	148 W River St	Providence	RI	02904	W Deffley	401-454-2600	R	8*	0.1
Roberts Automatic Products Inc	880 Lake Dr	Chanhassen	MN	55317	Walter G Roberts	612-949-1000	R	8	0.1
Supreme Lake Manufacturing	PO Box 19	Plantsville	CT	06479	Robert C Fazzone	203-621-8911	R	8	<0.1
Torco Inc	PO Box 4070	Marietta	GA	30061	Robert M Torras	404-427-3704	R	8	<0.1
Vallorbs Jewel Co	PO Box 958	Lancaster	PA	17608	Jeanette Steudler	717-392-3978	R	8	0.2
Revtech Inc	4288 SE Intl Way	Portland	OR	97222	Randy Todd	503-654-9543	R	8	0.1
BMC Bil-Mac Corp	2995 44th SW	Grandville	MI	49418	MR Bowen	616-538-1930	R	7	<0.1
Herker Screw Products Inc	PO Box 407	Butler	WI	53007	Robert Fancher	414-781-4220	R	7	<0.1
Milford Automatics Inc	1553 Boston Post	Milford	CT	06460	Steve Chernock Jr	203-878-7465	R	7*	<0.1
Ohio Screw Products Inc	PO Box 4027	Elyria	OH	44036	EN Imbrogno	216-322-6341	R	7*	<0.1
Pan American Metal Products	17401 NW 2nd Av	Miami	FL	33169	Omar A Vazquez	305-652-2400	R	7	<0.1
Rollin J Lobaugh Inc	240 Ryan Way	S San Francisco	CA	94080	J Corey	415-583-9682	R	7	<0.1
Swisstronics Inc	PO Box 76	Watertown	MA	02172	Frank Strangio	617-924-1050	R	7	<0.1
Delo Screw Products Co	PO Box 1203	Delaware	OH	43015	Margaret Russell	614-363-1971	R	7	<0.1
Air-Matic Products Company Inc	22218 Telegraph Rd	Southfield	MI	48034	WB Smolek	313-356-4200	R	6	<0.1
Bystrom Brothers Inc	2200 Snelling Av	Minneapolis	MN	55404	Michael Bystrom Sr	612-721-7511	R	6*	0.1
Fluidyne Ansonia	1 Riverside Dr	Ansonia	CT	06401	John Martineau	203-735-9311	S	6*	<0.1
H and L Tool Company Inc	32701 Dequindre	Madison H	MI	48071	Henry Brasza	810-585-7474	R	6*	<0.1
Hall Industries Inc	201 E Carson St	Pittsburgh	PA	15219	Jonathan C Hall	412-481-1100	R	6*	0.1
Hope Haven Inc	1800 19th St	Rock Valley	IA	51247	David Van Ningen	712-476-2737	R	6	0.1
Hy-Production Inc	PO Box 461	Valley City	OH	44280	D A Wildermuth	216-273-2400	R	6	<0.1
John J Steuby Co	6002 N Lindbergh	Hazelwood	MO	63042	John J Steuby	314-895-1000	R	6	0.1
LSC Co	100 Herrmann Rd	Pittsburgh	PA	15239	B A Woods III	412-795-6400	R	6	<0.1

Source: Ward's Business Directory of U.S. Private and Public Companies, Volumes 1 and 2, 1996. The company type code used is as follows: P - Public, R - Private, S - Subsidiary, D - Division, J - Joint Venture, A - Affiliate, G - Group. Sales are in millions of dollars, employees are in thousands. An asterisk (*) indicates an estimated sales volume. The symbol < stands for 'less than'. Company names and addresses are truncated, in some cases, to fit into the available space.

MATERIALS CONSUMED

Material	Quantity	Delivered Cost ($ million)
Materials, ingredients, containers, and supplies	(X)	1,050.1
Fabricated metal products, including forgings	(X)	42.5
Castings (rough and semifinished)	(X)	15.3
Steel bars and bar shapes	(X)	344.3
Steel sheet and strip, including tin plate	(X)	18.3
Steel wire and wire products	(X)	37.6
All other steel shapes and forms	(X)	10.1
Copper and copper-base alloy rod, bar, and bar shapes	(X)	136.6
Copper and copper-base alloy plate, sheet, and strip, including military cups and discs	(X)	0.6
All other copper and copper-base alloy shapes and forms	(X)	3.9
Aluminum and aluminum-base alloy extruded shapes, including extruded rod, bar, pipe, tube, etc.	(X)	41.5
All other aluminum and aluminum-base alloy shapes and forms	(X)	23.8
Other nonferrous shapes and forms	(X)	17.3
Paperboard containers, boxes, and corrugated paperboard	(X)	9.8
Special dies, tools, die sets, jigs, and fixtures, except cutting tools for machine tools	(X)	25.9
All other materials and components, parts, containers, and supplies	(X)	115.9
Materials, ingredients, containers, and supplies, nsk	(X)	206.7

Source: 1992 *Economic Census*. Explanation of symbols used: (D): Withheld to avoid disclosure of competitive data; na: Not available; (S): Withheld because statistical norms were not met; (X): Not applicable; (Z): Less than half the unit shown; nec: Not elsewhere classified; nsk: Not specified by kind; - : zero; * : 10-19 percent estimated; ** : 20-29 percent estimated.

PRODUCT SHARE DETAILS

Product or Product Class	% Share	Product or Product Class	% Share
Screw machine products	100.00	Screw machine products and turned parts for electric and electronic equipment, except household appliances (nonstandard items made from rod, bar, or tube stock)	15.58
Automotive screw machine products and turned parts (nonstandard items made from rod, bar, or tube stock)	30.05		
Other screw machine products and turned parts (nonstandard items made from rod, bar, or tube stock)	58.50	Screw machine products and turned parts for machinery (nonstandard items made from rod, bar, or tube stock)	18.98
Screw machine products and turned parts for aircraft (nonstandard items made from rod, bar, or tube stock)	8.97	Screw machine products and turned parts for all other end uses (nonstandard items made from rod, bar, or tube stock)	42.18
Screw machine products and turned parts for ordnance (nonstandard items made from rod, bar, or tube stock)	4.42	Other screw machine products and turned parts (nonstandard items made from rod, bar, or tube stock), nsk	2.30
Screw machine products and turned parts for household appliances, including radios and televisions (nonstandard items made from rod, bar, or tube stock)	7.56	Screw machine products, nsk	11.44

Source: 1992 *Economic Census*. The values shown are percent of total shipments in an industry. Values of indented subcategories are summed in the main heading. The symbol (D) appears when data are withheld to prevent disclosure of competitive information. The abbreviation nsk stands for 'not specified by kind' and nec for 'not elsewhere classified'.

INPUTS AND OUTPUTS FOR SCREW MACHINE AND RELATED PRODUCTS

Economic Sector or Industry Providing Inputs	%	Sector	Economic Sector or Industry Buying Outputs	%	Sector
Blast furnaces & steel mills	31.8	Manufg.	Motor vehicles & car bodies	5.7	Manufg.
Imports	19.2	Foreign	Motor vehicle parts & accessories	4.6	Manufg.
Wholesale trade	7.0	Trade	Fabricated structural metal	4.1	Manufg.
Copper rolling & drawing	3.0	Manufg.	Coal	3.2	Mining
Plating & polishing	3.0	Manufg.	Aircraft	3.2	Manufg.
Electric services (utilities)	2.9	Util.	Exports	2.9	Foreign
Advertising	2.8	Services	Pipe, valves, & pipe fittings	2.5	Manufg.
Special dies & tools & machine tool accessories	2.7	Manufg.	Radio & TV communication equipment	1.8	Manufg.
Banking	2.1	Fin/R.E.	Millwork	1.7	Manufg.
Aluminum rolling & drawing	1.9	Manufg.	Federal Government purchases, national defense	1.6	Fed Govt
Metal stampings, nec	1.7	Manufg.	Electronic computing equipment	1.5	Manufg.
Machinery, except electrical, nec	1.6	Manufg.	Farm machinery & equipment	1.5	Manufg.
Gas production & distribution (utilities)	1.4	Util.	Refrigeration & heating equipment	1.5	Manufg.
Screw machine and related products	1.2	Manufg.	Miscellaneous repair shops	1.5	Services
Motor freight transportation & warehousing	1.1	Util.	Electric services (utilities)	1.4	Util.
Equipment rental & leasing services	1.1	Services	Electronic components nec	1.3	Manufg.
Communications, except radio & TV	1.0	Util.	Telephone & telegraph apparatus	1.3	Manufg.
Real estate	1.0	Fin/R.E.	Communications, except radio & TV	1.3	Util.
Metal heat treating	0.8	Manufg.	Personal consumption expenditures	1.2	
Sanitary services, steam supply, irrigation	0.8	Util.	Aircraft & missile equipment, nec	1.2	Manufg.
Engineering, architectural, & surveying services	0.8	Services	Plumbing fixture fittings & trim	1.2	Manufg.
Eating & drinking places	0.7	Trade	Pumps & compressors	1.2	Manufg.
Metal coating & allied services	0.6	Manufg.	Wiring devices	1.2	Manufg.
Job training & related services	0.6	Services	Aircraft & missile engines & engine parts	1.1	Manufg.
Maintenance of nonfarm buildings nec	0.5	Constr.	Internal combustion engines, nec	1.1	Manufg.
Abrasive products	0.5	Manufg.	Job training & related services	1.1	Services
Iron & steel forgings	0.5	Manufg.	Automotive stampings	1.0	Manufg.
Paperboard containers & boxes	0.5	Manufg.	Blast furnaces & steel mills	1.0	Manufg.
Railroads & related services	0.5	Util.	Machinery, except electrical, nec	1.0	Manufg.

Continued on next page.

INPUTS AND OUTPUTS FOR SCREW MACHINE AND RELATED PRODUCTS - Continued

Economic Sector or Industry Providing Inputs	%	Sector	Economic Sector or Industry Buying Outputs	%	Sector
U.S. Postal Service	0.5	Gov't	Cold finishing of steel shapes	0.9	Manufg.
Machine tools, metal cutting types	0.4	Manufg.	Fabricated rubber products, nec	0.9	Manufg.
Nonferrous rolling & drawing, nec	0.4	Manufg.	Hardware, nec	0.9	Manufg.
Automotive repair shops & services	0.4	Services	Lighting fixtures & equipment	0.9	Manufg.
Petroleum refining	0.3	Manufg.	Logging camps & logging contractors	0.9	Manufg.
Detective & protective services	0.3	Services	Aluminum rolling & drawing	0.8	Manufg.
Management & consulting services & labs	0.3	Services	Mechanical measuring devices	0.8	Manufg.
Iron & steel foundries	0.2	Manufg.	Mobile homes	0.8	Manufg.
Miscellaneous plastics products	0.2	Manufg.	Motorcycles, bicycles, & parts	0.8	Manufg.
Nonferrous castings, nec	0.2	Manufg.	Prefabricated metal buildings	0.8	Manufg.
Nonferrous wire drawing & insulating	0.2	Manufg.	Sheet metal work	0.8	Manufg.
Plastics materials & resins	0.2	Manufg.	Ship building & repairing	0.8	Manufg.
Royalties	0.2	Fin/R.E.	Wood household furniture	0.8	Manufg.
Accounting, auditing & bookkeeping	0.2	Services	Construction machinery & equipment	0.7	Manufg.
Business/professional associations	0.2	Services	Fabricated plate work (boiler shops)	0.7	Manufg.
Computer & data processing services	0.2	Services	Games, toys, & children's vehicles	0.7	Manufg.
Legal services	0.2	Services	Switchgear & switchboard apparatus	0.7	Manufg.
Noncomparable imports	0.2	Foreign	Engine electrical equipment	0.6	Manufg.
Aluminum castings	0.1	Manufg.	Motors & generators	0.6	Manufg.
Hand saws & saw blades	0.1	Manufg.	Power transmission equipment	0.6	Manufg.
Lubricating oils & greases	0.1	Manufg.	Screw machine and related products	0.6	Manufg.
Manifold business forms	0.1	Manufg.	Steel pipe & tubes	0.6	Manufg.
Soap & other detergents	0.1	Manufg.	Architectural metal work	0.5	Manufg.
Air transportation	0.1	Util.	Carburetors, pistons, rings, & valves	0.5	Manufg.
Insurance carriers	0.1	Fin/R.E.	Electric housewares & fans	0.5	Manufg.
Laundry, dry cleaning, shoe repair	0.1	Services	Fabricated metal products, nec	0.5	Manufg.
			Household cooking equipment	0.5	Manufg.
			Industrial controls	0.5	Manufg.
			Metal stampings, nec	0.5	Manufg.
			Miscellaneous plastics products	0.5	Manufg.
			Blowers & fans	0.4	Manufg.
			General industrial machinery, nec	0.4	Manufg.
			Household laundry equipment	0.4	Manufg.
			Instruments to measure electricity	0.4	Manufg.
			Lawn & garden equipment	0.4	Manufg.
			Metal partitions & fixtures	0.4	Manufg.
			Nonferrous wire drawing & insulating	0.4	Manufg.
			Oil field machinery	0.4	Manufg.
			Photographic equipment & supplies	0.4	Manufg.
			Special dies & tools & machine tool accessories	0.4	Manufg.
			Wood products, nec	0.4	Manufg.
			Air transportation	0.4	Util.
			Nonferrous metal ores, except copper	0.3	Mining
			Electric utility facility construction	0.3	Constr.
			Conveyors & conveying equipment	0.3	Manufg.
			Electrical industrial apparatus, nec	0.3	Manufg.
			Environmental controls	0.3	Manufg.
			Furniture & fixtures, nec	0.3	Manufg.
			Guided missiles & space vehicles	0.3	Manufg.
			Machine tools, metal cutting types	0.3	Manufg.
			Manufacturing industries, nec	0.3	Manufg.
			Metal doors, sash, & trim	0.3	Manufg.
			Metal office furniture	0.3	Manufg.
			Optical instruments & lenses	0.3	Manufg.
			Power driven hand tools	0.3	Manufg.
			Radio & TV receiving sets	0.3	Manufg.
			Railroad equipment	0.3	Manufg.
			Service industry machines, nec	0.3	Manufg.
			Small arms	0.3	Manufg.
			Special industry machinery, nec	0.3	Manufg.
			Sporting & athletic goods, nec	0.3	Manufg.
			Steel wire & related products	0.3	Manufg.
			Surgical & medical instruments	0.3	Manufg.
			X-ray apparatus & tubes	0.3	Manufg.
			Automotive rental & leasing, without drivers	0.3	Services
			Automotive repair shops & services	0.3	Services
			S/L Govt. purch., other general government	0.3	S/L Govt
			Chemical & fertilizer mineral	0.2	Mining
			Copper ore	0.2	Mining
			Iron & ferroalloy ores	0.2	Mining
			Construction of educational buildings	0.2	Constr.
			Industrial buildings	0.2	Constr.
			Maintenance of electric utility facilities	0.2	Constr.
			Boat building & repairing	0.2	Manufg.
			Drapery hardware & blinds & shades	0.2	Manufg.

Continued on next page.

INPUTS AND OUTPUTS FOR SCREW MACHINE AND RELATED PRODUCTS - Continued

Economic Sector or Industry Providing Inputs	%	Sector	Economic Sector or Industry Buying Outputs	%	Sector
			Engineering & scientific instruments	0.2	Manufg.
			Household appliances, nec	0.2	Manufg.
			Household refrigerators & freezers	0.2	Manufg.
			Industrial trucks & tractors	0.2	Manufg.
			Iron & steel foundries	0.2	Manufg.
			Leather goods, nec	0.2	Manufg.
			Paints & allied products	0.2	Manufg.
			Printing trades machinery	0.2	Manufg.
			Public building furniture	0.2	Manufg.
			Semiconductors & related devices	0.2	Manufg.
			Structural wood members, nec	0.2	Manufg.
			Surgical appliances & supplies	0.2	Manufg.
			Textile machinery	0.2	Manufg.
			Travel trailers & campers	0.2	Manufg.
			Truck & bus bodies	0.2	Manufg.
			Truck trailers	0.2	Manufg.
			Turbines & turbine generator sets	0.2	Manufg.
			Typewriters & office machines, nec	0.2	Manufg.
			Wood pallets & skids	0.2	Manufg.
			Amusement & recreation services nec	0.2	Services
			Laundry, dry cleaning, shoe repair	0.2	Services
			Federal Government purchases, nondefense	0.2	Fed Govt
			Dimension, crushed & broken stone	0.1	Mining
			Maintenance of nonfarm buildings nec	0.1	Constr.
			Maintenance of railroads	0.1	Constr.
			Office buildings	0.1	Constr.
			Residential garden apartments	0.1	Constr.
			Sewer system facility construction	0.1	Constr.
			Automatic merchandising machines	0.1	Manufg.
			Ball & roller bearings	0.1	Manufg.
			Copper rolling & drawing	0.1	Manufg.
			Cutlery	0.1	Manufg.
			Elevators & moving stairways	0.1	Manufg.
			Heating equipment, except electric	0.1	Manufg.
			Household vacuum cleaners	0.1	Manufg.
			Industrial furnaces & ovens	0.1	Manufg.
			Metal sanitary ware	0.1	Manufg.
			Metalworking machinery, nec	0.1	Manufg.
			Mining machinery, except oil field	0.1	Manufg.
			Miscellaneous fabricated wire products	0.1	Manufg.
			Nonferrous rolling & drawing, nec	0.1	Manufg.
			Plating & polishing	0.1	Manufg.
			Prefabricated wood buildings	0.1	Manufg.
			Sawmills & planning mills, general	0.1	Manufg.
			Signs & advertising displays	0.1	Manufg.
			Tanks & tank components	0.1	Manufg.
			Transformers	0.1	Manufg.
			Watches, clocks, & parts	0.1	Manufg.
			Woodworking machinery	0.1	Manufg.

Source: Benchmark Input-Output Accounts for the U.S. Economy, 1982, U.S. Department of Commerce, Washington, D.C., July 1991. Data, as reported in the source, are organized by the 1977 SIC structure in use in 1982 but have been matched, as closely as is possible, to the 1987 SIC structure used in this book.

OCCUPATIONS EMPLOYED BY SIC 345 - SCREW MACHINE PRODUCTS, BOLTS, ETC.

Occupation	% of Total 1994	Change to 2005	Occupation	% of Total 1994	Change to 2005
Lathe & turning machine tool operators	9.4	-32.8	Metal & plastic machine workers nec	1.8	-25.5
Machinists	9.1	-16.0	Punching machine operators, metal & plastic	1.8	0.8
Machine tool cutting operators, metal & plastic	6.9	-13.9	Industrial production managers	1.7	-16.0
Machine tool cutting & forming etc. nec	4.9	-16.0	Industrial machinery mechanics	1.4	-7.5
Inspectors, testers, & graders, precision	4.9	-16.0	General office clerks	1.3	-28.4
Blue collar worker supervisors	4.6	-23.3	Bookkeeping, accounting, & auditing clerks	1.3	-37.0
Combination machine tool operators	4.5	-7.6	Secretaries, ex legal & medical	1.3	-23.5
General managers & top executives	3.4	-20.3	Helpers, laborers, & material movers nec	1.3	-16.0
Assemblers, fabricators, & hand workers nec	2.8	-16.0	Machine operators nec	1.2	-25.9
Drilling & boring machine tool workers	2.8	-16.0	Production, planning, & expediting clerks	1.1	13.4
Machine forming operators, metal & plastic	2.4	-16.0	Machine feeders & offbearers	1.1	-24.4
Traffic, shipping, & receiving clerks	2.4	-19.2	Hand packers & packagers	1.1	-28.0
Tool & die makers	2.3	-32.1	NC machine tool operators, metal & plastic	1.1	0.8
Sales & related workers nec	2.3	-16.0	Janitors & cleaners, incl maids	1.0	-32.8
Grinding machine operators, metal & plastic	2.2	0.8			

Source: Industry-Occupation Matrix, Bureau of Labor Statistics. These data relate to one or more 3-digit SIC industry groups rather than to a single 4-digit SIC. The change reported for each occupation to the year 2005 is a percent of growth or decline as estimated by the Bureau of Labor Statistics. The abbreviation nec stands for 'not elsewhere classified'.

LOCATION BY STATE AND REGIONAL CONCENTRATION

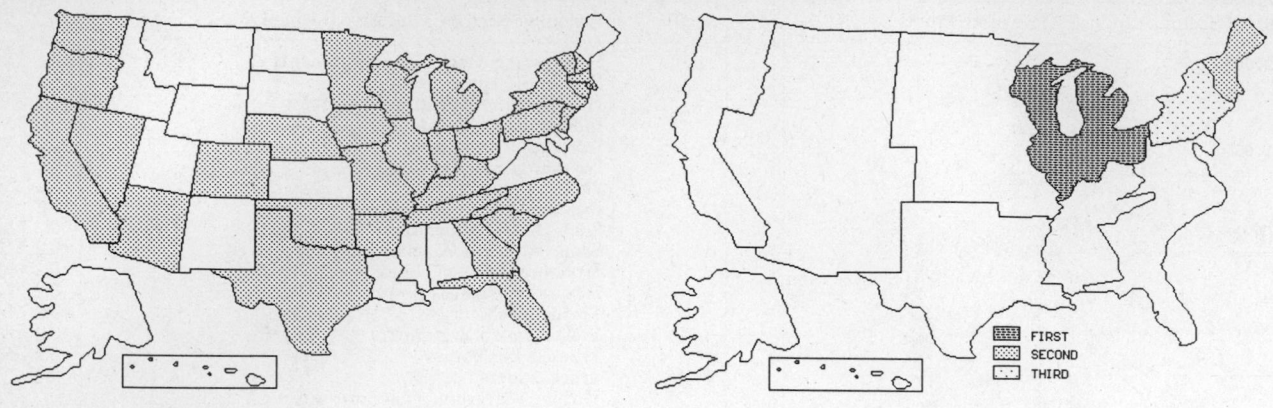

FIRST
SECOND
THIRD

INDUSTRY DATA BY STATE

| State | Establish-ments | Shipments | | | Employment | | | | Cost as % of Shipments | Investment per Employee ($) |
		Total ($ mil)	% of U.S.	Per Establ.	Total Number	% of U.S.	Per Establ.	Wages ($/hour)		
Michigan	213	728.7	19.0	3.4	7,000	15.1	33	11.97	44.5	3,086
Ohio	181	534.2	13.9	3.0	5,800	12.5	32	11.16	38.0	4,707
Illinois	188	379.9	9.9	2.0	5,100	11.0	27	11.19	36.2	2,980
New York	97	295.6	7.7	3.0	3,400	7.3	35	10.91	45.8	2,118
California	188	287.0	7.5	1.5	4,000	8.6	21	11.83	29.1	1,975
Indiana	70	175.5	4.6	2.5	2,300	5.0	33	11.00	36.8	3,609
Connecticut	125	163.7	4.3	1.3	2,400	5.2	19	11.87	33.7	1,208
Pennsylvania	64	146.7	3.8	2.3	1,900	4.1	30	10.94	35.0	2,684
Wisconsin	71	139.3	3.6	2.0	2,100	4.5	30	10.39	34.2	2,429
Missouri	45	127.5	3.3	2.8	1,500	3.2	33	12.40	35.0	2,067
Minnesota	48	119.8	3.1	2.5	1,500	3.2	31	12.33	29.1	4,467
Massachusetts	55	86.2	2.3	1.6	1,100	2.4	20	12.94	31.4	2,727
New Jersey	63	81.9	2.1	1.3	1,200	2.6	19	11.71	30.3	1,917
Florida	38	77.5	2.0	2.0	900	1.9	24	9.53	42.2	1,667
Tennessee	15	73.4	1.9	4.9	900	1.9	60	10.56	41.3	6,444
Arizona	17	64.9	1.7	3.8	500	1.1	29	12.00	36.2	-
Texas	35	39.0	1.0	1.1	600	1.3	17	9.30	36.2	1,333
Rhode Island	22	23.6	0.6	1.1	400	0.9	18	11.00	30.1	2,000
New Hampshire	17	20.2	0.5	1.2	200	0.4	12	10.75	28.7	3,500
South Carolina	9	19.0	0.5	2.1	300	0.6	33	9.75	34.7	-
Nebraska	7	19.0	0.5	2.7	300	0.6	43	9.00	29.5	1,333
Kentucky	12	18.9	0.5	1.6	200	0.4	17	12.33	41.8	1,500
Colorado	11	12.6	0.3	1.1	200	0.4	18	11.33	29.4	2,000
Georgia	7	6.6	0.2	0.9	100	0.2	14	10.00	25.8	1,000
Oregon	17	(D)	-	-	375 *	0.8	22	-	-	-
Oklahoma	12	(D)	-	-	175 *	0.4	15	-	-	-
North Carolina	11	(D)	-	-	375 *	0.8	34	-	-	2,133
Washington	10	(D)	-	-	175 *	0.4	18	-	-	-
Arkansas	9	(D)	-	-	175 *	0.4	19	-	-	-
Iowa	9	(D)	-	-	375 *	0.8	42	-	-	1,867
Nevada	7	(D)	-	-	175 *	0.4	25	-	-	1,143
Vermont	7	(D)	-	-	375 *	0.8	54	-	-	-

Source: 1992 *Economic Census*. The states are in descending order of shipments or establishments (if shipment data are missing for the majority). The symbol (D) appears when data are withheld to prevent disclosure of competitive information. States marked with (D) are sorted by number of establishments. A dash (-) indicates that the data element cannot be calculated; * indicates the midpoint of a range.

3452 - BOLTS, NUTS, RIVETS, & WASHERS

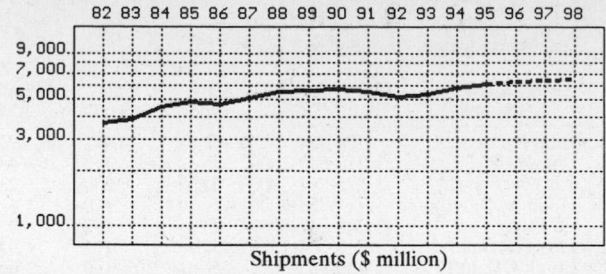

Shipments ($ million)

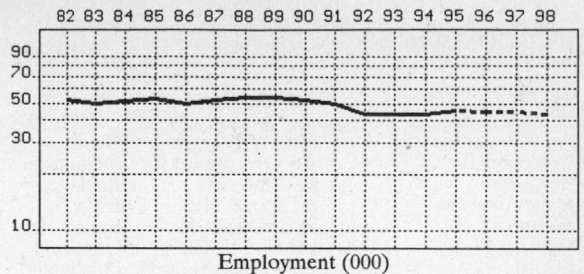

Employment (000)

GENERAL STATISTICS

Year	Companies	Establishments		Employment			Compensation		Production ($ million)			
		Total	with 20 or more employees	Total (000)	Production Workers (000)	Hours (Mil)	Payroll ($ mil)	Wages ($/hr)	Cost of Materials	Value Added by Manufacture	Value of Shipments	Capital Invest.
1982	781	903	450	52.2	37.2	69.1	1,008.1	9.33	1,588.1	1,981.2	3,661.3	109.8
1983		885	451	49.5	35.7	71.6	1,040.4	9.63	1,961.8	2,224.6	3,897.3	120.1
1984		867	452	51.0	38.5	78.2	1,139.0	9.79	2,001.1	2,605.8	4,568.7	124.1
1985		849	452	52.8	38.9	79.2	1,206.8	10.02	2,139.6	2,621.8	4,794.8	138.2
1986		833	434	49.7	36.6	74.2	1,197.0	10.79	2,061.7	2,640.9	4,713.6	144.5
1987	828	937	471	52.0	38.7	80.5	1,325.0	11.02	2,115.7	2,995.8	5,084.0	150.4
1988		923	501	54.3	40.4	85.9	1,426.8	11.11	2,299.4	3,219.2	5,483.7	155.8
1989		899	492	54.0	40.7	86.6	1,468.1	11.34	2,466.4	3,229.3	5,605.9	181.1
1990		910	494	52.7	39.0	84.4	1,489.5	11.39	2,474.3	3,194.4	5,688.6	166.6
1991		911	479	50.0	36.3	77.4	1,419.5	11.83	2,402.0	3,108.8	5,508.8	174.6
1992	806	930	436	44.0	31.8	66.4	1,350.0	12.80	2,309.2	2,866.4	5,183.1	150.9
1993		914	439	43.7	32.0	68.2	1,361.1	12.74	2,351.3	3,008.2	5,373.2	163.0
1994		927P	470P	43.4	31.9	68.5	1,413.9	12.76	2,546.2	3,276.4	5,793.7	157.3
1995		931P	472P	45.7P	33.7P	74.4P	1,536.1P	13.24P	2,670.0P	3,435.7P	6,075.3P	179.4P
1996		936P	473P	45.1P	33.3P	74.1P	1,570.4P	13.54P	2,735.8P	3,520.3P	6,225.0P	183.7P
1997		941P	474P	44.5P	32.9P	73.9P	1,604.8P	13.84P	2,801.6P	3,605.0P	6,374.7P	188.1P
1998		945P	475P	43.9P	32.5P	73.6P	1,639.1P	14.15P	2,867.3P	3,689.6P	6,524.4P	192.4P

Sources: 1982, 1987, 1992 *Economic Census*; *Annual Survey of Manufactures*, 83-86, 88-91, 93-94. Establishment counts for non-Census years are from *County Business Patterns*; establishment values for 83-84 are extrapolations. 'P's show projections by the editors. Industries reclassified in 87 will not have data for prior years.

INDICES OF CHANGE

Year	Companies	Establishments		Employment			Compensation		Production ($ million)			
		Total	with 20 or more employees	Total (000)	Production Workers (000)	Hours (Mil)	Payroll ($ mil)	Wages ($/hr)	Cost of Materials	Value Added by Manufacture	Value of Shipments	Capital Invest.
1982	97	97	103	119	117	104	75	73	69	69	71	73
1983		95	103	113	112	108	77	75	85	78	75	80
1984		93	104	116	121	118	84	76	87	91	88	82
1985		91	104	120	122	119	89	78	93	91	93	92
1986		90	100	113	115	112	89	84	89	92	91	96
1987	103	101	108	118	122	121	98	86	92	105	98	100
1988		99	115	123	127	129	106	87	100	112	106	103
1989		97	113	123	128	130	109	89	107	113	108	120
1990		98	113	120	123	127	110	89	107	111	110	110
1991		98	110	114	114	117	105	92	104	108	106	116
1992	100	100	100	100	100	100	100	100	100	100	100	100
1993		98	101	99	101	103	101	100	102	105	104	108
1994		100P	108P	99	100	103	105	100	110	114	112	104
1995		100P	108P	104P	106P	112P	114P	103P	116P	120P	117P	119P
1996		101P	108P	103P	105P	112P	116P	106P	118P	123P	120P	122P
1997		101P	109P	101P	103P	111P	119P	108P	121P	126P	123P	125P
1998		102P	109P	100P	102P	111P	121P	111P	124P	129P	126P	128P

Sources: Same as General Statistics. Values reflect change from the base year, 1992. Values above 100 mean greater than 92, values below 100 mean less than 92, and a value of 100 in the 82-91 or 93-98 period means same as 92. 'P's mark projections by the editors.

SELECTED RATIOS

For 1994	Avg. of All Manufact.	Analyzed Industry	Index	For 1994	Avg. of All Manufact.	Analyzed Industry	Index
Employees per Establishment	49	47	96	Value Added per Production Worker	134,084	102,708	77
Payroll per Establishment	1,500,273	1,525,542	102	Cost per Establishment	5,045,178	2,747,249	54
Payroll per Employee	30,620	32,578	106	Cost per Employee	102,970	58,668	57
Production Workers per Establishment	34	34	100	Cost per Production Worker	146,988	79,818	54
Wages per Establishment	853,319	943,076	111	Shipments per Establishment	9,576,895	6,251,172	65
Wages per Production Worker	24,861	27,400	110	Shipments per Employee	195,460	133,495	68
Hours per Production Worker	2,056	2,147	104	Shipments per Production Worker	279,017	181,621	65
Wages per Hour	12.09	12.76	106	Investment per Establishment	321,011	169,720	53
Value Added per Establishment	4,602,255	3,535,105	77	Investment per Employee	6,552	3,624	55
Value Added per Employee	93,930	75,493	80	Investment per Production Worker	9,352	4,931	53

Sources: Same as General Statistics. The 'Average of All Manufacturing' column represents the average of all manufacturing industries reported for the most recent complete year available. The Index shows the relationship between the Average and the Analyzed Industry. For example, 100 means that they are equal; 500 that the Analyzed Industry is five times the average; 50 means that the Analyzed Industry is half the national average. The abbreviation 'na' is used to show that data are 'not available'.

LEADING COMPANIES   Number shown: **74**   Total sales ($ mil): **6,590**   Total employment (000): **45.3**

Company Name	Address				CEO Name	Phone	Co. Type	Sales ($ mil)	Empl. (000)
Illinois Tool Works Inc	3600 W Lake Av	Glenview	IL	60025	John D Nichols	708-724-7500	P	3,461	19.5
SPS Technologies Inc	101 Greenwood Av	Jenkintown	PA	19046	Charles W Grigg	215-517-2000	P	349	4.0
Camcar Textron	600 18th Av	Rockford	IL	61104	J R MacGilvray	815-961-5000	D	230	2.0
Huck International Inc	6 Thomas St	Irvine	CA	92718	H George Faulkner	714-855-9000	S	170	1.1
ITW Shakeproof	St Charles Rd	Elgin	IL	60120	FS Ptak	708-741-7900	D	120	0.9
MNP Corp	PO Box 189002	Utica	MI	48318	Larry Berman	810-254-1320	R	115	0.8
Jennmar Corp	1330 Old Freeport	Pittsburgh	PA	15238	Frank Calandra Jr	412-963-9071	S	100	0.5
John Wagner Associates Inc	205 Mason Cir	Concord	CA	94520	Richard Holmberg	510-680-0777	R	98*	0.3
Rockford Products Corp	707 Harrison Av	Rockford	IL	61104	R Ray Wood	815-397-6000	R	90	0.9
Federal Screw Works	2400 Buhl Bldg	Detroit	MI	48226	W ZurSchmiede Jr	313-963-2323	P	81	0.5
Cold Heading Co	21777 Hoover Rd	Warren	MI	48089	Derek Stevens	810-497-7000	R	80	0.3
VSI Aerospace Fasteners	3000 Lomita Blv	Torrance	CA	90505	Joseph Hood	310-522-0700	D	65*	0.6
Lake Erie Screw Corp	13001 Athens Av	Lakewood	OH	44107	George Wasmer	216-521-1800	S	60	0.3
Bulldog Home Hardware	4131 Delp St	Memphis	TN	38118	Don Bergman	901-365-0479	D	57	0.3
Elastic Stop Nut	2330 Vauxhall Rd	Union	NJ	07083	Terry Margolis	908-686-6000	D	56*	0.7
Taptite Products	826 E Madison St	Belvidere	IL	61008	Hank Klotz	815-544-0331	D	56	0.3
Kaynar Technologies Inc	800 S State College	Fullerton	CA	92634	Jordan Law	714-871-1550	R	55	0.6
Medalist Industries Inc	10850 W Park Pl	Milwaukee	WI	53224	Edward D Hopkins	414-359-3000	P	55	1.0
Emco Industries Inc	PO Box 864	Des Moines	IA	50304	Daniel Ogden	515-265-6101	R	53	0.5
RB and W Corp	8341 Tyler Blv	Mentor	OH	44060	Ronald Leirvik	216-255-6511	D	50	0.4
Air Industries Corp	12570 Knott St	Garden Grove	CA	92641	Sam L Higgins	714-892-5571	R	49*	0.4
Construction Fasteners Inc	PO Box 6326	Wyomissing	PA	19610	Irvin Cohen	610-376-5751	R	45	0.4
ITW Fasteners	21555 S Harlem Av	Frankfort	IL	60423	Rick Kosick	708-720-2600	D	42	0.2
Amsco Products	345 E Marshall St	Wytheville	VA	24382	Stephen P Kline	703-228-8141	D	40	0.3
Metform Inc	2551 Wacker Rd	Savanna	IL	61074	Barry MacLean	815-273-2201	S	40	0.3
C-Tech Systems	10405 6th Av N	Plymouth	MN	55441		612-540-1700	D	35	0.3
Midwest Fastener Corp	665 W Armory Dr	South Holland	IL	60473	Lee M Loudermilk	708-331-1660	R	35	0.2
Huck International Inc	900 Watson Ctr Rd	Carson	CA	90745	Geo Faulkner	310-830-8200	D	33*	0.2
Accurate Threaded Fasteners	3550 W Pratt Av	Lincolnwood	IL	60645	Don Surber	708-677-1300	R	32	0.2
Alpha Bolt Co	1524 E 14 Mile Rd	Madison H	MI	48071	George Strumbos	313-585-6050	R	30	<0.1
Atlas Bolt and Screw Co	1628 Troy Rd	Ashland	OH	44805	Bob Moore	419-289-6171	S	30	0.2
Avdel Corp	50 Lackawanna Av	Parsippany	NJ	07054	J Kurtz	201-263-8100	D	30	0.2
Continental-Midland Inc	25000 S Western	Park Forest	IL	60466	RS Kaminski	708-747-1200	R	30	0.3
Landa Inc	13705 NE Airpt	Portland	OR	97230	Larry Linton	503-255-5980	R	30	0.2
Ohio Jacobson Co	941 Lake Rd	Medina	OH	44256	Paul M Parker	216-725-8853	D	30	0.1
Kendale Industries Inc	7600 Hub Pkwy	Valley View	OH	44125	D Honroth	216-524-5400	R	25	0.2
Key Manufacturing Group Inc	3200 W 14 Mile Rd	Royal Oak	MI	48073	JA Toth	313-280-2727	R	25	0.1
Milford Fastening Systems Inc	857 Bridgeport Av	Milford	CT	06460	David Melina	203-878-4631	R	25	0.3
VSI Screwcorp	135 N Unruh Av	City of Industry	CA	91744	Phil Anderson	818-968-3831	D	23*	0.4
Fastener Industries Inc	1 Berea Commons	Berea	OH	44017	R G Biernacki	216-243-0200	R	22	0.2
Heli Coil	510 River Rd	Shelton	CT	06484	Robert McCue	203-924-9341	D	22	0.4
Camloc Products	3016 Lomita Blv	Torrance	CA	90505	Eric Steiner	310-784-2600	D	21*	0.2
Palnut Co	152 Glen Rd	Mountainside	NJ	07092	S Wilson	908-233-3300	S	21	0.2
Pawtucket Fasteners LP	327 Pine St	Pawtucket	RI	02862	David M Hirsch	401-725-3880	R	21*	0.3
Federal Screw Works	34846 Goddard Rd	Romulus	MI	48174	R F Zurschmied	313-963-2477	D	20	0.1
Industrial Fasteners Corp	7 Harbor Park Dr	Pt Washington	NY	11050	Bernard Feldman	516-484-4900	R	20	0.1
Bristol Industries	PO Box 630	Brea	CA	92622	FA Klaus	714-990-4121	R	19	0.2
Amanda Bent Bolt Co	PO Box 1027	Logan	OH	43138	Donald Gruschow	614-385-9380	R	18*	0.2
Daniel Bolt and Gasket Inc	PO Box 292	Houston	TX	77001	Jerry Davis	713-224-5811	S	18	0.2
National Rivet & Mfg Co	PO Box 471	Waupun	WI	53963	JG Zeratsky	414-324-5511	R	18	0.3
Precision Form Inc	148 W Airport Rd	Lititz	PA	17543	WW Kopetz	717-560-7610	R	18	0.2
Ohio Nut and Bolt Co	33 1st Av	Berea	OH	44017	RG Biernacki	216-243-0200	D	17	0.1
Valley-Todeco Co	PO Box 4248	Sylmar	CA	91342	William G White	818-367-2261	D	17	0.3
Exemplar Manufacturing Co	800 Lowell St	Ypsilanti	MI	48197	Anthony Snoddy	313-483-5070	R	16	0.1
K-Tech Manufacturing Inc	288 Holbrook Dr	Wheeling	IL	60090	Donald Kuhns	708-459-6777	R	16	<0.1
Medalist Industries Inc	2700 York Rd	Elk Grove Vill	IL	60007	Bruce Woodward	708-766-9000	D	16*	0.3
TCR Corp	1600 67th Av N	Minneapolis	MN	55440	RH Bradley	612-560-2200	R	16	0.2
Terry Machine Co	PO Box 290487	Waterford	MI	48329	Robert Bego	313-623-0800	R	16	0.1
Commonwealth Bolt Inc	PO Box 328	Rich Creek	VA	24147	John Hale	703-726-2326	R	15*	<0.1
Gripco Fastener	111 E Broad St	South Whitley	IN	46787	Steve Slack	219-723-5111	D	15	0.3
ND Industries	1893 Barrett Rd	Troy	MI	48084	Richard M Wallace	810-362-1209	D	15	0.1
Oakland Corp	1893 Barrett Rd	Troy	MI	48084	Richard M Wallace	313-362-1200	R	15	0.1
Prestige Stamping Inc	PO Box 1086	Warren	MI	48090	Robert A Rink	810-773-2700	R	15*	0.1
Maynard Manufacturing Inc	50855 ER Schmidt	Chesterf Twnp	MI	48045	Don Artman	810-294-5830	R	14	<0.1
Shur-Lok Corp	PO Box 19584	Irvine	CA	92713	Peter Grefe	714-474-6000	R	14*	0.1
Parmatech Corp	2221 Pine View Way	Petaluma	CA	94954	Karl Zueger	707-778-2266	R	13*	0.1
Stafast Products Inc	505 Lakeshore Blv	Painesville	OH	44077	DS Selle	216-357-5546	R	13	<0.1
Assembly Fasteners Inc	6955 N Hamlin St	Lincolnwood	IL	60645	AJ Marinin	708-677-4644	R	12	0.1
Captive Fasteners Corp	115 Bauer Dr	Oakland	NJ	07436	Joseph Alderisio	201-337-6800	R	12*	0.1
Slidematic Products Co	4520 W Addison St	Chicago	IL	60641	David Magnuson	312-545-4213	R	12	<0.1
Voss Industries Inc	2168 W 25th St	Cleveland	OH	44113	William J Voss	216-771-7655	R	12	0.2
Automatic Screw Machine	PO Box 1608	Decatur	AL	35602	Stanley Belsky	205-353-1931	R	11*	0.2
Mid-Continent Screw Prod Co	3701 W Lunt Av	Lincolnwood	IL	60645	NC Couzin	708-679-3737	R	11	0.1
Keystone Screw Corp	PO Box V	Willow Grove	PA	19090	RD Wolf	215-657-7100	R	10	<0.1

Source: Ward's Business Directory of U.S. Private and Public Companies, Volumes 1 and 2, 1996. The company type code used is as follows: P - Public, R - Private, S - Subsidiary, D - Division, J - Joint Venture, A - Affiliate, G - Group. Sales are in millions of dollars, employees are in thousands. An asterisk (*) indicates an estimated sales volume. The symbol < stands for 'less than'. Company names and addresses are truncated, in some cases, to fit into the available space.

MATERIALS CONSUMED

Material	Quantity	Delivered Cost ($ million)
Materials, ingredients, containers, and supplies	(X)	1,721.8
Fabricated metal products, including forgings	(X)	96.1
Castings (rough and semifinished)	(X)	10.3
Steel bars and bar shapes	(X)	207.9
Steel sheet and strip, including tin plate	(X)	120.6
Steel wire and wire products	(X)	518.1
All other steel shapes and forms	(X)	67.3
Copper and copper-base alloy rod, bar, and bar shapes	(X)	16.0
Copper and copper-base alloy plate, sheet, and strip, including military cups and discs	(X)	13.4
All other copper and copper-base alloy shapes and forms	(X)	11.6
Aluminum and aluminum-base alloy extruded shapes, including extruded rod, bar, pipe, tube, etc.	(X)	5.7
All other aluminum and aluminum-base alloy shapes and forms	(X)	28.9
Other nonferrous shapes and forms	(X)	66.4
Paperboard containers, boxes, and corrugated paperboard	(X)	21.5
Special dies, tools, die sets, jigs, and fixtures, except cutting tools for machine tools	(X)	112.9
All other materials and components, parts, containers, and supplies	(X)	244.0
Materials, ingredients, containers, and supplies, nsk	(X)	181.2

Source: 1992 *Economic Census*. Explanation of symbols used: (D): Withheld to avoid disclosure of competitive data; na: Not available; (S): Withheld because statistical norms were not met; (X): Not applicable; (Z): Less than half the unit shown; nec: Not elsewhere classified; nsk: Not specified by kind; - : zero; * : 10-19 percent estimated; ** : 20-29 percent estimated.

PRODUCT SHARE DETAILS

Product or Product Class	% Share	Product or Product Class	% Share
Bolts, nuts, rivets, and washers	100.00	(meet specifications for flying vehicles)	15.84
Externally threaded metal fasteners, except aircraft types	41.57	Aircraft bolts, except plastics (including aerospace), less than 161 KSI tensile (meets specifications for flying vehicles)	15.01
Mine roof bolts	10.13		
Hex bolts, including heavy, tap-and-joint, excluding high-strength structural and aircraft	12.43	Aircraft bolts, except plastics (including aerospace), 161 KSI tensile or more (meets specifications for flying vehicles)	26.45
Other metal bolts, including square, round, plow, high-strength structural, and bent bolts (except aircraft types)	13.10	Aircraft screws and studs, except plastics (including aerospace) (meets specifications for flying vehicles)	5.17
Cap, set, machine, lag, flange, and self-locking screws, except aircraft types	21.32	Aircraft locknuts, except plastics (including aerospace), including flanged locknuts (meets specifications for flying vehicles)	18.78
Tapping screws (including fillister, flat, hex, oval, pan, and truss) and wood screws (including flat, oval, and round) (except aircraft types)	20.40	Other internally threaded aircraft fasteners, except plastics (including aerospace), including flanged nuts (all types except flanged locknuts), hex square nuts (all types) and sheet metal fasteners	13.36
Other externally threaded metal fasteners, including studs, except aircraft types	20.44		
Externally threaded metal fasteners, except aircraft types, nsk	2.20	Aircraft (including aerospace) washers, all types except plastics	1.09
Internally threaded metal fasteners, except aircraft types	11.49	Aircraft (including aerospace) rivets, all types except plastics	14.28
Hex nuts, including flanges, double chamfered, washer face, flat, jam, slotted, thick, castle, heavy, machine, and locking (except aircraft types)	51.20	Aircraft (including aerospace) pins, all types except plastics	5.77
Square nuts (including flat, washer, crowned, heavy, track, sleeve, and machine), sheet metal nuts, weld nuts, wing nuts, nut retainers, etc. (except aircraft types)	5.82	Aircraft (including aerospace) fasteners other than plastics (meet specifications for flying vehicles), nsk	0.08
Other internally threaded metal fasteners, including flanged nuts and locknuts (except aircraft types)	40.84	Other formed parts not made of plastics (made on fastener machines)	11.06
Internally threaded metal fasteners, except aircraft types, nsk	2.14	Other formed automotive parts not made of plastics (made on fastener machines)	53.69
Nonthreaded metal fasteners, except aircraft types	13.76	Other formed household appliance parts, including radio and television parts, not made of plastics (made on fastener machines)	5.77
Solid rivets, except aircraft types	13.77		
Tubular, split (including rivet caps) and blind rivets, except aircraft types	20.01	Other formed aircraft parts not made of plastics (made on fastener machines)	9.58
Washers, except aircraft types	24.76	All other formed parts, including ordnance parts, not made of plastics (made on fastener machines)	30.79
Other nonthreaded metal fasteners, including pins (except aircraft types)	41.32	Other formed parts not made of plastics (made on fastener machines), nsk	0.19
Nonthreaded metal fasteners, except aircraft types, nsk	0.13	Bolts, nuts, rivets, and washers, nsk	6.27
Aircraft (including aerospace) fasteners other than plastics			

Source: 1992 *Economic Census*. The values shown are percent of total shipments in an industry. Values of indented subcategories are summed in the main heading. The symbol (D) appears when data are withheld to prevent disclosure of competitive information. The abbreviation nsk stands for 'not specified by kind' and nec for 'not elsewhere classified'.

INPUTS AND OUTPUTS FOR SCREW MACHINE AND RELATED PRODUCTS

Economic Sector or Industry Providing Inputs	%	Sector	Economic Sector or Industry Buying Outputs	%	Sector
Blast furnaces & steel mills	31.8	Manufg.	Motor vehicles & car bodies	5.7	Manufg.
Imports	19.2	Foreign	Motor vehicle parts & accessories	4.6	Manufg.
Wholesale trade	7.0	Trade	Fabricated structural metal	4.1	Manufg.
Copper rolling & drawing	3.0	Manufg.	Coal	3.2	Mining
Plating & polishing	3.0	Manufg.	Aircraft	3.2	Manufg.
Electric services (utilities)	2.9	Util.	Exports	2.9	Foreign
Advertising	2.8	Services	Pipe, valves, & pipe fittings	2.5	Manufg.
Special dies & tools & machine tool accessories	2.7	Manufg.	Radio & TV communication equipment	1.8	Manufg.
Banking	2.1	Fin/R.E.	Millwork	1.7	Manufg.
Aluminum rolling & drawing	1.9	Manufg.	Federal Government purchases, national defense	1.6	Fed Govt
Metal stampings, nec	1.7	Manufg.	Electronic computing equipment	1.5	Manufg.
Machinery, except electrical, nec	1.6	Manufg.	Farm machinery & equipment	1.5	Manufg.
Gas production & distribution (utilities)	1.4	Util.	Refrigeration & heating equipment	1.5	Manufg.
Screw machine and related products	1.2	Manufg.	Miscellaneous repair shops	1.5	Services
Motor freight transportation & warehousing	1.1	Util.	Electric services (utilities)	1.4	Util.
Equipment rental & leasing services	1.1	Services	Electronic components nec	1.3	Manufg.
Communications, except radio & TV	1.0	Util.	Telephone & telegraph apparatus	1.3	Manufg.
Real estate	1.0	Fin/R.E.	Communications, except radio & TV	1.3	Util.
Metal heat treating	0.8	Manufg.	Personal consumption expenditures	1.2	
Sanitary services, steam supply, irrigation	0.8	Util.	Aircraft & missile equipment, nec	1.2	Manufg.
Engineering, architectural, & surveying services	0.8	Services	Plumbing fixture fittings & trim	1.2	Manufg.
Eating & drinking places	0.7	Trade	Pumps & compressors	1.2	Manufg.
Metal coating & allied services	0.6	Manufg.	Wiring devices	1.2	Manufg.
Job training & related services	0.6	Services	Aircraft & missile engines & engine parts	1.1	Manufg.
Maintenance of nonfarm buildings nec	0.5	Constr.	Internal combustion engines, nec	1.1	Manufg.
Abrasive products	0.5	Manufg.	Job training & related services	1.1	Services
Iron & steel forgings	0.5	Manufg.	Automotive stampings	1.0	Manufg.
Paperboard containers & boxes	0.5	Manufg.	Blast furnaces & steel mills	1.0	Manufg.
Railroads & related services	0.5	Util.	Machinery, except electrical, nec	1.0	Manufg.
U.S. Postal Service	0.5	Gov't	Cold finishing of steel shapes	0.9	Manufg.
Machine tools, metal cutting types	0.4	Manufg.	Fabricated rubber products, nec	0.9	Manufg.
Nonferrous rolling & drawing, nec	0.4	Manufg.	Hardware, nec	0.9	Manufg.
Automotive repair shops & services	0.4	Services	Lighting fixtures & equipment	0.9	Manufg.
Petroleum refining	0.3	Manufg.	Logging camps & logging contractors	0.9	Manufg.
Detective & protective services	0.3	Services	Aluminum rolling & drawing	0.8	Manufg.
Management & consulting services & labs	0.3	Services	Mechanical measuring devices	0.8	Manufg.
Iron & steel foundries	0.2	Manufg.	Mobile homes	0.8	Manufg.
Miscellaneous plastics products	0.2	Manufg.	Motorcycles, bicycles, & parts	0.8	Manufg.
Nonferrous castings, nec	0.2	Manufg.	Prefabricated metal buildings	0.8	Manufg.
Nonferrous wire drawing & insulating	0.2	Manufg.	Sheet metal work	0.8	Manufg.
Plastics materials & resins	0.2	Manufg.	Ship building & repairing	0.8	Manufg.
Royalties	0.2	Fin/R.E.	Wood household furniture	0.8	Manufg.
Accounting, auditing & bookkeeping	0.2	Services	Construction machinery & equipment	0.7	Manufg.
Business/professional associations	0.2	Services	Fabricated plate work (boiler shops)	0.7	Manufg.
Computer & data processing services	0.2	Services	Games, toys, & children's vehicles	0.7	Manufg.
Legal services	0.2	Services	Switchgear & switchboard apparatus	0.7	Manufg.
Noncomparable imports	0.2	Foreign	Engine electrical equipment	0.6	Manufg.
Aluminum castings	0.1	Manufg.	Motors & generators	0.6	Manufg.
Hand saws & saw blades	0.1	Manufg.	Power transmission equipment	0.6	Manufg.
Lubricating oils & greases	0.1	Manufg.	Screw machine and related products	0.6	Manufg.
Manifold business forms	0.1	Manufg.	Steel pipe & tubes	0.6	Manufg.
Soap & other detergents	0.1	Manufg.	Architectural metal work	0.5	Manufg.
Air transportation	0.1	Util.	Carburetors, pistons, rings, & valves	0.5	Manufg.
Insurance carriers	0.1	Fin/R.E.	Electric housewares & fans	0.5	Manufg.
Laundry, dry cleaning, shoe repair	0.1	Services	Fabricated metal products, nec	0.5	Manufg.
			Household cooking equipment	0.5	Manufg.
			Industrial controls	0.5	Manufg.
			Metal stampings, nec	0.5	Manufg.
			Miscellaneous plastics products	0.5	Manufg.
			Blowers & fans	0.4	Manufg.
			General industrial machinery, nec	0.4	Manufg.
			Household laundry equipment	0.4	Manufg.
			Instruments to measure electricity	0.4	Manufg.
			Lawn & garden equipment	0.4	Manufg.
			Metal partitions & fixtures	0.4	Manufg.
			Nonferrous wire drawing & insulating	0.4	Manufg.
			Oil field machinery	0.4	Manufg.
			Photographic equipment & supplies	0.4	Manufg.
			Special dies & tools & machine tool accessories	0.4	Manufg.
			Wood products, nec	0.4	Manufg.
			Air transportation	0.4	Util.
			Nonferrous metal ores, except copper	0.3	Mining
			Electric utility facility construction	0.3	Constr.
			Conveyors & conveying equipment	0.3	Manufg.
			Electrical industrial apparatus, nec	0.3	Manufg.
			Environmental controls	0.3	Manufg.
			Furniture & fixtures, nec	0.3	Manufg.
			Guided missiles & space vehicles	0.3	Manufg.

Continued on next page.

INPUTS AND OUTPUTS FOR SCREW MACHINE AND RELATED PRODUCTS - Continued

Economic Sector or Industry Providing Inputs	%	Sector	Economic Sector or Industry Buying Outputs	%	Sector
			Machine tools, metal cutting types	0.3	Manufg.
			Manufacturing industries, nec	0.3	Manufg.
			Metal doors, sash, & trim	0.3	Manufg.
			Metal office furniture	0.3	Manufg.
			Optical instruments & lenses	0.3	Manufg.
			Power driven hand tools	0.3	Manufg.
			Radio & TV receiving sets	0.3	Manufg.
			Railroad equipment	0.3	Manufg.
			Service industry machines, nec	0.3	Manufg.
			Small arms	0.3	Manufg.
			Special industry machinery, nec	0.3	Manufg.
			Sporting & athletic goods, nec	0.3	Manufg.
			Steel wire & related products	0.3	Manufg.
			Surgical & medical instruments	0.3	Manufg.
			X-ray apparatus & tubes	0.3	Manufg.
			Automotive rental & leasing, without drivers	0.3	Services
			Automotive repair shops & services	0.3	Services
			S/L Govt. purch., other general government	0.3	S/L Govt
			Chemical & fertilizer mineral	0.2	Mining
			Copper ore	0.2	Mining
			Iron & ferroalloy ores	0.2	Mining
			Construction of educational buildings	0.2	Constr.
			Industrial buildings	0.2	Constr.
			Maintenance of electric utility facilities	0.2	Constr.
			Boat building & repairing	0.2	Manufg.
			Drapery hardware & blinds & shades	0.2	Manufg.
			Engineering & scientific instruments	0.2	Manufg.
			Household appliances, nec	0.2	Manufg.
			Household refrigerators & freezers	0.2	Manufg.
			Industrial trucks & tractors	0.2	Manufg.
			Iron & steel foundries	0.2	Manufg.
			Leather goods, nec	0.2	Manufg.
			Paints & allied products	0.2	Manufg.
			Printing trades machinery	0.2	Manufg.
			Public building furniture	0.2	Manufg.
			Semiconductors & related devices	0.2	Manufg.
			Structural wood members, nec	0.2	Manufg.
			Surgical appliances & supplies	0.2	Manufg.
			Textile machinery	0.2	Manufg.
			Travel trailers & campers	0.2	Manufg.
			Truck & bus bodies	0.2	Manufg.
			Truck trailers	0.2	Manufg.
			Turbines & turbine generator sets	0.2	Manufg.
			Typewriters & office machines, nec	0.2	Manufg.
			Wood pallets & skids	0.2	Manufg.
			Amusement & recreation services nec	0.2	Services
			Laundry, dry cleaning, shoe repair	0.2	Services
			Federal Government purchases, nondefense	0.2	Fed Govt
			Dimension, crushed & broken stone	0.1	Mining
			Maintenance of nonfarm buildings nec	0.1	Constr.
			Maintenance of railroads	0.1	Constr.
			Office buildings	0.1	Constr.
			Residential garden apartments	0.1	Constr.
			Sewer system facility construction	0.1	Constr.
			Automatic merchandising machines	0.1	Manufg.
			Ball & roller bearings	0.1	Manufg.
			Copper rolling & drawing	0.1	Manufg.
			Cutlery	0.1	Manufg.
			Elevators & moving stairways	0.1	Manufg.
			Heating equipment, except electric	0.1	Manufg.
			Household vacuum cleaners	0.1	Manufg.
			Industrial furnaces & ovens	0.1	Manufg.
			Metal sanitary ware	0.1	Manufg.
			Metalworking machinery, nec	0.1	Manufg.
			Mining machinery, except oil field	0.1	Manufg.
			Miscellaneous fabricated wire products	0.1	Manufg.
			Nonferrous rolling & drawing, nec	0.1	Manufg.
			Plating & polishing	0.1	Manufg.
			Prefabricated wood buildings	0.1	Manufg.
			Sawmills & planning mills, general	0.1	Manufg.
			Signs & advertising displays	0.1	Manufg.
			Tanks & tank components	0.1	Manufg.
			Transformers	0.1	Manufg.
			Watches, clocks, & parts	0.1	Manufg.
			Woodworking machinery	0.1	Manufg.

Source: Benchmark Input-Output Accounts for the U.S. Economy, 1982, U.S. Department of Commerce, Washington, D.C., July 1991. Data, as reported in the source, are organized by the 1977 SIC structure in use in 1982 but have been matched, as closely as is possible, to the 1987 SIC structure used in this book.

OCCUPATIONS EMPLOYED BY SIC 345 - SCREW MACHINE PRODUCTS, BOLTS, ETC.

Occupation	% of Total 1994	Change to 2005	Occupation	% of Total 1994	Change to 2005
Lathe & turning machine tool operators	9.4	-32.8	Metal & plastic machine workers nec	1.8	-25.5
Machinists	9.1	-16.0	Punching machine operators, metal & plastic	1.8	0.8
Machine tool cutting operators, metal & plastic	6.9	-13.9	Industrial production managers	1.7	-16.0
Machine tool cutting & forming etc. nec	4.9	-16.0	Industrial machinery mechanics	1.4	-7.5
Inspectors, testers, & graders, precision	4.9	-16.0	General office clerks	1.3	-28.4
Blue collar worker supervisors	4.6	-23.3	Bookkeeping, accounting, & auditing clerks	1.3	-37.0
Combination machine tool operators	4.5	-7.6	Secretaries, ex legal & medical	1.3	-23.5
General managers & top executives	3.4	-20.3	Helpers, laborers, & material movers nec	1.3	-16.0
Assemblers, fabricators, & hand workers nec	2.8	-16.0	Machine operators nec	1.2	-25.9
Drilling & boring machine tool workers	2.8	-16.0	Production, planning, & expediting clerks	1.1	13.4
Machine forming operators, metal & plastic	2.4	-16.0	Machine feeders & offbearers	1.1	-24.4
Traffic, shipping, & receiving clerks	2.4	-19.2	Hand packers & packagers	1.1	-28.0
Tool & die makers	2.3	-32.1	NC machine tool operators, metal & plastic	1.1	0.8
Sales & related workers nec	2.3	-16.0	Janitors & cleaners, incl maids	1.0	-32.8
Grinding machine operators, metal & plastic	2.2	0.8			

Source: *Industry-Occupation Matrix*, Bureau of Labor Statistics. These data relate to one or more 3-digit SIC industry groups rather than to a single 4-digit SIC. The change reported for each occupation to the year 2005 is a percent of growth or decline as estimated by the Bureau of Labor Statistics. The abbreviation nec stands for 'not elsewhere classified'.

LOCATION BY STATE AND REGIONAL CONCENTRATION

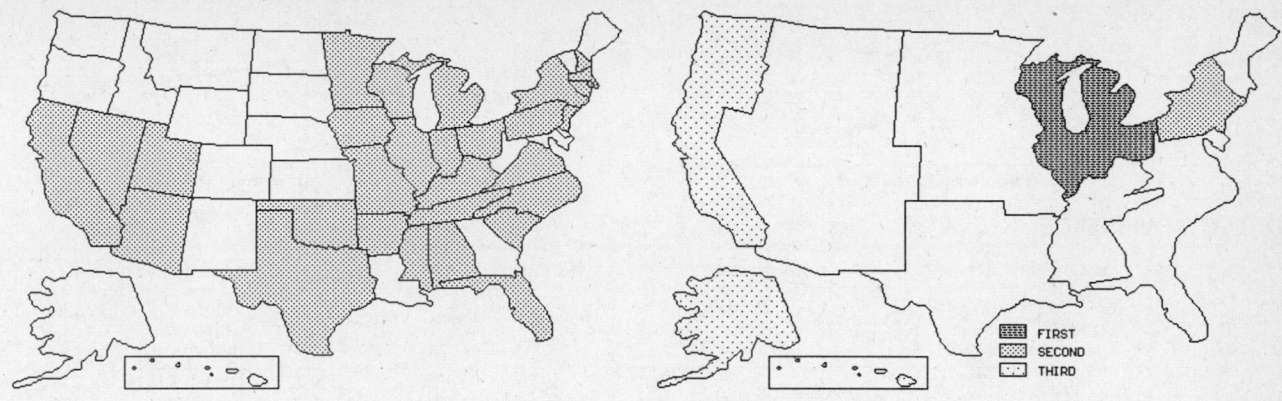

FIRST
SECOND
THIRD

INDUSTRY DATA BY STATE

State	Establish-ments	Shipments			Employment				Cost as % of Shipments	Investment per Employee ($)
		Total ($ mil)	% of U.S.	Per Establ.	Total Number	% of U.S.	Per Establ.	Wages ($/hour)		
Illinois	134	1,007.1	19.4	7.5	8,200	18.6	61	13.26	48.7	2,439
California	114	817.7	15.8	7.2	8,800	20.0	77	14.31	30.6	3,398
Michigan	98	640.9	12.4	6.5	3,900	8.9	40	13.27	51.9	6,103
Ohio	86	484.5	9.3	5.6	3,600	8.2	42	11.83	48.7	3,833
Pennsylvania	49	377.2	7.3	7.7	3,200	7.3	65	13.13	44.6	2,531
Connecticut	40	250.6	4.8	6.3	2,100	4.8	53	13.39	35.8	4,000
New York	54	214.3	4.1	4.0	2,000	4.5	37	11.69	46.7	2,100
New Jersey	40	175.9	3.4	4.4	1,700	3.9	43	13.42	32.1	3,353
Wisconsin	29	136.4	2.6	4.7	1,200	2.7	41	11.21	44.2	3,583
Texas	41	126.2	2.4	3.1	1,300	3.0	32	9.35	48.1	3,846
Massachusetts	29	101.5	2.0	3.5	900	2.0	31	10.92	54.7	2,444
Virginia	8	81.5	1.6	10.2	500	1.1	63	11.78	62.9	-
Kentucky	11	80.1	1.5	7.3	500	1.1	45	10.25	55.7	2,200
Tennessee	13	75.2	1.5	5.8	500	1.1	38	13.13	46.5	4,800
Alabama	12	68.5	1.3	5.7	500	1.1	42	10.00	56.4	2,000
Florida	16	38.9	0.8	2.4	500	1.1	31	11.25	40.6	1,400
Arizona	6	28.6	0.6	4.8	300	0.7	50	11.80	29.4	-
North Carolina	9	25.1	0.5	2.8	300	0.7	33	11.75	53.0	2,333
Missouri	12	25.1	0.5	2.1	300	0.7	25	10.00	42.2	1,667
Rhode Island	11	24.5	0.5	2.2	400	0.9	36	10.83	43.7	750
Minnesota	8	24.4	0.5	3.0	200	0.5	25	12.75	39.8	3,500
South Carolina	7	19.4	0.4	2.8	200	0.5	29	12.33	47.4	-
New Hampshire	6	13.4	0.3	2.2	200	0.5	33	12.50	36.6	2,000
Indiana	22	(D)	-	-	1,750 *	4.0	80	-	-	2,229
Nevada	6	(D)	-	-	175 *	0.4	29	-	-	2,857
Arkansas	5	(D)	-	-	175 *	0.4	35	-	-	-
Iowa	5	(D)	-	-	375 *	0.9	75	-	-	8,000
Mississippi	5	(D)	-	-	175 *	0.4	35	-	-	571
Oklahoma	5	(D)	-	-	375 *	0.9	75	-	-	-
Utah	5	(D)	-	-	175 *	0.4	35	-	-	-

Source: 1992 *Economic Census*. The states are in descending order of shipments or establishments (if shipment data are missing for the majority). The symbol (D) appears when data are withheld to prevent disclosure of competitive information. States marked with (D) are sorted by number of establishments. A dash (-) indicates that the data element cannot be calculated; * indicates the midpoint of a range.

3462 - IRON & STEEL FORGINGS

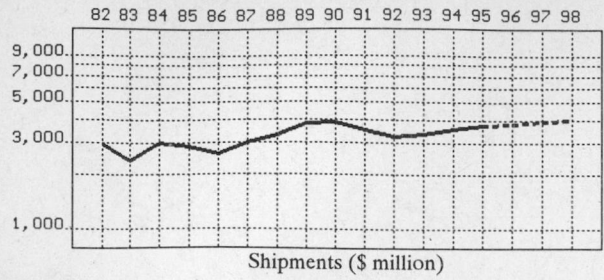

Shipments ($ million)

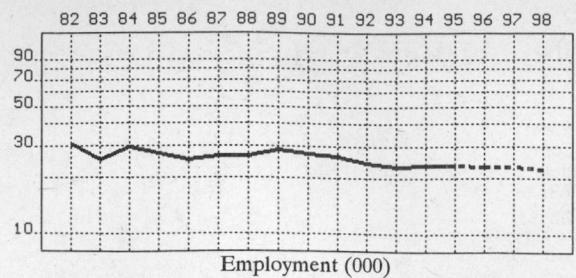

Employment (000)

GENERAL STATISTICS

Year	Companies	Establishments		Employment			Compensation		Production ($ million)			
		Total	with 20 or more employees	Total (000)	Production Workers (000)	Hours (Mil)	Payroll ($ mil)	Wages ($/hr)	Cost of Materials	Value Added by Manufacture	Value of Shipments	Capital Invest.
1982	337	381	221	30.9	22.5	40.9	692.5	11.35	1,524.6	1,352.6	2,952.5	158.4
1983		374	223	25.0	18.5	33.8	597.4	11.91	1,247.2	1,103.3	2,380.7	80.2
1984		367	225	29.8	22.0	42.0	754.7	12.36	1,531.3	1,386.1	3,014.8	97.3
1985		360	227	27.5	20.1	38.5	728.9	12.67	1,509.9	1,309.6	2,857.9	95.6
1986		357	210	25.2	18.3	35.0	675.6	12.84	1,368.6	1,247.1	2,602.1	74.4
1987	379	406	202	26.6	19.1	37.8	746.5	12.81	1,553.5	1,394.0	3,003.6	108.2
1988		395	202	26.4	19.6	40.5	751.4	12.42	1,757.9	1,587.8	3,284.6	105.3
1989		399	214	28.8	21.2	44.0	831.2	12.88	2,012.6	1,818.8	3,777.9	144.7
1990		384	214	27.3	21.3	43.0	861.2	13.45	2,088.3	1,746.9	3,858.8	130.1
1991		401	207	26.2	19.7	39.5	819.4	13.71	1,910.5	1,608.0	3,540.0	133.8
1992	367	403	199	24.2	18.0	36.8	765.9	13.58	1,631.7	1,586.3	3,232.3	114.4
1993		389	194	23.3	17.4	37.0	758.8	13.66	1,720.2	1,622.1	3,336.0	134.0
1994		403P	196P	23.9	18.8	40.3	828.7	14.34	1,816.2	1,807.4	3,608.4	88.9
1995		406P	193P	23.8P	18.3P	39.8P	847.0P	14.31P	1,893.6P	1,884.4P	3,762.1P	119.8P
1996		409P	191P	23.5P	18.1P	39.9P	860.1P	14.50P	1,934.8P	1,925.4P	3,844.1P	120.8P
1997		412P	189P	23.1P	17.9P	40.0P	873.3P	14.70P	1,976.1P	1,966.5P	3,926.0P	121.8P
1998		415P	186P	22.7P	17.7P	40.1P	886.5P	14.90P	2,017.3P	2,007.5P	4,008.0P	122.8P

Sources: 1982, 1987, 1992 *Economic Census*; *Annual Survey of Manufactures*, 83-86, 88-91, 93-94. Establishment counts for non-Census years are from *County Business Patterns*; establishment values for 83-84 are extrapolations. 'P's show projections by the editors. Industries reclassified in 87 will not have data for prior years.

INDICES OF CHANGE

Year	Companies	Establishments		Employment			Compensation		Production ($ million)			
		Total	with 20 or more employees	Total (000)	Production Workers (000)	Hours (Mil)	Payroll ($ mil)	Wages ($/hr)	Cost of Materials	Value Added by Manufacture	Value of Shipments	Capital Invest.
1982	92	95	111	128	125	111	90	84	93	85	91	138
1983		93	112	103	103	92	78	88	76	70	74	70
1984		91	113	123	122	114	99	91	94	87	93	85
1985		89	114	114	112	105	95	93	93	83	88	84
1986		89	106	104	102	95	88	95	84	79	81	65
1987	103	101	102	110	106	103	97	94	95	88	93	95
1988		98	102	109	109	110	98	91	108	100	102	92
1989		99	108	119	118	120	109	95	123	115	117	126
1990		95	108	113	118	117	112	99	128	110	119	114
1991		100	104	108	109	107	107	101	117	101	110	117
1992	100	100	100	100	100	100	100	100	100	100	100	100
1993		97	97	96	97	101	99	101	105	102	103	117
1994		100P	98P	99	104	110	108	106	111	114	112	78
1995		101P	97P	99P	102P	108P	111P	105P	116P	119P	116P	105P
1996		102P	96P	97P	101P	108P	112P	107P	119P	121P	119P	106P
1997		102P	95P	95P	100P	109P	114P	108P	121P	124P	121P	106P
1998		103P	94P	94P	98P	109P	116P	110P	124P	127P	124P	107P

Sources: Same as General Statistics. Values reflect change from the base year, 1992. Values above 100 mean greater than 92, values below 100 mean less than 92, and a value of 100 in the 82-91 or 93-98 period means same as 92. 'P's mark projections by the editors.

SELECTED RATIOS

For 1994	Avg. of All Manufact.	Analyzed Industry	Index	For 1994	Avg. of All Manufact.	Analyzed Industry	Index
Employees per Establishment	49	59	121	Value Added per Production Worker	134,084	96,138	72
Payroll per Establishment	1,500,273	2,054,551	137	Cost per Establishment	5,045,178	4,502,806	89
Payroll per Employee	30,620	34,674	113	Cost per Employee	102,970	75,992	74
Production Workers per Establishment	34	47	136	Cost per Production Worker	146,988	96,606	66
Wages per Establishment	853,319	1,432,761	168	Shipments per Establishment	9,576,895	8,946,110	93
Wages per Production Worker	24,861	30,739	124	Shipments per Employee	195,460	150,979	77
Hours per Production Worker	2,056	2,144	104	Shipments per Production Worker	279,017	191,936	69
Wages per Hour	12.09	14.34	119	Investment per Establishment	321,011	220,405	69
Value Added per Establishment	4,602,255	4,480,989	97	Investment per Employee	6,552	3,720	57
Value Added per Employee	93,930	75,623	81	Investment per Production Worker	9,352	4,729	51

Sources: Same as General Statistics. The 'Average of All Manufacturing' column represents the average of all manufacturing industries reported for the most recent complete year available. The Index shows the relationship between the Average and the Analyzed Industry. For example, 100 means that they are equal; 500 that the Analyzed Industry is five times the average; 50 means that the Analyzed Industry is half the national average. The abbreviation 'na' is used to show that data are 'not available'.

LEADING COMPANIES　　Number shown: **75**　　Total sales ($ mil): **2,502**　　Total employment (000): **18.2**

Company Name	Address				CEO Name	Phone	Co. Type	Sales ($ mil)	Empl. (000)
Ladish Company Inc	5481 S Packard Av	Cudahy	WI	53110	Kerry Woody	414-747-2611	R	175	1.6
Letts Industries Inc	1111 Bellevue St	Detroit	MI	48207	Charles E Letts Jr	313-579-1100	R	115	0.4
Griffin Wheel	200 W Monroe St	Chicago	IL	60606	Arthur W Goetschel	312-346-3300	D	100*	0.8
Scot Forge Co	PO Box 8	Spring Grove	IL	60081	J F McKinley Jr	708-587-1000	R	100	0.4
Interstate Forging Industries Inc	4051 N 27th St	Milwaukee	WI	53216	James Mitchell	414-444-0911	R	80	0.5
Keller Group Inc	1 Northfield Rd	Northfield	IL	60093	JP Keller	708-446-7550	R	80	0.4
FMC Corp Steel Products	2101 W 10th St	Anniston	AL	36201	Elmer L Doty	205-235-9683	D	75	0.5
National Forge Co	100 Front St	Irvine	PA	16329	RO Wilder	814-563-7522	R	74	0.7
Ellwood Group Inc	PO Box 31	Ellwood City	PA	16117	David Barensfeld	412-752-0055	R	70	0.6
Carlton Forge Works	7743 E Adams St	Paramount	CA	90723	Alan J Carlton Jr	310-633-1131	R	65*	0.2
Presrite Corp	3665 E 78th St	Cleveland	OH	44105	Don Diemer	216-441-5990	R	65	0.3
SIFCO Industries Inc	970 E 64th St	Cleveland	OH	44103	Jeffrey P Gotschall	216-881-8600	P	61	0.5
City Forge	PO Box 31	Ellwood City	PA	16117	Kevin J Handerhan	412-752-0055	D	60	0.3
Erie Forge and Steel Inc	1341 W 16th St	Erie	PA	16512	Allan Concoby	814-452-2300	R	52	0.3
BethForge Inc	1275 Daly Av	Bethlehem	PA	18016	Gregory F Paolini	610-694-2357	S	50	0.5
Meadville Forging Co	PO Box 459	Meadville	PA	16335	James R Martin	814-332-8200	S	50	0.3
T and D Metal Products Inc	PO Box 405	Watseka	IL	60970	Roger Dittrich	815-432-4938	R	50	0.8
Mercer Forge Co	PO Box 272	Mercer	PA	16137	Mark Clark	412-662-2750	D	45	0.2
Modern Drop Forge Co	PO Box 429	Blue Island	IL	60406	R Heim	708-388-1806	R	45	0.4
Kervick Enterprises Inc	40 Rockdale St	Worcester	MA	01606	Robert B Kervick	508-853-4500	R	41*	0.4
Jorgensen Forge Corp	8531 Marginal	Seattle	WA	98108	Jack Bunt	206-762-1100	R	40	0.3
McInnes Steel Co	441 E Main St	Corry	PA	16407	Steven J Mahoney	814-664-9664	R	40	0.3
Ovako Ajax Inc	PO Box 860	York	SC	29745	Anders Tenebeck	803-684-3133	S	40	0.1
Louisville Forge and Gear Works	PO Box 36426	Louisville	KY	40233	James L Peyton	502-367-3270	R	39*	0.3
Hermes Automotive Mfg Corp	2703 23rd St	Detroit	MI	48216	John Sinanis	313-897-6395	R	37	0.4
Colfor Inc	3255 Alliance NW	Malvern	OH	44644	WJ McCarthy	216-863-0404	S	35*	0.3
Komtek Inc	40 Rockdale St	Worcester	MA	01606	Robert Kervick	508-853-4500	S	31*	0.2
Dixie Industries	3510 N Orchard	Chattanooga	TN	37406	Zane Goggin	615-698-3323	D	30	0.2
Federal Forge Inc	2807 S Logan St	Lansing	MI	48910	Naga D Manohar	517-393-5300	R	30	0.2
International Crankshafts Inc	101 Carley Ct	Georgetown	KY	40324	Yasutaka Ishimatsu	502-868-0003	S	30	<0.1
Commercial Forged Products	5757 W 65th St	Bedford Park	IL	60638	Edward F Wozniak	708-458-1220	D	25	0.1
Crosby-LeBus Manufacturing Co	PO Box 271	Longview	TX	75606	Larry Postelwait	903-759-4424	S	25	0.3
Ellwood Texas Forge	PO Box 1477	Houston	TX	77251	Mike Henthorne	713-434-5100	D	25	0.1
Krupp Gerlach Co	PO Box 214	Danville	IL	61834	Joe Pvcz	217-431-0060	S	25	0.2
Mid-West Forge Corp	17301 St Clair Av	Cleveland	OH	44110	PC Gum	216-481-3030	R	25	0.1
Conley Frog Switch & Forge Co	PO Box 9188	Memphis	TN	38109	Edward M Lyons	901-948-4591	R	23	0.3
Cornell Forge Co	6666 W 66th St	Chicago	IL	60638	Bill Brewer	312-767-4242	R	20	0.3
Danville Metal Stamping	20 Oakwood Av	Danville	IL	61832	Judd C Peck	217-446-0647	R	20	0.4
ERI	10930 E 59th	Indianapolis	IN	46236	Peyton S Baker	317-823-4441	D	20*	0.2
T and W Forge Inc	562 Ely St	Alliance	OH	44601	Paul Schmidt	216-821-5740	S	20	0.2
Midland Forge	PO Box 1627	Cedar Rapids	IA	52406	Gary Eckley	319-362-1111	D	19*	0.2
Harris Thomas Drop Forge Co	1400 E 1st St	Dayton	OH	45403	Marvin L Billow	513-253-3152	R	18	0.1
Moroso Performance Products	80 Carter Dr	Guilford	CT	06437	RD Moroso	203-453-6571	R	18*	0.2
Tie Down Engineering Inc	5901 Wheaton Dr	Atlanta	GA	30336	Chuck Mackarvich	404-344-0000	R	18	0.1
Gunnebo Fastening Corp	1 Gunnebo Dr	Lonoke	AR	72086	Chuck Hicks	501-676-2222	S	17	0.2
Kropp Forge	5301 W Roosevelt	Chicago	IL	60650	Frank Greco	312-242-1900	D	16	0.2
W Pat Crow Forgings Inc	PO Box 1720	Fort Worth	TX	76101	Ron LeComte	817-536-2861	S	16	0.1
Hammond and Irving Inc	254 North St	Auburn	NY	13021	JD Taylor	315-253-6265	R	15	0.1
Gulf Forge Co	PO Box 2926	Houston	TX	77252	JW Brougher	713-683-0513	R	15	0.1
McWilliams Forge Company Inc	Franklin Rd	Rockaway	NJ	07866	A McWilliams	201-627-0200	R	15	0.1
Melling Forging Co	1709 Thompson St	Lansing	MI	48906	Richard B Horton	517-482-0791	S	15	0.2
National Flame and Forge Inc	PO Box 7385	Houston	TX	77248	William E Pielop Jr	713-869-5724	R	15	0.1
National Flange and Fitting Co	PO Box 924149	Houston	TX	77292	Alois O Keilers	713-688-2515	R	15	0.1
Forged Products Inc	6505 N	Houston	TX	77091	Charles McBride	713-462-3416	R	14	<0.1
Steel Industries Inc	12600 Beech Daly	Detroit	MI	48239	Paul Sakmar	313-531-1140	R	14	0.1
Uniflow Corp	PO Box 705	Novi	MI	48376	Corky MacKenzie	313-348-9370	S	14	0.2
Norforge and Machine Inc	195 N Dean St	Bushnell	IL	61422	Patricia Hayes	309-772-3124	R	13	0.1
Schaefer Manufacturing Inc	158 Duchaine Blv	New Bedford	MA	02745	R Smith	508-995-9511	R	13	0.2
Thermal Structures Inc	2362 Railroad St	Corona	CA	91720	Steven T Brauheim	909-736-9911	R	13	0.1
Berwick Forge & Fabricating	PO Box 188	Berwick	PA	18603	AC Bailey	717-752-2784	R	12	0.1
Clifford-Jacobs Forging Co	PO Box 830	Champaign	IL	61824	Paul J Hausmann	217-352-5172	R	12	<0.1
Coulter Steel and Forge Co	PO Box 8008	Emeryville	CA	94662	TM Coulter	510-420-3500	R	12	<0.1
CE Larson and Sons Inc	2645 N Keeler Av	Chicago	IL	60639	RL Larson	312-772-9700	R	12	<0.1
Forged Vessel Connections Inc	PO Box 38421	Houston	TX	77088	Charles Robinson	713-688-9705	R	12	<0.1
Moline Forge Inc	4101 4th Av	Moline	IL	61265	TG Getz	309-762-5506	R	12	<0.1
Rex Forge	335 Atwater St	Plantsville	CT	06479	Ronald Fontanella	203-628-0393	D	12*	0.2
Ogemaw Forge Co	PO Box 648	West Branch	MI	48661	Robert M Scibor	517-345-3850	R	11	0.1
Riverside Products	400 21st St	Moline	IL	61265	Claude D Robinson	309-764-2020	D	11	<0.1
Phoenix Forging Company Inc	PO Box 70	Catasauqua	PA	18032	Larry A Dildine	610-264-2861	R	10	0.2
Berkeley Forge and Tool Inc	1331 Eastshore Hwy	Berkeley	CA	94710	SF Bierwith	510-526-5034	R	10	<0.1
Dayton Forging	PO Box 1629	Dayton	OH	45401	Harlan H Todd	513-253-4126	R	10	0.1
Endicott Forging	1901 N North St	Endicott	NY	13760	Norm McDonald	607-785-3331	R	10*	0.1
Lefere Forge and Machine Co	665 Hupp Av	Jackson	MI	49203	MJ Lefere	517-784-7109	R	10	<0.1
Schaefer Equipment Inc	1590 Phoenix NE	Warren	OH	44483	Peter L Miller	216-372-4006	S	10	<0.1
SP Systems Armor	741 Flynn Rd	Camarillo	CA	93012	Don Carmichael	805-445-4542	R	10*	<0.1

Source: Ward's Business Directory of U.S. Private and Public Companies, Volumes 1 and 2, 1996. The company type code used is as follows: P - Public, R - Private, S - Subsidiary, D - Division, J - Joint Venture, A - Affiliate, G - Group. Sales are in millions of dollars, employees are in thousands. An asterisk (*) indicates an estimated sales volume. The symbol < stands for 'less than'. Company names and addresses are truncated, in some cases, to fit into the available space.

MATERIALS CONSUMED

Material	Quantity	Delivered Cost ($ million)
Materials, ingredients, containers, and supplies	(X)	1,313.2
Fabricated metal products (except castings and forgings)	(X)	18.3
Forgings ... 1,000 s tons	46.3*	81.7
Castings (rough and semifinished)	(X)	10.8
Steel ingot and semifinished shapes (blooms, billets, and slabs) ... 1,000 s tons	275.5**	259.6
Steel bars, bar shapes, and other shapes and forms ... 1,000 s tons	803.9*	476.0
Titanium and titanium-base alloy shapes and forms ... 1,000 s tons	2.7	38.2
Copper and copper-base alloy shapes and forms ... 1,000 s tons	(S)	0.7
Aluminum and aluminum-base alloy shapes and forms ... 1,000 s tons	3.0**	8.3
Nickel and nickel-base alloy, including nickel-copper alloys ... 1,000 s tons	15.7	52.2
All other nonferrous shapes and forms ... 1,000 s tons	(S)	1.3
Forging dies ... 1,000 s tons	(S)	17.8
All other materials and components, parts, containers, and supplies	(X)	161.4
Materials, ingredients, containers, and supplies, nsk	(X)	186.6

Source: 1992 *Economic Census*. Explanation of symbols used: (D): Withheld to avoid disclosure of competitive data; na: Not available; (S): Withheld because statistical norms were not met; (X): Not applicable; (Z): Less than half the unit shown; nec: Not elsewhere classified; nsk: Not specified by kind; - : zero; * : 10-19 percent estimated; ** : 20-29 percent estimated.

PRODUCT SHARE DETAILS

Product or Product Class	% Share	Product or Product Class	% Share
Iron and steel forgings	100.00	nsk	15.51
Hot impression die impact, press, and upset steel forgings	67.15	Seamless rolled ring forgings, ferrous (not made in steel mills)	6.24
Hot impression die impact, press, and upset carbon steel forgings	42.82	Seamless rolled ring carbon and alloy steel forgings, excluding stainless and hi-temp (not made in steel mills)	38.47
Hot impression die impact, press, and upset alloy steel forgings, excluding stainless and hi-temp	32.75	Seamless rolled ring stainless steel and hi-temp (iron, nickel, or cobalt-base alloy) forgings (not made in steel mills)	61.53
Hot impression die impact, press, and upset stainless steel forgings	4.91	Open die or smith forgings (hammer or press), ferrous (not made in steel mills)	10.39
Hot impression die impact, press, and upset hi-temp (iron, nickel, or cobalt-base alloy) steel forgings	12.81	Open die or smith carbon and alloy steel forgings (hammer or press), excluding stainless and hi-temp (not made in steel mills)	88.34
Hot impression die impact, press, and upset steel forgings, nsk	6.71	Open die or smith stainless steel and hi-temp (iron, nickel, or cobalt-base alloy) forgings (hammer or press) (not made in steel mills)	11.57
Cold impression die impact, press, and upset steel forgings	8.04	Open die or smith forgings (hammer or press), ferrous (not made in steel mills), nsk	0.09
Cold impression die impact, press, and upset carbon steel forgings	42.50	Iron and steel forgings, nsk	8.17
Cold impression die impact, press, and upset alloy steel forgings	39.95		
Cold impression die impact, press, and upset stainless steel and hi-temp (iron, nickel, or cobalt-base alloy) forgings	2.04		
Cold impression die impact, press, and upset steel forgings,			

Source: 1992 *Economic Census*. The values shown are percent of total shipments in an industry. Values of indented subcategories are summed in the main heading. The symbol (D) appears when data are withheld to prevent disclosure of competitive information. The abbreviation nsk stands for 'not specified by kind' and nec for 'not elsewhere classified'.

INPUTS AND OUTPUTS FOR IRON & STEEL FORGINGS

Economic Sector or Industry Providing Inputs	%	Sector	Economic Sector or Industry Buying Outputs	%	Sector
Blast furnaces & steel mills	47.2	Manufg.	Aircraft & missile engines & engine parts	15.3	Manufg.
Wholesale trade	9.1	Trade	Motor vehicle parts & accessories	14.4	Manufg.
Gas production & distribution (utilities)	5.6	Util.	Internal combustion engines, nec	10.4	Manufg.
Nonferrous rolling & drawing, nec	5.1	Manufg.	Construction machinery & equipment	7.8	Manufg.
Electric services (utilities)	4.1	Util.	Oil field machinery	7.0	Manufg.
Imports	3.9	Foreign	Turbines & turbine generator sets	6.5	Manufg.
Iron & steel forgings	2.1	Manufg.	Pipe, valves, & pipe fittings	4.5	Manufg.
Advertising	2.0	Services	Power transmission equipment	3.2	Manufg.
Pipe, valves, & pipe fittings	1.8	Manufg.	Railroad equipment	3.1	Manufg.
Special dies & tools & machine tool accessories	1.5	Manufg.	Exports	3.1	Foreign
Motor freight transportation & warehousing	1.5	Util.	Aircraft & missile equipment, nec	2.9	Manufg.
Petroleum refining	1.3	Manufg.	Farm machinery & equipment	2.8	Manufg.
Maintenance of nonfarm buildings nec	0.9	Constr.	Ball & roller bearings	2.5	Manufg.
Metal coating & allied services	0.9	Manufg.	Motor vehicles & car bodies	2.1	Manufg.
Miscellaneous plastics products	0.9	Manufg.	Pumps & compressors	1.5	Manufg.
Primary metal products, nec	0.8	Manufg.	Tanks & tank components	1.4	Manufg.
Railroads & related services	0.8	Util.	Iron & steel forgings	1.0	Manufg.
Fabricated metal products, nec	0.7	Manufg.	Copper ore	0.9	Mining
Metal heat treating	0.7	Manufg.	Hand & edge tools, nec	0.9	Manufg.
Machinery, except electrical, nec	0.6	Manufg.	Mining machinery, except oil field	0.9	Manufg.
Power transmission equipment	0.6	Manufg.	Refrigeration & heating equipment	0.8	Manufg.
Communications, except radio & TV	0.6	Util.	Iron & ferroalloy ores	0.5	Mining
Equipment rental & leasing services	0.6	Services	Aircraft	0.5	Manufg.
Eating & drinking places	0.5	Trade	Conveyors & conveying equipment	0.5	Manufg.

Continued on next page.

INPUTS AND OUTPUTS FOR IRON & STEEL FORGINGS - Continued

Economic Sector or Industry Providing Inputs	%	Sector	Economic Sector or Industry Buying Outputs	%	Sector
Banking	0.4	Fin/R.E.	Machine tools, metal cutting types	0.5	Manufg.
Business services nec	0.4	Services	Machine tools, metal forming types	0.5	Manufg.
Computer & data processing services	0.4	Services	Rolling mill machinery	0.5	Manufg.
Ball & roller bearings	0.3	Manufg.	Federal Government purchases, nondefense	0.5	Fed Govt
Paperboard containers & boxes	0.3	Manufg.	Screw machine and related products	0.4	Manufg.
Abrasive products	0.2	Manufg.	Small arms	0.4	Manufg.
Cyclic crudes and organics	0.2	Manufg.	Industrial trucks & tractors	0.3	Manufg.
Machine tools, metal cutting types	0.2	Manufg.	Surgical appliances & supplies	0.3	Manufg.
Machine tools, metal forming types	0.2	Manufg.	Cutlery	0.2	Manufg.
Air transportation	0.2	Util.	Fabricated metal products, nec	0.2	Manufg.
Real estate	0.2	Fin/R.E.	Hoists, cranes, & monorails	0.2	Manufg.
Engineering, architectural, & surveying services	0.2	Services	Nonferrous forgings	0.2	Manufg.
Management & consulting services & labs	0.2	Services	Special dies & tools & machine tool accessories	0.2	Manufg.
U.S. Postal Service	0.2	Gov't	Nonferrous metal ores, except copper	0.1	Mining
Coal	0.1	Mining	Motors & generators	0.1	Manufg.
Copper rolling & drawing	0.1	Manufg.	Ordnance & accessories nec	0.1	Manufg.
Fabricated rubber products, nec	0.1	Manufg.	Power driven hand tools	0.1	Manufg.
Lubricating oils & greases	0.1	Manufg.	Surgical & medical instruments	0.1	Manufg.
Primary aluminum	0.1	Manufg.	Transportation equipment, nec	0.1	Manufg.
Sanitary services, steam supply, irrigation	0.1	Util.	Federal Government purchases, national defense	0.1	Fed Govt
Royalties	0.1	Fin/R.E.			
Accounting, auditing & bookkeeping	0.1	Services			
Electrical repair shops	0.1	Services			
Legal services	0.1	Services			

Source: Benchmark Input-Output Accounts for the U.S. Economy, 1982, U.S. Department of Commerce, Washington, D.C., July 1991. Data, as reported in the source, are organized by the 1977 SIC structure in use in 1982 but have been matched, as closely as is possible, to the 1987 SIC structure used in this book.

OCCUPATIONS EMPLOYED BY SIC 346 - METAL FORGINGS AND STAMPINGS

Occupation	% of Total 1994	Change to 2005	Occupation	% of Total 1994	Change to 2005
Machine forming operators, metal & plastic	9.3	-65.9	Machine feeders & offbearers	1.8	-23.3
Tool & die makers	7.4	-11.4	Sheet metal workers & duct installers	1.7	-14.7
Machine tool cutting & forming etc. nec	6.4	10.9	Metal & plastic machine workers nec	1.5	-24.4
Assemblers, fabricators, & hand workers nec	4.7	-14.7	Helpers, laborers, & material movers nec	1.4	-14.7
Blue collar worker supervisors	4.6	-17.3	Industrial production managers	1.4	-14.7
Welding machine setters, operators	3.8	-23.2	Millwrights	1.3	2.3
Punching machine operators, metal & plastic	3.4	15.1	Traffic, shipping, & receiving clerks	1.3	-17.9
Inspectors, testers, & graders, precision	3.3	-14.7	Welders & cutters	1.3	-14.7
Industrial machinery mechanics	3.1	-6.2	Freight, stock, & material movers, hand	1.3	-31.8
Industrial truck & tractor operators	2.8	-14.7	Electricians	1.2	-20.0
Machine tool cutting operators, metal & plastic	2.6	-12.6	Janitors & cleaners, incl maids	1.1	-31.8
Machinists	2.4	-14.7	Sales & related workers nec	1.1	-14.7
Combination machine tool operators	2.1	27.9	Maintenance repairers, general utility	1.1	-23.3
General managers & top executives	1.8	-19.1			

Source: Industry-Occupation Matrix, Bureau of Labor Statistics. These data relate to one or more 3-digit SIC industry groups rather than to a single 4-digit SIC. The change reported for each occupation to the year 2005 is a percent of growth or decline as estimated by the Bureau of Labor Statistics. The abbreviation nec stands for 'not elsewhere classified'.

LOCATION BY STATE AND REGIONAL CONCENTRATION

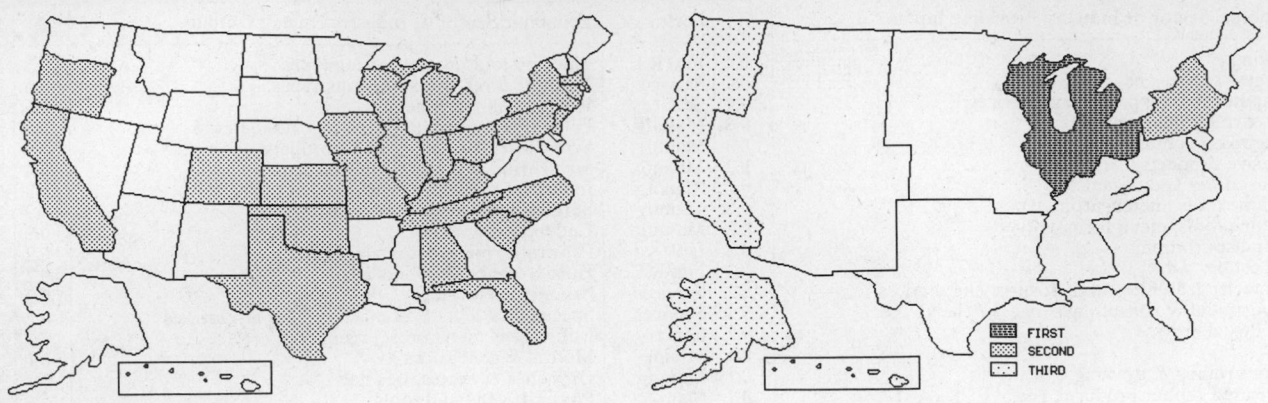

FIRST
SECOND
THIRD

INDUSTRY DATA BY STATE

| State | Establish-ments | Shipments | | | Employment | | | | Cost as % of Shipments | Investment per Employee ($) |
		Total ($ mil)	% of U.S.	Per Establ.	Total Number	% of U.S.	Per Establ.	Wages ($/hour)		
Ohio	51	528.9	16.4	10.4	3,500	14.5	69	13.35	54.6	4,914
Pennsylvania	34	390.4	12.1	11.5	2,900	12.0	85	13.77	46.4	5,724
Michigan	42	367.2	11.4	8.7	2,300	9.5	55	14.89	45.8	5,739
Texas	33	346.2	10.7	10.5	2,800	11.6	85	14.05	48.9	5,143
Wisconsin	14	324.6	10.0	23.2	2,700	11.2	193	17.82	47.9	4,556
Illinois	41	264.9	8.2	6.5	2,300	9.5	56	13.00	47.0	3,609
California	35	227.2	7.0	6.5	1,200	5.0	34	13.00	61.1	7,833
Indiana	13	100.1	3.1	7.7	900	3.7	69	12.15	52.2	-
Massachusetts	6	56.1	1.7	9.4	400	1.7	67	15.20	55.3	-
Connecticut	4	42.3	1.3	10.6	400	1.7	100	14.20	60.3	-
Florida	11	28.5	0.9	2.6	300	1.2	27	13.60	61.4	2,667
New York	18	(D)	-	-	750 *	3.1	42	-	-	3,067
Tennessee	9	(D)	-	-	750 *	3.1	83	-	-	5,600
Alabama	8	(D)	-	-	750 *	3.1	94	-	-	-
Kentucky	7	(D)	-	-	750 *	3.1	107	-	-	-
New Jersey	7	(D)	-	-	175 *	0.7	25	-	-	1,143
Oregon	7	(D)	-	-	175 *	0.7	25	-	-	-
Iowa	5	(D)	-	-	175 *	0.7	35	-	-	6,286
North Carolina	5	(D)	-	-	375 *	1.5	75	-	-	1,867
Kansas	4	(D)	-	-	175 *	0.7	44	-	-	-
Missouri	4	(D)	-	-	175 *	0.7	44	-	-	-
Oklahoma	3	(D)	-	-	175 *	0.7	58	-	-	-
Colorado	2	(D)	-	-	175 *	0.7	88	-	-	-
South Carolina	2	(D)	-	-	175 *	0.7	88	-	-	-

Source: 1992 *Economic Census*. The states are in descending order of shipments or establishments (if shipment data are missing for the majority). The symbol (D) appears when data are withheld to prevent disclosure of competitive information. States marked with (D) are sorted by number of establishments. A dash (-) indicates that the data element cannot be calculated; * indicates the midpoint of a range.

3463 - NONFERROUS FORGINGS

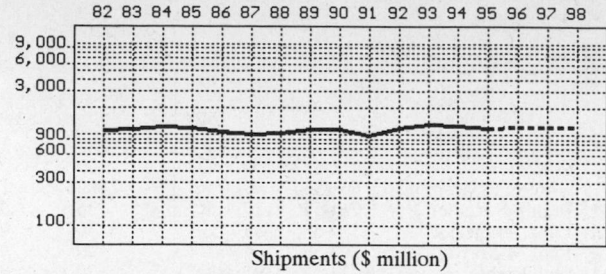

Shipments ($ million)

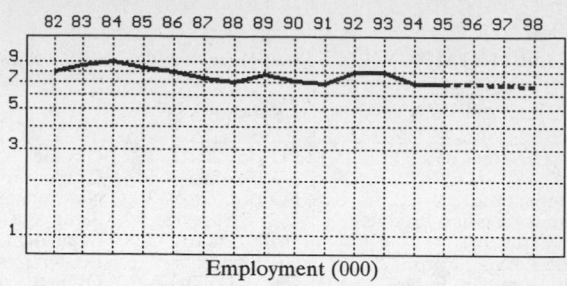

Employment (000)

GENERAL STATISTICS

Year	Com-panies	Establishments		Employment			Compensation		Production ($ million)			
		Total	with 20 or more employees	Total (000)	Production Workers (000)	Hours (Mil)	Payroll ($ mil)	Wages ($/hr)	Cost of Materials	Value Added by Manufacture	Value of Shipments	Capital Invest.
1982	59	64	46	7.9	5.7	10.6	200.7	12.14	569.3	462.1	1,093.8	100.3
1983		63	45	8.6	6.1	11.2	225.7	12.69	572.9	540.7	1,151.5	53.7
1984		62	44	9.1	6.6	12.7	258.8	13.37	733.5	516.2	1,237.4	45.5
1985		62	42	8.3	6.0	11.7	245.5	13.64	648.3	543.6	1,181.2	45.1
1986		60	40	7.9	5.6	10.9	247.2	14.80	586.1	497.8	1,086.1	43.8
1987	72	79	41	7.3	5.4	11.4	218.1	12.75	565.6	428.2	1,003.7	29.8
1988		85	42	7.0	5.3	11.6	222.8	13.10	664.2	404.0	1,029.5	33.5
1989		84	45	7.7	5.6	11.8	241.4	13.94	734.4	408.4	1,146.5	51.4
1990		80	45	7.2	5.5	11.4	249.7	14.83	649.5	495.2	1,159.1	58.1
1991		81	46	6.9	5.3	10.4	229.7	14.50	609.8	322.4	971.4	26.9
1992	72	78	44	7.9	5.9	12.4	276.2	14.15	622.0	588.3	1,196.5	53.0
1993		76	43	7.9	5.9	12.8	279.4	13.98	704.5	616.6	1,329.5	43.8
1994		86P	44P	6.9	5.2	11.8	256.8	14.13	706.4	602.1	1,290.1	33.8
1995		88P	44P	7.0P	5.4P	12.0P	267.7P	14.67P	659.5P	562.1P	1,204.5P	31.3P
1996		90P	44P	6.9P	5.3P	12.1P	271.3P	14.81P	664.2P	566.1P	1,213.1P	29.0P
1997		92P	44P	6.8P	5.3P	12.2P	274.9P	14.95P	668.9P	570.2P	1,221.7P	26.7P
1998		94P	44P	6.7P	5.2P	12.2P	278.5P	15.09P	673.6P	574.2P	1,230.2P	24.4P

Sources: 1982, 1987, 1992 *Economic Census*; *Annual Survey of Manufactures*, 83-86, 88-91, 93-94. Establishment counts for non-Census years are from *County Business Patterns*; establishment values for 83-84 are extrapolations. 'P's show projections by the editors. Industries reclassified in 87 will not have data for prior years.

INDICES OF CHANGE

Year	Com-panies	Establishments		Employment			Compensation		Production ($ million)			
		Total	with 20 or more employees	Total (000)	Production Workers (000)	Hours (Mil)	Payroll ($ mil)	Wages ($/hr)	Cost of Materials	Value Added by Manufacture	Value of Shipments	Capital Invest.
1982	82	82	105	100	97	85	73	86	92	79	91	189
1983		81	102	109	103	90	82	90	92	92	96	101
1984		79	100	115	112	102	94	94	118	88	103	86
1985		79	95	105	102	94	89	96	104	92	99	85
1986		77	91	100	95	88	90	105	94	85	91	83
1987	100	101	93	92	92	92	79	90	91	73	84	56
1988		109	95	89	90	94	81	93	107	69	86	63
1989		108	102	97	95	95	87	99	118	69	96	97
1990		103	102	91	93	92	90	105	104	84	97	110
1991		104	105	87	90	84	83	102	98	55	81	51
1992	100	100	100	100	100	100	100	100	100	100	100	100
1993		97	98	100	100	103	101	99	113	105	111	83
1994		110P	99P	87	88	95	93	100	114	102	108	64
1995		112P	99P	88P	91P	97P	97P	104P	106P	96P	101P	59P
1996		115P	99P	87P	90P	98P	98P	105P	107P	96P	101P	55P
1997		118P	99P	86P	89P	98P	100P	106P	108P	97P	102P	50P
1998		120P	99P	84P	88P	99P	101P	107P	108P	98P	103P	46P

Sources: Same as General Statistics. Values reflect change from the base year, 1992. Values above 100 mean greater than 92, values below 100 mean less than 92, and a value of 100 in the 82-91 or 93-98 period means same as 92. 'P's mark projections by the editors.

SELECTED RATIOS

For 1994	Avg. of All Manufact.	Analyzed Industry	Index	For 1994	Avg. of All Manufact.	Analyzed Industry	Index
Employees per Establishment	49	80	164	Value Added per Production Worker	134,084	115,788	86
Payroll per Establishment	1,500,273	2,995,017	200	Cost per Establishment	5,045,178	8,238,629	163
Payroll per Employee	30,620	37,217	122	Cost per Employee	102,970	102,377	99
Production Workers per Establishment	34	61	177	Cost per Production Worker	146,988	135,846	92
Wages per Establishment	853,319	1,944,592	228	Shipments per Establishment	9,576,895	15,046,227	157
Wages per Production Worker	24,861	32,064	129	Shipments per Employee	195,460	186,971	96
Hours per Production Worker	2,056	2,269	110	Shipments per Production Worker	279,017	248,096	89
Wages per Hour	12.09	14.13	117	Investment per Establishment	321,011	394,204	123
Value Added per Establishment	4,602,255	7,022,195	153	Investment per Employee	6,552	4,899	75
Value Added per Employee	93,930	87,261	93	Investment per Production Worker	9,352	6,500	70

Sources: Same as General Statistics. The 'Average of All Manufacturing' column represents the average of all manufacturing industries reported for the most recent complete year available. The Index shows the relationship between the Average and the Analyzed Industry. For example, 100 means that they are equal; 500 that the Analyzed Industry is five times the average; 50 means that the Analyzed Industry is half the national average. The abbreviation 'na' is used to show that data are 'not available'.

LEADING COMPANIES   Number shown: **21**   Total sales ($ mil): **809**   Total employment (000): **5.5**

Company Name	Address				CEO Name	Phone	Co. Type	Sales ($ mil)	Empl. (000)
Wyman-Gordon Co	PO Box 8001	Grafton	MA	01536	John M Nelson	508-756-5111	P	380	2.2
EST Co	PO Box 25	Grafton	WI	53024	G Kester	414-377-3270	D	100	0.6
Piper Impact Inc	PO Box 726-W	New Albany	MS	38652	Robert F Sammons	601-534-5046	R	100	1.0
Aluminum Precision Products	2621 S Susan St	Santa Ana	CA	92704	Phillip S Keeler	714-546-8125	R	36	0.3
Weber Metals Inc	16706 S Garfield	Paramount	CA	90723	Leon Kranz	310-636-1285	S	30	0.2
Accurate Forging Corp	201 Pine St	Bristol	CT	06010	Paul M Matyszyk	203-582-3169	S	20	0.1
Quality Aluminum Forge	430 W Collins Av	Orange	CA	92667	Gus Riehl	714-639-8191	D	20	0.2
Brass Forgings Co	PO Box 20129	Ferndale	MI	48220	Jeff Hillier	313-564-6831	R	15	0.1
Continental Forge Company Inc	PO Box 4789	Compton	CA	90224	Charles E Haueisen	213-774-3220	R	15	0.2
G and S Foundry Co	210 Kaskaskia St	Red Bud	IL	62278	Charles Wasem	618-282-4114	R	15*	<0.1
Consolidated Industries Inc	PO Box 280	Cheshire	CT	06410	JP Valentine	203-272-5371	R	12*	0.1
UNITED BRASS Manufacturers	PO Box 74095	Romulus	MI	48174	Richard Donahey	313-941-0700	R	11	0.1
Magnolia Metal Corp	PO Box 19110	Omaha	NE	68119	Adam M Koslosky	402-455-8760	R	10	<0.1
Wrayco Industries Inc	5010 Hudson Dr	Stow	OH	44224	Roberta Wray	216-688-5617	R	10*	<0.1
Independent Forge Company Inc	692 N Batavia St	Orange	CA	92668	RL Ruiz	714-997-7337	R	8	<0.1
D and E Industries Inc	PO Box 3089	Huntington	WV	25702	John J Klim Jr	304-736-3416	R	7	<0.1
Aerol Company Inc	3235 San Fernando	Los Angeles	CA	90065	Fred Siebert	213-254-2821	R	6	<0.1
Nonferrous Products Inc	PO Box 349	Franklin	IN	46131	D Brady	317-738-2558	R	5	<0.1
OBBCO Consolidated Industries	PO Box 7178	Odessa	TX	79760	John Holdridge	915-337-5341	R	5	0.1
Western Forge and Flange Co	PO Box 327	Santa Clara	CA	95054	Peter Zaklan	408-727-7000	R	3	<0.1
Premco Forge Inc	5200 Tweedy Blv	South Gate	CA	90280	Robert Pierson	213-564-7921	R	1	<0.1

Source: Ward's Business Directory of U.S. Private and Public Companies, Volumes 1 and 2, 1996. The company type code used is as follows: P - Public, R - Private, S - Subsidiary, D - Division, J - Joint Venture, A - Affiliate, G - Group. Sales are in millions of dollars, employees are in thousands. An asterisk (*) indicates an estimated sales volume. The symbol < stands for 'less than'. Company names and addresses are truncated, in some cases, to fit into the available space.

MATERIALS CONSUMED

Material	Quantity		Delivered Cost ($ million)
Materials, ingredients, containers, and supplies		(X)	474.0
Forgings	1,000 s tons	4.2	51.2
Steel ingot and semifinished shapes (blooms, billets, and slabs)	1,000 s tons	3.9	11.1
Steel bars, bar shapes, and other shapes and forms	1,000 s tons	(S)	1.6
Titanium and titanium-base alloy shapes and forms	1,000 s tons	229.6	57.2
Copper and copper-base alloy shapes and forms	1,000 s tons	(D)	(D)
Aluminum and aluminum-base alloy shapes and forms	1,000 s tons	118.3	171.9
Nickel and nickel-base alloy, including nickel-copper alloys	1,000 s tons	(D)	(D)
All other nonferrous shapes and forms	1,000 s tons	1.5	7.7
Forging dies	1,000 s tons	(S)	11.7
All other materials and components, parts, containers, and supplies		(X)	57.3
Materials, ingredients, containers, and supplies, nsk		(X)	34.1

Source: 1992 *Economic Census*. Explanation of symbols used: (D): Withheld to avoid disclosure of competitive data; na: Not available; (S): Withheld because statistical norms were not met; (X): Not applicable; (Z): Less than half the unit shown; nec: Not elsewhere classified; nsk: Not specified by kind; - : zero; * : 10-19 percent estimated; ** : 20-29 percent estimated.

PRODUCT SHARE DETAILS

Product or Product Class	% Share	Product or Product Class	% Share
Nonferrous forgings	100.00	nonferrous forgings	12.39
Hot impression die impact, press, and upset nonferrous forgings	80.90	Hot impression die impact, press, and upset nonferrous forgings, nsk	0.92
Hot impression die impact, press, and upset aluminum and aluminum alloy forgings	57.73	Other nonferrous forgings	11.04
Hot impression die impact, press, and upset titanium and titanium alloy forgings	19.99	Cold impression die impact, press, and upset nonferrous forgings	60.61
Hot impression die impact, press, and upset copper and copper-base alloy forgings	8.96	Seamless rolled ring nonferrous forgings	22.97
Other hot impression die impact, press, and upset		Open die or smith nonferrous forgings, hammer or press	14.90
		Other nonferrous forgings, nsk	1.52
		Nonferrous forgings, nsk	8.07

Source: 1992 *Economic Census*. The values shown are percent of total shipments in an industry. Values of indented subcategories are summed in the main heading. The symbol (D) appears when data are withheld to prevent disclosure of competitive information. The abbreviation nsk stands for 'not specified by kind' and nec for 'not elsewhere classified'.

INPUTS AND OUTPUTS FOR NONFERROUS FORGINGS

Economic Sector or Industry Providing Inputs	%	Sector	Economic Sector or Industry Buying Outputs	%	Sector
Nonferrous rolling & drawing, nec	23.6	Manufg.	Aircraft & missile engines & engine parts	33.9	Manufg.
Primary aluminum	15.4	Manufg.	Aircraft	10.5	Manufg.
Blast furnaces & steel mills	7.5	Manufg.	Aircraft & missile equipment, nec	8.2	Manufg.
Wholesale trade	5.8	Trade	Fabricated plate work (boiler shops)	5.7	Manufg.
Motor freight transportation & warehousing	5.3	Util.	Internal combustion engines, nec	5.3	Manufg.
Miscellaneous plastics products	4.8	Manufg.	Metal stampings, nec	4.6	Manufg.
Metal coating & allied services	4.6	Manufg.	Railroad equipment	4.3	Manufg.
Copper rolling & drawing	4.2	Manufg.	Automotive stampings	4.0	Manufg.
Aluminum rolling & drawing	3.6	Manufg.	Pumps & compressors	3.8	Manufg.
Gas production & distribution (utilities)	3.1	Util.	Tanks & tank components	3.8	Manufg.
Electric services (utilities)	3.0	Util.	Radio & TV communication equipment	2.5	Manufg.
Nonferrous forgings	2.9	Manufg.	Exports	2.2	Foreign
Advertising	2.8	Services	Ammunition, except for small arms, nec	1.8	Manufg.
Iron & steel forgings	1.3	Manufg.	Calculating & accounting machines	1.7	Manufg.
Cyclic crudes and organics	1.2	Manufg.	Nonferrous forgings	1.6	Manufg.
Fabricated metal products, nec	0.8	Manufg.	Turbines & turbine generator sets	1.5	Manufg.
Petroleum refining	0.8	Manufg.	Motor vehicle parts & accessories	1.4	Manufg.
Eating & drinking places	0.7	Trade	Hand & edge tools, nec	1.3	Manufg.
Sanitary services, steam supply, irrigation	0.6	Util.	Motor vehicles & car bodies	0.7	Manufg.
Maintenance of nonfarm buildings nec	0.5	Constr.	Boat building & repairing	0.4	Manufg.
Paperboard containers & boxes	0.5	Manufg.	Guided missiles & space vehicles	0.3	Manufg.
Communications, except radio & TV	0.5	Util.	Fabricated metal products, nec	0.1	Manufg.
Banking	0.5	Fin/R.E.			
Special dies & tools & machine tool accessories	0.4	Manufg.			
Railroads & related services	0.4	Util.			
Computer & data processing services	0.4	Services			
State & local government enterprises, nec	0.4	Gov't			
Mechanical measuring devices	0.3	Manufg.			
Metal heat treating	0.3	Manufg.			
Real estate	0.3	Fin/R.E.			
Detective & protective services	0.3	Services			
Equipment rental & leasing services	0.3	Services			
Machinery, except electrical, nec	0.2	Manufg.			
Air transportation	0.2	Util.			
Royalties	0.2	Fin/R.E.			
Engineering, architectural, & surveying services	0.2	Services			
U.S. Postal Service	0.2	Gov't			
Coal	0.1	Mining			
Chemical preparations, nec	0.1	Manufg.			
Industrial inorganic chemicals, nec	0.1	Manufg.			
Machine tools, metal cutting types	0.1	Manufg.			
Credit agencies other than banks	0.1	Fin/R.E.			
Insurance carriers	0.1	Fin/R.E.			
Automotive rental & leasing, without drivers	0.1	Services			
Business/professional associations	0.1	Services			

Source: Benchmark Input-Output Accounts for the U.S. Economy, 1982, U.S. Department of Commerce, Washington, D.C., July 1991. Data, as reported in the source, are organized by the 1977 SIC structure in use in 1982 but have been matched, as closely as is possible, to the 1987 SIC structure used in this book.

OCCUPATIONS EMPLOYED BY SIC 346 - METAL FORGINGS AND STAMPINGS

Occupation	% of Total 1994	Change to 2005	Occupation	% of Total 1994	Change to 2005
Machine forming operators, metal & plastic	9.3	-65.9	Machine feeders & offbearers	1.8	-23.3
Tool & die makers	7.4	-11.4	Sheet metal workers & duct installers	1.7	-14.7
Machine tool cutting & forming etc. nec	6.4	10.9	Metal & plastic machine workers nec	1.5	-24.4
Assemblers, fabricators, & hand workers nec	4.7	-14.7	Helpers, laborers, & material movers nec	1.4	-14.7
Blue collar worker supervisors	4.6	-17.3	Industrial production managers	1.4	-14.7
Welding machine setters, operators	3.8	-23.2	Millwrights	1.3	2.3
Punching machine operators, metal & plastic	3.4	15.1	Traffic, shipping, & receiving clerks	1.3	-17.9
Inspectors, testers, & graders, precision	3.3	-14.7	Welders & cutters	1.3	-14.7
Industrial machinery mechanics	3.1	-6.2	Freight, stock, & material movers, hand	1.3	-31.8
Industrial truck & tractor operators	2.8	-14.7	Electricians	1.2	-20.0
Machine tool cutting operators, metal & plastic	2.6	-12.6	Janitors & cleaners, incl maids	1.1	-31.8
Machinists	2.4	-14.7	Sales & related workers nec	1.1	-14.7
Combination machine tool operators	2.1	27.9	Maintenance repairers, general utility	1.1	-23.3
General managers & top executives	1.8	-19.1			

Source: Industry-Occupation Matrix, Bureau of Labor Statistics. These data relate to one or more 3-digit SIC industry groups rather than to a single 4-digit SIC. The change reported for each occupation to the year 2005 is a percent of growth or decline as estimated by the Bureau of Labor Statistics. The abbreviation nec stands for 'not elsewhere classified'.

LOCATION BY STATE AND REGIONAL CONCENTRATION

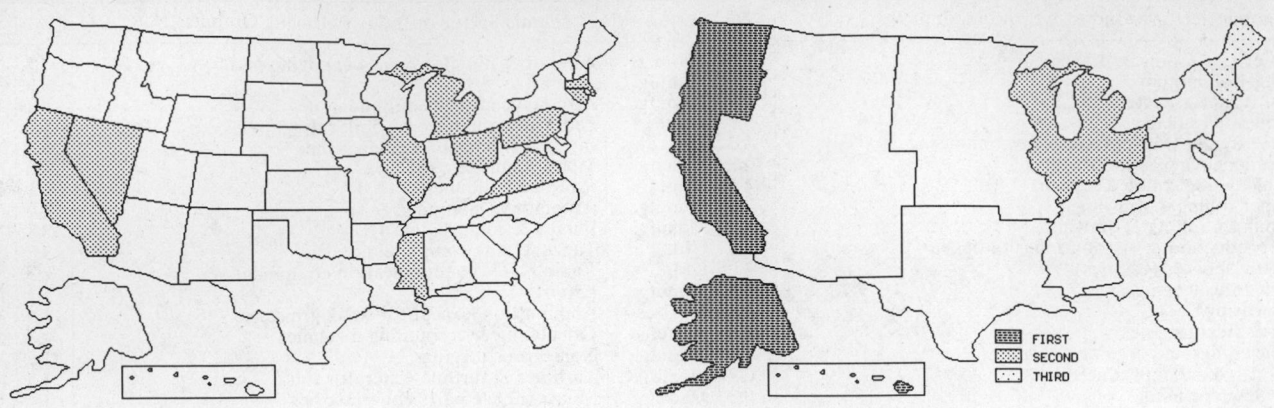

INDUSTRY DATA BY STATE

| State | Establish-ments | Shipments | | | Employment | | | | Cost as % of Shipments | Investment per Employee ($) |
		Total ($ mil)	% of U.S.	Per Establ.	Total Number	% of U.S.	Per Establ.	Wages ($/hour)		
California	20	232.3	19.4	11.6	2,200	27.8	110	13.72	53.0	2,682
Illinois	5	106.9	8.9	21.4	700	8.9	140	15.00	39.9	3,143
Connecticut	4	31.3	2.6	7.8	200	2.5	50	13.25	39.0	-
Michigan	6	(D)	-	-	375 *	4.7	63	-	-	-
Ohio	6	(D)	-	-	1,750 *	22.2	292	-	-	11,143
Pennsylvania	6	(D)	-	-	750 *	9.5	125	-	-	2,000
Massachusetts	3	(D)	-	-	750 *	9.5	250	-	-	-
Nevada	2	(D)	-	-	175 *	2.2	88	-	-	-
Virginia	2	(D)	-	-	175 *	2.2	88	-	-	-
Mississippi	1	(D)	-	-	750 *	9.5	750	-	-	-

Source: 1992 *Economic Census.* The states are in descending order of shipments or establishments (if shipment data are missing for the majority). The symbol (D) appears when data are withheld to prevent disclosure of competitive information. States marked with (D) are sorted by number of establishments. A dash (-) indicates that the data element cannot be calculated; * indicates the midpoint of a range.

3465 - AUTOMOTIVE STAMPINGS

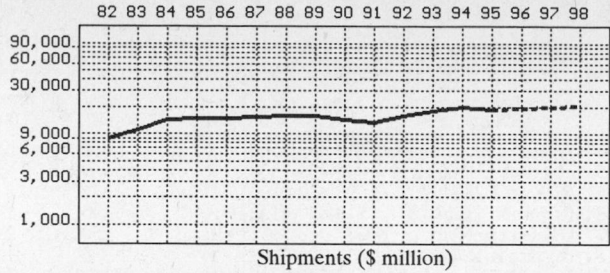

Shipments ($ million)

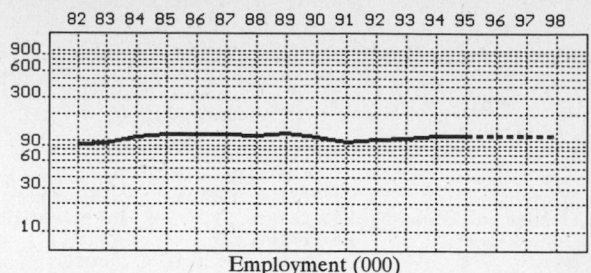

Employment (000)

GENERAL STATISTICS

Year	Com-panies	Establishments		Employment			Compensation		Production ($ million)			
		Total	with 20 or more employees	Total (000)	Production Workers (000)	Hours (Mil)	Payroll ($ mil)	Wages ($/hr)	Cost of Materials	Value Added by Manufacture	Value of Shipments	Capital Invest.
1982	566	668	434	90.5	74.5	145.2	2,292.6	12.34	4,621.6	4,114.4	8,777.4	465.0
1983		650	442	96.6	81.1	172.2	2,729.6	12.63	5,846.8	5,409.5	11,176.4	234.9
1984		632	450	112.4	95.2	203.8	3,368.9	13.33	7,594.8	6,631.8	14,137.4	462.3
1985		613	458	119.2	100.6	217.4	3,795.5	14.07	8,183.8	6,850.5	15,037.7	830.6
1986		607	454	119.1	99.9	216.5	3,942.7	14.61	8,310.9	6,521.2	14,826.5	1,183.0
1987	597	713	502	119.8	99.9	213.3	3,976.9	14.80	8,582.2	6,659.3	15,251.6	1,285.0
1988		707	509	117.1	98.3	211.9	4,164.0	15.78	8,863.7	7,064.5	15,874.5	947.2
1989		688	499	122.3	95.9	200.6	4,082.5	16.20	8,902.5	6,713.8	15,625.7	827.9
1990		670	489	111.5	91.3	189.0	3,887.6	16.34	8,282.8	6,300.2	14,544.5	912.4
1991		644	450	99.3	82.1	169.1	3,680.6	17.23	7,700.0	5,662.7	13,403.6	722.0
1992	585	700	506	105.2	87.1	184.0	4,098.9	17.76	8,598.1	7,241.8	15,821.4	519.9
1993		693	483	108.7	91.8	199.7	4,410.2	17.82	9,689.4	8,126.8	17,780.8	842.3
1994		696P	505P	114.0	95.4	212.5	4,882.2	18.42	10,641.5	9,074.9	19,652.8	1,201.1
1995		700P	510P	114.3P	94.5P	204.6P	4,813.0P	19.12P	10,074.3P	8,591.2P	18,605.3P	1,047.2P
1996		705P	515P	114.8P	94.8P	206.0P	4,958.6P	19.64P	10,371.6P	8,844.7P	19,154.3P	1,082.1P
1997		710P	520P	115.4P	95.2P	207.4P	5,104.3P	20.16P	10,668.9P	9,098.2P	19,703.3P	1,117.1P
1998		714P	525P	115.9P	95.6P	208.7P	5,250.0P	20.68P	10,966.1P	9,351.7P	20,252.3P	1,152.0P

Sources: 1982, 1987, 1992 *Economic Census*; *Annual Survey of Manufactures*, 83-86, 88-91, 93-94. Establishment counts for non-Census years are from *County Business Patterns*; establishment values for 83-84 are extrapolations. 'P's show projections by the editors. Industries reclassified in 87 will not have data for prior years.

INDICES OF CHANGE

Year	Com-panies	Establishments		Employment			Compensation		Production ($ million)			
		Total	with 20 or more employees	Total (000)	Production Workers (000)	Hours (Mil)	Payroll ($ mil)	Wages ($/hr)	Cost of Materials	Value Added by Manufacture	Value of Shipments	Capital Invest.
1982	97	95	86	86	86	79	56	69	54	57	55	89
1983		93	87	92	93	94	67	71	68	75	71	45
1984		90	89	107	109	111	82	75	88	92	89	89
1985		88	91	113	115	118	93	79	95	95	95	160
1986		87	90	113	115	118	96	82	97	90	94	228
1987	102	102	99	114	115	116	97	83	100	92	96	247
1988		101	101	111	113	115	102	89	103	98	100	182
1989		98	99	116	110	109	100	91	104	93	99	159
1990		96	97	106	105	103	95	92	96	87	92	175
1991		92	89	94	94	92	90	97	90	78	85	139
1992	100	100	100	100	100	100	100	100	100	100	100	100
1993		99	95	103	105	109	108	100	113	112	112	162
1994		99P	100P	108	110	115	119	104	124	125	124	231
1995		100P	101P	109P	108P	111P	117P	108P	117P	119P	118P	201P
1996		101P	102P	109P	109P	112P	121P	111P	121P	122P	121P	208P
1997		101P	103P	110P	109P	113P	125P	114P	124P	126P	125P	215P
1998		102P	104P	110P	110P	113P	128P	116P	128P	129P	128P	222P

Sources: Same as General Statistics. Values reflect change from the base year, 1992. Values above 100 mean greater than 92, values below 100 mean less than 92, and a value of 100 in the 82-91 or 93-98 period means same as 92. 'P's mark projections by the editors.

SELECTED RATIOS

For 1994	Avg. of All Manufact.	Analyzed Industry	Index	For 1994	Avg. of All Manufact.	Analyzed Industry	Index
Employees per Establishment	49	164	334	Value Added per Production Worker	134,084	95,125	71
Payroll per Establishment	1,500,273	7,018,016	468	Cost per Establishment	5,045,178	15,296,838	303
Payroll per Employee	30,620	42,826	140	Cost per Employee	102,970	93,346	91
Production Workers per Establishment	34	137	400	Cost per Production Worker	146,988	111,546	76
Wages per Establishment	853,319	5,626,617	659	Shipments per Establishment	9,576,895	28,250,311	295
Wages per Production Worker	24,861	41,030	165	Shipments per Employee	195,460	172,393	88
Hours per Production Worker	2,056	2,227	108	Shipments per Production Worker	279,017	206,004	74
Wages per Hour	12.09	18.42	152	Investment per Establishment	321,011	1,726,545	538
Value Added per Establishment	4,602,255	13,044,897	283	Investment per Employee	6,552	10,536	161
Value Added per Employee	93,930	79,604	85	Investment per Production Worker	9,352	12,590	135

Sources: Same as General Statistics. The 'Average of All Manufacturing' column represents the average of all manufacturing industries reported for the most recent complete year available. The Index shows the relationship between the Average and the Analyzed Industry. For example, 100 means that they are equal; 500 that the Analyzed Industry is five times the average; 50 means that the Analyzed Industry is half the national average. The abbreviation 'na' is used to show that data are 'not available'.

LEADING COMPANIES   Number shown: **75**   Total sales ($ mil): **6,329**   Total employment (000): **50.1**

Company Name	Address				CEO Name	Phone	Co. Type	Sales ($ mil)	Empl. (000)
Budd Co	PO Box 2601	Troy	MI	48007	Siegfried Buschmann	313-643-3500	S	1,080	10.0
Douglas and Lomason Co	24600 Hallwood Ct	Farmington Hls	MI	48335	Harry A Lomason II	810-478-7800	P	567	5.7
Varlen Corp	PO Box 3089	Naperville	IL	60566	Richard L Wellek	708-420-0400	P	342	2.5
Masco Technologies Stamping	PO Box 5011	Rochester Hills	MI	48308	Raymond Hunt	810-650-4100	S	290	1.6
Lobdell-Emery Corp	PO Box 129	Alma	MI	48801	DK Fesenmeyer	517-463-3151	R	250	2.0
Budd Co	12141 Charlevoix	Detroit	MI	48215	Robert R Kegerreis	313-823-9100	D	230	1.2
Tower Automotive Inc	4508 IDS Ctr	Minneapolis	MN	55402	Dugald K Campbell	612-342-2310	P	222	1.6
Randall Textron	750ephenson Hwy	Troy	MI	48080	Jane Warner	810-616-5100	S	210*	2.0
Benteler Industries Inc	50 Monroe Av NW	Grand Rapids	MI	49503	Lawrence A Abbott	616-247-3936	S	200	1.7
Lacks Enterprises Inc	5460 Cascade SE	Grand Rapids	MI	49546	Richard Lacks Sr	616-949-6570	R	200	1.5
Aetna Industries Inc	PO Box 3067	Center Line	MI	48015	Ueli Spring	313-759-2200	R	195	1.4
Checker Motors Corp	2016 N Pitcher St	Kalamazoo	MI	49007	David R Markin	616-343-6121	S	160	1.2
Yamakawa Mfg Corporation	PO Box 799	Portland	TN	37148	Toshihisa Ota	615-325-7311	S	125	0.5
Active Tool and Manufacturing	PO Box 722	Roseville	MI	48066	JA Blake	313-294-9220	R	100	1.4
EWI Inc	601 W Broadway	South Bend	IN	46601	George Hofmeister	219-287-7253	R	100	0.6
Trim Trends Inc	30665 Northwestern	Farmington Hls	MI	49334	David Stegemoller	313-626-4300	S	97	0.9
MPI International Inc	PO Box 1995	Rochester	MI	48308	Karl A Pfister	313-853-9010	S	90	0.7
RJ Tower Corp	PO Box 670	Greenville	MI	48838	DK Campbell	616-754-2211	S	90	0.6
Means Industries Inc	1860 S Jefferson St	Saginaw	MI	48601	William D Shaw	517-754-1433	S	80	0.4
Davis Industries Inc	14500 Sheldon Rd	Plymouth	MI	48170	Richard L Davis II	313-835-6000	R	75	0.5
L and W Engineering Company	6201 Haggerty Rd	Belleville	MI	48111	Wayne Jones	313-397-2212	R	75	0.4
Grand Haven Stamped Prod Co	1250 S Beechtree St	Grand Haven	MI	49417	Frank Nagy	616-842-5500	D	74	0.4
Production Stamping Inc	28175 Rosso	New Baltimore	MI	48047	William H John	313-949-1114	R	70	0.4
Franklin Aluminum Company	266 Bevis Rd	Franklin	GA	30217	James W Cranford	706-675-3341	R	65	0.4
Howell Industries Inc	17515 W 9 Mile Rd	Southfield	MI	48075	Morton Schiff	313-424-8220	P	63	0.4
Bliss Manufacturing Inc	PO Box 4266	Youngstown	OH	44515	William Adams	216-793-9801	S	60	0.4
Iroquois Die & Mfg Co	24400 Hoover Rd	Warren	MI	48089	Leo Deutschmann	313-756-6920	R	60	0.2
Ohi Automotive	1030 Hoover Blv	Frankfort	KY	40601	Tom Lingemon	502-695-4000	R	60	0.4
Steel Parts Corp	110 Berryman Pike	Tipton	IN	46072	BW Clark	317-675-2191	S	55	0.3
Davis Tool and Engineering Inc	19250 Plymouth Rd	Detroit	MI	48228	Lorne Greenwood	313-835-6000	S	53	0.3
Florida Production Engineering	225 Fentress Blv	Daytona Beach	FL	32114	Ernie Green	904-255-2566	R	52	0.4
Center Manufacturing Inc	PO Box 337	Byron Center	MI	49315	Roland E Johnson	616-878-3324	R	45	0.4
Dee Zee Inc	PO Box 3090	Des Moines	IA	50316	R Shivers	515-265-7331	S	45	0.4
Starboard Industries Inc	PO Box 217001	Auburn Hills	MI	48321	Ken Koroly	810-370-0020	R	43	0.4
Toledo Technologies	PO Box 596	Toledo	OH	43693	James R Terlizzi	419-661-1333	D	40	0.2
Modern Prototype Co	PO Box 4540	Troy	MI	48099	Dennis Cedar	313-589-7000	D	38	0.5
J-B Tool and Machine Inc	PO Box 387	Wapakoneta	OH	45895	Neil R Yantis	419-738-2177	R	37	0.3
AGS Metal System Inc	6640 Sterling Dr S	Sterling Hts	MI	48312	Ben Vergillio	810-939-3000	S	35*	0.2
Milford Fabricating Co	19200 Glendale St	Detroit	MI	48223	John Edwards	313-272-8400	S	34	0.5
Campbell Plastics	2900 Campbell Av	Schenectady	NY	12306	J Kaczynski	518-393-2167	D	28	0.3
BAE Industries Inc	24400 Sherwood	Center Line	MI	48015	EC Jones	313-754-3000	R	27	0.3
Wellington Industries Inc	39555 I-94 S Svc Dr	Belleville	MI	48111	Ron O'Dell	313-942-1060	R	26	0.2
ACEMCO INC	PO Box 857	Grand Haven	MI	49417	Jeffry Giangrande	616-842-5680	R	25	0.2
Active Products Corp	PO Box 479	Marion	IN	46952	Rick Miller	317-664-9084	S	25*	0.4
Manufacturers Products Co	26020 Sherwood	Warren	MI	48091	W C Fredericks Jr	313-961-4034	R	25	0.2
Paramount Fabricating Inc	13595 Helen St	Detroit	MI	48212	Vincent J Wood Sr	313-365-6600	R	25	0.2
Alofs Manufacturing Company	345 32nd SW	Grand Rapids	MI	49548	Thomas E Gleason	616-241-6405	R	22	0.3
Multi-Plex Inc	6505 N State Road 9	Howe	IN	46746	Merle Emery	219-562-2911	D	22*	0.1
Equity Resource Group	300 Madison Av	Toledo	OH	43604	Dan W Foy	419-241-1950	R	21*	0.2
Nelson Name Plate Company Inc	3191 Casitas Av	Los Angeles	CA	90039	David Lazier	213-663-3971	R	21*	0.2
Flint Manufacturing Co	PO Box 1630	Flint	MI	48529	James Lantz	313-742-8033	R	20	<0.1
Quality Metalcraft Inc	33355 Glendale St	Livonia	MI	48150	Alexander Chetcuti	313-261-6700	R	20	0.2
Wolverine Metal Specialties	PO Box 744	Jackson	MI	49204	Mark R Dailey	517-750-3414	S	20	<0.1
Decouper Industries Inc	21535 Hoover Rd	Warren	MI	48089	Thomas Robson	313-755-4488	R	18	0.2
Timco Manufacturing Company	27544 Groesbeck	Roseville	MI	48066	Carl Gilgallon	313-776-6720	R	18	0.1
Pullman Industries Inc	1228 Kirts Blv	Troy	MI	48084	E H Matthewson	810-244-8668	R	17	0.2
Thomas Die and Stamping Inc	2170 E Walton Blv	Auburn Hills	MI	48326	Thomas Robson	313-373-0388	D	17*	<0.1
Wauconda Tool and Engineering	821 W Algonquin	Algonquin	IL	60102	Robert Massi	708-658-4588	S	17	0.1
Fulton Industries Inc	PO Box 377	Wauseon	OH	43567	Richard G Volk	419-335-3015	R	15	0.2
LeMay Machine Company Inc	4725 Green Park Rd	St Louis	MO	63123	AA Suellentrop	314-892-8080	R	15	0.2
Manter Technologies	PO Box 308	Marine City	MI	48039	Willis C Manter	810-765-8000	R	15	0.1
Reichert Stamping Co	PO Box 351510	Toledo	OH	43635	Robert Reichert	419-841-8511	R	15*	0.1
Cleveland Metal Products Co	2019 Center St	Cleveland	OH	44113	Leon R Eiswerth	216-771-3888	S	13	0.1
Hofley Manufacturing Co	15500 12 Mile Rd	Roseville	MI	48066	N Hofley	810-778-5444	R	13	0.2
Lake Odessa Group	2820 29th St SE	Kentwood	MI	49512	Lawson K Smith	616-285-6850	S	13*	0.1
Midway Products Group Inc	PO Box 737	Monroe	MI	48161	George Hess	313-241-7242	R	13	0.1
Simpson Industries	131 W Harvest	Bluffton	IN	46714	Robert A Flynn	219-824-2360	D	13	0.1
Dickey-Grabler Co	10304 Madison Av	Cleveland	OH	44102	W J Primrose Jr	216-961-4172	R	12*	0.1
Snover Stamping Co	3279 W Snover St	Snover	MI	48472	Henry Sobell	313-672-9286	S	11*	0.1
American Metalcraft Co	218 Mechanic St	Waterville	OH	43566	Thomas Johnson	419-878-4015	S	10	<0.1
Banner Die Tool & Stamping Co	1308 Holly Av	Columbus	OH	43212	John E O'Brien III	614-291-3105	R	10*	<0.1
Central Manufacturing Inc	PO Box 508	Parker City	IN	47368	Vincent F Cimino	317-468-7133	R	10	<0.1
Clover Tool & Mfg Co	PO Box 407	Mount Clemens	MI	48046	Gerald Diez	810-468-0819	R	10*	0.1
Hy-Form Products Inc	35588 Veronica	Livonia	MI	48150	Stephen E Nash	313-464-3811	S	10	0.1
Blom Industries Inc	PO Box 486	Mount Clemens	MI	48046	Jack Blom	810-468-5600	R	9	<0.1

Source: Ward's Business Directory of U.S. Private and Public Companies, Volumes 1 and 2, 1996. The company type code used is as follows: P - Public, R - Private, S - Subsidiary, D - Division, J - Joint Venture, A - Affiliate, G - Group. Sales are in millions of dollars, employees are in thousands. An asterisk (*) indicates an estimated sales volume. The symbol < stands for 'less than'. Company names and addresses are truncated, in some cases, to fit into the available space.

MATERIALS CONSUMED

Material	Quantity	Delivered Cost ($ million)
Materials, ingredients, containers, and supplies	(X)	7,813.2
Metal bolts, nuts, screws, washers, rivets, and other screw machine products	(X)	210.7
Other fabricated metal products (except castings and forgings)	(X)	845.3
Iron and steel castings (rough and semifinished)	(X)	31.8
Nonferrous (aluminum, copper, etc.) castings (rough and semifinished)	(X)	10.0
Steel bars and bar shapes	(X)	67.7
Steel sheet and strip, including tin plate	1,000 s tons 6,831.4*	4,313.8
Steel plate	(X)	8.4
Steel wire and wire products	(X)	32.4
Steel tinplate, tin free steel, terneplate, and blackplate	1,000 s tons (S)	11.2
All other steel shapes and forms	(X)	326.2
Copper and copper-base alloy shapes and forms	(X)	25.5
Aluminum and aluminum-base alloy sheet, plate, foil, and welded tubing	(X)	85.2
Aluminum and aluminum-base alloy extruded shapes, including extruded rod, bar, pipe, tube, etc.	(X)	4.0
Other aluminum and aluminum-base alloy shapes and forms	(X)	13.4
Other nonferrous shapes and forms	(X)	16.5
Plastics resins consumed in the form of granules, pellets, powders, liquids, etc.	(X)	31.8
Paints, varnishes, lacquers, stains, shellacs, japans, enamels, and allied products	(X)	16.3
All other chemicals and allied products	(X)	20.1
Plastics products consumed in the form of sheets, rods, tubes, film, and other shapes	(X)	32.2
Paperboard containers, boxes, and corrugated paperboard	(X)	46.5
Other paper and paperboard products	(X)	9.4
Special dies, tools, die sets, jigs, and fixtures, except cutting tools for machine tools	(X)	287.6
All other materials and components, parts, containers, and supplies	(X)	980.0
Materials, ingredients, containers, and supplies, nsk	(X)	387.2

Source: 1992 *Economic Census*. Explanation of symbols used: (D): Withheld to avoid disclosure of competitive data; na: Not available; (S): Withheld because statistical norms were not met; (X): Not applicable; (Z): Less than half the unit shown; nec: Not elsewhere classified; nsk: Not specified by kind; - : zero; * : 10-19 percent estimated; ** : 20-29 percent estimated.

PRODUCT SHARE DETAILS

Product or Product Class	% Share	Product or Product Class	% Share
Automotive stampings	100.00		

Source: 1992 *Economic Census*. The values shown are percent of total shipments in an industry. Values of indented subcategories are summed in the main heading. The symbol (D) appears when data are withheld to prevent disclosure of competitive information. The abbreviation nsk stands for 'not specified by kind' and nec for 'not elsewhere classified'.

INPUTS AND OUTPUTS FOR AUTOMOTIVE STAMPINGS

Economic Sector or Industry Providing Inputs	%	Sector	Economic Sector or Industry Buying Outputs	%	Sector
Blast furnaces & steel mills	43.2	Manufg.	Motor vehicles & car bodies	55.1	Manufg.
Machinery, except electrical, nec	13.1	Manufg.	Automotive repair shops & services	27.6	Services
Wholesale trade	8.8	Trade	Exports	12.3	Foreign
Accounting, auditing & bookkeeping	7.4	Services	Motor vehicle parts & accessories	2.6	Manufg.
Imports	2.6	Foreign	S/L Govt. purch., elem. & secondary education	1.1	S/L Govt
Electric services (utilities)	2.3	Util.	Automotive stampings	0.4	Manufg.
Aluminum rolling & drawing	1.9	Manufg.	Truck & bus bodies	0.4	Manufg.
Motor freight transportation & warehousing	1.6	Util.	S/L Govt. purch., higher education	0.1	S/L Govt
Automotive & apparel trimmings	1.3	Manufg.			
Special dies & tools & machine tool accessories	1.3	Manufg.			
Banking	1.1	Fin/R.E.			
Screw machine and related products	1.0	Manufg.			
Nonferrous forgings	0.8	Manufg.			
Railroads & related services	0.8	Util.			
Automotive stampings	0.6	Manufg.			
Cyclic crudes and organics	0.6	Manufg.			
Plating & polishing	0.6	Manufg.			
Gas production & distribution (utilities)	0.6	Util.			
Maintenance of nonfarm buildings nec	0.5	Constr.			
Miscellaneous plastics products	0.5	Manufg.			
Fabricated metal products, nec	0.4	Manufg.			
Metal heat treating	0.4	Manufg.			
Paperboard containers & boxes	0.4	Manufg.			
Metal coating & allied services	0.3	Manufg.			
Paints & allied products	0.3	Manufg.			
Petroleum refining	0.3	Manufg.			
Primary zinc	0.3	Manufg.			
Communications, except radio & TV	0.3	Util.			
Eating & drinking places	0.3	Trade			
Advertising	0.3	Services			
Equipment rental & leasing services	0.3	Services			
Adhesives & sealants	0.2	Manufg.			

Continued on next page.

INPUTS AND OUTPUTS FOR AUTOMOTIVE STAMPINGS - Continued

Economic Sector or Industry Providing Inputs	%	Sector	Economic Sector or Industry Buying Outputs	%	Sector
Ball & roller bearings	0.2	Manufg.			
Copper rolling & drawing	0.2	Manufg.			
Fabricated rubber products, nec	0.2	Manufg.			
Hardware, nec	0.2	Manufg.			
Industrial controls	0.2	Manufg.			
Plastics materials & resins	0.2	Manufg.			
Automotive repair shops & services	0.2	Services			
Computer & data processing services	0.2	Services			
Detective & protective services	0.2	Services			
Engineering, architectural, & surveying services	0.2	Services			
U.S. Postal Service	0.2	Gov't			
Abrasive products	0.1	Manufg.			
Chemical preparations, nec	0.1	Manufg.			
Iron & steel foundries	0.1	Manufg.			
Machine tools, metal cutting types	0.1	Manufg.			
Miscellaneous fabricated wire products	0.1	Manufg.			
Nonferrous castings, nec	0.1	Manufg.			
Air transportation	0.1	Util.			
Sanitary services, steam supply, irrigation	0.1	Util.			
Water transportation	0.1	Util.			
Insurance carriers	0.1	Fin/R.E.			
Real estate	0.1	Fin/R.E.			
Royalties	0.1	Fin/R.E.			
Business/professional associations	0.1	Services			
Management & consulting services & labs	0.1	Services			

Source: Benchmark Input-Output Accounts for the U.S. Economy, 1982, U.S. Department of Commerce, Washington, D.C., July 1991. Data, as reported in the source, are organized by the 1977 SIC structure in use in 1982 but have been matched, as closely as is possible, to the 1987 SIC structure used in this book.

OCCUPATIONS EMPLOYED BY SIC 346 - METAL FORGINGS AND STAMPINGS

Occupation	% of Total 1994	Change to 2005	Occupation	% of Total 1994	Change to 2005
Machine forming operators, metal & plastic	9.3	-65.9	Machine feeders & offbearers	1.8	-23.3
Tool & die makers	7.4	-11.4	Sheet metal workers & duct installers	1.7	-14.7
Machine tool cutting & forming etc. nec	6.4	10.9	Metal & plastic machine workers nec	1.5	-24.4
Assemblers, fabricators, & hand workers nec	4.7	-14.7	Helpers, laborers, & material movers nec	1.4	-14.7
Blue collar worker supervisors	4.6	-17.3	Industrial production managers	1.4	-14.7
Welding machine setters, operators	3.8	-23.2	Millwrights	1.3	2.3
Punching machine operators, metal & plastic	3.4	15.1	Traffic, shipping, & receiving clerks	1.3	-17.9
Inspectors, testers, & graders, precision	3.3	-14.7	Welders & cutters	1.3	-14.7
Industrial machinery mechanics	3.1	-6.2	Freight, stock, & material movers, hand	1.3	-31.8
Industrial truck & tractor operators	2.8	-14.7	Electricians	1.2	-20.0
Machine tool cutting operators, metal & plastic	2.6	-12.6	Janitors & cleaners, incl maids	1.1	-31.8
Machinists	2.4	-14.7	Sales & related workers nec	1.1	-14.7
Combination machine tool operators	2.1	27.9	Maintenance repairers, general utility	1.1	-23.3
General managers & top executives	1.8	-19.1			

Source: Industry-Occupation Matrix, Bureau of Labor Statistics. These data relate to one or more 3-digit SIC industry groups rather than to a single 4-digit SIC. The change reported for each occupation to the year 2005 is a percent of growth or decline as estimated by the Bureau of Labor Statistics. The abbreviation nec stands for 'not elsewhere classified'.

LOCATION BY STATE AND REGIONAL CONCENTRATION

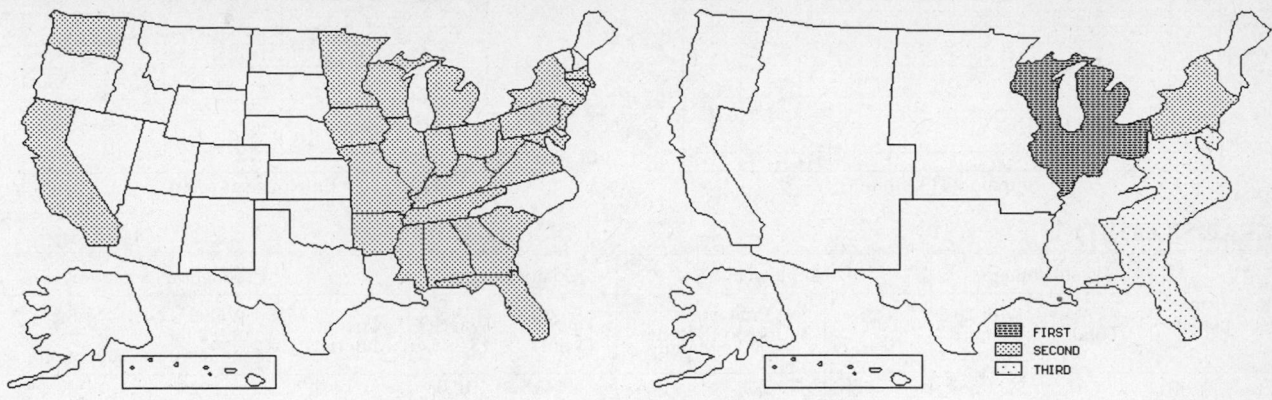

FIRST
SECOND
THIRD

INDUSTRY DATA BY STATE

State	Establish-ments	Shipments Total ($ mil)	Shipments % of U.S.	Shipments Per Establ.	Employment Total Number	Employment % of U.S.	Employment Per Establ.	Wages ($/hour)	Cost as % of Shipments	Investment per Employee ($)
Michigan	322	6,901.2	43.6	21.4	45,000	42.8	140	18.03	55.5	4,980
Ohio	134	3,753.1	23.7	28.0	25,200	24.0	188	19.68	58.1	3,476
Indiana	50	1,585.6	10.0	31.7	10,900	10.4	218	19.48	47.5	6,394
Illinois	32	615.1	3.9	19.2	3,400	3.2	106	20.08	54.1	-
Pennsylvania	20	404.3	2.6	20.2	3,300	3.1	165	18.11	43.6	8,697
Tennessee	11	343.4	2.2	31.2	2,300	2.2	209	11.02	61.3	-
Kentucky	6	325.4	2.1	54.2	800	0.8	133	11.24	52.2	14,750
South Carolina	7	148.8	0.9	21.3	1,000	1.0	143	10.69	54.8	5,900
Georgia	9	143.0	0.9	15.9	1,400	1.3	156	9.32	52.3	-
California	20	106.5	0.7	5.3	1,300	1.2	65	8.68	53.7	-
Connecticut	8	100.4	0.6	12.6	500	0.5	63	16.11	31.3	2,000
Missouri	6	70.0	0.4	11.7	600	0.6	100	13.22	50.0	2,667
Mississippi	3	65.8	0.4	21.9	600	0.6	200	9.60	62.6	2,667
Florida	7	45.4	0.3	6.5	500	0.5	71	7.13	58.4	-
New Jersey	6	29.5	0.2	4.9	300	0.3	50	11.20	50.8	-
Wisconsin	12	(D)	-	-	3,750 *	3.6	313	-	-	1,280
Iowa	7	(D)	-	-	1,750 *	1.7	250	-	-	-
Massachusetts	5	(D)	-	-	175 *	0.2	35	-	-	-
Maryland	4	(D)	-	-	375 *	0.4	94	-	-	-
New York	4	(D)	-	-	1,750 *	1.7	438	-	-	-
Alabama	2	(D)	-	-	175 *	0.2	88	-	-	-
Arkansas	2	(D)	-	-	375 *	0.4	188	-	-	-
Minnesota	2	(D)	-	-	375 *	0.4	188	-	-	-
Virginia	1	(D)	-	-	175 *	0.2	175	-	-	-
Washington	1	(D)	-	-	175 *	0.2	175	-	-	-
West Virginia	1	(D)	-	-	375 *	0.4	375	-	-	-

Source: 1992 *Economic Census.* The states are in descending order of shipments or establishments (if shipment data are missing for the majority). The symbol (D) appears when data are withheld to prevent disclosure of competitive information. States marked with (D) are sorted by number of establishments. A dash (-) indicates that the data element cannot be calculated, * indicates the midpoint of a range.

3466 - CROWNS & CLOSURES

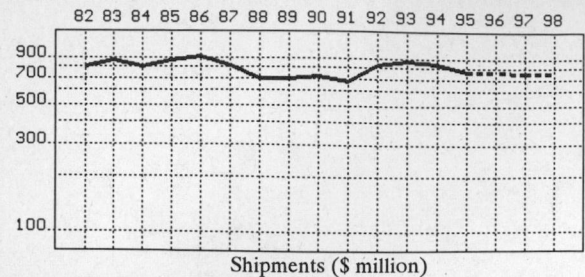

82 83 84 85 86 87 88 89 90 91 92 93 94 95 96 97 98

900.
700.
500.
300.
100.

Shipments ($ million)

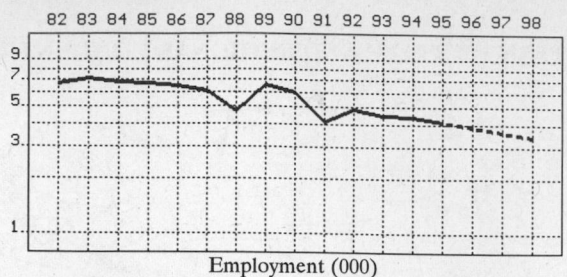

82 83 84 85 86 87 88 89 90 91 92 93 94 95 96 97 98

9.
7.
5.
3.
1.

Employment (000)

GENERAL STATISTICS

| Year | Com-panies | Establishments | | Employment | | | Compensation | | Production ($ million) | | | |
		Total	with 20 or more employees	Total (000)	Production Workers (000)	Hours (Mil)	Payroll ($ mil)	Wages ($/hr)	Cost of Materials	Value Added by Manufacture	Value of Shipments	Capital Invest.
1982	48	64	36	6.7	5.4	10.6	143.5	10.08	441.7	372.4	804.8	21.6
1983		61	36	7.1	5.7	11.4	161.1	10.67	494.9	373.0	866.5	19.8
1984		58	36	6.8	5.5	11.0	158.5	10.83	499.0	344.8	805.3	20.2
1985		55	35	6.7	5.3	10.1	161.6	11.96	509.7	353.7	862.5	22.0
1986		57	38	6.4	5.1	10.6	159.6	11.00	532.5	376.7	914.5	38.7
1987	44	57	34	6.1	4.9	10.3	161.6	11.63	443.1	382.1	819.8	26.5
1988		59	34	4.7	3.9	8.4	131.9	11.31	351.7	353.1	700.8	11.2
1989		53	34	6.6	3.7	7.6	128.6	11.91	363.6	328.8	702.4	18.8
1990		45	28	6.0	3.5	7.0	122.3	12.27	353.9	358.2	720.2	13.0
1991		43	29	4.1	3.3	6.8	121.1	12.10	352.6	324.5	673.8	12.3
1992	43	51	31	4.8	3.9	8.1	143.4	12.04	437.4	407.6	832.9	18.4
1993		50	27	4.5	3.6	8.1	143.4	11.94	467.1	382.4	859.7	22.6
1994		45P	28P	4.4	3.6	7.9	149.6	13.08	449.0	392.9	834.6	20.0
1995		44P	27P	4.1P	2.9P	6.6P	132.0P	12.85P	413.4P	361.7P	768.4P	16.9P
1996		42P	26P	3.9P	2.7P	6.3P	130.1P	13.03P	411.0P	359.6P	763.9P	16.4P
1997		41P	25P	3.6P	2.5P	5.9P	128.2P	13.21P	408.6P	357.5P	759.4P	15.9P
1998		39P	24P	3.4P	2.3P	5.5P	126.3P	13.38P	406.1P	355.4P	754.9P	15.4P

Sources: 1982, 1987, 1992 *Economic Census; Annual Survey of Manufactures*, 83-86, 88-91, 93-94. Establishment counts for non-Census years are from *County Business Patterns*; establishment values for 83-84 are extrapolations. 'P's show projections by the editors. Industries reclassified in 87 will not have data for prior years.

INDICES OF CHANGE

| Year | Com-panies | Establishments | | Employment | | | Compensation | | Production ($ million) | | | |
		Total	with 20 or more employees	Total (000)	Production Workers (000)	Hours (Mil)	Payroll ($ mil)	Wages ($/hr)	Cost of Materials	Value Added by Manufacture	Value of Shipments	Capital Invest.
1982	112	125	116	140	138	131	100	84	101	91	97	117
1983		120	116	148	146	141	112	89	113	92	104	108
1984		114	116	142	141	136	111	90	114	85	97	110
1985		108	113	140	136	125	113	99	117	87	104	120
1986		112	123	133	131	131	111	91	122	92	110	210
1987	102	112	110	127	126	127	113	97	101	94	98	144
1988		116	110	98	100	104	92	94	80	87	84	61
1989		104	110	137	95	94	90	99	83	81	84	102
1990		88	90	125	90	86	85	102	81	88	86	71
1991		84	94	85	85	84	84	100	81	80	81	67
1992	100	100	100	100	100	100	100	100	100	100	100	100
1993		98	87	94	92	100	100	99	107	94	103	123
1994		88P	89P	92	92	98	104	109	103	96	100	109
1995		86P	87P	86P	76P	82P	92P	107P	95P	89P	92P	92P
1996		83P	84P	81P	70P	77P	91P	108P	94P	88P	92P	89P
1997		80P	81P	76P	65P	73P	89P	110P	93P	88P	91P	86P
1998		77P	79P	71P	60P	68P	88P	111P	93P	87P	91P	84P

Sources: Same as General Statistics. Values reflect change from the base year, 1992. Values above 100 mean greater than 92, values below 100 mean less than 92, and a value of 100 in the 82-91 or 93-98 period means same as 92. 'P's mark projections by the editors.

SELECTED RATIOS

For 1994	Avg. of All Manufact.	Analyzed Industry	Index	For 1994	Avg. of All Manufact.	Analyzed Industry	Index
Employees per Establishment	49	98	199	Value Added per Production Worker	134,084	109,139	81
Payroll per Establishment	1,500,273	3,315,514	221	Cost per Establishment	5,045,178	9,950,974	197
Payroll per Employee	30,620	34,000	111	Cost per Employee	102,970	102,045	99
Production Workers per Establishment	34	80	232	Cost per Production Worker	146,988	124,722	85
Wages per Establishment	853,319	2,290,098	268	Shipments per Establishment	9,576,895	18,496,844	193
Wages per Production Worker	24,861	28,703	115	Shipments per Employee	195,460	189,682	97
Hours per Production Worker	2,056	2,194	107	Shipments per Production Worker	279,017	231,833	83
Wages per Hour	12.09	13.08	108	Investment per Establishment	321,011	443,251	138
Value Added per Establishment	4,602,255	8,707,656	189	Investment per Employee	6,552	4,545	69
Value Added per Employee	93,930	89,295	95	Investment per Production Worker	9,352	5,556	59

Sources: Same as General Statistics. The 'Average of All Manufacturing' column represents the average of all manufacturing industries reported for the most recent complete year available. The Index shows the relationship between the Average and the Analyzed Industry. For example, 100 means that they are equal; 500 that the Analyzed Industry is five times the average; 50 means that the Analyzed Industry is half the national average. The abbreviation 'na' is used to show that data are 'not available'.

LEADING COMPANIES   Number shown: 7   Total sales ($ mil): 707   Total employment (000): 5.5

Company Name	Address				CEO Name	Phone	Co. Type	Sales ($ mil)	Empl. (000)
Alltrista Corp	PO Box 5004	Muncie	IN	47307	Thomas B Clark	317-281-5000	P	296	1.5
Alcoa Closure Systems Intern	2485 Directors Row	Indianapolis	IN	46241	Timothy J Leveque	317-481-4203	D	210	2.5
Anchor Hocking Packaging Co	312 Elm St	Cincinnati	OH	45202	Larry Montanus	513-333-3400	S	150	1.0
American Flange & Mfg	PO Box 88688	Carol Stream	IL	60188	Ivan Signorelli	708-665-7900	S	25	0.2
Ferdinand Gutmann and Co	3611 14th Av	Brooklyn	NY	11218	Leonard Gutmann	718-436-6800	R	20	0.2
Allen Stevens	800 S Gilbert St	Danville	IL	61832	Richard Hird	217-442-4600	R	6	<0.1
Morgan Adhesives Co	4560 Darrow Rd	Stow	OH	44224	Lana Leggett	216-688-1111	D	1*	<0.1

Source: Ward's Business Directory of U.S. Private and Public Companies, Volumes 1 and 2, 1996. The company type code used is as follows: P - Public, R - Private, S - Subsidiary, D - Division, J - Joint Venture, A - Affiliate, G - Group. Sales are in millions of dollars, employees are in thousands. An asterisk (*) indicates an estimated sales volume. The symbol < stands for 'less than'. Company names and addresses are truncated, in some cases, to fit into the available space.

MATERIALS CONSUMED

Material		Quantity	Delivered Cost ($ million)
Materials, ingredients, containers, and supplies		(X)	409.6
Metal bolts, nuts, screws, washers, rivets, and other screw machine products		(X)	0.8
Other fabricated metal products (except castings and forgings)		(X)	(D)
Iron and steel castings (rough and semifinished)		(X)	(D)
Nonferrous (aluminum, copper, etc.) castings (rough and semifinished)		(X)	(D)
Steel bars and bar shapes		(X)	(D)
Steel sheet and strip, including tin plate	1,000 s tons	(S)	1.9
Steel tinplate, tin free steel, terneplate, and blackplate	1,000 s tons	(S)	209.6
All other steel shapes and forms		(X)	(D)
Aluminum and aluminum-base alloy sheet, plate, foil, and welded tubing		(X)	47.1
Aluminum and aluminum-base alloy extruded shapes, including extruded rod, bar, pipe, tube, etc.		(X)	(D)
Other aluminum and aluminum-base alloy shapes and forms		(X)	14.0
Plastics resins consumed in the form of granules, pellets, powders, liquids, etc.		(X)	23.1
Paints, varnishes, lacquers, stains, shellacs, japans, enamels, and allied products		(X)	28.2
All other chemicals and allied products		(X)	14.7
Plastics products consumed in the form of sheets, rods, tubes, film, and other shapes		(X)	(D)
Paperboard containers, boxes, and corrugated paperboard		(X)	10.0
Other paper and paperboard products		(X)	4.7
Special dies, tools, die sets, jigs, and fixtures, except cutting tools for machine tools		(X)	(D)
All other materials and components, parts, containers, and supplies		(X)	33.0
Materials, ingredients, containers, and supplies, nsk		(X)	10.1

Source: 1992 Economic Census. Explanation of symbols used: (D): Withheld to avoid disclosure of competitive data; na: Not available; (S): Withheld because statistical norms were not met; (X): Not applicable; (Z): Less than half the unit shown; nec: Not elsewhere classified; nsk: Not specified by kind; - : zero; * : 10-19 percent estimated; ** : 20-29 percent estimated.

PRODUCT SHARE DETAILS

Product or Product Class	% Share	Product or Product Class	% Share
Crowns and closures	100.00	Metal commercial closures and metal home-canning closures, except crowns, nsk.	1.16
Metal commercial closures and metal home-canning closures, except crowns	(D)	Metal crowns	(D)
Metal and metal-composite closures, including home-canning closures	78.36	Soft drink crowns	(D)
		All other metal crowns, including beer	(D)
Metal soft drink closures, including roll-ons	(D)	Crowns and closures, nsk	1.11
All other metal closures, including beer and roll-ons	(D)		

Source: 1992 Economic Census. The values shown are percent of total shipments in an industry. Values of indented subcategories are summed in the main heading. The symbol (D) appears when data are withheld to prevent disclosure of competitive information. The abbreviation nsk stands for 'not specified by kind' and nec for 'not elsewhere classified'.

INPUTS AND OUTPUTS FOR CROWNS & CLOSURES

Economic Sector or Industry Providing Inputs	%	Sector	Economic Sector or Industry Buying Outputs	%	Sector
Metal coating & allied services	26.8	Manufg.	Bottled & canned soft drinks	22.3	Manufg.
Aluminum rolling & drawing	20.2	Manufg.	Malt beverages	10.6	Manufg.
Wholesale trade	7.7	Trade	Pickles, sauces, & salad dressings	10.5	Manufg.
Blast furnaces & steel mills	5.8	Manufg.	Canned fruits & vegetables	7.7	Manufg.
Paints & allied products	4.1	Manufg.	Canned specialties	7.2	Manufg.
Motor freight transportation & warehousing	3.9	Util.	Drugs	6.3	Manufg.
Plastics materials & resins	3.8	Manufg.	Exports	6.0	Foreign
Adhesives & sealants	2.2	Manufg.	Food preparations, nec	4.3	Manufg.
Electric services (utilities)	2.0	Util.	Roasted coffee	3.8	Manufg.
Paperboard containers & boxes	1.9	Manufg.	Polishes & sanitation goods	3.7	Manufg.
Wood products, nec	1.6	Manufg.	Wines, brandy, & brandy spirits	3.0	Manufg.
Gas production & distribution (utilities)	1.6	Util.	Shortening & cooking oils	2.6	Manufg.
Imports	1.5	Foreign	Distilled liquor, except brandy	2.3	Manufg.
Miscellaneous plastics products	1.4	Manufg.	Personal consumption expenditures	1.9	
Paperboard mills	1.3	Manufg.	Flavoring extracts & syrups, nec	1.8	Manufg.
Special dies & tools & machine tool accessories	1.2	Manufg.	Toilet preparations	1.8	Manufg.
Chemical preparations, nec	1.1	Manufg.	Confectionery products	0.9	Manufg.
Advertising	0.9	Services	Glass containers	0.9	Manufg.
Fabricated rubber products, nec	0.7	Manufg.	Change in business inventories	0.8	In House
Machine tools, metal forming types	0.6	Manufg.	Fluid milk	0.3	Manufg.
Maintenance of nonfarm buildings nec	0.5	Constr.	S/L Govt. purch., health & hospitals	0.3	S/L Govt
Paper mills, except building paper	0.5	Manufg.	Soap & other detergents	0.2	Manufg.
Communications, except radio & TV	0.5	Util.	Agricultural chemicals, nec	0.1	Manufg.
Railroads & related services	0.5	Util.	Crowns & closures	0.1	Manufg.
Security & commodity brokers	0.5	Fin/R.E.	S/L Govt. purch., correction	0.1	S/L Govt
Metal heat treating	0.4	Manufg.	S/L Govt. purch., elem. & secondary education	0.1	S/L Govt
Royalties	0.4	Fin/R.E.	S/L Govt. purch., higher education	0.1	S/L Govt
Job training & related services	0.4	Services	S/L Govt. purch., natural resource & recreation.	0.1	S/L Govt
Ball & roller bearings	0.3	Manufg.			
Machinery, except electrical, nec	0.3	Manufg.			
Petroleum refining	0.3	Manufg.			
Eating & drinking places	0.3	Trade			
Banking	0.3	Fin/R.E.			
Real estate	0.3	Fin/R.E.			
Automotive repair shops & services	0.3	Services			
Personnel supply services	0.3	Services			
Crowns & closures	0.2	Manufg.			
Metal stampings, nec	0.2	Manufg.			
Sanitary services, steam supply, irrigation	0.2	Util.			
Engineering, architectural, & surveying services	0.2	Services			
Equipment rental & leasing services	0.2	Services			
Abrasive products	0.1	Manufg.			
Iron & steel foundries	0.1	Manufg.			
Lighting fixtures & equipment	0.1	Manufg.			
Machine tools, metal cutting types	0.1	Manufg.			
Primary zinc	0.1	Manufg.			
Screw machine and related products	0.1	Manufg.			
Water supply & sewage systems	0.1	Util.			
Credit agencies other than banks	0.1	Fin/R.E.			
Insurance carriers	0.1	Fin/R.E.			
Computer & data processing services	0.1	Services			
Laundry, dry cleaning, shoe repair	0.1	Services			
Legal services	0.1	Services			
Management & consulting services & labs	0.1	Services			
U.S. Postal Service	0.1	Gov't			

Source: Benchmark Input-Output Accounts for the U.S. Economy, 1982, U.S. Department of Commerce, Washington, D.C., July 1991. Data, as reported in the source, are organized by the 1977 SIC structure in use in 1982 but have been matched, as closely as is possible, to the 1987 SIC structure used in this book.

OCCUPATIONS EMPLOYED BY SIC 346 - METAL FORGINGS AND STAMPINGS

Occupation	% of Total 1994	Change to 2005	Occupation	% of Total 1994	Change to 2005
Machine forming operators, metal & plastic	9.3	-65.9	Machine feeders & offbearers	1.8	-23.3
Tool & die makers	7.4	-11.4	Sheet metal workers & duct installers	1.7	-14.7
Machine tool cutting & forming etc. nec	6.4	10.9	Metal & plastic machine workers nec	1.5	-24.4
Assemblers, fabricators, & hand workers nec	4.7	-14.7	Helpers, laborers, & material movers nec	1.4	-14.7
Blue collar worker supervisors	4.6	-17.3	Industrial production managers	1.4	-14.7
Welding machine setters, operators	3.8	-23.2	Millwrights	1.3	2.3
Punching machine operators, metal & plastic	3.4	15.1	Traffic, shipping, & receiving clerks	1.3	-17.9
Inspectors, testers, & graders, precision	3.3	-14.7	Welders & cutters	1.3	-14.7
Industrial machinery mechanics	3.1	-6.2	Freight, stock, & material movers, hand	1.3	-31.8
Industrial truck & tractor operators	2.8	-14.7	Electricians	1.2	-20.0
Machine tool cutting operators, metal & plastic	2.6	-12.6	Janitors & cleaners, incl maids	1.1	-31.8
Machinists	2.4	-14.7	Sales & related workers nec	1.1	-14.7
Combination machine tool operators	2.1	27.9	Maintenance repairers, general utility	1.1	-23.3
General managers & top executives	1.8	-19.1			

Source: Industry-Occupation Matrix, Bureau of Labor Statistics. These data relate to one or more 3-digit SIC industry groups rather than to a single 4-digit SIC. The change reported for each occupation to the year 2005 is a percent of growth or decline as estimated by the Bureau of Labor Statistics. The abbreviation nec stands for 'not elsewhere classified'.

LOCATION BY STATE AND REGIONAL CONCENTRATION

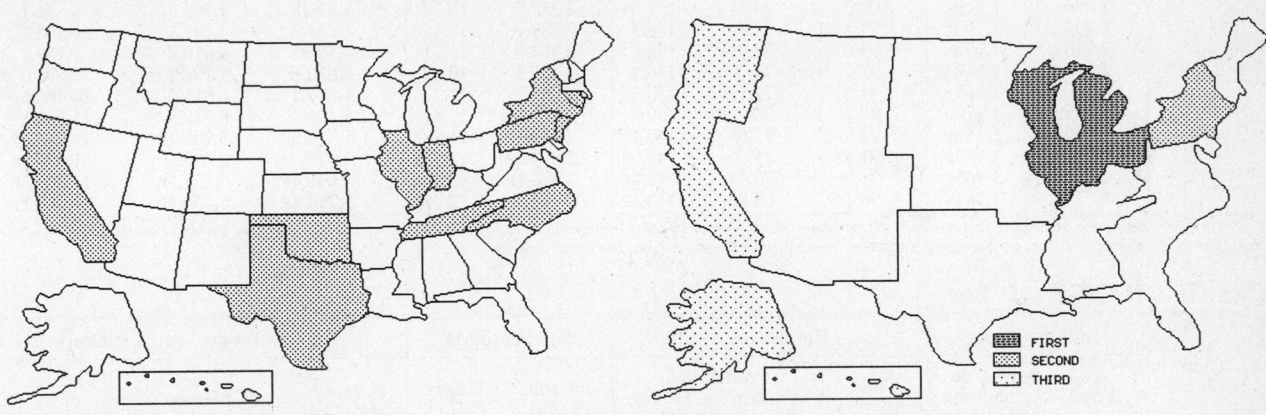

FIRST
SECOND
THIRD

INDUSTRY DATA BY STATE

State	Establish- ments	Shipments Total ($ mil)	Shipments % of U.S.	Shipments Per Establ.	Employment Total Number	Employment % of U.S.	Employment Per Establ.	Wages ($/hour)	Cost as % of Shipments	Investment per Employee ($)
Indiana	4	174.4	20.9	43.6	800	16.7	200	12.83	59.9	-
Illinois	7	127.3	15.3	18.2	800	16.7	114	11.79	40.4	-
Connecticut	6	48.3	5.8	8.1	400	8.3	67	14.43	38.7	1,750
California	6	47.4	5.7	7.9	300	6.3	50	14.67	48.5	-
Pennsylvania	6	(D)	-	-	750 *	15.6	125	-	-	3,733
New York	3	(D)	-	-	175 *	3.6	58	-	-	-
New Jersey	2	(D)	-	-	375 *	7.8	188	-	-	-
Oklahoma	2	(D)	-	-	175 *	3.6	88	-	-	-
North Carolina	1	(D)	-	-	175 *	3.6	175	-	-	-
Tennessee	1	(D)	-	-	175 *	3.6	175	-	-	-
Texas	1	(D)	-	-	175 *	3.6	175	-	-	-

Source: 1992 *Economic Census.* The states are in descending order of shipments or establishments (if shipment data are missing for the majority). The symbol (D) appears when data are withheld to prevent disclosure of competitive information. States marked with (D) are sorted by number of establishments. A dash (-) indicates that the data element cannot be calculated; * indicates the midpoint of a range.

3469 - METAL STAMPINGS, NEC

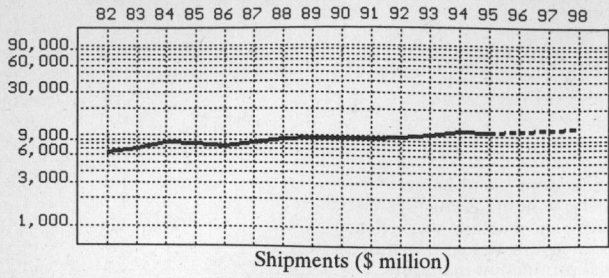

Shipments ($ million)

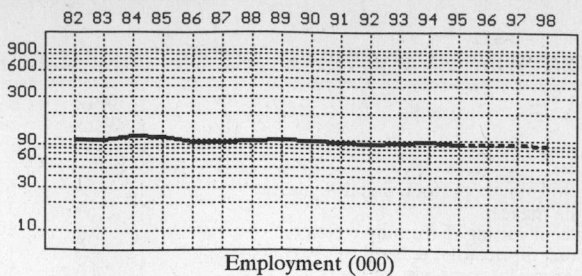

Employment (000)

GENERAL STATISTICS

Year	Companies	Establishments		Employment			Compensation		Production ($ million)			
		Total	with 20 or more employees	Total (000)	Production Workers (000)	Hours (Mil)	Payroll ($ mil)	Wages ($/hr)	Cost of Materials	Value Added by Manufacture	Value of Shipments	Capital Invest.
1982	2,718	2,843	1,156	100.4	75.6	146.3	1,782.7	7.89	2,986.4	3,414.2	6,437.7	200.1
1983		2,781	1,176	102.9	78.6	155.6	1,920.7	8.04	3,269.3	3,836.7	7,079.3	192.5
1984		2,719	1,196	110.5	85.7	167.4	2,193.5	8.65	4,062.4	4,436.0	8,410.2	290.9
1985		2,657	1,216	107.1	82.6	164.7	2,216.9	8.78	3,773.5	4,261.4	8,084.3	305.0
1986		2,573	1,183	95.9	73.3	142.4	2,067.6	9.58	3,584.1	4,032.4	7,625.5	256.9
1987	2,707	2,815	1,162	95.5	73.2	146.2	2,131.9	9.59	4,034.4	4,311.4	8,331.1	268.9
1988		2,730	1,211	97.1	74.2	153.1	2,259.5	9.65	4,461.7	4,670.5	9,042.7	221.4
1989		2,676	1,219	100.0	76.8	151.6	2,357.9	10.18	4,840.8	4,795.6	9,576.1	300.7
1990		2,703	1,193	97.1	74.7	148.7	2,382.9	10.43	4,634.2	4,747.3	9,380.2	277.7
1991		2,740	1,152	95.0	71.9	145.1	2,400.0	10.71	4,556.1	4,810.8	9,338.2	257.0
1992	2,635	2,748	1,142	92.7	70.7	146.3	2,487.5	10.88	4,520.5	5,192.2	9,695.0	298.1
1993		2,700	1,165	94.2	72.1	150.3	2,645.2	11.37	4,879.1	5,616.6	10,466.6	268.8
1994		2,695P	1,170P	97.7	76.9	159.7	2,765.0	11.32	5,423.9	6,019.5	11,361.5	408.3
1995		2,690P	1,169P	92.8P	71.6P	149.4P	2,743.1P	11.86P	5,320.9P	5,905.2P	11,145.8P	333.9P
1996		2,686P	1,167P	91.9P	71.0P	149.0P	2,809.6P	12.16P	5,478.7P	6,080.3P	11,476.2P	342.6P
1997		2,682P	1,165P	91.0P	70.4P	148.6P	2,876.1P	12.46P	5,636.4P	6,255.3P	11,806.6P	351.4P
1998		2,677P	1,164P	90.1P	69.8P	148.2P	2,942.5P	12.75P	5,794.1P	6,430.4P	12,137.1P	360.1P

Sources: 1982, 1987, 1992 *Economic Census*; *Annual Survey of Manufactures*, 83-86, 88-91, 93-94. Establishment counts for non-Census years are from *County Business Patterns*; establishment values for 83-84 are extrapolations. 'P's show projections by the editors. Industries reclassified in 87 will not have data for prior years.

INDICES OF CHANGE

Year	Companies	Establishments		Employment			Compensation		Production ($ million)			
		Total	with 20 or more employees	Total (000)	Production Workers (000)	Hours (Mil)	Payroll ($ mil)	Wages ($/hr)	Cost of Materials	Value Added by Manufacture	Value of Shipments	Capital Invest.
1982	103	103	101	108	107	100	72	73	66	66	66	67
1983		101	103	111	111	106	77	74	72	74	73	65
1984		99	105	119	121	114	88	80	90	85	87	98
1985		97	106	116	117	113	89	81	83	82	83	102
1986		94	104	103	104	97	83	88	79	78	79	86
1987	103	102	102	103	104	100	86	88	89	83	86	90
1988		99	106	105	105	105	91	89	99	90	93	74
1989		97	107	108	109	104	95	94	107	92	99	101
1990		98	104	105	106	102	96	96	103	91	97	93
1991		100	101	102	102	99	96	98	101	93	96	86
1992	100	100	100	100	100	100	100	100	100	100	100	100
1993		98	102	102	102	103	106	105	108	108	108	90
1994		98P	102P	105	109	109	111	104	120	116	117	137
1995		98P	102P	100P	101P	102P	110P	109P	118P	114P	115P	112P
1996		98P	102P	99P	100P	102P	113P	112P	121P	117P	118P	115P
1997		98P	102P	98P	100P	102P	116P	114P	125P	120P	122P	118P
1998		97P	102P	97P	99P	101P	118P	117P	128P	124P	125P	121P

Sources: Same as General Statistics. Values reflect change from the base year, 1992. Values above 100 mean greater than 92, values below 100 mean less than 92, and a value of 100 in the 82-91 or 93-98 period means same as 92. 'P's mark projections by the editors.

SELECTED RATIOS

For 1994	Avg. of All Manufact.	Analyzed Industry	Index	For 1994	Avg. of All Manufact.	Analyzed Industry	Index
Employees per Establishment	49	36	74	Value Added per Production Worker	134,084	78,277	58
Payroll per Establishment	1,500,273	1,026,009	68	Cost per Establishment	5,045,178	2,012,647	40
Payroll per Employee	30,620	28,301	92	Cost per Employee	102,970	55,516	54
Production Workers per Establishment	34	29	83	Cost per Production Worker	146,988	70,532	48
Wages per Establishment	853,319	670,822	79	Shipments per Establishment	9,576,895	4,215,912	44
Wages per Production Worker	24,861	23,509	95	Shipments per Employee	195,460	116,290	59
Hours per Production Worker	2,056	2,077	101	Shipments per Production Worker	279,017	147,744	53
Wages per Hour	12.09	11.32	94	Investment per Establishment	321,011	151,508	47
Value Added per Establishment	4,602,255	2,233,656	49	Investment per Employee	6,552	4,179	64
Value Added per Employee	93,930	61,612	66	Investment per Production Worker	9,352	5,309	57

Sources: Same as General Statistics. The 'Average of All Manufacturing' column represents the average of all manufacturing industries reported for the most recent complete year available. The Index shows the relationship between the Average and the Analyzed Industry. For example, 100 means that they are equal; 500 that the Analyzed Industry is five times the average; 50 means that the Analyzed Industry is half the national average. The abbreviation 'na' is used to show that data are 'not available'.

LEADING COMPANIES   Number shown: **75**   Total sales ($ mil): **4,267**   Total employment (000): **30.8**

Company Name	Address				CEO Name	Phone	Co. Type	Sales ($ mil)	Empl. (000)
Hexcel Corp	5794 W Las Positas	Pleasanton	CA	94588	John J Lee	510-847-9500	P	314	2.2
Tempel Steel Co	5215 Old Orchard	Skokie	IL	60077	Vincent Buonanno	708-581-9400	R	280	1.2
Ekco Group Inc	98 Spit Brook Rd	Nashua	NH	03062	Robert Stein	603-888-1212	P	246	1.5
Mirro-Foley Co	PO Box 1330	Manitowoc	WI	54221	Bill Frieder	414-684-4421	S	210*	2.0
Shiloh Industries Inc	1013 Centre Rd	Wilmington	DE	19805	Robert L Grissinger	302-998-0592	P	210	1.1
Ogihara America Corp	1480 W McPh	Howell	MI	48843	Tokio Ogihara	517-548-4900	S	170	0.5
Alpha Stamping Co	33375 Glendale St	Livonia	MI	48150	George Strumbos	313-523-1000	R	150	0.7
Western Industries Inc	1215 N 62nd St	Milwaukee	WI	53213	G Ronald Morris	414-771-7700	R	150*	1.2
Waterloo Industries Inc	PO Box 2095	Waterloo	IA	50704	John M Trebel	319-235-7131	S	130	1.2
MSI Corp	26269 Groesbeck	Warren	MI	48089	Fred Homann	810-773-0800	R	120	0.7
Vollrath Company Inc	PO Box 611	Sheboygan	WI	53081	TG Belot	414-457-4851	R	100	0.7
Delta Consolidated Industries	PO Box 41209	Raleigh	NC	27629	Markus Isenrich	919-821-0877	S	90	0.7
Spartanburg Steel Products Inc	PO Box 6428	Spartanburg	SC	29304	FW Schoen	803-585-5211	S	83	0.6
Sinter Metals Inc	50 Public Sq	Cleveland	OH	44113	Joseph W Carreras	216-771-6700	P	83	0.8
Midwest Stamping & Mfg Co	513 Napoleon Rd	Bowling Green	OH	43402	R L Thompson	419-353-5931	R	80	0.5
Brown Corporation of Ionia Inc	314 S Steele St	Ionia	MI	48846	RH Brown	616-527-1600	S	65	0.2
Pridgeon and Clay Inc	50 Cottage Gr	Grand Rapids	MI	49507	RE Clay	616-241-5675	R	63	0.3
Revere Ware Corp	PO Box 250	Clinton	IL	61727	Ashok Nayak	217-935-7200	S	57*	0.5
Griffiths Corp	3030 Hrbr Ln N	Plymouth	MN	55447	Harold Griffiths	612-557-8935	R	55*	0.5
Plews	6775 Shady Oak Rd	Eden Prairie	MN	55344	James Beiswanger	612-944-7255	D	55	0.4
California Industrial Products	11525 Shoemaker	Santa Fe Sprgs	CA	90670	Raymond Derby	310-941-3281	R	50	0.3
Eyelematic Manufacturing	1 Seemar Rd	Watertown	CT	06795	Henry F Seebach Jr	203-274-6791	R	50	0.4
Premium Allied Tool Inc	PO Box 1598	Owensboro	KY	42302	James Day	502-729-4242	R	46*	0.4
Instrument Specialties Company	PO Box A	Del Wtr Gap	PA	18327	William C Carson	717-424-8510	R	42	0.4
Bermo Inc	4501 Ball Rd NE	Circle Pines	MN	55014	D Berdass	612-786-7676	R	40*	0.4
Waggoner Corp	1400 Rochester Rd	Troy	MI	48083	DE Waggoner	313-588-2121	R	40	0.3
Small Parts Inc	PO Box 23	Logansport	IN	46947	Michael Jordan	219-753-6323	R	39	0.3
F and B Manufacturing Co	5480 N Northwest	Chicago	IL	60630	Milada F Anderson	312-774-6300	R	38*	0.6
American Metal and Plastics Inc	450 32nd SW	Grand Rapids	MI	49548	Thomas M Cook	616-452-6061	R	36	0.3
Cullman Products	PO Box 489	Cullman	AL	35056	George Pennington	205-734-4921	S	35	0.5
Trans-Matic Manufacturing Co	300 E 48th St	Holland	MI	49423	Patrick Thompson	616-396-1441	R	35	0.2
ER Wagner Manufacturing Co	4611 N 32nd St	Milwaukee	WI	53209	Frank Sterner	414-871-5080	R	33	0.3
Steel City Corp	PO Box 1227	Youngstown	OH	44501	C Kenneth Fibus	216-792-7663	R	33	<0.1
Parkview Metal Products Inc	4931 W Armitage	Chicago	IL	60639	CE Leutwiler	312-622-8414	R	32*	0.3
Alinabal Inc	28 Woodmont Rd	Milford	CT	06460	SS Bergami	203-877-3241	R	30	0.3
Dayton Rogers Mfg Co	2824 13th Av S	Minneapolis	MN	55407	JW Seeger	612-871-2471	R	30	0.4
Kennedy Manufacturing Co	PO Box 151	Van Wert	OH	45891	David Thompson	419-238-2442	R	30	0.4
Knaack Manufacturing Co	420 E Terra Cotta	Crystal Lake	IL	60014	Bob Ripley	815-459-6020	R	30	0.3
New Standard Corp	74 Commerce Way	York	PA	17406	M F Zifferer Jr	717-757-9450	R	30	0.4
Overland Bolling Co	9200 W Belmont	Franklin Park	IL	60131	Thomas Bolling	708-678-7950	R	30	0.2
Pacific Precision Metals Inc	601 S Vincent Av	Azusa	CA	91702	Michael S Bernath	818-334-0361	R	30	0.4
Wisconsin Tool and Stamp	9521 Ainslie St	Schiller Park	IL	60176	RE Ernst	708-678-7573	R	30	0.2
Hendrickson Stamping	PO Box 458	Joliet	IL	60434	Keith Stephenson	815-727-4031	D	28	0.1
ITW Fastex	195 Algonquin Rd	Des Plaines	IL	60016	Bob Williams	708-299-2222	D	28*	0.2
Burnside Manufacturing Co	PO Box 467	Spring Lake	MI	49456	Daniel H Burnside	616-798-3394	R	27*	0.1
Defiance Metal Products Co	PO Box 447	Defiance	OH	43512	Jon Zachrich	419-784-5332	R	27*	0.3
Highland Manufacturing Co	PO Box 1858	Waterbury	CT	06722	Michael Santogatta	203-574-3200	D	27	0.2
Drawform Inc	300 N Cen Av	Zeeland	MI	49464	Kirk Bush	616-772-1910	R	26	0.2
General Automotive Mfg Co	5801 W Franklin Dr	Franklin	WI	53132	N S Logarakis	414-421-6111	R	26	0.3
Commercial Tool and Die Co	1911 E 51st St	Los Angeles	CA	90058	DC Molinari	213-581-7151	D	25	0.2
Connecticut Fineblanking Corp	25 Forest Pkwy	Shelton	CT	06484	OR Teruzzi	203-925-0012	R	25	0.1
Freeway Corp	9301 Allen Dr	Cleveland	OH	44125	Jerry Grams	216-524-9700	R	25*	0.2
KI	501 Mayde Rd	Berea	KY	40403	Ralph Foster	606-986-1420	S	25	0.2
Punch Press Products Inc	1911 E 51st St	Los Angeles	CA	90058	DC Molinari	213-581-7151	R	25	0.2
United Screw and Bolt Corp	3636 W 58th St	Cleveland	OH	44102	JJ Tanis	216-651-1100	D	25	0.1
AJ Rose Manufacturing Co	3115 W 38th St	Cleveland	OH	44109	A Jay Rose	216-631-4645	R	24*	0.2
Diamond Manufacturing Co	PO Box 174	Wyoming	PA	18644	HF Flack III	717-693-0300	R	24	0.1
Erdle Perforating Co	100 Pixley Industrial	Rochester	NY	14624	WJ Erdle III	716-247-4700	R	24	<0.1
Winzeler Stamping Co	129 W Wabash	Montpelier	OH	43543	Robert C Winzeler	419-485-3147	R	24	0.2
Polar Ware Co	PO Box 211	Sheboygan	WI	53082	Walter Vollrath, III	414-458-3561	R	22*	0.1
Wainwright Industries Inc	PO Box 640	St Peters	MO	63376	WN Wainwright	314-278-5850	R	22	0.2
Cly-Del Manufacturing Co	PO Box 1367	Waterbury	CT	06721	RW Garthwait Sr	203-574-2100	R	21	0.2
Durex Inc	5 Stahuber Av	Union	NJ	07083	Robert Denholtz	908-688-0800	R	21	0.4
Fisher Corp	1625 W Maple Rd	Troy	MI	48084	CE Beckham	313-280-0808	D	21	0.2
Olson Metal Products Co	1001 Crossroads	Seguin	TX	78155	Don Cameron	210-379-7000	R	21	0.2
Plainfield Stamping Illinois Inc	1351 N Division St	Plainfield	IL	60544	C W Berglund	815-436-5671	R	21*	0.2
Precision Stamping	PO Box 660	Logansport	IN	46947	Larry L Dunn	219-722-5168	D	21	0.3
Stack On Products Co	PO Box 489	Wauconda	IL	60084	John Lynn	708-526-1611	R	21*	0.2
General Metal Products	3883 Delor St	St Louis	MO	63116	Anthony Vastardis	314-261-2687	D	20*	0.2
Stamco Industries Inc	26650 Lakeland	Euclid	OH	44132	William E Sopko	216-731-9333	R	20	0.1
Admiral Tool & Mfg Co	3700 N Talman Av	Chicago	IL	60618	ES Levine	312-477-4300	R	20	0.2
Chilton Metal Products	300 Breed St	Chilton	WI	53014	DG Schwalenberg	414-849-2381	D	20	0.4
Die-Matic Corp	201 Eastview Dr	Cleveland	OH	44131	Louie Zeitler	216-749-4656	R	20	0.1
Hendrick Manufacturing Co	7th Av & Clidco Dr	Carbondale	PA	18407	Michael Drake	717-282-1010	R	20	0.1
Hobson and Motzer Inc	PO Box 427	Durham	CT	06422	F Dworak	203-349-1756	R	20	0.1

Source: Ward's Business Directory of U.S. Private and Public Companies, Volumes 1 and 2, 1996. The company type code used is as follows: P - Public, R - Private, S - Subsidiary, D - Division, J - Joint Venture, A - Affiliate, G - Group. Sales are in millions of dollars, employees are in thousands. An asterisk (*) indicates an estimated sales volume. The symbol < stands for 'less than'. Company names and addresses are truncated, in some cases, to fit into the available space.

MATERIALS CONSUMED

Material	Quantity	Delivered Cost ($ million)
Materials, ingredients, containers, and supplies	(X)	3,802.1
Metal bolts, nuts, screws, washers, rivets, and other screw machine products	(X)	88.1
Other fabricated metal products (except castings and forgings)	(X)	309.1
Iron and steel castings (rough and semifinished)	(X)	46.6
Nonferrous (aluminum, copper, etc.) castings (rough and semifinished)	(X)	9.7
Forgings	(X)	1.1
Steel bars and bar shapes	(X)	35.0
Steel sheet and strip, including tin plate	1,000 s tons (S)	953.5
Steel plate	(X)	47.1
Steel wire and wire products	(X)	25.3
Steel tinplate, tin free steel, terneplate, and blackplate	1,000 s tons (S)	49.9
All other steel shapes and forms	(X)	244.5
Copper and copper-base alloy shapes and forms	(X)	105.0
Aluminum and aluminum-base alloy sheet, plate, foil, and welded tubing	(X)	193.4
Aluminum and aluminum-base alloy extruded shapes, including extruded rod, bar, pipe, tube, etc.	(X)	16.5
Other aluminum and aluminum-base alloy shapes and forms	(X)	141.7
Other nonferrous shapes and forms	(X)	57.4
Plastics resins consumed in the form of granules, pellets, powders, liquids, etc.	(X)	19.2
Paints, varnishes, lacquers, stains, shellacs, japans, enamels, and allied products	(X)	42.4
All other chemicals and allied products	(X)	21.8
Plastics products consumed in the form of sheets, rods, tubes, film, and other shapes	(X)	25.8
Paperboard containers, boxes, and corrugated paperboard	(X)	71.9
Other paper and paperboard products	(X)	20.4
Special dies, tools, die sets, jigs, and fixtures, except cutting tools for machine tools	(X)	40.8
All other materials and components, parts, containers, and supplies	(X)	527.3
Materials, ingredients, containers, and supplies, nsk	(X)	708.7

Source: 1992 *Economic Census*. Explanation of symbols used: (D): Withheld to avoid disclosure of competitive data; na: Not available; (S): Withheld because statistical norms were not met; (X): Not applicable; (Z): Less than half the unit shown; nec: Not elsewhere classified; nsk: Not specified by kind; - : zero; * : 10-19 percent estimated; ** : 20-29 percent estimated.

PRODUCT SHARE DETAILS

Product or Product Class	% Share	Product or Product Class	% Share
Metal stampings, nec	100.00	stamped and spun stainless steel utensils	7.98
Metal job stampings, except automotive	55.87	Other stamped and spun stainless steel cooking and kitchen utensils, including commercial, hospital, and outdoor cooking equipment	21.05
Recreational vehicle metal job stampings (motor homes, travel trailers, etc.)	1.99	Stamped and spun household tinware cooking and kitchen utensils	(D)
Motor and generator metal job stampings	8.85	Other stamped and spun tinware cooking and kitchen utensils, including commercial, hospital, and outdoor cooking equipment	(D)
Aviation metal job stampings	3.30	Vitreous enameled stamped and spun cooking and kitchen utensils	(D)
Agricultural equipment metal job stampings, including tractor	3.51	Other stamped and spun cooking and kitchen utensils, including copper	3.10
Computer metal job stampings	10.91	Metal spinning products, excluding cooking and kitchen utensils	1.22
Electrical appliance metal job stampings, except refrigeration and laundry equipment	5.45	Other stamped and pressed metal end products, including vitreous enameled products	20.15
Furniture metal job stampings	3.75	Stamped and pressed vitreous (porcelain) enameled metal architectural parts (exterior and interior), including store front and curtain wall components	1.34
Office machine metal job stampings, excluding computer	3.31	Other stamped and pressed vitreous (porcelain) enameled products (including refrigerator and laundry equipment parts and commercial and hospital utensils), except cooking and kitchen utensils	2.54
Radio and phonograph metal job stampings, except automotive	0.69	Stamped and pressed metal chemical milling products, milled contoured metal, and clad and bonded metal products	1.11
Television metal job stampings	2.29	Perforated metal end products	11.12
Refrigeration metal job stampings (residential, commercial, and industrial)	5.19	Stamped and pressed galvanized steel pails, ash cans, garbage cans, tubs, etc., excluding shipping containers	2.28
Stove, heater, and air-conditioner metal job stampings (residential, commercial, and industrial), except automotive	3.77	Other stamped and pressed metal pails, ash cans, garbage cans, tubs, etc., excluding shipping containers (including other grades of steel)	2.59
Laundry equipment metal job stampings (residential, commercial, and industrial)	1.22	Metal electronic enclosures (stamped and/or pressed), excluding computer stampings	19.84
Other industrial equipment metal job stampings	9.36	Stamped and pressed metal mailboxes (commercial and multiple unit residential)	3.41
Other metal job stampings, except automotive	30.90	Stamped and pressed metal toolboxes	18.61
Job stampings, except automotive, nsk	5.51	Other stamped and pressed metal end products, excluding spinning products	35.59
Stamped and spun utensils, cooking and kitchen, aluminum	6.10	Other stamped and pressed metal end products, including vitreous enameled products, nsk	1.57
Top of range household stamped and spun aluminum utensils (items generally used directly on top of source of heat), including pressure cookers	51.69	Metal stampings, nec, nsk	11.15
Bakeware, pantryware, and miscellaneous household stamped and spun aluminum utensils	34.51		
Camping and outdoor stamped and spun aluminum cooking equipment	1.55		
Other stamped and spun aluminum cooking and kitchen utensils, including commercial and hospital	12.18		
Stamped and spun utensils, cooking and kitchen, aluminum, nsk	0.07		
Stamped and spun utensils, cooking and kitchen, except aluminum	5.51		
Top of range household stamped and spun stainless steel utensils (items generally used directly on top of source of heat)	47.40		
Bakeware, pantryware, and miscellaneous household			

Source: 1992 *Economic Census.* The values shown are percent of total shipments in an industry. Values of indented subcategories are summed in the main heading. The symbol (D) appears when data are withheld to prevent disclosure of competitive information. The abbreviation nsk stands for 'not specified by kind' and nec for 'not elsewhere classified'.

INPUTS AND OUTPUTS FOR METAL STAMPINGS, NEC

Economic Sector or Industry Providing Inputs	%	Sector	Economic Sector or Industry Buying Outputs	%	Sector
Blast furnaces & steel mills	31.9	Manufg.	Personal consumption expenditures	13.1	
Wholesale trade	7.7	Trade	Fabricated structural metal	10.5	Manufg.
Aluminum rolling & drawing	5.8	Manufg.	Lighting fixtures & equipment	3.8	Manufg.
Communications, except radio & TV	5.6	Util.	Electronic components nec	2.5	Manufg.
Maintenance of nonfarm buildings nec	4.9	Constr.	Refrigeration & heating equipment	2.5	Manufg.
Imports	4.3	Foreign	Aircraft	2.0	Manufg.
Copper rolling & drawing	2.5	Manufg.	Optical instruments & lenses	2.0	Manufg.
Metal stampings, nec	2.4	Manufg.	Semiconductors & related devices	1.8	Manufg.
Electric services (utilities)	2.4	Util.	Surgical appliances & supplies	1.8	Manufg.
Nonferrous forgings	1.6	Manufg.	Surgical & medical instruments	1.7	Manufg.
Special dies & tools & machine tool accessories	1.6	Manufg.	Exports	1.7	Foreign
Motor freight transportation & warehousing	1.6	Util.	Electronic computing equipment	1.6	Manufg.
Paperboard containers & boxes	1.4	Manufg.	Communications, except radio & TV	1.5	Util.
Plating & polishing	1.4	Manufg.	Metal stampings, nec	1.3	Manufg.
Computer & data processing services	1.3	Services	Motors & generators	1.3	Manufg.
Gas production & distribution (utilities)	1.1	Util.	Photographic equipment & supplies	1.3	Manufg.
Real estate	1.1	Fin/R.E.	Radio & TV communication equipment	1.3	Manufg.
Paints & allied products	1.0	Manufg.	Fabricated rubber products, nec	1.2	Manufg.
Automotive repair shops & services	1.0	Services	Farm machinery & equipment	1.2	Manufg.
Adhesives & sealants	0.9	Manufg.	Miscellaneous plastics products	1.2	Manufg.
Glass & glass products, except containers	0.9	Manufg.	Polishes & sanitation goods	1.2	Manufg.
Metal heat treating	0.9	Manufg.	Blast furnaces & steel mills	1.1	Manufg.
Miscellaneous plastics products	0.9	Manufg.	Logging camps & logging contractors	1.1	Manufg.
Screw machine and related products	0.9	Manufg.	Radio & TV receiving sets	1.1	Manufg.
Cyclic crudes and organics	0.8	Manufg.	Telephone & telegraph apparatus	1.1	Manufg.
Machinery, except electrical, nec	0.7	Manufg.	Millwork	1.0	Manufg.

Continued on next page.

INPUTS AND OUTPUTS FOR METAL STAMPINGS, NEC - Continued

Economic Sector or Industry Providing Inputs	%	Sector	Economic Sector or Industry Buying Outputs	%	Sector
Railroads & related services	0.7	Util.	Prefabricated metal buildings	1.0	Manufg.
Primary metal products, nec	0.6	Manufg.	Primary batteries, dry & wet	1.0	Manufg.
Eating & drinking places	0.6	Trade	Wood products, nec	1.0	Manufg.
Equipment rental & leasing services	0.6	Services	Aircraft & missile engines & engine parts	0.9	Manufg.
Chemical preparations, nec	0.5	Manufg.	Aircraft & missile equipment, nec	0.9	Manufg.
Fabricated metal products, nec	0.5	Manufg.	Internal combustion engines, nec	0.9	Manufg.
Aluminum castings	0.4	Manufg.	Screw machine and related products	0.8	Manufg.
Ball & roller bearings	0.4	Manufg.	Hospitals	0.8	Services
Iron & steel foundries	0.4	Manufg.	Household laundry equipment	0.7	Manufg.
Plastics materials & resins	0.4	Manufg.	Metal office furniture	0.7	Manufg.
Banking	0.4	Fin/R.E.	Federal Government purchases, national defense	0.7	Fed Govt
Abrasive products	0.3	Manufg.	Architectural metal work	0.6	Manufg.
Metal coating & allied services	0.3	Manufg.	Conveyors & conveying equipment	0.6	Manufg.
Motors & generators	0.3	Manufg.	Electron tubes	0.6	Manufg.
Nonferrous castings, nec	0.3	Manufg.	Lawn & garden equipment	0.6	Manufg.
Petroleum refining	0.3	Manufg.	Pumps & compressors	0.6	Manufg.
Advertising	0.3	Services	Toilet preparations	0.6	Manufg.
Electrical repair shops	0.3	Services	Doctors & dentists	0.6	Services
Engineering, architectural, & surveying services	0.3	Services	Food products machinery	0.5	Manufg.
Brass, bronze, & copper castings	0.2	Manufg.	General industrial machinery, nec	0.5	Manufg.
Hardware, nec	0.2	Manufg.	Hardware, nec	0.5	Manufg.
Lubricating oils & greases	0.2	Manufg.	Household cooking equipment	0.5	Manufg.
Machine tools, metal cutting types	0.2	Manufg.	Mechanical measuring devices	0.5	Manufg.
Miscellaneous fabricated wire products	0.2	Manufg.	Metal doors, sash, & trim	0.5	Manufg.
Nonferrous wire drawing & insulating	0.2	Manufg.	Sheet metal work	0.5	Manufg.
Paperboard mills	0.2	Manufg.	Aluminum rolling & drawing	0.4	Manufg.
Welding apparatus, electric	0.2	Manufg.	Environmental controls	0.4	Manufg.
Wood products, nec	0.2	Manufg.	Glass & glass products, except containers	0.4	Manufg.
Accounting, auditing & bookkeeping	0.2	Services	Household refrigerators & freezers	0.4	Manufg.
Legal services	0.2	Services	Metal household furniture	0.4	Manufg.
Management & consulting services & labs	0.2	Services	Metal partitions & fixtures	0.4	Manufg.
Motor vehicle parts & accessories	0.1	Manufg.	Motorcycles, bicycles, & parts	0.4	Manufg.
Nonmetallic mineral products, nec	0.1	Manufg.	Small arms ammunition	0.4	Manufg.
Sawmills & planning mills, general	0.1	Manufg.	Transformers	0.4	Manufg.
Wiring devices	0.1	Manufg.	Typewriters & office machines, nec	0.4	Manufg.
Insurance carriers	0.1	Fin/R.E.	Watches, clocks, & parts	0.4	Manufg.
Royalties	0.1	Fin/R.E.	Wiring devices	0.4	Manufg.
U.S. Postal Service	0.1	Gov't	Wood kitchen cabinets	0.4	Manufg.
			Job training & related services	0.4	Services
			Residential high-rise apartments	0.3	Constr.
			Blowers & fans	0.3	Manufg.
			Electric housewares & fans	0.3	Manufg.
			Electrical equipment & supplies, nec	0.3	Manufg.
			Electrical industrial apparatus, nec	0.3	Manufg.
			Engine electrical equipment	0.3	Manufg.
			Fabricated metal products, nec	0.3	Manufg.
			Games, toys, & children's vehicles	0.3	Manufg.
			Guided missiles & space vehicles	0.3	Manufg.
			Iron & steel foundries	0.3	Manufg.
			Metalworking machinery, nec	0.3	Manufg.
			Miscellaneous metal work	0.3	Manufg.
			Pipe, valves, & pipe fittings	0.3	Manufg.
			Porcelain electrical supplies	0.3	Manufg.
			Sawmills & planning mills, general	0.3	Manufg.
			Truck trailers	0.3	Manufg.
			Wood partitions & fixtures	0.3	Manufg.
			X-ray apparatus & tubes	0.3	Manufg.
			Federal Government enterprises nec	0.3	Gov't
			Dairy farm products	0.2	Agric.
			Construction of stores & restaurants	0.2	Constr.
			Residential garden apartments	0.2	Constr.
			Brooms & brushes	0.2	Manufg.
			Chemical preparations, nec	0.2	Manufg.
			Furniture & fixtures, nec	0.2	Manufg.
			Hand & edge tools, nec	0.2	Manufg.
			Heating equipment, except electric	0.2	Manufg.
			Instruments to measure electricity	0.2	Manufg.
			Manufacturing industries, nec	0.2	Manufg.
			Marking devices	0.2	Manufg.
			Mattresses & bedsprings	0.2	Manufg.
			Metal sanitary ware	0.2	Manufg.
			Miscellaneous fabricated wire products	0.2	Manufg.
			Motor homes (made on purchased chassis)	0.2	Manufg.
			Nonferrous rolling & drawing, nec	0.2	Manufg.
			Plating & polishing	0.2	Manufg.

Continued on next page.

INPUTS AND OUTPUTS FOR METAL STAMPINGS, NEC - Continued

Economic Sector or Industry Providing Inputs	%	Sector	Economic Sector or Industry Buying Outputs	%	Sector
			Public building furniture	0.2	Manufg.
			Service industry machines, nec	0.2	Manufg.
			Small arms	0.2	Manufg.
			Special industry machinery, nec	0.2	Manufg.
			Switchgear & switchboard apparatus	0.2	Manufg.
			Truck & bus bodies	0.2	Manufg.
			Laundry, dry cleaning, shoe repair	0.2	Services
			Federal Government purchases, nondefense	0.2	Fed Govt
			Residential additions/alterations, nonfarm	0.1	Constr.
			Ammunition, except for small arms, nec	0.1	Manufg.
			Asbestos products	0.1	Manufg.
			Automatic merchandising machines	0.1	Manufg.
			Calculating & accounting machines	0.1	Manufg.
			Carburetors, pistons, rings, & valves	0.1	Manufg.
			Construction machinery & equipment	0.1	Manufg.
			Copper rolling & drawing	0.1	Manufg.
			Cutlery	0.1	Manufg.
			Drapery hardware & blinds & shades	0.1	Manufg.
			Lead pencils & art goods	0.1	Manufg.
			Machinery, except electrical, nec	0.1	Manufg.
			Oil field machinery	0.1	Manufg.
			Plumbing fixture fittings & trim	0.1	Manufg.
			Storage batteries	0.1	Manufg.
			Tanks & tank components	0.1	Manufg.
			Eating & drinking places	0.1	Trade
			Equipment rental & leasing services	0.1	Services
			S/L Govt. purch., elem. & secondary education	0.1	S/L Govt

Source: Benchmark Input-Output Accounts for the U.S. Economy, 1982, U.S. Department of Commerce, Washington, D.C., July 1991. Data, as reported in the source, are organized by the 1977 SIC structure in use in 1982 but have been matched, as closely as is possible, to the 1987 SIC structure used in this book.

OCCUPATIONS EMPLOYED BY SIC 346 - METAL FORGINGS AND STAMPINGS

Occupation	% of Total 1994	Change to 2005	Occupation	% of Total 1994	Change to 2005
Machine forming operators, metal & plastic	9.3	-65.9	Machine feeders & offbearers	1.8	-23.3
Tool & die makers	7.4	-11.4	Sheet metal workers & duct installers	1.7	-14.7
Machine tool cutting & forming etc. nec	6.4	10.9	Metal & plastic machine workers nec	1.5	-24.4
Assemblers, fabricators, & hand workers nec	4.7	-14.7	Helpers, laborers, & material movers nec	1.4	-14.7
Blue collar worker supervisors	4.6	-17.3	Industrial production managers	1.4	-14.7
Welding machine setters, operators	3.8	-23.2	Millwrights	1.3	2.3
Punching machine operators, metal & plastic	3.4	15.1	Traffic, shipping, & receiving clerks	1.3	-17.9
Inspectors, testers, & graders, precision	3.3	-14.7	Welders & cutters	1.3	-14.7
Industrial machinery mechanics	3.1	-6.2	Freight, stock, & material movers, hand	1.3	-31.8
Industrial truck & tractor operators	2.8	-14.7	Electricians	1.2	-20.0
Machine tool cutting operators, metal & plastic	2.6	-12.6	Janitors & cleaners, incl maids	1.1	-31.8
Machinists	2.4	-14.7	Sales & related workers nec	1.1	-14.7
Combination machine tool operators	2.1	27.9	Maintenance repairers, general utility	1.1	-23.3
General managers & top executives	1.8	-19.1			

Source: Industry-Occupation Matrix, Bureau of Labor Statistics. These data relate to one or more 3-digit SIC industry groups rather than to a single 4-digit SIC. The change reported for each occupation to the year 2005 is a percent of growth or decline as estimated by the Bureau of Labor Statistics. The abbreviation nec stands for 'not elsewhere classified'.

LOCATION BY STATE AND REGIONAL CONCENTRATION

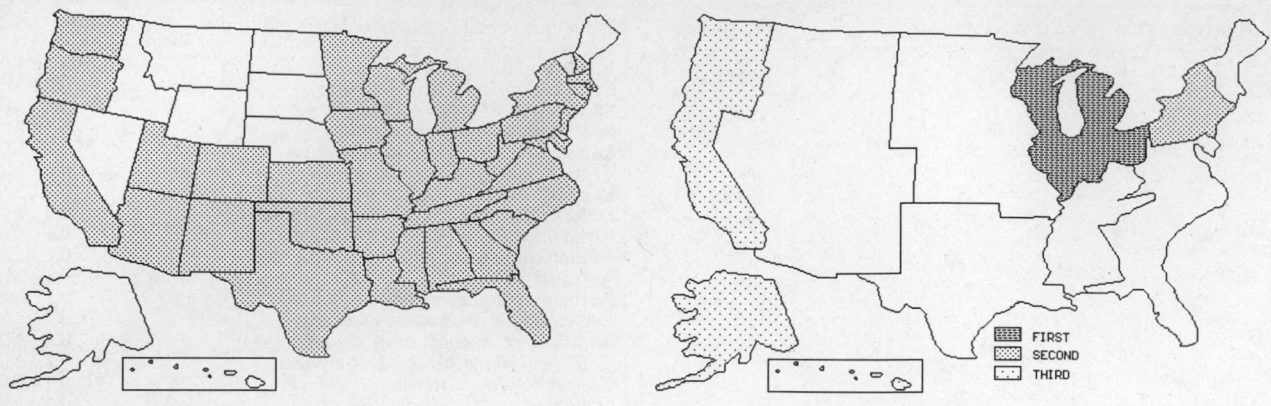

FIRST
SECOND
THIRD

INDUSTRY DATA BY STATE

State	Establish-ments	Shipments			Employment				Cost as % of Shipments	Investment per Employee ($)
		Total ($ mil)	% of U.S.	Per Establ.	Total Number	% of U.S.	Per Establ.	Wages ($/hour)		
Illinois	316	1,639.9	16.9	5.2	13,100	14.1	41	11.68	51.2	3,412
Ohio	251	1,011.4	10.4	4.0	9,000	9.7	36	10.88	47.4	3,067
Wisconsin	110	929.3	9.6	8.4	8,100	8.7	74	11.40	47.7	4,531
California	352	701.7	7.2	2.0	7,100	7.7	20	11.03	40.0	2,183
Pennsylvania	129	550.1	5.7	4.3	5,100	5.5	40	10.42	46.4	2,922
New York	206	544.1	5.6	2.6	6,300	6.8	31	10.84	49.4	1,984
Connecticut	122	495.5	5.1	4.1	4,300	4.6	35	11.31	42.5	4,395
Minnesota	91	412.4	4.3	4.5	4,600	5.0	51	11.82	40.3	2,609
Michigan	151	404.3	4.2	2.7	3,700	4.0	25	10.05	49.6	2,784
New Jersey	133	333.2	3.4	2.5	3,700	4.0	28	12.24	44.1	2,054
Massachusetts	107	308.2	3.2	2.9	3,400	3.7	32	11.10	43.7	2,500
Indiana	82	303.4	3.1	3.7	2,800	3.0	34	9.24	48.7	3,179
Kentucky	37	247.9	2.6	6.7	2,700	2.9	73	9.60	49.1	5,741
Missouri	52	178.7	1.8	3.4	1,300	1.4	25	11.59	46.1	3,231
Florida	62	138.1	1.4	2.2	1,500	1.6	24	9.67	42.5	1,600
Tennessee	43	112.8	1.2	2.6	1,300	1.4	30	8.62	46.9	2,615
North Carolina	41	105.3	1.1	2.6	1,100	1.2	27	10.05	44.9	2,364
Iowa	16	85.0	0.9	5.3	1,100	1.2	69	12.57	50.4	-
Colorado	30	71.3	0.7	2.4	700	0.8	23	9.60	37.7	4,000
Kansas	18	49.6	0.5	2.8	500	0.5	28	7.88	45.8	1,200
Oregon	29	45.3	0.5	1.6	500	0.5	17	9.57	41.3	1,600
Virginia	14	41.2	0.4	2.9	300	0.3	21	9.20	60.7	1,333
Oklahoma	16	38.1	0.4	2.4	300	0.3	19	8.60	45.9	3,000
Washington	20	28.0	0.3	1.4	300	0.3	15	13.40	47.1	4,000
Maryland	13	25.2	0.3	1.9	300	0.3	23	11.25	40.9	2,333
New Hampshire	15	20.1	0.2	1.3	300	0.3	20	9.25	37.8	3,333
South Carolina	11	18.6	0.2	1.7	200	0.2	18	13.00	55.4	-
New Mexico	6	13.8	0.1	2.3	200	0.2	33	11.67	35.5	2,000
Texas	86	(D)	-	-	1,750 *	1.9	20	-	-	-
Rhode Island	35	(D)	-	-	750 *	0.8	21	-	-	-
Georgia	32	(D)	-	-	1,750 *	1.9	55	-	-	5,943
Arizona	29	(D)	-	-	750 *	0.8	26	-	-	-
Alabama	21	(D)	-	-	750 *	0.8	36	-	-	-
Mississippi	15	(D)	-	-	750 *	0.8	50	-	-	-
Arkansas	11	(D)	-	-	1,750 *	1.9	159	-	-	-
West Virginia	9	(D)	-	-	375 *	0.4	42	-	-	-
Utah	8	(D)	-	-	175 *	0.2	22	-	-	-
Louisiana	6	(D)	-	-	175 *	0.2	29	-	-	571

Source: 1992 *Economic Census*. The states are in descending order of shipments or establishments (if shipment data are missing for the majority). The symbol (D) appears when data are withheld to prevent disclosure of competitive information. States marked with (D) are sorted by number of establishments. A dash (-) indicates that the data element cannot be calculated; * indicates the midpoint of a range.

3471 - PLATING & POLISHING

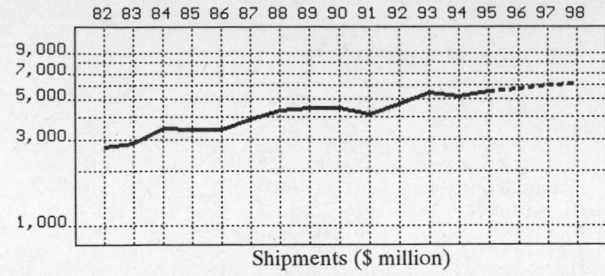

Shipments ($ million)

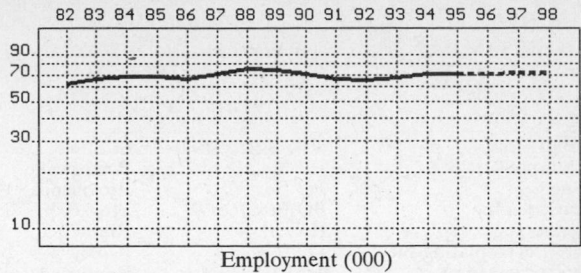

Employment (000)

GENERAL STATISTICS

Year	Companies	Establishments		Employment			Compensation		Production ($ million)			
		Total	with 20 or more employees	Total (000)	Production Workers (000)	Hours (Mil)	Payroll ($ mil)	Wages ($/hr)	Cost of Materials	Value Added by Manufacture	Value of Shipments	Capital Invest.
1982	3,367	3,450	898	61.9	49.7	98.1	919.0	6.44	1,038.4	1,693.9	2,731.4	109.2
1983		3,391	944	65.7	52.5	104.9	961.6	6.49	1,113.0	1,777.9	2,849.4	79.7
1984		3,332	990	68.7	55.7	112.6	1,143.4	7.05	1,254.8	2,207.8	3,457.4	184.5
1985		3,274	1,037	69.0	55.7	113.3	1,182.3	7.08	1,245.5	2,203.2	3,445.8	160.4
1986		3,204	978	65.8	51.5	104.6	1,179.1	7.50	1,149.2	2,252.3	3,408.3	137.7
1987	3,351	3,451	1,043	71.1	55.9	112.0	1,335.5	7.90	1,235.3	2,633.6	3,866.9	140.2
1988		3,278	1,065	76.3	58.2	118.5	1,530.2	8.27	1,523.7	2,810.7	4,324.0	148.6
1989		3,203	1,076	74.7	60.8	121.0	1,516.0	8.36	1,502.7	2,957.3	4,452.4	172.8
1990		3,180	1,013	70.6	57.9	115.8	1,534.4	8.77	1,538.4	2,981.0	4,513.3	167.1
1991		3,204	962	66.6	52.5	106.8	1,459.5	9.01	1,396.8	2,716.3	4,124.0	163.8
1992	3,161	3,296	950	65.4	50.3	103.9	1,563.5	9.51	1,567.0	3,192.8	4,725.7	207.9
1993		3,282	962	67.3	52.5	108.2	1,651.7	9.63	1,907.5	3,539.0	5,432.6	187.9
1994		3,199P	1,010P	70.6	54.5	113.3	1,740.5	9.52	1,598.1	3,649.1	5,232.5	170.1
1995		3,184P	1,013P	70.8P	55.0P	113.5P	1,816.0P	10.15P	1,679.3P	3,834.4P	5,498.2P	198.5P
1996		3,169P	1,016P	71.1P	55.1P	114.0P	1,880.7P	10.44P	1,742.7P	3,979.3P	5,706.0P	204.6P
1997		3,154P	1,018P	71.4P	55.2P	114.4P	1,945.5P	10.73P	1,806.2P	4,124.3P	5,913.9P	210.6P
1998		3,140P	1,021P	71.7P	55.3P	114.9P	2,010.2P	11.02P	1,869.7P	4,269.2P	6,121.7P	216.7P

Sources: 1982, 1987, 1992 *Economic Census*; *Annual Survey of Manufactures*, 83-86, 88-91, 93-94. Establishment counts for non-Census years are from *County Business Patterns*; establishment values for 83-84 are extrapolations. 'P's show projections by the editors. Industries reclassified in 87 will not have data for prior years.

INDICES OF CHANGE

Year	Companies	Establishments		Employment			Compensation		Production ($ million)			
		Total	with 20 or more employees	Total (000)	Production Workers (000)	Hours (Mil)	Payroll ($ mil)	Wages ($/hr)	Cost of Materials	Value Added by Manufacture	Value of Shipments	Capital Invest.
1982	107	105	95	95	99	94	59	68	66	53	58	53
1983		103	99	100	104	101	62	68	71	56	60	38
1984		101	104	105	111	108	73	74	80	69	73	89
1985		99	109	106	111	109	76	74	79	69	73	77
1986		97	103	101	102	101	75	79	73	71	72	66
1987	106	105	110	109	111	108	85	83	79	82	82	67
1988		99	112	117	116	114	98	87	97	88	91	71
1989		97	113	114	121	116	97	88	96	93	94	83
1990		96	107	108	115	111	98	92	98	93	96	80
1991		97	101	102	104	103	93	95	89	85	87	79
1992	100	100	100	100	100	100	100	100	100	100	100	100
1993		100	101	103	104	104	106	101	122	111	115	90
1994		97P	106P	108	108	109	111	100	102	114	111	82
1995		97P	107P	108P	109P	109P	116P	107P	107P	120P	116P	95P
1996		96P	107P	109P	110P	110P	120P	110P	111P	125P	121P	98P
1997		96P	107P	109P	110P	110P	124P	113P	115P	129P	125P	101P
1998		95P	107P	110P	110P	111P	129P	116P	119P	134P	130P	104P

Sources: Same as General Statistics. Values reflect change from the base year, 1992. Values above 100 mean greater than 92, values below 100 mean less than 92, and a value of 100 in the 82-91 or 93-98 period means same as 92. 'P's mark projections by the editors.

SELECTED RATIOS

For 1994	Avg. of All Manufact.	Analyzed Industry	Index	For 1994	Avg. of All Manufact.	Analyzed Industry	Index
Employees per Establishment	49	22	45	Value Added per Production Worker	134,084	66,956	50
Payroll per Establishment	1,500,273	544,087	36	Cost per Establishment	5,045,178	499,572	10
Payroll per Employee	30,620	24,653	81	Cost per Employee	102,970	22,636	22
Production Workers per Establishment	34	17	50	Cost per Production Worker	146,988	29,323	20
Wages per Establishment	853,319	337,179	40	Shipments per Establishment	9,576,895	1,635,698	17
Wages per Production Worker	24,861	19,791	80	Shipments per Employee	195,460	74,115	38
Hours per Production Worker	2,056	2,079	101	Shipments per Production Worker	279,017	96,009	34
Wages per Hour	12.09	9.52	79	Investment per Establishment	321,011	53,174	17
Value Added per Establishment	4,602,255	1,140,722	25	Investment per Employee	6,552	2,409	37
Value Added per Employee	93,930	51,687	55	Investment per Production Worker	9,352	3,121	33

Sources: Same as General Statistics. The 'Average of All Manufacturing' column represents the average of all manufacturing industries reported for the most recent complete year available. The Index shows the relationship between the Average and the Analyzed Industry. For example, 100 means that they are equal; 500 that the Analyzed Industry is five times the average; 50 means that the Analyzed Industry is half the national average. The abbreviation 'na' is used to show that data are 'not available'.

LEADING COMPANIES   Number shown: 75   Total sales ($ mil): 1,379   Total employment (000): 15.6

Company Name	Address				CEO Name	Phone	Co. Type	Sales ($ mil)	Empl. (000)
Siegel-Robert Inc	8645 S Broadway	St Louis	MO	63111	Bruce G Robert	314-638-8300	R	335	3.5
Plastene Supply Co	101 Meatte St	Portageville	MO	63873	Don Thompson	314-379-3857	D	57*	0.8
Technic Inc	PO Box 9650	Providence	RI	02906	Harant Shoushanian	401-781-6100	R	55*	0.2
Lorin Industries Inc	PO Box 766-W	Muskegon	MI	49443	RL Kersman	616-722-1631	R	50	0.2
S-R of Tennessee	Hwy 51 N	Ripley	TN	38063	Rod Schuh	901-635-3421	D	40	0.4
Tennessee Electroplating Inc	PO Box 233	Ripley	TN	38063	Tim Carter	901-635-0671	S	34	0.6
Lacks Enterprises Inc	4251 Brockton	Kentwood	MI	49512	Dan Savickas	616-698-2030	D	31	0.3
Handy & Harman	72 Elm St	N Attleboro	MA	02760	Allen E Molvar	508-695-1401	S	30	0.2
Aluminum Coil Anodizing Corp	501 E Lake St	Streamwood	IL	60107	Ronald L Rusch	708-837-4000	R	28	0.1
American Nickeloid Co	2900 W Main St	Peru	IL	61354	Ronald E Peterlin	815-223-0373	R	25	0.2
Custom Aluminum Products Inc	414 W Division St	South Elgin	IL	60177	JJ Castoro	708-741-6333	R	25	0.3
EC Industries Inc	PO Box 310	Martinez	CA	94550	Gary Garvens	510-372-3850	R	25	0.2
Pioneer Metal Finishing	1717 W River Rd N	Minneapolis	MN	55411	Delbert W Johnson	612-588-0855	D	25	0.4
Summit Corporation of America	1430 Waterbury Rd	Thomaston	CT	06787	Robert A Fumire	203-283-4391	R	25	0.2
Apollo Metals Ltd	PO Box 4045	Bethlehem	PA	18018	DE Kropp	215-867-5826	R	23	0.2
Applied Coating Technology Inc	12150 Techn Dr	Eden Prairie	MN	55344	PF Litchfield	612-941-4242	R	20	<0.1
Providence Metallizing Company	51 Fairlawn Av	Pawtucket	RI	02860	Harold Gadon	401-722-5300	R	20	0.4
US Chrome Corp	175 Garfield Av	Stratford	CT	06497	Nicholas R Spagnoli	203-378-9622	R	20	0.2
Metal Surfaces Inc	6060 Shull St	Bell Gardens	CA	90201	Don Brown	310-927-1331	R	17	0.3
Howard Plating Industry Inc	32565 Dequindre	Madison H	MI	48071	Paul A Sossi	810-588-5270	R	16	0.3
Embee Inc	2136 S Hathaway St	Santa Ana	CA	92705	John Dahlberg	714-546-9842	R	15	0.3
Lincoln Plating Co	600 W E St	Lincoln	NE	68522	Marc Lebaron	402-475-3671	R	15	0.3
SIFCO Selective Plating	5708 E Schaaf Rd	Independence	OH	44131	CA Graff	216-524-0099	D	15	0.1
Allied Finishing Inc	PO Box 3728	Grand Rapids	MI	49501	Robert W Corl Jr	616-698-7550	S	14	0.1
Amac Enterprises Inc	5909 W 130th St	Cleveland	OH	44130	Dean Chimples	216-362-1880	R	14	<0.1
MRC Inc	PO Box 3728	Grand Rapids	MI	49501	Robert Corl Jr	616-698-7550	R	14	0.2
Whyco Chromium Company Inc	670 Waterbury Rd	Thomaston	CT	06787	Mark S Hyner	203-283-5826	R	14	0.2
Capsco Inc	1101 W Blue Ridge	Greenville	SC	29609	William C Huffman	803-235-8000	R	11*	0.1
Empire Hard Chrome Inc	1615 S Kostner Av	Chicago	IL	60623	William G Horne Jr	312-762-3156	R	11*	0.2
Highland Plating Company Inc	1001 N Orange Dr	Los Angeles	CA	90038	Max Faeth	213-850-1020	R	11*	0.2
Horizons Inc	18531 S Miles Rd	Cleveland	OH	44128	HA Wainer	216-475-0555	R	11	0.1
Marsh Plating Corp	103 N Grove St	Ypsilanti	MI	48198	Matthew T Marsh	313-483-5767	R	11	0.2
Hytek Finishes Co	8127 S 216th	Kent	WA	98032	Douglas B Evans	206-872-7160	S	10*	0.1
Mansfield Industries Inc	PO Box 999	Mansfield	OH	44901	Otis M Cummins	419-524-1300	R	10	0.1
Master Finish Co	PO Box 7505	Grand Rapids	MI	49510	D Mulder	616-245-1228	R	10	0.1
Precision Plating Company Inc	4123 W Peterson	Chicago	IL	60646	James Belmonti	312-583-3333	R	10	<0.1
Superior Plating Inc	315 1st Av NE	Minneapolis	MN	55413	M P McMonagle	612-379-2121	R	10	0.1
SK Williams Co	4600 N 124th St	Wauwatosa	WI	53225	Robert Steuernagel	414-464-2220	R	10*	0.1
Acme Galvanizing Inc	PO Box 340050	Milwaukee	WI	53234	EJ Weiss	414-645-3250	R	9*	0.1
API Industries Inc	1250 Morris Av	Elk Grove Vill	IL	60007	Douglas T Walker	708-437-7474	R	9*	0.1
Fin-Clair Corp	4001 Gratiot St	St Louis	MO	63110	Robert C Mueller	314-535-7868	R	9*	0.1
Fountain Plating Company Inc	492 Prospect Av	W Springfield	MA	01089	LR Fountain	413-781-4651	R	9	0.1
George Industries	4116 Whiteside St	Los Angeles	CA	90063	GS Gering	213-264-6660	R	9*	0.2
United Plating Inc	PO Box 2046	Huntsville	AL	35804	J Caudle	205-852-8700	R	9	0.1
AACOA	2551 County Rd	Elkhart	IN	46514	Lou Gaspar	219-262-4685	R	8	0.1
EC Industries Inc	893 Carleton St	Berkeley	CA	94710	Tom Martin	510-849-4075	D	8*	0.1
Electronic Plating Service Inc	13021 S Budlong	Gardena	CA	90247	Richard P Brady	310-321-5747	S	8	<0.1
General Plating Inc	21841 Wyoming St	Oak Park	MI	48237	RP Guidebeck	313-546-9910	R	8	<0.1
Green Industries Corp	PO Box 41658	Cincinnati	OH	45262	Bernard Harris	513-769-4800	R	8	0.1
Modern Plating Co	PO Box 45007	Los Angeles	CA	90045	Leo Atimion	213-776-2440	R	8*	0.2
Production Anodizing Corp	PO Box 272	Adel	GA	31620	Zane Robinson	912-896-4531	R	8	0.2
Teer Plating Company Inc	6111 Wyche	Dallas	TX	75235	WM Teer Jr	214-637-1260	R	7	0.1
Amax Plating Inc	970 E Chicago St	Elgin	IL	60120	Paul Meyer	708-695-6100	R	7*	0.1
Bronson Plating Co	PO Box 69	Bronson	MI	49028	SR Welch	517-369-2885	R	7	0.1
CJ Saporito Plating Co	3119 S Austin Blv	Cicero	IL	60650	CJ Saporito Sr	312-242-2070	R	7	0.1
DV Industries Inc	2605 Industry Way	Lynwood	CA	90262	Peter J LaBarbera	213-563-1338	R	7	<0.1
Electro Machine and Engineering	PO Box 4998	Compton	CA	90224	Randall Turnbow	213-636-2391	R	7*	0.1
Liberty Industrial Finishing Corp	550 Suffolk Av	Brentwood	NY	11717	Lawrence Ripak Jr	516-273-4488	R	7*	0.1
Marlette National Corp	25 Rano St	Buffalo	NY	14207	EN Marlette Jr	716-874-1790	R	7	0.1
Modern Plating Corp	PO Box 838	Freeport	IL	61032	L Miller	815-235-3111	R	7*	0.1
P and H Plating Co	3416 W Belmont	Chicago	IL	60618	Jeff Pytlarz	312-463-2181	R	7*	0.1
Precision Anodizing and Plating	1601 N Miller St	Anaheim	CA	92806	John Rosser	714-996-1601	R	7*	<0.1
HM Quackenbush Inc	PO Box 429	Herkimer	NY	13350	Bronson Q Hager	315-866-3000	R	6	0.1
Almond Corp	PO Box 1387	Muskegon	MI	49443	Tammy Harley	616-798-7740	R	6	<0.1
AT and T Nassau Metals Corp	286 Richmond Val	Staten Island	NY	10307	Tom Kutyla	718-317-4400	S	6*	<0.1
Danco Metal Surfacing Inc	44 La Porte St	Arcadia	CA	91006	DL Tatge	818-447-8900	R	6	0.1
Erie Plating Co	656 W 12th St	Erie	PA	16512	Lewis T Briggs	814-453-7531	R	6	<0.1
Essex Machine Works Inc	PO Box 39	Essex	CT	06426	John S Johnson	203-767-8285	R	6	0.1
Michner Plating Co	520 N Mechanic St	Jackson	MI	49201	Walter S Michner	517-782-7151	R	6	0.1
South Holland Metal Finishing	143 W 154th St	South Holland	IL	60473	Robert Meagher	708-333-5292	R	6	<0.1
Varland Metal Service	3231 Fredonia Av	Cincinnati	OH	45229	Paul F Varland	513-861-0555	R	6	<0.1
Automation Plating Corp	927 Thompson Av	Glendale	CA	91201	William Wiggins	213-245-4951	R	6*	<0.1
Allegan Metal Finishing	PO Box 217	Allegan	MI	49010	WC Sosnowski	616-673-6604	R	5	0.1
Aluminum Finishing Corp	9850 E 30th St	Indianapolis	IN	46229	Dave Loner	317-897-9850	R	5	<0.1
Chromium Industries Inc	4645 W Chicago	Chicago	IL	60651	P Heidengren	312-287-3716	R	5	<0.1

Source: Ward's Business Directory of U.S. Private and Public Companies, Volumes 1 and 2, 1996. The company type code used is as follows: P - Public, R - Private, S - Subsidiary, D - Division, J - Joint Venture, A - Affiliate, G - Group. Sales are in millions of dollars, employees are in thousands. An asterisk (*) indicates an estimated sales volume. The symbol < stands for 'less than'. Company names and addresses are truncated, in some cases, to fit into the available space.

MATERIALS CONSUMED

Material	Quantity	Delivered Cost ($ million)
Materials, ingredients, containers, and supplies	(X)	1,200.4
Fabricated metal products, including forgings	(X)	105.6
Castings (rough and semifinished)	(X)	12.6
Steel shapes and forms	(X)	111.3
Nonferrous shapes and forms	(X)	9.8
Plastics materials and resins	(X)	38.8
Paints, varnishes, lacquers, stains, shellacs, japans, enamels, and allied products	(X)	19.6
Foundry chemicals, metal treating compounds, and plating compounds	(X)	201.8
Other chemicals and allied products	(X)	76.0
Grinding wheels and other abrasive products, except industrial diamonds	(X)	11.0
All other materials and components, parts, containers, and supplies	(X)	243.1
Materials, ingredients, containers, and supplies, nsk	(X)	370.9

Source: 1992 Economic Census. Explanation of symbols used: (D): Withheld to avoid disclosure of competitive data; na: Not available; (S): Withheld because statistical norms were not met; (X): Not applicable; (Z): Less than half the unit shown; nec: Not elsewhere classified; nsk: Not specified by kind; - : zero; * : 10-19 percent estimated; ** : 20-29 percent estimated.

PRODUCT SHARE DETAILS

Product or Product Class	% Share	Product or Product Class	% Share
Plating and polishing	100.00		

Source: 1992 Economic Census. The values shown are percent of total shipments in an industry. Values of indented subcategories are summed in the main heading. The symbol (D) appears when data are withheld to prevent disclosure of competitive information. The abbreviation nsk stands for 'not specified by kind' and nec for 'not elsewhere classified'.

INPUTS AND OUTPUTS FOR PLATING & POLISHING

Economic Sector or Industry Providing Inputs	%	Sector	Economic Sector or Industry Buying Outputs	%	Sector
Chemical preparations, nec	17.9	Manufg.	Electronic components nec	23.8	Manufg.
Communications, except radio & TV	15.3	Util.	Semiconductors & related devices	8.7	Manufg.
Wholesale trade	11.1	Trade	Hardware, nec	8.0	Manufg.
Primary copper	5.7	Manufg.	Motor vehicle parts & accessories	6.6	Manufg.
Electric services (utilities)	5.5	Util.	Wiring devices	3.6	Manufg.
Cyclic crudes and organics	3.6	Manufg.	Radio & TV communication equipment	3.4	Manufg.
Primary zinc	3.1	Manufg.	Screw machine and related products	3.4	Manufg.
Gas production & distribution (utilities)	3.0	Util.	Telephone & telegraph apparatus	3.0	Manufg.
Maintenance of nonfarm buildings nec	2.9	Constr.	Electronic computing equipment	2.7	Manufg.
Petroleum refining	2.5	Manufg.	Switchgear & switchboard apparatus	2.3	Manufg.
Motor freight transportation & warehousing	2.2	Util.	Aircraft & missile engines & engine parts	2.2	Manufg.
Real estate	1.6	Fin/R.E.	Plumbing fixture fittings & trim	2.0	Manufg.
Computer & data processing services	1.6	Services	Instruments to measure electricity	1.8	Manufg.
Abrasive products	1.4	Manufg.	Metal stampings, nec	1.8	Manufg.
Blast furnaces & steel mills	1.4	Manufg.	Hand & edge tools, nec	1.7	Manufg.
Machinery, except electrical, nec	1.4	Manufg.	Machinery, except electrical, nec	1.7	Manufg.
Eating & drinking places	1.3	Trade	Automotive stampings	1.5	Manufg.
Glass & glass products, except containers	1.1	Manufg.	Carburetors, pistons, rings, & valves	1.4	Manufg.
Paints & allied products	1.1	Manufg.	Aircraft	1.3	Manufg.
Metal stampings, nec	1.0	Manufg.	Aircraft & missile equipment, nec	1.3	Manufg.
Miscellaneous plastics products	1.0	Manufg.	Pipe, valves, & pipe fittings	1.3	Manufg.
Advertising	1.0	Services	Lighting fixtures & equipment	1.2	Manufg.
Special dies & tools & machine tool accessories	0.8	Manufg.	Industrial controls	1.1	Manufg.
Banking	0.8	Fin/R.E.	Motorcycles, bicycles, & parts	1.1	Manufg.
Plating & polishing	0.7	Manufg.	Typewriters & office machines, nec	1.1	Manufg.
Primary nonferrous metals, nec	0.7	Manufg.	Fabricated metal products, nec	0.9	Manufg.
Air transportation	0.7	Util.	Guided missiles & space vehicles	0.9	Manufg.
Railroads & related services	0.6	Util.	Internal combustion engines, nec	0.8	Manufg.
Engineering, architectural, & surveying services	0.6	Services	Special dies & tools & machine tool accessories	0.8	Manufg.
Equipment rental & leasing services	0.6	Services	Cutlery	0.7	Manufg.
Screw machine and related products	0.5	Manufg.	Environmental controls	0.7	Manufg.
Management & consulting services & labs	0.5	Services	Electron tubes	0.6	Manufg.
State & local government enterprises, nec	0.5	Gov't	Metal doors, sash, & trim	0.5	Manufg.
Industrial inorganic chemicals, nec	0.4	Manufg.	Small arms	0.5	Manufg.
Electrical repair shops	0.4	Services	Watches, clocks, & parts	0.5	Manufg.
Legal services	0.4	Services	Hand saws & saw blades	0.4	Manufg.
Lubricating oils & greases	0.3	Manufg.	Mechanical measuring devices	0.4	Manufg.
Metal heat treating	0.3	Manufg.	Motors & generators	0.4	Manufg.
Paperboard containers & boxes	0.3	Manufg.	Plating & polishing	0.4	Manufg.
Accounting, auditing & bookkeeping	0.3	Services	Special industry machinery, nec	0.4	Manufg.
Automotive rental & leasing, without drivers	0.3	Services	Ammunition, except for small arms, nec	0.3	Manufg.
Automotive repair shops & services	0.3	Services	Ball & roller bearings	0.3	Manufg.
Manifold business forms	0.2	Manufg.	Calculating & accounting machines	0.3	Manufg.
Sanitary services, steam supply, irrigation	0.2	Util.	General industrial machinery, nec	0.2	Manufg.
Water transportation	0.2	Util.	Metal sanitary ware	0.2	Manufg.
Insurance carriers	0.2	Fin/R.E.	Power driven hand tools	0.2	Manufg.
Business/professional associations	0.2	Services	Power transmission equipment	0.2	Manufg.
U.S. Postal Service	0.2	Gov't	Sheet metal work	0.2	Manufg.
Industrial gases	0.1	Manufg.	Small arms ammunition	0.2	Manufg.
Retail trade, except eating & drinking	0.1	Trade	Machine tools, metal cutting types	0.1	Manufg.
Hotels & lodging places	0.1	Services	Ordnance & accessories nec	0.1	Manufg.
Miscellaneous repair shops	0.1	Services	Scales & balances	0.1	Manufg.
Personnel supply services	0.1	Services			

Source: Benchmark Input-Output Accounts for the U.S. Economy, 1982, U.S. Department of Commerce, Washington, D.C., July 1991. Data, as reported in the source, are organized by the 1977 SIC structure in use in 1982 but have been matched, as closely as is possible, to the 1987 SIC structure used in this book.

OCCUPATIONS EMPLOYED BY SIC 347 - METAL SERVICES, NEC

Occupation	% of Total 1994	Change to 2005	Occupation	% of Total 1994	Change to 2005
Electrolytic plating machine workers	16.7	19.3	Traffic, shipping, & receiving clerks	2.0	14.8
Coating, painting, & spraying machine workers	8.1	19.3	Grinding machine operators, metal & plastic	2.0	19.3
Helpers, laborers, & material movers nec	6.4	19.3	Truck drivers light & heavy	2.0	23.0
Blue collar worker supervisors	5.1	9.7	Maintenance repairers, general utility	1.9	7.4
General managers & top executives	4.4	13.2	Industrial production managers	1.8	19.3
Metal & plastic machine workers nec	4.0	5.7	Bookkeeping, accounting, & auditing clerks	1.8	-10.5
Hand packers & packagers	3.8	2.3	General office clerks	1.8	1.8
Inspectors, testers, & graders, precision	3.4	19.3	Secretaries, ex legal & medical	1.6	8.6
Grinders & polishers, hand	3.2	-40.3	Freight, stock, & material movers, hand	1.5	-4.6
Assemblers, fabricators, & hand workers nec	2.9	19.3	Financial managers	1.2	19.4
Machine feeders & offbearers	2.5	7.4	Production, planning, & expediting clerks	1.0	19.3
Sales & related workers nec	2.2	19.3	Industrial truck & tractor operators	1.0	19.4

Source: Industry-Occupation Matrix, Bureau of Labor Statistics. These data relate to one or more 3-digit SIC industry groups rather than to a single 4-digit SIC. The change reported for each occupation to the year 2005 is a percent of growth or decline as estimated by the Bureau of Labor Statistics. The abbreviation nec stands for 'not elsewhere classified'.

LOCATION BY STATE AND REGIONAL CONCENTRATION

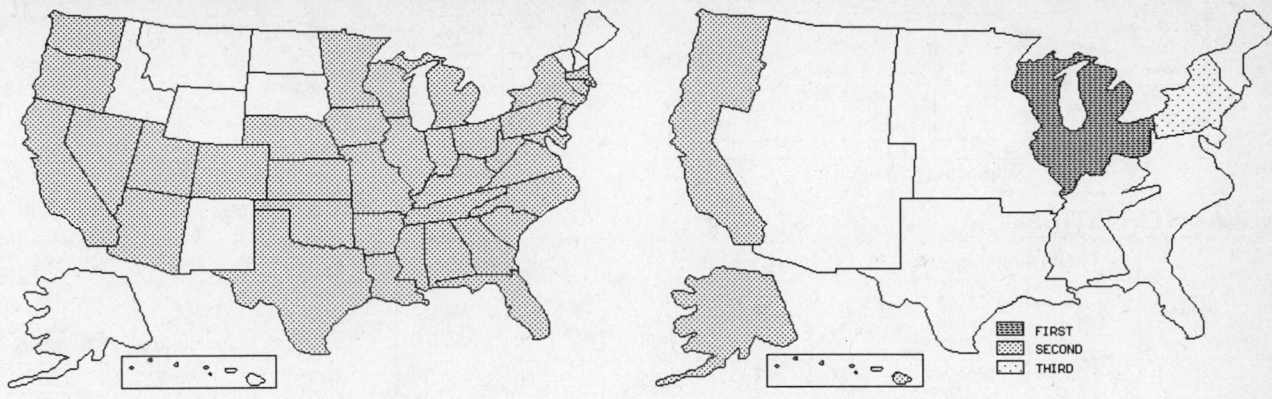

FIRST
SECOND
THIRD

INDUSTRY DATA BY STATE

| State | Establish-ments | Shipments | | | Employment | | | | Cost as % of Shipments | Investment per Employee ($) |
		Total ($ mil)	% of U.S.	Per Establ.	Total Number	% of U.S.	Per Establ.	Wages ($/hour)		
California	561	757.4	16.0	1.4	11,200	17.1	20	9.27	31.2	2,250
Ohio	275	478.1	10.1	1.7	6,300	9.6	23	9.59	27.8	2,556
Illinois	280	433.5	9.2	1.5	5,900	9.0	21	9.74	30.9	3,017
Michigan	252	415.4	8.8	1.6	5,800	8.9	23	9.29	31.2	2,810
Indiana	91	295.6	6.3	3.2	3,100	4.7	34	9.71	51.7	13,258
Pennsylvania	143	238.4	5.0	1.7	2,400	3.7	17	10.25	47.7	2,542
Connecticut	101	221.9	4.7	2.2	2,600	4.0	26	11.29	34.3	2,885
New York	167	172.6	3.7	1.0	2,700	4.1	16	8.84	30.9	1,296
Tennessee	61	172.4	3.6	2.8	2,100	3.2	34	8.88	44.1	4,619
Massachusetts	98	153.0	3.2	1.6	2,200	3.4	22	10.85	26.0	1,455
Texas	167	150.9	3.2	0.9	2,400	3.7	14	8.58	28.9	1,833
Minnesota	73	132.8	2.8	1.8	1,900	2.9	26	11.06	30.8	1,421
North Carolina	55	131.8	2.8	2.4	1,100	1.7	20	11.75	51.3	-
Rhode Island	101	121.2	2.6	1.2	1,900	2.9	19	8.90	29.9	3,158
New Jersey	116	119.2	2.5	1.0	1,700	2.6	15	10.12	33.6	1,941
Wisconsin	92	115.7	2.4	1.3	1,800	2.8	20	9.10	24.7	3,056
Missouri	66	109.1	2.3	1.7	1,700	2.6	26	8.93	31.9	2,176
Florida	93	58.7	1.2	0.6	900	1.4	10	9.29	26.1	2,333
Washington	51	39.7	0.8	0.8	700	1.1	14	10.30	24.9	2,714
Georgia	23	37.1	0.8	1.6	600	0.9	26	7.91	30.2	1,833
Kentucky	24	34.1	0.7	1.4	600	0.9	25	8.10	28.4	2,833
Arizona	43	32.9	0.7	0.8	700	1.1	16	8.82	29.5	2,571
Alabama	33	31.7	0.7	1.0	600	0.9	18	8.11	28.1	1,333
Oklahoma	36	26.3	0.6	0.7	500	0.8	14	7.71	20.9	800
West Virginia	10	21.5	0.5	2.2	300	0.5	30	11.80	25.6	-
Oregon	33	21.2	0.4	0.6	400	0.6	12	8.50	19.8	1,000
Colorado	30	21.0	0.4	0.7	400	0.6	13	8.17	28.6	1,250
Nebraska	7	17.6	0.4	2.5	200	0.3	29	9.25	21.6	1,500
Virginia	19	16.0	0.3	0.8	300	0.5	16	8.60	16.9	4,000
Maryland	26	14.6	0.3	0.6	300	0.5	12	9.25	25.3	1,000
Arkansas	17	14.0	0.3	0.8	200	0.3	12	8.00	45.7	3,500
Kansas	20	13.9	0.3	0.7	200	0.3	10	9.00	18.0	2,500
Louisiana	14	11.1	0.2	0.8	100	0.2	7	9.50	32.4	3,000
Nevada	14	10.3	0.2	0.7	200	0.3	14	8.67	21.4	1,500
Utah	9	7.7	0.2	0.9	100	0.2	11	10.00	11.7	2,000
Iowa	15	7.3	0.2	0.5	200	0.3	13	8.50	21.9	500
South Carolina	24	(D)	-	-	750 *	1.1	31	-	-	4,800
Mississippi	7	(D)	-	-	175 *	0.3	25	-	-	-

Source: 1992 *Economic Census.* The states are in descending order of shipments or establishments (if shipment data are missing for the majority). The symbol (D) appears when data are withheld to prevent disclosure of competitive information. States marked with (D) are sorted by number of establishments. A dash (-) indicates that the data element cannot be calculated; * indicates the midpoint of a range.

3479 - METAL COATING & ALLIED SERVICES

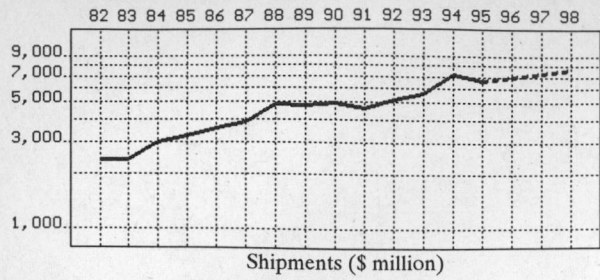

Shipments ($ million)

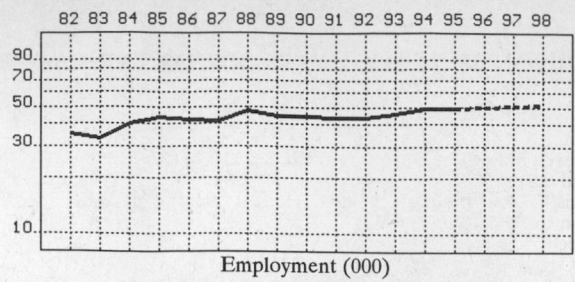

Employment (000)

GENERAL STATISTICS

Year	Companies	Establishments		Employment			Compensation		Production ($ million)			
		Total	with 20 or more employees	Total (000)	Production Workers (000)	Hours (Mil)	Payroll ($ mil)	Wages ($/hr)	Cost of Materials	Value Added by Manufacture	Value of Shipments	Capital Invest.
1982	1,524	1,620	507	35.0	27.5	54.1	599.1	7.46	1,195.9	1,202.6	2,393.4	91.0
1983		1,597	530	33.2	25.7	53.2	579.6	7.44	1,235.2	1,154.5	2,391.9	88.9
1984		1,574	553	40.6	32.5	65.1	715.5	7.36	1,525.8	1,502.7	3,008.3	111.3
1985		1,551	577	43.3	34.2	68.3	814.2	7.72	1,618.7	1,641.2	3,253.2	150.5
1986		1,533	563	42.7	32.8	66.1	824.2	8.26	1,824.2	1,735.0	3,573.7	137.9
1987	1,699	1,814	616	41.5	32.3	65.0	831.5	8.60	1,936.9	1,995.4	3,922.6	105.3
1988		1,741	607	47.0	36.3	75.3	984.2	8.70	2,230.7	2,647.1	4,867.4	128.2
1989		1,689	635	44.3	35.3	73.4	949.3	8.89	2,464.2	2,299.7	4,756.3	133.1
1990		1,746	636	44.1	34.7	70.5	999.9	9.77	2,503.6	2,429.3	4,928.7	158.8
1991		1,813	617	43.4	33.8	66.6	1,004.3	10.38	2,563.2	2,070.7	4,634.3	120.8
1992	1,821	1,945	640	43.7	33.8	69.5	1,093.0	10.33	2,680.7	2,504.7	5,160.7	159.3
1993		1,970	653	45.7	35.2	70.7	1,181.2	11.09	2,947.4	2,713.8	5,649.0	173.2
1994		1,944P	675P	48.6	37.8	79.0	1,286.0	10.50	3,779.2	3,316.8	7,081.8	223.0
1995		1,979P	687P	48.8P	37.8P	77.7P	1,284.7P	11.26P	3,529.4P	3,097.5P	6,613.7P	190.3P
1996		2,014P	700P	49.7P	38.5P	79.2P	1,337.9P	11.58P	3,707.4P	3,253.8P	6,947.2P	197.9P
1997		2,049P	712P	50.6P	39.2P	80.6P	1,391.1P	11.91P	3,885.4P	3,410.0P	7,280.8P	205.6P
1998		2,084P	725P	51.5P	39.8P	82.1P	1,444.3P	12.24P	4,063.4P	3,566.3P	7,614.4P	213.2P

Sources: 1982, 1987, 1992 *Economic Census*; *Annual Survey of Manufactures*, 83-86, 88-91, 93-94. Establishment counts for non-Census years are from *County Business Patterns*; establishment values for 83-84 are extrapolations. 'P's show projections by the editors. Industries reclassified in 87 will not have data for prior years.

INDICES OF CHANGE

Year	Companies	Establishments		Employment			Compensation		Production ($ million)			
		Total	with 20 or more employees	Total (000)	Production Workers (000)	Hours (Mil)	Payroll ($ mil)	Wages ($/hr)	Cost of Materials	Value Added by Manufacture	Value of Shipments	Capital Invest.
1982	84	83	79	80	81	78	55	72	45	48	46	57
1983		82	83	76	76	77	53	72	46	46	46	56
1984		81	86	93	96	94	65	71	57	60	58	70
1985		80	90	99	101	98	74	75	60	66	63	94
1986		79	88	98	97	95	75	80	68	69	69	87
1987	93	93	96	95	96	94	76	83	72	80	76	66
1988		90	95	108	107	108	90	84	83	106	94	80
1989		87	99	101	104	106	87	86	92	92	92	84
1990		90	99	101	103	101	91	95	93	97	96	100
1991		93	96	99	100	96	92	100	96	83	90	76
1992	100	100	100	100	100	100	100	100	100	100	100	100
1993		101	102	105	104	102	108	107	110	108	109	109
1994		100P	105P	111	112	114	118	102	141	132	137	140
1995		102P	107P	112P	112P	112P	118P	109P	132P	124P	128P	119P
1996		104P	109P	114P	114P	114P	122P	112P	138P	130P	135P	124P
1997		105P	111P	116P	116P	116P	127P	115P	145P	136P	141P	129P
1998		107P	113P	118P	118P	118P	132P	118P	152P	142P	148P	134P

Sources: Same as General Statistics. Values reflect change from the base year, 1992. Values above 100 mean greater than 92, values below 100 mean less than 92, and a value of 100 in the 82-91 or 93-98 period means same as 92. 'P's mark projections by the editors.

SELECTED RATIOS

For 1994	Avg. of All Manufact.	Analyzed Industry	Index	For 1994	Avg. of All Manufact.	Analyzed Industry	Index
Employees per Establishment	49	25	51	Value Added per Production Worker	134,084	87,746	65
Payroll per Establishment	1,500,273	661,548	44	Cost per Establishment	5,045,178	1,944,109	39
Payroll per Employee	30,620	26,461	86	Cost per Employee	102,970	77,761	76
Production Workers per Establishment	34	19	57	Cost per Production Worker	146,988	99,979	68
Wages per Establishment	853,319	426,714	50	Shipments per Establishment	9,576,895	3,643,043	38
Wages per Production Worker	24,861	21,944	88	Shipments per Employee	195,460	145,716	75
Hours per Production Worker	2,056	2,090	102	Shipments per Production Worker	279,017	187,349	67
Wages per Hour	12.09	10.50	87	Investment per Establishment	321,011	114,716	36
Value Added per Establishment	4,602,255	1,706,239	37	Investment per Employee	6,552	4,588	70
Value Added per Employee	93,930	68,247	73	Investment per Production Worker	9,352	5,899	63

Sources: Same as General Statistics. The 'Average of All Manufacturing' column represents the average of all manufacturing industries reported for the most recent complete year available. The Index shows the relationship between the Average and the Analyzed Industry. For example, 100 means that they are equal; 500 that the Analyzed Industry is five times the average; 50 means that the Analyzed Industry is half the national average. The abbreviation 'na' is used to show that data are 'not available'.

LEADING COMPANIES   Number shown: 75   Total sales ($ mil): **3,510**   Total employment (000): **17.1**

Company Name	Address				CEO Name	Phone	Co. Type	Sales ($ mil)	Empl. (000)
Engineered Materials	101 Wood Av S	Iselin	NJ	08830	Claude Azieres	908-205-5000	S	1,241	1.5
Material Sciences Corp	2300 E Pratt Blv	Elk Grove Vill	IL	60007	G Robert Evans	708-439-8270	P	228	0.9
Pre Finish Metals Inc	2300 E Pratt Blv	Elk Grove Vill	IL	60007	Gerald G Nadig	708-439-2210	S	171	0.8
Acheson Colloids Co	PO Box 611747	Port Huron	MI	48061	Micheal Porter	810-984-5581	R	130*	1.1
Metaltech	2400 2nd Av	Pittsburgh	PA	15219	Edwin H Gott	412-391-0483	R	130	<0.1
Sequa Corp Precoat Metals	4301 S Spring Av	St Louis	MO	63116	Gerald Dombek	314-352-8000	D	115	0.4
Optical Coating Laboratory Inc	2789 Northpoint	Santa Rosa	CA	95407	H M Dwight Jr	707-545-6440	D	110*	0.9
Metal Services Co	4100 W 42nd Pl	Chicago	IL	60632	Corry W Patton	312-523-0163	D	95	0.5
Metokote Corp	1340 Neubrecht Rd	Lima	OH	45801	James Blankemeyer	419-227-1100	R	80	1.0
Roll Coater Inc	PO Box 787	Greenfield	IN	46140	Don Ebert	317-462-7761	S	61	0.5
HI TecMetal Group Inc	1101 E 55th St	Cleveland	OH	44103	Terence Profughi	216-881-8100	R	50	0.5
Metal Prep Inc	501 N Greenwood	Houston	TX	77011	AR Ginn	713-921-7997	S	50	<0.1
Triumph Industries	8687 S 77th Av	Bridgeview	IL	60455	Thomas J Fournier	708-598-5100	D	45	<0.1
Walbridge Coatings	30610 E Broadway	Walbridge	OH	43465	Thomas E Moore	419-666-6130	D	42	0.2
Aztec Manufacturing Co	PO Box 668	Crowley	TX	76036	LC Martin	817-297-4361	P	41	0.3
Enamel Products and Plating Co	PO Box 159	McKeesport	PA	15134	KS Brand	412-461-2780	R	36*	0.3
I/N Kote	30755 Edison Rd	New Carlisle	IN	46552	Raymond Lepp	219-654-1000	R	36*	0.6
Unifoil Corp	217 Brook Av	Passaic Park	NJ	07055	EJ Alois	201-365-2000	R	35	0.1
Kinark Corp	PO Box 1499	Tulsa	OK	74101	Paul R Chastain	918-494-0964	P	35	0.5
Ceramx Corp	Hwy 14	Laurens	SC	29360	Peter Eichler	803-682-3215	R	30	0.4
E/M Corp	PO Box 2400	West Lafayette	IN	47906	LC Horwedel	317-497-6346	S	30	0.3
Hughes Brothers Inc	210 N 13th St	Seward	NE	68434	TR Hughes	402-643-2991	R	28	0.3
Multi-Arc Inc	200 Roundhill Dr	Rockaway	NJ	07866	Peter D Flood	201-625-3400	S	28	0.3
Alon Processing Inc	Grantham St	Tarentum	PA	15084	WP Heckel Jr	412-226-1677	S	25	<0.1
Eritech Inc	PO Box 487	Aberdeen	NC	28315	Jeff Church	910-944-3355	S	25	0.1
K and H Finishing Inc	2302 Trade Zone	San Jose	CA	95131	Edwin A Helwig	408-946-5440	R	25*	0.1
MATEC Corp	75 South St	Hopkinton	MA	01748	Robert B Gill	508-435-9039	P	24	0.3
Precoat Metals-Chicago	4800 S Kilbourn	Chicago	IL	60632	Raymond Druske	312-254-3400	D	24	0.1
Plating Technology Inc	PO Box 06236	Columbus	OH	43206	Dennis Goldman	614-228-2323	R	22*	0.2
Premier Coatings Inc	2250 Arthur Av	Elk Grove Vill	IL	60007	ER Andrus Jr	708-439-4200	R	20	<0.1
Witt Co	4454 Steel Pl	Cincinnati	OH	45209	Mary W Wydman	513-871-5700	R	20*	0.2
Berridge Manufacturing Co	1720 Maury St	Houston	TX	77026	Jack Berridge	713-223-4971	R	18*	0.2
TAFA Inc	146 Pembroke Rd	Concord	NH	03301	Walter Zanchuk	603-224-9585	S	18	<0.1
Chace Precision Metals Inc	1704 Barnes St	Reidsville	NC	27320	Louis E Skrabec	910-342-2381	S	17	0.2
Thermoclad Co	4690 Iroquois Av	Erie	PA	16511	AI Renkis	814-899-7628	R	17	<0.1
Gregory Galvanizing	4100 13th St SW	Canton	OH	44708	TS Gregory	216-477-4800	R	15	0.1
Linetec	725 S 75th Av	Wausau	WI	54401	Scott Platta	715-843-4100	D	15	0.3
Midwest Pipe Coating Inc	925 Kennedy Av	Schererville	IN	46375	Robert Theisen	219-322-4564	S	15*	0.1
Porcelain Industries Inc	20 Ceco Rd	Dickson	TN	37055	John M Walsh	615-446-7400	S	15	0.1
Porcelain Metals Corp	1400 S 13th St	Louisville	KY	40210	JH McBride	502-635-7421	R	15	0.1
Boyles Galvanizing Co	PO Box 187	Hurst	TX	76053	Harry D Jones	817-268-0713	S	15	0.2
Spraylat Corp	716 S Columbus	Mount Vernon	NY	10550	JE Borner	914-699-3030	R	14*	0.1
Eastern Etching & Mfg Co	Lower Grape St	Chicopee	MA	01013	Joseph A Lavallee	413-594-6601	R	13	0.2
A and A Coating Inc	PO Box 476	Lone Star	TX	75668	Kenneth J Clary	903-639-2595	R	12	0.1
Bayou Pipe Coating Co	PO Box 11010	New Iberia	LA	70562	Stewart Shea	318-369-3761	R	12*	0.1
Impact Label Corp	3434 S Burdick St	Kalamazoo	MI	49001	W Fogleson	616-381-4280	R	12	0.1
National Galvanizing LP	1500 Telb Rd	Monroe	MI	48161	Joe Rollins	313-243-1882	S	12*	0.1
Powder Coat Technology Inc	17024 Taft Rd	Spring Lake	MI	49456	Ken Goodsen	616-847-2000	S	12	0.1
Supracote Inc	11200 Arrow Rtc	R Cucamonga	CA	91730	Gerald E Smith	909-987-4711	R	12*	0.1
Texas Electric Cooperatives Inc	Bevel Loop Rd	Jasper	TX	75951	TJ Anthony	409-384-4633	D	12	0.1
Galvan Industries Inc	PO Box 369	Harrisburg	NC	28075	Laurens Willard	704-455-5102	R	11*	<0.1
Finish Co	1820 Hayes	Grand Haven	MI	49417	Don Wisner	616-846-6800	R	10	0.1
Monarch Wear Inc	PO Box 99	Algoma	WI	54201	Wes Drumm	414-487-5236	S	10	<0.1
Tandem Products Inc	3444 Dight Av S	Minneapolis	MN	55406	David Tweet	612-721-2911	R	10	<0.1
Universal Chemicals & Coatings	1975 Fox Ln	Elgin	IL	60123	YE Chin	708-931-1700	R	10	<0.1
General Magnaplate Corp	1331 US Rte 1	Linden	NJ	07036	Charles P Covino	908-862-6200	P	10	0.1
Roesch Inc	PO Box 328	Belleville	IL	62222	W Robert Voges	618-233-2760	R	9	0.2
Continuous Coating Corp	520 W Grove Av	Orange	CA	92665	RC Larrabee	714-637-4642	R	9	<0.1
Young Galvanizing Inc	Rte 551	Pulaski	PA	16143	Thomas E Litzinger	412-658-1666	R	9*	<0.1
Columbus Galvanizing Inc	1000 Buckeye Pk Rd	Columbus	OH	43207	Robert A Voigt	614-443-4621	S	8*	<0.1
Electro Static Finishing Inc	PO Box 8	Winsted	MN	55395	Jack Gustafson	612-781-6641	R	8*	0.1
Metal-Cladding Inc	230 S Niagara St	Lockport	NY	14094	Alexander F Robb	716-434-5513	R	8	0.1
Plasma Technology Inc	1754 Crenshaw Blv	Torrance	CA	90501	BE Fosket	310-320-3373	R	8	<0.1
Richter Precision Inc	1021 Commercial	E Petersburg	PA	17520	Hans Richter	717-560-9990	R	8*	<0.1
Voigt and Schweitzer Inc	1000 Buckeye Pk Rd	Columbus	OH	43207	Robert A Voigt	614-443-4621	R	8*	<0.1
Whitford Corp	PO Box 2347	West Chester	PA	19380	David P Willis Jr	610-296-3200	R	8*	<0.1
Wright Metal Processors	1700 W Cornell St	Milwaukee	WI	53209	Ronald G Jacob	414-263-1144	D	8*	<0.1
Alloy Surfaces Company Inc	100 Locke Rd	Wilmington	DE	19809	LS Bowers	302-762-8900	S	7	<0.1
Brainerd Industries	PO Box 77	Dayton	OH	45404	Edward Dudon	513-228-0488	D	7*	0.1
BL Downey Company Inc	2125 Gardner Rd	Broadview	IL	60153	Bernard L Downey	708-345-8000	R	7*	0.2
California Finished Metals Inc	9133 Center Av	R Cucamonga	CA	91730	Frank Molles	909-987-4681	R	7	0.1
Chemring Group Inc	100 Locke Rd	Wilmington	DE	19809	LS Bowers	302-762-8900	R	7	<0.1
ISPA Co	2915 Wilmarco Av	Baltimore	MD	21223	Ronald L Kaufman	410-644-4500	R	7	0.1
Metalade of PA Inc	PO Box 8287	Erie	PA	16505	Billy W Wallis	814-838-1911	S	7	0.1
Metallized Products Inc	37 East St	Winchester	MA	01890	Jason Weisman	617-729-8300	R	7	<0.1

Source: Ward's Business Directory of U.S. Private and Public Companies, Volumes 1 and 2, 1996. The company type code used is as follows: P - Public, R - Private, S - Subsidiary, D - Division, J - Joint Venture, A - Affiliate, G - Group. Sales are in millions of dollars, employees are in thousands. An asterisk (*) indicates an estimated sales volume. The symbol < stands for 'less than'. Company names and addresses are truncated, in some cases, to fit into the available space.

MATERIALS CONSUMED

Material	Quantity	Delivered Cost ($ million)
Materials, ingredients, containers, and supplies	(X)	2,416.2
Fabricated metal products, including forgings	(X)	112.0
Castings (rough and semifinished)	(X)	6.6
Steel shapes and forms	(X)	986.3
Nonferrous shapes and forms	(X)	40.5
Plastics materials and resins	(X)	61.6
Paints, varnishes, lacquers, stains, shellacs, japans, enamels, and allied products	(X)	302.1
Foundry chemicals, metal treating compounds, and plating compounds	(X)	104.1
Other chemicals and allied products	(X)	42.8
Grinding wheels and other abrasive products, except industrial diamonds	(X)	2.8
All other materials and components, parts, containers, and supplies	(X)	282.3
Materials, ingredients, containers, and supplies, nsk	(X)	475.2

Source: 1992 *Economic Census*. Explanation of symbols used: (D): Withheld to avoid disclosure of competitive data; na: Not available; (S): Withheld because statistical norms were not met; (X): Not applicable; (Z): Less than half the unit shown; nec: Not elsewhere classified; nsk: Not specified by kind; - : zero; * : 10-19 percent estimated; ** : 20-29 percent estimated.

PRODUCT SHARE DETAILS

Product or Product Class	% Share	Product or Product Class	% Share
Metal coating and allied services	100.00	(including organic coatings, enamels, lacquers, alkyds, plastics, etc.)	14.87
Electronic metal engraving, excluding metal nameplates	0.74		
Photo chemical metal etching, including machining (excluding metal nameplates)	2.37	Metal powder coating, including electrostatic and fluidized bed (including organic coatings, lacquers, alkyds, plastics, etc.)	11.64
Etching and engraving metal nameplates	1.54		
Other metal engraving and etching	1.45	All other metal coating, including curtain coating and wash coating (including organic coatings, enamels, lacquers, alkyds, plastics, etc.)	5.62
Metal galvanizing and other hot dip metal coating	26.02		
Metal coil coating (including organic coatings, enamels, lacquers, alkyds, plastics, etc.)	15.86	Inorganic metal coatings, including porcelain	5.70
Metal liquid spray coating, including electrostatic coating			

Source: 1992 *Economic Census*. The values shown are percent of total shipments in an industry. Values of indented subcategories are summed in the main heading. The symbol (D) appears when data are withheld to prevent disclosure of competitive information. The abbreviation nsk stands for 'not specified by kind' and nec for 'not elsewhere classified'.

INPUTS AND OUTPUTS FOR METAL COATING & ALLIED SERVICES

Economic Sector or Industry Providing Inputs	%	Sector	Economic Sector or Industry Buying Outputs	%	Sector
Blast furnaces & steel mills	13.4	Manufg.	Prefabricated metal buildings	10.0	Manufg.
Paints & allied products	11.9	Manufg.	Sheet metal work	6.2	Manufg.
Wholesale trade	10.8	Trade	Crowns & closures	5.4	Manufg.
Cyclic crudes and organics	10.6	Manufg.	Electronic components nec	5.1	Manufg.
Chemical preparations, nec	10.3	Manufg.	Fabricated structural metal	4.9	Manufg.
Primary zinc	7.2	Manufg.	Blast furnaces & steel mills	4.5	Manufg.
Metal coating & allied services	5.1	Manufg.	Metal cans	3.8	Manufg.
Advertising	3.9	Services	X-ray apparatus & tubes	3.7	Manufg.
Electric services (utilities)	3.2	Util.	Motor vehicles & car bodies	3.3	Manufg.
Gas production & distribution (utilities)	2.9	Util.	Metal coating & allied services	3.1	Manufg.
Miscellaneous plastics products	2.6	Manufg.	Nonferrous wire drawing & insulating	2.2	Manufg.
Motor freight transportation & warehousing	2.3	Util.	Fabricated plate work (boiler shops)	2.0	Manufg.
Petroleum refining	1.3	Manufg.	Aluminum rolling & drawing	1.9	Manufg.
Industrial inorganic chemicals, nec	1.2	Manufg.	Semiconductors & related devices	1.9	Manufg.
Real estate	1.2	Fin/R.E.	Hardware, nec	1.7	Manufg.
Railroads & related services	1.0	Util.	Metal barrels, drums, & pails	1.7	Manufg.
Banking	0.9	Fin/R.E.	Oil field machinery	1.6	Manufg.
Maintenance of nonfarm buildings nec	0.8	Constr.	Miscellaneous fabricated wire products	1.5	Manufg.
Machinery, except electrical, nec	0.8	Manufg.	Surgical appliances & supplies	1.5	Manufg.
Coal	0.7	Mining	Gaskets, packing & sealing devices	1.4	Manufg.
Air transportation	0.7	Util.	Motor vehicle parts & accessories	1.4	Manufg.
Eating & drinking places	0.7	Trade	Surgical & medical instruments	1.4	Manufg.
Abrasive products	0.5	Manufg.	Nonferrous forgings	1.3	Manufg.
Communications, except radio & TV	0.5	Util.	Iron & steel foundries	1.2	Manufg.
Water transportation	0.5	Util.	Cold finishing of steel shapes	0.9	Manufg.
Special dies & tools & machine tool accessories	0.4	Manufg.	Automotive stampings	0.8	Manufg.
Engineering, architectural, & surveying services	0.4	Services	Mobile homes	0.8	Manufg.
Equipment rental & leasing services	0.4	Services	Wiring devices	0.8	Manufg.
Management & consulting services & labs	0.3	Services	Elevators & moving stairways	0.7	Manufg.
Alkalies & chlorine	0.2	Manufg.	Measuring & dispensing pumps	0.7	Manufg.
Lubricating oils & greases	0.2	Manufg.	Radio & TV communication equipment	0.7	Manufg.
Metal heat treating	0.2	Manufg.	Screw machine and related products	0.7	Manufg.
Paperboard containers & boxes	0.2	Manufg.	Retail trade, except eating & drinking	0.7	Trade
Sanitary services, steam supply, irrigation	0.2	Util.	Electronic computing equipment	0.6	Manufg.
Accounting, auditing & bookkeeping	0.2	Services	Engine electrical equipment	0.6	Manufg.
Automotive rental & leasing, without drivers	0.2	Services	Iron & steel forgings	0.6	Manufg.

Continued on next page.

INPUTS AND OUTPUTS FOR METAL COATING & ALLIED SERVICES - Continued

Economic Sector or Industry Providing Inputs	%	Sector	Economic Sector or Industry Buying Outputs	%	Sector
Automotive repair shops & services	0.2	Services	Miscellaneous plastics products	0.6	Manufg.
Computer & data processing services	0.2	Services	Nonferrous rolling & drawing, nec	0.6	Manufg.
Legal services	0.2	Services	Steel pipe & tubes	0.6	Manufg.
U.S. Postal Service	0.2	Gov't	Telephone & telegraph apparatus	0.6	Manufg.
Manifold business forms	0.1	Manufg.	Aircraft & missile engines & engine parts	0.5	Manufg.
Soap & other detergents	0.1	Manufg.	Blowers & fans	0.5	Manufg.
Insurance carriers	0.1	Fin/R.E.	Conveyors & conveying equipment	0.5	Manufg.
Business/professional associations	0.1	Services	Copper rolling & drawing	0.5	Manufg.
Hotels & lodging places	0.1	Services	Fabricated metal products, nec	0.5	Manufg.
			Food products machinery	0.5	Manufg.
			Lighting fixtures & equipment	0.5	Manufg.
			Miscellaneous metal work	0.5	Manufg.
			Railroad equipment	0.5	Manufg.
			Switchgear & switchboard apparatus	0.5	Manufg.
			Equipment rental & leasing services	0.5	Services
			Hand & edge tools, nec	0.4	Manufg.
			Instruments to measure electricity	0.4	Manufg.
			Machinery, except electrical, nec	0.4	Manufg.
			Manufacturing industries, nec	0.4	Manufg.
			Metal stampings, nec	0.4	Manufg.
			Plumbing fixture fittings & trim	0.4	Manufg.
			Primary metal products, nec	0.4	Manufg.
			Service industry machines, nec	0.4	Manufg.
			Travel trailers & campers	0.4	Manufg.
			Aircraft	0.3	Manufg.
			Aircraft & missile equipment, nec	0.3	Manufg.
			Carburetors, pistons, rings, & valves	0.3	Manufg.
			Metal doors, sash, & trim	0.3	Manufg.
			Nonferrous castings, nec	0.3	Manufg.
			Pipe, valves, & pipe fittings	0.3	Manufg.
			Truck & bus bodies	0.3	Manufg.
			Truck trailers	0.3	Manufg.
			Dental equipment & supplies	0.2	Manufg.
			Electric housewares & fans	0.2	Manufg.
			Environmental controls	0.2	Manufg.
			Guided missiles & space vehicles	0.2	Manufg.
			Household appliances, nec	0.2	Manufg.
			Household laundry equipment	0.2	Manufg.
			Household refrigerators & freezers	0.2	Manufg.
			Industrial controls	0.2	Manufg.
			Internal combustion engines, nec	0.2	Manufg.
			Metal partitions & fixtures	0.2	Manufg.
			Motor homes (made on purchased chassis)	0.2	Manufg.
			Motorcycles, bicycles, & parts	0.2	Manufg.
			Pumps & compressors	0.2	Manufg.
			Refrigeration & heating equipment	0.2	Manufg.
			Special dies & tools & machine tool accessories	0.2	Manufg.
			Typewriters & office machines, nec	0.2	Manufg.
			Cutlery	0.1	Manufg.
			Electron tubes	0.1	Manufg.
			Sewing machines	0.1	Manufg.
			Small arms	0.1	Manufg.
			Watches, clocks, & parts	0.1	Manufg.

Source: Benchmark Input-Output Accounts for the U.S. Economy, 1982, U.S. Department of Commerce, Washington, D.C., July 1991. Data, as reported in the source, are organized by the 1977 SIC structure in use in 1982 but have been matched, as closely as is possible, to the 1987 SIC structure used in this book.

OCCUPATIONS EMPLOYED BY SIC 347 - METAL SERVICES, NEC

Occupation	% of Total 1994	Change to 2005	Occupation	% of Total 1994	Change to 2005
Electrolytic plating machine workers	16.7	19.3	Traffic, shipping, & receiving clerks	2.0	14.8
Coating, painting, & spraying machine workers	8.1	19.3	Grinding machine operators, metal & plastic	2.0	19.3
Helpers, laborers, & material movers nec	6.4	19.3	Truck drivers light & heavy	2.0	23.0
Blue collar worker supervisors	5.1	9.7	Maintenance repairers, general utility	1.9	7.4
General managers & top executives	4.4	13.2	Industrial production managers	1.8	19.3
Metal & plastic machine workers nec	4.0	5.7	Bookkeeping, accounting, & auditing clerks	1.8	-10.5
Hand packers & packagers	3.8	2.3	General office clerks	1.8	1.8
Inspectors, testers, & graders, precision	3.4	19.3	Secretaries, ex legal & medical	1.6	8.6
Grinders & polishers, hand	3.2	-40.3	Freight, stock, & material movers, hand	1.5	-4.6
Assemblers, fabricators, & hand workers nec	2.9	19.3	Financial managers	1.2	19.4
Machine feeders & offbearers	2.5	7.4	Production, planning, & expediting clerks	1.0	19.3
Sales & related workers nec	2.2	19.3	Industrial truck & tractor operators	1.0	19.4

Source: Industry-Occupation Matrix, Bureau of Labor Statistics. These data relate to one or more 3-digit SIC industry groups rather than to a single 4-digit SIC. The change reported for each occupation to the year 2005 is a percent of growth or decline as estimated by the Bureau of Labor Statistics. The abbreviation nec stands for 'not elsewhere classified'.

LOCATION BY STATE AND REGIONAL CONCENTRATION

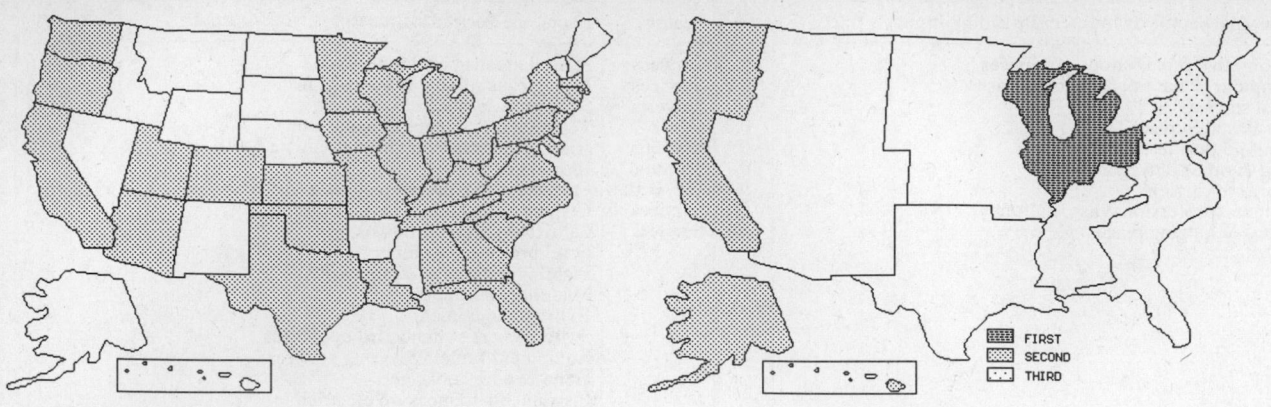

FIRST
SECOND
THIRD

INDUSTRY DATA BY STATE

State	Establish-ments	Shipments			Employment				Cost as % of Shipments	Investment per Employee ($)
		Total ($ mil)	% of U.S.	Per Establ.	Total Number	% of U.S.	Per Establ.	Wages ($/hour)		
Ohio	138	883.1	17.1	6.4	4,200	9.6	30	12.18	69.5	3,667
California	307	611.0	11.8	2.0	5,900	13.5	19	9.89	52.6	3,254
Illinois	121	474.0	9.2	3.9	3,300	7.6	27	11.72	58.2	3,788
Pennsylvania	103	430.2	8.3	4.2	2,200	5.0	21	11.33	60.7	3,364
Michigan	165	373.8	7.2	2.3	5,100	11.7	31	9.14	34.5	2,784
Texas	142	307.3	6.0	2.2	2,700	6.2	19	9.23	45.4	4,222
Indiana	66	220.5	4.3	3.3	2,200	5.0	33	10.38	44.5	5,818
New York	101	189.3	3.7	1.9	2,100	4.8	21	11.00	31.3	4,429
New Jersey	62	97.4	1.9	1.6	1,200	2.7	19	10.25	34.9	2,000
Wisconsin	52	95.5	1.9	1.8	1,500	3.4	29	9.08	35.7	2,867
Massachusetts	82	93.1	1.8	1.1	1,200	2.7	15	10.47	35.3	2,500
Connecticut	43	88.5	1.7	2.1	1,100	2.5	26	12.61	28.8	3,364
Missouri	37	76.9	1.5	2.1	700	1.6	19	11.69	25.5	4,571
Georgia	16	71.8	1.4	4.5	500	1.1	31	10.13	46.7	6,000
Oklahoma	26	68.3	1.3	2.6	500	1.1	19	12.40	39.5	9,400
Minnesota	43	67.8	1.3	1.6	1,100	2.5	26	9.06	33.5	1,636
Kentucky	14	59.2	1.1	4.2	500	1.1	36	9.57	65.2	5,000
Tennessee	25	55.2	1.1	2.2	700	1.6	28	7.42	39.3	9,143
Florida	56	51.8	1.0	0.9	500	1.1	9	8.89	37.6	2,800
Rhode Island	38	49.6	1.0	1.3	900	2.1	24	7.33	24.8	1,000
North Carolina	29	49.6	1.0	1.7	500	1.1	17	10.00	37.3	-
Washington	30	45.8	0.9	1.5	400	0.9	13	11.43	40.0	3,500
Louisiana	18	44.9	0.9	2.5	400	0.9	22	8.25	45.7	6,000
Colorado	27	40.6	0.8	1.5	300	0.7	11	10.00	56.9	3,000
Maryland	17	37.7	0.7	2.2	400	0.9	24	15.40	47.5	2,250
Alabama	17	35.7	0.7	2.1	500	1.1	29	9.57	53.5	2,400
Arizona	25	34.7	0.7	1.4	400	0.9	16	9.29	24.5	5,000
Oregon	25	30.1	0.6	1.2	400	0.9	16	9.00	32.6	2,000
Kansas	10	21.3	0.4	2.1	400	0.9	40	9.14	23.5	2,250
Iowa	9	19.9	0.4	2.2	300	0.7	33	7.17	31.7	667
Utah	14	16.3	0.3	1.2	200	0.5	14	10.33	45.4	2,000
Virginia	15	13.5	0.3	0.9	100	0.2	7	9.50	49.6	2,000
South Carolina	13	(D)	-	-	375 *	0.9	29	-	-	2,133
West Virginia	5	(D)	-	-	175 *	0.4	35	-	-	-

Source: 1992 *Economic Census*. The states are in descending order of shipments or establishments (if shipment data are missing for the majority). The symbol (D) appears when data are withheld to prevent disclosure of competitive information. States marked with (D) are sorted by number of establishments. A dash (-) indicates that the data element cannot be calculated; * indicates the midpoint of a range.

3482 - SMALL ARMS AMMUNITION

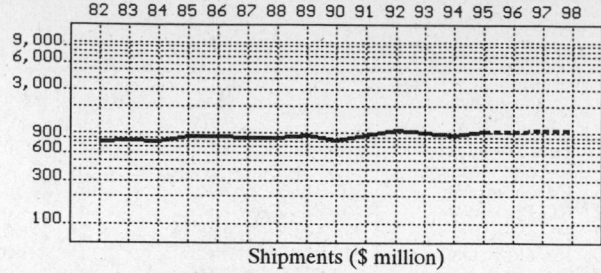

Shipments ($ million)

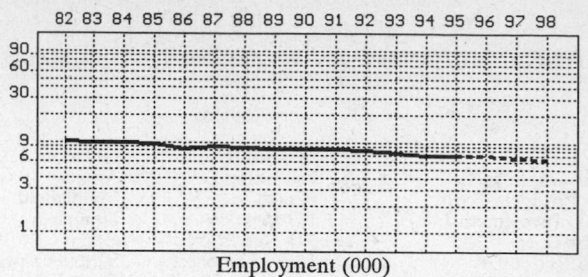

Employment (000)

GENERAL STATISTICS

Year	Companies	Establishments		Employment			Compensation		Production ($ million)			
		Total	with 20 or more employees	Total (000)	Production Workers (000)	Hours (Mil)	Payroll ($ mil)	Wages ($/hr)	Cost of Materials	Value Added by Manufacture	Value of Shipments	Capital Invest.
1982	59	66	19	10.1	7.4	14.1	221.6	10.12	265.6	518.4	799.8	23.3
1983		64	18	10.0	7.1	14.3	238.4	10.52	250.7	581.1	837.6	25.8
1984		62	17	9.8	7.0	13.8	258.6	11.58	261.7	561.7	811.4	32.9
1985		60	15	9.6	7.7	14.9	252.4	11.70	323.6	624.1	917.5	28.9
1986		65	16	8.4	6.8	13.4	222.7	12.21	303.0	604.2	918.7	34.3
1987	75	79	18	9.0	6.9	14.5	243.5	11.60	242.7	626.7	889.2	17.5
1988		80	18	8.7	6.5	14.0	240.8	11.61	304.3	608.7	883.3	20.0
1989		83	19	8.4	6.6	13.2	242.2	12.23	354.4	637.3	963.0	11.6
1990		79	21	8.3	6.3	12.5	242.9	13.19	344.5	535.9	844.1	29.4
1991		85	22	8.5	6.4	12.7	249.6	13.38	369.5	520.3	941.5	24.0
1992	101	103	19	8.1	6.1	12.6	258.6	14.04	397.6	694.5	1,088.6	31.4
1993		102	19	7.7	5.8	11.5	251.5	15.02	367.4	640.0	1,021.8	31.8
1994		101P	20P	7.2	5.5	11.6	241.7	14.77	338.1	628.4	963.4	35.4
1995		105P	20P	7.2P	5.6P	11.6P	251.8P	15.07P	361.3P	671.5P	1,029.5P	29.2P
1996		109P	20P	7.0P	5.4P	11.4P	253.0P	15.44P	367.1P	682.3P	1,046.1P	29.5P
1997		112P	21P	6.8P	5.3P	11.2P	254.1P	15.82P	372.9P	693.1P	1,062.6P	29.9P
1998		116P	21P	6.6P	5.2P	10.9P	255.3P	16.19P	378.7P	703.9P	1,079.1P	30.3P

Sources: 1982, 1987, 1992 *Economic Census*; *Annual Survey of Manufactures*, 83-86, 88-91, 93-94. Establishment counts for non-Census years are from *County Business Patterns*; establishment values for 83-84 are extrapolations. 'P's show projections by the editors. Industries reclassified in 87 will not have data for prior years.

INDICES OF CHANGE

Year	Companies	Establishments		Employment			Compensation		Production ($ million)			
		Total	with 20 or more employees	Total (000)	Production Workers (000)	Hours (Mil)	Payroll ($ mil)	Wages ($/hr)	Cost of Materials	Value Added by Manufacture	Value of Shipments	Capital Invest.
1982	58	64	100	125	121	112	86	72	67	75	73	74
1983		62	95	123	116	113	92	75	63	84	77	82
1984		60	89	121	115	110	100	82	66	81	75	105
1985		58	79	119	126	118	98	83	81	90	84	92
1986		63	84	104	111	106	86	87	76	87	84	109
1987	74	77	95	111	113	115	94	83	61	90	82	56
1988		78	95	107	107	111	93	83	77	88	81	64
1989		81	100	104	108	105	94	87	89	92	88	37
1990		77	111	102	103	99	94	94	87	77	78	94
1991		83	116	105	105	101	97	95	93	75	86	76
1992	100	100	100	100	100	100	100	100	100	100	100	100
1993		99	95	95	95	91	97	107	92	92	94	101
1994		98P	105P	89	90	92	93	105	85	90	88	113
1995		102P	106P	89P	92P	92P	97P	107P	91P	97P	95P	93P
1996		106P	108P	86P	89P	91P	98P	110P	92P	98P	96P	94P
1997		109P	109P	84P	87P	89P	98P	113P	94P	100P	98P	95P
1998		113P	110P	81P	85P	87P	99P	115P	95P	101P	99P	96P

Sources: Same as General Statistics. Values reflect change from the base year, 1992. Values above 100 mean greater than 92, values below 100 mean less than 92, and a value of 100 in the 82-91 or 93-98 period means same as 92. 'P's mark projections by the editors.

SELECTED RATIOS

For 1994	Avg. of All Manufact.	Analyzed Industry	Index	For 1994	Avg. of All Manufact.	Analyzed Industry	Index
Employees per Establishment	49	71	145	Value Added per Production Worker	134,084	114,255	85
Payroll per Establishment	1,500,273	2,384,128	159	Cost per Establishment	5,045,178	3,335,017	66
Payroll per Employee	30,620	33,569	110	Cost per Employee	102,970	46,958	46
Production Workers per Establishment	34	54	158	Cost per Production Worker	146,988	61,473	42
Wages per Establishment	853,319	1,690,018	198	Shipments per Establishment	9,576,895	9,502,974	99
Wages per Production Worker	24,861	31,151	125	Shipments per Employee	195,460	133,806	68
Hours per Production Worker	2,056	2,109	103	Shipments per Production Worker	279,017	175,164	63
Wages per Hour	12.09	14.77	122	Investment per Establishment	321,011	349,185	109
Value Added per Establishment	4,602,255	6,198,535	135	Investment per Employee	6,552	4,917	75
Value Added per Employee	93,930	87,278	93	Investment per Production Worker	9,352	6,436	69

Sources: Same as General Statistics. The 'Average of All Manufacturing' column represents the average of all manufacturing industries reported for the most recent complete year available. The Index shows the relationship between the Average and the Analyzed Industry. For example, 100 means that they are equal; 500 that the Analyzed Industry is five times the average; 50 means that the Analyzed Industry is half the national average. The abbreviation 'na' is used to show that data are 'not available'.

LEADING COMPANIES   Number shown: 7   Total sales ($ mil): 181   Total employment (000): 2.1

Company Name	Address				CEO Name	Phone	Co. Type	Sales ($ mil)	Empl. (000)
Blount Inc	PO Box 856	Lewiston	ID	83501	Tom Fruechtel	208-746-2351	D	84	0.9
Federal Cartridge Co	900 Ehlen Dr	Anoka	MN	55303	Ronald V Mason	612-323-2300	S	78	1.0
Lyman Products Corp	Rte 147	Middlefield	CT	06455	JM Thompson	203-349-3421	R	11	0.1
Three-D Investment Inc	PO Box J	Doniphan	NE	68832	RG Phelps	402-845-2285	R	4	<0.1
Murmur Corp	PO Box 224566	Dallas	TX	75222	Homer Kirby	214-630-5400	R	2*	<0.1
Buffalo Bullet Co	12637 Los Nietos	Santa Fe Sprgs	CA	90670	Ronald R Dahlitz	310-944-0322	R	1	<0.1
Kaswer Custom Inc	13 Surrey Dr	Brookfield	CT	06804	Stanley W Kaswer	203-775-0564	R	1	<0.1

Source: Ward's Business Directory of U.S. Private and Public Companies, Volumes 1 and 2, 1996. The company type code used is as follows: P - Public, R - Private, S - Subsidiary, D - Division, J - Joint Venture, A - Affiliate, G - Group. Sales are in millions of dollars, employees are in thousands. An asterisk (*) indicates an estimated sales volume. The symbol < stands for 'less than'. Company names and addresses are truncated, in some cases, to fit into the available space.

MATERIALS CONSUMED

Material	Quantity	Delivered Cost ($ million)
Materials, ingredients, containers, and supplies	(X)	374.7
Metal bolts, nuts, screws, washers, rivets, and other screw machine products	(X)	0.1
Other fabricated metal products (except castings and forgings)	(X)	65.9
Iron and steel castings (rough and semifinished)	(X)	(D)
Aluminum and aluminum-base alloy castings (rough and semifinished)	(X)	(D)
Other nonferrous castings (rough and semifinished)	(X)	2.8
Iron and steel forgings	(X)	(D)
Aluminum and aluminum-base alloy forgings	(X)	(D)
Steel shapes and forms	(X)	11.6
Copper and copper-base alloy shapes and forms	(X)	81.2
Aluminum and aluminum-base alloy shapes and forms	(X)	(D)
Other nonferrous shapes and forms	(X)	(D)
Smokeless powder	(X)	57.1
Plastics resins consumed in the form of granules, pellets, powders, liquids, etc.	(X)	5.7
Other chemicals and allied products	(X)	(D)
Rough and dressed lumber	(X)	(D)
Paperboard containers, boxes, and corrugated paperboard	(X)	22.4
Fabricated plastics products (except gaskets, hoses, and belting)	(X)	10.7
Machine tool accessories, including cutting tools	(X)	(D)
Electronic, hydraulic, and mechanical subassemblies	(X)	(D)
All other materials and components, parts, containers, and supplies	(X)	47.5
Materials, ingredients, containers, and supplies, nsk	(X)	14.5

Source: 1992 Economic Census. Explanation of symbols used: (D): Withheld to avoid disclosure of competitive data; na: Not available; (S): Withheld because statistical norms were not met; (X): Not applicable; (Z): Less than half the unit shown; nec: Not elsewhere classified; nsk: Not specified by kind; - : zero; * : 10-19 percent estimated; ** : 20-29 percent estimated.

PRODUCT SHARE DETAILS

Product or Product Class	% Share	Product or Product Class	% Share
Small arms ammunition	100.00	Small arms ammunition primers (30 mm or less, 1.18 in. or less)	2.68
Rimfire rifle/pistol cartridges (30 mm or less, 1.18 in. or less)	6.29	All other small arms ammunition components, including wads, shot cases (primed or unprimed), bullets, bullet jackets, and cases (30 mm or less, 1.18 in. or less)	13.94
Centerfire rifle cartridges (30 mm or less, 1.18 in. or less)	19.59		
Centerfire pistol cartridges, including cartridges interchangeable between rifles and pistols (30 mm or less, 1.18 in. or less)	13.76	Other small arms ammunition products, including industrial shells and cartridges, air gun ammunition, and percussion caps (30 mm or less, 1.18 in. or less)	22.44
Shotgun shells (30 mm or less, 1.18 in. or less)	18.38		

Source: 1992 Economic Census. The values shown are percent of total shipments in an industry. Values of indented subcategories are summed in the main heading. The symbol (D) appears when data are withheld to prevent disclosure of competitive information. The abbreviation nsk stands for 'not specified by kind' and nec for 'not elsewhere classified'.

INPUTS AND OUTPUTS FOR SMALL ARMS AMMUNITION

Economic Sector or Industry Providing Inputs	%	Sector	Economic Sector or Industry Buying Outputs	%	Sector
Primary lead	14.9	Manufg.	Personal consumption expenditures	39.0	
Wholesale trade	7.3	Trade	Federal Government purchases, national defense	20.6	Fed Govt
Explosives	7.2	Manufg.	Exports	10.5	Foreign
Metal stampings, nec	7.2	Manufg.	Cement, hydraulic	6.7	Manufg.
Copper rolling & drawing	7.1	Manufg.	S/L Govt. purch., police	5.5	S/L Govt
Advertising	6.9	Services	Forestry products	5.0	Agric.
Imports	4.8	Foreign	Blast furnaces & steel mills	4.3	Manufg.
Rubber & plastics hose & belting	3.5	Manufg.	Ready-mixed concrete	2.6	Manufg.
Blast furnaces & steel mills	2.8	Manufg.	S/L Govt. purch., correction	1.5	S/L Govt
Electric services (utilities)	2.8	Util.	Concrete block & brick	0.6	Manufg.
Motor freight transportation & warehousing	2.8	Util.	Lime	0.5	Manufg.
Gas production & distribution (utilities)	2.0	Util.	Federal Government purchases, nondefense	0.4	Fed Govt
Maintenance of nonfarm buildings nec	1.9	Constr.	Office buildings	0.3	Constr.
Aluminum rolling & drawing	1.9	Manufg.	Concrete products, nec	0.3	Manufg.
Paperboard containers & boxes	1.8	Manufg.	Gypsum products	0.3	Manufg.
Machinery, except electrical, nec	1.7	Manufg.	Construction of hospitals	0.2	Constr.
Plastics materials & resins	1.7	Manufg.	Construction of stores & restaurants	0.2	Constr.
Petroleum refining	1.6	Manufg.	Industrial buildings	0.2	Constr.
Primary metal products, nec	1.4	Manufg.	Small arms ammunition	0.2	Manufg.
Plating & polishing	1.3	Manufg.	Residential garden apartments	0.1	Constr.
Abrasive products	0.9	Manufg.			
Ordnance & accessories nec	0.9	Manufg.			
Special dies & tools & machine tool accessories	0.9	Manufg.			
Communications, except radio & TV	0.9	Util.			
Eating & drinking places	0.9	Trade			
Machine tools, metal cutting types	0.7	Manufg.			
Sanitary services, steam supply, irrigation	0.7	Util.			
Equipment rental & leasing services	0.7	Services			
Aluminum castings	0.5	Manufg.			
Miscellaneous plastics products	0.5	Manufg.			
Small arms ammunition	0.5	Manufg.			
Railroads & related services	0.5	Util.			
Hotels & lodging places	0.5	Services			
Legal services	0.5	Services			
Sawmills & planning mills, general	0.4	Manufg.			
Computer & data processing services	0.4	Services			
Ball & roller bearings	0.3	Manufg.			
Lubricating oils & greases	0.3	Manufg.			
Metal heat treating	0.3	Manufg.			
Motors & generators	0.3	Manufg.			
Banking	0.3	Fin/R.E.			
Real estate	0.3	Fin/R.E.			
Royalties	0.3	Fin/R.E.			
Automotive rental & leasing, without drivers	0.3	Services			
Engineering, architectural, & surveying services	0.3	Services			
Management & consulting services & labs	0.3	Services			
U.S. Postal Service	0.3	Gov't			
Iron & steel foundries	0.2	Manufg.			
Machine tools, metal forming types	0.2	Manufg.			
Manifold business forms	0.2	Manufg.			
Metal coating & allied services	0.2	Manufg.			
Insurance carriers	0.2	Fin/R.E.			
Accounting, auditing & bookkeeping	0.2	Services			
Automotive repair shops & services	0.2	Services			
Detective & protective services	0.2	Services			
Noncomparable imports	0.2	Foreign			
Security & commodity brokers	0.1	Fin/R.E.			
Business services nec	0.1	Services			
Electrical repair shops	0.1	Services			
Laundry, dry cleaning, shoe repair	0.1	Services			
Personnel supply services	0.1	Services			
Photofinishing labs, commercial photography	0.1	Services			

Source: Benchmark Input-Output Accounts for the U.S. Economy, 1982, U.S. Department of Commerce, Washington, D.C., July 1991. Data, as reported in the source, are organized by the 1977 SIC structure in use in 1982 but have been matched, as closely as is possible, to the 1987 SIC structure used in this book.

OCCUPATIONS EMPLOYED BY SIC 348 - ORDNANCE AND ACCESSORIES, NEC

Occupation	% of Total 1994	Change to 2005	Occupation	% of Total 1994	Change to 2005
Assemblers, fabricators, & hand workers nec	10.8	-5.4	General office clerks	1.7	-19.4
Inspectors, testers, & graders, precision	5.2	-5.4	Metal & plastic machine workers nec	1.7	-16.2
Blue collar worker supervisors	4.4	-10.2	Industrial machinery mechanics	1.7	4.0
Combination machine tool operators	3.9	4.0	Electricians	1.6	-11.2
Engineers nec	2.8	41.8	Electrical & electronic technicians,technologists	1.6	-5.5
Machinists	2.7	-5.5	Plant & system operators nec	1.5	-16.4
Guards	2.2	-14.9	Engineering, mathematical, & science managers	1.4	7.4
Machine operators nec	2.2	-16.7	Industrial engineers, ex safety engineers	1.4	4.0
Engineering technicians & technologists nec	2.2	-5.4	Janitors & cleaners, incl maids	1.4	-24.3
Mechanical engineers	2.2	4.1	NC machine tool operators, metal & plastic	1.3	13.4
Secretaries, ex legal & medical	2.1	-13.9	Tool & die makers	1.1	-23.6
Production, planning, & expediting clerks	2.0	27.6	Management support workers nec	1.1	-5.5
Professional workers nec	1.9	13.4	Industrial production managers	1.1	-5.4
Precision assemblers nec	1.9	-9.2	Freight, stock, & material movers, hand	1.1	-24.3
Machine tool cutting operators, metal & plastic	1.7	-21.2			

Source: Industry-Occupation Matrix, Bureau of Labor Statistics. These data relate to one or more 3-digit SIC industry groups rather than to a single 4-digit SIC. The change reported for each occupation to the year 2005 is a percent of growth or decline as estimated by the Bureau of Labor Statistics. The abbreviation nec stands for 'not elsewhere classified'.

LOCATION BY STATE AND REGIONAL CONCENTRATION

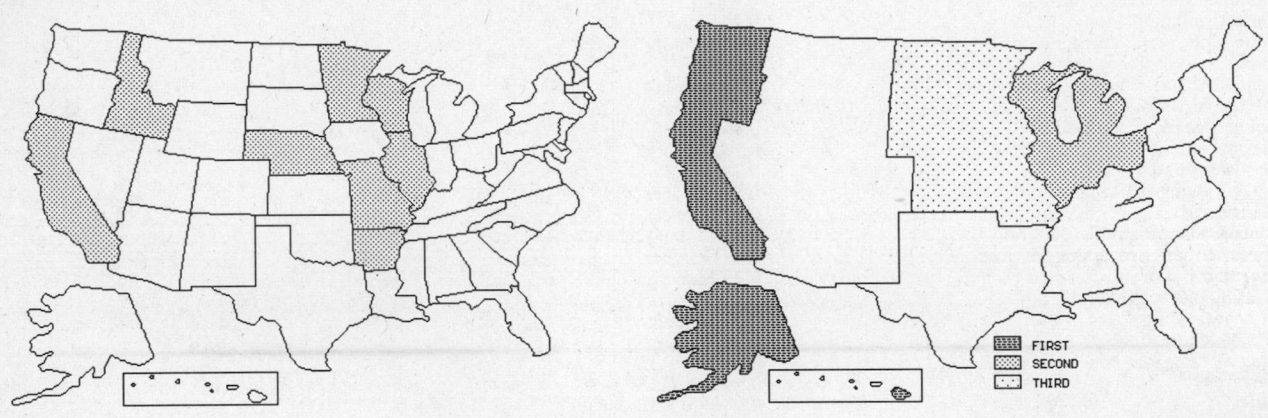

FIRST
SECOND
THIRD

INDUSTRY DATA BY STATE

State	Establish-ments	Shipments			Employment				Cost as % of Shipments	Investment per Employee ($)
		Total ($ mil)	% of U.S.	Per Establ.	Total Number	% of U.S.	Per Establ.	Wages ($/hour)		
California	14	(D)	-	-	375 *	4.6	27	-	-	6,133
Wisconsin	5	(D)	-	-	175 *	2.2	35	-	-	-
Illinois	4	(D)	-	-	1,750 *	21.6	438	-	-	-
Missouri	4	(D)	-	-	1,750 *	21.6	438	-	-	-
Idaho	3	(D)	-	-	750 *	9.3	250	-	-	-
Minnesota	2	(D)	-	-	750 *	9.3	375	-	-	-
Nebraska	2	(D)	-	-	175 *	2.2	88	-	-	-
Arkansas	1	(D)	-	-	750 *	9.3	750	-	-	-

Source: 1992 Economic Census. The states are in descending order of shipments or establishments (if shipment data are missing for the majority). The symbol (D) appears when data are withheld to prevent disclosure of competitive information. States marked with (D) are sorted by number of establishments. A dash (-) indicates that the data element cannot be calculated; * indicates the midpoint of a range.

3483 - AMMUNITION, EXCEPT FOR SMALL ARMS

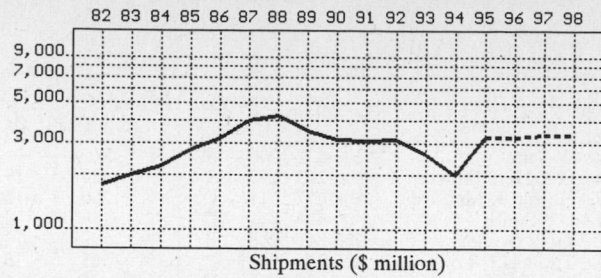

Shipments ($ million)

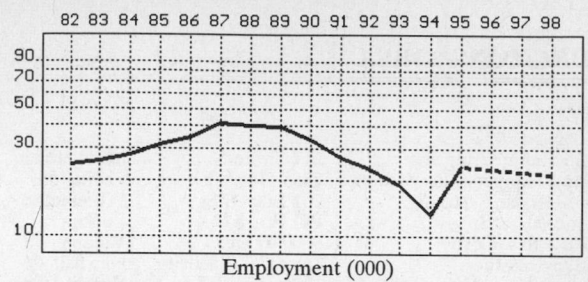

Employment (000)

GENERAL STATISTICS

Year	Companies	Establishments		Employment			Compensation		Production ($ million)			
		Total	with 20 or more employees	Total (000)	Production Workers (000)	Hours (Mil)	Payroll ($ mil)	Wages ($/hr)	Cost of Materials	Value Added by Manufacture	Value of Shipments	Capital Invest.
1982	63	74	53	24.4	15.9	29.1	497.5	9.11	625.2	1,125.0	1,757.1	56.2
1983		73	53	25.9	17.3	33.6	565.9	9.31	688.6	1,340.5	2,014.0	32.5
1984		72	53	27.8	18.1	34.0	640.5	10.35	822.4	1,499.4	2,249.8	74.0
1985		70	53	31.7	20.9	39.6	749.8	10.40	1,153.7	1,680.5	2,754.2	50.8
1986		71	55	34.8	22.7	44.0	849.3	10.22	1,185.4	2,052.0	3,204.6	56.6
1987	66	87	64	41.5	25.8	50.1	1,072.6	10.33	1,614.9	2,542.0	3,983.2	77.0
1988		86	66	39.9	24.2	49.2	1,061.9	10.14	1,900.3	2,340.4	4,290.8	83.6
1989		89	67	39.5	19.2	36.9	890.3	11.58	1,425.9	2,179.7	3,521.4	54.4
1990		86	65	33.5	14.6	27.8	807.0	12.55	1,178.1	1,908.9	3,128.6	39.0
1991		79	62	27.2	14.9	25.9	775.2	13.34	1,121.2	1,870.3	3,102.9	37.2
1992	56	70	53	23.4	13.5	24.0	719.5	12.19	1,224.4	1,889.8	3,136.5	35.2
1993		76	49	18.9	10.7	19.3	556.4	12.79	776.0	1,686.8	2,601.8	21.4
1994		82P	60P	13.0	7.8	15.2	397.7	12.31	621.6	1,377.1	2,008.2	17.5
1995		82P	61P	24.0P	11.9P	22.1P	717.2P	13.38P	999.2P	2,213.6P	3,228.1P	28.1P
1996		83P	61P	23.2P	11.2P	20.6P	714.3P	13.70P	1,013.5P	2,245.4P	3,274.4P	25.1P
1997		84P	62P	22.5P	10.4P	19.1P	711.4P	14.03P	1,027.9P	2,277.1P	3,320.7P	22.1P
1998		84P	62P	21.7P	9.6P	17.5P	708.6P	14.35P	1,042.2P	2,308.9P	3,367.0P	19.1P

Sources: 1982, 1987, 1992 *Economic Census*; *Annual Survey of Manufactures*, 83-86, 88-91, 93-94. Establishment counts for non-Census years are from *County Business Patterns*; establishment values for 83-84 are extrapolations. 'P's show projections by the editors. Industries reclassified in 87 will not have data for prior years.

INDICES OF CHANGE

Year	Companies	Establishments		Employment			Compensation		Production ($ million)			
		Total	with 20 or more employees	Total (000)	Production Workers (000)	Hours (Mil)	Payroll ($ mil)	Wages ($/hr)	Cost of Materials	Value Added by Manufacture	Value of Shipments	Capital Invest.
1982	113	106	100	104	118	121	69	75	51	60	56	160
1983		104	100	111	128	140	79	76	56	71	64	92
1984		103	100	119	134	142	89	85	67	79	72	210
1985		100	100	135	155	165	104	85	94	89	88	144
1986		101	104	149	168	183	118	84	97	109	102	161
1987	118	124	121	177	191	209	149	85	132	135	127	219
1988		123	125	171	179	205	148	83	155	124	137	238
1989		127	126	169	142	154	124	95	116	115	112	155
1990		123	123	143	108	116	112	103	96	101	100	111
1991		113	117	116	110	108	108	109	92	99	99	106
1992	100	100	100	100	100	100	100	100	100	100	100	100
1993		109	92	81	79	80	77	105	63	89	83	61
1994		117P	114P	56	58	63	55	101	51	73	64	50
1995		118P	115P	103P	88P	92P	100P	110P	82P	117P	103P	80P
1996		119P	116P	99P	83P	86P	99P	112P	83P	119P	104P	71P
1997		119P	116P	96P	77P	79P	99P	115P	84P	120P	106P	63P
1998		120P	117P	93P	71P	73P	98P	118P	85P	122P	107P	54P

Sources: Same as General Statistics. Values reflect change from the base year, 1992. Values above 100 mean greater than 92, values below 100 mean less than 92, and a value of 100 in the 82-91 or 93-98 period means same as 92. 'P's mark projections by the editors.

SELECTED RATIOS

For 1994	Avg. of All Manufact.	Analyzed Industry	Index	For 1994	Avg. of All Manufact.	Analyzed Industry	Index
Employees per Establishment	49	159	324	Value Added per Production Worker	134,084	176,551	132
Payroll per Establishment	1,500,273	4,863,480	324	Cost per Establishment	5,045,178	7,601,556	151
Payroll per Employee	30,620	30,592	100	Cost per Employee	102,970	47,815	46
Production Workers per Establishment	34	95	278	Cost per Production Worker	146,988	79,692	54
Wages per Establishment	853,319	2,288,196	268	Shipments per Establishment	9,576,895	24,558,310	256
Wages per Production Worker	24,861	23,989	96	Shipments per Employee	195,460	154,477	79
Hours per Production Worker	2,056	1,949	95	Shipments per Production Worker	279,017	257,462	92
Wages per Hour	12.09	12.31	102	Investment per Establishment	321,011	214,008	67
Value Added per Establishment	4,602,255	16,840,578	366	Investment per Employee	6,552	1,346	21
Value Added per Employee	93,930	105,931	113	Investment per Production Worker	9,352	2,244	24

Sources: Same as General Statistics. The 'Average of All Manufacturing' column represents the average of all manufacturing industries reported for the most recent complete year available. The Index shows the relationship between the Average and the Analyzed Industry. For example, 100 means that they are equal; 500 that the Analyzed Industry is five times the average; 50 means that the Analyzed Industry is half the national average. The abbreviation 'na' is used to show that data are 'not available'.

LEADING COMPANIES   Number shown: 11   Total sales ($ mil): 1,535   Total employment (000): 10.7

Company Name	Address				CEO Name	Phone	Co. Type	Sales ($ mil)	Empl. (000)
Raymond Engineering Inc	217 Smith St	Middletown	CT	06457	W Seitz	203-632-1000	S	560	0.4
Mason and Hanger-Silas Mason	2355 Harrodsburg	Lexington	KY	40504	D E Heffelbower	606-223-2277	S	420	5.5
Textron Defense Systems	201 Lowell St	Wilmington	MA	01887	Harold K McCard	508-657-5111	D	200	2.0
Lufkin Industries Inc	PO Box 849	Lufkin	TX	75901	DV Smith	409-634-2211	D	138	0.5
Allied Research Corp	8000 Towers	Vienna	VA	22182	Jar R Sculley	703-847-5268	P	70	0.5
Special Devices Inc	16830 W Placerita	Newhall	CA	91321	Thomas F Treinen	805-259-0753	P	65	0.9
Medico Industries Inc	1500 Hwy 315	Wilkes-Barre	PA	18711	Tom Medico	717-825-7711	R	30	0.2
Babcock & Wilcox Co	PO Box 11165	Lynchburg	VA	24506	DP Kohlhorst	804-522-6000	D	20	0.3
David B Lilly Company Inc	PO Box 10527	Wilmington	DE	19850	Philip Kadlecek	302-328-6675	R	19	0.2
Thiokol Corp	PO Box 30058	Shreveport	LA	71130	Steven J Shows	318-459-5501	D	11*	0.2
DP Packaging Inc	3410 NE 5th Av	Ft Lauderdale	FL	33334	Layne Dallett-Walls	305-565-8475	R	2	<0.1

Source: *Ward's Business Directory of U.S. Private and Public Companies*, Volumes 1 and 2, 1996. The company type code used is as follows: P - Public, R - Private, S - Subsidiary, D - Division, J - Joint Venture, A - Affiliate, G - Group. Sales are in millions of dollars, employees are in thousands. An asterisk (*) indicates an estimated sales volume. The symbol < stands for 'less than'. Company names and addresses are truncated, in some cases, to fit into the available space.

MATERIALS CONSUMED

Material	Quantity	Delivered Cost ($ million)
Materials, ingredients, containers, and supplies	(X)	1,126.7
Metal bolts, nuts, screws, washers, rivets, and other screw machine products	(X)	29.6
Other fabricated metal products (except castings and forgings)	(X)	69.5
Iron and steel castings (rough and semifinished)	(X)	40.6
Aluminum and aluminum-base alloy castings (rough and semifinished)	(X)	18.9
Other nonferrous castings (rough and semifinished)	(X)	4.3
Iron and steel forgings	(X)	7.7
Aluminum and aluminum-base alloy forgings	(X)	6.8
Other nonferrous forgings	(X)	2.2
Steel shapes and forms	(X)	81.8
Copper and copper-base alloy shapes and forms	(X)	13.0
Aluminum and aluminum-base alloy shapes and forms	(X)	38.2
Other nonferrous shapes and forms	(X)	6.6
Smokeless powder	(X)	(D)
Plastics resins consumed in the form of granules, pellets, powders, liquids, etc.	(X)	2.4
Other chemicals and allied products	(X)	32.0
Rough and dressed lumber	(X)	2.4
Paperboard containers, boxes, and corrugated paperboard	(X)	7.4
Fabricated plastics products (except gaskets, hoses, and belting)	(X)	17.6
Machine tool accessories, including cutting tools	(X)	7.1
Electronic, hydraulic, and mechanical subassemblies	(X)	19.9
Radio and electronic communication, navigation, search, detection systems	(X)	(D)
All other materials and components, parts, containers, and supplies	(X)	466.3
Materials, ingredients, containers, and supplies, nsk	(X)	9.2

Source: 1992 *Economic Census*. Explanation of symbols used: (D): Withheld to avoid disclosure of competitive data; na: Not available; (S): Withheld because statistical norms were not met; (X): Not applicable; (Z): Less than half the unit shown; nec: Not elsewhere classified; nsk: Not specified by kind; - : zero; * : 10-19 percent estimated; ** : 20-29 percent estimated.

PRODUCT SHARE DETAILS

Product or Product Class	% Share	Product or Product Class	% Share
Ammunition, except for small arms, nec	100.00	fuses, boosters, and nonsteel cases (more than 30 mm, more than 1.18 in.)	25.41
Artillery ammunition, more than 30 mm (or more than 1.18 in.)	43.55	Receipts for artillery ammunition loading and assembly (more than 30 mm, more than 1.18 in.)	6.48
Complete artillery rounds, loaded (more than 30 mm, more than 1.18 in.)	(D)	Artillery ammunition, more than 30 mm (or more than 1.18 inches), nsk	(D)
Artillery ammunition steel cases only (more than 30 mm, more than 1.18 in.)	6.90	Ammunition, except for small arms, nec	56.25
Artillery ammunition projectile metal parts (more than 30 mm, more than 1.18 in.)	17.40	Bombs, depth charges, mines, torpedoes, etc., and parts	99.04
Other artillery ammunition components, including primers,		Ammunition, except for small arms, nec, nsk	0.96
		Ammunition, except for small arms, nec, nsk	0.20

Source: 1992 *Economic Census*. The values shown are percent of total shipments in an industry. Values of indented subcategories are summed in the main heading. The symbol (D) appears when data are withheld to prevent disclosure of competitive information. The abbreviation nsk stands for 'not specified by kind' and nec for 'not elsewhere classified'.

INPUTS AND OUTPUTS FOR AMMUNITION, EXCEPT FOR SMALL ARMS, NEC

Economic Sector or Industry Providing Inputs	%	Sector	Economic Sector or Industry Buying Outputs	%	Sector
Ammunition, except for small arms, nec	12.9	Manufg.	Federal Government purchases, national defense	79.9	Fed Govt
Blast furnaces & steel mills	12.2	Manufg.	Exports	14.9	Foreign
Advertising	7.0	Services	Ammunition, except for small arms, nec	5.3	Manufg.
Wholesale trade	5.2	Trade			
Electronic components nec	4.6	Manufg.			
Aluminum rolling & drawing	4.3	Manufg.			
Gas production & distribution (utilities)	3.9	Util.			
Electric services (utilities)	3.6	Util.			
Detective & protective services	3.3	Services			
Semiconductors & related devices	3.2	Manufg.			
Nonferrous forgings	2.9	Manufg.			
Sawmills & planning mills, general	2.3	Manufg.			
Maintenance of nonfarm buildings nec	2.1	Constr.			
Machinery, except electrical, nec	2.0	Manufg.			
Petroleum refining	2.0	Manufg.			
Communications, except radio & TV	2.0	Util.			
Fabricated metal products, nec	1.4	Manufg.			
Imports	1.4	Foreign			
Miscellaneous fabricated wire products	1.2	Manufg.			
Abrasive products	1.1	Manufg.			
Metal stampings, nec	1.1	Manufg.			
Photographic equipment & supplies	1.1	Manufg.			
Special dies & tools & machine tool accessories	1.1	Manufg.			
Motor freight transportation & warehousing	1.0	Util.			
Eating & drinking places	1.0	Trade			
Plating & polishing	0.9	Manufg.			
Air transportation	0.9	Util.			
Paperboard containers & boxes	0.6	Manufg.			
Railroads & related services	0.6	Util.			
Real estate	0.6	Fin/R.E.			
Legal services	0.6	Services			
Copper rolling & drawing	0.5	Manufg.			
Credit agencies other than banks	0.5	Fin/R.E.			
Business/professional associations	0.5	Services			
Apparel made from purchased materials	0.4	Manufg.			
Iron & steel forgings	0.4	Manufg.			
Lubricating oils & greases	0.4	Manufg.			
Machine tools, metal cutting types	0.4	Manufg.			
Primary metal products, nec	0.4	Manufg.			
Screw machine and related products	0.4	Manufg.			
Security & commodity brokers	0.4	Fin/R.E.			
Hotels & lodging places	0.4	Services			
Management & consulting services & labs	0.4	Services			
Aluminum castings	0.3	Manufg.			
Manifold business forms	0.3	Manufg.			
Metal heat treating	0.3	Manufg.			
Plastics materials & resins	0.3	Manufg.			
Sanitary services, steam supply, irrigation	0.3	Util.			
Banking	0.3	Fin/R.E.			
Accounting, auditing & bookkeeping	0.3	Services			
Engineering, architectural, & surveying services	0.3	Services			
Equipment rental & leasing services	0.3	Services			
Explosives	0.2	Manufg.			
Metal coating & allied services	0.2	Manufg.			
Insurance carriers	0.2	Fin/R.E.			
Royalties	0.2	Fin/R.E.			
Automotive repair shops & services	0.2	Services			
Computer & data processing services	0.2	Services			
Laundry, dry cleaning, shoe repair	0.2	Services			
Photofinishing labs, commercial photography	0.2	Services			
Services to dwellings & other buildings	0.2	Services			
Noncomparable imports	0.2	Foreign			
Coal	0.1	Mining			
Fabricated rubber products, nec	0.1	Manufg.			
Iron & steel foundries	0.1	Manufg.			
Manufacturing industries, nec	0.1	Manufg.			
Miscellaneous plastics products	0.1	Manufg.			
Automotive rental & leasing, without drivers	0.1	Services			
Personnel supply services	0.1	Services			

Source: *Benchmark Input-Output Accounts for the U.S. Economy, 1982*, U.S. Department of Commerce, Washington, D.C., July 1991. Data, as reported in the source, are organized by the 1977 SIC structure in use in 1982 but have been matched, as closely as is possible, to the 1987 SIC structure used in this book.

OCCUPATIONS EMPLOYED BY SIC 348 - ORDNANCE AND ACCESSORIES, NEC

Occupation	% of Total 1994	Change to 2005	Occupation	% of Total 1994	Change to 2005
Assemblers, fabricators, & hand workers nec	10.8	-5.4	General office clerks	1.7	-19.4
Inspectors, testers, & graders, precision	5.2	-5.4	Metal & plastic machine workers nec	1.7	-16.2
Blue collar worker supervisors	4.4	-10.2	Industrial machinery mechanics	1.7	4.0
Combination machine tool operators	3.9	4.0	Electricians	1.6	-11.2
Engineers nec	2.8	41.8	Electrical & electronic technicians,technologists	1.6	-5.5
Machinists	2.7	-5.5	Plant & system operators nec	1.5	-16.4
Guards	2.2	-14.9	Engineering, mathematical, & science managers	1.4	7.4
Machine operators nec	2.2	-16.7	Industrial engineers, ex safety engineers	1.4	4.0
Engineering technicians & technologists nec	2.2	-5.4	Janitors & cleaners, incl maids	1.4	-24.3
Mechanical engineers	2.2	4.1	NC machine tool operators, metal & plastic	1.3	13.4
Secretaries, ex legal & medical	2.1	-13.9	Tool & die makers	1.1	-23.6
Production, planning, & expediting clerks	2.0	27.6	Management support workers nec	1.1	-5.5
Professional workers nec	1.9	13.4	Industrial production managers	1.1	-5.4
Precision assemblers nec	1.9	-9.2	Freight, stock, & material movers, hand	1.1	-24.3
Machine tool cutting operators, metal & plastic	1.7	-21.2			

Source: *Industry-Occupation Matrix*, Bureau of Labor Statistics. These data relate to one or more 3-digit SIC industry groups rather than to a single 4-digit SIC. The change reported for each occupation to the year 2005 is a percent of growth or decline as estimated by the Bureau of Labor Statistics. The abbreviation nec stands for 'not elsewhere classified'.

LOCATION BY STATE AND REGIONAL CONCENTRATION

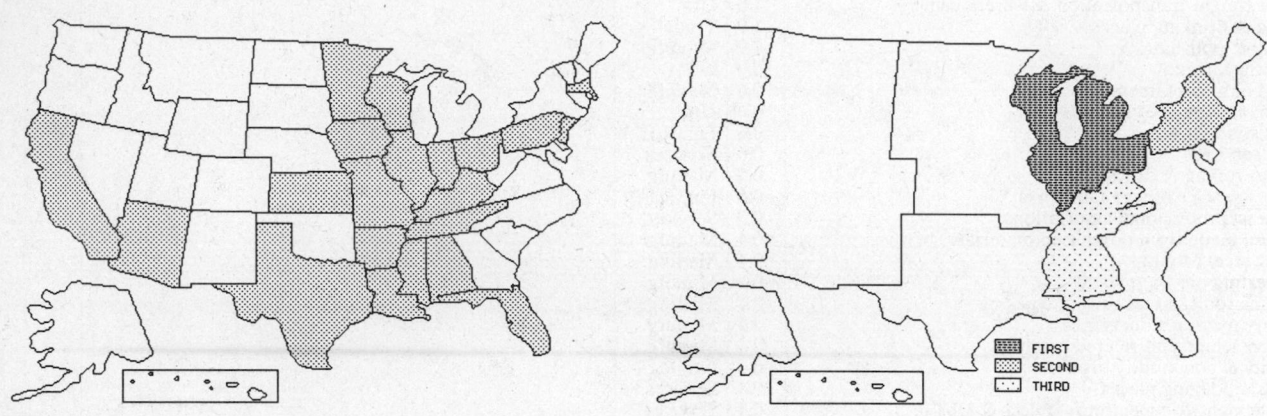

FIRST
SECOND
THIRD

INDUSTRY DATA BY STATE

State	Establish-ments	Shipments Total ($ mil)	% of U.S.	Per Establ.	Employment Total Number	% of U.S.	Per Establ.	Wages ($/hour)	Cost as % of Shipments	Investment per Employee ($)
Pennsylvania	8	222.3	7.1	27.8	2,300	9.8	288	11.16	39.1	-
Texas	5	184.4	5.9	36.9	3,000	12.8	600	13.67	35.5	2,100
California	6	170.8	5.4	28.5	1,300	5.6	217	13.76	42.9	3,846
Ohio	6	107.3	3.4	17.9	1,100	4.7	183	9.93	61.0	909
Illinois	6	(D)	-	-	375 *	1.6	63	-	-	-
Tennessee	5	(D)	-	-	1,750 *	7.5	350	-	-	-
New Jersey	4	(D)	-	-	375 *	1.6	94	-	-	-
Massachusetts	3	(D)	-	-	375 *	1.6	125	-	-	-
Wisconsin	3	(D)	-	-	750 *	3.2	250	-	-	-
Alabama	2	(D)	-	-	375 *	1.6	188	-	-	-
Arkansas	2	(D)	-	-	1,750 *	7.5	875	-	-	-
Florida	2	(D)	-	-	175 *	0.7	88	-	-	-
Indiana	2	(D)	-	-	750 *	3.2	375	-	-	-
Iowa	2	(D)	-	-	1,750 *	7.5	875	-	-	-
Minnesota	2	(D)	-	-	3,750 *	16.0	1,875	-	-	-
Missouri	2	(D)	-	-	750 *	3.2	375	-	-	-
Arizona	1	(D)	-	-	375 *	1.6	375	-	-	-
Kansas	1	(D)	-	-	750 *	3.2	750	-	-	-
Kentucky	1	(D)	-	-	175 *	0.7	175	-	-	-
Louisiana	1	(D)	-	-	1,750 *	7.5	1,750	-	-	-
Mississippi	1	(D)	-	-	375 *	1.6	375	-	-	-

Source: 1992 *Economic Census*. The states are in descending order of shipments or establishments (if shipment data are missing for the majority). The symbol (D) appears when data are withheld to prevent disclosure of competitive information. States marked with (D) are sorted by number of establishments. A dash (-) indicates that the data element cannot be calculated; * indicates the midpoint of a range.

3484 - SMALL ARMS

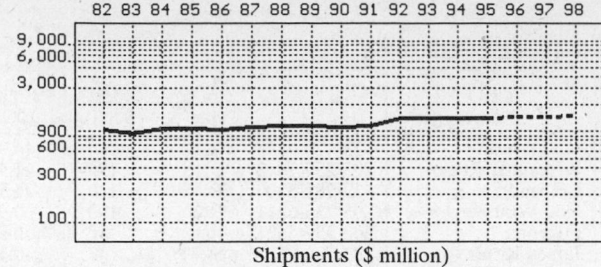

Shipments ($ million)

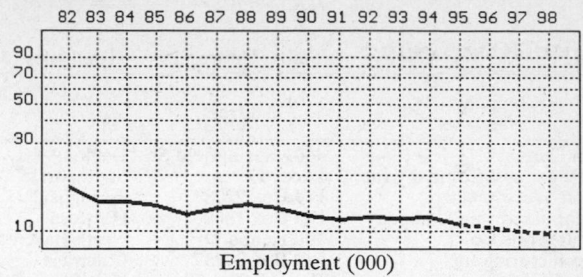

Employment (000)

GENERAL STATISTICS

Year	Companies	Establishments		Employment			Compensation		Production ($ million)			
		Total	with 20 or more employees	Total (000)	Production Workers (000)	Hours (Mil)	Payroll ($ mil)	Wages ($/hr)	Cost of Materials	Value Added by Manufacture	Value of Shipments	Capital Invest.
1982	136	144	44	17.4	13.0	25.0	338.2	8.41	339.9	715.7	1,032.1	34.3
1983		136	44	14.4	10.4	19.6	296.5	9.18	235.3	678.3	938.8	32.4
1984		128	44	14.3	10.6	21.4	326.0	9.52	290.4	763.0	1,063.0	28.5
1985		121	45	13.8	10.1	20.2	325.4	9.85	337.2	720.2	1,085.3	29.6
1986		116	42	12.3	8.5	17.6	308.9	10.35	298.4	711.3	1,030.8	32.3
1987	144	151	44	13.3	9.5	19.5	337.0	10.73	338.0	754.9	1,093.1	33.4
1988		141	46	13.9	10.3	20.2	364.9	11.76	359.6	843.6	1,168.6	35.8
1989		142	49	13.3	9.6	18.2	348.5	12.68	351.2	824.7	1,168.6	28.7
1990		136	56	12.1	9.4	18.4	348.6	13.17	317.8	859.9	1,108.8	25.7
1991		137	48	11.5	8.5	16.4	321.1	13.39	389.1	745.9	1,148.5	26.5
1992	177	184	60	11.8	9.2	18.1	334.9	12.78	401.1	939.8	1,384.0	32.0
1993		188	59	11.6	9.2	19.2	342.1	12.70	401.9	973.5	1,396.2	93.6
1994		169P	58P	11.8	9.1	19.3	358.4	13.04	383.3	987.9	1,384.1	33.5
1995		173P	59P	10.7P	8.3P	17.1P	352.5P	14.30P	384.6P	991.2P	1,388.8P	46.9P
1996		177P	61P	10.3P	8.1P	16.8P	355.0P	14.72P	393.9P	1,015.2P	1,422.3P	48.5P
1997		181P	62P	10.0P	7.9P	16.5P	357.5P	15.14P	403.2P	1,039.1P	1,455.8P	50.1P
1998		185P	64P	9.6P	7.7P	16.1P	360.1P	15.56P	412.5P	1,063.0P	1,489.4P	51.7P

Sources: 1982, 1987, 1992 *Economic Census*; *Annual Survey of Manufactures*, 83-86, 88-91, 93-94. Establishment counts for non-Census years are from *County Business Patterns*; establishment values for 83-84 are extrapolations. 'P's show projections by the editors. Industries reclassified in 87 will not have data for prior years.

INDICES OF CHANGE

Year	Companies	Establishments		Employment			Compensation		Production ($ million)			
		Total	with 20 or more employees	Total (000)	Production Workers (000)	Hours (Mil)	Payroll ($ mil)	Wages ($/hr)	Cost of Materials	Value Added by Manufacture	Value of Shipments	Capital Invest.
1982	77	78	73	147	141	138	101	66	85	76	75	107
1983		74	73	122	113	108	89	72	59	72	68	101
1984		70	73	121	115	118	97	74	72	81	77	89
1985		66	75	117	110	112	97	77	84	77	78	93
1986		63	70	104	92	97	92	81	74	76	74	101
1987	81	82	73	113	103	108	101	84	84	80	79	104
1988		77	77	118	112	112	109	92	90	90	84	112
1989		77	82	113	104	101	104	99	88	88	84	90
1990		74	93	103	102	102	104	103	79	91	80	80
1991		74	80	97	92	91	96	105	97	79	83	83
1992	100	100	100	100	100	100	100	100	100	100	100	100
1993		102	98	98	100	106	102	99	100	104	101	292
1994		92P	96P	100	99	107	107	102	96	105	100	105
1995		94P	99P	91P	91P	95P	105P	112P	96P	105P	100P	147P
1996		96P	101P	88P	88P	93P	106P	115P	98P	108P	103P	152P
1997		98P	104P	85P	86P	91P	107P	118P	101P	111P	105P	157P
1998		100P	106P	82P	84P	89P	108P	122P	103P	113P	108P	162P

Sources: Same as General Statistics. Values reflect change from the base year, 1992. Values above 100 mean greater than 92, values below 100 mean less than 92, and a value of 100 in the 82-91 or 93-98 period means same as 92. 'P's mark projections by the editors.

SELECTED RATIOS

For 1994	Avg. of All Manufact.	Analyzed Industry	Index	For 1994	Avg. of All Manufact.	Analyzed Industry	Index
Employees per Establishment	49	70	142	Value Added per Production Worker	134,084	108,560	81
Payroll per Establishment	1,500,273	2,118,621	141	Cost per Establishment	5,045,178	2,265,813	45
Payroll per Employee	30,620	30,373	99	Cost per Employee	102,970	32,483	32
Production Workers per Establishment	34	54	157	Cost per Production Worker	146,988	42,121	29
Wages per Establishment	853,319	1,487,716	174	Shipments per Establishment	9,576,895	8,181,872	85
Wages per Production Worker	24,861	27,656	111	Shipments per Employee	195,460	117,297	60
Hours per Production Worker	2,056	2,121	103	Shipments per Production Worker	279,017	152,099	55
Wages per Hour	12.09	13.04	108	Investment per Establishment	321,011	198,030	62
Value Added per Establishment	4,602,255	5,839,803	127	Investment per Employee	6,552	2,839	43
Value Added per Employee	93,930	83,720	89	Investment per Production Worker	9,352	3,681	39

Sources: Same as General Statistics. The 'Average of All Manufacturing' column represents the average of all manufacturing industries reported for the most recent complete year available. The Index shows the relationship between the Average and the Analyzed Industry. For example, 100 means that they are equal; 500 that the Analyzed Industry is five times the average; 50 means that the Analyzed Industry is half the national average. The abbreviation 'na' is used to show that data are 'not available'.

LEADING COMPANIES   Number shown: 24   Total sales ($ mil): 2,058   Total employment (000): 21.3

Company Name	Address				CEO Name	Phone	Co. Type	Sales ($ mil)	Empl. (000)
Tomkins Corp	4801 Springfield St	Dayton	OH	45431	Ian Duncan	513-476-0435	R	1,360*	14.0
Sturm, Ruger and Company Inc	Lacey Pl	Southport	CT	06490	William B Ruger	203-259-7843	P	196	1.9
Smith and Wesson Corp	PO Box 2208	Springfield	MA	01102	LE Shultz	413-781-8300	D	140	1.5
Colt's Manufacturing Company	PO Box 1868	Hartford	CT	06144	Ron Whitaker	203-236-6311	R	100	0.9
Marlin Firearms Co	100 Kenna Dr	North Haven	CT	06473	F Kenna	203-239-5621	R	50	0.5
FN Manufacturing Inc	PO Box 24257	Columbia	SC	29224	James Ritter	803-736-0522	S	47	0.4
Crosman Corp	Rte 5 & Rte 20	East Bloomfield	NY	14443	Kenneth R Scheele	716-657-6161	R	45	0.4
Savage Arms Inc	100 Springdale Rd	Westfield	MA	01085	Ronald Coburn	413-568-7001	S	26	0.2
OF Mossberg and Sons Inc	PO Box 497	North Haven	CT	06473	Alan I Mossberg	203-288-6491	R	20	0.3
KW Thompson Tool Co	PO Box 5002	Rochester	NH	03867	Robert L Gustafson	603-332-2333	R	18*	0.4
Benjamin Sheridan Corp	Rte 5 & Rte 20	East Bloomfield	NY	14443	Ken Scheele	716-657-6161	S	11	0.1
H and R 1871 Inc	Industrial Row	Gardner	MA	01440	James Garrison	508-632-9393	R	10	0.3
L and P Machine Inc	1340 Norman Av	Santa Clara	CA	95054	Raymond Leap	408-988-1720	R	9*	<0.1
Ram-Line Inc	545 31 Rd	Grand Junction	CO	81504	Jim Chesnut	303-434-4500	R	7*	<0.1
Charco Inc	26 Beaver St	Ansonia	CT	06401	Jeff Williams	203-735-4686	R	4	<0.1
Springfield Inc	420 W Main St	Geneseo	IL	61254	Thomas Reese	309-944-5631	R	3	<0.1
Wesson Firearms Company Inc	Maple Tree Indrial	Palmer	MA	01069	Seth K Wesson	413-267-4081	R	3	<0.1
Armscorp USA Inc	4424 John Av	Baltimore	MD	21227	John H Friese	410-247-6200	R	2	<0.1
A-Square Company Inc	1 Industrial Park	Bedford	KY	40006	Arthur B Alphin	502-255-7456	R	2	<0.1
Dakota Arms Inc	HC55 Box 326	Sturgis	SD	57785	Donald L Allen	605-347-4686	R	2	<0.1
CRL Inc	PO Box 111	Gladstone	MI	49837	Craig R Lauerman	906-428-3710	R	1	<0.1
Jarrett Rifles Inc	383 Brown Rd	Jackson	SC	29831	Kenny Jarrett	803-471-3616	R	1	<0.1
Ultra Light Arms Inc	PO Box 1270	Granville	WV	26534	Melvin D Forbes	304-599-5687	R	1	<0.1
Eagle Arms Inc	PO Box 457	Coal Valley	IL	61240	Mark Westrom	309-799-5619	R	1	<0.1

Source: *Ward's Business Directory of U.S. Private and Public Companies*, Volumes 1 and 2, 1996. The company type code used is as follows: P - Public, R - Private, S - Subsidiary, D - Division, J - Joint Venture, A - Affiliate, G - Group. Sales are in millions of dollars, employees are in thousands. An asterisk (*) indicates an estimated sales volume. The symbol < stands for 'less than'. Company names and addresses are truncated, in some cases, to fit into the available space.

MATERIALS CONSUMED

Material	Quantity	Delivered Cost ($ million)
Materials, ingredients, containers, and supplies	(X)	333.8
Metal bolts, nuts, screws, washers, rivets, and other screw machine products	(X)	18.8
Other fabricated metal products (except castings and forgings)	(X)	93.6
Iron and steel castings (rough and semifinished)	(X)	15.5
Aluminum and aluminum-base alloy castings (rough and semifinished)	(X)	14.9
Other nonferrous castings (rough and semifinished)	(X)	(D)
Iron and steel forgings	(X)	9.9
Aluminum and aluminum-base alloy forgings	(X)	(D)
Steel shapes and forms	(X)	17.2
Copper and copper-base alloy shapes and forms	(X)	0.4
Aluminum and aluminum-base alloy shapes and forms	(X)	2.1
Other nonferrous shapes and forms	(X)	5.7
Smokeless powder	(X)	(D)
Plastics resins consumed in the form of granules, pellets, powders, liquids, etc.	(X)	(D)
Other chemicals and allied products	(X)	0.9
Rough and dressed lumber	(X)	5.6
Paperboard containers, boxes, and corrugated paperboard	(X)	7.3
Fabricated plastics products (except gaskets, hoses, and belting)	(X)	5.5
Machine tool accessories, including cutting tools	(X)	10.5
Electronic, hydraulic, and mechanical subassemblies	(X)	(D)
All other materials and components, parts, containers, and supplies	(X)	101.9
Materials, ingredients, containers, and supplies, nsk	(X)	20.1

Source: 1992 *Economic Census*. Explanation of symbols used: (D): Withheld to avoid disclosure of competitive data; na: Not available; (S): Withheld because statistical norms were not met; (X): Not applicable; (Z): Less than half the unit shown; nec: Not elsewhere classified; nsk: Not specified by kind; - : zero; * : 10-19 percent estimated; ** : 20-29 percent estimated.

PRODUCT SHARE DETAILS

Product or Product Class	% Share	Product or Product Class	% Share
Small arms	100.00	Centerfire bolt repeater rifles (30 mm or less, 1.18 in. or less)	(D)
Machine guns, 30 mm or less (or 1.18 in. or less)	20.12	All other centerfire rifles (30 mm or less, 1.18 in. or less)	15.03
Small arms, 30 mm or less (or 1.18 in. or less)	74.58	Rimfire rifles (30 mm or less, 1.18 in. or less)	(D)
Centerfire pistols and revolvers, overall length 6 in. or less (30 mm or less, 1.18 in. or less)	12.17	Single barrel shotguns (30 mm or less, 1.18 in. or less)	17.98
All other centerfire pistols and revolvers, including units with interchangeable barrels (30 mm or less, 1.18 in. or less)	17.34	Other small firearms (30 mm or less, 1.18 in. or less), including double barrel shotguns and combination rifle-shotguns	3.36
Rimfire pistols and revolvers (30 mm or less, 1.18 in. or less)	5.64	Parts and attachments for small firearms (30 mm or less, 1.18 in. or less)	10.24
Centerfire semiautomatic rifles (30 mm or less, 1.18 in. or less)	5.67	Small arms, nsk	5.30

Source: 1992 *Economic Census*. The values shown are percent of total shipments in an industry. Values of indented subcategories are summed in the main heading. The symbol (D) appears when data are withheld to prevent disclosure of competitive information. The abbreviation nsk stands for 'not specified by kind' and nec for 'not elsewhere classified'.

INPUTS AND OUTPUTS FOR SMALL ARMS

Economic Sector or Industry Providing Inputs	%	Sector	Economic Sector or Industry Buying Outputs	%	Sector
Imports	20.7	Foreign	Personal consumption expenditures	57.7	
Semiconductors & related devices	10.5	Manufg.	Federal Government purchases, national defense	16.4	Fed Govt
Wholesale trade	8.2	Trade	Exports	13.0	Foreign
Explosives	4.8	Manufg.	Miscellaneous repair shops	5.9	Services
Blast furnaces & steel mills	4.7	Manufg.	Change in business inventories	2.9	In House
Advertising	4.3	Services	S/L Govt. purch., police	1.3	S/L Govt
Iron & steel foundries	4.2	Manufg.	Small arms	0.9	Manufg.
Aluminum rolling & drawing	2.4	Manufg.	S/L Govt. purch., other general government	0.3	S/L Govt
Miscellaneous plastics products	2.4	Manufg.	Wholesale trade	0.2	Trade
Iron & steel forgings	2.3	Manufg.	Detective & protective services	0.2	Services
Screw machine and related products	2.3	Manufg.	Federal Government purchases, nondefense	0.2	Fed Govt
Hardwood dimension & flooring mills	2.2	Manufg.	S/L Govt. purch., fire	0.2	S/L Govt
Electric services (utilities)	2.2	Util.	S/L Govt. purch., correction	0.1	S/L Govt
Metal stampings, nec	2.0	Manufg.			
Air transportation	2.0	Util.			
Motor freight transportation & warehousing	1.9	Util.			
Miscellaneous fabricated wire products	1.8	Manufg.			
Plating & polishing	1.8	Manufg.			
Machinery, except electrical, nec	1.4	Manufg.			
Small arms	1.4	Manufg.			
Maintenance of nonfarm buildings nec	0.9	Constr.			
Petroleum refining	0.9	Manufg.			
Business services nec	0.9	Services			
Aluminum castings	0.8	Manufg.			
Abrasive products	0.7	Manufg.			
Special dies & tools & machine tool accessories	0.7	Manufg.			
Communications, except radio & TV	0.7	Util.			
Eating & drinking places	0.7	Trade			
Sawmills & planning mills, general	0.5	Manufg.			
Electronic components nec	0.4	Manufg.			
Metal coating & allied services	0.4	Manufg.			
Nonferrous rolling & drawing, nec	0.4	Manufg.			
Paperboard containers & boxes	0.4	Manufg.			
Signs & advertising displays	0.4	Manufg.			
Gas production & distribution (utilities)	0.4	Util.			
Equipment rental & leasing services	0.4	Services			
Hotels & lodging places	0.4	Services			
Legal services	0.4	Services			
Machine tools, metal cutting types	0.3	Manufg.			
Railroads & related services	0.3	Util.			
Real estate	0.3	Fin/R.E.			
Management & consulting services & labs	0.3	Services			
U.S. Postal Service	0.3	Gov't			
Lubricating oils & greases	0.2	Manufg.			
Manifold business forms	0.2	Manufg.			
Metal heat treating	0.2	Manufg.			
Sanitary services, steam supply, irrigation	0.2	Util.			
Banking	0.2	Fin/R.E.			
Royalties	0.2	Fin/R.E.			
Accounting, auditing & bookkeeping	0.2	Services			
Computer & data processing services	0.2	Services			
Engineering, architectural, & surveying services	0.2	Services			
Noncomparable imports	0.2	Foreign			
Coal	0.1	Mining			
Ball & roller bearings	0.1	Manufg.			
Copper rolling & drawing	0.1	Manufg.			
Insurance carriers	0.1	Fin/R.E.			
Automotive rental & leasing, without drivers	0.1	Services			
Automotive repair shops & services	0.1	Services			
Detective & protective services	0.1	Services			
Photofinishing labs, commercial photography	0.1	Services			

Source: Benchmark Input-Output Accounts for the U.S. Economy, 1982, U.S. Department of Commerce, Washington, D.C., July 1991. Data, as reported in the source, are organized by the 1977 SIC structure in use in 1982 but have been matched, as closely as is possible, to the 1987 SIC structure used in this book.

OCCUPATIONS EMPLOYED BY SIC 348 - ORDNANCE AND ACCESSORIES, NEC

Occupation	% of Total 1994	Change to 2005	Occupation	% of Total 1994	Change to 2005
Assemblers, fabricators, & hand workers nec	10.8	-5.4	General office clerks	1.7	-19.4
Inspectors, testers, & graders, precision	5.2	-5.4	Metal & plastic machine workers nec	1.7	-16.2
Blue collar worker supervisors	4.4	-10.2	Industrial machinery mechanics	1.7	4.0
Combination machine tool operators	3.9	4.0	Electricians	1.6	-11.2
Engineers nec	2.8	41.8	Electrical & electronic technicians,technologists	1.6	-5.5
Machinists	2.7	-5.5	Plant & system operators nec	1.5	-16.4
Guards	2.2	-14.9	Engineering, mathematical, & science managers	1.4	7.4
Machine operators nec	2.2	-16.7	Industrial engineers, ex safety engineers	1.4	4.0
Engineering technicians & technologists nec	2.2	-5.4	Janitors & cleaners, incl maids	1.4	-24.3
Mechanical engineers	2.2	4.1	NC machine tool operators, metal & plastic	1.3	13.4
Secretaries, ex legal & medical	2.1	-13.9	Tool & die makers	1.1	-23.6
Production, planning, & expediting clerks	2.0	27.6	Management support workers nec	1.1	-5.5
Professional workers nec	1.9	13.4	Industrial production managers	1.1	-5.4
Precision assemblers nec	1.9	-9.2	Freight, stock, & material movers, hand	1.1	-24.3
Machine tool cutting operators, metal & plastic	1.7	-21.2			

Source: Industry-Occupation Matrix, Bureau of Labor Statistics. These data relate to one or more 3-digit SIC industry groups rather than to a single 4-digit SIC. The change reported for each occupation to the year 2005 is a percent of growth or decline as estimated by the Bureau of Labor Statistics. The abbreviation nec stands for 'not elsewhere classified'.

LOCATION BY STATE AND REGIONAL CONCENTRATION

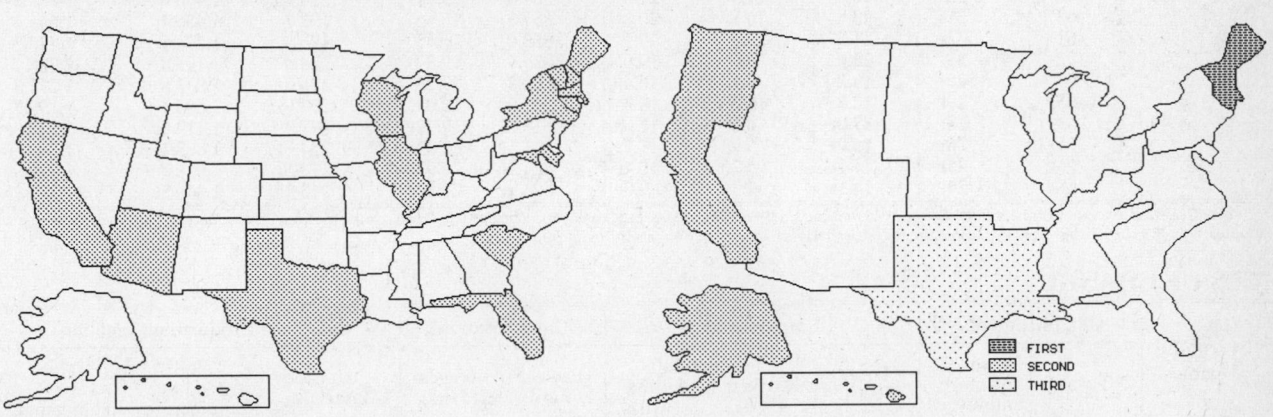

FIRST
SECOND
THIRD

INDUSTRY DATA BY STATE

State	Establish-ments	Shipments Total ($ mil)	Shipments % of U.S.	Shipments Per Establ.	Employment Total Number	Employment % of U.S.	Employment Per Establ.	Wages ($/hour)	Cost as % of Shipments	Investment per Employee ($)
Massachusetts	7	142.2	10.3	20.3	1,700	14.4	243	19.09	23.5	-
California	22	72.8	5.3	3.3	700	5.9	32	8.73	33.5	-
Florida	8	18.2	1.3	2.3	200	1.7	25	9.00	29.7	-
Texas	14	(D)	-	-	375 *	3.2	27	-	-	1,867
Connecticut	12	(D)	-	-	1,750 *	14.8	146	-	-	-
New York	10	(D)	-	-	1,750 *	14.8	175	-	-	-
Wisconsin	10	(D)	-	-	375 *	3.2	38	-	-	1,067
Arizona	6	(D)	-	-	375 *	3.2	63	-	-	-
Illinois	4	(D)	-	-	175 *	1.5	44	-	-	-
Maine	3	(D)	-	-	750 *	6.4	250	-	-	-
Maryland	3	(D)	-	-	375 *	3.2	125	-	-	-
New Hampshire	3	(D)	-	-	1,750 *	14.8	583	-	-	-
Vermont	3	(D)	-	-	750 *	6.4	250	-	-	-
South Carolina	1	(D)	-	-	750 *	6.4	750	-	-	-

Source: 1992 Economic Census. The states are in descending order of shipments or establishments (if shipment data are missing for the majority). The symbol (D) appears when data are withheld to prevent disclosure of competitive information. States marked with (D) are sorted by number of establishments. A dash (-) indicates that the data element cannot be calculated; * indicates the midpoint of a range.

3489 - ORDNANCE & ACCESSORIES, NEC

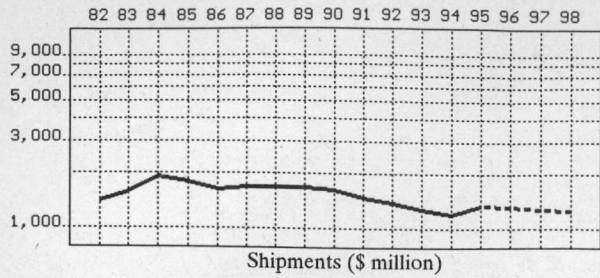

Shipments ($ million)

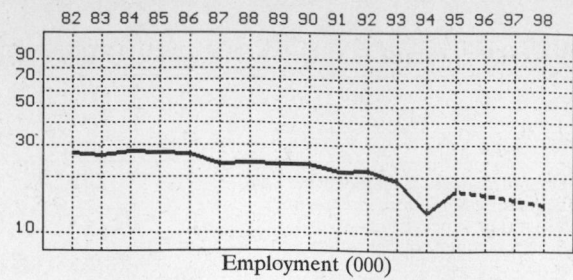

Employment (000)

GENERAL STATISTICS

Year	Companies	Establishments Total	Establishments with 20 or more employees	Employment Total (000)	Employment Production Workers (000)	Employment Hours (Mil)	Compensation Payroll ($ mil)	Compensation Wages ($/hr)	Production Cost of Materials	Production Value Added by Manufacture	Production Value of Shipments	Capital Invest.
1982	64	64	39	27.5	14.3	27.1	722.0	11.92	336.9	1,081.8	1,402.3	24.6
1983		64	41	26.7	14.3	27.7	716.2	12.40	298.4	1,361.9	1,575.8	17.3
1984		64	43	27.7	14.7	28.7	796.0	13.21	401.9	1,478.8	1,925.0	25.4
1985		65	45	27.3	14.6	27.5	811.4	13.59	296.2	1,512.8	1,790.5	28.9
1986		62	43	26.9	13.7	26.5	839.5	14.03	356.4	1,409.9	1,625.9	32.2
1987	59	59	33	23.9	12.1	23.9	805.1	15.23	218.4	1,425.4	1,678.1	50.0
1988		62	37	24.0	12.2	24.1	822.3	15.27	218.1	1,401.7	1,679.1	14.6
1989		60	33	23.6	11.5	21.8	807.8	15.71	210.8	1,476.0	1,688.6	19.7
1990		62	31	23.6	10.8	20.8	857.3	16.62	198.1	1,436.4	1,643.6	23.1
1991		61	33	21.5	9.3	19.1	858.6	16.45	196.3	1,241.5	1,480.2	13.2
1992	71	72	33	22.0	9.0	17.0	937.7	19.37	224.4	1,145.5	1,386.1	19.4
1993		67	29	19.3	8.5	16.4	882.8	19.33	356.5	873.8	1,272.0	903.0
1994		65P	29P	12.8	5.3	10.3	541.0	18.05	204.9	1,000.1	1,199.7	9.3
1995		65P	28P	17.0P	6.6P	13.0P	818.8P	19.72P	229.4P	1,119.6P	1,343.0P	253.0P
1996		65P	26P	16.0P	5.9P	11.7P	821.5P	20.32P	224.0P	1,093.1P	1,311.3P	276.2P
1997		66P	25P	15.1P	5.2P	10.4P	824.2P	20.93P	218.5P	1,066.7P	1,279.6P	299.4P
1998		66P	24P	14.1P	4.5P	9.0P	826.9P	21.54P	213.1P	1,040.2P	1,247.8P	322.5P

Sources: 1982, 1987, 1992 *Economic Census*; *Annual Survey of Manufactures*, 83-86, 88-91, 93-94. Establishment counts for non-Census years are from *County Business Patterns*; establishment values for 83-84 are extrapolations. 'P's show projections by the editors. Industries reclassified in 87 will not have data for prior years.

INDICES OF CHANGE

Year	Companies	Establishments Total	Establishments with 20 or more employees	Employment Total (000)	Employment Production Workers (000)	Employment Hours (Mil)	Compensation Payroll ($ mil)	Compensation Wages ($/hr)	Production Cost of Materials	Production Value Added by Manufacture	Production Value of Shipments	Capital Invest.
1982	90	89	118	125	159	159	77	62	150	94	101	127
1983		89	124	121	159	163	76	64	133	119	114	89
1984		89	130	126	163	169	85	68	179	129	139	131
1985		90	136	124	162	162	87	70	132	132	129	149
1986		86	130	122	152	156	90	72	159	123	117	166
1987	83	82	100	109	134	141	86	79	97	124	121	258
1988		86	112	109	136	142	88	79	97	122	121	75
1989		83	100	107	128	128	86	81	94	129	122	102
1990		86	94	107	120	122	91	86	88	125	119	119
1991		85	100	98	103	112	92	85	87	108	107	68
1992	100	100	100	100	100	100	100	100	100	100	100	100
1993		93	88	88	94	96	94	100	159	76	92	4,655
1994		90P	87P	58	59	61	58	93	91	87	87	48
1995		91P	83P	77P	74P	77P	87P	102P	102P	98P	97P	1,304P
1996		91P	80P	73P	66P	69P	88P	105P	100P	95P	95P	1,424P
1997		91P	76P	69P	58P	61P	88P	108P	97P	93P	92P	1,543P
1998		92P	72P	64P	50P	53P	88P	111P	95P	91P	90P	1,663P

Sources: Same as General Statistics. Values reflect change from the base year, 1992. Values above 100 mean greater than 92, values below 100 mean less than 92, and a value of 100 in the 82-91 or 93-98 period means same as 92. 'P's mark projections by the editors.

SELECTED RATIOS

For 1994	Avg. of All Manufact.	Analyzed Industry	Index	For 1994	Avg. of All Manufact.	Analyzed Industry	Index
Employees per Establishment	49	197	402	Value Added per Production Worker	134,084	188,698	141
Payroll per Establishment	1,500,273	8,323,077	555	Cost per Establishment	5,045,178	3,152,308	62
Payroll per Employee	30,620	42,266	138	Cost per Employee	102,970	16,008	16
Production Workers per Establishment	34	82	238	Cost per Production Worker	146,988	38,660	26
Wages per Establishment	853,319	2,860,231	335	Shipments per Establishment	9,576,895	18,456,923	193
Wages per Production Worker	24,861	35,078	141	Shipments per Employee	195,460	93,727	48
Hours per Production Worker	2,056	1,943	95	Shipments per Production Worker	279,017	226,358	81
Wages per Hour	12.09	18.05	149	Investment per Establishment	321,011	143,077	45
Value Added per Establishment	4,602,255	15,386,154	334	Investment per Employee	6,552	727	11
Value Added per Employee	93,930	78,133	83	Investment per Production Worker	9,352	1,755	19

Sources: Same as General Statistics. The 'Average of All Manufacturing' column represents the average of all manufacturing industries reported for the most recent complete year available. The Index shows the relationship between the Average and the Analyzed Industry. For example, 100 means that they are equal; 500 that the Analyzed Industry is five times the average; 50 means that the Analyzed Industry is half the national average. The abbreviation 'na' is used to show that data are 'not available'.

1442

LEADING COMPANIES   Number shown: 16   Total sales ($ mil): 1,520   Total employment (000): 10.6

Company Name	Address				CEO Name	Phone	Co. Type	Sales ($ mil)	Empl. (000)
Alliant Techsystems Inc	600 2nd St	Hopkins	MN	55343	Toby G Warson	612-931-6000	P	775	4.9
FMC Corp	4800 E River Rd	Minneapolis	MN	55421	James Ashton	612-571-9201	D	270	2.5
Alliant Techsystems Inc	600 2nd St	Hopkins	MN	55343	Don Cattell	612-931-6000	D	220•	1.0
Valentec International Corp	3190 Pullman St	Costa Mesa	CA	92626	Robert A Zummo	714-662-7756	R	65	0.8
Litton Industries Inc	2787 Orange	Apopka	FL	32703	Gregory M Misiak	407-297-4405	D	40	0.3
Loral Hycor Inc	10 Gill St	Woburn	MA	01801	James Larson	617-935-5950	S	30	0.3
Amron Corp	525 Progress Av	Waukesha	WI	53186	Phillip F Lange	414-547-1661	S	25	0.3
Lockley Manufacturing Group	PO Box 819	New Castle	PA	16103	George Frederick	412-658-1551	D	20	<0.1
Quantic Industries Inc	990 Commercial St	San Carlos	CA	94070	Charles Davis	415-595-1100	R	18	0.1
Hi-Shear Technology Corp	24225 Garnier St	Torrance	CA	90505	Thomas R Mooney	310-348-3769	P	15	0.1
MascoTech Inc	PO Box 856	Riverbank	CA	95367	Robert Jones	209-529-8100	D	15	0.1
Pachmayr Ltd	1875 S Mountain	Monrovia	CA	91016	Leslie H Whitney	818-357-7771	R	10	<0.1
ARES Inc	Erie Indstrial Park	Port Clinton	OH	43452	Karol N Wall	419-635-2175	R	5	<0.1
Hydroacoustics Inc	PO Box 23447	Rochester	NY	14692	John V Bouyoucos	716-359-1000	R	5	<0.1
Syn-Tech Systems Inc	PO Box 5258	Tallahassee	FL	32314	Douglas R Dunlap	904-878-2558	R	5	<0.1
Aerospace Design Inc	21200 S Figueora St	Carson	CA	90745	Loren Russakov	310-328-5175	R	1•	<0.1

Source: Ward's Business Directory of U.S. Private and Public Companies, Volumes 1 and 2, 1996. The company type code used is as follows: P - Public, R - Private, S - Subsidiary, D - Division, J - Joint Venture, A - Affiliate, G - Group. Sales are in millions of dollars, employees are in thousands. An asterisk (•) indicates an estimated sales volume. The symbol < stands for 'less than'. Company names and addresses are truncated, in some cases, to fit into the available space.

MATERIALS CONSUMED

Material	Quantity	Delivered Cost ($ million)
Materials, ingredients, containers, and supplies	(X)	164.2
Metal bolts, nuts, screws, washers, rivets, and other screw machine products	(X)	6.7
Other fabricated metal products (except castings and forgings)	(X)	13.5
Iron and steel castings (rough and semifinished)	(X)	1.1
Aluminum and aluminum-base alloy castings (rough and semifinished)	(X)	2.0
Other nonferrous castings (rough and semifinished)	(X)	(D)
Iron and steel forgings	(X)	(D)
Aluminum and aluminum-base alloy forgings	(X)	(D)
Other nonferrous forgings	(X)	(D)
Steel shapes and forms	(X)	22.8
Copper and copper-base alloy shapes and forms	(X)	(D)
Aluminum and aluminum-base alloy shapes and forms	(X)	0.9
Other nonferrous shapes and forms	(X)	(D)
Smokeless powder	(X)	(D)
Plastics resins consumed in the form of granules, pellets, powders, liquids, etc.	(X)	(D)
Other chemicals and allied products	(X)	0.7
Rough and dressed lumber	(X)	(D)
Paperboard containers, boxes, and corrugated paperboard	(X)	0.5
Fabricated plastics products (except gaskets, hoses, and belting)	(X)	1.2
Machine tool accessories, including cutting tools	(X)	1.3
Electronic, hydraulic, and mechanical subassemblies	(X)	(D)
Radio and electronic communication, navigation, search, detection systems	(X)	(D)
All other materials and components, parts, containers, and supplies	(X)	13.7
Materials, ingredients, containers, and supplies, nsk	(X)	7.3

Source: 1992 Economic Census. Explanation of symbols used: (D): Withheld to avoid disclosure of competitive data; na: Not available; (S): Withheld because statistical norms were not met; (X): Not applicable; (Z): Less than half the unit shown; nec: Not elsewhere classified; nsk: Not specified by kind; - : zero; • : 10-19 percent estimated; •• : 20-29 percent estimated.

PRODUCT SHARE DETAILS

Product or Product Class	% Share	Product or Product Class	% Share
Ordnance and accessories, nec.	100.00	Parts and other related equipment for heavy weapons (more than 30 mm, more than 1.18 in.)	(D)
Guns, howitzers, mortars, and related equipment, more than 30 mm (or more than 1.18 in.)	43.64	Guns, howitzers, mortars, and related equipment, more than 30 mm (or more than 1.18 inches), nsk	0.03
Guns, howitzers, and mortars, assembled and recoilless rifles (more than 30 mm, more than 1.18 in.)	55.69	Ordnance and accessories, nec (rocket projectors, line throwing guns, flame throwers, torpedo tubes, etc., and parts).	55.01
Turrets (except aircraft turrets and aircraft turret drives), mounts and carriages (except self-propelled) (more than 30 mm, more than 1.18 in.)	(D)	Ordnance and accessories, nec, nsk	1.35

Source: 1992 Economic Census. The values shown are percent of total shipments in an industry. Values of indented subcategories are summed in the main heading. The symbol (D) appears when data are withheld to prevent disclosure of competitive information. The abbreviation nsk stands for 'not specified by kind' and nec for 'not elsewhere classified'.

INPUTS AND OUTPUTS FOR ORDNANCE & ACCESSORIES NEC

Economic Sector or Industry Providing Inputs	%	Sector	Economic Sector or Industry Buying Outputs	%	Sector
Ordnance & accessories nec	9.0	Manufg.	Federal Government purchases, national defense	72.0	Fed Govt
Semiconductors & related devices	9.0	Manufg.	Exports	24.4	Foreign
Explosives	7.3	Manufg.	Ordnance & accessories nec	2.4	Manufg.
Wholesale trade	7.3	Trade	Change in business inventories	1.0	In House
Blast furnaces & steel mills	6.9	Manufg.	Small arms ammunition	0.2	Manufg.
Electric services (utilities)	6.1	Util.			
Advertising	4.7	Services			
Management & consulting services & labs	3.6	Services			
Air transportation	3.1	Util.			
Gas production & distribution (utilities)	2.9	Util.			
Motor freight transportation & warehousing	2.5	Util.			
Aluminum rolling & drawing	2.0	Manufg.			
Electronic components nec	1.9	Manufg.			
Internal combustion engines, nec	1.9	Manufg.			
Miscellaneous fabricated wire products	1.8	Manufg.			
Detective & protective services	1.8	Services			
Aircraft & missile equipment, nec	1.6	Manufg.			
Communications, except radio & TV	1.4	Util.			
U.S. Postal Service	1.4	Gov't			
Maintenance of nonfarm buildings nec	1.2	Constr.			
Equipment rental & leasing services	1.2	Services			
Iron & steel forgings	1.1	Manufg.			
Machinery, except electrical, nec	1.0	Manufg.			
Machine tools, metal cutting types	0.9	Manufg.			
Plating & polishing	0.9	Manufg.			
Fabricated metal products, nec	0.8	Manufg.			
Hand saws & saw blades	0.8	Manufg.			
Metal stampings, nec	0.8	Manufg.			
Miscellaneous plastics products	0.8	Manufg.			
Fabricated rubber products, nec	0.7	Manufg.			
Iron & steel foundries	0.7	Manufg.			
Hotels & lodging places	0.7	Services			
Screw machine and related products	0.6	Manufg.			
Personnel supply services	0.6	Services			
Imports	0.6	Foreign			
Abrasive products	0.5	Manufg.			
Aluminum castings	0.5	Manufg.			
Nonferrous rolling & drawing, nec	0.5	Manufg.			
Paperboard containers & boxes	0.5	Manufg.			
Petroleum refining	0.5	Manufg.			
Special dies & tools & machine tool accessories	0.5	Manufg.			
Sanitary services, steam supply, irrigation	0.5	Util.			
Eating & drinking places	0.5	Trade			
Noncomparable imports	0.5	Foreign			
Sawmills & planning mills, general	0.4	Manufg.			
Banking	0.4	Fin/R.E.			
Real estate	0.4	Fin/R.E.			
Nonferrous forgings	0.3	Manufg.			
Computer & data processing services	0.3	Services			
Legal services	0.3	Services			
Commercial printing	0.2	Manufg.			
Glass & glass products, except containers	0.2	Manufg.			
Lubricating oils & greases	0.2	Manufg.			
Metal coating & allied services	0.2	Manufg.			
Metal heat treating	0.2	Manufg.			
Railroads & related services	0.2	Util.			
Insurance carriers	0.2	Fin/R.E.			
Royalties	0.2	Fin/R.E.			
Business services nec	0.2	Services			
Engineering, architectural, & surveying services	0.2	Services			
Laundry, dry cleaning, shoe repair	0.2	Services			
State & local government enterprises, nec	0.2	Gov't			
Manifold business forms	0.1	Manufg.			
Accounting, auditing & bookkeeping	0.1	Services			
Automotive rental & leasing, without drivers	0.1	Services			
Automotive repair shops & services	0.1	Services			

Source: Benchmark Input-Output Accounts for the U.S. Economy, 1982, U.S. Department of Commerce, Washington, D.C., July 1991. Data, as reported in the source, are organized by the 1977 SIC structure in use in 1982 but have been matched, as closely as is possible, to the 1987 SIC structure used in this book.

OCCUPATIONS EMPLOYED BY SIC 348 - ORDNANCE AND ACCESSORIES, NEC

Occupation	% of Total 1994	Change to 2005	Occupation	% of Total 1994	Change to 2005
Assemblers, fabricators, & hand workers nec	10.8	-5.4	General office clerks	1.7	-19.4
Inspectors, testers, & graders, precision	5.2	-5.4	Metal & plastic machine workers nec	1.7	-16.2
Blue collar worker supervisors	4.4	-10.2	Industrial machinery mechanics	1.7	4.0
Combination machine tool operators	3.9	4.0	Electricians	1.6	-11.2
Engineers nec	2.8	41.8	Electrical & electronic technicians,technologists	1.6	-5.5
Machinists	2.7	-5.5	Plant & system operators nec	1.5	-16.4
Guards	2.2	-14.9	Engineering, mathematical, & science managers	1.4	7.4
Machine operators nec	2.2	-16.7	Industrial engineers, ex safety engineers	1.4	4.0
Engineering technicians & technologists nec	2.2	-5.4	Janitors & cleaners, incl maids	1.4	-24.3
Mechanical engineers	2.2	4.1	NC machine tool operators, metal & plastic	1.3	13.4
Secretaries, ex legal & medical	2.1	-13.9	Tool & die makers	1.1	-23.6
Production, planning, & expediting clerks	2.0	27.6	Management support workers nec	1.1	-5.5
Professional workers nec	1.9	13.4	Industrial production managers	1.1	-5.4
Precision assemblers nec	1.9	-9.2	Freight, stock, & material movers, hand	1.1	-24.3
Machine tool cutting operators, metal & plastic	1.7	-21.2			

Source: Industry-Occupation Matrix, Bureau of Labor Statistics. These data relate to one or more 3-digit SIC industry groups rather than to a single 4-digit SIC. The change reported for each occupation to the year 2005 is a percent of growth or decline as estimated by the Bureau of Labor Statistics. The abbreviation nec stands for 'not elsewhere classified'.

LOCATION BY STATE AND REGIONAL CONCENTRATION

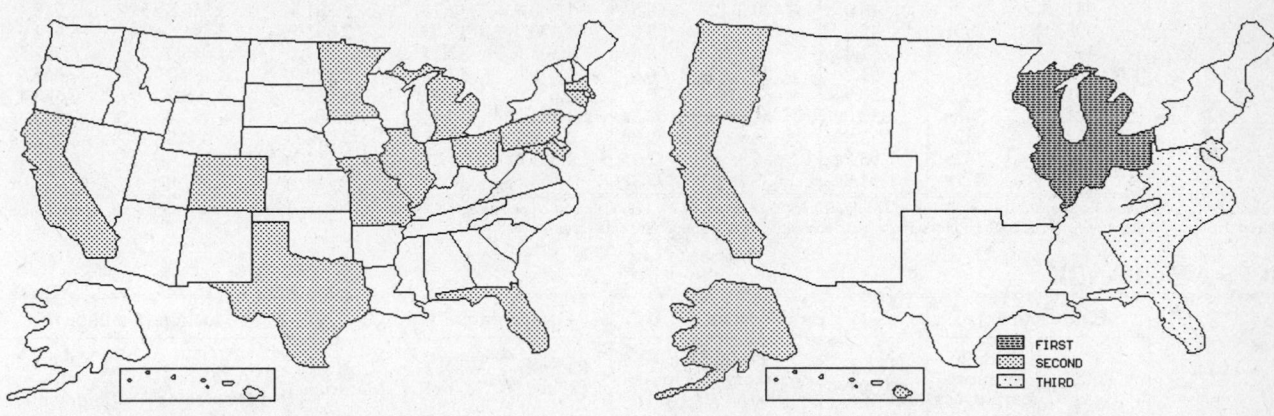

FIRST
SECOND
THIRD

INDUSTRY DATA BY STATE

State	Establish-ments	Shipments			Employment				Cost as % of Shipments	Investment per Employee ($)
		Total ($ mil)	% of U.S.	Per Establ.	Total Number	% of U.S.	Per Establ.	Wages ($/hour)		
Michigan	5	11.4	0.8	2.3	200	0.9	40	22.00	22.8	-
California	8	(D)	-	-	750*	3.4	94	-	-	-
Ohio	6	(D)	-	-	1,750*	8.0	292	-	-	-
Florida	5	(D)	-	-	750*	3.4	150	-	-	-
Texas	5	(D)	-	-	3,750*	17.0	750	-	-	27
Colorado	4	(D)	-	-	7,500*	34.1	1,875	-	-	-
Illinois	3	(D)	-	-	175*	0.8	58	-	-	-
Connecticut	2	(D)	-	-	375*	1.7	188	-	-	-
Maryland	2	(D)	-	-	175*	0.8	88	-	-	-
Minnesota	2	(D)	-	-	1,750*	8.0	875	-	-	-
Massachusetts	1	(D)	-	-	175*	0.8	175	-	-	-
Missouri	1	(D)	-	-	3,750*	17.0	3,750	-	-	-
Pennsylvania	1	(D)	-	-	375*	1.7	375	-	-	-

Source: 1992 *Economic Census.* The states are in descending order of shipments or establishments (if shipment data are missing for the majority). The symbol (D) appears when data are withheld to prevent disclosure of competitive information. States marked with (D) are sorted by number of establishments. A dash (-) indicates that the data element cannot be calculated; * indicates the midpoint of a range.

3491 - INDUSTRIAL VALVES

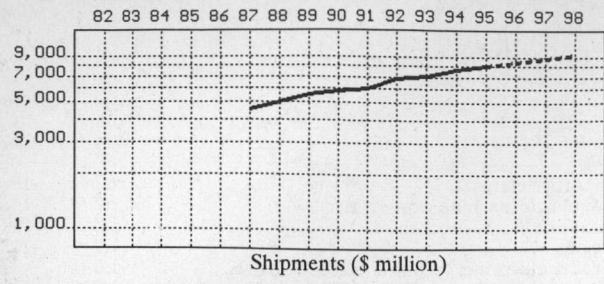

Shipments ($ million)

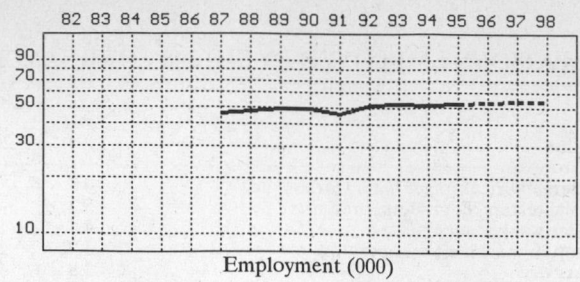

Employment (000)

GENERAL STATISTICS

Year	Com-panies	Establishments		Employment			Compensation		Production ($ million)			
		Total	with 20 or more employees	Total (000)	Production Workers (000)	Hours (Mil)	Payroll ($ mil)	Wages ($/hr)	Cost of Materials	Value Added by Manufacture	Value of Shipments	Capital Invest.
1982												
1983												
1984												
1985												
1986												
1987	314	385	273	46.2	29.8	59.2	1,199.3	10.94	1,803.9	2,793.2	4,595.7	150.5
1988		392	283	47.5	31.0	63.4	1,279.0	10.93	2,137.7	2,953.1	5,010.9	138.5
1989		380	279	48.6	31.7	63.7	1,364.0	11.45	2,343.2	3,246.9	5,501.2	162.6
1990		381	279	48.6	30.3	60.5	1,368.8	12.15	2,381.4	3,385.5	5,745.4	211.8
1991		395	285	45.9	29.6	58.8	1,430.3	12.73	2,392.1	3,591.8	5,972.6	191.0
1992	389	493	328	51.4	33.4	68.4	1,595.6	12.51	2,707.1	4,046.0	6,762.5	212.7
1993		504	331	51.7	33.7	66.6	1,619.0	12.93	2,820.9	4,189.2	7,060.5	202.8
1994		501P	333P	50.9	34.0	70.3	1,675.8	12.93	3,082.7	4,598.5	7,653.4	219.1
1995		521P	342P	52.0P	34.2P	69.5P	1,751.7P	13.55P	3,201.5P	4,775.7P	7,948.3P	236.0P
1996		542P	352P	52.8P	34.8P	70.8P	1,820.7P	13.88P	3,372.5P	5,030.8P	8,372.9P	247.1P
1997		562P	362P	53.5P	35.3P	72.1P	1,889.6P	14.21P	3,543.5P	5,285.9P	8,797.4P	258.2P
1998		583P	371P	54.2P	35.9P	73.3P	1,958.6P	14.54P	3,714.5P	5,541.0P	9,222.0P	269.3P

Sources: 1982, 1987, 1992 *Economic Census*; *Annual Survey of Manufactures*, 83-86, 88-91, 93-94. Establishment counts for non-Census years are from *County Business Patterns*; establishment values for 83-84 are extrapolations. 'P's show projections by the editors. Industries reclassified in 87 will not have data for prior years.

INDICES OF CHANGE

Year	Com-panies	Establishments		Employment			Compensation		Production ($ million)			
		Total	with 20 or more employees	Total (000)	Production Workers (000)	Hours (Mil)	Payroll ($ mil)	Wages ($/hr)	Cost of Materials	Value Added by Manufacture	Value of Shipments	Capital Invest.
1982												
1983												
1984												
1985												
1986												
1987	81	78	83	90	89	87	75	87	67	69	68	71
1988		80	86	92	93	93	80	87	79	73	74	65
1989		77	85	95	95	93	85	92	87	80	81	76
1990		77	85	95	91	88	86	97	88	84	85	100
1991		80	87	89	89	86	90	102	88	89	88	90
1992	100	100	100	100	100	100	100	100	100	100	100	100
1993		102	101	101	101	97	101	103	104	104	104	95
1994		102P	101P	99	102	103	105	103	114	114	113	103
1995		106P	104P	101P	102P	102P	110P	108P	118P	118P	118P	111P
1996		110P	107P	103P	104P	104P	114P	111P	125P	124P	124P	116P
1997		114P	110P	104P	106P	105P	118P	114P	131P	131P	130P	121P
1998		118P	113P	105P	108P	107P	123P	116P	137P	137P	136P	127P

Sources: Same as General Statistics. Values reflect change from the base year, 1992. Values above 100 mean greater than 92, values below 100 mean less than 92, and a value of 100 in the 82-91 or 93-98 period means same as 92. 'P's mark projections by the editors.

SELECTED RATIOS

For 1994	Avg. of All Manufact.	Analyzed Industry	Index	For 1994	Avg. of All Manufact.	Analyzed Industry	Index
Employees per Establishment	49	102	208	Value Added per Production Worker	134,084	135,250	101
Payroll per Establishment	1,500,273	3,347,774	223	Cost per Establishment	5,045,178	6,158,362	122
Payroll per Employee	30,620	32,923	108	Cost per Employee	102,970	60,564	59
Production Workers per Establishment	34	68	198	Cost per Production Worker	146,988	90,668	62
Wages per Establishment	853,319	1,815,883	213	Shipments per Establishment	9,576,895	15,289,326	160
Wages per Production Worker	24,861	26,735	108	Shipments per Employee	195,460	150,361	77
Hours per Production Worker	2,056	2,068	101	Shipments per Production Worker	279,017	225,100	81
Wages per Hour	12.09	12.93	107	Investment per Establishment	321,011	437,700	136
Value Added per Establishment	4,602,255	9,186,501	200	Investment per Employee	6,552	4,305	66
Value Added per Employee	93,930	90,344	96	Investment per Production Worker	9,352	6,444	69

Sources: Same as General Statistics. The 'Average of All Manufacturing' column represents the average of all manufacturing industries reported for the most recent complete year available. The Index shows the relationship between the Average and the Analyzed Industry. For example, 100 means that they are equal; 500 that the Analyzed Industry is five times the average; 50 means that the Analyzed Industry is half the national average. The abbreviation 'na' is used to show that data are 'not available'.

LEADING COMPANIES   Number shown: **73**   Total sales ($ mil): **2,546**   Total employment (000): **18.4**

Company Name	Address				CEO Name	Phone	Co. Type	Sales ($ mil)	Empl. (000)
AptarGroup Inc	475 W Terra Cotta	Crystal Lake	IL	60014	Ervin J LeCoque	815-477-0424	P	474	3.3
TIW Corp	PO Box 35729	Houston	TX	77235	SR Pearce	713-729-2110	S	400	0.3
Crane Valve	104 N Chicago St	Joliet	IL	60431	Bill Landholt	815-727-2600	D	250	2.2
Stockham Valves and Fittings	PO Box 10326	Birmingham	AL	35202	Larry Kinderman	205-592-6361	R	120	1.1
Anderson, Greenwood and Co	3950 Greenbriar	Stafford	TX	77477	Gregory E Hyland	713-274-4400	S	119	0.9
Vesuvius USA Corp	PO Box 4014	Champaign	IL	61824	Mark Fishler	217-351-5000	S	83*	0.8
Dresser Industries Indust Valve	PO Box 1430	Alexandria	LA	71309	John M Cooper	318-640-2250	D	71	0.7
FMC Corp	PO Box 3091	Houston	TX	77253	Robert Potter	713-591-4000	D	50	0.3
JM Huber Corp	14011 Park Dr	Tomball	TX	77375	Milton Hoff	713-351-0837	D	50	0.2
Harsco Corp	PO Box 790	Lockport	NY	14094	D Hathaway	716-433-3891	D	42	0.4
Engineered Controls Intern	PO Box 247	Elon College	NC	27244	Thomas R Darcy	919-449-7707	R	40	0.5
Leslie Controls Inc	12501 Telecom Dr	Tampa	FL	33637	Charles Wolley	813-978-1000	S	40	0.3
Suntec Industries Inc	PO Box 7010	Rockford	IL	61125	M Brown	815-226-3700	R	40	0.2
Orbit Valve Co	PO Box 193520	Little Rock	AR	72219	Ed Ligon	501-568-6000	R	39*	0.4
Ross Operating Valve Co	PO Box 7015	Troy	MI	48007	Russ J Cameron	810-362-1250	R	37	0.3
Control Components Inc	22591nida Empresa	R S Margari	CA	92688	Stuart Carson	714-858-1877	S	35	0.3
Precision Valve Corp	PO Box 309	Yonkers	NY	10702	R H Abplanalp	914-969-6500	R	35	0.5
Lonergan Valve	PO Box 2360	Fort Wayne	IN	46801	Randy Shoff	219-747-1077	D	30	0.3
Daniel Valve Co	PO Box 40421	Houston	TX	77240	Kenton Chickering	713-469-0550	D	30	0.2
Valcor Engineering Corp	2 Lawrence Rd	Springfield	NJ	07081	George G Landberg	201-467-8400	R	30*	0.3
Piedmont Manufacturing Co	PO Box 668	Altavista	VA	24517	Martin F Giudice	804-369-4741	D	28*	0.3
Wilkins	1747 Com Way	Paso Robles	CA	93446	A Marini	805-238-7100	D	28	0.1
Circle Seal Controls	2301 Wardlow Cir	Corona	CA	91720	Don Marshall	714-774-6110	R	27	0.3
General Valve Company Inc	800 Koomey Rd	Brookshire	TX	77423	Brian W Warren	713-934-6014	S	27	0.2
EBW Inc	2814 McCracken	Muskegon	MI	49441	Leo J Leblanc	616-755-1671	R	26	0.1
Kerotest Manufacturing Corp	2525 Liberty Av	Pittsburgh	PA	15222	Robert G Visalli	412-392-4300	R	23	0.2
Veriflo Corp	250 Canal Blv	Richmond	CA	94804	Thomas W Bates	510-235-9590	R	21	0.3
Parker Hannifin Corp	2651 Ala Hwy 21 N	Jacksonville	AL	36265	Ted Tabaka	205-435-2130	D	20	0.2
Cashco Inc	PO Box 6	Ellsworth	KS	67439	Philip G Rogers	913-472-4461	R	19	0.1
Hunt Valve Company Inc	1913 E State St	Salem	OH	44460	David B Huberfield	216-337-9535	D	17	0.2
Fleck Control Inc	20580 Enterprise	Brookfield	WI	53005	AJ Fleckenstein	414-784-4490	R	17	0.1
Precision General Inc	16101 Vallen Dr	Houston	TX	77041	Spencer Nimberger	713-466-0056	R	17*	0.2
Red Valve Co	700 N Bell Av	Carnegie	PA	15106	G Raftis	412-279-4660	R	17	0.2
Halkey Roberts Corp	11600 9th St N	St Petersburg	FL	33716	Charles S Gamble	813-577-1300	S	15	0.2
James Jones Co	PO Box 5428	El Monte	CA	91734	Kenneth A Davis	818-443-6191	S	15	0.1
Standard Machine & Mfg Co	10014 Big Bend	St Louis	MO	63122	Jack Deutsch	314-966-4500	R	15	0.4
Valco Instruments Company Inc	PO Box 55603	Houston	TX	77255	Stanley Stearns	713-688-9345	R	14	0.2
Sedco	PO Box 624	Adrian	MI	49221	RE Price	517-263-2220	D	13	<0.1
Value and Primer Corp	711 W 17th St	Costa Mesa	CA	92627	Chris M Dickson	714-631-2600	R	12*	<0.1
FETTEROLF/MG Corp	PO Box 103	Skippack	PA	19474	JW Williams IV	610-584-1500	R	10	<0.1
Newport News Industrial Corp	700 Thimble Shoal	Newport News	VA	23606	Tom Bond	804-380-7053	S	10	<0.1
Penberthy Inc	320 Locust St	Prophetstown	IL	61277	NM Simard	815-537-2311	S	9	0.2
Mazzella Wire Rope & Sling Co	14600 Brookpark	Cleveland	OH	44135	Anthony J Mazzella	216-362-4600	R	8	<0.1
Wright Components Inc	PO Box 160	Phelps	NY	14532	EJ Laskowski	315-548-9501	S	8	<0.1
GO Inc	305 S Acacia St	San Dimas	CA	91773	Philip Vaughan	909-599-6745	S	8	<0.1
Homestead Valve	160 W Walnut St	Allentown	PA	18102	James E Olson	610-770-1100	D	7	<0.1
High Vacuum Apparatus Mfg	PO Box 4764	Hayward	CA	94540	Arthur J Brenes	510-785-2744	R	7	<0.1
Wesco Valve & Mfg Co	PO Box J	Marshall	TX	75670	W Baker	903-938-9241	S	7	0.1
Girard Equipment Inc	1004 Hwy 1	Rahway	NJ	07065	JC Girard	908-382-4600	R	7	<0.1
C and S Valve Co	40 Chestnut Av	Westmont	IL	60559	Thomas Casale	708-789-8900	R	6	<0.1
Plattco Corp	18 White St	Plattsburgh	NY	12901	Peter B Guibord	518-563-4640	R	6	<0.1
Watson McDaniel Co	975 Madison Av	Norristown	PA	19403	Richard Picut	215-666-5711	R	6	<0.1
Atkomatic Valve Company Inc	PO Box 11258	Indianapolis	IN	46201	Wayne Rinker	317-357-8421	R	5	<0.1
Highfield Manufacturing Co	PO Box 549	Bridgeport	CT	06601	Timothy B Ely	203-384-2281	D	5*	<0.1
Romac Industries Inc	PO Box 580	Dallas	NC	28034	Joe Warner	704-824-4608	D	5	<0.1
Zimmermann and Jansen Inc	26535 Farmers Mkt	Magnolia	TX	77355	T Lynn Ash	713-931-3991	S	5	<0.1
Sem-Tec Inc	47 Lagrange St	Worcester	MA	01610	J M Krosoczka	508-798-8551	R	5	<0.1
Elliott Valve and Repair	5436 Clay St	Houston	TX	77023	Stan Bubien	713-926-8318	D	4*	<0.1
Ring-O-Valve Inc	510 Industrial Blv	Sugar Land	TX	77478	Joe Bocci	713-240-0711	R	4	<0.1
Sweeney Engineering Corp	17224 Gramercy Pl	Gardena	CA	90247	Stanley A Konin	310-324-4961	R	4*	<0.1
US Para Plate Corp	PO Box 611540	San Jose	CA	95161	Donald Gross	408-453-6311	R	4	<0.1
D/A Manufacturing Company	PO Box T	Tulia	TX	79088	Don Adams	806-995-3282	R	3*	<0.1
Superior Manufacturing Co	3200 Lynn Av	St Louis Park	MN	55416	Harry Rhodes	612-920-6474	R	3	<0.1
Ogontz Corp	PO Box 479	Willow Grove	PA	19090	TM Kenny	215-657-4770	R	3	<0.1
Hy-Matic Manufacturing Inc	PO Box 98	Kendallville	IN	46755	R Borger	219-347-3651	R	3	<0.1
Mirada Research & Mfg Co	125 Columbia Ct	Chaska	MN	55318	Lloyd Wass	612-448-3686	R	2*	<0.1
Redwood MicroSystems Corp	959 Hamilton Av	Menlo Park	CA	94025	Perry Constantine	415-617-1209	R	2*	<0.1
AMTECH Inc	3515 Stoneham	Houston	TX	77047	C Holmes	713-738-1166	R	2	<0.1
Alta Sales Inc	110 S 1200 W	Lindon	UT	84042	James Wilson	801-785-1114	R	1*	<0.1
Walworth Co	10190 Harwin Dr	Houston	TX	77036	John A Hotz	713-777-7788	R	1	<0.1
Rockwood Swendeman Corp	P O Box 2402	South Portland	ME	04106	Donald C Barratt	207-799-3341	R	1	<0.1
Savel Corp	PO Box 5151	Elm Grove	WI	53122	J Savignal	414-796-2860	R	1	<0.1
Hofer Valve and Supply Inc	130 Arlington Rd S	Jacksonville	FL	32216	Bob Waldeck	904-725-2172	R	0	<0.1

Source: Ward's Business Directory of U.S. Private and Public Companies, Volumes 1 and 2, 1996. The company type code used is as follows: P - Public, R - Private, S - Subsidiary, D - Division, J - Joint Venture, A - Affiliate, G - Group. Sales are in millions of dollars, employees are in thousands. An asterisk (*) indicates an estimated sales volume. The symbol < stands for 'less than'. Company names and addresses are truncated, in some cases, to fit into the available space.

MATERIALS CONSUMED

Material	Quantity	Delivered Cost ($ million)
Materials, ingredients, containers, and supplies	(X)	2,281.2
Metal bolts, nuts, screws, washers, rivets, and other screw machine products	(X)	111.8
Metal stampings	(X)	45.1
Valves, fittings, and couplings purchased for further assembly (except forgings)	(X)	178.4
Other fabricated metal products (except forgings)	(X)	114.3
Iron castings (rough and semifinished)	(X)	174.6
Steel castings (rough and semifinished)	(X)	252.9
Aluminum and aluminum-base alloy castings (rough and semifinished)	(X)	52.3
Copper and copper-base alloy castings (rough and semifinished)	(X)	79.1
Other nonferrous castings (rough and semifinished)	(X)	49.1
Iron and steel forgings	(X)	84.5
Nonferrous forgings	(X)	50.6
Steel bars, bar shapes, and plates	(X)	114.0
Steel sheet and strip, including tin plate	(X)	29.4
All other steel shapes and forms	(X)	51.8
Copper and copper-base alloy refinery shapes	(X)	46.9
Copper and copper-base alloy rod, bar, and mechanical wire, including extruded and/or drawn shapes	(X)	87.5
Copper and copper-base alloy pipe and tube	(X)	14.6
All other copper and copper-base alloy shapes and forms	(X)	2.4
Aluminum and aluminum-base alloy shapes and forms	(X)	19.7
Other nonferrous shapes and forms	(X)	8.9
Scrap, including iron, steel, aluminum and aluminum-base alloy (excluding home scrap)	(X)	39.0
Electric motors and generators less than 1 horsepower (less than 746 watts)	(X)	9.3
Paperboard containers, boxes, and corrugated paperboard	(X)	24.9
Hydraulic and pneumatic hose	(X)	1.1
Other rubber and plastics hose and belting	(X)	6.2
Gaskets (all types), and packing and sealing devices	(X)	44.4
Fabricated rubber products, except tires, tubes, hose, belting, and gaskets	(X)	28.7
Fabricated plastics products (except gaskets, hoses, and belting)	(X)	42.1
All other materials and components, parts, containers, and supplies	(X)	406.8
Materials, ingredients, containers, and supplies, nsk	(X)	110.7

Source: 1992 Economic Census. Explanation of symbols used: (D): Withheld to avoid disclosure of competitive data; na: Not available; (S): Withheld because statistical norms were not met; (X): Not applicable; (Z): Less than half the unit shown; nec: Not elsewhere classified; nsk: Not specified by kind; - : zero; * : 10-19 percent estimated; ** : 20-29 percent estimated.

PRODUCT SHARE DETAILS

Product or Product Class	% Share	Product or Product Class	% Share
Industrial valves	100.00	works and municipal equipment	10.56
Gates, globes, angles, straightway (Y-type) check, stop and check, cross, 3-and 4-way, and other industrial valves	19.00	Industrial AWWA butterfly valves (all pressures) for water works and municipal equipment	14.54
Iron body gates, globes, angles, straightway (Y-type) check, stop and check, cross, 3-and 4-way, and other industrial valves, including ductile or modular, all pressures (excl. IBBM, AWWA, and UL)	12.29	Parts for industrial valves for water works and municipal equipment (IBBW, AWWA, and UL)	9.58
		Valves for water works and municipal equipment (ibbw, awwa, and ul), nsk	1.46
Cast carbon steel gates, globes, angles, straightway (Y-type) check, stop and check, cross, 3-and 4-way, and other industrial valves	21.56	Industrial ball valves (all metals, pressures, and types), including manual and power-operated, on/off valves	15.37
Forged carbon steel gates, globes, angles, straightway (Y-type) check, stop and check, cross, 3-and 4-way, and other industrial valves	14.90	Industrial iron (including ductile) ball valves (all pressures and types), manual and power-operated, on/off valves	5.79
Alloy steel and other metal gates, globes, angles, straightway (Y-type) check, stop and check, cross, 3-and 4-way, and other industrial valves	20.77	Industrial brass and bronze ball valves (all pressures and types), including manual and power-operated, on/off valves	17.64
Brass and bronze (125 lb, w.s.p. or more) gates, globes, angles, straightway (Y-type) check, stop and check, cross, 3-and 4-way, and other industrial valves	21.85	Industrial carbon steel (cast and fabricated) ball valves (all pressures and types), including manual and power-operated, on/off valves	28.53
Actuators (power-operated, on/off mounted) for gates, globes, angles, straightway (Y-type) check, stop and check, cross, 3-and 4-way, and other industrial valves	0.92	Industrial alloy steel and other metal ball valves (all pressures and types), including manual and power-operated, on/off valves	29.54
Parts for gates, globes, angles, straightway (Y-type) check, stop and check, cross, 3-and 4-way, and other industrial valves	6.78	Actuators (power-operated, on/off mounted) for industrial ball valves (all metals, pressures, and types)	7.45
		Parts for industrial ball valves (all metals, pressures, and types)	10.98
Gates, globes, angles, straightway (y-type) checks, stop and check, cross, 3-and 4-way, etc., nsk	0.92	Ball valves (all metals, pressures, and types), including manual and power-operated, on/off valves, nsk	0.06
Industrial valves for water works and municipal equipment (IBBW, AWWA, and UL)	8.91	Industrial butterfly valves (all metals, pressures, and types), including manual and power-operated, on/off valves	6.01
Industrial IBBM gate line and tapping valves for water works and municipal equipment	29.97	Industrial iron (including ductile) butterfly valves, except high-pressure types, including elastomer and fluroplastics lined, manual and power-operated, on/off valves	24.34
Industrial UL check valves (all pressures) for water works and municipal equipment	2.78	Industrial brass and bronze butterfly valves, except high-pressure types, including elastomer and fluroplastics lined, manual and power-operated, on/off valves	5.76
All other industrial UL valves (all pressures), including pest indicators, for water works and municipal equipment	6.15	Industrial carbon steel (cast and fabricated) butterfly valves, except high-pressure types, including elastomer and fluroplastics lined, manual and power-operated, on/off valves	16.25
Tapping sleeves and crosses for industrial valves for water works and municipal equipment (IBBW, AWWA, and UL)	2.65		
Fire hydrants	22.32	Industrial alloy steel and other metal butterfly valves, except high-pressure types, including elastomer and	
Industrial AWWA check valves (all pressures) for water			

Continued on next page.

PRODUCT SHARE DETAILS - Continued

Product or Product Class	% Share	Product or Product Class	% Share
fluroplastics lined, manual and power-operated, on/off valves	5.05	Industrial diaphragm and pinch valves, including operators (all metals, pressures, and types), except automatic valves	7.90
Industrial high-pressure iron butterfly valves (shut-off to full ANSI class ratings), manual and power-operated, on/off valves	14.32	Industrial iron and steel pop safety valves and relief valves (more than 15 lb, w.s.p.)	22.68
Industrial high-pressure alloy steel and other metal butterfly valves (shut-off to full ANSI class ratings), manual and power-operated, on/off valves	14.18	Industrial brass and bronze pop safety valves and relief valves (more than 15 lb, w.s.p.)	9.33
		Industrial compressed gas cylinder valves	12.20
Industrial butterfly valve actuators (power-operated, on/off mounted)	13.56	Industrial valve steam traps (more than 15 lb, w.s.p.)	(D)
		Industrial thru conduit pipeline valves	(D)
Parts for industrial butterfly valves (all metals, pressures, and types), including manual and power-operated, on/off valves	5.87	Actuators for all other industrial valves, nec, sold separately (power-operated, on/off mounted)	5.05
Butterfly valves (all metals, pressures, and types), including manual and power-operated, on/off valves, nsk	0.65	Other industrial metal valves, excluding control valves, regulators, and solenoid valves	24.26
		Industrial valves, nec, nsk	9.74
Industrial plug valves (all metals, pressures, and types), such as lubricated, cylindrical eccentric, and sleeve-lined	4.48	Nuclear valves (N-stamp only)	1.65
Industrial iron (including ductile) plug valves (all pressures and types), such as lubricated, cylindrical eccentric, and sleeve-lined	29.97	Nuclear cast-carbon steel and low alloy gate, globe, and check valves (N-stamp only)	26.79
Industrial carbon steel plug valves (all pressures and types), such as lubricated, cylindrical eccentric, and sleeve-lined	32.19	Nuclear forged-carbon steel and low alloy gate, globe, and check valves (N-stamp only)	5.46
Industrial alloy steel and other metal plug valves (all pressures and types), such as lubricated, cylindrical eccentric, and sleeve-lined	29.75	Nuclear corrosion-resistant alloy steel gate, globe, and check valves (N-stamp only)	19.05
		Nuclear ball valves, butterfly valves, and plug valves (on/off, N-stamp only)	5.75
Industrial plug valve actuators (power-operated, on/off mounted)	(D)	Nuclear valve actuators (mounted power-operated, on/off) (N-stamp only)	8.53
		Nuclear automated control valves (N-stamp only)	4.56
Parts for industrial plug valves (all metals, pressures, and types), such as lubricated, cylindrical eccentric, and sleeve-lined	4.55	Parts for nuclear valves (N-stamp only)	27.18
		Nuclear valves (n-stamp only), nsk	2.68
Plug valves (all metals, pressures, and types), such as lubricated, cylindrical eccentric, and sleeve-lined, nsk	(D)	Automatic regulating and control valves and parts (except nuclear), power-operated, designed for modulating (throttling) service	24.98
Industrial valves, nec	10.48	Solenoid-operated valves and parts, except nuclear and fluid power transfer	6.42
Industrial valve cocks and stops (all metals, pressures, and types)	2.45	Industrial valves, nsk	2.71

Source: 1992 *Economic Census*. The values shown are percent of total shipments in an industry. Values of indented subcategories are summed in the main heading. The symbol (D) appears when data are withheld to prevent disclosure of competitive information. The abbreviation nsk stands for 'not specified by kind' and nec for 'not elsewhere classified'.

INPUTS AND OUTPUTS FOR PIPE, VALVES, & PIPE FITTINGS

Economic Sector or Industry Providing Inputs	%	Sector	Economic Sector or Industry Buying Outputs	%	Sector
Blast furnaces & steel mills	19.6	Manufg.	Industrial buildings	12.2	Constr.
Imports	13.8	Foreign	Gross private fixed investment	11.7	Cap Inv
Wholesale trade	8.1	Trade	Exports	7.8	Foreign
Iron & steel foundries	7.5	Manufg.	Maintenance of nonfarm buildings nec	5.5	Constr.
Pipe, valves, & pipe fittings	3.2	Manufg.	Crude petroleum & natural gas	4.0	Mining
Copper rolling & drawing	2.9	Manufg.	Cyclic crudes and organics	4.0	Manufg.
Advertising	2.8	Services	Sewer system facility construction	3.4	Constr.
Iron & steel forgings	2.5	Manufg.	Office buildings	2.8	Constr.
Screw machine and related products	2.4	Manufg.	Electric utility facility construction	2.7	Constr.
Electric services (utilities)	2.2	Util.	Automotive rental & leasing, without drivers	2.7	Services
Banking	1.9	Fin/R.E.	Water supply facility construction	2.0	Constr.
Fabricated structural metal	1.6	Manufg.	Maintenance of sewer facilities	1.8	Constr.
Aluminum castings	1.4	Manufg.	Pipe, valves, & pipe fittings	1.7	Manufg.
Brass, bronze, & copper castings	1.4	Manufg.	Turbines & turbine generator sets	1.3	Manufg.
Petroleum refining	1.4	Manufg.	Maintenance of water supply facilities	1.2	Constr.
Engineering, architectural, & surveying services	1.4	Services	Nonfarm residential structure maintenance	1.1	Constr.
Motor freight transportation & warehousing	1.2	Util.	Water transportation	1.1	Util.
Cyclic crudes and organics	1.1	Manufg.	Fabricated plate work (boiler shops)	1.0	Manufg.
Industrial patterns	1.1	Manufg.	Communications, except radio & TV	1.0	Util.
Primary copper	1.1	Manufg.	Motor freight transportation & warehousing	1.0	Util.
Nonferrous wire drawing & insulating	1.0	Manufg.	Sanitary services, steam supply, irrigation	1.0	Util.
Aluminum rolling & drawing	0.9	Manufg.	Water supply & sewage systems	1.0	Util.
Communications, except radio & TV	0.9	Util.	Blowers & fans	0.9	Manufg.
Equipment rental & leasing services	0.9	Services	Oil field machinery	0.9	Manufg.
Maintenance of nonfarm buildings nec	0.8	Constr.	Transit & bus transportation	0.9	Util.
Air transportation	0.8	Util.	Maintenance of electric utility facilities	0.8	Constr.
Gas production & distribution (utilities)	0.8	Util.	Residential 1-unit structures, nonfarm	0.8	Constr.
Machinery, except electrical, nec	0.7	Manufg.	Construction machinery & equipment	0.8	Manufg.
Business services nec	0.7	Services	Federal Government purchases, national defense	0.8	Fed Govt
Photofinishing labs, commercial photography	0.7	Services	Construction of stores & restaurants	0.7	Constr.
Plating & polishing	0.6	Manufg.	Maintenance of petroleum & natural gas wells	0.7	Constr.
Eating & drinking places	0.6	Trade	Ship building & repairing	0.7	Manufg.
Nonferrous castings, nec	0.5	Manufg.	Gas utility facility construction	0.6	Constr.
Primary nonferrous metals, nec	0.5	Manufg.	Warehouses	0.6	Constr.
Rubber & plastics hose & belting	0.5	Manufg.	Petroleum refining	0.6	Manufg.

Continued on next page.

INPUTS AND OUTPUTS FOR PIPE, VALVES, & PIPE FITTINGS - Continued

Economic Sector or Industry Providing Inputs	%	Sector	Economic Sector or Industry Buying Outputs	%	Sector
Real estate	0.5	Fin/R.E.	Pumps & compressors	0.6	Manufg.
Scrap	0.5	Scrap	Maintenance of gas utility facilities	0.5	Constr.
Fabricated rubber products, nec	0.4	Manufg.	Residential additions/alterations, nonfarm	0.5	Constr.
Industrial controls	0.4	Manufg.	Aircraft & missile engines & engine parts	0.5	Manufg.
Machine tools, metal cutting types	0.4	Manufg.	Machinery, except electrical, nec	0.5	Manufg.
Special dies & tools & machine tool accessories	0.4	Manufg.	Construction of hospitals	0.4	Constr.
Railroads & related services	0.4	Util.	Maintenance of farm service facilities	0.4	Constr.
Chemical preparations, nec	0.3	Manufg.	Blast furnaces & steel mills	0.4	Manufg.
Hardware, nec	0.3	Manufg.	Farm machinery & equipment	0.4	Manufg.
Metal stampings, nec	0.3	Manufg.	General industrial machinery, nec	0.4	Manufg.
Miscellaneous plastics products	0.3	Manufg.	Household cooking equipment	0.4	Manufg.
Paperboard containers & boxes	0.3	Manufg.	Household laundry equipment	0.4	Manufg.
Primary metal products, nec	0.3	Manufg.	Paper mills, except building paper	0.4	Manufg.
Automotive rental & leasing, without drivers	0.3	Services	Refrigeration & heating equipment	0.4	Manufg.
U.S. Postal Service	0.3	Gov't	Special industry machinery, nec	0.4	Manufg.
Abrasive products	0.2	Manufg.	Truck trailers	0.4	Manufg.
Gaskets, packing & sealing devices	0.2	Manufg.	Maintenance of petroleum pipelines	0.3	Constr.
Lubricating oils & greases	0.2	Manufg.	Maintenance, conservation & development facilities	0.3	Constr.
Metal heat treating	0.2	Manufg.	Petroleum & natural gas well drilling	0.3	Constr.
Miscellaneous fabricated wire products	0.2	Manufg.	Aircraft	0.3	Manufg.
Insurance carriers	0.2	Fin/R.E.	Household appliances, nec	0.3	Manufg.
Accounting, auditing & bookkeeping	0.2	Services	Internal combustion engines, nec	0.3	Manufg.
Automotive repair shops & services	0.2	Services	Motor vehicle parts & accessories	0.3	Manufg.
Computer & data processing services	0.2	Services	Paperboard mills	0.3	Manufg.
Legal services	0.2	Services	Polishes & sanitation goods	0.3	Manufg.
Management & consulting services & labs	0.2	Services	Sporting & athletic goods, nec	0.3	Manufg.
Noncomparable imports	0.2	Foreign	Toilet preparations	0.3	Manufg.
Electronic components nec	0.1	Manufg.	Railroads & related services	0.3	Util.
Fabricated metal products, nec	0.1	Manufg.	Construction of conservation facilities	0.2	Constr.
Industrial inorganic chemicals, nec	0.1	Manufg.	Construction of educational buildings	0.2	Constr.
Metal coating & allied services	0.1	Manufg.	Hotels & motels	0.2	Constr.
Motors & generators	0.1	Manufg.	Maintenance of farm residential buildings	0.2	Constr.
Credit agencies other than banks	0.1	Fin/R.E.	Maintenance of military facilities	0.2	Constr.
Security & commodity brokers	0.1	Fin/R.E.	Maintenance of nonbuilding facilities nec	0.2	Constr.
Hotels & lodging places	0.1	Services	Residential garden apartments	0.2	Constr.
			Industrial furnaces & ovens	0.2	Manufg.
			Industrial inorganic chemicals, nec	0.2	Manufg.
			Industrial trucks & tractors	0.2	Manufg.
			Iron & steel forgings	0.2	Manufg.
			Miscellaneous plastics products	0.2	Manufg.
			Nitrogenous & phosphatic fertilizers	0.2	Manufg.
			Paints & allied products	0.2	Manufg.
			Prepared feeds, nec	0.2	Manufg.
			Pulp mills	0.2	Manufg.
			Service industry machines, nec	0.2	Manufg.
			Farm service facilities	0.1	Constr.
			Maintenance of local transit facilities	0.1	Constr.
			Maintenance of railroads	0.1	Constr.
			Petroleum pipeline construction	0.1	Constr.
			Residential high-rise apartments	0.1	Constr.
			Agricultural chemicals, nec	0.1	Manufg.
			Aircraft & missile equipment, nec	0.1	Manufg.
			Bottled & canned soft drinks	0.1	Manufg.
			Cement, hydraulic	0.1	Manufg.
			Chemical preparations, nec	0.1	Manufg.
			Food products machinery	0.1	Manufg.
			Industrial gases	0.1	Manufg.
			Machine tools, metal cutting types	0.1	Manufg.
			Malt beverages	0.1	Manufg.
			Primary aluminum	0.1	Manufg.

Source: Benchmark Input-Output Accounts for the U.S. Economy, 1982, U.S. Department of Commerce, Washington, D.C., July 1991. Data, as reported in the source, are organized by the 1977 SIC structure in use in 1982 but have been matched, as closely as is possible, to the 1987 SIC structure used in this book.

OCCUPATIONS EMPLOYED BY SIC 349 - MISCELLANEOUS FABRICATED METAL PRODUCTS

Occupation	% of Total 1994	Change to 2005	Occupation	% of Total 1994	Change to 2005
Assemblers, fabricators, & hand workers nec	10.7	0.7	Industrial machinery mechanics	1.8	10.7
Blue collar worker supervisors	4.5	-7.4	Helpers, laborers, & material movers nec	1.8	0.7
Machinists	4.4	0.7	Hand packers & packagers	1.7	-13.7
Welding machine setters, operators	3.8	-9.4	Machine operators nec	1.6	-11.3
Machine tool cutting & forming etc. nec	3.7	20.8	Industrial production managers	1.5	0.7
Machine forming operators, metal & plastic	3.5	-9.4	Secretaries, ex legal & medical	1.4	-8.3
Sales & related workers nec	3.0	0.7	General office clerks	1.3	-14.2
Inspectors, testers, & graders, precision	3.0	0.7	Tool & die makers	1.3	-18.7
Metal & plastic machine workers nec	2.7	-28.7	Bookkeeping, accounting, & auditing clerks	1.3	-24.5
General managers & top executives	2.6	-4.5	Coating, painting, & spraying machine workers	1.3	0.7
Combination machine tool operators	2.4	51.0	Maintenance repairers, general utility	1.2	-9.4
Machine tool cutting operators, metal & plastic	2.4	41.9	Industrial truck & tractor operators	1.1	0.6
Lathe & turning machine tool operators	2.1	-9.4	Machine feeders & offbearers	1.1	-9.4
Welders & cutters	2.1	0.7	Production, planning, & expediting clerks	1.0	0.7
Traffic, shipping, & receiving clerks	2.0	-3.1	Mechanical engineers	1.0	-0.3

Source: *Industry-Occupation Matrix*, Bureau of Labor Statistics. These data relate to one or more 3-digit SIC industry groups rather than to a single 4-digit SIC. The change reported for each occupation to the year 2005 is a percent of growth or decline as estimated by the Bureau of Labor Statistics. The abbreviation nec stands for 'not elsewhere classified'.

LOCATION BY STATE AND REGIONAL CONCENTRATION

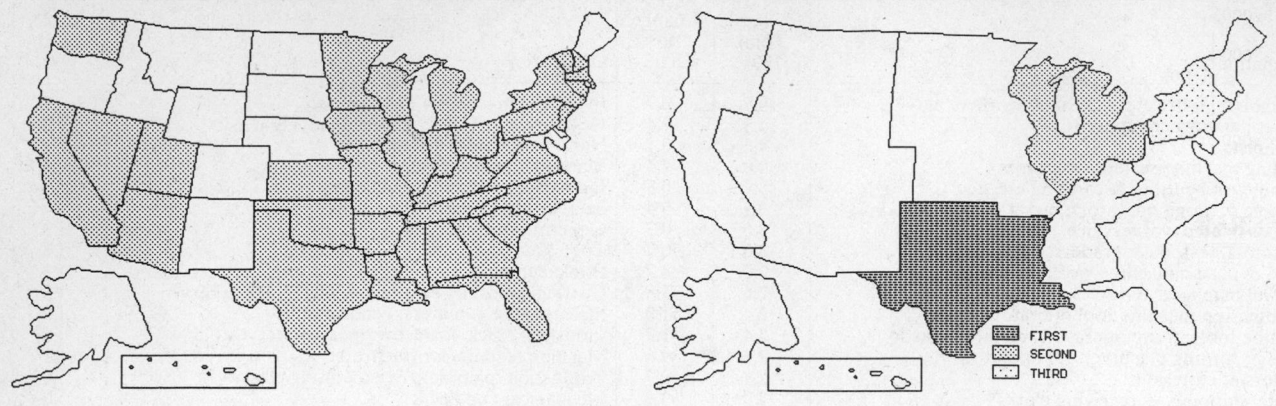

FIRST
SECOND
THIRD

INDUSTRY DATA BY STATE

State	Establish-ments	Shipments			Employment				Cost as % of Shipments	Investment per Employee ($)
		Total ($ mil)	% of U.S.	Per Establ.	Total Number	% of U.S.	Per Establ.	Wages ($/hour)		
Texas	73	898.7	13.3	12.3	6,300	12.3	86	12.67	42.1	5,667
California	54	579.7	8.6	10.7	4,000	7.8	74	12.77	41.3	2,750
Ohio	31	492.9	7.3	15.9	3,000	5.8	97	12.73	51.4	3,967
Iowa	10	370.8	5.5	37.1	3,100	6.0	310	15.82	27.3	-
Pennsylvania	45	365.0	5.4	8.1	3,100	6.0	69	14.38	33.7	3,290
Massachusetts	18	336.8	5.0	18.7	2,700	5.3	150	14.42	38.4	4,111
New Jersey	22	335.3	5.0	15.2	2,400	4.7	109	14.70	35.0	3,667
Illinois	25	314.6	4.7	12.6	2,200	4.3	88	13.53	44.4	4,955
North Carolina	14	255.4	3.8	18.2	1,400	2.7	100	9.61	40.8	3,857
Arkansas	6	230.2	3.4	38.4	2,200	4.3	367	9.57	43.4	2,773
Wisconsin	9	224.5	3.3	24.9	1,900	3.7	211	11.58	42.8	3,842
Missouri	12	210.8	3.1	17.6	1,500	2.9	125	11.76	38.9	4,067
Alabama	7	204.9	3.0	29.3	2,200	4.3	314	11.52	38.1	4,273
Tennessee	7	203.1	3.0	29.0	1,400	2.7	200	13.35	44.6	4,714
Oklahoma	25	197.1	2.9	7.9	1,400	2.7	56	12.29	43.6	5,714
South Carolina	7	177.5	2.6	25.4	1,500	2.9	214	9.00	20.8	-
New York	17	172.4	2.5	10.1	1,300	2.5	76	12.76	30.0	3,923
New Hampshire	8	142.7	2.1	17.8	800	1.6	100	12.54	40.7	-
Minnesota	6	136.3	2.0	22.7	1,200	2.3	200	16.08	42.0	4,750
Indiana	9	133.8	2.0	14.9	1,200	2.3	133	11.88	36.5	2,500
Louisiana	10	128.1	1.9	12.8	1,000	1.9	100	11.94	45.2	-
Connecticut	8	102.5	1.5	12.8	800	1.6	100	11.80	40.9	5,250
Michigan	13	86.0	1.3	6.6	800	1.6	62	14.80	31.3	3,250
Florida	10	72.5	1.1	7.3	700	1.4	70	10.44	36.6	3,143
Mississippi	4	39.1	0.6	9.8	400	0.8	100	10.20	53.5	-
Kentucky	3	28.5	0.4	9.5	300	0.6	100	10.75	41.8	-
Washington	5	20.2	0.3	4.0	200	0.4	40	16.00	33.2	1,000
Arizona	5	17.6	0.3	3.5	200	0.4	40	10.00	33.5	-
Kansas	5	16.6	0.2	3.3	200	0.4	40	9.50	61.4	-
Georgia	3	(D)	-	-	175 *	0.3	58	-	-	-
Nevada	3	(D)	-	-	175 *	0.3	58	-	-	-
Utah	3	(D)	-	-	750 *	1.5	250	-	-	-
Virginia	2	(D)	-	-	750 *	1.5	375	-	-	-
Vermont	1	(D)	-	-	175 *	0.3	175	-	-	-
West Virginia	1	(D)	-	-	375 *	0.7	375	-	-	-

Source: 1992 *Economic Census*. The states are in descending order of shipments or establishments (if shipment data are missing for the majority). The symbol (D) appears when data are withheld to prevent disclosure of competitive information. States marked with (D) are sorted by number of establishments. A dash (-) indicates that the data element cannot be calculated; * indicates the midpoint of a range.

3492 - FLUID POWER VALVES & HOSE FITTINGS

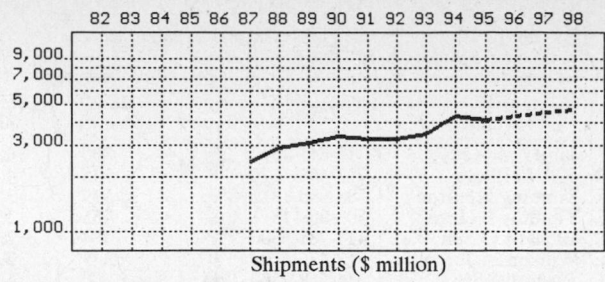

Shipments ($ million)

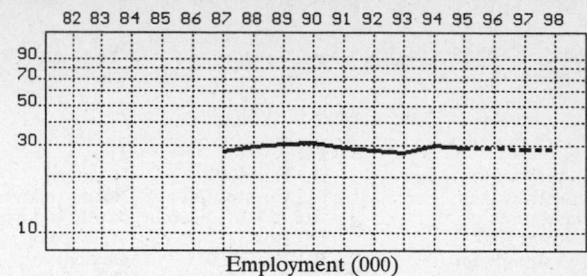

Employment (000)

GENERAL STATISTICS

Year	Companies	Establishments		Employment			Compensation		Production ($ million)			
		Total	with 20 or more employees	Total (000)	Production Workers (000)	Hours (Mil)	Payroll ($ mil)	Wages ($/hr)	Cost of Materials	Value Added by Manufacture	Value of Shipments	Capital Invest.
1982												
1983												
1984												
1985												
1986												
1987	331	386	203	27.9	18.4	37.0	686.1	10.89	1,001.8	1,445.4	2,451.5	77.9
1988		395	210	29.7	19.8	41.9	775.0	10.82	1,230.1	1,748.5	2,908.2	90.7
1989		378	213	30.5	20.0	40.6	781.4	11.62	1,229.9	1,851.4	3,063.7	94.0
1990		380	222	31.3	20.7	41.9	879.7	12.46	1,384.1	1,913.6	3,322.8	134.2
1991		393	212	29.0	19.1	38.6	856.0	12.95	1,405.6	1,825.7	3,214.3	90.5
1992	305	369	202	28.2	18.3	38.1	853.2	12.40	1,505.3	1,743.0	3,273.9	114.0
1993		385	201	27.5	18.5	39.3	857.9	12.44	1,550.1	1,941.2	3,460.8	117.1
1994		378P	206P	30.1	20.7	45.1	989.1	12.79	1,965.4	2,389.6	4,337.5	121.8
1995		377P	205P	29.0P	19.6P	42.1P	980.9P	13.34P	1,874.7P	2,279.3P	4,137.3P	129.4P
1996		375P	204P	29.0P	19.6P	42.5P	1,013.4P	13.63P	1,963.6P	2,387.5P	4,333.6P	134.9P
1997		374P	203P	28.9P	19.7P	42.9P	1,045.8P	13.92P	2,052.6P	2,495.6P	4,529.9P	140.3P
1998		372P	202P	28.8P	19.7P	43.2P	1,078.3P	14.21P	2,141.5P	2,603.7P	4,726.2P	145.7P

Sources: 1982, 1987, 1992 *Economic Census*; *Annual Survey of Manufactures*, 83-86, 88-91, 93-94. Establishment counts for non-Census years are from *County Business Patterns*; establishment values for 83-84 are extrapolations. 'P's show projections by the editors. Industries reclassified in 87 will not have data for prior years.

INDICES OF CHANGE

Year	Companies	Establishments		Employment			Compensation		Production ($ million)			
		Total	with 20 or more employees	Total (000)	Production Workers (000)	Hours (Mil)	Payroll ($ mil)	Wages ($/hr)	Cost of Materials	Value Added by Manufacture	Value of Shipments	Capital Invest.
1982												
1983												
1984												
1985												
1986												
1987	109	105	100	99	101	97	80	88	67	83	75	68
1988		107	104	105	108	110	91	87	82	100	89	80
1989		102	105	108	109	107	92	94	82	106	94	82
1990		103	110	111	113	110	103	100	92	110	101	118
1991		107	105	103	104	101	100	104	93	105	98	79
1992	100	100	100	100	100	100	100	100	100	100	100	100
1993		104	100	98	101	103	101	100	103	111	106	103
1994		102P	102P	107	113	118	116	103	131	137	132	107
1995		102P	101P	103P	107P	110P	115P	108P	125P	131P	126P	114P
1996		102P	101P	103P	107P	111P	119P	110P	130P	137P	132P	118P
1997		101P	101P	102P	107P	112P	123P	112P	136P	143P	138P	123P
1998		101P	100P	102P	108P	114P	126P	115P	142P	149P	144P	128P

Sources: Same as General Statistics. Values reflect change from the base year, 1992. Values above 100 mean greater than 92, values below 100 mean less than 92, and a value of 100 in the 82-91 or 93-98 period means same as 92. 'P's mark projections by the editors.

SELECTED RATIOS

For 1994	Avg. of All Manufact.	Analyzed Industry	Index	For 1994	Avg. of All Manufact.	Analyzed Industry	Index
Employees per Establishment	49	80	163	Value Added per Production Worker	134,084	115,440	86
Payroll per Establishment	1,500,273	2,616,667	174	Cost per Establishment	5,045,178	5,199,471	103
Payroll per Employee	30,620	32,860	107	Cost per Employee	102,970	65,296	63
Production Workers per Establishment	34	55	160	Cost per Production Worker	146,988	94,947	65
Wages per Establishment	853,319	1,526,003	179	Shipments per Establishment	9,576,895	11,474,868	120
Wages per Production Worker	24,861	27,866	112	Shipments per Employee	195,460	144,103	74
Hours per Production Worker	2,056	2,179	106	Shipments per Production Worker	279,017	209,541	75
Wages per Hour	12.09	12.79	106	Investment per Establishment	321,011	322,222	100
Value Added per Establishment	4,602,255	6,321,693	137	Investment per Employee	6,552	4,047	62
Value Added per Employee	93,930	79,389	85	Investment per Production Worker	9,352	5,884	63

Sources: Same as General Statistics. The 'Average of All Manufacturing' column represents the average of all manufacturing industries reported for the most recent complete year available. The Index shows the relationship between the Average and the Analyzed Industry. For example, 100 means that they are equal; 500 that the Analyzed Industry is five times the average; 50 means that the Analyzed Industry is half the national average. The abbreviation 'na' is used to show that data are 'not available'.

LEADING COMPANIES   Number shown: **71**   Total sales ($ mil): **4,320**   Total employment (000): **32.6**

Company Name	Address				CEO Name	Phone	Co. Type	Sales ($ mil)	Empl. (000)
Vickers Inc	PO Box 302	Troy	MI	48007	James McKee	313-641-4500	S	800	6.5
Keystone International Inc	PO Box 40010	Houston	TX	77040	R A LeBlanc	713-466-1176	P	535	4.2
Watts Industries Inc	815 Chestnut St	North Andover	MA	01845	Timothy P Horne	508-688-1811	P	519	3.1
Watts Regulator Co	Rte 114 & Chestnut	North Andover	MA	01845	Timothy P Horn	508-688-1811	S	465	2.3
Applied Power Inc	PO Box 325	Milwaukee	WI	53201	Richard G Sim	414-781-6600	P	434	2.7
Duriron Company Inc	PO Box 8820	Dayton	OH	45401	William M Jordan	513-476-6100	P	345	2.5
Alco Controls	PO Box 12700	St Louis	MO	63141	Richard J Schul	314-569-4500	D	100	1.7
HUSCO International Inc	PO Box 257	Waukesha	WI	53187	Agustin Ramirez	414-547-0261	R	100	0.5
Mobile Hydraulics	PO Box 394	Wooster	OH	44691	Michael D Bickel	216-263-3300	D	100	0.5
Parker Hannifin Corp	3885 Gateway Blv	Columbus	OH	43228	Don Gerosa	614-279-7070	D	56*	0.7
AVM Inc	Hwy 76 E	Marion	SC	29571	Bruce Walters	803-464-7823	D	50	0.5
Deltrol Corp	3001 Grant Av	Bellwood	IL	60104	Herbert Schwensohn	708-547-0500	R	50	0.5
Summit Packaging System Inc	16 Ammon Dr	Manchester	NH	03103	Gordon C Gilroy	603-669-5410	R	50	0.3
Milwaukee Valve Company Inc	2375 S Burrell St	Milwaukee	WI	53207	William Goglia	414-744-5240	R	45	0.5
Dixon Valve and Coupling Co	800 High St	Chestertown	MD	21620	Richard L Goodall	410-778-2000	R	44*	0.4
Hoke Inc	1 Tenakill Pk	Cresskill	NJ	07626	Walter H Jones	201-568-9100	R	42	0.8
Target Rock Corp	PO Box V	E Farmingdale	NY	11735	Martin R Benante	516-293-3800	S	30	0.2
HydraForce Inc	500 Barclay Blv	Lincolnshire	IL	60090	Jim Brizzolara	708-215-8710	R	30	0.2
Rexroth Corp	PO Box 13597	Lexington	KY	40512	Fred Chambers	606-254-8031	D	30*	0.3
Wilkerson Corp	PO Box 1237	Englewood	CO	80150	Richard Angelo	303-761-7601	S	30	0.3
Sun Hydraulics Corp	1500 University	Sarasota	FL	34243	RE Koski	813-355-2983	R	29	0.3
Barksdale Inc	3211 Fruitland Av	Los Angeles	CA	90058	Rudolf C Wolf	213-589-6181	D	25	0.2
Bijur Lubricating Corp	50 Kocher Dr	Bennington	VT	05201	D Preston	802-447-2174	R	25	0.3
Watts FluidAir	9 Cutts Rd	Kittery	ME	03904	Jerold Rudinsky	207-439-9511	S	22	0.2
Hoerbiger Corporation	PO Box 8888	Ft Lauderdale	FL	33310	Hubert Wagner	305-974-5700	D	21*	0.2
Deltrol Fluid Products	3001 Grant Av	Bellwood	IL	60104	Herbert Schwensohn	708-547-0500	D	20	0.1
Fluid Power Systems	511 Glenn Av	Wheeling	IL	60090	John A Tuzzolino	708-459-2800	D	20	0.1
Continental Hydraulics	12520 Quentin S	Savage	MN	55378	Michael Johnson	612-894-8900	D	19	0.1
Delta Power Hydraulic Co	4484 Boeing Dr	Rockford	IL	61109	John Fulton	815-397-6628	R	18	<0.1
Royal Brass and Hose	PO Box 51468	Knoxville	TN	37950	J Ingram	615-558-0224	R	18	<0.1
Fluidex	147 W Hoy Rd	Madison	MS	39110	Roger Riefler	601-853-7200	D	16	0.2
Catawissa Valve and Fittings Co	PO Box 157	Catawissa	PA	17820	Kevin Eagleton	717-356-2311	S	16	0.2
AAI/ACL Technologies Inc	1505 E Warner Av	Santa Ana	CA	92705	Thomas Wurzel	714-979-5500	R	15	0.1
DynaQuip Controls Corp	1645 Manufacturers	Fenton	MO	63026	J Rodney Bryan	314-343-1010	R	14	0.1
Faber Enterprises Inc	6606 Variel Av	Canoga Park	CA	91303	Ronald Spenser	818-999-1300	R	14	0.1
Spence Engineering Company	150 Coldenham Rd	Walden	NY	12586	S A Banyacski	914-778-5566	S	14	0.1
CMB Industries	PO Box 8070	Fresno	CA	93747	Kevin Coyne	209-252-0791	R	12*	0.1
Versa Products Company Inc	22 Spring Vally Rd	Paramus	NJ	07652	Karl Larsson	201-843-2400	R	12*	0.1
Sterling Hydraulics Inc	920 E State Pkwy	Schaumburg	IL	60173	Walter Braun	708-490-1333	S	11	<0.1
ITT Conoflow	PO Box 768	St George	SC	29477		803-563-9281	S	9	0.1
Bailey Polyjet	PO Box 8070	Fresno	CA	93747	John Brewer	209-252-4491	D	8	0.1
Campbell Fittings Inc	Front-Wash	Boyertown	PA	19512	Thomas J Paff	215-367-6916	R	8	<0.1
Catching Fluidpower Inc	1700 W 16th St	Broadview	IL	60153	Richard E Pyle	708-344-7272	R	8*	<0.1
Aerofit Products Inc	8531 Whitaker St	Buena Park	CA	90621	Robert R Peterjohn	714-521-5060	R	7	<0.1
Mead Fluid Dynamics Inc	4114 N Knox Av	Chicago	IL	60641	NS Framberg	312-685-6800	R	6*	<0.1
Aerodyne Controls Corp	30 Haynes Ct	Ronkonkoma	NY	11779	James Miller	516-737-1900	R	5	<0.1
Chem-Tec Equipment Co	234 SW 12th Av	Deerfield Bch	FL	33442	William Nolan	305-428-8259	R	5	<0.1
George W Dahl Company Inc	8439 Triad Dr	Greensboro	NC	27409	L T Nicolette	910-668-4444	S	5	<0.1
Morland Valve Co	1404 Tolland Tpk	Manchester	CT	06040	Paul Ruso	203-649-2893	S	5	<0.1
Nelson Aero Space Inc	3305 W Burbank	Burbank	CA	91505	Clemons Nelson	818-848-4414	R	5	<0.1
Pima Valve Inc	Box 5010	Chandler	AZ	85226	Allen J Link	602-796-1095	R	5	<0.1
Tylok International Inc	1061 E 260th St	Euclid	OH	44132	Carole Hahl	216-261-0400	R	5	<0.1
PJ Valves Inc	341 King St	Myerstown	PA	17067	RL Miller	717-866-5795	R	5	<0.1
Atchley Controls	21029 Osborne St	Canoga Park	CA	91304	Chuck Bernard	818-700-0770	D	4	<0.1
Brand Hydraulics Inc	2332 S 25th St	Omaha	NE	68105	Glen Brand	402-344-4434	R	4	<0.1
Lexair Inc	2025 Mercer Rd	Lexington	KY	40511	Clifford W Allen	606-255-5001	R	4	<0.1
Kay Pneumatics Inc	12 Austin Blv	Commack	NY	11725	Edmond Krauss	516-543-8200	R	3*	<0.1
Tavco Inc	20500 Prairie St	Chatsworth	CA	91311	Jeffrey E Jae	818-882-5411	R	3	<0.1
Warren Controls Corp	PO Box 509	Broadway	NJ	08808	C Alan Trent	908-689-1400	R	3	<0.1
Technical Prod & Precision Mfg	PO Box 904	Hatfield	PA	19440	R Edmonds Sr	215-822-3338	R	3	<0.1
JD Gould Company Inc	PO Box 18128	Indianapolis	IN	46218	John E Gould	317-547-5289	R	2	<0.1
Hydraulic Accessories Co	24301 Hoover Rd	Warren	MI	48089	John W Abar	810-757-6810	R	2	<0.1
Custom Hydraulic and Machine	22911 86th Av S	Kent	WA	98031	RR Wolfer	206-854-4666	R	2	<0.1
Diamond-U Products Inc	1429 Magnolia Av	Long Beach	CA	90813	DL Adams	310-436-8245	R	2	<0.1
MacKenzie Machine and Design	10 Industrial Pk Rd	Hingham	MA	02043	LS MacKenzie	617-749-1225	R	1	<0.1
Ace Aviation Service	7 N Main St	Pima	AZ	85543	Don Nicolai	602-485-2486	R	1	<0.1
Clearflow Valves Inc	631 Camelia St	Berkeley	CA	94710	Tom Impey	510-525-1700	R	1	<0.1
Control Mechanisms Inc	1036 Marlborough	Philadelphia	PA	19125	Melvin Ruttenberg	215-739-0822	R	1	<0.1
Westfield Hydraulics Inc	10275 Glenoaks	Pacoima	CA	91331	J Umbarger	818-896-6414	R	1	<0.1
ALE Hydraulic Machinery Co	6215 Airport Rd	Levittown	PA	19057	AR Zwiebel	215-547-3351	R	1*	<0.1
Uni-Seal Valve Co	1334 Callens Rd	Ventura	CA	93003	Keith Iveson	805-644-7238	S	1	<0.1

Source: Ward's Business Directory of U.S. Private and Public Companies, Volumes 1 and 2, 1996. The company type code used is as follows: P - Public, R - Private, S - Subsidiary, D - Division, J - Joint Venture, A - Affiliate, G - Group. Sales are in millions of dollars, employees are in thousands. An asterisk (*) indicates an estimated sales volume. The symbol < stands for 'less than'. Company names and addresses are truncated, in some cases, to fit into the available space.

MATERIALS CONSUMED

Material	Quantity	Delivered Cost ($ million)
Materials, ingredients, containers, and supplies	(X)	1,251.4
Metal bolts, nuts, screws, washers, rivets, and other screw machine products	(X)	194.3
Metal stampings	(X)	8.6
Valves, fittings, and couplings purchased for further assembly (except forgings)	(X)	112.8
Other fabricated metal products (except forgings)	(X)	57.7
Iron castings (rough and semifinished)	(X)	27.4
Steel castings (rough and semifinished)	(X)	14.0
Aluminum and aluminum-base alloy castings (rough and semifinished)	(X)	32.2
Copper and copper-base alloy castings (rough and semifinished)	(X)	4.6
Other nonferrous castings (rough and semifinished)	(X)	29.9
Iron and steel forgings	(X)	20.3
Nonferrous forgings	(X)	16.5
Steel bars, bar shapes, and plates	(X)	100.3
Steel sheet and strip, including tin plate	(X)	3.3
All other steel shapes and forms	(X)	35.4
Copper and copper-base alloy refinery shapes	(X)	(D)
Copper and copper-base alloy rod, bar, and mechanical wire, including extruded and/or drawn shapes	(X)	65.8
Copper and copper-base alloy pipe and tube	(X)	5.2
All other copper and copper-base alloy shapes and forms	(X)	(D)
Aluminum and aluminum-base alloy shapes and forms	(X)	28.6
Other nonferrous shapes and forms	(X)	8.3
Scrap, including iron, steel, aluminum and aluminum-base alloy (excluding home scrap)	(X)	2.6
Electric motors and generators less than 1 horsepower (less than 746 watts)	(X)	9.1
Paperboard containers, boxes, and corrugated paperboard	(X)	21.9
Hydraulic and pneumatic hose	(X)	104.8
Other rubber and plastics hose and belting	(X)	0.6
Gaskets (all types), and packing and sealing devices	(X)	10.7
Fabricated rubber products, except tires, tubes, hose, belting, and gaskets	(X)	5.5
Fabricated plastics products (except gaskets, hoses, and belting)	(X)	8.8
All other materials and components, parts, containers, and supplies	(X)	196.9
Materials, ingredients, containers, and supplies, nsk	(X)	107.8

Source: 1992 Economic Census. Explanation of symbols used: (D): Withheld to avoid disclosure of competitive data; na: Not available; (S): Withheld because statistical norms were not met; (X): Not applicable; (Z): Less than half the unit shown; nec: Not elsewhere classified; nsk: Not specified by kind; - : zero; * : 10-19 percent estimated; ** : 20-29 percent estimated.

PRODUCT SHARE DETAILS

Product or Product Class	% Share	Product or Product Class	% Share
Fluid power valves and hose fittings	100.00	Nonaerospace-type flared (metal) fittings, couplings for, and assemblies of tubing used in fluid power transfer systems	5.83
Aerospace-type hydraulic fluid power valves	10.89		
Aerospace-type pneumatic fluid power valves	3.31		
Nonaerospace-type hydraulic directional control valves	9.72	Nonaerospace-type flareless fittings and couplings (including nonmetal fittings) used in fluid power transfer systems	10.35
Nonaerospace-type hydraulic valves, except directional control	8.46		
Nonaerospace-type pneumatic directional control valves	9.90	Nonaerospace-type hydraulic and pneumatic fittings and couplings for hose	12.45
Nonaerospace-type pneumatic valves, except directional control	4.45	Nonaerospace-type hydraulic and pneumatic assemblies of hose	6.06
Parts for fluid power valves	4.17		
Aerospace-type hydraulic and pneumatic fluid power hose or tube end fittings and assemblies	10.94	Fluid power valves and hose fittings, nsk	3.48

Source: 1992 Economic Census. The values shown are percent of total shipments in an industry. Values of indented subcategories are summed in the main heading. The symbol (D) appears when data are withheld to prevent disclosure of competitive information. The abbreviation nsk stands for 'not specified by kind' and nec for 'not elsewhere classified'.

INPUTS AND OUTPUTS FOR PIPE, VALVES, & PIPE FITTINGS

Economic Sector or Industry Providing Inputs	%	Sector	Economic Sector or Industry Buying Outputs	%	Sector
Blast furnaces & steel mills	19.6	Manufg.	Industrial buildings	12.2	Constr.
Imports	13.8	Foreign	Gross private fixed investment	11.7	Cap Inv
Wholesale trade	8.1	Trade	Exports	7.8	Foreign
Iron & steel foundries	7.5	Manufg.	Maintenance of nonfarm buildings nec	5.5	Constr.
Pipe, valves, & pipe fittings	3.2	Manufg.	Crude petroleum & natural gas	4.0	Mining
Copper rolling & drawing	2.9	Manufg.	Cyclic crudes and organics	4.0	Manufg.
Advertising	2.8	Services	Sewer system facility construction	3.4	Constr.
Iron & steel forgings	2.5	Manufg.	Office buildings	2.8	Constr.
Screw machine and related products	2.4	Manufg.	Electric utility facility construction	2.7	Constr.
Electric services (utilities)	2.2	Util.	Automotive rental & leasing, without drivers	2.7	Services
Banking	1.9	Fin/R.E.	Water supply facility construction	2.0	Constr.
Fabricated structural metal	1.6	Manufg.	Maintenance of sewer facilities	1.8	Constr.
Aluminum castings	1.4	Manufg.	Pipe, valves, & pipe fittings	1.7	Manufg.
Brass, bronze, & copper castings	1.4	Manufg.	Turbines & turbine generator sets	1.3	Manufg.
Petroleum refining	1.4	Manufg.	Maintenance of water supply facilities	1.2	Constr.

Continued on next page.

INPUTS AND OUTPUTS FOR PIPE, VALVES, & PIPE FITTINGS - Continued

Economic Sector or Industry Providing Inputs	%	Sector	Economic Sector or Industry Buying Outputs	%	Sector
Engineering, architectural, & surveying services	1.4	Services	Nonfarm residential structure maintenance	1.1	Constr.
Motor freight transportation & warehousing	1.2	Util.	Water transportation	1.1	Util.
Cyclic crudes and organics	1.1	Manufg.	Fabricated plate work (boiler shops)	1.0	Manufg.
Industrial patterns	1.1	Manufg.	Communications, except radio & TV	1.0	Util.
Primary copper	1.1	Manufg.	Motor freight transportation & warehousing	1.0	Util.
Nonferrous wire drawing & insulating	1.0	Manufg.	Sanitary services, steam supply, irrigation	1.0	Util.
Aluminum rolling & drawing	0.9	Manufg.	Water supply & sewage systems	1.0	Util.
Communications, except radio & TV	0.9	Util.	Blowers & fans	0.9	Manufg.
Equipment rental & leasing services	0.9	Services	Oil field machinery	0.9	Manufg.
Maintenance of nonfarm buildings nec	0.8	Constr.	Transit & bus transportation	0.9	Util.
Air transportation	0.8	Util.	Maintenance of electric utility facilities	0.8	Constr.
Gas production & distribution (utilities)	0.8	Util.	Residential 1-unit structures, nonfarm	0.8	Constr.
Machinery, except electrical, nec	0.7	Manufg.	Construction machinery & equipment	0.8	Manufg.
Business services nec	0.7	Services	Federal Government purchases, national defense	0.8	Fed Govt
Photofinishing labs, commercial photography	0.7	Services	Construction of stores & restaurants	0.7	Constr.
Plating & polishing	0.6	Manufg.	Maintenance of petroleum & natural gas wells	0.7	Constr.
Eating & drinking places	0.6	Trade	Ship building & repairing	0.7	Manufg.
Nonferrous castings, nec	0.5	Manufg.	Gas utility facility construction	0.6	Constr.
Primary nonferrous metals, nec	0.5	Manufg.	Warehouses	0.6	Constr.
Rubber & plastics hose & belting	0.5	Manufg.	Petroleum refining	0.6	Manufg.
Real estate	0.5	Fin/R.E.	Pumps & compressors	0.6	Manufg.
Scrap	0.5	Scrap	Maintenance of gas utility facilities	0.5	Constr.
Fabricated rubber products, nec	0.4	Manufg.	Residential additions/alterations, nonfarm	0.5	Constr.
Industrial controls	0.4	Manufg.	Aircraft & missile engines & engine parts	0.5	Manufg.
Machine tools, metal cutting types	0.4	Manufg.	Machinery, except electrical, nec	0.5	Manufg.
Special dies & tools & machine tool accessories	0.4	Manufg.	Construction of hospitals	0.4	Constr.
Railroads & related services	0.4	Util.	Maintenance of farm service facilities	0.4	Constr.
Chemical preparations, nec	0.3	Manufg.	Blast furnaces & steel mills	0.4	Manufg.
Hardware, nec	0.3	Manufg.	Farm machinery & equipment	0.4	Manufg.
Metal stampings, nec	0.3	Manufg.	General industrial machinery, nec	0.4	Manufg.
Miscellaneous plastics products	0.3	Manufg.	Household cooking equipment	0.4	Manufg.
Paperboard containers & boxes	0.3	Manufg.	Household laundry equipment	0.4	Manufg.
Primary metal products, nec	0.3	Manufg.	Paper mills, except building paper	0.4	Manufg.
Automotive rental & leasing, without drivers	0.3	Services	Refrigeration & heating equipment	0.4	Manufg.
U.S. Postal Service	0.3	Gov't	Special industry machinery, nec	0.4	Manufg.
Abrasive products	0.2	Manufg.	Truck trailers	0.4	Manufg.
Gaskets, packing & sealing devices	0.2	Manufg.	Maintenance of petroleum pipelines	0.3	Constr.
Lubricating oils & greases	0.2	Manufg.	Maintenance, conservation & development facilities	0.3	Constr.
Metal heat treating	0.2	Manufg.	Petroleum & natural gas well drilling	0.3	Constr.
Miscellaneous fabricated wire products	0.2	Manufg.	Aircraft	0.3	Manufg.
Insurance carriers	0.2	Fin/R.E.	Household appliances, nec	0.3	Manufg.
Accounting, auditing & bookkeeping	0.2	Services	Internal combustion engines, nec	0.3	Manufg.
Automotive repair shops & services	0.2	Services	Motor vehicle parts & accessories	0.3	Manufg.
Computer & data processing services	0.2	Services	Paperboard mills	0.3	Manufg.
Legal services	0.2	Services	Polishes & sanitation goods	0.3	Manufg.
Management & consulting services & labs	0.2	Services	Sporting & athletic goods, nec	0.3	Manufg.
Noncomparable imports	0.2	Foreign	Toilet preparations	0.3	Manufg.
Electronic components nec	0.1	Manufg.	Railroads & related services	0.3	Util.
Fabricated metal products, nec	0.1	Manufg.	Construction of conservation facilities	0.2	Constr.
Industrial inorganic chemicals, nec	0.1	Manufg.	Construction of educational buildings	0.2	Constr.
Metal coating & allied services	0.1	Manufg.	Hotels & motels	0.2	Constr.
Motors & generators	0.1	Manufg.	Maintenance of farm residential buildings	0.2	Constr.
Credit agencies other than banks	0.1	Fin/R.E.	Maintenance of military facilities	0.2	Constr.
Security & commodity brokers	0.1	Fin/R.E.	Maintenance of nonbuilding facilities nec	0.2	Constr.
Hotels & lodging places	0.1	Services	Residential garden apartments	0.2	Constr.
			Industrial furnaces & ovens	0.2	Manufg.
			Industrial inorganic chemicals, nec	0.2	Manufg.
			Industrial trucks & tractors	0.2	Manufg.
			Iron & steel forgings	0.2	Manufg.
			Miscellaneous plastics products	0.2	Manufg.
			Nitrogenous & phosphatic fertilizers	0.2	Manufg.
			Paints & allied products	0.2	Manufg.
			Prepared feeds, nec	0.2	Manufg.
			Pulp mills	0.2	Manufg.
			Service industry machines, nec	0.2	Manufg.
			Farm service facilities	0.1	Constr.
			Maintenance of local transit facilities	0.1	Constr.
			Maintenance of railroads	0.1	Constr.
			Petroleum pipeline construction	0.1	Constr.
			Residential high-rise apartments	0.1	Constr.
			Agricultural chemicals, nec	0.1	Manufg.
			Aircraft & missile equipment, nec	0.1	Manufg.
			Bottled & canned soft drinks	0.1	Manufg.
			Cement, hydraulic	0.1	Manufg.
			Chemical preparations, nec	0.1	Manufg.

Continued on next page.

INPUTS AND OUTPUTS FOR PIPE, VALVES, & PIPE FITTINGS - Continued

Economic Sector or Industry Providing Inputs	%	Sector	Economic Sector or Industry Buying Outputs	%	Sector
			Food products machinery	0.1	Manufg.
			Industrial gases	0.1	Manufg.
			Machine tools, metal cutting types	0.1	Manufg.
			Malt beverages	0.1	Manufg.
			Primary aluminum	0.1	Manufg.

Source: Benchmark Input-Output Accounts for the U.S. Economy, 1982, U.S. Department of Commerce, Washington, D.C., July 1991. Data, as reported in the source, are organized by the 1977 SIC structure in use in 1982 but have been matched, as closely as is possible, to the 1987 SIC structure used in this book.

OCCUPATIONS EMPLOYED BY SIC 349 - MISCELLANEOUS FABRICATED METAL PRODUCTS

Occupation	% of Total 1994	Change to 2005	Occupation	% of Total 1994	Change to 2005
Assemblers, fabricators, & hand workers nec	10.7	0.7	Industrial machinery mechanics	1.8	10.7
Blue collar worker supervisors	4.5	-7.4	Helpers, laborers, & material movers nec	1.8	0.7
Machinists	4.4	0.7	Hand packers & packagers	1.7	-13.7
Welding machine setters, operators	3.8	-9.4	Machine operators nec	1.6	-11.3
Machine tool cutting & forming etc. nec	3.7	20.8	Industrial production managers	1.5	0.7
Machine forming operators, metal & plastic	3.5	-9.4	Secretaries, ex legal & medical	1.4	-8.3
Sales & related workers nec	3.0	0.7	General office clerks	1.3	-14.2
Inspectors, testers, & graders, precision	3.0	0.7	Tool & die makers	1.3	-18.7
Metal & plastic machine workers nec	2.7	-28.7	Bookkeeping, accounting, & auditing clerks	1.3	-24.5
General managers & top executives	2.6	-4.5	Coating, painting, & spraying machine workers	1.3	0.7
Combination machine tool operators	2.4	51.0	Maintenance repairers, general utility	1.2	-9.4
Machine tool cutting operators, metal & plastic	2.4	41.9	Industrial truck & tractor operators	1.1	0.6
Lathe & turning machine tool operators	2.1	-9.4	Machine feeders & offbearers	1.1	-9.4
Welders & cutters	2.1	0.7	Production, planning, & expediting clerks	1.0	0.7
Traffic, shipping, & receiving clerks	2.0	-3.1	Mechanical engineers	1.0	-0.3

Source: Industry-Occupation Matrix, Bureau of Labor Statistics. These data relate to one or more 3-digit SIC industry groups rather than to a single 4-digit SIC. The change reported for each occupation to the year 2005 is a percent of growth or decline as estimated by the Bureau of Labor Statistics. The abbreviation nec stands for 'not elsewhere classified'.

LOCATION BY STATE AND REGIONAL CONCENTRATION

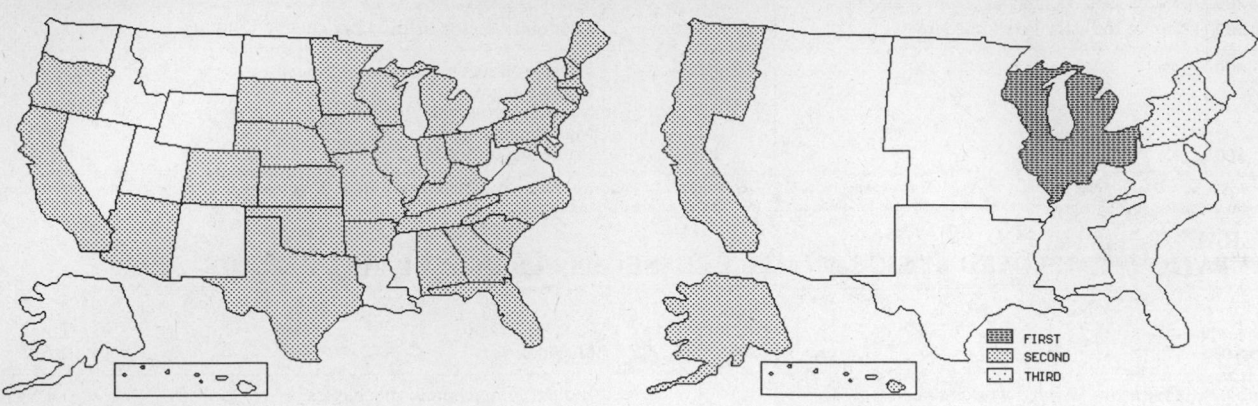

FIRST
SECOND
THIRD

INDUSTRY DATA BY STATE

State	Establish- ments	Shipments			Employment				Cost as % of Shipments	Investment per Employee ($)
		Total ($ mil)	% of U.S.	Per Establ.	Total Number	% of U.S.	Per Establ.	Wages ($/hour)		
Ohio	42	704.3	21.5	16.8	4,600	16.3	110	13.52	57.5	8,000
Michigan	47	415.5	12.7	8.8	3,700	13.1	79	12.89	49.9	3,324
California	65	380.0	11.6	5.8	3,700	13.1	57	14.73	36.6	3,189
Illinois	19	158.7	4.8	8.4	1,500	5.3	79	11.75	50.0	2,200
Wisconsin	12	155.8	4.8	13.0	1,300	4.6	108	13.53	38.3	5,538
Indiana	14	124.1	3.8	8.9	800	2.8	57	13.10	58.9	1,125
Florida	6	100.2	3.1	16.7	1,000	3.5	167	11.23	26.4	1,100
North Carolina	12	85.3	2.6	7.1	600	2.1	50	11.89	44.9	3,333
New York	12	78.6	2.4	6.6	800	2.8	67	15.78	35.0	3,125
New Jersey	9	78.1	2.4	8.7	600	2.1	67	13.83	28.6	–
Connecticut	9	77.4	2.4	8.6	600	2.1	67	15.00	37.9	6,333
Minnesota	14	72.4	2.2	5.2	800	2.8	57	11.78	48.5	3,875
Missouri	6	57.5	1.8	9.6	400	1.4	67	8.67	65.0	–
Texas	23	56.5	1.7	2.5	600	2.1	26	9.80	55.6	–
Arizona	8	55.5	1.7	6.9	700	2.5	88	12.25	45.4	2,429
Nebraska	4	25.2	0.8	6.3	300	1.1	75	11.50	40.9	–
Kansas	6	17.3	0.5	2.9	200	0.7	33	9.00	48.6	–
Pennsylvania	14	(D)	–	–	750 *	2.7	54	–	–	–
Massachusetts	6	(D)	–	–	375 *	1.3	63	–	–	–
Oklahoma	6	(D)	–	–	375 *	1.3	63	–	–	–
Colorado	5	(D)	–	–	1,750 *	6.2	350	–	–	–
Georgia	4	(D)	–	–	175 *	0.6	44	–	–	–
South Carolina	4	(D)	–	–	375 *	1.3	94	–	–	–
Maine	3	(D)	–	–	175 *	0.6	58	–	–	–
Alabama	2	(D)	–	–	750 *	2.7	375	–	–	–
Iowa	2	(D)	–	–	175 *	0.6	88	–	–	–
Kentucky	2	(D)	–	–	375 *	1.3	188	–	–	–
South Dakota	2	(D)	–	–	175 *	0.6	88	–	–	–
Tennessee	2	(D)	–	–	175 *	0.6	88	–	–	–
Arkansas	1	(D)	–	–	375 *	1.3	375	–	–	–
Maryland	1	(D)	–	–	375 *	1.3	375	–	–	–
New Hampshire	1	(D)	–	–	175 *	0.6	175	–	–	–
Oregon	1	(D)	–	–	175 *	0.6	175	–	–	–

Source: 1992 *Economic Census*. The states are in descending order of shipments or establishments (if shipment data are missing for the majority). The symbol (D) appears when data are withheld to prevent disclosure of competitive information. States marked with (D) are sorted by number of establishments. A dash (-) indicates that the data element cannot be calculated; * indicates the midpoint of a range.

3493 - STEEL SPRINGS, EXCEPT WIRE

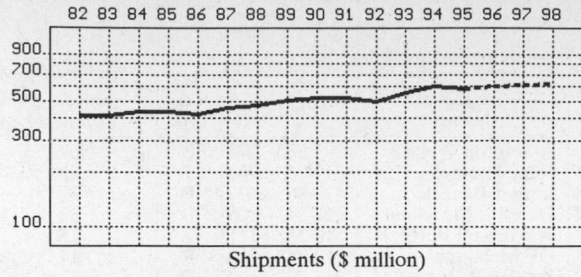

Shipments ($ million)

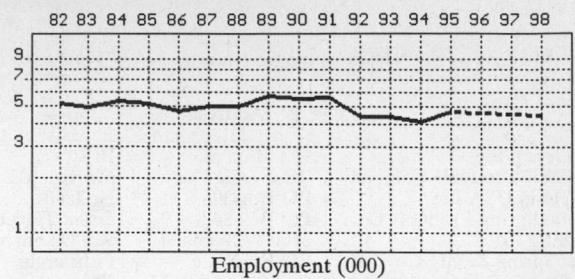

Employment (000)

GENERAL STATISTICS

Year	Com- panies	Establishments		Employment			Compensation		Production ($ million)			
		Total	with 20 or more employees	Total (000)	Production Workers (000)	Hours (Mil)	Payroll ($ mil)	Wages ($/hr)	Cost of Materials	Value Added by Manufacture	Value of Shipments	Capital Invest.
1982	123	137	55	5.2	3.8	6.8	103.5	9.90	204.8	204.5	413.5	7.3
1983		136	55	4.9	3.7	6.9	102.5	9.84	205.0	208.3	414.5	5.9
1984		135	55	5.3	4.0	7.6	118.4	10.54	223.9	219.7	432.7	9.4
1985		134	55	5.2	3.8	7.1	112.4	10.58	205.8	231.7	434.2	9.5
1986		128	57	4.7	3.5	6.6	109.5	10.83	191.6	221.6	419.3	15.4
1987	144	151	61	5.0	3.6	7.3	122.1	10.70	210.4	248.7	458.5	18.0
1988		146	60	5.0	3.7	7.5	127.5	11.07	213.5	256.1	468.3	11.2
1989		144	66	5.7	4.4	9.1	153.6	9.54	233.3	276.0	502.9	7.7
1990		147	62	5.5	4.1	9.0	148.4	9.90	229.6	286.4	524.7	7.3
1991		148	61	5.6	3.9	8.5	154.9	10.12	228.6	281.4	523.4	12.7
1992	107	114	46	4.4	3.1	5.9	118.9	12.41	244.8	247.3	495.8	17.5
1993		118	48	4.4	3.0	6.6	134.7	12.42	264.3	288.5	551.0	8.1
1994		132P	56P	4.1	3.1	6.3	126.8	13.60	305.5	297.6	600.5	30.2
1995		131P	55P	4.6P	3.3P	7.3P	146.4P	12.35P	294.6P	287.0P	579.2P	18.6P
1996		130P	55P	4.6P	3.3P	7.3P	149.3P	12.56P	301.9P	294.1P	593.3P	19.5P
1997		130P	55P	4.5P	3.2P	7.3P	152.3P	12.77P	309.1P	301.1P	607.5P	20.4P
1998		129P	55P	4.5P	3.2P	7.3P	155.3P	12.98P	316.3P	308.1P	621.7P	21.3P

Sources: 1982, 1987, 1992 *Economic Census*; *Annual Survey of Manufactures*, 83-86, 88-91, 93-94. Establishment counts for non-Census years are from *County Business Patterns*; establishment values for 83-84 are extrapolations. 'P's show projections by the editors. Industries reclassified in 87 will not have data for prior years.

INDICES OF CHANGE

Year	Com- panies	Establishments		Employment			Compensation		Production ($ million)			
		Total	with 20 or more employees	Total (000)	Production Workers (000)	Hours (Mil)	Payroll ($ mil)	Wages ($/hr)	Cost of Materials	Value Added by Manufacture	Value of Shipments	Capital Invest.
1982	115	120	120	118	123	115	87	80	84	83	83	42
1983		119	120	111	119	117	86	79	84	84	84	34
1984		118	120	120	129	129	100	85	91	89	87	54
1985		118	120	118	123	120	95	85	84	94	88	54
1986		112	124	107	113	112	92	87	78	90	85	88
1987	135	132	133	114	116	124	103	86	86	101	92	103
1988		128	130	114	119	127	107	89	87	104	94	64
1989		126	143	130	142	154	129	77	95	112	101	44
1990		129	135	125	132	153	125	80	94	116	106	42
1991		130	133	127	126	144	130	82	93	114	106	73
1992	100	100	100	100	100	100	100	100	100	100	100	100
1993		104	104	100	97	112	113	100	108	117	111	46
1994		116P	121P	93	100	107	107	110	125	120	121	173
1995		115P	120P	106P	107P	124P	123P	100P	120P	116P	117P	106P
1996		114P	120P	104P	106P	124P	126P	101P	123P	119P	120P	112P
1997		114P	119P	103P	104P	124P	128P	103P	126P	122P	123P	117P
1998		113P	119P	102P	102P	124P	131P	105P	129P	125P	125P	122P

Sources: Same as General Statistics. Values reflect change from the base year, 1992. Values above 100 mean greater than 92, values below 100 mean less than 92, and a value of 100 in the 82-91 or 93-98 period means same as 92. 'P's mark projections by the editors.

SELECTED RATIOS

For 1994	Avg. of All Manufact.	Analyzed Industry	Index	For 1994	Avg. of All Manufact.	Analyzed Industry	Index
Employees per Establishment	49	31	64	Value Added per Production Worker	134,084	96,000	72
Payroll per Establishment	1,500,273	962,263	64	Cost per Establishment	5,045,178	2,318,386	46
Payroll per Employee	30,620	30,927	101	Cost per Employee	102,970	74,512	72
Production Workers per Establishment	34	24	69	Cost per Production Worker	146,988	98,548	67
Wages per Establishment	853,319	650,210	76	Shipments per Establishment	9,576,895	4,557,089	48
Wages per Production Worker	24,861	27,639	111	Shipments per Employee	195,460	146,463	75
Hours per Production Worker	2,056	2,032	99	Shipments per Production Worker	279,017	193,710	69
Wages per Hour	12.09	13.60	112	Investment per Establishment	321,011	229,182	71
Value Added per Establishment	4,602,255	2,258,434	49	Investment per Employee	6,552	7,366	112
Value Added per Employee	93,930	72,585	77	Investment per Production Worker	9,352	9,742	104

Sources: Same as General Statistics. The 'Average of All Manufacturing' column represents the average of all manufacturing industries reported for the most recent complete year available. The Index shows the relationship between the Average and the Analyzed Industry. For example, 100 means that they are equal; 500 that the Analyzed Industry is five times the average; 50 means that the Analyzed Industry is half the national average. The abbreviation 'na' is used to show that data are 'not available'.

LEADING COMPANIES Number shown: 32 Total sales ($ mil): 827 Total employment (000): 6.6

Company Name	Address				CEO Name	Phone	Co. Type	Sales ($ mil)	Empl. (000)
Barnes Group Inc	123 Main St	Bristol	CT	06010	A Stanton Wells	203-583-7070	P	569	4.2
Detroit Steel Products Company	PO Box 285	Morristown	IN	46161	John Thomson	317-763-6089	S	40	0.4
Kern-Liebers USA Inc	PO Box 396	Holland	OH	43528	Lothar Bauerle	419-865-2437	R	30	0.2
Automatic Spring Products Co	PO Box 347	Grand Haven	MI	49417	Darell L Moreland	616-842-7800	R	22	0.2
Betts Spring Co	2100 Williams St	San Leandro	CA	94577	W Michael Betts	510-352-0111	R	15	0.2
H Miller Spring & Mfg Co	PO Box 7826	Pittsburgh	PA	15215	DF Melampy	412-782-0700	R	14	<0.1
Stanley Spring & Stamping Corp	5050 W Foster Av	Chicago	IL	60630	Stanley R Banas	312-777-2600	R	12	0.2
Union Spring & Mfg Corp	560 Epsilon Dr	Pittsburgh	PA	15238	Raymond Beacha	412-963-7703	R	12	0.1
Hasco Industries Inc	21648 Melrose Av	Southfield	MI	48075	Steve Aretakis	313-351-9450	R	11*	<0.1
Barber Manufacturing Company	PO Box 2454	Anderson	IN	46018	Jeffrey W Barber	317-643-6905	R	10	0.1
Toyoshima Indiana Inc	PO Box 765	Indianapolis	IN	46206	Toshi Osanai	317-638-3511	S	10*	0.1
Haven Steel Products Inc	PO Box 430	Haven	KS	67543	M Cohn	316-465-2573	R	9*	<0.1
Spring Dynamics Inc	35269 Cricklewood	New Baltimore	MI	48047	Joseph Doss	313-725-4447	R	8	<0.1
Pittsburgh Spring Inc	1 McCandless Av	Pittsburgh	PA	15201	Francis R Shuss	412-782-7300	R	8	<0.1
Rolex Co	385 Hillside Av	Hillside	NJ	07205	W Kramer	201-926-0900	R	6	<0.1
Service Spring	PO Box 765	Indianapolis	IN	46206	Toshi Osanai	317-638-3511	D	6	<0.1
Southern Spring and Stamping	401 Sub Station Rd	Venice	FL	34292	Dee Deaterly	813-488-2276	R	6*	<0.1
Wesco Spring Co	4501 S Knox Av	Chicago	IL	60632	Richard Chud	312-838-3350	R	6	0.1
Century Spring Corp	222 E 16th St	Los Angeles	CA	90015	William H Overfelt	213-749-1469	R	5*	<0.1
Arrow Manufacturing Co	16 Jeanette St	Bristol	CT	06011	Edward W Selnau	203-589-3900	R	4	<0.1
Colonial Spring Company Inc	PO Box 1598	Bristol	CT	06010		203-589-3231	R	4*	<0.1
Imperial Spring Company Inc	PO Box 457	Milldale	CT	06467	Larry Palazzo	203-628-9611	R	4	<0.1
Duluth Spring Co	300 Canal Park Dr	Duluth	MN	55802	B Tyacke	218-722-7938	R	3	<0.1
New England Spring Inc	PO Box 9068	Forestville	CT	06010	Robert Dabkowski	203-584-8688	R	3	<0.1
Bay State Auto Spring Mfg	83 Hampden St	Boston	MA	02119	E Kepnes	617-445-6692	R	3	<0.1
C and M Spring Engineering	PO Box 3566	S El Monte	CA	91733	Fred V Rowen	818-443-2205	R	2	<0.1
Benz Spring Co	700 S Forest St	Seattle	WA	98134	Alfred F Benz	206-624-7733	R	1*	<0.1
J and J Spring Company Inc	14100 23 Mile Rd	Shelby	MI	48315	Richard McGuire	810-566-7600	R	1	<0.1
Zenith Spring Company Inc	3116 W Michigan St	Duluth	MN	55806	Gary Tyacke	218-624-5779	R	1	<0.1
Spring Replacement Company	320 North Av	Bridgeport	CT	06606	Ron Martin	203-335-2138	R	1	<0.1
Laher Spring Corp	Hwy 78 W	New Albany	MS	38652	Tom Laher	601-534-2216	R	1	<0.1
Champ Spring Co	PO Box 7103	St Louis	MO	63177	NB Champ Jr	314-231-7570	R	1	<0.1

Source: Ward's Business Directory of U.S. Private and Public Companies, Volumes 1 and 2, 1996. The company type code used is as follows: P - Public, R - Private, S - Subsidiary, D - Division, J - Joint Venture, A - Affiliate, G - Group. Sales are in millions of dollars, employees are in thousands. An asterisk (*) indicates an estimated sales volume. The symbol < stands for 'less than'. Company names and addresses are truncated, in some cases, to fit into the available space.

MATERIALS CONSUMED

Material	Quantity	Delivered Cost ($ million)
Materials, ingredients, containers, and supplies	(X)	169.8
Metal bolts, nuts, screws, washers, rivets, and other screw machine products	(X)	7.3
Other fabricated metal products (except castings and forgings)	(X)	1.4
Castings (rough and semifinished)	(X)	(D)
Forgings	(X)	(D)
Steel bars and bar shapes	(X)	69.7
Steel sheet and strip, including tin plate	(X)	13.7
Steel wire and wire products	(X)	30.9
All other steel shapes and forms	(X)	(D)
Nonferrous shapes and forms	(X)	0.6
All other materials and components, parts, containers, and supplies	(X)	14.9
Materials, ingredients, containers, and supplies, nsk	(X)	16.6

Source: 1992 *Economic Census*. Explanation of symbols used: (D): Withheld to avoid disclosure of competitive data; na: Not available; (S): Withheld because statistical norms were not met; (X): Not applicable; (Z): Less than half the unit shown; nec: Not elsewhere classified; nsk: Not specified by kind; - : zero; * : 10-19 percent estimated; ** : 20-29 percent estimated.

PRODUCT SHARE DETAILS

Product or Product Class	% Share	Product or Product Class	% Share
Steel springs, except wire	100.00	Hot formed steel automotive (auto, truck, bus, trailer, etc.) leaf springs for shipment to U.S. motor vehicle manufacturers or their suppliers for use in original equipment	27.32
Hot formed steel springs, except wire	52.85		
Hot formed, hot wound, helical steel automobile coil springs for domestic replacement and shipments for export	24.60		
		Other hot formed steel springs, including torsion bar springs and leaf springs for tractors, farm equipment, locomotives, etc.	6.90
Hot formed, hot wound, helical steel automobile coil springs for shipment to U.S. motor vehicle manufacturers or their suppliers for use in original equipment	2.03	Hot formed springs, nsk	9.50
		Cold formed steel springs, except wire	42.36
Hot formed, hot wound locomotive, railroad car, and other helical steel springs	9.78	Cold formed flat steel springs made of sheet or strip	81.46
		Cold formed helical steel suspension springs	17.22
Hot formed steel automotive (auto, truck, bus, trailer, etc.) leaf springs for domestic replacement and shipments for export	19.85	Cold formed springs, nsk	1.32
		Steel springs, except wire, nsk	4.79

Source: 1992 Economic Census. The values shown are percent of total shipments in an industry. Values of indented subcategories are summed in the main heading. The symbol (D) appears when data are withheld to prevent disclosure of competitive information. The abbreviation nsk stands for 'not specified by kind' and nec for 'not elsewhere classified'.

INPUTS AND OUTPUTS FOR STEEL SPRINGS, EXCEPT WIRE

Economic Sector or Industry Providing Inputs	%	Sector	Economic Sector or Industry Buying Outputs	%	Sector
Imports	40.8	Foreign	Motor vehicles & car bodies	34.2	Manufg.
Blast furnaces & steel mills	25.5	Manufg.	Automotive rental & leasing, without drivers	31.3	Services
Wholesale trade	8.6	Trade	Automotive repair shops & services	9.0	Services
Iron & steel foundries	3.5	Manufg.	Exports	6.5	Foreign
Gas production & distribution (utilities)	2.3	Util.	Railroad equipment	5.3	Manufg.
Electric services (utilities)	1.7	Util.	Motor vehicle parts & accessories	3.0	Manufg.
Primary metal products, nec	1.3	Manufg.	Personal consumption expenditures	1.5	
Aluminum rolling & drawing	1.2	Manufg.	Farm machinery & equipment	1.5	Manufg.
Banking	1.1	Fin/R.E.	Construction machinery & equipment	1.1	Manufg.
Fabricated rubber products, nec	1.0	Manufg.	Local government passenger transit	1.1	Gov't
Steel springs, except wire	1.0	Manufg.	Truck trailers	1.0	Manufg.
Motor freight transportation & warehousing	1.0	Util.	Steel springs, except wire	0.6	Manufg.
Advertising	1.0	Services	Motor homes (made on purchased chassis)	0.4	Manufg.
Screw machine and related products	0.9	Manufg.	Miscellaneous repair shops	0.4	Services
Air transportation	0.8	Util.	Motor freight transportation & warehousing	0.3	Util.
Petroleum refining	0.7	Manufg.	Industrial trucks & tractors	0.2	Manufg.
Metal coating & allied services	0.6	Manufg.	Ship building & repairing	0.2	Manufg.
Maintenance of nonfarm buildings nec	0.5	Constr.	Retail trade, except eating & drinking	0.2	Trade
Machinery, except electrical, nec	0.5	Manufg.	Wholesale trade	0.2	Trade
Communications, except radio & TV	0.5	Util.	Truck & bus bodies	0.1	Manufg.
Railroads & related services	0.4	Util.	Membership organizations nec	0.1	Services
Eating & drinking places	0.4	Trade			
Noncomparable imports	0.4	Foreign			
Machine tools, metal forming types	0.3	Manufg.			
Special dies & tools & machine tool accessories	0.3	Manufg.			
Equipment rental & leasing services	0.3	Services			
Abrasive products	0.2	Manufg.			
Lubricating oils & greases	0.2	Manufg.			
Paperboard containers & boxes	0.2	Manufg.			
Automotive rental & leasing, without drivers	0.2	Services			
Automotive repair shops & services	0.2	Services			
Computer & data processing services	0.2	Services			
Legal services	0.2	Services			
Management & consulting services & labs	0.2	Services			
U.S. Postal Service	0.2	Gov't			
Copper rolling & drawing	0.1	Manufg.			
Manifold business forms	0.1	Manufg.			
Metal heat treating	0.1	Manufg.			
Paints & allied products	0.1	Manufg.			
Insurance carriers	0.1	Fin/R.E.			
Real estate	0.1	Fin/R.E.			
Accounting, auditing & bookkeeping	0.1	Services			
Hotels & lodging places	0.1	Services			

Source: Benchmark Input-Output Accounts for the U.S. Economy, 1982, U.S. Department of Commerce, Washington, D.C., July 1991. Data, as reported in the source, are organized by the 1977 SIC structure in use in 1982 but have been matched, as closely as is possible, to the 1987 SIC structure used in this book.

OCCUPATIONS EMPLOYED BY SIC 349 - MISCELLANEOUS FABRICATED METAL PRODUCTS

Occupation	% of Total 1994	Change to 2005	Occupation	% of Total 1994	Change to 2005
Assemblers, fabricators, & hand workers nec	10.7	0.7	Industrial machinery mechanics	1.8	10.7
Blue collar worker supervisors	4.5	-7.4	Helpers, laborers, & material movers nec	1.8	0.7
Machinists	4.4	0.7	Hand packers & packagers	1.7	-13.7
Welding machine setters, operators	3.8	-9.4	Machine operators nec	1.6	-11.3
Machine tool cutting & forming etc. nec	3.7	20.8	Industrial production managers	1.5	0.7
Machine forming operators, metal & plastic	3.5	-9.4	Secretaries, ex legal & medical	1.4	-8.3
Sales & related workers nec	3.0	0.7	General office clerks	1.3	-14.2
Inspectors, testers, & graders, precision	3.0	0.7	Tool & die makers	1.3	-18.7
Metal & plastic machine workers nec	2.7	-28.7	Bookkeeping, accounting, & auditing clerks	1.3	-24.5
General managers & top executives	2.6	-4.5	Coating, painting, & spraying machine workers	1.3	0.7
Combination machine tool operators	2.4	51.0	Maintenance repairers, general utility	1.2	-9.4
Machine tool cutting operators, metal & plastic	2.4	41.9	Industrial truck & tractor operators	1.1	0.6
Lathe & turning machine tool operators	2.1	-9.4	Machine feeders & offbearers	1.1	-9.4
Welders & cutters	2.1	0.7	Production, planning, & expediting clerks	1.0	0.7
Traffic, shipping, & receiving clerks	2.0	-3.1	Mechanical engineers	1.0	-0.3

Source: *Industry-Occupation Matrix*, Bureau of Labor Statistics. These data relate to one or more 3-digit SIC industry groups rather than to a single 4-digit SIC. The change reported for each occupation to the year 2005 is a percent of growth or decline as estimated by the Bureau of Labor Statistics. The abbreviation nec stands for 'not elsewhere classified'.

LOCATION BY STATE AND REGIONAL CONCENTRATION

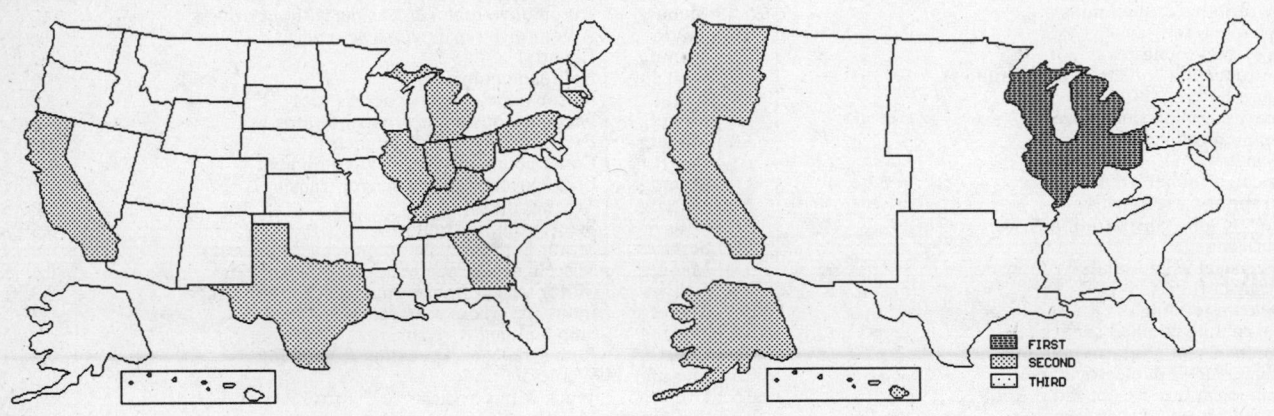

FIRST / SECOND / THIRD

INDUSTRY DATA BY STATE

State	Establish-ments	Shipments Total ($ mil)	% of U.S.	Per Establ.	Employment Total Number	% of U.S.	Per Establ.	Wages ($/hour)	Cost as % of Shipments	Investment per Employee ($)
Pennsylvania	13	138.3	27.9	10.6	900	20.5	69	13.36	59.4	5,000
Connecticut	8	48.9	9.9	6.1	500	11.4	63	14.14	31.3	1,000
Ohio	9	42.6	8.6	4.7	400	9.1	44	9.83	50.7	-
California	15	40.1	8.1	2.7	400	9.1	27	13.75	37.4	1,000
Illinois	10	39.1	7.9	3.9	400	9.1	40	11.80	47.8	-
Michigan	6	10.9	2.2	1.8	100	2.3	17	17.00	45.0	6,000
Indiana	8	(D)	-	-	750 *	17.0	94	-	-	1,600
Texas	5	(D)	-	-	175 *	4.0	35	-	-	-
Kentucky	2	(D)	-	-	175 *	4.0	88	-	-	-
Georgia	1	(D)	-	-	175 *	4.0	175	-	-	-

Source: 1992 *Economic Census*. The states are in descending order of shipments or establishments (if shipment data are missing for the majority). The symbol (D) appears when data are withheld to prevent disclosure of competitive information. States marked with (D) are sorted by number of establishments. A dash (-) indicates that the data element cannot be calculated; * indicates the midpoint of a range.

3494 - VALVES & PIPE FITTINGS, NEC

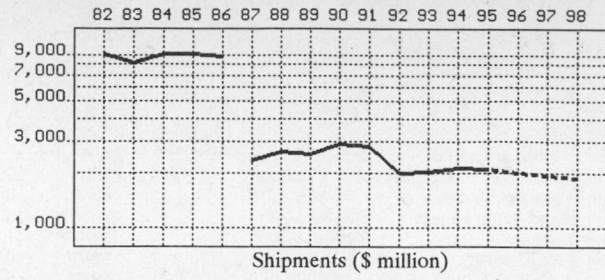

Shipments ($ million)

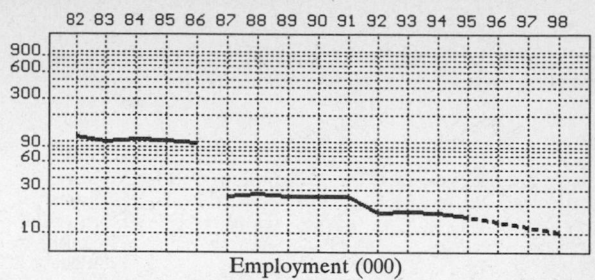

Employment (000)

GENERAL STATISTICS

Year	Companies	Establishments		Employment			Compensation		Production ($ million)			
		Total	with 20 or more employees	Total (000)	Production Workers (000)	Hours (Mil)	Payroll ($ mil)	Wages ($/hr)	Cost of Materials	Value Added by Manufacture	Value of Shipments	Capital Invest.
1982	945	1,161	726	115.2	75.3	144.2	2,334.6	9.32	3,734.8	5,219.8	9,039.9	363.9
1983		1,130	713	100.9	65.8	126.9	2,147.5	9.65	3,271.5	4,827.4	8,201.3	274.4
1984		1,099	700	106.8	71.3	141.0	2,404.7	9.84	3,789.3	5,317.3	9,075.7	267.5
1985		1,069	688	104.5	69.3	136.0	2,431.6	10.23	3,708.4	5,358.1	9,054.6	308.8
1986		1,063	674	99.2	65.2	128.7	2,375.7	10.42	3,636.8	5,089.8	8,823.4	260.1
1987*	371	416	231	25.1	17.6	34.9	570.4	10.11	1,002.0	1,366.5	2,377.3	62.4
1988		373	220	26.9	19.7	39.6	633.6	10.31	1,179.8	1,506.6	2,649.5	45.3
1989		360	227	25.9	17.5	35.2	611.4	10.53	1,163.3	1,392.9	2,571.4	63.7
1990		365	227	25.2	18.7	38.6	666.3	10.90	1,390.1	1,535.8	2,924.0	81.4
1991		372	229	25.0	18.1	36.7	640.5	11.02	1,292.6	1,519.3	2,803.0	55.5
1992	228	251	150	16.6	12.2	25.5	469.5	11.99	924.5	1,054.2	1,991.7	45.3
1993		245	148	17.3	12.9	26.0	490.7	12.33	936.7	1,101.6	2,041.5	54.3
1994		234P	149P	16.6	12.5	26.2	506.7	12.13	1,041.3	1,117.0	2,149.6	49.6
1995		207P	135P	15.1P	11.5P	24.3P	487.3P	12.70P	1,012.7P	1,086.3P	2,090.6P	50.5P
1996		181P	122P	13.4P	10.5P	22.4P	468.1P	13.05P	975.3P	1,046.2P	2,013.3P	49.0P
1997		154P	108P	11.8P	9.5P	20.5P	448.9P	13.39P	937.8P	1,006.0P	1,936.0P	47.5P
1998		127P	94P	10.2P	8.5P	18.6P	429.7P	13.73P	900.4P	965.8P	1,858.7P	46.0P

Sources: 1982, 1987, 1992 *Economic Census*; *Annual Survey of Manufactures*, 83-86, 88-91, 93-94. Establishment counts are from *County Business Patterns* for non-Census years; establishment counts for 83-84 are extrapolations. * indicates that industry content changed in 87; earlier years use 77 SICs. 'P's mark projections.

INDICES OF CHANGE

Year	Companies	Establishments		Employment			Compensation		Production ($ million)			
		Total	with 20 or more employees	Total (000)	Production Workers (000)	Hours (Mil)	Payroll ($ mil)	Wages ($/hr)	Cost of Materials	Value Added by Manufacture	Value of Shipments	Capital Invest.
1982	414	463	484	694	617	565	497	78	404	495	454	803
1983		450	475	608	539	498	457	80	354	458	412	606
1984		438	467	643	584	553	512	82	410	504	456	591
1985		426	459	630	568	533	518	85	401	508	455	682
1986		424	449	598	534	505	506	87	393	483	443	574
1987*	163	166	154	151	144	137	121	84	108	130	119	138
1988		149	147	162	161	155	135	86	128	143	133	100
1989		143	151	156	143	138	130	88	126	132	129	141
1990		145	151	152	153	151	142	91	150	146	147	180
1991		148	153	151	148	144	136	92	140	144	141	123
1992	100	100	100	100	100	100	100	100	100	100	100	100
1993		98	99	104	106	102	105	103	101	104	103	120
1994		93P	100P	100	102	103	108	101	113	106	108	109
1995		83P	90P	91P	95P	95P	104P	106P	110P	103P	105P	111P
1996		72P	81P	81P	86P	88P	100P	109P	105P	99P	101P	108P
1997		61P	72P	71P	78P	80P	96P	112P	101P	95P	97P	105P
1998		51P	63P	62P	69P	73P	92P	115P	97P	92P	93P	101P

Sources: Same as General Statistics. Values reflect change from the base year, 1992. Values above 100 mean greater than 92, values below 100 mean less than 92, and a value of 100 in the 82-91 or 93-98 period means same as 92. * indicates that industry content changed in 87. Data for earlier years are in 77 SIC format.

SELECTED RATIOS

For 1994	Avg. of All Manufact.	Analyzed Industry	Index	For 1994	Avg. of All Manufact.	Analyzed Industry	Index
Employees per Establishment	49	71	145	Value Added per Production Worker	134,084	89,360	67
Payroll per Establishment	1,500,273	2,166,707	144	Cost per Establishment	5,045,178	4,452,718	88
Payroll per Employee	30,620	30,524	100	Cost per Employee	102,970	62,729	61
Production Workers per Establishment	34	53	156	Cost per Production Worker	146,988	83,304	57
Wages per Establishment	853,319	1,358,975	159	Shipments per Establishment	9,576,895	9,191,936	96
Wages per Production Worker	24,861	25,424	102	Shipments per Employee	195,460	129,494	66
Hours per Production Worker	2,056	2,096	102	Shipments per Production Worker	279,017	171,968	62
Wages per Hour	12.09	12.13	100	Investment per Establishment	321,011	212,095	66
Value Added per Establishment	4,602,255	4,776,420	104	Investment per Employee	6,552	2,988	46
Value Added per Employee	93,930	67,289	72	Investment per Production Worker	9,352	3,968	42

Sources: Same as General Statistics. The 'Average of All Manufacturing' column represents the average of all manufacturing industries reported for the most recent complete year available. The Index shows the relationship between the Average and the Analyzed Industry. For example, 100 means that they are equal; 500 that the Analyzed Industry is five times the average; 50 means that the Analyzed Industry is half the national average. The abbreviation 'na' is used to show that data are 'not available'.

LEADING COMPANIES   Number shown: **75**   Total sales ($ mil): **4,820**   Total employment (000): **34.4**

Company Name	Address				CEO Name	Phone	Co. Type	Sales ($ mil)	Empl. (000)
Crane Co	100 1st Stamford Pl	Stamford	CT	06902	Robert S Evans	203-363-7300	P	1,654	10.7
Aeroquip Industrial Americas	1695 Indian	Maumee	OH	43537	ET Smith	419-891-7600	S	290•	2.5
Keystone Valve USA Inc	PO Box 40010	Houston	TX	77240	Tom Comstock	713-466-1176	S	206	0.5
Automatic Switch Co	50-60 Hanover Rd	Florham Park	NJ	07932	Randy P Smith	201-966-2000	S	180•	1.6
Westinghouse	Cheswick Av	Cheswick	PA	15024	Michael J Ferris	412-963-5000	D	135	0.7
Webster Valve	PO Box 431	Franklin	NH	03235	Jack Bishop	603-934-5110	D	100	0.6
Chicago Specialty Mfg Co	750 Northgate Pkwy	Wheeling	IL	60090	Albert D King	708-520-9080	S	85•	0.4
Dresser Manufacturing	41 Fisher Av	Bradford	PA	16701	JL Ward	814-362-9200	D	85	0.6
Senior Flexonics Inc	300 E Devon Av	Bartlett	IL	60103	S Perkins	708-837-1811	S	82•	0.7
Henry Vogt Machine Co	PO Box 1918	Louisville	KY	40201	Robert Campbell	502-634-1500	R	81	1.0
Valtek Inc	PO Box 2200	Springville	UT	84663	Mark Vernon	801-489-8611	S	80	0.5
Fluidmaster Inc	1800 Via Burton	Anaheim	CA	92806	Adolph Schoepe	714-774-1444	R	75	0.2
Yarway Corp	480 Norristown Rd	Blue Bell	PA	19422	Don S Cook	610-825-2100	S	62	0.4
Duriron Company Inc	PO Box 2609	Cookeville	TN	38501	Ronald L Hoffman	615-432-4021	D	60•	0.3
ITT Engineered Valves	33 Centerville Rd	Lancaster	PA	17603	James Greene	717-291-1901	D	60	0.5
Elkhart Products Corp	PO Box 1008	Elkhart	IN	46515	Dave Ewing	219-264-3181	S	55•	0.5
Parker Hannifin Corp	300 Parker Dr	Otsego	MI	49078	Tony Piscitello	616-694-9411	D	55•	0.5
Swagelok Co	31400 Aurora Rd	Solon	OH	44139	FJ Callahan	216-349-5934	R	55	0.5
Weatherhead Brake Prod	643 W Ellsworth St	Columbia City	IN	46725	Ray Frederick	219-481-3500	D	55	0.5
William Powell Co	PO Box 145434	Cincinnati	OH	45250	Randy Cowart	513-852-2000	R	52•	0.5
Armstrong International Inc	816 Maple St	Three Rivers	MI	49093	Rex Cheskaty	616-273-1415	R	50	0.5
Grove Valve and Regulator Co	6529 Hollis St	Emeryville	CA	94608	Kenneth G Banks	510-655-7700	R	50	0.1
Beck Manufacturing Inc	PO Box 510	Waynesboro	PA	17268	Stephen E Beck	717-762-9141	S	47	0.3
Bitrek Corp	PO Box 510	Waynesboro	PA	17268	Jacob E Beck Jr	717-762-9141	R	47	0.3
Kennedy Value	1021 E Water St	Elmira	NY	14901	Ray Schofield	607-734-2211	D	41•	0.5
Henry Valve Co	3215 North Av	Melrose Park	IL	60160	N Clark	708-344-1100	R	40	0.4
Lincoln Brass Works Inc	1161 Murfreesboro	Nashville	TN	37217	Douglas Fischer	615-367-9353	R	40	0.4
Bonney Forge Corp	PO Box 359	Allentown	PA	18105	John A Leone	215-435-9611	R	38•	0.4
Waterman Industries Inc	PO Box 458	Exeter	CA	93221	DF Appling	209-592-3174	R	38•	0.3
Fike Corp	PO Box 610	Blue Springs	MO	64015	LL Fike Jr	816-229-3405	R	36	0.3
Johnson Corp	805 Wood St	Three Rivers	MI	49093	R D Wiedenbeck	616-278-1715	R	35	0.1
Richards Industries Inc	3170 Wasson Rd	Cincinnati	OH	45209	Gilbert Richards	513-533-5600	R	35	0.2
TXT Inc	PO Box 60706	Houston	TX	77205	Stuart Brightman	713-443-7000	R	35	0.3
Kunkle Industries Inc	PO Box 1740	Fort Wayne	IN	46801	Richard Mattie	219-747-3405	S	31	0.3
Groth Corp	PO Box 15293	Houston	TX	77220	Ed Groth Jr	713-675-6151	R	30•	0.2
Humphrey Products Co	PO Box 2008	Kalamazoo	MI	49003	Randall M Webber	616-381-5500	R	30	0.4
Worcester Controls Corp	PO Box 538	Marlboro	MA	01752	Ronald Kozlowski	508-481-4800	S	30	0.2
Lee Co	PO Box 424	Westbrook	CT	06498	Leighton Lee III	203-399-6281	R	28•	0.5
Velan Valve Corp	Av C	Williston	VT	05495	Mike Parsons	802-863-2562	S	28•	0.2
Keystone Vanessa-Valvtron Inc	9600 W Gulf Bank	Houston	TX	77040	Terry Scanlan	713-937-5396	S	26	<0.1
Parker Hannifin Corp	2445 S 25th Av	Broadview	IL	60153	Paul G Bishop	708-681-6300	D	25	0.2
Stratoflex Aerospace	PO Box 10398	Fort Worth	TX	76114	Steve Hamilton	817-738-6543	D	25	0.3
Clayton Mark Inc	PO Box 248	Rogers	AR	72756	S Daniels	501-636-1800	D	23	0.1
IMI Cash Valve Inc	600 E Wabash Av	Decatur	IL	62523	TL Kobylarek	217-422-8574	R	22	0.2
Fresno Valves and Castings Inc	PO Box 40	Selma	CA	93662	OR Showalter Jr	209-834-2511	R	21•	0.2
Hansen Coupling	1000 W Bagley Rd	Berea	OH	44017	Melvin Fahnestock	216-252-3880	D	21	0.2
L and J Technologies Inc	5911 Butterfield Rd	Hillside	IL	60162	Lou Jannotta	708-236-6000	R	21•	0.2
Accord Industries	4001 Forsyth Rd	Winter Park	FL	32792	Edward W White	407-671-5200	R	20	0.2
Anchor-Darling Valve Co	PO Box 3428	Williamsport	PA	17701	JJ Chappell	717-327-4800	S	20	0.2
Pacific Pipe Co	PO Box 23711	Oakland	CA	94623	Ellis Jacobs	510-452-0122	R	20	0.1
Robert Mitchell Company Inc	PO Box 2008	Portland	ME	04104	Arthur Dubois	207-797-6771	R	20	0.1
Shafer Valve Co	PO Box 2507	Mansfield	OH	44906	Jerry L Kohler	419-529-4311	S	20	0.2
Specialty Manufacturing Co	5858 Centerville Rd	St Paul	MN	55127	B Lawin	612-653-0533	R	20	0.2
Flowline	PO Box 7027	New Castle	PA	16107	Roger S Brown Jr	412-658-3711	D	19•	0.2
JR Clarkson Co	650 Spice Island Dr	Sparks	NV	89431	CW Clarkson	702-359-4100	R	18	0.1
Michigan Hanger Company Inc	931 Summit Av	Niles	OH	44446	Ed Rumble	216-544-4700	R	17	0.1
Vacco Industries	10350 Vacco St	S El Monte	CA	91733	J Michael Conway	818-443-7121	S	17	0.2
Colder Products Co	1001 Westgate Dr	St Paul	MN	55114	Michael D Lyon	612-645-0091	R	16•	0.1
Ames Company Inc	PO Box 1387	Woodland	CA	95776	KD Powell	916-666-2493	R	15	0.1
MARPAC	400 Maple Av	Carpentersville	IL	60110	Rick Giannini	708-551-3388	S	15	<0.1
Masoneilan Montebello	1040 S Vail Av	Montebello	CA	90640	R R Williams	213-723-9351	D	15	<0.1
Robert Manufacturing Co	10667 Jersey Blv	R Cucamonga	CA	91730	Robert Hartwell	909-987-4654	R	15	0.2
Marshall Brass	450 Leggitt Rd	Marshall	MI	49068	RE Mell	616-781-3901	D	14•	0.1
Tube Forgings of America Inc	5200 NW Front Av	Portland	OR	97210	Jay N Zidell	503-228-8691	R	14•	0.1
Allenair Corp	255 E 2nd St	Mineola	NY	11501	AK Allen	516-747-5450	S	13	0.2
American Boa Inc	PO Box 1301	Cumming	GA	30130	F Berg	404-889-9400	S	13	<0.1
Primore Inc	PO Box 605	Adrian	MI	49221	RE Price	517-265-6168	R	13•	<0.1
Carpenter and Paterson Inc	225 Merrimac St	Woburn	MA	01801	Donald R Paterson	617-935-2950	R	12	<0.1
Tapco International Inc	PO Box 40472	Houston	TX	77240	John Arnoldy	713-466-0300	S	12	0.2
DPS Company Inc	PO Box 11324	Fort Wayne	IN	46857	DP Schenkel	219-747-9200	RS	11	0.2
L & L Fittings Mfg Co	PO Box 11324	Fort Wayne	IN	46857	David P Schenkel	219-747-9200	D	11	0.2
Automatic Machine Products Co	PO Box 1018	Attleboro	MA	02703	John S Holden Jr	508-222-2300	R	10	0.2
Automatic Valve Corp	PO Box 435	Novi	MI	48376	T Hutchins	313-474-6700	R	10	0.1
Byron Valve and Machine Co	PO Box 458	Siloam Springs	AR	72761	Thomas P Bevins II	501-524-3147	R	10	0.1
CM Industries Inc	330 Boston Post Rd	Old Saybrook	CT	06475	Paul Kowack	203-388-5747	R	10	<0.1

Source: Ward's Business Directory of U.S. Private and Public Companies, Volumes 1 and 2, 1996. The company type code used is as follows: P - Public, R - Private, S - Subsidiary, D - Division, J - Joint Venture, A - Affiliate, G - Group. Sales are in millions of dollars, employees are in thousands. An asterisk (*) indicates an estimated sales volume. The symbol < stands for 'less than'. Company names and addresses are truncated, in some cases, to fit into the available space.

MATERIALS CONSUMED

Material	Quantity	Delivered Cost ($ million)
Materials, ingredients, containers, and supplies	(X)	793.9
Metal bolts, nuts, screws, washers, rivets, and other screw machine products	(X)	37.7
Metal stampings	(X)	4.6
Valves, fittings, and couplings purchased for further assembly (except forgings)	(X)	24.7
Other fabricated metal products (except forgings)	(X)	10.9
Iron castings (rough and semifinished)	(X)	38.7
Steel castings (rough and semifinished)	(X)	14.3
Aluminum and aluminum-base alloy castings (rough and semifinished)	(X)	3.8
Copper and copper-base alloy castings (rough and semifinished)	(X)	18.9
Other nonferrous castings (rough and semifinished)	(X)	4.7
Iron and steel forgings	(X)	47.6
Nonferrous forgings	(X)	4.6
Steel bars, bar shapes, and plates	(X)	50.7
Steel sheet and strip, including tin plate	(X)	18.4
All other steel shapes and forms	(X)	54.3
Copper and copper-base alloy refinery shapes	(X)	(D)
Copper and copper-base alloy rod, bar, and mechanical wire, including extruded and/or drawn shapes	(X)	43.5
Copper and copper-base alloy pipe and tube	(X)	21.7
All other copper and copper-base alloy shapes and forms	(X)	33.5
Aluminum and aluminum-base alloy shapes and forms	(X)	2.3
Other nonferrous shapes and forms	(X)	6.3
Scrap, including iron, steel, aluminum and aluminum-base alloy (excluding home scrap)	(X)	(D)
Electric motors and generators less than 1 horsepower (less than 746 watts)	(X)	3.2
Paperboard containers, boxes, and corrugated paperboard	(X)	10.7
Hydraulic and pneumatic hose	(X)	(D)
Other rubber and plastics hose and belting	(X)	(D)
Gaskets (all types), and packing and sealing devices	(X)	12.3
Fabricated rubber products, except tires, tubes, hose, belting, and gaskets	(X)	3.9
Fabricated plastics products (except gaskets, hoses, and belting)	(X)	9.7
All other materials and components, parts, containers, and supplies	(X)	134.1
Materials, ingredients, containers, and supplies, nsk	(X)	122.2

Source: 1992 *Economic Census*. Explanation of symbols used: (D): Withheld to avoid disclosure of competitive data; na: Not available; (S): Withheld because statistical norms were not met; (X): Not applicable; (Z): Less than half the unit shown; nec: Not elsewhere classified; nsk: Not specified by kind; - : zero; * : 10-19 percent estimated; ** : 20-29 percent estimated.

PRODUCT SHARE DETAILS

Product or Product Class	% Share	Product or Product Class	% Share
Valves and pipe fittings, nec	100.00	Wrought copper and wrought copper alloy fittings, flanges, and unions for piping systems, including solder and threaded types	15.88
Plumbing and heating valves and specialties, except plumbers' brass goods	18.47	Cast carbon and alloy steel fittings, flanges, and unions for piping systems	2.19
Plumbing and heating safety and relief valves, except plumbers' brass goods	23.23	Forged carbon, alloy, and stainless steel fittings, flanges, and unions for piping systems, including socket-weld or threaded-type	13.96
Plumbing and heating check valves, except plumbers' brass goods	8.00	Forged carbon steel butt-welding type flanges for piping systems	4.17
All other plumbing and heating valves (less than 125 lb w.s.p.), except plumbers' brass goods	18.37	Forged alloy steel butt-welding type flanges for piping systems	1.23
Plumbing and heating valve specialties, except plumbers' brass goods	32.61	Forged stainless steel butt-welding type flanges for piping systems	2.20
Parts for plumbing and heating valves and specialties, except plumbers' brass goods	14.18	Forged carbon steel butt-welding type fittings for piping systems	2.82
Plumbing and heating valves and specialties, except plumbers' brass goods, nsk	3.61	Forged alloy steel butt-welding type fittings for piping systems	(D)
Metal fittings, flanges, and unions for piping systems	69.12	Forged stainless steel butt-welding type fittings for piping systems	4.22
Gray iron fittings, flanges, and unions for piping systems	6.35	Metal pipe hangers and supports (not including metal framing)	4.82
Gray iron grooved fittings and couplings for piping systems	0.73	Metal pipe couplings	7.28
Malleable iron fittings and flanges (including reducers, caps, etc.) for piping systems	7.47	Other metal fittings, flanges, and unions, including metal framing and fittings for mechanical and electrical supports	7.93
Malleable iron grooved fittings and couplings for piping systems	(D)	Metal fittings, flanges, and unions for piping systems, nsk	2.69
Malleable iron unions and union fittings for piping systems	2.40	Valves and pipe fittings, nec, nsk	12.42
Ductile iron fittings, flanges, and unions for piping systems	1.96		
Ductile iron grooved fittings and couplings for piping systems	1.63		
Cast brass or bronze fittings, flanges, and unions for piping systems, including solder and threaded types	7.93		

Source: 1992 *Economic Census*. The values shown are percent of total shipments in an industry. Values of indented subcategories are summed in the main heading. The symbol (D) appears when data are withheld to prevent disclosure of competitive information. The abbreviation nsk stands for 'not specified by kind' and nec for 'not elsewhere classified'.

INPUTS AND OUTPUTS FOR PIPE, VALVES, & PIPE FITTINGS

Economic Sector or Industry Providing Inputs	%	Sector	Economic Sector or Industry Buying Outputs	%	Sector
Blast furnaces & steel mills	19.6	Manufg.	Industrial buildings	12.2	Constr.
Imports	13.8	Foreign	Gross private fixed investment	11.7	Cap Inv
Wholesale trade	8.1	Trade	Exports	7.8	Foreign
Iron & steel foundries	7.5	Manufg.	Maintenance of nonfarm buildings nec	5.5	Constr.
Pipe, valves, & pipe fittings	3.2	Manufg.	Crude petroleum & natural gas	4.0	Mining
Copper rolling & drawing	2.9	Manufg.	Cyclic crudes and organics	4.0	Manufg.
Advertising	2.8	Services	Sewer system facility construction	3.4	Constr.
Iron & steel forgings	2.5	Manufg.	Office buildings	2.8	Constr.
Screw machine and related products	2.4	Manufg.	Electric utility facility construction	2.7	Constr.
Electric services (utilities)	2.2	Util.	Automotive rental & leasing, without drivers	2.7	Services
Banking	1.9	Fin/R.E.	Water supply facility construction	2.0	Constr.
Fabricated structural metal	1.6	Manufg.	Maintenance of sewer facilities	1.8	Constr.
Aluminum castings	1.4	Manufg.	Pipe, valves, & pipe fittings	1.7	Manufg.
Brass, bronze, & copper castings	1.4	Manufg.	Turbines & turbine generator sets	1.3	Manufg.
Petroleum refining	1.4	Manufg.	Maintenance of water supply facilities	1.2	Constr.
Engineering, architectural, & surveying services	1.4	Services	Nonfarm residential structure maintenance	1.1	Constr.
Motor freight transportation & warehousing	1.2	Util.	Water transportation	1.1	Util.
Cyclic crudes and organics	1.1	Manufg.	Fabricated plate work (boiler shops)	1.0	Manufg.
Industrial patterns	1.1	Manufg.	Communications, except radio & TV	1.0	Util.
Primary copper	1.1	Manufg.	Motor freight transportation & warehousing	1.0	Util.
Nonferrous wire drawing & insulating	1.0	Manufg.	Sanitary services, steam supply, irrigation	1.0	Util.
Aluminum rolling & drawing	0.9	Manufg.	Water supply & sewage systems	1.0	Util.
Communications, except radio & TV	0.9	Util.	Blowers & fans	0.9	Manufg.
Equipment rental & leasing services	0.9	Services	Oil field machinery	0.9	Manufg.
Maintenance of nonfarm buildings nec	0.8	Constr.	Transit & bus transportation	0.9	Util.
Air transportation	0.8	Util.	Maintenance of electric utility facilities	0.8	Constr.
Gas production & distribution (utilities)	0.8	Util.	Residential 1-unit structures, nonfarm	0.8	Constr.
Machinery, except electrical, nec	0.7	Manufg.	Construction machinery & equipment	0.8	Manufg.
Business services nec	0.7	Services	Federal Government purchases, national defense	0.8	Fed Govt
Photofinishing labs, commercial photography	0.7	Services	Construction of stores & restaurants	0.7	Constr.
Plating & polishing	0.6	Manufg.	Maintenance of petroleum & natural gas wells	0.7	Constr.
Eating & drinking places	0.6	Trade	Ship building & repairing	0.7	Manufg.
Nonferrous castings, nec	0.5	Manufg.	Gas utility facility construction	0.6	Constr.
Primary nonferrous metals, nec	0.5	Manufg.	Warehouses	0.6	Constr.
Rubber & plastics hose & belting	0.5	Manufg.	Petroleum refining	0.6	Manufg.
Real estate	0.5	Fin/R.E.	Pumps & compressors	0.6	Manufg.
Scrap	0.5	Scrap	Maintenance of gas utility facilities	0.5	Constr.
Fabricated rubber products, nec	0.4	Manufg.	Residential additions/alterations, nonfarm	0.5	Constr.
Industrial controls	0.4	Manufg.	Aircraft & missile engines & engine parts	0.5	Manufg.
Machine tools, metal cutting types	0.4	Manufg.	Machinery, except electrical, nec	0.5	Manufg.
Special dies & tools & machine tool accessories	0.4	Manufg.	Construction of hospitals	0.4	Constr.
Railroads & related services	0.4	Util.	Maintenance of farm service facilities	0.4	Constr.
Chemical preparations, nec	0.3	Manufg.	Blast furnaces & steel mills	0.4	Manufg.
Hardware, nec	0.3	Manufg.	Farm machinery & equipment	0.4	Manufg.
Metal stampings, nec	0.3	Manufg.	General industrial machinery, nec	0.4	Manufg.
Miscellaneous plastics products	0.3	Manufg.	Household cooking equipment	0.4	Manufg.
Paperboard containers & boxes	0.3	Manufg.	Household laundry equipment	0.4	Manufg.
Primary metal products, nec	0.3	Manufg.	Paper mills, except building paper	0.4	Manufg.
Automotive rental & leasing, without drivers	0.3	Services	Refrigeration & heating equipment	0.4	Manufg.
U.S. Postal Service	0.3	Gov't	Special industry machinery, nec	0.4	Manufg.
Abrasive products	0.2	Manufg.	Truck trailers	0.4	Manufg.
Gaskets, packing & sealing devices	0.2	Manufg.	Maintenance of petroleum pipelines	0.3	Constr.
Lubricating oils & greases	0.2	Manufg.	Maintenance, conservation & development facilities	0.3	Constr.
Metal heat treating	0.2	Manufg.	Petroleum & natural gas well drilling	0.3	Constr.
Miscellaneous fabricated wire products	0.2	Manufg.	Aircraft	0.3	Manufg.
Insurance carriers	0.2	Fin/R.E.	Household appliances, nec	0.3	Manufg.
Accounting, auditing & bookkeeping	0.2	Services	Internal combustion engines, nec	0.3	Manufg.
Automotive repair shops & services	0.2	Services	Motor vehicle parts & accessories	0.3	Manufg.
Computer & data processing services	0.2	Services	Paperboard mills	0.3	Manufg.
Legal services	0.2	Services	Polishes & sanitation goods	0.3	Manufg.
Management & consulting services & labs	0.2	Services	Sporting & athletic goods, nec	0.3	Manufg.
Noncomparable imports	0.2	Foreign	Toilet preparations	0.3	Manufg.
Electronic components nec	0.1	Manufg.	Railroads & related services	0.3	Util.
Fabricated metal products, nec	0.1	Manufg.	Construction of conservation facilities	0.2	Constr.
Industrial inorganic chemicals, nec	0.1	Manufg.	Construction of educational buildings	0.2	Constr.
Metal coating & allied services	0.1	Manufg.	Hotels & motels	0.2	Constr.
Motors & generators	0.1	Manufg.	Maintenance of farm residential buildings	0.2	Constr.
Credit agencies other than banks	0.1	Fin/R.E.	Maintenance of military facilities	0.2	Constr.
Security & commodity brokers	0.1	Fin/R.E.	Maintenance of nonbuilding facilities nec	0.2	Constr.
Hotels & lodging places	0.1	Services	Residential garden apartments	0.2	Constr.
			Industrial furnaces & ovens	0.2	Manufg.
			Industrial inorganic chemicals, nec	0.2	Manufg.
			Industrial trucks & tractors	0.2	Manufg.
			Iron & steel forgings	0.2	Manufg.
			Miscellaneous plastics products	0.2	Manufg.
			Nitrogenous & phosphatic fertilizers	0.2	Manufg.
			Paints & allied products	0.2	Manufg.
			Prepared feeds, nec	0.2	Manufg.

Continued on next page.

INPUTS AND OUTPUTS FOR PIPE, VALVES, & PIPE FITTINGS - Continued

Economic Sector or Industry Providing Inputs	%	Sector	Economic Sector or Industry Buying Outputs	%	Sector
			Pulp mills	0.2	Manufg.
			Service industry machines, nec	0.2	Manufg.
			Farm service facilities	0.1	Constr.
			Maintenance of local transit facilities	0.1	Constr.
			Maintenance of railroads	0.1	Constr.
			Petroleum pipeline construction	0.1	Constr.
			Residential high-rise apartments	0.1	Constr.
			Agricultural chemicals, nec	0.1	Manufg.
			Aircraft & missile equipment, nec	0.1	Manufg.
			Bottled & canned soft drinks	0.1	Manufg.
			Cement, hydraulic	0.1	Manufg.
			Chemical preparations, nec	0.1	Manufg.
			Food products machinery	0.1	Manufg.
			Industrial gases	0.1	Manufg.
			Machine tools, metal cutting types	0.1	Manufg.
			Malt beverages	0.1	Manufg.
			Primary aluminum	0.1	Manufg.

Source: Benchmark Input-Output Accounts for the U.S. Economy, 1982, U.S. Department of Commerce, Washington, D.C., July 1991. Data, as reported in the source, are organized by the 1977 SIC structure in use in 1982 but have been matched, as closely as is possible, to the 1987 SIC structure used in this book.

OCCUPATIONS EMPLOYED BY SIC 349 - MISCELLANEOUS FABRICATED METAL PRODUCTS

Occupation	% of Total 1994	Change to 2005	Occupation	% of Total 1994	Change to 2005
Assemblers, fabricators, & hand workers nec	10.7	0.7	Industrial machinery mechanics	1.8	10.7
Blue collar worker supervisors	4.5	-7.4	Helpers, laborers, & material movers nec	1.8	0.7
Machinists	4.4	0.7	Hand packers & packagers	1.7	-13.7
Welding machine setters, operators	3.8	-9.4	Machine operators nec	1.6	-11.3
Machine tool cutting & forming etc. nec	3.7	20.8	Industrial production managers	1.5	0.7
Machine forming operators, metal & plastic	3.5	-9.4	Secretaries, ex legal & medical	1.4	-8.3
Sales & related workers nec	3.0	0.7	General office clerks	1.3	-14.2
Inspectors, testers, & graders, precision	3.0	0.7	Tool & die makers	1.3	-18.7
Metal & plastic machine workers nec	2.7	-28.7	Bookkeeping, accounting, & auditing clerks	1.3	-24.5
General managers & top executives	2.6	-4.5	Coating, painting, & spraying machine workers	1.3	0.7
Combination machine tool operators	2.4	51.0	Maintenance repairers, general utility	1.2	-9.4
Machine tool cutting operators, metal & plastic	2.4	41.9	Industrial truck & tractor operators	1.1	0.6
Lathe & turning machine tool operators	2.1	-9.4	Machine feeders & offbearers	1.1	-9.4
Welders & cutters	2.1	0.7	Production, planning, & expediting clerks	1.0	0.7
Traffic, shipping, & receiving clerks	2.0	-3.1	Mechanical engineers	1.0	-0.3

Source: Industry-Occupation Matrix, Bureau of Labor Statistics. These data relate to one or more 3-digit SIC industry groups rather than to a single 4-digit SIC. The change reported for each occupation to the year 2005 is a percent of growth or decline as estimated by the Bureau of Labor Statistics. The abbreviation nec stands for 'not elsewhere classified'.

LOCATION BY STATE AND REGIONAL CONCENTRATION

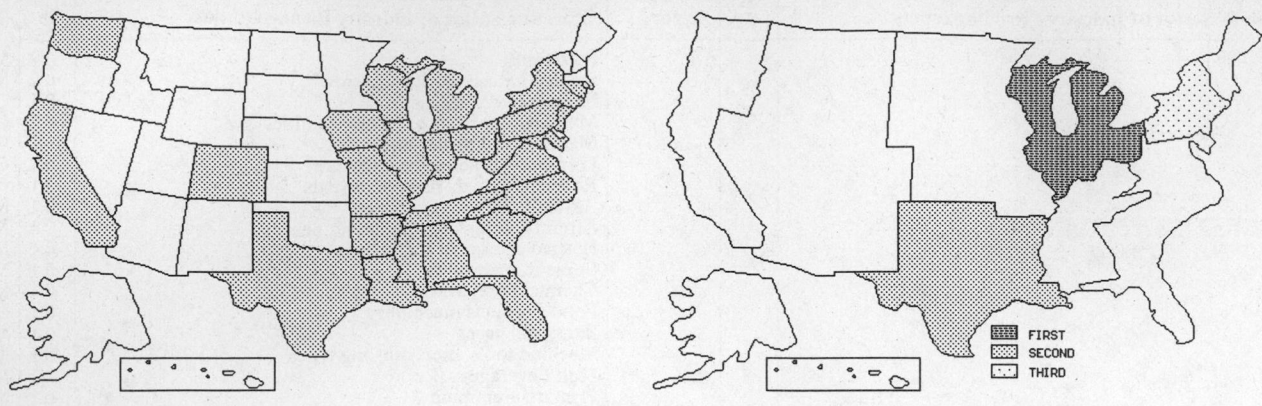

INDUSTRY DATA BY STATE

State	Establish-ments	Shipments			Employment				Cost as % of Shipments	Investment per Employee ($)
		Total ($ mil)	% of U.S.	Per Establ.	Total Number	% of U.S.	Per Establ.	Wages ($/hour)		
Pennsylvania	21	346.5	17.4	16.5	3,700	22.3	176	11.76	41.9	2,649
Ohio	29	214.0	10.7	7.4	1,600	9.6	55	12.88	47.5	2,750
Texas	47	191.8	9.6	4.1	1,600	9.6	34	10.62	47.2	2,500
Illinois	18	170.2	8.5	9.5	1,300	7.8	72	11.55	49.8	1,692
Indiana	8	136.3	6.8	17.0	1,100	6.6	138	13.94	47.3	2,818
Michigan	13	82.7	4.2	6.4	500	3.0	38	13.00	35.1	6,400
California	26	77.7	3.9	3.0	600	3.6	23	12.56	41.4	2,667
New York	8	40.8	2.0	5.1	400	2.4	50	14.67	52.5	2,250
Connecticut	5	36.6	1.8	7.3	200	1.2	40	13.50	45.1	3,500
New Jersey	9	34.5	1.7	3.8	400	2.4	44	9.00	52.2	1,250
Florida	7	21.7	1.1	3.1	200	1.2	29	9.00	49.3	6,000
Louisiana	5	(D)	-	-	750 *	4.5	150	-	-	-
Wisconsin	4	(D)	-	-	750 *	4.5	188	-	-	-
Alabama	3	(D)	-	-	375 *	2.3	125	-	-	-
Arkansas	3	(D)	-	-	375 *	2.3	125	-	-	-
Iowa	3	(D)	-	-	175 *	1.1	58	-	-	-
South Carolina	3	(D)	-	-	175 *	1.1	58	-	-	-
Tennessee	3	(D)	-	-	375 *	2.3	125	-	-	-
Washington	3	(D)	-	-	175 *	1.1	58	-	-	-
West Virginia	3	(D)	-	-	175 *	1.1	58	-	-	-
Missouri	2	(D)	-	-	375 *	2.3	188	-	-	-
North Carolina	2	(D)	-	-	375 *	2.3	188	-	-	-
Virginia	2	(D)	-	-	375 *	2.3	188	-	-	-
Colorado	1	(D)	-	-	175 *	1.1	175	-	-	-
Maryland	1	(D)	-	-	375 *	2.3	375	-	-	-
Mississippi	1	(D)	-	-	175 *	1.1	175	-	-	-

Source: 1992 *Economic Census*. The states are in descending order of shipments or establishments (if shipment data are missing for the majority). The symbol (D) appears when data are withheld to prevent disclosure of competitive information. States marked with (D) are sorted by number of establishments. A dash (-) indicates that the data element cannot be calculated; * indicates the midpoint of a range.

3495 - WIRE SPRINGS

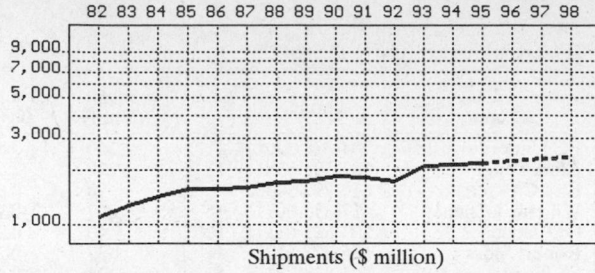

Shipments ($ million)

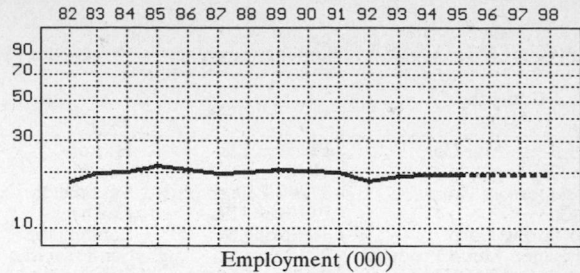

Employment (000)

GENERAL STATISTICS

Year	Companies	Establishments		Employment			Compensation		Production ($ million)			
		Total	with 20 or more employees	Total (000)	Production Workers (000)	Hours (Mil)	Payroll ($ mil)	Wages ($/hr)	Cost of Materials	Value Added by Manufacture	Value of Shipments	Capital Invest.
1982	322	432	236	17.9	14.5	27.5	294.1	7.59	498.7	596.8	1,104.3	31.9
1983		418	235	19.7	16.3	31.0	339.8	7.92	585.2	690.8	1,270.2	42.9
1984		404	234	20.2	16.6	32.4	388.6	8.13	649.2	804.1	1,441.2	71.2
1985		391	232	21.9	18.5	36.9	438.1	8.58	708.3	869.1	1,575.7	62.2
1986		386	222	21.0	17.6	35.4	438.5	9.10	696.9	882.3	1,574.0	60.7
1987	304	407	214	19.7	16.0	32.2	406.5	9.00	697.5	880.0	1,580.4	49.2
1988		403	212	20.1	16.3	31.8	425.2	9.81	797.7	908.7	1,702.6	49.8
1989		394	217	21.0	16.6	34.1	419.5	9.34	820.6	926.9	1,739.0	48.4
1990		386	207	20.4	16.4	32.9	437.2	10.07	885.8	974.7	1,843.9	54.9
1991		377	202	20.0	16.3	30.8	429.2	10.47	835.3	1,006.9	1,827.0	53.7
1992	313	399	208	18.0	13.9	27.6	443.4	11.20	792.2	953.1	1,743.5	61.5
1993		392	206	19.2	15.0	29.9	484.8	11.16	932.0	1,178.0	2,102.8	62.7
1994		381P	197P	19.4	15.7	32.3	496.4	11.89	984.9	1,162.9	2,129.3	85.3
1995		378P	194P	19.6P	15.4P	31.2P	500.9P	11.95P	1,000.6P	1,181.4P	2,163.2P	69.7P
1996		375P	191P	19.5P	15.3P	31.1P	512.7P	12.29P	1,033.5P	1,220.3P	2,234.5P	71.6P
1997		372P	188P	19.5P	15.2P	31.1P	524.5P	12.64P	1,066.5P	1,259.3P	2,305.7P	73.4P
1998		370P	184P	19.5P	15.1P	31.0P	536.2P	12.98P	1,099.5P	1,298.2P	2,377.0P	75.3P

Sources: 1982, 1987, 1992 *Economic Census*; *Annual Survey of Manufactures*, 83-86, 88-91, 93-94. Establishment counts for non-Census years are from *County Business Patterns*; establishment values for 83-84 are extrapolations. 'P's show projections by the editors. Industries reclassified in 87 will not have data for prior years.

INDICES OF CHANGE

Year	Companies	Establishments		Employment			Compensation		Production ($ million)			
		Total	with 20 or more employees	Total (000)	Production Workers (000)	Hours (Mil)	Payroll ($ mil)	Wages ($/hr)	Cost of Materials	Value Added by Manufacture	Value of Shipments	Capital Invest.
1982	103	108	113	99	104	100	66	68	63	63	63	52
1983		105	113	109	117	112	77	71	74	72	73	70
1984		101	113	112	119	117	88	73	82	84	83	116
1985		98	112	122	133	134	99	77	89	91	90	101
1986		97	107	117	127	128	99	81	88	93	90	99
1987	97	102	103	109	115	117	92	80	88	92	91	80
1988		101	102	112	117	115	96	88	101	95	98	81
1989		99	104	117	119	124	95	83	104	97	100	79
1990		97	100	113	118	119	99	90	112	102	106	89
1991		94	97	111	117	112	97	93	105	106	105	87
1992	100	100	100	100	100	100	100	100	100	100	100	100
1993		98	99	107	108	108	109	100	118	124	121	102
1994		95P	95P	108	113	117	112	106	124	122	122	139
1995		95P	93P	109P	111P	113P	113P	107P	126P	124P	124P	113P
1996		94P	92P	109P	110P	113P	116P	110P	130P	128P	128P	116P
1997		93P	90P	108P	109P	113P	118P	113P	135P	132P	132P	119P
1998		93P	89P	108P	109P	112P	121P	116P	139P	136P	136P	122P

Sources: Same as General Statistics. Values reflect change from the base year, 1992. Values above 100 mean greater than 92, values below 100 mean less than 92, and a value of 100 in the 82-91 or 93-98 period means same as 92. 'P's mark projections by the editors.

SELECTED RATIOS

For 1994	Avg. of All Manufact.	Analyzed Industry	Index	For 1994	Avg. of All Manufact.	Analyzed Industry	Index
Employees per Establishment	49	51	104	Value Added per Production Worker	134,084	74,070	55
Payroll per Establishment	1,500,273	1,303,613	87	Cost per Establishment	5,045,178	2,586,479	51
Payroll per Employee	30,620	25,588	84	Cost per Employee	102,970	50,768	49
Production Workers per Establishment	34	41	120	Cost per Production Worker	146,988	62,732	43
Wages per Establishment	853,319	1,008,559	118	Shipments per Establishment	9,576,895	5,591,827	58
Wages per Production Worker	24,861	24,462	98	Shipments per Employee	195,460	109,758	56
Hours per Production Worker	2,056	2,057	100	Shipments per Production Worker	279,017	135,624	49
Wages per Hour	12.09	11.89	98	Investment per Establishment	321,011	224,009	70
Value Added per Establishment	4,602,255	3,053,931	66	Investment per Employee	6,552	4,397	67
Value Added per Employee	93,930	59,943	64	Investment per Production Worker	9,352	5,433	58

Sources: Same as General Statistics. The 'Average of All Manufacturing' column represents the average of all manufacturing industries reported for the most recent complete year available. The Index shows the relationship between the Average and the Analyzed Industry. For example, 100 means that they are equal; 500 that the Analyzed Industry is five times the average; 50 means that the Analyzed Industry is half the national average. The abbreviation 'na' is used to show that data are 'not available'.

LEADING COMPANIES   Number shown: **56**   Total sales ($ mil): **1,275**   Total employment (000): **13.5**

Company Name	Address				CEO Name	Phone	Co. Type	Sales ($ mil)	Empl. (000)
Hickory Springs Mfg Co	PO Box 128	Hickory	NC	28603	PC Underdown Jr	704-328-2201	R	360•	3.8
Associated Spring	10 Main St	Bristol	CT	06010	Ali A Fadel	203-582-9581	D	230	2.0
Peterson American Corp	21200 Telegraph Rd	Southfield	MI	48034	Eric C Peterson	810-799-5400	R	90	0.8
Steadley Co	PO Box 419	Carthage	MO	64836	William J Cheney	417-358-6011	R	67•	0.7
Newcomb Spring Corp	10 Spring St	Southington	CT	06489	GD Jacobson Jr	203-621-0111	R	40	0.3
Connor Formed Metal Products	221 Main St	San Francisco	CA	94105	Robert Sloss	415-777-5277	R	32	0.2
Mid-West Spring Mfg Co	8 Greenwood Av	Romeoville	IL	60441	KE Scipta	815-838-7812	S	28•	0.4
Pathe Technologies Inc	17 Campton Rd	Irvington	NJ	07111	C Stephen Clegg	201-372-1112	P	28	0.4
Mid-West Express	8 Greenwood Av	Romeoville	IL	60441	Kenneth E Scipta	815-838-7812	D	28	0.4
Dudek & Bock Spring Mfg Co	5100 W Roosevelt	Chicago	IL	60650	W Dudek	312-379-4100	R	26	0.4
North Amer Spring & Stamping	346 Criss Cir	Elk Grove Vill	IL	60007	Mike McKee	708-437-1100	R	25	0.3
KL Spring and Stamping Inc	3323 W Addison St	Chicago	IL	60618	G Wilhite	312-583-7200	R	20	0.5
American Coil Spring Company	PO Box 388	Muskegon	MI	49443	Lee Palmer	616-726-4021	R	18	0.3
Twist Inc	PO Box 177	Jamestown	OH	45335	Joe Wright	513-675-9581	R	16•	0.2
Grand Rapids Spring & Wire	PO Box 141397	Grand Rapids	MI	49514	James Zawacki	616-453-4491	R	14	0.1
Spring Engineers Inc	PO Box 40446	Houston	TX	77240	Norman D Neeley	713-690-9488	R	13	0.1
Winamac Coil Spring Inc	PO Box 278	Kewanna	IN	46939	Joseph P Pesaresi	219-653-2186	R	13	0.2
Gilco Inc	16000 Common Rd	Roseville	MI	48066	Brian Gillum	810-779-5850	R	13	<0.1
Ark-Ell Springs Inc	PO Box 308	Houlka	MS	38850	WC Stewart Jr	601-568-3393	R	12	0.2
Mercer Spring and Wire Co	Mercer Rd	Townville	PA	16360	Richard P Krugle	814-967-2545	D	11	<0.1
Rowley Spring & Stamping Corp	210 Redstone Hill	Bristol	CT	06111	Stanley P Bitel	203-582-8175	R	11	0.2
Pa-Ted Spring Company Inc	137 Vincent P Kelly	Bristol	CT	06010	Fred Tedesco	203-582-6368	R	10	<0.1
John Evans Sons Inc	Spring Av	Lansdale	PA	19446	GD Bennett	215-368-7700	R	9	<0.1
General Spring Inc	PO Box 176	Hartsville	TN	37074	Marvin Hopper	615-374-9500	R	8	0.1
Helical Products Company Inc	PO Box 1069	Santa Maria	CA	93456	HL Merrell	805-928-3851	R	8	0.1
Monticello Spring Corp	PO Box 705	Monticello	IN	47960	Horst H Pimmler	219-583-8090	R	8•	0.1
O & G Spring & Wire	4407-31 W Divison	Chicago	IL	60651	Theodore Gregg	312-772-9331	R	8	<0.1
Rockford Spring Co	PO Box 5266	Rockford	IL	61125	John W Mink	815-397-2111	R	8•	0.1
Quincy Spring Group Inc	2652 W North Av	Chicago	IL	60647	Mark Andresik	312-342-2222	R	7•	0.1
Precision Coil Spring Co	PO Box 5450	El Monte	CA	91734	AH Goering	818-444-0561	R	7	<0.1
Atlantic Spring Company Inc	PO Box 650	Flemington	NJ	08822	Ed Reilly	908-788-5800	R	6•	<0.1
Bristol Spring Manufacturing Co	123 Whiting St	Plainville	CT	06062	BH Hittleman	203-747-5524	R	6	<0.1
Plymouth Spring Company Inc	281 Lake Av	Bristol	CT	06010	Richard Rubenstein	203-584-0594	R	6	<0.1
Trinacria Specialty Mfg	PO Box 469	Norwich	CT	06360	S Difrancesca	203-887-2856	R	6	0.1
Maryland Precision Spring	8900 Kelso Dr	Baltimore	MD	21221	Raymond Navarro	410-391-7400	S	5	<0.1
AJ Kay Co	5406 N Elston Av	Chicago	IL	60630	Robert C Schweda	312-545-5955	R	5	<0.1
Fairway Spring Company Inc	295 Hemlock St	Horseheads	NY	14845	Donald Peterson	607-739-3541	R	5	<0.1
Lee Spring Company Inc	1462 62nd St	Brooklyn	NY	11219	Peter G Giosa	718-236-2222	R	5	0.1
O'Hara Metal Products Co	185 Valley Dr	Brisbane	CA	94005	CV O'Hara	415-468-3350	R	5	<0.1
Sterling Spring Corp	5432 W 54th St	Chicago	IL	60638	Sam J Carrozza	312-582-6464	R	5	<0.1
Tru-Form Tool & Mfg Indust	14511 Anson Av	Santa Fe Sprgs	CA	90670	Vern Hildebrandt	310-802-2041	R	5	<0.1
Yost Superior Co	PO Box 1487	Springfield	OH	45501	Bert D Barnes	513-323-7591	R	5	<0.1
Hemphill Spring Co	4220 E Washington	Los Angeles	CA	90023	Betty Machlan	213-269-9276	R	4	<0.1
Ace Wire Spring and Form	1105 Thompson Av	McKees Rocks	PA	15136	Richard Froehlich	412-331-3353	R	4	<0.1
Dayon Manufacturing Company	PO Box 588	Farmington	CT	06034	Leslie R Dayon	203-677-8561	R	4	<0.1
Flex Metal Components Co	12304 McCann Dr	Santa Fe Sprgs	CA	90670	Thomas W Nickols	310-944-4444	R	4	<0.1
Paragon Spring Co	4435 W Rice St	Chicago	IL	60651	Scott C Whittle	312-489-6300	R	4	<0.1
James Spring and Wire Co	6 Bacton Hill Rd	Frazer	PA	19355	W R Krauss Jr	215-644-3450	R	4	<0.1
Holland Wire Products Inc	955 Brooks Av	Holland	MI	49423	Kyle Curtiss	616-392-8505	S	3•	<0.1
Precision Spring & Stamping Co	22617 85th Pl S	Kent	WA	98031	John T McHugh	206-852-6911	R	3	<0.1
Renton Coil Spring Co	PO Box 880	Renton	WA	98057	Charles F Pepka	206-255-1453	R	3	<0.1
Excel Spring and Stamping Inc	1080 Industrial Dr	Bensenville	IL	60106	Phillip B Matthaei	708-595-8585	R	3	<0.1
Hardware Products Co	191 Williams St	Chelsea	MA	02150	Henry Porter	617-884-9410	R	3	<0.1
Sanborn Wire Products Inc	3435 S Main St	Rock Creek	OH	44084	A David Morrow	216-563-3101	R	3	<0.1
Kokomo Spring Company Inc	329 E Firmin St	Kokomo	IN	46902	Douglas Bailey	317-459-5156	R	2•	<0.1
Duplex Inc	4080 Cheyenne Ct	Chino	CA	91710	Ralph Tollett	909-465-1011	R	1	<0.1

Source: Ward's Business Directory of U.S. Private and Public Companies, Volumes 1 and 2, 1996. The company type code used is as follows: P - Public, R - Private, S - Subsidiary, D - Division, J - Joint Venture, A - Affiliate, G - Group. Sales are in millions of dollars, employees are in thousands. An asterisk (•) indicates an estimated sales volume. The symbol < stands for 'less than'. Company names and addresses are truncated, in some cases, to fit into the available space.

MATERIALS CONSUMED

Material	Quantity	Delivered Cost ($ million)
Materials, ingredients, containers, and supplies	(X)	613.4
Metal bolts, nuts, screws, washers, rivets, and other screw machine products	(X)	14.5
Other fabricated metal products (except castings and forgings)	(X)	8.9
Castings (rough and semifinished)	(X)	(D)
Forgings	(X)	(D)
Steel bars and bar shapes	(X)	(D)
Steel sheet and strip, including tin plate	(X)	52.1
Steel wire and wire products	(X)	350.0
All other steel shapes and forms	(X)	42.7
Nonferrous shapes and forms	(X)	5.9
All other materials and components, parts, containers, and supplies	(X)	69.5
Materials, ingredients, containers, and supplies, nsk	(X)	66.2

Source: 1992 *Economic Census*. Explanation of symbols used: (D): Withheld to avoid disclosure of competitive data; na: Not available; (S): Withheld because statistical norms were not met; (X): Not applicable; (Z): Less than half the unit shown; nec: Not elsewhere classified; nsk: Not specified by kind; - : zero; * : 10-19 percent estimated; ** : 20-29 percent estimated.

PRODUCT SHARE DETAILS

Product or Product Class	% Share	Product or Product Class	% Share
Wire springs	100.00	Seat and back wire springs for motor vehicles (unassembled)	21.50
Precision mechanical wire springs	42.60	Wire spring units for box springs, innerspring mattresses, and dual-purpose sleep furniture (unassembled)	44.50
Precision mechanical compression-type wire springs, shipped to original equipment manufacturers	39.23	Wire spring units for upholstered furniture (unassembled)	4.21
Other precision mechanical compression-type wire spring shipments	11.54	Other upholstery and furniture wire springs (unassembled)	2.93
Precision mechanical extension-type wire springs	24.28	Wire valve springs	10.63
Precision mechanical torsion-type wire springs	21.48	Other wire springs	15.73
Precision mechanical springs, nsk	3.46	Other wire springs, nsk	0.49
Other wire springs	48.81	Wire springs, nsk	8.59

Source: 1992 *Economic Census*. The values shown are percent of total shipments in an industry. Values of indented subcategories are summed in the main heading. The symbol (D) appears when data are withheld to prevent disclosure of competitive information. The abbreviation nsk stands for 'not specified by kind' and nec for 'not elsewhere classified'.

INPUTS AND OUTPUTS FOR MISCELLANEOUS FABRICATED WIRE PRODUCTS

Economic Sector or Industry Providing Inputs	%	Sector	Economic Sector or Industry Buying Outputs	%	Sector
Blast furnaces & steel mills	32.6	Manufg.	Mattresses & bedsprings	5.8	Manufg.
Imports	18.5	Foreign	Logging camps & logging contractors	5.1	Manufg.
Wholesale trade	10.3	Trade	Concrete products, nec	4.0	Manufg.
Miscellaneous plastics products	9.5	Manufg.	Tires & inner tubes	3.2	Manufg.
Electric services (utilities)	1.8	Util.	Construction machinery & equipment	3.0	Manufg.
Metal coating & allied services	1.4	Manufg.	Personal consumption expenditures	2.9	
Paperboard containers & boxes	1.3	Manufg.	Exports	2.9	Foreign
Banking	1.3	Fin/R.E.	Office buildings	2.7	Constr.
Petroleum refining	1.2	Manufg.	Retail trade, except eating & drinking	2.5	Trade
Motor freight transportation & warehousing	1.2	Util.	Highway & street construction	2.4	Constr.
Miscellaneous fabricated wire products	1.1	Manufg.	Water transportation	2.3	Util.
Maintenance of nonfarm buildings nec	0.9	Constr.	Fabricated rubber products, nec	2.2	Manufg.
Copper rolling & drawing	0.9	Manufg.	Motor vehicles & car bodies	2.2	Manufg.
Machinery, except electrical, nec	0.9	Manufg.	Machinery, except electrical, nec	1.9	Manufg.
Wood products, nec	0.9	Manufg.	Oil field machinery	1.9	Manufg.
Gas production & distribution (utilities)	0.9	Util.	Upholstered household furniture	1.9	Manufg.
Eating & drinking places	0.8	Trade	Paper mills, except building paper	1.8	Manufg.
Sawmills & planning mills, general	0.7	Manufg.	Residential additions/alterations, nonfarm	1.7	Constr.
Advertising	0.7	Services	Residential 1-unit structures, nonfarm	1.6	Constr.
Fabricated rubber products, nec	0.6	Manufg.	Industrial buildings	1.4	Constr.
Metal stampings, nec	0.6	Manufg.	Electronic computing equipment	1.4	Manufg.
Air transportation	0.6	Util.	Coal	1.3	Mining
Communications, except radio & TV	0.6	Util.	Electric utility facility construction	1.3	Constr.
Railroads & related services	0.6	Util.	Motor vehicle parts & accessories	1.3	Manufg.
Real estate	0.6	Fin/R.E.	Nonfarm residential structure maintenance	1.2	Constr.
Nonferrous wire drawing & insulating	0.5	Manufg.	Construction of stores & restaurants	1.1	Constr.
Special dies & tools & machine tool accessories	0.5	Manufg.	Maintenance of farm service facilities	1.1	Constr.
Equipment rental & leasing services	0.5	Services	Farm service facilities	0.9	Constr.
Aluminum rolling & drawing	0.4	Manufg.	Household refrigerators & freezers	0.9	Manufg.
Hardware, nec	0.4	Manufg.	Metal doors, sash, & trim	0.9	Manufg.
Iron & steel foundries	0.4	Manufg.	Paperboard mills	0.9	Manufg.
Engineering, architectural, & surveying services	0.4	Services	Radio & TV communication equipment	0.9	Manufg.
U.S. Postal Service	0.4	Gov't	Feed grains	0.8	Agric.
Abrasive products	0.3	Manufg.	Construction of hospitals	0.8	Constr.
Ball & roller bearings	0.3	Manufg.	Telephone & telegraph apparatus	0.8	Manufg.
Screw machine and related products	0.3	Manufg.	Internal combustion engines, nec	0.7	Manufg.

Continued on next page.

1471

INPUTS AND OUTPUTS FOR MISCELLANEOUS FABRICATED WIRE PRODUCTS - Continued

Economic Sector or Industry Providing Inputs	%	Sector	Economic Sector or Industry Buying Outputs	%	Sector
Legal services	0.3	Services	Federal Government purchases, national defense	0.7	Fed Govt
Management & consulting services & labs	0.3	Services	Forestry products	0.6	Agric.
Lubricating oils & greases	0.2	Manufg.	Blankbooks & looseleaf binders	0.6	Manufg.
Machine tools, metal forming types	0.2	Manufg.	Job training & related services	0.6	Services
Metal heat treating	0.2	Manufg.	Hotels & motels	0.5	Constr.
Primary metal products, nec	0.2	Manufg.	Maintenance of nonbuilding facilities nec	0.5	Constr.
Welding apparatus, electric	0.2	Manufg.	Maintenance of nonfarm buildings nec	0.5	Constr.
Insurance carriers	0.2	Fin/R.E.	Blast furnaces & steel mills	0.5	Manufg.
Accounting, auditing & bookkeeping	0.2	Services	Electronic components nec	0.5	Manufg.
Automotive rental & leasing, without drivers	0.2	Services	Fabricated metal products, nec	0.5	Manufg.
Automotive repair shops & services	0.2	Services	Fabricated plate work (boiler shops)	0.5	Manufg.
Computer & data processing services	0.2	Services	Household cooking equipment	0.5	Manufg.
Hand saws & saw blades	0.1	Manufg.	Miscellaneous fabricated wire products	0.5	Manufg.
Manifold business forms	0.1	Manufg.	Railroads & related services	0.5	Util.
Sanitary services, steam supply, irrigation	0.1	Util.	Agricultural, forestry, & fishery services	0.4	Agric.
Royalties	0.1	Fin/R.E.	Construction of educational buildings	0.4	Constr.
Hotels & lodging places	0.1	Services	Residential garden apartments	0.4	Constr.
Noncomparable imports	0.1	Foreign	Sewer system facility construction	0.4	Constr.
			Warehouses	0.4	Constr.
			Miscellaneous plastics products	0.4	Manufg.
			Refrigeration & heating equipment	0.4	Manufg.
			Special industry machinery, nec	0.4	Manufg.
			Wiring devices	0.4	Manufg.
			Motor freight transportation & warehousing	0.4	Util.
			Wholesale trade	0.4	Trade
			Engineering, architectural, & surveying services	0.4	Services
			Maintenance of highways & streets	0.3	Constr.
			Nonbuilding facilities nec	0.3	Constr.
			Cold finishing of steel shapes	0.3	Manufg.
			Mineral wool	0.3	Manufg.
			Photographic equipment & supplies	0.3	Manufg.
			Prefabricated metal buildings	0.3	Manufg.
			Radio & TV receiving sets	0.3	Manufg.
			Meat animals	0.2	Agric.
			Amusement & recreation building construction	0.2	Constr.
			Construction of nonfarm buildings nec	0.2	Constr.
			Maintenance of military facilities	0.2	Constr.
			Maintenance of petroleum & natural gas wells	0.2	Constr.
			Resid. & other health facility construction	0.2	Constr.
			Residential high-rise apartments	0.2	Constr.
			Aircraft	0.2	Manufg.
			Aircraft & missile equipment, nec	0.2	Manufg.
			Aluminum rolling & drawing	0.2	Manufg.
			Ammunition, except for small arms, nec	0.2	Manufg.
			Blowers & fans	0.2	Manufg.
			Drugs	0.2	Manufg.
			Food products machinery	0.2	Manufg.
			Gaskets, packing & sealing devices	0.2	Manufg.
			Hardware, nec	0.2	Manufg.
			Lighting fixtures & equipment	0.2	Manufg.
			Machine tools, metal cutting types	0.2	Manufg.
			Mechanical measuring devices	0.2	Manufg.
			Metalworking machinery, nec	0.2	Manufg.
			Pipe, valves, & pipe fittings	0.2	Manufg.
			Semiconductors & related devices	0.2	Manufg.
			Small arms	0.2	Manufg.
			Switchgear & switchboard apparatus	0.2	Manufg.
			Turbines & turbine generator sets	0.2	Manufg.
			Wood products, nec	0.2	Manufg.
			Electric services (utilities)	0.2	Util.
			Pipelines, except natural gas	0.2	Util.
			Transit & bus transportation	0.2	Util.
			Real estate	0.2	Fin/R.E.
			Doctors & dentists	0.2	Services
			Membership organizations nec	0.2	Services
			Poultry & eggs	0.1	Agric.
			Crude petroleum & natural gas	0.1	Mining
			Construction of religious buildings	0.1	Constr.
			Garage & service station construction	0.1	Constr.
			Local transit facility construction	0.1	Constr.
			Maintenance of local transit facilities	0.1	Constr.
			Maintenance of railroads	0.1	Constr.
			Maintenance of telephone & telegraph facilities	0.1	Constr.
			Maintenance, conservation & development facilities	0.1	Constr.
			Residential 2-4 unit structures, nonfarm	0.1	Constr.

Continued on next page.

INPUTS AND OUTPUTS FOR MISCELLANEOUS FABRICATED WIRE PRODUCTS - Continued

Economic Sector or Industry Providing Inputs	%	Sector	Economic Sector or Industry Buying Outputs	%	Sector
			Automotive stampings	0.1	Manufg.
			Electric housewares & fans	0.1	Manufg.
			Environmental controls	0.1	Manufg.
			Games, toys, & children's vehicles	0.1	Manufg.
			Household laundry equipment	0.1	Manufg.
			Instruments to measure electricity	0.1	Manufg.
			Manufacturing industries, nec	0.1	Manufg.
			Metal barrels, drums, & pails	0.1	Manufg.
			Metal stampings, nec	0.1	Manufg.
			Ordnance & accessories nec	0.1	Manufg.
			Railroad equipment	0.1	Manufg.
			Ready-mixed concrete	0.1	Manufg.
			Surgical appliances & supplies	0.1	Manufg.
			Typewriters & office machines, nec	0.1	Manufg.
			Labor, civic, social, & fraternal associations	0.1	Services
			Legal services	0.1	Services
			S/L Govt. purch., higher education	0.1	S/L Govt

Source: *Benchmark Input-Output Accounts for the U.S. Economy, 1982*, U.S. Department of Commerce, Washington, D.C., July 1991. Data, as reported in the source, are organized by the 1977 SIC structure in use in 1982 but have been matched, as closely as is possible, to the 1987 SIC structure used in this book.

OCCUPATIONS EMPLOYED BY SIC 349 - MISCELLANEOUS FABRICATED METAL PRODUCTS

Occupation	% of Total 1994	Change to 2005	Occupation	% of Total 1994	Change to 2005
Assemblers, fabricators, & hand workers nec	10.7	0.7	Industrial machinery mechanics	1.8	10.7
Blue collar worker supervisors	4.5	-7.4	Helpers, laborers, & material movers nec	1.8	0.7
Machinists	4.4	0.7	Hand packers & packagers	1.7	-13.7
Welding machine setters, operators	3.8	-9.4	Machine operators nec	1.6	-11.3
Machine tool cutting & forming etc. nec	3.7	20.8	Industrial production managers	1.5	0.7
Machine forming operators, metal & plastic	3.5	-9.4	Secretaries, ex legal & medical	1.4	-8.3
Sales & related workers nec	3.0	0.7	General office clerks	1.3	-14.2
Inspectors, testers, & graders, precision	3.0	0.7	Tool & die makers	1.3	-18.7
Metal & plastic machine workers nec	2.7	-28.7	Bookkeeping, accounting, & auditing clerks	1.3	-24.5
General managers & top executives	2.6	-4.5	Coating, painting, & spraying machine workers	1.3	0.7
Combination machine tool operators	2.4	51.0	Maintenance repairers, general utility	1.2	-9.4
Machine tool cutting operators, metal & plastic	2.4	41.9	Industrial truck & tractor operators	1.1	0.6
Lathe & turning machine tool operators	2.1	-9.4	Machine feeders & offbearers	1.1	-9.4
Welders & cutters	2.1	0.7	Production, planning, & expediting clerks	1.0	0.7
Traffic, shipping, & receiving clerks	2.0	-3.1	Mechanical engineers	1.0	-0.3

Source: *Industry-Occupation Matrix*, Bureau of Labor Statistics. These data relate to one or more 3-digit SIC industry groups rather than to a single 4-digit SIC. The change reported for each occupation to the year 2005 is a percent of growth or decline as estimated by the Bureau of Labor Statistics. The abbreviation nec stands for 'not elsewhere classified'.

LOCATION BY STATE AND REGIONAL CONCENTRATION

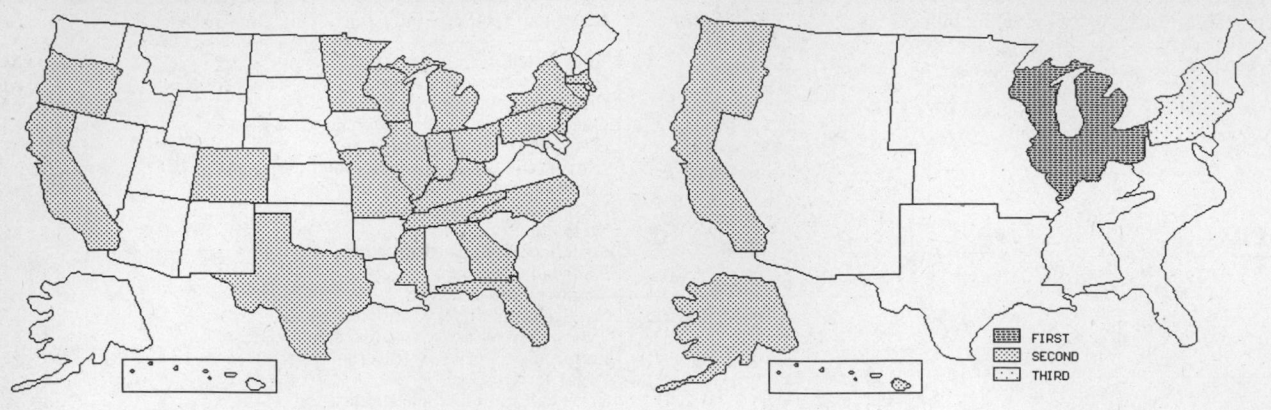

INDUSTRY DATA BY STATE

State	Establish-ments	Shipments			Employment				Cost as % of Shipments	Investment per Employee ($)
		Total ($ mil)	% of U.S.	Per Establ.	Total Number	% of U.S.	Per Establ.	Wages ($/hour)		
Illinois	51	201.1	11.5	3.9	2,200	12.2	43	11.30	44.2	3,682
Pennsylvania	21	150.0	8.6	7.1	1,200	6.7	57	15.24	40.7	3,667
Michigan	34	142.2	8.2	4.2	1,400	7.8	41	12.24	47.0	2,786
Ohio	36	133.1	7.6	3.7	1,700	9.4	47	10.38	41.2	6,176
Indiana	17	116.2	6.7	6.8	1,200	6.7	71	12.18	42.8	3,083
Tennessee	15	109.5	6.3	7.3	800	4.4	53	10.60	51.1	-
Texas	21	100.9	5.8	4.8	900	5.0	43	10.07	50.0	4,556
Kentucky	7	96.9	5.6	13.8	700	3.9	100	9.50	50.5	5,143
California	43	91.1	5.2	2.1	1,100	6.1	26	11.19	49.8	3,273
Missouri	11	86.8	5.0	7.9	1,000	5.6	91	8.53	45.3	-
Connecticut	30	75.1	4.3	2.5	1,200	6.7	40	12.35	32.8	2,083
Massachusetts	11	47.5	2.7	4.3	400	2.2	36	12.00	49.7	-
Wisconsin	7	37.0	2.1	5.3	600	3.3	86	12.67	31.1	833
New York	14	29.9	1.7	2.1	300	1.7	21	9.00	37.8	5,000
Colorado	6	24.6	1.4	4.1	200	1.1	33	10.00	48.8	-
Florida	10	22.3	1.3	2.2	200	1.1	20	10.33	45.3	-
Minnesota	7	16.0	0.9	2.3	100	0.6	14	12.50	46.3	-
Oregon	3	8.8	0.5	2.9	100	0.6	33	11.50	23.9	5,000
North Carolina	14	(D)	-	-	750 *	4.2	54	-	-	-
New Jersey	8	(D)	-	-	175 *	1.0	22	-	-	2,857
Georgia	5	(D)	-	-	375 *	2.1	75	-	-	-
Mississippi	1	(D)	-	-	750 *	4.2	750	-	-	-

Source: 1992 *Economic Census*. The states are in descending order of shipments or establishments (if shipment data are missing for the majority). The symbol (D) appears when data are withheld to prevent disclosure of competitive information. States marked with (D) are sorted by number of establishments. A dash (-) indicates that the data element cannot be calculated; * indicates the midpoint of a range.

3496 - MISC. FABRICATED WIRE PRODUCTS

82 83 84 85 86 87 88 89 90 91 92 93 94 95 96 97 98

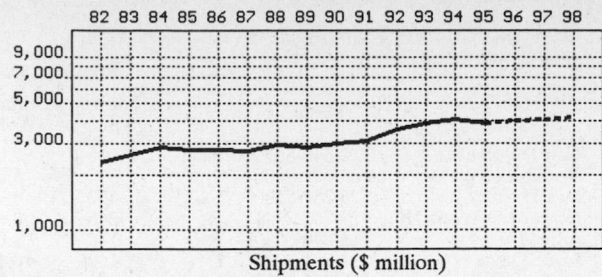

Shipments ($ million)

82 83 84 85 86 87 88 89 90 91 92 93 94 95 96 97 98

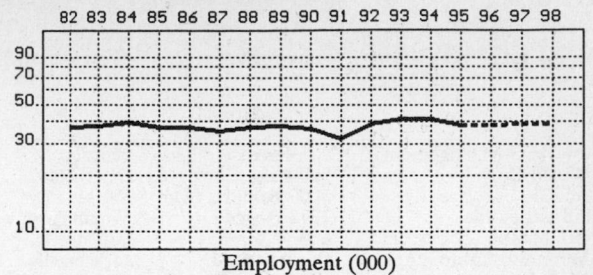

Employment (000)

GENERAL STATISTICS

Year	Companies	Establishments		Employment			Compensation		Production ($ million)			
		Total	with 20 or more employees	Total (000)	Production Workers (000)	Hours (Mil)	Payroll ($ mil)	Wages ($/hr)	Cost of Materials	Value Added by Manufacture	Value of Shipments	Capital Invest.
1982	1,108	1,181	445	36.8	27.8	53.2	580.2	7.05	1,171.7	1,164.8	2,357.3	66.7
1983		1,165	448	37.9	28.6	56.1	630.3	7.24	1,247.4	1,344.6	2,597.0	58.4
1984		1,149	451	39.7	30.4	60.0	682.9	7.28	1,364.6	1,503.5	2,837.5	72.3
1985		1,133	455	37.4	28.3	54.7	671.2	7.86	1,338.8	1,406.4	2,744.4	106.1
1986		1,113	474	37.2	28.3	55.7	678.6	7.71	1,337.1	1,447.1	2,775.5	68.7
1987	1,066	1,157	447	35.1	26.2	52.5	674.2	7.97	1,251.9	1,463.8	2,720.8	64.9
1988		1,122	450	36.9	28.0	55.2	723.2	8.14	1,422.4	1,555.1	2,962.9	56.5
1989		1,095	431	37.6	26.1	51.0	673.4	8.22	1,384.6	1,458.4	2,828.6	69.4
1990		1,059	428	36.7	24.9	50.0	681.0	8.44	1,447.3	1,552.1	2,999.7	89.7
1991		1,071	428	32.0	24.1	48.9	675.5	8.71	1,442.5	1,653.8	3,089.2	71.5
1992	1,091	1,165	448	38.8	29.0	59.6	878.8	9.04	1,701.1	1,857.5	3,553.0	102.9
1993		1,171	448	41.2	31.9	65.2	913.6	8.71	1,849.6	2,057.8	3,906.9	75.1
1994		1,106P	437P	41.1	31.6	64.0	921.3	8.92	1,960.0	2,112.6	4,063.6	100.7
1995		1,103P	436P	38.5P	28.6P	58.9P	885.8P	9.25P	1,857.2P	2,001.8P	3,850.5P	90.7P
1996		1,099P	435P	38.6P	28.7P	59.3P	909.3P	9.41P	1,913.5P	2,062.5P	3,967.2P	92.6P
1997		1,095P	433P	38.8P	28.8P	59.7P	932.7P	9.58P	1,969.8P	2,123.2P	4,083.9P	94.6P
1998		1,091P	432P	38.9P	28.9P	60.2P	956.1P	9.74P	2,026.1P	2,183.8P	4,200.6P	96.5P

Sources: 1982, 1987, 1992 *Economic Census*; *Annual Survey of Manufactures*, 83-86, 88-91, 93-94. Establishment counts for non-Census years are from *County Business Patterns*; establishment values for 83-84 are extrapolations. 'P's show projections by the editors. Industries reclassified in 87 will not have data for prior years.

INDICES OF CHANGE

Year	Companies	Establishments		Employment			Compensation		Production ($ million)			
		Total	with 20 or more employees	Total (000)	Production Workers (000)	Hours (Mil)	Payroll ($ mil)	Wages ($/hr)	Cost of Materials	Value Added by Manufacture	Value of Shipments	Capital Invest.
1982	102	101	99	95	96	89	66	78	69	63	66	65
1983		100	100	98	99	94	72	80	73	72	73	57
1984		99	101	102	105	101	78	81	80	81	80	70
1985		97	102	96	98	92	76	87	79	76	77	103
1986		96	106	96	98	93	77	85	79	78	78	67
1987	98	99	100	90	90	88	77	88	74	79	77	63
1988		96	100	95	97	93	82	90	84	84	83	55
1989		94	96	97	90	86	77	91	81	79	80	67
1990		91	96	95	86	84	77	93	85	84	84	87
1991		92	96	82	83	82	77	96	85	89	87	69
1992	100	100	100	100	100	100	100	100	100	100	100	100
1993		101	100	106	110	109	104	96	109	111	110	73
1994		95P	98P	106	109	107	105	99	115	114	114	98
1995		95P	97P	99P	99P	99P	101P	102P	109P	108P	108P	88P
1996		94P	97P	100P	99P	99P	103P	104P	112P	111P	112P	90P
1997		94P	97P	100P	99P	100P	106P	106P	116P	114P	115P	92P
1998		94P	96P	100P	100P	101P	109P	108P	119P	118P	118P	94P

Sources: Same as General Statistics. Values reflect change from the base year, 1992. Values above 100 mean greater than 92, values below 100 mean less than 92, and a value of 100 in the 82-91 or 93-98 period means same as 92. 'P's mark projections by the editors.

SELECTED RATIOS

For 1994	Avg. of All Manufact.	Analyzed Industry	Index	For 1994	Avg. of All Manufact.	Analyzed Industry	Index
Employees per Establishment	49	37	76	Value Added per Production Worker	134,084	66,854	50
Payroll per Establishment	1,500,273	832,694	56	Cost per Establishment	5,045,178	1,771,497	35
Payroll per Employee	30,620	22,416	73	Cost per Employee	102,970	47,689	46
Production Workers per Establishment	34	29	83	Cost per Production Worker	146,988	62,025	42
Wages per Establishment	853,319	515,976	60	Shipments per Establishment	9,576,895	3,672,783	38
Wages per Production Worker	24,861	18,066	73	Shipments per Employee	195,460	98,871	51
Hours per Production Worker	2,056	2,025	99	Shipments per Production Worker	279,017	128,595	46
Wages per Hour	12.09	8.92	74	Investment per Establishment	321,011	91,015	28
Value Added per Establishment	4,602,255	1,909,420	41	Investment per Employee	6,552	2,450	37
Value Added per Employee	93,930	51,401	55	Investment per Production Worker	9,352	3,187	34

Sources: Same as General Statistics. The 'Average of All Manufacturing' column represents the average of all manufacturing industries reported for the most recent complete year available. The Index shows the relationship between the Average and the Analyzed Industry. For example, 100 means that they are equal; 500 that the Analyzed Industry is five times the average; 50 means that the Analyzed Industry is half the national average. The abbreviation 'na' is used to show that data are 'not available'.

LEADING COMPANIES   Number shown: **75**   Total sales ($ mil): **5,232**   Total employment (000): **37.2**

Company Name	Address				CEO Name	Phone	Co. Type	Sales ($ mil)	Empl. (000)
Great American Management	2 N Riverside Plz	Chicago	IL	60606	Rod Dammeyer	312-648-5656	P	1,010	7.0
ACCO World Corp	500 Lake Cook Rd	Deerfield	IL	60015	Norman H Wesley	708-405-9000	S	977	5.0
GST Steel Co	7000 Roberts St	Kansas City	MO	64125	Jack D Stutz	816-242-5100	S	412	0.9
Insteel Industries Inc	1373 Boggs Dr	Mount Airy	NC	27030	Howard O Woltz III	919-786-2141	P	246	0.9
MMI Products Inc	515 W Greens	Houston	TX	77067	Julius Burns	713-876-0080	R	140	1.0
Florida Wire and Cable Inc	PO Box 6835	Jacksonville	FL	32236	Joyce Davis	904-781-8224	D	100	<0.1
Omega Wire Inc	PO Box 131	Camden	NY	13316	Rodney D Kent	315-964-2217	R	100	0.9
Axia Inc	2001 Spring Rd	Oak Brook	IL	60521	Dennis W Sheehan	708-571-3350	R	100	1.0
Master Halco Inc	110 E La Habra	La Habra	CA	90631	Barry Marrs	310-694-5064	R	91*	0.9
WS Tyler Inc	PO Box 8900	Gastonia	NC	28053	JB Cappio	704-629-2214	R	87	0.7
Nashville Wire Prod Mfg	PO Box 491	Nashville	TN	37202	David L Rollins	615-255-9874	R	71	0.7
Insteel Wire Products	PO Box 1122	Mount Airy	NC	27030	Jerry Noell	910-786-8336	S	70	0.3
Woven Electronics Corp	PO Box 189	Mauldin	SC	29662	JW Burnett III	803-963-5131	R	65	1.0
MacWhyte Co	2906 14th Av	Kenosha	WI	53140	RW Plaskett	414-654-5381	S	60	0.6
Gilbert & Bennett Mfg Co	PO Box 385	Georgetown	CT	06829	Paul Gossling	203-544-8323	R	55	0.4
Montrose/CDT	28 Sword St	Auburn	MA	01501	Joseph Merkawaz	508-791-3161	D	55*	0.2
Intern Staple and Machine Co	PO Box 629	Butler	PA	16003		412-287-7711	R	50	0.2
Koller Group	6800 W Calumet Rd	Milwaukee	WI	53223	J Muraski	414-354-6100	S	50	0.3
Metex Corp	970 New Durham	Edison	NJ	08818	Mason N Carter	908-287-0800	S	50	0.2
National Wire Product Corp	8203 Fischer Rd	Baltimore	MD	21222	Gerard A Seling	410-477-1700	S	50*	0.5
Richfield Iron Works Inc	3313 Richfield Rd	Flint	MI	48506	H W Campbell	810-736-2110	R	50	0.6
BIW Cable Systems Inc	22 Joseph E Warner	North Dighton	MA	02764	Dimitri Meistrellis	508-520-1200	S	46	0.2
Acco Chains & Lifting Prod	PO Box 792	York	PA	17405	Ron Weiskircher	717-741-4863	D	45	0.5
Bridon American Corp	PO Box 6000	Wilkes-Barre	PA	18773	William Hobbs	717-822-3349	S	45	0.4
Drives Inc	1009 1st St	Fulton	IL	61252	Ron W Den Besten	815-589-2211	R	45*	0.5
InterMetro Industries Corp	651 N Washington	Wilkes-Barre	PA	18705	R Maslow	717-825-2741	R	45*	1.0
Root Corp	152 N Main St	Mount Wolf	PA	17347	PT Root	717-266-5626	R	40	0.4
Union Wire Rope	2100 Manches	Kansas City	MO	64126	J Barclay	816-242-3191	D	40*	0.4
Semmerling Fence and Supply	700 N Wolf Rd	Wheeling	IL	60090	Larry Semmerling	708-537-3700	R	38	0.4
Woodstream Corp	69 N Locust St	Lititz	PA	17543	Harry E Whaley	717-626-2125	S	37	0.2
Peerless Chain Co	1416 E Sanborn St	Winona	MN	55987	J Van Osnabrugge	507-457-9100	R	36*	0.3
Laidlaw Corp	6625 N Scottsdale	Scottsdale	AZ	85250	JS Mueller	602-951-0003	R	35*	0.4
Hanover Wire Cloth	PO Box 473	Hanover	PA	17331	Richard Rinaldi	717-637-3795	D	34	0.3
Abbott and Co	1611 Cascade Dr	Marion	OH	43302	Roy Palmer	614-382-8212	R	32	0.8
Columbus McKinnon Corp	PO Box 110	Lexington	TN	38351	Earl L Loeswick	901-968-5271	D	30*	0.3
Flexible Material Handling	9501 Granger Rd	Cleveland	OH	44125	Donald Bachner	216-587-1575	D	30	<0.1
Kaspar Wire Works	PO Box 667	Shiner	TX	77984	DG Kaspar	512-594-3327	R	30	0.8
National Filtration Corp	PO Box 159	Star City	AR	71667	JA Pascale	501-628-4201	R	30	0.2
Western Wire Works Inc	PO Box 4382	Portland	OR	97208	Zanley F Galton	503-222-1644	R	30	0.2
Milwaukee Wire Products	4834 N 35th St	Milwaukee	WI	53209	Robert Melstrand	414-464-1350	D	28	0.2
Premier Manufacturing Corp	12117 Bennington	Cleveland	OH	44135	DC Dawson	216-941-9700	R	28	0.2
Amphenol Commercial	PO Box 4340	Hamden	CT	06514	Don Eisenberg	203-281-3200	D	27	0.1
Broderick and Bascom Rope	Oak Grove Indrial	Sedalia	MO	65301	RW Plaskett	816-827-3131	D	27	0.5
K-Lath	PO Box 489	Fontana	CA	92334	Frank Hoang	909-360-8288	D	25*	<0.1
Paulsen Wire Rope Corp	PO Box 192	Sunbury	PA	17801	FB Paulsen Jr	717-286-7141	R	25	0.3
Williamsport Wirerope Works	PO Box 3188	Williamsport	PA	17701	Larry Drummond	717-326-5146	R	25	0.3
Wayne Wire Cloth Products Inc	PO Box 550	Kalkaska	MI	49646	Edward Goossens	616-258-9187	R	22	0.3
Sommer Metalcraft Corp	PO Box 688	Crawfordsville	IN	47933	Jon W Sommer	317-362-6200	R	21	0.4
Abbott Industries Inc	95-25 149th St	Jamaica	NY	11435	LA Grossman	718-291-0800	R	20*	0.3
Adrian Fabricators Inc	412 W Beecher St	Adrian	MI	49221	GB Scully	517-263-4621	R	20	0.1
Amco Corp	901 N Kilpatrck	Chicago	IL	60651	Keith Jaffee	312-379-2100	R	20	0.2
Cablecraft	PO Box 11372	Tacoma	WA	98411	Keith Clarno	206-475-1080	D	20	0.2
Lift-All Company Inc	102 S Heintzelman	Manheim	PA	17545	Jeff Klibert	717-665-6821	R	20	0.2
Merchant Metals	PO Box 949	Statesville	NC	28687	Julius Burns	704-872-0926	D	20	0.2
Hoover Group Inc	446 Delaplain Rd	Georgetown	KY	40324	John P Kitchen	502-863-1500	D	18*	0.1
PPA Industries Inc	8222 Douglas Av	Dallas	TX	75225	David Crandall	214-373-1844	R	18	0.2
Straits Steel and Wire Co	PO Box 589	Ludington	MI	49431	Tom Curtis	616-843-3416	R	18*	0.2
Tote Cart Co	1802 Preston St	Rockford	IL	61102	Frank Mills Jr	815-963-3414	R	18	0.1
Angola Wire Products Inc	PO Box 247	Angola	IN	46703	MG Heroy	219-665-9447	R	18	0.2
Turner & Seymour Mfg Co	PO Box 358	Torrington	CT	06790	Thomas Pretak	203-489-9214	R	17	0.2
American Manufacturing Co	3600 N Hawthorne	Chattanooga	TN	37406	Everett C Warren	615-624-1191	R	16	0.2
Feldkircher Wire Fabricating	1015 W Kirkland	Nashville	TN	37216	Murray L Wakefield	615-262-0471	R	16	0.1
Precision Wire Products Inc	11215 S Wilmington	Los Angeles	CA	90059	Jon Ondrasik	213-569-8165	R	16*	0.2
Twin City Wire-MFI Inc	PO Box 21068	Eagan	MN	55121	Eugene J Lentsch	612-454-8835	R	16*	0.1
Anchor Wire Corp	425 Church St	Goodlettsville	TN	37072	Robert W Yeager	615-859-1306	S	15	0.1
Archer Wire International Corp	7300 S Narragansett	Bedford Park	IL	60638	John Svabek	708-563-1700	R	15*	0.2
Armbrust Corp	735 Allens Av	Providence	RI	02905	Howard Armburst	401-781-3300	R	15*	0.2
Cambridge Inc	105 Goodwill Rd	Cambridge	MD	21613	Theodore Dragich	410-228-3000	R	15	0.3
EH Baare Corp	500 Monroe St	Robinson	IL	62454	Harry Wong Jr	618-546-1575	S	15*	0.2
Spencer Products Inc	64 Main St	Spencer	MA	01562	George Krikorkian	508-885-3963	R	15*	0.2
Torpedo Wire and Strip Co	RD 2	Pittsfield	PA	16340	JB Lopez	814-563-7505	R	15	0.1
James Burn/American Inc	205 Cottage St	Poughkeepsie	NY	12602	N Comerford	914-454-8200	D	14	0.1
Bedford Industries Inc	PO Box 39	Worthington	MN	56187	Robert B Ludlow	507-376-4136	R	13	0.2
Nashville Wire Prod Mfg	PO Box 491	Nashville	TN	37210	ET White	615-255-6331	D	13	0.2
ALP Industries Inc	1229 W Lincoln	Coatesville	PA	19320	Reitzel Swaim	215-384-1300	R	12	0.1

Source: Ward's Business Directory of U.S. Private and Public Companies, Volumes 1 and 2, 1996. The company type code used is as follows: P - Public, R - Private, S - Subsidiary, D - Division, J - Joint Venture, A - Affiliate, G - Group. Sales are in millions of dollars, employees are in thousands. An asterisk (*) indicates an estimated sales volume. The symbol < stands for 'less than'. Company names and addresses are truncated, in some cases, to fit into the available space.

MATERIALS CONSUMED

Material	Quantity	Delivered Cost ($ million)
Materials, ingredients, containers, and supplies	(X)	1,475.3
Metal bolts, nuts, screws, washers, rivets, and other screw machine products	(X)	13.8
Other fabricated metal products (except castings and forgings)	(X)	44.6
Forgings	(X)	5.9
Castings (rough and semifinished)	(X)	5.9
Steel bars and bar shapes	(X)	39.3
Steel sheet and strip, including tin plate	(X)	61.8
Steel wire and wire products	(X)	689.1
All other steel shapes and forms	(X)	116.1
Aluminum and aluminum-base alloy shapes and forms	(X)	20.8
Other nonferrous shapes and forms	(X)	19.9
Plastics products consumed in the form of sheets, rods, tubes, film, and other shapes	(X)	24.2
All other materials and components, parts, containers, and supplies	(X)	251.7
Materials, ingredients, containers, and supplies, nsk	(X)	182.3

Source: 1992 *Economic Census*. Explanation of symbols used: (D): Withheld to avoid disclosure of competitive data; na: Not available; (S): Withheld because statistical norms were not met; (X): Not applicable; (Z): Less than half the unit shown; nec: Not elsewhere classified; nsk: Not specified by kind; - : zero; * : 10-19 percent estimated; ** : 20-29 percent estimated.

PRODUCT SHARE DETAILS

Product or Product Class	% Share	Product or Product Class	% Share
Miscellaneous fabricated wire products	100.00	fittings, not made in wiredrawing plants	7.84
Noninsulated ferrous wire rope, cable, and strand, not made in wiredrawing plants	12.92	Fencing and fence gates (not made in wiredrawing plants), nsk	2.21
Noninsulated ferrous wire rope and cable made from steel wire, excluding fabricated wire rope assemblies, not made in wiredrawing plants	21.28	Other fabricated ferrous wire products (except springs) not made in wiredrawing plants	67.02
Noninsulated fabricated ferrous wire rope assemblies, including lifting slings, not made in wiredrawing plants	21.66	Ferrous wire chain, including tire chain, stud-link chain, and welded link not made in wiredrawing plants	4.60
Noninsulated steel wire strand, including strand for prestressed concrete, composite wire strand, except ACSR, and guard rail cable, not made in wiredrawing plants	16.12	Barbed and twisted steel wire not made in wiredrawing plants	0.77
		Ferrous wire bale ties not made in wiredrawing plants	1.22
Other noninsulated ferrous wire rope, cable, and strand, including composite rope and cable and wire forms, not made in wiredrawing plants	40.95	Welded steel wire fabrics, including concrete reinforcing mesh, not made in wiredrawing plants	6.06
Ferrous wire cloth and other ferrous woven wire products not made in wiredrawing plants	5.18	Ferrous wire garment hangers not made in wiredrawing plants	4.00
Nonferrous wire cloth and other nonferrous woven wire products not made in nonferrous wiredrawing plants	4.21	Ferrous wire carts, including household, grocery, and industrial carts, not made in wiredrawing plants	5.66
		Steel wire cages not made in wiredrawing plants	4.64
Fabricated wire fencing and fence gates not made in wiredrawing plants	4.05	Ferrous wire baskets not made in wiredrawing plants	5.45
Chain link wire fencing, excluding posts, gates, and fittings (including galvanized and plastics coated wire), not made in wiredrawing plants	33.64	Ferrous wire shelving, including oven, refrigerator, closet, and barbeque grills, not made in wiredrawing plants	17.05
Woven and welded wire fencing, including galvanized and plastics coated, not made in wiredrawing plants	6.49	Ferrous wire racks, including shoe, bottle, display, and point of purchase, not made in wiredrawing plants	10.40
Fabricated wire fence gates, posts, and fittings, not made in wiredrawing plants	49.75	Other fabricated ferrous wire products, including guards, florists' designs, and paper clips, not made in wiredrawing plants	38.95
Ornamental wire lawn fence, excluding posts, gates, and		Other fabricated ferrous wire products (except springs) not made in wiredrawing plants), nsk	1.20
		Miscellaneous fabricated wire products, nsk	6.62

Source: 1992 *Economic Census*. The values shown are percent of total shipments in an industry. Values of indented subcategories are summed in the main heading. The symbol (D) appears when data are withheld to prevent disclosure of competitive information. The abbreviation nsk stands for 'not specified by kind' and nec for 'not elsewhere classified'.

INPUTS AND OUTPUTS FOR MISCELLANEOUS FABRICATED WIRE PRODUCTS

Economic Sector or Industry Providing Inputs	%	Sector	Economic Sector or Industry Buying Outputs	%	Sector
Blast furnaces & steel mills	32.6	Manufg.	Mattresses & bedsprings	5.8	Manufg.
Imports	18.5	Foreign	Logging camps & logging contractors	5.1	Manufg.
Wholesale trade	10.3	Trade	Concrete products, nec	4.0	Manufg.
Miscellaneous plastics products	9.5	Manufg.	Tires & inner tubes	3.2	Manufg.
Electric services (utilities)	1.8	Util.	Construction machinery & equipment	3.0	Manufg.
Metal coating & allied services	1.4	Manufg.	Personal consumption expenditures	2.9	
Paperboard containers & boxes	1.3	Manufg.	Exports	2.9	Foreign
Banking	1.3	Fin/R.E.	Office buildings	2.7	Constr.
Petroleum refining	1.2	Manufg.	Retail trade, except eating & drinking	2.5	Trade
Motor freight transportation & warehousing	1.2	Util.	Highway & street construction	2.4	Constr.
Miscellaneous fabricated wire products	1.1	Manufg.	Water transportation	2.3	Util.
Maintenance of nonfarm buildings nec	0.9	Constr.	Fabricated rubber products, nec	2.2	Manufg.
Copper rolling & drawing	0.9	Manufg.	Motor vehicles & car bodies	2.2	Manufg.
Machinery, except electrical, nec	0.9	Manufg.	Machinery, except electrical, nec	1.9	Manufg.
Wood products, nec	0.9	Manufg.	Oil field machinery	1.9	Manufg.

Continued on next page.

INPUTS AND OUTPUTS FOR MISCELLANEOUS FABRICATED WIRE PRODUCTS - Continued

Economic Sector or Industry Providing Inputs	%	Sector	Economic Sector or Industry Buying Outputs	%	Sector
Gas production & distribution (utilities)	0.9	Util.	Upholstered household furniture	1.9	Manufg.
Eating & drinking places	0.8	Trade	Paper mills, except building paper	1.8	Manufg.
Sawmills & planning mills, general	0.7	Manufg.	Residential additions/alterations, nonfarm	1.7	Constr.
Advertising	0.7	Services	Residential 1-unit structures, nonfarm	1.6	Constr.
Fabricated rubber products, nec	0.6	Manufg.	Industrial buildings	1.4	Constr.
Metal stampings, nec	0.6	Manufg.	Electronic computing equipment	1.4	Manufg.
Air transportation	0.6	Util.	Coal	1.3	Mining
Communications, except radio & TV	0.6	Util.	Electric utility facility construction	1.3	Constr.
Railroads & related services	0.6	Util.	Motor vehicle parts & accessories	1.3	Manufg.
Real estate	0.6	Fin/R.E.	Nonfarm residential structure maintenance	1.2	Constr.
Nonferrous wire drawing & insulating	0.5	Manufg.	Construction of stores & restaurants	1.1	Constr.
Special dies & tools & machine tool accessories	0.5	Manufg.	Maintenance of farm service facilities	1.1	Constr.
Equipment rental & leasing services	0.5	Services	Farm service facilities	0.9	Constr.
Aluminum rolling & drawing	0.4	Manufg.	Household refrigerators & freezers	0.9	Manufg.
Hardware, nec	0.4	Manufg.	Metal doors, sash, & trim	0.9	Manufg.
Iron & steel foundries	0.4	Manufg.	Paperboard mills	0.9	Manufg.
Engineering, architectural, & surveying services	0.4	Services	Radio & TV communication equipment	0.9	Manufg.
U.S. Postal Service	0.4	Gov't	Feed grains	0.8	Agric.
Abrasive products	0.3	Manufg.	Construction of hospitals	0.8	Constr.
Ball & roller bearings	0.3	Manufg.	Telephone & telegraph apparatus	0.8	Manufg.
Screw machine and related products	0.3	Manufg.	Internal combustion engines, nec	0.7	Manufg.
Legal services	0.3	Services	Federal Government purchases, national defense	0.7	Fed Govt
Management & consulting services & labs	0.3	Services	Forestry products	0.6	Agric.
Lubricating oils & greases	0.2	Manufg.	Blankbooks & looseleaf binders	0.6	Manufg.
Machine tools, metal forming types	0.2	Manufg.	Job training & related services	0.6	Services
Metal heat treating	0.2	Manufg.	Hotels & motels	0.5	Constr.
Primary metal products, nec	0.2	Manufg.	Maintenance of nonbuilding facilities nec	0.5	Constr.
Welding apparatus, electric	0.2	Manufg.	Maintenance of nonfarm buildings nec	0.5	Constr.
Insurance carriers	0.2	Fin/R.E.	Blast furnaces & steel mills	0.5	Manufg.
Accounting, auditing & bookkeeping	0.2	Services	Electronic components nec	0.5	Manufg.
Automotive rental & leasing, without drivers	0.2	Services	Fabricated metal products, nec	0.5	Manufg.
Automotive repair shops & services	0.2	Services	Fabricated plate work (boiler shops)	0.5	Manufg.
Computer & data processing services	0.2	Services	Household cooking equipment	0.5	Manufg.
Hand saws & saw blades	0.1	Manufg.	Miscellaneous fabricated wire products	0.5	Manufg.
Manifold business forms	0.1	Manufg.	Railroads & related services	0.5	Util.
Sanitary services, steam supply, irrigation	0.1	Util.	Agricultural, forestry, & fishery services	0.4	Agric.
Royalties	0.1	Fin/R.E.	Construction of educational buildings	0.4	Constr.
Hotels & lodging places	0.1	Services	Residential garden apartments	0.4	Constr.
Noncomparable imports	0.1	Foreign	Sewer system facility construction	0.4	Constr.
			Warehouses	0.4	Constr.
			Miscellaneous plastics products	0.4	Manufg.
			Refrigeration & heating equipment	0.4	Manufg.
			Special industry machinery, nec	0.4	Manufg.
			Wiring devices	0.4	Manufg.
			Motor freight transportation & warehousing	0.4	Util.
			Wholesale trade	0.4	Trade
			Engineering, architectural, & surveying services	0.4	Services
			Maintenance of highways & streets	0.3	Constr.
			Nonbuilding facilities nec	0.3	Constr.
			Cold finishing of steel shapes	0.3	Manufg.
			Mineral wool	0.3	Manufg.
			Photographic equipment & supplies	0.3	Manufg.
			Prefabricated metal buildings	0.3	Manufg.
			Radio & TV receiving sets	0.3	Manufg.
			Meat animals	0.2	Agric.
			Amusement & recreation building construction	0.2	Constr.
			Construction of nonfarm buildings nec	0.2	Constr.
			Maintenance of military facilities	0.2	Constr.
			Maintenance of petroleum & natural gas wells	0.2	Constr.
			Resid. & other health facility construction	0.2	Constr.
			Residential high-rise apartments	0.2	Constr.
			Aircraft	0.2	Manufg.
			Aircraft & missile equipment, nec	0.2	Manufg.
			Aluminum rolling & drawing	0.2	Manufg.
			Ammunition, except for small arms, nec	0.2	Manufg.
			Blowers & fans	0.2	Manufg.
			Drugs	0.2	Manufg.
			Food products machinery	0.2	Manufg.
			Gaskets, packing & sealing devices	0.2	Manufg.
			Hardware, nec	0.2	Manufg.
			Lighting fixtures & equipment	0.2	Manufg.
			Machine tools, metal cutting types	0.2	Manufg.
			Mechanical measuring devices	0.2	Manufg.
			Metalworking machinery, nec	0.2	Manufg.
			Pipe, valves, & pipe fittings	0.2	Manufg.

Continued on next page.

INPUTS AND OUTPUTS FOR MISCELLANEOUS FABRICATED WIRE PRODUCTS - Continued

Economic Sector or Industry Providing Inputs	%	Sector	Economic Sector or Industry Buying Outputs	%	Sector
			Semiconductors & related devices	0.2	Manufg.
			Small arms	0.2	Manufg.
			Switchgear & switchboard apparatus	0.2	Manufg.
			Turbines & turbine generator sets	0.2	Manufg.
			Wood products, nec	0.2	Manufg.
			Electric services (utilities)	0.2	Util.
			Pipelines, except natural gas	0.2	Util.
			Transit & bus transportation	0.2	Util.
			Real estate	0.2	Fin/R.E.
			Doctors & dentists	0.2	Services
			Membership organizations nec	0.2	Services
			Poultry & eggs	0.1	Agric.
			Crude petroleum & natural gas	0.1	Mining
			Construction of religious buildings	0.1	Constr.
			Garage & service station construction	0.1	Constr.
			Local transit facility construction	0.1	Constr.
			Maintenance of local transit facilities	0.1	Constr.
			Maintenance of railroads	0.1	Constr.
			Maintenance of telephone & telegraph facilities	0.1	Constr.
			Maintenance, conservation & development facilities	0.1	Constr.
			Residential 2-4 unit structures, nonfarm	0.1	Constr.
			Automotive stampings	0.1	Manufg.
			Electric housewares & fans	0.1	Manufg.
			Environmental controls	0.1	Manufg.
			Games, toys, & children's vehicles	0.1	Manufg.
			Household laundry equipment	0.1	Manufg.
			Instruments to measure electricity	0.1	Manufg.
			Manufacturing industries, nec	0.1	Manufg.
			Metal barrels, drums, & pails	0.1	Manufg.
			Metal stampings, nec	0.1	Manufg.
			Ordnance & accessories nec	0.1	Manufg.
			Railroad equipment	0.1	Manufg.
			Ready-mixed concrete	0.1	Manufg.
			Surgical appliances & supplies	0.1	Manufg.
			Typewriters & office machines, nec	0.1	Manufg.
			Labor, civic, social, & fraternal associations	0.1	Services
			Legal services	0.1	Services
			S/L Govt. purch., higher education	0.1	S/L Govt

Source: Benchmark Input-Output Accounts for the U.S. Economy, 1982, U.S. Department of Commerce, Washington, D.C., July 1991. Data, as reported in the source, are organized by the 1977 SIC structure in use in 1982 but have been matched, as closely as is possible, to the 1987 SIC structure used in this book.

OCCUPATIONS EMPLOYED BY SIC 349 - MISCELLANEOUS FABRICATED METAL PRODUCTS

Occupation	% of Total 1994	Change to 2005	Occupation	% of Total 1994	Change to 2005
Assemblers, fabricators, & hand workers nec	10.7	0.7	Industrial machinery mechanics	1.8	10.7
Blue collar worker supervisors	4.5	-7.4	Helpers, laborers, & material movers nec	1.8	0.7
Machinists	4.4	0.7	Hand packers & packagers	1.7	-13.7
Welding machine setters, operators	3.8	-9.4	Machine operators nec	1.6	-11.3
Machine tool cutting & forming etc. nec	3.7	20.8	Industrial production managers	1.5	0.7
Machine forming operators, metal & plastic	3.5	-9.4	Secretaries, ex legal & medical	1.4	-8.3
Sales & related workers nec	3.0	0.7	General office clerks	1.3	-14.2
Inspectors, testers, & graders, precision	3.0	0.7	Tool & die makers	1.3	-18.7
Metal & plastic machine workers nec	2.7	-28.7	Bookkeeping, accounting, & auditing clerks	1.3	-24.5
General managers & top executives	2.6	-4.5	Coating, painting, & spraying machine workers	1.3	0.7
Combination machine tool operators	2.4	51.0	Maintenance repairers, general utility	1.2	-9.4
Machine tool cutting operators, metal & plastic	2.4	41.9	Industrial truck & tractor operators	1.1	0.6
Lathe & turning machine tool operators	2.1	-9.4	Machine feeders & offbearers	1.1	-9.4
Welders & cutters	2.1	0.7	Production, planning, & expediting clerks	1.0	0.7
Traffic, shipping, & receiving clerks	2.0	-3.1	Mechanical engineers	1.0	-0.3

Source: Industry-Occupation Matrix, Bureau of Labor Statistics. These data relate to one or more 3-digit SIC industry groups rather than to a single 4-digit SIC. The change reported for each occupation to the year 2005 is a percent of growth or decline as estimated by the Bureau of Labor Statistics. The abbreviation nec stands for 'not elsewhere classified'.

LOCATION BY STATE AND REGIONAL CONCENTRATION

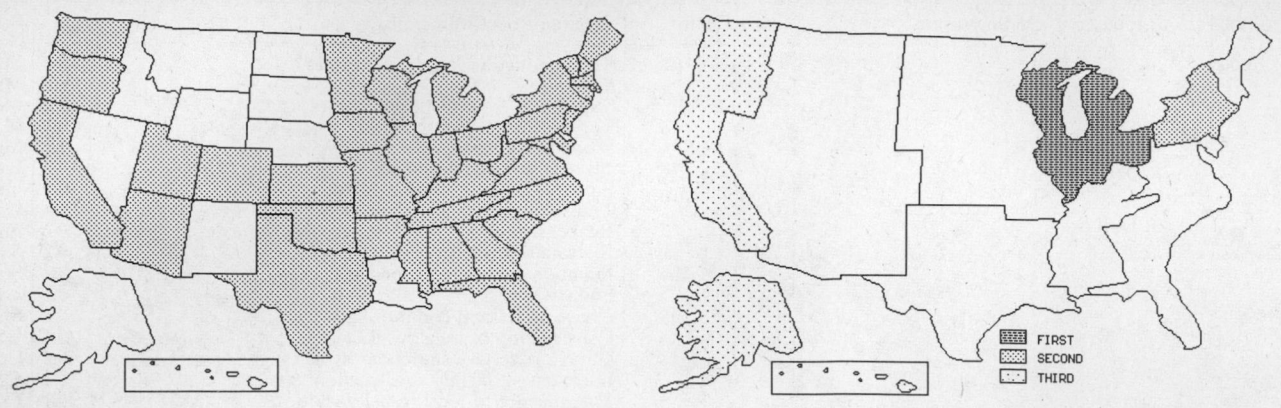

FIRST
SECOND
THIRD

INDUSTRY DATA BY STATE

State	Establish-ments	Shipments			Employment				Cost as % of Shipments	Investment per Employee ($)
		Total ($ mil)	% of U.S.	Per Establ.	Total Number	% of U.S.	Per Establ.	Wages ($/hour)		
Pennsylvania	71	331.2	9.3	4.7	2,700	7.0	38	10.61	47.9	3,556
Illinois	108	321.0	9.0	3.0	3,500	9.0	32	9.21	44.0	2,171
California	125	242.2	6.8	1.9	2,500	6.4	20	9.08	49.0	1,720
Missouri	25	229.9	6.5	9.2	3,100	8.0	124	7.41	42.1	3,581
Texas	70	198.0	5.6	2.8	2,500	6.4	36	8.49	54.5	4,400
Florida	35	174.4	4.9	5.0	1,300	3.4	37	8.53	54.4	2,846
Ohio	68	172.5	4.9	2.5	2,200	5.7	32	8.82	47.2	2,500
New Jersey	56	164.1	4.6	2.9	1,600	4.1	29	10.39	48.8	1,813
Tennessee	28	163.6	4.6	5.8	1,700	4.4	61	7.79	47.7	1,941
Indiana	37	141.4	4.0	3.8	2,000	5.2	54	9.48	48.5	6,000
Michigan	50	126.8	3.6	2.5	1,400	3.6	28	9.05	46.9	1,286
New York	69	114.8	3.2	1.7	1,700	4.4	25	9.25	45.7	1,765
North Carolina	26	97.6	2.7	3.8	1,200	3.1	46	8.40	50.4	2,083
Minnesota	25	95.0	2.7	3.8	800	2.1	32	12.36	43.4	1,625
Alabama	18	79.0	2.2	4.4	600	1.5	33	10.78	60.6	1,667
Maryland	14	74.9	2.1	5.3	800	2.1	57	11.18	51.5	2,500
Massachusetts	27	73.5	2.1	2.7	700	1.8	26	8.27	52.2	2,429
South Carolina	16	72.2	2.0	4.5	500	1.3	31	8.63	49.6	6,000
Arkansas	12	65.6	1.8	5.5	1,000	2.6	83	7.39	51.4	3,200
Wisconsin	30	63.7	1.8	2.1	1,000	2.6	33	7.93	40.8	1,300
Oklahoma	12	63.1	1.8	5.3	800	2.1	67	7.07	49.1	-
Connecticut	34	62.9	1.8	1.9	600	1.5	18	10.78	44.4	1,333
Kentucky	14	57.5	1.6	4.1	700	1.8	50	10.00	41.9	1,857
Georgia	21	55.0	1.5	2.6	500	1.3	24	10.83	51.1	3,000
Oregon	23	49.6	1.4	2.2	500	1.3	22	12.86	42.9	1,400
Iowa	13	35.2	1.0	2.7	400	1.0	31	10.57	41.2	-
Virginia	9	31.3	0.9	3.5	400	1.0	44	5.13	31.3	-
Kansas	13	24.9	0.7	1.9	300	0.8	23	9.00	55.4	1,000
West Virginia	4	24.1	0.7	6.0	300	0.8	75	8.20	60.6	2,333
Rhode Island	13	20.6	0.6	1.6	300	0.8	23	8.00	35.9	667
Maine	13	18.5	0.5	1.4	100	0.3	8	7.50	67.0	1,000
Washington	15	17.7	0.5	1.2	200	0.5	13	16.00	43.5	-
New Hampshire	8	14.7	0.4	1.8	200	0.5	25	10.50	35.4	1,000
Colorado	12	13.3	0.4	1.1	100	0.3	8	16.00	49.6	3,000
Arizona	11	8.9	0.3	0.8	100	0.3	9	10.00	60.7	1,000
Mississippi	6	(D)	-	-	175 *	0.5	29	-	-	571
Utah	4	(D)	-	-	175 *	0.5	44	-	-	-
Vermont	3	(D)	-	-	175 *	0.5	58	-	-	-

Source: 1992 *Economic Census*. The states are in descending order of shipments or establishments (if shipment data are missing for the majority). The symbol (D) appears when data are withheld to prevent disclosure of competitive information. States marked with (D) are sorted by number of establishments. A dash (-) indicates that the data element cannot be calculated; * indicates the midpoint of a range.

3497 - METAL FOIL & LEAF

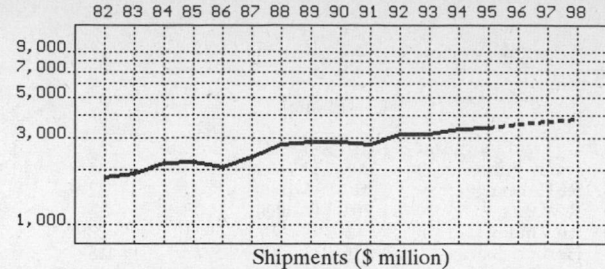

Shipments ($ million)

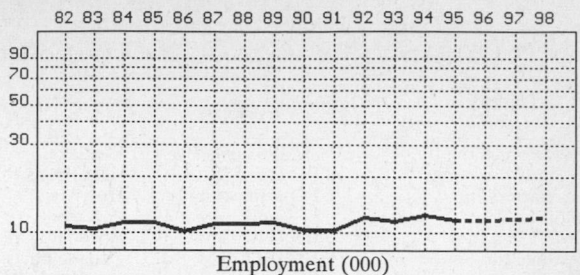

Employment (000)

GENERAL STATISTICS

Year	Com- panies	Establishments		Employment			Compensation		Production ($ million)			
		Total	with 20 or more employees	Total (000)	Production Workers (000)	Hours (Mil)	Payroll ($ mil)	Wages ($/hr)	Cost of Materials	Value Added by Manufacture	Value of Shipments	Capital Invest.
1982	79	96	67	10.8	8.0	15.9	241.6	10.77	1,156.7	660.6	1,831.3	40.1
1983		94	66	10.5	7.7	15.5	249.7	11.21	1,228.8	702.3	1,907.5	41.9
1984		92	65	11.4	8.5	17.4	287.3	11.16	1,445.2	762.8	2,191.1	58.4
1985		89	63	11.3	8.3	17.0	286.5	11.20	1,381.0	819.3	2,198.5	65.4
1986		93	63	10.1	7.4	15.5	270.8	11.63	1,279.4	764.8	2,056.9	50.9
1987	98	118	67	11.1	7.7	16.5	326.1	12.64	1,461.4	892.7	2,342.8	89.5
1988		116	65	11.1	8.2	17.4	332.5	13.05	1,828.4	971.9	2,739.0	95.2
1989		117	67	11.4	8.0	17.9	339.8	13.01	1,875.0	954.2	2,844.5	81.9
1990		115	68	10.3	7.9	18.3	354.0	13.35	1,916.6	938.4	2,845.8	95.5
1991		114	68	10.3	7.7	17.5	353.9	13.79	1,780.6	962.6	2,741.5	99.6
1992	104	121	73	12.0	8.7	18.5	405.9	14.25	1,839.9	1,274.8	3,118.5	90.6
1993		119	75	11.5	8.3	18.1	408.9	14.56	1,823.2	1,313.2	3,152.5	98.0
1994		126P	72P	12.4	9.3	19.8	451.9	15.17	2,034.9	1,330.9	3,336.1	85.2
1995		129P	73P	11.7P	8.6P	19.2P	443.6P	15.33P	2,094.6P	1,370.0P	3,434.0P	109.6P
1996		132P	73P	11.7P	8.6P	19.5P	459.6P	15.70P	2,170.6P	1,419.7P	3,558.6P	114.3P
1997		135P	74P	11.8P	8.7P	19.8P	475.6P	16.07P	2,246.6P	1,469.3P	3,683.1P	119.1P
1998		138P	75P	11.9P	8.7P	20.0P	491.6P	16.44P	2,322.6P	1,519.0P	3,807.7P	123.8P

Sources: 1982, 1987, 1992 *Economic Census*; *Annual Survey of Manufactures*, 83-86, 88-91, 93-94. Establishment counts for non-Census years are from *County Business Patterns*; establishment values for 83-84 are extrapolations. 'P's show projections by the editors. Industries reclassified in 87 will not have data for prior years.

INDICES OF CHANGE

Year	Com- panies	Establishments		Employment			Compensation		Production ($ million)			
		Total	with 20 or more employees	Total (000)	Production Workers (000)	Hours (Mil)	Payroll ($ mil)	Wages ($/hr)	Cost of Materials	Value Added by Manufacture	Value of Shipments	Capital Invest.
1982	76	79	92	90	92	86	60	76	63	52	59	44
1983		78	90	88	89	84	62	79	67	55	61	46
1984		76	89	95	98	94	71	78	79	60	70	64
1985		74	86	94	95	92	71	79	75	64	70	72
1986		77	86	84	85	84	67	82	70	60	66	56
1987	94	98	92	93	89	89	80	89	79	70	75	99
1988		96	89	93	94	94	82	92	99	76	88	105
1989		97	92	95	92	97	84	91	102	75	91	90
1990		95	93	86	91	99	87	94	104	74	91	105
1991		94	93	86	89	95	87	97	97	76	88	110
1992	100	100	100	100	100	100	100	100	100	100	100	100
1993		98	103	96	95	98	101	102	99	103	101	108
1994		104P	99P	103	107	107	111	106	111	104	107	94
1995		107P	100P	97P	98P	104P	109P	108P	114P	107P	110P	121P
1996		109P	101P	98P	99P	105P	113P	110P	118P	111P	114P	126P
1997		112P	102P	98P	100P	107P	117P	113P	122P	115P	118P	131P
1998		114P	103P	99P	100P	108P	121P	115P	126P	119P	122P	137P

Sources: Same as General Statistics. Values reflect change from the base year, 1992. Values above 100 mean greater than 92, values below 100 mean less than 92, and a value of 100 in the 82-91 or 93-98 period means same as 92. 'P's mark projections by the editors.

SELECTED RATIOS

For 1994	Avg. of All Manufact.	Analyzed Industry	Index	For 1994	Avg. of All Manufact.	Analyzed Industry	Index
Employees per Establishment	49	98	200	Value Added per Production Worker	134,084	143,108	107
Payroll per Establishment	1,500,273	3,577,474	238	Cost per Establishment	5,045,178	16,109,320	319
Payroll per Employee	30,620	36,444	119	Cost per Employee	102,970	164,105	159
Production Workers per Establishment	34	74	214	Cost per Production Worker	146,988	218,806	149
Wages per Establishment	853,319	2,377,852	279	Shipments per Establishment	9,576,895	26,410,291	276
Wages per Production Worker	24,861	32,297	130	Shipments per Employee	195,460	269,040	138
Hours per Production Worker	2,056	2,129	104	Shipments per Production Worker	279,017	358,720	129
Wages per Hour	12.09	15.17	125	Investment per Establishment	321,011	674,487	210
Value Added per Establishment	4,602,255	10,536,092	229	Investment per Employee	6,552	6,871	105
Value Added per Employee	93,930	107,331	114	Investment per Production Worker	9,352	9,161	98

Sources: Same as General Statistics. The 'Average of All Manufacturing' column represents the average of all manufacturing industries reported for the most recent complete year available. The Index shows the relationship between the Average and the Analyzed Industry. For example, 100 means that they are equal; 500 that the Analyzed Industry is five times the average; 50 means that the Analyzed Industry is half the national average. The abbreviation 'na' is used to show that data are 'not available'.

LEADING COMPANIES   Number shown: 13   Total sales ($ mil): 469   Total employment (000): 4.4

Company Name	Address				CEO Name	Phone	Co. Type	Sales ($ mil)	Empl. (000)
Gould Electronics Inc	35129 Curtis Blv	Eastlake	OH	44095	C David Ferguson	216-953-0525	S	280	3.0
Transfer Print Foils Inc	PO Box 538	East Brunswick	NJ	08816	Harry Parker	908-238-1800	R	35*	0.1
Hampden Papers Inc	PO Drawer 149	Holyoke	MA	01041	RK Fowler	413-536-1000	R	32	0.2
Crown Roll Leaf Inc	91 Illinois Av	Paterson	NJ	07503	R Waitts	201-742-4000	R	26	0.2
Circuit Foil USA Inc	US Hwy 130	Bordentown	NJ	08505	Alexander Cadron	609-298-4800	J	23*	0.2
Kurz-Hastings Inc	10901 Dutton Rd	Philadelphia	PA	19154	Herbert Kurz	215-632-2300	R	15	0.2
Oak-Mitsui Inc	PO Box 99	Hoosick Falls	NY	12090	Derek Carbin	518-686-4961	S	15*	0.2
M Swift and Sons Inc	10 Love Ln	Hartford	CT	06141	MA Swift	203-522-1181	R	10	0.1
Quick Roll Leaf Mfg Co	RD 2	Middletown	NY	10940	E Quick Jr	914-692-2500	R	8	<0.1
Sentinel Bag and Paper Company	219 36th St	Brooklyn	NY	11232	Peter M Nagler	718-788-2211	R	8*	<0.1
EKCO Products Inc	2100 Sanders Rd	Northbrook	IL	60065	Richard Wanbold	708-205-2300	D	7*	0.2
Ribbon Technology Corp	825 Taylor Station	Blacklick	OH	43004	LE Hackman	614-864-5444	R	6	<0.1
Delker Corporation Inc	PO Box 427	Branford	CT	06405	Fred R Weber	203-481-4277	R	4*	<0.1

Source: Ward's Business Directory of U.S. Private and Public Companies, Volumes 1 and 2, 1996. The company type code used is as follows: P - Public, R - Private, S - Subsidiary, D - Division, J - Joint Venture, A - Affiliate, G - Group. Sales are in millions of dollars, employees are in thousands. An asterisk (*) indicates an estimated sales volume. The symbol < stands for 'less than'. Company names and addresses are truncated, in some cases, to fit into the available space.

MATERIALS CONSUMED

Material		Quantity	Delivered Cost ($ million)
Materials, ingredients, containers, and supplies		(X)	1,672.9
Aluminum foil, converted (quantity represents metal content)	mil lb	386.8**	396.9
Other fabricated metal products (except forgings)	mil lb	(D)	(D)
Castings (rough and semifinished)		(X)	(D)
Aluminum and aluminum-base alloy sheet and plate	mil lb	132.2	106.4
Aluminum and aluminum-base alloy plain foil	mil lb	122.9	216.8
Other aluminum and aluminum-base alloy shapes and forms	mil lb	6.2	19.8
Copper and copper-base alloy shapes and forms	mil lb	(D)	(D)
Other nonferrous shapes and forms	mil lb	(D)	(D)
Glues and adhesives, including synthetic resin adhesives	mil lb	(S)	38.2
Printing ink	mil lb	28.5**	51.6
Other chemicals and allied products	mil lb	(S)	112.7
Plastics products consumed in the form of sheets, rods, tubes, film, and other shapes		(X)	217.6
Paper and paperboard containers, including shipping sacks and other paper packaging supplies		(X)	230.3
All other materials and components, parts, containers, and supplies		(X)	206.7
Materials, ingredients, containers, and supplies, nsk		(X)	11.8

Source: 1992 Economic Census. Explanation of symbols used: (D): Withheld to avoid disclosure of competitive data; na: Not available; (S): Withheld because statistical norms were not met; (X): Not applicable; (Z): Less than half the unit shown; nec: Not elsewhere classified; nsk: Not specified by kind; - : zero; * : 10-19 percent estimated; ** : 20-29 percent estimated.

PRODUCT SHARE DETAILS

Product or Product Class	% Share	Product or Product Class	% Share
Metal foil and leaf	100.00	sheets for flexible packaging uses	20.65
Converted unmounted aluminum foil packaging products	30.70	Laminated aluminum foil gift wrap	1.37
Converted unmounted aluminum household, institutional, and freezer foil	65.51	Laminated aluminum foil rolls and sheets for flexible packaging uses, nsk	0.26
Converted unmounted semirigid aluminum foil containers	19.98	Converted foil for nonpackaging applications and foil and leaf	20.37
Other converted unmounted aluminum foil flexible packaging products, including gift wrap	14.51	Converted aluminum foil, unmounted or coated, plain or printed, for nonpackaging applications	6.56
Laminated aluminum foil rolls and sheets for flexible packaging uses	48.46	Converted aluminum foil, laminated to other materials, for nonpackaging applications	35.98
Laminated aluminum film/foil (without paper) rolls and sheets for flexible packaging uses	31.19	Other converted foil, including composition (combination of two metals or more) for nonpackaging applications, and metal leaf (including aluminum leaf)	56.07
Extrusion laminated aluminum foil/paper combination rolls and sheets for flexible packaging uses	19.67	Converted foil for nonpackaging applications and foil and leaf, nsk	1.39
Adhesive or wax laminated aluminum foil/paper combination rolls and sheets for flexible packaging uses	26.86	Metal foil and leaf, nsk	0.47
Laminated aluminum foil/film/paper combination rolls and			

Source: 1992 Economic Census. The values shown are percent of total shipments in an industry. Values of indented subcategories are summed in the main heading. The symbol (D) appears when data are withheld to prevent disclosure of competitive information. The abbreviation nsk stands for 'not specified by kind' and nec for 'not elsewhere classified'.

INPUTS AND OUTPUTS FOR METAL FOIL & LEAF

Economic Sector or Industry Providing Inputs	%	Sector	Economic Sector or Industry Buying Outputs	%	Sector
Aluminum rolling & drawing	36.5	Manufg.	Personal consumption expenditures	23.4	
Paperboard containers & boxes	10.3	Manufg.	Frozen fruits, fruit juices & vegetables	13.2	Manufg.
Fabricated metal products, nec	10.2	Manufg.	Commercial printing	5.2	Manufg.
Imports	8.7	Foreign	Cigarettes	4.8	Manufg.
Motor freight transportation & warehousing	5.9	Util.	Food preparations, nec	4.4	Manufg.
Wholesale trade	5.7	Trade	Frozen specialties	4.2	Manufg.
Metal foil & leaf	5.1	Manufg.	Metal foil & leaf	3.3	Manufg.
Miscellaneous plastics products	3.5	Manufg.	Paperboard containers & boxes	3.1	Manufg.
Electric services (utilities)	3.4	Util.	Exports	2.9	Foreign
Advertising	1.8	Services	Cheese, natural & processed	2.8	Manufg.
Banking	1.5	Fin/R.E.	Miscellaneous plastics products	2.7	Manufg.
Primary lead	1.2	Manufg.	Confectionery products	2.6	Manufg.
Gas production & distribution (utilities)	1.0	Util.	Bread, cake, & related products	2.1	Manufg.
Air transportation	0.5	Util.	Eating & drinking places	2.0	Trade
Maintenance of nonfarm buildings nec	0.4	Constr.	Soap & other detergents	1.8	Manufg.
Petroleum refining	0.4	Manufg.	Malt beverages	1.7	Manufg.
Asbestos products	0.3	Manufg.	Drugs	1.5	Manufg.
Communications, except radio & TV	0.3	Util.	Toilet preparations	1.4	Manufg.
Railroads & related services	0.3	Util.	Dehydrated food products	1.3	Manufg.
Eating & drinking places	0.3	Trade	Mineral wool	1.2	Manufg.
Real estate	0.3	Fin/R.E.	Photographic equipment & supplies	1.2	Manufg.
Equipment rental & leasing services	0.3	Services	Retail trade, except eating & drinking	1.2	Trade
Soap & other detergents	0.1	Manufg.	Chewing gum	1.1	Manufg.
Transit & bus transportation	0.1	Util.	Chocolate & cocoa products	1.0	Manufg.
Water transportation	0.1	Util.	Surgical & medical instruments	1.0	Manufg.
Business services nec	0.1	Services	Cereal breakfast foods	0.9	Manufg.
Computer & data processing services	0.1	Services	Electronic components nec	0.8	Manufg.
			Cookies & crackers	0.6	Manufg.
			Blended & prepared flour	0.5	Manufg.
			Motor vehicles & car bodies	0.5	Manufg.
			Truck & bus bodies	0.5	Manufg.
			Bottled & canned soft drinks	0.4	Manufg.
			Chemical preparations, nec	0.4	Manufg.
			Guided missiles & space vehicles	0.4	Manufg.
			Shortening & cooking oils	0.4	Manufg.
			Aircraft	0.3	Manufg.
			Automotive stampings	0.3	Manufg.
			Book printing	0.3	Manufg.
			Chewing & smoking tobacco	0.3	Manufg.
			Distilled liquor, except brandy	0.3	Manufg.
			Fresh or frozen packaged fish	0.3	Manufg.
			Federal Government purchases, national defense	0.3	Fed Govt
			Bags, except textile	0.2	Manufg.
			Cigars	0.2	Manufg.
			Converted paper products, nec	0.2	Manufg.
			Paper coating & glazing	0.2	Manufg.
			Wines, brandy, & brandy spirits	0.2	Manufg.
			Condensed & evaporated milk	0.1	Manufg.
			Ice cream & frozen desserts	0.1	Manufg.

Source: Benchmark Input-Output Accounts for the U.S. Economy, 1982, U.S. Department of Commerce, Washington, D.C., July 1991. Data, as reported in the source, are organized by the 1977 SIC structure in use in 1982 but have been matched, as closely as is possible, to the 1987 SIC structure used in this book.

OCCUPATIONS EMPLOYED BY SIC 349 - MISCELLANEOUS FABRICATED METAL PRODUCTS

Occupation	% of Total 1994	Change to 2005	Occupation	% of Total 1994	Change to 2005
Assemblers, fabricators, & hand workers nec	10.7	0.7	Industrial machinery mechanics	1.8	10.7
Blue collar worker supervisors	4.5	-7.4	Helpers, laborers, & material movers nec	1.8	0.7
Machinists	4.4	0.7	Hand packers & packagers	1.7	-13.7
Welding machine setters, operators	3.8	-9.4	Machine operators nec	1.6	-11.3
Machine tool cutting & forming etc. nec	3.7	20.8	Industrial production managers	1.5	0.7
Machine forming operators, metal & plastic	3.5	-9.4	Secretaries, ex legal & medical	1.4	-8.3
Sales & related workers nec	3.0	0.7	General office clerks	1.3	-14.2
Inspectors, testers, & graders, precision	3.0	0.7	Tool & die makers	1.3	-18.7
Metal & plastic machine workers nec	2.7	-28.7	Bookkeeping, accounting, & auditing clerks	1.3	-24.5
General managers & top executives	2.6	-4.5	Coating, painting, & spraying machine workers	1.3	0.7
Combination machine tool operators	2.4	51.0	Maintenance repairers, general utility	1.2	-9.4
Machine tool cutting operators, metal & plastic	2.4	41.9	Industrial truck & tractor operators	1.1	0.6
Lathe & turning machine tool operators	2.1	-9.4	Machine feeders & offbearers	1.1	-9.4
Welders & cutters	2.1	0.7	Production, planning, & expediting clerks	1.0	0.7
Traffic, shipping, & receiving clerks	2.0	-3.1	Mechanical engineers	1.0	-0.3

Source: Industry-Occupation Matrix, Bureau of Labor Statistics. These data relate to one or more 3-digit SIC industry groups rather than to a single 4-digit SIC. The change reported for each occupation to the year 2005 is a percent of growth or decline as estimated by the Bureau of Labor Statistics. The abbreviation nec stands for 'not elsewhere classified'.

LOCATION BY STATE AND REGIONAL CONCENTRATION

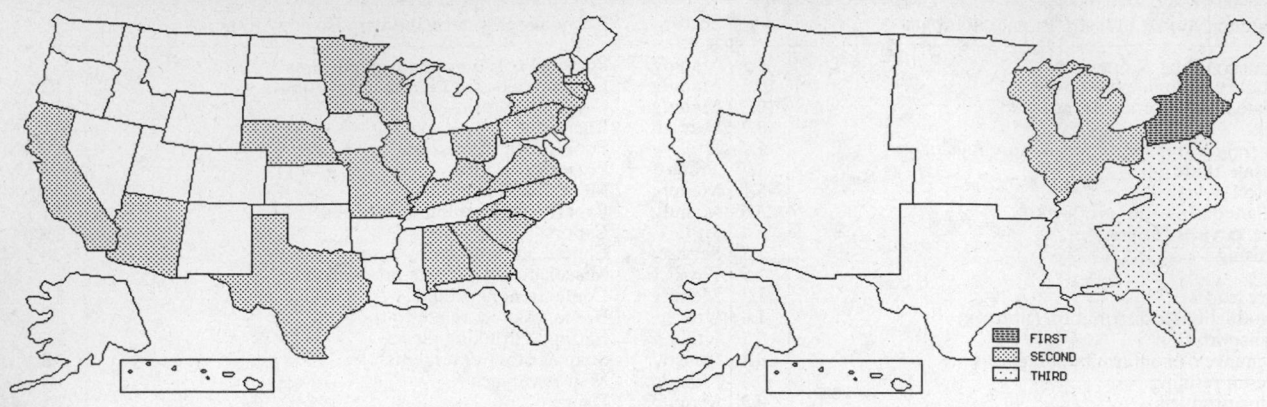

FIRST
SECOND
THIRD

INDUSTRY DATA BY STATE

| State | Establish-ments | Shipments | | | Employment | | | | Cost as % of Shipments | Investment per Employee ($) |
		Total ($ mil)	% of U.S.	Per Establ.	Total Number	% of U.S.	Per Establ.	Wages ($/hour)		
Kentucky	7	465.2	14.9	66.5	1,000	8.3	143	15.38	66.3	-
North Carolina	8	441.8	14.2	55.2	1,600	13.3	200	19.50	61.0	6,438
Ohio	9	270.2	8.7	30.0	1,300	10.8	144	13.35	59.4	-
New Jersey	16	262.7	8.4	16.4	1,400	11.7	88	13.80	56.9	11,357
Illinois	9	234.2	7.5	26.0	1,300	10.8	144	11.89	47.3	3,077
New York	6	85.3	2.7	14.2	400	3.3	67	15.67	54.6	-
California	13	58.9	1.9	4.5	400	3.3	31	12.20	53.1	2,500
Pennsylvania	5	(D)	-	-	375 *	3.1	75	-	-	-
Wisconsin	5	(D)	-	-	375 *	3.1	75	-	-	-
Georgia	4	(D)	-	-	175 *	1.5	44	-	-	1,714
Missouri	4	(D)	-	-	375 *	3.1	94	-	-	-
Alabama	3	(D)	-	-	175 *	1.5	58	-	-	-
South Carolina	3	(D)	-	-	375 *	3.1	125	-	-	-
Virginia	3	(D)	-	-	1,750 *	14.6	583	-	-	-
Arizona	2	(D)	-	-	175 *	1.5	88	-	-	-
Connecticut	2	(D)	-	-	175 *	1.5	88	-	-	-
Massachusetts	2	(D)	-	-	375 *	3.1	188	-	-	-
Minnesota	2	(D)	-	-	175 *	1.5	88	-	-	-
Texas	2	(D)	-	-	375 *	3.1	188	-	-	-
Nebraska	1	(D)	-	-	175 *	1.5	175	-	-	-

Source: 1992 *Economic Census*. The states are in descending order of shipments or establishments (if shipment data are missing for the majority). The symbol (D) appears when data are withheld to prevent disclosure of competitive information. States marked with (D) are sorted by number of establishments. A dash (-) indicates that the data element cannot be calculated; * indicates the midpoint of a range.

3498 - FABRICATED PIPE & FITTINGS

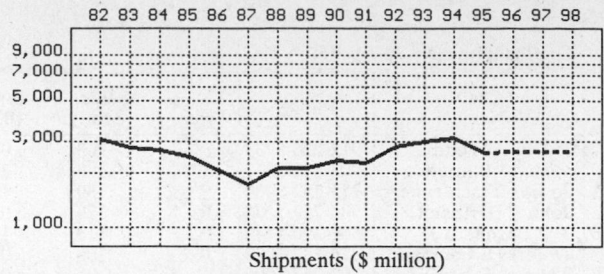

Shipments ($ million)

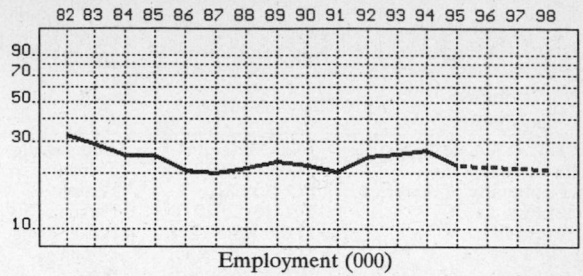

Employment (000)

GENERAL STATISTICS

Year	Com-panies	Establishments		Employment			Compensation		Production ($ million)			
		Total	with 20 or more employees	Total (000)	Production Workers (000)	Hours (Mil)	Payroll ($ mil)	Wages ($/hr)	Cost of Materials	Value Added by Manufacture	Value of Shipments	Capital Invest.
1982	704	778	347	32.7	23.8	46.4	651.7	9.16	1,674.3	1,440.4	3,115.8	137.7
1983		738	335	29.3	21.7	42.2	596.1	9.46	1,460.2	1,207.2	2,776.1	47.8
1984		698	323	25.2	18.9	36.7	563.9	10.36	1,552.9	1,079.8	2,679.8	52.7
1985		659	310	25.1	18.5	34.6	547.3	10.42	1,375.2	1,038.8	2,476.5	78.9
1986		661	295	20.9	15.1	28.8	473.8	10.55	1,122.1	926.4	2,071.1	47.0
1987	679	728	283	20.0	15.0	30.1	422.9	9.27	906.6	824.8	1,725.5	36.4
1988		697	276	21.4	16.2	34.3	471.6	9.01	1,058.2	1,111.8	2,142.0	33.9
1989		687	285	23.4	15.6	31.8	482.5	9.75	1,167.9	1,012.8	2,151.7	41.1
1990		661	284	22.3	16.1	33.3	529.5	10.36	1,291.5	1,027.0	2,333.8	35.2
1991		675	272	20.5	15.0	30.4	487.6	10.07	1,298.2	980.0	2,270.5	34.0
1992	810	856	315	24.8	17.8	37.9	647.7	10.68	1,487.7	1,303.6	2,794.3	59.3
1993		841	326	25.6	18.8	39.9	678.2	10.55	1,511.6	1,460.9	3,002.7	71.5
1994		761P	283P	27.2	19.7	42.1	747.5	11.17	1,591.7	1,574.2	3,142.9	91.1
1995		767P	280P	22.2P	15.9P	34.7P	612.0P	10.75P	1,313.9P	1,299.4P	2,594.3P	47.9P
1996		772P	276P	21.8P	15.6P	34.5P	619.2P	10.85P	1,319.7P	1,305.2P	2,605.8P	46.3P
1997		778P	273P	21.5P	15.3P	34.3P	626.5P	10.94P	1,325.5P	1,310.9P	2,617.3P	44.7P
1998		784P	270P	21.2P	15.0P	34.1P	633.7P	11.04P	1,331.3P	1,316.7P	2,628.7P	43.1P

Sources: 1982, 1987, 1992 *Economic Census*; *Annual Survey of Manufactures*, 83-86, 88-91, 93-94. Establishment counts for non-Census years are from *County Business Patterns*; establishment values for 83-84 are extrapolations. 'P's show projections by the editors. Industries reclassified in 87 will not have data for prior years.

INDICES OF CHANGE

Year	Com-panies	Establishments		Employment			Compensation		Production ($ million)			
		Total	with 20 or more employees	Total (000)	Production Workers (000)	Hours (Mil)	Payroll ($ mil)	Wages ($/hr)	Cost of Materials	Value Added by Manufacture	Value of Shipments	Capital Invest.
1982	87	91	110	132	134	122	101	86	113	110	112	232
1983		86	106	118	122	111	92	89	98	93	99	81
1984		82	103	102	106	97	87	97	104	83	96	89
1985		77	98	101	104	91	84	98	92	80	89	133
1986		77	94	84	85	76	73	99	75	71	74	79
1987	84	85	90	81	84	79	65	87	61	63	62	61
1988		81	88	86	91	91	73	84	71	85	77	57
1989		80	90	94	88	84	74	91	79	78	77	69
1990		77	90	90	90	88	82	97	87	79	84	59
1991		79	86	83	84	80	75	94	87	75	81	57
1992	100	100	100	100	100	100	100	100	100	100	100	100
1993		98	103	103	106	105	105	99	102	112	107	121
1994		89P	90P	110	111	111	115	105	107	121	112	154
1995		90P	89P	89P	89P	92P	94P	101P	88P	100P	93P	81P
1996		90P	88P	88P	88P	91P	96P	102P	89P	100P	93P	78P
1997		91P	87P	87P	86P	91P	97P	102P	89P	101P	94P	75P
1998		92P	86P	85P	84P	90P	98P	103P	89P	101P	94P	73P

Sources: Same as General Statistics. Values reflect change from the base year, 1992. Values above 100 mean greater than 92, values below 100 mean less than 92, and a value of 100 in the 82-91 or 93-98 period means same as 92. 'P's mark projections by the editors.

SELECTED RATIOS

For 1994	Avg. of All Manufact.	Analyzed Industry	Index	For 1994	Avg. of All Manufact.	Analyzed Industry	Index
Employees per Establishment	49	36	73	Value Added per Production Worker	134,084	79,909	60
Payroll per Establishment	1,500,273	982,554	65	Cost per Establishment	5,045,178	2,092,215	41
Payroll per Employee	30,620	27,482	90	Cost per Employee	102,970	58,518	57
Production Workers per Establishment	34	26	75	Cost per Production Worker	146,988	80,797	55
Wages per Establishment	853,319	618,131	72	Shipments per Establishment	9,576,895	4,131,194	43
Wages per Production Worker	24,861	23,871	96	Shipments per Employee	195,460	115,548	59
Hours per Production Worker	2,056	2,137	104	Shipments per Production Worker	279,017	159,538	57
Wages per Hour	12.09	11.17	92	Investment per Establishment	321,011	119,747	37
Value Added per Establishment	4,602,255	2,069,212	45	Investment per Employee	6,552	3,349	51
Value Added per Employee	93,930	57,875	62	Investment per Production Worker	9,352	4,624	49

Sources: Same as General Statistics. The 'Average of All Manufacturing' column represents the average of all manufacturing industries reported for the most recent complete year available. The Index shows the relationship between the Average and the Analyzed Industry. For example, 100 means that they are equal; 500 that the Analyzed Industry is five times the average; 50 means that the Analyzed Industry is half the national average. The abbreviation 'na' is used to show that data are 'not available'.

LEADING COMPANIES   Number shown: **75**   Total sales ($ mil): **1,992**   Total employment (000): **15.1**

Company Name	Address			CEO Name	Phone	Co. Type	Sales ($ mil)	Empl. (000)
Davis Water & Waste Industries	1820 Metcalf Av	Thomasville	GA 31792	R Doyle White	912-226-5733	P	203	0.7
Tyler Pipe Industry Inc	PO Box 2027	Tyler	TX 75710	Richard Barnett	903-882-5511	S	198	2.6
Victaulic Company of America	PO Box 31	Easton	PA 18042	Joseph Trachtenberg	215-252-6400	R	130•	1.2
Berg Steel Pipe Corp	PO Box 2029	Panama City	FL 32402	John P Berthelot	904-769-2273	R	120	0.2
Shaw Group Inc	11100 Mead Rd	Baton Rouge	LA 70816	J M Bernhard Jr	504-296-1140	P	113	1.2
PC Campana Inc	2115 W Park Dr	Lorain	OH 44053	Dolores S Campana	216-282-3803	R	100	0.6
Napa Pipe Corp	1025 Kaiser Rd	Napa	CA 94558	Thomas B Boklund	707-257-5000	S	90	0.3
FitzSimons Manufacturing Co	3775 E Outer Dr	Detroit	MI 48234	Joseph T Goerlich	313-891-4800	S	60	0.5
LaBarge Pipe and Steel Co	901 N 10th St	St Louis	MO 63101	PL LaBarge III	314-231-3400	R	59	0.2
Piping Companies Inc	PO Box 190	Sand Springs	OK 74063	Craig Bowman	918-245-6606	S	55•	0.6
Perfection Corp	222 Lake St	Madison	OH 44057	Joe Gregory	216-428-1171	S	45•	0.3
Picoma Industries Inc	PO Box 510	Waynesboro	PA 17268	Frank Muffeny	717-762-9141	S	40	0.4
Alloy Piping Products Company	PO Box 7368	Shreveport	LA 71137	Ron D Brown	318-226-9851	R	34•	0.2
International Piping Systems Ltd	PO Box 868	Port Allen	LA 70767	David Chapman	504-381-9422	S	34	0.3
Connex Pipe Systems Inc	1 Connex Way	Troutville	VA 24175	Drew Kershaw	703-992-1600	S	32•	0.2
Corrosion Materials	PO Box 666	Baker	LA 70714	Robert B Leonard	504-775-3675	D	32	0.1
H-P Products Inc	512 W Gorgas St	Louisville	OH 44641	PR Bishop	216-875-5556	R	30	0.3
Taracorp Industries Inc	1200 16th St	Granite City	IL 62040	Louis J Taratoott	618-451-4400	S	30	0.1
Thompson Pipe and Steel Co	6400 S Fiddler's	Englewood	CO 80111	Tim Phillips	303-290-9490	S	30	0.2
Word Industries Inc	PO Box 9615	Tulsa	OK 74157	Tom N Word III	918-446-6184	R	29•	0.3
Pioneer Pipe Inc	RR 4	Marietta	OH 45750	David Archer	614-374-3108	R	28•	0.3
Whitley Products Inc	PO Box 154	Pierceton	IN 46562	J Randolph Lowe	219-594-2112	R	25	0.3
Senior Boiler Tube Co	PO Box 517	Lyman	SC 29365	Mike Bird	803-439-0220	D	24	0.1
Douglas Brothers	475 Riverside	Portland	ME 04103	Arthur DuBois	207-797-6771	D	20	0.1
Ricwil Piping Systems LP	10100 Brecksville	Brecksville	OH 44141	G David Zeille	216-526-1600	R	20	< 0.1
Stark Manufacturing Inc	13300 S Elmira Rd	Russellville	AR 72801	Dick Dawson	501-968-8840	S	20	0.3
Curtis Products Inc	PO Box 1107	South Bend	IN 46624	CR Heckaman	219-289-4891	R	15	0.2
Lewis and Saunders Inc	PO Box 678	Laconia	NH 03247	Mark Celusniak	603-524-2064	S	15	0.2
Moeller Products Company Inc	PO Box 1736	Greenville	MS 38702	Douglas Estes	601-335-2325	R	15	< 0.1
Pipe Fabricating and Supply Co	9703 S Norwalk	Santa Fe Sprgs	CA 90670	AR Simmons	310-692-7226	R	15	0.2
Rovanco Corp	20535 SE Frontage	Joliet	IL 60436	Charles Ray	815-741-6700	R	15•	0.1
Stanley G Flagg Co	1020 W High St	Stowe	PA 19464	MA Cookman	610-326-9000	D	14	0.1
Carpenter Technology Corp	PO Box 609036	San Diego	CA 92160	DR Wozniak	619-448-1000	D	13	0.2
SSP Fittings Corp	8250 Boyle Pkwy	Twinsburg	OH 44087	R W King Jr	216-425-4250	R	13•	0.1
Industrial Piping Inc	PO Box 518	Pineville	NC 28134	Michael Jones	704-588-1100	R	12	0.1
NTF Inc	PO Box 10726	Fort Wayne	IN 46853	Edward Whipp	219-478-2363	R	10	0.2
Alloy Stainless Products	611 Union Blv	Totowa	NJ 07512		201-256-1616	R	10	< 0.1
Champion Furnace Pipe Co	PO Box 957	Peoria	IL 61653	Edward J Sepanik	309-676-0877	R	10•	< 0.1
HABCO Steel Service Inc	PO Box 13223	Memphis	TN 38113	Frank O Oakes Jr	901-775-1060	R	10	< 0.1
Inductoweld Tube Corp	1429 Ferris Pl	Bronx	NY 10461	Frank Rella Jr	718-828-4006	R	10	0.1
JL Allen Co	PO Box 347	Tuscola	IL 61953	ND Kellogg	217-253-3371	R	10	0.1
Mid-States Pipe Fabricating Inc	PO Box 1628	El Dorado	AR 71730	Jerry E Smith	501-862-5167	R	10	0.1
Pines Manufacturing Inc	30505 Clemens Rd	Westlake	OH 44145	Jeffrey Edmunds	216-835-5553	R	10	< 0.1
Arntzen Corp	1025 School St	Rockford	IL 61105	Richard Arntzen	815-964-9413	R	9	< 0.1
Flo-Bend Inc	8635 W 21st St	Tulsa	OK 74063	Russell McBroom	918-245-7501	S	9•	< 0.1
Spectrum Metals	37 Rogers St	Cambridge	MA 02142	Robert Leonard	617-491-1320	D	9	< 0.1
Bauer Welding	2159 Mustang Dr	St Paul	MN 55112	Gary A Bauer	612-786-6025	R	8	< 0.1
Dynamic Products Inc	16520 Peninsula	Houston	TX 77015	Ron Lindquist	713-457-3500	R	8•	< 0.1
Midland Pipe and Supply Co	2829 S 61st Ct	Chicago	IL 60650	LA Walsh	708-656-4200	R	8	< 0.1
Vacuum Barrier Corp	4 Barten Ln	Woburn	MA 01801	LB Thompson Jr	617-933-3570	R	8	< 0.1
A-1 Nipple Manufacturing Co	5404 Tweedy Pl	South Gate	CA 90280	Raul Gonzales	213-569-8151	S	7•	0.1
HLC Industries Inc	38880 Grand River	Farmington Hls	MI 48331	William K Morrow	810-477-9600	R	7•	< 0.1
Hydro Tube Corp	137 Artino St	Oberlin	OH 44074	Lawrence L Reining	216-774-1022	R	7	< 0.1
Woolf Aircraft Products Inc	6401 Cogswell	Romulus	MI 48174	Dan Woolf	313-721-5330	R	7	< 0.1
Cobra Pipe Supply Inc	PO Box 330475	Elmwood	CT 06133	Clifton O'Donnal	203-233-1231	R	6	< 0.1
Monona Tube and Welding Inc	5315 Paulson Rd	McFarland	WI 53558	Gene Henry	608-838-3188	R	6	< 0.1
Versatech Inc	PO Box 608	Export	PA 15632	Richard L Versaw	412-327-8324	R	6•	< 0.1
Anderson-Snow Corp	PO Box 2126	Schiller Park	IL 60176	Ted R Campbell	708-678-3823	R	5	< 0.1
Bassani Manufacturing Co	160 E La Jolla St	Placentia	CA 92670	Darryl Bassani	714-630-1821	R	5•	< 0.1
HB Larkin Corp	7155 Old Katy Rd	Houston	TX 77024	Kirk Weaver	713-866-6868	R	5	< 0.1
Hydraulic Tubes and Fittings Inc	3578 S Van Dyke	Almont	MI 48003	James Musser	810-798-8567	R	5•	0.1
Larkin Products Inc	PO Box 55289	Houston	TX 77255	Kirk Weaver	713-802-6785	S	5	< 0.1
Maass Manufacturing Inc	PO Box 547	Huntley	IL 60142	John J Surinak	708-669-5135	R	5•	< 0.1
Propipe Corp	1800 Clayton St	Middletown	OH 45042	James Bryson	513-424-5311	S	5	< 0.1
SL Piping Systems Inc	302 Falco Dr	Newport	DE 19804	Donald J Sipos	302-995-6136	S	5	< 0.1
Star Tubular Products Co	4747 S Richmond St	Chicago	IL 60632	WD Reed	312-523-8445	R	5	< 0.1
Edmund A Gray Company Inc	2277 E 15th St	Los Angeles	CA 90021	LC Gray Jr	213-625-2723	R	5	< 0.1
Engineering Tube Specialties Inc	PO Box 350	Ortonville	MI 48462	Valerie Oldenburg	810-627-2871	R	5	< 0.1
Humane Manufacturing Co	805 Moore St	Baraboo	WI 53913	Edwin C Sauey	608-356-8336	R	4	< 0.1
Standard Nipple Works Inc	PO Box 156	Garwood	NJ 07027	R Olesky	908-789-4747	R	4	< 0.1
Tulsa Tube Bending Co	PO Box 1017	Tulsa	OK 74101	Brad Frank	918-446-4461	R	4	< 0.1
Lindfor Inc	15600 32nd Av N	Plymouth	MN 55447	Jon Lindfors	612-559-2911	R	4	< 0.1
Crescent Metal Products Co	1303 Lincoln Rd	Allegan	MI 49010	Leroy Burgess	616-673-2151	R	4	< 0.1
Los Angeles Sleeve Co	12051 Rivera Rd	Santa Fe Sprgs	CA 90670	Gary N Metchkoff	310-945-7578	R	4	< 0.1
WA Call Manufacturing	1710 Rogers Av	San Jose	CA 95112	WA Call Jr	408-436-1450	R	2•	< 0.1

Source: Ward's Business Directory of U.S. Private and Public Companies, Volumes 1 and 2, 1996. The company type code used is as follows: P - Public, R - Private, S - Subsidiary, D - Division, J - Joint Venture, A - Affiliate, G - Group. Sales are in millions of dollars, employees are in thousands. An asterisk (•) indicates an estimated sales volume. The symbol < stands for 'less than'. Company names and addresses are truncated, in some cases, to fit into the available space.

MATERIALS CONSUMED

Material	Quantity	Delivered Cost ($ million)
Materials, ingredients, containers, and supplies	(X)	1,289.4
Metal fittings, flanges, and unions for piping systems (except forgings)	(X)	93.7
Metal stampings	(X)	5.3
Other fabricated metal products (except forgings)	(X)	39.0
Forgings	(X)	3.7
Iron and steel castings (rough and semifinished)	(X)	39.9
Aluminum and aluminum-base alloy castings (rough and semifinished)	(X)	1.1
Other nonferrous castings (rough and semifinished)	(X)	3.6
Steel bars and bar shapes	(X)	25.1
Steel sheet and strip, including tin plate	(X)	18.2
Steel plate	(X)	15.9
Steel structural shapes	(X)	25.5
Steel pipes	(X)	441.1
All other steel shapes and forms	(X)	74.4
Copper and copper-base alloy pipe and tube	(X)	39.2
All other copper and copper-base alloy shapes and forms	(X)	6.2
Aluminum and aluminum-base alloy sheet, plate, foil, and welded tubing	(X)	1.9
Aluminum and aluminum-base alloy extruded shapes, including extruded rod, bar, pipe, tube, etc.	(X)	12.9
Other aluminum and aluminum-base alloy shapes and forms	(X)	6.6
Other nonferrous shapes and forms	(X)	11.1
Metal powders	(X)	0.9
Plastics resins consumed in the form of granules, pellets, powders, liquids, etc.	(X)	12.0
All other materials and components, parts, containers, and supplies	(X)	194.5
Materials, ingredients, containers, and supplies, nsk	(X)	217.7

Source: 1992 *Economic Census*. Explanation of symbols used: (D): Withheld to avoid disclosure of competitive data; na: Not available; (S): Withheld because statistical norms were not met; (X): Not applicable; (Z): Less than half the unit shown; nec: Not elsewhere classified; nsk: Not specified by kind; - : zero; * : 10-19 percent estimated; ** : 20-29 percent estimated.

PRODUCT SHARE DETAILS

Product or Product Class	% Share	Product or Product Class	% Share
Fabricated pipe and fittings	100.00	Fabricated copper and copper-base alloy pipe and pipe fittings made from purchased pipe	3.23
Fabricated iron and steel pipe and pipe fittings made from purchased pipe	70.86	All other nonferrous fabricated pipe and pipe fittings made from purchased pipe	10.23
Fabricated aluminum and aluminum-base alloy pipe and pipe fittings made from purchased pipe	3.18		

Source: 1992 *Economic Census*. The values shown are percent of total shipments in an industry. Values of indented subcategories are summed in the main heading. The symbol (D) appears when data are withheld to prevent disclosure of competitive information. The abbreviation nsk stands for 'not specified by kind' and nec for 'not elsewhere classified'.

INPUTS AND OUTPUTS FOR PIPE, VALVES, & PIPE FITTINGS

Economic Sector or Industry Providing Inputs	%	Sector	Economic Sector or Industry Buying Outputs	%	Sector
Blast furnaces & steel mills	19.6	Manufg.	Industrial buildings	12.2	Constr.
Imports	13.8	Foreign	Gross private fixed investment	11.7	Cap Inv
Wholesale trade	8.1	Trade	Exports	7.8	Foreign
Iron & steel foundries	7.5	Manufg.	Maintenance of nonfarm buildings nec	5.5	Constr.
Pipe, valves, & pipe fittings	3.2	Manufg.	Crude petroleum & natural gas	4.0	Mining
Copper rolling & drawing	2.9	Manufg.	Cyclic crudes and organics	4.0	Manufg.
Advertising	2.8	Services	Sewer system facility construction	3.4	Constr.
Iron & steel forgings	2.5	Manufg.	Office buildings	2.8	Constr.
Screw machine and related products	2.4	Manufg.	Electric utility facility construction	2.7	Constr.
Electric services (utilities)	2.2	Util.	Automotive rental & leasing, without drivers	2.7	Services
Banking	1.9	Fin/R.E.	Water supply facility construction	2.0	Constr.
Fabricated structural metal	1.6	Manufg.	Maintenance of sewer facilities	1.8	Constr.
Aluminum castings	1.4	Manufg.	Pipe, valves, & pipe fittings	1.7	Manufg.
Brass, bronze, & copper castings	1.4	Manufg.	Turbines & turbine generator sets	1.3	Manufg.
Petroleum refining	1.4	Manufg.	Maintenance of water supply facilities	1.2	Constr.
Engineering, architectural, & surveying services	1.4	Services	Nonfarm residential structure maintenance	1.1	Constr.
Motor freight transportation & warehousing	1.2	Util.	Water transportation	1.1	Util.
Cyclic crudes and organics	1.1	Manufg.	Fabricated plate work (boiler shops)	1.0	Manufg.
Industrial patterns	1.1	Manufg.	Communications, except radio & TV	1.0	Util.
Primary copper	1.1	Manufg.	Motor freight transportation & warehousing	1.0	Util.
Nonferrous wire drawing & insulating	1.0	Manufg.	Sanitary services, steam supply, irrigation	1.0	Util.
Aluminum rolling & drawing	0.9	Manufg.	Water supply & sewage systems	1.0	Util.
Communications, except radio & TV	0.9	Util.	Blowers & fans	0.9	Manufg.
Equipment rental & leasing services	0.9	Services	Oil field machinery	0.9	Manufg.
Maintenance of nonfarm buildings nec	0.8	Constr.	Transit & bus transportation	0.9	Util.
Air transportation	0.8	Util.	Maintenance of electric utility facilities	0.8	Constr.
Gas production & distribution (utilities)	0.8	Util.	Residential 1-unit structures, nonfarm	0.8	Constr.
Machinery, except electrical, nec	0.7	Manufg.	Construction machinery & equipment	0.8	Manufg.
Business services nec	0.7	Services	Federal Government purchases, national defense	0.8	Fed Govt

Continued on next page.

INPUTS AND OUTPUTS FOR PIPE, VALVES, & PIPE FITTINGS - Continued

Economic Sector or Industry Providing Inputs	%	Sector	Economic Sector or Industry Buying Outputs	%	Sector
Photofinishing labs, commercial photography	0.7	Services	Construction of stores & restaurants	0.7	Constr.
Plating & polishing	0.6	Manufg.	Maintenance of petroleum & natural gas wells	0.7	Constr.
Eating & drinking places	0.6	Trade	Ship building & repairing	0.7	Manufg.
Nonferrous castings, nec	0.5	Manufg.	Gas utility facility construction	0.6	Constr.
Primary nonferrous metals, nec	0.5	Manufg.	Warehouses	0.6	Constr.
Rubber & plastics hose & belting	0.5	Manufg.	Petroleum refining	0.6	Manufg.
Real estate	0.5	Fin/R.E.	Pumps & compressors	0.6	Manufg.
Scrap	0.5	Scrap	Maintenance of gas utility facilities	0.5	Constr.
Fabricated rubber products, nec	0.4	Manufg.	Residential additions/alterations, nonfarm	0.5	Constr.
Industrial controls	0.4	Manufg.	Aircraft & missile engines & engine parts	0.5	Manufg.
Machine tools, metal cutting types	0.4	Manufg.	Machinery, except electrical, nec	0.5	Manufg.
Special dies & tools & machine tool accessories	0.4	Manufg.	Construction of hospitals	0.4	Constr.
Railroads & related services	0.4	Util.	Maintenance of farm service facilities	0.4	Constr.
Chemical preparations, nec	0.3	Manufg.	Blast furnaces & steel mills	0.4	Manufg.
Hardware, nec	0.3	Manufg.	Farm machinery & equipment	0.4	Manufg.
Metal stampings, nec	0.3	Manufg.	General industrial machinery, nec	0.4	Manufg.
Miscellaneous plastics products	0.3	Manufg.	Household cooking equipment	0.4	Manufg.
Paperboard containers & boxes	0.3	Manufg.	Household laundry equipment	0.4	Manufg.
Primary metal products, nec	0.3	Manufg.	Paper mills, except building paper	0.4	Manufg.
Automotive rental & leasing, without drivers	0.3	Services	Refrigeration & heating equipment	0.4	Manufg.
U.S. Postal Service	0.3	Gov't	Special industry machinery, nec	0.4	Manufg.
Abrasive products	0.2	Manufg.	Truck trailers	0.4	Manufg.
Gaskets, packing & sealing devices	0.2	Manufg.	Maintenance of petroleum pipelines	0.3	Constr.
Lubricating oils & greases	0.2	Manufg.	Maintenance, conservation & development facilities	0.3	Constr.
Metal heat treating	0.2	Manufg.	Petroleum & natural gas well drilling	0.3	Constr.
Miscellaneous fabricated wire products	0.2	Manufg.	Aircraft	0.3	Manufg.
Insurance carriers	0.2	Fin/R.E.	Household appliances, nec	0.3	Manufg.
Accounting, auditing & bookkeeping	0.2	Services	Internal combustion engines, nec	0.3	Manufg.
Automotive repair shops & services	0.2	Services	Motor vehicle parts & accessories	0.3	Manufg.
Computer & data processing services	0.2	Services	Paperboard mills	0.3	Manufg.
Legal services	0.2	Services	Polishes & sanitation goods	0.3	Manufg.
Management & consulting services & labs	0.2	Services	Sporting & athletic goods, nec	0.3	Manufg.
Noncomparable imports	0.2	Foreign	Toilet preparations	0.3	Manufg.
Electronic components nec	0.1	Manufg.	Railroads & related services	0.3	Util.
Fabricated metal products, nec	0.1	Manufg.	Construction of conservation facilities	0.2	Constr.
Industrial inorganic chemicals, nec	0.1	Manufg.	Construction of educational buildings	0.2	Constr.
Metal coating & allied services	0.1	Manufg.	Hotels & motels	0.2	Constr.
Motors & generators	0.1	Manufg.	Maintenance of farm residential buildings	0.2	Constr.
Credit agencies other than banks	0.1	Fin/R.E.	Maintenance of military facilities	0.2	Constr.
Security & commodity brokers	0.1	Fin/R.E.	Maintenance of nonbuilding facilities nec	0.2	Constr.
Hotels & lodging places	0.1	Services	Residential garden apartments	0.2	Constr.
			Industrial furnaces & ovens	0.2	Manufg.
			Industrial inorganic chemicals, nec	0.2	Manufg.
			Industrial trucks & tractors	0.2	Manufg.
			Iron & steel forgings	0.2	Manufg.
			Miscellaneous plastics products	0.2	Manufg.
			Nitrogenous & phosphatic fertilizers	0.2	Manufg.
			Paints & allied products	0.2	Manufg.
			Prepared feeds, nec	0.2	Manufg.
			Pulp mills	0.2	Manufg.
			Service industry machines, nec	0.2	Manufg.
			Farm service facilities	0.1	Constr.
			Maintenance of local transit facilities	0.1	Constr.
			Maintenance of railroads	0.1	Constr.
			Petroleum pipeline construction	0.1	Constr.
			Residential high-rise apartments	0.1	Constr.
			Agricultural chemicals, nec	0.1	Manufg.
			Aircraft & missile equipment, nec	0.1	Manufg.
			Bottled & canned soft drinks	0.1	Manufg.
			Cement, hydraulic	0.1	Manufg.
			Chemical preparations, nec	0.1	Manufg.
			Food products machinery	0.1	Manufg.
			Industrial gases	0.1	Manufg.
			Machine tools, metal cutting types	0.1	Manufg.
			Malt beverages	0.1	Manufg.
			Primary aluminum	0.1	Manufg.

Source: Benchmark Input-Output Accounts for the U.S. Economy, 1982, U.S. Department of Commerce, Washington, D.C., July 1991. Data, as reported in the source, are organized by the 1977 SIC structure in use in 1982 but have been matched, as closely as is possible, to the 1987 SIC structure used in this book.

OCCUPATIONS EMPLOYED BY SIC 349 - MISCELLANEOUS FABRICATED METAL PRODUCTS

Occupation	% of Total 1994	Change to 2005	Occupation	% of Total 1994	Change to 2005
Assemblers, fabricators, & hand workers nec	10.7	0.7	Industrial machinery mechanics	1.8	10.7
Blue collar worker supervisors	4.5	-7.4	Helpers, laborers, & material movers nec	1.8	0.7
Machinists	4.4	0.7	Hand packers & packagers	1.7	-13.7
Welding machine setters, operators	3.8	-9.4	Machine operators nec	1.6	-11.3
Machine tool cutting & forming etc. nec	3.7	20.8	Industrial production managers	1.5	0.7
Machine forming operators, metal & plastic	3.5	-9.4	Secretaries, ex legal & medical	1.4	-8.3
Sales & related workers nec	3.0	0.7	General office clerks	1.3	-14.2
Inspectors, testers, & graders, precision	3.0	0.7	Tool & die makers	1.3	-18.7
Metal & plastic machine workers nec	2.7	-28.7	Bookkeeping, accounting, & auditing clerks	1.3	-24.5
General managers & top executives	2.6	-4.5	Coating, painting, & spraying machine workers	1.3	0.7
Combination machine tool operators	2.4	51.0	Maintenance repairers, general utility	1.2	-9.4
Machine tool cutting operators, metal & plastic	2.4	41.9	Industrial truck & tractor operators	1.1	0.6
Lathe & turning machine tool operators	2.1	-9.4	Machine feeders & offbearers	1.1	-9.4
Welders & cutters	2.1	0.7	Production, planning, & expediting clerks	1.0	0.7
Traffic, shipping, & receiving clerks	2.0	-3.1	Mechanical engineers	1.0	-0.3

Source: *Industry-Occupation Matrix*, Bureau of Labor Statistics. These data relate to one or more 3-digit SIC industry groups rather than to a single 4-digit SIC. The change reported for each occupation to the year 2005 is a percent of growth or decline as estimated by the Bureau of Labor Statistics. The abbreviation nec stands for 'not elsewhere classified'.

LOCATION BY STATE AND REGIONAL CONCENTRATION

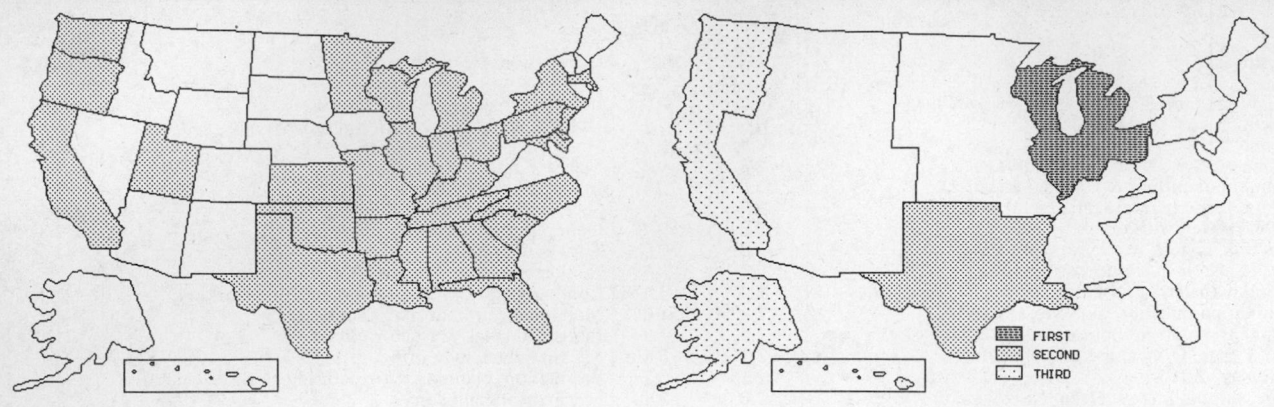

FIRST
SECOND
THIRD

INDUSTRY DATA BY STATE

| State | Establish-ments | Shipments | | | Employment | | | | Cost as % of Shipments | Investment per Employee ($) |
		Total ($ mil)	% of U.S.	Per Establ.	Total Number	% of U.S.	Per Establ.	Wages ($/hour)		
Texas	96	421.8	15.1	4.4	3,500	14.1	36	9.80	63.0	2,857
Michigan	69	239.2	8.6	3.5	2,600	10.5	38	9.66	48.9	2,731
Arkansas	11	203.6	7.3	18.5	1,100	4.4	100	10.60	54.8	3,636
Ohio	54	199.6	7.1	3.7	1,700	6.9	31	10.31	60.0	3,588
Louisiana	23	189.4	6.8	8.2	1,700	6.9	74	12.03	63.8	1,471
California	90	184.0	6.6	2.0	1,700	6.9	19	11.92	50.4	1,529
Oklahoma	37	159.4	5.7	4.3	1,600	6.5	43	12.91	42.6	2,000
Pennsylvania	47	149.5	5.4	3.2	1,500	6.0	32	12.35	38.5	2,067
Indiana	32	88.1	3.2	2.8	900	3.6	28	8.93	56.2	-
Tennessee	16	81.5	2.9	5.1	700	2.8	44	7.54	63.6	5,000
North Carolina	21	80.2	2.9	3.8	700	2.8	33	12.44	47.4	2,429
Illinois	35	74.4	2.7	2.1	800	3.2	23	9.75	50.8	-
New Jersey	30	65.8	2.4	2.2	700	2.8	23	10.20	44.7	1,286
Georgia	11	59.9	2.1	5.4	400	1.6	36	10.86	65.3	-
Wisconsin	22	54.8	2.0	2.5	600	2.4	27	10.60	50.0	2,167
Oregon	17	52.6	1.9	3.1	500	2.0	29	12.00	40.9	1,600
New York	36	51.3	1.8	1.4	500	2.0	14	11.00	45.8	3,000
Washington	14	41.7	1.5	3.0	300	1.2	21	11.75	60.7	3,000
Alabama	16	39.0	1.4	2.4	400	1.6	25	7.14	58.5	2,250
Utah	8	34.8	1.2	4.3	200	0.8	25	12.50	52.6	-
Minnesota	14	31.2	1.1	2.2	300	1.2	21	11.60	42.9	2,667
Missouri	17	23.9	0.9	1.4	300	1.2	18	8.50	52.7	2,000
Kansas	11	18.4	0.7	1.7	200	0.8	18	8.50	42.9	500
Florida	17	14.6	0.5	0.9	200	0.8	12	9.00	40.4	1,000
Mississippi	6	13.1	0.5	2.2	100	0.4	17	20.00	41.2	-
Massachusetts	16	12.4	0.4	0.8	100	0.4	6	9.50	59.7	2,000
Kentucky	11	(D)	-	-	175 *	0.7	16	-	-	2,286
South Carolina	8	(D)	-	-	375 *	1.5	47	-	-	-
Maryland	6	(D)	-	-	175 *	0.7	29	-	-	571
Delaware	4	(D)	-	-	175 *	0.7	44	-	-	1,714

Source: 1992 *Economic Census*. The states are in descending order of shipments or establishments (if shipment data are missing for the majority). The symbol (D) appears when data are withheld to prevent disclosure of competitive information. States marked with (D) are sorted by number of establishments. A dash (-) indicates that the data element cannot be calculated; * indicates the midpoint of a range.

3499 - FABRICATED METAL PRODUCTS, NEC

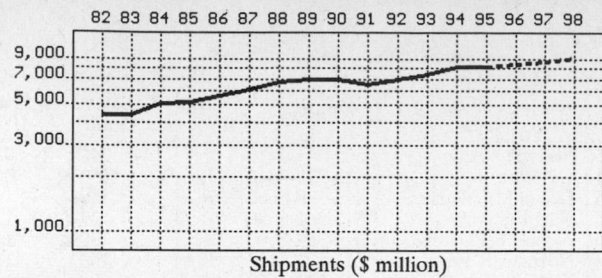

Shipments ($ million)

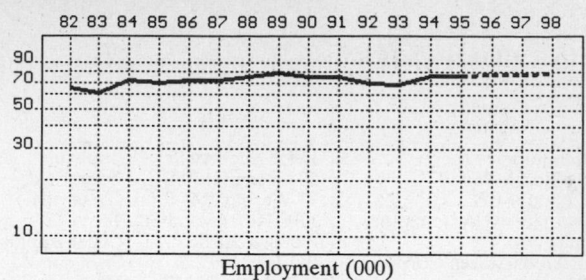

Employment (000)

GENERAL STATISTICS

Year	Companies	Establishments Total	Establishments with 20 or more employees	Employment Total (000)	Employment Production Workers (000)	Employment Hours (Mil)	Compensation Payroll ($ mil)	Compensation Wages ($/hr)	Production Cost of Materials	Production Value Added by Manufacture	Production Value of Shipments	Production Capital Invest.
1982	2,910	2,982	811	65.0	48.8	94.2	1,105.8	7.59	2,039.8	2,304.1	4,406.0	114.1
1983		2,773	822	61.1	46.5	89.5	1,061.5	7.83	2,025.5	2,382.9	4,399.8	97.3
1984		2,564	833	72.5	55.8	112.2	1,213.4	7.05	2,441.8	2,718.3	5,117.5	167.3
1985		2,355	843	70.0	53.5	107.0	1,277.8	7.81	2,395.4	2,809.1	5,216.2	172.4
1986		2,236	817	71.4	53.6	99.0	1,374.9	9.09	2,589.4	3,063.4	5,627.8	138.7
1987	3,720	3,782	849	72.5	52.7	105.6	1,525.9	8.92	2,778.0	3,380.9	6,148.6	152.5
1988		3,376	887	75.6	55.0	110.7	1,631.8	9.09	3,125.5	3,572.4	6,678.3	115.1
1989		3,100	908	78.7	56.2	115.2	1,718.9	9.30	3,173.8	3,768.9	6,947.0	148.3
1990		2,960	896	75.1	57.7	118.4	1,792.4	9.29	3,119.6	3,762.3	6,904.6	206.9
1991		2,938	830	75.6	53.7	111.0	1,785.6	9.74	2,976.3	6,540.0	6,517.7	148.3
1992	3,369	3,444	775	69.8	49.9	100.8	1,781.0	10.55	3,240.1	3,714.9	6,944.6	208.0
1993		2,993	771	68.8	50.4	103.5	1,779.9	10.52	3,366.0	3,953.0	7,311.8	181.6
1994		3,277P	830P	76.6	56.5	114.7	1,976.9	10.65	3,772.0	4,391.9	8,119.4	325.0
1995		3,325P	829P	76.7P	55.2P	114.3P	2,065.1P	11.05P	3,789.5P	4,412.3P	8,157.1P	240.8P
1996		3,374P	828P	77.4P	55.5P	115.4P	2,140.0P	11.34P	3,920.7P	4,565.1P	8,439.5P	251.3P
1997		3,423P	827P	78.1P	55.8P	116.6P	2,215.0P	11.62P	4,051.9P	4,717.8P	8,722.0P	261.8P
1998		3,472P	826P	78.8P	56.1P	117.7P	2,289.9P	11.91P	4,183.1P	4,870.6P	9,004.4P	272.3P

Sources: 1982, 1987, 1992 *Economic Census*; *Annual Survey of Manufactures*, 83-86, 88-91, 93-94. Establishment counts for non-Census years are from *County Business Patterns*; establishment values for 83-84 are extrapolations. 'P's show projections by the editors. Industries reclassified in 87 will not have data for prior years.

INDICES OF CHANGE

Year	Companies	Establishments Total	Establishments with 20 or more employees	Employment Total (000)	Employment Production Workers (000)	Employment Hours (Mil)	Compensation Payroll ($ mil)	Compensation Wages ($/hr)	Production Cost of Materials	Production Value Added by Manufacture	Production Value of Shipments	Production Capital Invest.
1982	86	87	105	93	98	93	62	72	63	62	63	55
1983		81	106	88	93	89	60	74	63	64	63	47
1984		74	107	104	112	111	68	67	75	73	74	80
1985		68	109	100	107	106	72	74	74	76	75	83
1986		65	105	102	107	98	77	86	80	82	81	67
1987	110	110	110	104	106	105	86	85	86	91	89	73
1988		98	114	108	110	110	92	86	96	96	96	55
1989		90	117	113	113	114	97	88	98	101	100	71
1990		86	116	108	116	117	101	88	96	101	99	99
1991		85	107	108	108	110	100	92	92	176	94	71
1992	100	100	100	100	100	100	100	100	100	100	100	100
1993		87	99	99	101	103	100	100	104	106	105	87
1994		95P	107P	110	113	114	111	101	116	118	117	156
1995		97P	107P	110P	111P	113P	116P	105P	117P	119P	117P	116P
1996		98P	107P	111P	111P	115P	120P	107P	121P	123P	122P	121P
1997		99P	107P	112P	112P	116P	124P	110P	125P	127P	126P	126P
1998		101P	107P	113P	112P	117P	129P	113P	129P	131P	130P	131P

Sources: Same as General Statistics. Values reflect change from the base year, 1992. Values above 100 mean greater than 92, values below 100 mean less than 92, and a value of 100 in the 82-91 or 93-98 period means same as 92. 'P's mark projections by the editors.

SELECTED RATIOS

For 1994	Avg. of All Manufact.	Analyzed Industry	Index	For 1994	Avg. of All Manufact.	Analyzed Industry	Index
Employees per Establishment	49	23	48	Value Added per Production Worker	134,084	77,733	58
Payroll per Establishment	1,500,273	603,354	40	Cost per Establishment	5,045,178	1,151,223	23
Payroll per Employee	30,620	25,808	84	Cost per Employee	102,970	49,243	48
Production Workers per Establishment	34	17	50	Cost per Production Worker	146,988	66,761	45
Wages per Establishment	853,319	372,821	44	Shipments per Establishment	9,576,895	2,478,060	26
Wages per Production Worker	24,861	21,620	87	Shipments per Employee	195,460	105,997	54
Hours per Production Worker	2,056	2,030	99	Shipments per Production Worker	279,017	143,706	52
Wages per Hour	12.09	10.65	88	Investment per Establishment	321,011	99,191	31
Value Added per Establishment	4,602,255	1,340,418	29	Investment per Employee	6,552	4,243	65
Value Added per Employee	93,930	57,336	61	Investment per Production Worker	9,352	5,752	62

Sources: Same as General Statistics. The 'Average of All Manufacturing' column represents the average of all manufacturing industries reported for the most recent complete year available. The Index shows the relationship between the Average and the Analyzed Industry. For example, 100 means that they are equal; 500 that the Analyzed Industry is five times the average; 50 means that the Analyzed Industry is half the national average. The abbreviation 'na' is used to show that data are 'not available'.

LEADING COMPANIES   Number shown: **75**   Total sales ($ mil): **3,342**   Total employment (000): **29.6**

Company Name	Address				CEO Name	Phone	Co. Type	Sales ($ mil)	Empl. (000)
Steel Technologies Inc	15415 Shelbyville	Louisville	KY	40245	Merwin J Ray	502-245-2110	P	241	0.5
Bete Fog Nozzle Inc	50 Greenfield St	Greenfield	MA	01301	DL Bete	413-772-0846	R	150	0.1
Werner Ladder Co	93 Werner Rd	Greenville	PA	16125	Donald M Werner	412-588-8600	R	150	1.6
Dura Automotive Systems Inc	1708 Northwood St	Troy	MI	48084	Karl F Storrie	313-362-8300	R	129	1.0
LeFebure Corp	PO Box 2028	Cedar Rapids	IA	52406	Joseph E Patton	319-369-5000	S	100*	1.1
Seymour Housewares Corp	PO Box 408	Seymour	IN	47274	Norman R Proulx	812-522-5130	R	100	1.2
Kurt Manufacturing Co	5280 Main St NE	Minneapolis	MN	55421	William G Kuban	612-572-1500	R	90	1.0
Mercury Aircraft Inc	17 Wheeler Av	Hammondsport	NY	14840	Joseph Meade III	607-569-4200	R	90	0.7
Pneumafil Corp	4500 Chesapeake	Charlotte	NC	28216	Bob Barbee	704-399-7441	R	90	0.6
Crenlo Inc	1600 4th Av NW	Rochester	MN	55901	Charles A Elliott	507-289-3371	R	89	0.9
Mayville Engineering Company	715 South St	Mayville	WI	53050	Carl N Bachhuber	414-387-4500	R	80	0.6
Sonoco Products Co	PO Box 668	Hartselle	AL	35640	Charles Reid	205-773-6581	D	80	0.9
DBA Sentry Group	900 Linden Av	Rochester	NY	14625	Douglas F Brush	716-381-4900	RA	74	0.6
Industrial Acoustics Company	1160 Commerce Av	Bronx	NY	10462	Martin Hirschorn	718-931-8000	P	72	0.7
Suspa Inc	3970 R B Chaffee	Grand Rapids	MI	49548	Erina Hanka	616-241-4200	R	64	0.2
MLX Corp	1000 Center Pl	Norcross	GA	30093	Brian R Esher	404-798-0677	P	61	0.6
Louisville Ladder Corp	1163 Algonquin	Louisville	KY	40208	DL Pringle	502-636-2811	S	60	0.3
Maysteel Corp	PO Box 1240	Menomonee Fls	WI	53052	AG Janos	414-387-5000	S	60	0.8
Sears Manufacturing Co	PO Box 3667	Davenport	IA	52802	I Weir Sears Jr	319-383-2800	R	60	0.5
Hitachi Magnetics Corp	7800 Neff Rd	Edmore	MI	48829	Karl Hiramoto	517-427-5151	S	58	0.5
Quamco Inc	18 Industrial Dr	Holden	MA	01520	John M Prosser	508-829-4491	R	57*	0.6
Unarco Industries Inc	701 16th Av E	Springfield	TN	37172	Bruce Wise	615-384-3531	S	56	0.7
Metalcraft of Mayville	1000 Metalcraft Dr	Mayville	WI	53050	EA Gallun Jr	414-387-3150	R	55	0.4
Arnold Engineering Co	300 N West St	Marengo	IL	60152	Walter T Benecki	815-568-2000	S	50	0.6
Walker Magnetics Group Inc	17 Rockdale St	Worcester	MA	01606	Hendrik Los	508-853-3232	R	48	0.6
Burgess-Norton Mfg Co	737 Peyton St	Geneva	IL	60134	CJ Brewer	708-232-4100	S	45*	0.8
Continental Metal Specialty Inc	PO Box 725	Richmond	KY	40475	George Hommel	606-623-7411	R	45	0.8
Energy Absorption Systems Inc	1 E Wacker Dr	Chicago	IL	60601	George D Ebersole	312-467-6750	S	44	0.4
Toyo Seat USA Corp	2155 S Almont Av	Imlay City	MI	48444	Michio Yamada	313-724-0300	R	40	0.1
Emson Inc	1100 Boston Av	Bridgeport	CT	06610	E Meshberg	203-366-4501	R	38*	0.5
FireKing International Inc	101 Security Pkwy	New Albany	IN	47150	Van G Carlisle	812-948-8400	R	37	0.4
GNC Industries Inc	2000 Pasadena Av	Los Angeles	CA	90031	Parviz Nazarian	213-223-1115	R	35	0.3
Zero-East	288 Main St	Monson	MA	01057	Louis Shew	413-267-5561	D	35*	0.4
Peerless Tube Co	58 Locust Av	Bloomfield	NJ	07003	F Remington Jr	201-743-5100	P	34	0.6
Xaloy Inc	102 Xaloy Way	Pulaski	VA	24301	W Cox	703-980-7560	R	32	0.3
Kunkle Industries Inc	PO Box 1740	Fort Wayne	IN	46801	Richard Mattie	219-747-3405	D	31	0.3
MAGNET Inc	7 Chamber Dr	Washington	MO	63090	Joseph Patane	314-239-5661	R	31	0.4
Samuel Strapping Systems	PO Box 32468	Columbus	OH	43232	AN Drechsel	614-864-3400	S	30	<0.1
Metal Powder Products Inc	10333 N Meridian	Indianapolis	IN	46290	Arlan Clayton	317-573-2420	R	29	0.3
Kay Home Products Inc	1971 W 85th St	Cleveland	OH	44102	Felix Tarorick	216-631-2400	S	27	0.3
Allied Safe and Vault Company	425 W 2nd Av	Spokane	WA	99204	TE Hunt	509-747-1123	R	26	0.3
American Security Products Co	11925 Pacific Av	Fontana	CA	92337	Frank Cademartori	909-685-9680	R	26	0.3
Hancock Manufacturing	PO Box 310	Toronto	OH	43964	Jack F Silcott	614-537-1581	S	25	0.2
Kooima Manufacturing	PO Box 178	Rock Valley	IA	51247	Wayne Groeneweg	712-476-5315	D	25	0.3
Maxcor Manufacturing Co	PO Box 7228	Co Springs	CO	80933	R Marold	719-598-4606	R	25	0.3
UGIMAG Inc	405 Elm St	Valparaiso	IN	46383	Antoine Darbois	219-462-3131	S	25	0.2
Supra Products Inc	PO Box 3167	Salem	OR	97302	Frank Consalvo	503-581-9101	R	21*	0.2
Cannon Equipment Southeast	PO Box 1446	Chattanooga	TN	37401	James A Bouldin	615-752-1000	S	20	0.2
P-W Industries Inc	801 W Street Rd	Feasterville	PA	19053	K Koundouriotis	215-364-3807	R	20	<0.1
Reelcraft Industries Inc	PO Box 248	Columbia City	IN	46725	Stan R Penn Jr	219-248-8188	R	20	0.1
Seiz Corp	PO Box 217	Perkasie	PA	18944	Frederick G Seiz	215-257-3600	D	20	0.1
Will-Burt Co	169 S Main St	Orrville	OH	44667	Dennis Donohue	216-682-7015	R	20*	0.3
Amcast Industrial Corp	PO Box 38	Geneva	IN	46740	Terry Garner	219-368-7246	D	19	0.2
DS Manufacturing Inc	67 5th St NE	Pine Island	MN	55963	Jim Priceler	507-356-8322	R	19*	0.2
National Service Industries	455 Academy Dr	Northbrook	IL	60062	James McClung	708-564-4550	D	19*	0.2
Stanco Metal Products Inc	PO Box 307	Grand Haven	MI	49417	W Stansberry III	616-842-5000	R	19	0.1
Amco Engineering Co	3801 N Rose St	Schiller Park	IL	60176	EV Anderson	708-671-6670	R	18	0.4
MDC Vacuum Products Corp	23842 Cabot Blv	Hayward	CA	94545	Joe Brownell	510-887-6100	R	18*	0.2
Plastic Dress-Up Co	11077 E Rush St	S El Monte	CA	91733	Myron Funk	818-442-7711	R	18*	0.4
Angus-Palm Industries Inc	PO Box 610	Watertown	SD	57201	John Calvin	605-886-5681	R	16	0.2
Jewel Case Corp	300 Niantic Av	Providence	RI	02907	Donald P Wolfe	401-943-1400	R	16	0.4
Mott Metallurgical Corp	84 Spring Ln	Farmington	CT	06032	Harry J Gray	203-677-7311	R	16	0.1
Standard Iron and Wire Works	207 Dundas Rd	Monticello	MN	55362	LT Demeules	612-295-8700	R	16	0.2
Jarvis Pemco	PO Box 1068	Kalamazoo	MI	49005	James L Campbell	616-349-9631	D	15	0.1
Dynamic Materials Corp	551 Aspen Ridge Dr	Lafayette	CO	80026	Paul Lange	303-666-6551	P	15	<0.1
Hansman Industries Inc	PO Box 0210	Stillwater	MN	55082	Ronald J Herold	612-439-7202	R	15	0.2
Lastad Properties Inc	11911 Industrial Av	South Gate	CA	90280	David Page	213-636-8124	R	15	<0.1
Martin Wheel Company Inc	PO Box 157	Tallmadge	OH	44278	John McCarthy	216-633-3278	R	15	<0.1
Metalfab Inc	401 Madison St	Beaver Dam	WI	53916	MJ Splaine	414-885-3381	R	15	0.3
Tech-Etch Inc	45 Aldrin Rd	Plymouth	MA	02360	George Keeler	508-747-0300	R	15	0.4
Thomas and Skinner Inc	PO Box 150-B	Indianapolis	IN	46206	Norris Krall	317-923-2501	R	15	0.2
Globe Electronic Hardware Inc	PO Box 770727	Woodside	NY	11377	Caroline Dennehy	718-457-0303	R	15	<0.1
Amer Grinding & Machine Co	2000 N Mango	Chicago	IL	60639	WE Kuchar	312-889-4343	R	14	0.1
Claude Sintz Inc	PO Box 167	Deshler	OH	43516	H William Gooding	419-278-1010	R	14	0.2
Dynamic Engineering Inc	703 Middle Ground	Newport News	VA	23606	David H Mullins	804-873-1344	R	14	0.2

Source: Ward's Business Directory of U.S. Private and Public Companies, Volumes 1 and 2, 1996. The company type code used is as follows: P - Public, R - Private, S - Subsidiary, D - Division, J - Joint Venture, A - Affiliate, G - Group. Sales are in millions of dollars, employees are in thousands. An asterisk (*) indicates an estimated sales volume. The symbol < stands for 'less than'. Company names and addresses are truncated, in some cases, to fit into the available space.

MATERIALS CONSUMED

Material	Quantity	Delivered Cost ($ million)
Materials, ingredients, containers, and supplies	(X)	2,685.7
Metal fittings, flanges, and unions for piping systems (except forgings)	(X)	5.4
Metal stampings	(X)	42.5
Other fabricated metal products (except forgings)	(X)	84.7
Forgings	(X)	11.7
Iron and steel castings (rough and semifinished)	(X)	190.8
Other nonferrous castings (rough and semifinished)	(X)	10.5
Steel bars and bar shapes	(X)	57.1
Steel sheet and strip, including tin plate	(X)	435.8
Steel plate	(X)	54.9
Steel structural shapes	(X)	24.9
Steel pipes	(X)	24.9
All other steel shapes and forms	(X)	111.9
Copper and copper-base alloy pipe and tube	(X)	1.4
All other copper and copper-base alloy shapes and forms	(X)	14.4
Aluminum and aluminum-base alloy sheet, plate, foil, and welded tubing	(X)	33.5
Aluminum and aluminum-base alloy extruded shapes, including extruded rod, bar, pipe, tube, etc.	(X)	45.1
Other aluminum and aluminum-base alloy shapes and forms	(X)	14.1
Other nonferrous shapes and forms	(X)	8.9
Metal powders	(X)	144.7
Plastics resins consumed in the form of granules, pellets, powders, liquids, etc.	(X)	31.8
All other materials and components, parts, containers, and supplies	(X)	836.1
Materials, ingredients, containers, and supplies, nsk	(X)	500.7

Source: 1992 Economic Census. Explanation of symbols used: (D): Withheld to avoid disclosure of competitive data; na: Not available; (S): Withheld because statistical norms were not met; (X): Not applicable; (Z): Less than half the unit shown; nec: Not elsewhere classified; nsk: Not specified by kind; - : zero; * : 10-19 percent estimated; ** : 20-29 percent estimated.

PRODUCT SHARE DETAILS

Product or Product Class	% Share	Product or Product Class	% Share
Fabricated metal products, nec	100.00	all cemented carbide parts	6.77
Fabricated metal safes and vaults (fire-resistive and burglary-resistive)	2.49	Iron and steel powder metallurgy parts, excluding bearings, gears, and machine cutting tools and all cemented carbide parts	52.66
Fabricated metal safes and vaults (fire-resistive and burglary-resistive)	78.25	Nickel-cobalt-base super alloy powder metallurgy parts, excluding bearings, gears, and machine cutting tools and all cemented carbide parts	2.74
Fabricated metal safe deposit boxes	7.60		
All other fabricated metal bank and security vaults and equipment (including bank security lockers, night depositories, etc.)	13.60	Tungsten metal and tungsten-base alloy powder metallurgy parts, excluding bearings, gears, and machine cutting tools and all cemented carbide parts	20.87
Safes and vaults, nsk	0.55	Other powder metallurgy materials, excluding bearings, gears, and machine cutting tools and all cemented carbide parts	12.04
Fabricated collapsible metal tubes	0.86		
Fabricated collapsible aluminum tubes	82.32	Powder metallurgy parts, excluding bearings, gears, and machine cutting tools and all cemented carbide parts, nsk	1.51
Other fabricated collapsible metal tubes, including tin, tin-coated, tin-lead alloy, and lead	15.36	All other fabricated metal products, nec	50.23
Collapsible tubes, nsk	2.32	Permanent magnets, except ceramic permanent magnets	7.14
Flat metal strapping	4.56	Fabricated metal assemblies of railroad frogs, switches, and crossings	2.08
Metal ladders	3.85		
Metal step and platform ladders	48.69	Fabricated steel boxes for packaging and shipping	0.80
Metal rung-type ladders (single, trestle, extension, sectional, etc.)	46.79	Fabricated steel boxes other than for shipping (ammunition boxes, jewelry cases, etc.)	5.10
Metal ladder-type step stools	0.95	Stamped metal wheels for golf carts, lawn mowers, etc., (disc type)	1.96
Ladder accessories (metal), including levelors, ladder feet, ladder jacks, roof hooks, bucket shelves, etc.	2.85	Metal spools and reels	1.27
Metal ladders, nsk	0.71	Other fabricated metal products, nec, including metal ironing boards, and metal memorial tablets and grave markers	78.06
Powder metallurgy parts, excluding bearings, gears, and machine cutting tools and all cemented carbide parts	13.50		
Aluminum and aluminum-base alloy powder metallurgy parts, excluding bearings, gears, and machine cutting tools and all cemented carbide parts	3.41	All other fabricated metal products, nec, nsk	3.60
Copper and copper-base alloy powder metallurgy parts, excluding bearings, gears, and machine cutting tools and		Fabricated metal products, nec, nsk	24.50

Source: 1992 Economic Census. The values shown are percent of total shipments in an industry. Values of indented subcategories are summed in the main heading. The symbol (D) appears when data are withheld to prevent disclosure of competitive information. The abbreviation nsk stands for 'not specified by kind' and nec for 'not elsewhere classified'.

INPUTS AND OUTPUTS FOR FABRICATED METAL PRODUCTS, NEC

Economic Sector or Industry Providing Inputs	%	Sector	Economic Sector or Industry Buying Outputs	%	Sector
Imports	23.5	Foreign	Gross private fixed investment	11.5	Cap Inv
Blast furnaces & steel mills	19.5	Manufg.	Exports	10.2	Foreign
Wholesale trade	9.1	Trade	Fabricated metal products, nec	3.6	Manufg.
Fabricated metal products, nec	5.6	Manufg.	Petroleum refining	3.1	Manufg.
Aluminum rolling & drawing	3.8	Manufg.	Federal Government purchases, national defense	3.0	Fed Govt
Internal combustion engines, nec	2.0	Manufg.	Personal consumption expenditures	2.8	
Electric services (utilities)	1.8	Util.	Photographic equipment & supplies	2.6	Manufg.
Motor freight transportation & warehousing	1.8	Util.	Primary batteries, dry & wet	2.4	Manufg.
Nonferrous wire drawing & insulating	1.7	Manufg.	Metal foil & leaf	2.3	Manufg.
Advertising	1.7	Services	Toilet preparations	2.2	Manufg.
Primary aluminum	1.5	Manufg.	Fabricated plate work (boiler shops)	1.9	Manufg.
Banking	1.2	Fin/R.E.	Maintenance of railroads	1.6	Constr.
Hardware, nec	1.1	Manufg.	Blast furnaces & steel mills	1.6	Manufg.
Petroleum refining	1.1	Manufg.	Manufacturing industries, nec	1.5	Manufg.
Maintenance of nonfarm buildings nec	1.0	Constr.	Electronic computing equipment	1.4	Manufg.
Primary metal products, nec	1.0	Manufg.	Miscellaneous plastics products	1.3	Manufg.
Iron & steel foundries	0.9	Manufg.	Crude petroleum & natural gas	1.2	Mining
Screw machine and related products	0.9	Manufg.	Lubricating oils & greases	1.2	Manufg.
Machinery, except electrical, nec	0.8	Manufg.	Machinery, except electrical, nec	1.2	Manufg.
Miscellaneous fabricated wire products	0.8	Manufg.	Radio & TV communication equipment	1.2	Manufg.
Nonferrous rolling & drawing, nec	0.8	Manufg.	Paper mills, except building paper	0.9	Manufg.
Gas production & distribution (utilities)	0.8	Util.	Typewriters & office machines, nec	0.9	Manufg.
Eating & drinking places	0.8	Trade	Local transit facility construction	0.8	Constr.
Plating & polishing	0.7	Manufg.	Drugs	0.8	Manufg.
Primary copper	0.7	Manufg.	Logging camps & logging contractors	0.8	Manufg.
Wood products, nec	0.7	Manufg.	Metal doors, sash, & trim	0.8	Manufg.
Air transportation	0.7	Util.	Metal office furniture	0.8	Manufg.
Real estate	0.7	Fin/R.E.	Paperboard mills	0.8	Manufg.
Communications, except radio & TV	0.6	Util.	Telephone & telegraph apparatus	0.8	Manufg.
Railroads & related services	0.6	Util.	Railroads & related services	0.8	Util.
Copper rolling & drawing	0.5	Manufg.	Water transportation	0.8	Util.
Glass & glass products, except containers	0.5	Manufg.	Aluminum rolling & drawing	0.7	Manufg.
Metal stampings, nec	0.5	Manufg.	Mechanical measuring devices	0.7	Manufg.
Paints & allied products	0.5	Manufg.	Millwork	0.7	Manufg.
Equipment rental & leasing services	0.5	Services	Prefabricated metal buildings	0.7	Manufg.
Cyclic crudes and organics	0.4	Manufg.	Engine electrical equipment	0.6	Manufg.
Gaskets, packing & sealing devices	0.4	Manufg.	Hand & edge tools, nec	0.6	Manufg.
Metal coating & allied services	0.4	Manufg.	Motor vehicles & car bodies	0.6	Manufg.
Miscellaneous plastics products	0.4	Manufg.	Nonferrous wire drawing & insulating	0.6	Manufg.
Paperboard containers & boxes	0.4	Manufg.	Oil field machinery	0.6	Manufg.
Primary nonferrous metals, nec	0.4	Manufg.	Hospitals	0.6	Services
Special dies & tools & machine tool accessories	0.4	Manufg.	Coal	0.5	Mining
Nonferrous castings, nec	0.3	Manufg.	Adhesives & sealants	0.5	Manufg.
Plastics materials & resins	0.3	Manufg.	Blankbooks & looseleaf binders	0.5	Manufg.
Engineering, architectural, & surveying services	0.3	Services	Iron & steel foundries	0.5	Manufg.
Management & consulting services & labs	0.3	Services	Motors & generators	0.5	Manufg.
Nonmetallic mineral services	0.2	Mining	Pens & mechanical pencils	0.5	Manufg.
Abrasive products	0.2	Manufg.	Railroad equipment	0.5	Manufg.
Aluminum castings	0.2	Manufg.	Retail trade, except eating & drinking	0.5	Trade
Ball & roller bearings	0.2	Manufg.	Wholesale trade	0.5	Trade
Iron & steel forgings	0.2	Manufg.	Aircraft	0.4	Manufg.
Metal cans	0.2	Manufg.	Aircraft & missile equipment, nec	0.4	Manufg.
Metal heat treating	0.2	Manufg.	Automotive stampings	0.4	Manufg.
Primary lead	0.2	Manufg.	Blowers & fans	0.4	Manufg.
Sawmills & planning mills, general	0.2	Manufg.	Engineering & scientific instruments	0.4	Manufg.
Wood pallets & skids	0.2	Manufg.	Heating equipment, except electric	0.4	Manufg.
Water transportation	0.2	Util.	Metal stampings, nec	0.4	Manufg.
Accounting, auditing & bookkeeping	0.2	Services	Paints & allied products	0.4	Manufg.
Automotive rental & leasing, without drivers	0.2	Services	Paperboard containers & boxes	0.4	Manufg.
Automotive repair shops & services	0.2	Services	Ship building & repairing	0.4	Manufg.
Business/professional associations	0.2	Services	Federal Government purchases, nondefense	0.4	Fed Govt
Computer & data processing services	0.2	Services	S/L Govt. purch., higher education	0.4	S/L Govt
Legal services	0.2	Services	Cold finishing of steel shapes	0.3	Manufg.
Noncomparable imports	0.2	Foreign	Electrical equipment & supplies, nec	0.3	Manufg.
Lubricating oils & greases	0.1	Manufg.	Elevators & moving stairways	0.3	Manufg.
Machine tools, metal cutting types	0.1	Manufg.	Hardware, nec	0.3	Manufg.
Machine tools, metal forming types	0.1	Manufg.	Lawn & garden equipment	0.3	Manufg.
Manifold business forms	0.1	Manufg.	Metal household furniture	0.3	Manufg.
Primary zinc	0.1	Manufg.	Motorcycles, bicycles, & parts	0.3	Manufg.
Insurance carriers	0.1	Fin/R.E.	Nonmetallic mineral products, nec	0.3	Manufg.
Hotels & lodging places	0.1	Services	Polishes & sanitation goods	0.3	Manufg.
U.S. Postal Service	0.1	Gov't	Radio & TV receiving sets	0.3	Manufg.
			Service industry machines, nec	0.3	Manufg.
			Special dies & tools & machine tool accessories	0.3	Manufg.
			Special industry machinery, nec	0.3	Manufg.
			Surgical & medical instruments	0.3	Manufg.
			Surgical appliances & supplies	0.3	Manufg.
			Wood products, nec	0.3	Manufg.

Continued on next page.

INPUTS AND OUTPUTS FOR FABRICATED METAL PRODUCTS, NEC - Continued

Economic Sector or Industry Providing Inputs	%	Sector	Economic Sector or Industry Buying Outputs	%	Sector
			Motor freight transportation & warehousing	0.3	Util.
			S/L Govt. purch., other general government	0.3	S/L Govt
			Agricultural, forestry, & fishery services	0.2	Agric.
			Maintenance of local transit facilities	0.2	Constr.
			Maintenance of nonfarm buildings nec	0.2	Constr.
			Office buildings	0.2	Constr.
			Railroad construction	0.2	Constr.
			Ammunition, except for small arms, nec	0.2	Manufg.
			Boat building & repairing	0.2	Manufg.
			Carbon paper & inked ribbons	0.2	Manufg.
			Chemical preparations, nec	0.2	Manufg.
			Copper rolling & drawing	0.2	Manufg.
			Electron tubes	0.2	Manufg.
			Electronic components nec	0.2	Manufg.
			Food products machinery	0.2	Manufg.
			Household refrigerators & freezers	0.2	Manufg.
			Iron & steel forgings	0.2	Manufg.
			Machine tools, metal cutting types	0.2	Manufg.
			Metal barrels, drums, & pails	0.2	Manufg.
			Metalworking machinery, nec	0.2	Manufg.
			Miscellaneous metal work	0.2	Manufg.
			Nonferrous rolling & drawing, nec	0.2	Manufg.
			Paving mixtures & blocks	0.2	Manufg.
			Signs & advertising displays	0.2	Manufg.
			Steel pipe & tubes	0.2	Manufg.
			Steel wire & related products	0.2	Manufg.
			Structural wood members, nec	0.2	Manufg.
			Wood household furniture	0.2	Manufg.
			Wood kitchen cabinets	0.2	Manufg.
			Wood partitions & fixtures	0.2	Manufg.
			Real estate	0.2	Fin/R.E.
			Doctors & dentists	0.2	Services
			Iron & ferroalloy ores	0.1	Mining
			Nonfarm residential structure maintenance	0.1	Constr.
			Agricultural chemicals, nec	0.1	Manufg.
			Brick & structural clay tile	0.1	Manufg.
			Cheese, natural & processed	0.1	Manufg.
			Cutlery	0.1	Manufg.
			Electric housewares & fans	0.1	Manufg.
			Fabricated structural metal	0.1	Manufg.
			Furniture & fixtures, nec	0.1	Manufg.
			Lead pencils & art goods	0.1	Manufg.
			Metal partitions & fixtures	0.1	Manufg.
			Mining machinery, except oil field	0.1	Manufg.
			Nonferrous castings, nec	0.1	Manufg.
			Optical instruments & lenses	0.1	Manufg.
			Pipe, valves, & pipe fittings	0.1	Manufg.
			Sawmills & planning mills, general	0.1	Manufg.
			Silverware & plated ware	0.1	Manufg.
			Truck & bus bodies	0.1	Manufg.
			Truck trailers	0.1	Manufg.
			Watches, clocks, & parts	0.1	Manufg.
			Legal services	0.1	Services

Source: Benchmark Input-Output Accounts for the U.S. Economy, 1982, U.S. Department of Commerce, Washington, D.C., July 1991. Data, as reported in the source, are organized by the 1977 SIC structure in use in 1982 but have been matched, as closely as is possible, to the 1987 SIC structure used in this book.

OCCUPATIONS EMPLOYED BY SIC 349 - MISCELLANEOUS FABRICATED METAL PRODUCTS

Occupation	% of Total 1994	Change to 2005	Occupation	% of Total 1994	Change to 2005
Assemblers, fabricators, & hand workers nec	10.7	0.7	Industrial machinery mechanics	1.8	10.7
Blue collar worker supervisors	4.5	-7.4	Helpers, laborers, & material movers nec	1.8	0.7
Machinists	4.4	0.7	Hand packers & packagers	1.7	-13.7
Welding machine setters, operators	3.8	-9.4	Machine operators nec	1.6	-11.3
Machine tool cutting & forming etc. nec	3.7	20.8	Industrial production managers	1.5	0.7
Machine forming operators, metal & plastic	3.5	-9.4	Secretaries, ex legal & medical	1.4	-8.3
Sales & related workers nec	3.0	0.7	General office clerks	1.3	-14.2
Inspectors, testers, & graders, precision	3.0	0.7	Tool & die makers	1.3	-18.7
Metal & plastic machine workers nec	2.7	-28.7	Bookkeeping, accounting, & auditing clerks	1.3	-24.5
General managers & top executives	2.6	-4.5	Coating, painting, & spraying machine workers	1.3	0.7
Combination machine tool operators	2.4	51.0	Maintenance repairers, general utility	1.2	-9.4
Machine tool cutting operators, metal & plastic	2.4	41.9	Industrial truck & tractor operators	1.1	0.6
Lathe & turning machine tool operators	2.1	-9.4	Machine feeders & offbearers	1.1	-9.4
Welders & cutters	2.1	0.7	Production, planning, & expediting clerks	1.0	0.7
Traffic, shipping, & receiving clerks	2.0	-3.1	Mechanical engineers	1.0	-0.3

Source: Industry-Occupation Matrix, Bureau of Labor Statistics. These data relate to one or more 3-digit SIC industry groups rather than to a single 4-digit SIC. The change reported for each occupation to the year 2005 is a percent of growth or decline as estimated by the Bureau of Labor Statistics. The abbreviation nec stands for 'not elsewhere classified'.

LOCATION BY STATE AND REGIONAL CONCENTRATION

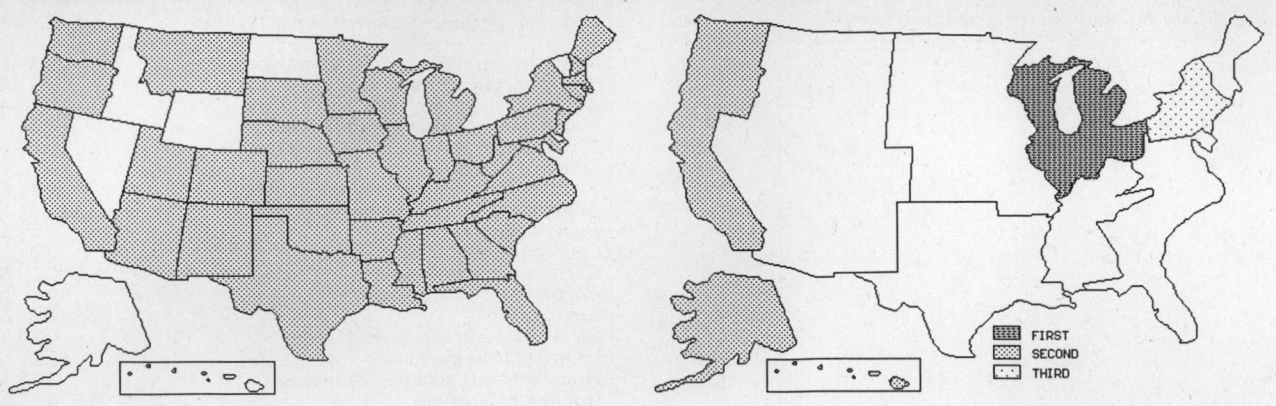

FIRST
SECOND
THIRD

INDUSTRY DATA BY STATE

State	Establish-ments	Shipments			Employment				Cost as % of Shipments	Investment per Employee ($)
		Total ($ mil)	% of U.S.	Per Establ.	Total Number	% of U.S.	Per Establ.	Wages ($/hour)		
Pennsylvania	211	766.2	11.0	3.6	7,900	11.3	37	11.63	48.2	3,228
Illinois	220	725.0	10.4	3.3	5,600	8.0	25	11.87	49.9	2,518
California	380	559.0	8.0	1.5	5,900	8.5	16	10.22	46.0	2,610
Michigan	257	474.7	6.8	1.8	4,800	6.9	19	10.45	42.0	2,375
Ohio	205	443.6	6.4	2.2	4,700	6.7	23	10.28	47.5	5,362
Indiana	137	344.1	5.0	2.5	3,900	5.6	28	10.24	41.5	3,949
Texas	190	297.8	4.3	1.6	2,900	4.2	15	11.42	42.4	3,897
New York	199	292.4	4.2	1.5	3,300	4.7	17	10.55	46.1	1,939
Connecticut	72	290.8	4.2	4.0	2,400	3.4	33	11.28	41.8	2,708
Wisconsin	95	240.7	3.5	2.5	2,300	3.3	24	10.53	46.4	4,217
Massachusetts	93	194.1	2.8	2.1	1,900	2.7	20	11.55	41.4	2,316
New Jersey	96	171.6	2.5	1.8	1,700	2.4	18	10.59	41.7	3,647
Florida	130	159.8	2.3	1.2	1,900	2.7	15	9.59	46.2	1,842
Kentucky	57	154.2	2.2	2.7	1,600	2.3	28	11.50	49.4	2,000
Tennessee	63	146.0	2.1	2.3	1,700	2.4	27	9.22	51.2	1,412
Minnesota	110	134.5	1.9	1.2	1,200	1.7	11	9.79	65.9	2,083
Nebraska	14	118.7	1.7	8.5	1,000	1.4	71	9.76	53.8	1,500
Virginia	40	110.6	1.6	2.8	700	1.0	18	12.33	30.2	3,429
Utah	34	103.0	1.5	3.0	1,100	1.6	32	9.63	34.4	2,091
Iowa	36	100.7	1.5	2.8	1,000	1.4	28	9.47	62.0	3,400
Oklahoma	53	98.5	1.4	1.9	900	1.3	17	9.15	45.0	2,000
North Carolina	64	93.6	1.3	1.5	1,300	1.9	20	10.21	37.5	3,385
Rhode Island	22	86.5	1.2	3.9	1,200	1.7	55	6.76	50.1	1,167
Georgia	50	80.4	1.2	1.6	900	1.3	18	8.54	50.0	2,222
Maryland	30	76.3	1.1	2.5	600	0.9	20	10.13	57.9	3,000
Washington	73	71.9	1.0	1.0	800	1.1	11	12.73	41.4	1,875
Arkansas	39	67.0	1.0	1.7	1,000	1.4	26	8.79	47.5	2,700
Colorado	53	66.6	1.0	1.3	700	1.0	13	7.88	45.9	3,857
Missouri	65	65.7	0.9	1.0	800	1.1	12	9.67	49.9	2,250
Alabama	39	64.0	0.9	1.6	500	0.7	13	10.57	58.4	2,400
Maine	8	58.5	0.8	7.3	500	0.7	63	15.13	31.5	-
Louisiana	48	50.0	0.7	1.0	600	0.9	13	8.88	46.8	1,500
South Carolina	34	46.7	0.7	1.4	800	1.1	24	9.70	64.0	4,750
New Hampshire	19	39.6	0.6	2.1	400	0.6	21	11.00	60.9	-
Oregon	53	33.6	0.5	0.6	400	0.6	8	10.40	51.2	1,250
West Virginia	10	26.8	0.4	2.7	100	0.1	10	12.50	76.5	3,000
Arizona	31	21.5	0.3	0.7	200	0.3	6	8.67	47.4	1,500
South Dakota	7	10.0	0.1	1.4	100	0.1	14	8.00	47.0	-
New Mexico	22	8.5	0.1	0.4	100	0.1	5	7.50	41.2	2,000
Mississippi	17	8.3	0.1	0.5	100	0.1	6	6.50	47.0	4,000
Kansas	14	(D)	-	-	175 *	0.3	13	-	-	1,714
Montana	5	(D)	-	-	175 *	0.3	35	-	-	571

Source: 1992 *Economic Census*. The states are in descending order of shipments or establishments (if shipment data are missing for the majority). The symbol (D) appears when data are withheld to prevent disclosure of competitive information. States marked with (D) are sorted by number of establishments. A dash (-) indicates that the data element cannot be calculated; * indicates the midpoint of a range.

3511 - TURBINES & TURBINE GENERATOR SETS

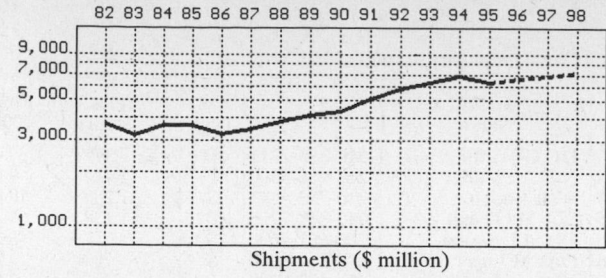

Shipments ($ million)

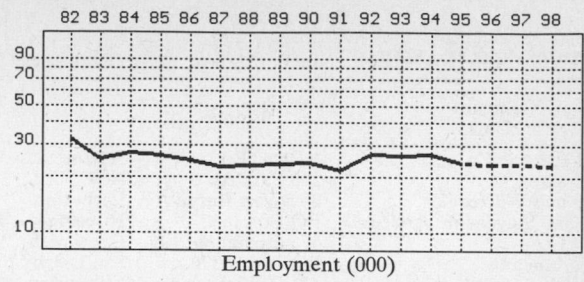

Employment (000)

GENERAL STATISTICS

Year	Com-panies	Establishments		Employment			Compensation		Production ($ million)			
		Total	with 20 or more employees	Total (000)	Production Workers (000)	Hours (Mil)	Payroll ($ mil)	Wages ($/hr)	Cost of Materials	Value Added by Manufacture	Value of Shipments	Capital Invest.
1982	71	88	49	32.4	18.9	37.4	870.8	11.54	1,515.0	2,152.9	3,676.3	127.9
1983		83	47	25.0	14.9	29.0	743.0	12.54	1,267.4	1,699.8	3,197.8	114.0
1984		78	45	27.5	14.6	29.0	868.8	13.82	1,346.1	2,149.0	3,624.0	116.3
1985		74	42	26.6	14.0	29.1	921.3	14.89	1,536.9	2,052.1	3,654.5	95.2
1986		74	44	24.7	13.3	27.6	870.2	14.87	1,416.8	1,814.9	3,219.9	82.1
1987	68	81	42	22.9	11.8	26.7	862.7	13.71	1,579.9	1,973.7	3,447.8	91.6
1988		75	44	23.0	11.9	23.3	826.6	16.63	1,676.2	2,204.2	3,790.6	137.3
1989		77	47	23.9	11.6	21.6	802.2	17.87	1,927.5	2,147.3	4,127.7	125.5
1990		74	42	24.1	12.5	25.3	858.8	16.97	2,328.7	2,259.2	4,356.7	150.9
1991		76	41	21.8	13.0	27.4	962.8	18.54	2,674.4	2,882.9	5,093.0	182.5
1992	64	79	48	27.1	15.0	30.1	1,106.5	18.38	2,690.9	2,952.8	5,842.6	312.0
1993		81	50	26.7	15.0	30.4	1,126.6	18.62	2,684.2	3,808.6	6,234.0	310.1
1994		76P	45P	27.1	15.1	30.0	1,237.1	19.67	3,049.8	3,947.2	6,801.6	246.5
1995		75P	45P	24.1P	13.0P	26.4P	1,123.9P	20.49P	2,831.9P	3,665.2P	6,315.7P	272.8P
1996		75P	45P	23.9P	12.8P	26.1P	1,152.0P	21.13P	2,955.3P	3,824.9P	6,590.9P	288.8P
1997		74P	45P	23.6P	12.7P	25.9P	1,180.1P	21.78P	3,078.7P	3,984.6P	6,866.0P	304.7P
1998		74P	45P	23.4P	12.6P	25.6P	1,208.1P	22.42P	3,202.1P	4,144.3P	7,141.2P	320.7P

Sources: 1982, 1987, 1992 *Economic Census*; *Annual Survey of Manufactures*, 83-86, 88-91, 93-94. Establishment counts for non-Census years are from *County Business Patterns*; establishment values for 83-84 are extrapolations. 'P's show projections by the editors. Industries reclassified in 87 will not have data for prior years.

INDICES OF CHANGE

Year	Com-panies	Establishments		Employment			Compensation		Production ($ million)			
		Total	with 20 or more employees	Total (000)	Production Workers (000)	Hours (Mil)	Payroll ($ mil)	Wages ($/hr)	Cost of Materials	Value Added by Manufacture	Value of Shipments	Capital Invest.
1982	111	111	102	120	126	124	79	63	56	73	63	41
1983		105	98	92	99	96	67	68	47	58	55	37
1984		99	94	101	97	96	79	75	50	73	62	37
1985		94	88	98	93	97	83	81	57	69	63	31
1986		94	92	91	89	92	79	81	53	61	55	26
1987	106	103	88	85	79	89	78	75	59	67	59	29
1988		95	92	85	79	77	75	90	62	75	65	44
1989		97	98	88	77	72	72	97	72	73	71	40
1990		94	88	89	83	84	78	92	87	77	75	48
1991		96	85	80	87	91	87	101	99	98	87	58
1992	100	100	100	100	100	100	100	100	100	100	100	100
1993		103	104	99	100	101	102	101	100	129	107	99
1994		96P	94P	100	101	100	112	107	113	134	116	79
1995		95P	94P	89P	87P	88P	102P	111P	105P	124P	108P	87P
1996		94P	94P	88P	86P	87P	104P	115P	110P	130P	113P	93P
1997		94P	94P	87P	85P	86P	107P	118P	114P	135P	118P	98P
1998		93P	94P	86P	84P	85P	109P	122P	119P	140P	122P	103P

Sources: Same as General Statistics. Values reflect change from the base year, 1992. Values above 100 mean greater than 92, values below 100 mean less than 92, and a value of 100 in the 82-91 or 93-98 period means same as 92. 'P's mark projections by the editors.

SELECTED RATIOS

For 1994	Avg. of All Manufact.	Analyzed Industry	Index	For 1994	Avg. of All Manufact.	Analyzed Industry	Index
Employees per Establishment	49	359	732	Value Added per Production Worker	134,084	261,404	195
Payroll per Establishment	1,500,273	16,382,143	1,092	Cost per Establishment	5,045,178	40,386,597	800
Payroll per Employee	30,620	45,649	149	Cost per Employee	102,970	112,539	109
Production Workers per Establishment	34	200	583	Cost per Production Worker	146,988	201,974	137
Wages per Establishment	853,319	7,814,326	916	Shipments per Establishment	9,576,895	90,069,342	940
Wages per Production Worker	24,861	39,079	157	Shipments per Employee	195,460	250,982	128
Hours per Production Worker	2,056	1,987	97	Shipments per Production Worker	279,017	450,437	161
Wages per Hour	12.09	19.67	163	Investment per Establishment	321,011	3,264,246	1,017
Value Added per Establishment	4,602,255	52,270,305	1,136	Investment per Employee	6,552	9,096	139
Value Added per Employee	93,930	145,653	155	Investment per Production Worker	9,352	16,325	175

Sources: Same as General Statistics. The 'Average of All Manufacturing' column represents the average of all manufacturing industries reported for the most recent complete year available. The Index shows the relationship between the Average and the Analyzed Industry. For example, 100 means that they are equal; 500 that the Analyzed Industry is five times the average; 50 means that the Analyzed Industry is half the national average. The abbreviation 'na' is used to show that data are 'not available'.

LEADING COMPANIES   Number shown: **37**   Total sales ($ mil): **14,578**   Total employment (000): **85.1**

Company Name	Address				CEO Name	Phone	Co. Type	Sales ($ mil)	Empl. (000)
General Electric Co	1 River Rd	Schenectady	NY	12345	DC G-Watling	518-385-2211	D	7,379	37.0
Sequa Corp	200 Park Av	New York	NY	10166	N E Alexander	212-986-5500	P	1,420	9.2
Babcock and Wilcox Co	20 S Van Buren Av	Barberton	OH	44203	Walt Boomer	216-753-4511	S	1,200	10.0
Stewart and Stevenson Services	PO Box 1637	Houston	TX	77251	Bob H O'Neal	713-868-7700	P	1,138	4.3
Howmet Corp	475 Steamboat Rd	Greenwich	CT	06836	David L Squier	203-661-4600	S	777	8.2
Solar Turbines Inc	PO Box 85376	San Diego	CA	92186	Don M Ings	619-544-5000	S	700	4.2
Jason Inc	411 E Wisconsin	Milwaukee	WI	53202	Vincent L Martin	414-277-9300	P	357	3.0
KENETECH Corp	500 Sansome St	San Francisco	CA	94111	Gerald R Alderson	415-398-3825	P	338	1.2
Elliott Turbomachinery Company	N 4th St	Jeannette	PA	15644	Paul R Smiy	412-527-2811	S	230	1.5
New Elliott Corp	N 4th St	Jeannette	PA	15644	Paul Smily	412-527-2811	R	230*	1.5
KENTECH Windpower Inc	500 Sansome St	San Francisco	CA	94111	Joel Canino	415-398-3825	S	130*	0.5
Toshiba International Corp	280 Utah Av	S San Francisco	CA	94080	Toshio Doshida	415-872-2722	S	110*	0.8
Turbo Power&Marine Systems	PO Box 611	Middletown	CT	06457	Randal Hogan	203-343-2000	S	100*	0.3
Voith Hydro Inc	PO Box 712	York	PA	17405	Wolfgang Heine	717-792-7000	S	95	0.6
Coppus Murray Group	Box 8000	Millbury	MA	01527	William F Jones	508-756-8391	D	50	0.4
Clayton Industries Inc	PO Box 5530	El Monte	CA	91734	W Clayton Jr	818-443-9381	R	40	0.8
European Gas Turbines Inc	15950 Park Row	Houston	TX	77084	John Paul	713-492-0222	S	40	0.2
Rotoflow Corporation Inc	540 E Rosecrans	Gardena	CA	90248	Frank Van Gogh	310-329-8447	S	38	0.2
US Turbine Corp	7685 S State Rte 48	Maineville	OH	45039	Larry D Davis	513-683-6100	S	32	0.1
Quabbin	2140 Westover Rd	Chicopee	MA	01022	Michael Corridarl	413-593-6746	D	31	0.2
Power Generation Inc	5309 Commonw	Midlothian	VA	23112	Alex Brnilivich	804-271-1261	D	23*	0.2
Metem Corp	700 Parsippany	Parsippany	NJ	07054	Stephen Chen	201-887-6635	R	20	0.2
Pacific Scientific Co	PO Box 1500	Santa Barbara	CA	93102	Dave Wightman	805-963-2055	D	12	0.1
Harco Laboratories Inc	186 Cedar St	Branford	CT	06405	Paul R Eden	203-483-3700	R	10	<0.1
Microturbo Inc	2707 Forum Dr	Grand Prairie	TX	75051	Robert P Schiller	214-660-5545	S	10*	<0.1
H and H Manufacturing	2 Horne Dr	Folcroft	PA	19032	Tom Tomei	215-532-8100	R	9*	<0.1
NoMac Energy Systems Inc	655 Deep Valley Dr	Rolling Hls Est	CA	90274	Jim Wensley	310-541-2528	R	9*	<0.1
Baker/MO Services Inc	PO Box 445	Cypress	TX	77429	Michael Gibbs	713-351-9847	S	8	0.1
Chromalloy	630 Anchors St NW	Ft Walton Bch	FL	32548	Jack B DeVria	904-244-7684	D	8	0.1
Preco Turbine Services Inc	17619 Aldine	Houston	TX	77073	Roy Goldberg	713-821-9620	R	7*	<0.1
GEA Power Cooling System Inc	5355 Mira Sorrento	San Diego	CA	92121	G E Hoffmeister	619-457-0086	S	6*	<0.1
Turbonetics Energy Inc	968 Albany-Shaker	Latham	NY	12110	Wayne Diesel	518-785-2211	S	6	<0.1
Jaycraft Corp	PO Box 1158	Spring Valley	CA	91979	P Van Vechten	619-670-3900	R	6	<0.1
Boyce Engineering Intern USA	10555 Rockley Rd	Houston	TX	77099	Meherwan P Boyce	713-933-7210	R	5	<0.1
American Motion Systems Inc	1221 Avnida Acaso	Camarillo	CA	93012	Daniel W McGee	805-482-0407	R	2	<0.1
Jarett Industries Inc	15 Saddle Rd	Cedar Knolls	NJ	07927	SH Weisman	201-539-4410	R	2*	<0.1
FD Contours Corp	175 Paularino Av	Costa Mesa	CA	92626	SV Folger	714-546-3030	R	1	<0.1

Source: Ward's Business Directory of U.S. Private and Public Companies, Volumes 1 and 2, 1996. The company type code used is as follows: P - Public, R - Private, S - Subsidiary, D - Division, J - Joint Venture, A - Affiliate, G - Group. Sales are in millions of dollars, employees are in thousands. An asterisk (*) indicates an estimated sales volume. The symbol < stands for 'less than'. Company names and addresses are truncated, in some cases, to fit into the available space.

MATERIALS CONSUMED

Material	Quantity	Delivered Cost ($ million)
Materials, ingredients, containers, and supplies	(X)	2,464.3
Fluid power pumps, motors, and hydrostatic transmissions (hydraulic and pneumatic)	(X)	(D)
Fluid power cylinders and rotary actuators (hydraulic and pneumatic)	(X)	2.3
Fluid power filters (hydraulic and pneumatic)	(X)	(D)
Fluid power hose or tube fittings and assemblies (hydraulic and pneumatic)	(X)	(D)
Fluid power valves (hydraulic and pneumatic)	(X)	30.7
Metal bolts, nuts, screws, washers, rivets, and other screw machine products	(X)	41.9
Metal stampings	(X)	(D)
Metal tanks, heat exchangers, steam condensers, and other boiler products	(X)	(D)
Fabricated structural metal products (except forgings)	(X)	(D)
All other fabricated metal products (except forgings)	(X)	161.2
Iron and steel forgings	(X)	(D)
Nonferrous forgings	(X)	(D)
Iron and steel castings (rough and semifinished)	(X)	330.6
Aluminum and aluminum-base alloy castings (rough and semifinished)	(X)	1.7
Other nonferrous castings (rough and semifinished)	(X)	(D)
Steel bars, bar shapes, and plates	(X)	70.9
Steel sheet and strip, including tin plate	(X)	24.7
All other steel shapes and forms	(X)	19.4
Nonferrous shapes and forms	(X)	34.0
Pistons, piston rings, carburetors, valves (intake and exhaust only)	(X)	(D)
Engine electrical equipment, including spark plugs, magnetos, generators, starters, etc.	(X)	(D)
Integral horsepower electric motors and generators (1 hp or more)	(X)	(D)
Ball bearings (mounted or unmounted)	(X)	4.5
Roller bearings (mounted or unmounted)	(X)	(D)
Plain bearings and bushings	(X)	12.6
Mechanical speed changers, gears, and industrial high-speed drives	(X)	30.1
Turbines purchased for incorporation into turbine generator sets	(X)	(D)
Generators purchased for incorporation into turbine generator sets	(X)	10.4
Gaskets (all types) and asbestos packing	(X)	3.3
Fabricated plastics products (except gaskets, hoses, and belting)	(X)	1.7
Rubber and plastics hose and belting	(X)	0.6
Cutting tools for machine tools	(X)	8.6
All other materials and components, parts, containers, and supplies	(X)	449.6
Materials, ingredients, containers, and supplies, nsk	(X)	184.9

Source: 1992 *Economic Census*. Explanation of symbols used: (D): Withheld to avoid disclosure of competitive data; na: Not available; (S): Withheld because statistical norms were not met; (X): Not applicable; (Z): Less than half the unit shown; nec: Not elsewhere classified; nsk: Not specified by kind; - : zero; * : 10-19 percent estimated; ** : 20-29 percent estimated.

PRODUCT SHARE DETAILS

Product or Product Class	% Share	Product or Product Class	% Share
Turbines, turbine generators, and turbine generator sets	100.00	(sold separately)	7.06
Turbine generator sets	30.65	Parts and accessories for gas turbines, except aircraft (sold separately)	7.11
Turbine generator parts and accessories (sold separately)	(D)	Parts and accessories for wind turbines (sold separately)	(D)
Steam turbines and other vapor turbines	4.05	Turbine generators	2.71
Hydraulic turbines (all sizes)	(D)	Turbine generator parts and accessories (sold separately)	(D)
Gas turbines, except aircraft (all sizes)	32.76		
Wind turbines	(D)		
Parts and accessories for steam and other vapor turbines			

Source: 1992 *Economic Census*. The values shown are percent of total shipments in an industry. Values of indented subcategories are summed in the main heading. The symbol (D) appears when data are withheld to prevent disclosure of competitive information. The abbreviation nsk stands for 'not specified by kind' and nec for 'not elsewhere classified'.

INPUTS AND OUTPUTS FOR TURBINES & TURBINE GENERATOR SETS

Economic Sector or Industry Providing Inputs	%	Sector	Economic Sector or Industry Buying Outputs	%	Sector
Iron & steel forgings	13.1	Manufg.	Gross private fixed investment	32.7	Cap Inv
Imports	11.3	Foreign	Exports	26.7	Foreign
Pipe, valves, & pipe fittings	9.4	Manufg.	Electric services (utilities)	25.3	Util.
Blast furnaces & steel mills	8.2	Manufg.	S/L Govt. purch., gas & electric utilities	3.9	S/L Govt
Wholesale trade	6.2	Trade	Federal Government purchases, nondefense	3.5	Fed Govt
Turbines & turbine generator sets	6.0	Manufg.	Federal Government purchases, national defense	2.6	Fed Govt
Iron & steel foundries	4.3	Manufg.	Turbines & turbine generator sets	2.5	Manufg.
Copper rolling & drawing	3.7	Manufg.	Federal electric utilities	2.4	Gov't
Industrial controls	2.5	Manufg.	State & local electric utilities	0.3	Gov't
Electric services (utilities)	2.2	Util.	Ship building & repairing	0.2	Manufg.
Internal combustion engines, nec	2.1	Manufg.			
Advertising	2.1	Services			
Fabricated structural metal	2.0	Manufg.			
Communications, except radio & TV	1.4	Util.			
Power transmission equipment	1.3	Manufg.			

Continued on next page.

INPUTS AND OUTPUTS FOR TURBINES & TURBINE GENERATOR SETS - Continued

Economic Sector or Industry Providing Inputs	%	Sector	Economic Sector or Industry Buying Outputs	%	Sector
Computer & data processing services	1.3	Services			
Maintenance of nonfarm buildings nec	1.1	Constr.			
Machinery, except electrical, nec	1.1	Manufg.			
Miscellaneous plastics products	1.1	Manufg.			
Motors & generators	1.1	Manufg.			
Nonferrous forgings	1.0	Manufg.			
Pumps & compressors	1.0	Manufg.			
Motor freight transportation & warehousing	1.0	Util.			
Petroleum refining	0.9	Manufg.			
Fabricated plate work (boiler shops)	0.8	Manufg.			
Nonferrous wire drawing & insulating	0.8	Manufg.			
Special dies & tools & machine tool accessories	0.7	Manufg.			
Engineering, architectural, & surveying services	0.7	Services			
Equipment rental & leasing services	0.7	Services			
Machine tools, metal cutting types	0.6	Manufg.			
Screw machine and related products	0.6	Manufg.			
Eating & drinking places	0.6	Trade			
Abrasive products	0.5	Manufg.			
Miscellaneous fabricated wire products	0.5	Manufg.			
Gas production & distribution (utilities)	0.5	Util.			
Banking	0.5	Fin/R.E.			
Real estate	0.5	Fin/R.E.			
Automotive repair shops & services	0.5	Services			
Ball & roller bearings	0.4	Manufg.			
Transformers	0.4	Manufg.			
U.S. Postal Service	0.4	Gov't			
Electrical equipment & supplies, nec	0.3	Manufg.			
Nonferrous castings, nec	0.3	Manufg.			
Electrical repair shops	0.3	Services			
Legal services	0.3	Services			
Noncomparable imports	0.3	Foreign			
Metal heat treating	0.2	Manufg.			
Nonferrous rolling & drawing, nec	0.2	Manufg.			
Air transportation	0.2	Util.			
Railroads & related services	0.2	Util.			
Royalties	0.2	Fin/R.E.			
Accounting, auditing & bookkeeping	0.2	Services			
Management & consulting services & labs	0.2	Services			
Aluminum castings	0.1	Manufg.			
Aluminum rolling & drawing	0.1	Manufg.			
Brass, bronze, & copper castings	0.1	Manufg.			
Lubricating oils & greases	0.1	Manufg.			
Manifold business forms	0.1	Manufg.			
Insurance carriers	0.1	Fin/R.E.			

Source: Benchmark Input-Output Accounts for the U.S. Economy, 1982, U.S. Department of Commerce, Washington, D.C., July 1991. Data, as reported in the source, are organized by the 1977 SIC structure in use in 1982 but have been matched, as closely as is possible, to the 1987 SIC structure used in this book.

OCCUPATIONS EMPLOYED BY SIC 351 - ENGINES AND TURBINES

Occupation	% of Total 1994	Change to 2005	Occupation	% of Total 1994	Change to 2005
Machine builders	7.8	30.8	Drilling & boring machine tool workers	1.5	-40.2
Assemblers, fabricators, & hand workers nec	5.9	-25.2	Secretaries, ex legal & medical	1.4	-31.9
Inspectors, testers, & graders, precision	5.4	-25.2	Precision metal workers nec	1.4	-25.2
Mechanical engineers	5.2	-17.7	Machine assemblers	1.3	-25.3
Combination machine tool operators	4.6	-17.8	Sales & related workers nec	1.3	-25.3
Blue collar worker supervisors	4.2	-23.0	Electricians	1.2	-29.8
Machinists	4.1	-25.2	Industrial engineers, ex safety engineers	1.2	-17.8
NC machine tool operators, metal & plastic	3.1	-10.3	Production, planning, & expediting clerks	1.2	-25.3
Metal & plastic machine workers nec	2.6	-33.7	Freight, stock, & material movers, hand	1.2	-40.3
Machine tool cutting operators, metal & plastic	2.1	-37.8	Tool & die makers	1.1	-39.6
Professional workers nec	2.1	-10.3	Helpers, laborers, & material movers nec	1.1	-25.3
Industrial machinery mechanics	2.0	-17.7	Managers & administrators nec	1.1	-25.3
Industrial truck & tractor operators	1.7	-25.3	Engineering, mathematical, & science managers	1.1	-15.1
Grinding machine operators, metal & plastic	1.6	-40.2	Drafters	1.0	-41.8
Lathe & turning machine tool operators	1.5	-40.2	Machine tool cutting & forming etc. nec	1.0	-25.3
Engineering technicians & technologists nec	1.5	-25.3			

Source: Industry-Occupation Matrix, Bureau of Labor Statistics. These data relate to one or more 3-digit SIC industry groups rather than to a single 4-digit SIC. The change reported for each occupation to the year 2005 is a percent of growth or decline as estimated by the Bureau of Labor Statistics. The abbreviation nec stands for 'not elsewhere classified'.

LOCATION BY STATE AND REGIONAL CONCENTRATION

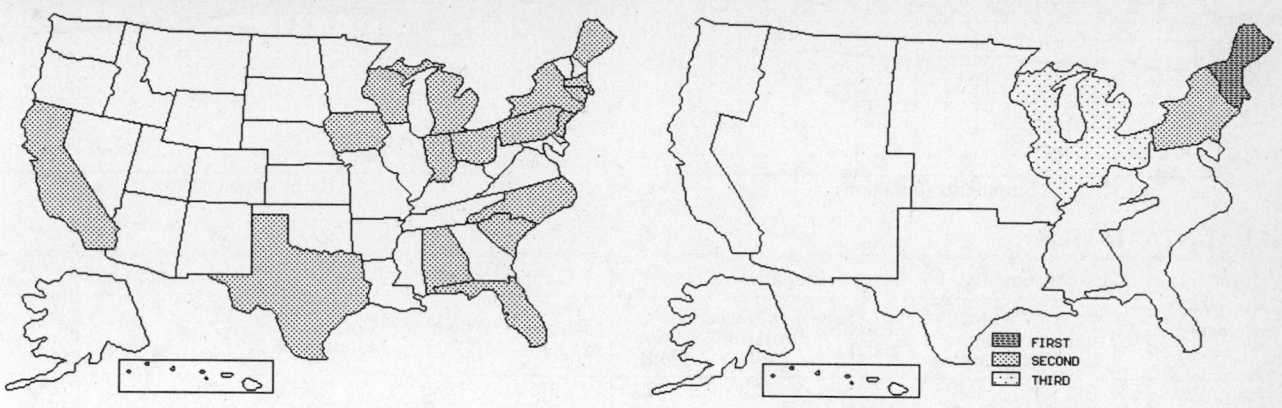

FIRST
SECOND
THIRD

INDUSTRY DATA BY STATE

| State | Establish-ments | Shipments | | | Employment | | | | Cost as % of Shipments | Investment per Employee ($) |
		Total ($ mil)	% of U.S.	Per Establ.	Total Number	% of U.S.	Per Establ.	Wages ($/hour)		
California	8	651.0	11.1	81.4	3,300	12.2	413	13.25	32.1	-
Massachusetts	10	(D)	-	-	1,750 *	6.5	175	-	-	-
Texas	9	(D)	-	-	1,750 *	6.5	194	-	-	-
New York	8	(D)	-	-	7,500 *	27.7	938	-	-	-
Connecticut	4	(D)	-	-	3,750 *	13.8	938	-	-	-
Florida	4	(D)	-	-	750 *	2.8	188	-	-	-
Pennsylvania	4	(D)	-	-	750 *	2.8	188	-	-	-
Wisconsin	4	(D)	-	-	750 *	2.8	188	-	-	-
North Carolina	3	(D)	-	-	1,750 *	6.5	583	-	-	-
Ohio	3	(D)	-	-	175 *	0.6	58	-	-	-
South Carolina	3	(D)	-	-	1,750 *	6.5	583	-	-	-
Indiana	2	(D)	-	-	375 *	1.4	188	-	-	-
Iowa	2	(D)	-	-	175 *	0.6	88	-	-	-
New Jersey	2	(D)	-	-	175 *	0.6	88	-	-	-
Alabama	1	(D)	-	-	175 *	0.6	175	-	-	-
Maine	1	(D)	-	-	375 *	1.4	375	-	-	-
Michigan	1	(D)	-	-	175 *	0.6	175	-	-	-

Source: 1992 *Economic Census*. The states are in descending order of shipments or establishments (if shipment data are missing for the majority). The symbol (D) appears when data are withheld to prevent disclosure of competitive information. States marked with (D) are sorted by number of establishments. A dash (-) indicates that the data element cannot be calculated; * indicates the midpoint of a range.

3519 - INTERNAL COMBUSTION ENGINES, NEC

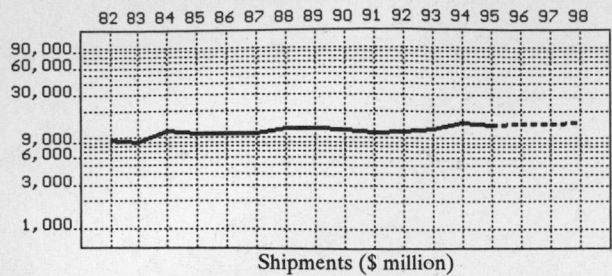

Shipments ($ million)

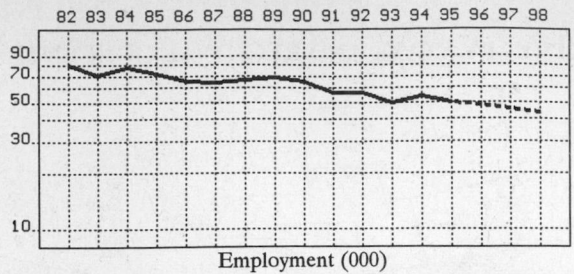

Employment (000)

GENERAL STATISTICS

Year	Companies	Establishments		Employment			Compensation		Production ($ million)			
		Total	with 20 or more employees	Total (000)	Production Workers (000)	Hours (Mil)	Payroll ($ mil)	Wages ($/hr)	Cost of Materials	Value Added by Manufacture	Value of Shipments	Capital Invest.
1982	200	252	158	79.5	52.4	95.2	1,976.3	12.82	5,129.3	3,911.1	9,355.6	599.5
1983		258	159	70.6	45.8	84.7	1,898.8	13.75	4,727.9	4,092.9	8,985.7	375.6
1984		264	160	78.2	55.2	108.9	2,369.0	14.47	6,298.9	5,692.6	11,869.7	417.3
1985		270	161	72.3	50.2	96.3	2,201.7	14.82	5,882.9	5,206.1	11,286.5	404.9
1986		274	167	65.5	44.9	86.9	2,054.6	15.07	5,719.8	5,142.8	10,896.2	338.0
1987	224	278	150	64.0	45.0	89.8	2,043.1	15.05	6,058.9	5,065.8	11,122.6	529.1
1988		263	154	66.8	47.6	99.4	2,294.0	15.42	7,121.5	5,506.5	12,432.4	467.3
1989		260	156	68.6	44.9	91.6	2,194.5	15.84	7,340.8	5,489.7	12,803.2	439.0
1990		252	152	64.4	42.9	85.7	2,117.6	16.01	7,250.4	4,899.8	12,224.2	524.6
1991		259	145	56.3	38.9	74.2	1,908.5	15.96	6,697.8	4,677.1	11,515.5	367.5
1992	250	294	135	56.6	37.9	77.0	2,071.6	16.86	6,996.1	4,794.2	11,826.9	461.1
1993		295	132	49.5	35.9	75.6	1,900.3	16.66	7,544.9	5,090.7	12,600.3	370.7
1994		282P	137P	54.2	39.9	85.3	2,145.0	16.61	8,888.5	6,291.5	14,952.6	400.4
1995		284P	135P	50.1P	35.8P	77.0P	2,060.7P	17.37P	8,236.5P	5,830.0P	13,855.8P	404.5P
1996		287P	132P	48.0P	34.5P	75.3P	2,056.4P	17.66P	8,421.1P	5,960.6P	14,166.3P	399.7P
1997		289P	130P	45.9P	33.3P	73.7P	2,052.2P	17.95P	8,605.6P	6,091.3P	14,476.7P	394.9P
1998		291P	127P	43.7P	32.0P	72.1P	2,048.0P	18.24P	8,790.2P	6,221.9P	14,787.2P	390.1P

Sources: 1982, 1987, 1992 *Economic Census; Annual Survey of Manufactures*, 83-86, 88-91, 93-94. Establishment counts for non-Census years are from *County Business Patterns*; establishment values for 83-84 are extrapolations. 'P's show projections by the editors. Industries reclassified in 87 will not have data for prior years.

INDICES OF CHANGE

Year	Companies	Establishments		Employment			Compensation		Production ($ million)			
		Total	with 20 or more employees	Total (000)	Production Workers (000)	Hours (Mil)	Payroll ($ mil)	Wages ($/hr)	Cost of Materials	Value Added by Manufacture	Value of Shipments	Capital Invest.
1982	80	86	117	140	138	124	95	76	73	82	79	130
1983		88	118	125	121	110	92	82	68	85	76	81
1984		90	119	138	146	141	114	86	90	119	100	91
1985		92	119	128	132	125	106	88	84	109	95	88
1986		93	124	116	118	113	99	89	82	107	92	73
1987	90	95	111	113	119	117	99	89	87	106	94	115
1988		89	114	118	126	129	111	91	102	115	105	101
1989		88	116	121	118	119	106	94	105	115	108	95
1990		86	113	114	113	111	102	95	104	102	103	114
1991		88	107	99	103	96	92	95	96	98	97	80
1992	100	100	100	100	100	100	100	100	100	100	100	100
1993		100	98	87	95	98	92	99	108	106	107	80
1994		96P	101P	96	105	111	104	99	127	131	126	87
1995		97P	100P	89P	95P	100P	99P	103P	118P	122P	117P	88P
1996		97P	98P	85P	91P	98P	99P	105P	120P	124P	120P	87P
1997		98P	96P	81P	88P	96P	99P	106P	123P	127P	122P	86P
1998		99P	94P	77P	84P	94P	99P	108P	126P	130P	125P	85P

Sources: Same as General Statistics. Values reflect change from the base year, 1992. Values above 100 mean greater than 92, values below 100 mean less than 92, and a value of 100 in the 82-91 or 93-98 period means same as 92. 'P's mark projections by the editors.

SELECTED RATIOS

For 1994	Avg. of All Manufact.	Analyzed Industry	Index	For 1994	Avg. of All Manufact.	Analyzed Industry	Index
Employees per Establishment	49	192	392	Value Added per Production Worker	134,084	157,682	118
Payroll per Establishment	1,500,273	7,600,258	507	Cost per Establishment	5,045,178	31,494,121	624
Payroll per Employee	30,620	39,576	129	Cost per Employee	102,970	163,994	159
Production Workers per Establishment	34	141	412	Cost per Production Worker	146,988	222,769	152
Wages per Establishment	853,319	5,020,185	588	Shipments per Establishment	9,576,895	52,980,705	553
Wages per Production Worker	24,861	35,510	143	Shipments per Employee	195,460	275,878	141
Hours per Production Worker	2,056	2,138	104	Shipments per Production Worker	279,017	374,752	134
Wages per Hour	12.09	16.61	137	Investment per Establishment	321,011	1,418,715	442
Value Added per Establishment	4,602,255	22,292,318	484	Investment per Employee	6,552	7,387	113
Value Added per Employee	93,930	116,079	124	Investment per Production Worker	9,352	10,035	107

Sources: Same as General Statistics. The 'Average of All Manufacturing' column represents the average of all manufacturing industries reported for the most recent complete year available. The Index shows the relationship between the Average and the Analyzed Industry. For example, 100 means that they are equal; 500 that the Analyzed Industry is five times the average; 50 means that the Analyzed Industry is half the national average. The abbreviation 'na' is used to show that data are 'not available'.

LEADING COMPANIES Number shown: 58 Total sales ($ mil): 12,121 Total employment (000): 63.9

Company Name	Address				CEO Name	Phone	Co. Type	Sales ($ mil)	Empl. (000)
Cummins Engine Company Inc	PO Box 3005	Columbus	IN	47202	Theodore M Solso	812-377-5000	D	2,423	12.0
Detroit Diesel Corp	13400 W Outer W	Detroit	MI	48239	Roger S Penske	313-592-5000	P	1,663	5.4
Briggs and Stratton Corp	PO Box 702	Milwaukee	WI	53201	F P Stratton Jr	414-259-5333	P	1,285	7.9
Mercury Marine	PO Box 1939	Fond du Lac	WI	54936	David D Jones Jr	414-929-5000	D	1,140	5.5
Outboard Marine Corp	100 Sea-Horse Dr	Waukegan	IL	60085	James C Chapman	708-689-6200	P	1,035	8.1
OMC SysteMatched	200 Sea Horse Dr	Waukegan	IL	60085	Michael M Potter	708-689-6200	D	1,000	6.0
Cooper Energy Services	N Sandusky St	Mount Vernon	OH	43050	K W Stevenson	614-393-8200	D	650	3.1
Consolidated Diesel Co	PO Box 670	Whitakers	NC	27891	Larry Moore	919-437-6611	S	520•	1.5
Tecumseh Prod Co	900 North St	Grafton	WI	53024	Harry Hans	414-377-2700	D	440	3.0
Westinghouse Electric Corp	401 E Hendy Av	Sunnyvale	CA	94088	Art Clark	408-735-3011	D	300	1.0
Mack Trucks Inc	13302 Pennsylvania	Hagerstown	MD	21742	Ross H Rhoads	301-790-5400	D	220	1.5
Dresser Industries Inc	1000 W St Paul Av	Waukesha	WI	53188	GP Tevis	414-547-3311	D	180	0.9
Navistar International	10400 W North Av	Melrose Park	IL	60160	Daniel Ustian	708-865-4276	D	170	1.2
Wartsila Diesel Inc	201 Defense Hwy	Annapolis	MD	21401	Terry Sirois	410-573-2100	S	110	0.3
Fairbanks Morse Engine	701 White Av	Beloit	WI	53511	Richard Dashnaw	608-364-4411	D	100	0.7
Interstate Co	2601 E 80th St	Minneapolis	MN	55425	Jeff Caswell	612-854-2044	R	100•	0.6
Onan Corp	PO Box 240004	Huntsville	AL	35824	Jim Renny	205-772-9671	D	98•	0.6
Interstate Detroit Diesel Inc	2501 E 80th St	Minneapolis	MN	55425	Gordan Galarneau	612-854-5511	S	72	0.6
Cummins Industrial Center	800 E 3rd St	Seymour	IN	47274	Charlie Graham	812-522-9366	S	68•	0.6
OMC Milwaukee	6101 N 64th St	Milwaukee	WI	53218	Terry Schneider	414-438-5097	D	51•	0.4
TANO Corp	3501 Jourdan Ct	New Orleans	LA	70129	James J Reiss Jr	504-243-2400	R	50	0.4
Hatch and Kirk Inc	5111 Leary NW	Seattle	WA	98107	R Van Hoozer	206-783-2766	R	40	0.2
Industrial Parts Depot Inc	23231 S Normandy	Torrance	CA	90501	Robert Rasmussen	310-530-1900	R	40	0.1
Central Detroit Diesel-Allison	PO Box 49	Liberty	MO	64068	Russell Redburn	816-781-8070	R	30	0.2
Mercury High Performance	N 7480 County UU	Fond du Lac	WI	54935	Fred C Kiekhaefer	414-921-5330	D	30	0.2
Western Diesel Services Inc	9290 W Floriss	St Louis	MO	63136	Hugh Scott III	314-868-8620	R	28	<0.1
Enterprise Engine Services	1351 Harbor Bay	Alameda	CA	94501	Robert Nimmo	510-614-7400	D	25	<0.1
Rochester Manufacturing Co	PO Box 1940	Rochester	MI	48308	Lawrence H Weting	810-652-2600	R	25	0.2
Dealers Manufacturing Co	5130 Main St NE	Minneapolis	MN	55421	David W Goodwin	612-572-1800	R	23	0.2
Indmar Products Inc	PO Box 27184	Memphis	TN	38127	Richard Rowe	901-353-9930	R	21	<0.1
Alaska Diesel Electric Inc	PO Box 70543	Seattle	WA	98107	Harold W Johnson	206-789-3880	S	18	<0.1
Chromium Corp	14643 Dallas Pkwy	Dallas	TX	75240	Mike Tatsch	214-851-0460	S	15	0.2
Detroit Diesel	13400 Outer Dr W	Detroit	MI	48239	Roger Penske	313-592-5246	S	13	<0.1
Crusader Engines	7100 E 15 Mile Rd	Sterling Hts	MI	48312	Chester Janssens	313-264-1200	D	12•	<0.1
Marine Power Inc	17506 Marien Power	Ponchatoula	LA	70454	WE Allbright	504-386-2081	R	12	<0.1
Peaker Services Inc	8080 Kensington Ct	Brighton	MI	48116	Richard Steele	810-437-4174	R	12	<0.1
Franklin Precision Industry Inc	PO Box 369	Franklin	KY	42135	Kunio Kadowaki	502-586-4450	S	11	<0.1
Gopher Motor Rebuilding	6401 Cambridge St	Minneapolis	MN	55426	JR Smith	612-929-0441	R	9	<0.1
Racing Head Service Inc	3416 Democrat Rd	Memphis	TN	38118	Ivars Smiltnicks	901-794-2830	R	8	<0.1
Richland Ltd	PO Box 489	Spring Green	WI	53588	Walter J Dallman	608-588-7779	R	7	<0.1
Rich Tool and Die Co	28 Pond View Dr	Scarborough	ME	04074	Allen C Estes	207-883-7424	R	7	<0.1
Custom Servo Motors Inc	2121 Bridge St	New Ulm	MN	56073	William Anderson	507-354-1616	R	6•	<0.1
Diesel America Inc	74257 Hwy 25	Covington	LA	70433	Mike Wittich	504-898-5800	R	6•	<0.1
MTU Detroit Diesel	10450 Corporate Dr	Sugar Land	TX	77478	Jeb Berg	713-240-4100	R	6•	<0.1
Beasley Manufacturing Inc	PO Box 113	Altoona	PA	16603	Joe Taylor	814-942-8538	R	5	<0.1
Benedict Manufacturing	PO Box 1117	Big Rapids	MI	49307	Jack R Benedict	616-796-8603	R	5	0.1
Start Master	959 Chaney Av	Marion	OH	43302	Paul Kestler	614-382-5771	D	4•	<0.1
Torque Engineering Corp	2932 Thorne Dr	Elkhart	IN	46514	Ray Wedel	219-264-2628	R	4	<0.1
AMW Cuyuna Engine Co	8 Schein Loop	Beaufort	SC	29902	Roger Worth	803-846-2167	R	3	<0.1
B and G Machine Inc	PO Box 80483	Seattle	WA	98108	John C Bianchi	206-767-3130	R	3	<0.1
Lingenfelter Performance Eng	1557 Winchester Rd	Decatur	IN	46733	John Lingenfelter	219-724-2552	R	3	<0.1
Pacific Rim Diesel Inc	3842 W Marginal	Seattle	WA	98106	Gwen Fraser	206-932-1800	R	3	<0.1
Vernay Products Inc	PO Box 3028	Thomasville	GA	31799	John R Ford	912-228-7653	S	3	<0.1
KBI Ltd	900 Pingree Rd	Lk in the Hls	IL	60102	James Burke	708-658-8561	R	3	<0.1
Lanco Engine Services	2732 S 3600 W	Salt Lake City	UT	84119	T Longenecker	801-964-9980	R	2	<0.1
Land and Sea Inc	PO Box 96	North Salem	NH	03073	Robert Bergeron	603-329-5645	R	2	<0.1
American Diesel Tube Corp	1240 Capital Dr	Addison	IL	60101	Kenneth J Pickering	708-628-1830	R	1	<0.1
Konrad Marine Inc	1421 Hanley Rd	Hudson	WI	54016	Ken Konrad	715-386-4203	R	1	<0.1

Source: *Ward's Business Directory of U.S. Private and Public Companies*, Volumes 1 and 2, 1996. The company type code used is as follows: P - Public, R - Private, S - Subsidiary, D - Division, J - Joint Venture, A - Affiliate, G - Group. Sales are in millions of dollars, employees are in thousands. An asterisk (*) indicates an estimated sales volume. The symbol < stands for 'less than'. Company names and addresses are truncated, in some cases, to fit into the available space.

MATERIALS CONSUMED

Material	Quantity	Delivered Cost ($ million)
Materials, ingredients, containers, and supplies	(X)	6,478.4
Fluid power pumps, motors, and hydrostatic transmissions (hydraulic and pneumatic)	(X)	164.1
Fluid power cylinders and rotary actuators (hydraulic and pneumatic)	(X)	26.7
Fluid power filters (hydraulic and pneumatic)	(X)	32.2
Fluid power hose or tube fittings and assemblies (hydraulic and pneumatic)	(X)	25.5
Fluid power valves (hydraulic and pneumatic)	(X)	(D)
Metal bolts, nuts, screws, washers, rivets, and other screw machine products	(X)	183.7
Metal stampings	(X)	111.7
Metal tanks, heat exchangers, steam condensers, and other boiler products	(X)	49.7
Fabricated structural metal products (except forgings)	(X)	(D)

Continued on next page.

MATERIALS CONSUMED - Continued

Material	Quantity	Delivered Cost ($ million)
All other fabricated metal products (except forgings)	(X)	307.0
Iron and steel forgings	(X)	494.9
Nonferrous forgings	(X)	(D)
Iron and steel castings (rough and semifinished)	(X)	987.5
Aluminum and aluminum-base alloy castings (rough and semifinished)	(X)	572.8
Other nonferrous castings (rough and semifinished)	(X)	21.3
Steel bars, bar shapes, and plates	(X)	70.6
Steel sheet and strip, including tin plate	(X)	39.5
All other steel shapes and forms	(X)	23.4
Nonferrous shapes and forms	(X)	18.7
Pistons, piston rings, carburetors, valves (intake and exhaust only)	(X)	570.1
Engine electrical equipment, including spark plugs, magnetos, generators, starters, etc.	(X)	236.4
Integral horsepower electric motors and generators (1 hp or more)	(X)	(D)
Ball bearings (mounted or unmounted)	(X)	16.9
Roller bearings (mounted or unmounted)	(X)	15.2
Plain bearings and bushings	(X)	45.1
Mechanical speed changers, gears, and industrial high-speed drives	(X)	70.8
Generators purchased for incorporation into turbine generator sets	(X)	(D)
Gaskets (all types) and asbestos packing	(X)	86.7
Fabricated plastics products (except gaskets, hoses, and belting)	(X)	47.7
Rubber and plastics hose and belting	(X)	43.3
Cutting tools for machine tools	(X)	38.1
All other materials and components, parts, containers, and supplies	(X)	1,802.4
Materials, ingredients, containers, and supplies, nsk	(X)	70.8

Source: 1992 *Economic Census*. Explanation of symbols used: (D): Withheld to avoid disclosure of competitive data; na: Not available; (S): Withheld because statistical norms were not met; (X): Not applicable; (Z): Less than half the unit shown; nec: Not elsewhere classified; nsk: Not specified by kind; - : zero; * : 10-19 percent estimated; ** : 20-29 percent estimated.

PRODUCT SHARE DETAILS

Product or Product Class	% Share	Product or Product Class	% Share
Internal combustion engines, nec	100.00	Fuel pumps, new, for internal combustion engines, except aircraft and gasoline automotive engines and gas turbines	(D)
Gasoline and gas-gasoline engines (except aircraft, automobile, highway truck, bus, tank, and outboard marine)	13.33	Water pumps, new, for internal combustion engines, except aircraft and gasoline automotive engines and gas turbines	0.79
Diesel, semidiesel, and dual-fuel engines (except automobile, highway truck, bus, and tank)	12.79	Engine blocks for internal combustion engines, except aircraft and gasoline automotive engines and gas turbines	3.07
Diesel, semidiesel, and dual-fuel engines for automobiles, highway trucks, and buses	27.21	Cylinder liners (sleeves) for internal combustion engines, except aircraft and gasoline automotive engines and gas turbines	4.12
Outboard motors (internal combustion)	(D)	Cylinder heads for internal combustion engines, except aircraft and gasoline automotive engines and gas turbines	3.31
Outboard motors (internal combustion)	(D)	Intake manifolds and exhaust manifolds for internal combustion engines, except aircraft and gasoline automotive engines and gas turbines	1.32
Piston-type natural gas engines, including LPG engines (excluding gas turbines)	1.51	Valve guides, seats, and tappets for internal combustion engines, except aircraft and gasoline automotive engines and gas turbines	0.40
Tank (except gas turbine) and converted internal combustion engines	(D)	Rocker arms and parts for internal combustion engines, except aircraft and gasoline automotive engines and gas turbines	0.65
Tank engines, except gas turbines	(D)	Fuel injection systems (multipoint) for internal combustion engines, except aircraft and gasoline automotive engines and gas turbines	11.89
Converted internal combustion engines (basic engines, short blocks purchased or intracompany transfer and converted to marine or other uses)	348.95	Engine speed governors for internal combustion engines, except aircraft and gasoline automotive engines and gas turbines	1.69
Parts and accessories for internal combustion engines, except aircraft and gasoline automotive engines and gas turbines	33.11	Stationary engine radiators for internal combustion engines, except aircraft and gasoline automotive engines and gas turbines	0.79
Connecting rods for internal combustion engines, except aircraft and gasoline automotive engines and gas turbines	1.23	Superchargers, including turbochargers, for internal combustion engines, except aircraft and gasoline automotive engines and gas turbines	3.98
Engine crankshafts for internal combustion engines, except aircraft and gasoline automotive engines and gas turbines	4.28	Other parts and accessories for internal combustion engines, except aircraft and gasoline automotive engines and gas turbines	26.09
Engine camshafts for internal combustion engines, except aircraft and gasoline automotive engines and gas turbines	2.42	Parts and accessories for internal combustion engines, except aircraft and gas automotive engines and turbines, nsk	26.96
Flywheels for internal combustion engines, except aircraft and gasoline automotive engines and gas turbines	0.77	Internal combustion engines, nec, nsk	1.57
Main (crankshaft) engine bearings (halves) for internal combustion engines, except aircraft and gasoline automotive engines and gas turbines	(D)		
Connecting rod bearings (halves) for internal combustion engines, except aircraft and gasoline automotive engines and gas turbines	0.58		
Other engine bearings (halves) (camshaft, balance shaft, etc.) for internal combustion engines, except aircraft and gasoline automotive engines and gas turbines	0.95		
Oil pumps, new, for internal combustion engines, except aircraft and gasoline automotive engines and gas turbines	0.50		

Source: 1992 *Economic Census*. The values shown are percent of total shipments in an industry. Values of indented subcategories are summed in the main heading. The symbol (D) appears when data is withheld to prevent disclosure of competitive information. The abbreviation nsk stands for 'not specified by kind' and nec for 'not elsewhere classified'.

INPUTS AND OUTPUTS FOR INTERNAL COMBUSTION ENGINES, NEC

Economic Sector or Industry Providing Inputs	%	Sector	Economic Sector or Industry Buying Outputs	%	Sector
Imports	18.5	Foreign	Exports	19.6	Foreign
Internal combustion engines, nec	13.7	Manufg.	Motor vehicles & car bodies	16.5	Manufg.
Iron & steel foundries	10.9	Manufg.	Internal combustion engines, nec	9.2	Manufg.
Wholesale trade	7.8	Trade	Gross private fixed investment	5.1	Cap Inv
Motors & generators	6.0	Manufg.	Farm machinery & equipment	4.0	Manufg.
Iron & steel forgings	5.5	Manufg.	Construction machinery & equipment	3.6	Manufg.
Carburetors, pistons, rings, & valves	4.0	Manufg.	Federal Government purchases, national defense	3.3	Fed Govt
Aluminum castings	3.6	Manufg.	Lawn & garden equipment	3.0	Manufg.
Blast furnaces & steel mills	2.4	Manufg.	Communications, except radio & TV	2.9	Util.
Engine electrical equipment	2.0	Manufg.	Coal	2.8	Mining
Power transmission equipment	1.6	Manufg.	Personal consumption expenditures	2.4	
Electric services (utilities)	1.5	Util.	Boat building & repairing	2.4	Manufg.
Primary aluminum	1.4	Manufg.	Water transportation	2.4	Util.
Pumps & compressors	1.1	Manufg.	Motorcycles, bicycles, & parts	1.7	Manufg.
Motor freight transportation & warehousing	1.1	Util.	Miscellaneous repair shops	1.6	Services
Screw machine and related products	1.0	Manufg.	Ship building & repairing	1.5	Manufg.
Nonferrous forgings	0.9	Manufg.	Electric services (utilities)	1.3	Util.
Metal stampings, nec	0.8	Manufg.	Motors & generators	1.1	Manufg.
Fabricated plate work (boiler shops)	0.7	Manufg.	Woodworking machinery	1.1	Manufg.
Gaskets, packing & sealing devices	0.7	Manufg.	Pumps & compressors	1.0	Manufg.
Special dies & tools & machine tool accessories	0.7	Manufg.	Fabricated metal products, nec	0.7	Manufg.
Maintenance of nonfarm buildings nec	0.6	Constr.	Oil field machinery	0.7	Manufg.
Machinery, except electrical, nec	0.6	Manufg.	Railroad equipment	0.7	Manufg.
Communications, except radio & TV	0.6	Util.	Automotive rental & leasing, without drivers	0.7	Services
Gas production & distribution (utilities)	0.6	Util.	Agricultural, forestry, & fishery services	0.6	Agric.
Advertising	0.6	Services	Machinery, except electrical, nec	0.6	Manufg.
Computer & data processing services	0.6	Services	Refrigeration & heating equipment	0.6	Manufg.
Ball & roller bearings	0.5	Manufg.	Motor freight transportation & warehousing	0.6	Util.
Fabricated structural metal	0.5	Manufg.	Industrial trucks & tractors	0.5	Manufg.
Industrial controls	0.5	Manufg.	Railroads & related services	0.5	Util.
Miscellaneous fabricated wire products	0.5	Manufg.	Chemical & fertilizer mineral	0.4	Mining
Pipe, valves, & pipe fittings	0.5	Manufg.	Sand & gravel	0.4	Mining
Rubber & plastics hose & belting	0.5	Manufg.	Blowers & fans	0.4	Manufg.
Banking	0.5	Fin/R.E.	Special industry machinery, nec	0.4	Manufg.
Equipment rental & leasing services	0.4	Services	Turbines & turbine generator sets	0.4	Manufg.
Abrasive products	0.3	Manufg.	Transit & bus transportation	0.4	Util.
Machine tools, metal cutting types	0.3	Manufg.	Local government passenger transit	0.4	Gov't
Plating & polishing	0.3	Manufg.	Crude petroleum & natural gas	0.3	Mining
Eating & drinking places	0.3	Trade	Iron & ferroalloy ores	0.3	Mining
Automotive repair shops & services	0.3	Services	Nonferrous metal ores, except copper	0.3	Mining
Scrap	0.3	Scrap	Motor vehicle parts & accessories	0.3	Manufg.
Copper rolling & drawing	0.2	Manufg.	Tanks & tank components	0.3	Manufg.
Miscellaneous plastics products	0.2	Manufg.	Truck & bus bodies	0.3	Manufg.
Motor vehicle parts & accessories	0.2	Manufg.	Pipelines, except natural gas	0.3	Util.
Nonferrous castings, nec	0.2	Manufg.	Clay, ceramic, & refractory minerals	0.2	Mining
Paperboard containers & boxes	0.2	Manufg.	Copper ore	0.2	Mining
Air transportation	0.2	Util.	Elevators & moving stairways	0.2	Manufg.
Railroads & related services	0.2	Util.	Iron & steel foundries	0.2	Manufg.
Legal services	0.2	Services	Miscellaneous plastics products	0.2	Manufg.
Noncomparable imports	0.2	Foreign	Motor homes (made on purchased chassis)	0.2	Manufg.
Aluminum rolling & drawing	0.1	Manufg.	Welding apparatus, electric	0.2	Manufg.
Hand & edge tools, nec	0.1	Manufg.	S/L Govt purch., other general government	0.2	S/L Govt
Metal heat treating	0.1	Manufg.	Commercial fishing	0.1	Agric.
Nonferrous rolling & drawing, nec	0.1	Manufg.	Nonmetallic mineral services	0.1	Mining
Nonferrous wire drawing & insulating	0.1	Manufg.	Transportation equipment, nec	0.1	Manufg.
Petroleum refining	0.1	Manufg.			
Credit agencies other than banks	0.1	Fin/R.E.			
Real estate	0.1	Fin/R.E.			
Electrical repair shops	0.1	Services			
Management & consulting services & labs	0.1	Services			
U.S. Postal Service	0.1	Gov't			

Source: Benchmark Input-Output Accounts for the U.S. Economy, 1982, U.S. Department of Commerce, Washington, D.C., July 1991. Data, as reported in the source, are organized by the 1977 SIC structure in use in 1982 but have been matched, as closely as is possible, to the 1987 SIC structure used in this book.

OCCUPATIONS EMPLOYED BY SIC 351 - ENGINES AND TURBINES

Occupation	% of Total 1994	Change to 2005	Occupation	% of Total 1994	Change to 2005
Machine builders	7.8	30.8	Drilling & boring machine tool workers	1.5	-40.2
Assemblers, fabricators, & hand workers nec	5.9	-25.2	Secretaries, ex legal & medical	1.4	-31.9
Inspectors, testers, & graders, precision	5.4	-25.2	Precision metal workers nec	1.4	-25.2
Mechanical engineers	5.2	-17.7	Machine assemblers	1.3	-25.3
Combination machine tool operators	4.6	-17.8	Sales & related workers nec	1.3	-25.3
Blue collar worker supervisors	4.2	-23.0	Electricians	1.2	-29.8
Machinists	4.1	-25.2	Industrial engineers, ex safety engineers	1.2	-17.8
NC machine tool operators, metal & plastic	3.1	-10.3	Production, planning, & expediting clerks	1.2	-25.3
Metal & plastic machine workers nec	2.6	-33.7	Freight, stock, & material movers, hand	1.2	-40.3
Machine tool cutting operators, metal & plastic	2.1	-37.8	Tool & die makers	1.1	-39.6
Professional workers nec	2.1	-10.3	Helpers, laborers, & material movers nec	1.1	-25.3
Industrial machinery mechanics	2.0	-17.7	Managers & administrators nec	1.1	-25.3
Industrial truck & tractor operators	1.7	-25.3	Engineering, mathematical, & science managers	1.1	-15.1
Grinding machine operators, metal & plastic	1.6	-40.2	Drafters	1.0	-41.8
Lathe & turning machine tool operators	1.5	-40.2	Machine tool cutting & forming etc. nec	1.0	-25.3
Engineering technicians & technologists nec	1.5	-25.3			

Source: *Industry-Occupation Matrix*, Bureau of Labor Statistics. These data relate to one or more 3-digit SIC industry groups rather than to a single 4-digit SIC. The change reported for each occupation to the year 2005 is a percent of growth or decline as estimated by the Bureau of Labor Statistics. The abbreviation nec stands for 'not elsewhere classified'.

LOCATION BY STATE AND REGIONAL CONCENTRATION

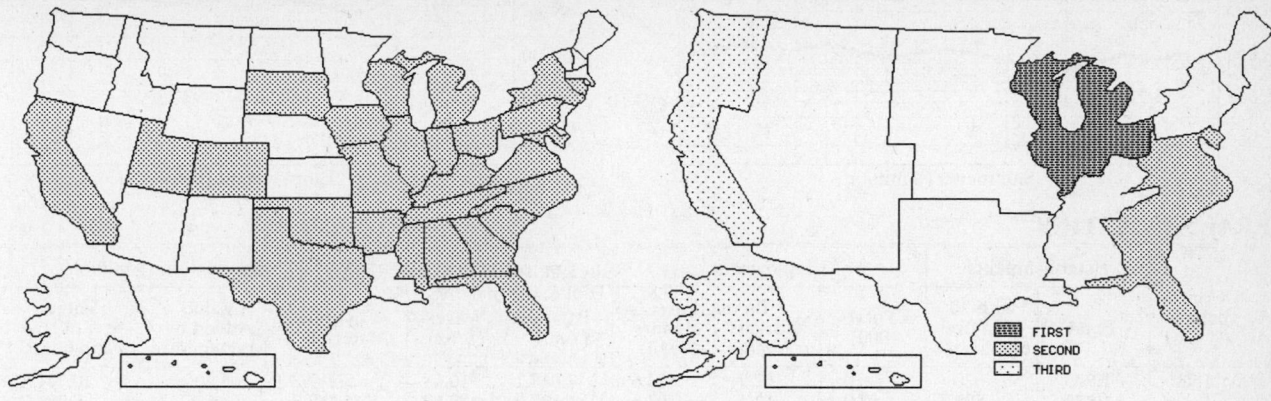

FIRST
SECOND
THIRD

INDUSTRY DATA BY STATE

State	Establish-ments	Shipments			Employment				Cost as % of Shipments	Investment per Employee ($)
		Total ($ mil)	% of U.S.	Per Establ.	Total Number	% of U.S.	Per Establ.	Wages ($/hour)		
Tennessee	9	320.0	2.7	35.6	1,500	2.7	167	9.11	65.8	-
California	31	267.7	2.3	8.6	3,300	5.8	106	34.85	43.1	-
Missouri	11	240.3	2.0	21.8	1,300	2.3	118	9.23	52.4	2,846
Florida	24	(D)	-	-	175 *	0.3	7	-	-	-
Illinois	23	(D)	-	-	7,500 *	13.3	326	-	-	-
Wisconsin	23	(D)	-	-	17,500 *	30.9	761	-	-	-
Michigan	19	(D)	-	-	3,750 *	6.6	197	-	-	-
Ohio	19	(D)	-	-	3,750 *	6.6	197	-	-	-
Texas	17	(D)	-	-	375 *	0.7	22	-	-	-
New York	12	(D)	-	-	1,750 *	3.1	146	-	-	-
North Carolina	10	(D)	-	-	3,750 *	6.6	375	-	-	-
Indiana	9	(D)	-	-	7,500 *	13.3	833	-	-	-
Pennsylvania	9	(D)	-	-	375 *	0.7	42	-	-	-
Oklahoma	6	(D)	-	-	1,750 *	3.1	292	-	-	-
Alabama	5	(D)	-	-	750 *	1.3	150	-	-	-
Georgia	5	(D)	-	-	375 *	0.7	75	-	-	-
Connecticut	4	(D)	-	-	1,750 *	3.1	438	-	-	-
Iowa	4	(D)	-	-	750 *	1.3	188	-	-	-
Mississippi	4	(D)	-	-	750 *	1.3	188	-	-	-
Arkansas	3	(D)	-	-	175 *	0.3	58	-	-	-
Utah	3	(D)	-	-	175 *	0.3	58	-	-	-
Colorado	2	(D)	-	-	1,750 *	3.1	875	-	-	-
Kentucky	2	(D)	-	-	750 *	1.3	375	-	-	-
Maryland	2	(D)	-	-	1,750 *	3.1	875	-	-	-
South Carolina	2	(D)	-	-	375 *	0.7	188	-	-	-
South Dakota	2	(D)	-	-	175 *	0.3	88	-	-	-
Virginia	2	(D)	-	-	175 *	0.3	88	-	-	-

Source: 1992 *Economic Census*. The states are in descending order of shipments or establishments (if shipment data are missing for the majority). The symbol (D) appears when data are withheld to prevent disclosure of competitive information. States marked with (D) are sorted by number of establishments. A dash (-) indicates that the data element cannot be calculated; * indicates the midpoint of a range.

3523 - FARM MACHINERY & EQUIPMENT

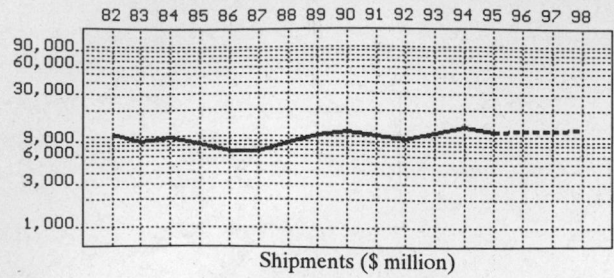

Shipments ($ million)

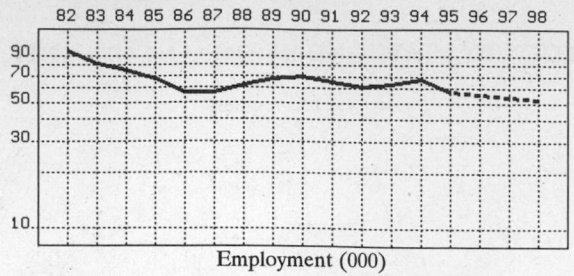

Employment (000)

GENERAL STATISTICS

| Year | Companies | Establishments | | Employment | | | Compensation | | Production ($ million) | | | |
		Total	with 20 or more employees	Total (000)	Production Workers (000)	Hours (Mil)	Payroll ($ mil)	Wages ($/hr)	Cost of Materials	Value Added by Manufacture	Value of Shipments	Capital Invest.
1982	1,786	1,903	620	96.0	62.9	113.6	2,067.1	10.61	5,217.2	5,200.0	10,741.8	341.1
1983		1,828	598	82.0	53.4	98.1	1,827.9	10.69	4,417.5	4,526.2	8,982.7	143.7
1984		1,753	576	75.4	51.5	97.1	1,794.6	11.06	4,741.9	5,115.5	9,858.1	158.7
1985		1,679	555	67.1	44.9	84.6	1,631.4	11.34	3,818.7	4,201.6	8,211.6	163.5
1986		1,593	523	57.1	38.2	74.3	1,384.2	10.46	3,217.7	3,414.2	6,745.4	138.9
1987	1,576	1,634	464	57.0	39.1	75.2	1,416.3	11.20	3,218.5	3,709.5	6,879.9	200.2
1988		1,580	499	62.4	43.6	88.0	1,604.4	11.38	4,218.2	4,691.5	8,731.7	179.7
1989		1,532	513	67.2	48.9	96.7	1,778.7	11.90	5,123.1	5,246.6	10,418.9	183.0
1990		1,538	525	69.3	49.9	97.6	1,882.6	12.31	5,517.4	5,978.5	11,546.2	210.1
1991		1,591	527	65.1	45.4	88.1	1,786.9	12.65	4,969.1	5,268.0	10,346.6	204.5
1992	1,578	1,631	465	61.4	42.5	83.2	1,784.4	13.18	4,435.8	5,165.0	9,617.0	196.2
1993		1,670	502	63.5	45.8	93.5	1,905.0	13.18	5,507.4	5,741.2	11,189.6	228.6
1994		1,515P	463P	67.1	48.1	99.5	2,110.9	13.81	6,542.4	6,855.3	13,217.7	259.7
1995		1,493P	452P	57.2P	42.3P	88.2P	1,860.8P	13.69P	5,662.6P	5,933.4P	11,440.2P	213.5P
1996		1,470P	442P	55.6P	41.6P	87.8P	1,874.2P	13.96P	5,783.5P	6,060.1P	11,684.6P	215.3P
1997		1,448P	431P	54.0P	40.9P	87.3P	1,887.6P	14.22P	5,904.5P	6,186.9P	11,928.9P	217.1P
1998		1,426P	421P	52.4P	40.2P	86.9P	1,901.0P	14.49P	6,025.4P	6,313.6P	12,173.3P	219.0P

Sources: 1982, 1987, 1992 *Economic Census*; *Annual Survey of Manufactures*, 83-86, 88-91, 93-94. Establishment counts for non-Census years are from *County Business Patterns*; establishment values for 83-84 are extrapolations. 'P's show projections by the editors. Industries reclassified in 87 will not have data for prior years.

INDICES OF CHANGE

| Year | Companies | Establishments | | Employment | | | Compensation | | Production ($ million) | | | |
		Total	with 20 or more employees	Total (000)	Production Workers (000)	Hours (Mil)	Payroll ($ mil)	Wages ($/hr)	Cost of Materials	Value Added by Manufacture	Value of Shipments	Capital Invest.
1982	113	117	133	156	148	137	116	81	118	101	112	174
1983		112	129	134	126	118	102	81	100	88	93	73
1984		107	124	123	121	117	101	84	107	99	103	81
1985		103	119	109	106	102	91	86	86	81	85	83
1986		98	112	93	90	89	78	79	73	66	70	71
1987	100	100	100	93	92	90	79	85	73	72	72	102
1988		97	107	102	103	106	90	86	95	91	91	92
1989		94	110	109	115	116	100	90	115	102	108	93
1990		94	113	113	117	117	106	93	124	116	120	107
1991		98	113	106	107	106	100	96	112	102	108	104
1992	100	100	100	100	100	100	100	100	100	100	100	100
1993		102	108	103	108	112	107	100	124	111	116	117
1994		93P	100P	109	113	120	118	105	147	133	137	132
1995		92P	97P	93P	100P	106P	104P	104P	128P	115P	119P	109P
1996		90P	95P	91P	98P	106P	105P	106P	130P	117P	121P	110P
1997		89P	93P	88P	96P	105P	106P	108P	133P	120P	124P	111P
1998		87P	91P	85P	95P	104P	107P	110P	136P	122P	127P	112P

Sources: Same as General Statistics. Values reflect change from the base year, 1992. Values above 100 mean greater than 92, values below 100 mean less than 92, and a value of 100 in the 82-91 or 93-98 period means same as 92. 'P's mark projections by the editors.

SELECTED RATIOS

For 1994	Avg. of All Manufact.	Analyzed Industry	Index	For 1994	Avg. of All Manufact.	Analyzed Industry	Index
Employees per Establishment	49	44	90	Value Added per Production Worker	134,084	142,522	106
Payroll per Establishment	1,500,273	1,393,083	93	Cost per Establishment	5,045,178	4,317,639	86
Payroll per Employee	30,620	31,459	103	Cost per Employee	102,970	97,502	95
Production Workers per Establishment	34	32	92	Cost per Production Worker	146,988	136,017	93
Wages per Establishment	853,319	906,830	106	Shipments per Establishment	9,576,895	8,722,984	91
Wages per Production Worker	24,861	28,567	115	Shipments per Employee	195,460	196,985	101
Hours per Production Worker	2,056	2,069	101	Shipments per Production Worker	279,017	274,796	98
Wages per Hour	12.09	13.81	114	Investment per Establishment	321,011	171,388	53
Value Added per Establishment	4,602,255	4,524,136	98	Investment per Employee	6,552	3,870	59
Value Added per Employee	93,930	102,165	109	Investment per Production Worker	9,352	5,399	58

Sources: Same as General Statistics. The 'Average of All Manufacturing' column represents the average of all manufacturing industries reported for the most recent complete year available. The Index shows the relationship between the Average and the Analyzed Industry. For example, 100 means that they are equal; 500 that the Analyzed Industry is five times the average; 50 means that the Analyzed Industry is half the national average. The abbreviation 'na' is used to show that data are 'not available'.

LEADING COMPANIES Number shown: **75** Total sales ($ mil): **35,149** Total employment (000): **161.9**

Company Name	Address				CEO Name	Phone	Co. Type	Sales ($ mil)	Empl. (000)
Tenneco Inc	PO Box 2511	Houston	TX	77252	Dana G Mead	713-757-2131	P	12,174	55.0
Deere and Co	John Deere Rd	Moline	IL	61265	Hans W Becherer	309-765-8000	P	9,030	33.0
Case Equipment Corp	700 State St	Racine	WI	53403	Jean-Pierre Rosso	414-636-6011	P	3,890	17.1
JI Case Co	700 State St	Racine	WI	53404	Dana Mead	414-636-6011	S	3,700	17.2
AGCO Corp	4830 River Green	Duluth	GA	30136	Robert J Ratliff	404-813-9200	P	1,319	5.8
Deere & Co	1100 13th Av	East Moline	IL	61244	Richard G Kleine	309-765-6321	D	720*	3.1
Butler Manufacturing Co	PO Box 419917	Kansas City	MO	64141	Donald H Pratt	816-968-3000	P	692	3.6
Ford New Holland Americas	PO Box 1895	New Holland	PA	17557	WT Kennedy	717-355-1121	S	460	3.8
TIC United Corp	4645 N Central	Dallas	TX	75205	S J Georgoulis	214-559-0580	R	250	2.0
Allied Products Corp	10 S Riverside Plz	Chicago	IL	60606	Richard A Drexler	312-454-1020	P	215	1.6
Deere & Co	PO Box 1595	Des Moines	IA	50306	Richard S White	515-289-1350	D	210*	1.5
Ag-Chem Equipment Company	5720 Smetana Dr	Minnetonka	MN	55343	Alvin E McQuinn	612-933-9006	P	183	1.0
James Hardie Irrigation Inc	27671 La Paz Rd	Laguna Niguel	CA	92656	Rick Parod	714-831-6000	S	160	0.9
Textron Inc	1721 Packard Av	Racine	WI	53403	Richard D Miller	414-637-6711	D	140	0.9
Woods Equipment Co	PO Box 1000	Oregon	IL	61061	Thomas J Laird	815-732-2141	R	120	1.0
Bush Hog Corp	PO Box 1039	Selma	AL	36701	Bobby Middlebrooks	334-872-6261	S	113	0.9
Lindsay Manufacturing Co	PO Box 156	Lindsay	NE	68644	Gary D Parker	402-428-2131	P	113	0.5
Hay and Forage Industries	PO Box 4000	Hesston	KS	67062	Charles Miller	316-327-6300	J	110	0.9
Grain Systems Inc	PO Box 20	Assumption	IL	62510	Craig Sloan	217-226-4421	R	100	0.3
Great Plains Manufacturing Inc	PO Box 218	Assaria	KS	67416	Roy Applequist	913-667-4755	R	90	1.0
Alamo Group Inc	1502 E Walnut	Seguin	TX	78155	Donald J Douglass	210-379-1480	P	89	1.0
Amerequip Corp	11033 N Towne Sq	Mequon	WI	53092	Richard H Lytle	414-241-6160	R	55*	0.4
Tyler LP	PO Box 249	Benson	MN	56215	Donald E McGrath	612-843-3333	R	50	0.2
AgEquipment Group	PO Box 1120	Lockney	TX	79241	John Tye	806-652-3367	R	47*	0.3
Cane Machine and Engineering	N 10th & Coulons	Thibodaux	LA	70301	Jacob Giordina	504-447-7285	R	47*	0.3
DMI Inc	PO Box 65	Goodfield	IL	61742	WH Schmidtgall	309-965-2233	R	35	0.3
Stock Equipment Co	16490 Chillicothe	Chagrin Falls	OH	44023	A Neil Small	216-543-6000	D	35	0.3
T-L Irrigation Co	PO Box 1047	Hastings	NE	68901	L Thom	402-462-4128	R	35	0.2
Long Mfg North Carolina	PO Box 1139	Tarboro	NC	27886	David Walsh	919-823-4151	R	32	0.3
French and Hecht	PO Box 739	Walcott	IA	52773	Maurice Taylor	319-284-5011	D	30	0.2
Reinke Manufacturing Company	PO Box 566	Deshler	NE	68340	Ron Schardt	402-365-7251	R	30	0.2
HCC Inc	1501 1st Av	Mendota	IL	61342	Carl E McNair	815-539-9371	R	29	0.3
Chick Master Incubator Co	PO Box 704	Medina	OH	44258	Robert Holzer	216-722-5591	R	28	<0.1
Ingersoll Products	1000 W 120th St	Chicago	IL	60643	Ed Budzinski	312-264-7800	D	28	0.2
J-Star Industries	801 Janesville Av	Fort Atkinson	WI	53538	John Neill	414-563-5521	R	27*	0.2
Yetter Manufacturing Company	PO Box 358	Colchester	IL	62326	Bernard Whalen	309-776-4111	R	27*	0.2
Chief Agri/Industrial	PO Box 848	Kearney	NE	68848	John Price	308-237-3186	D	26	0.1
Knight Manufacturing Corp	PO Box 167	Brodhead	WI	53520	WS Knight	608-897-2131	R	26	0.2
Bell Equipment USA Inc	2843 Hwy 80	Garden City	GA	31408	William C Rodgers	912-966-2615	S	25	<0.1
Diamond Automation Inc	23400 Haggerty Rd	Farmington Hls	MI	48335	James Nield	810-476-7100	R	25*	0.2
Wil-Rich Manufacturing	PO Box 1030	Wahpeton	ND	58074	John Kehrwald	701-642-2621	D	25*	0.1
Herschel Corp	1301 N 14th St	Indianola	IA	50125	John Annin	515-961-7481	S	25	0.2
Hiniker Co	Rte 5	Mankato	MN	56002	V Tomlonovic	507-625-6621	R	24	0.2
HD Hudson Manufacturing Co	500 N Michigan Av	Chicago	IL	60611	R C Hudson Jr	312-644-2830	R	23*	0.2
Tumac Industries Inc	1101 3rd Av	Grand Junction	CO	81501	JR McConnell	303-245-4400	R	23*	0.2
DuraTech Industries Intern	PO Box 1940	Jamestown	ND	58402	Robert Goff	701-252-4601	R	22	0.1
Kinze Manufacturing Inc	PO Box 806	Williamsburg	IA	52361	JE Kinzenbaw	319-668-1300	R	22*	0.3
Patz Sales Inc	PO Box 7	Pound	WI	54161	C Patz	414-897-2251	R	22	0.2
Art's-Way Manufacturing	PO Box 288	Armstrong	IA	50514	William S Garner	712-864-3131	P	20	0.2
CrustBuster/Speed King Inc	PO Box 1438	Dodge City	KS	67801	Don Hornung	316-227-7106	R	20	0.2
Lor-AL Products Inc	PO Box 265	Benson	MN	56215	AE McQuinn	612-843-4161	S	20	0.1
Unverferth Manufacturing	PO Box 357	Kalida	OH	45853	R A Unverferth	419-532-3121	R	20	0.2
Double L Manufacturing Inc	PO Box 597	American Falls	ID	83211	Melvin Grover	208-226-5592	S	19*	0.2
Mathews Co	PO Box 70	Crystal Lake	IL	60014	Ron Gillund	815-459-2210	R	19*	0.1
Shivvers Inc	614 W English St	Corydon	IA	50060	Steve Shivvers	515-872-1005	R	19*	0.1
Spudnik Equipment Co	PO Box 1045	Blackfoot	ID	83221	C Hobbs	208-785-0480	R	19	0.2
Hawkeye Steel Products Inc	PO Box 2000	Houghton	IA	52631	T Wenstrand	319-469-4141	R	18	0.2
Universal Cooperatives Inc	PO Box 115	Goshen	IN	46526		219-533-3131	R	18	0.3
XF Enterprises Inc	99 Locust Bend Rd	Ephrata	PA	17522	RH Klett	717-859-1166	R	18	<0.1
Badger Northland Inc	PO Box 1215	Kaukauna	WI	54130	Robert L Hartsock	414-766-4603	R	17	0.2
Byron Equipment Co	7275 Batavia Byron	Byron	NY	14422	Richard Glazier	716-548-2665	R	17	0.2
Nichols Tillage Tools Inc	312 Hereford Av	Sterling	CO	80751	John Nichols	303-522-8676	R	17	0.2
Orthman Manufacturing Inc	PO Box B	Lexington	NE	68850	William H Orthman	308-324-4654	R	16	0.2
Amadas Industries Inc	PO Box 1833	Suffolk	VA	23439	James C Adams	804-539-0231	R	16	0.1
Howard Price Turf Equipment	18155 Edison Av	Chesterfield	MO	63005	Howard Price	314-532-7000	R	15*	0.1
Howse Implement Company Inc	PO Box 86	Laurel	MS	39441	Ben T Howse	601-428-0841	R	15	0.3
Pixall LP	100 Bean St	Clear Lake	WI	54005	James T Glover	715-263-2112	R	15*	0.1
Sioux Steel Co	PO Drawer 1265	Sioux Falls	SD	57101	PM Rysdon	605-336-1750	R	15	0.2
Chick Master International Inc	120 Sylvan Av	Englewood Clfs	NJ	07632	Robert Holzer	201-947-8810	R	14*	<0.1
Conrad-American Inc	PO Box 88	Houghton	IA	52631	Marvin Bricker	319-469-4111	S	14*	0.1
Highway Equipment Co	616 D Av NW	Cedar Rapids	IA	52405	MW Rissi	319-363-8281	R	14	0.1
Hagie Manufacturing Co	721 Central Av W	Clarion	IA	50525	John Hagie	515-532-2861	R	13	0.1
Automatic Equipment Mfg Co	PO Box P	Pender	NE	68047	Jay Hesse	402-385-3051	R	13	0.2
Calkins Manufacturing Co	PO Box 14527	Spokane	WA	99214	JE Calkins	509-928-7420	R	13*	<0.1
Edstrom Industries Inc	819 Bakke Av	Waterford	WI	53185	WE Edstrom Sr	414-534-5181	R	13	0.2

Source: Ward's Business Directory of U.S. Private and Public Companies, Volumes 1 and 2, 1996. The company type code used is as follows: P - Public, R - Private, S - Subsidiary, D - Division, J - Joint Venture, A - Affiliate, G - Group. Sales are in millions of dollars, employees are in thousands. An asterisk (*) indicates an estimated sales volume. The symbol < stands for 'less than'. Company names and addresses are truncated, in some cases, to fit into the available space.

MATERIALS CONSUMED

Material	Quantity	Delivered Cost ($ million)
Materials, ingredients, containers, and supplies	(X)	3,975.4
Fluid power pumps, motors, and hydrostatic transmissions (hydraulic and pneumatic)	(X)	132.9
Fluid power cylinders and rotary actuators (hydraulic and pneumatic)	(X)	46.4
Fluid power filters (hydraulic and pneumatic)	(X)	5.3
Fluid power hose or tube fittings and assemblies (hydraulic and pneumatic)	(X)	40.4
Fluid power valves (hydraulic and pneumatic)	(X)	36.3
Metal bolts, nuts, screws, washers, rivets, and other screw machine products	(X)	89.5
Metal stampings	(X)	79.9
All other fabricated metal products (except forgings)	(X)	174.8
Iron and steel forgings	(X)	69.2
Nonferrous forgings	(X)	1.5
Iron and steel castings (rough and semifinished)	(X)	177.1
Aluminum and aluminum-base alloy castings (rough and semifinished)	(X)	22.8
Other nonferrous castings (rough and semifinished)	(X)	6.8
Steel bars, bar shapes, and plates	(X)	210.1
Steel sheet and strip, including tin plate	(X)	360.3
Steel structural shapes and sheet piling	(X)	53.2
All other steel shapes and forms	(X)	71.9
Nonferrous shapes and forms	(X)	3.3
Metal powders	(X)	8.5
Diesel and semidiesel engines	(X)	202.9
Gasoline and other carburetor engines	(X)	47.9
Engine electrical equipment, including spark plugs, magnetos, generators, starters, etc.	(X)	63.2
Electric motors and generators	(X)	30.9
Ball bearings (mounted or unmounted)	(X)	53.9
Roller bearings (mounted or unmounted)	(X)	36.3
Mechanical speed changers, gears, and industrial high-speed drives	(X)	116.4
Tires and inner tubes	(X)	135.0
Rubber and plastics hose and belting	(X)	61.5
Fabricated plastics products (except gaskets, hoses, and belting)	(X)	95.2
Paints, varnishes, lacquers, stains, shellacs, japans, enamels, and allied products	(X)	45.1
All other materials and components, parts, containers, and supplies	(X)	853.9
Materials, ingredients, containers, and supplies, nsk	(X)	643.0

Source: 1992 *Economic Census*. Explanation of symbols used: (D): Withheld to avoid disclosure of competitive data; na: Not available; (S): Withheld because statistical norms were not met; (X): Not applicable; (Z): Less than half the unit shown; nec: Not elsewhere classified; nsk: Not specified by kind; - : zero; * : 10-19 percent estimated; ** : 20-29 percent estimated.

PRODUCT SHARE DETAILS

Product or Product Class	% Share	Product or Product Class	% Share
Farm machinery and equipment	100.00	but including attachments	31.61
Farm dairy equipment, sprayers and dusters (except aerial types), farm elevators, farm blowers, and attachments	4.28	Parts for farm machinery, for sale separately	16.52
Planting, seeding, and fertilizing machinery, and attachments, excluding turf machinery	5.84	Parts for farm type wheel tractors (except operator cabs), for sale separately	25.27
Harvesting machinery (except hay and straw) and attachments	17.53	Other parts for farm machinery, (except for wheel tractors and operator cabs), for sale separately	67.27
Haying machinery and attachments	4.87	Operator cabs for farm equipment, for sale separately	2.85
Farm plows (including plowshares, primary tillage), harrows, rollers, pulverizers, and cultivators and weeders, and attachments	3.00	Parts for farm machinery, for sale separately, nsk	4.61
All other farm machinery and equipment, excluding parts,		Commercial turf and grounds care equipment, including parts and attachments	9.34
		Farm machinery and equipment, nsk	7.01

Source: 1992 *Economic Census*. The values shown are percent of total shipments in an industry. Values of indented subcategories are summed in the main heading. The symbol (D) appears when data are withheld to prevent disclosure of competitive information. The abbreviation nsk stands for 'not specified by kind' and nec for 'not elsewhere classified'.

INPUTS AND OUTPUTS FOR FARM MACHINERY & EQUIPMENT

Economic Sector or Industry Providing Inputs	%	Sector	Economic Sector or Industry Buying Outputs	%	Sector
Farm machinery & equipment	16.0	Manufg.	Gross private fixed investment	61.0	Cap Inv
Imports	15.7	Foreign	Exports	12.6	Foreign
Wholesale trade	12.0	Trade	Farm machinery & equipment	10.5	Manufg.
Blast furnaces & steel mills	9.0	Manufg.	Miscellaneous repair shops	2.9	Services
Internal combustion engines, nec	5.5	Manufg.	Feed grains	2.8	Agric.
Tires & inner tubes	3.7	Manufg.	Change in business inventories	1.9	In House
Iron & steel foundries	3.3	Manufg.	Meat animals	1.6	Agric.
Advertising	2.9	Services	State & local government enterprises, nec	1.3	Gov't
Carburetors, pistons, rings, & valves	2.3	Manufg.	Oil bearing crops	0.9	Agric.
Power transmission equipment	2.1	Manufg.	Food grains	0.6	Agric.
Machinery, except electrical, nec	1.7	Manufg.	S/L Govt. purch., higher education	0.5	S/L Govt
Pumps & compressors	1.5	Manufg.	Tobacco	0.4	Agric.
Iron & steel forgings	1.4	Manufg.	Vegetables	0.4	Agric.

Continued on next page.

INPUTS AND OUTPUTS FOR FARM MACHINERY & EQUIPMENT - Continued

Economic Sector or Industry Providing Inputs	%	Sector	Economic Sector or Industry Buying Outputs	%	Sector
Ball & roller bearings	1.2	Manufg.	S/L Govt. purch., natural resource & recreation.	0.4	S/L Govt
Screw machine and related products	1.2	Manufg.	Dairy farm products	0.3	Agric.
Electric services (utilities)	1.2	Util.	Fruits	0.3	Agric.
Maintenance of nonfarm buildings nec	1.1	Constr.	Cotton	0.2	Agric.
Metal stampings, nec	1.1	Manufg.	Miscellaneous livestock	0.2	Agric.
Motor freight transportation & warehousing	1.1	Util.	Federal Government purchases, national defense	0.2	Fed Govt
Air transportation	0.9	Util.	S/L Govt. purch., highways	0.2	S/L Govt
Engine electrical equipment	0.8	Manufg.	Forestry products	0.1	Agric.
Pipe, valves, & pipe fittings	0.7	Manufg.	Greenhouse & nursery products	0.1	Agric.
Gas production & distribution (utilities)	0.7	Util.	Poultry & eggs	0.1	Agric.
Motor vehicle parts & accessories	0.6	Manufg.	S/L Govt. purch., correction	0.1	S/L Govt
Engineering, architectural, & surveying services	0.6	Services			
Miscellaneous plastics products	0.5	Manufg.			
Paints & allied products	0.5	Manufg.			
Communications, except radio & TV	0.5	Util.			
Eating & drinking places	0.5	Trade			
Equipment rental & leasing services	0.5	Services			
Fabricated rubber products, nec	0.4	Manufg.			
Motors & generators	0.4	Manufg.			
Rubber & plastics hose & belting	0.4	Manufg.			
Veneer & plywood	0.4	Manufg.			
Banking	0.4	Fin/R.E.			
U.S. Postal Service	0.4	Gov't			
Lawn & garden equipment	0.3	Manufg.			
Machine tools, metal cutting types	0.3	Manufg.			
Special dies & tools & machine tool accessories	0.3	Manufg.			
Railroads & related services	0.3	Util.			
Computer & data processing services	0.3	Services			
Noncomparable imports	0.3	Foreign			
Aluminum castings	0.2	Manufg.			
Copper rolling & drawing	0.2	Manufg.			
Metal heat treating	0.2	Manufg.			
Petroleum refining	0.2	Manufg.			
Real estate	0.2	Fin/R.E.			
Legal services	0.2	Services			
Management & consulting services & labs	0.2	Services			
Scrap	0.2	Scrap			
Abrasive products	0.1	Manufg.			
Aluminum rolling & drawing	0.1	Manufg.			
Asbestos products	0.1	Manufg.			
Industrial gases	0.1	Manufg.			
Paperboard containers & boxes	0.1	Manufg.			
Sawmills & planning mills, general	0.1	Manufg.			
Steel springs, except wire	0.1	Manufg.			
Truck & bus bodies	0.1	Manufg.			
Insurance carriers	0.1	Fin/R.E.			
Accounting, auditing & bookkeeping	0.1	Services			
Automotive repair shops & services	0.1	Services			
Hotels & lodging places	0.1	Services			

Source: Benchmark Input-Output Accounts for the U.S. Economy, 1982, U.S. Department of Commerce, Washington, D.C., July 1991. Data, as reported in the source, are organized by the 1977 SIC structure in use in 1982 but have been matched, as closely as is possible, to the 1987 SIC structure used in this book.

OCCUPATIONS EMPLOYED BY SIC 352 - FARM AND GARDEN MACHINERY

Occupation	% of Total 1994	Change to 2005	Occupation	% of Total 1994	Change to 2005
Assemblers, fabricators, & hand workers nec	16.2	-15.3	Traffic, shipping, & receiving clerks	1.5	-18.4
Welders & cutters	5.7	-15.3	Helpers, laborers, & material movers nec	1.5	-15.2
Welding machine setters, operators	4.5	-23.7	Secretaries, ex legal & medical	1.4	-22.9
Blue collar worker supervisors	4.1	-22.4	Combination machine tool operators	1.4	-6.8
Machine tool cutting & forming etc. nec	2.8	-15.3	Industrial machinery mechanics	1.4	-6.8
Sales & related workers nec	2.7	-15.3	Stock clerks	1.4	-31.2
Industrial truck & tractor operators	2.3	-15.3	Freight, stock, & material movers, hand	1.3	-32.2
Machinists	2.2	-15.3	General office clerks	1.2	-27.8
Machine forming operators, metal & plastic	2.2	-15.3	Tool & die makers	1.2	-31.6
General managers & top executives	2.0	-19.6	Machine assemblers	1.2	-15.3
Coating, painting, & spraying machine workers	2.0	-15.3	Production, planning, & expediting clerks	1.1	-15.3
Inspectors, testers, & graders, precision	1.8	-15.3	Punching machine operators, metal & plastic	1.1	-32.2
Machine builders	1.8	1.7	Bookkeeping, accounting, & auditing clerks	1.1	-36.4
NC machine tool operators, metal & plastic	1.6	1.7	Drilling & boring machine tool workers	1.1	-32.2
Sheet metal workers & duct installers	1.6	-15.3	Mechanical engineers	1.1	-6.8
Machine tool cutting operators, metal & plastic	1.6	-29.4	Engineers nec	1.1	1.7

Source: Industry-Occupation Matrix, Bureau of Labor Statistics. These data relate to one or more 3-digit SIC industry groups rather than to a single 4-digit SIC. The change reported for each occupation to the year 2005 is a percent of growth or decline as estimated by the Bureau of Labor Statistics. The abbreviation nec stands for 'not elsewhere classified'.

LOCATION BY STATE AND REGIONAL CONCENTRATION

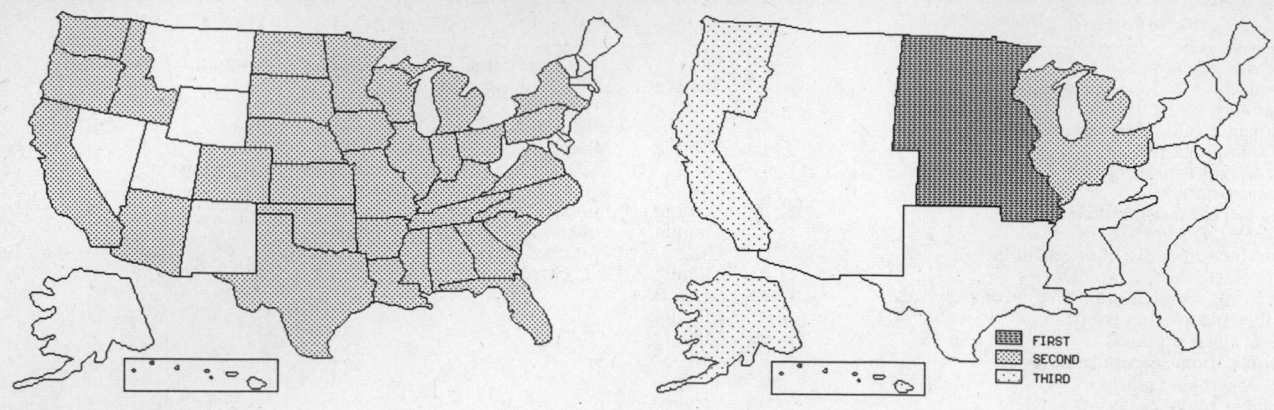

FIRST
SECOND
THIRD

INDUSTRY DATA BY STATE

| State | Establish-ments | Shipments | | | Employment | | | | Cost as % of Shipments | Investment per Employee ($) |
		Total ($ mil)	% of U.S.	Per Establ.	Total Number	% of U.S.	Per Establ.	Wages ($/hour)		
Illinois	97	1,867.6	19.4	19.3	8,600	14.0	89	18.45	41.0	5,058
Wisconsin	93	1,204.0	12.5	12.9	6,000	9.8	65	12.19	47.4	2,867
Kansas	79	384.3	4.0	4.9	4,000	6.5	51	10.31	53.8	-
Pennsylvania	41	269.9	2.8	6.6	1,600	2.6	39	13.52	47.4	1,250
Indiana	54	245.6	2.6	4.5	1,900	3.1	35	11.04	49.8	-
Missouri	39	200.5	2.1	5.1	1,200	2.0	31	11.47	44.2	-
Ohio	48	188.0	2.0	3.9	1,600	2.6	33	11.16	43.1	1,375
California	139	183.8	1.9	1.3	1,900	3.1	14	9.73	49.2	-
Texas	91	116.2	1.2	1.3	1,200	2.0	13	8.65	49.3	-
North Carolina	24	106.4	1.1	4.4	1,200	2.0	50	8.17	55.1	1,250
Florida	41	90.2	0.9	2.2	800	1.3	20	8.83	51.2	4,500
Michigan	34	80.1	0.8	2.4	500	0.8	15	11.86	48.6	-
Colorado	21	79.4	0.8	3.8	700	1.1	33	10.27	51.1	-
Idaho	37	64.0	0.7	1.7	800	1.3	22	8.09	49.1	-
Arkansas	38	61.9	0.6	1.6	600	1.0	16	9.71	53.0	-
South Dakota	25	51.6	0.5	2.1	500	0.8	20	9.00	44.0	1,600
Tennessee	18	37.7	0.4	2.1	400	0.7	22	9.00	50.1	1,000
South Carolina	5	17.5	0.2	3.5	200	0.3	40	7.00	40.0	500
Iowa	140	(D)	-	-	17,500 *	28.5	125	-	-	-
Minnesota	91	(D)	-	-	3,750 *	6.1	41	-	-	-
Nebraska	90	(D)	-	-	7,500 *	12.2	83	-	-	-
Georgia	59	(D)	-	-	1,750 *	2.9	30	-	-	-
North Dakota	45	(D)	-	-	1,750 *	2.9	39	-	-	-
Oklahoma	36	(D)	-	-	375 *	0.6	10	-	-	-
Washington	33	(D)	-	-	375 *	0.6	11	-	-	-
New York	31	(D)	-	-	750 *	1.2	24	-	-	-
Oregon	28	(D)	-	-	375 *	0.6	13	-	-	-
Alabama	26	(D)	-	-	1,750 *	2.9	67	-	-	-
Mississippi	20	(D)	-	-	750 *	1.2	38	-	-	-
Virginia	17	(D)	-	-	375 *	0.6	22	-	-	-
Kentucky	16	(D)	-	-	375 *	0.6	23	-	-	267
Louisiana	16	(D)	-	-	750 *	1.2	47	-	-	-
Arizona	11	(D)	-	-	175 *	0.3	16	-	-	-

Source: 1992 *Economic Census*. The states are in descending order of shipments or establishments (if shipment data are missing for the majority). The symbol (D) appears when data are withheld to prevent disclosure of competitive information. States marked with (D) are sorted by number of establishments. A dash (-) indicates that the data element cannot be calculated; * indicates the midpoint of a range.

3524 - LAWN & GARDEN EQUIPMENT

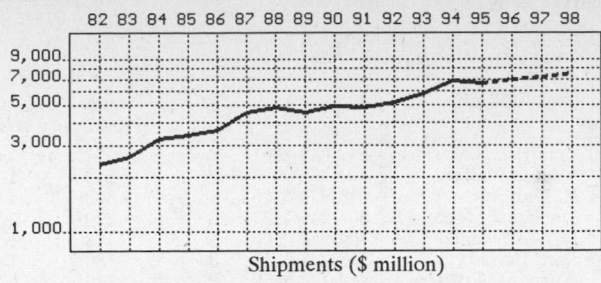

Shipments ($ million)

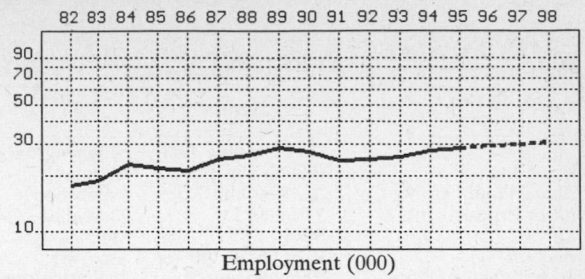

Employment (000)

GENERAL STATISTICS

| Year | Com-panies | Establishments | | Employment | | | Compensation | | Production ($ million) | | | |
		Total	with 20 or more employees	Total (000)	Production Workers (000)	Hours (Mil)	Payroll ($ mil)	Wages ($/hr)	Cost of Materials	Value Added by Manufacture	Value of Shipments	Capital Invest.
1982	151	175	84	17.8	12.4	24.6	318.1	8.08	1,375.7	945.5	2,365.2	51.1
1983		172	81	18.8	13.9	26.6	337.6	8.26	1,555.5	1,066.9	2,579.8	43.2
1984		169	78	23.2	17.9	34.1	443.0	9.04	2,049.3	1,246.3	3,239.8	65.9
1985		167	76	22.3	17.2	33.3	417.4	8.50	2,075.9	1,322.0	3,439.6	70.7
1986		169	76	21.5	16.5	30.6	429.9	9.42	2,220.0	1,438.3	3,647.0	54.0
1987	149	165	81	24.9	19.1	35.4	500.6	9.69	2,714.8	1,915.9	4,594.4	111.2
1988		169	82	26.1	19.9	38.1	537.0	9.55	2,995.9	1,916.1	4,828.4	97.4
1989		167	80	28.9	19.4	37.2	527.6	8.86	2,792.5	1,694.6	4,577.5	127.1
1990		155	79	27.4	18.3	36.0	526.6	9.82	3,006.1	2,006.5	4,910.0	82.2
1991		161	79	24.4	18.9	35.2	517.8	9.88	2,821.6	1,864.9	4,819.5	91.3
1992	127	144	73	24.8	19.8	36.3	568.9	10.71	3,080.1	2,153.4	5,168.7	125.1
1993		156	68	25.7	20.5	40.0	631.2	10.88	3,598.4	2,302.9	5,828.3	92.1
1994		151P	73P	27.8	22.4	44.4	682.7	10.64	4,194.0	2,778.4	6,819.2	125.3
1995		149P	72P	28.9P	22.4P	42.9P	675.3P	11.00P	4,044.2P	2,679.1P	6,575.6P	128.2P
1996		147P	72P	29.5P	23.0P	44.1P	701.0P	11.21P	4,237.9P	2,807.5P	6,890.6P	134.1P
1997		145P	71P	30.2P	23.6P	45.3P	726.7P	11.43P	4,431.7P	2,935.8P	7,205.6P	139.9P
1998		143P	70P	30.9P	24.2P	46.5P	752.4P	11.64P	4,625.4P	3,064.2P	7,520.6P	145.7P

Sources: 1982, 1987, 1992 *Economic Census*; *Annual Survey of Manufactures*, 83-86, 88-91, 93-94. Establishment counts for non-Census years are from *County Business Patterns*; establishment values for 83-84 are extrapolations. 'P's show projections by the editors. Industries reclassified in 87 will not have data for prior years.

INDICES OF CHANGE

| Year | Com-panies | Establishments | | Employment | | | Compensation | | Production ($ million) | | | |
		Total	with 20 or more employees	Total (000)	Production Workers (000)	Hours (Mil)	Payroll ($ mil)	Wages ($/hr)	Cost of Materials	Value Added by Manufacture	Value of Shipments	Capital Invest.
1982	119	122	115	72	63	68	56	75	45	44	46	41
1983		119	111	76	70	73	59	77	51	50	50	35
1984		117	107	94	90	94	78	84	67	58	63	53
1985		116	104	90	87	92	73	79	67	61	67	57
1986		117	104	87	83	84	76	88	72	67	71	43
1987	117	115	111	100	96	98	88	90	88	89	89	89
1988		117	112	105	101	105	94	89	97	89	93	78
1989		116	110	117	98	102	93	83	91	79	89	102
1990		108	108	110	92	99	93	92	98	93	95	66
1991		112	108	98	95	97	91	92	92	87	93	73
1992	100	100	100	100	100	100	100	100	100	100	100	100
1993		108	93	104	104	110	111	102	117	107	113	74
1994		105P	100P	112	113	122	120	99	136	129	132	100
1995		103P	99P	116P	113P	118P	119P	103P	131P	124P	127P	102P
1996		102P	98P	119P	116P	122P	123P	105P	138P	130P	133P	107P
1997		101P	97P	122P	119P	125P	128P	107P	144P	136P	139P	112P
1998		99P	96P	125P	122P	128P	132P	109P	150P	142P	146P	116P

Sources: Same as General Statistics. Values reflect change from the base year, 1992. Values above 100 mean greater than 92, values below 100 mean less than 92, and a value of 100 in the 82-91 or 93-98 period means same as 92. 'P's mark projections by the editors.

SELECTED RATIOS

For 1994	Avg. of All Manufact.	Analyzed Industry	Index	For 1994	Avg. of All Manufact.	Analyzed Industry	Index
Employees per Establishment	49	184	376	Value Added per Production Worker	134,084	124,036	93
Payroll per Establishment	1,500,273	4,523,461	302	Cost per Establishment	5,045,178	27,788,776	551
Payroll per Employee	30,620	24,558	80	Cost per Employee	102,970	150,863	147
Production Workers per Establishment	34	148	432	Cost per Production Worker	146,988	187,232	127
Wages per Establishment	853,319	3,130,153	367	Shipments per Establishment	9,576,895	45,182,933	472
Wages per Production Worker	24,861	21,090	85	Shipments per Employee	195,460	245,295	125
Hours per Production Worker	2,056	1,982	96	Shipments per Production Worker	279,017	304,429	109
Wages per Hour	12.09	10.64	88	Investment per Establishment	321,011	830,218	259
Value Added per Establishment	4,602,255	18,409,236	400	Investment per Employee	6,552	4,507	69
Value Added per Employee	93,930	99,942	106	Investment per Production Worker	9,352	5,594	60

Sources: Same as General Statistics. The 'Average of All Manufacturing' column represents the average of all manufacturing industries reported for the most recent complete year available. The Index shows the relationship between the Average and the Analyzed Industry. For example, 100 means that they are equal; 500 that the Analyzed Industry is five times the average; 50 means that the Analyzed Industry is half the national average. The abbreviation 'na' is used to show that data are 'not available'.

LEADING COMPANIES   Number shown: **44**   Total sales ($ mil): **3,502**   Total employment (000): **19.8**

Company Name	Address				CEO Name	Phone	Co. Type	Sales ($ mil)	Empl. (000)
Toro Co	8111 Lyndale Av S	Bloomington	MN	55420	Kendrick B Melrose	612-888-8801	P	794	3.4
Actava Group Inc	4900 G-Pacific	Atlanta	GA	30303	John D Phillips	404-658-9000	P	552	1.2
Murray Ohio Manufacturing Co	219 Franklin Rd	Brentwood	TN	37027	JL Duncan	615-373-6500	S	500*	4.0
WCI Outdoor Products Inc	1 Poulan Dr	Nashville	AR	71852	Michael J McCann	501-845-1234	S	250*	2.0
Snapper	PO Box 777	McDonough	GA	30253	Jerry Schweiner	404-954-2500	D	248	1.2
Garden Way Inc	102nd St & 9th Av	Troy	NY	12180	Jairo Estrada	518-235-6010	R	180*	1.5
Simplicity Manufacturing Inc	500 N Spring St	Pt Washington	WI	53074	Warner C Frazier	414-284-8669	R	120	0.5
True Temper Hardware Inc	PO Box 8859	Camp Hill	PA	17011	Brian T Schnabel	717-737-1500	S	100	0.8
Echo Inc	400 Oakwood Rd	Lake Zurich	IL	60047	Tak Sasaki	708-540-8400	R	78*	0.3
Ariens Co	655 W Ryan St	Brillion	WI	54110	Michael S Ariens	414-756-2141	R	75	0.5
Lawn Chief Manufacturing Co	201 E Brink St	Harvard	IL	60033	Bob Simmons	815-943-7419	D	70	0.4
Gilmour Manufacturing Co	Somerset Industrial	Somerset	PA	15501	M Zehnder	814-443-4802	S	63*	0.5
Easy Gardener Inc	PO Box 21025	Waco	TX	76702	Joe Owens	817-753-5353	R	50	<0.1
Honda Power Equipment Mfg	PO Box 37	Swepsonville	NC	27359	Pat Curtis	919-578-5300	S	31*	0.3
Rotary Corp	PO Box 947	Glennville	GA	30427	Ed Nelson	912-654-3433	R	29*	0.4
Shindaiwa Inc	PO Box 1090	Tualatin	OR	97062	Thomas Bunch	503-692-3070	S	29*	<0.1
Ingersoll Equipment Co	200 Ingersoll Rd	Winneconne	WI	54986	Thomas Lopina	414-582-4455	R	28	0.1
Yazoo Manufacturing Company	PO Box 4449	Jackson	MS	39296	Dan M Swain	601-366-6421	S	27	0.1
Agri-Fab Inc	PO Box 500	Sullivan	IL	61951	RD Harshman	217-728-8388	R	25	0.2
McLane Manufacturing Inc	7110 E Rosecrans	Paramount	CA	90723	Frank E McLane	310-633-8158	R	25	0.2
Southland Mower Corp	PO Box 347	Selma	AL	36701	Kenneth E Worrell	334-874-7405	S	25	0.1
Bunton Co	PO Box 33247	Louisville	KY	40232	Lawrence O'Connell	502-966-0550	R	20	0.3
Precision Products Inc	PO Box 2546	Springfield	IL	62708	JW Costa	217-528-1311	R	18	0.2
Grasshopper Co	PO Box 637	Moundridge	KS	67107	Stan Guyer	316-345-8621	R	17	0.1
Auburn Consolidated Industries	PO Box 350	Auburn	NE	68305	John S Skaggs	402-274-4911	R	15	0.2
Minuteman International Inc	111 S Rohlwing Rd	Addison	IL	60101	Jerry Rau	708-627-6900	R	14	0.1
Parker Sweeper Co	111 S Rohlwing Rd	Addison	IL	60101	Jerry Rau	708-627-6900	S	14	0.1
Earthway Products Inc	PO Box 547	Bristol	IN	46507	Doug Schrock	219-848-7491	R	12	0.1
Hoffco Inc	358 F St NW	Richmond	IN	47374	John Bratt	317-966-8161	R	12	0.2
Brinly-Hardy Co	340 E Main St	Louisville	KY	40202	Jane W Hardy	502-585-3351	R	10	0.1
Burgess Products Inc	23 Garden St	NY Mills	NY	13417	John F Romano	315-736-0037	R	10	<0.1
Frederick Manufacturing Corp	4840 E 12th St	Kansas City	MO	64127	P Lawler	816-231-5007	R	10*	0.1
FD Kees Manufacturing Co	700-800 Park Av	Beatrice	NE	68310	Mike Schaefer	402-223-2391	R	10	0.1
Studley Products Inc	210 Passaic St	Newark	NJ	07104	Mark W Howard	201-484-1100	S	10	0.1
Kut-Kwick Corp	PO Box 984	Brunswick	GA	31521	Robert Torras	912-265-1630	R	6	<0.1
Flink Company Inc	502 N Vermillion St	Streator	IL	61364	Lee Sprowl	815-673-4321	R	5	<0.1
Lamber New Corp	PO Box 278	Ansonia	OH	45303	Stephen K Lambert	513-337-3641	R	4*	<0.1
Orbex Inc	620 S 8th St	Minneapolis	MN	55404	Jeff Lawler	612-333-1208	R	4	<0.1
Evergreen International Inc	1001 South Ransdell	Lebanon	IN	46052	M Gramelspacher	317-482-1662	R	4	<0.1
MacKissic Inc	1189 Old Schuylkill	Parker Ford	PA	19457	Richard D Dhein	610-495-7181	S	3*	<0.1
Lindig Manufacturing Corp	PO Box 130130	Roseville	MN	55113	John F Lindig	612-633-3072	R	2*	<0.1
Lemont Industries Inc	PO Box 249	Coal City	IL	60416	Lawrence P Rybak	815-634-2214	R	2	<0.1
California Flexrake Corp	4335 Rowland Av	El Monte	CA	91731	GF Brock Jr	818-443-4026	R	1	<0.1
Hanson General Products	238 Charles St	South Beloit	IL	61080	John Wihlborg	815-389-5100	D	1*	<0.1

Source: Ward's Business Directory of U.S. Private and Public Companies, Volumes 1 and 2, 1996. The company type code used is as follows: P - Public, R - Private, S - Subsidiary, D - Division, J - Joint Venture, A - Affiliate, G - Group. Sales are in millions of dollars, employees are in thousands. An asterisk (*) indicates an estimated sales volume. The symbol < stands for 'less than'. Company names and addresses are truncated, in some cases, to fit into the available space.

MATERIALS CONSUMED

Material	Quantity	Delivered Cost ($ million)
Materials, ingredients, containers, and supplies	(X)	2,889.8
Fluid power pumps, motors, and hydrostatic transmissions (hydraulic and pneumatic)	(X)	44.3
Fluid power cylinders and rotary actuators (hydraulic and pneumatic)	(X)	2.4
Fluid power filters (hydraulic and pneumatic)	(X)	(D)
Fluid power hose or tube fittings and assemblies (hydraulic and pneumatic)	(X)	2.6
Fluid power valves (hydraulic and pneumatic)	(X)	(D)
Metal bolts, nuts, screws, washers, rivets, and other screw machine products	(X)	76.9
Metal stampings	(X)	137.3
All other fabricated metal products (except forgings)	(X)	116.9
Iron and steel forgings	(X)	(D)
Nonferrous forgings	(X)	(D)
Iron and steel castings (rough and semifinished)	(X)	11.9
Aluminum and aluminum-base alloy castings (rough and semifinished)	(X)	86.6
Other nonferrous castings (rough and semifinished)	(X)	(D)
Steel bars, bar shapes, and plates	(X)	42.0
Steel sheet and strip, including tin plate	(X)	150.4
Steel structural shapes and sheet piling	(X)	3.7
All other steel shapes and forms	(X)	19.1
Nonferrous shapes and forms	(X)	(D)
Metal powders	(X)	16.3
Diesel and semidiesel engines	(X)	(D)
Gasoline and other carburetor engines	(X)	914.7
Engine electrical equipment, including spark plugs, magnetos, generators, starters, etc.	(X)	59.6
Electric motors and generators	(X)	17.5

Continued on next page.

MATERIALS CONSUMED - Continued

Material	Quantity	Delivered Cost ($ million)
Ball bearings (mounted or unmounted)	(X)	28.0
Roller bearings (mounted or unmounted)	(X)	4.2
Mechanical speed changers, gears, and industrial high-speed drives	(X)	106.1
Tires and inner tubes	(X)	136.0
Rubber and plastics hose and belting	(X)	37.6
Fabricated plastics products (except gaskets, hoses, and belting)	(X)	132.1
Paints, varnishes, lacquers, stains, shellacs, japans, enamels, and allied products	(X)	40.5
Cabs purchased for installation on farm machinery	(X)	(D)
All other materials and components, parts, containers, and supplies	(X)	426.8
Materials, ingredients, containers, and supplies, nsk	(X)	241.8

Source: 1992 Economic Census. Explanation of symbols used: (D): Withheld to avoid disclosure of competitive data; na: Not available; (S): Withheld because statistical norms were not met; (X): Not applicable; (Z): Less than half the unit shown; nec: Not elsewhere classified; nsk: Not specified by kind; - : zero; * : 10-19 percent estimated; ** : 20-29 percent estimated.

PRODUCT SHARE DETAILS

Product or Product Class	% Share	Product or Product Class	% Share
Lawn and garden equipment	100.00	Parts and attachments for consumer lawn, garden, and snow equipment	14.65
Consumer nonriding lawn, garden, and snow equipment	44.75		
Consumer riding lawn, garden, and snow equipment	37.83	Lawn and garden equipment, nsk	2.77

Source: 1992 Economic Census. The values shown are percent of total shipments in an industry. Values of indented subcategories are summed in the main heading. The symbol (D) appears when data are withheld to prevent disclosure of competitive information. The abbreviation nsk stands for 'not specified by kind' and nec for 'not elsewhere classified'.

INPUTS AND OUTPUTS FOR LAWN & GARDEN EQUIPMENT

Economic Sector or Industry Providing Inputs	%	Sector	Economic Sector or Industry Buying Outputs	%	Sector
Internal combustion engines, nec	23.8	Manufg.	Gross private fixed investment	67.3	Cap Inv
Wholesale trade	13.8	Trade	Exports	6.5	Foreign
Blast furnaces & steel mills	8.3	Manufg.	Miscellaneous repair shops	6.4	Services
Tires & inner tubes	3.9	Manufg.	Owner-occupied dwellings	4.8	Fin/R.E.
Advertising	3.7	Services	Personal consumption expenditures	4.6	
Imports	3.3	Foreign	Change in business inventories	3.7	In House
Lawn & garden equipment	3.1	Manufg.	Lawn & garden equipment	1.7	Manufg.
Miscellaneous plastics products	3.0	Manufg.	Landscape & horticultural services	1.6	Agric.
Metal stampings, nec	2.9	Manufg.	Farm machinery & equipment	0.9	Manufg.
Power transmission equipment	2.8	Manufg.	Federal Government purchases, national defense	0.6	Fed Govt
Aluminum castings	2.5	Manufg.	Wholesale trade	0.4	Trade
Screw machine and related products	2.1	Manufg.	State & local government enterprises, nec	0.3	Gov't
Paperboard containers & boxes	1.5	Manufg.	S/L Govt. purch., natural resource & recreation.	0.3	S/L Govt
Fabricated metal products, nec	1.4	Manufg.	Agricultural, forestry, & fishery services	0.2	Agric.
Primary batteries, dry & wet	1.4	Manufg.	S/L Govt. purch., highways	0.2	S/L Govt
Maintenance of nonfarm buildings nec	1.1	Constr.	S/L Govt. purch., correction	0.1	S/L Govt
Machinery, except electrical, nec	1.1	Manufg.	S/L Govt. purch., higher education	0.1	S/L Govt
Electric services (utilities)	1.1	Util.			
Motor freight transportation & warehousing	1.1	Util.			
Iron & steel foundries	1.0	Manufg.			
Ball & roller bearings	0.9	Manufg.			
Pumps & compressors	0.9	Manufg.			
Air transportation	0.8	Util.			
Aluminum rolling & drawing	0.7	Manufg.			
Hand & edge tools, nec	0.7	Manufg.			
Primary metal products, nec	0.7	Manufg.			
Gas production & distribution (utilities)	0.7	Util.			
Engine electrical equipment	0.6	Manufg.			
Gaskets, packing & sealing devices	0.6	Manufg.			
Engineering, architectural, & surveying services	0.6	Services			
Fabricated rubber products, nec	0.5	Manufg.			
Motors & generators	0.5	Manufg.			
Nonferrous castings, nec	0.5	Manufg.			
Paints & allied products	0.5	Manufg.			
Rubber & plastics hose & belting	0.5	Manufg.			
Wiring devices	0.5	Manufg.			
Communications, except radio & TV	0.5	Util.			
Eating & drinking places	0.5	Trade			
Banking	0.5	Fin/R.E.			
Iron & steel forgings	0.3	Manufg.			
Special dies & tools & machine tool accessories	0.3	Manufg.			
Real estate	0.3	Fin/R.E.			
Computer & data processing services	0.3	Services			
Equipment rental & leasing services	0.3	Services			
U.S. Postal Service	0.3	Gov't			

Continued on next page.

INPUTS AND OUTPUTS FOR LAWN & GARDEN EQUIPMENT - Continued

Economic Sector or Industry Providing Inputs	%	Sector	Economic Sector or Industry Buying Outputs	%	Sector
Carburetors, pistons, rings, & valves	0.2	Manufg.			
Machine tools, metal cutting types	0.2	Manufg.			
Metal heat treating	0.2	Manufg.			
Pipe, valves, & pipe fittings	0.2	Manufg.			
Railroads & related services	0.2	Util.			
Legal services	0.2	Services			
Management & consulting services & labs	0.2	Services			
Scrap	0.2	Scrap			
Abrasive products	0.1	Manufg.			
Industrial gases	0.1	Manufg.			
Petroleum refining	0.1	Manufg.			
Insurance carriers	0.1	Fin/R.E.			
Accounting, auditing & bookkeeping	0.1	Services			
Noncomparable imports	0.1	Foreign			

Source: Benchmark Input-Output Accounts for the U.S. Economy, 1982, U.S. Department of Commerce, Washington, D.C., July 1991. Data, as reported in the source, are organized by the 1977 SIC structure in use in 1982 but have been matched, as closely as is possible, to the 1987 SIC structure used in this book.

OCCUPATIONS EMPLOYED BY SIC 352 - FARM AND GARDEN MACHINERY

Occupation	% of Total 1994	Change to 2005	Occupation	% of Total 1994	Change to 2005
Assemblers, fabricators, & hand workers nec	16.2	-15.3	Traffic, shipping, & receiving clerks	1.5	-18.4
Welders & cutters	5.7	-15.3	Helpers, laborers, & material movers nec	1.5	-15.2
Welding machine setters, operators	4.5	-23.7	Secretaries, ex legal & medical	1.4	-22.9
Blue collar worker supervisors	4.1	-22.4	Combination machine tool operators	1.4	-6.8
Machine tool cutting & forming etc. nec	2.8	-15.3	Industrial machinery mechanics	1.4	-6.8
Sales & related workers nec	2.7	-15.3	Stock clerks	1.4	-31.2
Industrial truck & tractor operators	2.3	-15.3	Freight, stock, & material movers, hand	1.3	-32.2
Machinists	2.2	-15.3	General office clerks	1.2	-27.8
Machine forming operators, metal & plastic	2.2	-15.3	Tool & die makers	1.2	-31.6
General managers & top executives	2.0	-19.6	Machine assemblers	1.2	-15.3
Coating, painting, & spraying machine workers	2.0	-15.3	Production, planning, & expediting clerks	1.1	-15.3
Inspectors, testers, & graders, precision	1.8	-15.3	Punching machine operators, metal & plastic	1.1	-32.2
Machine builders	1.8	1.7	Bookkeeping, accounting, & auditing clerks	1.1	-36.4
NC machine tool operators, metal & plastic	1.6	1.7	Drilling & boring machine tool workers	1.1	-32.2
Sheet metal workers & duct installers	1.6	-15.3	Mechanical engineers	1.1	-6.8
Machine tool cutting operators, metal & plastic	1.6	-29.4	Engineers nec	1.1	1.7

Source: Industry-Occupation Matrix, Bureau of Labor Statistics. These data relate to one or more 3-digit SIC industry groups rather than to a single 4-digit SIC. The change reported for each occupation to the year 2005 is a percent of growth or decline as estimated by the Bureau of Labor Statistics. The abbreviation nec stands for 'not elsewhere classified'.

LOCATION BY STATE AND REGIONAL CONCENTRATION

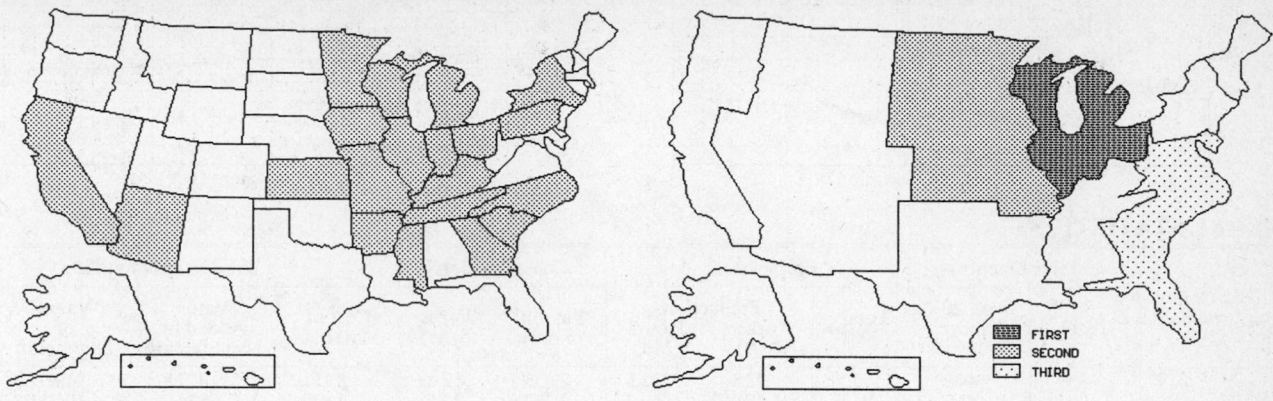

FIRST
SECOND
THIRD

INDUSTRY DATA BY STATE

| State | Establish-ments | Shipments | | | Employment | | | | Cost as % of Shipments | Investment per Employee ($) |
		Total ($ mil)	% of U.S.	Per Establ.	Total Number	% of U.S.	Per Establ.	Wages ($/hour)		
Tennessee	8	1,121.2	21.7	140.1	5,400	21.8	675	10.77	68.3	-
Wisconsin	10	916.6	17.7	91.7	3,000	12.1	300	16.97	49.6	6,900
Ohio	12	399.9	7.7	33.3	1,500	6.0	125	10.88	60.6	933
North Carolina	6	292.1	5.7	48.7	1,400	5.6	233	10.60	41.0	-
Illinois	9	110.5	2.1	12.3	700	2.8	78	9.50	57.8	1,286
Indiana	9	101.9	2.0	11.3	600	2.4	67	9.89	48.2	-
Pennsylvania	7	65.0	1.3	9.3	400	1.6	57	10.83	69.8	4,500
California	7	50.4	1.0	7.2	200	0.8	29	10.00	53.2	-
Kansas	4	31.5	0.6	7.9	300	1.2	75	9.00	45.1	-
Missouri	7	15.0	0.3	2.1	100	0.4	14	7.00	55.3	-
Minnesota	10	(D)	-	-	750 *	3.0	75	-	-	-
Georgia	9	(D)	-	-	1,750 *	7.1	194	-	-	-
Mississippi	4	(D)	-	-	1,750 *	7.1	438	-	-	-
Iowa	3	(D)	-	-	175 *	0.7	58	-	-	-
Kentucky	3	(D)	-	-	175 *	0.7	58	-	-	-
Arkansas	2	(D)	-	-	1,750 *	7.1	875	-	-	-
Michigan	2	(D)	-	-	750 *	3.0	375	-	-	-
South Carolina	2	(D)	-	-	1,750 *	7.1	875	-	-	-
Arizona	1	(D)	-	-	375 *	1.5	375	-	-	-
New York	1	(D)	-	-	750 *	3.0	750	-	-	-

Source: 1992 Economic Census. The states are in descending order of shipments or establishments (if shipment data are missing for the majority). The symbol (D) appears when data are withheld to prevent disclosure of competitive information. States marked with (D) are sorted by number of establishments. A dash (-) indicates that the data element cannot be calculated; * indicates the midpoint of a range.

3531 - CONSTRUCTION MACHINERY

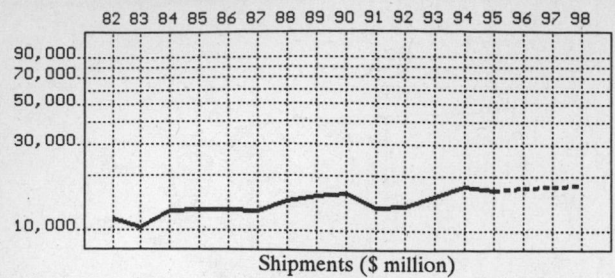

Shipments ($ million)

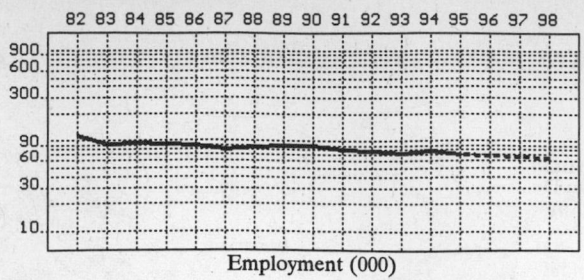

Employment (000)

GENERAL STATISTICS

Year	Companies	Establishments		Employment			Compensation		Production ($ million)			
		Total	with 20 or more employees	Total (000)	Production Workers (000)	Hours (Mil)	Payroll ($ mil)	Wages ($/hr)	Cost of Materials	Value Added by Manufacture	Value of Shipments	Capital Invest.
1982	817	939	444	115.5	72.8	122.6	2,653.1	12.42	6,151.6	5,477.8	11,657.9	419.4
1983		907	436	91.9	57.9	99.7	2,262.6	12.69	4,907.2	4,574.0	10,311.8	323.4
1984		875	428	95.8	63.7	121.1	2,618.7	13.33	6,775.1	5,801.3	12,692.7	248.1
1985		843	421	91.4	59.7	113.3	2,539.9	13.60	6,625.9	5,694.2	12,798.8	301.8
1986		826	412	87.9	57.7	109.4	2,503.5	13.55	6,978.7	5,902.9	12,987.1	296.0
1987	873	955	423	81.1	54.2	107.6	2,429.4	13.92	6,892.5	5,775.2	12,767.7	335.8
1988		903	434	85.2	57.7	115.3	2,634.9	14.18	7,677.2	6,980.3	14,476.8	462.8
1989		867	439	89.4	60.6	121.6	2,772.9	14.46	8,720.9	6,752.9	15,349.4	633.3
1990		863	426	88.3	60.7	118.9	2,878.1	14.96	9,178.0	6,797.3	16,069.6	638.1
1991		892	404	81.0	52.8	99.5	2,562.6	15.21	7,700.2	5,509.1	13,351.3	462.0
1992	864	944	399	77.1	49.6	93.8	2,571.4	15.41	7,581.2	5,828.9	13,451.8	424.5
1993		939	412	74.7	49.1	100.5	2,625.1	15.33	8,698.1	6,851.9	15,444.0	345.5
1994		910P	406P	77.9	52.9	103.9	2,705.8	14.84	10,153.1	7,673.8	17,697.1	359.4
1995		912P	404P	71.8P	48.9P	101.1P	2,716.0P	15.85P	9,564.5P	7,228.9P	16,671.1P	477.6P
1996		915P	401P	69.5P	47.6P	99.9P	2,733.0P	16.09P	9,802.0P	7,408.4P	17,085.0P	488.2P
1997		917P	399P	67.3P	46.4P	98.6P	2,750.1P	16.33P	10,039.4P	7,587.9P	17,499.0P	498.7P
1998		919P	396P	65.0P	45.1P	97.4P	2,767.1P	16.58P	10,276.9P	7,767.4P	17,912.9P	509.3P

Sources: 1982, 1987, 1992 *Economic Census*; *Annual Survey of Manufactures*, 83-86, 88-91, 93-94. Establishment counts for non-Census years are from *County Business Patterns*; establishment values for 83-84 are extrapolations. 'P's show projections by the editors. Industries reclassified in 87 will not have data for prior years.

INDICES OF CHANGE

Year	Companies	Establishments		Employment			Compensation		Production ($ million)			
		Total	with 20 or more employees	Total (000)	Production Workers (000)	Hours (Mil)	Payroll ($ mil)	Wages ($/hr)	Cost of Materials	Value Added by Manufacture	Value of Shipments	Capital Invest.
1982	95	99	111	150	147	131	103	81	81	94	87	99
1983		96	109	119	117	106	88	82	65	78	77	76
1984		93	107	124	128	129	102	87	89	100	94	58
1985		89	106	119	120	121	99	88	87	98	95	71
1986		88	103	114	116	117	97	88	92	101	97	70
1987	101	101	106	105	109	115	94	90	91	99	95	79
1988		96	109	111	116	123	102	92	101	120	108	109
1989		92	110	116	122	130	108	94	115	116	114	149
1990		91	107	115	122	127	112	97	121	117	119	150
1991		94	101	105	106	106	100	99	102	95	99	109
1992	100	100	100	100	100	100	100	100	100	100	100	100
1993		99	103	97	99	107	102	99	115	118	115	81
1994		96P	102P	101	107	111	105	96	134	132	132	85
1995		97P	101P	93P	99P	108P	106P	103P	126P	124P	124P	113P
1996		97P	101P	90P	96P	106P	106P	104P	129P	127P	127P	115P
1997		97P	100P	87P	93P	105P	107P	106P	132P	130P	130P	117P
1998		97P	99P	84P	91P	104P	108P	108P	136P	133P	133P	120P

Sources: Same as General Statistics. Values reflect change from the base year, 1992. Values above 100 mean greater than 92, values below 100 mean less than 92, and a value of 100 in the 82-91 or 93-98 period means same as 92. 'P's mark projections by the editors.

SELECTED RATIOS

For 1994	Avg. of All Manufact.	Analyzed Industry	Index	For 1994	Avg. of All Manufact.	Analyzed Industry	Index
Employees per Establishment	49	86	175	Value Added per Production Worker	134,084	145,062	108
Payroll per Establishment	1,500,273	2,972,615	198	Cost per Establishment	5,045,178	11,154,281	221
Payroll per Employee	30,620	34,734	113	Cost per Employee	102,970	130,335	127
Production Workers per Establishment	34	58	169	Cost per Production Worker	146,988	191,930	131
Wages per Establishment	853,319	1,693,918	199	Shipments per Establishment	9,576,895	19,442,183	203
Wages per Production Worker	24,861	29,147	117	Shipments per Employee	195,460	227,177	116
Hours per Production Worker	2,056	1,964	96	Shipments per Production Worker	279,017	334,539	120
Wages per Hour	12.09	14.84	123	Investment per Establishment	321,011	394,840	123
Value Added per Establishment	4,602,255	8,430,501	183	Investment per Employee	6,552	4,614	70
Value Added per Employee	93,930	98,508	105	Investment per Production Worker	9,352	6,794	73

Sources: Same as General Statistics. The 'Average of All Manufacturing' column represents the average of all manufacturing industries reported for the most recent complete year available. The Index shows the relationship between the Average and the Analyzed Industry. For example, 100 means that they are equal; 500 that the Analyzed Industry is five times the average; 50 means that the Analyzed Industry is half the national average. The abbreviation 'na' is used to show that data are 'not available'.

LEADING COMPANIES   Number shown: **75**   Total sales ($ mil): **5,110**   Total employment (000): **33.4**

Company Name	Address				CEO Name	Phone	Co. Type	Sales ($ mil)	Empl. (000)
Deere-Hitachi	PO Box 1187	Kernersville	NC	27285	James M Burns	910-996-8100	J	400	0.2
Komatsu Dresser Co	PO Box 1422	Lincolnshire	IL	60069	Arlie Tucker	708-367-2000	J	400	2.9
Grove North America	PO Box 21	Shady Grove	PA	17256	Robert C Stift	717-597-8121	D	330	3.0
Manitowoc Company Inc	PO Box 66	Manitowoc	WI	54221	Fred M Butler	414-684-6621	P	275	1.9
Vermeer Manufacturing Co	PO Box 200	Pella	IA	50219	Robert L Vermeer	515-628-3141	R	240*	1.9
Astec Industries Inc	PO Box 72787	Chattanooga	TN	37407	J Don Brock	615-867-4210	P	214	1.5
Haulpak	PO Box 240	Peoria	IL	61650	Edson McCord	309-672-7000	D	200	1.0
Melroe Co	PO Box 6019	Fargo	ND	58108	James Kertz	701-241-8700	S	200	1.6
Blount Inc	PO Box 568	Owatonna	MN	55060	Don Zorn	507-451-8654	D	178	1.0
VME Americas Inc	1 W Pack Sq	Asheville	NC	28801	Helmut Peters	704-257-2500	S	140*	1.0
CMI Corp	PO Box 1985	Oklahoma City	OK	73101	Bill Swisher	405-787-6020	P	128	1.2
Link-Belt Constr Equip Co	PO Box 13600	Lexington	KY	40583	T Izumi	606-263-5200	S	120	0.6
Charles Machine Works Inc	PO Box 66	Perry	OK	73077	Ed Malzahn	405-336-4404	R	110*	1.0
Cedarapids Inc	916 16th St NE	Cedar Rapids	IA	52402	Richard A Schwebel	319-363-3511	S	100	0.8
Hitachi Construction Machinery	20411 Imperial Val	Houston	TX	77073	Nick Stahl	713-821-2400	D	90	<0.1
Mark Industries	106 12th St SE	Waverly	IA	50677	Fil Filipo	319-352-3920	R	90	0.4
Snorkel	PO Box 4065	St Joseph	MO	64504	R Solon	913-989-4481	D	87	0.7
Ingersoll-Rand Co	7500 Shadwell Dr	Roanoke	VA	24019	ER Zimmerman	703-362-3321	D	73*	0.5
Blaw-Knox Constr Equip Corp	750 Broadway Av E	Mattoon	IL	61938	BJ Getz	217-234-8811	S	60	0.5
TRAK International Inc	369 W Western Av	Pt Washington	WI	53074	P Enoch Stiff	414-284-5571	R	60	0.3
Chemetron Railway Products	177 W Hintz Rd	Wheeling	IL	60090	PJ Cunningham	708-520-5454	R	59	0.2
Gencor Industries Inc	5201 Orange	Orlando	FL	32810	EJ Elliott	407-290-6000	P	58	0.5
Wacker Corp	PO Box 9007	Menomonee Fls	WI	53952	Larry O'Toole	414-255-0500	R	56*	0.4
Barber-Greene	12101 B-G	De Kalb	IL	60115	David Duffy	815-756-5600	D	53*	0.4
UpRight Inc	1775 Park St	Selma	CA	93662	David Sargent	209-896-5150	S	53	0.4
Rexworks Inc	PO Box 2037	Milwaukee	WI	53201	M C Hadjinian	414-747-7200	P	51	0.2
Dynapac	PO Box 615	Schertz	TX	78154	Art Kaplan	210-651-9700	S	50	<0.1
Eagle-Picher Industries Inc	1802 E 50th St	Lubbock	TX	79404	Robert Potter	806-747-4663	D	49*	0.4
AmClyde Engineered Products	240 E Plato Blv	St Paul	MN	55107	Wallace Fisk	612-293-4646	R	45	0.2
American Crane Corp	202 Raleigh St	Wilmington	NC	28412	David C Lewis	919-395-8500	R	45	0.2
Balderson Inc	PO Box 6	Wamego	KS	66547	SC Balderson	913-456-2224	S	45*	0.4
Gomaco Corp	PO Box 151	Ida Grove	IA	51445	Gary L Godbersen	712-364-3347	S	45*	0.3
McKees Rocks Forging Co	75 Nichol Av	McKees Rocks	PA	15136	JG Modic	412-778-2039	S	40	0.2
Ramsey Industries Inc	PO Box 581510	Tulsa	OK	74116	WW Ramsey	918-438-2760	R	40	0.3
Telsmith Inc	PO Box 539	Mequon	WI	53092	Robert G Stafford	414-242-6600	S	40	0.3
Mustang Manufacturing	PO Box 547	Owatonna	MN	55060	Bruce Collins	507-451-7112	R	36*	0.3
Paper, Calmenson and Co	PO Box 64432	St Paul	MN	55164	Willis M Forman	612-628-6301	R	35	0.3
Allied Construction Products	3900 Kelley Av	Cleveland	OH	44114	Leo Mathews	216-431-2600	R	34	<0.1
ED Etnyre and Co	1333 S Daysville Rd	Oregon	IL	61061	David H Abbott	815-732-2116	R	32	0.3
Dutton-Lainson Co	PO Box 729	Hastings	NE	68902	C Hermes	402-462-4141	R	30	0.4
LaBounty Manufacturing Inc	State Rd	Two Harbors	MN	55616	Roy LaBounty	218-834-2123	R	30*	0.2
Philadelphia Mixers Corp	1221 E Main St	Palmyra	PA	17078	Jack Sadler	717-838-1341	S	30	0.2
Portec Construction Equip	PO Box 20	Yankton	SD	57078	Walter G Lock	605-665-9311	D	30	0.3
Stone Construction Equipment	32 E Main St	Honeoye	NY	14471	R Fien	716-229-5141	R	28	0.2
Standard Havens	8800 E 63rd St	Kansas City	MO	64133	Jack Pasquariello	816-737-0400	D	27	0.2
Douglas Dynamics Inc	7777 N 73rd St	Milwaukee	WI	53223	James Janik	414-354-2310	S	25	0.3
Fulghum Industries Inc	PO Box 909	Wadley	GA	30477	Walter D Lampp III	912-252-5223	R	25	0.2
Fulton Performance Products	50 Indianhead Dr	Mosinee	WI	54455	Thomas Rudasics	715-693-1700	S	25	0.2
Ramsey Winch Company Inc	PO Box 581510	Tulsa	OK	74158	JE Henry	918-438-2760	S	25	0.2
Burkeen Manufacturing Co	11200 High Pt Cove	Olive Branch	MS	38654	James Burkeen	601-895-4150	R	24	0.2
Central Mine Equipment Co	4215 N Rider Trl	Earth City	MO	63045	Roberta Rassieur	314-291-7700	R	24*	0.2
Intern Pipe Machinery Corp	PO Box 1708	Sioux City	IA	51102	Al C Jensen	712-277-8111	R	24	0.2
Kress Corp	227 Illinois St	Brimfield	IL	61517	ES Kress	309-446-3395	R	24*	0.2
Reach All	PO Box 16047	Duluth	MN	55816	Robert J Adams	218-722-9200	R	24	<0.1
Smeal Manufacturing Co	Hwy 91 W	Snyder	NE	68664	Virgil Hunke	402-568-2221	R	24	<0.1
ABB Raymond	650 Warrenville Rd	Lisle	IL	60532	RR Rohlicek	708-971-2500	S	21*	0.2
Pierce Pacific Manufacturing	PO Box 1009	Tualatin	OR	97062	Jerry Clausen	503-620-9880	R	20	0.1
Simon-Ro Corp	550 Old Hwy 56	Olathe	KS	66061	Terry L Smith	913-782-1200	S	20*	0.2
Hendrix Manufacturing Company	PO Box 919	Mansfield	LA	71052	Calvin W Hall	318-872-1660	R	19*	0.1
Stow Manufacturing Co	PO Box 490	Binghamton	NY	13902	Mark Hotchkiss	607-723-6411	R	19	0.2
Bucyrus Blades Inc	PO Box 628	Bucyrus	OH	44820	Larry Huget	419-562-6015	S	18	0.2
Mobile Pulley & Machine Works	PO Box 1947	Mobile	AL	36633	Hannon S Hairston	205-432-7631	R	18	0.2
Racine Federated Inc	2200 South St	Racine	WI	53404	John E Erskine Jr	414-639-6770	R	18	<0.1
Ellicott Machine Corp Intern	1611 Bush St	Baltimore	MD	21230	Peter Bowe	410-837-7900	R	17	0.1
Allis Mineral Systems	PO Box 2219	Appleton	WI	54913	Robert S Morrison	414-734-9831	D	15*	0.3
Auto Crane Co	PO Box 581510	Tulsa	OK	74158	Joe E Henry	918-836-0463	S	15	0.1
Trencor Jetco Inc	3545 E Main St	Grand Prairie	TX	75050	Jerry Gilbert	214-264-0311	R	15	0.2
UEC Equipment Co	341 NW 122nd St	Oklahoma City	OK	73156	JV Neuberger	405-755-9703	S	15*	0.1
UEC Industries Inc	341 NW 122nd St	Oklahoma City	OK	73114	JV Neuberger	405-755-9703	R	15*	0.1
Valk Manufacturing Co	PO Box 218	Carlisle	PA	17013	Ted P Valk	717-766-0711	R	15	0.1
Western Products	PO Box 23045	Milwaukee	WI	53223	James Janik	414-354-2310	D	15	0.2
Young Corp	PO Box 3522	Seattle	WA	98124	RH Lindberg	206-624-1071	R	15	0.2
Besser-Appco	PO Box 1198	San Antonio	TX	78294	Raymond L Kinsel	210-333-1111	D	14	0.1
Foley Material Handling	1 Virginia Crane Dr	Ashland	VA	23005	Dale R Foley	804-798-1343	R	14*	0.1
Rose Industries Inc	Box 23907	Milwaukee	WI	53223	Gene E Lutz	414-352-2000	R	14	0.1

Source: Ward's Business Directory of U.S. Private and Public Companies, Volumes 1 and 2, 1996. The company type code used is as follows: P - Public, R - Private, S - Subsidiary, D - Division, J - Joint Venture, A - Affiliate, G - Group. Sales are in millions of dollars, employees are in thousands. An asterisk (*) indicates an estimated sales volume. The symbol < stands for 'less than'. Company names and addresses are truncated, in some cases, to fit into the available space.

MATERIALS CONSUMED

Material	Quantity	Delivered Cost ($ million)
Materials, ingredients, containers, and supplies	(X)	6,792.3
Fluid power pumps, motors, and hydrostatic transmissions (hydraulic and pneumatic)	(X)	266.7
All other pumps	(X)	21.5
Fluid power cylinders and rotary actuators (hydraulic and pneumatic)	(X)	158.6
Fluid power filters (hydraulic and pneumatic)	(X)	18.2
Fluid power hose or tube fittings and assemblies (hydraulic and pneumatic)	(X)	79.7
Fluid power valves (hydraulic and pneumatic)	(X)	100.0
Metal bolts, nuts, screws, washers, rivets, and other screw machine products	(X)	68.2
Fabricated structural metal products (except forgings)	(X)	240.9
Fabricated metal wire products (including wire rope, cable, springs, etc.)	(X)	29.2
All other fabricated metal products (except forgings)	(X)	97.0
Iron and steel forgings	(X)	252.3
Nonferrous forgings	(X)	1.6
Iron and steel castings (rough and semifinished)	(X)	550.9
Aluminum and aluminum-base alloy castings (rough and semifinished)	(X)	16.0
Other nonferrous castings (rough and semifinished)	(X)	18.0
Steel bars, bar shapes, and plates	(X)	446.9
Steel sheet and strip, including tin plate	(X)	262.8
Steel structural shapes and sheet piling	(X)	92.7
All other steel shapes and forms	(X)	149.9
Copper and copper-base alloy shapes and forms	(X)	8.8
All other nonferrous shapes and forms	(X)	7.8
Diesel and semidiesel engines	(X)	451.7
Gasoline and other carburetor engines	(X)	34.8
Integral horsepower electric motors and generators (1 hp or more)	(X)	35.2
Ball bearings (mounted or unmounted)	(X)	63.7
Roller bearings (mounted or unmounted)	(X)	64.6
Mechanical speed changers, gears, and industrial high-speed drives	(X)	154.4
Pneumatic tires	(X)	142.3
Rubber and plastics hose and belting	(X)	46.7
Fabricated plastics products (except gaskets, hoses, and belting)	(X)	36.3
Paints, varnishes, lacquers, stains, shellacs, japans, enamels, and allied products	(X)	37.8
Cabs purchased for installation on construction machinery	(X)	115.6
Cutting tools for machine tools	(X)	24.5
All other materials and components, parts, containers, and supplies	(X)	2,076.4
Materials, ingredients, containers, and supplies, nsk	(X)	620.6

Source: 1992 *Economic Census*. Explanation of symbols used: (D): Withheld to avoid disclosure of competitive data; na: Not available; (S): Withheld because statistical norms were not met; (X): Not applicable; (Z): Less than half the unit shown; nec: Not elsewhere classified; nsk: Not specified by kind; - : zero; * : 10-19 percent estimated; ** : 20-29 percent estimated.

PRODUCT SHARE DETAILS

Product or Product Class	% Share	Product or Product Class	% Share
Construction machinery	100.00	(excluding parts)	0.94
Wheel tractors, contractors' off-highway (2- and 4-wheel), rubber-tired dozers, and self-propelled wheeled log skidders	3.33	Electric winches, including marine use (excluding winches for tractor mounting and parts)	4.05
Crawler tractors, 20 net engine horsepower rating or more (sold with or without attachments) (excluding parts)	6.22	Other winches, including marine use (excluding winches for tractor mounting and parts)	4.38
Tractor shovel loaders (sold with or without attachments) (excluding parts)	14.82	Portable crushing plants, screening plants, washing plants, and combination plants (excluding parts)	4.85
Power cranes, draglines, and shovels (excavators) (including surface mining equipment and attachments) (excluding parts)	13.37	Handheld pavement breakers (excluding parts)	2.62
Mixers, pavers, and related equipment (excluding parts)	5.18	Rotary snow blowers, except residential (including integral units and attachments for mounting) (excluding parts)	0.66
Scrapers, graders, rollers, off-highway trucks and coal haulers, trailers, wagons, rough terrain forklifts, except parts	12.93	Snow clearing attachments for mounting on tractors or trucks (except rotary snow blowers), including snow plows, etc. (excluding parts)	5.11
Construction machinery for mounting on tractors and other prime movers (excluding parts and snow clearing attachments)	2.63	Personnel aerial work platforms (excluding parts)	31.78
Other construction machinery and equipment (excluding parts)	14.78	All other construction machinery and equipment, complete units, including well point systems (excluding parts)	18.07
Commercial brush, limb, and log chippers for waste wood reduction (excluding parts)	3.09	Other construction machinery and equipment (excluding parts), nsk	2.37
Log splitters (excluding parts)	0.92	Parts for construction machinery and equipment, sold separately	23.68
Dredging machinery, hydraulic and other types (excluding parts)	1.67	Parts for contractors' off-highway wheel tractors, crawler tractors, and tractor shovel loaders (sold separately)	40.65
Self-propelled continuous ditchers and trenchers (including ladder and wheel types) (integral units only) (excluding parts)	6.59	Parts for power cranes, draglines, and shovels (excavators) (including surface mining equipment) (sold separately)	12.96
Railway maintenance of way equipment (rail layers, ballast spreaders, etc.), except rail cars (excluding parts)	4.84	Parts for mixers, pavers, and related equipment (sold separately)	5.41
Vertical earth augers and power posthole diggers, excluding water well and blasthole drills (excluding parts)	1.17	Parts for scrapers, graders, rollers, off-highway trucks and coal haulers, trailers, wagons, rough terrain forklifts (sold separately)	9.77
Horizontal earth boring machines and accessories (excluding parts)	3.40	Parts for construction machinery for mounting on tractors and other prime movers (sold separately)	6.82
Digger-derricks (excluding parts)	3.51	Parts for other construction machinery and equipment listed in product class 3531P (sold separately)	22.05
Pile driving equipment (including air, steam, or diesel pile hammers and impact pile or vibratory driver extractors)		Parts for construction machinery and equipment, sold separately, nsk	2.33
		Construction machinery, nsk	3.05

Source: 1992 *Economic Census*. The values shown are percent of total shipments in an industry. Values of indented subcategories are summed in the main heading. The symbol (D) appears when data are withheld to prevent disclosure of competitive information. The abbreviation nsk stands for 'not specified by kind' and nec for 'not elsewhere classified'.

INPUTS AND OUTPUTS FOR CONSTRUCTION MACHINERY & EQUIPMENT

Economic Sector or Industry Providing Inputs	%	Sector	Economic Sector or Industry Buying Outputs	%	Sector
Imports	15.7	Foreign	Gross private fixed investment	41.2	Cap Inv
Wholesale trade	9.4	Trade	Exports	32.6	Foreign
Blast furnaces & steel mills	8.5	Manufg.	Coal	5.8	Mining
Construction machinery & equipment	7.4	Manufg.	Construction machinery & equipment	4.4	Manufg.
Iron & steel foundries	7.1	Manufg.	S/L Govt. purch., highways	2.1	S/L Govt
Internal combustion engines, nec	5.3	Manufg.	Ship building & repairing	2.0	Manufg.
Iron & steel forgings	4.1	Manufg.	Federal Government purchases, national defense	1.5	Fed Govt
Fabricated plate work (boiler shops)	3.6	Manufg.	Maintenance of nonfarm buildings nec	1.2	Constr.
Power transmission equipment	3.3	Manufg.	Industrial buildings	0.8	Constr.
Fabricated structural metal	2.5	Manufg.	Residential 1-unit structures, nonfarm	0.7	Constr.
Tires & inner tubes	2.5	Manufg.	Office buildings	0.6	Constr.
Machinery, except electrical, nec	2.3	Manufg.	S/L Govt. purch., other general government	0.6	S/L Govt
Miscellaneous fabricated wire products	2.2	Manufg.	S/L Govt. purch., sewerage	0.6	S/L Govt
Electric services (utilities)	1.9	Util.	Nonfarm residential structure maintenance	0.5	Constr.
Pumps & compressors	1.8	Manufg.	Electric utility facility construction	0.4	Constr.
Ball & roller bearings	1.5	Manufg.	S/L Govt. purch., water & air facilities	0.4	S/L Govt
Pipe, valves, & pipe fittings	1.4	Manufg.	Highway & street construction	0.3	Constr.
Maintenance of nonfarm buildings nec	1.3	Constr.	Maintenance of highways & streets	0.3	Constr.
Advertising	1.3	Services	Residential additions/alterations, nonfarm	0.3	Constr.
Motor freight transportation & warehousing	1.1	Util.	S/L Govt. purch., sanitation	0.3	S/L Govt
Communications, except radio & TV	0.8	Util.	S/L Govt. purch., water	0.3	S/L Govt
Gas production & distribution (utilities)	0.8	Util.	Construction of stores & restaurants	0.2	Constr.
Computer & data processing services	0.8	Services	Maintenance of electric utility facilities	0.2	Constr.
Rubber & plastics hose & belting	0.7	Manufg.	Telephone & telegraph facility construction	0.2	Constr.
Special dies & tools & machine tool accessories	0.7	Manufg.	S/L Govt. purch., gas & electric utilities	0.2	S/L Govt
Motors & generators	0.6	Manufg.	S/L Govt. purch., natural resource & recreation.	0.2	S/L Govt
Screw machine and related products	0.6	Manufg.	Construction of educational buildings	0.1	Constr.
Banking	0.6	Fin/R.E.	Construction of hospitals	0.1	Constr.
Equipment rental & leasing services	0.6	Services	Maintenance of railroads	0.1	Constr.
Petroleum refining	0.5	Manufg.	Maintenance of sewer facilities	0.1	Constr.
Eating & drinking places	0.5	Trade	Maintenance of telephone & telegraph facilities	0.1	Constr.
Noncomparable imports	0.5	Foreign	Residential garden apartments	0.1	Constr.
Industrial controls	0.4	Manufg.	Sewer system facility construction	0.1	Constr.

Continued on next page.

INPUTS AND OUTPUTS FOR CONSTRUCTION MACHINERY & EQUIPMENT - Continued

Economic Sector or Industry Providing Inputs	%	Sector	Economic Sector or Industry Buying Outputs	%	Sector
Air transportation	0.4	Util.	Federal Government purchases, nondefense	0.1	Fed Govt
U.S. Postal Service	0.4	Gov't			
Aluminum castings	0.3	Manufg.			
Gaskets, packing & sealing devices	0.3	Manufg.			
Paints & allied products	0.3	Manufg.			
Railroads & related services	0.3	Util.			
Real estate	0.3	Fin/R.E.			
Engineering, architectural, & surveying services	0.3	Services			
Abrasive products	0.2	Manufg.			
Aluminum rolling & drawing	0.2	Manufg.			
Asbestos products	0.2	Manufg.			
Machine tools, metal cutting types	0.2	Manufg.			
Metal heat treating	0.2	Manufg.			
Water transportation	0.2	Util.			
Electrical repair shops	0.2	Services			
Legal services	0.2	Services			
Management & consulting services & labs	0.2	Services			
Coal	0.1	Mining			
Blowers & fans	0.1	Manufg.			
Fabricated rubber products, nec	0.1	Manufg.			
Industrial gases	0.1	Manufg.			
Metal stampings, nec	0.1	Manufg.			
Miscellaneous plastics products	0.1	Manufg.			
Steel springs, except wire	0.1	Manufg.			
Sanitary services, steam supply, irrigation	0.1	Util.			
Insurance carriers	0.1	Fin/R.E.			
Accounting, auditing & bookkeeping	0.1	Services			
Hotels & lodging places	0.1	Services			
Scrap	0.1	Scrap			

Source: *Benchmark Input-Output Accounts for the U.S. Economy, 1982*, U.S. Department of Commerce, Washington, D.C., July 1991. Data, as reported in the source, are organized by the 1977 SIC structure in use in 1982 but have been matched, as closely as is possible, to the 1987 SIC structure used in this book.

OCCUPATIONS EMPLOYED BY SIC 353 - CONSTRUCTION AND RELATED MACHINERY

Occupation	% of Total 1994	Change to 2005	Occupation	% of Total 1994	Change to 2005
Welders & cutters	7.4	-9.1	NC machine tool operators, metal & plastic	1.6	-9.1
Assemblers, fabricators, & hand workers nec	6.3	-9.1	Industrial machinery mechanics	1.4	-0.0
Machinists	4.6	-9.1	Traffic, shipping, & receiving clerks	1.3	-12.5
Sales & related workers nec	3.9	-9.1	Freight, stock, & material movers, hand	1.3	-27.3
Blue collar worker supervisors	3.6	-13.9	Bookkeeping, accounting, & auditing clerks	1.3	-31.8
Welding machine setters, operators	2.9	-18.2	Production, planning, & expediting clerks	1.2	-9.1
Machine builders	2.3	-27.3	Coating, painting, & spraying machine workers	1.2	-9.1
General managers & top executives	2.3	-13.8	Sheet metal workers & duct installers	1.2	-9.1
Mechanical engineers	2.3	0.0	Engineering, mathematical, & science managers	1.2	3.2
Inspectors, testers, & graders, precision	2.1	-9.1	Machine tool cutting operators, metal & plastic	1.1	-24.3
Secretaries, ex legal & medical	2.0	-17.3	Purchasing agents, ex trade & farm products	1.1	9.1
Drafters	2.0	-29.2	Engineering technicians & technologists nec	1.1	-9.1
Machine assemblers	1.8	63.6	General office clerks	1.0	-22.5
Professional workers nec	1.8	9.1	Maintenance repairers, general utility	1.0	-18.2
Machine tool cutting & forming etc. nec	1.8	-9.1	Stock clerks	1.0	-26.1
Lathe & turning machine tool operators	1.6	-18.2	Drilling & boring machine tool workers	1.0	-54.6
Helpers, laborers, & material movers nec	1.6	-9.1	Industrial production managers	1.0	-9.1

Source: *Industry-Occupation Matrix*, Bureau of Labor Statistics. These data relate to one or more 3-digit SIC industry groups rather than to a single 4-digit SIC. The change reported for each occupation to the year 2005 is a percent of growth or decline as estimated by the Bureau of Labor Statistics. The abbreviation nec stands for 'not elsewhere classified'.

LOCATION BY STATE AND REGIONAL CONCENTRATION

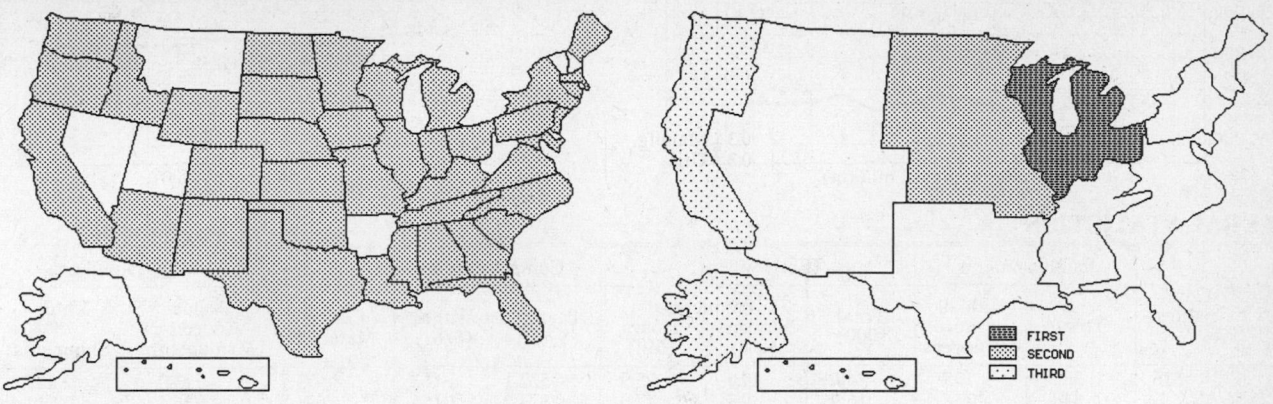

FIRST
SECOND
THIRD

INDUSTRY DATA BY STATE

| State | Establish-ments | Shipments | | | Employment | | | | Cost as % of Shipments | Investment per Employee ($) |
		Total ($ mil)	% of U.S.	Per Establ.	Total Number	% of U.S.	Per Establ.	Wages ($/hour)		
Iowa	28	1,948.5	14.5	69.6	7,900	10.2	282	17.52	57.6	4,494
Wisconsin	55	1,362.4	10.1	24.8	7,900	10.2	144	17.19	49.1	4,937
Pennsylvania	38	1,007.9	7.5	26.5	7,100	9.2	187	16.45	47.9	4,944
North Carolina	17	525.7	3.9	30.9	1,800	2.3	106	13.95	60.4	12,333
Minnesota	41	455.9	3.4	11.1	2,800	3.6	68	11.38	61.9	2,857
Ohio	46	435.8	3.2	9.5	2,700	3.5	59	12.94	49.5	1,963
Oklahoma	36	402.6	3.0	11.2	3,000	3.9	83	11.23	59.8	2,800
Texas	71	384.3	2.9	5.4	3,000	3.9	42	11.54	63.8	4,233
Kansas	24	339.0	2.5	14.1	1,800	2.3	75	12.55	61.2	6,444
Tennessee	16	272.8	2.0	17.0	1,100	1.4	69	12.75	52.9	2,818
Washington	35	218.2	1.6	6.2	1,300	1.7	37	13.87	60.3	2,231
California	75	192.4	1.4	2.6	1,500	1.9	20	11.89	51.7	1,933
Michigan	43	182.4	1.4	4.2	1,300	1.7	30	10.72	64.2	6,846
Missouri	17	147.5	1.1	8.7	900	1.2	53	12.60	75.9	4,444
Alabama	16	132.9	1.0	8.3	1,000	1.3	63	11.13	66.5	2,500
Virginia	11	119.6	0.9	10.9	800	1.0	73	11.80	54.5	1,125
Georgia	16	115.3	0.9	7.2	500	0.6	31	11.00	35.5	6,400
New York	33	109.0	0.8	3.3	900	1.2	27	10.30	53.7	1,111
Oregon	26	107.9	0.8	4.2	900	1.2	35	12.71	53.8	1,111
Indiana	22	102.8	0.8	4.7	800	1.0	36	11.50	55.0	1,750
Mississippi	8	97.9	0.7	12.2	900	1.2	113	10.13	55.9	1,000
Colorado	15	65.7	0.5	4.4	500	0.6	33	12.83	50.8	3,000
Nebraska	10	63.0	0.5	6.3	700	0.9	70	10.50	41.3	1,571
Connecticut	9	42.9	0.3	4.8	300	0.4	33	13.33	41.3	2,667
Florida	27	33.5	0.2	1.2	400	0.5	15	15.00	77.6	1,500
Massachusetts	16	32.6	0.2	2.0	300	0.4	19	11.67	44.8	1,000
Arizona	8	31.2	0.2	3.9	300	0.4	38	12.00	53.5	2,667
Louisiana	14	29.3	0.2	2.1	300	0.4	21	9.40	39.6	2,000
New Jersey	14	20.0	0.1	1.4	200	0.3	14	12.00	56.5	1,500
Idaho	6	18.6	0.1	3.1	100	0.1	17	11.00	62.4	2,000
New Mexico	6	12.1	0.1	2.0	100	0.1	17	11.00	53.7	3,000
Illinois	68	(D)	-	-	17,500*	22.7	257	-	-	-
South Dakota	12	(D)	-	-	1,750*	2.3	146	-	-	-
North Dakota	9	(D)	-	-	1,750*	2.3	194	-	-	-
South Carolina	7	(D)	-	-	750*	1.0	107	-	-	-
Kentucky	5	(D)	-	-	750*	1.0	150	-	-	-
Maine	5	(D)	-	-	175*	0.2	35	-	-	-
Maryland	4	(D)	-	-	175*	0.2	44	-	-	-
Wyoming	3	(D)	-	-	375*	0.5	125	-	-	-

Source: 1992 *Economic Census*. The states are in descending order of shipments or establishments (if shipment data are missing for the majority). The symbol (D) appears when data are withheld to prevent disclosure of competitive information. States marked with (D) are sorted by number of establishments. A dash (-) indicates that the data element cannot be calculated; * indicates the midpoint of a range.

3532 - MINING MACHINERY

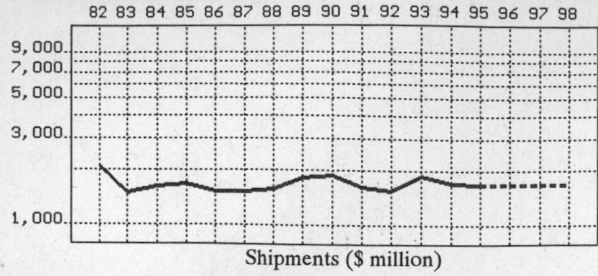

Shipments ($ million)

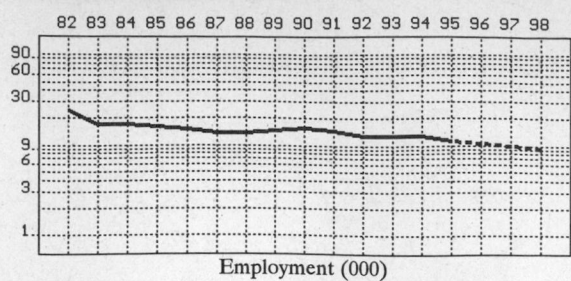

Employment (000)

GENERAL STATISTICS

| Year | Companies | Establishments | | Employment | | | Compensation | | Production ($ million) | | | |
		Total	with 20 or more employees	Total (000)	Production Workers (000)	Hours (Mil)	Payroll ($ mil)	Wages ($/hr)	Cost of Materials	Value Added by Manufacture	Value of Shipments	Capital Invest.
1982	316	369	175	24.6	14.3	25.9	522.1	10.63	991.7	1,113.0	2,109.3	66.0
1983		348	164	16.9	9.3	17.0	373.2	10.55	696.8	736.5	1,514.4	37.9
1984		327	153	16.9	9.9	18.6	391.5	10.65	808.4	816.0	1,628.4	30.7
1985		307	143	16.1	9.3	17.2	394.2	11.02	822.6	784.0	1,663.9	26.7
1986		284	123	14.7	8.5	16.0	374.8	11.33	725.9	783.2	1,532.5	32.8
1987	293	321	122	13.6	8.2	16.3	354.6	11.09	748.5	772.5	1,518.1	36.6
1988		309	130	13.6	8.5	16.7	359.8	11.25	801.0	787.1	1,568.9	29.7
1989		286	130	14.6	9.8	19.2	413.9	11.05	919.2	871.5	1,805.8	31.9
1990		290	128	15.3	9.7	19.4	415.6	10.92	940.4	912.8	1,865.5	38.2
1991		302	135	14.5	8.5	17.7	397.2	10.90	884.7	727.7	1,643.3	32.5
1992	268	295	116	12.6	7.4	15.9	393.7	11.89	811.8	729.6	1,557.5	33.0
1993		284	108	12.6	7.7	16.4	393.7	12.21	1,026.5	838.2	1,881.1	58.9
1994		272P	105P	12.8	8.1	16.3	402.6	12.96	925.8	849.1	1,695.0	29.5
1995		266P	100P	11.0P	7.1P	15.6P	381.5P	12.27P	922.7P	846.3P	1,689.4P	34.1P
1996		260P	96P	10.4P	6.8P	15.2P	379.0P	12.41P	922.6P	846.2P	1,689.1P	33.7P
1997		254P	91P	9.7P	6.5P	14.9P	376.5P	12.55P	922.5P	846.0P	1,688.9P	33.2P
1998		249P	86P	9.1P	6.2P	14.6P	373.9P	12.70P	922.3P	845.9P	1,688.6P	32.8P

Sources: 1982, 1987, 1992 *Economic Census*; *Annual Survey of Manufactures*, 83-86, 88-91, 93-94. Establishment counts for non-Census years are from *County Business Patterns*; establishment values for 83-84 are extrapolations. 'P's show projections by the editors. Industries reclassified in 87 will not have data for prior years.

INDICES OF CHANGE

| Year | Companies | Establishments | | Employment | | | Compensation | | Production ($ million) | | | |
		Total	with 20 or more employees	Total (000)	Production Workers (000)	Hours (Mil)	Payroll ($ mil)	Wages ($/hr)	Cost of Materials	Value Added by Manufacture	Value of Shipments	Capital Invest.
1982	118	125	151	195	193	163	133	89	122	153	135	200
1983		118	141	134	126	107	95	89	86	101	97	115
1984		111	132	134	134	117	99	90	100	112	105	93
1985		104	123	128	126	108	100	93	101	107	107	81
1986		96	106	117	115	101	95	95	89	107	98	99
1987	109	109	105	108	111	103	90	93	92	106	97	111
1988		105	112	108	115	105	91	95	99	108	101	90
1989		97	112	116	132	121	105	93	113	119	116	97
1990		98	110	121	131	122	106	92	116	125	120	116
1991		102	116	115	115	111	101	92	109	100	106	98
1992	100	100	100	100	100	100	100	100	100	100	100	100
1993		96	93	100	104	103	100	103	126	115	121	178
1994		92P	91P	102	109	103	102	109	114	116	109	89
1995		90P	87P	87P	96P	98P	97P	103P	114P	116P	108P	103P
1996		88P	83P	82P	92P	96P	96P	104P	114P	116P	108P	102P
1997		86P	78P	77P	88P	94P	96P	106P	114P	116P	108P	101P
1998		84P	74P	72P	84P	92P	95P	107P	114P	116P	108P	99P

Sources: Same as General Statistics. Values reflect change from the base year, 1992. Values above 100 mean greater than 92, values below 100 mean less than 92, and a value of 100 in the 82-91 or 93-98 period means same as 92. 'P's mark projections by the editors.

SELECTED RATIOS

For 1994	Avg. of All Manufact.	Analyzed Industry	Index	For 1994	Avg. of All Manufact.	Analyzed Industry	Index
Employees per Establishment	49	47	96	Value Added per Production Worker	134,084	104,827	78
Payroll per Establishment	1,500,273	1,479,982	99	Cost per Establishment	5,045,178	3,403,297	67
Payroll per Employee	30,620	31,453	103	Cost per Employee	102,970	72,328	70
Production Workers per Establishment	34	30	87	Cost per Production Worker	146,988	114,296	78
Wages per Establishment	853,319	776,561	91	Shipments per Establishment	9,576,895	6,230,923	65
Wages per Production Worker	24,861	26,080	105	Shipments per Employee	195,460	132,422	68
Hours per Production Worker	2,056	2,012	98	Shipments per Production Worker	279,017	209,259	75
Wages per Hour	12.09	12.96	107	Investment per Establishment	321,011	108,444	34
Value Added per Establishment	4,602,255	3,121,343	68	Investment per Employee	6,552	2,305	35
Value Added per Employee	93,930	66,336	71	Investment per Production Worker	9,352	3,642	39

Sources: Same as General Statistics. The 'Average of All Manufacturing' column represents the average of all manufacturing industries reported for the most recent complete year available. The Index shows the relationship between the Average and the Analyzed Industry. For example, 100 means that they are equal; 500 that the Analyzed Industry is five times the average; 50 means that the Analyzed Industry is half the national average. The abbreviation 'na' is used to show that data are 'not available'.

LEADING COMPANIES Number shown: 66 Total sales ($ mil): 3,824 Total employment (000): 27.7

Company Name	Address				CEO Name	Phone	Co. Type	Sales ($ mil)	Empl. (000)
Joy Technologies Inc	301 Grant St	Pittsburgh	PA	15219	Marc F Wray	412-562-4500	P	566	3.7
Jupiter Industries Inc	919 N Michigan Av	Chicago	IL	60611	Bill Jamison	312-642-6000	R	500*	7.0
INDRESCO Inc	PO Box 219022	Dallas	TX	75221	JL Jackson	214-953-4500	P	441	3.1
Harnischfeger Corp	PO Box 310	Milwaukee	WI	53201	Robert W Hale	414-671-4400	S	409	2.5
Bucyrus-Erie Co	PO Box 500	S Milwaukee	WI	53172	William B Winter	414-768-4400	P	270	1.0
Svedala Industries Inc	20965 Crossroads	Waukesha	WI	53186	John Platner	414-798-6200	S	250	1.1
Long-Airdox Co	227 W Maple Av	Oak Hill	WV	25901	WE Meador	304-469-3301	S	145	1.1
INDRESCO Inc	617 W Center St	Marion	OH	43302	Don Stout	614-383-5211	D	100	0.5
LeTourneau Inc	PO Box 2307	Longview	TX	75606	J Earl Beckman	903-237-7000	S	100	1.0
Oldenburg Group Inc	8600 W Bradley Rd	Milwaukee	WI	53224	Wayne Oldenburg	414-357-8600	R	100	0.3
Nordberg Inc	PO Box 383	Milwaukee	WI	53201	William O'Day	414-769-4300	S	90	0.5
Simmons-Rand Co	4201 Lee Hwy	Bristol	VA	24201	Bernard Simmons	703-669-9171	S	88*	0.7
Wagner Mining	PO Box 20307	Portland	OR	97220	Roderick J Brown	503-255-2863	S	75	0.5
Sandvik Rock Tools Inc	PO Box 40402	Houston	TX	77240	Olaf Lundblad	713-460-6200	S	65	0.3
California Pellet Mill Co	1114 E Wabash Av	Crawfordsville	IN	47933	Larry Pitsch	317-362-2600	S	50	0.5
EIMCO Coal Machinery Inc	210 Bland St	Bluefield	WV	24701	Lasse Hakoaho	304-327-0260	S	50	0.4
El Jay	PO Box 607	Eugene	OR	97440	Robert Hoitt	503-726-6541	D	45	0.3
JH Fletcher and Co	PO Box 2187	Huntington	WV	25722	J Fletcher	304-525-7811	R	35*	0.2
Tamrock/EJC USA Inc	860 Westlake Pkwy	Atlanta	GA	30336	Jim Ropp	404-346-6820	S	35	<0.1
Acrison Inc	20 Empire Blv	Moonachie	NJ	07074	Rocco Ricciardi	201-440-8300	R	30	0.2
Driltech Inc	PO Box 338	Alachua	FL	32615	Lorne Massel	904-462-4100	R	26*	0.2
Baker Hughes Inc Mining Tools	PO Box 531226	Grand Prairie	TX	75053	Robert W Thomas	214-988-3322	D	25	0.2
Stamler Corp	Main & Stamler St	Millersburg	KY	40348	C M Anderson	606-484-3431	S	25	0.2
Fairchild International	PO Box 300	Glen Lyn	VA	24093	Myrleen B Fairchild	703-726-2380	R	24	0.2
Line Power Manufacturing	PO Box 8200	Bristol	VA	24203	Scott Bullock	703-466-8200	D	24	0.2
Centrifugal & Mechanical Indust	201 President St	St Louis	MO	63118	N Andos	314-776-2848	D	21	0.1
Pennsylvania Crusher Corp	PO Box 100	Broomall	PA	19008	John D Whalen	610-544-7200	R	20	0.1
Goodman Equipment Corp	5430 W 70th Pl	Bedford Park	IL	60638	CA Campbell Jr	708-496-1188	R	19*	<0.1
California Pellet Mill Co	1114 E Wabash Av	Crawfordsville	IN	47933	Larry Pitsch	317-362-2600	D	16	0.2
Deister Machine Company Inc	PO Box 5188	Fort Wayne	IN	46895	Irwin F Deister Jr	219-426-7495	R	15	0.1
Eickoff Corp	PO Box 2000	Pittsburgh	PA	15230	Monika D Ludwig	412-788-1400	S	15	<0.1
Midwestern Industries Inc	PO Box 810	Massillon	OH	44648	LJ Riesbeck	216-837-4203	R	15	<0.1
Eagle Crusher Company Inc	4250 Rte 309	Galion	OH	44833	S Cobey	419-468-2288	R	11	<0.1
McLanahan Corp	200 Wall St	Hollidaysburg	PA	16648	M McLanahan	814-695-9807	R	10	0.2
Mega Corp	PO Box 26146	Albuquerque	NM	87125	William Schegel	505-345-2661	R	10*	<0.1
Prosser-Enpo Industries	PO Box 603	Piqua	OH	45356	Paul G Baldetti	513-773-2442	D	8	0.2
Williams Patent Crusher	813 Montgomery St	St Louis	MO	63102	RM Williams	314-621-3348	R	8	<0.1
Kue-Ken Corp	8383 Baldwin St	Oakland	CA	94621	RB Dediemar	510-569-8382	S	8	<0.1
Acker	PO Box 830	Scranton	PA	18501	Bob Heal	717-586-2061	D	7*	<0.1
McLellan Equipment Inc	251 Shaw Rd	S San Francisco	CA	94080	Dale C McLellan	415-873-8100	R	6*	<0.1
Bradley Pulverizer Co	PO Box 1318	Allentown	PA	18105	James J Fronheiser	610-434-5191	R	6	<0.1
American Pulverizer Co	5540 W Park Av	St Louis	MO	63110	H Chris Griesedieck	314-781-6100	R	5	<0.1
Epworth Manufacturing	1400 Kalamazoo St	South Haven	MI	49090	Roy A Nelson	616-637-2128	R	5	<0.1
Midwestern Machinery Company	PO Box 458	Joplin	MO	64802	GF Beechwood	417-624-2400	R	5	<0.1
Tug River Armature	PO Box 770	Williamson	WV	25661	TE Sheppard	304-235-5370	R	5*	0.1
Brookville Mining Equip Corp	PO Box 130	Brookville	PA	15825	Dalph McNeil	814-849-7321	R	4*	<0.1
Jadair Inc	PO Box 89	Pt Washington	WI	53074	DL Schmutzler	414-284-3411	R	4*	<0.1
Prox Company Inc	PO Box 1484	Terre Haute	IN	47808	Robert F Prox Jr	812-232-4324	R	4	<0.1
Amsat	PO Box 95	East Greenville	PA	18041	Greg Shemanski	215-679-5984	D	4	<0.1
Aggregate Machinery Inc	PO Box 17160	Salem	OR	97305	Edward Tompkins	503-390-6284	R	3	<0.1
B and D Machine Works Inc	307 Pinckneyville	Marissa	IL	62257	Donald Dickey	618-295-2112	R	3	<0.1
Balco Inc	PO Box 56	Blairsville	PA	15717	J Bartholow	412-459-6814	R	3	<0.1
BryDet Development Corp	PO Box 870	Coshocton	OH	43812	Paul Bryant	614-623-0455	R	3	<0.1
JR Hoe and Sons Inc	PO Box 1737	Middlesboro	KY	40965	Harry M Hoe	606-248-5560	R	3	<0.1
Salem Tool Inc	PO Box 760	London	KY	40743	Laird T Orr Sr	606-528-2963	R	3	<0.1
Superior Hydraulics Industries	PO Box 966	Morgantown	WV	26507	Giovanni Tarantini	304-292-6144	R	3	<0.1
Bowdil Co	PO Box 20470	Canton	OH	44701	Daniel Morrow	216-456-7176	R	3	<0.1
Construction & Tunneling Svcs	1609 S Central Av	Kent	WA	98032	Bruce Moulton	206-859-9724	R	2*	<0.1
California Graphite Machines	191 Granite St	Corona	CA	91719	Jerry W Thompson	909-734-1030	R	1	<0.1
Taconite Eng & Mfg Co	3800 W 5th Av	Hibbing	MN	55746	Karl Hnatko	218-262-3868	R	1	<0.1
WH Milroy and Company Inc	29 Washington Av	Hamden	CT	06518	WH Milroy Jr	203-248-4451	R	1	<0.1
Rel-Tek Corp	616 Beatty Rd	Monroeville	PA	15146	Albert Ketler	412-373-6700	R	1*	<0.1
Universal Road Machinery	27 Emerick St	Kingston	NY	12401	James J Hassett	914-331-8248	R	1*	<0.1
Atlas Copco Construction	3700 E 68th Av	Commerce City	CO	80022		303-286-8825	S	1	<0.1
BP Tracy Co	PO Box 518	Washington	PA	15301	NA Greig	412-228-2160	D	1	<0.1
Downard Hydraulics Inc	PO Box 1212	Princeton	WV	24740	DE Downard	304-487-1492	R	0*	0.1

Source: Ward's Business Directory of U.S. Private and Public Companies, Volumes 1 and 2, 1996. The company type code used is as follows: P - Public, R - Private, S - Subsidiary, D - Division, J - Joint Venture, A - Affiliate, G - Group. Sales are in millions of dollars, employees are in thousands. An asterisk (*) indicates an estimated sales volume. The symbol < stands for 'less than'. Company names and addresses are truncated, in some cases, to fit into the available space.

MATERIALS CONSUMED

Material	Quantity	Delivered Cost ($ million)
Materials, ingredients, containers, and supplies	(X)	682.1
Fluid power pumps, motors, and hydrostatic transmissions (hydraulic and pneumatic)	(X)	11.0
All other pumps	(X)	1.8
Fluid power cylinders and rotary actuators (hydraulic and pneumatic)	(X)	14.5
Fluid power filters (hydraulic and pneumatic)	(X)	3.0
Fluid power hose or tube fittings and assemblies (hydraulic and pneumatic)	(X)	6.1
Fluid power valves (hydraulic and pneumatic)	(X)	9.1
Metal bolts, nuts, screws, washers, rivets, and other screw machine products	(X)	7.1
Fabricated structural metal products (except forgings)	(X)	27.7
Fabricated metal wire products (including wire rope, cable, springs, etc.)	(X)	8.6
All other fabricated metal products (except forgings)	(X)	19.3
Iron and steel forgings	(X)	32.5
Nonferrous forgings	(X)	1.2
Iron and steel castings (rough and semifinished)	(X)	53.8
Aluminum and aluminum-base alloy castings (rough and semifinished)	(X)	0.4
Other nonferrous castings (rough and semifinished)	(X)	2.3
Steel bars, bar shapes, and plates	(X)	52.3
Steel sheet and strip, including tin plate	(X)	7.7
Steel structural shapes and sheet piling	(X)	4.9
All other steel shapes and forms	(X)	7.2
Copper and copper-base alloy shapes and forms	(X)	0.3
All other nonferrous shapes and forms	(X)	1.7
Diesel and semidiesel engines	(X)	8.5
Gasoline and other carburetor engines	(X)	1.4
Integral horsepower electric motors and generators (1 hp or more)	(X)	25.6
Ball bearings (mounted or unmounted)	(X)	7.0
Roller bearings (mounted or unmounted)	(X)	12.5
Mechanical speed changers, gears, and industrial high-speed drives	(X)	28.0
Pneumatic tires	(X)	2.9
Rubber and plastics hose and belting	(X)	5.6
Fabricated plastics products (except gaskets, hoses, and belting)	(X)	2.6
Paints, varnishes, lacquers, stains, shellacs, japans, enamels, and allied products	(X)	2.1
Cutting tools for machine tools	(X)	3.8
All other materials and components, parts, containers, and supplies	(X)	189.1
Materials, ingredients, containers, and supplies, nsk	(X)	120.5

Source: 1992 Economic Census. Explanation of symbols used: (D): Withheld to avoid disclosure of competitive data; na: Not available; (S): Withheld because statistical norms were not met; (X): Not applicable; (Z): Less than half the unit shown; nec: Not elsewhere classified; nsk: Not specified by kind; - : zero; * : 10-19 percent estimated; ** : 20-29 percent estimated.

PRODUCT SHARE DETAILS

Product or Product Class	% Share	Product or Product Class	% Share
Mining machinery	100.00	(sold separately)	42.79
Underground mining machinery (except parts sold separately)	23.65	Percussion rock mining drill bits (sold separately)	3.82
Mineral processing and beneficiation machinery (except parts sold separately)	5.37	Rock mining drill bits other than percussion (sold separately)	15.43
Crushing, pulverizing, and screening machinery (excluding portable combination plants), except parts sold separately	13.39	All other mining drill bits (sold separately)	11.63
Drills and other mining machinery, nec (except parts sold separately)	9.97	Parts and attachments for mining machinery and equipment (except drill bits) (sold separately)	61.41
Parts and attachments for mining machinery and equipment		Parts and attachments for mining machinery and equipment (sold separately), nsk	7.71
		Mining machinery, nsk	4.84

Source: 1992 Economic Census. The values shown are percent of total shipments in an industry. Values of indented subcategories are summed in the main heading. The symbol (D) appears when data are withheld to prevent disclosure of competitive information. The abbreviation nsk stands for 'not specified by kind' and nec for 'not elsewhere classified'.

INPUTS AND OUTPUTS FOR MINING MACHINERY, EXCEPT OIL FIELD

Economic Sector or Industry Providing Inputs	%	Sector	Economic Sector or Industry Buying Outputs	%	Sector
Imports	9.0	Foreign	Gross private fixed investment	35.4	Cap Inv
Blast furnaces & steel mills	8.5	Manufg.	Exports	17.8	Foreign
Mining machinery, except oil field	8.4	Manufg.	Coal	11.9	Mining
Wholesale trade	7.6	Trade	Mining machinery, except oil field	4.8	Manufg.
Iron & steel foundries	6.0	Manufg.	Dimension, crushed & broken stone	4.7	Mining
Fabricated plate work (boiler shops)	5.7	Manufg.	Chemical & fertilizer mineral	3.4	Mining
Power transmission equipment	5.3	Manufg.	Change in business inventories	3.3	In House
Motors & generators	3.4	Manufg.	Copper ore	2.7	Mining
Advertising	3.1	Services	Sand & gravel	2.3	Mining
Machinery, except electrical, nec	2.9	Manufg.	Clay, ceramic, & refractory minerals	2.2	Mining
Iron & steel forgings	2.8	Manufg.	Concrete products, nec	1.9	Manufg.
Primary metal products, nec	2.5	Manufg.	Ready-mixed concrete	1.8	Manufg.
Maintenance of nonfarm buildings nec	2.0	Constr.	Iron & ferroalloy ores	1.7	Mining

Continued on next page.

INPUTS AND OUTPUTS FOR MINING MACHINERY, EXCEPT OIL FIELD - Continued

Economic Sector or Industry Providing Inputs	%	Sector	Economic Sector or Industry Buying Outputs	%	Sector
Fabricated structural metal	1.9	Manufg.	Nonferrous metal ores, except copper	1.5	Mining
Ball & roller bearings	1.7	Manufg.	Miscellaneous repair shops	1.3	Services
Electric services (utilities)	1.7	Util.	Petroleum, gas, & solid mineral exploration	0.8	Constr.
Motor freight transportation & warehousing	1.6	Util.	Nonmetallic mineral services	0.5	Mining
Nonferrous rolling & drawing, nec	1.5	Manufg.	Cutstone & stone products	0.4	Manufg.
Noncomparable imports	1.5	Foreign	Gypsum products	0.4	Manufg.
Tires & inner tubes	1.3	Manufg.	Federal Government purchases, national defense	0.4	Fed Govt
Communications, except radio & TV	1.2	Util.	Access structures for mineral development	0.3	Constr.
Blowers & fans	1.0	Manufg.	Lime	0.3	Manufg.
Pumps & compressors	0.9	Manufg.	Federal Government purchases, nondefense	0.1	Fed Govt
Eating & drinking places	0.9	Trade			
Petroleum refining	0.8	Manufg.			
Pipe, valves, & pipe fittings	0.8	Manufg.			
Equipment rental & leasing services	0.8	Services			
Fabricated metal products, nec	0.7	Manufg.			
Sawmills & planning mills, general	0.7	Manufg.			
Pottery products, nec	0.6	Manufg.			
Screw machine and related products	0.6	Manufg.			
Gas production & distribution (utilities)	0.6	Util.			
Banking	0.6	Fin/R.E.			
Computer & data processing services	0.6	Services			
Internal combustion engines, nec	0.5	Manufg.			
Metal heat treating	0.4	Manufg.			
Nonferrous wire drawing & insulating	0.4	Manufg.			
Special dies & tools & machine tool accessories	0.4	Manufg.			
Air transportation	0.4	Util.			
Legal services	0.4	Services			
Abrasive products	0.3	Manufg.			
Fabricated rubber products, nec	0.3	Manufg.			
Industrial controls	0.3	Manufg.			
Machine tools, metal cutting types	0.3	Manufg.			
Miscellaneous fabricated wire products	0.3	Manufg.			
Paints & allied products	0.3	Manufg.			
Rubber & plastics hose & belting	0.3	Manufg.			
Railroads & related services	0.3	Util.			
Real estate	0.3	Fin/R.E.			
Accounting, auditing & bookkeeping	0.3	Services			
Management & consulting services & labs	0.3	Services			
U.S. Postal Service	0.3	Gov't			
Aluminum rolling & drawing	0.2	Manufg.			
Asbestos products	0.2	Manufg.			
Brass, bronze, & copper castings	0.2	Manufg.			
Copper rolling & drawing	0.2	Manufg.			
Industrial gases	0.2	Manufg.			
Manifold business forms	0.2	Manufg.			
Nonferrous castings, nec	0.2	Manufg.			
Signs & advertising displays	0.2	Manufg.			
Credit agencies other than banks	0.2	Fin/R.E.			
Aluminum castings	0.1	Manufg.			
General industrial machinery, nec	0.1	Manufg.			
Lubricating oils & greases	0.1	Manufg.			
Metal stampings, nec	0.1	Manufg.			
Miscellaneous plastics products	0.1	Manufg.			
Insurance carriers	0.1	Fin/R.E.			
Royalties	0.1	Fin/R.E.			
Automotive rental & leasing, without drivers	0.1	Services			
Automotive repair shops & services	0.1	Services			
Electrical repair shops	0.1	Services			
Hotels & lodging places	0.1	Services			

Source: Benchmark Input-Output Accounts for the U.S. Economy, 1982, U.S. Department of Commerce, Washington, D.C., July 1991. Data, as reported in the source, are organized by the 1977 SIC structure in use in 1982 but have been matched, as closely as is possible, to the 1987 SIC structure used in this book.

OCCUPATIONS EMPLOYED BY SIC 353 - CONSTRUCTION AND RELATED MACHINERY

Occupation	% of Total 1994	Change to 2005	Occupation	% of Total 1994	Change to 2005
Welders & cutters	7.4	-9.1	NC machine tool operators, metal & plastic	1.6	-9.1
Assemblers, fabricators, & hand workers nec	6.3	-9.1	Industrial machinery mechanics	1.4	-0.0
Machinists	4.6	-9.1	Traffic, shipping, & receiving clerks	1.3	-12.5
Sales & related workers nec	3.9	-9.1	Freight, stock, & material movers, hand	1.3	-27.3
Blue collar worker supervisors	3.6	-13.9	Bookkeeping, accounting, & auditing clerks	1.3	-31.8
Welding machine setters, operators	2.9	-18.2	Production, planning, & expediting clerks	1.2	-9.1
Machine builders	2.3	-27.3	Coating, painting, & spraying machine workers	1.2	-9.1
General managers & top executives	2.3	-13.8	Sheet metal workers & duct installers	1.2	-9.1
Mechanical engineers	2.3	0.0	Engineering, mathematical, & science managers	1.2	3.2
Inspectors, testers, & graders, precision	2.1	-9.1	Machine tool cutting operators, metal & plastic	1.1	-24.3
Secretaries, ex legal & medical	2.0	-17.3	Purchasing agents, ex trade & farm products	1.1	9.1
Drafters	2.0	-29.2	Engineering technicians & technologists nec	1.1	-9.1
Machine assemblers	1.8	63.6	General office clerks	1.0	-22.5
Professional workers nec	1.8	9.1	Maintenance repairers, general utility	1.0	-18.2
Machine tool cutting & forming etc. nec	1.8	-9.1	Stock clerks	1.0	-26.1
Lathe & turning machine tool operators	1.6	-18.2	Drilling & boring machine tool workers	1.0	-54.6
Helpers, laborers, & material movers nec	1.6	-9.1	Industrial production managers	1.0	-9.1

Source: *Industry-Occupation Matrix*, Bureau of Labor Statistics. These data relate to one or more 3-digit SIC industry groups rather than to a single 4-digit SIC. The change reported for each occupation to the year 2005 is a percent of growth or decline as estimated by the Bureau of Labor Statistics. The abbreviation nec stands for 'not elsewhere classified'.

LOCATION BY STATE AND REGIONAL CONCENTRATION

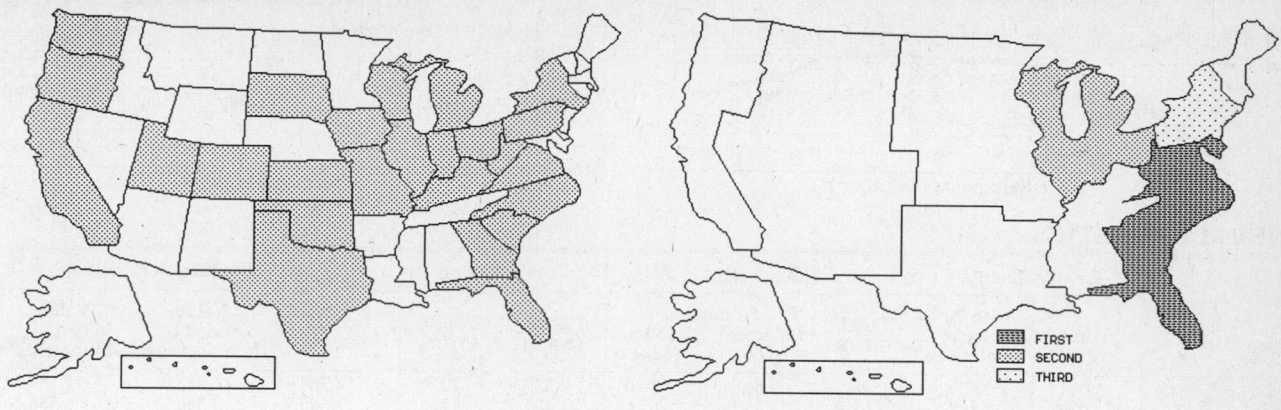

FIRST
SECOND
THIRD

INDUSTRY DATA BY STATE

State	Establish- ments	Shipments			Employment				Cost as % of Shipments	Investment per Employee ($)
		Total ($ mil)	% of U.S.	Per Establ.	Total Number	% of U.S.	Per Establ.	Wages ($/hour)		
Virginia	26	243.5	15.6	9.4	2,200	17.5	85	11.70	54.8	2,227
Pennsylvania	27	229.3	14.7	8.5	2,000	15.9	74	12.96	57.0	2,200
Wisconsin	5	129.9	8.3	26.0	800	6.3	160	12.50	71.2	-
West Virginia	41	122.1	7.8	3.0	1,000	7.9	24	12.83	49.2	3,900
Ohio	12	75.4	4.8	6.3	600	4.8	50	13.00	48.3	2,167
Kentucky	22	62.5	4.0	2.8	500	4.0	23	9.00	50.4	1,000
Illinois	12	58.6	3.8	4.9	400	3.2	33	12.67	42.7	-
California	14	43.6	2.8	3.1	300	2.4	21	10.67	47.9	2,667
Texas	14	43.2	2.8	3.1	500	4.0	36	13.20	48.4	-
Missouri	6	36.7	2.4	6.1	300	2.4	50	12.33	45.0	1,000
Colorado	11	27.1	1.7	2.5	200	1.6	18	15.50	35.4	-
Michigan	5	26.9	1.7	5.4	200	1.6	40	16.00	48.3	500
Indiana	6	24.0	1.5	4.0	200	1.6	33	11.00	48.3	-
Utah	9	22.9	1.5	2.5	200	1.6	22	11.67	52.8	500
Florida	7	15.8	1.0	2.3	200	1.6	29	8.67	55.1	1,000
Washington	7	13.5	0.9	1.9	100	0.8	14	16.00	51.9	6,000
Georgia	6	(D)	-	-	175 *	1.4	29	-	-	-
New York	6	(D)	-	-	175 *	1.4	29	-	-	-
Oklahoma	4	(D)	-	-	375 *	3.0	94	-	-	-
Oregon	4	(D)	-	-	750 *	6.0	188	-	-	-
North Carolina	2	(D)	-	-	175 *	1.4	88	-	-	-
South Carolina	2	(D)	-	-	375 *	3.0	188	-	-	-
Iowa	1	(D)	-	-	175 *	1.4	175	-	-	-
Kansas	1	(D)	-	-	175 *	1.4	175	-	-	-
South Dakota	1	(D)	-	-	375 *	3.0	375	-	-	-

Source: 1992 *Economic Census*. The states are in descending order of shipments or establishments (if shipment data are missing for the majority). The symbol (D) appears when data are withheld to prevent disclosure of competitive information. States marked with (D) are sorted by number of establishments. A dash (-) indicates that the data element cannot be calculated; * indicates the midpoint of a range.

3533 - OIL FIELD MACHINERY

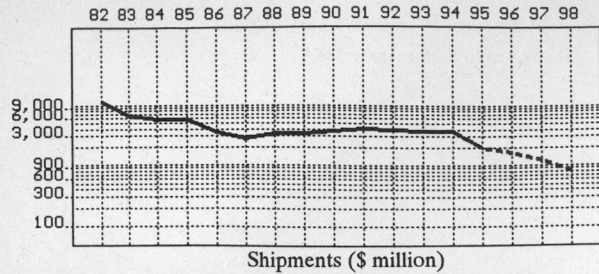

Shipments ($ million)

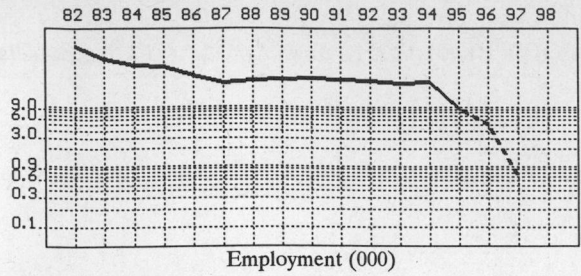

Employment (000)

GENERAL STATISTICS

| Year | Com-panies | Establishments | | Employment | | | Compensation | | Production ($ million) | | | |
		Total	with 20 or more employees	Total (000)	Production Workers (000)	Hours (Mil)	Payroll ($ mil)	Wages ($/hr)	Cost of Materials	Value Added by Manufacture	Value of Shipments	Capital Invest.
1982	848	1,011	499	98.5	60.0	120.0	2,339.9	10.76	4,784.3	6,542.0	11,189.5	902.9
1983		933	448	62.9	35.1	67.6	1,492.7	11.30	2,605.2	3,330.1	6,586.4	310.9
1984		855	397	49.2	28.7	57.7	1,308.5	12.10	2,273.1	3,484.0	5,803.5	243.6
1985		777	346	46.8	27.6	54.7	1,295.1	12.30	2,340.4	3,048.9	5,714.0	132.5
1986		702	285	32.6	15.7	29.9	877.9	13.53	1,524.9	1,701.3	3,577.0	72.7
1987	563	633	190	24.8	13.3	26.5	719.2	13.05	1,225.8	1,210.2	2,728.3	31.7
1988		601	212	27.0	15.5	31.1	807.4	13.11	1,545.2	1,963.3	3,400.6	46.9
1989		576	229	29.1	14.7	29.3	769.1	12.99	1,452.5	1,759.4	3,313.7	77.1
1990		572	228	30.1	15.3	32.1	841.8	13.11	1,593.3	2,040.9	3,634.7	72.2
1991		582	244	28.7	16.0	32.7	892.2	13.51	1,795.7	2,192.5	4,072.8	99.7
1992	473	537	203	27.0	15.4	31.2	876.5	13.62	1,707.7	2,104.2	3,917.9	104.6
1993		531	194	24.5	13.8	27.3	792.2	14.22	1,587.5	2,089.2	3,739.8	79.8
1994		415P	122P	25.4	14.7	27.8	845.4	15.59	1,610.7	2,082.5	3,752.7	77.6
1995		373P	96P	9.2P	4.1P	8.3P	472.6P	15.03P	862.4P	1,115.0P	2,009.3P	
1996		330P	71P	4.9P	1.5P	3.3P	387.8P	15.32P	695.9P	899.7P	1,621.3P	
1997		288P	45P	0.7P			303.0P	15.61P	529.3P	684.4P	1,233.3P	
1998		245P	19P				218.3P	15.89P	362.8P	469.1P	845.3P	

Sources: 1982, 1987, 1992 *Economic Census*; *Annual Survey of Manufactures*, 83-86, 88-91, 93-94. Establishment counts for non-Census years are from *County Business Patterns*; establishment values for 83-84 are extrapolations. 'P's show projections by the editors. Industries reclassified in 87 will not have data for prior years.

INDICES OF CHANGE

| Year | Com-panies | Establishments | | Employment | | | Compensation | | Production ($ million) | | | |
		Total	with 20 or more employees	Total (000)	Production Workers (000)	Hours (Mil)	Payroll ($ mil)	Wages ($/hr)	Cost of Materials	Value Added by Manufacture	Value of Shipments	Capital Invest.
1982	179	188	246	365	390	385	267	79	280	311	286	863
1983		174	221	233	228	217	170	83	153	158	168	297
1984		159	196	182	186	185	149	89	133	166	148	233
1985		145	170	173	179	175	148	90	137	145	146	127
1986		131	140	121	102	96	100	99	89	81	91	70
1987	119	118	94	92	86	85	82	96	72	58	70	30
1988		112	104	100	101	100	92	96	90	93	87	45
1989		107	113	108	95	94	88	95	85	84	85	74
1990		107	112	111	99	103	96	96	93	97	93	69
1991		108	120	106	104	105	102	99	105	104	104	95
1992	100	100	100	100	100	100	100	100	100	100	100	100
1993		99	96	91	90	87	90	104	93	99	95	76
1994		77P	60P	94	95	89	96	114	94	99	96	74
1995		69P	47P	34P	26P	27P	54P	110P	51P	53P	51P	
1996		61P	35P	18P	10P	10P	44P	112P	41P	43P	41P	
1997		54P	22P	2P			35P	115P	31P	33P	31P	
1998		46P	9P				25P	117P	21P	22P	22P	

Sources: Same as General Statistics. Values reflect change from the base year, 1992. Values above 100 mean greater than 92, values below 100 mean less than 92, and a value of 100 in the 82-91 or 93-98 period means same as 92. 'P's mark projections by the editors.

SELECTED RATIOS

For 1994	Avg. of All Manufact.	Analyzed Industry	Index	For 1994	Avg. of All Manufact.	Analyzed Industry	Index
Employees per Establishment	49	61	125	Value Added per Production Worker	134,084	141,667	106
Payroll per Establishment	1,500,273	2,034,880	136	Cost per Establishment	5,045,178	3,876,958	77
Payroll per Employee	30,620	33,283	109	Cost per Employee	102,970	63,413	62
Production Workers per Establishment	34	35	103	Cost per Production Worker	146,988	109,571	75
Wages per Establishment	853,319	1,043,200	122	Shipments per Establishment	9,576,895	9,032,757	94
Wages per Production Worker	24,861	29,483	119	Shipments per Employee	195,460	147,744	76
Hours per Production Worker	2,056	1,891	92	Shipments per Production Worker	279,017	255,286	91
Wages per Hour	12.09	15.59	129	Investment per Establishment	321,011	186,783	58
Value Added per Establishment	4,602,255	5,012,582	109	Investment per Employee	6,552	3,055	47
Value Added per Employee	93,930	81,988	87	Investment per Production Worker	9,352	5,279	56

Sources: Same as General Statistics. The 'Average of All Manufacturing' column represents the average of all manufacturing industries reported for the most recent complete year available. The Index shows the relationship between the Average and the Analyzed Industry. For example, 100 means that they are equal; 500 that the Analyzed Industry is five times the average; 50 means that the Analyzed Industry is half the national average. The abbreviation 'na' is used to show that data are 'not available'.

LEADING COMPANIES Number shown: **75** Total sales ($ mil): **6,776** Total employment (000): **46.5**

Company Name	Address				CEO Name	Phone	Co. Type	Sales ($ mil)	Empl. (000)
Baker Hughes Inc	PO Box 4740	Houston	TX	77210	James D Woods	713-439-8600	P	2,505	14.7
Smith International Inc	PO Box 60068	Houston	TX	77205	Doug Rock	713-443-3370	P	654	4.1
Camco International Inc	PO Box 14484	Houston	TX	77221	Gary D Nicholson	713-747-4000	P	590	4.3
Lone Star Technologies Inc	PO Box 803546	Dallas	TX	75380	John P Harbin	214-386-3981	P	357	1.6
Continental Emsco Company Inc	PO Box 469016	Garland	TX	75046	David E Althoff	214-487-3000	S	297	1.3
National-Oilwell	5555 San Felipe St	Houston	TX	77056	Joel Staff	713-960-5100	J	270	2.5
Varco International Inc	743 N Eckhoff St	Orange	CA	92668	George Boyadjieff	714-978-1900	P	224	1.4
Lufkin Industries Inc	PO Box 849	Lufkin	TX	75902	DV Smith	409-634-2211	P	217	2.0
Tuboscope Vetco Intern Corp	2835 Holmes Rd	Houston	TX	77051	William V Larkin Jr	713-799-5100	P	192	2.0
Baker Oil Tools	PO Box 3048	Houston	TX	77253	AG Avant	713-923-9351	D	130•	1.2
Hydril Co	PO Box 60458	Houston	TX	77205	Richard Seaver	713-449-2000	R	100	1.8
National Tank Co	2950 N Loop W	Houston	TX	77092	Nat Gregory	713-683-9292	S	100	0.4
Reed Tool Co	6501 Navigation	Houston	TX	77011	Roy Caldwell	713-924-5200	S	90	0.8
Crosby Group Inc	2801 Dawson Rd	Tulsa	OK	74110	Larry L Postelwait	918-834-4611	S	89	0.5
Martin Decker	1200 Cypress Creek	Cedar Park	TX	78613	Bob Gondek	512-331-0411	S	55	0.4
TD Williamson Inc	PO Box 2299	Tulsa	OK	74101	R B Williamson	918-254-9400	R	55	0.6
Varco BJ Oil Tools	PO Box 800457	Houston	TX	77280	M W Sutherlin	713-937-5500	D	53	0.1
Hycalog	9777 W Gulf Bank	Houston	TX	77040	Alex Newton	713-896-3600	D	40	0.3
Shaffer	PO Box 1473	Houston	TX	77251	Mark Merit	713-937-5000	D	40	0.2
Harbison-Fischer Manufacturing	PO Box 2477	Fort Worth	TX	76113	James A Burns	817-297-2211	R	36	0.3
Omsco Industries	PO Box 230589	Houston	TX	77223	Pete Samoff	713-926-7401	D	35	0.2
Shaw Resource Services Inc	10502 Fallstone Rd	Houston	TX	77099	Weems Turner	713-498-0600	S	35	0.3
Harrisburg/Woolley	PO Box 2108	Houston	TX	77252	DH Waldschmitt	713-462-4110	D	30	<0.1
Oil Dynamics Inc	PO Box 470446	Tulsa	OK	74147	John Hrncir	918-627-9021	J	30	0.3
Prideco Inc	6039 Thomas Rd	Houston	TX	77041	William Chunn	713-466-8161	S	29	0.2
Dickson GMP International Inc	PO Box 639	Belle Chasse	LA	70037	FE Gallander Jr	504-394-1890	R	27•	0.4
Brandt Co	PO Box 2327	Conroe	TX	77305	Dennis W Livesay	409-756-4800	D	26•	0.2
IRI International Corp	PO Box 1101	Pampa	TX	79066	VP Raymond	806-665-3701	J	25	0.4
ERC Industries Inc	2906 Holmes Rd	Houston	TX	77051	Richard H Rau	713-733-9301	P	23	0.3
Smith Industries Inc	PO Box 7398	Houston	TX	77248		713-869-1421	R	22•	0.2
Norriseal	PO Box 40525	Houston	TX	77240	Larry J Renaud	713-466-3552	D	21	0.2
Orbix Corp	PO Box 311	Clifton	TX	76634	Robert N Posey	817-675-8371	R	19	0.2
Schramm Inc	800 E Virginia Av	West Chester	PA	19380	RE Schramm	610-696-2500	R	18	0.1
US Synthetic Corp	744 S 100 E	Provo	UT	84606	Louis Pope	801-373-6311	R	18•	<0.1
Houston Engineers Inc	PO Box 567	Houston	TX	77001	Dwight E Beach Jr	713-237-3050	S	17	0.2
Baylor Technology Inc	500 Industrial Blv	Sugar Land	TX	77478	HB Payne	713-240-6111	R	16	0.2
Numa Tool Co	PO Box 348	Thompson	CT	06277	Roland R Leonard	203-923-9551	R	15	<0.1
US Enertek Inc	4901 E Main	Farmington	NM	87401	Rodney Heath	505-326-1151	R	15	0.2
George E Failing Co	PO Box 872	Enid	OK	73702	Jeff Smith	405-234-4141	D	14•	0.1
Titan Specialties Inc	PO Box 2316	Pampa	TX	79066	Robert B Echols	806-665-3781	R	14	<0.1
Southwest Oilfield Products Inc	PO Box 24068	Houston	TX	77229	John Leman	713-675-7541	R	13	0.1
Foster Valve Corp	6445 Burlington N	Houston	TX	77092	Dean Anderson	713-690-3553	S	12	0.1
Mark Products	10502 Fallstone Rd	Houston	TX	77099	Weems Turner	713-498-0600	D	12	0.1
McKissick Products Inc	PO Box 3128	Tulsa	OK	74101	Pat Gordon	918-834-4611	D	12•	0.2
Dresser Security	PO Box 210600	Dallas	TX	75211	EC Sharp	214-333-3211	D	11•	0.1
Daniel EN-FAB Systems Inc	PO Box 21361	Houston	TX	77223	Sandra Tripathy	713-225-4913	S	11•	0.1
Randolph Co	PO Box 130486	Houston	TX	77219	John Tatum Jr	713-526-2091	R	11•	0.1
Airtek Inc	76 Clair St	N Huntingdon	PA	15642	Robert C Whisner	412-863-1350	R	10•	<0.1
Ponder Industries Inc	PO Drawer 2229	Alice	TX	78333	Mack Ponder	512-664-5831	P	9	<0.1
Cooper Manufacturing Corp	PO Box 431	Brady	TX	76825	Joe Poe	915-597-0777	S	8	<0.1
Dreco Holding Co	PO Box 36619	Houston	TX	77236	Doug Frame	713-965-9122	R	8•	<0.1
Kremco Inc	PO Box 1624	Clearfield	UT	84016	Anthony J Colletti	801-773-9608	S	8	<0.1
Norton-Alcoa Proppants	5300 Gerber Rd	Fort Smith	AR	72904	PR Lemieux	501-782-2001	S	8	<0.1
Universal Equipment Inc	PO Box 51206	Lafayette	LA	70505	Eldon Adams	318-232-7905	R	8	<0.1
Winston Manufacturing Corp	PO Box 6440	Longview	TX	75608	John W Kinsel	903-757-7341	R	8	0.1
Jensen Mixers International Inc	PO Box 470368	Tulsa	OK	74147	Louis C Jensen	918-627-5770	R	7•	<0.1
Bolt Technology Corp	4 Duke Pl	Norwalk	CT	06854	Raymond M Soto	203-853-0700	P	7	<0.1
Nickles Machine Corp	600 S 1st	Ponca City	OK	74601	Clark Nickles	405-765-5557	R	6•	0.1
Partech Inc	4501 S County Rd	Odessa	TX	79765	Joe N Brown	915-563-2236	S	6	<0.1
Delta Corp	PO Box 2748	Conroe	TX	77305	Robert G Dipton Jr	409-760-4411	R	5	<0.1
Hendershot Tool Co	PO Box 94444	Oklahoma City	OK	73143	Elmer E Rubac	405-677-3386	R	5	<0.1
Specialty Machine and Supply	PO Box 606	Scott	LA	70583	JC Hisaw	318-232-8198	R	5•	<0.1
W-N Apache Corp	PO Box 2490	Wichita Falls	TX	76307	Clyde A Willis	817-723-0711	R	5	<0.1
Production Equipment Service	PO Box 52308	Lafayette	LA	70505	James W Young	318-232-0743	R	5	<0.1
NUMAR Corp	263 Great Val Pkwy	Malvern	PA	19355	Melvin N Miller	610-251-0116	P	4	<0.1
Mattco Manufacturing Inc	12000 Eastex Fwy	Houston	TX	77039	Robert W Hughes	713-449-0361	R	4	<0.1
Abasco Inc	PO Box 38573	Houston	TX	77238	HM Rhodes	713-931-4400	R	4	<0.1
American Rig Co	5400 Cedar Crest	Houston	TX	77087	PR Hampton	713-643-4321	R	3	<0.1
Dawson Enterprises	2853 Cherry Av	Long Beach	CA	90806	HW Dawson	310-424-8564	R	3•	<0.1
Maloney-Crawford Inc	PO Box 659	Tulsa	OK	74101	Frank Hadjcek	918-582-3461	S	3•	<0.1
Regional Fabricators Inc	PO Box 9037	New Iberia	LA	70562	L Berges	318-367-3488	R	3	<0.1
Southern Iowa Mfg Co	PO Box 448	Osceola	IA	50213	Robert M Hettinger	515-342-2166	S	3•	<0.1
Walker-Neer Corp	PO Box 2490	Wichita Falls	TX	76307	Clyde A Willis	817-723-0711	R	3	<0.1
WEDGE Group Inc	PO Box 130688	Houston	TX	77219	Richard E Blohm	713-620-7700	R	3•	<0.1
David Industries	4122 E Chapman	Orange	CA	92669	John M Rau	714-744-9234	R	2	<0.1

Source: Ward's Business Directory of U.S. Private and Public Companies, Volumes 1 and 2, 1996. The company type code used is as follows: P - Public, R - Private, S - Subsidiary, D - Division, J - Joint Venture, A - Affiliate, G - Group. Sales are in millions of dollars, employees are in thousands. An asterisk (*) indicates an estimated sales volume. The symbol < stands for 'less than'. Company names and addresses are truncated, in some cases, to fit into the available space.

MATERIALS CONSUMED

Material	Quantity	Delivered Cost ($ million)
Materials, ingredients, containers, and supplies	(X)	1,391.4
Fluid power (hydraulic and pneumatic) valves	(X)	15.8
All other valves	(X)	12.3
Fluid power pumps, motors, and hydrostatic transmissions (hydraulic and pneumatic)	(X)	17.8
All other pumps	(X)	10.0
Fluid power cylinders and rotary actuators (hydraulic and pneumatic)	(X)	12.9
Fluid power filters (hydraulic and pneumatic)	(X)	3.2
Fluid power hose or tube fittings and assemblies (hydraulic and pneumatic)	(X)	11.4
Fabricated metal products, except forgings	(X)	99.4
Iron and steel forgings	(X)	61.3
Nonferrous forgings	(X)	0.2
Iron and steel castings (rough and semifinished)	(X)	38.4
Nonferrous (aluminum, copper, etc.) castings (rough and semifinished)	(X)	14.4
Steel bars, bar shapes, and plates	(X)	166.0
Steel structural shapes	(X)	27.3
All other steel shapes and forms	(X)	55.3
Nonferrous shapes and forms	(X)	11.5
Metal powders	(X)	13.3
Diesel and semidiesel engines	(X)	21.0
Integral horsepower electric motors and generators (1 hp or more)	(X)	5.2
Ball and roller bearings (mounted or unmounted)	(X)	13.0
Fabricated rubber products, except tires, tubes, hose, belting, and gaskets	(X)	36.3
Fabricated plastics products (except gaskets, hoses, and belting)	(X)	8.0
Industrial diamonds	(X)	5.6
Cutting tools for machine tools	(X)	19.6
All other materials and components, parts, containers, and supplies	(X)	332.0
Materials, ingredients, containers, and supplies, nsk	(X)	380.2

Source: 1992 *Economic Census*. Explanation of symbols used: (D): Withheld to avoid disclosure of competitive data; na: Not available; (S): Withheld because statistical norms were not met; (X): Not applicable; (Z): Less than half the unit shown; nec: Not elsewhere classified; nsk: Not specified by kind; - : zero; * : 10-19 percent estimated; ** : 20-29 percent estimated.

PRODUCT SHARE DETAILS

Product or Product Class	% Share	Product or Product Class	% Share
Oil and gas field machinery	100.00	Parts for other oil and gas field drilling equipment, sold separately	20.59
Rotary oil and gas field drilling machinery and equipment	30.13	Oil and gas field production machinery and equipment (except pumps)	39.90
Blocks, crown and traveling rotary oil and gas field surface drilling equipment	(D)	Oil and gas field production well christmas tree assemblies (surface and subsurface, excluding subsea)	10.71
Rotary oil and gas field surface drilling draw works and accessories	(D)	Oil and gas field production well casings and tubing heads and supports (surface, subsurface, and subsea)	4.23
Rotary oil and gas field surface drilling tables	(D)	Oil and gas field production well chokes, manifolds, and other accessories (surface, subsurface, and subsea, excluding subsea manifolds and templates)	6.33
Rotary oil and gas field surface drilling elevators, spiders, slips, hooks, links, and connectors	2.34		
	3.22	Oil and gas field production well rodless oil lifting machinery and equipment (except pumps)	2.94
Rotary oil and gas field surface drilling well control equipment (blow-out preventers, etc.)	10.67	Oil and gas field production well subsea Christmas tree assemblies, manifolds, and templates	3.62
Other rotary oil and gas field surface drilling machinery and equipment, including kelly joints	7.63	Oil and gas field surface rod lifting pumping units and accessories, including back crank equipment	6.10
Rotary oil and gas field subsurface drilling bits, with working part of sintered metal carbide or cermets	18.40	Other oil and gas field surface rod lifting machinery and equipment	0.44
Rotary oil and gas field subsurface drilling bits, with working part of other material, including diamond	7.10	Oil and gas field subsurface rod lifting equipment (sucker rods), except pumps	3.63
Rotary oil and gas field subsurface drilling reamers and stabilizers	2.60	Oil and gas field packers	13.97
Rotary oil and gas field subsurface drilling coring equipment	(D)	Oil and gas field screens, tubing, catchers, etc.	3.99
Rotary oil and gas field subsurface drilling tool joints, subs, and connectors	7.99	Oil and gas field separating, metering, and treating equipment for use at the wellhead	8.32
Rotary oil and gas field subsurface drilling drill collars	3.03	Other oil and gas field production machinery and tools, except pumps	26.25
Rotary oil and gas field surface drilling fishing and cutting tools	4.94	Parts for oil and gas field production machinery and tools, except for pumps, sold separately	9.04
Rotary oil and gas field subsea drilling risers	1.69	Oil and gas field production machinery and equipment (except pumps), nsk	0.42
Other rotary oil and gas field subsurface drilling equipment	15.55	Portable oil and gas field drilling rigs and parts	11.35
Parts for rotary oil and gas field drilling equipment, sold separately (except for portable drilling rigs)	4.41	Portable oil and gas field drilling rigs (mounted and unmounted) used on the surface (above ground)	67.42
Rotary oil and gas field drilling machinery and equipment, nsk	2.66	Parts for portable oil and gas field drilling rigs used on the surface (above ground)	31.07
Other oil and gas field drilling machinery and equipment	6.16	Portable drilling rigs and parts, nsk	1.51
Cable tool oil and gas field drilling machinery and equipment, including both surface and subsurface equipment	0.49	Oil and gas field derricks and well surveying machinery	2.21
Oil and gas field drilling guide shoes, float collars, and combination guide and float shoes	16.61	Oil and gas field derricks, substructures, and accessories-- regular and portable	94.98
Other oil and gas field drilling cementing equipment	21.08	Oil and gas field derricks and well surveying machinery, nsk	4.87
Other oil and gas field drilling equipment, except rotary drilling equipment and portable drilling rigs	41.29	Oil and gas field machinery, nsk	10.25

Source: 1992 *Economic Census*. The values shown are percent of total shipments in an industry. Values of indented subcategories are summed in the main heading. The symbol (D) appears when data are withheld to prevent disclosure of competitive information. The abbreviation nsk stands for 'not specified by kind' and nec for 'not elsewhere classified'.

INPUTS AND OUTPUTS FOR OIL FIELD MACHINERY

Economic Sector or Industry Providing Inputs	%	Sector	Economic Sector or Industry Buying Outputs	%	Sector
Blast furnaces & steel mills	14.9	Manufg.	Exports	40.7	Foreign
Oil field machinery	12.3	Manufg.	Gross private fixed investment	30.3	Cap Inv
Wholesale trade	9.9	Trade	Petroleum & natural gas well drilling	13.1	Constr.
Iron & steel foundries	8.0	Manufg.	Crude petroleum & natural gas	6.2	Mining
Advertising	6.4	Services	Oil field machinery	5.8	Manufg.
Iron & steel forgings	5.3	Manufg.	Maintenance of petroleum & natural gas wells	1.0	Constr.
Fabricated rubber products, nec	3.6	Manufg.	Coal	0.8	Mining
Pipe, valves, & pipe fittings	2.3	Manufg.	Change in business inventories	0.8	In House
Electric services (utilities)	2.3	Util.	Federal Government purchases, national defense	0.4	Fed Govt
Miscellaneous fabricated wire products	2.1	Manufg.	Petroleum, gas, & solid mineral exploration	0.3	Constr.
Maintenance of nonfarm buildings nec	2.0	Constr.			
Machinery, except electrical, nec	2.0	Manufg.			
Internal combustion engines, nec	1.6	Manufg.			
Pumps & compressors	1.4	Manufg.			
Motor freight transportation & warehousing	1.3	Util.			
Primary metal products, nec	1.1	Manufg.			
Special dies & tools & machine tool accessories	1.1	Manufg.			
Communications, except radio & TV	1.1	Util.			
Gas production & distribution (utilities)	1.1	Util.			
Computer & data processing services	1.1	Services			
Ball & roller bearings	1.0	Manufg.			
Equipment rental & leasing services	1.0	Services			
Industrial controls	0.9	Manufg.			
Metal coating & allied services	0.8	Manufg.			
Eating & drinking places	0.8	Trade			
Fabricated metal products, nec	0.7	Manufg.			
Motors & generators	0.7	Manufg.			
Banking	0.7	Fin/R.E.			
Noncomparable imports	0.7	Foreign			
Gaskets, packing & sealing devices	0.6	Manufg.			
Machine tools, metal cutting types	0.6	Manufg.			
Power transmission equipment	0.6	Manufg.			
Real estate	0.6	Fin/R.E.			
Cyclic crudes and organics	0.5	Manufg.			
Rubber & plastics hose & belting	0.5	Manufg.			
Screw machine and related products	0.5	Manufg.			
Carburetors, pistons, rings, & valves	0.4	Manufg.			
Metal heat treating	0.4	Manufg.			
Nonferrous castings, nec	0.4	Manufg.			
Abrasive products	0.3	Manufg.			
Fabricated structural metal	0.3	Manufg.			
Petroleum refining	0.3	Manufg.			
Railroads & related services	0.3	Util.			
Royalties	0.3	Fin/R.E.			
Legal services	0.3	Services			
Management & consulting services & labs	0.3	Services			
Aluminum rolling & drawing	0.2	Manufg.			
Industrial gases	0.2	Manufg.			
Manifold business forms	0.2	Manufg.			
Metal stampings, nec	0.2	Manufg.			
Nonmetallic mineral products, nec	0.2	Manufg.			
Paints & allied products	0.2	Manufg.			
Power driven hand tools	0.2	Manufg.			
Storage batteries	0.2	Manufg.			
Air transportation	0.2	Util.			
Insurance carriers	0.2	Fin/R.E.			
Accounting, auditing & bookkeeping	0.2	Services			
Electrical repair shops	0.2	Services			
Aluminum castings	0.1	Manufg.			
Lubricating oils & greases	0.1	Manufg.			
Miscellaneous plastics products	0.1	Manufg.			
Sawmills & planning mills, general	0.1	Manufg.			
Switchgear & switchboard apparatus	0.1	Manufg.			
Security & commodity brokers	0.1	Fin/R.E.			

Source: Benchmark Input-Output Accounts for the U.S. Economy, 1982, U.S. Department of Commerce, Washington, D.C., July 1991. Data, as reported in the source, are organized by the 1977 SIC structure in use in 1982 but have been matched, as closely as is possible, to the 1987 SIC structure used in this book.

OCCUPATIONS EMPLOYED BY SIC 353 - CONSTRUCTION AND RELATED MACHINERY

Occupation	% of Total 1994	Change to 2005	Occupation	% of Total 1994	Change to 2005
Welders & cutters	7.4	-9.1	NC machine tool operators, metal & plastic	1.6	-9.1
Assemblers, fabricators, & hand workers nec	6.3	-9.1	Industrial machinery mechanics	1.4	-0.0
Machinists	4.6	-9.1	Traffic, shipping, & receiving clerks	1.3	-12.5
Sales & related workers nec	3.9	-9.1	Freight, stock, & material movers, hand	1.3	-27.3
Blue collar worker supervisors	3.6	-13.9	Bookkeeping, accounting, & auditing clerks	1.3	-31.8
Welding machine setters, operators	2.9	-18.2	Production, planning, & expediting clerks	1.2	-9.1
Machine builders	2.3	-27.3	Coating, painting, & spraying machine workers	1.2	-9.1
General managers & top executives	2.3	-13.8	Sheet metal workers & duct installers	1.2	-9.1
Mechanical engineers	2.3	0.0	Engineering, mathematical, & science managers	1.2	3.2
Inspectors, testers, & graders, precision	2.1	-9.1	Machine tool cutting operators, metal & plastic	1.1	-24.3
Secretaries, ex legal & medical	2.0	-17.3	Purchasing agents, ex trade & farm products	1.1	9.1
Drafters	2.0	-29.2	Engineering technicians & technologists nec	1.1	-9.1
Machine assemblers	1.8	63.6	General office clerks	1.0	-22.5
Professional workers nec	1.8	9.1	Maintenance repairers, general utility	1.0	-18.2
Machine tool cutting & forming etc. nec	1.8	-9.1	Stock clerks	1.0	-26.1
Lathe & turning machine tool operators	1.6	-18.2	Drilling & boring machine tool workers	1.0	-54.6
Helpers, laborers, & material movers nec	1.6	-9.1	Industrial production managers	1.0	-9.1

Source: Industry-Occupation Matrix, Bureau of Labor Statistics. These data relate to one or more 3-digit SIC industry groups rather than to a single 4-digit SIC. The change reported for each occupation to the year 2005 is a percent of growth or decline as estimated by the Bureau of Labor Statistics. The abbreviation nec stands for 'not elsewhere classified'.

LOCATION BY STATE AND REGIONAL CONCENTRATION

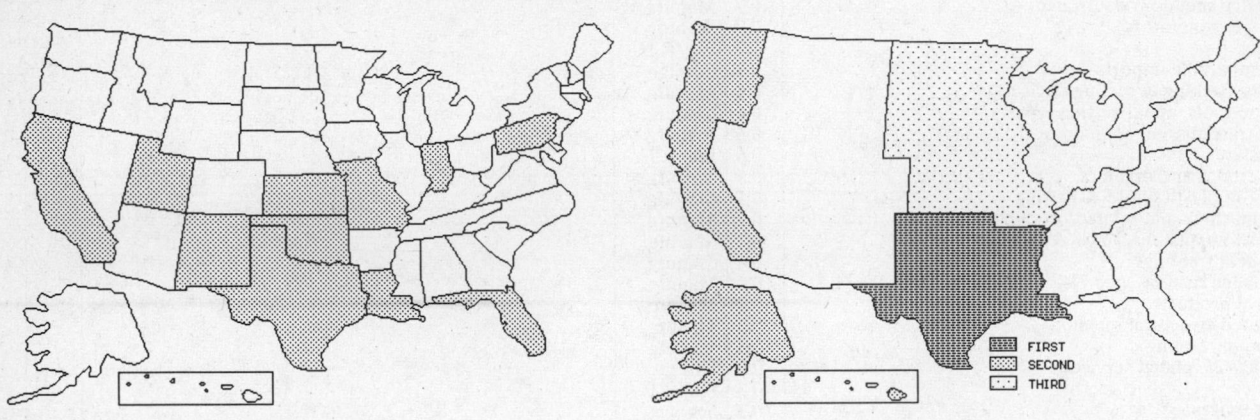

FIRST
SECOND
THIRD

INDUSTRY DATA BY STATE

State	Establish-ments	Shipments			Employment				Cost as % of Shipments	Investment per Employee ($)
		Total ($ mil)	% of U.S.	Per Establ.	Total Number	% of U.S.	Per Establ.	Wages ($/hour)		
Texas	280	2,642.9	67.5	9.4	19,000	70.4	68	14.00	44.8	4,332
Oklahoma	69	498.7	12.7	7.2	3,300	12.2	48	14.19	39.9	2,030
Louisiana	54	385.8	9.8	7.1	1,700	6.3	31	11.48	37.2	3,824
California	33	113.6	2.9	3.4	900	3.3	27	15.57	35.5	1,889
Pennsylvania	9	37.4	1.0	4.2	300	1.1	33	14.33	57.0	1,667
New Mexico	6	28.5	0.7	4.8	300	1.1	50	12.50	69.8	-
Indiana	4	24.9	0.6	6.2	200	0.7	50	13.00	47.0	1,500
Kansas	10	19.7	0.5	2.0	200	0.7	20	12.00	46.2	-
Florida	2	(D)	-	-	175 *	0.6	88	-	-	-
Missouri	1	(D)	-	-	175 *	0.6	175	-	-	-
Utah	1	(D)	-	-	175 *	0.6	175	-	-	-

Source: 1992 Economic Census. The states are in descending order of shipments or establishments (if shipment data are missing for the majority). The symbol (D) appears when data are withheld to prevent disclosure of competitive information. States marked with (D) are sorted by number of establishments. A dash (-) indicates that the data element cannot be calculated; * indicates the midpoint of a range.

3534 - ELEVATORS & MOVING STAIRWAYS

82 83 84 85 86 87 88 89 90 91 92 93 94 95 96 97 98

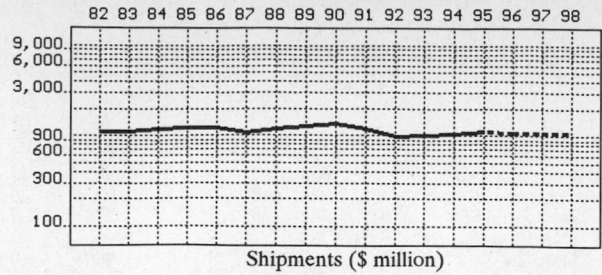

Shipments ($ million)

82 83 84 85 86 87 88 89 90 91 92 93 94 95 96 97 98

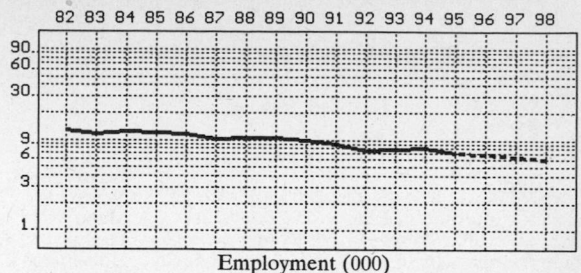

Employment (000)

GENERAL STATISTICS

Year	Companies	Establishments		Employment			Compensation		Production ($ million)			
		Total	with 20 or more employees	Total (000)	Production Workers (000)	Hours (Mil)	Payroll ($ mil)	Wages ($/hr)	Cost of Materials	Value Added by Manufacture	Value of Shipments	Capital Invest.
1982	148	165	83	13.0	7.7	15.1	270.7	9.07	557.6	589.4	1,120.7	31.2
1983		163	84	11.6	7.2	14.5	270.1	9.77	569.5	545.7	1,128.4	22.1
1984		161	85	12.2	7.5	15.2	284.3	10.33	604.3	566.0	1,171.6	20.4
1985		159	86	12.0	7.6	15.2	286.0	10.73	636.5	590.2	1,214.3	21.8
1986		156	87	11.7	6.7	13.4	286.4	10.96	620.7	597.3	1,224.3	16.4
1987	158	176	81	10.2	6.3	12.8	262.4	11.21	565.3	524.7	1,084.4	21.8
1988		173	79	10.5	6.7	13.3	273.1	11.50	679.2	544.7	1,197.0	28.9
1989		161	77	10.5	6.5	13.8	288.1	11.73	697.6	526.1	1,260.2	15.6
1990		165	80	9.9	6.2	13.5	269.1	11.83	772.4	556.5	1,343.1	14.5
1991		175	79	8.9	5.8	12.4	272.6	12.74	712.6	454.9	1,181.3	16.3
1992	162	179	62	7.7	4.8	10.4	233.7	12.19	591.1	386.8	975.6	20.2
1993		179	64	7.9	4.9	10.5	238.8	12.64	612.6	400.2	1,017.0	18.1
1994		178P	67P	8.0	4.9	10.1	248.5	13.21	625.4	433.2	1,053.4	17.4
1995		179P	66P	7.3P	4.6P	10.2P	247.2P	13.49P	651.1P	451.0P	1,096.6P	15.4P
1996		181P	64P	6.8P	4.4P	9.7P	244.2P	13.79P	646.4P	447.7P	1,088.8P	14.6P
1997		182P	62P	6.4P	4.1P	9.3P	241.2P	14.09P	641.7P	444.5P	1,080.9P	13.9P
1998		184P	60P	6.0P	3.9P	8.9P	238.2P	14.40P	637.1P	441.3P	1,073.0P	13.2P

Sources: 1982, 1987, 1992 Economic Census; Annual Survey of Manufactures, 83-86, 88-91, 93-94. Establishment counts for non-Census years are from County Business Patterns; establishment values for 83-84 are extrapolations. 'P's show projections by the editors. Industries reclassified in 87 will not have data for prior years.

INDICES OF CHANGE

Year	Companies	Establishments		Employment			Compensation		Production ($ million)			
		Total	with 20 or more employees	Total (000)	Production Workers (000)	Hours (Mil)	Payroll ($ mil)	Wages ($/hr)	Cost of Materials	Value Added by Manufacture	Value of Shipments	Capital Invest.
1982	91	92	134	169	160	145	116	74	94	152	115	154
1983		91	135	151	150	139	116	80	96	141	116	109
1984		90	137	158	156	146	122	85	102	146	120	101
1985		89	139	156	158	146	122	88	108	153	124	108
1986		87	140	152	140	129	123	90	105	154	125	81
1987	98	98	131	132	131	123	112	92	96	136	111	108
1988		97	127	136	140	128	117	94	115	141	123	143
1989		90	124	136	135	133	123	96	118	136	129	77
1990		92	129	129	129	130	115	97	131	144	138	72
1991		98	127	116	121	119	117	105	121	118	121	81
1992	100	100	100	100	100	100	100	100	100	100	100	100
1993		100	103	103	102	101	102	104	104	103	104	90
1994		99P	109P	104	102	97	106	108	106	112	108	86
1995		100P	106P	94P	96P	98P	106P	111P	110P	117P	112P	76P
1996		101P	103P	89P	91P	94P	104P	113P	109P	116P	112P	72P
1997		102P	100P	83P	86P	90P	103P	116P	109P	115P	111P	69P
1998		103P	97P	78P	81P	86P	102P	118P	108P	114P	110P	65P

Sources: Same as General Statistics. Values reflect change from the base year, 1992. Values above 100 mean greater than 92, values below 100 mean less than 92, and a value of 100 in the 82-91 or 93-98 period means same as 92. 'P's mark projections by the editors.

SELECTED RATIOS

For 1994	Avg. of All Manufact.	Analyzed Industry	Index	For 1994	Avg. of All Manufact.	Analyzed Industry	Index
Employees per Establishment	49	45	92	Value Added per Production Worker	134,084	88,408	66
Payroll per Establishment	1,500,273	1,399,045	93	Cost per Establishment	5,045,178	3,520,976	70
Payroll per Employee	30,620	31,063	101	Cost per Employee	102,970	78,175	76
Production Workers per Establishment	34	28	80	Cost per Production Worker	146,988	127,633	87
Wages per Establishment	853,319	751,155	88	Shipments per Establishment	9,576,895	5,930,598	62
Wages per Production Worker	24,861	27,229	110	Shipments per Employee	195,460	131,675	67
Hours per Production Worker	2,056	2,061	100	Shipments per Production Worker	279,017	214,980	77
Wages per Hour	12.09	13.21	109	Investment per Establishment	321,011	97,961	31
Value Added per Establishment	4,602,255	2,438,898	53	Investment per Employee	6,552	2,175	33
Value Added per Employee	93,930	54,150	58	Investment per Production Worker	9,352	3,551	38

Sources: Same as General Statistics. The 'Average of All Manufacturing' column represents the average of all manufacturing industries reported for the most recent complete year available. The Index shows the relationship between the Average and the Analyzed Industry. For example, 100 means that they are equal; 500 that the Analyzed Industry is five times the average; 50 means that the Analyzed Industry is half the national average. The abbreviation 'na' is used to show that data are 'not available'.

LEADING COMPANIES   Number shown: 37   Total sales ($ mil): 30,291   Total employment (000): 259.4

Company Name	Address				CEO Name	Phone	Co. Type	Sales ($ mil)	Empl. (000)
United Technologies Corp	U Techn	Hartford	CT	06101	Robert F Daniell	203-728-7000	P	20,801	171.5
Otis Elevator Co	10 Farm Springs	Farmington	CT	06032	J van Rooy	203-676-6000	S	4,400	47.0
Dover Corp	280 Park Av	New York	NY	10017	Thomas L Reece	212-922-1640	P	3,085	23.0
Dover Elevator International Inc	6750 Poplar Av	Memphis	TN	38138	John Apple	901-342-4300	S	778	7.5
Montgomery Elevator Co	1 Montgomery Ct	Moline	IL	61265	Dan Blount	309-764-6771	R	410	3.7
Total Energy Services Inc	15710 J F Kennedy	Houston	TX	77032	Dale Wood	713-987-3981	S	180*	<0.1
Dover Elevator Systems Inc	PO Box 2177	Memphis	TN	38101	LE Hamilton	601-393-2110	S	150	1.4
US Elevator Corp	10728 US Elevator	Spring Valley	CA	91978	John Ferguson	619-660-1000	S	100*	1.0
Access Industries Inc	4001 E 138th St	Grandview	MO	64030	Michael Mahoney	816-763-3100	R	51	0.5
Hogan Manufacturing Inc	PO Box 398	Escalon	CA	95320		209-838-7323	R	49	0.4
Ricon Corp	12450 Montague St	Pacoima	CA	91331	Jules Tremblay	818-899-7588	R	40	0.2
GAL Manufacturing Corp	50 E 153rd St	Bronx	NY	10451	HP Glaser	718-292-9000	R	36	0.3
Kelley Company Inc	PO Box 09993	Milwaukee	WI	53209	Larry Johnson	414-352-1000	R	36	0.3
Payne Elevator Co	665 Concord Av	Cambridge	MA	02138	John DeMartino	617-547-9000	R	25	0.3
Staley Elevator Company Inc	47-24 27th St	Long Island Ct	NY	11101	Kevin Leo	718-786-4300	R	15	0.1
American Stair-Glide	4001 E 138th St	Grandview	MO	64030	ME Mahoney	816-763-3100	D	12	0.1
Serge Elevator Company Inc	1 Industrial Rd	Wood Ridge	NJ	07075	Alex Sobel	201-438-8400	R	12	0.1
Oliver and Williams Elevator Inc	1411 Wilson St	Los Angeles	CA	90021	Lynn Park	213-746-1101	R	11*	<0.1
Beckwith Elevator Co	274 Southampton St	Boston	MA	02118	Norman C Bedford	617-427-5525	R	10	0.1
Courion Industries	3044 Lambdin Av	St Louis	MO	63115	Kevin O'Meara	314-533-5700	R	10	<0.1
Elevator Doors Inc	15 Jane St	Paterson	NJ	07522	Thomas Aveni	201-790-9100	R	9	<0.1
Inclinator Company of America	PO Box 1557	Harrisburg	PA	17105	PR Krum	717-234-8065	R	9*	<0.1
Bay State Elevator Co	PO Box 1210	Springfield	MA	01101	HF Potts Jr	413-736-2701	R	8*	<0.1
Schumacher Elevator Company	240 E Main St	Denver	IA	50622	MW Schumacher	319-984-5676	R	8	0.1
Matot Inc	1533 W Altgeld St	Chicago	IL	60614	Edward J Matot II	312-549-2177	R	7	<0.1
Elevator Equipment Corp	PO Box 39714	Los Angeles	CA	90039	Abe Salehpour	213-245-0147	R	6	<0.1
Payne Elevator Co New Haven	100 Sebethe Dr	Cromwell	CT	06416	John DeMartino	203-635-3377	D	6	<0.1
R and O Elevator Company Inc	8324 Pillsbury Av S	Bloomington	MN	55420	Lee Arnold	612-888-9255	R	6	<0.1
MD Knowlton Co	PO Box 29	Victor	NY	14564	KK Fellows	716-924-3230	R	4	<0.1
Mitsubishi Elevator Co	6360 Gateway Dr	Cypress	CA	90630	Davis Turner	714-220-4800	S	4*	<0.1
Elevator Systems Inc	207 Lawrence Av	Inwood	NY	11096	Ignatius Alcamo	516-239-4044	R	3*	<0.1
Advanced Elevator Eng & Mfg	3139 S Ridgeland	Berwyn	IL	60402	David Duncan Sr	708-484-7277	R	3	1.1
Detroit Elevator Co	1938 Franklin St	Detroit	MI	48207	Ralph McKian	313-259-3710	R	2*	<0.1
WE Palmer Company Inc	134 Southampton St	Boston	MA	02118	Sydney L Miller	617-445-1300	R	2	<0.1
Colley Elevator Co	226 Williams St	Bensenville	IL	60106	R E Zomchek	708-766-7230	R	1*	<0.1
Hoeck Metal Fabricators Inc	1184 Harrison St	San Francisco	CA	94103	Jeff Hoeck	415-621-4883	R	1*	<0.1
Handicaps Inc	4335 S Santa Fe Dr	Englewood	CO	80110	Jerry Kittle	303-781-2062	R	1	<0.1

Source: Ward's Business Directory of U.S. Private and Public Companies, Volumes 1 and 2, 1996. The company type code used is as follows: P - Public, R - Private, S - Subsidiary, D - Division, J - Joint Venture, A - Affiliate, G - Group. Sales are in millions of dollars, employees are in thousands. An asterisk (*) indicates an estimated sales volume. The symbol < stands for 'less than'. Company names and addresses are truncated, in some cases, to fit into the available space.

MATERIALS CONSUMED

Material	Quantity	Delivered Cost ($ million)
Materials, ingredients, containers, and supplies	(X)	540.7
Fluid power pumps, motors, and hydrostatic transmissions (hydraulic and pneumatic)	(X)	17.8
Fluid power cylinders and rotary actuators (hydraulic and pneumatic)	(X)	18.6
Fluid power hose or tube fittings and assemblies (hydraulic and pneumatic)	(X)	5.0
Fluid power valves (hydraulic and pneumatic)	(X)	8.0
Fabricated metal products, except forgings	(X)	31.2
Iron and steel forgings	(X)	1.9
Iron and steel castings (rough and semifinished)	(X)	10.4
Nonferrous (aluminum, copper, etc.) castings (rough and semifinished)	(X)	3.7
Steel bars, bar shapes, and plates	(X)	57.6
Steel sheet and strip, including tin plate	(X)	71.7
Steel structural shapes and sheet piling	(X)	22.4
All other steel shapes and forms	(X)	27.6
Copper and copper-base alloy shapes and forms	(X)	5.4
Aluminum and aluminum-base alloy shapes and forms	(X)	12.2
Other nonferrous shapes and forms	(X)	2.5
Fractional horsepower electric motors (less than 1 hp) (excluding timing motors)	(X)	5.0
Integral horsepower electric motors and generators (1 hp or more)	(X)	7.8
Electrical transmission, distribution, and control equipment	(X)	53.9
Storage batteries	(X)	0.5
Ball bearings (mounted or unmounted)	(X)	4.9
Roller bearings (mounted or unmounted)	(X)	4.0
Mechanical speed changers, gears, and industrial high-speed drives	(X)	4.6
Rubber and plastics hose and belting	(X)	4.1
All other materials and components, parts, containers, and supplies	(X)	103.7
Materials, ingredients, containers, and supplies, nsk	(X)	56.0

*Source: 1992 Economic Census. Explanation of symbols used: (D): Withheld to avoid disclosure of competitive data; na: Not available; (S): Withheld because statistical norms were not met; (X): Not applicable; (Z): Less than half the unit shown; nec: Not elsewhere classified; nsk: Not specified by kind; - : zero; * : 10-19 percent estimated; ** : 20-29 percent estimated.*

PRODUCT SHARE DETAILS

Product or Product Class	% Share	Product or Product Class	% Share
Elevators and moving stairways	100.00	Automobile lifts (service station and garage type)	13.81
Elevators and moving stairways	75.71	Moving stairways, escalators, and moving walkways	8.77
Geared electric passenger elevators (except farm, portable, and residential lifts)	17.06	Other nonfarm elevators, including sidewalk elevators, dumb waiters, man lifts, etc. (except portable elevator/ stackers)	9.16
Gearless electric passenger elevators (except farm, portable, and residential lifts)	6.07	Elevators and moving stairways, nsk	0.32
Hydraulic passenger elevators (except farm and portable)	36.52	Parts and attachments for elevators and moving stairways (sold separately)	19.41
Electric freight elevators (except farm and portable)	2.31	Elevators and moving stairways, nsk	4.87
Hydraulic freight elevators (except farm and portable)	5.96		

Source: 1992 Economic Census. The values shown are percent of total shipments in an industry. Values of indented subcategories are summed in the main heading. The symbol (D) appears when data are withheld to prevent disclosure of competitive information. The abbreviation nsk stands for 'not specified by kind' and nec for 'not elsewhere classified'.

INPUTS AND OUTPUTS FOR ELEVATORS & MOVING STAIRWAYS

Economic Sector or Industry Providing Inputs	%	Sector	Economic Sector or Industry Buying Outputs	%	Sector
Imports	9.1	Foreign	Office buildings	33.4	Constr.
Switchgear & switchboard apparatus	9.0	Manufg.	Nonfarm residential structure maintenance	12.7	Constr.
Blast furnaces & steel mills	7.5	Manufg.	Maintenance of nonfarm buildings nec	10.7	Constr.
Wholesale trade	7.0	Trade	Exports	6.4	Foreign
Sheet metal work	4.3	Manufg.	Construction of hospitals	5.5	Constr.
Motors & generators	3.9	Manufg.	Garage & service station construction	3.6	Constr.
Advertising	3.9	Services	Construction of stores & restaurants	3.3	Constr.
Iron & steel foundries	3.8	Manufg.	Industrial buildings	3.3	Constr.
Petroleum refining	3.6	Manufg.	Change in business inventories	2.3	In House
Internal combustion engines, nec	3.4	Manufg.	Hotels & motels	2.2	Constr.
Metal coating & allied services	3.0	Manufg.	Residential high-rise apartments	2.0	Constr.
Fabricated metal products, nec	2.6	Manufg.	Residential garden apartments	1.9	Constr.
Storage batteries	2.4	Manufg.	Residential additions/alterations, nonfarm	1.8	Constr.
Machinery, except electrical, nec	2.3	Manufg.	Federal Government purchases, national defense	1.7	Fed Govt
Aluminum rolling & drawing	1.9	Manufg.	Maintenance of petroleum & natural gas wells	1.6	Constr.
Maintenance of nonfarm buildings nec	1.8	Constr.	Amusement & recreation building construction	1.2	Constr.
Pumps & compressors	1.4	Manufg.	Construction of nonfarm buildings nec	1.2	Constr.
Screw machine and related products	1.4	Manufg.	Electric utility facility construction	1.1	Constr.
Electric services (utilities)	1.4	Util.	Resid. & other health facility construction	0.8	Constr.
Elevators & moving stairways	1.3	Manufg.	Construction of educational buildings	0.7	Constr.
Engineering, architectural, & surveying services	1.3	Services	Warehouses	0.6	Constr.
Communications, except radio & TV	1.2	Util.	Elevators & moving stairways	0.6	Manufg.
Motor freight transportation & warehousing	1.1	Util.	Local transit facility construction	0.4	Constr.
Nonferrous wire drawing & insulating	1.0	Manufg.	Dormitories & other group housing	0.3	Constr.
Power transmission equipment	1.0	Manufg.	Maintenance of electric utility facilities	0.2	Constr.
Banking	0.9	Fin/R.E.	Maintenance of railroads	0.2	Constr.
Automotive repair shops & services	0.9	Services	Maintenance of nonbuilding facilities nec	0.1	Constr.
Equipment rental & leasing services	0.9	Services			
Copper rolling & drawing	0.8	Manufg.			
Eating & drinking places	0.8	Trade			
Computer & data processing services	0.8	Services			
Detective & protective services	0.8	Services			
Ball & roller bearings	0.7	Manufg.			
Special dies & tools & machine tool accessories	0.7	Manufg.			
Tires & inner tubes	0.7	Manufg.			
Real estate	0.7	Fin/R.E.			
Automotive rental & leasing, without drivers	0.7	Services			
Transformers	0.6	Manufg.			
U.S. Postal Service	0.6	Gov't			
Fabricated rubber products, nec	0.5	Manufg.			
Gaskets, packing & sealing devices	0.5	Manufg.			
Pipe, valves, & pipe fittings	0.5	Manufg.			
Gas production & distribution (utilities)	0.5	Util.			
Aluminum castings	0.4	Manufg.			
Abrasive products	0.3	Manufg.			
Blankbooks & looseleaf binders	0.3	Manufg.			
Metal heat treating	0.3	Manufg.			
Retail trade, except eating & drinking	0.3	Trade			
Insurance carriers	0.3	Fin/R.E.			
Legal services	0.3	Services			
Management & consulting services & labs	0.3	Services			
Industrial gases	0.2	Manufg.			
Lubricating oils & greases	0.2	Manufg.			
Machine tools, metal cutting types	0.2	Manufg.			
Air transportation	0.2	Util.			
Royalties	0.2	Fin/R.E.			
Accounting, auditing & bookkeeping	0.2	Services			
Laundry, dry cleaning, shoe repair	0.2	Services			
Brass, bronze, & copper castings	0.1	Manufg.			
Cyclic crudes and organics	0.1	Manufg.			
Manifold business forms	0.1	Manufg.			
Miscellaneous fabricated wire products	0.1	Manufg.			
Motor vehicle parts & accessories	0.1	Manufg.			
Nonferrous castings, nec	0.1	Manufg.			
Paperboard containers & boxes	0.1	Manufg.			
Primary metal products, nec	0.1	Manufg.			
Electrical repair shops	0.1	Services			
Noncomparable imports	0.1	Foreign			

Source: Benchmark Input-Output Accounts for the U.S. Economy, 1982, U.S. Department of Commerce, Washington, D.C., July 1991. Data, as reported in the source, are organized by the 1977 SIC structure in use in 1982 but have been matched, as closely as is possible, to the 1987 SIC structure used in this book.

OCCUPATIONS EMPLOYED BY SIC 353 - CONSTRUCTION AND RELATED MACHINERY

Occupation	% of Total 1994	Change to 2005	Occupation	% of Total 1994	Change to 2005
Welders & cutters	7.4	-9.1	NC machine tool operators, metal & plastic	1.6	-9.1
Assemblers, fabricators, & hand workers nec	6.3	-9.1	Industrial machinery mechanics	1.4	-0.0
Machinists	4.6	-9.1	Traffic, shipping, & receiving clerks	1.3	-12.5
Sales & related workers nec	3.9	-9.1	Freight, stock, & material movers, hand	1.3	-27.3
Blue collar worker supervisors	3.6	-13.9	Bookkeeping, accounting, & auditing clerks	1.3	-31.8
Welding machine setters, operators	2.9	-18.2	Production, planning, & expediting clerks	1.2	-9.1
Machine builders	2.3	-27.3	Coating, painting, & spraying machine workers	1.2	-9.1
General managers & top executives	2.3	-13.8	Sheet metal workers & duct installers	1.2	-9.1
Mechanical engineers	2.3	0.0	Engineering, mathematical, & science managers	1.2	3.2
Inspectors, testers, & graders, precision	2.1	-9.1	Machine tool cutting operators, metal & plastic	1.1	-24.3
Secretaries, ex legal & medical	2.0	-17.3	Purchasing agents, ex trade & farm products	1.1	9.1
Drafters	2.0	-29.2	Engineering technicians & technologists nec	1.1	-9.1
Machine assemblers	1.8	63.6	General office clerks	1.0	-22.5
Professional workers nec	1.8	9.1	Maintenance repairers, general utility	1.0	-18.2
Machine tool cutting & forming etc. nec	1.8	-9.1	Stock clerks	1.0	-26.1
Lathe & turning machine tool operators	1.6	-18.2	Drilling & boring machine tool workers	1.0	-54.6
Helpers, laborers, & material movers nec	1.6	-9.1	Industrial production managers	1.0	-9.1

Source: Industry-Occupation Matrix, Bureau of Labor Statistics. These data relate to one or more 3-digit SIC industry groups rather than to a single 4-digit SIC. The change reported for each occupation to the year 2005 is a percent of growth or decline as estimated by the Bureau of Labor Statistics. The abbreviation nec stands for 'not elsewhere classified'.

LOCATION BY STATE AND REGIONAL CONCENTRATION

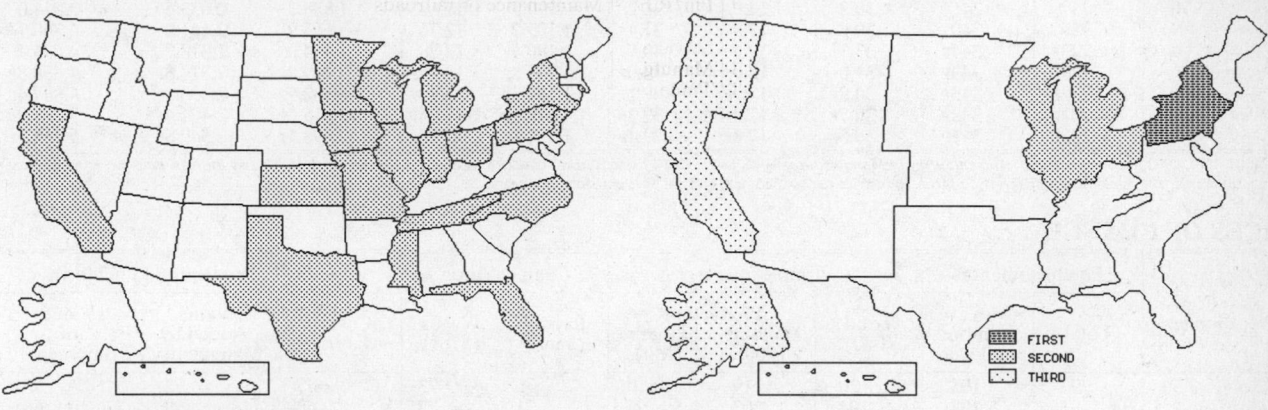

FIRST
SECOND
THIRD

INDUSTRY DATA BY STATE

State	Establish-ments	Shipments			Employment				Cost as % of Shipments	Investment per Employee ($)
		Total ($ mil)	% of U.S.	Per Establ.	Total Number	% of U.S.	Per Establ.	Wages ($/hour)		
Texas	8	114.1	11.7	14.3	500	6.5	63	9.29	78.2	1,600
Illinois	14	97.0	9.9	6.9	600	7.8	43	12.70	57.5	-
Pennsylvania	9	78.3	8.0	8.7	600	7.8	67	14.17	79.7	1,500
New York	22	74.0	7.6	3.4	700	9.1	32	11.17	42.8	857
Ohio	8	38.3	3.9	4.8	400	5.2	50	11.00	52.0	-
California	22	31.2	3.2	1.4	200	2.6	9	12.25	65.1	1,500
New Jersey	11	20.8	2.1	1.9	200	2.6	18	11.67	50.0	2,000
Florida	18	19.8	2.0	1.1	300	3.9	17	10.50	44.4	1,667
Minnesota	5	12.4	1.3	2.5	100	1.3	20	18.00	46.8	1,000
Wisconsin	6	(D)	-	-	175 *	2.3	29	-	-	1,143
Iowa	5	(D)	-	-	175 *	2.3	35	-	-	-
Indiana	4	(D)	-	-	1,750 *	22.7	438	-	-	-
Michigan	4	(D)	-	-	175 *	2.3	44	-	-	-
Kansas	3	(D)	-	-	175 *	2.3	58	-	-	-
Mississippi	3	(D)	-	-	750 *	9.7	250	-	-	-
Tennessee	3	(D)	-	-	375 *	4.9	125	-	-	-
Missouri	2	(D)	-	-	175 *	2.3	88	-	-	-
North Carolina	2	(D)	-	-	175 *	2.3	88	-	-	-
Rhode Island	2	(D)	-	-	175 *	2.3	88	-	-	-

Source: 1992 Economic Census. The states are in descending order of shipments or establishments (if shipment data are missing for the majority). The symbol (D) appears when data are withheld to prevent disclosure of competitive information. States marked with (D) are sorted by number of establishments. A dash (-) indicates that the data element cannot be calculated; * indicates the midpoint of a range.

3535 - CONVEYORS & CONVEYING EQUIPMENT

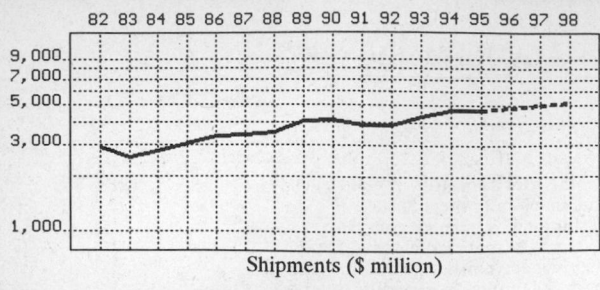

Shipments ($ million)

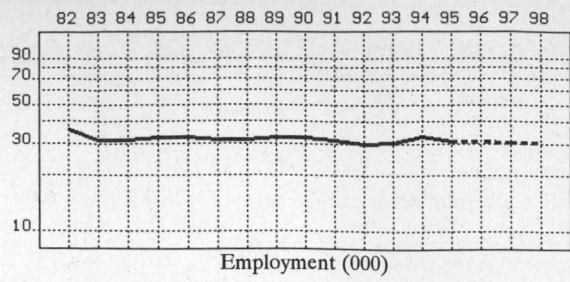

Employment (000)

GENERAL STATISTICS

Year	Companies	Establishments		Employment			Compensation		Production ($ million)			
		Total	with 20 or more employees	Total (000)	Production Workers (000)	Hours (Mil)	Payroll ($ mil)	Wages ($/hr)	Cost of Materials	Value Added by Manufacture	Value of Shipments	Capital Invest.
1982	645	698	362	36.4	20.3	39.7	756.0	8.71	1,451.1	1,465.8	2,936.0	57.9
1983		678	354	31.5	17.5	34.1	701.4	9.15	1,288.0	1,197.6	2,572.9	42.3
1984		658	346	31.8	18.3	37.3	735.0	9.22	1,367.8	1,395.6	2,742.5	45.1
1985		639	337	32.4	18.6	38.1	786.0	9.38	1,534.4	1,538.7	3,035.4	67.7
1986		644	339	32.7	18.3	38.2	836.0	9.71	1,668.8	1,650.5	3,343.6	49.3
1987	703	747	347	31.5	17.9	37.7	856.5	10.07	1,639.5	1,752.9	3,408.2	118.5
1988		723	361	31.7	18.0	37.6	871.8	10.28	1,753.8	1,784.0	3,490.1	59.1
1989		692	355	32.5	19.9	41.2	957.0	10.79	1,998.1	2,075.0	4,065.9	122.4
1990		672	359	32.8	19.3	39.8	957.2	11.18	2,046.9	2,066.3	4,089.9	97.8
1991		661	338	31.6	17.9	37.4	960.5	11.76	1,931.7	1,959.7	3,862.5	82.7
1992	707	747	346	30.3	17.0	36.2	990.5	11.88	1,888.3	2,033.2	3,914.0	55.3
1993		739	340	30.6	17.3	37.4	1,105.2	12.76	2,167.9	2,141.2	4,368.1	67.2
1994		723P	344P	33.1	19.1	40.7	1,180.6	12.78	2,454.0	2,318.7	4,696.7	89.4
1995		728P	343P	31.0P	18.0P	39.0P	1,147.8P	13.05P	2,449.9P	2,314.8P	4,688.8P	92.7P
1996		732P	343P	30.8P	17.9P	39.1P	1,183.3P	13.40P	2,532.7P	2,393.1P	4,847.4P	95.4P
1997		737P	342P	30.7P	17.8P	39.2P	1,218.7P	13.75P	2,615.6P	2,471.4P	5,005.9P	98.2P
1998		742P	341P	30.5P	17.8P	39.4P	1,254.2P	14.10P	2,698.4P	2,549.6P	5,164.5P	100.9P

Sources: 1982, 1987, 1992 *Economic Census*; *Annual Survey of Manufactures*, 83-86, 88-91, 93-94. Establishment counts for non-Census years are from *County Business Patterns*; establishment values for 83-84 are extrapolations. 'P's show projections by the editors. Industries reclassified in 87 will not have data for prior years.

INDICES OF CHANGE

Year	Companies	Establishments		Employment			Compensation		Production ($ million)			
		Total	with 20 or more employees	Total (000)	Production Workers (000)	Hours (Mil)	Payroll ($ mil)	Wages ($/hr)	Cost of Materials	Value Added by Manufacture	Value of Shipments	Capital Invest.
1982	91	93	105	120	119	110	76	73	77	72	75	105
1983		91	102	104	103	94	71	77	68	59	66	76
1984		88	100	105	108	103	74	78	72	69	70	82
1985		86	97	107	109	105	79	79	81	76	78	122
1986		86	98	108	108	106	84	82	88	81	85	89
1987	99	100	100	104	105	104	86	85	87	86	87	214
1988		97	104	105	106	104	88	87	93	88	89	107
1989		93	103	107	117	114	97	91	106	102	104	221
1990		90	104	108	114	110	97	94	108	102	104	177
1991		88	98	104	105	103	97	99	102	96	99	150
1992	100	100	100	100	100	100	100	100	100	100	100	100
1993		99	98	101	102	103	112	107	115	105	112	122
1994		97P	99P	109	112	112	119	108	130	114	120	162
1995		97P	99P	102P	106P	108P	116P	110P	130P	114P	120P	168P
1996		98P	99P	102P	105P	108P	119P	113P	134P	118P	124P	173P
1997		99P	99P	101P	105P	108P	123P	116P	139P	122P	128P	178P
1998		99P	99P	101P	105P	109P	127P	119P	143P	125P	132P	182P

Sources: Same as General Statistics. Values reflect change from the base year, 1992. Values above 100 mean greater than 92, values below 100 mean less than 92, and a value of 100 in the 82-91 or 93-98 period means same as 92. 'P's mark projections by the editors.

SELECTED RATIOS

For 1994	Avg. of All Manufact.	Analyzed Industry	Index	For 1994	Avg. of All Manufact.	Analyzed Industry	Index
Employees per Establishment	49	46	93	Value Added per Production Worker	134,084	121,398	91
Payroll per Establishment	1,500,273	1,633,329	109	Cost per Establishment	5,045,178	3,395,045	67
Payroll per Employee	30,620	35,668	116	Cost per Employee	102,970	74,139	72
Production Workers per Establishment	34	26	77	Cost per Production Worker	146,988	128,482	87
Wages per Establishment	853,319	719,608	84	Shipments per Establishment	9,576,895	6,497,761	68
Wages per Production Worker	24,861	27,233	110	Shipments per Employee	195,460	141,894	73
Hours per Production Worker	2,056	2,131	104	Shipments per Production Worker	279,017	245,901	88
Wages per Hour	12.09	12.78	106	Investment per Establishment	321,011	123,683	39
Value Added per Establishment	4,602,255	3,207,861	70	Investment per Employee	6,552	2,701	41
Value Added per Employee	93,930	70,051	75	Investment per Production Worker	9,352	4,681	50

Sources: Same as General Statistics. The 'Average of All Manufacturing' column represents the average of all manufacturing industries reported for the most recent complete year available. The Index shows the relationship between the Average and the Analyzed Industry. For example, 100 means that they are equal; 500 that the Analyzed Industry is five times the average; 50 means that the Analyzed Industry is half the national average. The abbreviation 'na' is used to show that data are 'not available'.

LEADING COMPANIES   Number shown: **75**   Total sales ($ mil): **3,694**   Total employment (000): **24.3**

Company Name	Address				CEO Name	Phone	Co. Type	Sales ($ mil)	Empl. (000)
Interlake Corp	550 Warrenville Rd	Lisle	IL	60532	W Robert Reum	708-852-8800	P	683	4.6
Rapistan Demag Corp	507 Plymouth NE	Grand Rapids	MI	49505	Pete Metros	616-451-6525	S	350	1.6
Jervis B Webb Company Inc	34375 W 12 Mile Rd	Farmington Hls	MI	48331	J M Hammond	810-553-1220	R	330	2.3
Alvey Inc	9301 Olive Blv	St Louis	MO	63132	Stephen J O'Neill	314-993-4700	S	100	0.7
Pinnacle Automation Inc	101 S Hanley	St Louis	MO	63105	William R Michaels	314-993-4700	R	100*	0.7
Western Atlas Mat Handling Syst	2300 Litton Ln	Hebron	KY	41048	John Paxton	606-334-2400	D	100	0.5
Buschman Co	10045 International	Cincinnati	OH	45246	Christopher C Cole	513-874-0788	S	80	0.6
Daifuku USA Inc	6700 Tussing Rd	Reynoldsburg	OH	43068	K Mabuchi	614-863-1888	R	80	0.1
Simplimatic Engineering Co	PO Box 11709	Lynchburg	VA	24506	DJ Jaisle	804-582-1200	S	80	0.5
Continental Conveyor	PO Box 400	Winfield	AL	35594	CE Bryant Jr	205-487-6492	S	70	0.5
Dearborn Fabricating	19440 Glendale Av	Detroit	MI	48223	J Wes Paisley	313-273-2800	S	70	0.1
IDAB Inc	PO Box 8157	Hampton	VA	23666	Ole B Rygh	804-825-2260	R	64*	0.5
Hytrol Conveyor Company Inc	2020 Hytrol	Jonesboro	AR	72401	Gregg Goodner	501-935-3700	R	60	0.7
McNally Wellman	4800 Grand Av	Neville Island	PA	15225	Ray Koper	412-269-5000	D	60	0.2
NKC of America Inc	1584 Brooks Rd E	Memphis	TN	38116	Yoshiaki Hata	901-396-5353	S	60	0.2
Acco Systems	12755 E 9 Mile Rd	Warren	MI	48089	William Salatino	313-755-7500	D	50	0.2
FMC Corp	7300 Presidents Dr	Orlando	FL	32809	WK Foster	407-851-3377	D	50	0.3
TransLogic Corp	10825 E 47th Av	Denver	CO	80239	JT Mahoney	303-371-7770	R	50	0.3
United Conveyor Corp	2100 Normanlv W	Waukegan	IL	60085	DN Basler	708-473-5900	R	50	0.4
Vulcan Engineering Co	PO Box 307	Helena	AL	35080	P Zettler	205-663-0732	R	50	0.3
Richards-Wilcox Mfg Co	600 S Lake St	Aurora	IL	60506	Manfred Haiderer	708-897-6951	S	48	0.4
Bosch Automation Products	816 E 3rd St	Buchanan	MI	49107	Dan Kelly	616-695-0151	D	40	0.1
MAC Equipment Inc	10741 Ambass	Kansas City	MO	64153	Rich Larson	816-891-9300	R	40	0.3
General Kinematics Corp	777 Lake Zurich Rd	Barrington	IL	60010	A Musschoot	708-381-2240	R	30	0.1
Integrated Material Handling Co	PO Box 2408	Oshkosh	WI	54903	Jim Streblow	414-235-5562	S	30	0.4
INTERROLL Corp	3000 Corporate Dr	Wilmington	NC	28405	C Agnoff	910-799-1100	R	30	0.3
Mathews Conveyor	PO Box 928	Danville	KY	40422	Harold E Davies	606-236-9400	D	30	0.5
Mayfran International	PO Box 43038	Cleveland	OH	44143	B Terry	216-461-4100	S	30	0.2
Predco Inc	PO Box 599	Pennsauken	NJ	08110	A Frascella	609-665-2644	R	30	0.3
W and H Systems Inc	120 Asia Pl	Carlstadt	NJ	07072	Alfred Iversen	201-933-7840	R	30	0.1
Webster Industries Inc	325 Hall St	Tiffin	OH	44883	FC Spurck	419-447-8232	R	30	0.3
SI Handling Systems Inc	600 Kuebler Rd	Easton	PA	18040	Leonard S Yurkovic	215-252-7321	P	29	0.2
TeleCom Corp	1545 Mockingb	Dallas	TX	75235	L W Schumann	214-638-0638	P	25	0.2
Semco Inc	13813 FM 529 Rd	Houston	TX	77041	CA Copenhaver	713-896-1825	R	25	<0.1
Shuttleworth Inc	10 Commercial Rd	Huntington	IN	46750	James Shuttleworth	219-356-8500	R	25	<0.1
Dorner Manufacturing Corp	PO Box 20	Hartland	WI	53029	Wolfgang Dorner	414-367-7600	R	24	0.2
Pevco Systems International Inc	9610 Pulaski Pk Dr	Baltimore	MD	21220	F M Valerino Sr	410-574-2800	R	22*	<0.1
Thomas Conveyor Company Inc	PO Box 2916	Fort Worth	TX	76113	BJ Hinterlong	817-295-7151	R	22	0.3
Control Engineering Co	8212 Harbor Spgs	Harbor Springs	MI	49740	AR Vokes	616-347-3931	S	21	0.2
Sasib Bakery North America	808 Stewart Av	Plano	TX	75074	Len Kilby	214-422-5808	S	21*	0.2
Southern Systems Inc	4101 Viscount Av	Memphis	TN	38118	Leon Linton	901-362-7340	R	21*	0.2
Dynamic Air Inc	1125 Wolters Blv	St Paul	MN	55110	JR Steele	612-484-2900	R	20	0.1
Ermanco Inc	PO Box 490	Spring Lake	MI	49456	Leon C Kirschner	616-798-4547	R	20	0.2
Hewitt-Robins Corp	40 Fairfield Pl	West Caldwell	NJ	07006	J Wrobel	201-777-5500	D	20	0.2
Key Handlings System Inc	210 S Newman St	Hackensack	NJ	07601	Charles Van Milis	201-342-6223	R	20	<0.1
Krupp Robins Inc	7730 E Belleview	Englewood	CO	80111	Ramsis Shehata	303-770-0808	S	20*	<0.1
Rapid Industries Inc	4003 Oaklawn Dr	Louisville	KY	40219	WG Sheets	502-968-3645	R	20	0.1
SCHEBLER Co	PO Box 1005	Bettendorf	IA	52722	George Kertesz	319-359-0110	S	20	0.2
Screw Conveyor Corp	700 Hoffman St	Hammond	IN	46327	Garry M Abraham	219-931-1450	R	20	0.3
Bristol Steel and Conveyor Corp	4144 Jimbo Dr	Burton	MI	48529	Raymond Oliver	313-743-8560	R	19	0.2
Ideas in Motion Inc	3470 Raleigh Av SE	Grand Rapids	MI	49512	Hank DiSalvatore	616-942-5488	R	19	0.1
Automotion Inc	11743 S Mayfield	Worth	IL	60482	Merle Davis	708-597-1910	R	18	0.1
Carrier Vibrating Equipment Inc	PO Box 37070	Louisville	KY	40233	Ken N Patel	502-969-3171	R	18	0.1
Nercon Eng & Manufacturing	PO Box 2288	Oshkosh	WI	54903	J L Nerenhausen	414-233-3268	R	18	0.3
Allor Manufacturing Inc	PO Box 647	Novi	MI	48376	Fred M Allor	313-348-2700	R	17*	0.1
Quipp Inc	4800 NW 157th St	Miami	FL	33014	James E Pruitt	305-623-8700	P	17	0.1
Quipp Systems Inc	4800 NW 157th St	Miami	FL	33014	Louis D Kipp	305-623-8700	S	17	0.1
Automated Conveyor Systems	3850 Southland Dr	West Memphis	AR	72301	Charles Dody	501-732-5050	R	16*	0.2
Martin Conveyor	PO Box 193	Mansfield	TX	76063	John Almarez	817-477-2104	D	16*	0.1
Wyard Machinery Group	907 SW 15th St	Forest Lake	MN	55025	Dennis Jaisle	612-464-4000	D	16	0.1
Automated Conveyor Systems	PO Box 2474	Lynchburg	VA	24501	Michael G Shenigo	804-528-3822	R	15	0.2
Jetstream Systems	4690 Joliet St	Denver	CO	80239	Phil Ostapowicz	303-371-9002	S	15	0.1
SandMold Systems Inc	PO Box 488	Newaygo	MI	49337	RJ Smith	616-652-1623	R	15	<0.1
Shick Tube Veyor Corp	4346 Clary Blv	Kansas City	MO	64130	Joseph T Ungashick	816-861-7224	R	15	0.2
Hoppmann Corp	14560 Lee Rd	Chantilly	VA	22021	Kurt Hoppmann	703-631-2700	R	14	0.3
KWS Manufacturing Company	PO Box 809	Joshua	TX	76058	Floyd Watkins	817-295-2247	R	14	0.1
Pulver Systems Inc	9999 Virginia Av	Chicago Ridge	IL	60415	Richard H Pulver	708-424-2500	R	14	0.1
Meyer Machine Co	PO Box 5460	San Antonio	TX	78201	Eugene W Teeter	210-736-1811	S	14	0.1
Kuka Welding	40675 Mound Rd	Sterling Hts	MI	48310	Richard Schmid	313-977-0100	S	13*	<0.1
Liftech Inc	124 Sylvania Pl	South Plainfield	NJ	07080	H Carrick	908-968-7100	R	13*	<0.1
Norfolk Conveyor	155 King St	Cohasset	MA	02025	Jim Ogburn	617-383-9400	D	13	<0.1
Beaumont Birch Co	3900 River Rd	Pennsauken	NJ	08110	JR Ford	609-663-6440	S	12	<0.1
Chain Supply Co	PO Box 1006	Wixom	MI	48393	JF Krempa	313-624-0500	R	12	<0.1
Garvey Corp	208 S Route 73	Hammonton	NJ	08037	Mark C Garvey	609-561-2450	R	12	<0.1
Overhead Conveyor Co	1330 Hilton Rd	Ferndale	MI	48220	M E Woodbeck Sr	313-547-3800	R	12	<0.1

Source: Ward's Business Directory of U.S. Private and Public Companies, Volumes 1 and 2, 1996. The company type code used is as follows: P - Public, R - Private, S - Subsidiary, D - Division, J - Joint Venture, A - Affiliate, G - Group. Sales are in millions of dollars, employees are in thousands. An asterisk (*) indicates an estimated sales volume. The symbol < stands for 'less than'. Company names and addresses are truncated, in some cases, to fit into the available space.

MATERIALS CONSUMED

Material	Quantity	Delivered Cost ($ million)
Materials, ingredients, containers, and supplies	(X)	1,431.9
Fluid power pumps, motors, and hydrostatic transmissions (hydraulic and pneumatic)	(X)	16.7
Fluid power cylinders and rotary actuators (hydraulic and pneumatic)	(X)	10.9
Fluid power filters (hydraulic and pneumatic)	(X)	2.3
Fluid power hose or tube fittings and assemblies (hydraulic and pneumatic)	(X)	8.3
Fluid power valves (hydraulic and pneumatic)	(X)	17.2
Fabricated metal products, except forgings	(X)	128.7
Iron and steel forgings	(X)	15.6
Nonferrous forgings	(X)	(D)
Iron and steel castings (rough and semifinished)	(X)	23.9
Nonferrous (aluminum, copper, etc.) castings (rough and semifinished)	(X)	5.4
Steel bars, bar shapes, and plates	(X)	71.0
Steel sheet and strip, including tin plate	(X)	107.4
Steel structural shapes and sheet piling	(X)	44.7
All other steel shapes and forms	(X)	45.6
Copper and copper-base alloy shapes and forms	(X)	0.4
Aluminum and aluminum-base alloy shapes and forms	(X)	12.7
Other nonferrous shapes and forms	(X)	5.4
Gasoline and other carburetor engines	(X)	(D)
Fractional horsepower electric motors (less than 1 hp) (excluding timing motors)	(X)	19.7
Integral horsepower electric motors and generators (1 hp or more)	(X)	32.7
Electrical transmission, distribution, and control equipment	(X)	56.0
Storage batteries	(X)	(Z)
Ball bearings (mounted or unmounted)	(X)	32.1
Roller bearings (mounted or unmounted)	(X)	21.0
Mechanical speed changers, gears, and industrial high-speed drives	(X)	47.4
Pneumatic tires and inner tubes	(X)	3.1
Rubber and plastics hose and belting	(X)	40.5
All other materials and components, parts, containers, and supplies	(X)	251.6
Materials, ingredients, containers, and supplies, nsk	(X)	407.8

Source: 1992 *Economic Census*. Explanation of symbols used: (D): Withheld to avoid disclosure of competitive data; na: Not available; (S): Withheld because statistical norms were not met; (X): Not applicable; (Z): Less than half the unit shown; nec: Not elsewhere classified; nsk: Not specified by kind; - : zero; * : 10-19 percent estimated; ** : 20-29 percent estimated.

PRODUCT SHARE DETAILS

Product or Product Class	% Share	Product or Product Class	% Share
Conveyors and conveying equipment	100.00	systems, except hoists and farm elevators	4.98
Unit handling conveyors and conveying systems, except hoists and farm elevators	49.98	Bulk material handling pneumatic conveyors and conveying systems, except hoists and farm elevators	12.23
Unit handling gravity conveyors and conveying systems (skate wheel and roller), except hoists and farm elevators	5.14	Bulk material handling portable conveyors and conveying systems, except hoists and farm elevators	2.32
Light-to medium-duty unit handling trolley (overhead) conveyors and conveying systems, except hoists and farm elevators	4.95	Bulk material handling en masse conveyors and conveying systems, except hoists and farm elevators	1.40
Heavy-duty unit handling trolley (overhead) conveyors and conveying systems, except hoists and farm elevators	13.43	Bulk material handling vibrating conveyors and conveying systems, except hoists and farm elevators	4.94
Unit handling tow conveyors and conveying systems (under floor systems), except hoists and farm elevators	1.70	All other bulk material handling conveyors and conveying systems, such as apron, flight, and drag conveyors, etc., except hoists and farm elevators	11.22
Light-to medium-duty unit handling powered conveyors and conveying systems (belt and roller), except hoists and farm elevators	27.55	Bulk material handling unloading and reclaiming vibrating feeders	4.67
Heavy-duty unit handling powered conveyors and conveying systems (belt and roller), except hoists and farm elevators	18.02	All other bulk material handling unloading and reclaiming systems, such as bins, apron feeders, gates, etc.	6.12
Unit handling pneumatic tube conveyors and conveying systems, except hoists and farm elevators	4.05	Bulk material handling loading and storing traveling stackers	1.88
Unit handling portable conveyors and conveying systems, except hoists and farm elevators	0.18	Other bulk material handling loading and storing systems, such as trippers, centrifugal throwers, etc.	0.69
Unit handling carousel conveyors and conveying systems, except hoists and farm elevators	1.07	Bulk material handling conveyors and conveying systems, except hoists and farm elevators, nsk	3.55
All other unit handling conveyors and conveying systems, such as pallet conveyors, etc., except hoists and farm elevators	11.18	Parts, attachments, and accessories for bulk material handling conveyors and conveying systems (sold separately)	10.36
Unit handling conveyors and conveying systems, except hoists and farm elevators, nsk	12.72	Belt conveyor idlers for bulk material handling conveyors and conveying systems (sold separately)	15.90
Parts, attachments, and accessories for unit handling conveyors and conveying systems (sold separately)	3.95	Belt conveyor pulleys for bulk material handling conveyors and conveying systems (sold separarely)	9.45
Bulk material handling conveyors and conveying systems, except hoists and farm elevators	27.28	All other parts, attachments, and accessories for bulk material handling conveyors and conveying systems (sold separately)	68.81
Bulk material handling belt conveyors and conveying systems, except hoists and farm elevators	37.93	Parts, attachments, and accessories for bulk material handling conveyors and conveying systems (sold separately), nsk	5.84
Bulk material handling screw conveyors and conveying systems, except hoists and farm elevators	8.06	Conveyors and conveying equipment, nsk	8.45
Bulk material handling bucket elevators and elevator			

Source: 1992 *Economic Census*. The values shown are percent of total shipments in an industry. Values of indented subcategories are summed in the main heading. The symbol (D) appears when data are withheld to prevent disclosure of competitive information. The abbreviation nsk stands for 'not specified by kind' and nec for 'not elsewhere classified'.

INPUTS AND OUTPUTS FOR CONVEYORS & CONVEYING EQUIPMENT

Economic Sector or Industry Providing Inputs	%	Sector	Economic Sector or Industry Buying Outputs	%	Sector
Blast furnaces & steel mills	12.1	Manufg.	Gross private fixed investment	70.4	Cap Inv
Conveyors & conveying equipment	10.2	Manufg.	Exports	6.2	Foreign
Imports	9.3	Foreign	Conveyors & conveying equipment	5.0	Manufg.
Wholesale trade	8.9	Trade	Miscellaneous repair shops	3.8	Services
Fabricated structural metal	4.2	Manufg.	Coal	3.1	Mining
Sheet metal work	4.2	Manufg.	Federal Government purchases, national defense	2.0	Fed Govt
Advertising	3.4	Services	Office buildings	1.7	Constr.
Motors & generators	3.3	Manufg.	Broadwoven fabric mills	1.0	Manufg.
Metal stampings, nec	3.1	Manufg.	Construction of hospitals	0.8	Constr.
Power transmission equipment	2.4	Manufg.	Dimension, crushed & broken stone	0.5	Mining
Ball & roller bearings	2.2	Manufg.	Electric utility facility construction	0.5	Constr.
Machinery, except electrical, nec	2.1	Manufg.	Chemical & fertilizer mineral	0.4	Mining
Iron & steel foundries	2.0	Manufg.	Copper ore	0.4	Mining
Switchgear & switchboard apparatus	1.9	Manufg.	Sand & gravel	0.4	Mining
Maintenance of nonfarm buildings nec	1.8	Constr.	Construction of stores & restaurants	0.4	Constr.
Electric services (utilities)	1.5	Util.	Industrial buildings	0.3	Constr.
Iron & steel forgings	1.4	Manufg.	Clay, ceramic, & refractory minerals	0.2	Mining
Screw machine and related products	1.4	Manufg.	Iron & ferroalloy ores	0.2	Mining
Communications, except radio & TV	1.4	Util.	Maintenance of nonfarm buildings nec	0.2	Constr.
Rubber & plastics hose & belting	1.3	Manufg.	Warehouses	0.2	Constr.
Hardware, nec	1.1	Manufg.	Special industry machinery, nec	0.2	Manufg.
Industrial controls	0.9	Manufg.	Federal Government purchases, nondefense	0.2	Fed Govt
Metal coating & allied services	0.9	Manufg.	Nonferrous metal ores, except copper	0.1	Mining
Real estate	0.9	Fin/R.E.	Residential high-rise apartments	0.1	Constr.
Aluminum rolling & drawing	0.8	Manufg.	Miscellaneous plastics products	0.1	Manufg.
Industrial trucks & tractors	0.8	Manufg.	Electric services (utilities)	0.1	Util.
Pipe, valves, & pipe fittings	0.8	Manufg.			
Motor freight transportation & warehousing	0.8	Util.			
Eating & drinking places	0.8	Trade			
Banking	0.8	Fin/R.E.			
Special dies & tools & machine tool accessories	0.7	Manufg.			
Gaskets, packing & sealing devices	0.6	Manufg.			
Gas production & distribution (utilities)	0.6	Util.			
Engineering, architectural, & surveying services	0.6	Services			
Equipment rental & leasing services	0.6	Services			
Petroleum refining	0.5	Manufg.			
Miscellaneous fabricated wire products	0.4	Manufg.			
Pumps & compressors	0.4	Manufg.			
Transformers	0.4	Manufg.			
Computer & data processing services	0.4	Services			
U.S. Postal Service	0.4	Gov't			
Aluminum castings	0.3	Manufg.			
Copper rolling & drawing	0.3	Manufg.			
Fabricated rubber products, nec	0.3	Manufg.			
Metal heat treating	0.3	Manufg.			
Miscellaneous plastics products	0.3	Manufg.			
Nonferrous wire drawing & insulating	0.3	Manufg.			
Tires & inner tubes	0.3	Manufg.			
Air transportation	0.3	Util.			
Railroads & related services	0.3	Util.			
Sanitary services, steam supply, irrigation	0.3	Util.			
Legal services	0.3	Services			
Management & consulting services & labs	0.3	Services			
Abrasive products	0.2	Manufg.			
Industrial gases	0.2	Manufg.			
Machine tools, metal cutting types	0.2	Manufg.			
Sawmills & planning mills, general	0.2	Manufg.			
Royalties	0.2	Fin/R.E.			
Accounting, auditing & bookkeeping	0.2	Services			
Noncomparable imports	0.2	Foreign			
Internal combustion engines, nec	0.1	Manufg.			
Lubricating oils & greases	0.1	Manufg.			
Manifold business forms	0.1	Manufg.			
Paints & allied products	0.1	Manufg.			
Insurance carriers	0.1	Fin/R.E.			
Security & commodity brokers	0.1	Fin/R.E.			
Hotels & lodging places	0.1	Services			

Source: Benchmark Input-Output Accounts for the U.S. Economy, 1982, U.S. Department of Commerce, Washington, D.C., July 1991. Data, as reported in the source, are organized by the 1977 SIC structure in use in 1982 but have been matched, as closely as is possible, to the 1987 SIC structure used in this book.

OCCUPATIONS EMPLOYED BY SIC 353 - CONSTRUCTION AND RELATED MACHINERY

Occupation	% of Total 1994	Change to 2005	Occupation	% of Total 1994	Change to 2005
Welders & cutters	7.4	-9.1	NC machine tool operators, metal & plastic	1.6	-9.1
Assemblers, fabricators, & hand workers nec	6.3	-9.1	Industrial machinery mechanics	1.4	-0.0
Machinists	4.6	-9.1	Traffic, shipping, & receiving clerks	1.3	-12.5
Sales & related workers nec	3.9	-9.1	Freight, stock, & material movers, hand	1.3	-27.3
Blue collar worker supervisors	3.6	-13.9	Bookkeeping, accounting, & auditing clerks	1.3	-31.8
Welding machine setters, operators	2.9	-18.2	Production, planning, & expediting clerks	1.2	-9.1
Machine builders	2.3	-27.3	Coating, painting, & spraying machine workers	1.2	-9.1
General managers & top executives	2.3	-13.8	Sheet metal workers & duct installers	1.2	-9.1
Mechanical engineers	2.3	0.0	Engineering, mathematical, & science managers	1.2	3.2
Inspectors, testers, & graders, precision	2.1	-9.1	Machine tool cutting operators, metal & plastic	1.1	-24.3
Secretaries, ex legal & medical	2.0	-17.3	Purchasing agents, ex trade & farm products	1.1	9.1
Drafters	2.0	-29.2	Engineering technicians & technologists nec	1.1	-9.1
Machine assemblers	1.8	63.6	General office clerks	1.0	-22.5
Professional workers nec	1.8	9.1	Maintenance repairers, general utility	1.0	-18.2
Machine tool cutting & forming etc. nec	1.8	-9.1	Stock clerks	1.0	-26.1
Lathe & turning machine tool operators	1.6	-18.2	Drilling & boring machine tool workers	1.0	-54.6
Helpers, laborers, & material movers nec	1.6	-9.1	Industrial production managers	1.0	-9.1

Source: *Industry-Occupation Matrix*, Bureau of Labor Statistics. These data relate to one or more 3-digit SIC industry groups rather than to a single 4-digit SIC. The change reported for each occupation to the year 2005 is a percent of growth or decline as estimated by the Bureau of Labor Statistics. The abbreviation nec stands for 'not elsewhere classified'.

LOCATION BY STATE AND REGIONAL CONCENTRATION

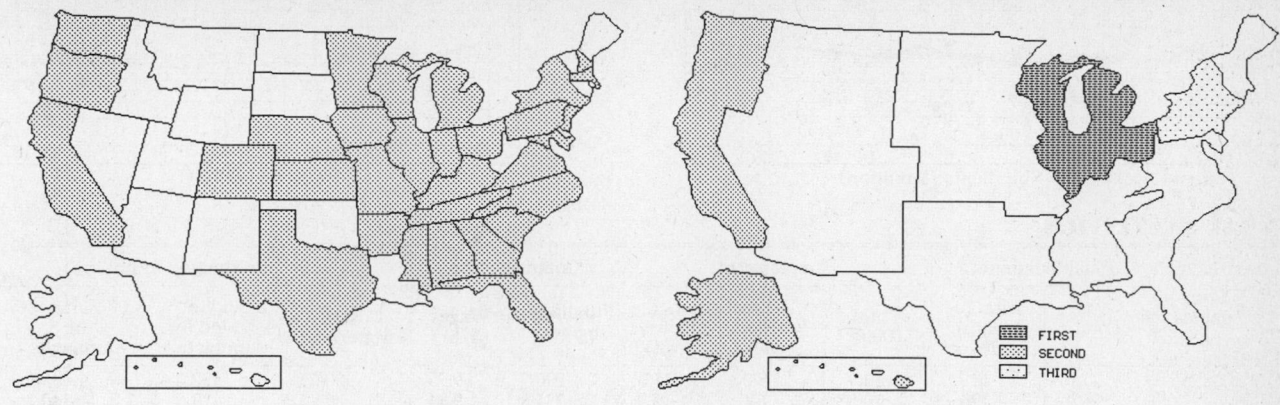

FIRST
SECOND
THIRD

INDUSTRY DATA BY STATE

State	Establish-ments	Shipments			Employment				Cost as % of Shipments	Investment per Employee ($)
		Total ($ mil)	% of U.S.	Per Establ.	Total Number	% of U.S.	Per Establ.	Wages ($/hour)		
Michigan	68	641.5	16.4	9.4	3,700	12.2	54	15.97	48.3	1,919
Ohio	59	299.9	7.7	5.1	2,700	8.9	46	11.29	43.8	1,704
Kentucky	19	299.2	7.6	15.7	2,100	6.9	111	13.29	48.7	1,619
Texas	45	273.2	7.0	6.1	2,200	7.3	49	10.50	43.3	1,591
Pennsylvania	33	216.6	5.5	6.6	1,500	5.0	45	13.00	44.8	2,000
Illinois	41	190.1	4.9	4.6	1,500	5.0	37	13.37	44.6	-
Tennessee	17	162.0	4.1	9.5	700	2.3	41	11.70	74.8	3,000
Virginia	16	145.3	3.7	9.1	1,000	3.3	63	9.00	47.9	1,300
California	67	145.2	3.7	2.2	1,300	4.3	19	12.41	46.4	1,615
Wisconsin	33	137.8	3.5	4.2	1,300	4.3	39	10.87	44.8	2,308
Kansas	12	137.3	3.5	11.4	800	2.6	67	11.00	63.9	2,375
Colorado	20	129.9	3.3	6.5	900	3.0	45	10.60	42.3	1,333
Missouri	23	117.4	3.0	5.1	700	2.3	30	9.60	50.5	857
New Jersey	32	109.5	2.8	3.4	1,000	3.3	31	12.89	50.1	2,700
Arkansas	8	107.9	2.8	13.5	1,200	4.0	150	10.06	49.6	-
Minnesota	26	88.1	2.3	3.4	800	2.6	31	10.36	48.0	2,000
Indiana	26	81.0	2.1	3.1	900	3.0	35	11.18	40.7	1,222
Florida	25	74.7	1.9	3.0	600	2.0	24	10.17	40.6	3,000
North Carolina	17	71.4	1.8	4.2	700	2.3	41	9.11	44.1	3,000
Alabama	11	56.3	1.4	5.1	600	2.0	55	9.38	51.7	833
Iowa	15	53.1	1.4	3.5	600	2.0	40	11.88	42.2	3,833
Oregon	15	52.3	1.3	3.5	600	2.0	40	12.29	44.7	1,000
New York	23	41.8	1.1	1.8	400	1.3	17	15.40	47.6	750
South Carolina	11	40.1	1.0	3.6	400	1.3	36	10.71	45.6	1,000
Georgia	13	33.1	0.8	2.5	300	1.0	23	14.67	46.5	667
Massachusetts	11	27.7	0.7	2.5	200	0.7	18	17.00	51.6	-
Washington	15	18.6	0.5	1.2	100	0.3	7	14.00	52.2	3,000
Mississippi	8	(D)	-	-	750 *	2.5	94	-	-	533
Nebraska	6	(D)	-	-	175 *	0.6	29	-	-	-
Maryland	4	(D)	-	-	175 *	0.6	44	-	-	-
New Hampshire	3	(D)	-	-	175 *	0.6	58	-	-	-

Source: 1992 *Economic Census*. The states are in descending order of shipments or establishments (if shipment data are missing for the majority). The symbol (D) appears when data are withheld to prevent disclosure of competitive information. States marked with (D) are sorted by number of establishments. A dash (-) indicates that the data element cannot be calculated; * indicates the midpoint of a range.

3536 - HOISTS, CRANES & MONORAILS

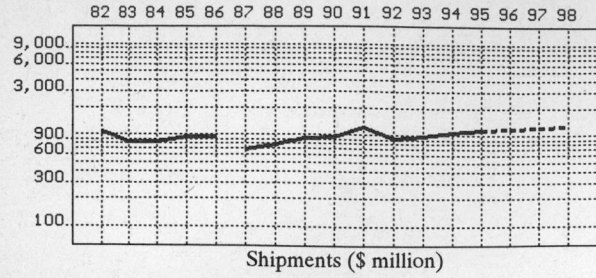

Shipments ($ million)

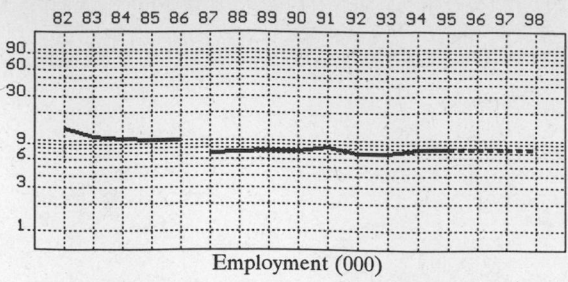

Employment (000)

GENERAL STATISTICS

Year	Companies	Establishments		Employment			Compensation		Production ($ million)			
		Total	with 20 or more employees	Total (000)	Production Workers (000)	Hours (Mil)	Payroll ($ mil)	Wages ($/hr)	Cost of Materials	Value Added by Manufacture	Value of Shipments	Capital Invest.
1982	255	276	127	13.7	8.2	15.3	290.7	9.95	504.1	532.2	1,085.6	20.0
1983		262	120	10.9	6.4	12.0	234.8	9.84	393.4	417.5	840.2	6.5
1984		248	113	10.2	6.4	12.7	216.6	9.31	412.1	408.2	840.9	9.7
1985		234	106	10.0	6.3	11.8	216.2	10.07	454.8	484.0	936.6	10.0
1986		239	107	9.8	6.2	11.9	217.3	9.66	476.2	438.2	940.4	6.3
1987*	165	175	81	7.0	4.2	8.1	175.3	11.11	318.7	358.1	675.4	7.0
1988		165	83	7.5	4.6	8.7	187.0	11.14	355.2	417.1	789.4	6.8
1989		159	85	7.7	4.2	8.6	201.3	11.52	432.9	483.5	912.4	11.7
1990		158	87	7.7	4.4	9.1	214.8	11.42	478.8	517.5	966.4	21.4
1991		160	82	8.4	5.0	10.3	262.8	13.45	519.0	738.9	1,210.3	28.5
1992	170	180	81	7.0	4.2	8.7	205.7	12.59	410.0	490.1	911.5	11.6
1993		169	81	7.2	4.2	8.8	213.3	12.34	426.1	548.3	974.8	10.2
1994		168P	82P	7.9	4.5	9.6	253.3	12.83	500.7	562.9	1,069.4	10.0
1995		169P	82P	7.7P	4.4P	9.7P	253.8P	13.30P	538.0P	604.8P	1,149.0P	15.8P
1996		169P	81P	7.8P	4.5P	9.8P	262.6P	13.57P	559.9P	629.4P	1,195.8P	16.3P
1997		170P	81P	7.8P	4.5P	10.0P	271.3P	13.85P	581.8P	654.0P	1,242.5P	16.9P
1998		170P	81P	7.9P	4.5P	10.1P	280.1P	14.13P	603.6P	678.6P	1,289.3P	17.4P

Sources: 1982, 1987, 1992 *Economic Census*; *Annual Survey of Manufactures*, 83-86, 88-91, 93-94. Establishment counts are from *County Business Patterns* for non-Census years; establishment counts for 83-84 are extrapolations. * indicates that industry content changed in 87; earlier years use 77 SICs. 'P's mark projections.

INDICES OF CHANGE

Year	Companies	Establishments		Employment			Compensation		Production ($ million)			
		Total	with 20 or more employees	Total (000)	Production Workers (000)	Hours (Mil)	Payroll ($ mil)	Wages ($/hr)	Cost of Materials	Value Added by Manufacture	Value of Shipments	Capital Invest.
1982	150	153	157	196	195	176	141	79	123	109	119	172
1983		146	148	156	152	138	114	78	96	85	92	56
1984		138	140	146	152	146	105	74	101	83	92	84
1985		130	131	143	150	136	105	80	111	99	103	86
1986		133	132	140	148	137	106	77	116	89	103	54
1987*	97	97	100	100	100	93	85	88	78	73	74	60
1988		92	102	107	110	100	91	88	87	85	87	59
1989		88	105	110	100	99	98	92	106	99	100	101
1990		88	107	110	105	105	104	91	117	106	106	184
1991		89	101	120	119	118	128	107	127	151	133	246
1992	100	100	100	100	100	100	100	100	100	100	100	100
1993		94	100	103	100	101	104	98	104	112	107	88
1994		94P	101P	113	107	110	123	102	122	115	117	86
1995		94P	101P	110P	106P	111P	123P	106P	131P	123P	126P	136P
1996		94P	100P	111P	106P	113P	128P	108P	137P	128P	131P	141P
1997		94P	100P	112P	106P	114P	132P	110P	142P	133P	136P	145P
1998		95P	100P	112P	107P	116P	136P	112P	147P	138P	141P	150P

Sources: Same as General Statistics. Values reflect change from the base year, 1992. Values above 100 mean greater than 92, values below 100 mean less than 92, and a value of 100 in the 82-91 or 93-98 period means same as 92. * indicates that industry content changed in 87. Data for earlier years are in 77 SIC format.

SELECTED RATIOS

For 1994	Avg. of All Manufact.	Analyzed Industry	Index	For 1994	Avg. of All Manufact.	Analyzed Industry	Index
Employees per Establishment	49	47	96	Value Added per Production Worker	134,084	125,089	93
Payroll per Establishment	1,500,273	1,503,902	100	Cost per Establishment	5,045,178	2,972,774	59
Payroll per Employee	30,620	32,063	105	Cost per Employee	102,970	63,380	62
Production Workers per Establishment	34	27	78	Cost per Production Worker	146,988	111,267	76
Wages per Establishment	853,319	731,277	86	Shipments per Establishment	9,576,895	6,349,279	66
Wages per Production Worker	24,861	27,371	110	Shipments per Employee	195,460	135,367	69
Hours per Production Worker	2,056	2,133	104	Shipments per Production Worker	279,017	237,644	85
Wages per Hour	12.09	12.83	106	Investment per Establishment	321,011	59,372	18
Value Added per Establishment	4,602,255	3,342,070	73	Investment per Employee	6,552	1,266	19
Value Added per Employee	93,930	71,253	76	Investment per Production Worker	9,352	2,222	24

Sources: Same as General Statistics. The 'Average of All Manufacturing' column represents the average of all manufacturing industries reported for the most recent complete year available. The Index shows the relationship between the Average and the Analyzed Industry. For example, 100 means that they are equal; 500 that the Analyzed Industry is five times the average; 50 means that the Analyzed Industry is half the national average. The abbreviation 'na' is used to show that data are 'not available'.

LEADING COMPANIES   Number shown: **50**   Total sales ($ mil): **997**   Total employment (000): **8.3**

Company Name	Address				CEO Name	Phone	Co. Type	Sales ($ mil)	Empl. (000)
JLG Industries Inc	JLG Dr	McConnellsburg	PA	17233	L David Black	717-485-5161	P	176	1.6
Columbus McKinnon Corp	140 Audubon	Amherst	NY	14228	Herbert P Ladds Jr	716-689-5400	R	120	1.2
Genie Industries	PO Box 69	Redmond	WA	98073	B Wilkerson	206-881-1800	R	55	0.5
Ingersoll-Rand Co	PO Box 24046	Seattle	WA	98124	Joe Kiah	206-624-0466	D	51	0.2
Kranco Crane Services Inc	PO Box 40400	Houston	TX	77240		713-466-7541	R	50*	0.4
Lift-Tech International Inc	414 W Broadway	Muskegon H	MI	49444	Larry Eggebrecht	616-733-0821	R	50	0.4
Whiting Corp	157th & Lathrop	Harvey	IL	60426	RE Gibson	708-468-9400	R	45	0.5
Iowa Mold Tooling Company Inc	PO Box 189	Garner	IA	50438	Richard Long	515-923-3711	S	44	0.4
Braden Carco Gearmatic	PO Box 547	Broken Arrow	OK	74012	GE Bowden	918-251-8511	D	37	0.3
Abell-Howe Co	7747 Van Buren St	Forest Park	IL	60130	MR Haeger	708-366-4900	R	35	0.2
Marine Travelift Inc	PO Box 66	Sturgeon Bay	WI	54235	Gerald Lamer	414-743-6202	R	34	0.3
Alliance Machine Co	1049 S Mahoning	Alliance	OH	44601	Christopher Sause	216-823-6120	D	32	0.3
Shepard Niles Inc	250 N Genesee	Montour Falls	NY	14865	J Underwood Jr	607-535-7111	R	30	0.3
Harsh International Inc	PO Box 7	Eaton	CO	80615	Andrew Brown	303-454-2291	R	21*	0.1
Philadelphia Tramrail Co	2207 E Ontario St	Philadelphia	PA	19134	RJ Riethmiller	215-533-5100	R	20	0.1
TC/American Monorail Inc	3839 County Rd	Hamel	MN	55340	James Brandt	612-478-6565	S	18	0.2
Westmont Industries Inc	10805 S Painter Av	Santa Fe Sprgs	CA	90670	Cliff Edey	213-723-3186	R	16	0.1
Ederer Inc	PO Box 24708	Seattle	WA	98124	DE Miller	206-622-4421	R	15	0.1
KONE-Landel Inc	7300 Chippewa Blv	Houston	TX	77086	Seppo Molsa	713-445-2225	S	15	0.1
Crane Mfg & Service Corp	PO Box 410	Cudahy	WI	53110	Richard A Corbett	414-769-8162	R	10*	0.1
North American Industries Inc	80 Holton St	Woburn	MA	01801	PH Levenson	617-721-4446	R	10	<0.1
Stanspec	1234 Washington St	Boston	MA	02118	Tony Polito	617-482-8383	D	10*	0.1
Western Hoist Inc	PO Box 1565	National City	CA	91951	Jon S Halstead	619-474-3361	R	10	<0.1
Bardex Corp	PO Box 1068	Goleta	CA	93117	JL Bartlett Jr	805-964-7747	R	8	<0.1
Anchor Crane and Hoist Service	2020 E Grauwyler	Irving	TX	75061	Laura Mays	214-438-5100	S	7*	<0.1
Smatco Industries Inc	PO Box 4036	Houma	LA	70361	John M Ledet	504-868-3927	R	7*	<0.1
Maasdam Pow'R-Pull Inc	PO Box 6130	Burbank	CA	91510	C C Maasdam	818-845-8769	R	6	<0.1
Progressive Crane Inc	13727 Bennington	Cleveland	OH	44135	WR Goforth	216-251-6126	R	6	<0.1
Nyman Marine Corp	1495 NW Gilman	Issaquah	WA	98027	Everett Johnson	206-391-1101	R	5*	<0.1
Downs Crane and Hoist	8827 Juniper St	Los Angeles	CA	90002	JW Downs Jr	213-589-6061	R	5	<0.1
General Hoist Corp	2772 Norton Av	Lynwood	CA	90262	Gary H Cox	213-636-0791	R	4	<0.1
Netec Inc	291 Eastern Av	Chelsea	MA	02150	William C Hoyt	617-884-4354	R	4*	<0.1
Overhead Crane and Service Inc	35171 Crane Rd	Romulus	MI	48174	S Yochum	313-941-3600	R	4	<0.1
Production Equipment Company	401 Liberty St	Meriden	CT	06450	Rebecca Davis	203-235-5795	R	4	<0.1
Continental Crane and Service	33681 Groesbeck	Fraser	MI	48026	E Dungan	810-294-7900	R	3	<0.1
Duct-O-Wire Co	PO Box 519	Corona	CA	91718	James L Holden	909-735-8220	R	3*	<0.1
Gardner Machinery Corp	PO Box 33818	Charlotte	NC	28233	SW Gardner Jr	704-372-3890	R	3	<0.1
Harding Co	6 Industrial Dr	Exeter	NH	03833	Andre J Demers	603-778-7070	D	3	<0.1
HECO Pacific Manufacturing	1510 Pacific St	Union City	CA	94587	MA Alarab	510-487-1155	R	3	<0.1
Industrial Crane and Equipment	4701 W Iowa St	Chicago	IL	60651	Gerald S Cole	312-378-0100	R	3*	<0.1
JC Renfroe and Sons Inc	1926 Spearing St	Jacksonville	FL	32205	Charles J Renfroe	904-356-4181	R	3*	<0.1
American Hook Lift Inc	1100 Denmill Rd	New Albany	MS	38652	Leslie Arthur Smart	601-534-9300	R	3	<0.1
Bushman Equipment Inc	4200 W Douglas	Milwaukee	WI	53209	Ralph C Deger	414-462-4380	R	2	<0.1
Walter Dankas and Co	1340 Underwood	San Francisco	CA	94124	Ernest Dankas	415-822-4010	R	2*	<0.1
Ratcliff Hoist Company Inc	1655 Old County	San Carlos	CA	94070	BE Ratcliff	415-595-3840	R	2	<0.1
Acme Marine Hoist Inc	690 Montauk Hwy	Bayport	NY	11705	Hugo Klingele	516-472-3030	R	1*	<0.1
Capital Material Handling Inc	435 E 11th St	Tacoma	WA	98421	GE Snyder	206-383-4367	R	1*	<0.1
Sasgen and Derrick Co	3101 W Grand Av	Chicago	IL	60622	MF Sasgen Jr	312-638-0800	R	1*	<0.1
American Crane	PO Box 14244	Houston	TX	77221	John Garland	713-433-5661	D	1	<0.1
Galloway Inc	320 Western Rd	Reno	NV	89506	JJ Galloway	702-329-0200	R	1	<0.1

Source: Ward's Business Directory of U.S. Private and Public Companies, Volumes 1 and 2, 1996. The company type code used is as follows: P - Public, R - Private, S - Subsidiary, D - Division, J - Joint Venture, A - Affiliate, G - Group. Sales are in millions of dollars, employees are in thousands. An asterisk (*) indicates an estimated sales volume. The symbol < stands for 'less than'. Company names and addresses are truncated, in some cases, to fit into the available space.

MATERIALS CONSUMED

Material	Quantity	Delivered Cost ($ million)
Materials, ingredients, containers, and supplies	(X)	352.4
Fluid power pumps, motors, and hydrostatic transmissions (hydraulic and pneumatic)	(X)	4.3
Fluid power cylinders and rotary actuators (hydraulic and pneumatic)	(X)	3.0
Fluid power filters (hydraulic and pneumatic)	(X)	0.1
Fluid power hose or tube fittings and assemblies (hydraulic and pneumatic)	(X)	1.2
Fluid power valves (hydraulic and pneumatic)	(X)	0.8
Fabricated structural metal products (except forgings)	(X)	24.8
Metal bolts, nuts, screws, washers, rivets, and other screw machine products	(X)	4.4
Metal stampings	(X)	2.8
Fabricated metal wire products (including wire rope, cable, springs, etc.)	(X)	10.9
All other fabricated metal products (except forgings)	(X)	23.6
Iron and steel forgings	(X)	8.4
Iron and steel castings (rough and semifinished)	(X)	14.3
Nonferrous (aluminum, copper, etc.) castings (rough and semifinished)	(X)	2.3
Steel bars, bar shapes, and plates	(X)	16.9
Steel sheet and strip, including tin plate	(X)	4.4
Steel structural shapes and sheet piling	(X)	7.0
All other steel shapes and forms	(X)	2.3
Aluminum and aluminum-base alloy shapes and forms	(X)	4.0
Other nonferrous shapes and forms	(X)	0.3
Integral horsepower electric motors and generators (1 hp or more)	(X)	21.3
Ball and roller bearings (mounted or unmounted)	(X)	6.1
Mechanical speed changers, gears, and industrial high-speed drives	(X)	7.2
Paints, varnishes, lacquers, stains, shellacs, japans, enamels, and allied products	(X)	1.3
All other materials and components, parts, containers, and supplies	(X)	59.1
Materials, ingredients, containers, and supplies, nsk	(X)	121.5

Source: 1992 Economic Census. Explanation of symbols used: (D): Withheld to avoid disclosure of competitive data; na: Not available; (S): Withheld because statistical norms were not met; (X): Not applicable; (Z): Less than half the unit shown; nec: Not elsewhere classified; nsk: Not specified by kind; - : zero; * : 10-19 percent estimated; ** : 20-29 percent estimated.

PRODUCT SHARE DETAILS

Product or Product Class	% Share	Product or Product Class	% Share
Hoists, cranes, and monorails	100.00	(except construction power cranes)	6.56
Hoists	46.94	Double top running bridge type overhead traveling cranes (except construction power cranes)	23.08
Chain hand hoists	3.97	Under running bridge type overhead traveling cranes (except construction power cranes)	9.01
Ratchet lever hand hoists	5.21		
Wire rope puller hand hoists	2.68	Gantry type overhead traveling cranes (except construction power cranes)	5.30
Electric (roller and link) chain hoists	15.83		
Air or other nonelectric chain hoists, except hand	5.55	Stacker/storage type overhead traveling cranes (except construction power cranes)	3.82
Electric wire rope hoists (excluding hand, mine shaft, and slope wire rope hoists)	17.96	Parts and attachments for overhead traveling cranes (sold separately)	15.89
Air or other nonelectric wire rope hoists (excluding hand, mine shaft, and slope wire rope hoists)	4.47	Monorail systems (manual and powered)	9.38
Other hoists	12.15	Parts and attachments for monorail systems (sold separately)	1.82
Parts and attachments for hoists (sold separately)	21.11		
Hoists, nsk	11.04	Overhead traveling cranes and monorail systems, nsk	25.14
Overhead traveling cranes and monorail systems	46.84	Hoists, cranes, and monorails, nsk	6.22
Single top running bridge type overhead traveling cranes			

Source: 1992 Economic Census. The values shown are percent of total shipments in an industry. Values of indented subcategories are summed in the main heading. The symbol (D) appears when data are withheld to prevent disclosure of competitive information. The abbreviation nsk stands for 'not specified by kind' and nec for 'not elsewhere classified'.

INPUTS AND OUTPUTS FOR HOISTS, CRANES, & MONORAILS

Economic Sector or Industry Providing Inputs	%	Sector	Economic Sector or Industry Buying Outputs	%	Sector
Blast furnaces & steel mills	14.8	Manufg.	Gross private fixed investment	38.5	Cap Inv
Imports	13.9	Foreign	Industrial buildings	29.2	Constr.
Hoists, cranes, & monorails	10.0	Manufg.	Exports	10.5	Foreign
Wholesale trade	8.7	Trade	Hoists, cranes, & monorails	5.7	Manufg.
Fabricated structural metal	5.3	Manufg.	Maintenance of nonfarm buildings nec	3.1	Constr.
Miscellaneous plastics products	3.5	Manufg.	Electric utility facility construction	1.9	Constr.
Motors & generators	3.3	Manufg.	Maintenance of petroleum & natural gas wells	1.2	Constr.
Iron & steel foundries	2.6	Manufg.	Dimension, crushed & broken stone	1.1	Mining
Fabricated plate work (boiler shops)	2.3	Manufg.	Construction of stores & restaurants	1.0	Constr.
Industrial controls	2.3	Manufg.	Chemical & fertilizer mineral	0.8	Mining
Power transmission equipment	2.3	Manufg.	Sand & gravel	0.8	Mining
Sheet metal work	2.3	Manufg.	Copper ore	0.6	Mining
Advertising	1.9	Services	Warehouses	0.6	Constr.
Maintenance of nonfarm buildings nec	1.6	Constr.	Water supply facility construction	0.6	Constr.
Ball & roller bearings	1.6	Manufg.	Clay, ceramic, & refractory minerals	0.5	Mining
Machinery, except electrical, nec	1.6	Manufg.	Iron & ferroalloy ores	0.4	Mining

Continued on next page.

INPUTS AND OUTPUTS FOR HOISTS, CRANES, & MONORAILS - Continued

Economic Sector or Industry Providing Inputs	%	Sector	Economic Sector or Industry Buying Outputs	%	Sector
Aluminum castings	1.5	Manufg.	Sewer system facility construction	0.4	Constr.
Electric services (utilities)	1.4	Util.	Nonferrous metal ores, except copper	0.3	Mining
Iron & steel forgings	1.3	Manufg.	Maintenance of electric utility facilities	0.3	Constr.
Communications, except radio & TV	1.1	Util.	Maintenance of sewer facilities	0.3	Constr.
Motor freight transportation & warehousing	0.9	Util.	Coal	0.2	Mining
Miscellaneous fabricated wire products	0.8	Manufg.	Petroleum, gas, & solid mineral exploration	0.2	Constr.
Gas production & distribution (utilities)	0.8	Util.	Motor vehicle parts & accessories	0.2	Manufg.
Pumps & compressors	0.7	Manufg.	Motor vehicles & car bodies	0.2	Manufg.
Eating & drinking places	0.7	Trade	Miscellaneous repair shops	0.2	Services
Metal stampings, nec	0.6	Manufg.	Federal Government purchases, national defense	0.2	Fed Govt
Nonferrous wire drawing & insulating	0.6	Manufg.	Nonmetallic mineral services	0.1	Mining
Screw machine and related products	0.6	Manufg.	Maintenance of water supply facilities	0.1	Constr.
Banking	0.6	Fin/R.E.	Nonbuilding facilities nec	0.1	Constr.
Real estate	0.6	Fin/R.E.			
Engineering, architectural, & surveying services	0.6	Services			
Equipment rental & leasing services	0.6	Services			
Gaskets, packing & sealing devices	0.5	Manufg.			
Computer & data processing services	0.5	Services			
Asbestos products	0.3	Manufg.			
Fabricated rubber products, nec	0.3	Manufg.			
Machine tools, metal cutting types	0.3	Manufg.			
Metal doors, sash, & trim	0.3	Manufg.			
Metal heat treating	0.3	Manufg.			
Paints & allied products	0.3	Manufg.			
Pipe, valves, & pipe fittings	0.3	Manufg.			
Special dies & tools & machine tool accessories	0.3	Manufg.			
Tires & inner tubes	0.3	Manufg.			
Air transportation	0.3	Util.			
Legal services	0.3	Services			
U.S. Postal Service	0.3	Gov't			
Abrasive products	0.2	Manufg.			
Aluminum rolling & drawing	0.2	Manufg.			
Petroleum refining	0.2	Manufg.			
Primary metal products, nec	0.2	Manufg.			
Railroads & related services	0.2	Util.			
Accounting, auditing & bookkeeping	0.2	Services			
Management & consulting services & labs	0.2	Services			
Noncomparable imports	0.2	Foreign			
Brass, bronze, & copper castings	0.1	Manufg.			
Industrial gases	0.1	Manufg.			
Internal combustion engines, nec	0.1	Manufg.			
Lubricating oils & greases	0.1	Manufg.			
Manifold business forms	0.1	Manufg.			
Insurance carriers	0.1	Fin/R.E.			
Hotels & lodging places	0.1	Services			

Source: Benchmark Input-Output Accounts for the U.S. Economy, 1982, U.S. Department of Commerce, Washington, D.C., July 1991. Data, as reported in the source, are organized by the 1977 SIC structure in use in 1982 but have been matched, as closely as is possible, to the 1987 SIC structure used in this book.

OCCUPATIONS EMPLOYED BY SIC 353 - CONSTRUCTION AND RELATED MACHINERY

Occupation	% of Total 1994	Change to 2005	Occupation	% of Total 1994	Change to 2005
Welders & cutters	7.4	-9.1	NC machine tool operators, metal & plastic	1.6	-9.1
Assemblers, fabricators, & hand workers nec	6.3	-9.1	Industrial machinery mechanics	1.4	-0.0
Machinists	4.6	-9.1	Traffic, shipping, & receiving clerks	1.3	-12.5
Sales & related workers nec	3.9	-9.1	Freight, stock, & material movers, hand	1.3	-27.3
Blue collar worker supervisors	3.6	-13.9	Bookkeeping, accounting, & auditing clerks	1.3	-31.8
Welding machine setters, operators	2.9	-18.2	Production, planning, & expediting clerks	1.2	-9.1
Machine builders	2.3	-27.3	Coating, painting, & spraying machine workers	1.2	-9.1
General managers & top executives	2.3	-13.8	Sheet metal workers & duct installers	1.2	-9.1
Mechanical engineers	2.3	0.0	Engineering, mathematical, & science managers	1.2	3.2
Inspectors, testers, & graders, precision	2.1	-9.1	Machine tool cutting operators, metal & plastic	1.1	-24.3
Secretaries, ex legal & medical	2.0	-17.3	Purchasing agents, ex trade & farm products	1.1	9.1
Drafters	2.0	-29.2	Engineering technicians & technologists nec	1.1	-9.1
Machine assemblers	1.8	63.6	General office clerks	1.0	-22.5
Professional workers nec	1.8	9.1	Maintenance repairers, general utility	1.0	-18.2
Machine tool cutting & forming etc. nec	1.8	-9.1	Stock clerks	1.0	-26.1
Lathe & turning machine tool operators	1.6	-18.2	Drilling & boring machine tool workers	1.0	-54.6
Helpers, laborers, & material movers nec	1.6	-9.1	Industrial production managers	1.0	-9.1

Source: Industry-Occupation Matrix, Bureau of Labor Statistics. These data relate to one or more 3-digit SIC industry groups rather than to a single 4-digit SIC. The change reported for each occupation to the year 2005 is a percent of growth or decline as estimated by the Bureau of Labor Statistics. The abbreviation nec stands for 'not elsewhere classified'.

LOCATION BY STATE AND REGIONAL CONCENTRATION

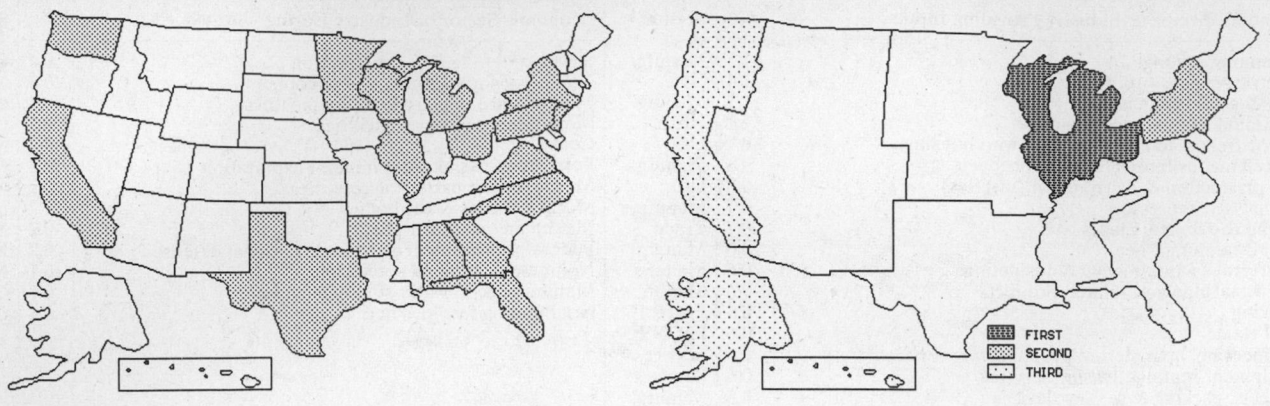

FIRST
SECOND
THIRD

INDUSTRY DATA BY STATE

| State | Establish-ments | Shipments | | | Employment | | | | Cost as % of Shipments | Investment per Employee ($) |
		Total ($ mil)	% of U.S.	Per Establ.	Total Number	% of U.S.	Per Establ.	Wages ($/hour)		
Texas	15	122.0	13.4	8.1	600	8.6	40	13.57	20.9	-
Ohio	17	121.1	13.3	7.1	1,000	14.3	59	13.00	60.0	2,200
Wisconsin	9	94.9	10.4	10.5	700	10.0	78	17.25	38.4	-
Michigan	17	69.7	7.6	4.1	600	8.6	35	11.71	43.9	1,667
Illinois	9	59.8	6.6	6.6	500	7.1	56	11.33	46.0	-
Pennsylvania	15	58.0	6.4	3.9	500	7.1	33	13.14	45.3	1,800
Minnesota	6	43.6	4.8	7.3	400	5.7	67	10.80	56.4	1,000
New York	7	42.6	4.7	6.1	400	5.7	57	12.00	41.5	750
California	19	36.3	4.0	1.9	300	4.3	16	11.67	50.4	1,333
Alabama	5	14.9	1.6	3.0	100	1.4	20	9.00	54.4	-
Georgia	5	12.1	1.3	2.4	100	1.4	20	11.50	46.3	-
Florida	9	(D)	-	-	175 *	2.5	19	-	-	571
Washington	6	(D)	-	-	175 *	2.5	29	-	-	-
Virginia	5	(D)	-	-	375 *	5.4	75	-	-	-
New Jersey	4	(D)	-	-	375 *	5.4	94	-	-	-
North Carolina	4	(D)	-	-	175 *	2.5	44	-	-	1,714
Arkansas	1	(D)	-	-	175 *	2.5	175	-	-	-

Source: 1992 *Economic Census.* The states are in descending order of shipments or establishments (if shipment data are missing for the majority). The symbol (D) appears when data are withheld to prevent disclosure of competitive information. States marked with (D) are sorted by number of establishments. A dash (-) indicates that the data element cannot be calculated; * indicates the midpoint of a range.

3537 - INDUSTRIAL TRUCKS & TRACTORS

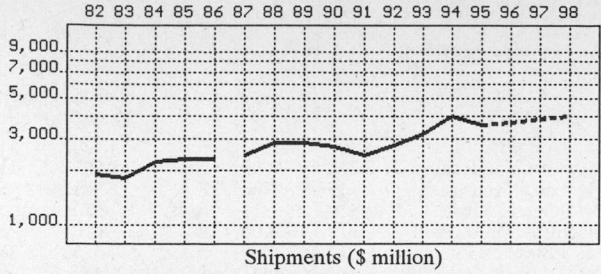

Shipments ($ million)

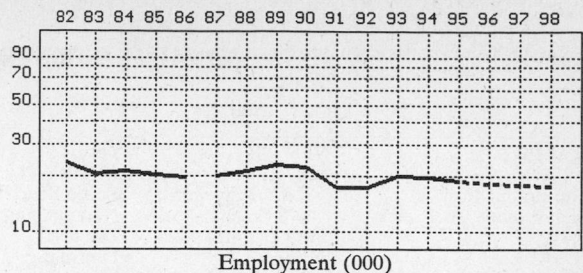

Employment (000)

GENERAL STATISTICS

Year	Com-panies	Establishments		Employment			Compensation		Production ($ million)			
		Total	with 20 or more employees	Total (000)	Production Workers (000)	Hours (Mil)	Payroll ($ mil)	Wages ($/hr)	Cost of Materials	Value Added by Manufacture	Value of Shipments	Capital Invest.
1982	463	489	175	24.0	14.3	26.2	494.7	9.55	1,109.7	722.4	1,922.2	80.3
1983		478	178	21.0	13.3	24.4	459.8	9.98	1,104.8	672.0	1,808.1	55.2
1984		467	181	21.6	14.7	27.1	492.9	9.99	1,327.8	917.5	2,255.8	36.3
1985		457	185	20.5	14.0	26.8	459.8	9.49	1,311.1	926.9	2,300.0	45.7
1986		445	180	19.9	13.3	24.2	439.5	10.17	1,430.6	867.0	2,330.2	30.8
1987*	448	467	173	20.1	13.0	25.0	476.1	10.01	1,479.5	953.7	2,440.2	37.5
1988		470	187	21.5	14.3	27.5	515.0	10.39	1,819.6	1,037.6	2,826.6	55.2
1989		451	193	23.2	13.6	27.1	528.1	10.94	1,645.9	1,194.3	2,841.3	53.0
1990		443	183	22.4	13.2	26.4	521.4	11.00	1,660.0	1,036.7	2,727.5	49.0
1991		477	187	17.3	11.4	22.5	471.5	11.73	1,458.0	947.9	2,405.6	46.6
1992	426	450	159	17.5	11.5	23.5	498.5	11.73	1,700.5	1,046.9	2,753.7	57.6
1993		454	167	20.3	14.1	28.1	592.2	12.29	1,983.6	1,204.7	3,199.7	43.3
1994		451P	167P	19.8	13.4	27.6	627.9	12.39	2,671.7	1,382.8	4,024.8	71.4
1995		449P	164P	18.6P	12.7P	26.3P	599.0P	12.88P	2,366.7P	1,224.9P	3,565.3P	61.8P
1996		447P	161P	18.3P	12.7P	26.4P	614.6P	13.23P	2,464.4P	1,275.5P	3,712.6P	64.1P
1997		446P	158P	17.9P	12.6P	26.5P	630.2P	13.57P	2,562.2P	1,326.1P	3,859.9P	66.3P
1998		444P	156P	17.6P	12.5P	26.5P	645.8P	13.92P	2,660.0P	1,376.7P	4,007.2P	68.6P

Sources: 1982, 1987, 1992 *Economic Census*; *Annual Survey of Manufactures*, 83-86, 88-91, 93-94. Establishment counts are from *County Business Patterns* for non-Census years; establishment counts for 83-84 are extrapolations. * indicates that industry content changed in 87; earlier years use 77 SICs. 'P's mark projections.

INDICES OF CHANGE

Year	Com-panies	Establishments		Employment			Compensation		Production ($ million)			
		Total	with 20 or more employees	Total (000)	Production Workers (000)	Hours (Mil)	Payroll ($ mil)	Wages ($/hr)	Cost of Materials	Value Added by Manufacture	Value of Shipments	Capital Invest.
1982	109	109	110	137	124	111	99	81	65	69	70	139
1983		106	112	120	116	104	92	85	65	64	66	96
1984		104	114	123	128	115	99	85	78	88	82	63
1985		102	116	117	122	114	92	81	77	89	84	79
1986		99	113	114	116	103	88	87	84	83	85	53
1987*	105	104	109	115	113	106	96	85	87	91	89	65
1988		104	118	123	124	117	103	89	107	99	103	96
1989		100	121	133	118	115	106	93	97	114	103	92
1990		98	115	128	115	112	105	94	98	99	99	85
1991		106	118	99	99	96	95	100	86	91	87	81
1992	100	100	100	100	100	100	100	100	100	100	100	100
1993		101	105	116	123	120	119	105	117	115	116	75
1994		100P	105P	113	117	117	126	106	157	132	146	124
1995		100P	103P	107P	111P	112P	120P	110P	139P	117P	129P	107P
1996		99P	101P	104P	110P	112P	123P	113P	145P	122P	135P	111P
1997		99P	100P	102P	109P	113P	126P	116P	151P	127P	140P	115P
1998		99P	98P	100P	109P	113P	130P	119P	156P	132P	146P	119P

Sources: Same as General Statistics. Values reflect change from the base year, 1992. Values above 100 mean greater than 92, values below 100 mean less than 92, and a value of 100 in the 82-91 or 93-98 period means same as 92. * indicates that industry content changed in 87. Data for earlier years are in 77 SIC format.

SELECTED RATIOS

For 1994	Avg. of All Manufact.	Analyzed Industry	Index	For 1994	Avg. of All Manufact.	Analyzed Industry	Index
Employees per Establishment	49	44	90	Value Added per Production Worker	134,084	103,194	77
Payroll per Establishment	1,500,273	1,391,358	93	Cost per Establishment	5,045,178	5,920,196	117
Payroll per Employee	30,620	31,712	104	Cost per Employee	102,970	134,934	131
Production Workers per Establishment	34	30	87	Cost per Production Worker	146,988	199,381	136
Wages per Establishment	853,319	757,755	89	Shipments per Establishment	9,576,895	8,918,519	93
Wages per Production Worker	24,861	25,520	103	Shipments per Employee	195,460	203,273	104
Hours per Production Worker	2,056	2,060	100	Shipments per Production Worker	279,017	300,358	108
Wages per Hour	12.09	12.39	102	Investment per Establishment	321,011	158,215	49
Value Added per Establishment	4,602,255	3,064,134	67	Investment per Employee	6,552	3,606	55
Value Added per Employee	93,930	69,838	74	Investment per Production Worker	9,352	5,328	57

Sources: Same as General Statistics. The 'Average of All Manufacturing' column represents the average of all manufacturing industries reported for the most recent complete year available. The Index shows the relationship between the Average and the Analyzed Industry. For example, 100 means that they are equal; 500 that the Analyzed Industry is five times the average; 50 means that the Analyzed Industry is half the national average. The abbreviation 'na' is used to show that data are 'not available'.

LEADING COMPANIES Number shown: 75   Total sales ($ mil): 21,614   Total employment (000): 96.6

Company Name	Address				CEO Name	Phone	Co. Type	Sales ($ mil)	Empl. (000)
Caterpillar Inc	100 NE Adams St	Peoria	IL	61629	Donald V Fites	309-675-1000	P	14,328	54.0
NACCO Industries Inc	5875 Landerbrook	Mayfield H	OH	44124	Alfred M Rankin Jr	216-449-9600	P	1,865	11.0
Clark Equipment Co	100 N Michigan St	South Bend	IN	46634	Leo J McKernan	219-239-0100	P	875	5.9
Crown Equipment Corp	40 S Washington St	New Bremen	OH	45869	James F Dicke II	419-629-2311	R	610*	3.5
Hyster Co	PO Box 847	Danville	IL	61834	Steve Finney	217-443-7000	D	610*	3.5
Hyster Co	PO Box 847	Danville	IL	61834	J Stephen Finney	217-443-7000	S	525	6.0
Mitsubishi Caterpillar Forklift	2011 W S Houston	Houston	TX	77043	T Okuna	713-365-1441	J	477	0.7
Clark Material Handling Co	333 W Vine St	Lexington	KY	40507	Martin M Dorio	606-288-1200	S	450	0.6
Nissan Forklift Corp	240 N Prospect St	Marengo	IL	60152	Kazuhide Yamada	815-568-0061	S	225	0.5
Cascade Corp	2020 SW 4th Av	Portland	OR	97201	Joseph J Barclay	503-227-0024	P	183	0.9
Raymond Corp	PO Box 130	Greene	NY	13778	Ross K Colquhoun	607-656-2311	P	172	1.2
TransTechnology Corp	700 Liberty Av	Union	NJ	07083	Michael J Berthelot	908-964-5666	P	122	1.1
Taylor Machine Works Inc	650 N Church St	Louisville	MS	39339	WA Taylor III	601-773-3421	R	110*	0.7
Eaton-Kenway Inc	515 E 100 S	Salt Lake City	UT	84102	Richard M Mooney	801-530-4000	S	88	0.4
Mobile Tool International Inc	PO Box 666	Westminster	CO	80080	Van J Walbridge	303-427-3700	R	70	0.5
Gradall Co	406 SW Mill Av	New Philad	OH	44663	Barry Phillips	216-339-2211	S	60	0.4
CIM Industrial Machinery Inc	7950 Blankenship	Houston	TX	77055	Jere L French	713-681-8888	S	50	<0.1
RA Industries Inc	PO Box 247	Lansdale	PA	19446	Allen Apter	215-699-0384	R	36*	0.2
Inter-American Transport Equip	3690 NW 62nd St	Miami	FL	33147	DR Suarez	305-633-0351	R	35*	0.2
Capacity of Texas Inc	401 Capacity Dr	Longview	TX	75604		903-759-0610	S	33	0.2
Breeze-Eastern	700 Liberty Av	Union	NJ	07083	Robert White	908-686-4000	D	30	0.2
Long Reach Holdings Inc	12300 Amelia Dr	Houston	TX	77045	DM Buchanan	713-433-9861	R	30*	0.3
Polar Tank Trailer Inc	12810 County Rd 17	Holdingford	MN	56340	James Jungels	612-746-2255	S	30	0.3
Accessory Control & Equip	805 Bloomfield Av	Windsor	CT	06095	Robert Barrack	203-688-4986	S	25	0.2
Material Handling Services Inc	315 E Fullerton Av	Carol Stream	IL	60188	Gerald Risch	708-665-7200	R	25*	0.1
Taylor-Dunn Manufacturing Co	2114 W Ball Rd	Anaheim	CA	92804	John Revere	714-956-4040	R	25*	0.2
Long Reach Manufacturing Co	PO Box 450069	Houston	TX	77245	DM Buchanan	713-433-9861	S	24*	0.3
Bright Cooperative Inc	PO Box 630730	Nacogdoches	TX	75961	Charles Bright	409-564-8378	R	22*	0.1
Garsite TSR Inc	539 S 10th St	Kansas City	KS	66105	Phil Hodes	913-342-5600	R	20	0.1
Magnum Resources Inc	2040 W Main	Rapid City	SD	57702		605-341-5660	R	20*	0.1
WB McGuire Company Inc	1 Hudson Av	Hudson	NY	12534	Tom DiSieno	518-828-7652	S	20	0.2
Tug Manufacturing Corp	PO Box 1447	Kennesaw	GA	30144	Don L Chapman	404-422-7230	R	19*	0.1
Kalmar AC Inc	777 Manor Park Dr	Columbus	OH	43228	Bengt Ljung	614-798-3600	R	18*	0.2
Perin Company Inc	21053 Alexander Ct	Hayward	CA	94541	RA Pattillos	510-887-0500	R	18	<0.1
Elwell-Parker Electric Co	4205 St Clair Av	Cleveland	OH	44103	LJ Wareham	216-881-6200	R	16	<0.1
Drexel Industries Inc	331 Maple Av	Horsham	PA	19044	Richard O McKerr	215-672-2200	R	15	0.1
New Castle Trailer Corp	PO Box 7257	New Castle	PA	16107	Robert Marinelli	412-658-5544	R	15	<0.1
Pettibone Michigan	PO Box 368	Baraga	MI	49908	Chris Yunkun	906-353-6611	D	15	<0.1
Harlo Products Corp	4210 Ferry St	Grandville	MI	49418	Dick Crooks	616-538-0550	R	14	0.1
Warren Manufacturing Inc	1008 37th St N	Birmingham	AL	35234	Russell Warren	205-591-3002	R	13*	<0.1
PathFinder Operations	1610 Fry Av	Canon City	CO	81212	Kevin C Rorke	719-269-1112	D	12	<0.1
Burtman Iron Works Inc	31 Industrial Dr	Boston	MA	02137	Charles Goodman	617-364-1200	R	12	0.1
Common Equipment Co	PO Box 988	Peoria	IL	61603	SJ Statler	309-672-9300	R	12	<0.1
Harlan Corp	27 Stanley Rd	Kansas City	KS	66115	James H Kaplan	913-342-5650	R	12*	0.1
K-D Manitou Inc	PO Box 154009	Waco	TX	76715	Serge Bosche	817-799-0232	S	12	0.1
Wesco Manufacturing Co	PO Box 47	Lansdale	PA	19446	Alan Apter	215-699-7031	D	12	0.1
Advance Lifts Inc	701 Kirk Rd	St Charles	IL	60174	David Clarke	708-584-9881	R	10	<0.1
Bruno Independent Living Aids	1780 Executive Dr	Oconomowoc	WI	53066	Mike Bruno	414-567-4990	R	10	0.1
Gannon Manufacturing	14821 Artesia Blv	La Mirada	CA	90638	Joe Kitchen	714-562-5353	D	10*	0.1
Hydra-Mac Inc	1110 Pennington	Thief River Fls	MN	56701	John Luoma	218-681-7130	R	10	<0.1
Schaeff Inc	PO Box 9700	Sioux City	IA	51102	Karl Schaeff	712-944-5111	R	10	0.1
CMF Inc	PO Box 89	Clare	MI	48617	Jeff Schwager	517-386-3231	R	9*	<0.1
Lift-A-Loft Corp	PO Box 2645	Muncie	IN	47307	RE Dennis	317-288-3691	R	9	0.1
Mohawk Resources Ltd	PO Box 110	Amsterdam	NY	12010	Rick Wells	518-842-1431	R	9	<0.1
Northwestern Motor Co	PO Box 265	Eau Claire	WI	54702	John Steingart	715-835-3151	R	9	<0.1
Wiggins Lift Company Inc	2571 Cortez St	Oxnard	CA	93030	Michael M Wiggins	805-485-7821	R	9	0.1
Materials Transportation Co	PO Box 1358	Temple	TX	76503	WA Jones III	817-778-1894	R	9	0.1
New Grand International Corp	PO Box 1867	Burleson	TX	76097	Ken Norris	817-478-1161	R	8	<0.1
United Tractor Co	116 N 15th St	Chesterton	IN	46304	Frank DeFino	219-926-1186	S	8	<0.1
Valley Craft Products Inc	2001 S Hwy 61	Lake City	MN	55041	Roger Hollman	612-345-3386	S	8	0.1
Dutro Co	PO Box 8447	Emeryville	CA	94662	William A Dutro	510-652-9130	R	8	<0.1
Darling Special Products Inc	PO Box 1000	Caruthersville	MO	63830	Jack Wilson	314-333-2070	S	7	<0.1
Master Craft Indust Equip Corp	Rte 2	Tifton	GA	31794	Jack Haswell	912-386-0610	R	7	<0.1
Skyhook Corp	1640 S Main St	Ottawa	KS	66067	Harold Sader	913-242-1584	R	7	<0.1
Autoquip Corp	PO Box 1058	Guthrie	OK	73044	L William Adel	405-282-5200	S	6	0.1
Power Machinery Center	PO Box 392	Oxnard	CA	93032	Richard Power	805-485-0577	R	6	<0.1
Baraga Products Inc	PO Box 248	Baraga	MI	49908	James Mayo	906-353-6675	R	5	<0.1
Roll-A-Way Conveyor Co	2335 Delaney Rd	Gurnee	IL	60031	W Chramosta	708-336-5033	R	5	<0.1
Air Technical Industries	7501 Clover Av	Mentor	OH	44060	P Novak	216-951-5191	R	5	0.1
JCM Industries Inc	125 E Selandia Ln	Carson	CA	90746	John D Krummell	310-217-0600	R	5	<0.1
Campbell International Inc	120 Kent Av	Wauconda	IL	60084	J M Campbell Jr	708-526-7300	R	4*	<0.1
Gregory Group Inc	15 Daniel Rd	Fairfield	NJ	07004	G Russell	201-808-8399	R	4	<0.1
Excel Inc	PO Box 459	Lincolnton	NC	28092	Charles W Eurey	704-735-6535	R	3	<0.1
Lubick Manufacturing Company	3101 S State St	Lockport	IL	60441	Ronald Lubick	815-722-6218	R	3	<0.1
Cavanaugh Machine Works Inc	220 E B St	Wilmington	CA	90744	JM Wells	310-834-5219	R	3	<0.1

Source: *Ward's Business Directory of U.S. Private and Public Companies*, Volumes 1 and 2, 1996. The company type code used is as follows: P - Public, R - Private, S - Subsidiary, D - Division, J - Joint Venture, A - Affiliate, G - Group. Sales are in millions of dollars, employees are in thousands. An asterisk (*) indicates an estimated sales volume. The symbol < stands for 'less than'. Company names and addresses are truncated, in some cases, to fit into the available space.

MATERIALS CONSUMED

Material	Quantity	Delivered Cost ($ million)
Materials, ingredients, containers, and supplies	(X)	1,524.7
Fluid power pumps, motors, and hydrostatic transmissions (hydraulic and pneumatic)	(X)	47.1
Fluid power cylinders and rotary actuators (hydraulic and pneumatic)	(X)	25.0
Fluid power filters (hydraulic and pneumatic)	(X)	1.7
Fluid power hose or tube fittings and assemblies (hydraulic and pneumatic)	(X)	18.8
Fluid power valves (hydraulic and pneumatic)	(X)	28.1
Fabricated metal products, except forgings	(X)	92.7
Iron and steel forgings	(X)	19.9
Nonferrous forgings	(X)	0.2
Iron and steel castings (rough and semifinished)	(X)	53.8
Nonferrous (aluminum, copper, etc.) castings (rough and semifinished)	(X)	1.4
Steel bars, bar shapes, and plates	(X)	133.3
Steel sheet and strip, including tin plate	(X)	31.3
Steel structural shapes and sheet piling	(X)	24.8
All other steel shapes and forms	(X)	30.4
Aluminum and aluminum-base alloy shapes and forms	(X)	7.5
Other nonferrous shapes and forms	(X)	3.3
Gasoline and other carburetor engines	(X)	56.9
Fractional horsepower electric motors (less than 1 hp) (excluding timing motors)	(X)	10.1
Integral horsepower electric motors and generators (1 hp or more)	(X)	29.2
Electrical transmission, distribution, and control equipment	(X)	39.0
Storage batteries	(X)	12.5
Ball bearings (mounted or unmounted)	(X)	21.0
Roller bearings (mounted or unmounted)	(X)	9.5
Mechanical speed changers, gears, and industrial high-speed drives	(X)	19.6
Pneumatic tires and inner tubes	(X)	19.8
Rubber and plastics hose and belting	(X)	11.2
All other materials and components, parts, containers, and supplies	(X)	521.6
Materials, ingredients, containers, and supplies, nsk	(X)	255.1

Source: 1992 *Economic Census*. Explanation of symbols used: (D): Withheld to avoid disclosure of competitive data; na: Not available; (S): Withheld because statistical norms were not met; (X): Not applicable; (Z): Less than half the unit shown; nec: Not elsewhere classified; nsk: Not specified by kind; - : zero; * : 10-19 percent estimated; ** : 20-29 percent estimated.

PRODUCT SHARE DETAILS

Product or Product Class	% Share	Product or Product Class	% Share
Industrial trucks and tractors	100.00	combustion engine or other nonelectric powered.	1.94
Industrial trucks, tractors, mobile straddle carriers and cranes, and automatic stacking machines	67.14	Work trucks and tractors not fitted with lifting or handling equipment, not self-propelled	3.17
Fork lift work trucks, operating riding, self-propelled, electric motor powered	27.21	Mobile straddle carriers and cranes	0.28
		Portable elevators/stackers (except farm type)	0.59
Other work trucks fitted with lifting or handling equipment, operating riding, self-propelled, electric motor powered	0.29	Palletizers and depalletizers (pallet loaders and unloaders)	3.22
Nonriding fork lift and other work trucks fitted with lifting or handling equipment, self-propelled, electric motor powered	10.42	Scissors type hydraulic lift tables (electrohydraulic lift platforms)	4.37
		Other hydraulic lift tables (electrohydraulic lift platforms)	1.52
Fork lift work trucks, operating riding, self-propelled, gasoline motor powered	5.00	Metal pallets and skids (excluding wood and metal combination)	0.71
Fork lift work trucks, operating riding, self-propelled, diesel motor powered	6.14	Dock boards (industrial loading ramps, hinged loading ramps)	2.03
		Automatic stacking machines	2.29
Fork lift work trucks, operating riding, self-propelled, LPG (liquid petroleum gas) motor powered	16.79	All other industrial trucks and tractors	4.98
Other work trucks fitted with lifting or handling equipment, operator riding, self-propelled, internal combustion engine or other nonelectric powered	2.71	Industrial trucks, tractors, mobile straddle carriers and cranes, and automatic stacking machines, nsk	2.67
Nonself-propelled fork lift and other work trucks fitted with lifting or handling equipment (hand lift, etc.)	1.83	Parts and attachments for industrial trucks and tractors (sold separately)	21.70
		Cabs for industrial trucks and tractors (sold separately)	1.58
Work trucks and tractors not fitted with lifting or handling equipment, operator riding, self-propelled, electric motor powered	1.87	All other parts and attachments for industrial trucks and tractors (sold separately)	94.80
Work trucks and tractors not fitted with lifting or handling equipment, operator riding, self-propelled, internal		Parts and attachments for industrial trucks and tractors (sold separately), nsk	3.61
		Industrial trucks and tractors, nsk	11.17

Source: 1992 *Economic Census*. The values shown are percent of total shipments in an industry. Values of indented subcategories are summed in the main heading. The symbol (D) appears when data are withheld to prevent disclosure of competitive information. The abbreviation nsk stands for 'not specified by kind' and nec for 'not elsewhere classified'.

INPUTS AND OUTPUTS FOR INDUSTRIAL TRUCKS & TRACTORS

Economic Sector or Industry Providing Inputs	%	Sector	Economic Sector or Industry Buying Outputs	%	Sector
Wholesale trade	13.4	Trade	Gross private fixed investment	67.2	Cap Inv
Imports	11.8	Foreign	Exports	10.0	Foreign
Blast furnaces & steel mills	7.2	Manufg.	Wholesale trade	5.6	Trade
Blowers & fans	6.0	Manufg.	Federal Government purchases, national defense	5.1	Fed Govt
Industrial trucks & tractors	6.0	Manufg.	Industrial trucks & tractors	3.9	Manufg.
Iron & steel foundries	4.3	Manufg.	Miscellaneous repair shops	3.8	Services
Pumps & compressors	4.2	Manufg.	Logging camps & logging contractors	1.1	Manufg.
Internal combustion engines, nec	3.6	Manufg.	Blast furnaces & steel mills	0.5	Manufg.
Machinery, except electrical, nec	3.2	Manufg.	Conveyors & conveying equipment	0.5	Manufg.
Power transmission equipment	2.6	Manufg.	Aluminum rolling & drawing	0.4	Manufg.
Motors & generators	2.5	Manufg.	Truck trailers	0.3	Manufg.
Hardware, nec	2.2	Manufg.	Electric services (utilities)	0.3	Util.
Public building furniture	2.2	Manufg.	Chemical & fertilizer mineral	0.1	Mining
Switchgear & switchboard apparatus	1.9	Manufg.	Dimension, crushed & broken stone	0.1	Mining
Maintenance of nonfarm buildings nec	1.7	Constr.	Motor vehicles & car bodies	0.1	Manufg.
Industrial controls	1.5	Manufg.	Nonferrous wire drawing & insulating	0.1	Manufg.
Pipe, valves, & pipe fittings	1.5	Manufg.	Primary aluminum	0.1	Manufg.
Electric services (utilities)	1.3	Util.			
Communications, except radio & TV	1.2	Util.			
Transformers	1.1	Manufg.			
Screw machine and related products	1.0	Manufg.			
Tires & inner tubes	1.0	Manufg.			
Motor freight transportation & warehousing	0.9	Util.			
Ball & roller bearings	0.8	Manufg.			
Iron & steel forgings	0.8	Manufg.			
Storage batteries	0.7	Manufg.			
Eating & drinking places	0.7	Trade			
Equipment rental & leasing services	0.7	Services			
Fabricated structural metal	0.6	Manufg.			
Special dies & tools & machine tool accessories	0.6	Manufg.			
Banking	0.6	Fin/R.E.			
Computer & data processing services	0.6	Services			
Aluminum rolling & drawing	0.5	Manufg.			
Veneer & plywood	0.5	Manufg.			
Gas production & distribution (utilities)	0.5	Util.			
Carburetors, pistons, rings, & valves	0.4	Manufg.			
Gaskets, packing & sealing devices	0.4	Manufg.			
Paints & allied products	0.4	Manufg.			
Petroleum refining	0.4	Manufg.			
Real estate	0.4	Fin/R.E.			
Engineering, architectural, & surveying services	0.4	Services			
Metal heat treating	0.3	Manufg.			
Rubber & plastics hose & belting	0.3	Manufg.			
Air transportation	0.3	Util.			
Legal services	0.3	Services			
Management & consulting services & labs	0.3	Services			
U.S. Postal Service	0.3	Gov't			
Noncomparable imports	0.3	Foreign			
Abrasive products	0.2	Manufg.			
Aluminum castings	0.2	Manufg.			
General industrial machinery, nec	0.2	Manufg.			
Industrial gases	0.2	Manufg.			
Machine tools, metal cutting types	0.2	Manufg.			
Metal stampings, nec	0.2	Manufg.			
Miscellaneous plastics products	0.2	Manufg.			
Nonferrous wire drawing & insulating	0.2	Manufg.			
Primary metal products, nec	0.2	Manufg.			
Sawmills & planning mills, general	0.2	Manufg.			
Signs & advertising displays	0.2	Manufg.			
Railroads & related services	0.2	Util.			
Accounting, auditing & bookkeeping	0.2	Services			
Asbestos products	0.1	Manufg.			
Copper rolling & drawing	0.1	Manufg.			
Fabricated metal products, nec	0.1	Manufg.			
Fabricated rubber products, nec	0.1	Manufg.			
Industrial patterns	0.1	Manufg.			
Lubricating oils & greases	0.1	Manufg.			
Manifold business forms	0.1	Manufg.			
Miscellaneous fabricated wire products	0.1	Manufg.			
Steel springs, except wire	0.1	Manufg.			
Insurance carriers	0.1	Fin/R.E.			
Electrical repair shops	0.1	Services			
Hotels & lodging places	0.1	Services			

Source: Benchmark Input-Output Accounts for the U.S. Economy, 1982, U.S. Department of Commerce, Washington, D.C., July 1991. Data, as reported in the source, are organized by the 1977 SIC structure in use in 1982 but have been matched, as closely as is possible, to the 1987 SIC structure used in this book.

OCCUPATIONS EMPLOYED BY SIC 353 - CONSTRUCTION AND RELATED MACHINERY

Occupation	% of Total 1994	Change to 2005	Occupation	% of Total 1994	Change to 2005
Welders & cutters	7.4	-9.1	NC machine tool operators, metal & plastic	1.6	-9.1
Assemblers, fabricators, & hand workers nec	6.3	-9.1	Industrial machinery mechanics	1.4	-0.0
Machinists	4.6	-9.1	Traffic, shipping, & receiving clerks	1.3	-12.5
Sales & related workers nec	3.9	-9.1	Freight, stock, & material movers, hand	1.3	-27.3
Blue collar worker supervisors	3.6	-13.9	Bookkeeping, accounting, & auditing clerks	1.3	-31.8
Welding machine setters, operators	2.9	-18.2	Production, planning, & expediting clerks	1.2	-9.1
Machine builders	2.3	-27.3	Coating, painting, & spraying machine workers	1.2	-9.1
General managers & top executives	2.3	-13.8	Sheet metal workers & duct installers	1.2	-9.1
Mechanical engineers	2.3	0.0	Engineering, mathematical, & science managers	1.2	3.2
Inspectors, testers, & graders, precision	2.1	-9.1	Machine tool cutting operators, metal & plastic	1.1	-24.3
Secretaries, ex legal & medical	2.0	-17.3	Purchasing agents, ex trade & farm products	1.1	9.1
Drafters	2.0	-29.2	Engineering technicians & technologists nec	1.1	-9.1
Machine assemblers	1.8	63.6	General office clerks	1.0	-22.5
Professional workers nec	1.8	9.1	Maintenance repairers, general utility	1.0	-18.2
Machine tool cutting & forming etc. nec	1.8	-9.1	Stock clerks	1.0	-26.1
Lathe & turning machine tool operators	1.6	-18.2	Drilling & boring machine tool workers	1.0	-54.6
Helpers, laborers, & material movers nec	1.6	-9.1	Industrial production managers	1.0	-9.1

Source: Industry-Occupation Matrix, Bureau of Labor Statistics. These data relate to one or more 3-digit SIC industry groups rather than to a single 4-digit SIC. The change reported for each occupation to the year 2005 is a percent of growth or decline as estimated by the Bureau of Labor Statistics. The abbreviation nec stands for 'not elsewhere classified'.

LOCATION BY STATE AND REGIONAL CONCENTRATION

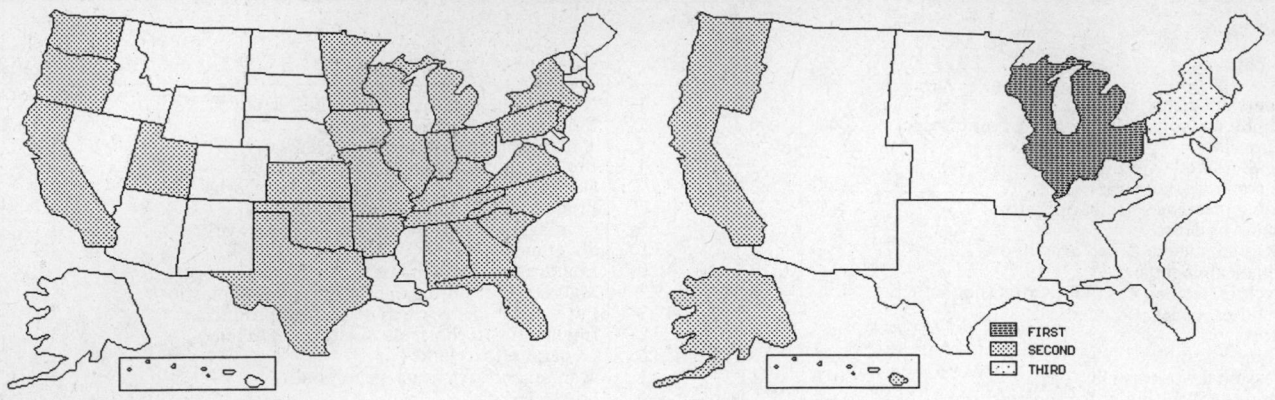

FIRST
SECOND
THIRD

INDUSTRY DATA BY STATE

| State | Establish-ments | Shipments | | | Employment | | | | Cost as % of Shipments | Investment per Employee ($) |
		Total ($ mil)	% of U.S.	Per Establ.	Total Number	% of U.S.	Per Establ.	Wages ($/hour)		
Ohio	29	399.8	14.5	13.8	2,700	15.4	93	13.18	50.6	-
Illinois	25	293.2	10.6	11.7	1,100	6.3	44	12.71	74.2	4,182
North Carolina	10	249.8	9.1	25.0	1,400	8.0	140	12.00	73.5	-
California	42	134.8	4.9	3.2	900	5.1	21	12.17	68.9	1,778
Indiana	20	132.9	4.8	6.6	700	4.0	35	10.56	65.4	-
Texas	16	112.1	4.1	7.0	800	4.6	50	10.77	56.4	2,250
Michigan	31	112.1	4.1	3.6	1,100	6.3	35	11.50	58.3	1,182
New York	24	100.9	3.7	4.2	1,100	6.3	46	10.14	53.1	2,545
Alabama	12	100.5	3.6	8.4	800	4.6	67	10.08	57.8	3,125
Wisconsin	18	93.9	3.4	5.2	400	2.3	22	11.25	53.4	2,000
Iowa	9	69.7	2.5	7.7	500	2.9	56	15.00	52.1	4,200
Florida	15	69.6	2.5	4.6	500	2.9	33	12.40	50.4	2,000
South Carolina	6	65.7	2.4	10.9	200	1.1	33	9.67	71.4	6,000
Pennsylvania	23	60.1	2.2	2.6	500	2.9	22	13.14	55.6	2,400
Oregon	16	57.4	2.1	3.6	400	2.3	25	15.25	41.3	-
Georgia	15	52.8	1.9	3.5	300	1.7	20	10.20	61.2	9,333
Washington	11	40.2	1.5	3.7	400	2.3	36	10.80	43.3	2,000
Minnesota	17	38.8	1.4	2.3	400	2.3	24	10.80	45.4	1,000
Arkansas	9	37.4	1.4	4.2	400	2.3	44	9.00	54.8	1,000
Virginia	7	34.0	1.2	4.9	200	1.1	29	13.00	55.6	-
Missouri	12	32.4	1.2	2.7	300	1.7	25	11.00	48.5	667
Kansas	7	26.9	1.0	3.8	400	2.3	57	8.60	46.1	-
New Jersey	9	16.4	0.6	1.8	100	0.6	11	11.00	48.2	4,000
Kentucky	9	(D)	-	-	1,750 *	10.0	194	-	-	-
Tennessee	9	(D)	-	-	175 *	1.0	19	-	-	571
Oklahoma	7	(D)	-	-	175 *	1.0	25	-	-	-
Utah	3	(D)	-	-	175 *	1.0	58	-	-	-

Source: 1992 *Economic Census*. The states are in descending order of shipments or establishments (if shipment data are missing for the majority). The symbol (D) appears when data are withheld to prevent disclosure of competitive information. States marked with (D) are sorted by number of establishments. A dash (-) indicates that the data element cannot be calculated; * indicates the midpoint of a range.

3541 - MACHINE TOOLS, METAL CUTTING TYPES

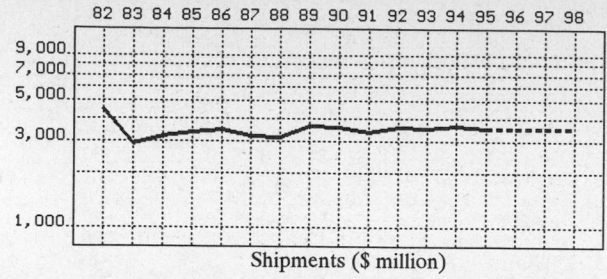

Shipments ($ million)

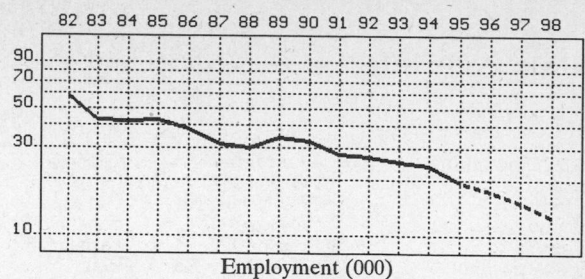

Employment (000)

GENERAL STATISTICS

Year	Com-panies	Establishments		Employment			Compensation		Production ($ million)			
		Total	with 20 or more employees	Total (000)	Production Workers (000)	Hours (Mil)	Payroll ($ mil)	Wages ($/hr)	Cost of Materials	Value Added by Manufacture	Value of Shipments	Capital Invest.
1982	864	940	293	58.1	33.7	63.4	1,331.7	11.15	1,589.5	2,554.5	4,440.2	157.1
1983		920	285	43.0	24.5	45.9	997.1	11.16	1,100.0	1,612.7	2,880.9	104.8
1984		900	277	42.4	24.7	46.5	1,053.9	11.65	1,369.5	1,820.0	3,211.6	108.8
1985		879	268	42.8	24.9	47.7	1,127.6	12.42	1,471.6	2,000.0	3,377.0	91.6
1986		867	271	38.6	22.0	43.6	1,088.8	12.78	1,538.0	1,823.8	3,456.1	62.7
1987	381	417	232	31.7	18.2	35.4	921.5	13.16	1,399.5	1,668.7	3,189.5	61.2
1988		454	247	29.9	17.7	35.8	940.2	13.52	1,404.2	1,828.1	3,137.8	64.5
1989		441	235	34.0	18.4	37.8	991.5	14.06	1,629.8	2,112.7	3,622.9	94.6
1990		496	236	32.7	18.2	37.8	1,002.2	14.13	1,682.8	1,890.3	3,606.8	84.2
1991		565	228	28.0	16.5	34.7	931.9	14.51	1,629.7	1,664.0	3,369.7	66.4
1992	391	422	220	27.0	15.2	32.2	982.1	15.48	1,762.0	1,840.9	3,567.8	82.4
1993		444	212	25.3	14.5	30.5	939.0	15.75	1,801.8	1,716.3	3,518.0	63.0
1994		295P	203P	24.0	13.7	30.0	945.9	16.31	1,781.6	1,894.6	3,658.1	105.9
1995		241P	196P	19.5P	10.9P	25.4P	881.7P	16.59P	1,699.0P	1,806.8P	3,488.6P	64.4P
1996		187P	189P	17.2P	9.6P	23.3P	862.0P	17.02P	1,700.7P	1,808.6P	3,492.1P	60.9P
1997		133P	182P	15.0P	8.3P	21.2P	842.3P	17.46P	1,702.4P	1,810.4P	3,495.5P	57.5P
1998		79P	174P	12.7P	6.9P	19.1P	822.6P	17.89P	1,704.1P	1,812.2P	3,499.0P	54.1P

Sources: 1982, 1987, 1992 *Economic Census*; *Annual Survey of Manufactures*, 83-86, 88-91, 93-94. Establishment counts for non-Census years are from *County Business Patterns*; establishment values for 83-84 are extrapolations. 'P's show projections by the editors. Industries reclassified in 87 will not have data for prior years.

INDICES OF CHANGE

Year	Com-panies	Establishments		Employment			Compensation		Production ($ million)			
		Total	with 20 or more employees	Total (000)	Production Workers (000)	Hours (Mil)	Payroll ($ mil)	Wages ($/hr)	Cost of Materials	Value Added by Manufacture	Value of Shipments	Capital Invest.
1982	221	223	133	215	222	197	136	72	90	139	124	191
1983		218	130	159	161	143	102	72	62	88	81	127
1984		213	126	157	163	144	107	75	78	99	90	132
1985		208	122	159	164	148	115	80	84	109	95	111
1986		205	123	143	145	135	111	83	87	99	97	76
1987	97	99	105	117	120	110	94	85	79	91	89	74
1988		108	112	111	116	111	96	87	80	99	88	78
1989		105	107	126	121	117	101	91	92	115	102	115
1990		118	107	121	120	117	102	91	96	103	101	102
1991		134	104	104	109	108	95	94	92	90	94	81
1992	100	100	100	100	100	100	100	100	100	100	100	100
1993		105	96	94	95	95	96	102	102	93	99	76
1994		70P	92P	89	90	93	96	105	101	103	103	129
1995		57P	89P	72P	72P	79P	90P	107P	96P	98P	98P	78P
1996		44P	86P	64P	63P	72P	88P	110P	97P	98P	98P	74P
1997		31P	83P	56P	54P	66P	86P	113P	97P	98P	98P	70P
1998		19P	79P	47P	46P	59P	84P	116P	97P	98P	98P	66P

Sources: Same as General Statistics. Values reflect change from the base year, 1992. Values above 100 mean greater than 92, values below 100 mean less than 92, and a value of 100 in the 82-91 or 93-98 period means same as 92. 'P's mark projections by the editors.

SELECTED RATIOS

For 1994	Avg. of All Manufact.	Analyzed Industry	Index	For 1994	Avg. of All Manufact.	Analyzed Industry	Index
Employees per Establishment	49	81	166	Value Added per Production Worker	134,084	138,292	103
Payroll per Establishment	1,500,273	3,211,554	214	Cost per Establishment	5,045,178	6,048,953	120
Payroll per Employee	30,620	39,412	129	Cost per Employee	102,970	74,233	72
Production Workers per Establishment	34	47	136	Cost per Production Worker	146,988	130,044	88
Wages per Establishment	853,319	1,661,289	195	Shipments per Establishment	9,576,895	12,420,114	130
Wages per Production Worker	24,861	35,715	144	Shipments per Employee	195,460	152,421	78
Hours per Production Worker	2,056	2,190	107	Shipments per Production Worker	279,017	267,015	96
Wages per Hour	12.09	16.31	135	Investment per Establishment	321,011	359,556	112
Value Added per Establishment	4,602,255	6,432,615	140	Investment per Employee	6,552	4,412	67
Value Added per Employee	93,930	78,942	84	Investment per Production Worker	9,352	7,730	83

Sources: Same as General Statistics. The 'Average of All Manufacturing' column represents the average of all manufacturing industries reported for the most recent complete year available. The Index shows the relationship between the Average and the Analyzed Industry. For example, 100 means that they are equal; 500 that the Analyzed Industry is five times the average; 50 means that the Analyzed Industry is half the national average. The abbreviation 'na' is used to show that data are 'not available'.

LEADING COMPANIES   Number shown: **75**   Total sales ($ mil): **7,304**   Total employment (000): **50.6**

Company Name	Address				CEO Name	Phone	Co. Type	Sales ($ mil)	Empl. (000)
Cincinnati Milacron Inc	4701 Marburg Av	Cincinnati	OH	45209	Daniel J Meyer	513-841-8100	P	1,197	8.4
Kennametal Inc	PO Box 231	Latrobe	PA	15650	R L McGeehan	412-539-5000	P	802	6.6
Sandvik Inc	1702 Nevins Rd	Fair Lawn	NJ	07410	James T Baker	201-794-5000	S	520	3.0
Giddings and Lewis Inc	PO Box 590	Fond du Lac	WI	54936	Joseph R Coppola	414-921-9400	P	518	3.7
Star Cutter Co	PO Box 376	Farmington	MI	48332	Brad Lawton	810-474-8200	R	400	0.6
Dover Industries Inc	675 Tollgate Rd	Elgin	IL	60123	LE Burns	708-468-0008	S	340	2.5
Precision Twist Drill Co	1 Precision Plz	Crystal Lake	IL	60014	Jim Beck	815-459-2040	R	300	1.0
Esterline Technologies Corp	10800 NE 8th St	Bellevue	WA	98004	Wendell P Hurlbut	206-453-9400	P	294	2.8
Okuma Machinery Inc	PO Box 7866	Charlotte	NC	28241	John Hendrick	704-588-7000	S	235	0.3
Thermodyne Industries Inc	101 S Hanley Rd	Clayton	MO	63105	James Mills	314-721-5573	R	140*	1.3
Giddings & Lewis	PO Box 590	Fond du Lac	WI	54936	Joseph R Coppola	414-921-9400	D	130	1.0
Gleason Corp	PO Box 22970	Rochester	NY	14692	James S Gleason	716-473-1000	P	128	1.1
Hardinge Inc	1 Hardinge Dr	Elmira	NY	14902	Robert E Agan	607-734-2281	P	117	1.0
General Broach & Engineering	13231 23-Mile Rd	Shelby	MI	48315	Tom Carter	810-726-8833	S	110	1.0
Acme-Cleveland Corp	PO Box 5639	Cleveland	OH	44101	David L Swift	216-432-5400	P	108	1.1
Bridgeport Machines Inc	PO Box 32	Bridgeport	CT	06601	Joseph E Clancy	203-367-3651	P	107	0.9
Flow International Corp	PO Box 97040	Kent	WA	98064	Ronald W Tarrant	206-850-3500	P	89	0.5
Monarch Machine Tool Co	615 N Oak Av	Sidney	OH	45365	Robert J Siewert	513-492-4111	P	76	0.7
Hurco Companies Inc	PO Box 68180	Indianapolis	IN	46268	B D McLaughlin	317-293-5309	P	73	0.4
Hurco Manufacturing Company	1 Technology Way	Indianapolis	IN	46268	James D Fabris	317-293-5309	D	70	0.3
Toyoda Machinery USA Inc	316 W University	Arlington H	IL	60004	K Jack Sakane	708-253-0340	S	65	0.2
DeVlieg-Bullard Inc	1 Gorham Island	Westport	CT	06880	William O Thomas	203-221-8201	P	64	0.6
Landis	20 E 6th St	Waynesboro	PA	17268	Charles E Hartle	717-762-2161	D	60	0.4
Talbot Holdings Ltd	31 S Harrison St	Easton	MD	21601	Timothy E Wyman	301-820-7676	R	57	0.5
Century Manufacturing Co	9231 Penn Av	Bloomington	MN	55431		612-884-3211	R	55*	0.5
Ann Arbor Machine Co	5800 Sibley Rd	Chelsea	MI	48118	Jim Breining	313-475-0505	R	50	0.2
Kingsbury Corp	PO Box 2020	Keene	NH	03431	JL Koontz	603-352-5212	R	50	0.5
Cargill Detroit Corp	1250 N Crooks Rd	Clawson	MI	48017	David Brown	313-435-3500	R	45	0.2
Kitamura Machinery of USA Inc	78 E Century Dr	Wheeling	IL	60090	Akihiro Kitamura	708-520-7755	S	42	<0.1
PMC Industries Inc	29100 Lakeland	Wickliffe	OH	44092	JF Roche	216-943-3300	R	37*	0.3
Hypertherm Inc	PO Box 5010	Hanover	NH	03755	R W Couch Jr	603-643-3441	R	36*	0.3
Delta Tooling Co	1350 Harmon Rd	Auburn Hills	MI	48326	Rudolph W Mozer	313-391-6800	R	35	0.3
Micromatic Textron Inc	345 E 48th St	Holland	MI	49423	Michael J Brennan	616-392-1461	S	35*	0.3
National Acme Co	170 E 131st St	Cleveland	OH	44108	Joe L Menger	216-268-4200	S	35	0.4
Carbide International Inc	5740 N Tripp Av	Chicago	IL	60646	Robert Britzke	312-583-6362	S	33*	0.3
Bryant Grinder Corp	PO Box 2002	Springfield	VT	05156	Richard Crossman	802-885-5161	R	30	0.4
Fellows Corp	PO Box 851	Springfield	VT	05156	Richard Crossman	802-886-8333	S	30	0.3
Regal Cutting Tools	PO Box 38	South Beloit	IL	61080	James L Packard	815-389-3461	D	30	0.9
Union Butterfield Corp	PO Box 50000	Asheville	NC	28813	Adrian M Waple	704-274-6070	R	30	0.2
US Group Inc	20580 Hoover Rd	Detroit	MI	48205	P Simon	313-372-7900	R	30	0.3
Western Atlas Inc	481 Gardner St	South Beloit	IL	61080	Jim Herrman	815-389-2251	D	30	0.3
National Broach&Machine Co	17500 23 Mile Rd	Macomb	MI	48044	Joseph W Back	810-263-0100	D	28	0.2
North American Products Corp	1180 Wernsing Rd	Jasper	IN	47546	Steve Segal	812-482-2000	R	27	0.5
Mattison Machine Works	545 Blackhawk Pk	Rockford	IL	61101	Arthur Mattison	815-962-5521	R	26	0.3
Moore Tool Co	PO Box 4088	Bridgeport	CT	06607	N M Marsilius III	203-366-3224	S	26*	0.2
Anorad Corp	110 Oser Av	Hauppauge	NY	11788	AK Chitayat	516-231-1995	R	25	0.3
Lou-Rich Inc	505 W Front St	Albert Lea	MN	56007	James A Anderson	507-377-8910	R	24	0.2
Edac Technologies Corp	1790 New Britain	Farmington	CT	06032	Robert T Whitty	203-677-2603	P	22	0.2
Saginaw Machine Systems Inc	301 Park St	Troy	MI	48083	Jeffrey A Clevenger	313-583-7200	R	22	0.2
Besly Products Corp	100 Dearborn Av	South Beloit	IL	61080	James Deeds	815-389-2231	R	21	0.2
Bohle Machine Tools Inc	38800 Grand River	Farmington Hls	MI	48335	Karl A Kreft	810-477-4500	R	21	<0.1
Motch Corp	1250 E 222nd St	Cleveland	OH	44117	Svend Weidemann	216-486-3600	S	21	0.2
Arland Tool and Manufacturing	PO Box 207	Sturbridge	MA	01566	Bill Gagnon	508-347-3368	R	20	0.2
Extrude Hone Corp	PO Box 527	Irwin	PA	15642	LJ Rhoades	412-863-5900	R	20	0.1
HR Krueger Machine Tool Inc	PO Box 310	Farmington	MI	48332	BL Moore	810-477-8400	R	20	0.1
Industrial Metal Products Corp	PO Box 10156	Lansing	MI	48901	Dave Houghton	517-484-9411	R	20	0.2
New York Twist Drill Inc	5368 E Rockton Rd	South Beloit	IL	61080		217-588-8800	D	20	0.1
Peddinghaus Corp	300 N Washington	Bradley	IL	60915	Tom Boyer	815-937-3800	R	20	0.1
S and S Machinery Co	140 53rd St	Brooklyn	NY	11232	Jed Srybnik	718-492-7400	R	20	0.1
Wasino Corp	6 Highpoint Dr	Wayne	NJ	07470	Ugo Boggio	201-696-7070	S	20	<0.1
AK Allen Inc	255 E 2nd St	Mineola	NY	11501	AK Allen	516-747-5450	R	19	0.2
American GFM Corp	1200 Cavalier Blv	Chesapeake	VA	23323	Robert Kralowetz	804-487-2442	R	18	0.2
Komo Machine Inc	11 Industrial Blv	Sauk Rapids	MN	56379	Bob Sexton	612-252-0580	R	18*	0.2
Kwik-Way Industries Inc	500 57th St	Marion	IA	52302	David Parks	319-377-9421	R	18	0.3
Schenck Turner Inc	100 Kay Indrial Dr	Orion	MI	48359	Jurgen Neugebaur	810-377-2100	S	18	<0.1
Waterbury Farrell Technologies	PO Box 400	Cheshire	CT	06410	Andre Nazarian	203-272-3271	S	18	<0.1
SE Huffman Corp	1050 Huffman Way	Clover	SC	29710	Roger Hayes	803-222-4561	S	18	<0.1
Sheldon Machine	PO Box 949	Fremont	OH	43420	Chuck Carnicom	419-334-8971	D	16*	0.2
Century Specialties Inc	2410 Aero Park Ct	Traverse City	MI	49684	William Janis	616-946-7500	R	15	0.2
Goldcrown Machinery Inc	4201 Malsbary Rd	Cincinnati	OH	45242	George Trenkamp	513-793-1500	R	15*	0.1
Republic Machinery Company	PO Box 5328	Carson	CA	90749	Norbert Toubes	310-518-1100	R	15	<0.1
Simmons Machine Tool Corp	1700 N Broadway	Albany	NY	12204	HJ Naumann	518-462-5431	R	15	0.1
South Bend Lathe Corp	400 W Sample St	South Bend	IN	46601	Norbert Toubes	219-289-7771	R	15	0.1
Standard Machine and Tool Co	29900 Hayes Rd	Roseville	MI	48066	George Volis	810-773-6800	R	15	<0.1
Strilich Technologies Inc	PO Box 210	Crown Point	IN	46307	David Strilich	219-663-9550	S	15	0.1

Source: Ward's Business Directory of U.S. Private and Public Companies, Volumes 1 and 2, 1996. The company type code used is as follows: P - Public, R - Private, S - Subsidiary, D - Division, J - Joint Venture, A - Affiliate, G - Group. Sales are in millions of dollars, employees are in thousands. An asterisk (*) indicates an estimated sales volume. The symbol < stands for 'less than'. Company names and addresses are truncated, in some cases, to fit into the available space.

MATERIALS CONSUMED

Material	Quantity	Delivered Cost ($ million)
Materials, ingredients, containers, and supplies	(X)	1,414.7
Fluid power pumps, motors, and hydrostatic transmissions (hydraulic and pneumatic)	(X)	33.1
Fluid power cylinders and rotary actuators (hydraulic and pneumatic)	(X)	15.1
Fluid power filters (hydraulic and pneumatic)	(X)	4.5
Fluid power hose or tube fittings and assemblies (hydraulic and pneumatic)	(X)	11.2
Fluid power valves (hydraulic and pneumatic)	(X)	13.3
Metal bolts, nuts, screws, washers, rivets, and other screw machine products	(X)	18.2
Other fabricated metal products (except fluid power products and forgings)	(X)	65.7
Iron and steel forgings	(X)	8.4
Nonferrous forgings	(X)	0.6
Iron and steel castings (rough and semifinished)	(X)	117.4
Aluminum and aluminum-base alloy castings (rough and semifinished)	(X)	6.6
Other nonferrous castings (rough and semifinished)	(X)	1.7
Steel bars, bar shapes, and plates	(X)	39.4
Steel sheet and strip, including tin plate	(X)	18.2
Steel structural shapes	(X)	10.4
All other steel shapes and forms	(X)	6.2
Aluminum and aluminum-base alloy shapes and forms	(X)	8.4
Other nonferrous shapes and forms	(X)	4.2
Fractional horsepower electric timing motors, synchronous and subsynchronous (less than 1 hp)	(X)	20.7
Other fractional horsepower electric motors (under 1 hp)	(X)	12.6
Integral horsepower electric motors and generators (1 hp or more)	(X)	19.7
Electrical transmission, distribution, and control equipment	(X)	37.2
Electrical industrial capacitors, resistors, rheostats, and coil windings	(X)	13.4
Numerical controls for metalworking machinery (except programmable)	(X)	46.5
Programmable controllers for metalworking machinery	(X)	67.7
Ball bearings (mounted or unmounted)	(X)	18.6
Roller bearings (mounted or unmounted)	(X)	7.3
Mechanical speed changers, gears, and industrial high-speed drives	(X)	40.5
Wood boxes, pallets, skids, and containers	(X)	5.9
Cutting tools for machine tools	(X)	23.0
All other materials and components, parts, containers, and supplies	(X)	295.5
Materials, ingredients, containers, and supplies, nsk	(X)	423.5

Source: 1992 Economic Census. Explanation of symbols used: (D): Withheld to avoid disclosure of competitive data; na: Not available; (S): Withheld because statistical norms were not met; (X): Not applicable; (Z): Less than half the unit shown; nec: Not elsewhere classified; nsk: Not specified by kind; - : zero; * : 10-19 percent estimated; ** : 20-29 percent estimated.

PRODUCT SHARE DETAILS

Product or Product Class	% Share	Product or Product Class	% Share
Machine tools, metal cutting types	100.00	for home workshops, labs, garages, etc., including crankshaft regrinding and valve grinding machines	6.24
Metal boring machines (excluding machining centers) and drilling machines (excluding machining centers)	4.96	Metal lathes designed primarily for home workshops, labs, garages, etc.	(D)
Metal gear cutting machines	2.86	Metal sawing and cut-off machines designed primarily for home workshops, labs, garages, etc.	(D)
Metal grinding, polishing, buffing, honing, and lapping machines, except gear-tooth grinding, lapping, polishing and buffing	12.31	Other metalworking (or primarily metalworking) machines designed primarily for home workshops, labs, garages, etc., including automotive cylinder reboring machines	54.52
Metal lathes (turning machines)	10.78		
Metal milling machines (excluding machining centers)	7.26	Machine tools designed primarily for home workshops, labs, etc. (metalworking and primarily metalworking), nsk	0.82
Metal machining centers (multifunction numerically controlled machines)	14.87	Parts for metal cutting machine tools (sold separately) and rebuilt metal cutting machine tools	19.85
Metal station type machines	15.36	Parts for metal cutting machine tools, sold separately	70.60
Other metal cutting machine tools (except those designed primarily for home workshops, laboratories, garages, etc.)	7.27	Rebuilt metal cutting machine tools	18.20
Machine tools designed primarily for home workshops, labs, garages, etc. (metalworking and primarily metalworking)	1.99	Parts for metal cutting machine tools (sold separately) and rebuilt metal cutting machine tools, nsk	11.18
Metal drilling machines designed primarily for home workshops, labs, garages, etc.	30.54	Machine tools, metal cutting types, nsk	2.47
Metal grinding and polishing machines designed primarily			

Source: 1992 Economic Census. The values shown are percent of total shipments in an industry. Values of indented subcategories are summed in the main heading. The symbol (D) appears when data are withheld to prevent disclosure of competitive information. The abbreviation nsk stands for 'not specified by kind' and nec for 'not elsewhere classified'.

INPUTS AND OUTPUTS FOR MACHINE TOOLS, METAL CUTTING TYPES

Economic Sector or Industry Providing Inputs	%	Sector	Economic Sector or Industry Buying Outputs	%	Sector
Imports	45.0	Foreign	Gross private fixed investment	73.1	Cap Inv
Wholesale trade	6.8	Trade	Exports	9.9	Foreign
Machine tools, metal cutting types	4.6	Manufg.	Machine tools, metal cutting types	2.6	Manufg.
Iron & steel foundries	4.3	Manufg.	Machinery, except electrical, nec	1.2	Manufg.
Industrial controls	4.0	Manufg.	Special dies & tools & machine tool accessories	0.6	Manufg.
Motors & generators	3.0	Manufg.	Aircraft & missile engines & engine parts	0.5	Manufg.
Machinery, except electrical, nec	2.8	Manufg.	Oil field machinery	0.5	Manufg.
Blast furnaces & steel mills	2.7	Manufg.	Pipe, valves, & pipe fittings	0.5	Manufg.
Fabricated plate work (boiler shops)	2.0	Manufg.	Federal Government purchases, national defense	0.5	Fed Govt
Special dies & tools & machine tool accessories	1.4	Manufg.	S/L Govt. purch., elem. & secondary education	0.5	S/L Govt
Electric services (utilities)	1.4	Util.	Aircraft & missile equipment, nec	0.4	Manufg.
Switchgear & switchboard apparatus	1.2	Manufg.	Internal combustion engines, nec	0.4	Manufg.
Abrasive products	1.1	Manufg.	Motor vehicle parts & accessories	0.4	Manufg.
Petroleum refining	1.1	Manufg.	Pumps & compressors	0.4	Manufg.
Communications, except radio & TV	1.0	Util.	Personal consumption expenditures	0.3	
Ball & roller bearings	0.8	Manufg.	Aircraft	0.3	Manufg.
Eating & drinking places	0.8	Trade	Construction machinery & equipment	0.3	Manufg.
Advertising	0.8	Services	Farm machinery & equipment	0.3	Manufg.
Noncomparable imports	0.8	Foreign	Power transmission equipment	0.3	Manufg.
Maintenance of nonfarm buildings nec	0.7	Constr.	Miscellaneous repair shops	0.3	Services
Iron & steel forgings	0.6	Manufg.	S/L Govt. purch., higher education	0.3	S/L Govt
Pipe, valves, & pipe fittings	0.6	Manufg.	Automotive stampings	0.2	Manufg.
Power transmission equipment	0.6	Manufg.	Electronic components nec	0.2	Manufg.
Banking	0.6	Fin/R.E.	Miscellaneous plastics products	0.2	Manufg.
Engineering, architectural, & surveying services	0.6	Services	Motor vehicles & car bodies	0.2	Manufg.
Equipment rental & leasing services	0.6	Services	Motors & generators	0.2	Manufg.
Screw machine and related products	0.5	Manufg.	Refrigeration & heating equipment	0.2	Manufg.
Air transportation	0.5	Util.	Screw machine and related products	0.2	Manufg.
Motor freight transportation & warehousing	0.5	Util.	Special industry machinery, nec	0.2	Manufg.
U.S. Postal Service	0.5	Gov't	Turbines & turbine generator sets	0.2	Manufg.
Cyclic crudes and organics	0.4	Manufg.	S/L Govt. purch., other general government	0.2	S/L Govt
Pumps & compressors	0.4	Manufg.	Carburetors, pistons, rings, & valves	0.1	Manufg.
Transformers	0.4	Manufg.	Electronic computing equipment	0.1	Manufg.
Gas production & distribution (utilities)	0.4	Util.	General industrial machinery, nec	0.1	Manufg.
Computer & data processing services	0.4	Services	Hand & edge tools, nec	0.1	Manufg.
Fabricated metal products, nec	0.3	Manufg.	Hardware, nec	0.1	Manufg.
Metal heat treating	0.3	Manufg.	Machine tools, metal forming types	0.1	Manufg.
Miscellaneous plastics products	0.3	Manufg.	Mechanical measuring devices	0.1	Manufg.
Real estate	0.3	Fin/R.E.	Metal stampings, nec	0.1	Manufg.
Legal services	0.3	Services	Ship building & repairing	0.1	Manufg.
Management & consulting services & labs	0.3	Services	Federal Government purchases, nondefense	0.1	Fed Govt
Metal stampings, nec	0.2	Manufg.	S/L Govt. purch., transit utilities	0.1	S/L Govt
Miscellaneous fabricated wire products	0.2	Manufg.			
Nonferrous wire drawing & insulating	0.2	Manufg.			
Accounting, auditing & bookkeeping	0.2	Services			
Automotive rental & leasing, without drivers	0.2	Services			
Automotive repair shops & services	0.2	Services			
Hotels & lodging places	0.2	Services			
Aluminum castings	0.1	Manufg.			
Aluminum rolling & drawing	0.1	Manufg.			
Lubricating oils & greases	0.1	Manufg.			
Manifold business forms	0.1	Manufg.			
Paints & allied products	0.1	Manufg.			
Plating & polishing	0.1	Manufg.			
Power driven hand tools	0.1	Manufg.			
Insurance carriers	0.1	Fin/R.E.			
Business/professional associations	0.1	Services			

Source: Benchmark Input-Output Accounts for the U.S. Economy, 1982, U.S. Department of Commerce, Washington, D.C., July 1991. Data, as reported in the source, are organized by the 1977 SIC structure in use in 1982 but have been matched, as closely as is possible, to the 1987 SIC structure used in this book.

OCCUPATIONS EMPLOYED BY SIC 354 - METALWORKING MACHINERY

Occupation	% of Total 1994	Change to 2005	Occupation	% of Total 1994	Change to 2005
Machinists	10.9	-17.4	Precision metal workers nec	1.9	10.1
Tool & die makers	10.6	-15.3	Mechanical engineers	1.9	1.0
General managers & top executives	3.7	-12.9	Inspectors, testers, & graders, precision	1.9	-8.2
Blue collar worker supervisors	3.4	-15.7	Secretaries, ex legal & medical	1.9	-16.4
Machine tool cutting operators, metal & plastic	2.9	-5.9	Drafters	1.8	-28.5
NC machine tool operators, metal & plastic	2.7	1.0	General office clerks	1.5	-21.7
Grinding machine operators, metal & plastic	2.7	-8.2	Machine forming operators, metal & plastic	1.4	-8.2
Sales & related workers nec	2.7	-8.2	Drilling & boring machine tool workers	1.3	-8.2
Assemblers, fabricators, & hand workers nec	2.5	-8.2	Bookkeeping, accounting, & auditing clerks	1.3	-31.1
Lathe & turning machine tool operators	2.3	-17.4	Industrial production managers	1.3	-8.2
Machine tool cutting & forming etc. nec	2.2	-54.1	Traffic, shipping, & receiving clerks	1.3	-11.7
Machine builders	2.0	1.0	Precision woodworkers nec	1.3	10.2
Combination machine tool operators	2.0	37.7	Janitors & cleaners, incl maids	1.2	-26.6

Source: *Industry-Occupation Matrix*, Bureau of Labor Statistics. These data relate to one or more 3-digit SIC industry groups rather than to a single 4-digit SIC. The change reported for each occupation to the year 2005 is a percent of growth or decline as estimated by the Bureau of Labor Statistics. The abbreviation nec stands for 'not elsewhere classified'.

LOCATION BY STATE AND REGIONAL CONCENTRATION

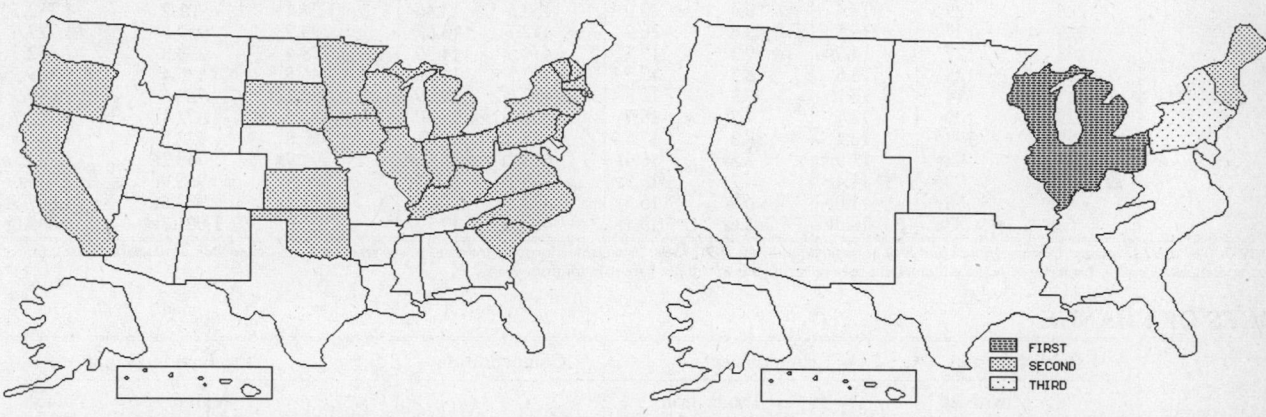

FIRST
SECOND
THIRD

INDUSTRY DATA BY STATE

State	Establish- ments	Shipments			Employment				Cost as % of Shipments	Investment per Employee ($)
		Total ($ mil)	% of U.S.	Per Establ.	Total Number	% of U.S.	Per Establ.	Wages ($/hour)		
Michigan	100	769.5	21.6	7.7	4,600	17.0	46	16.58	47.8	2,435
Ohio	36	659.5	18.5	18.3	4,600	17.0	128	16.67	44.5	2,174
Illinois	41	423.8	11.9	10.3	3,100	11.5	76	15.51	70.4	3,323
Wisconsin	28	326.4	9.1	11.7	2,600	9.6	93	14.71	34.9	1,423
New York	18	266.5	7.5	14.8	2,600	9.6	144	16.13	43.5	-
California	32	147.7	4.1	4.6	800	3.0	25	15.91	44.3	3,625
Connecticut	25	126.1	3.5	5.0	1,500	5.6	60	14.35	50.1	1,000
Pennsylvania	20	96.6	2.7	4.8	800	3.0	40	14.33	50.6	4,250
Minnesota	11	63.9	1.8	5.8	600	2.2	55	13.00	49.5	3,500
North Carolina	4	60.5	1.7	15.1	200	0.7	50	15.33	78.3	-
Massachusetts	18	58.6	1.6	3.3	600	2.2	33	11.83	47.4	1,000
Indiana	8	50.9	1.4	6.4	500	1.9	63	23.40	59.3	-
Oregon	7	19.0	0.5	2.7	200	0.7	29	12.00	50.0	2,000
New Jersey	9	18.1	0.5	2.0	200	0.7	22	13.67	34.8	-
Rhode Island	3	15.7	0.4	5.2	100	0.4	33	12.50	54.8	-
Kentucky	6	(D)	-	-	750 *	2.8	125	-	-	-
South Carolina	5	(D)	-	-	750 *	2.8	150	-	-	-
Vermont	4	(D)	-	-	750 *	2.8	188	-	-	-
Virginia	4	(D)	-	-	375 *	1.4	94	-	-	-
South Dakota	3	(D)	-	-	175 *	0.6	58	-	-	-
Iowa	2	(D)	-	-	175 *	0.6	88	-	-	-
Kansas	2	(D)	-	-	175 *	0.6	88	-	-	-
New Hampshire	2	(D)	-	-	750 *	2.8	375	-	-	-
Oklahoma	2	(D)	-	-	175 *	0.6	88	-	-	-

Source: 1992 *Economic Census*. The states are in descending order of shipments or establishments (if shipment data are missing for the majority). The symbol (D) appears when data are withheld to prevent disclosure of competitive information. States marked with (D) are sorted by number of establishments. A dash (-) indicates that the data element cannot be calculated; * indicates the midpoint of a range.

3542 - MACHINE TOOLS, METAL FORMING TYPES

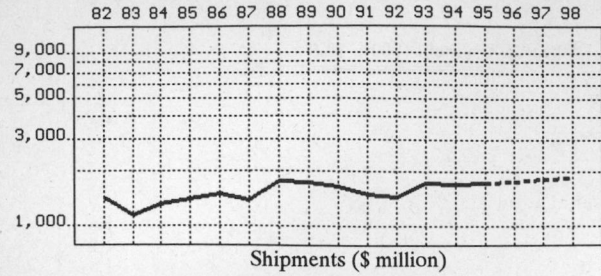

82 83 84 85 86 87 88 89 90 91 92 93 94 95 96 97 98

Shipments ($ million)

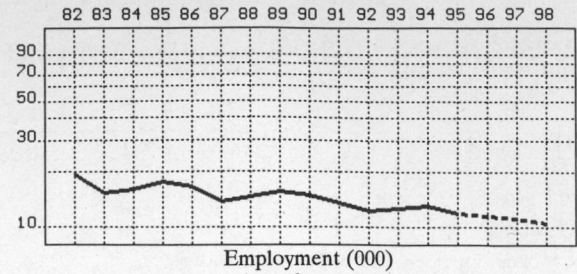

82 83 84 85 86 87 88 89 90 91 92 93 94 95 96 97 98

Employment (000)

GENERAL STATISTICS

Year	Companies	Establishments		Employment			Compensation		Production ($ million)			
		Total	with 20 or more employees	Total (000)	Production Workers (000)	Hours (Mil)	Payroll ($ mil)	Wages ($/hr)	Cost of Materials	Value Added by Manufacture	Value of Shipments	Capital Invest.
1982	434	452	162	19.5	12.1	23.2	438.1	10.57	574.6	777.8	1,428.7	43.4
1983		440	161	15.3	9.6	17.8	357.0	11.01	454.9	629.6	1,142.3	28.1
1984		428	160	16.1	10.5	20.7	407.0	11.72	535.4	775.9	1,308.7	31.0
1985		417	158	17.7	11.5	23.0	461.6	11.97	598.7	933.2	1,412.8	43.7
1986		412	157	16.6	10.7	21.4	440.5	12.02	713.2	765.7	1,507.9	33.9
1987	196	207	120	13.8	8.7	18.1	414.3	13.08	637.8	734.3	1,396.3	31.2
1988		226	129	14.6	9.4	20.0	458.6	13.46	824.4	921.2	1,752.7	36.5
1989		229	127	15.5	9.6	20.0	472.7	13.82	793.9	959.8	1,729.6	39.4
1990		248	127	14.7	9.3	19.5	474.5	14.30	778.4	853.8	1,652.7	33.3
1991		280	125	13.5	8.5	17.7	443.5	14.30	746.5	636.4	1,505.5	31.6
1992	211	217	114	12.2	7.8	15.8	417.6	14.99	700.1	728.6	1,450.9	40.4
1993		229	109	12.7	7.8	16.7	461.9	15.68	833.4	872.4	1,712.7	32.7
1994		159P	104P	13.1	8.3	17.5	475.3	15.91	769.8	922.0	1,694.3	25.4
1995		135P	98P	11.9P	7.5P	16.4P	473.4P	16.40P	791.7P	948.2P	1,742.5P	31.7P
1996		111P	93P	11.4P	7.2P	15.9P	478.1P	16.84P	806.5P	965.9P	1,775.0P	31.3P
1997		87P	88P	11.0P	6.9P	15.5P	482.8P	17.29P	821.2P	983.6P	1,807.5P	30.9P
1998		63P	83P	10.5P	6.6P	15.1P	487.6P	17.73P	836.0P	1,001.3P	1,840.0P	30.4P

Sources: 1982, 1987, 1992 *Economic Census*; *Annual Survey of Manufactures*, 83-86, 88-91, 93-94. Establishment counts for non-Census years are from *County Business Patterns*; establishment values for 83-84 are extrapolations. 'P's show projections by the editors. Industries reclassified in 87 will not have data for prior years.

INDICES OF CHANGE

Year	Companies	Establishments		Employment			Compensation		Production ($ million)			
		Total	with 20 or more employees	Total (000)	Production Workers (000)	Hours (Mil)	Payroll ($ mil)	Wages ($/hr)	Cost of Materials	Value Added by Manufacture	Value of Shipments	Capital Invest.
1982	206	208	142	160	155	147	105	71	82	107	98	107
1983		203	141	125	123	113	85	73	65	86	79	70
1984		197	140	132	135	131	97	78	76	106	90	77
1985		192	139	145	147	146	111	80	86	128	97	108
1986		190	138	136	137	135	105	80	102	105	104	84
1987	93	95	105	113	112	115	99	87	91	101	96	77
1988		104	113	120	121	127	110	90	118	126	121	90
1989		106	111	127	123	127	113	92	113	132	119	98
1990		114	111	120	119	123	114	95	111	117	114	82
1991		129	110	111	109	112	106	95	107	87	104	78
1992	100	100	100	100	100	100	100	100	100	100	100	100
1993		106	96	104	100	106	111	105	119	120	118	81
1994		73P	91P	107	106	111	114	106	110	127	117	63
1995		62P	86P	97P	96P	104P	113P	109P	113P	130P	120P	78P
1996		51P	82P	94P	92P	101P	114P	112P	115P	133P	122P	77P
1997		40P	77P	90P	88P	98P	116P	115P	117P	135P	125P	76P
1998		29P	73P	86P	84P	96P	117P	118P	119P	137P	127P	75P

Sources: Same as General Statistics. Values reflect change from the base year, 1992. Values above 100 mean greater than 92, values below 100 mean less than 92, and a value of 100 in the 82-91 or 93-98 period means same as 92. 'P's mark projections by the editors.

SELECTED RATIOS

For 1994	Avg. of All Manufact.	Analyzed Industry	Index	For 1994	Avg. of All Manufact.	Analyzed Industry	Index
Employees per Establishment	49	82	168	Value Added per Production Worker	134,084	111,084	83
Payroll per Establishment	1,500,273	2,984,473	199	Cost per Establishment	5,045,178	4,833,679	96
Payroll per Employee	30,620	36,282	118	Cost per Employee	102,970	58,763	57
Production Workers per Establishment	34	52	152	Cost per Production Worker	146,988	92,747	63
Wages per Establishment	853,319	1,748,268	205	Shipments per Establishment	9,576,895	10,638,740	111
Wages per Production Worker	24,861	33,545	135	Shipments per Employee	195,460	129,336	66
Hours per Production Worker	2,056	2,108	103	Shipments per Production Worker	279,017	204,133	73
Wages per Hour	12.09	15.91	132	Investment per Establishment	321,011	159,490	50
Value Added per Establishment	4,602,255	5,789,364	126	Investment per Employee	6,552	1,939	30
Value Added per Employee	93,930	70,382	75	Investment per Production Worker	9,352	3,060	33

Sources: Same as General Statistics. The 'Average of All Manufacturing' column represents the average of all manufacturing industries reported for the most recent complete year available. The Index shows the relationship between the Average and the Analyzed Industry. For example, 100 means that they are equal; 500 that the Analyzed Industry is five times the average; 50 means that the Analyzed Industry is half the national average. The abbreviation 'na' is used to show that data are 'not available'.

LEADING COMPANIES   Number shown: **75**   Total sales ($ mil): **1,247**   Total employment (000): **10.5**

Company Name	Address				CEO Name	Phone	Co. Type	Sales ($ mil)	Empl. (000)
National Machinery Co	161 Greenfield St	Tiffin	OH	44883	Paul Aley	419-447-5211	R	100*	0.8
Minster Machine Co	PO Box 120	Minster	OH	45865	John Winch	419-628-2331	R	87*	0.8
Dayton Reliable Tool & Mfg Co	PO Box 586	Dayton	OH	45409	Dennis T Casey	513-298-7391	R	56*	0.5
Tulip Corp	14963 E Salt Lake	City of Industry	CA	91746	Fred Teshinsky	818-968-0044	R	56*	0.5
Met-Coil Systems Corp	5486 6th St SW	Cedar Rapids	IA	52404	Ray Blakeman	319-363-6566	P	48	0.4
Cincinnati Inc	PO Box 11111	Cincinnati	OH	45211	CR Turner	513-367-7100	R	45	0.6
Trumpf Inc	Hyde Rd	Farmington	CT	06032	Daniel Dechamps	203-677-9741	S	45	0.2
Magnetic Metals Corp	PO Box 351	Camden	NJ	08101	Donald L Walsh	609-964-7842	S	40	0.4
Murata Wiedemann Inc	10510 Twin Lks	Charlotte	NC	28269	Joseph A Norwood	704-875-9280	S	40	0.2
Reed-Rico	18 Industrial Dr	Holden	MA	01520	John M Prosser	508-829-4491	D	40*	0.4
Fenn Manufacturing Co	300 Fenn Rd	Newington	CT	06111	FR Luszcz	203-666-2471	S	35	0.3
Citation Tool Inc	16660 E 13th Mile	Roseville	MI	48066	Ted Tsuda	313-773-4330	R	30	0.2
Littell Inc	145 Swift Rd	Addison	IL	60613	Sterling Stevenson	708-916-6667	S	30	0.2
Secom General Corp	37650 Prof Ctr	Livonia	MI	48154	Robert Clemente	313-953-3990	P	29	0.3
Efco Inc	1253 W 12th St	Erie	PA	16512	JA Currie Sr	814-455-3941	R	25	0.2
Lockformer Co	711 Ogden Av	Lisle	IL	60532	K John Del Vecchio	708-964-8000	S	25	0.2
Wagstaff Inc	N 3910 Flora Rd	Spokane	WA	99216	WG Wagstaff	509-922-1404	R	25	0.2
Cardinal American Corp	4259 E 49th St	Cleveland	OH	44125	SD Noll	216-883-3220	R	24	0.3
General Electro-Mechanical	785 Hertel Av	Buffalo	NY	14207	T H Speller Jr	716-876-9685	R	23	0.1
Royle Systems Group	1000 Cannonball Rd	Pompton Lakes	NJ	07442	JC Ramsey	201-839-8118	R	20	0.1
Williams, White and Co	600 River Dr	Moline	IL	61265	Thomas G Getz	309-797-7650	R	20	0.2
Header Products Inc	11850 Wayne Rd	Romulus	MI	48174	MF McManus Jr	313-941-2220	R	18	0.2
SMS Engineering Inc	100 Sandusky St	Pittsburgh	PA	15212	JD Butkus	412-231-2100	D	18	0.1
Wysong and Miles Company Inc	PO Box 21168	Greensboro	NC	27420	Russell F Hall III	919-621-3960	R	18	0.1
Pacific Press and Shear Inc	714 Walnut St	Mount Carmel	IL	62863	Mike Stein	618-262-8666	R	16	0.1
Koppy Corp	199 Kay Indrial Dr	Orion	MI	48359	Ronald Prater	313-373-5200	R	15	0.1
Roper Whitney of Rockford Inc	2833 Huffman Blv	Rockford	IL	61103	John Forlow	815-962-3011	R	14	0.1
Bonnot Co	1520 Corp Woods	Uniontown	OH	44895	BK Bain	216-896-6544	R	11*	0.1
Bruderer Inc	PO Box 208	Huntsville	AL	35804	Werner Vieh	205-859-4050	S	11*	0.1
Edmunds Manufacturing Co	PO Box 385	Farmington	CT	06032	RF Edmunds Jr	203-677-2813	R	11*	0.1
Feldmann Inc	4902 Hydraulic Dr	Rockford	IL	61109	MJ Pisano	815-874-2106	R	11	<0.1
Scotchman Industries Inc	PO Box 850	Philip	SD	57567	GA Carley	605-859-2542	R	11	<0.1
Welding Engineers Inc	1600 Union Meeting	Blue Bell	PA	19422	JG Hendrickson	215-643-6900	R	11	<0.1
Cerden and Son Manufacturing	105 N Park St	Frankton	IN	46044	Jack E Cerden	317-754-7577	R	10*	<0.1
Kinefac Corp	156 Goddard Mem	Worcester	MA	01603	HA Greis	508-754-6891	R	10	<0.1
Seaberg Precision Corp	165 Field St	West Babylon	NY	11704	Martha Seaberg	516-694-3871	R	10	<0.1
Anderson-Cook Inc	17650 15 Mile Rd	Fraser	MI	48026	William E Bogard	810-293-0800	R	9*	0.1
Suhner Industrial Products Corp	PO Box 1234	Rome	GA	30162	Paul Luthi	706-235-8046	R	9	0.1
Sesco Inc	7800 Dix Av	Detroit	MI	48209	Werner K Lehmann	313-843-7710	R	8	0.1
BTM Corp	300 Davis Rd	Marysville	MI	48040	Ed Sawdon	810-364-4567	R	8	<0.1
Dreis & Krump Mfg Co	7400 S Loomis Blv	Chicago	IL	60636	Rudy A Wolfer	312-874-1200	R	8	0.1
Grotnes Metalforming Systems	1025 W Thorndale	Itasca	IL	60143	Curtis Maas	312-769-1111	S	8	<0.1
AKH Inc	2405 Production Dr	Indianapolis	IN	46241	Gordon Goranson	317-243-5915	R	7	<0.1
Chicago Rivet and Machine Co	PO Box 8	Albia	IA	52531	Steve Larson	515-932-7107	D	7*	<0.1
Compumachine Inc	645 Main St	Wilmington	MA	01887	David Shaby	508-657-8440	R	7	<0.1
PRD Company Inc	1321 W Winton Av	Hayward	CA	94545	Robert F Miller	510-782-7242	R	7	<0.1
Tishken Products Co	13000 W 8 Mile Rd	Oak Park	MI	48237	GH Bahl Jr	313-399-9200	R	7	<0.1
Tools For Bending Inc	194 W Dakota Av	Denver	CO	80223	RR Stange	303-777-7170	R	7	<0.1
Abbey Etna Machine Co	11140 Avnue Rd	Perrysburg	OH	43551	Nelson Abbey III	419-874-4301	R	6*	<0.1
CJ Winter Machine Works	130 Albert St	Rochester	NY	14606	RJ Brinkman	716-429-5000	R	6	<0.1
LVD Corp	5 Northwest Dr	Plainville	CT	06062	P Rodin	203-747-4581	S	6	<0.1
Marks Machine and Tool Corp	PO Box 2095	Syracuse	NY	13220	Donald Etmanski	315-463-1751	R	6*	<0.1
PH Hydraulics and Automation	2365 Scioto Harper	Columbus	OH	43204	Charles Sherman	614-279-8877	S	6	<0.1
FH Peterson Machine Corp	PO Box 617	Stoughton	MA	02072	Stanley B Urban	617-341-4930	R	6	<0.1
Triana Industries Inc	500 6th St	Madison	AL	35758	George Malone	205-772-9304	R	5	0.1
Ajax Manufacturing Co	1441 Chardon Rd	Euclid	OH	44117	Ed Crawford	216-531-1010	R	5	<0.1
Eitel Presses Inc	PO Box 130	Orwigsburg	PA	17961	Ronald Schildge	717-366-0585	R	5	<0.1
Grob Inc	1731 10th Av	Grafton	WI	53024	B Grob	414-377-1400	R	5	<0.1
Rafter Equipment Corp	12430 Alameda Dr	Strongsville	OH	44136	Walter Krenz	216-572-3700	R	5	<0.1
Sonobond Ultrasonics Inc	887 S Matlack St	West Chester	PA	19382	Janet Devine	215-696-4710	S	5	<0.1
Threaded Rod Company Inc	1929 Columbia Av	Indianapolis	IN	46202	Harry Branson	317-921-3000	R	5	<0.1
Grant Assembly Technologies	PO Box 3345	Bridgeport	CT	06605	B W McNaugton	203-366-4557	R	4	<0.1
Mohawk Industries Inc	601 Amherst St	Buffalo	NY	14207	James Kenline	716-874-4371	R	4	<0.1
Multipress	560 Dublin Av	Columbus	OH	43215	James S Renald	614-228-0185	D	4	<0.1
O'Connell Machinery Co	175 Great Arrow	Buffalo	NY	14207	WC O'Connell Jr	716-877-3666	R	4	<0.1
Abel Automatics Inc	165 Aviador Rd	Camarillo	CA	93010	Steve Abel	805-484-8789	R	3	<0.1
Alva Allen Industries Inc	1001-15 N 3rd St	Clinton	MO	64735	Alva F Allen Jr	816-885-3331	R	3	<0.1
Container Tooling Corp	349 Progress Rd	Dayton	OH	45449	Steve Buck	513-859-5110	S	3	<0.1
Extek Inc	36B Cherry Hill Dr	Danvers	MA	01923	Bart Jones	508-762-6500	R	3	<0.1
Kard Corp	PO Box 6569	Orange	CA	92613	R Haupt	714-632-7611	R	3*	<0.1
National Diecasting Machinery	33 Plan Way	Warwick	RI	02886	Lyn Johnsen	401-737-3005	D	3	<0.1
New Deal Tool and Machine	245 Leo St	Dayton	OH	45404	Fred Ehrensberger	513-228-9109	D	3*	<0.1
Oak Products Inc	PO Box 840	Sturgis	MI	49091	NA Franks	616-651-8513	R	3	<0.1
Rapid Air Corp	4601 Kishwaukee St	Rockford	IL	61109	RD Nordlof	815-397-2578	S	3	<0.1
Walsh Press Co	1222 S Hannah Av	Forest Park	IL	60130	Jerome W Heyda	708-771-2480	S	3	<0.1

Source: Ward's Business Directory of U.S. Private and Public Companies, Volumes 1 and 2, 1996. The company type code used is as follows: P - Public, R - Private, S - Subsidiary, D - Division, J - Joint Venture, A - Affiliate, G - Group. Sales are in millions of dollars, employees are in thousands. An asterisk (*) indicates an estimated sales volume. The symbol < stands for 'less than'. Company names and addresses are truncated, in some cases, to fit into the available space.

MATERIALS CONSUMED

Material	Quantity	Delivered Cost ($ million)
Materials, ingredients, containers, and supplies	(X)	593.7
Fluid power pumps, motors, and hydrostatic transmissions (hydraulic and pneumatic)	(X)	17.6
Fluid power cylinders and rotary actuators (hydraulic and pneumatic)	(X)	8.2
Fluid power filters (hydraulic and pneumatic)	(X)	2.2
Fluid power hose or tube fittings and assemblies (hydraulic and pneumatic)	(X)	3.2
Fluid power valves (hydraulic and pneumatic)	(X)	8.9
Metal bolts, nuts, screws, washers, rivets, and other screw machine products	(X)	11.4
Other fabricated metal products (except fluid power products and forgings)	(X)	25.0
Iron and steel forgings	(X)	10.5
Nonferrous forgings	(X)	0.4
Iron and steel castings (rough and semifinished)	(X)	29.4
Aluminum and aluminum-base alloy castings (rough and semifinished)	(X)	4.0
Other nonferrous castings (rough and semifinished)	(X)	0.4
Steel bars, bar shapes, and plates	(X)	75.8
Steel sheet and strip, including tin plate	(X)	7.8
Steel structural shapes	(X)	19.1
All other steel shapes and forms	(X)	16.0
Aluminum and aluminum-base alloy shapes and forms	(X)	1.3
Other nonferrous shapes and forms	(X)	4.8
Fractional horsepower electric timing motors, synchronous and subsynchronous (less than 1 hp)	(X)	1.4
Other fractional horsepower electric motors (under 1 hp)	(X)	0.5
Integral horsepower electric motors and generators (1 hp or more)	(X)	14.9
Electrical transmission, distribution, and control equipment	(X)	16.3
Electrical industrial capacitors, resistors, rheostats, and coil windings	(X)	4.0
Numerical controls for metalworking machinery (except programmable)	(X)	3.7
Programmable controllers for metalworking machinery	(X)	22.2
Ball bearings (mounted or unmounted)	(X)	4.3
Roller bearings (mounted or unmounted)	(X)	2.9
Mechanical speed changers, gears, and industrial high-speed drives	(X)	6.9
Wood boxes, pallets, skids, and containers	(X)	2.8
Cutting tools for machine tools	(X)	16.6
All other materials and components, parts, containers, and supplies	(X)	141.6
Materials, ingredients, containers, and supplies, nsk	(X)	109.4

Source: 1992 *Economic Census*. Explanation of symbols used: (D): Withheld to avoid disclosure of competitive data; na: Not available; (S): Withheld because statistical norms were not met; (X): Not applicable; (Z): Less than half the unit shown; nec: Not elsewhere classified; nsk: Not specified by kind; - : zero; * : 10-19 percent estimated; ** : 20-29 percent estimated.

PRODUCT SHARE DETAILS

Product or Product Class	% Share	Product or Product Class	% Share
Machine tools, metal forming types	100.00	Parts for metal forming machine tools (sold separately) and rebuilt metal forming machine tools	27.06
Metal punching and shearing machines (including power and manual) and bending and forming machines (power only)	25.25	Parts for metal forming machine tools (sold separately)	83.54
Metalworking presses (except forging and die-stamping presses)	23.60	Rebuilt metal forming machine tools	10.27
Other metal forming machine tools, including forging and die-stamping machines (except metalworking presses)	22.75	Parts for metal forming machine tools (sold separately) and rebuilt metal forming machine tools, nsk	6.19
		Machine tools, metal forming types, nsk	1.35

Source: 1992 *Economic Census*. The values shown are percent of total shipments in an industry. Values of indented subcategories are summed in the main heading. The symbol (D) appears when data are withheld to prevent disclosure of competitive information. The abbreviation nsk stands for 'not specified by kind' and nec for 'not elsewhere classified'.

INPUTS AND OUTPUTS FOR MACHINE TOOLS, METAL FORMING TYPES

Economic Sector or Industry Providing Inputs	%	Sector	Economic Sector or Industry Buying Outputs	%	Sector
Imports	33.5	Foreign	Gross private fixed investment	62.5	Cap Inv
Blast furnaces & steel mills	8.8	Manufg.	Exports	20.7	Foreign
Wholesale trade	6.1	Trade	Miscellaneous repair shops	1.6	Services
Iron & steel foundries	5.0	Manufg.	Machine tools, metal forming types	1.4	Manufg.
Machinery, except electrical, nec	3.7	Manufg.	Federal Government purchases, national defense	1.2	Fed Govt
Machine tools, metal forming types	2.9	Manufg.	Fabricated plate work (boiler shops)	1.1	Manufg.
Iron & steel forgings	2.2	Manufg.	Nonferrous castings, nec	0.9	Manufg.
Motors & generators	2.1	Manufg.	Aircraft	0.5	Manufg.
Copper rolling & drawing	2.0	Manufg.	Miscellaneous fabricated wire products	0.4	Manufg.
Fabricated plate work (boiler shops)	2.0	Manufg.	Automotive stampings	0.3	Manufg.
Electric services (utilities)	2.0	Util.	Blast furnaces & steel mills	0.3	Manufg.
Petroleum refining	1.9	Manufg.	Electronic computing equipment	0.3	Manufg.
Ball & roller bearings	1.3	Manufg.	Fabricated metal products, nec	0.3	Manufg.
Communications, except radio & TV	1.3	Util.	Motor vehicles & car bodies	0.3	Manufg.
Industrial controls	1.2	Manufg.	Radio & TV communication equipment	0.3	Manufg.
Abrasive products	1.1	Manufg.	Switchgear & switchboard apparatus	0.3	Manufg.
Pumps & compressors	1.1	Manufg.	Aircraft & missile equipment, nec	0.2	Manufg.

Continued on next page.

INPUTS AND OUTPUTS FOR MACHINE TOOLS, METAL FORMING TYPES - Continued

Economic Sector or Industry Providing Inputs	%	Sector	Economic Sector or Industry Buying Outputs	%	Sector
Switchgear & switchboard apparatus	1.0	Manufg.	Crowns & closures	0.2	Manufg.
Power transmission equipment	0.9	Manufg.	Electronic components nec	0.2	Manufg.
Eating & drinking places	0.9	Trade	Household cooking equipment	0.2	Manufg.
Machine tools, metal cutting types	0.8	Manufg	Household refrigerators & freezers	0.2	Manufg.
Gas production & distribution (utilities)	0.8	Util.	Iron & steel forgings	0.2	Manufg.
Banking	0.8	Fin/R.E.	Lighting fixtures & equipment	0.2	Manufg.
Maintenance of nonfarm buildings nec	0.7	Constr.	Machinery, except electrical, nec	0.2	Manufg.
Special dies & tools & machine tool accessories	0.7	Manufg.	Metal doors, sash, & trim	0.2	Manufg.
Motor freight transportation & warehousing	0.7	Util.	Motor vehicle parts & accessories	0.2	Manufg.
Equipment rental & leasing services	0.7	Services	Motors & generators	0.2	Manufg.
Pipe, valves, & pipe fittings	0.6	Manufg.	Pipe, valves, & pipe fittings	0.2	Manufg.
Screw machine and related products	0.6	Manufg.	Prefabricated metal buildings	0.2	Manufg.
Air transportation	0.6	Util.	Rolling mill machinery	0.2	Manufg.
Advertising	0.6	Services	Special dies & tools & machine tool accessories	0.2	Manufg.
Computer & data processing services	0.6	Services	Telephone & telegraph apparatus	0.2	Manufg.
Brass, bronze, & copper castings	0.5	Manufg.	Wiring devices	0.2	Manufg.
Cyclic crudes and organics	0.5	Manufg.	Architectural metal work	0.1	Manufg.
Transformers	0.5	Manufg.	Blowers & fans	0.1	Manufg.
U.S. Postal Service	0.5	Gov't	Electric housewares & fans	0.1	Manufg.
Aluminum castings	0.4	Manufg.	Fabricated structural metal	0.1	Manufg.
Industrial patterns	0.4	Manufg.	Farm machinery & equipment	0.1	Manufg.
Metal heat treating	0.4	Manufg.	Heating equipment, except electric	0.1	Manufg.
Paints & allied products	0.4	Manufg.	Industrial controls	0.1	Manufg.
Real estate	0.4	Fin/R.E.	Metal stampings, nec	0.1	Manufg.
Engineering, architectural, & surveying services	0.4	Services	Semiconductors & related devices	0.1	Manufg.
Primary metal products, nec	0.3	Manufg.	Transformers	0.1	Manufg.
Automotive rental & leasing, without drivers	0.3	Services			
Automotive repair shops & services	0.3	Services			
Hotels & lodging places	0.3	Services			
Legal services	0.3	Services			
Management & consulting services & labs	0.3	Services			
Chemical preparations, nec	0.2	Manufg.			
Miscellaneous fabricated wire products	0.2	Manufg.			
Nonferrous rolling & drawing, nec	0.2	Manufg.			
Nonferrous wire drawing & insulating	0.2	Manufg.			
Sawmills & planning mills, general	0.2	Manufg.			
Insurance carriers	0.2	Fin/R.E.			
Accounting, auditing & bookkeeping	0.2	Services			
Noncomparable imports	0.2	Foreign			
Hand & edge tools, nec	0.1	Manufg.			
Lubricating oils & greases	0.1	Manufg.			
Manifold business forms	0.1	Manufg.			
Miscellaneous plastics products	0.1	Manufg.			
Paperboard containers & boxes	0.1	Manufg.			
Railroads & related services	0.1	Util.			
Sanitary services, steam supply, irrigation	0.1	Util.			
Electrical repair shops	0.1	Services			

Source: Benchmark Input-Output Accounts for the U.S. Economy, 1982, U.S. Department of Commerce, Washington, D.C., July 1991. Data, as reported in the source, are organized by the 1977 SIC structure in use in 1982 but have been matched, as closely as is possible, to the 1987 SIC structure used in this book.

OCCUPATIONS EMPLOYED BY SIC 354 - METALWORKING MACHINERY

Occupation	% of Total 1994	Change to 2005	Occupation	% of Total 1994	Change to 2005
Machinists	10.9	-17.4	Precision metal workers nec	1.9	10.1
Tool & die makers	10.6	-15.3	Mechanical engineers	1.9	1.0
General managers & top executives	3.7	-12.9	Inspectors, testers, & graders, precision	1.9	-8.2
Blue collar worker supervisors	3.4	-15.7	Secretaries, ex legal & medical	1.9	-16.4
Machine tool cutting operators, metal & plastic	2.9	-5.9	Drafters	1.8	-28.5
NC machine tool operators, metal & plastic	2.7	1.0	General office clerks	1.5	-21.7
Grinding machine operators, metal & plastic	2.7	-8.2	Machine forming operators, metal & plastic	1.4	-8.2
Sales & related workers nec	2.7	-8.2	Drilling & boring machine tool workers	1.3	-8.2
Assemblers, fabricators, & hand workers nec	2.5	-8.2	Bookkeeping, accounting, & auditing clerks	1.3	-31.1
Lathe & turning machine tool operators	2.3	-17.4	Industrial production managers	1.3	-8.2
Machine tool cutting & forming etc. nec	2.2	-54.1	Traffic, shipping, & receiving clerks	1.3	-11.7
Machine builders	2.0	1.0	Precision woodworkers nec	1.3	10.2
Combination machine tool operators	2.0	37.7	Janitors & cleaners, incl maids	1.2	-26.6

Source: Industry-Occupation Matrix, Bureau of Labor Statistics. These data relate to one or more 3-digit SIC industry groups rather than to a single 4-digit SIC. The change reported for each occupation to the year 2005 is a percent of growth or decline as estimated by the Bureau of Labor Statistics. The abbreviation nec stands for 'not elsewhere classified'.

LOCATION BY STATE AND REGIONAL CONCENTRATION

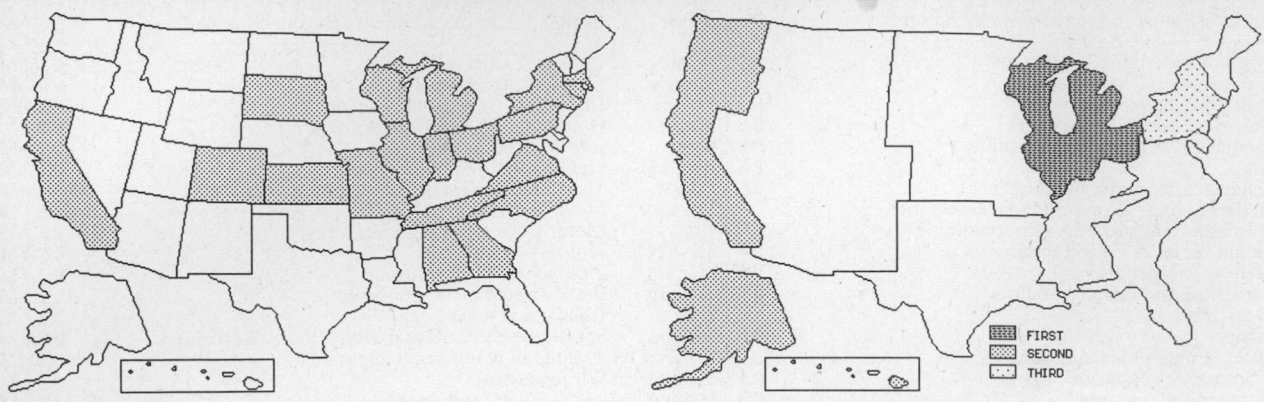

FIRST
SECOND
THIRD

INDUSTRY DATA BY STATE

State	Establish-ments	Shipments			Employment				Cost as % of Shipments	Investment per Employee ($)
		Total ($ mil)	% of U.S.	Per Establ.	Total Number	% of U.S.	Per Establ.	Wages ($/hour)		
Ohio	31	345.8	23.8	11.2	3,000	24.6	97	14.36	44.4	3,200
Illinois	32	341.0	23.5	10.7	2,700	22.1	84	17.64	61.3	2,704
Michigan	27	173.3	11.9	6.4	1,300	10.7	48	14.65	40.9	4,154
New York	10	117.6	8.1	11:8	1,100	9.0	110	15.25	46.0	-
California	21	113.1	7.8	5.4	800	6.6	38	14.67	47.9	2,875
Connecticut	7	46.5	3.2	6.6	400	3.3	57	16.00	42.4	-
Pennsylvania	8	45.6	3.1	5.7	500	4.1	63	15.17	61.0	2,200
Massachusetts	4	15.4	1.1	3.8	100	0.8	25	16.00	29.2	4,000
Wisconsin	6	13.5	0.9	2.3	100	0.8	17	13.00	46.7	1,000
Indiana	8	13.3	0.9	1.7	200	1.6	25	10.33	38.3	2,000
Missouri	6	12.9	0.9	2.2	100	0.8	17	13.00	41.1	1,000
Alabama	4	11.0	0.8	2.8	100	0.8	25	13.50	37.3	-
Tennessee	4	8.9	0.6	2.2	100	0.8	25	9.00	46.1	-
North Carolina	6	(D)	-	-	375 *	3.1	63	-	-	-
Georgia	4	(D)	-	-	175 *	1.4	44	-	-	1,714
South Dakota	3	(D)	-	-	175 *	1.4	58	-	-	-
Virginia	3	(D)	-	-	375 *	3.1	125	-	-	-
Colorado	2	(D)	-	-	175 *	1.4	88	-	-	-
Kansas	2	(D)	-	-	175 *	1.4	88	-	-	-

Source: 1992 *Economic Census*. The states are in descending order of shipments or establishments (if shipment data are missing for the majority). The symbol (D) appears when data are withheld to prevent disclosure of competitive information. States marked with (D) are sorted by number of establishments. A dash (-) indicates that the data element cannot be calculated; * indicates the midpoint of a range.

3543 - INDUSTRIAL PATTERNS

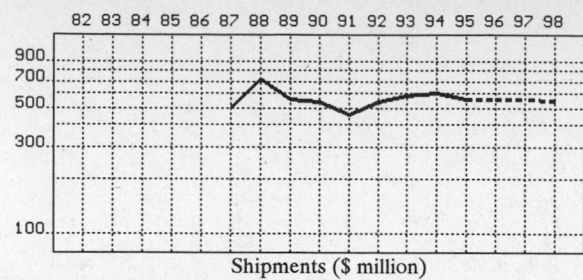

Shipments ($ million)

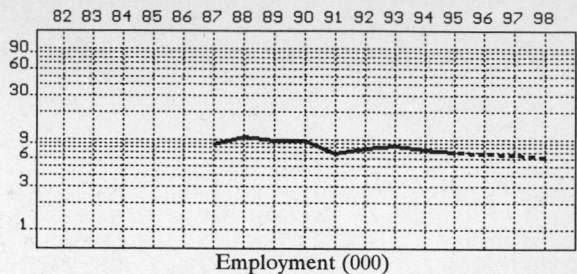

Employment (000)

GENERAL STATISTICS

Year	Com-panies	Establishments		Employment			Compensation		Production ($ million)			
		Total	with 20 or more employees	Total (000)	Production Workers (000)	Hours (Mil)	Payroll ($ mil)	Wages ($/hr)	Cost of Materials	Value Added by Manufacture	Value of Shipments	Capital Invest.
1982												
1983												
1984												
1985												
1986												
1987	812	813	85	8.6	6.9	13.8	237.5	13.51	112.5	389.3	499.4	18.5
1988		775	93	10.5	8.4	16.8	315.4	14.92	207.6	511.6	719.5	13.0
1989		738	98	9.7	7.5	15.0	271.2	14.28	132.6	436.2	556.9	15.7
1990		734	99	9.4	6.7	13.5	257.3	14.39	133.0	396.6	534.3	11.0
1991		735	92	6.9	5.6	10.5	223.0	16.08	117.5	340.5	461.7	6.0
1992	708	711	87	7.9	6.5	13.2	258.7	15.12	125.7	416.3	539.0	15.1
1993		677	97	8.3	7.1	14.3	249.8	14.31	123.7	459.3	576.5	16.3
1994		663P	96P	7.6	5.9	12.1	253.7	16.46	132.4	479.0	597.5	26.8
1995		644P	96P	7.2P	5.9P	11.9P	243.0P	16.05P	122.4P	442.8P	552.3P	18.9P
1996		625P	97P	6.9P	5.7P	11.5P	239.6P	16.31P	122.0P	441.3P	550.5P	19.7P
1997		606P	98P	6.6P	5.5P	11.1P	236.2P	16.57P	121.6P	439.8P	548.6P	20.5P
1998		586P	98P	6.3P	5.3P	10.7P	232.8P	16.83P	121.2P	438.4P	546.8P	21.4P

Sources: 1982, 1987, 1992 *Economic Census*; *Annual Survey of Manufactures*, 83-86, 88-91, 93-94. Establishment counts for non-Census years are from *County Business Patterns*; establishment values for 83-84 are extrapolations. 'P's show projections by the editors. Industries reclassified in 87 will not have data for prior years.

INDICES OF CHANGE

Year	Com-panies	Establishments		Employment			Compensation		Production ($ million)			
		Total	with 20 or more employees	Total (000)	Production Workers (000)	Hours (Mil)	Payroll ($ mil)	Wages ($/hr)	Cost of Materials	Value Added by Manufacture	Value of Shipments	Capital Invest.
1982												
1983												
1984												
1985												
1986												
1987	115	114	98	109	106	105	92	89	89	94	93	123
1988		109	107	133	129	127	122	99	165	123	133	86
1989		104	113	123	115	114	105	94	105	105	103	104
1990		103	114	119	103	102	99	95	106	95	99	73
1991		103	106	87	86	80	86	106	93	82	86	40
1992	100	100	100	100	100	100	100	100	100	100	100	100
1993		95	111	105	109	108	97	95	98	110	107	108
1994		93P	110P	96	91	92	98	109	105	115	111	177
1995		91P	111P	91P	90P	90P	94P	106P	97P	106P	102P	125P
1996		88P	111P	88P	87P	87P	93P	108P	97P	106P	102P	131P
1997		85P	112P	84P	84P	84P	91P	110P	97P	106P	102P	136P
1998		82P	113P	80P	81P	81P	90P	111P	96P	105P	101P	141P

Sources: Same as General Statistics. Values reflect change from the base year, 1992. Values above 100 mean greater than 92, values below 100 mean less than 92, and a value of 100 in the 82-91 or 93-98 period means same as 92. 'P's mark projections by the editors.

SELECTED RATIOS

For 1994	Avg. of All Manufact.	Analyzed Industry	Index	For 1994	Avg. of All Manufact.	Analyzed Industry	Index
Employees per Establishment	49	11	23	Value Added per Production Worker	134,084	81,186	61
Payroll per Establishment	1,500,273	382,407	25	Cost per Establishment	5,045,178	199,569	4
Payroll per Employee	30,620	33,382	109	Cost per Employee	102,970	17,421	17
Production Workers per Establishment	34	9	26	Cost per Production Worker	146,988	22,441	15
Wages per Establishment	853,319	300,207	35	Shipments per Establishment	9,576,895	900,624	9
Wages per Production Worker	24,861	33,757	136	Shipments per Employee	195,460	78,618	40
Hours per Production Worker	2,056	2,051	100	Shipments per Production Worker	279,017	101,271	36
Wages per Hour	12.09	16.46	136	Investment per Establishment	321,011	40,396	13
Value Added per Establishment	4,602,255	722,007	16	Investment per Employee	6,552	3,526	54
Value Added per Employee	93,930	63,026	67	Investment per Production Worker	9,352	4,542	49

Sources: Same as General Statistics. The 'Average of All Manufacturing' column represents the average of all manufacturing industries reported for the most recent complete year available. The Index shows the relationship between the Average and the Analyzed Industry. For example, 100 means that they are equal; 500 that the Analyzed Industry is five times the average; 50 means that the Analyzed Industry is half the national average. The abbreviation 'na' is used to show that data are 'not available'.

LEADING COMPANIES   Number shown: **22**   Total sales ($ mil): **203**   Total employment (000): **2.1**

Company Name	Address				CEO Name	Phone	Co. Type	Sales ($ mil)	Empl. (000)
Cole Pattern&Engineering Co	4912 Lima Rd	Fort Wayne	IN	46808	Byron J Cole	219-483-0382	R	77*	1.0
Progress Pattern Corp	21555 Telegraph Rd	Southfield	MI	48034	Al Goscinski	313-358-5100	S	25	0.2
D and F Corp	42455 Merrill Rd	Sterling Hts	MI	48314	Paul D Gard	810-854-5300	R	16	0.1
FAI Inc	PO Box 1588	Racine	WI	53401	Peter Christensen	414-637-9151	R	15*	0.1
Anderson Pattern Inc	PO Box 1088	Muskegon	MI	49443	JR McIntyre	616-733-2164	R	12	0.1
Southern Precision Corp	PO Box 100035	Birmingham	AL	35210	Robert McCulley	205-956-3556	S	12	<0.1
Binderline Development Inc	33100 Freeway Dr	St Clair Shores	MI	48082	Stephen E Nash	313-294-1620	S	10*	<0.1
Lake Erie Design Inc	1470 E 289th St	Wickliffe	OH	44092	JS Kryvicky	216-944-1880	R	8	0.1
Model Pattern Co	25 Leonard St	Grand Rapids	MI	49503	Otto Schlatter	616-456-5745	R	6	<0.1
Paragon Pattern & Mfg Co	2620 Park St	Muskegon H	MI	49444	LA Price	616-733-1582	R	5	<0.1
Watkins Pattern Company Inc	8420 220th West St	Lakeville	MN	55044	Chris May	612-469-4921	S	4	<0.1
Central Pattern Co	PO Box 10	Hazelwood	MO	63042	Mark C Petersen	314-524-3626	R	3	<0.1
Allen Pattern of Michigan Inc	202 McGrath Pl	Battle Creek	MI	49017	Wendell Allen	616-963-4131	R	2	<0.1
Acme Pattern Works Inc	PO Box 4	Chicago Hts	IL	60411	Edmund Heuberger	708-755-5613	R	2	<0.1
Gopher Pattern Works Inc	422 Roosevelt NE	Minneapolis	MN	55413	Floyd Jaehnert	612-331-5512	R	2	<0.1
Campbell Pattern Associates	300 N Orange Av	Brea	CA	92621	Bruce Bullock	714-990-5600	R	1	<0.1
Cunningham Pattern	PO Box 854	Columbus	IN	47202	J Cunningham	812-379-9571	R	1	<0.1
Eifel Pattern and Model	PO Box 190	Fraser	MI	48026	J Hecker	313-296-9640	R	1	<0.1
Production Pattern Shop Inc	4244 E 12th St	Oakland	CA	94601	Robert L Lambert	510-534-1133	R	1	<0.1
United Industries Inc	1899 Revere Beach	Everett	MA	02149	Richard B Adams	617-387-9500	R	1	<0.1
Lacy Foundries Inc	1602 Thames St	Baltimore	MD	21231	Joseph Lacy Sr	410-342-1148	R	1	<0.1
Peterson Pattern Works Inc	126 Spokane St	Seattle	WA	98134	Richard Peterson	206-622-4510	R	1	<0.1

Source: Ward's Business Directory of U.S. Private and Public Companies, Volumes 1 and 2, 1996. The company type code used is as follows: P - Public, R - Private, S - Subsidiary, D - Division, J - Joint Venture, A - Affiliate, G - Group. Sales are in millions of dollars, employees are in thousands. An asterisk (*) indicates an estimated sales volume. The symbol < stands for 'less than'. Company names and addresses are truncated, in some cases, to fit into the available space.

MATERIALS CONSUMED

Material	Quantity	Delivered Cost ($ million)
Materials, ingredients, containers, and supplies	(X)	99.4
Fabricated metal products (except castings and forgings)	(X)	2.4
Forgings	(X)	(Z)
Iron and steel castings (rough and semifinished)	(X)	8.9
Aluminum and aluminum-base alloy castings (rough and semifinished)	(X)	4.4
Other nonferrous castings (rough and semifinished)	(X)	1.1
Steel shapes and forms	(X)	0.8
Aluminum and aluminum-base alloy shapes and forms	(X)	1.1
Other nonferrous shapes and forms	(X)	0.1
Rough and dressed lumber	(X)	3.0
Plastics products consumed in the form of sheets, rods, tubes, film, and other shapes	(X)	1.1
All other materials and components, parts, containers, and supplies	(X)	27.1
Materials, ingredients, containers, and supplies, nsk	(X)	49.4

Source: 1992 Economic Census. Explanation of symbols used: (D): Withheld to avoid disclosure of competitive data; na: Not available; (S): Withheld because statistical norms were not met; (X): Not applicable; (Z): Less than half the unit shown; nec: Not elsewhere classified; nsk: Not specified by kind; - : zero; * : 10-19 percent estimated; ** : 20-29 percent estimated.

PRODUCT SHARE DETAILS

Product or Product Class	% Share	Product or Product Class	% Share
Industrial patterns	100.00	All other industrial patterns (except shoe patterns)	12.24
Foundry patterns	73.96		

Source: 1992 Economic Census. The values shown are percent of total shipments in an industry. Values of indented subcategories are summed in the main heading. The symbol (D) appears when data are withheld to prevent disclosure of competitive information. The abbreviation nsk stands for 'not specified by kind' and nec for 'not elsewhere classified'.

INPUTS AND OUTPUTS FOR INDUSTRIAL PATTERNS

Economic Sector or Industry Providing Inputs	%	Sector	Economic Sector or Industry Buying Outputs	%	Sector
Veneer & plywood	8.9	Manufg.	Miscellaneous repair shops	27.0	Services
Cyclic crudes and organics	8.3	Manufg.	Architectural metal work	21.4	Manufg.
Machinery, except electrical, nec	8.1	Manufg.	Pipe, valves, & pipe fittings	12.4	Manufg.
Wholesale trade	7.1	Trade	Iron & steel foundries	6.7	Manufg.
Sawmills & planning mills, general	5.5	Manufg.	Machinery, except electrical, nec	5.4	Manufg.
Iron & steel foundries	4.7	Manufg.	Special dies & tools & machine tool accessories	4.7	Manufg.
Blast furnaces & steel mills	3.9	Manufg.	Engineering & scientific instruments	3.5	Manufg.
Special dies & tools & machine tool accessories	3.5	Manufg.	Prefabricated metal buildings	3.2	Manufg.
Industrial patterns	3.0	Manufg.	Fabricated structural metal	1.7	Manufg.
Real estate	3.0	Fin/R.E.	Blowers & fans	1.6	Manufg.
Electric services (utilities)	2.8	Util.	Wiring devices	1.5	Manufg.
Miscellaneous plastics products	2.6	Manufg.	Fabricated plate work (boiler shops)	1.3	Manufg.
Maintenance of nonfarm buildings nec	2.3	Constr.	Aluminum castings	1.1	Manufg.
Aluminum rolling & drawing	2.2	Manufg.	Farm machinery & equipment	1.0	Manufg.
Eating & drinking places	2.2	Trade	Construction machinery & equipment	0.9	Manufg.
Engineering, architectural, & surveying services	2.2	Services	Cutlery	0.9	Manufg.
Aluminum castings	2.0	Manufg.	Industrial patterns	0.9	Manufg.
Gas production & distribution (utilities)	1.4	Util.	Motor vehicle parts & accessories	0.9	Manufg.
Brass, bronze, & copper castings	1.3	Manufg.	Railroad equipment	0.9	Manufg.
Air transportation	1.3	Util.	Carburetors, pistons, rings, & valves	0.8	Manufg.
Communications, except radio & TV	1.3	Util.	Machine tools, metal forming types	0.5	Manufg.
Banking	1.3	Fin/R.E.	Exports	0.5	Foreign
Abrasive products	1.2	Manufg.	Brass, bronze, & copper castings	0.3	Manufg.
Petroleum refining	1.2	Manufg.	Industrial trucks & tractors	0.3	Manufg.
Motor freight transportation & warehousing	1.2	Util.	Scales & balances	0.3	Manufg.
Sanitary services, steam supply, irrigation	1.2	Util.	Metalworking machinery, nec	0.1	Manufg.
Advertising	1.2	Services	Power driven hand tools	0.1	Manufg.
Nonferrous castings, nec	1.0	Manufg.			
Industrial inorganic chemicals, nec	0.9	Manufg.			
Railroads & related services	0.9	Util.			
Ball & roller bearings	0.8	Manufg.			
Legal services	0.8	Services			
Management & consulting services & labs	0.8	Services			
Copper rolling & drawing	0.7	Manufg.			
Metal heat treating	0.6	Manufg.			
Business/professional associations	0.6	Services			
Equipment rental & leasing services	0.6	Services			
Hotels & lodging places	0.6	Services			
Imports	0.6	Foreign			
Lubricating oils & greases	0.5	Manufg.			
Manifold business forms	0.5	Manufg.			
Accounting, auditing & bookkeeping	0.5	Services			
Computer & data processing services	0.5	Services			
U.S. Postal Service	0.4	Gov't			
General industrial machinery, nec	0.3	Manufg.			
Paperboard containers & boxes	0.3	Manufg.			
Alkalies & chlorine	0.2	Manufg.			
Gaskets, packing & sealing devices	0.2	Manufg.			
Industrial gases	0.2	Manufg.			
Machine tools, metal cutting types	0.2	Manufg.			
Miscellaneous fabricated wire products	0.2	Manufg.			
Screw machine and related products	0.2	Manufg.			
Transit & bus transportation	0.2	Util.			
Insurance carriers	0.2	Fin/R.E.			
Laundry, dry cleaning, shoe repair	0.2	Services			
Personnel supply services	0.2	Services			
Manufacturing industries, nec	0.1	Manufg.			
Mechanical measuring devices	0.1	Manufg.			
Photographic equipment & supplies	0.1	Manufg.			
Water transportation	0.1	Util.			
Automotive rental & leasing, without drivers	0.1	Services			
Automotive repair shops & services	0.1	Services			
Photofinishing labs, commercial photography	0.1	Services			
Services to dwellings & other buildings	0.1	Services			

Source: Benchmark Input-Output Accounts for the U.S. Economy, 1982, U.S. Department of Commerce, Washington, D.C., July 1991. Data, as reported in the source, are organized by the 1977 SIC structure in use in 1982 but have been matched, as closely as is possible, to the 1987 SIC structure used in this book.

OCCUPATIONS EMPLOYED BY SIC 354 - METALWORKING MACHINERY

Occupation	% of Total 1994	Change to 2005	Occupation	% of Total 1994	Change to 2005
Machinists	10.9	-17.4	Precision metal workers nec	1.9	10.1
Tool & die makers	10.6	-15.3	Mechanical engineers	1.9	1.0
General managers & top executives	3.7	-12.9	Inspectors, testers, & graders, precision	1.9	-8.2
Blue collar worker supervisors	3.4	-15.7	Secretaries, ex legal & medical	1.9	-16.4
Machine tool cutting operators, metal & plastic	2.9	-5.9	Drafters	1.8	-28.5
NC machine tool operators, metal & plastic	2.7	1.0	General office clerks	1.5	-21.7
Grinding machine operators, metal & plastic	2.7	-8.2	Machine forming operators, metal & plastic	1.4	-8.2
Sales & related workers nec	2.7	-8.2	Drilling & boring machine tool workers	1.3	-8.2
Assemblers, fabricators, & hand workers nec	2.5	-8.2	Bookkeeping, accounting, & auditing clerks	1.3	-31.1
Lathe & turning machine tool operators	2.3	-17.4	Industrial production managers	1.3	-8.2
Machine tool cutting & forming etc. nec	2.2	-54.1	Traffic, shipping, & receiving clerks	1.3	-11.7
Machine builders	2.0	1.0	Precision woodworkers nec	1.3	10.2
Combination machine tool operators	2.0	37.7	Janitors & cleaners, incl maids	1.2	-26.6

Source: *Industry-Occupation Matrix*, Bureau of Labor Statistics. These data relate to one or more 3-digit SIC industry groups rather than to a single 4-digit SIC. The change reported for each occupation to the year 2005 is a percent of growth or decline as estimated by the Bureau of Labor Statistics. The abbreviation nec stands for 'not elsewhere classified'.

LOCATION BY STATE AND REGIONAL CONCENTRATION

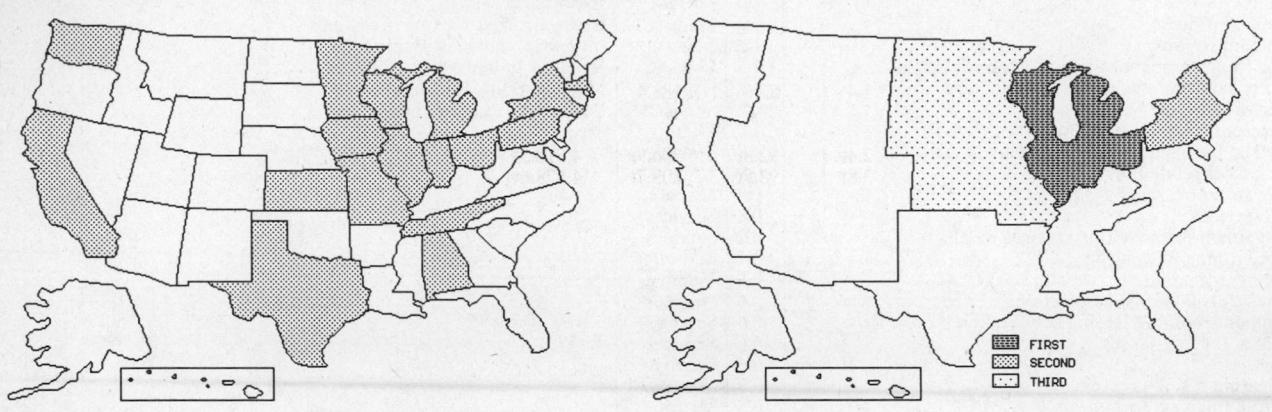

FIRST
SECOND
THIRD

INDUSTRY DATA BY STATE

State	Establish-ments	Shipments Total ($ mil)	Shipments % of U.S.	Shipments Per Establ.	Employment Total Number	Employment % of U.S.	Employment Per Establ.	Wages ($/hour)	Cost as % of Shipments	Investment per Employee ($)
Michigan	94	146.4	27.2	1.6	1,700	21.5	18	19.64	20.2	2,000
Ohio	94	68.1	12.6	0.7	1,100	13.9	12	13.79	24.7	1,273
Wisconsin	59	55.4	10.3	0.9	800	10.1	14	15.80	19.5	3,375
Indiana	29	30.4	5.6	1.0	500	6.3	17	14.38	22.7	1,800
Illinois	51	29.0	5.4	0.6	400	5.1	8	14.00	20.0	1,250
Tennessee	7	24.4	4.5	3.5	500	6.3	71	10.25	36.9	1,200
Alabama	21	17.4	3.2	0.8	200	2.5	10	15.00	39.7	2,000
Iowa	13	14.6	2.7	1.1	200	2.5	15	17.00	23.3	5,000
Minnesota	16	12.6	2.3	0.8	200	2.5	13	16.33	19.0	2,000
Missouri	15	9.7	1.8	0.6	100	1.3	7	15.50	24.7	5,000
Kansas	13	9.1	1.7	0.7	200	2.5	15	10.33	24.2	500
Texas	28	9.0	1.7	0.3	200	2.5	7	12.33	32.2	1,000
Washington	10	6.4	1.2	0.6	100	1.3	10	11.00	20.3	1,000
Pennsylvania	57	(D)	-	-	375 *	4.7	7	-	-	-
California	45	(D)	-	-	375 *	4.7	8	-	-	-
New York	28	(D)	-	-	175 *	2.2	6	-	-	1,714
Massachusetts	18	(D)	-	-	175 *	2.2	10	-	-	571

Source: 1992 *Economic Census*. The states are in descending order of shipments or establishments (if shipment data are missing for the majority). The symbol (D) appears when data are withheld to prevent disclosure of competitive information. States marked with (D) are sorted by number of establishments. A dash (-) indicates that the data element cannot be calculated; * indicates the midpoint of a range.

3544 - SPECIAL DIES, TOOLS, JIGS & FIXTURES

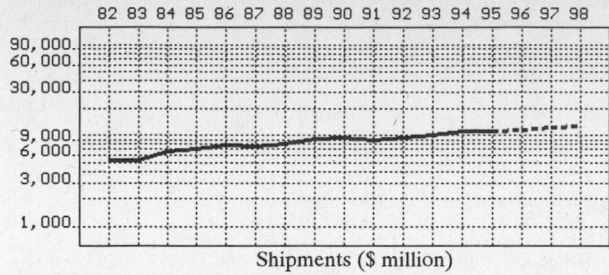

82 83 84 85 86 87 88 89 90 91 92 93 94 95 96 97 98

Shipments ($ million)

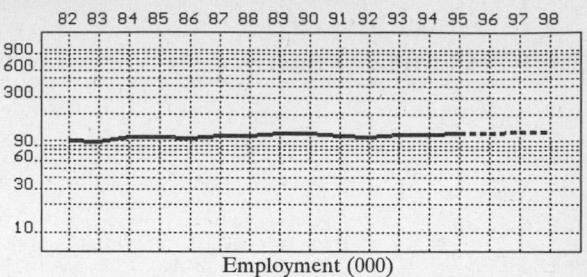

82 83 84 85 86 87 88 89 90 91 92 93 94 95 96 97 98

Employment (000)

GENERAL STATISTICS

| Year | Com-panies | Establishments | | Employment | | | Compensation | | Production ($ million) | | | |
		Total	with 20 or more employees	Total (000)	Production Workers (000)	Hours (Mil)	Payroll ($ mil)	Wages ($/hr)	Cost of Materials	Value Added by Manufacture	Value of Shipments	Capital Invest.
1982	7,131	7,255	1,322	102.9	81.7	164.9	2,293.3	10.22	1,535.4	3,780.5	5,374.9	232.9
1983		7,181	1,392	97.9	78.2	161.1	2,294.4	10.32	1,537.1	3,801.7	5,310.6	192.4
1984		7,107	1,462	111.2	89.7	188.1	2,737.8	10.75	2,010.7	4,681.5	6,614.7	398.8
1985		7,034	1,533	110.6	88.6	183.3	2,904.4	11.58	2,069.9	5,222.2	7,204.8	401.8
1986		7,070	1,584	108.9	85.7	179.4	3,000.1	12.07	2,158.1	5,582.2	7,725.1	333.6
1987	7,207	7,317	1,517	114.4	87.7	185.5	3,163.6	12.14	2,227.0	5,293.8	7,550.1	375.3
1988		7,106	1,582	115.7	89.7	193.3	3,357.8	12.58	2,606.7	5,528.6	8,078.3	272.9
1989		6,983	1,665	122.4	94.7	202.7	3,750.6	13.20	2,893.7	6,490.6	9,236.3	385.8
1990		7,040	1,659	123.0	92.2	200.5	3,804.8	13.56	2,952.5	6,525.4	9,487.2	412.8
1991		7,268	1,627	114.0	87.5	190.3	3,659.1	13.59	2,615.1	6,281.4	8,890.4	386.0
1992	7,227	7,350	1,521	111.4	85.3	188.4	3,898.3	14.57	2,690.7	6,644.7	9,309.8	370.1
1993		7,412	1,633	117.5	90.7	198.0	4,121.7	14.70	2,987.7	7,026.7	9,950.8	531.8
1994		7,266P	1,693P	119.5	92.4	204.1	4,352.7	15.00	3,296.9	8,090.0	11,195.0	541.2
1995		7,280P	1,716P	122.4P	92.8P	206.9P	4,510.5P	15.56P	3,296.4P	8,088.8P	11,193.3P	508.6P
1996		7,294P	1,740P	123.8P	93.5P	209.7P	4,678.6P	15.97P	3,424.5P	8,403.1P	11,628.3P	528.2P
1997		7,308P	1,763P	125.1P	94.2P	212.5P	4,846.7P	16.39P	3,552.6P	8,717.4P	12,063.3P	547.7P
1998		7,321P	1,786P	126.5P	94.9P	215.2P	5,014.8P	16.81P	3,680.7P	9,031.8P	12,498.3P	567.2P

Sources: 1982, 1987, 1992 *Economic Census*; *Annual Survey of Manufactures*, 83-86, 88-91, 93-94. Establishment counts for non-Census years are from *County Business Patterns*; establishment values for 83-84 are extrapolations. 'P's show projections by the editors. Industries reclassified in 87 will not have data for prior years.

INDICES OF CHANGE

| Year | Com-panies | Establishments | | Employment | | | Compensation | | Production ($ million) | | | |
		Total	with 20 or more employees	Total (000)	Production Workers (000)	Hours (Mil)	Payroll ($ mil)	Wages ($/hr)	Cost of Materials	Value Added by Manufacture	Value of Shipments	Capital Invest.
1982	99	99	87	92	96	88	59	70	57	57	58	63
1983		98	92	88	92	86	59	71	57	57	57	52
1984		97	96	100	105	100	70	74	75	70	71	108
1985		96	101	99	104	97	75	79	77	79	77	109
1986		96	104	98	100	95	77	83	80	84	83	90
1987	100	100	100	103	103	98	81	83	83	80	81	101
1988		97	104	104	105	103	86	86	97	83	87	74
1989		95	109	110	111	108	96	91	108	98	99	104
1990		96	109	110	108	106	98	93	110	98	102	112
1991		99	107	102	103	101	94	93	97	95	95	104
1992	100	100	100	100	100	100	100	100	100	100	100	100
1993		101	107	105	106	105	106	101	111	106	107	144
1994		99P	111P	107	108	108	112	103	123	122	120	146
1995		99P	113P	110P	109P	110P	116P	107P	123P	122P	120P	137P
1996		99P	114P	111P	110P	111P	120P	110P	127P	126P	125P	143P
1997		99P	116P	112P	110P	113P	124P	113P	132P	131P	130P	148P
1998		100P	117P	114P	111P	114P	129P	115P	137P	136P	134P	153P

Sources: Same as General Statistics. Values reflect change from the base year, 1992. Values above 100 mean greater than 92, values below 100 mean less than 92, and a value of 100 in the 82-91 or 93-98 period means same as 92. 'P's mark projections by the editors.

SELECTED RATIOS

For 1994	Avg. of All Manufact.	Analyzed Industry	Index	For 1994	Avg. of All Manufact.	Analyzed Industry	Index
Employees per Establishment	49	16	34	Value Added per Production Worker	134,084	87,554	65
Payroll per Establishment	1,500,273	599,025	40	Cost per Establishment	5,045,178	453,725	9
Payroll per Employee	30,620	36,424	119	Cost per Employee	102,970	27,589	27
Production Workers per Establishment	34	13	37	Cost per Production Worker	146,988	35,681	24
Wages per Establishment	853,319	421,328	49	Shipments per Establishment	9,576,895	1,540,673	16
Wages per Production Worker	24,861	33,133	133	Shipments per Employee	195,460	93,682	48
Hours per Production Worker	2,056	2,209	107	Shipments per Production Worker	279,017	121,158	43
Wages per Hour	12.09	15.00	124	Investment per Establishment	321,011	74,481	23
Value Added per Establishment	4,602,255	1,113,358	24	Investment per Employee	6,552	4,529	69
Value Added per Employee	93,930	67,699	72	Investment per Production Worker	9,352	5,857	63

Sources: Same as General Statistics. The 'Average of All Manufacturing' column represents the average of all manufacturing industries reported for the most recent complete year available. The Index shows the relationship between the Average and the Analyzed Industry. For example, 100 means that they are equal; 500 that the Analyzed Industry is five times the average; 50 means that the Analyzed Industry is half the national average. The abbreviation 'na' is used to show that data are 'not available'.

LEADING COMPANIES   Number shown: **75**   Total sales ($ mil): **2,269**   Total employment (000): **17.9**

Company Name	Address				CEO Name	Phone	Co. Type	Sales ($ mil)	Empl. (000)
Dynacast Inc	1401 Front St	Yorktown H	NY	10598	Keith Thompson	914-245-0064	R	200	0.8
D-M-E Co	29111ephenson Hwy	Madison H	MI	48071	Jerry Lirette	313-398-6000	D	145	0.8
Wright Industries Inc	707 Spence Ln	Nashville	TN	37217	B Robinson III	615-361-6600	R	85	0.5
Dayton Progress Corp	PO Box 39	Dayton	OH	45449	Steve Buck	513-859-5111	S	75	0.8
USM Corp	400 Research Dr	Wilmington	MA	01887	KC Cochrane	508-657-4700	S	70	0.2
A Finkl and Sons Co	2011 Southport Av	Chicago	IL	60614	Bruce Limatainen	312-975-2510	R	65	0.4
Danly Die Set	2115 S 54th Av	Chicago	IL	60650	Dave Lowum	708-780-5800	D	60	0.6
Jade Corp	3063 Philmont Av	Huntingdon Vl	PA	19006	JP McPartland	215-947-3333	S	60	0.5
Johnson Controls Inc	10501 Hwy M-52	Manchester	MI	48158	Kirt Deford	313-428-8371	D	58*	0.6
Phelps Tool and Die Company	4926 Lawn Av	Kansas City	MO	64130	MD Phelps	816-921-4373	R	46	0.4
Anchor Tool and Die Co	11830 Brookpark	Cleveland	OH	44130	Fred Pfaff	216-362-1850	R	40	0.3
Shelby Die Casting Co	PO Box 209	Shelby	MS	38774	Rives Neblett	601-398-5171	R	40	0.3
Zacova Industries Inc	16630 Eastland Dr	Roseville	MI	48066	Lester Sova	810-771-8580	R	40	0.2
Hess Industries Inc	191 Fir Rd	Niles	MI	49120	Fritz Kucklick	616-683-4182	R	37*	0.3
National Tool & Mfg Co	100 N 12th St	Kenilworth	NJ	07033	William Zeus Jr	908-276-1600	R	34	0.4
Penn United Technology Inc	PO Box 399	Saxonburg	PA	16056	Carl E Jones	412-352-1507	R	33	0.4
Demmer Corp	3525 Capital City	Lansing	MI	48906	John E Demmer	517-321-3600	R	30	0.3
Hercules Machine Tool	13920 E 10 Mile Rd	Warren	MI	48089	T Tsuda	313-778-4120	R	30	0.2
Hess Engineering Inc	191 Fir Rd	Niles	MI	49120	FC Kucklick	616-683-4182	S	30	0.2
Innovex Inc	1313 S 5th St	Hopkins	MN	55343	D Johnson	612-938-4155	D	30	0.4
Jasco Tools Inc	195 St Paul St	Rochester	NY	14606	John A Summers	716-546-1254	R	30	0.3
Select Tool and Die Corp	60 Heid Av	Dayton	OH	45404	RW Whited	513-233-9191	R	30	0.3
Belvac Production Machinery	PO Box 4276	Lynchburg	VA	24502	Jim Schneiders	804-239-0358	R	29*	0.3
Major Tool and Machine Inc	1458 E 19th St	Indianapolis	IN	46218	Steve Weyreter	317-636-6433	R	29*	0.3
Producto Machine Co	PO Box 780	Bridgeport	CT	06601	NM Maarsilius III	203-367-8675	R	29	0.3
R Olson Manufacturing	1820 W Grand Av	Chicago	IL	60622	Edward R Olson	312-738-6300	R	26	0.2
Capitol Technologies Inc	PO Box 3626	South Bend	IN	46619	David Steinbauer	219-232-3311	S	25	0.1
Elizabeth Carbide Die Company	601 Linden St	McKeesport	PA	15132	Richard A Pagliari	412-751-3000	R	25	0.2
Extrusion Dies Inc	911 Kurth Rd	Chippewa Falls	WI	54729	Robert Bariknecht	715-726-1201	R	25	0.2
Futuramic Tool & Eng Co	24680 Gibson Av	Warren	MI	48089	William Warner	313-758-2200	R	25	0.3
Hertel Cutting Technologies Inc	9041 Executive Pk	Knoxville	TN	37923	Jeff Turner	615-470-2300	S	25	<0.1
Jergens Inc	19520 Nottingham	Cleveland	OH	44110	JH Schron Jr	216-486-2100	R	25	0.3
Republic Die and Tool Co	PO Box 339	Belleville	MI	48111	J Lasko	313-699-3400	R	25	0.4
Sekely Industries Inc	PO Box 148	Salem	OH	44460	Richard J Sekely	216-337-3439	R	25*	0.2
Wirtz Manufacturing Company	PO Box 5006	Port Huron	MI	48060	JO Wirtz	313-987-4700	R	25	0.2
Rivera Prodion Tooling Group	5460 Executive S	Grand Rapids	MI	49512	Kenneth Rieth	616-698-2100	R	24	0.2
Teledyne Efficient Industries	5514 Old Brecksville	Cleveland	OH	44131	William Nordby	216-524-5250	D	24	0.2
United Tool and Engineering Co	PO Box 218	South Beloit	IL	61080	Rod C Meade	815-389-3021	R	24	<0.1
Fort Wayne Wire Die Inc	2424 American Way	Fort Wayne	IN	46809	Dwight Bieberich	219-747-1681	R	22*	0.2
MTD Technologies Inc	PO Box 2250	Pinellas Park	FL	34666	Dennis Ruppel	813-546-2446	R	21	0.3
Peko Precision Products Inc	1400 Emerson St	Rochester	NY	14606	John Oliveri	716-647-3010	R	21	0.2
Dieline Corp	3755 36th St SE	Kentwood	MI	49512	Roger L Schiefler	616-956-0081	R	20	0.2
Lamina Inc	14925 W 11 Mile Rd	Oak Park	MI	48237	Ronald E Smith	810-542-8341	S	20	0.2
Mate Punch and Die Co	PO Box 728	Anoka	MN	55303	Nils Sundquist	612-421-0230	R	20	0.2
Newton Tool & Mfg Co	Linden & Glassboro	Wenonah	NJ	08090	Robert Onraet	609-468-5595	S	20	<0.1
Paragon Die and Engineering Co	5225 33rd St SE	Grand Rapids	MI	49512	Ralph M Swain	616-949-2220	R	20	0.2
Autojectors Inc	1563 E State Rd 8	Albion	IN	46701	William Carteaux	219-636-2133	R	18	0.1
Caco Pacific Corp	PO Box 2369	Covina	CA	91722	M G Hoffmann	818-331-3361	R	18	0.2
Delaware Machinery and Tool	PO Box 2665	Muncie	IN	47307	Robert Haas Jr	317-284-3335	R	18	0.2
Howmet Tempcraft Inc	3960 S Marginal Rd	Cleveland	OH	44114	Elmer Miller Jr	216-391-3885	S	18	0.2
Lane Punch Corp	4985 Belleville Rd	Canton	MI	48187	W Porter	313-397-3200	R	18	0.2
Shawnee Plastics Inc	PO Box 280	Kuttawa	KY	42055	Dale Wendel	502-388-2253	R	18	0.3
Libbey-Owens-Ford Co	PO Box 779	Toledo	OH	43697	TF Hamstreet	419-729-9776	D	17	0.2
Lempco Industries Inc	5490 Dunham Rd	Cleveland	OH	44137	JJ Strnad	216-475-2400	R	16	0.2
Triangle Tool Corp	8609 W Port Av	Milwaukee	WI	53224	Leroy D Luther	414-357-7117	R	16*	0.1
Cybernetics Products Inc	180 Broad St	Carlstadt	NJ	07072	Joseph P Drier Jr	201-935-3000	P	16	<0.1
Hydro Carbide	PO Box 363	Latrobe	PA	15650	Thomas McLaren	412-539-9701	D	15	0.1
Link Inc	2066 Bristol Av NW	Grand Rapids	MI	49504	R Dennis Link	616-453-4431	R	15	0.1
Manchester Tool Co	5142 Manchester	Akron	OH	44319	Roger G Duffy	216-644-8853	S	15*	0.2
St Paul Metalcraft Inc	3737 Lexington N	St Paul	MN	55126	JT Walker	612-483-6641	R	15	0.2
WK Industries Inc	6120 Millette Av	Sterling Hts	MI	48312	W Kleinert	313-268-4090	R	15	0.1
Yorktown Tool and Die Corp	PO Box 218	Yorktown	IN	47396	JS Dunn	317-759-7767	R	15	0.2
Jet Die	PO Box 25066	Lansing	MI	48909	Stuart Kale	517-393-5110	D	15	0.1
City Machine Tool and Die Co	PO Box 2607	Muncie	IN	47305	John R Wagner	317-288-4431	R	14	0.2
Modineer Co	PO Box 640	Niles	MI	49120	Paul M Dreher	616-683-2550	R	13	0.1
MS Willett Inc	PO Box 266	Cockeysville	MD	21030	John W McCaughey	410-771-0460	R	13	0.1
Swift-Cor Tool Engineering Co	344 W 157th St	Gardena	CA	90248	R Scriba	310-538-2592	R	13*	0.2
Walker Tool and Die Inc	2411 Walker NW	Grand Rapids	MI	49504	G Hendricks	616-453-5471	R	13	0.1
Akromold Inc	1100 Main St	Cuyahoga Falls	OH	44221	WH Monteith Jr	216-929-3311	R	12	0.1
Birch Machinery Co	11160 Dixie Hwy	Birch Run	MI	48415	Harold Johnson	517-624-9373	R	12	<0.1
Ehlert Tool Company Inc	2500 S 162nd St	New Berlin	WI	53151	Stuart Meissner	414-784-5545	R	12	0.2
Hammill Manufacturing Co	PO Box 6680	Toledo	OH	43612	John Hammill	419-476-0789	R	12	0.2
Jaco Manufacturing Co	468 Geiger St	Berea	OH	44017	Lawrence Campbell	216-234-4000	R	12*	0.1
Mutual Tool and Die Inc	725 Lilac Av	Dayton	OH	45417	David A Dudon	513-268-6713	R	12	0.1
P and R Industries Inc	1524 N Clinton Av	Rochester	NY	14621	John Kress	716-266-6725	R	12	0.1

Source: Ward's Business Directory of U.S. Private and Public Companies, Volumes 1 and 2, 1996. The company type code used is as follows: P - Public, R - Private, S - Subsidiary, D - Division, J - Joint Venture, A - Affiliate, G - Group. Sales are in millions of dollars, employees are in thousands. An asterisk (*) indicates an estimated sales volume. The symbol < stands for 'less than'. Company names and addresses are truncated, in some cases, to fit into the available space.

MATERIALS CONSUMED

Material	Quantity	Delivered Cost ($ million)
Materials, ingredients, containers, and supplies	(X)	1,932.1
Metal bolts, nuts, screws, washers, rivets, and other screw machine products	(X)	39.7
Other fabricated metal products (except castings and forgings)	(X)	95.6
Forgings	(X)	13.1
Iron and steel castings (rough and semifinished)	(X)	85.8
Nonferrous (aluminum, copper, etc.) castings (rough and semifinished)	(X)	19.7
Steel bars, bar shapes, and plates	(X)	261.3
Steel sheet and strip, including tin plate	(X)	35.9
All other steel shapes and forms	(X)	38.4
Aluminum and aluminum-base alloy sheet, plate, foil, and welded tubing	(X)	25.6
All other aluminum and aluminum-base alloy shapes and forms	(X)	12.3
Other nonferrous shapes and forms	(X)	11.7
Tungsten carbide metal powders	(X)	20.4
All other metal powders	(X)	2.2
Iron and steel scrap (excluding home scrap)	(X)	2.0
Industrial diamonds	(X)	8.7
Electrical transmission, distribution, and control equipment	(X)	10.6
Grinding wheels and other abrasive products, except industrial diamonds	(X)	19.0
Fluid power products	(X)	23.7
Ceramic raw materials, including powders, chemicals, and fibers (excluding refractory uses)	(X)	3.7
Ceramic and ceramic composite parts, components, and accessories	(X)	1.3
All other materials and components, parts, containers, and supplies	(X)	345.6
Materials, ingredients, containers, and supplies, nsk	(X)	855.8

Source: 1992 *Economic Census*. Explanation of symbols used: (D): Withheld to avoid disclosure of competitive data; na: Not available; (S): Withheld because statistical norms were not met; (X): Not applicable; (Z): Less than half the unit shown; nec: Not elsewhere classified; nsk: Not specified by kind; - : zero; * : 10-19 percent estimated; ** : 20-29 percent estimated.

PRODUCT SHARE DETAILS

Product or Product Class	% Share	Product or Product Class	% Share
Special dies, tools, jigs, and fixtures	100.00	straightening dies	1.45
Special dies and tools, die sets, jigs, and fixtures	49.51	All other high-speed steel metalworking dies	1.10
Gauging and checking jigs and fixtures, less than 1,000 lb weight	4.14	All other carbide metalworking dies	1.24
Gauging and checking jigs and fixtures, 1,000 lb weight or more	0.93	All other metalworking dies	1.35
Other jigs and fixtures, less than 1,000 lb weight (holding, positioning, layout, assembly, etc.)	5.25	Standard and special metalworking die sets	1.86
		Standard steel punches for dies	3.72
Other jigs and fixtures, 1,000 lb weight or more (holding, positioning, layout, assembly, etc.)	4.60	Standard carbide punches for dies	0.53
		Other standard punches for dies	3.21
Standard catalog components and parts for jigs and fixtures, including drill bushings	2.68	Industrial models and prototypes	8.42
Press brake metalworking dies	0.70	Other specially designed tooling	9.74
		Special dies and tools, die sets, jigs, and fixtures, nsk	8.96
Metalworking forming and drawing dies, 500 lb weight or less	2.77	Industrial molds and mold boxes	36.53
Metalworking forming and drawing dies, 501 to 3,000 lb weight	2.97	Industrial molds made of metal, for low-pressure die-casting of metal or metal carbides (except ingot molds)	1.34
Metalworking forming and drawing dies, more than 3,000 lb weight	5.98	Industrial molds made of metal, for high-pressure die-casting of metal or metal carbides (except ingot molds)	7.56
Metalworking stamping dies (including lamination and blanking), progressive-type, high-speed steel	9.11	Industrial permanent molds made of metal for gravity casting of metal or metal carbides (except ingot molds)	0.85
Metalworking stamping dies (including lamination and blanking), progressive-type, carbide	1.62	Other industrial molds made of metal for metal or metal carbides (except ingot molds)	0.83
Other metalworking stamping dies (including lamination and blanking), progressive-type	3.72	Industrial molds made of metal for wax	1.87
		Industrial molds made of metal for mineral materials	0.17
All other metalworking stamping-type dies, including lamination and blanking dies (punch, trim, notch, pierce, perforate, etc.)	9.22	Industrial molds made of metal for glass	2.98
		Industrial injection or compression-type molds made of metal for rubber	4.91
Metalworking open-type forging dies, including cold forging and heading	0.81	Other industrial molds made of metal for rubber	1.44
Metalworking closed-type forging dies, including cold forging and heading	1.95	Industrial injection-type molds made of metal for plastics	57.82
Metalworking high-speed steel extrusion and wiredrawing and straightening dies	1.29	Industrial compression-type molds (including matched metal molds) made of metal for plastics	3.48
Metalworking carbide extrusion and wiredrawing and straightening dies	0.71	Other industrial molds (including transfer, plunger, and rotational molds) made of metal for plastics	4.21
		Industrial molds made of metal for other materials	1.47
Metalworking ceramic and ceramic composite extrusion and wiredrawing and straightening dies	0.01	Industrial mold bases made of metal	3.20
Other metalworking extrusion and wiredrawing and		Industrial molds made of materials other than metal	1.68
		Industrial mold boxes or flasks for use with patterns and sand molds in foundries	0.58
		Industrial molds and mold boxes, nsk	5.61
		Special dies, tools, jigs, and fixtures, nsk	13.96

Source: 1992 *Economic Census*. The values shown are percent of total shipments in an industry. Values of indented subcategories are summed in the main heading. The symbol (D) appears when data are withheld to prevent disclosure of competitive information. The abbreviation nsk stands for 'not specified by kind' and nec for 'not elsewhere classified'.

INPUTS AND OUTPUTS FOR SPECIAL DIES & TOOLS & MACHINE TOOL ACCESSORIES

Economic Sector or Industry Providing Inputs	%	Sector	Economic Sector or Industry Buying Outputs	%	Sector
Blast furnaces & steel mills	15.0	Manufg.	Gross private fixed investment	43.0	Cap Inv
Imports	10.8	Foreign	Motor vehicles & car bodies	4.6	Manufg.
Special dies & tools & machine tool accessories	9.7	Manufg.	Exports	4.4	Foreign
Wholesale trade	5.7	Trade	Aircraft	4.3	Manufg.
Machinery, except electrical, nec	5.1	Manufg.	Special dies & tools & machine tool accessories	3.6	Manufg.
Primary metal products, nec	4.4	Manufg.	Aluminum rolling & drawing	2.1	Manufg.
Advertising	3.5	Services	Machinery, except electrical, nec	1.9	Manufg.
Electric services (utilities)	3.3	Util.	Aluminum castings	1.6	Manufg.
Abrasive products	2.9	Manufg.	Steel wire & related products	1.6	Manufg.
Petroleum refining	2.2	Manufg.	Federal Government purchases, national defense	1.4	Fed Govt
Cyclic crudes and organics	1.8	Manufg.	Blast furnaces & steel mills	0.9	Manufg.
Motor freight transportation & warehousing	1.8	Util.	Motor vehicle parts & accessories	0.9	Manufg.
Eating & drinking places	1.8	Trade	Screw machine and related products	0.9	Manufg.
Noncomparable imports	1.5	Foreign	Automotive stampings	0.8	Manufg.
Communications, except radio & TV	1.4	Util.	Nonferrous rolling & drawing, nec	0.8	Manufg.
Iron & steel foundries	1.3	Manufg.	Refrigeration & heating equipment	0.8	Manufg.
Real estate	1.3	Fin/R.E.	Aircraft & missile engines & engine parts	0.7	Manufg.
Maintenance of nonfarm buildings nec	1.2	Constr.	Glass containers	0.7	Manufg.
Miscellaneous plastics products	1.2	Manufg.	Miscellaneous plastics products	0.7	Manufg.
Banking	1.2	Fin/R.E.	Metal stampings, nec	0.6	Manufg.
Machine tools, metal cutting types	1.0	Manufg.	Railroads & related services	0.6	Util.
Paperboard containers & boxes	1.0	Manufg.	Water transportation	0.6	Util.
Equipment rental & leasing services	1.0	Services	Construction machinery & equipment	0.5	Manufg.
Computer & data processing services	0.9	Services	Internal combustion engines, nec	0.5	Manufg.
Engineering, architectural, & surveying services	0.9	Services	Iron & steel foundries	0.5	Manufg.
Metal heat treating	0.8	Manufg.	Nonferrous castings, nec	0.5	Manufg.
Gas production & distribution (utilities)	0.8	Util.	Oil field machinery	0.5	Manufg.
Royalties	0.8	Fin/R.E.	Pumps & compressors	0.5	Manufg.
Aluminum rolling & drawing	0.7	Manufg.	Wholesale trade	0.5	Trade
Industrial patterns	0.7	Manufg.	Aircraft & missile equipment, nec	0.4	Manufg.
Screw machine and related products	0.7	Manufg.	Machine tools, metal cutting types	0.4	Manufg.
Air transportation	0.7	Util.	Manufacturing industries, nec	0.4	Manufg.
Management & consulting services & labs	0.7	Services	Special industry machinery, nec	0.4	Manufg.
Paints & allied products	0.6	Manufg.	Blowers & fans	0.3	Manufg.
Plating & polishing	0.6	Manufg.	Brass, bronze, & copper castings	0.3	Manufg.
Legal services	0.6	Services	Electronic components nec	0.3	Manufg.
Fabricated metal products, nec	0.5	Manufg.	Food products machinery	0.3	Manufg.
Nonmetallic mineral products, nec	0.4	Manufg.	General industrial machinery, nec	0.3	Manufg.
Primary lead	0.4	Manufg.	Guided missiles & space vehicles	0.3	Manufg.
Railroads & related services	0.4	Util.	Iron & steel forgings	0.3	Manufg.
Accounting, auditing & bookkeeping	0.4	Services	Power transmission equipment	0.3	Manufg.
Automotive rental & leasing, without drivers	0.4	Services	Radio & TV communication equipment	0.3	Manufg.
Industrial controls	0.3	Manufg.	Semiconductors & related devices	0.3	Manufg.
Lubricating oils & greases	0.3	Manufg.	Retail trade, except eating & drinking	0.3	Trade
Manifold business forms	0.3	Manufg.	Electronic computing equipment	0.2	Manufg.
Pumps & compressors	0.3	Manufg.	Farm machinery & equipment	0.2	Manufg.
Switchgear & switchboard apparatus	0.3	Manufg.	Glass & glass products, except containers	0.2	Manufg.
Insurance carriers	0.3	Fin/R.E.	Hardware, nec	0.2	Manufg.
Automotive repair shops & services	0.3	Services	Logging camps & logging contractors	0.2	Manufg.
Hotels & lodging places	0.3	Services	Paperboard containers & boxes	0.2	Manufg.
Aluminum castings	0.2	Manufg.	Petroleum refining	0.2	Manufg.
Ball & roller bearings	0.2	Manufg.	Photographic equipment & supplies	0.2	Manufg.
Brass, bronze, & copper castings	0.2	Manufg.	Pipe, valves, & pipe fittings	0.2	Manufg.
Copper rolling & drawing	0.2	Manufg.	Power driven hand tools	0.2	Manufg.
Industrial inorganic chemicals, nec	0.2	Manufg.	Printing trades machinery	0.2	Manufg.
Iron & steel forgings	0.2	Manufg.	Service industry machines, nec	0.2	Manufg.
Metal stampings, nec	0.2	Manufg.	Ship building & repairing	0.2	Manufg.
Motors & generators	0.2	Manufg.	Textile machinery	0.2	Manufg.
Business/professional associations	0.2	Services	Equipment rental & leasing services	0.2	Services
Electrical repair shops	0.2	Services	Miscellaneous repair shops	0.2	Services
Personnel supply services	0.2	Services	Federal Government purchases, nondefense	0.2	Fed Govt
Coal	0.1	Mining	Crude petroleum & natural gas	0.1	Mining
Hand & edge tools, nec	0.1	Manufg.	Ball & roller bearings	0.1	Manufg.
Machine tools, metal forming types	0.1	Manufg.	Broadwoven fabric mills	0.1	Manufg.
Metal coating & allied services	0.1	Manufg.	Electrical equipment & supplies, nec	0.1	Manufg.
Metalworking machinery, nec	0.1	Manufg.	Fabricated metal products, nec	0.1	Manufg.
Nonferrous castings, nec	0.1	Manufg.	Fabricated plate work (boiler shops)	0.1	Manufg.
Nonferrous rolling & drawing, nec	0.1	Manufg.	Industrial furnaces & ovens	0.1	Manufg.
Photographic equipment & supplies	0.1	Manufg.	Instruments to measure electricity	0.1	Manufg.
Transformers	0.1	Manufg.	Mechanical measuring devices	0.1	Manufg.
Transit & bus transportation	0.1	Util.	Metal cans	0.1	Manufg.
Retail trade, except eating & drinking	0.1	Trade	Miscellaneous fabricated wire products	0.1	Manufg.
Photofinishing labs, commercial photography	0.1	Services	Motors & generators	0.1	Manufg.
			Nonferrous wire drawing & insulating	0.1	Manufg.
			Paper industries machinery	0.1	Manufg.
			Paper mills, except building paper	0.1	Manufg.
			Plating & polishing	0.1	Manufg.
			Sheet metal work	0.1	Manufg.

Continued on next page.

INPUTS AND OUTPUTS FOR SPECIAL DIES & TOOLS & MACHINE TOOL ACCESSORIES - Continued

Economic Sector or Industry Providing Inputs	%	Sector	Economic Sector or Industry Buying Outputs	%	Sector
			Steel pipe & tubes	0.1	Manufg.
			Telephone & telegraph apparatus	0.1	Manufg.
			Turbines & turbine generator sets	0.1	Manufg.
			Wiring devices	0.1	Manufg.
			Electric services (utilities)	0.1	Util.
			Automotive repair shops & services	0.1	Services
			Management & consulting services & labs	0.1	Services
			S/L Govt. purch., other general government	0.1	S/L Govt

Source: Benchmark Input-Output Accounts for the U.S. Economy, 1982, U.S. Department of Commerce, Washington, D.C., July 1991. Data, as reported in the source, are organized by the 1977 SIC structure in use in 1982 but have been matched, as closely as is possible, to the 1987 SIC structure used in this book.

OCCUPATIONS EMPLOYED BY SIC 354 - METALWORKING MACHINERY

Occupation	% of Total 1994	Change to 2005	Occupation	% of Total 1994	Change to 2005
Machinists	10.9	-17.4	Precision metal workers nec	1.9	10.1
Tool & die makers	10.6	-15.3	Mechanical engineers	1.9	1.0
General managers & top executives	3.7	-12.9	Inspectors, testers, & graders, precision	1.9	-8.2
Blue collar worker supervisors	3.4	-15.7	Secretaries, ex legal & medical	1.9	-16.4
Machine tool cutting operators, metal & plastic	2.9	-5.9	Drafters	1.8	-28.5
NC machine tool operators, metal & plastic	2.7	1.0	General office clerks	1.5	-21.7
Grinding machine operators, metal & plastic	2.7	-8.2	Machine forming operators, metal & plastic	1.4	-8.2
Sales & related workers nec	2.7	-8.2	Drilling & boring machine tool workers	1.3	-8.2
Assemblers, fabricators, & hand workers nec	2.5	-8.2	Bookkeeping, accounting, & auditing clerks	1.3	-31.1
Lathe & turning machine tool operators	2.3	-17.4	Industrial production managers	1.3	-8.2
Machine tool cutting & forming etc. nec	2.2	-54.1	Traffic, shipping, & receiving clerks	1.3	-11.7
Machine builders	2.0	1.0	Precision woodworkers nec	1.3	10.2
Combination machine tool operators	2.0	37.7	Janitors & cleaners, incl maids	1.2	-26.6

Source: Industry-Occupation Matrix, Bureau of Labor Statistics. These data relate to one or more 3-digit SIC industry groups rather than to a single 4-digit SIC. The change reported for each occupation to the year 2005 is a percent of growth or decline as estimated by the Bureau of Labor Statistics. The abbreviation nec stands for 'not elsewhere classified'.

LOCATION BY STATE AND REGIONAL CONCENTRATION

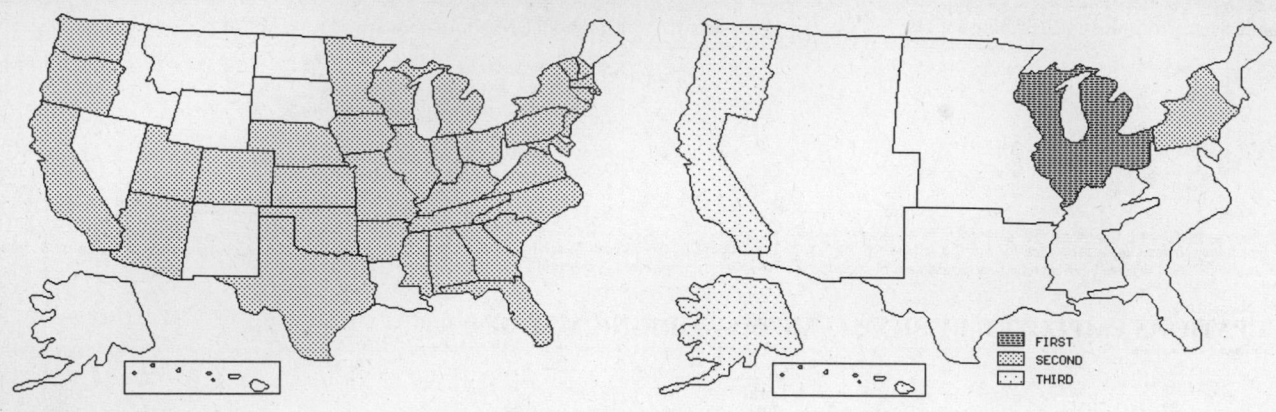

FIRST
SECOND
THIRD

INDUSTRY DATA BY STATE

State	Establish-ments	Shipments			Employment				Cost as % of Shipments	Investment per Employee ($)
		Total ($ mil)	% of U.S.	Per Establ.	Total Number	% of U.S.	Per Establ.	Wages ($/hour)		
Michigan	1,210	2,564.0	27.5	2.1	26,800	24.1	22	15.93	27.6	3,560
Ohio	772	1,239.1	13.3	1.6	15,300	13.7	20	14.05	27.9	3,111
Illinois	691	904.0	9.7	1.3	10,300	9.2	15	15.41	30.5	3,709
Pennsylvania	418	705.7	7.6	1.7	8,500	7.6	20	14.24	32.3	3,165
California	576	502.3	5.4	0.9	5,700	5.1	10	13.62	29.9	3,158
Indiana	407	468.2	5.0	1.2	6,400	5.7	16	13.60	27.0	2,922
Wisconsin	315	431.4	4.6	1.4	5,100	4.6	16	16.14	28.4	4,471
New York	338	288.2	3.1	0.9	4,000	3.6	12	12.94	29.8	2,850
Minnesota	184	241.2	2.6	1.3	2,800	2.5	15	14.93	26.9	4,964
New Jersey	249	231.8	2.5	0.9	2,800	2.5	11	13.90	28.9	2,250
Connecticut	251	215.0	2.3	0.9	2,500	2.2	10	15.52	29.3	3,040
Missouri	148	186.6	2.0	1.3	2,100	1.9	14	14.17	33.4	4,905
Massachusetts	199	177.2	1.9	0.9	2,200	2.0	11	14.73	27.6	3,182
Florida	177	116.2	1.2	0.7	1,600	1.4	9	12.68	30.3	2,438
Tennessee	148	110.3	1.2	0.7	1,700	1.5	11	12.33	29.9	2,294
Texas	198	109.8	1.2	0.6	1,800	1.6	9	12.21	28.9	2,444
North Carolina	107	94.7	1.0	0.9	1,300	1.2	12	13.10	25.9	3,308
Alabama	54	85.5	0.9	1.6	1,200	1.1	22	11.60	33.2	2,417
Iowa	63	74.3	0.8	1.2	1,000	0.9	16	14.00	27.5	3,000
Georgia	63	59.8	0.6	0.9	800	0.7	13	13.42	32.1	4,250
Kentucky	82	59.6	0.6	0.7	1,000	0.9	12	12.18	28.2	4,500
Arizona	86	57.5	0.6	0.7	800	0.7	9	13.77	32.7	3,875
South Carolina	46	42.9	0.5	0.9	600	0.5	13	11.50	29.6	2,333
Arkansas	69	40.5	0.4	0.6	600	0.5	9	11.10	29.9	2,667
Virginia	32	36.8	0.4	1.1	600	0.5	19	15.25	30.7	1,333
Colorado	55	35.9	0.4	0.7	500	0.4	9	13.25	29.2	2,200
Rhode Island	57	29.8	0.3	0.5	400	0.4	7	13.00	23.2	2,500
Oregon	60	28.5	0.3	0.5	400	0.4	7	15.67	31.6	1,500
Washington	46	28.1	0.3	0.6	500	0.4	11	14.00	30.6	2,200
Kansas	32	27.7	0.3	0.9	400	0.4	13	12.29	25.6	2,250
Mississippi	39	21.4	0.2	0.5	300	0.3	8	9.80	38.3	2,333
New Hampshire	33	19.5	0.2	0.6	300	0.3	9	13.60	21.5	3,000
Oklahoma	28	15.3	0.2	0.5	200	0.2	7	11.67	34.6	2,000
Nebraska	19	10.7	0.1	0.6	200	0.2	11	10.33	26.2	3,000
Utah	17	5.3	0.1	0.3	100	0.1	6	8.50	26.4	1,000
Maryland	24	(D)	-	-	175 *	0.2	7	-	-	1,143
Vermont	7	(D)	-	-	175 *	0.2	25	-	-	-

Source: 1992 *Economic Census*. The states are in descending order of shipments or establishments (if shipment data are missing for the majority). The symbol (D) appears when data are withheld to prevent disclosure of competitive information. States marked with (D) are sorted by number of establishments. A dash (-) indicates that the data element cannot be calculated; * indicates the midpoint of a range.

3545 - MACHINE TOOL ACCESSORIES

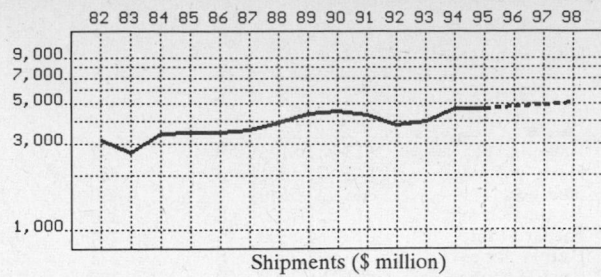

Shipments ($ million)

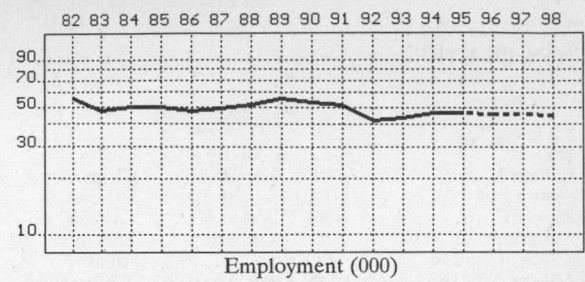

Employment (000)

GENERAL STATISTICS

| Year | Com-panies | Establishments | | Employment | | | Compensation | | Production ($ million) | | | |
		Total	with 20 or more employees	Total (000)	Production Workers (000)	Hours (Mil)	Payroll ($ mil)	Wages ($/hr)	Cost of Materials	Value Added by Manufacture	Value of Shipments	Capital Invest.
1982	1,444	1,620	562	55.1	38.8	73.5	1,069.4	9.39	979.9	2,162.9	3,163.9	144.0
1983		1,583	569	47.3	32.7	62.7	958.6	9.68	800.1	1,821.9	2,675.7	105.1
1984		1,546	576	49.5	36.2	72.8	1,087.9	9.77	996.4	2,432.7	3,426.3	168.8
1985		1,509	583	49.4	36.4	73.2	1,099.5	9.81	1,025.8	2,429.2	3,458.8	167.9
1986		1,496	574	47.0	34.0	68.3	1,113.0	10.55	1,002.7	2,423.7	3,452.3	108.5
1987	1,734	1,879	564	48.4	35.0	71.0	1,184.4	10.66	1,038.1	2,560.8	3,594.8	113.2
1988		1,779	550	50.7	36.7	75.2	1,263.9	10.72	1,163.2	2,813.6	3,948.5	102.0
1989		1,753	590	55.2	39.8	79.8	1,404.6	10.99	1,316.5	3,168.7	4,409.4	169.6
1990		1,748	570	52.9	39.6	79.9	1,445.0	11.14	1,425.7	3,072.4	4,550.4	182.8
1991		1,757	552	51.6	35.9	72.5	1,387.7	11.53	1,336.7	2,974.6	4,359.6	125.3
1992	1,759	1,866	485	42.7	30.1	62.4	1,243.9	12.00	1,113.9	2,653.6	3,786.3	142.2
1993		1,828	493	43.3	30.5	62.6	1,298.0	12.79	1,202.8	2,806.1	4,002.9	163.7
1994		1,883P	517P	46.6	32.7	68.8	1,429.5	12.49	1,425.6	3,283.5	4,665.0	188.1
1995		1,911P	511P	46.4P	33.1P	69.4P	1,469.3P	12.80P	1,431.4P	3,296.9P	4,684.0P	165.0P
1996		1,940P	505P	46.0P	32.7P	69.2P	1,503.5P	13.07P	1,469.7P	3,385.0P	4,809.2P	167.9P
1997		1,969P	499P	45.6P	32.4P	69.0P	1,537.7P	13.35P	1,508.0P	3,473.2P	4,934.5P	170.8P
1998		1,997P	493P	45.2P	32.1P	68.8P	1,572.0P	13.62P	1,546.2P	3,561.3P	5,059.7P	173.7P

Sources: 1982, 1987, 1992 *Economic Census*; *Annual Survey of Manufactures*, 83-86, 88-91, 93-94. Establishment counts for non-Census years are from *County Business Patterns*; establishment values for 83-84 are extrapolations. 'P's show projections by the editors. Industries reclassified in 87 will not have data for prior years.

INDICES OF CHANGE

| Year | Com-panies | Establishments | | Employment | | | Compensation | | Production ($ million) | | | |
		Total	with 20 or more employees	Total (000)	Production Workers (000)	Hours (Mil)	Payroll ($ mil)	Wages ($/hr)	Cost of Materials	Value Added by Manufacture	Value of Shipments	Capital Invest.
1982	82	87	116	129	129	118	86	78	88	82	84	101
1983		85	117	111	109	100	77	81	72	69	71	74
1984		83	119	116	120	117	87	81	89	92	90	119
1985		81	120	116	121	117	88	82	92	92	91	118
1986		80	118	110	113	109	89	88	90	91	91	76
1987	99	101	116	113	116	114	95	89	93	97	95	80
1988		95	113	119	122	121	102	89	104	106	104	72
1989		94	122	129	132	128	113	92	118	119	116	119
1990		94	118	124	132	128	116	93	128	116	120	129
1991		94	114	121	119	116	112	96	120	112	115	88
1992	100	100	100	100	100	100	100	100	100	100	100	100
1993		98	102	101	101	100	104	107	108	106	106	115
1994		101P	107P	109	109	110	115	104	128	124	123	132
1995		102P	105P	109P	110P	111P	118P	107P	129P	124P	124P	116P
1996		104P	104P	108P	109P	111P	121P	109P	132P	128P	127P	118P
1997		106P	103P	107P	108P	111P	124P	111P	135P	131P	130P	120P
1998		107P	102P	106P	107P	110P	126P	113P	139P	134P	134P	122P

Sources: Same as General Statistics. Values reflect change from the base year, 1992. Values above 100 mean greater than 92, values below 100 mean less than 92, and a value of 100 in the 82-91 or 93-98 period means same as 92. 'P's mark projections by the editors.

SELECTED RATIOS

For 1994	Avg. of All Manufact.	Analyzed Industry	Index	For 1994	Avg. of All Manufact.	Analyzed Industry	Index
Employees per Establishment	49	25	51	Value Added per Production Worker	134,084	100,413	75
Payroll per Establishment	1,500,273	759,216	51	Cost per Establishment	5,045,178	757,145	15
Payroll per Employee	30,620	30,676	100	Cost per Employee	102,970	30,592	30
Production Workers per Establishment	34	17	51	Cost per Production Worker	146,988	43,596	30
Wages per Establishment	853,319	456,386	53	Shipments per Establishment	9,576,895	2,477,609	26
Wages per Production Worker	24,861	26,279	106	Shipments per Employee	195,460	100,107	51
Hours per Production Worker	2,056	2,104	102	Shipments per Production Worker	279,017	142,661	51
Wages per Hour	12.09	12.49	103	Investment per Establishment	321,011	99,901	31
Value Added per Establishment	4,602,255	1,743,886	38	Investment per Employee	6,552	4,036	62
Value Added per Employee	93,930	70,461	75	Investment per Production Worker	9,352	5,752	62

Sources: Same as General Statistics. The 'Average of All Manufacturing' column represents the average of all manufacturing industries reported for the most recent complete year available. The Index shows the relationship between the Average and the Analyzed Industry. For example, 100 means that they are equal; 500 that the Analyzed Industry is five times the average; 50 means that the Analyzed Industry is half the national average. The abbreviation 'na' is used to show that data are 'not available'.

LEADING COMPANIES   Number shown: **75**   Total sales ($ mil): **4,896**   Total employment (000): **42.6**

Company Name	Address				CEO Name	Phone	Co. Type	Sales ($ mil)	Empl. (000)
Milacron Inc	4701 Marburg Av	Cincinnati	OH	45209	Raymond Ross	513-841-8100	R	1,197	8.5
Abrasive Industries Inc	3001 Executive Dr	Clearwater	FL	34622	Herb Quick	813-571-4501	R	730	7.0
Harbour Group Ltd	7701 Forsyth Blv	St Louis	MO	63105	Sam Fox	314-727-5550	R	330•	3.2
Black & Decker Corp	701 E Joppa Rd	Towson	MD	21286	Joseph Schmidt	410-527-7000	D	300•	1.0
Excellon Automation Co	24751 Crenshaw	Torrance	CA	90505	Richard Pinto	310-534-6300	S	294	0.4
Greenfield Industries Inc	PO Box 2587	Augusta	GA	30903	Paul W Jones	706-863-7708	P	272	3.8
Brown & Sharpe Mfg Co	200 Frenchtown Rd	N Kingstown	RI	02852	Fred M Stuber	401-886-2000	P	205	2.4
LS Starrett Co	121 Crescent St	Athol	MA	01331	Douglas R Starrett	508-249-3551	P	180	2.5
Commercial Cam Company Inc	1444 S Wolf Rd	Wheeling	IL	60090	Jim Lindemann	708-459-5200	S	100	0.3
International Knife and Saw Inc	1299 Cox Av	Erlanger	KY	41018	Jim Reed	606-371-0333	R	98	0.8
Cleveland Twist Drill Co	1242 E 49th St	Cleveland	OH	44114	Paul Jones	216-431-3120	S	75	1.1
Niagara Cutter Inc	889 Erie Av	N Tonawanda	NY	14120	Sherwood L Bollier	716-693-8400	R	50	0.3
Triumph Twist Drill Co	1 Precision Plz	Crystal Lake	IL	60014	Jim Beck	815-459-6250	S	46	0.9
Norton Construction Products	PO Box 2898	Gainesville	GA	30503	A John Siva	404-967-3954	D	40	0.2
Guhring Inc	PO Box 643	Brookfield	WI	53045	Greg Rees	414-784-6730	S	36	0.2
Varel Manufacturing Co	PO Box 540157	Dallas	TX	75354	DW Varel	214-351-6486	S	35	1.0
Sossner Tap and Tool Corp	676 E Fullerton Av	Glendale H	IL	60139	George Osawa	708-790-4445	S	33	0.3
Reynolds Machine & Tool Corp	2033 N 17th Av	Melrose Park	IL	60160	James P Reynolds	708-344-3280	R	32	<0.1
Ferguson Co	11820 Lackland Rd	St Louis	MO	63146	Dick King	314-567-3200	S	31•	0.1
Abraxas Inc	2015 Mitchell Blv	Schaumburg	IL	60193	KJ Stern	708-980-7400	S	30	<0.1
Greenleaf Corp	Greenleaf Dr	Saegertown	PA	16433	W J Greenleaf Jr	814-763-2915	R	30	0.5
SGS Tool Co	55 S Main St	Munroe Falls	OH	44262	JA Haag	216-688-6667	R	30	0.4
Stadco Corp	1931 Broadway Av	Los Angeles	CA	90031	Neil Kadisha	213-227-8888	S	30	0.2
Whessoe Varec Inc	10800 Val View Av	Cypress	CA	90630	N Mules	714-761-1300	S	25	0.2
KOMET of AMERICA Inc	2050 Mitchell Blv	Schaumburg	IL	60193	Rick Martin	708-924-8400	S	23•	0.2
Brubaker Tool Corp	200 Front St	Millersburg	PA	17061	Tim Wyman	717-692-2113	R	22•	0.4
Bachman Machine Co	4321 N Broadway	St Louis	MO	63147	W G Bachman Jr	314-231-4221	R	20	0.1
Newcomer Products Inc	PO Box 272	Latrobe	PA	15650	Robert S Jacobs	412-537-5531	S	20	0.2
Carmet Co	1 Tungsten Way	Duncan	SC	29334	Frank P Cyrill Jr	803-879-7621	R	18	0.3
Kwik-Way Manufacturing Co	500 57th St	Marion	IA	52302	David Parks	319-377-9421	S	18	0.3
OS Walker Company Inc	Rockdale St	Worcester	MA	01606	Charles Alesi	508-853-3232	S	18	0.2
Waukesha Cutting Tools Inc	1111 Sentry Dr	Waukesha	WI	53186	Charles P Brumder	414-542-4426	R	16	0.1
Tooling Systems	PO Box 590	Fond du Lac	WI	54936	JB Simon	414-921-6300	D	16	0.1
Tooling Systems	126 N Main St	Frankenmuth	MI	48734	Gerald Norton	517-652-9911	D	16	0.2
Cogsdill Tool Products Inc	PO Box 7007	Camden	SC	29020	W J Westerman	803-438-4000	R	15	0.2
IMT	7008 Northland Dr	Minneapolis	MN	55428	Hans P Deller	612-533-9990	D	15•	<0.1
ITW Woodworth	1300 E 9 Mile Rd	Ferndale	MI	48220	Leo Walterich	810-541-7500	D	15	0.1
NED Corp	18 Grafton St	Worcester	MA	01604	Peter F Wyatt	508-798-8546	R	15	<0.1
NED-Kut Diamond Products	18 Grafton St	Worcester	MA	01604	Peter F Wyatt	508-798-8546	D	15	<0.1
Viking Drill and Tool Inc	PO Box 65278	St Paul	MN	55165	John L Knight	612-227-8911	R	15	0.1
America Mine Tool	PO Box AG	Chilhowie	VA	24319	Carl Watson	703-646-8990	R	14•	0.1
O and E Machine Corp	PO Box 1836	Green Bay	WI	54305	Guy Meyerhoffer	414-437-6588	S	14•	0.1
Dearborn Gage Co	32330 Ford Rd	Garden City	MI	48135	Olof W Ellstrom Jr	313-422-8300	R	13	<0.1
Hougen Manufacturing Inc	PO Box 2005	Flint	MI	48501	Randall B Hougen	810-732-5840	R	13	0.1
Iowa Precision Industries Inc	5480 6th St SW	Cedar Rapids	IA	52404	Jim Heitt	319-364-9181	S	13	0.1
ITW Southern Gage	49 Midway Dr	Erin	TN	37061	Jim Osberg	615-289-4242	D	13	0.2
Michigan Drill Corp	PO Box 7012	Troy	MI	48083	Richard Kandarian	810-689-5050	R	13	0.1
Vermont Tap and Die Co	79 Main St	Lyndonville	VT	05851	Paul Jones	802-626-3331	S	13•	0.2
All American Products Company	PO Box 190	San Fernando	CA	91341	NO Shaw	818-361-0059	R	12	0.2
Buck Logansport Inc	PO Box 166	Logansport	IN	46947	GJ Duerr Jr	219-753-3104	R	12	0.1
E and E Engineering Inc	7200 Miller Dr	Warren	MI	48092	P Matthew Hirzel	810-978-3800	R	12•	<0.1
Fullerton Tool Co	PO Box 2008	Saginaw	MI	48605	Morgan Curry	517-799-4550	R	12	0.1
Garr Tool	PO Box 489	Alma	MI	48801	John C Leppien	517-463-6171	R	12•	0.1
Reiff and Nestor Company Inc	PO Box 147	Lykens	PA	17048	Patrick Savage	717-453-7113	R	12	0.3
Super-Cut Glass & Ceramic	84 O'Leary Dr	Bensenville	IL	60106	Carol E Spiller	312-286-5000	D	12•	0.2
Wisconsin Machine Tool Corp	445 S Curtis Rd	West Allis	WI	53214	William C Stevens	414-475-0900	R	12	0.1
PHB Machining	8150 W Ridge Rd	Fairview	PA	16415	David Parrish	814-474-1552	D	11	0.1
Alvord-Polk Inc	125 Gearhart St	Millersburg	PA	17061	RE Boyer	717-692-2128	R	10•	0.1
American Drill Bushing Co	2000 Camfield Av	City of Com	CA	90040	Albert Steele	213-725-1515	S	10•	0.1
Bronson and Bratton Inc	220 Shore Dr	Burr Ridge	IL	60521	HD Bronson	708-986-1815	R	10	0.1
CB Manufacturing and Sales	PO Box 37	West Carrollton	OH	45449	CS Biehn Jr	513-866-5986	R	10•	0.1
Citco	PO Box 84	Chardon	OH	44024	Robert J Hanslik	216-285-9181	D	10	0.2
Hannibal Carbide Tool Inc	PO Box 954	Hannibal	MO	63401	John Enander	314-221-2775	R	10	0.1
MA Ford Manufacturing	PO Box 3628	Davenport	IA	52808	S Morency	319-391-6220	R	10	0.3
Onsrud Cutter Inc	800 E Liberty Dr	Libertyville	IL	60048	Richard S O'Brien	708-362-1560	R	10	0.1
Russell T Gilman Inc	PO Box 5	Grafton	WI	53024	R T Gilman Sr	414-377-2434	R	10	0.1
Shred Pax Systems Inc	136 W Commercial	Wood Dale	IL	60191	Al Kaczmarek	708-595-8780	S	10•	<0.1
SP/Sheffer International Inc	1032 Woodmere	Traverse City	MI	49684	Richard E Slovenec	616-947-5755	D	10	<0.1
Vermont Precision Tools Inc	PO Box 182	Swanton	VT	05488	NC Leduc	802-868-4246	R	10	<0.1
Vlier	2333 Valley St	Burbank	CA	91505	Dave Wachtel	818-843-1922	D	10	0.1
H and H Machine Tool of Iowa	PO Box 568	Cedar Falls	IA	50613	Thomas Abbas	319-268-0181	R	10	<0.1
Sterling Die Operation	13811 Enterprise	Cleveland	OH	44135	Edmund T Wozniak	216-267-1300	S	10	0.1
S-T Industries Inc	301 Armstrong N	St James	MN	56081	LT Smith	507-375-3211	R	10	0.1
AVK Industrial Products	25323 Rye Canyon	Valencia	CA	91355		805-257-2329	D	9•	<0.1
Climax Portable Machine Tools	PO Box 230	Newberg	OR	97132	RL Benham	503-538-2185	R	9	<0.1

Source: Ward's Business Directory of U.S. Private and Public Companies, Volumes 1 and 2, 1996. The company type code used is as follows: P - Public, R - Private, S - Subsidiary, D - Division, J - Joint Venture, A - Affiliate, G - Group. Sales are in millions of dollars, employees are in thousands. An asterisk (*) indicates an estimated sales volume. The symbol < stands for 'less than'. Company names and addresses are truncated, in some cases, to fit into the available space.

MATERIALS CONSUMED

Material	Quantity	Delivered Cost ($ million)
Materials, ingredients, containers, and supplies	(X)	844.4
Metal bolts, nuts, screws, washers, rivets, and other screw machine products	(X)	20.4
Other fabricated metal products (except castings and forgings)	(X)	29.0
Forgings	(X)	1.6
Iron and steel castings (rough and semifinished)	(X)	9.7
Nonferrous (aluminum, copper, etc.) castings (rough and semifinished)	(X)	2.3
Steel bars, bar shapes, and plates	(X)	170.4
Steel sheet and strip, including tin plate	(X)	17.8
All other steel shapes and forms	(X)	12.8
Aluminum and aluminum-base alloy sheet, plate, foil, and welded tubing	(X)	2.7
All other aluminum and aluminum-base alloy shapes and forms	(X)	1.7
Other nonferrous shapes and forms	(X)	3.2
Tungsten carbide metal powders	(X)	123.9
All other metal powders	(X)	4.3
Iron and steel scrap (excluding home scrap)	(X)	0.1
Industrial diamonds	(X)	23.9
Electrical transmission, distribution, and control equipment	(X)	14.0
Grinding wheels and other abrasive products, except industrial diamonds	(X)	27.9
Fluid power products	(X)	11.4
Ceramic raw materials, including powders, chemicals, and fibers (excluding refractory uses)	(X)	1.0
Ceramic and ceramic composite parts, components, and accessories	(X)	6.4
All other materials and components, parts, containers, and supplies	(X)	171.1
Materials, ingredients, containers, and supplies, nsk	(X)	188.6

Source: 1992 *Economic Census*. Explanation of symbols used: (D): Withheld to avoid disclosure of competitive data; na: Not available; (S): Withheld because statistical norms were not met; (X): Not applicable; (Z): Less than half the unit shown; nec: Not elsewhere classified; nsk: Not specified by kind; - : zero; * : 10-19 percent estimated; ** : 20-29 percent estimated.

PRODUCT SHARE DETAILS

Product or Product Class	% Share	Product or Product Class	% Share
Machine tool accessories	100.00	complete) for machine tools and metalworking machinery	1.35
Small cutting tools for machine tools and metalworking machinery	56.69	High-speed steel milling cutters, nec, for machine tools and metalworking machinery.	2.90
Broaches (excluding holders and burnishing bars) for machine tools and metalworking machinery . . .	4.50	Solid and tipped carbide milling cutters, nec, for machine tools and metalworking macinery (excluding tips and blanks sold separately)	0.48
Carbon steel and high-speed steel taper shank twist drills for machine tools and metalworking machinery (excluding combined drills, countersinks, and gun drills)	0.76	Taps (excluding taps in threading sets and screw plates and inserted chaser types) for machine tools and metalworking machinery	5.62
Carbon steel and high-speed steel straight shank twist drills for machine tools and metalworking machinery (excluding combined drills, countersinks, and gun drills)	11.77	Dies, with two or more thread-forming edges integral with the body (excluding metalworking dies in product class 35441) for machine tools and metalworking machinery . .	0.63
Solid and tipped carbide twist drills for machine tools and metalworking machinery (excluding combined drills, countersinks, gun drills, tips and blanks sold separately, and masonry drills)	4.82	Chasers, single edge thread-cutting, circular blade and tangent types for mount in/on holders, die heads, and tap bodies on machine tools and metalworking machinery . .	1.32
Masonry twist drill bits for machinc tools and mctalworking machinery (excluding combined drills, countersinks, and gun drills)	1.90	Thread-rolling dies, including circular, flat, and planetary, for machine tools and metalworking machinery	1.95
Gun drills and gun reamers for machine tools and metalworking machinery.	0.98	Single and double point cutting tools for machine tools and metalworking machinery.	2.64
Combination drills and countersinks for machine tools and metalworking machinery.	0.77	Circular form cutting tools (including semifinished blanks) for machine tools and metalworking machinery	1.51
Countersinks (including port cutters, etc.) for machine tools and metalworking machinery (except combined drills-countersinks and pilots for interchangeable pilots) . . .	0.67	Molded blanks and tips (excluding pressed-to-size inserts) for machine tools and metalworking machinery	4.48
Counterbores (including spot facers, etc.) for machine tools and metalworking machinery (excluding pilots for interchangeable pilot types). .	0.58	Precision ground carbide indexible and throwaway inserts for machine tools and metalworking machinery	13.24
Carbon steel and high-speed steel reamers (all types, except gun reamers) for machine tools and metalworking machinery, including blades sold separately	1.34	Other carbide indexible and throwaway inserts for machine tools and metalworking machinery	6.34
Solid and tipped carbide reamers (all types, except gun reamers) for machine tools and metalworking machinery (including replaceable blades sold separately) (excluding tips and blanks sold separately)	1.76	Ceramic indexible and throwaway inserts for machine tools and metalworking machinery	1.55
Hobs (all types) for machine tools and metalworking machinery	1.45	Indexible and throwaway inserts other than carbide and ceramic for machine tools and metalworking machinery .	0.77
Gear shaper cutters and gear shaving cutters for machine tools and metalworking machinery	1.24	Carbide inserts, other than indexible and throwaway types, for machine tools and metalworking machinery	1.31
High-speed steel end mills (excluding inserted blade types and shell mills) for machine tools and metalworking machinery	4.44	Ceramic inserts, other than indexible and throwaway types, for machine tools and metalworking machinery	1.16
Solid and tipped carbide end mills (excluding inserted blade types, shell mills, and blades sold separately) for machine tools and metalworking machinery	4.32	Inserts, other than carbide and ceramic, other than indexible and throwaway, for machine tools and metalworking machinery.	0.90
Nonindexible inserted blade type milling cutters (all types, complete) for machine tools and metalworking machinery	0.43	Other carbon steel cutting tools for machine tools, nec (including rotary burrs, rotary files, and spade drills) . .	0.80
Indexible or throwaway insert type milling cutters (all types,		Other high-speed steel cutting tools for machine tools, nec (including rotary burrs, rotary files, and spade drills) . .	2.17
		Other solid and tipped carbide cutting tools for machine tools, nec (except tips and blanks sold separately) (including rotary burrs, rotary files, spade drills)	3.78
		Small cutting tools for machine tools and metalworking machinery, nsk	3.39
		Other attachments and accessories for machine tools and	

Continued on next page.

PRODUCT SHARE DETAILS - Continued

Product or Product Class	% Share	Product or Product Class	% Share
metalworking machinery	18.35	Other attachments and accessories for machine tools and metalworking machinery, nsk	2.03
Holders for turning tools, mechanically clamping for inserts and bits (except box tools and screw machine tool holders) for machine tools and metalworking machinery	12.50	Precision measuring tools (inspection, quality control, tool room, and machinists')	11.25
Holders for boring bars and heads for machine tools and metalworking machinery	7.51	Comparators (excluding optical) (inspection, quality control, tool room, and machinists' precision measuring tools)	1.20
Holders for drilling, reaming, and tapping chucks for machine tools and metalworking machinery	11.55	Fixture type fixed size precision measuring limit gauges (American Gauge Design Type C58-61) (inspection, quality control, tool room, and machinists')	10.53
Holders for special tooling and attachments for screw and automatic machines (box tools, tool holders, turrets, rollers, etc.)	3.63	Thread type fixed size precision measuring limit gauges (American Gauge Design Type C58-61) (inspection, quality control, tool room, and machinists')	9.26
Holders for die heads and tap bodies for chaser-type threading and thread-rolling heads (excluding hand-type die stocks)	1.52	Adjustable size precision measuring limit gauges	5.28
Other tool holders, including other chucks, drill heads, tool posts, turrets, sleeves, sockets, etc.	8.70	Precision measuring gauge blocks	1.42
		Precision measuring dial indicators	5.79
Rotary tables and indexing work holders, including numerically controlled	7.06	Precision measuring micrometers and calipers	8.57
Other work holding and positioning devices, including vises, mandrels, feeding fingers and collets, clamps, stops, etc.	24.38	Pneumatic and electronic precision measuring gauges (manual and automatic)	15.22
Tracer and tapering attachments, safety devices, centers, dogs, work rests, chutes, etc. for machine tools and metalworking machinery	2.23	Coordinate and contour precision measuring machines (inspection and gauging)	18.46
Lathe chucks for machine tools and metalworking machinery	4.28	Other machinists' precision measuring tools (including dividers, gear checking and surface texture measuring machines)	17.75
Tool room specialties (including levels, angle plates, parallels, sine bars, V-blocks, flats, etc.) for machine tools and metalworking machinery	0.74	Parts and accessories for machinists' precision measuring tools (sold separately)	4.44
Other attachments and accessories for machine tools and metalworking machinery	13.87	Precision measuring tools (inspection, quality control, tool room, and machinists'), nsk	2.04
		Machine tool accessories, nsk	13.71

Source: 1992 *Economic Census*. The values shown are percent of total shipments in an industry. Values of indented subcategories are summed in the main heading. The symbol (D) appears when data are withheld to prevent disclosure of competitive information. The abbreviation nsk stands for 'not specified by kind' and nec for 'not elsewhere classified'.

INPUTS AND OUTPUTS FOR SPECIAL DIES & TOOLS & MACHINE TOOL ACCESSORIES

Economic Sector or Industry Providing Inputs	%	Sector	Economic Sector or Industry Buying Outputs	%	Sector
Blast furnaces & steel mills	15.0	Manufg.	Gross private fixed investment	43.0	Cap Inv
Imports	10.8	Foreign	Motor vehicles & car bodies	4.6	Manufg.
Special dies & tools & machine tool accessories	9.7	Manufg.	Exports	4.4	Foreign
Wholesale trade	5.7	Trade	Aircraft	4.3	Manufg.
Machinery, except electrical, nec	5.1	Manufg.	Special dies & tools & machine tool accessories	3.6	Manufg.
Primary metal products, nec	4.4	Manufg.	Aluminum rolling & drawing	2.1	Manufg.
Advertising	3.5	Services	Machinery, except electrical, nec	1.9	Manufg.
Electric services (utilities)	3.3	Util.	Aluminum castings	1.6	Manufg.
Abrasive products	2.9	Manufg.	Steel wire & related products	1.6	Manufg.
Petroleum refining	2.2	Manufg.	Federal Government purchases, national defense	1.4	Fed Govt
Cyclic crudes and organics	1.8	Manufg.	Blast furnaces & steel mills	0.9	Manufg.
Motor freight transportation & warehousing	1.8	Util.	Motor vehicle parts & accessories	0.9	Manufg.
Eating & drinking places	1.8	Trade	Screw machine and related products	0.9	Manufg.
Noncomparable imports	1.5	Foreign	Automotive stampings	0.8	Manufg.
Communications, except radio & TV	1.4	Util.	Nonferrous rolling & drawing, nec	0.8	Manufg.
Iron & steel foundries	1.3	Manufg.	Refrigeration & heating equipment	0.8	Manufg.
Real estate	1.3	Fin/R.E.	Aircraft & missile engines & engine parts	0.7	Manufg.
Maintenance of nonfarm buildings nec	1.2	Constr.	Glass containers	0.7	Manufg.
Miscellaneous plastics products	1.2	Manufg.	Miscellaneous plastics products	0.7	Manufg.
Banking	1.2	Fin/R.E.	Metal stampings, nec	0.6	Manufg.
Machine tools, metal cutting types	1.0	Manufg.	Railroads & related services	0.6	Util.
Paperboard containers & boxes	1.0	Manufg.	Water transportation	0.6	Util.
Equipment rental & leasing services	1.0	Services	Construction machinery & equipment	0.5	Manufg.
Computer & data processing services	0.9	Services	Internal combustion engines, nec	0.5	Manufg.
Engineering, architectural, & surveying services	0.9	Services	Iron & steel foundries	0.5	Manufg.
Metal heat treating	0.8	Manufg.	Nonferrous castings, nec	0.5	Manufg.
Gas production & distribution (utilities)	0.8	Util.	Oil field machinery	0.5	Manufg.
Royalties	0.8	Fin/R.E.	Pumps & compressors	0.5	Manufg.
Aluminum rolling & drawing	0.7	Manufg.	Wholesale trade	0.5	Trade
Industrial patterns	0.7	Manufg.	Aircraft & missile equipment, nec	0.4	Manufg.
Screw machine and related products	0.7	Manufg.	Machine tools, metal cutting types	0.4	Manufg.
Air transportation	0.7	Util.	Manufacturing industries, nec	0.4	Manufg.
Management & consulting services & labs	0.7	Services	Special industry machinery, nec	0.4	Manufg.
Paints & allied products	0.6	Manufg.	Blowers & fans	0.3	Manufg.
Plating & polishing	0.6	Manufg.	Brass, bronze, & copper castings	0.3	Manufg.
Legal services	0.6	Services	Electronic components nec	0.3	Manufg.
Fabricated metal products, nec	0.5	Manufg.	Food products machinery	0.3	Manufg.
Nonmetallic mineral products, nec	0.4	Manufg.	General industrial machinery, nec	0.3	Manufg.
Primary lead	0.4	Manufg.	Guided missiles & space vehicles	0.3	Manufg.
Railroads & related services	0.4	Util.	Iron & steel forgings	0.3	Manufg.

Continued on next page.

INPUTS AND OUTPUTS FOR SPECIAL DIES & TOOLS & MACHINE TOOL ACCESSORIES - Continued

Economic Sector or Industry Providing Inputs	%	Sector	Economic Sector or Industry Buying Outputs	%	Sector
Accounting, auditing & bookkeeping	0.4	Services	Power transmission equipment	0.3	Manufg.
Automotive rental & leasing, without drivers	0.4	Services	Radio & TV communication equipment	0.3	Manufg.
Industrial controls	0.3	Manufg.	Semiconductors & related devices	0.3	Manufg.
Lubricating oils & greases	0.3	Manufg.	Retail trade, except eating & drinking	0.3	Trade
Manifold business forms	0.3	Manufg.	Electronic computing equipment	0.2	Manufg.
Pumps & compressors	0.3	Manufg.	Farm machinery & equipment	0.2	Manufg.
Switchgear & switchboard apparatus	0.3	Manufg.	Glass & glass products, except containers	0.2	Manufg.
Insurance carriers	0.3	Fin/R.E.	Hardware, nec	0.2	Manufg.
Automotive repair shops & services	0.3	Services	Logging camps & logging contractors	0.2	Manufg.
Hotels & lodging places	0.3	Services	Paperboard containers & boxes	0.2	Manufg.
Aluminum castings	0.2	Manufg.	Petroleum refining	0.2	Manufg.
Ball & roller bearings	0.2	Manufg.	Photographic equipment & supplies	0.2	Manufg.
Brass, bronze, & copper castings	0.2	Manufg.	Pipe, valves, & pipe fittings	0.2	Manufg.
Copper rolling & drawing	0.2	Manufg.	Power driven hand tools	0.2	Manufg.
Industrial inorganic chemicals, nec	0.2	Manufg.	Printing trades machinery	0.2	Manufg.
Iron & steel forgings	0.2	Manufg.	Service industry machines, nec	0.2	Manufg.
Metal stampings, nec	0.2	Manufg.	Ship building & repairing	0.2	Manufg.
Motors & generators	0.2	Manufg.	Textile machinery	0.2	Manufg.
Business/professional associations	0.2	Services	Equipment rental & leasing services	0.2	Services
Electrical repair shops	0.2	Services	Miscellaneous repair shops	0.2	Services
Personnel supply services	0.2	Services	Federal Government purchases, nondefense	0.2	Fed Govt
Coal	0.1	Mining	Crude petroleum & natural gas	0.1	Mining
Hand & edge tools, nec	0.1	Manufg.	Ball & roller bearings	0.1	Manufg.
Machine tools, metal forming types	0.1	Manufg.	Broadwoven fabric mills	0.1	Manufg.
Metal coating & allied services	0.1	Manufg.	Electrical equipment & supplies, nec	0.1	Manufg.
Metalworking machinery, nec	0.1	Manufg.	Fabricated metal products, nec	0.1	Manufg.
Nonferrous castings, nec	0.1	Manufg.	Fabricated plate work (boiler shops)	0.1	Manufg.
Nonferrous rolling & drawing, nec	0.1	Manufg.	Industrial furnaces & ovens	0.1	Manufg.
Photographic equipment & supplies	0.1	Manufg.	Instruments to measure electricity	0.1	Manufg.
Transformers	0.1	Manufg.	Mechanical measuring devices	0.1	Manufg.
Transit & bus transportation	0.1	Util.	Metal cans	0.1	Manufg.
Retail trade, except eating & drinking	0.1	Trade	Miscellaneous fabricated wire products	0.1	Manufg.
Photofinishing labs, commercial photography	0.1	Services	Motors & generators	0.1	Manufg.
			Nonferrous wire drawing & insulating	0.1	Manufg.
			Paper industries machinery	0.1	Manufg.
			Paper mills, except building paper	0.1	Manufg.
			Plating & polishing	0.1	Manufg.
			Sheet metal work	0.1	Manufg.
			Steel pipe & tubes	0.1	Manufg.
			Telephone & telegraph apparatus	0.1	Manufg.
			Turbines & turbine generator sets	0.1	Manufg.
			Wiring devices	0.1	Manufg.
			Electric services (utilities)	0.1	Util.
			Automotive repair shops & services	0.1	Services
			Management & consulting services & labs	0.1	Services
			S/L Govt. purch., other general government	0.1	S/L Govt

Source: Benchmark Input-Output Accounts for the U.S. Economy, 1982, U.S. Department of Commerce, Washington, D.C., July 1991. Data, as reported in the source, are organized by the 1977 SIC structure in use in 1982 but have been matched, as closely as is possible, to the 1987 SIC structure used in this book.

OCCUPATIONS EMPLOYED BY SIC 354 - METALWORKING MACHINERY

Occupation	% of Total 1994	Change to 2005	Occupation	% of Total 1994	Change to 2005
Machinists	10.9	-17.4	Precision metal workers nec	1.9	10.1
Tool & die makers	10.6	-15.3	Mechanical engineers	1.9	1.0
General managers & top executives	3.7	-12.9	Inspectors, testers, & graders, precision	1.9	-8.2
Blue collar worker supervisors	3.4	-15.7	Secretaries, ex legal & medical	1.9	-16.4
Machine tool cutting operators, metal & plastic	2.9	-5.9	Drafters	1.8	-28.5
NC machine tool operators, metal & plastic	2.7	1.0	General office clerks	1.5	-21.7
Grinding machine operators, metal & plastic	2.7	-8.2	Machine forming operators, metal & plastic	1.4	-8.2
Sales & related workers nec	2.7	-8.2	Drilling & boring machine tool workers	1.3	-8.2
Assemblers, fabricators, & hand workers nec	2.5	-8.2	Bookkeeping, accounting, & auditing clerks	1.3	-31.1
Lathe & turning machine tool operators	2.3	-17.4	Industrial production managers	1.3	-8.2
Machine tool cutting & forming etc. nec	2.2	-54.1	Traffic, shipping, & receiving clerks	1.3	-11.7
Machine builders	2.0	1.0	Precision woodworkers nec	1.3	10.2
Combination machine tool operators	2.0	37.7	Janitors & cleaners, incl maids	1.2	-26.6

Source: Industry-Occupation Matrix, Bureau of Labor Statistics. These data relate to one or more 3-digit SIC industry groups rather than to a single 4-digit SIC. The change reported for each occupation to the year 2005 is a percent of growth or decline as estimated by the Bureau of Labor Statistics. The abbreviation nec stands for 'not elsewhere classified'.

LOCATION BY STATE AND REGIONAL CONCENTRATION

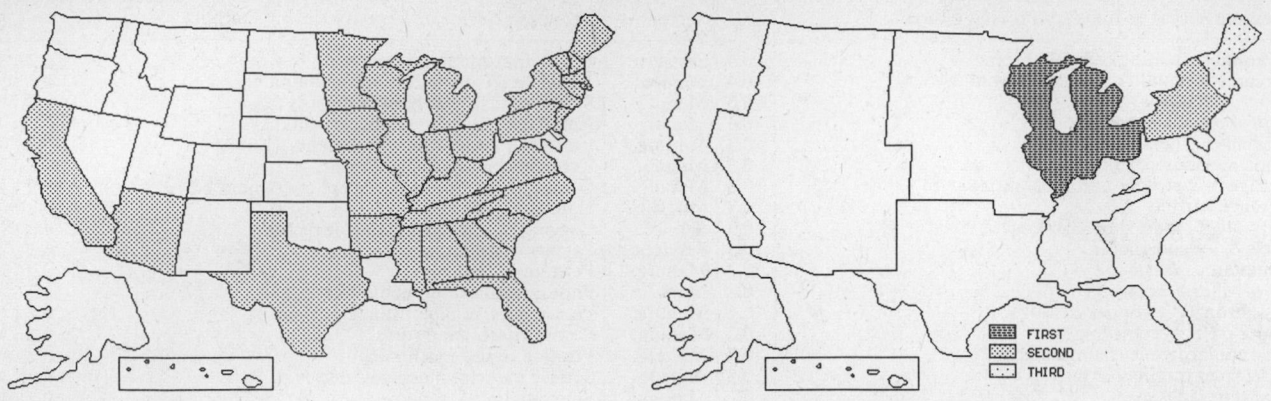

FIRST
SECOND
THIRD

INDUSTRY DATA BY STATE

State	Establish-ments	Shipments			Employment				Cost as % of Shipments	Investment per Employee ($)
		Total ($ mil)	% of U.S.	Per Establ.	Total Number	% of U.S.	Per Establ.	Wages ($/hour)		
Michigan	408	751.2	19.8	1.8	8,300	19.4	20	13.14	28.2	3,289
Ohio	170	515.3	13.6	3.0	4,900	11.5	29	12.46	23.5	3,490
Illinois	151	406.7	10.7	2.7	4,900	11.5	32	11.54	34.1	4,449
Massachusetts	67	232.1	6.1	3.5	2,600	6.1	39	13.17	28.2	2,962
South Carolina	34	193.4	5.1	5.7	1,800	4.2	53	10.21	32.2	4,389
California	192	191.8	5.1	1.0	2,500	5.9	13	10.65	31.1	2,560
Pennsylvania	79	171.1	4.5	2.2	2,200	5.2	28	12.35	33.3	3,955
Wisconsin	45	133.3	3.5	3.0	1,500	3.5	33	12.87	36.5	1,733
New York	100	123.9	3.3	1.2	1,500	3.5	15	12.80	29.9	4,267
Tennessee	24	118.6	3.1	4.9	1,100	2.6	46	13.56	26.8	4,545
Connecticut	93	118.5	3.1	1.3	1,500	3.5	16	12.52	32.8	1,467
North Carolina	25	91.2	2.4	3.6	900	2.1	36	10.79	21.9	7,333
Texas	50	69.2	1.8	1.4	900	2.1	18	11.85	26.7	2,778
Rhode Island	20	69.0	1.8	3.5	800	1.9	40	13.08	24.5	2,000
Minnesota	33	65.4	1.7	2.0	600	1.4	18	11.14	36.9	1,333
Indiana	45	65.2	1.7	1.4	900	2.1	20	9.93	25.2	2,222
New Jersey	54	51.6	1.4	1.0	500	1.2	9	13.00	29.1	4,000
Georgia	20	49.3	1.3	2.5	700	1.6	35	8.64	29.8	-
Florida	41	45.1	1.2	1.1	600	1.4	15	10.33	27.9	2,500
Vermont	14	38.3	1.0	2.7	700	1.6	50	11.20	26.6	1,286
Iowa	17	33.4	0.9	2.0	400	0.9	24	11.67	24.9	1,500
Missouri	19	32.6	0.9	1.7	400	0.9	21	10.00	39.0	1,500
New Hampshire	17	22.4	0.6	1.3	200	0.5	12	11.00	32.6	4,000
Kentucky	11	15.9	0.4	1.4	200	0.5	18	10.33	39.0	6,000
Virginia	10	9.8	0.3	1.0	200	0.5	20	11.67	31.6	-
Arizona	23	(D)	-	-	175 *	0.4	8	-	-	1,714
Arkansas	12	(D)	-	-	750 *	1.8	63	-	-	-
Alabama	5	(D)	-	-	175 *	0.4	35	-	-	-
Mississippi	4	(D)	-	-	175 *	0.4	44	-	-	-
Maine	3	(D)	-	-	175 *	0.4	58	-	-	-

Source: 1992 *Economic Census*. The states are in descending order of shipments or establishments (if shipment data are missing for the majority). The symbol (D) appears when data are withheld to prevent disclosure of competitive information. States marked with (D) are sorted by number of establishments. A dash (-) indicates that the data element cannot be calculated; * indicates the midpoint of a range.

3546 - POWER DRIVEN HANDTOOLS

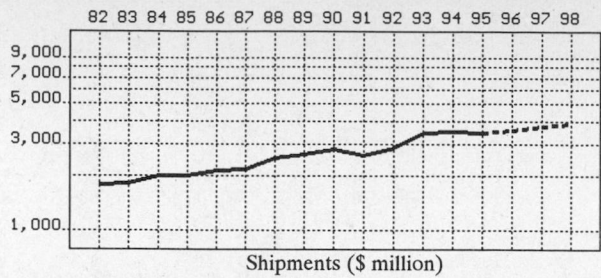

Shipments ($ million)

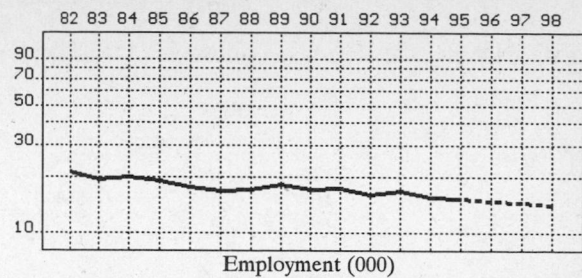

Employment (000)

GENERAL STATISTICS

Year	Companies	Establishments		Employment			Compensation		Production ($ million)			
		Total	with 20 or more employees	Total (000)	Production Workers (000)	Hours (Mil)	Payroll ($ mil)	Wages ($/hr)	Cost of Materials	Value Added by Manufacture	Value of Shipments	Capital Invest.
1982	180	203	74	21.6	14.5	25.9	393.6	8.83	791.5	940.3	1,795.3	68.1
1983		193	74	19.5	13.3	24.9	385.1	8.96	788.7	1,015.9	1,801.9	45.4
1984		183	74	20.0	14.2	26.4	392.2	9.28	965.9	1,088.8	2,016.3	50.6
1985		174	73	19.2	13.2	23.9	398.4	10.05	996.2	982.8	1,996.6	80.3
1986		165	76	17.9	12.5	24.0	383.8	9.55	998.3	1,121.5	2,142.4	72.1
1987	183	199	68	16.8	11.7	22.3	382.8	9.97	1,045.9	1,125.0	2,161.8	46.2
1988		201	76	17.1	12.0	23.3	391.2	9.69	1,241.0	1,275.7	2,505.0	59.0
1989		199	80	18.4	12.3	23.3	413.5	10.03	1,303.6	1,299.0	2,617.5	66.2
1990		199	76	17.1	12.6	24.4	446.9	10.18	1,344.1	1,471.8	2,805.8	98.4
1991		212	75	17.5	11.4	21.5	429.1	10.86	1,276.7	1,310.9	2,580.7	74.7
1992	214	226	69	16.1	10.6	21.7	440.6	11.08	1,359.9	1,506.8	2,872.5	72.3
1993		230	74	17.0	11.7	24.4	471.6	10.98	1,721.7	1,758.3	3,480.1	112.2
1994		222P	74P	15.6	10.9	23.5	449.0	11.19	1,734.8	1,786.0	3,495.2	103.9
1995		226P	74P	15.3P	10.5P	22.2P	460.1P	11.40P	1,719.9P	1,770.6P	3,465.1P	99.6P
1996		229P	74P	14.9P	10.2P	22.0P	466.8P	11.60P	1,789.5P	1,842.3P	3,605.5P	103.4P
1997		233P	74P	14.6P	10.0P	21.8P	473.4P	11.79P	1,859.2P	1,914.1P	3,745.9P	107.2P
1998		236P	74P	14.2P	9.7P	21.5P	480.0P	11.98P	1,928.9P	1,985.8P	3,886.2P	111.0P

Sources: 1982, 1987, 1992 *Economic Census*; *Annual Survey of Manufactures*, 83-86, 88-91, 93-94. Establishment counts for non-Census years are from *County Business Patterns*; establishment values for 83-84 are extrapolations. 'P's show projections by the editors. Industries reclassified in 87 will not have data for prior years.

INDICES OF CHANGE

Year	Companies	Establishments		Employment			Compensation		Production ($ million)			
		Total	with 20 or more employees	Total (000)	Production Workers (000)	Hours (Mil)	Payroll ($ mil)	Wages ($/hr)	Cost of Materials	Value Added by Manufacture	Value of Shipments	Capital Invest.
1982	84	90	107	134	137	119	89	80	58	62	62	94
1983		85	107	121	125	115	87	81	58	67	63	63
1984		81	107	124	134	122	89	84	71	72	70	70
1985		77	106	119	125	110	90	91	73	65	70	111
1986		73	110	111	118	111	87	86	73	74	75	100
1987	86	88	99	104	110	103	87	90	77	75	75	64
1988		89	110	106	113	107	89	87	91	85	87	82
1989		88	116	114	116	107	94	91	96	86	91	92
1990		88	110	106	119	112	101	92	99	98	98	136
1991		94	109	109	108	99	97	98	94	87	90	103
1992	100	100	100	100	100	100	100	100	100	100	100	100
1993		102	107	106	110	112	107	99	127	117	121	155
1994		98P	107P	97	103	108	102	101	128	119	122	144
1995		100P	107P	95P	99P	102P	104P	103P	126P	118P	121P	138P
1996		101P	107P	93P	97P	101P	106P	105P	132P	122P	126P	143P
1997		103P	107P	90P	94P	100P	107P	106P	137P	127P	130P	148P
1998		105P	107P	88P	92P	99P	109P	108P	142P	132P	135P	154P

Sources: Same as General Statistics. Values reflect change from the base year, 1992. Values above 100 mean greater than 92, values below 100 mean less than 92, and a value of 100 in the 82-91 or 93-98 period means same as 92. 'P's mark projections by the editors.

SELECTED RATIOS

For 1994	Avg. of All Manufact.	Analyzed Industry	Index	For 1994	Avg. of All Manufact.	Analyzed Industry	Index
Employees per Establishment	49	70	143	Value Added per Production Worker	134,084	163,853	122
Payroll per Establishment	1,500,273	2,022,661	135	Cost per Establishment	5,045,178	7,814,948	155
Payroll per Employee	30,620	28,782	94	Cost per Employee	102,970	111,205	108
Production Workers per Establishment	34	49	143	Cost per Production Worker	146,988	159,156	108
Wages per Establishment	853,319	1,184,608	139	Shipments per Establishment	9,576,895	15,745,219	164
Wages per Production Worker	24,861	24,125	97	Shipments per Employee	195,460	224,051	115
Hours per Production Worker	2,056	2,156	105	Shipments per Production Worker	279,017	320,661	115
Wages per Hour	12.09	11.19	93	Investment per Establishment	321,011	468,050	146
Value Added per Establishment	4,602,255	8,045,594	175	Investment per Employee	6,552	6,660	102
Value Added per Employee	93,930	114,487	122	Investment per Production Worker	9,352	9,532	102

Sources: Same as General Statistics. The 'Average of All Manufacturing' column represents the average of all manufacturing industries reported for the most recent complete year available. The Index shows the relationship between the Average and the Analyzed Industry. For example, 100 means that they are equal; 500 that the Analyzed Industry is five times the average; 50 means that the Analyzed Industry is half the national average. The abbreviation 'na' is used to show that data are 'not available'.

LEADING COMPANIES   Number shown: **54**   Total sales ($ mil): **10,491**   Total employment (000): **72.1**

Company Name	Address				CEO Name	Phone	Co. Type	Sales ($ mil)	Empl. (000)
Black and Decker Corp	701 E Joppa Rd	Towson	MD	21286	Nolan D Archibald	410-716-3900	P	5,248	35.8
North Amer Power Tools	701 E Joppa Rd	Towson	MD	21286	Gary T DiCamillo	410-716-3900	S	1,200	5.6
SPX Corp	PO Box 3301	Muskegon	MI	49443	Dale A Johnson	616-724-5000	P	1,093	8.2
Danaher Corp	1250 24th St NW	Washington	DC	20037	George M Sherman	202-828-0850	P	1,067	7.8
Milwaukee Electric Tool Corp	13135 W Lisbon Rd	Brookfield	WI	53005	Richard Grove	414-781-3600	S	370	2.2
Hilti Inc	PO Box 21148	Tulsa	OK	74121	Ewald Hoelker	918-252-6000	R	347	2.4
Porter-Cable Corp	PO Box 2468	Jackson	TN	38302	James A White	901-668-8600	S	150	1.0
Aro Fluid Products	1 Aro Ctr	Bryan	OH	43506	Carmine Bosco	419-636-4242	D	125	1.5
McCulloch Corp	PO Box 11990	Tucson	AZ	85734	Jonathan Miller	602-574-1311	R	95*	1.0
Simonds Industries Inc	PO Box 500	Fitchburg	MA	01420	R George	508-343-3731	R	90	0.6
INDRESCO Inc	PO Box 40430	Houston	TX	77240	TR Hurst	713-462-4521	D	69	0.6
Chemi-Trol Chemical Co	2776 County Rd 69	Gibsonburg	OH	43431	Arthur F Doust	419-665-2367	P	67	0.4
McCulloch Corp	900 Lake Havasu	Lk Havasu Ct	AZ	86403	Jonathan Miller	602-855-4171	D	49*	0.5
Shopsmith Inc	3931 Image Dr	Dayton	OH	45414	John R Folkerth	513-898-6070	P	48	0.1
Cooper Power Tools	PO Box 1410	Lexington	SC	29072	David Cartwright	803-359-1200	D	43*	0.5
Stihl Inc	536 Viking Dr	Virginia Beach	VA	23452	Fred J Whyte	804-486-9100	S	43	0.5
DESA International Inc	PO Box 90004	Bowling Green	KY	42102	D Vitale	502-781-9600	R	40*	0.4
Lisle Corp	PO Box 98	Clarinda	IA	51632	J Lisle	712-542-5101	R	38*	0.3
P and F Industries Inc	300 Smith St	Farmingdale	NY	11735	Sidney Horowitz	516-694-1800	P	36	0.3
Stanley Air Tools	700 Beta Dr	Cleveland	OH	44143	John E Turpin	216-461-5500	D	31*	0.3
Cushion Cut Inc	2565 W 237th St	Torrance	CA	90505	Mike Jagoe	310-325-5702	S	24	0.1
S-B Power Tool Co	4300 W Peterson	Chicago	IL	60646	G Thomas McKane	312-286-7330	J	24*	0.2
Multiquip Inc	PO Box 6254	Carson	CA	90746	Irving M Levine	310-537-3700	R	19*	0.2
Rotor Tool Co	26300 Lakeland	Cleveland	OH	44132	Robert W Zellner	216-731-8888	R	17	0.1
Dumore Corp	1030 Veteran St	Mauston	WI	53948	Bill Stout	608-847-6420	R	16	0.2
Deutsch Amer Pneumatic Tool	14710 Maple Av	Gardena	CA	90248	Lester Deutsch	310-538-2600	S	15	0.1
Blackstone Industries Inc	16 Stony Hill Rd	Bethel	CT	06801	Willard P Nelson	203-792-8622	R	13*	0.1
Dynabrade Inc	8989 Sheridan Dr	Clarence	NY	14031	Walter N Welsch	716-631-0100	R	12*	<0.1
Thomas C Wilson Inc	21-11 44th Av	Long Island Ct	NY	11101	Charles E Hanley	718-729-3360	R	12	0.1
ITW Heartland Components	1601 36th Av S	Alexandria	MN	56308	K Anderson	612-762-8138	D	10*	<0.1
Master Power Inc	909 Baltimore Blv	Westminster	MD	21157	Al Wordsworth	410-876-0076	R	9	0.1
Foredom Electric Co	16 Stony Hill Rd	Bethel	CT	06801	Willard P Nelson	203-792-8622	D	8	<0.1
Air Tool Service Co	7700 St Clair Av	Mentor	OH	44060	Steve Sabath	216-942-4475	R	6	<0.1
Fayscott Co	225 Spring St	Dexter	ME	04930	W Clukey	207-924-7331	R	6	0.1
Hutchins Manufacturing Co	49 N Lotus Av	Pasadena	CA	91107	AA Hutchins	818-792-8211	R	6	<0.1
United Grinding Technologies	510 Earl Blv	Miamisburg	OH	45342	Guergen Richter	513-859-1975	S	6*	<0.1
Wilde Tool Company Inc	PO Box 30	Hiawatha	KS	66434	Dan J Froeschl	913-742-7171	R	6	0.1
Delsteel Inc	PO Box 1268	Wilmington	DE	19899	RL Levy	302-764-2200	R	5	<0.1
Roseburrough Tool Inc	PO Box 1307	Orange	CA	92668	Jack Roseburrough	714-538-6015	R	4*	<0.1
Industrial Metalworking Services	1 Federick Rd	Warren	NJ	07059	Eugene Thomas	908-561-1706	S	3*	<0.1
Mechanic's Time Savers Inc	PO Box 531377	Grand Prairie	TX	75053	Jerry P Shaw	214-264-8170	R	3	<0.1
Stearns Prod Development Corp	5642 Borwick Av	South Gate	CA	90280	Eugene Raio	213-771-2270	R	3	<0.1
Cutawl Co	16 Stony Hill Rd	Bethel	CT	06801	Willard P Nelson	203-792-8622	D	2	<0.1
Danair Inc	PO Box 3898	Visalia	CA	93278	D Scilagyi	209-734-1961	R	2	<0.1
Ebbert Engineering Co	1925 W Maple Rd	Troy	MI	48084	Joe Basstee	810-649-9410	R	2	<0.1
Herr Manufacturing Company	17 Pearce Av	Tonawanda	NY	14150	Bruce McLean	716-874-5770	R	2	<0.1
Zircon Corp	1580 Dell Av	Campbell	CA	95008	John Stauss	408-866-8600	R	2	<0.1
Clipper Diamond Tool Co	47-16 Austell Pl	Long Island Ct	NY	11101	J Klipper	718-392-3671	R	1	<0.1
Hellyer Steel Parts Co	9858 Baldwin Pl	El Monte	CA	91734	R C Roeschlaub	818-443-0296	S	1*	<0.1
Hot Tools	PO Box 615	Marblehead	MA	01945	Charles Loutrel	617-639-1000	D	1*	<0.1
Racine Hydraulic Tools Inc	14272 Garwin	Menomonee Fls	WI	53051	Earl Rose	414-255-0999	R	1*	<0.1
W Rose Inc	PO Box 66	Sharon Hill	PA	19079	Glenn C King	215-583-4125	R	1*	<0.1
Air Turbine Technology Inc	1225 Broken Sound	Boca Raton	FL	33487	Gregory A Bowser	407-994-0500	R	1*	<0.1
Empire Technology Inc	3708 N Main St	Fort Worth	TX	76100	Bill Alexander	817-626-1504	R	0*	<0.1

Source: Ward's Business Directory of U.S. Private and Public Companies, Volumes 1 and 2, 1996. The company type code used is as follows: P - Public, R - Private, S - Subsidiary, D - Division, J - Joint Venture, A - Affiliate, G - Group. Sales are in millions of dollars, employees are in thousands. An asterisk (*) indicates an estimated sales volume. The symbol < stands for 'less than'. Company names and addresses are truncated, in some cases, to fit into the available space.

MATERIALS CONSUMED

Material	Quantity	Delivered Cost ($ million)
Materials, ingredients, containers, and supplies	(X)	1,079.3
Fluid power pumps, motors, and hydrostatic transmissions (hydraulic and pneumatic)	(X)	1.3
Fluid power filters (hydraulic and pneumatic)	(X)	0.1
Fluid power hose or tube fittings and assemblies (hydraulic and pneumatic)	(X)	0.5
Fluid power valves (hydraulic and pneumatic)	(X)	0.3
Other fluid power products (hydraulic and pneumatic)	(X)	0.2
Metal bolts, nuts, screws, washers, rivets, and other screw machine products	(X)	37.7
Other fabricated metal products (except fluid power products and forgings)	(X)	120.9
Forgings	(X)	1.8
Iron and steel castings (rough and semifinished)	(X)	29.2
Aluminum and aluminum-base alloy castings (rough and semifinished)	(X)	68.9
Other nonferrous castings (rough and semifinished)	(X)	7.4
Steel bars, bar shapes, and plates	(X)	51.1
All other steel shapes and forms	(X)	12.8

Continued on next page.

MATERIALS CONSUMED - Continued

Material	Quantity	Delivered Cost ($ million)
Nonferrous shapes and forms	(X)	5.3
Insulated copper wire and cable, except magnet wire	(X)	14.7
Magnet wire	(X)	20.6
Electric motors and generators	(X)	29.0
Storage batteries	(X)	43.9
Ball and roller bearings (mounted or unmounted)	(X)	46.0
Mechanical speed changers, gears, and industrial high-speed drives	(X)	18.9
Fabricated plastics products (except gaskets, hoses, and belting)	(X)	86.6
Paperboard containers, boxes, and corrugated paperboard	(X)	29.5
Packaging paper and plastics film, coated and laminated	(X)	4.0
All other materials and components, parts, containers, and supplies	(X)	319.8
Materials, ingredients, containers, and supplies, nsk	(X)	128.8

Source: 1992 *Economic Census*. Explanation of symbols used: (D): Withheld to avoid disclosure of competitive data; na: Not available; (S): Withheld because statistical norms were not met; (X): Not applicable; (Z): Less than half the unit shown; nec: Not elsewhere classified; nsk: Not specified by kind; - : zero; * : 10-19 percent estimated; ** : 20-29 percent estimated.

PRODUCT SHARE DETAILS

Product or Product Class	% Share	Product or Product Class	% Share
Power-driven handtools	100.00	other than sleeve bearings, 7 in. (177.80 mm) blade or less (except battery-powered)	8.86
Power-driven handtools, battery-powered (cordless)	12.07	Electric hand circular saws, armature mounted primarily on other than sleeve bearings, 8 in. (203.20 mm) blade or more (except battery-powered)	0.38
Battery-powered (cordless) driver/drills	33.97		
Other battery-powered (cordless) handtools	59.03	Electric hand jig and saber saws, armature mounted on ball bearings (except battery powered)	4.65
Parts, attachments, and accessories for battery-powered (cordless) handtools (sold separately)	7.00	Electric hand reciprocating saws (except battery-powered)	7.05
Power-driven handtools, electric (excluding battery-powered)	45.20	Electric hand chain saws (except battery-powered)	2.16
		Other electric-powered handtools (except battery-powered)	9.11
Electric hand drills (except battery-powered), armature mounted primarily on other than sleeve bearings, 5/16 in. (7.94 mm) to less than 1/2 in. (12.70 mm)	15.69	Parts, attachments, and accessories for electric-powered handtools (except battery powered) (sold separately)	8.24
Electric hand drills (except battery-powered), armature mounted primarily on other than sleeve bearings, 1/2 in. (12.70 mm) or more	5.32	Power-driven handtools, electric (excluding battery-powered), nsk	0.14
Electric screwdrivers and nut-runners (except battery-powered)	2.96	Power-driven handtools, pneumatic, hydraulic, and powder-actuated	26.27
Electric hammers, percussion and rotary, without drill chuck (except battery-powered)	3.34	Pneumatic drills, screwdrivers, and nut-runners	14.06
		Pneumatic percussion handtools (such as runners, riveters, chippers, scalers)	6.16
Electric hand impact wrenches (except battery-powered)	1.41	Pneumatic hand impact wrenches	11.46
Electric hand shears and nibblers (except battery-powered)	0.65	Pneumatic rotary hand grinders, polishers, and sanders	9.47
Electric hand planers (except battery-powered)	2.91	Other pneumatic hand grinders, polishers, and sanders	1.21
Electric hand routers, less than 1/2 inch collet size (maximum collet capacity) (except battery-powered)	3.33	Pneumatic hand staplers	7.30
		Pneumatic hand nailers	8.28
Electric hand routers, 1/2 inch collet size (maximum collet capacity) or more (except battery-powered)	2.03	Other pneumatic-powered handtools	7.01
Electric hand right angle polishers, circular sanders, and grinders, less than 7 inch wheel drive (except battery-powered)	3.79	Parts, attachments, and accessories for pneumatic-powered handtools (sold separately)	22.76
		Hydraulic hand chain saws, including pole	(D)
Electric hand right angle polishers, circular sanders, and grinders, 7 inch wheel drive or more (except battery-powered)	1.91	Other hydraulic-powered handtools	6.04
All other electric hand polishers, circular sanders, and grinders (including die grinders, but excluding bench) (except battery-powered)	1.81	Parts, attachments, and accessories for powder-actuated and hydraulic handtools (sold separately)	2.74
		Powder-actuated handtools	(D)
Electric hand belt sanders (except battery-powered)	3.32	Power-driven handtools, pneumatic, hydraulic, and powder-actuated, nsk	0.28
Electric hand oscillating, reciprocating, vibrating, and random orbit sanders (except battery powered)	5.28	Power-driven handtools, engine (internal combustion) driven	12.23
Electric hand circular saws, armature mounted primarily on sleeve bearings (except battery-powered)	5.66	Internal combustion engine driven chain saws	90.04
Electric hand circular saws, armature mounted primarily on		Other internal combustion engine-driven handtools, including cut-off saws and drills	9.93
		Power-driven handtools, nsk	4.23

Source: 1992 *Economic Census*. The values shown are percent of total shipments in an industry. Values of indented subcategories are summed in the main heading. The symbol (D) appears when data are withheld to prevent disclosure of competitive information. The abbreviation nsk stands for 'not specified by kind' and nec for 'not elsewhere classified'.

INPUTS AND OUTPUTS FOR POWER DRIVEN HAND TOOLS

Economic Sector or Industry Providing Inputs	%	Sector	Economic Sector or Industry Buying Outputs	%	Sector
Imports	27.1	Foreign	Gross private fixed investment	46.7	Cap Inv
Wholesale trade	9.0	Trade	Personal consumption expenditures	14.9	
Industrial controls	6.8	Manufg.	Exports	14.0	Foreign
Blast furnaces & steel mills	5.0	Manufg.	Logging camps & logging contractors	3.0	Manufg.
Fabricated plate work (boiler shops)	4.8	Manufg.	Meat animals	2.8	Agric.
Aluminum castings	3.2	Manufg.	Feed grains	2.2	Agric.
Advertising	2.8	Services	Miscellaneous repair shops	2.0	Services
Miscellaneous plastics products	2.5	Manufg.	Equipment rental & leasing services	1.3	Services
Nonferrous rolling & drawing, nec	2.4	Manufg.	Dairy farm products	1.1	Agric.
Machinery, except electrical, nec	2.1	Manufg.	Crude petroleum & natural gas	0.9	Mining
Petroleum refining	2.1	Manufg.	Federal Government purchases, national defense	0.9	Fed Govt
Special dies & tools & machine tool accessories	1.9	Manufg.	Oil bearing crops	0.8	Agric.
Primary metal products, nec	1.7	Manufg.	Food grains	0.6	Agric.
Ball & roller bearings	1.6	Manufg.	Poultry & eggs	0.6	Agric.
Electric services (utilities)	1.6	Util.	Millwork	0.6	Manufg.
Screw machine and related products	1.5	Manufg.	S/L Govt. purch., elem. & secondary education	0.6	S/L Govt
Nonferrous wire drawing & insulating	1.4	Manufg.	Vegetables	0.5	Agric.
Paperboard containers & boxes	0.9	Manufg.	Ship building & repairing	0.5	Manufg.
Power transmission equipment	0.9	Manufg.	Machinery, except electrical, nec	0.4	Manufg.
Communications, except radio & TV	0.9	Util.	Oil field machinery	0.4	Manufg.
Abrasive products	0.8	Manufg.	Fruits	0.3	Agric.
Iron & steel foundries	0.8	Manufg.	Greenhouse & nursery products	0.3	Agric.
Nonferrous castings, nec	0.8	Manufg.	Blast furnaces & steel mills	0.3	Manufg.
Eating & drinking places	0.8	Trade	Mobile homes	0.3	Manufg.
Banking	0.8	Fin/R.E.	Power driven hand tools	0.3	Manufg.
Semiconductors & related devices	0.7	Manufg.	Pumps & compressors	0.3	Manufg.
Motor freight transportation & warehousing	0.7	Util.	Electric services (utilities)	0.3	Util.
Equipment rental & leasing services	0.7	Services	Cotton	0.2	Agric.
Motors & generators	0.6	Manufg.	Tobacco	0.2	Agric.
Air transportation	0.6	Util.	Machine tools, metal cutting types	0.2	Manufg.
Royalties	0.6	Fin/R.E.	Miscellaneous livestock	0.1	Agric.
Engineering, architectural, & surveying services	0.6	Services	Nonfarm residential structure maintenance	0.1	Constr.
Maintenance of nonfarm buildings nec	0.5	Constr.	Office buildings	0.1	Constr.
Hand saws & saw blades	0.5	Manufg.	Residential 1-unit structures, nonfarm	0.1	Constr.
Machine tools, metal cutting types	0.5	Manufg.	Railroads & related services	0.1	Util.
Metal stampings, nec	0.5	Manufg.	S/L Govt. purch., higher education	0.1	S/L Govt
Plating & polishing	0.5	Manufg.			
Power driven hand tools	0.5	Manufg.			
Computer & data processing services	0.5	Services			
Brass, bronze, & copper castings	0.4	Manufg.			
Iron & steel forgings	0.4	Manufg.			
Wiring devices	0.4	Manufg.			
Automotive rental & leasing, without drivers	0.4	Services			
Automotive repair shops & services	0.4	Services			
Fabricated rubber products, nec	0.3	Manufg.			
Metal heat treating	0.3	Manufg.			
Sawmills & planning mills, general	0.3	Manufg.			
Gas production & distribution (utilities)	0.3	Util.			
Credit agencies other than banks	0.3	Fin/R.E.			
Legal services	0.3	Services			
Libraries, vocation education	0.3	Services			
Management & consulting services & labs	0.3	Services			
U.S. Postal Service	0.3	Gov't			
Pumps & compressors	0.2	Manufg.			
Insurance carriers	0.2	Fin/R.E.			
Real estate	0.2	Fin/R.E.			
Accounting, auditing & bookkeeping	0.2	Services			
Hotels & lodging places	0.2	Services			
Chemical preparations, nec	0.1	Manufg.			
Lubricating oils & greases	0.1	Manufg.			
Manifold business forms	0.1	Manufg.			
Metal coating & allied services	0.1	Manufg.			
Railroads & related services	0.1	Util.			
Noncomparable imports	0.1	Foreign			

Source: Benchmark Input-Output Accounts for the U.S. Economy, 1982, U.S. Department of Commerce, Washington, D.C., July 1991. Data, as reported in the source, are organized by the 1977 SIC structure in use in 1982 but have been matched, as closely as is possible, to the 1987 SIC structure used in this book.

OCCUPATIONS EMPLOYED BY SIC 354 - METALWORKING MACHINERY

Occupation	% of Total 1994	Change to 2005	Occupation	% of Total 1994	Change to 2005
Machinists	10.9	-17.4	Precision metal workers nec	1.9	10.1
Tool & die makers	10.6	-15.3	Mechanical engineers	1.9	1.0
General managers & top executives	3.7	-12.9	Inspectors, testers, & graders, precision	1.9	-8.2
Blue collar worker supervisors	3.4	-15.7	Secretaries, ex legal & medical	1.9	-16.4
Machine tool cutting operators, metal & plastic	2.9	-5.9	Drafters	1.8	-28.5
NC machine tool operators, metal & plastic	2.7	1.0	General office clerks	1.5	-21.7
Grinding machine operators, metal & plastic	2.7	-8.2	Machine forming operators, metal & plastic	1.4	-8.2
Sales & related workers nec	2.7	-8.2	Drilling & boring machine tool workers	1.3	-8.2
Assemblers, fabricators, & hand workers nec	2.5	-8.2	Bookkeeping, accounting, & auditing clerks	1.3	-31.1
Lathe & turning machine tool operators	2.3	-17.4	Industrial production managers	1.3	-8.2
Machine tool cutting & forming etc. nec	2.2	-54.1	Traffic, shipping, & receiving clerks	1.3	-11.7
Machine builders	2.0	1.0	Precision woodworkers nec	1.3	10.2
Combination machine tool operators	2.0	37.7	Janitors & cleaners, incl maids	1.2	-26.6

Source: Industry-Occupation Matrix, Bureau of Labor Statistics. These data relate to one or more 3-digit SIC industry groups rather than to a single 4-digit SIC. The change reported for each occupation to the year 2005 is a percent of growth or decline as estimated by the Bureau of Labor Statistics. The abbreviation nec stands for 'not elsewhere classified'.

LOCATION BY STATE AND REGIONAL CONCENTRATION

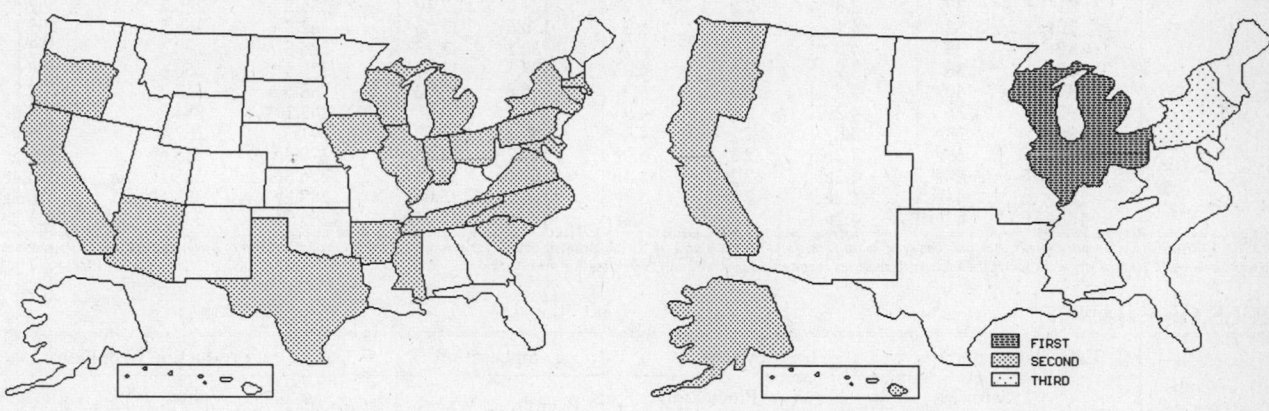

FIRST
SECOND
THIRD

INDUSTRY DATA BY STATE

State	Establish-ments	Shipments Total ($ mil)	Shipments % of U.S.	Shipments Per Establ.	Employment Total Number	Employment % of U.S.	Employment Per Establ.	Wages ($/hour)	Cost as % of Shipments	Investment per Employee ($)
South Carolina	4	234.3	8.2	58.6	2,000	12.4	500	10.58	51.1	1,450
Ohio	22	165.7	5.8	7.5	1,500	9.3	68	16.69	32.3	4,067
New York	12	136.9	4.8	11.4	1,000	6.2	83	16.67	44.6	3,800
Texas	10	50.5	1.8	5.1	500	3.1	50	19.75	24.4	800
Illinois	22	40.6	1.4	1.8	400	2.5	18	12.67	37.7	2,750
California	28	32.1	1.1	1.1	300	1.9	11	11.50	40.8	3,333
Massachusetts	10	31.0	1.1	3.1	300	1.9	30	10.00	49.0	2,667
Connecticut	12	16.3	0.6	1.4	200	1.2	17	13.00	50.9	3,000
Michigan	15	(D)	-	-	175 *	1.1	12	-	-	5,143
Pennsylvania	12	(D)	-	-	750 *	4.7	63	-	-	-
North Carolina	7	(D)	-	-	3,750 *	23.3	536	-	-	-
Oregon	6	(D)	-	-	375 *	2.3	63	-	-	-
Tennessee	5	(D)	-	-	750 *	4.7	150	-	-	-
Arizona	4	(D)	-	-	1,750 *	10.9	438	-	-	-
Arkansas	4	(D)	-	-	1,750 *	10.9	438	-	-	-
Maryland	4	(D)	-	-	750 *	4.7	188	-	-	-
Wisconsin	4	(D)	-	-	375 *	2.3	94	-	-	-
Indiana	3	(D)	-	-	175 *	1.1	58	-	-	-
Virginia	2	(D)	-	-	375 *	2.3	188	-	-	-
Iowa	1	(D)	-	-	175 *	1.1	175	-	-	-
Mississippi	1	(D)	-	-	375 *	2.3	375	-	-	-

Source: 1992 *Economic Census*. The states are in descending order of shipments or establishments (if shipment data are missing for the majority). The symbol (D) appears when data are withheld to prevent disclosure of competitive information. States marked with (D) are sorted by number of establishments. A dash (-) indicates that the data element cannot be calculated; * indicates the midpoint of a range.

3547 - ROLLING MILL MACHINERY

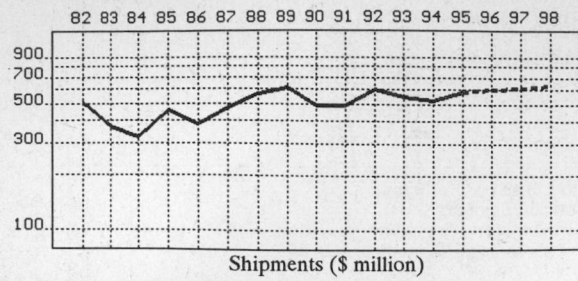

Shipments ($ million)

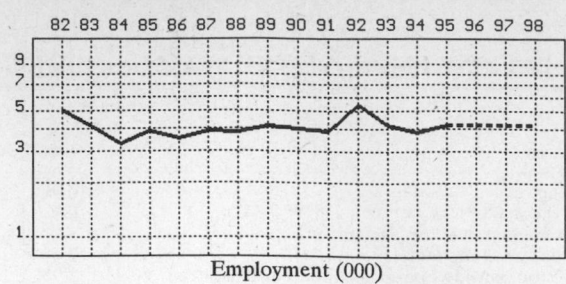

Employment (000)

GENERAL STATISTICS

| Year | Com-panies | Establishments | | Employment | | | Compensation | | Production ($ million) | | | |
		Total	with 20 or more employees	Total (000)	Production Workers (000)	Hours (Mil)	Payroll ($ mil)	Wages ($/hr)	Cost of Materials	Value Added by Manufacture	Value of Shipments	Capital Invest.
1982	58	63	32	5.1	3.3	6.1	125.1	11.61	246.8	276.4	502.9	14.8
1983		62	30	4.1	2.5	4.5	94.1	11.44	182.0	114.2	373.6	5.3
1984		61	28	3.3	2.0	3.8	89.3	11.97	168.5	192.8	323.1	6.0
1985		60	25	3.9	2.2	4.4	106.7	12.23	213.0	196.7	453.7	9.2
1986		60	24	3.5	1.9	3.8	101.1	12.34	209.7	169.6	380.8	3.6
1987	83	86	34	3.9	2.2	4.7	120.7	12.21	249.7	284.9	467.8	7.1
1988		81	37	3.8	2.2	4.9	121.6	12.10	335.3	246.2	561.4	5.2
1989		79	32	4.1	2.4	5.2	130.6	12.33	362.8	368.8	605.2	5.8
1990		82	33	4.0	2.2	4.7	130.1	13.13	226.8	173.3	483.4	9.9
1991		87	36	3.9	2.3	4.9	138.1	13.35	228.5	260.6	486.8	11.5
1992	87	89	40	5.4	3.0	6.3	186.8	13.52	288.6	313.2	602.8	13.7
1993		92	36	4.2	2.4	4.9	154.9	14.88	262.1	284.2	545.3	7.2
1994		96P	38P	3.8	2.1	4.4	139.3	14.59	241.9	282.4	521.3	9.9
1995		99P	39P	4.2P	2.3P	5.0P	162.2P	14.53P	270.3P	315.6P	582.5P	9.5P
1996		102P	40P	4.2P	2.2P	5.1P	167.4P	14.78P	276.8P	323.1P	596.4P	9.7P
1997		105P	41P	4.2P	2.2P	5.1P	172.6P	15.04P	283.2P	330.6P	610.3P	9.8P
1998		108P	42P	4.2P	2.2P	5.1P	177.8P	15.29P	289.7P	338.2P	624.2P	10.0P

Sources: 1982, 1987, 1992 *Economic Census*; *Annual Survey of Manufactures*, 83-86, 88-91, 93-94. Establishment counts for non-Census years are from *County Business Patterns*; establishment values for 83-84 are extrapolations. 'P's show projections by the editors. Industries reclassified in 87 will not have data for prior years.

INDICES OF CHANGE

| Year | Com-panies | Establishments | | Employment | | | Compensation | | Production ($ million) | | | |
		Total	with 20 or more employees	Total (000)	Production Workers (000)	Hours (Mil)	Payroll ($ mil)	Wages ($/hr)	Cost of Materials	Value Added by Manufacture	Value of Shipments	Capital Invest.
1982	67	71	80	94	110	97	67	86	86	88	83	108
1983		70	75	76	83	71	50	85	63	36	62	39
1984		69	70	61	67	60	48	89	58	62	54	44
1985		67	63	72	73	70	57	90	74	63	75	67
1986		67	60	65	63	60	54	91	73	54	63	26
1987	95	97	85	72	73	75	65	90	87	91	78	52
1988		91	93	70	73	78	65	89	116	79	93	38
1989		89	80	76	80	83	70	91	126	118	100	42
1990		92	83	74	73	75	70	97	79	55	80	72
1991		98	90	72	77	78	74	99	79	83	81	84
1992	100	100	100	100	100	100	100	100	100	100	100	100
1993		103	90	78	80	78	83	110	91	91	90	53
1994		108P	95P	70	70	70	75	108	84	90	86	72
1995		111P	97P	77P	75P	80P	87P	107P	94P	101P	97P	69P
1996		115P	100P	77P	75P	80P	90P	109P	96P	103P	99P	71P
1997		118P	102P	78P	74P	81P	92P	111P	98P	106P	101P	72P
1998		122P	104P	78P	74P	81P	95P	113P	100P	108P	104P	73P

Sources: Same as General Statistics. Values reflect change from the base year, 1992. Values above 100 mean greater than 92, values below 100 mean less than 92, and a value of 100 in the 82-91 or 93-98 period means same as 92. 'P's mark projections by the editors.

SELECTED RATIOS

For 1994	Avg. of All Manufact.	Analyzed Industry	Index	For 1994	Avg. of All Manufact.	Analyzed Industry	Index
Employees per Establishment	49	40	81	Value Added per Production Worker	134,084	134,476	100
Payroll per Establishment	1,500,273	1,454,715	97	Cost per Establishment	5,045,178	2,526,171	50
Payroll per Employee	30,620	36,658	120	Cost per Employee	102,970	63,658	62
Production Workers per Establishment	34	22	64	Cost per Production Worker	146,988	115,190	78
Wages per Establishment	853,319	670,401	79	Shipments per Establishment	9,576,895	5,443,956	57
Wages per Production Worker	24,861	30,570	123	Shipments per Employee	195,460	137,184	70
Hours per Production Worker	2,056	2,095	102	Shipments per Production Worker	279,017	248,238	89
Wages per Hour	12.09	14.59	121	Investment per Establishment	321,011	103,386	32
Value Added per Establishment	4,602,255	2,949,114	64	Investment per Employee	6,552	2,605	40
Value Added per Employee	93,930	74,316	79	Investment per Production Worker	9,352	4,714	50

Sources: Same as General Statistics. The 'Average of All Manufacturing' column represents the average of all manufacturing industries reported for the most recent complete year available. The Index shows the relationship between the Average and the Analyzed Industry. For example, 100 means that they are equal; 500 that the Analyzed Industry is five times the average; 50 means that the Analyzed Industry is half the national average. The abbreviation 'na' is used to show that data are 'not available'.

LEADING COMPANIES Number shown: 18 Total sales ($ mil): 228 Total employment (000): 1.5

Company Name	Address				CEO Name	Phone	Co. Type	Sales ($ mil)	Empl. (000)
Tippins Inc	435 Butler St	Pittsburgh	PA	15223	John E Thomas	412-781-7600	R	50	0.3
Integrated Industrial Systems Inc	475 Main St	Yalesville	CT	06492	Robert Herbst	203-265-5684	R	35	0.2
Hunter Engineering Company	6147 Rivercrest Dr	Riverside	CA	92507	Piero Bugnone	909-683-4010	S	25*	<0.1
BCO Industries Inc	PO Box 667	Moundridge	KS	67107	H David Bradbury	316-345-6394	R	21	0.2
Bradbury Company Inc	PO Box 667	Moundridge	KS	67107	H David Bradbury	316-345-6394	S	21	0.2
Weatherford-Pearland Mfg	PO Box 899	Pearland	TX	77588	Nelson A Byman	713-485-3264	D	15	0.1
T and H Machine Inc	55 W Laura Dr	Addison	IL	60101	Walter Heller	708-543-8880	R	12	<0.1
Magnat Rolls Inc	O'Neil St	Easthampton	MA	01027	William Daugherty	413-527-4256	R	11	<0.1
T Sendzimir Inc	269 Brookside Rd	Waterbury	CT	06721	M G Sendzimir	203-756-4617	R	10	<0.1
Rosemont Industries Inc	1700 West St	Cincinnati	OH	45215	Curt O Majors	513-733-4277	R	6	<0.1
Dahlstrom Industries Inc	9508 Winona Av	Schiller Park	IL	60176	James Williamson	708-678-5305	R	5*	<0.1
Hilman Rollers Inc	2604 Atlantic Av	Wall	NJ	07719	D Hill	908-528-0880	R	5	<0.1
Pecor Steel Technologies Co	12300 Perry Hwy	Wexford	PA	15090	Julie C Hereford	412-935-1553	R	3	<0.1
Zag Machine and Tool Co	99 John Downey Dr	New Britain	CT	06051	Adam Z Golas	203-224-7178	R	3	<0.1
Konrad Corp	1421 Hanley Rd	Hudson	WI	54016	K Konrad	715-386-4200	R	3	<0.1
Fanning-Schuett of New Jersey	3900 River Rd	Pennsauken	NJ	08110	JR Ford	609-663-6144	D	2	<0.1
Universal Tool and Engineering	3204 Hanover Dr	Johnson City	TN	37604	Manuel Abraham	615-282-0640	R	1	<0.1
UTE Straight-O-Matic	3204 Hanover Dr	Johnson City	TN	37604	Manuel Abraham	615-282-0640	S	1	<0.1

Source: Ward's Business Directory of U.S. Private and Public Companies, Volumes 1 and 2, 1996. The company type code used is as follows: P - Public, R - Private, S - Subsidiary, D - Division, J - Joint Venture, A - Affiliate, G - Group. Sales are in millions of dollars, employees are in thousands. An asterisk (*) indicates an estimated sales volume. The symbol < stands for 'less than'. Company names and addresses are truncated, in some cases, to fit into the available space.

MATERIALS CONSUMED

Material	Quantity	Delivered Cost ($ million)
Materials, ingredients, containers, and supplies	(X)	218.9
Fluid power pumps, motors, and hydrostatic transmissions (hydraulic and pneumatic)	(X)	4.8
Fluid power cylinders and rotary actuators (hydraulic and pneumatic)	(X)	2.1
Fluid power filters (hydraulic and pneumatic)	(X)	0.3
Fluid power hose or tube fittings and assemblies (hydraulic and pneumatic)	(X)	1.7
Fluid power valves (hydraulic and pneumatic)	(X)	0.7
Metal bolts, nuts, screws, washers, rivets, and other screw machine products	(X)	2.1
Other fabricated metal products (except fluid power products and forgings)	(X)	9.6
Iron and steel forgings	(X)	16.2
Nonferrous forgings	(X)	(D)
Iron and steel castings (rough and semifinished)	(X)	10.0
Aluminum and aluminum-base alloy castings (rough and semifinished)	(X)	(Z)
Other nonferrous castings (rough and semifinished)	(X)	0.6
Steel bars, bar shapes, and plates	(X)	16.9
Steel sheet and strip, including tin plate	(X)	1.5
Steel structural shapes	(X)	2.5
All other steel shapes and forms	(X)	5.1
Aluminum and aluminum-base alloy shapes and forms	(X)	0.6
Other nonferrous shapes and forms	(X)	0.2
Fractional horsepower electric timing motors, synchronous and subsynchronous (less than 1 hp)	(X)	1.4
Other fractional horsepower electric motors (under 1 hp)	(X)	(D)
Integral horsepower electric motors and generators (1 hp or more)	(X)	1.5
Electrical transmission, distribution, and control equipment	(X)	3.7
Electrical industrial capacitors, resistors, rheostats, and coil windings	(X)	(D)
Numerical controls for metalworking machinery (except programmable)	(X)	(D)
Programmable controllers for metalworking machinery	(X)	2.6
Ball bearings (mounted or unmounted)	(X)	(D)
Roller bearings (mounted or unmounted)	(X)	2.5
Mechanical speed changers, gears, and industrial high-speed drives	(X)	(D)
Wood boxes, pallets, skids, and containers	(X)	0.3
Cutting tools for machine tools	(X)	2.7
All other materials and components, parts, containers, and supplies	(X)	45.4
Materials, ingredients, containers, and supplies, nsk	(X)	68.4

Source: 1992 Economic Census. Explanation of symbols used: (D): Withheld to avoid disclosure of competitive data; na: Not available; (S): Withheld because statistical norms were not met; (X): Not applicable; (Z): Less than half the unit shown; nec: Not elsewhere classified; nsk: Not specified by kind; - : zero; * : 10-19 percent estimated; ** : 20-29 percent estimated.

PRODUCT SHARE DETAILS

Product or Product Class	% Share	Product or Product Class	% Share
Rolling mill machinery	100.00	Other rolling mill machinery (including tube mill machinery) and parts for all rolling mill machinery.	36.83
Hot rolling mill machinery (including combination hot and cold) (except tube rolling)	21.64	Tube rolling mill machinery.	12.97
Blooming and slabbing hot rolling mill machinery (including combination hot and cold) (except tube rolling)	47.11	Other rolling mill machinery and equipment, excluding parts.	21.80
Other hot rolling mill machinery and equipment (including hot and cold) (except tube rolling)	52.97	Machined rolls for rolling mills	45.29
Cold rolling mill machinery	11.92	Parts, excluding rolls, for rolling mill machinery (sold separately)	14.56
Tandem cold rolling mill machinery.	16.80	Other rolling mill machinery (including tube mill machinery) and parts for all rolling mill machinery, nsk	5.34
Single stand cold rolling mill machinery	56.55	Rolling mill machinery, nsk	29.63
Double stand cold rolling mill machinery	24.50		

Source: 1992 Economic Census. The values shown are percent of total shipments in an industry. Values of indented subcategories are summed in the main heading. The symbol (D) appears when data are withheld to prevent disclosure of competitive information. The abbreviation nsk stands for 'not specified by kind' and nec for 'not elsewhere classified'.

INPUTS AND OUTPUTS FOR ROLLING MILL MACHINERY

Economic Sector or Industry Providing Inputs	%	Sector	Economic Sector or Industry Buying Outputs	%	Sector
Imports	16.8	Foreign	Gross private fixed investment	75.4	Cap Inv
Wholesale trade	7.8	Trade	Exports	17.8	Foreign
Power transmission equipment	7.0	Manufg.	Change in business inventories	3.7	In House
Blast furnaces & steel mills	6.9	Manufg.	Rolling mill machinery	2.6	Manufg.
Iron & steel forgings	6.6	Manufg.	Federal Government purchases, national defense	0.1	Fed Govt
Rolling mill machinery	4.9	Manufg.			
Machinery, except electrical, nec	4.7	Manufg.			
Ball & roller bearings	4.2	Manufg.			
Advertising	3.1	Services			
Iron & steel foundries	2.7	Manufg.			
Motors & generators	2.7	Manufg.			
Electric services (utilities)	2.4	Util.			
Fabricated plate work (boiler shops)	2.2	Manufg.			
Pumps & compressors	1.8	Manufg.			
Switchgear & switchboard apparatus	1.4	Manufg.			
Machine tools, metal forming types	1.3	Manufg.			
Storage batteries	1.3	Manufg.			
Gas production & distribution (utilities)	1.3	Util.			
Abrasive products	1.2	Manufg.			
Metal stampings, nec	1.1	Manufg.			
Communications, except radio & TV	1.0	Util.			
Eating & drinking places	1.0	Trade			
Special dies & tools & machine tool accessories	0.8	Manufg.			
Motor freight transportation & warehousing	0.8	Util.			
Banking	0.8	Fin/R.E.			
Engineering, architectural, & surveying services	0.8	Services			
U.S. Postal Service	0.8	Gov't			
Maintenance of nonfarm buildings nec	0.7	Constr.			
Air transportation	0.7	Util.			
Royalties	0.7	Fin/R.E.			
Computer & data processing services	0.7	Services			
Equipment rental & leasing services	0.7	Services			
Industrial controls	0.6	Manufg.			
Machine tools, metal cutting types	0.6	Manufg.			
Pipe, valves, & pipe fittings	0.6	Manufg.			
Real estate	0.5	Fin/R.E.			
Metal heat treating	0.4	Manufg.			
Screw machine and related products	0.4	Manufg.			
Transformers	0.4	Manufg.			
Hotels & lodging places	0.4	Services			
Legal services	0.4	Services			
Management & consulting services & labs	0.4	Services			
Noncomparable imports	0.4	Foreign			
Brass, bronze, & copper castings	0.3	Manufg.			
Copper rolling & drawing	0.3	Manufg.			
Industrial inorganic chemicals, nec	0.3	Manufg.			
Petroleum refining	0.3	Manufg.			
Nonferrous wire drawing & insulating	0.2	Manufg.			
Sanitary services, steam supply, irrigation	0.2	Util.			
Accounting, auditing & bookkeeping	0.2	Services			
Automotive repair shops & services	0.2	Services			
Chemical preparations, nec	0.1	Manufg.			
Fabricated rubber products, nec	0.1	Manufg.			
Gaskets, packing & sealing devices	0.1	Manufg.			
Industrial patterns	0.1	Manufg.			
Lubricating oils & greases	0.1	Manufg.			
Manifold business forms	0.1	Manufg.			
Miscellaneous publishing	0.1	Manufg.			
Plating & polishing	0.1	Manufg.			
Electrical repair shops	0.1	Services			

Source: Benchmark Input-Output Accounts for the U.S. Economy, 1982, U.S. Department of Commerce, Washington, D.C., July 1991. Data, as reported in the source, are organized by the 1977 SIC structure in use in 1982 but have been matched, as closely as is possible, to the 1987 SIC structure used in this book.

OCCUPATIONS EMPLOYED BY SIC 354 - METALWORKING MACHINERY

Occupation	% of Total 1994	Change to 2005	Occupation	% of Total 1994	Change to 2005
Machinists	10.9	-17.4	Precision metal workers nec	1.9	10.1
Tool & die makers	10.6	-15.3	Mechanical engineers	1.9	1.0
General managers & top executives	3.7	-12.9	Inspectors, testers, & graders, precision	1.9	-8.2
Blue collar worker supervisors	3.4	-15.7	Secretaries, ex legal & medical	1.9	-16.4
Machine tool cutting operators, metal & plastic	2.9	-5.9	Drafters	1.8	-28.5
NC machine tool operators, metal & plastic	2.7	1.0	General office clerks	1.5	-21.7
Grinding machine operators, metal & plastic	2.7	-8.2	Machine forming operators, metal & plastic	1.4	-8.2
Sales & related workers nec	2.7	-8.2	Drilling & boring machine tool workers	1.3	-8.2
Assemblers, fabricators, & hand workers nec	2.5	-8.2	Bookkeeping, accounting, & auditing clerks	1.3	-31.1
Lathe & turning machine tool operators	2.3	-17.4	Industrial production managers	1.3	-8.2
Machine tool cutting & forming etc. nec	2.2	-54.1	Traffic, shipping, & receiving clerks	1.3	-11.7
Machine builders	2.0	1.0	Precision woodworkers nec	1.3	10.2
Combination machine tool operators	2.0	37.7	Janitors & cleaners, incl maids	1.2	-26.6

Source: Industry-Occupation Matrix, Bureau of Labor Statistics. These data relate to one or more 3-digit SIC industry groups rather than to a single 4-digit SIC. The change reported for each occupation to the year 2005 is a percent of growth or decline as estimated by the Bureau of Labor Statistics. The abbreviation nec stands for 'not elsewhere classified'.

LOCATION BY STATE AND REGIONAL CONCENTRATION

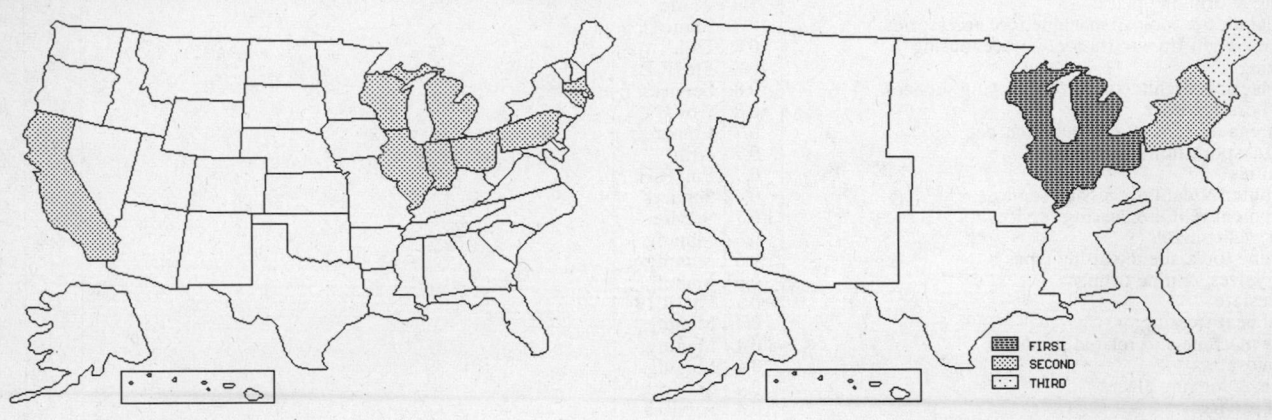

FIRST
SECOND
THIRD

INDUSTRY DATA BY STATE

State	Establish- ments	Shipments			Employment				Cost as % of Shipments	Investment per Employee ($)
		Total ($ mil)	% of U.S.	Per Establ.	Total Number	% of U.S.	Per Establ.	Wages ($/hour)		
Massachusetts	6	203.5	33.8	33.9	2,000	37.0	333	10.77	49.9	-
Ohio	16	161.7	26.8	10.1	1,300	24.1	81	15.50	47.7	2,308
Pennsylvania	10	87.1	14.4	8.7	700	13.0	70	15.17	52.8	1,714
Illinois	8	21.3	3.5	2.7	200	3.7	25	17.00	46.9	1,000
Michigan	7	(D)	-	-	375 *	6.9	54	-	-	800
California	5	(D)	-	-	175 *	3.2	35	-	-	-
Indiana	5	(D)	-	-	175 *	3.2	35	-	-	-
Wisconsin	4	(D)	-	-	175 *	3.2	44	-	-	571
Connecticut	2	(D)	-	-	175 *	3.2	88	-	-	-

Source: 1992 *Economic Census.* The states are in descending order of shipments or establishments (if shipment data are missing for the majority). The symbol (D) appears when data are withheld to prevent disclosure of competitive information. States marked with (D) are sorted by number of establishments. A dash (-) indicates that the data element cannot be calculated; * indicates the midpoint of a range.

3548 - WELDING & SOLDERING EQUIPMENT

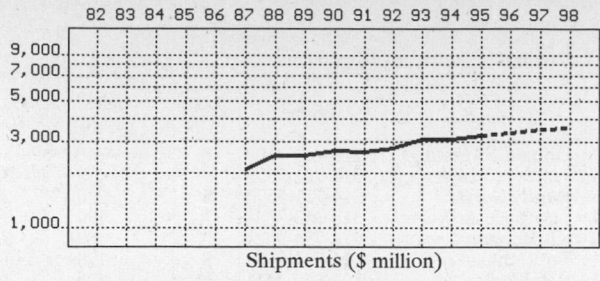

Shipments ($ million)

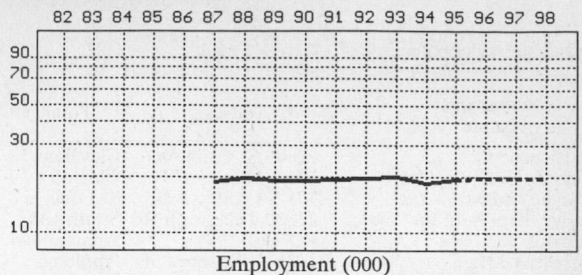

Employment (000)

GENERAL STATISTICS

| Year | Com-panies | Establishments | | Employment | | | Compensation | | Production ($ million) | | | |
		Total	with 20 or more employees	Total (000)	Production Workers (000)	Hours (Mil)	Payroll ($ mil)	Wages ($/hr)	Cost of Materials	Value Added by Manufacture	Value of Shipments	Capital Invest.
1982												
1983												
1984												
1985												
1986												
1987	203	225	130	18.7	11.5	24.2	541.5	12.10	1,062.1	1,084.4	2,104.6	45.4
1988		219	127	19.7	12.3	25.8	601.5	12.45	1,236.0	1,314.9	2,497.8	49.3
1989		215	129	19.3	11.6	24.3	578.0	12.67	1,223.4	1,274.9	2,520.5	59.1
1990		210	125	19.3	12.0	24.4	610.4	13.15	1,264.5	1,457.0	2,683.6	67.7
1991		216	126	19.5	11.8	24.5	620.8	13.07	1,295.1	1,344.1	2,651.2	50.5
1992	215	240	130	19.8	11.9	24.7	670.3	13.87	1,250.3	1,499.9	2,763.5	65.8
1993		244	132	20.3	13.1	28.3	736.2	14.76	1,445.9	1,722.5	3,100.7	80.6
1994		238P	130P	18.5	12.0	25.4	679.3	15.28	1,433.2	1,650.1	3,074.6	73.0
1995		242P	130P	19.6P	12.5P	26.4P	732.9P	15.42P	1,509.0P	1,737.3P	3,237.1P	80.3P
1996		246P	130P	19.6P	12.6P	26.7P	755.8P	15.86P	1,567.2P	1,804.4P	3,362.1P	84.5P
1997		249P	131P	19.6P	12.7P	26.9P	778.7P	16.31P	1,625.5P	1,871.5P	3,487.1P	88.7P
1998		253P	131P	19.7P	12.8P	27.2P	801.7P	16.75P	1,683.8P	1,938.6P	3,612.2P	92.9P

Sources: 1982, 1987, 1992 *Economic Census; Annual Survey of Manufactures*, 83-86, 88-91, 93-94. Establishment counts for non-Census years are from *County Business Patterns*; establishment values for 83-84 are extrapolations. 'P's show projections by the editors. Industries reclassified in 87 will not have data for prior years.

INDICES OF CHANGE

| Year | Com-panies | Establishments | | Employment | | | Compensation | | Production ($ million) | | | |
		Total	with 20 or more employees	Total (000)	Production Workers (000)	Hours (Mil)	Payroll ($ mil)	Wages ($/hr)	Cost of Materials	Value Added by Manufacture	Value of Shipments	Capital Invest.
1982												
1983												
1984												
1985												
1986												
1987	94	94	100	94	97	98	81	87	85	72	76	69
1988		91	98	99	103	104	90	90	99	88	90	75
1989		90	99	97	97	98	86	91	98	85	91	90
1990		88	96	97	101	99	91	95	101	97	97	103
1991		90	97	98	99	99	93	94	104	90	96	77
1992	100	100	100	100	100	100	100	100	100	100	100	100
1993		102	102	103	110	115	110	106	116	115	112	122
1994		99P	100P	93	101	103	101	110	115	110	111	111
1995		101P	100P	99P	105P	107P	109P	111P	121P	116P	117P	122P
1996		102P	100P	99P	106P	108P	113P	114P	125P	120P	122P	128P
1997		104P	101P	99P	106P	109P	116P	118P	130P	125P	126P	135P
1998		105P	101P	99P	107P	110P	120P	121P	135P	129P	131P	141P

Sources: Same as General Statistics. Values reflect change from the base year, 1992. Values above 100 mean greater than 92, values below 100 mean less than 92, and a value of 100 in the 82-91 or 93-98 period means same as 92. 'P's mark projections by the editors.

SELECTED RATIOS

For 1994	Avg. of All Manufact.	Analyzed Industry	Index	For 1994	Avg. of All Manufact.	Analyzed Industry	Index
Employees per Establishment	49	78	158	Value Added per Production Worker	134,084	137,508	103
Payroll per Establishment	1,500,273	2,849,071	190	Cost per Establishment	5,045,178	6,011,025	119
Payroll per Employee	30,620	36,719	120	Cost per Employee	102,970	77,470	75
Production Workers per Establishment	34	50	147	Cost per Production Worker	146,988	119,433	81
Wages per Establishment	853,319	1,627,791	191	Shipments per Establishment	9,576,895	12,895,267	135
Wages per Production Worker	24,861	32,343	130	Shipments per Employee	195,460	166,195	85
Hours per Production Worker	2,056	2,117	103	Shipments per Production Worker	279,017	256,217	92
Wages per Hour	12.09	15.28	126	Investment per Establishment	321,011	306,171	95
Value Added per Establishment	4,602,255	6,920,731	150	Investment per Employee	6,552	3,946	60
Value Added per Employee	93,930	89,195	95	Investment per Production Worker	9,352	6,083	65

Sources: Same as General Statistics. The 'Average of All Manufacturing' column represents the soverage of all manufacturing industries reported for the most recent complete year available. The Index shows the relationship between the Average and the Analyzed Industry. For example, 100 means that they are equal; 500 that the Analyzed Industry is five times the average; 50 means that the Analyzed Industry is half the national average. The abbreviation 'na' is used to show that data are 'not available'.

LEADING COMPANIES Number shown: **73** Total sales ($ mil): **3,189** Total employment (000): **25.2**

Company Name	Address				CEO Name	Phone	Co. Type	Sales ($ mil)	Empl. (000)
Lincoln Electric Co	22801 St Clair Av	Cleveland	OH	44117	Donald F Hastings	216-481-8100	P	846	6.0
Thermadyne Holdings Corp	101 S Hanley Rd	Clayton	MO	63105	James N Mills	314-721-5573	P	450	3.0
Thermadyne Industries Inc	101 S Hanley Rd	St Louis	MO	63105	Randy Curran	314-721-5573	S	220*	2.0
Progressive Tool & Indust Co	21000 Telegraph Rd	Southfield	MI	48034	Lawrence A Wisne	313-353-8888	R	190	1.7
Miller Group Inc	PO Box 1079	Appleton	WI	54912	KL Booher	414-734-9821	R	180*	1.6
Miller Electric Mfg Co	1635 W Spencer St	Appleton	WI	54914	Bob Jenkins,	414-734-9821	S	170*	1.5
ESAB Welding & Cutting Prod	PO Box 100545	Florence	SC	29501	Ray Hoglund	803-669-4411	D	120	1.1
Deloro Stellite Inc	101 S Hanley Rd	St Louis	MO	63105	RV Linn	314-862-2666	S	100	1.1
Weldmation Inc	31720ephenson Hwy	Madison H	MI	48071	A Kelsey	313-585-0010	R	100	0.4
Alloy Rods Corp	PO Box 517	Hanover	PA	17331	RC Hoglund	717-637-8911	S	70	0.5
Hughes Aircraft Co	2051 Palomar	Carlsbad	CA	92009	Gary E Gist	619-931-3000	S	57	0.6
Nippert Co	801 Pittsburgh Dr	Delaware	OH	43015	Russell A Nippert	614-363-1981	S	40	0.2
Nelson Stud Welding	7900 W Ridge Rd	Elyria	OH	44036	Ted J Woods	216-329-0400	D	37	0.4
Uniweld Products Inc	2850 Ravenswood	Ft Lauderdale	FL	33312	DS Pearl	305-584-2000	R	33	0.3
Forney Industries Inc	PO Box 563	Fort Collins	CO	80522	Ted G Anderson	303-482-7271	R	28	0.3
Thermatool Corp	PO Box 769	East Haven	CT	06512	Theodore J Morin	203-468-4100	S	28*	0.1
Newcor Bay City Inc	1846 Trumbull Dr	Bay City	MI	48707	John Kasuba	517-893-9505	D	27	0.2
Dimetrics Inc	PO Box 339	Davidson	NC	28036	Art Squicciarini	704-892-8872	S	26	0.3
Arc Machines Inc	10280 Glenoaks	Pacoima	CA	91331	M Gedgaudas	818-896-9556	R	25	0.2
Taylor-Winfield Corp	PO Box 500	Brookfield	OH	44403	David A Lynn	216-448-4464	S	25	0.3
Weltronic-Technitron Inc	150 E St Charles Rd	Carol Stream	IL	60188	Durrell Miller	708-462-8250	R	23	<0.1
National Torch Tip Company	50 Freeport Rd	Pittsburgh	PA	15215	David S Werner	412-781-4200	R	21*	0.2
Savair Inc	33200 Freeway Dr	St Clair Shores	MI	48082	E A Wolfbauer	313-296-7390	R	21	0.2
Techalloy Company Inc	2310 Chesapeake	Baltimore	MD	21222	David Daube	410-633-9300	D	20	0.1
Controls Corporation of America	1501 Harpers Rd	Virginia Beach	VA	23454	S Dukas	804-422-8330	R	19	0.2
Pertron Controls	PO Box 9247	Columbia	SC	29290	Dave Keaney	803-776-7500	D	18	0.2
Unitek Miyachi Corp	1820 S Myrtle Av	Monrovia	CA	91017	Jack D Lantz	818-303-5676	S	18	<0.1
Banner Welder Inc	PO Box 1008	Germantown	WI	53022	Robert J Thome	414-253-2900	R	15	0.1
Great Western Airgas Inc	1001 Dunn Av	Cheyenne	WY	82001	Dale Hess	307-632-0571	S	15	<0.1
Key Welder Corp	15686 Sturgeon Av	Roseville	MI	48066	N Ulmer	313-778-7700	R	15	<0.1
Genesis Systems Group	4821 Tremont Av	Davenport	IA	52807	Rich Litt	319-386-4034	R	14	<0.1
Tecnomatix Technologies Inc	39830 Grand River	Novi	MI	48375	Shlomo Dovrat	313-471-6140	S	13*	0.1
Smith Equipment	2601 Lockheed Av	Watertown	SD	57201	D Indahl	605-882-3200	D	12	0.2
Balaguer Corp	16 W Huron St	Pontiac	MI	48342	Richard Balaguer	810-338-6600	R	11*	0.1
Metcal Inc	1530 O'Brien Dr	Menlo Park	CA	94025	Keith Scott	415-325-3291	R	11	0.1
Joyal Products Inc	1233 W St George	Linden	NJ	07036	Allan Warner	908-486-6100	R	11	<0.1
American Torch Tip Co	6212 29th St E	Bradenton	FL	34203	John D Walters	813-753-7557	R	10	0.1
Arcos Alloys	1 Arcos Dr	Mount Carmel	PA	17851	Charles Perkey	717-339-5200	D	10	<0.1
MK Products Inc	16882 Armstrong	Irvine	CA	92714	Douglas M Kensrue	714-863-1234	R	10	<0.1
Laser Machining Inc	500 Laser Dr	Somerset	WI	54025	Noel Biebl	715-247-3285	R	9*	<0.1
Automation International Inc	1020 Bahls St	Danville	IL	61832	Bob Shutt	217-446-9500	R	7*	<0.1
NLC Inc	PO Box 348	Jackson	MO	63755	Kent Reese	314-243-3141	R	7	<0.1
Sonics and Materials Inc	Kenosia Av	Danbury	CT	06810	Robert Soloff	203-744-4400	R	7	<0.1
ABB Robotics Inc	4600 Innovation Dr	Fort Collins	CO	80525	Dave Prosser	303-225-7600	D	6*	<0.1
Atlas Welding Accessories Inc	PO Box 969	Troy	MI	48099	B Honhart	810-588-4666	R	6	<0.1
Craftmation Inc	1387 Piedmont Rd	Troy	MI	48083	Nils Karlson	810-689-8340	R	6*	<0.1
SEMTORQ Inc	PO Box 46406	Bedford	OH	44146	Joseph Seme Sr	216-232-4747	R	6	<0.1
Goss Inc	PO Box 57	Glenshaw	PA	15116	Jackie Goss	412-486-6100	R	5*	<0.1
Hexacon Electric Co	161 W Clay Av	Roselle Park	NJ	07204	Nick Rusignuolo	908-245-6200	R	5*	<0.1
Jetline Engineering Inc	15 Goodyear St	Irvine	CA	92718	Larry Russell	714-951-1515	R	5	<0.1
Spinweld Inc	PO Box 14485	Milwaukee	WI	53214	Richard B Leachy	414-327-0100	R	5	<0.1
IGM Robotic Systems Inc	W 133 N 5138	Menomonee Fls	WI	53051	William Heller	414-783-2720	S	5	<0.1
CK Systematics Inc	PO Box 2429	West Chester	PA	19380	J Conley	215-696-9040	R	4*	<0.1
Hercules Welding Products	11478 Timken St	Warren	MI	48090	Hiroaki Chiba	810-755-1250	D	4	<0.1
Seal-Seat Co	1200 Monterey	Monterey Park	CA	91754	Kathleen Trenschel	213-269-1311	D	4	<0.1
Tuffaloy Products Inc	2145 Crooks Rd	Troy	MI	48084	Michael S Simmons	313-643-9944	R	4	<0.1
CK Worldwide Inc	PO Box 1636	Auburn	WA	98071	Arthur Kleppen	206-854-5820	R	3	<0.1
Edsyn Inc	15958 Arminta St	Van Nuys	CA	91406	William S Fortune	818-989-2324	R	3*	<0.1
Entron Controls Inc	465 E Randy Rd	Carol Stream	IL	60188	Patricia S Adams	708-682-9600	R	3	<0.1
Maitlen and Bensen Inc	PO Box 4146	Long Beach	CA	90804	DL Maitlen	310-597-5594	R	3	<0.1
McDonald Welding and Machine	2337 Marshall Rd	McDonald	OH	44437	Nat A Gallo	216-530-9703	R	3*	<0.1
A-1 Welding and Fabricating Inc	1005 E 32nd St	Lorain	OH	44055	Daniel Balko	216-233-8474	R	2	<0.1
Dytron Corp	17000 Masonic Blv	Fraser	MI	48026	RC Donovan	810-296-9600	R	2	<0.1
Ogden Engineering Corp	372 W Division Av	Schererville	IN	46375	Ralph Ogden	219-322-5252	R	2	<0.1
Saturn Industries Inc	PO Box D	Hudson	NY	12534	K Werner	518-828-9956	R	2*	<0.1
Settle's Precision Manufacturing	Rte 1	Cuthbert	GA	31740	James L Hoover	912-732-3731	R	2	<0.1
Toddco General Inc	7888 Silverton Av	San Diego	CA	92126	Tom Todd	619-549-9229	R	2	<0.1
Ultrasonic Seal Co	368 Turner Indrial	Aston	PA	19014	Robert Soloff	215-497-5150	S	2*	<0.1
Stryco Industries Inc	40840 Onida Ct	Fremont	CA	94539	Joseph Borges Jr	510-373-1310	R	2	<0.1
Spot Weld Inc	2290 Wycliff St	St Paul	MN	55114	Dennis J Kilbane	612-646-1393	R	2	<0.1
High Frequency Technology	172-D Brook Av	Deer Park	NY	11729	Louis Amabile	516-242-3020	R	1	<0.1
Microflame Inc	14873 DeVeau Pl	Minnetonka	MN	55345	Richard Kurzeka	612-935-3777	S	1	<0.1
Sikama International Inc	118 E Gutierrez St	Santa Barbara	CA	93101	Sig Wathne	805-962-1000	R	1	<0.1

Source: Ward's Business Directory of U.S. Private and Public Companies, Volumes 1 and 2, 1996. The company type code used is as follows: P - Public, R - Private, S - Subsidiary, D - Division, J - Joint Venture, A - Affiliate, G - Group. Sales are in millions of dollars, employees are in thousands. An asterisk (*) indicates an estimated sales volume. The symbol < stands for 'less than'. Company names and addresses are truncated, in some cases, to fit into the available space.

MATERIALS CONSUMED

Material	Quantity	Delivered Cost ($ million)
Materials, ingredients, containers, and supplies	(X)	1,057.1
Metal bolts, nuts, screws, washers, rivets, and other screw machine products	(X)	11.2
Other fabricated metal products (except castings and forgings)	(X)	37.2
Forgings	(X)	2.8
Iron and steel castings (rough and semifinished)	(X)	4.2
Copper and copper-base alloy castings (rough and semifinished)	(X)	4.2
Other nonferrous castings (rough and semifinished)	(X)	2.7
Steel bars, bar shapes, and plates	(X)	19.7
Steel sheet and strip, including tin plate	(X)	55.0
Steel wire and wire products	(X)	91.1
All other steel shapes and forms	(X)	29.0
Copper and copper-base alloy shapes and forms	(X)	36.6
Aluminum and aluminum-base alloy shapes and forms	(X)	11.7
Other nonferrous shapes and forms	(X)	15.5
Insulated copper wire and cable, except magnet wire	(X)	9.4
Fractional horsepower electric timing motors, synchronous and subsynchronous (less than 1 hp)	(X)	2.3
Other fractional horsepower electric motors (less than 1 hp)	(X)	9.5
Integral horsepower electric motors and generators (1 hp or more)	(X)	9.0
Electrical transmission, distribution, and control equipment	(X)	47.7
Electrical industrial capacitors, resistors, rheostats, and coil windings	(X)	25.1
Pressure gauges	(X)	9.9
Ball bearings (mounted or unmounted)	(X)	4.3
Roller bearings (mounted or unmounted)	(X)	1.4
Paperboard containers, boxes, and corrugated paperboard	(X)	21.8
Industrial robots purchased for fabrication with welding equipment	(X)	6.7
All other materials and components, parts, containers, and supplies	(X)	400.1
Materials, ingredients, containers, and supplies, nsk	(X)	188.9

Source: 1992 Economic Census. Explanation of symbols used: (D): Withheld to avoid disclosure of competitive data; na: Not available; (S): Withheld because statistical norms were not met; (X): Not applicable; (Z): Less than half the unit shown; nec: Not elsewhere classified; nsk: Not specified by kind; - : zero; * : 10-19 percent estimated; ** : 20-29 percent estimated.

PRODUCT SHARE DETAILS

Product or Product Class	% Share	Product or Product Class	% Share
Welding apparatus	100.00	arc and inert gas welding, other than hard facing	2.07
Arc welding machines, components, and accessories (except electrodes), excluding stud welding equipment	36.07	Nonferrous metal coiled and spooled continuous solid wire electrodes for automatic arc and inert gas welding, other than hard facing	4.05
Alternating current transformer arc welding machines, except stud welding equipment	5.03	Coiled and spooled continuous cored metal wire electrodes for automatic arc and inert gas welding, other than hard facing	24.63
Direct current arc welding generators, except stud welding equipment	14.97	Arc welding electrodes, metal, nsk	1.35
Rectifier-type direct current arc welders, including ac/dc, except stud welding equipment	21.23	Resistance welders, components, accessories, and electrodes	11.15
Complete direct current arc welding units, except stud welding equipment	2.94	Spot and projection resistance welders, single electrode	12.21
		Spot and projection resistance welders, multielectrode	21.91
Automatic and semiautomatic wire drive apparatus and related accessories for arc welding machines (except electrodes and stud welding equipment)	6.81	Seam resistance welders	6.74
Automatic and semiautomatic welding torches, guns and cables, and related accessories for arc welding machines (except electrodes and stud welding equipment)	9.98	Other resistance welders, including flash, upset, and butt welders	9.03
		Resistance welder transformers (sold separately)	11.21
Special purpose automatic arc welding apparatus (except electrodes and stud welding equipment)	1.97	Resistance welder electrodes	8.03
Circuit welding accessories (including electrode holders, ground clamps, cable connectors, cables sold separately, etc.) (except electrodes and stud welding equipment)	6.67	Resistance welder components and accessories, including electrode holders, etc.	22.69
Positioning and manipulating arc welding equipment, including turn rolls, head and tail stock, weld head manipulators, seamers, etc. (except electrodes and stud welding equipment)	3.40	Resistance welders, components, accessories, and electrodes, nsk	8.22
		Gas welding and cutting equipment, parts, attachments, and accessories	10.44
All other components and accessories for arc welding machinery, excluding welding rods, electrodes, and stud welding equipment	7.78	Gas welding and cutting torches (including gas air torches)	14.64
		Gas cutting machines and carriages, stationary and portable	10.05
Arc welding machines, components, and accessories (except electrodes), excluding stud welding equipment, nsk	19.21	Other gas welding and cutting equipment, excluding pressure containers	19.46
Arc welding electrodes, metal	26.08	Tips for gas welding and cutting equipment (sold separately)	12.30
Hard facing metal arc welding stick electrodes (including solid, cored, covered, and bare electrodes)	3.06	Pressure regulators for gas welding and cutting equipment (sold separately)	18.24
Mild steel arc welding stick electrodes (including solid, cored, covered, and bare electrodes), other than hard facing	22.87	All other spare parts, accessories, attachments, adaptors, etc. for gas welding and cutting equipment (sold separately)	14.52
Low alloy steel arc welding stick electrodes (including solid, cored, covered, and bare electrodes), other than hard facing	9.88	Gas welding and cutting equipment, parts, attachments, and accessories, nsk	10.76
Nonferrous metal arc welding stick electrodes (including solid, cored, covered, and bare electrodes), other than hard facing	4.10	Other welding equipment, components, and accessories (excluding arc, resistance, and gas)	10.01
		Stud welding equipment (except arc, resistance, and gas)	12.86
Hard facing coiled and spooled continuous metal wire electrodes for automatic arc and inert gas welding	4.59	Plasma welding and cutting equipment	47.40
Mild steel coiled and spooled continuous solid wire electrodes for automatic arc and inert gas welding, other than hard facing	23.40	All other welding equipment (excluding laser, electron beam, and ultrasonic equipment, and arc, resistance, and gas welding equipment)	13.56
		Soldering equipment (except hand and ultrasonic)	9.65
Stainless steel (chromium, 4 percent or more) coiled and spooled continuous solid wire electrodes for automatic		Components and accessories for all other welding equipment, excluding arc, resistance, and gas welding equipment	15.54
		Other welding equipment, components, and accessories (excluding arc, resistance, and gas), nsk	0.99
		Welding apparatus, nsk	6.25

Source: 1992 *Economic Census*. The values shown are percent of total shipments in an industry. Values of indented subcategories are summed in the main heading. The symbol (D) appears when data are withheld to prevent disclosure of competitive information. The abbreviation nsk stands for 'not specified by kind' and nec for 'not elsewhere classified'.

INPUTS AND OUTPUTS FOR METALWORKING MACHINERY, NEC

Economic Sector or Industry Providing Inputs	%	Sector	Economic Sector or Industry Buying Outputs	%	Sector
Blast furnaces & steel mills	9.5	Manufg.	Gross private fixed investment	67.5	Cap Inv
Wholesale trade	9.0	Trade	Exports	11.5	Foreign
Fabricated plate work (boiler shops)	6.0	Manufg.	Machinery, except electrical, nec	5.1	Manufg.
Metalworking machinery, nec	5.8	Manufg.	Personal consumption expenditures	4.0	
Petroleum refining	4.2	Manufg.	Fabricated plate work (boiler shops)	2.6	Manufg.
Motors & generators	4.0	Manufg.	Metalworking machinery, nec	2.5	Manufg.
Switchgear & switchboard apparatus	3.9	Manufg.	Blast furnaces & steel mills	1.7	Manufg.
Advertising	3.6	Services	Metal partitions & fixtures	1.5	Manufg.
Machinery, except electrical, nec	3.4	Manufg.	Aircraft & missile engines & engine parts	1.0	Manufg.
Metal stampings, nec	3.0	Manufg.	Federal Government purchases, national defense	0.7	Fed Govt
Nonferrous rolling & drawing, nec	2.9	Manufg.	Automotive repair shops & services	0.5	Services
Miscellaneous plastics products	2.6	Manufg.	Special dies & tools & machine tool accessories	0.3	Manufg.
Electric services (utilities)	2.4	Util.	Iron & steel foundries	0.2	Manufg.
Copper rolling & drawing	2.1	Manufg.	Motor vehicles & car bodies	0.2	Manufg.
Iron & steel foundries	1.8	Manufg.	Motor freight transportation & warehousing	0.2	Util.
Communications, except radio & TV	1.8	Util.			
Fabricated metal products, nec	1.6	Manufg.			
Miscellaneous fabricated wire products	1.5	Manufg.			
Power transmission equipment	1.4	Manufg.			
Screw machine and related products	1.3	Manufg.			

Continued on next page.

INPUTS AND OUTPUTS FOR METALWORKING MACHINERY, NEC - Continued

Economic Sector or Industry Providing Inputs	%	Sector	Economic Sector or Industry Buying Outputs	%	Sector
Banking	1.2	Fin/R.E.			
Aluminum castings	1.1	Manufg.			
Pipe, valves, & pipe fittings	1.1	Manufg.			
Ball & roller bearings	1.0	Manufg.			
Equipment rental & leasing services	1.0	Services			
Maintenance of nonfarm buildings nec	0.9	Constr.			
Abrasive products	0.9	Manufg.			
Paperboard containers & boxes	0.9	Manufg.			
Pumps & compressors	0.9	Manufg.			
Automotive rental & leasing, without drivers	0.9	Services			
Motor freight transportation & warehousing	0.8	Util.			
Real estate	0.8	Fin/R.E.			
Royalties	0.8	Fin/R.E.			
Automotive repair shops & services	0.8	Services			
Computer & data processing services	0.8	Services			
Engineering, architectural, & surveying services	0.8	Services			
Aluminum rolling & drawing	0.7	Manufg.			
Industrial controls	0.7	Manufg.			
Transformers	0.7	Manufg.			
Gas production & distribution (utilities)	0.7	Util.			
Eating & drinking places	0.7	Trade			
Imports	0.7	Foreign			
Machine tools, metal cutting types	0.6	Manufg.			
Special dies & tools & machine tool accessories	0.6	Manufg.			
Air transportation	0.6	Util.			
Iron & steel forgings	0.4	Manufg.			
Insurance carriers	0.4	Fin/R.E.			
Brass, bronze, & copper castings	0.3	Manufg.			
Metal heat treating	0.3	Manufg.			
Rubber & plastics hose & belting	0.3	Manufg.			
Wiring devices	0.3	Manufg.			
Detective & protective services	0.3	Services			
Hotels & lodging places	0.3	Services			
Legal services	0.3	Services			
Management & consulting services & labs	0.3	Services			
Noncomparable imports	0.3	Foreign			
Fabricated rubber products, nec	0.2	Manufg.			
Lubricating oils & greases	0.2	Manufg.			
Motor vehicle parts & accessories	0.2	Manufg.			
Nonferrous wire drawing & insulating	0.2	Manufg.			
Plating & polishing	0.2	Manufg.			
Primary metal products, nec	0.2	Manufg.			
Tires & inner tubes	0.2	Manufg.			
Retail trade, except eating & drinking	0.2	Trade			
Accounting, auditing & bookkeeping	0.2	Services			
U.S. Postal Service	0.2	Gov't			
Machine tools, metal forming types	0.1	Manufg.			
Manifold business forms	0.1	Manufg.			
Nonferrous castings, nec	0.1	Manufg.			
Electrical repair shops	0.1	Services			

Source: Benchmark Input-Output Accounts for the U.S. Economy, 1982, U.S. Department of Commerce, Washington, D.C., July 1991. Data, as reported in the source, are organized by the 1977 SIC structure in use in 1982 but have been matched, as closely as is possible, to the 1987 SIC structure used in this book.

OCCUPATIONS EMPLOYED BY SIC 354 - METALWORKING MACHINERY

Occupation	% of Total 1994	Change to 2005	Occupation	% of Total 1994	Change to 2005
Machinists	10.9	-17.4	Precision metal workers nec	1.9	10.1
Tool & die makers	10.6	-15.3	Mechanical engineers	1.9	1.0
General managers & top executives	3.7	-12.9	Inspectors, testers, & graders, precision	1.9	-8.2
Blue collar worker supervisors	3.4	-15.7	Secretaries, ex legal & medical	1.9	-16.4
Machine tool cutting operators, metal & plastic	2.9	-5.9	Drafters	1.8	-28.5
NC machine tool operators, metal & plastic	2.7	1.0	General office clerks	1.5	-21.7
Grinding machine operators, metal & plastic	2.7	-8.2	Machine forming operators, metal & plastic	1.4	-8.2
Sales & related workers nec	2.7	-8.2	Drilling & boring machine tool workers	1.3	-8.2
Assemblers, fabricators, & hand workers nec	2.5	-8.2	Bookkeeping, accounting, & auditing clerks	1.3	-31.1
Lathe & turning machine tool operators	2.3	-17.4	Industrial production managers	1.3	-8.2
Machine tool cutting & forming etc. nec	2.2	-54.1	Traffic, shipping, & receiving clerks	1.3	-11.7
Machine builders	2.0	1.0	Precision woodworkers nec	1.3	10.2
Combination machine tool operators	2.0	37.7	Janitors & cleaners, incl maids	1.2	-26.6

Source: Industry-Occupation Matrix, Bureau of Labor Statistics. These data relate to one or more 3-digit SIC industry groups rather than to a single 4-digit SIC. The change reported for each occupation to the year 2005 is a percent of growth or decline as estimated by the Bureau of Labor Statistics. The abbreviation nec stands for 'not elsewhere classified'.

LOCATION BY STATE AND REGIONAL CONCENTRATION

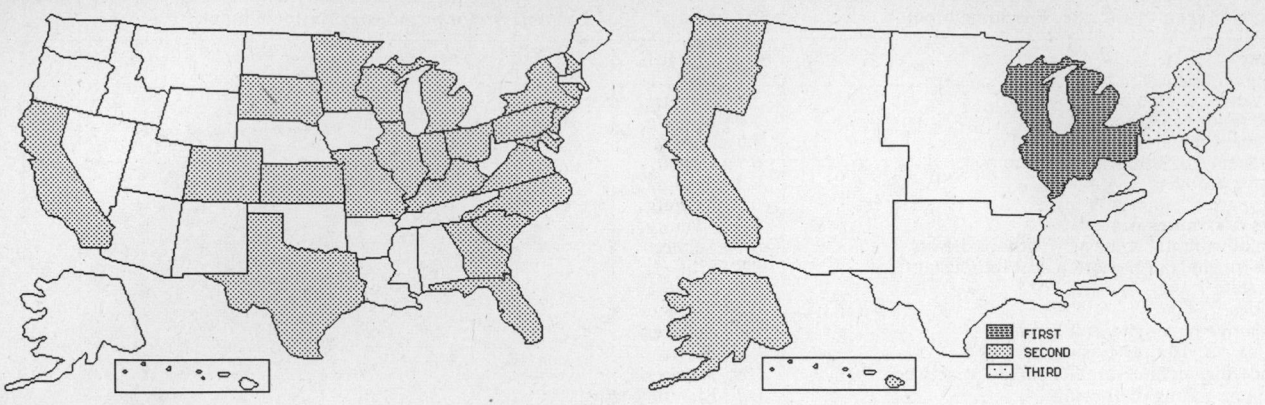

FIRST
SECOND
THIRD

INDUSTRY DATA BY STATE

State	Establish-ments	Shipments			Employment				Cost as % of Shipments	Investment per Employee ($)
		Total ($ mil)	% of U.S.	Per Establ.	Total Number	% of U.S.	Per Establ.	Wages ($/hour)		
Ohio	24	798.1	28.9	33.3	4,900	24.7	204	16.26	42.5	4,939
Michigan	46	482.5	17.5	10.5	3,000	15.2	65	15.00	59.7	3,900
Wisconsin	10	292.3	10.6	29.2	2,100	10.6	210	14.38	50.5	-
Pennsylvania	13	141.0	5.1	10.8	900	4.5	69	14.00	32.7	1,667
New York	8	124.4	4.5	15.6	800	4.0	100	18.67	44.0	-
California	28	120.0	4.3	4.3	1,000	5.1	36	12.27	41.0	3,200
North Carolina	4	57.6	2.1	14.4	500	2.5	125	8.50	35.4	3,000
Illinois	12	54.7	2.0	4.6	500	2.5	42	14.17	40.4	1,400
Missouri	7	37.8	1.4	5.4	500	2.5	71	11.20	25.7	-
Florida	8	33.3	1.2	4.2	500	2.5	63	7.13	37.5	800
Connecticut	5	27.1	1.0	5.4	200	1.0	40	13.00	38.7	-
Indiana	5	25.3	0.9	5.1	300	1.5	60	13.00	32.0	2,667
Georgia	4	22.5	0.8	5.6	200	1.0	50	13.00	76.4	-
New Jersey	5	19.2	0.7	3.8	200	1.0	40	15.50	43.8	2,000
Texas	9	(D)	-	-	750 *	3.8	83	-	-	-
Colorado	6	(D)	-	-	375 *	1.9	63	-	-	-
Minnesota	5	(D)	-	-	375 *	1.9	75	-	-	-
New Hampshire	5	(D)	-	-	750 *	3.8	150	-	-	-
South Carolina	4	(D)	-	-	750 *	3.8	188	-	-	-
Kansas	3	(D)	-	-	375 *	1.9	125	-	-	-
Kentucky	3	(D)	-	-	175 *	0.9	58	-	-	-
Maryland	2	(D)	-	-	175 *	0.9	88	-	-	-
South Dakota	1	(D)	-	-	175 *	0.9	175	-	-	-
Virginia	1	(D)	-	-	175 *	0.9	175	-	-	-

Source: 1992 *Economic Census*. The states are in descending order of shipments or establishments (if shipment data are missing for the majority). The symbol (D) appears when data are withheld to prevent disclosure of competitive information. States marked with (D) are sorted by number of establishments. A dash (-) indicates that the data element cannot be calculated; * indicates the midpoint of a range.

3549 - METALWORKING MACHINERY, NEC

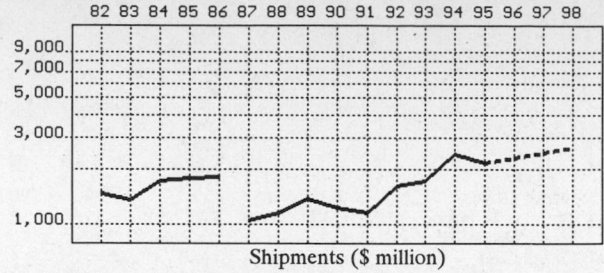

Shipments ($ million)

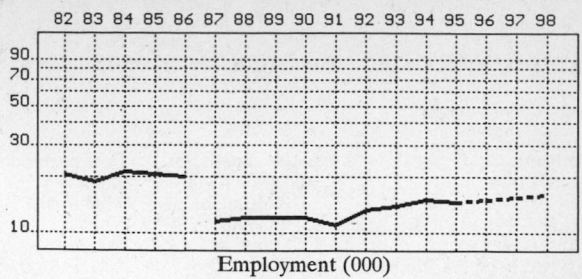

Employment (000)

GENERAL STATISTICS

Year	Com-panies	Establishments		Employment			Compensation		Production ($ million)			
		Total	with 20 or more employees	Total (000)	Production Workers (000)	Hours (Mil)	Payroll ($ mil)	Wages ($/hr)	Cost of Materials	Value Added by Manufacture	Value of Shipments	Capital Invest.
1982	421	446	197	20.7	13.0	25.4	431.5	9.19	600.0	830.3	1,471.5	52.4
1983		435	197	18.8	11.9	22.6	409.8	9.54	550.0	763.9	1,356.5	31.4
1984		424	197	21.5	14.1	27.9	495.6	9.74	765.7	1,003.2	1,727.5	56.2
1985		414	198	20.8	13.8	27.6	505.1	10.26	778.8	1,063.1	1,788.2	48.6
1986		397	191	20.0	12.9	26.5	497.0	10.39	823.2	938.9	1,811.8	35.1
1987*	293	302	147	11.5	7.0	14.7	338.8	12.37	451.6	593.3	1,055.0	21.4
1988		286	143	12.0	7.3	15.6	370.6	12.38	507.8	673.1	1,164.5	26.8
1989		285	146	12.0	7.6	16.8	393.9	12.19	664.2	756.6	1,385.3	29.3
1990		293	145	11.9	6.7	15.5	384.4	12.45	584.7	675.1	1,231.1	19.6
1991		297	143	10.9	6.5	14.8	358.9	12.41	505.1	702.0	1,150.8	32.5
1992	389	400	167	13.2	7.9	17.6	487.5	14.05	687.6	942.7	1,618.3	27.6
1993		390	171	13.9	8.6	18.9	514.8	13.77	739.8	1,003.3	1,711.4	17.2
1994		394P	168P	15.1	9.8	22.5	593.7	13.88	1,147.9	1,400.4	2,411.9	48.5
1995		412P	173P	14.6P	9.1P	21.0P	578.2P	14.17P	1,025.4P	1,251.0P	2,154.5P	35.9P
1996		430P	177P	15.0P	9.4P	21.8P	611.1P	14.45P	1,098.2P	1,339.8P	2,307.5P	37.7P
1997		448P	181P	15.4P	9.7P	22.7P	643.9P	14.72P	1,171.0P	1,428.6P	2,460.5P	39.4P
1998		466P	185P	15.9P	10.1P	23.6P	676.8P	15.00P	1,243.8P	1,517.4P	2,613.5P	41.2P

Sources: 1982, 1987, 1992 Economic Census; Annual Survey of Manufactures, 83-86, 88-91, 93-94. Establishment counts are from County Business Patterns for non-Census years; establishment counts for 83-84 are extrapolations. * indicates that industry content changed in 87; earlier years use 77 SICs. 'P's mark projections.

INDICES OF CHANGE

Year	Com-panies	Establishments		Employment			Compensation		Production ($ million)			
		Total	with 20 or more employees	Total (000)	Production Workers (000)	Hours (Mil)	Payroll ($ mil)	Wages ($/hr)	Cost of Materials	Value Added by Manufacture	Value of Shipments	Capital Invest.
1982	108	111	118	157	165	144	89	65	87	88	91	190
1983		109	118	142	151	128	84	68	80	81	84	114
1984		106	118	163	178	159	102	69	111	106	107	204
1985		103	119	158	175	157	104	73	113	113	110	176
1986		99	114	152	163	151	102	74	120	100	112	127
1987*	75	76	88	87	89	84	69	88	66	63	65	78
1988		72	86	91	92	89	76	88	74	71	72	97
1989		71	87	91	96	95	81	87	97	80	86	106
1990		73	87	90	85	88	79	89	85	72	76	71
1991		74	86	83	82	84	74	88	73	74	71	118
1992	100	100	100	100	100	100	100	100	100	100	100	100
1993		98	102	105	109	107	106	98	108	106	106	62
1994		98P	101P	114	124	128	122	99	167	149	149	176
1995		103P	103P	110P	115P	119P	119P	101P	149P	133P	133P	130P
1996		107P	106P	114P	119P	124P	125P	103P	160P	142P	143P	136P
1997		112P	108P	117P	123P	129P	132P	105P	170P	152P	152P	143P
1998		116P	111P	120P	127P	134P	139P	107P	181P	161P	161P	149P

Sources: Same as General Statistics. Values reflect change from the base year, 1992. Values above 100 mean greater than 92, values below 100 mean less than 92, and a value of 100 in the 82-91 or 93-98 period means same as 92. * indicates that industry content changed in 87. Data for earlier years are in 77 SIC format.

SELECTED RATIOS

For 1994	Avg. of All Manufact.	Analyzed Industry	Index	For 1994	Avg. of All Manufact.	Analyzed Industry	Index
Employees per Establishment	49	38	78	Value Added per Production Worker	134,084	142,898	107
Payroll per Establishment	1,500,273	1,507,399	100	Cost per Establishment	5,045,178	2,914,509	58
Payroll per Employee	30,620	39,318	128	Cost per Employee	102,970	76,020	74
Production Workers per Establishment	34	25	72	Cost per Production Worker	146,988	117,133	80
Wages per Establishment	853,319	792,927	93	Shipments per Establishment	9,576,895	6,123,794	64
Wages per Production Worker	24,861	31,867	128	Shipments per Employee	195,460	159,728	82
Hours per Production Worker	2,056	2,296	112	Shipments per Production Worker	279,017	246,112	88
Wages per Hour	12.09	13.88	115	Investment per Establishment	321,011	123,141	38
Value Added per Establishment	4,602,255	3,555,604	77	Investment per Employee	6,552	3,212	49
Value Added per Employee	93,930	92,742	99	Investment per Production Worker	9,352	4,949	53

Sources: Same as General Statistics. The 'Average of All Manufacturing' column represents the average of all manufacturing industries reported for the most recent complete year available. The Index shows the relationship between the Average and the Analyzed Industry. For example, 100 means that they are equal; 500 that the Analyzed Industry is five times the average; 50 means that the Analyzed Industry is half the national average. The abbreviation 'na' is used to show that data are 'not available'.

LEADING COMPANIES   Number shown: 56   Total sales ($ mil): 841   Total employment (000): 6.0

Company Name	Address				CEO Name	Phone	Co. Type	Sales ($ mil)	Empl. (000)
Ingersoll-Rand Co	23400 Halsted St	Farmington Hls	MI	48335	Robert Seccombe	810-477-0800	D	85	0.3
Giddings & Lewis	305 W Delavan Dr	Janesville	WI	53547	PN Ciarlo	608-756-1211	D	70	0.5
Ristance Corp	1718 Home St	Mishawaka	IN	46545	Frank Hayes	219-259-9903	D	68	0.4
Cummins Mid-America Inc	3527 Gardner St	Kansas City	MO	64120	Thomas H Payne	816-483-6313	R	47	0.2
Clemco Industries Corp	1 Cable Car Dr	Washington	MO	63090	Mark W Cleary	314-239-0300	R	38*	0.3
Fargo Assembly of PA Inc	PO Box 550	Norristown	PA	19404	Dennis A Rees	610-272-6850	R	37*	0.3
Atlas Technologies Inc	201 S Alloy Dr	Fenton	MI	48430	Ronald Prime	810-629-6663	R	30	0.2
Spirol International Corp	30 Rock Av	Danielson	CT	06239	James C Shaw	203-774-8571	S	30	0.3
Spirol Intern Holding Corp	30 Rock Av	Danielson	CT	06239	James C Shaw	203-774-8571	R	30	0.3
Stamco	125 S Herman St	New Bremen	OH	45869	Robert J Kindt	419-629-2061	D	30	0.2
Bodine Corp	PO Box 3245	Bridgeport	CT	06605	David Bodine	203-334-3100	R	27*	0.2
Delta Brands Inc	2204 Century Ctr	Irving	TX	75062	Samuel Savariego	214-438-7150	R	26	0.1
ADS Machinery Corp	PO Box 1027	Warren	OH	44482	Carl H Minton	216-399-3601	R	20	<0.1
Zero Products	1 Cable Car Dr	Washington	MO	63090	Mark W Cleary	314-239-0300	D	20	0.2
Amistar Corp	237 Via Vera Cruz	San Marcos	CA	92069	Stuart C Baker	619-471-1700	P	16	0.2
Speedring Inc	PO Box 1588	Cullman	AL	35056	Arnold Scheu	205-737-5200	S	16*	0.1
Winona Van-Norman	4730 W Service Dr	Winona	MN	55987	CE Mieras	507-454-4330	D	16	<0.1
Merrill Tool and Machine Inc	100 E Mahoney	Merrill	MI	48637	Gary J Yackel	517-643-7214	R	15	0.2
Schumag-Kieserling Machinery	155 Hudson Av	Norwood	NJ	07648	Dennis P Capolete	201-767-6850	S	15	<0.1
Aidlin Automation Corp	PO Box 13125	Sarasota	FL	34278	Stephen H Aidlin	813-756-0641	R	14	0.1
John Dusenbery Company Inc	220 Franklin Rd	Randolph	NJ	07869	John D Wilkes	201-366-7500	R	12	0.2
Muller-Ray Corp	805 Housatonic Av	Bridgeport	CT	06604	Monica Ray	203-367-6910	R	12	<0.1
Vanguard Automation Inc	10900 N Stallard Pl	Tucson	AZ	85737	Bill Orinski	602-297-2621	R	12	0.2
GWI Engineering Inc	1411 Michigan NE	Grand Rapids	MI	49503	Peter A Cordes	616-459-8274	R	11	0.1
CMI Equipment and Engineering	533 N Court St	Au Gres	MI	48703	Ken McKibben	517-876-7161	S	10	<0.1
ESAB Automation Inc	4600 Innovation Dr	Fort Collins	CO	80525	Dave Prosser	303-225-7600	S	10*	<0.1
Ahaus Tool and Engineering Inc	PO Box 280	Richmond	IN	47374	Fredric A Ahaus	317-962-3571	R	9	<0.1
Advex Corp	121 Thompson	Hampton	VA	23666	Bennie Barnett	804-865-0920	R	9	0.1
Abrasive Eng & Manufacturing	540 E Hwy 56	Olathe	KS	66061	Ray Vold	913-764-6040	S	8	0.1
Syncro Machine Co	611 Sayre Av	Perth Amboy	NJ	08861	DP Johnson	908-442-5500	R	8	<0.1
Visi-Trol Engineering Co	12720 Burt Rd	Detroit	MI	48223	Larry Galarowic	313-535-4140	R	8	<0.1
Bee Line Co	2700-62 Street Ct	Bettendorf	IA	52722		319-332-4066	D	7	0.1
Bachi LP	1201 Ardmore Av	Itasca	IL	60143	Neil Tiffin	708-773-5600	R	7	0.1
80/20 Inc	2570 Commercial	Fort Wayne	IN	46809	DF Wood	219-478-8020	R	6	<0.1
Bracker Corp	PO Box 441	Carnegie	PA	15106	F Boesch	412-276-4400	S	6	<0.1
Dixon Automatic Tool Inc	2300 23rd Av	Rockford	IL	61104	John H Dixon	815-226-3000	R	6*	<0.1
Ferranti Holdings Inc	PO Box 3025	Lancaster	PA	17604	James Shinehause	717-285-3113	R	6*	<0.1
Watson Machinery International	74-102 Railroad Av	Paterson	NJ	07509	Ian M McLaughlin	201-684-3700	R	6	<0.1
Braner USA Inc	9301 W Bernice	Schiller Park	IL	60176	Douglas Matsunaga	708-671-6210	R	5*	<0.1
Red Bud Industries Inc	200 B and E Indrial	Red Bud	IL	62278	K Voges	618-282-3801	R	5	<0.1
Amacoil Inc	PO Box 2228	Aston	PA	19014	H Zavaleta	610-485-8300	R	4	<0.1
Ohlinger Industries Inc	PO Box 42268	Phoenix	AZ	85080	HF Ohlinger	602-285-0911	R	4	<0.1
Beta Instrument Inc	125 J Hancock Rd	Taunton	MA	02780	Michael Conners	508-880-0771	S	3*	<0.1
New Production Machinery Inc	8500 Station St	Mentor	OH	44060	Norman Ferguson	216-255-3437	S	3*	<0.1
Air-Hydraulics Inc	PO Box 831	Jackson	MI	49204	R Michael Clark	517-787-9444	S	2	<0.1
American Steel Line Co	255 76th St	Grand Rapids	MI	49548	BW Fox	616-455-3000	R	2	<0.1
Shuster-Mettler Corp	PO Box 883	New Haven	CT	06504	William L DeSenti	203-562-3178	R	2	<0.1
Wyrepak Industries Inc	136 James St	Bridgeport	CT	06604	Raymond E Browne	203-334-4274	R	2	<0.1
American Best Tool Corp	PO Box 1149	Apple Valley	CA	92308	WT Martin	619-247-0102	R	1	<0.1
MacBee Engineering Corp	8613 Helms Av	R Cucamonga	CA	91730	W McMillan	909-466-6021	R	1*	<0.1
Robinson Engineering Corp	13063 Park St	Santa Fe Sprgs	CA	90670	P Robinson	310-946-4461	R	1	<0.1
Tubetronics	6030nida Encinas	Carlsbad	CA	92009	Paul Howard	619-438-5322	S	1*	<0.1
Roberts Metal Mfg Co	6301 Maywood Av	Huntington Pk	CA	90255	James M Roberts	213-585-1296	R	1*	<0.1
CE Cox Co	2415 S Broadway	Santa Ana	CA	92707	Ina Cox	714-540-2444	R	1	<0.1
Ettaco Inc	1829 W Drake Dr	Tempe	AZ	85283	David Foster	602-831-7445	R	1	<0.1
G-T Energy Concepts Inc	500 Griswold St	Detroit	MI	48226	Steven Ewing	313-256-6752	S	1	<0.1

Source: Ward's Business Directory of U.S. Private and Public Companies, Volumes 1 and 2, 1996. The company type code used is as follows: P - Public, R - Private, S - Subsidiary, D - Division, J - Joint Venture, A - Affiliate, G - Group. Sales are in millions of dollars, employees are in thousands. An asterisk (*) indicates an estimated sales volume. The symbol < stands for 'less than'. Company names and addresses are truncated, in some cases, to fit into the available space.

MATERIALS CONSUMED

Material	Quantity	Delivered Cost ($ million)
Materials, ingredients, containers, and supplies	(X)	552.4
Fluid power pumps, motors, and hydrostatic transmissions (hydraulic and pneumatic)	(X)	10.0
Fluid power cylinders and rotary actuators (hydraulic and pneumatic)	(X)	6.5
Fluid power filters (hydraulic and pneumatic)	(X)	1.1
Fluid power hose or tube fittings and assemblies (hydraulic and pneumatic)	(X)	3.2
Fluid power valves (hydraulic and pneumatic)	(X)	5.0
Metal bolts, nuts, screws, washers, rivets, and other screw machine products	(X)	6.7
Other fabricated metal products (except fluid power products and forgings)	(X)	18.1
Iron and steel forgings	(X)	0.3
Iron and steel castings (rough and semifinished)	(X)	8.7
Aluminum and aluminum-base alloy castings (rough and semifinished)	(X)	4.3
Other nonferrous castings (rough and semifinished)	(X)	0.6

Continued on next page.

MATERIALS CONSUMED - Continued

Material	Quantity	Delivered Cost ($ million)
Steel bars, bar shapes, and plates	(X)	23.4
Steel sheet and strip, including tin plate	(X)	5.1
Steel structural shapes	(X)	4.2
All other steel shapes and forms	(X)	4.0
Aluminum and aluminum-base alloy shapes and forms	(X)	2.5
Other nonferrous shapes and forms	(X)	0.8
Fractional horsepower electric timing motors, synchronous and subsynchronous (less than 1 hp)	(X)	2.1
Other fractional horsepower electric motors (under 1 hp)	(X)	2.5
Integral horsepower electric motors and generators (1 hp or more)	(X)	12.4
Electrical transmission, distribution, and control equipment	(X)	40.6
Electrical industrial capacitors, resistors, rheostats, and coil windings	(X)	3.0
Numerical controls for metalworking machinery (except programmable)	(X)	1.4
Programmable controllers for metalworking machinery	(X)	12.1
Ball bearings (mounted or unmounted)	(X)	4.3
Roller bearings (mounted or unmounted)	(X)	3.4
Mechanical speed changers, gears, and industrial high-speed drives	(X)	8.4
Wood boxes, pallets, skids, and containers	(X)	1.6
Cutting tools for machine tools	(X)	2.7
All other materials and components, parts, containers, and supplies	(X)	180.4
Materials, ingredients, containers, and supplies, nsk	(X)	172.7

Source: 1992 *Economic Census*. Explanation of symbols used: (D): Withheld to avoid disclosure of competitive data; na: Not available; (S): Withheld because statistical norms were not met; (X): Not applicable; (Z): Less than half the unit shown; nec: Not elsewhere classified; nsk: Not specified by kind; - : zero; * : 10-19 percent estimated; ** : 20-29 percent estimated.

PRODUCT SHARE DETAILS

Product or Product Class	% Share	Product or Product Class	% Share
Metalworking machinery, nec	100.00	(except dies and handheld)	4.33
Assembly machines	54.63	Wire rope or wire cable making machines (except handheld)	2.73
Rotary transfer metalworking assembly machines (dial or rotary, trunnion, center column)	10.52	Other metalworking machines for working wire (except handheld)	12.66
Synchronous in-line transfer metalworking assembly machines	12.47	Cut-to-length coil handling lines (conversion or straightening), except handheld	9.72
Nonsynchronous in-line transfer metalworking assembly machines	18.78	Slitting coil handling lines (conversion or straightening), except handheld	8.53
Special-purpose and other types of metalworking assembly machines	53.08	Other metalworking machinery (except handheld or ultrasonic)	42.81
Parts and attachments for metalworking assembly machines (sold separately)	4.00	Parts and attachments for other metalworking machinery (except handheld or ultrasonic) (sold separately)	16.51
Assembly machines, nsk	1.15	Other metalworking machinery (except handheld and ultrasonic), nsk	2.69
Other metalworking machinery (except handheld and ultrasonic)	33.05		
Metalworking draw benches and wiredrawing machines		Metalworking machinery, nec, nsk	12.32

Source: 1992 *Economic Census*. The values shown are percent of total shipments in an industry. Values of indented subcategories are summed in the main heading. The symbol (D) appears when data are withheld to prevent disclosure of competitive information. The abbreviation nsk stands for 'not specified by kind' and nec for 'not elsewhere classified'.

INPUTS AND OUTPUTS FOR METALWORKING MACHINERY, NEC

Economic Sector or Industry Providing Inputs	%	Sector	Economic Sector or Industry Buying Outputs	%	Sector
Blast furnaces & steel mills	9.5	Manufg.	Gross private fixed investment	67.5	Cap Inv
Wholesale trade	9.0	Trade	Exports	11.5	Foreign
Fabricated plate work (boiler shops)	6.0	Manufg.	Machinery, except electrical, nec	5.1	Manufg.
Metalworking machinery, nec	5.8	Manufg.	Personal consumption expenditures	4.0	
Petroleum refining	4.2	Manufg.	Fabricated plate work (boiler shops)	2.6	Manufg.
Motors & generators	4.0	Manufg.	Metalworking machinery, nec	2.5	Manufg.
Switchgear & switchboard apparatus	3.9	Manufg.	Blast furnaces & steel mills	1.7	Manufg.
Advertising	3.6	Services	Metal partitions & fixtures	1.5	Manufg.
Machinery, except electrical, nec	3.4	Manufg.	Aircraft & missile engines & engine parts	1.0	Manufg.
Metal stampings, nec	3.0	Manufg.	Federal Government purchases, national defense	0.7	Fed Govt
Nonferrous rolling & drawing, nec	2.9	Manufg.	Automotive repair shops & services	0.5	Services
Miscellaneous plastics products	2.6	Manufg.	Special dies & tools & machine tool accessories	0.3	Manufg.
Electric services (utilities)	2.4	Util.	Iron & steel foundries	0.2	Manufg.
Copper rolling & drawing	2.1	Manufg.	Motor vehicles & car bodies	0.2	Manufg.
Iron & steel foundries	1.8	Manufg.	Motor freight transportation & warehousing	0.2	Util.
Communications, except radio & TV	1.8	Util.			
Fabricated metal products, nec	1.6	Manufg.			
Miscellaneous fabricated wire products	1.5	Manufg.			
Power transmission equipment	1.4	Manufg.			
Screw machine and related products	1.3	Manufg.			
Banking	1.2	Fin/R.E.			
Aluminum castings	1.1	Manufg.			

Continued on next page.

INPUTS AND OUTPUTS FOR METALWORKING MACHINERY, NEC - Continued

Economic Sector or Industry Providing Inputs	%	Sector	Economic Sector or Industry Buying Outputs	%	Sector
Pipe, valves, & pipe fittings	1.1	Manufg.			
Ball & roller bearings	1.0	Manufg.			
Equipment rental & leasing services	1.0	Services			
Maintenance of nonfarm buildings nec	0.9	Constr.			
Abrasive products	0.9	Manufg.			
Paperboard containers & boxes	0.9	Manufg.			
Pumps & compressors	0.9	Manufg.			
Automotive rental & leasing, without drivers	0.9	Services			
Motor freight transportation & warehousing	0.8	Util.			
Real estate	0.8	Fin/R.E.			
Royalties	0.8	Fin/R.E.			
Automotive repair shops & services	0.8	Services			
Computer & data processing services	0.8	Services			
Engineering, architectural, & surveying services	0.8	Services			
Aluminum rolling & drawing	0.7	Manufg.			
Industrial controls	0.7	Manufg.			
Transformers	0.7	Manufg.			
Gas production & distribution (utilities)	0.7	Util.			
Eating & drinking places	0.7	Trade			
Imports	0.7	Foreign			
Machine tools, metal cutting types	0.6	Manufg.			
Special dies & tools & machine tool accessories	0.6	Manufg.			
Air transportation	0.6	Util.			
Iron & steel forgings	0.4	Manufg.			
Insurance carriers	0.4	Fin/R.E.			
Brass, bronze, & copper castings	0.3	Manufg.			
Metal heat treating	0.3	Manufg.			
Rubber & plastics hose & belting	0.3	Manufg.			
Wiring devices	0.3	Manufg.			
Detective & protective services	0.3	Services			
Hotels & lodging places	0.3	Services			
Legal services	0.3	Services			
Management & consulting services & labs	0.3	Services			
Noncomparable imports	0.3	Foreign			
Fabricated rubber products, nec	0.2	Manufg.			
Lubricating oils & greases	0.2	Manufg.			
Motor vehicle parts & accessories	0.2	Manufg.			
Nonferrous wire drawing & insulating	0.2	Manufg.			
Plating & polishing	0.2	Manufg.			
Primary metal products, nec	0.2	Manufg.			
Tires & inner tubes	0.2	Manufg.			
Retail trade, except eating & drinking	0.2	Trade			
Accounting, auditing & bookkeeping	0.2	Services			
U.S. Postal Service	0.2	Gov't			
Machine tools, metal forming types	0.1	Manufg.			
Manifold business forms	0.1	Manufg.			
Nonferrous castings, nec	0.1	Manufg.			
Electrical repair shops	0.1	Services			

Source: Benchmark Input-Output Accounts for the U.S. Economy, 1982, U.S. Department of Commerce, Washington, D.C., July 1991. Data, as reported in the source, are organized by the 1977 SIC structure in use in 1982 but have been matched, as closely as is possible, to the 1987 SIC structure used in this book.

OCCUPATIONS EMPLOYED BY SIC 354 - METALWORKING MACHINERY

Occupation	% of Total 1994	Change to 2005	Occupation	% of Total 1994	Change to 2005
Machinists	10.9	-17.4	Precision metal workers nec	1.9	10.1
Tool & die makers	10.6	-15.3	Mechanical engineers	1.9	1.0
General managers & top executives	3.7	-12.9	Inspectors, testers, & graders, precision	1.9	-8.2
Blue collar worker supervisors	3.4	-15.7	Secretaries, ex legal & medical	1.9	-16.4
Machine tool cutting operators, metal & plastic	2.9	-5.9	Drafters	1.8	-28.5
NC machine tool operators, metal & plastic	2.7	1.0	General office clerks	1.5	-21.7
Grinding machine operators, metal & plastic	2.7	-8.2	Machine forming operators, metal & plastic	1.4	-8.2
Sales & related workers nec	2.7	-8.2	Drilling & boring machine tool workers	1.3	-8.2
Assemblers, fabricators, & hand workers nec	2.5	-8.2	Bookkeeping, accounting, & auditing clerks	1.3	-31.1
Lathe & turning machine tool operators	2.3	-17.4	Industrial production managers	1.3	-8.2
Machine tool cutting & forming etc. nec	2.2	-54.1	Traffic, shipping, & receiving clerks	1.3	-11.7
Machine builders	2.0	1.0	Precision woodworkers nec	1.3	10.2
Combination machine tool operators	2.0	37.7	Janitors & cleaners, incl maids	1.2	-26.6

Source: Industry-Occupation Matrix, Bureau of Labor Statistics. These data relate to one or more 3-digit SIC industry groups rather than to a single 4-digit SIC. The change reported for each occupation to the year 2005 is a percent of growth or decline as estimated by the Bureau of Labor Statistics. The abbreviation nec stands for 'not elsewhere classified'.

LOCATION BY STATE AND REGIONAL CONCENTRATION

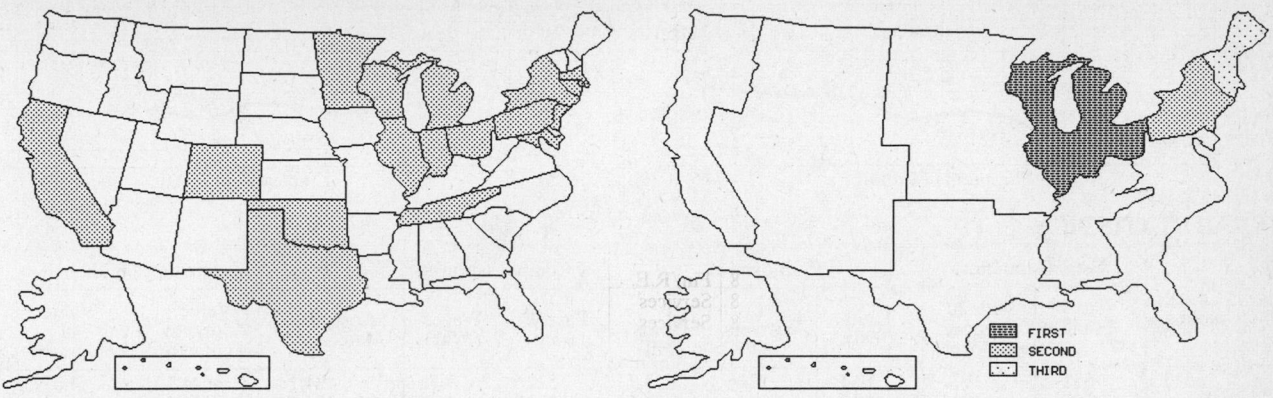

FIRST
SECOND
THIRD

INDUSTRY DATA BY STATE

| State | Establish-ments | Shipments | | | Employment | | | | Cost as % of Shipments | Investment per Employee ($) |
		Total ($ mil)	% of U.S.	Per Establ.	Total Number	% of U.S.	Per Establ.	Wages ($/hour)		
Michigan	65	456.3	28.2	7.0	2,900	22.0	45	16.10	49.9	2,034
Ohio	43	222.6	13.8	5.2	1,900	14.4	44	13.75	36.7	2,316
Illinois	40	158.2	9.8	4.0	1,400	10.6	35	14.76	44.8	2,714
Wisconsin	15	124.0	7.7	8.3	1,000	7.6	67	14.50	23.3	1,000
Pennsylvania	24	85.0	5.3	3.5	900	6.8	38	12.00	40.6	2,333
New York	21	76.9	4.8	3.7	800	6.1	38	13.11	45.1	750
Connecticut	23	59.8	3.7	2.6	500	3.8	22	17.14	30.3	1,000
New Jersey	17	46.7	2.9	2.7	400	3.0	24	12.75	48.0	750
Indiana	11	44.8	2.8	4.1	500	3.8	45	14.50	39.1	1,800
California	33	44.1	2.7	1.3	500	3.8	15	11.57	38.1	1,200
Massachusetts	16	26.4	1.6	1.6	300	2.3	19	10.75	42.8	1,333
Texas	9	25.3	1.6	2.8	200	1.5	22	9.00	57.7	-
Minnesota	8	21.9	1.4	2.7	200	1.5	25	16.00	41.6	2,500
Rhode Island	6	21.4	1.3	3.6	200	1.5	33	11.33	35.5	-
Maryland	5	13.4	0.8	2.7	100	0.8	20	12.00	56.7	-
Colorado	5	(D)	-	-	175 *	1.3	35	-	-	-
Tennessee	5	(D)	-	-	375 *	2.8	75	-	-	-
Oklahoma	4	(D)	-	-	175 *	1.3	44	-	-	-

Source: 1992 *Economic Census*. The states are in descending order of shipments or establishments (if shipment data are missing for the majority). The symbol (D) appears when data are withheld to prevent disclosure of competitive information. States marked with (D) are sorted by number of establishments. A dash (-) indicates that the data element cannot be calculated; * indicates the midpoint of a range.

3552 - TEXTILE MACHINERY

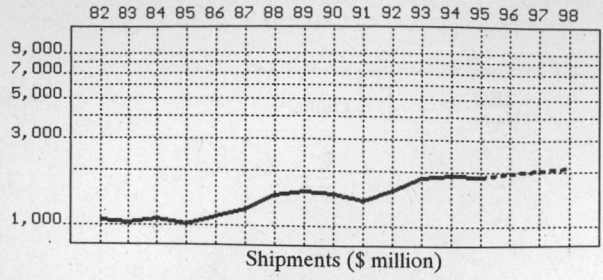

Shipments ($ million)

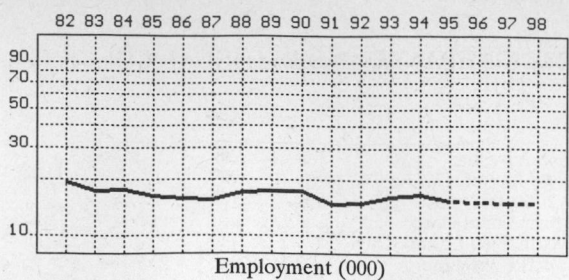

Employment (000)

GENERAL STATISTICS

Year	Com- panies	Establishments		Employment			Compensation		Production ($ million)			
		Total	with 20 or more employees	Total (000)	Production Workers (000)	Hours (Mil)	Payroll ($ mil)	Wages ($/hr)	Cost of Materials	Value Added by Manufacture	Value of Shipments	Capital Invest.
1982	512	551	197	19.4	12.8	23.9	322.9	7.72	424.1	641.4	1,059.0	46.1
1983		537	192	17.5	12.0	23.4	324.3	8.05	454.7	534.5	1,038.2	35.4
1984		523	187	17.7	11.9	23.4	340.1	8.35	485.6	627.8	1,089.3	24.3
1985		510	183	16.2	10.9	21.5	316.9	8.60	451.1	543.0	1,019.8	29.2
1986		495	173	15.8	10.7	21.6	331.7	9.04	478.4	669.3	1,132.2	22.1
1987	475	506	172	15.6	10.6	22.2	352.4	9.32	547.1	712.1	1,240.7	33.2
1988		493	182	17.1	11.7	23.3	388.2	9.87	642.8	766.0	1,487.4	40.6
1989		491	182	17.3	11.5	23.9	428.7	10.26	710.7	889.0	1,561.0	37.4
1990		500	181	17.4	11.0	22.7	452.7	10.85	687.3	814.9	1,505.1	46.6
1991		500	168	14.9	10.1	20.9	437.1	11.42	621.8	776.8	1,395.8	32.9
1992	480	506	158	15.0	9.6	19.8	424.5	11.04	644.4	908.1	1,565.2	39.5
1993		496	169	16.5	10.1	21.2	486.0	10.96	845.2	1,025.7	1,853.4	46.9
1994		484P	162P	16.7	10.6	21.9	494.0	11.92	818.5	1,090.3	1,908.3	71.4
1995		480P	160P	15.5P	9.8P	20.9P	501.9P	12.24P	808.0P	1,076.3P	1,883.9P	51.8P
1996		476P	157P	15.3P	9.6P	20.7P	517.6P	12.59P	839.3P	1,118.0P	1,956.8P	53.6P
1997		472P	155P	15.2P	9.4P	20.5P	533.3P	12.94P	870.6P	1,159.6P	2,029.7P	55.4P
1998		468P	152P	15.0P	9.2P	20.3P	549.0P	13.29P	901.8P	1,201.3P	2,102.6P	57.3P

Sources: 1982, 1987, 1992 *Economic Census*; *Annual Survey of Manufactures*, 83-86, 88-91, 93-94. Establishment counts for non-Census years are from *County Business Patterns*; establishment values for 83-84 are extrapolations. 'P's show projections by the editors. Industries reclassified in 87 will not have data for prior years.

INDICES OF CHANGE

Year	Com- panies	Establishments		Employment			Compensation		Production ($ million)			
		Total	with 20 or more employees	Total (000)	Production Workers (000)	Hours (Mil)	Payroll ($ mil)	Wages ($/hr)	Cost of Materials	Value Added by Manufacture	Value of Shipments	Capital Invest.
1982	107	109	125	129	133	121	76	70	66	71	68	117
1983		106	122	117	125	118	76	73	71	59	66	90
1984		103	118	118	124	118	80	76	75	69	70	62
1985		101	116	108	114	109	75	78	70	60	65	74
1986		98	109	105	111	109	78	82	74	74	72	56
1987	99	100	109	104	110	112	83	84	85	78	79	84
1988		97	115	114	122	118	91	89	100	84	95	103
1989		97	115	115	120	121	101	93	110	98	100	95
1990		99	115	116	115	115	107	98	107	90	96	118
1991		99	106	99	105	106	103	103	96	86	89	83
1992	100	100	100	100	100	100	100	100	100	100	100	100
1993		98	107	110	105	107	114	99	131	113	118	119
1994		96P	103P	111	110	111	116	108	127	120	122	181
1995		95P	101P	103P	102P	106P	118P	111P	125P	119P	120P	131P
1996		94P	100P	102P	100P	105P	122P	114P	130P	123P	125P	136P
1997		93P	98P	101P	98P	104P	126P	117P	135P	128P	130P	140P
1998		92P	96P	100P	96P	103P	129P	120P	140P	132P	134P	145P

Sources: Same as General Statistics. Values reflect change from the base year, 1992. Values above 100 mean greater than 92, values below 100 mean less than 92, and a value of 100 in the 82-91 or 93-98 period means same as 92. 'P's mark projections by the editors.

SELECTED RATIOS

For 1994	Avg. of All Manufact.	Analyzed Industry	Index	For 1994	Avg. of All Manufact.	Analyzed Industry	Index
Employees per Establishment	49	35	70	Value Added per Production Worker	134,084	102,858	77
Payroll per Establishment	1,500,273	1,021,621	68	Cost per Establishment	5,045,178	1,692,705	34
Payroll per Employee	30,620	29,581	97	Cost per Employee	102,970	49,012	48
Production Workers per Establishment	34	22	64	Cost per Production Worker	146,988	77,217	53
Wages per Establishment	853,319	539,862	63	Shipments per Establishment	9,576,895	3,946,475	41
Wages per Production Worker	24,861	24,627	99	Shipments per Employee	195,460	114,269	58
Hours per Production Worker	2,056	2,066	100	Shipments per Production Worker	279,017	180,028	65
Wages per Hour	12.09	11.92	99	Investment per Establishment	321,011	147,659	46
Value Added per Establishment	4,602,255	2,254,804	49	Investment per Employee	6,552	4,275	65
Value Added per Employee	93,930	65,287	70	Investment per Production Worker	9,352	6,736	72

Sources: Same as General Statistics. The 'Average of All Manufacturing' column represents the average of all manufacturing industries reported for the most recent complete year available. The Index shows the relationship between the Average and the Analyzed Industry. For example, 100 means that they are equal; 500 that the Analyzed Industry is five times the average; 50 means that the Analyzed Industry is half the national average. The abbreviation 'na' is used to show that data are 'not available'.

LEADING COMPANIES   Number shown: **74**   Total sales ($ mil): **1,044**   Total employment (000): **9.1**

Company Name	Address				CEO Name	Phone	Co. Type	Sales ($ mil)	Empl. (000)
Day International Inc	PO Box 338	Dayton	OH	45401	Dennis R Walter	513-226-5866	S	80*	0.8
Steel Heddle Manufacturing Co	PO Box 1867	Greenville	SC	29602	BG Team	803-244-4110	R	70*	0.9
Speizman Industries Inc	508 W 5th St	Charlotte	NC	28202	Robert S Speizman	704-372-3751	P	69	<0.1
Hirsch International Corp	355 Marcus Blv	Hauppauge	NY	11788	Henry Arnberg	516-436-7100	P	50	0.1
Hollingsworth Saco Lowell Corp	PO Box 2327	Greenville	SC	29602	J D Hollingsworth	803-859-3211	S	42	0.4
Tubular Textile Machinery	PO Box 2097	Lexington	NC	27293	Richard Bruce	704-956-6444	S	40	0.2
Vanguard Supreme	PO Box 5009	Monroe	NC	28110	Irving Bienstock	704-283-8171	D	40	0.3
Setco Sales Co	5880 Hillside Av	Cincinnati	OH	45233	WA Ferguson III	513-941-5110	R	36	0.3
Gaston Cty Dyeing Machine	PO Box 308	Stanley	NC	28164	HM Craig, Jr	704-822-5000	R	35	0.5
Morrison Textile Machinery Co	PO Box 1	Fort Lawn	SC	29714	JJ O'Neill	803-872-4401	R	35	0.2
Eastman Machine Co	779 Washington St	Buffalo	NY	14203	RL Stevenson	716-856-2200	R	33	0.3
Marshall and Williams Co	PO Box 17268	Greenville	SC	29606	SR MacDonald	803-242-6750	R	33*	0.3
Precision Screen Machines Inc	44 Utter Av	Hawthorne	NJ	07506	Richard Hoffman	201-427-5100	R	30*	0.2
Atlantic Zeiser Co	15 Patton Dr	West Caldwell	NJ	07006	Walter Klingher	201-228-0800	R	27*	0.3
Whitin-Roberts Co	PO Box 250	Sanford	NC	27330	DA Smith	919-775-7321	R	25*	0.3
Kusters Corp	PO Box 6128	Spartanburg	SC	29304	Blas Miyares	803-576-0660	R	24	0.1
West Point Foundry	PO Box 151	West Point	GA	31833	AD Cotney	706-643-2127	S	22	0.3
Barudan America Inc	29500 Fountain	Solon	OH	44139	Yoshio Shibata	216-248-8770	S	20	<0.1
Carter Traveler Co	PO Box 518	Gastonia	NC	28053	Rick Craig	704-865-1201	S	20	0.4
Tuftco Corp	2318 Holtzclaw Av	Chattanooga	TN	37404	Jack L Frost	615-698-8601	R	18*	0.2
GHM Industries Inc	41 Fremont St	Worcester	MA	01603	Michael Wilson	508-793-7000	R	16	<0.1
Greenville Machinery Corp	PO Box 1209	Greenville	SC	29602	Wolf Stromberg	803-879-3011	S	15	0.1
Wardwell Braiding Machine Co	1211 High St	Central Falls	RI	02863	JK Farnum	401-724-8800	R	14	0.2
Morrison Berkshire Inc	PO Box 958	North Adams	MA	01247	James S White	413-663-6501	R	13	<0.1
Bahan Machine & Foundry Corp	PO Box 1908	Greenville	SC	29602	Marshall Croy	803-244-4220	R	12*	<0.1
Tuftco Finishing System Inc	PO Box 704	Dalton	GA	30720	Frank Johnson	706-277-1110	S	12*	0.1
Harco Graphic Products Inc	101 Garden St	Grand Rapids	MI	49507	CW Harpold	616-452-3400	R	10	<0.1
Spencer Wright Industries Inc	1731 Kimberly Pk	Dalton	GA	30720	Spencer H Wright	706-278-1857	R	10	0.2
Mount Hope Machine Co	15 5th St	Taunton	MA	02780	NA August	508-824-6994	S	9	0.1
Jenkins Metal Corp	PO Box 2089	Gastonia	NC	28053	Robert B Jenkins Jr	704-867-6394	R	8	0.1
Belmont Textile Machinery	1212 W Catawba	Mount Holly	NC	28120	Walter Rhyne	704-827-5836	R	7*	<0.1
Bowman Dunn Mfg Co	PO Box 19249	Charlotte	NC	28219	Thomas H Bowman	704-374-1500	R	7	<0.1
Cronland Warp Roll Co	PO Box 574	Lincolnton	NC	28093	B Cronland	704-735-6564	R	7	<0.1
DR Kenyon and Son Inc	PO Box 6200	Bridgewater	NJ	08807	Loren W Jones	908-722-0001	R	7*	<0.1
Parks and Woolson Machine Co	PO Box 859	Springfield	VT	05156	Thomas J Koledo	802-885-2102	R	7	<0.1
Spuhl-Anderson Machine Co	1610 Parallel St	Chaska	MN	55318	Marlin Schutte	612-448-2676	S	7	<0.1
BF Perkins	939 Chicopee St	Chicopee	MA	01013	Paul Staszko	413-536-1311	D	6	<0.1
H Maimin Company Inc	PO Box 549	Kent	CT	06757	Daniel J Bangser	203-927-4601	R	6*	0.1
Louis P Batson Co	PO Box 3978	Greenville	SC	29608	HE Batson	803-242-5262	R	6*	<0.1
Mayer Textile Machine Corp	310 Brighton Rd	Clifton	NJ	07012	Gert Rohmert	201-773-3350	R	6*	<0.1
McCoy-Ellison Inc	1101 Curtis St	Monroe	NC	28110	Bruce McCoy	704-289-5413	R	6*	<0.1
Robert A Main and Sons Inc	555 Goffle Rd	Wyckoff	NJ	07481	Robert A Main Jr	201-447-3700	R	6	<0.1
Wolf Machine Company Inc	5570 Creek Rd	Cincinnati	OH	45242	George L Andre	513-791-5194	R	6	0.1
Elliott Metal Works Inc	PO Box 8675	Greenville	SC	29604	OS Elliott Jr	803-269-8930	R	5*	<0.1
Sew Simple Systems Inc	PO Box 68	Fountain Inn	SC	29644	Cecil Eggert	803-862-4252	S	5*	<0.1
Silkcraft of Oregon	11120 SW Industrial	Tualatin	OR	97062	Marshall A Stevens	503-692-8286	R	5	0.1
Parkinson Machinery	PO Box 17099	Esmond	RI	02917	Ernest Abrahamson	401-231-7100	R	5	<0.1
Tapistron International Inc	PO Box 1067	Ringgold	GA	30736	Larry Pierce	706-965-9300	P	4	<0.1
Baxter Corp	PO Box 1766	Shelby	NC	28150	Andrew Price	704-482-2476	R	4	<0.1
Curtin-Hebert Company Inc	PO Box 511	Gloversville	NY	12078	James A Curtin	518-725-7157	R	4	<0.1
Fiber Controls Corp	PO Box 1358	Gastonia	NC	28053	Bill Tench	704-864-9911	R	4	<0.1
Gunter and Cooke	PO Box 15220	Durham	NC	27704	Jon L Hasbrouck	910-852-4200	D	4	<0.1
Judelshon Industries Inc	220 Franklin Rd	Randolph	NJ	07869	John D Wilkes	201-366-7500	S	4	<0.1
Lawson-Hemphill Inc	PO Box 759	Pawtucket	RI	02862	Avischai Nevel	401-724-7130	R	4*	<0.1
Rando Machine Corp	PO Box 614	Macedon	NY	14502	M Wilder	315-986-2761	R	4*	<0.1
Thermal Engineering of Arizona	2250 W Wetmore	Tucson	AZ	85705	R O Kaufmann	520-888-4000	R	4	<0.1
WH Bagshaw Co	PO Box 766	Nashua	NH	03061	David A Bagshaw	603-883-7758	R	4	<0.1
Dallas Machine Company Inc	PO Box 12606	Gastonia	NC	28053	Cecil C Bumgarner	704-865-8545	R	4*	<0.1
Fletcher Industries Inc	PO Box 1359	Southern Pines	NC	28388	Edward T Taws Jr	910-692-7133	R	3	<0.1
Lever Manufacturing Corp	150 E 7th St	Paterson	NJ	07524	Irving Gerstein	201-684-5000	R	3	<0.1
Sims Machinery Company Inc	Huguley Industrial	Lanett	AL	36863	Lynn Duncan	205-576-2101	R	3	<0.1
Transcolor East Inc	PO Box 4170	Albany	GA	31706	Beverly Hanna	912-883-8300	S	3	<0.1
American Wax Inc	PO Box 504	Heath Springs	SC	29058	Dave Eudy	803-273-9492	R	2	<0.1
Haskell-Dawes Inc	2231 E Ontario St	Philadelphia	PA	19134	A Allen Woodruff	215-743-1432	R	2*	<0.1
MEG US Inc	401 Central Av	E Rutherford	NJ	07073	Donald Dionne	201-939-6600	S	2	<0.1
North Amer Textile Machinery	PO Box 26936	Philadelphia	PA	19134	RW Farley	215-739-9107	R	2	<0.1
Textile Parts and Machine	PO Box 12305	Gastonia	NC	28052	MB Stewart	704-865-8564	R	2*	<0.1
Advanced Graphics Inc	920 Honeyspot Rd	Stratford	CT	06497	John Alesvich	203-378-0471	R	1	<0.1
Barney Knitting Machinery	1140 Seneca Av	Ridgewood	NY	11385	Harold Gray	718-497-2266	R	1	<0.1
GEM Textile Company Inc	PO Box 12567	Gastonia	NC	28052	J R Franklin Jr	704-629-3181	R	1	<0.1
Pacific Fabric Reels Inc	3401 Etiwanda Av	Mira Loma	CA	91752	Remy O'Neill	909-681-2993	R	1*	<0.1
Parts and Systems Company Inc	PO Box 5468	Asheville	NC	28813	John M Crook	704-684-7070	R	1*	<0.1
Whitinsville Spinning Rings	PO Box 518	Gastonia	NC	28053	Buddy Brock	704-864-7846	D	1*	<0.1
HF Livermore Corp	Endicott St	Norwood	MA	02062	William Seller	617-782-2800	R	1	<0.1

Source: Ward's Business Directory of U.S. Private and Public Companies, Volumes 1 and 2, 1996. The company type code used is as follows: P - Public, R - Private, S - Subsidiary, D - Division, J - Joint Venture, A - Affiliate, G - Group. Sales are in millions of dollars, employees are in thousands. An asterisk (*) indicates an estimated sales volume. The symbol < stands for 'less than'. Company names and addresses are truncated, in some cases, to fit into the available space.

MATERIALS CONSUMED

Material	Quantity	Delivered Cost ($ million)
Materials, ingredients, containers, and supplies	(X)	509.1
Fluid power products	(X)	25.8
Fabricated metal products, except forgings	(X)	31.4
Forgings	(X)	6.4
Iron and steel castings (rough and semifinished)	(X)	14.8
Aluminum and aluminum-base alloy castings (rough and semifinished)	(X)	4.8
Other nonferrous castings (rough and semifinished)	(X)	1.1
Steel bars, bar shapes, and plates	(X)	24.7
Steel sheet and strip, including tin plate	(X)	10.4
Steel structural shapes and sheet piling	(X)	4.3
Steel wire and wire products	(X)	26.5
All other steel shapes and forms	(X)	11.8
Aluminum and aluminum-base alloy shapes and forms	(X)	9.2
Other nonferrous shapes and forms	(X)	3.1
Integral horsepower electric motors and generators (1 hp or more)	(X)	8.9
Ball bearings (mounted or unmounted)	(X)	6.4
Roller bearings (mounted or unmounted)	(X)	1.8
Speed reducers, gears, drives, and other mechanical power transmission equipment, except bearings	(X)	15.2
Fabricated plastics products (except gaskets, hoses, and belting)	(X)	4.5
Electrical transmission, distribution, and control equipment	(X)	39.6
All other materials and components, parts, containers, and supplies	(X)	142.2
Materials, ingredients, containers, and supplies, nsk	(X)	116.2

Source: 1992 Economic Census. Explanation of symbols used: (D): Withheld to avoid disclosure of competitive data; na: Not available; (S): Withheld because statistical norms were not met; (X): Not applicable; (Z): Less than half the unit shown; nec: Not elsewhere classified; nsk: Not specified by kind; - : zero; * : 10-19 percent estimated; ** : 20-29 percent estimated.

PRODUCT SHARE DETAILS

Product or Product Class	% Share	Product or Product Class	% Share
Textile machinery	100.00	Textile bleaching, mercerizing, and dyeing machinery (except parts, attachments, and accessories)	10.56
Textile machinery, except parts, attachments, and accessories	52.50	Textile printing machinery (except parts, attachments, and accessories)	4.46
Cleaning and opening fiber-to-fabric textile machinery, including picker, garnetting, and other (except parts, attachments, and accessories)	3.90	Textile finishing machinery, including calendering or rolling machines (except parts, attachments, and accessories)	9.45
Carding and combing fiber-to-fabric textile machines (except parts, attachments, and accessories)	0.85	Textile machinery for drying stocks, yarns, cloth, carpet, nonwoven, etc. (except parts, attachments, and accessories)	8.09
Drawing and roving fiber-to-fabric frames (except parts, attachments, and accessories)	(D)	Other textile industries machinery, nec (except parts, attachments, and accessories)	24.84
Spinning and twisting fiber-to-fabric textile frames (except parts, attachments, and accessories)	(D)	Textile machinery, except parts, attachments, and accessories, nsk	2.34
Fiber-to-fabric yarn winding machines (skein, spool, bobbin, quill, cone, etc.) (except parts, attachments, and accessories)	3.63	Parts and attachments for textile machinery	35.69
Other fiber-to-fabric yarn preparing machines (beaming, warping, warp tying, warp drawing in, slashing, etc.) (except parts, attachments, and accessories)	2.61	Textile machinery turnings and shapes (bobbins, shuttles, spools, picker sticks, cops, etc.)	9.75
Other fiber-to-fabric machinery, including machines for extruding, drawing, or cutting manmade textile fibers (except parts, attachments, and accessories)	6.33	Parts and attachments for fiber-to-fabric card clothing machinery	4.81
Fabric weaving machinery, power driven (including machinery for broad and narrow fabrics), shuttle and shuttleless (except parts, attachments, and accessories)	1.43	Parts and attachments for other fiber-to-fabric machinery, except card clothing	17.96
Fabric knitting machinery (except parts, attachments, and accessories)	(D)	Parts and attachments for weaving machines, including broad and narrow fabrics	16.34
Other fabric machinery, including lace, embroidery, braiding, and tufting machinery and hand looms (except parts, attachments, and accessories)	9.93	Parts and attachments for knitting machines, excluding needles	6.65
		Parts and attachments for finishing machinery	7.01
		Parts and attachments for other textile machinery, including printing	34.34
		Parts and attachments for textile machinery, nsk	3.12
		Textile machinery, nsk	11.82

Source: 1992 Economic Census. The values shown are percent of total shipments in an industry. Values of indented subcategories are summed in the main heading. The symbol (D) appears when data are withheld to prevent disclosure of competitive information. The abbreviation nsk stands for 'not specified by kind' and nec for 'not elsewhere classified'.

INPUTS AND OUTPUTS FOR TEXTILE MACHINERY

Economic Sector or Industry Providing Inputs	%	Sector	Economic Sector or Industry Buying Outputs	%	Sector
Imports	51.1	Foreign	Gross private fixed investment	52.5	Cap Inv
Wholesale trade	6.5	Trade	Exports	15.0	Foreign
Blast furnaces & steel mills	5.5	Manufg.	Broadwoven fabric mills	8.4	Manufg.
Machinery, except electrical, nec	2.7	Manufg.	Floor coverings	5.4	Manufg.
Textile machinery	2.3	Manufg.	Yarn mills & finishing of textiles, nec	4.8	Manufg.
Iron & steel foundries	2.0	Manufg.	Knit fabric mills	2.7	Manufg.
Switchgear & switchboard apparatus	1.7	Manufg.	Textile machinery	1.8	Manufg.
Aluminum rolling & drawing	1.5	Manufg.	Knit outerwear mills	1.5	Manufg.
Power transmission equipment	1.5	Manufg.	Nonwoven fabrics	0.9	Manufg.
Advertising	1.3	Services	Women's hosiery, except socks	0.9	Manufg.
Special dies & tools & machine tool accessories	1.2	Manufg.	Hosiery, nec	0.8	Manufg.
Electric services (utilities)	1.1	Util.	Tire cord & fabric	0.8	Manufg.
Ball & roller bearings	1.0	Manufg.	Coated fabrics, not rubberized	0.7	Manufg.
Aluminum castings	0.9	Manufg.	Canvas & related products	0.5	Manufg.
Industrial controls	0.9	Manufg.	Glass & glass products, except containers	0.5	Manufg.
Motors & generators	0.9	Manufg.	Knit underwear mills	0.5	Manufg.
Screw machine and related products	0.9	Manufg.	Thread mills	0.4	Manufg.
Engineering, architectural, & surveying services	0.9	Services	House slippers	0.3	Manufg.
Maintenance of nonfarm buildings nec	0.8	Constr.	Processed textile waste	0.3	Manufg.
Eating & drinking places	0.7	Trade	Felt goods, nec	0.2	Manufg.
Banking	0.7	Fin/R.E.	Narrow fabric mills	0.2	Manufg.
Fabricated rubber products, nec	0.6	Manufg.	Padding & upholstery filling	0.2	Manufg.
Petroleum refining	0.6	Manufg.	Textile goods, nec	0.2	Manufg.
Pipe, valves, & pipe fittings	0.6	Manufg.	Cordage & twine	0.1	Manufg.
Rubber & plastics hose & belting	0.6	Manufg.	Schiffli machine embroideries	0.1	Manufg.
Air transportation	0.6	Util.			
Communications, except radio & TV	0.6	Util.			
Noncomparable imports	0.6	Foreign			
Nonferrous rolling & drawing, nec	0.5	Manufg.			
Pumps & compressors	0.5	Manufg.			
Real estate	0.5	Fin/R.E.			
U.S. Postal Service	0.5	Gov't			
Abrasive products	0.4	Manufg.			
Miscellaneous plastics products	0.4	Manufg.			
Nonferrous castings, nec	0.4	Manufg.			
Sawmills & planning mills, general	0.4	Manufg.			
Gas production & distribution (utilities)	0.4	Util.			
Motor freight transportation & warehousing	0.4	Util.			
Machine tools, metal cutting types	0.3	Manufg.			
Transformers	0.3	Manufg.			
Legal services	0.3	Services			
Management & consulting services & labs	0.3	Services			
Gaskets, packing & sealing devices	0.2	Manufg.			
Hardwood dimension & flooring mills	0.2	Manufg.			
Metal heat treating	0.2	Manufg.			
Primary nonferrous metals, nec	0.2	Manufg.			
Accounting, auditing & bookkeeping	0.2	Services			
Computer & data processing services	0.2	Services			
Detective & protective services	0.2	Services			
Equipment rental & leasing services	0.2	Services			
Hotels & lodging places	0.2	Services			
Copper rolling & drawing	0.1	Manufg.			
Lubricating oils & greases	0.1	Manufg.			
Manifold business forms	0.1	Manufg.			
Royalties	0.1	Fin/R.E.			
Automotive rental & leasing, without drivers	0.1	Services			
Automotive repair shops & services	0.1	Services			

Source: Benchmark Input-Output Accounts for the U.S. Economy, 1982, U.S. Department of Commerce, Washington, D.C., July 1991. Data, as reported in the source, are organized by the 1977 SIC structure in use in 1982 but have been matched, as closely as is possible, to the 1987 SIC structure used in this book.

OCCUPATIONS EMPLOYED BY SIC 355 - SPECIAL INDUSTRY MACHINERY

Occupation	% of Total 1994	Change to 2005	Occupation	% of Total 1994	Change to 2005
Machinists	8.3	-1.6	Bookkeeping, accounting, & auditing clerks	1.5	-26.2
Machine builders	4.3	18.1	Welding machine setters, operators	1.5	-11.4
Sales & related workers nec	4.3	-1.6	Lathe & turning machine tool operators	1.4	-21.3
Assemblers, fabricators, & hand workers nec	3.9	-1.6	Engineering technicians & technologists nec	1.4	-1.6
Blue collar worker supervisors	3.7	-9.0	Machine tool cutting operators, metal & plastic	1.3	-18.0
General managers & top executives	3.3	-6.6	Industrial production managers	1.3	-1.6
Mechanical engineers	3.1	8.3	Electrical & electronic equipment assemblers	1.3	8.3
Welders & cutters	2.6	-1.6	NC machine tool operators, metal & plastic	1.3	18.1
Secretaries, ex legal & medical	2.5	-10.4	Purchasing agents, ex trade & farm products	1.3	-1.6
Drafters	2.3	-23.4	General office clerks	1.3	-16.1
Machine assemblers	2.1	-1.6	Stock clerks	1.2	-20.0
Sheet metal workers & duct installers	2.0	-1.6	Industrial machinery mechanics	1.2	8.3
Machine tool cutting & forming etc. nec	1.7	-1.6	Managers & administrators nec	1.1	-1.6
Electrical & electronics engineers	1.7	4.8	Financial managers	1.1	-1.6
Inspectors, testers, & graders, precision	1.6	-1.6	Production, planning, & expediting clerks	1.0	-1.6
Engineering, mathematical, & science managers	1.6	11.8	Electrical & electronic assemblers	1.0	-1.6
Traffic, shipping, & receiving clerks	1.6	-5.3	Electrical & electronic technicians,technologists	1.0	-1.6

Source: *Industry-Occupation Matrix*, Bureau of Labor Statistics. These data relate to one or more 3-digit SIC industry groups rather than to a single 4-digit SIC. The change reported for each occupation to the year 2005 is a percent of growth or decline as estimated by the Bureau of Labor Statistics. The abbreviation nec stands for 'not elsewhere classified'.

LOCATION BY STATE AND REGIONAL CONCENTRATION

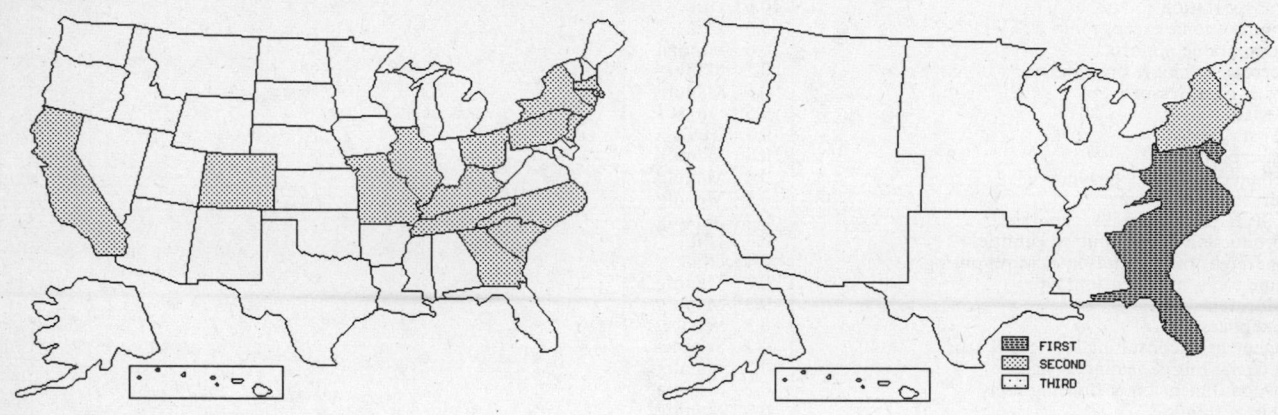

FIRST
SECOND
THIRD

INDUSTRY DATA BY STATE

State	Establish-ments	Shipments			Employment				Cost as % of Shipments	Investment per Employee ($)
		Total ($ mil)	% of U.S.	Per Establ.	Total Number	% of U.S.	Per Establ.	Wages ($/hour)		
North Carolina	124	517.5	33.1	4.2	4,400	29.3	35	10.67	41.7	3,477
South Carolina	68	371.6	23.7	5.5	3,900	26.0	57	10.90	41.4	2,154
Georgia	55	104.5	6.7	1.9	1,200	8.0	22	9.94	44.1	1,417
New York	33	93.1	5.9	2.8	800	5.3	24	11.73	43.3	2,125
Pennsylvania	23	47.0	3.0	2.0	500	3.3	22	11.67	42.1	3,200
Tennessee	14	45.8	2.9	3.3	400	2.7	29	11.83	46.5	2,500
New Jersey	25	43.7	2.8	1.7	500	3.3	20	13.50	39.4	600
Massachusetts	21	36.2	2.3	1.7	400	2.7	19	11.67	49.4	3,500
California	18	22.7	1.5	1.3	300	2.0	17	9.00	43.2	1,000
Illinois	8	18.6	1.2	2.3	200	1.3	25	15.50	35.5	-
Kentucky	7	13.1	0.8	1.9	200	1.3	29	8.67	24.4	1,000
Rhode Island	14	7.7	0.5	0.6	100	0.7	7	14.00	37.7	1,000
Connecticut	8	(D)	-	-	750 *	5.0	94	-	-	-
Ohio	8	(D)	-	-	175 *	1.2	22	-	-	-
Missouri	6	(D)	-	-	175 *	1.2	29	-	-	2,857
Colorado	2	(D)	-	-	175 *	1.2	88	-	-	-

Source: 1992 *Economic Census*. The states are in descending order of shipments or establishments (if shipment data are missing for the majority). The symbol (D) appears when data are withheld to prevent disclosure of competitive information. States marked with (D) are sorted by number of establishments. A dash (-) indicates that the data element cannot be calculated; * indicates the midpoint of a range.

3553 - WOODWORKING MACHINERY

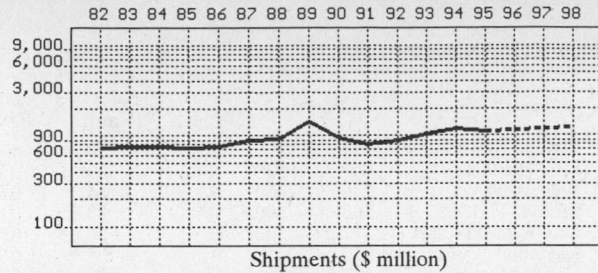

82 83 84 85 86 87 88 89 90 91 92 93 94 95 96 97 98

Shipments ($ million)

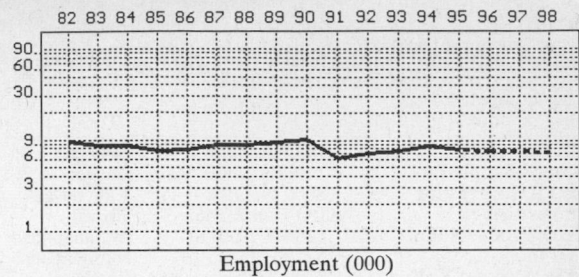

82 83 84 85 86 87 88 89 90 91 92 93 94 95 96 97 98

Employment (000)

GENERAL STATISTICS

Year	Companies	Establishments		Employment			Compensation		Production ($ million)			
		Total	with 20 or more employees	Total (000)	Production Workers (000)	Hours (Mil)	Payroll ($ mil)	Wages ($/hr)	Cost of Materials	Value Added by Manufacture	Value of Shipments	Capital Invest.
1982	267	279	96	9.7	5.9	10.9	173.7	8.38	332.2	387.4	730.7	20.2
1983		273	97	8.8	5.5	10.2	172.1	9.39	352.4	394.1	756.1	9.5
1984		267	98	8.6	5.5	10.7	164.5	8.44	371.6	405.2	753.3	12.9
1985		262	99	7.7	4.9	10.0	157.8	9.13	387.9	326.2	725.8	10.1
1986		261	100	7.9	5.0	10.5	172.9	9.41	379.2	380.8	760.5	8.9
1987	280	292	97	8.9	5.8	11.9	205.3	9.72	423.7	464.3	884.3	17.0
1988		284	102	8.9	5.8	11.3	208.2	10.16	469.7	451.5	921.1	12.5
1989		284	105	9.7	6.4	12.8	258.5	10.20	563.5	590.4	1,441.4	18.5
1990		284	106	10.1	5.4	10.6	226.2	11.25	439.8	477.4	936.6	14.2
1991		291	97	6.2	3.9	8.1	184.7	11.15	368.9	420.3	811.4	7.2
1992	277	289	90	7.2	4.7	9.5	206.1	11.56	439.5	458.8	894.6	19.5
1993		291	93	7.6	5.1	10.7	221.0	11.62	507.0	548.8	1,028.2	20.0
1994		294P	97P	8.6	5.4	12.0	264.2	12.43	522.4	675.3	1,184.3	30.7
1995		296P	97P	7.8P	5.0P	10.7P	247.1P	12.46P	500.0P	646.3P	1,133.5P	21.1P
1996		298P	97P	7.7P	4.9P	10.7P	253.7P	12.77P	514.1P	664.5P	1,165.4P	21.9P
1997		300P	97P	7.6P	4.8P	10.7P	260.2P	13.09P	528.2P	682.7P	1,197.3P	22.7P
1998		302P	97P	7.5P	4.8P	10.7P	266.8P	13.41P	542.2P	701.0P	1,229.3P	23.5P

Sources: 1982, 1987, 1992 *Economic Census*; *Annual Survey of Manufactures*, 83-86, 88-91, 93-94. Establishment counts for non-Census years are from *County Business Patterns*; establishment values for 83-84 are extrapolations. 'P's show projections by the editors. Industries reclassified in 87 will not have data for prior years.

INDICES OF CHANGE

Year	Companies	Establishments		Employment			Compensation		Production ($ million)			
		Total	with 20 or more employees	Total (000)	Production Workers (000)	Hours (Mil)	Payroll ($ mil)	Wages ($/hr)	Cost of Materials	Value Added by Manufacture	Value of Shipments	Capital Invest.
1982	96	97	107	135	126	115	84	72	76	84	82	104
1983		94	108	122	117	107	84	81	80	86	85	49
1984		92	109	119	117	113	80	73	85	88	84	66
1985		91	110	107	104	105	77	79	88	71	81	52
1986		90	111	110	106	111	84	81	86	83	85	46
1987	101	101	108	124	123	125	100	84	96	101	99	87
1988		98	113	124	123	119	101	88	107	98	103	64
1989		98	117	135	136	135	125	88	128	129	161	95
1990		98	118	140	115	112	110	97	100	104	105	73
1991		101	108	86	83	85	90	96	84	92	91	37
1992	100	100	100	100	100	100	100	100	100	100	100	100
1993		101	103	106	109	113	107	101	115	120	115	103
1994		102P	108P	119	115	126	128	108	119	147	132	157
1995		102P	108P	108P	105P	113P	120P	108P	114P	141P	127P	108P
1996		103P	108P	107P	104P	113P	123P	111P	117P	145P	130P	112P
1997		104P	107P	105P	103P	113P	126P	113P	120P	149P	134P	116P
1998		105P	107P	104P	102P	113P	129P	116P	123P	153P	137P	120P

Sources: Same as General Statistics. Values reflect change from the base year, 1992. Values above 100 mean greater than 92, values below 100 mean less than 92, and a value of 100 in the 82-91 or 93-98 period means same as 92. 'P's mark projections by the editors.

SELECTED RATIOS

For 1994	Avg. of All Manufact.	Analyzed Industry	Index	For 1994	Avg. of All Manufact.	Analyzed Industry	Index
Employees per Establishment	49	29	60	Value Added per Production Worker	134,084	125,056	93
Payroll per Establishment	1,500,273	899,474	60	Cost per Establishment	5,045,178	1,778,521	35
Payroll per Employee	30,620	30,721	100	Cost per Employee	102,970	60,744	59
Production Workers per Establishment	34	18	54	Cost per Production Worker	146,988	96,741	66
Wages per Establishment	853,319	507,818	60	Shipments per Establishment	9,576,895	4,031,972	42
Wages per Production Worker	24,861	27,622	111	Shipments per Employee	195,460	137,709	70
Hours per Production Worker	2,056	2,222	108	Shipments per Production Worker	279,017	219,315	79
Wages per Hour	12.09	12.43	103	Investment per Establishment	321,011	104,519	33
Value Added per Establishment	4,602,255	2,299,071	50	Investment per Employee	6,552	3,570	54
Value Added per Employee	93,930	78,523	84	Investment per Production Worker	9,352	5,685	61

Sources: Same as General Statistics. The 'Average of All Manufacturing' column represents the average of all manufacturing industries reported for the most recent complete year available. The Index shows the relationship between the Average and the Analyzed Industry. For example, 100 means that they are equal; 500 that the Analyzed Industry is five times the average; 50 means that the Analyzed Industry is half the national average. The abbreviation 'na' is used to show that data are 'not available'.

LEADING COMPANIES   Number shown: **57**   Total sales ($ mil): **928**   Total employment (000): **7.0**

Company Name	Address				CEO Name	Phone	Co. Type	Sales ($ mil)	Empl. (000)
Mitek Industries Inc	PO Box 7359	St Louis	MO	63177	Eugene M Toombs	314-434-1200	R	170	0.9
Woodworker's Supply Inc	1108 N Glenn Rd	Casper	WY	82601	John Wirth Jr	307-237-5528	R	130*	0.2
US Natural Resources Inc	8000 NE Pkway Dr	Vancouver	WA	98662	Richard H Ward	206-892-2650	R	84	0.8
Cole Manufacturing Co	PO Box 23366	Tigard	OR	97223	Fred W Fields	503-684-2600	R	80	0.8
Wood-Mizer Products Inc	8180 W 10th St	Indianapolis	IN	46214	Don R Laskowski	317-271-1542	R	50	0.5
Timesavers Inc	5270 Hanson Ct	Minneapolis	MN	55429	Raymond S Vold	612-537-3611	S	35	0.2
Pacific/Hoe, Saw and Knife Co	2700 SE Tacoma St	Portland	OR	97282	George Jacobson	503-234-9501	R	30	0.3
Bandit Industries Inc	6750 Millbrook Rd	Remus	MI	49340	Terry Morey	517-561-2270	R	22	0.1
Precision Husky Corp	PO Drawer 507	Leeds	AL	35094	Bob Smith	205-640-5181	R	22	0.2
Irvington-Moore West	PO Box 23038	Portland	OR	97223	Gary Hogue	503-620-0800	D	21	0.3
Nicholson Manufacturing	3670 Marginal	Seattle	WA	98134	TW Nicholson	206-682-2752	R	20	0.2
Schurman Machine	PO Box 310	Woodland	WA	98674	Gary Hogue	206-225-8267	D	19	0.2
Powermatic	607 Morrison St	McMinnville	TN	37110	George W Delaney	615-473-5551	D	18	0.2
Newman Machine Company Inc	PO Box 5467	Greensboro	NC	27435	Frank W York	910-273-8261	R	16	0.1
Mereen-Johnson Machine Co	4401 Lyndale Av N	Minneapolis	MN	55412	Russ McBroom	612-529-7791	R	11*	0.2
Industrial Woodworking Machine	PO Box 461466	Garland	TX	75046	Gile Cromeens	214-272-4521	R	10	0.1
Corley Manufacturing Co	PO Box 471	Chattanooga	TN	37401	HA Corley III	615-698-0284	R	10*	0.1
Mill Services and Manufacturing	PO Box 17077	Hattiesburg	MS	39404	Donald D Rake	601-268-2300	S	10	0.1
Salem Equipment Inc	PO Box 947	Salem	OR	97308	Lewis Judson	503-581-8411	R	10*	0.1
Viking Eng & Development	5750 NE Main	Fridley	MN	55432	William Hanneman	612-571-2400	R	10*	0.1
Pauli and Griffin Co	907 Cotting Ln	Vacaville	CA	95688	R Pauli	707-447-7000	R	8	<0.1
Wisconsin Automated Machinery	PO Box 30008	Oshkosh	WI	54903	Paul Ehrlich	414-231-4100	R	8	0.1
Voorwood Co	PO Box 1127	Anderson	CA	96007	Jerry W Voorhees	916-365-3315	R	8	<0.1
Hi-Tech Engineering Inc	PO Box 3591	Hot Springs	AR	71914	Russell Kennedy	501-760-1100	R	7*	<0.1
KVAL Inc	PO Drawer A	Petaluma	CA	94953	I Kvalheim	707-762-7367	R	7	<0.1
Kimwood Corp	PO Box 97	Cottage Grove	OR	97424	Dave Evans	503-942-4401	R	6	<0.1
Kohler-General Corp	100 Clark St	Sheboygan Fls	WI	53085	PG Kohler	414-467-4674	R	6*	<0.1
Norfield Industries	PO Drawer 688	Chico	CA	95927	B Norlie	916-891-4214	R	6	<0.1
Western Saw Manufacturing Inc	1000 Del Norte Blv	Oxnard	CA	93030	Kevin Baron	805-981-0999	R	6	<0.1
Premier Gear & Machine Works	1700 NW Thurman	Portland	OR	97209	Allen Cole	503-227-3514	R	6	<0.1
Biesemeyer Manufacturing Corp	216 S Alma School	Mesa	AZ	85210	William Biesemeyer	602-835-9300	R	5	<0.1
Evans Machinery Inc	PO Box 1406	Glendale	AZ	85311	R Perez	602-934-7294	R	5	<0.1
McDonough Manufacturing Co	PO Box 510	Eau Claire	WI	54702	BJ Wathke	715-834-7755	R	5	<0.1
Black Bros Co	PO Box 410	Mendota	IL	61342	James S Carroll	815-539-7451	R	4	<0.1
Hahn Machinery Inc	PO Box 220	Two Harbors	MN	55616	Gary Olsen	218-834-2156	R	4	<0.1
Mattison Woodworking Co	545 Blackhawk Pk	Rockford	IL	61101	PL Mattison	815-962-5521	D	4*	<0.1
Safety Speed Cut Mfg Co	13460 N Hwy 65	Anoka	MN	55304	John R Hammett	612-755-1600	R	4	<0.1
Southern California Machinery	15531 Arrow Hwy	Irwindale	CA	91706	Wayne D Anderson	818-960-5374	R	4	<0.1
Tannewitz Inc	794 Chicago Dr	Jenison	MI	49428	Morry Pysarchik	616-457-5999	R	4*	<0.1
Tyler Machinery Company Inc	621 S Detroit St	Warsaw	IN	46580	David Tyler	219-267-3530	R	4	<0.1
James L Taylor Mfg Co	PO Box 712	Poughkeepsie	NY	12602	Michael C Burdis	914-452-3780	R	4	<0.1
Mellott Manufacturing Company	13156 Long Ln	Mercersburg	PA	17236	HR Mellott	717-369-3125	R	4	<0.1
Soule' Steam Feed Works	PO Box 5757	Meridian	MS	39301	G Robert Soule	601-693-1982	R	4	<0.1
Midwest Automation Inc	3530 E 28th St	Minneapolis	MN	55406	Kenneth P Holley	612-721-5347	R	3	<0.1
BM Root Co	PO Box 1226	York	PA	17405	Patrick Lever	717-848-2356	S	3	<0.1
Cornell Manufacturing Inc	RD 2	Laceyville	PA	18623	W E Griffin Jr	717-869-1227	R	3	<0.1
Filer and Stowell Sales	147 E Becher St	Milwaukee	WI	53207	Charles Read	414-744-6170	D	3	<0.1
Northfield Foundry	PO Drawer 140	Northfield	MN	55057	JP Machacek	507-645-5641	R	3	<0.1
Accu-Router Inc	634 Mountainview	Morrison	TN	37357	Todd Herzog	615-668-7127	R	3	<0.1
Carter Products Company Inc	437 Spring St NE	Grand Rapids	MI	49503	Peter Perez	616-451-2928	R	2	<0.1
Holzma-US	1200 Tulip Dr	Gastonia	NC	28052	Bill Pitt	704-861-8239	D	2*	<0.1
Industrial Saws Inc	PO Box C-14118	Seattle	WA	98114	David Court	206-329-1050	R	2	<0.1
Lewis Controls Inc	PO Box 526	Cornelius	OR	97113	Richard Girouard	503-648-9119	S	2	<0.1
Jackson Lumber Harvester	Hwy 37 N	Mondovi	WI	54755	Thomas F Meis	715-926-3816	R	1	<0.1
Multiscore Inc	1143 NW 51st St	Seattle	WA	98107	Hugh Lade	206-784-7640	R	1	<0.1
Sprunger Corp	PO Box 1621	Elkhart	IN	46515	DD Fahlbeck	219-262-2476	R	1	<0.1
Cabinet Shop Machinery Inc	717 Fountain Av	Lancaster	PA	17601	Gerald Cohen	717-393-5831	R	0*	<0.1

Source: Ward's Business Directory of U.S. Private and Public Companies, Volumes 1 and 2, 1996. The company type code used is as follows: P - Public, R - Private, S - Subsidiary, D - Division, J - Joint Venture, A - Affiliate, G - Group. Sales are in millions of dollars, employees are in thousands. An asterisk (*) indicates an estimated sales volume. The symbol < stands for 'less than'. Company names and addresses are truncated, in some cases, to fit into the available space.

MATERIALS CONSUMED

Material	Quantity	Delivered Cost ($ million)
Materials, ingredients, containers, and supplies	(X)	334.0
Fluid power pumps, motors, and hydrostatic transmissions (hydraulic and pneumatic)	(X)	5.8
All other pumps	(X)	0.2
Fluid power valves (hydraulic and pneumatic)	(X)	3.1
Fluid power hose or tube fittings and assemblies (hydraulic and pneumatic)	(X)	1.4
Fluid power cylinders and rotary actuators (hydraulic and pneumatic)	(X)	4.5
Fluid power filters (hydraulic and pneumatic)	(X)	1.0
Other fluid power products (hydraulic and pneumatic)	(X)	1.4
Fabricated metal products, except forgings	(X)	14.6
Forgings	(X)	0.5
Iron and steel castings (rough and semifinished)	(X)	23.5

Continued on next page.

MATERIALS CONSUMED - Continued

Material	Quantity	Delivered Cost ($ million)
Aluminum and aluminum-base alloy castings (rough and semifinished)	(X)	6.9
Steel bars, bar shapes, and plates	(X)	33.5
Steel sheet and strip, including tin plate	(X)	9.5
All other steel shapes and forms	(X)	4.9
Nonferrous shapes and forms	(X)	2.3
Other fractional horsepower electric motors (under 1 hp)	(X)	14.2
Integral horsepower electric motors and generators (1 hp or more)	(X)	33.5
Ball bearings (mounted or unmounted)	(X)	7.2
Roller bearings (mounted or unmounted)	(X)	2.3
Fabricated plastics products (except gaskets, hoses, and belting)	(X)	4.9
Electronic computing equipment and parts	(X)	6.2
Electrical transmission, distribution, and control equipment	(X)	11.6
Numerical controls for woodworking machinery and equipment	(X)	2.2
Paperboard containers, boxes, and corrugated paperboard	(X)	5.9
All other materials and components, parts, containers, and supplies	(X)	49.7
Materials, ingredients, containers, and supplies, nsk	(X)	83.3

Source: 1992 *Economic Census*. Explanation of symbols used: (D): Withheld to avoid disclosure of competitive data; na: Not available; (S): Withheld because statistical norms were not met; (X): Not applicable; (Z): Less than half the unit shown; nec: Not elsewhere classified; nsk: Not specified by kind; - : zero; * : 10-19 percent estimated; ** : 20-29 percent estimated.

PRODUCT SHARE DETAILS

Product or Product Class	% Share	Product or Product Class	% Share
Woodworking machinery	100.00	Parts, attachments, and accessories for woodworking machinery (sold separately), excluding saw blades and cutting tools	25.07
Woodworking machinery, including parts, attachments and accessories	61.78	Woodworking machinery, including parts, attachments and accessories, nsk	1.06
Woodworking sawmill circular saws (head rigs)	3.70	Woodworking machinery for home workshops, garages, and service shops	23.42
Woodworking sawmill band saws (head rigs)	9.16		
Other woodworking sawmill equipment, except saws (head rigs)	15.80	Woodworking saws (including circular and band saws, excluding chain saws) for home workshops, garages, and service shops	69.27
Woodworking sawing machines, except sawmill equipment	9.33	Other woodworking machines and equipment for home workshops, garages, and service shops, except power driven handtools	19.05
Woodworking planing machinery, including single and double planers, facers, jointers, and abrasive planers	2.38		
Woodworking sanding machines	5.20	Parts, attachments, and accessories for woodworking machinery for home workshops, garages, and service shops (sold separately)	10.35
Woodworking boring machines	0.58		
Woodworking lathes or turning machines	0.73		
Woodworking routers	5.59	Woodworking machinery for home workshops, garages, service shops, nsk	1.32
Woodworking shapers and profilers	2.05	Woodworking machinery, nsk	14.79
Woodworking assembling, gluing, laminating, and finishing machines	4.82		
Multifunction woodworking machines	0.94		
Other woodworking machines and equipment, including moulders	13.55		

Source: 1992 *Economic Census*. The values shown are percent of total shipments in an industry. Values of indented subcategories are summed in the main heading. The symbol (D) appears when data are withheld to prevent disclosure of competitive information. The abbreviation nsk stands for 'not specified by kind' and nec for 'not elsewhere classified'.

INPUTS AND OUTPUTS FOR WOODWORKING MACHINERY

Economic Sector or Industry Providing Inputs	%	Sector	Economic Sector or Industry Buying Outputs	%	Sector
Internal combustion engines, nec	19.1	Manufg.	Gross private fixed investment	47.7	Cap Inv
Imports	13.1	Foreign	Personal consumption expenditures	29.1	
Wholesale trade	10.4	Trade	Exports	11.9	Foreign
Blast furnaces & steel mills	9.4	Manufg.	Sawmills & planning mills, general	1.4	Manufg.
Iron & steel foundries	4.0	Manufg.	Woodworking machinery	1.4	Manufg.
Machinery, except electrical, nec	3.3	Manufg.	Wood household furniture	1.3	Manufg.
Aluminum castings	2.8	Manufg.	S/L Govt. purch., elem. & secondary education	1.2	S/L Govt
Fabricated plate work (boiler shops)	2.5	Manufg.	Logging camps & logging contractors	1.0	Manufg.
Motors & generators	2.1	Manufg.	S/L Govt. purch., higher education	0.8	S/L Govt
Ball & roller bearings	2.0	Manufg.	Veneer & plywood	0.7	Manufg.
Advertising	2.0	Services	Wood products, nec	0.6	Manufg.
Woodworking machinery	1.7	Manufg.	Millwork	0.5	Manufg.
Nonferrous castings, nec	1.6	Manufg.	Mobile homes	0.5	Manufg.
Paperboard containers & boxes	1.6	Manufg.	Wood kitchen cabinets	0.4	Manufg.
Miscellaneous plastics products	1.4	Manufg.	Hardwood dimension & flooring mills	0.3	Manufg.
Switchgear & switchboard apparatus	1.2	Manufg.	Wood pallets & skids	0.3	Manufg.
Electric services (utilities)	1.2	Util.	Wood containers	0.2	Manufg.
Industrial controls	1.1	Manufg.	Federal Government purchases, national defense	0.2	Fed Govt
Screw machine and related products	1.1	Manufg.	Prefabricated wood buildings	0.1	Manufg.
Special dies & tools & machine tool accessories	1.1	Manufg.	Structural wood members, nec	0.1	Manufg.
Communications, except radio & TV	1.0	Util.	Wood preserving	0.1	Manufg.
Banking	1.0	Fin/R.E.			

Continued on next page.

INPUTS AND OUTPUTS FOR WOODWORKING MACHINERY - Continued

Economic Sector or Industry Providing Inputs	%	Sector	Economic Sector or Industry Buying Outputs	%	Sector
Pipe, valves, & pipe fittings	0.8	Manufg.			
Air transportation	0.8	Util.			
Engineering, architectural, & surveying services	0.8	Services			
Maintenance of nonfarm buildings nec	0.7	Constr.			
Sawmills & planning mills, general	0.7	Manufg.			
Eating & drinking places	0.7	Trade			
Real estate	0.7	Fin/R.E.			
Petroleum refining	0.6	Manufg.			
Motor freight transportation & warehousing	0.6	Util.			
Equipment rental & leasing services	0.6	Services			
Fabricated rubber products, nec	0.5	Manufg.			
Pumps & compressors	0.5	Manufg.			
Gas production & distribution (utilities)	0.5	Util.			
Abrasive products	0.4	Manufg.			
Rubber & plastics hose & belting	0.4	Manufg.			
U.S. Postal Service	0.4	Gov't			
Gaskets, packing & sealing devices	0.3	Manufg.			
General industrial machinery, nec	0.3	Manufg.			
Machine tools, metal cutting types	0.3	Manufg.			
Transformers	0.3	Manufg.			
Computer & data processing services	0.3	Services			
Noncomparable imports	0.3	Foreign			
Aluminum rolling & drawing	0.2	Manufg.			
Metal heat treating	0.2	Manufg.			
Railroads & related services	0.2	Util.			
Sanitary services, steam supply, irrigation	0.2	Util.			
Royalties	0.2	Fin/R.E.			
Accounting, auditing & bookkeeping	0.2	Services			
Hotels & lodging places	0.2	Services			
Legal services	0.2	Services			
Management & consulting services & labs	0.2	Services			
Lubricating oils & greases	0.1	Manufg.			
Manifold business forms	0.1	Manufg.			
Nonferrous rolling & drawing, nec	0.1	Manufg.			
Automotive rental & leasing, without drivers	0.1	Services			

Source: *Benchmark Input-Output Accounts for the U.S. Economy, 1982*, U.S. Department of Commerce, Washington, D.C., July 1991. Data, as reported in the source, are organized by the 1977 SIC structure in use in 1982 but have been matched, as closely as is possible, to the 1987 SIC structure used in this book.

OCCUPATIONS EMPLOYED BY SIC 355 - SPECIAL INDUSTRY MACHINERY

Occupation	% of Total 1994	Change to 2005	Occupation	% of Total 1994	Change to 2005
Machinists	8.3	-1.6	Bookkeeping, accounting, & auditing clerks	1.5	-26.2
Machine builders	4.3	18.1	Welding machine setters, operators	1.5	-11.4
Sales & related workers nec	4.3	-1.6	Lathe & turning machine tool operators	1.4	-21.3
Assemblers, fabricators, & hand workers nec	3.9	-1.6	Engineering technicians & technologists nec	1.4	-1.6
Blue collar worker supervisors	3.7	-9.0	Machine tool cutting operators, metal & plastic	1.3	-18.0
General managers & top executives	3.3	-6.6	Industrial production managers	1.3	-1.6
Mechanical engineers	3.1	8.3	Electrical & electronic equipment assemblers	1.3	8.3
Welders & cutters	2.6	-1.6	NC machine tool operators, metal & plastic	1.3	18.1
Secretaries, ex legal & medical	2.5	-10.4	Purchasing agents, ex trade & farm products	1.3	-1.6
Drafters	2.3	-23.4	General office clerks	1.3	-16.1
Machine assemblers	2.1	-1.6	Stock clerks	1.2	-20.0
Sheet metal workers & duct installers	2.0	-1.6	Industrial machinery mechanics	1.2	8.3
Machine tool cutting & forming etc. nec	1.7	-1.6	Managers & administrators nec	1.1	-1.6
Electrical & electronics engineers	1.7	4.8	Financial managers	1.1	-1.6
Inspectors, testers, & graders, precision	1.6	-1.6	Production, planning, & expediting clerks	1.0	-1.6
Engineering, mathematical, & science managers	1.6	11.8	Electrical & electronic assemblers	1.0	-1.6
Traffic, shipping, & receiving clerks	1.6	-5.3	Electrical & electronic technicians, technologists	1.0	-1.6

Source: *Industry-Occupation Matrix*, Bureau of Labor Statistics. These data relate to one or more 3-digit SIC industry groups rather than to a single 4-digit SIC. The change reported for each occupation to the year 2005 is a percent of growth or decline as estimated by the Bureau of Labor Statistics. The abbreviation nec stands for 'not elsewhere classified'.

LOCATION BY STATE AND REGIONAL CONCENTRATION

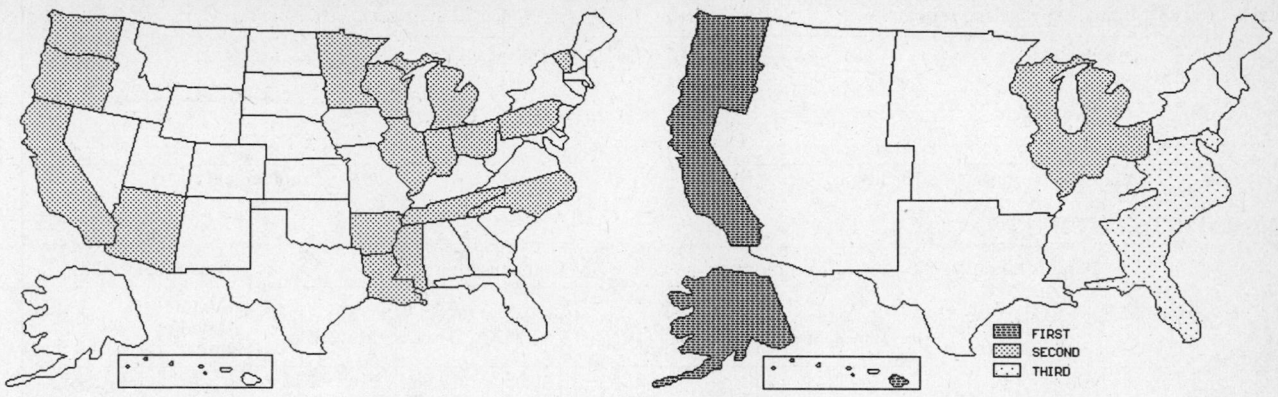

INDUSTRY DATA BY STATE

| State | Establish-ments | Shipments | | | Employment | | | | Cost as % of Shipments | Investment per Employee ($) |
		Total ($ mil)	% of U.S.	Per Establ.	Total Number	% of U.S.	Per Establ.	Wages ($/hour)		
Tennessee	6	129.0	14.4	21.5	700	9.7	117	10.33	70.5	3,571
Oregon	33	119.0	13.3	3.6	1,100	15.3	33	13.31	48.7	2,636
North Carolina	21	80.8	9.0	3.8	500	6.9	24	10.57	51.5	1,000
Mississippi	8	78.0	8.7	9.8	500	6.9	63	12.25	46.2	-
Indiana	11	54.3	6.1	4.9	500	6.9	45	12.80	35.9	3,000
California	26	53.0	5.9	2.0	400	5.6	15	11.14	46.6	2,750
Minnesota	5	28.8	3.2	5.8	300	4.2	60	14.00	49.7	333
Arkansas	5	21.9	2.4	4.4	200	2.8	40	10.67	45.2	3,000
Illinois	13	21.5	2.4	1.7	300	4.2	23	9.00	34.9	-
Pennsylvania	10	13.8	1.5	1.4	200	2.8	20	12.50	29.0	500
Louisiana	4	10.8	1.2	2.7	100	1.4	25	8.50	44.4	3,000
Arizona	3	6.8	0.8	2.3	100	1.4	33	13.00	42.6	1,000
Michigan	16	(D)	-	-	175 *	2.4	11	-	-	3,429
Wisconsin	15	(D)	-	-	375 *	5.2	25	-	-	-
Washington	13	(D)	-	-	375 *	5.2	29	-	-	800
Ohio	7	(D)	-	-	375 *	5.2	54	-	-	-
Vermont	3	(D)	-	-	175 *	2.4	58	-	-	-

Source: 1992 *Economic Census*. The states are in descending order of shipments or establishments (if shipment data are missing for the majority). The symbol (D) appears when data are withheld to prevent disclosure of competitive information. States marked with (D) are sorted by number of establishments. A dash (-) indicates that the data element cannot be calculated; * indicates the midpoint of a range.

3554 - PAPER INDUSTRIES MACHINERY

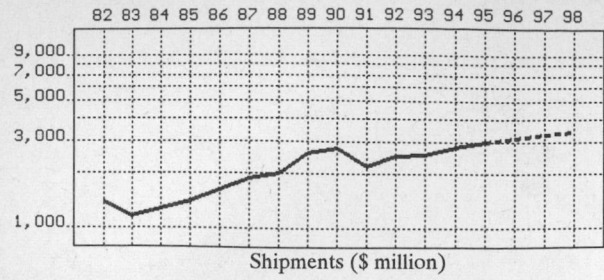

Shipments ($ million)

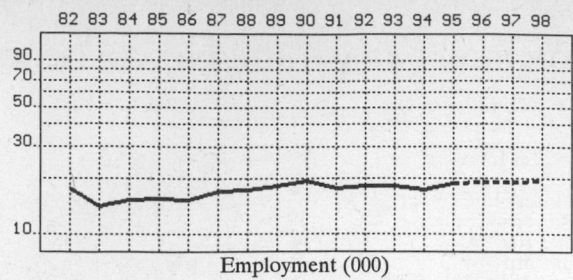

Employment (000)

GENERAL STATISTICS

Year	Com-panies	Establishments		Employment			Compensation		Production ($ million)			
		Total	with 20 or more employees	Total (000)	Production Workers (000)	Hours (Mil)	Payroll ($ mil)	Wages ($/hr)	Cost of Materials	Value Added by Manufacture	Value of Shipments	Capital Invest.
1982	230	253	126	17.8	10.0	20.2	405.8	9.98	622.5	762.6	1,387.4	29.4
1983		245	126	14.1	7.1	13.7	352.0	12.18	497.5	596.4	1,150.8	15.7
1984		237	126	15.4	8.5	16.9	392.8	11.29	566.2	731.7	1,274.3	22.6
1985		229	125	15.6	8.9	17.6	442.6	12.58	654.3	792.9	1,408.2	40.6
1986		230	123	15.4	8.7	17.4	448.9	12.41	860.6	706.0	1,630.7	37.8
1987	256	278	127	17.1	9.6	19.1	502.2	12.51	868.8	1,027.0	1,867.1	45.6
1988		272	132	17.3	10.4	22.2	538.2	12.68	1,064.6	1,022.5	2,012.1	60.0
1989		282	134	18.2	11.7	24.7	624.8	13.74	1,410.1	1,208.8	2,579.8	84.6
1990		285	134	19.5	12.2	26.3	648.3	13.19	1,591.2	1,118.7	2,770.4	72.0
1991		292	141	17.8	9.6	20.8	606.6	14.83	1,200.0	987.8	2,206.3	62.6
1992	298	333	153	18.2	10.2	21.3	651.8	14.76	1,209.8	1,273.8	2,524.2	65.4
1993		332	140	18.1	9.8	19.6	667.2	15.24	1,330.5	1,206.3	2,528.9	55.9
1994		329P	146P	17.4	9.7	20.2	691.3	16.11	1,413.0	1,407.5	2,812.5	49.3
1995		337P	148P	18.8P	10.9P	23.1P	741.7P	16.10P	1,509.8P	1,503.9P	3,005.1P	74.9P
1996		346P	150P	19.0P	11.0P	23.5P	771.0P	16.51P	1,581.1P	1,574.9P	3,147.0P	78.6P
1997		355P	152P	19.3P	11.2P	24.0P	800.3P	16.93P	1,652.3P	1,645.9P	3,288.9P	82.2P
1998		363P	154P	19.5P	11.4P	24.4P	829.7P	17.34P	1,723.6P	1,716.9P	3,430.8P	85.9P

Sources: 1982, 1987, 1992 *Economic Census*; *Annual Survey of Manufactures*, 83-86, 88-91, 93-94. Establishment counts for non-Census years are from *County Business Patterns*; establishment values for 83-84 are extrapolations. 'P's show projections by the editors. Industries reclassified in 87 will not have data for prior years.

INDICES OF CHANGE

Year	Com-panies	Establishments		Employment			Compensation		Production ($ million)			
		Total	with 20 or more employees	Total (000)	Production Workers (000)	Hours (Mil)	Payroll ($ mil)	Wages ($/hr)	Cost of Materials	Value Added by Manufacture	Value of Shipments	Capital Invest.
1982	77	76	82	98	98	95	62	68	51	60	55	45
1983		74	82	77	70	64	54	83	41	47	46	24
1984		71	82	85	83	79	60	76	47	57	50	35
1985		69	82	86	87	83	68	85	54	62	56	62
1986		69	80	85	85	82	69	84	71	55	65	58
1987	86	83	83	94	94	90	77	85	72	81	74	70
1988		82	86	95	102	104	83	86	88	80	80	92
1989		85	88	100	115	116	96	93	117	95	102	129
1990		86	88	107	120	123	99	89	132	88	110	110
1991		88	92	98	94	98	93	100	99	78	87	96
1992	100	100	100	100	100	100	100	100	100	100	100	100
1993		100	92	99	96	92	102	103	110	95	100	85
1994		99P	95P	96	95	95	106	109	117	110	111	75
1995		101P	96P	103P	107P	108P	114P	109P	125P	118P	119P	115P
1996		104P	98P	105P	108P	110P	118P	112P	131P	124P	125P	120P
1997		106P	99P	106P	110P	112P	123P	115P	137P	129P	130P	126P
1998		109P	100P	107P	111P	115P	127P	117P	142P	135P	136P	131P

Sources: Same as General Statistics. Values reflect change from the base year, 1992. Values above 100 mean greater than 92, values below 100 mean less than 92, and a value of 100 in the 82-91 or 93-98 period means same as 92. 'P's mark projections by the editors.

SELECTED RATIOS

For 1994	Avg. of All Manufact.	Analyzed Industry	Index	For 1994	Avg. of All Manufact.	Analyzed Industry	Index
Employees per Establishment	49	53	108	Value Added per Production Worker	134,084	145,103	108
Payroll per Establishment	1,500,273	2,103,735	140	Cost per Establishment	5,045,178	4,299,982	85
Payroll per Employee	30,620	39,730	130	Cost per Employee	102,970	81,207	79
Production Workers per Establishment	34	30	86	Cost per Production Worker	146,988	145,670	99
Wages per Establishment	853,319	990,310	116	Shipments per Establishment	9,576,895	8,558,880	89
Wages per Production Worker	24,861	33,549	135	Shipments per Employee	195,460	161,638	83
Hours per Production Worker	2,056	2,082	101	Shipments per Production Worker	279,017	289,948	104
Wages per Hour	12.09	16.11	133	Investment per Establishment	321,011	150,028	47
Value Added per Establishment	4,602,255	4,283,244	93	Investment per Employee	6,552	2,833	43
Value Added per Employee	93,930	80,891	86	Investment per Production Worker	9,352	5,082	54

Sources: Same as General Statistics. The 'Average of All Manufacturing' column represents the average of all manufacturing industries reported for the most recent complete year available. The Index shows the relationship between the Average and the Analyzed Industry. For example, 100 means that they are equal; 500 that the Analyzed Industry is five times the average; 50 means that the Analyzed Industry is half the national average. The abbreviation 'na' is used to show that data are 'not available'.

LEADING COMPANIES   Number shown: **62**   Total sales ($ mil): **3,300**   Total employment (000): **27.6**

Company Name	Address				CEO Name	Phone	Co. Type	Sales ($ mil)	Empl. (000)
Harnischfeger Industries Inc	PO Box 554	Milwaukee	WI	53201	Jeffery T Grade	414-671-4400	P	1,117	11.2
Beloit Corp	1 St Lawrence Av	Beloit	WI	53511	John A McKay	608-365-3311	S	713	6.5
Eimco Process Equipment Co	PO Box 300	Salt Lake City	UT	84110	Tim Probert	801-526-2000	D	200	0.7
Paper Converting Machine Co	PO Box 19005	Green Bay	WI	54307	Rick Baer	414-494-5601	R	120*	1.0
Sulzer Papertec Middletown Inc	PO Box 509	Middletown	OH	45042	O L Heissenberger	513-423-9281	S	100	0.5
Langston Corp	111 Woodcrest Rd	Cherry Hill	NJ	08034	L P Maynes	609-795-7100	S	75	0.4
Enterprises International Inc	PO Box 293	Hoquiam	WA	98550	Isabelle S Lamb	206-533-6222	R	64	0.2
Sunds Defibrator Inc	2900 Courtyards Dr	Norcross	GA	30071	Aron Mikaelsson	404-263-7863	S	60	0.2
Jones	401 South St	Dalton	MA	01226	LK Swift	413-443-5621	D	50	0.4
Lamb-Grays Harbor Co	PO Box 359	Hoquiam	WA	98550	Jack B Sparks	360-532-1000	R	50	0.4
Impco	150 Burke St	Nashua	NH	03061	Larry H Pitsch	603-882-2711	D	48*	0.5
Rader Cos	PO Box 20128	Portland	OR	97220	Garvin Pitney	503-255-5330	D	48	0.2
Valmet-Appleton Inc	2111 N Sandra St	Appleton	WI	54915	Bob Lemke	414-733-7361	S	46*	0.4
Voith Sulzer Paper	2620 E Glendale	Appleton	WI	54911	Werner Kade	414-731-7724	S	46	0.6
Sandusky International Inc	PO Box 5012	Sandusky	OH	44871	CW Rainger	419-626-5340	R	45	0.2
Beloit Lenox	PO Box 846	Lenox	MA	01240	Richard Stile	413-637-2424	D	35*	0.3
Ward Machinery Co	10615 Beaver Dam	Cockeysville	MD	21030	T M Scanlan Jr	410-584-7700	R	33	0.4
Entwistle Co	Bigelow St	Hudson	MA	01749	Herbert I Corkin	508-481-4000	R	31	0.4
CG Bretting Manufacturing	PO Box 113	Ashland	WI	54806	HL Bretting	715-682-5231	R	30	0.4
Shartle	605 Clark St	Middletown	OH	45042	WJ Fondow	513-424-7400	D	28*	0.2
AES Engineered Systems	PO Box 7010	Queensbury	NY	12804	Jan-Eric Bergstedt	518-793-8801	D	25	0.1
BTG Inc	2364 Pk Central	Decatur	GA	30035	Randy Linville	404-981-3998	S	21	0.1
Langston Staley Corp	11110 Pepper Rd	Hunt Valley	MD	21031	Leo P Maynes	410-785-1550	S	21*	0.1
FMP/Rauma Co	104 Inverness Ctr Pl	Birmingham	AL	35242	GL Albertson	205-995-0190	R	20	0.1
Geo M Martin Co	1250 67th St	Emeryville	CA	94608	Merrill D Martin	510-652-2200	R	17*	0.1
International Paper Box Machine	PO Box 787	Nashua	NH	03061	LC Chagnon	603-889-6651	R	17	0.2
Baum Folder Corp	1660 Campbell Rd	Sidney	OH	45365	Michael Grauel	513-492-1281	R	16	0.1
Agnati America Inc	371 Gees Mill	Conyers	GA	30208	Edward Bollinger	404-922-5870	S	15	<0.1
Green Bay Packaging Inc	PO Box 398	Birmingham	AL	35201	Richard Buchanan	205-324-7511	D	15	<0.1
Black Clawson Co	103 Pearl St	Watertown	NY	13601	Peter Tecori	315-788-2000	D	13	<0.1
Butler Automatic Inc	480 Neponset St	Canton	MA	02021	Richard A Butler Jr	617-828-5450	R	13	0.1
Sherwood Tool Inc	PO Box 7118	Kensington	CT	06037	Richard F Varano	203-828-4161	S	13*	0.1
Faustel Inc	PO Box 1000	Germantown	WI	53022	Lee J Hartenstein	414-253-3333	R	12	0.1
Baumfolder Corp	1660 Campbell Rd	Sidney	OH	45365	Michael Grauel	513-492-1281	R	11*	0.1
Meridian Machine Works Inc	PO Box 5393	Meridian	MS	39302	Harold W Johnson	601-483-9283	S	11	0.1
CA Lawton Co	PO Box 330	De Pere	WI	54115	RW Lawton	414-337-2460	R	10	0.1
SWF Machinery	PO Box 337	Sanger	CA	93657	Thomas Berardino	209-875-2545	R	10	<0.1
Herman Schwabe Inc	147 Prince St	Brooklyn	NY	11201	Jerrold M Schwabe	718-237-1700	R	9	<0.1
Double E Company Inc	PO Box 574	W Bridgewater	MA	02379	RE Flagg	508-588-8099	R	7	<0.1
Kempsmith Machine Co	PO Box 14336	West Allis	WI	53214	RE Burris	414-256-8160	R	7	<0.1
Renard Machine	1367 Reber St	Green Bay	WI	54302	Gary Steiner	414-432-8412	D	7	<0.1
Standard Paper Box Machine Co	347 Coster St	Bronx	NY	10474	Bruce Adams	718-328-3300	R	7	0.1
Montague Machine Company	PO Box 777	Turners Falls	MA	01376	Sylvester J Pierce	413-863-4301	R	6*	<0.1
Thomson National Press Co	115 Dean Av	Franklin	MA	02038	AF St Andre	508-528-2000	R	6	<0.1
SCM Container Machinery Inc	PO Box 405	Agawam	MA	01001	Peter Horton	413-786-3366	S	5	<0.1
Holyoke Machine Co	PO Box 988	Holyoke	MA	01041	I Sagalyn	413-534-5612	R	5	<0.1
Bar-Plate Manufacturing Co	506 Boston Post Rd	Orange	CT	06477	Dennis J Garrity	203-799-2345	R	4	<0.1
Hycorr Machine Corp	5801 E Kilgore	Kalamazoo	MI	49001	Fred Harrison	616-381-5905	R	4	<0.1
MAN Roland Inc	PO Box 578	Salem	IL	62881	RW Sherman	618-548-2600	D	4*	<0.1
Simon-LG Industries Inc	PO Box 329	Wagontown	PA	19376	Ronald L Constein	610-384-3100	S	4*	<0.1
Associated Pacific Machine Corp	724 Via Alondra	Camarillo	CA	93012	Robert Wax	213-749-6401	R	3	<0.1
Bolton-Emerson Americas Inc	9 Osgood St	Lawrence	MA	01842	Sandra Krug	508-686-3961	R	3*	<0.1
JH Horne and Sons Inc	PO Box 128	Lawrence	MA	01843	Byron Cleveland Jr	508-683-2463	R	3*	<0.1
MBO America	400 Highland Dr	Mount Holly	NJ	08060	Hans Max	609-267-2900	S	3	<0.1
Rice Barton Corp	PO Box 15006	Worcester	MA	01615	RL Couture	508-752-2821	R	3	<0.1
Zink Manufacturing Co	PO Box 550	Fulton	NY	13069	Stanley C Zink	315-592-4254	R	3	<0.1
Paco Winders Manufacturing Inc	2040 Bennett Rd	Philadelphia	PA	19116	SR Kiss	215-673-6265	R	3	<0.1
Allegheny Paper Shredders Corp	PO Box 80	Delmont	PA	15626	John Wagner	412-468-4300	R	2	<0.1
Wallace Co	PO Box 91746	Pasadena	CA	91109	R Little Jr	818-564-8777	R	2	<0.1
B and R Machine Inc	PO Box 2003	Camden	AR	71701	Robert W Barnwell	501-231-4307	R	1	<0.1
Schaefer-Ross Corp	82 Main St E	Webster	NY	14580	Brendan Hanna	716-265-9800	R	1	<0.1
Donahue and Associates Intern	2002 Ford Cir	Milford	OH	45150	John L Donahue	513-831-2770	R	1*	<0.1

Source: Ward's Business Directory of U.S. Private and Public Companies, Volumes 1 and 2, 1996. The company type code used is as follows: P - Public, R - Private, S - Subsidiary, D - Division, J - Joint Venture, A - Affiliate, G - Group. Sales are in millions of dollars, employees are in thousands. An asterisk (*) indicates an estimated sales volume. The symbol < stands for 'less than'. Company names and addresses are truncated, in some cases, to fit into the available space.

MATERIALS CONSUMED

Material	Quantity	Delivered Cost ($ million)
Materials, ingredients, containers, and supplies	(X)	968.1
Fluid power pumps, motors, and hydrostatic transmissions (hydraulic and pneumatic)	(X)	8.0
All other pumps	(X)	0.4
Fluid power valves (hydraulic and pneumatic)	(X)	9.5
Fluid power hose or tube fittings and assemblies (hydraulic and pneumatic)	(X)	7.6
Fluid power cylinders and rotary actuators (hydraulic and pneumatic)	(X)	9.6
Fluid power filters (hydraulic and pneumatic)	(X)	0.5
Other fluid power products (hydraulic and pneumatic)	(X)	4.6
Fabricated metal products, except forgings	(X)	46.9
Forgings	(X)	7.3
Iron and steel castings (rough and semifinished)	(X)	35.2
Nonferrous (aluminum, copper, etc.) castings (rough and semifinished)	(X)	15.3
Steel bars, bar shapes, and plates	(X)	65.8
Steel sheet and strip, including tin plate	(X)	19.1
All other steel shapes and forms	(X)	13.0
Nonferrous shapes and forms	(X)	7.9
Integral horsepower electric motors and generators (1 hp or more)	(X)	14.2
Ball bearings (mounted or unmounted)	(X)	8.2
Roller bearings (mounted or unmounted)	(X)	7.8
Plain bearings and bushings	(X)	7.0
Mechanical speed changers, gears, and industrial high-speed drives	(X)	28.1
Electrical transmission, distribution, and control equipment	(X)	53.3
Fabricated rubber products, except tires, tubes, hose, belting, and gaskets	(X)	9.9
All other materials and components, parts, containers, and supplies	(X)	478.8
Materials, ingredients, containers, and supplies, nsk	(X)	110.0

Source: 1992 *Economic Census*. Explanation of symbols used: (D): Withheld to avoid disclosure of competitive data; na: Not available; (S): Withheld because statistical norms were not met; (X): Not applicable; (Z): Less than half the unit shown; nec: Not elsewhere classified; nsk: Not specified by kind; - : zero; * : 10-19 percent estimated; ** : 20-29 percent estimated.

PRODUCT SHARE DETAILS

Product or Product Class	% Share	Product or Product Class	% Share
Paper industries machinery	100.00	envelopes	(D)
Paper industries machinery	61.68	Paper and paperboard corrugated box making machines	12.95
Paper industries wood preparation equipment, including debarkers, chippers, knotters, splitters, chipscreens, etc.	3.06	Machines for making paper and paperboard cartons, boxes, cases, tubes, drums, or similar containers, except by molding	(D)
Paper industries pulp mill grinders and refiners (TMP) for the manufacture of mechanical pulp	(D)	Machines for molding articles in paper pulp, paper, or paperboard	1.00
Paper industries pulp mill digesters and other equipment for the manufacture of chemical pulp	(D)	Other paper and paperboard converting equipment (folding, gluing, laminating, gumming, scoring, sandpapering, etc.)	12.52
Paper industries pulp mill deckers, thickeners, wet lap machines, bleaching equipment, pulp screens, washers, and save-alls	5.28	Paper industries machinery, nsk	2.45
Other paper industries pulp mill machinery	2.50	Parts and attachments for paper industries machinery (sold separately)	30.98
Paper mill stock preparation equipment, including refiners (chip, conical, deflaker, disk, etc.), pulpers, beaters, jordans, etc.	5.15	Parts and attachments for paper industries wood preparation equipment (sold separately)	2.30
Paper mill paper making machines, including headbox forming area, presses, dryers, and reels	16.77	Parts and attachments for paper industries pulp mill machinery (sold separately)	10.08
Paper mill paper coating machines, including equipment for applying sizing or pigment coating to paper	2.50	Parts and attachments for paper industry machines for finishing paper	10.15
Paper mill paper calendering and similar rolling machines for finishing paper	2.16	Parts and attachments for paper mill machinery, except machines for finishing paper	40.34
Paper mill machines for finishing paper, except calendering or similar rolling	9.13	Parts and attachments for paper and paperboard converting equipment	34.91
Other paper mill paper machines	8.21	Parts and attachments for paper industries machinery (sold separately), nsk	2.24
Paper and paperboard cutting machines, except sheeters and winders	2.37	Paper industries machinery, nsk	7.34
Machines for making paper and paperboard bags, sacks, or			

Source: 1992 *Economic Census*. The values shown are percent of total shipments in an industry. Values of indented subcategories are summed in the main heading. The symbol (D) appears when data are withheld to prevent disclosure of competitive information. The abbreviation nsk stands for 'not specified by kind' and nec for 'not elsewhere classified'.

INPUTS AND OUTPUTS FOR PAPER INDUSTRIES MACHINERY

Economic Sector or Industry Providing Inputs	%	Sector	Economic Sector or Industry Buying Outputs	%	Sector
Imports	25.8	Foreign	Gross private fixed investment	66.2	Cap Inv
Blast furnaces & steel mills	12.4	Manufg.	Exports	19.7	Foreign
Wholesale trade	7.8	Trade	Paper industries machinery	3.3	Manufg.
Iron & steel foundries	5.9	Manufg.	Paperboard containers & boxes	2.6	Manufg.
Paper industries machinery	5.8	Manufg.	Paper mills, except building paper	2.3	Manufg.
Machinery, except electrical, nec	3.6	Manufg.	Paperboard mills	1.2	Manufg.
Copper rolling & drawing	3.4	Manufg.	Sanitary paper products	1.0	Manufg.
Power transmission equipment	2.6	Manufg.	Bags, except textile	0.8	Manufg.
Motors & generators	2.2	Manufg.	Paper coating & glazing	0.8	Manufg.
Advertising	2.1	Services	Pulp mills	0.4	Manufg.
Ball & roller bearings	2.0	Manufg.	Converted paper products, nec	0.3	Manufg.
Industrial controls	1.6	Manufg.	Die-cut paper & board	0.2	Manufg.
Switchgear & switchboard apparatus	1.6	Manufg.	Envelopes	0.2	Manufg.
Electric services (utilities)	1.6	Util.	Retail trade, except eating & drinking	0.2	Trade
Special dies & tools & machine tool accessories	1.4	Manufg.	Wholesale trade	0.2	Trade
Banking	1.3	Fin/R.E.	Stationery products	0.1	Manufg.
Maintenance of nonfarm buildings nec	1.0	Constr.			
Communications, except radio & TV	0.9	Util.			
Eating & drinking places	0.9	Trade			
Engineering, architectural, & surveying services	0.9	Services			
Fabricated rubber products, nec	0.7	Manufg.			
Pipe, valves, & pipe fittings	0.7	Manufg.			
Air transportation	0.7	Util.			
Motor freight transportation & warehousing	0.7	Util.			
Machine tools, metal cutting types	0.6	Manufg.			
Miscellaneous fabricated wire products	0.6	Manufg.			
Petroleum refining	0.6	Manufg.			
Scrap	0.6	Scrap			
Abrasive products	0.5	Manufg.			
Pumps & compressors	0.5	Manufg.			
Transformers	0.5	Manufg.			
Gas production & distribution (utilities)	0.5	Util.			
Equipment rental & leasing services	0.5	Services			
U.S. Postal Service	0.5	Gov't			
Brass, bronze, & copper castings	0.4	Manufg.			
Gaskets, packing & sealing devices	0.4	Manufg.			
Miscellaneous plastics products	0.4	Manufg.			
Rubber & plastics hose & belting	0.4	Manufg.			
Computer & data processing services	0.4	Services			
Aluminum castings	0.3	Manufg.			
Metal heat treating	0.3	Manufg.			
Sawmills & planning mills, general	0.3	Manufg.			
Screw machine and related products	0.3	Manufg.			
Hotels & lodging places	0.3	Services			
Legal services	0.3	Services			
Management & consulting services & labs	0.3	Services			
Noncomparable imports	0.3	Foreign			
Railroads & related services	0.2	Util.			
Real estate	0.2	Fin/R.E.			
Royalties	0.2	Fin/R.E.			
Accounting, auditing & bookkeeping	0.2	Services			
Lubricating oils & greases	0.1	Manufg.			
Manifold business forms	0.1	Manufg.			
Sanitary services, steam supply, irrigation	0.1	Util.			
Insurance carriers	0.1	Fin/R.E.			
Automotive rental & leasing, without drivers	0.1	Services			

Source: Benchmark Input-Output Accounts for the U.S. Economy, 1982, U.S. Department of Commerce, Washington, D.C., July 1991. Data, as reported in the source, are organized by the 1977 SIC structure in use in 1982 but have been matched, as closely as is possible, to the 1987 SIC structure used in this book.

OCCUPATIONS EMPLOYED BY SIC 355 - SPECIAL INDUSTRY MACHINERY

Occupation	% of Total 1994	Change to 2005	Occupation	% of Total 1994	Change to 2005
Machinists	8.3	-1.6	Bookkeeping, accounting, & auditing clerks	1.5	-26.2
Machine builders	4.3	18.1	Welding machine setters, operators	1.5	-11.4
Sales & related workers nec	4.3	-1.6	Lathe & turning machine tool operators	1.4	-21.3
Assemblers, fabricators, & hand workers nec	3.9	-1.6	Engineering technicians & technologists nec	1.4	-1.6
Blue collar worker supervisors	3.7	-9.0	Machine tool cutting operators, metal & plastic	1.3	-18.0
General managers & top executives	3.3	-6.6	Industrial production managers	1.3	-1.6
Mechanical engineers	3.1	8.3	Electrical & electronic equipment assemblers	1.3	8.3
Welders & cutters	2.6	-1.6	NC machine tool operators, metal & plastic	1.3	18.1
Secretaries, ex legal & medical	2.5	-10.4	Purchasing agents, ex trade & farm products	1.3	-1.6
Drafters	2.3	-23.4	General office clerks	1.3	-16.1
Machine assemblers	2.1	-1.6	Stock clerks	1.2	-20.0
Sheet metal workers & duct installers	2.0	-1.6	Industrial machinery mechanics	1.2	8.3
Machine tool cutting & forming etc. nec	1.7	-1.6	Managers & administrators nec	1.1	-1.6
Electrical & electronics engineers	1.7	4.8	Financial managers	1.1	-1.6
Inspectors, testers, & graders, precision	1.6	-1.6	Production, planning, & expediting clerks	1.0	-1.6
Engineering, mathematical, & science managers	1.6	11.8	Electrical & electronic assemblers	1.0	-1.6
Traffic, shipping, & receiving clerks	1.6	-5.3	Electrical & electronic technicians, technologists	1.0	-1.6

Source: *Industry-Occupation Matrix*, Bureau of Labor Statistics. These data relate to one or more 3-digit SIC industry groups rather than to a single 4-digit SIC. The change reported for each occupation to the year 2005 is a percent of growth or decline as estimated by the Bureau of Labor Statistics. The abbreviation nec stands for 'not elsewhere classified'.

LOCATION BY STATE AND REGIONAL CONCENTRATION

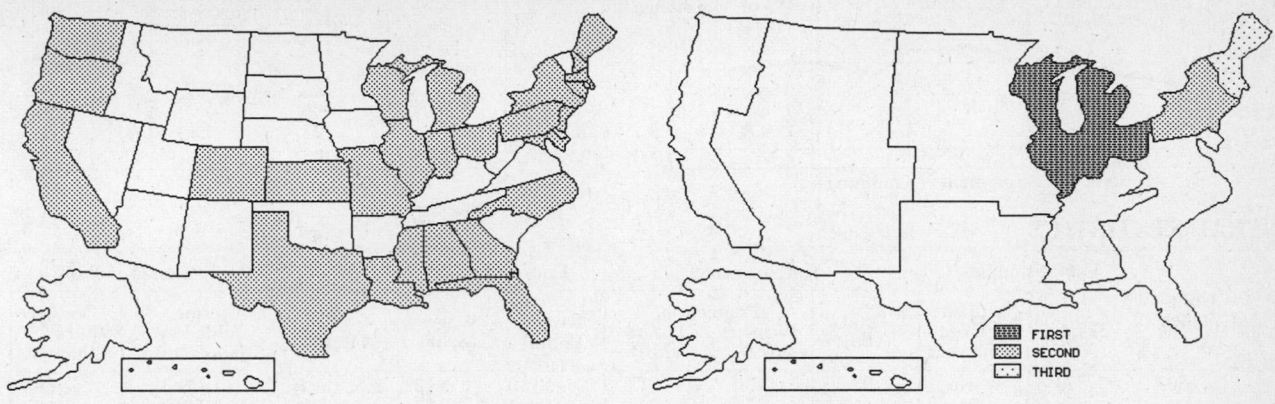

INDUSTRY DATA BY STATE

| State | Establish-ments | Shipments | | | Employment | | | | Cost as % of Shipments | Investment per Employee ($) |
		Total ($ mil)	% of U.S.	Per Establ.	Total Number	% of U.S.	Per Establ.	Wages ($/hour)		
Wisconsin	40	831.8	33.0	20.8	6,000	33.0	150	16.00	40.3	5,317
New Jersey	14	262.6	10.4	18.8	1,100	6.0	79	18.08	64.6	5,545
Massachusetts	28	185.3	7.3	6.6	1,400	7.7	50	12.37	44.8	3,143
Ohio	20	153.0	6.1	7.7	1,100	6.0	55	14.54	59.3	2,273
Pennsylvania	20	144.0	5.7	7.2	1,400	7.7	70	13.06	38.6	2,000
New York	30	115.2	4.6	3.8	1,000	5.5	33	14.25	43.7	2,300
New Hampshire	10	99.3	3.9	9.9	900	4.9	90	18.82	61.7	3,000
Washington	14	92.6	3.7	6.6	600	3.3	43	15.71	56.5	500
North Carolina	6	80.0	3.2	13.3	400	2.2	67	14.67	35.6	2,000
Maryland	7	70.4	2.8	10.1	400	2.2	57	16.83	54.7	2,000
Georgia	9	56.8	2.3	6.3	300	1.6	33	11.75	58.5	3,333
Oregon	7	36.1	1.4	5.2	400	2.2	57	14.75	37.7	750
Indiana	9	32.1	1.3	3.6	300	1.6	33	13.25	42.4	-
California	18	31.3	1.2	1.7	200	1.1	11	12.67	39.9	1,000
Florida	7	27.7	1.1	4.0	200	1.1	29	8.75	51.6	-
Alabama	13	27.4	1.1	2.1	300	1.6	23	10.75	47.1	1,333
Michigan	11	25.3	1.0	2.3	200	1.1	18	8.67	57.3	2,500
Connecticut	4	20.8	0.8	5.2	200	1.1	50	15.50	31.3	-
Illinois	11	20.3	0.8	1.8	200	1.1	18	13.00	59.1	1,000
Missouri	6	17.8	0.7	3.0	200	1.1	33	15.50	37.1	-
Kansas	3	13.5	0.5	4.5	100	0.5	33	18.00	60.7	-
Texas	9	(D)	-	-	175 *	1.0	19	-	-	-
Maine	4	(D)	-	-	175 *	1.0	44	-	-	-
Louisiana	3	(D)	-	-	175 *	1.0	58	-	-	-
Mississippi	2	(D)	-	-	175 *	1.0	88	-	-	-
Colorado	1	(D)	-	-	175 *	1.0	175	-	-	-

Source: 1992 *Economic Census*. The states are in descending order of shipments or establishments (if shipment data are missing for the majority). The symbol (D) appears when data are withheld to prevent disclosure of competitive information. States marked with (D) are sorted by number of establishments. A dash (-) indicates that the data element cannot be calculated; * indicates the midpoint of a range.

3555 - PRINTING TRADES MACHINERY

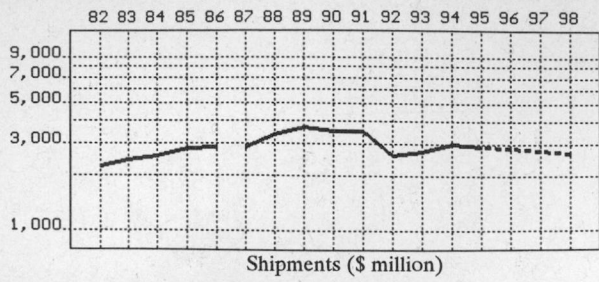

Shipments ($ million)

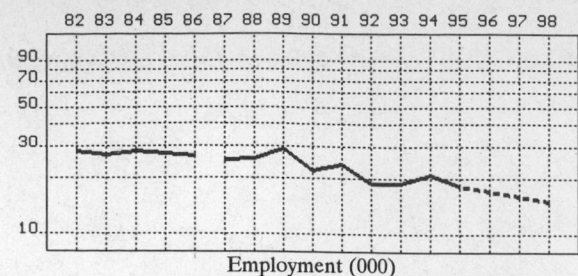

Employment (000)

GENERAL STATISTICS

| Year | Com-panies | Establishments | | Employment | | | Compensation | | Production ($ million) | | | |
		Total	with 20 or more employees	Total (000)	Production Workers (000)	Hours (Mil)	Payroll ($ mil)	Wages ($/hr)	Cost of Materials	Value Added by Manufacture	Value of Shipments	Capital Invest.
1982	507	570	215	28.2	15.5	31.2	597.0	9.42	974.1	1,231.3	2,256.9	72.2
1983		558	214	26.8	13.7	26.8	620.1	10.32	1,010.7	1,448.3	2,447.9	55.3
1984		546	213	28.3	14.7	30.1	709.2	10.49	1,135.5	1,454.7	2,575.3	107.2
1985		534	212	27.4	14.8	29.6	739.3	10.94	1,292.7	1,570.8	2,798.1	121.4
1986		529	211	26.7	14.0	28.7	747.3	11.24	1,262.1	1,534.5	2,834.3	74.3
1987*	408	438	194	25.0	12.1	25.3	751.6	12.10	1,315.1	1,606.6	2,857.8	114.2
1988		431	206	25.6	13.1	27.0	806.1	12.81	1,600.1	1,739.8	3,313.1	86.3
1989		428	207	29.3	13.6	28.6	833.9	12.77	1,820.4	1,948.7	3,691.9	99.5
1990		430	201	22.1	12.9	27.7	821.2	13.07	1,752.8	1,808.2	3,538.2	89.6
1991		443	190	24.0	12.1	25.5	799.9	13.29	1,706.2	1,850.1	3,538.1	74.7
1992	461	506	180	18.7	10.3	22.0	655.9	13.65	1,267.6	1,266.4	2,591.9	59.6
1993		514	175	18.9	10.3	21.3	683.8	14.59	1,334.5	1,393.9	2,727.2	52.9
1994		512P	175P	21.1	11.3	24.0	810.9	14.27	1,471.3	1,573.3	3,015.4	70.7
1995		526P	171P	18.2P	10.3P	22.0P	730.1P	14.76P	1,407.5P	1,505.0P	2,884.6P	48.5P
1996		540P	166P	17.1P	10.0P	21.3P	721.2P	15.08P	1,377.7P	1,473.2P	2,823.5P	41.3P
1997		554P	162P	16.1P	9.6P	20.6P	712.2P	15.40P	1,347.9P	1,441.4P	2,762.5P	34.0P
1998		568P	157P	15.0P	9.3P	19.9P	703.3P	15.72P	1,318.1P	1,409.5P	2,701.5P	26.8P

Sources: 1982, 1987, 1992 *Economic Census*; *Annual Survey of Manufactures*, 83-86, 88-91, 93-94. Establishment counts are from *County Business Patterns* for non-Census years; establishment counts for 83-84 are extrapolations. * indicates that industry content changed in 87; earlier years use 77 SICs. 'P's mark projections.

INDICES OF CHANGE

| Year | Com-panies | Establishments | | Employment | | | Compensation | | Production ($ million) | | | |
		Total	with 20 or more employees	Total (000)	Production Workers (000)	Hours (Mil)	Payroll ($ mil)	Wages ($/hr)	Cost of Materials	Value Added by Manufacture	Value of Shipments	Capital Invest.
1982	110	113	119	151	150	142	91	69	77	97	87	121
1983		110	119	143	133	122	95	76	80	114	94	93
1984		108	118	151	143	137	108	77	90	115	99	180
1985		106	118	147	144	135	113	80	102	124	108	204
1986		105	117	143	136	130	114	82	100	121	109	125
1987*	89	87	108	134	117	115	115	89	104	127	110	192
1988		85	114	137	127	123	123	94	126	137	128	145
1989		85	115	157	132	130	127	94	144	154	142	167
1990		85	112	118	125	126	125	96	138	143	137	150
1991		88	106	128	117	116	122	97	135	146	137	125
1992	100	100	100	100	100	100	100	100	100	100	100	100
1993		102	97	101	100	97	104	107	105	110	105	89
1994		101P	97P	113	110	109	124	105	116	124	116	119
1995		104P	95P	97P	100P	100P	111P	108P	111P	119P	111P	81P
1996		107P	92P	92P	97P	97P	110P	111P	109P	116P	109P	69P
1997		109P	90P	86P	93P	93P	109P	113P	106P	114P	107P	57P
1998		112P	87P	80P	90P	90P	107P	115P	104P	111P	104P	45P

Sources: Same as General Statistics. Values reflect change from the base year, 1992. Values above 100 mean greater than 92, values below 100 mean less than 92, and a value of 100 in the 82-91 or 93-98 period means same as 92. * indicates that industry content changed in 87. Data for earlier years are in 77 SIC format.

SELECTED RATIOS

For 1994	Avg. of All Manufact.	Analyzed Industry	Index	For 1994	Avg. of All Manufact.	Analyzed Industry	Index
Employees per Establishment	49	41	84	Value Added per Production Worker	134,084	139,230	104
Payroll per Establishment	1,500,273	1,584,231	106	Cost per Establishment	5,045,178	2,874,435	57
Payroll per Employee	30,620	38,431	126	Cost per Employee	102,970	69,730	68
Production Workers per Establishment	34	22	64	Cost per Production Worker	146,988	130,204	89
Wages per Establishment	853,319	669,093	78	Shipments per Establishment	9,576,895	5,891,097	62
Wages per Production Worker	24,861	30,308	122	Shipments per Employee	195,460	142,910	73
Hours per Production Worker	2,056	2,124	103	Shipments per Production Worker	279,017	266,850	96
Wages per Hour	12.09	14.27	118	Investment per Establishment	321,011	138,124	43
Value Added per Establishment	4,602,255	3,073,709	67	Investment per Employee	6,552	3,351	51
Value Added per Employee	93,930	74,564	79	Investment per Production Worker	9,352	6,257	67

Sources: Same as General Statistics. The 'Average of All Manufacturing' column represents the average of all manufacturing industries reported for the most recent complete year available. The Index shows the relationship between the Average and the Analyzed Industry. For example, 100 means that they are equal; 500 that the Analyzed Industry is five times the average; 50 means that the Analyzed Industry is half the national average. The abbreviation 'na' is used to show that data are 'not available'.

LEADING COMPANIES Number shown: 75 Total sales ($ mil): 3,908 Total employment (000): 16.7

Company Name	Address				CEO Name	Phone	Co. Type	Sales ($ mil)	Empl. (000)
Rockwell Graphic Systems	700 Oakmont Ln	Westmont	IL	60559	Robert Switt	708-850-5600	D	750	3.5
AB Dick Co	5700 W Touhy Av	Niles	IL	60714	Ron Peterson	708-763-1900	S	400	0.2
AM Graphics	PO Box 14468	Dayton	OH	45414	Richard J Bonnie	513-278-2651	D	370	0.9
Heidelberg Harris Inc	121 Broadway Av	Dover	NH	03820	Robert A Brown	603-749-6600	S	300	1.0
Sequa Can Machinery	401 Central Av	E Rutherford	NJ	07073	John P Skrypek	201-933-1200	D	210	0.3
Markem Corp	150 Congress St	Keene	NH	03431	Tom Putnam	603-352-1130	R	200	1.4
Baldwin Technology Company	65 Rowayton Av	Rowayton	CT	06853	Wendell M Smith	203-838-7470	P	198	1.1
MAN Roland Inc	PO Box 280	N Stonington	CT	06359	Leif Reslow	203-599-7000	S	189	0.4
Materials Research Corp	560 Rte 303	Orangeburg	NY	10962	T Marmen	914-359-4200	S	151	0.8
Etec Systems Inc	26460 Corporate	Hayward	CA	94545	Steve Cooper	510-783-9210	R	90*	0.5
DS America Inc	5110 Tollview Dr	Rolling Mdws	IL	60008	Kenneth D Newton	708-870-7400	S	76	0.2
Delphax Systems	5 Campanelli Cir	Canton	MA	02021	Alex Cimochowski	617-828-9017	J	70	0.4
Zebra Technologies Corp	333 Corp Woods	Vernon Hills	IL	60061	Edward L Kaplan	708-634-6700	P	64	0.4
Longer Life Products Inc	PO Box 2657	Darien	CT	06820	K W Grossberndt	203-655-5177	R	62	0.2
Hamilton-Stevens Group Inc	851 Walnut St	Hamilton	OH	45012	Andrew J Jones	513-863-1200	S	55*	0.3
Mark Andy Inc	PO Box 1023	Chesterfield	MO	63006	John Eulich	314-532-4433	R	50	0.3
PrePress Solutions Inc	11 Mt Pleasant Av	East Hanover	NJ	07936	R H Trenkamp Jr	201-887-8000	R	49*	0.3
M and R Printing Equipment Inc	1 N 372 Main St	Glen Ellyn	IL	60137	Richard Hoffman	708-858-6101	R	48	0.4
Varitronic Systems Inc	300 Interchange N	Minneapolis	MN	55426	Scott F Drill	612-542-1500	P	45	0.2
New Hermes Inc	535 Connecticut Av	Norwalk	CT	06854	Jon C Estes	203-852-7000	R	43	0.3
Autologic Inc	1050 R Conejo	Thousand Oaks	CA	91320	Dennis Doolittle	805-498-9611	S	37	0.3
Kidder-Stacy Inc	270 Main St	Agawam	MA	01001	R Berger	413-786-8692	R	31	0.2
KBA-Motter Corp	PO Box 1562	York	PA	17405	Scott R Smith	717-755-1071	S	30	0.2
Max Daetwyler Corp	13420 Reese Blvd W	Huntersville	NC	28078	Peter Daetwyler	704-875-1200	S	25	0.1
Southern Litho Plate Inc	PO Box 9400	Wake Forest	NC	27588	E A Casson Jr	919-556-9400	R	21*	0.2
Martin Automatic Inc	1661 Northrock Ct	Rockford	IL	61103	David A Wright	815-654-4800	R	20	0.2
McCain Manufacturing Corp	6200 W 60th St	Chicago	IL	60638	John S Iverson	312-586-6200	R	20	0.2
Hyphen Inc	181 Ballordvale St	Wilmington	MA	01887	Gianni Smaniotto	508-988-0880	S	15	<0.1
Ryco Graphic Manufacturing Inc	2181 S Foster Av	Wheeling	IL	60090	Anthony J Magro	708-259-3330	R	14*	<0.1
King Press Corp	PO Box 21	Joplin	MO	64802	Evans Kostas	417-781-3700	S	13*	0.1
Brandtjen and Kluge Inc	539 Bl Woods	St Croix Falls	WI	54024	Hank Brandtjen Jr	715-483-3265	R	12	<0.1
Franklin Manufacturing Corp	692 Pleasant St	Norwood	MA	02062	J Olsen	617-769-5800	R	12	<0.1
Presstek Inc	8 Commercial St	Hudson	NH	03051	Lawrence Howard	603-595-7000	P	12	<0.1
Birmy Graphics Corp	255 East Dr	Melbourne	FL	32904	Joe Birmingham	407-768-6766	R	11	<0.1
Baldwin Stobb	1351 E Riverview	San Bernardino	CA	92408	John St John	909-799-9950	D	10	<0.1
Electro Sprayer Systems Inc	1090 Fargo Av	Elk Grove Vill	IL	60007	Harold D Versten	708-439-9292	R	10	<0.1
K and F Manufacturing Co	12633 Industrial Dr	Granger	IN	46530	Alex Kocsis	219-272-9950	R	10	<0.1
Burgess Industries Inc	2700 Campus Dr	Plymouth	MN	55441	RJ Burgess	612-553-7800	R	9*	<0.1
Trident Inc	1114 Federal Rd	Brookfield	CT	06804	R Hugh VanBrimer	203-740-9333	S	9	<0.1
Dahlgren USA Inc	PO Box 115140	Carrollton	TX	75011	Brian Bargenquest	214-245-0035	R	8*	<0.1
Ternes Register System	2361 W Hwy 36	St Paul	MN	55113	Clifford Allen	612-633-2361	R	8	<0.1
Troy	2331 S Pullman St	Santa Ana	CA	92705	Patrick J Dirk	714-250-3280	D	8*	<0.1
JetFill Inc	5902 Sovereign Dr	Houston	TX	77036	Charles Schwarz	713-779-9393	R	7	<0.1
Arthur J Evers Corp	221 Banker St	Brooklyn	NY	11222	Eileen E Carlson	718-383-7191	R	7*	<0.1
Lith-O-Roll Corp	PO Box 5328	El Monte	CA	91734	Eldon V Swanson	818-579-0340	R	7	0.1
Precision Rubber Plate Company	5620 Elmwood Av	Indianapolis	IN	46203	Manuel S Green	317-783-3226	R	7	<0.1
Scheffer Inc	1565 E 91st Av	Merrillville	IN	46410	Bruce Scheffer	219-736-6200	R	7	<0.1
Wisconsin Automated Machinery	PO Box 3008	Oshkosh	WI	54903	Paul Ehrlich	414-231-4100	R	7	0.1
Web Press Corp	22023 68th Av S	Kent	WA	98032	W R Marcouiller	206-395-3343	P	7	<0.1
Eltron International Inc	21617 Nordhoff St	Chatsworth	CA	91311	Donald K Skinner	818-885-6484	P	7	<0.1
Baldwin Web Controls	1051-B W Main St	Lombard	IL	60148	Pete Anselmo	708-261-9180	D	6*	<0.1
FP Rosback Co	PO Box 899	Benton Harbor	MI	49023	LR Bowman	616-983-2582	R	6	<0.1
Deluxe Stitcher Company Inc	6639 W Irving Park	Chicago	IL	60634	Frank V Bocchieri	312-777-6500	R	6	<0.1
Chicago Manifold Products Co	171 E Marquardt	Wheeling	IL	60090	IB Harmon	708-459-6000	R	6	<0.1
Adolph Gottscho Company Inc	835 Lehigh Av	Union	NJ	07083	Eva Gottscho	908-688-2400	R	5	<0.1
Matthews International Corp	5555 Fresca Dr	La Palma	CA	90623	William Hauber	714-523-5511	D	5	<0.1
Mosstype Corp	150 Franklin Tpk	Waldwick	NJ	07463	Lester Moss	201-444-8000	R	5	0.1
RapidTec Inc	9126 Industrial Blv	Covington	GA	30209	KJ Holmes	404-787-5080	S	5	<0.1
Luminite Products Corp	115 Rochester St	Salamanca	NY	14779	John F Vosbuerg Jr	716-945-2270	R	5	<0.1
Craftsmen Machinery Co	PO Box 38	Millis	MA	02054	SJ Marks	508-376-2001	R	4	<0.1
National Printing Plate Company	1415 Drover St	Indianapolis	IN	46221	William B Ruch	317-630-2355	R	4*	<0.1
DICO Group Inc	PO Box 539	Monterey Park	CA	91754	Fern Haberman	213-264-2000	R	3*	<0.1
Mekatronics Inc	85 Channel Dr	Pt Washington	NY	11050	Jack Bendror	516-883-6805	R	3*	<0.1
Van Dam Machine Corp	20 Andrews Dr	West Paterson	NJ	07424	Anthony Hooimeijer	201-785-4444	S	3	<0.1
Xitron Inc	1428 E Ellsworth	Ann Arbor	MI	48108	Wendy Darland	313-971-8530	S	3*	<0.1
Carl G Wiklander Co	365 Criss Cir	Elk Grove Vill	IL	60007	NJ Douglas	708-593-6800	R	2	<0.1
Autographic Services Inc	31 Industrial Av	Mahwah	NJ	07430	Walter C Shoup	201-825-0400	R	2	<0.1
Exxtra Corp	55 Cabot Ct	Hauppauge	NY	11788	Walter Hansen	516-231-3998	R	2	<0.1
Graphics LX Corp	260 Eliot St	Ashland	MA	01721	Laurence Murray	508-881-8207	R	2	<0.1
Hydraulic & Pneumatic	633 Meriwether Av	Louisville	KY	40217	Steven Sherer	502-636-4030	R	2	<0.1
Imperial Stamp and Engraving	1280 Kyle Ct	Wauconda	IL	60084	WS Mackey	708-487-2400	R	2*	<0.1
International Composites Corp	14413 NE 10th Av	Vancouver	WA	98685	W L Thompson	206-574-5394	R	2*	<0.1
Lembo Corp	235 McLean Blv	Paterson	NJ	07504	Frank P Lembo	201-345-5555	R	2*	<0.1
Mosstype Corporation of Illinois	150 Scott St	Elk Grove Vill	IL	60007	Joseph G Rogers	708-437-1300	D	2	<0.1
Pamarco Inc	234 E 11th Av	Roselle	NJ	07203	M Baldasare Jr	908-241-1200	S	2	<0.1

Source: Ward's Business Directory of U.S. Private and Public Companies, Volumes 1 and 2, 1996. The company type code used is as follows: P - Public, R - Private, S - Subsidiary, D - Division, J - Joint Venture, A - Affiliate, G - Group. Sales are in millions of dollars, employees are in thousands. An asterisk (*) indicates an estimated sales volume. The symbol < stands for 'less than'. Company names and addresses are truncated, in some cases, to fit into the available space.

MATERIALS CONSUMED

Material	Quantity	Delivered Cost ($ million)
Materials, ingredients, containers, and supplies	(X)	1,078.7
Fluid power products	(X)	29.7
Fabricated metal products, except forgings	(X)	98.3
Forgings	(X)	5.2
Iron and steel castings (rough and semifinished)	(X)	47.7
Nonferrous (aluminum, copper, etc.) castings (rough and semifinished)	(X)	4.7
Steel bars, bar shapes, and plates	(X)	47.0
All other steel shapes and forms	(X)	16.4
Aluminum and aluminum-base alloy shapes and forms	(X)	13.1
Other nonferrous shapes and forms	(X)	3.0
Other fractional horsepower electric motors (under 1 hp)	(X)	12.5
Integral horsepower electric motors and generators (1 hp or more)	(X)	15.2
Ball bearings (mounted or unmounted)	(X)	16.6
Roller bearings (mounted or unmounted)	(X)	7.8
Mechanical speed changers, gears, and industrial high-speed drives	(X)	66.0
Electrical transmission, distribution, and control equipment	(X)	80.7
All other materials and components, parts, containers, and supplies	(X)	443.4
Materials, ingredients, containers, and supplies, nsk	(X)	171.5

Source: 1992 *Economic Census*. Explanation of symbols used: (D): Withheld to avoid disclosure of competitive data; na: Not available; (S): Withheld because statistical norms were not met; (X): Not applicable; (Z): Less than half the unit shown; nec: Not elsewhere classified; nsk: Not specified by kind; - : zero; * : 10-19 percent estimated; ** : 20-29 percent estimated.

PRODUCT SHARE DETAILS

Product or Product Class	% Share	Product or Product Class	% Share
Printing trades machinery	100.00	Printing trades paper cutting machines	9.77
Printing presses, offset lithographic	26.79	Printing trades collating and/or gathering machines (sold separately)	13.16
Small sheet-fed offset lithographic printing presses, less than 14 inches	2.27	Printing trades folding machines	10.64
Other sheet-fed offset lithographic printing presses, 14 inches or more	2.14	Printing trades newspaper inserting equipment	25.67
Offset roll-fed (web-fed) lithographic newspaper printing presses	48.94	Other printing trades binding machinery and equipment, including stitchers and trimmers	12.75
Offset roll-fed (web-fed) lithographic business form printing presses	7.15	Printing trades machinery, nec	35.98
Offset roll-fed (web-fed) lithographic commercial (including heat-set) printing presses	36.51	Digital electronic pre-press systems components and elements (excluding typesetting equipment and cameras)	3.88
All other roll-fed (web-fed) offset lithographic printing presses	1.72	Digital pre-press proofing devices, miscellaneous digital electronic pre-press systems (excluding typesetting equipment and cameras)	4.35
Printing presses, offset lithographic, nsk	1.30	Printing trades engravers' materials and equipment, including metal plates, etc.	7.98
Printing presses, other than lithographic	13.13	Other printing trades machinery and equipment, including platens, except typewriters	18.93
Sheet-fed and web-fed flexographic printing presses, less than 16 inches	23.63	Parts, attachments, and accessories for printing presses, including flying pasters, dryers, folders, and reels	47.78
Sheet-fed and web-fed flexographic printing presses, 16 inches or more	20.25	Parts, attachments, and accessories for typesetting machines	0.21
Other printing presses, including letterpress, gravure, metal decorating, proof, screen, pad printing, and rebuilt (except lithographic)	53.74	Parts, attachments, and accessories for bindery equipment	3.67
Printing presses, other than lithographic, nsk	2.38	Parts, attachments, and accessories for pre-press preparatory equipment, excluding typesetting and camera parts	1.80
Typesetting machinery, excluding justifying typewriters	5.80	Parts, attachments, and accessories for other printing trades machinery and equipment	9.36
Printing trades binding machinery and equipment, including paper cutting and collating or gathering machines	7.43	Printing trades machinery, nec, nsk	2.07
Saddle, perfect/adhesive, and hard case (edition) printing trades binding equipment	27.95	Printing trades machinery, nsk	10.87

Source: 1992 *Economic Census*. The values shown are percent of total shipments in an industry. Values of indented subcategories are summed in the main heading. The symbol (D) appears when data are withheld to prevent disclosure of competitive information. The abbreviation nsk stands for 'not specified by kind' and nec for 'not elsewhere classified'.

INPUTS AND OUTPUTS FOR PRINTING TRADES MACHINERY

Economic Sector or Industry Providing Inputs	%	Sector	Economic Sector or Industry Buying Outputs	%	Sector
Imports	22.6	Foreign	Gross private fixed investment	60.9	Cap Inv
Wholesale trade	8.3	Trade	Exports	18.6	Foreign
Blast furnaces & steel mills	7.0	Manufg.	Commercial printing	4.7	Manufg.
Printing trades machinery	5.2	Manufg.	Printing trades machinery	3.9	Manufg.
Switchgear & switchboard apparatus	5.2	Manufg.	Newspapers	2.1	Manufg.
Industrial controls	4.6	Manufg.	Miscellaneous repair shops	1.7	Services
Machinery, except electrical, nec	4.4	Manufg.	Lithographic platemaking & services	1.5	Manufg.
Iron & steel foundries	3.7	Manufg.	Typesetting	1.4	Manufg.
Pumps & compressors	2.8	Manufg.	Periodicals	1.0	Manufg.
Motors & generators	2.6	Manufg.	Manifold business forms	0.6	Manufg.
Aluminum rolling & drawing	1.8	Manufg.	Blankbooks & looseleaf binders	0.5	Manufg.

Continued on next page.

INPUTS AND OUTPUTS FOR PRINTING TRADES MACHINERY - Continued

Economic Sector or Industry Providing Inputs	%	Sector	Economic Sector or Industry Buying Outputs	%	Sector
Fabricated plate work (boiler shops)	1.8	Manufg.	Book printing	0.5	Manufg.
Power transmission equipment	1.8	Manufg.	S/L Govt. purch., higher education	0.4	S/L Govt
Rubber & plastics hose & belting	1.7	Manufg.	Bookbinding & related work	0.3	Manufg.
Advertising	1.6	Services	Engraving & plate printing	0.3	Manufg.
Transformers	1.4	Manufg.	Greeting card publishing	0.3	Manufg.
Ball & roller bearings	1.2	Manufg.	Miscellaneous publishing	0.3	Manufg.
Copper rolling & drawing	1.2	Manufg.	Book publishing	0.2	Manufg.
Primary lead	1.2	Manufg.	Federal Government purchases, national defense	0.2	Fed Govt
Special dies & tools & machine tool accessories	1.2	Manufg.	S/L Govt. purch., elem. & secondary education	0.2	S/L Govt
Communications, except radio & TV	1.1	Util.	Equipment rental & leasing services	0.1	Services
Electric services (utilities)	1.1	Util.			
Aluminum castings	1.0	Manufg.			
Banking	1.0	Fin/R.E.			
General industrial machinery, nec	0.9	Manufg.			
Maintenance of nonfarm buildings nec	0.8	Constr.			
Fabricated rubber products, nec	0.8	Manufg.			
Motor freight transportation & warehousing	0.7	Util.			
Eating & drinking places	0.7	Trade			
Engineering, architectural, & surveying services	0.7	Services			
Petroleum refining	0.6	Manufg.			
Screw machine and related products	0.6	Manufg.			
Air transportation	0.6	Util.			
Electronic components nec	0.5	Manufg.			
Pipe, valves, & pipe fittings	0.5	Manufg.			
Real estate	0.5	Fin/R.E.			
Equipment rental & leasing services	0.5	Services			
Abrasive products	0.4	Manufg.			
Nonferrous wire drawing & insulating	0.4	Manufg.			
Computer & data processing services	0.4	Services			
Gaskets, packing & sealing devices	0.3	Manufg.			
Machine tools, metal cutting types	0.3	Manufg.			
Miscellaneous plastics products	0.3	Manufg.			
Gas production & distribution (utilities)	0.3	Util.			
Legal services	0.3	Services			
Management & consulting services & labs	0.3	Services			
Noncomparable imports	0.3	Foreign			
Metal heat treating	0.2	Manufg.			
Nonferrous rolling & drawing, nec	0.2	Manufg.			
Sawmills & planning mills, general	0.2	Manufg.			
Railroads & related services	0.2	Util.			
Royalties	0.2	Fin/R.E.			
Accounting, auditing & bookkeeping	0.2	Services			
Hotels & lodging places	0.2	Services			
U.S. Postal Service	0.2	Gov't			
Lubricating oils & greases	0.1	Manufg.			
Manifold business forms	0.1	Manufg.			
Miscellaneous fabricated wire products	0.1	Manufg.			
Automotive rental & leasing, without drivers	0.1	Services			

Source: Benchmark Input-Output Accounts for the U.S. Economy, 1982, U.S. Department of Commerce, Washington, D.C., July 1991. Data, as reported in the source, are organized by the 1977 SIC structure in use in 1982 but have been matched, as closely as is possible, to the 1987 SIC structure used in this book.

OCCUPATIONS EMPLOYED BY SIC 355 - SPECIAL INDUSTRY MACHINERY

Occupation	% of Total 1994	Change to 2005	Occupation	% of Total 1994	Change to 2005
Machinists	8.3	-1.6	Bookkeeping, accounting, & auditing clerks	1.5	-26.2
Machine builders	4.3	18.1	Welding machine setters, operators	1.5	-11.4
Sales & related workers nec	4.3	-1.6	Lathe & turning machine tool operators	1.4	-21.3
Assemblers, fabricators, & hand workers nec	3.9	-1.6	Engineering technicians & technologists nec	1.4	-1.6
Blue collar worker supervisors	3.7	-9.0	Machine tool cutting operators, metal & plastic	1.3	-18.0
General managers & top executives	3.3	-6.6	Industrial production managers	1.3	-1.6
Mechanical engineers	3.1	8.3	Electrical & electronic equipment assemblers	1.3	8.3
Welders & cutters	2.6	-1.6	NC machine tool operators, metal & plastic	1.3	18.1
Secretaries, ex legal & medical	2.5	-10.4	Purchasing agents, ex trade & farm products	1.3	-1.6
Drafters	2.3	-23.4	General office clerks	1.3	-16.1
Machine assemblers	2.1	-1.6	Stock clerks	1.2	-20.0
Sheet metal workers & duct installers	2.0	-1.6	Industrial machinery mechanics	1.2	8.3
Machine tool cutting & forming etc. nec	1.7	-1.6	Managers & administrators nec	1.1	-1.6
Electrical & electronics engineers	1.7	4.8	Financial managers	1.1	-1.6
Inspectors, testers, & graders, precision	1.6	-1.6	Production, planning, & expediting clerks	1.0	-1.6
Engineering, mathematical, & science managers	1.6	11.8	Electrical & electronic assemblers	1.0	-1.6
Traffic, shipping, & receiving clerks	1.6	-5.3	Electrical & electronic technicians, technologists	1.0	-1.6

Source: Industry-Occupation Matrix, Bureau of Labor Statistics. These data relate to one or more 3-digit SIC industry groups rather than to a single 4-digit SIC. The change reported for each occupation to the year 2005 is a percent of growth or decline as estimated by the Bureau of Labor Statistics. The abbreviation nec stands for 'not elsewhere classified'.

LOCATION BY STATE AND REGIONAL CONCENTRATION

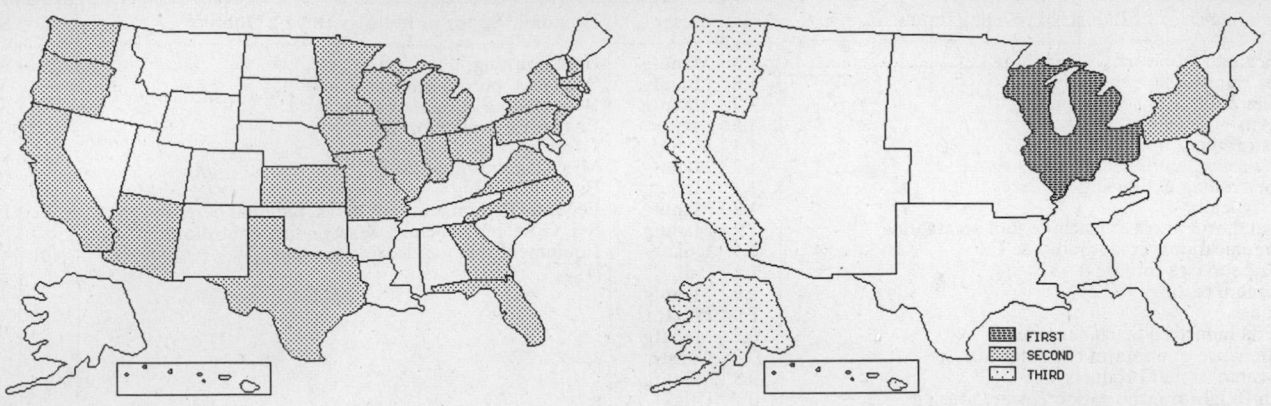

FIRST
SECOND
THIRD

INDUSTRY DATA BY STATE

| State | Establish-ments | Shipments | | | Employment | | | | Cost as % of Shipments | Investment per Employee ($) |
		Total ($ mil)	% of U.S.	Per Establ.	Total Number	% of U.S.	Per Establ.	Wages ($/hour)		
New Jersey	34	293.3	11.3	8.6	1,600	8.6	47	16.46	60.8	2,750
Ohio	35	292.6	11.3	8.4	2,100	11.2	60	14.59	41.0	2,524
Illinois	59	215.9	8.3	3.7	1,800	9.6	31	12.54	40.8	1,889
Pennsylvania	23	185.9	7.2	8.1	1,100	5.9	48	14.08	61.7	7,727
Massachusetts	25	161.6	6.2	6.5	1,200	6.4	48	14.00	43.5	3,500
California	49	149.6	5.8	3.1	1,200	6.4	24	12.31	34.6	2,167
Wisconsin	20	120.1	4.6	6.0	1,100	5.9	55	13.93	39.5	2,636
New York	50	105.6	4.1	2.1	1,000	5.3	20	10.27	42.5	1,300
Missouri	18	89.9	3.5	5.0	800	4.3	44	12.00	44.2	3,500
Michigan	12	69.0	2.7	5.8	600	3.2	50	14.25	57.8	13,500
Connecticut	13	68.5	2.6	5.3	500	2.7	38	16.50	52.3	1,000
Kansas	8	59.4	2.3	7.4	600	3.2	75	11.75	53.9	1,333
Texas	21	58.3	2.2	2.8	600	3.2	29	13.13	41.7	-
Minnesota	15	49.7	1.9	3.3	400	2.1	27	15.00	33.2	3,500
Florida	19	43.6	1.7	2.3	400	2.1	21	11.25	47.0	-
Virginia	8	38.6	1.5	4.8	300	1.6	38	16.25	43.5	-
North Carolina	9	38.0	1.5	4.2	300	1.6	33	10.00	32.6	3,000
Washington	7	19.0	0.7	2.7	100	0.5	14	13.00	53.7	-
Georgia	12	15.1	0.6	1.3	100	0.5	8	11.50	43.0	5,000
Indiana	9	(D)	-	-	375 *	2.0	42	-	-	533
Oregon	5	(D)	-	-	175 *	0.9	35	-	-	1,714
Iowa	4	(D)	-	-	750 *	4.0	188	-	-	-
Arizona	3	(D)	-	-	175 *	0.9	58	-	-	-
New Hampshire	3	(D)	-	-	750 *	4.0	250	-	-	-

Source: 1992 *Economic Census*. The states are in descending order of shipments or establishments (if shipment data are missing for the majority). The symbol (D) appears when data are withheld to prevent disclosure of competitive information. States marked with (D) are sorted by number of establishments. A dash (-) indicates that the data element cannot be calculated; * indicates the midpoint of a range.

3556 - FOOD PRODUCTS MACHINERY

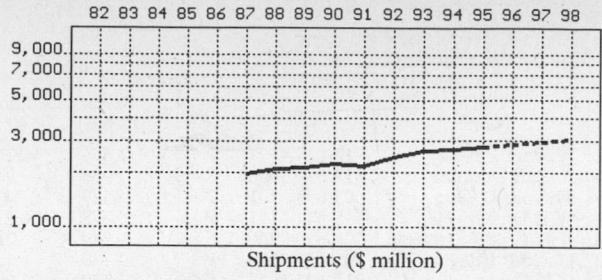

Shipments ($ million)

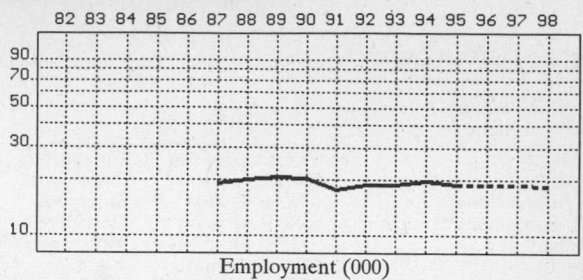

Employment (000)

GENERAL STATISTICS

| Year | Com-panies | Establishments | | Employment | | | Compensation | | Production ($ million) | | | |
		Total	with 20 or more employees	Total (000)	Production Workers (000)	Hours (Mil)	Payroll ($ mil)	Wages ($/hr)	Cost of Materials	Value Added by Manufacture	Value of Shipments	Capital Invest.
1982												
1983												
1984												
1985												
1986												
1987	483	512	219	19.2	11.8	22.8	487.9	11.32	832.7	1,140.5	1,971.4	47.4
1988		503	236	20.0	12.2	24.1	525.7	11.41	866.8	1,249.0	2,091.8	33.7
1989		480	227	20.8	11.1	22.0	531.5	11.50	886.4	1,253.2	2,125.9	63.6
1990		466	237	20.4	11.2	22.0	566.3	12.07	1,008.5	1,266.3	2,260.9	45.8
1991		488	216	17.9	10.5	21.0	570.5	12.70	996.6	1,183.9	2,193.1	42.1
1992	495	518	228	18.8	11.1	22.2	605.5	13.06	1,008.4	1,392.1	2,407.2	46.8
1993		516	229	18.8	11.3	22.8	621.8	13.23	1,069.3	1,569.5	2,630.2	48.8
1994		505P	228P	19.8	12.0	24.3	649.7	13.05	1,126.6	1,580.7	2,673.5	48.5
1995		507P	228P	18.9P	11.2P	22.8P	668.4P	13.71P	1,156.0P	1,622.0P	2,743.3P	48.6P
1996		508P	228P	18.8P	11.2P	22.9P	690.3P	14.03P	1,198.1P	1,681.0P	2,843.1P	49.0P
1997		510P	228P	18.7P	11.1P	22.9P	712.2P	14.34P	1,240.1P	1,740.0P	2,942.9P	49.3P
1998		512P	228P	18.5P	11.1P	23.0P	734.1P	14.66P	1,282.2P	1,799.0P	3,042.7P	49.7P

Sources: 1982, 1987, 1992 *Economic Census*; *Annual Survey of Manufactures*, 83-86, 88-91, 93-94. Establishment counts for non-Census years are from *County Business Patterns*; establishment values for 83-84 are extrapolations. 'P's show projections by the editors. Industries reclassified in 87 will not have data for prior years.

INDICES OF CHANGE

| Year | Com-panies | Establishments | | Employment | | | Compensation | | Production ($ million) | | | |
		Total	with 20 or more employees	Total (000)	Production Workers (000)	Hours (Mil)	Payroll ($ mil)	Wages ($/hr)	Cost of Materials	Value Added by Manufacture	Value of Shipments	Capital Invest.
1982												
1983												
1984												
1985												
1986												
1987	98	99	96	102	106	103	81	87	83	82	82	101
1988		97	104	106	110	109	87	87	86	90	87	72
1989		93	100	111	100	99	88	88	88	90	88	136
1990		90	104	109	101	99	94	92	100	91	94	98
1991		94	95	95	95	95	94	97	99	85	91	90
1992	100	100	100	100	100	100	100	100	100	100	100	100
1993		100	100	100	102	103	103	101	106	113	109	104
1994		97P	100P	105	108	109	107	100	112	114	111	104
1995		98P	100P	101P	101P	103P	110P	105P	115P	117P	114P	104P
1996		98P	100P	100P	100P	103P	114P	107P	119P	121P	118P	105P
1997		98P	100P	99P	100P	103P	118P	110P	123P	125P	122P	105P
1998		99P	100P	99P	100P	103P	121P	112P	127P	129P	126P	106P

Sources: Same as General Statistics. Values reflect change from the base year, 1992. Values above 100 mean greater than 92, values below 100 mean less than 92, and a value of 100 in the 82-91 or 93-98 period means same as 92. 'P's mark projections by the editors.

SELECTED RATIOS

For 1994	Avg. of All Manufact.	Analyzed Industry	Index	For 1994	Avg. of All Manufact.	Analyzed Industry	Index
Employees per Establishment	49	39	80	Value Added per Production Worker	134,084	131,725	98
Payroll per Establishment	1,500,273	1,287,263	86	Cost per Establishment	5,045,178	2,232,154	44
Payroll per Employee	30,620	32,813	107	Cost per Employee	102,970	56,899	55
Production Workers per Establishment	34	24	69	Cost per Production Worker	146,988	93,883	64
Wages per Establishment	853,319	628,306	74	Shipments per Establishment	9,576,895	5,297,056	55
Wages per Production Worker	24,861	26,426	106	Shipments per Employee	195,460	135,025	69
Hours per Production Worker	2,056	2,025	98	Shipments per Production Worker	279,017	222,792	80
Wages per Hour	12.09	13.05	108	Investment per Establishment	321,011	96,094	30
Value Added per Establishment	4,602,255	3,131,871	68	Investment per Employee	6,552	2,449	37
Value Added per Employee	93,930	79,833	85	Investment per Production Worker	9,352	4,042	43

Sources: Same as General Statistics. The 'Average of All Manufacturing' column represents the average of all manufacturing industries reported for the most recent complete year available. The Index shows the relationship between the Average and the Analyzed Industry. For example, 100 means that they are equal; 500 that the Analyzed Industry is five times the average; 50 means that the Analyzed Industry is half the national average. The abbreviation 'na' is used to show that data are 'not available'.

LEADING COMPANIES   Number shown: **75**   Total sales ($ mil): **5,632**   Total employment (000): **44.1**

Company Name	Address				CEO Name	Phone	Co. Type	Sales ($ mil)	Empl. (000)
Newell Co	29 E Stephenson St	Freeport	IL	61032	William P Sovey	815-235-4171	P	1,645	11.0
PMI Food Equipment Group	701 Ridge Av	Troy	OH	45374	Joseph Deering	513-332-3000	S	990	9.2
Standex International Corp	6 Manor Pkwy	Salem	NH	03079	Thomas L King	603-893-9701	P	529	4.5
Specialty Equipment Companies	6581 Revlon Dr	Belvidere	IL	61008	W E Dotterweich	815-544-5111	P	372	2.2
APV Crepaco Inc	9525 W Bryn Mawr	Rosemont	IL	60018	Jens-Erik Kristensen	708-678-4300	S	240	2.2
Middleby Corp	1400 Toastmaster	Elgin	IL	60120	David P Riley	708-741-3300	P	130	1.0
Taylor Co	750 Blackhawk Blv	Rockton	IL	61072	D B Greenwood	815-624-8333	D	120	0.7
Delfield Co	980 S Isabella Rd	Mount Pleasant	MI	48858	Kevin E McCrone	517-773-7981	D	90	0.7
Heat and Control Inc	225 Shaw Rd	S San Francisco	CA	94080	CK Benson	415-871-9234	R	90	0.6
APV Baker Inc	3200 Fruit Ridge	Grand Rapids	MI	49504	RH Rander	616-784-3111	S	87	0.5
Lincoln Foodservice Products	PO Box 1229	Fort Wayne	IN	46801	William A Thomas	219-432-9511	P	58	0.4
Waukesha Pumps	611 Sugar Creek Rd	Delavan	WI	53115	Jim Dahlke	414-548-5800	D	48*	0.5
Chicago Metallic Products Inc	800 Ela Rd	Lake Zurich	IL	60047	Geoffrey C Fear	708-438-2171	R	42*	0.3
APV Gaulin	500 Research Dr	Wilmington	MA	01887	RN Cooper	508-988-9300	D	40	0.3
Johnson Food Equipment Inc	PO Box 15300	Kansas City	KS	66115	D Franken	913-621-3366	R	40	0.2
Stein Inc	PO Box 5001	Sandusky	OH	44871	Richard J Dibbs	419-626-0304	S	40	0.3
Cincinnati Butchers Supply Co	PO Box 16098	Cincinnati	OH	45216	MW Schmidt	513-242-1535	R	35*	0.2
Gold Medal Products Co	2001 Dalton Av	Cincinnati	OH	45214	Dan Kroeger	513-381-1313	R	35	0.3
Berkel Inc	1 Berkel Dr	La Porte	IN	46350	Richard G Graf	219-324-2938	S	33	0.3
Key Technology Inc	150 Avry St	Walla Walla	WA	99362	Thomas C Madsen	509-529-2161	P	31	0.3
Automated Machinery Systems	PO Box 9168	Richmond	VA	23227	Harry Franze	804-355-7961	R	30	0.5
BWI KartridgPak	PO Box 3848	Davenport	IA	52808	Barry Shoulders	319-391-1100	D	27	0.2
FMC Corp	PO Box 1708	Lakeland	FL	33802	Robert Blackwell	813-683-5411	D	27*	0.3
ICEE-USA Corp	4701 Airport Dr	Ontario	CA	91761	Gerald Shrieber	909-467-4233	S	27*	0.3
Perlick Corp	PO Box 23098	Milwaukee	WI	53223	RD Perlick	414-353-7060	R	27*	0.3
Hosokawa Bepex Corp	333 NE Taft St	Minneapolis	MN	55413	Gordon Ettie	612-331-4370	S	25	0.2
Lee Industries Inc	514 W Pine St	Philipsburg	PA	16866	Robert W Montler	814-342-0461	R	25	0.2
Oliver Products Co	455 6th St NW	Grand Rapids	MI	49504	VP Tuthill	616-456-7711	R	25	0.2
Par-Way Group	15635 Alton Pkwy	Irvine	CA	92718	Ken Lowrey	714-453-8820	R	25	<0.1
Reading Pretzel Machinery Corp	380 Old W Penn	Robesonia	PA	19551	ET Groff	610-693-5816	R	25	0.1
Rheon USA	13400 Reese Blvd W	Huntersville	NC	28078	Makoto Nakagawa	704-875-9191	S	24	0.2
Littleford Group Inc	PO Box 128	Florence	KY	41022	Donald D Steedman	606-525-7600	R	23	0.2
Brown International Corp	633 N Barranca St	Covina	CA	91723	LB Alexander	818-966-8361	R	21	0.2
Stoelting Inc	502 Hwy 67	Kiel	WI	53042	Marion D Huggins	414-894-2293	R	21	0.2
Baxter Manufacturing Company	PO Box 729	Orting	WA	98360	Marlin Palmer	206-893-5554	R	20*	0.2
Bettcher Industries Inc	PO Box 336	Vermilion	OH	44089	L Bettcher	216-965-4422	R	20*	0.2
FMC Corp Food and Machinery	PO Box 5710	Riverside	CA	92517	Richard L Houtzer	909-222-2300	D	20*	0.2
M-E-C Co	PO Box 330	Neodesha	KS	66757	Dave M Parker	316-325-2673	R	20	0.2
Urschel Laboratories Inc	PO Box 2200	Valparaiso	IN	46384	Robert Urschel	219-464-4811	R	20*	0.2
Servolift-Eastern Corp	266 Hancock St	Boston	MA	02125	IL Kaplan	617-825-9000	R	19	0.2
Damrow Company Inc	196 Western Av	Fond du Lac	WI	54936	George F Manninen	414-922-1500	R	18	0.2
Howden Food Equipment Inc	72 Santa Felicia St	Santa Barbara	CA	93117	Merlin E Rossow	805-968-7444	S	18	0.2
Crescent Metal Products	12825 Taft Av	Cleveland	OH	44108	George E Baggott	216-851-6800	R	17*	0.2
Henny Penny Corp	PO Box 60	Eaton	OH	45320	JL Cobb	513-456-8400	R	17*	0.4
Tomlinson Industries	13700 Broadway	Cleveland	OH	44125	John Chernak	216-587-3400	R	17	0.1
Schlueter Company Inc	PO Box 548	Janesville	WI	53547	Bradley Losching	608-755-0740	R	16	<0.1
Aeroglide Corp	PO Box 29505	Raleigh	NC	27626	J Fredrick Kelly Jr	919-851-2000	R	15	0.1
AMF Bakery Systems	2115 W Laburnum	Richmond	VA	23227	Harry Franze	804-355-7961	D	15*	0.3
Belshaw Brothers Inc	1750 22nd Av S	Seattle	WA	98144	Louis J Agathos	206-322-5474	S	15	0.2
Buffalo Technologies Corp	PO Box 1041	Buffalo	NY	14240	John B Nemcek	716-895-2100	R	15	0.1
Food Automation	905 Honeyspot Rd	Stratford	CT	06497	Bernard G Koether	203-377-4414	R	15	0.2
HC Duke and Son Inc	2116 8th Av	East Moline	IL	61244	Dan Duke	309-755-4553	R	15	0.1
Jarvis Products Corp	33 Anderson Rd	Middletown	CT	06457	VR Volpe	203-347-7271	R	15	0.1
National Drying Machinery Co	2190 Hornig Rd	Philadelphia	PA	19116	Richard B Parkes	215-464-6070	R	15*	0.1
SaniServ	2020 Production Dr	Indianapolis	IN	46241	Carlton Curry	317-247-0460	R	15	<0.1
Silver Engineering Works Inc	14800 E Moncrieff	Aurora	CO	80011	Derrald Houston	303-373-2311	S	15	0.2
Cardwell Machine Co	PO Box 34588	Richmond	VA	23234	Haldun M Turgay	804-275-1471	R	14	0.1
Nieco Corp	PO Box 4506	Burlingame	CA	94011	ED Baker	415-697-7335	R	13*	<0.1
Thomas L Green and Company	202 N Miley Av	Indianapolis	IN	46222	Thomas R Lugar	317-263-6935	R	13*	<0.1
Weiler and Company Inc	1116 E Main St	Whitewater	WI	53190	NJ Lesar	414-473-5254	R	13	0.1
C Cretors and Co	3243 N California	Chicago	IL	60618	Charles D Cretors	312-588-1690	R	12	<0.1
Hartness International Inc	PO Box 26509	Greenville	SC	29616	Bern McPheely	803-297-1200	R	12	0.2
Lang Manufacturing Co	PO Box 905	Redmond	WA	98073	David Lawrence	206-885-4045	R	12	0.1
United Bakery Equipment	18408 Laurel Pk Rd	Compton	CA	90220	FJ Bastasch	310-635-8121	D	12	<0.1
Anderson International Corp	6200 Harvard Av	Cleveland	OH	44105	JC Ansley	216-641-1112	R	11	<0.1
Grindmaster Corp	PO Box 35020	Louisville	KY	40232	Karl D Kuiper	502-499-0770	R	11*	0.1
Server Products Inc	PO Box 530	Menomonee Fls	WI	53052	Paul Wickesberg	414-251-7100	R	11	0.1
APV Crepaco Inc	3009 Industrial Ter	Austin	TX	78759	Randy Campbell	512-836-0920	D	10	<0.1
Automatic Mach & Electronics	PO Box 713	Winter Haven	FL	33882	Preston Troutman	813-299-2111	S	10	<0.1
Crippen Manufacturing Company	PO Box 128	Alma	MI	48801	Jim Gascho	517-463-2119	R	10	<0.1
Demaco	46-25 Metropolitan	Ridgewood	NY	11385	Leonard DeFrancisci	718-456-6600	D	10	<0.1
Duchess Bakers' Machinery Co	1101 John Av	Superior	WI	54880	L Donald Tenerelli	715-394-2387	S	10	<0.1
Flohr Metal Fabricators Inc	PO Box 70469	Seattle	WA	98107	B Feldt	206-633-2222	R	10	0.1
Fulton Ferracute Indust Intern	3844 Walsh St	St Louis	MO	63116	Ramon Marin	314-752-2400	R	10	<0.1
Juice Tree Inc	7300 Bolsa Av	Westminster	CA	92683	James Beck	714-891-4425	R	10	0.1

Source: Ward's Business Directory of U.S. Private and Public Companies, Volumes 1 and 2, 1996. The company type code used is as follows: P - Public, R - Private, S - Subsidiary, D - Division, J - Joint Venture, A - Affiliate, G - Group. Sales are in millions of dollars, employees are in thousands. An asterisk (*) indicates an estimated sales volume. The symbol < stands for 'less than'. Company names and addresses are truncated, in some cases, to fit into the available space.

MATERIALS CONSUMED

Material	Quantity	Delivered Cost ($ million)
Materials, ingredients, containers, and supplies	(X)	834.3
Fluid power pumps, motors, and hydrostatic transmissions (hydraulic and pneumatic)	(X)	9.2
All other pumps	(X)	9.9
Fluid power valves (hydraulic and pneumatic)	(X)	9.9
Fluid power cylinders and rotary actuators (hydraulic and pneumatic)	(X)	3.8
Fluid power hose or tube fittings and assemblies (hydraulic and pneumatic)	(X)	4.7
Fluid power filters (hydraulic and pneumatic)	(X)	2.5
Other fluid power products (hydraulic and pneumatic)	(X)	1.4
Metal tanks, heat exchangers, steam condensers, and other boiler products	(X)	13.7
All other fabricated metal products, except forgings	(X)	35.1
Forgings	(X)	2.6
Iron and steel castings (rough and semifinished)	(X)	27.5
Aluminum and aluminum-base alloy castings (rough and semifinished)	(X)	10.9
Other nonferrous castings (rough and semifinished)	(X)	5.6
Steel bars, bar shapes, and plates	(X)	47.8
Steel sheet and strip, including tin plate	(X)	52.2
All other steel shapes and forms	(X)	20.6
Aluminum and aluminum-base alloy sheet, plate, foil, and welded tubing	(X)	16.2
All other aluminum and aluminum-base alloy shapes and forms	(X)	8.0
Other nonferrous shapes and forms	(X)	11.4
Other fractional horsepower electric motors (under 1 hp)	(X)	7.5
Integral horsepower electric motors and generators (1 hp or more)	(X)	24.2
Ball and roller bearings (mounted or unmounted)	(X)	12.5
Mechanical speed changers, gears, and industrial high-speed drives	(X)	20.0
Electrical transmission, distribution, and control equipment	(X)	33.1
Wood boxes, pallets, skids, and containers	(X)	3.8
All other materials and components, parts, containers, and supplies	(X)	264.3
Materials, ingredients, containers, and supplies, nsk	(X)	175.7

Source: 1992 *Economic Census*. Explanation of symbols used: (D): Withheld to avoid disclosure of competitive data; na: Not available; (S): Withheld because statistical norms were not met; (X): Not applicable; (Z): Less than half the unit shown; nec: Not elsewhere classified; nsk: Not specified by kind; - : zero; * : 10-19 percent estimated; ** : 20-29 percent estimated.

PRODUCT SHARE DETAILS

Product or Product Class	% Share	Product or Product Class	% Share
Food products machinery	100.00	Other industrial bakery machinery and equipment, including pastry rolling machines, except packaging machinery and food cooking and warming equipment	10.96
Dairy and milk products plant machinery and equipment, except bottling and packaging.	12.38	Other commercial food preparation machines, including tenderizers (power driven), except packaging machinery and food cooking and warming equipment	13.21
Dairy plant ice cream freezers, except bottling and packaging machinery and equipment	(D)	Parts and attachments for commercial food preparation machines, except packaging machinery and food cooking and warming equipment	15.75
Butter and cheese processing plant machinery and equipment, except bottling and packaging machinery and equipment	5.92	Commercial food products machinery, except packaging machinery and food cooking and warming equipment, nsk	4.66
Dairy and milk products plant homogenizers.	(D)	Industrial machinery and equipment for manufacturing or processing foods, beverages, and animal or fowl feed	38.54
Other dairy and milk products plant machinery and equipment, including cream separators, pasteurizers, sterilizers, etc., except bottling and packaging machinery and equipment	37.52	Industrial machinery for sorting, grading, or cleaning fruits, vegetables, or eggs	13.13
Parts and attachments for dairy and milk products plant machinery and equipment, except for bottling and packaging equipment	12.07	Industrial meat and poultry processing and preparation machinery and equipment (killing, dehairing, stuffing, cooking, rendering)	17.41
Dairy and milk products plant machinery and equipment, except bottling and packaging, nsk	0.38	Industrial machinery, nec, for the preparation of fruits, vegetables, and nuts	7.80
Commercial food products machinery, except packaging machinery and food cooking and warming equipment	35.53	Other industrial food and feed products machinery	41.33
Commercial food products slicers.	14.80	Parts and attachments for industrial food products machinery	18.19
Commercial food products choppers, grinders, cutters, dicers, and similar machines	15.18	Industrial machinery and equipment for manufacturing or processing foods, beverages, and animal or fowl feed, nsk	2.15
Commercial food products mixers and whippers, except drink mixers	10.71	Food products machinery, nsk	13.55
Industrial bakery dough mixers, dividers, and molders	8.11		
Industrial bakery bake ovens, including traveling tray	6.65		

Source: 1992 *Economic Census*. The values shown are percent of total shipments in an industry. Values of indented subcategories are summed in the main heading. The symbol (D) appears when data are withheld to prevent disclosure of competitive information. The abbreviation nsk stands for 'not specified by kind' and nec for 'not elsewhere classified'.

INPUTS AND OUTPUTS FOR FOOD PRODUCTS MACHINERY

Economic Sector or Industry Providing Inputs	%	Sector	Economic Sector or Industry Buying Outputs	%	Sector
Imports	29.1	Foreign	Gross private fixed investment	65.8	Cap Inv
Blast furnaces & steel mills	10.4	Manufg.	Exports	16.7	Foreign
Wholesale trade	8.9	Trade	Eating & drinking places	4.5	Trade
Machinery, except electrical, nec	3.9	Manufg.	Miscellaneous repair shops	2.0	Services
Motors & generators	3.7	Manufg.	Food products machinery	1.5	Manufg.
Iron & steel foundries	2.6	Manufg.	Federal Government purchases, national defense	1.5	Fed Govt
Food products machinery	2.3	Manufg.	Retail trade, except eating & drinking	1.0	Trade
Power transmission equipment	1.9	Manufg.	Malt beverages	0.7	Manufg.
Advertising	1.9	Services	Bottled & canned soft drinks	0.6	Manufg.
Aluminum rolling & drawing	1.7	Manufg.	Bread, cake, & related products	0.6	Manufg.
Metal stampings, nec	1.7	Manufg.	Food preparations, nec	0.4	Manufg.
Transformers	1.6	Manufg.	Wet corn milling	0.4	Manufg.
Ball & roller bearings	1.5	Manufg.	Canned fruits & vegetables	0.3	Manufg.
Special dies & tools & machine tool accessories	1.5	Manufg.	Cookies & crackers	0.3	Manufg.
Industrial controls	1.3	Manufg.	Fluid milk	0.3	Manufg.
Miscellaneous plastics products	1.1	Manufg.	Meat packing plants	0.3	Manufg.
Electric services (utilities)	1.1	Util.	Cereal breakfast foods	0.2	Manufg.
Banking	1.1	Fin/R.E.	Cheese, natural & processed	0.2	Manufg.
Maintenance of nonfarm buildings nec	1.0	Constr.	Confectionery products	0.2	Manufg.
Aluminum castings	1.0	Manufg.	Dog, cat, & other pet food	0.2	Manufg.
Fabricated plate work (boiler shops)	1.0	Manufg.	Prepared feeds, nec	0.2	Manufg.
Nonferrous wire drawing & insulating	1.0	Manufg.	Soybean oil mills	0.2	Manufg.
Pipe, valves, & pipe fittings	0.9	Manufg.	Change in business inventories	0.2	In House
Eating & drinking places	0.9	Trade	Animal & marine fats & oils	0.1	Manufg.
Air transportation	0.8	Util.	Chocolate & cocoa products	0.1	Manufg.
Communications, except radio & TV	0.8	Util.	Condensed & evaporated milk	0.1	Manufg.
Motor freight transportation & warehousing	0.8	Util.	Dehydrated food products	0.1	Manufg.
Miscellaneous fabricated wire products	0.7	Manufg.	Distilled liquor, except brandy	0.1	Manufg.
Engineering, architectural, & surveying services	0.7	Services	Flavoring extracts & syrups, nec	0.1	Manufg.
Brass, bronze, & copper castings	0.6	Manufg.	Fresh or frozen packaged fish	0.1	Manufg.
Metal coating & allied services	0.6	Manufg.	Frozen fruits, fruit juices & vegetables	0.1	Manufg.
Petroleum refining	0.6	Manufg.	Frozen specialties	0.1	Manufg.
Real estate	0.6	Fin/R.E.	Pickles, sauces, & salad dressings	0.1	Manufg.
Abrasive products	0.5	Manufg.	Poultry dressing plants	0.1	Manufg.
Copper rolling & drawing	0.5	Manufg.	Sausages & other prepared meats	0.1	Manufg.
Fabricated metal products, nec	0.5	Manufg.	Shortening & cooking oils	0.1	Manufg.
Nonferrous castings, nec	0.5	Manufg.	Wines, brandy, & brandy spirits	0.1	Manufg.
Gaskets, packing & sealing devices	0.4	Manufg.			
Pumps & compressors	0.4	Manufg.			
Rubber & plastics hose & belting	0.4	Manufg.			
Gas production & distribution (utilities)	0.4	Util.			
Equipment rental & leasing services	0.4	Services			
U.S. Postal Service	0.4	Gov't			
Machine tools, metal cutting types	0.3	Manufg.			
Metal heat treating	0.3	Manufg.			
Refrigeration & heating equipment	0.3	Manufg.			
Sawmills & planning mills, general	0.3	Manufg.			
Screw machine and related products	0.3	Manufg.			
Hotels & lodging places	0.3	Services			
Legal services	0.3	Services			
Management & consulting services & labs	0.3	Services			
Noncomparable imports	0.3	Foreign			
Fabricated rubber products, nec	0.2	Manufg.			
Lubricating oils & greases	0.2	Manufg.			
Manifold business forms	0.2	Manufg.			
Nonferrous rolling & drawing, nec	0.2	Manufg.			
Paperboard containers & boxes	0.2	Manufg.			
Railroads & related services	0.2	Util.			
Royalties	0.2	Fin/R.E.			
Accounting, auditing & bookkeeping	0.2	Services			
Computer & data processing services	0.2	Services			
Wood pallets & skids	0.1	Manufg.			
Insurance carriers	0.1	Fin/R.E.			
Automotive rental & leasing, without drivers	0.1	Services			

Source: Benchmark Input-Output Accounts for the U.S. Economy, 1982, U.S. Department of Commerce, Washington, D.C., July 1991. Data, as reported in the source, are organized by the 1977 SIC structure in use in 1982 but have been matched, as closely as is possible, to the 1987 SIC structure used in this book.

OCCUPATIONS EMPLOYED BY SIC 355 - SPECIAL INDUSTRY MACHINERY

Occupation	% of Total 1994	Change to 2005	Occupation	% of Total 1994	Change to 2005
Machinists	8.3	-1.6	Bookkeeping, accounting, & auditing clerks	1.5	-26.2
Machine builders	4.3	18.1	Welding machine setters, operators	1.5	-11.4
Sales & related workers nec	4.3	-1.6	Lathe & turning machine tool operators	1.4	-21.3
Assemblers, fabricators, & hand workers nec	3.9	-1.6	Engineering technicians & technologists nec	1.4	-1.6
Blue collar worker supervisors	3.7	-9.0	Machine tool cutting operators, metal & plastic	1.3	-18.0
General managers & top executives	3.3	-6.6	Industrial production managers	1.3	-1.6
Mechanical engineers	3.1	8.3	Electrical & electronic equipment assemblers	1.3	8.3
Welders & cutters	2.6	-1.6	NC machine tool operators, metal & plastic	1.3	18.1
Secretaries, ex legal & medical	2.5	-10.4	Purchasing agents, ex trade & farm products	1.3	-1.6
Drafters	2.3	-23.4	General office clerks	1.3	-16.1
Machine assemblers	2.1	-1.6	Stock clerks	1.2	-20.0
Sheet metal workers & duct installers	2.0	-1.6	Industrial machinery mechanics	1.2	8.3
Machine tool cutting & forming etc. nec	1.7	-1.6	Managers & administrators nec	1.1	-1.6
Electrical & electronics engineers	1.7	4.8	Financial managers	1.1	-1.6
Inspectors, testers, & graders, precision	1.6	-1.6	Production, planning, & expediting clerks	1.0	-1.6
Engineering, mathematical, & science managers	1.6	11.8	Electrical & electronic assemblers	1.0	-1.6
Traffic, shipping, & receiving clerks	1.6	-5.3	Electrical & electronic technicians, technologists	1.0	-1.6

Source: Industry-Occupation Matrix, Bureau of Labor Statistics. These data relate to one or more 3-digit SIC industry groups rather than to a single 4-digit SIC. The change reported for each occupation to the year 2005 is a percent of growth or decline as estimated by the Bureau of Labor Statistics. The abbreviation nec stands for 'not elsewhere classified'.

LOCATION BY STATE AND REGIONAL CONCENTRATION

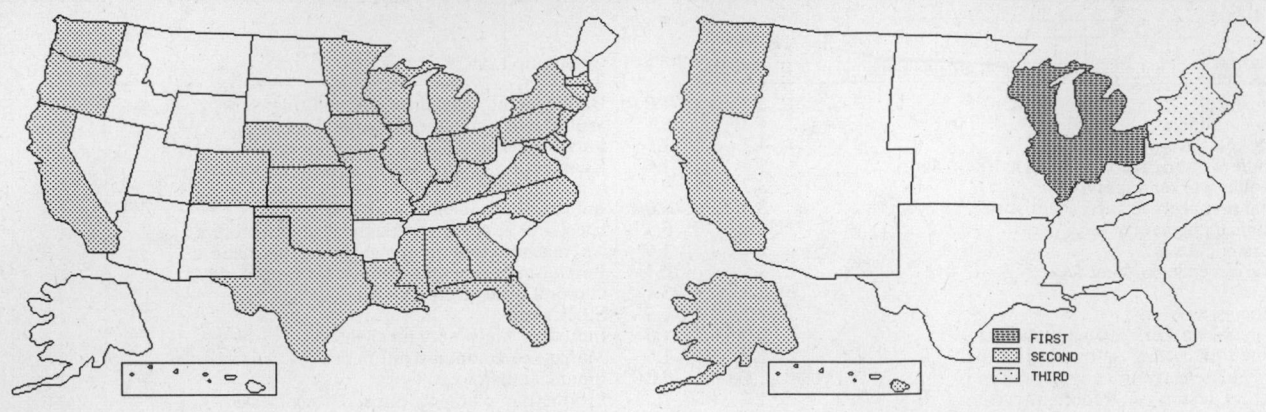

FIRST
SECOND
THIRD

INDUSTRY DATA BY STATE

| State | Establish-ments | Shipments | | | Employment | | | | Cost as % of Shipments | Investment per Employee ($) |
		Total ($ mil)	% of U.S.	Per Establ.	Total Number	% of U.S.	Per Establ.	Wages ($/hour)		
Illinois	47	290.5	12.1	6.2	2,000	10.6	43	16.00	44.3	3,350
Ohio	31	253.8	10.5	8.2	1,500	8.0	48	14.37	31.2	4,133
Wisconsin	34	219.6	9.1	6.5	2,100	11.2	62	14.24	50.7	2,619
California	77	191.5	8.0	2.5	1,900	10.1	25	11.12	44.9	2,158
Georgia	23	139.0	5.8	6.0	900	4.8	39	12.00	43.3	1,667
North Carolina	11	112.6	4.7	10.2	800	4.3	73	11.10	46.9	1,875
Iowa	15	108.9	4.5	7.3	800	4.3	53	14.00	38.6	-
Indiana	9	105.0	4.4	11.7	500	2.7	56	18.29	39.7	3,200
Virginia	13	80.5	3.3	6.2	600	3.2	46	12.14	37.6	1,667
Pennsylvania	20	69.6	2.9	3.5	500	2.7	25	11.14	41.7	1,600
Michigan	13	69.0	2.9	5.3	500	2.7	38	11.83	42.2	-
New York	30	68.0	2.8	2.3	700	3.7	23	14.38	37.1	2,286
Kansas	12	67.0	2.8	5.6	600	3.2	50	11.80	48.7	2,000
Minnesota	23	62.4	2.6	2.7	600	3.2	26	13.25	43.8	1,667
Kentucky	7	56.5	2.3	8.1	400	2.1	57	12.67	56.1	-
Florida	15	53.0	2.2	3.5	600	3.2	40	11.00	26.2	4,333
Washington	14	42.4	1.8	3.0	300	1.6	21	15.75	42.2	1,333
New Jersey	18	38.0	1.6	2.1	300	1.6	17	13.33	42.4	-
Texas	14	34.4	1.4	2.5	400	2.1	29	10.00	44.5	-
Missouri	9	31.0	1.3	3.4	300	1.6	33	12.00	47.7	-
Massachusetts	9	23.5	1.0	2.6	100	0.5	11	12.00	43.0	7,000
Nebraska	4	18.7	0.8	4.7	100	0.5	25	10.50	47.6	3,000
Oregon	13	17.1	0.7	1.3	200	1.1	15	13.50	38.0	-
Maryland	5	16.3	0.7	3.3	200	1.1	40	12.50	45.4	-
Mississippi	3	15.2	0.6	5.1	100	0.5	33	7.50	48.0	-
Louisiana	6	13.2	0.5	2.2	200	1.1	33	14.00	34.8	1,500
Oklahoma	5	10.0	0.4	2.0	100	0.5	20	9.00	43.0	1,000
Colorado	4	(D)	-	-	175 *	0.9	44	-	-	-
Alabama	3	(D)	-	-	375 *	2.0	125	-	-	-
Connecticut	2	(D)	-	-	375 *	2.0	188	-	-	-

Source: 1992 *Economic Census*. The states are in descending order of shipments or establishments (if shipment data are missing for the majority). The symbol (D) appears when data are withheld to prevent disclosure of competitive information. States marked with (D) are sorted by number of establishments. A dash (-) indicates that the data element cannot be calculated; * indicates the midpoint of a range.

3559 - SPECIAL INDUSTRY MACHINERY, NEC

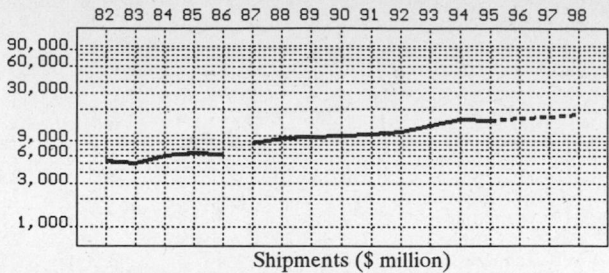

Shipments ($ million)

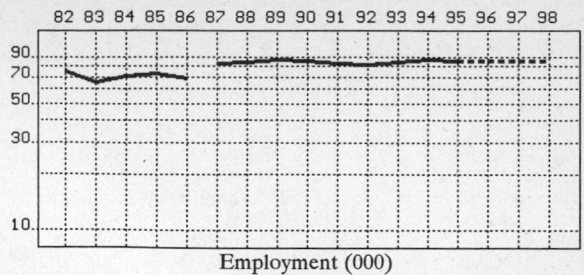

Employment (000)

GENERAL STATISTICS

| Year | Com-panies | Establishments | | Employment | | | Compensation | | Production ($ million) | | | |
		Total	with 20 or more employees	Total (000)	Production Workers (000)	Hours (Mil)	Payroll ($ mil)	Wages ($/hr)	Cost of Materials	Value Added by Manufacture	Value of Shipments	Capital Invest.
1982	1,754	1,826	693	74.9	43.7	84.5	1,557.6	9.40	2,263.2	3,015.1	5,351.1	173.4
1983		1,745	683	65.0	37.8	73.4	1,469.0	9.84	2,154.5	2,782.9	5,031.4	98.0
1984		1,664	673	70.6	41.7	84.3	1,703.8	10.18	2,587.2	3,631.5	6,072.5	194.2
1985		1,583	664	73.1	41.8	82.7	1,829.0	10.88	2,768.1	3,685.9	6,495.1	204.3
1986		1,556	653	67.8	36.8	74.3	1,796.6	11.58	2,692.8	3,541.5	6,254.4	172.5
1987*	2,438	2,531	817	83.3	46.3	92.8	2,285.6	11.40	3,593.4	4,704.1	8,275.3	217.3
1988		2,446	817	84.6	47.7	95.7	2,444.8	11.65	4,066.0	5,414.9	9,338.2	189.1
1989		2,380	852	86.1	47.9	97.3	2,580.7	12.11	4,480.9	5,450.7	9,853.8	297.7
1990		2,286	846	85.4	47.3	95.2	2,709.8	12.62	4,724.2	5,517.0	10,247.1	311.7
1991		2,313	813	83.0	45.7	91.9	2,778.4	13.33	4,882.9	5,657.4	10,542.5	285.6
1992	2,469	2,557	814	81.9	44.0	89.8	2,903.9	13.02	5,146.9	6,226.2	11,297.2	348.1
1993		2,443	837	84.3	45.1	90.2	3,103.3	13.83	5,878.1	7,096.6	12,917.2	310.9
1994		2,407P	830P	87.7	49.1	102.2	3,473.0	14.69	6,805.8	8,748.1	15,191.7	637.7
1995		2,403P	831P	85.3P	46.3P	95.1P	3,462.2P	14.83P	6,611.5P	8,498.3P	14,758.0P	521.7P
1996		2,399P	831P	85.5P	46.2P	95.2P	3,612.7P	15.28P	6,989.8P	8,984.6P	15,602.5P	565.5P
1997		2,395P	832P	85.6P	46.1P	95.4P	3,763.2P	15.72P	7,368.1P	9,470.9P	16,446.9P	609.3P
1998		2,391P	832P	85.8P	46.0P	95.5P	3,913.7P	16.17P	7,746.5P	9,957.2P	17,291.4P	653.1P

Sources: 1982, 1987, 1992 Economic Census; Annual Survey of Manufactures, 83-86, 88-91, 93-94. Establishment counts are from County Business Patterns for non-Census years; establishment counts for 83-84 are extrapolations. * indicates that industry content changed in 87; earlier years use 77 SICs. 'P's mark projections.

INDICES OF CHANGE

| Year | Com-panies | Establishments | | Employment | | | Compensation | | Production ($ million) | | | |
		Total	with 20 or more employees	Total (000)	Production Workers (000)	Hours (Mil)	Payroll ($ mil)	Wages ($/hr)	Cost of Materials	Value Added by Manufacture	Value of Shipments	Capital Invest.
1982	71	71	85	91	99	94	54	72	44	48	47	50
1983		68	84	79	86	82	51	76	42	45	45	28
1984		65	83	86	95	94	59	78	50	58	54	56
1985		62	82	89	95	92	63	84	54	59	57	59
1986		61	80	83	84	83	62	89	52	57	55	50
1987*	99	99	100	102	105	103	79	88	70	76	73	62
1988		96	100	103	108	107	84	89	79	87	83	54
1989		93	105	105	109	108	89	93	87	88	87	86
1990		89	104	104	107	106	93	97	92	89	91	90
1991		90	100	101	104	102	96	102	95	91	93	82
1992	100	100	100	100	100	100	100	100	100	100	100	100
1993		96	103	103	102	100	107	106	114	114	114	89
1994		94P	102P	107	112	114	120	113	132	141	134	183
1995		94P	102P	104P	105P	106P	119P	114P	128P	136P	131P	150P
1996		94P	102P	104P	105P	106P	124P	117P	136P	144P	138P	162P
1997		94P	102P	105P	105P	106P	130P	121P	143P	152P	146P	175P
1998		94P	102P	105P	105P	106P	135P	124P	151P	160P	153P	188P

Sources: Same as General Statistics. Values reflect change from the base year, 1992. Values above 100 mean greater than 92, values below 100 mean less than 92, and a value of 100 in the 82-91 or 93-98 period means same as 92. * indicates that industry content changed in 87. Data for earlier years are in 77 SIC format.

SELECTED RATIOS

For 1994	Avg. of All Manufact.	Analyzed Industry	Index	For 1994	Avg. of All Manufact.	Analyzed Industry	Index
Employees per Establishment	49	36	74	Value Added per Production Worker	134,084	178,169	133
Payroll per Establishment	1,500,273	1,443,046	96	Cost per Establishment	5,045,178	2,827,839	56
Payroll per Employee	30,620	39,601	129	Cost per Employee	102,970	77,603	75
Production Workers per Establishment	34	20	59	Cost per Production Worker	146,988	138,611	94
Wages per Establishment	853,319	623,804	73	Shipments per Establishment	9,576,895	6,312,216	66
Wages per Production Worker	24,861	30,577	123	Shipments per Employee	195,460	173,223	89
Hours per Production Worker	2,056	2,081	101	Shipments per Production Worker	279,017	309,403	111
Wages per Hour	12.09	14.69	121	Investment per Establishment	321,011	264,967	83
Value Added per Establishment	4,602,255	3,634,873	79	Investment per Employee	6,552	7,271	111
Value Added per Employee	93,930	99,750	106	Investment per Production Worker	9,352	12,988	139

Sources: Same as General Statistics. The 'Average of All Manufacturing' column represents the average of all manufacturing industries reported for the most recent complete year available. The Index shows the relationship between the Average and the Analyzed Industry. For example, 100 means that they are equal; 500 that the Analyzed Industry is five times the average; 50 means that the Analyzed Industry is half the national average. The abbreviation 'na' is used to show that data are 'not available'.

LEADING COMPANIES   Number shown: 75   Total sales ($ mil): 8,794   Total employment (000): 46.0

Company Name	Address				CEO Name	Phone	Co. Type	Sales ($ mil)	Empl. (000)
Applied Materials Inc	PO Box 58039	Santa Clara	CA	95052	James C Morgan	408-727-5555	P	1,660	4.7
Cincinnati Plastics Machinery	4165 Half Acre Rd	Batavia	OH	45103	Harold Faig	513-536-2000	D	503	2.4
Lam Research Corp	4650 Cushing Pkwy	Fremont	CA	94538	Roger D Emerick	510-659-0200	P	493	2.2
Engineered Support Systems Inc	1270 N Price Rd	St Louis	MO	63132	M F Shanahan Jr	314-993-5880	P	440	0.3
John Brown Co	PO Box 720421	Houston	TX	77272	Alastair Gowan	713-988-2002	S	370*	2.9
Watkins-Johnson Co	3333 Hillview Av	Palo Alto	CA	94304	WK Kennedy	415-493-4141	P	333	2.2
Silicon Valley Group Inc	2240 Ringwood Av	San Jose	CA	95131	P S Der Torossian	408-434-0500	P	320	1.9
Gerber Scientific Inc	83 Gerber Rd W	South Windsor	CT	06074	H Joseph Gerber	203-644-1551	P	261	1.5
ASM Lithography	2315 W Fairmont	Tempe	AZ	85282	C Douglas Marsh	602-438-0559	S	250	0.5
Koch Engineering Company Inc	PO Box 8127	Wichita	KS	67208	Vern E Griffith	316-832-5110	S	250*	2.0
Novellus Systems Inc	81 Vista Montana	San Jose	CA	95134	Richard S Hill	408-943-9700	P	225	0.5
Davis-Standard Co	1 Extrusion Dr	Pawcatuck	CT	06379	Robert W Ackley	203-599-1010	S	200	1.0
Van Dorn Demag Corp	11792 Alameda Dr	Strongsville	OH	44136	William G Pryor	216-238-8960	S	180	0.8
Van Dorn Plastic Machinery	11792 Alameda Dr	Strongsville	OH	44136	WG Pryor	216-238-8960	D	180	0.8
Kulicke and Soffa Industries Inc	2101 Blair Mill Rd	Willow Grove	PA	19090	C Scott Kulicke	215-784-6000	P	173	1.3
Amdura Corp	PO Box 870	Southbury	CT	06488	William F Andrews	203-262-0570	P	130	1.5
Conair Group Inc	PO Box 790	Franklin	PA	16323	E Niles Kenyon	814-437-6861	R	120*	1.0
Newcor Inc	1825 S Woodward	Bloomfield Hls	MI	48302	W John Weinhardt	313-253-2400	P	115	0.9
DT Industries Inc	441 W Elm St	Lebanon	MO	65536	Stephen J Gore	417-532-2141	P	108	0.9
Haden Schweitzer Corp	1399 Pacific Dr	Auburn Hills	MI	48326	Kenneth C Dargatz	810-475-5000	S	100*	0.2
FSI International Inc	322 Lake Hazeltine	Chaska	MN	55318	Joel A Elftmann	612-448-5440	P	94	0.5
Hunter Engineering Co	11250 Hunter Dr	Bridgeton	MO	63044	Stephen Brauer	314-731-3020	R	92	0.6
Kent-Moore Tool Group	28635 Mound Rd	Warren	MI	48092	Peter F Schneider	810-574-2332	D	85	0.5
Watkins-Johnson Co	440 Kings Village	Scotts Valley	CA	95066	James L Schram	408-438-2100	D	81	0.4
HPM Corp	820 Marion Rd	Mount Gilead	OH	43338	WT Flickinger	419-946-0222	R	75	0.8
Pangborn Corp	PO Box #380	Hagerstown	MD	21740	Ed R Reuschling	301-739-3500	R	75	0.4
Emhart-Hartford	123 Day Hill Rd	Windsor	CT	06095	EH Pemberton	203-688-8551	D	70	0.3
GaSonics International Corp	2730 Junction Av	San Jose	CA	95134	Monte M Toole	408-944-0212	P	67	0.3
FMC Corp	PO Box 3000	Conway	AR	72032	James R Collins	501-327-4433	D	65	0.4
Genus Inc	1139 Karlstad Dr	Sunnyvale	CA	94089	William WR Elder	408-747-7120	P	64	0.3
Intertec Corp	3400 Executive	Toledo	OH	43606	George Seifreid	419-537-9711	R	63*	0.5
Battenfeld Gloucester Eng	PO Box 900	Gloucester	MA	01930	Harold F Wrede	508-281-1800	S	60	0.5
Ingersoll Cutting Tool Co	505 Fulton Av	Rockford	IL	61103	Merle Clewett	815-987-6600	S	60*	0.5
Semitool Inc	655 W Reserve Dr	Kalispell	MT	59901	R F Thompson	406-752-2107	P	56	0.6
Detroit Tool&Engineering Co	PO Box 232	Lebanon	MO	65536	Stephen Gore	417-532-2141	S	55	0.5
Besser Co	PO Box 336	Alpena	MI	49707	James C Park	517-354-4111	R	50	0.6
Burr Oak Tool and Gauge	PO Box 338	Sturgis	MI	49091	LA Franks	616-651-9393	R	50	0.4
Columbia Machine Inc	PO Box 8950	Vancouver	WA	98668	Gerry O'Meara	206-694-1501	R	50	0.5
Veeco Instruments Inc	Terminal Dr	Plainview	NY	11803	Edward H Braun	516-349-8300	P	49	0.2
Continental Eagle Corp	PO Box 1000	Prattville	AL	36067	Joseph C Fermon	205-365-8811	R	46*	0.4
ARTOS Engineering Co	PO Box 1650	Waukesha	WI	53187	Jon Olsen	414-782-3300	R	45	0.3
Braswell Services Group Inc	60 Braswell St	Charleston	SC	29405	ES Braswell	803-577-4692	R	45	0.5
Luwa Bahnson Inc	PO Box 10458	Winston-Salem	NC	27108	Timothy J Whitener	919-760-3111	S	45	0.8
Waterjet Cutting	23629 Industrial Pk	Farmington Hls	MI	48335	Dick Demming	313-826-9274	D	45	0.1
BTU International Inc	23 Esquire Rd	North Billerica	MA	01862	P v der Wansem	508-667-4111	P	43	0.3
AG Associates Inc	1325 Borregas Av	Sunnyvale	CA	94089	Arnon Gat	408-745-1790	P	40	0.2
Chemineer Inc	PO Box 1123	Dayton	OH	45401	JG Fenic	513-454-3200	S	40	0.3
Integrated Solutions Inc	836 North St	Tewksbury	MA	01876	Ellery R Buchanan	508-640-1400	R	40	0.2
Morgen Manufacturing Co	PO Box 160	Yankton	SD	57078	Jim Cope	605-665-9654	S	39	0.1
CRC-Evans Pipeline Intern	PO Box 3227	Houston	TX	77253	C Paul Evans	713-460-2900	S	38*	0.3
Akron Equipment Co	PO Box 990	Akron	OH	44309	EL McCartt	216-762-9361	R	30	0.2
Dupps Co	548 N Cherry St	Germantown	OH	45327	JA Dupps Jr	513-855-6555	R	30	0.2
Globe Products Inc	5051 Kitridge Rd	Dayton	OH	45424	James E Kroencke	513-233-0233	R	30	0.2
Lynch Machinery-Miller Hydro	601 Independent St	Bainbridge	GA	31717	Robert T Pando	912-246-2244	S	30	0.2
Niro Hudson Inc	1600 O'Keefe Rd	Hudson	WI	54016	Carl Dyrbye	715-386-9371	S	30	0.3
Resources Conservation Co	3006 Northup Way	Bellevue	WA	98004	RC Vandenberg	206-828-2400	D	30	<0.1
Welex Inc	850 Jolly Rd	Blue Bell	PA	19422	Frank R Nissel	215-542-8000	R	30	0.1
Brown Machine	PO Box 434	Beaverton	MI	48612	R Stewart	517-435-7741	D	28	0.2
Hansford Manufacturing Corp	3111 Winton Rd S	Rochester	NY	14623	VN Hansford Jr	716-427-0660	R	28*	0.2
Mechanical Equipment Company	861 Carondelet St	New Orleans	LA	70130	LR McMillan II	504-599-4000	R	28	0.3
Powers Manufacturing Inc	1140 Sullivan St	Elmira	NY	14902	Joseph F Laundry	607-734-3671	S	28	0.2
Brunner-Hildebrand Lumber	7532 Little Av	Charlotte	NC	28226	Rein Juergen	704-543-7121	R	25*	0.2
Cosmodyne Inc	2920 Columbia St	Torrance	CA	90503	Ross Brown	310-320-5650	R	25*	0.3
Cumberland Engineering	100 Roddy Av	S Attleboro	MA	02703	A Cohen	508-399-6400	D	25	0.2
Graco-LTI	31 Volunteer Dr	Hendersonville	TN	37075	Valarie Lapinski	615-824-3634	D	25	0.2
Newbury Industries Inc	10975 Kinsman Rd	Newbury	OH	44065	Sudhir J Amin	216-564-2285	S	25	0.1
VHC Ltd	601 Clearwater Pk	W Palm Beach	FL	33401	J F Bradway Jr	407-659-7770	R	25	0.2
Brooks Automation Inc	41 Wellman St	Lowell	MA	01851	Robert J Therrien	508-453-1112	P	24	0.2
Forward Technology Industries	13500 County Rd 6	Minneapolis	MN	55441	David A Buckland	612-559-1785	S	24	0.2
McCracken Concrete Pipe Mach	PO Box 1708	Sioux City	IA	51102	JC Dannenberg	712-277-8111	D	24	0.2
Production Experts Inc	4259 E 49th St	Cleveland	OH	44125	SD Noll	216-883-3220	S	24	0.3
Kona Corp	PO Box 1227	Gloucester	MA	01930	Paul Swenson	508-281-3810	S	23	0.2
FEI Co	7451 NE Evergreen	Hillsboro	OR	97124	L W Swanson	503-640-7500	P	22	0.1
Athens Corp	1922nida del Oro	Oceanside	CA	92056	Robert L Corey	619-758-0994	R	22	<0.1
Automatic Feed Co	476 E Riverview	Napoleon	OH	43545	Kim Beck	419-592-0050	R	22	0.1

Source: Ward's Business Directory of U.S. Private and Public Companies, Volumes 1 and 2, 1996. The company type code used is as follows: P - Public, R - Private, S - Subsidiary, D - Division, J - Joint Venture, A - Affiliate, G - Group. Sales are in millions of dollars, employees are in thousands. An asterisk (*) indicates an estimated sales volume. The symbol < stands for 'less than'. Company names and addresses are truncated, in some cases, to fit into the available space.

MATERIALS CONSUMED

Material	Quantity	Delivered Cost ($ million)
Materials, ingredients, containers, and supplies	(X)	4,299.3
Electrical transmission, distribution, and control equipment	(X)	226.9
Fluid power pumps, motors, and hydrostatic transmissions (hydraulic and pneumatic)	(X)	88.2
Other pumps and pump parts, except fluid power (complete assemblies)	(X)	71.7
Fluid power valves (hydraulic and pneumatic)	(X)	55.5
Fluid power cylinders and rotary actuators (hydraulic and pneumatic)	(X)	40.9
Fluid power hose or tube fittings and assemblies (hydraulic and pneumatic)	(X)	25.7
Fluid power filters (hydraulic and pneumatic)	(X)	11.0
Other fluid power products (hydraulic and pneumatic)	(X)	27.6
Metal bolts, nuts, screws, washers, rivets, and other screw machine products	(X)	70.2
Metal tanks, heat exchangers, steam condensers, and other boiler products	(X)	92.9
Metal pipe, valves, and pipe fittings (except forgings)	(X)	57.6
Other fabricated metal products (except fluid power products and forgings)	(X)	205.8
Forgings	(X)	16.3
Iron and steel castings (rough and semifinished)	(X)	185.3
Aluminum and aluminum-base alloy castings (rough and semifinished)	(X)	28.4
Other nonferrous castings (rough and semifinished)	(X)	29.1
Steel bars, bar shapes, and plates	(X)	169.1
Steel sheet and strip, including tin plate	(X)	106.5
Steel structural shapes and sheet piling	(X)	34.9
All other steel shapes and forms	(X)	61.5
Aluminum and aluminum-base alloy sheet, plate, foil, and welded tubing	(X)	62.0
All other aluminum and aluminum-base alloy shapes and forms	(X)	55.8
Other nonferrous shapes and forms	(X)	29.0
Other fractional horsepower electric motors (under 1 hp)	(X)	36.7
Integral horsepower electric motors and generators (1 hp or more)	(X)	69.8
Ball and roller bearings (mounted or unmounted)	(X)	42.4
Mechanical speed changers, gears, and industrial high-speed drives	(X)	55.0
Air and gas compressors except refrigeration compressors	(X)	4.0
Filter paper	(X)	113.8
All other materials and components, parts, containers, and supplies	(X)	1,099.4
Materials, ingredients, containers, and supplies, nsk	(X)	1,126.4

Source: 1992 *Economic Census.* Explanation of symbols used: (D): Withheld to avoid disclosure of competitive data; na: Not available; (S): Withheld because statistical norms were not met; (X): Not applicable; (Z): Less than half the unit shown; nec: Not elsewhere classified; nsk: Not specified by kind; - : zero; * : 10-19 percent estimated; ** : 20-29 percent estimated.

PRODUCT SHARE DETAILS

Product or Product Class	% Share	Product or Product Class	% Share
Special industry machinery, nec	100.00	Chemical manufacturing distilling, rectifying, or fractionating machinery and equipment	4.60
Chemical manufacturing machinery, equipment, and parts	7.80		

Source: 1992 *Economic Census.* The values shown are percent of total shipments in an industry. Values of indented subcategories are summed in the main heading. The symbol (D) appears when data are withheld to prevent disclosure of competitive information. The abbreviation nsk stands for 'not specified by kind' and nec for 'not elsewhere classified'.

INPUTS AND OUTPUTS FOR SPECIAL INDUSTRY MACHINERY, NEC

Economic Sector or Industry Providing Inputs	%	Sector	Economic Sector or Industry Buying Outputs	%	Sector
Wholesale trade	11.3	Trade	Gross private fixed investment	79.1	Cap Inv
Imports	9.7	Foreign	Exports	10.0	Foreign
Blast furnaces & steel mills	8.9	Manufg.	Cyclic crudes and organics	4.2	Manufg.
Cyclic crudes and organics	5.9	Manufg.	Industrial inorganic chemicals, nec	1.9	Manufg.
Machinery, except electrical, nec	3.2	Manufg.	Miscellaneous plastics products	1.6	Manufg.
Switchgear & switchboard apparatus	2.9	Manufg.	Special industry machinery, nec	0.9	Manufg.
Motors & generators	2.4	Manufg.	Federal Government purchases, national defense	0.8	Fed Govt
Advertising	2.4	Services	Alkalies & chlorine	0.7	Manufg.
Industrial controls	2.2	Manufg.	Inorganic pigments	0.5	Manufg.
Iron & steel foundries	2.2	Manufg.	Plastics materials & resins	0.2	Manufg.
Electric services (utilities)	2.2	Util.	Tires & inner tubes	0.1	Manufg.
Fabricated plate work (boiler shops)	2.0	Manufg.	Federal Government purchases, nondefense	0.1	Fed Govt
Pumps & compressors	2.0	Manufg.			
Sheet metal work	1.8	Manufg.			
Special industry machinery, nec	1.7	Manufg.			
Pipe, valves, & pipe fittings	1.6	Manufg.			
Power transmission equipment	1.5	Manufg.			
Communications, except radio & TV	1.5	Util.			
Banking	1.5	Fin/R.E.			
Internal combustion engines, nec	1.4	Manufg.			
Aluminum rolling & drawing	1.3	Manufg.			
Maintenance of nonfarm buildings nec	1.2	Constr.			
Nonferrous wire drawing & insulating	1.2	Manufg.			
Special dies & tools & machine tool accessories	1.2	Manufg.			

Continued on next page.

INPUTS AND OUTPUTS FOR SPECIAL INDUSTRY MACHINERY, NEC - Continued

Economic Sector or Industry Providing Inputs	%	Sector	Economic Sector or Industry Buying Outputs	%	Sector
Copper rolling & drawing	1.1	Manufg.			
Petroleum refining	1.1	Manufg.			
Engineering, architectural, & surveying services	1.1	Services			
Ball & roller bearings	1.0	Manufg.			
Air transportation	1.0	Util.			
Glass & glass products, except containers	0.9	Manufg.			
Real estate	0.9	Fin/R.E.			
Equipment rental & leasing services	0.9	Services			
Aluminum castings	0.8	Manufg.			
Motor freight transportation & warehousing	0.8	Util.			
Industrial inorganic chemicals, nec	0.7	Manufg.			
Screw machine and related products	0.7	Manufg.			
Transformers	0.7	Manufg.			
Eating & drinking places	0.7	Trade			
Fabricated metal products, nec	0.6	Manufg.			
Gaskets, packing & sealing devices	0.6	Manufg.			
Miscellaneous fabricated wire products	0.6	Manufg.			
Miscellaneous plastics products	0.6	Manufg.			
Nonferrous rolling & drawing, nec	0.6	Manufg.			
Rubber & plastics hose & belting	0.6	Manufg.			
Gas production & distribution (utilities)	0.6	Util.			
Brass, bronze, & copper castings	0.5	Manufg.			
Fabricated rubber products, nec	0.5	Manufg.			
Metal stampings, nec	0.5	Manufg.			
Nonferrous castings, nec	0.5	Manufg.			
Computer & data processing services	0.5	Services			
U.S. Postal Service	0.5	Gov't			
Abrasive products	0.4	Manufg.			
Electronic components nec	0.4	Manufg.			
Machine tools, metal cutting types	0.4	Manufg.			
Plating & polishing	0.4	Manufg.			
General industrial machinery, nec	0.3	Manufg.			
Sawmills & planning mills, general	0.3	Manufg.			
Railroads & related services	0.3	Util.			
Royalties	0.3	Fin/R.E.			
Detective & protective services	0.3	Services			
Hotels & lodging places	0.3	Services			
Legal services	0.3	Services			
Management & consulting services & labs	0.3	Services			
Conveyors & conveying equipment	0.2	Manufg.			
Metal heat treating	0.2	Manufg.			
Paperboard containers & boxes	0.2	Manufg.			
Insurance carriers	0.2	Fin/R.E.			
Accounting, auditing & bookkeeping	0.2	Services			
Automotive rental & leasing, without drivers	0.2	Services			
Automotive repair shops & services	0.2	Services			
Alkalies & chlorine	0.1	Manufg.			
Lubricating oils & greases	0.1	Manufg.			
Manifold business forms	0.1	Manufg.			
Primary metal products, nec	0.1	Manufg.			
Primary nonferrous metals, nec	0.1	Manufg.			
Sanitary services, steam supply, irrigation	0.1	Util.			
Noncomparable imports	0.1	Foreign			
Scrap	0.1	Scrap			

Source: Benchmark Input-Output Accounts for the U.S. Economy, 1982, U.S. Department of Commerce, Washington, D.C., July 1991. Data, as reported in the source, are organized by the 1977 SIC structure in use in 1982 but have been matched, as closely as is possible, to the 1987 SIC structure used in this book.

OCCUPATIONS EMPLOYED BY SIC 355 - SPECIAL INDUSTRY MACHINERY

Occupation	% of Total 1994	Change to 2005	Occupation	% of Total 1994	Change to 2005
Machinists	8.3	-1.6	Bookkeeping, accounting, & auditing clerks	1.5	-26.2
Machine builders	4.3	18.1	Welding machine setters, operators	1.5	-11.4
Sales & related workers nec	4.3	-1.6	Lathe & turning machine tool operators	1.4	-21.3
Assemblers, fabricators, & hand workers nec	3.9	-1.6	Engineering technicians & technologists nec	1.4	-1.6
Blue collar worker supervisors	3.7	-9.0	Machine tool cutting operators, metal & plastic	1.3	-18.0
General managers & top executives	3.3	-6.6	Industrial production managers	1.3	-1.6
Mechanical engineers	3.1	8.3	Electrical & electronic equipment assemblers	1.3	8.3
Welders & cutters	2.6	-1.6	NC machine tool operators, metal & plastic	1.3	18.1
Secretaries, ex legal & medical	2.5	-10.4	Purchasing agents, ex trade & farm products	1.3	-1.6
Drafters	2.3	-23.4	General office clerks	1.3	-16.1
Machine assemblers	2.1	-1.6	Stock clerks	1.2	-20.0
Sheet metal workers & duct installers	2.0	-1.6	Industrial machinery mechanics	1.2	8.3
Machine tool cutting & forming etc. nec	1.7	-1.6	Managers & administrators nec	1.1	-1.6
Electrical & electronics engineers	1.7	4.8	Financial managers	1.1	-1.6
Inspectors, testers, & graders, precision	1.6	-1.6	Production, planning, & expediting clerks	1.0	-1.6
Engineering, mathematical, & science managers	1.6	11.8	Electrical & electronic assemblers	1.0	-1.6
Traffic, shipping, & receiving clerks	1.6	-5.3	Electrical & electronic technicians,technologists	1.0	-1.6

Source: *Industry-Occupation Matrix*, Bureau of Labor Statistics. These data relate to one or more 3-digit SIC industry groups rather than to a single 4-digit SIC. The change reported for each occupation to the year 2005 is a percent of growth or decline as estimated by the Bureau of Labor Statistics. The abbreviation nec stands for 'not elsewhere classified'.

LOCATION BY STATE AND REGIONAL CONCENTRATION

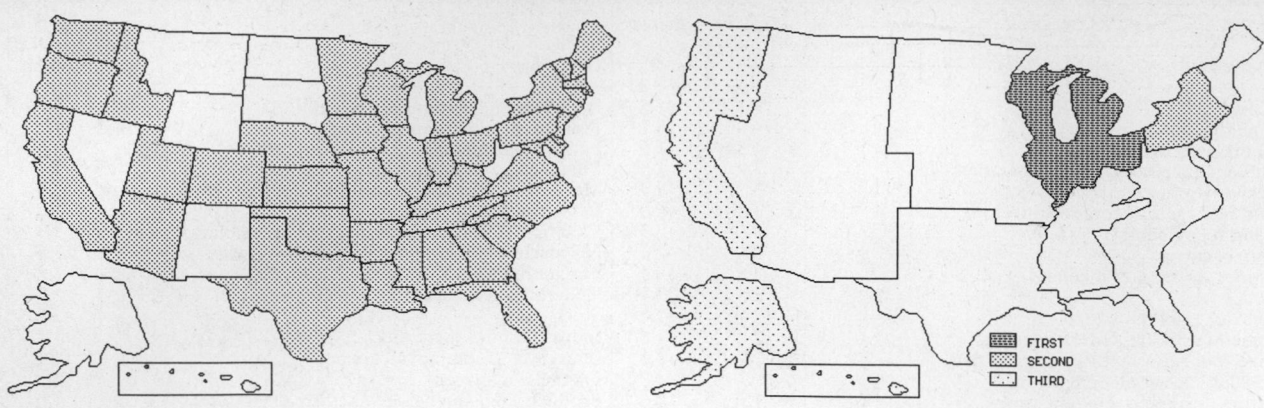

FIRST
SECOND
THIRD

INDUSTRY DATA BY STATE

| State | Establish-ments | Shipments | | | Employment | | | | Cost as % of Shipments | Investment per Employee ($) |
		Total ($ mil)	% of U.S.	Per Establ.	Total Number	% of U.S.	Per Establ.	Wages ($/hour)		
California	320	2,193.4	19.4	6.9	12,500	15.3	39	13.01	38.6	8,768
Ohio	210	1,118.3	9.9	5.3	8,200	10.0	39	13.27	51.0	3,341
Michigan	188	872.6	7.7	4.6	5,900	7.2	31	14.75	53.0	1,932
Massachusetts	117	855.5	7.6	7.3	5,700	7.0	49	15.96	42.9	5,333
Pennsylvania	134	622.2	5.5	4.6	4,800	5.9	36	12.51	51.9	2,625
Illinois	127	563.1	5.0	4.4	4,500	5.5	35	14.00	42.6	2,400
New York	137	545.0	4.8	4.0	3,800	4.6	28	15.28	44.7	4,421
New Jersey	141	431.8	3.8	3.1	3,300	4.0	23	14.60	44.8	1,788
Texas	137	423.0	3.7	3.1	3,500	4.3	26	10.36	46.4	2,200
Connecticut	65	401.7	3.6	6.2	2,800	3.4	43	15.59	43.1	2,607
Indiana	79	217.0	1.9	2.7	1,800	2.2	23	12.48	55.4	3,111
Oklahoma	40	212.7	1.9	5.3	2,000	2.4	50	11.60	40.2	12,750
Wisconsin	66	209.1	1.9	3.2	1,800	2.2	27	12.95	49.2	2,444
Minnesota	59	193.4	1.7	3.3	1,900	2.3	32	10.90	45.1	2,579
Georgia	51	161.4	1.4	3.2	1,600	2.0	31	10.55	57.2	4,813
North Carolina	69	161.2	1.4	2.3	1,400	1.7	20	11.53	43.2	2,929
Washington	37	160.7	1.4	4.3	1,400	1.7	38	13.37	50.7	2,429
Florida	69	156.1	1.4	2.3	1,300	1.6	19	9.69	44.7	2,000
Arizona	28	148.2	1.3	5.3	1,000	1.2	36	13.20	48.5	3,200
Tennessee	38	148.0	1.3	3.9	900	1.1	24	11.00	47.8	3,111
Missouri	41	143.8	1.3	3.5	1,000	1.2	24	12.10	35.5	3,200
Kansas	22	120.2	1.1	5.5	900	1.1	41	11.91	35.4	2,778
Arkansas	16	108.1	1.0	6.8	800	1.0	50	11.38	58.6	2,250
South Carolina	33	101.2	0.9	3.1	800	1.0	24	11.60	51.9	2,625
Virginia	25	96.1	0.9	3.8	900	1.1	36	12.27	41.0	1,667
Maryland	19	92.0	0.8	4.8	700	0.9	37	12.71	37.3	-
Mississippi	12	65.9	0.6	5.5	600	0.7	50	9.11	36.6	-
Alabama	18	61.1	0.5	3.4	600	0.7	33	10.14	61.0	2,333
Kentucky	24	54.5	0.5	2.3	600	0.7	25	10.43	53.8	-
Louisiana	13	49.8	0.4	3.8	500	0.6	38	12.33	43.8	1,800
Colorado	32	44.7	0.4	1.4	500	0.6	16	10.17	52.3	3,400
Rhode Island	26	43.1	0.4	1.7	400	0.5	15	10.00	39.0	4,500
Maine	9	38.9	0.3	4.3	200	0.2	22	16.00	58.4	500
New Hampshire	15	35.9	0.3	2.4	400	0.5	27	14.00	41.8	-
Vermont	9	31.9	0.3	3.5	200	0.2	22	10.00	46.7	-
Oregon	32	31.7	0.3	1.0	300	0.4	9	10.00	48.6	7,000
Iowa	36	(D)	-	-	750 *	0.9	21	-	-	7,600
Nebraska	10	(D)	-	-	175 *	0.2	18	-	-	-
Utah	9	(D)	-	-	750 *	0.9	83	-	-	-
Idaho	7	(D)	-	-	375 *	0.5	54	-	-	-
Delaware	4	(D)	-	-	175 *	0.2	44	-	-	-

Source: 1992 *Economic Census.* The states are in descending order of shipments or establishments (if shipment data are missing for the majority). The symbol (D) appears when data are withheld to prevent disclosure of competitive information. States marked with (D) are sorted by number of establishments. A dash (-) indicates that the data element cannot be calculated; * indicates the midpoint of a range.

3561 - PUMPS & PUMPING EQUIPMENT

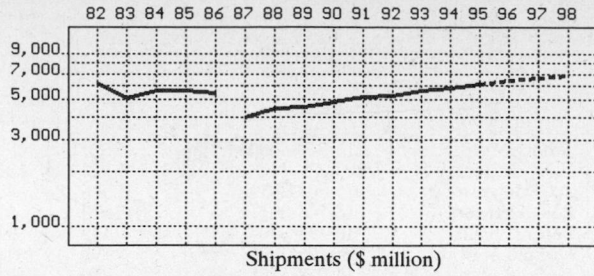

Shipments ($ million)

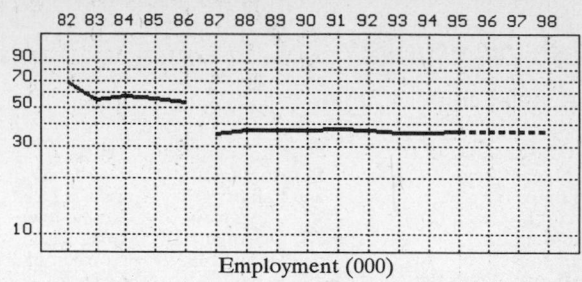

Employment (000)

GENERAL STATISTICS

Year	Com-panies	Establishments		Employment			Compensation		Production ($ million)			
		Total	with 20 or more employees	Total (000)	Production Workers (000)	Hours (Mil)	Payroll ($ mil)	Wages ($/hr)	Cost of Materials	Value Added by Manufacture	Value of Shipments	Capital Invest.
1982	518	626	325	68.5	39.1	74.7	1,484.7	10.33	2,742.4	3,336.8	6,198.3	227.5
1983		607	322	54.4	30.7	57.9	1,275.7	11.01	2,161.7	2,796.6	5,076.0	151.1
1984		588	319	57.5	33.7	65.4	1,403.9	10.94	2,509.2	3,206.9	5,680.1	175.5
1985		569	316	55.7	32.5	63.4	1,401.2	11.27	2,563.8	3,085.3	5,617.5	184.5
1986		544	307	52.8	29.7	58.7	1,391.3	11.74	2,421.7	3,012.0	5,433.6	142.9
1987*	333	405	226	35.2	19.7	38.4	969.9	11.95	1,837.2	2,154.5	3,998.3	95.8
1988		393	233	36.9	20.7	41.0	1,040.2	12.35	2,126.2	2,436.1	4,497.9	102.2
1989		370	226	37.0	20.4	40.6	1,077.1	12.71	2,179.2	2,379.6	4,520.0	99.8
1990		374	224	36.8	20.5	40.1	1,110.4	13.15	2,311.9	2,552.8	4,830.3	146.2
1991		401	232	37.8	20.8	40.5	1,168.1	13.42	2,501.7	2,706.0	5,218.1	140.1
1992	354	430	231	36.9	20.9	41.0	1,225.3	14.20	2,473.1	2,746.0	5,268.4	155.0
1993		465	233	36.0	20.5	40.4	1,234.8	14.50	2,558.8	3,049.2	5,594.3	148.9
1994		446P	233P	35.8	21.1	42.8	1,272.6	14.36	2,692.2	3,192.6	5,851.0	188.6
1995		456P	233P	36.6P	21.1P	42.2P	1,329.8P	15.06P	2,807.6P	3,329.4P	6,101.8P	190.4P
1996		467P	234P	36.6P	21.3P	42.5P	1,372.6P	15.45P	2,923.1P	3,466.4P	6,352.8P	202.8P
1997		477P	235P	36.6P	21.4P	42.9P	1,415.4P	15.83P	3,038.6P	3,603.4P	6,603.8P	215.3P
1998		487P	236P	36.6P	21.5P	43.2P	1,458.2P	16.22P	3,154.1P	3,740.3P	6,854.8P	227.7P

Sources: 1982, 1987, 1992 *Economic Census*; *Annual Survey of Manufactures*, 83-86, 88-91, 93-94. Establishment counts are from *County Business Patterns* for non-Census years; establishment counts for 83-84 are extrapolations. * indicates that industry content changed in 87; earlier years use 77 SICs. 'P's mark projections.

INDICES OF CHANGE

Year	Com-panies	Establishments		Employment			Compensation		Production ($ million)			
		Total	with 20 or more employees	Total (000)	Production Workers (000)	Hours (Mil)	Payroll ($ mil)	Wages ($/hr)	Cost of Materials	Value Added by Manufacture	Value of Shipments	Capital Invest.
1982	146	146	141	186	187	182	121	73	111	122	118	147
1983		141	139	147	147	141	104	78	87	102	96	97
1984		137	138	156	161	160	115	77	101	117	108	113
1985		132	137	151	156	155	114	79	104	112	107	119
1986		127	133	143	142	143	114	83	98	110	103	92
1987*	94	94	98	95	94	94	79	84	74	78	76	62
1988		91	101	100	99	100	85	87	86	89	85	66
1989		86	98	100	98	99	88	90	88	87	86	64
1990		87	97	100	98	98	91	93	93	93	92	94
1991		93	100	102	100	99	95	95	101	99	99	90
1992	100	100	100	100	100	100	100	100	100	100	100	100
1993		108	101	98	98	99	101	102	103	111	106	96
1994		104P	101P	97	101	104	104	101	109	116	111	122
1995		106P	101P	99P	101P	103P	109P	106P	114P	121P	116P	123P
1996		108P	101P	99P	102P	104P	112P	109P	118P	126P	121P	131P
1997		111P	102P	99P	102P	105P	116P	112P	123P	131P	125P	139P
1998		113P	102P	99P	103P	105P	119P	114P	128P	136P	130P	147P

Sources: Same as General Statistics. Values reflect change from the base year, 1992. Values above 100 mean greater than 92, values below 100 mean less than 92, and a value of 100 in the 82-91 or 93-98 period means same as 92. * indicates that industry content changed in 87. Data for earlier years are in 77 SIC format.

SELECTED RATIOS

For 1994	Avg. of All Manufact.	Analyzed Industry	Index	For 1994	Avg. of All Manufact.	Analyzed Industry	Index
Employees per Establishment	49	80	164	Value Added per Production Worker	134,084	151,308	113
Payroll per Establishment	1,500,273	2,852,450	190	Cost per Establishment	5,045,178	6,034,390	120
Payroll per Employee	30,620	35,547	116	Cost per Employee	102,970	75,201	73
Production Workers per Establishment	34	47	138	Cost per Production Worker	146,988	127,592	87
Wages per Establishment	853,319	1,377,604	161	Shipments per Establishment	9,576,895	13,114,633	137
Wages per Production Worker	24,861	29,128	117	Shipments per Employee	195,460	163,436	84
Hours per Production Worker	2,056	2,028	99	Shipments per Production Worker	279,017	277,299	99
Wages per Hour	12.09	14.36	119	Investment per Establishment	321,011	422,735	132
Value Added per Establishment	4,602,255	7,156,004	155	Investment per Employee	6,552	5,268	80
Value Added per Employee	93,930	89,179	95	Investment per Production Worker	9,352	8,938	96

Sources: Same as General Statistics. The 'Average of All Manufacturing' column represents the average of all manufacturing industries reported for the most recent complete year available. The Index shows the relationship between the Average and the Analyzed Industry. For example, 100 means that they are equal; 500 that the Analyzed Industry is five times the average; 50 means that the Analyzed Industry is half the national average. The abbreviation 'na' is used to show that data are 'not available'.

LEADING COMPANIES   Number shown: **75**   Total sales ($ mil): **12,155**   Total employment (000): **81.2**

Company Name	Address				CEO Name	Phone	Co. Type	Sales ($ mil)	Empl. (000)
Dresser Industries Inc	2001 Ross Av	Dallas	TX	75201	John J Murphy	214-740-6000	P	5,331	29.2
General Signal Corp	PO Box 10010	Stamford	CT	06904	E M Carpenter	203-329-4100	P	1,528	12.2
Goulds Pumps Inc	240 Fall St	Seneca Falls	NY	13148	T C McDermott	315-568-2811	P	586	4.2
Commercial Intertech Corp	1775 Logan Av	Youngstown	OH	44505	Paul J Powers	216-746-8011	P	517	4.5
BWIP Holding Inc	200 Oceangate Blv	Long Beach	CA	90802	Peter C Valli	310-435-3700	P	449	3.0
IDEX Corp	630 Dundee Rd	Northbrook	IL	60062	Donald N Boyce	708-498-7070	P	399	3.0
Sta-Rite Indust Water Syst	293 Wright St	Delavan	WI	53115	George Wardeberg	414-728-5551	D	260	0.8
Tuthill Corp	908 N Elm St	Hinsdale	IL	60521	JG Tuthill Jr	708-655-2266	R	200	1.8
ITT Bell and Gossett	8200 N Austin Av	Morton Grove	IL	60053	Pat DePalma	708-966-3700	D	150	0.8
Roper Industries Inc	PO Box 550	Bogart	GA	30622	Derrick N Key	706-335-5551	P	148	0.8
Gorman-Rupp Co	PO Box 1217	Mansfield	OH	44901	James C Gorman	419-755-1011	P	138	0.9
Robbins and Myers Inc	1400 Kettering Twr	Dayton	OH	45423	Daniel W Duval	513-222-2610	P	122	2.2
FE Myers Co	1101 Myers Pkwy	Ashland	OH	44805	Fred C Lavender	419-289-1144	S	114	0.6
Centrilift	200 W Stuart Roosa	Claremore	OK	74017	Joseph F Brady	918-341-9600	D	100	1.2
Sulzer Bingham Pumps Inc	2800 NW Front Av	Portland	OR	97210	Robert L Ayers	503-226-5200	S	100	0.9
Wheatly TXT Corp	6750 S 57th W Av	Tulsa	OK	74131	Gary L Rosenthal	918-446-4551	P	98	1.1
Viking Pump Inc	406 State St	Cedar Falls	IA	50613	Frank Hansen	319-266-1741	S	91	0.8
Wilden Pump & Engineering Co	22069 Van Buren St	Colton	CA	92324	Roark Moudy	909-422-1730	R	68	0.2
Hale Products Inc	700 Spring Mill Av	Conshohocken	PA	19428	Peter J Andrews	215-825-6300	R	65*	0.6
Waukesha Fluid Handling	611 Sugar Creek Rd	Delavan	WI	53115	James S Dahlke	414-728-1900	S	65*	0.5
Axelson Inc	PO Box 2427	Longview	TX	75606	Jim B McMichael	903-757-6650	S	60	0.5
Fairbanks Morse Pump Corp	PO Box 6999	Kansas City	KS	66106	WJ Letts	913-371-5000	R	60	0.4
Taco Inc	1160 Cranston St	Cranston	RI	02920	JH White	401-942-8000	R	60	0.4
CTI-Cryogenics	9 Hampshire St	Mansfield	MA	02048	RJ Lepofsky	508-337-5000	D	55	0.3
Pac-Fab Inc	1620 Hawkins Av	Sanford	NC	27330	Douglas Brittellel	919-774-4151	S	52	0.3
Ingersoll-Dresser Pump Co	PO Box 91	Taneytown	MD	21787	JE Pienat	410-756-2602	D	50	0.2
Waterous Co	300 JE Carroll Av	South St Paul	MN	55075	Donald J Haugen	612-450-5000	S	50	0.3
Calmar Inc	PO Box 1203	City of Industry	CA	91745	Richard Hartl	818-330-3161	R	48	0.5
Shurflo	12650 Westminster	Santa Ana	CA	92706	JW Casey	714-554-7709	R	47*	0.4
Johnston Pump Co	800 Koomey Rd	Brookshire	TX	77423	Robert K Elders	713-934-6012	R	46	0.3
Paco Pumps Inc	800 Koomey Rd	Brookshire	TX	77423	Robert K Elders	713-934-6012	R	46	0.3
Patterson Pump Co	PO Box 790	Toccoa	GA	30577	Albert Huber	706-886-2101	S	45	0.4
Haskel International Inc	100 E Graham Pl	Burbank	CA	91502	Maury S Friedman	818-843-4000	P	41	0.3
Gast Manufacturing Corp	PO Box 97	Benton Harbor	MI	49023	Warren Gast	616-926-6171	R	40	0.6
Hypro Corp	375 5th Av NW	New Brighton	MN	55112	WT Dudley	612-633-9300	R	40	0.1
Jacuzzi Brothers	PO Box 8903	Little Rock	AR	72219	D Van Duinen	501-455-1234	D	40	0.2
Little Giant Pump Co	PO Box 12010	Oklahoma City	OK	73157	Gabe Zablatnik	405-947-2511	S	40	0.3
Simer Pump Co	800 E 101st Ter	Kansas City	MO	64131	T Manning	816-943-4100	D	36	0.1
Roper Pump Co	PO Box 269	Commerce	GA	30529	Ron Bridgers	706-335-5551	D	35	0.3
Milton Roy Co	201 Ivyland Rd	Ivyland	PA	18974	Bob Mulholland	215-441-0800	D	33	0.2
Floway Pumps Inc	2494 S Railroad Av	Fresno	CA	93707	Raymond W Dunn	209-442-4000	S	30	0.2
Warren Pumps Inc	82 Bridges Av	Warren	MA	01083	Kevin Powers	413-436-7711	S	30	0.3
Zoeller Co	PO Box 16347	Louisville	KY	40216	RF Zoeller	502-778-2731	R	30	0.1
Hazleton Pumps Inc	225 N Cedar St	Hazleton	PA	18201	Mick Von Bergen	717-455-7711	R	28	0.2
GIW Industries Inc	5000 Wrightsboro	Grovetown	GA	30813	Thomas W Hagler	706-863-1011	S	27*	0.3
Marcum Natural Gas Services	1675 Broadway	Denver	CO	80202	W Phillip Marcum	303-592-5555	P	27	0.3
Barnes Pumps Inc	1485 Lexington Av	Mansfield	OH	44907	Paul Baldetti	419-774-1511	D	25*	0.2
Berkeley Pumps	293 Wright St	Delavan	WI	53115	George Wardenerg	414-728-7551	D	25	0.1
Layne and Bowler	800 Airport Rd	North Aurora	IL	60542	RE Goodwill Jr	708-000-0000	S	25*	0.2
Serfilco Ltd	1777 Shermer Rd	Northbrook	IL	60062	Jack H Berg	708-559-1777	R	25	0.3
Flint and Walling Industries Inc	95 N Oak St	Kendallville	IN	46755	B Stuckey	219-347-1600	D	24	0.1
Goulds Pumps Vertical Products	3951 Capitol Av	City of Industry	CA	91749	Bill Wigley	310-949-2113	D	24*	0.1
Aermotor Pumps Inc	PO Box 1364	Conway	AR	72032	Darryl Stearle	501-329-9811	S	22	0.2
Beckett Corp	2521 Willowbrook	Dallas	TX	75220	Wingate Sung	214-352-4200	R	21*	0.2
ASM Industries Inc	1 Lark Av	Leola	PA	17540	J Berg	717-656-2161	S	20	0.2
Ruthman Pump and Engineering	1212 Streng St	Cincinnati	OH	45223	Thomas R Ruthman	513-559-1900	R	20	0.2
Western Land Rollers Co	1341 W 2nd St	Hastings	NE	68901	DM Giovannini	402-463-1306	D	20	0.2
G/H Products Corp	PO Box 1199	Kenosha	WI	53141	Ole Petersen	414-694-1010	S	20	<0.1
American Machine and Tool	PO Box 70	Royersford	PA	19468	H Pollak	610-948-3800	R	18	<0.1
Pierce Co	PO Box 2000	Upland	IN	46989	Scott L Bowser	317-998-2712	S	18	0.2
Chas S Lewis and Company Inc	8625 Grant Rd	St Louis	MO	63123	Robert G Hanssen	314-843-4437	D	18	0.1
Warren Rupp Inc	PO Box 1568	Mansfield	OH	44901	David Fix	419-524-8388	D	17*	0.2
Graymills Corp	3705 N Lincoln Av	Chicago	IL	60613	GN Shields	312-248-6825	R	16	0.1
ACD Inc	2321 S Pullman St	Santa Ana	CA	92705	Robert E Crowl	714-261-7533	S	15	<0.1
Cornell Pump Co	2323 SE Harvester	Portland	OR	97222	Glenn O Tribe	503-653-0330	D	15	0.1
Gorman-Rupp Industries	180 Hines Av	Bellville	OH	44813	Leroy J Sargent	419-886-3001	D	15	0.1
Goulds Pumps Inc	E Center St	Ashland	PA	17921	Thomas McDermott	717-875-2660	D	15*	0.4
Gusher Pumps Inc	1212 Streng St	Cincinnati	OH	45223	Thomas R Ruthman	513-559-1900	S	15	0.2
Jensen International Inc	PO Box 1509	Coffeyville	KS	67337	JB Jensen	316-251-5700	S	15	0.1
Sturm Engineered Products Inc	PO Box 277	Barboursville	WV	25504	Joseph F Markus	304-736-3476	R	15	0.1
Weil Pump Company Inc	PO Box 887	Cedarburg	WI	53012	RI Bratt	414-377-1399	R	15	0.1
Buffalo Pumps Inc	874 Oliver St	N Tonawanda	NY	14120	CR Kistner	716-693-1850	S	14	0.1
Micropump Corp	PO Box 8975	Vancouver	WA	98668	W Ross	206-253-2008	R	14*	0.1
Procon Products	910 Ridgely Rd	Murfreesboro	TN	37129	John Zavisa	615-890-5710	D	14*	0.1
Environment/One Corp	2773 Balltown Rd	Schenectady	NY	12309	Angelo Dounoucos	518-346-6161	P	14	0.1

Source: Ward's Business Directory of U.S. Private and Public Companies, Volumes 1 and 2, 1996. The company type code used is as follows: P - Public, R - Private, S - Subsidiary, D - Division, J - Joint Venture, A - Affiliate, G - Group. Sales are in millions of dollars, employees are in thousands. An asterisk (*) indicates an estimated sales volume. The symbol < stands for 'less than'. Company names and addresses are truncated, in some cases, to fit into the available space.

MATERIALS CONSUMED

Material	Quantity	Delivered Cost ($ million)
Materials, ingredients, containers, and supplies	(X)	2,129.4
Fluid power products	(X)	186.0
Metal bolts, nuts, screws, washers, rivets, and other screw machine products	(X)	43.4
Metal stampings	(X)	27.1
Metal pipe, valves, and pipe fittings, except plumbers' and forgings	(X)	25.0
Metal tanks, heat exchangers, steam condensers, and other boiler products	(X)	24.8
All other fabricated metal products (except forgings)	(X)	52.4
Iron and steel forgings	(X)	28.6
Nonferrous forgings	(X)	5.7
Iron and steel castings (rough and semifinished)	(X)	298.1
Aluminum and aluminum-base alloy castings (rough and semifinished)	(X)	39.4
Other nonferrous castings (rough and semifinished)	(X)	70.8
Steel bars, bar shapes, and plates	(X)	109.4
Steel sheet and strip, including tin plate	(X)	16.6
All other steel shapes and forms	(X)	10.4
Copper and copper-base alloy shapes and forms	(X)	11.3
Aluminum and aluminum-base alloy shapes and forms	(X)	4.4
Other nonferrous shapes and forms	(X)	12.9
Engines (diesel, semidiesel, gasoline, and other carburetor)	(X)	56.7
Other fractional horsepower electric motors (under 1 hp)	(X)	176.0
Integral horsepower electric motors and generators (1 hp or more)	(X)	169.3
Ball and roller bearings (mounted or unmounted)	(X)	19.9
Paperboard containers, boxes, and corrugated paperboard	(X)	19.5
Fabricated rubber products, except gaskets	(X)	23.0
Fabricated plastics products (except gaskets, hoses, and belting)	(X)	46.7
Gaskets (all types), and packing and sealing devices	(X)	33.4
Electrical transmission, distribution, and control equipment	(X)	55.1
Paints, varnishes, stains, lacquers, shellacs, japans, enamels, and allied products	(X)	4.9
All other materials and components, parts, containers, and supplies	(X)	293.3
Materials, ingredients, containers, and supplies, nsk	(X)	265.4

Source: 1992 Economic Census. Explanation of symbols used: (D): Withheld to avoid disclosure of competitive data; na: Not available; (S): Withheld because statistical norms were not met; (X): Not applicable; (Z): Less than half the unit shown; nec: Not elsewhere classified; nsk: Not specified by kind; - : zero; * : 10-19 percent estimated; ** : 20-29 percent estimated.

PRODUCT SHARE DETAILS

Product or Product Class	% Share	Product or Product Class	% Share
Pumps and pumping equipment	100.00	Oil-well and oil-field pumps, except boiler feed (including the value of the driver if shipped as a complete unit)	29.82
Industrial pumps, except hydraulic fluid power pumps, automotive circulating pumps, and measuring and dispensing pumps	51.93	Other pumps, except automotive circulating pumps and measuring and dispensing pumps	48.57
Domestic water systems (pumps for farm and home use), excluding irrigation pumps	6.90	Parts and attachments for pumps and pumping equipment (except for hydraulic fluid power, and air and gas compressors)	23.98
Pumps, nec	15.15		
Domestic sump pumps (1 hp or less) (including the value of the driver if shipped as a complete unit)	21.62	Pumps and pumping equipment, nsk	2.04

Source: 1992 Economic Census. The values shown are percent of total shipments in an industry. Values of indented subcategories are summed in the main heading. The symbol (D) appears when data are withheld to prevent disclosure of competitive information. The abbreviation nsk stands for 'not specified by kind' and nec for 'not elsewhere classified'.

INPUTS AND OUTPUTS FOR PUMPS & COMPRESSORS

Economic Sector or Industry Providing Inputs	%	Sector	Economic Sector or Industry Buying Outputs	%	Sector
Imports	10.8	Foreign	Gross private fixed investment	37.9	Cap Inv
Wholesale trade	9.4	Trade	Exports	17.2	Foreign
Iron & steel foundries	8.2	Manufg.	Petroleum & natural gas well drilling	3.7	Constr.
Motors & generators	8.1	Manufg.	Miscellaneous repair shops	3.2	Services
Blast furnaces & steel mills	7.5	Manufg.	Coal	2.7	Mining
Pumps & compressors	4.1	Manufg.	Pumps & compressors	2.2	Manufg.
Advertising	3.3	Services	Federal Government purchases, national defense	2.2	Fed Govt
Machinery, except electrical, nec	2.6	Manufg.	Water transportation	1.6	Util.
Fabricated plate work (boiler shops)	2.1	Manufg.	Construction machinery & equipment	1.3	Manufg.
Power transmission equipment	2.1	Manufg.	Crude petroleum & natural gas	1.2	Mining
Internal combustion engines, nec	2.0	Manufg.	Blast furnaces & steel mills	1.2	Manufg.
Ball & roller bearings	1.9	Manufg.	Farm machinery & equipment	1.2	Manufg.
Electric services (utilities)	1.8	Util.	Soap & other detergents	1.0	Manufg.
Pipe, valves, & pipe fittings	1.5	Manufg.	Residential 1-unit structures, nonfarm	0.9	Constr.
Screw machine and related products	1.5	Manufg.	Nonfarm residential structure maintenance	0.8	Constr.
Aluminum castings	1.4	Manufg.	Office buildings	0.8	Constr.
Copper rolling & drawing	1.3	Manufg.	Explosives	0.8	Manufg.
Communications, except radio & TV	1.2	Util.	Internal combustion engines, nec	0.8	Manufg.
Iron & steel forgings	1.1	Manufg.	Residential garden apartments	0.7	Constr.

Continued on next page.

INPUTS AND OUTPUTS FOR PUMPS & COMPRESSORS - Continued

Economic Sector or Industry Providing Inputs	%	Sector	Economic Sector or Industry Buying Outputs	%	Sector
Special dies & tools & machine tool accessories	1.1	Manufg.	Oil field machinery	0.7	Manufg.
Miscellaneous plastics products	1.0	Manufg.	Pipelines, except natural gas	0.7	Util.
Maintenance of nonfarm buildings nec	0.9	Constr.	Aircraft & missile engines & engine parts	0.6	Manufg.
Nonferrous forgings	0.9	Manufg.	General industrial machinery, nec	0.6	Manufg.
Nonferrous wire drawing & insulating	0.9	Manufg.	Household refrigerators & freezers	0.6	Manufg.
Switchgear & switchboard apparatus	0.9	Manufg.	Industrial trucks & tractors	0.6	Manufg.
Air transportation	0.9	Util.	Printing trades machinery	0.6	Manufg.
Banking	0.9	Fin/R.E.	Ship building & repairing	0.6	Manufg.
Equipment rental & leasing services	0.9	Services	Special industry machinery, nec	0.6	Manufg.
Fabricated rubber products, nec	0.8	Manufg.	S/L Govt. purch., water	0.6	S/L Govt
General industrial machinery, nec	0.8	Manufg.	Electric utility facility construction	0.5	Constr.
Metal stampings, nec	0.8	Manufg.	Drugs	0.5	Manufg.
Paperboard containers & boxes	0.8	Manufg.	Plastics materials & resins	0.5	Manufg.
Petroleum refining	0.8	Manufg.	Truck & bus bodies	0.5	Manufg.
Motor freight transportation & warehousing	0.8	Util.	Aircraft	0.4	Manufg.
Aluminum rolling & drawing	0.7	Manufg.	Cold finishing of steel shapes	0.4	Manufg.
Brass, bronze, & copper castings	0.7	Manufg.	Condensed & evaporated milk	0.4	Manufg.
Gaskets, packing & sealing devices	0.7	Manufg.	Construction of hospitals	0.3	Constr.
Eating & drinking places	0.7	Trade	Maintenance of nonfarm buildings nec	0.3	Constr.
Computer & data processing services	0.7	Services	Primary aluminum	0.3	Manufg.
Engineering, architectural, & surveying services	0.7	Services	Steel wire & related products	0.3	Manufg.
Industrial controls	0.6	Manufg.	Copper ore	0.2	Mining
Gas production & distribution (utilities)	0.6	Util.	Iron & ferroalloy ores	0.2	Mining
Electronic computing equipment	0.5	Manufg.	Construction of stores & restaurants	0.2	Constr.
Nonferrous castings, nec	0.5	Manufg.	Industrial buildings	0.2	Constr.
Rubber & plastics hose & belting	0.5	Manufg.	Residential 2-4 unit structures, nonfarm	0.2	Constr.
Abrasive products	0.4	Manufg.	Carbon black	0.2	Manufg.
Machine tools, metal cutting types	0.4	Manufg.	Distilled liquor, except brandy	0.2	Manufg.
Sawmills & planning mills, general	0.4	Manufg.	Fabricated plate work (boiler shops)	0.2	Manufg.
Transformers	0.4	Manufg.	Fertilizers, mixing only	0.2	Manufg.
Real estate	0.4	Fin/R.E.	Gum & wood chemicals	0.2	Manufg.
Nonferrous rolling & drawing, nec	0.3	Manufg.	Machine tools, metal cutting types	0.2	Manufg.
Primary metal products, nec	0.3	Manufg.	Nitrogenous & phosphatic fertilizers	0.2	Manufg.
Hotels & lodging places	0.3	Services	Nonferrous wire drawing & insulating	0.2	Manufg.
Legal services	0.3	Services	Paperboard mills	0.2	Manufg.
Management & consulting services & labs	0.3	Services	Railroad equipment	0.2	Manufg.
U.S. Postal Service	0.3	Gov't	Service industry machines, nec	0.2	Manufg.
Electronic components nec	0.2	Manufg.	Sheet metal work	0.2	Manufg.
Metal heat treating	0.2	Manufg.	Synthetic rubber	0.2	Manufg.
Primary copper	0.2	Manufg.	Turbines & turbine generator sets	0.2	Manufg.
Railroads & related services	0.2	Util.	Electric services (utilities)	0.2	Util.
Insurance carriers	0.2	Fin/R.E.	Commercial fishing	0.1	Agric.
Royalties	0.2	Fin/R.E.	Dimension, crushed & broken stone	0.1	Mining
Accounting, auditing & bookkeeping	0.2	Services	Hotels & motels	0.1	Constr.
Business/professional associations	0.2	Services	Residential additions/alterations, nonfarm	0.1	Constr.
Scrap	0.2	Scrap	Guided missiles & space vehicles	0.1	Manufg.
Lubricating oils & greases	0.1	Manufg.	Industrial gases	0.1	Manufg.
Manifold business forms	0.1	Manufg.	Iron & steel foundries	0.1	Manufg.
Power driven hand tools	0.1	Manufg.	Lawn & garden equipment	0.1	Manufg.
Sanitary services, steam supply, irrigation	0.1	Util.	Malt beverages	0.1	Manufg.
Water transportation	0.1	Util.	Mining machinery, except oil field	0.1	Manufg.
Automotive rental & leasing, without drivers	0.1	Services	Nonferrous rolling & drawing, nec	0.1	Manufg.
Automotive repair shops & services	0.1	Services	Special dies & tools & machine tool accessories	0.1	Manufg.
Electrical repair shops	0.1	Services	Steel pipe & tubes	0.1	Manufg.
Noncomparable imports	0.1	Foreign	Tires & inner tubes	0.1	Manufg.

Source: Benchmark Input-Output Accounts for the U.S. Economy, 1982, U.S. Department of Commerce, Washington, D.C., July 1991. Data, as reported in the source, are organized by the 1977 SIC structure in use in 1982 but have been matched, as closely as is possible, to the 1987 SIC structure used in this book.

OCCUPATIONS EMPLOYED BY SIC 356 - GENERAL INDUSTRIAL MACHINERY

Occupation	% of Total 1994	Change to 2005	Occupation	% of Total 1994	Change to 2005
Assemblers, fabricators, & hand workers nec	7.9	-2.9	Machine tool cutting & forming etc. nec	1.5	-2.9
Machinists	4.9	6.9	Machine assemblers	1.5	-2.8
Sales & related workers nec	3.8	-2.9	Traffic, shipping, & receiving clerks	1.5	-6.5
Blue collar worker supervisors	3.8	-9.0	Combination machine tool operators	1.5	6.9
Inspectors, testers, & graders, precision	3.1	-2.9	Welding machine setters, operators	1.2	-12.6
Welders & cutters	2.8	31.2	Engineering, mathematical, & science managers	1.2	10.4
Mechanical engineers	2.8	6.9	Machine operators nec	1.2	-14.4
General managers & top executives	2.5	-7.8	Industrial production managers	1.2	-2.8
NC machine tool operators, metal & plastic	2.3	45.7	Bookkeeping, accounting, & auditing clerks	1.2	-27.2
Machine builders	2.3	31.1	Stock clerks	1.1	-21.0
Lathe & turning machine tool operators	2.2	-36.8	Engineering technicians & technologists nec	1.1	-2.9
Secretaries, ex legal & medical	2.1	-11.6	Machine forming operators, metal & plastic	1.1	-2.9
Grinding machine operators, metal & plastic	2.1	-12.6	General office clerks	1.1	-17.2
Drafters	2.0	-24.3	Production, planning, & expediting clerks	1.1	-2.8
Machine tool cutting operators, metal & plastic	1.8	-19.1	Helpers, laborers, & material movers nec	1.1	-2.8
Sheet metal workers & duct installers	1.6	-2.8	Drilling & boring machine tool workers	1.1	-51.4
Industrial machinery mechanics	1.6	6.9	Purchasing agents, ex trade & farm products	1.0	6.9

Source: *Industry-Occupation Matrix*, Bureau of Labor Statistics. These data relate to one or more 3-digit SIC industry groups rather than to a single 4-digit SIC. The change reported for each occupation to the year 2005 is a percent of growth or decline as estimated by the Bureau of Labor Statistics. The abbreviation nec stands for 'not elsewhere classified'.

LOCATION BY STATE AND REGIONAL CONCENTRATION

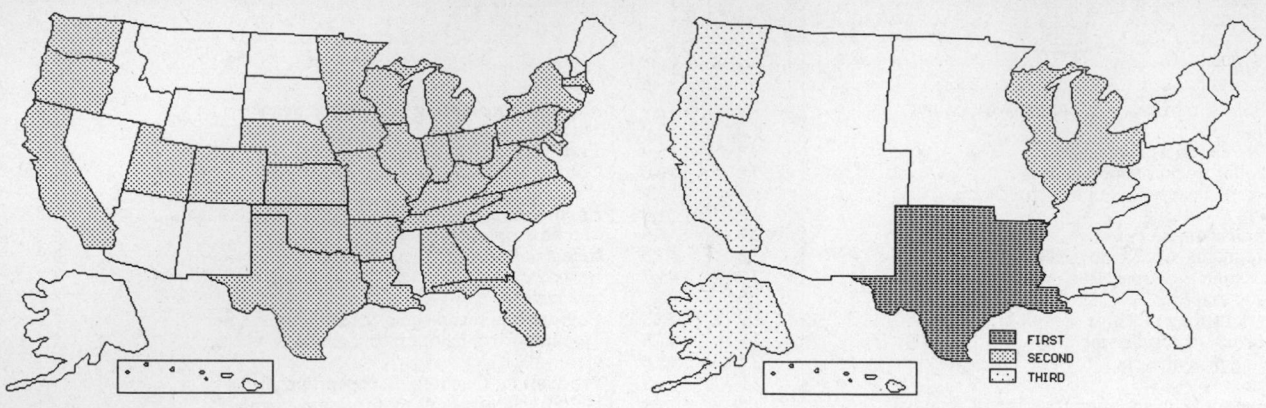

INDUSTRY DATA BY STATE

State	Establish-ments	Shipments			Employment				Cost as % of Shipments	Investment per Employee ($)
		Total ($ mil)	% of U.S.	Per Establ.	Total Number	% of U.S.	Per Establ.	Wages ($/hour)		
California	61	644.8	12.2	10.6	4,000	10.8	66	13.67	53.6	4,200
Ohio	25	597.7	11.3	23.9	4,200	11.4	168	13.88	44.9	3,786
Texas	56	430.8	8.2	7.7	3,100	8.4	55	13.80	52.0	2,806
New York	16	415.1	7.9	25.9	2,200	6.0	138	16.17	52.0	7,955
Oklahoma	24	377.3	7.2	15.7	2,800	7.6	117	15.38	46.5	3,750
Pennsylvania	21	375.4	7.1	17.9	3,100	8.4	148	17.50	39.5	3,355
Illinois	23	291.2	5.5	12.7	2,200	6.0	96	13.81	39.9	3,636
Minnesota	5	281.5	5.3	56.3	1,500	4.1	300	20.56	35.8	-
New Jersey	23	262.0	5.0	11.4	2,000	5.4	87	14.39	44.9	5,150
Wisconsin	11	174.8	3.3	15.9	1,300	3.5	118	13.87	57.2	4,385
Michigan	9	141.5	2.7	15.7	1,200	3.3	133	11.17	33.3	4,833
Indiana	8	139.2	2.6	17.4	800	2.2	100	11.20	65.6	3,000
Georgia	14	123.1	2.3	8.8	1,100	3.0	79	10.07	48.8	9,545
Massachusetts	7	74.1	1.4	10.6	600	1.6	86	24.25	36.6	3,000
Kansas	7	71.5	1.4	10.2	600	1.6	86	13.00	58.9	2,667
Tennessee	7	65.3	1.2	9.3	400	1.1	57	12.40	45.0	2,000
Nebraska	6	58.5	1.1	9.8	400	1.1	67	9.00	63.1	1,250
North Carolina	7	53.6	1.0	7.7	400	1.1	57	12.50	41.2	4,000
Florida	23	49.2	0.9	2.1	500	1.4	22	11.83	47.2	3,200
Washington	4	31.6	0.6	7.9	100	0.3	25	15.50	31.0	-
Louisiana	13	19.4	0.4	1.5	200	0.5	15	10.67	43.3	2,000
Alabama	4	10.1	0.2	2.5	100	0.3	25	15.00	30.7	2,000
Missouri	6	(D)	-	-	175 *	0.5	29	-	-	4,000
Arkansas	5	(D)	-	-	375 *	1.0	75	-	-	1,067
Maryland	4	(D)	-	-	175 *	0.5	44	-	-	2,286
Oregon	4	(D)	-	-	750 *	2.0	188	-	-	-
Colorado	3	(D)	-	-	375 *	1.0	125	-	-	-
Iowa	3	(D)	-	-	750 *	2.0	250	-	-	-
Kentucky	3	(D)	-	-	375 *	1.0	125	-	-	-
Virginia	3	(D)	-	-	175 *	0.5	58	-	-	-
Rhode Island	2	(D)	-	-	375 *	1.0	188	-	-	-
Utah	2	(D)	-	-	375 *	1.0	188	-	-	-
West Virginia	2	(D)	-	-	175 *	0.5	88	-	-	-

Source: 1992 *Economic Census*. The states are in descending order of shipments or establishments (if shipment data are missing for the majority). The symbol (D) appears when data are withheld to prevent disclosure of competitive information. States marked with (D) are sorted by number of establishments. A dash (-) indicates that the data element cannot be calculated; * indicates the midpoint of a range.

3562 - BALL & ROLLER BEARINGS

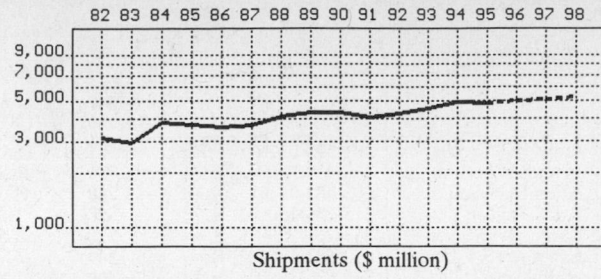

Shipments ($ million)

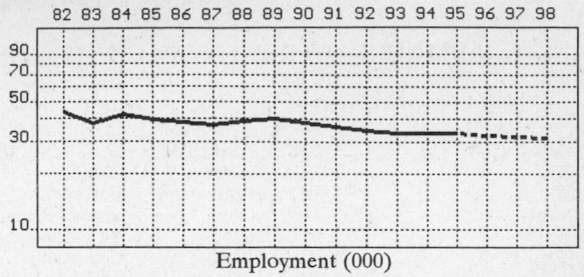

Employment (000)

GENERAL STATISTICS

Year	Companies	Establishments Total	Establishments with 20 or more employees	Employment Total (000)	Employment Production Workers (000)	Employment Hours (Mil)	Compensation Payroll ($ mil)	Compensation Wages ($/hr)	Production Cost of Materials	Production Value Added by Manufacture	Production Value of Shipments	Production Capital Invest.
1982	107	161	115	43.7	33.5	62.6	907.0	10.32	1,215.8	1,839.2	3,135.8	164.5
1983		164	115	37.7	29.9	58.4	863.7	10.88	1,272.0	1,655.3	2,964.6	114.7
1984		167	115	42.4	34.0	66.7	1,026.5	11.43	1,623.4	2,222.0	3,775.8	126.8
1985		169	115	39.6	31.1	60.8	969.6	11.76	1,410.2	2,220.4	3,679.3	138.7
1986		169	118	38.4	30.3	60.6	961.2	11.71	1,402.3	2,159.9	3,597.3	173.0
1987	113	169	116	36.9	29.2	60.0	949.9	11.99	1,511.5	2,203.3	3,723.7	154.7
1988		176	121	38.8	31.4	67.2	1,077.7	12.18	1,818.6	2,361.2	4,143.7	196.1
1989		172	120	39.9	32.1	66.9	1,108.3	12.70	1,828.9	2,575.5	4,327.3	271.2
1990		171	120	38.3	32.1	64.2	1,102.1	13.14	1,790.8	2,481.7	4,306.3	363.9
1991		181	126	36.6	29.8	57.3	1,054.0	13.85	1,540.3	2,451.9	4,051.2	305.7
1992	122	183	122	34.9	28.2	57.5	1,091.2	14.34	1,717.4	2,546.7	4,287.9	206.5
1993		185	126	33.8	27.3	57.2	1,081.5	14.48	1,778.0	2,746.6	4,557.4	205.5
1994		185P	126P	33.4	27.2	56.4	1,188.0	16.17	1,974.3	2,947.5	4,890.5	172.8
1995		187P	127P	33.5P	27.7P	58.3P	1,172.6P	15.56P	1,965.1P	2,933.7P	4,867.7P	269.6P
1996		189P	128P	32.9P	27.3P	57.9P	1,193.1P	15.97P	2,017.6P	3,012.2P	4,997.8P	279.6P
1997		191P	129P	32.2P	26.9P	57.4P	1,213.6P	16.38P	2,070.1P	3,090.6P	5,127.9P	289.6P
1998		193P	130P	31.6P	26.5P	57.0P	1,234.1P	16.79P	2,122.6P	3,169.0P	5,258.0P	299.7P

Sources: 1982, 1987, 1992 *Economic Census*; *Annual Survey of Manufactures*, 83-86, 88-91, 93-94. Establishment counts for non-Census years are from *County Business Patterns*; establishment values for 83-84 are extrapolations. 'P's show projections by the editors. Industries reclassified in 87 will not have data for prior years.

INDICES OF CHANGE

Year	Companies	Establishments Total	Establishments with 20 or more employees	Employment Total (000)	Employment Production Workers (000)	Employment Hours (Mil)	Compensation Payroll ($ mil)	Compensation Wages ($/hr)	Production Cost of Materials	Production Value Added by Manufacture	Production Value of Shipments	Production Capital Invest.
1982	88	88	94	125	119	109	83	72	71	72	73	80
1983		90	94	108	106	102	79	76	74	65	69	56
1984		91	94	121	121	116	94	80	95	87	88	61
1985		92	94	113	110	106	89	82	82	87	86	67
1986		92	97	110	107	105	88	82	82	85	84	84
1987	93	92	95	106	104	104	87	84	88	87	87	75
1988		96	99	111	111	117	99	85	106	93	97	95
1989		94	98	114	114	116	102	89	106	101	101	131
1990		93	98	110	114	112	101	92	104	97	100	176
1991		99	103	105	106	100	97	97	90	96	94	148
1992	100	100	100	100	100	100	100	100	100	100	100	100
1993		101	103	97	97	99	99	101	104	108	106	100
1994		101P	103P	96	96	98	109	113	115	116	114	84
1995		102P	104P	96P	98P	101P	107P	108P	114P	115P	114P	131P
1996		103P	105P	94P	97P	101P	109P	111P	117P	118P	117P	135P
1997		104P	106P	92P	96P	100P	111P	114P	121P	121P	120P	140P
1998		105P	107P	90P	94P	99P	113P	117P	124P	124P	123P	145P

Sources: Same as General Statistics. Values reflect change from the base year, 1992. Values above 100 mean greater than 92, values below 100 mean less than 92, and a value of 100 in the 82-91 or 93-98 period means same as 92. 'P's mark projections by the editors.

SELECTED RATIOS

For 1994	Avg. of All Manufact.	Analyzed Industry	Index	For 1994	Avg. of All Manufact.	Analyzed Industry	Index
Employees per Establishment	49	181	369	Value Added per Production Worker	134,084	108,364	81
Payroll per Establishment	1,500,273	6,423,200	428	Cost per Establishment	5,045,178	10,674,515	212
Payroll per Employee	30,620	35,569	116	Cost per Employee	102,970	59,111	57
Production Workers per Establishment	34	147	428	Cost per Production Worker	146,988	72,585	49
Wages per Establishment	853,319	4,930,876	578	Shipments per Establishment	9,576,895	26,441,632	276
Wages per Production Worker	24,861	33,529	135	Shipments per Employee	195,460	146,422	75
Hours per Production Worker	2,056	2,074	101	Shipments per Production Worker	279,017	179,798	64
Wages per Hour	12.09	16.17	134	Investment per Establishment	321,011	934,284	291
Value Added per Establishment	4,602,255	15,936,348	346	Investment per Employee	6,552	5,174	79
Value Added per Employee	93,930	88,249	94	Investment per Production Worker	9,352	6,353	68

Sources: Same as General Statistics. The 'Average of All Manufacturing' column represents the average of all manufacturing industries reported for the most recent complete year available. The Index shows the relationship between the Average and the Analyzed Industry. For example, 100 means that they are equal; 500 that the Analyzed Industry is five times the average; 50 means that the Analyzed Industry is half the national average. The abbreviation 'na' is used to show that data are 'not available'.

LEADING COMPANIES   Number shown: **71**   Total sales ($ mil): **10,919**   Total employment (000): **86.8**

Company Name	Address				CEO Name	Phone	Co. Type	Sales ($ mil)	Empl. (000)
Ingersoll-Rand Co	200 Chestnut Ridge	Woodcliff Lake	NJ	07675	James E Perrella	201-573-0123	P	4,507	35.0
Timken Co	1835 Dueber SW	Canton	OH	44706	Joseph F Toot Jr	216-438-3000	P	1,930	16.2
Torrington Co	59 Field St	Torrington	CT	06790	Allen M Nixon	203-482-9511	S	1,260•	11.0
SKF USA Inc	1100 1st Av	King of Prussia	PA	19406	R B Langton	215-265-1900	S	800	6.0
NSK Corp	PO Box 1507	Ann Arbor	MI	48106	Larry McPherson	313-761-9500	S	320	1.5
Kaydon Corp	19345 US Hwy 19 N	Clearwater	FL	34624	Lawrence J Cawley	813-531-1101	P	244	1.7
American Koyo Corp	29570 Clemens Rd	Westlake	OH	44145	H Shiose	216-835-1000	S	200	0.8
MPB Corp	PO Box 547	Keene	NH	03431	Scott Mathot	603-352-0310	S	140	1.5
MRC Bearings	402 Chandler St	Jamestown	NY	14701	Bengt Nilsson	716-661-2600	D	140	1.6
INA Bearing Company Inc	308 Springhill Farm	Fort Mill	SC	29715	Bruce Warmbold	803-548-8500	R	130	1.2
Brenco Inc	PO Box 389	Petersburg	VA	23804	N B Whitfield	804-732-0202	P	118	0.9
New Hampshire Ball Bearings	Rte 202 S	Peterborough	NH	03458	Gary Yomantas	603-924-3311	S	110•	1.1
McGill Manufacturing Company	909 N Lafayette St	Valparaiso	IN	46383	RE Swinehart	219-465-2200	S	100	1.4
Spherical Roller Bearing	525 Fame Av	Hanover	PA	17331	Inge Rahmquist	717-637-8981	D	75	0.4
Hoover Precision Products Inc	PO Box 899	Cumming	GA	30130	Carl Pfizenmaier	404-889-9223	S	72	0.6
Peer Bearing Co	241 W Palatine Rd	Wheeling	IL	60090	L W Spungen	708-870-3300	R	60	0.2
Roller Bearing of America	PO Box 1237	Newtown	PA	18940	Michael Hartnett	215-579-4300	R	50•	0.6
Thomson Precision Ball	PO Box 68	Unionville	CT	06085	Robert Lawson	203-673-2534	S	50•	0.1
Astro	155 Lexington Dr	Laconia	NH	03246	Gary Yomantas	603-524-0004	S	45	0.4
Nachi America Inc	223 Veterans Blv	Carlstadt	NJ	07072	T Hisazawa	201-935-1417	S	40	<0.1
Barden Corp	PO Box 2449	Danbury	CT	06813	John McCloskey	203-744-2211	S	32•	0.5
Kamatics Corp	1330 Blue Hills Av	Bloomfield	CT	06002	Alan A Whitfield	203-243-9704	S	32•	0.3
TransDigm Corp	26380 C Wright	Richmond H	OH	44143	Nick Howley	216-289-8900	S	32•	0.3
Bearing Service of Pennsylvania	500 Dargan St	Pittsburgh	PA	15224	Jacob W Banks	412-621-7300	R	30	0.1
Nucor Bearing Products Inc	PO Box 370	Wilson	NC	27893	J Downing	919-237-8181	D	30	0.3
Hartford Bearing Co	1022 Elm St	Rocky Hill	CT	06067	Laura Thompson	203-573-3602	R	25	0.1
Hartford Ball Co	1022 Elm St	Rocky Hill	CT	06067	George Gigon	203-563-0111	D	21•	0.2
Powder Metal Products Inc	PO Box 580	St Marys	PA	15857	Frances Samick	814-834-7261	R	21•	0.2
Margoil Bearing	15 Belmont St	Worcester	MA	01605	Scott Salter	508-755-6111	D	20	<0.1
Universal Bearings Inc	PO Box 38	Bremen	IN	46506	Wook Yang	219-546-2261	R	20	0.1
AeroControlex	26380 C Wright	Richmond H	OH	44143	Nick Howley	216-289-8900	D	17•	0.2
Frost Inc	2020 Bristol Av NW	Grand Rapids	MI	49504	Chad Frost	616-453-7781	R	16	0.1
Standard Locknut Inc	PO Box 662	Westfield	IN	46074	Robert F Waddell	317-867-0100	R	16•	0.2
Twentieth Century Machine Co	6070 E 18 Mile Rd	Sterling Hts	MI	48314	Ericka Dabringhaus	313-264-4641	R	16	0.1
Du Pont Tribon Composites Inc	5581 W 164th St	Cleveland	OH	44142	Terry L Stimeling	216-267-4100	S	14	0.1
Kahr Bearing	5675 W Burlingame	Tucson	AZ	85743	Donald Tarquin	602-744-1000	D	14•	0.1
Aetna Bearing Co	4600 W Schubert	Chicago	IL	60639	Patrick Balson	312-227-2410	R	12	0.2
Industrial Tectonics Inc	7222 Huron River	Dexter	MI	48130	Joanna Sutton	313-426-4681	S	12	<0.1
Industrial Tectonics Bearings	18301 Santa Fe Av	R Dominguez	CA	90221	Michael J Hartnett	310-537-3750	S	12•	<0.1
Oiles America Corp	14941 Cleat St	Plymouth	MI	48170	T Shirosaki	313-459-2940	S	12	<0.1
National Bearings Inc	1596 Manheim Pike	Lancaster	PA	17601	Jessica H May	717-569-0485	R	11	0.1
Winsted Precision Ball Co	PO Box 679	Winsted	CT	06098	Robert Davis	203-379-2788	D	10	0.1
Stein Seal Company Inc	PO Box 316	Kulpsville	PA	19443	PC Stein Jr	215-256-0201	R	9•	0.1
Green Ball Bearing Co	9801 Harvard Av	Cleveland	OH	44105	B Green	216-883-7800	R	8•	<0.1
IKS American Corp	1555 W Rosecrans	Gardena	CA	90249	Toshi Noma	213-770-2700	S	8	<0.1
Networks Electronic Corp	9750 De Soto Av	Chatsworth	CA	91311	Mahai D Patrichi	818-341-0440	P	7	<0.1
Randall Bearings Inc	PO Box 1258	Lima	OH	45802	BG Dickerson	419-223-1075	P	6	<0.1
Ball Screws and Actuators	3616 Snell Av	San Jose	CA	95136	W Scott Davis	408-629-1132	S	6•	<0.1
Schatz Bearing Corp	PO Box 1191	Poughkeepsie	NY	12602	Steve Pomeroy	914-452-6000	R	6•	<0.1
Schneeberger Inc	11 De Angelo Dr	Bedford	MA	01730	A Fischer	617-271-0140	S	5	<0.1
Barnes Industries Inc	1161 E 11 Mile Rd	Madison H	MI	48071	GR Barnes	313-541-2333	R	4	<0.1
Linear Industries Ltd	1850 Ent Way	Monrovia	CA	91016	Tony Angelica	818-303-1130	R	4	<0.1
Pacamor Kubar Bearings Inc	165 Jordan Rd	Troy	NY	12180	Augustine Sperraza	518-283-8003	R	4	<0.1
Ball and Roller Bearings Co	566 Danbury Rd	New Milford	CT	06776	H Nohe	203-355-4161	R	3	<0.1
Berliss Bearing Co	PO Box 45	Livingston	NJ	07039	FM Rubinstein	201-992-4242	R	3	<0.1
Messinger Bearings Corp	PO Box 9570	Philadelphia	PA	19124	Robert F Matthews	215-739-6880	R	3	<0.1
New England Mini Ball Co	PO Box 585	Norfolk	CT	06058	B Lawrence	203-542-5543	R	3	<0.1
PE-Del Mar Inc	625 W 18th St	Hialeah	FL	33010	Richard Riechelson	305-885-1911	S	3	<0.1
Quality Bearing Service	14570 Mayer	Fontana	CA	92336	Gino Ackerman	909-357-0267	S	3•	<0.1
US Bearing	9750 De Soto Av	Chatsworth	CA	91311	David Wachtel	818-341-0440	D	3	<0.1
Quality Bearing	PO Box 37	Gonzales	LA	70707	L Conner	504-673-6103	R	2•	<0.1
Royersford Foundry&Machine	808 Township Line	Phoenixville	PA	19460	Kurt B Deisher	215-935-7200	R	2•	<0.1
Seal Tech Inc	PO Box 389	Petersburg	VA	23804	Craig Rice	804-732-0202	S	2•	<0.1
Auburn Ball Bearing	PO Box 29	Victor	NY	14564	KK Fellows	716-924-3230	D	2	<0.1
Radial Bearing Corp	250 Taylor St	Danbury	CT	06810	S Papish	203-744-0323	R	2	<0.1
Bal-Tec	1550 E Slauson Av	Los Angeles	CA	90011	E A Gleason Jr	213-582-7348	D	1	<0.1
Lutco Bearings	130 Higgins St	Worcester	MA	01606	JC Stowe	508-853-3590	D	1•	<0.1
Micro Surface Engineering Inc	1550 E Slauson Av	Los Angeles	CA	90011	Eugene Gleason Jr	213-582-7348	R	1	<0.1
Moline Bearing Co	PO Box 509	St Charles	IL	60174	Dick Vanthournout	708-584-4600	R	1	<0.1
RH Little Co	4434 Southway SW	Canton	OH	44706	David Little	216-477-3455	R	1	<0.1
American Roller Bearing Co	150 Gamma Dr	Pittsburgh	PA	15238	Lawrence N Succop	412-781-1190	R	0	<0.1

Source: Ward's Business Directory of U.S. Private and Public Companies, Volumes 1 and 2, 1996. The company type code used is as follows: P - Public, R - Private, S - Subsidiary, D - Division, J - Joint Venture, A - Affiliate, G - Group. Sales are in millions of dollars, employees are in thousands. An asterisk (*) indicates an estimated sales volume. The symbol < stands for 'less than'. Company names and addresses are truncated, in some cases, to fit into the available space.

MATERIALS CONSUMED

Material	Quantity	Delivered Cost ($ million)
Materials, ingredients, containers, and supplies	(X)	1,499.0
Metal bolts, nuts, screws, washers, rivets, and other screw machine products	(X)	27.0
Other fabricated metal products (except castings and forgings)	(X)	18.0
Cold iron and steel forgings	(X)	26.5
Other iron and steel forgings	(X)	150.8
Iron and steel castings (rough and semifinished)	(X)	44.2
Nonferrous (aluminum, copper, etc.) castings (rough and semifinished)	(X)	6.5
Steel bars, bar shapes, and plates	(X)	184.0
Steel sheet and strip, including tin plate	(X)	59.2
All other steel shapes and forms	(X)	128.0
All other nonferrous shapes and forms	(X)	7.7
Ball bearings (mounted or unmounted)	(X)	16.0
Roller bearings (mounted or unmounted)	(X)	17.6
Balls, rollers, cages, collars, races, and other antifriction bearing components and parts	(X)	376.0
Clutches, couplings, shafts, sprockets, and other mechanical power transmission equipment	(X)	4.5
Electric motors, generators, and parts	(X)	5.7
Paperboard containers, boxes, and corrugated paperboard	(X)	12.6
All other materials and components, parts, containers, and supplies	(X)	363.9
Materials, ingredients, containers, and supplies, nsk	(X)	50.8

Source: 1992 Economic Census. Explanation of symbols used: (D): Withheld to avoid disclosure of competitive data; na: Not available; (S): Withheld because statistical norms were not met; (X): Not applicable; (Z): Less than half the unit shown; nec: Not elsewhere classified; nsk: Not specified by kind; - : zero; * : 10-19 percent estimated; ** : 20-29 percent estimated.

PRODUCT SHARE DETAILS

Product or Product Class	% Share	Product or Product Class	% Share
Ball and roller bearings	100.00	Mounted bearings, except plain	8.68
Ball bearings, complete, unmounted	37.58	Parts and components for ball and roller bearings, except cups and cones (including ball and rollers sold separately)	10.38
Tapered roller bearings (including cups and cones), unmounted	22.62	Ball and roller bearings, nsk	1.16
Roller bearings, except tapered, unmounted	19.58		

Source: 1992 Economic Census. The values shown are percent of total shipments in an industry. Values of indented subcategories are summed in the main heading. The symbol (D) appears when data are withheld to prevent disclosure of competitive information. The abbreviation nsk stands for 'not specified by kind' and nec for 'not elsewhere classified'.

INPUTS AND OUTPUTS FOR BALL & ROLLER BEARINGS

Economic Sector or Industry Providing Inputs	%	Sector	Economic Sector or Industry Buying Outputs	%	Sector
Imports	25.8	Foreign	Motor vehicle parts & accessories	7.9	Manufg.
Blast furnaces & steel mills	17.3	Manufg.	Exports	7.1	Foreign
Wholesale trade	10.0	Trade	Federal Government purchases, national defense	6.9	Fed Govt
Ball & roller bearings	9.3	Manufg.	Ball & roller bearings	5.3	Manufg.
Iron & steel forgings	4.5	Manufg.	Machinery, except electrical, nec	4.0	Manufg.
Machinery, except electrical, nec	3.7	Manufg.	Communications, except radio & TV	3.7	Util.
Electric services (utilities)	3.5	Util.	Coal	3.6	Mining
Iron & steel foundries	2.2	Manufg.	Construction machinery & equipment	3.0	Manufg.
Gas production & distribution (utilities)	1.7	Util.	Pumps & compressors	2.8	Manufg.
Abrasive products	1.3	Manutg.	Water transportation	2.8	Util.
Maintenance of nonfarm buildings nec	1.1	Constr.	Railroads & related services	2.7	Util.
Petroleum refining	1.1	Manufg.	Blast furnaces & steel mills	2.6	Manufg.
Eating & drinking places	1.0	Trade	Farm machinery & equipment	2.6	Manufg.
Air transportation	0.9	Util.	Aircraft & missile engines & engine parts	2.2	Manufg.
Motor freight transportation & warehousing	0.9	Util.	Miscellaneous repair shops	1.8	Services
Computer & data processing services	0.9	Services	Power transmission equipment	1.7	Manufg.
Engineering, architectural, & surveying services	0.9	Services	Refrigeration & heating equipment	1.6	Manufg.
Brass, bronze, & copper castings	0.8	Manufg.	Motor vehicles & car bodies	1.4	Manufg.
Power transmission equipment	0.8	Manufg.	Oil field machinery	1.4	Manufg.
Special dies & tools & machine tool accessories	0.7	Manufg.	Transit & bus transportation	1.4	Util.
Communications, except radio & TV	0.7	Util.	Motors & generators	1.3	Manufg.
Banking	0.7	Fin/R.E.	Nonferrous wire drawing & insulating	1.2	Manufg.
Advertising	0.7	Services	Aircraft & missile equipment, nec	1.0	Manufg.
Paperboard containers & boxes	0.6	Manufg.	Aluminum rolling & drawing	1.0	Manufg.
Copper rolling & drawing	0.5	Manufg.	Internal combustion engines, nec	1.0	Manufg.
Equipment rental & leasing services	0.5	Services	Crude petroleum & natural gas	0.9	Mining
U.S. Postal Service	0.5	Gov't	Conveyors & conveying equipment	0.9	Manufg.
Nonferrous castings, nec	0.4	Manufg.	Electronic computing equipment	0.9	Manufg.
Plating & polishing	0.4	Manufg.	Railroad equipment	0.9	Manufg.
Screw machine and related products	0.4	Manufg.	Pipelines, except natural gas	0.9	Util.
Legal services	0.4	Services	Food products machinery	0.8	Manufg.
Management & consulting services & labs	0.4	Services	Nonferrous rolling & drawing, nec	0.8	Manufg.
Metal heat treating	0.3	Manufg.	Special industry machinery, nec	0.8	Manufg.
Motors & generators	0.3	Manufg.	Steel wire & related products	0.8	Manufg.

Continued on next page.

INPUTS AND OUTPUTS FOR BALL & ROLLER BEARINGS - Continued

Economic Sector or Industry Providing Inputs	%	Sector	Economic Sector or Industry Buying Outputs	%	Sector
Railroads & related services	0.3	Util.	Blowers & fans	0.7	Manufg.
Hotels & lodging places	0.3	Services	Machine tools, metal cutting types	0.7	Manufg.
Lubricating oils & greases	0.2	Manufg.	Printing trades machinery	0.7	Manufg.
Machine tools, metal cutting types	0.2	Manufg.	Cold finishing of steel shapes	0.6	Manufg.
Manifold business forms	0.2	Manufg.	Fabricated plate work (boiler shops)	0.6	Manufg.
Nonmetallic mineral products, nec	0.2	Manufg.	Fabricated structural metal	0.6	Manufg.
Insurance carriers	0.2	Fin/R.E.	Fluid milk	0.6	Manufg.
Accounting, auditing & bookkeeping	0.2	Services	Household laundry equipment	0.6	Manufg.
Automotive rental & leasing, without drivers	0.2	Services	Mining machinery, except oil field	0.6	Manufg.
Electrical repair shops	0.2	Services	Power driven hand tools	0.6	Manufg.
Automotive repair shops & services	0.1	Services	Copper ore	0.5	Mining
Noncomparable imports	0.1	Foreign	Aircraft	0.5	Manufg.
Scrap	0.1	Scrap	General industrial machinery, nec	0.5	Manufg.
			Paper industries machinery	0.5	Manufg.
			Automotive stampings	0.4	Manufg.
			Industrial furnaces & ovens	0.4	Manufg.
			Metal office furniture	0.4	Manufg.
			Metal stampings, nec	0.4	Manufg.
			Service industry machines, nec	0.4	Manufg.
			Steel pipe & tubes	0.4	Manufg.
			Textile machinery	0.4	Manufg.
			X-ray apparatus & tubes	0.4	Manufg.
			Dimension, crushed & broken stone	0.3	Mining
			Bottled & canned soft drinks	0.3	Manufg.
			Engine electrical equipment	0.3	Manufg.
			Hoists, cranes, & monorails	0.3	Manufg.
			Industrial trucks & tractors	0.3	Manufg.
			Iron & steel foundries	0.3	Manufg.
			Lawn & garden equipment	0.3	Manufg.
			Machine tools, metal forming types	0.3	Manufg.
			Rolling mill machinery	0.3	Manufg.
			Ship building & repairing	0.3	Manufg.
			Woodworking machinery	0.3	Manufg.
			Chemical & fertilizer mineral	0.2	Mining
			Clay, ceramic, & refractory minerals	0.2	Mining
			Nonferrous metal ores, except copper	0.2	Mining
			Nonmetallic mineral services	0.2	Mining
			Sand & gravel	0.2	Mining
			Boat building & repairing	0.2	Manufg.
			Engineering & scientific instruments	0.2	Manufg.
			Environmental controls	0.2	Manufg.
			Fabricated metal products, nec	0.2	Manufg.
			Iron & steel forgings	0.2	Manufg.
			Knit fabric mills	0.2	Manufg.
			Mechanical measuring devices	0.2	Manufg.
			Metal doors, sash, & trim	0.2	Manufg.
			Metalworking machinery, nec	0.2	Manufg.
			Miscellaneous fabricated wire products	0.2	Manufg.
			Optical instruments & lenses	0.2	Manufg.
			Special dies & tools & machine tool accessories	0.2	Manufg.
			Transportation equipment, nec	0.2	Manufg.
			Turbines & turbine generator sets	0.2	Manufg.
			Typewriters & office machines, nec	0.2	Manufg.
			Welding apparatus, electric	0.2	Manufg.
			Electric services (utilities)	0.2	Util.
			Calculating & accounting machines	0.1	Manufg.
			Copper rolling & drawing	0.1	Manufg.
			Elevators & moving stairways	0.1	Manufg.
			Switchgear & switchboard apparatus	0.1	Manufg.
			Federal Government enterprises nec	0.1	Gov't
			Federal Government purchases, nondefense	0.1	Fed Govt

Source: Benchmark Input-Output Accounts for the U.S. Economy, 1982, U.S. Department of Commerce, Washington, D.C., July 1991. Data, as reported in the source, are organized by the 1977 SIC structure in use in 1982 but have been matched, as closely as is possible, to the 1987 SIC structure used in this book.

OCCUPATIONS EMPLOYED BY SIC 356 - GENERAL INDUSTRIAL MACHINERY

Occupation	% of Total 1994	Change to 2005	Occupation	% of Total 1994	Change to 2005
Assemblers, fabricators, & hand workers nec	7.9	-2.9	Machine tool cutting & forming etc. nec	1.5	-2.9
Machinists	4.9	6.9	Machine assemblers	1.5	-2.8
Sales & related workers nec	3.8	-2.9	Traffic, shipping, & receiving clerks	1.5	-6.5
Blue collar worker supervisors	3.8	-9.0	Combination machine tool operators	1.5	6.9
Inspectors, testers, & graders, precision	3.1	-2.9	Welding machine setters, operators	1.2	-12.6
Welders & cutters	2.8	31.2	Engineering, mathematical, & science managers	1.2	10.4
Mechanical engineers	2.8	6.9	Machine operators nec	1.2	-14.4
General managers & top executives	2.5	-7.8	Industrial production managers	1.2	-2.8
NC machine tool operators, metal & plastic	2.3	45.7	Bookkeeping, accounting, & auditing clerks	1.2	-27.2
Machine builders	2.3	31.1	Stock clerks	1.1	-21.0
Lathe & turning machine tool operators	2.2	-36.8	Engineering technicians & technologists nec	1.1	-2.9
Secretaries, ex legal & medical	2.1	-11.6	Machine forming operators, metal & plastic	1.1	-2.9
Grinding machine operators, metal & plastic	2.1	-12.6	General office clerks	1.1	-17.2
Drafters	2.0	-24.3	Production, planning, & expediting clerks	1.1	-2.8
Machine tool cutting operators, metal & plastic	1.8	-19.1	Helpers, laborers, & material movers nec	1.1	-2.8
Sheet metal workers & duct installers	1.6	-2.8	Drilling & boring machine tool workers	1.1	-51.4
Industrial machinery mechanics	1.6	6.9	Purchasing agents, ex trade & farm products	1.0	6.9

Source: Industry-Occupation Matrix, Bureau of Labor Statistics. These data relate to one or more 3-digit SIC industry groups rather than to a single 4-digit SIC. The change reported for each occupation to the year 2005 is a percent of growth or decline as estimated by the Bureau of Labor Statistics. The abbreviation nec stands for 'not elsewhere classified'.

LOCATION BY STATE AND REGIONAL CONCENTRATION

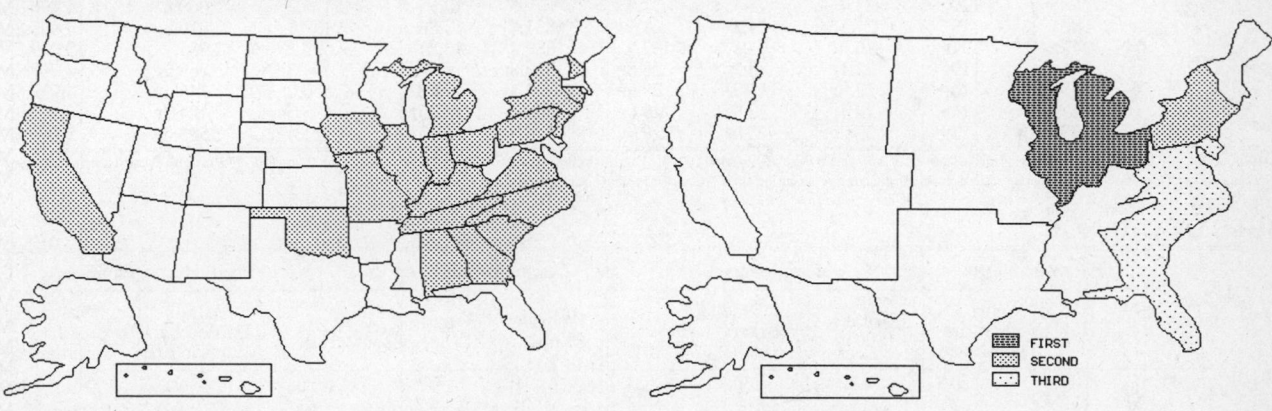

INDUSTRY DATA BY STATE

State	Establish-ments	Shipments Total ($ mil)	Shipments % of U.S.	Shipments Per Establ.	Employment Total Number	Employment % of U.S.	Employment Per Establ.	Wages ($/hour)	Cost as % of Shipments	Investment per Employee ($)
South Carolina	14	627.4	14.6	44.8	5,000	14.3	357	11.22	38.1	7,660
Pennsylvania	14	321.4	7.5	23.0	1,800	5.2	129	14.57	35.8	6,389
Connecticut	17	262.6	6.1	15.4	3,300	9.5	194	17.65	39.6	5,091
Indiana	10	255.1	5.9	25.5	2,100	6.0	210	13.71	33.0	2,905
Illinois	13	246.9	5.8	19.0	1,900	5.4	146	14.08	43.3	10,737
Tennessee	8	245.0	5.7	30.6	1,600	4.6	200	13.04	33.8	4,500
Georgia	9	242.7	5.7	27.0	2,100	6.0	233	11.18	40.1	3,667
New York	16	211.4	4.9	13.2	2,300	6.6	144	12.54	36.0	3,609
North Carolina	9	178.7	4.2	19.9	1,400	4.0	156	12.96	48.1	7,286
Michigan	10	152.0	3.5	15.2	1,000	2.9	100	17.27	35.3	8,500
New Jersey	8	50.8	1.2	6.3	400	1.1	50	11.83	45.3	-
California	11	47.8	1.1	4.3	400	1.1	36	14.00	34.5	-
Ohio	13	(D)	-	-	7,500 *	21.5	577	-	-	-
New Hampshire	5	(D)	-	-	1,750 *	5.0	350	-	-	-
Kentucky	4	(D)	-	-	750 *	2.1	188	-	-	3,067
Oklahoma	4	(D)	-	-	175 *	0.5	44	-	-	-
Missouri	2	(D)	-	-	375 *	1.1	188	-	-	-
Virginia	2	(D)	-	-	750 *	2.1	375	-	-	-
Alabama	1	(D)	-	-	375 *	1.1	375	-	-	-
Iowa	1	(D)	-	-	375 *	1.1	375	-	-	-

Source: 1992 *Economic Census.* The states are in descending order of shipments or establishments (if shipment data are missing for the majority). The symbol (D) appears when data are withheld to prevent disclosure of competitive information. States marked with (D) are sorted by number of establishments. A dash (-) indicates that the data element cannot be calculated; * indicates the midpoint of a range.

3563 - AIR & GAS COMPRESSORS

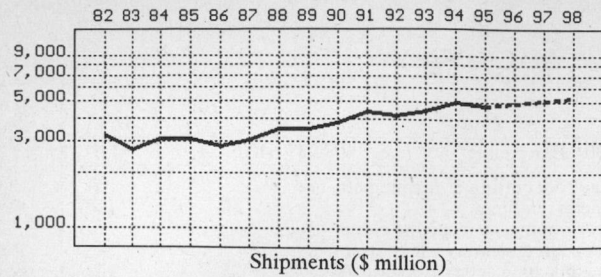

Shipments ($ million)

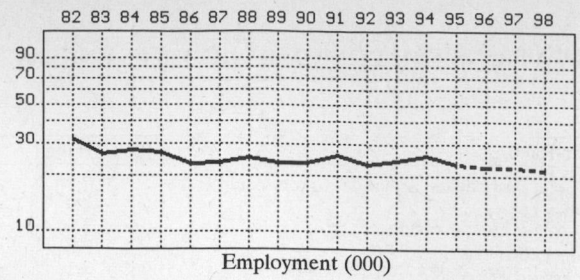

Employment (000)

GENERAL STATISTICS

| Year | Com-panies | Establishments | | Employment | | | Compensation | | Production ($ million) | | | |
		Total	with 20 or more employees	Total (000)	Production Workers (000)	Hours (Mil)	Payroll ($ mil)	Wages ($/hr)	Cost of Materials	Value Added by Manufacture	Value of Shipments	Capital Invest.
1982	239	282	144	32.1	17.5	34.1	709.3	10.09	1,698.3	1,470.6	3,270.0	118.1
1983		277	146	26.2	13.8	25.9	606.1	7.19	1,328.2	1,347.3	2,683.4	94.8
1984		272	148	27.8	15.3	29.9	695.0	11.23	1,507.3	1,586.8	3,108.9	115.3
1985		268	149	26.8	14.7	28.3	681.3	11.49	1,481.4	1,557.1	3,077.5	110.3
1986		262	149	23.3	12.1	23.7	640.1	12.21	1,419.0	1,392.7	2,817.5	68.0
1987	223	259	136	23.8	12.4	24.5	651.8	12.19	1,609.8	1,415.1	3,050.9	68.8
1988		252	134	25.2	14.0	28.4	735.4	12.68	1,879.6	1,606.4	3,485.7	70.7
1989		241	131	23.6	13.1	26.8	742.9	13.25	1,936.1	1,600.3	3,537.3	49.2
1990		234	138	23.7	13.5	27.4	768.1	13.38	2,057.6	1,769.9	3,806.9	60.3
1991		244	137	26.1	14.8	30.1	837.0	13.80	2,305.0	2,016.5	4,389.7	96.8
1992	220	258	120	23.4	13.5	27.4	777.5	13.95	2,120.2	2,069.8	4,170.3	138.0
1993		283	118	24.5	14.2	28.9	831.6	13.88	2,137.8	2,310.8	4,493.2	122.2
1994		247P	121P	26.0	16.3	32.9	886.2	13.81	2,593.2	2,385.1	5,019.2	129.4
1995		245P	119P	23.1P	13.9P	28.8P	863.8P	15.20P	2,464.3P	2,266.5P	4,769.7P	104.0P
1996		243P	116P	22.8P	13.9P	28.9P	882.1P	15.63P	2,550.0P	2,345.4P	4,935.6P	105.2P
1997		241P	114P	22.4P	13.8P	29.0P	900.4P	16.05P	2,635.7P	2,424.2P	5,101.5P	106.4P
1998		239P	112P	22.0P	13.8P	29.0P	918.7P	16.47P	2,721.4P	2,503.0P	5,267.4P	107.6P

Sources: 1982, 1987, 1992 *Economic Census*; *Annual Survey of Manufactures*, 83-86, 88-91, 93-94. Establishment counts for non-Census years are from *County Business Patterns*; establishment values for 83-84 are extrapolations. 'P's show projections by the editors. Industries reclassified in 87 will not have data for prior years.

INDICES OF CHANGE

| Year | Com-panies | Establishments | | Employment | | | Compensation | | Production ($ million) | | | |
		Total	with 20 or more employees	Total (000)	Production Workers (000)	Hours (Mil)	Payroll ($ mil)	Wages ($/hr)	Cost of Materials	Value Added by Manufacture	Value of Shipments	Capital Invest.
1982	109	109	120	137	130	124	91	72	80	71	78	86
1983		107	122	112	102	95	78	52	63	65	64	69
1984		105	123	119	113	109	89	81	71	77	75	84
1985		104	124	115	109	103	88	82	70	75	74	80
1986		102	124	100	90	86	82	88	67	67	68	49
1987	101	100	113	102	92	89	84	87	76	68	73	50
1988		98	112	108	104	104	95	91	89	78	84	51
1989		93	109	101	97	98	96	95	91	77	85	36
1990		91	115	101	100	100	99	96	97	86	91	44
1991		95	114	112	110	110	108	99	109	97	105	70
1992	100	100	100	100	100	100	100	100	100	100	100	100
1993		110	98	105	105	105	107	99	101	112	108	89
1994		96P	101P	111	121	120	114	99	122	115	120	94
1995		95P	99P	99P	103P	105P	111P	109P	116P	110P	114P	75P
1996		94P	97P	97P	103P	105P	113P	112P	120P	113P	118P	76P
1997		93P	95P	96P	102P	106P	116P	115P	124P	117P	122P	77P
1998		93P	93P	94P	102P	106P	118P	118P	128P	121P	126P	78P

Sources: Same as General Statistics. Values reflect change from the base year, 1992. Values above 100 mean greater than 92, values below 100 mean less than 92, and a value of 100 in the 82-91 or 93-98 period means same as 92. 'P's mark projections by the editors.

SELECTED RATIOS

For 1994	Avg. of All Manufact.	Analyzed Industry	Index	For 1994	Avg. of All Manufact.	Analyzed Industry	Index
Employees per Establishment	49	105	214	Value Added per Production Worker	134,084	146,325	109
Payroll per Establishment	1,500,273	3,581,264	239	Cost per Establishment	5,045,178	10,479,500	208
Payroll per Employee	30,620	34,085	111	Cost per Employee	102,970	99,738	97
Production Workers per Establishment	34	66	192	Cost per Production Worker	146,988	159,092	108
Wages per Establishment	853,319	1,836,091	215	Shipments per Establishment	9,576,895	20,283,321	212
Wages per Production Worker	24,861	27,874	112	Shipments per Employee	195,460	193,046	99
Hours per Production Worker	2,056	2,018	98	Shipments per Production Worker	279,017	307,926	110
Wages per Hour	12.09	13.81	114	Investment per Establishment	321,011	522,924	163
Value Added per Establishment	4,602,255	9,638,538	209	Investment per Employee	6,552	4,977	76
Value Added per Employee	93,930	91,735	98	Investment per Production Worker	9,352	7,939	85

Sources: Same as General Statistics. The 'Average of All Manufacturing' column represents the average of all manufacturing industries reported for the most recent complete year available. The Index shows the relationship between the Average and the Analyzed Industry. For example, 100 means that they are equal; 500 that the Analyzed Industry is five times the average; 50 means that the Analyzed Industry is half the national average. The abbreviation 'na' is used to show that data are 'not available'.

LEADING COMPANIES

Number shown: **75** Total sales ($ mil): **4,818** Total employment (000): **32.5**

Company Name	Address				CEO Name	Phone	Co. Type	Sales ($ mil)	Empl. (000)
Dresser-Rand Co	1 Baron Steuben Pl	Corning	NY	14830	Ben Stuart	607-937-6400	J	1,070	7.1
Thomas Industries Inc	PO Box 35120	Louisville	KY	40232	Timothy C Brown	502-893-4600	P	457	3.2
Binks Manufacturing Co	9201 W Belmont	Franklin Park	IL	60131	Burke B Roche	708-671-3000	P	244	1.7
DeVilbiss Industrial Coating	1724 Indian	Maumee	OH	43537	James Farrell	419-891-8100	S	220	2.0
CRL Industries Inc	2345 Waukegan Rd	Bannockburn	IL	60015	Thomas W Carroll	708-940-1500	R	210*	1.6
Gardner Denver Machinery Inc	PO Box 4024	Quincy	IL	62305	Ross J Centanni	217-222-5400	P	158	1.2
Compression Services	20602 E 81st St	Broken Arrow	OK	74014	Tom Gamble	918-251-8571	D	150	0.8
Dresser-Rand Co	PO Box 560	Olean	NY	14760	DW Norton	716-375-3000	D	140*	1.5
Spraying Systems Co	PO Box 7900	Wheaton	IL	60188	James E Bramsen	708-665-5000	R	130*	1.0
Metco	1101 Prospect Av	Westbury	NY	11590	Andrew B Mazzone	516-334-1300	D	130	0.4
Ansul Inc	1 Stanton St	Marinette	WI	54143	Karl Kinkead	715-735-7411	S	110	0.6
Atlas Copco Compressors Inc	PO Box 431	Holyoke	MA	01041	A Limongelli	413-536-0600	S	105	0.2
Sullair Corp	3700 E Michigan	Michigan City	IN	46360	Joseph Weifiger	219-879-5451	S	100*	0.8
Sanborn Manufacturing Co	118 W Rock St	Springfield	MN	56087	Steffen Magnell	507-723-6211	R	90	0.4
Ingersoll-Rand Co	PO Box 868	Mocksville	NC	27028	Chris Vasiloff	704-634-3561	D	78	0.7
Wahlco Environmental Systems	3600 W Segerstrom	Santa Ana	CA	92704	Henry Huta	714-979-7300	P	70	0.8
Wagner Spray Tech Corp	1770 Fernbrook Ln	Minneapolis	MN	55447	Sean James	612-553-0759	R	66*	0.3
Energy Industries Inc	PO Box 1979	Corpus Christi	TX	78467	P Holt	512-853-9933	S	65	0.4
Helix Technology Corp	9 Hampshire St	Mansfield	MA	02048	Robert J Lepofsky	508-337-5500	P	64	0.3
KCI Inc	6105 W 68th St	Tulsa	OK	74131	Neal Cartwright	918-446-1801	R	60	0.1
Power Application & Mfg Co	10777 E 45th Av	Denver	CO	80239	Thomas Gamble	303-371-0330	S	54	0.2
Thomas Industries Inc	1419 Illinois Av	Sheboygan	WI	53082	Bernard Berntson	414-457-4891	D	51*	0.4
Ariel Corp	35 Blackjack Rd	Mount Vernon	OH	43050	JP Buchwald	614-397-0311	R	50	0.2
Compressor Systems Inc	PO Box 60760	Midland	TX	79711	Johhny R Warren	915-563-1170	R	50	0.2
Leybold Vacuum Products Inc	5700 Mellon Rd	Export	PA	15632	RT Heglin	412-327-5700	S	50	0.3
Mattson Spray Equipment Inc	PO Box 132	Rice Lake	WI	54868	Mark Johnson	715-234-1617	R	50	<0.1
Graham Corp	PO Box 719	Batavia	NY	14021	F D Berkeley	716-343-2216	P	47	0.4
Graham Manufacturing	PO Box 719	Batavia	NY	14021	F D Berkeley	716-343-2216	P	47	0.4
Dresser Industries Inc	900 W Mount St	Connersville	IN	47331	MD Pope	317-827-9200	D	45	0.7
Aerosol Systems Inc	9100 Val View Rd	Macedonia	OH	44056	John H Elhert	216-468-1380	S	42*	0.2
LeROI International Inc	PO Box 90	Sidney	OH	45365	Mike Toal	513-492-2500	S	40	0.3
Stewert and Stevenson	PO Box 1181	Houston	TX	77251	Don Wallin	713-461-7867	R	40*	<0.1
Varian Associates Inc	121 Hartwell Av	Lexington	MA	02173	Peter Frasso	617-861-7200	D	30	0.2
Champion Pneumatic Machinery	1301 N Euclid Av	Princeton	IL	61356	Gilbert Williamson	815-875-3321	S	26*	0.2
Kinney Vacuum Co	495 Turnpike St	Canton	MA	02021	Kurt Bramer	617-828-9500	D	26	0.2
Rogers Machinery Company Inc	PO Box 23279	Portland	OR	97281	James D Kirkmire	503-639-6151	R	26	0.2
Cooper Industries Inc	3101 Broadway	Cheektowaga	NY	14225	EF Minter	716-896-6600	D	25	0.4
Emglo Products Corp	303 Indrial Pk Rd	Johnstown	PA	15904	Jerry N Fetter	814-269-1000	S	25	0.2
ABB Flakt Inc	1250 Brown Rd	Auburn Hills	MI	48326	Tomas E Mark	810-391-9000	D	21*	0.2
Wahlco Inc	3600 W Segerstrom	Santa Ana	CA	92704	VF Middleton	714-979-7300	S	20	<0.1
CompAir Kellogg	PO Box 737	Independence	VA	24348	Stephen A Lang	703-773-3100	S	18	<0.1
Howden Airdynamics	2616 Research Dr	Corona	CA	91720	Jeffery A Logan	909-734-0070	S	18*	<0.1
Curtis Toledo Inc	1905 Kienlen Av	St Louis	MO	63133	Kenneth Carpenter	314-383-1300	R	16	0.3
Sullivan Industries Inc	River Rd	Claremont	NH	03743	David Pollock	603-543-3131	R	15	<0.1
Cherco Compressors Inc	PO Box 7516	Longview	TX	75607	D Mike Stewart	903-753-4488	S	14	<0.1
Sharpe Manufacturing Co	8750 Pioneer Blv	Santa Fe Sprgs	CA	90670	Michael McCourt	310-908-6800	R	14	<0.1
Thermoform Plastics Inc	4221 Otter Lake Rd	St Paul	MN	55110	Curt Zamec	612-426-7319	S	14	0.2
Wittemann Company Inc	2 Commerce Blv	Palm Coast	FL	32164	W Geiger	904-445-4200	S	14	<0.1
Compressor Engineering Corp	5440 Alder St	Houston	TX	77081	Ernest G Hotze	713-664-7333	R	13	0.1
Panhandle Industrial Company	PO Box 702	Pampa	TX	79066	Loyd McKnight	806-665-1648	S	13*	0.1
Norwalk Company Inc	PO Box 548	South Norwalk	CT	06856	Arthur McCauley	203-838-4766	R	12	<0.1
Zeks Air Drier Corp	PO Box 396	Malvern	PA	19355	Ralph Clark	610-647-1600	R	12	<0.1
Grimmer-Schmidt Corp	PO Box 489	Franklin	IN	46131	Michael D Carter	317-736-8416	R	11	0.1
Sames Electrostatic Inc	11998 Merriman Rd	Livonia	MI	48150	Steve Mathers	313-261-5970	S	11	<0.1
Dover Corp	2605 W 42nd St	Odessa	TX	79764	Wayne Elder	915-367-7786	D	10	0.1
Mass Oxygen Equipment	PO Box 897	Westborough	MA	01581	John Finn	508-366-8361	R	10	0.1
NEAC Compressor Service USA	191 Howard St	Franklin	PA	16323	Andre Schmitz	814-437-3711	S	10	<0.1
Neuman and Esser USA Inc	1035 Dairy Ashford	Houston	TX	77079	Harinder S Gujral	713-497-5113	S	10	0.1
Bauer Compressors Inc	1328 Azalea Garden	Norfolk	VA	23502	Jan Von Dobeneck	804-855-6006	R	9*	0.1
Hardie-Tynes Manufacturing Co	PO Box 12166	Birmingham	AL	35202	Gordon L Flynn	205-252-5191	R	9	0.1
Howden Compressors Inc	23 Old Windsor Rd	Bloomfield	CT	06002	George W York	203-242-7351	S	9	<0.1
Graco Robotics Inc	1250 Brown Rd	Auburn Hills	MI	48326	Richard Armbrust	313-391-9000	S	8	<0.1
Sihi Pumps Inc	303 Industrial Blv	Grand Island	NY	14072	Robert H Mallette	716-773-6450	R	8	0.1
Earl E Knox Co	PO Box 1248	Erie	PA	16512	Robert E Knox	814-459-2754	R	8	0.1
Santa Fe Systems Co	PO Box 866	Atwood	CA	92601	GL Gates	310-944-6265	R	7	<0.1
Saylor Beall Manufacturing Co	PO Box 40	St Johns	MI	48879	Robert McFee	517-224-2371	R	7	<0.1
Air-Vac Engineering Co	PO Box 522	Milford	CT	06460	RA Duhaime	203-874-2541	R	6*	<0.1
ITW Ransburg Electrostatic Syst	320 Phillips Av	Toledo	OH	43612	Wade Hickam	419-470-2000	D	5	<0.1
Kremlin Inc	211 S Lombard Rd	Addison	IL	60101	John Patry	708-543-0708	S	5	<0.1
Exxel Container Inc	33 School House Rd	Somerset	NJ	08873	Ronald L Lemke	908-560-3655	R	4*	<0.1
EL Smith Air Compressors	63 Nottingham Rd	Deerfield	NH	03037	Frank Hodgman III	603-463-8311	D	4	<0.1
Gas Drying Inc	PO Box D6	Wharton	NJ	07885	G Behrens	201-361-2212	R	4	<0.1
Processall Inc	10596 Springfield Pk	Cincinnati	OH	45215	Albert J Shohet	513-771-2266	R	4	<0.1
Smith Air Compressors	1535 Old Louisville	Bowling Green	KY	42101	Jerry Fetler	502-842-1689	D	3	<0.1
Combined Fluid Products Co	805 Oakwood Rd	Lake Zurich	IL	60047	Bill Kist	708-540-0054	R	3	<0.1

Source: Ward's Business Directory of U.S. Private and Public Companies, Volumes 1 and 2, 1996. The company type code used is as follows: P - Public, R - Private, S - Subsidiary, D - Division, J - Joint Venture, A - Affiliate, G - Group. Sales are in millions of dollars, employees are in thousands. An asterisk (*) indicates an estimated sales volume. The symbol < stands for 'less than'. Company names and addresses are truncated, in some cases, to fit into the available space.

MATERIALS CONSUMED

Material	Quantity	Delivered Cost ($ million)
Materials, ingredients, containers, and supplies	(X)	1,629.9
Fluid power products	(X)	63.3
Metal bolts, nuts, screws, washers, rivets, and other screw machine products	(X)	39.4
Metal stampings	(X)	12.6
Metal pipe, valves, and pipe fittings, except plumbers' and forgings	(X)	39.1
Metal tanks, heat exchangers, steam condensers, and other boiler products	(X)	64.7
All other fabricated metal products (except forgings)	(X)	84.1
Iron and steel forgings	(X)	42.3
Nonferrous forgings	(X)	1.1
Iron and steel castings (rough and semifinished)	(X)	130.2
Aluminum and aluminum-base alloy castings (rough and semifinished)	(X)	28.1
Other nonferrous castings (rough and semifinished)	(X)	5.0
Steel bars, bar shapes, and plates	(X)	49.3
Steel sheet and strip, including tin plate	(X)	29.4
All other steel shapes and forms	(X)	10.9
Copper and copper-base alloy shapes and forms	(X)	5.2
Aluminum and aluminum-base alloy shapes and forms	(X)	3.8
Other nonferrous shapes and forms	(X)	8.9
Engines (diesel, semidiesel, gasoline, and other carburetor)	(X)	92.6
Other fractional horsepower electric motors (under 1 hp)	(X)	52.9
Integral horsepower electric motors and generators (1 hp or more)	(X)	83.2
Ball and roller bearings (mounted or unmounted)	(X)	19.5
Paperboard containers, boxes, and corrugated paperboard	(X)	14.2
Fabricated rubber products, except gaskets	(X)	6.5
Fabricated plastics products (except gaskets, hoses, and belting)	(X)	27.8
Gaskets (all types), and packing and sealing devices	(X)	28.6
Electrical transmission, distribution, and control equipment	(X)	68.2
Paints, varnishes, stains, lacquers, shellacs, japans, enamels, and allied products	(X)	7.2
All other materials and components, parts, containers, and supplies	(X)	370.2
Materials, ingredients, containers, and supplies, nsk	(X)	241.5

Source: 1992 *Economic Census.* Explanation of symbols used: (D): Withheld to avoid disclosure of competitive data; na: Not available; (S): Withheld because statistical norms were not met; (X): Not applicable; (Z): Less than half the unit shown; nec: Not elsewhere classified; nsk: Not specified by kind; - : zero; * : 10-19 percent estimated; ** : 20-29 percent estimated.

PRODUCT SHARE DETAILS

Product or Product Class	% Share	Product or Product Class	% Share
Air and gas compressors	100.00	Parts and attachments for air and gas compressors, except for refrigeration, ice making, and air-conditioning equipment	20.77
Air and gas compressors and vacuum pumps	58.01	Industrial spraying equipment	19.91
Air and gas vacuum pumps (compressors) (including value of the driver if shipped as a complete unit), except laboratory	8.59	Industrial power paint spraying outfits and other liquid power sprayers, except agricultural	92.23
Air and gas compressors, except compressors for ice making, refrigeration, or air-conditioning equipment, and air motors	91.22	Industrial hand sprayers, except agricultural and flame	7.19
		Industrial spraying equipment, nsk	0.58
Air and gas compressors and vacuum pumps, nsk	0.19	Air and gas compressors, nsk	1.31

Source: 1992 *Economic Census.* The values shown are percent of total shipments in an industry. Values of indented subcategories are summed in the main heading. The symbol (D) appears when data are withheld to prevent disclosure of competitive information. The abbreviation nsk stands for 'not specified by kind' and nec for 'not elsewhere classified'.

INPUTS AND OUTPUTS FOR PUMPS & COMPRESSORS

Economic Sector or Industry Providing Inputs	%	Sector	Economic Sector or Industry Buying Outputs	%	Sector
Imports	10.8	Foreign	Gross private fixed investment	37.9	Cap Inv
Wholesale trade	9.4	Trade	Exports	17.2	Foreign
Iron & steel foundries	8.2	Manufg.	Petroleum & natural gas well drilling	3.7	Constr.
Motors & generators	8.1	Manufg.	Miscellaneous repair shops	3.2	Services
Blast furnaces & steel mills	7.5	Manufg.	Coal	2.7	Mining
Pumps & compressors	4.1	Manufg.	Pumps & compressors	2.2	Manufg.
Advertising	3.3	Services	Federal Government purchases, national defense	2.2	Fed Govt
Machinery, except electrical, nec	2.6	Manufg.	Water transportation	1.6	Util.
Fabricated plate work (boiler shops)	2.1	Manufg.	Construction machinery & equipment	1.3	Manufg.
Power transmission equipment	2.1	Manufg.	Crude petroleum & natural gas	1.2	Mining
Internal combustion engines, nec	2.0	Manufg.	Blast furnaces & steel mills	1.2	Manufg.
Ball & roller bearings	1.9	Manufg.	Farm machinery & equipment	1.2	Manufg.
Electric services (utilities)	1.8	Util.	Soap & other detergents	1.0	Manufg.
Pipe, valves, & pipe fittings	1.5	Manufg.	Residential 1-unit structures, nonfarm	0.9	Constr.
Screw machine and related products	1.5	Manufg.	Nonfarm residential structure maintenance	0.8	Constr.
Aluminum castings	1.4	Manufg.	Office buildings	0.8	Constr.
Copper rolling & drawing	1.3	Manufg.	Explosives	0.8	Manufg.
Communications, except radio & TV	1.2	Util.	Internal combustion engines, nec	0.8	Manufg.
Iron & steel forgings	1.1	Manufg.	Residential garden apartments	0.7	Constr.

Continued on next page.

INPUTS AND OUTPUTS FOR PUMPS & COMPRESSORS - Continued

Economic Sector or Industry Providing Inputs	%	Sector	Economic Sector or Industry Buying Outputs	%	Sector
Special dies & tools & machine tool accessories	1.1	Manufg.	Oil field machinery	0.7	Manufg.
Miscellaneous plastics products	1.0	Manufg.	Pipelines, except natural gas	0.7	Util.
Maintenance of nonfarm buildings nec	0.9	Constr.	Aircraft & missile engines & engine parts	0.6	Manufg.
Nonferrous forgings	0.9	Manufg.	General industrial machinery, nec	0.6	Manufg.
Nonferrous wire drawing & insulating	0.9	Manufg.	Household refrigerators & freezers	0.6	Manufg.
Switchgear & switchboard apparatus	0.9	Manufg.	Industrial trucks & tractors	0.6	Manufg.
Air transportation	0.9	Util.	Printing trades machinery	0.6	Manufg.
Banking	0.9	Fin/R.E.	Ship building & repairing	0.6	Manufg.
Equipment rental & leasing services	0.9	Services	Special industry machinery, nec	0.6	Manufg.
Fabricated rubber products, nec	0.8	Manufg.	S/L Govt. purch., water	0.6	S/L Govt
General industrial machinery, nec	0.8	Manufg.	Electric utility facility construction	0.5	Constr.
Metal stampings, nec	0.8	Manufg.	Drugs	0.5	Manufg.
Paperboard containers & boxes	0.8	Manufg.	Plastics materials & resins	0.5	Manufg.
Petroleum refining	0.8	Manufg.	Truck & bus bodies	0.5	Manufg.
Motor freight transportation & warehousing	0.8	Util.	Aircraft	0.4	Manufg.
Aluminum rolling & drawing	0.7	Manufg.	Cold finishing of steel shapes	0.4	Manufg.
Brass, bronze, & copper castings	0.7	Manufg.	Condensed & evaporated milk	0.4	Manufg.
Gaskets, packing & sealing devices	0.7	Manufg.	Construction of hospitals	0.3	Constr.
Eating & drinking places	0.7	Trade	Maintenance of nonfarm buildings nec	0.3	Constr.
Computer & data processing services	0.7	Services	Primary aluminum	0.3	Manufg.
Engineering, architectural, & surveying services	0.7	Services	Steel wire & related products	0.3	Manufg.
Industrial controls	0.6	Manufg.	Copper ore	0.2	Mining
Gas production & distribution (utilities)	0.6	Util.	Iron & ferroalloy ores	0.2	Mining
Electronic computing equipment	0.5	Manufg.	Construction of stores & restaurants	0.2	Constr.
Nonferrous castings, nec	0.5	Manufg.	Industrial buildings	0.2	Constr.
Rubber & plastics hose & belting	0.5	Manufg.	Residential 2-4 unit structures, nonfarm	0.2	Constr.
Abrasive products	0.4	Manufg.	Carbon black	0.2	Manufg.
Machine tools, metal cutting types	0.4	Manufg.	Distilled liquor, except brandy	0.2	Manufg.
Sawmills & planning mills, general	0.4	Manufg.	Fabricated plate work (boiler shops)	0.2	Manufg.
Transformers	0.4	Manufg.	Fertilizers, mixing only	0.2	Manufg.
Real estate	0.4	Fin/R.E.	Gum & wood chemicals	0.2	Manufg.
Nonferrous rolling & drawing, nec	0.3	Manufg.	Machine tools, metal cutting types	0.2	Manufg.
Primary metal products, nec	0.3	Manufg.	Nitrogenous & phosphatic fertilizers	0.2	Manufg.
Hotels & lodging places	0.3	Services	Nonferrous wire drawing & insulating	0.2	Manufg.
Legal services	0.3	Services	Paperboard mills	0.2	Manufg.
Management & consulting services & labs	0.3	Services	Railroad equipment	0.2	Manufg.
U.S. Postal Service	0.3	Gov't	Service industry machines, nec	0.2	Manufg.
Electronic components nec	0.2	Manufg.	Sheet metal work	0.2	Manufg.
Metal heat treating	0.2	Manufg.	Synthetic rubber	0.2	Manufg.
Primary copper	0.2	Manufg.	Turbines & turbine generator sets	0.2	Manufg.
Railroads & related services	0.2	Util.	Electric services (utilities)	0.2	Util.
Insurance carriers	0.2	Fin/R.E.	Commercial fishing	0.1	Agric.
Royalties	0.2	Fin/R.E.	Dimension, crushed & broken stone	0.1	Mining
Accounting, auditing & bookkeeping	0.2	Services	Hotels & motels	0.1	Constr.
Business/professional associations	0.2	Services	Residential additions/alterations, nonfarm	0.1	Constr.
Scrap	0.2	Scrap	Guided missiles & space vehicles	0.1	Manufg.
Lubricating oils & greases	0.1	Manufg.	Industrial gases	0.1	Manufg.
Manifold business forms	0.1	Manufg.	Iron & steel foundries	0.1	Manufg.
Power driven hand tools	0.1	Manufg.	Lawn & garden equipment	0.1	Manufg.
Sanitary services, steam supply, irrigation	0.1	Util.	Malt beverages	0.1	Manufg.
Water transportation	0.1	Util.	Mining machinery, except oil field	0.1	Manufg.
Automotive rental & leasing, without drivers	0.1	Services	Nonferrous rolling & drawing, nec	0.1	Manufg.
Automotive repair shops & services	0.1	Services	Special dies & tools & machine tool accessories	0.1	Manufg.
Electrical repair shops	0.1	Services	Steel pipe & tubes	0.1	Manufg.
Noncomparable imports	0.1	Foreign	Tires & inner tubes	0.1	Manufg.

Source: Benchmark Input-Output Accounts for the U.S. Economy, 1982, U.S. Department of Commerce, Washington, D.C., July 1991. Data, as reported in the source, are organized by the 1977 SIC structure in use in 1982 but have been matched, as closely as is possible, to the 1987 SIC structure used in this book.

OCCUPATIONS EMPLOYED BY SIC 356 - GENERAL INDUSTRIAL MACHINERY

Occupation	% of Total 1994	Change to 2005	Occupation	% of Total 1994	Change to 2005
Assemblers, fabricators, & hand workers nec	7.9	-2.9	Machine tool cutting & forming etc. nec	1.5	-2.9
Machinists	4.9	6.9	Machine assemblers	1.5	-2.8
Sales & related workers nec	3.8	-2.9	Traffic, shipping, & receiving clerks	1.5	-6.5
Blue collar worker supervisors	3.8	-9.0	Combination machine tool operators	1.5	6.9
Inspectors, testers, & graders, precision	3.1	-2.9	Welding machine setters, operators	1.2	-12.6
Welders & cutters	2.8	31.2	Engineering, mathematical, & science managers	1.2	10.4
Mechanical engineers	2.8	6.9	Machine operators nec	1.2	-14.4
General managers & top executives	2.5	-7.8	Industrial production managers	1.2	-2.8
NC machine tool operators, metal & plastic	2.3	45.7	Bookkeeping, accounting, & auditing clerks	1.2	-27.2
Machine builders	2.3	31.1	Stock clerks	1.1	-21.0
Lathe & turning machine tool operators	2.2	-36.8	Engineering technicians & technologists nec	1.1	-2.9
Secretaries, ex legal & medical	2.1	-11.6	Machine forming operators, metal & plastic	1.1	-2.9
Grinding machine operators, metal & plastic	2.1	-12.6	General office clerks	1.1	-17.2
Drafters	2.0	-24.3	Production, planning, & expediting clerks	1.1	-2.8
Machine tool cutting operators, metal & plastic	1.8	-19.1	Helpers, laborers, & material movers nec	1.1	-2.8
Sheet metal workers & duct installers	1.6	-2.8	Drilling & boring machine tool workers	1.1	-51.4
Industrial machinery mechanics	1.6	6.9	Purchasing agents, ex trade & farm products	1.0	6.9

Source: Industry-Occupation Matrix, Bureau of Labor Statistics. These data relate to one or more 3-digit SIC industry groups rather than to a single 4-digit SIC. The change reported for each occupation to the year 2005 is a percent of growth or decline as estimated by the Bureau of Labor Statistics. The abbreviation nec stands for 'not elsewhere classified'.

LOCATION BY STATE AND REGIONAL CONCENTRATION

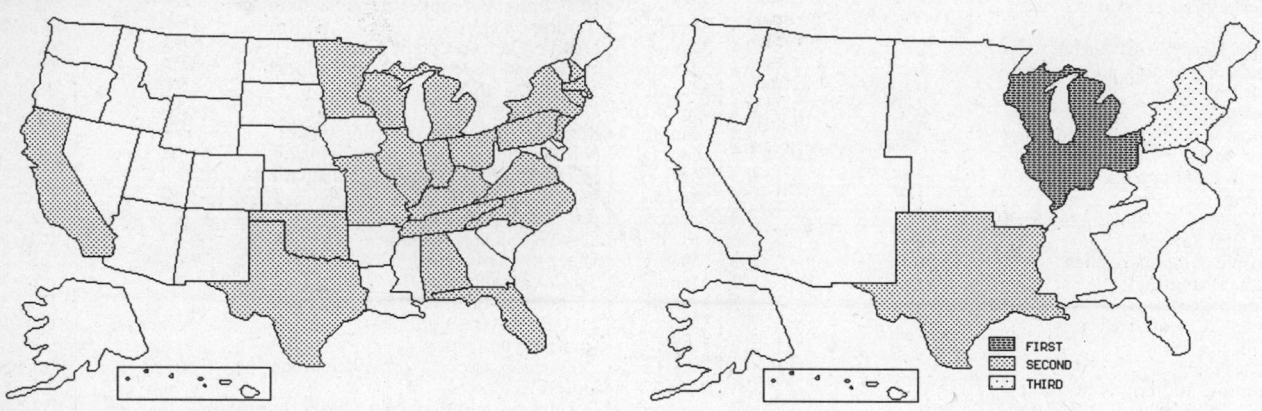

INDUSTRY DATA BY STATE

State	Establish-ments	Shipments			Employment				Cost as % of Shipments	Investment per Employee ($)
		Total ($ mil)	% of U.S.	Per Establ.	Total Number	% of U.S.	Per Establ.	Wages ($/hour)		
New York	13	772.4	18.5	59.4	3,700	15.8	285	14.37	53.5	-
Pennsylvania	17	485.2	11.6	28.5	3,200	13.7	188	15.16	38.8	5,219
Ohio	13	367.5	8.8	28.3	2,300	9.8	177	16.04	36.3	-
Illinois	24	363.7	8.7	15.2	3,100	13.2	129	14.58	34.7	3,968
Kentucky	7	212.2	5.1	30.3	1,200	5.1	171	13.19	56.8	-
Texas	33	182.6	4.4	5.5	1,200	5.1	36	12.87	52.4	1,833
Indiana	13	163.6	3.9	12.6	800	3.4	62	10.30	60.7	2,250
New Jersey	9	140.1	3.4	15.6	900	3.8	100	18.25	36.3	-
Missouri	5	94.9	2.3	19.0	300	1.3	60	14.60	35.5	-
California	25	87.6	2.1	3.5	700	3.0	28	11.71	48.2	3,714
Oklahoma	10	86.5	2.1	8.6	400	1.7	40	10.40	67.5	7,500
Virginia	3	40.8	1.0	13.6	200	0.9	67	8.67	59.1	3,500
New Hampshire	4	20.4	0.5	5.1	200	0.9	50	11.00	71.6	1,500
Florida	11	14.8	0.4	1.3	100	0.4	9	17.00	36.5	2,000
Michigan	9	(D)	-	-	750 *	3.2	83	-	-	-
Minnesota	8	(D)	-	-	750 *	3.2	94	-	-	-
North Carolina	6	(D)	-	-	1,750 *	7.5	292	-	-	-
Wisconsin	6	(D)	-	-	750 *	3.2	125	-	-	-
Alabama	4	(D)	-	-	175 *	0.7	44	-	-	571
Connecticut	4	(D)	-	-	750 *	3.2	188	-	-	533
Tennessee	4	(D)	-	-	375 *	1.6	94	-	-	-
Massachusetts	3	(D)	-	-	375 *	1.6	125	-	-	-

Source: 1992 Economic Census. The states are in descending order of shipments or establishments (if shipment data are missing for the majority). The symbol (D) appears when data are withheld to prevent disclosure of competitive information. States marked with (D) are sorted by number of establishments. A dash (-) indicates that the data element cannot be calculated; * indicates the midpoint of a range.

3564 - BLOWERS & FANS

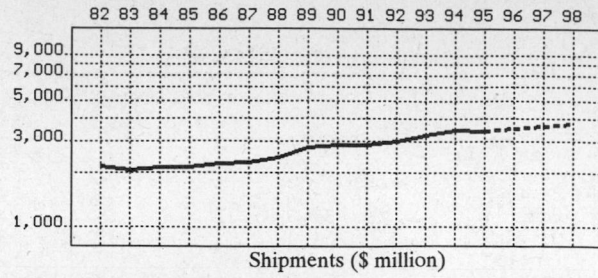

Shipments ($ million)

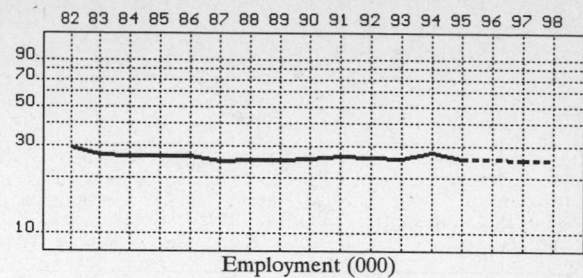

Employment (000)

GENERAL STATISTICS

Year	Companies	Establishments		Employment			Compensation		Production ($ million)			
		Total	with 20 or more employees	Total (000)	Production Workers (000)	Hours (Mil)	Payroll ($ mil)	Wages ($/hr)	Cost of Materials	Value Added by Manufacture	Value of Shipments	Capital Invest.
1982	450	502	240	29.8	19.1	37.2	553.6	8.23	999.8	1,160.0	2,173.5	57.1
1983		488	238	26.8	16.9	32.6	528.8	8.60	915.8	1,131.6	2,055.9	34.7
1984		474	236	26.4	17.8	34.4	533.7	8.89	955.9	1,169.7	2,119.4	47.9
1985		460	233	26.2	17.7	33.4	538.0	9.03	947.5	1,198.2	2,149.7	42.6
1986	445	445	225	26.3	17.5	34.5	575.6	9.43	1,034.5	1,155.5	2,239.1	53.6
1987	445	507	242	24.8	16.0	32.6	548.6	9.48	996.7	1,282.4	2,272.4	46.8
1988		494	245	25.2	16.9	33.9	576.0	9.60	1,097.6	1,365.6	2,441.0	42.1
1989		471	234	25.0	17.3	34.7	674.7	10.22	1,254.1	1,528.8	2,760.4	55.1
1990		482	244	25.5	16.8	33.9	724.8	10.72	1,341.7	1,519.6	2,850.1	56.9
1991		516	244	26.6	16.4	32.8	700.2	10.73	1,293.9	1,557.7	2,863.5	46.0
1992	517	587	257	26.0	17.5	34.9	723.9	11.30	1,339.0	1,647.9	3,000.9	60.7
1993		614	269	25.7	17.0	34.5	733.1	11.30	1,505.9	1,750.0	3,251.0	61.5
1994		562P	257P	27.8	18.8	38.9	783.0	11.21	1,569.1	1,905.2	3,459.5	54.7
1995		571P	259P	25.6P	17.1P	35.3P	786.8P	11.80P	1,541.1P	1,871.2P	3,397.8P	58.3P
1996		580P	261P	25.5P	17.1P	35.4P	809.2P	12.08P	1,593.6P	1,935.0P	3,513.5P	59.4P
1997		589P	263P	25.4P	17.0P	35.5P	831.6P	12.35P	1,646.1P	1,998.7P	3,629.3P	60.4P
1998		598P	265P	25.3P	17.0P	35.6P	853.9P	12.62P	1,698.6P	2,062.5P	3,745.0P	61.5P

Sources: 1982, 1987, 1992 *Economic Census*; *Annual Survey of Manufactures*, 83-86, 88-91, 93-94. Establishment counts for non-Census years are from *County Business Patterns*; establishment values for 83-84 are extrapolations. 'P's show projections by the editors. Industries reclassified in 87 will not have data for prior years.

INDICES OF CHANGE

Year	Companies	Establishments		Employment			Compensation		Production ($ million)			
		Total	with 20 or more employees	Total (000)	Production Workers (000)	Hours (Mil)	Payroll ($ mil)	Wages ($/hr)	Cost of Materials	Value Added by Manufacture	Value of Shipments	Capital Invest.
1982	87	86	93	115	109	107	76	73	75	70	72	94
1983		83	93	103	97	93	73	76	68	69	69	57
1984		81	92	102	102	99	74	79	71	71	71	79
1985		78	91	101	101	96	74	80	71	73	72	70
1986	86	76	88	101	100	99	80	83	77	70	75	88
1987	86	86	94	95	91	93	76	84	74	78	76	77
1988		84	95	97	97	97	80	85	82	83	81	69
1989		80	91	96	99	99	93	90	94	93	92	91
1990		82	95	98	96	97	100	95	100	92	95	94
1991		88	95	102	94	94	97	95	97	95	95	76
1992	100	100	100	100	100	100	100	100	100	100	100	100
1993		105	105	99	97	99	101	100	112	106	108	101
1994		96P	100P	107	107	111	108	99	117	116	115	90
1995		97P	101P	98P	98P	101P	109P	104P	115P	114P	113P	96P
1996		99P	102P	98P	98P	101P	112P	107P	119P	117P	117P	98P
1997		100P	102P	98P	97P	102P	115P	109P	123P	121P	121P	100P
1998		102P	103P	97P	97P	102P	118P	112P	127P	125P	125P	101P

Sources: Same as General Statistics. Values reflect change from the base year, 1992. Values above 100 mean greater than 92, values below 100 mean less than 92, and a value of 100 in the 82-91 or 93-98 period means same as 92. 'P's mark projections by the editors.

SELECTED RATIOS

For 1994	Avg. of All Manufact.	Analyzed Industry	Index	For 1994	Avg. of All Manufact.	Analyzed Industry	Index
Employees per Establishment	49	49	101	Value Added per Production Worker	134,084	101,340	76
Payroll per Establishment	1,500,273	1,392,638	93	Cost per Establishment	5,045,178	2,790,789	55
Payroll per Employee	30,620	28,165	92	Cost per Employee	102,970	56,442	55
Production Workers per Establishment	34	33	97	Cost per Production Worker	146,988	83,463	57
Wages per Establishment	853,319	775,589	91	Shipments per Establishment	9,576,895	6,153,040	64
Wages per Production Worker	24,861	23,195	93	Shipments per Employee	195,460	124,442	64
Hours per Production Worker	2,056	2,069	101	Shipments per Production Worker	279,017	184,016	66
Wages per Hour	12.09	11.21	93	Investment per Establishment	321,011	97,289	30
Value Added per Establishment	4,602,255	3,388,574	74	Investment per Employee	6,552	1,968	30
Value Added per Employee	93,930	68,532	73	Investment per Production Worker	9,352	2,910	31

Sources: Same as General Statistics. The 'Average of All Manufacturing' column represents the average of all manufacturing industries reported for the most recent complete year available. The Index shows the relationship between the Average and the Analyzed Industry. For example, 100 means that they are equal; 500 that the Analyzed Industry is five times the average; 50 means that the Analyzed Industry is half the national average. The abbreviation 'na' is used to show that data are 'not available'.

LEADING COMPANIES Number shown: 75   Total sales ($ mil): **4,379**   Total employment (000): **35.5**

Company Name	Address				CEO Name	Phone	Co. Type	Sales ($ mil)	Empl. (000)
Eagle Industries Inc	2 N Riverside Plz	Chicago	IL	60606	William K Hall	312-906-8700	P	1,142	9.4
Air & Water Technologies Corp	PO Box 1500	Somerville	NJ	08876	Arthur L Glenn	908-685-4600	P	523	3.6
Fasco Motors Group	500 Chesterfield Ctr	Chesterfield	MO	63017	Joseph L Palazzi	314-532-3505	S	470	4.0
National Safety Associates Inc	PO Box 18603	Memphis	TN	38181	Jay Martin	901-366-9288	R	150	0.4
NSA International Inc	4260 E Raines Rd	Memphis	TN	38118	A Jay Martin	901-541-1223	P	113	0.3
Farr Co	2221 Park Pl	El Segundo	CA	90245	HJ Meany	310-536-6300	P	107	1.2
Clarkson Industries Inc	2 Corporate Dr	Shelton	CT	06484	George H Bagnall	203-322-3990	S	96	1.0
PrecipTech Inc	8800 E 63rd St	Kansas City	MO	64133	James Lund	816-356-8400	S	88	0.6
Flair Corp	4647 SW 40th Av	Ocala	FL	34474	Richard A Bearse	904-237-1220	P	84	0.9
Environmental Elements Corp	PO Box 1318	Baltimore	MD	21203	F Bradford Smith	410-368-7000	P	73	0.3
Lakewood Eng & Mfg Co	501 N Sacramento	Chicago	IL	60612	Carl W Krauss	312-722-4300	R	70*	0.7
Lau Industrial	4509 Springfield St	Dayton	OH	45401	Will Jones	513-476-6500	D	65	0.7
Pollenex Corp	800 E 101st Ter	Kansas City	MO	64131	Tom Manning	816-943-4100	S	60	0.6
Twin City Fan and Blower Co	5959 Trenton Ln	Minneapolis	MN	55442	C Barry	612-551-7600	R	60	0.6
Acme Eng & Mfg Corp	PO Box 978	Muskogee	OK	74402	Edward Buddrus	918-682-7791	R	50	0.5
Fasco Consumer Products	810 Gillespie St	Fayetteville	NC	28306	Michael T Fuller	919-483-0421	D	50	0.5
New York Blower Co	7660 Quincy St	Willowbrook	IL	60521	Edward J Blondell	708-655-4881	R	48	0.5
Daw Technologies Inc	2700 S 900 W	Salt Lake City	UT	84119	Ronald W Daw	801-977-3100	P	48	0.5
Brookside Group Inc	PO Box 38	McCordsville	IN	46055	Claude A Forth Jr	317-335-2101	R	47*	0.4
Loren Cook Co	PO Box 4047	Springfield	MO	65808	GA Cook	417-869-6474	R	45	0.3
Smith Engineering Co	2837 E Cedar St	Ontario	CA	91761	J Archibald	909-923-3331	S	41	0.1
Penn Ventilator Company Inc	Red Lion & Gantry	Philadelphia	PA	19115	Donald A Silver	215-464-8900	R	36*	0.6
Revcor Inc	251 Edwards St	Carpentersville	IL	60110	JH Reichwein	708-428-4411	R	36	0.4
GE Environmental Systems	200 N 7th St	Lebanon	PA	17042	James Geurts	717-274-7000	D	35*	0.4
Joy-Green Fan	338 S Broadway	New Philad	OH	44663	James Miller	216-339-1111	D	35	0.2
Flair Pneumatic Products Corp	4647 SW 40th Av	Ocala	FL	32674	Richard Bearse	904-237-1220	S	35	0.3
Trion Inc	PO Box 760	Sanford	NC	27331	Steven L Schneider	919-775-2201	P	34	0.3
Spencer Turbine Co	600 Day Hill Rd	Windsor	CT	06095	R Rayve	203-688-8361	R	34	0.4
Chicago Blower Corp	1675 Glen Ellyn Rd	Glendale H	IL	60139	J Dubeck	708-858-2600	R	30	0.2
Deltech Engineering LP	PO Box 667	New Castle	DE	19720	TD Duffy	302-328-1345	S	30	0.3
Lamson Corp	PO Box 4857	Syracuse	NY	13221	Stephen R Beason	315-433-5500	S	30	0.2
Somerset Technologies Inc	PO Box 791	New Brunswick	NJ	08903	Frank Linsalata	908-356-6000	R	30	0.2
United Air Specialists Inc	4440 Creek Rd	Cincinnati	OH	45242	Durwood G Rorie	513-891-0400	P	29	0.3
Howden Fan Co	1 Westinghouse Plz	Hyde Park	MA	02136	HGY Lincoln	617-361-3700	S	27	0.2
Flanders Filters Inc	531 Flanders Filters	Washington	NC	27889	Robert Amerson	919-946-8081	R	24	0.3
TCF Aerovent Co	5959 Trenton Ln	Plymouth	MN	55442	Zika Srejovic	612-551-7500	S	22	0.2
American Fan Co	2933 Symmes Rd	Fairfield	OH	45014	Richard Niehaus	513-874-2400	R	21	0.2
Howard Industries Inc	1 N Dixie Hwy	Milford	IL	60953	Chuck McLaughlin	815-889-4105	R	21*	0.3
ABB Garden City Fan	PO Box 760	Niles	MI	49120	Doug Niemeyer	616-683-1150	S	20*	0.1
Hartzell Fan Inc	PO Box 919	Piqua	OH	45356	RW Wallace	513-773-7411	R	20	0.2
King Co	PO Box 287	Owatonna	MN	55060	Jerry Jungquist	507-451-3770	R	20	0.2
Hoffman Air & Filtration Syst	PO Box 548	East Syracuse	NY	13057	Stephen J Demello	315-432-8600	D	19*	0.2
Kooltronic Inc	PO Box 300	Hopewell	NJ	08525	Anne L Freedman	609-466-3400	R	19*	0.2
Airmaster Fan Co	1300 Falahee Rd	Jackson	MI	49203	Robert H LaZebnik	517-764-2300	R	16	0.1
Mikropul Envir Syst	20 Chatham Rd	Summit	NJ	07901	Louis A Luedtke	201-606-5900	D	16*	<0.1
Barry Blower	PO Box 1551	Minneapolis	MN	55440	CJ Tambornino	612-571-5200	D	15	0.2
Blocksom and Co	406 Center St	Michigan City	IN	46360	Clay T Barnes	219-874-3231	R	15*	0.2
Envirco Corp	6701 Jefferson NE	Albuquerque	NM	87109	David M Schlegel	505-345-3561	R	15	0.1
Ventline	PO Box 629	Bristol	IN	46507	Dan Alt	219-848-4491	D	15	0.3
Air Systems Clarage Fans	PO Box 2206	Birmingham	AL	35201	Karl L Strick	205-942-2211	D	14	0.1
American Coolair Corp	PO Box 2300	Jacksonville	FL	32203	Harry Graves Jr	904-389-3646	R	14	0.2
Aerotech Inc	929 Terminal Rd	Lansing	MI	48906	R Mitchell	517-323-2930	R	13	<0.1
World Dryer Corp	5700 McDermott Dr	Berkeley	IL	60163	Randy M Cordova	708-449-6950	D	13*	<0.1
Metal-Fab Inc	PO Box 1138	Wichita	KS	67201	Ken Shannon	316-943-2351	R	12*	0.2
Andersen 2000 Inc	306 Dividend Dr	Peachtree City	GA	30269	Jack Brady	404-997-2000	S	11	<0.1
Able Corp	23695 Via Del Rio	Yorba Linda	CA	92687	EF Ward	714-692-0200	R	10	<0.1
Air Kontrol Inc	PO Box 597	Batesville	MS	38606	Joe B Reed	601-563-4736	R	10*	<0.1
Airfoil Impellers Corp	PO Box 4669	Bryan	TX	77801	Dennis Anderholm	409-822-6418	R	10*	<0.1
Busch Co	904 Mount Royal	Pittsburgh	PA	15223	AM Halapin	412-487-4131	R	10	<0.1
Essick Air Products Inc	5800 Murray St	Little Rock	AR	72209	David Pritchard	501-562-1094	S	10	<0.1
Honeywell Envir Air Control	747 Bowman Av	Hagerstown	MD	21740	William O'Hara	301-797-9700	S	10	<0.1
Laminaire Corp	960 E Hazelwood	Rahway	NJ	07065	CJ Garay	908-381-8200	R	10	0.1
Paxton Products Inc	1260 Calle Suerte	Camarillo	CA	93012	Joseph Granatelli	805-987-5555	R	10	<0.1
Tek Air Systems Inc	157 Veterans Dr	Northvale	NJ	07647	Kenneth Kolkebeck	201-784-8700	R	10	<0.1
Engineered Cooling System Inc	201 W Carmel Dr	Carmel	IN	46032	DG Pieper	317-846-3438	R	10	0.1
Glasfloss Industries Inc	PO Box 427	Millersport	OH	43046	Scott Lange	614-467-2010	R	9	0.1
Steelcraft Corp	PO Box 12748	Memphis	TN	38182	Lewis Ford	901-452-5200	R	9	0.1
Air-Cure Technologies Inc	PO Box 811630	Cleveland	OH	44181	John R Cummings	216-243-0700	D	8*	<0.1
RP Fedder Corp	PO Box 92	Rochester	NY	14601	RP Fedder	716-288-1600	R	8*	<0.1
Aercology Inc	Custom Dr	Old Saybrook	CT	06475	Earle M Hamilton	203-399-7941	R	8	<0.1
Chelsea Fans and Blowers	PO Box 767	Jackson	MI	49204	Daniel Gregorich	517-764-6200	D	7	0.1
Hepa Corp	3071 E Coronado St	Anaheim	CA	92806	Richard J Braman	714-630-5700	R	7	<0.1
Sly Inc	PO Box 5939	Cleveland	OH	44101	TB Kurz	216-891-3200	R	7	<0.1
Aget Manufacturing Co	PO Box 248	Adrian	MI	49221	Ray Wakefield	517-263-5781	R	6	<0.1
Air Quality Engineering Inc	3340 Winpark Dr	Minneapolis	MN	55427	Heidi Oas	612-544-4426	R	6	<0.1

Source: Ward's Business Directory of U.S. Private and Public Companies, Volumes 1 and 2, 1996. The company type code used is as follows: P - Public, R - Private, S - Subsidiary, D - Division, J - Joint Venture, A - Affiliate, G - Group. Sales are in millions of dollars, employees are in thousands. An asterisk (*) indicates an estimated sales volume. The symbol < stands for 'less than'. Company names and addresses are truncated, in some cases, to fit into the available space.

MATERIALS CONSUMED

Material	Quantity	Delivered Cost ($ million)
Materials, ingredients, containers, and supplies	(X)	1,189.8
Metal bolts, nuts, screws, washers, rivets, and other screw machine products	(X)	16.6
Other fabricated metal products (except electrical enclosures and forgings)	(X)	66.8
Forgings	(X)	9.7
Iron and steel castings (rough and semifinished)	(X)	30.1
Aluminum and aluminum-base alloy castings (rough and semifinished)	(X)	16.3
Other nonferrous castings (rough and semifinished)	(X)	1.5
Steel bars, bar shapes, and plates	(X)	44.1
Steel sheet and strip, including tin plate	(X)	98.2
Steel structural shapes and sheet piling	(X)	11.9
All other steel shapes and forms	(X)	19.7
Copper and copper-base alloy shapes and forms	(X)	2.1
Aluminum and aluminum-base alloy sheet, plate, foil, and welded tubing	(X)	27.0
All other aluminum and aluminum-base alloy shapes and forms	(X)	9.3
Other nonferrous shapes and forms	(X)	4.6
Fractional horsepower electric motors (less than 1 hp)	(X)	48.3
Integral horsepower electric motors and generators (1 hp or more)	(X)	69.6
Ball and roller bearings (mounted or unmounted)	(X)	20.7
Electrical enclosures (metal and plastics)	(X)	7.5
Air filtration systems and parts	(X)	30.2
Paperboard containers, boxes, and corrugated paperboard	(X)	19.9
Flexible packaging materials	(X)	1.6
All other materials and components, parts, containers, and supplies	(X)	412.1
Materials, ingredients, containers, and supplies, nsk	(X)	222.0

Source: 1992 Economic Census. Explanation of symbols used: (D): Withheld to avoid disclosure of competitive data; na: Not available; (S): Withheld because statistical norms were not met; (X): Not applicable; (Z): Less than half the unit shown; nec: Not elsewhere classified; nsk: Not specified by kind; - : zero; * : 10-19 percent estimated; ** : 20-29 percent estimated.

PRODUCT SHARE DETAILS

Product or Product Class	% Share	Product or Product Class	% Share
Blowers and fans	100.00	parts	7.06
Centrifugal fans and blowers	24.73	Centrifugal type power roof ventilators, except parts	13.93
Centrifugal blower-filter units	13.47	Parts for power roof ventilators	1.38
Centrifugal classes I and II fans (more than 1 1/2 in. to 6 3/4 in. maximum total pressure)	9.94	Propeller fans and accessories, axial fans and power roof ventilators, and parts, nsk	1.13
Centrifugal classes III and IV fans (more than 6 3/4 in. maximum total pressure)	9.95	Dust collection and other air purification equipment for cleaning incoming air	31.97
Industrial centrifugal fans, excluding blowers, turboblowers, and multistage blowers	20.13	Air washers (purification equipment for cleaning incoming air), except parts	6.00
Positive displacement centrifugal blowers, excluding turboblowers	13.80	Electrostatic precipitation dust collection and air purification equipment, except parts	8.34
Multistage centrifugal blowers	7.78	Air filters for air-conditioners and furnaces, etc., of 2400 CFM or less, except parts	33.45
Small housed centrifugal blowers (utility sets)	2.10		
Other centrifugal fans and blowers (including furnace blowers, lightweight air-conditioning blowers, and turboblowers)	21.62	Other dust collection and other air purification equipment (including air filters for air-conditioners and furnaces), except parts	44.60
Centrifugal fans and blowers, nsk	1.23	Parts for dust collection and air purification equipment	4.72
Propeller fans and accessories, axial fans and power roof ventilators, and parts	20.72	Dust collection and other air purification equipment for cleaning incoming air, nsk	2.89
Axial fans directly connected to driver	22.21	Dust collection and other air purification equipment for industrial gas cleaning systems (for cleaning outgoing air)	18.17
Belt-driven axial fans	5.69		
Industrial propeller fans and accessories directly connected to driver	13.88	Dust collection and other air purification equipment for industrial gas cleaning systems (for cleaning outgoing air), except parts	97.10
Industrial belt-driven propeller fans and accessories	11.18		
Propeller fan penthouses, shutters, guards, and other accessories	3.54	Parts for industrial air purification equipment	1.96
Parts for fans and blowers	20.01	Dust collection and other air purification equipment for industrial gas cleaning systems, nsk	0.94
Axial and propeller type power roof ventilators, except		Blowers and fans, nsk	4.42

Source: 1992 Economic Census. The values shown are percent of total shipments in an industry. Values of indented subcategories are summed in the main heading. The symbol (D) appears when data are withheld to prevent disclosure of competitive information. The abbreviation nsk stands for 'not specified by kind' and nec for 'not elsewhere classified'.

INPUTS AND OUTPUTS FOR BLOWERS & FANS

Economic Sector or Industry Providing Inputs	%	Sector	Economic Sector or Industry Buying Outputs	%	Sector
Imports	21.8	Foreign	Gross private fixed investment	22.4	Cap Inv
Blast furnaces & steel mills	11.0	Manufg.	Exports	7.2	Foreign
Wholesale trade	8.0	Trade	Office buildings	5.0	Constr.
Pipe, valves, & pipe fittings	5.7	Manufg.	Water transportation	4.8	Util.
Blowers & fans	5.4	Manufg.	Blowers & fans	4.5	Manufg.
Motors & generators	4.4	Manufg.	Ship building & repairing	3.8	Manufg.
Nonferrous wire drawing & insulating	4.0	Manufg.	Refrigeration & heating equipment	3.4	Manufg.
Power transmission equipment	3.3	Manufg.	Industrial trucks & tractors	3.3	Manufg.
Machinery, except electrical, nec	2.7	Manufg.	Miscellaneous repair shops	3.2	Services
Internal combustion engines, nec	2.1	Manufg.	Railroad equipment	3.1	Manufg.
Nonwoven fabrics	1.9	Manufg.	Construction of hospitals	2.9	Constr.
Aluminum rolling & drawing	1.8	Manufg.	Coal	2.7	Mining
Iron & steel foundries	1.6	Manufg.	Fabricated plate work (boiler shops)	2.5	Manufg.
Copper rolling & drawing	1.4	Manufg.	Blast furnaces & steel mills	2.3	Manufg.
Ball & roller bearings	1.2	Manufg.	General industrial machinery, nec	2.2	Manufg.
Special dies & tools & machine tool accessories	1.2	Manufg.	Railroads & related services	2.1	Util.
Screw machine and related products	1.1	Manufg.	Cold finishing of steel shapes	2.0	Manufg.
Metal stampings, nec	1.0	Manufg.	Construction of stores & restaurants	1.9	Constr.
Electric services (utilities)	1.0	Util.	Heating equipment, except electric	1.8	Manufg.
Aluminum castings	0.9	Manufg.	Service industry machines, nec	1.6	Manufg.
Fabricated metal products, nec	0.9	Manufg.	Electric utility facility construction	1.4	Constr.
Felt goods, nec	0.9	Manufg.	Motor freight transportation & warehousing	1.2	Util.
Motor freight transportation & warehousing	0.9	Util.	Construction of educational buildings	1.1	Constr.
Maintenance of nonfarm buildings nec	0.8	Constr.	Iron & ferroalloy ores	0.8	Mining
Environmental controls	0.7	Manufg.	Amusement & recreation building construction	0.8	Constr.
Eating & drinking places	0.7	Trade	Maintenance of nonfarm buildings nec	0.8	Constr.
Rubber & plastics hose & belting	0.6	Manufg.	Residential 1-unit structures, nonfarm	0.8	Constr.
Air transportation	0.6	Util.	Federal Government purchases, national defense	0.8	Fed Govt
Communications, except radio & TV	0.6	Util.	Electric services (utilities)	0.7	Util.
Banking	0.6	Fin/R.E.	Industrial buildings	0.6	Constr.
Engineering, architectural, & surveying services	0.6	Services	Residential garden apartments	0.6	Constr.
Metal coating & allied services	0.5	Manufg.	Sewer system facility construction	0.6	Constr.
Petroleum refining	0.5	Manufg.	Credit agencies other than banks	0.6	Fin/R.E.
Real estate	0.5	Fin/R.E.	Hotels & motels	0.5	Constr.
Advertising	0.5	Services	Nonfarm residential structure maintenance	0.5	Constr.
Abrasive products	0.4	Manufg.	Residential high-rise apartments	0.5	Constr.
Brass, bronze, & copper castings	0.4	Manufg.	Telephone & telegraph facility construction	0.5	Constr.
Industrial patterns	0.4	Manufg.	Mining machinery, except oil field	0.5	Manufg.
Miscellaneous fabricated wire products	0.4	Manufg.	Construction of nonfarm buildings nec	0.4	Constr.
Nonferrous castings, nec	0.4	Manufg.	Construction machinery & equipment	0.4	Manufg.
Paints & allied products	0.4	Manufg.	Warehouses	0.3	Constr.
Paperboard containers & boxes	0.4	Manufg.	Household refrigerators & freezers	0.3	Manufg.
Gas production & distribution (utilities)	0.4	Util.	Industrial furnaces & ovens	0.3	Manufg.
Gaskets, packing & sealing devices	0.3	Manufg.	Real estate	0.3	Fin/R.E.
Sawmills & planning mills, general	0.3	Manufg.	Resid. & other health facility construction	0.2	Constr.
Computer & data processing services	0.3	Services	Boat building & repairing	0.2	Manufg.
Equipment rental & leasing services	0.3	Services	Iron & steel foundries	0.2	Manufg.
Legal services	0.3	Services	Gas utility facility construction	0.1	Constr.
Management & consulting services & labs	0.3	Services	Maintenance of nonbuilding facilities nec	0.1	Constr.
Industrial controls	0.2	Manufg.	Residential 2-4 unit structures, nonfarm	0.1	Constr.
Metal heat treating	0.2	Manufg.	Residential additions/alterations, nonfarm	0.1	Constr.
Railroads & related services	0.2	Util.			
Accounting, auditing & bookkeeping	0.2	Services			
Business/professional associations	0.2	Services			
Hotels & lodging places	0.2	Services			
U.S. Postal Service	0.2	Gov't			
Lubricating oils & greases	0.1	Manufg.			
Manifold business forms	0.1	Manufg.			
Transformers	0.1	Manufg.			
Security & commodity brokers	0.1	Fin/R.E.			
Automotive rental & leasing, without drivers	0.1	Services			

Source: Benchmark Input-Output Accounts for the U.S. Economy, 1982, U.S. Department of Commerce, Washington, D.C., July 1991. Data, as reported in the source, are organized by the 1977 SIC structure in use in 1982 but have been matched, as closely as is possible, to the 1987 SIC structure used in this book.

OCCUPATIONS EMPLOYED BY SIC 356 - GENERAL INDUSTRIAL MACHINERY

Occupation	% of Total 1994	Change to 2005	Occupation	% of Total 1994	Change to 2005
Assemblers, fabricators, & hand workers nec	7.9	-2.9	Machine tool cutting & forming etc. nec	1.5	-2.9
Machinists	4.9	6.9	Machine assemblers	1.5	-2.8
Sales & related workers nec	3.8	-2.9	Traffic, shipping, & receiving clerks	1.5	-6.5
Blue collar worker supervisors	3.8	-9.0	Combination machine tool operators	1.5	6.9
Inspectors, testers, & graders, precision	3.1	-2.9	Welding machine setters, operators	1.2	-12.6
Welders & cutters	2.8	31.2	Engineering, mathematical, & science managers	1.2	10.4
Mechanical engineers	2.8	6.9	Machine operators nec	1.2	-14.4
General managers & top executives	2.5	-7.8	Industrial production managers	1.2	-2.8
NC machine tool operators, metal & plastic	2.3	45.7	Bookkeeping, accounting, & auditing clerks	1.2	-27.2
Machine builders	2.3	31.1	Stock clerks	1.1	-21.0
Lathe & turning machine tool operators	2.2	-36.8	Engineering technicians & technologists nec	1.1	-2.9
Secretaries, ex legal & medical	2.1	-11.6	Machine forming operators, metal & plastic	1.1	-2.9
Grinding machine operators, metal & plastic	2.1	-12.6	General office clerks	1.1	-17.2
Drafters	2.0	-24.3	Production, planning, & expediting clerks	1.1	-2.8
Machine tool cutting operators, metal & plastic	1.8	-19.1	Helpers, laborers, & material movers nec	1.1	-2.8
Sheet metal workers & duct installers	1.6	-2.8	Drilling & boring machine tool workers	1.1	-51.4
Industrial machinery mechanics	1.6	6.9	Purchasing agents, ex trade & farm products	1.0	6.9

Source: *Industry-Occupation Matrix*, Bureau of Labor Statistics. These data relate to one or more 3-digit SIC industry groups rather than to a single 4-digit SIC. The change reported for each occupation to the year 2005 is a percent of growth or decline as estimated by the Bureau of Labor Statistics. The abbreviation nec stands for 'not elsewhere classified'.

LOCATION BY STATE AND REGIONAL CONCENTRATION

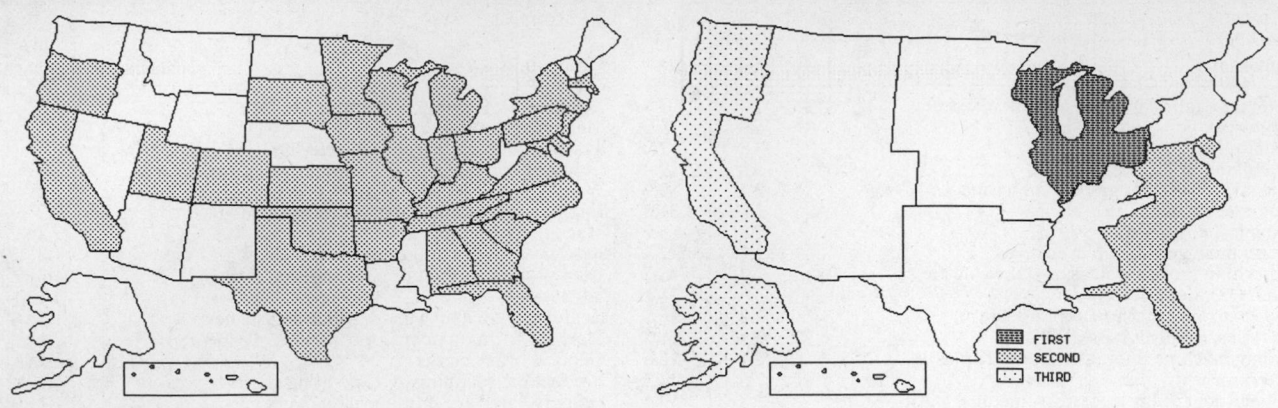

FIRST
SECOND
THIRD

INDUSTRY DATA BY STATE

| State | Establish-ments | Shipments | | | Employment | | | | Cost as % of Shipments | Investment per Employee ($) |
		Total ($ mil)	% of U.S.	Per Establ.	Total Number	% of U.S.	Per Establ.	Wages ($/hour)		
Ohio	38	314.4	10.5	8.3	2,700	10.4	71	11.30	48.9	1,889
California	72	243.8	8.1	3.4	2,000	7.7	28	12.60	39.7	1,300
Illinois	46	225.6	7.5	4.9	2,000	7.7	43	10.81	45.4	2,450
New York	20	199.9	6.7	10.0	1,700	6.5	85	14.72	39.8	3,118
Pennsylvania	32	184.3	6.1	5.8	1,400	5.4	44	11.65	47.9	1,786
North Carolina	29	163.8	5.5	5.6	1,900	7.3	66	8.47	43.1	895
Wisconsin	14	155.5	5.2	11.1	1,100	4.2	79	15.36	40.8	2,455
Indiana	15	128.4	4.3	8.6	1,400	5.4	93	14.47	33.5	1,929
Maryland	9	124.5	4.1	13.8	900	3.5	100	16.86	42.5	3,444
Texas	44	115.8	3.9	2.6	1,300	5.0	30	6.95	50.3	1,769
Michigan	27	111.7	3.7	4.1	1,100	4.2	41	11.00	47.4	1,727
Kentucky	12	111.6	3.7	9.3	900	3.5	75	11.15	44.3	1,556
Virginia	7	104.5	3.5	14.9	600	2.3	86	8.22	33.8	1,667
Kansas	9	86.2	2.9	9.6	500	1.9	56	11.71	47.7	-
Florida	25	76.6	2.6	3.1	800	3.1	32	11.18	48.7	1,875
Georgia	10	74.5	2.5	7.4	400	1.5	40	6.60	74.0	4,250
New Jersey	13	70.2	2.3	5.4	400	1.5	31	11.40	52.1	2,750
Missouri	10	61.6	2.1	6.2	600	2.3	60	10.44	38.5	-
Alabama	12	49.6	1.7	4.1	600	2.3	50	10.14	47.8	1,000
Minnesota	19	44.1	1.5	2.3	400	1.5	21	11.80	41.7	1,250
Massachusetts	14	38.3	1.3	2.7	200	0.8	14	14.67	44.6	5,500
Tennessee	14	34.6	1.2	2.5	400	1.5	29	11.00	49.4	1,000
Arkansas	3	31.4	1.0	10.5	300	1.2	100	9.33	47.8	2,000
Colorado	11	12.6	0.4	1.1	100	0.4	9	9.50	52.4	2,000
Iowa	7	12.3	0.4	1.8	100	0.4	14	7.00	40.7	1,000
South Carolina	7	11.4	0.4	1.6	200	0.8	29	10.00	49.1	3,000
Utah	4	8.6	0.3	2.2	100	0.4	25	8.00	43.0	-
Oklahoma	13	(D)	-	-	750 *	2.9	58	-	-	667
Oregon	10	(D)	-	-	750 *	2.9	75	-	-	-
South Dakota	2	(D)	-	-	375 *	1.4	188	-	-	-

Source: 1992 *Economic Census*. The states are in descending order of shipments or establishments (if shipment data are missing for the majority). The symbol (D) appears when data are withheld to prevent disclosure of competitive information. States marked with (D) are sorted by number of establishments. A dash (-) indicates that the data element cannot be calculated; * indicates the midpoint of a range.

3565 - PACKAGING MACHINERY

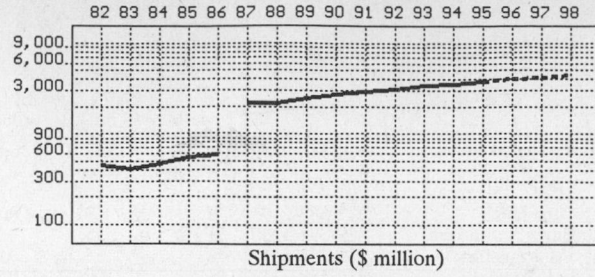

Shipments ($ million)

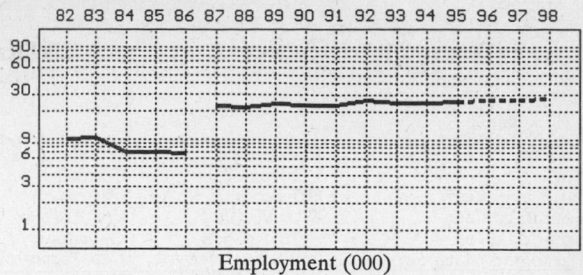

Employment (000)

GENERAL STATISTICS

Year	Com-panies	Establishments		Employment			Compensation		Production ($ million)			
		Total	with 20 or more employees	Total (000)	Production Workers (000)	Hours (Mil)	Payroll ($ mil)	Wages ($/hr)	Cost of Materials	Value Added by Manufacture	Value of Shipments	Capital Invest.
1982	994	996	105	9.8	8.1	14.8	217.6	11.72	97.6	347.6	452.3	15.8
1983		955	112	10.2	8.3	14.9	218.9	11.89	112.2	311.8	417.2	14.4
1984		914	119	7.0	5.8	11.1	173.7	12.34	186.3	312.2	461.1	11.4
1985		873	126	7.1	6.0	12.0	188.1	12.72	216.1	349.0	549.3	15.6
1986		812	114	6.9	5.8	12.3	191.7	12.63	216.8	384.9	590.4	14.1
1987*	414	439	231	22.6	13.4	26.6	631.9	12.30	785.1	1,406.8	2,189.9	54.4
1988		431	218	21.7	12.5	25.1	650.8	12.60	841.9	1,362.0	2,185.9	39.4
1989		449	232	24.4	14.0	28.4	699.2	12.65	950.9	1,553.6	2,497.8	67.1
1990		449	224	23.8	13.3	27.1	762.2	13.61	1,080.3	1,735.4	2,762.2	73.5
1991		478	241	23.9	13.7	28.1	797.2	13.99	1,222.4	1,627.0	2,879.9	68.4
1992	590	631	264	26.2	15.4	31.0	894.4	13.79	1,252.4	1,913.5	3,126.9	70.1
1993		633	265	24.9	14.2	29.2	899.5	14.29	1,378.4	2,010.0	3,418.2	90.2
1994		646P	268P	24.8	14.9	30.2	895.4	14.52	1,563.2	2,120.2	3,620.9	98.6
1995		682P	276P	26.0P	15.2P	31.1P	977.5P	14.96P	1,644.5P	2,230.5P	3,809.3P	100.6P
1996		718P	283P	26.5P	15.5P	31.8P	1,021.7P	15.29P	1,738.0P	2,357.3P	4,025.8P	107.4P
1997		754P	290P	26.9P	15.8P	32.4P	1,065.8P	15.62P	1,831.5P	2,484.0P	4,242.3P	114.1P
1998		790P	297P	27.3P	16.0P	33.1P	1,110.0P	15.95P	1,924.9P	2,610.8P	4,458.7P	120.9P

Sources: 1982, 1987, 1992 *Economic Census*; *Annual Survey of Manufactures*, 83-86, 88-91, 93-94. Establishment counts are from *County Business Patterns* for non-Census years; establishment counts for 83-84 are extrapolations. * indicates that industry content changed in 87; earlier years use 77 SICs. 'P's mark projections.

INDICES OF CHANGE

Year	Com-panies	Establishments		Employment			Compensation		Production ($ million)			
		Total	with 20 or more employees	Total (000)	Production Workers (000)	Hours (Mil)	Payroll ($ mil)	Wages ($/hr)	Cost of Materials	Value Added by Manufacture	Value of Shipments	Capital Invest.
1982	168	158	40	37	53	48	24	85	8	18	14	23
1983		151	42	39	54	48	24	86	9	16	13	21
1984		145	45	27	38	36	19	89	15	16	15	16
1985		138	48	27	39	39	21	92	17	18	18	22
1986		129	43	26	38	40	21	92	17	20	19	20
1987*	70	70	88	86	87	86	71	89	63	74	70	78
1988		68	83	83	81	81	73	91	67	71	70	56
1989		71	88	93	91	92	78	92	76	81	80	96
1990		71	85	91	86	87	85	99	86	91	88	105
1991		76	91	91	89	91	89	101	98	85	92	98
1992	100	100	100	100	100	100	100	100	100	100	100	100
1993		100	100	95	92	94	101	104	110	105	109	129
1994		102P	102P	95	97	97	100	105	125	111	116	141
1995		108P	104P	99P	99P	100P	109P	108P	131P	117P	122P	144P
1996		114P	107P	101P	100P	103P	114P	111P	139P	123P	129P	153P
1997		120P	110P	103P	102P	105P	119P	113P	146P	130P	136P	163P
1998		125P	113P	104P	104P	107P	124P	116P	154P	136P	143P	172P

Sources: Same as General Statistics. Values reflect change from the base year, 1992. Values above 100 mean greater than 92, values below 100 mean less than 92, and a value of 100 in the 82-91 or 93-98 period means same as 92. * indicates that industry content changed in 87. Data for earlier years are in 77 SIC format.

SELECTED RATIOS

For 1994	Avg. of All Manufact.	Analyzed Industry	Index	For 1994	Avg. of All Manufact.	Analyzed Industry	Index
Employees per Establishment	49	38	78	Value Added per Production Worker	134,084	142,295	106
Payroll per Establishment	1,500,273	1,386,375	92	Cost per Establishment	5,045,178	2,420,349	48
Payroll per Employee	30,620	36,105	118	Cost per Employee	102,970	63,032	61
Production Workers per Establishment	34	23	67	Cost per Production Worker	146,988	104,913	71
Wages per Establishment	853,319	678,949	80	Shipments per Establishment	9,576,895	5,606,348	59
Wages per Production Worker	24,861	29,430	118	Shipments per Employee	195,460	146,004	75
Hours per Production Worker	2,056	2,027	99	Shipments per Production Worker	279,017	243,013	87
Wages per Hour	12.09	14.52	120	Investment per Establishment	321,011	152,665	48
Value Added per Establishment	4,602,255	3,282,769	71	Investment per Employee	6,552	3,976	61
Value Added per Employee	93,930	85,492	91	Investment per Production Worker	9,352	6,617	71

Sources: Same as General Statistics. The 'Average of All Manufacturing' column represents the average of all manufacturing industries reported for the most recent complete year available. The Index shows the relationship between the Average and the Analyzed Industry. For example, 100 means that they are equal; 500 that the Analyzed Industry is five times the average; 50 means that the Analyzed Industry is half the national average. The abbreviation 'na' is used to show that data are 'not available'.

LEADING COMPANIES Number shown: **75** Total sales ($ mil): **1,586** Total employment (000): **13.4**

Company Name	Address				CEO Name	Phone	Co. Type	Sales ($ mil)	Empl. (000)
Liqui-Box Corp	PO Box 494	Worthington	OH	43085	Samuel B Davis	614-888-9280	P	148	0.8
Krones Inc	PO Box 32100	Franklin	WI	53132	Rudolph Zeus	414-421-5650	S	80	0.5
Permanent Label Corp	801 Bloomfield Av	Clifton	NJ	07012	M A Contreras	201-471-6617	R	80	0.5
Angelus Sanitary Can	4900 Pacific Blv	Los Angeles	CA	90058	Otto Heck	213-583-2171	R	63*	0.5
BW International Inc	PO Box 3848	Davenport	IA	52808	Barry Shoulders	319-391-1100	S	55	0.5
Doboy Packaging Machinery	869 S Knowles Av	New Richmond	WI	54017	JM Johnston	715-246-6511	S	55	0.5
Hayssen	PO Box 571	Sheboygan	WI	53082	D L Jones	414-458-2111	S	50*	0.7
Kliklok Corp	5224 Snapf	Decatur	GA	30035	PE Black	404-981-5200	R	50	0.4
RA Jones and Company Inc	2701 Crescent Spgs	Covington	KY	41017	AW Koehlinger	606-341-0400	R	50*	0.5
Evergreen Packaging Equipment	PO Box 3000	Cedar Rapids	IA	52406	Allan F Lips	319-399-3200	D	48	0.5
Lantech Inc	11000 Bluegrass	Louisville	KY	40299	PR Lancaster III	502-267-4200	R	45	0.3
Alloyd Company Inc	1401 Pleasant St	De Kalb	IL	60115	Dale Meyer	815-756-8451	R	40	0.4
Douglas Machine Corp	3404 Iowa St	Alexandria	MN	56308	Vernon J Anderson	612-763-6587	S	37	0.4
Weldotron Corp	1532 S Washington	Piscataway	NJ	08855	William L Remley	908-752-6700	P	30	0.2
Package Machinery Co	24 Scitico Rd	Somersville	CT	06072	Katherine E Putnam	203-749-0000	P	28	0.1
Loveshaw Corp	10 Fleetwood Ct	Ronkonkoma	NY	11779	A Di Russo, Jr	516-471-4090	R	25	0.1
Minnesota Automation	PO Box 190	Crosby	MN	56441	Greg Mangan	218-546-2222	D	25	0.2
Standard-Knapp Inc	127 Main St	Portland	CT	06480	Arthur A Tanner	203-342-1100	R	25*	0.2
Bemis Packaging Machinery Co	315 27th Av	Minneapolis	MN	55418	Larry Smith	612-782-1200	D	22*	0.2
Cannon Conveyor Systems Inc	1001 Johnson Pkwy	St Paul	MN	55106	Arnie Eliason	612-776-8501	S	20	0.2
Cloud Corp	424 Howard Av	Des Plaines	IL	60018	Doug Robison	708-390-6170	R	20	0.2
Liberty Industries Inc	PO Box 525	Girard	OH	44420	Ronald C Ringness	216-539-4744	R	20*	<0.1
Paxall Group Inc	7515 N Linder Av	Skokie	IL	60077	George Freer	708-677-7800	S	19*	0.2
Salwasser Manufacturing	PO Box 548	Reedley	CA	93654	M Salwasser	209-638-8484	R	19	0.2
General Devices Company Inc	PO Box 39100	Indianapolis	IN	46239	Maxwell S Fall	317-897-7000	R	18	0.3
Label-Aire Inc	550 Burning Tree	Fullerton	CA	92633	RW Riley	714-441-0700	R	18	0.1
Pure-Pak Inc	30000 S Hill Rd	New Hudson	MI	48165	Frank J Torrens	810-486-4600	R	18*	0.2
RA Pearson Company Inc	W 8120 Sunset Hwy	Spokane	WA	99204	Pamela Senske	509-838-6226	R	18	0.2
Electronic Liquid Fillers Inc	1535 S Hwy 39	La Porte	IN	46350	Sim Ake	219-393-5541	R	16	0.1
Acma USA Inc	501 Southlake Blv	Richmond	VA	23236	Giuseppe Venturi	804-794-6688	S	16*	0.1
Barry Wehmiller Packaging Syst	5320 140th Av N	Clearwater	FL	34620	Carl Goodwin	813-535-4100	S	15	0.1
Busse Inc	PO Drawer B	Randolph	WI	53956	Tom Young	414-326-3131	S	15	0.1
Mateer-Burt Company Inc	434 Devon Park Dr	Wayne	PA	19087	Anthony J Izzi	610-293-0100	R	15	0.1
Shanklin Corp	100 Westford Rd	Ayer	MA	01432	FG Shanklin	508-772-3200	R	15	0.2
Woodman Company Inc	5224 Snapf	Decatur	GA	30035	PE Black	404-981-5200	S	15	0.2
Brenton Engineering Co	4750 County 13 N	Alexandria	MN	56308	Buck Cody	612-852-7705	S	13*	0.1
Renco Machine Co	1421 Eastman Av	Green Bay	WI	54302	Donald J Renard	414-448-8000	R	13	0.1
Allied Gear and Machine Co	1101 Research Blv	St Louis	MO	63132	David E Bouchein	314-991-5900	R	12	0.2
Can Lines Inc	9839 Downey	Downey	CA	90241	Don Koplien	213-773-5676	R	12*	0.1
Exact Equipment Corp	PO Box 666	Levittown	PA	19058	Robert C Enichen	215-750-9090	R	12	0.1
Klockner Medipak Inc	14501 58th St N	Clearwater	FL	34620	Robert W Singleton	813-538-4644	D	12*	0.1
Southern Tool Company Inc	PO Box 457	West Monroe	LA	71294	Bill Meuwly	318-387-2263	R	12	0.2
United Bakery Equipment	15815 W 110th St	Lenexa	KS	66219	Frank Bastasch	913-541-8700	R	12*	0.1
Fowler Products Co	PO Box 80268	Athens	GA	30608	CP Wyllie	706-549-3300	R	11	<0.1
Ro-An Industries Corp	6420 Admiral Av	Middle Village	NY	11379	Angelo Cervera	718-821-1115	R	11	0.1
Van Doren Sales Inc	PO Box 1746	Wenatchee	WA	98801	Louis Van Doren	509-662-6197	R	11	<0.1
Potdevin Machine Co	200 North St	Teterboro	NJ	07608	RA Potdevin	201-288-1941	R	11	<0.1
Accraply Inc	15410 Mintka	Minnetonka	MN	55345	Gregory J Tschida	612-933-0800	D	10	<0.1
B and H Manufacturing Co	PO Box 247	Ceres	CA	95307	Lyn Bright	209-537-5785	R	10	0.1
Scandia Packaging Machinery Co	180 Brighton Rd	Clifton	NJ	07012	W Bronander III	201-473-6100	R	10	0.1
WA Lane Inc	998 S Sierra Way	San Bernardino	CA	92408	William A Lane	909-885-0715	R	10*	<0.1
TL Systems Corp	8700 Wyoming N	Minneapolis	MN	55445	Don DeMorett	612-493-6770	R	9	0.1
A-B-C Packaging Machine	811 Live Oak St	Tarpon Springs	FL	34689	JL Neal	813-937-5144	R	9	0.1
Klockner Packaging Machinery	N8 W22455 Johnson	Waukesha	WI	53186	Hans Dubell	414-521-9988	S	9*	<0.1
Paxall Clybourn Machinery	7515 N Linder Av	Skokie	IL	60077	Earl Ritchie	708-677-7800	D	9	0.1
Roberts Systems Inc	8500 S Tryon St	Charlotte	NC	28273	John Robert	704-588-2950	R	9*	<0.1
National Instrument Company	4119 Fordleigh Rd	Baltimore	MD	21215	S Rosen	410-764-0900	R	9	0.1
Catty Corp	PO Box 187	Huntley	IL	60142	RG Scott	708-669-5161	R	8	<0.1
Packaging Systems International	4990 Acoma St	Denver	CO	80216	CD McCrorie	303-296-4445	R	8*	<0.1
Tri-Pak Machinery Inc	PO Box 1228	Harlingen	TX	78551	AC Fitzgerald	210-423-5140	R	8	0.1
Zed Industries Inc	PO Box 458	Vandalia	OH	45377	David R Zelnick	513-667-8407	R	8	<0.1
New England Machinery Inc	6204 29th St E	Bradenton	FL	34203	GE Bankuty	813-755-5550	R	8	<0.1
Superior Packaging Equip Corp	625 Gotham Pkwy	Carlstadt	NJ	07072	A Farnow	201-438-4500	R	8	<0.1
Acraloc Corp	PO Box 4129	Oak Ridge	TN	37831	L George Andre	615-483-1368	R	7	0.1
Lakso Co	PO Box 929	Leominster	MA	01453	Robert Marston	508-537-8534	D	7	<0.1
Gabilan Manufacturing Inc	PO Box 2027	Salinas	CA	93902	Paul E Bickel	408-422-7824	R	7	<0.1
Greene Line Mfg Corp	2703 W 9th St	Marion	IN	46953	Jerry W Greene	317-662-9881	R	6*	<0.1
Heisler Industries Inc	224 Passaic Av	Fairfield	NJ	07004	Joseph W Reilly	201-227-6300	R	6	<0.1
Union Standard Equipment Co	801 E 141st St	Bronx	NY	10454	Richard Greenberg	718-585-0200	S	6	0.1
Sabel Engineering Corp	PO Box 1223	Sonoma	CA	95476	Herbert J Sabel	707-938-4771	R	6	<0.1
Stapling Machines Co	41 Pine St	Rockaway	NJ	07866	William Gillespie	201-627-4400	S	6	0.1
John R Nalbach Engineering	6139 W Ogden Av	Cicero	IL	60650	John C Nalbach	708-652-8900	R	5	<0.1
Flarpak Corp	PO Box 7409	Grand Rapids	MI	49510	LH Flaherty	616-245-8631	S	5	<0.1
Labeljet	4901 NE Parkway	Fort Worth	TX	76106		817-624-2600	D	5*	<0.1
Luthi Machine and Engineering	PO Box 2679	Gardena	CA	90247	Fred Avers	213-321-1337	R	5	<0.1

Source: Ward's Business Directory of U.S. Private and Public Companies, Volumes 1 and 2, 1996. The company type code used is as follows: P - Public, R - Private, S - Subsidiary, D - Division, J - Joint Venture, A - Affiliate, G - Group. Sales are in millions of dollars, employees are in thousands. An asterisk (*) indicates an estimated sales volume. The symbol < stands for 'less than'. Company names and addresses are truncated, in some cases, to fit into the available space.

MATERIALS CONSUMED

Material	Quantity	Delivered Cost ($ million)
Materials, ingredients, containers, and supplies	(X)	1,062.6
Fluid power pumps, motors, and hydrostatic transmissions (hydraulic and pneumatic)	(X)	8.0
All other pumps	(X)	4.8
Fluid power products	(X)	22.5
Mechanical speed changers, gears, and industrial high-speed drives	(X)	22.8
Ball and roller bearings (mounted or unmounted)	(X)	17.0
Other mechanical power transmission equipment	(X)	25.4
Metal bolts, nuts, screws, washers, rivets, and other screw machine products	(X)	23.7
Other fabricated metal products (except fluid power products and forgings)	(X)	96.5
Forgings	(X)	1.3
Iron and steel castings (rough and semifinished)	(X)	20.7
Aluminum and aluminum-base alloy castings (rough and semifinished)	(X)	12.4
Other nonferrous castings (rough and semifinished)	(X)	3.9
Stainless steel shapes and forms	(X)	29.4
All other steel shapes and forms	(X)	28.5
Aluminum and aluminum-base alloy shapes and forms	(X)	9.9
Other nonferrous shapes and forms	(X)	4.8
Relays and industrial controls for drives, clutches, brakes, motors, etc.	(X)	56.1
Transformers and related equipment	(X)	12.8
Other fractional horsepower electric motors (under 1 hp)	(X)	16.3
Integral horsepower electric motors and generators (1 hp or more)	(X)	10.4
Purchased filler devices (volumetric and others)	(X)	5.7
Purchased weighing machines and scales	(X)	3.1
Purchased coding, code dating, and marking devices (including imprinting and labeling)	(X)	4.5
Purchased detection devices (including checkweighing & metal detection)	(X)	4.5
Purchased adhesive devices (hot melt and cold glue)	(X)	10.4
Purchased conveyors and conveying equipment	(X)	6.4
Purchased accumulating, collating, feeding, and unscrambling machinery	(X)	3.8
All other materials and components, parts, containers, and supplies	(X)	322.6
Materials, ingredients, containers, and supplies, nsk	(X)	274.5

Source: 1992 *Economic Census*. Explanation of symbols used: (D): Withheld to avoid disclosure of competitive data; na: Not available; (S): Withheld because statistical norms were not met; (X): Not applicable; (Z): Less than half the unit shown; nec: Not elsewhere classified; nsk: Not specified by kind; - : zero; * : 10-19 percent estimated; ** : 20-29 percent estimated.

PRODUCT SHARE DETAILS

Product or Product Class	% Share	Product or Product Class	% Share
Packaging machinery	100.00	canning machinery, and parts)	3.62
Packing, packaging, and bottling machinery, except parts	77.84	Labeling machinery (all types of applications and methods), except parts	12.30
Cartoning, multipacking, and leaflet/coupon placing machinery, except parts	12.32	Coding, dating, imprinting, jet printing, marking, and stamping machinery, except parts	2.56
Thermoforming, blister, and skin machinery, including carded display machinery, except parts	1.39	Corrugated and solid fibre case and tray forming, loading, and sealing machinery, except parts	7.18
Bag (pre-form) opening, filling, and closing machinery and systems, except parts	3.20	Accumulating, collating, feeding, and unscrambling machinery, except parts	1.25
Vacuum, gas, and other modified atmosphere laminating and bagging machines, except parts	0.42	Testing, inspecting, detecting, checkweighing, and other quality control devices, except parts	4.34
Horizontal bag or pouch form, fill, and seal machinery (performs all three functions)	3.49	Paper, film, and foil wrapping machines (all types, except shrink and stretch film equipment, and parts)	1.38
Vertical bag or pouch form, fill, and seal machinery (performs all three functions)	3.61	Shrink and stretch film overwrapping, banding, and bundling machinery (excluding pallet unitizing and parts)	6.12
Machinery for cleaning or drying bottles or other containers, except parts	1.02	Palletizing, depalletizing, and pallet unitizing machinery with stretch film, adhesive, or strapping, except parts	2.04
Bottling and canning machinery (including fillers, all types of closers, and accessory equipment), except parts	3.93	Other packing, packaging, and bottling machinery or systems and combination of equipment nec, except parts	17.88
Dry products filling machinery (free and nonfree flowing), including by count machinery, except bags and parts	2.61	Packing, packaging, and bottling machinery, except parts, nsk	4.51
Liquids and viscous products filling machinery (very heavy liquids, slurries, and pumpable semisolids), except parts	4.86	Parts for packing, packaging, and bottling machinery	15.08
Glass or plastics container or can capping, sealing, and lidding machinery (excluding all filling, bottling and		Packaging machinery, nsk	7.09

Source: 1992 *Economic Census*. The values shown are percent of total shipments in an industry. Values of indented subcategories are summed in the main heading. The symbol (D) appears when data are withheld to prevent disclosure of competitive information. The abbreviation nsk stands for 'not specified by kind' and nec for 'not elsewhere classified'.

INPUTS AND OUTPUTS FOR SPECIAL INDUSTRY MACHINERY, NEC

Economic Sector or Industry Providing Inputs	%	Sector	Economic Sector or Industry Buying Outputs	%	Sector
Wholesale trade	11.3	Trade	Gross private fixed investment	79.1	Cap Inv
Imports	9.7	Foreign	Exports	10.0	Foreign
Blast furnaces & steel mills	8.9	Manufg.	Cyclic crudes and organics	4.2	Manufg.
Cyclic crudes and organics	5.9	Manufg.	Industrial inorganic chemicals, nec	1.9	Manufg.
Machinery, except electrical, nec	3.2	Manufg.	Miscellaneous plastics products	1.6	Manufg.
Switchgear & switchboard apparatus	2.9	Manufg.	Special industry machinery, nec	0.9	Manufg.
Motors & generators	2.4	Manufg.	Federal Government purchases, national defense	0.8	Fed Govt
Advertising	2.4	Services	Alkalies & chlorine	0.7	Manufg.
Industrial controls	2.2	Manufg.	Inorganic pigments	0.5	Manufg.
Iron & steel foundries	2.2	Manufg.	Plastics materials & resins	0.2	Manufg.
Electric services (utilities)	2.2	Util.	Tires & inner tubes	0.1	Manufg.
Fabricated plate work (boiler shops)	2.0	Manufg.	Federal Government purchases, nondefense	0.1	Fed Govt
Pumps & compressors	2.0	Manufg.			
Sheet metal work	1.8	Manufg.			
Special industry machinery, nec	1.7	Manufg.			
Pipe, valves, & pipe fittings	1.6	Manufg.			
Power transmission equipment	1.5	Manufg.			
Communications, except radio & TV	1.5	Util.			
Banking	1.5	Fin/R.E.			
Internal combustion engines, nec	1.4	Manufg.			
Aluminum rolling & drawing	1.3	Manufg.			
Maintenance of nonfarm buildings nec	1.2	Constr.			
Nonferrous wire drawing & insulating	1.2	Manufg.			
Special dies & tools & machine tool accessories	1.2	Manufg.			
Copper rolling & drawing	1.1	Manufg.			
Petroleum refining	1.1	Manufg.			
Engineering, architectural, & surveying services	1.1	Services			
Ball & roller bearings	1.0	Manufg.			
Air transportation	1.0	Util.			
Glass & glass products, except containers	0.9	Manufg.			
Real estate	0.9	Fin/R.E.			
Equipment rental & leasing services	0.9	Services			
Aluminum castings	0.8	Manufg.			
Motor freight transportation & warehousing	0.8	Util.			
Industrial inorganic chemicals, nec	0.7	Manufg.			
Screw machine and related products	0.7	Manufg.			
Transformers	0.7	Manufg.			
Eating & drinking places	0.7	Trade			
Fabricated metal products, nec	0.6	Manufg.			
Gaskets, packing & sealing devices	0.6	Manufg.			
Miscellaneous fabricated wire products	0.6	Manufg.			
Miscellaneous plastics products	0.6	Manufg.			
Nonferrous rolling & drawing, nec	0.6	Manufg.			
Rubber & plastics hose & belting	0.6	Manufg.			
Gas production & distribution (utilities)	0.6	Util.			
Brass, bronze, & copper castings	0.5	Manufg.			
Fabricated rubber products, nec	0.5	Manufg.			
Metal stampings, nec	0.5	Manufg.			
Nonferrous castings, nec	0.5	Manufg.			
Computer & data processing services	0.5	Services			
U.S. Postal Service	0.5	Gov't			
Abrasive products	0.4	Manufg.			
Electronic components nec	0.4	Manufg.			
Machine tools, metal cutting types	0.4	Manufg.			
Plating & polishing	0.4	Manufg.			
General industrial machinery, nec	0.3	Manufg.			
Sawmills & planning mills, general	0.3	Manufg.			
Railroads & related services	0.3	Util.			
Royalties	0.3	Fin/R.E.			
Detective & protective services	0.3	Services			
Hotels & lodging places	0.3	Services			
Legal services	0.3	Services			
Management & consulting services & labs	0.3	Services			
Conveyors & conveying equipment	0.2	Manufg.			
Metal heat treating	0.2	Manufg.			
Paperboard containers & boxes	0.2	Manufg.			
Insurance carriers	0.2	Fin/R.E.			
Accounting, auditing & bookkeeping	0.2	Services			
Automotive rental & leasing, without drivers	0.2	Services			
Automotive repair shops & services	0.2	Services			
Alkalies & chlorine	0.1	Manufg.			
Lubricating oils & greases	0.1	Manufg.			
Manifold business forms	0.1	Manufg.			
Primary metal products, nec	0.1	Manufg.			
Primary nonferrous metals, nec	0.1	Manufg.			
Sanitary services, steam supply, irrigation	0.1	Util.			
Noncomparable imports	0.1	Foreign			
Scrap	0.1	Scrap			

Source: *Benchmark Input-Output Accounts for the U.S. Economy, 1982*, U.S. Department of Commerce, Washington, D.C., July 1991. Data, as reported in the source, are organized by the 1977 SIC structure in use in 1982 but have been matched, as closely as is possible, to the 1987 SIC structure used in this book.

OCCUPATIONS EMPLOYED BY SIC 356 - GENERAL INDUSTRIAL MACHINERY

Occupation	% of Total 1994	Change to 2005	Occupation	% of Total 1994	Change to 2005
Assemblers, fabricators, & hand workers nec	7.9	-2.9	Machine tool cutting & forming etc. nec	1.5	-2.9
Machinists	4.9	6.9	Machine assemblers	1.5	-2.8
Sales & related workers nec	3.8	-2.9	Traffic, shipping, & receiving clerks	1.5	-6.5
Blue collar worker supervisors	3.8	-9.0	Combination machine tool operators	1.5	6.9
Inspectors, testers, & graders, precision	3.1	-2.9	Welding machine setters, operators	1.2	-12.6
Welders & cutters	2.8	31.2	Engineering, mathematical, & science managers	1.2	10.4
Mechanical engineers	2.8	6.9	Machine operators nec	1.2	-14.4
General managers & top executives	2.5	-7.8	Industrial production managers	1.2	-2.8
NC machine tool operators, metal & plastic	2.3	45.7	Bookkeeping, accounting, & auditing clerks	1.2	-27.2
Machine builders	2.3	31.1	Stock clerks	1.1	-21.0
Lathe & turning machine tool operators	2.2	-36.8	Engineering technicians & technologists nec	1.1	-2.9
Secretaries, ex legal & medical	2.1	-11.6	Machine forming operators, metal & plastic	1.1	-2.9
Grinding machine operators, metal & plastic	2.1	-12.6	General office clerks	1.1	-17.2
Drafters	2.0	-24.3	Production, planning, & expediting clerks	1.1	-2.8
Machine tool cutting operators, metal & plastic	1.8	-19.1	Helpers, laborers, & material movers nec	1.1	-2.8
Sheet metal workers & duct installers	1.6	-2.8	Drilling & boring machine tool workers	1.1	-51.4
Industrial machinery mechanics	1.6	6.9	Purchasing agents, ex trade & farm products	1.0	6.9

Source: *Industry-Occupation Matrix*, Bureau of Labor Statistics. These data relate to one or more 3-digit SIC industry groups rather than to a single 4-digit SIC. The change reported for each occupation to the year 2005 is a percent of growth or decline as estimated by the Bureau of Labor Statistics. The abbreviation nec stands for 'not elsewhere classified'.

LOCATION BY STATE AND REGIONAL CONCENTRATION

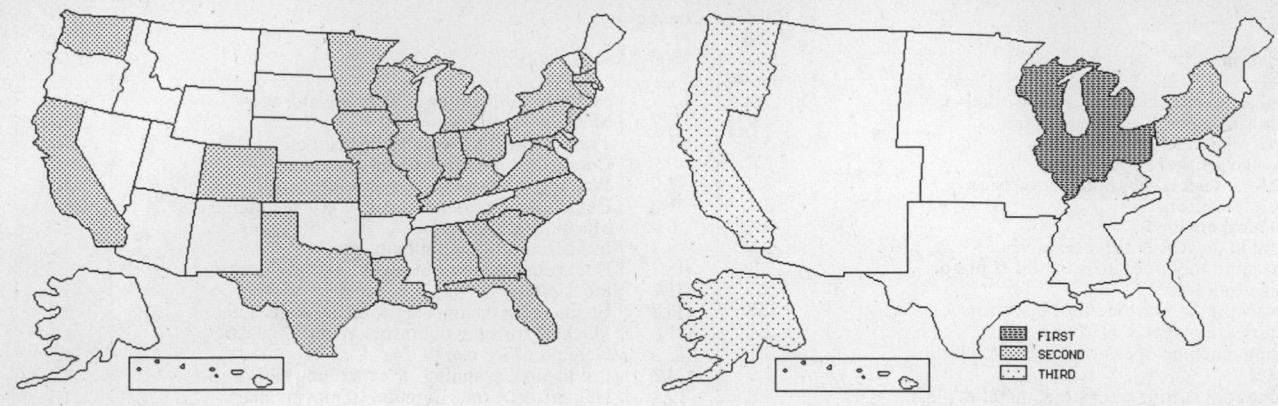

FIRST
SECOND
THIRD

INDUSTRY DATA BY STATE

| State | Establish-ments | Shipments | | | Employment | | | | Cost as % of Shipments | Investment per Employee ($) |
		Total ($ mil)	% of U.S.	Per Establ.	Total Number	% of U.S.	Per Establ.	Wages ($/hour)		
Ohio	43	434.9	13.9	10.1	2,700	10.3	63	13.32	34.6	3,111
Illinois	58	403.5	12.9	7.0	2,900	11.1	50	13.55	42.1	2,345
California	87	339.1	10.8	3.9	3,000	11.5	34	14.86	34.9	2,067
Wisconsin	41	313.9	10.0	7.7	2,800	10.7	68	13.55	43.3	4,929
Minnesota	22	155.3	5.0	7.1	1,500	5.7	68	12.78	50.3	2,600
New Jersey	44	146.4	4.7	3.3	1,300	5.0	30	14.07	39.7	2,154
Florida	31	133.1	4.3	4.3	1,200	4.6	39	11.69	38.7	2,083
Georgia	18	111.3	3.6	6.2	800	3.1	44	13.40	42.9	3,000
New York	36	100.8	3.2	2.8	1,000	3.8	28	12.23	39.7	900
Pennsylvania	30	100.3	3.2	3.3	800	3.1	27	13.70	39.2	1,375
South Carolina	10	97.7	3.1	9.8	700	2.7	70	14.25	64.3	2,286
Massachusetts	18	90.3	2.9	5.0	1,000	3.8	56	12.50	34.0	1,800
Maryland	6	87.1	2.8	14.5	700	2.7	117	14.70	36.1	1,714
Washington	22	71.1	2.3	3.2	800	3.1	36	12.56	36.7	2,125
Texas	24	61.7	2.0	2.6	700	2.7	29	12.00	37.1	1,143
Missouri	10	47.2	1.5	4.7	400	1.5	40	11.00	59.1	-
North Carolina	15	45.2	1.4	3.0	300	1.1	20	18.50	32.7	5,667
Connecticut	14	44.7	1.4	3.2	400	1.5	29	14.40	50.3	-
New Hampshire	5	28.1	0.9	5.6	200	0.8	40	15.50	46.6	1,000
Louisiana	4	23.0	0.7	5.8	200	0.8	50	13.50	33.5	-
Virginia	6	14.2	0.5	2.4	100	0.4	17	13.00	26.1	-
Kansas	4	12.9	0.4	3.2	100	0.4	25	10.00	37.2	1,000
Alabama	5	8.6	0.3	1.7	100	0.4	20	9.50	37.2	1,000
Michigan	20	(D)	-	-	750 *	2.9	38	-	-	-
Indiana	11	(D)	-	-	175 *	0.7	16	-	-	-
Iowa	7	(D)	-	-	750 *	2.9	107	-	-	-
Colorado	5	(D)	-	-	175 *	0.7	35	-	-	-
Kentucky	5	(D)	-	-	750 *	2.9	150	-	-	-

Source: 1992 *Economic Census*. The states are in descending order of shipments or establishments (if shipment data are missing for the majority). The symbol (D) appears when data are withheld to prevent disclosure of competitive information. States marked with (D) are sorted by number of establishments. A dash (-) indicates that the data element cannot be calculated; * indicates the midpoint of a range.

3566 - SPEED CHANGERS, DRIVES & GEARS

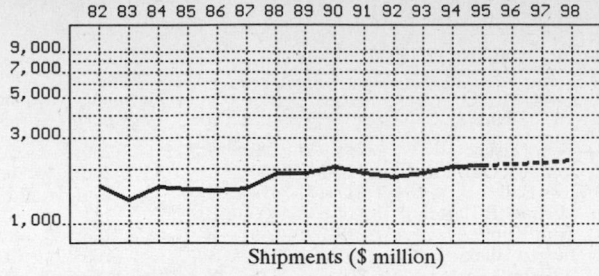

Shipments ($ million)

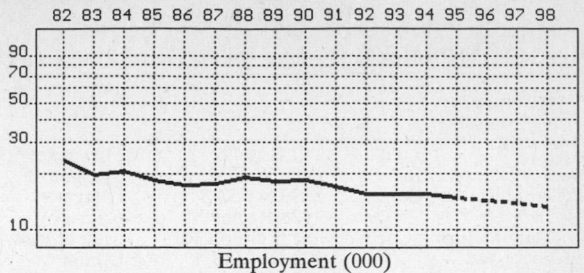

Employment (000)

GENERAL STATISTICS

Year	Com-panies	Establishments		Employment			Compensation		Production ($ million)			
		Total	with 20 or more employees	Total (000)	Production Workers (000)	Hours (Mil)	Payroll ($ mil)	Wages ($/hr)	Cost of Materials	Value Added by Manufacture	Value of Shipments	Capital Invest.
1982	282	309	180	24.1	15.8	29.7	503.9	10.05	552.7	1,020.9	1,631.6	93.2
1983		299	176	19.6	12.8	24.2	436.2	10.21	476.4	857.3	1,363.7	79.7
1984		289	172	20.6	13.8	28.1	505.0	10.89	586.7	1,029.6	1,609.0	73.6
1985		280	168	18.6	12.7	25.8	478.1	11.53	552.1	992.2	1,555.6	71.3
1986		276	163	17.4	11.7	23.7	474.3	11.95	527.8	986.6	1,529.8	65.6
1987	250	276	157	17.9	11.9	23.8	474.0	12.14	555.4	1,004.4	1,569.0	65.0
1988		266	158	19.3	13.3	26.5	530.0	12.30	667.3	1,260.8	1,916.9	63.4
1989		259	161	18.4	12.6	25.5	524.2	12.97	727.1	1,166.5	1,911.6	59.0
1990		267	166	18.5	12.6	26.0	551.0	13.01	734.9	1,353.0	2,055.7	81.1
1991		266	156	17.2	11.5	23.1	509.5	12.84	703.8	1,194.1	1,916.5	63.2
1992	256	287	145	15.7	10.4	20.6	495.8	13.77	646.3	1,160.5	1,823.1	69.8
1993		277	141	15.7	10.6	22.1	526.0	14.33	696.4	1,229.5	1,921.5	55.4
1994		262P	143P	15.6	10.9	23.2	541.1	14.66	710.0	1,361.3	2,054.0	72.8
1995		260P	140P	14.8P	10.2P	21.7P	539.7P	14.92P	722.9P	1,386.0P	2,091.3P	60.3P
1996		257P	137P	14.3P	9.9P	21.2P	544.8P	15.29P	739.3P	1,417.5P	2,138.8P	58.9P
1997		254P	134P	13.8P	9.6P	20.8P	549.9P	15.65P	755.8P	1,449.0P	2,186.4P	57.5P
1998		252P	131P	13.3P	9.3P	20.3P	555.1P	16.02P	772.2P	1,480.6P	2,234.0P	56.0P

Sources: 1982, 1987, 1992 *Economic Census*; *Annual Survey of Manufactures*, 83-86, 88-91, 93-94. Establishment counts for non-Census years are from *County Business Patterns*; establishment values for 83-84 are extrapolations. 'P's show projections by the editors. Industries reclassified in 87 will not have data for prior years.

INDICES OF CHANGE

Year	Com-panies	Establishments		Employment			Compensation		Production ($ million)			
		Total	with 20 or more employees	Total (000)	Production Workers (000)	Hours (Mil)	Payroll ($ mil)	Wages ($/hr)	Cost of Materials	Value Added by Manufacture	Value of Shipments	Capital Invest.
1982	110	108	124	154	152	144	102	73	86	88	89	134
1983		104	121	125	123	117	88	74	74	74	75	114
1984		101	119	131	133	136	102	79	91	89	88	105
1985		98	116	118	122	125	96	84	85	85	85	102
1986		96	112	111	113	115	96	87	82	85	84	94
1987	98	96	108	114	114	116	96	88	86	87	86	93
1988		93	109	123	128	129	107	89	103	109	105	91
1989		90	111	117	121	124	106	94	113	101	105	85
1990		93	114	118	121	126	111	94	114	117	113	116
1991		93	108	110	111	112	103	93	109	103	105	91
1992	100	100	100	100	100	100	100	100	100	100	100	100
1993		97	97	100	102	107	106	104	108	106	105	79
1994		91P	99P	99	105	113	109	106	110	117	113	104
1995		90P	97P	94P	98P	105P	109P	108P	112P	119P	115P	86P
1996		90P	95P	91P	95P	103P	110P	111P	114P	122P	117P	84P
1997		89P	93P	88P	93P	101P	111P	114P	117P	125P	120P	82P
1998		88P	91P	85P	90P	99P	112P	116P	119P	128P	123P	80P

Sources: Same as General Statistics. Values reflect change from the base year, 1992. Values above 100 mean greater than 92, values below 100 mean less than 92, and a value of 100 in the 82-91 or 93-98 period means same as 92. 'P's mark projections by the editors.

SELECTED RATIOS

For 1994	Avg. of All Manufact.	Analyzed Industry	Index	For 1994	Avg. of All Manufact.	Analyzed Industry	Index
Employees per Establishment	49	59	121	Value Added per Production Worker	134,084	124,890	93
Payroll per Establishment	1,500,273	2,063,120	138	Cost per Establishment	5,045,178	2,707,106	54
Payroll per Employee	30,620	34,686	113	Cost per Employee	102,970	45,513	44
Production Workers per Establishment	34	42	121	Cost per Production Worker	146,988	65,138	44
Wages per Establishment	853,319	1,296,788	152	Shipments per Establishment	9,576,895	7,831,542	82
Wages per Production Worker	24,861	31,203	126	Shipments per Employee	195,460	131,667	67
Hours per Production Worker	2,056	2,128	104	Shipments per Production Worker	279,017	188,440	68
Wages per Hour	12.09	14.66	121	Investment per Establishment	321,011	277,574	86
Value Added per Establishment	4,602,255	5,190,399	113	Investment per Employee	6,552	4,667	71
Value Added per Employee	93,930	87,263	93	Investment per Production Worker	9,352	6,679	71

Sources: Same as General Statistics. The 'Average of All Manufacturing' column represents the average of all manufacturing industries reported for the most recent complete year available. The Index shows the relationship between the Average and the Analyzed Industry. For example, 100 means that they are equal; 500 that the Analyzed Industry is five times the average; 50 means that the Analyzed Industry is half the national average. The abbreviation 'na' is used to show that data are 'not available'.

LEADING COMPANIES   Number shown: **75**   Total sales ($ mil): **1,972**   Total employment (000): **14.4**

Company Name	Address				CEO Name	Phone	Co. Type	Sales ($ mil)	Empl. (000)
Falk Corp	PO Box 492	Milwaukee	WI	53201	Tom Misiak	414-937-4140	S	220	1.4
Allen-Bradley Company Inc	6400 W Enterprise	Mequon	WI	53092	John McDermott	414-242-8200	D	200*	0.9
Fairfield Manufacturing Co	US Hwy 52 S	Lafayette	IN	47903	Jess Ball	317-474-3474	S	150	1.0
Twin Disc Inc	1328 Racine St	Racine	WI	53403	Michael E Batten	414-638-4000	P	141	1.1
American Manufacturing Corp	181 S Gulph Rd	King of Prussia	PA	19406	Tom Kling	215-265-3000	R	120*	1.0
SEW-Eurodrive Inc	PO Box 518	Lyman	SC	29365	Jurgen Blickle	803-439-7537	S	100	0.4
Sumitomo Machinery Corp	4200 Holland Blv	Chesapeake	VA	23323	William Lechler	804-485-3355	S	80	0.3
Boston Gear	14 Hayward St	Quincy	MA	02171	RA Pennycock	617-328-3300	D	62	0.7
Eaton Corp	3122 14th Av	Kenosha	WI	53141	Peter F Krol	414-656-4011	D	50	0.3
Von Weise Gear Co	PO Box 228	St Clair	MO	63077	Leigh Ewell	314-629-1010	R	48	0.5
Hub City Inc	2914 Industrial Av	Aberdeen	SD	57402	Jim Campbell	605-225-0360	S	47	0.5
Flender Corp	PO Box 1449	Elgin	IL	60121	Gottfried Versock	708-931-1990	S	40	0.2
Ross Hill Controls Corp	1530 Houston	Houston	TX	77043	Ross K Hill	713-935-8000	S	40	0.3
Philadelphia Gear Corp	181 S Gulph Rd	King of Prussia	PA	19406	Ramon T Tork	215-265-3000	S	37*	0.4
Penn Machine Co	106 Station St	Johnstown	PA	15905	H Karl Wiegand	814-288-1547	D	36	0.2
Auburn Gear Inc	400 E Auburn Dr	Auburn	IN	46706	GE Callas	219-925-3200	R	35	0.3
Bedford Gear	6160 Cochran Rd	Solon	OH	44139	Ray Albertini	216-248-7970	D	33	0.1
Horsburgh and Scott Co	5114 Hamilton Av	Cleveland	OH	44114	Jeff Echko	216-431-3900	R	32*	0.3
Foote-Jones/Illinois Gear	2102 N Natchez Av	Chicago	IL	60635	JC Drecoll	312-622-8000	D	31	0.3
Cone Drive Operations Inc	PO Box 272	Traverse City	MI	49685	John Melvin	616-946-8410	S	30	0.3
Rexnord Corp Link-Belt	2045 W Hunting Pk	Philadelphia	PA	19140	Richard McGovern	215-225-6000	D	25	0.1
Milwaukee Gear Company Inc	PO Box 17615	Milwaukee	WI	53217	Harold Trusky	414-962-3532	R	24*	0.2
Cincinnati Gear Co	5657 Wooster Pike	Cincinnati	OH	45227	W Rye	513-271-7700	R	23	0.3
Bison Gear and Engineering Inc	2424 Wisconsin Av	Downers Grove	IL	60515	RD Bullock	708-968-6400	R	22	0.2
Cleveland Gear Inc	3249 E 80th St	Cleveland	OH	44104	Dana Lynch	216-641-9000	S	20	0.1
Amarillo Gear Co	PO Box 1789	Amarillo	TX	79105	Glen McCain	806-622-1273	S	19	0.1
Allied Devices Corp	PO Box 503	Baldwin	NY	11510	P Bartow	516-623-6300	R	16	0.2
Stock Drive Products	Box 5416	New Hyde Park	NY	11042	Frank Buchsbaum	516-328-3300	D	16	0.2
Brook Hansen Inc	PO Box 710	Branford	CT	06405	TP Harold	203-488-6396	S	15	<0.1
Grant Gear Inc	921 Providence Hwy	Norwood	MA	02062	RJ Hurley	617-769-7200	R	15	0.1
Peerless-Winsmith	172 Eaton St	Springville	NY	14141	D McCann	716-592-9311	D	13	0.1
Overton Gear and Tool Corp	530 Westgate Dr	Addison	IL	60101	Carl Overton	708-543-9570	R	12	0.1
Delroyd Worm Gear	PO Box 8633	Trenton	NJ	08650	Donald K Farrar	609-890-6800	D	11	0.1
Zero-Max Inc	13200 6th Av N	Minneapolis	MN	55441	S Vogel	612-546-4300	R	10	<0.1
David Brown Radicon Corp	755 N Rte 83	Bensenville	IL	60106	Gary Combs	708-860-7719	R	10*	<0.1
Schafer Gear Works Inc	814-20 S Main St	South Bend	IN	46624	Bipin Doshi	219-234-4116	R	10	<0.1
Acme Gear Company Inc	129 Coolidge Av	Englewood	NJ	07631	Joseph Geeles	201-568-2245	R	8	<0.1
Chicago Gear-DO James Corp	2823 W Fulton St	Chicago	IL	60612	Wayne H Wellman	312-638-0508	R	8*	<0.1
Gear Motions Inc	1750 Milton Av	Syracuse	NY	13209	SR Haines II	315-488-0100	R	8	<0.1
Gear Works, Seattle Inc	500 S Portland St	Seattle	WA	98108	Roland Ramberg	206-762-3333	R	8	0.1
Leedy Manufacturing Company	PO Box 1946	Grand Rapids	MI	49507	Harold Leedy Jr	616-245-0519	R	8*	0.1
Precision Industrial	PO Box 1004	Middlebury	CT	06762	J Herz	203-758-8272	S	8	0.1
Joliet Equipment Corp	PO Box 114	Joliet	IL	60434	J Keck	815-727-6606	R	7	<0.1
Chicago Gear Works Inc	1805 S 55th Av	Cicero	IL	60650	W Richard Jones	708-863-2700	R	6	<0.1
Nord Gear Corp	800 Nord Dr	Waunakee	WI	53597	Gary K Miller	608-849-7300	S	6	<0.1
Nordex Inc	50 Newtown Rd	Danbury	CT	06810	JG Agius	203-792-9050	R	6	<0.1
Sewall Gear Manufacturing Co	705 Raymond Av	St Paul	MN	55114	John R Bataglia	612-645-7721	R	6*	<0.1
Geartronics Industries Inc	PO Box 376	North Billerica	MA	01862	RL Duffy	508-663-6566	R	6	<0.1
Speed Selector Inc	17050 Munn Rd	Chagrin Falls	OH	44023	George C Wick Jr	216-543-8233	R	6	<0.1
Adams Co	100 E 4th St	Dubuque	IA	52001	John Hendry	319-583-3591	R	5*	<0.1
Dalton Gear Co	212 Colfax Av N	Minneapolis	MN	55405	ER Wood	612-374-2150	R	5	<0.1
Franke Gear Works Inc	4401 N Ravenswood	Chicago	IL	60640	CR Mason	312-561-0950	R	5*	<0.1
Gear Research Inc	PO Box 8837	Grand Rapids	MI	49518	G Barry Lawrence	616-241-3411	S	5*	<0.1
Globe Gear Co	550 Virginia Dr	Ft Washington	PA	19034	GR McGann	215-542-9000	R	5*	<0.1
Skidmore Gear Co	5507 Avion Park Dr	Cleveland	OH	44143	Frank Sirianni	216-442-6670	S	5	<0.1
St Louis Gear Company Inc	PO Box 880	Keokuk	IA	52632	DW Hodges	319-524-5042	R	5*	<0.1
Hi Lo Manufacturing Co	1700-G Freeway	Minneapolis	MN	55430	VG Nordley	612-566-2510	R	5	<0.1
Sterling Instrument	2101 Jericho Tpk	New Hyde Park	NY	11040	Martin Hoffman	516-328-3300	D	5	<0.1
Productigear Company Inc	1900 W 34th St	Chicago	IL	60608	Richard Wieker	312-847-4506	R	4	<0.1
Nixon Gear Inc	1750 Milton Av	Syracuse	NY	13209	Samuel R Haines	315-488-0100	S	4*	<0.1
Euclid Universal Corp	7280 Wright Av	Bedford	OH	44146	R K Formanek	216-439-6970	R	3	<0.1
Advance Gear & Machine Corp	PO Box 2378	Gardena	CA	90247	C E Breneman	213-770-1951	R	3	<0.1
Advanced Mechanical Techn	176 Waltham St	Watertown	MA	02172	Walter Syniuta	617-926-6700	R	3	<0.1
O'Brien Gear Company Inc	2396 Skokie Val Rd	Highland Park	IL	60035	George O'Brien	708-433-3580	R	3	<0.1
Randolph Manufacturing Co	PO Box 2067	Lubbock	TX	79408	John Osteen	806-765-5583	R	3	<0.1
Walter Machine Company Inc	PO Box 7700	Jersey City	NJ	07307	DR Chatrnuck	201-656-5654	R	3	<0.1
XTEK/Arizona	1710 W Broadway	Tempe	AZ	85282	George W Corfield	602-968-6141	D	3	<0.1
Hadley Gear Manufacturing Co	4444 W Roosevelt	Chicago	IL	60624	John A Davey	312-722-1030	R	3	<0.1
Charles Bond Co	1035 Louis Dr	Warminster	PA	18971	CC Bond Jr	215-329-1700	R	3	<0.1
American Chain and Gear Co	2451 E 57th St	Los Angeles	CA	90058	John Kyle	213-581-9131	R	2	<0.1
Diefendorf Gear Corp	PO Box 6489	Syracuse	NY	13217	A A Diefendorf	315-422-2281	R	2	<0.1
Star Tool and Engineering	PO Box 7130	Redwood City	CA	94063	R Wantin	415-366-8235	R	2*	<0.1
Christiana Machine Co	PO Box 105	Christiana	PA	17509	WR Bond	610-593-5171	D	2	<0.1
America Precision Gear	PO Box 906	San Carlos	CA	94070	Dennis Regan	415-595-3664	R	1	<0.1
Lamont Gear Co	208 1st Av	Rock Falls	IL	61071	Rolla M Montee	815-625-7542	R	1	<0.1

Source: Ward's Business Directory of U.S. Private and Public Companies, Volumes 1 and 2, 1996. The company type code used is as follows: P - Public, R - Private, S - Subsidiary, D - Division, J - Joint Venture, A - Affiliate, G - Group. Sales are in millions of dollars, employees are in thousands. An asterisk (*) indicates an estimated sales volume. The symbol < stands for 'less than'. Company names and addresses are truncated, in some cases, to fit into the available space.

MATERIALS CONSUMED

Material	Quantity	Delivered Cost ($ million)
Materials, ingredients, containers, and supplies	(X)	539.6
Metal bolts, nuts, screws, washers, rivets, and other screw machine products	(X)	17.1
Other fabricated metal products (except castings and forgings)	(X)	18.3
Cold iron and steel forgings	(X)	16.3
Other iron and steel forgings	(X)	31.9
Nonferrous forgings	(X)	2.6
Iron and steel castings (rough and semifinished)	(X)	60.0
Nonferrous (aluminum, copper, etc.) castings (rough and semifinished)	(X)	12.4
Steel bars, bar shapes, and plates	(X)	46.5
Steel sheet and strip, including tin plate	(X)	0.8
All other steel shapes and forms	(X)	3.6
Copper and copper-base alloy shapes and forms	(X)	5.5
All other nonferrous shapes and forms	(X)	4.3
Ball bearings (mounted or unmounted)	(X)	16.2
Roller bearings (mounted or unmounted)	(X)	22.9
Balls, rollers, cages, collars, races, and other antifriction bearing components and parts	(X)	1.4
Clutches, couplings, shafts, sprockets, and other mechanical power transmission equipment	(X)	44.6
Electric motors, generators, and parts	(X)	47.4
Paperboard containers, boxes, and corrugated paperboard	(X)	4.0
All other materials and components, parts, containers, and supplies	(X)	107.4
Materials, ingredients, containers, and supplies, nsk	(X)	76.4

Source: 1992 *Economic Census*. Explanation of symbols used: (D): Withheld to avoid disclosure of competitive data; na: Not available; (S): Withheld because statistical norms were not met; (X): Not applicable; (Z): Less than half the unit shown; nec: Not elsewhere classified; nsk: Not specified by kind; - : zero; * : 10-19 percent estimated; ** : 20-29 percent estimated.

PRODUCT SHARE DETAILS

Product or Product Class	% Share	Product or Product Class	% Share
Speed changers, drives, and gears	100.00	and scoop mount, less than 1 horsepower (746.0w)	3.86
Mechanical nonhydraulic variable speed changers and parts, excluding value of drivers	6.70	Worm gearmotors, sold with motors, including "C" flange and scoop mount, 1 horsepower (746.0w) or more	1.06
Industrial high-speed drives, fixed ratio (pitch line velocity of 5,000 feet (1,525 meters) per minute or more)	3.44	Helical, herringbone, spur, and spiral bevel gearmotors, sold with motors, including "C" flange and scoop mount, less than 1 horsepower (746.0w)	4.55
Worm gear speed reducers, fixed ratio, enclosed, excluding gear motor, including "C" flange or scoop mount, 6 in. (15.24 cm) centers or more	1.51	Helical, herringbone, spur, and spiral bevel gearmotors, sold with motors, including "C" flange and scoop mount, 1 horsepower (746.0w) or more	3.25
Worm gear speed reducers, fixed ratio, enclosed, excluding gear motor, including "C" flange or scoop mount, 3 in. (7.62 cm) to 5.99 in. (15.22 cm) centers	3.97	Fine pitch loose gears, pinions, and racks (19.99 diametral pitch and finer), excluding spare parts for reducers	3.32
Worm gear speed reducers, fixed ratio, enclosed, excluding gear motor, including "C" flange or scoop mount, less than 3 in. (7.62 cm) centers	4.42	Coarse pitch (less than 19.99 diametral pitch) loose helical, herringbone, and spur gears, 24 in. (60.96 cm) or less, excluding spare parts for reducers	8.64
Shaft mounted speed reducers and screw conveyor drives, fixed ratio, enclosed, excluding gear motor, hollow shaft diameter, 2 1/2 in. (6.35 cm) or less, including repair parts	3.07	Coarse pitch (less than 19.99 diametral pitch) loose helical, herringbone, and spur gears, more than 24 in. (60.96 cm) diameter through 72 in. (182.88 cm) diameter, excluding spare parts for reducers	9.69
Shaft mounted speed reducers and screw conveyor drives, fixed ratio, enclosed, excluding gear motor, hollow shaft diameter, more than 2 1/2 in. (6.35 cm), including repair parts	2.71	Coarse pitch (less than 19.99 diametral pitch) loose helical, herringbone, and spur gears, more than 72 in. (182.88 cm) diameter, excluding spare parts for reducers	0.76
Helical, herringbone, spur, and spiral bevel speed reducers, fixed ratio, enclosed, excluding gear motor, more than 15 in. (38.10 cm) low-speed center	6.08	Coarse pitch (less than 19.99 diametral pitch) loose worms and worm gearing, excluding spare parts for reducers	4.10
Helical, herringbone, spur, and spiral bevel speed reducers, fixed ratio, enclosed, excluding gear motor, 15 in. (38.10 cm) low-speed center or less	7.48	Other coarse pitch (less than 19.99 diametral pitch) loose gears, pinions, and racks, including bevel gears, excluding spare parts for reducers	5.13
Worm gearmotors, sold with motors, including "C" flange		Other parts and components for speed changers, including housings, shafts, pins, and spacers	5.89

Source: 1992 *Economic Census*. The values shown are percent of total shipments in an industry. Values of indented subcategories are summed in the main heading. The symbol (D) appears when data are withheld to prevent disclosure of competitive information. The abbreviation nsk stands for 'not specified by kind' and nec for 'not elsewhere classified'.

INPUTS AND OUTPUTS FOR POWER TRANSMISSION EQUIPMENT

Economic Sector or Industry Providing Inputs	%	Sector	Economic Sector or Industry Buying Outputs	%	Sector
Imports	13.2	Foreign	Motor vehicle parts & accessories	10.0	Manufg.
Blast furnaces & steel mills	11.8	Manufg.	Exports	7.5	Foreign
Power transmission equipment	11.6	Manufg.	Power transmission equipment	6.3	Manufg.
Wholesale trade	10.0	Trade	Construction machinery & equipment	6.1	Manufg.
Iron & steel forgings	5.6	Manufg.	Farm machinery & equipment	3.9	Manufg.
Iron & steel foundries	5.6	Manufg.	Motorcycles, bicycles, & parts	3.8	Manufg.
Motors & generators	3.2	Manufg.	Aluminum rolling & drawing	3.1	Manufg.
Machinery, except electrical, nec	3.0	Manufg.	Internal combustion engines, nec	2.8	Manufg.
Ball & roller bearings	2.8	Manufg.	Pumps & compressors	2.7	Manufg.
Electric services (utilities)	2.4	Util.	Crude petroleum & natural gas	2.5	Mining
Screw machine and related products	1.7	Manufg.	Machinery, except electrical, nec	2.5	Manufg.
Brass, bronze, & copper castings	1.3	Manufg.	Ship building & repairing	2.5	Manufg.
Special dies & tools & machine tool accessories	1.3	Manufg.	Federal Government purchases, national defense	2.4	Fed Govt
Maintenance of nonfarm buildings nec	1.2	Constr.	Tanks & tank components	2.3	Manufg.
Communications, except radio & TV	1.2	Util.	Coal	2.2	Mining
Computer & data processing services	1.2	Services	Iron & steel foundries	2.0	Manufg.
Copper rolling & drawing	1.1	Manufg.	Railroad equipment	1.9	Manufg.
Nonferrous castings, nec	1.1	Manufg.	Nonferrous wire drawing & insulating	1.8	Manufg.
Aluminum castings	1.0	Manufg.	Blast furnaces & steel mills	1.7	Manufg.
Air transportation	1.0	Util.	Blowers & fans	1.7	Manufg.
Gas production & distribution (utilities)	1.0	Util.	Logging camps & logging contractors	1.7	Manufg.
Motor freight transportation & warehousing	1.0	Util.	Mining machinery, except oil field	1.6	Manufg.
Equipment rental & leasing services	1.0	Services	Secondary nonferrous metals	1.5	Manufg.
Abrasive products	0.9	Manufg.	Special industry machinery, nec	1.1	Manufg.
Nonferrous rolling & drawing, nec	0.9	Manufg.	Miscellaneous repair shops	1.1	Services
Banking	0.9	Fin/R.E.	Printing trades machinery	1.0	Manufg.
Engineering, architectural, & surveying services	0.9	Services	Steel pipe & tubes	1.0	Manufg.
Machine tools, metal cutting types	0.8	Manufg.	Copper rolling & drawing	0.9	Manufg.
Petroleum refining	0.8	Manufg.	Food products machinery	0.9	Manufg.
Eating & drinking places	0.8	Trade	General industrial machinery, nec	0.9	Manufg.
Advertising	0.8	Services	Industrial trucks & tractors	0.9	Manufg.
Primary copper	0.7	Manufg.	Lawn & garden equipment	0.9	Manufg.
Real estate	0.6	Fin/R.E.	Refrigeration & heating equipment	0.9	Manufg.
Miscellaneous plastics products	0.5	Manufg.	Aircraft & missile equipment, nec	0.8	Manufg.
Fabricated rubber products, nec	0.4	Manufg.	Conveyors & conveying equipment	0.8	Manufg.
Gaskets, packing & sealing devices	0.4	Manufg.	Household laundry equipment	0.8	Manufg.
Paperboard containers & boxes	0.4	Manufg.	Cold finishing of steel shapes	0.7	Manufg.
Primary metal products, nec	0.4	Manufg.	Oil field machinery	0.7	Manufg.
Hotels & lodging places	0.4	Services	Furniture & fixtures, nec	0.6	Manufg.
U.S. Postal Service	0.4	Gov't	Industrial furnaces & ovens	0.6	Manufg.
Electrical repair shops	0.3	Services	Paper industries machinery	0.6	Manufg.
Legal services	0.3	Services	Service industry machines, nec	0.6	Manufg.
Management & consulting services & labs	0.3	Services	Turbines & turbine generator sets	0.6	Manufg.
Scrap	0.3	Scrap	Fabricated plate work (boiler shops)	0.5	Manufg.
Asbestos products	0.2	Manufg.	Machine tools, metal cutting types	0.5	Manufg.
Lubricating oils & greases	0.2	Manufg.	Nonferrous rolling & drawing, nec	0.5	Manufg.
Metal heat treating	0.2	Manufg.	Rolling mill machinery	0.5	Manufg.
Plating & polishing	0.2	Manufg.	Steel wire & related products	0.5	Manufg.
Railroads & related services	0.2	Util.	Textile machinery	0.5	Manufg.
Accounting, auditing & bookkeeping	0.2	Services	Ball & roller bearings	0.4	Manufg.
Automotive rental & leasing, without drivers	0.2	Services	Hoists, cranes, & monorails	0.4	Manufg.
Manifold business forms	0.1	Manufg.	Motors & generators	0.4	Manufg.
Paints & allied products	0.1	Manufg.	Boat building & repairing	0.3	Manufg.
Insurance carriers	0.1	Fin/R.E.	Engine electrical equipment	0.3	Manufg.
Automotive repair shops & services	0.1	Services	Power driven hand tools	0.3	Manufg.
			Truck & bus bodies	0.3	Manufg.
			Chemical & fertilizer mineral	0.2	Mining
			Dimension, crushed & broken stone	0.2	Mining
			Calculating & accounting machines	0.2	Manufg.
			Fabricated structural metal	0.2	Manufg.
			Iron & steel forgings	0.2	Manufg.
			Machine tools, metal forming types	0.2	Manufg.
			Metalworking machinery, nec	0.2	Manufg.
			Truck trailers	0.2	Manufg.
			Clay, ceramic, & refractory minerals	0.1	Mining
			Copper ore	0.1	Mining
			Sand & gravel	0.1	Mining
			Elevators & moving stairways	0.1	Manufg.
			Gypsum products	0.1	Manufg.
			Motor homes (made on purchased chassis)	0.1	Manufg.
			Photographic equipment & supplies	0.1	Manufg.
			Primary metal products, nec	0.1	Manufg.
			Local government passenger transit	0.1	Gov't

Source: Benchmark Input-Output Accounts for the U.S. Economy, 1982, U.S. Department of Commerce, Washington, D.C., July 1991. Data, as reported in the source, are organized by the 1977 SIC structure in use in 1982 but have been matched, as closely as is possible, to the 1987 SIC structure used in this book.

OCCUPATIONS EMPLOYED BY SIC 356 - GENERAL INDUSTRIAL MACHINERY

Occupation	% of Total 1994	Change to 2005	Occupation	% of Total 1994	Change to 2005
Assemblers, fabricators, & hand workers nec	7.9	-2.9	Machine tool cutting & forming etc. nec	1.5	-2.9
Machinists	4.9	.6.9	Machine assemblers	1.5	-2.8
Sales & related workers nec	3.8	-2.9	Traffic, shipping, & receiving clerks	1.5	-6.5
Blue collar worker supervisors	3.8	-9.0	Combination machine tool operators	1.5	6.9
Inspectors, testers, & graders, precision	3.1	-2.9	Welding machine setters, operators	1.2	-12.6
Welders & cutters	2.8	31.2	Engineering, mathematical, & science managers	1.2	10.4
Mechanical engineers	2.8	6.9	Machine operators nec	1.2	-14.4
General managers & top executives	2.5	-7.8	Industrial production managers	1.2	-2.8
NC machine tool operators, metal & plastic	2.3	45.7	Bookkeeping, accounting, & auditing clerks	1.2	-27.2
Machine builders	2.3	31.1	Stock clerks	1.1	-21.0
Lathe & turning machine tool operators	2.2	-36.8	Engineering technicians & technologists nec	1.1	-2.9
Secretaries, ex legal & medical	2.1	-11.6	Machine forming operators, metal & plastic	1.1	-2.9
Grinding machine operators, metal & plastic	2.1	-12.6	General office clerks	1.1	-17.2
Drafters	2.0	-24.3	Production, planning, & expediting clerks	1.1	-2.8
Machine tool cutting operators, metal & plastic	1.8	-19.1	Helpers, laborers, & material movers nec	1.1	-2.8
Sheet metal workers & duct installers	1.6	-2.8	Drilling & boring machine tool workers	1.1	-51.4
Industrial machinery mechanics	1.6	6.9	Purchasing agents, ex trade & farm products	1.0	6.9

Source: *Industry-Occupation Matrix*, Bureau of Labor Statistics. These data relate to one or more 3-digit SIC industry groups rather than to a single 4-digit SIC. The change reported for each occupation to the year 2005 is a percent of growth or decline as estimated by the Bureau of Labor Statistics. The abbreviation nec stands for 'not elsewhere classified'.

LOCATION BY STATE AND REGIONAL CONCENTRATION

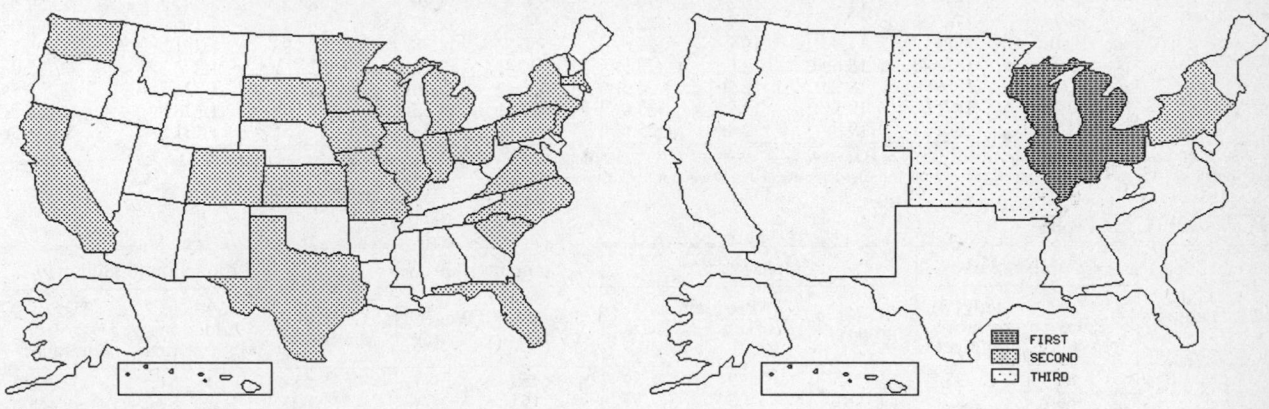

FIRST
SECOND
THIRD

INDUSTRY DATA BY STATE

State	Establish-ments	Shipments			Employment				Cost as % of Shipments	Investment per Employee ($)
		Total ($ mil)	% of U.S.	Per Establ.	Total Number	% of U.S.	Per Establ.	Wages ($/hour)		
Wisconsin	17	272.7	15.0	16.0	2,100	13.4	124	15.21	27.9	4,476
Illinois	35	231.1	12.7	6.6	2,200	14.0	63	13.79	37.8	3,773
Pennsylvania	16	107.1	5.9	6.7	900	5.7	56	12.82	35.9	5,111
California	27	95.9	5.3	3.6	700	4.5	26	15.75	41.2	2,286
Michigan	20	91.7	5.0	4.6	800	5.1	40	11.92	28.9	3,500
North Carolina	10	85.0	4.7	8.5	600	3.8	60	11.63	26.0	4,667
New York	19	75.2	4.1	4.0	1,000	6.4	53	11.31	29.3	3,600
Missouri	12	58.9	3.2	4.9	600	3.8	50	10.50	48.9	4,167
New Jersey	12	46.9	2.6	3.9	400	2.5	33	14.80	40.1	1,500
Massachusetts	14	44.2	2.4	3.2	500	3.2	36	13.50	28.5	3,800
Texas	12	43.8	2.4	3.7	300	1.9	25	10.60	40.6	9,333
Minnesota	12	43.2	2.4	3.6	400	2.5	33	13.33	34.7	9,250
Florida	7	16.0	0.9	2.3	200	1.3	29	9.67	36.9	2,000
Iowa	5	11.8	0.6	2.4	200	1.3	40	13.00	44.1	-
Ohio	26	(D)	-	-	750 *	4.8	29	-	-	3,600
Indiana	5	(D)	-	-	1,750 *	11.1	350	-	-	-
Kansas	4	(D)	-	-	375 *	2.4	94	-	-	-
South Carolina	3	(D)	-	-	375 *	2.4	125	-	-	-
Virginia	2	(D)	-	-	175 *	1.1	88	-	-	-
Colorado	1	(D)	-	-	175 *	1.1	175	-	-	-
South Dakota	1	(D)	-	-	375 *	2.4	375	-	-	-
Washington	1	(D)	-	-	175 *	1.1	175	-	-	-

Source: 1992 *Economic Census*. The states are in descending order of shipments or establishments (if shipment data are missing for the majority). The symbol (D) appears when data are withheld to prevent disclosure of competitive information. States marked with (D) are sorted by number of establishments. A dash (-) indicates that the data element cannot be calculated; * indicates the midpoint of a range.

3567 - INDUSTRIAL FURNACES & OVENS

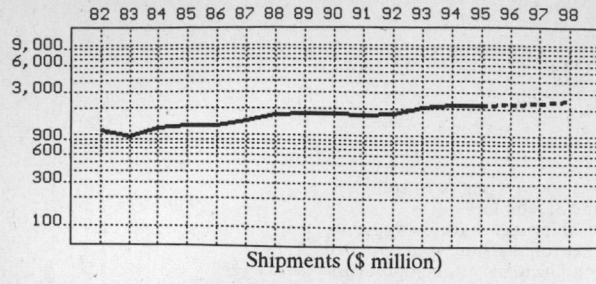

Shipments ($ million)

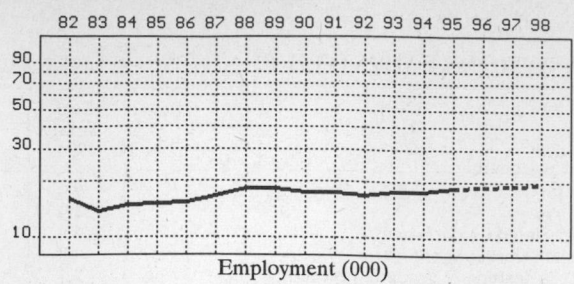

Employment (000)

GENERAL STATISTICS

| Year | Companies | Establishments | | Employment | | | Compensation | | Production ($ million) | | | |
		Total	with 20 or more employees	Total (000)	Production Workers (000)	Hours (Mil)	Payroll ($ mil)	Wages ($/hr)	Cost of Materials	Value Added by Manufacture	Value of Shipments	Capital Invest.
1982	321	353	172	16.1	9.2	17.8	312.6	8.38	465.2	624.9	1,102.2	21.1
1983		345	167	13.8	8.2	15.5	267.7	8.46	383.2	582.3	954.2	12.2
1984		337	162	14.7	8.7	16.5	306.0	8.99	519.6	702.7	1,197.2	26.5
1985		328	157	15.2	8.6	15.7	327.1	9.75	596.5	686.9	1,288.5	27.4
1986		327	157	15.4	8.4	15.9	357.7	10.10	580.6	687.3	1,291.7	21.9
1987	342	370	170	16.6	9.9	19.1	401.1	10.25	623.6	821.6	1,434.8	27.2
1988		358	181	18.2	11.1	21.6	453.4	9.81	744.5	970.9	1,697.6	39.4
1989		362	188	18.1	12.5	25.0	508.6	10.35	798.2	1,009.1	1,778.7	46.1
1990		358	179	17.5	11.7	23.6	500.0	10.43	791.4	902.5	1,766.1	40.2
1991		353	171	17.5	10.9	22.1	472.3	10.40	714.3	940.0	1,679.6	29.0
1992	377	409	181	17.0	10.3	20.1	529.4	11.49	764.2	982.1	1,757.7	27.6
1993		407	176	17.7	11.4	22.5	550.6	11.99	869.1	1,190.5	2,033.0	36.7
1994		394P	182P	17.9	10.8	20.8	550.3	11.38	903.9	1,261.9	2,163.4	40.8
1995		399P	183P	18.6P	12.0P	23.9P	601.4P	12.00P	912.1P	1,273.3P	2,183.0P	42.2P
1996		404P	185P	18.9P	12.2P	24.5P	626.4P	12.26P	949.9P	1,326.1P	2,273.5P	43.9P
1997		410P	187P	19.2P	12.5P	25.0P	651.5P	12.53P	987.7P	1,378.9P	2,364.0P	45.6P
1998		415P	188P	19.5P	12.8P	25.6P	676.6P	12.79P	1,025.5P	1,431.7P	2,454.5P	47.2P

Sources: 1982, 1987, 1992 *Economic Census*; *Annual Survey of Manufactures*, 83-86, 88-91, 93-94. Establishment counts for non-Census years are from *County Business Patterns*; establishment values for 83-84 are extrapolations. 'P's show projections by the editors. Industries reclassified in 87 will not have data for prior years.

INDICES OF CHANGE

| Year | Companies | Establishments | | Employment | | | Compensation | | Production ($ million) | | | |
		Total	with 20 or more employees	Total (000)	Production Workers (000)	Hours (Mil)	Payroll ($ mil)	Wages ($/hr)	Cost of Materials	Value Added by Manufacture	Value of Shipments	Capital Invest.
1982	85	86	95	95	89	89	59	73	61	64	63	76
1983		84	92	81	80	77	51	74	50	59	54	44
1984		82	90	86	84	82	58	78	68	72	68	96
1985		80	87	89	83	78	62	85	78	70	73	99
1986		80	87	91	82	79	68	88	76	70	73	79
1987	91	90	94	98	96	95	76	89	82	84	82	99
1988		88	100	107	108	107	86	85	97	99	97	143
1989		89	104	106	121	124	96	90	104	103	101	167
1990		88	99	103	114	117	94	91	104	92	100	146
1991		86	94	103	106	110	89	91	93	96	96	105
1992	100	100	100	100	100	100	100	100	100	100	100	100
1993		100	97	104	111	112	104	104	114	121	116	133
1994		96P	101P	105	105	103	104	99	118	128	123	148
1995		98P	101P	109P	116P	119P	114P	104P	119P	130P	124P	153P
1996		99P	102P	111P	119P	122P	118P	107P	124P	135P	129P	159P
1997		100P	103P	113P	121P	125P	123P	109P	129P	140P	134P	165P
1998		101P	104P	114P	124P	128P	128P	111P	134P	146P	140P	171P

Sources: Same as General Statistics. Values reflect change from the base year, 1992. Values above 100 mean greater than 92, values below 100 mean less than 92, and a value of 100 in the 82-91 or 93-98 period means same as 92. 'P's mark projections by the editors.

SELECTED RATIOS

For 1994	Avg. of All Manufact.	Analyzed Industry	Index	For 1994	Avg. of All Manufact.	Analyzed Industry	Index
Employees per Establishment	49	45	93	Value Added per Production Worker	134,084	116,843	87
Payroll per Establishment	1,500,273	1,398,206	93	Cost per Establishment	5,045,178	2,296,635	46
Payroll per Employee	30,620	30,743	100	Cost per Employee	102,970	50,497	49
Production Workers per Establishment	34	27	80	Cost per Production Worker	146,988	83,694	57
Wages per Establishment	853,319	601,419	70	Shipments per Establishment	9,576,895	5,496,782	57
Wages per Production Worker	24,861	21,917	88	Shipments per Employee	195,460	120,860	62
Hours per Production Worker	2,056	1,926	94	Shipments per Production Worker	279,017	200,315	72
Wages per Hour	12.09	11.38	94	Investment per Establishment	321,011	103,665	32
Value Added per Establishment	4,602,255	3,206,244	70	Investment per Employee	6,552	2,279	35
Value Added per Employee	93,930	70,497	75	Investment per Production Worker	9,352	3,778	40

Sources: Same as General Statistics. The 'Average of All Manufacturing' column represents the average of all manufacturing industries reported for the most recent complete year available. The Index shows the relationship between the Average and the Analyzed Industry. For example, 100 means that they are equal; 500 that the Analyzed Industry is five times the average; 50 means that the Analyzed Industry is half the national average. The abbreviation 'na' is used to show that data are 'not available'.

LEADING COMPANIES   Number shown: 75   Total sales ($ mil): 3,767   Total employment (000): 24.9

Company Name	Address				CEO Name	Phone	Co. Type	Sales ($ mil)	Empl. (000)
Ogden Projects Inc	PO Box 2615	Fairfield	NJ	07007	Scott Mackin	201-882-9000	P	681	0.5
Inductotherm Industries Inc	10 Indel Av	Rancocas	NJ	08073	Henry M Rowan Jr	609-267-9000	R	550	4.3
George Koch Sons Inc	10 S 11th Av	Evansville	IN	47744	Robert L Koch II	812-465-9600	R	340	2.0
Wiegand Industrial	641 Alpha Dr	Pittsburgh	PA	15238	John Stoops	412-967-3800	D	200	2.1
Watlow Electric Mfg Co	12001 Lackland Rd	St Louis	MO	63146	Gary Neal	314-878-4600	R	150	1.8
Salem Corp	PO Box 2222	Pittsburgh	PA	15230	AA Fornataro	412-923-2200	P	128	0.9
Thermo Process Systems Inc	12068 Market St	Livonia	MI	48150	John P Appleton	313-591-1000	P	110	1.1
Gallagher-Kaiser Corp	13710 Mount Elliot	Detroit	MI	48212	Joseph P Kaiser Jr	313-368-3100	R	98	0.3
Glasstech Inc	995 4th St	Perrysburg	OH	43551	MD Christman	419-661-9500	R	90	0.4
Inductoheat Inc	32251 N Avis Dr	Madison H	MI	48071	BL Taylor	313-585-9393	S	75	0.5
Selas Corporation of America	PO Box 200	Dresher	PA	19025	Stephen F Ryan	215-646-6600	P	74	0.5
Ajax Magnethermic Corp	1745 NE Overland	Warren	OH	44482	Frank Spalla	216-372-8511	S	67	0.6
Abar Ipsen Industries	3260 Tillman Dr	Bensalem	PA	19020	Mac Moore	215-244-4900	S	65	0.3
Continental Materials Corp	325 N Wells St	Chicago	IL	60610	James G Gidwitz	312-661-7200	P	63	0.8
Inductotherm Corp	10 Indel Av	Rancocas	NJ	08073	John H Mortimer	609-267-9000	S	60	0.4
Despatch Industries Inc	63 St Anthony Pkwy	Minneapolis	MN	55440	Sid Johnston	612-781-5363	R	42	0.5
Coen Company Inc	1510 Rollins Rd	Burlingame	CA	94010	JH White	415-697-0440	R	40	0.3
Drever Co	PO Box 98	Huntingdon Vl	PA	19006	F Johnson	215-947-3400	S	40*	0.2
Templeton Coal Company Inc	501 Merch	Terre Haute	IN	47807	John A Templeton	812-232-7037	R	40	0.4
United Dominion Industries	470 Beauty Spot E	Bennettsville	SC	29512	Dennis Porzio	803-479-4006	S	37*	0.4
Callidus Technologies Inc	7130 S Lewis St	Tulsa	OK	74136	William P Bartlett	918-496-7599	S	30	0.2
Detroit Stoker Co	PO Box 732	Monroe	MI	48161	J M Ballentine Jr	313-241-9500	S	30	0.1
Seco Warwick Corp	527 Mercer St	Meadville	PA	16335	PL Huber	814-724-1400	R	30	0.2
Tocco Inc	PO Box 447	Boaz	AL	35957	Denis J Liederbach	205-593-7770	S	27*	0.3
Pillar Corp	330 E Kilbourn Av	Milwaukee	WI	53202	EC Goggio	414-223-5100	R	26*	0.3
Pillar Industries	15800 Megal	Menomonee Fls	WI	53051	William Blackmore	414-255-6470	S	25	0.2
Swindell-Dressler Intern Co	PO Box 15541	Pittsburgh	PA	15244	Richard V Wright	412-788-7100	R	25	<0.1
Ogden Manufacturing Co	719 W Algonquin	Arlington H	IL	60005	G Grendys	708-593-8050	R	24*	0.3
Advantage Engineering Inc	PO Box 407	Greenwood	IN	46142	Philip Oswalt	317-887-0729	R	23	0.1
Holcroft	12068 Market St	Livonia	MI	48150	Francis A Ragone	313-591-1000	D	23*	0.1
Thermo Electron Wisconsin Inc	PO Box 7001	Kaukauna	WI	54130	Robert Rabuck	414-766-7200	S	23	0.2
Oven Systems Inc	4697 W Greenfield	Milwaukee	WI	53214	ES Napoleon	414-672-7700	R	22	0.2
Watlow Industries	6 Indrial Loop Dr	Hannibal	MO	63401	John E Kuhfahl	314-221-2816	D	22	0.3
Wellman Thermal Systems Inc	1 Progress Rd	Shelbyville	IN	46176	JH Cherry	317-398-4411	R	22	0.2
Industrial Airsystems Inc	PO Box 8066	St Paul	MN	55108		612-646-9631	S	20	<0.1
Hi-Temp Inc	75 E Lake St	Northlake	IL	60164	A Lukowicz	708-345-5800	R	18*	0.2
Kolene Corp	12890 Westwood	Detroit	MI	48223	Thomas F McCardle	313-273-9220	R	18	<0.1
Thermal Engineering Corp	PO Box 868	Columbia	SC	29202	WH Best	803-783-0756	R	17	0.1
Consarc Corp	100 Indel Av	Rancocas	NJ	08073	Raymond J Roberts	609-267-8000	S	15	<0.1
Lepel Corp	50 Heartland Blv	Edgewood	NY	11717	John Stoll	516-586-3300	S	15	<0.1
Moco Thermal Industries Inc	1 Oven Pl	Romulus	MI	48174	Curtis B Peterson	313-728-6800	R	15	0.1
Vulcan Iron Works Inc	1050 M Bank Ctr	Wilkes-Barre	PA	18701	Ernst G Klein	717-822-2161	R	15	<0.1
Atmosphere Furnace Co	PO Box 317	Wixom	MI	48393	William M Keough	810-624-8191	R	13	0.2
GC Broach Co	7667 E 46th Pl	Tulsa	OK	74145	GC Broach	918-664-7420	R	13	0.1
Salem Furnace Inc	PO Box 2222	Pittsburgh	PA	15230	Lee A Weaver	412-276-5700	R	13	0.1
Surface Combustion Inc	PO Box 428	Maumee	OH	43537	DG Orzechowski	419-891-7150	R	13	0.1
Born Inc	PO Box 102	Tulsa	OK	74101	Jim Master	918-582-2186	R	12*	<0.1
Centorr/Vacuum Industries Inc	542 Amherst St	Nashua	NH	03063	James S Kellogg	603-595-7233	S	12	<0.1
NGE Inc	2937 Tanager Av	Los Angeles	CA	90040	T Haldeman	213-685-8340	R	12	<0.1
Sherwood Industries Inc	PO Box 317	Wixom	MI	48393	WM Keough	313-624-8191	D	12	<0.1
Thermal Technology Inc	90 Airport Rd	Concord	NH	03301	George M Johnston	603-225-6605	R	12	<0.1
FECO Engineered Systems Inc	5855 Grant Av	Cleveland	OH	44105	K A Krismanth	216-441-2400	R	11	0.1
Amer Induction Heating Corp	PO Box 248	Fraser	MI	48026	Paul N Lavins	810-294-1700	S	10	<0.1
Bickley Inc	PO Box 369	Bensalem	PA	19020	Vincent B Harris	215-638-4500	R	10	<0.1
Blue M Electric	304 Hart St	Watertown	WI	53094	Pete Walter	414-261-0819	D	10*	0.4
EW Bowman Inc	PO Box 849	Uniontown	PA	15401	JA Ulmer	412-438-0503	R	10*	0.1
Frank W Schaefer Inc	PO Box 1508	Dayton	OH	45401	Richard Schaefer	513-253-3342	R	10*	0.1
Fulton Boiler Works Inc	PO Box 257	Pulaski	NY	13142	RB Palm	315-298-5121	R	10	0.2
Grieve Corp	500 Hart Rd	Round Lake	IL	60073	PJ Calabrese	708-546-8225	R	10	<0.1
Harper Electric Furnace	W Drullard Av	Lancaster	NY	14086	Guenter Feucht	716-684-7400	R	10*	<0.1
Hotwatt Inc	128 Maple St	Danvers	MA	01923	RS Lee	508-777-0070	R	10	0.2
Retech Inc	PO Box 997	Ukiah	CA	95482	Max P Schlienger	707-462-6522	R	10	0.1
Salem Engelhard	245 S Mill Rd	South Lyon	MI	48178	Michael Thompson	313-437-1400	R	10*	0.1
Tempco Electric Heater Corp	607 N Central Av	Wood Dale	IL	60191	Fermin Adames	708-350-2252	R	10	0.2
Vacuum-Atmospheres Co	PO Box 1043	Hawthorne	CA	90250	Terry Sweem	310-644-0255	D	10	<0.1
CI Hayes Inc	800 Wellington Av	Cranston	RI	02910	Carl I Hayes	401-467-5200	R	9*	<0.1
Procedyne Corp	11 Industrial Dr	New Brunswick	NJ	08901	H Kenneth Staffin	908-249-8347	R	9	0.1
Heatron Inc	PO Box 45	Leavenworth	KS	66048	Mike Keenan	913-651-4420	R	8	0.2
Granco-Clark Inc	7298 N Storey Rd	Belding	MI	48809	C Gentry	616-794-2600	R	8	<0.1
Kayex Corp	1000 Millstead Way	Rochester	NY	14624	James L Stanton	716-235-2524	S	8	<0.1
Process Combustion Corp	PO Box 12866	Pittsburgh	PA	15241	JE Johns	412-655-0955	S	8*	<0.1
Thermcraft Inc	PO Box 12037	Winston-Salem	NC	27117	Morris Crafton	910-784-4800	R	8	<0.1
Advanced Vacuum Systems Inc	60 Fitchburg Rd	Ayer	MA	01432	Norman R Buck	508-772-0712	R	7*	<0.1
Gladd Industries Inc	15450 Dale St	Detroit	MI	48223	Andrew Gladd	313-537-2800	R	7*	<0.1
Thorpe Technologies Inc	9905 Painter Av	Whittier	CA	90605	JE Allen	310-903-8230	R	7	<0.1

Source: Ward's Business Directory of U.S. Private and Public Companies, Volumes 1 and 2, 1996. The company type code used is as follows: P - Public, R - Private, S - Subsidiary, D - Division, J - Joint Venture, A - Affiliate, G - Group. Sales are in millions of dollars, employees are in thousands. An asterisk (*) indicates an estimated sales volume. The symbol < stands for 'less than'. Company names and addresses are truncated, in some cases, to fit into the available space.

MATERIALS CONSUMED

Material	Quantity	Delivered Cost ($ million)
Materials, ingredients, containers, and supplies	(X)	673.4
Fluid power products	(X)	21.6
Metal stampings	(X)	4.0
All other fabricated metal products (except fluid power products and forgings)	(X)	48.6
Forgings	(X)	1.0
Iron and steel castings (rough and semifinished)	(X)	4.3
Nonferrous (aluminum, copper, etc.) castings (rough and semifinished)	(X)	3.4
Steel bars, bar shapes, and plates	(X)	18.1
Steel sheet and strip, including tin plate	(X)	24.7
Steel structural shapes and sheet piling	(X)	8.2
All other steel shapes and forms	(X)	15.6
Copper and copper-base alloy shapes and forms	(X)	5.3
Aluminum and aluminum-base alloy sheet, plate, foil, and welded tubing	(X)	7.0
All other aluminum and aluminum-base alloy shapes and forms	(X)	1.9
Other nonferrous shapes and forms	(X)	6.9
Electrical transmission, distribution, and control equipment	(X)	67.3
Electric heating elements for industrial furnaces, ovens, and kilns	(X)	18.5
All other materials and components, parts, containers, and supplies	(X)	266.2
Materials, ingredients, containers, and supplies, nsk	(X)	150.9

Source: 1992 Economic Census. Explanation of symbols used: (D): Withheld to avoid disclosure of competitive data; na: Not available; (S): Withheld because statistical norms were not met; (X): Not applicable; (Z): Less than half the unit shown; nec: Not elsewhere classified; nsk: Not specified by kind; - : zero; * : 10-19 percent estimated; ** : 20-29 percent estimated.

PRODUCT SHARE DETAILS

Product or Product Class	% Share	Product or Product Class	% Share
Industrial furnaces and ovens	100.00	All other industrial electric heating units and devices, including strip heaters, ring heaters, water and oil immersion heaters, glue and compound pots, etc. (except soldering irons)	62.47
Electric industrial furnaces, ovens, and kilns	23.78	Parts and attachments for electrical industrial space heaters	2.95
Electric industrial metal melting furnaces (excluding induction)	7.50	Parts and attachments for other electrical industrial furnaces and ovens	21.38
Electric industrial metal processing and heat treating furnaces (such as annealing, hardening, carburizing, and porcelain enameling furnaces, excluding induction)	31.84	Fuel-fired industrial furnaces, ovens, and kilns	20.81
Other electric industrial furnaces, excluding induction	36.71	Fuel-fired industrial metal melting furnaces, including blast furnaces and cupolas, except parts and attachments	2.11
Electric industrial ovens and kilns, including infrared	23.97	Fuel-fired industrial metal processing and heat treating furnaces (such as annealing, hardening, carburizing, and porcelain enameling furnaces), except parts and attachments	40.54
High-frequency induction and dielectric heating equipment	9.11		
Radio frequency type (including spark gap) and line and motor-generator set frequency type furnaces and ovens, (induction or dielectric), except metal melting	11.82	Other fuel-fired industrial furnaces (including hot rolling, forging, forming, and extruding), except parts and attachments	14.34
High-frequency metal melting induction furnaces	63.46	Fuel-fired industrial ovens and kilns (except cement, wood, and chemical), except parts and attachments	30.84
Other high-frequency induction or dielectric furnaces and ovens	12.40	Parts and attachments for industrial fuel-fired furnaces, ovens, and kilns	11.61
Other high-frequency induction or dielectric heating equipment	12.07	Fuel-fired industrial furnaces, ovens, and kilns, nsk	0.56
High-frequency induction and dielectric heating equipment, nsk	0.32	Industrial furnaces and ovens, nsk	8.55
Electrical heating equipment for industrial use, nec, (except soldering iron) and parts and attachments	37.75		
Industrial electric tubular heaters	10.08		
Industrial electric space heaters	3.15		

Source: 1992 Economic Census. The values shown are percent of total shipments in an industry. Values of indented subcategories are summed in the main heading. The symbol (D) appears when data are withheld to prevent disclosure of competitive information. The abbreviation nsk stands for 'not specified by kind' and nec for 'not elsewhere classified'.

INPUTS AND OUTPUTS FOR INDUSTRIAL FURNACES & OVENS

Economic Sector or Industry Providing Inputs	%	Sector	Economic Sector or Industry Buying Outputs	%	Sector
Blast furnaces & steel mills	13.9	Manufg.	Gross private fixed investment	76.1	Cap Inv
Wholesale trade	9.3	Trade	Exports	16.3	Foreign
Industrial furnaces & ovens	8.8	Manufg.	Industrial furnaces & ovens	5.9	Manufg.
Imports	8.8	Foreign	Federal Government purchases, national defense	0.8	Fed Govt
Switchgear & switchboard apparatus	6.0	Manufg.	Machinery, except electrical, nec	0.5	Manufg.
Pipe, valves, & pipe fittings	3.8	Manufg.	Blast furnaces & steel mills	0.2	Manufg.
Machinery, except electrical, nec	3.7	Manufg.	Primary metal products, nec	0.1	Manufg.
Copper rolling & drawing	3.5	Manufg.			
Fabricated plate work (boiler shops)	3.3	Manufg.			
Power transmission equipment	3.0	Manufg.			
Transformers	2.5	Manufg.			
Electronic components nec	2.3	Manufg.			
Ball & roller bearings	2.0	Manufg.			
Aluminum rolling & drawing	1.9	Manufg.			
Special dies & tools & machine tool accessories	1.6	Manufg.			

Continued on next page.

INPUTS AND OUTPUTS FOR INDUSTRIAL FURNACES & OVENS - Continued

Economic Sector or Industry Providing Inputs	%	Sector	Economic Sector or Industry Buying Outputs	%	Sector
Electric services (utilities)	1.4	Util.			
Iron & steel foundries	1.3	Manufg.			
Industrial controls	1.2	Manufg.			
Maintenance of nonfarm buildings nec	1.1	Constr.			
Communications, except radio & TV	1.1	Util.			
Fabricated rubber products, nec	1.0	Manufg.			
Eating & drinking places	1.0	Trade			
Blowers & fans	0.9	Manufg.			
Pumps & compressors	0.9	Manufg.			
Screw machine and related products	0.9	Manufg.			
Nonferrous rolling & drawing, nec	0.8	Manufg.			
Air transportation	0.8	Util.			
Motor freight transportation & warehousing	0.8	Util.			
Advertising	0.8	Services			
Engineering, architectural, & surveying services	0.8	Services			
Abrasive products	0.7	Manufg.			
Banking	0.7	Fin/R.E.			
Miscellaneous plastics products	0.6	Manufg.			
Gas production & distribution (utilities)	0.6	Util.			
Real estate	0.6	Fin/R.E.			
Equipment rental & leasing services	0.5	Services			
U.S. Postal Service	0.5	Gov't			
Die-cut paper & board	0.4	Manufg.			
Sawmills & planning mills, general	0.4	Manufg.			
Business/professional associations	0.4	Services			
Legal services	0.4	Services			
Management & consulting services & labs	0.4	Services			
Metal heat treating	0.3	Manufg.			
Motors & generators	0.3	Manufg.			
Petroleum refining	0.3	Manufg.			
Sheet metal work	0.3	Manufg.			
Railroads & related services	0.3	Util.			
Computer & data processing services	0.3	Services			
Hotels & lodging places	0.3	Services			
Gaskets, packing & sealing devices	0.2	Manufg.			
Lubricating oils & greases	0.2	Manufg.			
Manifold business forms	0.2	Manufg.			
Paperboard containers & boxes	0.2	Manufg.			
Accounting, auditing & bookkeeping	0.2	Services			
Insurance carriers	0.1	Fin/R.E.			
Noncomparable imports	0.1	Foreign			

Source: Benchmark Input-Output Accounts for the U.S. Economy, 1982, U.S. Department of Commerce, Washington, D.C., July 1991. Data, as reported in the source, are organized by the 1977 SIC structure in use in 1982 but have been matched, as closely as is possible, to the 1987 SIC structure used in this book.

OCCUPATIONS EMPLOYED BY SIC 356 - GENERAL INDUSTRIAL MACHINERY

Occupation	% of Total 1994	Change to 2005	Occupation	% of Total 1994	Change to 2005
Assemblers, fabricators, & hand workers nec	7.9	-2.9	Machine tool cutting & forming etc. nec	1.5	-2.9
Machinists	4.9	6.9	Machine assemblers	1.5	-2.8
Sales & related workers nec	3.8	-2.9	Traffic, shipping, & receiving clerks	1.5	-6.5
Blue collar worker supervisors	3.8	-9.0	Combination machine tool operators	1.5	6.9
Inspectors, testers, & graders, precision	3.1	-2.9	Welding machine setters, operators	1.2	-12.6
Welders & cutters	2.8	31.2	Engineering, mathematical, & science managers	1.2	10.4
Mechanical engineers	2.8	6.9	Machine operators nec	1.2	-14.4
General managers & top executives	2.5	-7.8	Industrial production managers	1.2	-2.8
NC machine tool operators, metal & plastic	2.3	45.7	Bookkeeping, accounting, & auditing clerks	1.2	-27.2
Machine builders	2.3	31.1	Stock clerks	1.1	-21.0
Lathe & turning machine tool operators	2.2	-36.8	Engineering technicians & technologists nec	1.1	-2.9
Secretaries, ex legal & medical	2.1	-11.6	Machine forming operators, metal & plastic	1.1	-2.9
Grinding machine operators, metal & plastic	2.1	-12.6	General office clerks	1.1	-17.2
Drafters	2.0	-24.3	Production, planning, & expediting clerks	1.1	-2.8
Machine tool cutting operators, metal & plastic	1.8	-19.1	Helpers, laborers, & material movers nec	1.1	-2.8
Sheet metal workers & duct installers	1.6	-2.8	Drilling & boring machine tool workers	1.1	-51.4
Industrial machinery mechanics	1.6	6.9	Purchasing agents, ex trade & farm products	1.0	6.9

Source: Industry-Occupation Matrix, Bureau of Labor Statistics. These data relate to one or more 3-digit SIC industry groups rather than to a single 4-digit SIC. The change reported for each occupation to the year 2005 is a percent of growth or decline as estimated by the Bureau of Labor Statistics. The abbreviation nec stands for 'not elsewhere classified'.

LOCATION BY STATE AND REGIONAL CONCENTRATION

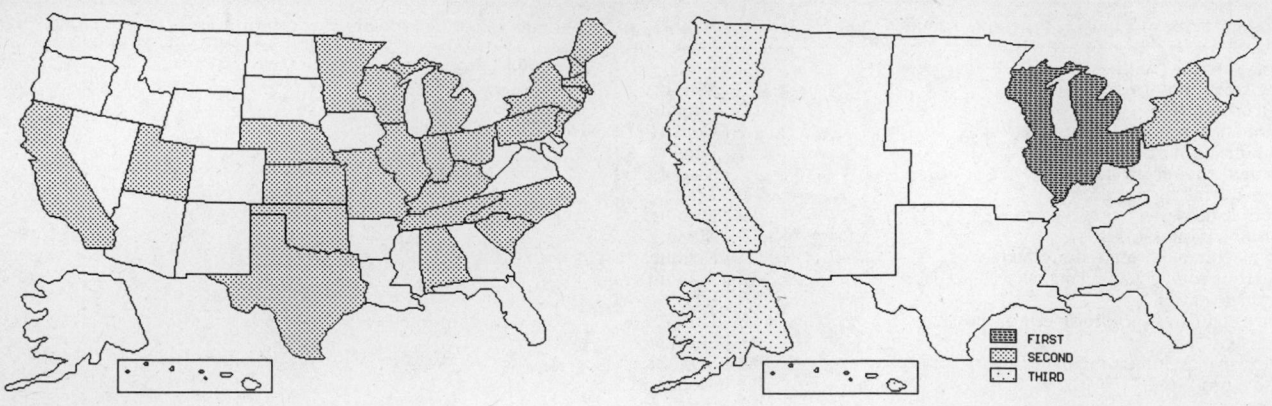

INDUSTRY DATA BY STATE

| State | Establish-ments | Shipments | | | Employment | | | | Cost as % of Shipments | Investment per Employee ($) |
		Total ($ mil)	% of U.S.	Per Establ.	Total Number	% of U.S.	Per Establ.	Wages ($/hour)		
California	48	237.0	13.5	4.9	2,000	11.8	42	12.19	48.4	-
Pennsylvania	34	164.2	9.3	4.8	1,300	7.6	38	11.07	50.1	1,231
Michigan	40	157.9	9.0	3.9	1,400	8.2	35	13.47	41.9	2,214
Ohio	38	143.6	8.2	3.8	1,200	7.1	32	15.77	49.5	917
New Jersey	24	134.5	7.7	5.6	1,000	5.9	42	13.89	40.7	2,400
Illinois	29	110.8	6.3	3.8	1,300	7.6	45	10.47	42.5	769
Missouri	15	100.8	5.7	6.7	1,600	9.4	107	9.14	23.7	-
Massachusetts	15	91.1	5.2	6.1	900	5.3	60	10.90	46.4	1,222
Wisconsin	17	86.2	4.9	5.1	900	5.3	53	12.36	37.2	2,222
Minnesota	8	58.0	3.3	7.3	600	3.5	75	14.00	46.0	833
New York	16	50.8	2.9	3.2	500	2.9	31	11.67	39.8	-
Oklahoma	8	36.4	2.1	4.6	500	2.9	63	13.20	63.2	-
Indiana	10	31.3	1.8	3.1	400	2.4	40	12.00	37.7	2,000
New Hampshire	5	26.5	1.5	5.3	200	1.2	40	12.00	39.2	1,500
Kansas	7	25.2	1.4	3.6	300	1.8	43	7.75	45.2	667
Alabama	4	23.7	1.3	5.9	200	1.2	50	12.50	48.1	-
Texas	15	21.6	1.2	1.4	200	1.2	13	9.67	45.8	1,000
Connecticut	7	17.9	1.0	2.6	200	1.2	29	9.00	46.9	-
South Carolina	5	11.1	0.6	2.2	100	0.6	20	12.00	47.7	-
North Carolina	9	(D)	-	-	375 *	2.2	42	-	-	-
Tennessee	5	(D)	-	-	750 *	4.4	150	-	-	-
Rhode Island	4	(D)	-	-	175 *	1.0	44	-	-	-
Kentucky	2	(D)	-	-	175 *	1.0	88	-	-	-
Nebraska	2	(D)	-	-	175 *	1.0	88	-	-	-
Maine	1	(D)	-	-	175 *	1.0	175	-	-	-
Utah	1	(D)	-	-	375 *	2.2	375	-	-	-

Source: 1992 *Economic Census*. The states are in descending order of shipments or establishments (if shipment data are missing for the majority). The symbol (D) appears when data are withheld to prevent disclosure of competitive information. States marked with (D) are sorted by number of establishments. A dash (-) indicates that the data element cannot be calculated; * indicates the midpoint of a range.

3568 - POWER TRANSMISSION EQUIPMENT, NEC

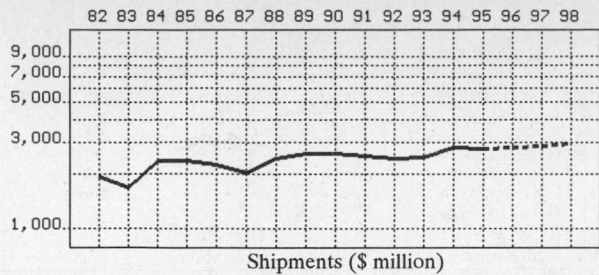

82 83 84 85 86 87 88 89 90 91 92 93 94 95 96 97 98

Shipments ($ million)

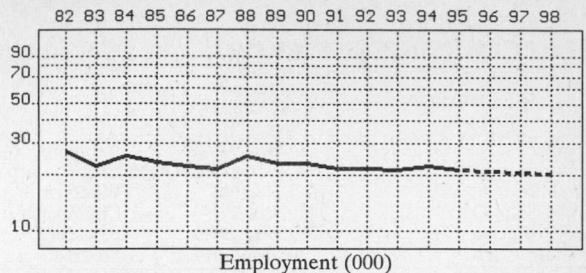

82 83 84 85 86 87 88 89 90 91 92 93 94 95 96 97 98

Employment (000)

GENERAL STATISTICS

| Year | Com-panies | Establishments | | Employment | | | Compensation | | Production ($ million) | | | |
		Total	with 20 or more employees	Total (000)	Production Workers (000)	Hours (Mil)	Payroll ($ mil)	Wages ($/hr)	Cost of Materials	Value Added by Manufacture	Value of Shipments	Capital Invest.
1982	243	294	195	27.1	18.1	33.3	556.4	10.13	767.4	1,153.9	1,940.5	75.0
1983		288	193	22.7	15.8	29.3	497.0	10.33	686.5	980.0	1,672.8	48.4
1984		282	191	25.7	18.2	36.2	623.3	10.89	1,019.4	1,366.2	2,357.5	83.6
1985		276	188	23.7	16.3	33.1	611.6	11.32	996.3	1,325.5	2,343.0	81.8
1986		268	175	22.4	15.3	30.9	599.2	11.71	922.4	1,295.8	2,221.3	60.0
1987	262	308	183	22.0	15.0	29.6	562.6	11.86	776.0	1,258.6	2,041.1	61.5
1988		299	178	25.5	17.7	35.3	676.7	11.52	1,016.6	1,438.9	2,409.1	63.1
1989		295	182	23.3	17.2	34.7	643.7	11.48	1,092.8	1,527.7	2,598.9	57.6
1990		282	180	23.3	16.6	33.7	638.2	11.62	1,102.0	1,503.3	2,596.5	68.1
1991		290	179	21.7	14.8	29.8	606.0	11.84	1,049.8	1,442.4	2,479.1	61.9
1992	270	311	171	21.8	14.6	29.6	679.0	13.05	922.0	1,493.0	2,411.4	72.5
1993		306	167	21.3	14.6	30.3	693.5	13.53	1,016.6	1,462.6	2,462.3	72.6
1994		303P	168P	22.7	15.5	32.5	734.9	13.31	1,162.4	1,648.1	2,796.3	72.3
1995		305P	166P	21.3P	14.8P	31.2P	717.8P	13.46P	1,145.9P	1,624.8P	2,756.7P	68.1P
1996		306P	164P	21.0P	14.6P	31.1P	731.1P	13.70P	1,171.1P	1,660.4P	2,817.2P	68.1P
1997		308P	161P	20.8P	14.4P	30.9P	744.4P	13.95P	1,196.2P	1,696.1P	2,877.7P	68.2P
1998		310P	159P	20.5P	14.2P	30.8P	757.7P	14.19P	1,221.4P	1,731.8P	2,938.2P	68.3P

Sources: 1982, 1987, 1992 *Economic Census*; *Annual Survey of Manufactures*, 83-86, 88-91, 93-94. Establishment counts for non-Census years are from *County Business Patterns*; establishment values for 83-84 are extrapolations. 'P's show projections by the editors. Industries reclassified in 87 will not have data for prior years.

INDICES OF CHANGE

| Year | Com-panies | Establishments | | Employment | | | Compensation | | Production ($ million) | | | |
		Total	with 20 or more employees	Total (000)	Production Workers (000)	Hours (Mil)	Payroll ($ mil)	Wages ($/hr)	Cost of Materials	Value Added by Manufacture	Value of Shipments	Capital Invest.
1982	90	95	114	124	124	113	82	78	83	77	80	103
1983		93	113	104	108	99	73	79	74	66	69	67
1984		91	112	118	125	122	92	83	111	92	98	115
1985		89	110	109	112	112	90	87	108	89	97	113
1986		86	102	103	105	104	88	90	100	87	92	83
1987	97	99	107	101	103	100	83	91	84	84	85	85
1988		96	104	117	121	119	100	88	110	96	100	87
1989		95	106	107	118	117	95	88	119	102	108	79
1990		91	105	107	114	114	94	89	120	101	108	94
1991		93	105	100	101	101	89	91	114	97	103	85
1992	100	100	100	100	100	100	100	100	100	100	100	100
1993		98	98	98	100	102	102	104	110	98	102	100
1994		97P	98P	104	106	110	108	102	126	110	116	100
1995		98P	97P	98P	101P	105P	106P	103P	124P	109P	114P	94P
1996		99P	96P	97P	100P	105P	108P	105P	127P	111P	117P	94P
1997		99P	94P	95P	98P	104P	110P	107P	130P	114P	119P	94P
1998		100P	93P	94P	97P	104P	112P	109P	132P	116P	122P	94P

Sources: Same as General Statistics. Values reflect change from the base year, 1992. Values above 100 mean greater than 92, values below 100 mean less than 92, and a value of 100 in the 82-91 or 93-98 period means same as 92. 'P's mark projections by the editors.

SELECTED RATIOS

For 1994	Avg. of All Manufact.	Analyzed Industry	Index	For 1994	Avg. of All Manufact.	Analyzed Industry	Index
Employees per Establishment	49	75	153	Value Added per Production Worker	134,084	106,329	79
Payroll per Establishment	1,500,273	2,426,383	162	Cost per Establishment	5,045,178	3,837,839	76
Payroll per Employee	30,620	32,374	106	Cost per Employee	102,970	51,207	50
Production Workers per Establishment	34	51	149	Cost per Production Worker	146,988	74,994	51
Wages per Establishment	853,319	1,428,212	167	Shipments per Establishment	9,576,895	9,232,406	96
Wages per Production Worker	24,861	27,908	112	Shipments per Employee	195,460	123,185	63
Hours per Production Worker	2,056	2,097	102	Shipments per Production Worker	279,017	180,406	65
Wages per Hour	12.09	13.31	110	Investment per Establishment	321,011	238,709	74
Value Added per Establishment	4,602,255	5,441,451	118	Investment per Employee	6,552	3,185	49
Value Added per Employee	93,930	72,604	77	Investment per Production Worker	9,352	4,665	50

Sources: Same as General Statistics. The 'Average of All Manufacturing' column represents the average of all manufacturing industries reported for the most recent complete year available. The Index shows the relationship between the Average and the Analyzed Industry. For example, 100 means that they are equal; 500 that the Analyzed Industry is five times the average; 50 means that the Analyzed Industry is half the national average. The abbreviation 'na' is used to show that data are 'not available'.

LEADING COMPANIES Number shown: **70**   Total sales ($ mil): **1,281**   Total employment (000): **12.0**

Company Name	Address				CEO Name	Phone	Co. Type	Sales ($ mil)	Empl. (000)
Martin Sprocket and Gear Inc	PO Box 91588	Arlington	TX	76015	Gary C Martin	817-465-6377	R	170•	1.7
Glacier Vandervell Inc	50 W Big Beaver Rd	Troy	MI	48084	James Connor	313-524-0600	S	120	1.1
US Tsubaki Inc	PO Box 665	Wheeling	IL	60090	H Miyazaki	708-459-9500	S	100	0.9
TB Wood's Sons Co	440 5th Av	Chambersburg	PA	17201	Michael L Hurt	717-264-7161	S	65	0.7
Rockford Powertrain Inc	PO Box 2908	Rockford	IL	61132	GR Welch	815-633-7460	R	56	0.6
Jeffrey Chain Corp	2307 Maden Dr	Morristown	TN	37814	Gerd Krohn	615-586-1951	R	53	0.3
Formsprag-Warren	PO Box 778	Warren	MI	48090	Bill Wilson	313-758-5000	D	52	0.3
Airflex	9919 Clinton Rd	Cleveland	OH	44144	James W Fisher	216-281-2211	D	34	0.2
Horton Manufacturing Co	1170 15th Av SE	Minneapolis	MN	55414	Hugh K Schilling Jr	612-331-5931	S	31	0.3
Lovejoy Inc	2655 Wisconsin Av	Downers Grove	IL	60515	M W Hennessy	708-852-0500	R	31	0.3
Weasler Engineering Inc	PO Box 558	West Bend	WI	53095	JT Hawkins	414-338-2161	R	30	0.3
Regal-Beloit Corp	PO Box 298	Beloit	WI	53512	Carroll V Stein	608-365-2563	D	27	0.2
Designatronics Inc	PO Box 5416	New Hyde Park	NY	11040	Martin Hoffman	516-328-3300	P	25	0.3
RBC Transport Dynamics Corp	PO Box 1953	Santa Ana	CA	92702	M Hartnett	714-546-3131	R	25•	0.3
Hilliard Corp	100 W 4th St	Elmira	NY	14902	GF Schichtel	607-733-7121	R	23	0.2
Renold Inc	PO Box A	Westfield	NY	14787	JL Ellison	716-326-3121	S	23	0.1
Morse Industrial	S Aurora St	Ithaca	NY	14850	Chris Smith	607-272-7220	D	22	0.6
Acme Chain	821 Main St	Holyoke	MA	01040	Kenji Ohara	413-536-1576	D	21•	0.2
Marquette Coppersmithing Inc	PO Box 4584	Philadelphia	PA	19131	Chester M Heller	215-877-9362	R	20	0.1
PSI	PO Box 118	Pitman	NJ	08071	Peter Skidmore	609-589-0815	R	20•	0.2
Zurn Industries Inc	PO Box 13801	Erie	PA	16514	NJ Anderson	814-871-6150	D	20	0.2
Capitol Stampings Corp	PO Box 12365	Milwaukee	WI	53212	Paul A Cadorin	414-963-3500	R	18	<0.1
Orion Corp	PO Box 84	Grafton	WI	53024	Bruce L Boegel	414-377-2210	R	18•	0.2
Sparks Belting Co	3800 Stahl Dr SE	Grand Rapids	MI	49546	J Wardrop	616-949-2750	D	16	0.1
BW Elliott Manufacturing	PO Box 773	Binghamton	NY	13904	GM Scherer	607-772-0404	R	14	0.2
A-1 Carbide Corp	PO Box 1387	Placentia	CA	92670	V Russell Benedict	714-630-9422	R	12	0.1
Bishop-Wisecarver Corp	PO Box 1109	Pittsburg	CA	94565	W R Wisecarver	510-439-8272	R	12	<0.1
EC Styberg Engineering	PO Box 788	Racine	WI	53401	EC Styberg Jr	414-637-9301	R	12	0.2
Linn Gear Co	PO Box 397	Lebanon	OR	97355	Gene Hartl	503-259-1211	R	12	0.2
Graetz Manufacturing Inc	W 11094 Hwy 64	Pound	WI	54161	Aulden Graetz	414-897-4041	R	11•	0.2
Midwest Control Products Corp	PO Box 299	Bushnell	IL	61422	Mark Rauschert	309-772-3163	R	11	0.2
Avon Bearings Corp	1500 Nagle Rd	Avon	OH	44011	John W Walsh	216-871-2500	R	10•	0.1
Cotta Transmission Co	2210 Harrison Av	Rockford	IL	61104	William Sims	815-394-7400	R	10•	0.1
Industrial Sales Co	1200 W Hamburg St	Baltimore	MD	21230	Henry Schloss	410-727-0665	R	10•	<0.1
Nim-Cor Inc	PO Box K	Nashua	NH	03061	JM Smethurst	603-889-2153	S	9	<0.1
Dover Instrument Corp	PO Box 200	Westborough	MA	01581	Steve Hero	508-366-1456	R	8	<0.1
Dynacorp	5173 26th Av	Rockford	IL	61109	David Dyckman	815-229-3190	D	8	0.1
Star Linear Systems Co	9432 Southern Pine	Charlotte	NC	28273	Dennis J Barnes	704-523-2088	S	8	<0.1
Force Control Industries Inc	PO Box 18366	Fairfield	OH	45018	James Besl	513-868-0900	R	6•	<0.1
Marland Clutch	PO Box 308	La Grange	IL	60525	Thomas L Werking	708-352-3330	D	6	<0.1
American Metal Bearing Co	7191 Acacia Av	Garden Grove	CA	92641	Dominic Negri	714-892-5527	R	5	<0.1
Beckett Bronze Co	PO Box 2425	Muncie	IN	47307	LR Dixon	317-282-2261	R	5	<0.1
Centric Clutch	PO Box 668	Woodbridge	NJ	07095	Lee C Fantone	908-634-1761	D	5	<0.1
Don Dye Company Inc	PO Box 107	Kingman	KS	67068	Don D Dye	316-532-3131	R	5	<0.1
Highway Machine Company Inc	RR 1	Princeton	IN	47670	Robert J Smith III	812-385-3639	R	5	<0.1
Industrial Clutch Corp	PO Box 118	Waukesha	WI	53187	David J Pfeffer	414-547-3357	R	5	<0.1
Liquid Drive Corp	10799 Plaza Dr	Whitmore Lake	MI	48189	Robert G Nelson	313-449-4443	R	5	<0.1
Ramsey Products Corp	PO Box 668827	Charlotte	NC	28266	SG Hall	704-394-0322	R	5	<0.1
Ruland Manufacturing Company	380 Pleasant St	Watertown	MA	02172	Robert Ruland	617-924-8000	R	5•	<0.1
SafeWay Hydraulics Inc	4040 Norex Dr	Chaska	MN	55378	Steven J Berkey	612-448-2600	R	5•	<0.1
Orlandi Gear Co	6566 Sterling Dr S	Sterling Hts	MI	48312	Ralph A Orlandi	810-264-6700	R	5	<0.1
ADSCO Manufacturing Corp	4979 Lake Av	Buffalo	NY	14219	Gustav Linda	716-827-5450	R	4	<0.1
Alves Precision Engineered Prod	PO Box 518	Watertown	CT	06795	Alexander R Alves	203-274-6756	R	4	0.1
Milwaukee Bearing & Machining	9532 W Carmen	Milwaukee	WI	53225	K Chybowski	414-438-1100	R	4	<0.1
North American Clutch Corp	3131 W Mill Rd	Milwaukee	WI	53209	W Hargarten Jr	414-352-9727	R	4•	<0.1
Wheeler Manufacturing Corp	155 Brookline	Cambridge	MA	02139	William J McTighe	617-547-6016	R	4	<0.1
Cannon Bronze Corp	PO Box 289	Keithsburg	IL	61442	George E Mugford	309-374-2211	R	3	<0.1
Graseby Controls Inc	100 Bonita Dr	Greensboro	NC	27405	Tim O'Brien	910-375-7444	S	3	<0.1
HydraMechanica Corp	6625 Cobb St	Sterling Hts	MI	48312	Bernd Werkhausen	313-939-0620	S	3	<0.1
KTR Corp	PO Box 9065	Michigan City	IN	46361	Jerrold S Elenz	219-872-9100	R	3	<0.1
St Louis Bearing Company Inc	333 E B St	Wilmington	CA	90744	Luz Huseman	310-834-8506	R	3•	<0.1
Avcon Inc	5210 Lewis Rd	Agoura Hills	CA	91301	Crawford R Meeks	818-865-0250	R	2	<0.1
Roscommon Manufacturing Co	PO Box 239	Roscommon	MI	48653	WR Waterman	517-275-5126	R	2	<0.1
SEPAC Inc	453 E Clinton St	Elmira	NY	14901	Gary R Packard	607-732-2030	R	2	<0.1
Wisconsin Ordnance Works Ltd	3375 County Rd A	Oshkosh	WI	54901	Lloyd Zellmer	414-426-1977	R	1	<0.1
Brown Bearing Supply Co	7 Deer Park Dr	Monmouth Jnc	NJ	08852	Jeff Brown	908-274-2244	R	1•	<0.1
Gray and Prior Machine Co	95 Granby St	Bloomfield	CT	06002	RW Gray III	203-243-8381	R	1	<0.1
Sanders Company Inc	PO Box 324	Elizabeth City	NC	27907	Clarence T Sanders	919-338-3995	R	1•	<0.1
Placid Industries Inc	100 River St	Lake Placid	NY	12946	Margaret D Pedu	518-523-2422	R	1	<0.1
Applied Mechanical Energy	6401 Carlton Av	Seattle	WA	98108	Rich DeGroen	206-767-6963	R	1	<0.1

Source: Ward's Business Directory of U.S. Private and Public Companies, Volumes 1 and 2, 1996. The company type code used is as follows: P - Public, R - Private, S - Subsidiary, D - Division, J - Joint Venture, A - Affiliate, G - Group. Sales are in millions of dollars, employees are in thousands. An asterisk (*) indicates an estimated sales volume. The symbol < stands for 'less than'. Company names and addresses are truncated, in some cases, to fit into the available space.

MATERIALS CONSUMED

Material	Quantity	Delivered Cost ($ million)
Materials, ingredients, containers, and supplies	(X)	740.5
Metal bolts, nuts, screws, washers, rivets, and other screw machine products	(X)	28.2
Other fabricated metal products (except castings and forgings)	(X)	33.9
Cold iron and steel forgings	(X)	15.8
Other iron and steel forgings	(X)	13.9
Nonferrous forgings	(X)	0.2
Iron and steel castings (rough and semifinished)	(X)	60.4
Nonferrous (aluminum, copper, etc.) castings (rough and semifinished)	(X)	23.5
Steel bars, bar shapes, and plates	(X)	109.5
Steel sheet and strip, including tin plate	(X)	43.7
All other steel shapes and forms	(X)	22.7
Copper and copper-base alloy shapes and forms	(X)	8.9
All other nonferrous shapes and forms	(X)	9.3
Scrap, including iron, steel, aluminum and aluminum-base alloy (excluding home scrap)	(X)	(D)
Ball bearings (mounted or unmounted)	(X)	10.4
Roller bearings (mounted or unmounted)	(X)	5.5
Balls, rollers, cages, collars, races, and other antifriction bearing components and parts	(X)	(D)
Clutches, couplings, shafts, sprockets, and other mechanical power transmission equipment	(X)	15.8
Electric motors, generators, and parts	(X)	12.4
Paperboard containers, boxes, and corrugated paperboard	(X)	6.9
All other materials and components, parts, containers, and supplies	(X)	148.0
Materials, ingredients, containers, and supplies, nsk	(X)	154.5

Source: 1992 Economic Census. Explanation of symbols used: (D): Withheld to avoid disclosure of competitive data; na: Not available; (S): Withheld because statistical norms were not met; (X): Not applicable; (Z): Less than half the unit shown; nec: Not elsewhere classified; nsk: Not specified by kind; - : zero; * : 10-19 percent estimated; ** : 20-29 percent estimated.

PRODUCT SHARE DETAILS

Product or Product Class	% Share	Product or Product Class	% Share
Power transmission equipment, nec	100.00	ASA standard roller chains for sprocket drives	5.66
Plain bearings and bushings	14.88	Other chains for sprocket drives	8.96
Plain bearings and bushings, unmounted, machined	78.53	Sprockets	6.54
Plain bearings, mounted (except engine)	20.12	Pulleys	3.37
Plain bearings and bushings, nsk	1.33	Single drive sheaves	2.27
Mechanical power transmission equipment, except speed		Multiple drive sheaves	2.93
changers, drives, and gears, nec	80.10	Inboard marine propulsion gear transmissions, including	
Friction-type clutches and brakes	12.05	reversing, speed changing, and turbine driven gear drives	1.99
Hydraulic-type clutches and brakes, including hydraulic		Other mechanical power transmission equipment, nec,	
couplings	2.20	except speed changers, drives, and gears	18.44
All other clutches and brakes	6.58	Other parts for mechanical power transmission equipment,	
Gear-type flexible couplings	3.41	except speed changers, drives, and gears	9.06
Flexible couplings other than gear-type	9.58	Mechanical power transmission equipment, except speed	
Nonflexible couplings	0.29	changers, drives, and gears, nec, nsk	2.32
Universal joints	3.36	Power transmission equipment, nec, nsk	5.02
Drive shafts, except flexible shafts	0.98		

Source: 1992 Economic Census. The values shown are percent of total shipments in an industry. Values of indented subcategories are summed in the main heading. The symbol (D) appears when data are withheld to prevent disclosure of competitive information. The abbreviation nsk stands for 'not specified by kind' and nec for 'not elsewhere classified'.

INPUTS AND OUTPUTS FOR POWER TRANSMISSION EQUIPMENT

Economic Sector or Industry Providing Inputs	%	Sector	Economic Sector or Industry Buying Outputs	%	Sector
Imports	13.2	Foreign	Motor vehicle parts & accessories	10.0	Manufg.
Blast furnaces & steel mills	11.8	Manufg.	Exports	7.5	Foreign
Power transmission equipment	11.6	Manufg.	Power transmission equipment	6.3	Manufg.
Wholesale trade	10.0	Trade	Construction machinery & equipment	6.1	Manufg.
Iron & steel forgings	5.6	Manufg.	Farm machinery & equipment	3.9	Manufg.
Iron & steel foundries	5.6	Manufg.	Motorcycles, bicycles, & parts	3.8	Manufg.
Motors & generators	3.2	Manufg.	Aluminum rolling & drawing	3.1	Manufg.
Machinery, except electrical, nec	3.0	Manufg.	Internal combustion engines, nec	2.8	Manufg.
Ball & roller bearings	2.8	Manufg.	Pumps & compressors	2.7	Manufg.
Electric services (utilities)	2.4	Util.	Crude petroleum & natural gas	2.5	Mining
Screw machine and related products	1.7	Manufg.	Machinery, except electrical, nec	2.5	Manufg.
Brass, bronze, & copper castings	1.3	Manufg.	Ship building & repairing	2.5	Manufg.
Special dies & tools & machine tool accessories	1.3	Manufg.	Federal Government purchases, national defense	2.4	Fed Govt
Maintenance of nonfarm buildings nec	1.2	Constr.	Tanks & tank components	2.3	Manufg.
Communications, except radio & TV	1.2	Util.	Coal	2.2	Mining
Computer & data processing services	1.2	Services	Iron & steel foundries	2.0	Manufg.
Copper rolling & drawing	1.1	Manufg.	Railroad equipment	1.9	Manufg.
Nonferrous castings, nec	1.1	Manufg.	Nonferrous wire drawing & insulating	1.8	Manufg.
Aluminum castings	1.0	Manufg.	Blast furnaces & steel mills	1.7	Manufg.
Air transportation	1.0	Util.	Blowers & fans	1.7	Manufg.

Continued on next page.

INPUTS AND OUTPUTS FOR POWER TRANSMISSION EQUIPMENT - Continued

Economic Sector or Industry Providing Inputs	%	Sector	Economic Sector or Industry Buying Outputs	%	Sector
Gas production & distribution (utilities)	1.0	Util.	Logging camps & logging contractors	1.7	Manufg.
Motor freight transportation & warehousing	1.0	Util.	Mining machinery, except oil field	1.6	Manufg.
Equipment rental & leasing services	1.0	Services	Secondary nonferrous metals	1.5	Manufg.
Abrasive products	0.9	Manufg.	Special industry machinery, nec	1.1	Manufg.
Nonferrous rolling & drawing, nec	0.9	Manufg.	Miscellaneous repair shops	1.1	Services
Banking	0.9	Fin/R.E.	Printing trades machinery	1.0	Manufg.
Engineering, architectural, & surveying services	0.9	Services	Steel pipe & tubes	1.0	Manufg.
Machine tools, metal cutting types	0.8	Manufg.	Copper rolling & drawing	0.9	Manufg.
Petroleum refining	0.8	Manufg.	Food products machinery	0.9	Manufg.
Eating & drinking places	0.8	Trade	General industrial machinery, nec	0.9	Manufg.
Advertising	0.8	Services	Industrial trucks & tractors	0.9	Manufg.
Primary copper	0.7	Manufg.	Lawn & garden equipment	0.9	Manufg.
Real estate	0.6	Fin/R.E.	Refrigeration & heating equipment	0.9	Manufg.
Miscellaneous plastics products	0.5	Manufg.	Aircraft & missile equipment, nec	0.8	Manufg.
Fabricated rubber products, nec	0.4	Manufg.	Conveyors & conveying equipment	0.8	Manufg.
Gaskets, packing & sealing devices	0.4	Manufg.	Household laundry equipment	0.8	Manufg.
Paperboard containers & boxes	0.4	Manufg.	Cold finishing of steel shapes	0.7	Manufg.
Primary metal products, nec	0.4	Manufg.	Oil field machinery	0.7	Manufg.
Hotels & lodging places	0.4	Services	Furniture & fixtures, nec	0.6	Manufg.
U.S. Postal Service	0.4	Gov't	Industrial furnaces & ovens	0.6	Manufg.
Electrical repair shops	0.3	Services	Paper industries machinery	0.6	Manufg.
Legal services	0.3	Services	Service industry machines, nec	0.6	Manufg.
Management & consulting services & labs	0.3	Services	Turbines & turbine generator sets	0.6	Manufg.
Scrap	0.3	Scrap	Fabricated plate work (boiler shops)	0.5	Manufg.
Asbestos products	0.2	Manufg.	Machine tools, metal cutting types	0.5	Manufg.
Lubricating oils & greases	0.2	Manufg.	Nonferrous rolling & drawing, nec	0.5	Manufg.
Metal heat treating	0.2	Manufg.	Rolling mill machinery	0.5	Manufg.
Plating & polishing	0.2	Manufg.	Steel wire & related products	0.5	Manufg.
Railroads & related services	0.2	Util.	Textile machinery	0.5	Manufg.
Accounting, auditing & bookkeeping	0.2	Services	Ball & roller bearings	0.4	Manufg.
Automotive rental & leasing, without drivers	0.2	Services	Hoists, cranes, & monorails	0.4	Manufg.
Manifold business forms	0.1	Manufg.	Motors & generators	0.4	Manufg.
Paints & allied products	0.1	Manufg.	Boat building & repairing	0.3	Manufg.
Insurance carriers	0.1	Fin/R.E.	Engine electrical equipment	0.3	Manufg.
Automotive repair shops & services	0.1	Services	Power driven hand tools	0.3	Manufg.
			Truck & bus bodies	0.3	Manufg.
			Chemical & fertilizer mineral	0.2	Mining
			Dimension, crushed & broken stone	0.2	Mining
			Calculating & accounting machines	0.2	Manufg.
			Fabricated structural metal	0.2	Manufg.
			Iron & steel forgings	0.2	Manufg.
			Machine tools, metal forming types	0.2	Manufg.
			Metalworking machinery, nec	0.2	Manufg.
			Truck trailers	0.2	Manufg.
			Clay, ceramic, & refractory minerals	0.1	Mining
			Copper ore	0.1	Mining
			Sand & gravel	0.1	Mining
			Elevators & moving stairways	0.1	Manufg.
			Gypsum products	0.1	Manufg.
			Motor homes (made on purchased chassis)	0.1	Manufg.
			Photographic equipment & supplies	0.1	Manufg.
			Primary metal products, nec	0.1	Manufg.
			Local government passenger transit	0.1	Gov't

Source: Benchmark Input-Output Accounts for the U.S. Economy, 1982, U.S. Department of Commerce, Washington, D.C., July 1991. Data, as reported in the source, are organized by the 1977 SIC structure in use in 1982 but have been matched, as closely as is possible, to the 1987 SIC structure used in this book.

OCCUPATIONS EMPLOYED BY SIC 356 - GENERAL INDUSTRIAL MACHINERY

Occupation	% of Total 1994	Change to 2005	Occupation	% of Total 1994	Change to 2005
Assemblers, fabricators, & hand workers nec	7.9	-2.9	Machine tool cutting & forming etc. nec	1.5	-2.9
Machinists	4.9	6.9	Machine assemblers	1.5	-2.8
Sales & related workers nec	3.8	-2.9	Traffic, shipping, & receiving clerks	1.5	-6.5
Blue collar worker supervisors	3.8	-9.0	Combination machine tool operators	1.5	6.9
Inspectors, testers, & graders, precision	3.1	-2.9	Welding machine setters, operators	1.2	-12.6
Welders & cutters	2.8	31.2	Engineering, mathematical, & science managers	1.2	10.4
Mechanical engineers	2.8	6.9	Machine operators nec	1.2	-14.4
General managers & top executives	2.5	-7.8	Industrial production managers	1.2	-2.8
NC machine tool operators, metal & plastic	2.3	45.7	Bookkeeping, accounting, & auditing clerks	1.2	-27.2
Machine builders	2.3	31.1	Stock clerks	1.1	-21.0
Lathe & turning machine tool operators	2.2	-36.8	Engineering technicians & technologists nec	1.1	-2.9
Secretaries, ex legal & medical	2.1	-11.6	Machine forming operators, metal & plastic	1.1	-2.9
Grinding machine operators, metal & plastic	2.1	-12.6	General office clerks	1.1	-17.2
Drafters	2.0	-24.3	Production, planning, & expediting clerks	1.1	-2.8
Machine tool cutting operators, metal & plastic	1.8	-19.1	Helpers, laborers, & material movers nec	1.1	-2.8
Sheet metal workers & duct installers	1.6	-2.8	Drilling & boring machine tool workers	1.1	-51.4
Industrial machinery mechanics	1.6	6.9	Purchasing agents, ex trade & farm products	1.0	6.9

Source: *Industry-Occupation Matrix*, Bureau of Labor Statistics. These data relate to one or more 3-digit SIC industry groups rather than to a single 4-digit SIC. The change reported for each occupation to the year 2005 is a percent of growth or decline as estimated by the Bureau of Labor Statistics. The abbreviation nec stands for 'not elsewhere classified'.

LOCATION BY STATE AND REGIONAL CONCENTRATION

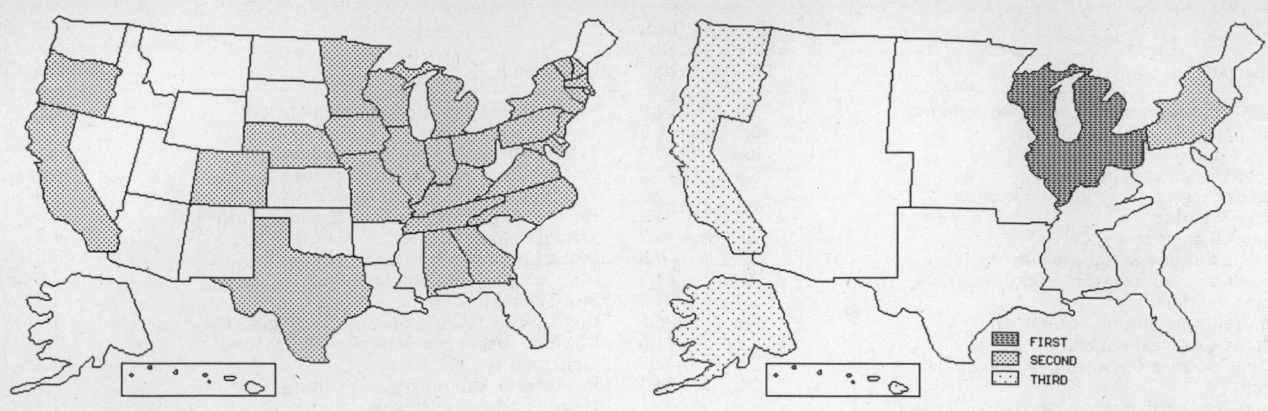

FIRST
SECOND
THIRD

INDUSTRY DATA BY STATE

State	Establish-ments	Shipments			Employment				Cost as % of Shipments	Investment per Employee ($)
		Total ($ mil)	% of U.S.	Per Establ.	Total Number	% of U.S.	Per Establ.	Wages ($/hour)		
Illinois	29	345.2	14.3	11.9	3,200	14.7	110	13.37	35.0	5,313
Wisconsin	23	294.2	12.2	12.8	2,200	10.1	96	16.48	36.5	4,682
Ohio	29	249.0	10.3	8.6	1,900	8.7	66	13.84	42.2	2,632
Michigan	25	161.8	6.7	6.5	1,300	6.0	52	15.17	45.4	3,308
Indiana	12	121.8	5.1	10.1	1,500	6.9	125	13.33	31.0	1,467
New York	19	118.5	4.9	6.2	1,200	5.5	63	12.50	32.7	2,500
Texas	21	103.9	4.3	4.9	1,100	5.0	52	8.39	42.5	727
Pennsylvania	11	103.2	4.3	9.4	1,000	4.6	91	15.00	27.9	6,200
North Carolina	9	86.7	3.6	9.6	800	3.7	89	11.00	42.9	3,375
California	24	71.6	3.0	3.0	800	3.7	33	11.73	34.5	1,000
Connecticut	12	63.8	2.6	5.3	700	3.2	58	11.63	30.6	1,571
Tennessee	8	57.2	2.4	7.2	600	2.8	75	10.29	41.3	-
New Jersey	15	50.5	2.1	3.4	500	2.3	33	13.17	35.2	1,400
Massachusetts	11	40.5	1.7	3.7	600	2.8	55	18.67	40.2	2,500
Nebraska	5	39.1	1.6	7.8	300	1.4	60	12.00	48.3	2,667
Minnesota	6	29.7	1.2	4.9	300	1.4	50	12.80	40.7	7,000
Oregon	4	18.6	0.8	4.7	200	0.9	50	10.25	37.1	2,000
Georgia	5	(D)	-	-	175 *	0.8	35	-	-	1,714
Missouri	5	(D)	-	-	175 *	0.8	35	-	-	-
Iowa	4	(D)	-	-	375 *	1.7	94	-	-	-
New Hampshire	3	(D)	-	-	175 *	0.8	58	-	-	-
Alabama	2	(D)	-	-	175 *	0.8	88	-	-	-
Colorado	2	(D)	-	-	375 *	1.7	188	-	-	-
Kentucky	1	(D)	-	-	1,750 *	8.0	1,750	-	-	-
Maryland	1	(D)	-	-	375 *	1.7	375	-	-	-
Vermont	1	(D)	-	-	175 *	0.8	175	-	-	-
Virginia	1	(D)	-	-	375 *	1.7	375	-	-	-

Source: 1992 *Economic Census*. The states are in descending order of shipments or establishments (if shipment data are missing for the majority). The symbol (D) appears when data are withheld to prevent disclosure of competitive information. States marked with (D) are sorted by number of establishments. A dash (-) indicates that the data element cannot be calculated; * indicates the midpoint of a range.

3569 - GENERAL INDUSTRIAL MACHINERY, NEC

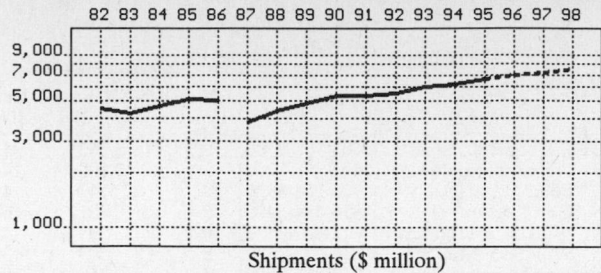

Shipments ($ million)

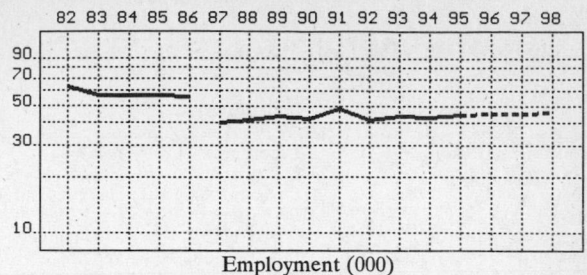

Employment (000)

GENERAL STATISTICS

Year	Com-panies	Establishments Total	Establishments with 20 or more employees	Employment Total (000)	Employment Production Workers (000)	Employment Hours (Mil)	Compensation Payroll ($ mil)	Compensation Wages ($/hr)	Production Cost of Materials	Production Value Added by Manufacture	Production Value of Shipments	Production Capital Invest.
1982	1,391	1,460	565	63.1	37.4	73.7	1,256.0	8.43	1,893.0	2,623.9	4,566.5	132.3
1983		1,418	571	56.8	33.2	64.2	1,178.8	8.88	1,762.8	2,458.0	4,281.0	88.4
1984		1,376	577	56.9	34.3	67.7	1,241.1	9.16	1,952.9	2,757.0	4,681.7	144.3
1985		1,334	582	56.5	33.7	65.6	1,314.0	9.73	2,152.2	2,922.0	5,086.9	146.4
1986		1,333	602	55.4	32.5	65.3	1,342.4	9.75	2,148.1	2,851.3	5,030.1	137.6
1987*	1,157	1,219	444	40.6	23.6	47.5	1,013.7	9.84	1,614.6	2,236.0	3,840.4	105.2
1988		1,158	430	41.4	24.8	49.7	1,072.7	10.20	1,872.4	2,621.1	4,420.5	100.2
1989		1,138	462	43.8	27.1	58.0	1,249.2	10.15	2,169.7	2,818.5	4,886.4	164.4
1990		1,113	454	42.3	27.3	56.6	1,353.1	10.70	2,423.0	2,992.8	5,364.7	171.7
1991		1,128	458	48.1	28.3	55.9	1,382.8	11.28	2,496.6	2,897.0	5,331.1	305.2
1992	965	1,028	426	41.5	23.3	47.1	1,308.4	11.65	2,316.8	3,229.4	5,526.1	180.1
1993		1,049	445	43.8	24.7	49.9	1,402.7	11.87	2,578.2	3,359.6	5,924.6	161.3
1994		1,008P	444P	42.8	25.1	51.3	1,441.1	12.52	2,718.4	3,517.0	6,181.4	198.4
1995		980P	444P	44.4P	25.5P	51.7P	1,537.7P	12.75P	2,887.6P	3,735.9P	6,566.2P	234.3P
1996		952P	444P	44.8P	25.5P	51.6P	1,595.5P	13.13P	3,022.6P	3,910.6P	6,873.2P	247.9P
1997		924P	443P	45.1P	25.5P	51.6P	1,653.2P	13.52P	3,157.7P	4,085.3P	7,180.3P	261.4P
1998		896P	443P	45.4P	25.5P	51.5P	1,710.9P	13.90P	3,292.7P	4,260.0P	7,487.3P	275.0P

Sources: 1982, 1987, 1992 *Economic Census*; *Annual Survey of Manufactures*, 83-86, 88-91, 93-94. Establishment counts are from *County Business Patterns* for non-Census years; establishment counts for 83-84 are extrapolations. * indicates that industry content changed in 87; earlier years use 77 SICs. 'P's mark projections.

INDICES OF CHANGE

Year	Com-panies	Establishments Total	Establishments with 20 or more employees	Employment Total (000)	Employment Production Workers (000)	Employment Hours (Mil)	Compensation Payroll ($ mil)	Compensation Wages ($/hr)	Production Cost of Materials	Production Value Added by Manufacture	Production Value of Shipments	Production Capital Invest.
1982	144	142	133	152	161	156	96	72	82	81	83	73
1983		138	134	137	142	136	90	76	76	76	77	49
1984		134	135	137	147	144	95	79	84	85	85	80
1985		130	137	136	145	139	100	84	93	90	92	81
1986		130	141	133	139	139	103	84	93	88	91	76
1987*	120	119	104	98	101	101	77	84	70	69	69	58
1988		113	101	100	106	106	82	88	81	81	80	56
1989		111	108	106	116	123	95	87	94	87	88	91
1990		108	107	102	117	120	103	92	105	93	97	95
1991		110	108	116	121	119	106	97	108	90	96	169
1992	100	100	100	100	100	100	100	100	100	100	100	100
1993		102	104	106	106	106	107	102	111	104	107	90
1994		98P	104P	103	108	109	110	107	117	109	112	110
1995		95P	104P	107P	109P	110P	118P	109P	125P	116P	119P	130P
1996		93P	104P	108P	109P	110P	122P	113P	130P	121P	124P	138P
1997		90P	104P	109P	109P	109P	126P	116P	136P	127P	130P	145P
1998		87P	104P	109P	109P	109P	131P	119P	142P	132P	135P	153P

Sources: Same as General Statistics. Values reflect change from the base year, 1992. Values above 100 mean greater than 92, values below 100 mean less than 92, and a value of 100 in the 82-91 or 93-98 period means same as 92. * indicates that industry content changed in 87. Data for earlier years are in 77 SIC format.

SELECTED RATIOS

For 1994	Avg. of All Manufact.	Analyzed Industry	Index	For 1994	Avg. of All Manufact.	Analyzed Industry	Index
Employees per Establishment	49	42	87	Value Added per Production Worker	134,084	140,120	105
Payroll per Establishment	1,500,273	1,430,271	95	Cost per Establishment	5,045,178	2,697,972	53
Payroll per Employee	30,620	33,671	110	Cost per Employee	102,970	63,514	62
Production Workers per Establishment	34	25	73	Cost per Production Worker	146,988	108,303	74
Wages per Establishment	853,319	637,450	75	Shipments per Establishment	9,576,895	6,134,950	64
Wages per Production Worker	24,861	25,589	103	Shipments per Employee	195,460	144,425	74
Hours per Production Worker	2,056	2,044	99	Shipments per Production Worker	279,017	246,271	88
Wages per Hour	12.09	12.52	104	Investment per Establishment	321,011	196,909	61
Value Added per Establishment	4,602,255	3,490,571	76	Investment per Employee	6,552	4,636	71
Value Added per Employee	93,930	82,173	87	Investment per Production Worker	9,352	7,904	85

Sources: Same as General Statistics. The 'Average of All Manufacturing' column represents the average of all manufacturing industries reported for the most recent complete year available. The Index shows the relationship between the Average and the Analyzed Industry. For example, 100 means that they are equal; 500 that the Analyzed Industry is five times the average; 50 means that the Analyzed Industry is half the national average. The abbreviation 'na' is used to show that data are 'not available'.

LEADING COMPANIES   Number shown: **75**   Total sales ($ mil): **12,770**   Total employment (000): **93.4**

Company Name	Address				CEO Name	Phone	Co. Type	Sales ($ mil)	Empl. (000)
Tyco International Ltd	1 Tyco Park	Exeter	NH	03833	L Dennis Kozlowski	603-778-9700	P	3,263	24.0
Grinnell Corp	3 Tyco Park	Exeter	NH	03833	L Dennis Kozlowski	603-778-9200	S	3,100•	24.0
AEG Corp	180 Mount Airy Rd	Basking Ridge	NJ	07920	Peter Westrick	908-204-8900	S	747	4.0
Pall Corp	2200 Northern Blv	East Hills	NY	11548	Eric Krasnoff	516-484-5400	P	701	6.2
Donaldson Company Inc	PO Box 1299	Minneapolis	MN	55440	William A Hodder	612-887-3131	P	593	4.4
Nordson Corp	28601 Clemens Rd	Westlake	OH	44145	William P Madar	216-892-1580	P	507	3.3
FANUC Robotics Corp	2000 S Adams Rd	Auburn Hills	MI	48326	Eric Mittelstadt	313-377-7000	S	260	0.6
FANUC Robotics North America	2000 S Adams Rd	Auburn Hills	MI	48326	Eric Mittelstadt	810-377-7000	S	260	0.6
Parker Hannifin Corp	17325 Euclid Av	Cleveland	OH	44112	Warren E McHale	216-531-3000	D	240•	3.6
Dorr-Oliver Inc	PO Box 3819	Milford	CT	06460	Cornelius J Barton	203-876-5400	R	170	0.7
Thermo Fibertek Inc	PO Box 9046	Waltham	MA	02254	William A Rainville	617-622-1000	P	163	1.0
Pall Corp	2200 Northern Blv	East Hills	NY	11548	Eric Krasnoff	516-484-5400	D	158	1.1
United States Filter Corp	PO Box 560	Rockford	IL	61105	R J Heckmann	815-877-3041	P	148	1.5
Central Sprinkler Corp	451 N Cannon Av	Lansdale	PA	19446	George G Meyer	215-362-0700	P	116	0.7
Duff-Norton Co	PO Box 7010	Charlotte	NC	28241	James E Boyd	704-588-0510	S	100	0.7
Lincoln Industrial	1 Lincoln Way	St Louis	MO	63120	John Little	314-679-4200	S	99•	0.8
Osmonics Inc	5951 Clearwater Dr	Minnetonka	MN	55343	D Dean Spatz	612-933-2277	P	96	0.9
Gelman Sciences Inc	PO Box 1448	Ann Arbor	MI	48106	Kim A Davis	313-665-0651	P	95	0.7
ISI Robotics Inc	PO Box 220	Fraser	MI	48026	LD Blatt	810-294-9500	R	90	0.6
Pall Trinity Micro	PO Box 2030	Cortland	NY	13045	Paul Kohn	607-753-6041	S	82	0.8
Sunnen Products Co	7910 Manchester	St Louis	MO	63143	Jim Berthold	314-781-2100	R	75	0.7
Continental Air Systems	PO Box 400	Winfield	AL	35594	CE Bryant Jr	205-487-6492	D	70	0.5
Durr Industries Inc	PO Box 2129	Plymouth	MI	48170	R J Mulholland	313-459-6800	S	68•	0.6
Munters Corp	PO Box 6428	Fort Myers	FL	33911	Sven Lundin	813-936-1555	S	55	0.4
Eriez Magnetics	PO Box 10608	Erie	PA	16514	CF Giermak	814-835-6000	R	50	0.3
Hankison	1000 Philadelphia St	Canonsburg	PA	15317	Doanld R Ranalli	412-745-1555	D	50	0.4
Met-Pro Corp	PO Box 144	Harleysville	PA	19438	William L Kacin	215-723-6751	P	50	0.4
PTI Technologies Inc	PO Box 2000	Newbury Park	CA	91319	J Michael Conway	805-499-2661	S	50	0.3
Pyrotek Inc	E 9601 Montgomery	Spokane	WA	99206	DC Swanson	509-926-6212	R	50	<0.1
Reliable Automatic Sprinkler	525 N MacQueston	Mount Vernon	NY	10552	FJ Fee III	914-592-1414	R	50	0.4
Racor	PO Box 3208	Modesto	CA	95353	Peter Popoff	209-521-7860	D	49•	0.4
Chief Automotive Systems Inc	1924 E 4th St	Grand Island	NE	68801	James E Aylward	308-384-9747	S	45	0.3
Air-Maze Corp	115 Eeels Corners	Stow	OH	44224	John F Cummings	216-928-4100	R	37	0.3
Foley-Belsaw Co	3300 5th St NE	Minneapolis	MN	55418	Richard J Hentges	612-789-8831	R	37•	0.3
Perry Equipment Corp	PO Box 640	Mineral Wells	TX	76068	D Perry Jr	817-325-2575	R	37	0.4
Steel King Industries Inc	2700 Chambers St	Stevens Point	WI	54481	Frederic Anderson	715-341-3120	R	36	0.3
Burke E Porter Machinery Co	730 Plymouth NE	Grand Rapids	MI	49505	Andrew D Murch	616-459-9531	R	35	0.1
Cuno Inc	400 Research Pkwy	Meriden	CT	06450	Mark Kachur	203-237-5541	S	35•	0.7
RDI Inc	1025 W Thorndale	Itasca	IL	60143	Curtis N Maas	708-773-2500	R	35	0.1
Verteq Inc	PO Box 35033	Santa Ana	CA	92705	Keith Norby	714-708-0330	R	35	0.3
Viking Corp	210 N Industrial Pk	Hastings	MI	49058	Thomas T Groos	616-945-9501	R	35	0.3
Fibercor	14605 28th Av N	Minneapolis	MN	55447	Louis C Cosentino	612-553-3300	D	34•	0.3
Bird Machine Co	100 Neponset St	South Walpole	MA	02071	Tim Davis	508-668-0400	S	32•	0.5
Parker Compumotor	5500 Business Pk	Rohnert Park	CA	94928	Roy Glassett	707-584-7558	D	32	0.2
Aqua-Aerobic Systems Inc	PO Box 2026	Rockford	IL	61130	JD Brubaker	815-654-2501	R	30	0.1
Automatic Handling Inc	360 Lavoy Rd	Erie	MI	48133	David Pienta	313-847-0633	R	30	0.3
Kawasaki Robotics	28059 Ctr Oaks Ct	Wixom	MI	48393	Michiharu Tanimoto	810-305-7610	S	30	<0.1
Saulsbury Fire Equipment Corp	PO Box 690	Tully	NY	13159	AR Saulsbury	315-696-8909	R	28	0.2
Hosokawa Micron Powder Syst	10 Chatham Rd	Summit	NJ	07901	F Sawamura	908-273-6360	S	27	0.2
Peerless Manufacturing Co	PO Box 540667	Dallas	TX	75354	Sherrill Stone	214-357-6181	P	26	0.2
JWI Inc	2155 112th Av	Holland	MI	49424	David Spyker	616-772-9011	R	25	0.1
Komline-Sanderson Engineering	12 Holland Av	Peapack	NJ	07977	Russell M Komline	908-234-1000	R	25•	0.3
Lubriquip Inc	18901 Cranwood	Cleveland	OH	44128	Mark Baker	216-581-2000	S	25	0.3
Read Corp	25 Wareham St	Middleboro	MA	02346	J Read	508-946-1200	R	25	<0.1
Universal Dynamics Inc	PO Box X	Woodbridge	VA	22194	Donald Rainville	703-491-2191	S	25•	0.2
Robotic Vision Systems Inc	425 Rabro Dr E	Hauppauge	NY	11788	Pat V Costa	516-273-9700	P	25	0.1
Joy Environmental Technologies	10700 N Fwy	Houston	TX	77037	Marc F Wray	713-878-1000	S	24	0.2
Dollinger Corp	3951 Westerre Way	Richmond	VA	23233	Anthony M Vincent	804-965-2700	S	23	0.2
Clifford B Hannay and Son	600 E Main St	Westerlo	NY	12193	George A Hannay	518-797-3791	R	22	0.2
Western States Machine Co	PO Box 327	Hamilton	OH	45012	John Wake	513-863-4758	R	22	0.1
Hirata Corporation of America	3901 Industrial Blv	Indianapolis	IN	46254	Steve Race	317-299-8800	S	21	<0.1
Parker Hannifin Corp	PO Box 901	Richland	MI	49083	WH Long	616-629-5000	D	21	0.3
Byers Industries Inc	PO Box 13097	Portland	OR	97213	Edwin J Fackler	503-281-0069	R	20	0.1
Cochrane Environmental Systems	800 3rd Av	King of Prussia	PA	19406	AP Roy Choudhury	215-265-5050	D	20	<0.1
Micrion Corp	1 Corporation Way	Peabody	MA	01960	Nicholas Economou	508-531-6464	P	20	0.1
Netzsch Inc	119 Pickering Way	Exton	PA	19341	Hanno W Spranger	610-363-8010	R	20	0.1
Sweco Inc	7120 N Buff	Florence	KY	41022	John Kaiser	606-727-5147	D	20	0.3
Thomas Engineering Inc	575 W Central Rd	Hoffman Est	IL	60195	Brian Casey	708-358-5800	R	20	0.1
Schroeder Industries	PO Box 72	McKees Rocks	PA	15136	AO Schroeder	412-771-4810	D	19	0.2
Thiele Engineering Co	7225 Bush Lake Rd	Minneapolis	MN	55439	PNY Pan	612-835-2290	R	19•	0.2
Boart Longyear	PO Box 1959	Stone Mt	GA	30086	Bruce Wood	404-469-2720	D	18•	0.2
Elkhart Brass Manufacturing	PO Box 1127	Elkhart	IN	46514	RG Ashbaugh Jr	219-295-8330	R	18•	0.2
Kaydon Corp	1571 Lukken Indrial	La Grange	GA	30240	TA Bushar	706-884-3041	D	18	0.1
Templeton, Kenly and Company	2525 Gardner Rd	Broadview	IL	60153	Robert Spath	708-865-1500	R	18	0.1
Huntington Mechanical Labs	1040 L'Avenida	Mountain View	CA	94043	Vince D'Amato	415-964-3323	R	17	<0.1

Source: Ward's Business Directory of U.S. Private and Public Companies, Volumes 1 and 2, 1996. The company type code used is as follows: P - Public, R - Private, S - Subsidiary, D - Division, J - Joint Venture, A - Affiliate, G - Group. Sales are in millions of dollars, employees are in thousands. An asterisk (•) indicates an estimated sales volume. The symbol < stands for 'less than'. Company names and addresses are truncated, in some cases, to fit into the available space.

MATERIALS CONSUMED

Material	Quantity	Delivered Cost ($ million)
Materials, ingredients, containers, and supplies	(X)	1,951.6
Electrical transmission, distribution, and control equipment	(X)	40.9
Fluid power pumps, motors, and hydrostatic transmissions (hydraulic and pneumatic)	(X)	21.7
Other pumps and pump parts, except fluid power (complete assemblies)	(X)	19.0
Fluid power valves (hydraulic and pneumatic)	(X)	14.1
Fluid power cylinders and rotary actuators (hydraulic and pneumatic)	(X)	12.2
Fluid power hose or tube fittings and assemblies (hydraulic and pneumatic)	(X)	10.2
Fluid power filters (hydraulic and pneumatic)	(X)	49.9
Other fluid power products (hydraulic and pneumatic)	(X)	5.1
Metal bolts, nuts, screws, washers, rivets, and other screw machine products	(X)	28.5
Metal tanks, heat exchangers, steam condensers, and other boiler products	(X)	27.5
Metal pipe, valves, and pipe fittings (except forgings)	(X)	31.1
Other fabricated metal products (except fluid power products and forgings)	(X)	125.8
Forgings	(X)	3.5
Iron and steel castings (rough and semifinished)	(X)	77.2
Aluminum and aluminum-base alloy castings (rough and semifinished)	(X)	17.7
Other nonferrous castings (rough and semifinished)	(X)	17.2
Steel bars, bar shapes, and plates	(X)	65.1
Steel sheet and strip, including tin plate	(X)	52.0
Steel structural shapes and sheet piling	(X)	13.6
All other steel shapes and forms	(X)	30.0
Aluminum and aluminum-base alloy sheet, plate, foil, and welded tubing	(X)	20.8
All other aluminum and aluminum-base alloy shapes and forms	(X)	6.4
Other nonferrous shapes and forms	(X)	21.5
Other fractional horsepower electric motors (under 1 hp)	(X)	8.0
Integral horsepower electric motors and generators (1 hp or more)	(X)	25.1
Ball and roller bearings (mounted or unmounted)	(X)	13.0
Mechanical speed changers, gears, and industrial high-speed drives	(X)	16.0
Air and gas compressors except refrigeration compressors	(X)	5.4
Filter paper	(X)	97.3
All other materials and components, parts, containers, and supplies	(X)	691.3
Materials, ingredients, containers, and supplies, nsk	(X)	384.4

Source: 1992 *Economic Census*. Explanation of symbols used: (D): Withheld to avoid disclosure of competitive data; na: Not available; (S): Withheld because statistical norms were not met; (X): Not applicable; (Z): Less than half the unit shown; nec: Not elsewhere classified; nsk: Not specified by kind; - : zero; * : 10-19 percent estimated; ** : 20-29 percent estimated.

PRODUCT SHARE DETAILS

Product or Product Class	% Share	Product or Product Class	% Share
General industrial machinery, nec	100.00	General industrial gas generating equipment	1.60
Filters and strainers, except fluid power	35.54	General industrial gas separating equipment	5.01
Containment filters and strainers (housing devices) for water, except parts and accessories	31.29	General industrial steam and vapor separators	0.36
Containment filters and strainers (housing devices) for beverages other than water, except parts and accessories	0.78	General industrial compressed air and gas refrigerated dryers	2.02
Containment filters and strainers (housing devices) for other fluids, except for fluid power, and except parts and accessories	19.85	General industrial compressed air and gas disiccant dryers	1.44
Parts and accessories for containment filters and strainers (housing devices), sold separately	5.09	General industrial compressed air and gas deliquescent dryers	(D)
Reusable (cleanable) media filters and strainers, except for fluid power	13.56	Other general industrial compressed air and gas dryers	(D)
Nonreusable media filters and strainers, including disposable (throw away) litter cartridges, except for fluid power	27.86	Mixers for general industrial processes, solids or liquids	8.06
		Lubricating systems, industrial, centralized and automatic	2.93
Filters and strainers, except fluid power, nsk	1.58	General industrial sifting and screening machines	1.09
Filters for hydraulic fluid power systems, non-aerospace	4.39	General industrial metal baling presses	1.76
Filters for pneumatic fluid power systems, non-aerospace	1.41	General industrial centrifugals and separators (except cream, grain, and berry)	7.10
Filters for hydraulic and pneumatic fluid power systems, aerospace	1.39	General industrial automatic fire sprinklers	4.96
		General industrial pneumatic jacks	(D)
General industrial machinery, nec	52.47	General industrial hydraulic jacks	1.75
General industrial robots, attachments and parts	14.89	General industrial screwjacks (except automotive)	1.17
		Other general industrial machinery	35.50
		Parts for general industrial equipment, nec	8.92
		General industrial machinery, nec, nsk	0.82
		General industrial machinery, nec, nsk	4.80

Source: 1992 *Economic Census*. The values shown are percent of total shipments in an industry. Values of indented subcategories are summed in the main heading. The symbol (D) appears when data are withheld to prevent disclosure of competitive information. The abbreviation nsk stands for 'not specified by kind' and nec for 'not elsewhere classified'.

INPUTS AND OUTPUTS FOR GENERAL INDUSTRIAL MACHINERY, NEC

Economic Sector or Industry Providing Inputs	%	Sector	Economic Sector or Industry Buying Outputs	%	Sector
Imports	41.1	Foreign	Gross private fixed investment	54.4	Cap Inv
Wholesale trade	6.3	Trade	Exports	29.3	Foreign
Blast furnaces & steel mills	5.5	Manufg.	General industrial machinery, nec	2.4	Manufg.
General industrial machinery, nec	3.7	Manufg.	Federal Government purchases, national defense	2.2	Fed Govt
Nonferrous wire drawing & insulating	2.4	Manufg.	Radio & TV communication equipment	0.8	Manufg.
Machinery, except electrical, nec	2.3	Manufg.	Miscellaneous repair shops	0.7	Services
Miscellaneous plastics products	2.2	Manufg.	Industrial buildings	0.6	Constr.
Iron & steel foundries	1.6	Manufg.	Pumps & compressors	0.6	Manufg.
Switchgear & switchboard apparatus	1.5	Manufg.	Office buildings	0.5	Constr.
Aluminum rolling & drawing	1.4	Manufg.	Crude petroleum & natural gas	0.4	Mining
Blowers & fans	1.4	Manufg.	Feed grains	0.3	Agric.
Motors & generators	1.4	Manufg.	Meat animals	0.3	Agric.
Pipe, valves, & pipe fittings	1.4	Manufg.	Plastics materials & resins	0.3	Manufg.
Pumps & compressors	1.4	Manufg.	Printing trades machinery	0.3	Manufg.
Industrial controls	1.2	Manufg.	Coal	0.2	Mining
Nonwoven fabrics	1.2	Manufg.	Maintenance of nonfarm buildings nec	0.2	Constr.
Electric services (utilities)	1.2	Util.	Nonfarm residential structure maintenance	0.2	Constr.
Fabricated plate work (boiler shops)	1.1	Manufg.	Fabricated plate work (boiler shops)	0.2	Manufg.
Transformers	1.1	Manufg.	Motor vehicles & car bodies	0.2	Manufg.
Power transmission equipment	0.9	Manufg.	Sporting & athletic goods, nec	0.2	Manufg.
Metal stampings, nec	0.8	Manufg.	S/L Govt. purch., fire	0.2	S/L Govt
Special dies & tools & machine tool accessories	0.8	Manufg.	Oil bearing crops	0.1	Agric.
Communications, except radio & TV	0.8	Util.	Construction of stores & restaurants	0.1	Constr.
Banking	0.7	Fin/R.E.	Hotels & motels	0.1	Constr.
Advertising	0.7	Services	Ship building & repairing	0.1	Manufg.
Maintenance of nonfarm buildings nec	0.6	Constr.	Special industry machinery, nec	0.1	Manufg.
Aluminum castings	0.6	Manufg.	Electric services (utilities)	0.1	Util.
Brass, bronze, & copper castings	0.6	Manufg.			
Copper rolling & drawing	0.6	Manufg.			
Petroleum refining	0.6	Manufg.			
Screw machine and related products	0.6	Manufg.			
Air transportation	0.6	Util.			
Personnel supply services	0.6	Services			
Electronic components nec	0.5	Manufg.			
Felt goods, nec	0.5	Manufg.			
Motor freight transportation & warehousing	0.5	Util.			
Eating & drinking places	0.5	Trade			
Real estate	0.5	Fin/R.E.			
Engineering, architectural, & surveying services	0.5	Services			
Abrasive products	0.4	Manufg.			
Ball & roller bearings	0.4	Manufg.			
Paperboard containers & boxes	0.4	Manufg.			
Plastics materials & resins	0.4	Manufg.			
Equipment rental & leasing services	0.4	Services			
Fabricated rubber products, nec	0.3	Manufg.			
Gaskets, packing & sealing devices	0.3	Manufg.			
Nonferrous rolling & drawing, nec	0.3	Manufg.			
Gas production & distribution (utilities)	0.3	Util.			
Business services nec	0.3	Services			
Computer & data processing services	0.3	Services			
U.S. Postal Service	0.3	Gov't			
Machine tools, metal cutting types	0.2	Manufg.			
Nonferrous castings, nec	0.2	Manufg.			
Paper mills, except building paper	0.2	Manufg.			
Railroads & related services	0.2	Util.			
Hotels & lodging places	0.2	Services			
Legal services	0.2	Services			
Management & consulting services & labs	0.2	Services			
Noncomparable imports	0.2	Foreign			
Converted paper products, nec	0.1	Manufg.			
Lubricating oils & greases	0.1	Manufg.			
Metal heat treating	0.1	Manufg.			
Plating & polishing	0.1	Manufg.			
Sawmills & planning mills, general	0.1	Manufg.			
Accounting, auditing & bookkeeping	0.1	Services			
Business/professional associations	0.1	Services			
Scrap	0.1	Scrap			

Source: Benchmark Input-Output Accounts for the U.S. Economy, 1982, U.S. Department of Commerce, Washington, D.C., July 1991. Data, as reported in the source, are organized by the 1977 SIC structure in use in 1982 but have been matched, as closely as is possible, to the 1987 SIC structure used in this book.

OCCUPATIONS EMPLOYED BY SIC 356 - GENERAL INDUSTRIAL MACHINERY

Occupation	% of Total 1994	Change to 2005	Occupation	% of Total 1994	Change to 2005
Assemblers, fabricators, & hand workers nec	7.9	-2.9	Machine tool cutting & forming etc. nec	1.5	-2.9
Machinists	4.9	6.9	Machine assemblers	1.5	-2.8
Sales & related workers nec	3.8	-2.9	Traffic, shipping, & receiving clerks	1.5	-6.5
Blue collar worker supervisors	3.8	-9.0	Combination machine tool operators	1.5	6.9
Inspectors, testers, & graders, precision	3.1	-2.9	Welding machine setters, operators	1.2	-12.6
Welders & cutters	2.8	31.2	Engineering, mathematical, & science managers	1.2	10.4
Mechanical engineers	2.8	6.9	Machine operators nec	1.2	-14.4
General managers & top executives	2.5	-7.8	Industrial production managers	1.2	-2.8
NC machine tool operators, metal & plastic	2.3	45.7	Bookkeeping, accounting, & auditing clerks	1.2	-27.2
Machine builders	2.3	31.1	Stock clerks	1.1	-21.0
Lathe & turning machine tool operators	2.2	-36.8	Engineering technicians & technologists nec	1.1	-2.9
Secretaries, ex legal & medical	2.1	-11.6	Machine forming operators, metal & plastic	1.1	-2.9
Grinding machine operators, metal & plastic	2.1	-12.6	General office clerks	1.1	-17.2
Drafters	2.0	-24.3	Production, planning, & expediting clerks	1.1	-2.8
Machine tool cutting operators, metal & plastic	1.8	-19.1	Helpers, laborers, & material movers nec	1.1	-2.8
Sheet metal workers & duct installers	1.6	-2.8	Drilling & boring machine tool workers	1.1	-51.4
Industrial machinery mechanics	1.6	6.9	Purchasing agents, ex trade & farm products	1.0	6.9

Source: *Industry-Occupation Matrix*, Bureau of Labor Statistics. These data relate to one or more 3-digit SIC industry groups rather than to a single 4-digit SIC. The change reported for each occupation to the year 2005 is a percent of growth or decline as estimated by the Bureau of Labor Statistics. The abbreviation nec stands for 'not elsewhere classified'.

LOCATION BY STATE AND REGIONAL CONCENTRATION

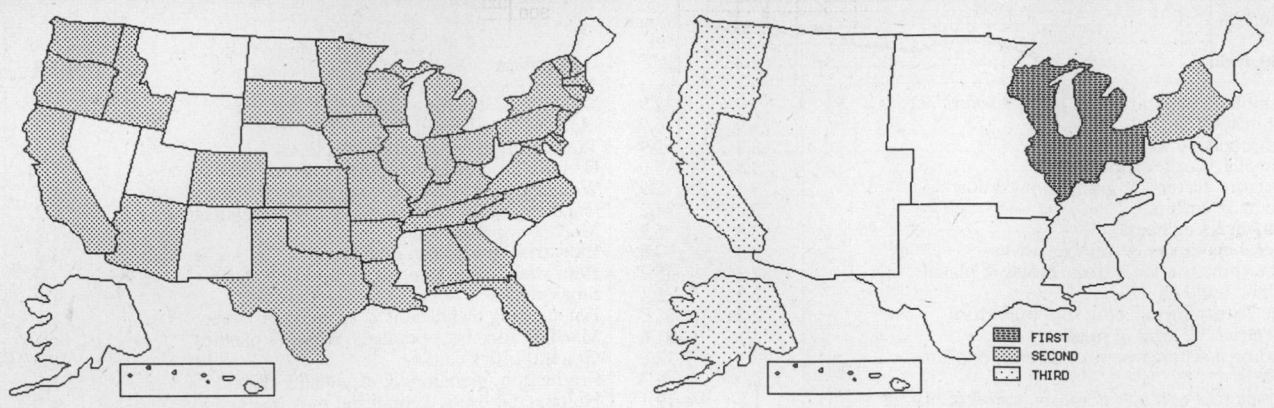

FIRST
SECOND
THIRD

INDUSTRY DATA BY STATE

| State | Establish-ments | Shipments | | | Employment | | | | Cost as % of Shipments | Investment per Employee ($) |
		Total ($ mil)	% of U.S.	Per Establ.	Total Number	% of U.S.	Per Establ.	Wages ($/hour)		
Michigan	90	547.5	9.9	6.1	3,600	8.7	40	12.14	42.5	3,167
California	111	445.4	8.1	4.0	3,700	8.9	33	11.53	42.3	4,270
Pennsylvania	52	437.7	7.9	8.4	2,800	6.7	54	14.38	46.4	8,214
New York	48	431.9	7.8	9.0	4,400	10.6	92	13.94	43.9	8,250
Illinois	75	300.0	5.4	4.0	2,500	6.0	33	10.18	47.2	3,640
Massachusetts	33	255.1	4.6	7.7	1,300	3.1	39	15.07	44.0	4,154
Florida	38	244.2	4.4	6.4	1,900	4.6	50	12.45	35.4	4,579
Ohio	66	217.8	3.9	3.3	1,600	3.9	24	10.87	46.6	1,688
New Jersey	47	207.3	3.8	4.4	1,700	4.1	36	12.68	44.2	2,118
Texas	63	195.7	3.5	3.1	1,800	4.3	29	10.20	48.0	1,833
Wisconsin	36	185.7	3.4	5.2	1,500	3.6	42	11.82	41.7	2,667
Indiana	32	184.9	3.3	5.8	1,600	3.9	50	9.82	44.1	2,313
North Carolina	27	173.5	3.1	6.4	1,400	3.4	52	10.11	39.9	3,429
New Hampshire	6	173.0	3.1	28.8	1,200	2.9	200	11.50	25.7	-
Minnesota	27	171.9	3.1	6.4	1,200	2.9	44	13.58	35.2	6,167
Connecticut	33	138.2	2.5	4.2	1,200	2.9	36	11.29	35.5	6,167
Kentucky	11	84.0	1.5	7.6	900	2.2	82	7.14	52.3	2,000
Georgia	15	70.5	1.3	4.7	600	1.4	40	10.14	52.2	1,000
Washington	16	68.8	1.2	4.3	500	1.2	31	12.75	33.9	-
Virginia	20	60.5	1.1	3.0	600	1.4	30	10.43	44.5	5,333
Oklahoma	18	41.3	0.7	2.3	400	1.0	22	11.50	45.0	2,500
Oregon	16	38.5	0.7	2.4	300	0.7	19	11.00	26.5	4,333
Missouri	11	38.4	0.7	3.5	300	0.7	27	9.75	40.9	-
Iowa	5	36.2	0.7	7.2	300	0.7	60	9.25	56.4	1,333
Alabama	9	35.1	0.6	3.9	300	0.7	33	13.00	38.7	-
Kansas	9	28.6	0.5	3.2	400	1.0	44	10.60	36.0	1,250
Rhode Island	8	19.0	0.3	2.4	200	0.5	25	10.00	36.3	-
Arizona	6	18.5	0.3	3.1	200	0.5	33	7.67	36.8	-
Louisiana	9	18.0	0.3	2.0	200	0.5	22	10.50	47.2	500
Tennessee	19	(D)	-	-	750 *	1.8	39	-	-	5,867
Maryland	12	(D)	-	-	750 *	1.8	63	-	-	-
Colorado	7	(D)	-	-	375 *	0.9	54	-	-	-
Arkansas	6	(D)	-	-	750 *	1.8	125	-	-	-
Vermont	6	(D)	-	-	175 *	0.4	29	-	-	-
Idaho	3	(D)	-	-	175 *	0.4	58	-	-	-
Delaware	2	(D)	-	-	175 *	0.4	88	-	-	-
South Dakota	2	(D)	-	-	175 *	0.4	88	-	-	-

Source: 1992 *Economic Census*. The states are in descending order of shipments or establishments (if shipment data are missing for the majority). The symbol (D) appears when data are withheld to prevent disclosure of competitive information. States marked with (D) are sorted by number of establishments. A dash (-) indicates that the data element cannot be calculated; * indicates the midpoint of a range.

3571 - ELECTRONIC COMPUTERS

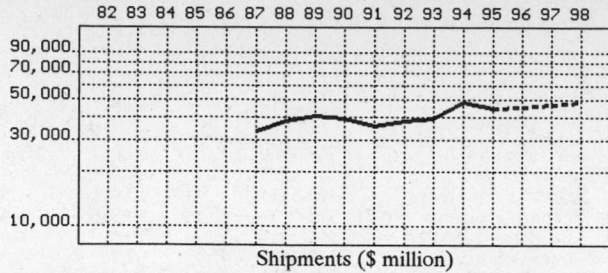

82 83 84 85 86 87 88 89 90 91 92 93 94 95 96 97 98

Shipments ($ million)

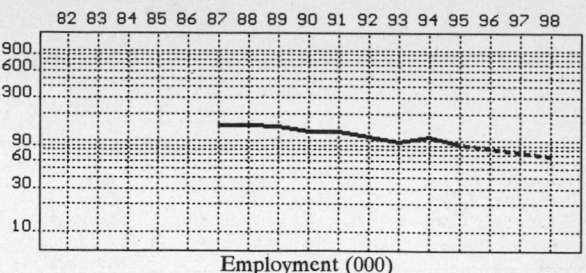

82 83 84 85 86 87 88 89 90 91 92 93 94 95 96 97 98

Employment (000)

GENERAL STATISTICS

| Year | Companies | Establishments | | Employment | | | Compensation | | Production ($ million) | | | |
		Total	with 20 or more employees	Total (000)	Production Workers (000)	Hours (Mil)	Payroll ($ mil)	Wages ($/hr)	Cost of Materials	Value Added by Manufacture	Value of Shipments	Capital Invest.
1982												
1983												
1984												
1985												
1986												
1987	938	974	394	151.9	54.7	105.8	4,953.0	10.64	15,222.9	18,322.2	33,626.5	1,223.3
1988		983	462	150.6	54.7	109.2	5,513.9	11.09	18,090.6	19,702.7	37,683.3	1,304.0
1989		829	407	143.7	49.7	100.3	5,454.8	12.05	20,307.8	21,052.3	40,550.7	1,304.5
1990		783	374	126.0	44.0	90.0	5,264.5	12.45	19,355.7	19,666.3	39,293.6	1,222.3
1991		766	377	126.0	38.1	78.7	5,343.7	12.80	19,078.4	16,877.5	35,572.9	1,153.5
1992	803	834	289	110.8	31.2	65.5	4,855.9	12.51	21,388.9	16,137.4	38,205.9	1,242.4
1993		747	278	99.5	28.8	60.2	4,514.7	13.49	23,498.6	16,218.4	39,176.8	1,042.9
1994		696P	265P	109.4	30.4	62.7	5,073.9	13.54	27,046.1	21,066.4	48,546.9	932.7
1995		659P	239P	92.3P	22.1P	48.6P	4,807.5P	14.14P	24,792.1P	19,310.8P	44,501.1P	985.6P
1996		622P	214P	84.6P	17.8P	40.7P	4,737.6P	14.55P	25,463.0P	19,833.3P	45,705.3P	942.8P
1997		585P	188P	76.8P	13.5P	32.8P	4,667.8P	14.95P	26,133.9P	20,355.9P	46,909.6P	900.0P
1998		547P	162P	69.0P	9.2P	24.9P	4,597.9P	15.36P	26,804.8P	20,878.5P	48,113.8P	857.2P

Sources: 1982, 1987, 1992 *Economic Census*; *Annual Survey of Manufactures*, 83-86, 88-91, 93-94. Establishment counts for non-Census years are from *County Business Patterns*; establishment values for 83-84 are extrapolations. 'P's show projections by the editors. Industries reclassified in 87 will not have data for prior years.

INDICES OF CHANGE

| Year | Companies | Establishments | | Employment | | | Compensation | | Production ($ million) | | | |
		Total	with 20 or more employees	Total (000)	Production Workers (000)	Hours (Mil)	Payroll ($ mil)	Wages ($/hr)	Cost of Materials	Value Added by Manufacture	Value of Shipments	Capital Invest.
1982												
1983												
1984												
1985												
1986												
1987	117	117	136	137	175	162	102	85	71	114	88	98
1988		118	160	136	175	167	114	89	85	122	99	105
1989		99	141	130	159	153	112	96	95	130	106	105
1990		94	129	114	141	137	108	100	90	122	103	98
1991		92	130	114	122	120	110	102	89	105	93	93
1992	100	100	100	100	100	100	100	100	100	100	100	100
1993		90	96	90	92	92	93	108	110	101	103	84
1994		83P	92P	99	97	96	104	108	126	131	127	75
1995		79P	83P	83P	71P	74P	99P	113P	116P	120P	116P	79P
1996		75P	74P	76P	57P	62P	98P	116P	119P	123P	120P	76P
1997		70P	65P	69P	43P	50P	96P	120P	122P	126P	123P	72P
1998		66P	56P	62P	30P	38P	95P	123P	125P	129P	126P	69P

Sources: Same as General Statistics. Values reflect change from the base year, 1992. Values above 100 mean greater than 92, values below 100 mean less than 92, and a value of 100 in the 82-91 or 93-98 period means same as 92. 'P's mark projections by the editors.

SELECTED RATIOS

For 1994	Avg. of All Manufact.	Analyzed Industry	Index	For 1994	Avg. of All Manufact.	Analyzed Industry	Index
Employees per Establishment	49	157	321	Value Added per Production Worker	134,084	692,974	517
Payroll per Establishment	1,500,273	7,287,095	486	Cost per Establishment	5,045,178	38,843,394	770
Payroll per Employee	30,620	46,379	151	Cost per Employee	102,970	247,222	240
Production Workers per Establishment	34	44	127	Cost per Production Worker	146,988	889,674	605
Wages per Establishment	853,319	1,219,267	143	Shipments per Establishment	9,576,895	69,722,671	728
Wages per Production Worker	24,861	27,926	112	Shipments per Employee	195,460	443,756	227
Hours per Production Worker	2,056	2,063	100	Shipments per Production Worker	279,017	1,596,938	572
Wages per Hour	12.09	13.54	112	Investment per Establishment	321,011	1,339,536	417
Value Added per Establishment	4,602,255	30,255,396	657	Investment per Employee	6,552	8,526	130
Value Added per Employee	93,930	192,563	205	Investment per Production Worker	9,352	30,681	328

Sources: Same as General Statistics. The 'Average of All Manufacturing' column represents the average of all manufacturing industries reported for the most recent complete year available. The Index shows the relationship between the Average and the Analyzed Industry. For example, 100 means that they are equal; 500 that the Analyzed Industry is five times the average; 50 means that the Analyzed Industry is half the national average. The abbreviation 'na' is used to show that data are 'not available'.

LEADING COMPANIES   Number shown: **75**   Total sales ($ mil): **166,050**   Total employment (000): **639.4**

Company Name	Address				CEO Name	Phone	Co. Type	Sales ($ mil)	Empl. (000)
Intern Business Machines Corp	Old Orchard Rd	Armonk	NY	10504	L V Gerstner Jr	914-765-1900	P	64,052	219.8
Hewlett-Packard Co	3000 Hanover St	Palo Alto	CA	94304	Lewis E Platt	415-857-1501	P	24,991	98.4
Digital Equipment Corp	146 Main St	Maynard	MA	01754	Robert B Palmer	508-493-5111	P	13,451	77.8
Compaq Computer Corp	PO Box 692000	Houston	TX	77269	Eckhard Pfeiffer	713-370-0670	P	10,866	14.4
Apple Computer Inc	1 Infinite Loop	Cupertino	CA	95014	Michael H Spindler	408-996-1010	P	9,189	14.6
Unisys Corp	PO Box 500	Blue Bell	PA	19424	James A Unruh	215-542-4011	P	7,400	46.3
Sun Microsystems Inc	2550 Garcia Av	Mountain View	CA	94043	Scott G McNealy	415-960-1300	P	4,690	13.3
Litton Industries Inc	21240 Burbank Blv	Woodland Hills	CA	91367	John M Leonis	310-859-2026	P	3,446	32.3
AST Research Inc	PO Box 57005	Irvine	CA	92619	Safi U Qureshey	714-727-4141	P	2,367	6.5
Federal Systems Co	6600 Rockledge Dr	Bethesda	MD	20817	Gerald Ebker	301-493-1500	D	2,300	10.0
Advanced Micro Devices Inc	PO Box 3453	Sunnyvale	CA	94088	WJ Sanders III	408-732-2400	P	2,135	11.8
Tandem Computers Inc	19333 Vallco Pkwy	Cupertino	CA	95014	James G Treybig	408-725-6000	P	2,031	10.0
Amdahl Corp	1250 E Arques Av	Sunnyvale	CA	94088	E Joseph Zemke	408-746-6000	P	1,639	5.6
DR Holdings Incorporated	100 Crosby Dr	Bedford	MA	01730	L E Landrigan	617-275-1800	R	1,590	7.0
Silicon Graphics Inc	PO Box 7311	Mountain View	CA	94039	E R McCracken	415-960-1980	P	1,482	4.4
Data General Corp	4400 Computer Dr	Westboro	MA	01581	Ronald L Skates	508-898-5000	P	1,121	5.8
Intergraph Corp	289 Dunlap Blv	Huntsville	AL	35824	James W Meadlock	205-730-2000	P	1,050	9.5
Memorex Telex Corp	545 J Carpenter Fwy	Irving	TX	75062	Marcelo Gumucio	214-444-3500	S	1,000	5.0
Zenith Datasystems Inc	2150 E Lake Cook	Buffalo Grove	IL	60089	Jacques Noels	708-808-5000	S	1,000	1.5
Packard Bell Electronics Inc	31717 La Tienda Dr	Westlake Vil	CA	90362	Beny Alagem	818-865-1555	R	930	1.2
Cray Research Inc	655A Lone Oak Dr	Eagan	MN	55121	John F Carlson	612-683-7100	P	895	4.9
Acer America Corp	2641 Orchard Pkwy	San Jose	CA	95134	Ronald Chwang	408-432-6200	S	858	0.9
Control Data Systems Inc	4201 Lexington N	Arden Hills	MN	55126	James E Ousley	612-482-2401	P	524	2.9
Sequent Computer Systems Inc	15450 SW Koll	Beaverton	OR	97006	Karl C Powell Jr	503-626-5700	P	451	1.8
Litton Guidance & Control Syst	19601 Nordhoff	Northridge	CA	91324	Darwin D Beckel	818-886-2211	D	400	2.8
Intel Products Group	5200 Young	Hillsboro	OR	97124	Frank Gill	503-681-8080	D	350	2.1
Standard Microsystems Corp	80 Arkay Dr	Hauppauge	NY	11788	Victor F Trizzino	516-273-3100	P	323	0.7
Sun Microsystems Federal Inc	2550 Garcia Av	Mountain View	CA	94043	John Marselle	415-960-1300	S	300	0.2
Supercom Inc	410 S Abbott Av	Milpitas	CA	95035	James Fang	408-456-8888	R	300	0.3
Telxon Corp	PO Box 5582	Akron	OH	44334	Robert F Meyerson	216-867-3700	P	296	1.6
Pyramid Technology Corp	3860 N 1st St	San Jose	CA	95134	Richard H Lussier	408-428-9000	P	219	0.8
Bolt Beranek and Newman Inc	150 CambridgePark	Cambridge	MA	02139	George H Conrades	617-873-2000	P	196	1.7
Norand Corp	550 2nd St SE	Cedar Rapids	IA	52401	N Robert Hammer	319-369-3100	P	193	0.8
Northgate Computer Systems	PO Box 59080	Minneapolis	MN	55459	Khaled Ibrahim	612-943-8181	P	190	0.4
Advanced Logic Research Inc	9401 Jeronimo Rd	Irvine	CA	92718	Gene Lu	714-581-6770	P	183	0.5
Concurrent Computer Corp	2 Crescent Pl	Oceanport	NJ	07757	John T Stihl	908-870-4500	P	179	1.3
Datapoint Corp	8400 Datapoint Dr	San Antonio	TX	78229	Asher B Edelman	210-593-7000	P	173	1.4
PAXAR Corp	275 N Middletown	Pearl River	NY	10965	Arthur Hershaft	914-735-9200	P	167	1.9
CONVEX Computer Corp	PO Box 83351	Richardson	TX	75083	Robert J Paluck	214-497-4000	P	144	0.9
Evans & Sutherland	600 Komas Dr	Salt Lake City	UT	84108	Rodney S Rougleot	801-582-5847	P	142	1.2
Titan Corp	3033 Science Pk Rd	San Diego	CA	92121	Gene W Ray	619-552-9500	P	136	1.0
Encore Computer Corp	PO Box 9148	Ft Lauderdale	FL	33310	Kenneth G Fisher	305-587-2900	P	131	1.3
Lockheed Sanders Inc	PO Box 0868	Nashua	NH	03061	Stevan I Feldman	603-885-5431	D	125	0.3
Twinhead Corp	1537 Ctr Point Dr	Milpitas	CA	95035	John Lin	408-945-0808	S	110*	<0.1
American Research Corp	1101 Monterey	Monterey Park	CA	91754	Alex Ho	213-265-0835	R	100	0.3
Fora Inc	48434 Milmont Dr	Fremont	CA	94538	Victor Wu	510-438-0233	R	100*	0.2
Intern Data Products Corp	20 Firstfield Rd	Gaithersburg	MD	20878	George Fuster	301-590-8100	R	100	0.2
Toshiba America Info Systems	9740 Irvine Blv	Irvine	CA	92718	Michael Winkler	714-583-3000	D	100	0.4
Unisys Corp	41100 Plymouth Rd	Plymouth	MI	48170	James Mavel	313-451-4000	D	100	1.0
Comverse Technology Inc	170 Crossways Pk	Woodbury	NY	11797	Kobi Alexander	516-677-7200	P	99	0.7
Epson America Inc	20770 Madrona Av	Torrance	CA	90503	Norio Niwa	310-782-0770	R	94*	0.6
Thinking Machines Corp	245 1st St	Cambridge	MA	02142	Richard Fishman	617-876-1111	R	92	0.2
NetFRAME Systems Inc	1545 Barber Ln	Milpitas	CA	95035	Robert L Puette	408-944-0600	P	89	0.3
Osicom Technologies Inc	198 Green Pond	Rockaway	NJ	07866	Parvinder S Chadha	201-586-2550	P	84	0.2
Franklin Electronic Publishers	122 Burrs Rd	Mount Holly	NJ	08060	Morton E David	609-261-4800	P	83	0.2
Government Micro Resources	7203 Gateway Ct	Manassas	VA	22110	Humberto A Pujals	703-330-1199	R	80	0.2
Micros Systems Inc	12000 Baltimore	Beltsville	MD	20705	AL Giannopoulos	301-210-6000	P	79	0.4
PCC Group Inc	640 Puente St	Brea	CA	92621	Jack Wen	714-256-5000	P	75	<0.1
Micron Electronics Inc	900 E Karcher Rd	Nampa	ID	83687	Chase Mart	208-465-3434	S	75*	0.5
Computer Sales Professional Ltd	400B Pierce St	Somerset	NJ	08873	Ted Hsu	908-560-0113	R	72*	0.4
WIN Laboratories Ltd	11090 Industrial Rd	Manassas	VA	22110	Winfred Wu	703-690-4227	R	72	0.1
APAQ Technology Inc	45250 Northport Dr	Fremont	CA	94538	Ming Hsu	510-226-7800	R	70	0.1
Caliber Computer Corp	1500 McCandless	Milpitas	CA	95035	Eddie Wang	408-942-1220	R	70	<0.1
Everex Systems Inc	5020 Brandin Ct	Fremont	CA	94538	Cher Wang	510-498-1111	R	70	0.2
Megatron Computer Systems	4992 E Hunter Av	Anaheim	CA	92807	Jason Pang	714-777-6166	R	70	<0.1
Solbourne Computer Inc	1900 Pike Rd	Longmont	CO	80501	Carl Hermann	303-772-3400	R	66	0.2
Comptek Research Inc	2732 Transit Rd	Buffalo	NY	14224	John R Cummings	716-677-4070	P	63	0.7
Victron Inc	2530 Zanker Rd	San Jose	CA	95131	Todd Yun	408-943-9141	R	63	0.4
Insight Direct Inc	1912 W 4th St	Tempe	AZ	85281	Eric J Crown	602-902-1000	S	61	0.4
Austin Computer Systems Inc	10300 Metric Blv	Austin	TX	78758	David Scull	512-339-3500	R	60	0.2
Dynamic Decisions Inc	53 Murray St	New York	NY	10007	Peter Phame	212-227-9888	R	50	0.1
Eltech Research Inc	48890 Milmont Dr	Fremont	CA	94538	Theresa Lam	510-438-0990	R	50	<0.1
Industrial Computer Source	9950 Barnes	San Diego	CA	92121	Steve Peltier	619-271-9340	S	50	0.1
Magnetic Data Inc	6754 Shady Oak Rd	Eden Prairie	MN	55344	Ron Nelsen	612-941-0453	S	50	4.1
Swan Technologies	PO Box 1006	State College	PA	16801	Jack Conrad	814-238-1820	R	50	0.1

Source: *Ward's Business Directory of U.S. Private and Public Companies*, Volumes 1 and 2, 1996. The company type code used is as follows: P - Public, R - Private, S - Subsidiary, D - Division, J - Joint Venture, A - Affiliate, G - Group. Sales are in millions of dollars, employees are in thousands. An asterisk (*) indicates an estimated sales volume. The symbol < stands for 'less than'. Company names and addresses are truncated, in some cases, to fit into the available space.

MATERIALS CONSUMED

Material	Quantity	Delivered Cost ($ million)
Materials, ingredients, containers, and supplies	(X)	19,849.3
Cathode ray tubes (CRT'S)	(X)	70.2
Printed circuit boards (without inserted components) for electronic circuitry	(X)	327.9
Printed memory boards for electronic circuitry	(X)	2,013.8
Printed peripheral controllers (graphic boards, drive controllers, etc.) for electronic circuitry	(X)	456.9
Printed computer processors (system boards, array processors, etc.) for electronic circuitry	(X)	1,411.3
Printed communication boards (LAN, D/A, A/D converters, etc.) with fax capability	(X)	(D)
Other printed communication boards (LAN, D/A, A/D converters, etc.)	(X)	(D)
Other printed circuit boards (loaded boards, subassemblies, and modules) for electronic circuitry	(X)	178.5
Semiconductors, including transistors, diodes, rectifiers, and integrated circuits for electronic circuitry	(X)	789.1
Capacitors for electronic circuitry	(X)	42.8
Resistors for electronic circuitry	(X)	35.6
Connectors for electronic circuitry	(X)	59.4
Battery packs for electronic circuitry	(X)	462.8
Other power supply units for electronic circuitry	(X)	168.2
Other components and accessories for electric circuitry	(X)	388.2
Electrical transmission, distribution, and control equipment	(X)	28.8
Steel, aluminum, and other metal electronic enclosurse	(X)	204.4
Plastics electronic enclosures	(X)	114.0
Sheet metal products (including stampings), except enclosures	(X)	88.5
All other fabricated metal products (except forgings)	(X)	32.0
Forgings	(X)	(D)
Castings (rough and semifinished)	(X)	3.4
Metal shapes and forms, except castings, forgings, and fabricated metal products	(X)	1.3
Insulated copper wire and cable (including magnet wire)	(X)	149.4
Fabricated plastics products, except enclosures	(X)	9.3
Purchased software	(X)	194.4
Appliance outlets, switches, lampholders, and other current-carrying wiring devices	(X)	30.4
Electric motors and generators (all types)	(X)	5.0
Paper and paperboard products including paperboard boxes, containers, and corrugated paperboard	(X)	60.2
Purchased computers	(X)	4,671.2
Purchased peripheral storage devices	(X)	1,387.0
Purchased computer terminals	(X)	774.9
Purchased peripheral input devices, including keyboards, mouse devices, trackballs, etc.	(X)	430.0
Purchased peripheral printers	(X)	76.3
Other purchased electronic computing and peripheral equipment	(X)	121.3
All other materials and components, parts, containers, and supplies	(X)	1,059.5
Materials, ingredients, containers, and supplies, nsk	(X)	3,904.2

Source: 1992 *Economic Census*. Explanation of symbols used: (D): Withheld to avoid disclosure of competitive data; na: Not available; (S): Withheld because statistical norms were not met; (X): Not applicable; (Z): Less than half the unit shown; nec: Not elsewhere classified; nsk: Not specified by kind; - : zero; * : 10-19 percent estimated; ** : 20-29 percent estimated.

PRODUCT SHARE DETAILS

Product or Product Class	% Share	Product or Product Class	% Share
Electronic computers	100.00	peripherals, and parts	55.29
Large-scale general purpose digital computer processing equipment (64 megabytes or more in MINIMUM main memory configuration), excluding word processors, peripherals, and parts	17.81	General purpose portable digital computers (machines with attached display), excluding word processors, peripherals, and parts	5.72
Medium-scale and small-scale general purpose digital computer processing equipment (less than 64 megabytes in MINIMUM main memory configuration), excluding word processors, peripherals, and parts	10.40	Other general purpose digital processing units, excluding word processors, peripherals, and parts	1.54
General prupose personal digital computers and workstations, excluding portables, word processors,		Other computers, including array, database, and image processors, and other analog, hybrid or special purpose computers	2.77
		Electronic computers, nsk	6.47

Source: 1992 *Economic Census*. The values shown are percent of total shipments in an industry. Values of indented subcategories are summed in the main heading. The symbol (D) appears when data are withheld to prevent disclosure of competitive information. The abbreviation nsk stands for 'not specified by kind' and nec for 'not elsewhere classified'.

INPUTS AND OUTPUTS FOR ELECTRONIC COMPUTING EQUIPMENT

Economic Sector or Industry Providing Inputs	%	Sector	Economic Sector or Industry Buying Outputs	%	Sector
Electronic computing equipment	29.4	Manufg.	Gross private fixed investment	45.6	Cap Inv
Wholesale trade	11.9	Trade	Exports	21.6	Foreign
Imports	10.2	Foreign	Electronic computing equipment	17.8	Manufg.
Electronic components nec	9.0	Manufg.	Federal Government purchases, national defense	4.6	Fed Govt
Semiconductors & related devices	5.0	Manufg.	Federal Government purchases, nondefense	1.6	Fed Govt
Industrial controls	2.8	Manufg.	Computer & data processing services	1.4	Services
Miscellaneous plastics products	2.7	Manufg.	Change in business inventories	1.3	In House
Air transportation	2.1	Util.	Personal consumption expenditures	1.0	
Banking	1.6	Fin/R.E.	Radio & TV communication equipment	0.8	Manufg.
Hotels & lodging places	1.5	Services	Typewriters & office machines, nec	0.5	Manufg.
Communications, except radio & TV	1.4	Util.	S/L Govt. purch., elem. & secondary education	0.4	S/L Govt
Electric services (utilities)	1.3	Util.	S/L Govt. purch., higher education	0.4	S/L Govt
Sheet metal work	1.2	Manufg.	S/L Govt. purch., other general government	0.4	S/L Govt
Eating & drinking places	1.2	Trade	Calculating & accounting machines	0.3	Manufg.
Real estate	1.2	Fin/R.E.	Banking	0.3	Fin/R.E.
Noncomparable imports	1.1	Foreign	Accounting, auditing & bookkeeping	0.3	Services
Petroleum refining	1.0	Manufg.	Electronic components nec	0.2	Manufg.
Paperboard containers & boxes	0.9	Manufg.	Instruments to measure electricity	0.2	Manufg.
Motors & generators	0.8	Manufg.	Insurance carriers	0.2	Fin/R.E.
Switchgear & switchboard apparatus	0.8	Manufg.	Mechanical measuring devices	0.1	Manufg.
Legal services	0.7	Services	Retail trade, except eating & drinking	0.1	Trade
Maintenance of nonfarm buildings nec	0.5	Constr.	Wholesale trade	0.1	Trade
Gaskets, packing & sealing devices	0.5	Manufg.			
Metal stampings, nec	0.5	Manufg.			
Wiring devices	0.5	Manufg.			
Management & consulting services & labs	0.5	Services			
Aluminum castings	0.4	Manufg.			
Electron tubes	0.4	Manufg.			
Screw machine and related products	0.4	Manufg.			
Advertising	0.4	Services			
Equipment rental & leasing services	0.4	Services			
Aluminum rolling & drawing	0.3	Manufg.			
Blast furnaces & steel mills	0.3	Manufg.			
Fabricated metal products, nec	0.3	Manufg.			
Iron & steel foundries	0.3	Manufg.			
Miscellaneous fabricated wire products	0.3	Manufg.			
Nonferrous wire drawing & insulating	0.3	Manufg.			
Plating & polishing	0.3	Manufg.			
Motor freight transportation & warehousing	0.3	Util.			
Accounting, auditing & bookkeeping	0.3	Services			
Computer & data processing services	0.3	Services			
Electrical industrial apparatus, nec	0.2	Manufg.			
Machinery, except electrical, nec	0.2	Manufg.			
Manifold business forms	0.2	Manufg.			
Transformers	0.2	Manufg.			
Insurance carriers	0.2	Fin/R.E.			
Royalties	0.2	Fin/R.E.			
Security & commodity brokers	0.2	Fin/R.E.			
Automotive rental & leasing, without drivers	0.2	Services			
Ball & roller bearings	0.1	Manufg.			
Die-cut paper & board	0.1	Manufg.			
Nonferrous rolling & drawing, nec	0.1	Manufg.			
Gas production & distribution (utilities)	0.1	Util.			
Automotive repair shops & services	0.1	Services			
Photofinishing labs, commercial photography	0.1	Services			
U.S. Postal Service	0.1	Gov't			

Source: Benchmark Input-Output Accounts for the U.S. Economy, 1982, U.S. Department of Commerce, Washington, D.C., July 1991. Data, as reported in the source, are organized by the 1977 SIC structure in use in 1982 but have been matched, as closely as is possible, to the 1987 SIC structure used in this book.

OCCUPATIONS EMPLOYED BY SIC 357 - COMPUTER AND OFFICE EQUIPMENT

Occupation	% of Total 1994	Change to 2005	Occupation	% of Total 1994	Change to 2005
Computer engineers	8.4	4.1	Engineering technicians & technologists nec	1.8	-29.7
Electrical & electronics engineers	5.4	-10.2	General managers & top executives	1.7	-33.3
Electrical & electronic technicians,technologists	4.9	-29.7	General office clerks	1.6	-40.1
Electrical & electronic equipment assemblers	4.8	-36.8	Purchasing agents, ex trade & farm products	1.4	5.4
Electrical & electronic assemblers	4.4	-15.7	Engineers nec	1.4	-15.7
Sales & related workers nec	3.9	-29.7	Industrial production managers	1.4	-36.8
Computer programmers	3.9	-43.1	Professional workers nec	1.4	-15.7
Engineering, mathematical, & science managers	3.5	-20.2	Industrial engineers, ex safety engineers	1.3	-38.2
Secretaries, ex legal & medical	3.3	-36.0	Electromechanical equipment assemblers	1.3	-86.0
Inspectors, testers, & graders, precision	2.7	-29.7	Accountants & auditors	1.3	-29.7
Production, planning, & expediting clerks	2.6	-29.7	Traffic, shipping, & receiving clerks	1.2	-32.4
Management support workers nec	2.4	-29.7	Mechanical engineers	1.2	-22.7
Systems analysts	2.2	12.4	Financial managers	1.1	-29.7
Marketing, advertising, & PR managers	2.0	-29.7	Bookkeeping, accounting, & auditing clerks	1.1	-47.3
Assemblers, fabricators, & hand workers nec	2.0	-29.7	Stock clerks	1.1	-42.9
Managers & administrators nec	1.9	-29.8	Writers & editors, incl technical writers	1.1	5.4
Blue collar worker supervisors	1.8	-32.6	Clerical supervisors & managers	1.0	-28.1

Source: *Industry-Occupation Matrix*, Bureau of Labor Statistics. These data relate to one or more 3-digit SIC industry groups rather than to a single 4-digit SIC. The change reported for each occupation to the year 2005 is a percent of growth or decline as estimated by the Bureau of Labor Statistics. The abbreviation nec stands for 'not elsewhere classified'.

LOCATION BY STATE AND REGIONAL CONCENTRATION

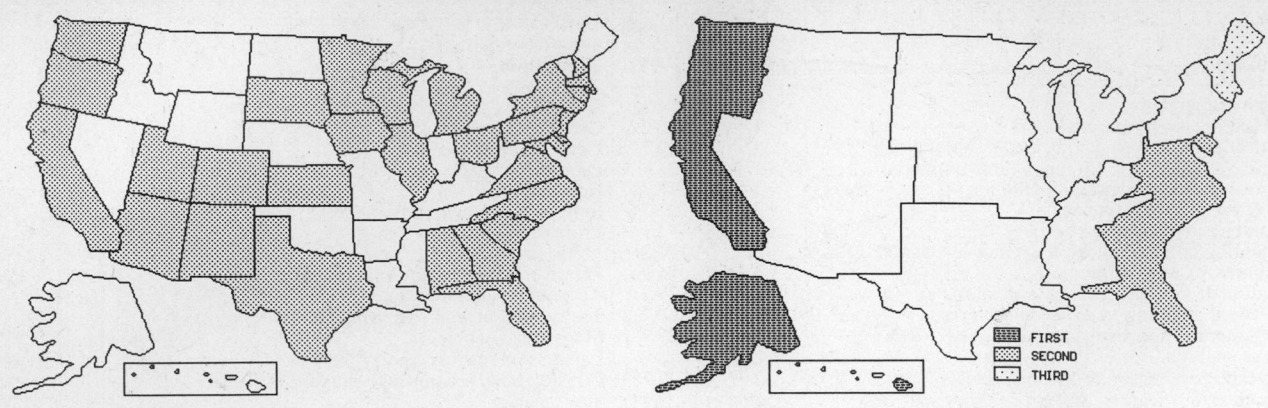

INDUSTRY DATA BY STATE

State	Establish- ments	Shipments			Employment				Cost as % of Shipments	Investment per Employee ($)
		Total ($ mil)	% of U.S.	Per Establ.	Total Number	% of U.S.	Per Establ.	Wages ($/hour)		
California	249	14,891.3	39.0	59.8	43,700	39.4	176	13.15	40.9	15,332
Texas	55	5,320.8	13.9	96.7	7,700	6.9	140	12.22	55.7	14,260
Massachusetts	52	1,893.3	5.0	36.4	6,100	5.5	117	15.08	56.8	7,164
Minnesota	23	704.7	1.8	30.6	5,300	4.8	230	12.33	87.2	4,000
Michigan	24	597.2	1.6	24.9	2,600	2.3	108	12.57	79.2	4,346
Oregon	20	410.6	1.1	20.5	1,600	1.4	80	17.63	50.3	-
New Jersey	21	326.4	0.9	15.5	1,000	0.9	48	13.10	43.3	-
Florida	37	307.6	0.8	8.3	2,000	1.8	54	12.08	41.8	3,250
Georgia	14	183.3	0.5	13.1	600	0.5	43	10.20	66.1	4,333
Utah	18	147.0	0.4	8.2	1,200	1.1	67	9.50	28.1	-
Connecticut	20	108.1	0.3	5.4	500	0.5	25	10.25	70.8	-
Kansas	6	98.0	0.3	16.3	500	0.5	83	10.50	56.5	4,800
Virginia	11	91.1	0.2	8.3	400	0.4	36	9.50	66.5	2,750
Pennsylvania	19	89.9	0.2	4.7	400	0.4	21	15.75	48.1	7,000
Illinois	25	60.5	0.2	2.4	500	0.5	20	9.25	54.2	2,200
Washington	18	58.9	0.2	3.3	700	0.6	39	7.67	47.4	-
Maryland	16	36.3	0.1	2.3	200	0.2	13	11.00	57.9	4,500
Ohio	16	29.9	0.1	1.9	300	0.3	19	10.00	50.2	2,667
Arizona	15	18.7	0.0	1.2	200	0.2	13	9.00	42.2	4,000
New York	43	(D)	-	-	7,500 *	6.8	174	-	-	1,240
Colorado	21	(D)	-	-	750 *	0.7	36	-	-	-
North Carolina	18	(D)	-	-	17,500 *	15.8	972	-	-	-
New Hampshire	15	(D)	-	-	1,750 *	1.6	117	-	-	1,657
Wisconsin	13	(D)	-	-	3,750 *	3.4	288	-	-	-
South Carolina	7	(D)	-	-	1,750 *	1.6	250	-	-	-
Alabama	4	(D)	-	-	7,500 *	6.8	1,875	-	-	-
Iowa	4	(D)	-	-	175 *	0.2	44	-	-	-
New Mexico	4	(D)	-	-	750 *	0.7	188	-	-	-
Delaware	2	(D)	-	-	750 *	0.7	375	-	-	-
South Dakota	2	(D)	-	-	1,750 *	1.6	875	-	-	-
Rhode Island	1	(D)	-	-	175 *	0.2	175	-	-	-

Source: 1992 *Economic Census*. The states are in descending order of shipments or establishments (if shipment data are missing for the majority). The symbol (D) appears when data are withheld to prevent disclosure of competitive information. States marked with (D) are sorted by number of establishments. A dash (-) indicates that the data element cannot be calculated; * indicates the midpoint of a range.

3572 - COMPUTER STORAGE DEVICES

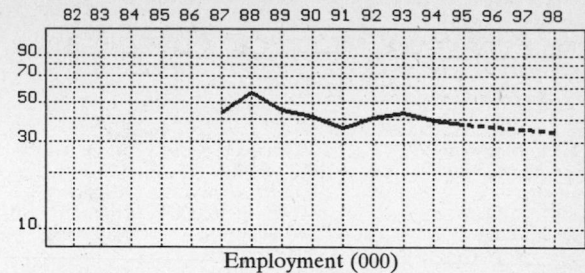

82 83 84 85 86 87 88 89 90 91 92 93 94 95 96 97 98

Shipments ($ million)

82 83 84 85 86 87 88 89 90 91 92 93 94 95 96 97 98

Employment (000)

GENERAL STATISTICS

Year	Com-panies	Establishments		Employment			Compensation		Production ($ million)			
		Total	with 20 or more employees	Total (000)	Production Workers (000)	Hours (Mil)	Payroll ($ mil)	Wages ($/hr)	Cost of Materials	Value Added by Manufacture	Value of Shipments	Capital Invest.
1982												
1983												
1984												
1985												
1986												
1987	100	106	64	43.3	15.0	31.0	1,442.6	10.58	3,252.8	3,268.5	6,394.8	347.0
1988		125	74	56.1	20.2	42.6	1,975.9	10.44	5,470.7	4,208.6	9,543.9	404.4
1989		114	70	44.7	16.6	34.7	1,684.3	11.31	4,087.8	3,666.7	7,612.5	432.4
1990		118	67	41.7	15.4	32.1	1,490.1	11.12	4,368.9	4,359.0	8,751.1	426.5
1991		126	71	36.0	12.2	27.1	1,444.1	13.01	3,674.2	3,600.2	7,188.6	392.8
1992	163	179	96	40.8	15.2	31.3	1,795.4	13.64	4,991.7	4,658.8	9,544.3	455.6
1993		178	91	43.8	17.6	34.9	1,796.0	13.72	6,110.0	4,294.8	10,395.3	557.7
1994		183P	94P	39.8	19.7	44.0	1,650.5	14.83	6,759.4	4,364.6	11,004.5	521.9
1995		195P	99P	37.7P	17.2P	36.7P	1,705.0P	15.28P	6,749.2P	4,358.0P	10,987.8P	550.9P
1996		207P	103P	36.5P	17.3P	37.2P	1,715.1P	15.93P	7,047.2P	4,550.4P	11,473.0P	575.0P
1997		219P	108P	35.3P	17.5P	37.6P	1,725.1P	16.59P	7,345.2P	4,742.9P	11,958.3P	599.1P
1998		231P	112P	34.0P	17.6P	38.0P	1,735.1P	17.24P	7,643.3P	4,935.3P	12,443.5P	623.2P

Sources: 1982, 1987, 1992 *Economic Census*; *Annual Survey of Manufactures*, 83-86, 88-91, 93-94. Establishment counts for non-Census years are from *County Business Patterns*; establishment values for 83-84 are extrapolations. 'P's show projections by the editors. Industries reclassified in 87 will not have data for prior years.

INDICES OF CHANGE

Year	Com-panies	Establishments		Employment			Compensation		Production ($ million)			
		Total	with 20 or more employees	Total (000)	Production Workers (000)	Hours (Mil)	Payroll ($ mil)	Wages ($/hr)	Cost of Materials	Value Added by Manufacture	Value of Shipments	Capital Invest.
1982												
1983												
1984												
1985												
1986												
1987	61	59	67	106	99	99	80	78	65	70	67	76
1988		70	77	138	133	136	110	77	110	90	100	89
1989		64	73	110	109	111	94	83	82	79	80	95
1990		66	70	102	101	103	83	82	88	94	92	94
1991		70	74	88	80	87	80	95	74	77	75	86
1992	100	100	100	100	100	100	100	100	100	100	100	100
1993		99	95	107	116	112	100	101	122	92	109	122
1994		102P	98P	98	130	141	92	109	135	94	115	115
1995		109P	103P	92P	113P	117P	95P	112P	135P	94P	115P	121P
1996		116P	107P	89P	114P	119P	96P	117P	141P	98P	120P	126P
1997		122P	112P	86P	115P	120P	96P	122P	147P	102P	125P	132P
1998		129P	117P	83P	116P	122P	97P	126P	153P	106P	130P	137P

Sources: Same as General Statistics. Values reflect change from the base year, 1992. Values above 100 mean greater than 92, values below 100 mean less than 92, and a value of 100 in the 82-91 or 93-98 period means same as 92. 'P's mark projections by the editors.

SELECTED RATIOS

For 1994	Avg. of All Manufact.	Analyzed Industry	Index	For 1994	Avg. of All Manufact.	Analyzed Industry	Index
Employees per Establishment	49	217	444	Value Added per Production Worker	134,084	221,553	165
Payroll per Establishment	1,500,273	9,012,090	601	Cost per Establishment	5,045,178	36,907,800	732
Payroll per Employee	30,620	41,470	135	Cost per Employee	102,970	169,834	165
Production Workers per Establishment	34	108	313	Cost per Production Worker	146,988	343,117	233
Wages per Establishment	853,319	3,562,902	418	Shipments per Establishment	9,576,895	60,086,973	627
Wages per Production Worker	24,861	33,123	133	Shipments per Employee	195,460	276,495	141
Hours per Production Worker	2,056	2,234	109	Shipments per Production Worker	279,017	558,604	200
Wages per Hour	12.09	14.83	123	Investment per Establishment	321,011	2,849,688	888
Value Added per Establishment	4,602,255	23,831,669	518	Investment per Employee	6,552	13,113	200
Value Added per Employee	93,930	109,663	117	Investment per Production Worker	9,352	26,492	283

Sources: Same as General Statistics. The 'Average of All Manufacturing' column represents the average of all manufacturing industries reported for the most recent complete year available. The Index shows the relationship between the Average and the Analyzed Industry. For example, 100 means that they are equal; 500 that the Analyzed Industry is five times the average; 50 means that the Analyzed Industry is half the national average. The abbreviation 'na' is used to show that data are 'not available'.

LEADING COMPANIES Number shown: **75**   Total sales ($ mil): **19,734**   Total employment (000): **135.3**

Company Name	Address				CEO Name	Phone	Co. Type	Sales ($ mil)	Empl. (000)
Seagate Technology Inc	920 Disc Dr	Scotts Valley	CA	95066	Alan F Shugart	408-438-6550	P	3,500	53.0
Conner Peripherals Inc	3081 Zanker Rd	San Jose	CA	95134	Finis F Conner	408-456-4500	P	2,152	9.1
Quantum Corp	500 McCarthy Blv	Milpitas	CA	95035	William J Miller	408-894-4000	P	2,131	3.0
Western Digital Corp	8105 Irvine Ctr Dr	Irvine	CA	92718	Charles A Haggerty	714-932-5000	P	1,540	6.6
Storage Technology Corp	2270 S 88th St	Louisville	CO	80028	Ryal R Poppa	303-673-5151	P	1,405	10.1
Intern Business Machines Corp	5600 Cottle Rd	San Jose	CA	95193	Ed Zschau	408-256-1600	D	1,240*	5.1
Maxtor Corp	211 River Oaks	San Jose	CA	95134	L R Hootnick	408-432-1700	P	1,153	8.8
EMC Corp	171 South St	Hopkinton	MA	01748	M C Ruettgers	508-435-1000	P	783	2.5
TBG Inc	565 5th Av	New York	NY	10017	GH Thyssen	212-850-8500	S	690*	6.0
Kingston Technology Corp	17600 Newhope St	Fountain Val	CA	92708	John Tu	714-435-2600	R	500*	0.3
Micropolis Corp	21211 Nordhoff St	Chatsworth	CA	91311	Stuart P Mabon	818-709-3300	P	346	2.2
Applied Magnetics Corp	75 Robin Hill Rd	Goleta	CA	93117	Craig D Crisman	805-683-5353	P	276	5.5
SyQuest Technology Inc	47071 Bayside Pkwy	Fremont	CA	94538	Syed H Iftikar	510-226-4000	P	221	1.4
REXON Inc	1 Progress Plz	St Petersburg	FL	33701	Robert C Genesi	813-896-9609	P	205	0.7
Ampex Systems Corp	401 Broadway St	Redwood City	CA	94063	Edward Bramson	415-367-2011	S	170	1.1
Digital Equip Corp	301 Rockrimmon S	Co Springs	CO	80919	Robert Palmer	719-260-2820	D	150	2.5
Smart Modular Technologies Inc	45531 Npt Loop	Fremont	CA	94538	Ajay Shah	510-623-1231	R	150	0.2
IOMEGA Corp	1821 Iomega	Roy	UT	84067	Kim B Edwards	801-778-1000	P	141	0.9
Wangtek Inc	41 Moreland Rd	Simi Valley	CA	93065	Robert Genesi	805-582-3300	S	140*	0.6
Ampex Corp	401 Broadway	Redwood City	CA	94063	Edward J Bramson	415-367-2011	P	127	0.6
Sequel Inc	2300 Central Expwy	Santa Clara	CA	95054	Michael Haltom	408-987-1000	R	125	0.6
MTI Technology Corp	4905 E La Palma	Anaheim	CA	92807	Steven J Hamerslag	714-970-0300	P	123	0.6
MTI Inc	4905 E La Palma	Anaheim	CA	92807	Steve Hamerslag	714-970-0300	R	120	0.5
Multi-Tech Systems Inc	2205 Woodale Dr	Mounds View	MN	55112	Raghu Sharma	612-785-3500	R	103	0.3
La Cie Ltd	8700 SW Creekside	Beaverton	OR	97005	Joel Kamerman	503-520-9000	S	100	0.9
Intern Components Techn Corp	2360 Zanker Rd	San Jose	CA	95131	Hirosi Akita	408-435-8780	R	90*	0.7
AmeriQuest Technologies Inc	2722 Michelson Dr	Irvine	CA	92715	Harold L Clark	714-222-6000	P	88	0.4
Seagate Technologies	PO Box 12313	Oklahoma City	OK	73157	BA Carballo	405-324-3040	D	85	2.0
Tallgrass Technologies Corp	11100 W 82nd St	Lenexa	KS	66214	Ernest Wassmann	913-492-6002	S	85	<0.1
Odetics Inc	1515 S Manchester	Anaheim	CA	92802	Joel Slutzky	714-774-5000	P	84	0.6
Odetics Inc	240 E Palais Rd	Anaheim	CA	92805	Kevin Daly	714-774-6900	D	83	0.6
Storage Dimensions Inc	1656 MCarthy Blv	Milpitas	CA	95035	David A Eeg	408-954-0710	R	82	0.2
Intel Corp	5200 Elam Young	Hillsboro	OR	97124	Jim Johnson	503-629-7402	D	81	0.7
Kennedy Co	9292 Jeronimo Rd	Irvine	CA	92718	Dennis Narlinger	714-770-1100	S	69*	0.6
Tecmar	6225 Cochran Rd	Solon	OH	44139	Robert Werbicki	216-349-0600	S	66*	0.2
Plextor Inc	4255 Burton Dr	Santa Clara	CA	95054	N Takahashi	408-980-1838	S	64*	0.6
Emulex Corp	3535 Harbor Blv	Costa Mesa	CA	92626	Paul F Folino	714-662-5600	P	62	0.4
Metrum Information Storage	4800 E Dry Creek	Littleton	CO	80122	John Brenan	303-773-4700	D	59	0.3
Formation Inc	121 Whittendale Dr	Moorestown	NJ	08057	AD Beard	609-234-5020	R	57	0.3
Philips Laser Magnetic Storage	4425 Arrowswest Dr	Co Springs	CO	80907	Charles Johnston	719-593-7900	S	55	0.5
Falcon Systems Inc	1417 W North Mkt	Sacramento	CA	95834	Craig Caudill	916-929-9255	R	50	<0.1
Procom Technology Inc	2181 Dupont Dr	Irvine	CA	92715	Alex Razmjoo	714-852-1000	R	50	0.1
Network Imaging Systems Corp	600 Huntmar Pk Dr	Herndon	VA	22070	Robert Bernardi	703-478-2260	P	49	0.2
MPM	15285 Alton Pkwy	Irvine	CA	92718	Christ Zomaya	714-753-1200	D	43	<0.1
Datatape Inc	PO Box 7014	Pasadena	CA	91109	Dominic Saccacio	818-796-9381	S	41*	0.4
Delta Tango Inc	110 El Paso	Santa Barbara	CA	93101	M Stewart Millar	805-965-6453	R	41	0.4
Cambex Corp	360 2nd Av	Waltham	MA	02154	Joseph F Kruy	617-890-6000	P	41	0.2
CORE International Inc	7171 N Federal Hwy	Boca Raton	FL	33487	Hal Prewitt	407-997-6055	R	40*	0.2
Pinnacle Micro Inc	19 Technology Dr	Irvine	CA	92718	William F Blum	714-727-3300	P	39	<0.1
Information Storage Devices Inc	2045 Hamilton Av	San Jose	CA	95125	David L Angel	408-369-2400	P	39	<0.1
Mega Drive Systems Inc	489 S Robertson	Beverly Hills	CA	90211	Paul Bloch	310-247-0006	R	38*	0.2
Peripheral Land Inc	47421 Bayside Pkwy	Fremont	CA	94538	Leo Berenguel	510-657-2211	R	36	<0.1
Alliance Peripheral Systems Inc	6131 Deramus	Kansas City	MO	64120	Paul Mandel	816-483-1600	R	35*	<0.1
Great Valley Products Inc	600 Clark Av	King of Prussia	PA	19406	Gerard Bucas	215-337-8770	R	35	<0.1
M4 Data Inc	3815 N US 1	Cocoa	FL	32926	Duke Ebenezer	407-639-6487	R	35	0.2
Spectrum Engineering Inc	7803 E Osie St	Wichita	KS	67207	James Wiebe	316-685-4904	R	35	<0.1
SyDOS	6501 Pk of Com	Boca Raton	FL	33487	Tim Mahoney	407-998-5450	D	30	<0.1
UltraStor Corp	310 Hammond Av	Fremont	CA	94539	Steve A Roberts	510-623-8955	R	30	<0.1
MountainGate Data Systems Inc	9393 Gateway Dr	Reno	NV	89511	Glen T Williamson	702-851-9393	S	29*	0.1
Anthem Technology Systems	1160 Ridder Pk Dr	San Jose	CA	95131	Michael A Rynas	408-441-7177	S	26	<0.1
Focus Enhancements Inc	800 Cummings	Woburn	MA	01801	Thomas L Massie	617-938-8088	P	24	<0.1
Sankyo Seiki America Inc	3191 Airport Loop	Costa Mesa	CA	92626		714-751-5959	S	23*	0.2
TEAC America Inc	7733 Telegraph Rd	Montebello	CA	90640	Norio Tamura	213-726-0303	D	23*	0.2
Advanced Digital Info Corp	PO Box 97057	Redmond	WA	98073	Peter H van Oppen	206-881-8004	S	22	<0.1
Array Technology Corp	4775 Walnut St	Boulder	CO	80301	David W Gordon	303-444-9300	S	21*	0.1
Sanyo/Icon	18301 Von Karman	Irvine	CA	92715	John Bonne	714-263-3777	S	20	<0.1
Micro Memory Inc	9540 Vassar Av	Chatsworth	CA	91311	Robert E Lepore	818-998-0070	R	19	<0.1
Zitel Corp	47211 Bayside Pkwy	Fremont	CA	94538	Jack H King	510-440-9600	P	18	0.1
Nordigo Peripherals	127 M LaPorte St	Arcadia	CA	91006	Edmond Chan	818-445-2091	R	16	0.6
ACS Computer	260 E Grand Av	S San Francisco	CA	94080	Yong Chew	415-875-6633	A,S	15	<0.1
Optima Technology Corp	17526 Von Karman	Irvine	CA	92714	HR Assadian	714-476-0515	R	15	<0.1
Pacific Micro Data Inc	3002 Dow Av	Tustin	CA	92663	Ame Gur	714-838-8900	R	15	<0.1
Transitional Technology Inc	5401 E La Palma	Anaheim	CA	92807	M T Goldbach	714-693-1133	R	15*	<0.1
Cognitronics Corp	3 Corporate Dr	Danbury	CT	06810	Brian J Kelley	203-830-3400	P	15	<0.1
EBSCO Publishing	PO Box 2250	Peabody	MA	01960	Tim Collins	508-535-8500	D	14*	0.1

Source: Ward's Business Directory of U.S. Private and Public Companies, Volumes 1 and 2, 1996. The company type code used is as follows: P - Public, R - Private, S - Subsidiary, D - Division, J - Joint Venture, A - Affiliate, G - Group. Sales are in millions of dollars, employees are in thousands. An asterisk (*) indicates an estimated sales volume. The symbol < stands for 'less than'. Company names and addresses are truncated, in some cases, to fit into the available space.

MATERIALS CONSUMED

Material	Quantity	Delivered Cost ($ million)
Materials, ingredients, containers, and supplies	(X)	3,612.3
Printed circuit boards (without inserted components) for electronic circuitry	(X)	29.6
Printed memory boards for electronic circuitry	(X)	281.2
Semiconductors, including transistors, diodes, rectifiers, and integrated circuits for electronic circuitry	(X)	120.0
Capacitors for electronic circuitry	(X)	10.6
Resistors for electronic circuitry	(X)	5.1
Connectors for electronic circuitry	(X)	15.9
Battery packs for electronic circuitry	(X)	(D)
Other components and accessories for electric circuitry	(X)	36.2
Electrical transmission, distribution, and control equipment	(X)	(D)
Steel, aluminum, and other metal electronic enclosurse	(X)	16.0
Plastics electronic enclosures	(X)	(D)
Sheet metal products (including stampings), except enclosures	(X)	31.2
All other fabricated metal products (except forgings)	(X)	35.7
Forgings	(X)	(D)
Castings (rough and semifinished)	(X)	29.6
Metal shapes and forms, except castings, forgings, and fabricated metal products	(X)	27.5
Insulated copper wire and cable (including magnet wire)	(X)	23.4
Fabricated plastics products, except enclosures	(X)	15.1
Purchased software	(X)	(D)
Appliance outlets, switches, lampholders, and other current-carrying wiring devices	(X)	(D)
Electric motors and generators (all types)	(X)	85.6
Paper and paperboard products including paperboard boxes, containers, and corrugated paperboard	(X)	12.4
Purchased computers	(X)	1,303.5
All other materials and components, parts, containers, and supplies	(X)	805.7
Materials, ingredients, containers, and supplies, nsk	(X)	667.8

Source: 1992 *Economic Census*. Explanation of symbols used: (D): Withheld to avoid disclosure of competitive data; na: Not available; (S): Withheld because statistical norms were not met; (X): Not applicable; (Z): Less than half the unit shown; nec: Not elsewhere classified; nsk: Not specified by kind; - : zero; * : 10-19 percent estimated; ** : 20-29 percent estimated.

PRODUCT SHARE DETAILS

Product or Product Class	% Share	Product or Product Class	% Share
Computer storage devices	100.00	Parts, attachments, and accessories for computer storage devices	16.29
Computer storage devices (except parts, attachments, and accessories)	78.42	Computer storage devices, nsk	5.29

Source: 1992 *Economic Census*. The values shown are percent of total shipments in an industry. Values of indented subcategories are summed in the main heading. The symbol (D) appears when data are withheld to prevent disclosure of competitive information. The abbreviation nsk stands for 'not specified by kind' and nec for 'not elsewhere classified'.

INPUTS AND OUTPUTS FOR ELECTRONIC COMPUTING EQUIPMENT

Economic Sector or Industry Providing Inputs	%	Sector	Economic Sector or Industry Buying Outputs	%	Sector
Electronic computing equipment	29.4	Manufg.	Gross private fixed investment	45.6	Cap Inv
Wholesale trade	11.9	Trade	Exports	21.6	Foreign
Imports	10.2	Foreign	Electronic computing equipment	17.8	Manufg.
Electronic components nec	9.0	Manufg.	Federal Government purchases, national defense	4.6	Fed Govt
Semiconductors & related devices	5.0	Manufg.	Federal Government purchases, nondefense	1.6	Fed Govt
Industrial controls	2.8	Manufg.	Computer & data processing services	1.4	Services
Miscellaneous plastics products	2.7	Manufg.	Change in business inventories	1.3	In House
Air transportation	2.1	Util.	Personal consumption expenditures	1.0	
Banking	1.6	Fin/R.E.	Radio & TV communication equipment	0.8	Manufg.
Hotels & lodging places	1.5	Services	Typewriters & office machines, nec	0.5	Manufg.
Communications, except radio & TV	1.4	Util.	S/L Govt. purch., elem. & secondary education	0.4	S/L Govt
Electric services (utilities)	1.3	Util.	S/L Govt. purch., higher education	0.4	S/L Govt
Sheet metal work	1.2	Manufg.	S/L Govt. purch., other general government	0.4	S/L Govt
Eating & drinking places	1.2	Trade	Calculating & accounting machines	0.3	Manufg.
Real estate	1.2	Fin/R.E.	Banking	0.3	Fin/R.E.
Noncomparable imports	1.1	Foreign	Accounting, auditing & bookkeeping	0.3	Services
Petroleum refining	1.0	Manufg.	Electronic components nec	0.2	Manufg.
Paperboard containers & boxes	0.9	Manufg.	Instruments to measure electricity	0.2	Manufg.
Motors & generators	0.8	Manufg.	Insurance carriers	0.2	Fin/R.E.
Switchgear & switchboard apparatus	0.8	Manufg.	Mechanical measuring devices	0.1	Manufg.
Legal services	0.7	Services	Retail trade, except eating & drinking	0.1	Trade
Maintenance of nonfarm buildings nec	0.5	Constr.	Wholesale trade	0.1	Trade
Gaskets, packing & sealing devices	0.5	Manufg.			
Metal stampings, nec	0.5	Manufg.			
Wiring devices	0.5	Manufg.			
Management & consulting services & labs	0.5	Services			
Aluminum castings	0.4	Manufg.			
Electron tubes	0.4	Manufg.			
Screw machine and related products	0.4	Manufg.			

Continued on next page.

INPUTS AND OUTPUTS FOR ELECTRONIC COMPUTING EQUIPMENT - Continued

Economic Sector or Industry Providing Inputs	%	Sector	Economic Sector or Industry Buying Outputs	%	Sector
Advertising	0.4	Services			
Equipment rental & leasing services	0.4	Services			
Aluminum rolling & drawing	0.3	Manufg.			
Blast furnaces & steel mills	0.3	Manufg.			
Fabricated metal products, nec	0.3	Manufg.			
Iron & steel foundries	0.3	Manufg.			
Miscellaneous fabricated wire products	0.3	Manufg.			
Nonferrous wire drawing & insulating	0.3	Manufg.			
Plating & polishing	0.3	Manufg.			
Motor freight transportation & warehousing	0.3	Util.			
Accounting, auditing & bookkeeping	0.3	Services			
Computer & data processing services	0.3	Services			
Electrical industrial apparatus, nec	0.2	Manufg.			
Machinery, except electrical, nec	0.2	Manufg.			
Manifold business forms	0.2	Manufg.			
Transformers	0.2	Manufg.			
Insurance carriers	0.2	Fin/R.E.			
Royalties	0.2	Fin/R.E.			
Security & commodity brokers	0.2	Fin/R.E.			
Automotive rental & leasing, without drivers	0.2	Services			
Ball & roller bearings	0.1	Manufg.			
Die-cut paper & board	0.1	Manufg.			
Nonferrous rolling & drawing, nec	0.1	Manufg.			
Gas production & distribution (utilities)	0.1	Util.			
Automotive repair shops & services	0.1	Services			
Photofinishing labs, commercial photography	0.1	Services			
U.S. Postal Service	0.1	Gov't			

Source: Benchmark Input-Output Accounts for the U.S. Economy, 1982, U.S. Department of Commerce, Washington, D.C., July 1991. Data, as reported in the source, are organized by the 1977 SIC structure in use in 1982 but have been matched, as closely as is possible, to the 1987 SIC structure used in this book.

OCCUPATIONS EMPLOYED BY SIC 357 - COMPUTER AND OFFICE EQUIPMENT

Occupation	% of Total 1994	Change to 2005	Occupation	% of Total 1994	Change to 2005
Computer engineers	8.4	4.1	Engineering technicians & technologists nec	1.8	-29.7
Electrical & electronics engineers	5.4	-10.2	General managers & top executives	1.7	-33.3
Electrical & electronic technicians,technologists	4.9	-29.7	General office clerks	1.6	-40.1
Electrical & electronic equipment assemblers	4.8	-36.8	Purchasing agents, ex trade & farm products	1.4	5.4
Electrical & electronic assemblers	4.4	-15.7	Engineers nec	1.4	-15.7
Sales & related workers nec	3.9	-29.7	Industrial production managers	1.4	-36.8
Computer programmers	3.9	-43.1	Professional workers nec	1.4	-15.7
Engineering, mathematical, & science managers	3.5	-20.2	Industrial engineers, ex safety engineers	1.3	-38.2
Secretaries, ex legal & medical	3.3	-36.0	Electromechanical equipment assemblers	1.3	-86.0
Inspectors, testers, & graders, precision	2.7	-29.7	Accountants & auditors	1.3	-29.7
Production, planning, & expediting clerks	2.6	-29.7	Traffic, shipping, & receiving clerks	1.2	-32.4
Management support workers nec	2.4	-29.7	Mechanical engineers	1.2	-22.7
Systems analysts	2.2	12.4	Financial managers	1.1	-29.7
Marketing, advertising, & PR managers	2.0	-29.7	Bookkeeping, accounting, & auditing clerks	1.1	-47.3
Assemblers, fabricators, & hand workers nec	2.0	-29.7	Stock clerks	1.1	-42.9
Managers & administrators nec	1.9	-29.8	Writers & editors, incl technical writers	1.1	5.4
Blue collar worker supervisors	1.8	-32.6	Clerical supervisors & managers	1.0	-28.1

Source: Industry-Occupation Matrix, Bureau of Labor Statistics. These data relate to one or more 3-digit SIC industry groups rather than to a single 4-digit SIC. The change reported for each occupation to the year 2005 is a percent of growth or decline as estimated by the Bureau of Labor Statistics. The abbreviation nec stands for 'not elsewhere classified'.

LOCATION BY STATE AND REGIONAL CONCENTRATION

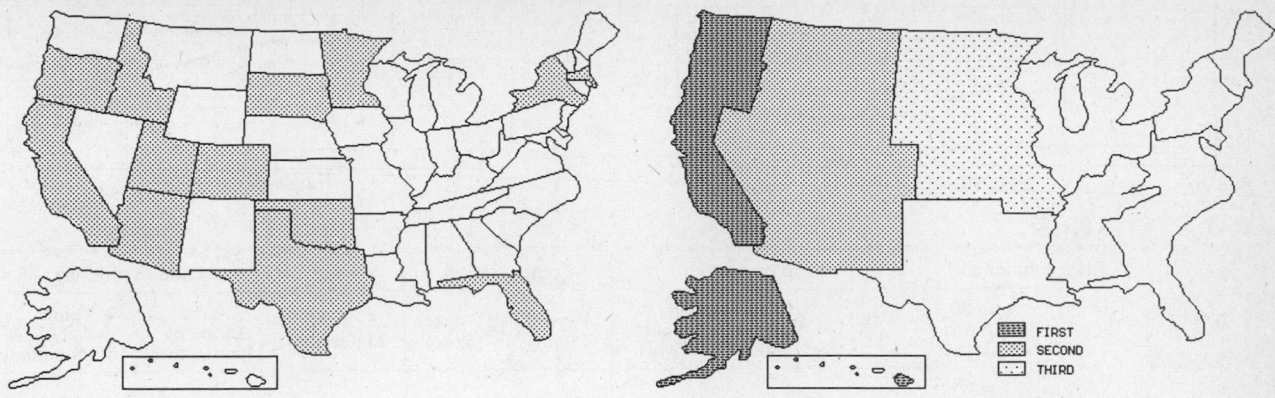

FIRST
SECOND
THIRD

INDUSTRY DATA BY STATE

State	Establish-ments	Shipments			Employment				Cost as % of Shipments	Investment per Employee ($)
		Total ($ mil)	% of U.S.	Per Establ.	Total Number	% of U.S.	Per Establ.	Wages ($/hour)		
California	68	4,486.5	47.0	66.0	14,500	35.5	213	14.30	60.4	14,372
Colorado	14	1,960.5	20.5	140.0	9,600	23.5	686	23.33	38.6	14,458
Minnesota	14	512.4	5.4	36.6	6,000	14.7	429	9.97	57.5	-
Massachusetts	13	282.2	3.0	21.7	1,100	2.7	85	14.77	59.0	-
New York	5	16.0	0.2	3.2	100	0.2	20	15.00	30.0	-
Florida	7	(D)	-	-	750 *	1.8	107	-	-	2,400
Texas	5	(D)	-	-	375 *	0.9	75	-	-	-
Arizona	4	(D)	-	-	175 *	0.4	44	-	-	1,143
Oregon	3	(D)	-	-	750 *	1.8	250	-	-	-
Idaho	2	(D)	-	-	3,750 *	9.2	1,875	-	-	-
Oklahoma	2	(D)	-	-	1,750 *	4.3	875	-	-	-
South Dakota	1	(D)	-	-	750 *	1.8	750	-	-	-
Utah	1	(D)	-	-	750 *	1.8	750	-	-	-

Source: 1992 *Economic Census.* The states are in descending order of shipments or establishments (if shipment data are missing for the majority). The symbol (D) appears when data are withheld to prevent disclosure of competitive information. States marked with (D) are sorted by number of establishments. A dash (-) indicates that the data element cannot be calculated; * indicates the midpoint of a range.

3575 - COMPUTER TERMINALS

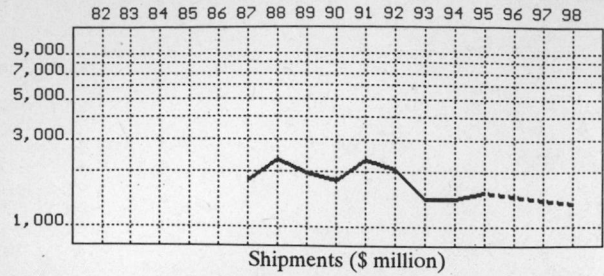

Shipments ($ million)

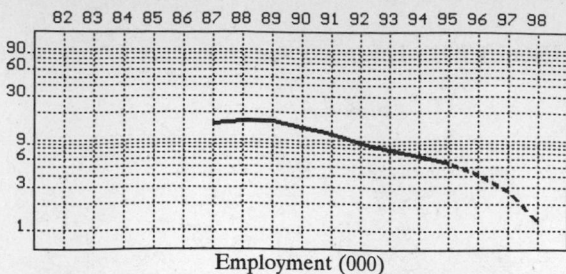

Employment (000)

GENERAL STATISTICS

| Year | Companies | Establishments | | Employment | | | Compensation | | Production ($ million) | | | |
		Total	with 20 or more employees	Total (000)	Production Workers (000)	Hours (Mil)	Payroll ($ mil)	Wages ($/hr)	Cost of Materials	Value Added by Manufacture	Value of Shipments	Capital Invest.
1982												
1983												
1984												
1985												
1986												
1987	122	121	74	15.0	5.5	10.7	441.7	11.60	742.0	1,014.3	1,799.0	58.0
1988		129	67	15.8	5.9	11.2	451.5	10.96	1,159.2	1,210.3	2,332.6	42.7
1989		141	79	15.8	6.3	12.9	450.5	11.47	1,219.1	726.1	1,969.2	69.3
1990		138	72	13.5	4.8	9.8	412.9	11.96	1,035.5	728.5	1,790.0	45.4
1991		154	70	11.7	5.2	10.3	386.1	11.99	1,429.5	865.0	2,326.5	42.4
1992	186	190	63	9.3	4.2	8.0	344.1	12.84	1,288.2	800.2	2,070.7	44.4
1993		177	61	7.9	3.5	6.8	307.2	12.25	807.8	615.8	1,435.0	29.3
1994		193P	61P	6.6	2.7	5.2	260.4	12.85	894.7	516.2	1,427.5	41.7
1995		204P	59P	5.5P	2.8P	5.4P	256.6P	13.03P	977.2P	563.8P	1,559.1P	32.8P
1996		215P	57P	4.1P	2.3P	4.5P	228.8P	13.26P	930.6P	536.9P	1,484.7P	29.7P
1997		226P	55P	2.7P	1.9P	3.6P	201.0P	13.49P	884.0P	510.0P	1,410.4P	26.6P
1998		237P	53P	1.3P	1.4P	2.7P	173.2P	13.72P	837.4P	483.1P	1,336.0P	23.5P

Sources: 1982, 1987, 1992 *Economic Census*; *Annual Survey of Manufactures*, 83-86, 88-91, 93-94. Establishment counts for non-Census years are from *County Business Patterns*; establishment values for 83-84 are extrapolations. 'P's show projections by the editors. Industries reclassified in 87 will not have data for prior years.

INDICES OF CHANGE

| Year | Companies | Establishments | | Employment | | | Compensation | | Production ($ million) | | | |
		Total	with 20 or more employees	Total (000)	Production Workers (000)	Hours (Mil)	Payroll ($ mil)	Wages ($/hr)	Cost of Materials	Value Added by Manufacture	Value of Shipments	Capital Invest.
1982												
1983												
1984												
1985												
1986												
1987	66	64	117	161	131	134	128	90	58	127	87	131
1988		68	106	170	140	140	131	85	90	151	113	96
1989		74	125	170	150	161	131	89	95	91	95	156
1990		73	114	145	114	123	120	93	80	91	86	102
1991		81	111	126	124	129	112	93	111	108	112	95
1992	100	100	100	100	100	100	100	100	100	100	100	100
1993		93	97	85	83	85	89	95	63	77	69	66
1994		102P	98P	71	64	65	76	100	69	65	69	94
1995		107P	94P	60P	66P	67P	75P	101P	76P	70P	75P	74P
1996		113P	91P	44P	55P	56P	66P	103P	72P	67P	72P	67P
1997		119P	88P	29P	44P	45P	58P	105P	69P	64P	68P	60P
1998		125P	85P	14P	34P	34P	50P	107P	65P	60P	65P	53P

Sources: Same as General Statistics. Values reflect change from the base year, 1992. Values above 100 mean greater than 92, values below 100 mean less than 92, and a value of 100 in the 82-91 or 93-98 period means same as 92. 'P's mark projections by the editors.

SELECTED RATIOS

For 1994	Avg. of All Manufact.	Analyzed Industry	Index	For 1994	Avg. of All Manufact.	Analyzed Industry	Index
Employees per Establishment	49	34	70	Value Added per Production Worker	134,084	191,185	143
Payroll per Establishment	1,500,273	1,347,228	90	Cost per Establishment	5,045,178	4,628,899	92
Payroll per Employee	30,620	39,455	129	Cost per Employee	102,970	135,561	132
Production Workers per Establishment	34	14	41	Cost per Production Worker	146,988	331,370	225
Wages per Establishment	853,319	345,706	41	Shipments per Establishment	9,576,895	7,385,440	77
Wages per Production Worker	24,861	24,748	100	Shipments per Employee	195,460	216,288	111
Hours per Production Worker	2,056	1,926	94	Shipments per Production Worker	279,017	528,704	189
Wages per Hour	12.09	12.85	106	Investment per Establishment	321,011	215,743	67
Value Added per Establishment	4,602,255	2,670,658	58	Investment per Employee	6,552	6,318	96
Value Added per Employee	93,930	78,212	83	Investment per Production Worker	9,352	15,444	165

Sources: Same as General Statistics. The 'Average of All Manufacturing' column represents the average of all manufacturing industries reported for the most recent complete year available. The Index shows the relationship between the Average and the Analyzed Industry. For example, 100 means that they are equal; 500 that the Analyzed Industry is five times the average; 50 means that the Analyzed Industry is half the national average. The abbreviation 'na' is used to show that data are 'not available'.

LEADING COMPANIES   Number shown: **62**   Total sales ($ mil): **1,667**   Total employment (000): **7.2**

Company Name	Address				CEO Name	Phone	Co. Type	Sales ($ mil)	Empl. (000)
Wyse Technology Inc	3471 N 1st St	San Jose	CA	95134	C Daniel Wu	408-473-1200	R	290•	1.8
Compac Microelectronics Inc	3797 Spinnaker Ct	Fremont	CA	94538	Bob Huang	510-656-3333	S	200	0.2
Network Computing Devices Inc	350 N Bernardo	Mountain View	CA	94043	Edward Marinaro	415-694-0650	P	161	0.4
Radius Inc	215 Moffett Park Dr	Sunnyvale	CA	94089	Charles Berger	408-541-6100	P	135	0.3
VTech Computers Inc	800 Church St	Lake Zurich	IL	60047	David Gish	708-540-8086	S	99•	0.2
DATAMAX Corp	4501 Pkway Com	Orlando	FL	32808	R C Strandberg	407-578-8007	R	90	0.3
Applied Digital Data Systems	100 Marcus Blv	Hauppauge	NY	11788	Gerald Youngblood	516-342-7400	S	80•	0.4
Planar Systems Inc	1400 NW Compton	Beaverton	OR	97006	James M Hurd	503-690-1100	P	60	0.5
BancTec Systems Inc	4435 Spring Val Rd	Dallas	TX	75244	G N Clark Jr	214-450-7700	S	50	0.3
CliniCom Inc	4720 Walnut St	Boulder	CO	80301	William H Brehm	303-443-9660	P	35	0.2
Electronic Associates Inc	185 Monmouth	W Long Branch	NJ	07764	Joseph R Spalliero	908-229-1100	P	30	0.3
Falco Data Products Inc	440 Potrero Av	Sunnyvale	CA	94086	Joseph D'Alessandro	408-745-7123	R	30	<0.1
Human Designed Systems Inc	421 Feheley Dr	King of Prussia	PA	19406	Mark Gelberg	215-277-8300	R	30	<0.1
GCH Systems Inc	777 E Middlefield	Mountain View	CA	94043	George Huang	415-968-3400	R	25	0.1
International Totalizator Systems	2131 Faraday Av	Carlsbad	CA	92008	Frederick A Brunn	619-931-4000	P	24	0.3
Mag-Tek Inc	20725 S Annalee	Carson	CA	90746	Thomas C McGeary	310-631-8602	R	23	0.1
Intecolor Corp	2150 Boggs Rd	Duluth	GA	30136	David M Deans	404-623-9145	R	20	0.2
Zynk Industrial Corp	41650 Christy St	Fremont	CA	94538	Frank Yao	510-490-6611	R	20	0.2
Lannet Inc	17942 Cowan Av	Irvine	CA	92714	Bill Atkinson	714-752-6638	P	18	<0.1
Digital F/X Inc	755 Ravendale Dr	Mountain View	CA	94043	Rolando Esteverena	415-961-2800	R	16	0.1
Chase Research Inc	545 Marriott Dr	Nashville	TN	37214	Jeff Pack	615-872-0770	R	15	<0.1
Lynk Corp	101 Queens Dr	King of Prussia	PA	19406	James Beisty	215-265-3550	R	15	<0.1
Ultimate Technology Corp	6280 Rte 96 E	Victor	NY	14564	Dennis Lewis	716-924-9500	S	15	<0.1
TeleVideo Systems Inc	PO Box 49048	San Jose	CA	95161	K Philip Hwang	408-954-8333	P	13	<0.1
General Parametrics Corp	1250 9th St	Berkeley	CA	94710	Herbert B Baskin	510-524-3950	P	12	<0.1
GraphOn Corp	544 Division St	Campbell	CA	95008	Walt Keller	408-370-4080	R	12	<0.1
Motif Inc	27700 A SW Pkwy	Wilsonville	OR	97070	Paul Gulick	503-682-7700	J	12•	<0.1
ADI Systems Inc	2115 Ringwood Av	San Jose	CA	95131	Raymond Hou	408-944-0100	R	11	<0.1
Sherwood Kimtron Corp	4181 Business Ctr	Fremont	CA	94538	Phillip Graham Sr	510-623-8900	S	11	<0.1
Suntronic Technology Group Inc	6711 Sands Rd	Crystal Lake	IL	60014	Michael P Nicholas	815-459-1959	R	10	<0.1
Interaction Systems Inc	10 Commerce Way	Woburn	MA	01801	Michael J Marino	617-923-6001	R	8•	<0.1
Phase X Systems Inc	19545 Neumann	Beaverton	OR	97006	Chong Lee	503-531-2400	R	8•	<0.1
Visentech Systems Inc	1825 E Plano Pkwy	Plano	TX	75074	HC Ham	214-423-1677	R	8	<0.1
Control Transaction Corp	130 Clinton Rd	Fairfield	NJ	07004	Arthur Rush	201-575-9100	R	7	<0.1
Elite Products Co	8324 Veterans Hwy	Millersville	MD	21108	Gershon Hoffer	410-987-3048	R	6	<0.1
Mercury Minnesota Inc	901 Hulett Av	Faribault	MN	55021	Gordon Adamek	507-334-5513	S	6	0.1
Comtrex Systems Corp	102 Executive Dr	Moorestown	NJ	08057	Jeffrey Rice	609-778-0090	P	6	<0.1
Termiflex Corp	316 Daniel Webster	Merrimack	NH	03054	Velton Casler	603-424-3700	P	6	<0.1
Photonics Systems Inc	6975 Wales Rd	Northwood	OH	43619	Ray A Stoller	419-666-6325	R	5	<0.1
Graphics Technology Company	2113 Wells Branch	Austin	TX	78728	Gary L Barrett	512-990-9700	R	4•	<0.1
Informer Computer Systems Inc	12833 Monarch St	Garden Grove	CA	92641	Edward Dailey	714-891-1112	R	4	<0.1
Linx Data Terminals Inc	625 Digital Dr	Plano	TX	75075	David Willis	214-964-7090	R	4•	<0.1
Candes Systems Inc	3131 Detwiler Rd	Harleysville	PA	19438	Daniel Signore	215-256-4130	R	3	<0.1
Cumulus Technology Corp	725 N Shoreline	Mountain View	CA	94403	John Darke	415-960-1200	S	3	<0.1
SunRiver Corp	9430 Research Blv	Austin	TX	78758	G F Youngblood	512-835-8001	R	3•	<0.1
Granite Communications Inc	9 Townsend W	Nashua	NH	03063	Harry Klein	603-881-8666	R	3	<0.1
Digix America Corp	10430 NW 31st	Miami	FL	33172	Arnon I Schreiber	305-593-8070	R	2•	<0.1
IICON Corp	16040 Caputo Dr	Morgan Hill	CA	95037	Robert Steinberg	408-779-7466	R	2	<0.1
Jupiter Systems Inc	3073 Teagarden St	San Leandro	CA	94577	Eric Wogsberg	510-523-9000	R	2	<0.1
Kangaroo Technologies Corp	601 Gateway Blv	S San Francisco	CA	94080	Gary Kench	415-588-0715	R	2	<0.1
Peripheral Systems Inc	150 Wright Brothers	Salt Lake City	UT	84116	Pat Volz	801-521-0383	R	2	<0.1
TransTechnology Syst & Services	30777 Schoolcraft	Livonia	MI	48150	Steven Holthen	313-458-8649	D	2•	<0.1
Bel-Air Technologies Inc	555 N Mathilda Av	Sunnyvale	CA	94086	Saeed Kazmi	408-720-4424	R	1•	<0.1
CheckOutPlus Inc	221 Broadway	Amityville	NY	11701	Anthony R Santoro	516-691-3020	R	1•	<0.1
Durasys Corp	PO Box 814	Dover	NH	03821	George Perrine	603-742-7363	R	1	<0.1
Electronic Systems International	23531 Ridge Route	Laguna Hills	CA	92653	Michael T Doyle	714-770-3246	R	1•	<0.1
Industrial Data Entry	2362 E Lake Rd	Skaneateles	NY	13152	John Hattersley	315-685-8311	R	1	<0.1
NetData Solutions Inc	3390 W 11th Av	Eugene	OR	97440	John Braymer	503-683-2110	R	1•	<0.1
AccuData Inc	3007 SW Temple	Salt Lake City	UT	84115	C Burton Pugh	801-485-7400	R	1	<0.1
TriAm Inc	5445 Oceanus Dr	Huntington Bch	CA	92649	SH Jeong	714-890-5332	R	1•	<0.1
EASI Computer Systems Inc	3405 Army St	San Francisco	CA	94110	L Louis Chu	415-285-4096	R	0	<0.1
Safe Technologies Corp	1950 NE 208 Ter	N Miami Beach	FL	33179	George Lechter	305-933-2026	R	0	<0.1

Source: Ward's Business Directory of U.S. Private and Public Companies, Volumes 1 and 2, 1996. The company type code used is as follows: P - Public, R - Private, S - Subsidiary, D - Division, J - Joint Venture, A - Affiliate, G - Group. Sales are in millions of dollars, employees are in thousands. An asterisk (•) indicates an estimated sales volume. The symbol < stands for 'less than'. Company names and addresses are truncated, in some cases, to fit into the available space.

MATERIALS CONSUMED

Material	Quantity	Delivered Cost ($ million)
Materials, ingredients, containers, and supplies	(X)	1,233.3
Printed circuit boards (without inserted components) for electronic circuitry	(X)	22.2
Printed peripheral controllers (graphic boards, drive controllers, etc.) for electronic circuitry	(X)	3.3
Capacitors for electronic circuitry	(X)	5.7
Resistors for electronic circuitry	(X)	4.2
Battery packs for electronic circuitry	(X)	(D)
Other components and accessories for electric circuitry	(X)	337.2
Electrical transmission, distribution, and control equipment	(X)	(D)
Steel, aluminum, and other metal electronic enclosurse	(X)	6.7
Plastics electronic enclosures	(X)	26.4
Sheet metal products (including stampings), except enclosures	(X)	(D)
All other fabricated metal products (except forgings)	(X)	(D)
Castings (rough and semifinished)	(X)	(D)
Metal shapes and forms, except castings, forgings, and fabricated metal products	(X)	(D)
Insulated copper wire and cable (including magnet wire)	(X)	2.1
Fabricated plastics products, except enclosures	(X)	2.0
Purchased software	(X)	(D)
Appliance outlets, switches, lampholders, and other current-carrying wiring devices	(X)	(D)
Electric motors and generators (all types)	(X)	(D)
Paper and paperboard products including paperboard boxes, containers, and corrugated paperboard	(X)	(D)
Purchased computers	(X)	133.0
All other materials and components, parts, containers, and supplies	(X)	222.8
Materials, ingredients, containers, and supplies, nsk	(X)	410.9

Source: 1992 *Economic Census*. Explanation of symbols used: (D): Withheld to avoid disclosure of competitive data; na: Not available; (S): Withheld because statistical norms were not met; (X): Not applicable; (Z): Less than half the unit shown; nec: Not elsewhere classified; nsk: Not specified by kind; - : zero; * : 10-19 percent estimated; ** : 20-29 percent estimated.

PRODUCT SHARE DETAILS

Product or Product Class	% Share	Product or Product Class	% Share
Computer terminals	100.00	Parts, attachments, and accessories for computer terminals (excluding point-of-sale and funds-transfer devices)	8.50
Computer terminals (excluding point-of-sale and funds-transfer devices, and parts, attachments, and accessories)	78.22	Computer terminals, nsk	13.28

Source: 1992 *Economic Census*. The values shown are percent of total shipments in an industry. Values of indented subcategories are summed in the main heading. The symbol (D) appears when data are withheld to prevent disclosure of competitive information. The abbreviation nsk stands for 'not specified by kind' and nec for 'not elsewhere classified'.

INPUTS AND OUTPUTS FOR ELECTRONIC COMPUTING EQUIPMENT

Economic Sector or Industry Providing Inputs	%	Sector	Economic Sector or Industry Buying Outputs	%	Sector
Electronic computing equipment	29.4	Manufg.	Gross private fixed investment	45.6	Cap Inv
Wholesale trade	11.9	Trade	Exports	21.6	Foreign
Imports	10.2	Foreign	Electronic computing equipment	17.8	Manufg.
Electronic components nec	9.0	Manufg.	Federal Government purchases, national defense	4.6	Fed Govt
Semiconductors & related devices	5.0	Manufg.	Federal Government purchases, nondefense	1.6	Fed Govt
Industrial controls	2.8	Manufg.	Computer & data processing services	1.4	Services
Miscellaneous plastics products	2.7	Manufg.	Change in business inventories	1.3	In House
Air transportation	2.1	Util.	Personal consumption expenditures	1.0	
Banking	1.6	Fin/R.E.	Radio & TV communication equipment	0.8	Manufg.
Hotels & lodging places	1.5	Services	Typewriters & office machines, nec	0.5	Manufg.
Communications, except radio & TV	1.4	Util.	S/L Govt. purch., elem. & secondary education	0.4	S/L Govt
Electric services (utilities)	1.3	Util.	S/L Govt. purch., higher education	0.4	S/L Govt
Sheet metal work	1.2	Manufg.	S/L Govt. purch., other general government	0.4	S/L Govt
Eating & drinking places	1.2	Trade	Calculating & accounting machines	0.3	Manufg.
Real estate	1.2	Fin/R.E.	Banking	0.3	Fin/R.E.
Noncomparable imports	1.1	Foreign	Accounting, auditing & bookkeeping	0.3	Services
Petroleum refining	1.0	Manufg.	Electronic components nec	0.2	Manufg.
Paperboard containers & boxes	0.9	Manufg.	Instruments to measure electricity	0.2	Manufg.
Motors & generators	0.8	Manufg.	Insurance carriers	0.2	Fin/R.E.
Switchgear & switchboard apparatus	0.8	Manufg.	Mechanical measuring devices	0.1	Manufg.
Legal services	0.7	Services	Retail trade, except eating & drinking	0.1	Trade
Maintenance of nonfarm buildings nec	0.5	Constr.	Wholesale trade	0.1	Trade
Gaskets, packing & sealing devices	0.5	Manufg.			
Metal stampings, nec	0.5	Manufg.			
Wiring devices	0.5	Manufg.			
Management & consulting services & labs	0.5	Services			
Aluminum castings	0.4	Manufg.			
Electron tubes	0.4	Manufg.			
Screw machine and related products	0.4	Manufg.			
Advertising	0.4	Services			
Equipment rental & leasing services	0.4	Services			
Aluminum rolling & drawing	0.3	Manufg.			

Continued on next page.

INPUTS AND OUTPUTS FOR ELECTRONIC COMPUTING EQUIPMENT - Continued

Economic Sector or Industry Providing Inputs	%	Sector	Economic Sector or Industry Buying Outputs	%	Sector
Blast furnaces & steel mills	0.3	Manufg.			
Fabricated metal products, nec	0.3	Manufg.			
Iron & steel foundries	0.3	Manufg.			
Miscellaneous fabricated wire products	0.3	Manufg.			
Nonferrous wire drawing & insulating	0.3	Manufg.			
Plating & polishing	0.3	Manufg.			
Motor freight transportation & warehousing	0.3	Util.			
Accounting, auditing & bookkeeping	0.3	Services			
Computer & data processing services	0.3	Services			
Electrical industrial apparatus, nec	0.2	Manufg.			
Machinery, except electrical, nec	0.2	Manufg.			
Manifold business forms	0.2	Manufg.			
Transformers	0.2	Manufg.			
Insurance carriers	0.2	Fin/R.E.			
Royalties	0.2	Fin/R.E.			
Security & commodity brokers	0.2	Fin/R.E.			
Automotive rental & leasing, without drivers	0.2	Services			
Ball & roller bearings	0.1	Manufg.			
Die-cut paper & board	0.1	Manufg.			
Nonferrous rolling & drawing, nec	0.1	Manufg.			
Gas production & distribution (utilities)	0.1	Util.			
Automotive repair shops & services	0.1	Services			
Photofinishing labs, commercial photography	0.1	Services			
U.S. Postal Service	0.1	Gov't			

Source: Benchmark Input-Output Accounts for the U.S. Economy, 1982, U.S. Department of Commerce, Washington, D.C., July 1991. Data, as reported in the source, are organized by the 1977 SIC structure in use in 1982 but have been matched, as closely as is possible, to the 1987 SIC structure used in this book.

OCCUPATIONS EMPLOYED BY SIC 357 - COMPUTER AND OFFICE EQUIPMENT

Occupation	% of Total 1994	Change to 2005	Occupation	% of Total 1994	Change to 2005
Computer engineers	8.4	4.1	Engineering technicians & technologists nec	1.8	-29.7
Electrical & electronics engineers	5.4	-10.2	General managers & top executives	1.7	-33.3
Electrical & electronic technicians,technologists	4.9	-29.7	General office clerks	1.6	-40.1
Electrical & electronic equipment assemblers	4.8	-36.8	Purchasing agents, ex trade & farm products	1.4	5.4
Electrical & electronic assemblers	4.4	-15.7	Engineers nec	1.4	-15.7
Sales & related workers nec	3.9	-29.7	Industrial production managers	1.4	-36.8
Computer programmers	3.9	-43.1	Professional workers nec	1.4	-15.7
Engineering, mathematical, & science managers	3.5	-20.2	Industrial engineers, ex safety engineers	1.3	-38.2
Secretaries, ex legal & medical	3.3	-36.0	Electromechanical equipment assemblers	1.3	-86.0
Inspectors, testers, & graders, precision	2.7	-29.7	Accountants & auditors	1.3	-29.7
Production, planning, & expediting clerks	2.6	-29.7	Traffic, shipping, & receiving clerks	1.2	-32.4
Management support workers nec	2.4	-29.7	Mechanical engineers	1.2	-22.7
Systems analysts	2.2	12.4	Financial managers	1.1	-29.7
Marketing, advertising, & PR managers	2.0	-29.7	Bookkeeping, accounting, & auditing clerks	1.1	-47.3
Assemblers, fabricators, & hand workers nec	2.0	-29.7	Stock clerks	1.1	-42.9
Managers & administrators nec	1.9	-29.8	Writers & editors, incl technical writers	1.1	5.4
Blue collar worker supervisors	1.8	-32.6	Clerical supervisors & managers	1.0	-28.1

Source: Industry-Occupation Matrix, Bureau of Labor Statistics. These data relate to one or more 3-digit SIC industry groups rather than to a single 4-digit SIC. The change reported for each occupation to the year 2005 is a percent of growth or decline as estimated by the Bureau of Labor Statistics. The abbreviation nec stands for 'not elsewhere classified'.

LOCATION BY STATE AND REGIONAL CONCENTRATION

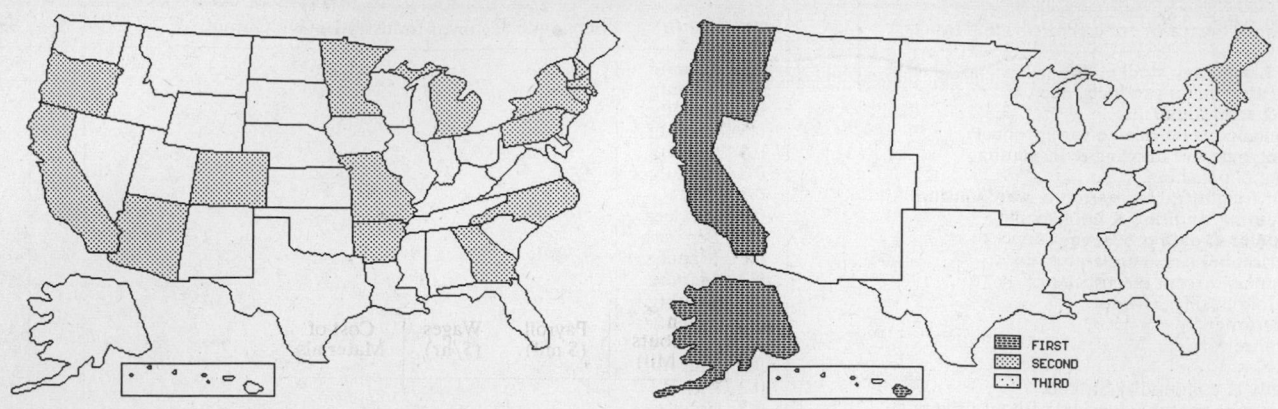

FIRST
SECOND
THIRD

INDUSTRY DATA BY STATE

State	Establish-ments	Shipments			Employment				Cost as % of Shipments	Investment per Employee ($)
		Total ($ mil)	% of U.S.	Per Establ.	Total Number	% of U.S.	Per Establ.	Wages ($/hour)		
California	61	596.4	28.8	9.8	3,100	33.3	51	12.00	56.4	3,387
New York	11	144.7	7.0	13.2	900	9.7	82	16.14	43.7	-
Massachusetts	15	53.0	2.6	3.5	500	5.4	33	14.33	71.7	2,000
Pennsylvania	7	32.3	1.6	4.6	300	3.2	43	13.00	48.6	1,667
Missouri	5	19.5	0.9	3.9	100	1.1	20	7.00	63.1	2,000
Georgia	6	(D)	-	-	750 *	8.1	125	-	-	-
Michigan	5	(D)	-	-	175 *	1.9	35	-	-	-
Minnesota	5	(D)	-	-	175 *	1.9	35	-	-	571
New Hampshire	5	(D)	-	-	750 *	8.1	150	-	-	-
Oregon	5	(D)	-	-	175 *	1.9	35	-	-	-
Connecticut	4	(D)	-	-	175 *	1.9	44	-	-	-
Colorado	3	(D)	-	-	175 *	1.9	58	-	-	-
North Carolina	3	(D)	-	-	750 *	8.1	250	-	-	-
Arizona	2	(D)	-	-	750 *	8.1	375	-	-	-
Arkansas	2	(D)	-	-	750 *	8.1	375	-	-	-

Source: 1992 *Economic Census*. The states are in descending order of shipments or establishments (if shipment data are missing for the majority). The symbol (D) appears when data are withheld to prevent disclosure of competitive information. States marked with (D) are sorted by number of establishments. A dash (-) indicates that the data element cannot be calculated; * indicates the midpoint of a range.

3577 - COMPUTER PERIPHERAL EQUIPMENT, NEC

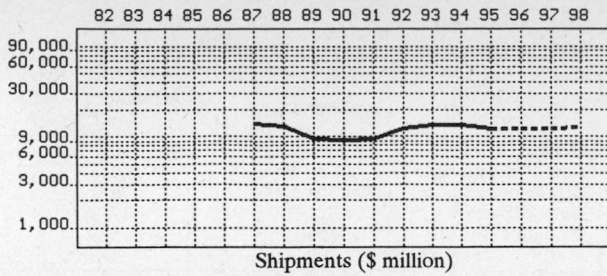

Shipments ($ million)

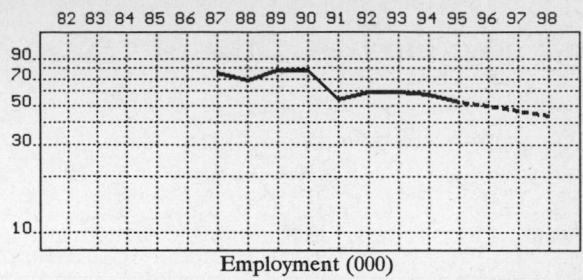

Employment (000)

GENERAL STATISTICS

| Year | Com-panies | Establishments | | Employment | | | Compensation | | Production ($ million) | | | |
		Total	with 20 or more employees	Total (000)	Production Workers (000)	Hours (Mil)	Payroll ($ mil)	Wages ($/hr)	Cost of Materials	Value Added by Manufacture	Value of Shipments	Capital Invest.
1982												
1983												
1984												
1985												
1986												
1987	520	549	252	76.2	26.2	58.7	2,625.4	9.87	7,107.3	6,918.1	13,965.5	391.2
1988		663	289	67.8	24.0	46.8	2,330.6	11.00	7,761.3	5,928.0	13,213.3	461.5
1989		616	280	77.2	24.2	46.8	1,992.2	11.22	4,909.0	4,726.6	9,625.2	341.6
1990		597	267	78.1	25.4	49.9	1,857.5	10.75	5,239.3	3,923.0	9,146.3	299.1
1991		651	285	53.5	20.7	42.3	1,963.9		5,779.9	3,748.4	9,614.7	223.8
1992	748	772	275	59.3	23.5	48.4	2,175.7	11.00	7,166.4	5,034.8	12,156.5	393.0
1993		786	272	59.3	24.0	49.5	2,260.4	11.31	8,184.2	5,182.6	13,366.6	415.1
1994		800P	280P	57.6	23.1	48.4	2,383.8	12.91	8,293.2	5,422.4	13,665.9	410.8
1995		834P	281P	52.7P	22.4P	45.6P	2,124.5P		7,406.6P	4,842.7P	12,204.9P	366.2P
1996		869P	282P	49.7P	22.0P	44.8P	2,108.0P		7,455.2P	4,874.5P	12,285.0P	366.0P
1997		903P	284P	46.7P	21.7P	44.1P	2,091.5P		7,503.8P	4,906.3P	12,365.1P	365.8P
1998		937P	285P	43.7P	21.3P	43.4P	2,075.0P		7,552.5P	4,938.1P	12,445.3P	365.6P

Sources: 1982, 1987, 1992 *Economic Census*; *Annual Survey of Manufactures*, 83-86, 88-91, 93-94. Establishment counts for non-Census years are from *County Business Patterns*; establishment values for 83-84 are extrapolations. 'P's show projections by the editors. Industries reclassified in 87 will not have data for prior years.

INDICES OF CHANGE

| Year | Com-panies | Establishments | | Employment | | | Compensation | | Production ($ million) | | | |
		Total	with 20 or more employees	Total (000)	Production Workers (000)	Hours (Mil)	Payroll ($ mil)	Wages ($/hr)	Cost of Materials	Value Added by Manufacture	Value of Shipments	Capital Invest.
1982												
1983												
1984												
1985												
1986												
1987	70	71	92	128	111	121	121	90	99	137	115	100
1988		86	105	114	102	97	107	100	108	118	109	117
1989		80	102	130	103	97	92	102	69	94	79	87
1990		77	97	132	108	103	85	98	73	78	75	76
1991		84	104	90	88	87	90		81	74	79	57
1992	100	100	100	100	100	100	100	100	100	100	100	100
1993		102	99	100	102	102	104	103	114	103	110	106
1994		104P	102P	97	98	100	110	117	116	108	112	105
1995		108P	102P	89P	95P	94P	98P		103P	96P	100P	93P
1996		113P	103P	84P	94P	93P	97P		104P	97P	101P	93P
1997		117P	103P	79P	92P	91P	96P		105P	97P	102P	93P
1998		121P	104P	74P	91P	90P	95P		105P	98P	102P	93P

Sources: Same as General Statistics. Values reflect change from the base year, 1992. Values above 100 mean greater than 92, values below 100 mean less than 92, and a value of 100 in the 82-91 or 93-98 period means same as 92. 'P's mark projections by the editors.

SELECTED RATIOS

For 1994	Avg. of All Manufact.	Analyzed Industry	Index	For 1994	Avg. of All Manufact.	Analyzed Industry	Index
Employees per Establishment	49	72	147	Value Added per Production Worker	134,084	234,736	175
Payroll per Establishment	1,500,273	2,980,815	199	Cost per Establishment	5,045,178	10,370,204	206
Payroll per Employee	30,620	41,385	135	Cost per Employee	102,970	143,979	140
Production Workers per Establishment	34	29	84	Cost per Production Worker	146,988	359,013	244
Wages per Establishment	853,319	781,334	92	Shipments per Establishment	9,576,895	17,088,478	178
Wages per Production Worker	24,861	27,050	109	Shipments per Employee	195,460	237,255	121
Hours per Production Worker	2,056	2,095	102	Shipments per Production Worker	279,017	591,597	212
Wages per Hour	12.09	12.91	107	Investment per Establishment	321,011	513,683	160
Value Added per Establishment	4,602,255	6,780,422	147	Investment per Employee	6,552	7,132	109
Value Added per Employee	93,930	94,139	100	Investment per Production Worker	9,352	17,784	190

Sources: Same as General Statistics. The 'Average of All Manufacturing' column represents the average of all manufacturing industries reported for the most recent complete year available. The Index shows the relationship between the Average and the Analyzed Industry. For example, 100 means that they are equal; 500 that the Analyzed Industry is five times the average; 50 means that the Analyzed Industry is half the national average. The abbreviation 'na' is used to show that data are 'not available'.

LEADING COMPANIES   Number shown: 75   Total sales ($ mil): 19,459   Total employment (000): 72.5

Company Name	Address				CEO Name	Phone	Co. Type	Sales ($ mil)	Empl. (000)
Canon USA Inc	1 Canon Plz	Lake Success	NY	11042	Hajime Mitarai	516-488-6700	S	6,000	8.2
Cisco Systems Inc	170 W Tasman Dr	San Jose	CA	95134	John P Morgridge	408-526-4000	P	1,243	2.4
3Com Corp	5400 Bayfront Plz	Santa Clara	CA	95052	Eric A Benhamou	408-764-5000	P	827	2.3
CalComp Inc	2411 W La Palma	Anaheim	CA	92801	Gary R Long	714-821-2000	S	750*	1.1
Stratus Computer Inc	55 Fairbanks Blv	Marlboro	MA	01752	William E Foster	508-460-2000	P	576	2.9
NMB	9730 Independence	Chatsworth	CA	91311	Marty Yamanaka	818-709-1770	S	500	2.8
AM International Inc	9399 W Higgins Rd	Rosemont	IL	60018	Jerome D Brady	708-818-1294	P	422	3.4
Okidata Group	532 Fellowship Rd	Mount Laurel	NJ	08054	Dennis P Flannagan	609-235-2600	D	400	0.5
Adaptec Inc	691 S Milpitas Blv	Milpitas	CA	95035	John G Adler	408-945-8600	P	372	1.8
National Computer Systems Inc	11000 Prairie Lks	Eden Prairie	MN	55344	Russell A Gullotti	612-829-3000	P	306	2.6
Sato America Inc	545 Weddell Dr	Sunnyvale	CA	94089	T Fujita	408-745-1300	S	300	1.2
QMS Inc	1 Magnum Pass	Mobile	AL	36618	James L Busby	205-633-4300	P	293	1.1
Logitech Inc	6505 Kaiser Dr	Fremont	CA	94555	Pierlugi Zappacosta	510-795-8500	S	292	2.1
Aztech Labs Inc	47811 Warm Spgs	Fremont	CA	94539	Michael Mun	510-623-8988	P	280	0.8
BancTec Inc	4435 Spring Val Rd	Dallas	TX	75244	G N Clark Jr	214-450-7700	P	247	2.4
Hutchinson Technology Inc	40 W Highland Park	Hutchinson	MN	55350	Jeffrey W Green	612-587-3797	P	239	4.6
GENICOM Corp	14800 Conference	Chantilly	VA	22021	Paul T Winn	703-802-9200	P	234	2.6
XES Inc	5853 Rue Ferrari	San Jose	CA	95138	Wilbur Pittman	408-225-2800	S	227	1.3
TeleSec	1 Corporate Dr	Clearwater	FL	34622	James C Garrett	813-573-0330	D	226	1.7
Recognition International Inc	PO Box 660204	Dallas	TX	75266	Robert A Vanourek	214-579-6000	P	219	1.5
Monarch Marking Systems Inc	PO Box 608	Dayton	OH	45401	Dakiel B Teich	513-865-2123	S	200	1.9
FileNet Corp	3565 Harbor Blv	Costa Mesa	CA	92626	Theodore J Smith	714-966-3400	P	180	0.9
Videojet Systems International	1500 Mittel Blv	Wood Dale	IL	60191	Henry J Bode	708-860-7300	S	170*	1.1
LANpoint Systems	6550 S Bay Colony	Tucson	AZ	85706	Thomas Brown	602-741-4209	D	163	1.4
Key Tronic Corp	PO Box 14687	Spokane	WA	99214	Stanley Hiller Jr	509-928-8000	P	159	2.2
Chipcom Corp	118 Turnpike Rd	Southborough	MA	01772	J Robert Held	508-460-8900	P	150	0.6
Hayes Microcomputer Products	PO Box 105203	Atlanta	GA	30310	Dennis C Hayes	404-840-9200	R	150*	1.0
NMB Technologies Inc	9730 Independence	Chatsworth	CA	91311	Myron Jones	818-341-3355	S	150	0.1
Best Power Technology Inc	PO Box 280	Necedah	WI	54646	William L Paul	608-565-7200	P	149	1.1
Pyxis Corp	9380 Carroll Pk Dr	San Diego	CA	92121	Gerald E Forth	619-625-3300	P	142	0.6
Xircom Inc	2300 Corporate Ctr	Thousand Oaks	CA	91320	Dirk I Gates	805-376-9300	P	132	0.3
Digi International Inc	6400 Flying Cloud	Eden Prairie	MN	55344	Ervin F Kamm Jr	612-943-9020	P	131	0.4
Dataproducts Corp	6219 De Soto Av	Woodland Hills	CA	91367	Irvin Maloney	818-887-8000	S	130	0.9
IDEXX Corp	1716 Orange Av	Costa Mesa	CA	92627	David E Shaw	714-548-7574	P	126	0.5
Allied Telesis Inc	575 E Middlefield	Mountain View	CA	94043	Tony Russo	415-964-2771	R	120	0.3
Olicom USA Inc	900 E Park Blv	Plano	TX	75074	Max Jensen	214-423-7560	P	114	0.2
Primax Electronics	254 E Hacienda Av	Campbell	CA	95008	Ray Sun	408-364-2800	S	110	1.3
Decision Data	10230 W 70th St	Eden Prairie	MN	55344	Thomas S Bednarik	612-828-0400	S	107	0.2
LaserMaster Corp	7156 Shady Oak Rd	Eden Prairie	MN	55344	Robert J Wenzel	612-941-8687	S	106	0.5
Microdyne Corp	3601 Eisenhower	Alexandria	VA	22304	P T Cunningham	703-739-0500	P	101	0.5
CTX International Inc	20530 Earlgate St	Walnut	CA	91789	YC Liu	909-595-6146	S	100	<0.1
Marubeni Intern Electr Corp	20 William St	Wellesley	MA	02181	Mitch Noda	617-237-2115	S	100	<0.1
Siemens Nixdorf Printing Syst	5500 Broken Sound	Boca Raton	FL	33487	H Werner Krause	407-997-3100	J	100*	0.7
TM Digital Solutions Inc	11205 Knott Av	Cypress	CA	90630	Craig McHugh	714-373-9989	S	96	0.6
Catalina Marketing Corp	11300 9th St N	St Petersburg	FL	33716	Tommy D Greer	813-579-5000	P	91	0.4
Bourns Networks Inc	1400 N 1000 West St	Logan	UT	84321	Mike Ehman	801-750-7200	S	89	0.6
American Megatrends Inc	6145-F Northbelt	Norcross	GA	30071	S Shankar	404-263-8181	R	87	0.2
Optical Data Systems Inc	1101 E Arapaho Rd	Richardson	TX	75081	G Ward Paxton	214-234-6400	P	87	0.3
MicroNet Technology Inc	80 Technology	Irvine	CA	92718	Dennis Bradshaw	714-453-6000	R	86	0.1
DH Print	860 College View	Riverton	WY	82501	Bernard Masson	307-856-4821	D	85	0.3
DH Technology Inc	15070 Av of Science	San Diego	CA	92128	William H Gibbs	619-451-3485	P	85	0.3
Auspex Systems Inc	5200 Great America	Santa Clara	CA	95054	L B Boucher	408-986-2000	P	83	0.3
NAI Technologies Inc	60 Plant Av	Hauppauge	NY	11788	Robert A Carlson	516-582-6500	P	81	0.4
Asante Technologies Inc	821 Fox Ln	San Jose	CA	95131	Jeff Lin	408-435-8388	P	80	0.2
Xyplex Inc	330 Codman Hill	Boxborough	MA	01719	Peter J Nesbeda	508-264-9900	P	80	0.4
Computer Network Techn Corp	605 N Hwy 169	Minneapolis	MN	55441	CM Lewis	612-797-6000	P	80	0.3
RasterOps Corp	2500 Walsh Av	Santa Clara	CA	95051	Keith E Sorenson	408-562-4200	P	79	0.2
Telematics International Inc	1201 W Cypress	Ft Lauderdale	FL	33309	W A Hightower	305-772-3070	S	78	0.4
Tricord Systems Inc	3750 Annapolis Ln	Plymouth	MN	55447	James D Edwards	612-557-9005	P	77	0.2
Telebit Corp	1315 Chesap	Sunnyvale	CA	94089	James D Norrod	408-734-4333	P	76	0.3
Miltope Group Inc	500 Richardson S	Hope Hull	AL	36043	George K Webster	334-284-8665	P	76	0.3
Mextel Inc	159 Beeline Dr	Bensenville	IL	60106	Vedran Skulic	708-595-4146	R	71	<0.1
Cornerstone Imaging Inc	1710 Fortune Dr	San Jose	CA	95131	T T van Overbeek	408-435-8900	P	70	0.2
Jaton Corp	556 S Milpitas Blv	Milpitas	CA	95035	JS Hong	408-942-9888	P	70	0.1
STB Systems Inc	1651 N Glenville	Richardson	TX	75081	Bill Ogle	214-234-8750	R	70	0.1
Trident Microsystems Inc	189 N Bernardo	Mountain View	CA	94043	Frank C Lin	415-691-9211	P	69	0.1
Number Nine Visual	18 Hartwell Av	Lexington	MA	02173	Andrew Najda	617-674-0009	P	66	0.1
Iris Graphics Inc	6 Crosby Dr	Bedford	MA	01730	Alphonse Lucchese	617-275-8777	S	65	0.3
Boca Research Inc	6413 Congress Av	Boca Raton	FL	33487	Timothy Farris	407-997-6227	P	65	0.3
Summagraphics Corp	8500 Cameron Rd	Austin	TX	78754	Michael S Bennett	512-835-0900	P	65	0.3
Madge Networks Inc	2310 N 1st St	San Jose	CA	95131	Robert Madge	408-955-0700	S	63	0.1
Mylex Corp	34551 Ardenwood	Fremont	CA	94555	Albert E Montross	510-796-6100	P	63	0.1
Sony Electronics Inc	655 River Oaks	San Jose	CA	95134	Ted Matsumoto	408-432-1600	D	62*	0.4
Data I/O Corp	PO Box 97046	Redmond	WA	98073	William C Erxleben	206-881-6444	P	62	0.4
Buffalo Inc	2805 19th St SE	Salem	OR	97302	Richard J Stanczak	503-585-3414	S	60	0.1

Source: Ward's Business Directory of U.S. Private and Public Companies, Volumes 1 and 2, 1996. The company type code used is as follows: P - Public, R - Private, S - Subsidiary, D - Division, J - Joint Venture, A - Affiliate, G - Group. Sales are in millions of dollars, employees are in thousands. An asterisk (*) indicates an estimated sales volume. The symbol < stands for 'less than'. Company names and addresses are truncated, in some cases, to fit into the available space.

MATERIALS CONSUMED

Material	Quantity	Delivered Cost ($ million)
Materials, ingredients, containers, and supplies	(X)	6,377.5
Cathode ray tubes (CRTs)	(X)	14.9
Printed circuit boards (without inserted components) for electronic circuitry	(X)	137.7
Printed peripheral controllers (graphic boards, drive controllers, etc.) for electronic circuitry	(X)	67.5
Printed computer processors (system boards, array processors, etc.) for electronic circuitry	(X)	144.8
Printed communication boards (LAN, D/A, A/D converters, etc.) with fax capability	(X)	39.5
Other printed circuit boards (loaded boards, subassemblies, and modules) for electronic circuitry	(X)	300.9
Semiconductors, including transistors, diodes, rectifiers, and integrated circuits for electronic circuitry	(X)	526.1
Capacitors for electronic circuitry	(X)	36.2
Resistors for electronic circuitry	(X)	17.4
Connectors for electronic circuitry	(X)	65.2
Battery packs for electronic circuitry	(X)	37.8
Other power supply units for electronic circuitry	(X)	87.3
Other components and accessories for electric circuitry	(X)	244.4
Electrical transmission, distribution, and control equipment	(X)	52.1
Steel, aluminum, and other metal electronic enclousrse	(X)	33.2
Plastics electronic enclosures	(X)	128.4
Sheet metal products (including stampings), except enclosures	(X)	(D)
All other fabricated metal products (except forgings)	(X)	100.9
Forgings	(X)	(D)
Castings (rough and semifinished)	(X)	22.9
Metal shapes and forms, except castings, forgings, and fabricated metal products	(X)	47.7
Insulated copper wire and cable (including magnet wire)	(X)	32.3
Fabricated plastics products, except enclosures	(X)	152.0
Purchased software	(X)	29.8
Appliance outlets, switches, lampholders, and other current-carrying wiring devices	(X)	18.5
Electric motors and generators (all types)	(X)	(D)
Paper and paperboard products including paperboard boxes, containers, and corrugated paperboard	(X)	47.3
Purchased computers	(X)	9.7
Purchased peripheral input devices, including keyboards, mouse devices, trackballs, etc.	(X)	171.0
Purchased peripheral printers	(X)	253.2
Other purchased electronic computing and peripheral equipment	(X)	415.4
All other materials and components, parts, containers, and supplies	(X)	1,264.3
Materials, ingredients, containers, and supplies, nsk	(X)	945.1

Source: 1992 Economic Census. Explanation of symbols used: (D): Withheld to avoid disclosure of competitive data; na: Not available; (S): Withheld because statistical norms were not met; (X): Not applicable; (Z): Less than half the unit shown; nec: Not elsewhere classified; nsk: Not specified by kind; - : zero; * : 10-19 percent estimated; ** : 20-29 percent estimated.

PRODUCT SHARE DETAILS

Product or Product Class	% Share	Product or Product Class	% Share
Computer peripheral equipment, nec	100.00	Parts, attachments, and accessories for computer peripheral (input/output) equipment, nec	24.51
Computer peripheral (input/output) equipment, nec, except parts, attachments, and accessories	67.93	Computer peripheral equipment, nec, nsk	7.56

Source: 1992 Economic Census. The values shown are percent of total shipments in an industry. Values of indented subcategories are summed in the main heading. The symbol (D) appears when data are withheld to prevent disclosure of competitive information. The abbreviation nsk stands for 'not specified by kind' and nec for 'not elsewhere classified'.

INPUTS AND OUTPUTS FOR ELECTRONIC COMPUTING EQUIPMENT

Economic Sector or Industry Providing Inputs	%	Sector	Economic Sector or Industry Buying Outputs	%	Sector
Electronic computing equipment	29.4	Manufg.	Gross private fixed investment	45.6	Cap Inv
Wholesale trade	11.9	Trade	Exports	21.6	Foreign
Imports	10.2	Foreign	Electronic computing equipment	17.8	Manufg.
Electronic components nec	9.0	Manufg.	Federal Government purchases, national defense	4.6	Fed Govt
Semiconductors & related devices	5.0	Manufg.	Federal Government purchases, nondefense	1.6	Fed Govt
Industrial controls	2.8	Manufg.	Computer & data processing services	1.4	Services
Miscellaneous plastics products	2.7	Manufg.	Change in business inventories	1.3	In House
Air transportation	2.1	Util.	Personal consumption expenditures	1.0	
Banking	1.6	Fin/R.E.	Radio & TV communication equipment	0.8	Manufg.
Hotels & lodging places	1.5	Services	Typewriters & office machines, nec	0.5	Manufg.
Communications, except radio & TV	1.4	Util.	S/L Govt. purch., elem. & secondary education	0.4	S/L Govt
Electric services (utilities)	1.3	Util.	S/L Govt. purch., higher education	0.4	S/L Govt
Sheet metal work	1.2	Manufg.	S/L Govt. purch., other general government	0.4	S/L Govt
Eating & drinking places	1.2	Trade	Calculating & accounting machines	0.3	Manufg.
Real estate	1.2	Fin/R.E.	Banking	0.3	Fin/R.E.
Noncomparable imports	1.1	Foreign	Accounting, auditing & bookkeeping	0.3	Services
Petroleum refining	1.0	Manufg.	Electronic components nec	0.2	Manufg.
Paperboard containers & boxes	0.9	Manufg.	Instruments to measure electricity	0.2	Manufg.
Motors & generators	0.8	Manufg.	Insurance carriers	0.2	Fin/R.E.
Switchgear & switchboard apparatus	0.8	Manufg.	Mechanical measuring devices	0.1	Manufg.
Legal services	0.7	Services	Retail trade, except eating & drinking	0.1	Trade

Continued on next page.

INPUTS AND OUTPUTS FOR ELECTRONIC COMPUTING EQUIPMENT - Continued

Economic Sector or Industry Providing Inputs	%	Sector	Economic Sector or Industry Buying Outputs	%	Sector
Maintenance of nonfarm buildings nec	0.5	Constr.	Wholesale trade	0.1	Trade
Gaskets, packing & sealing devices	0.5	Manufg.			
Metal stampings, nec	0.5	Manufg.			
Wiring devices	0.5	Manufg.			
Management & consulting services & labs	0.5	Services			
Aluminum castings	0.4	Manufg.			
Electron tubes	0.4	Manufg.			
Screw machine and related products	0.4	Manufg.			
Advertising	0.4	Services			
Equipment rental & leasing services	0.4	Services			
Aluminum rolling & drawing	0.3	Manufg.			
Blast furnaces & steel mills	0.3	Manufg.			
Fabricated metal products, nec	0.3	Manufg.			
Iron & steel foundries	0.3	Manufg.			
Miscellaneous fabricated wire products	0.3	Manufg.			
Nonferrous wire drawing & insulating	0.3	Manufg.			
Plating & polishing	0.3	Manufg.			
Motor freight transportation & warehousing	0.3	Util.			
Accounting, auditing & bookkeeping	0.3	Services			
Computer & data processing services	0.3	Services			
Electrical industrial apparatus, nec	0.2	Manufg.			
Machinery, except electrical, nec	0.2	Manufg.			
Manifold business forms	0.2	Manufg.			
Transformers	0.2	Manufg.			
Insurance carriers	0.2	Fin/R.E.			
Royalties	0.2	Fin/R.E.			
Security & commodity brokers	0.2	Fin/R.E.			
Automotive rental & leasing, without drivers	0.2	Services			
Ball & roller bearings	0.1	Manufg.			
Die-cut paper & board	0.1	Manufg.			
Nonferrous rolling & drawing, nec	0.1	Manufg.			
Gas production & distribution (utilities)	0.1	Util.			
Automotive repair shops & services	0.1	Services			
Photofinishing labs, commercial photography	0.1	Services			
U.S. Postal Service	0.1	Gov't			

Source: Benchmark Input-Output Accounts for the U.S. Economy, 1982, U.S. Department of Commerce, Washington, D.C., July 1991. Data, as reported in the source, are organized by the 1977 SIC structure in use in 1982 but have been matched, as closely as is possible, to the 1987 SIC structure used in this book.

OCCUPATIONS EMPLOYED BY SIC 357 - COMPUTER AND OFFICE EQUIPMENT

Occupation	% of Total 1994	Change to 2005	Occupation	% of Total 1994	Change to 2005
Computer engineers	8.4	4.1	Engineering technicians & technologists nec	1.8	-29.7
Electrical & electronics engineers	5.4	-10.2	General managers & top executives	1.7	-33.3
Electrical & electronic technicians,technologists	4.9	-29.7	General office clerks	1.6	-40.1
Electrical & electronic equipment assemblers	4.8	-36.8	Purchasing agents, ex trade & farm products	1.4	5.4
Electrical & electronic assemblers	4.4	-15.7	Engineers nec	1.4	-15.7
Sales & related workers nec	3.9	-29.7	Industrial production managers	1.4	-36.8
Computer programmers	3.9	-43.1	Professional workers nec	1.4	-15.7
Engineering, mathematical, & science managers	3.5	-20.2	Industrial engineers, ex safety engineers	1.3	-38.2
Secretaries, ex legal & medical	3.3	-36.0	Electromechanical equipment assemblers	1.3	-86.0
Inspectors, testers, & graders, precision	2.7	-29.7	Accountants & auditors	1.3	-29.7
Production, planning, & expediting clerks	2.6	-29.7	Traffic, shipping, & receiving clerks	1.2	-32.4
Management support workers nec	2.4	-29.7	Mechanical engineers	1.2	-22.7
Systems analysts	2.2	12.4	Financial managers	1.1	-29.7
Marketing, advertising, & PR managers	2.0	-29.7	Bookkeeping, accounting, & auditing clerks	1.1	-47.3
Assemblers, fabricators, & hand workers nec	2.0	-29.7	Stock clerks	1.1	-42.9
Managers & administrators nec	1.9	-29.8	Writers & editors, incl technical writers	1.1	5.4
Blue collar worker supervisors	1.8	-32.6	Clerical supervisors & managers	1.0	-28.1

Source: Industry-Occupation Matrix, Bureau of Labor Statistics. These data relate to one or more 3-digit SIC industry groups rather than to a single 4-digit SIC. The change reported for each occupation to the year 2005 is a percent of growth or decline as estimated by the Bureau of Labor Statistics. The abbreviation nec stands for 'not elsewhere classified'.

LOCATION BY STATE AND REGIONAL CONCENTRATION

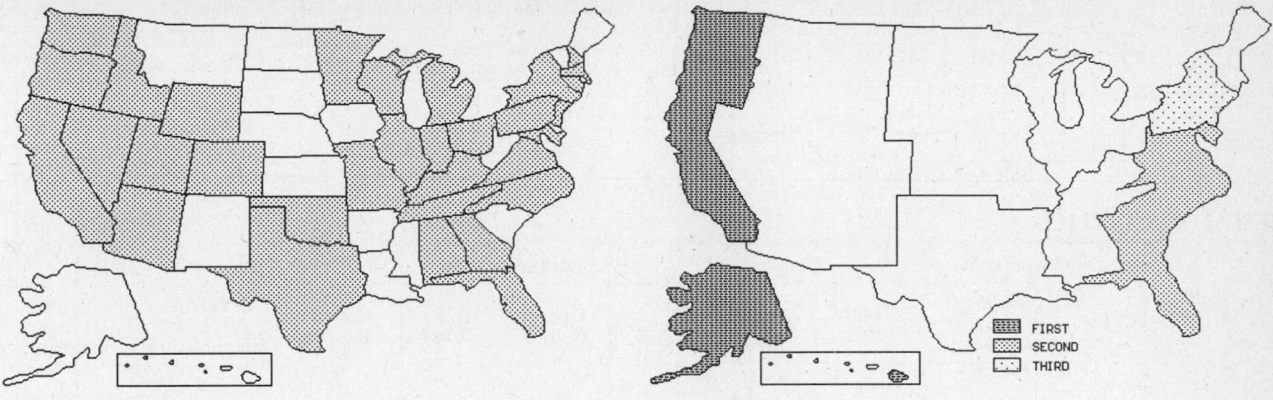

FIRST
SECOND
THIRD

INDUSTRY DATA BY STATE

State	Establish-ments	Shipments			Employment				Cost as % of Shipments	Investment per Employee ($)
		Total ($ mil)	% of U.S.	Per Establ.	Total Number	% of U.S.	Per Establ.	Wages ($/hour)		
California	248	2,842.5	23.4	11.5	13,500	22.8	54	13.59	56.3	5,889
Washington	33	801.0	6.6	24.3	4,500	7.6	136	9.79	55.1	5,289
Texas	43	568.2	4.7	13.2	3,100	5.2	72	6.38	74.5	3,839
Oregon	11	478.7	3.9	43.5	2,500	4.2	227	9.52	64.2	8,800
New York	42	475.8	3.9	11.3	3,300	5.6	79	9.04	28.9	5,576
Massachusetts	42	286.6	2.4	6.8	2,000	3.4	48	13.07	41.4	4,900
Colorado	24	221.0	1.8	9.2	1,100	1.9	46	13.50	51.9	3,909
Florida	32	186.4	1.5	5.8	2,000	3.4	63	9.22	51.8	1,850
Connecticut	14	186.2	1.5	13.3	1,100	1.9	79	14.13	31.0	5,636
Illinois	26	181.2	1.5	7.0	800	1.3	31	11.50	58.6	5,875
Ohio	16	116.9	1.0	7.3	800	1.3	50	10.75	36.3	1,750
Arizona	18	104.8	0.9	5.8	800	1.3	44	7.14	50.1	3,500
Pennsylvania	29	86.8	0.7	3.0	800	1.3	28	12.71	47.5	2,000
North Carolina	15	82.4	0.7	5.5	800	1.3	53	18.25	66.7	1,125
Michigan	11	57.9	0.5	5.3	500	0.8	45	8.25	53.2	3,800
Maryland	12	43.0	0.4	3.6	300	0.5	25	8.50	49.8	3,000
New Jersey	10	39.1	0.3	3.9	400	0.7	40	11.25	39.4	1,250
New Hampshire	11	30.5	0.3	2.8	200	0.3	18	8.67	63.6	-
Georgia	13	23.2	0.2	1.8	200	0.3	15	4.33	44.4	6,000
Oklahoma	4	19.1	0.2	4.8	200	0.3	50	4.50	57.6	2,500
Nevada	4	13.6	0.1	3.4	200	0.3	50	9.00	58.1	-
Indiana	7	13.1	0.1	1.9	100	0.2	14	5.50	66.4	1,000
Minnesota	25	(D)	-	-	7,500 *	12.6	300	-	-	-
Utah	14	(D)	-	-	175 *	0.3	13	-	-	3,429
Virginia	13	(D)	-	-	3,750 *	6.3	288	-	-	-
Wisconsin	9	(D)	-	-	175 *	0.3	19	-	-	-
Idaho	7	(D)	-	-	1,750 *	3.0	250	-	-	-
Missouri	7	(D)	-	-	175 *	0.3	25	-	-	-
Alabama	5	(D)	-	-	1,750 *	3.0	350	-	-	-
Tennessee	3	(D)	-	-	175 *	0.3	58	-	-	-
Kentucky	2	(D)	-	-	3,750 *	6.3	1,875	-	-	-
Wyoming	1	(D)	-	-	375 *	0.6	375	-	-	-

Source: 1992 *Economic Census*. The states are in descending order of shipments or establishments (if shipment data are missing for the majority). The symbol (D) appears when data are withheld to prevent disclosure of competitive information. States marked with (D) are sorted by number of establishments. A dash (-) indicates that the data element cannot be calculated; * indicates the midpoint of a range.

3578 - CALCULATING AND ACCOUNTING EQUIPMENT

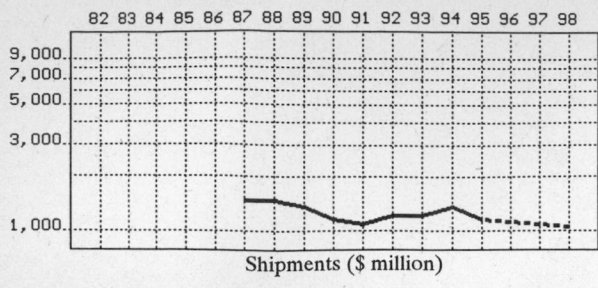

82 83 84 85 86 87 88 89 90 91 92 93 94 95 96 97 98

Shipments ($ million)

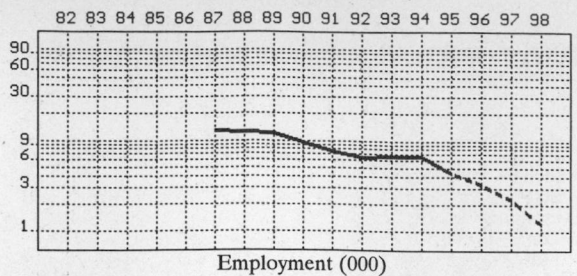

82 83 84 85 86 87 88 89 90 91 92 93 94 95 96 97 98

Employment (000)

GENERAL STATISTICS

| Year | Com-panies | Establishments | | Employment | | | Compensation | | Production ($ million) | | | |
		Total	with 20 or more employees	Total (000)	Production Workers (000)	Hours (Mil)	Payroll ($ mil)	Wages ($/hr)	Cost of Materials	Value Added by Manufacture	Value of Shipments	Capital Invest.
1982												
1983												
1984												
1985												
1986												
1987	93	98	53	12.8	6.0	12.4	380.7	11.51	560.6	937.3	1,486.8	40.2
1988		94	56	12.3	5.2	10.4	348.4	9.25	568.5	897.7	1,469.1	50.7
1989		87	55	11.8	3.8	7.7	332.3	9.99	535.6	798.2	1,352.1	46.0
1990		84	51	9.4	4.0	8.1	214.1	10.44	544.7	620.4	1,170.2	101.3
1991		84	48	7.6	3.6	7.3	211.9	9.44	482.4	617.5	1,086.6	15.7
1992	91	95	46	6.5	2.7	5.3	191.2	10.66	443.6	768.5	1,206.5	25.5
1993		100	48	6.6	3.1	6.0	198.4	11.80	458.7	772.6	1,229.4	29.9
1994		92P	45P	6.6	2.9	6.3	226.5	10.52	592.8	774.3	1,346.8	34.8
1995		93P	43P	4.4P	2.0P	4.0P	142.1P	10.82P	505.7P	660.5P	1,148.9P	27.5P
1996		93P	42P	3.3P	1.6P	3.2P	115.3P	10.90P	491.5P	642.0P	1,116.7P	24.1P
1997		93P	41P	2.3P	1.1P	2.3P	88.5P	10.98P	477.4P	623.6P	1,084.6P	20.7P
1998		93P	39P	1.2P	0.7P	1.4P	61.6P	11.06P	463.2P	605.1P	1,052.5P	17.2P

Sources: 1982, 1987, 1992 *Economic Census*; *Annual Survey of Manufactures*, 83-86, 88-91, 93-94. Establishment counts for non-Census years are from *County Business Patterns*; establishment values for 83-84 are extrapolations. 'P's show projections by the editors. Industries reclassified in 87 will not have data for prior years.

INDICES OF CHANGE

| Year | Com-panies | Establishments | | Employment | | | Compensation | | Production ($ million) | | | |
		Total	with 20 or more employees	Total (000)	Production Workers (000)	Hours (Mil)	Payroll ($ mil)	Wages ($/hr)	Cost of Materials	Value Added by Manufacture	Value of Shipments	Capital Invest.
1982												
1983												
1984												
1985												
1986												
1987	102	103	115	197	222	234	199	108	126	122	123	158
1988		99	122	189	193	196	182	87	128	117	122	199
1989		92	120	182	141	145	174	94	121	104	112	180
1990		88	111	145	148	153	112	98	123	81	97	397
1991		88	104	117	133	138	111	89	109	80	90	62
1992	100	100	100	100	100	100	100	100	100	100	100	100
1993		105	104	102	115	113	104	111	103	101	102	117
1994		97P	98P	102	107	119	118	99	134	101	112	136
1995		97P	95P	68P	74P	76P	74P	101P	114P	86P	95P	108P
1996		98P	91P	51P	58P	60P	60P	102P	111P	84P	93P	94P
1997		98P	88P	35P	42P	44P	46P	103P	108P	81P	90P	81P
1998		98P	85P	18P	26P	27P	32P	104P	104P	79P	87P	68P

Sources: Same as General Statistics. Values reflect change from the base year, 1992. Values above 100 mean greater than 92, values below 100 mean less than 92, and a value of 100 in the 82-91 or 93-98 period means same as 92. 'P's mark projections by the editors.

SELECTED RATIOS

For 1994	Avg. of All Manufact.	Analyzed Industry	Index	For 1994	Avg. of All Manufact.	Analyzed Industry	Index
Employees per Establishment	49	71	146	Value Added per Production Worker	134,084	267,000	199
Payroll per Establishment	1,500,273	2,450,541	163	Cost per Establishment	5,045,178	6,413,601	127
Payroll per Employee	30,620	34,318	112	Cost per Employee	102,970	89,818	87
Production Workers per Establishment	34	31	91	Cost per Production Worker	146,988	204,414	139
Wages per Establishment	853,319	717,051	84	Shipments per Establishment	9,576,895	14,571,252	152
Wages per Production Worker	24,861	22,854	92	Shipments per Employee	195,460	204,061	104
Hours per Production Worker	2,056	2,172	106	Shipments per Production Worker	279,017	464,414	166
Wages per Hour	12.09	10.52	87	Investment per Establishment	321,011	376,507	117
Value Added per Establishment	4,602,255	8,377,280	182	Investment per Employee	6,552	5,273	80
Value Added per Employee	93,930	117,318	125	Investment per Production Worker	9,352	12,000	128

Sources: Same as General Statistics. The 'Average of All Manufacturing' column represents the average of all manufacturing industries reported for the most recent complete year available. The Index shows the relationship between the Average and the Analyzed Industry. For example, 100 means that they are equal; 500 that the Analyzed Industry is five times the average; 50 means that the Analyzed Industry is half the national average. The abbreviation 'na' is used to show that data are 'not available'.

LEADING COMPANIES   Number shown: 29   Total sales ($ mil): 1,763   Total employment (000): 11.4

Company Name	Address				CEO Name	Phone	Co. Type	Sales ($ mil)	Empl. (000)
Diebold Inc	PO Box 8230	Canton	OH	44711	Robert W Mahoney	216-489-4000	P	760	4.7
Symbol Technologies Inc	116 Wilbur Pl	Bohemia	NY	11716	Jerome Swartz	516-563-2400	P	465	2.4
Monroe Systems for Business	1000 The American	Morris Plains	NJ	07950	Jeffry Picower	201-993-2000	R	290	2.4
Stores Automated Systems Inc	311 Sinclair St	Bristol	PA	19007	Bernard Greenberg	215-785-4321	R	45	0.3
Innovax Concepts Corp	1303 Walnut Hill Ln	Irving	TX	75038	John Young	214-550-8371	R	30*	0.3
Remanco International Inc	260 Fordham Rd	Wilmington	MA	01887	Frank J Hughes Jr	508-658-7400	R	25	0.2
Geac/Fasfax	Simon and Ledge	Nashua	NH	03060	Charles V Lee	603-889-5152	S	20	0.3
Pegasystems Inc	101 Main St	Cambridge	MA	02142	Alan Trefler	617-576-3580	R	18*	0.2
Cyberdata Corp	2700 Garden Rd	Monterey	CA	93940	Phil Lembo	408-373-2601	R	14	<0.1
AMT Industries Inc	5847 San Felipe	Houston	TX	77057	James T Rash	713-783-8200	P	13	<0.1
APG	5250 Industrial Blv	Minneapolis	MN	55421	Mark J Olson	612-560-1440	D	12	<0.1
InterBold	PO 3091	North Canton	OH	44720	Patrick Green	216-497-5099	J	12*	0.1
Coin Bill Validator Inc	425B Oser Av	Hauppauge	NY	11788	William H Wood	516-231-1177	P	10	<0.1
Trendar Corp	1655 Murfreesboro	Nashville	TN	37217	Ernest Betancourt	615-367-1000	R	9	<0.1
Birum Corp	Rte 29	Frenchtown	NJ	08825	Steve Fiala	908-996-3113	R	8	0.1
Harco Industries Inc	2362 W Shangri La	Phoenix	AZ	85029	Phillip Jalowiec	602-944-1565	R	6	<0.1
CECORP	8 Chrysler	Irvine	CA	92718	Geatano Cimo	714-583-0792	R	4*	<0.1
Hospitality Systems Inc	6401 Congress Av	Boca Raton	FL	33487	Jim Carlson	407-241-9998	R	4	<0.1
Datacap Systems Inc	212-A Progress Dr	Montgomeryv	PA	18936	Terry Ziegler	215-699-7051	R	3*	<0.1
MacSema Inc	94 SE Wilson St	Bend	OR	97702	Don Rowden	503-389-1122	R	3	<0.1
Progress Wire Products Inc	3535 W 140th St	Cleveland	OH	44111	T Harman	216-251-2181	R	3	<0.1
BankCard Services Corp	4940 Peachtree	Norcross	GA	30071	Richard Hicks	404-446-8891	R	2	<0.1
CompuRegister Corp	1213 Bittersweet Rd	Lake Ozark	MO	65049	Ted Ave-Lallemant	314-365-2050	R	2	<0.1
Smart Mortgage Access Inc	123 E Ogden Av	Hinsdale	IL	60521	Weston E Edwards	708-654-9700	R	2*	<0.1
RJP Electronics Inc	656-A Monterey	Monterey Park	CA	91754	Ray Lee	818-293-8458	R	1*	<0.1
Kush Industries	55 Messina Dr	Braintree	MA	02184	Al Ewing	617-848-9398	R	1	<0.1
Peregrin Technologies Inc	1400 NW Compton	Beaverton	OR	97006	Sam Bosch	503-690-1111	R	1	<0.1
CommStar Inc	6440 Flying Cloud	Eden Prairie	MN	55344	Verne L Severson	612-473-4284	R	0*	<0.1
POS Technology LC	2526 Manana St	Dallas	TX	75220	Mike LeSieur	214-357-4435	R	0*	<0.1

Source: Ward's Business Directory of U.S. Private and Public Companies, Volumes 1 and 2, 1996. The company type code used is as follows: P - Public, R - Private, S - Subsidiary, D - Division, J - Joint Venture, A - Affiliate, G - Group. Sales are in millions of dollars, employees are in thousands. An asterisk (*) indicates an estimated sales volume. The symbol < stands for 'less than'. Company names and addresses are truncated, in some cases, to fit into the available space.

MATERIALS CONSUMED

Material	Quantity	Delivered Cost ($ million)
Materials, ingredients, containers, and supplies	(X)	419.9
Cathode ray tubes (CRTs)	(X)	4.4
Printed circuit boards (without inserted components) for electronic circuitry	(X)	13.2
Printed peripheral controllers (graphic boards, drive controllers, etc.) for electronic circuitry	(X)	5.9
Printed computer processors (system boards, array processors, etc.) for electronic circuitry	(X)	8.1
Other printed circuit boards (loaded boards, subassemblies, and modules) for electronic circuitry	(X)	11.9
Semiconductors, including transistors, diodes, rectifiers, and integrated circuits for electronic circuitry	(X)	26.3
Capacitors for electronic circuitry	(X)	2.0
Resistors for electronic circuitry	(X)	1.9
Connectors for electronic circuitry	(X)	6.5
Battery packs for electronic circuitry	(X)	(D)
Other power supply units for electronic circuitry	(X)	11.9
Other components and accessories for electric circuitry	(X)	4.2
Electrical transmission, distribution, and control equipment	(X)	(D)
Steel, aluminum, and other metal electronic enclosurse	(X)	13.7
Plastics electronic enclosures	(X)	9.5
Sheet metal products (including stampings), except enclosures	(X)	29.4
All other fabricated metal products (except forgings)	(X)	14.7
Castings (rough and semifinished)	(X)	(D)
Metal shapes and forms, except castings, forgings, and fabricated metal products	(X)	2.5
Insulated copper wire and cable (including magnet wire)	(X)	(D)
Fabricated plastics products, except enclosures	(X)	9.7
Purchased software	(X)	(D)
Appliance outlets, switches, lampholders, and other current-carrying wiring devices	(X)	(D)
Electric motors and generators (all types)	(X)	7.0
Paper and paperboard products including paperboard boxes, containers, and corrugated paperboard	(X)	3.8
Purchased peripheral input devices, including keyboards, mouse devices, trackballs, etc.	(X)	19.4
Purchased peripheral printers	(X)	15.7
Other purchased electronic computing and peripheral equipment	(X)	18.9
All other materials and components, parts, containers, and supplies	(X)	54.1
Materials, ingredients, containers, and supplies, nsk	(X)	101.0

Source: 1992 *Economic Census.* Explanation of symbols used: (D): Withheld to avoid disclosure of competitive data; na: Not available; (S): Withheld because statistical norms were not met; (X): Not applicable; (Z): Less than half the unit shown; nec: Not elsewhere classified; nsk: Not specified by kind; - : zero; * : 10-19 percent estimated; ** : 20-29 percent estimated.

PRODUCT SHARE DETAILS

Product or Product Class	% Share	Product or Product Class	% Share
Calculating and accounting equipment	100.00	Parts and attachments for calculating, accounting machines, cash registers, funds-transfer and point-of-sale terminals. .	5.46
Accounting machines and cash registers, including calculator, funds-transfer terminals, and point-of-sale terminals, except parts and attachments	83.97	Calculating and accounting equipment, nsk	10.58

Source: 1992 *Economic Census*. The values shown are percent of total shipments in an industry. Values of indented subcategories are summed in the main heading. The symbol (D) appears when data are withheld to prevent disclosure of competitive information. The abbreviation nsk stands for 'not specified by kind' and nec for 'not elsewhere classified'.

INPUTS AND OUTPUTS FOR CALCULATING & ACCOUNTING MACHINES

Economic Sector or Industry Providing Inputs	%	Sector	Economic Sector or Industry Buying Outputs	%	Sector
Imports	31.2	Foreign	Gross private fixed investment	61.0	Cap Inv
Electronic components nec	13.7	Manufg.	Exports	13.2	Foreign
Wholesale trade	8.4	Trade	Accounting, auditing & bookkeeping	7.8	Services
Electronic computing equipment	6.3	Manufg.	Calculating & accounting machines	4.3	Manufg.
Calculating & accounting machines	4.6	Manufg.	Personal consumption expenditures	3.6	
Semiconductors & related devices	4.6	Manufg.	S/L Govt. purch., elem. & secondary education	2.2	S/L Govt
Primary batteries, dry & wet	2.7	Manufg.	Federal Government purchases, national defense	1.5	Fed Govt
Noncomparable imports	2.6	Foreign	Change in business inventories	1.5	In House
Miscellaneous publishing	2.2	Manufg.	S/L Govt. purch., higher education	1.5	S/L Govt
Banking	1.6	Fin/R.E.	Electrical repair shops	0.9	Services
Miscellaneous plastics products	1.3	Manufg.	S/L Govt. purch., public assistance & relief	0.9	S/L Govt
Nonferrous forgings	1.1	Manufg.	S/L Govt. purch., other general government	0.8	S/L Govt
Paperboard containers & boxes	1.0	Manufg.	Federal Government purchases, nondefense	0.3	Fed Govt
Commercial printing	0.8	Manufg.	S/L Govt. purch., highways	0.1	S/L Govt
Typewriters & office machines, nec	0.8	Manufg.			
Plastics materials & resins	0.7	Manufg.			
Wiring devices	0.7	Manufg.			
Communications, except radio & TV	0.7	Util.			
Eating & drinking places	0.7	Trade			
Motor freight transportation & warehousing	0.6	Util.			
Aluminum castings	0.5	Manufg.			
Blast furnaces & steel mills	0.5	Manufg.			
Petroleum refining	0.5	Manufg.			
Plating & polishing	0.5	Manufg.			
Die-cut paper & board	0.4	Manufg.			
Metal stampings, nec	0.4	Manufg.			
Paints & allied products	0.4	Manufg.			
Switchgear & switchboard apparatus	0.4	Manufg.			
Electric services (utilities)	0.4	Util.			
Real estate	0.4	Fin/R.E.			
Engineering, architectural, & surveying services	0.4	Services			
Iron & steel foundries	0.3	Manufg.			
Power transmission equipment	0.3	Manufg.			
Sheet metal work	0.3	Manufg.			
Signs & advertising displays	0.3	Manufg.			
Air transportation	0.3	Util.			
Advertising	0.3	Services			
Equipment rental & leasing services	0.3	Services			
Legal services	0.3	Services			
Management & consulting services & labs	0.3	Services			
Maintenance of nonfarm buildings nec	0.2	Constr.			
Aluminum rolling & drawing	0.2	Manufg.			
Ball & roller bearings	0.2	Manufg.			
Electron tubes	0.2	Manufg.			
Fabricated metal products, nec	0.2	Manufg.			
Gaskets, packing & sealing devices	0.2	Manufg.			
Industrial controls	0.2	Manufg.			
Instruments to measure electricity	0.2	Manufg.			
Machinery, except electrical, nec	0.2	Manufg.			
Miscellaneous fabricated wire products	0.2	Manufg.			
Motors & generators	0.2	Manufg.			
Nonferrous rolling & drawing, nec	0.2	Manufg.			
Screw machine and related products	0.2	Manufg.			
Gas production & distribution (utilities)	0.2	Util.			
Security & commodity brokers	0.2	Fin/R.E.			
Accounting, auditing & bookkeeping	0.2	Services			
U.S. Postal Service	0.2	Gov't			
Manifold business forms	0.1	Manufg.			
Nonferrous wire drawing & insulating	0.1	Manufg.			
Special dies & tools & machine tool accessories	0.1	Manufg.			
Royalties	0.1	Fin/R.E.			
Automotive rental & leasing, without drivers	0.1	Services			
Computer & data processing services	0.1	Services			

Source: *Benchmark Input-Output Accounts for the U.S. Economy, 1982*, U.S. Department of Commerce, Washington, D.C., July 1991. Data, as reported in the source, are organized by the 1977 SIC structure in use in 1982 but have been matched, as closely as is possible, to the 1987 SIC structure used in this book.

OCCUPATIONS EMPLOYED BY SIC 357 - COMPUTER AND OFFICE EQUIPMENT

Occupation	% of Total 1994	Change to 2005	Occupation	% of Total 1994	Change to 2005
Computer engineers	8.4	4.1	Engineering technicians & technologists nec	1.8	-29.7
Electrical & electronics engineers	5.4	-10.2	General managers & top executives	1.7	-33.3
Electrical & electronic technicians,technologists	4.9	-29.7	General office clerks	1.6	-40.1
Electrical & electronic equipment assemblers	4.8	-36.8	Purchasing agents, ex trade & farm products	1.4	5.4
Electrical & electronic assemblers	4.4	-15.7	Engineers nec	1.4	-15.7
Sales & related workers nec	3.9	-29.7	Industrial production managers	1.4	-36.8
Computer programmers	3.9	-43.1	Professional workers nec	1.4	-15.7
Engineering, mathematical, & science managers	3.5	-20.2	Industrial engineers, ex safety engineers	1.3	-38.2
Secretaries, ex legal & medical	3.3	-36.0	Electromechanical equipment assemblers	1.3	-86.0
Inspectors, testers, & graders, precision	2.7	-29.7	Accountants & auditors	1.3	-29.7
Production, planning, & expediting clerks	2.6	-29.7	Traffic, shipping, & receiving clerks	1.2	-32.4
Management support workers nec	2.4	-29.7	Mechanical engineers	1.2	-22.7
Systems analysts	2.2	12.4	Financial managers	1.1	-29.7
Marketing, advertising, & PR managers	2.0	-29.7	Bookkeeping, accounting, & auditing clerks	1.1	-47.3
Assemblers, fabricators, & hand workers nec	2.0	-29.7	Stock clerks	1.1	-42.9
Managers & administrators nec	1.9	-29.8	Writers & editors, incl technical writers	1.1	5.4
Blue collar worker supervisors	1.8	-32.6	Clerical supervisors & managers	1.0	-28.1

Source: *Industry-Occupation Matrix*, Bureau of Labor Statistics. These data relate to one or more 3-digit SIC industry groups rather than to a single 4-digit SIC. The change reported for each occupation to the year 2005 is a percent of growth or decline as estimated by the Bureau of Labor Statistics. The abbreviation nec stands for 'not elsewhere classified'.

LOCATION BY STATE AND REGIONAL CONCENTRATION

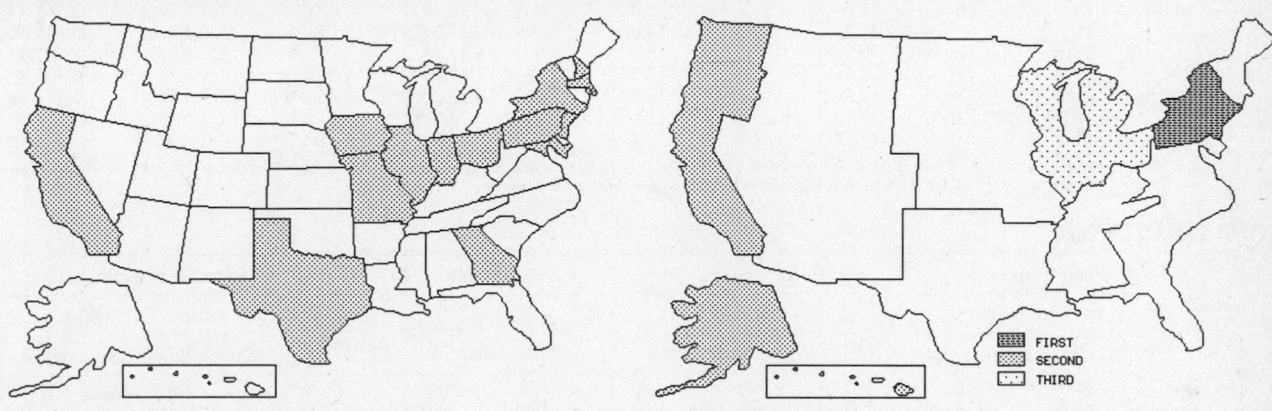

FIRST
SECOND
THIRD

INDUSTRY DATA BY STATE

State	Establish-ments	Shipments Total ($ mil)	Shipments % of U.S.	Shipments Per Establ.	Employment Total Number	Employment % of U.S.	Employment Per Establ.	Wages ($/hour)	Cost as % of Shipments	Investment per Employee ($)
New York	6	99.1	8.2	16.5	900	13.8	150	11.60	48.2	3,667
California	12	78.4	6.5	6.5	600	9.2	50	8.75	27.7	5,000
Pennsylvania	9	56.4	4.7	6.3	500	7.7	56	7.60	44.9	2,200
Massachusetts	5	22.6	1.9	4.5	200	3.1	40	19.00	49.6	-
New Jersey	5	20.5	1.7	4.1	200	3.1	40	9.00	49.8	2,500
Texas	9	(D)	-	-	175 *	2.7	19	-	-	1,714
Ohio	6	(D)	-	-	750 *	11.5	125	-	-	-
Maryland	4	(D)	-	-	375 *	5.8	94	-	-	-
Illinois	3	(D)	-	-	375 *	5.8	125	-	-	-
Indiana	3	(D)	-	-	175 *	2.7	58	-	-	-
Georgia	2	(D)	-	-	750 *	11.5	375	-	-	-
Rhode Island	2	(D)	-	-	375 *	5.8	188	-	-	-
Iowa	1	(D)	-	-	375 *	5.8	375	-	-	-
Missouri	1	(D)	-	-	175 *	2.7	175	-	-	-
New Hampshire	1	(D)	-	-	175 *	2.7	175	-	-	-

Source: 1992 *Economic Census*. The states are in descending order of shipments or establishments (if shipment data are missing for the majority). The symbol (D) appears when data are withheld to prevent disclosure of competitive information. States marked with (D) are sorted by number of establishments. A dash (-) indicates that the data element cannot be calculated; * indicates the midpoint of a range.

3579 - OFFICE MACHINES, NEC

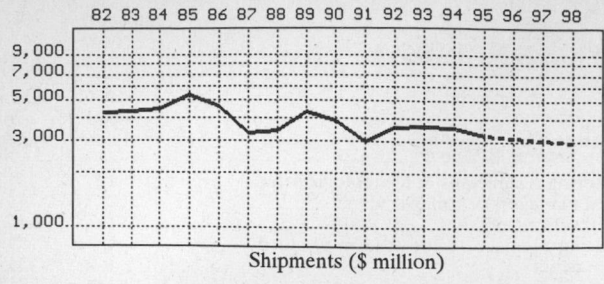

Shipments ($ million)

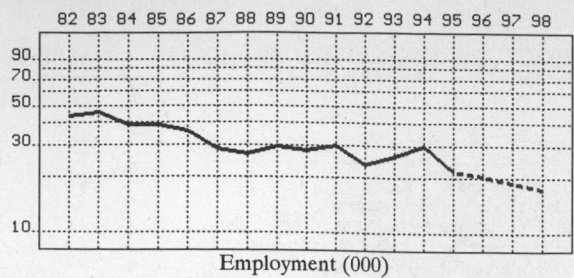

Employment (000)

GENERAL STATISTICS

| Year | Com-panies | Establishments | | Employment | | | Compensation | | Production ($ million) | | | |
		Total	with 20 or more employees	Total (000)	Production Workers (000)	Hours (Mil)	Payroll ($ mil)	Wages ($/hr)	Cost of Materials	Value Added by Manufacture	Value of Shipments	Capital Invest.
1982	203	232	125	43.7	21.2	40.1	963.7	8.97	1,572.7	2,689.3	4,256.7	198.5
1983		220	121	45.9	24.2	46.7	1,060.0	9.47	1,691.9	2,641.6	4,346.3	206.7
1984		208	117	38.7	17.2	34.2	1,037.7	10.27	2,193.7	2,382.5	4,449.5	281.0
1985		195	112	38.8	16.5	31.7	1,091.9	10.44	2,347.5	3,147.9	5,383.5	256.6
1986		191	107	35.9	15.8	30.1	1,018.3	10.70	1,906.8	2,652.0	4,630.9	164.9
1987	191	204	102	28.5	13.5	26.3	824.8	10.30	1,448.4	1,869.8	3,297.2	125.1
1988		207	103	27.2	12.6	25.0	888.2	9.76	1,640.8	1,743.3	3,401.7	121.4
1989		193	92	29.6	13.8	27.4	1,053.2	11.37	2,273.9	2,208.6	4,316.0	153.1
1990		202	94	28.3	12.7	26.6	1,029.7	11.76	1,732.7	1,986.2	3,922.1	150.3
1991		206	92	30.3	10.5	20.6	732.1	11.45	1,273.3	1,709.9	2,966.2	110.4
1992	143	157	88	23.8	11.9	22.3	781.1	10.98	1,681.3	1,844.8	3,562.8	115.4
1993		152	92	26.2	10.3	21.6	798.5	11.24	1,728.6	1,969.6	3,645.4	113.6
1994		165P	82P	30.0	9.0	18.3	808.0	11.91	1,565.9	2,036.8	3,581.5	117.6
1995		160P	78P	22.1P	7.3P	15.3P	772.2P	12.03P	1,425.4P	1,854.1P	3,260.2P	83.7P
1996		155P	75P	20.5P	6.3P	13.4P	749.7P	12.23P	1,380.4P	1,795.5P	3,157.2P	72.4P
1997		150P	72P	19.0P	5.3P	11.6P	727.2P	12.42P	1,335.3P	1,736.9P	3,054.2P	61.1P
1998		145P	68P	17.5P	4.2P	9.7P	704.7P	12.62P	1,290.3P	1,678.3P	2,951.1P	49.9P

Sources: 1982, 1987, 1992 *Economic Census*; *Annual Survey of Manufactures*, 83-86, 88-91, 93-94. Establishment counts for non-Census years are from *County Business Patterns*; establishment values for 83-84 are extrapolations. 'P's show projections by the editors. Industries reclassified in 87 will not have data for prior years.

INDICES OF CHANGE

| Year | Com-panies | Establishments | | Employment | | | Compensation | | Production ($ million) | | | |
		Total	with 20 or more employees	Total (000)	Production Workers (000)	Hours (Mil)	Payroll ($ mil)	Wages ($/hr)	Cost of Materials	Value Added by Manufacture	Value of Shipments	Capital Invest.
1982	142	148	142	184	178	180	123	82	94	146	119	172
1983		140	138	193	203	209	136	86	101	143	122	179
1984		132	133	163	145	153	133	94	130	129	125	244
1985		124	127	163	139	142	140	95	140	171	151	222
1986		122	122	151	133	135	130	97	113	144	130	143
1987	134	130	116	120	113	118	106	94	86	101	93	108
1988		132	117	114	106	112	114	89	98	94	95	105
1989		123	105	124	116	123	135	104	135	120	121	133
1990		129	107	119	107	119	132	107	103	108	110	130
1991		131	105	127	88	92	94	104	76	93	83	96
1992	100	100	100	100	100	100	100	100	100	100	100	100
1993		97	105	110	87	97	102	102	103	107	102	98
1994		105P	93P	126	76	82	103	108	93	110	101	102
1995		102P	89P	93P	62P	69P	99P	110P	85P	101P	92P	73P
1996		99P	85P	86P	53P	60P	96P	111P	82P	97P	89P	63P
1997		96P	81P	80P	44P	52P	93P	113P	79P	94P	86P	53P
1998		93P	77P	73P	36P	43P	90P	115P	77P	91P	83P	43P

Sources: Same as General Statistics. Values reflect change from the base year, 1992. Values above 100 mean greater than 92, values below 100 mean less than 92, and a value of 100 in the 82-91 or 93-98 period means same as 92. 'P's mark projections by the editors.

SELECTED RATIOS

For 1994	Avg. of All Manufact.	Analyzed Industry	Index	For 1994	Avg. of All Manufact.	Analyzed Industry	Index
Employees per Establishment	49	182	371	Value Added per Production Worker	134,084	226,311	169
Payroll per Establishment	1,500,273	4,895,621	326	Cost per Establishment	5,045,178	9,487,689	188
Payroll per Employee	30,620	26,933	88	Cost per Employee	102,970	52,197	51
Production Workers per Establishment	34	55	159	Cost per Production Worker	146,988	173,989	118
Wages per Establishment	853,319	1,320,563	155	Shipments per Establishment	9,576,895	21,700,083	227
Wages per Production Worker	24,861	24,217	97	Shipments per Employee	195,460	119,383	61
Hours per Production Worker	2,056	2,033	99	Shipments per Production Worker	279,017	397,944	143
Wages per Hour	12.09	11.91	98	Investment per Establishment	321,011	712,531	222
Value Added per Establishment	4,602,255	12,340,843	268	Investment per Employee	6,552	3,920	60
Value Added per Employee	93,930	67,893	72	Investment per Production Worker	9,352	13,067	140

Sources: Same as General Statistics. The 'Average of All Manufacturing' column represents the average of all manufacturing industries reported for the most recent complete year available. The Index shows the relationship between the Average and the Analyzed Industry. For example, 100 means that they are equal; 500 that the Analyzed Industry is five times the average; 50 means that the Analyzed Industry is half the national average. The abbreviation 'na' is used to show that data are 'not available'.

LEADING COMPANIES   Number shown: 45   Total sales ($ mil): 7,263   Total employment (000): 59.9

Company Name	Address				CEO Name	Phone	Co. Type	Sales ($ mil)	Empl. (000)
Pitney Bowes Inc	1 Elmcroft Rd	Stamford	CT	06926	George B Harvey	203-356-5000	P	3,271	30.8
Lexmark International Inc	55 Railroad Av	Greenwich	CT	06836	Marvin Mann	203-629-6700	R	640	5.0
Simplex Time Recorder Co	1 Simplex Plz	Gardner	MA	01441	Edward G Watkins	508-632-2500	R	530	3.5
General Binding Corp	1 GBC Plz	Northbrook	IL	60062	Govi C Reddy	708-272-3700	P	420	3.2
ElectroCom Automation Inc	PO Box 95080	Arlington	TX	76005	G Dan Thompson	817-640-5690	P	414	1.0
Dictaphone Corp	3191 Broadbridge	Stratford	CT	06497	Marc Breslawsky	203-381-7000	S	306	3.3
AM Multigraphics	1800 W Central Rd	Mt Prospect	IL	60056	Jerome D Brady	708-398-1900	D	290•	2.4
Smith Corona Corp	65 Locust Av	New Canaan	CT	06840	G Lee Thompson	203-972-1471	P	279	3.0
Toshiba America Info Systems	9740 Irvine Blv	Irvine	CA	92718	Atsutoshi Nishida	714-583-3000	S	240•	1.6
Brother Industries USA Inc	2950 Brother Blv	Bartlett	TN	38133	Sam Matsumoto	901-377-7777	S	160	0.5
Friden Neopost	30955 Huntwood	Hayward	CA	94544	Neil Mahlstedt	510-489-6800	R	130	0.9
Kronos Inc	400 5th Av	Waltham	MA	02154	Mark S Ain	617-890-3232	P	93	0.9
Ascom Hasler Mailing Systems	19 Forest Pkwy	Shelton	CT	06484	Michael A Allocca	203-926-1087	S	60	0.3
Bates Manufacturing Co	36 Newburgh Rd	Hackettstown	NJ	07840	Thomas M Williams	908-852-9300	R	50	0.3
Cummins-Allison Corp	891 Feehanville Dr	Mt Prospect	IL	60056	JE Jones	708-299-9550	R	35•	0.5
Stenograph Corp	1500 Bishop Ct	Mt Prospect	IL	60056	M J LaForte Jr	708-803-1400	R	35	0.2
United Barcode Industries Inc	12240 Indian Cr	Beltsville	MD	20705	Jeremy Meta	301-210-3000	R	30	0.2
Kroy Inc	PO Box 12279	Scottsdale	AZ	85260	Kenneth Cleveland	602-948-2222	R	29	0.1
Check Technology Corp	1284 Corporate Ctr	St Paul	MN	55121	Jay A Herman	612-454-9300	P	23	0.2
MBM Corp	PO Box 40249	Charleston	SC	29423	W Golde	803-552-2700	R	20	<0.1
Newbold Corp	510 Weaver St NE	Rocky Mount	VA	24151	Dean Whaley	703-489-4400	R	20	0.2
Ace Fasteners Co	1100 Hicks Rd	Rolling Mdws	IL	60008	Mark Wilson	708-259-1620	D	18	0.1
Atalla Corp	2304 Zanker Rd	San Jose	CA	95131	Robert Gargus	408-435-8850	S	17	<0.1
Spiral Binding Company Inc	PO Box 286	Totowa	NJ	07511	Robert M Roth	201-256-0666	R	15	0.3
Martin-Yale Industries Inc	251 Wedcor Dr	Wabash	IN	46992	C William Reed	219-563-0641	S	15	0.1
Dragon Systems Inc	320 Nevada St	Newton	MA	02160	James Baker	617-965-5200	R	13•	<0.1
Citifax Corp	28427 N Ballard Dr	Lake Forest	IL	60045	Michael Einarsen	708-362-3300	R	12	0.1
Omron Systems Inc	55 E Commerce Dr	Schaumburg	IL	60173	H Patrick Green	708-843-0515	S	12	<0.1
Security Engineered Machinery	PO Box 1045	Westborough	MA	01581	L Rosen	508-366-1488	R	11	<0.1
COPE Inc	2425 E Medina Rd	Tucson	AZ	85706	David Harcort	602-746-3241	S	10•	0.2
Master Products Manufacturing	PO Box 23985	Los Angeles	CA	90023	Jerome A Sixel	213-265-8042	S	10	0.1
Rapidprint Inc	2055 S Main St	Middletown	CT	06457	Donald Bidwell	203-346-9283	R	9•	<0.1
EMF Corp	15110 NE 95th St	Redmond	WA	98052	David Isett	206-883-0045	S	8	<0.1
Documail Systems	6802 Brummel Pl	Evanston	IL	60645	Gerry Loftis	708-475-6100	D	7•	<0.1
Sudbury Systems Inc	490 Boston Post Rd	Sudbury	MA	01776	Gerald Delaney	508-443-1100	R	7•	<0.1
EPE Corp	540 N Commercial	Manchester	NH	03101	Jim Bell	603-669-9181	R	5	<0.1
R Funk and Company Inc	825 N Easton Rd	Doylestown	PA	18901	M Thorton	215-348-8181	R	5	<0.1
Eastern Engraving	355 Warren Av	Stirling	NJ	07980	Thomas Garrett	908-647-3300	S	3	<0.1
McGill Inc	131 E Prairie St	Marengo	IL	60152	Wayne Schwartzman	815-568-7244	R	3	<0.1
Numberall Stamp and Tool	PO Box 187	Sangerville	ME	04479	Herman Bayerdoffer	207-876-3541	R	3•	<0.1
Parking Products Inc	PO Box S	Willow Grove	PA	19090	Dieter E Niebisch	215-657-7500	R	3	<0.1
Shuffle Master Inc	10921 Val View Rd	Eden Prairie	MN	55344	John G Breeding	612-943-1951	P	3	<0.1
Addressing Machines and Supply	940 Virginia Av	Indianapolis	IN	46203	Martin D Cole	317-637-8537	R	1	<0.1
Electronic Voting Systems Inc	475 10th Av	New York	NY	10018	Herbert F Kozlov	212-629-5777	S	1	<0.1
Office Systems Inc	111 N Wabash	Chicago	IL	60602	Roger Coleman	312-873-5000	R	1	<0.1

Source: Ward's Business Directory of U.S. Private and Public Companies, Volumes 1 and 2, 1996. The company type code used is as follows: P - Public, R - Private, S - Subsidiary, D - Division, J - Joint Venture, A - Affiliate, G - Group. Sales are in millions of dollars, employees are in thousands. An asterisk (•) indicates an estimated sales volume. The symbol < stands for 'less than'. Company names and addresses are truncated, in some cases, to fit into the available space.

MATERIALS CONSUMED

Material	Quantity	Delivered Cost ($ million)
Materials, ingredients, containers, and supplies	(X)	1,270.3
Cathode ray tubes (CRTs)	(X)	(D)
Printed circuit boards (without inserted components) for electronic circuitry	(X)	13.1
Printed computer processors (system boards, array processors, etc.) for electronic circuitry . .	(X)	5.3
Other printed communication boards (LAN, D/A, A/D converters, etc.)	(X)	7.7
Other printed circuit boards (loaded boards, subassemblies, and modules) for electronic circuitry .	(X)	3.4
Semiconductors, including transistors, diodes, rectifiers, and integrated circuits for electronic circuitry	(X)	63.5
Capacitors for electronic circuitry	(X)	3.6
Resistors for electronic circuitry	(X)	2.7
Connectors for electronic circuitry	(X)	7.8
Battery packs for electronic circuitry	(X)	(D)
Other power supply units for electronic circuitry	(X)	2.8
Other components and accessories for electric circuitry	(X)	4.4
Electrical transmission, distribution, and control equipment	(X)	4.0
Steel, aluminum, and other metal electronic enclousrse	(X)	(D)
Plastics electronic enclosures	(X)	11.1
Sheet metal products (including stampings), except enclosures	(X)	23.0
All other fabricated metal products (except forgings)	(X)	60.8
Forgings	(X)	(D)
Castings (rough and semifinished)	(X)	0.9
Metal shapes and forms, except castings, forgings, and fabricated metal products	(X)	3.6
Insulated copper wire and cable (including magnet wire)	(X)	5.0
Fabricated plastics products, except enclosures	(X)	21.6

Continued on next page.

MATERIALS CONSUMED - Continued

Material	Quantity	Delivered Cost ($ million)
Purchased software	(X)	2.2
Appliance outlets, switches, lampholders, and other current-carrying wiring devices	(X)	1.0
Electric motors and generators (all types)	(X)	20.8
Paper and paperboard products including paperboard boxes, containers, and corrugated paperboard	(X)	17.6
Purchased computers	(X)	24.9
Purchased peripheral storage devices	(X)	37.5
Purchased peripheral printers	(X)	2.9
Other purchased electronic computing and peripheral equipment	(X)	16.8
All other materials and components, parts, containers, and supplies	(X)	326.7
Materials, ingredients, containers, and supplies, nsk	(X)	503.3

Source: 1992 *Economic Census*. Explanation of symbols used: (D): Withheld to avoid disclosure of competitive data; na: Not available; (S): Withheld because statistical norms were not met; (X): Not applicable; (Z): Less than half the unit shown; nec: Not elsewhere classified; nsk: Not specified by kind; - : zero; * : 10-19 percent estimated; ** : 20-29 percent estimated.

PRODUCT SHARE DETAILS

Product or Product Class	% Share	Product or Product Class	% Share
Office machines, nec	100.00	machines, and all other office machines, nec, except parts and attachments	(D)
Automatic typing and word processing machines, parts, and attachments	16.15	Standard typewriters, dictating, transcribing, and recording machines, and all other office machines, nec, except parts and attachments	(D)
Duplicating machines, except parts and attachments	(D)		
Duplicating machines, except parts and attachments	(D)	Parts and attachments for standard typewriters, and dictating, duplicating, and other office machines, nec	9.21
Mailing, letter handling, and addressing machines, except parts and attachments	39.34	Office machines, nec, nsk	13.55
Standard typewriters, dictating, transcribing, and recording			

Source: 1992 *Economic Census*. The values shown are percent of total shipments in an industry. Values of indented subcategories are summed in the main heading. The symbol (D) appears when data are withheld to prevent disclosure of competitive information. The abbreviation nsk stands for 'not specified by kind' and nec for 'not elsewhere classified'.

INPUTS AND OUTPUTS FOR TYPEWRITERS & OFFICE MACHINES, NEC

Economic Sector or Industry Providing Inputs	%	Sector	Economic Sector or Industry Buying Outputs	%	Sector
Imports	28.0	Foreign	Gross private fixed investment	45.2	Cap Inv
Miscellaneous plastics products	16.2	Manufg.	Exports	10.6	Foreign
Wholesale trade	11.3	Trade	Electrical repair shops	7.8	Services
Electronic computing equipment	4.9	Manufg.	Miscellaneous repair shops	6.5	Services
Typewriters & office machines, nec	3.9	Manufg.	Personal consumption expenditures	3.9	
Noncomparable imports	2.9	Foreign	Typewriters & office machines, nec	3.3	Manufg.
Business services nec	2.7	Services	S/L Govt. purch., elem. & secondary education	3.1	S/L Govt
Electronic components nec	1.8	Manufg.	Change in business inventories	2.0	In House
Paints & allied products	1.6	Manufg.	S/L Govt. purch., other general government	1.9	S/L Govt
Banking	1.6	Fin/R.E.	S/L Govt. purch., higher education	1.7	S/L Govt
Semiconductors & related devices	1.5	Manufg.	Federal Government purchases, nondefense	1.3	Fed Govt
Switchgear & switchboard apparatus	1.5	Manufg.	Accounting, auditing & bookkeeping	1.1	Services
Blast furnaces & steel mills	1.3	Manufg.	Retail trade, except eating & drinking	1.0	Trade
Fabricated metal products, nec	1.2	Manufg.	Membership organizations nec	1.0	Services
Motors & generators	1.1	Manufg.	Colleges, universities, & professional schools	0.9	Services
Plastics materials & resins	1.1	Manufg.	Federal Government purchases, national defense	0.8	Fed Govt
Die-cut paper & board	1.0	Manufg.	Wholesale trade	0.7	Trade
Communications, except radio & TV	0.8	Util.	S/L Govt. purch., public assistance & relief	0.7	S/L Govt
Electric services (utilities)	0.8	Util.	Motor freight transportation & warehousing	0.4	Util.
Metal stampings, nec	0.7	Manufg.	Doctors & dentists	0.4	Services
Plating & polishing	0.7	Manufg.	S/L Govt. purch., other education & libraries	0.4	S/L Govt
Motor freight transportation & warehousing	0.7	Util.	Blast furnaces & steel mills	0.3	Manufg.
Eating & drinking places	0.7	Trade	Calculating & accounting machines	0.3	Manufg.
Real estate	0.7	Fin/R.E.	Real estate	0.3	Fin/R.E.
Paperboard containers & boxes	0.6	Manufg.	Commercial printing	0.2	Manufg.
Sheet metal work	0.6	Manufg.	Arrangement of passenger transportation	0.2	Util.
Personnel supply services	0.6	Services	Electric services (utilities)	0.2	Util.
Paper mills, except building paper	0.5	Manufg.	Freight forwarders	0.2	Util.
Equipment rental & leasing services	0.5	Services	Banking	0.2	Fin/R.E.
Aluminum rolling & drawing	0.4	Manufg.	Credit agencies other than banks	0.2	Fin/R.E.
Wiring devices	0.4	Manufg.	Security & commodity brokers	0.2	Fin/R.E.
Advertising	0.4	Services	Elementary & secondary schools	0.2	Services
Maintenance of nonfarm buildings nec	0.3	Constr.	Labor, civic, social, & fraternal associations	0.2	Services
Electron tubes	0.3	Manufg.	Legal services	0.2	Services
Gaskets, packing & sealing devices	0.3	Manufg.	S/L Govt. purch., health & hospitals	0.2	S/L Govt
Iron & steel foundries	0.3	Manufg.	S/L Govt. purch., highways	0.2	S/L Govt
Nonferrous rolling & drawing, nec	0.3	Manufg.	Eating & drinking places	0.1	Trade
Screw machine and related products	0.3	Manufg.	Computer & data processing services	0.1	Services
Legal services	0.3	Services	Management & consulting services & labs	0.1	Services
Aluminum castings	0.2	Manufg.	S/L Govt. purch., natural resource & recreation.	0.1	S/L Govt

Continued on next page.

INPUTS AND OUTPUTS FOR TYPEWRITERS & OFFICE MACHINES, NEC - Continued

Economic Sector or Industry Providing Inputs	%	Sector	Economic Sector or Industry Buying Outputs	%	Sector
Ball & roller bearings	0.2	Manufg.			
Machinery, except electrical, nec	0.2	Manufg.			
Miscellaneous fabricated wire products	0.2	Manufg.			
Transformers	0.2	Manufg.			
Air transportation	0.2	Util.			
Gas production & distribution (utilities)	0.2	Util.			
Royalties	0.2	Fin/R.E.			
Security & commodity brokers	0.2	Fin/R.E.			
Computer & data processing services	0.2	Services			
Management & consulting services & labs	0.2	Services			
U.S. Postal Service	0.2	Gov't			
Fabricated rubber products, nec	0.1	Manufg.			
Industrial controls	0.1	Manufg.			
Manifold business forms	0.1	Manufg.			
Metal coating & allied services	0.1	Manufg.			
Nonferrous wire drawing & insulating	0.1	Manufg.			
Signs & advertising displays	0.1	Manufg.			
Special dies & tools & machine tool accessories	0.1	Manufg.			
Accounting, auditing & bookkeeping	0.1	Services			

Source: Benchmark Input-Output Accounts for the U.S. Economy, 1982, U.S. Department of Commerce, Washington, D.C., July 1991. Data, as reported in the source, are organized by the 1977 SIC structure in use in 1982 but have been matched, as closely as is possible, to the 1987 SIC structure used in this book.

OCCUPATIONS EMPLOYED BY SIC 357 - COMPUTER AND OFFICE EQUIPMENT

Occupation	% of Total 1994	Change to 2005	Occupation	% of Total 1994	Change to 2005
Computer engineers	8.4	4.1	Engineering technicians & technologists nec	1.8	-29.7
Electrical & electronics engineers	5.4	-10.2	General managers & top executives	1.7	-33.3
Electrical & electronic technicians,technologists	4.9	-29.7	General office clerks	1.6	-40.1
Electrical & electronic equipment assemblers	4.8	-36.8	Purchasing agents, ex trade & farm products	1.4	5.4
Electrical & electronic assemblers	4.4	-15.7	Engineers nec	1.4	-15.7
Sales & related workers nec	3.9	-29.7	Industrial production managers	1.4	-36.8
Computer programmers	3.9	-43.1	Professional workers nec	1.4	-15.7
Engineering, mathematical, & science managers	3.5	-20.2	Industrial engineers, ex safety engineers	1.3	-38.2
Secretaries, ex legal & medical	3.3	-36.0	Electromechanical equipment assemblers	1.3	-86.0
Inspectors, testers, & graders, precision	2.7	-29.7	Accountants & auditors	1.3	-29.7
Production, planning, & expediting clerks	2.6	-29.7	Traffic, shipping, & receiving clerks	1.2	-32.4
Management support workers nec	2.4	-29.7	Mechanical engineers	1.2	-22.7
Systems analysts	2.2	12.4	Financial managers	1.1	-29.7
Marketing, advertising, & PR managers	2.0	-29.7	Bookkeeping, accounting, & auditing clerks	1.1	-47.3
Assemblers, fabricators, & hand workers nec	2.0	-29.7	Stock clerks	1.1	-42.9
Managers & administrators nec	1.9	-29.8	Writers & editors, incl technical writers	1.1	5.4
Blue collar worker supervisors	1.8	-32.6	Clerical supervisors & managers	1.0	-28.1

Source: Industry-Occupation Matrix, Bureau of Labor Statistics. These data relate to one or more 3-digit SIC industry groups rather than to a single 4-digit SIC. The change reported for each occupation to the year 2005 is a percent of growth or decline as estimated by the Bureau of Labor Statistics. The abbreviation nec stands for 'not elsewhere classified'.

LOCATION BY STATE AND REGIONAL CONCENTRATION

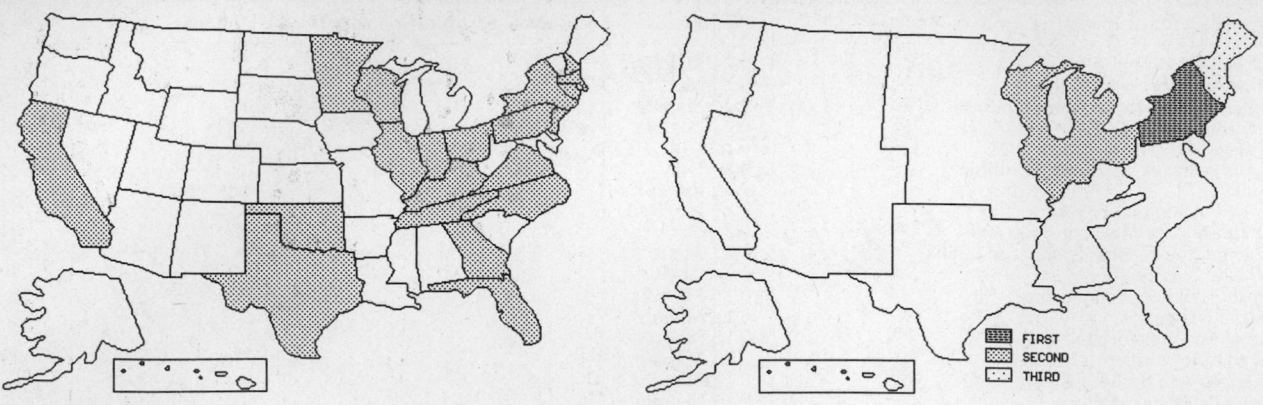

INDUSTRY DATA BY STATE

| State | Establish-ments | Shipments | | | Employment | | | | Cost as % of Shipments | Investment per Employee ($) |
		Total ($ mil)	% of U.S.	Per Establ.	Total Number	% of U.S.	Per Establ.	Wages ($/hour)		
Illinois	20	595.7	16.7	29.8	2,700	11.3	135	12.23	50.9	4,444
California	13	375.5	10.5	28.9	2,400	10.1	185	10.29	45.2	5,208
Massachusetts	10	142.5	4.0	14.2	800	3.4	80	13.20	34.6	7,375
New Jersey	10	47.0	1.3	4.7	600	2.5	60	10.33	42.1	1,833
Georgia	6	26.7	0.7	4.4	300	1.3	50	10.40	43.1	3,333
New York	12	(D)	-	-	1,750 *	7.4	146	-	-	2,514
Connecticut	10	(D)	-	-	7,500 *	31.5	750	-	-	-
Minnesota	10	(D)	-	-	750 *	3.2	75	-	-	4,133
Texas	7	(D)	-	-	1,750 *	7.4	250	-	-	-
Pennsylvania	6	(D)	-	-	750 *	3.2	125	-	-	-
Florida	5	(D)	-	-	750 *	3.2	150	-	-	-
North Carolina	5	(D)	-	-	750 *	3.2	150	-	-	-
Kentucky	3	(D)	-	-	750 *	3.2	250	-	-	-
Ohio	3	(D)	-	-	375 *	1.6	125	-	-	-
Oklahoma	3	(D)	-	-	375 *	1.6	125	-	-	-
Tennessee	3	(D)	-	-	750 *	3.2	250	-	-	-
Virginia	3	(D)	-	-	375 *	1.6	125	-	-	1,333
Wisconsin	3	(D)	-	-	375 *	1.6	125	-	-	-
Indiana	1	(D)	-	-	175 *	0.7	175	-	-	-
New Hampshire	1	(D)	-	-	175 *	0.7	175	-	-	-

Source: 1992 *Economic Census*. The states are in descending order of shipments or establishments (if shipment data are missing for the majority). The symbol (D) appears when data are withheld to prevent disclosure of competitive information. States marked with (D) are sorted by number of establishments. A dash (-) indicates that the data element cannot be calculated; * indicates the midpoint of a range.

3581 - AUTOMATIC VENDING MACHINES

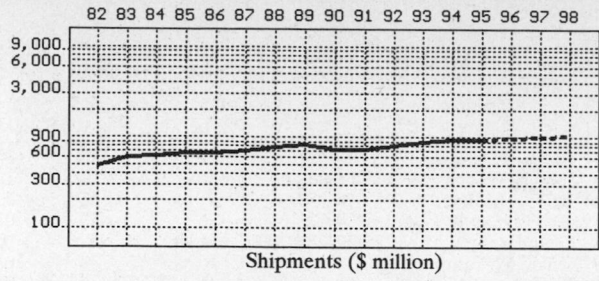

Shipments ($ million)

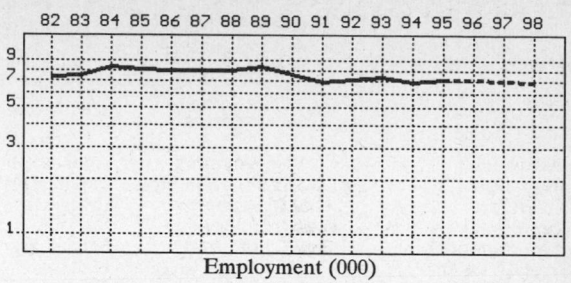

Employment (000)

GENERAL STATISTICS

Year	Companies	Establishments		Employment			Compensation		Production ($ million)			
		Total	with 20 or more employees	Total (000)	Production Workers (000)	Hours (Mil)	Payroll ($ mil)	Wages ($/hr)	Cost of Materials	Value Added by Manufacture	Value of Shipments	Capital Invest.
1982	88	92	36	7.4	5.1	9.8	123.3	7.38	267.1	215.0	479.4	13.1
1983		94	37	7.5	5.1	9.8	132.9	7.39	345.0	257.1	596.0	25.3
1984		96	38	8.3	6.0	11.6	156.8	8.91	349.8	279.3	611.3	23.5
1985		97	39	8.1	5.9	11.4	163.4	9.09	372.5	306.2	664.8	13.4
1986		95	33	8.0	5.6	10.7	176.6	9.22	363.1	306.3	668.6	27.0
1987	97	98	33	7.9	5.7	11.1	185.6	9.79	372.7	344.0	714.6	25.9
1988		97	34	7.9	5.5	10.6	179.0	9.24	406.0	366.8	776.1	21.9
1989		99	33	8.3	5.7	11.0	188.9	9.74	462.7	381.0	828.6	21.6
1990		89	32	7.6	5.2	10.1	181.5	9.62	399.8	338.1	741.7	80.2
1991		90	30	6.9	4.8	9.2	164.9	8.99	385.4	339.2	733.7	10.4
1992	105	107	38	7.2	5.1	10.1	181.2	8.84	422.7	384.7	815.7	13.4
1993		115	38	7.3	5.4	10.9	183.3	9.17	504.7	385.4	887.2	20.3
1994		104P	34P	6.9	5.1	10.8	197.5	9.29	503.3	426.7	942.0	21.3
1995		105P	34P	7.2P	5.2P	10.5P	201.6P	9.76P	502.1P	425.7P	939.8P	27.4P
1996		106P	33P	7.1P	5.1P	10.4P	206.1P	9.88P	518.3P	439.4P	970.1P	27.8P
1997		107P	33P	7.0P	5.1P	10.4P	210.6P	9.99P	534.5P	453.2P	1,000.4P	28.2P
1998		108P	33P	7.0P	5.1P	10.4P	215.0P	10.10P	550.7P	466.9P	1,030.8P	28.6P

Sources: 1982, 1987, 1992 *Economic Census; Annual Survey of Manufactures*, 83-86, 88-91, 93-94. Establishment counts for non-Census years are from *County Business Patterns*; establishment values for 83-84 are extrapolations. 'P's show projections by the editors. Industries reclassified in 87 will not have data for prior years.

INDICES OF CHANGE

Year	Companies	Establishments		Employment			Compensation		Production ($ million)			
		Total	with 20 or more employees	Total (000)	Production Workers (000)	Hours (Mil)	Payroll ($ mil)	Wages ($/hr)	Cost of Materials	Value Added by Manufacture	Value of Shipments	Capital Invest.
1982	84	86	95	103	100	97	68	83	63	56	59	98
1983		88	97	104	100	97	73	84	82	67	73	189
1984		90	100	115	118	115	87	101	83	73	75	175
1985		91	103	113	116	113	90	103	88	80	82	100
1986		89	87	111	110	106	97	104	86	80	82	201
1987	92	92	87	110	112	110	102	111	88	89	88	193
1988		91	89	110	108	105	99	105	96	95	95	163
1989		93	87	115	112	109	104	110	109	99	102	161
1990		83	84	106	102	100	100	109	95	88	91	599
1991		84	79	96	94	91	91	102	91	88	90	78
1992	100	100	100	100	100	100	100	100	100	100	100	100
1993		107	100	101	106	108	101	104	119	100	109	151
1994		97P	89P	96	100	107	109	105	119	111	115	159
1995		98P	88P	99P	101P	104P	111P	110P	119P	111P	115P	204P
1996		99P	88P	99P	101P	103P	114P	112P	123P	114P	119P	207P
1997		100P	87P	98P	100P	103P	116P	113P	126P	118P	123P	211P
1998		101P	87P	97P	99P	103P	119P	114P	130P	121P	126P	214P

Sources: Same as General Statistics. Values reflect change from the base year, 1992. Values above 100 mean greater than 92, values below 100 mean less than 92, and a value of 100 in the 82-91 or 93-98 period means same as 92. 'P's mark projections by the editors.

SELECTED RATIOS

For 1994	Avg. of All Manufact.	Analyzed Industry	Index	For 1994	Avg. of All Manufact.	Analyzed Industry	Index
Employees per Establishment	49	66	135	Value Added per Production Worker	134,084	83,667	62
Payroll per Establishment	1,500,273	1,895,173	126	Cost per Establishment	5,045,178	4,829,573	96
Payroll per Employee	30,620	28,623	93	Cost per Employee	102,970	72,942	71
Production Workers per Establishment	34	49	143	Cost per Production Worker	146,988	98,686	67
Wages per Establishment	853,319	962,767	113	Shipments per Establishment	9,576,895	9,039,256	94
Wages per Production Worker	24,861	19,673	79	Shipments per Employee	195,460	136,522	70
Hours per Production Worker	2,056	2,118	103	Shipments per Production Worker	279,017	184,706	66
Wages per Hour	12.09	9.29	77	Investment per Establishment	321,011	204,391	64
Value Added per Establishment	4,602,255	4,094,533	89	Investment per Employee	6,552	3,087	47
Value Added per Employee	93,930	61,841	66	Investment per Production Worker	9,352	4,176	45

Sources: Same as General Statistics. The 'Average of All Manufacturing' column represents the average of all manufacturing industries reported for the most recent complete year available. The Index shows the relationship between the Average and the Analyzed Industry. For example, 100 means that they are equal; 500 that the Analyzed Industry is five times the average; 50 means that the Analyzed Industry is half the national average. The abbreviation 'na' is used to show that data are 'not available'.

LEADING COMPANIES   Number shown: **30**   Total sales ($ mil): **656**   Total employment (000): **7.8**

Company Name	Address				CEO Name	Phone	Co. Type	Sales ($ mil)	Empl. (000)
IMI Cornelius Inc	1 Cornelius Pl	Anoka	MN	55303	Richard Barkley	612-421-6120	S	140	1.5
Rowe International Inc	27 Druid Hill Dr	Parsippany	NJ	07054	James Gang	201-887-0400	R	99	1.2
National Vendors	12955 Enterpr	Bridgeton	MO	63044	Robert Muller	314-383-3500	D	65	0.9
Coin Acceptors Inc	300 Hunter Av	St Louis	MO	63124	JE Thomas Jr	314-725-0100	R	50	0.8
SerVend International Inc	2100 Future Dr	Sellersburg	IN	47172	Greg Fischer	812-246-7000	R	39	0.3
Booth Crystal Tips Inc	2007 Royal Ln	Dallas	TX	75229	David W Campbell	214-488-1030	S	38	0.2
WICO Corp	6400 W Gross Pt Rd	Niles	IL	60714	Ed Sokolofski	708-647-7500	R	29	0.2
Gross-Given Manufacturing	75 W Plato Blv	St Paul	MN	55107	Alan Suitor	612-224-4391	R	28	0.3
Automated Custom Food Svcs	7700 Brookhollow	Dallas	TX	75235	Steve Errico	214-631-7040	R	25	0.3
Polyvend Inc	700 S German Ln	Conway	AR	72032	JS Stoltz	501-327-1301	R	22	0.4
Mountain View Fabricating	Hwy 60	Mountain View	MO	65548	Frank Blood	417-934-2048	S	21	0.4
H and R Electronics Co	PO Box 196	High Ridge	MO	63049	Bill Peterson	314-677-3377	S	15*	0.2
Cavalier Corp	1105 E 10th St	Chattanooga	TN	37403	Kevin Ward	615-267-6671	R	13	0.1
Water Point Systems	14500 Trinity Blv	Fort Worth	TX	76155	Mark Hope	817-545-0809	R	12*	0.1
A and A Co	2301 York St	Timonium	MD	21093	Eugene Lipman	410-252-1020	D	11*	<0.1
Federal Machine Corp	PO Box 1779	Des Moines	IA	50306	L A Kershbaumer	515-274-1555	R	9*	<0.1
International Lottery Inc	6665 Creek Rd	Cincinnati	OH	45242	L Rogers Wells Jr	513-792-7000	P	6	<0.1
Scribe International	790 Maple Ln	Bensenville	IL	60106	Charlie Flubacker	708-860-6500	D	6	<0.1
Singer Data Products Inc	790 Maple Ln	Bensenville	IL	60106	Ted Singer	708-860-6500	R	6*	<0.1
Wilshire Corp	PO Box 1639	Torrington	CT	06790	William May	203-489-6748	R	6	0.4
Seaga Manufacturing Inc	PO Box 47	Shannon	IL	61078	Steven Chesney	815-864-2600	R	4	<0.1
Nouveau International Inc	2546 Armistead	Norristown	PA	19403	Gary W Black Sr	215-630-8007	R	3*	<0.1
Amer Business Computers	451 Kennedy Rd	Akron	OH	44305	Joseph W Shannon	216-733-2841	P	2	<0.1
Advance Manufacturing Co	11760 Roscoe Blv	Sun Valley	CA	91352	Don C Lemke	818-767-5466	R	2*	<0.1
Honor Gard	720 Bonnie Ln	Elk Grove Vill	IL	60007	Jerome Remien	708-806-0888	S	1	<0.1
Jerome Remien Corp	720 Bonnie Ln	Elk Grove Vill	IL	60007	Jerome Remien	708-806-0888	R	1	<0.1
Fun Industries Inc	627 15th Av	East Moline	IL	61244	Bud Johnston	309-755-5021	R	1	<0.1
Video Dome Inc	1500 Hwy 315	Wilkes-Barre	PA	18711	Lawrence P Medico	717-825-8833	R	1	<0.1
Cafe Quick Enterprises Inc	2023 E Shady Grove	Irving	TX	75080	Bill Taylor	214-721-9496	R	0*	<0.1
Associated Vending Co	6735 32nd St	N Highlands	CA	95660	J Duke Marlowe	916-349-8910	R	0	<0.1

Source: *Ward's Business Directory of U.S. Private and Public Companies*, Volumes 1 and 2, 1996. The company type code used is as follows: P - Public, R - Private, S - Subsidiary, D - Division, J - Joint Venture, A - Affiliate, G - Group. Sales are in millions of dollars, employees are in thousands. An asterisk (*) indicates an estimated sales volume. The symbol < stands for 'less than'. Company names and addresses are truncated, in some cases, to fit into the available space.

MATERIALS CONSUMED

Material	Quantity	Delivered Cost ($ million)
Materials, ingredients, containers, and supplies	(X)	384.1
Fractional horsepower electric motors and generators (less than 1 hp)	(X)	21.3
Refrigeration compressors, compressor units, condensing units, and other heat transfer equipment	(X)	18.3
Current-carrying wiring devices	(X)	22.0
Electrical enclosures (metal and plastics)	(X)	1.0
Metal bolts, nuts, screws, washers, rivets, and other screw machine products	(X)	8.6
Metal stampings	(X)	6.5
Other fabricated metal products (except electrical enclosures and forgings)	(X)	7.7
Castings (rough and semifinished)	(X)	4.1
Steel shapes and forms	(X)	39.1
Nonferrous shapes and forms	(X)	1.3
Plastics products consumed in the form of sheets, rods, tubes, film, and other shapes	(X)	23.4
Paints, varnishes, stains, lacquers, shellacs, japans, enamels, and allied products	(X)	7.9
Paper and paperboard containers, including shipping sacks and other paper packaging supplies	(X)	6.7
All other materials and components, parts, containers, and supplies	(X)	140.3
Materials, ingredients, containers, and supplies, nsk	(X)	75.7

Source: 1992 *Economic Census*. Explanation of symbols used: (D): Withheld to avoid disclosure of competitive data; na: Not available; (S): Withheld because statistical norms were not met; (X): Not applicable; (Z): Less than half the unit shown; nec: Not elsewhere classified; nsk: Not specified by kind; - : zero; * : 10-19 percent estimated; ** : 20-29 percent estimated.

PRODUCT SHARE DETAILS

Product or Product Class	% Share	Product or Product Class	% Share
Automatic merchandising machines	100.00	separately).	70.49
Automatic merchandising machines, coin-operated (vending), excluding money changing machines, coin-operated mechanisms, and parts	73.83	Parts for automatic merchandising machines, except coin-operated mechanisms.	29.33
Coin-operated mechanisms and parts for automatic merchandising machines	21.64	Coin-operated mechanisms and parts for automatic merchandising machines, nsk	0.24
Coin-operated mechanisms for vending machines (for sale		Automatic merchandising machines, nsk	4.53

Source: 1992 *Economic Census*. The values shown are percent of total shipments in an industry. Values of indented subcategories are summed in the main heading. The symbol (D) appears when data are withheld to prevent disclosure of competitive information. The abbreviation nsk stands for 'not specified by kind' and nec for 'not elsewhere classified'.

INPUTS AND OUTPUTS FOR AUTOMATIC MERCHANDISING MACHINES

Economic Sector or Industry Providing Inputs	%	Sector	Economic Sector or Industry Buying Outputs	%	Sector
Wholesale trade	10.5	Trade	Gross private fixed investment	68.7	Cap Inv
Sheet metal work	9.2	Manufg.	Exports	8.9	Foreign
Refrigeration & heating equipment	7.9	Manufg.	Retail trade, except eating & drinking	6.7	Trade
Blast furnaces & steel mills	5.2	Manufg.	Manufacturing industries, nec	6.5	Manufg.
Machinery, except electrical, nec	4.8	Manufg.	Amusement & recreation services nec	3.6	Services
Motors & generators	4.7	Manufg.	Radio & TV receiving sets	1.7	Manufg.
Nonferrous wire drawing & insulating	4.4	Manufg.	U.S. Postal Service	0.9	Gov't
Imports	3.2	Foreign	Household laundry equipment	0.8	Manufg.
Screw machine and related products	2.8	Manufg.	Commercial laundry equipment	0.7	Manufg.
Transformers	2.7	Manufg.	Laundry, dry cleaning, shoe repair	0.7	Services
Miscellaneous plastics products	2.4	Manufg.	Change in business inventories	0.4	In House
Pipe, valves, & pipe fittings	2.4	Manufg.	Portrait, photographic studios	0.3	Services
Metal stampings, nec	2.2	Manufg.	Automatic merchandising machines	0.1	Manufg.
Special dies & tools & machine tool accessories	2.1	Manufg.			
Industrial controls	1.9	Manufg.			
Hardware, nec	1.7	Manufg.			
Electric services (utilities)	1.5	Util.			
Maintenance of nonfarm buildings nec	1.4	Constr.			
Eating & drinking places	1.3	Trade			
Engineering, architectural, & surveying services	1.3	Services			
Paints & allied products	1.2	Manufg.			
Communications, except radio & TV	1.2	Util.			
Aluminum rolling & drawing	1.1	Manufg.			
Wiring devices	1.1	Manufg.			
Advertising	1.1	Services			
Abrasive products	1.0	Manufg.			
Paperboard containers & boxes	1.0	Manufg.			
Air transportation	1.0	Util.			
Plastics materials & resins	0.9	Manufg.			
Power transmission equipment	0.9	Manufg.			
Nonferrous castings, nec	0.8	Manufg.			
Petroleum refining	0.8	Manufg.			
Signs & advertising displays	0.8	Manufg.			
U.S. Postal Service	0.8	Gov't			
Gas production & distribution (utilities)	0.7	Util.			
Motor freight transportation & warehousing	0.7	Util.			
Real estate	0.7	Fin/R.E.			
Carbon & graphite products	0.6	Manufg.			
Fabricated rubber products, nec	0.6	Manufg.			
Nonferrous rolling & drawing, nec	0.6	Manufg.			
Rubber & plastics hose & belting	0.6	Manufg.			
Computer & data processing services	0.6	Services			
Equipment rental & leasing services	0.6	Services			
Management & consulting services & labs	0.5	Services			
Metal heat treating	0.4	Manufg.			
Banking	0.4	Fin/R.E.			
Hotels & lodging places	0.4	Services			
Legal services	0.4	Services			
Lubricating oils & greases	0.3	Manufg.			
Manifold business forms	0.3	Manufg.			
Switchgear & switchboard apparatus	0.3	Manufg.			
Wood containers	0.3	Manufg.			
Sanitary services, steam supply, irrigation	0.3	Util.			
Security & commodity brokers	0.3	Fin/R.E.			
Accounting, auditing & bookkeeping	0.3	Services			
Automatic merchandising machines	0.2	Manufg.			
Machine tools, metal forming types	0.2	Manufg.			
Insurance carriers	0.2	Fin/R.E.			
Automotive rental & leasing, without drivers	0.1	Services			
Electrical repair shops	0.1	Services			

Source: Benchmark Input-Output Accounts for the U.S. Economy, 1982, U.S. Department of Commerce, Washington, D.C., July 1991. Data, as reported in the source, are organized by the 1977 SIC structure in use in 1982 but have been matched, as closely as is possible, to the 1987 SIC structure used in this book.

OCCUPATIONS EMPLOYED BY SIC 358 - REFRIGERATION AND SERVICE MACHINERY

Occupation	% of Total 1994	Change to 2005	Occupation	% of Total 1994	Change to 2005
Assemblers, fabricators, & hand workers nec	20.6	0.4	Machine tool cutting operators, metal & plastic	1.3	-16.4
Machine assemblers	5.5	50.6	Drafters	1.3	-21.8
Blue collar worker supervisors	3.7	-4.4	Combination machine tool operators	1.3	10.4
Sheet metal workers & duct installers	3.2	0.4	Industrial machinery mechanics	1.2	10.4
Inspectors, testers, & graders, precision	2.7	0.4	Helpers, laborers, & material movers nec	1.2	0.4
Sales & related workers nec	2.6	0.4	Traffic, shipping, & receiving clerks	1.2	-3.4
Welders & cutters	2.5	0.4	Solderers & brazers	1.1	0.4
Industrial truck & tractor operators	2.1	0.4	Stock clerks	1.1	-18.4
Machine tool cutting & forming etc. nec	2.1	0.4	Freight, stock, & material movers, hand	1.1	-19.7
Machine forming operators, metal & plastic	1.7	0.4	Engineering technicians & technologists nec	1.1	0.4
General managers & top executives	1.7	-4.8	Metal & plastic machine workers nec	1.0	-11.1
Welding machine setters, operators	1.7	-9.7	Mechanical engineers	1.0	10.5
Machine builders	1.4	20.4	Machinists	1.0	0.4
Secretaries, ex legal & medical	1.4	-8.6			

Source: Industry-Occupation Matrix, Bureau of Labor Statistics. These data relate to one or more 3-digit SIC industry groups rather than to a single 4-digit SIC. The change reported for each occupation to the year 2005 is a percent of growth or decline as estimated by the Bureau of Labor Statistics. The abbreviation nec stands for 'not elsewhere classified'.

LOCATION BY STATE AND REGIONAL CONCENTRATION

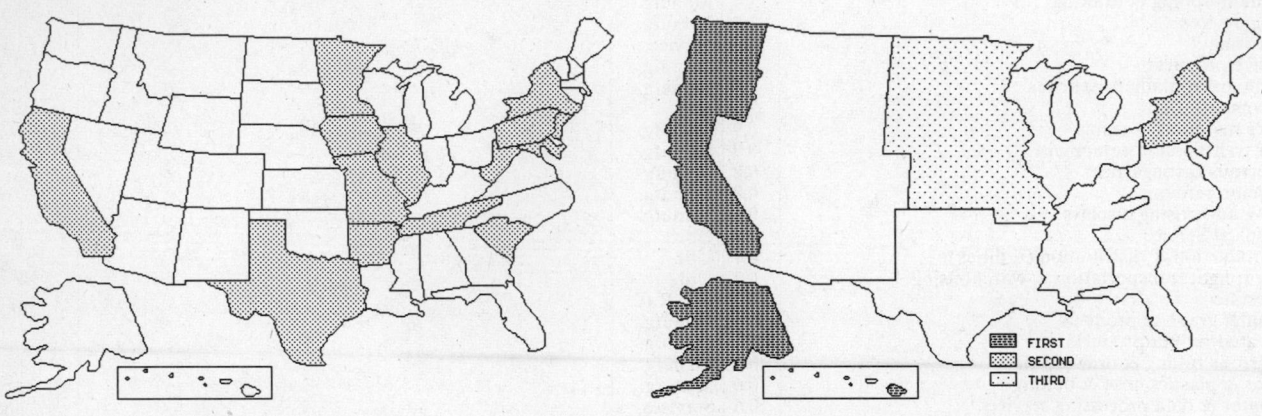

FIRST
SECOND
THIRD

INDUSTRY DATA BY STATE

State	Establish-ments	Shipments			Employment				Cost as % of Shipments	Investment per Employee ($)
		Total ($ mil)	% of U.S.	Per Establ.	Total Number	% of U.S.	Per Establ.	Wages ($/hour)		
Illinois	11	39.0	4.8	3.5	300	4.2	27	8.00	50.0	-
New York	8	22.5	2.8	2.8	300	4.2	38	9.50	48.0	667
California	20	(D)	-	-	1,750 *	24.3	88	-	-	857
Minnesota	6	(D)	-	-	375 *	5.2	63	-	-	-
Pennsylvania	5	(D)	-	-	750 *	10.4	150	-	-	-
Texas	4	(D)	-	-	750 *	10.4	188	-	-	-
Arkansas	3	(D)	-	-	375 *	5.2	125	-	-	-
Iowa	3	(D)	-	-	375 *	5.2	125	-	-	-
Missouri	2	(D)	-	-	1,750 *	24.3	875	-	-	-
Maryland	1	(D)	-	-	175 *	2.4	175	-	-	-
New Jersey	1	(D)	-	-	375 *	5.2	375	-	-	-
South Carolina	1	(D)	-	-	750 *	10.4	750	-	-	-
Tennessee	1	(D)	-	-	175 *	2.4	175	-	-	-
West Virginia	1	(D)	-	-	375 *	5.2	375	-	-	-

Source: 1992 *Economic Census.* The states are in descending order of shipments or establishments (if shipment data are missing for the majority). The symbol (D) appears when data are withheld to prevent disclosure of competitive information. States marked with (D) are sorted by number of establishments. A dash (-) indicates that the data element cannot be calculated; * indicates the midpoint of a range.

3582 - COMMERCIAL LAUNDRY EQUIPMENT

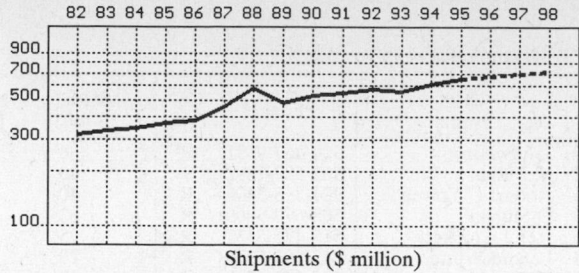

Shipments ($ million)

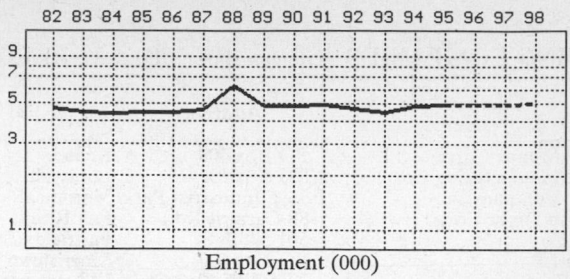

Employment (000)

GENERAL STATISTICS

Year	Com-panies	Establishments Total	Establishments with 20 or more employees	Employment Total (000)	Employment Production Workers (000)	Employment Hours (Mil)	Compensation Payroll ($ mil)	Compensation Wages ($/hr)	Production ($ million) Cost of Materials	Production ($ million) Value Added by Manufacture	Production ($ million) Value of Shipments	Capital Invest.
1982	86	87	45	4.7	3.2	6.4	83.0	7.75	147.8	170.0	321.0	4.9
1983		87	44	4.5	3.1	6.5	85.6	8.12	160.0	177.5	338.5	6.0
1984		87	43	4.4	3.1	6.5	89.5	8.52	159.1	191.5	346.0	9.3
1985		86	41	4.5	3.1	6.5	92.9	8.58	186.1	193.3	373.5	6.8
1986		81	42	4.5	3.1	6.6	100.2	9.06	184.5	196.0	381.6	6.9
1987	81	82	41	4.6	3.4	7.0	107.4	9.53	213.0	247.5	457.0	6.6
1988		79	40	6.3	4.3	9.0	138.3	8.87	308.2	294.1	586.7	9.2
1989		74	35	4.8	3.5	7.2	111.8	8.96	283.7	181.2	478.9	13.4
1990		75	40	4.8	3.5	7.1	120.3	9.49	277.6	240.4	526.6	13.2
1991		75	37	4.9	3.3	7.3	125.9	9.53	287.6	247.0	542.3	11.1
1992	74	77	33	4.7	3.2	6.7	129.3	10.61	281.5	286.7	570.3	10.8
1993		75	35	4.5	3.0	6.3	122.9	10.48	285.5	273.5	555.1	6.2
1994		72P	33P	4.8	3.4	7.0	131.1	10.77	329.3	285.4	614.1	8.8
1995		70P	32P	4.9P	3.4P	7.2P	140.6P	10.84P	345.3P	299.2P	643.9P	11.1P
1996		69P	31P	4.9P	3.4P	7.2P	144.8P	11.07P	358.7P	310.9P	668.9P	11.5P
1997		68P	30P	5.0P	3.5P	7.3P	149.1P	11.30P	372.1P	322.5P	693.9P	11.8P
1998		66P	29P	5.0P	3.5P	7.3P	153.4P	11.53P	385.5P	334.1P	719.0P	12.2P

Sources: 1982, 1987, 1992 *Economic Census*; *Annual Survey of Manufactures*, 83-86, 88-91, 93-94. Establishment counts for non-Census years are from *County Business Patterns*; establishment values for 83-84 are extrapolations. 'P's show projections by the editors. Industries reclassified in 87 will not have data for prior years.

INDICES OF CHANGE

Year	Com-panies	Establishments Total	Establishments with 20 or more employees	Employment Total (000)	Employment Production Workers (000)	Employment Hours (Mil)	Compensation Payroll ($ mil)	Compensation Wages ($/hr)	Production ($ million) Cost of Materials	Production ($ million) Value Added by Manufacture	Production ($ million) Value of Shipments	Capital Invest.
1982	116	113	136	100	100	96	64	73	53	59	56	45
1983		113	133	96	97	97	66	77	57	62	59	56
1984		113	130	94	97	97	69	80	57	67	61	86
1985		112	124	96	97	97	72	81	66	67	65	63
1986		105	127	96	97	99	77	85	66	68	67	64
1987	109	106	124	98	106	104	83	90	76	86	80	61
1988		103	121	134	134	134	107	84	109	103	103	85
1989		96	106	102	109	107	86	84	101	63	84	124
1990		97	121	102	109	106	93	89	99	84	92	122
1991		97	112	104	103	109	97	90	102	86	95	103
1992	100	100	100	100	100	100	100	100	100	100	100	100
1993		97	106	96	94	94	95	99	101	95	97	57
1994		93P	101P	102	106	104	101	102	117	100	108	81
1995		91P	98P	105P	107P	107P	109P	102P	123P	104P	113P	103P
1996		90P	95P	105P	107P	108P	112P	104P	127P	108P	117P	106P
1997		88P	92P	105P	108P	109P	115P	107P	132P	112P	122P	109P
1998		86P	89P	106P	108P	109P	119P	109P	137P	117P	126P	113P

Sources: Same as General Statistics. Values reflect change from the base year, 1992. Values above 100 mean greater than 92, values below 100 mean less than 92, and a value of 100 in the 82-91 or 93-98 period means same as 92. 'P's mark projections by the editors.

SELECTED RATIOS

For 1994	Avg. of All Manufact.	Analyzed Industry	Index	For 1994	Avg. of All Manufact.	Analyzed Industry	Index
Employees per Establishment	49	67	137	Value Added per Production Worker	134,084	83,941	63
Payroll per Establishment	1,500,273	1,829,302	122	Cost per Establishment	5,045,178	4,594,884	91
Payroll per Employee	30,620	27,313	89	Cost per Employee	102,970	68,604	67
Production Workers per Establishment	34	47	138	Cost per Production Worker	146,988	96,853	66
Wages per Establishment	853,319	1,051,953	123	Shipments per Establishment	9,576,895	8,568,837	89
Wages per Production Worker	24,861	22,174	89	Shipments per Employee	195,460	127,938	65
Hours per Production Worker	2,056	2,059	100	Shipments per Production Worker	279,017	180,618	65
Wages per Hour	12.09	10.77	89	Investment per Establishment	321,011	122,791	38
Value Added per Establishment	4,602,255	3,982,326	87	Investment per Employee	6,552	1,833	28
Value Added per Employee	93,930	59,458	63	Investment per Production Worker	9,352	2,588	28

Sources: Same as General Statistics. The 'Average of All Manufacturing' column represents the average of all manufacturing industries reported for the most recent complete year available. The Index shows the relationship between the Average and the Analyzed Industry. For example, 100 means that they are equal; 500 that the Analyzed Industry is five times the average; 50 means that the Analyzed Industry is half the national average. The abbreviation 'na' is used to show that data are 'not available'.

LEADING COMPANIES   Number shown: 19   Total sales ($ mil): 322   Total employment (000): 3.1

Company Name	Address				CEO Name	Phone	Co. Type	Sales ($ mil)	Empl. (000)
Pellerin Milnor Corp	PO Box 400	Kenner	LA	70063	JW Pellerin	504-467-9591	R	86*	0.9
Cissell Manufacturing Co	831 S 1st St	Louisville	KY	40203	M Milam	502-587-1292	S	40	0.4
Unimac Company Inc	3595 Industrial Park	Marianna	FL	32446	Robert C Cowen	904-526-3405	R	40	0.4
American Dryer Corp	88 Currant Rd	Fall River	MA	02720	D Slutsky	508-678-9000	R	33*	0.3
Dexter Co	501 N 8th St	Fairfield	IA	52556	Rex J Crockett	515-472-5131	R	25	0.4
Forenta LP	PO Box 607	Morristown	TN	37814	Leland White	615-586-5370	R	20	0.2
Jensen Corp	2775 NW 63rd Ct	Ft Lauderdale	FL	33309	Howard Eglowstein	305-974-6300	S	17	<0.1
Bermil Industries Corp	461 Doughty Blv	Inwood	NY	11696	B Milch	516-371-4400	R	10	<0.1
Challenge Industries	633 Commerce Dr	Bryan	OH	43506	Larry D Talkington	419-636-3111	S	10	0.1
Chicago Dryer Co	2210 N Pulaski Rd	Chicago	IL	60639	Bruce W Johnson	312-235-4430	R	10*	0.1
Hoyt Corp	251 Forge Rd	Westport	MA	02790	Steven D Rooney	508-636-8811	R	8*	<0.1
Ludell Manufacturing Co	5200 W State St	Milwaukee	WI	53208	Gerald Roszak	414-476-9934	R	6	<0.1
MMC Systems	5922 San Pedro Av	San Antonio	TX	78212	Joe Hemmi	210-344-8551	R	4	<0.1
Edro Corp	PO Box 308	East Berlin	CT	06023	E S Kirejczyk, Jr	203-828-0311	R	4	<0.1
Easy Card Corp	100 Hoods Ln	Marblehead	MA	01945	John Hooper	617-639-4000	R	3*	<0.1
Sioux Steam Cleaner Corp	1 Sioux Plz	Beresford	SD	57004	JF Finger	605-763-2776	R	3	<0.1
Bishop Freeman Co	1600 Foster St	Evanston	IL	60204	Steven Davis	708-328-5200	R	2	<0.1
Rototherm Corp	30 Laurel Pl	Howell	NJ	07731	Ben D Herschel	908-370-0695	R	1	<0.1
Web Systems Inc	1354 Linden Dr	Boulder	CO	80304	Edward Nowaczek	303-440-4868	R	1	<0.1

Source: *Ward's Business Directory of U.S. Private and Public Companies*, Volumes 1 and 2, 1996. The company type code used is as follows: P - Public, R - Private, S - Subsidiary, D - Division, J - Joint Venture, A - Affiliate, G - Group. Sales are in millions of dollars, employees are in thousands. An asterisk (*) indicates an estimated sales volume. The symbol < stands for 'less than'. Company names and addresses are truncated, in some cases, to fit into the available space.

MATERIALS CONSUMED

Material	Quantity	Delivered Cost ($ million)
Materials, ingredients, containers, and supplies	(X)	255.7
Metal stampings	(X)	10.8
Metal pipe, valves, and pipe fittings, except plumbers' and forgings	(X)	7.9
Other fabricated metal products (except forgings)	(X)	19.8
Forgings	(X)	0.2
Castings (rough and semifinished)	(X)	6.6
Steel sheet and strip, including tin plate	(X)	40.7
All other steel shapes and forms	(X)	8.2
Nonferrous shapes and forms	(X)	8.4
Other fractional horsepower electric motors (under 1 hp)	(X)	18.4
Integral horsepower electric motors and generators (1 hp or more)	(X)	10.1
Rubber and plastics hose and belting	(X)	4.3
Plain bearings and bushings	(X)	15.0
Mechanical speed changers, gears, and industrial high-speed drives	(X)	8.9
Current-carrying wiring devices	(X)	7.9
Electrical transmission, distribution, and control equipment	(X)	20.5
Wood boxes, pallets, skids, and containers	(X)	2.2
All other materials and components, parts, containers, and supplies	(X)	46.6
Materials, ingredients, containers, and supplies, nsk	(X)	19.1

Source: 1992 *Economic Census*. Explanation of symbols used: (D): Withheld to avoid disclosure of competitive data; na: Not available; (S): Withheld because statistical norms were not met; (X): Not applicable; (Z): Less than half the unit shown; nec: Not elsewhere classified; nsk: Not specified by kind; - : zero; * : 10-19 percent estimated; ** : 20-29 percent estimated.

PRODUCT SHARE DETAILS

Product or Product Class	% Share	Product or Product Class	% Share
Commercial laundry equipment	100.00	Commercial laundry presses, more than 10 kg (22 lb) load capacity, except parts, attachments, and accessories	7.84
Commercial laundry washers (only), more than 10 kg (22 lb) load capacity, except parts, attachments, and accessories	0.62	Other commercial laundry equipment (more than 10 kg (22 lb) load capacity), including extractors only, flatwork ironers, etc., except parts, attachments, and accessories	15.42
Commercial laundry washer-extractor combination machines, more than 10 kg (22 lb) load capacity, except parts, attachments, and accessories	26.39	Parts, attachments, and accessories for commercial laundry equipment and presses, more than 10 kg (22 lb) load capacity	11.56
Commercial coin-operated laundry drying tumblers, more than 10 kg (22 lb) load capacity, except parts, attachments, and accessories	13.04	Commercial dry-cleaning units, presses (including garment manufacturers' (needle trades) presses), reclaiming units, etc., except parts, attachments, and accessories	7.40
Commercial laundry drying tumblers, other than coin-operated, more than 10 kg (22 lb) load capacity, except parts, attachments, and accessories	6.83	Parts, attachments, and accessories for commercial dry-cleaning equipment and clothing presses	3.13

Source: 1992 *Economic Census*. The values shown are percent of total shipments in an industry. Values of indented subcategories are summed in the main heading. The symbol (D) appears when data are withheld to prevent disclosure of competitive information. The abbreviation nsk stands for 'not specified by kind' and nec for 'not elsewhere classified'.

INPUTS AND OUTPUTS FOR COMMERCIAL LAUNDRY EQUIPMENT

Economic Sector or Industry Providing Inputs	%	Sector	Economic Sector or Industry Buying Outputs	%	Sector
Imports	13.1	Foreign	Gross private fixed investment	60.6	Cap Inv
Petroleum refining	11.9	Manufg.	Laundry, dry cleaning, shoe repair	17.5	Services
Blast furnaces & steel mills	9.8	Manufg.	Exports	12.0	Foreign
Wholesale trade	9.1	Trade	S/L Govt. purch., higher education	4.1	S/L Govt
Motors & generators	6.0	Manufg.	Federal Government purchases, national defense	2.3	Fed Govt
Miscellaneous plastics products	4.7	Manufg.	Commercial laundry equipment	1.3	Manufg.
Machinery, except electrical, nec	3.0	Manufg.	S/L Govt. purch., natural resource & recreation.	1.0	S/L Govt
Automotive rental & leasing, without drivers	2.5	Services	S/L Govt. purch., health & hospitals	0.8	S/L Govt
Switchgear & switchboard apparatus	2.2	Manufg.	S/L Govt. purch., correction	0.4	S/L Govt
Commercial laundry equipment	2.1	Manufg.			
Advertising	2.0	Services			
Automotive repair shops & services	1.8	Services			
Power transmission equipment	1.6	Manufg.			
Special dies & tools & machine tool accessories	1.3	Manufg.			
Transformers	1.3	Manufg.			
Automatic merchandising machines	1.2	Manufg.			
Iron & steel foundries	1.2	Manufg.			
Nonferrous castings, nec	1.1	Manufg.			
Pipe, valves, & pipe fittings	1.1	Manufg.			
Die-cut paper & board	1.0	Manufg.			
Electric services (utilities)	1.0	Util.			
Engineering, architectural, & surveying services	1.0	Services			
Sheet metal work	0.9	Manufg.			
Screw machine and related products	0.8	Manufg.			
Eating & drinking places	0.8	Trade			
Maintenance of nonfarm buildings nec	0.7	Constr.			
Aluminum rolling & drawing	0.7	Manufg.			
Metal stampings, nec	0.7	Manufg.			
Refrigeration & heating equipment	0.7	Manufg.			
Wood containers	0.7	Manufg.			
Air transportation	0.7	Util.			
Motor freight transportation & warehousing	0.7	Util.			
Real estate	0.7	Fin/R.E.			
Security & commodity brokers	0.7	Fin/R.E.			
Abrasive products	0.6	Manufg.			
Wiring devices	0.6	Manufg.			
Insurance carriers	0.6	Fin/R.E.			
Motor vehicle parts & accessories	0.5	Manufg.			
Rubber & plastics hose & belting	0.5	Manufg.			
Tires & inner tubes	0.5	Manufg.			
Communications, except radio & TV	0.5	Util.			
Retail trade, except eating & drinking	0.5	Trade			
Aluminum castings	0.4	Manufg.			
Gaskets, packing & sealing devices	0.4	Manufg.			
Pumps & compressors	0.4	Manufg.			
Gas production & distribution (utilities)	0.4	Util.			
Banking	0.4	Fin/R.E.			
Lubricating oils & greases	0.3	Manufg.			
Nonferrous wire drawing & insulating	0.3	Manufg.			
Paperboard containers & boxes	0.3	Manufg.			
Legal services	0.3	Services			
Management & consulting services & labs	0.3	Services			
U.S. Postal Service	0.3	Gov't			
Manifold business forms	0.2	Manufg.			
Metal heat treating	0.2	Manufg.			
Plastics materials & resins	0.2	Manufg.			
Accounting, auditing & bookkeeping	0.2	Services			
Detective & protective services	0.2	Services			
Hotels & lodging places	0.2	Services			
State & local government enterprises, nec	0.2	Gov't			
Pipelines, except natural gas	0.1	Util.			
Railroads & related services	0.1	Util.			
Water transportation	0.1	Util.			

Source: Benchmark Input-Output Accounts for the U.S. Economy, 1982, U.S. Department of Commerce, Washington, D.C., July 1991. Data, as reported in the source, are organized by the 1977 SIC structure in use in 1982 but have been matched, as closely as is possible, to the 1987 SIC structure used in this book.

OCCUPATIONS EMPLOYED BY SIC 358 - REFRIGERATION AND SERVICE MACHINERY

Occupation	% of Total 1994	Change to 2005	Occupation	% of Total 1994	Change to 2005
Assemblers, fabricators, & hand workers nec	20.6	0.4	Machine tool cutting operators, metal & plastic	1.3	-16.4
Machine assemblers	5.5	50.6	Drafters	1.3	-21.8
Blue collar worker supervisors	3.7	-4.4	Combination machine tool operators	1.3	10.4
Sheet metal workers & duct installers	3.2	0.4	Industrial machinery mechanics	1.2	10.4
Inspectors, testers, & graders, precision	2.7	0.4	Helpers, laborers, & material movers nec	1.2	0.4
Sales & related workers nec	2.6	0.4	Traffic, shipping, & receiving clerks	1.2	-3.4
Welders & cutters	2.5	0.4	Solderers & brazers	1.1	0.4
Industrial truck & tractor operators	2.1	0.4	Stock clerks	1.1	-18.4
Machine tool cutting & forming etc. nec	2.1	0.4	Freight, stock, & material movers, hand	1.1	-19.7
Machine forming operators, metal & plastic	1.7	0.4	Engineering technicians & technologists nec	1.1	0.4
General managers & top executives	1.7	-4.8	Metal & plastic machine workers nec	1.0	-11.1
Welding machine setters, operators	1.7	-9.7	Mechanical engineers	1.0	10.5
Machine builders	1.4	20.4	Machinists	1.0	0.4
Secretaries, ex legal & medical	1.4	-8.6			

Source: *Industry-Occupation Matrix*, Bureau of Labor Statistics. These data relate to one or more 3-digit SIC industry groups rather than to a single 4-digit SIC. The change reported for each occupation to the year 2005 is a percent of growth or decline as estimated by the Bureau of Labor Statistics. The abbreviation nec stands for 'not elsewhere classified'.

LOCATION BY STATE AND REGIONAL CONCENTRATION

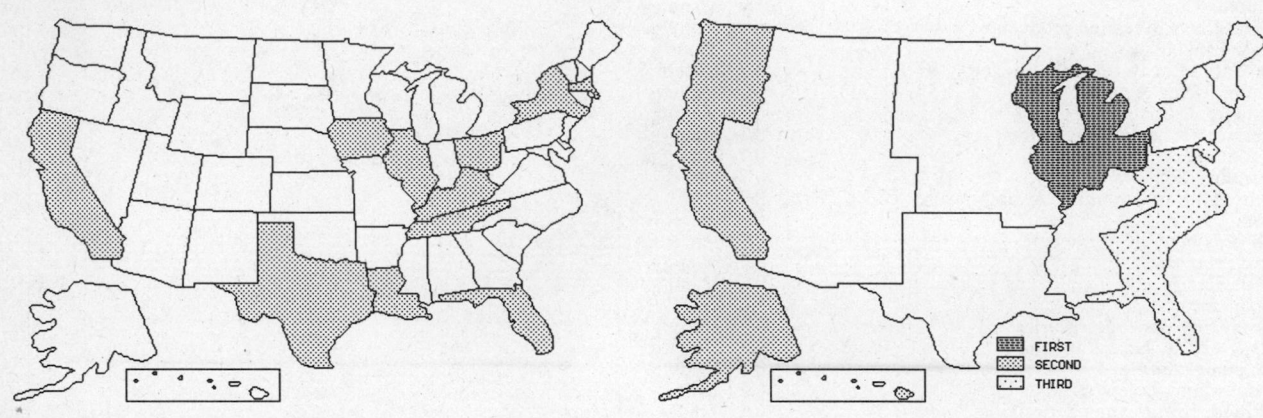

FIRST
SECOND
THIRD

INDUSTRY DATA BY STATE

State	Establish-ments	Shipments Total ($ mil)	Shipments % of U.S.	Shipments Per Establ.	Employment Total Number	Employment % of U.S.	Employment Per Establ.	Wages ($/hour)	Cost as % of Shipments	Investment per Employee ($)
Florida	7	93.8	16.4	13.4	600	12.8	86	8.38	59.1	-
Illinois	12	(D)	-	-	375 *	8.0	31	-	-	-
California	10	(D)	-	-	175 *	3.7	18	-	-	1,143
Ohio	5	(D)	-	-	375 *	8.0	75	-	-	-
Massachusetts	4	(D)	-	-	375 *	8.0	94	-	-	-
Texas	4	(D)	-	-	375 *	8.0	94	-	-	-
Iowa	2	(D)	-	-	375 *	8.0	188	-	-	-
Kentucky	2	(D)	-	-	750 *	16.0	375	-	-	-
New York	2	(D)	-	-	375 *	8.0	188	-	-	-
Tennessee	2	(D)	-	-	175 *	3.7	88	-	-	-
Louisiana	1	(D)	-	-	750 *	16.0	750	-	-	-

Source: 1992 *Economic Census*. The states are in descending order of shipments or establishments (if shipment data are missing for the majority). The symbol (D) appears when data are withheld to prevent disclosure of competitive information. States marked with (D) are sorted by number of establishments. A dash (-) indicates that the data element cannot be calculated; * indicates the midpoint of a range.

3585 - REFRIGERATION & HEATING EQUIPMENT

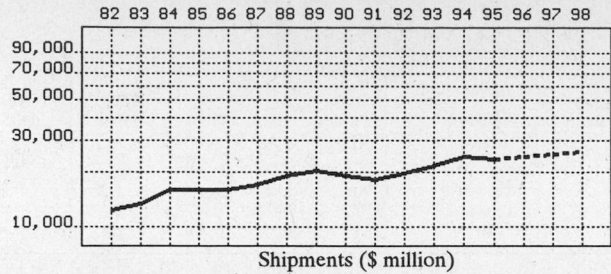

Shipments ($ million)

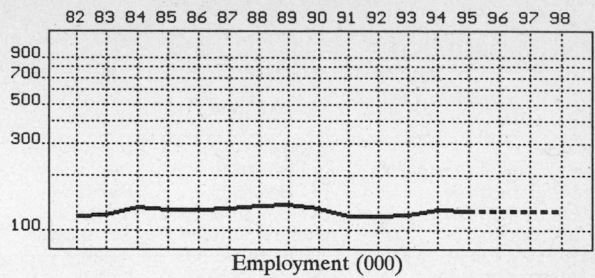

Employment (000)

GENERAL STATISTICS

| Year | Com-panies | Establishments | | Employment | | | Compensation | | Production ($ million) | | | |
		Total	with 20 or more employees	Total (000)	Production Workers (000)	Hours (Mil)	Payroll ($ mil)	Wages ($/hr)	Cost of Materials	Value Added by Manufacture	Value of Shipments	Capital Invest.
1982	730	865	496	120.5	85.0	159.3	2,393.2	9.62	6,190.6	6,049.4	12,390.3	329.3
1983		852	496	122.1	88.5	167.3	2,565.0	10.04	6,966.7	6,373.1	13,375.2	225.6
1984		839	496	134.2	97.7	189.6	3,021.0	10.46	8,576.8	7,409.1	15,775.5	418.9
1985		827	495	130.4	94.2	183.3	3,102.3	11.03	8,608.2	7,249.5	15,836.4	436.5
1986		806	497	128.9	91.8	181.3	3,194.5	11.40	8,484.8	7,212.6	15,826.2	493.1
1987	736	892	511	133.3	96.4	192.8	3,355.1	11.26	9,128.8	8,051.9	17,035.4	433.3
1988		914	536	135.2	99.5	200.2	3,456.7	11.35	10,373.1	8,822.6	18,963.6	452.4
1989		896	526	138.8	101.5	204.3	3,683.1	12.01	11,161.7	9,340.7	20,249.1	534.2
1990		885	519	131.2	92.4	183.9	3,507.2	12.30	10,425.0	8,339.5	19,043.2	472.4
1991		905	511	119.0	87.1	173.4	3,377.1	12.70	9,867.4	8,186.2	18,223.8	443.2
1992	769	913	512	120.5	88.9	179.8	3,590.2	13.27	10,341.2	9,413.3	19,697.0	556.1
1993		918	503	121.9	90.1	183.9	3,721.4	13.36	11,363.6	10,124.9	21,529.7	517.9
1994		925P	521P	130.6	99.1	201.6	4,103.1	13.76	13,359.6	11,285.5	24,397.0	779.0
1995		933P	523P	127.5P	94.9P	195.6P	4,086.0P	14.05P	12,847.2P	10,852.7P	23,461.3P	652.8P
1996		941P	525P	127.4P	95.1P	197.2P	4,196.4P	14.38P	13,284.4P	11,222.0P	24,259.7P	679.1P
1997		948P	527P	127.2P	95.3P	198.7P	4,306.8P	14.71P	13,721.6P	11,591.3P	25,058.1P	705.4P
1998		956P	529P	127.1P	95.6P	200.3P	4,417.2P	15.05P	14,158.8P	11,960.6P	25,856.5P	731.7P

Sources: 1982, 1987, 1992 *Economic Census*; *Annual Survey of Manufactures*, 83-86, 88-91, 93-94. Establishment counts for non-Census years are from *County Business Patterns*; establishment values for 83-84 are extrapolations. 'P's show projections by the editors. Industries reclassified in 87 will not have data for prior years.

INDICES OF CHANGE

| Year | Com-panies | Establishments | | Employment | | | Compensation | | Production ($ million) | | | |
		Total	with 20 or more employees	Total (000)	Production Workers (000)	Hours (Mil)	Payroll ($ mil)	Wages ($/hr)	Cost of Materials	Value Added by Manufacture	Value of Shipments	Capital Invest.
1982	95	95	97	100	96	89	67	72	60	64	63	59
1983		93	97	101	100	93	71	76	67	68	68	41
1984		92	97	111	110	105	84	79	83	79	80	75
1985		91	97	108	106	102	86	83	83	77	80	78
1986		88	97	107	103	101	89	86	82	77	80	89
1987	96	98	100	111	108	107	93	85	88	86	86	78
1988		100	105	112	112	111	96	86	100	94	96	81
1989		98	103	115	114	114	103	91	108	99	103	96
1990		97	101	109	104	102	98	93	101	89	97	85
1991		99	100	99	98	96	94	96	95	87	93	80
1992	100	100	100	100	100	100	100	100	100	100	100	100
1993		101	98	101	101	102	104	101	110	108	109	93
1994		101P	102P	108	111	112	114	104	129	120	124	140
1995		102P	102P	106P	107P	109P	114P	106P	124P	115P	119P	117P
1996		103P	102P	106P	107P	110P	117P	108P	128P	119P	123P	122P
1997		104P	103P	106P	107P	111P	120P	111P	133P	123P	127P	127P
1998		105P	103P	106P	108P	111P	123P	113P	137P	127P	131P	132P

Sources: Same as General Statistics. Values reflect change from the base year, 1992. Values above 100 mean greater than 92, values below 100 mean less than 92, and a value of 100 in the 82-91 or 93-98 period means same as 92. 'P's mark projections by the editors.

SELECTED RATIOS

For 1994	Avg. of All Manufact.	Analyzed Industry	Index	For 1994	Avg. of All Manufact.	Analyzed Industry	Index
Employees per Establishment	49	141	288	Value Added per Production Worker	134,084	113,880	85
Payroll per Establishment	1,500,273	4,433,605	296	Cost per Establishment	5,045,178	14,435,717	286
Payroll per Employee	30,620	31,417	103	Cost per Employee	102,970	102,294	99
Production Workers per Establishment	34	107	312	Cost per Production Worker	146,988	134,809	92
Wages per Establishment	853,319	2,997,463	351	Shipments per Establishment	9,576,895	26,362,181	275
Wages per Production Worker	24,861	27,992	113	Shipments per Employee	195,460	186,807	96
Hours per Production Worker	2,056	2,034	99	Shipments per Production Worker	279,017	246,186	88
Wages per Hour	12.09	13.76	114	Investment per Establishment	321,011	841,749	262
Value Added per Establishment	4,602,255	12,194,548	265	Investment per Employee	6,552	5,965	91
Value Added per Employee	93,930	86,413	92	Investment per Production Worker	9,352	7,861	84

Sources: Same as General Statistics. The 'Average of All Manufacturing' column represents the average of all manufacturing industries reported for the most recent complete year available. The Index shows the relationship between the Average and the Analyzed Industry. For example, 100 means that they are equal; 500 that the Analyzed Industry is five times the average; 50 means that the Analyzed Industry is half the national average. The abbreviation 'na' is used to show that data are 'not available'.

LEADING COMPANIES Number shown: 75 Total sales ($ mil): 28,394 Total employment (000): 206.0

Company Name	Address				CEO Name	Phone	Co. Type	Sales ($ mil)	Empl. (000)
Carrier Corp	1 Carrier Pl	Farmington	CT	06034	William S Frago	203-674-3000	S	4,500	27.0
American Standard Companies	PO Box 6820	Piscataway	NJ	08855	E A Kampouris	908-980-6000	P	4,457	38.0
American Standard Inc	PO Box 6820	Piscataway	NJ	08855	E A Kampouris	908-980-6000	S	3,831	38.5
York International Corp	PO Box 1592-364M	York	PA	17405	R N Pokelwaldt	717-771-7890	P	2,032	13.8
Tecumseh Products Co	General Office	Tecumseh	MI	49286	Todd W Herrick	517-423-8411	P	1,533	14.4
ABB Flexible Automation Inc	7000 Central NE	Atlanta	GA	30328	John Camardella	404-393-6130	S	1,200	3.6
Lennox International Inc	PO Box 799900	Dallas	TX	75379	Tom Keefe	214-497-5000	R	1,047	8.0
Pace Industries Inc	405 Lexington Av	New York	NY	10174	William J Bisset	212-916-8199	S	950	5.0
Rheem Manufacturing Co	405 Lexington Av	New York	NY	10174	Gary Tapella	212-916-8100	S	950	5.0
Hussmann Corp	12999 Charles Rock	Bridgeton	MO	63044	J Larry Vowell	314-291-2000	S	860	6.7
Copeland Corp	1675 Campbell Rd	Sidney	OH	45365	Dean Ruwe	513-498-3011	S	780*	4.5
Bristol Compressors Inc	15185 Industrial Pk	Bristol	VA	24202	Joseph Loprete	703-466-4121	S	440*	3.0
Rheem Air Conditioning	PO Box 17010	Fort Smith	AR	72917	Ross Willis	501-646-4311	D	350*	1.9
McQuay International	PO Box 1551	Minneapolis	MN	55440	Charles Tamborino	612-553-5330	D	330	2.5
Kysor Industrial Corp	1 Madison Av	Cadillac	MI	49601	George R Kempton	616-779-2200	P	314	2.0
Heatcraft Inc	PO Box 948	Grenada	MS	38901	Dennis Custance	601-226-3421	S	300	2.4
Scotsman Industries Inc	775 Corp Woods	Vernon Hills	IL	60061	Richard C Osborne	708-215-4500	P	267	0.9
Fedders Corp	PO Box 813	Liberty Corner	NJ	07938	S Giordano Jr	908-234-2100	P	232	1.8
Fedders North America Inc	415 Wabash Av	Effingham	IL	62401	NW Swartz	217-342-3901	S	190*	1.2
Sanden International	601 S Sanden Blv	Wylie	TX	75098	Michitake Naka	214-442-8400	S	185	0.5
Baltimore Aircoil Company Inc	PO Box 7322	Baltimore	MD	21227	M J McKenna	410-799-6200	S	180*	1.5
Nordyne Inc	1801 Park 270 Dr	St Louis	MO	63416	R EG Ractliffe	314-878-6200	S	180	1.0
Tyler Refrigeration Corp	1329 Lake St	Niles	MI	49120	James E Mack	616-683-2000	R	160	1.3
Evcon Industries Inc	PO Box 19014	Wichita	KS	67204	Michael Young	316-832-6300	R	150	0.8
Dunham-Bush Inc	175 South St	West Hartford	CT	06110	Thomas A Zacaroli	203-249-8671	R	130	1.1
Sanyo E and E Corp	2001 Sanyo Av	San Diego	CA	92173	Shigeru Otsuka	619-661-1134	S	121	0.3
Heatcraft Inc	PO Box 1699	Atlanta	GA	30371	Ed French	404-939-4450	D	120	0.6
Beverage-Air	700 Ruffington Rd	Spartanburg	SC	29304	W W Robertson	803-582-8111	D	119	0.4
AdobeAir Inc	500 S 15th St	Phoenix	AZ	85084	Ron Rosin	602-257-0060	R	100	1.0
Evapco Inc	PO Box 1300	Westminster	MD	21157	JW Bowles	410-756-2600	R	100	0.6
Frick Co	PO Box 997	Waynesboro	PA	17268	James E Erdly	717-762-2121	D	100	0.6
Wynn's Climate Systems Inc	SE Loop 820	Fort Worth	TX	76140	John J Halenda	817-293-4600	S	100	0.5
Manitowoc Equipment Works	PO Box 1720	Manitowoc	WI	54221	Terry Growcock	414-682-0161	D	93	0.3
Thermo Power Corp	PO Box 9046	Waltham	MA	02254	M J Armstrong	617-622-1111	P	89	0.4
Jordon Coml Refrigeration Co	2200 Kennedy St	Philadelphia	PA	19137	William Fogel	215-535-8300	R	80	1.0
Kolpak	PO Box 137	River Falls	WI	54022	Gary Hainky	715-425-6741	D	80	0.7
Master-Bilt Products	Hwy 15 N	New Albany	MS	38652	Michael Stewart	601-534-9061	D	80	0.9
Shannon Group	PO Box 137	River Falls	WI	54022	Gary Hainley	715-425-6741	D	80	0.7
AAON Inc	2425 S Yukon Av	Tulsa	OK	74107	N H Asbjornson	918-583-2266	P	80	0.4
Nycor Inc	287 Childs Rd	Basking Ridge	NJ	07920	Joseph Giordano	908-953-8200	P	75	0.8
Russell Coil Co	PO Box 1030	Brea	CA	92622	WJ Wilson	714-529-1935	S	70	0.4
Reznor	McKinley Av	Mercer	PA	16137	Dick McCullough	412-662-4400	D	65	0.3
Zurn Balcke-Durr Inc	405 N Reo St	Tampa	FL	33609	Jack T Eunson	813-289-1516	J	65	0.3
Signet Systems Inc	551 Tapp Rd	Harrodsburg	KY	40330	LE Mentzer	606-734-7711	S	60	0.4
Phoenix Refrigeration Systems	709 Sigman Rd NE	Conyers	GA	30208	Grant Brown	404-388-0706	S	60	0.4
Rotorex Conpnay Inc	PO Box 1168	Frederick	MD	21701		301-898-7011	S	60*	0.5
Revco/Limberg	275 Aiken Rd	Asheville	NC	28804	DB Dawley	704-658-2711	S	53*	0.4
Vilter Manufacturing Corp	2217 S 1st St	Milwaukee	WI	53207	Paul Szymaszek	414-744-0111	R	53*	0.4
Heat Controller Inc	PO Box 1089	Jackson	MI	49204	JA Knight	517-787-2100	R	50	<0.1
Parker Hannifin Corp	100 Dunn Rd	Lyons	NY	14489	Robert Havrilla	315-946-4891	D	50	0.5
Sunroc Corp	300 S Pennell Rd	Media	PA	19063	A A Salamone	215-459-1100	R	50	0.4
Friedrich Air Conditioning Co	PO Box 1540	San Antonio	TX	78295	Charles M Marino	210-225-2000	D	49	0.4
Ice-O-Matic	PO Box 39487	Denver	CO	80239	Frank Thomas	303-371-3737	S	47*	0.3
Engineered Air Systems Inc	1270 N Price Rd	St Louis	MO	63132	MF Shanahan	314-993-5880	S	44	0.2
RECO York International	5680 E Houston St	San Antonio	TX	78220	Terry Hobson	210-662-5700	D	41	0.3
Frigette Corp	PO Box 40577	Fort Worth	TX	76140	Holt Hickman	817-293-5313	R	40	0.5
Gem Products Inc	12472 Edison Way	Garden Grove	CA	92641	Kenny Pritchett	714-898-2788	S	40	0.2
Unitech	2720 US Hwy 22	Union	NJ	07083	Roland Flores	908-964-2400	D	40	0.1
Commercial Environmental Syst	101 W 82nd St	Chaska	MN	55318	Dave Huntley	612-361-2711	S	36*	0.3
NESLAB Instruments Inc	PO Box 1178	Portsmouth	NH	03802	Andrew Bebbington	603-436-9444	S	33	0.3
Flex Technologies Inc	PO Box 400	Midvale	OH	44653	Glenn E Burket	614-922-5992	R	32	0.8
Mammoth Inc	101 W 82nd St	Chaska	MN	55318	DJ Huntley	612-361-2711	S	31*	0.3
White-Rodgers Inc	131 Godfrey St	Logansport	IN	46947	Jim George	219-753-0600	R	31*	0.3
Bergstrom Manufacturing Co	PO Box 6007	Rockford	IL	61125	DR Rydell	815-874-7821	R	30	0.4
Governair Corp	4841 N Sewell	Oklahoma City	OK	73118	Walter P Mecozzi	405-525-6546	S	30	0.2
Keco Industries Inc	7375 Industrial Rd	Florence	KY	41042	George W Andrews	606-525-2102	R	30*	0.4
Peerless of America Inc	PO Box 850	Lincolnshire	IL	60064	R Paulman	708-634-7500	R	30	0.3
White Industries LLC	8804 Bash St	Indianapolis	IN	46256	Stan Hirschseld	317-849-6830	R	30*	0.1
Kramer Co	3075 N Lanier Pkwy	Decatur	GA	30034	Will Wilson	404-244-8004	R	27*	0.2
Astro Air Inc	PO Box 1988	Jacksonville	TX	75766	Rex Dacus	903-586-3691	R	26	0.2
Mile High Equipment Co	11100 E 45th Av	Denver	CO	80239	Frank Thomas	303-371-3737	S	26	0.2
Nuclear Cooling Inc	PO Box 171	High Ridge	MO	63049	David G Ault	314-677-6600	R	26	0.2
Standard Refrigeration Co	2050 N Ruby St	Melrose Park	IL	60160	R Levin	708-345-5400	R	25	0.3
Anemostat Products	888 N Keyser Av	Scranton	PA	18501	Robert Yarnchak	717-346-6586	D	25	0.3
Duro Dyne Corp	PO Box 117	Farmingdale	NY	11735	M Hinden	516-249-9000	S	25	0.1

Source: Ward's Business Directory of U.S. Private and Public Companies, Volumes 1 and 2, 1996. The company type code used is as follows: P - Public, R - Private, S - Subsidiary, D - Division, J - Joint Venture, A - Affiliate, G - Group. Sales are in millions of dollars, employees are in thousands. An asterisk (*) indicates an estimated sales volume. The symbol < stands for 'less than'. Company names and addresses are truncated, in some cases, to fit into the available space.

MATERIALS CONSUMED

Material	Quantity	Delivered Cost ($ million)
Materials, ingredients, containers, and supplies	(X)	9,607.0
Refrigeration compressors, compressor units, condensing units, and other heat transfer equipment	(X)	1,497.5
Fractional horsepower electric timing motors, synchronous and subsynchronous (less than 1 hp)	(X)	210.0
Other fractional horsepower electric motors (less than 1 hp)	(X)	316.7
Integral horsepower electric motors and generators (1 hp or more)	(X)	507.5
Fans and blowers	(X)	136.5
Electrical transmission, distribution, and control equipment	(X)	341.8
Current-carrying wiring devices	(X)	114.5
Automatic temperature controls (thermostats, regulators, etc.)	(X)	386.6
Ball and roller bearings (mounted or unmounted)	(X)	46.0
Metal bolts, nuts, screws, washers, rivets, and other screw machine products	(X)	178.7
Metal stampings	(X)	231.1
Metal hardware, including hinges, handles, locks, casters, etc. (except castings and forgings)	(X)	55.8
Metal pipe, valves, and pipe fittings (except forgings)	(X)	162.0
All other fabricated metal products (except forgings)	(X)	236.5
Forgings	(X)	22.1
Iron and steel castings (rough and semifinished)	(X)	300.9
Aluminum and aluminum-base alloy castings (rough and semifinished)	(X)	182.2
Other nonferrous castings (rough and semifinished)	(X)	5.7
Steel bars, bar shapes, and plates	(X)	101.1
Steel sheet and strip, including tin plate	(X)	549.8
Steel wire and wire products	(X)	45.7
All other steel shapes and forms	(X)	111.7
Copper and copper-base alloy pipe and tube	(X)	384.6
All other copper and copper-base alloy shapes and forms	(X)	115.5
Aluminum and aluminum-base alloy sheet, plate, foil, and welded tubing	(X)	410.8
All other aluminum and aluminum-base alloy shapes and forms	(X)	93.4
Other nonferrous shapes and forms	(X)	28.0
Paper and paperboard containers, including shipping sacks and other paper packaging supplies	(X)	128.0
Wooden containers, complete (including combination woood and paperboard)	(X)	29.9
Paints, varnishes, lacquers, stains, shellacs, japans, enamels, and allied products	(X)	55.0
Refrigerant gases and other synthetic organic chemicals	(X)	55.8
Plastics resins consumed in the form of granules, pellets, powders, liquids, etc.	(X)	52.5
Plastics products consumed in the form of sheets, rods, tubes, film, and other shapes	(X)	53.1
Fabricated plastics products (except gaskets, hoses, and belting)	(X)	226.4
Rubber and plastics hose and belting	(X)	104.3
Gaskets (all types) and asbestos packing	(X)	66.2
Mineral wool insulation (fibrous glass, rock wool, etc.)	(X)	62.6
All other materials and components, parts, containers, and supplies	(X)	1,317.1
Materials, ingredients, containers, and supplies, nsk	(X)	683.9

Source: 1992 *Economic Census*. Explanation of symbols used: (D): Withheld to avoid disclosure of competitive data; na: Not available; (S): Withheld because statistical norms were not met; (X): Not applicable; (Z): Less than half the unit shown; nec: Not elsewhere classified; nsk: Not specified by kind; - : zero; * : 10-19 percent estimated; ** : 20-29 percent estimated.

PRODUCT SHARE DETAILS

Product or Product Class	% Share	Product or Product Class	% Share
Refrigeration and heating equipment	100.00	Commercial mechanical refrigerated bulk beverage dispensers, including malt dispensers and precooler cabinets, except coin-operated	7.90
Heat transfer equipment (except electrically operated dehumidifiers), mechanically refrigerated, self-contained	24.42	Other commercial refrigerators and related equipment	11.40
Unitary air-conditioners	21.03	Commercial refrigerators and related equipment, nsk	0.75
Commercial refrigerators and related equipment	10.08	Compressors and compressor units, all refrigerants	17.04
Commercial refrigerated sectional coolers or cooling rooms of the prefabricated (factory produced) type, including self-contained and remote units	16.87	Refrigeration condensing units, all refrigerants, except ammonia (complete)	1.16
Commercial reach-in refrigerators and reach-in verticle display cabinets for normal temperature applications (not intended for frozen foods, ice cream, etc.), incl. self-contained and remote units	12.30	Room air-conditioners and dehumidifiers, except portable dehumidifiers	4.86
Commercial reach-in refrigerators and reach-in type verticle display cabinets for low temperature application, including self-contained and remote units	12.05	Refrigeration and air-conditioning equipment, nec	3.84
Commercial closed refrigerated display cases, operated at normal temperatures, including self-contained and remote units	2.39	Soda fountain refrigeration equipment (cooler box, fountainette, and similar equipment)	24.26
		Beer dispensing refrigeration equipment	3.44
Commercial open, one level, self-service refrigerated display cases, operated at normal temperatures, including self-contained and remote units	7.81	Evaporative air coolers	23.26
		Other refrigeration machinery and air-conditioning equipment	46.34
Commercial open, multilevel, self-service refrigerated display cases, operated at normal temperatures, including self-contained and remote units	8.33	Refrigeration and air-conditioning equipment, nec, nsk	2.71
		Warm air furnaces, including duct furnaces and humidifiers, and electric comfort heating equipment	7.23
Commercial open, self-service refrigerated frozen food display cases, including self-contained and remote units	3.39	Parts and accessories for air-conditioning and heat transfer equipment	8.12
Commercial closed, refrigerated frozen food cabinets, other than reach-in type, including self-contained and remote units	2.10	Parts for heat transfer equipment, including parts for air-conditioning condensing units	42.87
Other commercial refrigerated display cases operated at low temperatures, including self-contained and remote units	2.17	Parts for unitary air-conditioners	19.43
		Parts for commercial refrigeration and related equipment	9.52
Commercial mechanical refrigerated drinking water coolers	11.32	Parts for compressors and compressor units	13.59
Commercial mechanical refrigerated bottled beverage coolers, dry and wet types, except coin-operated	1.23	Parts for condensing units, excluding air-conditioning condensing units	0.46
		Parts for dehumidifiers and room air-conditioners	5.48
		Parts for warm air furnaces, including duct furnaces (excluding complete humidifiers)	5.06
		Parts and accessories for air-conditioning and heat transfer equipment, nsk	3.59
		Refrigeration and heating equipment, nsk	2.22

Source: 1992 *Economic Census*. The values shown are percent of total shipments in an industry. Values of indented subcategories are summed in the main heading. The symbol (D) appears when data are withheld to prevent disclosure of competitive information. The abbreviation nsk stands for 'not specified by kind' and nec for 'not elsewhere classified'.

INPUTS AND OUTPUTS FOR REFRIGERATION & HEATING EQUIPMENT

Economic Sector or Industry Providing Inputs	%	Sector	Economic Sector or Industry Buying Outputs	%	Sector
Refrigeration & heating equipment	15.8	Manufg.	Gross private fixed investment	17.7	Cap Inv
Wholesale trade	13.5	Trade	Motor vehicles & car bodies	13.6	Manufg.
Motors & generators	8.9	Manufg.	Refrigeration & heating equipment	9.7	Manufg.
Blast furnaces & steel mills	6.7	Manufg.	Exports	9.1	Foreign
Imports	4.3	Foreign	Personal consumption expenditures	6.7	
Aluminum rolling & drawing	4.0	Manufg.	Maintenance of nonfarm buildings nec	6.2	Constr.
Copper rolling & drawing	2.9	Manufg.	Office buildings	5.2	Constr.
Switchgear & switchboard apparatus	2.6	Manufg.	Industrial buildings	3.6	Constr.
Advertising	2.5	Services	Nonfarm residential structure maintenance	3.3	Constr.
Machinery, except electrical, nec	2.3	Manufg.	Household refrigerators & freezers	3.2	Manufg.
Metal stampings, nec	2.2	Manufg.	Residential 1-unit structures, nonfarm	2.9	Constr.
Environmental controls	1.9	Manufg.	Automotive repair shops & services	2.9	Services
Electric services (utilities)	1.7	Util.	Residential garden apartments	1.7	Constr.
Iron & steel foundries	1.5	Manufg.	Construction of hospitals	1.2	Constr.
Screw machine and related products	1.3	Manufg.	Residential additions/alterations, nonfarm	1.1	Constr.
Aluminum castings	1.1	Manufg.	Motor vehicle parts & accessories	1.0	Manufg.
Blowers & fans	1.1	Manufg.	S/L Govt. purch., elem. & secondary education	1.0	S/L Govt
Industrial controls	1.1	Manufg.	Truck trailers	0.9	Manufg.
Miscellaneous plastics products	1.0	Manufg.	Construction of educational buildings	0.8	Constr.
Special dies & tools & machine tool accessories	1.0	Manufg.	Construction of stores & restaurants	0.7	Constr.
Wiring devices	1.0	Manufg.	Automotive rental & leasing, without drivers	0.7	Services
Air transportation	1.0	Util.	Electric utility facility construction	0.5	Constr.
Internal combustion engines, nec	0.9	Manufg.	Warehouses	0.4	Constr.
Motor freight transportation & warehousing	0.9	Util.	Heating equipment, except electric	0.4	Manufg.
Paperboard containers & boxes	0.8	Manufg.	Federal Government purchases, national defense	0.4	Fed Govt
Maintenance of nonfarm buildings nec	0.7	Constr.	Hotels & motels	0.3	Constr.
Ball & roller bearings	0.7	Manufg.	Residential 2-4 unit structures, nonfarm	0.3	Constr.
Pipe, valves, & pipe fittings	0.7	Manufg.	Cyclic crudes and organics	0.3	Manufg.
Communications, except radio & TV	0.7	Util.	Retail trade, except eating & drinking	0.3	Trade
Cyclic crudes and organics	0.6	Manufg.	Wholesale trade	0.3	Trade
Gas production & distribution (utilities)	0.6	Util.	Amusement & recreation building construction	0.2	Constr.
Eating & drinking places	0.6	Trade	Construction of nonfarm buildings nec	0.2	Constr.
Engineering, architectural, & surveying services	0.6	Services	Automatic merchandising machines	0.2	Manufg.
Abrasive products	0.5	Manufg.	Ship building & repairing	0.2	Manufg.
Plastics materials & resins	0.5	Manufg.	Travel trailers & campers	0.2	Manufg.

Continued on next page.

INPUTS AND OUTPUTS FOR REFRIGERATION & HEATING EQUIPMENT - Continued

Economic Sector or Industry Providing Inputs	%	Sector	Economic Sector or Industry Buying Outputs	%	Sector
Power transmission equipment	0.5	Manufg.	Eating & drinking places	0.2	Trade
Banking	0.5	Fin/R.E.	Farm service facilities	0.1	Constr.
Equipment rental & leasing services	0.5	Services	Resid. & other health facility construction	0.1	Constr.
Iron & steel forgings	0.4	Manufg.	Residential high-rise apartments	0.1	Constr.
Nonferrous rolling & drawing, nec	0.4	Manufg.	Household laundry equipment	0.1	Manufg.
Paints & allied products	0.4	Manufg.	Mobile homes	0.1	Manufg.
Transformers	0.4	Manufg.	Prefabricated wood buildings	0.1	Manufg.
Computer & data processing services	0.4	Services	Truck & bus bodies	0.1	Manufg.
Fabricated rubber products, nec	0.3	Manufg.			
Mineral wool	0.3	Manufg.			
Miscellaneous fabricated wire products	0.3	Manufg.			
Petroleum refining	0.3	Manufg.			
Rubber & plastics hose & belting	0.3	Manufg.			
Wood pallets & skids	0.3	Manufg.			
Sanitary services, steam supply, irrigation	0.3	Util.			
Real estate	0.3	Fin/R.E.			
Hotels & lodging places	0.3	Services			
Brass, bronze, & copper castings	0.2	Manufg.			
Glass & glass products, except containers	0.2	Manufg.			
Hardware, nec	0.2	Manufg.			
Heating equipment, except electric	0.2	Manufg.			
Metal heat treating	0.2	Manufg.			
Nonferrous wire drawing & insulating	0.2	Manufg.			
Sawmills & planning mills, general	0.2	Manufg.			
Signs & advertising displays	0.2	Manufg.			
Wood containers	0.2	Manufg.			
Railroads & related services	0.2	Util.			
Colleges, universities, & professional schools	0.2	Services			
Legal services	0.2	Services			
Management & consulting services & labs	0.2	Services			
Gaskets, packing & sealing devices	0.1	Manufg.			
Lubricating oils & greases	0.1	Manufg.			
Machine tools, metal cutting types	0.1	Manufg.			
Manifold business forms	0.1	Manufg.			
Primary metal products, nec	0.1	Manufg.			
Veneer & plywood	0.1	Manufg.			
Insurance carriers	0.1	Fin/R.E.			
Royalties	0.1	Fin/R.E.			
Accounting, auditing & bookkeeping	0.1	Services			

Source: Benchmark Input-Output Accounts for the U.S. Economy, 1982, U.S. Department of Commerce, Washington, D.C., July 1991. Data, as reported in the source, are organized by the 1977 SIC structure in use in 1982 but have been matched, as closely as is possible, to the 1987 SIC structure used in this book.

OCCUPATIONS EMPLOYED BY SIC 358 - REFRIGERATION AND SERVICE MACHINERY

Occupation	% of Total 1994	Change to 2005	Occupation	% of Total 1994	Change to 2005
Assemblers, fabricators, & hand workers nec	20.6	0.4	Machine tool cutting operators, metal & plastic	1.3	-16.4
Machine assemblers	5.5	50.6	Drafters	1.3	-21.8
Blue collar worker supervisors	3.7	-4.4	Combination machine tool operators	1.3	10.4
Sheet metal workers & duct installers	3.2	0.4	Industrial machinery mechanics	1.2	10.4
Inspectors, testers, & graders, precision	2.7	0.4	Helpers, laborers, & material movers nec	1.2	0.4
Sales & related workers nec	2.6	0.4	Traffic, shipping, & receiving clerks	1.2	-3.4
Welders & cutters	2.5	0.4	Solderers & brazers	1.1	0.4
Industrial truck & tractor operators	2.1	0.4	Stock clerks	1.1	-18.4
Machine tool cutting & forming etc. nec	2.1	0.4	Freight, stock, & material movers, hand	1.1	-19.7
Machine forming operators, metal & plastic	1.7	0.4	Engineering technicians & technologists nec	1.1	0.4
General managers & top executives	1.7	-4.8	Metal & plastic machine workers nec	1.0	-11.1
Welding machine setters, operators	1.7	-9.7	Mechanical engineers	1.0	10.5
Machine builders	1.4	20.4	Machinists	1.0	0.4
Secretaries, ex legal & medical	1.4	-8.6			

Source: Industry-Occupation Matrix, Bureau of Labor Statistics. These data relate to one or more 3-digit SIC industry groups rather than to a single 4-digit SIC. The change reported for each occupation to the year 2005 is a percent of growth or decline as estimated by the Bureau of Labor Statistics. The abbreviation nec stands for 'not elsewhere classified'.

LOCATION BY STATE AND REGIONAL CONCENTRATION

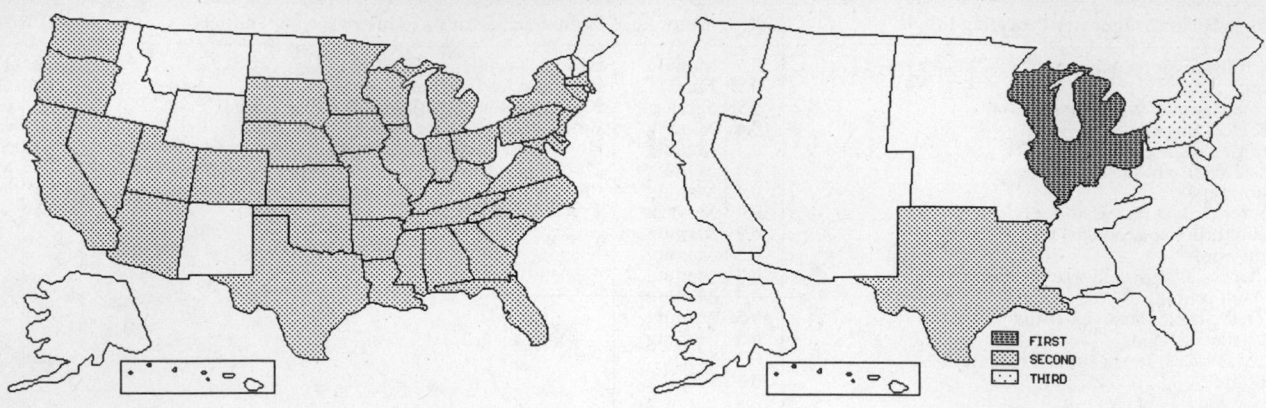

FIRST
SECOND
THIRD

INDUSTRY DATA BY STATE

State	Establish-ments	Shipments			Employment				Cost as % of Shipments	Investment per Employee ($)
		Total ($ mil)	% of U.S.	Per Establ.	Total Number	% of U.S.	Per Establ.	Wages ($/hour)		
Ohio	44	2,220.5	11.3	50.5	12,200	10.1	277	17.78	52.7	7,508
Tennessee	25	2,170.2	11.0	86.8	9,900	8.2	396	11.47	54.6	5,717
Texas	118	1,759.8	8.9	14.9	11,700	9.7	99	10.29	61.5	3,897
New York	64	1,673.3	8.5	26.1	11,300	9.4	177	20.21	46.4	4,726
Indiana	29	1,420.9	7.2	49.0	6,800	5.6	234	15.40	46.6	4,618
Pennsylvania	51	885.0	4.5	17.4	5,200	4.3	102	12.87	41.2	2,827
Georgia	27	753.6	3.8	27.9	4,700	3.9	174	10.80	59.0	5,532
Illinois	39	721.1	3.7	18.5	4,400	3.7	113	11.06	42.4	4,705
California	99	633.4	3.2	6.4	3,800	3.2	38	11.26	57.3	2,579
Arkansas	15	626.9	3.2	41.8	3,700	3.1	247	12.19	50.4	3,270
New Jersey	20	582.2	3.0	29.1	3,700	3.1	185	12.09	49.2	3,351
Wisconsin	24	535.0	2.7	22.3	4,000	3.3	167	14.13	52.0	2,900
Michigan	30	522.9	2.7	17.4	3,700	3.1	123	13.06	66.5	-
Virginia	12	510.9	2.6	42.6	3,800	3.2	317	12.76	58.8	-
Missouri	19	504.7	2.6	26.6	4,200	3.5	221	10.40	52.3	2,619
Kentucky	14	494.2	2.5	35.3	3,300	2.7	236	11.46	52.0	2,212
Alabama	11	492.7	2.5	44.8	2,400	2.0	218	12.30	61.0	7,042
Minnesota	24	464.8	2.4	19.4	3,000	2.5	125	14.06	57.7	2,267
Oklahoma	15	320.5	1.6	21.4	1,900	1.6	127	10.52	55.0	3,263
Iowa	11	295.8	1.5	26.9	1,400	1.2	127	13.10	34.5	3,000
North Carolina	18	291.3	1.5	16.2	1,900	1.6	106	10.03	50.1	14,842
South Carolina	8	224.9	1.1	28.1	1,600	1.3	200	9.28	50.9	2,750
Kansas	8	202.2	1.0	25.3	1,100	0.9	138	14.23	60.4	1,727
Arizona	17	129.7	0.7	7.6	1,100	0.9	65	8.53	52.7	2,273
Maryland	16	127.5	0.6	8.0	1,400	1.2	88	15.46	43.6	2,071
Florida	49	125.1	0.6	2.6	1,300	1.1	27	8.71	54.4	1,077
Connecticut	12	104.3	0.5	8.7	700	0.6	58	11.80	48.9	2,143
Massachusetts	15	85.6	0.4	5.7	700	0.6	47	12.00	47.4	3,143
Nevada	10	22.4	0.1	2.2	200	0.2	20	9.33	54.0	2,500
Washington	11	(D)	-	-	750 *	0.6	68	-	-	2,133
Oregon	9	(D)	-	-	175 *	0.1	19	-	-	-
Mississippi	8	(D)	-	-	3,750 *	3.1	469	-	-	-
Colorado	7	(D)	-	-	750 *	0.6	107	-	-	-
South Dakota	4	(D)	-	-	375 *	0.3	94	-	-	-
Delaware	3	(D)	-	-	175 *	0.1	58	-	-	-
Louisiana	3	(D)	-	-	175 *	0.1	58	-	-	-
Nebraska	3	(D)	-	-	375 *	0.3	125	-	-	-
Utah	3	(D)	-	-	175 *	0.1	58	-	-	-

Source: 1992 *Economic Census*. The states are in descending order of shipments or establishments (if shipment data are missing for the majority). The symbol (D) appears when data are withheld to prevent disclosure of competitive information. States marked with (D) are sorted by number of establishments. A dash (-) indicates that the data element cannot be calculated; * indicates the midpoint of a range.

3586 - MEASURING & DISPENSING PUMPS

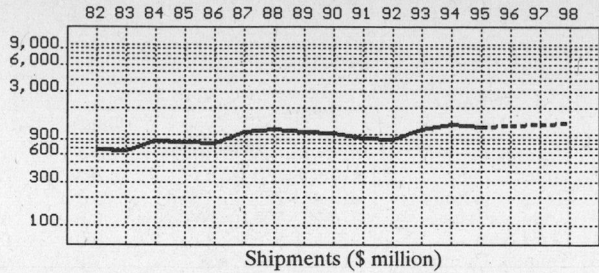

Shipments ($ million)

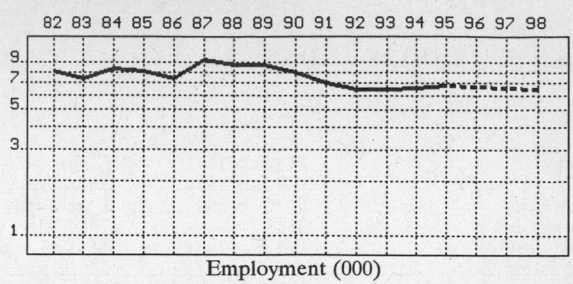

Employment (000)

GENERAL STATISTICS

| Year | Com-panies | Establishments | | Employment | | | Compensation | | Production ($ million) | | | |
		Total	with 20 or more employees	Total (000)	Production Workers (000)	Hours (Mil)	Payroll ($ mil)	Wages ($/hr)	Cost of Materials	Value Added by Manufacture	Value of Shipments	Capital Invest.
1982	58	61	35	8.1	5.2	9.9	160.0	8.78	299.7	365.1	676.2	12.8
1983		59	35	7.4	4.7	8.7	157.6	9.21	292.1	364.3	668.2	13.1
1984		57	35	8.4	5.6	11.0	182.3	9.40	396.2	465.7	858.8	13.1
1985		56	36	8.2	5.3	10.2	185.1	9.51	423.7	417.9	840.3	21.3
1986		62	37	7.3	4.7	9.1	172.8	10.01	413.5	392.8	813.1	16.4
1987	70	83	44	9.4	5.8	12.0	249.0	11.07	470.3	617.8	1,068.7	24.4
1988		81	44	8.8	5.7	11.2	230.4	11.91	490.3	641.7	1,135.7	26.8
1989		81	44	8.8	5.2	10.0	229.5	12.35	505.1	544.3	1,071.0	29.7
1990		75	42	7.9	4.9	9.5	221.1	12.74	492.9	519.1	1,029.5	21.1
1991		73	40	7.0	4.5	8.1	192.8	11.93	426.4	493.5	927.7	22.2
1992	71	77	41	6.5	4.1	8.2	196.6	12.02	451.4	422.0	896.3	27.1
1993		76	41	6.4	4.7	9.2	208.4	13.79	582.4	585.5	1,137.6	31.8
1994		83P	44P	6.6	4.8	10.1	223.3	14.36	679.7	635.4	1,295.0	27.5
1995		86P	45P	6.8P	4.6P	9.2P	231.1P	14.42P	640.8P	599.0P	1,220.9P	31.9P
1996		88P	46P	6.7P	4.5P	9.1P	235.5P	14.87P	660.7P	617.7P	1,258.8P	33.3P
1997		90P	46P	6.5P	4.5P	9.0P	239.8P	15.31P	680.6P	636.3P	1,296.8P	34.7P
1998		92P	47P	6.4P	4.4P	9.0P	244.2P	15.76P	700.6P	654.9P	1,334.7P	36.1P

Sources: 1982, 1987, 1992 *Economic Census*; *Annual Survey of Manufactures*, 83-86, 88-91, 93-94. Establishment counts for non-Census years are from *County Business Patterns*; establishment values for 83-84 are extrapolations. 'P's show projections by the editors. Industries reclassified in 87 will not have data for prior years.

INDICES OF CHANGE

| Year | Com-panies | Establishments | | Employment | | | Compensation | | Production ($ million) | | | |
		Total	with 20 or more employees	Total (000)	Production Workers (000)	Hours (Mil)	Payroll ($ mil)	Wages ($/hr)	Cost of Materials	Value Added by Manufacture	Value of Shipments	Capital Invest.
1982	82	79	85	125	127	121	81	73	66	87	75	47
1983		77	85	114	115	106	80	77	65	86	75	48
1984		74	85	129	137	134	93	78	88	110	96	48
1985		73	88	126	129	124	94	79	94	99	94	79
1986		81	90	112	115	111	88	83	92	93	91	61
1987	99	108	107	145	141	146	127	92	104	146	119	90
1988		105	107	135	139	137	117	99	109	152	127	99
1989		105	107	135	127	122	117	103	112	129	119	110
1990		97	102	122	120	116	112	106	109	123	115	78
1991		95	98	108	110	99	98	99	94	117	104	82
1992	100	100	100	100	100	100	100	100	100	100	100	100
1993		99	100	98	115	112	106	115	129	139	127	117
1994		108P	108P	102	117	123	114	119	151	151	144	101
1995		111P	110P	105P	112P	112P	118P	120P	142P	142P	136P	118P
1996		114P	111P	103P	111P	111P	120P	124P	146P	146P	140P	123P
1997		116P	113P	101P	109P	110P	122P	127P	151P	151P	145P	128P
1998		119P	115P	98P	108P	109P	124P	131P	155P	155P	149P	133P

Sources: Same as General Statistics. Values reflect change from the base year, 1992. Values above 100 mean greater than 92, values below 100 mean less than 92, and a value of 100 in the 82-91 or 93-98 period means same as 92. 'P's mark projections by the editors.

SELECTED RATIOS

For 1994	Avg. of All Manufact.	Analyzed Industry	Index	For 1994	Avg. of All Manufact.	Analyzed Industry	Index
Employees per Establishment	49	79	161	Value Added per Production Worker	134,084	132,375	99
Payroll per Establishment	1,500,273	2,675,222	178	Cost per Establishment	5,045,178	8,143,075	161
Payroll per Employee	30,620	33,833	110	Cost per Employee	102,970	102,985	100
Production Workers per Establishment	34	58	168	Cost per Production Worker	146,988	141,604	96
Wages per Establishment	853,319	1,737,589	204	Shipments per Establishment	9,576,895	15,514,612	162
Wages per Production Worker	24,861	30,216	122	Shipments per Employee	195,460	196,212	100
Hours per Production Worker	2,056	2,104	102	Shipments per Production Worker	279,017	269,792	97
Wages per Hour	12.09	14.36	119	Investment per Establishment	321,011	329,461	103
Value Added per Establishment	4,602,255	7,612,343	165	Investment per Employee	6,552	4,167	64
Value Added per Employee	93,930	96,273	102	Investment per Production Worker	9,352	5,729	61

Sources: Same as General Statistics. The 'Average of All Manufacturing' column represents the average of all manufacturing industries reported for the most recent complete year available. The Index shows the relationship between the Average and the Analyzed Industry. For example, 100 means that they are equal; 500 that the Analyzed Industry is five times the average; 50 means that the Analyzed Industry is half the national average. The abbreviation 'na' is used to show that data are 'not available'.

LEADING COMPANIES   Number shown: 14   Total sales ($ mil): 1,200   Total employment (000): 6.8

Company Name	Address				CEO Name	Phone	Co. Type	Sales ($ mil)	Empl. (000)
Graco Inc	PO Box 1441	Minneapolis	MN	55440	David A Koch	612-623-6000	P	360	2.1
Wayne	PO Box 1859	Salisbury	MD	21802	James C Hilton	410-546-6600	D	283	0.9
Tokheim Corp	PO Box 360	Fort Wayne	IN	46801	Douglas K Pinner	219-423-2552	P	202	1.9
Pulsafeeder Inc	2883	Rochester	NY	14623	Rodney L Usher	716-292-8000	S	140	0.3
Marley Pump Co	PO Box 2973	Mission	KS	66201	Leif Lomo	913-831-5700	D	100	0.5
Bennett Pump Co	PO Box 597	Muskegon	MI	49443	Thomas Thompson	616-733-1302	R	40	0.2
Blackmer	1809 Century SW	Grand Rapids	MI	49509	Raymon Pilch	616-241-1611	D	23*	0.3
Schlumberger	PO Drawer 280	Bonham	TX	75418	Malcome Unsworth	903-583-3134	D	18*	0.2
O'Day Equipment Inc	PO Box 2706	Fargo	ND	58108	J O'Day	701-282-9260	R	12	<0.1
National-Spencer Inc	PO Box 57	Wichita	KS	67201	OB Elliott	316-265-5601	R	9	0.1
Robotics Inc	2421 Rte 9	Ballston Spa	NY	12020	John E Soron	518-899-4211	R	5	<0.1
Coen Company Inc	1510 Tanforan Av	Woodland	CA	95776	EG Deane	916-661-6127	D	4	0.1
DVCO Fuel Systems Inc	702 W 48th Av	Denver	CO	80216	A Bradley Gabbard	303-296-9666	S	2	<0.1
Par-Tee Company Inc	PO Box 69	Spencerville	IN	46788	A Ronald Kohart	219-238-4483	R	1	<0.1

Source: Ward's Business Directory of U.S. Private and Public Companies, Volumes 1 and 2, 1996. The company type code used is as follows: P - Public, R - Private, S - Subsidiary, D - Division, J - Joint Venture, A - Affiliate, G - Group. Sales are in millions of dollars, employees are in thousands. An asterisk (*) indicates an estimated sales volume. The symbol < stands for 'less than'. Company names and addresses are truncated, in some cases, to fit into the available space.

MATERIALS CONSUMED

Material	Quantity	Delivered Cost ($ million)
Materials, ingredients, containers, and supplies	(X)	424.1
Fluid power products	(X)	10.9
Metal bolts, nuts, screws, washers, rivets, and other screw machine products	(X)	16.6
Metal stampings	(X)	6.9
Metal pipe, valves, and pipe fittings, except plumbers' and forgings	(X)	5.6
Metal tanks, heat exchangers, steam condensers, and other boiler products	(X)	1.2
Iron and steel castings (rough and semifinished)	(X)	10.0
Aluminum and aluminum-base alloy castings (rough and semifinished)	(X)	16.6
Other nonferrous castings (rough and semifinished)	(X)	3.4
Steel sheet and strip, including tin plate	(X)	16.7
All other steel shapes and forms	(X)	4.7
Aluminum and aluminum-base alloy shapes and forms	(X)	4.9
Other nonferrous shapes and forms	(X)	2.4
Other fractional horsepower electric motors (under 1 hp)	(X)	13.1
Integral horsepower electric motors and generators (1 hp or more)	(X)	2.5
Ball and roller bearings (mounted or unmounted)	(X)	2.1
Paperboard containers, boxes, and corrugated paperboard	(X)	5.1
Fabricated rubber products, except gaskets	(X)	5.8
Fabricated plastics products (except gaskets, hoses, and belting)	(X)	31.8
Gaskets (all types), and packing and sealing devices	(X)	7.5
Electrical transmission, distribution, and control equipment	(X)	8.5
Paints, varnishes, stains, lacquers, shellacs, japans, enamels, and allied products	(X)	1.4
All other materials and components, parts, containers, and supplies	(X)	214.8
Materials, ingredients, containers, and supplies, nsk	(X)	31.7

Source: 1992 Economic Census. Explanation of symbols used: (D): Withheld to avoid disclosure of competitive data; na: Not available; (S): Withheld because statistical norms were not met; (X): Not applicable; (Z): Less than half the unit shown; nec: Not elsewhere classified; nsk: Not specified by kind; - : zero; * : 10-19 percent estimated; ** : 20-29 percent estimated.

PRODUCT SHARE DETAILS

Product or Product Class	% Share	Product or Product Class	% Share
Measuring and dispensing pumps	100.00	parts and attachments	4.04
Single unit gasoline dispensing pumps, computing type (filling station type), with suction pumping unit, except parts and attachments	7.88	Multiple unit gasoline dispensing pumps, computing type (filling station type), without suction pumping unit, except parts and attachments	26.63
Single unit gasoline dispensing pumps, computing type (filling station type), without suction pumping unit, except parts and attachments	2.11	Lubricating oil measuring and dispensing pumps, including barrel pumps, except parts and attachments	1.72
Multiple unit gasoline dispensing pumps, computing type (filling station type), with suction pumping unit, except		Measuring and dispensing grease guns, except parts and attachments	32.71
		Parts and attachments for measuring and dispensing pumps	21.97

Source: 1992 Economic Census. The values shown are percent of total shipments in an industry. Values of indented subcategories are summed in the main heading. The symbol (D) appears when data are withheld to prevent disclosure of competitive information. The abbreviation nsk stands for 'not specified by kind' and nec for 'not elsewhere classified'.

INPUTS AND OUTPUTS FOR MEASURING & DISPENSING PUMPS

Economic Sector or Industry Providing Inputs	%	Sector	Economic Sector or Industry Buying Outputs	%	Sector
Mechanical measuring devices	17.1	Manufg.	Gross private fixed investment	89.1	Cap Inv
Blast furnaces & steel mills	10.5	Manufg.	Exports	9.7	Foreign
Wholesale trade	7.5	Trade	Federal Government purchases, national defense	0.7	Fed Govt
Metal coating & allied services	6.7	Manufg.	Retail trade, except eating & drinking	0.3	Trade
Motors & generators	5.2	Manufg.	Federal Government purchases, nondefense	0.1	Fed Govt
Iron & steel foundries	4.6	Manufg.			
Machinery, except electrical, nec	4.3	Manufg.			
Advertising	4.1	Services			
Aluminum castings	3.6	Manufg.			
Electric services (utilities)	2.2	Util.			
Imports	2.0	Foreign			
Communications, except radio & TV	1.8	Util.			
Security & commodity brokers	1.6	Fin/R.E.			
Maintenance of nonfarm buildings nec	1.4	Constr.			
Engineering, architectural, & surveying services	1.4	Services			
Pipe, valves, & pipe fittings	1.3	Manufg.			
Rubber & plastics hose & belting	1.3	Manufg.			
Gas production & distribution (utilities)	1.3	Util.			
Eating & drinking places	1.2	Trade			
Pumps & compressors	1.1	Manufg.			
Air transportation	1.1	Util.			
Computer & data processing services	1.1	Services			
Miscellaneous plastics products	1.0	Manufg.			
Abrasive products	0.9	Manufg.			
Paperboard containers & boxes	0.9	Manufg.			
Equipment rental & leasing services	0.9	Services			
Screw machine and related products	0.7	Manufg.			
Machine tools, metal cutting types	0.6	Manufg.			
Motor freight transportation & warehousing	0.6	Util.			
Banking	0.6	Fin/R.E.			
U.S. Postal Service	0.6	Gov't			
Ball & roller bearings	0.5	Manufg.			
Metal stampings, nec	0.5	Manufg.			
Nonferrous castings, nec	0.5	Manufg.			
Paints & allied products	0.5	Manufg.			
Power transmission equipment	0.5	Manufg.			
Aluminum rolling & drawing	0.4	Manufg.			
Sanitary services, steam supply, irrigation	0.4	Util.			
Hotels & lodging places	0.4	Services			
Legal services	0.4	Services			
Management & consulting services & labs	0.4	Services			
Fabricated rubber products, nec	0.3	Manufg.			
Lubricating oils & greases	0.3	Manufg.			
Metal heat treating	0.3	Manufg.			
Special dies & tools & machine tool accessories	0.3	Manufg.			
Water transportation	0.3	Util.			
Credit agencies other than banks	0.3	Fin/R.E.			
Real estate	0.3	Fin/R.E.			
Accounting, auditing & bookkeeping	0.3	Services			
Business/professional associations	0.3	Services			
Fabricated plate work (boiler shops)	0.2	Manufg.			
Gaskets, packing & sealing devices	0.2	Manufg.			
Manifold business forms	0.2	Manufg.			
Measuring & dispensing pumps	0.2	Manufg.			
Petroleum refining	0.2	Manufg.			
Transformers	0.2	Manufg.			
Insurance carriers	0.2	Fin/R.E.			
Royalties	0.2	Fin/R.E.			
Electrical repair shops	0.2	Services			
Industrial gases	0.1	Manufg.			
Machine tools, metal forming types	0.1	Manufg.			
Miscellaneous fabricated wire products	0.1	Manufg.			
Railroads & related services	0.1	Util.			
Laundry, dry cleaning, shoe repair	0.1	Services			

Source: Benchmark Input-Output Accounts for the U.S. Economy, 1982, U.S. Department of Commerce, Washington, D.C., July 1991. Data, as reported in the source, are organized by the 1977 SIC structure in use in 1982 but have been matched, as closely as is possible, to the 1987 SIC structure used in this book.

OCCUPATIONS EMPLOYED BY SIC 358 - REFRIGERATION AND SERVICE MACHINERY

Occupation	% of Total 1994	Change to 2005	Occupation	% of Total 1994	Change to 2005
Assemblers, fabricators, & hand workers nec	20.6	0.4	Machine tool cutting operators, metal & plastic	1.3	-16.4
Machine assemblers	5.5	50.6	Drafters	1.3	-21.8
Blue collar worker supervisors	3.7	-4.4	Combination machine tool operators	1.3	10.4
Sheet metal workers & duct installers	3.2	0.4	Industrial machinery mechanics	1.2	10.4
Inspectors, testers, & graders, precision	2.7	0.4	Helpers, laborers, & material movers nec	1.2	0.4
Sales & related workers nec	2.6	0.4	Traffic, shipping, & receiving clerks	1.2	-3.4
Welders & cutters	2.5	0.4	Solderers & brazers	1.1	0.4
Industrial truck & tractor operators	2.1	0.4	Stock clerks	1.1	-18.4
Machine tool cutting & forming etc. nec	2.1	0.4	Freight, stock, & material movers, hand	1.1	-19.7
Machine forming operators, metal & plastic	1.7	0.4	Engineering technicians & technologists nec	1.1	0.4
General managers & top executives	1.7	-4.8	Metal & plastic machine workers nec	1.0	-11.1
Welding machine setters, operators	1.7	-9.7	Mechanical engineers	1.0	10.5
Machine builders	1.4	20.4	Machinists	1.0	0.4
Secretaries, ex legal & medical	1.4	-8.6			

Source: Industry-Occupation Matrix, Bureau of Labor Statistics. These data relate to one or more 3-digit SIC industry groups rather than to a single 4-digit SIC. The change reported for each occupation to the year 2005 is a percent of growth or decline as estimated by the Bureau of Labor Statistics. The abbreviation nec stands for 'not elsewhere classified'.

LOCATION BY STATE AND REGIONAL CONCENTRATION

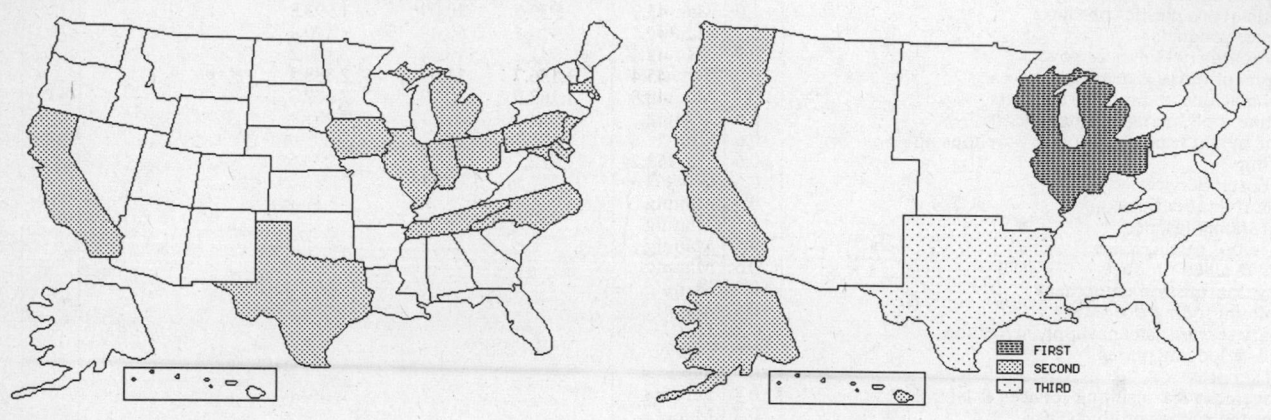

FIRST
SECOND
THIRD

INDUSTRY DATA BY STATE

State	Establish-ments	Shipments			Employment				Cost as % of Shipments	Investment per Employee ($)
		Total ($ mil)	% of U.S.	Per Establ.	Total Number	% of U.S.	Per Establ.	Wages ($/hour)		
Indiana	7	137.9	15.4	19.7	1,100	16.9	157	9.73	59.8	2,727
Ohio	6	65.5	7.3	10.9	400	6.2	67	13.20	40.3	4,750
California	11	58.6	6.5	5.3	400	6.2	36	11.50	44.4	3,000
Texas	11	42.6	4.8	3.9	400	6.2	36	9.60	43.0	1,250
Michigan	7	(D)	-	-	375 *	5.8	54	-	-	-
Illinois	3	(D)	-	-	375 *	5.8	125	-	-	-
North Carolina	3	(D)	-	-	1,750 *	26.9	583	-	-	-
Tennessee	3	(D)	-	-	375 *	5.8	125	-	-	-
Iowa	2	(D)	-	-	375 *	5.8	188	-	-	-
Massachusetts	2	(D)	-	-	175 *	2.7	88	-	-	-
Maryland	1	(D)	-	-	750 *	11.5	750	-	-	-
New Jersey	1	(D)	-	-	175 *	2.7	175	-	-	-
Pennsylvania	1	(D)	-	-	375 *	5.8	375	-	-	-

Source: 1992 Economic Census. The states are in descending order of shipments or establishments (if shipment data are missing for the majority). The symbol (D) appears when data are withheld to prevent disclosure of competitive information. States marked with (D) are sorted by number of establishments. A dash (-) indicates that the data element cannot be calculated; * indicates the midpoint of a range.

3589 - SERVICE INDUSTRY MACHINERY, NEC

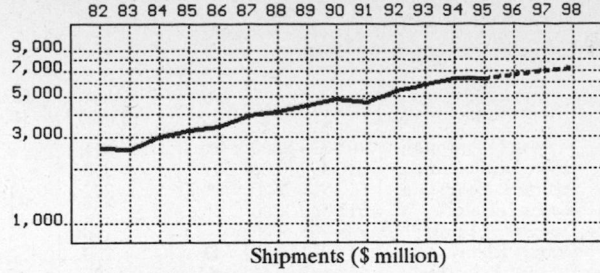

Shipments ($ million)

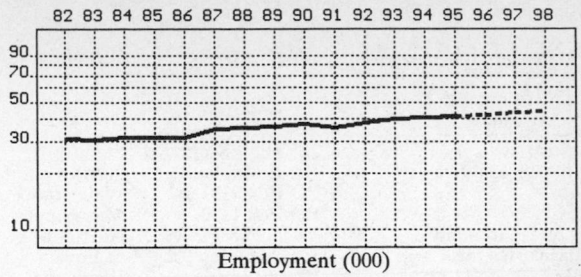

Employment (000)

GENERAL STATISTICS

| Year | Companies | Establishments | | Employment | | | Compensation | | Production ($ million) | | | |
		Total	with 20 or more employees	Total (000)	Production Workers (000)	Hours (Mil)	Payroll ($ mil)	Wages ($/hr)	Cost of Materials	Value Added by Manufacture	Value of Shipments	Capital Invest.
1982	791	832	287	31.0	17.8	33.5	584.0	8.45	1,235.4	1,321.3	2,583.0	54.6
1983		809	291	30.5	17.9	33.6	589.5	8.55	1,245.5	1,306.6	2,570.1	50.0
1984		786	295	31.5	19.4	38.1	676.9	8.82	1,429.2	1,587.5	2,991.3	80.2
1985		763	298	31.6	19.5	37.9	694.5	8.90	1,568.6	1,669.2	3,214.7	65.0
1986		771	294	31.7	19.1	36.5	736.4	9.47	1,611.7	1,815.4	3,428.3	58.0
1987	895	949	333	35.2	21.5	43.3	854.8	9.21	1,765.5	2,200.5	3,960.4	88.9
1988		924	355	35.9	22.2	43.0	874.2	9.78	1,977.6	2,169.6	4,116.6	100.3
1989		904	357	36.5	22.4	45.2	932.5	10.09	1,993.8	2,592.9	4,428.1	104.0
1990		914	359	37.8	23.0	46.3	1,018.1	10.39	2,170.4	2,721.4	4,877.2	32.5
1991		944	366	35.9	21.2	42.6	971.2	10.64	2,130.2	2,483.7	4,636.2	60.0
1992	1,046	1,113	391	38.5	22.4	45.4	1,136.7	11.09	2,388.3	2,965.1	5,344.6	109.6
1993		1,109	393	40.5	24.1	48.8	1,163.9	10.99	2,639.7	3,171.5	5,743.6	85.1
1994		1,084P	405P	40.9	25.6	52.3	1,240.0	11.29	3,085.1	3,245.2	6,289.2	130.0
1995		1,112P	416P	41.5P	25.2P	51.8P	1,271.7P	11.60P	3,085.0P	3,245.1P	6,289.0P	112.7P
1996		1,140P	426P	42.4P	25.8P	53.2P	1,327.3P	11.85P	3,233.6P	3,401.5P	6,592.0P	117.1P
1997		1,168P	437P	43.3P	26.4P	54.6P	1,382.9P	12.11P	3,382.3P	3,557.8P	6,895.0P	121.5P
1998		1,196P	448P	44.2P	26.9P	56.0P	1,438.5P	12.36P	3,530.9P	3,714.2P	7,198.1P	125.8P

Sources: 1982, 1987, 1992 Economic Census; Annual Survey of Manufactures, 83-86, 88-91, 93-94. Establishment counts for non-Census years are from County Business Patterns; establishment values for 83-84 are extrapolations. 'P's show projections by the editors. Industries reclassified in 87 will not have data for prior years.

INDICES OF CHANGE

| Year | Companies | Establishments | | Employment | | | Compensation | | Production ($ million) | | | |
		Total	with 20 or more employees	Total (000)	Production Workers (000)	Hours (Mil)	Payroll ($ mil)	Wages ($/hr)	Cost of Materials	Value Added by Manufacture	Value of Shipments	Capital Invest.
1982	76	75	73	81	79	74	51	76	52	45	48	50
1983		73	74	79	80	74	52	77	52	44	48	46
1984		71	75	82	87	84	60	80	60	54	56	73
1985		69	76	82	87	83	61	80	66	56	60	59
1986		69	75	82	85	80	65	85	67	61	64	53
1987	86	85	85	91	96	95	75	83	74	74	74	81
1988		83	91	93	99	95	77	88	83	73	77	92
1989		81	91	95	100	100	82	91	83	87	83	95
1990		82	92	98	103	102	90	94	91	92	91	75
1991		85	94	93	95	94	85	96	89	84	87	55
1992	100	100	100	100	100	100	100	100	100	100	100	100
1993		100	101	105	108	107	102	99	111	107	107	78
1994		97P	104P	106	114	115	109	102	129	109	118	119
1995		100P	106P	108P	113P	114P	112P	105P	129P	109P	118P	103P
1996		102P	109P	110P	115P	117P	117P	107P	135P	115P	123P	107P
1997		105P	112P	112P	118P	120P	122P	109P	142P	120P	129P	111P
1998		107P	115P	115P	120P	123P	127P	111P	148P	125P	135P	115P

Sources: Same as General Statistics. Values reflect change from the base year, 1992. Values above 100 mean greater than 92, values below 100 mean less than 92, and a value of 100 in the 82-91 or 93-98 period means same as 92. 'P's mark projections by the editors.

SELECTED RATIOS

For 1994	Avg. of All Manufact.	Analyzed Industry	Index	For 1994	Avg. of All Manufact.	Analyzed Industry	Index
Employees per Establishment	49	38	77	Value Added per Production Worker	134,084	126,766	95
Payroll per Establishment	1,500,273	1,144,199	76	Cost per Establishment	5,045,178	2,846,749	56
Payroll per Employee	30,620	30,318	99	Cost per Employee	102,970	75,430	73
Production Workers per Establishment	34	24	69	Cost per Production Worker	146,988	120,512	82
Wages per Establishment	853,319	544,848	64	Shipments per Establishment	9,576,895	5,803,305	61
Wages per Production Worker	24,861	23,065	93	Shipments per Employee	195,460	153,770	79
Hours per Production Worker	2,056	2,043	99	Shipments per Production Worker	279,017	245,672	88
Wages per Hour	12.09	11.29	93	Investment per Establishment	321,011	119,956	37
Value Added per Establishment	4,602,255	2,994,480	65	Investment per Employee	6,552	3,178	49
Value Added per Employee	93,930	79,345	84	Investment per Production Worker	9,352	5,078	54

Sources: Same as General Statistics. The 'Average of All Manufacturing' column represents the average of all manufacturing industries reported for the most recent complete year available. The Index shows the relationship between the Average and the Analyzed Industry. For example, 100 means that they are equal; 500 that the Analyzed Industry is five times the average; 50 means that the Analyzed Industry is half the national average. The abbreviation 'na' is used to show that data are 'not available'.

LEADING COMPANIES   Number shown: **75**   Total sales ($ mil): **3,667**   Total employment (000): **23.9**

Company Name	Address				CEO Name	Phone	Co. Type	Sales ($ mil)	Empl. (000)
Welbilt Corp	225 High Ridge Rd	Stamford	CT	06905	Marion H Antonini	203-325-8300	P	411	2.1
Hoover North America	101 E Maple St	North Canton	OH	44720	Brian Girdlestone	216-499-9200	D	350	2.7
Ionics Inc	PO Box 9131	Watertown	MA	02272	Arthur L Goldstein	617-926-2500	P	222	1.2
Tennant Co	PO Box 1452	Minneapolis	MN	55440	Roger L Hale	612-540-1200	P	221	1.4
Culligan International Co	1 Culligan Pkwy	Northbrook	IL	60062	Douglas A Pertz	708-205-6000	S	190*	1.5
Clarke Industries Inc	101 S Hanley Rd	St Louis	MO	63105	Robert Elkin	314-721-7255	D	100	0.6
Envirex Inc	1901 S Prairie Av	Waukesha	WI	53186	Edward P Saffran	414-547-0141	S	95	0.7
Frymaster Corp	PO Box 51000	Shreveport	LA	71135	David E Mosteller	318-865-1711	S	70	0.4
Ryko Manufacturing Co	PO Box 38	Grimes	IA	50111	James A Nelson	515-986-3700	R	65	0.6
Cleveland Range Co	1333 E 179th St	Cleveland	OH	44110	Stephen Amos	216-481-4900	D	60	0.3
Wallace and Tiernan Inc	25 Main St	Belleville	NJ	07109	G Rufenacht	201-759-8000	D	60	0.7
Seco Products Corp	PO Box 187	Washington	MO	63090	Richard C Runyan	314-239-4788	R	58*	0.5
Marathon Equipment Co	PO Box 1798	Vernon	AL	35592	Grant Fenner	205-695-9105	S	55	0.5
Middleby Cooking Systems	1400 Toastmaster	Elgin	IL	60120	David Riley	708-741-3300	D	55	0.3
Aladdin Synergetics Inc	555 Marriot Dr	Nashville	TN	37214	Bob Garda	615-748-3600	S	52	0.4
Pitco Frialator Inc	PO Box 501	Concord	NH	03302	Robert A Nerbonne	603-225-6684	S	52*	0.4
Aqua-Chem Inc	PO Box 421	Milwaukee	WI	53201	J H Creaghead	414-359-0600	D	50	0.4
Infilco Degremont Inc	PO Box 71390	Richmond	VA	23255	Yves M Moyne	804-756-7600	S	50	0.2
Parkson Corp	PO Box 408399	Ft Lauderdale	FL	33340	G Parks Souther	305-974-6610	S	50	0.2
Solid Waste Equipment	500 Sherman St	Galion	OH	44833	Dale Schenian	419-468-2120	D	48	0.4
Garland Commercial Industries	185 East South St	Freeland	PA	18224	Frank A Radice	717-636-1000	S	45	0.4
Windsor Industries Inc	1351 W Stanford	Englewood	CO	80110	TM Francis	303-762-1800	R	45	0.2
Andritz Ruthner Inc	1010 Commercial S	Arlington	TX	76017	Tim Ryan	817-465-5611	S	40	0.1
Graver Co	2720 US Hwy 22	Union	NJ	07083	Randall Hansen	908-964-2400	S	40	0.1
Prince Castle Inc	355 E Kehoe Blv	Carol Stream	IL	60186	Norman Terry	708-462-8800	R	40	0.2
Vactor Manufacturing Inc	1621 S Illinois St	Streator	IL	61364	William A Gaff	815-672-3171	S	40	0.3
Hako Minuteman Inc	111 S Rohlwing Rd	Addison	IL	60101	Jerome E Rau	708-627-6900	P	38	0.2
Haviland Enterprises Inc	421 Ann St NW	Grand Rapids	MI	49504	H Richard Garner	616-361-6691	R	38	0.1
Alto-Shaam Inc	PO Box 450	Menomonee Fls	WI	53052	Jerry Maahs	414-251-3800	R	35	0.3
Dorian International Inc	2 Gannett Dr	White Plains	NY	10604	Edward S Dorian Jr	914-697-9800	R	35	<0.1
Hanna Car Wash International	2000 SE Hanna Dr	Milwaukie	OR	97222	Ted Graves	503-659-0361	R	34*	0.2
FilmTec Corp	7200 Ohms Ln	Minneapolis	MN	55439	James Cederna	612-897-4200	S	32	0.3
Smith and Loveless Inc	14040 San Fe	Lenexa	KS	66215	Robert L Rebori	913-888-5201	R	32*	0.3
Ashbrook Corp	116 E Hardy	Houston	TX	77093	Robert J Wimmer	713-449-0322	S	30	0.2
Fluid Systems Corp	10054 Old Grove	San Diego	CA	92131	Philip Turnock	619-695-3840	S	30	0.2
Hatco Corp	635 S 28th St	Milwaukee	WI	53215	David G Hatch	414-671-6350	R	30*	0.3
Kent Co	2310 Indrial Pkwy	Elkhart	IN	46515	Phil Hayes	219-293-8661	S	30	0.2
Merco/Savory Inc	725 Vassar Av	Lakewood	NJ	08701	Michael Auld	908-364-9600	S	28*	0.3
AAR PowerBoss Inc	PO Box 1227	Aberdeen	NC	28315	David Webb	910-944-2105	S	25	0.1
FM Manufacturing Inc	35 Garvey St	Everett	MA	02149	Harold Hamilton	617-387-4100	R	25	0.2
McNish Corp	840 N Russell Av	Aurora	IL	60506	J McNish	708-892-7921	R	25	0.2
National Super Service Co	3115 Frenchmens	Toledo	OH	43607	Mark Bevington	419-531-2121	R	25	0.1
Pure Solutions	4101 E Wood St	Phoenix	AZ	85040	Bob Ritz	602-437-1355	D	25	<0.1
Martin Engineering	One Martin Pl	Neponset	IL	61345	RT Swinderman	309-594-2384	R	24	0.2
Wolf Range Co	19600 S Alameda St	Compton	CA	90224	Stanley Waldman	310-637-3737	D	24	0.3
Southbend	1100 Old Honeycutt	Fuquay-Varina	NC	27526	Johnie Cooper	919-552-9161	D	23	0.2
Brewmatic Co	3828 S Main St	Los Angeles	CA	90037	Roy F Farmer	213-233-8204	D	22	0.1
Carter-Hoffmann Corp	1551 McCormick	Mundelein	IL	60060	Philip Hartung	708-362-5500	R	22	0.2
Dean/US Range Inc	14501 S Broadway	Gardena	CA	90248		310-353-5000	S	22	0.2
Pioneer-Eclipse Corp	PO Box 909	Sparta	NC	28675	WH Wilson	910-372-8080	S	21	0.1
Breuer/Tornado Corp	7401 W Lawrence	Chicago	IL	60656	Linda Breuer	708-867-5100	R	20	0.1
FB Leopold Company Inc	227 S Division St	Zelienople	PA	16063	Marvin Brown	412-452-6300	R	20	0.1
Kinetico Inc	PO Box 193	Newbury	OH	44065	William C Prior	216-564-9111	R	20	0.2
Westech Engineering Inc	3605 W Temple St	Salt Lake City	UT	84115	James Larsen	801-265-1000	R	20	<0.1
Clements National Co	2150 W 16th St	Broadview	IL	60153	RW Barrett	708-681-4330	R	19*	0.2
Racine Industries Inc	PO Box 1648	Racine	WI	53401	J Fritz Rench	414-637-4491	R	19*	0.2
Southern Equipment Co	PO Box 7115	St Louis	MO	63117	Robert E Niemeyier	314-481-0660	R	19	0.1
Cecilware Corp	43-05 20th Av	Long Island Ct	NY	11105	Fran Kaplan	718-932-1414	R	19	0.2
Biothane Corp	2500 Broadway	Camden	NJ	08104	Robert I Sax	609-541-3500	S	18	<0.1
Permutit Company Inc	30 Technology Dr	Warren	NJ	07059	L Olejar	908-668-1700	S	18	<0.1
Shop Vac Corp	2323 Reach Rd	Williamsport	PA	17701	J Miller	717-326-0502	R	18	0.3
Silver King	1600 Xenium Ln N	Minneapolis	MN	55441	Harold Rubin	612-553-1881	S	18*	0.1
General Filter Co	600 Arrasmith Trail	Ames	IA	50010	Charles Biskner	515-232-4121	S	17	0.2
Lodal Inc	PO Box 2315	Kingsford	MI	49801	Bernard Leger	906-779-1700	R	16	<0.1
Bruner Corp	500 W Oklahoma	Milwaukee	WI	53207	Joseph Prochot	414-747-3700	R	16	0.1
Matt-Son Inc	28W005 Indrial Av	Barrington	IL	60010	Robert B Oleskow	708-628-8766	R	16	<0.1
Aeration Industries International	PO Box 59144	Minneapolis	MN	55459	Daniel Durda	612-448-6789	R	15	<0.1
American Wyott Corp	PO Box 1829	Cheyenne	WY	82001	Brian Rosenbloom	307-634-5801	S	15*	0.2
Dorian America	2 Gannett Dr	White Plains	NY	10604	Edward S Dorian Jr	914-697-9800	D	15*	<0.1
Hydranautics Inc	8444 Miralani Dr	San Diego	CA	92126	Mickey Ohishi	619-536-2500	S	15*	0.1
Sewer Rodding Equipment Co	PO Box 2957	Culver City	CA	90231	Patrick Crane	310-390-4444	R	15*	0.1
Von Schrader Co	1600 Junction Av	Racine	WI	53403	Quentin H Rench	414-634-1956	R	15	<0.1
Vulcan Peroxidation Systems Inc	5151 E Broadway	Tucson	AZ	85711	Don Hager	602-790-8383	R	15	0.1
Walker Process Equipment	840 N Russell Av	Aurora	IL	60506	James A McNish	708-892-7921	D	15	0.1
Champion Industries Inc	PO Box 4149	Winston-Salem	NC	27115	H Holt	919-661-1556	R	14	0.2

Source: Ward's Business Directory of U.S. Private and Public Companies, Volumes 1 and 2, 1996. The company type code used is as follows: P - Public, R - Private, S - Subsidiary, D - Division, J - Joint Venture, A - Affiliate, G - Group. Sales are in millions of dollars, employees are in thousands. An asterisk (*) indicates an estimated sales volume. The symbol < stands for 'less than'. Company names and addresses are truncated, in some cases, to fit into the available space.

MATERIALS CONSUMED

Material	Quantity	Delivered Cost ($ million)
Materials, ingredients, containers, and supplies	(X)	2,053.3
Fluid power pumps, motors, and hydrostatic transmissions (hydraulic and pneumatic)	(X)	41.1
All other pumps	(X)	24.8
Fluid power valves (hydraulic and pneumatic)	(X)	35.2
Fluid power cylinders and rotary actuators (hydraulic and pneumatic)	(X)	11.9
Fluid power hose or tube fittings and assemblies (hydraulic and pneumatic)	(X)	23.5
Fluid power filters (hydraulic and pneumatic)	(X)	15.7
Other fluid power products (hydraulic and pneumatic)	(X)	11.1
Metal bolts, nuts, screws, washers, rivets, and other screw machine products	(X)	49.0
Metal pipe, valves, and pipe fittings, except plumbers' and forgings	(X)	44.4
Other fabricated metal products, except electrical enclosures and forgings	(X)	90.0
Forgings	(X)	3.5
Iron and steel castings (rough and semifinished)	(X)	33.3
Nonferrous (aluminum, copper, etc.) castings (rough and semifinished)	(X)	22.8
Steel bars, bar shapes, and plates	(X)	49.0
Steel sheet and strip, including tin plate	(X)	148.7
Steel structural shapes and sheet piling	(X)	15.0
All other steel shapes and forms	(X)	33.9
Aluminum and aluminum-base alloy shapes and forms	(X)	24.6
Other nonferrous shapes and forms	(X)	8.6
Fractional horsepower electric motors and generators (less than 1 hp)	(X)	43.3
Integral horsepower electric motors and generators (1 hp or more)	(X)	55.9
Plastics resins consumed in the form of granules, pellets, powders, liquids, etc.	(X)	52.3
Plastics products consumed in the form of sheets, rods, tubes, film, and other shapes	(X)	70.6
Mechanical speed changers, gears, and industrial high-speed drives	(X)	18.1
Electrical enclosures (metal and plastics)	(X)	19.3
Current-carrying wiring devices	(X)	31.3
Electrical transmission, distribution, and control equipment	(X)	79.2
Paper and paperboard containers, including shipping sacks and other paper packaging supplies	(X)	35.6
All other materials and components, parts, containers, and supplies	(X)	606.7
Materials, ingredients, containers, and supplies, nsk	(X)	354.9

Source: 1992 *Economic Census.* Explanation of symbols used: (D): Withheld to avoid disclosure of competitive data; na: Not available; (S): Withheld because statistical norms were not met; (X): Not applicable; (Z): Less than half the unit shown; nec: Not elsewhere classified; nsk: Not specified by kind; - : zero; * : 10-19 percent estimated; ** : 20-29 percent estimated.

PRODUCT SHARE DETAILS

Product or Product Class	% Share	Product or Product Class	% Share
Service industry machinery, nec	100.00	Other commercial and industrial floor and carpet cleaning machines, including waxing and polishing machines, except vacuum cleaners	5.37
Commercial cooking and food-warming equipment	25.80	Parts and attachments for commercial and industrial floor and carpet cleaning equipment	2.09
Commercial ranges, ovens, and broilers (except electric)	22.13	Conveyor type commercial dishwashing machines	2.04
Commercial deep-fat fryers (except electric)	9.32	All other commercial dishwashing machines	2.67
Other commercial cooking equipment (griddles, toasters, coffee urns, pressure cookers-steam, etc.), except electric	5.87	Parts for commercial dishwashing machines	1.31
Commercial food-warming equipment, including steam tables (except electric)	1.82	Sewer pipe and drain cleaning equipment	2.45
Commercial electric ranges, ovens, and broilers	12.11	High-pressure (more than 1,000 p.s.i.) cleaning and blasting machinery and equipment (except foundry)	4.91
Commercial electric deep-fat fryers	5.84	Commercial car, truck, and bus washing machinery and equipment	3.96
Other commercial electric cooking equipment, including griddles, toasters, coffee makers, coffee urns, etc.	17.67	Service industry trash and garbage compactors	2.75
Commercial electric food-warming equipment, including hot-food server units and steam tables	14.10	Service industry sewage treatment equipment	18.07
		Other service industry equipment	24.22
Parts and accessories for commercial cooking and food-warming equipment	10.86	Parts for other service industry machines	4.78
Commercial cooking and food-warming equipment, nsk	0.28	Service industry machines and parts, nsk	1.65
Service industry machines and parts	59.56	Commercial and industrial vacuum cleaners, including parts and attachments	6.88
Instantaneous service industry water heaters, including parts	1.05	Commercial and industrial portable vacuum cleaners	54.64
All others service industry water heaters (including parts) with more than 120 gallons (454.2 liters capacity)	2.85	Parts and attachments for commercial and industrial portable vacuum cleaners	33.17
Industrial water softeners	1.71	Commercial and industrial central vacuum cleaner systems, including parts and attachments	9.92
Farm, household and commercial water softeners	7.27	Commercial and industrial vacuum cleaners, including parts and attachments, nsk	2.27
Parts for water softeners (excluding tanks)	1.41	Service industry machinery, nec, nsk	7.75
Commercial and industrial floor sanding and scrubbing machines	2.48		
Commercial and industrial carpet cleaning equipment, including sweepers (except vacuum cleaners)	6.95		

Source: 1992 *Economic Census.* The values shown are percent of total shipments in an industry. Values of indented subcategories are summed in the main heading. The symbol (D) appears when data are withheld to prevent disclosure of competitive information. The abbreviation nsk stands for 'not specified by kind' and nec for 'not elsewhere classified'.

INPUTS AND OUTPUTS FOR SERVICE INDUSTRY MACHINES, NEC

Economic Sector or Industry Providing Inputs	%	Sector	Economic Sector or Industry Buying Outputs	%	Sector
Wholesale trade	11.9	Trade	Gross private fixed investment	54.0	Cap Inv
Fabricated plate work (boiler shops)	8.2	Manufg.	Sewer system facility construction	13.2	Constr.
Blast furnaces & steel mills	7.7	Manufg.	Electrical repair shops	6.1	Services
Motors & generators	5.5	Manufg.	S/L Govt. purch., elem. & secondary education	3.3	S/L Govt
Machinery, except electrical, nec	3.3	Manufg.	S/L Govt. purch., higher education	2.8	S/L Govt
Blowers & fans	2.9	Manufg.	Exports	2.0	Foreign
Advertising	2.6	Services	Business services nec	1.9	Services
Communications, except radio & TV	2.4	Util.	Wholesale trade	1.7	Trade
Chemical & fertilizer mineral	2.2	Mining	Miscellaneous repair shops	1.5	Services
Miscellaneous plastics products	2.2	Manufg.	Office buildings	1.2	Constr.
Pipe, valves, & pipe fittings	2.2	Manufg.	S/L Govt. purch., sewerage	1.2	S/L Govt
Plastics materials & resins	2.2	Manufg.	S/L Govt. purch., sanitation	1.1	S/L Govt
Sheet metal work	2.2	Manufg.	Residential 1-unit structures, nonfarm	0.9	Constr.
Switchgear & switchboard apparatus	2.2	Manufg.	Construction of stores & restaurants	0.7	Constr.
Maintenance of nonfarm buildings nec	2.1	Constr.	Nonfarm residential structure maintenance	0.7	Constr.
Motor freight transportation & warehousing	2.1	Util.	Federal Government purchases, national defense	0.7	Fed Govt
Cyclic crudes and organics	1.7	Manufg.	Maintenance of sewer facilities	0.5	Constr.
Iron & steel foundries	1.6	Manufg.	Residential garden apartments	0.5	Constr.
Paperboard containers & boxes	1.6	Manufg.	Service industry machines, nec	0.5	Manufg.
Power transmission equipment	1.6	Manufg.	Ship building & repairing	0.5	Manufg.
Special dies & tools & machine tool accessories	1.5	Manufg.	Construction of hospitals	0.4	Constr.
Aluminum rolling & drawing	1.3	Manufg.	Maintenance of nonfarm buildings nec	0.4	Constr.
Screw machine and related products	1.3	Manufg.	Motor freight transportation & warehousing	0.4	Util.
Paints & allied products	1.2	Manufg.	Services to dwellings & other buildings	0.4	Services
Electric services (utilities)	1.2	Util.	Industrial buildings	0.3	Constr.
Fabricated metal products, nec	1.1	Manufg.	Residential additions/alterations, nonfarm	0.3	Constr.
Nonferrous rolling & drawing, nec	1.1	Manufg.	S/L Govt. purch., correction	0.3	S/L Govt
Pumps & compressors	1.1	Manufg.	Construction of educational buildings	0.2	Constr.
Wiring devices	1.1	Manufg.	Local government passenger transit	0.2	Gov't
Air transportation	1.0	Util.	State & local government enterprises, nec	0.2	Gov't
Real estate	1.0	Fin/R.E.	Federal Government purchases, nondefense	0.2	Fed Govt
Ball & roller bearings	0.9	Manufg.	S/L Govt. purch., health & hospitals	0.2	S/L Govt
Service industry machines, nec	0.9	Manufg.	S/L Govt. purch., natural resource & recreation.	0.2	S/L Govt
Eating & drinking places	0.9	Trade	S/L Govt. purch., water	0.2	S/L Govt
Engineering, architectural, & surveying services	0.9	Services	Hotels & motels	0.1	Constr.
Aluminum castings	0.8	Manufg.	Resid. & other health facility construction	0.1	Constr.
Metal stampings, nec	0.8	Manufg.	Residential 2-4 unit structures, nonfarm	0.1	Constr.
Abrasive products	0.7	Manufg.	Blast furnaces & steel mills	0.1	Manufg.
Metal coating & allied services	0.7	Manufg.	Federal Government enterprises nec	0.1	Gov't
Gaskets, packing & sealing devices	0.6	Manufg.			
Hardware, nec	0.6	Manufg.			
Petroleum refining	0.6	Manufg.			
Railroads & related services	0.6	Util.			
Computer & data processing services	0.6	Services			
Fabricated rubber products, nec	0.5	Manufg.			
Refrigeration & heating equipment	0.5	Manufg.			
Gas production & distribution (utilities)	0.5	Util.			
Banking	0.5	Fin/R.E.			
Equipment rental & leasing services	0.5	Services			
Brass, bronze, & copper castings	0.4	Manufg.			
Nonferrous wire drawing & insulating	0.4	Manufg.			
Rubber & plastics hose & belting	0.4	Manufg.			
Transformers	0.4	Manufg.			
Wood containers	0.4	Manufg.			
Metal heat treating	0.3	Manufg.			
Water transportation	0.3	Util.			
Hotels & lodging places	0.3	Services			
Legal services	0.3	Services			
Management & consulting services & labs	0.3	Services			
U.S. Postal Service	0.3	Gov't			
General industrial machinery, nec	0.2	Manufg.			
Industrial inorganic chemicals, nec	0.2	Manufg.			
Lubricating oils & greases	0.2	Manufg.			
Manifold business forms	0.2	Manufg.			
Accounting, auditing & bookkeeping	0.2	Services			
Coal	0.1	Mining			
Machine tools, metal cutting types	0.1	Manufg.			
Machine tools, metal forming types	0.1	Manufg.			
Insurance carriers	0.1	Fin/R.E.			
Royalties	0.1	Fin/R.E.			
Automotive rental & leasing, without drivers	0.1	Services			
Electrical repair shops	0.1	Services			

Source: Benchmark Input-Output Accounts for the U.S. Economy, 1982, U.S. Department of Commerce, Washington, D.C., July 1991. Data, as reported in the source, are organized by the 1977 SIC structure in use in 1982 but have been matched, as closely as is possible, to the 1987 SIC structure used in this book.

OCCUPATIONS EMPLOYED BY SIC 358 - REFRIGERATION AND SERVICE MACHINERY

Occupation	% of Total 1994	Change to 2005	Occupation	% of Total 1994	Change to 2005
Assemblers, fabricators, & hand workers nec	20.6	0.4	Machine tool cutting operators, metal & plastic	1.3	-16.4
Machine assemblers	5.5	50.6	Drafters	1.3	-21.8
Blue collar worker supervisors	3.7	-4.4	Combination machine tool operators	1.3	10.4
Sheet metal workers & duct installers	3.2	0.4	Industrial machinery mechanics	1.2	10.4
Inspectors, testers, & graders, precision	2.7	0.4	Helpers, laborers, & material movers nec	1.2	0.4
Sales & related workers nec	2.6	0.4	Traffic, shipping, & receiving clerks	1.2	-3.4
Welders & cutters	2.5	0.4	Solderers & brazers	1.1	0.4
Industrial truck & tractor operators	2.1	0.4	Stock clerks	1.1	-18.4
Machine tool cutting & forming etc. nec	2.1	0.4	Freight, stock, & material movers, hand	1.1	-19.7
Machine forming operators, metal & plastic	1.7	0.4	Engineering technicians & technologists nec	1.1	0.4
General managers & top executives	1.7	-4.8	Metal & plastic machine workers nec	1.0	-11.1
Welding machine setters, operators	1.7	-9.7	Mechanical engineers	1.0	10.5
Machine builders	1.4	20.4	Machinists	1.0	0.4
Secretaries, ex legal & medical	1.4	-8.6			

Source: *Industry-Occupation Matrix*, Bureau of Labor Statistics. These data relate to one or more 3-digit SIC industry groups rather than to a single 4-digit SIC. The change reported for each occupation to the year 2005 is a percent of growth or decline as estimated by the Bureau of Labor Statistics. The abbreviation nec stands for 'not elsewhere classified'.

LOCATION BY STATE AND REGIONAL CONCENTRATION

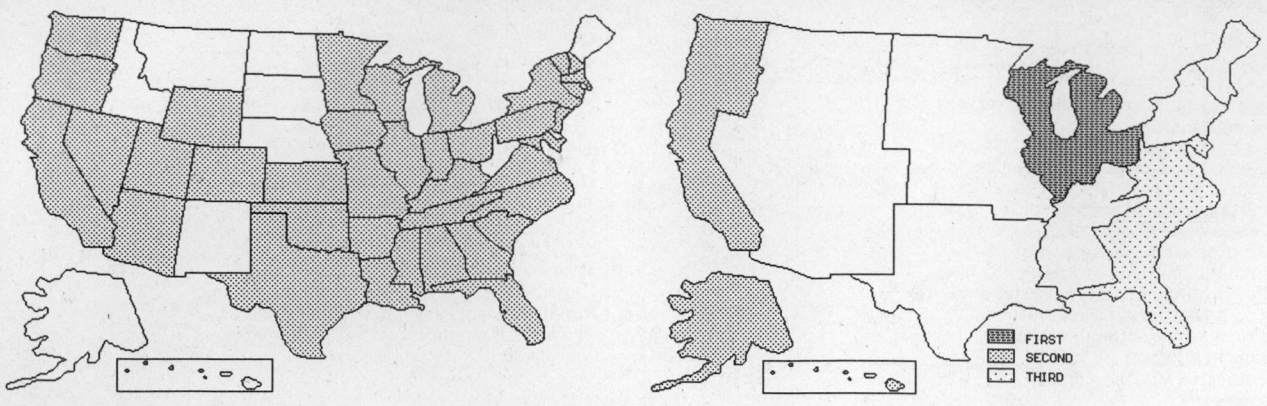

INDUSTRY DATA BY STATE

| State | Establish-ments | Shipments | | | Employment | | | | Cost as % of Shipments | Investment per Employee ($) |
		Total ($ mil)	% of U.S.	Per Establ.	Total Number	% of U.S.	Per Establ.	Wages ($/hour)		
Illinois	77	666.4	12.5	8.7	4,800	12.5	62	11.83	49.2	2,146
Ohio	65	646.5	12.1	9.9	3,700	9.6	57	10.86	41.3	2,432
California	189	575.7	10.8	3.0	4,400	11.4	23	11.15	44.5	2,318
Michigan	41	421.4	7.9	10.3	2,400	6.2	59	10.41	45.1	6,083
Minnesota	40	325.0	6.1	8.1	2,400	6.2	60	14.48	42.6	-
Texas	58	164.1	3.1	2.8	1,200	3.1	21	10.08	39.8	2,333
Pennsylvania	46	163.1	3.1	3.5	1,300	3.4	28	10.67	46.7	-
Florida	73	152.0	2.8	2.1	1,300	3.4	18	10.94	41.7	1,538
New Jersey	34	139.9	2.6	4.1	1,400	3.6	41	13.67	42.7	714
Iowa	17	116.9	2.2	6.9	900	2.3	53	12.75	46.8	1,556
Indiana	20	105.2	2.0	5.3	800	2.1	40	10.58	52.0	2,500
New York	35	104.0	1.9	3.0	1,000	2.6	29	9.21	54.6	1,400
Massachusetts	25	100.2	1.9	4.0	700	1.8	28	16.17	45.4	-
Tennessee	19	86.0	1.6	4.5	700	1.8	37	9.71	43.6	3,143
Alabama	9	83.4	1.6	9.3	700	1.8	78	9.88	41.1	-
Washington	20	71.3	1.3	3.6	600	1.6	30	12.63	34.9	-
Colorado	19	61.8	1.2	3.3	300	0.8	16	10.00	46.0	2,667
Oregon	13	53.7	1.0	4.1	500	1.3	38	8.50	45.8	1,800
Arizona	18	46.3	0.9	2.6	400	1.0	22	9.25	54.9	1,250
Nevada	10	41.3	0.8	4.1	300	0.8	30	11.25	42.4	667
Georgia	21	39.0	0.7	1.9	600	1.6	29	9.67	60.5	2,333
Utah	16	28.2	0.5	1.8	200	0.5	13	9.00	53.5	1,500
Wisconsin	46	(D)	-	-	1,750 *	4.5	38	-	-	-
Missouri	26	(D)	-	-	375 *	1.0	14	-	-	1,600
North Carolina	23	(D)	-	-	750 *	1.9	33	-	-	-
Louisiana	17	(D)	-	-	750 *	1.9	44	-	-	-
Arkansas	15	(D)	-	-	750 *	1.9	50	-	-	1,600
Kansas	15	(D)	-	-	375 *	1.0	25	-	-	1,067
Oklahoma	14	(D)	-	-	175 *	0.5	13	-	-	-
South Carolina	13	(D)	-	-	375 *	1.0	29	-	-	-
Virginia	10	(D)	-	-	375 *	1.0	38	-	-	-
Kentucky	9	(D)	-	-	375 *	1.0	42	-	-	-
Connecticut	8	(D)	-	-	375 *	1.0	47	-	-	-
New Hampshire	7	(D)	-	-	375 *	1.0	54	-	-	-
Maryland	4	(D)	-	-	375 *	1.0	94	-	-	-
Mississippi	4	(D)	-	-	175 *	0.5	44	-	-	-
Vermont	2	(D)	-	-	375 *	1.0	188	-	-	-
Wyoming	2	(D)	-	-	175 *	0.5	88	-	-	-

Source: 1992 *Economic Census*. The states are in descending order of shipments or establishments (if shipment data are missing for the majority). The symbol (D) appears when data are withheld to prevent disclosure of competitive information. States marked with (D) are sorted by number of establishments. A dash (-) indicates that the data element cannot be calculated; * indicates the midpoint of a range.

3592 - CARBURETORS, PISTONS, RINGS, VALVES

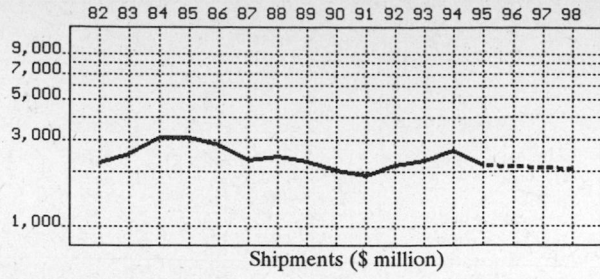

Shipments ($ million)

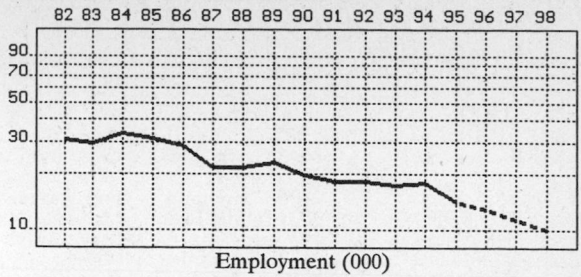

Employment (000)

GENERAL STATISTICS

Year	Companies	Establishments		Employment			Compensation		Production ($ million)			
		Total	with 20 or more employees	Total (000)	Production Workers (000)	Hours (Mil)	Payroll ($ mil)	Wages ($/hr)	Cost of Materials	Value Added by Manufacture	Value of Shipments	Capital Invest.
1982	149	171	94	31.2	24.0	46.0	691.9	10.70	795.6	1,385.1	2,224.5	106.0
1983		168	93	29.8	23.2	46.7	745.5	11.65	928.4	1,542.5	2,485.0	71.1
1984		165	92	33.4	26.6	54.3	892.9	12.38	1,234.7	1,888.4	3,096.4	96.9
1985		162	92	31.6	24.9	49.8	870.9	12.93	1,286.8	1,774.5	3,091.9	188.0
1986		157	85	28.8	22.3	46.3	838.4	13.27	1,243.7	1,585.6	2,818.2	244.4
1987	132	155	86	21.7	17.3	36.7	648.2	13.22	1,088.7	1,198.0	2,287.4	104.9
1988		144	80	22.0	17.8	37.1	653.1	13.45	1,221.3	1,200.2	2,408.6	128.9
1989		147	81	23.2	17.5	35.6	652.4	13.83	1,094.3	1,160.1	2,265.5	71.8
1990		142	76	19.7	16.5	31.6	601.1	14.30	980.7	1,045.8	2,042.4	87.9
1991		145	80	18.4	14.7	28.9	558.6	14.25	916.9	997.0	1,920.6	79.3
1992	117	138	68	18.4	14.6	29.6	586.6	14.82	981.6	1,146.5	2,155.4	76.8
1993		136	68	17.4	14.0	29.6	582.3	14.79	1,011.0	1,335.6	2,330.4	99.5
1994		131p	67p	18.0	14.8	32.6	653.4	15.19	1,129.7	1,544.2	2,657.1	134.7
1995		128p	65p	14.2p	11.7p	25.1p	548.9p	15.72p	924.5p	1,263.8p	2,174.5p	97.8p
1996		125p	62p	12.8p	10.7p	23.1p	528.7p	16.04p	908.1p	1,241.3p	2,135.9p	95.4p
1997		121p	60p	11.4p	9.6p	21.1p	508.5p	16.36p	891.7p	1,218.9p	2,097.3p	93.0p
1998		118p	57p	10.0p	8.6p	19.2p	488.2p	16.69p	875.3p	1,196.4p	2,058.7p	90.5p

Sources: 1982, 1987, 1992 *Economic Census*; *Annual Survey of Manufactures*, 83-86, 88-91, 93-94. Establishment counts for non-Census years are from *County Business Patterns*; establishment values for 83-84 are extrapolations. 'P's show projections by the editors. Industries reclassified in 87 will not have data for prior years.

INDICES OF CHANGE

Year	Companies	Establishments		Employment			Compensation		Production ($ million)			
		Total	with 20 or more employees	Total (000)	Production Workers (000)	Hours (Mil)	Payroll ($ mil)	Wages ($/hr)	Cost of Materials	Value Added by Manufacture	Value of Shipments	Capital Invest.
1982	127	124	138	170	164	155	118	72	81	121	103	138
1983		122	137	162	159	158	127	79	95	135	115	93
1984		120	135	182	182	183	152	84	126	165	144	126
1985		117	135	172	171	168	148	87	131	155	143	245
1986		114	125	157	153	156	143	90	127	138	131	318
1987	113	112	126	118	118	124	111	89	111	104	106	137
1988		104	118	120	122	125	111	91	124	105	112	168
1989		107	119	126	120	120	111	93	111	101	105	93
1990		103	112	107	113	107	102	96	100	91	95	114
1991		105	118	100	101	98	95	96	93	87	89	103
1992	100	100	100	100	100	100	100	100	100	100	100	100
1993		99	100	95	96	100	99	100	103	116	108	130
1994		95p	99p	98	101	110	111	102	115	135	123	175
1995		93p	95p	77p	80p	85p	94p	106p	94p	110p	101p	127p
1996		90p	92p	70p	73p	78p	90p	108p	93p	108p	99p	124p
1997		88p	88p	62p	66p	71p	87p	110p	91p	106p	97p	121p
1998		86p	85p	54p	59p	65p	83p	113p	89p	104p	96p	118p

Sources: Same as General Statistics. Values reflect change from the base year, 1992. Values above 100 mean greater than 92, values below 100 mean less than 92, and a value of 100 in the 82-91 or 93-98 period means same as 92. 'P's mark projections by the editors.

SELECTED RATIOS

For 1994	Avg. of All Manufact.	Analyzed Industry	Index	For 1994	Avg. of All Manufact.	Analyzed Industry	Index
Employees per Establishment	49	137	280	Value Added per Production Worker	134,084	104,338	78
Payroll per Establishment	1,500,273	4,979,148	332	Cost per Establishment	5,045,178	8,608,729	171
Payroll per Employee	30,620	36,300	119	Cost per Employee	102,970	62,761	61
Production Workers per Establishment	34	113	329	Cost per Production Worker	146,988	76,331	52
Wages per Establishment	853,319	3,773,560	442	Shipments per Establishment	9,576,895	20,248,078	211
Wages per Production Worker	24,861	33,459	135	Shipments per Employee	195,460	147,617	76
Hours per Production Worker	2,056	2,203	107	Shipments per Production Worker	279,017	179,534	64
Wages per Hour	12.09	15.19	126	Investment per Establishment	321,011	1,026,463	320
Value Added per Establishment	4,602,255	11,767,371	256	Investment per Employee	6,552	7,483	114
Value Added per Employee	93,930	85,789	91	Investment per Production Worker	9,352	9,101	97

Sources: Same as General Statistics. The 'Average of All Manufacturing' column represents the average of all manufacturing industries reported for the most recent complete year available. The Index shows the relationship between the Average and the Analyzed Industry. For example, 100 means that they are equal; 500 that the Analyzed Industry is five times the average; 50 means that the Analyzed Industry is half the national average. The abbreviation 'na' is used to show that data are 'not available'.

LEADING COMPANIES   Number shown: 24   Total sales ($ mil): **981**   Total employment (000): **9.1**

Company Name	Address				CEO Name	Phone	Co. Type	Sales ($ mil)	Empl. (000)
Walbro Corp	6242 Garfield St	Cass City	MI	48726	Lambert E Althaver	517-872-2131	P	325	3.2
Mahle Inc	PO Box 748	Morristown	TN	37815	HD Jehle	615-581-6603	S	113	1.0
Zollner Co LP	2425 Coliseum S	Fort Wayne	IN	46803	Gary Riley	219-426-8081	R	95	0.8
Karl Schmidt Unisia Inc	1731 Indrial Pkwy	Marinette	WI	54143	Stephen Demster	715-732-0181	D	73*	0.7
AE Goetze North American	PO Box 456	Lake City	MN	55041	David D Krohn	612-345-4541	D	55	0.5
KSG Industries Inc	151 S Warner Rd	Wayne	PA	19087	Jerry McGrath	610-964-8590	S	50	0.4
Kaydon Ring and Seal Inc	PO Box 626	Baltimore	MD	21203	Arthur Ridler	410-547-7700	S	45	0.3
Metal Leve Inc	560 Avis Dr	Ann Arbor	MI	48108	Durand Mahrus	313-930-1590	R	40	0.3
AirSensors Inc	16804 Gridley Pl	Cerritos	CA	90701	Robert M Stemmler	310-860-6666	P	36	0.3
IMPCO Technologies Inc	16804 Gridley Pl	Cerritos	CA	90701	Bertram R Martin	310-860-6666	S	32	0.2
AE Goetze-Wausau	PO Box 620	Schofield	WI	54476	Russel A Duggan	715-359-3111	D	30	0.3
Tomco Auto Products Inc	4330 E 26th St	Los Angeles	CA	90023	Victor Moss	213-268-4830	R	22	0.4
Pacific Piston Ring Company Inc	PO Box 987	Culver City	CA	90232	F Shannon	310-836-3322	R	9*	<0.1
Allied Ring Corp	916 W State St	St Johns	MI	48879	RC Shunta	517-224-2384	J	8	<0.1
Zenith Fuel Systems Inc	14570 Industrial Pk	Bristol	VA	24202	Rick Hamill	703-669-5555	D	8	<0.1
Precision Rings Inc	PO Box 418187	Indianapolis	IN	46241	JS Crannell	317-247-4786	R	7	<0.1
Starflo Corp	940 Crosscreek Rd	Orangeburg	SC	29115	Frank Parker	803-536-9660	S	7*	<0.1
Ertel Manufacturing Corp	PO Box 21	New Albany	MS	38652	Douglas Wolfe	601-534-7601	D	6	<0.1
FRON Corp	1360 N Jefferson	Anaheim	CA	92807	FT Grady	714-996-0050	R	6	<0.1
Double Seal Ring Company Inc	PO Box 566	Fort Worth	TX	76101	H Green Jr	817-738-6581	R	4	<0.1
Safety Seal Piston Ring Co	PO Box 106	Fort Worth	TX	76101	Stanley Baker Jr	817-283-1573	R	4	<0.1
Katty Industries Inc	PO Box 14290	Odessa	TX	79768	Ken Hutson	915-362-0303	R	4	<0.1
Carburetion-J and S Inc	PO Box 4778	Dallas	TX	75208	JT Vail	214-741-3980	R	1*	<0.1
Jahns Quality Pistons	1360 N Jefferson St	Anaheim	CA	92807	Frank Grady	714-579-3795	S	1	<0.1

Source: *Ward's Business Directory of U.S. Private and Public Companies*, Volumes 1 and 2, 1996. The company type code used is as follows: P - Public, R - Private, S - Subsidiary, D - Division, J - Joint Venture, A - Affiliate, G - Group. Sales are in millions of dollars, employees are in thousands. An asterisk (*) indicates an estimated sales volume. The symbol < stands for 'less than'. Company names and addresses are truncated, in some cases, to fit into the available space.

MATERIALS CONSUMED

Material	Quantity	Delivered Cost ($ million)
Materials, ingredients, containers, and supplies	(X)	907.6
Metal bolts, nuts, screws, washers, rivets, and other screw machine products	(X)	38.0
Other fabricated metal products (except castings and forgings)	(X)	49.1
Forgings	(X)	(D)
Iron and steel castings (rough and semifinished)	(X)	68.2
Aluminum and aluminum-base alloy castings (rough and semifinished)	(X)	157.6
Other nonferrous castings (rough and semifinished)	(X)	4.3
Steel bars, bar shapes, and plates	(X)	65.6
Steel sheet and strip, including tin plate	(X)	41.1
All other steel shapes and forms	(X)	(D)
Copper and copper-base alloy shapes and forms	(X)	(D)
Aluminum and aluminum-base alloy shapes and forms	(X)	45.7
Other nonferrous shapes and forms	(X)	4.1
Paperboard containers, boxes, and corrugated paperboard	(X)	36.6
Flexible packaging materials	(X)	0.4
Fabricated plastics products (except gaskets, hoses, and belting)	(X)	(D)
Gaskets (all types), and packing and sealing devices	(X)	8.5
All other materials and components, parts, containers, and supplies	(X)	294.1
Materials, ingredients, containers, and supplies, nsk	(X)	36.4

Source: 1992 *Economic Census*. Explanation of symbols used: (D): Withheld to avoid disclosure of competitive data; na: Not available; (S): Withheld because statistical norms were not met; (X): Not applicable; (Z): Less than half the unit shown; nec: Not elsewhere classified; nsk: Not specified by kind; - : zero; * : 10-19 percent estimated; ** : 20-29 percent estimated.

PRODUCT SHARE DETAILS

Product or Product Class	% Share	Product or Product Class	% Share
Carburetors, pistons, rings, and valves	100.00	Oil type piston rings for motor vehicle engines (passenger car, truck, and bus), all types	11.83
Carburetors, new and rebuilt (all types)	39.87	All other oil type engine piston rings	7.12
Carburators for motor vehicle engines (passenger car, truck, and bus), new, all types	62.48	Compression type piston rings for motor vehicle engines (passenger car, truck, and bus), all types	22.56
Carburetors, rebuilt, all types	9.60	All other compression type engine piston rings	5.53
Parts for carburetors (excluding gaskets and screw machine products)	25.99	Engine piston pins	10.64
Carburetors, new and rebuilt (all types), nsk	1.95	Valves (engine intake and exhaust)	19.31
Pistons, piston rings, and piston pins (engine)	38.08	Intake and exhaust valves for motor vehicle engines (passenger car, truck, and bus)	80.68
Pistons for motor vehicle engines (passenger car, truck, and bus), all types (machined), except rough castings	32.60	Intake and exhaust valves for other engines	19.32
All other engine pistons, all types (machined), except rough castings	9.73	Carburetors, pistons, rings, and valves, nsk	2.74

Source: 1992 *Economic Census*. The values shown are percent of total shipments in an industry. Values of indented subcategories are summed in the main heading. The symbol (D) appears when data are withheld to prevent disclosure of competitive information. The abbreviation nsk stands for 'not specified by kind' and nec for 'not elsewhere classified'.

INPUTS AND OUTPUTS FOR CARBURETORS, PISTONS, RINGS, & VALVES

Economic Sector or Industry Providing Inputs	%	Sector	Economic Sector or Industry Buying Outputs	%	Sector
Blast furnaces & steel mills	8.2	Manufg.	Motor vehicle parts & accessories	34.3	Manufg.
Carburetors, pistons, rings, & valves	6.5	Manufg.	Motor vehicles & car bodies	16.9	Manufg.
Wholesale trade	5.6	Trade	Internal combustion engines, nec	14.3	Manufg.
Machinery, except electrical, nec	5.5	Manufg.	Farm machinery & equipment	8.7	Manufg.
Electric services (utilities)	5.2	Util.	Automotive repair shops & services	8.2	Services
Automotive repair shops & services	5.2	Services	Carburetors, pistons, rings, & valves	3.0	Manufg.
Plating & polishing	4.3	Manufg.	Miscellaneous repair shops	2.3	Services
Iron & steel foundries	4.2	Manufg.	Truck & bus bodies	2.1	Manufg.
Maintenance of nonfarm buildings nec	4.1	Constr.	Personal consumption expenditures	1.5	
Aluminum castings	4.1	Manufg.	Feed grains	1.3	Agric.
Screw machine and related products	3.4	Manufg.	Oil field machinery	0.9	Manufg.
Advertising	3.3	Services	State & local government enterprises, nec	0.9	Gov't
Aluminum rolling & drawing	2.9	Manufg.	Federal Government purchases, national defense	0.9	Fed Govt
Computer & data processing services	2.8	Services	Meat animals	0.7	Agric.
Nonferrous castings, nec	2.5	Manufg.	Oil bearing crops	0.4	Agric.
Miscellaneous plastics products	1.6	Manufg.	Food grains	0.3	Agric.
Paperboard containers & boxes	1.5	Manufg.	Industrial trucks & tractors	0.3	Manufg.
Petroleum refining	1.5	Manufg.	Motor freight transportation & warehousing	0.3	Util.
Gas production & distribution (utilities)	1.5	Util.	S/L Govt. purch., elem. & secondary education	0.3	S/L Govt
Eating & drinking places	1.5	Trade	Tobacco	0.2	Agric.
Banking	1.4	Fin/R.E.	Vegetables	0.2	Agric.
Engineering, architectural, & surveying services	1.3	Services	Dairy farm products	0.1	Agric.
Air transportation	1.2	Util.	Fruits	0.1	Agric.
Motor freight transportation & warehousing	1.2	Util.	Lawn & garden equipment	0.1	Manufg.
Abrasive products	1.1	Manufg.	Retail trade, except eating & drinking	0.1	Trade
Equipment rental & leasing services	1.1	Services	Wholesale trade	0.1	Trade
Copper rolling & drawing	1.0	Manufg.	S/L Govt. purch., higher education	0.1	S/L Govt
Communications, except radio & TV	0.9	Util.	S/L Govt. purch., other general government	0.1	S/L Govt
Used & secondhand goods	0.9	Scrap			
Metal coating & allied services	0.8	Manufg.			
Metal stampings, nec	0.8	Manufg.			
Special dies & tools & machine tool accessories	0.8	Manufg.			
Gaskets, packing & sealing devices	0.7	Manufg.			
Machine tools, metal cutting types	0.7	Manufg.			
Electrical repair shops	0.7	Services			
Brass, bronze, & copper castings	0.5	Manufg.			
Industrial patterns	0.5	Manufg.			
Miscellaneous fabricated wire products	0.5	Manufg.			
Royalties	0.5	Fin/R.E.			
Legal services	0.5	Services			
Management & consulting services & labs	0.5	Services			
Fabricated rubber products, nec	0.4	Manufg.			
Metal heat treating	0.4	Manufg.			
Real estate	0.4	Fin/R.E.			
Accounting, auditing & bookkeeping	0.4	Services			
Hotels & lodging places	0.4	Services			
U.S. Postal Service	0.4	Gov't			
Lubricating oils & greases	0.3	Manufg.			
Rubber & plastics hose & belting	0.3	Manufg.			
Retail trade, except eating & drinking	0.3	Trade			
Insurance carriers	0.3	Fin/R.E.			
Automotive rental & leasing, without drivers	0.3	Services			
State & local government enterprises, nec	0.3	Gov't			
Hand & edge tools, nec	0.2	Manufg.			
Manifold business forms	0.2	Manufg.			
Nonferrous rolling & drawing, nec	0.2	Manufg.			
Railroads & related services	0.2	Util.			
Security & commodity brokers	0.2	Fin/R.E.			
Miscellaneous repair shops	0.2	Services			
Industrial gases	0.1	Manufg.			
Photographic equipment & supplies	0.1	Manufg.			
Sanitary services, steam supply, irrigation	0.1	Util.			
Transit & bus transportation	0.1	Util.			
Laundry, dry cleaning, shoe repair	0.1	Services			
Personnel supply services	0.1	Services			

Source: Benchmark Input-Output Accounts for the U.S. Economy, 1982, U.S. Department of Commerce, Washington, D.C., July 1991. Data, as reported in the source, are organized by the 1977 SIC structure in use in 1982 but have been matched, as closely as is possible, to the 1987 SIC structure used in this book.

OCCUPATIONS EMPLOYED BY SIC 359 - INDUSTRIAL MACHINERY, NEC

Occupation	% of Total 1994	Change to 2005	Occupation	% of Total 1994	Change to 2005
Machinists	20.3	-5.6	Sales & related workers nec	1.8	-5.6
NC machine tool operators, metal & plastic	4.9	41.6	Tool & die makers	1.6	-67.3
General managers & top executives	4.6	-10.4	Bookkeeping, accounting, & auditing clerks	1.6	-29.2
Blue collar worker supervisors	3.8	-13.3	Industrial production managers	1.6	3.8
Assemblers, fabricators, & hand workers nec	3.5	-5.6	Welding machine setters, operators	1.6	-15.0
Machine tool cutting operators, metal & plastic	3.5	-27.4	Drilling & boring machine tool workers	1.4	-52.8
Inspectors, testers, & graders, precision	2.7	-5.6	Traffic, shipping, & receiving clerks	1.3	-9.1
Welders & cutters	2.6	-5.6	Janitors & cleaners, incl maids	1.2	-24.5
Lathe & turning machine tool operators	2.6	-52.8	Industrial machinery mechanics	1.2	3.9
Combination machine tool operators	2.5	-5.6	Mechanical engineers	1.2	14.3
Grinding machine operators, metal & plastic	2.2	-5.6	Financial managers	1.1	-5.6
Secretaries, ex legal & medical	2.0	-14.1	Machine builders	1.1	-24.5
General office clerks	2.0	-19.5	Clerical supervisors & managers	1.0	-3.4
Machine tool cutting & forming etc. nec	1.9	-24.5	Grinders & polishers, hand	1.0	51.0

Source: Industry-Occupation Matrix, Bureau of Labor Statistics. These data relate to one or more 3-digit SIC industry groups rather than to a single 4-digit SIC. The change reported for each occupation to the year 2005 is a percent of growth or decline as estimated by the Bureau of Labor Statistics. The abbreviation nec stands for 'not elsewhere classified'.

LOCATION BY STATE AND REGIONAL CONCENTRATION

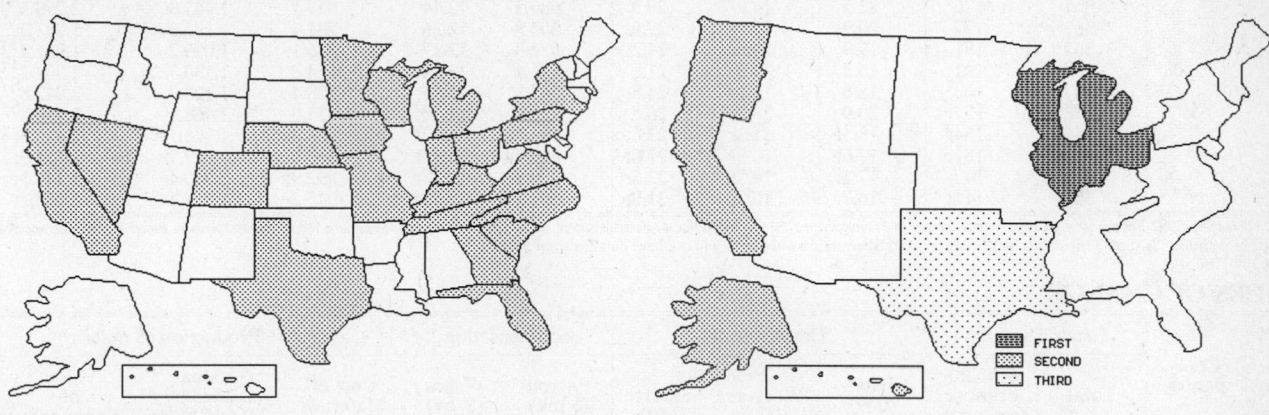

FIRST
SECOND
THIRD

INDUSTRY DATA BY STATE

State	Establish-ments	Shipments			Employment				Cost as % of Shipments	Investment per Employee ($)
		Total ($ mil)	% of U.S.	Per Establ.	Total Number	% of U.S.	Per Establ.	Wages ($/hour)		
Michigan	13	815.2	37.8	62.7	4,000	21.7	308	20.88	56.0	-
Indiana	7	175.4	8.1	25.1	2,000	10.9	286	14.24	33.6	3,800
Pennsylvania	7	64.5	3.0	9.2	900	4.9	129	9.65	29.3	1,222
California	27	61.6	2.9	2.3	900	4.9	33	8.63	39.4	2,222
Texas	14	23.0	1.1	1.6	400	2.2	29	9.67	32.2	1,250
Ohio	7	(D)	-	-	1,750 *	9.5	250	-	-	-
New York	6	(D)	-	-	1,750 *	9.5	292	-	-	-
Wisconsin	6	(D)	-	-	1,750 *	9.5	292	-	-	-
Tennessee	4	(D)	-	-	1,750 *	9.5	438	-	-	-
Iowa	3	(D)	-	-	750 *	4.1	250	-	-	-
Minnesota	3	(D)	-	-	375 *	2.0	125	-	-	-
Missouri	3	(D)	-	-	375 *	2.0	125	-	-	-
South Carolina	3	(D)	-	-	175 *	1.0	58	-	-	-
Virginia	3	(D)	-	-	175 *	1.0	58	-	-	-
Florida	2	(D)	-	-	375 *	2.0	188	-	-	-
Kentucky	2	(D)	-	-	750 *	4.1	375	-	-	-
Nebraska	2	(D)	-	-	750 *	4.1	375	-	-	-
Colorado	1	(D)	-	-	175 *	1.0	175	-	-	-
Georgia	1	(D)	-	-	175 *	1.0	175	-	-	-
Nevada	1	(D)	-	-	375 *	2.0	375	-	-	-
North Carolina	1	(D)	-	-	750 *	4.1	750	-	-	-

Source: 1992 Economic Census. The states are in descending order of shipments or establishments (if shipment data are missing for the majority). The symbol (D) appears when data are withheld to prevent disclosure of competitive information. States marked with (D) are sorted by number of establishments. A dash (-) indicates that the data element cannot be calculated; * indicates the midpoint of a range.

3593 - FLUID POWER CYLINDERS & ACTUATORS

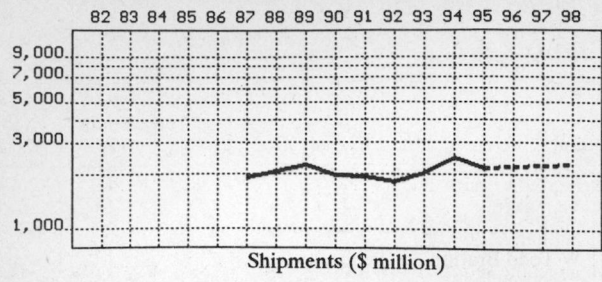

Shipments ($ million)

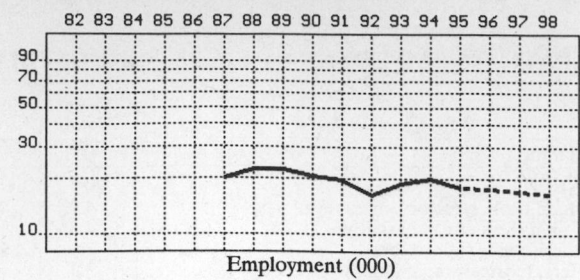

Employment (000)

GENERAL STATISTICS

Year	Com-panies	Establishments		Employment			Compensation		Production ($ million)			
		Total	with 20 or more employees	Total (000)	Production Workers (000)	Hours (Mil)	Payroll ($ mil)	Wages ($/hr)	Cost of Materials	Value Added by Manufacture	Value of Shipments	Capital Invest.
1982												
1983												
1984												
1985												
1986												
1987	330	362	168	20.2	13.2	26.1	602.0	12.79	637.5	1,254.5	1,896.6	81.1
1988		385	180	22.5	15.1	29.4	660.8	12.61	706.6	1,340.6	2,061.7	73.2
1989		361	176	22.5	14.9	29.9	686.7	12.44	819.1	1,447.3	2,238.6	57.9
1990		362	177	20.9	13.0	27.0	603.8	12.26	781.4	1,195.3	1,981.9	56.9
1991		369	181	19.9	11.9	24.2	572.8	12.17	836.1	1,093.2	1,931.7	69.4
1992	308	348	141	16.5	10.2	21.0	531.1	13.34	664.8	1,130.7	1,804.2	54.2
1993		367	163	18.8	12.1	24.8	645.8	14.85	928.4	1,123.7	2,042.8	71.5
1994		358P	157P	20.0	13.1	26.8	707.5	14.73	1,022.0	1,468.4	2,458.8	90.9
1995		356P	154P	18.1P	11.3P	23.6P	635.2P	14.62P	908.3P	1,305.1P	2,185.3P	72.7P
1996		354P	151P	17.6P	10.9P	23.0P	637.2P	14.94P	920.6P	1,322.7P	2,214.9P	73.4P
1997		352P	147P	17.2P	10.5P	22.5P	639.1P	15.27P	932.9P	1,340.4P	2,244.5P	74.1P
1998		350P	144P	16.7P	10.2P	21.9P	641.1P	15.59P	945.2P	1,358.1P	2,274.1P	74.9P

Sources: 1982, 1987, 1992 *Economic Census*; *Annual Survey of Manufactures*, 83-86, 88-91, 93-94. Establishment counts for non-Census years are from *County Business Patterns*; establishment values for 83-84 are extrapolations. 'P's show projections by the editors. Industries reclassified in 87 will not have data for prior years.

INDICES OF CHANGE

Year	Com-panies	Establishments		Employment			Compensation		Production ($ million)			
		Total	with 20 or more employees	Total (000)	Production Workers (000)	Hours (Mil)	Payroll ($ mil)	Wages ($/hr)	Cost of Materials	Value Added by Manufacture	Value of Shipments	Capital Invest.
1982												
1983												
1984												
1985												
1986												
1987	107	104	119	122	129	124	113	96	96	111	105	150
1988		111	128	136	148	140	124	95	106	119	114	135
1989		104	125	136	146	142	129	93	123	128	124	107
1990		104	126	127	127	129	114	92	118	106·	110	105
1991		106	128	121	117	115	108	91	126	97	107	128
1992	100	100	100	100	100	100	100	100	100	100	100	100
1993		105	116	114	119	118	122	111	140	99	113	132
1994		103P	111P	121	128	128	133	110	154	130	136	168
1995		102P	109P	110P	111P	112P	120P	110P	137P	115P	121P	134P
1996		102P	107P	107P	107P	110P	120P	112P	138P	117P	123P	135P
1997		101P	105P	104P	103P	107P	120P	114P	140P	119P	124P	137P
1998		101P	102P	101P	100P	104P	121P	117P	142P	120P	126P	138P

Sources: Same as General Statistics. Values reflect change from the base year, 1992. Values above 100 mean greater than 92, values below 100 mean less than 92, and a value of 100 in the 82-91 or 93-98 period means same as 92. 'P's mark projections by the editors.

SELECTED RATIOS

For 1994	Avg. of All Manufact.	Analyzed Industry	Index	For 1994	Avg. of All Manufact.	Analyzed Industry	Index
Employees per Establishment	49	56	114	Value Added per Production Worker	134,084	112,092	84
Payroll per Establishment	1,500,273	1,978,626	132	Cost per Establishment	5,045,178	2,858,170	57
Payroll per Employee	30,620	35,375	116	Cost per Employee	102,970	51,100	50
Production Workers per Establishment	34	37	107	Cost per Production Worker	146,988	78,015	53
Wages per Establishment	853,319	1,104,014	129	Shipments per Establishment	9,576,895	6,876,388	72
Wages per Production Worker	24,861	30,135	121	Shipments per Employee	195,460	122,940	63
Hours per Production Worker	2,056	2,046	100	Shipments per Production Worker	279,017	187,695	67
Wages per Hour	12.09	14.73	122	Investment per Establishment	321,011	254,215	79
Value Added per Establishment	4,602,255	4,106,592	89	Investment per Employee	6,552	4,545	69
Value Added per Employee	93,930	73,420	78	Investment per Production Worker	9,352	6,939	74

Sources: Same as General Statistics. The 'Average of All Manufacturing' column represents the average of all manufacturing industries reported for the most recent complete year available. The Index shows the relationship between the Average and the Analyzed Industry. For example, 100 means that they are equal; 500 that the Analyzed Industry is five times the average; 50 means that the Analyzed Industry is half the national average. The abbreviation 'na' is used to show that data are 'not available'.

LEADING COMPANIES   Number shown: **71**   Total sales ($ mil): **3,839**   Total employment (000): **38.6**

Company Name	Address				CEO Name	Phone	Co. Type	Sales ($ mil)	Empl. (000)
Parker Hannifin Corp	17325 Euclid Av	Cleveland	OH	44112	Duane E Collins	216-531-3000	P	2,576	26.7
Rexroth Corp	PO Box 25407	Lehigh Valley	PA	18001	AJ Krug	610-694-8300	S	200•	1.0
Curtiss-Wright Corp	1200 Wall St W	Lyndhurst	NJ	07071	David Lasky	201-896-8400	P	166	1.5
Parker Hannifin Corp	500 S Wolf Rd	Des Plaines	IL	60016	John K Oelslager	708-298-2400	D	80•	0.7
Galveston Houston Co	PO Box 2207	Houston	TX	77252	Nathan M Avery	713-966-2500	P	63	0.7
Bettis Corp	PO Box 508	Waller	TX	77484	W Todd Bratton	713-463-5100	P	53	0.4
BW/IP International Inc	7500 Tyrone Av	Van Nuys	CA	91409	Woodrow Lane	818-781-4000	D	36	0.3
McCanna Inc	400 Maple Av	Carpentersville	IL	60110	Rick Giannini	708-426-4100	S	35	0.2
Limitorque Corp	PO Box 11318	Lynchburg	VA	24506	T Mignogna	804-528-4400	S	30	0.5
Prince Manufacturing Corp	PO Box 537	Sioux City	IA	51102	Roland D Junck	712-277-4061	R	30•	0.3
PHD Inc	PO Box 9070	Fort Wayne	IN	46899	Joe Oberlin	219-747-6151	R	28	0.4
Monarch Hydraulics Inc	PO Box 1764	Grand Rapids	MI	49501	GA Jackoboice	616-458-1306	R	25	0.2
Sheffer Corp	6990 Cornell Rd	Cincinnati	OH	45242	Bob Warner	513-489-9770	R	25	0.3
Cross Manufacturing Inc	11011 King St	Overland Park	KS	66210	JH Cross	913-451-1233	R	22	0.3
RHM Fluid Power Inc	375 Manufacturers	Westland	MI	48185	W W Tulloch III	313-326-5400	R	21	<0.1
Oildyne	4301 Quebec Av N	Minneapolis	MN	55428	David Roberts	612-533-1600	D	20	<0.1
Norris Cylinder Co	PO Box 7486	Longview	TX	75607	Ed McSweeney	903-757-7633	S	19•	0.2
Thomson Saginaw Ball Screw	628 N Hamilton St	Saginaw	MI	48602	William A Pauwels	517-776-5111	S	19•	0.2
Atlas Cylinder	PO Box 2248	Eugene	OR	97402	Frank Bishop	503-689-9111	D	16	0.1
Bimba Manufacturing Co	PO Box 68	Monee	IL	60449	PJ Ormsby	708-534-8544	R	15•	0.4
Clippard Instrument Laboratory	7390 Colerain Av	Cincinnati	OH	45239	William Clippard III	513-521-4261	R	15	0.2
Milwaukee Cylinder	5877 S Pennsylvania	Cudahy	WI	53110	David J McKendrey	414-769-9700	D	15	0.1
Seitz Manufacturing Company	3645 W Elm St	Milwaukee	WI	53209	Richard S Melrose	414-352-1700	R	15	0.1
Victor Fluid Power Inc	1123 Hwy 212 W	Granite Falls	MN	56241	Ed Olson	612-564-2311	D	15	0.2
Iowa Industrial Hydraulics	Industrial Park Rd	Pocahontas	IA	50574	Don D Clover	712-335-3311	D	15	0.2
Hanna Corp	1765 N Elston Av	Chicago	IL	60622	Robert Rakstang	312-384-7000	R	14	<0.1
Remco Hydraulics Inc	934 S Main St	Willits	CA	95490	Guy Madden	707-459-5301	D	14	<0.1
Tol-O-Matic Inc	3800 County Rd 116	Hamel	MN	55340	William Toles	612-478-8000	R	13•	0.1
Hol-Mac Corp	PO Box 349	Bay Springs	MS	39422	CB Holder	601-764-4121	R	12	0.2
CM Smillie Co	1200 Woodwrd	Ferndale	MI	48220	CM Smillie III	810-544-3100	R	11	0.1
Mosier Industries Inc	PO Box 189	Brookville	OH	45309	KC Mosier II	513-833-4033	R	11•	0.2
Automax Inc	11444 Deerfield Rd	Cincinnati	OH	45242	Joseph Stigler	513-489-7800	S	10•	0.1
Best Metal Products Co	PO Box 888440	Grand Rapids	MI	49588	David J Faasse	616-942-7141	R	10	0.1
C and C Manufacturing Inc	1330 Anvil Rd	Rockford	IL	61115	Michael Vinski	815-633-8897	S	10	<0.1
Eckel Manufacturing Company	PO Box 1375	Odessa	TX	79760	Terry L Eckel	915-362-4336	R	10	<0.1
Green Manufacturing Inc	PO Box 408	Bowling Green	OH	43402	JF Snook	419-352-9484	R	10•	0.1
Hannon Hydraulics Inc	PO Box 36197	Dallas	TX	75235	HW Reed	214-438-2870	R	10•	0.1
Ralph A Hiller Co	6005 Enterprise Dr	Export	PA	15632	J Randolf Hiller	412-325-1200	R	10•	<0.1
WABASH MPI	PO Box 298	Wabash	IN	46992	Tim Shively	219-563-1184	S	10	<0.1
Carter Machine Company Inc	820 Edwards St	Galion	OH	44833	T Carter	419-468-3530	R	9	0.2
Fabco-Air Inc	3716 NE 49 Rd	Gainesville	FL	32609	W Schmidt	904-373-3578	R	8	0.1
Omhaline Hydraulic Co	PO Box 19	N Sioux City	SD	57049	George Sully	605-232-9060	S	8•	<0.1
Ortman Fluid Power	19 143rd St	Hammond	IN	46327	Ernest Kuhnen	219-931-1710	S	8	<0.1
Production Saw and Machine Co	3211 Gregory Rd	Jackson	MI	49202	J M Vancalbergh	517-787-9262	R	8	0.1
Daman Products Company Inc	3622 N Home St	Mishawaka	IN	46545	Jack D Davis	219-259-7841	R	7•	<0.1
Jarp Industries Inc	PO Box 923	Wausau	WI	54402	John Kraft	715-359-4241	R	7	<0.1
Columbus Hydraulic Co	PO Box 250	Columbus	NE	68602	Emanuel C Cimpl	402-564-8544	R	6	<0.1
General Engineering Co	PO Box 549	Abingdon	VA	24210	D W Tuckwiller	703-628-6068	R	6	<0.1
Helac Corp	PO Box 398	Enumclaw	WA	98022	Paul Weyer	360-825-1601	R	6	<0.1
Hydranamics	820 Edwards St	Galion	OH	44833	T Carter	419-468-3530	D	6•	0.1
Young Corp	PO Box 3522	Seattle	WA	98124	RH Lindberg	206-624-1071	D	6	<0.1
American Cylinder Company Inc	481 Governors Hwy	Peotone	IL	60468	Joseph White	708-258-3935	R	5	<0.1
Cunningham Manufacturing Co	318 S Webster St	Seattle	WA	98108	Scott Ericksen	206-767-3713	R	5	<0.1
Flo-Tork Inc	PO Box 68	Orrville	OH	44667	A Fejes	216-682-0010	R	5	0.1
Origa Corp	928 Oaklawn Av	Elmhurst	IL	60126	Joseph M Hughes	708-832-4321	S	4•	<0.1
Aurelius Manufacturing	PO Box 508	Braham	MN	55006	Conrad L Nelson	612-396-3343	R	3	<0.1
Hypress Technologies Inc	340 S Hearst Dr	Oxnard	CA	93030	Walter Moe	805-485-4060	R	3	<0.1
ICT Manufacturing Inc	4601 N Tyler Rd	Wichita	KS	67205	Jerry James	316-722-0265	R	3	<0.1
Star Dynamics Inc	PO Box 893	Washington	PA	15301	Ronald J Jordan	412-228-3700	R	3	<0.1
Advance Automation Company	3526 N Elston Av	Chicago	IL	60618	Joseph Hanley	312-539-7633	R	2	<0.1
Hydraulic Drives Inc	898 Commercial St	San Jose	CA	95112	Andrew A Weady	408-453-5010	R	2•	<0.1
Kemp Industries Inc	280 Marshall Hill	West Milford	NJ	07480	Anthony Mirti	201-728-8181	R	2•	<0.1
PFA Inc	15885 W Overland	New Berlin	WI	53151	Jerry Klimowicz	414-785-1869	S	2	<0.1
Chant Engineering Company Inc	3331 County Line	Chalfont	PA	18914	L James Chant	215-822-6048	R	2	<0.1
Catching Engineering Inc	1733 N 25th Av	Melrose Park	IL	60160	Inderjit S Sundal	708-344-2334	R	1	<0.1
Davis Welding & Mfg Co	511 W 8th St	Gibson City	IL	60936	Robert G Davis	217-784-5480	R	1	<0.1
Downey Manufacturing Inc	11421 S Downey	Downey	CA	90241	Bill J Read	213-773-2434	R	1	<0.1
REXA Corp	427 Turnpike St	Canton	MA	02021	Kevin Hynes	617-575-1199	R	1	<0.1
ETREMA Products Inc	2500 N Loop Dr	Ames	IA	50010	Larry Larson	515-296-8030	S	1	<0.1
Teknocraft Inc	1340 Clearmont NE	Palm Bay	FL	32905	Sam Kumar	407-729-9634	R	1	<0.1
Beltac Corp	711 First Parish Rd	Scituate	MA	02066	James M Denker	617-545-2617	R	0	<0.1

Source: Ward's Business Directory of U.S. Private and Public Companies, Volumes 1 and 2, 1996. The company type code used is as follows: P - Public, R - Private, S - Subsidiary, D - Division, J - Joint Venture, A - Affiliate, G - Group. Sales are in millions of dollars, employees are in thousands. An asterisk (•) indicates an estimated sales volume. The symbol < stands for 'less than'. Company names and addresses are truncated, in some cases, to fit into the available space.

MATERIALS CONSUMED

Material	Quantity	Delivered Cost ($ million)
Materials, ingredients, containers, and supplies	(X)	526.7
Fluid power pumps purchased for incorporation into products of this establishment	(X)	22.7
Metal bolts, nuts, screws, washers, rivets, and other screw machine products	(X)	34.1
Metal pipe, valves, and pipe fittings (except forgings)	(X)	9.4
Other fabricated metal products (except forgings)	(X)	37.4
Forgings	(X)	3.0
Iron and steel castings (rough and semifinished)	(X)	27.9
Aluminum and aluminum-base alloy castings (rough and semifinished)	(X)	22.7
Other nonferrous castings (rough and semifinished)	(X)	5.6
Steel bars, bar shapes, and plates	(X)	83.2
Steel sheet and strip, including tin plate	(X)	3.1
Steel tubing	(X)	45.5
All other steel shapes and forms	(X)	44.6
Aluminum and aluminum-base alloy extruded shapes, including extruded rod, bar, pipe, tube, etc.	(X)	11.8
Other aluminum and aluminum-base alloy shapes and forms	(X)	10.8
Other nonferrous shapes and forms	(X)	3.2
Metal powders	(X)	0.3
Lubricating and cutting oils	(X)	3.1
Paints, varnishes, lacquers, stains, shellacs, japans, enamels, and allied products	(X)	1.5
Electric motors and generators	(X)	4.0
Ball bearings (mounted or unmounted)	(X)	1.1
Roller bearings (mounted or unmounted)	(X)	0.8
Plain bearings and bushings	(X)	2.2
Mechanical speed changers, gears, and industrial high-speed drives	(X)	0.2
Gaskets (all types), and packing and sealing devices	(X)	12.3
Cutting tools for machine tools	(X)	10.9
All other materials and components, parts, containers, and supplies	(X)	76.1
Materials, ingredients, containers, and supplies, nsk	(X)	49.1

Source: 1992 *Economic Census*. Explanation of symbols used: (D): Withheld to avoid disclosure of competitive data; na: Not available; (S): Withheld because statistical norms were not met; (X): Not applicable; (Z): Less than half the unit shown; nec: Not elsewhere classified; nsk: Not specified by kind; - : zero; * : 10-19 percent estimated; ** : 20-29 percent estimated.

PRODUCT SHARE DETAILS

Product or Product Class	% Share	Product or Product Class	% Share
Fluid power cylinders and actuators	100.00	Aerospace type fluid power cylinders and actuators, hydraulic and pneumatic	26.42
Non-aerospace type hydraulic fluid power cylinders and actuators, linear and rotary	33.92	Parts for hydraulic and pneumatic fluid power cylinders and actuators, including accumulators, cushions, etc.	12.93
Non-aerospace type pneumatic fluid power cylinders and actuators, linear and rotary	18.71	Fluid power cylinders and actuators, nsk	8.03

Source: 1992 *Economic Census*. The values shown are percent of total shipments in an industry. Values of indented subcategories are summed in the main heading. The symbol (D) appears when data are withheld to prevent disclosure of competitive information. The abbreviation nsk stands for 'not specified by kind' and nec for 'not elsewhere classified'.

INPUTS AND OUTPUTS FOR MACHINERY, EXCEPT ELECTRICAL, NEC

Economic Sector or Industry Providing Inputs	%	Sector	Economic Sector or Industry Buying Outputs	%	Sector
Machinery, except electrical, nec	14.8	Manufg.	Machinery, except electrical, nec	6.9	Manufg.
Blast furnaces & steel mills	8.2	Manufg.	Exports	6.9	Foreign
Wholesale trade	7.6	Trade	Automotive stampings	6.4	Manufg.
Advertising	3.7	Services	Federal Government purchases, national defense	6.2	Fed Govt
Banking	3.4	Fin/R.E.	Automotive rental & leasing, without drivers	4.8	Services
Special dies & tools & machine tool accessories	3.1	Manufg.	Motor vehicle parts & accessories	4.7	Manufg.
Electric services (utilities)	2.5	Util.	Eating & drinking places	4.7	Trade
Business services nec	2.4	Services	State & local government enterprises, nec	3.7	Gov't
Ball & roller bearings	2.3	Manufg.	Aircraft	2.9	Manufg.
Maintenance of nonfarm buildings nec	2.0	Constr.	Water transportation	2.9	Util.
Aluminum castings	2.0	Manufg.	Coal	1.9	Mining
Eating & drinking places	1.8	Trade	Special dies & tools & machine tool accessories	1.4	Manufg.
Real estate	1.8	Fin/R.E.	Refrigeration & heating equipment	1.3	Manufg.
Iron & steel foundries	1.6	Manufg.	Construction machinery & equipment	1.2	Manufg.
Miscellaneous fabricated wire products	1.6	Manufg.	Logging camps & logging contractors	1.2	Manufg.
Power transmission equipment	1.6	Manufg.	Railroads & related services	1.1	Util.
Aluminum rolling & drawing	1.5	Manufg.	Aircraft & missile engines & engine parts	1.0	Manufg.
Mineral wool	1.5	Manufg.	Farm machinery & equipment	1.0	Manufg.
Abrasive products	1.4	Manufg.	Miscellaneous plastics products	1.0	Manufg.
Detective & protective services	1.4	Services	Pumps & compressors	1.0	Manufg.
Sheet metal work	1.3	Manufg.	Wholesale trade	1.0	Trade
Metalworking machinery, nec	1.2	Manufg.	Blast furnaces & steel mills	0.8	Manufg.
Engineering, architectural, & surveying services	1.2	Services	General industrial machinery, nec	0.7	Manufg.
Equipment rental & leasing services	1.2	Services	Machine tools, metal cutting types	0.7	Manufg.
Photofinishing labs, commercial photography	1.2	Services	Manufacturing industries, nec	0.7	Manufg.

Continued on next page.

INPUTS AND OUTPUTS FOR MACHINERY, EXCEPT ELECTRICAL, NEC - Continued

Economic Sector or Industry Providing Inputs	%	Sector	Economic Sector or Industry Buying Outputs	%	Sector
Fabricated metal products, nec	1.1	Manufg.	Oil field machinery	0.7	Manufg.
Internal combustion engines, nec	1.1	Manufg.	Printing trades machinery	0.7	Manufg.
Machine tools, metal cutting types	1.1	Manufg.	Special industry machinery, nec	0.7	Manufg.
Petroleum refining	1.1	Manufg.	Aircraft & missile equipment, nec	0.6	Manufg.
Pipe, valves, & pipe fittings	1.1	Manufg.	Ball & roller bearings	0.6	Manufg.
Paperboard containers & boxes	1.0	Manufg.	Motor vehicles & car bodies	0.6	Manufg.
Screw machine and related products	1.0	Manufg.	Ship building & repairing	0.6	Manufg.
Communications, except radio & TV	1.0	Util.	Retail trade, except eating & drinking	0.6	Trade
Motor freight transportation & warehousing	1.0	Util.	Food products machinery	0.5	Manufg.
Air transportation	0.9	Util.	Power transmission equipment	0.5	Manufg.
Plating & polishing	0.8	Manufg.	Transformers	0.5	Manufg.
Computer & data processing services	0.8	Services	Federal Government purchases, nondefense	0.5	Fed Govt
Nonferrous castings, nec	0.7	Manufg.	Blowers & fans	0.4	Manufg.
Gas production & distribution (utilities)	0.7	Util.	Carburetors, pistons, rings, & valves	0.4	Manufg.
Legal services	0.7	Services	Electronic components nec	0.4	Manufg.
Management & consulting services & labs	0.7	Services	Guided missiles & space vehicles	0.4	Manufg.
Electrical equipment & supplies, nec	0.6	Manufg.	Radio & TV communication equipment	0.4	Manufg.
Gaskets, packing & sealing devices	0.6	Manufg.	Screw machine and related products	0.4	Manufg.
Transformers	0.6	Manufg.	Service industry machines, nec	0.4	Manufg.
Royalties	0.6	Fin/R.E.	S/L Govt. purch., higher education	0.4	S/L Govt
Copper rolling & drawing	0.5	Manufg.	Electronic computing equipment	0.3	Manufg.
Felt goods, nec	0.5	Manufg.	Glass & glass products, except containers	0.3	Manufg.
Industrial patterns	0.5	Manufg.	Industrial trucks & tractors	0.3	Manufg.
Metal heat treating	0.5	Manufg.	Internal combustion engines, nec	0.3	Manufg.
Services to dwellings & other buildings	0.5	Services	Iron & steel foundries	0.3	Manufg.
Aircraft & missile engines & engine parts	0.4	Manufg.	Mining machinery, except oil field	0.3	Manufg.
Brass, bronze, & copper castings	0.4	Manufg.	Mobile homes	0.3	Manufg.
Accounting, auditing & bookkeeping	0.4	Services	Petroleum refining	0.3	Manufg.
Business/professional associations	0.4	Services	Pipe, valves, & pipe fittings	0.3	Manufg.
Lubricating oils & greases	0.3	Manufg.	Semiconductors & related devices	0.3	Manufg.
Manifold business forms	0.3	Manufg.	Textile machinery	0.3	Manufg.
Miscellaneous plastics products	0.3	Manufg.	Electric services (utilities)	0.3	Util.
Hotels & lodging places	0.3	Services	Accounting, auditing & bookkeeping	0.3	Services
U.S. Postal Service	0.3	Gov't	Portrait, photographic studios	0.3	Services
Industrial gases	0.2	Manufg.	S/L Govt. purch., other general government	0.3	S/L Govt
Metal stampings, nec	0.2	Manufg.	Feed grains	0.2	Agric.
Nonferrous rolling & drawing, nec	0.2	Manufg.	Crude petroleum & natural gas	0.2	Mining
Primary nonferrous metals, nec	0.2	Manufg.	Apparel made from purchased materials	0.2	Manufg.
Railroads & related services	0.2	Util.	Broadwoven fabric mills	0.2	Manufg.
Insurance carriers	0.2	Fin/R.E.	Conveyors & conveying equipment	0.2	Manufg.
Automotive rental & leasing, without drivers	0.2	Services	Cyclic crudes and organics	0.2	Manufg.
Automotive repair shops & services	0.2	Services	Electrical equipment & supplies, nec	0.2	Manufg.
Metal coating & allied services	0.1	Manufg.	Fabricated metal products, nec	0.2	Manufg.
Paints & allied products	0.1	Manufg.	Fabricated plate work (boiler shops)	0.2	Manufg.
Photographic equipment & supplies	0.1	Manufg.	Glass containers	0.2	Manufg.
Power driven hand tools	0.1	Manufg.	Hardware, nec	0.2	Manufg.
Welding apparatus, electric	0.1	Manufg.	Industrial furnaces & ovens	0.2	Manufg.
Transit & bus transportation	0.1	Util.	Instruments to measure electricity	0.2	Manufg.
Water transportation	0.1	Util.	Machine tools, metal forming types	0.2	Manufg.
Electrical repair shops	0.1	Services	Mechanical measuring devices	0.2	Manufg.
Personnel supply services	0.1	Services	Metal stampings, nec	0.2	Manufg.
Noncomparable imports	0.1	Foreign	Metalworking machinery, nec	0.2	Manufg.
			Miscellaneous fabricated wire products	0.2	Manufg.
			Motors & generators	0.2	Manufg.
			Nonferrous wire drawing & insulating	0.2	Manufg.
			Paper industries machinery	0.2	Manufg.
			Paper mills, except building paper	0.2	Manufg.
			Paperboard containers & boxes	0.2	Manufg.
			Photographic equipment & supplies	0.2	Manufg.
			Plating & polishing	0.2	Manufg.
			Power driven hand tools	0.2	Manufg.
			Turbines & turbine generator sets	0.2	Manufg.
			Woodworking machinery	0.2	Manufg.
			Air transportation	0.2	Util.
			Automotive repair shops & services	0.2	Services
			Management & consulting services & labs	0.2	Services
			Miscellaneous repair shops	0.2	Services
			Gross private fixed investment	0.2	Cap Inv
			Meat animals	0.1	Agric.
			Clay, ceramic, & refractory minerals	0.1	Mining
			Aluminum rolling & drawing	0.1	Manufg.
			Ammunition, except for small arms, nec	0.1	Manufg.
			Automatic merchandising machines	0.1	Manufg.
			Commercial printing	0.1	Manufg.
			Engineering & scientific instruments	0.1	Manufg.

Continued on next page.

INPUTS AND OUTPUTS FOR MACHINERY, EXCEPT ELECTRICAL, NEC - Continued

Economic Sector or Industry Providing Inputs	%	Sector	Economic Sector or Industry Buying Outputs	%	Sector
			Fabricated rubber products, nec	0.1	Manufg.
			Fabricated structural metal	0.1	Manufg.
			Hand & edge tools, nec	0.1	Manufg.
			Industrial inorganic chemicals, nec	0.1	Manufg.
			Industrial patterns	0.1	Manufg.
			Lawn & garden equipment	0.1	Manufg.
			Metal cans	0.1	Manufg.
			Metal doors, sash, & trim	0.1	Manufg.
			Newspapers	0.1	Manufg.
			Organic fibers, noncellulosic	0.1	Manufg.
			Plastics materials & resins	0.1	Manufg.
			Rolling mill machinery	0.1	Manufg.
			Sheet metal work	0.1	Manufg.
			Surgical & medical instruments	0.1	Manufg.
			Surgical appliances & supplies	0.1	Manufg.
			Telephone & telegraph apparatus	0.1	Manufg.
			Travel trailers & campers	0.1	Manufg.
			Wiring devices	0.1	Manufg.

Source: Benchmark Input-Output Accounts for the U.S. Economy, 1982, U.S. Department of Commerce, Washington, D.C., July 1991. Data, as reported in the source, are organized by the 1977 SIC structure in use in 1982 but have been matched, as closely as is possible, to the 1987 SIC structure used in this book.

OCCUPATIONS EMPLOYED BY SIC 359 - INDUSTRIAL MACHINERY, NEC

Occupation	% of Total 1994	Change to 2005	Occupation	% of Total 1994	Change to 2005
Machinists	20.3	-5.6	Sales & related workers nec	1.8	-5.6
NC machine tool operators, metal & plastic	4.9	41.6	Tool & die makers	1.6	-67.3
General managers & top executives	4.6	-10.4	Bookkeeping, accounting, & auditing clerks	1.6	-29.2
Blue collar worker supervisors	3.8	-13.3	Industrial production managers	1.6	3.8
Assemblers, fabricators, & hand workers nec	3.5	-5.6	Welding machine setters, operators	1.6	-15.0
Machine tool cutting operators, metal & plastic	3.5	-27.4	Drilling & boring machine tool workers	1.4	-52.8
Inspectors, testers, & graders, precision	2.7	-5.6	Traffic, shipping, & receiving clerks	1.3	-9.1
Welders & cutters	2.6	-5.6	Janitors & cleaners, incl maids	1.2	-24.5
Lathe & turning machine tool operators	2.6	-52.8	Industrial machinery mechanics	1.2	3.9
Combination machine tool operators	2.5	-5.6	Mechanical engineers	1.2	14.3
Grinding machine operators, metal & plastic	2.2	-5.6	Financial managers	1.1	-5.6
Secretaries, ex legal & medical	2.0	-14.1	Machine builders	1.1	-24.5
General office clerks	2.0	-19.5	Clerical supervisors & managers	1.0	-3.4
Machine tool cutting & forming etc. nec	1.9	-24.5	Grinders & polishers, hand	1.0	51.0

Source: Industry-Occupation Matrix, Bureau of Labor Statistics. These data relate to one or more 3-digit SIC industry groups rather than to a single 4-digit SIC. The change reported for each occupation to the year 2005 is a percent of growth or decline as estimated by the Bureau of Labor Statistics. The abbreviation nec stands for 'not elsewhere classified'.

LOCATION BY STATE AND REGIONAL CONCENTRATION

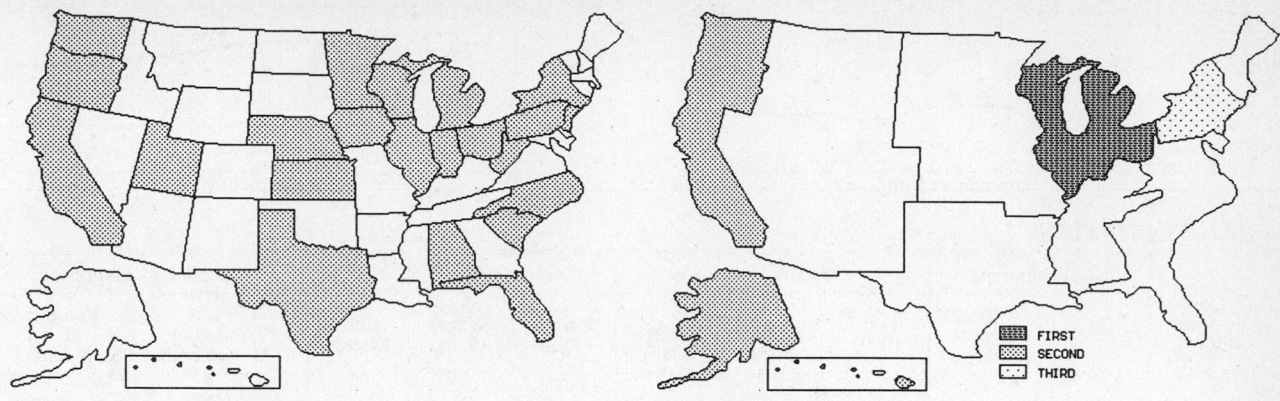

FIRST
SECOND
THIRD

INDUSTRY DATA BY STATE

State	Establish-ments	Shipments			Employment				Cost as % of Shipments	Investment per Employee ($)
		Total ($ mil)	% of U.S.	Per Establ.	Total Number	% of U.S.	Per Establ.	Wages ($/hour)		
California	53	363.0	20.1	6.8	3,300	20.0	62	17.53	30.8	1,576
New York	20	341.2	18.9	17.1	3,200	19.4	160	17.97	27.3	5,031
Illinois	37	165.7	9.2	4.5	1,700	10.3	46	11.38	36.9	6,000
Ohio	33	141.0	7.8	4.3	1,400	8.5	42	12.15	35.4	2,857
Michigan	23	137.5	7.6	6.0	1,000	6.1	43	12.46	53.8	2,100
Wisconsin	19	121.6	6.7	6.4	700	4.2	37	11.90	46.3	-
Indiana	12	86.5	4.8	7.2	600	3.6	50	12.90	35.8	6,000
Texas	12	46.6	2.6	3.9	300	1.8	25	11.00	42.3	3,333
Minnesota	7	33.9	1.9	4.8	300	1.8	43	13.75	44.5	1,667
Alabama	7	28.5	1.6	4.1	300	1.8	43	9.00	45.6	2,667
New Jersey	7	14.7	0.8	2.1	200	1.2	29	9.33	34.7	2,000
Pennsylvania	10	11.0	0.6	1.1	100	0.6	10	11.00	56.4	1,000
Washington	8	8.2	0.5	1.0	100	0.6	13	8.00	25.6	7,000
Florida	11	(D)	-	-	175 *	1.1	16	-	-	-
Iowa	10	(D)	-	-	750 *	4.5	75	-	-	-
Oregon	9	(D)	-	-	175 *	1.1	19	-	-	-
North Carolina	5	(D)	-	-	175 *	1.1	35	-	-	-
Kansas	3	(D)	-	-	375 *	2.3	125	-	-	-
South Carolina	3	(D)	-	-	175 *	1.1	58	-	-	-
Utah	3	(D)	-	-	375 *	2.3	125	-	-	-
West Virginia	3	(D)	-	-	175 *	1.1	58	-	-	-
Nebraska	2	(D)	-	-	175 *	1.1	88	-	-	-

Source: 1992 *Economic Census*. The states are in descending order of shipments or establishments (if shipment data are missing for the majority). The symbol (D) appears when data are withheld to prevent disclosure of competitive information. States marked with (D) are sorted by number of establishments. A dash (-) indicates that the data element cannot be calculated; * indicates the midpoint of a range.

3594 - FLUID POWER PUMPS & MOTORS

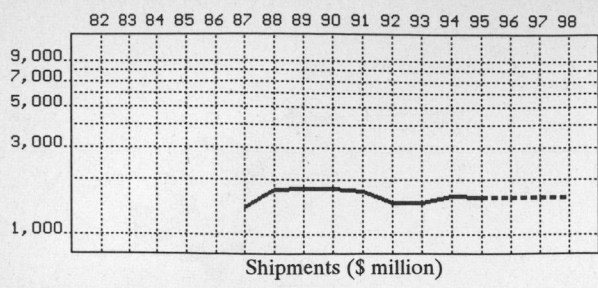
Shipments ($ million)

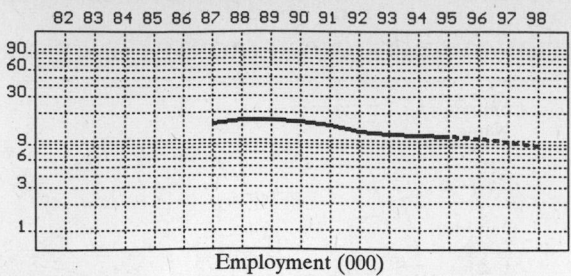
Employment (000)

GENERAL STATISTICS

| Year | Companies | Establishments | | Employment | | | Compensation | | Production ($ million) | | | |
		Total	with 20 or more employees	Total (000)	Production Workers (000)	Hours (Mil)	Payroll ($ mil)	Wages ($/hr)	Cost of Materials	Value Added by Manufacture	Value of Shipments	Capital Invest.
1982												
1983												
1984												
1985												
1986												
1987	133	150	67	14.8	9.0	18.9	428.9	12.71	559.6	830.6	1,404.4	58.5
1988		173	86	16.2	10.0	20.8	484.6	13.14	679.9	1,110.4	1,752.7	48.7
1989		155	72	16.4	9.6	19.2	452.3	13.35	734.9	1,050.2	1,788.2	64.3
1990		157	75	15.8	9.4	18.6	464.2	13.76	784.7	1,004.1	1,798.6	70.9
1991		161	72	14.3	8.6	17.5	457.8	13.66	693.4	1,002.9	1,732.6	53.3
1992	158	176	68	12.4	8.1	17.2	430.3	14.14	590.6	915.1	1,506.2	48.5
1993		174	66	11.7	7.5	16.1	406.2	14.17	591.8	922.1	1,515.6	36.0
1994		176P	67P	11.3	8.0	16.8	400.7	14.89	714.4	915.7	1,622.1	69.6
1995		179P	65P	10.9P	7.4P	15.7P	405.2P	14.94P	708.8P	908.5P	1,609.3P	53.5P
1996		182P	64P	10.2P	7.2P	15.2P	397.3P	15.21P	705.8P	904.6P	1,602.5P	52.9P
1997		185P	63P	9.4P	6.9P	14.6P	389.4P	15.48P	702.8P	900.8P	1,595.7P	52.3P
1998		188P	61P	8.7P	6.6P	14.1P	381.5P	15.75P	699.7P	896.9P	1,588.8P	51.7P

Sources: 1982, 1987, 1992 *Economic Census*; *Annual Survey of Manufactures*, 83-86, 88-91, 93-94. Establishment counts for non-Census years are from *County Business Patterns*; establishment values for 83-84 are extrapolations. 'P's show projections by the editors. Industries reclassified in 87 will not have data for prior years.

INDICES OF CHANGE

| Year | Companies | Establishments | | Employment | | | Compensation | | Production ($ million) | | | |
		Total	with 20 or more employees	Total (000)	Production Workers (000)	Hours (Mil)	Payroll ($ mil)	Wages ($/hr)	Cost of Materials	Value Added by Manufacture	Value of Shipments	Capital Invest.
1982												
1983												
1984												
1985												
1986												
1987	84	85	99	119	111	110	100	90	95	91	93	121
1988		98	126	131	123	121	113	93	115	121	116	100
1989		88	106	132	119	112	105	94	124	115	119	133
1990		89	110	127	116	108	108	97	133	110	119	146
1991		91	106	115	106	102	106	97	117	110	115	110
1992	100	100	100	100	100	100	100	100	100	100	100	100
1993		99	97	94	93	94	94	100	100	101	101	74
1994		100P	98P	91	99	98	93	105	121	100	108	144
1995		102P	96P	88P	92P	91P	94P	106P	120P	99P	107P	110P
1996		103P	94P	82P	88P	88P	92P	108P	119P	99P	106P	109P
1997		105P	92P	76P	85P	85P	90P	109P	119P	98P	106P	108P
1998		107P	90P	70P	81P	82P	89P	111P	118P	98P	105P	107P

Sources: Same as General Statistics. Values reflect change from the base year, 1992. Values above 100 mean greater than 92, values below 100 mean less than 92, and a value of 100 in the 82-91 or 93-98 period means same as 92. 'P's mark projections by the editors.

SELECTED RATIOS

For 1994	Avg. of All Manufact.	Analyzed Industry	Index	For 1994	Avg. of All Manufact.	Analyzed Industry	Index
Employees per Establishment	49	64	131	Value Added per Production Worker	134,084	114,463	85
Payroll per Establishment	1,500,273	2,280,407	152	Cost per Establishment	5,045,178	4,065,691	81
Payroll per Employee	30,620	35,460	116	Cost per Employee	102,970	63,221	61
Production Workers per Establishment	34	46	133	Cost per Production Worker	146,988	89,300	61
Wages per Establishment	853,319	1,423,629	167	Shipments per Establishment	9,576,895	9,231,463	96
Wages per Production Worker	24,861	31,269	126	Shipments per Employee	195,460	143,549	73
Hours per Production Worker	2,056	2,100	102	Shipments per Production Worker	279,017	202,762	73
Wages per Hour	12.09	14.89	123	Investment per Establishment	321,011	396,098	123
Value Added per Establishment	4,602,255	5,211,301	113	Investment per Employee	6,552	6,159	94
Value Added per Employee	93,930	81,035	86	Investment per Production Worker	9,352	8,700	93

Sources: Same as General Statistics. The 'Average of All Manufacturing' column represents the average of all manufacturing industries reported for the most recent complete year available. The Index shows the relationship between the Average and the Analyzed Industry. For example, 100 means that they are equal; 500 that the Analyzed Industry is five times the average; 50 means that the Analyzed Industry is half the national average. The abbreviation 'na' is used to show that data are 'not available'.

LEADING COMPANIES   Number shown: **43**   Total sales ($ mil): **5,868**   Total employment (000): **44.7**

Company Name	Address				CEO Name	Phone	Co. Type	Sales ($ mil)	Empl. (000)
TRINOVA Corp	3000 Strayer Rd	Maumee	OH	43537	Darryl F Allen	419-867-2200	P	1,795	15.0
Mannesmann Capital Corp	450 Park Av	New York	NY	10022	P P Wittgenstein	212-826-0040	S	1,750	7.0
ITT Fluid Technology Corp	445 Godwin Av	Midland Park	NJ	07432	Bertil T Nilsson	201-444-6030	S	1,125	8.0
Emerson Power Transmission	620 S Aurora St	Ithaca	NY	14850	Eugene A Yarussi	607-272-7220	S	450	4.5
Sta-Rite Industries Inc	293 Wright St	Delavan	WI	53115	James Donnelly	414-728-5551	S	140*	1.4
Sauer-Sundstrand	2800 E 13th St	Ames	IA	50010	David Pfeifle	515-239-6000	R	110	1.0
Oilgear Co	PO Box 343924	Milwaukee	WI	53234	Otto E Klieve	414-327-1700	P	70	0.8
Danfoss Fluid Power	8635 Washington	Racine	WI	53406	James C White	414-884-7400	D	69*	0.6
Danfoss Inc	8635 Washington	Racine	WI	53406	Ronald G Wickline	414-884-7400	S	55	0.5
John S Barnes Corp	PO Box 6166	Rockford	IL	61125	John M Pepe	815-398-4400	S	40	0.3
Florida Pneumatic Mfg Corp	851 Jupiter Park Ln	Jupiter	FL	33458	Charles B Swank	407-744-9500	S	35*	<0.1
Robert Bosch Fluid Power Corp	PO Box 2025	Racine	WI	53401	Mike Hadfield	414-554-7100	S	34	0.3
Denison Hydraulics Inc	14249 Indrial Pkwy	Marysville	OH	43040	Robert J Naples	513-644-3915	S	27*	0.2
Best Equipment Co	PO Box 7095	Tyler	TX	75711	RL Wall	903-931-8363	R	22	0.1
Dynex/Rivett Inc	770 Capitol Dr	Pewaukee	WI	53072	Raymond M Warell	414-691-0300	R	12	0.1
Micro Motors Inc	151 E Columbine	Santa Ana	CA	92707	Ronald G Coss	714-546-4045	R	12	0.1
Brett Aqualine Inc	5221 Oceanus Dr	Huntington Bch	CA	92649	James Brett	714-891-7211	R	11	<0.1
Dynamic Technology	2608 Nordic Rd	Dayton	OH	45414	M Schlater	513-274-3007	R	10	<0.1
M and W Pump Corp	PO Drawer E	Deerfield Bch	FL	33443	J David Eller	305-426-1500	R	10*	0.2
Applied Energy Company Inc	11431 Chairman Dr	Dallas	TX	75243	James E Steedly Jr	214-349-1171	S	9	<0.1
Fluid Regulators Corp	313 Gillette St	Painesville	OH	44077	Richard Schreiner	216-352-6182	R	8	<0.1
Aqua-Dyne Inc	3620 W 11th St	Houston	TX	77008	GJ Rankin	713-864-6929	R	8	<0.1
Erie Press Systems Inc	1253 W 12th St	Erie	PA	16512	GE Lunger	814-455-3941	D	7	0.1
Flow Products Inc	2626 W Addison St	Chicago	IL	60618	HM Burgh	312-528-2000	R	7	<0.1
Duke's Inc	9060 Winnetka Av	Northridge	CA	91324	RL Huffman	818-998-9811	R	5*	<0.1
Edwards Engineering	3401 E Randol Mill	Arlington	TX	76011	Karl Brown	817-640-0673	D	5	<0.1
LPI Corp	54 N Lively Blv	Elk Grove Vill	IL	60007	Magnus Mankert	708-593-1585	R	5	<0.1
Eaton Corp	15151 Hwy 5	Eden Prairie	MN	55344	Arthur J Warburton	612-937-9800	D	4	3.0
United Supply Company Inc	3700 S Broadway Pl	Los Angeles	CA	90007	Richard Barkley	213-233-8241	R	4*	<0.1
CRS Service Inc	1986 Rochester	Rochester Hills	MI	48309	RS Solgan	313-652-9940	R	4	<0.1
Wisconsin Hydraulics Inc	1666 S Johnson Rd	New Berlin	WI	53146	Kenneth Kersten	414-547-8550	R	4	<0.1
Anko Products Inc	3007 29th Av E	Bradenton	FL	34208	David S Anderson	813-749-1960	R	3	<0.1
Lucker Manufacturing	444 Henderson Rd	King of Prussia	PA	19406	Luke Franovich	215-337-0444	D	3*	<0.1
Newton Manufacturing Company	4249 Delemere Blv	Royal Oak	MI	48073	Noel R Cook	313-549-9600	R	3	<0.1
Star Hydraulics Inc	2727 Clinton St	River Grove	IL	60171	John F Tindall	708-453-3238	R	3	<0.1
La-Man Corp	7450 S Homestead	Hamilton	IN	46742	Richard W Coffman	219-488-3511	P	3	<0.1
Krogh Pump Co	531 Getty Ct	Benicia	CA	94510	Charles J O'Brien	707-747-7585	R	2*	<0.1
Reidville Hydraulics & Mfg Co	87 Sharon Rd	Waterbury	CT	06705	Peter G Heller	203-757-1268	R	2*	<0.1
American Machine & Hydraulics	294 Dawson Dr	Camarillo	CA	93012	Sam Grimaldo	805-388-2082	R	1	0.4
Accudriv Inc	PO Box 2341	York	PA	17405	Robert Vaughn	717-848-3088	R	1	<0.1
RDA Corp	10760 Shady Trail	Dallas	TX	75220	RD Arnett Jr	214-357-7165	R	1	<0.1
Kings Screw Machine Products	7289 Hwy 291	Tumtum	WA	99034	Gary Kingsbury	509-276-7999	R	1	<0.1
Pugh Manufacturing Co	5052 Alhambra Av	Los Angeles	CA	90032	William M Pugh	213-225-5691	R	0*	<0.1

Source: Ward's Business Directory of U.S. Private and Public Companies, Volumes 1 and 2, 1996. The company type code used is as follows: P - Public, R - Private, S - Subsidiary, D - Division, J - Joint Venture, A - Affiliate, G - Group. Sales are in millions of dollars, employees are in thousands. An asterisk (*) indicates an estimated sales volume. The symbol < stands for 'less than'. Company names and addresses are truncated, in some cases, to fit into the available space.

MATERIALS CONSUMED

Material	Quantity	Delivered Cost ($ million)
Materials, ingredients, containers, and supplies	(X)	501.5
Fluid power pumps purchased for incorporation into products of this establishment	(X)	9.7
Metal bolts, nuts, screws, washers, rivets, and other screw machine products	(X)	38.0
Metal pipe, valves, and pipe fittings (except forgings)	(X)	6.5
Other fabricated metal products (except forgings)	(X)	46.6
Forgings	(X)	9.7
Iron and steel castings (rough and semifinished)	(X)	51.6
Aluminum and aluminum-base alloy castings (rough and semifinished)	(X)	14.1
Other nonferrous castings (rough and semifinished)	(X)	1.8
Steel bars, bar shapes, and plates	(X)	19.1
Steel sheet and strip, including tin plate	(X)	6.6
All other steel shapes and forms	(X)	8.2
Aluminum and aluminum-base alloy extruded shapes, including extruded rod, bar, pipe, tube, etc.	(X)	1.2
Other aluminum and aluminum-base alloy shapes and forms	(X)	3.4
Other nonferrous shapes and forms	(X)	7.2
Metal powders	(X)	8.4
Lubricating and cutting oils	(X)	4.7
Paints, varnishes, lacquers, stains, shellacs, japans, enamels, and allied products	(X)	1.3
Electric motors and generators	(X)	25.1
Ball bearings (mounted or unmounted)	(X)	5.3
Roller bearings (mounted or unmounted)	(X)	13.0
Plain bearings and bushings	(X)	3.4
Mechanical speed changers, gears, and industrial high-speed drives	(X)	12.2
Gaskets (all types), and packing and sealing devices	(X)	8.3
Cutting tools for machine tools	(X)	11.0
All other materials and components, parts, containers, and supplies	(X)	120.4
Materials, ingredients, containers, and supplies, nsk	(X)	64.5

Source: 1992 *Economic Census.* Explanation of symbols used: (D): Withheld to avoid disclosure of competitive data; na: Not available; (S): Withheld because statistical norms were not met; (X): Not applicable; (Z): Less than half the unit shown; nec: Not elsewhere classified; nsk: Not specified by kind; - : zero; * : 10-19 percent estimated; ** : 20-29 percent estimated.

PRODUCT SHARE DETAILS

Product or Product Class	% Share	Product or Product Class	% Share
Fluid power pumps, motors, and hydrostatic transmission components	100.00	Aerospace type fluid power pumps and motors	10.63
Non-aerospace type reciprocating fluid power pumps	17.20	Parts for fluid power pumps, motors, and hydrostatic transmissions	23.35
Non-aerospace type rotary and other fluid power pumps	28.07	Fluid power pumps, motors, and hydrostatic transmission components, nsk	7.29
Non-aeropsace type fluid power motors	13.48		

Source: 1992 *Economic Census.* The values shown are percent of total shipments in an industry. Values of indented subcategories are summed in the main heading. The symbol (D) appears when data are withheld to prevent disclosure of competitive information. The abbreviation nsk stands for 'not specified by kind' and nec for 'not elsewhere classified'.

INPUTS AND OUTPUTS FOR MACHINERY, EXCEPT ELECTRICAL, NEC

Economic Sector or Industry Providing Inputs	%	Sector	Economic Sector or Industry Buying Outputs	%	Sector
Machinery, except electrical, nec	14.8	Manufg.	Machinery, except electrical, nec	6.9	Manufg.
Blast furnaces & steel mills	8.2	Manufg.	Exports	6.9	Foreign
Wholesale trade	7.6	Trade	Automotive stampings	6.4	Manufg.
Advertising	3.7	Services	Federal Government purchases, national defense	6.2	Fed Govt
Banking	3.4	Fin/R.E.	Automotive rental & leasing, without drivers	4.8	Services
Special dies & tools & machine tool accessories	3.1	Manufg.	Motor vehicle parts & accessories	4.7	Manufg.
Electric services (utilities)	2.5	Util.	Eating & drinking places	4.7	Trade
Business services nec	2.4	Services	State & local government enterprises, nec	3.7	Gov't
Ball & roller bearings	2.3	Manufg.	Aircraft	2.9	Manufg.
Maintenance of nonfarm buildings nec	2.0	Constr.	Water transportation	2.9	Util.
Aluminum castings	2.0	Manufg.	Coal	1.9	Mining
Eating & drinking places	1.8	Trade	Special dies & tools & machine tool accessories	1.4	Manufg.
Real estate	1.8	Fin/R.E.	Refrigeration & heating equipment	1.3	Manufg.
Iron & steel foundries	1.6	Manufg.	Construction machinery & equipment	1.2	Manufg.
Miscellaneous fabricated wire products	1.6	Manufg.	Logging camps & logging contractors	1.2	Manufg.
Power transmission equipment	1.6	Manufg.	Railroads & related services	1.1	Util.
Aluminum rolling & drawing	1.5	Manufg.	Aircraft & missile engines & engine parts	1.0	Manufg.
Mineral wool	1.5	Manufg.	Farm machinery & equipment	1.0	Manufg.
Abrasive products	1.4	Manufg.	Miscellaneous plastics products	1.0	Manufg.
Detective & protective services	1.4	Services	Pumps & compressors	1.0	Manufg.
Sheet metal work	1.3	Manufg.	Wholesale trade	1.0	Trade
Metalworking machinery, nec	1.2	Manufg.	Blast furnaces & steel mills	0.8	Manufg.
Engineering, architectural, & surveying services	1.2	Services	General industrial machinery, nec	0.7	Manufg.
Equipment rental & leasing services	1.2	Services	Machine tools, metal cutting types	0.7	Manufg.
Photofinishing labs, commercial photography	1.2	Services	Manufacturing industries, nec	0.7	Manufg.
Fabricated metal products, nec	1.1	Manufg.	Oil field machinery	0.7	Manufg.

Continued on next page.

INPUTS AND OUTPUTS FOR MACHINERY, EXCEPT ELECTRICAL, NEC - Continued

Economic Sector or Industry Providing Inputs	%	Sector	Economic Sector or Industry Buying Outputs	%	Sector
Internal combustion engines, nec	1.1	Manufg.	Printing trades machinery	0.7	Manufg.
Machine tools, metal cutting types	1.1	Manufg.	Special industry machinery, nec	0.7	Manufg.
Petroleum refining	1.1	Manufg.	Aircraft & missile equipment, nec	0.6	Manufg.
Pipe, valves, & pipe fittings	1.1	Manufg.	Ball & roller bearings	0.6	Manufg.
Paperboard containers & boxes	1.0	Manufg.	Motor vehicles & car bodies	0.6	Manufg.
Screw machine and related products	1.0	Manufg.	Ship building & repairing	0.6	Manufg.
Communications, except radio & TV	1.0	Util.	Retail trade, except eating & drinking	0.6	Trade
Motor freight transportation & warehousing	1.0	Util.	Food products machinery	0.5	Manufg.
Air transportation	0.9	Util.	Power transmission equipment	0.5	Manufg.
Plating & polishing	0.8	Manufg.	Transformers	0.5	Manufg.
Computer & data processing services	0.8	Services	Federal Government purchases, nondefense	0.5	Fed Govt
Nonferrous castings, nec	0.7	Manufg.	Blowers & fans	0.4	Manufg.
Gas production & distribution (utilities)	0.7	Util.	Carburetors, pistons, rings, & valves	0.4	Manufg.
Legal services	0.7	Services	Electronic components nec	0.4	Manufg.
Management & consulting services & labs	0.7	Services	Guided missiles & space vehicles	0.4	Manufg.
Electrical equipment & supplies, nec	0.6	Manufg.	Radio & TV communication equipment	0.4	Manufg.
Gaskets, packing & sealing devices	0.6	Manufg.	Screw machine and related products	0.4	Manufg.
Transformers	0.6	Manufg.	Service industry machines, nec	0.4	Manufg.
Royalties	0.6	Fin/R.E.	S/L Govt. purch., higher education	0.4	S/L Govt
Copper rolling & drawing	0.5	Manufg.	Electronic computing equipment	0.3	Manufg.
Felt goods, nec	0.5	Manufg.	Glass & glass products, except containers	0.3	Manufg.
Industrial patterns	0.5	Manufg.	Industrial trucks & tractors	0.3	Manufg.
Metal heat treating	0.5	Manufg.	Internal combustion engines, nec	0.3	Manufg.
Services to dwellings & other buildings	0.5	Services	Iron & steel foundries	0.3	Manufg.
Aircraft & missile engines & engine parts	0.4	Manufg.	Mining machinery, except oil field	0.3	Manufg.
Brass, bronze, & copper castings	0.4	Manufg.	Mobile homes	0.3	Manufg.
Accounting, auditing & bookkeeping	0.4	Services	Petroleum refining	0.3	Manufg.
Business/professional associations	0.4	Services	Pipe, valves, & pipe fittings	0.3	Manufg.
Lubricating oils & greases	0.3	Manufg.	Semiconductors & related devices	0.3	Manufg.
Manifold business forms	0.3	Manufg.	Textile machinery	0.3	Manufg.
Miscellaneous plastics products	0.3	Manufg.	Electric services (utilities)	0.3	Util.
Hotels & lodging places	0.3	Services	Accounting, auditing & bookkeeping	0.3	Services
U.S. Postal Service	0.3	Gov't	Portrait, photographic studios	0.3	Services
Industrial gases	0.2	Manufg.	S/L Govt. purch., other general government	0.3	S/L Govt
Metal stampings, nec	0.2	Manufg.	Feed grains	0.2	Agric.
Nonferrous rolling & drawing, nec	0.2	Manufg.	Crude petroleum & natural gas	0.2	Mining
Primary nonferrous metals, nec	0.2	Manufg.	Apparel made from purchased materials	0.2	Manufg.
Railroads & related services	0.2	Util.	Broadwoven fabric mills	0.2	Manufg.
Insurance carriers	0.2	Fin/R.E.	Conveyors & conveying equipment	0.2	Manufg.
Automotive rental & leasing, without drivers	0.2	Services	Cyclic crudes and organics	0.2	Manufg.
Automotive repair shops & services	0.2	Services	Electrical equipment & supplies, nec	0.2	Manufg.
Metal coating & allied services	0.1	Manufg.	Fabricated metal products, nec	0.2	Manufg.
Paints & allied products	0.1	Manufg.	Fabricated plate work (boiler shops)	0.2	Manufg.
Photographic equipment & supplies	0.1	Manufg.	Glass containers	0.2	Manufg.
Power driven hand tools	0.1	Manufg.	Hardware, nec	0.2	Manufg.
Welding apparatus, electric	0.1	Manufg.	Industrial furnaces & ovens	0.2	Manufg.
Transit & bus transportation	0.1	Util.	Instruments to measure electricity	0.2	Manufg.
Water transportation	0.1	Util.	Machine tools, metal forming types	0.2	Manufg.
Electrical repair shops	0.1	Services	Mechanical measuring devices	0.2	Manufg.
Personnel supply services	0.1	Services	Metal stampings, nec	0.2	Manufg.
Noncomparable imports	0.1	Foreign	Metalworking machinery, nec	0.2	Manufg.
			Miscellaneous fabricated wire products	0.2	Manufg.
			Motors & generators	0.2	Manufg.
			Nonferrous wire drawing & insulating	0.2	Manufg.
			Paper industries machinery	0.2	Manufg.
			Paper mills, except building paper	0.2	Manufg.
			Paperboard containers & boxes	0.2	Manufg.
			Photographic equipment & supplies	0.2	Manufg.
			Plating & polishing	0.2	Manufg.
			Power driven hand tools	0.2	Manufg.
			Turbines & turbine generator sets	0.2	Manufg.
			Woodworking machinery	0.2	Manufg.
			Air transportation	0.2	Util.
			Automotive repair shops & services	0.2	Services
			Management & consulting services & labs	0.2	Services
			Miscellaneous repair shops	0.2	Services
			Gross private fixed investment	0.2	Cap Inv
			Meat animals	0.1	Agric.
			Clay, ceramic, & refractory minerals	0.1	Mining
			Aluminum rolling & drawing	0.1	Manufg.
			Ammunition, except for small arms, nec	0.1	Manufg.
			Automatic merchandising machines	0.1	Manufg.
			Commercial printing	0.1	Manufg.
			Engineering & scientific instruments	0.1	Manufg.
			Fabricated rubber products, nec	0.1	Manufg.

Continued on next page.

INPUTS AND OUTPUTS FOR MACHINERY, EXCEPT ELECTRICAL, NEC - Continued

Economic Sector or Industry Providing Inputs	%	Sector	Economic Sector or Industry Buying Outputs	%	Sector
			Fabricated structural metal	0.1	Manufg.
			Hand & edge tools, nec	0.1	Manufg.
			Industrial inorganic chemicals, nec	0.1	Manufg.
			Industrial patterns	0.1	Manufg.
			Lawn & garden equipment	0.1	Manufg.
			Metal cans	0.1	Manufg.
			Metal doors, sash, & trim	0.1	Manufg.
			Newspapers	0.1	Manufg.
			Organic fibers, noncellulosic	0.1	Manufg.
			Plastics materials & resins	0.1	Manufg.
			Rolling mill machinery	0.1	Manufg.
			Sheet metal work	0.1	Manufg.
			Surgical & medical instruments	0.1	Manufg.
			Surgical appliances & supplies	0.1	Manufg.
			Telephone & telegraph apparatus	0.1	Manufg.
			Travel trailers & campers	0.1	Manufg.
			Wiring devices	0.1	Manufg.

Source: Benchmark Input-Output Accounts for the U.S. Economy, 1982, U.S. Department of Commerce, Washington, D.C., July 1991. Data, as reported in the source, are organized by the 1977 SIC structure in use in 1982 but have been matched, as closely as is possible, to the 1987 SIC structure used in this book.

OCCUPATIONS EMPLOYED BY SIC 359 - INDUSTRIAL MACHINERY, NEC

Occupation	% of Total 1994	Change to 2005	Occupation	% of Total 1994	Change to 2005
Machinists	20.3	-5.6	Sales & related workers nec	1.8	-5.6
NC machine tool operators, metal & plastic	4.9	41.6	Tool & die makers	1.6	-67.3
General managers & top executives	4.6	-10.4	Bookkeeping, accounting, & auditing clerks	1.6	-29.2
Blue collar worker supervisors	3.8	-13.3	Industrial production managers	1.6	3.8
Assemblers, fabricators, & hand workers nec	3.5	-5.6	Welding machine setters, operators	1.6	-15.0
Machine tool cutting operators, metal & plastic	3.5	-27.4	Drilling & boring machine tool workers	1.4	-52.8
Inspectors, testers, & graders, precision	2.7	-5.6	Traffic, shipping, & receiving clerks	1.3	-9.1
Welders & cutters	2.6	-5.6	Janitors & cleaners, incl maids	1.2	-24.5
Lathe & turning machine tool operators	2.6	-52.8	Industrial machinery mechanics	1.2	3.9
Combination machine tool operators	2.5	-5.6	Mechanical engineers	1.2	14.3
Grinding machine operators, metal & plastic	2.2	-5.6	Financial managers	1.1	-5.6
Secretaries, ex legal & medical	2.0	-14.1	Machine builders	1.1	-24.5
General office clerks	2.0	-19.5	Clerical supervisors & managers	1.0	-3.4
Machine tool cutting & forming etc. nec	1.9	-24.5	Grinders & polishers, hand	1.0	51.0

Source: Industry-Occupation Matrix, Bureau of Labor Statistics. These data relate to one or more 3-digit SIC industry groups rather than to a single 4-digit SIC. The change reported for each occupation to the year 2005 is a percent of growth or decline as estimated by the Bureau of Labor Statistics. The abbreviation nec stands for 'not elsewhere classified'.

LOCATION BY STATE AND REGIONAL CONCENTRATION

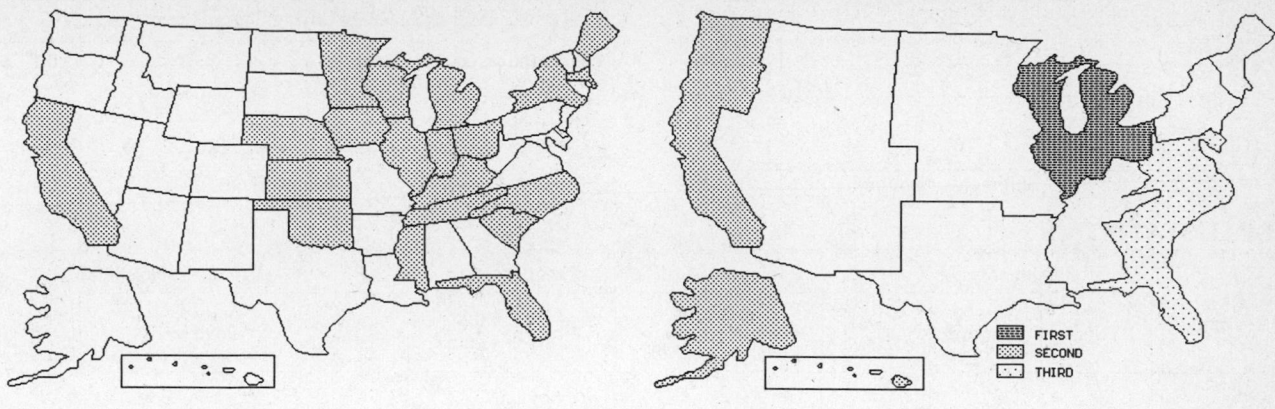

FIRST
SECOND
THIRD

INDUSTRY DATA BY STATE

State	Establish-ments	Shipments			Employment				Cost as % of Shipments	Investment per Employee ($)
		Total ($ mil)	% of U.S.	Per Establ.	Total Number	% of U.S.	Per Establ.	Wages ($/hour)		
Ohio	12	130.7	8.7	10.9	1,000	8.1	83	13.12	40.6	4,700
California	21	124.9	8.3	5.9	900	7.3	43	21.27	38.3	3,444
Minnesota	8	119.3	7.9	14.9	1,200	9.7	150	12.23	30.7	3,500
Illinois	14	117.0	7.8	8.4	1,100	8.9	79	12.75	47.0	3,727
South Carolina	6	77.6	5.2	12.9	500	4.0	83	12.33	61.9	4,400
Wisconsin	9	62.9	4.2	7.0	600	4.8	67	13.71	42.8	3,833
Florida	6	46.5	3.1	7.8	400	3.2	67	15.50	41.7	2,250
Tennessee	6	45.0	3.0	7.5	400	3.2	67	13.20	41.8	2,000
New York	7	9.9	0.7	1.4	100	0.8	14	9.00	31.3	-
Michigan	11	(D)	-	-	375 *	3.0	34	-	-	-
Indiana	5	(D)	-	-	175 *	1.4	35	-	-	2,286
North Carolina	5	(D)	-	-	375 *	3.0	75	-	-	-
Oklahoma	4	(D)	-	-	375 *	3.0	94	-	-	-
Kansas	3	(D)	-	-	750 *	6.0	250	-	-	-
Mississippi	3	(D)	-	-	1,750 *	14.1	583	-	-	-
Nebraska	3	(D)	-	-	1,750 *	14.1	583	-	-	-
Iowa	2	(D)	-	-	750 *	6.0	375	-	-	-
Kentucky	2	(D)	-	-	175 *	1.4	88	-	-	-
Maine	2	(D)	-	-	375 *	3.0	188	-	-	-
Massachusetts	1	(D)	-	-	175 *	1.4	175	-	-	-

Source: 1992 *Economic Census*. The states are in descending order of shipments or establishments (if shipment data are missing for the majority). The symbol (D) appears when data are withheld to prevent disclosure of competitive information. States marked with (D) are sorted by number of establishments. A dash (-) indicates that the data element cannot be calculated; * indicates the midpoint of a range.

3596 - SCALES & BALANCES EXCEPT LABORATORY

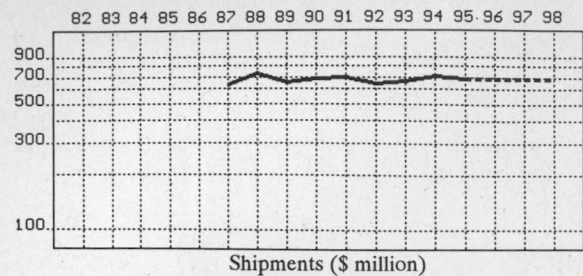

Shipments ($ million)

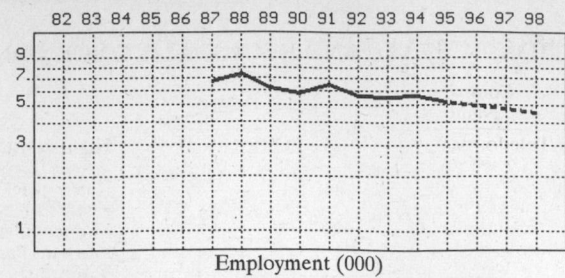

Employment (000)

GENERAL STATISTICS

| Year | Companies | Establishments | | Employment | | | Compensation | | Production ($ million) | | | |
		Total	with 20 or more employees	Total (000)	Production Workers (000)	Hours (Mil)	Payroll ($ mil)	Wages ($/hr)	Cost of Materials	Value Added by Manufacture	Value of Shipments	Capital Invest.
1982												
1983												
1984												
1985												
1986												
1987	118	134	57	6.7	3.8	8.0	146.1	8.21	273.8	355.6	633.0	11.5
1988		127	55	7.3	4.1	8.9	162.2	8.67	334.7	397.2	726.0	6.5
1989		112	58	6.2	3.5	7.7	149.2	8.35	322.0	327.8	648.4	8.3
1990		111	51	5.8	3.3	6.9	152.7	8.94	348.0	336.4	680.0	13.3
1991		119	52	6.4	3.3	7.1	161.2	9.24	375.4	322.2	697.0	15.9
1992	114	128	54	5.6	3.0	6.6	154.8	9.35	296.4	349.1	641.1	17.2
1993		119	51	5.5	3.0	6.6	155.1	9.42	296.7	371.6	659.2	11.3
1994		116P	50P	5.6	3.1	6.6	167.5	10.41	360.5	348.0	704.8	20.5
1995		115P	49P	5.2P	2.8P	6.0P	163.6P	10.28P	349.1P	337.0P	682.5P	19.3P
1996		114P	48P	5.0P	2.6P	5.7P	165.2P	10.54P	350.1P	337.9P	684.4P	20.7P
1997		112P	48P	4.8P	2.5P	5.4P	166.9P	10.81P	351.1P	338.9P	686.4P	22.1P
1998		111P	47P	4.5P	2.3P	5.1P	168.6P	11.08P	352.1P	339.9P	688.3P	23.4P

Sources: 1982, 1987, 1992 *Economic Census*; *Annual Survey of Manufactures*, 83-86, 88-91, 93-94. Establishment counts for non-Census years are from *County Business Patterns*; establishment values for 83-84 are extrapolations. 'P's show projections by the editors. Industries reclassified in 87 will not have data for prior years.

INDICES OF CHANGE

| Year | Companies | Establishments | | Employment | | | Compensation | | Production ($ million) | | | |
		Total	with 20 or more employees	Total (000)	Production Workers (000)	Hours (Mil)	Payroll ($ mil)	Wages ($/hr)	Cost of Materials	Value Added by Manufacture	Value of Shipments	Capital Invest.
1982												
1983												
1984												
1985												
1986												
1987	104	105	106	120	127	121	94	88	92	102	99	67
1988		99	102	130	137	135	105	93	113	114	113	38
1989		88	107	111	117	117	96	89	109	94	101	48
1990		87	94	104	110	105	99	96	117	96	106	77
1991		93	96	114	110	108	104	99	127	92	109	92
1992	100	100	100	100	100	100	100	100	100	100	100	100
1993		93	94P	98	100	100	100	101	100	106	103	66
1994		91P	93P	100	103	100	108	111	122	100	110	119
1995		90P	91P	92P	92P	91P	106P	110P	118P	97P	106P	112P
1996		89P	90P	89P	87P	86P	107P	113P	118P	97P	107P	120P
1997		88P	88P	85P	82P	82P	108P	116P	118P	97P	107P	128P
1998		87P	86P	81P	78P	78P	109P	118P	119P	97P	107P	136P

Sources: Same as General Statistics. Values reflect change from the base year, 1992. Values above 100 mean greater than 92, values below 100 mean less than 92, and a value of 100 in the 82-91 or 93-98 period means same as 92. 'P's mark projections by the editors.

SELECTED RATIOS

For 1994	Avg. of All Manufact.	Analyzed Industry	Index	For 1994	Avg. of All Manufact.	Analyzed Industry	Index
Employees per Establishment	49	48	98	Value Added per Production Worker	134,084	112,258	84
Payroll per Establishment	1,500,273	1,440,418	96	Cost per Establishment	5,045,178	3,100,123	61
Payroll per Employee	30,620	29,911	98	Cost per Employee	102,970	64,375	63
Production Workers per Establishment	34	27	78	Cost per Production Worker	146,988	116,290	79
Wages per Establishment	853,319	590,838	69	Shipments per Establishment	9,576,895	6,060,934	63
Wages per Production Worker	24,861	22,163	89	Shipments per Employee	195,460	125,857	64
Hours per Production Worker	2,056	2,129	104	Shipments per Production Worker	279,017	227,355	81
Wages per Hour	12.09	10.41	86	Investment per Establishment	321,011	176,290	55
Value Added per Establishment	4,602,255	2,992,629	65	Investment per Employee	6,552	3,661	56
Value Added per Employee	93,930	62,143	66	Investment per Production Worker	9,352	6,613	71

Sources: Same as General Statistics. The 'Average of All Manufacturing' column represents the average of all manufacturing industries reported for the most recent complete year available. The Index shows the relationship between the Average and the Analyzed Industry. For example, 100 means that they are equal; 500 that the Analyzed Industry is five times the average; 50 means that the Analyzed Industry is half the national average. The abbreviation 'na' is used to show that data are 'not available'.

LEADING COMPANIES   Number shown: 37   Total sales ($ mil): 783   Total employment (000): 6.2

Company Name	Address				CEO Name	Phone	Co. Type	Sales ($ mil)	Empl. (000)
Mettler-Toledo Inc	PO Box 71	Hightstown	NJ	08520	MP Knapp	609-448-3000	S	190	1.4
Chronos Richardson Inc	435 Hamburg Tpk	Wayne	NJ	07470	Mark Swift	201-595-7979	S	85	0.1
Fairbanks Inc	821 Locust St	Kansas City	MO	64106	FA Norden	816-471-0231	R	75*	0.8
Health o meter Products Inc	24700 Miles Rd	Bedford Hts	OH	44146	P C McCHowell	216-464-4000	P	69	1.0
Ramsey Technology Inc	501 90th Av NW	Minneapolis	MN	55433	Lew Ribich	612-783-2500	S	60	0.2
Ohaus Corp	29 Hanover Rd	Florham Park	NJ	07932	James G Ohaus	201-377-9000	S	52*	0.4
Weigh-Tronix Inc	PO Box 1000	Fairmont	MN	56031	Randy Newlin	507-238-4461	S	49*	0.4
Cardinal Scale Mfg Co	PO Box 151	Webb City	MO	64870	WH Perry	417-673-4631	R	35	0.5
Counselor Co	2107 Kishwaukee St	Rockford	IL	61104	Don Bergman	815-968-9621	D	25	0.3
Pelouze Scale Co	7400 W 100th Pl	Bridgeview	IL	60455	Lawrence Zawsky	708-430-8330	S	20	<0.1
John Chatillon and Sons Inc	PO Box 35668	Greensboro	NC	27425	Kenneth Stenton	910-668-0841	S	13	0.1
Emery Winslow Scale Co	73 Cogwheel Ln	Seymour	CT	06483	WM Young	203-881-9333	R	10	0.1
Structural Instrumentation Inc	4611 S 134th Pl	Seattle	WA	98168	Rick A Beets	206-244-6100	P	8	<0.1
Exact Equipment Corp	Rte 2	Whiteville	NC	28472	Robert C Enichen	910-642-5913	D	8*	<0.1
Metrigraphics	60 Concord St	Wilmington	MA	01887	Chester Ju	508-658-6100	D	7*	<0.1
Metro Corp	PO Box 1240	Las Cruces	NM	88004	Eric L Hanssen	505-526-0944	R	7	<0.1
Structural Instrumentation Mfg	4611 S 134th Pl	Tukwila	WA	98168	Rick Beets	206-244-6100	S	7*	<0.1
Thurman Manufacturing Co	1939 Refugee Rd	Columbus	OH	43207	Millard Cummins	614-443-9741	R	7	<0.1
Thurman Scale Co	1939 Refugee Rd	Columbus	OH	43207	Millard Cummins	614-443-9741	S	7	<0.1
Micro General Corp	1740 E Wilshire Av	Santa Ana	CA	92705	George D O'Leary	714-667-0557	P	5	<0.1
Control Systems Engineering Inc	12500 US Hwy 64	Eads	TN	38028	Layne Carruth	901-867-8500	R	5	<0.1
Pennsylvania Scale Co	21 Graybill Rd	Leola	PA	17540	Wayne W Johnson	717-656-2653	R	5	<0.1
Scale-Tronix Inc	200 E Post Rd	White Plains	NY	10601	DC Hale	914-948-8117	R	5	<0.1
Triner Scale and Manufacturing	2842 Sanderwood	Memphis	TN	38118	John B Wendt	901-795-0746	R	5	<0.1
CST/Auto Weigh Inc	PO Box 4017	Modesto	CA	95352	Gerald Berger	209-526-3557	R	3	<0.1
General Electrodynamics Corp	8000 Calendar Rd	Arlington	TX	76017	Dick Davis	817-572-0366	R	3	<0.1
Sterling Scale Company Inc	20950 Boening Dr	Southfield	MI	48075	ED Dixon	313-358-0590	R	3	<0.1
Division Scale Company Inc	3957 E Raines Rd	Memphis	TN	38118	Steven Muccillo	901-366-4220	R	3	<0.1
Evergreen Weigh Inc	19023 36th Av W	Lynnwood	WA	98036	Carl R Harris	206-771-4545	R	3	<0.1
Toroid Corp	PO Box 1435	Huntsville	AL	35807	A Paelian	205-837-7510	R	2	<0.1
Circuits and Systems Inc	Foot of 2nd St	East Rockaway	NY	11518	Arnold Gordon	516-593-4301	R	2	<0.1
West Weigh Scale Inc	3990 Hicock St	San Diego	CA	92110	VJ Muccillo Jr	619-291-8231	R	2	<0.1
Kernco Instruments Company	420 Kenazo Av	El Paso	TX	79927	John P Kelly	915-852-3375	R	1	<0.1
HSI Scales Inc	61 Endicott St	Norwood	MA	02062	EM Farrow	617-762-7540	R	1*	<0.1
Ormond Inc	12030 Rivera Rd	Santa Fe Sprgs	CA	90670	AN Ormond	310-945-1425	R	1*	<0.1
Terraillon Corp	700 Canal St	Stamford	CT	06902	Richard W Agresta	203-328-3715	S	1*	<0.1
J and L Metrology Company Inc	PO Box 851	Springfield	VT	05156	Richard Crossman	802-886-8333	S	0	<0.1

Source: Ward's Business Directory of U.S. Private and Public Companies, Volumes 1 and 2, 1996. The company type code used is as follows: P - Public, R - Private, S - Subsidiary, D - Division, J - Joint Venture, A - Affiliate, G - Group. Sales are in millions of dollars, employees are in thousands. An asterisk (*) indicates an estimated sales volume. The symbol < stands for 'less than'. Company names and addresses are truncated, in some cases, to fit into the available space.

MATERIALS CONSUMED

Material	Quantity	Delivered Cost ($ million)
Materials, ingredients, containers, and supplies	(X)	269.9
Sheet metal products, except stampings	(X)	12.9
Metal bolts, nuts, screws, washers, rivets, and other screw machine products	(X)	6.3
All other fabricated metal products (except forgings)	(X)	19.7
Forgings	(X)	0.1
Iron and steel castings (rough and semifinished)	(X)	7.6
Nonferrous (aluminum, copper, etc.) castings (rough and semifinished)	(X)	3.9
Steel bars, bar shapes, and plates	(X)	13.4
Steel structural shapes and sheet piling	(X)	4.4
All other steel shapes and forms	(X)	4.6
Semiconductors, including transistors, diodes, rectifiers, and integrated circuits for electronic circuitry	(X)	27.6
Resistors, capacitors, transformers, transducers, and other electronic-type components	(X)	31.4
Appliance outlets, switches, lampholders, and other current-carrying wiring devices	(X)	5.8
Electrical transmission, distribution, and control equipment	(X)	7.9
Paperboard containers, boxes, and corrugated paperboard	(X)	4.9
Fabricated plastics products (except gaskets, hoses, and belting)	(X)	3.8
All other materials and components, parts, containers, and supplies	(X)	97.4
Materials, ingredients, containers, and supplies, nsk	(X)	18.3

Source: 1992 Economic Census. Explanation of symbols used: (D): Withheld to avoid disclosure of competitive data; na: Not available; (S): Withheld because statistical norms were not met; (X): Not applicable; (Z): Less than half the unit shown; nec: Not elsewhere classified; nsk: Not specified by kind; - : zero; * : 10-19 percent estimated; ** : 20-29 percent estimated.

PRODUCT SHARE DETAILS

Product or Product Class	% Share	Product or Product Class	% Share
Scales and balances, except laboratory	100.00	Household and person-weighing scales (including bathroom, coin-operated, free weighing, kitchen, and baby scales)	48.80
Vehicle and industrial scales	41.91		
Motor truck scales	19.31	Mailing and parcel post scales (including handheld scales)	22.49
Railroad track scales	1.24	Balances with or without weights (of all sensitivities), except laboratory	9.97
Industrial bench and portable scales	9.91		
Industrial floor scales, dormant, pitless	9.84	Parts, attachments, and accessories for scales and balances, except laboratory	22.11
Industrial automatic checkweigher scales	12.43		
Industrial over-under (predetermined weight) scales	1.76	Printers for scales and balances, except laboratory (sold separately)	49.69
Miscellaneous industrial scales (crane, suspension, tank, hopper, force measuring devices, bulk conveyor, etc.)	41.28	Other accessories and attachments for scales and balances, except laboratory (sold separately)	20.51
Industrial counting scales	3.62	Parts for scales and balances, except laboratory (sold for assembly elsewhere, repair, service, etc.)	29.73
Vehicle and industrial scales, nsk	0.66		
Retail, commercial, household, and mailing scales	30.59	Parts, attachments, and accessories for scales and balances, nsk	0.07
Retail and commercial delicatessen scales	4.31		
Retail and commercial checkstand scales	(D)	Scales and balances, except laboratory, nsk	5.41
Retail and commercial automatic prepack scales	(D)		
Other retail and commercial scales	(D)		

Source: 1992 *Economic Census*. The values shown are percent of total shipments in an industry. Values of indented subcategories are summed in the main heading. The symbol (D) appears when data are withheld to prevent disclosure of competitive information. The abbreviation nsk stands for 'not specified by kind' and nec for 'not elsewhere classified'.

INPUTS AND OUTPUTS FOR SCALES & BALANCES

Economic Sector or Industry Providing Inputs	%	Sector	Economic Sector or Industry Buying Outputs	%	Sector
Imports	11.7	Foreign	Gross private fixed investment	55.3	Cap Inv
Scales & balances	10.5	Manufg.	Personal consumption expenditures	12.9	
Electronic components nec	8.7	Manufg.	Exports	12.5	Foreign
Blast furnaces & steel mills	8.5	Manufg.	Business services nec	6.7	Services
Plastics materials & resins	6.6	Manufg.	Scales & balances	6.3	Manufg.
Wholesale trade	6.0	Trade	S/L Govt. purch., highways	1.6	S/L Govt
Iron & steel foundries	3.9	Manufg.	Federal Government purchases, national defense	1.2	Fed Govt
Semiconductors & related devices	3.4	Manufg.	Wholesale trade	0.8	Trade
Computer & data processing services	3.3	Services	S/L Govt. purch., elem. & secondary education	0.8	S/L Govt
Petroleum refining	3.1	Manufg.	Federal Government purchases, nondefense	0.7	Fed Govt
Electronic computing equipment	2.2	Manufg.	Management & consulting services & labs	0.4	Services
Banking	1.9	Fin/R.E.	U.S. Postal Service	0.3	Gov't
Switchgear & switchboard apparatus	1.8	Manufg.	Equipment rental & leasing services	0.2	Services
Aluminum castings	1.5	Manufg.	S/L Govt. purch., health & hospitals	0.2	S/L Govt
Eating & drinking places	1.4	Trade	S/L Govt. purch., higher education	0.2	S/L Govt
Communications, except radio & TV	1.2	Util.	Blast furnaces & steel mills	0.1	Manufg.
Noncomparable imports	1.2	Foreign			
Plating & polishing	1.0	Manufg.			
Sheet metal work	1.0	Manufg.			
Electric services (utilities)	1.0	Util.			
Miscellaneous plastics products	0.9	Manufg.			
Transformers	0.9	Manufg.			
Paperboard containers & boxes	0.8	Manufg.			
Screw machine and related products	0.8	Manufg.			
Motor freight transportation & warehousing	0.8	Util.			
Ball & roller bearings	0.7	Manufg.			
Wiring devices	0.7	Manufg.			
Aluminum rolling & drawing	0.6	Manufg.			
Automotive rental & leasing, without drivers	0.6	Services			
Legal services	0.6	Services			
Copper rolling & drawing	0.5	Manufg.			
Industrial patterns	0.5	Manufg.			
Motors & generators	0.5	Manufg.			
Real estate	0.5	Fin/R.E.			
Automotive repair shops & services	0.5	Services			
Management & consulting services & labs	0.5	Services			
U.S. Postal Service	0.5	Gov't			
Maintenance of nonfarm buildings nec	0.4	Constr.			
Electron tubes	0.4	Manufg.			
Gaskets, packing & sealing devices	0.4	Manufg.			
Machinery, except electrical, nec	0.4	Manufg.			
Advertising	0.4	Services			
Equipment rental & leasing services	0.4	Services			
Metal stampings, nec	0.3	Manufg.			
Paints & allied products	0.3	Manufg.			
Air transportation	0.3	Util.			
Gas production & distribution (utilities)	0.3	Util.			
Accounting, auditing & bookkeeping	0.3	Services			
Abrasive products	0.2	Manufg.			
Lubricating oils & greases	0.2	Manufg.			
Manifold business forms	0.2	Manufg.			
Metal coating & allied services	0.2	Manufg.			
Miscellaneous fabricated wire products	0.2	Manufg.			

Continued on next page.

INPUTS AND OUTPUTS FOR SCALES & BALANCES - Continued

Economic Sector or Industry Providing Inputs	%	Sector	Economic Sector or Industry Buying Outputs	%	Sector
Nonferrous wire drawing & insulating	0.2	Manufg.			
Special dies & tools & machine tool accessories	0.2	Manufg.			
Railroads & related services	0.2	Util.			
Sanitary services, steam supply, irrigation	0.2	Util.			
Credit agencies other than banks	0.2	Fin/R.E.			
Insurance carriers	0.2	Fin/R.E.			
Machine tools, metal cutting types	0.1	Manufg.			
Mechanical measuring devices	0.1	Manufg.			
Metal heat treating	0.1	Manufg.			
Motor vehicle parts & accessories	0.1	Manufg.			
Photographic equipment & supplies	0.1	Manufg.			
Tires & inner tubes	0.1	Manufg.			
Retail trade, except eating & drinking	0.1	Trade			
Royalties	0.1	Fin/R.E.			
Security & commodity brokers	0.1	Fin/R.E.			
Photofinishing labs, commercial photography	0.1	Services			

Source: Benchmark Input-Output Accounts for the U.S. Economy, 1982, U.S. Department of Commerce, Washington, D.C., July 1991. Data, as reported in the source, are organized by the 1977 SIC structure in use in 1982 but have been matched, as closely as is possible, to the 1987 SIC structure used in this book.

OCCUPATIONS EMPLOYED BY SIC 359 - INDUSTRIAL MACHINERY, NEC

Occupation	% of Total 1994	Change to 2005	Occupation	% of Total 1994	Change to 2005
Machinists	20.3	-5.6	Sales & related workers nec	1.8	-5.6
NC machine tool operators, metal & plastic	4.9	41.6	Tool & die makers	1.6	-67.3
General managers & top executives	4.6	-10.4	Bookkeeping, accounting, & auditing clerks	1.6	-29.2
Blue collar worker supervisors	3.8	-13.3	Industrial production managers	1.6	3.8
Assemblers, fabricators, & hand workers nec	3.5	-5.6	Welding machine setters, operators	1.6	-15.0
Machine tool cutting operators, metal & plastic	3.5	-27.4	Drilling & boring machine tool workers	1.4	-52.8
Inspectors, testers, & graders, precision	2.7	-5.6	Traffic, shipping, & receiving clerks	1.3	-9.1
Welders & cutters	2.6	-5.6	Janitors & cleaners, incl maids	1.2	-24.5
Lathe & turning machine tool operators	2.6	-52.8	Industrial machinery mechanics	1.2	3.9
Combination machine tool operators	2.5	-5.6	Mechanical engineers	1.2	14.3
Grinding machine operators, metal & plastic	2.2	-5.6	Financial managers	1.1	-5.6
Secretaries, ex legal & medical	2.0	-14.1	Machine builders	1.1	-24.5
General office clerks	2.0	-19.5	Clerical supervisors & managers	1.0	-3.4
Machine tool cutting & forming etc. nec	1.9	-24.5	Grinders & polishers, hand	1.0	51.0

Source: Industry-Occupation Matrix, Bureau of Labor Statistics. These data relate to one or more 3-digit SIC industry groups rather than to a single 4-digit SIC. The change reported for each occupation to the year 2005 is a percent of growth or decline as estimated by the Bureau of Labor Statistics. The abbreviation nec stands for 'not elsewhere classified'.

LOCATION BY STATE AND REGIONAL CONCENTRATION

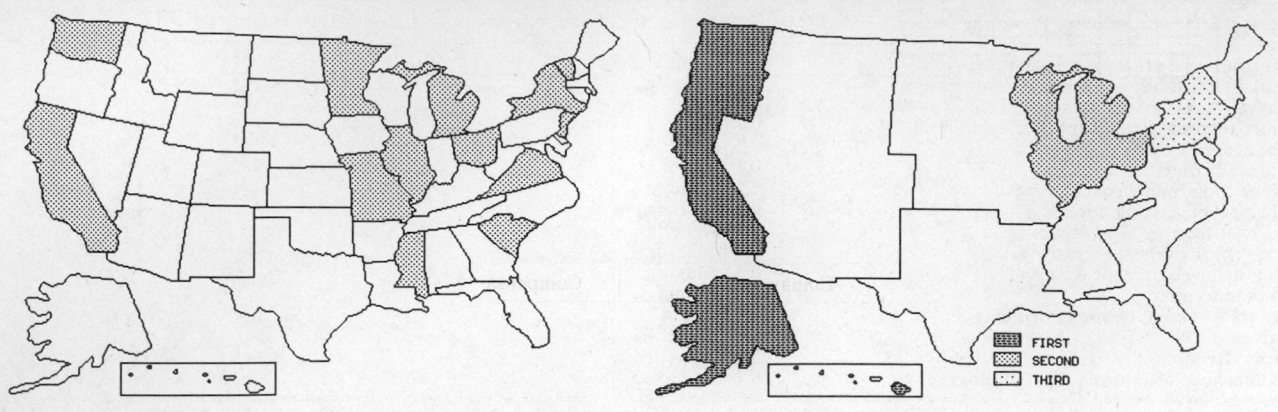

FIRST
SECOND
THIRD

INDUSTRY DATA BY STATE

State	Establish-ments	Shipments			Employment				Cost as % of Shipments	Investment per Employee ($)
		Total ($ mil)	% of U.S.	Per Establ.	Total Number	% of U.S.	Per Establ.	Wages ($/hour)		
Illinois	8	83.5	13.0	10.4	700	12.5	88	7.46	47.5	-
California	21	76.8	12.0	3.7	800	14.3	38	12.14	51.2	2,625
Washington	6	22.7	3.5	3.8	200	3.6	33	9.00	28.6	500
Ohio	8	(D)	-	-	750 *	13.4	94	-	-	-
Minnesota	6	(D)	-	-	375 *	6.7	63	-	-	-
New Jersey	6	(D)	-	-	375 *	6.7	63	-	-	1,067
New York	6	(D)	-	-	175 *	3.1	29	-	-	-
Michigan	3	(D)	-	-	175 *	3.1	58	-	-	-
Mississippi	3	(D)	-	-	375 *	6.7	125	-	-	533
Missouri	2	(D)	-	-	375 *	6.7	188	-	-	-
Vermont	2	(D)	-	-	175 *	3.1	88	-	-	-
Virginia	2	(D)	-	-	175 *	3.1	88	-	-	-
South Carolina	1	(D)	-	-	375 *	6.7	375	-	-	-

Source: 1992 Economic Census. The states are in descending order of shipments or establishments (if shipment data are missing for the majority). The symbol (D) appears when data are withheld to prevent disclosure of competitive information. States marked with (D) are sorted by number of establishments. A dash (-) indicates that the data element cannot be calculated; * indicates the midpoint of a range.

3599 - INDUSTRIAL MACHINERY, NEC

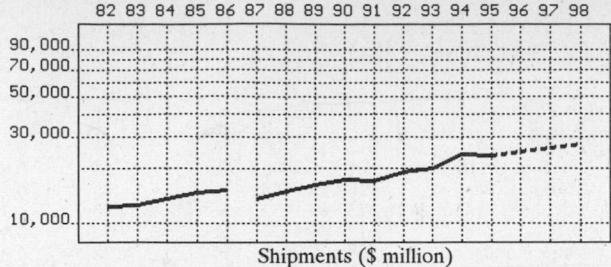

Shipments ($ million)

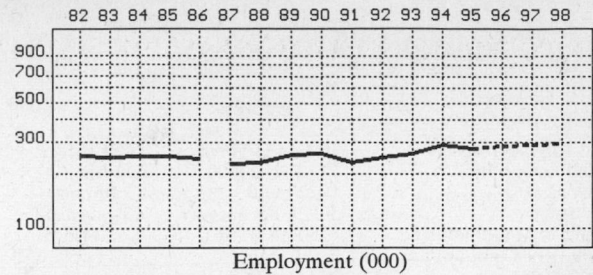

Employment (000)

GENERAL STATISTICS

Year	Companies	Establishments Total	Establishments with 20 or more employees	Employment Total (000)	Employment Production Workers (000)	Employment Hours (Mil)	Compensation Payroll ($ mil)	Compensation Wages ($/hr)	Production Cost of Materials	Production Value Added by Manufacture	Production Value of Shipments	Production Capital Invest.
1982	21,903	22,075	3,099	250.9	196.8	390.3	4,543.1	8.20	3,820.3	8,420.5	12,271.4	660.7
1983		21,619	3,183	247.2	194.7	387.1	4,618.2	8.23	3,939.3	8,485.6	12,483.2	543.3
1984		21,163	3,267	249.9	195.1	384.9	4,913.3	8.86	4,486.0	9,121.3	13,560.2	559.5
1985		20,707	3,350	252.6	197.4	394.2	5,316.2	9.28	4,903.9	9,976.9	14,765.1	701.9
1986		20,525	3,333	244.2	185.1	373.1	5,439.0	9.74	4,940.3	10,271.6	15,167.0	572.7
1987*	21,412	21,545	2,985	228.4	177.5	366.3	5,118.7	9.50	4,170.7	9,516.3	13,692.2	539.4
1988		20,610	3,193	231.8	181.8	380.1	5,449.8	9.84	4,696.7	10,295.9	14,926.0	267.3
1989		20,309	3,492	257.9	192.1	405.7	6,067.7	10.18	5,150.6	11,521.6	16,287.8	643.5
1990		20,890	3,475	259.2	190.3	401.8	6,311.6	10.61	5,629.6	11,741.2	17,184.1	690.2
1991		21,514	3,395	233.0	178.7	380.9	6,226.7	11.10	5,729.9	11,205.2	16,924.8	610.6
1992	22,591	22,756	3,356	248.2	188.7	396.2	7,020.8	11.71	6,207.0	12,845.9	19,057.7	683.9
1993		22,835	3,468	258.8	198.7	413.7	7,284.6	11.69	7,010.1	13,090.4	19,997.4	702.2
1994		22,832P	3,577P	288.8	221.2	474.1	8,285.5	11.70	7,761.1	16,487.9	23,955.3	1,120.9
1995		23,167P	3,637P	277.7P	210.9P	449.1P	8,298.3P	12.38P	7,578.4P	16,099.9P	23,391.5P	994.0P
1996		23,501P	3,697P	283.7P	215.3P	459.5P	8,704.5P	12.74P	7,984.4P	16,962.2P	24,644.5P	1,068.9P
1997		23,836P	3,757P	289.6P	219.6P	469.9P	9,110.6P	13.09P	8,390.3P	17,824.6P	25,897.4P	1,143.7P
1998		24,170P	3,817P	295.6P	224.0P	480.3P	9,516.7P	13.45P	8,796.3P	18,687.0P	27,150.4P	1,218.6P

Sources: 1982, 1987, 1992 *Economic Census*; *Annual Survey of Manufactures*, 83-86, 88-91, 93-94. Establishment counts are from *County Business Patterns* for non-Census years; establishment counts for 83-84 are extrapolations. * indicates that industry content changed in 87; earlier years use 77 SICs. 'P's mark projections.

INDICES OF CHANGE

Year	Companies	Establishments Total	Establishments with 20 or more employees	Employment Total (000)	Employment Production Workers (000)	Employment Hours (Mil)	Compensation Payroll ($ mil)	Compensation Wages ($/hr)	Production Cost of Materials	Production Value Added by Manufacture	Production Value of Shipments	Production Capital Invest.
1982	97	97	92	101	104	99	65	70	62	66	64	97
1983		95	95	100	103	98	66	70	63	66	66	79
1984		93	97	101	103	97	70	76	72	71	71	82
1985		91	100	102	105	99	76	79	79	78	77	103
1986		90	99	98	98	94	77	83	80	80	80	84
1987*	95	95	89	92	94	92	73	81	67	74	72	79
1988		91	95	93	96	96	78	84	76	80	78	39
1989		89	104	104	102	102	86	87	83	90	85	94
1990		92	104	104	101	101	90	91	91	91	90	101
1991		95	101	94	95	96	89	95	92	87	89	89
1992	100	100	100	100	100	100	100	100	100	100	100	100
1993		100	103	104	105	104	104	100	113	102	105	103
1994		100P	107P	116	117	120	118	100	125	128	126	164
1995		102P	108P	112P	112P	113P	118P	106P	122P	125P	123P	145P
1996		103P	110P	114P	114P	116P	124P	109P	129P	132P	129P	156P
1997		105P	112P	117P	116P	119P	130P	112P	135P	139P	136P	167P
1998		106P	114P	119P	119P	121P	136P	115P	142P	145P	142P	178P

Sources: Same as General Statistics. Values reflect change from the base year, 1992. Values above 100 mean greater than 92, values below 100 mean less than 92, and a value of 100 in the 82-91 or 93-98 period means same as 92. * indicates that industry content changed in 87. Data for earlier years are in 77 SIC format.

SELECTED RATIOS

For 1994	Avg. of All Manufact.	Analyzed Industry	Index	For 1994	Avg. of All Manufact.	Analyzed Industry	Index
Employees per Establishment	49	13	26	Value Added per Production Worker	134,084	74,538	56
Payroll per Establishment	1,500,273	362,885	24	Cost per Establishment	5,045,178	339,918	7
Payroll per Employee	30,620	28,689	94	Cost per Employee	102,970	26,874	26
Production Workers per Establishment	34	10	28	Cost per Production Worker	146,988	35,086	24
Wages per Establishment	853,319	242,944	28	Shipments per Establishment	9,576,895	1,049,185	11
Wages per Production Worker	24,861	25,077	101	Shipments per Employee	195,460	82,948	42
Hours per Production Worker	2,056	2,143	104	Shipments per Production Worker	279,017	108,297	39
Wages per Hour	12.09	11.70	97	Investment per Establishment	321,011	49,093	15
Value Added per Establishment	4,602,255	722,131	16	Investment per Employee	6,552	3,881	59
Value Added per Employee	93,930	57,091	61	Investment per Production Worker	9,352	5,067	54

Sources: Same as General Statistics. The 'Average of All Manufacturing' column represents the average of all manufacturing industries reported for the most recent complete year available. The Index shows the relationship between the Average and the Analyzed Industry. For example, 100 means that they are equal; 500 that the Analyzed Industry is five times the average; 50 means that the Analyzed Industry is half the national average. The abbreviation 'na' is used to show that data are 'not available'.

LEADING COMPANIES Number shown: **75** Total sales ($ mil): **1,584** Total employment (000): **13.7**

Company Name	Address				CEO Name	Phone	Co. Type	Sales ($ mil)	Empl. (000)
Remmele Engineering Inc	10 Old Highway	New Brighton	MN	55112	T Moore	612-635-4100	R	77	0.5
Diversified Group Inc	PO Box 23890	Harahan	LA	70123	Herbert D Hughes	504-733-2800	R	55*	0.6
Phillips Corp	10220 Old Columbia	Columbia	MD	21046	Margie Lewis	410-995-6148	R	45	0.1
PBM Intermet	50925 Richard W	Chesterfield	MI	48051	Michael J Mahoney	313-949-1433	S	41*	0.3
Motek Eng & Manufacturing	625 2nd St SE	Cambridge	MN	55008	D Ricke	612-689-1333	R	40	0.3
Cam Fran Tool Company Inc	737 Fargo	Elk Grove Vill	IL	60007	Mark Knudson	708-640-2382	R	38	0.2
Oberg Industries Inc	PO Box 368	Freeport	PA	16229	David O Shondeck	412-295-2121	R	37*	0.4
Royal Oak Industries Inc	PO Box 127	Lake Orion	MI	48361	Patrick J Carroll	313-340-9200	R	36	0.4
Chance Industries Inc	PO Box 12328	Wichita	KS	67277	Richard G Chance	316-942-7411	R	35	0.4
UNC Johnson Technology Inc	2034 Latimer Dr	Muskegon	MI	49442	Ron Fredrick	616-777-2685	S	35	0.3
Leybold Inficon Inc	Two Technology Pl	East Syracuse	NY	13057	JL Brissenden	315-434-1100	S	33	0.2
Hypro Inc	PO Box 30	Waterford	WI	53185	Robert Schildt	414-534-5141	R	30	0.3
Modern Industries Inc	PO Box 399	Erie	PA	16512	Herbert S Sweny	814-455-8061	R	30	0.3
Rite-On Industries Inc	12540 Beech-Daly	Redford	MI	48239	Gary S Muscat	313-937-2000	R	30	0.2
Velcon Filters Inc	4525 Cen Blvd	Co Springs	CO	80919	David C Taylor	719-531-5855	R	26	0.2
Belfab	PO Box 9370	Daytona Beach	FL	32120	DD Fockler	904-253-0628	D	25	0.2
Michigan Wheel Corp	1501 Buchanan SW	Grand Rapids	MI	49507	Stan Heide	616-452-6941	R	25	0.2
Mid-State Machine Products Inc	1501 Verti Dr	Winslow	ME	04901	Douglas Sukeforth	207-873-6136	R	25	0.2
Multiplex Company Inc	250 Old Ballwin Rd	Ballwin	MO	63021	J Walter Kisling	314-256-7777	R	25	0.2
RMS Co	8600 Evergreen	Minneapolis	MN	55433	Art Mouyard	612-786-1520	S	25	0.3
Schuler Industries Inc	PO Box 5366	Birmingham	AL	35207	Jerry D Hart	205-252-5010	R	25	0.3
Piqua Engineering Inc	234 1st St	Piqua	OH	45356	John Scarbrough Sr	513-773-8966	R	23	0.2
RM Kerner Co	2208 E 33rd St	Erie	PA	16510	RM Kerner	814-898-2000	R	23*	0.3
Superior Machine	492 N Cashua Rd	Florence	SC	29501	Bryan Jackson	803-664-3001	R	23	0.3
Waltco Engineering Company	401 W Redondo	Gardena	CA	90248	JF Brannan	310-538-0321	R	22	0.2
Ace Controls Inc	23435 Industrial Pk	Farmington	MI	48335	WJ Chorkey	810-476-0213	R	20	0.2
Dreison International Inc	6501 Barberton Av	Cleveland	OH	44102	Theodore Berger	216-281-7810	R	20	0.2
Globe Machine Mfg Co	PO Box 2274	Tacoma	WA	98401	Calvin Bamford	206-383-2584	R	20	0.2
Komar Industries Inc	4425 Marketing Pl	Groveport	OH	43125	Larry Koenig	614-836-2366	R	20	<0.1
Midwest Irrigation Co	PO Box 516	Henderson	NE	68371	Carl C Buller	402-723-5374	R	20	<0.1
Votaw Precision Technologies	13153 Lakeland Rd	Santa Fe Sprgs	CA	90670	Stuart Gordon	310-944-0661	S	20	0.1
Tibor Machine Products	6350 W Birmingham	Chicago Ridge	IL	60415	Larry Overbey	708-499-3700	R	19	0.2
Chance Rides Inc	PO Box 12328	Wichita	KS	67277	Scott Culbertson	316-942-7411	S	18	0.4
James Machine Works Inc	PO Box 1752	Monroe	LA	71210	JW Neal	318-322-6104	R	18	0.2
Littleford Day Inc	PO Box 128	Florence	KY	41022	DL Steedman	606-525-7600	S	18	0.2
Loral American Beryllium Corp	PO Box 1087	Tallevast	FL	34270	George Allen	813-355-5105	S	18	<0.1
Nationwide Precision Prod Corp	200 Tech Park Dr	Rochester	NY	14623	Michael R Nuccitelli	716-272-7100	R	18	0.2
Robot Aided Mfg Center	5140 Moundview Dr	Red Wing	MN	55066	Larry Lautt	612-388-1821	R	18	<0.1
Foundry and Steel Inc	PO Box 349	Anderson	SC	29622	Ray Hopkins	803-226-0381	R	17*	0.2
Advance Manufacturing	Turnpike Indust	Westfield	MA	01085	Anthony E Amante	413-568-2411	R	16	0.2
Biddle Precision Components	701 S Main St	Sheridan	IN	46069	Dale McCullough	317-758-4451	R	16	0.2
Husky Injection Molding	303 Washington St	Auburn	MA	01501	AH Robinson	508-832-5911	S	16	<0.1
Pennfield Precision Inc	PO Box 380	Sellersville	PA	18960	JF Matczak	215-257-5191	R	16	0.2
Acromil Corp	18421 Railroad St	City of Industry	CA	91744	Victor Bowman	818-964-2522	R	15	<0.1
Adapto Inc	PO Box 280	Litchfield Park	AZ	85340	TL Schoaf	602-935-2681	R	15	0.3
Columbus Industries Inc	PO Box 257	Ashville	OH	43103	H Pontius	614-983-2552	R	15	0.2
Elox Corp	PO Box 220	Davidson	NC	28036	Gabriele G Carinci	704-892-8011	S	15	<0.1
Excel Recycling	PO Box 31118	Amarillo	TX	79120	Matt Garth	806-335-3737	R	15	<0.1
Mac Corp	201 E Shady Grove	Grand Prairie	TX	75050	Glen Newton	214-790-7800	R	15	<0.1
Modern Manufacturing Inc	2900 Lind Av SW	Renton	WA	98055	Dennis Speer	206-251-1515	S	15*	0.1
Precision Metal Products Inc	307 Pepe's Farm Rd	Milford	CT	06460	Joseph Martino	203-877-4258	R	15	0.1
Primeway Tool and Engineering	32033 Edward St	Madison H	MI	48071	Bob Ianitelli	810-588-6614	R	15*	0.2
Roberds-Johnson Industries Inc	PO Box 472	Galena Park	TX	77547	W Monteleone	713-676-2636	R	15	0.2
Seabee Corp	PO Box 457	Hampton	IA	50441	D Yadon	515-456-4871	R	15	0.3
Arrowsmith Tool and Die Inc	PO Box 407	Southfield	MI	48037	Mark E Greenbury	810-357-4400	R	14	0.1
Porter Precision Products Inc	2734 Banning Rd	Cincinnati	OH	45239	John F Cipriani Sr	513-923-3777	R	14	0.2
Steward Machine Company Inc	PO Box 11008	Birmingham	AL	35202	W Debardeleben	205-841-6461	R	14	0.2
Ultimate Precision Inc	29370ephenson Hwy	Madison H	MI	48071	Donald J Chinn	313-543-7670	R	14	<0.1
Ver-Sa-Til Associates Inc	18400 W 77 St	Chanhassen	MN	55317	Brion Roberts	612-929-0626	R	14	0.2
Continental Machine & Eng Co	PO Box 270	East Chicago	IN	46312	A Sakelaris	219-398-7300	R	13*	0.2
Lindquist Machine Corp	PO Box 2327	Green Bay	WI	54306	PD Mancuso	414-499-0831	R	13*	0.2
TMI Industries	2001 W Melinda Ln	Phoenix	AZ	85027	Bill Schultz	602-581-8030	D	13	0.3
Universal Machine Co	525 W Vine St	Stowe	PA	19464	Richard F Francis	215-323-1810	R	13*	0.2
Bay Cast Inc	PO Box 126	Bay City	MI	48707	Scott L Holman	517-892-0511	R	12*	<0.1
Benton Harbor Engineering Inc	PO Box 367	Benton Harbor	MI	49023	LJ Derubbo	616-925-7081	R	12	0.1
Floturn Inc	4236 Thunderb	Fairfield	OH	45014	Robert V Glutting	513-860-8040	R	12	<0.1
Guarantee Specialties Inc	9401 Carr Av	Cleveland	OH	44108	Ed Pages	216-451-9744	R	12	<0.1
Lakeside Machine Inc	PO Box 151	Gladstone	MI	49837	Gregory Hansen	906-428-2333	R	12	<0.1
McNally Industries Inc	PO Box 129	Grantsburg	WI	54840	Al L Scheideler	715-463-5311	R	12	0.1
Nelmor Company Inc	PO Box 328	North Uxbridge	MA	01538		508-278-5584	S	12*	<0.1
Sonfarrel Inc	3000 E La Jolla St	Anaheim	CA	92806	Frank Power	714-630-7280	R	12*	0.1
Aline Components Inc	PO Box 263	Kulpsville	PA	19443	H Davis	215-368-0300	R	11	<0.1
UIP Engineered Products	5501 W Grand Av	Chicago	IL	60639	Chuck Wilken	312-237-6004	S	11	0.1
Atols Tool and Mold Corp	3828 River Rd	Schiller Park	IL	60176	JM Atols	708-671-4420	R	11*	0.1
Com-Tal Machine & Engineering	1239 Wolters Blv	St Paul	MN	55110	John Melquist	612-483-2611	R	11	<0.1

Source: Ward's Business Directory of U.S. Private and Public Companies, Volumes 1 and 2, 1996. The company type code used is as follows: P - Public, R - Private, S - Subsidiary, D - Division, J - Joint Venture, A - Affiliate, G - Group. Sales are in millions of dollars, employees are in thousands. An asterisk (*) indicates an estimated sales volume. The symbol < stands for 'less than'. Company names and addresses are truncated, in some cases, to fit into the available space.

MATERIALS CONSUMED

Material	Quantity	Delivered Cost ($ million)
Materials, ingredients, containers, and supplies	(X)	4,802.4
Metal bolts, nuts, screws, washers, rivets, and other screw machine products	(X)	79.2
Other fabricated metal products (except forgings)	(X)	159.3
Forgings	(X)	66.3
Iron and steel castings (rough and semifinished)	(X)	312.6
Aluminum and aluminum-base alloy castings (rough and semifinished)	(X)	180.5
Other nonferrous castings (rough and semifinished)	(X)	51.9
Steel bars, bar shapes, and plates	(X)	394.2
Steel sheet and strip, including tin plate	(X)	127.4
Steel structural shapes and sheet piling	(X)	27.5
All other steel shapes and forms	(X)	55.2
Copper and copper-base alloy rod, bar, and mechanical wire, including extruded and/or drawn shapes	(X)	24.0
Aluminum and aluminum-base alloy sheet, plate, foil, and welded tubing	(X)	80.5
All other aluminum and aluminum-base alloy shapes and forms	(X)	81.1
Other nonferrous shapes and forms	(X)	40.1
Cutting tools for machine tools	(X)	91.4
Fluid power products	(X)	76.5
All other materials and components, parts, containers, and supplies	(X)	714.9
Materials, ingredients, containers, and supplies, nsk	(X)	2,239.8

Source: 1992 Economic Census. Explanation of symbols used: (D): Withheld to avoid disclosure of competitive data; na: Not available; (S): Withheld because statistical norms were not met; (X): Not applicable; (Z): Less than half the unit shown; nec: Not elsewhere classified; nsk: Not specified by kind; - : zero; * : 10-19 percent estimated; ** : 20-29 percent estimated.

PRODUCT SHARE DETAILS

Product or Product Class	% Share	Product or Product Class	% Share
Machinery, except electrical, nec	100.00	Carnival and amusement park equipment (ferris wheels, merry-go-rounds, etc.), excluding electric equipment, and coin-operated amusement machines	2.70
Miscellaneous machinery products, except electrical (including flexible metal hose and tubing, metal bellows, etc.)	9.99	All other miscellaneous machinery products, except electrical	70.16
Flexible copper and copper-base alloy hose and tubing	2.22	Miscellaneous machinery products (including flexible metal hose and tubing, metal bellows, etc.), nsk	9.46
Flexible aluminum and aluminum-base alloy hose and tubing	1.81	Receipts for machine shop job work and job order repairs	68.41
Flexible stainless steel hose and tubing	6.13	Machinery, except electrical, nec, nsk	21.60
Other flexible metal hose and tubing	2.08		
Metal bellows	5.42		

Source: 1992 Economic Census. The values shown are percent of total shipments in an industry. Values of indented subcategories are summed in the main heading. The symbol (D) appears when data are withheld to prevent disclosure of competitive information. The abbreviation nsk stands for 'not specified by kind' and nec for 'not elsewhere classified'.

INPUTS AND OUTPUTS FOR GENERAL INDUSTRIAL MACHINERY, NEC

Economic Sector or Industry Providing Inputs	%	Sector	Economic Sector or Industry Buying Outputs	%	Sector
Imports	41.1	Foreign	Gross private fixed investment	54.4	Cap Inv
Wholesale trade	6.3	Trade	Exports	29.3	Foreign
Blast furnaces & steel mills	5.5	Manufg.	General industrial machinery, nec	2.4	Manufg.
General industrial machinery, nec	3.7	Manufg.	Federal Government purchases, national defense	2.2	Fed Govt
Nonferrous wire drawing & insulating	2.4	Manufg.	Radio & TV communication equipment	0.8	Manufg.
Machinery, except electrical, nec	2.3	Manufg.	Miscellaneous repair shops	0.7	Services
Miscellaneous plastics products	2.2	Manufg.	Industrial buildings	0.6	Constr.
Iron & steel foundries	1.6	Manufg.	Pumps & compressors	0.6	Manufg.
Switchgear & switchboard apparatus	1.5	Manufg.	Office buildings	0.5	Constr.
Aluminum rolling & drawing	1.4	Manufg.	Crude petroleum & natural gas	0.4	Mining
Blowers & fans	1.4	Manufg.	Feed grains	0.3	Agric.
Motors & generators	1.4	Manufg.	Meat animals	0.3	Agric.
Pipe, valves, & pipe fittings	1.4	Manufg.	Plastics materials & resins	0.3	Manufg.
Pumps & compressors	1.4	Manufg.	Printing trades machinery	0.3	Manufg.
Industrial controls	1.2	Manufg.	Coal	0.2	Mining
Nonwoven fabrics	1.2	Manufg.	Maintenance of nonfarm buildings nec	0.2	Constr.
Electric services (utilities)	1.2	Util.	Nonfarm residential structure maintenance	0.2	Constr.
Fabricated plate work (boiler shops)	1.1	Manufg.	Fabricated plate work (boiler shops)	0.2	Manufg.
Transformers	1.1	Manufg.	Motor vehicles & car bodies	0.2	Manufg.
Power transmission equipment	0.9	Manufg.	Sporting & athletic goods, nec	0.2	Manufg.
Metal stampings, nec	0.8	Manufg.	S/L Govt. purch., fire	0.2	S/L Govt
Special dies & tools & machine tool accessories	0.8	Manufg.	Oil bearing crops	0.1	Agric.
Communications, except radio & TV	0.8	Util.	Construction of stores & restaurants	0.1	Constr.
Banking	0.7	Fin/R.E.	Hotels & motels	0.1	Constr.
Advertising	0.7	Services	Ship building & repairing	0.1	Manufg.
Maintenance of nonfarm buildings nec	0.6	Constr.	Special industry machinery, nec	0.1	Manufg.
Aluminum castings	0.6	Manufg.	Electric services (utilities)	0.1	Util.
Brass, bronze, & copper castings	0.6	Manufg.			
Copper rolling & drawing	0.6	Manufg.			

Continued on next page.

INPUTS AND OUTPUTS FOR GENERAL INDUSTRIAL MACHINERY, NEC - Continued

Economic Sector or Industry Providing Inputs	%	Sector	Economic Sector or Industry Buying Outputs	%	Sector
Petroleum refining	0.6	Manufg.			
Screw machine and related products	0.6	Manufg.			
Air transportation	0.6	Util.			
Personnel supply services	0.6	Services			
Electronic components nec	0.5	Manufg.			
Felt goods, nec	0.5	Manufg.			
Motor freight transportation & warehousing	0.5	Util.			
Eating & drinking places	0.5	Trade			
Real estate	0.5	Fin/R.E.			
Engineering, architectural, & surveying services	0.5	Services			
Abrasive products	0.4	Manufg.			
Ball & roller bearings	0.4	Manufg.			
Paperboard containers & boxes	0.4	Manufg.			
Plastics materials & resins	0.4	Manufg.			
Equipment rental & leasing services	0.4	Services			
Fabricated rubber products, nec	0.3	Manufg.			
Gaskets, packing & sealing devices	0.3	Manufg.			
Nonferrous rolling & drawing, nec	0.3	Manufg.			
Gas production & distribution (utilities)	0.3	Util.			
Business services nec	0.3	Services			
Computer & data processing services	0.3	Services			
U.S. Postal Service	0.3	Gov't			
Machine tools, metal cutting types	0.2	Manufg.			
Nonferrous castings, nec	0.2	Manufg.			
Paper mills, except building paper	0.2	Manufg.			
Railroads & related services	0.2	Util.			
Hotels & lodging places	0.2	Services			
Legal services	0.2	Services			
Management & consulting services & labs	0.2	Services			
Noncomparable imports	0.2	Foreign			
Converted paper products, nec	0.1	Manufg.			
Lubricating oils & greases	0.1	Manufg.			
Metal heat treating	0.1	Manufg.			
Plating & polishing	0.1	Manufg.			
Sawmills & planning mills, general	0.1	Manufg.			
Accounting, auditing & bookkeeping	0.1	Services			
Business/professional associations	0.1	Services			
Scrap	0.1	Scrap			

Source: Benchmark Input-Output Accounts for the U.S. Economy, 1982, U.S. Department of Commerce, Washington, D.C., July 1991. Data, as reported in the source, are organized by the 1977 SIC structure in use in 1982 but have been matched, as closely as is possible, to the 1987 SIC structure used in this book.

OCCUPATIONS EMPLOYED BY SIC 359 - INDUSTRIAL MACHINERY, NEC

Occupation	% of Total 1994	Change to 2005	Occupation	% of Total 1994	Change to 2005
Machinists	20.3	-5.6	Sales & related workers nec	1.8	-5.6
NC machine tool operators, metal & plastic	4.9	41.6	Tool & die makers	1.6	-67.3
General managers & top executives	4.6	-10.4	Bookkeeping, accounting, & auditing clerks	1.6	-29.2
Blue collar worker supervisors	3.8	-13.3	Industrial production managers	1.6	3.8
Assemblers, fabricators, & hand workers nec	3.5	-5.6	Welding machine setters, operators	1.6	-15.0
Machine tool cutting operators, metal & plastic	3.5	-27.4	Drilling & boring machine tool workers	1.4	-52.8
Inspectors, testers, & graders, precision	2.7	-5.6	Traffic, shipping, & receiving clerks	1.3	-9.1
Welders & cutters	2.6	-5.6	Janitors & cleaners, incl maids	1.2	-24.5
Lathe & turning machine tool operators	2.6	-52.8	Industrial machinery mechanics	1.2	3.9
Combination machine tool operators	2.5	-5.6	Mechanical engineers	1.2	14.3
Grinding machine operators, metal & plastic	2.2	-5.6	Financial managers	1.1	-5.6
Secretaries, ex legal & medical	2.0	-14.1	Machine builders	1.1	-24.5
General office clerks	2.0	-19.5	Clerical supervisors & managers	1.0	-3.4
Machine tool cutting & forming etc. nec	1.9	-24.5	Grinders & polishers, hand	1.0	51.0

Source: Industry-Occupation Matrix, Bureau of Labor Statistics. These data relate to one or more 3-digit SIC industry groups rather than to a single 4-digit SIC. The change reported for each occupation to the year 2005 is a percent of growth or decline as estimated by the Bureau of Labor Statistics. The abbreviation nec stands for 'not elsewhere classified'.

LOCATION BY STATE AND REGIONAL CONCENTRATION

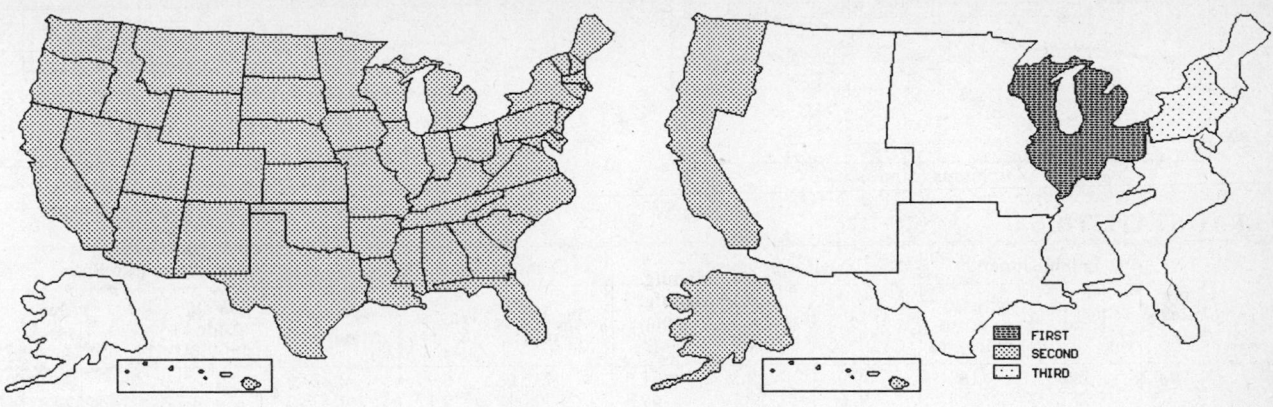

FIRST
SECOND
THIRD

INDUSTRY DATA BY STATE

| State | Establish-ments | Shipments | | | Employment | | | | Cost as % of Shipments | Investment per Employee ($) |
		Total ($ mil)	% of U.S.	Per Establ.	Total Number	% of U.S.	Per Establ.	Wages ($/hour)		
California	3,391	2,404.4	12.6	0.7	29,600	11.9	9	12.64	29.1	2,294
Ohio	1,597	1,555.5	8.2	1.0	20,500	8.3	13	11.58	32.3	2,532
Michigan	1,395	1,504.2	7.9	1.1	17,100	6.9	12	12.36	36.8	3,497
Illinois	1,235	1,331.6	7.0	1.1	16,500	6.6	13	11.80	33.4	2,721
Pennsylvania	1,197	1,256.1	6.6	1.0	17,000	6.8	14	11.66	29.6	2,476
Texas	1,566	1,047.6	5.5	0.7	14,300	5.8	9	11.27	34.1	2,049
Wisconsin	666	789.4	4.1	1.2	9,500	3.8	14	11.01	36.6	3,905
New York	956	772.9	4.1	0.8	10,200	4.1	11	11.97	32.4	2,745
Minnesota	545	700.0	3.7	1.3	8,400	3.4	15	12.40	33.4	4,298
Massachusetts	698	686.1	3.6	1.0	8,100	3.3	12	13.41	29.0	2,580
Connecticut	537	557.5	2.9	1.0	6,500	2.6	12	12.51	35.4	2,354
Indiana	600	554.8	2.9	0.9	7,600	3.1	13	11.09	33.2	2,974
New Jersey	700	501.1	2.6	0.7	6,300	2.5	9	12.65	31.2	2,365
Florida	599	353.6	1.9	0.6	5,200	2.1	9	10.48	32.9	2,558
Washington	444	335.8	1.8	0.8	4,400	1.8	10	12.97	32.6	3,159
Tennessee	361	315.4	1.7	0.9	4,400	1.8	12	11.08	32.7	2,841
North Carolina	551	308.9	1.6	0.6	4,700	1.9	9	10.35	32.3	3,702
Georgia	374	294.8	1.5	0.8	4,100	1.7	11	10.39	36.3	4,098
Alabama	363	289.2	1.5	0.8	3,900	1.6	11	9.63	36.1	2,769
Virginia	319	280.0	1.5	0.9	4,000	1.6	13	12.27	29.2	2,875
Missouri	388	274.6	1.4	0.7	3,900	1.6	10	11.72	32.3	2,410
Arizona	305	247.2	1.3	0.8	3,400	1.4	11	12.24	32.0	2,500
Kansas	235	238.1	1.2	1.0	3,100	1.2	13	11.11	34.7	4,871
Oklahoma	339	201.3	1.1	0.6	3,200	1.3	9	9.85	31.2	-
Oregon	327	195.8	1.0	0.6	2,700	1.1	8	11.60	31.3	2,333
Colorado	302	195.8	1.0	0.6	2,600	1.0	9	11.31	35.1	2,962
Kentucky	227	172.6	0.9	0.8	2,600	1.0	11	10.47	31.4	3,000
South Carolina	264	155.2	0.8	0.6	2,500	1.0	9	10.64	32.3	1,880
Iowa	213	143.2	0.8	0.7	2,100	0.8	10	9.97	31.3	2,905
West Virginia	143	134.5	0.7	0.9	1,800	0.7	13	10.69	39.6	2,278
Mississippi	129	65.5	0.3	0.5	1,000	0.4	8	10.19	34.7	1,800
Nebraska	117	56.3	0.3	0.5	1,000	0.4	9	9.38	36.8	1,400
Delaware	41	39.7	0.2	1.0	500	0.2	12	12.25	29.2	2,800
Wyoming	40	24.0	0.1	0.6	300	0.1	8	11.60	44.2	1,667
North Dakota	28	15.9	0.1	0.6	300	0.1	11	9.25	35.8	-
Hawaii	19	8.5	0.0	0.4	100	0.0	5	10.50	20.0	2,000
Louisiana	276	(D)	-	-	3,750 *	1.5	14	-	-	-
Maryland	198	(D)	-	-	1,750 *	0.7	9	-	-	-
New Hampshire	183	(D)	-	-	1,750 *	0.7	10	-	-	-
Arkansas	161	(D)	-	-	1,750 *	0.7	11	-	-	-
Utah	143	(D)	-	-	1,750 *	0.7	12	-	-	2,343
Rhode Island	117	(D)	-	-	750 *	0.3	6	-	-	-
New Mexico	99	(D)	-	-	750 *	0.3	8	-	-	-
Maine	95	(D)	-	-	750 *	0.3	8	-	-	-
Idaho	70	(D)	-	-	750 *	0.3	11	-	-	2,800
Nevada	53	(D)	-	-	375 *	0.2	7	-	-	-
Montana	50	(D)	-	-	375 *	0.2	8	-	-	-
Vermont	44	(D)	-	-	375 *	0.2	9	-	-	-
South Dakota	40	(D)	-	-	375 *	0.2	9	-	-	-

Source: 1992 *Economic Census*. The states are in descending order of shipments or establishments (if shipment data are missing for the majority). The symbol (D) appears when data are withheld to prevent disclosure of competitive information. States marked with (D) are sorted by number of establishments. A dash (-) indicates that the data element cannot be calculated; * indicates the midpoint of a range.

3612 - TRANSFORMERS

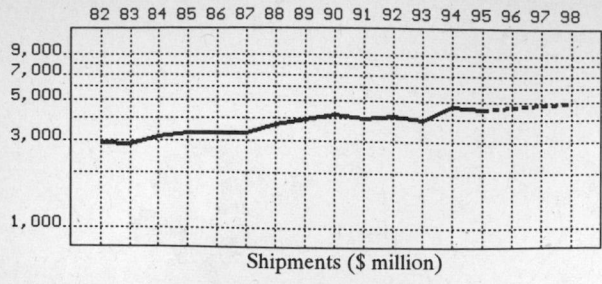

Shipments ($ million)

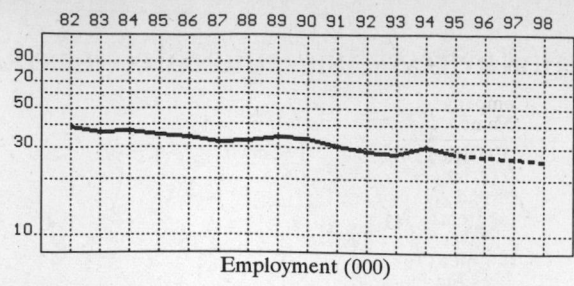

Employment (000)

GENERAL STATISTICS

Year	Companies	Establishments		Employment			Compensation		Production ($ million)			
		Total	with 20 or more employees	Total (000)	Production Workers (000)	Hours (Mil)	Payroll ($ mil)	Wages ($/hr)	Cost of Materials	Value Added by Manufacture	Value of Shipments	Capital Invest.
1982	244	293	162	39.0	28.5	51.3	732.1	8.99	1,421.8	1,439.0	2,916.0	76.5
1983		287	163	36.6	27.1	49.8	708.0	9.17	1,425.3	1,397.3	2,867.0	63.2
1984		281	164	37.1	28.3	52.7	761.4	9.52	1,686.1	1,534.1	3,170.3	82.8
1985		275	166	35.3	26.6	49.5	752.1	9.95	1,721.4	1,558.4	3,288.5	79.1
1986		272	165	34.3	25.9	48.4	746.7	10.22	1,674.8	1,591.8	3,298.4	66.2
1987	239	286	154	32.2	24.4	46.7	720.4	10.07	1,656.4	1,615.4	3,289.5	66.0
1988		283	165	32.9	25.1	49.6	753.3	10.11	1,928.9	1,798.9	3,669.9	66.8
1989		271	162	34.0	25.8	50.3	774.8	10.16	2,094.4	1,844.0	3,933.5	78.0
1990		275	163	33.5	24.8	49.2	816.9	10.89	2,239.4	1,892.3	4,177.8	109.0
1991		286	164	30.6	23.2	45.6	772.8	11.17	2,124.1	1,855.0	3,995.2	114.6
1992	223	276	146	28.9	21.9	43.7	777.1	11.64	2,052.7	2,010.1	4,096.3	85.1
1993		290	151	27.9	21.1	41.3	754.7	12.00	2,094.4	1,848.3	3,939.5	77.8
1994		279P	154P	30.7	21.7	45.4	874.9	11.80	2,311.9	2,380.7	4,672.7	165.5
1995		279P	153P	27.9P	20.8P	43.3P	819.2P	12.15P	2,259.9P	2,327.2P	4,567.6P	118.5P
1996		278P	152P	27.2P	20.2P	42.7P	827.0P	12.40P	2,325.5P	2,394.7P	4,700.2P	123.0P
1997		278P	151P	26.4P	19.6P	42.0P	834.7P	12.64P	2,391.1P	2,462.3P	4,832.8P	127.5P
1998		277P	150P	25.6P	19.1P	41.3P	842.5P	12.89P	2,456.7P	2,529.8P	4,965.4P	132.0P

Sources: 1982, 1987, 1992 *Economic Census*; *Annual Survey of Manufactures*, 83-86, 88-91, 93-94. Establishment counts for non-Census years are from *County Business Patterns*; establishment values for 83-84 are extrapolations. 'P's show projections by the editors. Industries reclassified in 87 will not have data for prior years.

INDICES OF CHANGE

Year	Companies	Establishments		Employment			Compensation		Production ($ million)			
		Total	with 20 or more employees	Total (000)	Production Workers (000)	Hours (Mil)	Payroll ($ mil)	Wages ($/hr)	Cost of Materials	Value Added by Manufacture	Value of Shipments	Capital Invest.
1982	109	106	111	135	130	117	94	77	69	72	71	90
1983		104	112	127	124	114	91	79	69	70	70	74
1984		102	112	128	129	121	98	82	82	76	77	97
1985		100	114	122	121	113	97	85	84	78	80	93
1986		99	113	119	118	111	96	88	82	79	81	78
1987	107	104	105	111	111	107	93	87	81	80	80	78
1988		103	113	114	115	114	97	87	94	89	90	78
1989		98	111	118	118	115	100	87	102	92	96	92
1990		100	112	116	113	113	105	94	109	94	102	128
1991		104	112	106	106	104	99	96	103	92	98	135
1992	100	100	100	100	100	100	100	100	100	100	100	100
1993		105	103	97	96	95	97	103	102	92	96	91
1994		101P	105P	106	99	104	113	101	113	118	114	194
1995		101P	105P	97P	95P	99P	105P	104P	110P	116P	112P	139P
1996		101P	104P	94P	92P	98P	106P	107P	113P	119P	115P	145P
1997		101P	103P	91P	90P	96P	107P	109P	116P	122P	118P	150P
1998		101P	103P	89P	87P	95P	108P	111P	120P	126P	121P	155P

Sources: Same as General Statistics. Values reflect change from the base year, 1992. Values above 100 mean greater than 92, values below 100 mean less than 92, and a value of 100 in the 82-91 or 93-98 period means same as 92. 'P's mark projections by the editors.

SELECTED RATIOS

For 1994	Avg. of All Manufact.	Analyzed Industry	Index	For 1994	Avg. of All Manufact.	Analyzed Industry	Index
Employees per Establishment	49	110	225	Value Added per Production Worker	134,084	109,710	82
Payroll per Establishment	1,500,273	3,136,864	209	Cost per Establishment	5,045,178	8,289,081	164
Payroll per Employee	30,620	28,498	93	Cost per Employee	102,970	75,306	73
Production Workers per Establishment	34	78	227	Cost per Production Worker	146,988	106,539	72
Wages per Establishment	853,319	1,920,769	225	Shipments per Establishment	9,576,895	16,753,488	175
Wages per Production Worker	24,861	24,688	99	Shipments per Employee	195,460	152,205	78
Hours per Production Worker	2,056	2,092	102	Shipments per Production Worker	279,017	215,332	77
Wages per Hour	12.09	11.80	98	Investment per Establishment	321,011	593,383	185
Value Added per Establishment	4,602,255	8,535,756	185	Investment per Employee	6,552	5,391	82
Value Added per Employee	93,930	77,547	83	Investment per Production Worker	9,352	7,627	82

Sources: Same as General Statistics. The 'Average of All Manufacturing' column represents the average of all manufacturing industries reported for the most recent complete year available. The Index shows the relationship between the Average and the Analyzed Industry. For example, 100 means that they are equal; 500 that the Analyzed Industry is five times the average; 50 means that the Analyzed Industry is half the national average. The abbreviation 'na' is used to show that data are 'not available'.

LEADING COMPANIES   Number shown: **75**   Total sales ($ mil): **17,686**   Total employment (000): **163.2**

Company Name	Address				CEO Name	Phone	Co. Type	Sales ($ mil)	Empl. (000)
Westinghouse Electric Corp	11 Stanwix St	Pittsburgh	PA	15222	Michael H Jordan	412-244-2000	P	8,848	84.4
Cooper Industries Inc	PO Box 4446	Houston	TX	77210	H John Riley	713-739-5400	P	4,588	40.8
Siemens Energy and Automation	PO Box 89000	Atlanta	GA	30356	TJ Malott	404-751-2000	S	999	7.0
Cooper Power Systems	PO Box 1640	Waukesha	WI	53187	William D Brewer	414-524-3300	D	612	5.3
MagneTek Inc	200 Robin Rd	Paramus	NJ	07652	Bill Landry	201-967-7600	D	460*	4.0
Advance Transformer Co	10275 W Higgins Rd	Rosemont	IL	60018	Ted Filson	708-390-5000	D	250	3.0
Howard Industries Inc	PO Box 1588	Laurel	MS	39441	Billy W Howard	601-425-3151	R	170*	1.5
General Electric Co	1935 Redmond Cir	Rome	GA	30165	B Daugherty	706-291-3000	D	110	1.0
Sola Electric	1717 Busse Rd	Elk Grove Vill	IL	60007	Carlo Gorla	708-439-2800	D	101	1.0
Soca/Hevi-Duty Electric	PO Box 268	Goldsboro	NC	27530	Mike Cheshire	919-734-8900	D	100	0.9
Valmont Electric Inc	1770 Com Pk Dr	El Paso	TX	79912	Richard Galberaith	915-877-4380	S	100*	0.5
MagneTek Electric Inc	400 S Prairie Av	Waukesha	WI	53186	Nick Aversa	414-547-0121	S	90	0.5
Central Moloney Inc	PO Box 6608	Pine Bluff	AR	71611	EW Copeland	501-534-5332	R	75	0.7
North American Transformer	1200 Piper Dr	Milpitas	CA	95035	Harral Robin	408-262-7000	S	75	0.4
EPE Technologies Inc	1660 Scenic Av	Costa Mesa	CA	92626	Dave Petratis	714-557-1636	S	68*	0.4
Kentucky Association	PO Box 32170	Louisville	KY	40232	Ron Sheets	502-459-4011	R	51	0.2
VRN International	PO Box 44000	St Petersburg	FL	33743	Joe Lebas	813-347-2181	D	43	0.5
TII Industries Inc	1385 Akron St	Copiague	NY	11726	Alfred J Roach	516-789-5000	P	40	1.0
Magnetek Ohio	1776 Constitution	Louisville	OH	44641	Rich Cummings	216-875-3333	S	40*	0.2
Scott Fetzer Co	PO Box 300	Fairview	TN	37062	Phillip Jones	615-799-0551	D	40	0.7
LeeMAH Electronics Inc	1088 Sansome St	San Francisco	CA	94133	BH Mah	415-394-1288	R	39*	0.6
Deltec Corp	2727 Kurtz St	San Diego	CA	92110	Ray Meyer	619-291-4211	S	36	0.4
Intern Power Machines Corp	2975 Miller Park N	Garland	TX	75042	John J Mulcair	214-272-8000	P	36	0.2
Kuhlman Corp	101 Porter St	Crystal Springs	MS	39059	Paul Acheson	601-892-4661	D	32*	0.2
SNC Manufacturing Company	101 Waukau Av	Oshkosh	WI	54901	John L Vette III	414-231-7370	R	32	0.9
Pauwels Transformers Inc	PO Box 189	Washington	MO	63090	F Robberechts	314-239-6783	S	30	0.2
Robertson Transformer	13611 Thornton Rd	Blue Island	IL	60406	Margerie Pelino	708-388-2315	R	28*	0.6
Joslyn Electronic Systems Corp	PO Box 817	Goleta	CA	93116	Russell Gray	805-968-3551	S	26*	0.2
Del Electronics Corp	1 Commerce Park	Valhalla	NY	10595	L A Trugman	914-686-3600	P	24	0.3
Shepherd Electric Company Inc	7401 Pulaski Hwy	Baltimore	MD	21237	CC Vogel III	410-866-6000	R	24	<0.1
EG and G Power Systems	1330 E Cypress St	Covina	CA	91724	Richard Brownhill	818-967-9521	D	23*	0.2
Rapid Power Technologies Inc	18 Graysbridge Rd	Brookfield	CT	06804	Ronald J Viola	203-775-0411	R	23*	0.2
Knopp Inc	1307 66th St	Emeryville	CA	94608	Alex Finlay	510-653-1661	R	20	<0.1
Power Sentry Inc	6271 Bury Dr	Eden Prairie	MN	55344	Bob Lovett	612-949-1100	R	20	0.2
Shape Electronics Inc	901 N Dupage Av	Lombard	IL	60148	Richard Ryan	708-620-8394	S	20	0.2
Trine Products Co	1430 Ferris Pl	Bronx	NY	10461	FH Schildwachter	718-829-4796	S	20	0.1
NWL Transformers Inc	PO Box 358	Bordentown	NJ	08505	J David Seitz	609-298-7300	R	19	0.2
OB Systems and Mining Inc	PO Box 880	Barboursville	WV	25705	Oliver Fearing	304-736-8933	S	17*	0.2
Altronic Inc	712 Trumbull Av	Girard	OH	44420	Bruce R Beeghly	216-545-9768	R	16*	0.1
Computer Power Inc	124 W Main St	High Bridge	NJ	08829	Roger N Love	908-638-8000	P	16	0.2
Staco Energy Products Co	301 Gaddis Blv	Dayton	OH	45403	Leroy Carver	513-253-1191	S	15	0.2
Stocker and Yale Inc	PO Box 493	Beverly	MA	01915	Mark Blodgett	508-927-3940	R	14	<0.1
Associated Engineering Co	101 Kuhlman Blv	Versailles	KY	40383		606-879-2920	S	13	0.1
Deltona Transformer Corp	801 US Hwy 92 E	Deland	FL	32724	Michael G Prelec	904-736-7900	R	13	0.3
Foster Transformer Co	3820 Colerain Av	Cincinnati	OH	45223	Herman A Harrison	513-681-2420	S	13	0.2
Virginia Transformer Corp	220 Glade View NE	Roanoke	VA	24012	Prab Jain	703-345-9892	R	13*	0.2
Johnson Electric Coil Co	821 Watson St	Antigo	WI	54409	William H Bockes	715-627-4367	R	12	0.2
Neco Hammond Inc	1100 Lake St	Baraboo	WI	53913	Randy Ekern	608-356-3921	S	12*	0.2
T and R Electric Supply	PO Box 180	Colman	SD	57017	JR Thompson	605-534-3555	R	12	0.1
Ensign Corp	7960 S Madison St	Burr Ridge	IL	60521	John Ensign	708-325-9343	R	11*	0.1
Cramer Coil and Transformer	401 Progress Dr	Saukville	WI	53080	Terrence Wilkinson	414-268-2150	R	10	0.1
Grand Transformers Inc	PO Box 799	Grand Haven	MI	49417	Gerry V Retzlaff	616-842-5430	R	10	0.2
Hitran Corp	362 Hwy 31	Flemington	NJ	08822	JC Hindle Jr	908-782-5525	R	10	0.1
ITW Linx	201 Scott St	Elk Grove Vill	IL	60007	Ron Schmidt	708-952-8844	D	10*	<0.1
MTE Corp	PO Box 9013	Menomonee Fls	WI	53052	F Lewis	414-253-8200	R	10	0.1
Superior Technology Inc	PO Box 24590	Canton	OH	44701	S Mort Zimmerman	216-452-4681	S	10*	<0.1
Transformer Manufacturers Inc	7051 W Wilson Av	Chicago	IL	60656	A A Gianaras	708-457-1200	R	10	0.2
Van Tran Industries Inc	PO Box 20128	Waco	TX	76702	A Waterman	817-772-9740	R	9	<0.1
Electronic Ballast Technology	23600 Telo Av	Torrance	CA	90505	Peter Shen	310-784-2000	S	8*	<0.1
Electronic Components Corp	793 Winthrop Av	Addison	IL	60101	George W Banser	708-543-7720	R	8*	0.4
Micron Industries Corp	4250 Madison St	Hillside	IL	60162	Donald Clark	708-547-0900	R	8	0.1
Niagara Transformer Corp	PO Box 233	Buffalo	NY	14225	Fred Darby	716-896-6500	R	8*	0.1
Polyphase Corp	175 Commerce Dr	Ft Washington	PA	19034	Paul A Tanner	215-643-6950	P	7	0.1
American Monarch Corp	2801 37th Av NE	Minneapolis	MN	55421	Jess Barber	612-788-9161	R	7	<0.1
Cin-Tran Inc	8950 Rossash Rd	Cincinnati	OH	45236	Dennie Meeks	513-793-1100	R	7	0.2
Coil-Tran Corp	160 S Illinois St	Hobart	IN	46342	Gary Kriadis	219-942-8511	R	7*	<0.1
RE Uptegraff Mfg Co	PO Box 182	Scottdale	PA	15683	Susan U Endersbe	412-887-7700	R	7	0.1
Stangenes Industries Inc	1052 E Meadow Cir	Palo Alto	CA	94303	M Stangenes	415-493-0814	R	7	<0.1
WPI Power Systems Inc	PO Box 267	Warner	NH	03278	David J Edwards	603-456-3111	S	7	<0.1
Superior Magnetics Inc	3401 Texoma Dr	Denison	TX	75020	Gabriel Prieto	903-415-2850	S	6	0.1
Carson Manufacturing Co	PO Box 20464	Indianapolis	IN	46220	William H Carson	317-257-3191	R	6	<0.1
Electronic Transformer Corp	PO Box 487	Paterson	NJ	07544	M Gorman	201-942-2222	R	6	0.1
ISOREG	PO Box 486	Littleton	MA	01460	Thomas Castle	508-486-9483	D	6	<0.1
Memco Manufacturing Company	PO Box 380	Commack	NY	11725	Norman L Hasel	516-499-2442	R	6*	<0.1
Ward Transformer Co	PO Box 30009	Raleigh	NC	27622	RE Ward III	919-787-3553	R	6	<0.1

Source: Ward's Business Directory of U.S. Private and Public Companies, Volumes 1 and 2, 1996. The company type code used is as follows: P - Public, R - Private, S - Subsidiary, D - Division, J - Joint Venture, A - Affiliate, G - Group. Sales are in millions of dollars, employees are in thousands. An asterisk (*) indicates an estimated sales volume. The symbol < stands for 'less than'. Company names and addresses are truncated, in some cases, to fit into the available space.

MATERIALS CONSUMED

Material	Quantity	Delivered Cost ($ million)
Materials, ingredients, containers, and supplies	(X)	1,885.4
Metal bolts, nuts, screws, washers, rivets, and other screw machine products	(X)	18.5
Other fabricated metal products (except castings and forgings)	(X)	54.5
Castings (rough and semifinished)	(X)	2.9
Steel bars, bar shapes, and plates	(X)	17.1
Steel sheet and strip, including tin plate	(X)	298.9
Steel structural shapes and sheet piling	(X)	7.3
All other steel shapes and forms	(X)	74.1
Aluminum and aluminum-base alloy sheet, plate, foil, and welded tubing	(X)	28.8
All other aluminum and aluminum-base alloy shapes and forms	(X)	33.9
Copper and copper-base alloy bare wire for electrical conduction only	(X)	29.9
Copper and copper-base alloy rod, bar, and mechanical wire, including extruded and/or drawn shapes	(X)	53.6
Other nonferrous shapes and forms	(X)	0.8
Magnet wire	(X)	117.0
Insulated wire and cable, except magnet wire	(X)	40.9
Refined petroleum products (transformer oils, lubricating oils and greases, etc.)	(X)	59.6
Porcelain, steatite, and other ceramic electrical products	(X)	45.2
Paper and paperboard products except paperboard boxes, containers, and corrugated paperboard	(X)	53.6
Paints, varnishes, lacquers, stains, shellacs, japans, enamels, and allied products	(X)	16.8
Electrical industrial capacitors, resistors, rheostats, and coil windings	(X)	95.0
Current-carrying wiring devices	(X)	41.1
All other materials and components, parts, containers, and supplies	(X)	669.0
Materials, ingredients, containers, and supplies, nsk	(X)	126.9

Source: 1992 *Economic Census*. Explanation of symbols used: (D): Withheld to avoid disclosure of competitive data; na: Not available; (S): Withheld because statistical norms were not met; (X): Not applicable; (Z): Less than half the unit shown; nec: Not elsewhere classified; nsk: Not specified by kind; - : zero; * : 10-19 percent estimated; ** : 20-29 percent estimated.

PRODUCT SHARE DETAILS

Product or Product Class	% Share	Product or Product Class	% Share
Transformers, except electronic	100.00	tap-changing, 30,001 kVA, OA (50,001 kVA, top FOA) to 100,000 kVA, OA (167,000 kVA, top FOA)	3.09
Power and distribution transformers, except parts	53.45	Large liquid-immersed power transformers with load-tap-changing, 100,001 kVA, OA (167,001 kVA, top FOA) and larger	2.64
Distribution transformers, except parts, overhead type, single-phase, liquid-immersed; 500 kVA and smaller (excluding general purpose)	21.63	Large liquid-immersed power transformers without load-tap-changing, 100,001 kVA, OA (167,001 kVA, top FOA) and larger	7.12
Distribution transformers, except parts, compartmentalized pad-mounted, single-phase, liquid-immersed; 500 kVA and smaller (excluding general purpose)	10.04	Power and distribution transformers, except parts, nsk . .	0.56
Distribution transformers, except parts, subsurface and subway types, single-phase, liquid-immersed; 500 kVA and smaller (excluding general purpose)	0.72	Specialty transformers, except flourescent lamp ballast . . .	10.80
Distribution three-phase transformers, except parts, 500 kVA and smaller, liquid-immersed (all voltages) (excluding general purpose)	6.38	Open core and coil units, excluding machine tool control transformers and all units end-bell enclosed (250 VA and under).	6.78
		Machine tool control transformers	6.23
Distribution network transformers, except parts, all ratings, excluding network protectors (excluding general purpose)	1.85	Indoor and outdoor current instrument transformers . .	11.79
Distribution transformers, except parts, single-phase and three-phase, pad-mounted (dry); 500 kVA and smaller (excluding general purpose)	(D)	Indoor and outdoor voltage instrument transformers . . .	8.00
		All other specialty transformers, excluding internal combustion engine ignition	62.21
Small power transformers, single- and three-phase, all voltages, compartmentalized pad-mounted and subsurface underground and conventional subway type 501 kVA through 2500 kVA liquid-immersed	3.53	Specialty transformers, except fluorescent lamp ballast, nsk .	5.01
		Fluorescent lamp ballasts	19.73
Small conventional transformers and autotransformers, single-and three-phase, all voltages, primary unit substation and single circuit unit substations 501 kVA through 2500 kVA liquid-immersed	(D)	Commercial, institutional, and industrial general-purpose transformers, all voltages	6.77
Small power transformers, single- and three-phase, all voltages, liquid-immersed conventionals, primary unit and single circuit unit substations, 2501 kVA through 10,000 kVA	4.99	Commercial, institutional, and industrial general-purpose transformers, single-and three-phase, 3 kVA and below, all voltages	21.57
Dry-type small power transformers, conventional, primary unit substation, and core and coil units, single-and three-phase, all voltages	4.45	Commercial, institutional, and industrial general-purpose transformers, single-and three-phase, 3.01 kVA through 15 kVA, all voltages	9.46
Secondary unit substation power transformers, liquid-immersed, all kVA ratings	(D)	Commercial, institutional, and industrial general-purpose transformers, single-and three-phase, 15.01 kVA through 100 kVA, all voltages	15.92
Secondary unit substation power transformers, dry-type, all kVA ratings	2.59	Commercial, institutional, and industrial general-purpose transformers, single-and three-phase, 100.01 kVA and above, all voltages	9.02
Large power transformers with load-tap-changing, 10,001 kVA, OA to 30,000 kVA, OA (50,000 kVA, top FOA), liquid-immersed	8.81	Commercial, institutional, and industrial general-purpose transformers, voltage regulating transformers, except transmission and distribution voltage regulators . . .	37.97
Large power transformers without load-tap-changing, 10,001 kVA, OA to 30,000 kVA, OA (50,000 kVA, top FOA), liquid-immersed	3.93	Commercial, institutional, and industrial general-purpose transformers, all voltages, nsk	6.05
		Power regulators, boosters, and other transformers and parts for all transformers	6.65
Large liquid-immersed power transformers with load-tap-changing, 30,001 kVA, OA (50,001 kVA, top FOA) to 100,000 kVA, OA (167,000 kVA, top FOA)	3.05	Power regulators, boosters, and other transformers . . .	52.84
Large liquid-immersed power transformers without load-		Parts, including renewal and repair parts, subassemblies and accessories for all transformers	37.38
		Power regulators, boosters, and other transformers and parts for all transformers, nsk	9.78
		Transformers, except electronic, nsk	2.60

Source: 1992 *Economic Census*. The values shown are percent of total shipments in an industry. Values of indented subcategories are summed in the main heading. The symbol (D) appears when data are withheld to prevent disclosure of competitive information. The abbreviation nsk stands for 'not specified by kind' and nec for 'not elsewhere classified'.

INPUTS AND OUTPUTS FOR TRANSFORMERS

Economic Sector or Industry Providing Inputs	%	Sector	Economic Sector or Industry Buying Outputs	%	Sector
Blast furnaces & steel mills	15.7	Manufg.	Gross private fixed investment	54.6	Cap Inv
Wholesale trade	12.1	Trade	Lighting fixtures & equipment	9.3	Manufg.
Imports	10.5	Foreign	Exports	6.6	Foreign
Nonferrous wire drawing & insulating	8.2	Manufg.	Federal Government purchases, national defense	5.0	Fed Govt
Lubricating oils & greases	4.1	Manufg.	Household cooking equipment	3.5	Manufg.
Machinery, except electrical, nec	3.4	Manufg.	Electronic computing equipment	1.7	Manufg.
Advertising	3.4	Services	General industrial machinery, nec	1.4	Manufg.
Industrial controls	3.3	Manufg.	Signs & advertising displays	1.3	Manufg.
Aluminum rolling & drawing	3.1	Manufg.	Machinery, except electrical, nec	1.2	Manufg.
Electric services (utilities)	2.3	Util.	Refrigeration & heating equipment	1.1	Manufg.
Banking	2.2	Fin/R.E.	Food products machinery	1.0	Manufg.
Fabricated plate work (boiler shops)	1.8	Manufg.	Printing trades machinery	0.9	Manufg.
Air transportation	1.8	Util.	Mechanical measuring devices	0.7	Manufg.
Copper rolling & drawing	1.7	Manufg.	Pumps & compressors	0.7	Manufg.
Paper mills, except building paper	1.7	Manufg.	Industrial furnaces & ovens	0.6	Manufg.
Porcelain electrical supplies	1.7	Manufg.	Radio & TV communication equipment	0.6	Manufg.
Metal stampings, nec	1.4	Manufg.	Special industry machinery, nec	0.6	Manufg.
Communications, except radio & TV	1.3	Util.	Crude petroleum & natural gas	0.5	Mining
Miscellaneous plastics products	1.1	Manufg.	Industrial trucks & tractors	0.5	Manufg.
Wiring devices	1.1	Manufg.	Switchgear & switchboard apparatus	0.5	Manufg.
Motor freight transportation & warehousing	1.1	Util.	S/L Govt. purch., gas & electric utilities	0.5	S/L Govt
Hotels & lodging places	1.1	Services	Electrical industrial apparatus, nec	0.4	Manufg.

Continued on next page.

INPUTS AND OUTPUTS FOR TRANSFORMERS - Continued

Economic Sector or Industry Providing Inputs	%	Sector	Economic Sector or Industry Buying Outputs	%	Sector
Maintenance of nonfarm buildings nec	0.9	Constr.	Machine tools, metal cutting types	0.4	Manufg.
Electrical industrial apparatus, nec	0.9	Manufg.	Motors & generators	0.4	Manufg.
Paper coating & glazing	0.9	Manufg.	Welding apparatus, electric	0.4	Manufg.
Petroleum refining	0.9	Manufg.	Automatic merchandising machines	0.3	Manufg.
Eating & drinking places	0.9	Trade	Federal Government purchases, nondefense	0.3	Fed Govt
Paints & allied products	0.7	Manufg.	Cold finishing of steel shapes	0.2	Manufg.
Paperboard containers & boxes	0.7	Manufg.	Conveyors & conveying equipment	0.2	Manufg.
Cyclic crudes and organics	0.6	Manufg.	Electric housewares & fans	0.2	Manufg.
Sawmills & planning mills, general	0.6	Manufg.	Electronic components nec	0.2	Manufg.
Gas production & distribution (utilities)	0.6	Util.	Environmental controls	0.2	Manufg.
Computer & data processing services	0.6	Services	Industrial controls	0.2	Manufg.
Equipment rental & leasing services	0.6	Services	Service industry machines, nec	0.2	Manufg.
Screw machine and related products	0.5	Manufg.	Ship building & repairing	0.2	Manufg.
Railroads & related services	0.5	Util.	Transformers	0.2	Manufg.
Legal services	0.5	Services	Turbines & turbine generator sets	0.2	Manufg.
Transformers	0.4	Manufg.	Typewriters & office machines, nec	0.2	Manufg.
Management & consulting services & labs	0.4	Services	Wiring devices	0.2	Manufg.
Special dies & tools & machine tool accessories	0.3	Manufg.	Electric services (utilities)	0.2	Util.
U.S. Postal Service	0.3	Gov't	Personal consumption expenditures	0.1	
Abrasive products	0.2	Manufg.	Commercial laundry equipment	0.1	Manufg.
Gaskets, packing & sealing devices	0.2	Manufg.	Elevators & moving stairways	0.1	Manufg.
Manifold business forms	0.2	Manufg.	Instruments to measure electricity	0.1	Manufg.
Credit agencies other than banks	0.2	Fin/R.E.	Machine tools, metal forming types	0.1	Manufg.
Insurance carriers	0.2	Fin/R.E.	Metalworking machinery, nec	0.1	Manufg.
Real estate	0.2	Fin/R.E.	Paper industries machinery	0.1	Manufg.
Accounting, auditing & bookkeeping	0.2	Services	Photographic equipment & supplies	0.1	Manufg.
Coal	0.1	Mining	Scales & balances	0.1	Manufg.
Machine tools, metal forming types	0.1	Manufg.	Special dies & tools & machine tool accessories	0.1	Manufg.
Royalties	0.1	Fin/R.E.	Textile machinery	0.1	Manufg.
Automotive rental & leasing, without drivers	0.1	Services			
Automotive repair shops & services	0.1	Services			
Electrical repair shops	0.1	Services			
Photofinishing labs, commercial photography	0.1	Services			
Noncomparable imports	0.1	Foreign			

Source: Benchmark Input-Output Accounts for the U.S. Economy, 1982, U.S. Department of Commerce, Washington, D.C., July 1991. Data, as reported in the source, are organized by the 1977 SIC structure in use in 1982 but have been matched, as closely as is possible, to the 1987 SIC structure used in this book.

OCCUPATIONS EMPLOYED BY SIC 361 - ELECTRIC DISTRIBUTING EQUIPMENT

Occupation	% of Total 1994	Change to 2005	Occupation	% of Total 1994	Change to 2005
Electrical & electronic assemblers	11.7	-19.4	Mechanical engineers	1.5	8.3
Electrical & electronic equipment assemblers	11.0	-10.5	Secretaries, ex legal & medical	1.4	-18.6
Assemblers, fabricators, & hand workers nec	8.3	-10.5	Sheet metal workers & duct installers	1.4	-10.4
Coil winders, tapers, & finishers	5.0	-28.4	Industrial production managers	1.2	-10.5
Electrical & electronics engineers	3.6	-4.7	Machine tool cutting & forming etc. nec	1.1	-28.4
Blue collar worker supervisors	3.4	-24.8	Engineering, mathematical, & science managers	1.1	1.6
Inspectors, testers, & graders, precision	3.4	-37.3	Stock clerks	1.1	-27.2
Electromechanical equipment assemblers	2.5	-1.6	Traffic, shipping, & receiving clerks	1.1	-13.9
Sales & related workers nec	2.1	-10.5	Freight, stock, & material movers, hand	1.1	-28.4
Drafters	2.0	-30.3	Industrial engineers, ex safety engineers	1.0	-1.5
Electrical & electronic technicians,technologists	1.8	-10.5	Coating, painting, & spraying machine workers	1.0	-10.4
General managers & top executives	1.7	-15.1	Welding machine setters, operators	1.0	-19.5

Source: Industry-Occupation Matrix, Bureau of Labor Statistics. These data relate to one or more 3-digit SIC industry groups rather than to a single 4-digit SIC. The change reported for each occupation to the year 2005 is a percent of growth or decline as estimated by the Bureau of Labor Statistics. The abbreviation nec stands for 'not elsewhere classified'.

LOCATION BY STATE AND REGIONAL CONCENTRATION

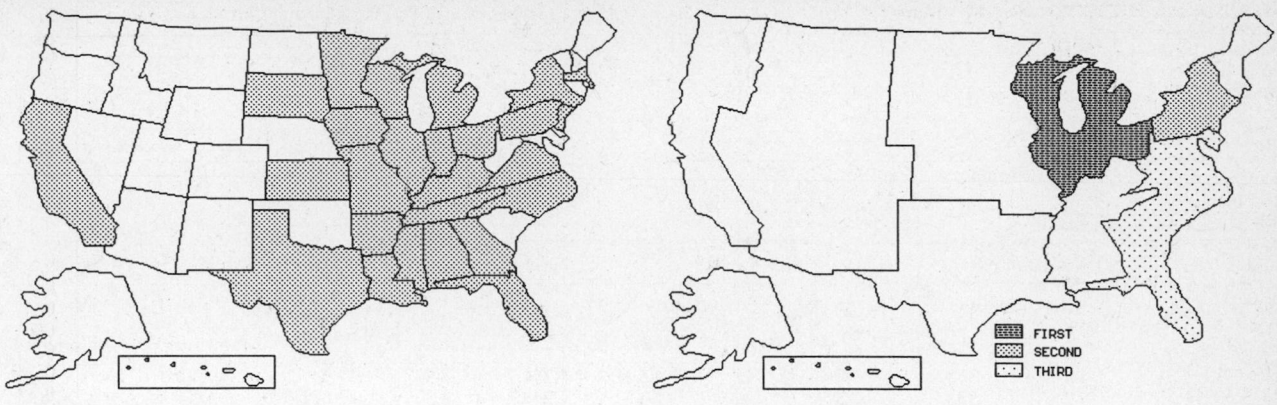

FIRST
SECOND
THIRD

INDUSTRY DATA BY STATE

State	Establish-ments	Shipments			Employment				Cost as % of Shipments	Investment per Employee ($)
		Total ($ mil)	% of U.S.	Per Establ.	Total Number	% of U.S.	Per Establ.	Wages ($/hour)		
Illinois	24	275.4	6.7	11.5	1,800	6.2	75	10.46	57.9	1,000
North Carolina	11	262.4	6.4	23.9	2,100	7.3	191	14.43	49.1	5,048
California	28	160.5	3.9	5.7	1,000	3.5	36	13.67	44.4	1,600
Virginia	10	157.8	3.9	15.8	1,200	4.2	120	14.00	48.1	-
Pennsylvania	13	136.2	3.3	10.5	1,300	4.5	100	9.55	41.3	462
New Jersey	19	115.6	2.8	6.1	600	2.1	32	11.33	49.7	3,500
Tennessee	8	108.9	2.7	13.6	800	2.8	100	8.83	49.0	1,875
New York	13	99.6	2.4	7.7	800	2.8	62	10.82	57.8	1,625
Florida	13	68.3	1.7	5.3	800	2.8	62	7.62	29.4	1,000
Texas	20	62.3	1.5	3.1	500	1.7	25	10.86	57.0	400
Ohio	9	50.1	1.2	5.6	500	1.7	56	9.22	41.5	-
Michigan	6	28.6	0.7	4.8	200	0.7	33	6.50	60.5	5,500
Minnesota	8	9.4	0.2	1.2	100	0.3	13	7.50	44.7	-
Massachusetts	6	8.4	0.2	1.4	100	0.3	17	7.00	52.4	1,000
Wisconsin	13	(D)	-	-	3,750 *	13.0	288	-	-	2,853
Indiana	9	(D)	-	-	1,750 *	6.1	194	-	-	-
Mississippi	7	(D)	-	-	3,750 *	13.0	536	-	-	-
Georgia	6	(D)	-	-	1,750 *	6.1	292	-	-	1,657
Missouri	6	(D)	-	-	1,750 *	6.1	292	-	-	-
Arkansas	5	(D)	-	-	1,750 *	6.1	350	-	-	-
Kansas	5	(D)	-	-	175 *	0.6	35	-	-	-
Alabama	4	(D)	-	-	375 *	1.3	94	-	-	-
Kentucky	4	(D)	-	-	750 *	2.6	188	-	-	3,467
Iowa	2	(D)	-	-	175 *	0.6	88	-	-	-
Louisiana	1	(D)	-	-	750 *	2.6	750	-	-	-
South Dakota	1	(D)	-	-	175 *	0.6	175	-	-	-

Source: 1992 *Economic Census*. The states are in descending order of shipments or establishments (if shipment data are missing for the majority). The symbol (D) appears when data are withheld to prevent disclosure of competitive information. States marked with (D) are sorted by number of establishments. A dash (-) indicates that the data element cannot be calculated; * indicates the midpoint of a range.

3613 - SWITCHGEAR & SWITCHBOARD APPARATUS

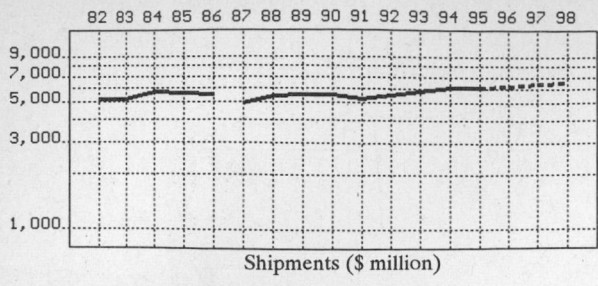

Shipments ($ million)

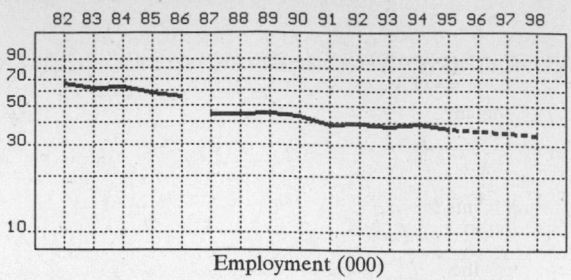

Employment (000)

GENERAL STATISTICS

Year	Com-panies	Establishments		Employment			Compensation		Production ($ million)			
		Total	with 20 or more employees	Total (000)	Production Workers (000)	Hours (Mil)	Payroll ($ mil)	Wages ($/hr)	Cost of Materials	Value Added by Manufacture	Value of Shipments	Capital Invest.
1982	512	646	334	66.0	44.5	83.4	1,256.7	8.59	1,947.2	3,133.0	5,172.8	142.6
1983		625	333	62.3	42.3	80.3	1,248.2	8.80	2,001.5	3,151.2	5,158.7	137.0
1984		604	332	62.8	42.3	81.9	1,321.6	9.30	2,184.6	3,559.2	5,696.2	151.7
1985		582	330	58.1	39.6	76.4	1,273.5	9.74	2,254.7	3,388.1	5,676.3	155.0
1986		574	326	55.6	37.6	72.6	1,272.5	10.01	2,195.4	3,292.9	5,504.7	178.7
1987*	370	474	256	44.4	29.4	57.1	1,057.8	10.42	1,969.0	2,888.7	4,907.3	120.5
1988		457	248	44.3	29.3	57.4	1,114.9	11.05	2,268.5	3,232.1	5,427.3	119.2
1989		453	248	45.2	29.7	57.6	1,162.4	11.32	2,147.0	3,315.9	5,466.8	118.9
1990		463	252	44.0	28.2	55.2	1,128.8	11.11	2,224.7	3,314.4	5,550.8	110.9
1991		480	245	39.4	26.2	51.3	1,100.4	11.52	2,137.7	3,084.2	5,280.6	114.5
1992	441	533	241	39.5	26.3	52.8	1,167.0	12.15	2,240.6	3,277.7	5,527.7	118.4
1993		522	237	38.5	25.7	50.0	1,185.9	12.89	2,396.1	3,443.6	5,849.0	120.2
1994		529P	236P	39.5	27.2	54.5	1,250.9	13.25	2,552.4	3,614.9	6,140.2	145.7
1995		541P	234P	37.3P	25.3P	50.6P	1,236.7P	13.42P	2,531.2P	3,584.9P	6,089.3P	130.9P
1996		552P	231P	36.3P	24.8P	49.7P	1,256.8P	13.80P	2,584.0P	3,659.6P	6,216.1P	133.1P
1997		564P	228P	35.3P	24.2P	48.8P	1,276.9P	14.18P	2,636.7P	3,734.2P	6,342.9P	135.2P
1998		575P	226P	34.3P	23.7P	47.9P	1,297.1P	14.56P	2,689.4P	3,808.9P	6,469.7P	137.4P

Sources: 1982, 1987, 1992 *Economic Census; Annual Survey of Manufactures,* 83-86, 88-91, 93-94. Establishment counts are from *County Business Patterns* for non-Census years; establishment counts for 83-84 are extrapolations. * indicates that industry content changed in 87; earlier years use 77 SICs. 'P's mark projections.

INDICES OF CHANGE

Year	Com-panies	Establishments		Employment			Compensation		Production ($ million)			
		Total	with 20 or more employees	Total (000)	Production Workers (000)	Hours (Mil)	Payroll ($ mil)	Wages ($/hr)	Cost of Materials	Value Added by Manufacture	Value of Shipments	Capital Invest.
1982	116	121	139	167	169	158	108	71	87	96	94	120
1983		117	138	158	161	152	107	72	89	96	93	116
1984		113	138	159	161	155	113	77	98	109	103	128
1985		109	137	147	151	145	109	80	101	103	103	131
1986		108	135	141	143	138	109	82	98	100	100	151
1987*	84	89	106	112	112	108	91	86	88	88	89	102
1988		86	103	112	111	109	96	91	101	99	98	101
1989		85	103	114	113	109	100	93	96	101	99	100
1990		87	105	111	107	105	97	91	99	101	100	94
1991		90	102	100	100	97	94	95	95	94	96	97
1992	100	100	100	100	100	100	100	100	100	100	100	100
1993		98	98	97	98	95	102	106	107	105	106	102
1994		99P	98P	100	103	103	107	109	114	110	111	123
1995		101P	97P	94P	96P	96P	106P	110P	113P	109P	110P	111P
1996		104P	96P	92P	94P	94P	108P	114P	115P	112P	112P	112P
1997		106P	95P	89P	92P	92P	109P	117P	118P	114P	115P	114P
1998		108P	94P	87P	90P	91P	111P	120P	120P	116P	117P	116P

Sources: Same as General Statistics. Values reflect change from the base year, 1992. Values above 100 mean greater than 92, values below 100 mean less than 92, and a value of 100 in the 82-91 or 93-98 period means same as 92. * indicates that industry content changed in 87. Data for earlier years are in 77 SIC format.

SELECTED RATIOS

For 1994	Avg. of All Manufact.	Analyzed Industry	Index	For 1994	Avg. of All Manufact.	Analyzed Industry	Index
Employees per Establishment	49	75	152	Value Added per Production Worker	134,084	132,901	99
Payroll per Establishment	1,500,273	2,363,374	158	Cost per Establishment	5,045,178	4,822,348	96
Payroll per Employee	30,620	31,668	103	Cost per Employee	102,970	64,618	63
Production Workers per Establishment	34	51	150	Cost per Production Worker	146,988	93,838	64
Wages per Establishment	853,319	1,364,339	160	Shipments per Establishment	9,576,895	11,600,918	121
Wages per Production Worker	24,861	26,549	107	Shipments per Employee	195,460	155,448	80
Hours per Production Worker	2,056	2,004	97	Shipments per Production Worker	279,017	225,743	81
Wages per Hour	12.09	13.25	110	Investment per Establishment	321,011	275,277	86
Value Added per Establishment	4,602,255	6,829,771	148	Investment per Employee	6,552	3,689	56
Value Added per Employee	93,930	91,516	97	Investment per Production Worker	9,352	5,357	57

Sources: Same as General Statistics. The 'Average of All Manufacturing' column represents the average of all manufacturing industries reported for the most recent complete year available. The Index shows the relationship between the Average and the Analyzed Industry. For example, 100 means that they are equal; 500 that the Analyzed Industry is five times the average; 50 means that the Analyzed Industry is half the national average. The abbreviation 'na' is used to show that data are 'not available'.

LEADING COMPANIES Number shown: **75** Total sales ($ mil): **3,272** Total employment (000): **30.7**

Company Name	Address				CEO Name	Phone	Co. Type	Sales ($ mil)	Empl. (000)
Micro Switch	11 W Spring St	Freeport	IL	61032	Ray A Alvarez	815-235-6600	D	440	4.5
Bussmann	PO Box 14460	St Louis	MO	63178	John Monter	314-394-2877	D	370•	3.4
C and K Components Inc	57 Stanley Av	Watertown	MA	02172	James E Walsh	617-926-6400	R	220	2.0
S and C Electric Co	6601 N Ridge Blv	Chicago	IL	60626	John R Conrad	312-338-1000	R	200	1.5
Littelfuse Inc	800 E Northwest	Des Plaines	IL	60016	Howard B Witt	708-824-1188	P	194	2.3
Powell Industries Inc	PO Box 12818	Houston	TX	77217	Thomas W Powell	713-944-6900	P	152	0.9
Liebert Corp	1050 Dearborn Dr	Columbus	OH	43085	Walden O'Dell	614-888-0246	S	150•	1.4
Sentrol Inc	12345 SW Leveton	Tualatin	OR	97062	John W Hakanson	503-620-8540	R	130•	1.2
Lutron Electronics Company Inc	7200 Suter Rd	Coopersburg	PA	18036	Joel S Spira	215-282-3800	R	83•	0.8
Powell Electrical Mfg Co	PO Box 12818	Houston	TX	77217	M Gus Zeller	713-944-6900	S	83	0.5
ABB Power T and D Company	201 Hickman Dr	Sanford	FL	32711	Norbert Hagenhoff	407-323-8220	D	70	0.4
Hughes Corp	16900 Foltz Indrial	Cleveland	OH	44136	David Hughes Sr	216-238-2550	R	68	0.2
Airpax Protector Products	PO Box 520	Cambridge	MD	21613	Dennis Karr	410-228-4600	R	52	1.0
Heinmann Products	PO Box 13	Salisbury	MD	21803	Steven I Burack	410-546-9778	D	50•	0.5
Telenex Corp	13000 Midlantic Dr	Mount Laurel	NJ	08054	Robert Coackley	609-234-7900	S	47•	0.4
Bel Fuse Inc	198 Van Vorst St	Jersey City	NJ	07302	Elliot Bernstein	201-432-0463	P	46	0.7
Electric and Gas Technology Inc	13636 Neutron Rd	Dallas	TX	75244	S Mort Zimmerman	214-934-8797	P	43	0.6
ITE Electrical Products	PO Box 1612	Spartanburg	SC	29304	Roger Kroes	803-576-6510	D	42•	0.3
Russelectric Inc	S Shore Park	Hingham	MA	02043	RG Russell	617-749-6000	R	40	0.3
Powercon Corp	PO Box 477	Severn	MD	21144	Ralph Siegel	410-551-6500	R	35•	0.4
SPD Technologies	13500 Roosevelt	Philadelphia	PA	19116	LA Colangelo	215-677-4900	R	35•	0.4
Appliance Control Technology	1431 Jeffrey Dr	Addison	IL	60101	Bill Stafford	708-916-0900	R	30	0.3
Electroswitch Corp	775-1 Pleasant St	Weymouth	MA	02189	Frank Meissner	617-335-1195	R	30	0.3
Oneac Corp	27944 N Bradley Rd	Libertyville	IL	60048	Charles W Pearson	708-816-6000	R	30	0.2
On-Line Power Corp	5701 Smithway St	Commerce	CA	90040	Abbie Gougerchian	213-721-5017	R	30	0.2
G and W Electric Co	3500 W 127th St	Blue Island	IL	60406	John D Mueller	708-388-5010	R	25	0.3
Mechanical Products Inc	PO Box 729	Jackson	MI	49204	LR Matzen	517-782-0391	S	25	0.5
Southern States Inc	30 Georgia Av	Hampton	GA	30228	T W McGarity	404-946-4562	R	25	0.2
USD Products	7300 W Wilson Av	Chicago	IL	60656	H Ray Ege	708-867-4600	D	25	0.2
Strand Lighting Inc	18111 S Santa Fe	R Dominguez	UT	84115	Jean Griffith	801-487-9861	R	25	0.2
Anderson Power Products	PO Box 579	Sterling	MA	01564	Ronald Robinson	617-787-5880	D	23	0.2
American Circuit Breaker Corp	PO Box 1308	Albemarle	NC	28002	Nelson Park	704-463-7361	R	21•	0.2
Mitsubishi Electric Power Prod	512 Keystone Dr	Warrendale	PA	15086	Roger L Barna	412-772-2555	S	20	<0.1
International Switchboard Corp	PO Box 2001	Sugar Land	TX	77487	LL Voyles	713-565-7022	S	20	0.2
Thore Inc	PO Box 2001	Sugar Land	TX	77487	LL Voyles	713-565-5464	R	20	0.2
Circle AW Products Co	PO Box 1171	Modesto	CA	95353	Tom Ruddy	209-524-5281	R	19	0.2
Hartman Electrical Mfg	175 N Diamond St	Mansfield	OH	44902	James R Mikesell	419-524-1411	D	16•	0.2
MicroENERGY Inc	350 Randy Rd	Carol Stream	IL	60188	Robert G Gatza	708-653-5900	P	16	0.2
Central Electric Co	PO Box 310	Fulton	MO	65251	CW James	314-642-6811	R	15•	0.1
Durham Co	PO Box 908	Lebanon	MO	65536	Doug Russell	417-532-7121	R	15	0.1
Joslyn Hi Voltage Corp	4000 E 116th St	Cleveland	OH	44105	George W Diehl	216-271-6600	S	15	0.2
M and I Electric Industries Inc	PO Box 1792	Beaumont	TX	77704	A Dauber	409-838-0441	R	15	0.2
Point Eight Power Inc	1510 Engineers Rd	Belle Chasse	LA	70037	Carl J Boudreaux	504-394-6100	R	15	<0.1
Switching Power Inc	3601 Veterans Hwy	Ronkonkoma	NY	11779	Mel Cravitz	516-981-5353	R	13	0.1
Control Concepts Corp	328 Water St	Binghamton	NY	13902	Pat V Gillette	607-724-2484	S	11	<0.1
Linemaster Switch Corp	29 Plaine Hill Rd	Woodstock	CT	06281	NB Blakely	203-974-1000	R	11	0.2
Atkinson Industries Inc	PO Box 268	Pittsburg	KS	66762	LC Martin	316-231-6900	S	10	<0.1
Connecticut Electric	PO Box 50130	Mobile	AL	36605	Charles J Beamish	205-438-3701	D	10	<0.1
Electro-Mechanical Products Inc	1200ephenson Hwy	Troy	MI	48083	Timothy J Marker	810-589-1510	R	10	0.5
General Switch Co	PO Box 1308	Albemarle	NC	28001	Nathan Mazurek	704-463-7361	D	10	0.2
Hendry Telephone Products	PO Box 998	Goleta	CA	93117	JJ Keenan	805-968-5511	R	10	0.2
Kilovac Corp	PO Box 4422	Santa Barbara	CA	93140	Doug Campbell	805-684-4560	R	10	0.1
Usco Power Equipment Corp	PO Box 10023	Birmingham	AL	35202	BT Lankford	205-592-7241	R	10	<0.1
Edison Fuse Gear Inc	11939 Manchester	Des Peres	MO	63131	Scott Saunders	314-527-1394	S	9•	0.1
Golden Gate Switchboard Co	PO Box 389	Napa	CA	94558	Christopher Connors	707-255-4261	R	9•	<0.1
Clary Corp	1960 S Walker Av	Monrovia	CA	91016	Donald G Ash	818-287-6111	P	9	<0.1
Panlmatic Co	79 Bond St	Elk Grove Vill	IL	60007	Gerald Luc	708-439-4030	R	9	<0.1
Anello Corp	2641 Walnut Av	Tustin	CA	92680	Peter J Anello	714-669-9940	R	8	<0.1
Electrical Design and Control	2545 Industrial Row	Troy	MI	48084	Eric Monson	313-280-0630	R	8•	<0.1
Powell-Esco Co	PO Box 1039	Greenville	TX	75403	Glenn Aver	903-455-6234	S	8•	<0.1
Capco Inc	1328 Winters Av	Grand Junction	CO	81501	Steve Wood	303-243-8750	R	7	0.1
Stacoswitch Inc	1139 Baker St	Costa Mesa	CA	92626	JF Gust	714-549-3041	S	7	<0.1
Digital Power Corp	41920 Christy St	Fremont	CA	94538	Bob Smith	510-657-2635	R	6	<0.1
Control Assemblies Co	15400 Medina Rd	Minneapolis	MN	55447	Garber Trambley	612-544-3364	R	6	<0.1
Frank Electric Corp	PO Box 69	York	PA	17405	Fred Le Page	717-764-5959	R	6	<0.1
Morpac Industries Inc	117 Frontage Rd N	Pacific	WA	98047	Lloyd Morgan	206-735-8922	R	6	<0.1
Silver Cloud Manufacturing Co	525 Orange St	Millville	NJ	08332	Harry E Cloud	609-825-8900	R	6•	<0.1
Electric Switchboard Company	185 3rd Av	Brooklyn	NY	11217	PC Walsh	718-643-1105	R	6	<0.1
Metropolitan Electric Mfg Co	200 Dixon Av	Amityville	NY	11701	JJ Shelley	516-842-4555	R	6	<0.1
Federal Pacific	PO Box 8200	Bristol	VA	24203	Frank L Leonard	703-669-4084	D	5	0.2
General Equipment & Mfg	PO Box 37290	Louisville	KY	40213	C Rob Marcum	502-969-8000	R	5	<0.1
Kearney K-P-F Electric	PO Box 8485	Stockton	CA	95208	John Rose	209-464-8381	S	5	<0.1
Protection Controls Inc	PO Box 287	Skokie	IL	60076	J Yates	708-674-7676	R	5•	<0.1
States Electric Mfg Co	650 Ottawa Av N	Minneapolis	MN	55422	GH Shallbetter	612-588-0536	R	5•	<0.1
Turner Electric Corp	PO Box 3158	Fairview Hts	IL	62208	A Jack Hoppenjans	618-397-1865	S	5	<0.1

Source: Ward's Business Directory of U.S. Private and Public Companies, Volumes 1 and 2, 1996. The company type code used is as follows: P - Public, R - Private, S - Subsidiary, D - Division, J - Joint Venture, A - Affiliate, G - Group. Sales are in millions of dollars, employees are in thousands. An asterisk (*) indicates an estimated sales volume. The symbol < stands for 'less than'. Company names and addresses are truncated, in some cases, to fit into the available space.

MATERIALS CONSUMED

Material	Quantity	Delivered Cost ($ million)
Materials, ingredients, containers, and supplies	(X)	1,955.3
Sheet metal products, except stampings	(X)	41.5
Metal bolts, nuts, screws, washers, rivets, and other screw machine products	(X)	70.6
All other fabricated metal products (except forgings)	(X)	89.1
Forgings	(X)	5.3
Castings (rough and semifinished)	(X)	38.2
Steel bars, bar shapes, and plates	(X)	22.0
Steel sheet and strip, including tin plate	(X)	132.0
Steel structural shapes and sheet piling	(X)	5.5
All other steel shapes and forms	(X)	13.0
Nonferrous metal smelter and refinery shapes	(X)	12.0
Aluminum and aluminum-base alloy shapes and forms	(X)	35.9
Copper and copper-base alloy shapes and forms	(X)	124.0
All other nonferrous shapes and forms	(X)	13.8
Nonferrous wire and cable, including magnet wire, bare or insulated wire, etc.	(X)	50.6
Industrial electrical control equipment purchased from other companies	(X)	209.8
Industrial electrical control equipment received from other plants of the same company	(X)	165.2
Porcelain, steatite, and other ceramic electrical products	(X)	57.3
Plastics products consumed in the form of sheets, rods, tubes, film, and other shapes	(X)	48.6
Plastics resins consumed in the form of granules, pellets, powders, liquids, etc.	(X)	36.7
Switches, except snap, toggle and push, and circuit breakers	(X)	92.4
Resistors, capacitors, transformers, and other electronic-type components	(X)	99.6
All other materials and components, parts, containers, and supplies	(X)	348.5
Materials, ingredients, containers, and supplies, nsk	(X)	243.5

Source: 1992 Economic Census. Explanation of symbols used: (D): Withheld to avoid disclosure of competitive data; na: Not available; (S): Withheld because statistical norms were not met; (X): Not applicable; (Z): Less than half the unit shown; nec: Not elsewhere classified; nsk: Not specified by kind; - : zero; * : 10-19 percent estimated; ** : 20-29 percent estimated.

PRODUCT SHARE DETAILS

Product or Product Class	% Share	Product or Product Class	% Share
Switchgear and switchboard apparatus	100.00	Molded case circuit breakers, 1000 volts or less	17.02
Power circuit breakers, all voltages	8.27	Duct, including plug-in units and accessories, 1000 volts or less	3.44
Low voltage panelboards and distribution boards and other switching and interrupting devices, 1000 volts or less	29.43	Switchgear, except ducts and relays	29.95
Fuses and fuse equipment, less than 2300 volts, except power distribution cut-outs	7.98	Switchgear and switchboard apparatus, nsk	3.92

Source: 1992 Economic Census. The values shown are percent of total shipments in an industry. Values of indented subcategories are summed in the main heading. The symbol (D) appears when data are withheld to prevent disclosure of competitive information. The abbreviation nsk stands for 'not specified by kind' and nec for 'not elsewhere classified'.

INPUTS AND OUTPUTS FOR SWITCHGEAR & SWITCHBOARD APPARATUS

Economic Sector or Industry Providing Inputs	%	Sector	Economic Sector or Industry Buying Outputs	%	Sector
Imports	13.9	Foreign	Gross private fixed investment	21.4	Cap Inv
Industrial controls	13.3	Manufg.	Exports	6.7	Foreign
Wholesale trade	10.0	Trade	Office buildings	5.6	Constr.
Primary copper	8.6	Manufg.	Maintenance of nonfarm buildings nec	5.1	Constr.
Advertising	6.2	Services	Industrial buildings	4.6	Constr.
Blast furnaces & steel mills	4.6	Manufg.	Residential 1-unit structures, nonfarm	4.2	Constr.
Electronic components nec	4.0	Manufg.	Refrigeration & heating equipment	3.7	Manufg.
Copper rolling & drawing	3.3	Manufg.	Electronic computing equipment	3.6	Manufg.
Banking	2.2	Fin/R.E.	Nonfarm residential structure maintenance	3.5	Constr.
Plating & polishing	2.0	Manufg.	Communications, except radio & TV	3.3	Util.
Switchgear & switchboard apparatus	2.0	Manufg.	Residential additions/alterations, nonfarm	3.0	Constr.
Air transportation	1.7	Util.	Printing trades machinery	2.1	Manufg.
Maintenance of nonfarm buildings nec	1.6	Constr.	Construction of educational buildings	1.7	Constr.
Electric services (utilities)	1.6	Util.	Special industry machinery, nec	1.6	Manufg.
Screw machine and related products	1.4	Manufg.	Residential garden apartments	1.5	Constr.
Aluminum rolling & drawing	1.3	Manufg.	Construction of hospitals	1.4	Constr.
Porcelain electrical supplies	1.2	Manufg.	Electric utility facility construction	1.4	Constr.
Communications, except radio & TV	1.1	Util.	Sewer system facility construction	1.4	Constr.
Miscellaneous plastics products	1.0	Manufg.	Switchgear & switchboard apparatus	1.2	Manufg.
Plastics materials & resins	1.0	Manufg.	Typewriters & office machines, nec	1.2	Manufg.
Hotels & lodging places	1.0	Services	Hotels & motels	1.1	Constr.
Nonferrous wire drawing & insulating	0.9	Manufg.	General industrial machinery, nec	1.1	Manufg.
Petroleum refining	0.7	Manufg.	Railroad equipment	1.1	Manufg.
Sheet metal work	0.6	Manufg.	Maintenance of electric utility facilities	1.0	Constr.
Wiring devices	0.6	Manufg.	Elevators & moving stairways	0.9	Manufg.
Eating & drinking places	0.6	Trade	Industrial furnaces & ovens	0.9	Manufg.
Engineering, architectural, & surveying services	0.6	Services	Pumps & compressors	0.9	Manufg.
Equipment rental & leasing services	0.6	Services	Personal consumption expenditures	0.8	

Continued on next page.

INPUTS AND OUTPUTS FOR SWITCHGEAR & SWITCHBOARD APPARATUS - Continued

Economic Sector or Industry Providing Inputs	%	Sector	Economic Sector or Industry Buying Outputs	%	Sector
Aluminum castings	0.5	Manufg.	Construction of stores & restaurants	0.8	Constr.
Gaskets, packing & sealing devices	0.5	Manufg.	Warehouses	0.7	Constr.
Metal stampings, nec	0.5	Manufg.	Industrial controls	0.7	Manufg.
Paperboard containers & boxes	0.5	Manufg.	Machine tools, metal cutting types	0.7	Manufg.
Motor freight transportation & warehousing	0.5	Util.	Service industry machines, nec	0.6	Manufg.
Real estate	0.5	Fin/R.E.	Maintenance of farm service facilities	0.5	Constr.
Machinery, except electrical, nec	0.4	Manufg.	Residential high-rise apartments	0.5	Constr.
Metal coating & allied services	0.4	Manufg.	Conveyors & conveying equipment	0.5	Manufg.
Paints & allied products	0.4	Manufg.	Industrial trucks & tractors	0.5	Manufg.
Primary aluminum	0.4	Manufg.	Metalworking machinery, nec	0.5	Manufg.
Transformers	0.4	Manufg.	Construction of nonfarm buildings nec	0.4	Constr.
Gas production & distribution (utilities)	0.4	Util.	Household laundry equipment	0.4	Manufg.
Computer & data processing services	0.4	Services	Motors & generators	0.4	Manufg.
Brass, bronze, & copper castings	0.3	Manufg.	Textile machinery	0.4	Manufg.
Fabricated rubber products, nec	0.3	Manufg.	Federal Government purchases, national defense	0.4	Fed Govt
Instruments to measure electricity	0.3	Manufg.	Federal Government purchases, nondefense	0.4	Fed Govt
Miscellaneous fabricated wire products	0.3	Manufg.	S/L Govt. purch., gas & electric utilities	0.4	S/L Govt
Motors & generators	0.3	Manufg.	Amusement & recreation building construction	0.3	Constr.
Signs & advertising displays	0.3	Manufg.	Farm service facilities	0.3	Constr.
Veneer & plywood	0.3	Manufg.	Highway & street construction	0.3	Constr.
Railroads & related services	0.3	Util.	Resid. & other health facility construction	0.3	Constr.
Legal services	0.3	Services	Residential 2-4 unit structures, nonfarm	0.3	Constr.
Management & consulting services & labs	0.3	Services	Paper industries machinery	0.3	Manufg.
U.S. Postal Service	0.3	Gov't	Local transit facility construction	0.2	Constr.
Ball & roller bearings	0.2	Manufg.	Telephone & telegraph facility construction	0.2	Constr.
Iron & steel foundries	0.2	Manufg.	Electric housewares & fans	0.2	Manufg.
Nonferrous castings, nec	0.2	Manufg.	Machine tools, metal forming types	0.2	Manufg.
Special dies & tools & machine tool accessories	0.2	Manufg.	Special dies & tools & machine tool accessories	0.2	Manufg.
Royalties	0.2	Fin/R.E.	S/L Govt. purch., higher education	0.2	S/L Govt
Accounting, auditing & bookkeeping	0.2	Services	S/L Govt. purch., transit utilities	0.2	S/L Govt
Abrasive products	0.1	Manufg.	Construction of religious buildings	0.1	Constr.
Fabricated metal products, nec	0.1	Manufg.	Maintenance of telephone & telegraph facilities	0.1	Constr.
Machine tools, metal forming types	0.1	Manufg.	Calculating & accounting machines	0.1	Manufg.
Manifold business forms	0.1	Manufg.	Commercial laundry equipment	0.1	Manufg.
Primary nonferrous metals, nec	0.1	Manufg.	Household cooking equipment	0.1	Manufg.
Insurance carriers	0.1	Fin/R.E.	Instruments to measure electricity	0.1	Manufg.
Automotive rental & leasing, without drivers	0.1	Services	Oil field machinery	0.1	Manufg.
Noncomparable imports	0.1	Foreign	Scales & balances	0.1	Manufg.
			Woodworking machinery	0.1	Manufg.

Source: Benchmark Input-Output Accounts for the U.S. Economy, 1982, U.S. Department of Commerce, Washington, D.C., July 1991. Data, as reported in the source, are organized by the 1977 SIC structure in use in 1982 but have been matched, as closely as is possible, to the 1987 SIC structure used in this book.

OCCUPATIONS EMPLOYED BY SIC 361 - ELECTRIC DISTRIBUTING EQUIPMENT

Occupation	% of Total 1994	Change to 2005	Occupation	% of Total 1994	Change to 2005
Electrical & electronic assemblers	11.7	-19.4	Mechanical engineers	1.5	8.3
Electrical & electronic equipment assemblers	11.0	-10.5	Secretaries, ex legal & medical	1.4	-18.6
Assemblers, fabricators, & hand workers nec	8.3	-10.5	Sheet metal workers & duct installers	1.4	-10.4
Coil winders, tapers, & finishers	5.0	-28.4	Industrial production managers	1.2	-10.5
Electrical & electronics engineers	3.6	-4.7	Machine tool cutting & forming etc. nec	1.1	-28.4
Blue collar worker supervisors	3.4	-24.8	Engineering, mathematical, & science managers	1.1	1.6
Inspectors, testers, & graders, precision	3.4	-37.3	Stock clerks	1.1	-27.2
Electromechanical equipment assemblers	2.5	-1.6	Traffic, shipping, & receiving clerks	1.1	-13.9
Sales & related workers nec	2.1	-10.5	Freight, stock, & material movers, hand	1.1	-28.4
Drafters	2.0	-30.3	Industrial engineers, ex safety engineers	1.0	-1.5
Electrical & electronic technicians, technologists	1.8	-10.5	Coating, painting, & spraying machine workers	1.0	-10.4
General managers & top executives	1.7	-15.1	Welding machine setters, operators	1.0	-19.5

Source: Industry-Occupation Matrix, Bureau of Labor Statistics. These data relate to one or more 3-digit SIC industry groups rather than to a single 4-digit SIC. The change reported for each occupation to the year 2005 is a percent of growth or decline as estimated by the Bureau of Labor Statistics. The abbreviation nec stands for 'not elsewhere classified'.

LOCATION BY STATE AND REGIONAL CONCENTRATION

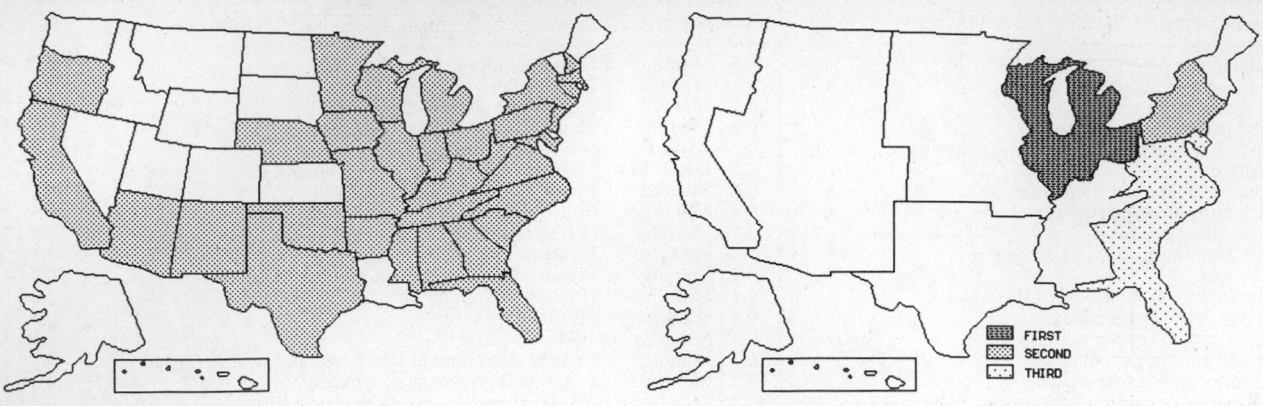

FIRST
SECOND
THIRD

INDUSTRY DATA BY STATE

| State | Establish- ments | Shipments | | | Employment | | | | Cost as % of Shipments | Investment per Employee ($) |
		Total ($ mil)	% of U.S.	Per Establ.	Total Number	% of U.S.	Per Establ.	Wages ($/hour)		
Illinois	47	657.1	11.9	14.0	5,800	14.7	123	12.89	37.1	3,328
North Carolina	18	410.5	7.4	22.8	2,100	5.3	117	10.07	43.0	3,095
Ohio	40	358.1	6.5	9.0	2,800	7.1	70	13.13	40.9	3,000
Pennsylvania	29	331.3	6.0	11.4	2,100	5.3	72	13.45	45.5	3,571
California	53	226.1	4.1	4.3	1,900	4.8	36	12.39	42.1	1,316
Texas	36	214.9	3.9	6.0	1,700	4.3	47	13.09	67.4	1,588
Tennessee	9	204.8	3.7	22.8	1,100	2.8	122	13.37	45.2	3,636
Massachusetts	12	141.5	2.6	11.8	1,200	3.0	100	10.33	28.1	1,750
New Jersey	24	103.7	1.9	4.3	1,100	2.8	46	12.15	37.8	1,909
Michigan	29	95.9	1.7	3.3	900	2.3	31	10.46	50.4	1,222
Virginia	10	82.0	1.5	8.2	1,100	2.8	110	8.87	55.0	-
Florida	14	62.8	1.1	4.5	500	1.3	36	11.50	80.4	3,400
New York	24	47.4	0.9	2.0	400	1.0	17	5.00	42.0	1,000
Alabama	8	26.1	0.5	3.3	300	0.8	38	9.50	50.2	1,000
Connecticut	17	(D)	-	-	1,750 *	4.4	103	-	-	-
Missouri	14	(D)	-	-	1,750 *	4.4	125	-	-	-
Oregon	14	(D)	-	-	375 *	0.9	27	-	-	-
Georgia	12	(D)	-	-	1,750 *	4.4	146	-	-	686
Maryland	10	(D)	-	-	1,750 *	4.4	175	-	-	-
Wisconsin	10	(D)	-	-	750 *	1.9	75	-	-	-
Indiana	8	(D)	-	-	750 *	1.9	94	-	-	-
Kentucky	8	(D)	-	-	1,750 *	4.4	219	-	-	3,257
Minnesota	8	(D)	-	-	375 *	0.9	47	-	-	-
Oklahoma	7	(D)	-	-	375 *	0.9	54	-	-	1,600
South Carolina	6	(D)	-	-	1,750 *	4.4	292	-	-	-
Arizona	5	(D)	-	-	175 *	0.4	35	-	-	-
Iowa	5	(D)	-	-	1,750 *	4.4	350	-	-	-
Mississippi	5	(D)	-	-	1,750 *	4.4	350	-	-	-
Nebraska	4	(D)	-	-	750 *	1.9	188	-	-	-
New Hampshire	4	(D)	-	-	375 *	0.9	94	-	-	-
Arkansas	3	(D)	-	-	175 *	0.4	58	-	-	-
West Virginia	2	(D)	-	-	175 *	0.4	88	-	-	-
New Mexico	1	(D)	-	-	375 *	0.9	375	-	-	-

Source: 1992 *Economic Census*. The states are in descending order of shipments or establishments (if shipment data are missing for the majority). The symbol (D) appears when data are withheld to prevent disclosure of competitive information. States marked with (D) are sorted by number of establishments. A dash (-) indicates that the data element cannot be calculated; * indicates the midpoint of a range.

3621 - MOTORS & GENERATORS

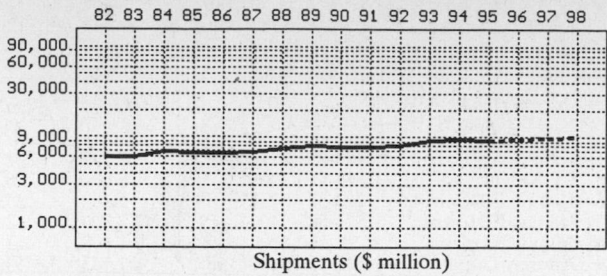

Shipments ($ million)

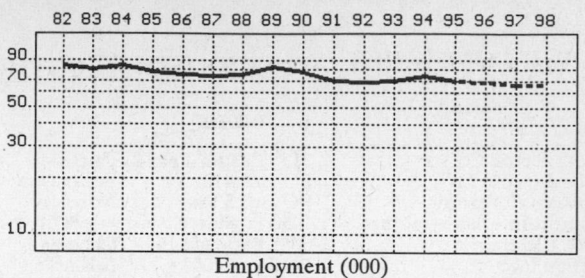

Employment (000)

GENERAL STATISTICS

| Year | Com-panies | Establishments | | Employment | | | Compensation | | Production ($ million) | | | |
		Total	with 20 or more employees	Total (000)	Production Workers (000)	Hours (Mil)	Payroll ($ mil)	Wages ($/hr)	Cost of Materials	Value Added by Manufacture	Value of Shipments	Capital Invest.
1982	348	471	324	84.1	61.0	114.3	1,545.7	8.61	2,525.9	3,434.3	6,058.1	275.4
1983		463	323	80.4	60.1	114.4	1,555.0	8.77	2,647.7	3,372.5	6,002.0	204.6
1984		455	322	84.2	63.3	124.5	1,743.5	9.22	2,915.7	3,929.3	6,760.5	244.9
1985		448	322	77.6	57.8	112.1	1,648.9	9.80	2,888.7	3,668.1	6,583.6	262.5
1986		432	305	75.6	55.5	108.1	1,681.7	10.01	2,921.1	3,623.4	6,608.1	232.7
1987	349	462	302	74.6	56.5	110.2	1,663.5	9.98	2,962.9	3,815.2	6,753.1	201.6
1988		475	312	75.7	58.0	115.0	1,785.8	10.36	3,504.1	4,182.5	7,601.4	205.2
1989		467	307	82.8	58.5	116.8	1,815.3	10.35	3,802.5	4,271.8	8,072.8	215.5
1990		466	312	78.5	55.1	108.5	1,751.1	10.62	3,634.5	4,005.3	7,672.2	238.8
1991		472	311	69.7	51.9	103.5	1,741.8	10.97	3,683.6	4,037.2	7,673.8	238.1
1992	370	470	274	67.9	52.0	104.1	1,764.6	11.23	3,812.9	4,244.3	8,039.7	242.0
1993		463	273	69.2	53.6	109.7	1,834.1	11.33	4,279.4	4,915.4	9,181.8	249.3
1994		469P	282P	74.0	58.1	118.6	1,914.2	10.90	4,705.5	4,777.2	9,429.3	290.5
1995		470P	278P	69.1P	52.7P	108.5P	1,890.4P	11.69P	4,622.4P	4,692.9P	9,262.9P	248.4P
1996		471P	274P	68.1P	52.1P	108.0P	1,913.8P	11.91P	4,753.9P	4,826.4P	9,526.4P	249.8P
1997		472P	270P	67.0P	51.5P	107.5P	1,937.2P	12.13P	4,885.5P	4,959.9P	9,789.9P	251.2P
1998		473P	266P	66.0P	50.9P	106.9P	1,960.6P	12.35P	5,017.0P	5,093.4P	10,053.4P	252.6P

Sources: 1982, 1987, 1992 *Economic Census*; *Annual Survey of Manufactures*, 83-86, 88-91, 93-94. Establishment counts for non-Census years are from *County Business Patterns*; establishment values for 83-84 are extrapolations. 'P's show projections by the editors. Industries reclassified in 87 will not have data for prior years.

INDICES OF CHANGE

| Year | Com-panies | Establishments | | Employment | | | Compensation | | Production ($ million) | | | |
		Total	with 20 or more employees	Total (000)	Production Workers (000)	Hours (Mil)	Payroll ($ mil)	Wages ($/hr)	Cost of Materials	Value Added by Manufacture	Value of Shipments	Capital Invest.
1982	94	100	118	124	117	110	88	77	66	81	75	114
1983		99	118	118	116	110	88	78	69	79	75	85
1984		97	118	124	122	120	99	82	76	93	84	101
1985		95	118	114	111	108	93	87	76	86	82	108
1986		92	111	111	107	104	95	89	77	85	82	96
1987	94	98	110	110	109	106	94	89	78	90	84	83
1988		101	114	111	112	110	101	92	92	99	95	85
1989		99	112	122	113	112	103	92	100	101	100	89
1990		99	114	116	106	104	99	95	95	94	95	99
1991		100	114	103	100	99	99	98	97	95	95	98
1992	100	100	100	100	100	100	100	100	100	100	100	100
1993		99	100	102	103	105	104	101	112	116	114	103
1994		100P	103P	109	112P	114	108	97	123	113	117	120
1995		100P	101P	102P	101P	104P	107P	104P	121P	111P	115P	103P
1996		100P	100P	100P	100P	104P	108P	106P	125P	114P	118P	103P
1997		100P	99P	99P	99P	103P	110P	108P	128P	117P	122P	104P
1998		101P	97P	97P	98P	103P	111P	110P	132P	120P	125P	104P

Sources: Same as General Statistics. Values reflect change from the base year, 1992. Values above 100 mean greater than 92, values below 100 mean less than 92, and a value of 100 in the 82-91 or 93-98 period means same as 92. 'P's mark projections by the editors.

SELECTED RATIOS

For 1994	Avg. of All Manufact.	Analyzed Industry	Index	For 1994	Avg. of All Manufact.	Analyzed Industry	Index
Employees per Establishment	49	158	322	Value Added per Production Worker	134,084	82,224	61
Payroll per Establishment	1,500,273	4,082,637	272	Cost per Establishment	5,045,178	10,035,967	199
Payroll per Employee	30,620	25,868	84	Cost per Employee	102,970	63,588	62
Production Workers per Establishment	34	124	361	Cost per Production Worker	146,988	80,990	55
Wages per Establishment	853,319	2,757,177	323	Shipments per Establishment	9,576,895	20,110,965	210
Wages per Production Worker	24,861	22,250	89	Shipments per Employee	195,460	127,423	65
Hours per Production Worker	2,056	2,041	99	Shipments per Production Worker	279,017	162,294	58
Wages per Hour	12.09	10.90	90	Investment per Establishment	321,011	619,583	193
Value Added per Establishment	4,602,255	10,188,890	221	Investment per Employee	6,552	3,926	60
Value Added per Employee	93,930	64,557	69	Investment per Production Worker	9,352	5,000	53

Sources: Same as General Statistics. The 'Average of All Manufacturing' column represents the average of all manufacturing industries reported for the most recent complete year available. The Index shows the relationship between the Average and the Analyzed Industry. For example, 100 means that they are equal; 500 that the Analyzed Industry is five times the average; 50 means that the Analyzed Industry is half the national average. The abbreviation 'na' is used to show that data are 'not available'.

LEADING COMPANIES   Number shown: 75   Total sales ($ mil): 85,781   Total employment (000): 425.0

Company Name	Address				CEO Name	Phone	Co. Type	Sales ($ mil)	Empl. (000)
General Electric Co	3135 Easton Tpk	Fairfield	CT	06431	John F Welch Jr	203-373-2211	P	60,109	221.0
Emerson Electric Co	PO Box 4100	St Louis	MO	63136	CF Knight	314-553-2000	P	8,607	76.0
Asea Brown Boveri Inc	PO Box 5308	Norwalk	CT	06856	Robert E Donovan	203-329-8771	S	5,100	25.0
McDermott International Inc	1450 Poydras St	New Orleans	LA	70112	Robert E Howson	504-587-5400	P	3,060	22.5
Reliance Electric Co	6065 Parkland Blv	Cleveland	OH	44124	John C Morley	216-266-5800	P	1,608	13.8
MagneTek Incc	PO Box 290159	Nashville	TN	37229	Andrew G Galef	615-316-5100	P	1,133	14.3
GE Motors	PO Box 2205	Fort Wayne	IN	46801	James W Rogers	219-439-2000	D	630*	3.0
Onan Corp	1400 73rd Av NE	Minneapolis	MN	55432	Richard B Stoner	612-574-5000	S	575	3.6
Baldor Electric Co	PO Box 2400	Fort Smith	AR	72902	RL Qualls	501-646-4711	P	418	3.4
Philips Technologies	PO Box 868	Cheshire	CT	06410	Robert T Hamilton	203-271-6000	S	250	3.2
AO Smith Corp	531 N 4th St	Tipp City	OH	45371	John Bertrand	513-667-2431	D	242	3.0
Franklin Electric Company Inc	400 E Spring St	Bluffton	IN	46714	William H Lawson	219-824-2900	P	241	2.5
Magnetek Century Inc	1881 Pine St	St Louis	MO	63103	Brian R Dundon	314-436-7800	S	240	2.3
Pacific Scientific Co	620 Newport Ctr Dr	Newport Beach	CA	92660	Edgar S Brower	714-720-1714	P	235	1.8
United Techn Motor Systems	PO Box 2228	Columbus	MS	39704	David Harris	601-328-4150	D	210	2.0
Emerson Electric Co US	8100 W Florissant	St Louis	MO	63136	Dave Wathen	314-553-2000	D	200	2.0
Kollmorgen Corp	1601 Trapelo Rd	Waltham	MA	02154	Gideon Argov	617-890-5655	P	192	1.6
Lamb Electric	PO Box 1599	Kent	OH	44240	Walter Sliva	216-673-3451	D	172	1.4
GS Electric	1700 Ritner Hwy	Carlisle	PA	17013	Mike Jacqmin	717-243-4041	D	160*	1.0
Genie Co	22790 Lake Pk Blvd	Alliance	OH	44601	John Gray	216-821-5360	R	120*	0.9
Leeson Electric Corp	PO Box 241	Grafton	WI	53024	CL Doerr	414-377-8810	R	113	1.0
Siemens Energy and Automation	4620 Forest Av	Norwood	OH	45212	Larry Ray	513-841-3100	D	110	0.6
Toshiba International Corp	PO Box 40906	Houston	TX	77240	Toshio Doshida	713-466-0277	D	110	0.8
Coleman Powermate Inc	PO Box 6001	Kearney	NE	68848	Gerry Brown	308-237-2181	S	90	0.3
Magnetek National	800 King Av	Columbus	OH	43216	Robert G Barton	614-488-1151	D	90	1.0
Pacific Scientific Co	PO Box 106	Rockford	IL	61105	Ronald Nelson	815-226-3100	D	90	0.6
Peerless-Winsmith Inc	1700 E Market St	Warren	OH	44483	RT Groner	216-395-1010	S	80	1.0
Morganite Inc	1 Morganite Dr	Dunn	NC	28334	DB Muller	910-892-8081	S	75	0.8
General Industries Co	PO Box 4002	Elyria	OH	44036	Jim Lynam	216-323-3136	S	70	0.8
Industrial General Corp	PO Box 4002	Elyria	OH	44036	Jim Lynam	216-323-3136	R	70	0.8
Vernitron Corp	645 Madison Av	New York	NY	10022	Stephen W Bershad	212-593-7900	P	68	0.7
Motor Coils Manufacturing Co	100 Talbot Av	Braddock	PA	15104	MJ Farrell	412-273-4900	S	60	0.6
Superior Electric	383 Middle St	Bristol	CT	06010	Willis J Kanaley	203-585-4500	S	60	0.6
Kato Engineering	PO Box 8447	Mankato	MN	56002	Jim Holderge	507-625-4011	D	55	0.6
Bodine Electric Co	2500 W Bradley Pl	Chicago	IL	60618	John R Bodine	312-478-3515	R	50	0.7
Dresser-Rand Electric Machinery	800 Central Av	Minneapolis	MN	55413	G Johnson	612-378-8000	D	50*	0.5
Emerson Motor Co	957 W Mullins St	Humboldt	TN	38343	Rich Graff	901-784-3611	S	50	0.5
Merkle-Korff Industries	1776 Winthrop Dr	Des Plaines	IL	60018	JD Simms	708-296-8800	R	50	0.3
Buehler Products Inc	PO Box 33400	Raleigh	NC	27636	L LaFreniere	919-469-8522	R	45*	0.5
General Industries Co	PO Box 260	Bald Knob	AR	72010	J Lynam	501-724-3227	D	45	0.3
Imperial Electric Co	PO Box 309	Akron	OH	44309	Francis A Collins	216-253-9126	R	33	0.2
Lucas Aerospace Electro Prod	610 Neptune Av	Brea	CA	92621	Tom Yearian	714-671-4500	D	33	0.2
Hansen Corp	PO Box 23	Princeton	IN	47670	WK Poyner	812-385-3415	S	31	0.4
Libby Corp	5800 Stillwell St	Kansas City	MO	64120	Hugh L Libby	816-231-6039	R	31	0.4
Yaskawa Electric America Inc	3160 MacArthur	Northbrook	IL	60062	D Tonimoto	708-291-2340	S	31*	0.2
General Dynamics Corp	150 Avnel St	Avenel	NJ	07001	Donald R Sisk	908-636-9100	D	30	0.3
Mamco Corp	8630 Industrial Dr	Franksville	WI	53126	WH Meltzer	414-886-9069	R	30	0.2
Morrill Motors Inc	3685 Northrop St	Fort Wayne	IN	46805	Giles W Morrill	219-484-1519	R	30	0.6
Eaton Technologies Inc	402 E Haven St	Eaton Rapids	MI	48827	Larry Kipp	517-663-2161	S	29*	0.4
Comet Industries Inc	4800 Deramus	Kansas City	MO	64120	Edwin Johnson	816-245-9400	R	28	0.3
Ideal Electric Co	330 E 1st St	Mansfield	OH	44903	M Vucelic	419-522-3611	R	28	0.3
DMT Corp	2494 N Hwy 164	Waukesha	WI	53186	Bradley J Harring	414-549-0014	R	27	<0.1
Kirkwood Commutator Co	4855 W 130th St	Cleveland	OH	44135	W C DeLorenzo	216-267-6200	D	27*	0.5
WA Kraft Corp	PO Box 2189	Woburn	MA	01888	Owen M Duffy	617-938-9100	R	26	0.1
Pittman	PO Box 3	Harleysville	PA	19438	K A Swanstrom	215-256-6601	D	25	0.2
Hurst Manufacturing	PO Box 326	Princeton	IN	47670	Dennis M Hurst	812-385-2564	D	25	0.3
Power Convertibles Corp	3450 S Broadmont	Tucson	AZ	85713	Larry McDonald	602-628-8292	S	25	0.6
Unitron Inc	10925 Miller Rd	Dallas	TX	75238	RL Beutel	214-340-8600	R	25	0.3
AO Smith Corp	PO Box 368	Mebane	NC	27302	Carl Steinbecker	919-563-9100	D	24*	0.3
Stature Electric Inc	22543 Fisher Rd	Watertown	NY	13601	LP Huntsinger	315-782-5910	R	23	0.2
Barber-Colman Co	PO Box 7040	Rockford	IL	61125	Peter C Spaulding	815-397-7400	D	22	0.3
Indramat	255 Mittel Dr	Wood Dale	IL	60191	Robert Rickert	708-860-1010	D	21*	0.1
Electro-Tec Corp	1501 N Main St	Blacksburg	VA	24060	Sam St Amour	703-552-2111	S	20	0.2
Everson Electric Co	2000 City Line Rd	Bethlehem	PA	18017	Dave Everson Jr	215-264-8611	R	20	0.2
Indiana General Motor Products	1168 Barranca Dr	El Paso	TX	79935		915-593-1621	D	20	0.3
Midwestern Power	5194 NE 17th	Des Moines	IA	50313	William Hanley	515-264-1650	D	20	<0.1
Molon Motor and Coil Inc	3737 Industrial Av	Rolling Mdws	IL	60008	EF Moloney	708-253-6000	R	20	0.3
Sterling Electric Inc	16752 Armstrong	Irvine	CA	92714	RE Helton	714-474-0520	R	20	0.2
Rotor Clip Company Inc	187 Davidson Av	Somerset	NJ	08875	Robert Slass	908-469-7333	R	19	0.2
Electro Sales Company Inc	100 Fellsway W	Somerville	MA	02145	Stewart D Berg	617-666-0500	R	18	<0.1
Micro Mo Electronics Inc	742 2nd Av S	St Petersburg	FL	33701	F Faulhaber	813-822-2529	R	18*	<0.1
Skurka Engineering Co	PO Box 2869	Camarillo	CA	93011	M Skurka	805-484-8884	R	18*	0.1
Winco Inc	225 S Cordova Av	Le Center	MN	56057	Ralph I Call	612-357-6821	R	18	0.1
Ohio Electric Motors Inc	PO Box 168	Barnardsville	NC	28709	Ken Simmons	704-626-2901	S	17	0.2
DURHAM Products	PO Box 15730	Durham	NC	27704	Ernest Wendell	919-471-4488	D	16	0.1

Source: *Ward's Business Directory of U.S. Private and Public Companies*, Volumes 1 and 2, 1996. The company type code used is as follows: P - Public, R - Private, S - Subsidiary, D - Division, J - Joint Venture, A - Affiliate, G - Group. Sales are in millions of dollars, employees are in thousands. An asterisk (*) indicates an estimated sales volume. The symbol < stands for 'less than'. Company names and addresses are truncated, in some cases, to fit into the available space.

MATERIALS CONSUMED

Material	Quantity	Delivered Cost ($ million)
Materials, ingredients, containers, and supplies	(X)	3,502.8
Metal stampings	(X)	231.5
Metal bolts, nuts, screws, washers, rivets, and other screw machine products	(X)	77.1
All other fabricated metal products (except forgings)	(X)	81.3
Forgings	(X)	3.0
Iron and steel castings (rough and semifinished)	(X)	85.8
Aluminum and aluminum-base alloy castings (rough and semifinished)	(X)	90.1
Other nonferrous castings (rough and semifinished)	(X)	49.8
Steel bars, bar shapes, and plates	(X)	80.2
Steel sheet and strip, including tin plate	(X)	374.9
Steel wire and wire products	(X)	3.7
All other steel shapes and forms	(X)	93.0
Copper and copper-base alloy bare wire for electrical conduction only	(X)	10.3
Copper and copper-base alloy rod, bar, and mechanical wire, including extruded and/or drawn shapes	(X)	3.0
All other copper and copper-base alloy shapes and forms	(X)	49.2
Aluminum and aluminum-base alloy sheet, plate, foil, and welded tubing	(X)	5.4
All other aluminum and aluminum-base alloy shapes and forms	(X)	31.3
Other nonferrous shapes and forms	(X)	3.6
Magnet wire	(X)	355.6
Insulated copper wire and cable, except magnet wire	(X)	35.1
Primary aluminum and aluminum-base alloy refinery shapes	(X)	42.0
Diesel and semidiesel engines	(X)	188.1
Gasoline and other carburetor engines	(X)	60.1
Fractional horsepower electric motors (less than 1 hp)	(X)	24.6
Integral horsepower electric motors and generators (1 hp or more)	(X)	35.9
Ball and roller bearings (mounted or unmounted)	(X)	62.1
Plain bearings and bushings	(X)	46.6
Mechanical speed changers, gears, and industrial high-speed drives	(X)	9.6
Semiconductors, including transistors, diodes, rectifiers, and integrated circuits for electronic circuitry	(X)	65.4
Carbon brushes	(X)	12.4
Ceramic magnets (ferrite)	(X)	47.4
Plastics products consumed in the form of sheets, rods, tubes, film, and other shapes	(X)	135.2
Plastics resins consumed in the form of granules, pellets, powders, liquids, etc.	(X)	24.9
Paints, varnishes, lacquers, stains, shellacs, japans, enamels, and allied products	(X)	40.2
Electrical industrial capacitors, resistors, rheostats, and coil windings	(X)	110.9
Electrical transmission, distribution, and control equipment	(X)	36.3
Paperboard containers, boxes, and corrugated paperboard	(X)	38.2
All other materials and components, parts, containers, and supplies	(X)	513.7
Materials, ingredients, containers, and supplies, nsk	(X)	345.3

Source: 1992 *Economic Census*. Explanation of symbols used: (D): Withheld to avoid disclosure of competitive data; na: Not available; (S): Withheld because statistical norms were not met; (X): Not applicable; (Z): Less than half the unit shown; nec: Not elsewhere classified; nsk: Not specified by kind; - : zero; * : 10-19 percent estimated; ** : 20-29 percent estimated.

PRODUCT SHARE DETAILS

Product or Product Class	% Share	Product or Product Class	% Share
Motors and generators	100.00	turbine	13.60
Fractional horsepower motors (rated at less than 746 watts) (excluding hermetics)	45.39	Fractional motor generator sets and other rotating equipment, including hermetics	3.88
Integral horsepower motors and generators other than for land transportation equipment (rated at 746 watts or more)	20.76	Integral motor generator sets and other rotating equipment, including hermetics	5.19
Land transportation motors, generators, and control equipment, excluding parts	2.22	Parts, supplies for motors, generators, generator sets, and other rotating equipment, excluding motors for built-in jobs	6.99
Prime mover generator sets, except steam or hydraulic		Motors and generators, nsk	1.96

Source: 1992 *Economic Census*. The values shown are percent of total shipments in an industry. Values of indented subcategories are summed in the main heading. The symbol (D) appears when data are withheld to prevent disclosure of competitive information. The abbreviation nsk stands for 'not specified by kind' and nec for 'not elsewhere classified'.

INPUTS AND OUTPUTS FOR MOTORS & GENERATORS

Economic Sector or Industry Providing Inputs	%	Sector	Economic Sector or Industry Buying Outputs	%	Sector
Imports	17.2	Foreign	Exports	16.1	Foreign
Blast furnaces & steel mills	11.1	Manufg.	Gross private fixed investment	11.9	Cap Inv
Wholesale trade	9.9	Trade	Refrigeration & heating equipment	8.7	Manufg.
Nonferrous wire drawing & insulating	7.1	Manufg.	Local government passenger transit	7.8	Gov't
Motors & generators	6.4	Manufg.	Internal combustion engines, nec	5.5	Manufg.
Advertising	4.4	Services	Pumps & compressors	5.4	Manufg.
Internal combustion engines, nec	2.9	Manufg.	Motors & generators	3.4	Manufg.
Metal stampings, nec	2.1	Manufg.	Electronic computing equipment	2.5	Manufg.
Electric services (utilities)	2.0	Util.	Electric housewares & fans	2.0	Manufg.
Iron & steel foundries	1.9	Manufg.	Household laundry equipment	2.0	Manufg.

Continued on next page.

INPUTS AND OUTPUTS FOR MOTORS & GENERATORS - Continued

Economic Sector or Industry Providing Inputs	%	Sector	Economic Sector or Industry Buying Outputs	%	Sector
Banking	1.9	Fin/R.E.	Crude petroleum & natural gas	1.6	Mining
Air transportation	1.6	Util.	Federal Government purchases, national defense	1.5	Fed Govt
Industrial controls	1.4	Manufg.	Machine tools, metal cutting types	1.3	Manufg.
Aluminum castings	1.2	Manufg.	Blowers & fans	1.2	Manufg.
Ball & roller bearings	1.2	Manufg.	Hardware, nec	1.0	Manufg.
Communications, except radio & TV	1.2	Util.	Household refrigerators & freezers	1.0	Manufg.
Miscellaneous plastics products	1.0	Manufg.	Service industry machines, nec	1.0	Manufg.
Petroleum refining	1.0	Manufg.	Power transmission equipment	0.9	Manufg.
Primary aluminum	1.0	Manufg.	Special industry machinery, nec	0.9	Manufg.
Screw machine and related products	1.0	Manufg.	Personal consumption expenditures	0.8	
Hotels & lodging places	1.0	Services	Food products machinery	0.8	Manufg.
Motor freight transportation & warehousing	0.9	Util.	Photographic equipment & supplies	0.8	Manufg.
Eating & drinking places	0.9	Trade	Railroad equipment	0.8	Manufg.
Paperboard containers & boxes	0.8	Manufg.	General industrial machinery, nec	0.7	Manufg.
Noncomparable imports	0.8	Foreign	Household appliances, nec	0.7	Manufg.
Maintenance of nonfarm buildings nec	0.7	Constr.	Printing trades machinery	0.7	Manufg.
Aluminum rolling & drawing	0.7	Manufg.	Construction machinery & equipment	0.6	Manufg.
Fabricated metal products, nec	0.7	Manufg.	Conveyors & conveying equipment	0.6	Manufg.
Gas production & distribution (utilities)	0.7	Util.	Motor vehicle parts & accessories	0.6	Manufg.
Equipment rental & leasing services	0.7	Services	Radio & TV communication equipment	0.6	Manufg.
Nonmetallic mineral products, nec	0.6	Manufg.	Typewriters & office machines, nec	0.6	Manufg.
Switchgear & switchboard apparatus	0.6	Manufg.	Household cooking equipment	0.5	Manufg.
Copper rolling & drawing	0.5	Manufg.	Household vacuum cleaners	0.5	Manufg.
Gaskets, packing & sealing devices	0.5	Manufg.	Mining machinery, except oil field	0.5	Manufg.
Machinery, except electrical, nec	0.5	Manufg.	Coal	0.4	Mining
Rubber & plastics hose & belting	0.5	Manufg.	Farm machinery & equipment	0.4	Manufg.
Computer & data processing services	0.5	Services	Heating equipment, except electric	0.4	Manufg.
Electrical industrial apparatus, nec	0.4	Manufg.	Industrial controls	0.4	Manufg.
Electronic components nec	0.4	Manufg.	Industrial trucks & tractors	0.4	Manufg.
Paints & allied products	0.4	Manufg.	Motor vehicles & car bodies	0.4	Manufg.
Plastics materials & resins	0.4	Manufg.	Oil field machinery	0.4	Manufg.
Porcelain electrical supplies	0.4	Manufg.	Ship building & repairing	0.4	Manufg.
Power transmission equipment	0.4	Manufg.	Federal Government purchases, nondefense	0.4	Fed Govt
Legal services	0.4	Services	S/L Govt. purch., gas & electric utilities	0.4	S/L Govt
Management & consulting services & labs	0.4	Services	S/L Govt. purch., higher education	0.4	S/L Govt
U.S. Postal Service	0.4	Gov't	Elevators & moving stairways	0.3	Manufg.
Fabricated rubber products, nec	0.3	Manufg.	Engineering & scientific instruments	0.3	Manufg.
Plating & polishing	0.3	Manufg.	Hoists, cranes, & monorails	0.3	Manufg.
Special dies & tools & machine tool accessories	0.3	Manufg.	Mechanical measuring devices	0.3	Manufg.
Transformers	0.3	Manufg.	Metalworking machinery, nec	0.3	Manufg.
Railroads & related services	0.3	Util.	Turbines & turbine generator sets	0.3	Manufg.
Real estate	0.3	Fin/R.E.	Motor freight transportation & warehousing	0.3	Util.
Abrasive products	0.2	Manufg.	Transit & bus transportation	0.3	Util.
Carbon & graphite products	0.2	Manufg.	Automotive repair shops & services	0.3	Services
Die-cut paper & board	0.2	Manufg.	S/L Govt. purch., elem. & secondary education	0.3	S/L Govt
Lighting fixtures & equipment	0.2	Manufg.	Feed grains	0.2	Agric.
Machine tools, metal cutting types	0.2	Manufg.	Aircraft	0.2	Manufg.
Manifold business forms	0.2	Manufg.	Aircraft & missile equipment, nec	0.2	Manufg.
Mineral wool	0.2	Manufg.	Automatic merchandising machines	0.2	Manufg.
Nonferrous castings, nec	0.2	Manufg.	Commercial laundry equipment	0.2	Manufg.
Primary nonferrous metals, nec	0.2	Manufg.	Fabricated plate work (boiler shops)	0.2	Manufg.
Semiconductors & related devices	0.2	Manufg.	Furniture & fixtures, nec	0.2	Manufg.
Wood pallets & skids	0.2	Manufg.	Logging camps & logging contractors	0.2	Manufg.
Insurance carriers	0.2	Fin/R.E.	Machine tools, metal forming types	0.2	Manufg.
Royalties	0.2	Fin/R.E.	Measuring & dispensing pumps	0.2	Manufg.
Accounting, auditing & bookkeeping	0.2	Services	Paper industries machinery	0.2	Manufg.
Automotive rental & leasing, without drivers	0.2	Services	Sheet metal work	0.2	Manufg.
Brass, bronze, & copper castings	0.1	Manufg.	Tanks & tank components	0.2	Manufg.
Felt goods, nec	0.1	Manufg.	Textile machinery	0.2	Manufg.
Hand & edge tools, nec	0.1	Manufg.	Woodworking machinery	0.2	Manufg.
Iron & steel forgings	0.1	Manufg.	Electric services (utilities)	0.2	Util.
Lubricating oils & greases	0.1	Manufg.	Railroads & related services	0.2	Util.
Paper coating & glazing	0.1	Manufg.	Electrical repair shops	0.2	Services
Automotive repair shops & services	0.1	Services	Management & consulting services & labs	0.2	Services
Electrical repair shops	0.1	Services	Boat building & repairing	0.1	Manufg.
Photofinishing labs, commercial photography	0.1	Services	Electronic components nec	0.1	Manufg.
			Instruments to measure electricity	0.1	Manufg.
			Metal stampings, nec	0.1	Manufg.
			Pipe, valves, & pipe fittings	0.1	Manufg.
			Sewing machines	0.1	Manufg.
			Switchgear & switchboard apparatus	0.1	Manufg.
			Welding apparatus, electric	0.1	Manufg.
			Computer & data processing services	0.1	Services

Source: Benchmark Input-Output Accounts for the U.S. Economy, 1982, U.S. Department of Commerce, Washington, D.C., July 1991. Data, as reported in the source, are organized by the 1977 SIC structure in use in 1982 but have been matched, as closely as is possible, to the 1987 SIC structure used in this book.

OCCUPATIONS EMPLOYED BY SIC 362 - ELECTRICAL INDUSTRIAL APPARATUS

Occupation	% of Total 1994	Change to 2005	Occupation	% of Total 1994	Change to 2005
Electrical & electronic assemblers	16.8	-28.3	Secretaries, ex legal & medical	1.4	-27.5
Assemblers, fabricators, & hand workers nec	8.0	-20.3	Drafters	1.3	-38.0
Electrical & electronic equipment assemblers	4.3	-20.3	Stock clerks	1.3	-35.3
Inspectors, testers, & graders, precision	4.2	-44.2	Machine assemblers	1.3	-28.3
Electrical & electronics engineers	3.3	-15.2	Machine tool cutting & forming etc. nec	1.2	-36.3
Blue collar worker supervisors	3.2	-33.5	Lathe & turning machine tool operators	1.2	-48.2
Coil winders, tapers, & finishers	3.0	-36.3	Traffic, shipping, & receiving clerks	1.1	-23.4
Electrical & electronic technicians,technologists	2.3	-20.4	Industrial production managers	1.1	-20.3
Machinists	2.2	-40.3	Engineering, mathematical, & science managers	1.1	-9.5
Sales & related workers nec	1.8	-20.3	Industrial machinery mechanics	1.1	7.5
Electromechanical equipment assemblers	1.5	-12.4	Production, planning, & expediting clerks	1.0	-20.3
General managers & top executives	1.5	-24.4			

Source: Industry-Occupation Matrix, Bureau of Labor Statistics. These data relate to one or more 3-digit SIC industry groups rather than to a single 4-digit SIC. The change reported for each occupation to the year 2005 is a percent of growth or decline as estimated by the Bureau of Labor Statistics. The abbreviation nec stands for 'not elsewhere classified'.

LOCATION BY STATE AND REGIONAL CONCENTRATION

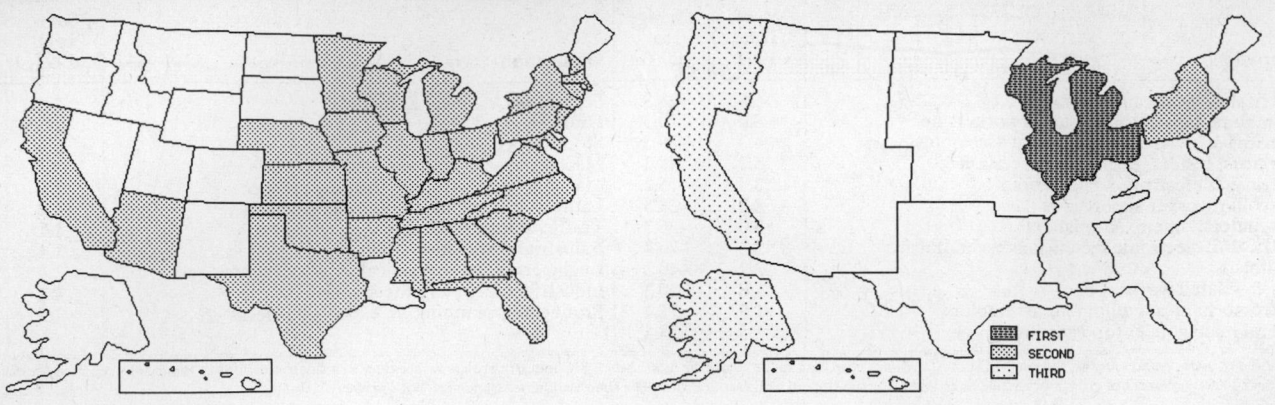

FIRST
SECOND
THIRD

INDUSTRY DATA BY STATE

| State | Establish-ments | Shipments | | | Employment | | | | Cost as % of Shipments | Investment per Employee ($) |
		Total ($ mil)	% of U.S.	Per Establ.	Total Number	% of U.S.	Per Establ.	Wages ($/hour)		
Wisconsin	32	776.7	9.7	24.3	6,200	9.1	194	10.63	52.2	4,323
Arkansas	12	683.8	8.5	57.0	5,600	8.2	467	10.37	44.5	3,857
Missouri	19	638.2	7.9	33.6	6,100	9.0	321	9.87	39.8	3,541
Ohio	34	608.0	7.6	17.9	5,400	8.0	159	11.17	41.4	3,463
Tennessee	16	591.5	7.4	37.0	5,200	7.7	325	9.88	51.6	3,423
Minnesota	15	534.2	6.6	35.6	3,000	4.4	200	18.39	44.1	4,333
Pennsylvania	27	378.0	4.7	14.0	3,500	5.2	130	11.88	38.3	3,600
Kentucky	8	357.4	4.4	44.7	2,500	3.7	313	9.21	54.3	6,200
Illinois	40	350.0	4.4	8.8	3,100	4.6	78	8.86	51.9	1,677
Mississippi	6	341.5	4.2	56.9	3,100	4.6	517	10.27	50.2	-
Indiana	24	313.2	3.9	13.1	4,200	6.2	175	11.18	43.1	2,405
North Carolina	15	272.3	3.4	18.2	2,900	4.3	193	9.64	51.9	6,103
Connecticut	14	176.5	2.2	12.6	1,800	2.7	129	10.52	40.1	2,389
Texas	16	176.1	2.2	11.0	1,200	1.8	75	13.44	57.9	5,583
California	50	162.5	2.0	3.3	1,800	2.7	36	12.56	34.5	3,222
Virginia	9	114.1	1.4	12.7	1,300	1.9	144	9.19	33.2	-
Oklahoma	7	111.3	1.4	15.9	900	1.3	129	8.91	45.4	-
Michigan	14	102.4	1.3	7.3	1,200	1.8	86	11.65	59.2	2,833
New York	22	(D)	-	-	3,750 *	5.5	170	-	-	-
New Jersey	16	(D)	-	-	750 *	1.1	47	-	-	400
Florida	14	(D)	-	-	175 *	0.3	13	-	-	-
Massachusetts	8	(D)	-	-	375 *	0.6	47	-	-	-
South Carolina	5	(D)	-	-	175 *	0.3	35	-	-	-
Alabama	4	(D)	-	-	375 *	0.6	94	-	-	-
Iowa	4	(D)	-	-	175 *	0.3	44	-	-	-
Kansas	4	(D)	-	-	750 *	1.1	188	-	-	-
Arizona	3	(D)	-	-	175 *	0.3	58	-	-	-
Georgia	3	(D)	-	-	750 *	1.1	250	-	-	-
New Hampshire	3	(D)	-	-	175 *	0.3	58	-	-	-
Maryland	1	(D)	-	-	175 *	0.3	175	-	-	-
Nebraska	1	(D)	-	-	375 *	0.6	375	-	-	-

Source: 1992 *Economic Census*. The states are in descending order of shipments or establishments (if shipment data are missing for the majority). The symbol (D) appears when data are withheld to prevent disclosure of competitive information. States marked with (D) are sorted by number of establishments. A dash (-) indicates that the data element cannot be calculated; * indicates the midpoint of a range.

3624 - CARBON & GRAPHITE PRODUCTS

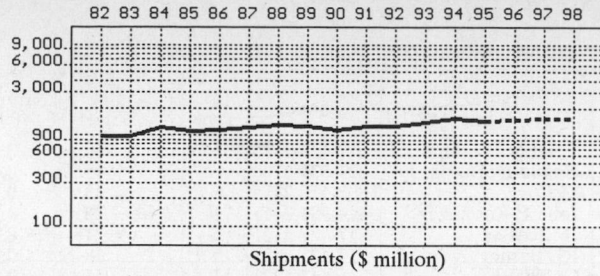

Shipments ($ million)

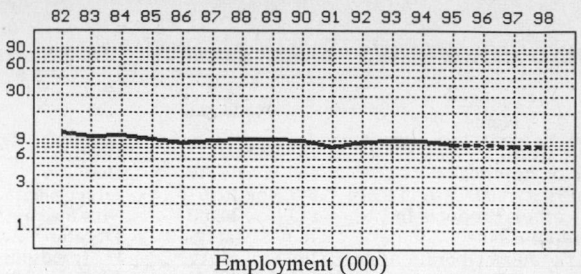

Employment (000)

GENERAL STATISTICS

Year	Com- panies	Establishments		Employment			Compensation		Production ($ million)			
		Total	with 20 or more employees	Total (000)	Production Workers (000)	Hours (Mil)	Payroll ($ mil)	Wages ($/hr)	Cost of Materials	Value Added by Manufacture	Value of Shipments	Capital Invest.
1982	71	91	60	12.1	8.5	16.8	260.7	9.88	467.4	554.9	980.4	151.9
1983		90	60	10.8	7.5	15.2	246.1	9.96	446.9	476.9	994.6	125.0
1984		89	60	11.4	8.3	17.2	276.5	10.58	655.9	584.6	1,222.8	68.9
1985		88	60	10.2	7.3	15.2	260.8	11.09	633.8	379.4	1,120.6	52.3
1986		89	58	9.3	6.6	13.5	240.3	11.26	587.7	508.9	1,136.3	38.5
1987	77	95	64	9.8	7.0	14.8	262.7	11.21	613.1	604.4	1,224.5	47.3
1988		98	64	10.3	7.4	16.1	278.8	11.42	662.2	618.7	1,303.2	62.4
1989		93	62	10.2	7.1	15.0	278.5	12.35	657.4	579.5	1,253.5	82.8
1990		92	61	9.9	6.3	13.5	271.9	13.34	559.3	586.8	1,166.9	110.9
1991		97	59	8.4	6.0	13.0	279.9	13.68	585.9	684.8	1,266.9	66.0
1992	86	110	67	9.6	7.0	14.4	286.6	12.85	561.2	724.2	1,276.4	43.6
1993		108	64	9.9	7.4	15.9	300.9	12.69	612.5	788.7	1,397.7	40.3
1994		105P	64P	10.0	7.3	15.5	329.9	13.87	699.6	844.4	1,527.0	50.4
1995		107P	65P	9.1P	6.5P	14.2P	308.2P	14.16P	665.7P	803.5P	1,453.1P	37.2P
1996		109P	65P	8.9P	6.4P	14.1P	313.0P	14.49P	680.9P	821.9P	1,486.3P	32.2P
1997		110P	65P	8.8P	6.4P	14.0P	317.7P	14.81P	696.1P	840.2P	1,519.4P	27.2P
1998		112P	66P	8.6P	6.3P	13.9P	322.5P	15.14P	711.3P	858.6P	1,552.6P	22.2P

Sources: 1982, 1987, 1992 *Economic Census; Annual Survey of Manufactures*, 83-86, 88-91, 93-94. Establishment counts for non-Census years are from *County Business Patterns*; establishment values for 83-84 are extrapolations. 'P's show projections by the editors. Industries reclassified in 87 will not have data for prior years.

INDICES OF CHANGE

Year	Com- panies	Establishments		Employment			Compensation		Production ($ million)			
		Total	with 20 or more employees	Total (000)	Production Workers (000)	Hours (Mil)	Payroll ($ mil)	Wages ($/hr)	Cost of Materials	Value Added by Manufacture	Value of Shipments	Capital Invest.
1982	83	83	90	126	121	117	91	77	83	77	77	348
1983		82	90	113	107	106	86	78	80	66	78	287
1984		81	90	119	119	119	96	82	117	81	96	158
1985		80	90	106	104	106	91	86	113	52	88	120
1986		81	87	97	94	94	84	88	105	70	89	88
1987	90	86	96	102	100	103	92	87	109	83	96	108
1988		89	96	107	106	112	97	89	118	85	102	143
1989		85	93	106	101	104	97	96	117	80	98	190
1990		84	91	103	90	94	95	104	100	81	91	254
1991		88	88	87	86	90	98	106	104	95	99	151
1992	100	100	100	100	100	100	100	100	100	100	100	100
1993		98	96	103	106	110	105	99	109	109	110	92
1994		96P	96P	104	104	108	115	108	125	117	120	116
1995		97P	96P	94P	93P	99P	108P	110P	119P	111P	114P	85P
1996		99P	97P	93P	92P	98P	109P	113P	121P	113P	116P	74P
1997		100P	98P	91P	91P	97P	111P	115P	124P	116P	119P	62P
1998		102P	98P	90P	89P	96P	113P	118P	127P	119P	122P	51P

Sources: Same as General Statistics. Values reflect change from the base year, 1992. Values above 100 mean greater than 92, values below 100 mean less than 92, and a value of 100 in the 82-91 or 93-98 period means same as 92. 'P's mark projections by the editors.

SELECTED RATIOS

For 1994	Avg. of All Manufact.	Analyzed Industry	Index	For 1994	Avg. of All Manufact.	Analyzed Industry	Index
Employees per Establishment	49	95	194	Value Added per Production Worker	134,084	115,671	86
Payroll per Establishment	1,500,273	3,129,711	209	Cost per Establishment	5,045,178	6,636,999	132
Payroll per Employee	30,620	32,990	108	Cost per Employee	102,970	69,960	68
Production Workers per Establishment	34	69	202	Cost per Production Worker	146,988	95,836	65
Wages per Establishment	853,319	2,039,530	239	Shipments per Establishment	9,576,895	14,486,417	151
Wages per Production Worker	24,861	29,450	118	Shipments per Employee	195,460	152,700	78
Hours per Production Worker	2,056	2,123	103	Shipments per Production Worker	279,017	209,178	75
Wages per Hour	12.09	13.87	115	Investment per Establishment	321,011	478,137	149
Value Added per Establishment	4,602,255	8,010,694	174	Investment per Employee	6,552	5,040	77
Value Added per Employee	93,930	84,440	90	Investment per Production Worker	9,352	6,904	74

Sources: Same as General Statistics. The 'Average of All Manufacturing' column represents the average of all manufacturing industries reported for the most recent complete year available. The Index shows the relationship between the Average and the Analyzed Industry. For example, 100 means that they are equal; 500 that the Analyzed Industry is five times the average; 50 means that the Analyzed Industry is half the national average. The abbreviation 'na' is used to show that data are 'not available'.

LEADING COMPANIES   Number shown: 21   Total sales ($ mil): 514   Total employment (000): 5.1

Company Name	Address				CEO Name	Phone	Co. Type	Sales ($ mil)	Empl. (000)
Stackpole Corp	133 Federal St	Boston	MA	02110	J Samuel Parkhill	617-423-3520	R	260	2.5
Carbone of America	400 Myrtle Av	Boonton	NJ	07005	Emilio DeBarnardo	201-334-0700	S	45	0.4
National Electrical Carbon Corp	PO Box 1056	Greenville	SC	29602	M Cox	803-458-7777	S	45	0.7
Helwig Carbon Products Inc	PO Box 24400	Milwaukee	WI	53224	John E Koenitzer	414-354-2411	R	23	0.3
Poco Graphite Inc	1601 S State St	Decatur	TX	76234	JF Beasley	817-627-2121	S	21*	0.2
Carbon Products Operation Inc	100 Stokes Av	E Stroudsburg	PA	18301	RL Barney	717-421-9921	S	18	0.2
Graphite Sales Inc	PO Box 185	Chagrin Falls	OH	44022	GC Hanna	216-543-8221	R	15	0.1
Zoltek Companies Inc	3101 McKelvey Rd	Bridgeton	MO	63044	Zsolt Rumy	314-291-5110	P	15	0.1
US Graphite Inc	1621 E Holland Av	Saginaw	MI	48601	Alan Vandall	517-755-0441	R	13	0.2
Metallized Carbon Corp	PO Box 1198	Ossining	NY	10562	Bruce Neri	914-941-3738	R	10	0.1
Pacific Combining Corp	3055 E Fruitland	Los Angeles	CA	90058	Samuel R Goodson	213-583-4231	S	10	<0.1
JS McCormick Co	318 Oliver Av	Pittsburgh	PA	15222	JB Snyder	412-471-7246	R	8	<0.1
Zoltek Companies Inc	11 Mo Res	St Louis	MO	63304	Zsolt Rumy	314-926-9999	D	8	<0.1
Advance Carbon Products Inc	171 Industrial Way	Brisbane	CA	94005	William J Crader Jr	415-468-1670	R	6*	<0.1
Graphite Metallizing Corp	1050 Nepperhan	Yonkers	NY	10702	Eben Walker	914-968-8400	R	6	<0.1
EDM Supplies Inc	9806 Everest St	Downey	CA	90242	David Muhs	310-803-6563	R	5	<0.1
Micro Mech Inc	PO Box 229	Ipswich	MA	01938	Akira Hashimoto	508-356-2966	S	3	<0.1
Pacific Fibre and Rope Company	PO Box 187	Wilmington	CA	90748	Allan Goldman	310-834-4567	R	1	<0.1
Ohio Carbon Co	705 S US Rte 224	Nova	OH	44859	Leland P Reineke	419-736-3610	R	1	<0.1
Graphite Systems Inc	PO Box 8798	Fort Worth	TX	76124	Robert B Trask	817-457-1851	R	1*	<0.1
Carbospheres Inc	PO Box 8116	Fredericksburg	VA	22404	Robert G Shaver	703-898-4040	R	0	<0.1

Source: Ward's Business Directory of U.S. Private and Public Companies, Volumes 1 and 2, 1996. The company type code used is as follows: P - Public, R - Private, S - Subsidiary, D - Division, J - Joint Venture, A - Affiliate, G - Group. Sales are in millions of dollars, employees are in thousands. An asterisk (*) indicates an estimated sales volume. The symbol < stands for 'less than'. Company names and addresses are truncated, in some cases, to fit into the available space.

MATERIALS CONSUMED

Material	Quantity	Delivered Cost ($ million)
Materials, ingredients, containers, and supplies	(X)	434.7
Pitch	(X)	33.5
Coke, petroleum coke, metallurgical coke, calcined coke, foundry coke, etc. used as raw material	(X)	91.2
Natural graphite	(X)	22.4
Artificial graphite	(X)	28.1
Carbon, ground or treated	(X)	13.1
Metal powders	(X)	2.7
All other materials and components, parts, containers, and supplies	(X)	205.3
Materials, ingredients, containers, and supplies, nsk	(X)	38.3

Source: 1992 Economic Census. Explanation of symbols used: (D): Withheld to avoid disclosure of competitive data; na: Not available; (S): Withheld because statistical norms were not met; (X): Not applicable; (Z): Less than half the unit shown; nec: Not elsewhere classified; nsk: Not specified by kind; - : zero; * : 10-19 percent estimated; ** : 20-29 percent estimated.

PRODUCT SHARE DETAILS

Product or Product Class	% Share	Product or Product Class	% Share
Carbon and graphite products	100.00	illuminating carbons, battery, except silver or other metal contacts	10.99
Electrodes.	42.30	All other carbon and graphite products, except refractories,	
Carbon electrodes for electric furnaces and electrolytic cell use	24.70	for mechanical uses, rotor vanes, and other uses where motion is between two parts, except metallic oilless	
Graphite electrodes for electric furnaces and electrolytic cell use	74.83	bearings	17.94
Electrodes, nsk	0.46	All other carbon and graphite products, except refractories, for aerospace uses, including unmachined stock and	
All other carbon and graphite products	55.22	machined items not included elsewhere	1.86
Carbon and graphite automotive brushes, including replacement brushes and those that are coded, except automobile accessory brushes	9.57	All other carbon and graphite products, except refractories, for all other uses (including chemical, metallurgical, etc.), paste	7.05
Other carbon and graphite industrial brushes and contacts (brushes more than 1/4 sq in. in cross section and more than 1 1/2 in. long)	9.31	All other carbon and graphite products, except refractories, for all other uses (including chemical, metallurgical, etc.), other	11.11
Carbon and graphite brush plates	4.39	All other carbon and graphite products, nsk	0.35
Carbon and graphite fibers	27.45	Carbon and graphite products, nsk	2.47
All other carbon and graphite products, except refractories, for electrical uses, including welding products,			

Source: 1992 Economic Census. The values shown are percent of total shipments in an industry. Values of indented subcategories are summed in the main heading. The symbol (D) appears when data are withheld to prevent disclosure of competitive information. The abbreviation nsk stands for 'not specified by kind' and nec for 'not elsewhere classified'.

INPUTS AND OUTPUTS FOR CARBON & GRAPHITE PRODUCTS

Economic Sector or Industry Providing Inputs	%	Sector	Economic Sector or Industry Buying Outputs	%	Sector
Imports	24.3	Foreign	Blast furnaces & steel mills	21.1	Manufg.
Miscellaneous plastics products	10.2	Manufg.	Exports	13.5	Foreign
Primary copper	10.0	Manufg.	X-ray apparatus & tubes	12.6	Manufg.
Electric services (utilities)	7.1	Util.	Secondary nonferrous metals	6.3	Manufg.
Products of petroleum & coal, nec	6.8	Manufg.	Primary aluminum	6.1	Manufg.
Carbon & graphite products	6.6	Manufg.	Semiconductors & related devices	5.4	Manufg.
Wholesale trade	4.8	Trade	Carbon & graphite products	4.6	Manufg.
Gas production & distribution (utilities)	4.0	Util.	Electronic components nec	3.6	Manufg.
Motor freight transportation & warehousing	4.0	Util.	Iron & steel foundries	3.5	Manufg.
Advertising	2.7	Services	Cyclic crudes and organics	2.4	Manufg.
Minerals, ground or treated	2.3	Manufg.	Glass & glass products, except containers	2.1	Manufg.
Railroads & related services	2.0	Util.	Primary batteries, dry & wet	2.1	Manufg.
Banking	1.4	Fin/R.E.	Electric services (utilities)	1.6	Util.
Cyclic crudes and organics	1.2	Manufg.	Engine electrical equipment	1.5	Manufg.
Air transportation	1.2	Util.	Electrometallurgical products	1.2	Manufg.
Industrial inorganic chemicals, nec	0.9	Manufg.	Change in business inventories	1.2	In House
Petroleum refining	0.8	Manufg.	Fabricated plate work (boiler shops)	1.1	Manufg.
Communications, except radio & TV	0.8	Util.	Management & consulting services & labs	1.1	Services
Hotels & lodging places	0.8	Services	Federal Government purchases, national defense	1.1	Fed Govt
Computer & data processing services	0.6	Services	Motor vehicle parts & accessories	1.0	Manufg.
Eating & drinking places	0.5	Trade	Crude petroleum & natural gas	0.9	Mining
Maintenance of nonfarm buildings nec	0.4	Constr.	Steel pipe & tubes	0.9	Manufg.
Electronic components nec	0.4	Manufg.	Motors & generators	0.8	Manufg.
Primary metal products, nec	0.4	Manufg.	Transit & bus transportation	0.8	Util.
Water transportation	0.4	Util.	Nonferrous wire drawing & insulating	0.7	Manufg.
Aluminum rolling & drawing	0.3	Manufg.	Railroads & related services	0.7	Util.
Copper rolling & drawing	0.3	Manufg.	Abrasive products	0.3	Manufg.
Machinery, except electrical, nec	0.3	Manufg.	Air transportation	0.3	Util.
Porcelain electrical supplies	0.3	Manufg.	Automatic merchandising machines	0.2	Manufg.
Real estate	0.3	Fin/R.E.	Storage batteries	0.2	Manufg.
Equipment rental & leasing services	0.3	Services	Telephone & telegraph apparatus	0.2	Manufg.
Noncomparable imports	0.3	Foreign	Water transportation	0.2	Util.
Alkalies & chlorine	0.2	Manufg.	Miscellaneous repair shops	0.2	Services
Nonferrous wire drawing & insulating	0.2	Manufg.	Coal	0.1	Mining
Special dies & tools & machine tool accessories	0.2	Manufg.			
Sanitary services, steam supply, irrigation	0.2	Util.			
Legal services	0.2	Services			
Management & consulting services & labs	0.2	Services			
Paperboard containers & boxes	0.1	Manufg.			
Insurance carriers	0.1	Fin/R.E.			
Royalties	0.1	Fin/R.E.			
Accounting, auditing & bookkeeping	0.1	Services			
Automotive rental & leasing, without drivers	0.1	Services			
Automotive repair shops & services	0.1	Services			
Electrical repair shops	0.1	Services			
U.S. Postal Service	0.1	Gov't			

Source: Benchmark Input-Output Accounts for the U.S. Economy, 1982, U.S. Department of Commerce, Washington, D.C., July 1991. Data, as reported in the source, are organized by the 1977 SIC structure in use in 1982 but have been matched, as closely as is possible, to the 1987 SIC structure used in this book.

OCCUPATIONS EMPLOYED BY SIC 362 - ELECTRICAL INDUSTRIAL APPARATUS

Occupation	% of Total 1994	Change to 2005	Occupation	% of Total 1994	Change to 2005
Electrical & electronic assemblers	16.8	-28.3	Secretaries, ex legal & medical	1.4	-27.5
Assemblers, fabricators, & hand workers nec	8.0	-20.3	Drafters	1.3	-38.0
Electrical & electronic equipment assemblers	4.3	-20.3	Stock clerks	1.3	-35.3
Inspectors, testers, & graders, precision	4.2	-44.2	Machine assemblers	1.3	-28.3
Electrical & electronics engineers	3.3	-15.2	Machine tool cutting & forming etc. nec	1.2	-36.3
Blue collar worker supervisors	3.2	-33.5	Lathe & turning machine tool operators	1.2	-48.2
Coil winders, tapers, & finishers	3.0	-36.3	Traffic, shipping, & receiving clerks	1.1	-23.4
Electrical & electronic technicians, technologists	2.3	-20.4	Industrial production managers	1.1	-20.3
Machinists	2.2	-40.3	Engineering, mathematical, & science managers	1.1	-9.5
Sales & related workers nec	1.8	-20.3	Industrial machinery mechanics	1.1	7.5
Electromechanical equipment assemblers	1.5	-12.4	Production, planning, & expediting clerks	1.0	-20.3
General managers & top executives	1.5	-24.4			

Source: Industry-Occupation Matrix, Bureau of Labor Statistics. These data relate to one or more 3-digit SIC industry groups rather than to a single 4-digit SIC. The change reported for each occupation to the year 2005 is a percent of growth or decline as estimated by the Bureau of Labor Statistics. The abbreviation nec stands for 'not elsewhere classified'.

LOCATION BY STATE AND REGIONAL CONCENTRATION

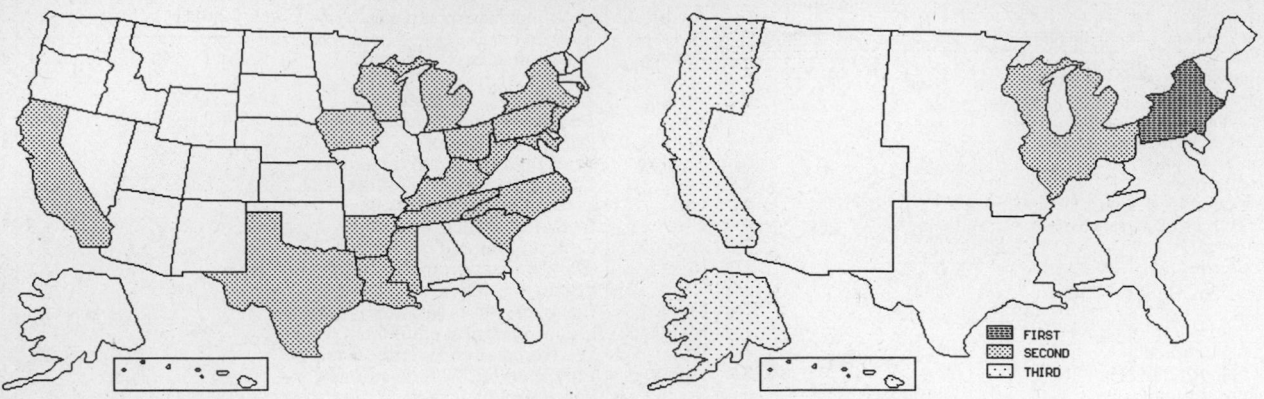

INDUSTRY DATA BY STATE

| State | Establish-ments | Shipments | | | Employment | | | | Cost as % of Shipments | Investment per Employee ($) |
		Total ($ mil)	% of U.S.	Per Establ.	Total Number	% of U.S.	Per Establ.	Wages ($/hour)		
South Carolina	5	202.5	15.9	40.5	1,100	11.5	220	13.27	42.9	-
Pennsylvania	12	181.2	14.2	15.1	1,900	19.8	158	13.24	44.9	3,737
New York	7	138.8	10.9	19.8	700	7.3	100	15.60	50.9	2,143
Ohio	14	86.3	6.8	6.2	700	7.3	50	12.91	46.3	3,571
California	11	66.2	5.2	6.0	500	5.2	45	10.14	49.8	5,400
Michigan	5	16.2	1.3	3.2	200	2.1	40	13.67	29.6	1,000
New Jersey	4	(D)	-	-	375 *	3.9	94	-	-	1,867
Texas	4	(D)	-	-	375 *	3.9	94	-	-	-
Kentucky	3	(D)	-	-	175 *	1.8	58	-	-	-
Tennessee	3	(D)	-	-	750 *	7.8	250	-	-	-
Wisconsin	3	(D)	-	-	375 *	3.9	125	-	-	-
Arkansas	2	(D)	-	-	175 *	1.8	88	-	-	-
North Carolina	2	(D)	-	-	1,750 *	18.2	875	-	-	-
West Virginia	2	(D)	-	-	375 *	3.9	188	-	-	-
Iowa	1	(D)	-	-	175 *	1.8	175	-	-	-
Louisiana	1	(D)	-	-	175 *	1.8	175	-	-	-
Maryland	1	(D)	-	-	175 *	1.8	175	-	-	-
Mississippi	1	(D)	-	-	175 *	1.8	175	-	-	-
Rhode Island	1	(D)	-	-	175 *	1.8	175	-	-	-

Source: 1992 *Economic Census.* The states are in descending order of shipments or establishments (if shipment data are missing for the majority). The symbol (D) appears when data are withheld to prevent disclosure of competitive information. States marked with (D) are sorted by number of establishments. A dash (-) indicates that the data element cannot be calculated; * indicates the midpoint of a range.

3625 - RELAYS & INDUSTRIAL CONTROLS

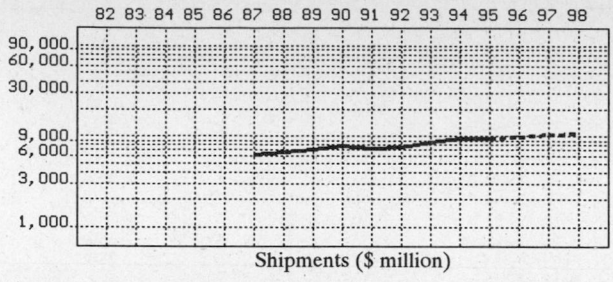

Shipments ($ million)

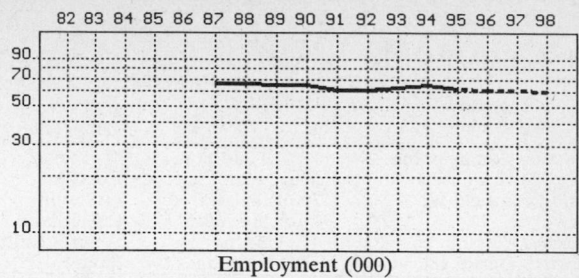

Employment (000)

GENERAL STATISTICS

| Year | Com-panies | Establishments | | Employment | | | Compensation | | Production ($ million) | | | |
		Total	with 20 or more employees	Total (000)	Production Workers (000)	Hours (Mil)	Payroll ($ mil)	Wages ($/hr)	Cost of Materials	Value Added by Manufacture	Value of Shipments	Capital Invest.
1982												
1983												
1984												
1985												
1986												
1987	1,100	1,168	439	66.6	37.6	72.7	1,630.7	9.59	2,292.7	3,781.2	6,100.5	165.5
1988		1,137	449	65.8	38.4	73.2	1,656.7	9.77	2,499.3	4,173.5	6,652.1	167.9
1989		1,090	437	64.6	39.4	73.7	1,694.5	10.01	2,931.8	4,322.1	7,222.2	213.9
1990		1,055	453	65.2	38.9	76.5	1,818.6	10.06	3,149.5	4,688.4	7,854.2	221.2
1991		1,074	438	61.2	35.3	72.3	1,798.7	10.37	3,072.8	4,289.2	7,378.6	196.4
1992	1,168	1,242	448	61.6	35.2	66.7	1,846.1	10.96	3,025.8	4,548.6	7,573.1	218.5
1993		1,221	450	63.1	37.2	68.3	1,928.8	11.58	3,442.2	5,275.5	8,726.7	217.0
1994		1,191P	449P	64.8	37.6	73.3	2,121.8	12.15	3,955.3	5,559.7	9,473.4	260.5
1995		1,204P	451P	62.0P	36.3P	69.7P	2,092.3P	12.18P	3,955.6P	5,560.1P	9,474.1P	255.8P
1996		1,217P	452P	61.6P	36.0P	69.1P	2,154.6P	12.53P	4,127.4P	5,801.5P	9,885.5P	266.5P
1997		1,229P	453P	61.1P	35.7P	68.6P	2,216.9P	12.89P	4,299.1P	6,043.0P	10,296.9P	277.2P
1998		1,242P	454P	60.6P	35.5P	68.0P	2,279.2P	13.25P	4,470.9P	6,284.5P	10,708.4P	287.9P

Sources: 1982, 1987, 1992 *Economic Census*; *Annual Survey of Manufactures*, 83-86, 88-91, 93-94. Establishment counts for non-Census years are from *County Business Patterns*; establishment values for 83-84 are extrapolations. 'P's show projections by the editors. Industries reclassified in 87 will not have data for prior years.

INDICES OF CHANGE

| Year | Com-panies | Establishments | | Employment | | | Compensation | | Production ($ million) | | | |
		Total	with 20 or more employees	Total (000)	Production Workers (000)	Hours (Mil)	Payroll ($ mil)	Wages ($/hr)	Cost of Materials	Value Added by Manufacture	Value of Shipments	Capital Invest.
1982												
1983												
1984												
1985												
1986												
1987	94	94	98	108	107	109	88	87	76	83	81	76
1988		92	100	107	109	110	90	89	83	92	88	77
1989		88	98	105	112	110	92	91	97	95	95	98
1990		85	101	106	111	115	99	92	104	103	104	101
1991		86	98	99	100	108	97	95	102	94	97	90
1992	100	100	100	100	100	100	100	100	100	100	100	100
1993		98	100	102	106	102	104	106	114	116	115	99
1994		96P	100P	105	107	110	115	111	131	122	125	119
1995		97P	101P	101P	103P	104P	113P	111P	131P	122P	125P	117P
1996		98P	101P	100P	102P	104P	117P	114P	136P	128P	131P	122P
1997		99P	101P	99P	102P	103P	120P	118P	142P	133P	136P	127P
1998		100P	101P	98P	101P	102P	123P	121P	148P	138P	141P	132P

Sources: Same as General Statistics. Values reflect change from the base year, 1992. Values above 100 mean greater than 92, values below 100 mean less than 92, and a value of 100 in the 82-91 or 93-98 period means same as 92. 'P's mark projections by the editors.

SELECTED RATIOS

For 1994	Avg. of All Manufact.	Analyzed Industry	Index	For 1994	Avg. of All Manufact.	Analyzed Industry	Index
Employees per Establishment	49	54	111	Value Added per Production Worker	134,084	147,864	110
Payroll per Establishment	1,500,273	1,780,887	119	Cost per Establishment	5,045,178	3,319,796	66
Payroll per Employee	30,620	32,744	107	Cost per Employee	102,970	61,039	59
Production Workers per Establishment	34	32	92	Cost per Production Worker	146,988	105,194	72
Wages per Establishment	853,319	747,502	88	Shipments per Establishment	9,576,895	7,951,295	83
Wages per Production Worker	24,861	23,686	95	Shipments per Employee	195,460	146,194	75
Hours per Production Worker	2,056	1,949	95	Shipments per Production Worker	279,017	251,952	90
Wages per Hour	12.09	12.15	100	Investment per Establishment	321,011	218,645	68
Value Added per Establishment	4,602,255	4,666,415	101	Investment per Employee	6,552	4,020	61
Value Added per Employee	93,930	85,798	91	Investment per Production Worker	9,352	6,928	74

Sources: Same as General Statistics. The 'Average of All Manufacturing' column represents the average of all manufacturing industries reported for the most recent complete year available. The Index shows the relationship between the Average and the Analyzed Industry. For example, 100 means that they are equal; 500 that the Analyzed Industry is five times the average; 50 means that the Analyzed Industry is half the national average. The abbreviation 'na' is used to show that data are 'not available'.

LEADING COMPANIES   Number shown: **75**   Total sales ($ mil): **5,180**   Total employment (000): **43.6**

Company Name	Address				CEO Name	Phone	Co. Type	Sales ($ mil)	Empl. (000)
Allen-Bradley Company Inc	1201 S 2nd St	Milwaukee	WI	53204	Jodie Glore	414-382-2000	S	1,500	11.0
Woodward Governor Co	5001 N 2nd St	Rockford	IL	61125	John A Halbrook	815-877-7441	P	333	3.4
Potter and Brumfield Inc	200 Richland	Princeton	IN	47671	Rolf Cousin	812-386-1000	S	330	3.0
Measurex Corp	1 Results Way	Cupertino	CA	95014	David A Bossen	408-255-1500	P	254	2.3
Dana Warner Electric	449 Gardner St	South Beloit	IL	61080	Bob Johnson	815-389-3771	D	250	2.1
Hella North America Inc	1101 Vincennes Av	Flora	IL	62839	Michael J Buford	618-662-4402	S	250	2.0
Kone Holdings Inc	12540 Westport Rd	Louisville	KY	40245	Pekka Herlin	502-339-7900	S	183	1.0
GE Control Products	W Wall St	Morrison	IL	61270	Ron Sansom	815-772-1100	D	160	1.0
Johnson Yokogawa Corp	4 Dart Rd	Newnan	GA	30265	George Doig	404-254-0400	J	160	0.6
Mallory Controls	2831 Waterfront	Indianapolis	IN	46214	Lowell Kilgus	317-328-4000	D	140	1.6
Aromat Corp	629 Central Av	New Providence	NJ	07974	Phil Yamamoto	908-464-3550	A	130	0.5
Furnas Electric Co	1000 McKee St	Batavia	IL	60510	RW Hansen	708-879-6000	R	125	1.4
Paragon Electric Company Inc	PO Box 28	Two Rivers	WI	54241	Richard DeMarle	414-793-1161	S	61*	0.6
Hi-Ram Inc	323 Skeels St	Coopersville	MI	49404	David Tenniswood	616-837-9711	S	55*	0.5
Woodward Governor Co	PO Box 1519	Fort Collins	CO	80522	John A Halbrook	303-482-5811	D	55	0.7
Leach Corp	6900 Orangethorpe	Buena Park	CA	90622	Larry Parker	714-739-0770	D	52	0.5
Magnecraft Electric Co	211 Waukegan Rd	Northfield	IL	60093	JA Steinback	708-441-2525	R	52*	0.5
Cincinnati Milacron Inc	1151 Morrow	Lebanon	OH	45036	Kenneth Schubeler	513-494-1200	D	50	0.5
Hamlin Inc	612 E Lake St	Lake Mills	WI	53551	MJ Fitzpatrick	414-648-3000	S	50	1.0
Cleveland Machine Controls Inc	7550 Hub Pkwy	Valley View	OH	44125	Jerome C Nunn	216-524-8800	R	45	0.4
Chatham Corp	350 Barclay Blv	Lincolnshire	IL	60069	Garry Brainin	708-634-5506	R	42	0.4
Minarik Corp	PO Box 25033	Glendale	CA	91201	CE Carson	818-502-1528	R	40*	0.2
Struthers-Dunn Inc	PO Box 901	Pitman	NJ	08071	CK Rivard	609-589-7500	S	40	0.3
Halmar Robicon Group	100 Sagamore Hill	Pittsburgh	PA	15239	N Antonuccio	412-327-7000	S	36	0.3
Amer Electronic Components	PO Box 280	Elkhart	IN	46515	David D Webster	219-264-1116	P	30	0.6
Gerhardt's Inc	PO Box 10161	Jefferson	LA	70181	Bruce Gerhardt	504-733-2500	R	27	0.2
Signal Control Co	2430 McGilchrist S	Salem	OR	97302	R d'Alessandro	503-371-1032	S	27*	0.2
Scientific Technologies Inc	31069 Genstar Rd	Hayward	CA	94544	Joseph J Lazzara	510-471-9717	P	26	0.2
Patriot Sensors and Controls	1080 N Crooks Rd	Clawson	MI	48017	Dick Baumhauer	313-435-0700	R	26	0.2
Fincor Electronics	3750 E Market St	York	PA	17402	A Samuelsen	717-751-4200	D	25	0.2
Klockner-Moeller Corp	25 Forge Pkwy	Franklin	MA	02038	N Dudas	508-520-7080	S	25	0.2
Namco Controls Corp	5335 Avion Park Dr	Highland H	OH	44143	Jon Slaybaugh	216-946-9900	S	25	0.2
Turck Inc	3000 Campus Dr	Plymouth	MN	55441	W A Schneider	612-553-9224	R	25	0.2
Giddings & Lewis	666 S Military Rd	Fond du Lac	WI	54935	Jody E Kurtzhalts	414-921-7100	D	22	0.2
Aerotech Inc	101 Zeta Dr	Pittsburgh	PA	15238	SJ Botos	412-963-7470	R	20	0.2
Automatic Timing & Controls	PO Box 491	Marion	OH	43302	Paul Kesler	614-387-8827	D	20*	0.2
Enercon Engineering Inc	1 Altorfer Ln	East Peoria	IL	61611	Edward Tangel	309-694-1418	R	20*	0.1
Guardian Electric Mfg Co	1425 Lake Av	Woodstock	IL	60098	Robert F Heyne	815-337-0050	S	20	0.2
Hubbell Industrial Controls Inc	50 Edwards St	Madison	OH	44057	Gary N Amato	216-428-1161	S	20	0.2
Joslyn Jennings Corp	970 McLaughlin Av	San Jose	CA	95122	JE Berkeland	408-292-4025	S	20	0.3
ITT General Controls	900 Turnbull	City of Industry	CA	91749	Roger McDonough	818-961-2547	D	19	<0.1
Avtron Manufacturing Inc	7900 E Pleasant	Independence	OH	44131	Robert J Fritz	216-642-1230	R	18	0.3
Deltrol Controls	2740 S 20th St	Milwaukee	WI	53215	Robert Oster	414-671-6800	D	17	0.3
Happ Controls Inc	106 Garlisch Dr	Elk Grove Vill	IL	60007	Frank Happ	708-593-6130	R	17	0.1
Coto Wabash	55 DuPont Dr	Providence	RI	02907	Gerry Labutti	401-943-2686	S	16	0.2
Tech/Ops Sevcon Inc	1 Beacon St	Boston	MA	02108	Bernard F Start	617-523-2030	P	16	0.2
Advanced Grade Technology Inc	396 Earhart Way	Livermore	CA	94550	R W Davidson	510-443-8161	S	15	<0.1
Midtex	9-B2 Butterfld	El Paso	TX	79906	Ted Anderson	915-772-1061	D	15	0.5
SSAC Inc	PO Box 1000	Baldwinsville	NY	13027	R Shutt	315-638-1300	R	15	0.3
Unico Inc	3725 Nicholson Rd	Franksville	WI	53126	Tom L Beck	414-886-5678	R	14	0.2
Ormec Systems Corp	19 Linden Park	Rochester	NY	14625	Gordon Presher	716-385-3520	R	13	<0.1
Banner Engineering Corp	9714 10th Av N	Minneapolis	MN	55441	R Fayfield	612-544-3164	R	12	0.2
Control Techniques Inc	4 Blackstone Val Pl	Lincoln	RI	02865	Joseph Raposa	401-333-3331	S	12*	<0.1
Inertia Dynamics Inc	PO Box 300	Collinsville	CT	06022	Steve Nyquist	203-693-0231	S	12	0.1
MSD Inc	700 Orange St	Darlington	SC	29532	James Steinbeck	803-393-5421	S	12*	0.3
Sensor Engineering Co	2155 State St	Hamden	CT	06517	Brian C Clarke	203-777-7444	D	12	0.1
Clemar Manufacturing Corp	7590 Brittan Ct	San Diego	CA	92173	AW La Fetra	818-963-9311	R	11*	0.1
Jordan Controls Inc	5607 W Douglas	Milwaukee	WI	53218	Robert Seidell	414-461-9200	R	11	<0.1
Telemotive	6470 W Courtland	Chicago	IL	60635	B Tyler	312-889-9035	D	11*	0.1
Autocon Technologies Inc	38455 Hills Tech Dr	Farmington Hls	MI	48331	John George	313-488-0440	S	10	<0.1
Electroid Co	45 Fadem Rd	Springfield	NJ	07081	Haim Loran	201-467-8100	D	10	0.1
Electromotive Systems Inc	PO Box 13615	Milwaukee	WI	53213	Frederick C Lach	414-783-3500	R	10	<0.1
KB Electronics Inc	73 Wortman Av	Brooklyn	NY	11207	G Knauer	718-257-3300	R	10*	0.2
KJ Law Engineers Inc	42300 W 9 Mile Rd	Novi	MI	48375	Kenneth J Law	810-347-3300	R	10	0.1
Uticor Technology Inc	4140 Utica Ridge	Bettendorf	IA	52722	Donald E Henry	319-359-7501	R	10	0.1
Beckwith Electric Co	6190 118th Av N	Largo	FL	34643	Robert Pettigrew	813-535-3408	R	9*	<0.1
Simplex Inc	1139 N MacArthur	Springfield	IL	62702	James L Debrey	217-528-3130	R	9	<0.1
Precision Multiple Controls Inc	33 Greenwood Av	Midland Park	NJ	07432	PH Zecher	201-444-0600	R	9	0.1
American Precision Industries	4401 Genesee St	Buffalo	NY	14225	Kurt Wiedenhaupt	716-631-9800	D	8	<0.1
Optical Associates Inc	1425 McCandless	Milpitas	CA	95035	Charles D Turk	408-263-4944	R	8	<0.1
Oregon Micro Systems Inc	1800 NW 169th Pl	Beaverton	OR	97006	Wayne Hunter	503-629-8081	R	8	<0.1
Schweitzer Eng Laboratories	2350 Hopkins	Pullman	WA	99163	E O Schweitzer III	509-332-1890	R	8*	<0.1
Vista Controls Corp	27825 Fremont Ct	Valencia	CA	91355	Ron Rambin	805-257-4430	R	8	<0.1
Solidyne Corp	1202 Carnegie St	Rolling Mdws	IL	60008	Baha Erturk	708-394-3333	R	8	<0.1
Control Chief Holdings Inc	PO Box 141	Bradford	PA	16701	J F Lamendola	814-368-4132	P	7	<0.1

Source: Ward's Business Directory of U.S. Private and Public Companies, Volumes 1 and 2, 1996. The company type code used is as follows: P - Public, R - Private, S - Subsidiary, D - Division, J - Joint Venture, A - Affiliate, G - Group. Sales are in millions of dollars, employees are in thousands. An asterisk (*) indicates an estimated sales volume. The symbol < stands for 'less than'. Company names and addresses are truncated, in some cases, to fit into the available space.

MATERIALS CONSUMED

Material	Quantity	Delivered Cost ($ million)
Materials, ingredients, containers, and supplies	(X)	2,427.4
Sheet metal products, except stampings	(X)	86.5
Metal bolts, nuts, screws, washers, rivets, and other screw machine products	(X)	45.7
All other fabricated metal products (except forgings)	(X)	81.7
Forgings	(X)	3.6
Castings (rough and semifinished)	(X)	21.3
Steel bars, bar shapes, and plates	(X)	8.6
Steel sheet and strip, including tin plate	(X)	35.4
Steel structural shapes and sheet piling	(X)	3.9
All other steel shapes and forms	(X)	8.3
Nonferrous metal smelter and refinery shapes	(X)	19.8
Aluminum and aluminum-base alloy shapes and forms	(X)	10.9
Copper and copper-base alloy shapes and forms	(X)	30.8
All other nonferrous shapes and forms	(X)	6.0
Nonferrous wire and cable, including magnet wire, bare or insulated wire, etc.	(X)	32.4
Industrial electrical control equipment purchased from other companies	(X)	247.8
Industrial electrical control equipment received from other plants of the same company	(X)	257.8
Porcelain, steatite, and other ceramic electrical products	(X)	7.7
Plastics products consumed in the form of sheets, rods, tubes, film, and other shapes	(X)	44.4
Plastics resins consumed in the form of granules, pellets, powders, liquids, etc.	(X)	22.1
Switches, except snap, toggle and push, and circuit breakers	(X)	31.5
Resistors, capacitors, transformers, and other electronic-type components	(X)	404.2
All other materials and components, parts, containers, and supplies	(X)	674.1
Materials, ingredients, containers, and supplies, nsk	(X)	342.9

Source: 1992 Economic Census. Explanation of symbols used: (D): Withheld to avoid disclosure of competitive data; na: Not available; (S): Withheld because statistical norms were not met; (X): Not applicable; (Z): Less than half the unit shown; nec: Not elsewhere classified; nsk: Not specified by kind; - : zero; * : 10-19 percent estimated; ** : 20-29 percent estimated.

PRODUCT SHARE DETAILS

Product or Product Class	% Share	Product or Product Class	% Share
Relays and industrial controls	100.00	General purpose industrial controls	40.38
Relays for electronic circuitry, industrial control, overload,		Parts for industrial controls and motor-control accessories	6.91
and switchgear type	11.82	Relays and industrial controls, nsk	9.00
Specific purpose industrial controls	31.89		

Source: 1992 Economic Census. The values shown are percent of total shipments in an industry. Values of indented subcategories are summed in the main heading. The symbol (D) appears when data are withheld to prevent disclosure of competitive information. The abbreviation nsk stands for 'not specified by kind' and nec for 'not elsewhere classified'.

INPUTS AND OUTPUTS FOR INDUSTRIAL CONTROLS

Economic Sector or Industry Providing Inputs	%	Sector	Economic Sector or Industry Buying Outputs	%	Sector
Electronic components nec	16.0	Manufg.	Electronic computing equipment	13.9	Manufg.
Wholesale trade	12.8	Trade	Gross private fixed investment	10.5	Cap Inv
Communications, except radio & TV	12.4	Util.	Switchgear & switchboard apparatus	9.3	Manufg.
Industrial controls	12.1	Manufg.	Industrial controls	6.9	Manufg.
Imports	4.5	Foreign	Instruments to measure electricity	6.0	Manufg.
Advertising	3.5	Services	Radio & TV communication equipment	3.1	Manufg.
Maintenance of nonfarm buildings nec	2.1	Constr.	Management & consulting services & labs	3.1	Services
Banking	2.1	Fin/R.E.	Machine tools, metal cutting types	2.8	Manufg.
Blast furnaces & steel mills	1.7	Manufg.	Crude petroleum & natural gas	2.7	Mining
Copper rolling & drawing	1.6	Manufg.	Exports	2.7	Foreign
Electric services (utilities)	1.6	Util.	Printing trades machinery	2.1	Manufg.
Air transportation	1.5	Util.	Electric services (utilities)	1.9	Util.
Switchgear & switchboard apparatus	1.4	Manufg.	Local government passenger transit	1.9	Gov't
Sheet metal work	1.3	Manufg.	Power driven hand tools	1.8	Manufg.
Plating & polishing	1.2	Manufg.	Refrigeration & heating equipment	1.8	Manufg.
Nonferrous wire drawing & insulating	1.1	Manufg.	Motor vehicles & car bodies	1.6	Manufg.
Primary aluminum	1.1	Manufg.	Special industry machinery, nec	1.4	Manufg.
Screw machine and related products	1.1	Manufg.	Blast furnaces & steel mills	1.3	Manufg.
Miscellaneous plastics products	1.0	Manufg.	Ship building & repairing	1.3	Manufg.
Motors & generators	1.0	Manufg.	Transformers	1.3	Manufg.
Equipment rental & leasing services	1.0	Services	Motors & generators	1.2	Manufg.
Primary copper	0.9	Manufg.	Automotive repair shops & services	1.2	Services
Instruments to measure electricity	0.8	Manufg.	Coal	1.0	Mining
Petroleum refining	0.8	Manufg.	General industrial machinery, nec	1.0	Manufg.
Eating & drinking places	0.8	Trade	Oil field machinery	1.0	Manufg.
Hotels & lodging places	0.8	Services	Turbines & turbine generator sets	1.0	Manufg.
Noncomparable imports	0.8	Foreign	Federal Government purchases, national defense	0.9	Fed Govt
Paperboard containers & boxes	0.7	Manufg.	Internal combustion engines, nec	0.8	Manufg.
Plastics materials & resins	0.7	Manufg.	Aircraft	0.7	Manufg.
Wiring devices	0.7	Manufg.	Aluminum rolling & drawing	0.7	Manufg.

Continued on next page.

INPUTS AND OUTPUTS FOR INDUSTRIAL CONTROLS - Continued

Economic Sector or Industry Providing Inputs	%	Sector	Economic Sector or Industry Buying Outputs	%	Sector
Aluminum rolling & drawing	0.6	Manufg.	Pumps & compressors	0.7	Manufg.
Real estate	0.6	Fin/R.E.	Construction machinery & equipment	0.6	Manufg.
Aluminum castings	0.5	Manufg.	Pipe, valves, & pipe fittings	0.6	Manufg.
Gaskets, packing & sealing devices	0.5	Manufg.	Food products machinery	0.5	Manufg.
Machinery, except electrical, nec	0.5	Manufg.	Household refrigerators & freezers	0.5	Manufg.
Gas production & distribution (utilities)	0.5	Util.	Railroad equipment	0.5	Manufg.
Motor freight transportation & warehousing	0.5	Util.	Motor freight transportation & warehousing	0.5	Util.
Fabricated rubber products, nec	0.4	Manufg.	Cold finishing of steel shapes	0.4	Manufg.
Paints & allied products	0.4	Manufg.	Copper rolling & drawing	0.4	Manufg.
Computer & data processing services	0.4	Services	Industrial trucks & tractors	0.4	Manufg.
Legal services	0.4	Services	Iron & steel foundries	0.4	Manufg.
Iron & steel foundries	0.3	Manufg.	Mechanical measuring devices	0.4	Manufg.
Paper coating & glazing	0.3	Manufg.	Nonferrous wire drawing & insulating	0.4	Manufg.
Special dies & tools & machine tool accessories	0.3	Manufg.	Welding apparatus, electric	0.4	Manufg.
Management & consulting services & labs	0.3	Services	Railroads & related services	0.4	Util.
Manifold business forms	0.2	Manufg.	Transit & bus transportation	0.4	Util.
Metal coating & allied services	0.2	Manufg.	Automotive stampings	0.3	Manufg.
Miscellaneous fabricated wire products	0.2	Manufg.	Conveyors & conveying equipment	0.3	Manufg.
Nonferrous castings, nec	0.2	Manufg.	Hoists, cranes, & monorails	0.3	Manufg.
Transformers	0.2	Manufg.	Nonferrous rolling & drawing, nec	0.3	Manufg.
Accounting, auditing & bookkeeping	0.2	Services	Paper industries machinery	0.3	Manufg.
U.S. Postal Service	0.2	Gov't	Textile machinery	0.3	Manufg.
Abrasive products	0.1	Manufg.	X-ray apparatus & tubes	0.3	Manufg.
Lubricating oils & greases	0.1	Manufg.	Chemical & fertilizer mineral	0.2	Mining
Machine tools, metal cutting types	0.1	Manufg.	Dimension, crushed & broken stone	0.2	Mining
Railroads & related services	0.1	Util.	Heating equipment, except electric	0.2	Manufg.
Insurance carriers	0.1	Fin/R.E.	Industrial furnaces & ovens	0.2	Manufg.
Royalties	0.1	Fin/R.E.	Machine tools, metal forming types	0.2	Manufg.
Automotive rental & leasing, without drivers	0.1	Services	Signs & advertising displays	0.2	Manufg.
Photofinishing labs, commercial photography	0.1	Services	Special dies & tools & machine tool accessories	0.2	Manufg.
			Steel pipe & tubes	0.2	Manufg.
			S/L Govt. purch., higher education	0.2	S/L Govt
			Copper ore	0.1	Mining
			Nonferrous metal ores, except copper	0.1	Mining
			Sand & gravel	0.1	Mining
			Automatic merchandising machines	0.1	Manufg.
			Steel wire & related products	0.1	Manufg.
			Typewriters & office machines, nec	0.1	Manufg.
			Woodworking machinery	0.1	Manufg.
			Pipelines, except natural gas	0.1	Util.
			Federal Government purchases, nondefense	0.1	Fed Govt

Source: Benchmark Input-Output Accounts for the U.S. Economy, 1982, U.S. Department of Commerce, Washington, D.C., July 1991. Data, as reported in the source, are organized by the 1977 SIC structure in use in 1982 but have been matched, as closely as is possible, to the 1987 SIC structure used in this book.

OCCUPATIONS EMPLOYED BY SIC 362 - ELECTRICAL INDUSTRIAL APPARATUS

Occupation	% of Total 1994	Change to 2005	Occupation	% of Total 1994	Change to 2005
Electrical & electronic assemblers	16.8	-28.3	Secretaries, ex legal & medical	1.4	-27.5
Assemblers, fabricators, & hand workers nec	8.0	-20.3	Drafters	1.3	-38.0
Electrical & electronic equipment assemblers	4.3	-20.3	Stock clerks	1.3	-35.3
Inspectors, testers, & graders, precision	4.2	-44.2	Machine assemblers	1.3	-28.3
Electrical & electronics engineers	3.3	-15.2	Machine tool cutting & forming etc. nec	1.2	-36.3
Blue collar worker supervisors	3.2	-33.5	Lathe & turning machine tool operators	1.2	-48.2
Coil winders, tapers, & finishers	3.0	-36.3	Traffic, shipping, & receiving clerks	1.1	-23.4
Electrical & electronic technicians,technologists	2.3	-20.4	Industrial production managers	1.1	-20.3
Machinists	2.2	-40.3	Engineering, mathematical, & science managers	1.1	-9.5
Sales & related workers nec	1.8	-20.3	Industrial machinery mechanics	1.1	7.5
Electromechanical equipment assemblers	1.5	-12.4	Production, planning, & expediting clerks	1.0	-20.3
General managers & top executives	1.5	-24.4			

Source: Industry-Occupation Matrix, Bureau of Labor Statistics. These data relate to one or more 3-digit SIC industry groups rather than to a single 4-digit SIC. The change reported for each occupation to the year 2005 is a percent of growth or decline as estimated by the Bureau of Labor Statistics. The abbreviation nec stands for 'not elsewhere classified'.

LOCATION BY STATE AND REGIONAL CONCENTRATION

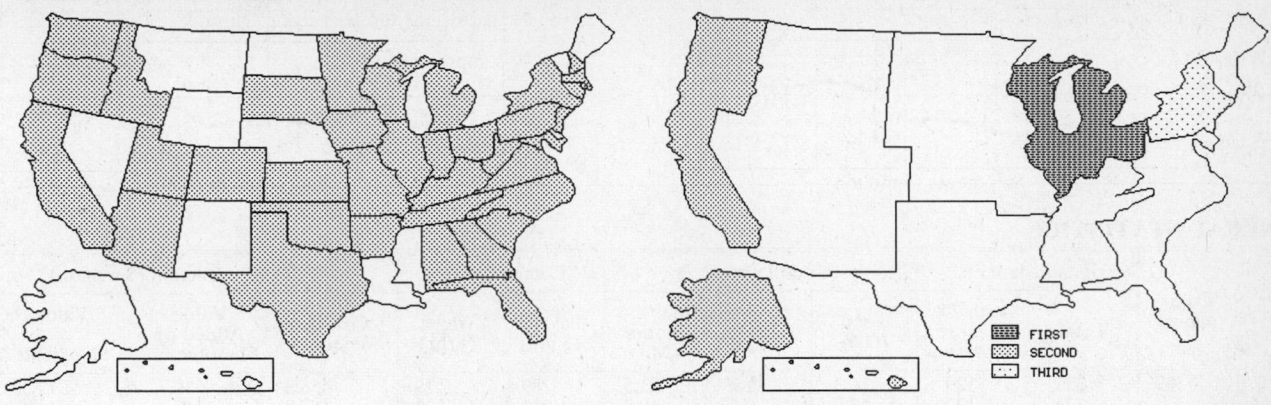

FIRST
SECOND
THIRD

INDUSTRY DATA BY STATE

| State | Establish-ments | Shipments | | | Employment | | | | Cost as % of Shipments | Investment per Employee ($) |
		Total ($ mil)	% of U.S.	Per Establ.	Total Number	% of U.S.	Per Establ.	Wages ($/hour)		
Wisconsin	62	1,099.5	14.5	17.7	8,000	13.0	129	10.74	36.4	4,875
North Carolina	31	712.6	9.4	23.0	4,500	7.3	145	12.37	43.6	3,933
Ohio	76	688.2	9.1	9.1	4,900	8.0	64	10.82	28.7	4,776
California	153	609.9	8.1	4.0	5,300	8.6	35	10.59	36.6	2,642
Illinois	89	477.7	6.3	5.4	4,100	6.7	46	9.94	46.9	3,049
Pennsylvania	76	402.7	5.3	5.3	3,100	5.0	41	13.00	47.7	-
Massachusetts	45	303.8	4.0	6.8	2,600	4.2	58	14.00	39.8	3,731
New York	71	291.4	3.8	4.1	3,300	5.4	46	9.87	40.8	2,636
Florida	48	273.0	3.6	5.7	2,000	3.2	42	9.68	33.7	3,500
Texas	60	240.9	3.2	4.0	2,000	3.2	33	9.05	52.6	1,700
Minnesota	39	178.1	2.4	4.6	1,700	2.8	44	9.65	41.2	-
Kentucky	8	166.1	2.2	20.8	1,300	2.1	163	18.44	41.3	-
Michigan	68	154.2	2.0	2.3	1,500	2.4	22	10.33	40.9	1,933
South Carolina	15	138.6	1.8	9.2	1,600	2.6	107	10.05	54.8	2,938
Indiana	35	125.0	1.7	3.6	1,500	2.4	43	10.38	42.5	-
Connecticut	40	124.4	1.6	3.1	1,200	1.9	30	10.00	40.0	1,750
New Jersey	49	110.9	1.5	2.3	1,600	2.6	33	9.79	35.5	1,563
Maryland	9	74.5	1.0	8.3	700	1.1	78	9.20	36.9	-
Washington	32	59.8	0.8	1.9	600	1.0	19	10.67	41.5	5,667
Tennessee	14	51.4	0.7	3.7	400	0.6	29	8.00	44.6	-
Missouri	18	34.1	0.5	1.9	400	0.6	22	9.50	34.3	-
Colorado	13	32.3	0.4	2.5	300	0.5	23	8.00	39.0	2,000
Alabama	14	22.6	0.3	1.6	400	0.6	29	10.00	38.9	1,250
Oregon	17	13.7	0.2	0.8	100	0.2	6	11.00	51.8	2,000
West Virginia	7	11.9	0.2	1.7	200	0.3	29	17.00	40.3	-
Georgia	24	(D)	-	-	1,750 *	2.8	73	-	-	1,771
Virginia	20	(D)	-	-	3,750 *	6.1	188	-	-	-
Oklahoma	15	(D)	-	-	175 *	0.3	12	-	-	1,143
Kansas	14	(D)	-	-	375 *	0.6	27	-	-	-
Iowa	12	(D)	-	-	750 *	1.2	63	-	-	-
New Hampshire	12	(D)	-	-	375 *	0.6	31	-	-	-
Arizona	11	(D)	-	-	175 *	0.3	16	-	-	-
Rhode Island	6	(D)	-	-	750 *	1.2	125	-	-	667
Utah	6	(D)	-	-	175 *	0.3	29	-	-	-
Arkansas	5	(D)	-	-	175 *	0.3	35	-	-	-
Idaho	5	(D)	-	-	175 *	0.3	35	-	-	-
South Dakota	2	(D)	-	-	175 *	0.3	88	-	-	-

Source: 1992 *Economic Census*. The states are in descending order of shipments or establishments (if shipment data are missing for the majority). The symbol (D) appears when data are withheld to prevent disclosure of competitive information. States marked with (D) are sorted by number of establishments. A dash (-) indicates that the data element cannot be calculated; * indicates the midpoint of a range.

3629 - ELECTRICAL INDUSTRIAL APPARATUS, NEC

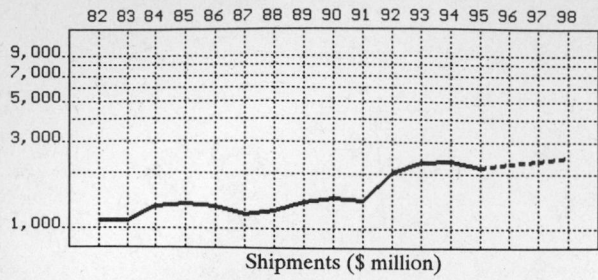

82 83 84 85 86 87 88 89 90 91 92 93 94 95 96 97 98

Shipments ($ million)

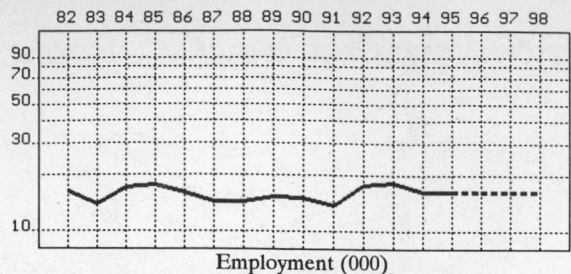

82 83 84 85 86 87 88 89 90 91 92 93 94 95 96 97 98

Employment (000)

GENERAL STATISTICS

| Year | Com-panies | Establishments | | Employment | | | Compensation | | Production ($ million) | | | |
		Total	with 20 or more employees	Total (000)	Production Workers (000)	Hours (Mil)	Payroll ($ mil)	Wages ($/hr)	Cost of Materials	Value Added by Manufacture	Value of Shipments	Capital Invest.
1982	309	323	120	16.3	11.3	21.5	280.4	7.19	456.3	644.7	1,111.3	29.3
1983		323	127	13.9	9.7	18.8	257.6	7.60	465.8	641.0	1,112.8	35.2
1984		323	134	17.2	11.3	21.8	342.6	8.69	573.3	765.1	1,318.0	33.0
1985		324	141	17.7	11.3	21.9	362.3	8.67	585.9	767.1	1,353.4	31.5
1986		315	140	16.0	10.6	19.8	355.2	9.34	564.8	762.5	1,327.6	29.4
1987	472	481	106	14.5	8.9	17.7	325.1	9.03	483.8	698.4	1,188.2	31.9
1988		449	110	14.3	9.1	18.0	327.4	9.15	533.8	749.4	1,262.2	24.2
1989		411	116	15.4	9.1	17.5	351.8	9.29	570.4	831.2	1,406.5	22.1
1990		387	118	15.1	8.9	18.2	365.6	9.34	622.6	846.3	1,465.4	19.7
1991		376	121	13.8	8.7	17.6	345.7	9.42	619.1	812.4	1,434.8	25.0
1992	342	355	151	17.6	11.7	22.9	473.5	9.55	951.5	1,096.9	2,039.3	65.3
1993		330	148	18.2	12.0	24.9	503.8	9.57	1,175.5	1,180.3	2,332.9	40.5
1994		396P	133P	16.4	11.4	23.3	484.0	10.30	1,117.0	1,258.8	2,353.9	68.5
1995		401P	134P	16.3P	10.4P	21.4P	481.7P	10.34P	1,032.9P	1,164.0P	2,176.6P	48.2P
1996		405P	135P	16.4P	10.4P	21.6P	498.0P	10.53P	1,077.7P	1,214.5P	2,271.0P	50.1P
1997		410P	136P	16.4P	10.4P	21.8P	514.4P	10.71P	1,122.5P	1,264.9P	2,365.4P	52.0P
1998		414P	137P	16.5P	10.5P	21.9P	530.7P	10.90P	1,167.2P	1,315.4P	2,459.8P	53.8P

Sources: 1982, 1987, 1992 *Economic Census*; *Annual Survey of Manufactures*, 83-86, 88-91, 93-94. Establishment counts for non-Census years are from *County Business Patterns*; establishment values for 83-84 are extrapolations. 'P's show projections by the editors. Industries reclassified in 87 will not have data for prior years.

INDICES OF CHANGE

| Year | Com-panies | Establishments | | Employment | | | Compensation | | Production ($ million) | | | |
		Total	with 20 or more employees	Total (000)	Production Workers (000)	Hours (Mil)	Payroll ($ mil)	Wages ($/hr)	Cost of Materials	Value Added by Manufacture	Value of Shipments	Capital Invest.
1982	90	91	79	93	97	94	59	75	48	59	54	45
1983		91	84	79	83	82	54	80	49	58	55	54
1984		91	89	98	97	95	72	91	60	70	65	51
1985		91	93	101	97	96	77	91	62	70	66	48
1986		89	93	91	91	86	75	98	59	70	65	45
1987	138	135	70	82	76	77	69	95	51	64	58	49
1988		126	73	81	78	79	69	96	56	68	62	37
1989		116	77	87	78	76	74	97	60	76	69	34
1990		109	78	86	76	79	77	98	65	77	72	30
1991		106	80	78	74	77	73	99	65	74	70	38
1992	100	100	100	100	100	100	100	100	100	100	100	100
1993		93	98	103	103	109	106	100	124	108	114	62
1994		112P	88P	93	97	102	102	108	117	115	115	105
1995		113P	89P	93P	89P	94P	102P	108P	109P	106P	107P	74P
1996		114P	89P	93P	89P	94P	105P	110P	113P	111P	111P	77P
1997		115P	90P	93P	89P	95P	109P	112P	118P	115P	116P	80P
1998		117P	91P	94P	89P	96P	112P	114P	123P	120P	121P	82P

Sources: Same as General Statistics. Values reflect change from the base year, 1992. Values above 100 mean greater than 92, values below 100 mean less than 92, and a value of 100 in the 82-91 or 93-98 period means same as 92. 'P's mark projections by the editors.

SELECTED RATIOS

For 1994	Avg. of All Manufact.	Analyzed Industry	Index	For 1994	Avg. of All Manufact.	Analyzed Industry	Index
Employees per Establishment	49	41	84	Value Added per Production Worker	134,084	110,421	82
Payroll per Establishment	1,500,273	1,221,848	81	Cost per Establishment	5,045,178	2,819,844	56
Payroll per Employee	30,620	29,512	96	Cost per Employee	102,970	68,110	66
Production Workers per Establishment	34	29	84	Cost per Production Worker	146,988	97,982	67
Wages per Establishment	853,319	605,850	71	Shipments per Establishment	9,576,895	5,942,373	62
Wages per Production Worker	24,861	21,052	85	Shipments per Employee	195,460	143,530	73
Hours per Production Worker	2,056	2,044	99	Shipments per Production Worker	279,017	206,482	74
Wages per Hour	12.09	10.30	85	Investment per Establishment	321,011	172,927	54
Value Added per Establishment	4,602,255	3,177,815	69	Investment per Employee	6,552	4,177	64
Value Added per Employee	93,930	76,756	82	Investment per Production Worker	9,352	6,009	64

Sources: Same as General Statistics. The 'Average of All Manufacturing' column represents the average of all manufacturing industries reported for the most recent complete year available. The Index shows the relationship between the Average and the Analyzed Industry. For example, 100 means that they are equal; 500 that the Analyzed Industry is five times the average; 50 means that the Analyzed Industry is half the national average. The abbreviation 'na' is used to show that data are 'not available'.

LEADING COMPANIES Number shown: 74 Total sales ($ mil): 1,739 Total employment (000): 13.2

Company Name	Address				CEO Name	Phone	Co. Type	Sales ($ mil)	Empl. (000)
Exide Electronics Group Inc	8521 Six Forks Rd	Raleigh	NC	27615	James A Risher	919-872-3020	P	327	1.4
ZEXEL USA Corp	625 Southside Dr	Decatur	IL	62521	A Tanaka	217-362-2300	S	170	0.8
International Components Corp	420 N May St	Chicago	IL	60622	Jim Gaza	312-829-7101	R	150	0.9
Exide Electronics Corp	8521 Six Forks Rd	Raleigh	NC	27615	James A Risher	919-872-3020	S	120*	1.1
Maxwell Laboratories Inc	8888 Balboa Av	San Diego	CA	92123	Alan C Kolb	619-279-5100	P	85	0.7
Vicor Corp	23 Frontage Rd	Andover	MA	01810	Patrizio Vinciarelli	508-470-2900	P	84	0.7
Acme Electric Corp	400 Quaker Rd	East Aurora	NY	14052	G Wayne Hawk	716-655-3800	P	79	0.7
Thermon Manufacturing Co	PO Box 609	San Marcos	TX	78667	Richard L Burdick	512-396-5801	R	60	0.3
Schumacher Electric Corp	7474 N Rogers Av	Chicago	IL	60626	Donald Schumacher	312-973-1600	R	46*	0.8
Midwest Electric Products Inc	Hwy 22 N	Mankato	MN	56002	Mary Frances-Cox	507-625-4414	S	43	0.2
International Fuel Cells Corp	PO Box 739	South Windsor	CT	06074	H David Ramm	203-727-2200	S	40	0.4
AGIE USA Ltd	PO Box 220	Davidson	NC	28036	Gabriele G Carinci	704-892-8011	S	30	0.1
Siemens Solar Industries Inc	PO Box 6032	Camarillo	CA	93010	George Roland	805-482-6800	S	28*	0.3
Mupac Corp	10 Mupac Dr	Brockton	MA	02401	Frank Angelis	508-588-6110	S	25	0.2
Robotron Corp	PO Box 5090	Southfield	MI	48086	L J Brzozowski	810-350-1444	R	22	0.2
Tripp Lite	500 N Orleans St	Chicago	IL	60610	Ebert Howell	312-329-1777	D	21	0.2
La Marche Manufacturing Co	106 Bradrock Dr	Des Plaines	IL	60018	John J Walsh	708-299-1188	R	20	0.2
Lester Electrical of Nebraska	625 W A St	Lincoln	NE	68522	JL Carrier	402-477-8988	R	20	0.3
M-C Power Corp	8040 S Madison St	Burr Ridge	IL	60521	Paul B Tarman	708-986-8040	S	20	<0.1
Ault Inc	7300 Boone Av N	Minneapolis	MN	55428	Frederick M Green	612-493-1900	P	18	0.3
Associated Equipment Corp	5043 Farlin Av	St Louis	MO	63115	B Drake	314-385-5178	R	15	0.1
Cyberex Inc	7171 Industrial Park	Mentor	OH	44060	E Yohman	216-946-1783	R	15	0.1
Para Systems Inc	1455 LeMay Dr	Dallas	TX	75007	Steve Lam	214-446-7363	S	15	<0.1
Astrosystems Inc	6 Nevada Dr	Lake Success	NY	11042	Seymour Barth	516-328-1600	P	13	0.1
WPI Group Inc	PO Box 267	Warner	NH	03278	Michael H Foster	603-456-3111	P	12	0.1
Delta-Unibus Corp	11323 W Franklin	Franklin Park	IL	60131	Adam Janas	708-451-1982	S	12	<0.1
Good-All Electric Inc	3725 Canal Dr	Fort Collins	CO	80524	Ron Bladon	303-484-3080	S	12	<0.1
Marlow Industries Inc	10451 Vista Park Rd	Dallas	TX	75238	Raymond Marlow	214-340-4900	R	12	0.2
Japler Aquisition Co	4500 Alpine Av	Cincinnati	OH	45242	Arthur P Hopmeier	513-791-3030	R	11	0.2
Pauluhn Electric Mfg Co	PO Box 53	Pearland	TX	77588	JP Russo	713-485-4311	S	11	<0.1
Ascom Warren Inc	12 Executive Dr	Hudson	NH	03051	John Toomey	603-598-4700	R	10	<0.1
Insul-8 Corp	1417 Indrial Pkwy	Harlan	IA	51537	Donald Brockley	712-755-3050	R	10	<0.1
Teledyne Isotopes Inc	10707 Gilroy Rd	Hunt Valley	MD	21031	William C Kincaidi	410-771-8600	S	10*	0.1
Pemco Corp	PO Box 1319	Bluefield	VA	24605	WL Sowers	703-326-2611	R	10	<0.1
Heart Interface Corp	21440 68th Av S	Kent	WA	98032	Warren Stokes	206-872-7225	S	9*	<0.1
Simco Company Inc	2257 N Penn Rd	Hatfield	PA	19440	Gary Swink	215-822-2171	S	9	0.1
Industrial Devices Corp	64 Digital Dr	Novato	CA	94949	Scott Johnson	415-883-2094	R	8	<0.1
Lortec Power Systems Inc	145 Keep Ct	Elyria	OH	44035	John Mulcair	216-327-5050	S	8	<0.1
Energy Conversion Devices Inc	1675 W Maple Rd	Troy	MI	48084	S R Ovshinsky	313-280-1900	P	7	0.2
Associated Ceramics	PO Box 144	Sarver	PA	16055	RC Lassinger	412-353-1585	R	7	0.1
Briskheat Corp	PO Box 628	Columbus	OH	43216	Richard E Jacob	614-294-3376	R	7	<0.1
Ekstrom Industries Inc	23847 Industrial Pk	Farmington Hls	MI	48335	J Roessner	313-477-0040	S	7	<0.1
Blue Grass Manufacturing	673 Kennedy Rd	Lexington	KY	40511	D Bundy	606-233-7445	R	7	<0.1
Arnold Magnetics Corp	4000 Via Pescador	Camarillo	CA	93012	Robert Deininger	805-484-4221	R	6	<0.1
Bitrode Corp	1642 Manufacturers	Fenton	MO	63026	Don Brandt	314-343-6112	R	6	<0.1
Able Coil and Electronics Co	PO Box 9127	Bolton	CT	06043	K D Rockefeller	203-646-5686	R	6	<0.1
Component Research Company	1655 26th St	Santa Monica	CA	90404	Hillel Kellerman	310-829-3615	R	6	0.2
Hardy Instruments Inc	3860 Fortunada	San Diego	CA	92123	David Ness	619-278-2900	S	6	0.1
EMF Corp	505 Pokagon Trl	Angola	IN	46703	Richard Poe	219-665-9541	R	5	0.2
Vanner Weldon Inc	4282 Reynolds Dr	Hilliard	OH	43026	Simon Russell	614-771-2718	S	5*	<0.1
Universal Voltronics	27 Radio Circle Dr	Mount Kisco	NY	10549	James W Wood Jr	914-241-1300	D	5	<0.1
DLS	510 E 2nd St	Bremen	IN	46506	WL Dickerhoff	219-546-4717	D	5	<0.1
Elma Engineering	1066 E Meadow Cir	Palo Alto	CA	94303	KW Widl	415-494-7303	R	4	<0.1
Hot Shot Products Company Inc	5441 W 125th St	Savage	MN	55378	William Bartel Sr	612-890-3520	R	4	<0.1
Technipower Inc	PO Box 222	Danbury	CT	06810	John D Familetti	203-748-7001	S	4*	<0.1
Electrofilm Manufacturing Co	PO Box 55669	Valencia	CA	91385	Myron Tomikawa	805-257-2242	R	4	<0.1
GMH Electronics Corp	PO Box 1194	Roxboro	NC	27573	Philip Greenspan	910-599-1011	R	3*	<0.1
Howell Corp	1180 Stratford Rd	Stratford	CT	06497	Robin A Clarke	203-375-5651	R	3	<0.1
Willow Manufacturing Inc	PO Box 17	Parkville	MN	55773	Reynold Herzog	218-741-0073	R	3	<0.1
HV Component Associates Inc	PO Box 2484	Farmingdale	NJ	07727	Dennis R Dean	908-938-4499	R	3	<0.1
Oryx Technology Corp	47341 Bayside Pkwy	Fremont	CA	94538	Kailash Joshi	510-249-1144	P	2	<0.1
Cable Service Technologies Inc	1140 Pearl St	Boulder	CO	80302	Gary Beyson	303-442-6200	R	2*	<0.1
Clinton Power Co	PO Box 1597	Northbrook	IL	60062	Bruce Benjamin	708-498-4200	R	2*	<0.1
Stored Energy Systems	1840 Industrial Cir	Longmont	CO	80501	Herb Kaewert	303-678-7500	R	2*	<0.1
MIL Electronics Inc	1 Chesnut St	Nashua	NH	03060	Robert Maynard	603-882-3200	R	2	<0.1
Tribotech	PO Box 5030	Napa	CA	94581	JA Carlson	707-643-2148	R	2	<0.1
Eastron Corp	15 Hale St	Haverhill	MA	01830	William E Slusher	508-373-3824	R	1*	<0.1
Electronic Power Technology Inc	6400 Atlantic Blv	Norcross	GA	30071	Karen A Robinson	404-449-1104	R	1*	<0.1
Lightning Master Corp	PO Box 6017	Clearwater	FL	34618	Bruce A Kaiser	813-447-6800	R	1	<0.1
Lind Electronic Design Inc	6414 Cambridge St	Minneapolis	MN	55426	Leroy Lind	612-927-6303	R	1*	<0.1
Superconductivity Inc	PO Box 56074	Madison	WI	53705	Paul F Koeppe	608-831-5773	R	1*	<0.1
AB Tool and Manufacturing Inc	1350 Balata St	Easton	PA	18042	David F Mowen	215-258-7678	R	1	<0.1
Crown Industrial Products Co	PO Box 350	Hebron	IL	60034	John Strett	815-648-2427	R	1*	<0.1
Humbug Mountain Research	PO Box 1380	Duarte	CA	91009	Alan Vetter	818-303-2400	R	0*	<0.1

Source: Ward's Business Directory of U.S. Private and Public Companies, Volumes 1 and 2, 1996. The company type code used is as follows: P - Public, R - Private, S - Subsidiary, D - Division, J - Joint Venture, A - Affiliate, G - Group. Sales are in millions of dollars, employees are in thousands. An asterisk (*) indicates an estimated sales volume. The symbol < stands for 'less than'. Company names and addresses are truncated, in some cases, to fit into the available space.

MATERIALS CONSUMED

Material	Quantity	Delivered Cost ($ million)
Materials, ingredients, containers, and supplies	(X)	830.0
Metal bolts, nuts, screws, washers, rivets, and other screw machine products	(X)	14.0
Other fabricated metal products (except castings and forgings)	(X)	16.3
Castings (rough and semifinished)	(X)	2.0
Steel shapes and forms	(X)	13.5
Copper and copper-base alloy shapes and forms	(X)	6.8
Aluminum and aluminum-base alloy sheet, plate, foil, and welded tubing	(X)	4.8
All other aluminum and aluminum-base alloy shapes and forms	(X)	3.6
Other nonferrous shapes and forms	(X)	4.4
Magnet wire	(X)	24.7
Insulated wire and cable, except magnet wire	(X)	17.1
Porcelain, steatite, and other ceramic electrical products	(X)	0.7
Industrial electrical control equipment	(X)	58.1
Plastics products consumed in the form of sheets, rods, tubes, film, and other shapes	(X)	9.1
Resistors, capacitors, transformers, and other electronic-type components	(X)	70.0
Paperboard containers, boxes, and corrugated paperboard	(X)	5.4
Flexible packaging materials	(X)	1.5
All other materials and components, parts, containers, and supplies	(X)	436.6
Materials, ingredients, containers, and supplies, nsk	(X)	141.4

Source: 1992 Economic Census. Explanation of symbols used: (D): Withheld to avoid disclosure of competitive data; na: Not available; (S): Withheld because statistical norms were not met; (X): Not applicable; (Z): Less than half the unit shown; nec: Not elsewhere classified; nsk: Not specified by kind; - : zero; * : 10-19 percent estimated; ** : 20-29 percent estimated.

PRODUCT SHARE DETAILS

Product or Product Class	% Share	Product or Product Class	% Share
Electrical industrial apparatus, nec	100.00	supplies more than 100 kW	4.78
Capacitors for industrial use (except for electronic circuitry)	14.68	All other ac to dc semiconductor power conversion apparatus, including computer supplies	20.17
Shunt and series power capacitors, units, and equipment, 1/2 kVA or more, and accessories for industrial use (except for electronic circuitry)	64.65	Other rectifying (power conversion) apparatus	44.08
		Rectifying apparatus, nsk	2.81
Other capacitors (except electrolytic) incl. ac, general purpose for motors, controls, high intensity discharge lighting for industrial use (except for electronic circuitry)	32.84	Other electrical equipment for industrial use, except for electronic circuitry	21.53
		Electrical equipment for industrial use, except for electronic circuitry, coil windings, electrical	7.68
Capacitors for industrial use (except for electronic circuitry), nsk	2.50	Electrical equipment for industrial use, except for electronic circuitry, solenoids (except solenoid-actuated regulating valves)	12.06
Rectifying apparatus	55.85		
Automotive rectifying semiconductor power conversion apparatus, except for electronic circuitry, battery chargers	7.36	Electrical equipment for industrial use, except for electronic circuitry, surge suppressors	9.72
Industrial and railroad rectifying semiconductor power conversion apparatus, except for electronic circuitry, battery chargers	7.64	Electrical equipment for industrial use, except for electronic circuitry, cathodic protection equipment	1.78
Rectifying power conversion apparatus, except for electronic circuitry, semiconductor high-voltage power supplies in excess of 2 kV, 100 kW or less	13.16	Other miscellaneous electrical equipment for industrial use, nec, including electrical discharge equipment	67.35
Rectifying power conversion apparatus, except for electronic circuitry, semiconductor high-voltage power		Other electrical equipment for industrial use, except for electronic circuitry, nsk	1.41
		Electrical industrial apparatus, nec, nsk	7.95

Source: 1992 Economic Census. The values shown are percent of total shipments in an industry. Values of indented subcategories are summed in the main heading. The symbol (D) appears when data are withheld to prevent disclosure of competitive information. The abbreviation nsk stands for 'not specified by kind' and nec for 'not elsewhere classified'.

INPUTS AND OUTPUTS FOR ELECTRICAL INDUSTRIAL APPARATUS, NEC

Economic Sector or Industry Providing Inputs	%	Sector	Economic Sector or Industry Buying Outputs	%	Sector
Die-cut paper & board	7.8	Manufg.	Gross private fixed investment	47.2	Cap Inv
Wholesale trade	7.7	Trade	Federal Government purchases, national defense	29.2	Fed Govt
Advertising	6.7	Services	Federal Government purchases, nondefense	10.5	Fed Govt
Miscellaneous plastics products	6.3	Manufg.	Automotive repair shops & services	3.8	Services
Noncomparable imports	5.4	Foreign	Electronic computing equipment	3.4	Manufg.
Aircraft & missile engines & engine parts	5.2	Manufg.	Transformers	1.6	Manufg.
Nonferrous wire drawing & insulating	4.8	Manufg.	Motors & generators	1.5	Manufg.
Electronic components nec	4.3	Manufg.	Communications, except radio & TV	1.1	Util.
Metal stampings, nec	3.3	Manufg.	S/L Govt. purch., higher education	0.7	S/L Govt
Screw machine and related products	3.0	Manufg.	S/L Govt. purch., elem. & secondary education	0.4	S/L Govt
Communications, except radio & TV	3.0	Util.	Electrical industrial apparatus, nec	0.3	Manufg.
Banking	3.0	Fin/R.E.	Personal consumption expenditures	0.1	
Paperboard containers & boxes	2.8	Manufg.	S/L Govt. purch., gas & electric utilities	0.1	S/L Govt
Electric services (utilities)	2.7	Util.			
Paper mills, except building paper	2.3	Manufg.			
Transformers	2.2	Manufg.			
Air transportation	2.1	Util.			
Semiconductors & related devices	1.9	Manufg.			

Continued on next page.

INPUTS AND OUTPUTS FOR ELECTRICAL INDUSTRIAL APPARATUS, NEC - Continued

Economic Sector or Industry Providing Inputs	%	Sector	Economic Sector or Industry Buying Outputs	%	Sector
Maintenance of nonfarm buildings nec	1.6	Constr.			
Aircraft	1.3	Manufg.			
Aircraft & missile equipment, nec	1.3	Manufg.			
Petroleum refining	1.3	Manufg.			
Motor freight transportation & warehousing	1.3	Util.			
Hotels & lodging places	1.2	Services			
Copper rolling & drawing	1.1	Manufg.			
Engineering, architectural, & surveying services	1.1	Services			
Aluminum rolling & drawing	1.0	Manufg.			
Paints & allied products	1.0	Manufg.			
Computer & data processing services	0.9	Services			
Eating & drinking places	0.7	Trade			
Equipment rental & leasing services	0.7	Services			
Cyclic crudes and organics	0.6	Manufg.			
Electrical industrial apparatus, nec	0.6	Manufg.			
Gaskets, packing & sealing devices	0.6	Manufg.			
Wood pallets & skids	0.6	Manufg.			
Gas production & distribution (utilities)	0.6	Util.			
Real estate	0.6	Fin/R.E.			
Detective & protective services	0.6	Services			
Nonferrous rolling & drawing, nec	0.5	Manufg.			
Machinery, except electrical, nec	0.4	Manufg.			
Porcelain electrical supplies	0.4	Manufg.			
Switchgear & switchboard apparatus	0.4	Manufg.			
U.S. Postal Service	0.4	Gov't			
Sanitary services, steam supply, irrigation	0.3	Util.			
Royalties	0.3	Fin/R.E.			
Legal services	0.3	Services			
Management & consulting services & labs	0.3	Services			
Lubricating oils & greases	0.2	Manufg.			
Manifold business forms	0.2	Manufg.			
Special dies & tools & machine tool accessories	0.2	Manufg.			
Accounting, auditing & bookkeeping	0.2	Services			
Automotive rental & leasing, without drivers	0.2	Services			
Automotive repair shops & services	0.2	Services			
Electrical repair shops	0.2	Services			
Laundry, dry cleaning, shoe repair	0.2	Services			
Abrasive products	0.1	Manufg.			
Hand & edge tools, nec	0.1	Manufg.			
Industrial inorganic chemicals, nec	0.1	Manufg.			

Source: Benchmark Input-Output Accounts for the U.S. Economy, 1982, U.S. Department of Commerce, Washington, D.C., July 1991. Data, as reported in the source, are organized by the 1977 SIC structure in use in 1982 but have been matched, as closely as is possible, to the 1987 SIC structure used in this book.

OCCUPATIONS EMPLOYED BY SIC 362 - ELECTRICAL INDUSTRIAL APPARATUS

Occupation	% of Total 1994	Change to 2005	Occupation	% of Total 1994	Change to 2005
Electrical & electronic assemblers	16.8	-28.3	Secretaries, ex legal & medical	1.4	-27.5
Assemblers, fabricators, & hand workers nec	8.0	-20.3	Drafters	1.3	-38.0
Electrical & electronic equipment assemblers	4.3	-20.3	Stock clerks	1.3	-35.3
Inspectors, testers, & graders, precision	4.2	-44.2	Machine assemblers	1.3	-28.3
Electrical & electronics engineers	3.3	-15.2	Machine tool cutting & forming etc. nec	1.2	-36.3
Blue collar worker supervisors	3.2	-33.5	Lathe & turning machine tool operators	1.2	-48.2
Coil winders, tapers, & finishers	3.0	-36.3	Traffic, shipping, & receiving clerks	1.1	-23.4
Electrical & electronic technicians,technologists	2.3	-20.4	Industrial production managers	1.1	-20.3
Machinists	2.2	-40.3	Engineering, mathematical, & science managers	1.1	-9.5
Sales & related workers nec	1.8	-20.3	Industrial machinery mechanics	1.1	7.5
Electromechanical equipment assemblers	1.5	-12.4	Production, planning, & expediting clerks	1.0	-20.3
General managers & top executives	1.5	-24.4			

Source: Industry-Occupation Matrix, Bureau of Labor Statistics. These data relate to one or more 3-digit SIC industry groups rather than to a single 4-digit SIC. The change reported for each occupation to the year 2005 is a percent of growth or decline as estimated by the Bureau of Labor Statistics. The abbreviation nec stands for 'not elsewhere classified'.

LOCATION BY STATE AND REGIONAL CONCENTRATION

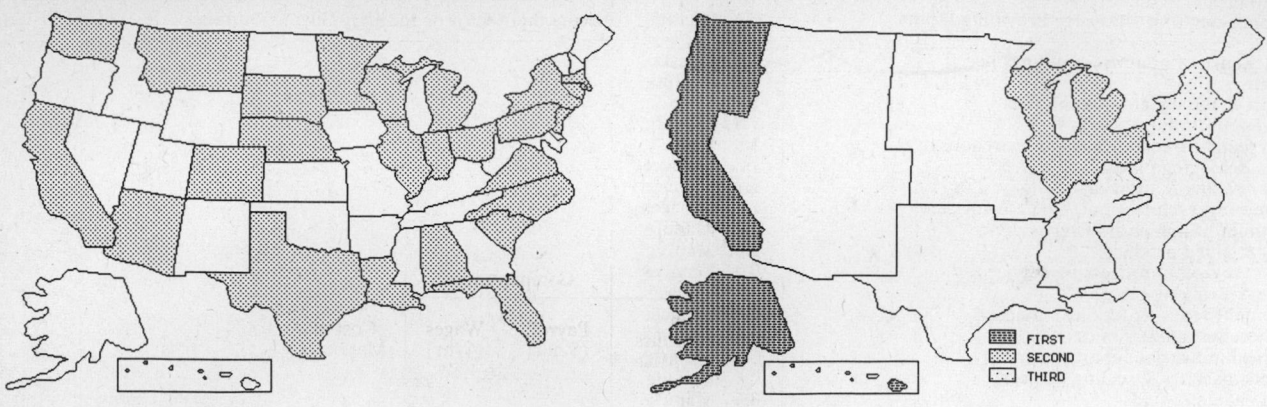

FIRST
SECOND
THIRD

INDUSTRY DATA BY STATE

| State | Establish-ments | Shipments | | | Employment | | | | Cost as % of Shipments | Investment per Employee ($) |
		Total ($ mil)	% of U.S.	Per Establ.	Total Number	% of U.S.	Per Establ.	Wages ($/hour)		
California	72	297.8	14.6	4.1	3,000	17.0	42	10.76	41.8	1,800
Ohio	16	166.7	8.2	10.4	1,400	8.0	88	8.94	39.4	7,214
Illinois	22	133.7	6.6	6.1	1,600	9.1	73	8.17	48.6	1,688
Massachusetts	12	99.5	4.9	8.3	1,200	6.8	100	8.11	50.2	-
Pennsylvania	13	78.3	3.8	6.0	500	2.8	38	13.33	38.6	-
Connecticut	13	65.6	3.2	5.0	600	3.4	46	11.25	45.4	3,167
New Jersey	14	64.1	3.1	4.6	800	4.5	57	7.08	43.7	1,250
Florida	11	43.9	2.2	4.0	500	2.8	45	8.50	47.8	2,800
Arizona	12	32.3	1.6	2.7	400	2.3	33	12.50	27.9	3,500
Michigan	11	17.5	0.9	1.6	200	1.1	18	8.50	45.7	2,000
Colorado	4	14.4	0.7	3.6	200	1.1	50	11.00	38.2	-
New York	20	(D)	-	-	750 *	4.3	38	-	-	-
Texas	20	(D)	-	-	750 *	4.3	38	-	-	-
Indiana	17	(D)	-	-	750 *	4.3	44	-	-	-
Wisconsin	11	(D)	-	-	750 *	4.3	68	-	-	-
North Carolina	9	(D)	-	-	750 *	4.3	83	-	-	-
Washington	8	(D)	-	-	375 *	2.1	47	-	-	-
Minnesota	5	(D)	-	-	375 *	2.1	75	-	-	-
Alabama	4	(D)	-	-	375 *	2.1	94	-	-	-
Louisiana	4	(D)	-	-	175 *	1.0	44	-	-	-
Virginia	4	(D)	-	-	175 *	1.0	44	-	-	-
Rhode Island	3	(D)	-	-	750 *	4.3	250	-	-	-
Montana	2	(D)	-	-	175 *	1.0	88	-	-	-
Nebraska	2	(D)	-	-	375 *	2.1	188	-	-	-
South Carolina	2	(D)	-	-	175 *	1.0	88	-	-	-
South Dakota	2	(D)	-	-	175 *	1.0	88	-	-	-

Source: 1992 *Economic Census*. The states are in descending order of shipments or establishments (if shipment data are missing for the majority). The symbol (D) appears when data are withheld to prevent disclosure of competitive information. States marked with (D) are sorted by number of establishments. A dash (-) indicates that the data element cannot be calculated; * indicates the midpoint of a range.

3631 - HOUSEHOLD COOKING EQUIPMENT

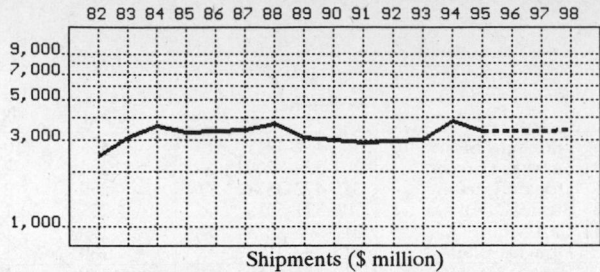

Shipments ($ million)

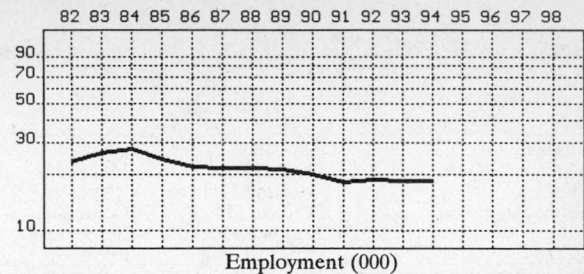

Employment (000)

GENERAL STATISTICS

Year	Companies	Establishments		Employment			Compensation		Production ($ million)			
		Total	with 20 or more employees	Total (000)	Production Workers (000)	Hours (Mil)	Payroll ($ mil)	Wages ($/hr)	Cost of Materials	Value Added by Manufacture	Value of Shipments	Capital Invest.
1982	71	88	59	23.7	17.0	31.4	402.1	8.11	1,408.3	941.4	2,414.9	51.5
1983		87	58	26.5	19.7	39.4	488.7	8.22	1,862.0	1,276.9	3,075.8	69.4
1984		86	57	27.7	21.5	42.9	544.8	8.64	2,161.4	1,528.8	3,578.5	86.3
1985		85	56	24.2	18.5	35.9	492.4	9.02	2,006.7	1,236.5	3,297.3	76.6
1986		82	55	22.2	16.9	32.6	461.9	9.51	2,089.8	1,305.3	3,328.7	88.6
1987	65	78	47	21.9	16.9	33.3	475.8	9.72	2,118.3	1,267.9	3,395.8	79.6
1988		75	52			33.3	463.8		2,382.3	1,404.7	3,699.4	
1989		77	48	21.5	15.9	31.3	440.7	9.77	1,940.0	1,087.6	3,094.5	63.2
1990		79	48	20.2	15.3	30.2	442.2	10.03	1,821.7	1,138.9	2,994.0	84.7
1991		90	53	18.4	14.8	28.0	401.5	9.95	1,801.0	1,091.2	2,890.7	95.7
1992	80	89	43	18.8	15.0	29.9	437.0	9.81	1,811.7	1,141.4	2,950.0	82.9
1993		98	42	18.6	14.9	29.6	461.1	10.34	1,741.5	1,330.7	3,010.2	82.6
1994		87P	42P	18.6	15.0	30.2	491.2	10.69	1,922.1	1,908.0	3,813.9	101.6
1995		87P	41P			27.6P	447.1P		1,675.7P	1,663.4P	3,325.0P	
1996		88P	39P			26.8P	445.0P		1,685.0P	1,672.6P	3,343.4P	
1997		88P	38P			26.1P	442.9P		1,694.3P	1,681.9P	3,361.9P	
1998		88P	37P			25.3P	440.8P		1,703.6P	1,691.1P	3,380.4P	

Sources: 1982, 1987, 1992 *Economic Census*; *Annual Survey of Manufactures*, 83-86, 88-91, 93-94. Establishment counts for non-Census years are from *County Business Patterns*; establishment values for 83-84 are extrapolations. 'P's show projections by the editors. Industries reclassified in 87 will not have data for prior years.

INDICES OF CHANGE

Year	Companies	Establishments		Employment			Compensation		Production ($ million)			
		Total	with 20 or more employees	Total (000)	Production Workers (000)	Hours (Mil)	Payroll ($ mil)	Wages ($/hr)	Cost of Materials	Value Added by Manufacture	Value of Shipments	Capital Invest.
1982	89	99	137	126	113	105	92	83	78	82	82	62
1983		98	135	141	131	132	112	84	103	112	104	84
1984		97	133	147	143	143	125	88	119	134	121	104
1985		96	130	129	123	120	113	92	111	108	112	92
1986		92	128	118	113	109	106	97	115	114	113	107
1987	81	88	109	116	113	111	109	99	117	111	115	96
1988		84	121			111	106		131	123	125	
1989		87	112	114	106	105	101	100	107	95	105	76
1990		89	112	107	102	101	101	102	101	100	101	102
1991		101	123	98	99	94	92	101	99	96	98	115
1992	100	100	100	100	100	100	100	100	100	100	100	100
1993		110	98	99	99	99	106	105	96	117	102	100
1994		98P	98P	99	100	101	112	109	106	167	129	123
1995		98P	95P			92P	102P		92P	146P	113P	
1996		99P	92P			90P	102P		93P	147P	113P	
1997		99P	88P			87P	101P		94P	147P	114P	
1998		99P	85P			85P	101P		94P	148P	115P	

Sources: Same as General Statistics. Values reflect change from the base year, 1992. Values above 100 mean greater than 92, values below 100 mean less than 92, and a value of 100 in the 82-91 or 93-98 period means same as 92. 'P's mark projections by the editors.

SELECTED RATIOS

For 1994	Avg. of All Manufact.	Analyzed Industry	Index	For 1994	Avg. of All Manufact.	Analyzed Industry	Index
Employees per Establishment	49	214	437	Value Added per Production Worker	134,084	127,200	95
Payroll per Establishment	1,500,273	5,648,928	377	Cost per Establishment	5,045,178	22,104,652	438
Payroll per Employee	30,620	26,409	86	Cost per Employee	102,970	103,339	100
Production Workers per Establishment	34	173	503	Cost per Production Worker	146,988	128,140	87
Wages per Establishment	853,319	3,712,721	435	Shipments per Establishment	9,576,895	43,860,847	458
Wages per Production Worker	24,861	21,523	87	Shipments per Employee	195,460	205,048	105
Hours per Production Worker	2,056	2,013	98	Shipments per Production Worker	279,017	254,260	91
Wages per Hour	12.09	10.69	88	Investment per Establishment	321,011	1,168,427	364
Value Added per Establishment	4,602,255	21,942,499	477	Investment per Employee	6,552	5,462	83
Value Added per Employee	93,930	102,581	109	Investment per Production Worker	9,352	6,773	72

Sources: Same as General Statistics. The 'Average of All Manufacturing' column represents the average of all manufacturing industries reported for the most recent complete year available. The Index shows the relationship between the Average and the Analyzed Industry. For example, 100 means that they are equal; 500 that the Analyzed Industry is five times the average; 50 means that the Analyzed Industry is half the national average. The abbreviation 'na' is used to show that data are 'not available'.

LEADING COMPANIES   Number shown: 25   Total sales ($ mil): 9,172   Total employment (000): 22.7

Company Name	Address				CEO Name	Phone	Co. Type	Sales ($ mil)	Empl. (000)
GE Appliances	Appliance Park	Louisville	KY	40225	J Richard Stonesifer	502-452-4311	D	5,555	10.0
Sharp Electronics Corp	PO Box 650	Mahwah	NJ	07430	Sueyuki Hiroka	201-529-8200	S	2,200•	2.2
Magic Chef Co	740 King Edward	Cleveland	TN	37311	Donald Lorton	615-472-3371	D	250	2.0
WC Bradley Co	PO Box 140	Columbus	GA	31902	Steve Butler	706-571-6055	R	215	1.4
Jenn-Air Co	3035 N Shadeland	Indianapolis	IN	46226	Carl Moe	317-545-2271	D	190	1.3
Thermos Co	1555 Rte 75 E	Freeport	IL	61032	Douglas Blair	815-232-2111	S	180	1.0
Sunbeam Outdoor Products Inc	4101 H Bush Dr	Neosho	MO	64850	Rick D Davidson	417-451-4550	S	130•	1.3
Unaka Company Inc	PO Box 877	Greeneville	TN	37744	Gerald Jaynes	615-639-1163	R	100	1.2
Whirlpool Corp	927 Whirlpool Dr	Oxford	MS	38655	Edward Messal	601-234-3131	D	91	0.5
Brinkmann Corp	4215 McEwen Rd	Dallas	TX	75244	JB Brinkmann	214-387-4939	R	68•	0.5
Alternative Pioneering Systems	PO Box 159	Chaska	MN	55318	David Dornbush	612-448-4400	R	45•	0.2
Peerless Premier Appliance Co	PO Box 387	Belleville	IL	62222	Joseph E Geary	618-233-0475	R	30	0.3
Barbeques Galore Inc	15041 Bake Pkwy	Irvine	CA	92718	Sidney Selati	714-522-0660	R	23	0.2
Viking Range Corp	111 Front St	Greenwood	MS	38930	Fred E Carl Jr	601-455-1200	R	20	0.1
Jensen Industries Inc	1946 E 46th St	Los Angeles	CA	90058		213-235-6800	S	15	0.1
Maytag Clarence	PO Box 70	Clarence	MO	63437	Gary Hicks	816-699-2156	D	13	0.2
Dwyer Products Corp	418 N Calumet Av	Michigan City	IN	46360	Charles Linski	219-874-5236	R	10	<0.1
Helman Holding Inc	1701 Pacific #160	Oxnard	CA	93033	Barry Helman	805-487-7772	R	10	<0.1
Vitantonio Manufacturing Co	34355 Vokes Dr	Eastlake	OH	44095	Louis Vitantonio	216-946-1661	R	9•	<0.1
Charcoal Companion Inc	7955 Edgewater Dr	Oakland	CA	94621	Chuck Adams	510-632-2100	R	6	<0.1
Camp Chef	PO Box 88447	Emeryville	CA	94662	William A Dutro	510-652-9130	D	5	<0.1
Quadlux Inc	47817 Fremont Blv	Fremont	CA	94538	Robert Beaver	510-498-4200	R	3•	<0.1
Town Food Service Equip Co	351 Bowery St	New York	NY	10003	Charles Suss	212-473-8355	R	3	<0.1
Pyromid Inc	3292 S Hwy 97	Redmond	OR	97756	Paul Hait	503-548-1041	R	1	<0.1
Regency VSA Appliances Ltd	1442 Irvine St	Tustin	CA	92680	Becky Moulton	714-544-3530	R	1•	<0.1

Source: Ward's Business Directory of U.S. Private and Public Companies, Volumes 1 and 2, 1996. The company type code used is as follows: P - Public, R - Private, S - Subsidiary, D - Division, J - Joint Venture, A - Affiliate, G - Group. Sales are in millions of dollars, employees are in thousands. An asterisk (*) indicates an estimated sales volume. The symbol < stands for 'less than'. Company names and addresses are truncated, in some cases, to fit into the available space.

MATERIALS CONSUMED

Material	Quantity	Delivered Cost ($ million)
Materials, ingredients, containers, and supplies	(X)	1,482.0
Metal stampings	(X)	98.9
Metal wire racks, grills, springs, and other fabricated nonelectric wire products (except forgings)	(X)	36.5
Metal bolts, nuts, screws, washers, rivets, and other screw machine products	(X)	20.5
All other fabricated metal products (except forgings)	(X)	29.9
Forgings	(X)	(D)
Iron and steel castings (rough and semifinished)	(X)	12.6
Aluminum and aluminum-base alloy castings (rough and semifinished)	(X)	65.4
Steel bars, bar shapes, and plates	(X)	(D)
Steel sheet and strip, including tin plate	(X)	103.2
All other steel shapes and forms	(X)	49.9
Aluminum and aluminum-base alloy shapes and forms	(X)	8.7
Nonferrous wire and cable, including magnet wire, bare or insulated wire, etc.	(X)	4.7
Fractional horsepower electric timing motors, synchronous and subsynchronous (less than 1 hp)	(X)	13.2
Paper and paperboard containers, including shipping sacks and other paper packaging supplies	(X)	58.8
Electrical transmission, distribution, and control equipment	(X)	49.0
Current-carrying wiring devices	(X)	34.0
Timing mechanisms, except microprocessors	(X)	12.8
Automatic temperature controls (thermostats, regulators, etc.)	(X)	28.8
Paints, varnishes, lacquers, stains, shellacs, japans, enamels, and allied products	(X)	27.8
Fabricated rubber products, except tires, tubes, hose, belting, and gaskets	(X)	4.6
Rubber and plastics hose and belting	(X)	6.0
Plastics products consumed in the form of sheets, rods, tubes, film, and other shapes	(X)	28.8
Plastics resins consumed in the form of granules, pellets, powders, liquids, etc.	(X)	3.3
Complete flexible cord sets	(X)	(D)
Resistors, capacitors, transformers, electron tubes, semiconductors, and other electronic components	(X)	53.9
Mineral wool insulation (fibrous glass, rock wool, etc.)	(X)	12.2
All other materials and components, parts, containers, and supplies	(X)	361.0
Materials, ingredients, containers, and supplies, nsk	(X)	341.7

Source: 1992 Economic Census. Explanation of symbols used: (D): Withheld to avoid disclosure of competitive data; na: Not available; (S): Withheld because statistical norms were not met; (X): Not applicable; (Z): Less than half the unit shown; nec: Not elsewhere classified; nsk: Not specified by kind; - : zero; * : 10-19 percent estimated; ** : 20-29 percent estimated.

PRODUCT SHARE DETAILS

Product or Product Class	% Share	Product or Product Class	% Share
Household cooking equipment	100.00	Gas household ranges, ovens, surface cooking units, and equipment	94.80
Electric household cooking equipment, including microwave, ranges, ovens, surface cooking units and equipment parts and accessories	53.53	Parts and accessories for gas household ranges and ovens, such as burners, rotisseries, oven racks, broiler pans, etc.	5.20
Electric household ranges, ovens, surface cooking units, and equipment	92.59	Other household ranges and cooking equipment	25.39
Parts and accessories for electric household ranges and ovens, such as burners, rotisseries, oven racks, broiler pans	7.41	Other household ranges and cooking equipment (except gas and electric), and outdoor cooking equipment	89.40
Gas household ranges, ovens, surface cooking units, and equipment parts and accessories	20.41	Parts and accessories for outdoor and other cooking equipment, sold separately	10.60
		Household cooking equipment, nsk	0.68

Source: 1992 *Economic Census.* The values shown are percent of total shipments in an industry. Values of indented subcategories are summed in the main heading. The symbol (D) appears when data are withheld to prevent disclosure of competitive information. The abbreviation nsk stands for 'not specified by kind' and nec for 'not elsewhere classified'.

INPUTS AND OUTPUTS FOR HOUSEHOLD COOKING EQUIPMENT

Economic Sector or Industry Providing Inputs	%	Sector	Economic Sector or Industry Buying Outputs	%	Sector
Imports	19.4	Foreign	Personal consumption expenditures	63.0	
Blast furnaces & steel mills	11.1	Manufg.	Gross private fixed investment	23.9	Cap Inv
Wholesale trade	10.8	Trade	Exports	5.3	Foreign
Transformers	5.8	Manufg.	Electrical repair shops	3.8	Services
Environmental controls	4.2	Manufg.	Mobile homes	1.3	Manufg.
Hardware, nec	3.7	Manufg.	Federal Government purchases, national defense	0.9	Fed Govt
Glass & glass products, except containers	3.3	Manufg.	Travel trailers & campers	0.8	Manufg.
Paperboard containers & boxes	2.9	Manufg.	S/L Govt. purch., health & hospitals	0.4	S/L Govt
Pipe, valves, & pipe fittings	2.7	Manufg.	Household cooking equipment	0.2	Manufg.
Wiring devices	2.5	Manufg.	S/L Govt. purch., elem. & secondary education	0.2	S/L Govt
Motors & generators	2.3	Manufg.			
Aluminum castings	2.2	Manufg.			
Metal stampings, nec	1.8	Manufg.			
Screw machine and related products	1.6	Manufg.			
Advertising	1.6	Services			
Miscellaneous fabricated wire products	1.5	Manufg.			
Paints & allied products	1.5	Manufg.			
Watches, clocks, & parts	1.5	Manufg.			
Electric services (utilities)	1.3	Util.			
Chemical preparations, nec	1.1	Manufg.			
Aluminum rolling & drawing	1.0	Manufg.			
Mineral wool	1.0	Manufg.			
Gas production & distribution (utilities)	1.0	Util.			
Motor freight transportation & warehousing	1.0	Util.			
Miscellaneous plastics products	0.8	Manufg.			
Banking	0.8	Fin/R.E.			
Communications, except radio & TV	0.7	Util.			
Nonferrous wire drawing & insulating	0.6	Manufg.			
Air transportation	0.6	Util.			
Equipment rental & leasing services	0.6	Services			
Maintenance of nonfarm buildings nec	0.5	Constr.			
Detective & protective services	0.5	Services			
U.S. Postal Service	0.5	Gov't			
Fabricated rubber products, nec	0.4	Manufg.			
Eating & drinking places	0.4	Trade			
Personnel supply services	0.4	Services			
Electric housewares & fans	0.3	Manufg.			
Household cooking equipment	0.3	Manufg.			
Paper coating & glazing	0.3	Manufg.			
Switchgear & switchboard apparatus	0.3	Manufg.			
Computer & data processing services	0.3	Services			
Hotels & lodging places	0.3	Services			
Gaskets, packing & sealing devices	0.2	Manufg.			
Machine tools, metal forming types	0.2	Manufg.			
Machinery, except electrical, nec	0.2	Manufg.			
Nonferrous castings, nec	0.2	Manufg.			
Rubber & plastics hose & belting	0.2	Manufg.			
Semiconductors & related devices	0.2	Manufg.			
Special dies & tools & machine tool accessories	0.2	Manufg.			
Railroads & related services	0.2	Util.			
Credit agencies other than banks	0.2	Fin/R.E.			
Real estate	0.2	Fin/R.E.			
Abrasive products	0.1	Manufg.			
Electric lamps	0.1	Manufg.			
Iron & steel foundries	0.1	Manufg.			
Nonferrous rolling & drawing, nec	0.1	Manufg.			
Petroleum refining	0.1	Manufg.			
Signs & advertising displays	0.1	Manufg.			
Automotive repair shops & services	0.1	Services			
Legal services	0.1	Services			
Management & consulting services & labs	0.1	Services			

Source: Benchmark Input-Output Accounts for the U.S. Economy, 1982, U.S. Department of Commerce, Washington, D.C., July 1991. Data, as reported in the source, are organized by the 1977 SIC structure in use in 1982 but have been matched, as closely as is possible, to the 1987 SIC structure used in this book.

OCCUPATIONS EMPLOYED BY SIC 363 - HOUSEHOLD APPLIANCES

Occupation	% of Total 1994	Change to 2005	Occupation	% of Total 1994	Change to 2005
Assemblers, fabricators, & hand workers nec	31.2	-16.9	Electrical & electronic equipment assemblers	1.7	-16.9
Helpers, laborers, & material movers nec	4.1	-16.9	Machine operators nec	1.5	-41.4
Machine assemblers	3.9	-25.2	Welding machine setters, operators	1.5	-25.2
Inspectors, testers, & graders, precision	3.1	-41.9	Sales & related workers nec	1.5	-17.0
Blue collar worker supervisors	3.0	-25.4	Maintenance repairers, general utility	1.5	-25.2
Electrical & electronic assemblers	2.7	-25.3	Janitors & cleaners, incl maids	1.2	-33.6
Industrial truck & tractor operators	2.6	-16.9	Tool & die makers	1.2	-32.9
Plastic molding machine workers	2.6	-0.3	Material moving equipment operators nec	1.2	-16.9
Machine forming operators, metal & plastic	2.5	-58.5	General managers & top executives	1.1	-21.2
Freight, stock, & material movers, hand	1.9	-33.6	Machine tool cutting & forming etc. nec	1.1	-33.6

Source: Industry-Occupation Matrix, Bureau of Labor Statistics. These data relate to one or more 3-digit SIC industry groups rather than to a single 4-digit SIC. The change reported for each occupation to the year 2005 is a percent of growth or decline as estimated by the Bureau of Labor Statistics. The abbreviation nec stands for 'not elsewhere classified'.

LOCATION BY STATE AND REGIONAL CONCENTRATION

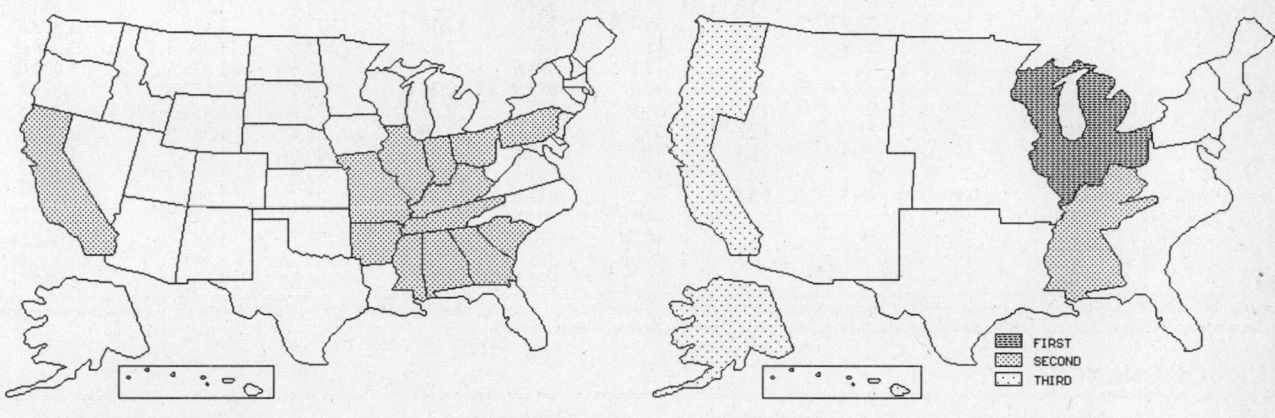

FIRST
SECOND
THIRD

INDUSTRY DATA BY STATE

State	Establish-ments	Shipments Total ($ mil)	Shipments % of U.S.	Shipments Per Establ.	Employment Total Number	Employment % of U.S.	Employment Per Establ.	Wages ($/hour)	Cost as % of Shipments	Investment per Employee ($)
Tennessee	9	850.4	28.8	94.5	4,500	23.9	500	9.21	63.8	4,844
Illinois	10	264.2	9.0	26.4	2,300	12.2	230	10.24	62.8	-
California	11	(D)	-	-	750 *	4.0	68	-	-	-
Ohio	6	(D)	-	-	1,750 *	9.3	292	-	-	-
Georgia	5	(D)	-	-	3,750 *	19.9	750	-	-	4,773
Indiana	5	(D)	-	-	1,750 *	9.3	350	-	-	-
Missouri	4	(D)	-	-	1,750 *	9.3	438	-	-	-
Pennsylvania	4	(D)	-	-	175 *	0.9	44	-	-	-
Arkansas	3	(D)	-	-	375 *	2.0	125	-	-	-
Mississippi	3	(D)	-	-	750 *	4.0	250	-	-	-
Alabama	2	(D)	-	-	750 *	4.0	375	-	-	-
Kentucky	2	(D)	-	-	750 *	4.0	375	-	-	-
South Carolina	1	(D)	-	-	750 *	4.0	750	-	-	-

Source: 1992 Economic Census. The states are in descending order of shipments or establishments (if shipment data are missing for the majority). The symbol (D) appears when data are withheld to prevent disclosure of competitive information. States marked with (D) are sorted by number of establishments. A dash (-) indicates that the data element cannot be calculated; * indicates the midpoint of a range.

3632 - HOUSEHOLD REFRIGERATORS & FREEZERS

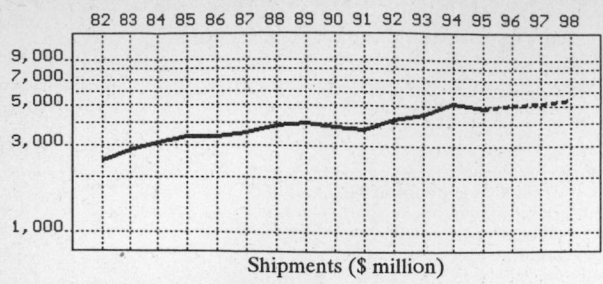

Shipments ($ million)

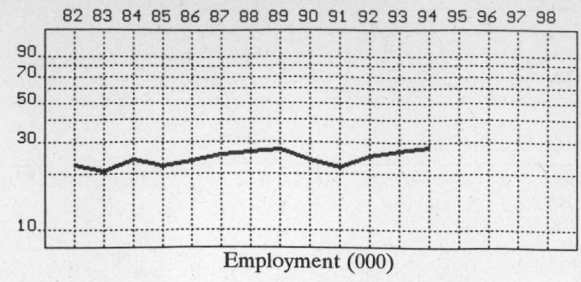

Employment (000)

GENERAL STATISTICS

Year	Companies	Establishments		Employment			Compensation		Production ($ million)			
		Total	with 20 or more employees	Total (000)	Production Workers (000)	Hours (Mil)	Payroll ($ mil)	Wages ($/hr)	Cost of Materials	Value Added by Manufacture	Value of Shipments	Capital Invest.
1982	39	50	30	22.4	18.4	33.1	441.2	10.28	1,460.5	1,032.3	2,470.7	79.1
1983		49	30	21.0	17.5	33.4	460.2	10.75	1,653.7	1,146.2	2,821.2	53.7
1984		48	30	24.0	20.4	37.9	526.5	10.96	2,052.6	1,152.5	3,088.0	56.3
1985		48	30	22.3	18.4	34.1	514.5	11.53	1,889.5	1,352.2	3,340.5	64.3
1986		48	29	23.9	20.1	38.4	594.7	12.14	1,977.8	1,427.2	3,352.8	84.9
1987	40	49	26	25.7	21.6	41.5	650.7	12.49	2,161.6	1,450.4	3,518.9	100.3
1988		47	25			43.1	695.6		2,382.4	1,538.6	3,902.0	
1989		52	28	27.4	20.7	42.0	683.9	12.75	2,444.1	1,653.4	4,015.4	172.6
1990		49	21	24.1	19.2	36.6	609.2	12.58	2,193.9	1,464.3	3,799.8	101.9
1991		48	22	21.8	17.8	35.8	605.5	12.85	2,238.4	1,386.9	3,721.3	168.7
1992	52	58	19	25.4	21.4	41.5	719.0	12.81	2,596.6	1,629.1	4,232.4	187.4
1993		54	22	27.1	22.5	42.6	741.5	13.21	2,883.7	1,602.4	4,463.1	139.9
1994		53P	19P	28.3	23.9	46.3	817.6	13.43	3,329.7	1,909.7	5,149.1	170.5
1995		54P	18P			44.4P	803.5P		3,162.2P	1,813.6P	4,890.0P	
1996		54P	17P			45.2P	829.7P		3,273.7P	1,877.6P	5,062.5P	
1997		55P	16P			45.9P	855.9P		3,385.2P	1,941.6P	5,235.0P	
1998		55P	15P			46.7P	882.1P		3,496.8P	2,005.5P	5,407.5P	

Sources: 1982, 1987, 1992 *Economic Census*; *Annual Survey of Manufactures*, 83-86, 88-91, 93-94. Establishment counts for non-Census years are from *County Business Patterns*; establishment values for 83-84 are extrapolations. 'P's show projections by the editors. Industries reclassified in 87 will not have data for prior years.

INDICES OF CHANGE

Year	Companies	Establishments		Employment			Compensation		Production ($ million)			
		Total	with 20 or more employees	Total (000)	Production Workers (000)	Hours (Mil)	Payroll ($ mil)	Wages ($/hr)	Cost of Materials	Value Added by Manufacture	Value of Shipments	Capital Invest.
1982	75	86	158	88	86	80	61	80	56	63	58	42
1983		84	158	83	82	80	64	84	64	70	67	29
1984		83	158	94	95	91	73	86	79	71	73	30
1985		83	158	88	86	82	72	90	73	83	79	34
1986		83	153	94	94	93	83	95	76	88	79	45
1987	77	84	137	101	101	100	91	98	83	89	83	54
1988		81	132			104	97		92	94	92	
1989		90	147	108	97	101	95	100	94	101	95	92
1990		84	111	95	90	88	85	98	84	90	90	54
1991		83	116	86	83	86	84	100	86	85	88	90
1992	100	100	100	100	100	100	100	100	100	100	100	100
1993		93	116	107	105	103	103	103	111	98	105	75
1994		92P	102P	111	112	112	114	105	128	117	122	91
1995		93P	97P			107P	112P		122P	111P	116P	
1996		93P	91P			109P	115P		126P	115P	120P	
1997		94P	86P			111P	119P		130P	119P	124P	
1998		95P	80P			113P	123P		135P	123P	128P	

Sources: Same as General Statistics. Values reflect change from the base year, 1992. Values above 100 mean greater than 92, values below 100 mean less than 92, and a value of 100 in the 82-91 or 93-98 period means same as 92. 'P's mark projections by the editors.

SELECTED RATIOS

For 1994	Avg. of All Manufact.	Analyzed Industry	Index	For 1994	Avg. of All Manufact.	Analyzed Industry	Index
Employees per Establishment	49	532	1,086	Value Added per Production Worker	134,084	79,904	60
Payroll per Establishment	1,500,273	15,373,675	1,025	Cost per Establishment	5,045,178	62,609,744	1,241
Payroll per Employee	30,620	28,890	94	Cost per Employee	102,970	117,657	114
Production Workers per Establishment	34	449	1,309	Cost per Production Worker	146,988	139,318	95
Wages per Establishment	853,319	11,692,135	1,370	Shipments per Establishment	9,576,895	96,820,684	1,011
Wages per Production Worker	24,861	26,017	105	Shipments per Employee	195,460	181,947	93
Hours per Production Worker	2,056	1,937	94	Shipments per Production Worker	279,017	215,444	77
Wages per Hour	12.09	13.43	111	Investment per Establishment	321,011	3,205,983	999
Value Added per Establishment	4,602,255	35,908,889	780	Investment per Employee	6,552	6,025	92
Value Added per Employee	93,930	67,481	72	Investment per Production Worker	9,352	7,134	76

Sources: Same as General Statistics. The 'Average of All Manufacturing' column represents the average of all manufacturing industries reported for the most recent complete year available. The Index shows the relationship between the Average and the Analyzed Industry. For example, 100 means that they are equal; 500 that the Analyzed Industry is five times the average; 50 means that the Analyzed Industry is half the national average. The abbreviation 'na' is used to show that data are 'not available'.

LEADING COMPANIES   Number shown: 10   Total sales ($ mil): **9,316**   Total employment (000): **50.0**

Company Name	Address				CEO Name	Phone	Co. Type	Sales ($ mil)	Empl. (000)
Whirlpool Corp	2000 N M-63	Benton Harbor	MI	49022	David R Whitwam	616-926-5000	P	7,949	39.0
Amana Refrigeration Inc	Hwy 20	Amana	IA	52204	Robert L Swam	319-622-5511	S	590	5.5
Galesburg Refrigeration Prod	1 Admiral Dr	Galesburg	IL	61401	Rick Foltz	309-343-0181	D	360	2.3
Frigidaire Refrigerator Products	635 W Charles St	Greenville	MI	48838	Steve Black	616-754-7131	D	260	2.0
Sub Zero Freezer Co	PO Box 44130	Madison	WI	53744	James J Bakke	608-271-2233	R	75	0.5
Norcold	PO Box 180	Sidney	OH	45365	Ron Riethman	513-492-1111	D	40	0.2
MicroFridge Inc	1 Merchant St	Sharon	MA	02067	Robert P Bennett	617-784-1713	R	15	<0.1
Arctic Industries Inc	8207 NW 74th Av	Miami	FL	33166	Donald Goodstein	305-883-5581	R	10	<0.1
Northland Corp	PO Box 400	Greenville	MI	48838	G Stauffer	616-754-5601	R	9	0.2
King Refrigerator Corp	7602 Woodhaven	Glendale	NY	11385	Andrew Koeppel	718-897-2200	R	8	0.2

Source: Ward's Business Directory of U.S. Private and Public Companies, Volumes 1 and 2, 1996. The company type code used is as follows: P - Public, R - Private, S - Subsidiary, D - Division, J - Joint Venture, A - Affiliate, G - Group. Sales are in millions of dollars, employees are in thousands. An asterisk (*) indicates an estimated sales volume. The symbol < stands for 'less than'. Company names and addresses are truncated, in some cases, to fit into the available space.

MATERIALS CONSUMED

Material	Quantity	Delivered Cost ($ million)
Materials, ingredients, containers, and supplies	(X)	2,375.7
Refrigeration compressors, compressor units, condensing units, and other heat transfer equipment	(X)	343.7
Fractional horsepower electric timing motors, synchronous and subsynchronous (less than 1 hp)	(X)	87.8
Fans and blowers	(X)	(D)
Electrical transmission, distribution, and control equipment	(X)	30.6
Current-carrying wiring devices	(X)	73.0
Automatic temperature controls (thermostats, regulators, etc.)	(X)	48.1
Ball and roller bearings (mounted or unmounted)	(X)	(D)
Metal bolts, nuts, screws, washers, rivets, and other screw machine products	(X)	59.1
Metal stampings	(X)	30.2
Metal hardware, including hinges, handles, locks, casters, etc. (except castings and forgings)	(X)	37.8
Metal pipe, valves, and pipe fittings (except forgings)	(X)	19.5
Iron and steel castings (rough and semifinished)	(X)	30.8
Steel bars, bar shapes, and plates	(X)	(Z)
Steel sheet and strip, including tin plate	(X)	249.5
Steel wire and wire products	(X)	100.6
Copper and copper-base alloy pipe and tube	(X)	19.4
All other copper and copper-base alloy shapes and forms	(X)	15.2
Aluminum and aluminum-base alloy sheet, plate, foil, and welded tubing	(X)	20.6
All other aluminum and aluminum-base alloy shapes and forms	(X)	15.4
Paper and paperboard containers, including shipping sacks and other paper packaging supplies	(X)	48.3
Wooden containers, complete (including combination woood and paperboard)	(X)	6.3
Paints, varnishes, lacquers, stains, shellacs, japans, enamels, and allied products	(X)	29.1
Refrigerant gases and other synthetic organic chemicals	(X)	90.2
Plastics resins consumed in the form of granules, pellets, powders, liquids, etc.	(X)	139.7
Plastics products consumed in the form of sheets, rods, tubes, film, and other shapes	(X)	44.6
Fabricated plastics products (except gaskets, hoses, and belting)	(X)	209.1
Rubber and plastics hose and belting	(X)	(D)
Gaskets (all types) and asbestos packing	(X)	41.6
Mineral wool insulation (fibrous glass, rock wool, etc.)	(X)	24.3
All other materials and components, parts, containers, and supplies	(X)	192.0
Materials, ingredients, containers, and supplies, nsk	(X)	343.6

Source: 1992 Economic Census. Explanation of symbols used: (D): Withheld to avoid disclosure of competitive data; na: Not available; (S): Withheld because statistical norms were not met; (X): Not applicable; (Z): Less than half the unit shown; nec: Not elsewhere classified; nsk: Not specified by kind; - : zero; * : 10-19 percent estimated; ** : 20-29 percent estimated.

PRODUCT SHARE DETAILS

Product or Product Class	% Share	Product or Product Class	% Share
Household refrigerators and freezers	100.00	(excluding compressors, condensing units, and ice making machines)	1.91
Food freezers, complete units, household type	97.67		
Parts, attachments for household refrigerators and freezers		Household refrigerators and freezers, nsk	0.42

Source: 1992 Economic Census. The values shown are percent of total shipments in an industry. Values of indented subcategories are summed in the main heading. The symbol (D) appears when data are withheld to prevent disclosure of competitive information. The abbreviation nsk stands for 'not specified by kind' and nec for 'not elsewhere classified'.

INPUTS AND OUTPUTS FOR HOUSEHOLD REFRIGERATORS & FREEZERS

Economic Sector or Industry Providing Inputs	%	Sector	Economic Sector or Industry Buying Outputs	%	Sector
Refrigeration & heating equipment	19.7	Manufg.	Personal consumption expenditures	61.2	
Wholesale trade	15.5	Trade	Gross private fixed investment	25.3	Cap Inv
Blast furnaces & steel mills	9.7	Manufg.	Exports	6.9	Foreign
Miscellaneous plastics products	4.9	Manufg.	Electrical repair shops	2.6	Services
Imports	4.7	Foreign	Mobile homes	1.9	Manufg.
Motors & generators	3.9	Manufg.	Travel trailers & campers	0.9	Manufg.
Pumps & compressors	2.8	Manufg.	Federal Government purchases, national defense	0.3	Fed Govt
Plastics materials & resins	2.6	Manufg.	S/L Govt. purch., elem. & secondary education	0.3	S/L Govt
Miscellaneous fabricated wire products	2.5	Manufg.	S/L Govt. purch., health & hospitals	0.3	S/L Govt
Paperboard containers & boxes	1.9	Manufg.	S/L Govt. purch., higher education	0.1	S/L Govt
Wiring devices	1.9	Manufg.			
Aluminum rolling & drawing	1.6	Manufg.			
Environmental controls	1.6	Manufg.			
Veneer & plywood	1.5	Manufg.			
Banking	1.5	Fin/R.E.			
Advertising	1.5	Services			
Electric services (utilities)	1.3	Util.			
Cyclic crudes and organics	1.2	Manufg.			
Metal stampings, nec	1.2	Manufg.			
Industrial controls	1.1	Manufg.			
Motor freight transportation & warehousing	1.0	Util.			
Paints & allied products	0.9	Manufg.			
Gaskets, packing & sealing devices	0.8	Manufg.			
Screw machine and related products	0.8	Manufg.			
Mineral wool	0.7	Manufg.			
Air transportation	0.7	Util.			
U.S. Postal Service	0.7	Gov't			
Copper rolling & drawing	0.6	Manufg.			
Fabricated metal products, nec	0.6	Manufg.			
Hardware, nec	0.6	Manufg.			
Maintenance of nonfarm buildings nec	0.5	Constr.			
Narrow fabric mills	0.5	Manufg.			
Gas production & distribution (utilities)	0.5	Util.			
Aluminum castings	0.4	Manufg.			
Blowers & fans	0.4	Manufg.			
Chemical preparations, nec	0.4	Manufg.			
Communications, except radio & TV	0.4	Util.			
Railroads & related services	0.4	Util.			
Eating & drinking places	0.4	Trade			
Equipment rental & leasing services	0.4	Services			
Hotels & lodging places	0.4	Services			
Metal coating & allied services	0.3	Manufg.			
Nonferrous castings, nec	0.3	Manufg.			
Asphalt felts & coatings	0.2	Manufg.			
Fabricated rubber products, nec	0.2	Manufg.			
Glass & glass products, except containers	0.2	Manufg.			
Iron & steel foundries	0.2	Manufg.			
Machinery, except electrical, nec	0.2	Manufg.			
Paper coating & glazing	0.2	Manufg.			
Paper mills, except building paper	0.2	Manufg.			
Rubber & plastics hose & belting	0.2	Manufg.			
Special dies & tools & machine tool accessories	0.2	Manufg.			
Sanitary services, steam supply, irrigation	0.2	Util.			
Computer & data processing services	0.2	Services			
Legal services	0.2	Services			
Coal	0.1	Mining			
Abrasive products	0.1	Manufg.			
Industrial inorganic chemicals, nec	0.1	Manufg.			
Machine tools, metal forming types	0.1	Manufg.			
Sawmills & planning mills, general	0.1	Manufg.			
Real estate	0.1	Fin/R.E.			
Automotive repair shops & services	0.1	Services			
Business/professional associations	0.1	Services			
Management & consulting services & labs	0.1	Services			

Source: Benchmark Input-Output Accounts for the U.S. Economy, 1982, U.S. Department of Commerce, Washington, D.C., July 1991. Data, as reported in the source, are organized by the 1977 SIC structure in use in 1982 but have been matched, as closely as is possible, to the 1987 SIC structure used in this book.

OCCUPATIONS EMPLOYED BY SIC 363 - HOUSEHOLD APPLIANCES

Occupation	% of Total 1994	Change to 2005	Occupation	% of Total 1994	Change to 2005
Assemblers, fabricators, & hand workers nec	31.2	-16.9	Electrical & electronic equipment assemblers	1.7	-16.9
Helpers, laborers, & material movers nec	4.1	-16.9	Machine operators nec	1.5	-41.4
Machine assemblers	3.9	-25.2	Welding machine setters, operators	1.5	-25.2
Inspectors, testers, & graders, precision	3.1	-41.9	Sales & related workers nec	1.5	-17.0
Blue collar worker supervisors	3.0	-25.4	Maintenance repairers, general utility	1.5	-25.2
Electrical & electronic assemblers	2.7	-25.3	Janitors & cleaners, incl maids	1.2	-33.6
Industrial truck & tractor operators	2.6	-16.9	Tool & die makers	1.2	-32.9
Plastic molding machine workers	2.6	-0.3	Material moving equipment operators nec	1.2	-16.9
Machine forming operators, metal & plastic	2.5	-58.5	General managers & top executives	1.1	-21.2
Freight, stock, & material movers, hand	1.9	-33.6	Machine tool cutting & forming etc. nec	1.1	-33.6

Source: *Industry-Occupation Matrix*, Bureau of Labor Statistics. These data relate to one or more 3-digit SIC industry groups rather than to a single 4-digit SIC. The change reported for each occupation to the year 2005 is a percent of growth or decline as estimated by the Bureau of Labor Statistics. The abbreviation nec stands for 'not elsewhere classified'.

LOCATION BY STATE AND REGIONAL CONCENTRATION

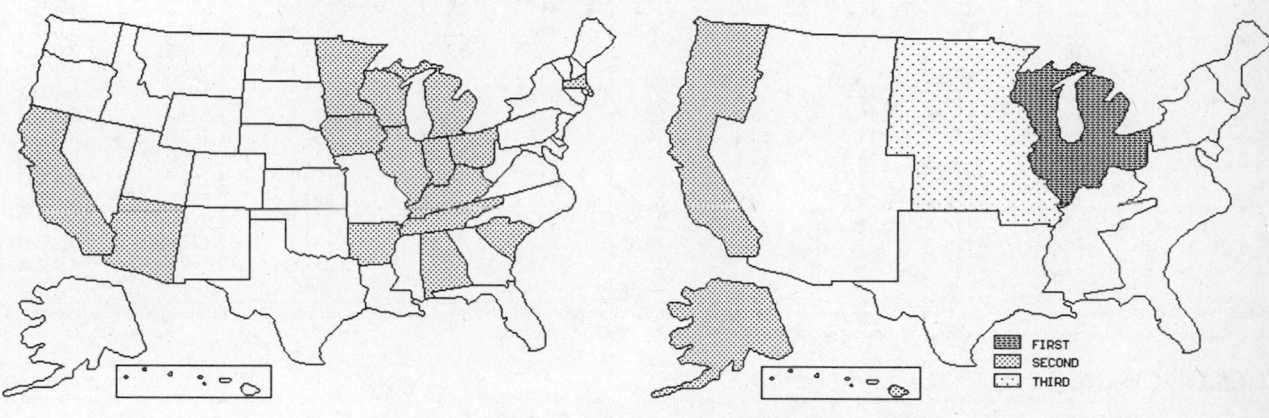

FIRST
SECOND
THIRD

INDUSTRY DATA BY STATE

State	Establish-ments	Shipments Total ($ mil)	Shipments % of U.S.	Shipments Per Establ.	Employment Total Number	Employment % of U.S.	Employment Per Establ.	Wages ($/hour)	Cost as % of Shipments	Investment per Employee ($)
California	10	(D)	-	-	375 *	1.5	38	-	-	-
Michigan	4	(D)	-	-	1,750 *	6.9	438	-	-	-
Ohio	4	(D)	-	-	375 *	1.5	94	-	-	-
Indiana	3	(D)	-	-	7,500 *	29.5	2,500	-	-	-
Massachusetts	3	(D)	-	-	375 *	1.5	125	-	-	-
Illinois	2	(D)	-	-	1,750 *	6.9	875	-	-	-
Minnesota	2	(D)	-	-	1,750 *	6.9	875	-	-	-
South Carolina	2	(D)	-	-	750 *	3.0	375	-	-	-
Wisconsin	2	(D)	-	-	375 *	1.5	188	-	-	-
Alabama	1	(D)	-	-	1,750 *	6.9	1,750	-	-	-
Arizona	1	(D)	-	-	175 *	0.7	175	-	-	-
Arkansas	1	(D)	-	-	3,750 *	14.8	3,750	-	-	-
Iowa	1	(D)	-	-	3,750 *	14.8	3,750	-	-	-
Kentucky	1	(D)	-	-	3,750 *	14.8	3,750	-	-	-
Tennessee	1	(D)	-	-	175 *	0.7	175	-	-	-

Source: 1992 *Economic Census*. The states are in descending order of shipments or establishments (if shipment data are missing for the majority). The symbol (D) appears when data are withheld to prevent disclosure of competitive information. States marked with (D) are sorted by number of establishments. A dash (-) indicates that the data element cannot be calculated; * indicates the midpoint of a range.

3633 - HOUSEHOLD LAUNDRY EQUIPMENT

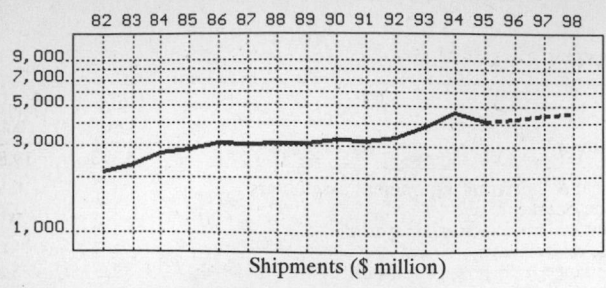

82 83 84 85 86 87 88 89 90 91 92 93 94 95 96 97 98

Shipments ($ million)

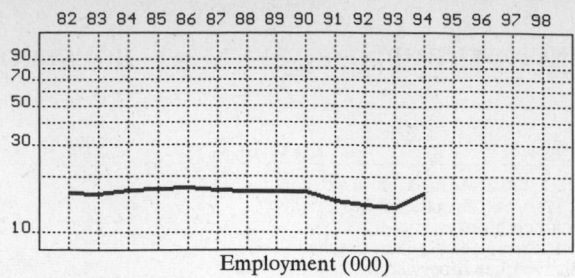

82 83 84 85 86 87 88 89 90 91 92 93 94 95 96 97 98

Employment (000)

GENERAL STATISTICS

| Year | Com-panies | Establishments | | Employment | | | Compensation | | Production ($ million) | | | |
		Total	with 20 or more employees	Total (000)	Production Workers (000)	Hours (Mil)	Payroll ($ mil)	Wages ($/hr)	Cost of Materials	Value Added by Manufacture	Value of Shipments	Capital Invest.
1982	15	25	20	16.5	13.4	23.4	335.4	11.16	1,116.2	984.1	2,122.2	35.1
1983		24	20	16.1	13.6	25.4	361.9	11.35	1,290.5	1,114.8	2,369.4	36.9
1984		23	20	16.7	14.1	27.0	401.2	11.83	1,435.1	1,292.5	2,715.6	69.9
1985		22	21	17.2	14.4	27.1	415.5	12.19	1,450.9	1,375.5	2,838.4	81.4
1986		21	19	17.4	14.7	28.7	452.3	12.51	1,558.4	1,520.4	3,062.9	69.3
1987	11	18	16	16.7	14.1	27.7	465.8	13.48	1,649.0	1,406.2	3,034.8	73.4
1988		17	15			26.5	449.1		1,727.4	1,398.1	3,117.9	
1989		16	14	16.6	14.2	27.7	453.3	13.22	1,657.0	1,460.6	3,103.8	59.4
1990		17	15	16.6	13.9	27.4	439.5	13.17	1,726.1	1,543.1	3,234.4	54.6
1991		18	14	14.7	12.4	24.5	413.3	13.65	1,654.3	1,517.6	3,205.7	82.2
1992	10	17	15	14.2	12.1	24.8	423.1	13.71	1,721.2	1,545.2	3,328.5	93.6
1993		17	13	13.8	11.9	24.2	440.7	14.76	2,170.0	1,761.9	3,871.3	73.5
1994		14P	12P	16.2	12.9	25.3	537.2	15.15	2,327.2	2,290.0	4,612.1	79.6
1995		14P	11P			25.6P	493.2P		2,089.1P	2,055.8P	4,140.3P	
1996		13P	11P			25.5P	502.3P		2,162.4P	2,127.8P	4,285.5P	
1997		12P	10P			25.4P	511.3P		2,235.6P	2,199.9P	4,430.6P	
1998		11P	9P			25.4P	520.4P		2,308.8P	2,271.9P	4,575.7P	

Sources: 1982, 1987, 1992 *Economic Census*; *Annual Survey of Manufactures*, 83-86, 88-91, 93-94. Establishment counts for non-Census years are from *County Business Patterns*; establishment values for 83-84 are extrapolations. 'P's show projections by the editors. Industries reclassified in 87 will not have data for prior years.

INDICES OF CHANGE

| Year | Com-panies | Establishments | | Employment | | | Compensation | | Production ($ million) | | | |
		Total	with 20 or more employees	Total (000)	Production Workers (000)	Hours (Mil)	Payroll ($ mil)	Wages ($/hr)	Cost of Materials	Value Added by Manufacture	Value of Shipments	Capital Invest.
1982	150	147	133	116	111	94	79	81	65	64	64	38
1983		141	133	113	112	102	86	83	75	72	71	39
1984		135	133	118	117	109	95	86	83	84	82	75
1985		129	140	121	119	109	98	89	84	89	85	87
1986		124	127	123	121	116	107	91	91	98	92	74
1987	110	106	107	118	117	112	110	98	96	91	91	78
1988		100	100			107	106		100	90	94	
1989		94	93	117	117	112	107	96	96	95	93	63
1990		100	100	117	115	110	104	96	100	100	97	58
1991		106	93	104	102	99	98	100	96	98	96	88
1992	100	100	100	100	100	100	100	100	100	100	100	100
1993		100	87	97	98	98	104	108	126	114	116	79
1994		85P	80P	114	107	102	127	111	135	148	139	85
1995		80P	76P			103P	117P		121P	133P	124P	
1996		76P	71P			103P	119P		126P	138P	129P	
1997		71P	66P			103P	121P		130P	142P	133P	
1998		66P	61P			102P	123P		134P	147P	137P	

Sources: Same as General Statistics. Values reflect change from the base year, 1992. Values above 100 mean greater than 92, values below 100 mean less than 92, and a value of 100 in the 82-91 or 93-98 period means same as 92. 'P's mark projections by the editors.

SELECTED RATIOS

For 1994	Avg. of All Manufact.	Analyzed Industry	Index	For 1994	Avg. of All Manufact.	Analyzed Industry	Index
Employees per Establishment	49	1,123	2,292	Value Added per Production Worker	134,084	177,519	132
Payroll per Establishment	1,500,273	37,242,857	2,482	Cost per Establishment	5,045,178	161,339,496	3,198
Payroll per Employee	30,620	33,160	108	Cost per Employee	102,970	143,654	140
Production Workers per Establishment	34	894	2,606	Cost per Production Worker	146,988	180,403	123
Wages per Establishment	853,319	26,572,973	3,114	Shipments per Establishment	9,576,895	319,746,429	3,339
Wages per Production Worker	24,861	29,713	120	Shipments per Employee	195,460	284,698	146
Hours per Production Worker	2,056	1,961	95	Shipments per Production Worker	279,017	357,527	128
Wages per Hour	12.09	15.15	125	Investment per Establishment	321,011	5,518,487	1,719
Value Added per Establishment	4,602,255	158,760,504	3,450	Investment per Employee	6,552	4,914	75
Value Added per Employee	93,930	141,358	150	Investment per Production Worker	9,352	6,171	66

Sources: Same as General Statistics. The 'Average of All Manufacturing' column represents the average of all manufacturing industries reported for the most recent complete year available. The Index shows the relationship between the Average and the Analyzed Industry. For example, 100 means that they are equal; 500 that the Analyzed Industry is five times the average; 50 means that the Analyzed Industry is half the national average. The abbreviation 'na' is used to show that data are 'not available'.

LEADING COMPANIES   Number shown: **4**   Total sales ($ mil): **4,798**   Total employment (000): **27.1**

Company Name	Address				CEO Name	Phone	Co. Type	Sales ($ mil)	Empl. (000)
Maytag Corp	403 W 4th St N	Newton	IA	50208	Leonard A Hadley	515-792-8000	P	3,373	19.8
Whirlpool Corp	119 Birdseye St	Clyde	OH	43410	Kevin M Cooney	419-547-7711	D	600	3.3
Speed Queen Co	PO Box 990	Ripon	WI	54971	James L Bennett	414-748-3121	S	450	2.5
Frigidaire Laundry	400 Des Moines St	Webster City	IA	50595	Jenny Senion	515-832-5334	D	375	1.5

Source: Ward's Business Directory of U.S. Private and Public Companies, Volumes 1 and 2, 1996. The company type code used is as follows: P - Public, R - Private, S - Subsidiary, D - Division, J - Joint Venture, A - Affiliate, G - Group. Sales are in millions of dollars, employees are in thousands. An asterisk (*) indicates an estimated sales volume. The symbol < stands for 'less than'. Company names and addresses are truncated, in some cases, to fit into the available space.

MATERIALS CONSUMED

Material	Quantity	Delivered Cost ($ million)
Materials, ingredients, containers, and supplies	(X)	1,531.7
Metal stampings	(X)	42.8
Metal wire racks, grills, springs, and other fabricated nonelectric wire products (except forgings)	(X)	8.8
Metal bolts, nuts, screws, washers, rivets, and other screw machine products	(X)	40.1
All other fabricated metal products (except forgings)	(X)	34.6
Iron and steel castings (rough and semifinished)	(X)	134.8
Steel bars, bar shapes, and plates	(X)	19.1
Steel sheet and strip, including tin plate	(X)	293.4
Aluminum and aluminum-base alloy shapes and forms	(X)	(D)
Other nonferrous shapes and forms	(X)	(D)
Nonferrous wire and cable, including magnet wire, bare or insulated wire, etc.	(X)	(D)
Fractional horsepower electric timing motors, synchronous and subsynchronous (less than 1 hp)	(X)	226.0
Integral horsepower electric motors and generators (1 hp or more)	(X)	29.9
Paper and paperboard containers, including shipping sacks and other paper packaging supplies	(X)	37.7
Electrical transmission, distribution, and control equipment	(X)	49.4
Current-carrying wiring devices	(X)	(D)
Timing mechanisms, except microprocessors	(X)	72.9
Automatic temperature controls (thermostats, regulators, etc.)	(X)	42.5
Paints, varnishes, lacquers, stains, shellacs, japans, enamels, and allied products	(X)	46.3
Fabricated rubber products, except tires, tubes, hose, belting, and gaskets	(X)	(D)
Rubber and plastics hose and belting	(X)	46.2
Plastics products consumed in the form of sheets, rods, tubes, film, and other shapes	(X)	67.6
Plastics resins consumed in the form of granules, pellets, powders, liquids, etc.	(X)	29.3
Complete flexible cord sets	(X)	(D)
Resistors, capacitors, transformers, electron tubes, semiconductors, and other electronic components	(X)	(D)
Mineral wool insulation (fibrous glass, rock wool, etc.)	(X)	(D)
All other materials and components, parts, containers, and supplies	(X)	142.4
Materials, ingredients, containers, and supplies, nsk	(X)	4.5

Source: 1992 *Economic Census*. Explanation of symbols used: (D): Withheld to avoid disclosure of competitive data; na: Not available; (S): Withheld because statistical norms were not met; (X): Not applicable; (Z): Less than half the unit shown; nec: Not elsewhere classified; nsk: Not specified by kind; - : zero; * : 10-19 percent estimated; ** : 20-29 percent estimated.

PRODUCT SHARE DETAILS

Product or Product Class	% Share	Product or Product Class	% Share
Household laundry equipment.	100.00	combinations	95.01
Household laundry machines, including both coin- and non-coin-operated washing machines, dryers, and		Parts, accessories, and attachments for household laundry equipment, sold separately.	4.76

Source: 1992 *Economic Census*. The values shown are percent of total shipments in an industry. Values of indented subcategories are summed in the main heading. The symbol (D) appears when data are withheld to prevent disclosure of competitive information. The abbreviation nsk stands for 'not specified by kind' and nec for 'not elsewhere classified'.

INPUTS AND OUTPUTS FOR HOUSEHOLD LAUNDRY EQUIPMENT

Economic Sector or Industry Providing Inputs	%	Sector	Economic Sector or Industry Buying Outputs	%	Sector
Blast furnaces & steel mills	15.3	Manufg.	Personal consumption expenditures	78.7	
Wholesale trade	12.1	Trade	Exports	6.1	Foreign
Motors & generators	11.7	Manufg.	Laundry, dry cleaning, shoe repair	4.5	Services
Pipe, valves, & pipe fittings	3.9	Manufg.	Gross private fixed investment	2.8	Cap Inv
Watches, clocks, & parts	3.5	Manufg.	State & local government enterprises, nec	2.3	Gov't
Metal stampings, nec	3.2	Manufg.	Real estate	2.1	Fin/R.E.
Advertising	3.0	Services	Mobile homes	1.6	Manufg.
Paperboard containers & boxes	2.6	Manufg.	S/L Govt. purch., higher education	0.5	S/L Govt
Environmental controls	2.4	Manufg.	Electrical repair shops	0.4	Services
Glass containers	2.4	Manufg.	Household laundry equipment	0.3	Manufg.
Power transmission equipment	2.4	Manufg.	S/L Govt. purch., correction	0.3	S/L Govt
Wiring devices	2.1	Manufg.	Federal Government purchases, national defense	0.1	Fed Gov't
Iron & steel foundries	1.9	Manufg.	S/L Govt. purch., health & hospitals	0.1	S/L Govt
Rubber & plastics hose & belting	1.8	Manufg.	S/L Govt. purch., natural resource & recreation.	0.1	S/L Govt
Screw machine and related products	1.8	Manufg.			
Electric services (utilities)	1.8	Util.			
Switchgear & switchboard apparatus	1.7	Manufg.			
Miscellaneous plastics products	1.6	Manufg.			
Ball & roller bearings	1.5	Manufg.			
Motor freight transportation & warehousing	1.4	Util.			
Aluminum castings	1.3	Manufg.			
Gas production & distribution (utilities)	1.3	Util.			
Paints & allied products	1.2	Manufg.			
Plastics materials & resins	1.2	Manufg.			
Imports	1.0	Foreign			
Refrigeration & heating equipment	0.9	Manufg.			
Banking	0.9	Fin/R.E.			
Air transportation	0.8	Util.			
U.S. Postal Service	0.8	Gov't			
Aluminum rolling & drawing	0.7	Manufg.			
Maintenance of nonfarm buildings nec	0.6	Constr.			
Fabricated rubber products, nec	0.6	Manufg.			
Mineral wool	0.6	Manufg.			
Signs & advertising displays	0.6	Manufg.			
Household laundry equipment	0.5	Manufg.			
Metal coating & allied services	0.5	Manufg.			
Miscellaneous fabricated wire products	0.5	Manufg.			
Nonferrous castings, nec	0.5	Manufg.			
Primary metal products, nec	0.5	Manufg.			
Eating & drinking places	0.5	Trade			
Communications, except radio & TV	0.4	Util.			
Hotels & lodging places	0.4	Services			
Automatic merchandising machines	0.3	Manufg.			
Gaskets, packing & sealing devices	0.3	Manufg.			
Glass & glass products, except containers	0.3	Manufg.			
Nonferrous wire drawing & insulating	0.3	Manufg.			
Paper coating & glazing	0.3	Manufg.			
Railroads & related services	0.3	Util.			
Copper rolling & drawing	0.2	Manufg.			
Machinery, except electrical, nec	0.2	Manufg.			
Special dies & tools & machine tool accessories	0.2	Manufg.			
Sanitary services, steam supply, irrigation	0.2	Util.			
Computer & data processing services	0.2	Services			
Equipment rental & leasing services	0.2	Services			
Legal services	0.2	Services			
Management & consulting services & labs	0.2	Services			
Abrasive products	0.1	Manufg.			
Machine tools, metal forming types	0.1	Manufg.			
Petroleum refining	0.1	Manufg.			
Wood pallets & skids	0.1	Manufg.			
Insurance carriers	0.1	Fin/R.E.			
Security & commodity brokers	0.1	Fin/R.E.			
Accounting, auditing & bookkeeping	0.1	Services			
Automotive repair shops & services	0.1	Services			

Source: Benchmark Input-Output Accounts for the U.S. Economy, 1982, U.S. Department of Commerce, Washington, D.C., July 1991. Data, as reported in the source, are organized by the 1977 SIC structure in use in 1982 but have been matched, as closely as is possible, to the 1987 SIC structure used in this book.

OCCUPATIONS EMPLOYED BY SIC 363 - HOUSEHOLD APPLIANCES

Occupation	% of Total 1994	Change to 2005	Occupation	% of Total 1994	Change to 2005
Assemblers, fabricators, & hand workers nec	31.2	-16.9	Electrical & electronic equipment assemblers	1.7	-16.9
Helpers, laborers, & material movers nec	4.1	-16.9	Machine operators nec	1.5	-41.4
Machine assemblers	3.9	-25.2	Welding machine setters, operators	1.5	-25.2
Inspectors, testers, & graders, precision	3.1	-41.9	Sales & related workers nec	1.5	-17.0
Blue collar worker supervisors	3.0	-25.4	Maintenance repairers, general utility	1.5	-25.2
Electrical & electronic assemblers	2.7	-25.3	Janitors & cleaners, incl maids	1.2	-33.6
Industrial truck & tractor operators	2.6	-16.9	Tool & die makers	1.2	-32.9
Plastic molding machine workers	2.6	-0.3	Material moving equipment operators nec	1.2	-16.9
Machine forming operators, metal & plastic	2.5	-58.5	General managers & top executives	1.1	-21.2
Freight, stock, & material movers, hand	1.9	-33.6	Machine tool cutting & forming etc. nec	1.1	-33.6

Source: *Industry-Occupation Matrix*, Bureau of Labor Statistics. These data relate to one or more 3-digit SIC industry groups rather than to a single 4-digit SIC. The change reported for each occupation to the year 2005 is a percent of growth or decline as estimated by the Bureau of Labor Statistics. The abbreviation nec stands for 'not elsewhere classified'.

LOCATION BY STATE AND REGIONAL CONCENTRATION

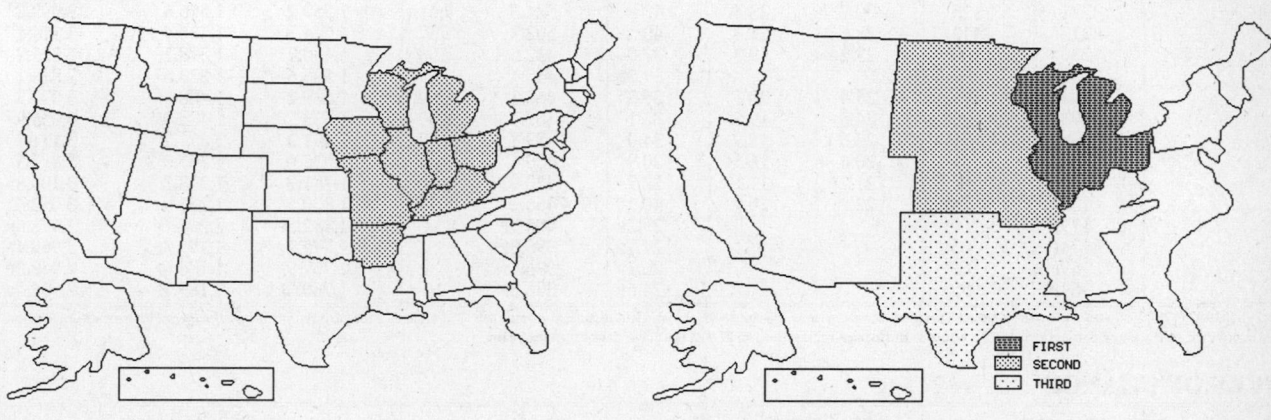

FIRST
SECOND
THIRD

INDUSTRY DATA BY STATE

State	Establish-ments	Shipments Total ($ mil)	Shipments % of U.S.	Shipments Per Establ.	Employment Total Number	Employment % of U.S.	Employment Per Establ.	Wages ($/hour)	Cost as % of Shipments	Investment per Employee ($)
Iowa	4	(D)	-	-	3,750 *	26.4	938	-	-	-
Illinois	2	(D)	-	-	750 *	5.3	375	-	-	-
Indiana	2	(D)	-	-	750 *	5.3	375	-	-	-
Ohio	2	(D)	-	-	7,500 *	52.8	3,750	-	-	-
Arkansas	1	(D)	-	-	375 *	2.6	375	-	-	-
Kentucky	1	(D)	-	-	1,750 *	12.3	1,750	-	-	-
Michigan	1	(D)	-	-	375 *	2.6	375	-	-	-
Missouri	1	(D)	-	-	175 *	1.2	175	-	-	-
Wisconsin	1	(D)	-	-	1,750 *	12.3	1,750	-	-	-

Source: 1992 *Economic Census*. The states are in descending order of shipments or establishments (if shipment data are missing for the majority). The symbol (D) appears when data are withheld to prevent disclosure of competitive information. States marked with (D) are sorted by number of establishments. A dash (-) indicates that the data element cannot be calculated; * indicates the midpoint of a range.

3634 - ELECTRIC HOUSEWARES & FANS

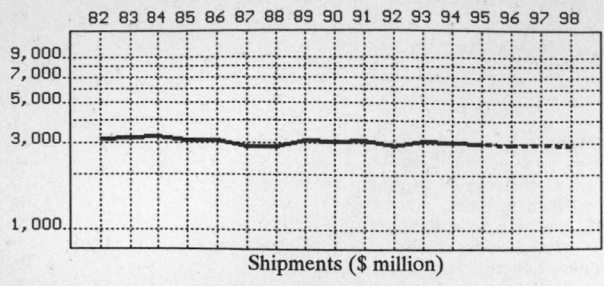

82 83 84 85 86 87 88 89 90 91 92 93 94 95 96 97 98

Shipments ($ million)

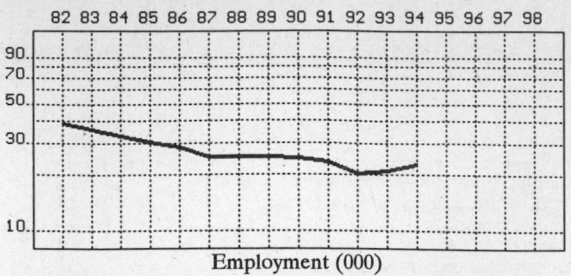

82 83 84 85 86 87 88 89 90 91 92 93 94 95 96 97 98

Employment (000)

GENERAL STATISTICS

Year	Com-panies	Establishments		Employment			Compensation		Production ($ million)			
		Total	with 20 or more employees	Total (000)	Production Workers (000)	Hours (Mil)	Payroll ($ mil)	Wages ($/hr)	Cost of Materials	Value Added by Manufacture	Value of Shipments	Capital Invest.
1982	218	263	134	38.7	29.3	54.4	539.1	6.28	1,556.4	1,538.5	3,155.7	73.0
1983		249	131	36.0	28.1	51.0	543.3	6.84	1,548.7	1,639.6	3,186.2	66.1
1984		235	128	33.1	25.7	47.1	522.6	7.18	1,607.9	1,669.7	3,238.6	64.9
1985		221	125	30.7	23.4	43.8	524.3	7.61	1,559.2	1,546.6	3,099.6	68.5
1986		212	113	28.6	21.3	40.5	502.2	7.57	1,644.3	1,436.4	3,116.1	65.7
1987	201	230	100	25.2	19.3	37.0	432.5	7.58	1,430.7	1,378.1	2,825.7	57.0
1988		208	92			37.3	445.4		1,506.6	1,322.6	2,828.1	
1989		194	95	25.4	20.5	37.3	463.1	8.00	1,639.2	1,488.4	3,077.9	44.5
1990		193	99	24.8	19.5	35.1	462.1	8.50	1,633.7	1,425.0	3,055.9	48.9
1991		206	95	23.6	18.9	34.0	453.8	8.69	1,651.2	1,371.7	3,111.8	47.1
1992	189	209	81	20.4	16.5	30.9	400.6	8.45	1,525.0	1,387.0	2,897.5	46.3
1993		199	85	21.2	17.5	32.7	420.9	8.56	1,781.2	1,328.2	3,105.8	59.1
1994		185P	74P	22.9	20.1	40.1	466.2	8.14	1,830.5	1,236.9	3,052.6	66.4
1995		179P	69P			29.2P	405.9P		1,782.5P	1,204.5P	2,972.6P	
1996		174P	65P			27.7P	396.0P		1,775.2P	1,199.5P	2,960.4P	
1997		169P	60P			26.1P	386.1P		1,767.9P	1,194.6P	2,948.2P	
1998		164P	55P			24.6P	376.3P		1,760.6P	1,189.7P	2,936.0P	

Sources: 1982, 1987, 1992 *Economic Census*; *Annual Survey of Manufactures*, 83-86, 88-91, 93-94. Establishment counts for non-Census years are from *County Business Patterns*; establishment values for 83-84 are extrapolations. 'P's show projections by the editors. Industries reclassified in 87 will not have data for prior years.

INDICES OF CHANGE

Year	Com-panies	Establishments		Employment			Compensation		Production ($ million)			
		Total	with 20 or more employees	Total (000)	Production Workers (000)	Hours (Mil)	Payroll ($ mil)	Wages ($/hr)	Cost of Materials	Value Added by Manufacture	Value of Shipments	Capital Invest.
1982	115	126	165	190	178	176	135	74	102	111	109	158
1983		119	162	176	170	165	136	81	102	118	110	143
1984		112	158	162	156	152	130	85	105	120	112	140
1985		106	154	150	142	142	131	90	102	112	107	148
1986		101	140	140	129	131	125	90	108	104	108	142
1987	106	110	123	124	117	120	108	90	94	99	98	123
1988		100	114			121	111		99	95	98	
1989		93	117	125	124	121	116	95	107	107	106	96
1990		92	122	122	118	114	115	101	107	103	105	106
1991		99	117	116	115	110	113	103	108	99	107	102
1992	100	100	100	100	100	100	100	100	100	100	100	100
1993		95	105	104	106	106	105	101	117	96	107	128
1994		88P	92P	112	122	130	116	96	120	89	105	143
1995		86P	86P			95P	101P		117P	87P	103P	
1996		83P	80P			90P	99P		116P	86P	102P	
1997		81P	74P			85P	96P		116P	86P	102P	
1998		78P	67P			80P	94P		115P	86P	101P	

Sources: Same as General Statistics. Values reflect change from the base year, 1992. Values above 100 mean greater than 92, values below 100 mean less than 92, and a value of 100 in the 82-91 or 93-98 period means same as 92. 'P's mark projections by the editors.

SELECTED RATIOS

For 1994	Avg. of All Manufact.	Analyzed Industry	Index	For 1994	Avg. of All Manufact.	Analyzed Industry	Index
Employees per Establishment	49	124	253	Value Added per Production Worker	134,084	61,537	46
Payroll per Establishment	1,500,273	2,526,207	168	Cost per Establishment	5,045,178	9,918,966	197
Payroll per Employee	30,620	20,358	66	Cost per Employee	102,970	79,934	78
Production Workers per Establishment	34	109	317	Cost per Production Worker	146,988	91,070	62
Wages per Establishment	853,319	1,768,746	207	Shipments per Establishment	9,576,895	16,541,182	173
Wages per Production Worker	24,861	16,240	65	Shipments per Employee	195,460	133,301	68
Hours per Production Worker	2,056	1,995	97	Shipments per Production Worker	279,017	151,871	54
Wages per Hour	12.09	8.14	67	Investment per Establishment	321,011	359,803	112
Value Added per Establishment	4,602,255	6,702,414	146	Investment per Employee	6,552	2,900	44
Value Added per Employee	93,930	54,013	58	Investment per Production Worker	9,352	3,303	35

Sources: Same as General Statistics. The 'Average of All Manufacturing' column represents the average of all manufacturing industries reported for the most recent complete year available. The Index shows the relationship between the Average and the Analyzed Industry. For example, 100 means that they are equal; 500 that the Analyzed Industry is five times the average; 50 means that the Analyzed Industry is half the national average. The abbreviation 'na' is used to show that data are 'not available'.

LEADING COMPANIES Number shown: 62 Total sales ($ mil): 5,059 Total employment (000): 43.1

Company Name	Address				CEO Name	Phone	Co. Type	Sales ($ mil)	Empl. (000)
Sunbeam-Oster Company Inc	200 E Las Olas Blvd	Ft Lauderdale	FL	33301	Roger W Schipke	305-767-2100	P	1,198	12.0
Sunbeam-Oster Household Prod	PO Box 247	Laurel	MS	39440	R L Boynton Jr	601-425-7800	S	600	6.0
Hamilton Beach/Proctor-Silex	4421 Waterfront Dr	Glen Allen	VA	23060	George C Nebel	804-273-9777	S	378	4.1
Rival Co	800 E 101st Ter	Kansas City	MO	64131	Thomas K Manning	816-943-4100	P	229	2.1
NuTone Inc	Madison	Cincinnati	OH	45227	Greg Lawton	513-527-5100	S	210	1.5
Broan Manufacturing Company	PO Box 140	Hartford	WI	53027	JG Santowski	414-673-4340	S	200	0.7
Rival Manufacturing Co	800 E 101st Ter	Kansas City	MO	64131	Thomas K Manning	816-943-4100	S	180	1.4
Toastmaster Inc	1801 Nadium Blvd	Columbia	MO	65202	Robert H Deming	314-445-8666	P	164	1.4
Teledyne Water Pik	1730 E Prospect Rd	Fort Collins	CO	80525	Melvin E Cruger	303-484-1352	S	150*	1.1
West Bend Co	400 Washington St	West Bend	WI	53095	Thomas Kieckhafer	414-334-2311	S	150	1.2
Duracraft Corp	355 Main St	Whitinsville	MA	01588	Bernard K Chiu	508-234-4600	P	143	2.2
Conair Corp	1 Cummings Pt Rd	Stamford	CT	06904	Lee P Rizzuto	203-351-9000	R	140*	1.0
National Presto Industries Inc	3925 N Hastings	Eau Claire	WI	54703	Maryjo Cohen	715-839-2121	P	128	0.7
Helen of Troy Corp	6827 Market Av	El Paso	TX	79915	Gerald J Rubin	915-779-6363	S	123	0.3
Dazey Corp	1 Dazey Cir	Indust Apt	KS	66031	Henry Talge	913-782-7500	R	110	0.5
Hunter Fan Co	PO Box 14775	Memphis	TN	38114	R E Beasley Jr	901-743-1360	R	100	0.3
Bunn-O-Matic Corp	1400 Stevenson Dr	Springfield	IL	62708	Arthur Bunn	217-529-6601	R	97*	0.7
TPI Corp	PO Box 4973-CRS	Johnson City	TN	37602	RE Henry	615-477-4131	R	81	0.8
Encon Industries Inc	6901 Snowden Rd	Fort Worth	TX	76140	HW Markwardt	817-293-7400	R	65	0.1
Salton/Maxim Housewares Inc	550 Business Ctr Dr	Mt Prospect	IL	60056	Leonhard Dreimann	708-803-4600	P	52	<0.1
Waring Products	283 Main St	New Hartford	CT	06057	Bruno M Valbona	203-379-0731	D	48	0.5
Rival Manufacturing Co	217 E 16th St	Sedalia	MO	65301	Carol Bottcher	816-826-6600	D	41*	0.3
Lasko Metal Products Inc	820 Lincoln Av	West Chester	PA	19380	Oscar Lasko	215-692-7400	R	39	0.4
Tutco Inc	500 Gould Dr	Cookeville	TN	38501	Michael C Mahoney	615-432-4141	R	37	0.4
Cooperation Opportunity Inc	PO Box 39108	Minneapolis	MN	55439	JS Braun	612-941-5600	S	35	0.5
Casablanca Fan Co	450 N Baldwin Park	City of Industry	CA	91746	Thomas Laird	818-369-6441	R	33*	0.2
Norelco Consumer Products	PO Box 120015	Stamford	CT	06912	Patrick Dinley	203-973-0200	D	31	0.2
Creative Technologies Corp	170 53rd St	Brooklyn	NY	11232	Richard Helfman	718-492-8400	P	30	0.1
Captive-Aire Systems Inc	112 Wheaton Dr	Youngsville	NC	27629	Robert L Luddy	919-554-2410	R	30	0.2
Research Products Corp	PO Box 1467	Madison	WI	53701	Vern J Hellenbrand	608-257-8801	R	25*	0.3
Metal Ware Corp	PO Box 237	Two Rivers	WI	54241	WC Drumm	414-793-1368	R	20	0.1
Vita-Mix Corp	8615 Usher Rd	Cleveland	OH	44138	W Grover Barnard	216-235-4840	R	20	0.1
Penn Ventilator Company Inc	9995 Gantry Rd	Philadelphia	PA	19115	Louis G Malissa	215-464-8900	D	19*	0.1
King Electrical Mfg Co	9131 10th Av S	Seattle	WA	98108	Robert E Wilson	206-762-0400	R	15	0.1
Cadet Manufacturing Co	PO Box 1675	Vancouver	WA	98668	Richard Anderson	206-693-2505	R	14	0.1
Vornado Air Circulation Systems	550 N 159th St E	Wichita	KS	67230	Michael C Coup	316-733-0035	R	12	<0.1
Air Technology Systems Inc	1572 Tilco Dr	Frederick	MD	21701	James A Garrett	301-620-2033	R	10	0.1
Hankscraft Motors	PO Box 190	Reedsburg	WI	53959	Steve Royster	608-524-4341	D	10	<0.1
Hartman Products Inc	4949 W 147th St	Hawthorne	CA	90250	M Dorfman	310-676-7700	R	10	<0.1
Hudson Standard Corp	90 South St	Newark	NJ	07102	T Pearlman	201-589-6140	R	10	0.1
Takka	170 53rd St	Brooklyn	NY	11232	Richard Halfman	718-492-8400	D	9*	<0.1
Air-Dry Corporation of America	5297 Maureen Ln	Moorpark	CA	93021	Bob Schneider	805-529-6226	D	8	<0.1
Homedics Inc	2240 Greer St	Keego Harbor	MI	48320	Ron Ferber	810-681-9600	R	8*	<0.1
Waugh Controls Corp	733 Lakefield Rd	Westlake Vil	CA	91361	R Lucas	805-495-7111	R	6*	0.1
Dynamic Cooking Systems Inc	10850 Portal Dr	Los Alamitos	CA	90720	Surgit Kalsi	714-220-9505	R	5*	<0.1
Stewart Manufacturing Co	1280 N Senate Av	Indianapolis	IN	46202	Tom Brennan	317-634-8585	R	5*	<0.1
Thermal Circuits Inc	4 Jefferson Av	Salem	MA	01970	Anthony A Klein	508-745-1162	R	4*	<0.1
WB Marvin Manufacturing Co	PO Box 230	Urbana	OH	43078	Angus Randolph	513-653-7131	R	4	<0.1
Airflow By Casablanca	16175 Stephens St	City of Industry	CA	91745	John Rattan	818-333-8313	R	3*	<0.1
Marpac Corp	PO Box 3098	Wilmington	NC	28406	James Buckwalter	910-763-7861	R	3*	<0.1
Rama Corp	600 W Esplanade	San Jacinto	CA	92583	MM Renshaw	909-654-7351	R	3	0.2
Ramfan Corp	2746 Orange	Spring Valley	CA	91978	Wayne Allen	619-670-9590	R	3	<0.1
American Electrical Heater Co	6110 Cass Av	Detroit	MI	48202	Robert W Kuhn	313-875-2505	R	2*	<0.1
Tops Manufacturing Company	83 Salisbury Rd	Darien	CT	06820	Mitch Himmel	203-655-9367	R	2	<0.1
Appliance Science Corp	PO Box 566	Southport	CT	06490	Robert Wilson	203-255-2402	R	2	<0.1
Erincraft Manufacturing Co	27 Pine Lake Av	La Porte	IN	46350	Sam Verma	219-324-2789	R	1	<0.1
Aqua-Mist Inc	PO Box 4558	Winston-Salem	NC	27115	Robert B Morrow	910-767-2201	R	1	<0.1
Sonex International Corp	PO Box 533	Brewster	NY	10509	Robert Bock	914-279-7048	R	1*	<0.1
PAJ America Inc	11 Martine Av	White Plains	NY	10606	Peter O Bodnar	914-997-1100	R	1	<0.1
Memphis Metal Manufacturing	PO Box 11271	Memphis	TN	38111	William B Mason Jr	901-276-6363	R	1	<0.1
Medisonic USA Inc	9600 Main St	Clarence	NY	14031	Don Trainer	716-759-7213	R	0	<0.1
Phoenix Housewares Corp	PO Box 643	Wilton	CT	06897	Yvonne Cosacow	203-762-7799	R	0*	<0.1

Source: Ward's Business Directory of U.S. Private and Public Companies, Volumes 1 and 2, 1996. The company type code used is as follows: P - Public, R - Private, S - Subsidiary, D - Division, J - Joint Venture, A - Affiliate, G - Group. Sales are in millions of dollars, employees are in thousands. An asterisk (*) indicates an estimated sales volume. The symbol < stands for 'less than'. Company names and addresses are truncated, in some cases, to fit into the available space.

MATERIALS CONSUMED

Material	Quantity	Delivered Cost ($ million)
Materials, ingredients, containers, and supplies	(X)	1,278.2
Metal stampings	(X)	31.8
Metal wire racks, grills, springs, and other fabricated nonelectric wire products (except forgings)	(X)	21.3
Metal bolts, nuts, screws, washers, rivets, and other screw machine products	(X)	21.6
All other fabricated metal products (except forgings)	(X)	16.0
Iron and steel castings (rough and semifinished)	(X)	5.2
Aluminum and aluminum-base alloy castings (rough and semifinished)	(X)	25.5
Other nonferrous castings (rough and semifinished)	(X)	13.0
Steel sheet and strip, including tin plate	(X)	86.9
All other steel shapes and forms	(X)	5.8
Aluminum and aluminum-base alloy shapes and forms	(X)	25.7
Other nonferrous shapes and forms	(X)	7.8
Nonferrous wire and cable, including magnet wire, bare or insulated wire, etc.	(X)	9.5
Fractional horsepower electric timing motors, synchronous and subsynchronous (less than 1 hp)	(X)	140.8
Integral horsepower electric motors and generators (1 hp or more)	(X)	4.3
Paper and paperboard containers, including shipping sacks and other paper packaging supplies	(X)	72.1
Electrical transmission, distribution, and control equipment	(X)	20.0
Current-carrying wiring devices	(X)	24.8
Timing mechanisms, except microprocessors	(X)	(D)
Automatic temperature controls (thermostats, regulators, etc.)	(X)	61.4
Paints, varnishes, lacquers, stains, shellacs, japans, enamels, and allied products	(X)	21.7
Fabricated rubber products, except tires, tubes, hose, belting, and gaskets	(X)	6.5
Rubber and plastics hose and belting	(X)	(D)
Plastics products consumed in the form of sheets, rods, tubes, film, and other shapes	(X)	74.5
Plastics resins consumed in the form of granules, pellets, powders, liquids, etc.	(X)	50.2
Complete flexible cord sets	(X)	39.1
Resistors, capacitors, transformers, electron tubes, semiconductors, and other electronic components	(X)	32.5
Mineral wool insulation (fibrous glass, rock wool, etc.)	(X)	(D)
All other materials and components, parts, containers, and supplies	(X)	286.6
Materials, ingredients, containers, and supplies, nsk	(X)	164.3

Source: 1992 *Economic Census*. Explanation of symbols used: (D): Withheld to avoid disclosure of competitive data; na: Not available; (S): Withheld because statistical norms were not met; (X): Not applicable; (Z): Less than half the unit shown; nec: Not elsewhere classified; nsk: Not specified by kind; - : zero; * : 10-19 percent estimated; ** : 20-29 percent estimated.

PRODUCT SHARE DETAILS

Product or Product Class	% Share	Product or Product Class	% Share
Electric housewares and fans	100.00	appliances	5.78
Electric fans, except industrial type	18.81	Parts and attachments for household electric fans	13.05
Small electric household appliances, except fans	72.21	Parts for other small household electric appliances	86.95
Parts and attachments for small household electric		Electric housewares and fans, nsk	3.20

Source: 1992 *Economic Census*. The values shown are percent of total shipments in an industry. Values of indented subcategories are summed in the main heading. The symbol (D) appears when data are withheld to prevent disclosure of competitive information. The abbreviation nsk stands for 'not specified by kind' and nec for 'not elsewhere classified'.

INPUTS AND OUTPUTS FOR ELECTRIC HOUSEWARES & FANS

Economic Sector or Industry Providing Inputs	%	Sector	Economic Sector or Industry Buying Outputs	%	Sector
Imports	23.6	Foreign	Personal consumption expenditures	69.5	
Wholesale trade	7.4	Trade	Exports	6.3	Foreign
Motors & generators	7.3	Manufg.	Gross private fixed investment	4.1	Cap Inv
Blast furnaces & steel mills	5.6	Manufg.	Residential 1-unit structures, nonfarm	4.0	Constr.
Plastics materials & resins	3.9	Manufg.	Electrical repair shops	3.5	Services
Wood products, nec	3.4	Manufg.	Nonfarm residential structure maintenance	2.1	Constr.
Paperboard containers & boxes	3.3	Manufg.	Residential additions/alterations, nonfarm	1.8	Constr.
Miscellaneous plastics products	3.1	Manufg.	Maintenance of nonfarm buildings nec	1.4	Constr.
Advertising	2.8	Services	Beauty & barber shops	0.9	Services
Nonferrous wire drawing & insulating	2.5	Manufg.	Industrial buildings	0.5	Constr.
Glass & glass products, except containers	2.3	Manufg.	Residential garden apartments	0.5	Constr.
Aluminum rolling & drawing	2.0	Manufg.	Federal Government purchases, national defense	0.5	Fed Govt
Wiring devices	1.8	Manufg.	Mobile homes	0.4	Manufg.
Screw machine and related products	1.6	Manufg.	S/L Govt. purch., higher education	0.4	S/L Govt
Electric services (utilities)	1.6	Util.	S/L Govt. purch., other general government	0.4	S/L Govt
Nonferrous castings, nec	1.5	Manufg.	Residential 2-4 unit structures, nonfarm	0.3	Constr.
Environmental controls	1.4	Manufg.	Residential high-rise apartments	0.3	Constr.
Hardware, nec	1.4	Manufg.	S/L Govt. purch., elem. & secondary education	0.3	S/L Govt
Noncomparable imports	1.4	Foreign	Electric housewares & fans	0.2	Manufg.
Nonferrous rolling & drawing, nec	1.3	Manufg.	S/L Govt. purch., correction	0.2	S/L Govt
Motor freight transportation & warehousing	1.1	Util.	S/L Govt. purch., public assistance & relief	0.2	S/L Govt
Broadwoven fabric mills	1.0	Manufg.	Household cooking equipment	0.1	Manufg.
Metal stampings, nec	1.0	Manufg.	Hotels & lodging places	0.1	Services
Mineral wool	1.0	Manufg.	Services to dwellings & other buildings	0.1	Services

Continued on next page.

INPUTS AND OUTPUTS FOR ELECTRIC HOUSEWARES & FANS - Continued

Economic Sector or Industry Providing Inputs	%	Sector	Economic Sector or Industry Buying Outputs	%	Sector
Maintenance of nonfarm buildings nec	0.8	Constr.	Federal Government purchases, nondefense	0.1	Fed Govt
Communications, except radio & TV	0.8	Util.	S/L Govt. purch., health & hospitals	0.1	S/L Govt
Banking	0.8	Fin/R.E.	S/L Govt. purch., highways	0.1	S/L Govt
Aluminum castings	0.7	Manufg.			
Air transportation	0.7	Util.			
Gas production & distribution (utilities)	0.6	Util.			
Eating & drinking places	0.6	Trade			
Equipment rental & leasing services	0.6	Services			
U.S. Postal Service	0.6	Gov't			
Paints & allied products	0.5	Manufg.			
Computer & data processing services	0.5	Services			
Electric housewares & fans	0.4	Manufg.			
Fabricated metal products, nec	0.4	Manufg.			
Signs & advertising displays	0.4	Manufg.			
Switchgear & switchboard apparatus	0.4	Manufg.			
Real estate	0.4	Fin/R.E.			
Chemical preparations, nec	0.3	Manufg.			
Machinery, except electrical, nec	0.3	Manufg.			
Miscellaneous fabricated wire products	0.3	Manufg.			
Primary metal products, nec	0.3	Manufg.			
Rubber & plastics hose & belting	0.3	Manufg.			
Special dies & tools & machine tool accessories	0.3	Manufg.			
Transformers	0.3	Manufg.			
Watches, clocks, & parts	0.3	Manufg.			
Railroads & related services	0.3	Util.			
Hotels & lodging places	0.3	Services			
Job training & related services	0.3	Services			
Abrasive products	0.2	Manufg.			
Fabricated rubber products, nec	0.2	Manufg.			
Metal coating & allied services	0.2	Manufg.			
Automotive repair shops & services	0.2	Services			
Legal services	0.2	Services			
Management & consulting services & labs	0.2	Services			
Coal	0.1	Mining			
Commercial printing	0.1	Manufg.			
Copper rolling & drawing	0.1	Manufg.			
Gaskets, packing & sealing devices	0.1	Manufg.			
Lubricating oils & greases	0.1	Manufg.			
Machine tools, metal forming types	0.1	Manufg.			
Manifold business forms	0.1	Manufg.			
Metal heat treating	0.1	Manufg.			
Petroleum refining	0.1	Manufg.			
Semiconductors & related devices	0.1	Manufg.			
Accounting, auditing & bookkeeping	0.1	Services			
Electrical repair shops	0.1	Services			
Laundry, dry cleaning, shoe repair	0.1	Services			

Source: Benchmark Input-Output Accounts for the U.S. Economy, 1982, U.S. Department of Commerce, Washington, D.C., July 1991. Data, as reported in the source, are organized by the 1977 SIC structure in use in 1982 but have been matched, as closely as is possible, to the 1987 SIC structure used in this book.

OCCUPATIONS EMPLOYED BY SIC 363 - HOUSEHOLD APPLIANCES

Occupation	% of Total 1994	Change to 2005	Occupation	% of Total 1994	Change to 2005
Assemblers, fabricators, & hand workers nec	31.2	-16.9	Electrical & electronic equipment assemblers	1.7	-16.9
Helpers, laborers, & material movers nec	4.1	-16.9	Machine operators nec	1.5	-41.4
Machine assemblers	3.9	-25.2	Welding machine setters, operators	1.5	-25.2
Inspectors, testers, & graders, precision	3.1	-41.9	Sales & related workers nec	1.5	-17.0
Blue collar worker supervisors	3.0	-25.4	Maintenance repairers, general utility	1.5	-25.2
Electrical & electronic assemblers	2.7	-25.3	Janitors & cleaners, incl maids	1.2	-33.6
Industrial truck & tractor operators	2.6	-16.9	Tool & die makers	1.2	-32.9
Plastic molding machine workers	2.6	-0.3	Material moving equipment operators nec	1.2	-16.9
Machine forming operators, metal & plastic	2.5	-58.5	General managers & top executives	1.1	-21.2
Freight, stock, & material movers, hand	1.9	-33.6	Machine tool cutting & forming etc. nec	1.1	-33.6

Source: Industry-Occupation Matrix, Bureau of Labor Statistics. These data relate to one or more 3-digit SIC industry groups rather than to a single 4-digit SIC. The change reported for each occupation to the year 2005 is a percent of growth or decline as estimated by the Bureau of Labor Statistics. The abbreviation nec stands for 'not elsewhere classified'.

LOCATION BY STATE AND REGIONAL CONCENTRATION

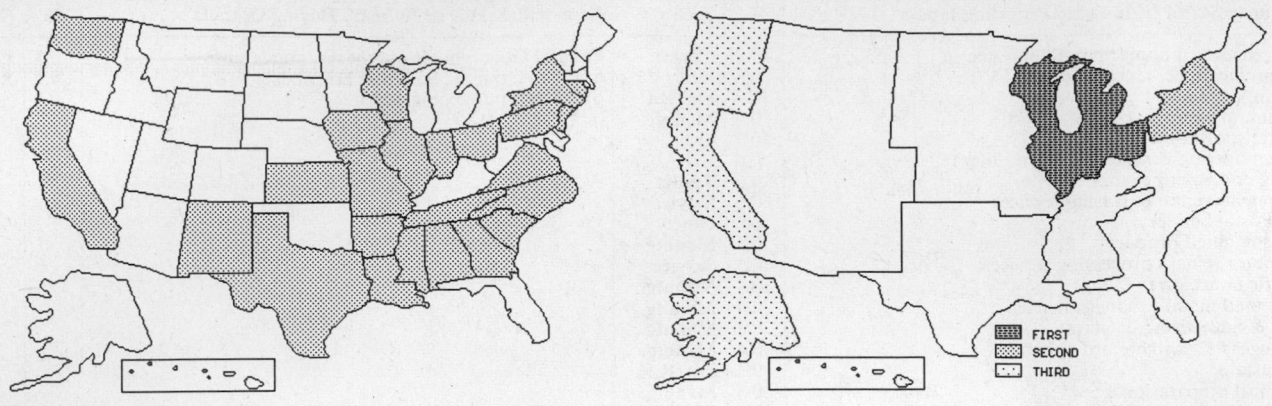

INDUSTRY DATA BY STATE

State	Establish-ments	Shipments			Employment				Cost as % of Shipments	Investment per Employee ($)
		Total ($ mil)	% of U.S.	Per Establ.	Total Number	% of U.S.	Per Establ.	Wages ($/hour)		
Ohio	13	406.4	14.0	31.3	2,000	9.8	154	9.38	49.8	3,300
Tennessee	7	160.6	5.5	22.9	1,300	6.4	186	9.27	70.5	-
California	20	125.5	4.3	6.3	600	2.9	30	9.00	56.0	2,833
Texas	7	104.5	3.6	14.9	800	3.9	114	6.54	55.4	1,500
Indiana	8	56.7	2.0	7.1	600	2.9	75	8.33	61.0	7,167
New York	18	41.3	1.4	2.3	300	1.5	17	6.67	42.6	2,333
New Jersey	5	9.9	0.3	2.0	100	0.5	20	5.50	60.6	-
Illinois	15	(D)	-	-	1,750 *	8.6	117	-	-	114
Wisconsin	9	(D)	-	-	1,750 *	8.6	194	-	-	5,714
Missouri	8	(D)	-	-	3,750 *	18.4	469	-	-	-
North Carolina	8	(D)	-	-	3,750 *	18.4	469	-	-	1,627
Pennsylvania	8	(D)	-	-	375 *	1.8	47	-	-	-
Mississippi	7	(D)	-	-	1,750 *	8.6	250	-	-	-
Washington	6	(D)	-	-	175 *	0.9	29	-	-	-
Alabama	5	(D)	-	-	375 *	1.8	75	-	-	267
Georgia	4	(D)	-	-	175 *	0.9	44	-	-	-
Louisiana	4	(D)	-	-	375 *	1.8	94	-	-	-
Arkansas	3	(D)	-	-	175 *	0.9	58	-	-	-
Connecticut	3	(D)	-	-	750 *	3.7	250	-	-	-
Iowa	3	(D)	-	-	175 *	0.9	58	-	-	-
Kansas	3	(D)	-	-	375 *	1.8	125	-	-	-
Virginia	3	(D)	-	-	175 *	0.9	58	-	-	-
South Carolina	2	(D)	-	-	375 *	1.8	188	-	-	-
New Mexico	1	(D)	-	-	375 *	1.8	375	-	-	-

Source: 1992 *Economic Census*. The states are in descending order of shipments or establishments (if shipment data are missing for the majority). The symbol (D) appears when data are withheld to prevent disclosure of competitive information. States marked with (D) are sorted by number of establishments. A dash (-) indicates that the data element cannot be calculated; * indicates the midpoint of a range.

3635 - HOUSEHOLD VACUUM CLEANERS

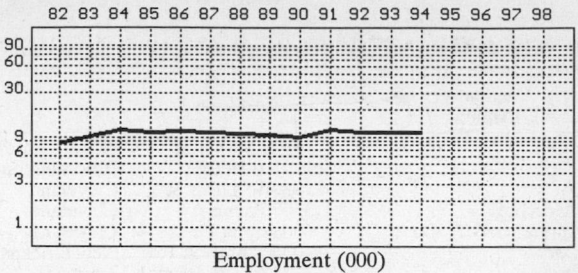

82 83 84 85 86 87 88 89 90 91 92 93 94 95 96 97 98

Shipments ($ million)

82 83 84 85 86 87 88 89 90 91 92 93 94 95 96 97 98

Employment (000)

GENERAL STATISTICS

Year	Companies	Establishments		Employment			Compensation		Production ($ million)			
		Total	with 20 or more employees	Total (000)	Production Workers (000)	Hours (Mil)	Payroll ($ mil)	Wages ($/hr)	Cost of Materials	Value Added by Manufacture	Value of Shipments	Capital Invest.
1982	29	31	18	8.4	5.2	9.5	151.1	9.96	294.9	463.6	775.7	17.1
1983		30	18	10.1	6.7	12.4	189.4	9.84	423.6	556.7	970.6	23.4
1984		29	18	11.9	8.2	15.5	237.5	10.12	610.9	746.5	1,307.3	43.4
1985		29	17	11.2	7.5	14.3	236.4	10.48	592.9	691.9	1,301.9	35.3
1986		29	19	11.5	7.5	14.9	242.4	10.22	626.6	702.8	1,337.3	46.7
1987	28	31	21	11.3	7.6	14.4	247.1	11.30	622.5	718.2	1,324.2	59.6
1988		35	24			14.4	256.7		712.6	795.6	1,473.0	
1989		35	23	10.6	8.3	15.9	272.7	10.89	773.0	835.8	1,605.0	32.1
1990		33	20	9.8	8.6	17.2	295.6	10.85	860.4	997.8	1,860.1	33.8
1991		39	22	11.9	8.2	15.8	282.5	11.20	827.4	961.8	1,804.5	53.3
1992	35	43	22	11.3	7.7	15.2	278.7	11.32	859.8	1,054.6	1,905.3	66.8
1993		40	21	11.4	7.9	16.4	300.8	10.87	990.2	1,102.6	2,095.6	72.2
1994		41P	23P	11.0	7.5	15.3	299.6	11.63	914.7	1,057.5	1,932.8	58.9
1995		42P	24P			17.2P	325.5P		1,040.8P	1,203.3P	2,199.2P	
1996		43P	24P			17.5P	335.9P		1,087.1P	1,256.8P	2,297.0P	
1997		45P	24P			17.9P	346.2P		1,133.3P	1,310.3P	2,394.8P	
1998		46P	25P			18.2P	356.6P		1,179.6P	1,363.8P	2,492.6P	

Sources: 1982, 1987, 1992 *Economic Census*; *Annual Survey of Manufactures*, 83-86, 88-91, 93-94. Establishment counts for non-Census years are from *County Business Patterns*; establishment values for 83-84 are extrapolations. 'P's show projections by the editors. Industries reclassified in 87 will not have data for prior years.

INDICES OF CHANGE

Year	Companies	Establishments		Employment			Compensation		Production ($ million)			
		Total	with 20 or more employees	Total (000)	Production Workers (000)	Hours (Mil)	Payroll ($ mil)	Wages ($/hr)	Cost of Materials	Value Added by Manufacture	Value of Shipments	Capital Invest.
1982	83	72	82	74	68	63	54	88	34	44	41	26
1983		70	82	89	87	82	68	87	49	53	51	35
1984		67	82	105	106	102	85	89	71	71	69	65
1985		67	77	99	97	94	85	93	69	66	68	53
1986		67	86	102	97	98	87	90	73	67	70	70
1987	80	72	95	100	99	95	89	100	72	68	70	89
1988		81	109			95	92		83	75	77	
1989		81	105	94	108	105	98	96	90	79	84	48
1990		77	91	87	112	113	106	96	100	95	98	51
1991		91	100	105	106	104	101	99	96	91	95	80
1992	100	100	100	100	100	100	100	100	100	100	100	100
1993		93	95	101	103	108	108	96	115	105	110	108
1994		96P	105P	97	97	101	107	103	106	100	101	88
1995		98P	107P			113P	117P		121P	114P	115P	
1996		101P	109P			115P	121P		126P	119P	121P	
1997		104P	111P			118P	124P		132P	124P	126P	
1998		106P	113P			120P	128P		137P	129P	131P	

Sources: Same as General Statistics. Values reflect change from the base year, 1992. Values above 100 mean greater than 92, values below 100 mean less than 92, and a value of 100 in the 82-91 or 93-98 period means same as 92. 'P's mark projections by the editors.

SELECTED RATIOS

For 1994	Avg. of All Manufact.	Analyzed Industry	Index	For 1994	Avg. of All Manufact.	Analyzed Industry	Index
Employees per Establishment	49	268	546	Value Added per Production Worker	134,084	141,000	105
Payroll per Establishment	1,500,273	7,285,777	486	Cost per Establishment	5,045,178	22,243,994	441
Payroll per Employee	30,620	27,236	89	Cost per Employee	102,970	83,155	81
Production Workers per Establishment	34	182	531	Cost per Production Worker	146,988	121,960	83
Wages per Establishment	853,319	4,327,183	507	Shipments per Establishment	9,576,895	47,002,506	491
Wages per Production Worker	24,861	23,725	95	Shipments per Employee	195,460	175,709	90
Hours per Production Worker	2,056	2,040	99	Shipments per Production Worker	279,017	257,707	92
Wages per Hour	12.09	11.63	96	Investment per Establishment	321,011	1,432,351	446
Value Added per Establishment	4,602,255	25,716,654	559	Investment per Employee	6,552	5,355	82
Value Added per Employee	93,930	96,136	102	Investment per Production Worker	9,352	7,853	84

Sources: Same as General Statistics. The 'Average of All Manufacturing' column represents the average of all manufacturing industries reported for the most recent complete year available. The Index shows the relationship between the Average and the Analyzed Industry. For example, 100 means that they are equal; 500 that the Analyzed Industry is five times the average; 50 means that the Analyzed Industry is half the national average. The abbreviation 'na' is used to show that data are 'not available'.

LEADING COMPANIES   Number shown: 13   Total sales ($ mil): 2,836   Total employment (000): 18.2

Company Name	Address				CEO Name	Phone	Co. Type	Sales ($ mil)	Empl. (000)
Hoover Co	101 E Maple St	North Canton	OH	44720	Brian A Girdlestone	216-499-9200	S	1,500*	10.6
Eureka Co	1201 E Bell St	Bloomington	IL	61701	Gilbert L Dorsey	309-828-2367	S	630*	3.5
Kirby Co	1920 W 114th St	Cleveland	OH	44102	Gene Windfeldt	216-228-2400	D	200	0.8
Health-Mor Inc	3500 Payne Av	Cleveland	OH	44114	Kirk W Foley	216-432-1990	P	135	1.2
Royal Appliance Mfg Co	650 Alpha Dr	Cleveland	OH	44143	John A Balch	216-449-6150	P	120	0.7
Oreck Corp	100 Plantation Rd	New Orleans	LA	70123	David Oreck	504-733-6983	R	80	0.3
Rexair Inc	3221 W Big Beaver	Troy	MI	48084	D Cunningham	313-643-7222	S	72*	0.4
Rexair Manufacturing Inc	230 7th St	Cadillac	MI	49601	Paul E Dryburgh	616-775-3413	S	72	0.4
Douglas Products	118 E Douglas	Walnut Ridge	AR	72476	C Nathan Howard	501-886-6774	D	10	0.2
Blue Lustre Products Inc	7950 Castleway Dr	Indianapolis	IN	46250	George Feldman	317-842-0820	R	9*	<0.1
Metropolitan Vacuum Cleaner	PO Box 149	Suffern	NY	10901	Israel Stern	914-357-1600	R	5	<0.1
Central Vac International	3133 E 12th St	Los Angeles	CA	90023	JD Hanks	213-268-1135	R	2*	<0.1
Sequoia Vacuum Systems	164 Jefferson Dr	Menlo Park	CA	94025	Michael D White	415-322-7281	R	2	<0.1

Source: Ward's Business Directory of U.S. Private and Public Companies, Volumes 1 and 2, 1996. The company type code used is as follows: P - Public, R - Private, S - Subsidiary, D - Division, J - Joint Venture, A - Affiliate, G - Group. Sales are in millions of dollars, employees are in thousands. An asterisk (*) indicates an estimated sales volume. The symbol < stands for 'less than'. Company names and addresses are truncated, in some cases, to fit into the available space.

MATERIALS CONSUMED

Material	Quantity	Delivered Cost ($ million)
Materials, ingredients, containers, and supplies	(X)	804.4
Metal stampings	(X)	9.3
Metal wire racks, grills, springs, and other fabricated nonelectric wire products (except forgings)	(X)	3.6
Metal bolts, nuts, screws, washers, rivets, and other screw machine products	(X)	14.4
All other fabricated metal products (except forgings)	(X)	8.2
Aluminum and aluminum-base alloy castings (rough and semifinished)	(X)	16.8
Steel bars, bar shapes, and plates	(X)	16.2
Aluminum and aluminum-base alloy shapes and forms	(X)	2.6
Nonferrous wire and cable, including magnet wire, bare or insulated wire, etc.	(X)	9.9
Fractional horsepower electric timing motors, synchronous and subsynchronous (less than 1 hp)	(X)	54.7
Integral horsepower electric motors and generators (1 hp or more)	(X)	36.4
Paper and paperboard containers, including shipping sacks and other paper packaging supplies	(X)	39.0
Electrical transmission, distribution, and control equipment	(X)	(D)
Current-carrying wiring devices	(X)	13.4
Automatic temperature controls (thermostats, regulators, etc.)	(X)	(D)
Paints, varnishes, lacquers, stains, shellacs, japans, enamels, and allied products	(X)	0.7
Fabricated rubber products, except tires, tubes, hose, belting, and gaskets	(X)	3.3
Rubber and plastics hose and belting	(X)	15.7
Plastics products consumed in the form of sheets, rods, tubes, film, and other shapes	(X)	(D)
Plastics resins consumed in the form of granules, pellets, powders, liquids, etc.	(X)	115.5
Complete flexible cord sets	(X)	19.5
Resistors, capacitors, transformers, electron tubes, semiconductors, and other electronic components	(X)	(D)
All other materials and components, parts, containers, and supplies	(X)	184.5
Materials, ingredients, containers, and supplies, nsk	(X)	195.6

Source: 1992 Economic Census. Explanation of symbols used: (D): Withheld to avoid disclosure of competitive data; na: Not available; (S): Withheld because statistical norms were not met; (X): Not applicable; (Z): Less than half the unit shown; nec: Not elsewhere classified; nsk: Not specified by kind; - : zero; * : 10-19 percent estimated; ** : 20-29 percent estimated.

PRODUCT SHARE DETAILS

Product or Product Class	% Share	Product or Product Class	% Share
Household vacuum cleaners	100.00	Household vacuum cleaners, cannister/tank type	23.41
Household vacuum cleaners, complete power units, central system type	1.28	Attachments and cleaning tools for household vacuum cleaners, including central system attachments	4.61
Household vacuum cleaners, hand type	9.37	Parts for household type vacuum cleaners, including central system parts	6.91
Household vacuum cleaners, upright/stick type	52.56		

Source: 1992 Economic Census. The values shown are percent of total shipments in an industry. Values of indented subcategories are summed in the main heading. The symbol (D) appears when data are withheld to prevent disclosure of competitive information. The abbreviation nsk stands for 'not specified by kind' and nec for 'not elsewhere classified'.

INPUTS AND OUTPUTS FOR HOUSEHOLD VACUUM CLEANERS

Economic Sector or Industry Providing Inputs	%	Sector	Economic Sector or Industry Buying Outputs	%	Sector
Motors & generators	10.7	Manufg.	Personal consumption expenditures	71.8	
Wholesale trade	10.2	Trade	Exports	8.1	Foreign
Plastics materials & resins	7.2	Manufg.	Electrical repair shops	5.4	Services
Imports	7.0	Foreign	Office buildings	4.0	Constr.
Miscellaneous plastics products	6.8	Manufg.	Gross private fixed investment	3.8	Cap Inv
Blast furnaces & steel mills	5.9	Manufg.	Construction of stores & restaurants	1.1	Constr.
Advertising	5.1	Services	Residential garden apartments	1.1	Constr.
Paperboard containers & boxes	4.6	Manufg.	Residential 1-unit structures, nonfarm	1.0	Constr.
Signs & advertising displays	3.9	Manufg.	Residential additions/alterations, nonfarm	0.9	Constr.
Banking	3.4	Fin/R.E.	Industrial buildings	0.7	Constr.
Nonferrous wire drawing & insulating	2.6	Manufg.	Construction of hospitals	0.3	Constr.
Screw machine and related products	2.2	Manufg.	Hotels & motels	0.3	Constr.
Electric services (utilities)	2.1	Util.	Construction of educational buildings	0.2	Constr.
Environmental controls	1.9	Manufg.	S/L Govt. purch., elem. & secondary education	0.2	S/L Govt
Watches, clocks, & parts	1.7	Manufg.	S/L Govt. purch., health & hospitals	0.2	S/L Govt
Aluminum rolling & drawing	1.6	Manufg.	Nonfarm residential structure maintenance	0.1	Constr.
Communications, except radio & TV	1.5	Util.	Federal Government purchases, national defense	0.1	Fed Govt
Motor freight transportation & warehousing	1.5	Util.			
Aluminum castings	1.3	Manufg.			
Equipment rental & leasing services	1.2	Services			
Rubber & plastics hose & belting	1.1	Manufg.			
U.S. Postal Service	1.1	Gov't			
Wiring devices	1.0	Manufg.			
Maintenance of nonfarm buildings nec	0.9	Constr.			
Metal stampings, nec	0.9	Manufg.			
Air transportation	0.8	Util.			
Fabricated rubber products, nec	0.7	Manufg.			
Gas production & distribution (utilities)	0.7	Util.			
Eating & drinking places	0.7	Trade			
Real estate	0.5	Fin/R.E.			
Hotels & lodging places	0.5	Services			
Computer & data processing services	0.4	Services			
Machinery, except electrical, nec	0.3	Manufg.			
Miscellaneous fabricated wire products	0.3	Manufg.			
Paints & allied products	0.3	Manufg.			
Special dies & tools & machine tool accessories	0.3	Manufg.			
Switchgear & switchboard apparatus	0.3	Manufg.			
Railroads & related services	0.3	Util.			
Automotive repair shops & services	0.3	Services			
Engineering, architectural, & surveying services	0.3	Services			
Job training & related services	0.3	Services			
Legal services	0.3	Services			
Abrasive products	0.2	Manufg.			
Copper rolling & drawing	0.2	Manufg.			
Gaskets, packing & sealing devices	0.2	Manufg.			
Household vacuum cleaners	0.2	Manufg.			
Iron & steel foundries	0.2	Manufg.			
Lubricating oils & greases	0.2	Manufg.			
Machine tools, metal forming types	0.2	Manufg.			
Manifold business forms	0.2	Manufg.			
Metal heat treating	0.2	Manufg.			
Nonferrous castings, nec	0.2	Manufg.			
Paper mills, except building paper	0.2	Manufg.			
Petroleum refining	0.2	Manufg.			
Sanitary services, steam supply, irrigation	0.2	Util.			
Water transportation	0.2	Util.			
Accounting, auditing & bookkeeping	0.2	Services			
Laundry, dry cleaning, shoe repair	0.2	Services			
Management & consulting services & labs	0.2	Services			
State & local government enterprises, nec	0.2	Gov't			
Bags, except textile	0.1	Manufg.			
Brooms & brushes	0.1	Manufg.			
Insurance carriers	0.1	Fin/R.E.			
Security & commodity brokers	0.1	Fin/R.E.			

Source: Benchmark Input-Output Accounts for the U.S. Economy, 1982, U.S. Department of Commerce, Washington, D.C., July 1991. Data, as reported in the source, are organized by the 1977 SIC structure in use in 1982 but have been matched, as closely as is possible, to the 1987 SIC structure used in this book.

OCCUPATIONS EMPLOYED BY SIC 363 - HOUSEHOLD APPLIANCES

Occupation	% of Total 1994	Change to 2005	Occupation	% of Total 1994	Change to 2005
Assemblers, fabricators, & hand workers nec	31.2	-16.9	Electrical & electronic equipment assemblers	1.7	-16.9
Helpers, laborers, & material movers nec	4.1	-16.9	Machine operators nec	1.5	-41.4
Machine assemblers	3.9	-25.2	Welding machine setters, operators	1.5	-25.2
Inspectors, testers, & graders, precision	3.1	-41.9	Sales & related workers nec	1.5	-17.0
Blue collar worker supervisors	3.0	-25.4	Maintenance repairers, general utility	1.5	-25.2
Electrical & electronic assemblers	2.7	-25.3	Janitors & cleaners, incl maids	1.2	-33.6
Industrial truck & tractor operators	2.6	-16.9	Tool & die makers	1.2	-32.9
Plastic molding machine workers	2.6	-0.3	Material moving equipment operators nec	1.2	-16.9
Machine forming operators, metal & plastic	2.5	-58.5	General managers & top executives	1.1	-21.2
Freight, stock, & material movers, hand	1.9	-33.6	Machine tool cutting & forming etc. nec	1.1	-33.6

Source: Industry-Occupation Matrix, Bureau of Labor Statistics. These data relate to one or more 3-digit SIC industry groups rather than to a single 4-digit SIC. The change reported for each occupation to the year 2005 is a percent of growth or decline as estimated by the Bureau of Labor Statistics. The abbreviation nec stands for 'not elsewhere classified'.

LOCATION BY STATE AND REGIONAL CONCENTRATION

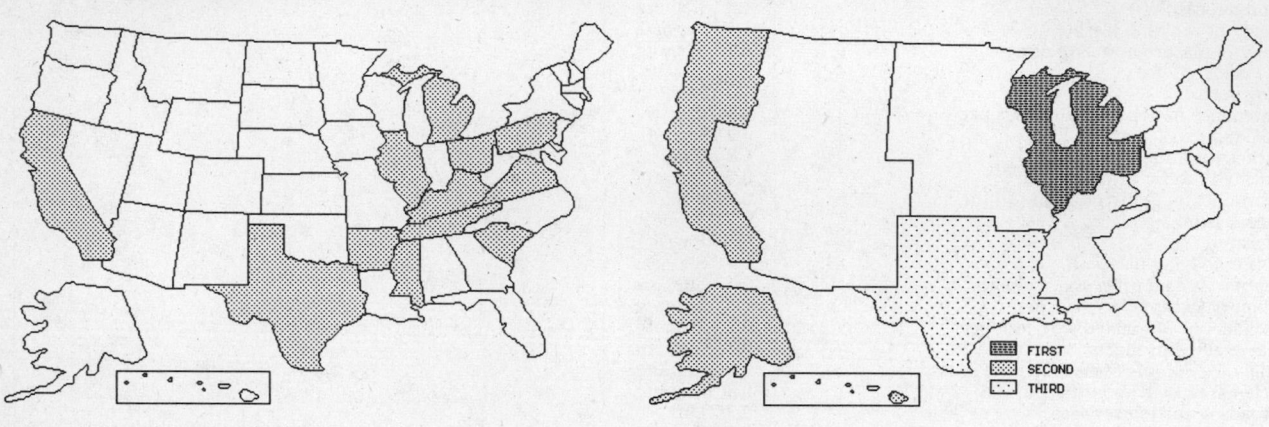

FIRST
SECOND
THIRD

INDUSTRY DATA BY STATE

State	Establish-ments	Shipments Total ($ mil)	% of U.S.	Per Establ.	Employment Total Number	% of U.S.	Per Establ.	Wages ($/hour)	Cost as % of Shipments	Investment per Employee ($)
Ohio	11	736.3	38.6	66.9	3,900	34.5	355	13.84	33.6	-
California	5	(D)	-	-	175 *	1.5	35	-	-	-
Texas	4	(D)	-	-	750 *	6.6	188	-	-	-
Illinois	3	(D)	-	-	1,750 *	15.5	583	-	-	-
Michigan	2	(D)	-	-	375 *	3.3	188	-	-	-
Arkansas	1	(D)	-	-	175 *	1.5	175	-	-	-
Kentucky	1	(D)	-	-	750 *	6.6	750	-	-	-
Mississippi	1	(D)	-	-	750 *	6.6	750	-	-	-
Pennsylvania	1	(D)	-	-	375 *	3.3	375	-	-	-
South Carolina	1	(D)	-	-	1,750 *	15.5	1,750	-	-	-
Tennessee	1	(D)	-	-	175 *	1.5	175	-	-	-
Virginia	1	(D)	-	-	750 *	6.6	750	-	-	-

Source: 1992 *Economic Census.* The states are in descending order of shipments or establishments (if shipment data are missing for the majority). The symbol (D) appears when data are withheld to prevent disclosure of competitive information. States marked with (D) are sorted by number of establishments. A dash (-) indicates that the data element cannot be calculated; * indicates the midpoint of a range.

3639 - HOUSEHOLD APPLIANCES, NEC

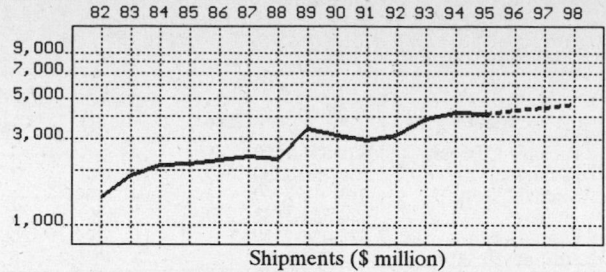

Shipments ($ million)

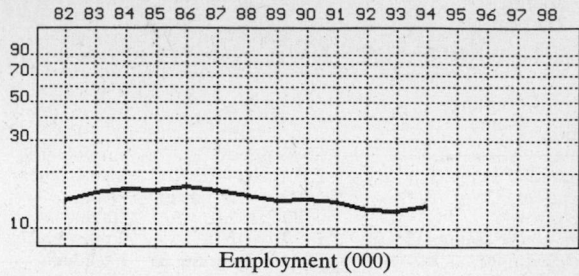

Employment (000)

GENERAL STATISTICS

| Year | Com-panies | Establishments | | Employment | | | Compensation | | Production ($ million) | | | |
		Total	with 20 or more employees	Total (000)	Production Workers (000)	Hours (Mil)	Payroll ($ mil)	Wages ($/hr)	Cost of Materials	Value Added by Manufacture	Value of Shipments	Capital Invest.
1982	70	83	49	14.2	11.4	21.7	252.4	8.30	754.9	673.8	1,432.0	40.6
1983		81	47	15.7	13.0	25.9	313.9	9.22	1,026.0	869.7	1,879.0	36.6
1984		79	45	16.2	13.6	27.8	338.3	9.43	1,157.6	1,025.5	2,150.7	48.5
1985		78	44	16.0	13.2	26.7	337.1	9.71	1,147.7	1,018.9	2,175.3	71.9
1986		79	45	16.7	13.9	27.1	373.5	10.54	1,247.1	1,053.8	2,280.1	44.9
1987	61	75	38	16.0	13.1	26.0	370.2	10.63	1,208.2	1,215.5	2,398.3	60.2
1988		73	39			24.6	361.7		1,215.8	1,103.6	2,312.0	
1989		66	36	14.2	12.2	24.8	359.5	11.11	1,959.0	1,647.7	3,438.0	113.5
1990		66	39	14.5	11.8	23.5	355.9	11.71	1,874.0	1,266.9	3,124.9	66.5
1991		69	36	14.0	11.7	23.4	340.1	11.29	1,771.3	1,083.5	2,958.2	49.4
1992	61	68	34	12.7	10.5	21.0	329.7	12.07	1,954.5	1,061.3	3,169.1	79.2
1993		68	28	12.4	10.3	21.1	355.9	12.74	2,302.1	1,619.5	3,889.2	54.0
1994		63P	30P	13.3	11.1	22.1	379.7	12.37	2,320.5	1,903.2	4,233.3	39.6
1995		62P	28P			21.7P	378.4P		2,253.6P	1,848.3P	4,111.2P	
1996		60P	26P			21.3P	383.4P		2,362.0P	1,937.2P	4,309.0P	
1997		59P	25P			21.0P	388.3P		2,470.5P	2,026.2P	4,506.9P	
1998		57P	23P			20.6P	393.3P		2,578.9P	2,115.2P	4,704.8P	

Sources: 1982, 1987, 1992 *Economic Census*; *Annual Survey of Manufactures*, 83-86, 88-91, 93-94. Establishment counts for non-Census years are from *County Business Patterns*; establishment values for 83-84 are extrapolations. 'P's show projections by the editors. Industries reclassified in 87 will not have data for prior years.

INDICES OF CHANGE

| Year | Com-panies | Establishments | | Employment | | | Compensation | | Production ($ million) | | | |
		Total	with 20 or more employees	Total (000)	Production Workers (000)	Hours (Mil)	Payroll ($ mil)	Wages ($/hr)	Cost of Materials	Value Added by Manufacture	Value of Shipments	Capital Invest.
1982	115	122	144	112	109	103	77	69	39	63	45	51
1983		119	138	124	124	123	95	76	52	82	59	46
1984		116	132	128	130	132	103	78	59	97	68	61
1985		115	129	126	126	127	102	80	59	96	69	91
1986		116	132	131	132	129	113	87	64	99	72	57
1987	100	110	112	126	125	124	112	88	62	115	76	76
1988		107	115			117	110		62	104	73	
1989		97	106	112	116	118	109	92	100	155	108	143
1990		97	115	114	112	112	108	97	96	119	99	84
1991		101	106	110	111	111	103	94	91	102	93	62
1992	100	100	100	100	100	100	100	100	100	100	100	100
1993		100	82	98	98	100	108	106	118	153	123	68
1994		93P	87P	105	106	105	115	102	119	179	134	50
1995		91P	82P			103P	115P		115P	174P	130P	
1996		89P	77P			102P	116P		121P	183P	136P	
1997		86P	73P			100P	118P		126P	191P	142P	
1998		84P	68P			98P	119P		132P	199P	148P	

Sources: Same as General Statistics. Values reflect change from the base year, 1992. Values above 100 mean greater than 92, values below 100 mean less than 92, and a value of 100 in the 82-91 or 93-98 period means same as 92. 'P's mark projections by the editors.

SELECTED RATIOS

For 1994	Avg. of All Manufact.	Analyzed Industry	Index	For 1994	Avg. of All Manufact.	Analyzed Industry	Index
Employees per Establishment	49	210	428	Value Added per Production Worker	134,084	171,459	128
Payroll per Establishment	1,500,273	5,983,811	399	Cost per Establishment	5,045,178	36,569,484	725
Payroll per Employee	30,620	28,549	93	Cost per Employee	102,970	174,474	169
Production Workers per Establishment	34	175	510	Cost per Production Worker	146,988	209,054	142
Wages per Establishment	853,319	4,308,234	505	Shipments per Establishment	9,576,895	66,713,897	697
Wages per Production Worker	24,861	24,629	99	Shipments per Employee	195,460	318,293	163
Hours per Production Worker	2,056	1,991	97	Shipments per Production Worker	279,017	381,378	137
Wages per Hour	12.09	12.37	102	Investment per Establishment	321,011	624,069	194
Value Added per Establishment	4,602,255	29,993,123	652	Investment per Employee	6,552	2,977	45
Value Added per Employee	93,930	143,098	152	Investment per Production Worker	9,352	3,568	38

Sources: Same as General Statistics. The 'Average of All Manufacturing' column represents the average of all manufacturing industries reported for the most recent complete year available. The Index shows the relationship between the Average and the Analyzed Industry. For example, 100 means that they are equal; 500 that the Analyzed Industry is five times the average; 50 means that the Analyzed Industry is half the national average. The abbreviation 'na' is used to show that data are 'not available'.

LEADING COMPANIES   Number shown: 13   Total sales ($ mil): 1,729   Total employment (000): 10.2

Company Name	Address				CEO Name	Phone	Co. Type	Sales ($ mil)	Empl. (000)
Nortek Inc	50 Kennedy Plz	Providence	RI	02903	Richard L Bready	401-751-1600	P	737	5.3
Rheem Ruud	PO Box 244020	Montgomery	AL	36124	IS Farwell	334-260-1500	D	400	2.0
Water Products Co	600 E J Carpenter	Irving	TX	75062	Michael Watts	214-719-5900	D	240*	1.0
In-Sink-Erator	4700 21st St	Racine	WI	53406		414-554-5432	D	190	0.9
American Water Heater Group	PO Box 1378	Johnson City	TN	37605	Doug Booth	615-434-1500	R	120*	0.7
Distinctive Appliances Inc	950 S Raymond Av	Pasadena	CA	91109	Anthony Joseph	818-799-1000	R	15*	0.2
Kemco Systems Inc	11500 47th St	Clearwater	FL	34622	Lee R Kemberling	813-573-2323	R	10	<0.1
Refuse Compactor Service Inc	12776 Foothill Blv	Sylmar	CA	91342	Art Nevill	818-365-6868	R	6*	<0.1
Multi-Pak Corp	400 Railroad Av	Hackensack	NJ	07601	Philip Cahill	201-342-7474	R	3*	<0.1
M Joseph Machine Company	5287 Washington St	Boston	MA	02132	Nicholas Joseph	617-325-6818	R	3	<0.1
Chronomite Laboratories Inc	21011 S Figueroa St	Carson	CA	90745	Robert G Russell	310-320-9452	R	2*	<0.1
Compax Systems Inc	650 Albert Dr	Brookville	OH	45309	KC Mosier II	513-833-3148	R	2*	<0.1
Weigh-In Inc	5798 Cimarron	Tucson	AZ	85715	Manny Homen	602-299-9353	R	1*	<0.1

Source: *Ward's Business Directory of U.S. Private and Public Companies*, Volumes 1 and 2, 1996. The company type code used is as follows: P - Public, R - Private, S - Subsidiary, D - Division, J - Joint Venture, A - Affiliate, G - Group. Sales are in millions of dollars, employees are in thousands. An asterisk (*) indicates an estimated sales volume. The symbol < stands for 'less than'. Company names and addresses are truncated, in some cases, to fit into the available space.

MATERIALS CONSUMED

Material	Quantity	Delivered Cost ($ million)
Materials, ingredients, containers, and supplies	(X)	1,230.4
Metal stampings	(X)	30.0
Metal wire racks, grills, springs, and other fabricated nonelectric wire products (except forgings)	(X)	23.2
Metal bolts, nuts, screws, washers, rivets, and other screw machine products	(X)	23.4
All other fabricated metal products (except forgings)	(X)	30.8
Iron and steel castings (rough and semifinished)	(X)	18.6
Steel sheet and strip, including tin plate	(X)	146.7
All other steel shapes and forms	(X)	39.6
Aluminum and aluminum-base alloy shapes and forms	(X)	21.1
Nonferrous wire and cable, including magnet wire, bare or insulated wire, etc.	(X)	7.5
Fractional horsepower electric timing motors, synchronous and subsynchronous (less than 1 hp)	(X)	57.2
Integral horsepower electric motors and generators (1 hp or more)	(X)	2.5
Paper and paperboard containers, including shipping sacks and other paper packaging supplies	(X)	43.8
Electrical transmission, distribution, and control equipment	(X)	55.4
Current-carrying wiring devices	(X)	27.8
Timing mechanisms, except microprocessors	(X)	23.5
Automatic temperature controls (thermostats, regulators, etc.)	(X)	85.7
Paints, varnishes, lacquers, stains, shellacs, japans, enamels, and allied products	(X)	16.7
Fabricated rubber products, except tires, tubes, hose, belting, and gaskets	(X)	5.2
Rubber and plastics hose and belting	(X)	11.4
Plastics products consumed in the form of sheets, rods, tubes, film, and other shapes	(X)	77.7
Plastics resins consumed in the form of granules, pellets, powders, liquids, etc.	(X)	79.2
Complete flexible cord sets	(X)	0.4
Resistors, capacitors, transformers, electron tubes, semiconductors, and other electronic components	(X)	3.7
Mineral wool insulation (fibrous glass, rock wool, etc.)	(X)	18.6
All other materials and components, parts, containers, and supplies	(X)	215.1
Materials, ingredients, containers, and supplies, nsk	(X)	165.8

Source: 1992 *Economic Census*. Explanation of symbols used: (D): Withheld to avoid disclosure of competitive data; na: Not available; (S): Withheld because statistical norms were not met; (X): Not applicable; (Z): Less than half the unit shown; nec: Not elsewhere classified; nsk: Not specified by kind; - : zero; * : 10-19 percent estimated; ** : 20-29 percent estimated.

PRODUCT SHARE DETAILS

Product or Product Class	% Share	Product or Product Class	% Share
Household appliances, nec	100.00	Other major household appliances, nec, excluding parts	92.07
Household water heaters, electric, for permanent installation	18.98	Parts and accessories for other household appliances, nec	7.79
Household water heaters, except electric	24.94	Household appliances, nec, and parts for household appliances, nec, nsk	0.15
Household appliances, nec, and parts for household appliances, nec	54.03	Household appliances, nec, nsk	2.04

Source: 1992 *Economic Census*. The values shown are percent of total shipments in an industry. Values of indented subcategories are summed in the main heading. The symbol (D) appears when data are withheld to prevent disclosure of competitive information. The abbreviation nsk stands for 'not specified by kind' and nec for 'not elsewhere classified'.

INPUTS AND OUTPUTS FOR HOUSEHOLD APPLIANCES, NEC

Economic Sector or Industry Providing Inputs	%	Sector	Economic Sector or Industry Buying Outputs	%	Sector
Blast furnaces & steel mills	21.3	Manufg.	Personal consumption expenditures	26.3	
Wholesale trade	10.6	Trade	Nonfarm residential structure maintenance	19.1	Constr.
Environmental controls	7.1	Manufg.	Gross private fixed investment	10.1	Cap Inv
Motors & generators	6.5	Manufg.	Residential 1-unit structures, nonfarm	8.0	Constr.
Pipe, valves, & pipe fittings	4.6	Manufg.	Office buildings	5.6	Constr.
Noncomparable imports	3.4	Foreign	Maintenance of nonfarm buildings nec	5.3	Constr.
Imports	3.2	Foreign	Electrical repair shops	5.3	Services
Paperboard containers & boxes	3.0	Manufg.	Exports	4.0	Foreign
Hardware, nec	2.9	Manufg.	Construction of stores & restaurants	2.2	Constr.
Advertising	2.7	Services	Residential additions/alterations, nonfarm	2.1	Constr.
Wiring devices	1.9	Manufg.	Mobile homes	2.0	Manufg.
Electric services (utilities)	1.9	Util.	Construction of hospitals	1.9	Constr.
Paints & allied products	1.7	Manufg.	Residential garden apartments	1.8	Constr.
Gas production & distribution (utilities)	1.6	Util.	Industrial buildings	1.5	Constr.
Plastics materials & resins	1.5	Manufg.	Hotels & motels	0.7	Constr.
Motor freight transportation & warehousing	1.5	Util.	Residential high-rise apartments	0.6	Constr.
Aluminum rolling & drawing	1.4	Manufg.	Travel trailers & campers	0.6	Manufg.
Chemical preparations, nec	1.4	Manufg.	Construction of nonfarm buildings nec	0.5	Constr.
Glass & glass products, except containers	1.4	Manufg.	Amusement & recreation building construction	0.3	Constr.
Fabricated rubber products, nec	1.3	Manufg.	Resid. & other health facility construction	0.3	Constr.
Screw machine and related products	1.2	Manufg.	Residential 2-4 unit structures, nonfarm	0.3	Constr.
Nonferrous wire drawing & insulating	0.9	Manufg.	S/L Govt. purch., health & hospitals	0.3	S/L Govt
Communications, except radio & TV	0.9	Util.	Construction of educational buildings	0.2	Constr.
Banking	0.9	Fin/R.E.	Nonbuilding facilities nec	0.2	Constr.
Equipment rental & leasing services	0.9	Services	Warehouses	0.2	Constr.
U.S. Postal Service	0.9	Gov't	Federal Government purchases, national defense	0.2	Fed Govt
Air transportation	0.8	Util.	Dormitories & other group housing	0.1	Constr.
Nonferrous castings, nec	0.7	Manufg.	Maintenance of farm residential buildings	0.1	Constr.
Maintenance of nonfarm buildings nec	0.6	Constr.			
Mineral wool	0.6	Manufg.			
Copper rolling & drawing	0.5	Manufg.			
Iron & steel foundries	0.5	Manufg.			
Metal coating & allied services	0.5	Manufg.			
Miscellaneous plastics products	0.5	Manufg.			
Nonferrous rolling & drawing, nec	0.5	Manufg.			
Eating & drinking places	0.5	Trade			
Computer & data processing services	0.5	Services			
Metal stampings, nec	0.4	Manufg.			
Signs & advertising displays	0.4	Manufg.			
Railroads & related services	0.4	Util.			
Hotels & lodging places	0.4	Services			
Job training & related services	0.4	Services			
Electric housewares & fans	0.3	Manufg.			
Hardwood dimension & flooring mills	0.3	Manufg.			
Petroleum refining	0.3	Manufg.			
Wood partitions & fixtures	0.3	Manufg.			
Real estate	0.3	Fin/R.E.			
Aluminum castings	0.2	Manufg.			
Gaskets, packing & sealing devices	0.2	Manufg.			
Machinery, except electrical, nec	0.2	Manufg.			
Rubber & plastics hose & belting	0.2	Manufg.			
Semiconductors & related devices	0.2	Manufg.			
Special dies & tools & machine tool accessories	0.2	Manufg.			
Automotive repair shops & services	0.2	Services			
Legal services	0.2	Services			
Management & consulting services & labs	0.2	Services			
Abrasive products	0.1	Manufg.			
Machine tools, metal forming types	0.1	Manufg.			
Miscellaneous fabricated wire products	0.1	Manufg.			
Insurance carriers	0.1	Fin/R.E.			
Accounting, auditing & bookkeeping	0.1	Services			

Source: Benchmark Input-Output Accounts for the U.S. Economy, 1982, U.S. Department of Commerce, Washington, D.C., July 1991. Data, as reported in the source, are organized by the 1977 SIC structure in use in 1982 but have been matched, as closely as is possible, to the 1987 SIC structure used in this book.

OCCUPATIONS EMPLOYED BY SIC 363 - HOUSEHOLD APPLIANCES

Occupation	% of Total 1994	Change to 2005	Occupation	% of Total 1994	Change to 2005
Assemblers, fabricators, & hand workers nec	31.2	-16.9	Electrical & electronic equipment assemblers	1.7	-16.9
Helpers, laborers, & material movers nec	4.1	-16.9	Machine operators nec	1.5	-41.4
Machine assemblers	3.9	-25.2	Welding machine setters, operators	1.5	-25.2
Inspectors, testers, & graders, precision	3.1	-41.9	Sales & related workers nec	1.5	-17.0
Blue collar worker supervisors	3.0	-25.4	Maintenance repairers, general utility	1.5	-25.2
Electrical & electronic assemblers	2.7	-25.3	Janitors & cleaners, incl maids	1.2	-33.6
Industrial truck & tractor operators	2.6	-16.9	Tool & die makers	1.2	-32.9
Plastic molding machine workers	2.6	-0.3	Material moving equipment operators nec	1.2	-16.9
Machine forming operators, metal & plastic	2.5	-58.5	General managers & top executives	1.1	-21.2
Freight, stock, & material movers, hand	1.9	-33.6	Machine tool cutting & forming etc. nec	1.1	-33.6

Source: Industry-Occupation Matrix, Bureau of Labor Statistics. These data relate to one or more 3-digit SIC industry groups rather than to a single 4-digit SIC. The change reported for each occupation to the year 2005 is a percent of growth or decline as estimated by the Bureau of Labor Statistics. The abbreviation nec stands for 'not elsewhere classified'.

LOCATION BY STATE AND REGIONAL CONCENTRATION

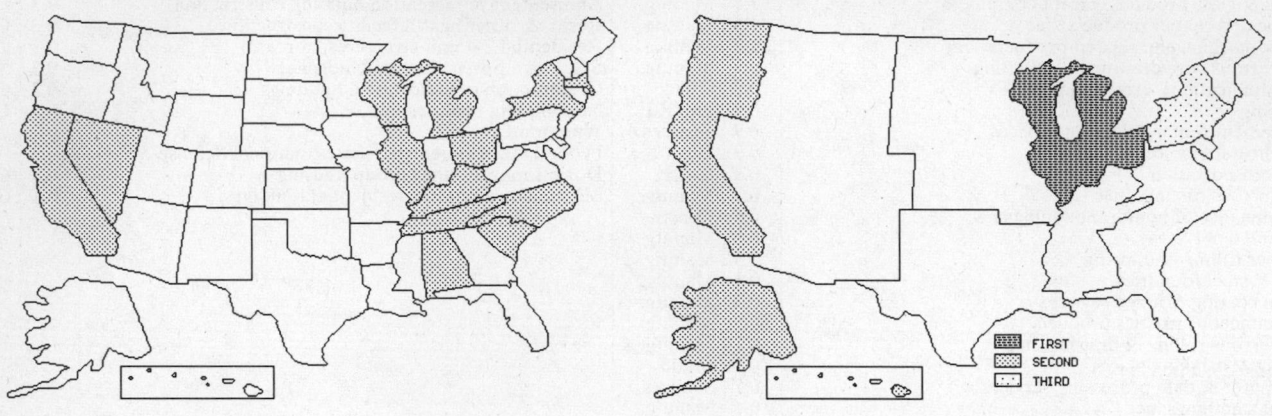

FIRST
SECOND
THIRD

INDUSTRY DATA BY STATE

State	Establish-ments	Shipments			Employment				Cost as % of Shipments	Investment per Employee ($)
		Total ($ mil)	% of U.S.	Per Establ.	Total Number	% of U.S.	Per Establ.	Wages ($/hour)		
California	13	114.2	3.6	8.8	800	6.3	62	11.33	62.1	1,375
Massachusetts	5	20.3	0.6	4.1	200	1.6	40	8.00	57.1	-
New York	9	(D)	-	-	175 *	1.4	19	-	-	-
Wisconsin	6	(D)	-	-	1,750 *	13.8	292	-	-	-
Ohio	5	(D)	-	-	1,750 *	13.8	350	-	-	-
Tennessee	4	(D)	-	-	3,750 *	29.5	938	-	-	-
North Carolina	3	(D)	-	-	750 *	5.9	250	-	-	-
Illinois	2	(D)	-	-	175 *	1.4	88	-	-	-
Kentucky	2	(D)	-	-	1,750 *	13.8	875	-	-	-
South Carolina	2	(D)	-	-	1,750 *	13.8	875	-	-	-
Alabama	1	(D)	-	-	750 *	5.9	750	-	-	-
Michigan	1	(D)	-	-	750 *	5.9	750	-	-	-
Nevada	1	(D)	-	-	175 *	1.4	175	-	-	-

Source: 1992 *Economic Census.* The states are in descending order of shipments or establishments (if shipment data are missing for the majority). The symbol (D) appears when data are withheld to prevent disclosure of competitive information. States marked with (D) are sorted by number of establishments. A dash (-) indicates that the data element cannot be calculated; * indicates the midpoint of a range.

3641 - ELECTRIC LAMP BULBS & TUBES

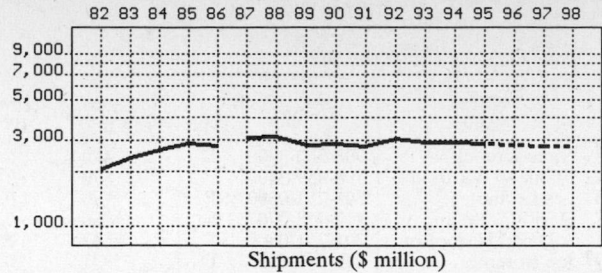

82 83 84 85 86 87 88 89 90 91 92 93 94 95 96 97 98

Shipments ($ million)

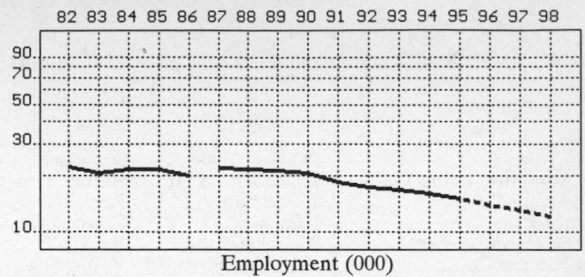

82 83 84 85 86 87 88 89 90 91 92 93 94 95 96 97 98

Employment (000)

GENERAL STATISTICS

| Year | Com-panies | Establishments | | Employment | | | Compensation | | Production ($ million) | | | |
		Total	with 20 or more employees	Total (000)	Production Workers (000)	Hours (Mil)	Payroll ($ mil)	Wages ($/hr)	Cost of Materials	Value Added by Manufacture	Value of Shipments	Capital Invest.
1982	115	149	61	22.4	18.9	33.5	396.9	9.26	792.7	1,283.8	2,072.6	60.5
1983		150	63	20.6	17.2	31.9	418.4	10.17	896.8	1,512.5	2,388.8	61.3
1984		151	65	22.0	19.1	35.9	462.7	10.48	1,036.9	1,622.5	2,633.6	94.1
1985		151	67	21.8	18.7	34.8	492.6	11.32	1,039.0	1,841.6	2,849.1	122.9
1986		141	65	20.1	17.2	32.5	455.1	11.11	846.0	1,877.3	2,743.6	128.2
1987*	93	127	67	22.2	19.0	36.8	526.8	11.44	977.8	2,125.5	3,096.7	101.9
1988		121	69	21.9	18.7	37.1	546.1	11.89	957.7	2,201.7	3,168.9	122.8
1989		120	71	21.3	16.9	35.1	513.1	11.86	924.9	1,861.2	2,793.5	96.3
1990		121	74	20.9	16.7	35.6	521.8	11.77	969.3	1,862.5	2,830.9	107.5
1991		128	75	18.6	15.6	33.0	486.8	11.68	898.4	1,863.0	2,772.9	105.4
1992	76	105	66	17.5	14.7	29.3	510.7	13.86	955.6	2,067.5	3,026.9	99.6
1993		103	65	16.8	14.2	26.8	514.9	15.59	981.5	1,911.3	2,900.6	115.7
1994		104P	68P	16.0	13.5	28.1	514.8	14.89	1,006.0	1,902.0	2,915.0	110.1
1995		101P	68P	15.0P	12.5P	25.6P	501.8P	15.47P	977.6P	1,848.2P	2,832.6P	109.0P
1996		97P	68P	14.0P	11.7P	24.1P	498.4P	16.05P	969.5P	1,832.9P	2,809.1P	109.4P
1997		94P	68P	13.0P	10.8P	22.5P	495.0P	16.63P	961.4P	1,817.6P	2,785.6P	109.7P
1998		90P	67P	12.0P	10.0P	20.9P	491.7P	17.21P	953.3P	1,802.3P	2,762.2P	110.1P

Sources: 1982, 1987, 1992 *Economic Census*; *Annual Survey of Manufactures*, 83-86, 88-91, 93-94. Establishment counts are from *County Business Patterns* for non-Census years; establishment counts for 83-84 are extrapolations. * indicates that industry content changed in 87; earlier years use 77 SICs. 'P's mark projections.

INDICES OF CHANGE

| Year | Com-panies | Establishments | | Employment | | | Compensation | | Production ($ million) | | | |
		Total	with 20 or more employees	Total (000)	Production Workers (000)	Hours (Mil)	Payroll ($ mil)	Wages ($/hr)	Cost of Materials	Value Added by Manufacture	Value of Shipments	Capital Invest.
1982	151	142	92	128	129	114	78	67	83	62	68	61
1983		143	95	118	117	109	82	73	94	73	79	62
1984		144	98	126	130	123	91	76	109	78	87	94
1985		144	102	125	127	119	96	82	109	89	94	123
1986		134	98	115	117	111	89	80	89	91	91	129
1987*	122	121	102	127	129	126	103	83	102	103	102	102
1988		115	105	125	127	127	107	86	100	106	105	123
1989		114	108	122	115	120	100	86	97	90	92	97
1990		115	112	119	114	122	102	85	101	90	94	108
1991		122	114	106	106	113	95	84	94	90	92	106
1992	100	100	100	100	100	100	100	100	100	100	100	100
1993		98	98	96	97	91	101	112	103	92	96	116
1994		99P	104P	91	92	96	101	107	105	92	96	111
1995		96P	103P	86P	85P	87P	98P	112P	102P	89P	94P	109P
1996		93P	103P	80P	79P	82P	98P	116P	101P	89P	93P	110P
1997		89P	102P	74P	74P	77P	97P	120P	101P	88P	92P	110P
1998		86P	102P	69P	68P	71P	96P	124P	100P	87P	91P	111P

Sources: Same as General Statistics. Values reflect change from the base year, 1992. Values above 100 mean greater than 92, values below 100 mean less than 92, and a value of 100 in the 82-91 or 93-98 period means same as 92. * indicates that industry content changed in 87. Data for earlier years are in 77 SIC format.

SELECTED RATIOS

For 1994	Avg. of All Manufact.	Analyzed Industry	Index	For 1994	Avg. of All Manufact.	Analyzed Industry	Index
Employees per Establishment	49	154	314	Value Added per Production Worker	134,084	140,889	105
Payroll per Establishment	1,500,273	4,943,210	329	Cost per Establishment	5,045,178	9,659,808	191
Payroll per Employee	30,620	32,175	105	Cost per Employee	102,970	62,875	61
Production Workers per Establishment	34	130	378	Cost per Production Worker	146,988	74,519	51
Wages per Establishment	853,319	4,017,645	471	Shipments per Establishment	9,576,895	27,990,398	292
Wages per Production Worker	24,861	30,993	125	Shipments per Employee	195,460	182,187	93
Hours per Production Worker	2,056	2,081	101	Shipments per Production Worker	279,017	215,926	77
Wages per Hour	12.09	14.89	123	Investment per Establishment	321,011	1,057,202	329
Value Added per Establishment	4,602,255	18,263,374	397	Investment per Employee	6,552	6,881	105
Value Added per Employee	93,930	118,875	127	Investment per Production Worker	9,352	8,156	87

Sources: Same as General Statistics. The 'Average of All Manufacturing' column represents the average of all manufacturing industries reported for the most recent complete year available. The Index shows the relationship between the Average and the Analyzed Industry. For example, 100 means that they are equal; 500 that the Analyzed Industry is five times the average; 50 means that the Analyzed Industry is half the national average. The abbreviation 'na' is used to show that data are 'not available'.

LEADING COMPANIES   Number shown: 33   Total sales ($ mil): 1,038   Total employment (000): 10.5

Company Name	Address				CEO Name	Phone	Co. Type	Sales ($ mil)	Empl. (000)
Philips Lighting Co	PO Box 6800	Somerset	NJ	08873	Nicco Bruijel	908-563-3000	D	490	6.0
Chicago Miniature Lamp Inc	1080 Johnson Dr	Buffalo Grove	IL	60089	Frank M Ward	708-459-3400	P	70	0.7
Fusion Systems Corp	7600 Standish Pl	Rockville	MD	20855	Les Levine	301-251-0300	P	70	0.4
Alsy Lighting Inc	1 Early St	Ellwood City	PA	16117	David Grolman	412-758-0707	S	50	0.5
Fusion UV Curing Systems	7600 Standish Pl	Rockville	MD	20855	A David Harbourne	301-251-0300	S	42	0.3
General Electric Co	280 N Meridian Rd	Youngstown	OH	44509	Joe Borenko	216-793-3911	D	31	0.3
Voltarc Technologies Inc	186 Linwood Av	Fairfield	CT	06430	V Mehta	203-255-2633	R	29	0.3
Frederick Cooper Lamps Inc	2545 W Diversey	Chicago	IL	60647	Peter Gershanov	312-384-0800	R	28	0.2
Trojan Inc	PO Box 850	Mount Sterling	KY	40353	Edward F Duzyk	606-498-0526	R	25	0.3
Venture Lighting International	32000 Aurora Rd	Solon	OH	44139	Wayne Hellman	216-248-3510	R	24	0.1
Marvel Lighting Corp	25 E Spring Valley	Maywood	NJ	07607	Paul Greenberg	201-368-8015	R	20	0.2
Tensor Corp	100 Justin Dr	Chelsea	MA	02150	William Sherman	617-884-7744	R	20	<0.1
Eltrex Industries Inc	65 Sullivan St	Rochester	NY	14605	Matthew Augustine	716-454-6100	R	14	0.1
Tomar Electronics Inc	2100 W Obispo	Gilbert	AZ	85233	Tom Sikora	602-497-4400	R	14•	0.1
Carley Inc	1502 W 228th St	Torrance	CA	90501	James Carley	310-534-3860	R	12	<0.1
Unity Manufacturing Co	1260 N Clybourn	Chicago	IL	60610	LE Gross	312-943-5200	R	12	0.2
Beacon Light Products Inc	723 W Taylor Av	Meridian	ID	83642	Ron Porter	208-888-5905	R	10	<0.1
American Power Products Inc	14040 Central Av	Chino	CA	91710	Afzar Hussain	909-627-0886	R	9•	<0.1
LiteTronics Inc	4101 W 123rd St	Alsip	IL	60658	Robert Sorenson	708-389-8000	R	9	<0.1
Pace Control Technologies Inc	14040 Central Av	Chino	CA	91710	Afzar Hussain	909-627-0886	S	9•	<0.1
Elscott Corp	Rte 1	Gouldsboro	ME	04607	Ed H Soper	207-422-6747	R	7•	<0.1
UVP Inc	2066 W 11th St	Upland	CA	91786	Paul Warren	909-946-3197	R	7	<0.1
Electrix Inc	45 Spring St	New Haven	CT	06519	H Shwisha	203-776-5577	R	6	<0.1
Pennsylvania Illuminating Corp	526 Ash St	Scranton	PA	18509	JT Schofield	717-346-7356	R	6	0.1
Eximco Manufacturing Co	5311 N Kedzie Av	Chicago	IL	60625	R Ramsden	312-463-1470	R	5	<0.1
Electric M and R Inc	PO Box 326	Bethel Park	PA	15102	Gretchen Oswald	412-831-6101	P	5	<0.1
Poly-Optical Products Inc	17475 Gillette Av	Irvine	CA	92714	Joe Atchison	714-250-8557	S	3	<0.1
EMA Capital Inc	1816 W 135th St	Gardena	CA	90249	Daniel S Edelist	310-323-4310	R	2•	<0.1
Leeazanne	635 Regal Row	Dallas	TX	75247	Lam Lee	214-631-6611	D	2	<0.1
Mercury Techn Intern Corp	30677 Huntwood	Hayward	CA	94544	William Erhardt	510-429-1129	R	2•	<0.1
Precision Filaments Inc	89 Bannard St	Freehold	NJ	07728	Donald L Wilcox	908-462-3755	R	2	<0.1
Aero-Tech Light Bulb Co	534 Pratt Av N	Schaumburg	IL	60193	Ray M Schlosser	708-351-4900	R	2	<0.1
Sloan Co	7704 San Fernando	Sun Valley	CA	91352	Ellen Sloan	818-767-1729	R	1•	<0.1

Source: Ward's Business Directory of U.S. Private and Public Companies, Volumes 1 and 2, 1996. The company type code used is as follows: P - Public, R - Private, S - Subsidiary, D - Division, J - Joint Venture, A - Affiliate, G - Group. Sales are in millions of dollars, employees are in thousands. An asterisk (•) indicates an estimated sales volume. The symbol < stands for 'less than'. Company names and addresses are truncated, in some cases, to fit into the available space.

MATERIALS CONSUMED

Material	Quantity	Delivered Cost ($ million)
Materials, ingredients, containers, and supplies	(X)	848.4
Glass and glass products (including lamp bulb blanks)	(X)	257.8
Paper and paperboard containers, including shipping sacks and other paper packaging supplies	(X)	89.7
Industrial inorganic chemicals	(X)	33.2
Nonferrous metal wire	(X)	124.3
Electric lamp (bulb) bases	(X)	92.0
All other materials and components, parts, containers, and supplies	(X)	151.8
Materials, ingredients, containers, and supplies, nsk	(X)	99.8

Source: 1992 Economic Census. Explanation of symbols used: (D): Withheld to avoid disclosure of competitive data; na: Not available; (S): Withheld because statistical norms were not met; (X): Not applicable; (Z): Less than half the unit shown; nec: Not elsewhere classified; nsk: Not specified by kind; - : zero; • : 10-19 percent estimated; •• : 20-29 percent estimated.

PRODUCT SHARE DETAILS

Product or Product Class	% Share	Product or Product Class	% Share
Electric lamps (bulbs and tubes)	100.00	supports, lead-in, filaments, etc. but excluding lamp bulb blanks)	7.21
Electric lamp bulbs and tubes (including sealed beam lamp bulbs)	91.51	Electric lamps (bulbs and tubes), nsk	1.28
Electric lamp (bulbs and tubes) components (bases,			

Source: 1992 Economic Census. The values shown are percent of total shipments in an industry. Values of indented subcategories are summed in the main heading. The symbol (D) appears when data are withheld to prevent disclosure of competitive information. The abbreviation nsk stands for 'not specified by kind' and nec for 'not elsewhere classified'.

INPUTS AND OUTPUTS FOR ELECTRIC LAMPS

Economic Sector or Industry Providing Inputs	%	Sector	Economic Sector or Industry Buying Outputs	%	Sector
Glass & glass products, except containers	19.4	Manufg.	Personal consumption expenditures	40.2	
Wholesale trade	17.0	Trade	S/L Govt. purch., elem. & secondary education	6.1	S/L Govt
Imports	15.4	Foreign	Exports	6.0	Foreign
Nonferrous wire drawing & insulating	6.4	Manufg.	Hotels & lodging places	3.5	Services
Electrical equipment & supplies, nec	5.7	Manufg.	Hospitals	3.3	Services
Paperboard containers & boxes	5.0	Manufg.	Services to dwellings & other buildings	3.1	Services
Advertising	3.7	Services	Electric services (utilities)	3.0	Util.
Cyclic crudes and organics	3.4	Manufg.	Motor vehicles & car bodies	2.3	Manufg.
Electric services (utilities)	2.5	Util.	State & local electric utilities	2.1	Gov't
Banking	2.5	Fin/R.E.	Lighting fixtures & equipment	1.9	Manufg.
Industrial inorganic chemicals, nec	2.2	Manufg.	Retail trade, except eating & drinking	1.5	Trade
Air transportation	2.2	Util.	Real estate	1.4	Fin/R.E.
Gas production & distribution (utilities)	1.4	Util.	S/L Govt. purch., health & hospitals	1.1	S/L Govt
Hotels & lodging places	1.3	Services	S/L Govt. purch., higher education	1.1	S/L Govt
Motor freight transportation & warehousing	1.0	Util.	S/L Govt. purch., other general government	1.1	S/L Govt
Petroleum refining	0.9	Manufg.	Water transportation	1.0	Util.
Communications, except radio & TV	0.9	Util.	S/L Govt. purch., natural resource & recreation.	1.0	S/L Govt
Eating & drinking places	0.9	Trade	Eating & drinking places	0.9	Trade
Maintenance of nonfarm buildings nec	0.8	Constr.	Colleges, universities, & professional schools	0.9	Services
Computer & data processing services	0.7	Services	Nursing & personal care facilities	0.9	Services
Machinery, except electrical, nec	0.5	Manufg.	Elementary & secondary schools	0.8	Services
Royalties	0.5	Fin/R.E.	Federal Government purchases, nondefense	0.8	Fed Govt
Legal services	0.4	Services	U.S. Postal Service	0.7	Gov't
Management & consulting services & labs	0.4	Services	Photographic equipment & supplies	0.5	Manufg.
Special dies & tools & machine tool accessories	0.3	Manufg.	Wholesale trade	0.5	Trade
Railroads & related services	0.3	Util.	Accounting, auditing & bookkeeping	0.5	Services
Equipment rental & leasing services	0.3	Services	Office buildings	0.4	Constr.
Abrasive products	0.2	Manufg.	Electrical equipment & supplies, nec	0.4	Manufg.
Manifold business forms	0.2	Manufg.	Banking	0.4	Fin/R.E.
Miscellaneous fabricated wire products	0.2	Manufg.	Business/professional associations	0.4	Services
Insurance carriers	0.2	Fin/R.E.	Religious organizations	0.4	Services
Accounting, auditing & bookkeeping	0.2	Services	State & local government enterprises, nec	0.4	Gov't
Automotive rental & leasing, without drivers	0.2	Services	Federal Government purchases, national defense	0.4	Fed Govt
Automotive repair shops & services	0.2	Services	Industrial buildings	0.3	Constr.
Electrical repair shops	0.2	Services	Residential 1-unit structures, nonfarm	0.3	Constr.
U.S. Postal Service	0.2	Gov't	Engineering, architectural, & surveying services	0.3	Services
Lubricating oils & greases	0.1	Manufg.	Change in business inventories	0.3	In House
Sanitary services, steam supply, irrigation	0.1	Util.	Feed grains	0.2	Agric.
Real estate	0.1	Fin/R.E.	Meat animals	0.2	Agric.
Photofinishing labs, commercial photography	0.1	Services	Maintenance of nonfarm buildings nec	0.2	Constr.
Noncomparable imports	0.1	Foreign	Nonfarm residential structure maintenance	0.2	Constr.
			Residential additions/alterations, nonfarm	0.2	Constr.
			Signs & advertising displays	0.2	Manufg.
			Motor freight transportation & warehousing	0.2	Util.
			Amusement & recreation services nec	0.2	Services
			Doctors & dentists	0.2	Services
			Laundry, dry cleaning, shoe repair	0.2	Services
			Membership sports & recreation clubs	0.2	Services
			S/L Govt. purch., highways	0.2	S/L Govt
			S/L Govt. purch., public assistance & relief	0.2	S/L Govt
			Construction of hospitals	0.1	Constr.
			Blast furnaces & steel mills	0.1	Manufg.
			Household cooking equipment	0.1	Manufg.
			Miscellaneous plastics products	0.1	Manufg.
			Motor vehicle parts & accessories	0.1	Manufg.
			Air transportation	0.1	Util.
			Communications, except radio & TV	0.1	Util.
			Insurance carriers	0.1	Fin/R.E.
			Beauty & barber shops	0.1	Services
			Job training & related services	0.1	Services
			Labor, civic, social, & fraternal associations	0.1	Services
			S/L Govt. purch., correction	0.1	S/L Govt
			S/L Govt. purch., police	0.1	S/L Govt

Source: Benchmark Input-Output Accounts for the U.S. Economy, 1982, U.S. Department of Commerce, Washington, D.C., July 1991. Data, as reported in the source, are organized by the 1977 SIC structure in use in 1982 but have been matched, as closely as is possible, to the 1987 SIC structure used in this book.

OCCUPATIONS EMPLOYED BY SIC 364 - ELECTRIC LIGHTING AND WIRING EQUIPMENT

Occupation	% of Total 1994	Change to 2005	Occupation	% of Total 1994	Change to 2005
Assemblers, fabricators, & hand workers nec	16.0	-7.4	Tool & die makers	1.5	-25.2
Electrical & electronic assemblers	7.6	-16.7	Hand packers & packagers	1.5	-20.6
Blue collar worker supervisors	3.7	-20.3	Sheet metal workers & duct installers	1.5	-7.4
Inspectors, testers, & graders, precision	3.2	-35.2	Machinists	1.5	-30.5
Plastic molding machine workers	3.2	11.1	Freight, stock, & material movers, hand	1.4	-25.9
Machine operators nec	2.5	-34.7	Industrial truck & tractor operators	1.3	-7.4
Industrial machinery mechanics	2.5	25.0	Machine tool cutting & forming etc. nec	1.2	-25.9
Sales & related workers nec	2.1	-7.4	Production, planning, & expediting clerks	1.2	-7.4
Electrical & electronic equipment assemblers	1.9	-7.4	Secretaries, ex legal & medical	1.2	-15.7
General managers & top executives	1.9	-12.2	Maintenance repairers, general utility	1.1	-16.6
Machine feeders & offbearers	1.8	-16.7	Industrial production managers	1.0	-7.4
Traffic, shipping, & receiving clerks	1.7	-10.9	Coating, painting, & spraying machine workers	1.0	-7.4
Helpers, laborers, & material movers nec	1.6	-7.4			

Source: *Industry-Occupation Matrix*, Bureau of Labor Statistics. These data relate to one or more 3-digit SIC industry groups rather than to a single 4-digit SIC. The change reported for each occupation to the year 2005 is a percent of growth or decline as estimated by the Bureau of Labor Statistics. The abbreviation nec stands for 'not elsewhere classified'.

LOCATION BY STATE AND REGIONAL CONCENTRATION

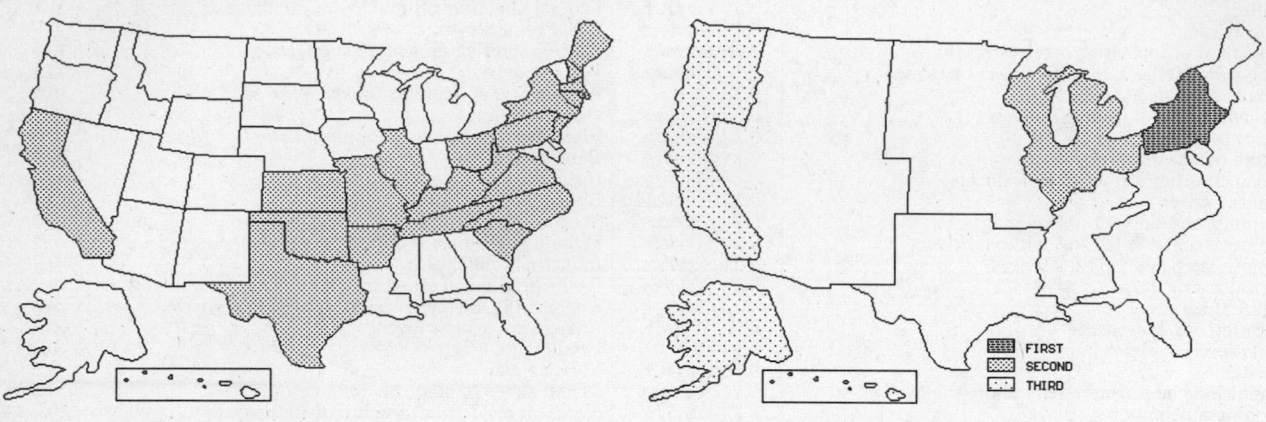

FIRST
SECOND
THIRD

INDUSTRY DATA BY STATE

State	Establish- ments	Shipments			Employment				Cost as % of Shipments	Investment per Employee ($)
		Total ($ mil)	% of U.S.	Per Establ.	Total Number	% of U.S.	Per Establ.	Wages ($/hour)		
New Jersey	10	80.4	2.7	8.0	700	4.0	70	11.00	33.6	857
Tennessee	5	76.9	2.5	15.4	900	5.1	180	12.59	25.2	-
Illinois	8	74.3	2.5	9.3	700	4.0	88	12.54	25.0	-
California	12	29.0	1.0	2.4	300	1.7	25	8.50	39.7	3,333
Ohio	12	(D)	-	-	3,750 *	21.4	313	-	-	-
Pennsylvania	8	(D)	-	-	1,750 *	10.0	219	-	-	-
New York	7	(D)	-	-	750 *	4.3	107	-	-	-
Kentucky	6	(D)	-	-	1,750 *	10.0	292	-	-	4,343
Maine	4	(D)	-	-	375 *	2.1	94	-	-	-
Connecticut	3	(D)	-	-	750 *	4.3	250	-	-	-
Texas	3	(D)	-	-	175 *	1.0	58	-	-	-
Kansas	2	(D)	-	-	750 *	4.3	375	-	-	-
Massachusetts	2	(D)	-	-	750 *	4.3	375	-	-	-
New Hampshire	2	(D)	-	-	1,750 *	10.0	875	-	-	-
Oklahoma	2	(D)	-	-	175 *	1.0	88	-	-	-
South Carolina	2	(D)	-	-	375 *	2.1	188	-	-	-
Virginia	2	(D)	-	-	375 *	2.1	188	-	-	-
Arkansas	1	(D)	-	-	750 *	4.3	750	-	-	-
Missouri	1	(D)	-	-	375 *	2.1	375	-	-	-
North Carolina	1	(D)	-	-	175 *	1.0	175	-	-	-
Rhode Island	1	(D)	-	-	175 *	1.0	175	-	-	-
West Virginia	1	(D)	-	-	750 *	4.3	750	-	-	-

Source: 1992 *Economic Census*. The states are in descending order of shipments or establishments (if shipment data are missing for the majority). The symbol (D) appears when data are withheld to prevent disclosure of competitive information. States marked with (D) are sorted by number of establishments. A dash (-) indicates that the data element cannot be calculated; * indicates the midpoint of a range.

3643 - CURRENT-CARRYING WIRING DEVICES

82 83 84 85 86 87 88 89 90 91 92 93 94 95 96 97 98

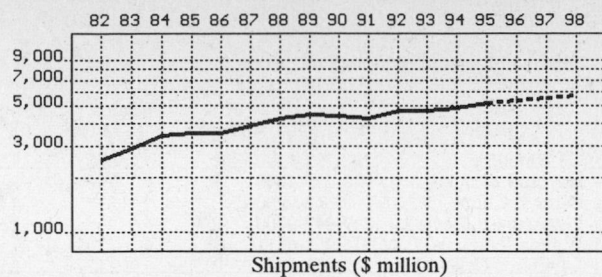

Shipments ($ million)

82 83 84 85 86 87 88 89 90 91 92 93 94 95 96 97 98

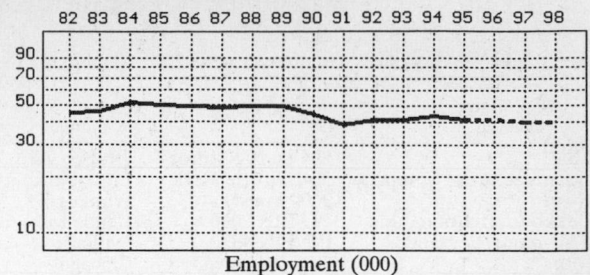

Employment (000)

GENERAL STATISTICS

| Year | Companies | Establishments | | Employment | | | Compensation | | Production ($ million) | | | |
		Total	with 20 or more employees	Total (000)	Production Workers (000)	Hours (Mil)	Payroll ($ mil)	Wages ($/hr)	Cost of Materials	Value Added by Manufacture	Value of Shipments	Capital Invest.
1982	361	415	222	44.5	31.9	57.2	699.8	7.37	996.7	1,464.2	2,510.3	88.8
1983		410	229	45.9	34.1	63.6	786.5	7.60	1,210.2	1,742.5	2,914.0	85.5
1984		405	236	50.7	37.5	69.3	905.5	7.76	1,446.1	2,042.1	3,415.1	100.3
1985		401	242	49.8	36.5	66.1	920.3	8.20	1,416.1	2,089.9	3,518.2	109.0
1986		402	241	48.3	34.6	65.0	941.3	8.30	1,421.7	2,147.5	3,537.0	101.2
1987	365	430	246	47.9	34.5	66.3	963.8	8.42	1,518.2	2,330.8	3,848.1	111.9
1988		451	273	48.7	36.2	69.8	1,019.2	8.66	1,768.3	2,542.0	4,262.6	120.4
1989		421	264	48.8	32.7	63.4	1,011.4	9.20	1,767.8	2,663.6	4,434.1	128.5
1990		415	252	44.4	31.8	62.0	1,019.4	9.40	1,739.5	2,642.1	4,404.0	155.2
1991		445	261	39.4	28.3	55.5	961.9	9.84	1,692.8	2,560.7	4,260.2	98.8
1992	406	475	264	41.1	28.9	57.7	1,071.4	10.28	1,795.3	2,881.7	4,668.1	124.2
1993		477	270	41.5	29.4	56.7	1,079.0	10.35	1,884.7	2,783.0	4,681.0	122.6
1994		467P	276P	43.3	30.6	62.1	1,156.8	10.24	1,965.8	2,889.7	4,874.7	145.7
1995		473P	281P	41.7P	29.1P	59.1P	1,164.2P	10.78P	2,097.5P	3,083.3P	5,201.3P	142.3P
1996		479P	285P	41.1P	28.6P	58.6P	1,192.8P	11.05P	2,169.7P	3,189.4P	5,380.3P	146.3P
1997		485P	289P	40.5P	28.0P	58.1P	1,221.3P	11.31P	2,241.9P	3,295.5P	5,559.3P	150.2P
1998		491P	293P	39.9P	27.5P	57.6P	1,249.9P	11.58P	2,314.1P	3,401.7P	5,738.3P	154.1P

Sources: 1982, 1987, 1992 *Economic Census*; *Annual Survey of Manufactures*, 83-86, 88-91, 93-94. Establishment counts for non-Census years are from *County Business Patterns*; establishment values for 83-84 are extrapolations. 'P's show projections by the editors. Industries reclassified in 87 will not have data for prior years.

INDICES OF CHANGE

| Year | Companies | Establishments | | Employment | | | Compensation | | Production ($ million) | | | |
		Total	with 20 or more employees	Total (000)	Production Workers (000)	Hours (Mil)	Payroll ($ mil)	Wages ($/hr)	Cost of Materials	Value Added by Manufacture	Value of Shipments	Capital Invest.
1982	89	87	84	108	110	99	65	72	56	51	54	71
1983		86	87	112	118	110	73	74	67	60	62	69
1984		85	89	123	130	120	85	75	81	71	73	81
1985		84	92	121	126	115	86	80	79	73	75	88
1986		85	91	118	120	113	88	81	79	75	76	81
1987	90	91	93	117	119	115	90	82	85	81	82	90
1988		95	103	118	125	121	95	84	98	88	91	97
1989		89	100	119	113	110	94	89	98	92	95	103
1990		87	95	108	110	107	95	91	97	92	94	125
1991		94	99	96	98	96	90	96	94	89	91	80
1992	100	100	100	100	100	100	100	100	100	100	100	100
1993		100	102	101	102	98	101	101	105	97	100	99
1994		98P	105P	105	106	108	108	100	109	100	104	117
1995		100P	106P	101P	101P	102P	109P	105P	117P	107P	111P	115P
1996		101P	108P	100P	99P	102P	111P	107P	121P	111P	115P	118P
1997		102P	109P	99P	97P	101P	114P	110P	125P	114P	119P	121P
1998		103P	111P	97P	95P	100P	117P	113P	129P	118P	123P	124P

Sources: Same as General Statistics. Values reflect change from the base year, 1992. Values above 100 mean greater than 92, values below 100 mean less than 92, and a value of 100 in the 82-91 or 93-98 period means same as 92. 'P's mark projections by the editors.

SELECTED RATIOS

For 1994	Avg. of All Manufact.	Analyzed Industry	Index	For 1994	Avg. of All Manufact.	Analyzed Industry	Index
Employees per Establishment	49	93	189	Value Added per Production Worker	134,084	94,435	70
Payroll per Establishment	1,500,273	2,474,759	165	Cost per Establishment	5,045,178	4,205,465	83
Payroll per Employee	30,620	26,716	87	Cost per Employee	102,970	45,400	44
Production Workers per Establishment	34	65	191	Cost per Production Worker	146,988	64,242	44
Wages per Establishment	853,319	1,360,399	159	Shipments per Establishment	9,576,895	10,428,518	109
Wages per Production Worker	24,861	20,781	84	Shipments per Employee	195,460	112,580	58
Hours per Production Worker	2,056	2,029	99	Shipments per Production Worker	279,017	159,304	57
Wages per Hour	12.09	10.24	85	Investment per Establishment	321,011	311,698	97
Value Added per Establishment	4,602,255	6,181,978	134	Investment per Employee	6,552	3,365	51
Value Added per Employee	93,930	66,737	71	Investment per Production Worker	9,352	4,761	51

Sources: Same as General Statistics. The 'Average of All Manufacturing' column represents the average of all manufacturing industries reported for the most recent complete year available. The Index shows the relationship between the Average and the Analyzed Industry. For example, 100 means that they are equal; 500 that the Analyzed Industry is five times the average; 50 means that the Analyzed Industry is half the national average. The abbreviation 'na' is used to show that data are 'not available'.

LEADING COMPANIES   Number shown: **75**   Total sales ($ mil): **5,255**   Total employment (000): **48.0**

Company Name	Address				CEO Name	Phone	Co. Type	Sales ($ mil)	Empl. (000)
Hubbell Inc	584 Derby Milford	Orange	CT	06477	GJ Ratcliffe	203-799-4100	P	832	5.9
Leviton Manufacturing Company	59-25 Little Neck	Little Neck	NY	11362	Harold Leviton	718-229-4040	R	790	8.0
Cherry Corp	3600 Sunset Av	Waukegan	IL	60087	Peter B Cherry	708-662-9200	P	339	4.0
Group Dekko International Inc	PO Box 2000	Kendallville	IN	46755	Lawrence P Doyle	219-347-0700	R	270	2.5
Pass and Seymour/Legrand	PO Box 4822	Syracuse	NY	13221	William Nuckols	315-468-6211	S	250	1.0
Joslyn Corp	30 S Wacker Dr	Chicago	IL	60606	Lawrence G Wolski	312-454-2900	P	217	2.0
Panduit Corp	17301 Ridgeland	Tinley Park	IL	60477	DM Henderson	708-532-1800	R	200	1.0
Eagle Electric Manufacturing	45-31 Court Sq	Long Island Ct	NY	11101	Neal Kluger	718-937-8000	R	180	1.9
Carlingswitch Inc	60 Johnson Av	Plainville	CT	06062	RW Sorenson	203-793-9281	R	110*	1.2
Woodhead Industries Inc	2150 E Lake Cook	Buffalo Grove	IL	60089	C Mark DeWinter	708-465-8300	P	106	1.1
EF Johnson Co	438 Gateway Blv	Burnsville	MN	55337	William Weksel	612-882-5500	R	100*	1.0
Gould Inc	374 Merrimac St	Newburyport	MA	01950	William A Trotman	508-462-6662	D	100	0.5
Hubbell Inc	PO Box 3999	Bridgeport	CT	06605	Wesson M Brown	203-333-1181	D	100*	0.8
Wiremold Co	60 Woodlawn St	West Hartford	CT	06110	Art Byrne	203-233-6251	R	100*	0.4
Pacific Electricord Co	747 W Redondo	Gardena	CA	90247	Edwin Kanner	310-532-6600	S	85	0.7
SL Industries Inc	520 Fellowship Rd	Mount Laurel	NJ	08054	Owen Farren	609-727-1500	P	77	1.3
Samtec Inc	PO Box 1147	New Albany	IN	47151	John Shine	812-944-6733	R	70	0.3
Bardes Corp	4730 Madison Rd	Cincinnati	OH	45227	DJ FitzGibbon	513-533-6200	R	68	0.7
Robinson Nugent Inc	PO Box 1208	New Albany	IN	47151	Larry W Burke	812-945-0211	P	68	0.6
Erico Inc	34600 Solon Rd	Solon	OH	44139	RB Savage	216-248-0100	S	60	0.3
Ilsco	4730 Madison Rd	Cincinnati	OH	45227	DJ Fitzgibbon	513-533-6200	D	60	0.5
Tri-Star Electronics International	2201 Rosecrans Av	El Segundo	CA	90245	J DeCrane	310-536-0444	S	60	0.3
G and H Technology Inc	750 W Ventura Blvd	Camarillo	CA	93010	Thomas Cleary	805-484-0543	S	52	0.4
Medinan Voltage	PO Box 341	Bloomington	IN	47402	Ake Almgren	812-335-4459	D	40	0.3
Ohio Brass Co	1850 Richland Av E	Aiken	SC	29801	Mike Dalton	803-648-8386	S	40	0.2
Brooks Electronics Inc	4001 N American St	Philadelphia	PA	19140	Gary J Brooks	215-228-4433	S	35*	0.1
Southern Devices Inc	PO Drawer 68	Morganton	NC	28655	Harold Leviton	704-584-1611	S	32	0.5
Iowa Assemblies Inc	3330 W McLane St	Osceola	IA	50213	Phil Salisbury	515-342-6559	R	30*	0.5
Deringer Manufacturing Co	1250 Town Line Rd	Mundelein	IL	60060	RW Lamm	708-566-4100	R	29*	0.3
Lucerne Products Inc	7600 Old 8th Rd	Hudson	OH	44236	Blair Hamilton	216-653-6661	R	29*	0.3
Penn-Union Corp	229 Waterford St	Edinboro	PA	16412	J Tramontano	814-734-1631	S	29	0.3
Fusite	6000 Fernview Av	Cincinnati	OH	45212	Earle L Weaver	513-731-2020	D	27	0.3
American Electric Cordsets LP	PO Box 802	Bensenville	IL	60106	TD McKee	708-595-3900	R	25	0.4
Crouse-Hinds Molded Products	Rte 4	La Grange	NC	28551	Gregory Quinton	919-566-3014	D	24	0.2
Autosplice Inc	10121 Barnes	San Diego	CA	92121	Peter Zahn	619-535-0077	R	23	0.2
SL Waber Inc	520 Fellowship Rd	Mount Laurel	NJ	08054	Ronald Mazik	609-866-8888	S	23*	0.2
Continental Wirt Electr Corp	130 James Way	Southampton	PA	18966	Kalman Lifson	215-355-7080	R	22*	0.3
ITT Aerospace Controls	28150 Industry Dr	Valencia	CA	91355	Jack Rubin	805-295-4000	D	22	0.4
Otto Engineering Inc	2 E Main St	Carpentersville	IL	60110	Jack O Roeser	708-428-7171	R	21	0.2
Central Industries of Indiana Inc	1325 E Virginia	Evansville	IN	47711	Terry Silver	812-421-0231	R	20	0.4
Dearborn Wire and Cable LP	250 W Carpenter St	Wheeling	IL	60090	B Greene	708-459-1000	S	20	0.3
Lyall Assemblies Inc	PO Box 110	Albion	IN	46701	Dave Schinbeckler	219-636-2551	R	20	0.4
NTT Inc	632 Arch St	Meadville	PA	16335	Gary Gerhold	814-724-6440	R	20	0.4
Pyle-National Co	1334 N Kostner Av	Chicago	IL	60651	Michael Carroccia	312-342-6300	D	20	0.1
Sine Companies Inc	25325 Joy Blv	Mount Clemens	MI	48046	Jerry Zaccardelli	313-465-6570	R	20	0.3
Jay-El Products Inc	PO Box 6240	Carson	CA	90749	Robert Borlet	310-513-7200	S	19	0.2
Methode Electrique	1700 Hicks Rd	Rolling Mdws	IL	60008	William J McGinley	708-392-3500	D	19*	0.2
Mac Products Inc	60 Penn Av	Kearny	NJ	07032	Edward Gollob	201-344-0700	R	18	0.1
Pent Products Inc	PO Box 246	Kendallville	IN	46755	Randy Sutton Sr	219-347-5831	S	18*	0.2
Clayton/Unimax	8182 US 70 W	Clayton	NC	27520	Ted McGowan	919-553-3131	D	17	0.3
CMW Inc	PO Box 2266	Indianapolis	IN	46206	HD Johnston	317-634-8884	D	17*	0.2
McGill Electric Switch Prod	1002 Campbell Av	Valparaiso	IN	46383	Peter Banks	219-465-2200	D	17*	0.2
Condor DC Power Supplies Inc	2311 Statham Pkwy	Oxnard	CA	93033	Thomas Ingman	805-486-4565	S	16	0.5
Superior Manufacturing Co	3133 W Harvard St	Santa Ana	CA	92704	R Fung	714-540-4605	R	16	0.2
Techdyne Inc	2230 W 77th St	Hialeah	FL	33016	Thomas K Langbein	305-556-9210	P	15	0.3
Advanced Circuit Technology	118 Northeastern	Nashua	NH	03062	Joseph A Roberts	603-880-6000	R	15	0.2
Charles E Gillman Co	519 Taft Dr	South Holland	IL	60473	Charles E Gillman	708-333-4900	R	15	0.4
Citel America Inc	1111 Pkcentre Blvd	Miami	FL	33169	Fabrice Larmier	305-621-0022	S	14*	0.2
Eldre Corp	1500 Jefferson Rd	Rochester	NY	14623	Harvey B Erdle	716-427-7280	R	14*	0.2
Phoenix Contact Inc	PO Box 4100	Harrisburg	PA	17111	Don Springer	717-944-1300	R	14	0.2
Specialty Connector Co	PO Box 547	Franklin	IN	46131	Scott Kelley	317-738-2800	S	13	0.2
Aircraft & Electronic Specialties	PO Box 248	Plainfield	IN	46168	NB Higbie	317-272-2551	R	12	0.3
Beau Interconnect Systems	PO Box 10	Laconia	NH	03246	Tony Peleckis	603-524-5101	D	12	0.2
Haydon Switch and Instrument	1500 Meriden Rd	Waterbury	CT	06705	B Dubois	203-756-7441	R	12	0.1
Marquardt Switches Inc	2711 Rte 20 E	Cazenovia	NY	13035	Gerald Groff	315-655-8050	R	11*	0.1
Perma Power Electronics Inc	5601 W H Av	Niles	IL	60714	Norman Ackerman	312-763-0763	S	11	0.1
BIW Connector Systems Inc	500 Tesconi Cir	Santa Rosa	CA	95401	Tom Fitzsimmons	707-523-2300	S	10	0.1
Midcon Cables	PO Box 1786	Joplin	MO	64802	C Wheeler	417-781-4331	D	10	<0.1
DG O'Brien Inc	PO Box 159	Seabrook	NH	03874	DG O'Brien	603-474-5571	R	8	0.1
Green Technologies Inc	5490 Spine Rd	Boulder	CO	80301	John Hay	303-581-9600	R	8*	<0.1
Heaters Engineering Inc	PO Box 337	North Webster	IN	46555	Jeff Thornburgh	219-834-2818	R	8	0.2
Hydra-Electric Co	3151 Kenwood St	Burbank	CA	91505	Henry Acuff	818-843-6211	R	8	0.1
Rowe Industries Inc	PO Box 6877	Toledo	OH	43612	Haywood Bower	419-729-9761	S	8	<0.1
Lucas Products	PO Box 7	Lucas	IA	50151	Lyle Persels	515-766-6121	D	8	0.1
Aerospace Systems Co	PO Box 998	Fairmont	MN	56031	Robert A Jensen	507-235-3355	D	7	<0.1

Source: Ward's Business Directory of U.S. Private and Public Companies, Volumes 1 and 2, 1996. The company type code used is as follows: P - Public, R - Private, S - Subsidiary, D - Division, J - Joint Venture, A - Affiliate, G - Group. Sales are in millions of dollars, employees are in thousands. An asterisk (*) indicates an estimated sales volume. The symbol < stands for 'less than'. Company names and addresses are truncated, in some cases, to fit into the available space.

MATERIALS CONSUMED

Material	Quantity	Delivered Cost ($ million)
Materials, ingredients, containers, and supplies	(X)	1,539.0
Metal stampings	(X)	77.1
Metal bolts, nuts, screws, washers, rivets, and other screw machine products	(X)	71.1
All other fabricated metal products (except forgings)	(X)	19.4
Forgings	(X)	2.2
Iron and steel castings (rough and semifinished)	(X)	12.3
Aluminum and aluminum-base alloy castings (rough and semifinished)	(X)	18.0
Copper and copper-base alloy castings (rough and semifinished)	(X)	18.2
Other nonferrous castings (rough and semifinished)	(X)	2.4
Steel bars, bar shapes, and plates	(X)	7.3
Steel sheet and strip, including tin plate	(X)	26.0
All other steel shapes and forms	(X)	1.6
Copper and copper-base alloy rod, bar, and bar shapes	(X)	13.2
Copper and copper-base alloy plate, sheet, and strip, including military cups and discs	(X)	95.3
All other copper and copper-base alloy mill shapes and forms	(X)	20.7
Aluminum and aluminum-base alloy shapes and forms	(X)	25.8
Other nonferrous shapes and forms	(X)	18.5
Plastics resins consumed in the form of granules, pellets, powders, liquids, etc.	(X)	107.0
Plastics products consumed in the form of sheets, rods, tubes, film, and other shapes	(X)	64.0
Current-carrying wiring devices	(X)	76.9
Precious metals (gold, platinum, etc.), all forms	(X)	54.6
Insulated wire and cable, except magnet wire	(X)	45.6
Semiconductors, including transistors, diodes, rectifiers, and integrated circuits for electronic circuitry	(X)	74.3
Paper and paperboard containers, including shipping sacks and other paper packaging supplies	(X)	21.8
All other materials and components, parts, containers, and supplies	(X)	341.4
Materials, ingredients, containers, and supplies, nsk	(X)	324.2

Source: 1992 *Economic Census*. Explanation of symbols used: (D): Withheld to avoid disclosure of competitive data; na: Not available; (S): Withheld because statistical norms were not met; (X): Not applicable; (Z): Less than half the unit shown; nec: Not elsewhere classified; nsk: Not specified by kind; - : zero; * : 10-19 percent estimated; ** : 20-29 percent estimated.

PRODUCT SHARE DETAILS

Product or Product Class	% Share	Product or Product Class	% Share
Current-carrying wiring devices	100.00	Current-carrying wiring devices, metal contacts, precious and other	5.72
Current-carrying wiring devices, lampholders	4.65		
Current-carrying wiring devices, convenience and power outlets, both general and special purpose (excluding pin-and-sleeve type)	7.77	Current-carrying wiring devices, wire connectors for electrical circuitry.	23.46
		Other current-carrying wiring devices, nec (attachments, plug caps, connector bodies, lightning arrestors, etc.)	22.45
Current-carrying wiring devices, switches for electrical circuitry (including vehicular switches)	32.17	Current-carrying wiring devices, nsk	3.77

Source: 1992 *Economic Census*. The values shown are percent of total shipments in an industry. Values of indented subcategories are summed in the main heading. The symbol (D) appears when data are withheld to prevent disclosure of competitive information. The abbreviation nsk stands for 'not specified by kind' and nec for 'not elsewhere classified'.

INPUTS AND OUTPUTS FOR WIRING DEVICES

Economic Sector or Industry Providing Inputs	%	Sector	Economic Sector or Industry Buying Outputs	%	Sector
Imports	21.5	Foreign	Electric utility facility construction	15.6	Constr.
Blast furnaces & steel mills	10.4	Manufg.	Exports	11.2	Foreign
Wholesale trade	9.5	Trade	Maintenance of nonfarm buildings nec	8.3	Constr.
Wiring devices	6.6	Manufg.	Residential 1-unit structures, nonfarm	5.9	Constr.
Plastics materials & resins	5.6	Manufg.	Residential additions/alterations, nonfarm	4.8	Constr.
Plating & polishing	3.1	Manufg.	Nonfarm residential structure maintenance	4.2	Constr.
Copper rolling & drawing	2.9	Manufg.	Industrial buildings	3.8	Constr.
Electric services (utilities)	2.5	Util.	Wiring devices	3.8	Manufg.
Screw machine and related products	2.3	Manufg.	Maintenance of electric utility facilities	3.6	Constr.
Banking	2.0	Fin/R.E.	Office buildings	2.7	Constr.
Primary nonferrous metals, nec	1.9	Manufg.	Telephone & telegraph facility construction	2.2	Constr.
Advertising	1.8	Services	Electronic computing equipment	2.0	Manufg.
Air transportation	1.4	Util.	Construction of hospitals	1.7	Constr.
Communications, except radio & TV	1.4	Util.	Refrigeration & heating equipment	1.4	Manufg.
Aluminum rolling & drawing	1.3	Manufg.	Residential garden apartments	1.3	Constr.
Miscellaneous plastics products	1.2	Manufg.	Lighting fixtures & equipment	1.2	Manufg.
Nonferrous wire drawing & insulating	1.2	Manufg.	Radio & TV communication equipment	1.1	Manufg.
Paperboard containers & boxes	1.1	Manufg.	Radio & TV receiving sets	1.1	Manufg.
Petroleum refining	1.1	Manufg.	Telephone & telegraph apparatus	1.1	Manufg.
Aluminum castings	1.0	Manufg.	Construction of stores & restaurants	1.0	Constr.
Cyclic crudes and organics	1.0	Manufg.	Mobile homes	1.0	Manufg.
Eating & drinking places	1.0	Trade	Warehouses	0.8	Constr.
Maintenance of nonfarm buildings nec	0.9	Constr.	Household cooking equipment	0.8	Manufg.
Motor freight transportation & warehousing	0.9	Util.	Construction of educational buildings	0.7	Constr.
Hotels & lodging places	0.9	Services	Farm service facilities	0.7	Constr.

Continued on next page.

INPUTS AND OUTPUTS FOR WIRING DEVICES - Continued

Economic Sector or Industry Providing Inputs	%	Sector	Economic Sector or Industry Buying Outputs	%	Sector
Metal stampings, nec	0.8	Manufg.	Maintenance of railroads	0.7	Constr.
Fabricated rubber products, nec	0.7	Manufg.	Electric housewares & fans	0.7	Manufg.
Gas production & distribution (utilities)	0.7	Util.	Electronic components nec	0.7	Manufg.
Equipment rental & leasing services	0.7	Services	Household refrigerators & freezers	0.7	Manufg.
Iron & steel foundries	0.6	Manufg.	Maintenance of telephone & telegraph facilities	0.6	Constr.
Machinery, except electrical, nec	0.6	Manufg.	Engine electrical equipment	0.6	Manufg.
Metal coating & allied services	0.6	Manufg.	Construction of nonfarm buildings nec	0.5	Constr.
Miscellaneous fabricated wire products	0.6	Manufg.	Highway & street construction	0.5	Constr.
Nonferrous castings, nec	0.6	Manufg.	Hotels & motels	0.5	Constr.
Semiconductors & related devices	0.6	Manufg.	Maintenance of farm service facilities	0.5	Constr.
Computer & data processing services	0.6	Services	Sewer system facility construction	0.5	Constr.
Real estate	0.5	Fin/R.E.	Electrical equipment & supplies, nec	0.5	Manufg.
Legal services	0.5	Services	Household laundry equipment	0.5	Manufg.
Glass & glass products, except containers	0.4	Manufg.	Mechanical measuring devices	0.5	Manufg.
Primary zinc	0.4	Manufg.	Amusement & recreation building construction	0.4	Constr.
Special dies & tools & machine tool accessories	0.4	Manufg.	Local transit facility construction	0.4	Constr.
Railroads & related services	0.4	Util.	Maintenance of local transit facilities	0.4	Constr.
Management & consulting services & labs	0.4	Services	Resid. & other health facility construction	0.4	Constr.
U.S. Postal Service	0.4	Gov't	Residential 2-4 unit structures, nonfarm	0.4	Constr.
Industrial patterns	0.3	Manufg.	Instruments to measure electricity	0.4	Manufg.
Primary aluminum	0.3	Manufg.	Transformers	0.4	Manufg.
Accounting, auditing & bookkeeping	0.3	Services	Crude petroleum & natural gas	0.3	Mining
Abrasive products	0.2	Manufg.	Maintenance of highways & streets	0.3	Constr.
Brass, bronze, & copper castings	0.2	Manufg.	Maintenance of military facilities	0.3	Constr.
Chemical preparations, nec	0.2	Manufg.	Residential high-rise apartments	0.3	Constr.
Manifold business forms	0.2	Manufg.	Water supply facility construction	0.3	Constr.
Nonferrous rolling & drawing, nec	0.2	Manufg.	Household appliances, nec	0.3	Manufg.
Transformers	0.2	Manufg.	Industrial controls	0.3	Manufg.
Royalties	0.2	Fin/R.E.	Service industry machines, nec	0.3	Manufg.
Automotive rental & leasing, without drivers	0.2	Services	Switchgear & switchboard apparatus	0.3	Manufg.
Noncomparable imports	0.2	Foreign	Typewriters & office machines, nec	0.3	Manufg.
Hand & edge tools, nec	0.1	Manufg.	Communications, except radio & TV	0.3	Util.
Industrial inorganic chemicals, nec	0.1	Manufg.	Construction of religious buildings	0.2	Constr.
Lubricating oils & greases	0.1	Manufg.	Maintenance of farm residential buildings	0.2	Constr.
Machine tools, metal cutting types	0.1	Manufg.	Maintenance of nonbuilding facilities nec	0.2	Constr.
Machine tools, metal forming types	0.1	Manufg.	Maintenance of sewer facilities	0.2	Constr.
Primary copper	0.1	Manufg.	Nonbuilding facilities nec	0.2	Constr.
Insurance carriers	0.1	Fin/R.E.	Calculating & accounting machines	0.2	Manufg.
Automotive repair shops & services	0.1	Services	Travel trailers & campers	0.2	Manufg.
Electrical repair shops	0.1	Services	Federal Government purchases, national defense	0.2	Fed Govt
Photofinishing labs, commercial photography	0.1	Services	Gross private fixed investment	0.2	Cap Inv
			Dormitories & other group housing	0.1	Constr.
			Farm housing units & additions & alterations	0.1	Constr.
			Maintenance of water supply facilities	0.1	Constr.
			Railroad construction	0.1	Constr.
			Environmental controls	0.1	Manufg.
			Lawn & garden equipment	0.1	Manufg.
			Signs & advertising displays	0.1	Manufg.
			Vitreous plumbing fixtures	0.1	Manufg.

Source: Benchmark Input-Output Accounts for the U.S. Economy, 1982, U.S. Department of Commerce, Washington, D.C., July 1991. Data, as reported in the source, are organized by the 1977 SIC structure in use in 1982 but have been matched, as closely as is possible, to the 1987 SIC structure used in this book.

OCCUPATIONS EMPLOYED BY SIC 364 - ELECTRIC LIGHTING AND WIRING EQUIPMENT

Occupation	% of Total 1994	Change to 2005	Occupation	% of Total 1994	Change to 2005
Assemblers, fabricators, & hand workers nec	16.0	-7.4	Tool & die makers	1.5	-25.2
Electrical & electronic assemblers	7.6	-16.7	Hand packers & packagers	1.5	-20.6
Blue collar worker supervisors	3.7	-20.3	Sheet metal workers & duct installers	1.5	-7.4
Inspectors, testers, & graders, precision	3.2	-35.2	Machinists	1.5	-30.5
Plastic molding machine workers	3.2	11.1	Freight, stock, & material movers, hand	1.4	-25.9
Machine operators nec	2.5	-34.7	Industrial truck & tractor operators	1.3	-7.4
Industrial machinery mechanics	2.5	25.0	Machine tool cutting & forming etc. nec	1.2	-25.9
Sales & related workers nec	2.1	-7.4	Production, planning, & expediting clerks	1.2	-7.4
Electrical & electronic equipment assemblers	1.9	-7.4	Secretaries, ex legal & medical	1.2	-15.7
General managers & top executives	1.9	-12.2	Maintenance repairers, general utility	1.1	-16.6
Machine feeders & offbearers	1.8	-16.7	Industrial production managers	1.0	-7.4
Traffic, shipping, & receiving clerks	1.7	-10.9	Coating, painting, & spraying machine workers	1.0	-7.4
Helpers, laborers, & material movers nec	1.6	-7.4			

Source: Industry-Occupation Matrix, Bureau of Labor Statistics. These data relate to one or more 3-digit SIC industry groups rather than to a single 4-digit SIC. The change reported for each occupation to the year 2005 is a percent of growth or decline as estimated by the Bureau of Labor Statistics. The abbreviation nec stands for 'not elsewhere classified'.

LOCATION BY STATE AND REGIONAL CONCENTRATION

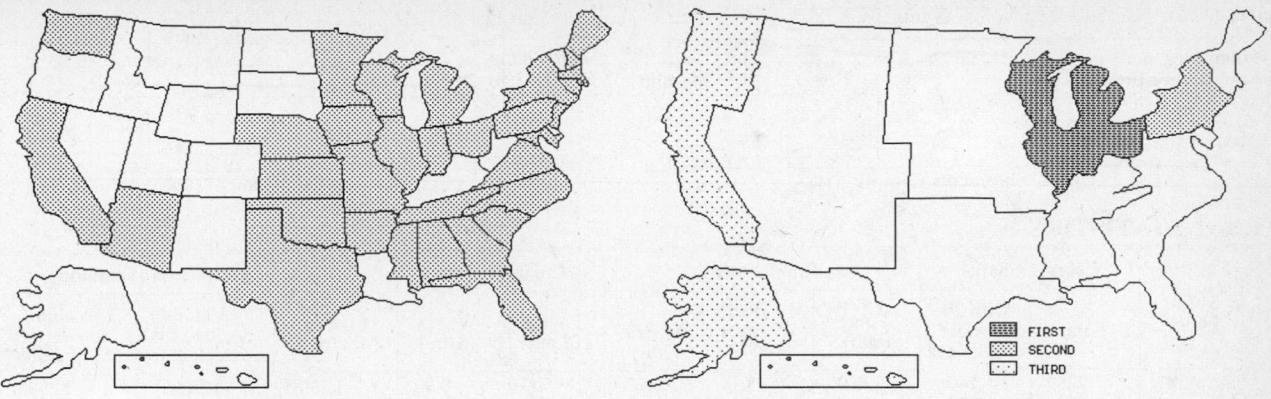

FIRST
SECOND
THIRD

INDUSTRY DATA BY STATE

State	Establish-ments	Shipments			Employment				Cost as % of Shipments	Investment per Employee ($)
		Total ($ mil)	% of U.S.	Per Establ.	Total Number	% of U.S.	Per Establ.	Wages ($/hour)		
Illinois	44	795.8	17.0	18.1	8,300	20.2	189	10.56	35.8	4,337
Pennsylvania	41	434.3	9.3	10.6	3,000	7.3	73	11.40	41.2	3,300
New York	35	374.8	8.0	10.7	4,100	10.0	117	9.49	35.1	1,293
North Carolina	15	301.6	6.5	20.1	2,300	5.6	153	8.68	20.4	2,174
Ohio	28	228.2	4.9	8.1	3,000	7.3	107	9.27	48.2	1,633
Connecticut	15	219.2	4.7	14.6	1,500	3.6	100	12.80	35.5	4,667
California	64	191.4	4.1	3.0	2,100	5.1	33	10.14	33.4	2,190
Massachusetts	21	163.9	3.5	7.8	1,500	3.6	71	13.11	38.5	2,867
New Jersey	21	162.8	3.5	7.8	1,300	3.2	62	11.50	36.4	6,538
Michigan	18	149.3	3.2	8.3	1,600	3.9	89	9.48	47.8	3,188
Indiana	14	109.3	2.3	7.8	1,000	2.4	71	10.60	38.6	2,700
Alabama	8	103.4	2.2	12.9	1,300	3.2	163	10.16	38.9	-
South Carolina	5	95.3	2.0	19.1	800	1.9	160	7.57	53.9	2,875
Florida	32	88.2	1.9	2.8	1,200	2.9	38	9.44	34.5	4,500
Wisconsin	8	70.9	1.5	8.9	400	1.0	50	10.80	47.5	-
Texas	19	67.5	1.4	3.6	1,000	2.4	53	8.40	26.7	1,400
Missouri	10	48.4	1.0	4.8	600	1.5	60	10.13	50.6	3,167
Tennessee	4	34.8	0.7	8.7	600	1.5	150	8.44	48.0	-
Arizona	7	16.4	0.4	2.3	300	0.7	43	7.50	42.1	1,000
Georgia	6	10.8	0.2	1.8	200	0.5	33	8.00	46.3	500
Minnesota	8	(D)	-	-	750 *	1.8	94	-	-	-
Kansas	6	(D)	-	-	375 *	0.9	63	-	-	1,067
Rhode Island	5	(D)	-	-	1,750 *	4.3	350	-	-	-
Maryland	4	(D)	-	-	375 *	0.9	94	-	-	-
Virginia	4	(D)	-	-	175 *	0.4	44	-	-	-
Arkansas	3	(D)	-	-	175 *	0.4	58	-	-	-
Iowa	3	(D)	-	-	175 *	0.4	58	-	-	-
Maine	3	(D)	-	-	750 *	1.8	250	-	-	-
Mississippi	2	(D)	-	-	375 *	0.9	188	-	-	-
New Hampshire	2	(D)	-	-	375 *	0.9	188	-	-	-
Washington	2	(D)	-	-	175 *	0.4	88	-	-	-
Nebraska	1	(D)	-	-	175 *	0.4	175	-	-	-
Oklahoma	1	(D)	-	-	175 *	0.4	175	-	-	-

Source: 1992 *Economic Census*. The states are in descending order of shipments or establishments (if shipment data are missing for the majority). The symbol (D) appears when data are withheld to prevent disclosure of competitive information. States marked with (D) are sorted by number of establishments. A dash (-) indicates that the data element cannot be calculated; * indicates the midpoint of a range.

3644 - NONCURRENT-CARRYING WIRING DEVICES

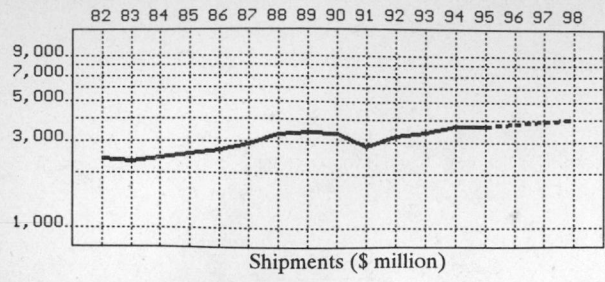

Shipments ($ million)

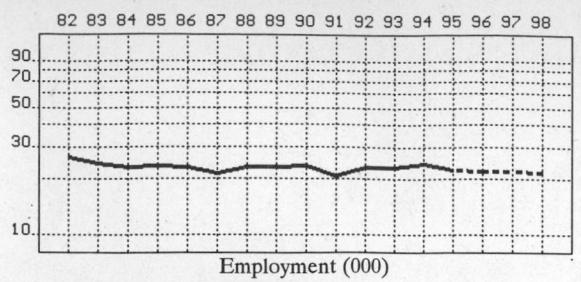

Employment (000)

GENERAL STATISTICS

Year	Companies	Establishments Total	Establishments with 20 or more employees	Employment Total (000)	Employment Production Workers (000)	Employment Hours (Mil)	Compensation Payroll ($ mil)	Compensation Wages ($/hr)	Production ($ million) Cost of Materials	Production ($ million) Value Added by Manufacture	Production ($ million) Value of Shipments	Production ($ million) Capital Invest.
1982	187	226	140	26.3	18.1	35.3	509.9	8.90	1,130.5	1,218.7	2,399.6	69.1
1983		220	137	24.5	17.1	33.9	511.2	9.33	1,132.0	1,194.3	2,309.4	47.5
1984		214	134	23.3	16.5	32.8	515.7	9.87	1,171.3	1,289.0	2,426.2	59.1
1985		207	132	23.6	16.7	33.6	541.2	10.06	1,139.8	1,390.0	2,541.5	79.1
1986		199	131	23.4	16.2	31.6	557.7	10.76	1,177.1	1,483.1	2,660.3	62.8
1987	177	209	124	21.5	15.3	30.7	550.3	11.62	1,340.8	1,588.9	2,903.2	73.8
1988		209	134	23.1	16.1	31.9	607.4	11.93	1,550.5	1,782.4	3,283.8	75.6
1989		207	133	23.0	16.2	33.9	618.7	11.30	1,606.4	1,784.3	3,392.7	83.4
1990		201	131	23.6	15.8	31.8	618.6	11.44	1,418.6	1,919.5	3,346.1	145.4
1991		204	126	20.6	14.6	29.5	565.8	11.69	1,162.3	1,629.9	2,824.6	81.8
1992	177	220	134	23.2	17.1	35.4	643.0	11.52	1,344.8	1,871.7	3,223.7	80.7
1993		209	130	23.4	17.3	35.8	665.2	11.91	1,471.6	1,921.4	3,443.2	85.9
1994		204P	128P	24.4	18.4	37.3	697.2	12.11	1,558.2	2,200.8	3,732.2	93.6
1995		204P	127P	22.4P	16.5P	34.2P	687.3P	12.68P	1,550.2P	2,189.5P	3,713.0P	103.2P
1996		203P	127P	22.3P	16.5P	34.4P	702.0P	12.92P	1,595.1P	2,252.9P	3,820.5P	106.6P
1997		202P	126P	22.2P	16.5P	34.5P	716.6P	13.17P	1,639.9P	2,316.3P	3,928.0P	109.9P
1998		201P	125P	22.0P	16.5P	34.6P	731.3P	13.41P	1,684.8P	2,379.6P	4,035.5P	113.3P

Sources: 1982, 1987, 1992 *Economic Census; Annual Survey of Manufactures*, 83-86, 88-91, 93-94. Establishment counts for non-Census years are from *County Business Patterns*; establishment values for 83-84 are extrapolations. 'P's show projections by the editors. Industries reclassified in 87 will not have data for prior years.

INDICES OF CHANGE

Year	Companies	Establishments Total	Establishments with 20 or more employees	Employment Total (000)	Employment Production Workers (000)	Employment Hours (Mil)	Compensation Payroll ($ mil)	Compensation Wages ($/hr)	Production ($ million) Cost of Materials	Production ($ million) Value Added by Manufacture	Production ($ million) Value of Shipments	Production ($ million) Capital Invest.
1982	106	103	104	113	106	100	79	77	84	65	74	86
1983		100	102	106	100	96	80	81	84	64	72	59
1984		97	100	100	96	93	80	86	87	69	75	73
1985		94	99	102	98	95	84	87	85	74	79	98
1986		90	98	101	95	89	87	93	88	79	83	78
1987	100	95	93	93	89	87	86	101	100	85	90	91
1988		95	100	100	94	90	94	104	115	95	102	94
1989		94	99	99	95	96	96	98	119	95	105	103
1990		91	98	102	92	90	96	99	105	103	104	180
1991		93	94	89	85	83	88	101	86	87	88	101
1992	100	100	100	100	100	100	100	100	100	100	100	100
1993		95	97	101	101	101	103	103	109	103	107	106
1994		93P	96P	105	108	105	108	105	116	118	116	116
1995		93P	95P	97P	97P	97P	107P	110P	115P	117P	115P	128P
1996		92P	95P	96P	97P	97P	109P	112P	119P	120P	119P	132P
1997		92P	94P	96P	97P	97P	111P	114P	122P	124P	122P	136P
1998		91P	94P	95P	97P	98P	114P	116P	125P	127P	125P	140P

Sources: Same as General Statistics. Values reflect change from the base year, 1992. Values above 100 mean greater than 92, values below 100 mean less than 92, and a value of 100 in the 82-91 or 93-98 period means same as 92. 'P's mark projections by the editors.

SELECTED RATIOS

For 1994	Avg. of All Manufact.	Analyzed Industry	Index	For 1994	Avg. of All Manufact.	Analyzed Industry	Index
Employees per Establishment	49	119	244	Value Added per Production Worker	134,084	119,609	89
Payroll per Establishment	1,500,273	3,410,302	227	Cost per Establishment	5,045,178	7,621,819	151
Payroll per Employee	30,620	28,574	93	Cost per Employee	102,970	63,861	62
Production Workers per Establishment	34	90	262	Cost per Production Worker	146,988	84,685	58
Wages per Establishment	853,319	2,209,471	259	Shipments per Establishment	9,576,895	18,255,777	191
Wages per Production Worker	24,861	24,549	99	Shipments per Employee	195,460	152,959	78
Hours per Production Worker	2,056	2,027	99	Shipments per Production Worker	279,017	202,837	73
Wages per Hour	12.09	12.11	100	Investment per Establishment	321,011	457,837	143
Value Added per Establishment	4,602,255	10,765,049	234	Investment per Employee	6,552	3,836	59
Value Added per Employee	93,930	90,197	96	Investment per Production Worker	9,352	5,087	54

Sources: Same as General Statistics. The 'Average of All Manufacturing' column represents the average of all manufacturing industries reported for the most recent complete year available. The Index shows the relationship between the Average and the Analyzed Industry. For example, 100 means that they are equal; 500 that the Analyzed Industry is five times the average; 50 means that the Analyzed Industry is half the national average. The abbreviation 'na' is used to show that data are 'not available'.

LEADING COMPANIES   Number shown: **40**   Total sales ($ mil): **1,568**   Total employment (000): **9.7**

Company Name	Address				CEO Name	Phone	Co. Type	Sales ($ mil)	Empl. (000)
Lamson and Sessions Co	25701 Science Pk Dr	Beachwood	OH	44122	John B Schulze	216-464-3400	P	301	1.9
Allied Tube and Conduit Co	16100 S Lathrop	Harvey	IL	60426	Robert P Mead	708-339-1610	D	270	0.9
Hoffman Engineering Co	900 Ehlen Dr	Anoka	MN	55303	Richard Ingman	612-421-2240	S	240	1.7
Appleton Electric Co	1701 W Wellington	Chicago	IL	60657	Gary Karnes	312-327-7200	S	200•	1.4
Western Tube and Conduit Co	PO Box 2720	Long Beach	CA	90801	T Nakajima	310-537-6300	S	100	0.2
Superior Manufacturing	PO Box 936	Georgetown	SC	29442	Tony Johnson	803-546-2516	D	65•	0.5
Bowers Manufacturing Corp	8685 Bowers Av	South Gate	CA	90280	Warren Hanelsman	213-566-2111	D	40	0.3
EM Wiegmann and Company	501 W Apple St	Freeburg	IL	62243	Wayne Plaisance	618-539-3193	R	40	0.4
Bridgeport Fittings Inc	705 Lordship Blv	Stratford	CT	06497	Delbert L Auray	203-377-5944	R	29	0.3
Chase Corp	50 Braintree Hill Pk	Braintree	MA	02184	Peter R Chase	617-848-2810	P	29	0.1
Neer Manufacturing Company	PO Box 3089	Lexington	OH	44904	Robert Lape	419-884-2274	R	27•	0.1
M Stephens Manufacturing Inc	8420 S Atlantic Av	Cudahy	CA	90201	Sam W Friedman	213-560-8301	R	24	0.3
McDATA Corp	310 Interlocken	Broomfield	CO	80021	Jack McDonnell	303-460-9200	R	20•	0.1
Westinghouse Electric Corp	PO Box 657	Bedford	PA	15522	Richard Carmody	814-623-9014	D	19•	<0.1
Electroline Manufacturing Co	18681 S Miles Indstl	Cleveland	OH	44128	Daniel Morgenstern	216-475-9044	R	18	<0.1
Robroy Industries-Texas Inc	PO Box 1828	Gilmer	TX	75644	Lewis Hardt	903-843-5591	S	17•	0.2
TJ Cope Inc	PO Box 991	Collegeville	PA	19426	CE Humphreys	215-489-4200	R	12	0.1
Electri-Flex Co	222 W Central Av	Roselle	IL	60172	EJ Marinelli	708-529-2920	R	10	0.1
Myers Electric Products	1130 S Vail Av	Montebello	CA	90640	Maury Markowitz	213-724-0450	R	10	<0.1
International Metal Hose Co	520 Goodrich Rd	Bellevue	OH	44811	Michael D Allen	419-483-7690	R	8•	<0.1
McKinstry Inc	285 McKinstry Av	Chicopee	MA	01013	Beatrice Levine	413-592-7716	R	8•	<0.1
Custom Materials Inc	16865 Pk Circle Dr	Chagrin Falls	OH	44023	G C Robinson	216-543-8284	R	8	0.1
US Samica Corp	PO Box 848	Rutland	VT	05071	Frank Burnham	802-775-5528	S	8	<0.1
Icore International Inc	180 N Wolfe Rd	Sunnyvale	CA	94086	Keith J Paterson	408-732-5400	S	7•	<0.1
Sherman and Reilly Inc	400 W 33rd St	Chattanooga	TN	37410	James Reilly	615-756-5300	R	7	<0.1
Minerallac Electric Co	466 Vista Av	Addison	IL	60101	John W Alton	708-543-7080	R	6	<0.1
Universal Manufacturing Co	PO Box 381220	Duncanville	TX	75138	D Vassallo	214-298-0531	R	6•	<0.1
Kortick Manufacturing Co	5600 3rd St	San Francisco	CA	94124	CM Hutchison	415-822-6660	R	5	<0.1
County Insulation Co	461 Churchmans	New Castle	DE	19720	James W Betley	302-322-8946	R	5	<0.1
Bo-Witt Products Inc	500 N Walnut	Edinburgh	IN	46124	William Bobbs	812-526-5561	R	4	<0.1
IBC Corp	27 Belmont St	South Easton	MA	02375	A Stikeleather	508-238-7941	R	4•	<0.1
Utilities Service Co	PO Box 627	Allentown	PA	18105	RJ Guidera	215-434-9541	R	4•	<0.1
Delta Metal Products Company	476 Flushing Av	Brooklyn	NY	11205	Mark Beer	718-855-4200	R	3•	<0.1
Hilec Inc	PO Box 157	Arcade	NY	14009	Kevin J Maguire	716-492-2212	S	3	<0.1
Metalform Company Inc	555 J Downey Dr	New Britain	CT	06051	Derek A Dibble	203-224-2630	R	3	<0.1
Queen Products Company Inc	1234 Rowan St	Louisville	KY	40203	Joan L Junghaene	502-585-5071	R	3	<0.1
KNS Metals	5301 Tacony St	Philadelphia	PA	19137	William Langendorf	215-744-5930	R	2•	<0.1
Lyn-Tron Inc	6001 S T Mallen	Spokane	WA	99204	Donald E Lynn	509-456-4545	R	2•	<0.1
Industrial Dielectrics Inc	PO Box 1099	San Bernardino	CA	92402	Wayne Nicely	909-381-4734	D	2	<0.1
Airflex Industries Inc	2422 Merced Av	S El Monte	CA	91733	M Joobits	818-579-7060	R	1•	<0.1

Source: Ward's Business Directory of U.S. Private and Public Companies, Volumes 1 and 2, 1996. The company type code used is as follows: P - Public, R - Private, S - Subsidiary, D - Division, J - Joint Venture, A - Affiliate, G - Group. Sales are in millions of dollars, employees are in thousands. An asterisk (*) indicates an estimated sales volume. The symbol < stands for 'less than'. Company names and addresses are truncated, in some cases, to fit into the available space.

MATERIALS CONSUMED

Material	Quantity	Delivered Cost ($ million)
Materials, ingredients, containers, and supplies	(X)	1,218.7
Metal stampings	(X)	18.1
Metal bolts, nuts, screws, washers, rivets, and other screw machine products	(X)	38.0
All other fabricated metal products (except forgings)	(X)	8.4
Forgings	(X)	1.3
Iron and steel castings (rough and semifinished)	(X)	46.7
Aluminum and aluminum-base alloy castings (rough and semifinished)	(X)	48.2
Copper and copper-base alloy castings (rough and semifinished)	(X)	3.8
Other nonferrous castings (rough and semifinished)	(X)	9.2
Steel bars, bar shapes, and plates	(X)	52.9
Steel sheet and strip, including tin plate	(X)	225.1
All other steel shapes and forms	(X)	156.9
Copper and copper-base alloy rod, bar, and bar shapes	(X)	19.9
Aluminum and aluminum-base alloy shapes and forms	(X)	20.6
Other nonferrous shapes and forms	(X)	10.9
Plastics resins consumed in the form of granules, pellets, powders, liquids, etc.	(X)	186.7
Plastics products consumed in the form of sheets, rods, tubes, film, and other shapes	(X)	15.6
Current-carrying wiring devices	(X)	4.4
Precious metals (gold, platinum, etc.), all forms	(X)	15.1
Insulated wire and cable, except magnet wire	(X)	10.9
Semiconductors, including transistors, diodes, rectifiers, and integrated circuits for electronic circuitry	(X)	2.6
Paper and paperboard containers, including shipping sacks and other paper packaging supplies	(X)	21.8
All other materials and components, parts, containers, and supplies	(X)	252.1
Materials, ingredients, containers, and supplies, nsk	(X)	49.3

Source: 1992 Economic Census. Explanation of symbols used: (D): Withheld to avoid disclosure of competitive data; na: Not available; (S): Withheld because statistical norms were not met; (X): Not applicable; (Z): Less than half the unit shown; nec: Not elsewhere classified; nsk: Not specified by kind; - : zero; * : 10-19 percent estimated; ** : 20-29 percent estimated.

PRODUCT SHARE DETAILS

Product or Product Class	% Share	Product or Product Class	% Share
Noncurrent-carrying wiring devices	100.00	fittings	48.09
Noncurrent-carrying wiring devices, pole and transmission line hardware	18.17	Other noncurrent-carrying wiring devices and supplies (boxes, covers, bar hangers, etc.)	32.29
Noncurrent-carrying wiring devices, electrical conduit and conduit fittings, including plastics conduit and conduit		Noncurrent-carrying wiring devices, nsk	1.45

Source: 1992 *Economic Census*. The values shown are percent of total shipments in an industry. Values of indented subcategories are summed in the main heading. The symbol (D) appears when data are withheld to prevent disclosure of competitive information. The abbreviation nsk stands for 'not specified by kind' and nec for 'not elsewhere classified'.

INPUTS AND OUTPUTS FOR WIRING DEVICES

Economic Sector or Industry Providing Inputs	%	Sector	Economic Sector or Industry Buying Outputs	%	Sector
Imports	21.5	Foreign	Electric utility facility construction	15.6	Constr.
Blast furnaces & steel mills	10.4	Manufg.	Exports	11.2	Foreign
Wholesale trade	9.5	Trade	Maintenance of nonfarm buildings nec	8.3	Constr.
Wiring devices	6.6	Manufg.	Residential 1-unit structures, nonfarm	5.9	Constr.
Plastics materials & resins	5.6	Manufg.	Residential additions/alterations, nonfarm	4.8	Constr.
Plating & polishing	3.1	Manufg.	Nonfarm residential structure maintenance	4.2	Constr.
Copper rolling & drawing	2.9	Manufg.	Industrial buildings	3.8	Constr.
Electric services (utilities)	2.5	Util.	Wiring devices	3.8	Manufg.
Screw machine and related products	2.3	Manufg.	Maintenance of electric utility facilities	3.6	Constr.
Banking	2.0	Fin/R.E.	Office buildings	2.7	Constr.
Primary nonferrous metals, nec	1.9	Manufg.	Telephone & telegraph facility construction	2.2	Constr.
Advertising	1.8	Services	Electronic computing equipment	2.0	Manufg.
Air transportation	1.4	Util.	Construction of hospitals	1.7	Constr.
Communications, except radio & TV	1.4	Util.	Refrigeration & heating equipment	1.4	Manufg.
Aluminum rolling & drawing	1.3	Manufg.	Residential garden apartments	1.3	Constr.
Miscellaneous plastics products	1.2	Manufg.	Lighting fixtures & equipment	1.2	Manufg.
Nonferrous wire drawing & insulating	1.2	Manufg.	Radio & TV communication equipment	1.1	Manufg.
Paperboard containers & boxes	1.1	Manufg.	Radio & TV receiving sets	1.1	Manufg.
Petroleum refining	1.1	Manufg.	Telephone & telegraph apparatus	1.1	Manufg.
Aluminum castings	1.0	Manufg.	Construction of stores & restaurants	1.0	Constr.
Cyclic crudes and organics	1.0	Manufg.	Mobile homes	1.0	Manufg.
Eating & drinking places	1.0	Trade	Warehouses	0.8	Constr.
Maintenance of nonfarm buildings nec	0.9	Constr.	Household cooking equipment	0.8	Manufg.
Motor freight transportation & warehousing	0.9	Util.	Construction of educational buildings	0.7	Constr.
Hotels & lodging places	0.9	Services	Farm service facilities	0.7	Constr.
Metal stampings, nec	0.8	Manufg.	Maintenance of railroads	0.7	Constr.
Fabricated rubber products, nec	0.7	Manufg.	Electric housewares & fans	0.7	Manufg.
Gas production & distribution (utilities)	0.7	Util.	Electronic components nec	0.7	Manufg.
Equipment rental & leasing services	0.7	Services	Household refrigerators & freezers	0.7	Manufg.
Iron & steel foundries	0.6	Manufg.	Maintenance of telephone & telegraph facilities	0.6	Constr.
Machinery, except electrical, nec	0.6	Manufg.	Engine electrical equipment	0.6	Manufg.
Metal coating & allied services	0.6	Manufg.	Construction of nonfarm buildings nec	0.5	Constr.
Miscellaneous fabricated wire products	0.6	Manufg.	Highway & street construction	0.5	Constr.
Nonferrous castings, nec	0.6	Manufg.	Hotels & motels	0.5	Constr.
Semiconductors & related devices	0.6	Manufg.	Maintenance of farm service facilities	0.5	Constr.
Computer & data processing services	0.6	Services	Sewer system facility construction	0.5	Constr.
Real estate	0.5	Fin/R.E.	Electrical equipment & supplies, nec	0.5	Manufg.
Legal services	0.5	Services	Household laundry equipment	0.5	Manufg.
Glass & glass products, except containers	0.4	Manufg.	Mechanical measuring devices	0.5	Manufg.
Primary zinc	0.4	Manufg.	Amusement & recreation building construction	0.4	Constr.
Special dies & tools & machine tool accessories	0.4	Manufg.	Local transit facility construction	0.4	Constr.
Railroads & related services	0.4	Util.	Maintenance of local transit facilities	0.4	Constr.
Management & consulting services & labs	0.4	Services	Resid. & other health facility construction	0.4	Constr.
U.S. Postal Service	0.4	Gov't	Residential 2-4 unit structures, nonfarm	0.4	Constr.
Industrial patterns	0.3	Manufg.	Instruments to measure electricity	0.4	Manufg.
Primary aluminum	0.3	Manufg.	Transformers	0.4	Manufg.
Accounting, auditing & bookkeeping	0.3	Services	Crude petroleum & natural gas	0.3	Mining
Abrasive products	0.2	Manufg.	Maintenance of highways & streets	0.3	Constr.
Brass, bronze, & copper castings	0.2	Manufg.	Maintenance of military facilities	0.3	Constr.
Chemical preparations, nec	0.2	Manufg.	Residential high-rise apartments	0.3	Constr.
Manifold business forms	0.2	Manufg.	Water supply facility construction	0.3	Constr.
Nonferrous rolling & drawing, nec	0.2	Manufg.	Household appliances, nec	0.3	Manufg.
Transformers	0.2	Manufg.	Industrial controls	0.3	Manufg.
Royalties	0.2	Fin/R.E.	Service industry machines, nec	0.3	Manufg.
Automotive rental & leasing, without drivers	0.2	Services	Switchgear & switchboard apparatus	0.3	Manufg.
Noncomparable imports	0.2	Foreign	Typewriters & office machines, nec	0.3	Manufg.
Hand & edge tools, nec	0.1	Manufg.	Communications, except radio & TV	0.3	Util.
Industrial inorganic chemicals, nec	0.1	Manufg.	Construction of religious buildings	0.2	Constr.
Lubricating oils & greases	0.1	Manufg.	Maintenance of farm residential buildings	0.2	Constr.
Machine tools, metal cutting types	0.1	Manufg.	Maintenance of nonbuilding facilities nec	0.2	Constr.
Machine tools, metal forming types	0.1	Manufg.	Maintenance of sewer facilities	0.2	Constr.
Primary copper	0.1	Manufg.	Nonbuilding facilities nec	0.2	Constr.
Insurance carriers	0.1	Fin/R.E.	Calculating & accounting machines	0.2	Manufg.
Automotive repair shops & services	0.1	Services	Travel trailers & campers	0.2	Manufg.
Electrical repair shops	0.1	Services	Federal Government purchases, national defense	0.2	Fed Govt

Continued on next page.

INPUTS AND OUTPUTS FOR WIRING DEVICES - Continued

Economic Sector or Industry Providing Inputs	%	Sector	Economic Sector or Industry Buying Outputs	%	Sector
Photofinishing labs, commercial photography	0.1	Services	Gross private fixed investment	0.2	Cap Inv
			Dormitories & other group housing	0.1	Constr.
			Farm housing units & additions & alterations	0.1	Constr.
			Maintenance of water supply facilities	0.1	Constr.
			Railroad construction	0.1	Constr.
			Environmental controls	0.1	Manufg.
			Lawn & garden equipment	0.1	Manufg.
			Signs & advertising displays	0.1	Manufg.
			Vitreous plumbing fixtures	0.1	Manufg.

Source: Benchmark Input-Output Accounts for the U.S. Economy, 1982, U.S. Department of Commerce, Washington, D.C., July 1991. Data, as reported in the source, are organized by the 1977 SIC structure in use in 1982 but have been matched, as closely as is possible, to the 1987 SIC structure used in this book.

OCCUPATIONS EMPLOYED BY SIC 364 - ELECTRIC LIGHTING AND WIRING EQUIPMENT

Occupation	% of Total 1994	Change to 2005	Occupation	% of Total 1994	Change to 2005
Assemblers, fabricators, & hand workers nec	16.0	-7.4	Tool & die makers	1.5	-25.2
Electrical & electronic assemblers	7.6	-16.7	Hand packers & packagers	1.5	-20.6
Blue collar worker supervisors	3.7	-20.3	Sheet metal workers & duct installers	1.5	-7.4
Inspectors, testers, & graders, precision	3.2	-35.2	Machinists	1.5	-30.5
Plastic molding machine workers	3.2	11.1	Freight, stock, & material movers, hand	1.4	-25.9
Machine operators nec	2.5	-34.7	Industrial truck & tractor operators	1.3	-7.4
Industrial machinery mechanics	2.5	25.0	Machine tool cutting & forming etc. nec	1.2	-25.9
Sales & related workers nec	2.1	-7.4	Production, planning, & expediting clerks	1.2	-7.4
Electrical & electronic equipment assemblers	1.9	-7.4	Secretaries, ex legal & medical	1.2	-15.7
General managers & top executives	1.9	-12.2	Maintenance repairers, general utility	1.1	-16.6
Machine feeders & offbearers	1.8	-16.7	Industrial production managers	1.0	-7.4
Traffic, shipping, & receiving clerks	1.7	-10.9	Coating, painting, & spraying machine workers	1.0	-7.4
Helpers, laborers, & material movers nec	1.6	-7.4			

Source: Industry-Occupation Matrix, Bureau of Labor Statistics. These data relate to one or more 3-digit SIC industry groups rather than to a single 4-digit SIC. The change reported for each occupation to the year 2005 is a percent of growth or decline as estimated by the Bureau of Labor Statistics. The abbreviation nec stands for 'not elsewhere classified'.

LOCATION BY STATE AND REGIONAL CONCENTRATION

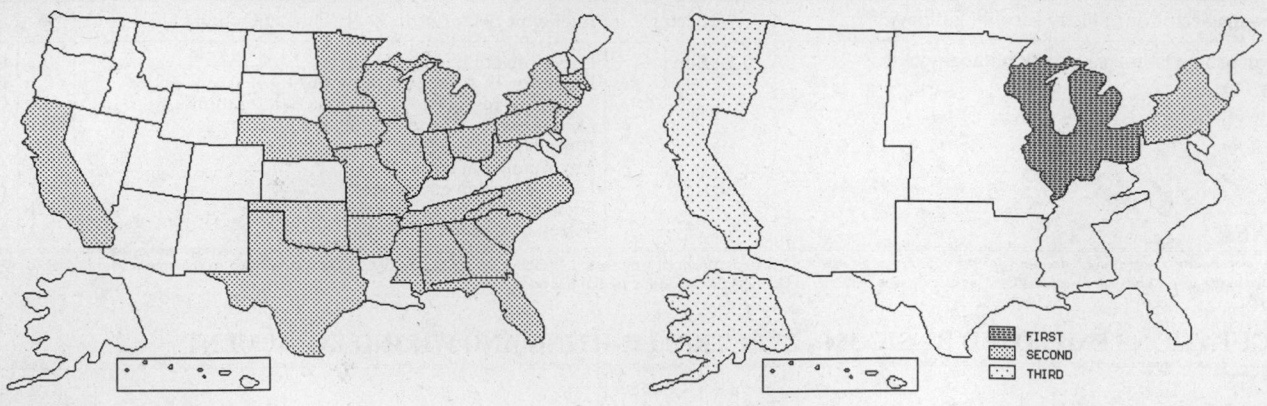

FIRST
SECOND
THIRD

INDUSTRY DATA BY STATE

State	Establish-ments	Shipments			Employment				Cost as % of Shipments	Investment per Employee ($)
		Total ($ mil)	% of U.S.	Per Establ.	Total Number	% of U.S.	Per Establ.	Wages ($/hour)		
Illinois	20	311.4	9.7	15.6	2,200	9.5	110	11.25	38.4	4,000
California	33	278.6	8.6	8.4	1,600	6.9	48	11.30	55.2	3,625
Pennsylvania	18	275.7	8.6	15.3	1,700	7.3	94	11.89	53.6	5,824
Connecticut	9	270.8	8.4	30.1	1,700	7.3	189	12.76	20.8	3,000
Ohio	17	151.4	4.7	8.9	1,200	5.2	71	12.35	49.8	5,333
Tennessee	4	130.4	4.0	32.6	800	3.4	200	9.36	37.3	-
Texas	15	118.2	3.7	7.9	700	3.0	47	9.75	58.7	3,571
New Jersey	6	90.1	2.8	15.0	600	2.6	100	14.62	32.4	1,500
North Carolina	5	76.1	2.4	15.2	1,600	6.9	320	9.21	32.6	-
Alabama	9	68.0	2.1	7.6	500	2.2	56	9.43	53.5	5,600
Georgia	6	63.9	2.0	10.6	200	0.9	33	9.00	60.1	-
Florida	9	54.5	1.7	6.1	500	2.2	56	9.71	55.8	3,200
South Carolina	4	38.9	1.2	9.7	500	2.2	125	7.00	46.8	-
New York	13	(D)	-	-	1,750 *	7.5	135	-	-	-
Missouri	8	(D)	-	-	1,750 *	7.5	219	-	-	3,714
Iowa	4	(D)	-	-	375 *	1.6	94	-	-	-
Massachusetts	4	(D)	-	-	375 *	1.6	94	-	-	-
Minnesota	3	(D)	-	-	1,750 *	7.5	583	-	-	-
Oklahoma	3	(D)	-	-	175 *	0.8	58	-	-	-
Wisconsin	3	(D)	-	-	375 *	1.6	125	-	-	1,867
Arkansas	2	(D)	-	-	375 *	1.6	188	-	-	-
Indiana	2	(D)	-	-	750 *	3.2	375	-	-	-
Michigan	2	(D)	-	-	175 *	0.8	88	-	-	-
Mississippi	1	(D)	-	-	375 *	1.6	375	-	-	-
Nebraska	1	(D)	-	-	750 *	3.2	750	-	-	-
West Virginia	1	(D)	-	-	375 *	1.6	375	-	-	-

Source: 1992 *Economic Census*. The states are in descending order of shipments or establishments (if shipment data are missing for the majority). The symbol (D) appears when data are withheld to prevent disclosure of competitive information. States marked with (D) are sorted by number of establishments. A dash (-) indicates that the data element cannot be calculated; * indicates the midpoint of a range.

3645 - RESIDENTIAL LIGHTING FIXTURES

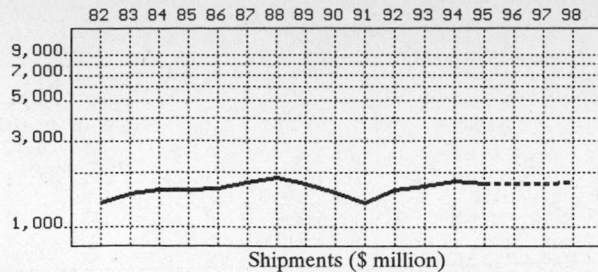

Shipments ($ million)

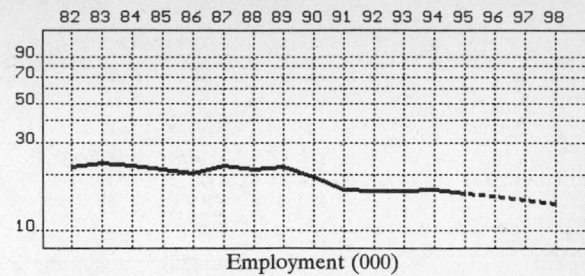

Employment (000)

GENERAL STATISTICS

Year	Com- panies	Establishments		Employment			Compensation		Production ($ million)			
		Total	with 20 or more employees	Total (000)	Production Workers (000)	Hours (Mil)	Payroll ($ mil)	Wages ($/hr)	Cost of Materials	Value Added by Manufacture	Value of Shipments	Capital Invest.
1982	610	643	250	22.2	16.9	31.5	308.8	6.18	634.5	708.7	1,351.8	17.9
1983		616	243	23.4	18.4	34.9	328.9	6.01	729.9	818.3	1,529.5	15.7
1984		589	236	22.6	17.9	34.2	344.5	6.45	783.6	841.7	1,606.7	28.8
1985		563	229	21.5	16.9	30.7	349.6	7.28	758.1	856.4	1,607.5	31.3
1986		545	235	20.5	15.8	30.2	353.7	7.30	748.3	888.9	1,628.5	20.2
1987	551	580	220	22.5	17.5	34.1	398.8	7.45	790.9	1,005.5	1,772.6	24.7
1988		549	218	21.6	16.4	32.0	392.5	7.51	822.2	1,032.0	1,862.9	26.4
1989		523	219	22.3	16.2	31.8	381.0	7.35	739.8	981.3	1,725.8	27.0
1990		522	198	19.4	13.8	26.8	344.3	7.76	721.4	826.0	1,561.3	24.3
1991		498	185	16.6	13.0	25.1	314.8	7.63	651.3	678.4	1,352.1	13.0
1992	511	526	176	16.5	12.6	23.9	329.3	7.85	757.7	857.9	1,610.9	21.4
1993		499	159	16.2	12.3	24.1	344.2	8.05	779.3	916.0	1,680.1	20.9
1994		479P	165P	16.6	13.1	25.8	351.3	8.10	862.2	959.0	1,792.8	20.0
1995		467P	157P	15.9P	11.9P	23.6P	354.4P	8.42P	825.6P	918.3P	1,716.7P	21.1P
1996		455P	150P	15.3P	11.4P	22.8P	355.1P	8.58P	832.1P	925.5P	1,730.2P	20.9P
1997		444P	142P	14.7P	10.9P	21.9P	355.8P	8.75P	838.6P	932.8P	1,743.8P	20.7P
1998		432P	134P	14.0P	10.4P	21.1P	356.5P	8.91P	845.2P	940.0P	1,757.4P	20.5P

Sources: 1982, 1987, 1992 *Economic Census*; *Annual Survey of Manufactures*, 83-86, 88-91, 93-94. Establishment counts for non-Census years are from *County Business Patterns*; establishment values for 83-84 are extrapolations. 'P's show projections by the editors. Industries reclassified in 87 will not have data for prior years.

INDICES OF CHANGE

Year	Com- panies	Establishments		Employment			Compensation		Production ($ million)			
		Total	with 20 or more employees	Total (000)	Production Workers (000)	Hours (Mil)	Payroll ($ mil)	Wages ($/hr)	Cost of Materials	Value Added by Manufacture	Value of Shipments	Capital Invest.
1982	119	122	142	135	134	132	94	79	84	83	84	84
1983		117	138	142	146	146	100	77	96	95	95	73
1984		112	134	137	142	143	105	82	103	98	100	135
1985		107	130	130	134	128	106	93	100	100	100	146
1986		104	134	124	125	126	107	93	99	104	101	94
1987	108	110	125	136	139	143	121	95	104	117	110	115
1988		104	124	131	130	134	119	96	109	120	116	123
1989		99	124	135	129	133	116	94	98	114	107	126
1990		99	113	118	110	112	105	99	95	96	97	114
1991		95	105	101	103	105	96	97	86	79	84	61
1992	100	100	100	100	100	100	100	100	100	100	100	100
1993		95	90	98	98	101	105	103	103	107	104	98
1994		91P	94P	101	104	108	107	103	114	112	111	93
1995		89P	89P	96P	95P	99P	108P	107P	109P	107P	107P	98P
1996		87P	85P	93P	91P	95P	108P	109P	110P	108P	107P	98P
1997		84P	81P	89P	87P	92P	108P	111P	111P	109P	108P	97P
1998		82P	76P	85P	83P	88P	108P	113P	112P	110P	109P	96P

Sources: Same as General Statistics. Values reflect change from the base year, 1992. Values above 100 mean greater than 92, values below 100 mean less than 92, and a value of 100 in the 82-91 or 93-98 period means same as 92. 'P's mark projections by the editors.

SELECTED RATIOS

For 1994	Avg. of All Manufact.	Analyzed Industry	Index	For 1994	Avg. of All Manufact.	Analyzed Industry	Index
Employees per Establishment	49	35	71	Value Added per Production Worker	134,084	73,206	55
Payroll per Establishment	1,500,273	733,914	49	Cost per Establishment	5,045,178	1,801,253	36
Payroll per Employee	30,620	21,163	69	Cost per Employee	102,970	51,940	50
Production Workers per Establishment	34	27	80	Cost per Production Worker	146,988	65,817	45
Wages per Establishment	853,319	436,588	51	Shipments per Establishment	9,576,895	3,745,404	39
Wages per Production Worker	24,861	15,953	64	Shipments per Employee	195,460	108,000	55
Hours per Production Worker	2,056	1,969	96	Shipments per Production Worker	279,017	136,855	49
Wages per Hour	12.09	8.10	67	Investment per Establishment	321,011	41,783	13
Value Added per Establishment	4,602,255	2,003,482	44	Investment per Employee	6,552	1,205	18
Value Added per Employee	93,930	57,771	62	Investment per Production Worker	9,352	1,527	16

Sources: Same as General Statistics. The 'Average of All Manufacturing' column represents the average of all manufacturing industries reported for the most recent complete year available. The Index shows the relationship between the Average and the Analyzed Industry. For example, 100 means that they are equal; 500 that the Analyzed Industry is five times the average; 50 means that the Analyzed Industry is half the national average. The abbreviation 'na' is used to show that data are 'not available'.

LEADING COMPANIES   Number shown: **75**   Total sales ($ mil): **1,965**   Total employment (000): **15.7**

Company Name	Address				CEO Name	Phone	Co. Type	Sales ($ mil)	Empl. (000)
Genlyte Group Inc	100 Lighting Way	Secaucus	NJ	07096	Larry K Powers	201-864-3000	P	425	3.0
Metalux Lighting Co	PO Box 1207	Americus	GA	31709	Dudley Gatewood	912-924-8000	S	272	1.3
Lightolier Inc	631 Airport Rd	Fall River	MA	02720	Zia Eftekar	508-679-8131	D	198	1.3
Catalina Lighting Inc	18191 NW 68th Av	Miami	FL	33015	Robert Hersh	305-558-4777	P	113	0.2
Progress Lighting	Box 5704	Spartanburg	SC	29304	Scott Muse	803-599-6000	S	100	1.1
Spartus Corp	24 Richmond Hill	Stamford	CT	06901	Kevin P O'Malley	203-359-1007	S	68	0.7
Sea Gull Lighting Products Inc	301 W Washington	Riverside	NJ	08075	Edwin N Solomon	609-764-0500	R	60	0.5
Harris Marcus Group	3757 S Ashland Av	Chicago	IL	60609	David Marcus	312-247-7500	R	58	0.6
American Lantern Co	4344 Hwy 67 N	Newport	AR	72112	Donald B Little	501-523-2705	R	45	0.5
LD Kichler Co	7711 E Pleasant	Cleveland	OH	44114	Barry Minoff	216-573-1000	R	43	0.4
Toro Co	8111 Lindale Av S	Bloomington	MN	55420	David McIntosh	612-888-8801	D	40•	0.4
Quoizel Inc	325 Kennedy Dr	Hauppauge	NY	11788	I Phillips	516-273-2700	R	35	0.2
SOI Industries Inc	1051 E 24th St	Hialeah	FL	33013	Donald E Courtney	305-835-2214	P	34	0.7
Coronet Manufacturing	1620 S Avalon Blv	Gardena	CA	90248	David Smith	310-327-6700	R	34	0.3
Harris Lamps Co	3757 S Ashland Av	Chicago	IL	60609	David Marcus	312-247-7500	D	30•	0.3
Rangaire Co	PO Box 177	Cleburne	TX	76033	Joe McKenzie	817-556-6500	R	30	0.3
Sterner Lighting Systems Inc	351 W Lewis Av	Winsted	MN	55395	Jack Desalliers	612-473-1251	S	30	0.3
Stiffel Co	700 N Kingsbury St	Chicago	IL	60610	Jim Cunningham	312-664-9200	R	28	0.4
Wood River Industries	PO Box 329	Riverside	NJ	08075	EN Solomon	609-764-0500	D	22	0.2
Specialty Lighting Inc	PO Box 1680	Shelby	NC	28151	HM Schellpfeffer	704-482-3416	S	19	0.2
Prestigeline Inc	5 Inez Dr	Brentwood	NY	11717	Scott Roth	516-273-3636	R	18	<0.1
Kensington Lamp Co	PO Box 165	Youngwood	PA	15697	Walter Tymoczko	412-925-1312	S	15•	<0.1
Millstein Industries	4th & Gaskill Av	Jeannette	PA	15644	Jack Millstein Jr	412-523-5531	R	15	<0.1
Duray Fluorescent Mfg Co	2050 W Balmoral	Chicago	IL	60625	Bernard R Meyer	312-271-2800	R	12	0.1
Uspar Enterprises Inc	13404 M Vista	Chino	CA	91710	HE Parekh	909-591-7506	R	12	<0.1
Natalie Lamp and Shade Corp	220 Straight St	Paterson	NJ	07509	George F Reisman	201-278-8800	R	11•	0.2
Swivelier Company Inc	PO Box 619	Nanuet	NY	10954	Michael Schwartz	914-623-3471	R	11•	0.1
Adjusta-Post Manufacturing Co	PO Box 71	Norton	OH	44203	Patrick Ruggles	216-745-1692	S	10	0.1
Mar-Kel Lighting Inc	PO Box 190	Paris	TN	38242	HL Kosser	901-642-7190	R	10•	0.2
Wildwood Lamps Co	PO Box 672	Rocky Mount	NC	27802	William Kincheloe	919-446-3266	R	10•	<0.1
George Kovacs Lighting Inc	67-25 Otto Rd	Glendale	NY	11385	George Kovacs	718-628-5201	R	9	0.2
Hinkley Lighting Inc	12600 Berea Rd	Cleveland	OH	44111	RA Wiedemer Jr	216-671-3300	R	9•	0.1
McEnroe Inc	105 N Bouchelle St	Morganton	NC	28655	Bill Kinchloe	704-437-3711	S	9•	<0.1
Edwards Lamp and Shade Co	PO Box 4564	P Verdes Est	CA	90274	BL Edwards	213-583-6474	R	8•	<0.1
Old Forge Lamp and Shade Inc	500 Hillcrest Dr	Old Forge	PA	18518	George Cohen	717-344-1235	R	8•	<0.1
Cambridge Lamps Inc	2605 W 8th Av	Hialeah	FL	33010	Stuart Schiller	305-885-3800	R	7	<0.1
Boyd Lighting Fixture Co	56 12th St	San Francisco	CA	94103	JS Sweet Jr	415-431-4300	R	6•	<0.1
Mario Industries of Virginia Inc	PO Box 3190	Roanoke	VA	24015	Louis M Scutellaro	703-342-1111	R	6	0.1
Tempo Lighting Inc	1051 E 24th St	Hialeah	FL	33013	Kevin B Halter	305-835-2214	S	5	<0.1
Adelphia Lamp and Shade Inc	5000 Paschall Av	Philadelphia	PA	19143	F Denenberg	215-729-2600	R	5	<0.1
Kenton Industries Inc	PO Box 878	Westminster	CA	92684	MP Nyssen	714-892-6681	R	5	<0.1
Nessen Lighting Inc	420 Railroad Way	Mamaroneck	NY	10543	Robert N Haidinger	914-698-7799	S	5	<0.1
Pieri Creations Inc	100 W Oxford St	Philadelphia	PA	19122	GJ Pieri	215-634-0700	R	5	<0.1
Rejuvenation Inc	1100 SE Grand Av	Portland	OR	97214	James O Kelly	503-231-1900	R	5	<0.1
Sempre Lighting Group Inc	14225 Telephone	Chino	CA	91710	Laura L Ditte	909-902-1296	R	5	<0.1
Texas Lamp Manufacturers Inc	3419 E Kiest Blv	Dallas	TX	75203	FJ Pinnell	214-943-4663	R	5	<0.1
Reisco Inc	220 Straight St	Paterson	NJ	07501	George Reisman	201-278-8800	D	4	0.1
Remington Lamp Co	5000 Paschall Av	Philadelphia	PA	19143	Alfred Bennenberg	215-729-2600	R	4	<0.1
Austin Innovations Inc	2600 McHale Ct	Austin	TX	78758	Joseph E Marischen	512-339-6765	R	4	<0.1
Bauer Lamp Co	PO Box 10385	Riviera Beach	FL	33419	D Eric Bauer	407-848-0828	R	3	<0.1
Dinico Products Inc	123 S Newman St	Hackensack	NJ	07601	R Dini	201-488-5700	R	3	<0.1
Flos Inc	200 Mckay Rd	Huntington St	NY	11743	Tom DeNapili	516-549-2745	R	3•	<0.1
Flute Inc	1500 S Western Av	Chicago	IL	60608	Gerald Garfinkle	312-738-0622	R	3	<0.1
Galaxy Electrical Manufacturing	5295 NW 163rd St	Hialeah	FL	33014	Samon Kao	305-625-0088	R	3•	<0.1
Lightcraft Corp	PO Box 259	Jeannette	PA	15644	Jack Millstein Jr	412-523-5531	S	3	<0.1
Reliance Lamp Company Inc	125 Laser Ct	Hauppauge	NY	11788	M Magun	516-434-1120	R	3•	<0.1
Signature Lighting Corp	1667 E 40th St	Cleveland	OH	44103	Tom Gruber	216-361-4354	R	3•	<0.1
Royal Haeger Lamp Co	PO Box 218	Macomb	IL	61455	Nicholas H Estes	309-837-9966	R	3	<0.1
Active Specialty Corp	106 Pierces Rd	Newburgh	NY	12550	David Litman	914-565-3635	R	2	<0.1
Lawrin Lighting Inc	PO Box 728	Kosciusko	MS	39090	Keith Thweatt	601-289-1711	R	2	<0.1
Olympia Lighting Inc	2215 NW 30th Pl	Pompano Bch	FL	33069	Paul Murch	305-979-1300	R	2	<0.1
Solar Outdoor Lighting Inc	3131 SE Waaler St	Stuart	FL	34997	Alan Hurst	407-286-9461	R	2	<0.1
KBD Inc	20 Kenton Lands	Erlanger	KY	41018	James Shepherd	606-331-0800	R	1	<0.1
Scientific NRG Inc	2651 Dow Av	Tustin	CA	92680	Malcolm L Fickel	714-730-3555	P	1	<0.1
Dansk Lights Inc	2000 N Dixie Hwy	Ft Lauderdale	FL	33305	Stark Holekamp	305-565-0003	R	1	<0.1
KB Lighting Manufacturing	2101 Byberry Rd	Philadelphia	PA	19116	Gary Kessel	215-673-6400	R	1•	<0.1
Lamptek Co	1229 Hyde Park Av	Boston	MA	02136	C Edwards	617-364-3600	R	1•	<0.1
Norman Perry Co	501 W Green Dr	High Point	NC	27260		919-841-5222	R	1	<0.1
Nautilus Lamp Company Inc	PO Box 588	Orange	CT	06477	Martin Feldman	203-795-1178	R	1•	<0.1
Janelle Products Inc	PO Box 198	Jackson	WI	53037	John Wundrock	414-774-4008	S	1	<0.1
Roctronics Lighting	100 WBD Roct	Pembroke	MA	02359	Richard Iacobucci	617-826-8888	R	0	<0.1
Victor Illuminating Inc	1115 W Sunset Blv	Rocklin	CA	95765	Rudy Mortensen	916-645-9663	R	0	<0.1
A Schonbek and Company Inc	61 Industrial Blv	Plattsburgh	NY	12901	Andrew J Schonbek	518-563-7500	S	0•	0.2
Dizzy Designs	69 Mill Rd	Jersey City	NJ	07302	Alan Slesinger	201-434-8519	R	0	<0.1
Jet-Lite Products Inc	618-13th St	Highland	IL	62249	John D Kutz Jr	618-654-6366	R	0•	<0.1

Source: Ward's Business Directory of U.S. Private and Public Companies, Volumes 1 and 2, 1996. The company type code used is as follows: P - Public, R - Private, S - Subsidiary, D - Division, J - Joint Venture, A - Affiliate, G - Group. Sales are in millions of dollars, employees are in thousands. An asterisk (•) indicates an estimated sales volume. The symbol < stands for 'less than'. Company names and addresses are truncated, in some cases, to fit into the available space.

MATERIALS CONSUMED

Material	Quantity	Delivered Cost ($ million)
Materials, ingredients, containers, and supplies	(X)	635.5
Specialty transformers and fluorescent ballasts	(X)	7.2
Current-carrying wiring devices	(X)	21.6
Electric lamp bulbs	(X)	11.2
Flat glass (plate, float, and sheet)	(X)	16.6
Plastics resins consumed in the form of granules, pellets, powders, liquids, etc.	(X)	3.6
Plastics products consumed in the form of sheets, rods, tubes, film, and other shapes	(X)	5.1
Fabricated plastics products (except gaskets, hoses, and belting)	(X)	4.7
Insulated wire and cable, including magnet wire	(X)	8.5
Paperboard containers, boxes, and corrugated paperboard	(X)	25.7
Metal bolts, nuts, screws, washers, rivets, and other screw machine products	(X)	12.4
Metal poles	(X)	2.8
All other fabricated metal products (except forgings)	(X)	28.4
Forgings	(X)	0.3
Iron and steel castings (rough and semifinished)	(X)	4.5
Aluminum and aluminum-base alloy castings (rough and semifinished)	(X)	6.7
Other nonferrous castings (rough and semifinished)	(X)	8.2
Steel sheet and strip, including tin plate	(X)	15.6
Steel wire and wire products	(X)	1.7
All other steel mill shapes and forms (except castings and forgings)	(X)	5.5
Aluminum and aluminum-base alloy sheet, plate, foil, and welded tubing	(X)	4.2
Copper and copper-base alloy shapes and forms	(X)	8.0
Other nonferrous shapes and forms	(X)	6.3
Lamp shades	(X)	22.1
All other materials and components, parts, containers, and supplies	(X)	258.6
Materials, ingredients, containers, and supplies, nsk	(X)	146.0

Source: 1992 *Economic Census*. Explanation of symbols used: (D): Withheld to avoid disclosure of competitive data; na: Not available; (S): Withheld because statistical norms were not met; (X): Not applicable; (Z): Less than half the unit shown; nec: Not elsewhere classified; nsk: Not specified by kind; - : zero; * : 10-19 percent estimated; ** : 20-29 percent estimated.

PRODUCT SHARE DETAILS

Product or Product Class	% Share	Product or Product Class	% Share
Residential lighting fixtures	100.00	Other portable residential lighting fixtures (including parts and accessories), (including boudoir and desk), incandescent complete with shade	11.05
Residential type electric lighting fixtures (except portable, including parts and accessories)	50.61		
Portable residential lighting fixtures (including parts and accessories for portable residential lighting fixtures)	36.20	Portable residential lighting fixtures (including parts and accessories) lamps sold without shades; including floor, table, etc., complete incandescent	3.30
Portable residential lighting fixtures (including parts and accessories), Floor lamps, incandescent complete with shade	16.65	Portable residential lighting fixtures (including parts and accessories) lamps (desks, etc.), fluorescent	3.24
Portable residential lighting fixtures (including parts and accessories), wall lamps (including adjustable types), incandescent complete with shade	4.76	Parts and accessories for portable residential lighting fixtures	5.38
Portable residential lighting fixtures (including parts and accessories), table lamps (excluding desk), incandescent complete with shade	47.03	Portable residential lighting fixtures (incl. parts and accessories for portable residential lighting fixtures), nsk	8.61
		Residential lighting fixtures, nsk	13.19

Source: 1992 *Economic Census*. The values shown are percent of total shipments in an industry. Values of indented subcategories are summed in the main heading. The symbol (D) appears when data are withheld to prevent disclosure of competitive information. The abbreviation nsk stands for 'not specified by kind' and nec for 'not elsewhere classified'.

INPUTS AND OUTPUTS FOR LIGHTING FIXTURES & EQUIPMENT

Economic Sector or Industry Providing Inputs	%	Sector	Economic Sector or Industry Buying Outputs	%	Sector
Wholesale trade	11.2	Trade	Personal consumption expenditures	16.4	
Transformers	8.4	Manufg.	Office buildings	9.3	Constr.
Glass & glass products, except containers	8.0	Manufg.	Motor vehicles & car bodies	5.6	Manufg.
Metal stampings, nec	7.5	Manufg.	Residential 1-unit structures, nonfarm	5.0	Constr.
Imports	7.1	Foreign	Maintenance of nonfarm buildings nec	4.8	Constr.
Blast furnaces & steel mills	6.4	Manufg.	Gross private fixed investment	4.6	Cap Inv
Aluminum rolling & drawing	4.0	Manufg.	Electric utility facility construction	4.3	Constr.
Miscellaneous plastics products	2.5	Manufg.	Industrial buildings	4.3	Constr.
Paperboard containers & boxes	2.5	Manufg.	Exports	4.2	Foreign
Lighting fixtures & equipment	2.4	Manufg.	Nonfarm residential structure maintenance	3.2	Constr.
Plastics materials & resins	2.4	Manufg.	Residential additions/alterations, nonfarm	2.9	Constr.
Advertising	2.0	Services	Automotive repair shops & services	2.8	Services
Wiring devices	1.9	Manufg.	Construction of stores & restaurants	2.2	Constr.
Banking	1.9	Fin/R.E.	Construction of hospitals	1.9	Constr.
Screw machine and related products	1.7	Manufg.	Highway & street construction	1.9	Constr.
Electric services (utilities)	1.6	Util.	Construction of educational buildings	1.7	Constr.
Air transportation	1.5	Util.	Lighting fixtures & equipment	1.5	Manufg.
Aluminum castings	1.4	Manufg.	Eating & drinking places	1.5	Trade

Continued on next page.

INPUTS AND OUTPUTS FOR LIGHTING FIXTURES & EQUIPMENT - Continued

Economic Sector or Industry Providing Inputs	%	Sector	Economic Sector or Industry Buying Outputs	%	Sector
Electric lamps	1.3	Manufg.	Hospitals	1.2	Services
Nonferrous wire drawing & insulating	1.2	Manufg.	Miscellaneous plastics products	1.1	Manufg.
Paints & allied products	1.2	Manufg.	Automobile parking & car washes	1.0	Services
Motor freight transportation & warehousing	1.2	Util.	Maintenance of highways & streets	0.9	Constr.
Plating & polishing	1.0	Manufg.	Warehouses	0.9	Constr.
Cyclic crudes and organics	0.9	Manufg.	Hotels & motels	0.7	Constr.
Communications, except radio & TV	0.9	Util.	Residential garden apartments	0.7	Constr.
Hotels & lodging places	0.9	Services	S/L Govt. purch., elem. & secondary education	0.7	S/L Govt
Manufacturing industries, nec	0.8	Manufg.	Coal	0.5	Mining
Semiconductors & related devices	0.8	Manufg.	Resid. & other health facility construction	0.5	Constr.
Electrical equipment & supplies, nec	0.7	Manufg.	Colleges, universities, & professional schools	0.5	Services
Petroleum refining	0.7	Manufg.	Amusement & recreation building construction	0.4	Constr.
Real estate	0.7	Fin/R.E.	Construction of nonfarm buildings nec	0.4	Constr.
Gas production & distribution (utilities)	0.6	Util.	Maintenance of farm service facilities	0.4	Constr.
Eating & drinking places	0.6	Trade	Telephone & telegraph facility construction	0.4	Constr.
Equipment rental & leasing services	0.6	Services	Mobile homes	0.4	Manufg.
Maintenance of nonfarm buildings nec	0.5	Constr.	Electric services (utilities)	0.4	Util.
Copper rolling & drawing	0.5	Manufg.	Elementary & secondary schools	0.4	Services
Iron & steel foundries	0.5	Manufg.	Crude petroleum & natural gas	0.3	Mining
Nonferrous castings, nec	0.5	Manufg.	Construction of religious buildings	0.3	Constr.
Primary zinc	0.5	Manufg.	Maintenance of electric utility facilities	0.3	Constr.
Wood products, nec	0.5	Manufg.	Maintenance of nonbuilding facilities nec	0.3	Constr.
Engineering, architectural, & surveying services	0.5	Services	Residential 2-4 unit structures, nonfarm	0.3	Constr.
Broadwoven fabric mills	0.4	Manufg.	Photographic equipment & supplies	0.3	Manufg.
Electronic components nec	0.4	Manufg.	Porcelain electrical supplies	0.3	Manufg.
Metal coating & allied services	0.4	Manufg.	Communications, except radio & TV	0.3	Util.
Miscellaneous fabricated wire products	0.4	Manufg.	Nursing & personal care facilities	0.3	Services
Gaskets, packing & sealing devices	0.3	Manufg.	Photofinishing labs, commercial photography	0.3	Services
Machinery, except electrical, nec	0.3	Manufg.	Theatrical producers, bands, entertainers	0.3	Services
Pottery products, nec	0.3	Manufg.	S/L Govt. purch., higher education	0.3	S/L Govt
Railroads & related services	0.3	Util.	S/L Govt. purch., other general government	0.3	S/L Govt
Computer & data processing services	0.3	Services	Nonbuilding facilities nec	0.2	Constr.
Legal services	0.3	Services	Residential high-rise apartments	0.2	Constr.
U.S. Postal Service	0.3	Gov't	Logging camps & logging contractors	0.2	Manufg.
Brass, bronze, & copper castings	0.2	Manufg.	Motor vehicle parts & accessories	0.2	Manufg.
Fabricated rubber products, nec	0.2	Manufg.	Petroleum refining	0.2	Manufg.
Hardwood dimension & flooring mills	0.2	Manufg.	Travel trailers & campers	0.2	Manufg.
Primary aluminum	0.2	Manufg.	Railroads & related services	0.2	Util.
Sawmills & planning mills, general	0.2	Manufg.	Water supply & sewage systems	0.2	Util.
Special dies & tools & machine tool accessories	0.2	Manufg.	Real estate	0.2	Fin/R.E.
Royalties	0.2	Fin/R.E.	Libraries, vocation education	0.2	Services
Accounting, auditing & bookkeeping	0.2	Services	S/L Govt. purch., health & hospitals	0.2	S/L Govt
Management & consulting services & labs	0.2	Services	S/L Govt. purch., natural resource & recreation.	0.2	S/L Govt
Machine tools, metal forming types	0.1	Manufg.	Farm service facilities	0.1	Constr.
Manifold business forms	0.1	Manufg.	Maintenance of military facilities	0.1	Constr.
Insurance carriers	0.1	Fin/R.E.	Sewer system facility construction	0.1	Constr.
Automotive rental & leasing, without drivers	0.1	Services	Automotive stampings	0.1	Manufg.
Automotive repair shops & services	0.1	Services	Blast furnaces & steel mills	0.1	Manufg.
Business/professional associations	0.1	Services	Boat building & repairing	0.1	Manufg.
			Motorcycles, bicycles, & parts	0.1	Manufg.
			Motors & generators	0.1	Manufg.
			Radio & TV communication equipment	0.1	Manufg.
			Gas production & distribution (utilities)	0.1	Util.
			Motor freight transportation & warehousing	0.1	Util.
			Credit agencies other than banks	0.1	Fin/R.E.
			Labor, civic, social, & fraternal associations	0.1	Services
			S/L Govt. purch., correction	0.1	S/L Govt
			S/L Govt. purch., public assistance & relief	0.1	S/L Govt

Source: Benchmark Input-Output Accounts for the U.S. Economy, 1982, U.S. Department of Commerce, Washington, D.C., July 1991. Data, as reported in the source, are organized by the 1977 SIC structure in use in 1982 but have been matched, as closely as is possible, to the 1987 SIC structure used in this book.

OCCUPATIONS EMPLOYED BY SIC 364 - ELECTRIC LIGHTING AND WIRING EQUIPMENT

Occupation	% of Total 1994	Change to 2005	Occupation	% of Total 1994	Change to 2005
Assemblers, fabricators, & hand workers nec	16.0	-7.4	Tool & die makers	1.5	-25.2
Electrical & electronic assemblers	7.6	-16.7	Hand packers & packagers	1.5	-20.6
Blue collar worker supervisors	3.7	-20.3	Sheet metal workers & duct installers	1.5	-7.4
Inspectors, testers, & graders, precision	3.2	-35.2	Machinists	1.5	-30.5
Plastic molding machine workers	3.2	11.1	Freight, stock, & material movers, hand	1.4	-25.9
Machine operators nec	2.5	-34.7	Industrial truck & tractor operators	1.3	-7.4
Industrial machinery mechanics	2.5	25.0	Machine tool cutting & forming etc. nec	1.2	-25.9
Sales & related workers nec	2.1	-7.4	Production, planning, & expediting clerks	1.2	-7.4
Electrical & electronic equipment assemblers	1.9	-7.4	Secretaries, ex legal & medical	1.2	-15.7
General managers & top executives	1.9	-12.2	Maintenance repairers, general utility	1.1	-16.6
Machine feeders & offbearers	1.8	-16.7	Industrial production managers	1.0	-7.4
Traffic, shipping, & receiving clerks	1.7	-10.9	Coating, painting, & spraying machine workers	1.0	-7.4
Helpers, laborers, & material movers nec	1.6	-7.4			

Source: Industry-Occupation Matrix, Bureau of Labor Statistics. These data relate to one or more 3-digit SIC industry groups rather than to a single 4-digit SIC. The change reported for each occupation to the year 2005 is a percent of growth or decline as estimated by the Bureau of Labor Statistics. The abbreviation nec stands for 'not elsewhere classified'.

LOCATION BY STATE AND REGIONAL CONCENTRATION

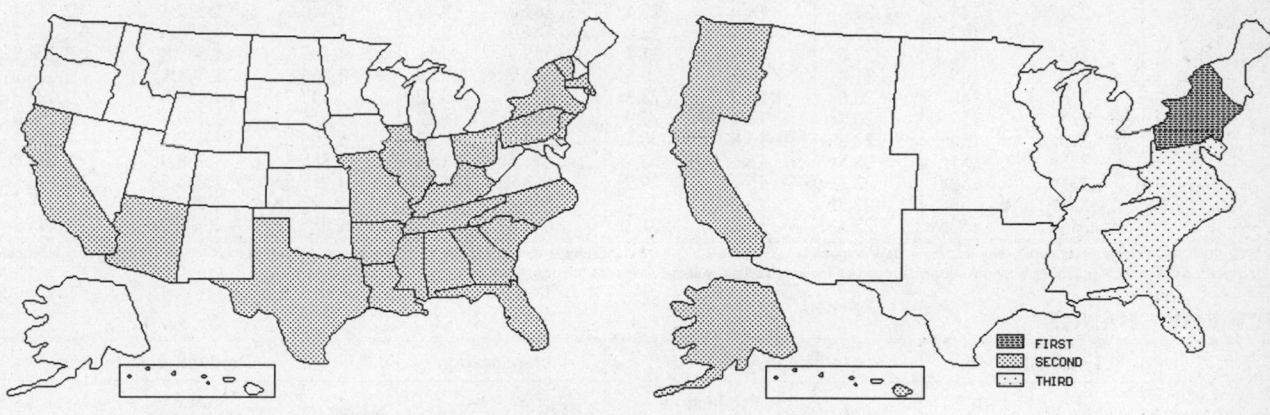

FIRST
SECOND
THIRD

INDUSTRY DATA BY STATE

State	Establish-ments	Shipments Total ($ mil)	Shipments % of U.S.	Shipments Per Establ.	Employment Total Number	Employment % of U.S.	Employment Per Establ.	Wages ($/hour)	Cost as % of Shipments	Investment per Employee ($)
California	119	266.8	16.6	2.2	3,300	20.0	28	7.47	44.5	1,152
New York	69	186.8	11.6	2.7	1,500	9.1	22	9.86	49.0	867
Pennsylvania	40	182.5	11.3	4.6	1,300	7.9	33	7.06	40.2	-
Ohio	16	139.7	8.7	8.7	900	5.5	56	8.75	45.1	2,222
Illinois	26	122.9	7.6	4.7	1,500	9.1	58	9.14	41.8	-
New Jersey	29	115.9	7.2	4.0	800	4.8	28	8.27	48.1	875
Arkansas	12	99.6	6.2	8.3	1,000	6.1	83	6.22	55.2	800
Massachusetts	14	71.9	4.5	5.1	600	3.6	43	8.89	43.4	-
North Carolina	20	55.5	3.4	2.8	700	4.2	35	7.00	47.4	571
Florida	31	47.4	2.9	1.5	800	4.8	26	6.36	43.5	500
Kentucky	9	36.5	2.3	4.1	500	3.0	56	8.13	67.7	-
Georgia	9	16.3	1.0	1.8	300	1.8	33	8.50	60.7	333
Arizona	11	9.6	0.6	0.9	200	1.2	18	7.50	43.8	500
Texas	23	(D)	-	-	750 *	4.5	33	-	-	667
Alabama	8	(D)	-	-	175 *	1.1	22	-	-	-
Louisiana	6	(D)	-	-	175 *	1.1	29	-	-	-
Tennessee	6	(D)	-	-	175 *	1.1	29	-	-	-
Mississippi	5	(D)	-	-	750 *	4.5	150	-	-	-
Missouri	5	(D)	-	-	175 *	1.1	35	-	-	-
South Carolina	5	(D)	-	-	375 *	2.3	75	-	-	-
Vermont	4	(D)	-	-	175 *	1.1	44	-	-	-
Rhode Island	3	(D)	-	-	175 *	1.1	58	-	-	-

Source: 1992 Economic Census. The states are in descending order of shipments or establishments (if shipment data are missing for the majority). The symbol (D) appears when data are withheld to prevent disclosure of competitive information. States marked with (D) are sorted by number of establishments. A dash (-) indicates that the data element cannot be calculated; * indicates the midpoint of a range.

3646 - COMMERCIAL LIGHTING FIXTURES

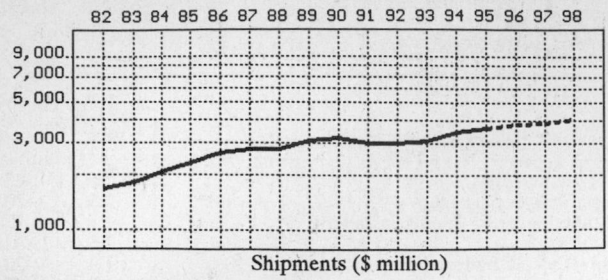

Shipments ($ million)

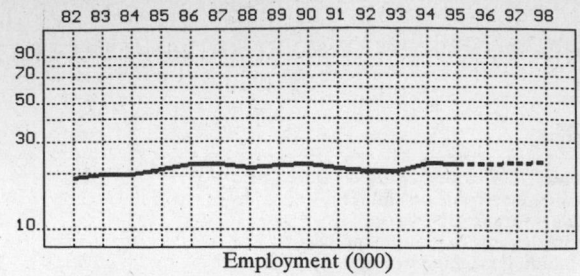

Employment (000)

GENERAL STATISTICS

| Year | Companies | Establishments | | Employment | | | Compensation | | Production ($ million) | | | |
		Total	with 20 or more employees	Total (000)	Production Workers (000)	Hours (Mil)	Payroll ($ mil)	Wages ($/hr)	Cost of Materials	Value Added by Manufacture	Value of Shipments	Capital Invest.
1982	219	243	140	18.9	13.4	25.7	313.3	7.17	892.3	774.8	1,671.9	32.0
1983		243	141	19.6	14.0	27.5	342.1	7.39	938.3	871.6	1,800.0	37.5
1984		243	142	19.7	14.6	28.4	362.3	7.83	1,068.3	987.5	2,051.6	47.5
1985		243	144	21.1	15.6	29.9	411.6	8.15	1,228.0	1,104.1	2,321.2	53.0
1986		236	144	22.5	16.0	31.3	456.7	8.62	1,340.0	1,282.6	2,623.7	56.4
1987	242	271	148	22.7	15.6	31.3	478.2	8.89	1,375.1	1,367.4	2,739.9	60.6
1988		274	148	21.6	15.0	29.4	482.0	9.44	1,440.1	1,289.4	2,743.4	60.8
1989		259	149	22.2	15.5	32.0	544.9	9.35	1,563.6	1,533.9	3,083.7	63.0
1990		262	146	22.6	15.3	31.7	559.9	9.50	1,585.8	1,609.8	3,208.9	64.4
1991		280	145	21.4	14.2	29.5	536.7	9.89	1,466.8	1,541.1	3,020.9	63.3
1992	305	336	148	20.6	14.0	28.3	530.5	10.02	1,547.7	1,479.8	3,023.3	57.7
1993		321	150	20.7	14.1	27.9	536.3	10.22	1,616.5	1,479.2	3,105.0	43.2
1994		316P	150P	22.8	15.8	30.7	612.0	10.83	1,799.6	1,678.5	3,469.9	50.4
1995		323P	151P	22.5P	15.1P	30.7P	631.4P	11.04P	1,885.8P	1,758.9P	3,636.2P	61.9P
1996		331P	152P	22.7P	15.2P	30.9P	653.9P	11.32P	1,956.5P	1,824.9P	3,772.5P	63.1P
1997		338P	153P	22.9P	15.2P	31.1P	676.3P	11.61P	2,027.2P	1,890.8P	3,908.8P	64.4P
1998		345P	153P	23.1P	15.2P	31.3P	698.8P	11.90P	2,097.9P	1,956.8P	4,045.2P	65.6P

Sources: 1982, 1987, 1992 *Economic Census*; *Annual Survey of Manufactures*, 83-86, 88-91, 93-94. Establishment counts for non-Census years are from *County Business Patterns*; establishment values for 83-84 are extrapolations. 'P's show projections by the editors. Industries reclassified in 87 will not have data for prior years.

INDICES OF CHANGE

| Year | Companies | Establishments | | Employment | | | Compensation | | Production ($ million) | | | |
		Total	with 20 or more employees	Total (000)	Production Workers (000)	Hours (Mil)	Payroll ($ mil)	Wages ($/hr)	Cost of Materials	Value Added by Manufacture	Value of Shipments	Capital Invest.
1982	72	72	95	92	96	91	59	72	58	52	55	55
1983		72	95	95	100	97	64	74	61	59	60	65
1984		72	96	96	104	100	68	78	69	67	68	82
1985		72	97	102	111	106	78	81	79	75	77	92
1986		70	97	109	114	111	86	86	87	87	87	98
1987	79	81	100	110	111	111	90	89	89	92	91	105
1988		82	100	105	107	104	91	94	93	87	91	105
1989		77	101	108	111	113	103	93	101	104	102	109
1990		78	99	110	109	112	106	95	102	109	106	112
1991		83	98	104	101	104	101	99	95	104	100	110
1992	100	100	100	100	100	100	100	100	100	100	100	100
1993		96	101	100	101	99	101	102	104	100	103	75
1994		94P	102P	111	113	108	115	108	116	113	115	87
1995		96P	102P	109P	108P	109P	119P	110P	122P	119P	120P	107P
1996		98P	103P	110P	108P	109P	123P	113P	126P	123P	125P	109P
1997		101P	103P	111P	109P	110P	127P	116P	131P	128P	129P	112P
1998		103P	104P	112P	109P	110P	132P	119P	136P	132P	134P	114P

Sources: Same as General Statistics. Values reflect change from the base year, 1992. Values above 100 mean greater than 92, values below 100 mean less than 92, and a value of 100 in the 82-91 or 93-98 period means same as 92. 'P's mark projections by the editors.

SELECTED RATIOS

For 1994	Avg. of All Manufact.	Analyzed Industry	Index	For 1994	Avg. of All Manufact.	Analyzed Industry	Index
Employees per Establishment	49	72	147	Value Added per Production Worker	134,084	106,234	79
Payroll per Establishment	1,500,273	1,938,010	129	Cost per Establishment	5,045,178	5,698,762	113
Payroll per Employee	30,620	26,842	88	Cost per Employee	102,970	78,930	77
Production Workers per Establishment	34	50	146	Cost per Production Worker	146,988	113,899	77
Wages per Establishment	853,319	1,052,862	123	Shipments per Establishment	9,576,895	10,988,072	115
Wages per Production Worker	24,861	21,043	85	Shipments per Employee	195,460	152,189	78
Hours per Production Worker	2,056	1,943	95	Shipments per Production Worker	279,017	219,614	79
Wages per Hour	12.09	10.83	90	Investment per Establishment	321,011	159,601	50
Value Added per Establishment	4,602,255	5,315,277	115	Investment per Employee	6,552	2,211	34
Value Added per Employee	93,930	73,618	78	Investment per Production Worker	9,352	3,190	34

Sources: Same as General Statistics. The 'Average of All Manufacturing' column represents the average of all manufacturing industries reported for the most recent complete year available. The Index shows the relationship between the Average and the Analyzed Industry. For example, 100 means that they are equal; 500 that the Analyzed Industry is five times the average; 50 means that the Analyzed Industry is half the national average. The abbreviation 'na' is used to show that data are 'not available'.

LEADING COMPANIES Number shown: 75   Total sales ($ mil): 6,252   Total employment (000): 58.0

Company Name	Address				CEO Name	Phone	Co. Type	Sales ($ mil)	Empl. (000)
National Service Industries Inc	1420 Peachtree NE	Atlanta	GA	30309	D Raymond Riddle	404-853-1000	P	1,882	22.0
Osram Sylvania Inc	100 Endicott St	Danvers	MA	01923	Dean Langford	508-777-1900	S	1,250	12.5
Lithonia Lighting Co	PO Box A	Conyers	GA	30207	James H McClung	404-922-9000	D	764	4.5
Cooper Lighting	400 Busse Rd	Elk Grove Vill	IL	60007	Daniel Thomson	708-956-8400	D	330	3.5
USI Lighting Inc	1251 Doolittle Dr	San Leandro	CA	94577	K White	510-562-3500	S	300	2.2
Hubbell Lighting Inc	2000 Electric Way	Christiansburg	VA	24073	Glenn Grunewald	703-382-6111	S	195	1.4
Holophane Corp	250 E Broad St	Columbus	OH	43215	John R DallePezze	614-224-3134	P	151	1.2
Lightolier/Fall River	631 Airport Rd	Fall River	MA	02720	Richard Kurtz	508-679-8131	D	110*	0.3
JJI Lighting Group Inc	PO Box 4207	Greenwich	CT	06830	Robert N Haidinger	203-869-9330	R	98	1.0
Lights of America Inc	611 Reyes Dr	Walnut	CA	91789	Usman U Vakil	909-594-7883	R	98	1.0
LSI Industries Inc	PO Box 42728	Cincinnati	OH	45242	Robert J Ready	513-793-3200	P	94	0.7
Genlyte Group Inc	PO Box 129	Union	NJ	07083	Chuck Havers	908-964-7000	D	83	0.3
Columbia Lighting Inc	PO Box 2787	Spokane	WA	99220	Robert H Ingram	509-924-7000	S	70*	0.5
SIMKAR Lighting Fixture Co	601 E Cayuga St	Philadelphia	PA	19120	Frank W Tobin	215-831-7700	D	70	0.5
Kenall Manufacturing Co	1020 Lakeside Dr	Gurnee	IL	60031	James W Hawkins	708-360-8200	R	54	<0.1
Entergy Systems and Service Inc	4740 Shelby Dr	Memphis	TN	38141	Paul E Williams	901-367-2880	S	44*	0.5
Kim Lighting Inc	PO Box 1275	City of Industry	CA	91749	Lou Goren	818-968-5666	S	40	0.3
Luminator Mass Transit Dive	1200 E Plano Pkwy	Plano	TX	75074	RG Strickland	214-424-6511	D	40	0.4
High End Systems	2217 W Braker Ln	Austin	TX	78758	Lowell Fowler	512-836-2242	R	35	0.3
Capri Lighting Inc	6430 E Slauson Av	Los Angeles	CA	90040	Charles Harris	213-726-1800	S	33	0.2
Williams Inc	PO Box 837	Carthage	MO	64836	Ronald Snyder	417-358-4065	R	33	0.3
Strand Lighting Inc	18111 S Santa Fe	R Dominguez	CA	90221	Gene Griffith	310-637-7500	S	30	0.2
American Fluorescent Corp	2345 Krueger	Waukegan	IL	60079	William Solomon	708-249-5970	R	24	0.2
Litecontrol Corp	100 Hawks Av	Hanson	MA	02341	Robert Danforth	617-294-0100	R	22	0.2
Chloride Systems	126 Chloride Rd	Burgaw	NC	28425	Fred Sturm	919-259-1000	D	20	0.3
Wide-Lite	PO Box 606	San Marcos	TX	78666	George O'Donnell	512-392-5821	D	20	0.2
Moldcast Lighting	1251 Doolittle Dr	San Leandro	CA	94577	J Kiernan White	510-562-3500	D	19	0.3
Edison Price Inc	409 E 60th St	New York	NY	10022	Emma Price	212-838-5212	R	18	0.1
Mid-West Chandelier Co	100 Funston Rd	Kansas City	KS	66115	S Lefkovitz	913-281-1100	R	18	0.2
Peerless Lighting Corp	PO Box 2556	Berkeley	CA	94702	DJ Herst	510-845-2760	R	17*	0.2
Prudential Lighting Corp	PO Box 58736	Los Angeles	CA	90058	JM Ellis	213-746-0360	R	15	<0.1
Devine Lighting Inc	1 Design Dr	N Kansas City	MO	64116	WF Budnovitch	816-221-9440	R	14	0.1
Flex-O-Lite Inc	16330 Phoebe Av	La Mirada	CA	90638	Mark Fernandez	714-994-3880	D	14*	0.2
Loctite Luminescent Systems Inc	101 Etna Rd	Lebanon	NH	03766	Jeffrey Piccolomini	603-448-3444	S	14	0.1
Neo-Ray Lighting Products Inc	537 Johnson Av	Brooklyn	NY	11237	Leon Conn	718-456-7400	R	14	0.1
Fiberstars Inc	2883 Bayview Dr	Fremont	CA	94538	David N Ruckert	510-490-0719	P	14	<0.1
NL Corp	14901 Broadway	Cleveland	OH	44137	Lawrence A Terkel	216-662-2080	R	13	0.2
Indy Lighting Inc	12001 Exit Five	Fishers	IN	46038	Jacques P LeFevre	317-849-1233	S	12	0.1
Lumax Industries Inc	PO Box 991	Altoona	PA	16601	Donald E Snyder	814-944-2537	R	12	0.1
Parke Industries Inc	2246 Lindsay Way	Glendora	CA	91740	Dan Parke	909-599-1204	R	12	0.1
Guth Lighting	PO Box 7079	St Louis	MO	63177	Walter Coleman	314-533-3200	D	10	<0.1
Lightron of Cornwall Inc	PO Box 4270	New Windsor	NY	12553	E Littman	914-562-5500	S	10*	<0.1
Mark Lighting Fixture Company	25 Knickerbocker	Moonachie	NJ	07074	Carl Coppola	201-939-0880	R	10*	0.1
Durel Corp	645 W 24th St	Tempe	AZ	85282	Bob Krafcik	602-731-6200	J	8	<0.1
Luxo Corp	PO Box 951	Port Chester	NY	10573	Edgar R Connors	914-937-4433	R	8	<0.1
Steel Craft Fluorescent Inc	191 Murray St	Newark	NJ	07114	Myron Melster	201-824-5871	R	8	0.1
E-L FlexKey Technologies Inc	77 Olean Rd	East Aurora	NY	14052	Peter Gunderman	716-655-0800	D	8	<0.1
Louis Baldinger and Son Inc	1902 Steinway St	Astoria	NY	11105	Daniel Baldinger	718-204-5700	R	7	<0.1
Paramount Industries Inc	304 N Howard St	Croswell	MI	48422	BR Bailey	313-679-2551	R	7	<0.1
Nu-Art Lighting & Mfg	160 N 400 W	North Salt Lake	UT	84054	Norman Schoepf	801-298-1600	R	6	0.1
Borden/Reaves Inc	1141 Marina Way S	Richmond	CA	94804	J Randy Borden	510-234-2370	R	5	<0.1
House-O-Lite Corp	13603 S Halsted St	Riverdale	IL	60627	Susan M Larson	708-841-3800	R	5	<0.1
LC Doane Co	55 Plains Rd	Essex	CT	06426	Margaret P Eagan	203-767-8295	R	5	<0.1
Mason Candlelight Co	PO Box 367	Middlesex	NJ	08846	Richard J Kane	908-469-4212	D	5	<0.1
Rambusch Decorating Co	40 W 13th St	New York	NY	10011	Viggo B Rambusch	212-675-0400	R	5*	<0.1
Dazor Manufacturing Corp	4483 Duncan Av	St Louis	MO	63110	Mark Hogrebe	314-652-2400	R	4	<0.1
DeVoe Lighting Corp	527 Main St	Belleville	NJ	07109	M Goldstone	201-450-4200	R	4	<0.1
Hasco Electric Corp	84 S Water St	Greenwich	CT	06830	Samuel Benkel	203-531-9400	R	4	<0.1
Legion Lighting Company Inc	221 Glenmore Av	Brooklyn	NY	11207	Sheldon Bellovin	718-498-1770	R	4	<0.1
RAB Electric Manufacturing	PO Box 970	Northvale	NJ	07647	RA Barna	201-784-8600	R	4	<0.1
Warner Technologies Inc	11859 Wilshire Blv	Los Angeles	CA	90025	Thomas S Hathaway	310-444-0488	R	4	<0.1
RA Manning Company Inc	PO Box 1063	Sheboygan	WI	53082	Thomas Manning	414-458-2184	R	3	<0.1
Challenger Lighting Company	2420-V E Oakton St	Arlington H	IL	60005	Woodrow Paradis	708-364-9100	R	3	<0.1
CW Cole and Co	2560 Rosemead	S El Monte	CA	91733	SW Cole	818-443-2473	R	3	<0.1
Loran Inc	1705 E Colton Av	Redlands	CA	92374	William J Locklin	909-794-2121	R	3*	0.2
McPhilben	170 Rodeo Dr	Edgewood	NY	11717	Joe Gillin	516-254-4400	D	3*	<0.1
National Electric Mfg Corp	6361 Chalet Dr	Commerce	CA	90040	James M Galvez	310-928-8488	R	3*	<0.1
Fuller Ultraviolet Corp	9416 Gulf Stream	Frankfort	IL	60423	William R Eckstrom	815-469-3301	R	2*	<0.1
Moffatt Products Inc	222 Cessna	Watertown	SD	57201	David Moffatt	605-886-5700	R	2	<0.1
Roxter Manufacturing Corp	10-11 40th Av	Long Island Ct	NY	11101	Alan Hochster	718-392-5060	R	2	<0.1
Envel Design Corp	5740 Corsa Av	Westlake Vil	CA	91362	Quinn B Mayer	818-865-8111	R	1	<0.1
Lite Center California Inc	306 Elizabeth Ln	Corona	CA	91720	Ray Wetzel Jr	909-273-7380	R	1*	<0.1
Hastings Lighting Company Inc	1206 Long Beach	Los Angeles	CA	90021	J Culbertson	213-622-2009	R	1	<0.1
Nemco Electric Co	207 S Horton St	Seattle	WA	98134	A Larson	206-622-1551	R	1	<0.1
Novo Products Inc	977 Irvin St	Plymouth	MI	48170	Don Gaines	313-451-2011	R	0*	<0.1

Source: Ward's Business Directory of U.S. Private and Public Companies, Volumes 1 and 2, 1996. The company type code used is as follows: P - Public, R - Private, S - Subsidiary, D - Division, J - Joint Venture, A - Affiliate, G - Group. Sales are in millions of dollars, employees are in thousands. An asterisk (*) indicates an estimated sales volume. The symbol < stands for 'less than'. Company names and addresses are truncated, in some cases, to fit into the available space.

MATERIALS CONSUMED

Material	Quantity	Delivered Cost ($ million)
Materials, ingredients, containers, and supplies	(X)	1,443.8
Specialty transformers and fluorescent ballasts	(X)	464.4
Current-carrying wiring devices	(X)	57.2
Electric lamp bulbs	(X)	41.1
Flat glass (plate, float, and sheet)	(X)	8.9
Plastics resins consumed in the form of granules, pellets, powders, liquids, etc.	(X)	24.5
Plastics products consumed in the form of sheets, rods, tubes, film, and other shapes	(X)	38.7
Fabricated plastics products (except gaskets, hoses, and belting)	(X)	45.4
Insulated wire and cable, including magnet wire	(X)	6.3
Paperboard containers, boxes, and corrugated paperboard	(X)	56.1
Metal bolts, nuts, screws, washers, rivets, and other screw machine products	(X)	38.2
Metal poles	(X)	1.1
All other fabricated metal products (except forgings)	(X)	52.3
Iron and steel castings (rough and semifinished)	(X)	14.3
Aluminum and aluminum-base alloy castings (rough and semifinished)	(X)	28.0
Other nonferrous castings (rough and semifinished)	(X)	2.0
Steel sheet and strip, including tin plate	(X)	172.9
Steel wire and wire products	(X)	4.8
All other steel mill shapes and forms (except castings and forgings)	(X)	17.0
Aluminum and aluminum-base alloy sheet, plate, foil, and welded tubing	(X)	63.8
Aluminum and aluminum-base alloy extruded shapes, including extruded rod, bar, pipe, tube, etc.	(X)	19.8
Other aluminum and aluminum-base alloy shapes and forms	(X)	0.4
Copper and copper-base alloy shapes and forms	(X)	0.9
All other materials and components, parts, containers, and supplies	(X)	134.1
Materials, ingredients, containers, and supplies, nsk	(X)	151.4

Source: 1992 *Economic Census*. Explanation of symbols used: (D): Withheld to avoid disclosure of competitive data; na: Not available; (S): Withheld because statistical norms were not met; (X): Not applicable; (Z): Less than half the unit shown; nec: Not elsewhere classified; nsk: Not specified by kind; - : zero; * : 10-19 percent estimated; ** : 20-29 percent estimated.

PRODUCT SHARE DETAILS

Product or Product Class	% Share	Product or Product Class	% Share
Commercial, industrial, and institutional electric lighting fixtures	100.00	Industrial type electric lighting fixtures, including parts and accessories	15.44
Commercial and institutional-type electric lighting fixtures, including parts and accessories	78.99	Commercial, industrial, and institutional electric lighting fixtures, nsk	5.57

Source: 1992 *Economic Census*. The values shown are percent of total shipments in an industry. Values of indented subcategories are summed in the main heading. The symbol (D) appears when data are withheld to prevent disclosure of competitive information. The abbreviation nsk stands for 'not specified by kind' and nec for 'not elsewhere classified'.

INPUTS AND OUTPUTS FOR LIGHTING FIXTURES & EQUIPMENT

Economic Sector or Industry Providing Inputs	%	Sector	Economic Sector or Industry Buying Outputs	%	Sector
Wholesale trade	11.2	Trade	Personal consumption expenditures	16.4	
Transformers	8.4	Manufg.	Office buildings	9.3	Constr.
Glass & glass products, except containers	8.0	Manufg.	Motor vehicles & car bodies	5.6	Manufg.
Metal stampings, nec	7.5	Manufg.	Residential 1-unit structures, nonfarm	5.0	Constr.
Imports	7.1	Foreign	Maintenance of nonfarm buildings nec	4.8	Constr.
Blast furnaces & steel mills	6.4	Manufg.	Gross private fixed investment	4.6	Cap Inv
Aluminum rolling & drawing	4.0	Manufg.	Electric utility facility construction	4.3	Constr.
Miscellaneous plastics products	2.5	Manufg.	Industrial buildings	4.3	Constr.
Paperboard containers & boxes	2.5	Manufg.	Exports	4.2	Foreign
Lighting fixtures & equipment	2.4	Manufg.	Nonfarm residential structure maintenance	3.2	Constr.
Plastics materials & resins	2.4	Manufg.	Residential additions/alterations, nonfarm	2.9	Constr.
Advertising	2.0	Services	Automotive repair shops & services	2.8	Services
Wiring devices	1.9	Manufg.	Construction of stores & restaurants	2.2	Constr.
Banking	1.9	Fin/R.E.	Construction of hospitals	1.9	Constr.
Screw machine and related products	1.7	Manufg.	Highway & street construction	1.9	Constr.
Electric services (utilities)	1.6	Util.	Construction of educational buildings	1.7	Constr.
Air transportation	1.5	Util.	Lighting fixtures & equipment	1.5	Manufg.
Aluminum castings	1.4	Manufg.	Eating & drinking places	1.5	Trade
Electric lamps	1.3	Manufg.	Hospitals	1.2	Services
Nonferrous wire drawing & insulating	1.2	Manufg.	Miscellaneous plastics products	1.1	Manufg.
Paints & allied products	1.2	Manufg.	Automobile parking & car washes	1.0	Services
Motor freight transportation & warehousing	1.2	Util.	Maintenance of highways & streets	0.9	Constr.
Plating & polishing	1.0	Manufg.	Warehouses	0.9	Constr.
Cyclic crudes and organics	0.9	Manufg.	Hotels & motels	0.7	Constr.
Communications, except radio & TV	0.9	Util.	Residential garden apartments	0.7	Constr.
Hotels & lodging places	0.9	Services	S/L Govt. purch., elem. & secondary education	0.7	S/L Govt
Manufacturing industries, nec	0.8	Manufg.	Coal	0.5	Mining
Semiconductors & related devices	0.8	Manufg.	Resid. & other health facility construction	0.5	Constr.
Electrical equipment & supplies, nec	0.7	Manufg.	Colleges, universities, & professional schools	0.5	Services

Continued on next page.

INPUTS AND OUTPUTS FOR LIGHTING FIXTURES & EQUIPMENT - Continued

Economic Sector or Industry Providing Inputs	%	Sector	Economic Sector or Industry Buying Outputs	%	Sector
Petroleum refining	0.7	Manufg.	Amusement & recreation building construction	0.4	Constr.
Real estate	0.7	Fin/R.E.	Construction of nonfarm buildings nec	0.4	Constr.
Gas production & distribution (utilities)	0.6	Util.	Maintenance of farm service facilities	0.4	Constr.
Eating & drinking places	0.6	Trade	Telephone & telegraph facility construction	0.4	Constr.
Equipment rental & leasing services	0.6	Services	Mobile homes	0.4	Manufg.
Maintenance of nonfarm buildings nec	0.5	Constr.	Electric services (utilities)	0.4	Util.
Copper rolling & drawing	0.5	Manufg.	Elementary & secondary schools	0.4	Services
Iron & steel foundries	0.5	Manufg.	Crude petroleum & natural gas	0.3	Mining
Nonferrous castings, nec	0.5	Manufg.	Construction of religious buildings	0.3	Constr.
Primary zinc	0.5	Manufg.	Maintenance of electric utility facilities	0.3	Constr.
Wood products, nec	0.5	Manufg.	Maintenance of nonbuilding facilities nec	0.3	Constr.
Engineering, architectural, & surveying services	0.5	Services	Residential 2-4 unit structures, nonfarm	0.3	Constr.
Broadwoven fabric mills	0.4	Manufg.	Photographic equipment & supplies	0.3	Manufg.
Electronic components nec	0.4	Manufg.	Porcelain electrical supplies	0.3	Manufg.
Metal coating & allied services	0.4	Manufg.	Communications, except radio & TV	0.3	Util.
Miscellaneous fabricated wire products	0.4	Manufg.	Nursing & personal care facilities	0.3	Services
Gaskets, packing & sealing devices	0.3	Manufg.	Photofinishing labs, commercial photography	0.3	Services
Machinery, except electrical, nec	0.3	Manufg.	Theatrical producers, bands, entertainers	0.3	Services
Pottery products, nec	0.3	Manufg.	S/L Govt. purch., higher education	0.3	S/L Govt
Railroads & related services	0.3	Util.	S/L Govt. purch., other general government	0.3	S/L Govt
Computer & data processing services	0.3	Services	Nonbuilding facilities nec	0.2	Constr.
Legal services	0.3	Services	Residential high-rise apartments	0.2	Constr.
U.S. Postal Service	0.3	Gov't	Logging camps & logging contractors	0.2	Manufg.
Brass, bronze, & copper castings	0.2	Manufg.	Motor vehicle parts & accessories	0.2	Manufg.
Fabricated rubber products, nec	0.2	Manufg.	Petroleum refining	0.2	Manufg.
Hardwood dimension & flooring mills	0.2	Manufg.	Travel trailers & campers	0.2	Manufg.
Primary aluminum	0.2	Manufg.	Railroads & related services	0.2	Util.
Sawmills & planning mills, general	0.2	Manufg.	Water supply & sewage systems	0.2	Util.
Special dies & tools & machine tool accessories	0.2	Manufg.	Real estate	0.2	Fin/R.E.
Royalties	0.2	Fin/R.E.	Libraries, vocation education	0.2	Services
Accounting, auditing & bookkeeping	0.2	Services	S/L Govt. purch., health & hospitals	0.2	S/L Govt
Management & consulting services & labs	0.2	Services	S/L Govt. purch., natural resource & recreation.	0.2	S/L Govt
Machine tools, metal forming types	0.1	Manufg.	Farm service facilities	0.1	Constr.
Manifold business forms	0.1	Manufg.	Maintenance of military facilities	0.1	Constr.
Insurance carriers	0.1	Fin/R.E.	Sewer system facility construction	0.1	Constr.
Automotive rental & leasing, without drivers	0.1	Services	Automotive stampings	0.1	Manufg.
Automotive repair shops & services	0.1	Services	Blast furnaces & steel mills	0.1	Manufg.
Business/professional associations	0.1	Services	Boat building & repairing	0.1	Manufg.
			Motorcycles, bicycles, & parts	0.1	Manufg.
			Motors & generators	0.1	Manufg.
			Radio & TV communication equipment	0.1	Manufg.
			Gas production & distribution (utilities)	0.1	Util.
			Motor freight transportation & warehousing	0.1	Util.
			Credit agencies other than banks	0.1	Fin/R.E.
			Labor, civic, social, & fraternal associations	0.1	Services
			S/L Govt. purch., correction	0.1	S/L Govt
			S/L Govt. purch., public assistance & relief	0.1	S/L Govt

Source: Benchmark Input-Output Accounts for the U.S. Economy, 1982, U.S. Department of Commerce, Washington, D.C., July 1991. Data, as reported in the source, are organized by the 1977 SIC structure in use in 1982 but have been matched, as closely as is possible, to the 1987 SIC structure used in this book.

OCCUPATIONS EMPLOYED BY SIC 364 - ELECTRIC LIGHTING AND WIRING EQUIPMENT

Occupation	% of Total 1994	Change to 2005	Occupation	% of Total 1994	Change to 2005
Assemblers, fabricators, & hand workers nec	16.0	-7.4	Tool & die makers	1.5	-25.2
Electrical & electronic assemblers	7.6	-16.7	Hand packers & packagers	1.5	-20.6
Blue collar worker supervisors	3.7	-20.3	Sheet metal workers & duct installers	1.5	-7.4
Inspectors, testers, & graders, precision	3.2	-35.2	Machinists	1.5	-30.5
Plastic molding machine workers	3.2	11.1	Freight, stock, & material movers, hand	1.4	-25.9
Machine operators nec	2.5	-34.7	Industrial truck & tractor operators	1.3	-7.4
Industrial machinery mechanics	2.5	25.0	Machine tool cutting & forming etc. nec	1.2	-25.9
Sales & related workers nec	2.1	-7.4	Production, planning, & expediting clerks	1.2	-7.4
Electrical & electronic equipment assemblers	1.9	-7.4	Secretaries, ex legal & medical	1.2	-15.7
General managers & top executives	1.9	-12.2	Maintenance repairers, general utility	1.1	-16.6
Machine feeders & offbearers	1.8	-16.7	Industrial production managers	1.0	-7.4
Traffic, shipping, & receiving clerks	1.7	-10.9	Coating, painting, & spraying machine workers	1.0	-7.4
Helpers, laborers, & material movers nec	1.6	-7.4			

Source: Industry-Occupation Matrix, Bureau of Labor Statistics. These data relate to one or more 3-digit SIC industry groups rather than to a single 4-digit SIC. The change reported for each occupation to the year 2005 is a percent of growth or decline as estimated by the Bureau of Labor Statistics. The abbreviation nec stands for 'not elsewhere classified'.

LOCATION BY STATE AND REGIONAL CONCENTRATION

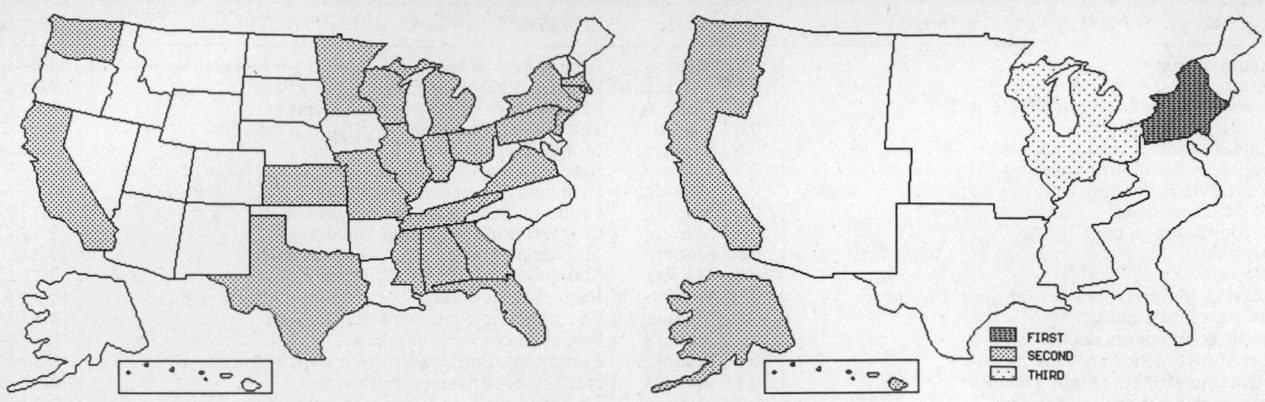

FIRST
SECOND
THIRD

INDUSTRY DATA BY STATE

State	Establish-ments	Shipments			Employment				Cost as % of Shipments	Investment per Employee ($)
		Total ($ mil)	% of U.S.	Per Establ.	Total Number	% of U.S.	Per Establ.	Wages ($/hour)		
California	65	406.5	13.4	6.3	2,800	13.6	43	9.69	43.4	2,143
Illinois	23	352.9	11.7	15.3	2,400	11.7	104	9.04	39.1	4,625
Pennsylvania	25	204.7	6.8	8.2	1,500	7.3	60	8.24	66.4	5,133
Massachusetts	13	200.5	6.6	15.4	1,100	5.3	85	12.67	38.4	3,364
New York	43	157.6	5.2	3.7	1,400	6.8	33	10.37	43.7	1,214
New Jersey	22	157.0	5.2	7.1	1,000	4.9	45	11.54	53.4	500
Ohio	17	146.2	4.8	8.6	900	4.4	53	10.38	49.0	2,222
Tennessee	7	118.2	3.9	16.9	700	3.4	100	9.18	56.9	4,143
Texas	13	83.7	2.8	6.4	600	2.9	46	11.00	51.9	2,000
Missouri	5	42.9	1.4	8.6	400	1.9	80	10.20	54.3	-
Connecticut	6	39.6	1.3	6.6	300	1.5	50	11.67	44.9	-
Michigan	11	30.6	1.0	2.8	300	1.5	27	6.75	38.6	1,333
Wisconsin	5	28.0	0.9	5.6	200	1.0	40	10.67	36.8	2,500
Florida	11	27.4	0.9	2.5	200	1.0	18	5.50	81.0	1,500
Minnesota	10	15.8	0.5	1.6	200	1.0	20	13.00	50.6	500
Rhode Island	3	13.9	0.5	4.6	100	0.5	33	10.50	41.0	-
Georgia	8	(D)	-	-	3,750 *	18.2	469	-	-	2,827
Indiana	7	(D)	-	-	750 *	3.6	107	-	-	-
Washington	7	(D)	-	-	750 *	3.6	107	-	-	-
Virginia	4	(D)	-	-	175 *	0.8	44	-	-	-
Alabama	2	(D)	-	-	375 *	1.8	188	-	-	-
Kansas	2	(D)	-	-	175 *	0.8	88	-	-	-
Mississippi	1	(D)	-	-	750 *	3.6	750	-	-	-

Source: 1992 *Economic Census*. The states are in descending order of shipments or establishments (if shipment data are missing for the majority). The symbol (D) appears when data are withheld to prevent disclosure of competitive information. States marked with (D) are sorted by number of establishments. A dash (-) indicates that the data element cannot be calculated; * indicates the midpoint of a range.

3647 - VEHICULAR LIGHTING EQUIPMENT

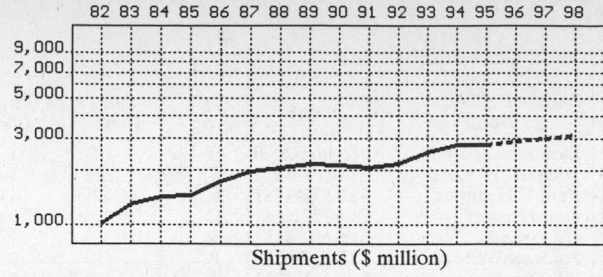

Shipments ($ million)

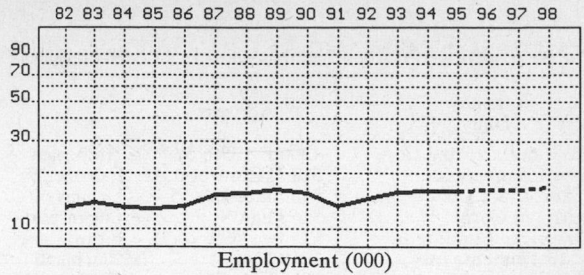

Employment (000)

GENERAL STATISTICS

Year	Com-panies	Establishments		Employment			Compensation		Production ($ million)			
		Total	with 20 or more employees	Total (000)	Production Workers (000)	Hours (Mil)	Payroll ($ mil)	Wages ($/hr)	Cost of Materials	Value Added by Manufacture	Value of Shipments	Capital Invest.
1982	77	84	54	12.9	9.6	18.8	281.6	10.29	411.2	602.4	1,013.6	43.8
1983		81	54	13.9	10.7	21.6	344.1	11.57	501.4	790.9	1,282.1	34.4
1984		78	54	13.0	9.9	20.0	382.3	13.58	602.0	832.4	1,418.6	39.9
1985		76	53	12.8	9.6	19.4	393.8	14.60	659.0	798.6	1,453.1	87.7
1986		71	50	13.3	9.8	20.0	439.0	15.49	870.7	873.9	1,729.7	150.6
1987	68	72	49	15.4	11.6	23.4	508.4	15.78	1,025.6	957.1	1,969.6	75.7
1988		72	53	15.6	11.5	23.1	519.1	15.98	1,053.9	989.8	2,051.6	67.9
1989		70	55	16.2	11.6	22.5	491.5	14.91	1,125.6	1,051.7	2,179.1	79.8
1990		73	52	15.6	10.9	20.9	480.8	15.32	1,095.6	1,025.1	2,121.7	103.5
1991		79	53	13.2	9.7	19.3	460.7	16.13	1,074.2	990.6	2,061.9	74.4
1992	90	96	58	14.6	10.8	21.3	499.0	16.17	1,024.0	1,143.4	2,190.4	61.3
1993		98	60	15.8	12.0	23.8	537.3	15.66	1,312.4	1,213.0	2,517.5	84.5
1994		85P	56P	16.0	12.2	26.2	578.1	15.81	1,376.1	1,386.4	2,747.1	91.5
1995		86P	57P	16.1P	11.8P	23.9P	588.9P	17.30P	1,386.4P	1,396.7P	2,767.6P	95.5P
1996		87P	57P	16.3P	12.0P	24.3P	608.0P	17.67P	1,448.2P	1,459.1P	2,891.6P	98.2P
1997		88P	57P	16.5P	12.2P	24.6P	627.1P	18.04P	1,510.1P	1,521.4P	3,014.7P	100.9P
1998		89P	58P	16.7P	12.3P	24.9P	646.2P	18.41P	1,572.0P	1,583.8P	3,138.2P	103.6P

Sources: 1982, 1987, 1992 Economic Census; Annual Survey of Manufactures, 83-86, 88-91, 93-94. Establishment counts for non-Census years are from County Business Patterns; establishment values for 83-84 are extrapolations. 'P's show projections by the editors. Industries reclassified in 87 will not have data for prior years.

INDICES OF CHANGE

Year	Com-panies	Establishments		Employment			Compensation		Production ($ million)			
		Total	with 20 or more employees	Total (000)	Production Workers (000)	Hours (Mil)	Payroll ($ mil)	Wages ($/hr)	Cost of Materials	Value Added by Manufacture	Value of Shipments	Capital Invest.
1982	86	88	93	88	89	88	56	64	40	53	46	71
1983		84	93	95	99	101	69	72	49	69	59	56
1984		81	93	89	92	94	77	84	59	73	65	65
1985		79	91	88	89	91	79	90	64	70	66	143
1986		74	86	91	91	94	88	96	85	76	79	246
1987	76	75	84	105	107	110	102	98	100	84	90	123
1988		75	91	107	106	108	104	99	103	87	94	111
1989		73	95	111	107	106	98	92	110	92	99	130
1990		76	90	107	101	98	96	95	107	90	97	169
1991		82	91	90	90	91	92	100	105	87	94	121
1992	100	100	100	100	100	100	100	100	100	100	100	100
1993		102	103	108	111	112	108	97	128	106	115	138
1994		89P	97P	110	113	123	116	98	134	121	125	149
1995		90P	98P	110P	110P	112P	118P	107P	135P	122P	126P	156P
1996		91P	98P	112P	111P	114P	122P	109P	141P	128P	132P	160P
1997		92P	99P	113P	113P	115P	126P	112P	147P	133P	138P	165P
1998		93P	100P	115P	114P	117P	130P	114P	154P	139P	143P	169P

Sources: Same as General Statistics. Values reflect change from the base year, 1992. Values above 100 mean greater than 92, values below 100 mean less than 92, and a value of 100 in the 82-91 or 93-98 period means same as 92. 'P's mark projections by the editors.

SELECTED RATIOS

For 1994	Avg. of All Manufact.	Analyzed Industry	Index	For 1994	Avg. of All Manufact.	Analyzed Industry	Index
Employees per Establishment	49	187	382	Value Added per Production Worker	134,084	113,639	85
Payroll per Establishment	1,500,273	6,762,602	451	Cost per Establishment	5,045,178	16,097,590	319
Payroll per Employee	30,620	36,131	118	Cost per Employee	102,970	86,006	84
Production Workers per Establishment	34	143	416	Cost per Production Worker	146,988	112,795	77
Wages per Establishment	853,319	4,845,560	568	Shipments per Establishment	9,576,895	32,135,519	336
Wages per Production Worker	24,861	33,953	137	Shipments per Employee	195,460	171,694	88
Hours per Production Worker	2,056	2,148	104	Shipments per Production Worker	279,017	225,172	81
Wages per Hour	12.09	15.81	131	Investment per Establishment	321,011	1,070,365	333
Value Added per Establishment	4,602,255	16,218,079	352	Investment per Employee	6,552	5,719	87
Value Added per Employee	93,930	86,650	92	Investment per Production Worker	9,352	7,500	80

Sources: Same as General Statistics. The 'Average of All Manufacturing' column represents the average of all manufacturing industries reported for the most recent complete year available. The Index shows the relationship between the Average and the Analyzed Industry. For example, 100 means that they are equal; 500 that the Analyzed Industry is five times the average; 50 means that the Analyzed Industry is half the national average. The abbreviation 'na' is used to show that data are 'not available'.

LEADING COMPANIES   Number shown: 13   Total sales ($ mil): 530   Total employment (000): 6.4

Company Name	Address				CEO Name	Phone	Co. Type	Sales ($ mil)	Empl. (000)
Peterson Manufacturing Co	4200 E 135th St	Grandview	MO	64030	Don Armacost Jr	816-765-2000	R	107*	1.2
Truck-Lite Company Inc	310 E Elmwood Av	Falconer	NY	14733	Richard W Kotsi	716-665-6214	S	102	1.1
Grimes Aerospace Co	550 State Rte 55	Urbana	OH	43078	Paul E Gralnick	513-652-1431	R	95*	1.2
Federal-Mogul Corp	516 High St	Logansport	IN	46947	Ken Dresser	219-722-6141	D	71*	0.6
North American Lighting Inc	PO Box 499	Flora	IL	62839	Ed Grenda	618-662-4483	S	63	1.1
Concord Instruments Inc	1910 Elm St	Cincinnati	OH	45210	Andrew L Stone	513-621-4211	R	19	0.3
KD Lamp Co	1910 Elm St	Cincinnati	OH	45210	AL Stone	513-621-4211	S	19	0.2
Sate-Lite Manufacturing	6230 Grosse Pt Rd	Niles	IL	60714	Larry Michelson	708-647-1515	S	18*	0.2
Luminator Aircraft Products	1200 E Plano Pkwy	Plano	TX	75074	John Hartzler	214-424-6511	D	16	0.1
JW Speaker Corp	PO Box 489	Germantown	WI	53022	JA Speaker	414-251-6660	R	7	0.2
Progressive Dynamics Inc	507 Industrial Rd	Marshall	MI	49068	Eugene L Kilbourn	616-781-4241	R	7	<0.1
Soderberg Manufacturing	20821 Currier Rd	Walnut	CA	91789	BW Soderberg	909-595-1291	R	5	<0.1
Griffin Lamp Co	PO Box 66	Shelby	MS	38774	Robert D Gray	601-398-5131	R	2*	<0.1

Source: Ward's Business Directory of U.S. Private and Public Companies, Volumes 1 and 2, 1996. The company type code used is as follows: P - Public, R - Private, S - Subsidiary, D - Division, J - Joint Venture, A - Affiliate, G - Group. Sales are in millions of dollars, employees are in thousands. An asterisk (*) indicates an estimated sales volume. The symbol < stands for 'less than'. Company names and addresses are truncated, in some cases, to fit into the available space.

MATERIALS CONSUMED

Material	Quantity	Delivered Cost ($ million)
Materials, ingredients, containers, and supplies	(X)	911.8
Specialty transformers and fluorescent ballasts	(X)	6.6
Current-carrying wiring devices	(X)	13.5
Electric lamp bulbs	(X)	78.5
Flat glass (plate, float, and sheet)	(X)	(D)
Plastics resins consumed in the form of granules, pellets, powders, liquids, etc.	(X)	119.6
Plastics products consumed in the form of sheets, rods, tubes, film, and other shapes	(X)	14.5
Fabricated plastics products (except gaskets, hoses, and belting)	(X)	142.1
Insulated wire and cable, including magnet wire	(X)	10.2
Paperboard containers, boxes, and corrugated paperboard	(X)	20.0
Metal bolts, nuts, screws, washers, rivets, and other screw machine products	(X)	22.2
Metal poles	(X)	(D)
All other fabricated metal products (except forgings)	(X)	16.5
Iron and steel castings (rough and semifinished)	(X)	5.9
Aluminum and aluminum-base alloy castings (rough and semifinished)	(X)	2.0
Other nonferrous castings (rough and semifinished)	(X)	(D)
Steel sheet and strip, including tin plate	(X)	5.2
Steel wire and wire products	(X)	(D)
All other steel mill shapes and forms (except castings and forgings)	(X)	1.1
Aluminum and aluminum-base alloy sheet, plate, foil, and welded tubing	(X)	(D)
Aluminum and aluminum-base alloy extruded shapes, including extruded rod, bar, pipe, tube, etc.	(X)	(D)
Other aluminum and aluminum-base alloy shapes and forms	(X)	(D)
Copper and copper-base alloy shapes and forms	(X)	1.3
Other nonferrous shapes and forms	(X)	0.6
Lamp shades	(X)	(D)
All other materials and components, parts, containers, and supplies	(X)	320.9
Materials, ingredients, containers, and supplies, nsk	(X)	70.0

Source: 1992 Economic Census. Explanation of symbols used: (D): Withheld to avoid disclosure of competitive data; na: Not available; (S): Withheld because statistical norms were not met; (X): Not applicable; (Z): Less than half the unit shown; nec: Not elsewhere classified; nsk: Not specified by kind; - : zero; * : 10-19 percent estimated; ** : 20-29 percent estimated.

PRODUCT SHARE DETAILS

Product or Product Class	% Share	Product or Product Class	% Share
Vehicular lighting equipment	100.00		

Source: 1992 Economic Census. The values shown are percent of total shipments in an industry. Values of indented subcategories are summed in the main heading. The symbol (D) appears when data are withheld to prevent disclosure of competitive information. The abbreviation nsk stands for 'not specified by kind' and nec for 'not elsewhere classified'.

INPUTS AND OUTPUTS FOR LIGHTING FIXTURES & EQUIPMENT

Economic Sector or Industry Providing Inputs	%	Sector	Economic Sector or Industry Buying Outputs	%	Sector
Wholesale trade	11.2	Trade	Personal consumption expenditures	16.4	
Transformers	8.4	Manufg.	Office buildings	9.3	Constr.
Glass & glass products, except containers	8.0	Manufg.	Motor vehicles & car bodies	5.6	Manufg.
Metal stampings, nec	7.5	Manufg.	Residential 1-unit structures, nonfarm	5.0	Constr.
Imports	7.1	Foreign	Maintenance of nonfarm buildings nec	4.8	Constr.
Blast furnaces & steel mills	6.4	Manufg.	Gross private fixed investment	4.6	Cap Inv
Aluminum rolling & drawing	4.0	Manufg.	Electric utility facility construction	4.3	Constr.
Miscellaneous plastics products	2.5	Manufg.	Industrial buildings	4.3	Constr.
Paperboard containers & boxes	2.5	Manufg.	Exports	4.2	Foreign
Lighting fixtures & equipment	2.4	Manufg.	Nonfarm residential structure maintenance	3.2	Constr.
Plastics materials & resins	2.4	Manufg.	Residential additions/alterations, nonfarm	2.9	Constr.
Advertising	2.0	Services	Automotive repair shops & services	2.8	Services
Wiring devices	1.9	Manufg.	Construction of stores & restaurants	2.2	Constr.
Banking	1.9	Fin/R.E.	Construction of hospitals	1.9	Constr.
Screw machine and related products	1.7	Manufg.	Highway & street construction	1.9	Constr.
Electric services (utilities)	1.6	Util.	Construction of educational buildings	1.7	Constr.
Air transportation	1.5	Util.	Lighting fixtures & equipment	1.5	Manufg.
Aluminum castings	1.4	Manufg.	Eating & drinking places	1.5	Trade
Electric lamps	1.3	Manufg.	Hospitals	1.2	Services
Nonferrous wire drawing & insulating	1.2	Manufg.	Miscellaneous plastics products	1.1	Manufg.
Paints & allied products	1.2	Manufg.	Automobile parking & car washes	1.0	Services
Motor freight transportation & warehousing	1.2	Util.	Maintenance of highways & streets	0.9	Constr.
Plating & polishing	1.0	Manufg.	Warehouses	0.9	Constr.
Cyclic crudes and organics	0.9	Manufg.	Hotels & motels	0.7	Constr.
Communications, except radio & TV	0.9	Util.	Residential garden apartments	0.7	Constr.
Hotels & lodging places	0.9	Services	S/L Govt. purch., elem. & secondary education	0.7	S/L Govt
Manufacturing industries, nec	0.8	Manufg.	Coal	0.5	Mining
Semiconductors & related devices	0.8	Manufg.	Resid. & other health facility construction	0.5	Constr.
Electrical equipment & supplies, nec	0.7	Manufg.	Colleges, universities, & professional schools	0.5	Services
Petroleum refining	0.7	Manufg.	Amusement & recreation building construction	0.4	Constr.
Real estate	0.7	Fin/R.E.	Construction of nonfarm buildings nec	0.4	Constr.
Gas production & distribution (utilities)	0.6	Util.	Maintenance of farm service facilities	0.4	Constr.
Eating & drinking places	0.6	Trade	Telephone & telegraph facility construction	0.4	Constr.
Equipment rental & leasing services	0.6	Services	Mobile homes	0.4	Manufg.
Maintenance of nonfarm buildings nec	0.5	Constr.	Electric services (utilities)	0.4	Util.
Copper rolling & drawing	0.5	Manufg.	Elementary & secondary schools	0.4	Services
Iron & steel foundries	0.5	Manufg.	Crude petroleum & natural gas	0.3	Mining
Nonferrous castings, nec	0.5	Manufg.	Construction of religious buildings	0.3	Constr.
Primary zinc	0.5	Manufg.	Maintenance of electric utility facilities	0.3	Constr.
Wood products, nec	0.5	Manufg.	Maintenance of nonbuilding facilities nec	0.3	Constr.
Engineering, architectural, & surveying services	0.5	Services	Residential 2-4 unit structures, nonfarm	0.3	Constr.
Broadwoven fabric mills	0.4	Manufg.	Photographic equipment & supplies	0.3	Manufg.
Electronic components nec	0.4	Manufg.	Porcelain electrical supplies	0.3	Manufg.
Metal coating & allied services	0.4	Manufg.	Communications, except radio & TV	0.3	Util.
Miscellaneous fabricated wire products	0.4	Manufg.	Nursing & personal care facilities	0.3	Services
Gaskets, packing & sealing devices	0.3	Manufg.	Photofinishing labs, commercial photography	0.3	Services
Machinery, except electrical, nec	0.3	Manufg.	Theatrical producers, bands, entertainers	0.3	Services
Pottery products, nec	0.3	Manufg.	S/L Govt. purch., higher education	0.3	S/L Govt
Railroads & related services	0.3	Util.	S/L Govt. purch., other general government	0.3	S/L Govt
Computer & data processing services	0.3	Services	Nonbuilding facilities nec	0.2	Constr.
Legal services	0.3	Services	Residential high-rise apartments	0.2	Constr.
U.S. Postal Service	0.3	Gov't	Logging camps & logging contractors	0.2	Manufg.
Brass, bronze, & copper castings	0.2	Manufg.	Motor vehicle parts & accessories	0.2	Manufg.
Fabricated rubber products, nec	0.2	Manufg.	Petroleum refining	0.2	Manufg.
Hardwood dimension & flooring mills	0.2	Manufg.	Travel trailers & campers	0.2	Manufg.
Primary aluminum	0.2	Manufg.	Railroads & related services	0.2	Util.
Sawmills & planning mills, general	0.2	Manufg.	Water supply & sewage systems	0.2	Util.
Special dies & tools & machine tool accessories	0.2	Manufg.	Real estate	0.2	Fin/R.E.
Royalties	0.2	Fin/R.E.	Libraries, vocation education	0.2	Services
Accounting, auditing & bookkeeping	0.2	Services	S/L Govt. purch., health & hospitals	0.2	S/L Govt
Management & consulting services & labs	0.2	Services	S/L Govt. purch., natural resource & recreation.	0.2	S/L Govt
Machine tools, metal forming types	0.1	Manufg.	Farm service facilities	0.1	Constr.
Manifold business forms	0.1	Manufg.	Maintenance of military facilities	0.1	Constr.
Insurance carriers	0.1	Fin/R.E.	Sewer system facility construction	0.1	Constr.
Automotive rental & leasing, without drivers	0.1	Services	Automotive stampings	0.1	Manufg.
Automotive repair shops & services	0.1	Services	Blast furnaces & steel mills	0.1	Manufg.
Business/professional associations	0.1	Services	Boat building & repairing	0.1	Manufg.
			Motorcycles, bicycles, & parts	0.1	Manufg.
			Motors & generators	0.1	Manufg.
			Radio & TV communication equipment	0.1	Manufg.
			Gas production & distribution (utilities)	0.1	Util.
			Motor freight transportation & warehousing	0.1	Util.
			Credit agencies other than banks	0.1	Fin/R.E.
			Labor, civic, social, & fraternal associations	0.1	Services
			S/L Govt. purch., correction	0.1	S/L Govt
			S/L Govt. purch., public assistance & relief	0.1	S/L Govt

Source: Benchmark Input-Output Accounts for the U.S. Economy, 1982, U.S. Department of Commerce, Washington, D.C., July 1991. Data, as reported in the source, are organized by the 1977 SIC structure in use in 1982 but have been matched, as closely as is possible, to the 1987 SIC structure used in this book.

OCCUPATIONS EMPLOYED BY SIC 364 - ELECTRIC LIGHTING AND WIRING EQUIPMENT

Occupation	% of Total 1994	Change to 2005	Occupation	% of Total 1994	Change to 2005
Assemblers, fabricators, & hand workers nec	16.0	-7.4	Tool & die makers	1.5	-25.2
Electrical & electronic assemblers	7.6	-16.7	Hand packers & packagers	1.5	-20.6
Blue collar worker supervisors	3.7	-20.3	Sheet metal workers & duct installers	1.5	-7.4
Inspectors, testers, & graders, precision	3.2	-35.2	Machinists	1.5	-30.5
Plastic molding machine workers	3.2	11.1	Freight, stock, & material movers, hand	1.4	-25.9
Machine operators nec	2.5	-34.7	Industrial truck & tractor operators	1.3	-7.4
Industrial machinery mechanics	2.5	25.0	Machine tool cutting & forming etc. nec	1.2	-25.9
Sales & related workers nec	2.1	-7.4	Production, planning, & expediting clerks	1.2	-7.4
Electrical & electronic equipment assemblers	1.9	-7.4	Secretaries, ex legal & medical	1.2	-15.7
General managers & top executives	1.9	-12.2	Maintenance repairers, general utility	1.1	-16.6
Machine feeders & offbearers	1.8	-16.7	Industrial production managers	1.0	-7.4
Traffic, shipping, & receiving clerks	1.7	-10.9	Coating, painting, & spraying machine workers	1.0	-7.4
Helpers, laborers, & material movers nec	1.6	-7.4			

Source: *Industry-Occupation Matrix*, Bureau of Labor Statistics. These data relate to one or more 3-digit SIC industry groups rather than to a single 4-digit SIC. The change reported for each occupation to the year 2005 is a percent of growth or decline as estimated by the Bureau of Labor Statistics. The abbreviation nec stands for 'not elsewhere classified'.

LOCATION BY STATE AND REGIONAL CONCENTRATION

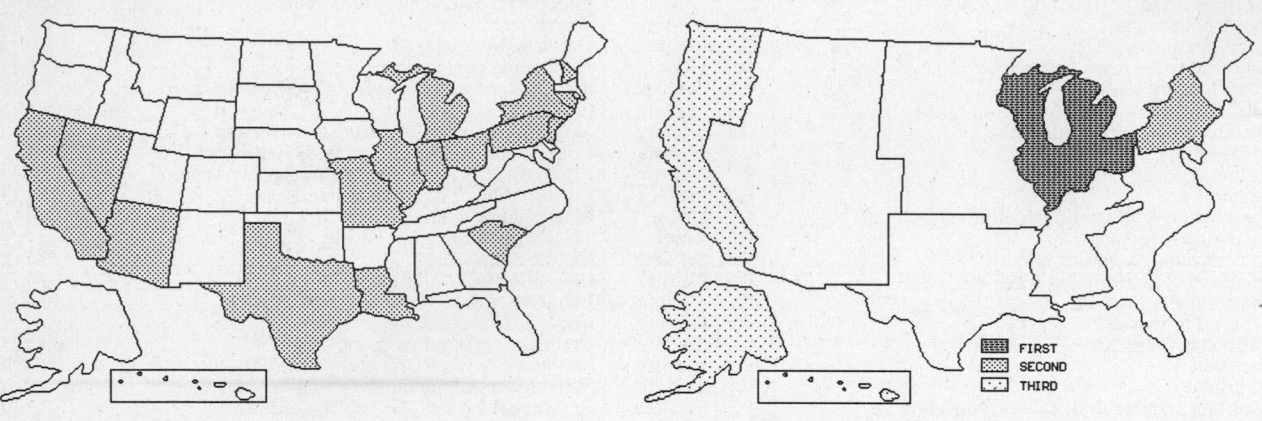

FIRST
SECOND
THIRD

INDUSTRY DATA BY STATE

State	Establish-ments	Shipments			Employment				Cost as % of Shipments	Investment per Employee ($)
		Total ($ mil)	% of U.S.	Per Establ.	Total Number	% of U.S.	Per Establ.	Wages ($/hour)		
Indiana	8	717.2	32.7	89.6	5,700	39.0	713	18.75	44.2	2,333
Illinois	8	155.0	7.1	19.4	1,200	8.2	150	9.78	50.3	-
Connecticut	5	65.7	3.0	13.1	400	2.7	80	10.00	42.9	3,500
Texas	5	55.8	2.5	11.2	600	4.1	120	10.83	41.9	-
Michigan	6	23.6	1.1	3.9	200	1.4	33	8.33	47.5	3,000
California	12	20.2	0.9	1.7	200	1.4	17	16.50	45.5	2,500
Ohio	9	(D)	-	-	1,750 *	12.0	194	-	-	3,657
New York	8	(D)	-	-	750 *	5.1	94	-	-	-
Missouri	5	(D)	-	-	750 *	5.1	150	-	-	-
New Jersey	3	(D)	-	-	375 *	2.6	125	-	-	1,867
Pennsylvania	3	(D)	-	-	375 *	2.6	125	-	-	-
Arizona	2	(D)	-	-	175 *	1.2	88	-	-	-
New Hampshire	2	(D)	-	-	175 *	1.2	88	-	-	-
Louisiana	1	(D)	-	-	750 *	5.1	750	-	-	-
Nevada	1	(D)	-	-	375 *	2.6	375	-	-	-
South Carolina	1	(D)	-	-	375 *	2.6	375	-	-	-

Source: 1992 *Economic Census*. The states are in descending order of shipments or establishments (if shipment data are missing for the majority). The symbol (D) appears when data are withheld to prevent disclosure of competitive information. States marked with (D) are sorted by number of establishments. A dash (-) indicates that the data element cannot be calculated; * indicates the midpoint of a range.

3648 - LIGHTING EQUIPMENT, NEC

82 83 84 85 86 87 88 89 90 91 92 93 94 95 96 97 98

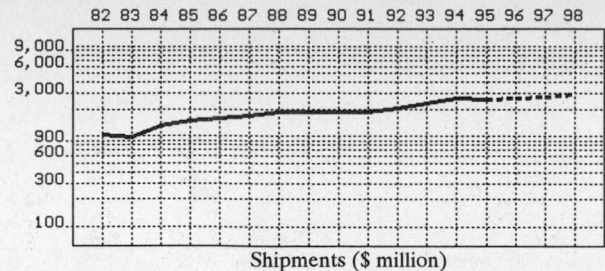

Shipments ($ million)

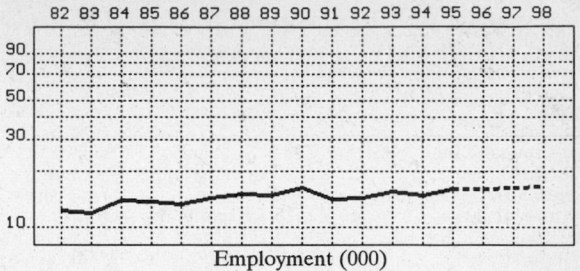

Employment (000)

GENERAL STATISTICS

Year	Companies	Establishments		Employment			Compensation		Production ($ million)			
		Total	with 20 or more employees	Total (000)	Production Workers (000)	Hours (Mil)	Payroll ($ mil)	Wages ($/hr)	Cost of Materials	Value Added by Manufacture	Value of Shipments	Capital Invest.
1982	222	233	113	12.2	8.5	16.1	207.9	7.53	454.4	566.4	1,028.0	28.0
1983		230	113	11.8	8.4	15.8	210.5	7.58	458.6	539.0	992.9	17.1
1984		227	113	13.9	9.8	18.9	251.3	7.61	614.1	693.7	1,303.5	33.9
1985		224	113	13.6	9.4	18.6	272.7	8.24	670.3	843.9	1,515.5	43.4
1986		226	111	13.2	9.1	17.5	270.6	8.58	681.2	922.9	1,607.3	32.5
1987	246	262	128	14.4	9.9	19.3	304.4	8.63	801.5	884.7	1,673.9	31.3
1988		251	127	15.1	10.1	19.9	344.8	9.39	917.2	952.3	1,863.1	32.1
1989		251	130	14.7	9.9	19.5	337.4	9.33	962.2	933.6	1,889.3	48.5
1990		248	130	16.4	9.7	19.2	347.6	9.26	958.2	883.6	1,849.5	46.8
1991		249	131	14.1	9.3	18.6	350.4	9.89	938.2	897.0	1,838.5	33.5
1992	272	293	132	14.5	9.3	18.7	394.3	10.21	1,018.0	1,073.8	2,083.5	50.9
1993		289	123	15.5	10.1	20.1	434.1	10.26	1,170.3	1,118.2	2,290.1	49.2
1994		283P	134P	14.9	9.5	17.9	425.4	11.80	1,235.8	1,470.3	2,672.7	52.9
1995		288P	136P	15.9P	10.0P	19.8P	450.7P	11.27P	1,180.1P	1,404.0P	2,552.3P	53.6P
1996		294P	138P	16.2P	10.1P	20.0P	469.5P	11.58P	1,233.8P	1,467.9P	2,668.4P	55.8P
1997		299P	140P	16.4P	10.1P	20.2P	488.2P	11.89P	1,287.5P	1,531.9P	2,784.6P	58.0P
1998		304P	142P	16.7P	10.2P	20.4P	507.0P	12.20P	1,341.3P	1,595.8P	2,900.8P	60.1P

Sources: 1982, 1987, 1992 *Economic Census*; *Annual Survey of Manufactures*, 83-86, 88-91, 93-94. Establishment counts for non-Census years are from *County Business Patterns*; establishment values for 83-84 are extrapolations. 'P's show projections by the editors. Industries reclassified in 87 will not have data for prior years.

INDICES OF CHANGE

Year	Companies	Establishments		Employment			Compensation		Production ($ million)			
		Total	with 20 or more employees	Total (000)	Production Workers (000)	Hours (Mil)	Payroll ($ mil)	Wages ($/hr)	Cost of Materials	Value Added by Manufacture	Value of Shipments	Capital Invest.
1982	82	80	86	84	91	86	53	74	45	53	49	55
1983		78	86	81	90	84	53	74	45	50	48	34
1984		77	86	96	105	101	64	75	60	65	63	67
1985		76	86	94	101	99	69	81	66	79	73	85
1986		77	84	91	98	94	69	84	67	86	77	64
1987	90	89	97	99	106	103	77	85	79	82	80	61
1988		86	96	104	109	106	87	92	90	89	89	63
1989		86	98	101	106	104	86	91	95	87	91	95
1990		85	98	113	104	103	88	91	94	82	89	92
1991		85	99	97	100	99	89	97	92	84	88	66
1992	100	100	100	100	100	100	100	100	100	100	100	100
1993		99	93	107	109	107	110	100	115	104	110	97
1994		97P	102P	103	102	96	108	116	121P	137	128P	104
1995		98P	103P	110P	107P	106P	114P	110P	116P	131P	122P	105P
1996		100P	105P	112P	108P	107P	119P	113P	121P	137P	128P	110P
1997		102P	106P	113P	109P	108P	124P	116P	126P	143P	134P	114P
1998		104P	108P	115P	110P	109P	129P	120P	132P	149P	139P	118P

Sources: Same as General Statistics. Values reflect change from the base year, 1992. Values above 100 mean greater than 92, values below 100 mean less than 92, and a value of 100 in the 82-91 or 93-98 period means same as 92. 'P's mark projections by the editors.

SELECTED RATIOS

For 1994	Avg. of All Manufact.	Analyzed Industry	Index	For 1994	Avg. of All Manufact.	Analyzed Industry	Index
Employees per Establishment	49	53	107	Value Added per Production Worker	134,084	154,768	115
Payroll per Establishment	1,500,273	1,502,376	100	Cost per Establishment	5,045,178	4,364,448	87
Payroll per Employee	30,620	28,550	93	Cost per Employee	102,970	82,940	81
Production Workers per Establishment	34	34	98	Cost per Production Worker	146,988	130,084	88
Wages per Establishment	853,319	745,961	87	Shipments per Establishment	9,576,895	9,439,116	99
Wages per Production Worker	24,861	22,234	89	Shipments per Employee	195,460	179,376	92
Hours per Production Worker	2,056	1,884	92	Shipments per Production Worker	279,017	281,337	101
Wages per Hour	12.09	11.80	98	Investment per Establishment	321,011	186,826	58
Value Added per Establishment	4,602,255	5,192,626	113	Investment per Employee	6,552	3,550	54
Value Added per Employee	93,930	98,678	105	Investment per Production Worker	9,352	5,568	60

Sources: Same as General Statistics. The 'Average of All Manufacturing' column represents the average of all manufacturing industries reported for the most recent complete year available. The Index shows the relationship between the Average and the Analyzed Industry. For example, 100 means that they are equal; 500 that the Analyzed Industry is five times the average; 50 means that the Analyzed Industry is half the national average. The abbreviation 'na' is used to show that data are 'not available'.

LEADING COMPANIES   Number shown: 55   Total sales ($ mil): 1,764   Total employment (000): 10.7

Company Name	Address				CEO Name	Phone	Co. Type	Sales ($ mil)	Empl. (000)
Coleman Company Inc	250 N St Francis	Wichita	KS	67202	Michael N Hammes	316-261-3211	P	752	4.8
AMSCO Healthcare	112 Washington Pl	Pittsburgh	PA	15219	Daniel R Barry	412-338-6500	D	216	<0.1
Juno Lighting Inc	PO Box 5065	Des Plaines	IL	60017	Robert S Fremont	708-827-9880	P	127	0.9
Mag Instrument Inc	1635 S Sacramento	Ontario	CA	91761	Tony Maglica	909-947-1006	R	100	0.5
Publicker Industries Inc	1445 E Putnam Av	Old Greenwich	CT	06870	James J Weis	203-637-4500	P	76	1.0
Garrity Industries	14 New Rd	Madison	CT	06443	Kevin S Garrity	203-245-8383	R	75	0.3
Dual-Lite Inc	Simm Ln	Newtown	CT	06470	JR Carson	203-426-8011	S	50*	0.4
ADB Alnaco Inc	PO Box 30829	Columbus	OH	43230	Steve Rauch	614-861-1304	S	33*	0.1
Lightalarms Electronics Corp	1170 Atlantic Av	Baldwin	NY	11510	Michael Kaufman	516-379-1000	S	25	0.1
Colortran Inc	1015 Chestnut St	Burbank	CA	91506	Robert Sherman	818-843-1200	S	23	0.2
Bright Image Corp	4900 Harrison St	Hillside	IL	60162	Atiq Jilani	708-449-5656	R	20	<0.1
ALM Surgical Equipment Inc	1820 N Lemon St	Anaheim	CA	92801	George Crispin	714-578-1234	S	16*	<0.1
Pelican Products Inc	23215 Early Av	Torrance	CA	90505	David Parker	310-326-3700	R	16	0.2
Amida Industries Inc	PO Box 3147	Rock Hill	SC	29732	Irvin Plowden Sr	803-324-3011	R	15	0.1
Paraflex Industries Inc	PO Box 920	Beacon	NY	12508	Michael A James	914-831-9000	R	14	<0.1
Ledtronics Inc	4009 Pacific Coast	Torrance	CA	90505		310-534-1505	R	12	0.1
North Starlighting Inc	2150 W 16th St	Broadview	IL	60153	Len J Blaszak	708-681-4330	S	12*	<0.1
Bright Star Industries Inc	380 Stewart Rd	Wilkes-Barre	PA	18706	Lee Kirwan	717-825-1900	S	11*	<0.1
Gem Electric Manufacturing	390 Vanderbilt	Hauppauge	NY	11788	Harvey Cooper	516-273-2230	R	11	0.1
Automatic Power Inc	PO Box 230738	Houston	TX	77223	Steve Trenchard	713-228-5208	D	10	<0.1
Spectronics Corp	956 Brush Hollow	Westbury	NY	11590	BW Cooper	516-333-4840	R	10	0.2
Spring City Electrical Mfg Co	PO Drawer A	Spring City	PA	19475	John A Miller	215-948-4000	R	10	0.2
Foremost Manufacturing	941 Ball Av	Union	NJ	07083	Herbert Schiller	908-687-4646	R	9	0.1
Koehler Manufacturing Co	123 Felton St	Marlboro	MA	01752	MW Powning	508-485-1000	R	9	0.1
Bieber Lighting Corp	970 W Manchester	Inglewood	CA	90301	Lawrence Bieber	213-776-4744	R	8*	<0.1
Multi-Electric Manufacturing Inc	4223 W Lake St	Chicago	IL	60624	Mike A Mongoven	312-722-1900	R	8*	<0.1
Radiant Illumination Inc	7121 Case Av	N Hollywood	CA	91605	Ernest Silva	818-982-0160	S	8*	<0.1
Haggerty Enterprises Inc	2321 N Keystone	Chicago	IL	60639	Jack F Mundy	312-342-5700	R	7	0.1
American Ultraviolet Co	562 Central Av	Murray Hill	NJ	07974	MC Stines	908-665-2211	R	6	<0.1
LAM Lighting Systems Inc	2930 S Fairview St	Santa Ana	CA	92704	Gary Gulden	714-549-9765	S	6	<0.1
Norman Enterprises Inc	2601 Empire Av	Burbank	CA	91504	William Norman	818-843-6811	S	6*	<0.1
Norwell Manufacturing Inc	82 Stevens St	East Taunton	MA	02718	Robert Cannon	508-822-5854	R	6*	<0.1
Highline Products Corp	800 South St	Waltham	MA	02154	Michael Brown	617-736-0002	R	5	<0.1
LSI Lectro Science Inc	6410 W Ridge Rd	Erie	PA	16506	Floyd Devroy	814-833-6487	R	5	<0.1
Siltron Illumination Inc	301 Doubleday Av	Ontario	CA	91761	Brad Kies	909-941-3500	S	5	<0.1
Western Lighting Standards Inc	3151 Airway Av	Costa Mesa	CA	92626	John Miller	714-586-7031	S	5	0.1
General Metalworks Corp	10245 N Enterprise	Mequon	WI	53092	Jerry Dubiel	414-242-2800	D	4	<0.1
Silverlight Corp	16 W 151 Shore Ct	Burr Ridge	IL	60521	James W Zarlenga	708-986-1651	S	4	<0.1
Gilbert Industries Inc	5611 Krueger Dr	Jonesboro	AR	72401	David W Gilbert II	501-932-6070	R	3	<0.1
Hipwell Manufacturing Co	831 W North Av	Pittsburgh	PA	15233	Harry H Hipwell Sr	412-231-7310	R	3	<0.1
Lava-Simplex Internationale	2321 N Keystone	Chicago	IL	60639	JF Mundy	312-342-5700	D	3	<0.1
Prestige Products Inc	1250 Arthur Av	Elk Grove Vill	IL	60007	Dan Mazzie	708-595-8000	R	3	<0.1
Stewart R Browne Mfg Co	1165 Htower	Atlanta	GA	30350	Robert Browne	404-993-9600	R	3	<0.1
United Lighting and Ceiling Co	513 Independent Rd	Oakland	CA	94621	Robert A Schwartz	510-569-6700	D	3	<0.1
Kilowatt Saver Inc	924 Sligh Blv	Orlando	FL	32806	Ronald Flaherty	407-872-0143	R	2	<0.1
Dynaray Inc	PO Box 969	Westbrook	CT	06498	WJ Simpson Sr	203-388-3007	S	2	<0.1
Guyco Corp	2751 NW 75th St	Miami	FL	33147	Michael Turner	305-836-1215	R	1	<0.1
ACO Inc	5656 Northwest N	Chicago	IL	60646	WR Nordlof	312-774-5200	R	1	<0.1
Machinery Products Co	1893 Com Pk E	Lancaster	PA	17601	David Gammache	717-295-2422	R	1	<0.1
Syracuse Safety-Lites Inc	6397 Joy Rd	East Syracuse	NY	13057	Peter Low	315-437-8423	R	1	<0.1
Natale Machine and Tool	339 13th St	Carlstadt	NJ	07072	Dominick Natale	201-933-5500	R	1*	<0.1
Peak Beam Systems Inc	523 E Bill Williams	Williams	AZ	86046	Robert Brunson	602-635-2695	R	1	<0.1
Cantek Metatron Corp	19 W Water St	Canonsburg	PA	15317	Lucian J Spalla	412-745-6760	R	1	<0.1
Prasse Enterprise	1315 N Bolton Rd	Freeport	IL	61032	Dave Prasse	815-233-1966	R	1	<0.1
Bardwell and McAlister Inc	10314 Norris Av	Pazoima	CA	91331	James Kalisz	818-834-6136	S	0*	<0.1

Source: *Ward's Business Directory of U.S. Private and Public Companies*, Volumes 1 and 2, 1996. The company type code used is as follows: P - Public, R - Private, S - Subsidiary, D - Division, J - Joint Venture, A - Affiliate, G - Group. Sales are in millions of dollars, employees are in thousands. An asterisk (*) indicates an estimated sales volume. The symbol < stands for 'less than'. Company names and addresses are truncated, in some cases, to fit into the available space.

MATERIALS CONSUMED

Material	Quantity	Delivered Cost ($ million)
Materials, ingredients, containers, and supplies	(X)	889.8
Specialty transformers and fluorescent ballasts	(X)	66.8
Current-carrying wiring devices	(X)	57.2
Electric lamp bulbs	(X)	50.8
Flat glass (plate, float, and sheet)	(X)	11.5
Plastics resins consumed in the form of granules, pellets, powders, liquids, etc.	(X)	9.7
Plastics products consumed in the form of sheets, rods, tubes, film, and other shapes	(X)	14.5
Fabricated plastics products (except gaskets, hoses, and belting)	(X)	36.5
Insulated wire and cable, including magnet wire	(X)	15.0
Paperboard containers, boxes, and corrugated paperboard	(X)	28.3
Metal bolts, nuts, screws, washers, rivets, and other screw machine products	(X)	22.1
Metal poles	(X)	18.7
All other fabricated metal products (except forgings)	(X)	30.9

Continued on next page.

MATERIALS CONSUMED - Continued

Material	Quantity	Delivered Cost ($ million)
Iron and steel castings (rough and semifinished)	(X)	3.7
Aluminum and aluminum-base alloy castings (rough and semifinished)	(X)	58.1
Other nonferrous castings (rough and semifinished)	(X)	4.1
Steel sheet and strip, including tin plate	(X)	24.1
Steel wire and wire products	(X)	4.5
All other steel mill shapes and forms (except castings and forgings)	(X)	8.6
Aluminum and aluminum-base alloy sheet, plate, foil, and welded tubing	(X)	19.2
Aluminum and aluminum-base alloy extruded shapes, including extruded rod, bar, pipe, tube, etc.	(X)	41.7
Other aluminum and aluminum-base alloy shapes and forms	(X)	7.3
Copper and copper-base alloy shapes and forms	(X)	2.7
Other nonferrous shapes and forms	(X)	1.8
Lamp shades	(X)	7.7
All other materials and components, parts, containers, and supplies	(X)	157.9
Materials, ingredients, containers, and supplies, nsk	(X)	186.3

Source: 1992 *Economic Census*. Explanation of symbols used: (D): Withheld to avoid disclosure of competitive data; na: Not available; (S): Withheld because statistical norms were not met; (X): Not applicable; (Z): Less than half the unit shown; nec: Not elsewhere classified; nsk: Not specified by kind; - : zero; * : 10-19 percent estimated; ** : 20-29 percent estimated.

PRODUCT SHARE DETAILS

Product or Product Class	% Share	Product or Product Class	% Share
Lighting equipment, nec	100.00	Other electric lighting equipment, including electrical discharge, such as mercury vapor, sodium vapor, etc.	7.13
Outdoor lighting equipment (including parts and accessories)	57.23	Ultraviolet and infrared health lamp fixtures (excluding lamp bulbs sold separately)	(D)
Electric and nonelectric lighting equipment, nec, including hand portable, and parts and accessories	37.81	Parts and accessories for other electric lighting fixtures, nec.	3.25
Rechargeable battery-operated incandescent hand portable lighting equipment, and parts and accessories	4.52	Nonelectric lighting equipment (including parts), lamps and lanterns (including kerosene, gasoline, propane, butane, etc.)	(D)
Incandescent hand portable lighting equipment, other than rechargeable battery-operated, flashlights and flashlight lanterns (one to five cells)	35.44	Other nonelectric lighting fixtures and equipment, complete units (including carbide lamps of all types)	(D)
Other incandescent hand portable lighting equipment, other than recargeable battery-operated, such as miners' lights, emergency warning lights, generator flashlights, etc.	3.13	Parts and accessories for nonelectric lighting equipment, including reflectors and fittings, incandescent mantles, etc.)	3.01
Other incandescent electric lighting equipment (including marine markers or beacons)	4.72	Electric and nonelectric lighting equipment, nec, including hand portable, and parts and accessories, nsk	7.69
Other fluorescent lighting equipment, complete units, including processing and technical equipment	16.51	Lighting equipment, nec, nsk	4.97

Source: 1992 *Economic Census*. The values shown are percent of total shipments in an industry. Values of indented subcategories are summed in the main heading. The symbol (D) appears when data are withheld to prevent disclosure of competitive information. The abbreviation nsk stands for 'not specified by kind' and nec for 'not elsewhere classified'.

INPUTS AND OUTPUTS FOR LIGHTING FIXTURES & EQUIPMENT

Economic Sector or Industry Providing Inputs	%	Sector	Economic Sector or Industry Buying Outputs	%	Sector
Wholesale trade	11.2	Trade	Personal consumption expenditures	16.4	
Transformers	8.4	Manufg.	Office buildings	9.3	Constr.
Glass & glass products, except containers	8.0	Manufg.	Motor vehicles & car bodies	5.6	Manufg.
Metal stampings, nec	7.5	Manufg.	Residential 1-unit structures, nonfarm	5.0	Constr.
Imports	7.1	Foreign	Maintenance of nonfarm buildings nec	4.8	Constr.
Blast furnaces & steel mills	6.4	Manufg.	Gross private fixed investment	4.6	Cap Inv
Aluminum rolling & drawing	4.0	Manufg.	Electric utility facility construction	4.3	Constr.
Miscellaneous plastics products	2.5	Manufg.	Industrial buildings	4.3	Constr.
Paperboard containers & boxes	2.5	Manufg.	Exports	4.2	Foreign
Lighting fixtures & equipment	2.4	Manufg.	Nonfarm residential structure maintenance	3.2	Constr.
Plastics materials & resins	2.4	Manufg.	Residential additions/alterations, nonfarm	2.9	Constr.
Advertising	2.0	Services	Automotive repair shops & services	2.8	Services
Wiring devices	1.9	Manufg.	Construction of stores & restaurants	2.2	Constr.
Banking	1.9	Fin/R.E.	Construction of hospitals	1.9	Constr.
Screw machine and related products	1.7	Manufg.	Highway & street construction	1.9	Constr.
Electric services (utilities)	1.6	Util.	Construction of educational buildings	1.7	Constr.
Air transportation	1.5	Util.	Lighting fixtures & equipment	1.5	Manufg.
Aluminum castings	1.4	Manufg.	Eating & drinking places	1.5	Trade
Electric lamps	1.3	Manufg.	Hospitals	1.2	Services
Nonferrous wire drawing & insulating	1.2	Manufg.	Miscellaneous plastics products	1.1	Manufg.
Paints & allied products	1.2	Manufg.	Automobile parking & car washes	1.0	Services
Motor freight transportation & warehousing	1.2	Util.	Maintenance of highways & streets	0.9	Constr.
Plating & polishing	1.0	Manufg.	Warehouses	0.9	Constr.
Cyclic crudes and organics	0.9	Manufg.	Hotels & motels	0.7	Constr.
Communications, except radio & TV	0.9	Util.	Residential garden apartments	0.7	Constr.
Hotels & lodging places	0.9	Services	S/L Govt. purch., elem. & secondary education	0.7	S/L Govt
Manufacturing industries, nec	0.8	Manufg.	Coal	0.5	Mining

Continued on next page.

INPUTS AND OUTPUTS FOR LIGHTING FIXTURES & EQUIPMENT - Continued

Economic Sector or Industry Providing Inputs	%	Sector	Economic Sector or Industry Buying Outputs	%	Sector
Semiconductors & related devices	0.8	Manufg.	Resid. & other health facility construction	0.5	Constr.
Electrical equipment & supplies, nec	0.7	Manufg.	Colleges, universities, & professional schools	0.5	Services
Petroleum refining	0.7	Manufg.	Amusement & recreation building construction	0.4	Constr.
Real estate	0.7	Fin/R.E.	Construction of nonfarm buildings nec	0.4	Constr.
Gas production & distribution (utilities)	0.6	Util.	Maintenance of farm service facilities	0.4	Constr.
Eating & drinking places	0.6	Trade	Telephone & telegraph facility construction	0.4	Constr.
Equipment rental & leasing services	0.6	Services	Mobile homes	0.4	Manufg.
Maintenance of nonfarm buildings nec	0.5	Constr.	Electric services (utilities)	0.4	Util.
Copper rolling & drawing	0.5	Manufg.	Elementary & secondary schools	0.4	Services
Iron & steel foundries	0.5	Manufg.	Crude petroleum & natural gas	0.3	Mining
Nonferrous castings, nec	0.5	Manufg.	Construction of religious buildings	0.3	Constr.
Primary zinc	0.5	Manufg.	Maintenance of electric utility facilities	0.3	Constr.
Wood products, nec	0.5	Manufg.	Maintenance of nonbuilding facilities nec	0.3	Constr.
Engineering, architectural, & surveying services	0.5	Services	Residential 2-4 unit structures, nonfarm	0.3	Constr.
Broadwoven fabric mills	0.4	Manufg.	Photographic equipment & supplies	0.3	Manufg.
Electronic components nec	0.4	Manufg.	Porcelain electrical supplies	0.3	Manufg.
Metal coating & allied services	0.4	Manufg.	Communications, except radio & TV	0.3	Util.
Miscellaneous fabricated wire products	0.4	Manufg.	Nursing & personal care facilities	0.3	Services
Gaskets, packing & sealing devices	0.3	Manufg.	Photofinishing labs, commercial photography	0.3	Services
Machinery, except electrical, nec	0.3	Manufg.	Theatrical producers, bands, entertainers	0.3	Services
Pottery products, nec	0.3	Manufg.	S/L Govt. purch., higher education	0.3	S/L Govt
Railroads & related services	0.3	Util.	S/L Govt. purch., other general government	0.3	S/L Govt
Computer & data processing services	0.3	Services	Nonbuilding facilities nec	0.2	Constr.
Legal services	0.3	Services	Residential high-rise apartments	0.2	Constr.
U.S. Postal Service	0.3	Gov't	Logging camps & logging contractors	0.2	Manufg.
Brass, bronze, & copper castings	0.2	Manufg.	Motor vehicle parts & accessories	0.2	Manufg.
Fabricated rubber products, nec	0.2	Manufg.	Petroleum refining	0.2	Manufg.
Hardwood dimension & flooring mills	0.2	Manufg.	Travel trailers & campers	0.2	Manufg.
Primary aluminum	0.2	Manufg.	Railroads & related services	0.2	Util.
Sawmills & planning mills, general	0.2	Manufg.	Water supply & sewage systems	0.2	Util.
Special dies & tools & machine tool accessories	0.2	Manufg.	Real estate	0.2	Fin/R.E.
Royalties	0.2	Fin/R.E.	Libraries, vocation education	0.2	Services
Accounting, auditing & bookkeeping	0.2	Services	S/L Govt. purch., health & hospitals	0.2	S/L Govt
Management & consulting services & labs	0.2	Services	S/L Govt. purch., natural resource & recreation.	0.2	S/L Govt
Machine tools, metal forming types	0.1	Manufg.	Farm service facilities	0.1	Constr.
Manifold business forms	0.1	Manufg.	Maintenance of military facilities	0.1	Constr.
Insurance carriers	0.1	Fin/R.E.	Sewer system facility construction	0.1	Constr.
Automotive rental & leasing, without drivers	0.1	Services	Automotive stampings	0.1	Manufg.
Automotive repair shops & services	0.1	Services	Blast furnaces & steel mills	0.1	Manufg.
Business/professional associations	0.1	Services	Boat building & repairing	0.1	Manufg.
			Motorcycles, bicycles, & parts	0.1	Manufg.
			Motors & generators	0.1	Manufg.
			Radio & TV communication equipment	0.1	Manufg.
			Gas production & distribution (utilities)	0.1	Util.
			Motor freight transportation & warehousing	0.1	Util.
			Credit agencies other than banks	0.1	Fin/R.E.
			Labor, civic, social, & fraternal associations	0.1	Services
			S/L Govt. purch., correction	0.1	S/L Govt
			S/L Govt. purch., public assistance & relief	0.1	S/L Govt

Source: Benchmark Input-Output Accounts for the U.S. Economy, 1982, U.S. Department of Commerce, Washington, D.C., July 1991. Data, as reported in the source, are organized by the 1977 SIC structure in use in 1982 but have been matched, as closely as is possible, to the 1987 SIC structure used in this book.

OCCUPATIONS EMPLOYED BY SIC 364 - ELECTRIC LIGHTING AND WIRING EQUIPMENT

Occupation	% of Total 1994	Change to 2005	Occupation	% of Total 1994	Change to 2005
Assemblers, fabricators, & hand workers nec	16.0	-7.4	Tool & die makers	1.5	-25.2
Electrical & electronic assemblers	7.6	-16.7	Hand packers & packagers	1.5	-20.6
Blue collar worker supervisors	3.7	-20.3	Sheet metal workers & duct installers	1.5	-7.4
Inspectors, testers, & graders, precision	3.2	-35.2	Machinists	1.5	-30.5
Plastic molding machine workers	3.2	11.1	Freight, stock, & material movers, hand	1.4	-25.9
Machine operators nec	2.5	-34.7	Industrial truck & tractor operators	1.3	-7.4
Industrial machinery mechanics	2.5	25.0	Machine tool cutting & forming etc. nec	1.2	-25.9
Sales & related workers nec	2.1	-7.4	Production, planning, & expediting clerks	1.2	-7.4
Electrical & electronic equipment assemblers	1.9	-7.4	Secretaries, ex legal & medical	1.2	-15.7
General managers & top executives	1.9	-12.2	Maintenance repairers, general utility	1.1	-16.6
Machine feeders & offbearers	1.8	-16.7	Industrial production managers	1.0	-7.4
Traffic, shipping, & receiving clerks	1.7	-10.9	Coating, painting, & spraying machine workers	1.0	-7.4
Helpers, laborers, & material movers nec	1.6	-7.4			

Source: Industry-Occupation Matrix, Bureau of Labor Statistics. These data relate to one or more 3-digit SIC industry groups rather than to a single 4-digit SIC. The change reported for each occupation to the year 2005 is a percent of growth or decline as estimated by the Bureau of Labor Statistics. The abbreviation nec stands for 'not elsewhere classified'.

LOCATION BY STATE AND REGIONAL CONCENTRATION

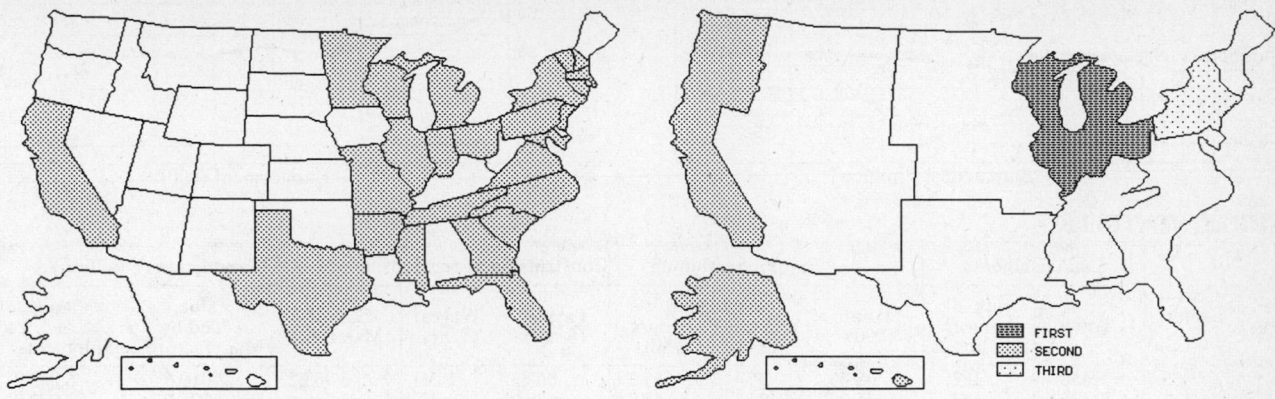

FIRST
SECOND
THIRD

INDUSTRY DATA BY STATE

| State | Establish-ments | Shipments | | | Employment | | | | Cost as % of Shipments | Investment per Employee ($) |
		Total ($ mil)	% of U.S.	Per Establ.	Total Number	% of U.S.	Per Establ.	Wages ($/hour)		
California	63	382.9	18.4	6.1	2,900	20.0	46	9.51	46.2	3,069
Connecticut	9	94.3	4.5	10.5	400	2.8	44	10.00	62.1	-
Texas	13	94.2	4.5	7.2	600	4.1	46	9.29	50.5	2,833
Wisconsin	12	85.3	4.1	7.1	600	4.1	50	7.57	36.6	-
Illinois	24	83.1	4.0	3.5	700	4.8	29	8.91	46.6	6,714
New York	23	66.8	3.2	2.9	500	3.4	22	14.17	44.9	2,600
Indiana	5	40.0	1.9	8.0	300	2.1	60	12.00	46.8	667
Florida	18	26.5	1.3	1.5	400	2.8	22	10.75	58.1	2,000
Michigan	5	19.4	0.9	3.9	100	0.7	20	11.50	51.0	5,000
Ohio	20	(D)	-	-	750 *	5.2	38	-	-	-
Pennsylvania	15	(D)	-	-	750 *	5.2	50	-	-	2,000
New Jersey	12	(D)	-	-	375 *	2.6	31	-	-	1,867
Massachusetts	8	(D)	-	-	375 *	2.6	47	-	-	3,200
North Carolina	7	(D)	-	-	1,750 *	12.1	250	-	-	-
Minnesota	4	(D)	-	-	375 *	2.6	94	-	-	-
Missouri	3	(D)	-	-	175 *	1.2	58	-	-	-
New Hampshire	3	(D)	-	-	175 *	1.2	58	-	-	-
South Carolina	3	(D)	-	-	175 *	1.2	58	-	-	-
Tennessee	3	(D)	-	-	175 *	1.2	58	-	-	-
Vermont	3	(D)	-	-	175 *	1.2	58	-	-	-
Arkansas	2	(D)	-	-	375 *	2.6	188	-	-	-
Georgia	2	(D)	-	-	375 *	2.6	188	-	-	-
Virginia	2	(D)	-	-	750 *	5.2	375	-	-	-
Maryland	1	(D)	-	-	375 *	2.6	375	-	-	-
Mississippi	1	(D)	-	-	750 *	5.2	750	-	-	-

Source: 1992 *Economic Census*. The states are in descending order of shipments or establishments (if shipment data are missing for the majority). The symbol (D) appears when data are withheld to prevent disclosure of competitive information. States marked with (D) are sorted by number of establishments. A dash (-) indicates that the data element cannot be calculated; * indicates the midpoint of a range.

3651 - HOUSEHOLD AUDIO & VIDEO EQUIPMENT

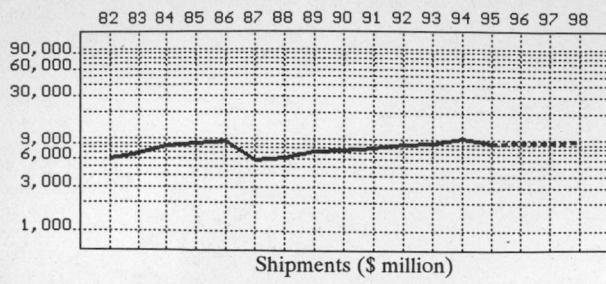

Shipments ($ million)

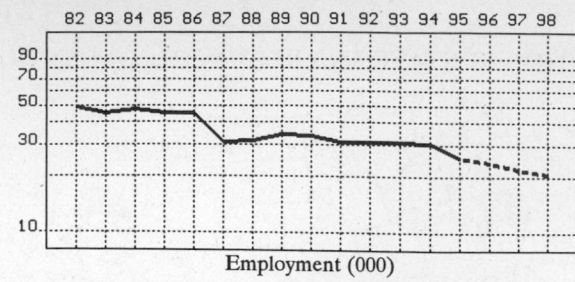

Employment (000)

GENERAL STATISTICS

Year	Companies	Establishments		Employment			Compensation		Production ($ million)			
		Total	with 20 or more employees	Total (000)	Production Workers (000)	Hours (Mil)	Payroll ($ mil)	Wages ($/hr)	Cost of Materials	Value Added by Manufacture	Value of Shipments	Capital Invest.
1982	435	458	182	48.4	35.4	65.3	862.3	8.36	3,967.2	2,010.6	6,063.9	140.9
1983		430	173	45.3	32.5	64.8	935.9	9.17	4,648.4	2,096.6	6,772.8	166.7
1984		402	164	47.5	35.9	68.9	1,038.0	9.85	5,642.5	2,868.0	8,216.9	255.9
1985		374	156	45.1	33.7	64.6	1,033.2	10.47	6,333.9	2,323.1	8,888.1	252.9
1986		354	156	44.9	33.7	65.6	1,067.9	10.48	6,994.6	2,539.2	9,363.9	240.5
1987	352	378	150	30.9	23.6	45.2	583.9	8.02	4,247.8	1,702.1	5,911.2	124.9
1988		368	149	31.6	24.2	47.1	659.2	8.54	4,709.3	1,553.4	6,326.8	127.7
1989		376	164	34.0	24.9	50.1	704.1	8.54	5,532.1	1,904.3	7,360.2	139.0
1990		395	173	33.7	22.5	45.2	704.3	9.00	5,592.9	1,892.0	7,520.5	255.7
1991		401	170	31.1	21.7	41.9	732.4	9.36	5,893.8	2,122.4	7,993.6	277.5
1992	400	427	163	31.2	22.3	44.3	736.6	9.52	6,444.2	2,280.1	8,769.3	252.9
1993		439	164	31.2	22.6	45.0	774.8	9.60	6,596.7	2,567.4	9,159.3	211.5
1994		398P	161P	30.5	23.3	46.6	798.2	9.64	7,617.4	2,756.3	10,285.6	225.8
1995		398P	160P	25.6P	18.4P	37.5P	667.6P	9.38P	6,830.9P	2,471.7P	9,223.7P	237.8P
1996		398P	160P	24.0P	17.1P	35.3P	646.1P	9.39P	6,971.5P	2,522.6P	9,413.5P	242.4P
1997		397P	159P	22.3P	15.9P	33.0P	624.6P	9.41P	7,112.1P	2,573.5P	9,603.3P	247.1P
1998		397P	159P	20.6P	14.6P	30.7P	603.2P	9.42P	7,252.7P	2,624.3P	9,793.2P	251.7P

Sources: 1982, 1987, 1992 *Economic Census*; *Annual Survey of Manufactures*, 83-86, 88-91, 93-94. Establishment counts for non-Census years are from *County Business Patterns*; establishment values for 83-84 are extrapolations. 'P's show projections by the editors. Industries reclassified in 87 will not have data for prior years.

INDICES OF CHANGE

Year	Companies	Establishments		Employment			Compensation		Production ($ million)			
		Total	with 20 or more employees	Total (000)	Production Workers (000)	Hours (Mil)	Payroll ($ mil)	Wages ($/hr)	Cost of Materials	Value Added by Manufacture	Value of Shipments	Capital Invest.
1982	109	107	112	155	159	147	117	88	62	88	69	56
1983		101	106	145	146	146	127	96	72	92	77	66
1984		94	101	152	161	156	141	103	88	126	94	101
1985		88	96	145	151	146	140	110	98	102	101	100
1986		83	96	144	151	148	145	110	109	111	107	95
1987	88	89	92	99	106	102	79	84	66	75	67	49
1988		86	91	101	109	106	89	90	73	68	72	50
1989		88	101	109	112	113	96	90	86	84	84	55
1990		93	106	108	101	102	96	95	87	83	86	101
1991		94	104	100	97	95	99	98	91	93	91	110
1992	100	100	100	100	100	100	100	100	100	100	100	100
1993		103	101	100	101	102	105	101	102	113	104	84
1994		93P	98P	98	104	105	108	101	118	121	117	89
1995		93P	98P	82P	83P	85P	91P	99P	106P	108P	105P	94P
1996		93P	98P	77P	77P	80P	88P	99P	108P	111P	107P	96P
1997		93P	98P	71P	71P	74P	85P	99P	110P	113P	110P	98P
1998		93P	97P	66P	65P	69P	82P	99P	113P	115P	112P	100P

Sources: Same as General Statistics. Values reflect change from the base year, 1992. Values above 100 mean greater than 92, values below 100 mean less than 92, and a value of 100 in the 82-91 or 93-98 period means same as 92. 'P's mark projections by the editors.

SELECTED RATIOS

For 1994	Avg. of All Manufact.	Analyzed Industry	Index	For 1994	Avg. of All Manufact.	Analyzed Industry	Index
Employees per Establishment	49	77	156	Value Added per Production Worker	134,084	118,296	88
Payroll per Establishment	1,500,273	2,004,002	134	Cost per Establishment	5,045,178	19,124,635	379
Payroll per Employee	30,620	26,170	85	Cost per Employee	102,970	249,751	243
Production Workers per Establishment	34	58	170	Cost per Production Worker	146,988	326,927	222
Wages per Establishment	853,319	1,127,845	132	Shipments per Establishment	9,576,895	25,823,554	270
Wages per Production Worker	24,861	19,280	78	Shipments per Employee	195,460	337,233	173
Hours per Production Worker	2,056	2,000	97	Shipments per Production Worker	279,017	441,442	158
Wages per Hour	12.09	9.64	80	Investment per Establishment	321,011	566,905	177
Value Added per Establishment	4,602,255	6,920,108	150	Investment per Employee	6,552	7,403	113
Value Added per Employee	93,930	90,370	96	Investment per Production Worker	9,352	9,691	104

Sources: Same as General Statistics. The 'Average of All Manufacturing' column represents the average of all manufacturing industries reported for the most recent complete year available. The Index shows the relationship between the Average and the Analyzed Industry. For example, 100 means that they are equal; 500 that the Analyzed Industry is five times the average; 50 means that the Analyzed Industry is half the national average. The abbreviation 'na' is used to show that data are 'not available'.

LEADING COMPANIES   Number shown: **75**   Total sales ($ mil): **21,811**   Total employment (000): **151.3**

Company Name	Address				CEO Name	Phone	Co. Type	Sales ($ mil)	Empl. (000)
Philips Electric	100 E 42nd St	New York	NY	10017	Stephen Tumminello	212-850-5000	S	6,000	40.0
Thomson Consumer Electronics	600 N Sherman Dr	Indianapolis	IN	46201	Alain Prestat	317-267-5000	S	6,000	54.0
Mitsubishi Electronics America	PO Box 6007	Cypress	CA	90630	Takeo Iinuma	714-220-2500	S	2,200	1.3
Philips Consumer Electronics Co	PO Box 14810	Knoxville	TN	37914	Robert Minkhorst	615-521-4316	S	1,800	8.0
Zenith Electronics Corp	1000 Milwaukee Av	Glenview	IL	60025	Jerry K Pearlman	708-391-7000	P	1,469	22.5
Harman International Industries	1101 Pennsylvania	Washington	DC	20004	Sidney Harman	202-393-1101	P	862	6.8
Sanyo Fisher	PO Box 2329	Chatsworth	CA	91311	Toshiaki Taguchi	818-998-7322	D	580	0.4
Bose Corp	The Mountain	Framingham	MA	01701	Sherwin Greenblatt	508-879-7330	R	500	3.0
International Jensen Inc	25 Tri-State	Lincolnshire	IL	60069	Robert G Shaw	708-317-3700	P	253	1.8
Peavey Electronics Corp	PO Box 2898	Meridian	MS	39302	Melia Peavey	601-483-5365	R	210*	1.9
Gold Star of America Inc	PO Box 6126	Huntsville	AL	35824	Man S Kim	205-772-0623	S	162	0.2
Universal Electronics Inc	1864 Enterprise	Twinsburg	OH	44087	David M Gabrielsen	216-487-1110	P	96	0.5
Sparkomatic Corp	PO Box 277	Milford	PA	18337	Edward Anchel	717-296-6444	R	90	0.4
Infinity Systems Inc	20630 Nordhoff St	Chatsworth	CA	91311	Hank Suerth	818-709-9400	S	80	0.1
Rockford Corp	546 S Rockford Dr	Tempe	AZ	85281	Gary Suttle	602-967-3565	R	65	0.4
DOD Electronics Corp	8670 Sandy Pkwy	Salt Lake City	UT	84070	John Johnson	801-566-8800	S	64	0.3
Mitsubishi Consumer	2001 E Carnegie	Santa Ana	CA	92705	Mikio Omaru	714-261-3200	S	62*	0.6
MTX	4545 E Baseline Rd	Phoenix	AZ	85044	Loyd Ivey	602-438-4545	R	61	0.4
Oxford International Ltd	4237 W 42nd Pl	Chicago	IL	60632	Michael J Oslac	312-927-3715	R	60	0.8
Legacy Storage Systems Inc	25 South St	Hopkinton	MA	01748	David Killins	508-881-6442	R	50	<0.1
Mackie Designs Inc	16220 Woodingville	Woodinville	WA	98072	Greg C Mackie	206-487-4333	R	48*	0.2
Cerwin-Vega Inc	555 E Easy St	Simi Valley	CA	93065	Connie Czerwinski	805-584-9332	R	45	0.5
MPO Videotronics Inc	1167 Lawrence Dr	Newbury Park	CA	91320	Larry Kaiser	805-499-8513	R	45	0.2
JBL Inc	8500 Balboa Blv	Northridge	CA	91329	Bernard Girod	818-893-8411	S	42*	0.3
Oxford Speakers Co	4237 W 42nd Pl	Chicago	IL	60632	M Oslac	312-927-3715	D	38*	0.4
Home Theater Products Intern	1620 S Lewis St	Anaheim	CA	92805	Stanley Snyder	714-937-9300	P	36	0.2
Koss Corp	4129 N Port	Milwaukee	WI	53212	Michael J Koss	414-964-5000	P	36	0.2
Eminence Speaker Corp	PO Box 207	Eminence	KY	40019	RA Gault	502-845-5622	R	35	0.2
Boston Acoustics Inc	1 Bio-Logic Plz	Mundelein	IL	60060	Francis L Reed	708-949-5200	P	34	0.2
Polk Audio Inc	5601 Metro Dr	Baltimore	MD	21215	George M Klopfer	410-358-3600	P	34	0.2
Videonics Inc	1370 Dell Av	Campbell	CA	95008	Michael L D'Addio	408-866-8300	P	32	<0.1
Go-Video Inc	14455 N Hayden Rd	Scottsdale	AZ	85260	R Terren Dunlap	602-998-3400	P	31	<0.1
Allsop Inc	PO Box 23	Bellingham	WA	98227	I Allsop	206-734-9090	R	30	0.2
Lavcon Inc	11131 Dora St	Sun Valley	CA	91352	Lavere Lund	818-767-2843	R	30	0.3
Pyle Industries Inc	501 Center St	Huntington	IN	46750	F Pyle Jr	219-356-1200	S	30	0.3
Royal Sound Company Inc	6 Industrial Way W	Eatontown	NJ	07724	Mervin Dayan	908-542-8400	R	30	<0.1
Analog and Digital Systems Inc	1 Progress Way	Wilmington	MA	01887	Karien Jacob	508-658-5100	R	28	0.2
QSC Audio Products Inc	1675 McArthur Blv	Costa Mesa	CA	92626	Barry Andrews	714-754-6175	R	28*	0.3
Crest Audio Inc	100 Eisenhower Dr	Paramus	NJ	07652	John V Lee	201-909-8700	R	25	0.3
Sencore Inc	3200 Sencore Dr	Sioux Falls	SD	57107	Al Bowden	605-339-0100	R	25	0.2
AMX Corp	11995 Forestgate Dr	Dallas	TX	75243	Scott Miller	214-644-3048	R	24	0.1
Carver Corp	PO Box 1237	Lynnwood	WA	98036	Robert A Fulton	206-775-1202	P	22	0.1
ASA Corp	PO Box 4009	Elkhart	IN	46514	Thomas Irions	219-264-3135	R	22*	<0.1
Gemini Sound Products Corp	1100 Milik St	Carteret	NJ	07008	Ike Cabasso	908-969-9000	R	20	<0.1
Kinyo Company Inc	14235 Lomitas Av	La Puente	CA	91746	Jacob Chang	818-333-3711	R	20	<0.1
KLH Research & Dev Corp	11131 Dora St	Sun Valley	CA	91352	Lavere Lund	818-767-2843	S	20*	0.3
Cambridge SoundWorks Inc	311 Needham St	Newton	MA	02164	Thomas J DeVesto	617-332-5936	P	19	0.2
PTS Electronics Corp	PO Box 272	Bloomington	IN	47402	Jack D Craig	812-824-9331	PTS	19*	0.2
J and R Film Company Inc	1135 N Mansfield	Hollywood	CA	90038	Joseph Paskal	213-467-3107	R	18	0.2
Jasco Products Company Inc	311 NW 122nd St	Oklahoma City	OK	73114	Steve Trice	405-752-0710	R	17*	0.2
Phoenix Gold International Inc	9300 N Decatur St	Portland	OR	97203	Keith A Peterson	503-288-2008	P	16	0.2
Dynalec Corp	87 W Main St	Sodus	NY	14551	Craig Cuvelier	315-483-6923	R	15*	0.1
Teleview Research Inc	900 Hansen Way	Palo Alto	CA	94304	Wayne Catlett	415-813-6481	R	15	<0.1
Horizon Music Inc	230 N Spring St	Cape Girardeau	MO	63701	Ernie Eudy	314-651-6500	R	14	0.2
Southern Audio Services Inc	15049 Florida Blv	Baton Rouge	LA	70819	Jon Jordan	504-272-7135	R	14	<0.1
JBL Consumer Products	80 Crossways Pk Dr	Woodbury	NY	11797	Thomas Jacoby	516-496-3400	D	13*	0.1
Nu-Way Speaker Products Inc	945 Anita Av	Antioch	IL	60002	Dennis Smith	708-395-5141	R	13	0.3
Quam-Nichols Co	234 E Marquette	Chicago	IL	60637	WG Little	312-488-5800	R	12	0.1
Rane Corp	10802 47th Av W	Mukilteo	WA	98275	Linda Arink	206-355-6000	R	12*	0.1
Accom Inc	1490 O'Brien Dr	Menlo Park	CA	94025	Junaid Sheikh	415-328-3818	R	11*	0.1
Apogee Sound Inc	1150 Industrial Av	Petaluma	CA	94952	Ken Deloria	707-778-8887	R	11	0.1
Samson Technologies Corp	PO Box 9031	Syosset	NY	11791	Scott Goodman	516-364-2244	S	11*	<0.1
Meyer Sound Laboratories Inc	2832 San Pablo Av	Berkeley	CA	94702	John Meyer	510-486-1166	R	10	<0.1
Sima Products Corp	6153 W Mulford St	Niles	IL	60714	Steven Breslau	708-679-7462	R	10	<0.1
Alphasonik Inc	701 Heinz Av	Berkeley	CA	94710	Henry Eberle	510-548-4005	R	9	<0.1
Naki Electronics	10100 Santa Monica	Los Angeles	CA	90067	Herschel Naghi	818-890-9955	R	9	<0.1
Pioneer New Media Technologies	2265 E 220th St	Long Beach	CA	90810	Tom Haga	310-952-2111	S	9*	<0.1
Larcan-TTC Inc	650 S Taylor Av	Louisville	CO	80027	Dirk B Freeman	303-665-8000	P	9	<0.1
BGW Systems Inc	13130 Yukon Av	Hawthorne	CA	90250	Brian G Wachner	310-973-8090	R	8	<0.1
Coustic Co	4260 Charter St	Vernon	CA	90058	David Kwang	213-582-2832	D	8*	<0.1
United Speaker Systems	6400 Youngerman	Jacksonville	FL	32244	William Hecht	904-777-0700	R	7*	<0.1
Eastern Acoustic Works	1 Main St	Whitinsville	MA	01588	Kenneth P Berger	508-234-6158	R	7	<0.1
Ramsa Pro Audio	6550 Katella Av	Cypress	CA	90630	Steve Wooley	714-373-7277	D	7	<0.1
Zenasia International Corp	1210 E 223rd St	Carson	CA	90745	M Ito	310-518-3335	R	7	<0.1
Anchor Audio Inc	913 W 223rd St	Torrance	CA	90502	David Jacobs	310-533-5984	R	6	<0.1

Source: Ward's Business Directory of U.S. Private and Public Companies, Volumes 1 and 2, 1996. The company type code used is as follows: P - Public, R - Private, S - Subsidiary, D - Division, J - Joint Venture, A - Affiliate, G - Group. Sales are in millions of dollars, employees are in thousands. An asterisk (*) indicates an estimated sales volume. The symbol < stands for 'less than'. Company names and addresses are truncated, in some cases, to fit into the available space.

MATERIALS CONSUMED

Material	Quantity	Delivered Cost ($ million)
Materials, ingredients, containers, and supplies	(X)	5,999.0
Cabinets (wood, metal, and plastics)	(X)	510.4
Tuners	(X)	183.2
Speakers and speaker systems	(X)	159.3
Cathode ray picture tubes	(X)	1,593.3
Printed circuit boards (without inserted components) for electronic circuitry	(X)	143.5
Printed circuit assemblies, loaded boards or modules	(X)	751.2
Semiconductors, including transistors, diodes, rectifiers, and integrated circuits for electronic circuitry	(X)	110.3
Capacitors for electronic circuitry	(X)	41.6
Resistors for electronic circuitry	(X)	23.8
Other components and accessories for electronic circuitry, nec, except tubes	(X)	314.6
Plastics products consumed in the form of sheets, rods, tubes, film, and other shapes	(X)	98.9
Paper and paperboard containers, including shipping sacks and other paper packaging supplies	(X)	84.0
Current-carrying wiring devices	(X)	112.6
Metal stampings	(X)	88.7
Metal bolts, nuts, screws, washers, rivets, and other screw machine products	(X)	28.0
All other fabricated metal products (except forgings)	(X)	73.2
Forgings	(X)	(D)
Castings (rough and semifinished)	(X)	(D)
Steel shapes and forms	(X)	19.6
Nonferrous shapes and forms	(X)	25.5
Insulated wire and cable, including magnet wire	(X)	75.8
All other materials and components, parts, containers, and supplies	(X)	661.4
Materials, ingredients, containers, and supplies, nsk	(X)	874.6

Source: 1992 Economic Census. Explanation of symbols used: (D): Withheld to avoid disclosure of competitive data; na: Not available; (S): Withheld because statistical norms were not met; (X): Not applicable; (Z): Less than half the unit shown; nec: Not elsewhere classified; nsk: Not specified by kind; - : zero; * : 10-19 percent estimated; ** : 20-29 percent estimated.

PRODUCT SHARE DETAILS

Product or Product Class	% Share	Product or Product Class	% Share
Radio receivers, television sets, phonographs, speakers, and related equipment	100.00	sold separately, and commercial sound equipment)	21.04
Home, portable, and automobile radios and radio phonograph-tape recorder combinations	7.70	Other consumer high fidelity components (including audio and video recorders and players, camcorders)	2.98
Television receivers (including combination models)	63.81	Radio receivers, television sets, phonographs, speakers, and related equipment, nsk	4.46
Speakers (including loudspeaker systems and loudspeakers			

Source: 1992 Economic Census. The values shown are percent of total shipments in an industry. Values of indented subcategories are summed in the main heading. The symbol (D) appears when data are withheld to prevent disclosure of competitive information. The abbreviation nsk stands for 'not specified by kind' and nec for 'not elsewhere classified'.

INPUTS AND OUTPUTS FOR RADIO & TV RECEIVING SETS

Economic Sector or Industry Providing Inputs	%	Sector	Economic Sector or Industry Buying Outputs	%	Sector
Imports	51.2	Foreign	Personal consumption expenditures	77.3	
Electronic components nec	11.5	Manufg.	Radio & TV receiving sets	6.4	Manufg.
Wholesale trade	7.4	Trade	Exports	5.7	Foreign
Radio & TV receiving sets	6.4	Manufg.	Motor vehicles & car bodies	3.6	Manufg.
Electron tubes	5.6	Manufg.	Gross private fixed investment	2.9	Cap Inv
Electrical repair shops	2.7	Services	Automotive rental & leasing, without drivers	0.5	Services
Wood TV & radio cabinets	2.0	Manufg.	Retail trade, except eating & drinking	0.3	Trade
Semiconductors & related devices	1.6	Manufg.	Child day care services	0.3	Services
Computer & data processing services	1.0	Services	Electrical repair shops	0.3	Services
Miscellaneous plastics products	0.9	Manufg.	S/L Govt. purch., elem. & secondary education	0.3	S/L Govt
Noncomparable imports	0.8	Foreign	Motor vehicle parts & accessories	0.2	Manufg.
Metal stampings, nec	0.7	Manufg.	Radio & TV communication equipment	0.2	Manufg.
Household furniture, nec	0.6	Manufg.	S/L Govt. purch., higher education	0.2	S/L Govt
Advertising	0.6	Services	Wholesale trade	0.1	Trade
Wiring devices	0.5	Manufg.	Insurance carriers	0.1	Fin/R.E.
Paperboard containers & boxes	0.4	Manufg.	Real estate	0.1	Fin/R.E.
Air transportation	0.4	Util.	S/L Govt. purch., public assistance & relief	0.1	S/L Govt
Blast furnaces & steel mills	0.3	Manufg.			
Hardware, nec	0.3	Manufg.			
Electric services (utilities)	0.3	Util.			
Die-cut paper & board	0.2	Manufg.			
Fabricated metal products, nec	0.2	Manufg.			
Miscellaneous fabricated wire products	0.2	Manufg.			
Petroleum refining	0.2	Manufg.			
Plastics materials & resins	0.2	Manufg.			
Screw machine and related products	0.2	Manufg.			
Veneer & plywood	0.2	Manufg.			
Communications, except radio & TV	0.2	Util.			

Continued on next page.

INPUTS AND OUTPUTS FOR RADIO & TV RECEIVING SETS - Continued

Economic Sector or Industry Providing Inputs	%	Sector	Economic Sector or Industry Buying Outputs	%	Sector
Motor freight transportation & warehousing	0.2	Util.			
Eating & drinking places	0.2	Trade			
Banking	0.2	Fin/R.E.			
Engineering, architectural, & surveying services	0.2	Services			
Equipment rental & leasing services	0.2	Services			
Metal household furniture	0.1	Manufg.			
Real estate	0.1	Fin/R.E.			
Colleges, universities, & professional schools	0.1	Services			
Legal services	0.1	Services			
U.S. Postal Service	0.1	Gov't			

Source: Benchmark Input-Output Accounts for the U.S. Economy, 1982, U.S. Department of Commerce, Washington, D.C., July 1991. Data, as reported in the source, are organized by the 1977 SIC structure in use in 1982 but have been matched, as closely as is possible, to the 1987 SIC structure used in this book.

OCCUPATIONS EMPLOYED BY SIC 365 - HOUSEHOLD AUDIO AND VIDEO EQUIPMENT

Occupation	% of Total 1994	Change to 2005	Occupation	% of Total 1994	Change to 2005
Assemblers, fabricators, & hand workers nec	9.3	-35.8	Electrical & electronic technicians,technologists	1.5	-35.8
Electrical & electronic assemblers	8.1	-42.2	Hand packers & packagers	1.5	-45.0
Inspectors, testers, & graders, precision	4.7	-55.0	Stock clerks	1.5	-47.8
Blue collar worker supervisors	4.1	-41.8	Engineering technicians & technologists nec	1.4	-35.8
Engineers nec	4.0	-3.7	Electrical & electronics engineers	1.4	-31.6
Machine operators nec	3.7	-54.7	Managers & administrators nec	1.3	-35.8
Metal & plastic machine workers nec	2.4	-31.7	Bookkeeping, accounting, & auditing clerks	1.3	-51.8
Electrical & electronic equipment assemblers	2.4	-35.8	Helpers, laborers, & material movers nec	1.3	-35.7
Secretaries, ex legal & medical	2.2	-41.5	Machine feeders & offbearers	1.2	-42.2
Packaging & filling machine operators	2.1	-35.7	Freight, stock, & material movers, hand	1.1	-48.6
Traffic, shipping, & receiving clerks	2.1	-38.2	Marketing, advertising, & PR managers	1.1	-35.8
Combination machine tool operators	2.1	-35.8	General office clerks	1.1	-45.2
Sales & related workers nec	2.0	-35.8	Production, planning, & expediting clerks	1.0	-35.8
Plastic molding machine workers	1.9	-23.0	Janitors & cleaners, incl maids	1.0	-48.6
General managers & top executives	1.8	-39.1	Industrial truck & tractor operators	1.0	-35.8
Professional workers nec	1.7	-22.9	Management support workers nec	1.0	-35.8

Source: Industry-Occupation Matrix, Bureau of Labor Statistics. These data relate to one or more 3-digit SIC industry groups rather than to a single 4-digit SIC. The change reported for each occupation to the year 2005 is a percent of growth or decline as estimated by the Bureau of Labor Statistics. The abbreviation nec stands for 'not elsewhere classified'.

LOCATION BY STATE AND REGIONAL CONCENTRATION

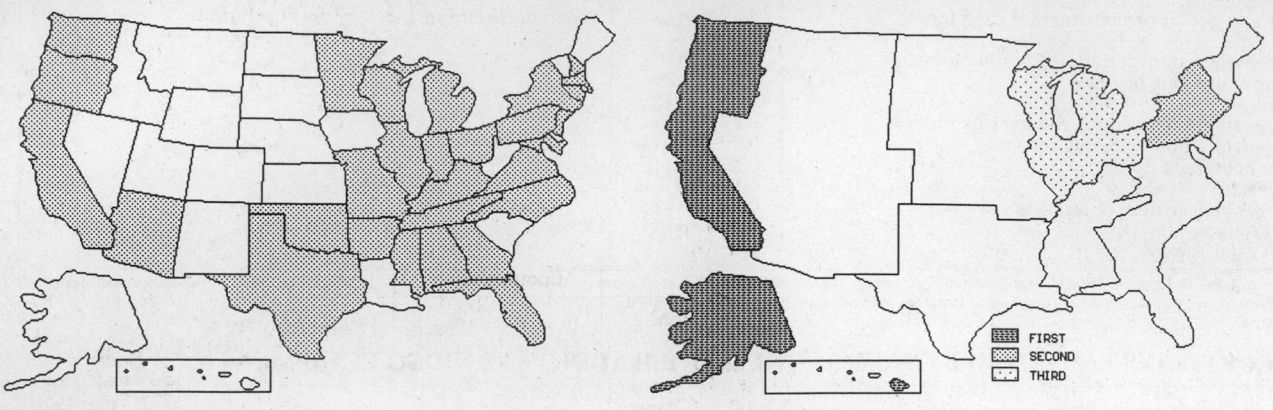

		FIRST
		SECOND
		THIRD

INDUSTRY DATA BY STATE

| State | Establish- ments | Shipments | | | Employment | | | | Cost as % of Shipments | Investment per Employee ($) |
		Total ($ mil)	% of U.S.	Per Establ.	Total Number	% of U.S.	Per Establ.	Wages ($/hour)		
California	100	2,237.8	25.5	22.4	7,000	22.4	70	10.49	73.8	8,014
Tennessee	9	1,543.6	17.6	171.5	4,100	13.1	456	9.60	78.0	13,683
Illinois	28	450.5	5.1	16.1	1,900	6.1	68	9.65	40.6	4,105
Massachusetts	17	264.7	3.0	15.6	1,000	3.2	59	12.08	33.5	2,600
New Jersey	14	248.9	2.8	17.8	600	1.9	43	9.11	81.3	-
Washington	12	208.6	2.4	17.4	600	1.9	50	8.89	77.9	6,000
Michigan	11	167.1	1.9	15.2	900	2.9	82	12.31	48.6	1,000
New York	42	72.5	0.8	1.7	600	1.9	14	8.50	65.4	4,333
Arizona	7	56.9	0.6	8.1	400	1.3	57	11.75	46.2	4,250
Texas	13	41.6	0.5	3.2	400	1.3	31	7.33	60.1	3,250
Connecticut	5	27.6	0.3	5.5	300	1.0	60	10.67	47.8	1,333
Florida	16	25.2	0.3	1.6	200	0.6	13	8.50	59.5	2,000
Virginia	7	22.9	0.3	3.3	300	1.0	43	8.33	49.3	-
Minnesota	10	13.2	0.2	1.3	100	0.3	10	8.00	48.5	3,000
Indiana	15	(D)	-	-	3,750 *	12.0	250	-	-	-
Pennsylvania	14	(D)	-	-	750 *	2.4	54	-	-	-
Oregon	12	(D)	-	-	375 *	1.2	31	-	-	-
Kentucky	8	(D)	-	-	750 *	2.4	94	-	-	2,000
North Carolina	8	(D)	-	-	1,750 *	5.6	219	-	-	-
Wisconsin	8	(D)	-	-	750 *	2.4	94	-	-	-
Arkansas	6	(D)	-	-	750 *	2.4	125	-	-	-
New Hampshire	6	(D)	-	-	175 *	0.6	29	-	-	1,714
Ohio	6	(D)	-	-	175 *	0.6	29	-	-	-
Oklahoma	5	(D)	-	-	375 *	1.2	75	-	-	-
Georgia	4	(D)	-	-	750 *	2.4	188	-	-	-
Maryland	3	(D)	-	-	175 *	0.6	58	-	-	-
Missouri	3	(D)	-	-	1,750 *	5.6	583	-	-	-
Alabama	2	(D)	-	-	175 *	0.6	88	-	-	-
Mississippi	2	(D)	-	-	1,750 *	5.6	875	-	-	-

Source: 1992 *Economic Census*. The states are in descending order of shipments or establishments (if shipment data are missing for the majority). The symbol (D) appears when data are withheld to prevent disclosure of competitive information. States marked with (D) are sorted by number of establishments. A dash (-) indicates that the data element cannot be calculated; * indicates the midpoint of a range.

3652 - PRERECORDED RECORDS & TAPES

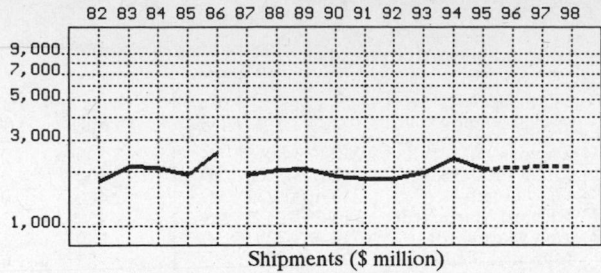

Shipments ($ million)

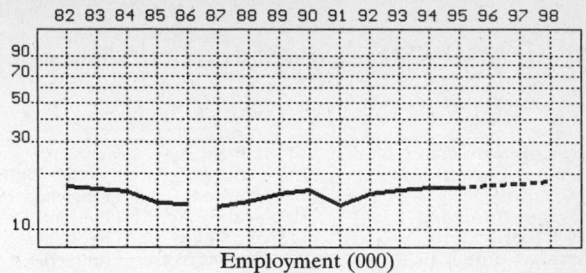

Employment (000)

GENERAL STATISTICS

Year	Com-panies	Establishments		Employment			Compensation		Production ($ million)			
		Total	with 20 or more employees	Total (000)	Production Workers (000)	Hours (Mil)	Payroll ($ mil)	Wages ($/hr)	Cost of Materials	Value Added by Manufacture	Value of Shipments	Capital Invest.
1982	548	574	131	17.1	11.8	23.6	292.0	6.75	578.8	1,189.5	1,768.9	36.4
1983		544	120	16.6	11.3	22.2	298.8	6.85	647.5	1,492.1	2,151.6	34.8
1984		514	109	16.0	10.9	21.3	302.6	7.65	655.5	1,447.5	2,092.5	31.4
1985		485	98	14.0	9.5	19.0	283.1	7.92	665.8	1,223.2	1,896.7	48.8
1986		479	106	13.6	9.5	18.6	300.0	8.70	791.8	1,789.9	2,574.0	76.3
1987*	449	476	94	13.3	10.2	20.4	265.7	7.86	567.4	1,362.2	1,921.6	50.7
1988		475	100	14.1	10.6	19.5	311.8	9.51	696.9	1,346.5	2,033.7	86.7
1989		414	96	15.7	10.5	20.7	286.0	8.54	583.0	1,498.1	2,060.9	97.0
1990		423	102	16.3	10.6	21.3	294.5	9.17	583.7	1,257.9	1,856.1	90.0
1991		415	104	13.5	10.5	20.8	294.8	9.69	550.2	1,247.4	1,829.0	70.6
1992	402	419	95	15.7	12.1	24.5	382.4	10.24	631.5	1,181.6	1,808.2	182.6
1993		424	94	16.2	12.5	25.3	403.9	10.40	720.0	1,241.5	1,966.3	138.1
1994		397P	98P	17.0	13.3	30.4	442.9	9.71	829.0	1,523.1	2,342.5	164.4
1995		387P	98P	17.0P	13.2P	28.8P	441.9P	10.62P	734.3P	1,349.2P	2,075.0P	179.1P
1996		378P	97P	17.4P	13.6P	30.1P	465.6P	10.90P	742.0P	1,363.3P	2,096.7P	194.5P
1997		368P	97P	17.8P	14.1P	31.4P	489.3P	11.17P	749.7P	1,377.4P	2,118.4P	209.9P
1998		359P	97P	18.2P	14.5P	32.7P	513.0P	11.45P	757.4P	1,391.5P	2,140.2P	225.2P

Sources: 1982, 1987, 1992 *Economic Census*; *Annual Survey of Manufactures*, 83-86, 88-91, 93-94. Establishment counts are from *County Business Patterns* for non-Census years; establishment counts for 83-84 are extrapolations. * indicates that industry content changed in 87; earlier years use 77 SICs. 'P's mark projections.

INDICES OF CHANGE

Year	Com-panies	Establishments		Employment			Compensation		Production ($ million)			
		Total	with 20 or more employees	Total (000)	Production Workers (000)	Hours (Mil)	Payroll ($ mil)	Wages ($/hr)	Cost of Materials	Value Added by Manufacture	Value of Shipments	Capital Invest.
1982	136	137	138	109	98	96	76	66	92	101	98	20
1983		130	126	106	93	91	78	67	103	126	119	19
1984		123	115	102	90	87	79	75	104	123	116	17
1985		116	103	89	79	78	74	77	105	104	105	27
1986		114	112	87	79	76	78	85	125	151	142	42
1987*	112	114	99	85	84	83	69	77	90	115	106	28
1988		113	105	90	88	80	82	93	110	114	112	47
1989		99	101	100	87	84	75	83	92	127	114	53
1990		101	107	104	88	87	77	90	92	106	103	49
1991		99	109	86	87	85	77	95	87	106	101	39
1992	100	100	100	100	100	100	100	100	100	100	100	100
1993		101	99	103	103	103	106	102	114	105	109	76
1994		95P	103P	108	110	124	116	95	131	129	130	90
1995		92P	103P	108P	109P	117P	116P	104P	116P	114P	115P	98P
1996		90P	103P	111P	113P	123P	122P	106P	118P	115P	116P	107P
1997		88P	102P	114P	116P	128P	128P	109P	119P	117P	117P	115P
1998		86P	102P	116P	120P	133P	134P	112P	120P	118P	118P	123P

Sources: Same as General Statistics. Values reflect change from the base year, 1992. Values above 100 mean greater than 92, values below 100 mean less than 92, and a value of 100 in the 82-91 or 93-98 period means same as 92. * indicates that industry content changed in 87. Data for earlier years are in 77 SIC format.

SELECTED RATIOS

For 1994	Avg. of All Manufact.	Analyzed Industry	Index	For 1994	Avg. of All Manufact.	Analyzed Industry	Index
Employees per Establishment	49	43	87	Value Added per Production Worker	134,084	114,519	85
Payroll per Establishment	1,500,273	1,115,617	74	Cost per Establishment	5,045,178	2,088,161	41
Payroll per Employee	30,620	26,053	85	Cost per Employee	102,970	48,765	47
Production Workers per Establishment	34	34	98	Cost per Production Worker	146,988	62,331	42
Wages per Establishment	853,319	743,537	87	Shipments per Establishment	9,576,895	5,900,504	62
Wages per Production Worker	24,861	22,194	89	Shipments per Employee	195,460	137,794	70
Hours per Production Worker	2,056	2,286	111	Shipments per Production Worker	279,017	176,128	63
Wages per Hour	12.09	9.71	80	Investment per Establishment	321,011	414,106	129
Value Added per Establishment	4,602,255	3,836,524	83	Investment per Employee	6,552	9,671	148
Value Added per Employee	93,930	89,594	95	Investment per Production Worker	9,352	12,361	132

Sources: Same as General Statistics. The 'Average of All Manufacturing' column represents the average of all manufacturing industries reported for the most recent complete year available. The Index shows the relationship between the Average and the Analyzed Industry. For example, 100 means that they are equal; 500 that the Analyzed Industry is five times the average; 50 means that the Analyzed Industry is half the national average. The abbreviation 'na' is used to show that data are 'not available'.

LEADING COMPANIES   Number shown: 49   Total sales ($ mil): 3,626   Total employment (000): 18.9

Company Name	Address				CEO Name	Phone	Co. Type	Sales ($ mil)	Empl. (000)
Sony Music Entertainment Inc	550 Madison Av	New York	NY	10022	Thomas D Mottola	212-833-8000	S	1,500*	10.0
Better Quality Cassettes Inc	2101 S 35th St	Council Bluffs	IA	51501	Linda Frederiksen	712-328-8060	R	500	0.3
Capitol Records Inc	1750 N Vine St	Hollywood	CA	90028	Gary Gersh	213-462-6252	S	480	0.3
WEA Manufacturing Inc	210 N Valley Av	Olyphant	PA	18447	Richard Marquardt	717-383-2471	S	450	3.0
Specialty Records Corp	PO Box 1400	Olyphant	PA	18447	Ann Marquardt	717-383-3291	D	180	1.8
Disc Manufacturing Inc	4905 Moores Mill	Huntsville	AL	35811	Myron R Shain	205-859-9042	S	83	0.7
Motown Record Company LP	5750 Wilshire Blv	Los Angeles	CA	90036	Jheryl Busby	213-634-3500	R	65	0.2
HMG Digital Technologies Corp	30 Gilpin Av	Hauppauge	NY	11788	George N Fishman	516-234-0200	P	62	0.7
Cinram Inc	1600 Rich Rd	Richmond	IN	47374	Dave Rubenstein	317-962-9511	S	53*	0.4
A and M Records Inc	1416 N La Brea Av	Los Angeles	CA	90028	Al Cafaro	310-333-8000	S	31*	0.2
Allied Record Co	6110 Peachtree St	Los Angeles	CA	90040	Richard Marquardt	213-725-6900	S	25	0.3
Rykodisc Inc	27 Concord St	Salem	MA	01970	Don Rose	508-744-7678	R	20*	<0.1
intouch group inc	333 Bryant St	San Francisco	CA	94107	Joshua Kaplan	415-974-5000	R	18	<0.1
Metacom Inc	5353 Nathan Ln	Plymouth	MN	55442	Phillip Levin	612-553-2000	R	17*	0.1
PILZ America Inc	PO Box 220	Concordville	PA	19331	Martin J Mair	610-459-5035	R	15	0.1
Dove Audio Inc	301 N Canon Dr	Beverly Hills	CA	90210	Michael Viner	310-273-7722	P	12	<0.1
ASR Recording Services	8960 Eton Av	Canoga Park	CA	91304	Conrad Enricas	818-341-1124	R	11	<0.1
Intersound Entertainment Inc	PO Box 1724	Roswell	GA	30077	Don Johnson	404-664-9262	R	11*	<0.1
Original Sound Records	7120 Sunset Blv	Los Angeles	CA	90046	A Laboe	213-851-2500	R	11	<0.1
MMO Music Group Inc	50 Executive Blv	Elmsford	NY	10523	I Kratka	914-592-1188	R	10	<0.1
Rainbo Record Mfg Corp	1738 Berkeley St	Santa Monica	CA	90404	Jack G Brown	310-829-3476	R	10	0.1
K-Tel International	15535 Medina Rd	Plymouth	MN	55447	Mickey Elfenbein	612-559-6800	S	9	<0.1
Hollywood Records	500 S Buena Vista	Burbank	CA	91521	Bob Pfeifer	818-560-1000	D	7*	<0.1
MacKenzie Laboratories Inc	PO Box 1416	Glendora	CA	91740	Nagy Khattar	909-394-9007	R	6	<0.1
Higher Octave Music Inc	23715 W Malibu Rd	Malibu	CA	90065	Matt Marshall	310-589-1515	R	5	<0.1
Educational Activities Inc	1937 Grand Av	Baldwin	NY	11510	Alfred S Harris	516-223-4666	R	4*	<0.1
Profile Records Inc	740 Broadway	New York	NY	10003	Steve Plotnicki	212-529-2600	R	4*	<0.1
Audio Renaissance Tapes	5858 Wilshire Blv	Los Angeles	CA	90036	Bill Hartley	213-939-1840	D	3	<0.1
Transco Products Co	PO Box 1025	Linden	NJ	07036	FL Buehler	908-862-0030	R	3	<0.1
Antone's Records and Tapes	500 San Marcos	Austin	TX	78702	Clifford Antone	512-322-0617	R	3*	<0.1
Hold Company Inc	540 Township Line	Blue Bell	PA	19422	Anthony Stagliano	215-643-0700	R	2	<0.1
Koss Classics Ltd	4129 N Port WA	Milwaukee	WI	53212	Michael Koss	414-964-5000	S	2	<0.1
RWM Associates	PO Box 1324	Bethesda	MD	20817	Robert W Magee	301-299-7817	R	2*	<0.1
QCA Inc	2832 Spring Grove	Cincinnati	OH	45225	James Bosken	513-681-8400	R	2	<0.1
Alberti Record Mfg Co	312 Monterey	Monterey Park	CA	91754	John Alberti Sr	818-282-5181	R	1*	<0.1
Dixie Record Pressing Inc	631 Hamilton Av	Nashville	TN	37203	James Gann	615-256-0922	R	1	<0.1
Flyte Tyme Productions Inc	4100 W 76th St	Edina	MN	55435	Jimmy Harris	612-897-3901	R	1*	<0.1
Justice Record Co	PO Box 980369	Houston	TX	77098	Randall Jamail	713-520-6669	R	1	<0.1
Red House Records Inc	PO Box 4044	St Paul	MN	55104	Bob Feldman	612-379-1089	R	1	<0.1
Wayzata Technology Inc	2515 E Highway 2	Grand Rapids	MN	55744	Mark Engelhardt	218-326-0597	R	1*	<0.1
Publishing Mills Inc	1680 N Vine St	Los Angeles	CA	90028	Jessica Kaye	213-467-7831	R	1	<0.1
Audio Literature Inc	3800 P Verdes	S San Francisco	CA	94080	John Hunt	415-952-3400	S	1	<0.1
Pacific Audio Recording	2477 Orangeth	Fullerton	CA	92631	James Campbell	714-441-0782	R	1	<0.1
Smarty Pants Audio and Video	15104 Detroit Av	Lakewood	OH	44107	S Tirk	216-221-5300	S	1	<0.1
Wakefield Lauber Co	PO Box 22555-180	Tempe	AZ	85282	K A Wakefield	602-252-5644	R	1	<0.1
California Magnetics	7898 Ostrow St	San Diego	CA	92111	Don Nuzzo	619-576-0291	R	0	<0.1
New Albion Records Inc	584 Castro St	San Francisco	CA	94114	Foster Reed	415-621-5757	R	0	<0.1
River City Sound Productions	PO Box 750786	Memphis	TN	38175	Bob Pierce	901-274-7277	R	0*	<0.1
Audio Computer Information	PO Box 216	Spring Grove	MN	55974	John Socha	507-498-3279	R	0	<0.1

Source: Ward's Business Directory of U.S. Private and Public Companies, Volumes 1 and 2, 1996. The company type code used is as follows: P - Public, R - Private, S - Subsidiary, D - Division, J - Joint Venture, A - Affiliate, G - Group. Sales are in millions of dollars, employees are in thousands. An asterisk (*) indicates an estimated sales volume. The symbol < stands for 'less than'. Company names and addresses are truncated, in some cases, to fit into the available space.

MATERIALS CONSUMED

Material	Quantity	Delivered Cost ($ million)
Materials, ingredients, containers, and supplies	(X)	576.7
Unrecorded audio-range magnetic tape, with or without cassettes or cartridges	(X)	123.0
Record blanks, audio	(X)	8.8
Compact disc blanks for audio and computer use	(X)	37.3
Empty tape cassettes and cartridges	(X)	33.4
Plastics products consumed in the form of sheets, rods, tubes, film, and other shapes	(X)	35.8
Plastics resins consumed in the form of granules, pellets, powders, liquids, etc.	(X)	36.8
Paper and paperboard products (including album covers, sleeves, etc.)	(X)	33.8
All other materials and components, parts, containers, and supplies	(X)	42.8
Materials, ingredients, containers, and supplies, nsk	(X)	225.0

Source: 1992 Economic Census. Explanation of symbols used: (D): Withheld to avoid disclosure of competitive data; na: Not available; (S): Withheld because statistical norms were not met; (X): Not applicable; (Z): Less than half the unit shown; nec: Not elsewhere classified; nsk: Not specified by kind; - : zero; * : 10-19 percent estimated; ** : 20-29 percent estimated.

PRODUCT SHARE DETAILS

Product or Product Class	% Share	Product or Product Class	% Share
Prerecorded records and tapes	100.00	Other audio discs or records (including compact discs prerecorded), including digitally mastered records for consumer use, and master records used to press commercial records	1.20
Audio discs or records (including compact discs prerecorded) long playing (LP), excluding digitally mastered records for consumer use	2.10	Audio tapes, prerecorded cassette singles/maxisingles	4.22
Audio discs or records (including compact discs prerecorded) compact disc (CD) singles/maxisingles	1.39	Audio tapes, prerecorded cassette full-length	26.81
Audio discs or records (including compact discs prerecorded) compact disc (CD) full-length	46.58	Other audio tapes, prerecorded (including 8 track and DAT)	0.06

Source: 1992 *Economic Census*. The values shown are percent of total shipments in an industry. Values of indented subcategories are summed in the main heading. The symbol (D) appears when data are withheld to prevent disclosure of competitive information. The abbreviation nsk stands for 'not specified by kind' and nec for 'not elsewhere classified'.

INPUTS AND OUTPUTS FOR PHONOGRAPH RECORDS & TAPES

Economic Sector or Industry Providing Inputs	%	Sector	Economic Sector or Industry Buying Outputs	%	Sector
Phonograph records & tapes	26.0	Manufg.	Personal consumption expenditures	66.3	
Wholesale trade	10.8	Trade	Phonograph records & tapes	15.4	Manufg.
Electronic components nec	10.0	Manufg.	Exports	7.2	Foreign
Plastics materials & resins	8.0	Manufg.	Change in business inventories	2.0	In House
Commercial printing	7.4	Manufg.	Colleges, universities, & professional schools	1.8	Services
Imports	6.7	Foreign	Libraries, vocation education	1.7	Services
Blankbooks & looseleaf binders	6.5	Manufg.	Business services nec	1.6	Services
Advertising	5.7	Services	Elementary & secondary schools	1.2	Services
Miscellaneous plastics products	3.7	Manufg.	Radio & TV broadcasting	1.1	Util.
U.S. Postal Service	2.3	Gov't	Federal Government purchases, national defense	0.4	Fed Govt
Electric services (utilities)	1.5	Util.	Child day care services	0.3	Services
Communications, except radio & TV	1.2	Util.	Labor, civic, social, & fraternal associations	0.2	Services
Gas production & distribution (utilities)	1.2	Util.	S/L Govt. purch., elem. & secondary education	0.2	S/L Govt
Real estate	0.8	Fin/R.E.	S/L Govt. purch., higher education	0.2	S/L Govt
Equipment rental & leasing services	0.8	Services	S/L Govt. purch., public assistance & relief	0.2	S/L Govt
Air transportation	0.7	Util.	Air transportation	0.1	Util.
Motor freight transportation & warehousing	0.5	Util.			
Eating & drinking places	0.5	Trade			
Computer & data processing services	0.5	Services			
Engineering, architectural, & surveying services	0.5	Services			
Maintenance of nonfarm buildings nec	0.4	Constr.			
Royalties	0.4	Fin/R.E.			
Petroleum refining	0.3	Manufg.			
Banking	0.3	Fin/R.E.			
Legal services	0.3	Services			
Noncomparable imports	0.3	Foreign			
Machinery, except electrical, nec	0.2	Manufg.			
Paper coating & glazing	0.2	Manufg.			
Railroads & related services	0.2	Util.			
Accounting, auditing & bookkeeping	0.2	Services			
Colleges, universities, & professional schools	0.2	Services			
Management & consulting services & labs	0.2	Services			
Manifold business forms	0.1	Manufg.			
Special dies & tools & machine tool accessories	0.1	Manufg.			
Insurance carriers	0.1	Fin/R.E.			
Electrical repair shops	0.1	Services			
Hotels & lodging places	0.1	Services			
Photofinishing labs, commercial photography	0.1	Services			

Source: Benchmark Input-Output Accounts for the U.S. Economy, 1982, U.S. Department of Commerce, Washington, D.C., July 1991. Data, as reported in the source, are organized by the 1977 SIC structure in use in 1982 but have been matched, as closely as is possible, to the 1987 SIC structure used in this book.

OCCUPATIONS EMPLOYED BY SIC 365 - HOUSEHOLD AUDIO AND VIDEO EQUIPMENT

Occupation	% of Total 1994	Change to 2005	Occupation	% of Total 1994	Change to 2005
Assemblers, fabricators, & hand workers nec	9.3	-35.8	Electrical & electronic technicians,technologists	1.5	-35.8
Electrical & electronic assemblers	8.1	-42.2	Hand packers & packagers	1.5	-45.0
Inspectors, testers, & graders, precision	4.7	-55.0	Stock clerks	1.5	-47.8
Blue collar worker supervisors	4.1	-41.8	Engineering technicians & technologists nec	1.4	-35.8
Engineers nec	4.0	-3.7	Electrical & electronics engineers	1.4	-31.6
Machine operators nec	3.7	-54.7	Managers & administrators nec	1.3	-35.8
Metal & plastic machine workers nec	2.4	-31.7	Bookkeeping, accounting, & auditing clerks	1.3	-51.8
Electrical & electronic equipment assemblers	2.4	-35.8	Helpers, laborers, & material movers nec	1.3	-35.7
Secretaries, ex legal & medical	2.2	-41.5	Machine feeders & offbearers	1.2	-42.2
Packaging & filling machine operators	2.1	-35.7	Freight, stock, & material movers, hand	1.1	-48.6
Traffic, shipping, & receiving clerks	2.1	-38.2	Marketing, advertising, & PR managers	1.1	-35.8
Combination machine tool operators	2.1	-35.8	General office clerks	1.1	-45.2
Sales & related workers nec	2.0	-35.8	Production, planning, & expediting clerks	1.0	-35.8
Plastic molding machine workers	1.9	-23.0	Janitors & cleaners, incl maids	1.0	-48.6
General managers & top executives	1.8	-39.1	Industrial truck & tractor operators	1.0	-35.8
Professional workers nec	1.7	-22.9	Management support workers nec	1.0	-35.8

Source: *Industry-Occupation Matrix*, Bureau of Labor Statistics. These data relate to one or more 3-digit SIC industry groups rather than to a single 4-digit SIC. The change reported for each occupation to the year 2005 is a percent of growth or decline as estimated by the Bureau of Labor Statistics. The abbreviation nec stands for 'not elsewhere classified'.

LOCATION BY STATE AND REGIONAL CONCENTRATION

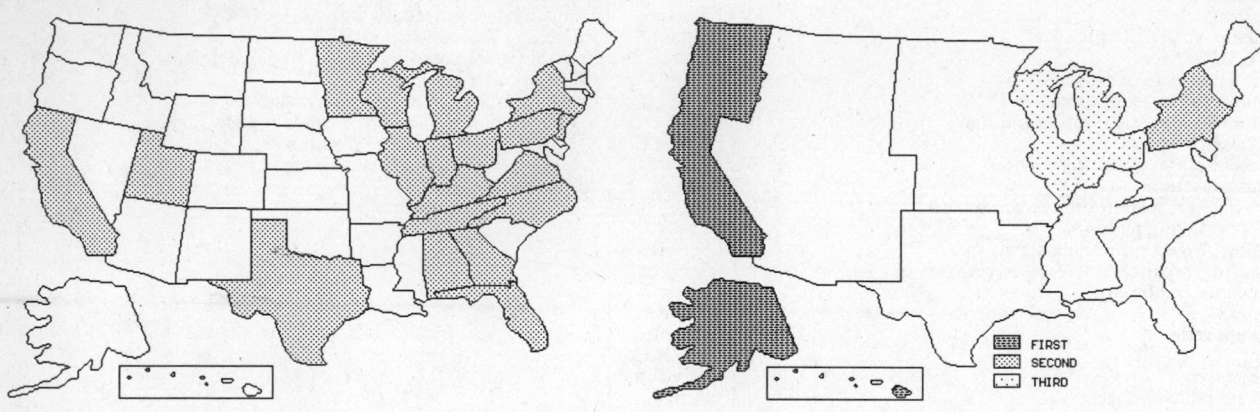

FIRST
SECOND
THIRD

INDUSTRY DATA BY STATE

State	Establish-ments	Shipments Total ($ mil)	Shipments % of U.S.	Shipments Per Establ.	Employment Total Number	Employment % of U.S.	Employment Per Establ.	Wages ($/hour)	Cost as % of Shipments	Investment per Employee ($)
Illinois	16	216.0	11.9	13.5	1,300	8.3	81	8.71	42.4	6,923
New York	61	202.0	11.2	3.3	1,400	8.9	23	12.05	38.8	7,071
California	98	189.3	10.5	1.9	1,500	9.6	15	9.68	31.7	8,800
North Carolina	10	164.4	9.1	16.4	1,600	10.2	160	11.28	33.7	-
New Jersey	20	112.7	6.2	5.6	1,000	6.4	50	11.88	26.5	-
Tennessee	30	59.8	3.3	2.0	500	3.2	17	9.38	47.3	8,000
Minnesota	7	45.4	2.5	6.5	300	1.9	43	11.75	26.4	7,000
Texas	22	43.9	2.4	2.0	500	3.2	23	9.67	31.4	4,200
Florida	14	34.7	1.9	2.5	400	2.5	29	9.67	39.8	4,000
Michigan	8	15.3	0.8	1.9	200	1.3	25	7.00	29.4	2,500
Ohio	10	(D)	-	-	375 *	2.4	38	-	-	-
Pennsylvania	10	(D)	-	-	3,750 *	23.9	375	-	-	-
Alabama	7	(D)	-	-	750 *	4.8	107	-	-	-
Georgia	7	(D)	-	-	1,750 *	11.1	250	-	-	-
Indiana	5	(D)	-	-	175 *	1.1	35	-	-	-
Utah	3	(D)	-	-	175 *	1.1	58	-	-	-
Virginia	3	(D)	-	-	375 *	2.4	125	-	-	-
Wisconsin	3	(D)	-	-	175 *	1.1	58	-	-	-
Kentucky	2	(D)	-	-	375 *	2.4	188	-	-	-

Source: 1992 *Economic Census*. The states are in descending order of shipments or establishments (if shipment data are missing for the majority). The symbol (D) appears when data are withheld to prevent disclosure of competitive information. States marked with (D) are sorted by number of establishments. A dash (-) indicates that the data element cannot be calculated; * indicates the midpoint of a range.

3661 - TELEPHONE & TELEGRAPH APPARATUS

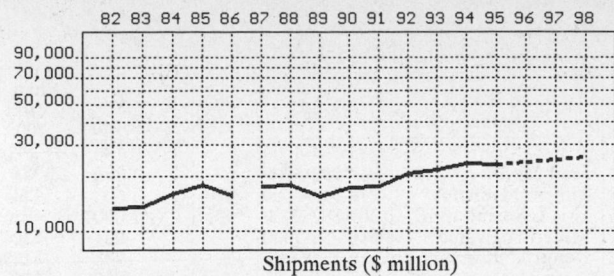

Shipments ($ million)

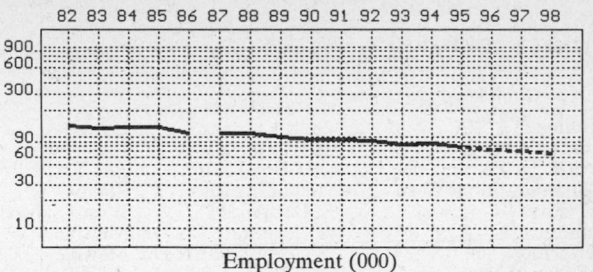

Employment (000)

GENERAL STATISTICS

Year	Companies	Establishments		Employment			Compensation		Production ($ million)			
		Total	with 20 or more employees	Total (000)	Production Workers (000)	Hours (Mil)	Payroll ($ mil)	Wages ($/hr)	Cost of Materials	Value Added by Manufacture	Value of Shipments	Capital Invest.
1982	259	333	210	136.5	85.6	158.4	3,021.2	10.19	6,357.8	7,120.8	13,394.4	513.1
1983		342	219	128.0	81.3	149.5	3,124.4	10.84	6,789.6	6,725.8	13,527.2	592.1
1984		351	228	132.5	82.3	155.1	3,547.1	12.09	8,563.0	7,695.9	15,783.2	866.1
1985		359	237	130.5	75.6	143.3	3,524.9	12.08	8,928.3	8,502.9	17,775.0	725.7
1986		359	233	110.9	58.5	110.7	3,019.2	12.60	7,409.5	7,780.7	15,687.8	629.6
1987*	390	469	297	112.3	58.6	109.4	3,178.4	12.82	7,956.3	9,588.0	17,582.5	552.0
1988		448	284	111.7	57.2	107.5	3,458.4	13.37	8,756.7	9,195.1	17,901.1	625.1
1989		449	278	101.7	49.0	93.9	3,312.6	14.51	6,975.0	8,326.5	15,467.0	573.1
1990		433	273	94.0	46.5	91.8	3,421.2	14.48	7,606.3	9,619.4	17,297.3	592.7
1991		468	281	94.1	46.9	89.7	3,468.1	14.47	7,838.7	9,502.5	17,424.9	458.7
1992	479	544	315	91.0	44.7	87.4	3,741.4	14.95	8,153.5	12,463.1	20,498.3	614.8
1993		532	303	84.9	38.5	75.1	3,731.1	15.34	8,002.8	13,589.9	21,539.8	594.6
1994		535P	302P	86.1	41.2	78.6	3,576.6	15.46	9,638.6	14,435.7	23,471.8	781.3
1995		549P	305P	78.3P	35.6P	70.3P	3,779.8P	16.01P	9,402.3P	14,081.8P	22,896.4P	676.4P
1996		563P	308P	74.1P	32.9P	65.5P	3,845.1P	16.37P	9,767.2P	14,628.3P	23,785.0P	693.6P
1997		578P	311P	69.9P	30.2P	60.8P	3,910.3P	16.72P	10,132.1P	15,174.8P	24,673.6P	710.8P
1998		592P	314P	65.8P	27.5P	56.0P	3,975.6P	17.07P	10,497.0P	15,721.3P	25,562.1P	727.9P

Sources: 1982, 1987, 1992 *Economic Census*; *Annual Survey of Manufactures*, 83-86, 88-91, 93-94. Establishment counts are from *County Business Patterns* for non-Census years; establishment counts for 83-84 are extrapolations. * indicates that industry content changed in 87; earlier years use 77 SICs. 'P's mark projections.

INDICES OF CHANGE

Year	Companies	Establishments		Employment			Compensation		Production ($ million)			
		Total	with 20 or more employees	Total (000)	Production Workers (000)	Hours (Mil)	Payroll ($ mil)	Wages ($/hr)	Cost of Materials	Value Added by Manufacture	Value of Shipments	Capital Invest.
1982	54	61	67	150	191	181	81	68	78	57	65	83
1983		63	70	141	182	171	84	73	83	54	66	96
1984		65	72	146	184	177	95	81	105	62	77	141
1985		66	75	143	169	164	94	81	110	68	87	118
1986		66	74	122	131	127	81	84	91	62	77	102
1987*	81	86	94	123	131	125	85	86	98	77	86	90
1988		82	90	123	128	123	92	89	107	74	87	102
1989		83	88	112	110	107	89	97	86	67	75	93
1990		80	87	103	104	105	91	97	93	77	84	96
1991		86	89	103	105	103	93	97	96	76	85	75
1992	100	100	100	100	100	100	100	100	100	100	100	100
1993		98	96	93	86	86	100	103	98	109	105	97
1994		98P	96P	95	92	90	96	103	118	116	115	127
1995		101P	97P	86P	80P	80P	101P	107P	115P	113P	112P	110P
1996		104P	98P	81P	74P	75P	103P	109P	120P	117P	116P	113P
1997		106P	99P	77P	68P	70P	105P	112P	124P	122P	120P	116P
1998		109P	100P	72P	61P	64P	106P	114P	129P	126P	125P	118P

Sources: Same as General Statistics. Values reflect change from the base year, 1992. Values above 100 mean greater than 92, values below 100 mean less than 92, and a value of 100 in the 82-91 or 93-98 period means same as 92. * indicates that industry content changed in 87. Data for earlier years are in 77 SIC format.

SELECTED RATIOS

For 1994	Avg. of All Manufact.	Analyzed Industry	Index	For 1994	Avg. of All Manufact.	Analyzed Industry	Index
Employees per Establishment	49	161	329	Value Added per Production Worker	134,084	350,381	261
Payroll per Establishment	1,500,273	6,688,806	446	Cost per Establishment	5,045,178	18,025,701	357
Payroll per Employee	30,620	41,540	136	Cost per Employee	102,970	111,947	109
Production Workers per Establishment	34	77	224	Cost per Production Worker	146,988	233,947	159
Wages per Establishment	853,319	2,272,533	266	Shipments per Establishment	9,576,895	43,895,966	458
Wages per Production Worker	24,861	29,494	119	Shipments per Employee	195,460	272,611	139
Hours per Production Worker	2,056	1,908	93	Shipments per Production Worker	279,017	569,704	204
Wages per Hour	12.09	15.46	128	Investment per Establishment	321,011	1,461,154	455
Value Added per Establishment	4,602,255	26,997,034	587	Investment per Employee	6,552	9,074	139
Value Added per Employee	93,930	167,662	178	Investment per Production Worker	9,352	18,964	203

Sources: Same as General Statistics. The 'Average of All Manufacturing' column represents the average of all manufacturing industries reported for the most recent complete year available. The Index shows the relationship between the Average and the Analyzed Industry. For example, 100 means that they are equal; 500 that the Analyzed Industry is five times the average; 50 means that the Analyzed Industry is half the national average. The abbreviation 'na' is used to show that data are 'not available'.

LEADING COMPANIES Number shown: 75 Total sales ($ mil): 24,559 Total employment (000): 126.9

Company Name	Address				CEO Name	Phone	Co. Type	Sales ($ mil)	Empl. (000)
Northern Telecom Inc	200 Athens Way	Nashville	TN	37228	Jean C Monty	615-734-4000	S	8,150	45.0
Northern Telecom Inc	PO Box 13010	Res Tri Pk	NC	27709	Jean Monty	919-992-5000	D	1,850	10.0
DSC Communications Corp	1000 Coit Rd	Plano	TX	75075	James L Donald	214-519-3000	P	1,003	5.4
RICOH Corp	5 Dedrick Pl	West Caldwell	NJ	07006	Eric L Steenburgh	201-882-2000	S	1,000	1.5
NEC America Inc	8 Corporate Ctr Dr	Melville	NY	11747	Mineo Sugiyama	516-753-7000	S	991	2.4
NEC USA Inc	8 Corporate Ctr Dr	Melville	NY	11747	Kenjiro Nitta	516-753-7000	S	991	2.5
Siemens Stromberg-Carlson	900 Broken Sound	Boca Raton	FL	33487	Anton Hasholzner	407-955-5000	S	750*	3.8
Racal-Datacom Inc	PO Box 407044	Ft Lauderdale	FL	33340	James Norman	305-846-4811	S	553	2.4
Tellabs Operations Inc	4951 Indiana Av	Lisle	IL	60532	Michael J Birck	708-969-8800	P	494	2.4
Dynatech Corp	3 New England	Burlington	MA	01803	John F Reno	617-272-6100	P	489	2.9
ADC Telecommunications Inc	4900 W 78th St	Minneapolis	MN	55403	William J Cadogan	612-938-8080	P	449	2.6
Octel Communications Corp	1001 Murphy Ranch	Milpitas	CA	95035	Robert Cohn	408-321-2000	P	406	2.4
California Microwave Inc	985 Almanor Av	Sunnyvale	CA	94086	Philip F Otto	408-732-4000	P	369	1.9
Hughes Network Systems Inc	11717 Expl Ln	Germantown	MD	20876	Jack A Shaw	301-428-5500	S	310*	1.6
Cypress Semiconductor Corp	3901 N 1st St	San Jose	CA	95134	TJ Rodgers	408-943-2600	P	304	1.3
Network Equipment Techn	800 Saginaw Dr	Redwood City	CA	94063	J J Francesconi	415-366-4400	P	284	1.2
Gilbert Associates Inc	PO Box 1498	Reading	PA	19603	Timothy S Cobb	610-775-5900	P	283	3.4
EXECUTONE Info Systems	478 Wheelers Farm	Milford	CT	06460	Alan Kessman	203-876-7600	P	282	2.4
Network Systems Corp	7600 Boone Av N	Minneapolis	MN	55428	Lyle D Altman	612-424-4888	P	216	1.6
General DataComm Industries	1579 Straits Tpk	Middlebury	CT	06762	Charles P Johnson	203-574-1118	P	211	1.8
Litton Data Systems	29851 Agoura Av	Agoura Hills	CA	91301	Allen Powers	818-902-4000	D	200	1.6
Mitel Inc	11921 Freedom Dr	Reston	VA	22090	Greg Spierkel	703-318-7020	S	200	0.3
Murata/Muratec	5560 Tennyson	Plano	TX	75024	R Michael Franz	214-403-3300	S	200	0.2
Fujitsu Business Comm Syst	7776 S Pte Pkwy W	Phoenix	AZ	85044	Anthony V Carollo	602-921-5900	S	199*	1.0
Fujitsu Network Syst	2801 Telecom Pkwy	Richardson	TX	75082	K Inoue	214-690-6000	S	190*	1.0
US Robotics Inc	8100 N McCormick	Skokie	IL	60076	Casey G Cowell	708-982-5010	P	189	0.8
Plantronics Inc	PO Box 1802	Santa Cruz	CA	95061	Robert S Cecil	408-426-6060	P	170	1.1
IPC Information Systems Inc	88 Pine St	New York	NY	10005	R P Kleinknecht	212-825-9060	P	164	0.9
StrataCom Inc	1400 Parkmoor Av	San Jose	CA	95126	Richard M Moley	408-294-7600	P	154	0.6
Rockwell International Corp	1431 Opus Pl	Downers Grove	IL	60515	Steven Panyko	708-960-8000	D	150*	0.8
TRW Inc	18901 S Wilmington	Carson	CA	90746	Jack R Distaso	310-764-6000	D	140*	0.8
TIE communications Inc	4 Progress Av	Seymour	CT	06483	G N Benjamin III	203-888-8000	P	125	1.1
ADTRAN Inc	901 Explorer Blv	Huntsville	AL	35806	Mark C Smith	205-971-8000	P	123	0.5
R-Tec Systems	2100 Reliance Pkwy	Bedford	TX	76021	Pat Welker	817-267-3141	D	110*	0.7
Aspect Telecomm Corp	1730 Fox Dr	San Jose	CA	95131	James K Carreker	408-441-2200	P	107	0.5
Telco Systems Inc	63 Nahatan St	Norwood	MA	02062	John A Ruggiero	617-551-0300	P	100	0.4
CIDCO Inc	220 Cochrane Cir	Morgan Hill	CA	95037	Paul G Locklin	408-779-1162	P	100	0.3
CMC Industries Inc	PO Box 831	Corinth	MS	38834	Steve J Fry	601-287-3771	P	100	1.2
ComStream Corp	10180 Barnes	San Diego	CA	92121	Patrick Courtin	619-458-1800	S	100	0.5
Hypercom Inc	2851 W Kathleen	Phoenix	AZ	85023	Al Irato	602-866-5399	R	100	0.2
Symmetricom Inc	85 W Tasman Dr	San Jose	CA	95134	William D Rasdal	408-943-9403	P	98	0.6
Charles Industries Ltd	5600 Apollo Dr	Rolling Mdws	IL	60008	JT Charles	708-806-6300	R	95	1.0
VMX Inc	1001 Murphy Ranch	Milpitas	CA	95035	Patrick S Howard	408-321-2000	P	91	0.6
Boston Technology Inc	100 Quannapowitt	Wakefield	MA	01880	John C Taylor	617-246-9000	P	89	0.4
Intecom	5057 Keller Spgs Rd	Dallas	TX	75248	George C Platt	214-447-9000	S	88	0.7
Comdial Corp	PO Box 7266	Charlottesville	VA	22906	William G Mustain	804-978-2200	P	77	0.8
Hughes LAN Systems Inc	1225 Charleston Rd	Mountain View	CA	94043	Joseph Kennedy	415-966-7300	S	75	0.4
Communications Systems Inc	213 S Main St	Hector	MN	55342	Curtis A Sampson	612-848-6231	P	74	1.3
Penril Datability Networks Inc	1300 Quince	Gaithersburg	MD	20878	Henry D Epstein	301-417-0552	P	74	0.4
EF Data Corp	2105 W 5th Pl	Tempe	AZ	85281	Steve Eymann	602-968-0447	P	73*	0.4
MaxTech Corp	13915 Cerritos	Cerritos	CA	90703	Gary Fan	310-921-1698	R	70	0.2
MICOM Communications Corp	4100 Los Angeles	Simi Valley	CA	93063	Barry Phelps	805-583-8600	S	69	0.4
Zoom Telephonics Inc	207 South St	Boston	MA	02111	Frank B Manning	617-423-1072	P	68	0.2
Dynatel Systems	PO Box 2963	Austin	TX	78769	Roger H D Lacey	512-984-3400	D	67	0.4
Electronic Information Systems	1351 Washington	Stamford	CT	06902	Joseph J Porfeli	203-358-0764	P	65	0.3
ADC Fibermux Corp	21415 Plummer St	Chatsworth	CA	91311	Steve Kim	818-709-6000	S	60	0.3
Coil Sales and Manufacturing Co	5600 Apollo Dr	Rolling Mdws	IL	60008	JT Charles	708-806-6300	S	60	0.4
Harris Corp	PO Box 1188	Novato	CA	94948	Gerald L Doyle	415-382-5000	D	60	0.5
Superior TeleTec Inc	150 Interstate N	Atlanta	GA	30339	Justin F Deedy Jr	404-953-8338	S	60*	0.5
PairGain Technologies Inc	12921 E 166th St	Cerritos	CA	90701	Charles S Strauch	310-404-8811	P	59	0.3
Microcom Inc	500 River Ridge Dr	Norwood	MA	02062	Roland D Pampel	617-551-1000	P	57	0.3
IDB WorldCom Services Inc	380 Madison Av	New York	NY	10017	Jeffrey Sudikoff	212-478-6100	S	55*	0.3
OST Inc	14225 Sullyfield Cir	Chantilly	VA	22021	Thao Lane	703-817-0400	R	55	0.4
VTEL Corp	108 Wild Basin Rd	Austin	TX	78746	FH Moeller	512-314-2700	P	54	0.3
Megahertz Corp	PO Box 16020	Salt Lake City	UT	84116	Spencer F Kirk	801-272-6000	S	53	0.4
Vystar Group Inc	4505 S Wasatch	Salt Lake City	UT	84124	Spencer F Kirk	801-272-6000	P	53	0.4
Digital Systems International Inc	6464 185th Av NE	Redmond	WA	98052	Pat Howard	206-881-7544	P	53	0.3
XEL Communications Inc	17600 E Exposition	Aurora	CO	80017	William J Sanko	303-369-7000	R	52	0.3
ECI Telecom Inc	927 Fern St	Altamonte Sp	FL	32701	JR Kennedy	407-331-5500	S	50	<0.1
GVC Technologies Inc	376 Lafayette Rd	Sparta	NJ	07871	Charles Lee	201-579-3630	S	50	<0.1
Unisys	8230 Montgomery	Cincinnati	OH	45236	Paul Rodwick	513-745-0500	D	50	0.2
Teltrend Inc	620 Stetson Av	St Charles	IL	60174	Howard L Kirby Jr	708-377-1700	P	49	0.3
Teltrend Subsidiary Inc	620 Stetson Av	St Charles	IL	60174	Howard L Kirby Jr	708-377-1700	S	49	0.3
TI Holdings Inc	620 Stetson Av	St Charles	IL	60174	Howard L Kirby Jr	708-377-1700	S	49	0.3
Protel Inc	4150 Kidron Rd	Lakeland	FL	33811	Jerry Yachabach	813-644-5558	S	49*	0.3

Source: Ward's Business Directory of U.S. Private and Public Companies, Volumes 1 and 2, 1996. The company type code used is as follows: P - Public, R - Private, S - Subsidiary, D - Division, J - Joint Venture, A - Affiliate, G - Group. Sales are in millions of dollars, employees are in thousands. An asterisk (*) indicates an estimated sales volume. The symbol < stands for 'less than'. Company names and addresses are truncated, in some cases, to fit into the available space.

MATERIALS CONSUMED

Material	Quantity	Delivered Cost ($ million)
Materials, ingredients, containers, and supplies	(X)	7,446.9
Printed circuit boards (without inserted components) for electronic circuitry	(X)	630.7
Printed circuit assemblies, loaded boards or modules	(X)	729.8
Semiconductors, including transistors, diodes, rectifiers, and integrated circuits for electronic circuitry	(X)	1,500.6
Capacitors for electronic circuitry	(X)	161.7
Resistors for electronic circuitry	(X)	156.8
Other components and accessories for electronic circuitry, nec, except tubes	(X)	403.6
Electronic communication equipment	(X)	486.8
Electrical instrument mechanisms and meter movements (including instrument relays)	(X)	20.8
Electronic computing equipment	(X)	123.8
Current-carrying wiring devices	(X)	49.3
Insulated wire and cable, including magnet wire	(X)	139.0
Loudspeakers, microphones, and tuners (all types)	(X)	9.2
Fractional horsepower electric motors (less than 1 hp)	(X)	0.8
Silicon, hyperpure	(X)	(D)
Plastics resins consumed in the form of granules, pellets, powders, liquids, etc.	(X)	43.4
Fabricated plastics products (except gaskets, hoses, and belting)	(X)	46.8
Sheet metal products, except stampings	(X)	92.4
Metal stampings	(X)	19.8
Metal bolts, nuts, screws, washers, rivets, and other screw machine products	(X)	22.8
Other fabricated metal products (except forgings)	(X)	63.1
Forgings	(X)	(D)
Castings (rough and semifinished)	(X)	12.1
Steel shapes and forms	(X)	7.0
Aluminum and aluminum-base alloy shapes and forms	(X)	10.0
Other nonferrous shapes and forms	(X)	(D)
Paper and paperboard containers, including shipping sacks and other paper packaging supplies	(X)	29.4
All other materials and components, parts, containers, and supplies	(X)	1,099.8
Materials, ingredients, containers, and supplies, nsk	(X)	1,572.2

Source: 1992 Economic Census. Explanation of symbols used: (D): Withheld to avoid disclosure of competitive data; na: Not available; (S): Withheld because statistical norms were not met; (X): Not applicable; (Z): Less than half the unit shown; nec: Not elsewhere classified; nsk: Not specified by kind; - : zero; * : 10-19 percent estimated; ** : 20-29 percent estimated.

PRODUCT SHARE DETAILS

Product or Product Class	% Share	Product or Product Class	% Share
Telephone and telegraph apparatus	100.00	equipment) and modems, including auxiliary sets	26.91
Telephone switching and switchboard equipment	41.01	Other telephone and telegraph (wire) apparatus, including	
Telephone and telegraph apparatus, carrier line equipment		telephone sets, telephone answering, and fax machines	29.95
(office and line repeaters and line terminating carrier		Telephone and telegraph apparatus, nsk	2.13

Source: 1992 Economic Census. The values shown are percent of total shipments in an industry. Values of indented subcategories are summed in the main heading. The symbol (D) appears when data are withheld to prevent disclosure of competitive information. The abbreviation nsk stands for 'not specified by kind' and nec for 'not elsewhere classified'.

INPUTS AND OUTPUTS FOR TELEPHONE & TELEGRAPH APPARATUS

Economic Sector or Industry Providing Inputs	%	Sector	Economic Sector or Industry Buying Outputs	%	Sector
Electronic components nec	23.2	Manufg.	Gross private fixed investment	58.0	Cap Inv
Miscellaneous plastics products	13.3	Manufg.	Communications, except radio & TV	22.2	Util.
Wholesale trade	10.9	Trade	Exports	6.1	Foreign
Telephone & telegraph apparatus	8.7	Manufg.	Telephone & telegraph apparatus	5.9	Manufg.
Imports	8.5	Foreign	Personal consumption expenditures	4.4	
Semiconductors & related devices	3.8	Manufg.	Federal Government purchases, national defense	2.1	Fed Govt
Nonferrous rolling & drawing, nec	2.9	Manufg.	Change in business inventories	0.5	In House
Advertising	2.7	Services	Federal Government purchases, nondefense	0.2	Fed Govt
Primary nonferrous metals, nec	1.8	Manufg.	S/L Govt. purch., elem. & secondary education	0.2	S/L Govt
Communications, except radio & TV	1.2	Util.	S/L Govt. purch., health & hospitals	0.2	S/L Govt
Electric services (utilities)	1.0	Util.	S/L Govt. purch., higher education	0.1	S/L Govt
Business services nec	1.0	Services			
Plating & polishing	0.9	Manufg.			
Screw machine and related products	0.9	Manufg.			
Banking	0.9	Fin/R.E.			
Electron tubes	0.8	Manufg.			
Metal stampings, nec	0.8	Manufg.			
Nonferrous wire drawing & insulating	0.8	Manufg.			
Plastics materials & resins	0.8	Manufg.			
Wiring devices	0.7	Manufg.			
Air transportation	0.7	Util.			
Motor freight transportation & warehousing	0.7	Util.			
Equipment rental & leasing services	0.7	Services			
Noncomparable imports	0.7	Foreign			
Adhesives & sealants	0.6	Manufg.			

Continued on next page.

INPUTS AND OUTPUTS FOR TELEPHONE & TELEGRAPH APPARATUS - Continued

Economic Sector or Industry Providing Inputs	%	Sector	Economic Sector or Industry Buying Outputs	%	Sector
Maintenance of nonfarm buildings nec	0.5	Constr.			
Blast furnaces & steel mills	0.5	Manufg.			
Fabricated metal products, nec	0.5	Manufg.			
Hardware, nec	0.5	Manufg.			
Miscellaneous fabricated wire products	0.5	Manufg.			
Paints & allied products	0.5	Manufg.			
Eating & drinking places	0.5	Trade			
Engineering, architectural, & surveying services	0.5	Services			
Real estate	0.4	Fin/R.E.			
Computer & data processing services	0.4	Services			
Die-cut paper & board	0.3	Manufg.			
Paperboard containers & boxes	0.3	Manufg.			
Gas production & distribution (utilities)	0.3	Util.			
Legal services	0.3	Services			
Chemical preparations, nec	0.2	Manufg.			
Copper rolling & drawing	0.2	Manufg.			
Glass & glass products, except containers	0.2	Manufg.			
Hand & edge tools, nec	0.2	Manufg.			
Machinery, except electrical, nec	0.2	Manufg.			
Metal coating & allied services	0.2	Manufg.			
Petroleum refining	0.2	Manufg.			
Royalties	0.2	Fin/R.E.			
Management & consulting services & labs	0.2	Services			
U.S. Postal Service	0.2	Gov't			
Aluminum rolling & drawing	0.1	Manufg.			
Electrical equipment & supplies, nec	0.1	Manufg.			
Fabricated rubber products, nec	0.1	Manufg.			
Manifold business forms	0.1	Manufg.			
Sawmills & planning mills, general	0.1	Manufg.			
Special dies & tools & machine tool accessories	0.1	Manufg.			
Railroads & related services	0.1	Util.			
Insurance carriers	0.1	Fin/R.E.			
Accounting, auditing & bookkeeping	0.1	Services			
Business/professional associations	0.1	Services			
Hotels & lodging places	0.1	Services			

Source: Benchmark Input-Output Accounts for the U.S. Economy, 1982, U.S. Department of Commerce, Washington, D.C., July 1991. Data, as reported in the source, are organized by the 1977 SIC structure in use in 1982 but have been matched, as closely as is possible, to the 1987 SIC structure used in this book.

OCCUPATIONS EMPLOYED BY SIC 366 - COMMUNICATIONS EQUIPMENT

Occupation	% of Total 1994	Change to 2005	Occupation	% of Total 1994	Change to 2005
Electrical & electronic assemblers	9.3	-22.9	Engineers nec	1.6	28.5
Electrical & electronics engineers	7.6	9.6	General managers & top executives	1.6	-18.7
Electrical & electronic equipment assemblers	7.2	-14.3	Managers & administrators nec	1.5	-14.4
Electrical & electronic technicians,technologists	5.0	-14.3	Mechanical engineers	1.4	3.7
Inspectors, testers, & graders, precision	4.5	-40.0	Production, planning, & expediting clerks	1.4	-14.3
Assemblers, fabricators, & hand workers nec	4.0	-14.3	Systems analysts	1.3	37.0
Sales & related workers nec	3.0	-14.3	Traffic, shipping, & receiving clerks	1.3	-17.6
Blue collar worker supervisors	2.5	-25.1	Stock clerks	1.3	-30.4
Secretaries, ex legal & medical	2.4	-22.0	Plant & system operators nec	1.2	-24.3
Computer engineers	2.4	26.9	Drafters	1.2	-33.3
Engineering technicians & technologists nec	2.1	-14.3	General office clerks	1.2	-27.0
Engineering, mathematical, & science managers	2.1	-2.7	Industrial production managers	1.2	-14.4
Management support workers nec	2.0	-14.3	Marketing, advertising, & PR managers	1.2	-14.4
Electromechanical equipment assemblers	1.7	-5.8	Purchasing agents, ex trade & farm products	1.1	-14.3

Source: Industry-Occupation Matrix, Bureau of Labor Statistics. These data relate to one or more 3-digit SIC industry groups rather than to a single 4-digit SIC. The change reported for each occupation to the year 2005 is a percent of growth or decline as estimated by the Bureau of Labor Statistics. The abbreviation nec stands for 'not elsewhere classified'.

LOCATION BY STATE AND REGIONAL CONCENTRATION

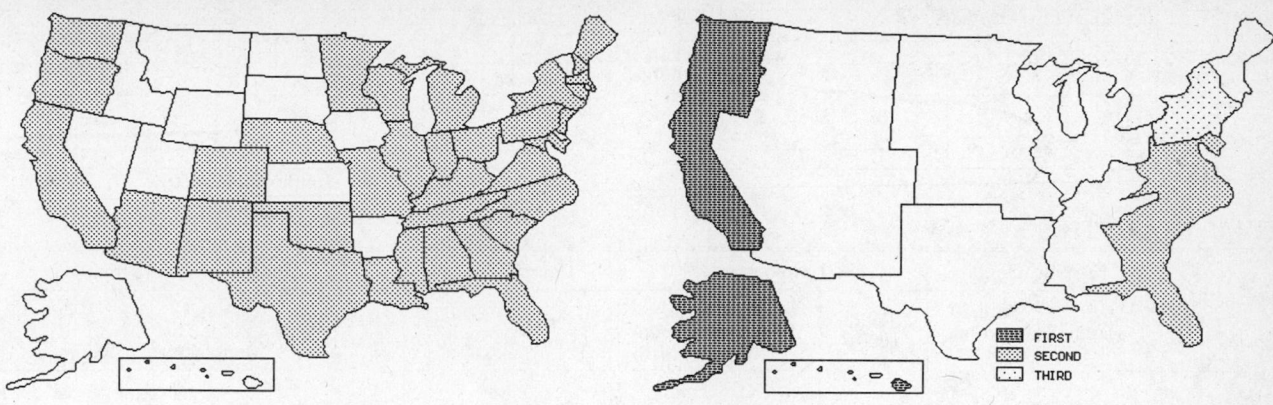

FIRST
SECOND
THIRD

INDUSTRY DATA BY STATE

State	Establish- ments	Shipments			Employment				Cost as % of Shipments	Investment per Employee ($)
		Total ($ mil)	% of U.S.	Per Establ.	Total Number	% of U.S.	Per Establ.	Wages ($/hour)		
California	122	4,031.4	19.7	33.0	16,900	18.6	139	12.80	34.0	8,976
Massachusetts	28	2,531.1	12.3	90.4	10,700	11.8	382	18.85	39.0	9,411
Texas	43	1,841.3	9.0	42.8	8,100	8.9	188	14.65	39.7	5,815
Florida	32	937.1	4.6	29.3	5,900	6.5	184	13.00	24.3	4,373
Illinois	37	863.2	4.2	23.3	6,500	7.1	176	16.84	26.2	-
New Jersey	26	567.5	2.8	21.8	3,000	3.3	115	20.93	54.2	1,733
Minnesota	16	442.2	2.2	27.6	2,400	2.6	150	9.88	29.4	5,750
Georgia	17	315.9	1.5	18.6	1,700	1.9	100	14.36	51.0	4,882
Alabama	13	282.1	1.4	21.7	2,000	2.2	154	9.63	35.7	7,600
Virginia	15	242.4	1.2	16.2	1,900	2.1	127	10.63	45.8	5,211
Connecticut	11	190.0	0.9	17.3	1,100	1.2	100	8.13	42.1	8,000
New York	25	184.4	0.9	7.4	1,800	2.0	72	12.31	46.4	1,889
Maryland	10	106.3	0.5	10.6	700	0.8	70	13.80	37.3	3,714
Washington	16	89.7	0.4	5.6	800	0.9	50	7.88	37.6	3,000
Pennsylvania	15	69.1	0.3	4.6	700	0.8	47	9.25	37.5	-
Wisconsin	5	55.5	0.3	11.1	500	0.5	100	9.00	40.7	-
Michigan	5	30.6	0.1	6.1	300	0.3	60	13.00	37.9	-
North Carolina	16	(D)	-	-	7,500 *	8.2	469	-	-	-
Colorado	10	(D)	-	-	1,750 *	1.9	175	-	-	-
Ohio	9	(D)	-	-	3,750 *	4.1	417	-	-	-
Missouri	8	(D)	-	-	175 *	0.2	22	-	-	2,286
New Hampshire	8	(D)	-	-	750 *	0.8	94	-	-	-
Tennessee	7	(D)	-	-	175 *	0.2	25	-	-	-
Oregon	6	(D)	-	-	750 *	0.8	125	-	-	-
Arizona	5	(D)	-	-	175 *	0.2	35	-	-	-
Kentucky	4	(D)	-	-	375 *	0.4	94	-	-	533
Maine	4	(D)	-	-	750 *	0.8	188	-	-	-
Nebraska	4	(D)	-	-	1,750 *	1.9	438	-	-	-
Louisiana	3	(D)	-	-	1,750 *	1.9	583	-	-	-
Mississippi	3	(D)	-	-	750 *	0.8	250	-	-	-
Oklahoma	3	(D)	-	-	3,750 *	4.1	1,250	-	-	-
South Carolina	3	(D)	-	-	375 *	0.4	125	-	-	-
Indiana	2	(D)	-	-	175 *	0.2	88	-	-	-
New Mexico	2	(D)	-	-	375 *	0.4	188	-	-	-

Source: 1992 *Economic Census*. The states are in descending order of shipments or establishments (if shipment data are missing for the majority). The symbol (D) appears when data are withheld to prevent disclosure of competitive information. States marked with (D) are sorted by number of establishments. A dash (-) indicates that the data element cannot be calculated; * indicates the midpoint of a range.

3663 - RADIO & TV COMMUNICATIONS EQUIPMENT

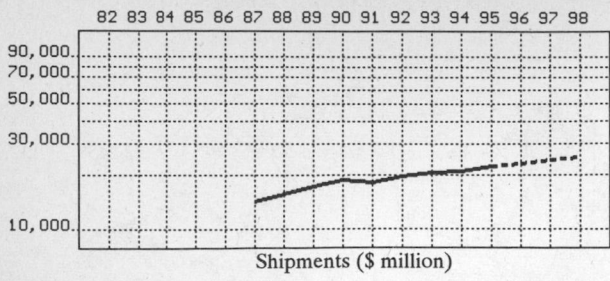

Shipments ($ million)

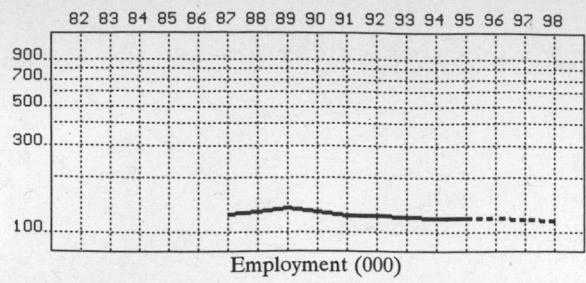

Employment (000)

GENERAL STATISTICS

| Year | Com-panies | Establishments | | Employment | | | Compensation | | Production ($ million) | | | |
		Total	with 20 or more employees	Total (000)	Production Workers (000)	Hours (Mil)	Payroll ($ mil)	Wages ($/hr)	Cost of Materials	Value Added by Manufacture	Value of Shipments	Capital Invest.
1982												
1983												
1984												
1985												
1986												
1987	535	655	455	126.0	58.5	115.2	3,775.6	11.17	5,265.0	9,060.8	14,228.6	622.7
1988		605	406	129.4	58.5	115.3	4,080.4	11.55	6,052.1	10,054.3	15,693.1	704.5
1989		657	445	135.8	62.1	125.2	4,624.4	12.10	6,434.4	10,899.3	17,286.5	697.5
1990		657	446	129.2	62.4	125.0	4,651.9	11.90	7,840.0	11,278.0	18,759.3	751.8
1991		681	443	123.4	53.7	110.4		12.51	7,638.7	10,339.3	18,164.9	592.4
1992	861	948	492	124.4	58.3	115.1	4,700.8	13.81	7,311.6	12,246.9	19,472.3	700.3
1993		907	478	122.6	56.0	108.4	4,630.4	13.66	9,186.0	11,929.8	20,649.9	850.6
1994		939P	486P	120.2	54.5	111.1	4,836.7	13.91	8,892.4	12,354.4	20,877.2	799.5
1995		992P	495P	120.2P	54.8P	109.9P		14.48P	9,490.7P	13,185.6P	22,281.9P	812.3P
1996		1,044P	503P	118.9P	54.0P	108.6P		14.90P	9,882.6P	13,730.1P	23,201.9P	833.9P
1997		1,096P	512P	117.5P	53.3P	107.3P		15.32P	10,274.5P	14,274.6P	24,122.0P	855.5P
1998		1,149P	520P	116.1P	52.6P	106.1P		15.74P	10,666.4P	14,819.1P	25,042.1P	877.2P

Sources: 1982, 1987, 1992 *Economic Census*; *Annual Survey of Manufactures*, 83-86, 88-91, 93-94. Establishment counts for non-Census years are from *County Business Patterns*; establishment values for 83-84 are extrapolations. 'P's show projections by the editors. Industries reclassified in 87 will not have data for prior years.

INDICES OF CHANGE

| Year | Com-panies | Establishments | | Employment | | | Compensation | | Production ($ million) | | | |
		Total	with 20 or more employees	Total (000)	Production Workers (000)	Hours (Mil)	Payroll ($ mil)	Wages ($/hr)	Cost of Materials	Value Added by Manufacture	Value of Shipments	Capital Invest.
1982												
1983												
1984												
1985												
1986												
1987	62	69	92	101	100	100	80	81	72	74	73	89
1988		64	83	104	100	100	87	84	83	82	81	101
1989		69	90	109	107	109	98	88	88	89	89	100
1990		69	91	104	107	109	99	86	107	92	96	107
1991		72	90	99	92	96		91	104	84	93	85
1992	100	100	100	100	100	100	100	100	100	100	100	100
1993		96	97	99	96	94	99	99	126	97	106	121
1994		99P	99P	97	93	97	103	101	122	101	107	114
1995		105P	101P	97P	94P	96P		105P	130P	108P	114P	116P
1996		110P	102P	96P	93P	94P		108P	135P	112P	119P	119P
1997		116P	104P	94P	91P	93P		111P	141P	117P	124P	122P
1998		121P	106P	93P	90P	92P		114P	146P	121P	129P	125P

Sources: Same as General Statistics. Values reflect change from the base year, 1992. Values above 100 mean greater than 92, values below 100 mean less than 92, and a value of 100 in the 82-91 or 93-98 period means same as 92. 'P's mark projections by the editors.

SELECTED RATIOS

For 1994	Avg. of All Manufact.	Analyzed Industry	Index	For 1994	Avg. of All Manufact.	Analyzed Industry	Index
Employees per Establishment	49	128	261	Value Added per Production Worker	134,084	226,686	169
Payroll per Establishment	1,500,273	5,148,555	343	Cost per Establishment	5,045,178	9,465,754	188
Payroll per Employee	30,620	40,239	131	Cost per Employee	102,970	73,980	72
Production Workers per Establishment	34	58	169	Cost per Production Worker	146,988	163,163	111
Wages per Establishment	853,319	1,645,044	193	Shipments per Establishment	9,576,895	22,223,297	232
Wages per Production Worker	24,861	28,356	114	Shipments per Employee	195,460	173,687	89
Hours per Production Worker	2,056	2,039	99	Shipments per Production Worker	279,017	383,068	137
Wages per Hour	12.09	13.91	115	Investment per Establishment	321,011	851,049	265
Value Added per Establishment	4,602,255	13,150,973	286	Investment per Employee	6,552	6,651	102
Value Added per Employee	93,930	102,782	109	Investment per Production Worker	9,352	14,670	157

Sources: Same as General Statistics. The 'Average of All Manufacturing' column represents the average of all manufacturing industries reported for the most recent complete year available. The Index shows the relationship between the Average and the Analyzed Industry. For example, 100 means that they are equal; 500 that the Analyzed Industry is five times the average; 50 means that the Analyzed Industry is half the national average. The abbreviation 'na' is used to show that data are 'not available'.

LEADING COMPANIES   Number shown: 74   Total sales ($ mil): 32,522   Total employment (000): 205.0

Company Name	Address				CEO Name	Phone	Co. Type	Sales ($ mil)	Empl. (000)
Motorola Inc	1303 E Algonquin	Schaumburg	IL	60196	Gary L Tooker	708-576-5000	P	22,245	132.0
Space Systems/Loral Inc	1111 J Davis	Arlington	VA	22202	Robert E Berry	703-685-5500	S	2,130*	17.0
GTE Government Systems Corp	77 A St	Needham	MA	02194	Frances A Gicca	617-449-2000	S	1,590	9.5
Scientific-Atlanta Inc	PO Box 105600	Atlanta	GA	30348	James F McDonald	404-903-5000	P	812	4.0
E-Systems Inc	PO Box 12248	St Petersburg	FL	33733	Jim Garrett	813-381-2000	D	340*	2.1
Ascom Timeplex Inc	400 Chestnut Ridge	Woodcliff Lake	NJ	07675	W Y O'Connor	201-391-1111	S	280*	2.1
Ball Aerospace	PO Box 1235	Broomfield	CO	80038	Donovan Hicks	303-939-4000	D	268	2.0
California Microwave	171 N Covington Dr	Bloomingdale	IL	60108	Michael Terhune	708-307-5900	R	267	0.4
Cubic Corp	PO Box 85587	San Diego	CA	92186	Walter J Zable	619-277-6780	P	261	2.7
Allen Telecom Group Inc	30500 Bruce Indrial	Cleveland	OH	44139	Erik Van der Kaay	216-349-8400	S	200	1.2
RF Communications	1680 University Av	Rochester	NY	14610	Bruce Fennie	716-244-5830	D	197	1.6
AlliedSignal	400 N Rogers Rd	Olathe	KS	66062	Greg Summe	913-782-0400	D	180*	3.1
Grass Valley Group Inc	PO Box 1114	Grass Valley	CA	95945	Dan Castles	916-478-3000	S	160	1.0
Compression Labs Inc	2860 Junction Av	San Jose	CA	95134	John E Tyson	408-435-3000	P	157	0.5
American Lightwave Systems Inc	999 Research Pkwy	Meriden	CT	06450		203-630-5700	S	150	0.5
Aydin Corp	PO Box 349	Horsham	PA	19044	Ayhan Hakimoglu	215-657-7510	P	142	1.5
Antenna Specialists Co	30500 Bruce Indrial	Cleveland	OH	44139	Erik van der Kaay	216-349-8400	D	135	0.7
Telex Communications Inc	9600 Aldrich Av S	Minneapolis	MN	55420	John Hale	612-884-4051	R	130	1.4
Radiation Systems Inc	1501 Moran Rd	Sterling	VA	20166	Richard E Thomas	703-450-5680	P	122	0.9
Farinon	1691 Bayport Av	San Carlos	CA	94070	Ricardo Diaz	415-594-3000	D	120*	1.0
Burle Industries Inc	1000 New Holland	Lancaster	PA	17601	Eric Burlefinger	717-295-6000	R	117	1.0
Digital Microwave Corp	170 Rose	San Jose	CA	95134	M M Michigami	408-943-0777	P	108	0.5
McDonnell Douglas	700 Royal Oaks Dr	Monrovia	CA	91016	F Pat Daher	818-303-9000	D	100*	0.8
Philips Broadband Network Inc	100 Fairgrounds Dr	Manlius	NY	13104	Dieter B Brauer	315-682-9105	S	87*	0.7
Wiltron Co	490 Jarvis Dr	Morgan Hill	CA	95037	DE Dunwoodie	408-778-2000	D	77	0.8
Toko America Inc	1250 Feehanville Dr	Mt Prospect	IL	60056	Kent Kitano	708-297-0070	S	76	0.2
C-Cor Electronics Inc	60 Decibel Rd	State College	PA	16801	Richard E Perry	814-238-2461	P	75	0.5
AT and T Tridom	840 Franklin Ct	Marietta	GA	30067	E Mc-Raisch	404-426-4261	S	73	0.4
General Railway Signal Corp	PO Box 20600	Rochester	NY	14602	Stuart A Brown	716-783-2000	S	72*	0.8
ADC Kentrox Inc	14375 NW Science	Portland	OR	97229	Richard S Gilbert	503-643-1681	S	70	0.3
Sony Trans Com Inc	PO Box 19713	Irvine	CA	92713	Douglas Cline	714-252-0600	S	70	0.4
Aeroflex Inc	35 S Service Rd	Plainview	NY	11803	Harvey R Blau	516-694-6700	P	66	0.7
Datron Systems Inc	304 Enterprise St	Escondido	CA	92029	David A Derby	619-747-3734	P	66	0.3
LXE Inc	PO Box 926000	Norcross	GA	30092	Thomas E Sharon	404-447-4224	P	63	0.4
Spectrian Corp	550 Ellis St	Mountain View	CA	94043	C Woodrow Rea Jr	415-961-1473	P	63	0.4
Vertex Communications Corp	PO Box 1277	Kilgore	TX	75663	J Rex Vardeman	903-984-0555	P	57	0.5
MPD Inc	316 E 9th St	Owensboro	KY	42303	Gary J Braswell	502-685-6200	R	55	0.6
Decibel Products	PO Box 569610	Dallas	TX	75356	Peter Mailandt	214-631-0310	D	54*	0.6
Telecom Solutions	85 W Tasman Dr	San Jose	CA	95134	D Ronald Duren	408-433-0910	S	53	0.4
Telecom Solutions	85 W Tasman Dr	San Jose	CA	95134	D Ronald Duren	408-943-9403	D	52	0.3
TV/COM International	16516 Via Esprillo	San Diego	CA	92127	Henk Hanselaar	619-451-1500	R	50	0.3
GTE Government Systems Corp	100 1st Av	Waltham	MA	02254	Frances Gicca	617-890-9200	D	47*	0.4
Vicon Industries Inc	525 Broad Hollow	Melville	NY	11747	Kenneth M Darby	516-293-2200	P	46	0.2
Racal Communications Inc	5 Research Pl	Rockville	MD	20850	Mark Lipp	301-948-4420	S	44*	0.3
Loral Randtron Systems	130 Constitution Dr	Menlo Park	CA	94025	Bill Gates	415-326-9500	S	43*	0.2
Gai-Tronics Corp	PO Box 31	Reading	PA	19603	G Edward Smith	215-372-5151	S	42	0.4
LoJack Corp	333 Elm St	Dedham	MA	02026	C Michael Daley	617-326-4700	P	41	0.2
California Amplifier Inc	460 Calle San Pablo	Camarillo	CA	93012	Barry W Hall	805-987-9000	P	41	0.4
Chaparral Communications Inc	2450 N First St	San Jose	CA	95131	Robert Taggart	408-435-1530	R	40	0.3
Flightline Electronics Inc	7500 Main St	Fishers	NY	14453	William Blossom	716-924-4000	R	40	0.2
M/A-COM Inc	110 Haverhill Rd	Amesbury	MA	01913	Ken Andrews	508-388-5210	D	40	0.4
Eagle Comtronics Inc	4562 Waterhouse	Clay	NY	13041	Alan Devendorf	315-622-3402	R	38	0.7
RL Drake Co	PO Box 3006	Miamisburg	OH	45343	Ronald E Wysong	513-866-2421	R	38	0.4
American Microwave Technology	550 Ellis St	Mountain View	CA	94043	William P Clark	415-961-1473	S	38	0.3
Texscan Corp	5-D Butterf	El Paso	TX	79906	William H Lambert	915-772-4400	S	38	0.6
Alpha Technologies Inc	3767 Alpha Way	Bellingham	WA	98226	Fred Kaiser	360-647-2360	R	35*	0.4
Microwave Radio Corp	20 Alpha Rd	Chelmsford	MA	01824	Robert Morrill	508-250-1110	S	34*	0.2
Wandel and Goltermann Inc	PO Box 13585	Res Tri Pk	NC	27709	Robert B Davidson	919-941-5730	S	33	0.2
Wandel & Goltermann Techn	PO Box 13585	Res Tri Pk	NC	27709	Robert B Davidson	919-941-5730	P	33	0.2
Telemobile Inc	19840 Hamilton Av	Torrance	CA	90502	William Thomas	310-538-5100	R	33	0.2
Cohu Inc	5755 Kearny Villa	San Diego	CA	92123	James Barnes	619-277-6700	D	30	0.2
Telesensory Corp	PO Box 7455	Mountain View	CA	94039	James W Morrell	415-960-0920	R	30	0.2
Pico Products Inc	12500 Foothill Blv	Lakeview Ter	CA	91342	Everett T Keech	818-897-0028	P	30	0.3
SSE Technologies Inc	47823 Westinghouse	Fremont	CA	94539	F C Toombs	510-657-9815	S	29	0.2
SSE Telecom Inc	1430 Spring Hill Rd	McLean	VA	22102	Frank S Trumbower	703-790-0250	P	29	0.2
LNR Communications Inc	180 Marcus Blv	Hauppauge	NY	11788	S Okwit	516-273-7111	R	29	0.2
American Dynamics Inc	10 Corporate Dr	Orangeburg	NY	10962	Glenn Waehner	914-365-1000	S	27*	0.2
Dorne and Margolin Inc	2950 Veterans Mem	Bohemia	NY	11716	Robert J McInerney	516-585-4000	S	27	0.2
Information Transmission	375 Val Brook Rd	McMurray	PA	15317	Robert M Unetich	412-941-1500	R	27*	0.2
Mackay Communications Inc	PO Box 58649	Raleigh	NC	27658	Francis Neary	919-850-3000	R	27	0.2
Maxon America Inc	10828 NW Airworld	Kansas City	MO	64153	Daniel Devling	816-891-1093	S	27*	0.2
MCL Inc	501 S Woodcreek	Bolingbrook	IL	60440	FP Morgan	708-759-9500	R	27	0.1
Maxon Systems Inc	10828 NW Airworld	Kansas City	MO	64153	Robert D Haler	816-891-1093	S	26*	0.2
TCI International Inc	222 Caspian Dr	Sunnyvale	CA	94089	John W Ballard	408-747-6100	P	26	0.1

Source: Ward's Business Directory of U.S. Private and Public Companies, Volumes 1 and 2, 1996. The company type code used is as follows: P - Public, R - Private, S - Subsidiary, D - Division, J - Joint Venture, A - Affiliate, G - Group. Sales are in millions of dollars, employees are in thousands. An asterisk (*) indicates an estimated sales volume. The symbol < stands for 'less than'. Company names and addresses are truncated, in some cases, to fit into the available space.

MATERIALS CONSUMED

Material	Quantity	Delivered Cost ($ million)
Materials, ingredients, containers, and supplies	(X)	5,977.9
Printed circuit boards (without inserted components) for electronic circuitry	(X)	226.3
Printed circuit assemblies, loaded boards or modules	(X)	185.0
Semiconductors, including transistors, diodes, rectifiers, and integrated circuits for electronic circuitry	(X)	590.7
Capacitors for electronic circuitry	(X)	147.3
Resistors for electronic circuitry	(X)	92.1
Other components and accessories for electronic circuitry, nec, except tubes	(X)	1,296.3
Electronic communication equipment	(X)	385.1
Electrical instrument mechanisms and meter movements (including instrument relays)	(X)	59.7
Electronic computing equipment	(X)	45.5
Current-carrying wiring devices	(X)	(D)
Insulated wire and cable, including magnet wire	(X)	110.1
Loudspeakers, microphones, and tuners (all types)	(X)	42.2
Fractional horsepower electric motors (less than 1 hp)	(X)	11.6
Silicon, hyperpure	(X)	0.9
Plastics resins consumed in the form of granules, pellets, powders, liquids, etc.	(X)	18.7
Fabricated plastics products (except gaskets, hoses, and belting)	(X)	69.9
Sheet metal products, except stampings	(X)	109.3
Metal stampings	(X)	160.6
Metal bolts, nuts, screws, washers, rivets, and other screw machine products	(X)	96.1
Other fabricated metal products (except forgings)	(X)	63.3
Forgings	(X)	(D)
Castings (rough and semifinished)	(X)	(D)
Steel shapes and forms	(X)	25.1
Aluminum and aluminum-base alloy shapes and forms	(X)	46.2
Other nonferrous shapes and forms	(X)	26.2
Paper and paperboard containers, including shipping sacks and other paper packaging supplies	(X)	46.7
All other materials and components, parts, containers, and supplies	(X)	678.9
Materials, ingredients, containers, and supplies, nsk	(X)	1,129.6

Source: 1992 *Economic Census*. Explanation of symbols used: (D): Withheld to avoid disclosure of competitive data; na: Not available; (S): Withheld because statistical norms were not met; (X): Not applicable; (Z): Less than half the unit shown; nec: Not elsewhere classified; nsk: Not specified by kind; - : zero; * : 10-19 percent estimated; ** : 20-29 percent estimated.

PRODUCT SHARE DETAILS

Product or Product Class	% Share	Product or Product Class	% Share
Radio and tv communications equipment	100.00	Broadcast, studio, and related electronic equipment	11.19
Communication systems and equipment, except broadcast, but including microwave equipment, and space satellites	86.49	Radio and television communications equipment, nsk	2.32

Source: 1992 *Economic Census*. The values shown are percent of total shipments in an industry. Values of indented subcategories are summed in the main heading. The symbol (D) appears when data are withheld to prevent disclosure of competitive information. The abbreviation nsk stands for 'not specified by kind' and nec for 'not elsewhere classified'.

INPUTS AND OUTPUTS FOR RADIO & TV COMMUNICATION EQUIPMENT

Economic Sector or Industry Providing Inputs	%	Sector	Economic Sector or Industry Buying Outputs	%	Sector
Imports	21.2	Foreign	Gross private fixed investment	42.4	Cap Inv
Electronic components nec	20.1	Manufg.	Federal Government purchases, national defense	32.4	Fed Govt
Wholesale trade	7.6	Trade	Exports	6.3	Foreign
Semiconductors & related devices	6.3	Manufg.	Federal Government purchases, nondefense	2.7	Fed Govt
Radio & TV communication equipment	4.1	Manufg.	Aircraft	1.9	Manufg.
Advertising	3.8	Services	Guided missiles & space vehicles	1.7	Manufg.
Communications, except radio & TV	2.5	Util.	Radio & TV communication equipment	1.6	Manufg.
Electric services (utilities)	2.4	Util.	Industrial buildings	1.4	Constr.
Electronic computing equipment	1.9	Manufg.	Personal consumption expenditures	1.0	
Real estate	1.3	Fin/R.E.	Residential 1-unit structures, nonfarm	0.8	Constr.
Equipment rental & leasing services	1.3	Services	Residential additions/alterations, nonfarm	0.8	Constr.
Electron tubes	1.2	Manufg.	Highway & street construction	0.7	Constr.
Banking	1.2	Fin/R.E.	Maintenance of nonfarm buildings nec	0.7	Constr.
Sheet metal work	1.1	Manufg.	Change in business inventories	0.7	In House
Industrial controls	1.0	Manufg.	Office buildings	0.6	Constr.
Maintenance of nonfarm buildings nec	0.9	Constr.	Warehouses	0.5	Constr.
Eating & drinking places	0.9	Trade	S/L Govt. purch., higher education	0.5	S/L Govt
Screw machine and related products	0.8	Manufg.	Construction of educational buildings	0.3	Constr.
Air transportation	0.8	Util.	Maintenance of railroads	0.3	Constr.
Engineering, architectural, & surveying services	0.8	Services	Construction of hospitals	0.2	Constr.
U.S. Postal Service	0.8	Gov't	Construction of stores & restaurants	0.2	Constr.
Aluminum rolling & drawing	0.7	Manufg.	Electric utility facility construction	0.2	Constr.
Blast furnaces & steel mills	0.6	Manufg.	Maintenance of electric utility facilities	0.2	Constr.
Metal stampings, nec	0.6	Manufg.	Maintenance of highways & streets	0.2	Constr.
Nonferrous wire drawing & insulating	0.6	Manufg.	Nonfarm residential structure maintenance	0.2	Constr.
Plating & polishing	0.6	Manufg.	Ship building & repairing	0.2	Manufg.

Continued on next page.

INPUTS AND OUTPUTS FOR RADIO & TV COMMUNICATION EQUIPMENT - Continued

Economic Sector or Industry Providing Inputs	%	Sector	Economic Sector or Industry Buying Outputs	%	Sector
Aluminum castings	0.5	Manufg.	Maintenance of nonbuilding facilities nec	0.1	Constr.
Engineering & scientific instruments	0.5	Manufg.	Nonbuilding facilities nec	0.1	Constr.
Instruments to measure electricity	0.5	Manufg.	Residential garden apartments	0.1	Constr.
Miscellaneous plastics products	0.5	Manufg.			
Computer & data processing services	0.5	Services			
Legal services	0.5	Services			
Fabricated metal products, nec	0.4	Manufg.			
General industrial machinery, nec	0.4	Manufg.			
Petroleum refining	0.4	Manufg.			
Wiring devices	0.4	Manufg.			
Wood TV & radio cabinets	0.4	Manufg.			
Motor freight transportation & warehousing	0.4	Util.			
Management & consulting services & labs	0.4	Services			
Noncomparable imports	0.4	Foreign			
Chemical preparations, nec	0.3	Manufg.			
Machinery, except electrical, nec	0.3	Manufg.			
Miscellaneous fabricated wire products	0.3	Manufg.			
Motors & generators	0.3	Manufg.			
Paperboard containers & boxes	0.3	Manufg.			
Gas production & distribution (utilities)	0.3	Util.			
Royalties	0.3	Fin/R.E.			
Accounting, auditing & bookkeeping	0.3	Services			
Fabricated rubber products, nec	0.2	Manufg.			
Fabricated structural metal	0.2	Manufg.			
Iron & steel foundries	0.2	Manufg.			
Manifold business forms	0.2	Manufg.			
Nonferrous forgings	0.2	Manufg.			
Optical instruments & lenses	0.2	Manufg.			
Primary nonferrous metals, nec	0.2	Manufg.			
Radio & TV receiving sets	0.2	Manufg.			
Special dies & tools & machine tool accessories	0.2	Manufg.			
Storage batteries	0.2	Manufg.			
Insurance carriers	0.2	Fin/R.E.			
Photofinishing labs, commercial photography	0.2	Services			
Metal coating & allied services	0.1	Manufg.			
Miscellaneous metal work	0.1	Manufg.			
Synthetic rubber	0.1	Manufg.			
Transformers	0.1	Manufg.			
Credit agencies other than banks	0.1	Fin/R.E.			
Electrical repair shops	0.1	Services			
Hotels & lodging places	0.1	Services			

Source: Benchmark Input-Output Accounts for the U.S. Economy, 1982, U.S. Department of Commerce, Washington, D.C., July 1991. Data, as reported in the source, are organized by the 1977 SIC structure in use in 1982 but have been matched, as closely as is possible, to the 1987 SIC structure used in this book.

OCCUPATIONS EMPLOYED BY SIC 366 - COMMUNICATIONS EQUIPMENT

Occupation	% of Total 1994	Change to 2005	Occupation	% of Total 1994	Change to 2005
Electrical & electronic assemblers	9.3	-22.9	Engineers nec	1.6	28.5
Electrical & electronics engineers	7.6	9.6	General managers & top executives	1.6	-18.7
Electrical & electronic equipment assemblers	7.2	-14.3	Managers & administrators nec	1.5	-14.4
Electrical & electronic technicians,technologists	5.0	-14.3	Mechanical engineers	1.4	3.7
Inspectors, testers, & graders, precision	4.5	-40.0	Production, planning, & expediting clerks	1.4	-14.3
Assemblers, fabricators, & hand workers nec	4.0	-14.3	Systems analysts	1.3	37.0
Sales & related workers nec	3.0	-14.3	Traffic, shipping, & receiving clerks	1.3	-17.6
Blue collar worker supervisors	2.5	-25.1	Stock clerks	1.3	-30.4
Secretaries, ex legal & medical	2.4	-22.0	Plant & system operators nec	1.2	-24.3
Computer engineers	2.4	26.9	Drafters	1.2	-33.3
Engineering technicians & technologists nec	2.1	-14.3	General office clerks	1.2	-27.0
Engineering, mathematical, & science managers	2.1	-2.7	Industrial production managers	1.2	-14.4
Management support workers nec	2.0	-14.3	Marketing, advertising, & PR managers	1.2	-14.4
Electromechanical equipment assemblers	1.7	-5.8	Purchasing agents, ex trade & farm products	1.1	-14.3

Source: Industry-Occupation Matrix, Bureau of Labor Statistics. These data relate to one or more 3-digit SIC industry groups rather than to a single 4-digit SIC. The change reported for each occupation to the year 2005 is a percent of growth or decline as estimated by the Bureau of Labor Statistics. The abbreviation nec stands for 'not elsewhere classified'.

LOCATION BY STATE AND REGIONAL CONCENTRATION

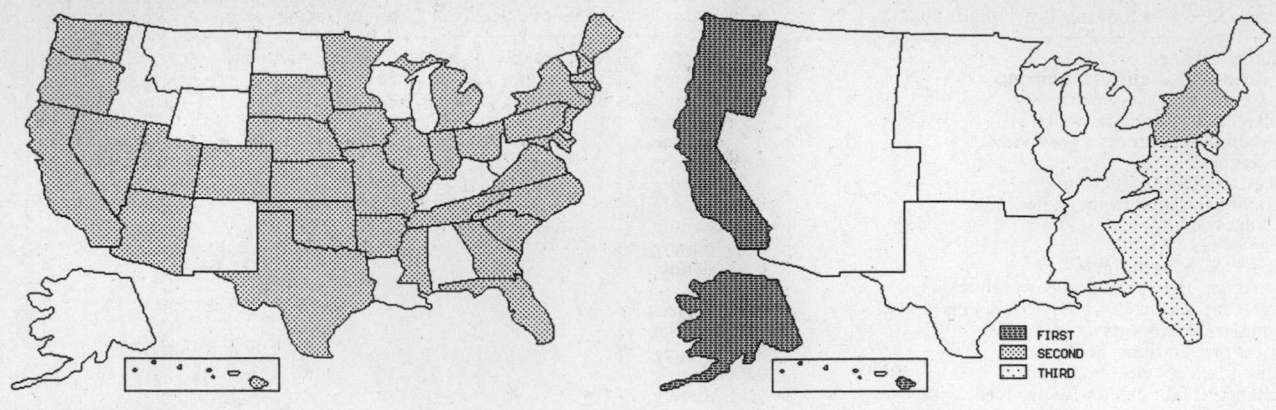

FIRST
SECOND
THIRD

INDUSTRY DATA BY STATE

State	Establish-ments	Shipments			Employment				Cost as % of Shipments	Investment per Employee ($)
		Total ($ mil)	% of U.S.	Per Establ.	Total Number	% of U.S.	Per Establ.	Wages ($/hour)		
Florida	61	3,380.8	17.4	55.4	19,500	15.7	320	11.61	31.0	6,990
California	235	3,033.3	15.6	12.9	19,400	15.6	83	16.38	29.4	5,000
Massachusetts	37	1,597.8	8.2	43.2	10,000	8.0	270	21.14	40.7	-
Texas	57	970.6	5.0	17.0	6,700	5.4	118	11.34	41.0	4,687
Georgia	20	962.4	4.9	48.1	3,700	3.0	185	12.36	56.3	5,351
New York	72	794.0	4.1	11.0	6,400	5.1	89	12.05	51.2	3,875
Virginia	26	787.4	4.0	30.3	4,700	3.8	181	17.20	45.8	5,553
New Jersey	47	626.2	3.2	13.3	3,700	3.0	79	13.67	21.4	5,514
Maryland	20	525.7	2.7	26.3	2,700	2.2	135	16.74	41.3	6,444
Pennsylvania	44	454.1	2.3	10.3	4,000	3.2	91	11.25	43.3	1,750
Arizona	21	441.5	2.3	21.0	4,400	3.5	210	19.53	21.5	-
Minnesota	13	258.0	1.3	19.8	2,300	1.8	177	9.31	41.9	5,783
Colorado	18	237.3	1.2	13.2	1,300	1.0	72	12.18	52.0	2,308
Utah	18	153.2	0.8	8.5	1,100	0.9	61	7.27	58.3	-
Washington	27	114.7	0.6	4.2	1,100	0.9	41	11.40	37.9	3,000
Oregon	12	39.9	0.2	3.3	300	0.2	25	12.00	38.3	-
Missouri	6	20.9	0.1	3.5	200	0.2	33	12.50	56.9	-
Tennessee	10	16.3	0.1	1.6	300	0.2	30	8.00	41.7	2,333
Michigan	11	11.2	0.1	1.0	100	0.1	9	8.50	45.5	-
Illinois	39	(D)	-	-	17,500 *	14.1	449	-	-	-
Ohio	27	(D)	-	-	3,750 *	3.0	139	-	-	5,120
Connecticut	14	(D)	-	-	375 *	0.3	27	-	-	-
New Hampshire	14	(D)	-	-	375 *	0.3	27	-	-	-
North Carolina	12	(D)	-	-	750 *	0.6	63	-	-	-
Indiana	10	(D)	-	-	7,500 *	6.0	750	-	-	2,053
Oklahoma	10	(D)	-	-	175 *	0.1	18	-	-	-
South Carolina	7	(D)	-	-	375 *	0.3	54	-	-	-
Nebraska	6	(D)	-	-	750 *	0.6	125	-	-	-
Iowa	5	(D)	-	-	3,750 *	3.0	750	-	-	-
Maine	5	(D)	-	-	175 *	0.1	35	-	-	-
Nevada	5	(D)	-	-	175 *	0.1	35	-	-	-
Hawaii	4	(D)	-	-	175 *	0.1	44	-	-	-
Arkansas	3	(D)	-	-	175 *	0.1	58	-	-	-
Kansas	3	(D)	-	-	750 *	0.6	250	-	-	-
Mississippi	3	(D)	-	-	175 *	0.1	58	-	-	-
South Dakota	1	(D)	-	-	375 *	0.3	375	-	-	-

Source: 1992 *Economic Census*. The states are in descending order of shipments or establishments (if shipment data are missing for the majority). The symbol (D) appears when data are withheld to prevent disclosure of competitive information. States marked with (D) are sorted by number of establishments. A dash (-) indicates that the data element cannot be calculated; * indicates the midpoint of a range.

3669 - COMMUNICATIONS EQUIPMENT, NEC

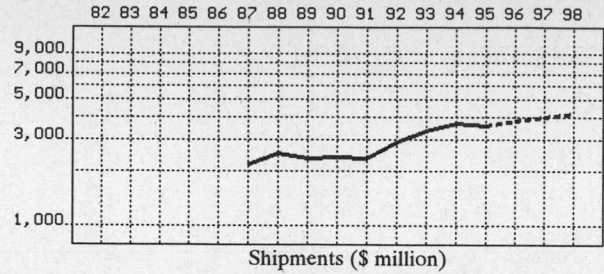

Shipments ($ million)

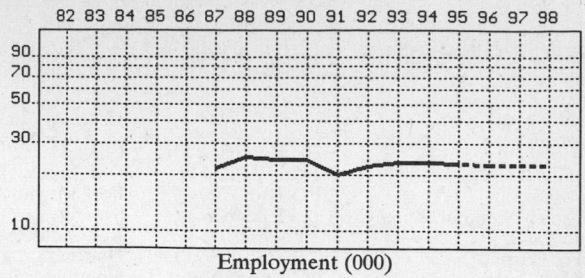

Employment (000)

GENERAL STATISTICS

Year	Com-panies	Establishments Total	Establishments with 20 or more employees	Employment Total (000)	Employment Production Workers (000)	Employment Hours (Mil)	Compensation Payroll ($ mil)	Compensation Wages ($/hr)	Production Cost of Materials	Production Value Added by Manufacture	Production Value of Shipments	Production Capital Invest.
1982												
1983												
1984												
1985												
1986												
1987	363	382	162	21.9	11.1	21.5	533.0	9.21	870.9	1,316.7	2,189.6	64.1
1988		403	169	25.4	12.3	24.4	647.9	9.19	999.0	1,509.8	2,498.5	58.7
1989		411	178	24.4	10.6	20.7	566.2	9.41	942.4	1,423.3	2,364.1	69.6
1990		402	178	24.2	10.5	21.7	587.9	9.30	945.0	1,452.3	2,395.1	71.6
1991		421	177	20.0	8.8	18.2	530.7	9.72	926.2	1,444.6	2,356.1	63.3
1992	502	517	174	22.5	10.8	21.3	650.2	10.36	1,167.5	1,767.0	2,923.9	65.0
1993		523	185	23.7	11.7	22.6	698.4	10.19	1,412.5	1,939.5	3,349.7	83.8
1994		531P	186P	23.9	12.3	23.0	745.3	10.44	1,492.8	2,241.0	3,703.4	99.6
1995		555P	189P	23.0P	11.2P	21.7P	723.5P	10.63P	1,453.6P	2,182.1P	3,606.1P	90.8P
1996		579P	191P	23.0P	11.3P	21.7P	746.5P	10.83P	1,532.7P	2,300.9P	3,802.4P	95.0P
1997		602P	194P	22.9P	11.3P	21.7P	769.6P	11.03P	1,611.9P	2,419.7P	3,998.8P	99.2P
1998		626P	197P	22.9P	11.4P	21.7P	792.6P	11.23P	1,691.0P	2,538.6P	4,195.1P	103.4P

Sources: 1982, 1987, 1992 *Economic Census*; *Annual Survey of Manufactures*, 83-86, 88-91, 93-94. Establishment counts for non-Census years are from *County Business Patterns*; establishment values for 83-84 are extrapolations. 'P's show projections by the editors. Industries reclassified in 87 will not have data for prior years.

INDICES OF CHANGE

Year	Com-panies	Establishments Total	Establishments with 20 or more employees	Employment Total (000)	Employment Production Workers (000)	Employment Hours (Mil)	Compensation Payroll ($ mil)	Compensation Wages ($/hr)	Production Cost of Materials	Production Value Added by Manufacture	Production Value of Shipments	Production Capital Invest.
1982												
1983												
1984												
1985												
1986												
1987	72	74	93	97	103	101	82	89	75	75	75	99
1988		78	97	113	114	115	100	89	86	85	85	90
1989		79	102	108	98	97	87	91	81	81	81	107
1990		78	102	108	97	102	90	90	81	82	82	110
1991		81	102	89	81	85	82	94	79	82	81	97
1992	100	100	100	100	100	100	100	100	100	100	100	100
1993		101	106	105	108	106	107	98	121	110	115	129
1994		103P	107P	106	114	108	115	101	128	127	127	153
1995		107P	108P	102P	104P	102P	111P	103P	125P	123P	123P	140P
1996		112P	110P	102P	105P	102P	115P	105P	131P	130P	130P	146P
1997		116P	112P	102P	105P	102P	118P	107P	138P	137P	137P	153P
1998		121P	113P	102P	106P	102P	122P	108P	145P	144P	143P	159P

Sources: Same as General Statistics. Values reflect change from the base year, 1992. Values above 100 mean greater than 92, values below 100 mean less than 92, and a value of 100 in the 82-91 or 93-98 period means same as 92. 'P's mark projections by the editors.

SELECTED RATIOS

For 1994	Avg. of All Manufact.	Analyzed Industry	Index	For 1994	Avg. of All Manufact.	Analyzed Industry	Index
Employees per Establishment	49	45	92	Value Added per Production Worker	134,084	182,195	136
Payroll per Establishment	1,500,273	1,402,446	93	Cost per Establishment	5,045,178	2,809,032	56
Payroll per Employee	30,620	31,184	102	Cost per Employee	102,970	62,460	61
Production Workers per Establishment	34	23	67	Cost per Production Worker	146,988	121,366	83
Wages per Establishment	853,319	451,839	53	Shipments per Establishment	9,576,895	6,968,763	73
Wages per Production Worker	24,861	19,522	79	Shipments per Employee	195,460	154,954	79
Hours per Production Worker	2,056	1,870	91	Shipments per Production Worker	279,017	301,089	108
Wages per Hour	12.09	10.44	86	Investment per Establishment	321,011	187,419	58
Value Added per Establishment	4,602,255	4,216,935	92	Investment per Employee	6,552	4,167	64
Value Added per Employee	93,930	93,766	100	Investment per Production Worker	9,352	8,098	87

Sources: Same as General Statistics. The 'Average of All Manufacturing' column represents the average of all manufacturing industries reported for the most recent complete year available. The Index shows the relationship between the Average and the Analyzed Industry. For example, 100 means that they are equal; 500 that the Analyzed Industry is five times the average; 50 means that the Analyzed Industry is half the national average. The abbreviation 'na' is used to show that data are 'not available'.

LEADING COMPANIES   Number shown: **75**   Total sales ($ mil): **6,110**   Total employment (000): **44.8**

Company Name	Address				CEO Name	Phone	Co. Type	Sales ($ mil)	Empl. (000)
Pittway Corp	200 S Wacker Dr	Chicago	IL	60606	King Harris	312-831-1070	P	778	5.4
Federal Signal Corp	1415 W 22nd St	Oak Brook	IL	60521	Joseph J Ross	708-954-2000	P	677	5.2
ITT Aerospace	PO Box 3700	Fort Wayne	IN	46801	Marvin R Sambur	219-487-6000	D	425	2.5
Alarm Device Manufacturing Co	165 Eileen Way	Syosset	NY	11791	Leo A Guthart	516-921-6704	D	300	1.5
VeriFone Inc	3 Lagoon Dr	Redwood City	CA	94065	Hatim A Tyabji	415-591-6500	P	259	1.8
PictureTel Corp	222 Rosewood Dr	Danvers	MA	01923	Norman E Gaut	508-762-5000	P	255	0.8
Mosler Inc	1561 Grand Blv	Hamilton	OH	45012	Angelo M Marzano	513-867-4000	R	250	2.0
First Alert Inc	780 McClure Rd	Aurora	IL	60504	Malcolm Candlish	708-851-7330	P	248	2.1
BRK Brands Inc	780 McClure Rd	Aurora	IL	60504	Malcolm Candlish	708-851-7330	S	158	2.0
Edwards Company Inc	PO Box 310	Pittsfield	ME	04967	Timothy J Mellen	207-487-3104	D	130	0.5
Sentrol Life Safety Corp	12345 SW Leveton	Tualatin	OR	97062	Ken Boyda	503-692-4052	S	130•	1.2
Checkpoint Systems Inc	PO Box 188	Thorofare	NJ	08086	Albert E Wolf	609-848-1800	P	128	1.8
Itron Inc	2818 N Sullivan Rd	Spokane	WA	99216	J M Humphreys	509-924-9900	P	121	0.7
Harmon Industries Inc	1300 Jefferson Ct	Blue Springs	MO	64015	Bjorn E Olsson	816-229-3345	P	120	1.0
Telephonics Corp	815 Broad Hollow	Farmingdale	NY	11735	Z Papazissimos	516-755-7000	S	104	1.1
Poly-Scientific	1213 N Main St	Blacksburg	VA	24060	E Wade	703-552-3011	D	100	0.6
Federal Signal Corp	2645 Federal Signal	University Park	IL	60466	Richard G Gibb	708-534-3400	D	90	0.7
Knogo Corp	350 Wireless Blv	Hauppauge	NY	11788	Thomas A Nicolette	516-232-2100	P	89	0.8
Ultrak Inc	1220 Champion Cir	Carrollton	TX	75006	George K Broady	214-280-9675	P	79	0.2
Dukane Corp	2900 Dukane Dr	St Charles	IL	60174	JM Stone Jr	708-584-2300	R	75	0.8
Maple Chase Co	2820 Thatcher Rd	Downers Grove	IL	60515	James McCrink	708-963-1550	R	75	1.1
Thorn Automated Systems Inc	835 Sharon Dr	Westlake	OH	44145	Charles M Jones	216-871-9900	S	65	0.4
MILCOM Systems Corp	532 Viking Dr	Virginia Beach	VA	23452	William Fleming Jr	804-463-2800	R	64	0.8
Directed Electronics Inc	2560 Progress St	Vista	CA	92083	Darrell E Issa	619-598-6200	R	63	0.1
Interactive Technologies Inc	2266 N 2nd St	North St Paul	MN	55109	TL Auth	612-777-2690	P	60	0.4
Norment Industries WSA Inc	PO Drawer 6129	Montgomery	AL	36106	Dennis A Flynn	205-281-8440	S	60	0.3
Scientific-Atlanta Inc	13112 Evening	San Diego	CA	92128	Phillip L Heltman	619-679-6000	D	60•	0.4
Moose Products	PO Box 2904	Hickory	NC	28603	Cliff Licko	704-322-2333	D	57•	0.5
Code-Alarm Inc	950 E Whitcomb	Madison H	MI	48071	Rand W Mueller	313-583-9620	P	50	0.4
Radionics Inc	1800 Abbott St	Salinas	CA	93901	Glen Greer	408-757-8877	R	50	0.4
Westinghouse Security Electr	5452 Betsy Ross Dr	Santa Clara	CA	95054	William Prevost	408-727-5170	D	50	0.3
Winegard Co	3000 Kirkwood St	Burlington	IA	52601	Randy Winegard	319-754-0600	R	50	0.3
Peek Traffic Inc	1500 N Washington	Sarasota	FL	34236	William Sowell	813-366-8770	S	48•	0.3
Napco Security Systems Inc	333 Bayview Av	Amityville	NY	11701	Kenneth Rosenberg	516-842-9400	P	47	1.1
Condor Systems Inc	2133 Samaritan Dr	San Jose	CA	95124	Robert E Young II	408-371-9580	R	46	0.2
Larscom Inc	4600 Patrick Henry	Santa Clara	CA	95054	Deborah M Soon	408-988-6600	S	40	0.2
Rauland-Borg Corp	3450 W Oakton St	Skokie	IL	60076	William Krucks	708-679-0900	R	40	0.3
ICI Security Systems	PO Box 819	Valley Forge	PA	19482	Thomas Oxenfeld	610-666-8600	D	37	0.3
Cardkey Systems Inc	101 W Cochran St	Simi Valley	CA	93065	Clas Thelin	805-522-5555	R	35	0.2
Stewart-Decatur Security Systems	PO Box 18700	Erlanger	KY	41018	James K Ramsey	606-371-6000	R	33•	0.2
CASI-RUSCO Inc	1155 Broken Sound	Boca Raton	FL	33487	David A Schuldt	407-998-6100	R	28	0.2
Andrew SciComm Inc	2908 National Dr	Garland	TX	75041	Bill Shockley	214-840-4900	S	26	0.2
Econolite Control Products Inc	3360 E La Palma	Anaheim	CA	92806	Mike Doyle	714-630-3700	R	25•	0.2
FiberCom Inc	PO Box 11966	Roanoke	VA	24022	Albert D Bender	703-342-6700	R	25	0.3
SYSTECH Corp	6465 Nancy Ridge	San Diego	CA	92121	Mark Fowler	619-453-8970	R	25	<0.1
Grinnell Systems Company Inc	PO Box 128	Cleveland	NC	27013	Don Alcorn	704-278-2221	D	24•	0.2
IGYS Inc	4462 Corporate Ctr	Los Alamitos	CA	90720	Arthur Van Leuven	714-220-2040	R	22•	0.1
New Bedford Panoramex Corp	1037 W 9th St	Upland	CA	91786	Robert L Ozuna	909-982-9806	R	22	0.1
Jetronic Industries Inc	4200 Mitchell St	Philadelphia	PA	19128	Daniel R Kursman	215-482-7660	P	22	0.2
Robot Research Inc	5636 Ruffin Rd	San Diego	CA	92123	Thomas E Cashman	619-279-9430	R	20•	0.1
WLI Industries Inc	PO Box 7050	Villa Park	IL	60181	James Van De Velde	708-932-4600	R	19•	0.1
MERET Optical Commun	1800 Stewart St	Santa Monica	CA	90404	Xin Cheng	310-828-7496	R	18•	0.1
Faraday Inc	805 S Maumee St	Tecumseh	MI	49286	Robert F Bell	517-423-2111	R	17	0.2
Custom Cable Industries Inc	3221 Cherry Palm	Tampa	FL	33619	Rick Watson	813-623-2232	R	15	0.2
Cubix Corp	2800 Lockheed Way	Carson City	NV	89706	D Lehr	702-883-7611	R	15	0.1
National Security Systems Inc	511 M Wood	Manhasset	NY	11030	Jay Barron	516-627-2222	R	15	<0.1
Western-Cullen Hayes Inc	2700 W 36th Pl	Chicago	IL	60632	R L MacDaniel	312-254-9600	R	15	0.2
Secure Computing Corp	2675 Long Lake Rd	Roseville	MN	55113	Kermit M Beseke	612-628-2700	R	15	0.2
Clifford Electronics Inc	20750 Lassen St	Chatsworth	CA	91311	Ze'ev Drori	818-709-7551	R	14•	0.1
Mason Electric Company Inc	PO Box 311	San Fernando	CA	91341	William Southern	818-361-3366	S	14	0.1
Protex International Corp	180 Keyland Ct	Bohemia	NY	11716	David Wachsman	516-563-4250	R	14•	0.1
Silent Knight Security Systems	7550 Merdian Cir	Maple Grove	MN	55369	John Ellis	612-493-6400	R	14•	<0.1
Telecorp Systems Inc	1000 Holcomb	Roswell	GA	30076	J Lawrence Bradner	404-587-0700	R	13	<0.1
Network Communications Corp	5501 Green Val Dr	Bloomington	MN	55437	George Wood	612-844-1003	R	12•	<0.1
ST Research Corp	8419 Terminal Rd	Newington	VA	22122	SR Perrino	703-550-7000	R	12•	0.1
TyLink Corp	10 Commerce Way	Norton	MA	02766	Robert Degan	508-285-0033	R	12	<0.1
CompuDyne Corp	90 State House Sq	Hartford	CT	06103	Norman Silberdick	203-247-7611	P	12	0.1
Firecom Case/Acme	39-27 59th St	Woodside	NY	11377	Paul Mendez	718-899-6100	P	12	0.1
Falcon Safety Products Inc	PO Box 1299	Somerville	NJ	08876	Phil M Lapin	908-707-4900	R	11	0.1
Garrett Metal Detectors	1881 W State St	Garland	TX	75042	CL Garrett	214-494-6151	R	10	<0.1
Habitec Security	3790 Hauck Rd	Cincinnati	OH	45241	Jim Smythe	513-554-3660	R	10	<0.1
Quantum Group Inc	11211 Sorrento Val	San Diego	CA	92121	Mark Goldstein	619-457-3048	R	10	0.1
Continental Instruments Corp	70 Hopper St	Westbury	NY	11590	John Banks	516-334-0900	S	9	<0.1
Firetector Inc	575 Underhill Blv	Syosset	NY	11791	Richard H Axelsen	516-921-3400	R	9	<0.1
Sonicraft Inc	8859 S Greenwood	Chicago	IL	60619	JT Jones	312-933-9200	R	9•	<0.1

Source: *Ward's Business Directory of U.S. Private and Public Companies*, Volumes 1 and 2, 1996. The company type code used is as follows: P - Public, R - Private, S - Subsidiary, D - Division, J - Joint Venture, A - Affiliate, G - Group. Sales are in millions of dollars, employees are in thousands. An asterisk (•) indicates an estimated sales volume. The symbol < stands for 'less than'. Company names and addresses are truncated, in some cases, to fit into the available space.

MATERIALS CONSUMED

Material	Quantity	Delivered Cost ($ million)
Materials, ingredients, containers, and supplies	(X)	989.5
Printed circuit boards (without inserted components) for electronic circuitry	(X)	46.0
Printed circuit assemblies, loaded boards or modules	(X)	35.1
Semiconductors, including transistors, diodes, rectifiers, and integrated circuits for electronic circuitry	(X)	73.8
Capacitors for electronic circuitry	(X)	17.9
Resistors for electronic circuitry	(X)	14.0
Other components and accessories for electronic circuitry, nec, except tubes	(X)	64.3
Electronic communication equipment	(X)	22.7
Electrical instrument mechanisms and meter movements (including instrument relays)	(X)	24.8
Electronic computing equipment	(X)	3.3
Current-carrying wiring devices	(X)	9.1
Insulated wire and cable, including magnet wire	(X)	12.5
Loudspeakers, microphones, and tuners (all types)	(X)	6.7
Fractional horsepower electric motors (less than 1 hp)	(X)	7.3
Silicon, hyperpure	(X)	(D)
Plastics resins consumed in the form of granules, pellets, powders, liquids, etc.	(X)	7.2
Fabricated plastics products (except gaskets, hoses, and belting)	(X)	38.2
Sheet metal products, except stampings	(X)	29.4
Metal stampings	(X)	17.2
Metal bolts, nuts, screws, washers, rivets, and other screw machine products	(X)	10.0
Other fabricated metal products (except forgings)	(X)	(D)
Forgings	(X)	(D)
Castings (rough and semifinished)	(X)	22.1
Steel shapes and forms	(X)	6.2
Aluminum and aluminum-base alloy shapes and forms	(X)	6.4
Other nonferrous shapes and forms	(X)	1.7
Paper and paperboard containers, including shipping sacks and other paper packaging supplies	(X)	12.6
All other materials and components, parts, containers, and supplies	(X)	125.0
Materials, ingredients, containers, and supplies, nsk	(X)	330.1

Source: 1992 *Economic Census*. Explanation of symbols used: (D): Withheld to avoid disclosure of competitive data; na: Not available; (S): Withheld because statistical norms were not met; (X): Not applicable; (Z): Less than half the unit shown; nec: Not elsewhere classified; nsk: Not specified by kind; - : zero; * : 10-19 percent estimated; ** : 20-29 percent estimated.

PRODUCT SHARE DETAILS

Product or Product Class	% Share	Product or Product Class	% Share
Communications equipment, nec	100.00	Intercommunications systems, including inductive paging systems (selective paging), except telephone and telegraph	11.76
Alarm systems, including electric sirens, and horns	54.45	Communications equipment, nec, nsk	11.05
Vehicular and pedestrian traffic control equipment, electric railway signals and attachments	22.74		

Source: 1992 *Economic Census*. The values shown are percent of total shipments in an industry. Values of indented subcategories are summed in the main heading. The symbol (D) appears when data are withheld to prevent disclosure of competitive information. The abbreviation nsk stands for 'not specified by kind' and nec for 'not elsewhere classified'.

INPUTS AND OUTPUTS FOR RADIO & TV COMMUNICATION EQUIPMENT

Economic Sector or Industry Providing Inputs	%	Sector	Economic Sector or Industry Buying Outputs	%	Sector
Imports	21.2	Foreign	Gross private fixed investment	42.4	Cap Inv
Electronic components nec	20.1	Manufg.	Federal Government purchases, national defense	32.4	Fed Govt
Wholesale trade	7.6	Trade	Exports	6.3	Foreign
Semiconductors & related devices	6.3	Manufg.	Federal Government purchases, nondefense	2.7	Fed Govt
Radio & TV communication equipment	4.1	Manufg.	Aircraft	1.9	Manufg.
Advertising	3.8	Services	Guided missiles & space vehicles	1.7	Manufg.
Communications, except radio & TV	2.5	Util.	Radio & TV communication equipment	1.6	Manufg.
Electric services (utilities)	2.4	Util.	Industrial buildings	1.4	Constr.
Electronic computing equipment	1.9	Manufg.	Personal consumption expenditures	1.0	
Real estate	1.3	Fin/R.E.	Residential 1-unit structures, nonfarm	0.8	Constr.
Equipment rental & leasing services	1.3	Services	Residential additions/alterations, nonfarm	0.8	Constr.
Electron tubes	1.2	Manufg.	Highway & street construction	0.7	Constr.
Banking	1.2	Fin/R.E.	Maintenance of nonfarm buildings nec	0.7	Constr.
Sheet metal work	1.1	Manufg.	Change in business inventories	0.7	In House
Industrial controls	1.0	Manufg.	Office buildings	0.6	Constr.
Maintenance of nonfarm buildings nec	0.9	Constr.	Warehouses	0.5	Constr.
Eating & drinking places	0.9	Trade	S/L Govt. purch., higher education	0.5	S/L Govt
Screw machine and related products	0.8	Manufg.	Construction of educational buildings	0.3	Constr.
Air transportation	0.8	Util.	Maintenance of railroads	0.3	Constr.
Engineering, architectural, & surveying services	0.8	Services	Construction of hospitals	0.2	Constr.
U.S. Postal Service	0.8	Gov't	Construction of stores & restaurants	0.2	Constr.
Aluminum rolling & drawing	0.7	Manufg.	Electric utility facility construction	0.2	Constr.
Blast furnaces & steel mills	0.6	Manufg.	Maintenance of electric utility facilities	0.2	Constr.
Metal stampings, nec	0.6	Manufg.	Maintenance of highways & streets	0.2	Constr.
Nonferrous wire drawing & insulating	0.6	Manufg.	Nonfarm residential structure maintenance	0.2	Constr.

Continued on next page.

INPUTS AND OUTPUTS FOR RADIO & TV COMMUNICATION EQUIPMENT - Continued

Economic Sector or Industry Providing Inputs	%	Sector	Economic Sector or Industry Buying Outputs	%	Sector
Plating & polishing	0.6	Manufg.	Ship building & repairing	0.2	Manufg.
Aluminum castings	0.5	Manufg.	Maintenance of nonbuilding facilities nec	0.1	Constr.
Engineering & scientific instruments	0.5	Manufg.	Nonbuilding facilities nec	0.1	Constr.
Instruments to measure electricity	0.5	Manufg.	Residential garden apartments	0.1	Constr.
Miscellaneous plastics products	0.5	Manufg.			
Computer & data processing services	0.5	Services			
Legal services	0.5	Services			
Fabricated metal products, nec	0.4	Manufg.			
General industrial machinery, nec	0.4	Manufg.			
Petroleum refining	0.4	Manufg.			
Wiring devices	0.4	Manufg.			
Wood TV & radio cabinets	0.4	Manufg.			
Motor freight transportation & warehousing	0.4	Util.			
Management & consulting services & labs	0.4	Services			
Noncomparable imports	0.4	Foreign			
Chemical preparations, nec	0.3	Manufg.			
Machinery, except electrical, nec	0.3	Manufg.			
Miscellaneous fabricated wire products	0.3	Manufg.			
Motors & generators	0.3	Manufg.			
Paperboard containers & boxes	0.3	Manufg.			
Gas production & distribution (utilities)	0.3	Util.			
Royalties	0.3	Fin/R.E.			
Accounting, auditing & bookkeeping	0.3	Services			
Fabricated rubber products, nec	0.2	Manufg.			
Fabricated structural metal	0.2	Manufg.			
Iron & steel foundries	0.2	Manufg.			
Manifold business forms	0.2	Manufg.			
Nonferrous forgings	0.2	Manufg.			
Optical instruments & lenses	0.2	Manufg.			
Primary nonferrous metals, nec	0.2	Manufg.			
Radio & TV receiving sets	0.2	Manufg.			
Special dies & tools & machine tool accessories	0.2	Manufg.			
Storage batteries	0.2	Manufg.			
Insurance carriers	0.2	Fin/R.E.			
Photofinishing labs, commercial photography	0.2	Services			
Metal coating & allied services	0.1	Manufg.			
Miscellaneous metal work	0.1	Manufg.			
Synthetic rubber	0.1	Manufg.			
Transformers	0.1	Manufg.			
Credit agencies other than banks	0.1	Fin/R.E.			
Electrical repair shops	0.1	Services			
Hotels & lodging places	0.1	Services			

Source: Benchmark Input-Output Accounts for the U.S. Economy, 1982, U.S. Department of Commerce, Washington, D.C., July 1991. Data, as reported in the source, are organized by the 1977 SIC structure in use in 1982 but have been matched, as closely as is possible, to the 1987 SIC structure used in this book.

OCCUPATIONS EMPLOYED BY SIC 366 - COMMUNICATIONS EQUIPMENT

Occupation	% of Total 1994	Change to 2005	Occupation	% of Total 1994	Change to 2005
Electrical & electronic assemblers	9.3	-22.9	Engineers nec	1.6	28.5
Electrical & electronics engineers	7.6	9.6	General managers & top executives	1.6	-18.7
Electrical & electronic equipment assemblers	7.2	-14.3	Managers & administrators nec	1.5	-14.4
Electrical & electronic technicians,technologists	5.0	-14.3	Mechanical engineers	1.4	3.7
Inspectors, testers, & graders, precision	4.5	-40.0	Production, planning, & expediting clerks	1.4	-14.3
Assemblers, fabricators, & hand workers nec	4.0	-14.3	Systems analysts	1.3	37.0
Sales & related workers nec	3.0	-14.3	Traffic, shipping, & receiving clerks	1.3	-17.6
Blue collar worker supervisors	2.5	-25.1	Stock clerks	1.3	-30.4
Secretaries, ex legal & medical	2.4	-22.0	Plant & system operators nec	1.2	-24.3
Computer engineers	2.4	26.9	Drafters	1.2	-33.3
Engineering technicians & technologists nec	2.1	-14.3	General office clerks	1.2	-27.0
Engineering, mathematical, & science managers	2.1	-2.7	Industrial production managers	1.2	-14.4
Management support workers nec	2.0	-14.3	Marketing, advertising, & PR managers	1.2	-14.4
Electromechanical equipment assemblers	1.7	-5.8	Purchasing agents, ex trade & farm products	1.1	-14.3

Source: Industry-Occupation Matrix, Bureau of Labor Statistics. These data relate to one or more 3-digit SIC industry groups rather than to a single 4-digit SIC. The change reported for each occupation to the year 2005 is a percent of growth or decline as estimated by the Bureau of Labor Statistics. The abbreviation nec stands for 'not elsewhere classified'.

LOCATION BY STATE AND REGIONAL CONCENTRATION

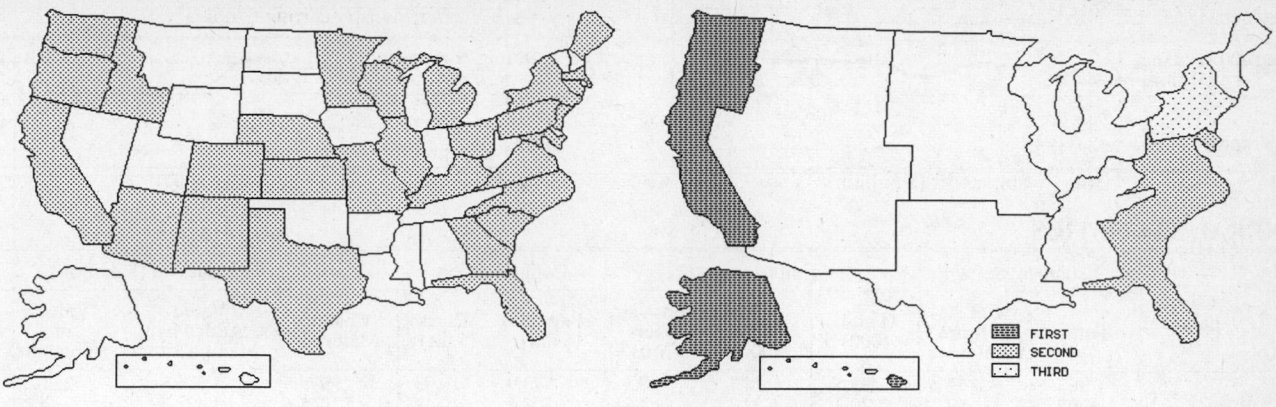

INDUSTRY DATA BY STATE

| State | Establish- ments | Shipments | | | Employment | | | | Cost as % of Shipments | Investment per Employee ($) |
		Total ($ mil)	% of U.S.	Per Establ.	Total Number	% of U.S.	Per Establ.	Wages ($/hour)		
California	106	479.8	16.4	4.5	4,000	17.8	38	10.34	42.3	2,400
Massachusetts	15	431.7	14.8	28.8	1,800	8.0	120	10.75	18.9	-
New York	36	352.7	12.1	9.8	2,900	12.9	81	11.24	32.9	2,966
Minnesota	14	137.9	4.7	9.9	1,000	4.4	71	13.82	40.8	4,000
Florida	40	129.7	4.4	3.2	1,100	4.9	28	9.67	36.2	2,818
New Jersey	23	122.1	4.2	5.3	1,000	4.4	43	11.30	49.5	3,100
Michigan	15	80.0	2.7	5.3	600	2.7	40	10.14	45.7	1,667
Missouri	9	79.9	2.7	8.9	800	3.6	89	12.00	36.8	-
North Carolina	11	70.4	2.4	6.4	600	2.7	55	7.00	54.0	-
Texas	29	65.4	2.2	2.3	600	2.7	21	7.57	49.1	2,167
Pennsylvania	14	47.1	1.6	3.4	500	2.2	36	9.75	41.0	-
Georgia	10	31.0	1.1	3.1	300	1.3	30	9.00	46.1	2,333
Maryland	8	20.7	0.7	2.6	200	0.9	25	14.50	27.5	1,500
Virginia	19	16.8	0.6	0.9	300	1.3	16	8.00	36.9	1,000
Kansas	6	16.4	0.6	2.7	200	0.9	33	11.00	54.9	1,000
Washington	15	15.4	0.5	1.0	200	0.9	13	12.00	38.3	2,000
Nebraska	5	11.2	0.4	2.2	100	0.4	20	8.00	38.4	1,000
Illinois	30	(D)	-	-	1,750 *	7.8	58	-	-	-
Colorado	17	(D)	-	-	375 *	1.7	22	-	-	-
Oregon	10	(D)	-	-	750 *	3.3	75	-	-	-
Arizona	9	(D)	-	-	175 *	0.8	19	-	-	-
Ohio	8	(D)	-	-	375 *	1.7	47	-	-	-
Connecticut	7	(D)	-	-	375 *	1.7	54	-	-	-
South Carolina	5	(D)	-	-	375 *	1.7	75	-	-	-
Wisconsin	5	(D)	-	-	175 *	0.8	35	-	-	-
Kentucky	4	(D)	-	-	175 *	0.8	44	-	-	-
Idaho	3	(D)	-	-	175 *	0.8	58	-	-	-
New Mexico	3	(D)	-	-	375 *	1.7	125	-	-	-
Maine	1	(D)	-	-	375 *	1.7	375	-	-	-

Source: 1992 *Economic Census*. The states are in descending order of shipments or establishments (if shipment data are missing for the majority). The symbol (D) appears when data are withheld to prevent disclosure of competitive information. States marked with (D) are sorted by number of establishments. A dash (-) indicates that the data element cannot be calculated; * indicates the midpoint of a range.

3671 - ELECTRON TUBES

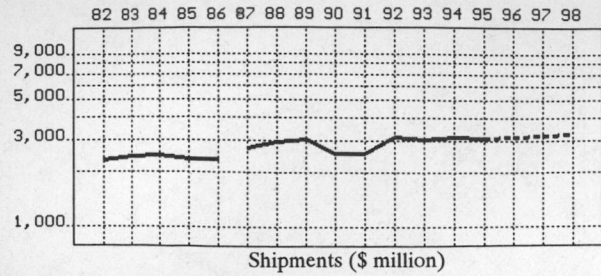

Shipments ($ million)

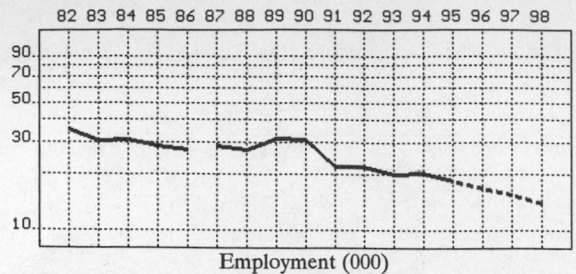

Employment (000)

GENERAL STATISTICS

Year	Companies	Establishments		Employment			Compensation		Production ($ million)			
		Total	with 20 or more employees	Total (000)	Production Workers (000)	Hours (Mil)	Payroll ($ mil)	Wages ($/hr)	Cost of Materials	Value Added by Manufacture	Value of Shipments	Capital Invest.
1982	85	102	61	35.5	24.7	47.5	773.0	10.03	979.5	1,306.8	2,302.3	89.0
1983		97	59	30.6	22.0	43.5	703.5	10.43	965.6	1,437.3	2,417.7	88.8
1984		92	57	31.3	22.5	45.2	746.1	10.97	1,005.5	1,510.7	2,466.7	134.2
1985		88	56	28.6	20.9	42.1	718.8	10.96	993.4	1,325.5	2,362.9	160.2
1986		83	52	27.4	20.0	39.2	722.1	11.67	1,054.8	1,341.0	2,361.0	109.0
1987*	102	121	64	28.4	20.5	41.2	792.4	12.18	1,234.7	1,513.0	2,735.4	103.4
1988		164	86	27.4	19.4	37.7	810.6	13.64	1,330.8	1,629.3	2,942.7	193.9
1989		173	101	31.6	20.0	40.5	813.1	12.53	1,443.4	1,606.8	3,038.8	201.9
1990		172	104	31.3	17.5	34.1	653.3	12.58	1,258.8	1,317.8	2,570.4	169.7
1991		185	100	22.1	16.3	32.2	630.3	12.90	1,454.6	1,131.0	2,568.3	77.3
1992	174	189	69	22.2	16.8	33.5	677.4	13.21	1,883.6	1,280.4	3,144.9	61.7
1993		195	73	20.2	15.3	30.9	628.9	13.73	1,912.9	1,135.8	3,051.5	85.5
1994		212p	84p	20.4	15.8	31.7	629.8	13.25	1,798.6	1,357.3	3,148.1	132.3
1995		222p	84p	18.5p	14.3p	28.6p	571.8p	13.55p	1,771.6p	1,336.9p	3,100.9p	82.5p
1996		232p	84p	17.0p	13.5p	27.1p	542.3p	13.68p	1,797.1p	1,356.2p	3,145.5p	72.4p
1997		242p	83p	15.4p	12.7p	25.7p	512.8p	13.80p	1,822.6p	1,375.4p	3,190.1p	62.2p
1998		252p	83p	13.9p	12.0p	24.2p	483.3p	13.92p	1,848.1p	1,394.7p	3,234.8p	52.1p

Sources: 1982, 1987, 1992 *Economic Census*; *Annual Survey of Manufactures*, 83-86, 88-91, 93-94. Establishment counts are from *County Business Patterns* for non-Census years; establishment counts for 83-84 are extrapolations. * indicates that industry content changed in 87; earlier years use 77 SICs. 'P's mark projections.

INDICES OF CHANGE

Year	Companies	Establishments		Employment			Compensation		Production ($ million)			
		Total	with 20 or more employees	Total (000)	Production Workers (000)	Hours (Mil)	Payroll ($ mil)	Wages ($/hr)	Cost of Materials	Value Added by Manufacture	Value of Shipments	Capital Invest.
1982	49	54	88	160	147	142	114	76	52	102	73	144
1983		51	86	138	131	130	104	79	51	112	77	144
1984		49	83	141	134	135	110	83	53	118	78	218
1985		47	81	129	124	126	106	83	53	104	75	260
1986		44	75	123	119	117	107	88	56	105	75	177
1987*	59	64	93	128	122	123	117	92	66	118	87	168
1988		87	125	123	115	113	120	103	71	127	94	314
1989		92	146	142	119	121	120	95	77	125	97	327
1990		91	151	141	104	102	96	95	67	103	82	275
1991		98	145	100	97	96	93	98	77	88	82	125
1992	100	100	100	100	100	100	100	100	100	100	100	100
1993		103	106	91	91	92	93	104	102	89	97	139
1994		112p	122p	92	94	95	93	100	95	106	100	214
1995		117p	122p	83p	85p	85p	84p	103p	94p	104p	99p	134p
1996		123p	121p	76p	80p	81p	80p	104p	95p	106p	100p	117p
1997		128p	121p	70p	76p	77p	76p	104p	97p	107p	101p	101p
1998		134p	120p	63p	71p	72p	71p	105p	98p	109p	103p	84p

Sources: Same as General Statistics. Values reflect change from the base year, 1992. Values above 100 mean greater than 92, values below 100 mean less than 92, and a value of 100 in the 82-91 or 93-98 period means same as 92. * indicates that industry content changed in 87. Data for earlier years are in 77 SIC format.

SELECTED RATIOS

For 1994	Avg. of All Manufact.	Analyzed Industry	Index	For 1994	Avg. of All Manufact.	Analyzed Industry	Index
Employees per Establishment	49	96	197	Value Added per Production Worker	134,084	85,905	64
Payroll per Establishment	1,500,273	2,972,758	198	Cost per Establishment	5,045,178	8,489,683	168
Payroll per Employee	30,620	30,873	101	Cost per Employee	102,970	88,167	86
Production Workers per Establishment	34	75	217	Cost per Production Worker	146,988	113,835	77
Wages per Establishment	853,319	1,982,586	232	Shipments per Establishment	9,576,895	14,859,541	155
Wages per Production Worker	24,861	26,584	107	Shipments per Employee	195,460	154,319	79
Hours per Production Worker	2,056	2,006	98	Shipments per Production Worker	279,017	199,247	71
Wages per Hour	12.09	13.25	110	Investment per Establishment	321,011	624,477	195
Value Added per Establishment	4,602,255	6,406,676	139	Investment per Employee	6,552	6,485	99
Value Added per Employee	93,930	66,534	71	Investment per Production Worker	9,352	8,373	90

Sources: Same as General Statistics. The 'Average of All Manufacturing' column represents the average of all manufacturing industries reported for the most recent complete year available. The Index shows the relationship between the Average and the Analyzed Industry. For example, 100 means that they are equal; 500 that the Analyzed Industry is five times the average; 50 means that the Analyzed Industry is half the national average. The abbreviation 'na' is used to show that data are 'not available'.

LEADING COMPANIES   Number shown: 26   Total sales ($ mil): 3,141   Total employment (000): 17.1

Company Name	Address				CEO Name	Phone	Co. Type	Sales ($ mil)	Empl. (000)
Varian Associates Inc	3050 Hansen Way	Palo Alto	CA	94304	J Tracy O'Rourke	415-493-4000	P	1,552	8.1
Schlumberger Technologies Inc	277 Park Av	New York	NY	10172	Clermont Matten	212-350-9400	S	300*	1.6
Philips Display Components Co	1600 Huron Pkwy	Ann Arbor	MI	48105	Iva M Wilson	313-996-9400	S	270	1.8
Litton Industries Inc	960 Industrial Rd	San Carlos	CA	94070	John Niddaugh	415-591-8411	D	190*	1.3
Richardson Electronics Ltd	40W267 Keslinger	Lafox	IL	60147	E J Richardson	708-208-2200	P	172	0.7
Varian Power Grid Tube Prod	301 Industrial Way	San Carlos	CA	94070	Frederick Kohler	415-592-1221	D	150	0.4
Hughes Aircraft Co	PO Box 2999	Torrance	CA	90509	Tim T Fong	310-517-6000	D	100	0.9
Ni-Tec	3414 Hermann Dr	Garland	TX	75041	Steve Lambert	214-840-5600	D	100	0.5
Thomson Components	406 Commerce Way	Totowa	NJ	07511	Ernest L Stern	201-812-9000	S	90	<0.1
Video Display Corp	1868 Tucker Indrial	Tucker	GA	30084	Ronald D Ordway	404-938-2080	P	53	0.3
ITT Corp	PO Box 100	Easton	PA	18044	Travis Gerould	610-252-7331	D	40	0.4
Thomas Electronics Inc	100 Riverview Dr	Wayne	NJ	07470	Harold A Ketchum	201-696-5200	R	20*	0.3
Beam-Stream Inc	1635 Magda Dr	Montpelier	OH	43543	Leonard H Howell	419-485-3166	R	17	0.3
Esprit Systems Inc	2115 Ringwood Av	San Jose	CA	95131	Raymond Hou	408-954-9900	R	16	<0.1
Dotronix Inc	160 1st St SE	New Brighton	MN	55112	William S Sadler	612-633-1742	P	15	0.2
EMR Photoelectric	PO Box 44	Princeton	NJ	08542	Robert R Davis	609-799-1000	D	12	<0.1
Incom Inc	PO Drawer G	Southbridge	MA	01550	A M Detarando	508-765-9151	R	11*	<0.1
LogiMetrics Inc	121-03 Dupont St	Plainview	NY	11803	M H Feigenbaum	516-349-1700	P	7	<0.1
Syntronic Instruments Inc	100 Industrial Rd	Addison	IL	60101	GN Marcy	708-543-6444	R	6	<0.1
Ultra Clean Techn Syst & Service	150 Independence	Menlo Park	CA	94025	Hisayoshi Kobayashi	415-323-4100	S	6*	<0.1
Spectra-Mat Inc	100 Westgate Dr	Watsonville	CA	95076	James Abendshan	408-722-4116	R	5	<0.1
Teltron Technologies Inc	2 Riga Ln	Birdsboro	PA	19508	Arthur H Mengel	610-582-9450	S	3	<0.1
TELvision Laboratories	PO Box 519	Wauconda	IL	60084	WA Schwalm	708-526-2511	R	2	<0.1
Apex Electronics Inc	100 8th St	Passaic	NJ	07055	Michael Dorota	201-773-1220	S	1	<0.1
Coloray Display Corp	1045 Mission Ct	Fremont	CA	94539	Charles Antony	510-623-3300	R	1*	<0.1
CRT Scientific Corp	14746 Raymer St	Van Nuys	CA	91406	Kenneth Keller	818-989-4610	R	1*	<0.1

Source: Ward's Business Directory of U.S. Private and Public Companies, Volumes 1 and 2, 1996. The company type code used is as follows: P - Public, R - Private, S - Subsidiary, D - Division, J - Joint Venture, A - Affiliate, G - Group. Sales are in millions of dollars, employees are in thousands. An asterisk (*) indicates an estimated sales volume. The symbol < stands for 'less than'. Company names and addresses are truncated, in some cases, to fit into the available space.

MATERIALS CONSUMED

Material	Quantity	Delivered Cost ($ million)
Materials, ingredients, containers, and supplies	(X)	1,710.5
Tube blanks	(X)	(D)
Printed circuit boards (without inserted components) for electronic circuitry	(X)	5.0
Semiconductors, including transistors, diodes, rectifiers, and integrated circuits for electronic circuitry	(X)	3.9
Capacitors for electronic circuitry	(X)	3.1
Resistors for electronic circuitry	(X)	2.1
Other components and accessories for electronic circuitry, nec, except tubes	(X)	29.4
Silicon, hyperpure	(X)	(D)
Gold and other precious metals, all forms (including ingot, sheet, strip, solder, plating, electrodes, etc.)	(X)	8.6
Doped chemicals, and other doped materials for electronic use	(X)	(D)
Ferrites (powder and paste)	(X)	(D)
Metal powders	(X)	7.9
Electronic computing equipment	(X)	(D)
Current-carrying wiring devices	(X)	(D)
Electronic communication equipment	(X)	(D)
Electrical instrument mechanisms and meter movements (including instrument relays)	(X)	(D)
Optical instruments and lenses (except sighting, tracking, and fire control)	(X)	(D)
Plastics resins consumed in the form of granules, pellets, powders, liquids, etc.	(X)	(D)
Fabricated plastics products (except gaskets, hoses, and belting)	(X)	1.8
Sheet metal products, except stampings	(X)	65.6
Metal stampings	(X)	57.9
Metal bolts, nuts, screws, washers, rivets, and other screw machine products	(X)	5.4
Other fabricated metal products (except forgings)	(X)	28.1
Forgings	(X)	(D)
Castings (rough and semifinished)	(X)	(D)
Steel shapes and forms	(X)	1.3
Copper and copper-base alloy shapes and forms	(X)	2.1
Aluminum and aluminum-base alloy shapes and forms	(X)	0.8
Other nonferrous shapes and forms	(X)	3.1
Insulated wire and cable, including magnet wire	(X)	6.6
Paper and paperboard containers, including shipping sacks and other paper packaging supplies	(X)	2.0
All other materials and components, parts, containers, and supplies	(X)	240.2
Materials, ingredients, containers, and supplies, nsk	(X)	833.8

Source: 1992 *Economic Census.* Explanation of symbols used: (D): Withheld to avoid disclosure of competitive data; na: Not available; (S): Withheld because statistical norms were not met; (X): Not applicable; (Z): Less than half the unit shown; nec: Not elsewhere classified; nsk: Not specified by kind; - : zero; * : 10-19 percent estimated; ** : 20-29 percent estimated.

PRODUCT SHARE DETAILS

Product or Product Class	% Share	Product or Product Class	% Share
Electron tubes	100.00	and rebuilt).	63.29
Transmittal, industrial, and special purpose electron tubes,		Electron tube parts	4.90
except x-ray	27.35	Electron tubes, nsk	4.45
Receiving-type electron tubes (including cathode ray) (new			

Source: 1992 *Economic Census.* The values shown are percent of total shipments in an industry. Values of indented subcategories are summed in the main heading. The symbol (D) appears when data are withheld to prevent disclosure of competitive information. The abbreviation nsk stands for 'not specified by kind' and nec for 'not elsewhere classified'.

INPUTS AND OUTPUTS FOR ELECTRON TUBES

Economic Sector or Industry Providing Inputs	%	Sector	Economic Sector or Industry Buying Outputs	%	Sector
Glass & glass products, except containers	18.9	Manufg.	Radio & TV receiving sets	29.4	Manufg.
Electronic components nec	18.6	Manufg.	Federal Government purchases, national defense	12.8	Fed Govt
Imports	13.3	Foreign	Exports	11.6	Foreign
Wholesale trade	9.7	Trade	Electrical repair shops	10.6	Services
Banking	3.6	Fin/R.E.	Radio & TV broadcasting	10.1	Util.
Electric services (utilities)	3.5	Util.	Radio & TV communication equipment	8.4	Manufg.
Metal stampings, nec	3.0	Manufg.	Electronic computing equipment	4.6	Manufg.
Air transportation	1.7	Util.	Telephone & telegraph apparatus	3.6	Manufg.
Cyclic crudes and organics	1.5	Manufg.	X-ray apparatus & tubes	2.3	Manufg.
Petroleum refining	1.5	Manufg.	Personal consumption expenditures	0.9	
Electron tubes	1.4	Manufg.	Electron tubes	0.9	Manufg.
Communications, except radio & TV	1.4	Util.	Computer & data processing services	0.8	Services
Plating & polishing	1.2	Manufg.	Federal Government purchases, nondefense	0.7	Fed Govt
Equipment rental & leasing services	1.1	Services	Instruments to measure electricity	0.5	Manufg.
Gas production & distribution (utilities)	1.0	Util.	Typewriters & office machines, nec	0.5	Manufg.
Eating & drinking places	1.0	Trade	Colleges, universities, & professional schools	0.5	Services
Hotels & lodging places	1.0	Services	Miscellaneous plastics products	0.3	Manufg.
Maintenance of nonfarm buildings nec	0.9	Constr.	Calculating & accounting machines	0.2	Manufg.
Copper rolling & drawing	0.9	Manufg.	Electronic components nec	0.2	Manufg.
Nonferrous rolling & drawing, nec	0.9	Manufg.	Mechanical measuring devices	0.2	Manufg.
Blast furnaces & steel mills	0.7	Manufg.	S/L Govt. purch., elem. & secondary education	0.2	S/L Govt
Advertising	0.7	Services	S/L Govt. purch., fire	0.2	S/L Govt
Fabricated metal products, nec	0.6	Manufg.	S/L Govt. purch., highways	0.1	S/L Govt
Machinery, except electrical, nec	0.6	Manufg.	S/L Govt. purch., police	0.1	S/L Govt
Paperboard containers & boxes	0.6	Manufg.			
Motor freight transportation & warehousing	0.6	Util.			
Computer & data processing services	0.6	Services			
U.S. Postal Service	0.6	Gov't			
Primary nonferrous metals, nec	0.5	Manufg.			
Legal services	0.5	Services			
Noncomparable imports	0.5	Foreign			
Household furniture, nec	0.4	Manufg.			
Semiconductors & related devices	0.4	Manufg.			
Special dies & tools & machine tool accessories	0.4	Manufg.			
Real estate	0.4	Fin/R.E.			
Management & consulting services & labs	0.4	Services			
Nonmetallic mineral products, nec	0.3	Manufg.			
Screw machine and related products	0.3	Manufg.			
Accounting, auditing & bookkeeping	0.3	Services			
Abrasive products	0.2	Manufg.			
Industrial inorganic chemicals, nec	0.2	Manufg.			
Manifold business forms	0.2	Manufg.			
Metal coating & allied services	0.2	Manufg.			
Miscellaneous plastics products	0.2	Manufg.			
Nonferrous wire drawing & insulating	0.2	Manufg.			
Insurance carriers	0.2	Fin/R.E.			
Security & commodity brokers	0.2	Fin/R.E.			
Automotive rental & leasing, without drivers	0.2	Services			
Automotive repair shops & services	0.2	Services			
Lubricating oils & greases	0.1	Manufg.			
Sheet metal work	0.1	Manufg.			
Railroads & related services	0.1	Util.			
Sanitary services, steam supply, irrigation	0.1	Util.			
Electrical repair shops	0.1	Services			
Photofinishing labs, commercial photography	0.1	Services			
State & local government enterprises, nec	0.1	Gov't			

Source: Benchmark Input-Output Accounts for the U.S. Economy, 1982, U.S. Department of Commerce, Washington, D.C., July 1991. Data, as reported in the source, are organized by the 1977 SIC structure in use in 1982 but have been matched, as closely as is possible, to the 1987 SIC structure used in this book.

OCCUPATIONS EMPLOYED BY SIC 367 - ELECTRONIC COMPONENTS AND ACCESSORIES

Occupation	% of Total 1994	Change to 2005	Occupation	% of Total 1994	Change to 2005
Electrical & electronic assemblers	9.8	-8.4	Secretaries, ex legal & medical	1.6	-7.4
Electrical & electronic equipment assemblers	5.7	1.8	Industrial production managers	1.5	-28.8
Electrical & electronic technicians,technologists	5.4	22.1	Electromechanical equipment assemblers	1.4	11.9
Inspectors, testers, & graders, precision	5.4	-28.8	Coil winders, tapers, & finishers	1.3	-18.6
Electronic semiconductor processors	5.3	-1.4	Production, planning, & expediting clerks	1.3	1.8
Electrical & electronics engineers	4.4	29.8	Industrial engineers, ex safety engineers	1.3	12.0
Assemblers, fabricators, & hand workers nec	4.0	1.8	Electrolytic plating machine workers	1.2	11.9
Blue collar worker supervisors	3.4	-14.1	Traffic, shipping, & receiving clerks	1.2	-2.1
Sales & related workers nec	2.2	1.8	Freight, stock, & material movers, hand	1.2	-18.6
Computer engineers	2.2	50.7	Precision assemblers nec	1.1	62.9
Engineering, mathematical, & science managers	1.8	15.6	Stock clerks	1.0	-17.3
Engineering technicians & technologists nec	1.8	1.8	Engineers nec	1.0	52.6
General managers & top executives	1.6	-3.4	General office clerks	1.0	-13.2

Source: Industry-Occupation Matrix, Bureau of Labor Statistics. These data relate to one or more 3-digit SIC industry groups rather than to a single 4-digit SIC. The change reported for each occupation to the year 2005 is a percent of growth or decline as estimated by the Bureau of Labor Statistics. The abbreviation nec stands for 'not elsewhere classified'.

LOCATION BY STATE AND REGIONAL CONCENTRATION

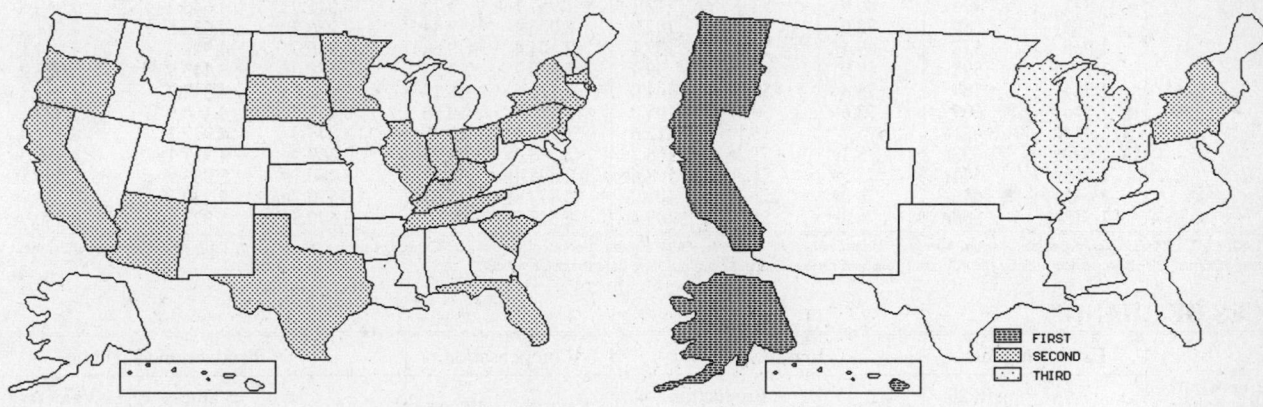

FIRST
SECOND
THIRD

INDUSTRY DATA BY STATE

State	Establish-ments	Shipments			Employment				Cost as % of Shipments	Investment per Employee ($)
		Total ($ mil)	% of U.S.	Per Establ.	Total Number	% of U.S.	Per Establ.	Wages ($/hour)		
California	54	427.5	13.6	7.9	4,000	18.0	74	19.41	34.3	1,500
Pennsylvania	8	364.7	11.6	45.6	1,900	8.6	238	13.69	54.7	-
Massachusetts	11	192.1	6.1	17.5	1,800	8.1	164	16.46	31.4	1,333
New Jersey	13	53.8	1.7	4.1	600	2.7	46	11.86	36.4	3,167
Arizona	6	53.3	1.7	8.9	500	2.3	83	10.14	49.2	1,600
Ohio	10	(D)	-	-	3,750 *	16.9	375	-	-	-
Texas	10	(D)	-	-	750 *	3.4	75	-	-	-
Florida	8	(D)	-	-	175 *	0.8	22	-	-	1,714
Illinois	8	(D)	-	-	3,750 *	16.9	469	-	-	-
New York	6	(D)	-	-	1,750 *	7.9	292	-	-	-
Minnesota	5	(D)	-	-	175 *	0.8	35	-	-	571
Oregon	5	(D)	-	-	175 *	0.8	35	-	-	-
Indiana	2	(D)	-	-	1,750 *	7.9	875	-	-	-
South Carolina	2	(D)	-	-	375 *	1.7	188	-	-	-
Tennessee	2	(D)	-	-	375 *	1.7	188	-	-	-
Kentucky	1	(D)	-	-	175 *	0.8	175	-	-	-
Rhode Island	1	(D)	-	-	375 *	1.7	375	-	-	-
South Dakota	1	(D)	-	-	175 *	0.8	175	-	-	-

Source: 1992 Economic Census. The states are in descending order of shipments or establishments (if shipment data are missing for the majority). The symbol (D) appears when data are withheld to prevent disclosure of competitive information. States marked with (D) are sorted by number of establishments. A dash (-) indicates that the data element cannot be calculated; * indicates the midpoint of a range.

3672 - PRINTED CIRCUIT BOARDS

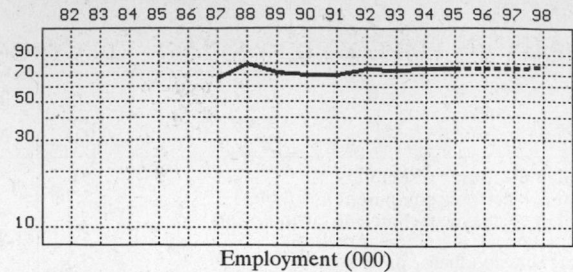

82 83 84 85 86 87 88 89 90 91 92 93 94 95 96 97 98

Shipments ($ million)

Employment (000)

GENERAL STATISTICS

Year	Com-panies	Establishments		Employment			Compensation		Production ($ million)			
		Total	with 20 or more employees	Total (000)	Production Workers (000)	Hours (Mil)	Payroll ($ mil)	Wages ($/hr)	Cost of Materials	Value Added by Manufacture	Value of Shipments	Capital Invest.
1982												
1983												
1984												
1985												
1986												
1987	950	1,009	566	66.6	48.6	100.0	1,378.5	8.36	2,023.8	2,683.1	4,672.6	239.3
1988		871	496	80.9	53.9	112.1	1,957.1	9.12	3,041.9	4,643.2	7,960.6	336.8
1989		982	589	72.0	52.4	107.9	2,006.4	9.44	2,776.2	4,623.5	7,354.4	372.8
1990		1,060	578	69.4	51.0	109.1	2,104.4	9.68	2,886.7	4,997.2	7,844.1	405.1
1991		1,187	598	69.9	47.1	99.4	1,920.7	9.92	2,678.0	3,443.9	6,352.9	311.1
1992	1,261	1,324	589	75.8	50.8	104.8	2,110.6	10.17	2,972.8	4,348.3	7,311.8	316.8
1993		1,346	609	73.6	50.5	105.0	2,129.4	10.58	3,151.2	4,160.0	7,377.6	282.8
1994		1,414P	621P	75.7	53.1	112.6	2,162.6	10.49	3,376.9	4,947.3	8,262.4	365.9
1995		1,490P	633P	75.1P	51.2P	108.2P	2,318.3P	11.04P	3,369.9P	4,937.1P	8,245.3P	347.8P
1996		1,566P	644P	75.5P	51.3P	108.6P	2,395.4P	11.33P	3,470.1P	5,083.9P	8,490.5P	352.0P
1997		1,642P	656P	76.0P	51.4P	109.0P	2,472.6P	11.63P	3,570.3P	5,230.7P	8,735.7P	356.2P
1998		1,718P	668P	76.5P	51.4P	109.4P	2,549.7P	11.92P	3,670.5P	5,377.5P	8,980.9P	360.5P

Sources: 1982, 1987, 1992 *Economic Census*; *Annual Survey of Manufactures*, 83-86, 88-91, 93-94. Establishment counts for non-Census years are from *County Business Patterns*; establishment values for 83-84 are extrapolations. 'P's show projections by the editors. Industries reclassified in 87 will not have data for prior years.

INDICES OF CHANGE

Year	Com-panies	Establishments		Employment			Compensation		Production ($ million)			
		Total	with 20 or more employees	Total (000)	Production Workers (000)	Hours (Mil)	Payroll ($ mil)	Wages ($/hr)	Cost of Materials	Value Added by Manufacture	Value of Shipments	Capital Invest.
1982												
1983												
1984												
1985												
1986												
1987	75	76	96	88	96	95	65	82	68	62	64	76
1988		66	84	107	106	107	93	90	102	107	109	106
1989		74	100	95	103	103	95	93	93	106	101	118
1990		80	98	92	100	104	100	95	97	115	107	128
1991		90	102	92	93	95	91	98	90	79	87	98
1992	100	100	100	100	100	100	100	100	100	100	100	100
1993		102	103	97	99	100	101	104	106	96	101	89
1994		107P	105P	100	105	107	102	103	114	114	113	115
1995		113P	107P	99P	101P	103P	110P	109P	113P	114P	113P	110P
1996		118P	109P	100P	101P	104P	113P	111P	117P	117P	116P	111P
1997		124P	111P	100P	101P	104P	117P	114P	120P	120P	119P	112P
1998		130P	113P	101P	101P	104P	121P	117P	123P	124P	123P	114P

Sources: Same as General Statistics. Values reflect change from the base year, 1992. Values above 100 mean greater than 92, values below 100 mean less than 92, and a value of 100 in the 82-91 or 93-98 period means same as 92. 'P's mark projections by the editors.

SELECTED RATIOS

For 1994	Avg. of All Manufact.	Analyzed Industry	Index	For 1994	Avg. of All Manufact.	Analyzed Industry	Index
Employees per Establishment	49	54	109	Value Added per Production Worker	134,084	93,169	69
Payroll per Establishment	1,500,273	1,528,957	102	Cost per Establishment	5,045,178	2,387,466	47
Payroll per Employee	30,620	28,568	93	Cost per Employee	102,970	44,609	43
Production Workers per Establishment	34	38	109	Cost per Production Worker	146,988	63,595	43
Wages per Establishment	853,319	835,089	98	Shipments per Establishment	9,576,895	5,841,511	61
Wages per Production Worker	24,861	22,244	89	Shipments per Employee	195,460	109,147	56
Hours per Production Worker	2,056	2,121	103	Shipments per Production Worker	279,017	155,601	56
Wages per Hour	12.09	10.49	87	Investment per Establishment	321,011	258,691	81
Value Added per Establishment	4,602,255	3,497,738	76	Investment per Employee	6,552	4,834	74
Value Added per Employee	93,930	65,354	70	Investment per Production Worker	9,352	6,891	74

Sources: Same as General Statistics. The 'Average of All Manufacturing' column represents the average of all manufacturing industries reported for the most recent complete year available. The Index shows the relationship between the Average and the Analyzed Industry. For example, 100 means that they are equal; 500 that the Analyzed Industry is five times the average; 50 means that the Analyzed Industry is half the national average. The abbreviation 'na' is used to show that data are 'not available'.

LEADING COMPANIES Number shown: 75   Total sales ($ mil): **6,474**   Total employment (000): **46.2**

Company Name	Address				CEO Name	Phone	Co. Type	Sales ($ mil)	Empl. (000)
Solectron Corp	777 Gibraltar Dr	Milpitas	CA	95035	Koichi Nishimura	408-957-8500	P	1,457	6.6
Jabil Circuit Inc	10800 Roosevelt	St Petersburg	FL	33716	William D Morean	813-577-9749	P	376	1.0
Photocircuits Corp	31 Sea Cliff Av	Glen Cove	NY	11542	John Endee	516-674-1000	R	237	2.1
Hadco Corp	12A Manor Pkwy	Salem	NH	03079	Patrick Sweeney	603-898-8000	P	222	2.0
Diamond Multimedia Systems	2880 Junction Av	San Jose	CA	95134	William J Schroeder	408-325-7000	P	203	0.3
Lucas Body Systems	14241 Fenton Rd	Fenton	MI	48430	Tom Schilling	810-750-7510	S	200	2.0
Micronics Computers Inc	232 E Warren Av	Fremont	CA	94539	Steven P Kitrosser	510-651-2300	P	182	0.3
Solectron Technology Inc	PO Box 562148	Charlotte	NC	28256	Hank Ewert	704-598-3300	S	170•	0.8
AMP-AKZO Co	200 Fairforest Way	Greenville	SC	29607	Javad K Hassan	803-297-4100	S	160	1.1
Bairnco Corp	2251 Lucien Way	Maitland	FL	32751	Luke E Fichthorn	407-875-2222	P	145	0.9
Advance Circuits Inc	15102 Minnetonka	Minnetonka	MN	55345	Robert W Heller	612-930-8000	P	143	1.3
Zycon Corp	445 El Camino Real	Santa Clara	CA	95050	Ronald Donati	408-241-9900	R	140	1.6
IEC Electronics Corp	PO Box 271	Newark	NY	14513	Roger E Main	315-331-7742	P	130	1.9
Comptronix Corp	3 Maryland Farms	Brentwood	TN	37027	E Townes Duncan	615-377-3330	P	120	0.7
VIP Computer Inc	16851 Knots Av	La Mirada	CA	90638	Chris Yang	714-562-6999	R	120	0.8
Sanmina Corp	355 Trimble Rd	San Jose	CA	95131	Jure Sola	408-435-8444	P	115	0.9
LaserMaster Systems	7156 Shady Oak Rd	Eden Prairie	MN	55344	Melvin L Masters	612-941-8687	P	106	0.5
Altron Inc	1 Jewel Dr	Wilmington	MA	01887	Samuel Altschuler	508-658-5800	P	104	0.9
Benchmark Electronics Inc	3000 Technology Dr	Angleton	TX	77515	Donald E Nigbor	409-849-6550	P	98	0.4
Diceon Electronics Inc	18522 Von Karman	Irvine	CA	92715	Milan Mandaric	714-833-0870	P	96	1.3
Integrated Circuit Systems Inc	PO Box 968	Valley Forge	PA	19482	Edward H Arnold	610-630-5300	P	94	0.3
K-Byte Manufacturing	14201 McCormick	Tampa	FL	33626	Patrick J Flynn	813-854-2000	S	90	0.7
ACT Manufacturing Inc	108 Forest Av	Hudson	MA	01749	John A Pino	508-562-1200	P	86	0.5
Continental Circuits Corp	3502 E Roeser Rd	Phoenix	AZ	85040	F G McNamee III	602-268-3461	P	80	0.9
ADFlex Solutions Inc	2001 W Chandler	Chandler	AZ	85224	R C Esteverena	602-963-4584	P	78	2.7
Rockwell International Corp	6 Butterfield Trail	El Paso	TX	79906	T Woo	915-775-2100	D	74•	0.8
Toppan West Inc	7770 Miramar Rd	San Diego	CA	92126	Steve Schlepp	619-695-2222	S	70	0.5
Circuit Systems Inc	2350 E Lunt Av	Elk Grove Vill	IL	60007	Dahyabhai S Patel	708-439-1999	P	60	0.6
Litton Industries Inc	PO Box 2847	Springfield	MO	65801	Robert Schutz	417-862-0751	D	60	0.6
Data-Design Laboratories Inc	1270 NW 167th Pl	Beaverton	OR	97006	William E Cook	503-645-3807	P	58	0.7
Advanced Quick Circuits LP	245 East Dr	Melbourne	FL	32904	CA Rossi	407-768-9901	R	55	0.6
CMAC of America Inc	1601 Hill Av	W Palm Beach	FL	33407	Dennis Wood	407-845-8455	S	51•	0.4
Xylogics Inc	53 3rd Av	Burlington	MA	01803	Bruce I Sachs	617-272-8140	P	50	0.2
Universal Circuits Inc	9240 Fountain	Menomonee Fls	WI	53051	Raymond Esser	414-255-0802	R	50	0.7
Ferranti Technologies Inc	PO Box 3025	Lancaster	PA	17604	Nathan C Blackwell	717-285-7151	S	45•	0.4
Riverhead Circuits	PO Box 700	Aquebogue	NY	11931	Robert Emgee	516-722-4100	D	45	0.5
Cuplex Inc	1500 E Hwy 66	Garland	TX	75040	Ron P Ryno	214-276-0333	R	44	0.5
Automata Inc	1200 Severn Way	Sterling	VA	20166	Mohamed El Ezaby	703-450-2605	R	40	0.4
Sigma Circuits Inc	393 Mathew St	Santa Clara	CA	95050	B Kevin Kelly	408-727-9169	P	39	0.4
SigmaTron International Inc	2201 Landmeier Rd	Elk Grove Vill	IL	60007	Gary R Fairhead	708-956-8000	P	37	0.7
Herco Technology Corp	13330 Evening	San Diego	CA	92128	Robert Herring Sr	619-679-2800	R	34•	0.4
Xetel Corp	2525 Brockton Dr	Austin	TX	78758	Richard N Winter	512-834-2266	R	34•	0.4
HF Henderson Industries	45 Fairfield Pl	West Caldwell	NJ	07006	HF Henderson Jr	201-227-9250	R	31	0.2
Inter-Pak Electronics	2500 Airport Com	Springfield	MO	65803	Robert J Schutz	417-862-0751	D	30	0.1
LION Computers Inc	1751 McCarthy Blv	Milpitas	CA	95035	KM Chow	408-954-8070	S	30	<0.1
Calidad Electronics Inc	1920 SE Industrial	Edinburg	TX	78539	Tina Rolsten	210-381-0909	S	28•	0.3
Advanced Flex Inc	15115 Minnetonka	Minnetonka	MN	55345	Larry Bergman	612-930-4800	R	27	0.3
Philway Products Inc	701 Virginia Av	Ashland	OH	44805	Mehendra Patel	419-281-7777	R	27	0.3
Kalmus and Associates	2424 S 25th Av	Broadview	IL	60153	HJ Kalmus Jr	708-343-7004	R	25	0.3
Targ-It-Tronics Inc	7100 Technology Dr	W Melbourne	FL	32904	Larry Groves	407-725-6993	S	25•	0.3
Velie Circuits Inc	1267 Logan Av	Costa Mesa	CA	92626	Larry Velie	714-751-4994	R	25	0.3
Capitol Circuits Corp	24 Denby Rd	Boston	MA	02134	Dominic Emello	617-787-2030	R	23	0.2
Hauppauge Computer Works	91 Cabot Ct	Hauppauge	NY	11788	Ken Plotkin	516-434-1600	R	22	<0.1
Excel Inc	PO Box 327	Sagamore Bch	MA	02562	Bob Madonna	508-833-1144	R	22	0.2
Lytton Inc	1784 Stanley Av	Dayton	OH	45404	Lytton F Crossley	513-898-9800	R	21	0.2
Electronic Instrumentation	108 Carpenter Dr	Sterling	VA	20164	Joe T May	703-478-0700	R	20	0.2
Methode of California	9334 Mason Av	Chatsworth	CA	91311	Thomas J Beaven	818-886-2272	S	20•	0.2
Printed Circuit Corp	10 Micro Dr	Woburn	MA	01801	Peter Sarmonian	617-935-9570	R	20	0.2
Antares Group Inc	10052 Mesa Ridge	San Diego	CA	92121	Cliff Cooke	619-457-2111	S	19•	0.2
Area Electronics Systems Inc	330 E Orangethrope	Placentia	CA	92670	William Huang	714-993-0300	R	18	<0.1
Diagnostic Instrument Corp	4 Copeland Dr	Ayer	MA	01432	James Hashem	508-772-4572	R	18	0.1
Dynaco Acquisition Corp	1000 S Preist Rd	Tempe	AZ	85281	Mauri Needham	602-968-2000	D	18	0.2
Holaday Circuits Inc	11126 Bren Rd W	Minnetonka	MN	55343	Marshall Lewis	612-933-3303	R	18	0.3
Sky Computers Inc	27 Industrial Av	Chelmsford	MA	01824	Bruce R Rusch	508-250-1920	S	18•	<0.1
Stallion Technologies Inc	60 Penny Ln	Watsonville	CA	95076	William Kiely	408-761-9499	S	18•	<0.1
H-R Industries Inc	1302 E Collins Blv	Richardson	TX	75081	Paul R Dickinson	214-301-6620	R	16•	0.2
Micom Corp	475 NW 8th Av	New Brighton	MN	55112	Edwin Walhof	612-636-5616	R	16	0.3
Star Gate Technologies Inc	29300 Aurora Rd	Solon	OH	44139	Raymond Wymer	216-349-1860	S	16•	<0.1
Acsist Associates Inc	3965 Meadowbrook	Minneapolis	MN	55426	Frank Voight	612-931-1300	R	15	0.2
Advanced Controls Corporation	16901 Jamboree	Irvine	CA	92714	RW Quest	714-863-9300	R	15	0.1
Electro-Etch Circuits Inc	339 S Isis Av	Inglewood	CA	90301	John Mayer	310-649-2411	R	15•	0.2
Laminating Company of America	7311 Doig Dr	Garden Grove	CA	92641	JP Block	714-891-3581	R	15	0.1
McCurdy Circuits Inc	1739 N Case St	Orange	CA	92665	Scott G McCurdy	714-974-0401	R	15	0.1
Solid State Circuits Inc	3300 S Golden Av	Springfield	MO	65807	Jim Vieth	417-887-0823	R	15•	<0.1
Tingstol Co	1340 W Fullerton	Chicago	IL	60614	John P Zopp Jr	312-935-4422	R	15	0.2

Source: Ward's Business Directory of U.S. Private and Public Companies, Volumes 1 and 2, 1996. The company type code used is as follows: P - Public, R - Private, S - Subsidiary, D - Division, J - Joint Venture, A - Affiliate, G - Group. Sales are in millions of dollars, employees are in thousands. An asterisk (*) indicates an estimated sales volume. The symbol < stands for 'less than'. Company names and addresses are truncated, in some cases, to fit into the available space.

MATERIALS CONSUMED

Material	Quantity	Delivered Cost ($ million)
Materials, ingredients, containers, and supplies	(X)	2,701.8
Tube blanks	(X)	(D)
Printed circuit boards (without inserted components) for electronic circuitry	(X)	293.1
Semiconductors, including transistors, diodes, rectifiers, and integrated circuits for electronic circuitry	(X)	67.0
Capacitors for electronic circuitry	(X)	11.1
Resistors for electronic circuitry	(X)	6.7
Other components and accessories for electronic circuitry, nec, except tubes	(X)	259.5
Silicon, hyperpure	(X)	(D)
Gold and other precious metals, all forms (including ingot, sheet, strip, solder, plating, electrodes, etc.)	(X)	56.7
Doped chemicals, and other doped materials for electronic use	(X)	50.7
Ferrites (powder and paste)	(X)	(D)
Metal powders	(X)	(D)
Electronic computing equipment	(X)	0.4
Current-carrying wiring devices	(X)	2.4
Electronic communication equipment	(X)	(D)
Electrical instrument mechanisms and meter movements (including instrument relays)	(X)	0.5
Optical instruments and lenses (except sighting, tracking, and fire control)	(X)	0.3
Plastics resins consumed in the form of granules, pellets, powders, liquids, etc.	(X)	3.8
Fabricated plastics products (except gaskets, hoses, and belting)	(X)	24.0
Sheet metal products, except stampings	(X)	9.8
Metal stampings	(X)	2.2
Metal bolts, nuts, screws, washers, rivets, and other screw machine products	(X)	5.2
Other fabricated metal products (except forgings)	(X)	91.0
Castings (rough and semifinished)	(X)	0.2
Steel shapes and forms	(X)	0.5
Copper and copper-base alloy shapes and forms	(X)	42.7
Aluminum and aluminum-base alloy shapes and forms	(X)	5.6
Other nonferrous shapes and forms	(X)	6.7
Insulated wire and cable, including magnet wire	(X)	9.8
Paper and paperboard containers, including shipping sacks and other paper packaging supplies	(X)	6.5
All other materials and components, parts, containers, and supplies	(X)	877.8
Materials, ingredients, containers, and supplies, nsk	(X)	824.9

Source: 1992 *Economic Census*. Explanation of symbols used: (D): Withheld to avoid disclosure of competitive data; na: Not available; (S): Withheld because statistical norms were not met; (X): Not applicable; (Z): Less than half the unit shown; nec: Not elsewhere classified; nsk: Not specified by kind; - : zero; * : 10-19 percent estimated; ** : 20-29 percent estimated.

PRODUCT SHARE DETAILS

Product or Product Class	% Share	Product or Product Class	% Share
Printed circuit boards	100.00		

Source: 1992 *Economic Census*. The values shown are percent of total shipments in an industry. Values of indented subcategories are summed in the main heading. The symbol (D) appears when data are withheld to prevent disclosure of competitive information. The abbreviation nsk stands for 'not specified by kind' and nec for 'not elsewhere classified'.

INPUTS AND OUTPUTS FOR SEMICONDUCTORS & RELATED DEVICES

Economic Sector or Industry Providing Inputs	%	Sector	Economic Sector or Industry Buying Outputs	%	Sector
Imports	31.9	Foreign	Exports	21.5	Foreign
Semiconductors & related devices	23.6	Manufg.	Semiconductors & related devices	20.9	Manufg.
Wholesale trade	6.5	Trade	Electronic computing equipment	7.0	Manufg.
Nonferrous rolling & drawing, nec	5.0	Manufg.	Photographic equipment & supplies	6.6	Manufg.
Cyclic crudes and organics	4.2	Manufg.	Federal Government purchases, national defense	6.3	Fed Govt
Electronic components nec	3.0	Manufg.	Radio & TV communication equipment	5.7	Manufg.
Nonferrous wire drawing & insulating	2.6	Manufg.	Electrical repair shops	4.6	Services
Electric services (utilities)	1.8	Util.	Communications, except radio & TV	3.8	Util.
Plating & polishing	1.7	Manufg.	Aircraft & missile equipment, nec	2.5	Manufg.
Banking	1.4	Fin/R.E.	Management & consulting services & labs	2.3	Services
Miscellaneous plastics products	1.3	Manufg.	Telephone & telegraph apparatus	2.1	Manufg.
Communications, except radio & TV	1.2	Util.	Electronic components nec	1.5	Manufg.
Primary nonferrous metals, nec	1.1	Manufg.	Computer & data processing services	1.3	Services
Chemical preparations, nec	1.0	Manufg.	Aircraft	1.1	Manufg.
Industrial inorganic chemicals, nec	0.8	Manufg.	Radio & TV receiving sets	1.1	Manufg.
Metal stampings, nec	0.8	Manufg.	Instruments to measure electricity	1.0	Manufg.
Air transportation	0.8	Util.	Motor vehicles & car bodies	1.0	Manufg.
Advertising	0.8	Services	Business services nec	0.9	Services
Maintenance of nonfarm buildings nec	0.6	Constr.	Electrical equipment & supplies, nec	0.6	Manufg.
Noncomparable imports	0.6	Foreign	Change in business inventories	0.6	In House
Eating & drinking places	0.5	Trade	Calculating & accounting machines	0.5	Manufg.
Real estate	0.5	Fin/R.E.	Mechanical measuring devices	0.5	Manufg.
Engineering, architectural, & surveying services	0.5	Services	Small arms	0.5	Manufg.
Equipment rental & leasing services	0.5	Services	X-ray apparatus & tubes	0.5	Manufg.
Carbon & graphite products	0.4	Manufg.	Optical instruments & lenses	0.4	Manufg.

Continued on next page.

INPUTS AND OUTPUTS FOR SEMICONDUCTORS & RELATED DEVICES - Continued

Economic Sector or Industry Providing Inputs	%	Sector	Economic Sector or Industry Buying Outputs	%	Sector
Motor freight transportation & warehousing	0.4	Util.	Typewriters & office machines, nec	0.4	Manufg.
Business services nec	0.4	Services	Detective & protective services	0.4	Services
Machinery, except electrical, nec	0.3	Manufg.	Miscellaneous repair shops	0.4	Services
Metal coating & allied services	0.3	Manufg.	Photofinishing labs, commercial photography	0.4	Services
Gas production & distribution (utilities)	0.3	Util.	Engine electrical equipment	0.3	Manufg.
Computer & data processing services	0.3	Services	Motor vehicle parts & accessories	0.3	Manufg.
Hotels & lodging places	0.3	Services	Watches, clocks, & parts	0.3	Manufg.
Legal services	0.3	Services	Services to dwellings & other buildings	0.3	Services
Fabricated rubber products, nec	0.2	Manufg.	Ammunition, except for small arms, nec	0.2	Manufg.
Glass & glass products, except containers	0.2	Manufg.	Engineering & scientific instruments	0.2	Manufg.
Plastics materials & resins	0.2	Manufg.	Guided missiles & space vehicles	0.2	Manufg.
Special dies & tools & machine tool accessories	0.2	Manufg.	Lighting fixtures & equipment	0.2	Manufg.
Railroads & related services	0.2	Util.	Ordnance & accessories nec	0.2	Manufg.
Colleges, universities, & professional schools	0.2	Services	Personnel supply services	0.2	Services
Management & consulting services & labs	0.2	Services	Wiring devices	0.1	Manufg.
Electronic computing equipment	0.1	Manufg.	Air transportation	0.1	Util.
Accounting, auditing & bookkeeping	0.1	Services	Equipment rental & leasing services	0.1	Services
Business/professional associations	0.1	Services			

Source: Benchmark Input-Output Accounts for the U.S. Economy, 1982, U.S. Department of Commerce, Washington, D.C., July 1991. Data, as reported in the source, are organized by the 1977 SIC structure in use in 1982 but have been matched, as closely as is possible, to the 1987 SIC structure used in this book.

OCCUPATIONS EMPLOYED BY SIC 367 - ELECTRONIC COMPONENTS AND ACCESSORIES

Occupation	% of Total 1994	Change to 2005	Occupation	% of Total 1994	Change to 2005
Electrical & electronic assemblers	9.8	-8.4	Secretaries, ex legal & medical	1.6	-7.4
Electrical & electronic equipment assemblers	5.7	1.8	Industrial production managers	1.5	-28.8
Electrical & electronic technicians,technologists	5.4	22.1	Electromechanical equipment assemblers	1.4	11.9
Inspectors, testers, & graders, precision	5.4	-28.8	Coil winders, tapers, & finishers	1.3	-18.6
Electronic semiconductor processors	5.3	-1.4	Production, planning, & expediting clerks	1.3	1.8
Electrical & electronics engineers	4.4	29.8	Industrial engineers, ex safety engineers	1.3	12.0
Assemblers, fabricators, & hand workers nec	4.0	1.8	Electrolytic plating machine workers	1.2	11.9
Blue collar worker supervisors	3.4	-14.1	Traffic, shipping, & receiving clerks	1.2	-2.1
Sales & related workers nec	2.2	1.8	Freight, stock, & material movers, hand	1.2	-18.6
Computer engineers	2.2	50.7	Precision assemblers nec	1.1	62.9
Engineering, mathematical, & science managers	1.8	15.6	Stock clerks	1.0	-17.3
Engineering technicians & technologists nec	1.8	1.8	Engineers nec	1.0	52.6
General managers & top executives	1.6	-3.4	General office clerks	1.0	-13.2

Source: Industry-Occupation Matrix, Bureau of Labor Statistics. These data relate to one or more 3-digit SIC industry groups rather than to a single 4-digit SIC. The change reported for each occupation to the year 2005 is a percent of growth or decline as estimated by the Bureau of Labor Statistics. The abbreviation nec stands for 'not elsewhere classified'.

LOCATION BY STATE AND REGIONAL CONCENTRATION

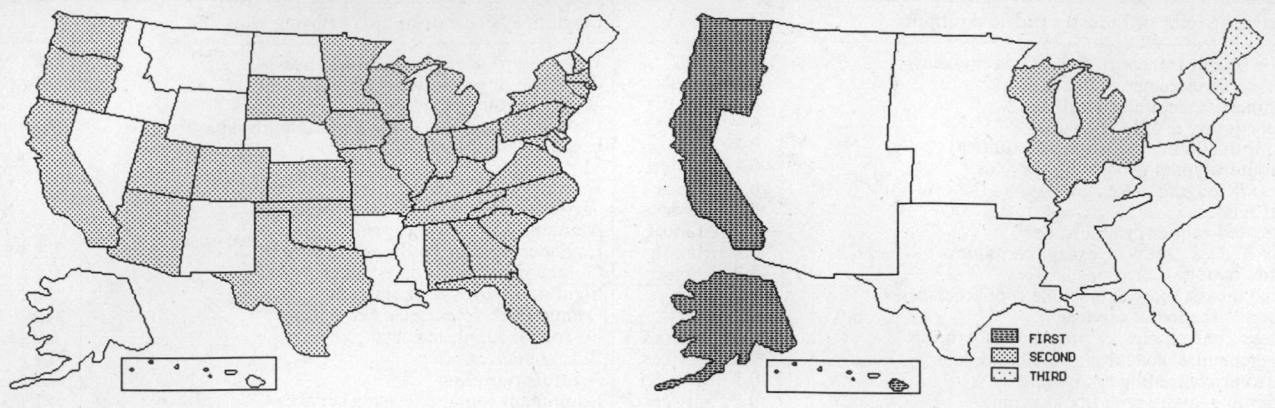

FIRST
SECOND
THIRD

INDUSTRY DATA BY STATE

State	Establish-ments	Shipments			Employment				Cost as % of Shipments	Investment per Employee ($)
		Total ($ mil)	% of U.S.	Per Establ.	Total Number	% of U.S.	Per Establ.	Wages ($/hour)		
New York	60	1,964.7	26.9	32.7	13,700	18.1	228	11.61	33.1	5,212
California	417	1,666.4	22.8	4.0	18,000	23.7	43	10.50	40.7	4,511
Minnesota	47	383.5	5.2	8.2	5,200	6.9	111	10.12	41.5	4,442
Illinois	72	356.0	4.9	4.9	4,200	5.5	58	8.29	41.7	2,643
Massachusetts	64	343.2	4.7	5.4	3,900	5.1	61	10.05	43.5	3,077
Arizona	39	260.9	3.6	6.7	2,800	3.7	72	10.72	36.6	4,500
New Hampshire	31	213.7	2.9	6.9	2,100	2.8	68	10.43	42.1	2,619
Florida	59	187.3	2.6	3.2	2,400	3.2	41	9.02	48.0	3,208
Connecticut	34	148.3	2.0	4.4	1,500	2.0	44	10.96	34.1	2,267
Texas	83	137.7	1.9	1.7	2,000	2.6	24	9.55	43.1	1,950
Alabama	11	103.7	1.4	9.4	1,500	2.0	136	11.00	52.3	-
Pennsylvania	37	103.7	1.4	2.8	1,200	1.6	32	7.29	44.8	2,083
Ohio	41	95.7	1.3	2.3	1,300	1.7	32	9.45	39.0	3,231
Oregon	22	92.7	1.3	4.2	1,200	1.6	55	9.08	38.4	5,583
Colorado	37	86.7	1.2	2.3	1,200	1.6	32	9.11	49.1	7,667
Washington	21	76.7	1.0	3.7	900	1.2	43	11.25	36.9	1,889
Georgia	20	71.7	1.0	3.6	800	1.1	40	8.50	51.9	6,875
New Jersey	30	68.9	0.9	2.3	800	1.1	27	10.00	42.2	4,125
Michigan	27	68.1	0.9	2.5	900	1.2	33	7.77	45.5	1,667
Wisconsin	16	64.3	0.9	4.0	1,000	1.3	63	7.71	41.7	4,200
North Carolina	17	39.2	0.5	2.3	500	0.7	29	8.86	38.0	4,000
Indiana	15	32.7	0.4	2.2	500	0.7	33	8.29	41.6	2,000
Maryland	15	28.4	0.4	1.9	400	0.5	27	10.60	29.9	4,500
Kansas	9	8.2	0.1	0.9	200	0.3	22	7.00	42.7	500
Utah	18	(D)	-	-	750 *	1.0	42	-	-	4,267
Virginia	13	(D)	-	-	3,750 *	4.9	288	-	-	-
Kentucky	8	(D)	-	-	175 *	0.2	22	-	-	571
Missouri	7	(D)	-	-	750 *	1.0	107	-	-	-
Oklahoma	7	(D)	-	-	175 *	0.2	25	-	-	571
Tennessee	7	(D)	-	-	750 *	1.0	107	-	-	2,800
Iowa	5	(D)	-	-	175 *	0.2	35	-	-	-
South Carolina	5	(D)	-	-	1,750 *	2.3	350	-	-	-
Rhode Island	4	(D)	-	-	175 *	0.2	44	-	-	-
South Dakota	2	(D)	-	-	175 *	0.2	88	-	-	-

Source: 1992 *Economic Census.* The states are in descending order of shipments or establishments (if shipment data are missing for the majority). The symbol (D) appears when data are withheld to prevent disclosure of competitive information. States marked with (D) are sorted by number of establishments. A dash (-) indicates that the data element cannot be calculated; * indicates the midpoint of a range.

3674 - SEMICONDUCTORS & RELATED DEVICES

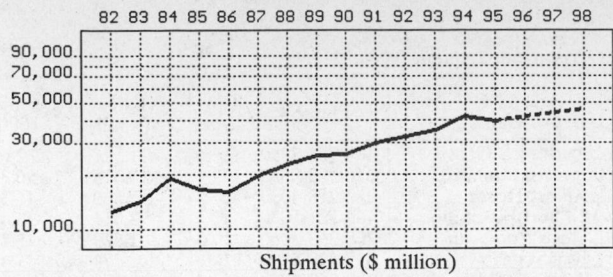

Shipments ($ million)

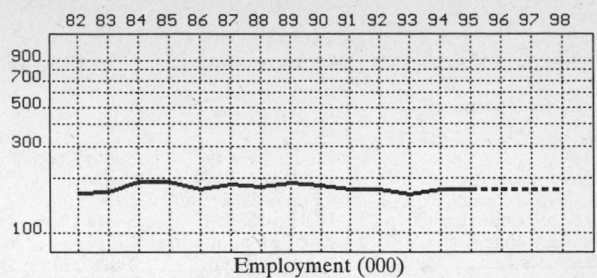

Employment (000)

GENERAL STATISTICS

| Year | Com-panies | Establishments | | Employment | | | Compensation | | Production ($ million) | | | |
		Total	with 20 or more employees	Total (000)	Production Workers (000)	Hours (Mil)	Payroll ($ mil)	Wages ($/hr)	Cost of Materials	Value Added by Manufacture	Value of Shipments	Capital Invest.
1982	685	766	352	166.5	81.3	154.4	3,785.0	8.51	4,196.0	8,356.6	12,429.9	1,723.8
1983		774	373	169.3	84.1	161.7	4,287.1	9.06	4,640.7	9,991.3	14,339.4	1,831.6
1984		782	394	192.3	96.1	186.6	5,216.0	9.85	6,039.5	13,414.3	19,134.5	2,817.6
1985		791	415	190.4	91.8	177.5	5,016.3	10.35	5,427.6	10,872.1	16,487.3	2,831.7
1986		804	438	172.9	79.2	155.4	4,967.1	10.86	4,784.5	10,910.9	15,785.0	2,220.2
1987	757	853	437	184.6	87.4	175.4	5,494.8	10.57	6,462.5	13,429.3	19,794.9	1,920.8
1988		831	458	179.4	86.5	170.7	5,899.4	11.38	7,248.8	15,694.1	22,596.6	
1989		837	457	188.8	90.5	176.0	6,314.1	12.02	7,956.3	17,819.8	25,707.7	3,132.0
1990		858	457	181.3	87.7	174.9	6,532.4	12.58	8,197.3	17,855.5	25,977.3	3,439.3
1991		901	449	175.0	86.2	177.4	6,490.8	12.69	9,197.7	20,151.9	29,668.1	2,945.0
1992	823	921	438	171.9	84.7	172.2	6,879.8	13.55	9,823.3	22,299.7	32,157.0	3,118.0
1993		930	436	162.5	82.2	167.2	6,770.4	14.08	8,937.5	26,465.2	35,151.5	3,838.5
1994		937P	475P	173.6	89.1	177.5	7,464.2	14.48	10,388.9	32,083.6	42,252.1	5,697.8
1995		952P	483P	173.8P	86.5P	177.0P	7,682.9P	14.91P	9,682.8P	29,902.9P	39,380.2P	
1996		967P	490P	173.3P	86.5P	177.8P	7,955.0P	15.39P	10,224.4P	31,575.6P	41,583.1P	
1997		983P	498P	172.7P	86.4P	178.6P	8,227.1P	15.87P	10,766.1P	33,248.3P	43,786.0P	
1998		998P	505P	172.2P	86.4P	179.4P	8,499.2P	16.35P	11,307.7P	34,921.1P	45,988.9P	

Sources: 1982, 1987, 1992 *Economic Census*; *Annual Survey of Manufactures*, 83-86, 88-91, 93-94. Establishment counts for non-Census years are from *County Business Patterns*; establishment values for 83-84 are extrapolations. 'P's show projections by the editors. Industries reclassified in 87 will not have data for prior years.

INDICES OF CHANGE

| Year | Com-panies | Establishments | | Employment | | | Compensation | | Production ($ million) | | | |
		Total	with 20 or more employees	Total (000)	Production Workers (000)	Hours (Mil)	Payroll ($ mil)	Wages ($/hr)	Cost of Materials	Value Added by Manufacture	Value of Shipments	Capital Invest.
1982	83	83	80	97	96	90	55	63	43	37	39	55
1983		84	85	98	99	94	62	67	47	45	45	59
1984		85	90	112	113	108	76	73	61	60	60	90
1985		86	95	111	108	103	73	76	55	49	51	91
1986		87	100	101	94	90	72	80	49	49	49	71
1987	92	93	100	107	103	102	80	78	66	60	62	62
1988		90	105	104	102	99	86	84	74	70	70	
1989		91	104	110	107	102	92	89	81	80	80	100
1990		93	104	105	104	102	95	93	83	80	81	110
1991		98	103	102	102	103	94	94	94	90	92	94
1992	100	100	100	100	100	100	100	100	100	100	100	100
1993		101	100	95	97	97	98	104	91	119	109	123
1994		102P	108P	101	105	103	108	107	106	144	131	183
1995		103P	110P	101P	102P	103P	112P	110P	99P	134P	122P	
1996		105P	112P	101P	102P	103P	116P	114P	104P	142P	129P	
1997		107P	114P	100P	102P	104P	120P	117P	110P	149P	136P	
1998		108P	115P	100P	102P	104P	124P	121P	115P	157P	143P	

Sources: Same as General Statistics. Values reflect change from the base year, 1992. Values above 100 mean greater than 92, values below 100 mean less than 92, and a value of 100 in the 82-91 or 93-98 period means same as 92. 'P's mark projections by the editors.

SELECTED RATIOS

For 1994	Avg. of All Manufact.	Analyzed Industry	Index	For 1994	Avg. of All Manufact.	Analyzed Industry	Index
Employees per Establishment	49	185	378	Value Added per Production Worker	134,084	360,085	269
Payroll per Establishment	1,500,273	7,968,639	531	Cost per Establishment	5,045,178	11,090,993	220
Payroll per Employee	30,620	42,997	140	Cost per Employee	102,970	59,844	58
Production Workers per Establishment	34	95	277	Cost per Production Worker	146,988	116,598	79
Wages per Establishment	853,319	2,743,897	322	Shipments per Establishment	9,576,895	45,107,544	471
Wages per Production Worker	24,861	28,846	116	Shipments per Employee	195,460	243,388	125
Hours per Production Worker	2,056	1,992	97	Shipments per Production Worker	279,017	474,210	170
Wages per Hour	12.09	14.48	120	Investment per Establishment	321,011	6,082,864	1,895
Value Added per Establishment	4,602,255	34,251,846	744	Investment per Employee	6,552	32,821	501
Value Added per Employee	93,930	184,813	197	Investment per Production Worker	9,352	63,948	684

Sources: Same as General Statistics. The 'Average of All Manufacturing' column represents the average of all manufacturing industries reported for the most recent complete year available. The Index shows the relationship between the Average and the Analyzed Industry. For example, 100 means that they are equal; 500 that the Analyzed Industry is five times the average; 50 means that the Analyzed Industry is half the national average. The abbreviation 'na' is used to show that data are 'not available'.

LEADING COMPANIES Number shown: **75**   Total sales ($ mil): **45,485**   Total employment (000): **237.9**

Company Name	Address				CEO Name	Phone	Co. Type	Sales ($ mil)	Empl. (000)
Intel Corp	PO Box 58119	Santa Clara	CA	95052	Andrew S Grove	408-765-8080	P	11,521	32.6
Texas Instruments Inc	PO Box 655474	Dallas	TX	75265	Jerry R Junkins	214-995-3333	P	10,315	56.3
Siemens Corp	1301 of Americas	New York	NY	10019	Albert Hoser	212-258-4000	S	7,300	46.3
National Semiconductor Corp	PO Box 58090	Santa Clara	CA	95020	Gilbert F Amelio	408-721-5000	P	2,014	23.4
Micron Technology Inc	2805 E Columbia	Boise	ID	83706	Steve Appleton	208-368-4000	P	1,629	5.4
Analog Devices Inc	PO Box 9106	Norwood	MA	02062	Ray Stata	617-329-4700	P	774	5.4
Micron Semiconductor Inc	2805 E Columbia	Boise	ID	83706	Steve Appleton	208-368-4000	S	670•	4.0
LSI Logic Corp	1551 McCarthy Blv	Milpitas	CA	95035	Wilfred J Corrigan	408-433-8000	P	655	4.4
Harris Corp	PO Box 883	Melbourne	FL	32901	John C Garrett	407-724-7000	D	638	8.5
VLSI Technology Inc	1109 McKay Dr	San Jose	CA	95131	Alfred J Stein	408-434-3000	P	587	2.7
Cirrus Logic Inc	3100 W Warren Av	Fremont	CA	94538	M L Hackworth	510-623-8300	P	544	1.8
Fujitsu Microelectronics Inc	3545 N 1st St	San Jose	CA	95134	Ken Katashiba	408-922-9000	S	544	1.1
Amkor Electronics Inc	1345 Enterprise Dr	West Chester	PA	19380	James J Kim	215-431-9600	R	500	0.2
Atmel Corp	2125 O'Nel Dr	San Jose	CA	95131	George Perlegos	408-441-0311	P	375	1.2
Xilinx Inc	2100 Logic Dr	San Jose	CA	95124	B V Vonderschmitt	408-559-7778	P	355	0.9
M/A-COM Inc	401 Edgewater Pl	Wakefield	MA	01880	Allan L Rayfield	617-224-5600	P	342	3.9
Integrated Device Technology	PO Box 58015	Santa Clara	CA	95052	Leonard C Perham	408-727-6116	P	331	2.6
General Instrument Corp	10 Melville Park Rd	Melville	NY	11747	Ronald A Ostertag	516-847-3000	D	316	3.3
International Rectifier Corp	233 Kansas St	El Segundo	CA	90245	Eric Lidow	310-322-3331	P	282	3.0
AT & T Global Info Solutions	2001 Danfield Ct	Fort Collins	CO	80525	Gene Patterson	303-223-5100	D	275	1.4
Group Technologies Corp	10901 McKinley	Tampa	FL	33612	Carl P McCormick	813-972-6426	R	250	2.2
Cyrix Corp	2703 N Central	Richardson	TX	75080	Gerald D Rogers	214-994-8388	P	246	0.3
Zilog Inc	210 E Hacienda Av	Campbell	CA	95008	Edgar A Sack	408-370-8000	P	223	1.4
Linear Technology Corp	1630 McCarthy Blv	Milpitas	CA	95035	R H Swanson Jr	408-432-1900	P	201	1.0
AEG Capital Corp	180 Mount Airy Rd	Basking Ridge	NJ	07920	Peter Westrick	908-722-9800	S	197	1.2
Siliconix Inc	PO Box 54951	Santa Clara	CA	95056	Richard J Kulle	408-988-8000	P	197	1.2
Burr-Brown Corp	PO Box 11400	Tucson	AZ	85734	Syrus P Madavi	602-746-1111	P	194	1.8
Sumitomo Sitix Silicon Inc	537 Grandin Rd	Maineville	OH	45039	George Renfeldt	513-583-2600	S	180	0.5
American Microsystems Inc	2300 Buckskin Rd	Pocatello	ID	83201	C W Wredberg	208-233-4690	S	171	1.3
Exar Corp	PO Box 49007	San Jose	CA	95161	George D Wells	408-434-6400	P	161	0.5
Viking Components	11 Columbia	Laguna Hills	CA	92656	Glenn McCusker	714-643-7255	R	160	<0.1
Maxim Integrated Products Inc	120 San Gabriel Dr	Sunnyvale	CA	94086	John F Gifford	408-737-7600	P	154	1.0
Omron Electronics Inc	1 E Commerce Dr	Schaumburg	IL	60173	Noboru Sano	708-843-7900	S	150	0.2
OPTi Inc	2525 Walsh Av	Santa Clara	CA	95051	Jerry Chang	408-980-8178	P	134	0.2
Lattice Semiconductor Corp	5555 NE Moore Ct	Hillsboro	OR	97124	Cyrus Y Tsui	503-681-0118	P	126	0.4
Wacker Siltronic Corp	PO Box 83180	Portland	OR	97283	Jim Ellis	503-243-2020	S	120•	1.0
Microsemi Corp	PO Box 26890	Santa Ana	CA	92799	Philip Frey Jr	714-979-8220	P	119	2.0
Elitegroup Computer Systems	45225 Northport Ct	Fremont	CA	94538	Ben Chan	510-226-7333	R	111	0.1
Brooktree Corp	9868 Scranton Rd	San Diego	CA	92121	James Bixby	619-452-7580	P	109	0.6
Sierra Semiconductor Corp	2075 N Capitol Av	San Jose	CA	95132	James V Diller	408-263-9300	P	109	0.5
Xicor Inc	1511 Buckeye Dr	Milpitas	CA	95035	Raphael Klein	408-432-8888	P	103	0.9
SVG Lithography Systems Inc	77 Danbury Rd	Wilton	CT	06897	R J Richardson	203-761-4000	S	87•	0.6
Three-Five Systems Inc	1600 N Desert Dr	Phoenix	AZ	85281	David R Buchanan	602-496-0035	P	86	<0.1
Micro-Rel	2343 W 10th Pl	Tempe	AZ	85281	William Murray	602-968-6411	D	85	0.9
PMI	1500 Space Park Dr	Santa Clara	CA	95054	Jock Ochiltree	408-727-9222	D	85•	0.6
Tseng Labs Inc	6 Terry Dr	Newtown	PA	18940	Jack Tseng	215-968-0502	P	81	<0.1
Kelly Micro Systems Inc	25 Musick	Irvine	CA	92718	Rick Scherle	714-859-3900	S	80	<0.1
Mitsubishi Chemical America	401 Volvo Pkwy	Chesapeake	VA	23320	T Imai	804-547-5050	D	80	0.3
Dataram Corp	PO Box 7528	Princeton	NJ	08543	Robert V Tarantino	609-799-0071	P	80	0.2
Curtis Mathes Holding Corp	2855 Marquis Dr	Garland	TX	75042	Patrick A Custer	214-494-6411	P	79	<0.1
Altera Corp	2610 Orchard Pkwy	San Jose	CA	95134	Rodney Smith	408-894-7000	P	78	0.7
Cherry Semiconductor Corp	2000 S C Trail	East Greenwich	RI	02818	AS Budnick	401-885-3600	S	77	0.9
Actel Corp	955 E Arques Av	Sunnyvale	CA	94086	John C East	408-739-1010	P	76	0.2
Powerex Inc	Hillis St	Youngwood	PA	15697	SR Hunt	412-925-7272	R	74	0.7
Chips and Technologies Inc	2950 Zanker Rd	San Jose	CA	95134	James F Stafford	408-434-0600	P	73	0.2
Alpha Industries Inc	20 Sylvan Rd	Woburn	MA	01801	Martin J Reid	617-935-5150	P	70	0.8
Integrated Silicon Solution Inc	680 Almanor Dr	Sunnyvale	CA	94086	Jimmy SM Lee	408-733-4774	P	61	0.2
ILC Industries Inc	105 Wilbur Pl	Bohemia	NY	11716	Cliff Lane	516-567-5600	R	60	0.6
Aurora Electronics Inc	2030 Main St	Irvine	CA	92714	James Cowart	714-660-1232	P	58	0.3
Interpoint Corp	PO Box 97005	Redmond	WA	98073	Peter H van Oppen	206-882-3100	P	55	0.6
Alliance Semiconductor Corp	1930 Zanker Rd	San Jose	CA	95112	N Damodar Reddy	408-383-4900	P	55	<0.1
Catalyst Semiconductor Inc	2231 Calle de Luna	Santa Clara	CA	95054	C Michael Powell	408-748-7700	P	54	<0.1
Zing Technologies Inc	115 Stevens Av	Valhalla	NY	10595	Robert E Schrader	914-747-7474	P	53	0.1
Semi-Alloys Co	888 S Columbus	Mount Vernon	NY	10550	David Lloyd	914-664-2800	R	50	0.3
IMP Inc	2830 N 1st St	San Jose	CA	95134	Barry M Carrington	408-432-9100	P	48	0.4
Cypress Semiconductor	17 Cypress Blv	Round Rock	TX	78664	Frank Digesualdo	512-244-7789	S	47•	0.3
Level One Communications Inc	9750 Goethe Rd	Sacramento	CA	95827	Robert S Pepper	916-855-5000	P	47	0.2
Applied Micro Circuits Corp	6195 Lusk Blv	San Diego	CA	92121	Albert Martinez	619-450-9333	R	46	0.3
Orbit Semiconductor Inc	1215 Bordeaux Dr	Sunnyvale	CA	94089	Gary P Kennedy	408-744-1800	P	43	0.2
Donnelly Applied Films Corp	6797 Winchester Cir	Boulder	CO	80301	Cecil Van Alsburg	303-530-1411	J	43	0.3
Oak Technology Inc	139 Kifer Ct	Sunnyvale	CA	94086	David D Tsang	408-737-0888	P	43	0.2
Nartron Corp	5000 N US Hwy 131	Reed City	MI	49677	Norman Rautiola	616-832-5525	P	40	0.2
Linfinity Microelectronics Inc	11861 Western Av	Garden Grove	CA	92641	Brad P Whitney	714-898-8121	S	39	0.2
Tosoh SMD Inc	3600 Gantz Rd	Grove City	OH	43123	Raymond Kidner	614-875-7912	S	38	0.2
Micro-C	9477 Waples St	San Diego	CA	92121	Robert Allison	619-552-1213	D	36•	0.2

Source: Ward's Business Directory of U.S. Private and Public Companies, Volumes 1 and 2, 1996. The company type code used is as follows: P - Public, R - Private, S - Subsidiary, D - Division, J - Joint Venture, A - Affiliate, G - Group. Sales are in millions of dollars, employees are in thousands. An asterisk (•) indicates an estimated sales volume. The symbol < stands for 'less than'. Company names and addresses are truncated, in some cases, to fit into the available space.

MATERIALS CONSUMED

Material	Quantity	Delivered Cost ($ million)
Materials, ingredients, containers, and supplies	(X)	6,687.2
Tube blanks	(X)	2.5
Printed circuit boards (without inserted components) for electronic circuitry	(X)	44.5
Printed circuit assemblies, loaded boards or modules	(X)	98.0
Semiconductors, including transistors, diodes, rectifiers, and integrated circuits for electronic circuitry	(X)	1,341.8
Capacitors for electronic circuitry	(X)	15.9
Resistors for electronic circuitry	(X)	4.6
Other components and accessories for electronic circuitry, nec, except tubes	(X)	195.7
Silicon, hyperpure	(X)	675.7
Gold and other precious metals, all forms (including ingot, sheet, strip, solder, plating, electrodes, etc.)	(X)	63.7
Doped chemicals, and other doped materials for electronic use	(X)	133.9
Ferrites (powder and paste)	(X)	(D)
Metal powders	(X)	6.4
Electronic computing equipment	(X)	38.6
Current-carrying wiring devices	(X)	6.5
Electronic communication equipment	(X)	7.8
Electrical instrument mechanisms and meter movements (including instrument relays)	(X)	(D)
Optical instruments and lenses (except sighting, tracking, and fire control)	(X)	(D)
Plastics resins consumed in the form of granules, pellets, powders, liquids, etc.	(X)	29.8
Fabricated plastics products (except gaskets, hoses, and belting)	(X)	30.7
Sheet metal products, except stampings	(X)	10.8
Metal stampings	(X)	49.9
Metal bolts, nuts, screws, washers, rivets, and other screw machine products	(X)	3.0
Other fabricated metal products (except forgings)	(X)	41.6
Forgings	(X)	(D)
Castings (rough and semifinished)	(X)	1.0
Steel shapes and forms	(X)	4.4
Copper and copper-base alloy shapes and forms	(X)	3.4
Aluminum and aluminum-base alloy shapes and forms	(X)	11.4
Other nonferrous shapes and forms	(X)	7.7
Insulated wire and cable, including magnet wire	(X)	6.3
Paper and paperboard containers, including shipping sacks and other paper packaging supplies	(X)	44.4
All other materials and components, parts, containers, and supplies	(X)	1,508.4
Materials, ingredients, containers, and supplies, nsk	(X)	2,233.7

Source: 1992 *Economic Census*. Explanation of symbols used: (D): Withheld to avoid disclosure of competitive data; na: Not available; (S): Withheld because statistical norms were not met; (X): Not applicable; (Z): Less than half the unit shown; nec: Not elsewhere classified; nsk: Not specified by kind; - : zero; * : 10-19 percent estimated; ** : 20-29 percent estimated.

PRODUCT SHARE DETAILS

Product or Product Class	% Share	Product or Product Class	% Share
Semiconductors and related devices	100.00	Diodes and rectifiers	2.43
Integrated microcircuits (including semiconductor networks, microprocessors, and MOS memories)	70.68	Other semiconductor devices (including semiconductor parts such as chips, wafers, and heat sinks)	20.71
Transistors	2.42	Semiconductors and related devices, nsk	3.76

Source: 1992 *Economic Census*. The values shown are percent of total shipments in an industry. Values of indented subcategories are summed in the main heading. The symbol (D) appears when data are withheld to prevent disclosure of competitive information. The abbreviation nsk stands for 'not specified by kind' and nec for 'not elsewhere classified'.

INPUTS AND OUTPUTS FOR SEMICONDUCTORS & RELATED DEVICES

Economic Sector or Industry Providing Inputs	%	Sector	Economic Sector or Industry Buying Outputs	%	Sector
Imports	31.9	Foreign	Exports	21.5	Foreign
Semiconductors & related devices	23.6	Manufg.	Semiconductors & related devices	20.9	Manufg.
Wholesale trade	6.5	Trade	Electronic computing equipment	7.0	Manufg.
Nonferrous rolling & drawing, nec	5.0	Manufg.	Photographic equipment & supplies	6.6	Manufg.
Cyclic crudes and organics	4.2	Manufg.	Federal Government purchases, national defense	6.3	Fed Govt
Electronic components nec	3.0	Manufg.	Radio & TV communication equipment	5.7	Manufg.
Nonferrous wire drawing & insulating	2.6	Manufg.	Electrical repair shops	4.6	Services
Electric services (utilities)	1.8	Util.	Communications, except radio & TV	3.8	Util.
Plating & polishing	1.7	Manufg.	Aircraft & missile equipment, nec	2.5	Manufg.
Banking	1.4	Fin/R.E.	Management & consulting services & labs	2.3	Services
Miscellaneous plastics products	1.3	Manufg.	Telephone & telegraph apparatus	2.1	Manufg.
Communications, except radio & TV	1.2	Util.	Electronic components nec	1.5	Manufg.
Primary nonferrous metals, nec	1.1	Manufg.	Computer & data processing services	1.3	Services
Chemical preparations, nec	1.0	Manufg.	Aircraft	1.1	Manufg.
Industrial inorganic chemicals, nec	0.8	Manufg.	Radio & TV receiving sets	1.1	Manufg.
Metal stampings, nec	0.8	Manufg.	Instruments to measure electricity	1.0	Manufg.
Air transportation	0.8	Util.	Motor vehicles & car bodies	1.0	Manufg.
Advertising	0.8	Services	Business services nec	0.9	Services
Maintenance of nonfarm buildings nec	0.6	Constr.	Electrical equipment & supplies, nec	0.6	Manufg.
Noncomparable imports	0.6	Foreign	Change in business inventories	0.6	In House

Continued on next page.

INPUTS AND OUTPUTS FOR SEMICONDUCTORS & RELATED DEVICES - Continued

Economic Sector or Industry Providing Inputs	%	Sector	Economic Sector or Industry Buying Outputs	%	Sector
Eating & drinking places	0.5	Trade	Calculating & accounting machines	0.5	Manufg.
Real estate	0.5	Fin/R.E.	Mechanical measuring devices	0.5	Manufg.
Engineering, architectural, & surveying services	0.5	Services	Small arms	0.5	Manufg.
Equipment rental & leasing services	0.5	Services	X-ray apparatus & tubes	0.5	Manufg.
Carbon & graphite products	0.4	Manufg.	Optical instruments & lenses	0.4	Manufg.
Motor freight transportation & warehousing	0.4	Util.	Typewriters & office machines, nec	0.4	Manufg.
Business services nec	0.4	Services	Detective & protective services	0.4	Services
Machinery, except electrical, nec	0.3	Manufg.	Miscellaneous repair shops	0.4	Services
Metal coating & allied services	0.3	Manufg.	Photofinishing labs, commercial photography	0.4	Services
Gas production & distribution (utilities)	0.3	Util.	Engine electrical equipment	0.3	Manufg.
Computer & data processing services	0.3	Services	Motor vehicle parts & accessories	0.3	Manufg.
Hotels & lodging places	0.3	Services	Watches, clocks, & parts	0.3	Manufg.
Legal services	0.3	Services	Services to dwellings & other buildings	0.3	Services
Fabricated rubber products, nec	0.2	Manufg.	Ammunition, except for small arms, nec	0.2	Manufg.
Glass & glass products, except containers	0.2	Manufg.	Engineering & scientific instruments	0.2	Manufg.
Plastics materials & resins	0.2	Manufg.	Guided missiles & space vehicles	0.2	Manufg.
Special dies & tools & machine tool accessories	0.2	Manufg.	Lighting fixtures & equipment	0.2	Manufg.
Railroads & related services	0.2	Util.	Ordnance & accessories nec	0.2	Manufg.
Colleges, universities, & professional schools	0.2	Services	Personnel supply services	0.2	Services
Management & consulting services & labs	0.2	Services	Wiring devices	0.1	Manufg.
Electronic computing equipment	0.1	Manufg.	Air transportation	0.1	Util.
Accounting, auditing & bookkeeping	0.1	Services	Equipment rental & leasing services	0.1	Services
Business/professional associations	0.1	Services			

Source: Benchmark Input-Output Accounts for the U.S. Economy, 1982, U.S. Department of Commerce, Washington, D.C., July 1991. Data, as reported in the source, are organized by the 1977 SIC structure in use in 1982 but have been matched, as closely as is possible, to the 1987 SIC structure used in this book.

OCCUPATIONS EMPLOYED BY SIC 367 - ELECTRONIC COMPONENTS AND ACCESSORIES

Occupation	% of Total 1994	Change to 2005	Occupation	% of Total 1994	Change to 2005
Electrical & electronic assemblers	9.8	-8.4	Secretaries, ex legal & medical	1.6	-7.4
Electrical & electronic equipment assemblers	5.7	1.8	Industrial production managers	1.5	-28.8
Electrical & electronic technicians,technologists	5.4	22.1	Electromechanical equipment assemblers	1.4	11.9
Inspectors, testers, & graders, precision	5.4	-28.8	Coil winders, tapers, & finishers	1.3	-18.6
Electronic semiconductor processors	5.3	-1.4	Production, planning, & expediting clerks	1.3	1.8
Electrical & electronics engineers	4.4	29.8	Industrial engineers, ex safety engineers	1.3	12.0
Assemblers, fabricators, & hand workers nec	4.0	1.8	Electrolytic plating machine workers	1.2	11.9
Blue collar worker supervisors	3.4	-14.1	Traffic, shipping, & receiving clerks	1.2	-2.1
Sales & related workers nec	2.2	1.8	Freight, stock, & material movers, hand	1.2	-18.6
Computer engineers	2.2	50.7	Precision assemblers nec	1.1	62.9
Engineering, mathematical, & science managers	1.8	15.6	Stock clerks	1.0	-17.3
Engineering technicians & technologists nec	1.8	1.8	Engineers nec	1.0	52.6
General managers & top executives	1.6	-3.4	General office clerks	1.0	-13.2

Source: Industry-Occupation Matrix, Bureau of Labor Statistics. These data relate to one or more 3-digit SIC industry groups rather than to a single 4-digit SIC. The change reported for each occupation to the year 2005 is a percent of growth or decline as estimated by the Bureau of Labor Statistics. The abbreviation nec stands for 'not elsewhere classified'.

LOCATION BY STATE AND REGIONAL CONCENTRATION

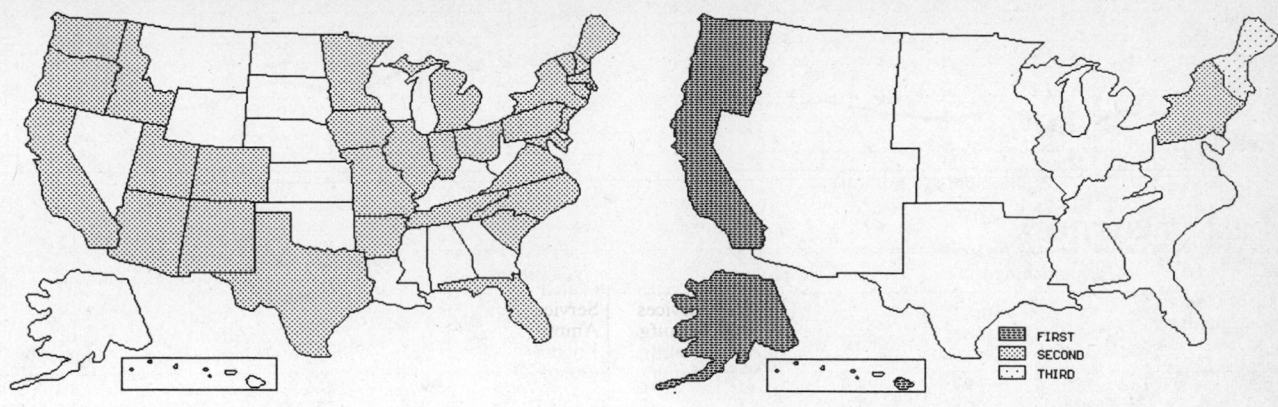

FIRST
SECOND
THIRD

INDUSTRY DATA BY STATE

| State | Establish-ments | Shipments | | | Employment | | | | Cost as % of Shipments | Investment per Employee ($) |
		Total ($ mil)	% of U.S.	Per Establ.	Total Number	% of U.S.	Per Establ.	Wages ($/hour)		
California	343	9,082.5	28.2	26.5	50,600	29.4	148	14.24	28.6	12,937
Texas	60	5,183.6	16.1	86.4	25,300	14.7	422	13.18	20.7	21,542
Arizona	37	2,558.2	8.0	69.1	16,700	9.7	451	10.68	13.0	11,647
Pennsylvania	50	1,995.2	6.2	39.9	8,100	4.7	162	19.48	23.8	-
Oregon	22	1,598.8	5.0	72.7	7,900	4.6	359	13.00	25.5	22,367
Massachusetts	76	1,230.3	3.8	16.2	11,200	6.5	147	13.16	45.4	9,821
Idaho	5	581.4	1.8	116.3	6,200	3.6	1,240	15.65	32.8	14,371
Colorado	20	423.3	1.3	21.2	3,200	1.9	160	13.35	31.7	18,281
Florida	21	356.2	1.1	17.0	3,300	1.9	157	12.47	35.4	-
Washington	13	298.7	0.9	23.0	2,200	1.3	169	12.47	40.4	17,591
North Carolina	13	251.0	0.8	19.3	1,400	0.8	108	9.11	55.3	-
Minnesota	15	152.5	0.5	10.2	1,400	0.8	93	12.73	21.7	16,929
New Jersey	40	136.5	0.4	3.4	1,400	0.8	35	10.65	30.6	10,143
New Hampshire	8	128.1	0.4	16.0	1,000	0.6	125	10.54	32.4	16,800
Illinois	25	89.5	0.3	3.6	500	0.3	20	6.83	62.8	15,400
Connecticut	14	37.4	0.1	2.7	400	0.2	29	13.60	31.0	-
Maryland	7	35.6	0.1	5.1	400	0.2	57	12.00	42.7	15,250
Michigan	14	30.6	0.1	2.2	300	0.2	21	9.60	35.9	8,667
Virginia	13	10.9	0.0	0.8	100	0.1	8	14.00	36.7	10,000
Iowa	5	6.0	0.0	1.2	100	0.1	20	9.00	33.3	-
New York	41	(D)	-	-	17,500 *	10.2	427	-	-	-
Ohio	10	(D)	-	-	1,750 *	1.0	175	-	-	3,943
New Mexico	8	(D)	-	-	3,750 *	2.2	469	-	-	-
Missouri	7	(D)	-	-	1,750 *	1.0	250	-	-	-
Maine	5	(D)	-	-	1,750 *	1.0	350	-	-	-
Vermont	5	(D)	-	-	7,500 *	4.4	1,500	-	-	-
Indiana	4	(D)	-	-	375 *	0.2	94	-	-	-
Tennessee	4	(D)	-	-	175 *	0.1	44	-	-	-
Rhode Island	3	(D)	-	-	375 *	0.2	125	-	-	-
Utah	3	(D)	-	-	1,750 *	1.0	583	-	-	-
Arkansas	2	(D)	-	-	175 *	0.1	88	-	-	-
South Carolina	2	(D)	-	-	750 *	0.4	375	-	-	-

Source: 1992 *Economic Census*. The states are in descending order of shipments or establishments (if shipment data are missing for the majority). The symbol (D) appears when data are withheld to prevent disclosure of competitive information. States marked with (D) are sorted by number of establishments. A dash (-) indicates that the data element cannot be calculated; * indicates the midpoint of a range.

3675 - ELECTRONIC CAPACITORS

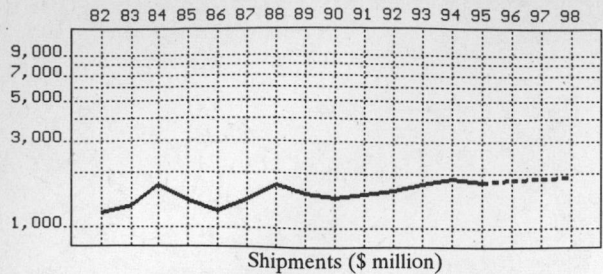

Shipments ($ million)

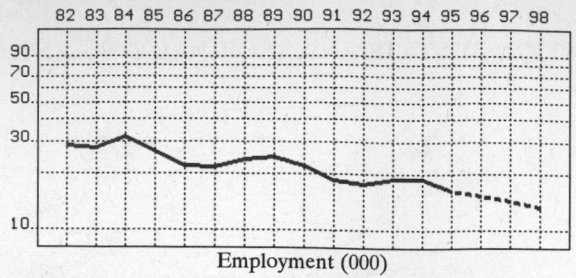

Employment (000)

GENERAL STATISTICS

Year	Companies	Establishments		Employment			Compensation		Production ($ million)			
		Total	with 20 or more employees	Total (000)	Production Workers (000)	Hours (Mil)	Payroll ($ mil)	Wages ($/hr)	Cost of Materials	Value Added by Manufacture	Value of Shipments	Capital Invest.
1982	93	130	99	28.9	21.6	40.7	387.9	5.99	466.9	728.4	1,188.9	72.3
1983		130	101	27.9	21.0	40.5	403.1	6.63	539.0	813.8	1,330.6	68.4
1984		130	103	32.2	25.1	50.4	501.8	6.73	696.5	1,059.0	1,702.0	111.9
1985		130	104	26.5	20.2	38.3	417.9	7.38	536.1	851.6	1,410.0	125.2
1986		119	88	22.1	17.0	33.1	380.0	7.36	483.5	747.2	1,241.4	75.1
1987	116	148	105	21.7	16.4	31.7	392.8	7.86	551.3	892.0	1,440.1	78.2
1988		138	97	24.0	18.1	36.3	465.0	8.17	663.3	1,044.2	1,723.9	95.3
1989		135	101	24.8	15.7	30.6	408.1	8.10	614.9	937.4	1,563.2	55.6
1990		130	92	22.1	14.4	28.2	398.4	8.37	609.8	848.5	1,471.6	52.1
1991		125	94	18.5	13.5	27.2	417.8	8.82	618.6	926.3	1,546.1	57.1
1992	99	117	80	17.9	13.4	26.3	415.7	9.43	703.0	930.0	1,630.1	64.0
1993		115	81	18.7	13.7	27.9	438.7	9.37	788.2	979.0	1,762.8	63.9
1994		123P	85P	18.8	14.2	29.4	474.2	9.99	879.8	1,002.3	1,871.9	78.5
1995		122P	83P	16.3P	11.3P	23.5P	438.7P	10.13P	844.7P	962.3P	1,797.2P	59.4P
1996		121P	81P	15.3P	10.5P	22.0P	440.9P	10.44P	862.7P	982.8P	1,835.5P	57.0P
1997		120P	80P	14.3P	9.7P	20.5P	443.1P	10.74P	880.7P	1,003.3P	1,873.8P	54.5P
1998		119P	78P	13.2P	8.8P	19.0P	445.3P	11.04P	898.7P	1,023.8P	1,912.0P	52.0P

Sources: 1982, 1987, 1992 *Economic Census*; *Annual Survey of Manufactures*, 83-86, 88-91, 93-94. Establishment counts for non-Census years are from *County Business Patterns*; establishment values for 83-84 are extrapolations. 'P's show projections by the editors. Industries reclassified in 87 will not have data for prior years.

INDICES OF CHANGE

Year	Companies	Establishments		Employment			Compensation		Production ($ million)			
		Total	with 20 or more employees	Total (000)	Production Workers (000)	Hours (Mil)	Payroll ($ mil)	Wages ($/hr)	Cost of Materials	Value Added by Manufacture	Value of Shipments	Capital Invest.
1982	94	111	124	161	161	155	93	64	66	78	73	113
1983		111	126	156	157	154	97	70	77	88	82	107
1984		111	129	180	187	192	121	71	99	114	104	175
1985		111	130	148	151	146	101	78	76	92	86	196
1986		102	110	123	127	126	91	78	69	80	76	117
1987	117	126	131	121	122	121	94	83	78	96	88	122
1988		118	121	134	135	138	112	87	94	112	106	149
1989		115	126	139	117	116	98	86	87	101	96	87
1990		111	115	123	107	107	96	89	87	91	90	81
1991		107	117	103	101	103	101	94	88	100	95	89
1992	100	100	100	100	100	100	100	100	100	100	100	100
1993		98	101	104	102	106	106	99	112	105	108	100
1994		105P	106P	105	106	112	114	106	125	108	115	123
1995		104P	104P	91P	85P	89P	106P	107P	120P	103P	110P	93P
1996		103P	101P	85P	78P	84P	106P	111P	123P	106P	113P	89P
1997		102P	99P	80P	72P	78P	107P	114P	125P	108P	115P	85P
1998		101P	97P	74P	66P	72P	107P	117P	128P	110P	117P	81P

Sources: Same as General Statistics. Values reflect change from the base year, 1992. Values above 100 mean greater than 92, values below 100 mean less than 92, and a value of 100 in the 82-91 or 93-98 period means same as 92. 'P's mark projections by the editors.

SELECTED RATIOS

For 1994	Avg. of All Manufact.	Analyzed Industry	Index	For 1994	Avg. of All Manufact.	Analyzed Industry	Index
Employees per Establishment	49	153	313	Value Added per Production Worker	134,084	70,585	53
Payroll per Establishment	1,500,273	3,868,628	258	Cost per Establishment	5,045,178	7,177,602	142
Payroll per Employee	30,620	25,223	82	Cost per Employee	102,970	46,798	45
Production Workers per Establishment	34	116	338	Cost per Production Worker	146,988	61,958	42
Wages per Establishment	853,319	2,396,118	281	Shipments per Establishment	9,576,895	15,271,372	159
Wages per Production Worker	24,861	20,684	83	Shipments per Employee	195,460	99,569	51
Hours per Production Worker	2,056	2,070	101	Shipments per Production Worker	279,017	131,824	47
Wages per Hour	12.09	9.99	83	Investment per Establishment	321,011	640,420	200
Value Added per Establishment	4,602,255	8,176,984	178	Investment per Employee	6,552	4,176	64
Value Added per Employee	93,930	53,314	57	Investment per Production Worker	9,352	5,528	59

Sources: Same as General Statistics. The 'Average of All Manufacturing' column represents the average of all manufacturing industries reported for the most recent complete year available. The Index shows the relationship between the Average and the Analyzed Industry. For example, 100 means that they are equal; 500 that the Analyzed Industry is five times the average; 50 means that the Analyzed Industry is half the national average. The abbreviation 'na' is used to show that data are 'not available'.

LEADING COMPANIES   Number shown: 27   Total sales ($ mil): 1,598   Total employment (000): 23.9

Company Name	Address				CEO Name	Phone	Co. Type	Sales ($ mil)	Empl. (000)
KEMET Corp	2835 Kemet Way	Simpsonville	SC	29681	David E Maguire	803-963-6300	P	473	8.4
Kemet Electronics Corp	2835 Kemet Way	Simpsonville	SC	29681	David E Maguire	803-963-6300	S	385	7.2
General Electric Co	381 Broadway	Fort Edward	NY	12828	Joseph McSweeney	518-746-5561	D	225	2.0
Aerovox Inc	370 Faunce Corner	N Dartmouth	MA	02747	Clifford H Tuttle	508-995-8000	P	126	1.9
Vitramon Inc	PO Box 544	Bridgeport	CT	06601	Donald Alfson	203-268-6261	S	100	1.5
Spectrum Control Inc	6000 W Ridge Rd	Erie	PA	16506	John L Johnston	814-835-4000	P	44	0.6
Elpac Electronics Inc	1562 Reynolds Av	Irvine	CA	92714	FG Benhard	714-476-6070	R	40*	0.1
North American Capacitor Co	7545 Rockville Rd	Indianapolis	IN	46214	Thomas Arnold	317-273-0090	R	35	0.2
Aerovox Aero M Inc	20 Aberdeen Dr	Glasgow	KY	42141	Gerald Challender	502-651-8301	S	25	0.2
Amer Technical Ceramics Corp	17 Stepar Pl	Huntington St	NY	11746	Victor Insetta	516-547-5700	P	23	0.3
Maida Development Co	PO Box 3529	Hampton	VA	23663	Edward T Maida	804-723-0785	R	22	0.4
Cera-Mite Corp	PO Box 166	Grafton	WI	53024	JS Sarnowski	414-377-3500	R	15	0.3
Illinois Capacitor Inc	3757 W Touhy Av	Lincolnwood	IL	60645	CJ Kurland	708-675-1760	R	15	<0.1
Johanson Manufacturing Corp	Rockaway Val Rd	Boonton	NJ	07005	Nancy Johanson	201-334-2676	R	15	0.2
ITW Paktron	PO Box 4539	Lynchburg	VA	24502	Ian Clelland	804-239-6941	D	10	<0.1
Circuit Components Inc	2400 S Roosevelt St	Tempe	AZ	85282	N L Greenman	602-967-0624	R	9	<0.1
Center Engineering Inc	2820 E College Av	State College	PA	16801	James McCrea	814-237-0321	R	8*	0.1
STK Electronics Inc	2747 Rte 20 E	Cazenovia	NY	13035	Peter Kip	315-655-2530	R	6	<0.1
ASC Industries Inc	8967 Pleasantwood	North Canton	OH	44720	Ted Swaldo	216-499-1210	R	5*	<0.1
Custom Electronics Inc	12 Browne St	Oneonta	NY	13820	PS Dokuchitz	607-432-3880	R	5	<0.1
Myron Zucker Inc	315 E Parent St	Royal Oak	MI	48067	William Zobel	313-543-2277	R	3	<0.1
CSI Technologies Inc	810 Rancheros Dr	San Marcos	CA	92069	R Testut	619-747-4000	R	3	<0.1
Plastic Capacitors Inc	2623 N Pulaski Rd	Chicago	IL	60639	CT Pavey	312-489-2229	R	3	<0.1
Radio Materials Corp	PO Box 339	Attica	IN	47918	Joseph F Riley Jr	317-762-2491	R	2	<0.1
Oren Elliott Products	128 W Vine St	Edgerton	OH	43517	Oren Elliott	419-298-2306	R	1	<0.1
S and EI Manufacturing Inc	PO Box 280832	Northridge	CA	91328	KP Ellenberger	818-349-4111	R	1*	<0.1
Arco Electronics Inc	5310-J Derry Av	Agoura Hills	CA	91301	Jerald Aarons	818-707-6465	R	1	<0.1

Source: Ward's Business Directory of U.S. Private and Public Companies, Volumes 1 and 2, 1996. The company type code used is as follows: P - Public, R - Private, S - Subsidiary, D - Division, J - Joint Venture, A - Affiliate, G - Group. Sales are in millions of dollars, employees are in thousands. An asterisk (*) indicates an estimated sales volume. The symbol < stands for 'less than'. Company names and addresses are truncated, in some cases, to fit into the available space.

MATERIALS CONSUMED

Material	Quantity	Delivered Cost ($ million)
Materials, ingredients, containers, and supplies	(X)	472.9
Printed circuit boards (without inserted components) for electronic circuitry	(X)	0.2
Semiconductors, including transistors, diodes, rectifiers, and integrated circuits for electronic circuitry	(X)	0.7
Capacitors for electronic circuitry	(X)	20.2
Other components and accessories for electronic circuitry, nec, except tubes	(X)	0.9
Silicon, hyperpure	(X)	(D)
Gold and other precious metals, all forms (including ingot, sheet, strip, solder, plating, electrodes, etc.)	(X)	43.0
Doped chemicals, and other doped materials for electronic use	(X)	(D)
Ferrites (powder and paste)	(X)	(D)
Metal powders	(X)	(D)
Electronic computing equipment	(X)	(D)
Electronic communication equipment	(X)	(D)
Electrical instrument mechanisms and meter movements (including instrument relays)	(X)	(D)
Optical instruments and lenses (except sighting, tracking, and fire control)	(X)	(D)
Plastics resins consumed in the form of granules, pellets, powders, liquids, etc.	(X)	13.0
Fabricated plastics products (except gaskets, hoses, and belting)	(X)	2.1
Sheet metal products, except stampings	(X)	(D)
Metal stampings	(X)	(D)
Metal bolts, nuts, screws, washers, rivets, and other screw machine products	(X)	3.0
Other fabricated metal products (except forgings)	(X)	3.8
Castings (rough and semifinished)	(X)	(D)
Steel shapes and forms	(X)	(D)
Copper and copper-base alloy shapes and forms	(X)	(D)
Aluminum and aluminum-base alloy shapes and forms	(X)	17.7
Other nonferrous shapes and forms	(X)	(D)
Insulated wire and cable, including magnet wire	(X)	0.8
Paper and paperboard containers, including shipping sacks and other paper packaging supplies	(X)	3.9
All other materials and components, parts, containers, and supplies	(X)	114.3
Materials, ingredients, containers, and supplies, nsk	(X)	151.1

Source: 1992 Economic Census. Explanation of symbols used: (D): Withheld to avoid disclosure of competitive data; na: Not available; (S): Withheld because statistical norms were not met; (X): Not applicable; (Z): Less than half the unit shown; nec: Not elsewhere classified; nsk: Not specified by kind; - : zero; * : 10-19 percent estimated; ** : 20-29 percent estimated.

PRODUCT SHARE DETAILS

Product or Product Class	% Share	Product or Product Class	% Share
Capacitors for electronic circuitry.	100.00		

Source: 1992 *Economic Census*. The values shown are percent of total shipments in an industry. Values of indented subcategories are summed in the main heading. The symbol (D) appears when data are withheld to prevent disclosure of competitive information. The abbreviation nsk stands for 'not specified by kind' and nec for 'not elsewhere classified'.

INPUTS AND OUTPUTS FOR ELECTRONIC COMPONENTS NEC

Economic Sector or Industry Providing Inputs	%	Sector	Economic Sector or Industry Buying Outputs	%	Sector
Miscellaneous plastics products	16.2	Manufg.	Radio & TV communication equipment	13.5	Manufg.
Imports	15.7	Foreign	Exports	9.8	Foreign
Electronic components nec	15.5	Manufg.	Telephone & telegraph apparatus	9.6	Manufg.
Wholesale trade	10.9	Trade	Electronic computing equipment	9.4	Manufg.
Plating & polishing	5.2	Manufg.	Electronic components nec	9.1	Manufg.
Primary nonferrous metals, nec	2.6	Manufg.	Radio & TV receiving sets	5.8	Manufg.
Electric services (utilities)	2.0	Util.	Guided missiles & space vehicles	3.2	Manufg.
Semiconductors & related devices	1.9	Manufg.	Personal consumption expenditures	2.9	
Advertising	1.5	Services	X-ray apparatus & tubes	2.6	Manufg.
Communications, except radio & TV	1.4	Util.	Aircraft	2.2	Manufg.
Metal stampings, nec	1.3	Manufg.	Instruments to measure electricity	2.2	Manufg.
Banking	1.2	Fin/R.E.	Aircraft & missile equipment, nec	2.0	Manufg.
Aluminum rolling & drawing	1.1	Manufg.	Semiconductors & related devices	2.0	Manufg.
Copper rolling & drawing	1.1	Manufg.	Industrial controls	1.9	Manufg.
Nonferrous wire drawing & insulating	1.1	Manufg.	Optical instruments & lenses	1.7	Manufg.
Air transportation	1.1	Util.	Computer & data processing services	1.7	Services
Metal coating & allied services	1.0	Manufg.	Federal Government purchases, national defense	1.7	Fed Govt
Blast furnaces & steel mills	0.9	Manufg.	Games, toys, & children's vehicles	1.6	Manufg.
Maintenance of nonfarm buildings nec	0.8	Constr.	Mechanical measuring devices	1.3	Manufg.
Plastics materials & resins	0.8	Manufg.	Calculating & accounting machines	1.2	Manufg.
Motor freight transportation & warehousing	0.8	Util.	Electron tubes	1.2	Manufg.
Real estate	0.8	Fin/R.E.	Photographic equipment & supplies	1.1	Manufg.
Chemical preparations, nec	0.7	Manufg.	Radio & TV broadcasting	1.0	Util.
Petroleum refining	0.7	Manufg.	Change in business inventories	1.0	In House
Screw machine and related products	0.7	Manufg.	Communications, except radio & TV	0.8	Util.
Eating & drinking places	0.7	Trade	Federal Government purchases, nondefense	0.7	Fed Govt
Equipment rental & leasing services	0.6	Services	Engine electrical equipment	0.6	Manufg.
Hotels & lodging places	0.6	Services	Musical instruments	0.6	Manufg.
Cyclic crudes and organics	0.5	Manufg.	Surgical appliances & supplies	0.6	Manufg.
Electronic computing equipment	0.5	Manufg.	Switchgear & switchboard apparatus	0.6	Manufg.
Paperboard containers & boxes	0.5	Manufg.	Phonograph records & tapes	0.5	Manufg.
Sheet metal work	0.5	Manufg.	Management & consulting services & labs	0.5	Services
Adhesives & sealants	0.4	Manufg.	Engineering & scientific instruments	0.4	Manufg.
Machinery, except electrical, nec	0.4	Manufg.	Surgical & medical instruments	0.4	Manufg.
Nonferrous rolling & drawing, nec	0.4	Manufg.	Banking	0.4	Fin/R.E.
Gas production & distribution (utilities)	0.4	Util.	Environmental controls	0.3	Manufg.
Computer & data processing services	0.4	Services	Typewriters & office machines, nec	0.3	Manufg.
Legal services	0.4	Services	Libraries, vocation education	0.3	Services
Carbon & graphite products	0.3	Manufg.	Aircraft & missile engines & engine parts	0.2	Manufg.
Fabricated rubber products, nec	0.3	Manufg.	Ammunition, except for small arms, nec	0.2	Manufg.
Primary metal products, nec	0.3	Manufg.	Electrical equipment & supplies, nec	0.2	Manufg.
Special dies & tools & machine tool accessories	0.3	Manufg.	Detective & protective services	0.2	Services
Wiring devices	0.3	Manufg.	Elementary & secondary schools	0.2	Services
Management & consulting services & labs	0.3	Services	Gross private fixed investment	0.2	Cap Inv
Noncomparable imports	0.3	Foreign	Electrical industrial apparatus, nec	0.1	Manufg.
Aluminum castings	0.2	Manufg.	Motor vehicle parts & accessories	0.1	Manufg.
Instruments to measure electricity	0.2	Manufg.	Motor vehicles & car bodies	0.1	Manufg.
Miscellaneous fabricated wire products	0.2	Manufg.	Scales & balances	0.1	Manufg.
Sanitary services, steam supply, irrigation	0.2	Util.	S/L Govt. purch., higher education	0.1	S/L Govt
Accounting, auditing & bookkeeping	0.2	Services			
U.S. Postal Service	0.2	Gov't			
Abrasive products	0.1	Manufg.			
Glass & glass products, except containers	0.1	Manufg.			
Industrial inorganic chemicals, nec	0.1	Manufg.			
Lubricating oils & greases	0.1	Manufg.			
Manifold business forms	0.1	Manufg.			
Metal foil & leaf	0.1	Manufg.			
Paper coating & glazing	0.1	Manufg.			
Railroads & related services	0.1	Util.			
Insurance carriers	0.1	Fin/R.E.			
Automotive rental & leasing, without drivers	0.1	Services			
Colleges, universities, & professional schools	0.1	Services			

Source: Benchmark Input-Output Accounts for the U.S. Economy, 1982, U.S. Department of Commerce, Washington, D.C., July 1991. Data, as reported in the source, are organized by the 1977 SIC structure in use in 1982 but have been matched, as closely as is possible, to the 1987 SIC structure used in this book.

OCCUPATIONS EMPLOYED BY SIC 367 - ELECTRONIC COMPONENTS AND ACCESSORIES

Occupation	% of Total 1994	Change to 2005	Occupation	% of Total 1994	Change to 2005
Electrical & electronic assemblers	9.8	-8.4	Secretaries, ex legal & medical	1.6	-7.4
Electrical & electronic equipment assemblers	5.7	1.8	Industrial production managers	1.5	-28.8
Electrical & electronic technicians,technologists	5.4	22.1	Electromechanical equipment assemblers	1.4	11.9
Inspectors, testers, & graders, precision	5.4	-28.8	Coil winders, tapers, & finishers	1.3	-18.6
Electronic semiconductor processors	5.3	-1.4	Production, planning, & expediting clerks	1.3	1.8
Electrical & electronics engineers	4.4	29.8	Industrial engineers, ex safety engineers	1.3	12.0
Assemblers, fabricators, & hand workers nec	4.0	1.8	Electrolytic plating machine workers	1.2	11.9
Blue collar worker supervisors	3.4	-14.1	Traffic, shipping, & receiving clerks	1.2	-2.1
Sales & related workers nec	2.2	1.8	Freight, stock, & material movers, hand	1.2	-18.6
Computer engineers	2.2	50.7	Precision assemblers nec	1.1	62.9
Engineering, mathematical, & science managers	1.8	15.6	Stock clerks	1.0	-17.3
Engineering technicians & technologists nec	1.8	1.8	Engineers nec	1.0	52.6
General managers & top executives	1.6	-3.4	General office clerks	1.0	-13.2

Source: *Industry-Occupation Matrix*, Bureau of Labor Statistics. These data relate to one or more 3-digit SIC industry groups rather than to a single 4-digit SIC. The change reported for each occupation to the year 2005 is a percent of growth or decline as estimated by the Bureau of Labor Statistics. The abbreviation nec stands for 'not elsewhere classified'.

LOCATION BY STATE AND REGIONAL CONCENTRATION

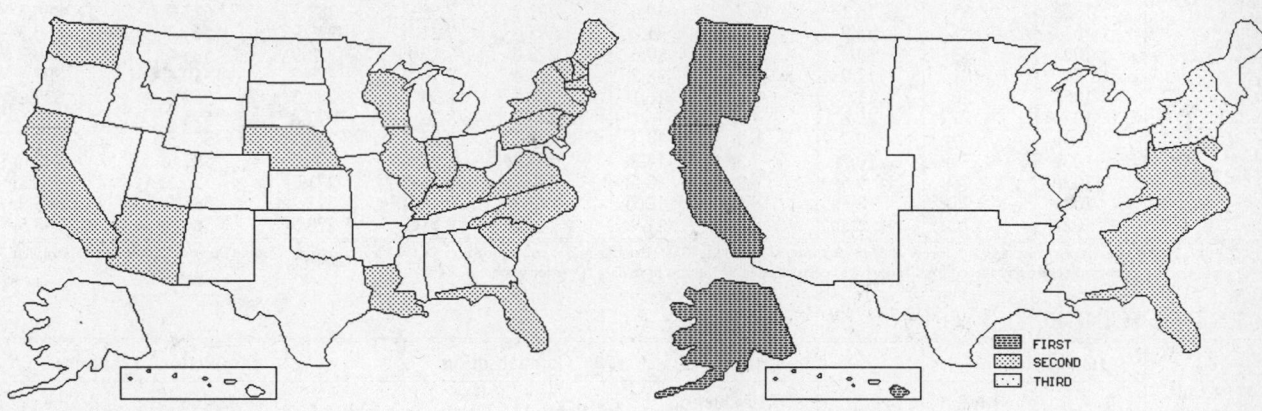

FIRST
SECOND
THIRD

INDUSTRY DATA BY STATE

State	Establish-ments	Shipments			Employment				Cost as % of Shipments	Investment per Employee ($)
		Total ($ mil)	% of U.S.	Per Establ.	Total Number	% of U.S.	Per Establ.	Wages ($/hour)		
South Carolina	10	493.9	30.3	49.4	5,600	31.3	560	9.79	45.9	-
New York	7	176.3	10.8	25.2	1,300	7.3	186	12.42	40.4	-
North Carolina	5	155.6	9.5	31.1	1,500	8.4	300	10.21	44.0	1,133
California	26	124.3	7.6	4.8	1,600	8.9	62	8.61	40.5	3,125
Massachusetts	6	56.0	3.4	9.3	600	3.4	100	10.25	43.0	833
New Jersey	5	38.3	2.3	7.7	400	2.2	80	11.67	36.8	-
Virginia	3	36.1	2.2	12.0	400	2.2	133	7.75	25.2	-
Florida	7	35.8	2.2	5.1	600	3.4	86	7.10	35.5	1,333
Connecticut	5	34.3	2.1	6.9	600	3.4	120	8.20	38.8	-
Vermont	3	20.9	1.3	7.0	300	1.7	100	8.40	45.5	-
Pennsylvania	9	(D)	-	-	1,750 *	9.8	194	-	-	-
Illinois	7	(D)	-	-	175 *	1.0	25	-	-	1,143
Arizona	2	(D)	-	-	175 *	1.0	88	-	-	-
Indiana	2	(D)	-	-	175 *	1.0	88	-	-	-
Maine	2	(D)	-	-	1,750 *	9.8	875	-	-	-
Wisconsin	2	(D)	-	-	175 *	1.0	88	-	-	-
Kentucky	1	(D)	-	-	175 *	1.0	175	-	-	-
Louisiana	1	(D)	-	-	175 *	1.0	175	-	-	-
Nebraska	1	(D)	-	-	175 *	1.0	175	-	-	-
New Hampshire	1	(D)	-	-	175 *	1.0	175	-	-	-
Washington	1	(D)	-	-	750 *	4.2	750	-	-	-

Source: 1992 *Economic Census*. The states are in descending order of shipments or establishments (if shipment data are missing for the majority). The symbol (D) appears when data are withheld to prevent disclosure of competitive information. States marked with (D) are sorted by number of establishments. A dash (-) indicates that the data element cannot be calculated; * indicates the midpoint of a range.

3676 - ELECTRONIC RESISTORS

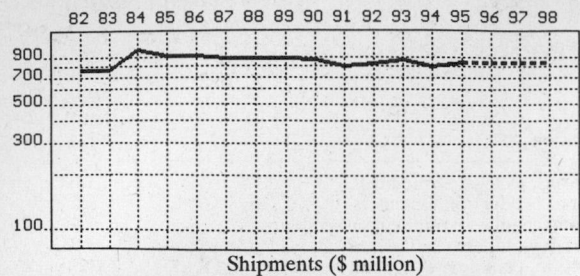

82 83 84 85 86 87 88 89 90 91 92 93 94 95 96 97 98

Shipments ($ million)

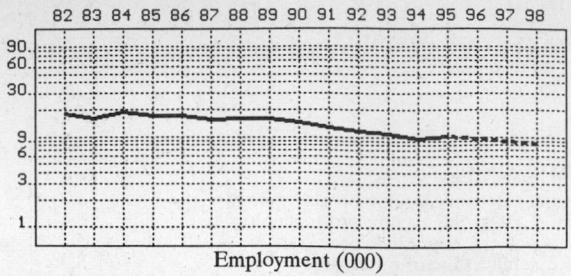

82 83 84 85 86 87 88 89 90 91 92 93 94 95 96 97 98

Employment (000)

GENERAL STATISTICS

Year	Companies	Establishments		Employment			Compensation		Production ($ million)			
		Total	with 20 or more employees	Total (000)	Production Workers (000)	Hours (Mil)	Payroll ($ mil)	Wages ($/hr)	Cost of Materials	Value Added by Manufacture	Value of Shipments	Capital Invest.
1982	77	103	85	18.3	12.4	23.8	258.4	5.84	238.4	524.0	765.8	28.9
1983		103	86	15.8	11.0	20.7	246.7	6.06	214.3	543.0	761.6	38.6
1984		103	87	18.5	13.0	26.9	294.3	5.93	269.6	724.1	983.0	
1985		104	88	17.3	11.8	22.8	297.2	7.01	280.9	643.8	908.5	88.0
1986		107	88	17.2	11.7	22.7	297.2	6.86	276.2	640.7	918.9	69.3
1987	87	118	88	15.7	10.9	21.4	293.4	7.02	281.3	602.1	882.7	43.5
1988		106	85	16.1	10.7	19.6	291.2	7.58	283.5	613.9	888.3	48.9
1989		101	85	15.9	11.7	21.0	285.4	7.21	323.9	561.3	890.3	56.3
1990		102	83	14.4	10.1	19.6	273.0	7.13	328.5	535.4	862.7	53.3
1991		99	81	12.9	8.6	16.2	244.4	7.16	277.7	510.4	797.1	33.6
1992	87	105	80	11.7	8.3	16.1	258.7	8.34	258.3	562.6	827.2	21.3
1993		102	77	10.8	7.9	16.1	249.9	8.76	277.8	598.7	870.5	25.7
1994		103P	79P	9.7	6.9	14.3	237.2	9.26	258.7	558.5	806.0	53.0
1995		103P	79P	10.2P	7.3P	14.3P	253.3P	8.97P	272.2P	587.6P	848.0P	
1996		103P	78P	9.6P	6.9P	13.5P	250.7P	9.21P	271.7P	586.6P	846.5P	
1997		102P	77P	8.9P	6.5P	12.7P	248.1P	9.46P	271.2P	585.5P	845.0P	
1998		102P	76P	8.2P	6.0P	11.9P	245.6P	9.71P	270.7P	584.5P	843.5P	

Sources: 1982, 1987, 1992 *Economic Census*; *Annual Survey of Manufactures*, 83-86, 88-91, 93-94. Establishment counts for non-Census years are from *County Business Patterns*; establishment values for 83-84 are extrapolations. 'P's show projections by the editors. Industries reclassified in 87 will not have data for prior years.

INDICES OF CHANGE

Year	Companies	Establishments		Employment			Compensation		Production ($ million)			
		Total	with 20 or more employees	Total (000)	Production Workers (000)	Hours (Mil)	Payroll ($ mil)	Wages ($/hr)	Cost of Materials	Value Added by Manufacture	Value of Shipments	Capital Invest.
1982	89	98	106	156	149	148	100	70	92	93	93	136
1983		98	108	135	133	129	95	73	83	97	92	181
1984		98	109	158	157	167	114	71	104	129	119	
1985		99	110	148	142	142	115	84	109	114	110	413
1986		102	110	147	141	141	115	82	107	114	111	325
1987	100	112	110	134	131	133	113	84	109	107	107	204
1988		101	106	138	129	122	113	91	110	109	107	230
1989		96	106	136	141	130	110	86	125	100	108	264
1990		97	104	123	122	122	106	85	127	95	104	250
1991		94	101	110	104	101	94	86	108	91	96	158
1992	100	100	100	100	100	100	100	100	100	100	100	100
1993		97	96	92	95	100	97	105	108	106	105	121
1994		98P	99P	83	83	89	92	111	100	99	97	249
1995		98P	98P	88P	88P	89P	98P	108P	105P	104P	103P	
1996		98P	97P	82P	83P	84P	97P	110P	105P	104P	102P	
1997		98P	96P	76P	78P	79P	96P	113P	105P	104P	102P	
1998		97P	95P	70P	73P	74P	95P	116P	105P	104P	102P	

Sources: Same as General Statistics. Values reflect change from the base year, 1992. Values above 100 mean greater than 92, values below 100 mean less than 92, and a value of 100 in the 82-91 or 93-98 period means same as 92. 'P's mark projections by the editors.

SELECTED RATIOS

For 1994	Avg. of All Manufact.	Analyzed Industry	Index	For 1994	Avg. of All Manufact.	Analyzed Industry	Index
Employees per Establishment	49	94	192	Value Added per Production Worker	134,084	80,942	60
Payroll per Establishment	1,500,273	2,302,235	153	Cost per Establishment	5,045,178	2,510,912	50
Payroll per Employee	30,620	24,454	80	Cost per Employee	102,970	26,670	26
Production Workers per Establishment	34	67	195	Cost per Production Worker	146,988	37,493	26
Wages per Establishment	853,319	1,285,234	151	Shipments per Establishment	9,576,895	7,822,941	82
Wages per Production Worker	24,861	19,191	77	Shipments per Employee	195,460	83,093	43
Hours per Production Worker	2,056	2,072	101	Shipments per Production Worker	279,017	116,812	42
Wages per Hour	12.09	9.26	77	Investment per Establishment	321,011	514,412	160
Value Added per Establishment	4,602,255	5,420,735	118	Investment per Employee	6,552	5,464	83
Value Added per Employee	93,930	57,577	61	Investment per Production Worker	9,352	7,681	82

Sources: Same as General Statistics. The 'Average of All Manufacturing' column represents the average of all manufacturing industries reported for the most recent complete year available. The Index shows the relationship between the Average and the Analyzed Industry. For example, 100 means that they are equal; 500 that the Analyzed Industry is five times the average; 50 means that the Analyzed Industry is half the national average. The abbreviation 'na' is used to show that data are 'not available'.

LEADING COMPANIES   Number shown: 30   Total sales ($ mil): 765   Total employment (000): 13.2

Company Name	Address				CEO Name	Phone	Co. Type	Sales ($ mil)	Empl. (000)
Bourns Inc	1200 Columbia Av	Riverside	CA	92507	Gordon Bourns	909-781-5690	R	250	4.5
Dale Electronics Inc	PO Box 609	Columbus	NE	68601	Felix Zandman	402-564-3131	S	186	4.3
IRC Inc	PO Box 1860	Boone	NC	28607	Robert P Rhen	704-264-8861	S	50*	1.0
RCD Components Inc	520 E Industrial Pk	Manchester	NH	03109	LJ Arcidy	603-669-0054	R	50	0.5
K and M Electronics Inc	11 Interstate Dr	W Springfield	MA	01089	James L Knak	413-781-1350	S	48	0.4
CTS Corp	406 Parr Rd	Berne	IN	46711	Bing Harding	219-589-3111	D	29*	0.5
Allen-Bradley Company Inc	1414 Allen-Bradley	El Paso	TX	79936	Bruce I Womer	915-592-4888	D	28	0.2
SEI Electronics Inc	PO Box 58789	Raleigh	NC	27604	Charles J Hill	919-850-9500	R	20	0.2
Ohmtek Inc	2160 Liberty Dr	Niagara Falls	NY	14304	Carl Fritz	716-283-4025	D	15	0.2
Post Glover Resistors Inc	PO Box 18666	Erlanger	KY	41018	N Gambow	606-283-0778	R	11	<0.1
State of the Art Inc	2470 Fox Hill Rd	State College	PA	16803	Donald W Hamer	814-355-8004	R	10	0.1
Prime Technology Inc	PO Box 185	North Branford	CT	06471	Raymon S Sterman	203-481-5721	R	6*	0.1
Rodan	2900 E Blue Star St	Anaheim	CA	92806	Jonathan Smith	714-630-0081	D	6	<0.1
Victory Engineering Corp	118 Victory Rd	Springfield	NJ	07081	Frank Mascuch	201-379-5900	R	6	0.1
IRC Shallcross	US Hwy 70 E	Smithfield	NC	27577	Jerry August	919-934-5181	S	6	0.1
Angstrohm Precision Inc	PO Box 1827	Hagerstown	MD	21740	Robert E Black	301-739-8722	S	5	<0.1
Techno Components Inc	7803 Lemona Av	Van Nuys	CA	91405	Peter Huber	818-781-1642	S	5	0.1
Interlink Electronics Inc	1110 Mark Av	Carpinteria	CA	93013	E Michael Thoben	805-684-2100	R	4	<0.1
Milwaukee Resistor Corp	PO Box 24200	Milwaukee	WI	53224	MA Loewi	414-362-8900	R	4	<0.1
Tepro Florida Inc	PO Box 1260	Clearwater	FL	34617	Roger C Mayo	813-796-1044	R	4	<0.1
Ultronix Inc	PO Box 1090	Grand Junction	CO	81502	Craig Marsh	303-242-0810	S	4*	0.1
Huntington Electric Inc	PO Box 366	Huntington	IN	46750	Michael Khorshid	219-356-0756	R	4	<0.1
Colber Corp	26 Buffington St	Irvington	NJ	07111	Anthony Collett	201-371-9500	R	3	<0.1
Pacific Resistor Co	18300 Oxnard St	Tarzana	CA	91356	Steve C Trewhitt Jr	818-345-7811	R	3	<0.1
Twinpoint Inc	11917 County Rd	Delta	OH	43515	Rudy Lapoint	419-923-7525	R	3	<0.1
Micro Ohm Corp	1088 Hamilton Rd	Duarte	CA	91010	BH Ritchey	818-357-5377	R	2	<0.1
Vamistor Corp	144 River Bend Dr	Sevierville	TN	37876	John Boatman	615-453-0001	R	2	<0.1
Mills Resistor Co	3840 Catalina	Los Alamitos	CA	90720	BT Mills	213-598-2454	R	1	0.1
RF Power Components Inc	125 Wilbur Pl	Bohemia	NY	11716	Paul Davidson	516-563-5050	R	1*	<0.1
Microplex Inc	1977 S State College	Anaheim	CA	92806	C Kucenas	714-634-1535	R	1	<0.1

Source: *Ward's Business Directory of U.S. Private and Public Companies*, Volumes 1 and 2, 1996. The company type code used is as follows: P - Public, R - Private, S - Subsidiary, D - Division, J - Joint Venture, A - Affiliate, G - Group. Sales are in millions of dollars, employees are in thousands. An asterisk (*) indicates an estimated sales volume. The symbol < stands for 'less than'. Company names and addresses are truncated, in some cases, to fit into the available space.

MATERIALS CONSUMED

Material	Quantity	Delivered Cost ($ million)
Materials, ingredients, containers, and supplies	(X)	190.1
Tube blanks	(X)	(D)
Printed circuit boards (without inserted components) for electronic circuitry	(X)	0.5
Printed circuit assemblies, loaded boards or modules	(X)	0.3
Resistors for electronic circuitry	(X)	8.4
Other components and accessories for electronic circuitry, nec, except tubes	(X)	35.5
Silicon, hyperpure	(X)	(D)
Gold and other precious metals, all forms (including ingot, sheet, strip, solder, plating, electrodes, etc.)	(X)	6.6
Doped chemicals, and other doped materials for electronic use	(X)	(D)
Ferrites (powder and paste)	(X)	2.2
Metal powders	(X)	0.7
Electronic computing equipment	(X)	(D)
Current-carrying wiring devices	(X)	0.7
Electrical instrument mechanisms and meter movements (including instrument relays)	(X)	(Z)
Optical instruments and lenses (except sighting, tracking, and fire control)	(X)	0.4
Plastics resins consumed in the form of granules, pellets, powders, liquids, etc.	(X)	4.7
Fabricated plastics products (except gaskets, hoses, and belting)	(X)	3.8
Sheet metal products, except stampings	(X)	0.2
Metal stampings	(X)	14.2
Metal bolts, nuts, screws, washers, rivets, and other screw machine products	(X)	2.5
Other fabricated metal products (except forgings)	(X)	3.7
Castings (rough and semifinished)	(X)	(D)
Steel shapes and forms	(X)	1.5
Copper and copper-base alloy shapes and forms	(X)	1.5
Aluminum and aluminum-base alloy shapes and forms	(X)	0.6
Other nonferrous shapes and forms	(X)	(D)
Insulated wire and cable, including magnet wire	(X)	3.7
Paper and paperboard containers, including shipping sacks and other paper packaging supplies	(X)	2.1
All other materials and components, parts, containers, and supplies	(X)	41.4
Materials, ingredients, containers, and supplies, nsk	(X)	46.9

Source: 1992 *Economic Census*. Explanation of symbols used: (D): Withheld to avoid disclosure of competitive data; na: Not available; (S): Withheld because statistical norms were not met; (X): Not applicable; (Z): Less than half the unit shown; nec: Not elsewhere classified; nsk: Not specified by kind; - : zero; * : 10-19 percent estimated; ** : 20-29 percent estimated.

PRODUCT SHARE DETAILS

Product or Product Class	% Share	Product or Product Class	% Share
Resistors for electronic circuitry	100.00		

Source: 1992 Economic Census. The values shown are percent of total shipments in an industry. Values of indented subcategories are summed in the main heading. The symbol (D) appears when data are withheld to prevent disclosure of competitive information. The abbreviation nsk stands for 'not specified by kind' and nec for 'not elsewhere classified'.

INPUTS AND OUTPUTS FOR ELECTRONIC COMPONENTS NEC

Economic Sector or Industry Providing Inputs	%	Sector	Economic Sector or Industry Buying Outputs	%	Sector
Miscellaneous plastics products	16.2	Manufg.	Radio & TV communication equipment	13.5	Manufg.
Imports	15.7	Foreign	Exports	9.8	Foreign
Electronic components nec	15.5	Manufg.	Telephone & telegraph apparatus	9.6	Manufg.
Wholesale trade	10.9	Trade	Electronic computing equipment	9.4	Manufg.
Plating & polishing	5.2	Manufg.	Electronic components nec	9.1	Manufg.
Primary nonferrous metals, nec	2.6	Manufg.	Radio & TV receiving sets	5.8	Manufg.
Electric services (utilities)	2.0	Util.	Guided missiles & space vehicles	3.2	Manufg.
Semiconductors & related devices	1.9	Manufg.	Personal consumption expenditures	2.9	
Advertising	1.5	Services	X-ray apparatus & tubes	2.6	Manufg.
Communications, except radio & TV	1.4	Util.	Aircraft	2.2	Manufg.
Metal stampings, nec	1.3	Manufg.	Instruments to measure electricity	2.2	Manufg.
Banking	1.2	Fin/R.E.	Aircraft & missile equipment, nec	2.0	Manufg.
Aluminum rolling & drawing	1.1	Manufg.	Semiconductors & related devices	2.0	Manufg.
Copper rolling & drawing	1.1	Manufg.	Industrial controls	1.9	Manufg.
Nonferrous wire drawing & insulating	1.1	Manufg.	Optical instruments & lenses	1.7	Manufg.
Air transportation	1.1	Util.	Computer & data processing services	1.7	Services
Metal coating & allied services	1.0	Manufg.	Federal Government purchases, national defense	1.7	Fed Govt
Blast furnaces & steel mills	0.9	Manufg.	Games, toys, & children's vehicles	1.6	Manufg.
Maintenance of nonfarm buildings nec	0.8	Constr.	Mechanical measuring devices	1.3	Manufg.
Plastics materials & resins	0.8	Manufg.	Calculating & accounting machines	1.2	Manufg.
Motor freight transportation & warehousing	0.8	Util.	Electron tubes	1.2	Manufg.
Real estate	0.8	Fin/R.E.	Photographic equipment & supplies	1.1	Manufg.
Chemical preparations, nec	0.7	Manufg.	Radio & TV broadcasting	1.0	Util.
Petroleum refining	0.7	Manufg.	Change in business inventories	1.0	In House
Screw machine and related products	0.7	Manufg.	Communications, except radio & TV	0.8	Util.
Eating & drinking places	0.7	Trade	Federal Government purchases, nondefense	0.7	Fed Govt
Equipment rental & leasing services	0.6	Services	Engine electrical equipment	0.6	Manufg.
Hotels & lodging places	0.6	Services	Musical instruments	0.6	Manufg.
Cyclic crudes and organics	0.5	Manufg.	Surgical appliances & supplies	0.6	Manufg.
Electronic computing equipment	0.5	Manufg.	Switchgear & switchboard apparatus	0.6	Manufg.
Paperboard containers & boxes	0.5	Manufg.	Phonograph records & tapes	0.5	Manufg.
Sheet metal work	0.5	Manufg.	Management & consulting services & labs	0.5	Services
Adhesives & sealants	0.4	Manufg.	Engineering & scientific instruments	0.4	Manufg.
Machinery, except electrical, nec	0.4	Manufg.	Surgical & medical instruments	0.4	Manufg.
Nonferrous rolling & drawing, nec	0.4	Manufg.	Banking	0.4	Fin/R.E.
Gas production & distribution (utilities)	0.4	Util.	Environmental controls	0.3	Manufg.
Computer & data processing services	0.4	Services	Typewriters & office machines, nec	0.3	Manufg.
Legal services	0.4	Services	Libraries, vocation education	0.3	Services
Carbon & graphite products	0.3	Manufg.	Aircraft & missile engines & engine parts	0.2	Manufg.
Fabricated rubber products, nec	0.3	Manufg.	Ammunition, except for small arms, nec	0.2	Manufg.
Primary metal products, nec	0.3	Manufg.	Electrical equipment & supplies, nec	0.2	Manufg.
Special dies & tools & machine tool accessories	0.3	Manufg.	Detective & protective services	0.2	Services
Wiring devices	0.3	Manufg.	Elementary & secondary schools	0.2	Services
Management & consulting services & labs	0.3	Services	Gross private fixed investment	0.2	Cap Inv
Noncomparable imports	0.3	Foreign	Electrical industrial apparatus, nec	0.1	Manufg.
Aluminum castings	0.2	Manufg.	Motor vehicle parts & accessories	0.1	Manufg.
Instruments to measure electricity	0.2	Manufg.	Motor vehicles & car bodies	0.1	Manufg.
Miscellaneous fabricated wire products	0.2	Manufg.	Scales & balances	0.1	Manufg.
Sanitary services, steam supply, irrigation	0.2	Util.	S/L Govt. purch., higher education	0.1	S/L Govt
Accounting, auditing & bookkeeping	0.2	Services			
U.S. Postal Service	0.2	Gov't			
Abrasive products	0.1	Manufg.			
Glass & glass products, except containers	0.1	Manufg.			
Industrial inorganic chemicals, nec	0.1	Manufg.			
Lubricating oils & greases	0.1	Manufg.			
Manifold business forms	0.1	Manufg.			
Metal foil & leaf	0.1	Manufg.			
Paper coating & glazing	0.1	Manufg.			
Railroads & related services	0.1	Util.			
Insurance carriers	0.1	Fin/R.E.			
Automotive rental & leasing, without drivers	0.1	Services			
Colleges, universities, & professional schools	0.1	Services			

Source: Benchmark Input-Output Accounts for the U.S. Economy, 1982, U.S. Department of Commerce, Washington, D.C., July 1991. Data, as reported in the source, are organized by the 1977 SIC structure in use in 1982 but have been matched, as closely as is possible, to the 1987 SIC structure used in this book.

OCCUPATIONS EMPLOYED BY SIC 367 - ELECTRONIC COMPONENTS AND ACCESSORIES

Occupation	% of Total 1994	Change to 2005	Occupation	% of Total 1994	Change to 2005
Electrical & electronic assemblers	9.8	-8.4	Secretaries, ex legal & medical	1.6	-7.4
Electrical & electronic equipment assemblers	5.7	1.8	Industrial production managers	1.5	-28.8
Electrical & electronic technicians,technologists	5.4	22.1	Electromechanical equipment assemblers	1.4	11.9
Inspectors, testers, & graders, precision	5.4	-28.8	Coil winders, tapers, & finishers	1.3	-18.6
Electronic semiconductor processors	5.3	-1.4	Production, planning, & expediting clerks	1.3	1.8
Electrical & electronics engineers	4.4	29.8	Industrial engineers, ex safety engineers	1.3	12.0
Assemblers, fabricators, & hand workers nec	4.0	1.8	Electrolytic plating machine workers	1.2	11.9
Blue collar worker supervisors	3.4	-14.1	Traffic, shipping, & receiving clerks	1.2	-2.1
Sales & related workers nec	2.2	1.8	Freight, stock, & material movers, hand	1.2	-18.6
Computer engineers	2.2	50.7	Precision assemblers nec	1.1	62.9
Engineering, mathematical, & science managers	1.8	15.6	Stock clerks	1.0	-17.3
Engineering technicians & technologists nec	1.8	1.8	Engineers nec	1.0	52.6
General managers & top executives	1.6	-3.4	General office clerks	1.0	-13.2

Source: Industry-Occupation Matrix, Bureau of Labor Statistics. These data relate to one or more 3-digit SIC industry groups rather than to a single 4-digit SIC. The change reported for each occupation to the year 2005 is a percent of growth or decline as estimated by the Bureau of Labor Statistics. The abbreviation nec stands for 'not elsewhere classified'.

LOCATION BY STATE AND REGIONAL CONCENTRATION

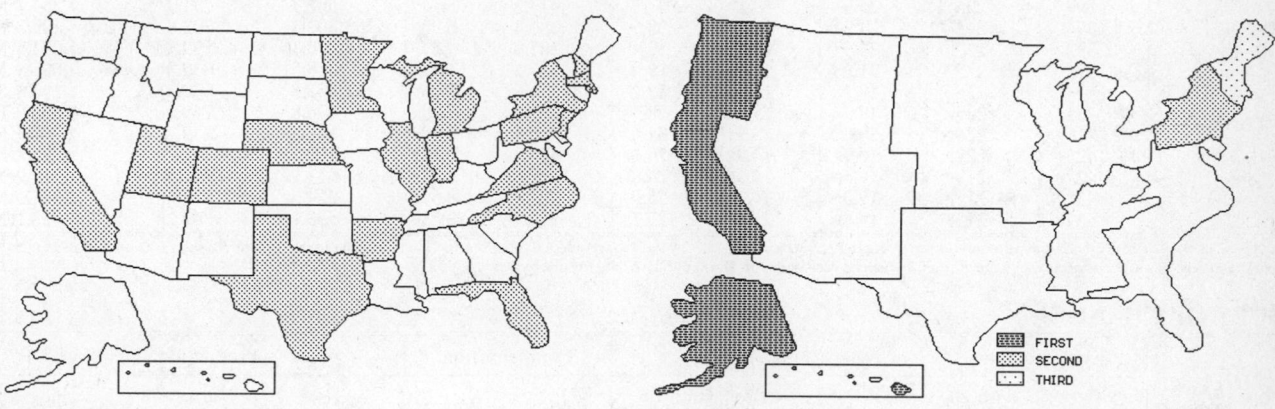

FIRST
SECOND
THIRD

INDUSTRY DATA BY STATE

State	Establish-ments	Shipments			Employment				Cost as % of Shipments	Investment per Employee ($)
		Total ($ mil)	% of U.S.	Per Establ.	Total Number	% of U.S.	Per Establ.	Wages ($/hour)		
California	17	172.3	20.8	10.1	1,500	12.8	88	11.00	38.9	2,733
Texas	7	75.4	9.1	10.8	1,100	9.4	157	8.79	33.7	2,091
New Hampshire	4	54.1	6.5	13.5	800	6.8	200	8.85	34.6	2,375
Pennsylvania	6	47.0	5.7	7.8	600	5.1	100	9.50	20.6	-
New York	5	38.0	4.6	7.6	600	5.1	120	8.50	29.7	-
Florida	7	36.3	4.4	5.2	900	7.7	129	7.00	30.9	-
Massachusetts	6	31.8	3.8	5.3	500	4.3	83	8.57	25.2	1,000
New Jersey	5	28.1	3.4	5.6	400	3.4	80	9.40	35.2	1,000
Rhode Island	3	18.1	2.2	6.0	200	1.7	67	9.67	33.1	3,000
Illinois	6	(D)	-	-	375 *	3.2	63	-	-	-
Indiana	4	(D)	-	-	750 *	6.4	188	-	-	-
North Carolina	3	(D)	-	-	750 *	6.4	250	-	-	-
Arkansas	2	(D)	-	-	175 *	1.5	88	-	-	-
Colorado	2	(D)	-	-	175 *	1.5	88	-	-	-
Michigan	2	(D)	-	-	175 *	1.5	88	-	-	-
Minnesota	2	(D)	-	-	375 *	3.2	188	-	-	-
Nebraska	2	(D)	-	-	1,750 *	15.0	875	-	-	-
Utah	2	(D)	-	-	750 *	6.4	375	-	-	-
Virginia	1	(D)	-	-	375 *	3.2	375	-	-	-

Source: 1992 Economic Census. The states are in descending order of shipments or establishments (if shipment data are missing for the majority). The symbol (D) appears when data are withheld to prevent disclosure of competitive information. States marked with (D) are sorted by number of establishments. A dash (-) indicates that the data element cannot be calculated; * indicates the midpoint of a range.

3677 - ELECTRONIC COILS & TRANSFORMERS

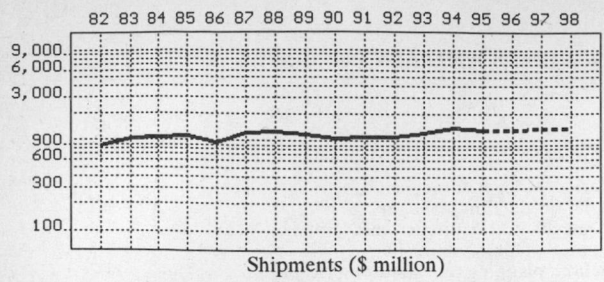

82 83 84 85 86 87 88 89 90 91 92 93 94 95 96 97 98

Shipments ($ million)

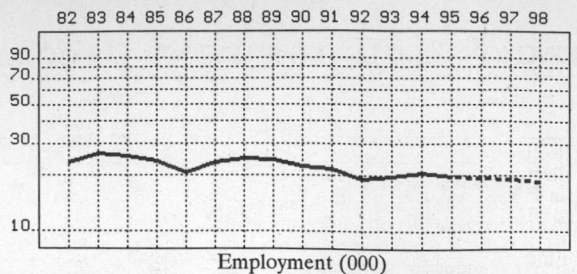

82 83 84 85 86 87 88 89 90 91 92 93 94 95 96 97 98

Employment (000)

GENERAL STATISTICS

| Year | Companies | Establishments | | Employment | | | Compensation | | Production ($ million) | | | |
		Total	with 20 or more employees	Total (000)	Production Workers (000)	Hours (Mil)	Payroll ($ mil)	Wages ($/hr)	Cost of Materials	Value Added by Manufacture	Value of Shipments	Capital Invest.
1982	353	386	245	23.7	18.7	35.2	293.4	5.21	327.6	537.6	863.3	36.5
1983		383	245	26.2	21.2	39.6	346.3	5.54	415.2	625.9	1,029.4	31.8
1984		380	245	25.5	21.1	41.9	339.2	5.55	446.0	653.2	1,085.2	
1985		377	245	24.0	19.5	35.9	334.5	6.00	437.3	682.6	1,130.2	19.7
1986		375	242	20.9	16.7	31.5	300.4	6.02	373.5	567.6	952.4	15.8
1987	393	416	246	23.9	18.3	34.8	396.3	6.84	476.1	747.9	1,228.4	29.1
1988		399	236	24.9	19.1	37.3	413.2	6.74	502.7	754.2	1,254.7	17.5
1989		394	242	24.3	18.7	33.5	382.4	6.96	502.2	702.0	1,199.3	28.8
1990		392	227	22.5	17.4	30.4	365.8	7.27	470.1	599.1	1,074.6	25.3
1991		391	225	21.8	16.6	29.6	355.4	6.97	458.5	639.0	1,098.8	21.1
1992	401	423	228	19.2	14.1	27.2	374.0	7.66	452.3	680.6	1,133.8	20.1
1993		431	225	19.7	15.2	28.9	388.5	7.78	476.8	745.2	1,222.9	21.9
1994		419P	224P	20.8	16.5	31.6	410.1	7.57	538.5	849.7	1,380.3	31.8
1995	·	423P	222P	19.9P	14.9P	27.6P	408.8P	8.14P	505.5P	797.7P	1,295.8P	
1996		427P	219P	19.4P	14.5P	26.8P	415.6P	8.35P	514.9P	812.5P	1,319.9P	
1997		430P	217P	19.0P	14.1P	25.9P	422.3P	8.57P	524.3P	827.4P	1,344.0P	
1998		434P	215P	18.6P	13.6P	25.1P	429.1P	8.78P	533.7P	842.2P	1,368.1P	

Sources: 1982, 1987, 1992 *Economic Census*; *Annual Survey of Manufactures*, 83-86, 88-91, 93-94. Establishment counts for non-Census years are from *County Business Patterns*; establishment values for 83-84 are extrapolations. 'P's show projections by the editors. Industries reclassified in 87 will not have data for prior years.

INDICES OF CHANGE

| Year | Companies | Establishments | | Employment | | | Compensation | | Production ($ million) | | | |
		Total	with 20 or more employees	Total (000)	Production Workers (000)	Hours (Mil)	Payroll ($ mil)	Wages ($/hr)	Cost of Materials	Value Added by Manufacture	Value of Shipments	Capital Invest.
1982	88	91	107	123	133	129	78	68	72	79	76	182
1983		91	107	136	150	146	93	72	92	92	91	158
1984		90	107	133	150	154	91	72	99	96	96	
1985		89	107	125	138	132	89	78	97	100	100	98
1986		89	106	109	118	116	80	79	83	83	84	79
1987	98	98	108	124	130	128	106	89	105	110	108	145
1988		94	104	130	135	137	110	88	111	111	111	87
1989		93	106	127	133	123	102	91	111	103	106	143
1990		93	100	117	123	112	98	95	104	88	95	126
1991		92	99	114	118	109	95	91	101	94	97	105
1992	100	100	100	100	100	100	100	100	100	100	100	100
1993		102	99	103	108	106	104	102	105	109	108	109
1994		99P	98P	108	117	116	110	99	119	125	122	158
1995		100P	97P	104P	106P	102P	109P	106P	112P	117P	114P	
1996		101P	96P	101P	103P	98P	111P	109P	114P	119P	116P	
1997		102P	95P	99P	100P	95P	113P	112P	116P	122P	119P	
1998		103P	94P	97P	97P	92P	115P	115P	118P	124P	121P	

Sources: Same as General Statistics. Values reflect change from the base year, 1992. Values above 100 mean greater than 92, values below 100 mean less than 92, and a value of 100 in the 82-91 or 93-98 period means same as 92. 'P's mark projections by the editors.

SELECTED RATIOS

For 1994	Avg. of All Manufact.	Analyzed Industry	Index	For 1994	Avg. of All Manufact.	Analyzed Industry	Index
Employees per Establishment	49	50	101	Value Added per Production Worker	134,084	51,497	38
Payroll per Establishment	1,500,273	977,875	65	Cost per Establishment	5,045,178	1,284,042	25
Payroll per Employee	30,620	19,716	64	Cost per Employee	102,970	25,889	25
Production Workers per Establishment	34	39	115	Cost per Production Worker	146,988	32,636	22
Wages per Establishment	853,319	570,396	67	Shipments per Establishment	9,576,895	3,291,297	34
Wages per Production Worker	24,861	14,498	58	Shipments per Employee	195,460	66,361	34
Hours per Production Worker	2,056	1,915	93	Shipments per Production Worker	279,017	83,655	30
Wages per Hour	12.09	7.57	63	Investment per Establishment	321,011	75,826	24
Value Added per Establishment	4,602,255	2,026,092	44	Investment per Employee	6,552	1,529	23
Value Added per Employee	93,930	40,851	43	Investment per Production Worker	9,352	1,927	21

Sources: Same as General Statistics. The 'Average of All Manufacturing' column represents the average of all manufacturing industries reported for the most recent complete year available. The Index shows the relationship between the Average and the Analyzed Industry. For example, 100 means that they are equal; 500 that the Analyzed Industry is five times the average; 50 means that the Analyzed Industry is half the national average. The abbreviation 'na' is used to show that data are 'not available'.

LEADING COMPANIES   Number shown: 75   Total sales ($ mil): 929   Total employment (000): 20.4

Company Name	Address				CEO Name	Phone	Co. Type	Sales ($ mil)	Empl. (000)
Valor Electronics Inc	9715 Business Pk	San Diego	CA	92131	Richard Barron	619-537-2500	S	103	7.0
Basler Electric Co	Rte 143	Highland	IL	62249	WL Basler	618-654-2341	R	100	1.8
American Precision Industries	2777 Walden Av	Buffalo	NY	14225	Kurt Wiedenhaupt	716-684-9700	P	65	0.9
Products Unlimited Corp	PO Box 413	Sterling	IL	61081	Gary Schreiner	815-626-0300	R	57	0.7
Delevan	270 Quaker Rd	East Aurora	NY	14052	James Bingel	716-652-3600	D	50	0.2
Midcom Inc	PO Box 1330	Watertown	SD	57201	Rich Lowe	605-886-4385	R	50	1.5
Signal Transformer Company	500 Bayview Av	Inwood	NY	11696	Jim Oberlender	516-239-5777	S	25	0.5
Robinson-Halpern Products	PO Box 9011	Valley Forge	PA	19485	Robert Carlson	215-539-4400	D	22	0.2
Kepco Inc	131-38 Sanford Av	Flushing	NY	11352	Max Kupferberg	718-461-7000	R	17	0.3
TNI Inc	1001eeple Square Ct	Knightdale	NC	27545	J Strathmeyer	919-266-4411	R	17	0.3
KSC Industries Inc	1138 E 6th St	Corona	CA	91719	Jeff King Sr	909-371-4140	R	16	0.2
RF Power Products Inc	520 G-M	Voorhees	NJ	08043	Joseph Stach	609-751-0033	P	16	0.1
OPT Industries Inc	300 Red School Ln	Phillipsburg	NJ	08865	Christian Hughes	908-454-2600	S	16	0.3
Espey Mfg & Electronics Corp	PO Box 422	Saratoga Sp	NY	12866	Sol Pinsley	518-584-4100	P	15	0.2
Schott Corp	1000 Pkers Lake Rd	Wayzata	MN	55391	Owen W Schott	612-475-1173	R	15	0.3
MagneTek	PO Box 490	Goodland	IN	47948	John Waldron	219-297-3111	D	14	0.3
Alcoils	802 E Short St	Columbia City	IN	46725	Bary Lloyd	219-244-6183	D	13	0.1
Deltron Inc	290 Wissahickon	North Wales	PA	19454	Aaron Anton	215-699-9261	R	11	0.2
Olsun Electrics Corp	PO Box 1	Richmond	IL	60071	AF Asta	815-678-2421	R	11	0.1
Axel Electronics Inc	19060 S Dominguez	R Dominguez	CA	90220	John S McGovern	310-884-5200	S	10	0.2
BH Electronics Inc	12219 Wood Lake	Burnsville	MN	55337	Richard Jackson	612-894-9590	R	10	0.1
Bicron Electronics Co	Barlow St	Canaan	CT	06018	PB Kent	203-824-5125	R	10	0.1
Coiltronics Inc	6000 Pk of Com	Boca Raton	FL	33487	Lynn Hayden	407-241-7876	R	10*	0.3
Electrical Windings Inc	2015 N Kolmar Av	Chicago	IL	60639	DR Murphy	312-235-3360	R	10	0.3
Multi Products International	250 Lackawanna Dr	West Paterson	NJ	07424	Howard Longin	201-890-1344	R	10	<0.1
Quality Coils Inc	PO Box 1480	Bristol	CT	06011	KA Gibson	203-584-0927	R	10*	0.2
Sigmapower Inc	19060 S Dominguez	R Dominguez	CA	90220	John J McGovern	310-884-5200	S	10	0.1
Dial Products Co	PO Box 456	Bayonne	NJ	07002	M Krull	201-437-0720	R	9*	<0.1
Climco Coils Co	400 Oakwood Dr	Morrison	IL	61270	Scott Selman	815-772-2107	D	8	0.1
GFS Manufacturing Company	140 Crosby Rd	Dover	NH	03820	Janet Sylvester	603-742-4375	R	8	0.1
Gowanda Electronics Corp	PO Box 111	Gowanda	NY	14070	D Schaack	716-532-2234	R	8*	0.2
Industrial Coils Inc	PO Box 170	Baraboo	WI	53913	James Kieffer	608-356-6601	R	8	0.2
RFI Corp	100 Pine Aire Dr	Bay Shore	NY	11706	Seymour Rubin	516-231-6400	S	8*	0.1
Tur-bo Jet Products Company	PO Box 677	Rosemead	CA	91770	Richard L Bloom	818-285-1294	R	8	0.1
Warsaw Coil Company Inc	PO Box 1057	Warsaw	IN	46581	T Joyner	219-267-6041	R	8	0.1
EWC Inc	385 Hwy 33	Englishtown	NJ	07726	Christ G Hiotis	908-446-3110	R	7	0.1
Northlake Engineering Inc	PO Box 370	Bristol	WI	53104	WA Hardt	414-857-9600	R	7	0.1
Coast Magnetics	1207 N La Brea Av	Inglewood	CA	90302	Dev Dosha	310-673-3245	R	6	<0.1
Elgin E2	5533 New Perry	Erie	PA	16509	William Mosconi	814-864-4921	R	6*	<0.1
Merrimack Magnetics Corp	121 Hale St	Lowell	MA	01851	Carol A Mariano	508-458-1487	R	6	<0.1
Lepco Inc	85 Industrial Dr	Brownsville	TX	78521	RW Smith	210-546-1625	R	5	0.2
Daykin Electric Corp	34425 Schoolcraft St	Livonia	MI	48150	JP Williams	313-261-3310	R	5*	<0.1
ESC Electronics Corp	534 Bergen Blv	Palisades Park	NJ	07650	L Hammer	201-947-0400	R	5	0.1
Hisonic Inc	PO Box 1130	Olathe	KS	66051	Jack O Cooper III	913-782-0012	R	5	0.1
Sag Harbor Industries Inc	1668 S Harbor	Sag Harbor	NY	11963	P Scheerer	516-725-0440	R	5	0.2
Trinetics Inc	55807 Currant Rd	Mishawaka	IN	46545	NL Walters	219-259-8535	R	5	<0.1
Wabash Transformer Corp	411 E South St	Tipton	IA	52772	Mahboob Khan	319-886-6086	S	5	<0.1
Electronic Coil Corp	125 Old Iron Ore	Bloomfield	CT	06002	Raymond B Gorski	203-243-5233	R	5	0.1
Hytronics Corp	PO Box 4050	Clearwater	FL	34618	Robert J Enersen	813-536-7861	R	5	0.2
Microtran Company Inc	PO Box 236	Valley Stream	NY	11582	AJ Eisenberg	516-561-6050	R	5	<0.1
Airborne Power Supply	1231 W 23rd St	Tempe	AZ	85282	James A Boyd	602-967-8604	D	4	<0.1
Coiltronics International Corp	6000 Pk of Com	Boca Raton	FL	33487	Lynn A Hayden	407-241-7876	S	4*	<0.1
Communication Coil Inc	9601 Soreng Av	Schiller Park	IL	60176	Elliot H Goldman	708-671-1333	R	4	0.1
Discom Inc	334 Littleton Rd	Westford	MA	01886	WC Curry	508-692-6000	S	4	<0.1
Magnetic Coils Inc	411 Manhattan Av	North Babylon	NY	11704	JH Williams	516-587-0510	R	4	<0.1
Raycom Electronics Inc	PO Box 250	Dover	PA	17315	R Ford	717-292-3641	R	4	0.1
South Haven Coil Inc	PO Box 409	South Haven	MI	49090	Virgil Brambaugh	616-637-5201	S	4*	<0.1
ATR Coil Company Inc	PO Box 2089	Bloomington	IN	47402	David Wiley	812-336-5096	R	4	<0.1
Custom Magnetics Inc	801 W Main St	N Manchester	IN	46962	Kirti Shah	219-982-8508	R	3	<0.1
Frequency Devices Inc	25 Locust St	Haverhill	MA	01830	G T Anderson	508-374-0761	R	3	<0.1
Lark Engineering Company Inc	27282 Calle Arroyo	S J Capistrano	CA	92675	Fred Baier	714-240-1233	R	3*	<0.1
Polara Engineering Inc	4115 W Artesia Av	Fullerton	CA	92633	John McGaffey	714-521-0900	R	3	<0.1
Precision Inc	3415 48th Av N	Minneapolis	MN	55429	David J Anderson	612-537-9340	R	3	<0.1
Corona Magnetics Inc	PO Box 1355	Corona	CA	91718	UK Paasch	909-735-7558	R	3	<0.1
Neshaminy Transformer Corp	40 Indian Dr	Ivyland	PA	18974	John Ralston	215-322-2727	R	3	<0.1
American Trans-Coil Corp	PO Box 357	Richmond Hill	NY	11419	WB Rogers	718-441-5207	R	2*	<0.1
Atlantic Magnetics Inc	1441 SW 30th Av	Pompano Bch	FL	33069	Edward Cammarata	305-979-7920	R	2	<0.1
Coil Specialty Company Inc	PO Box 978	State College	PA	16804	Janice Hoffman	814-234-7044	R	2	<0.1
Hermetic Coil Inc	PO Box 216	Bicknell	IN	47512	Dwayne Davis	812-735-2400	R	2*	<0.1
Keytronics Inc	707 North St	Endicott	NY	13760	HH Horton	607-754-5405	R	2	<0.1
Sartron Inc	114 N Main St	Newberg	OR	97132	Timothy James	503-538-3191	R	2	<0.1
MC Davis Company Inc	PO Box 2266	Arizona City	AZ	85223	J Tooley	602-466-5151	R	2	<0.1
Teco Corp	PO Box A	Winnisquam	NH	03289	A Costa	603-524-1998	R	2	<0.1
AH and R Industries Inc	21366 S Alameda St	Long Beach	CA	90810		310-549-5060	R	1	<0.1
Electro Engineering Works	401 Preda St	San Leandro	CA	94577	W M Niederjohn	510-569-3326	R	1*	<0.1

Source: Ward's Business Directory of U.S. Private and Public Companies, Volumes 1 and 2, 1996. The company type code used is as follows: P - Public, R - Private, S - Subsidiary, D - Division, J - Joint Venture, A - Affiliate, G - Group. Sales are in millions of dollars, employees are in thousands. An asterisk (*) indicates an estimated sales volume. The symbol < stands for 'less than'. Company names and addresses are truncated, in some cases, to fit into the available space.

MATERIALS CONSUMED

Material	Quantity	Delivered Cost ($ million)
Materials, ingredients, containers, and supplies	(X)	413.8
Tube blanks	(X)	0.2
Printed circuit boards (without inserted components) for electronic circuitry	(X)	1.2
Printed circuit assemblies, loaded boards or modules	(X)	1.4
Semiconductors, including transistors, diodes, rectifiers, and integrated circuits for electronic circuitry	(X)	1.4
Capacitors for electronic circuitry	(X)	1.2
Resistors for electronic circuitry	(X)	0.8
Other components and accessories for electronic circuitry, nec, except tubes	(X)	11.3
Silicon, hyperpure	(X)	1.3
Gold and other precious metals, all forms (including ingot, sheet, strip, solder, plating, electrodes, etc.)	(X)	1.9
Doped chemicals, and other doped materials for electronic use	(X)	0.3
Ferrites (powder and paste)	(X)	12.1
Metal powders	(X)	3.4
Electronic computing equipment	(X)	0.3
Current-carrying wiring devices	(X)	1.0
Electronic communication equipment	(X)	0.1
Electrical instrument mechanisms and meter movements (including instrument relays)	(X)	1.5
Optical instruments and lenses (except sighting, tracking, and fire control)	(X)	(D)
Plastics resins consumed in the form of granules, pellets, powders, liquids, etc.	(X)	4.4
Fabricated plastics products (except gaskets, hoses, and belting)	(X)	5.8
Sheet metal products, except stampings	(X)	7.6
Metal stampings	(X)	14.1
Metal bolts, nuts, screws, washers, rivets, and other screw machine products	(X)	4.1
Other fabricated metal products (except forgings)	(X)	2.4
Forgings	(X)	(D)
Castings (rough and semifinished)	(X)	(D)
Steel shapes and forms	(X)	8.1
Copper and copper-base alloy shapes and forms	(X)	2.9
Aluminum and aluminum-base alloy shapes and forms	(X)	0.6
Other nonferrous shapes and forms	(X)	1.1
Insulated wire and cable, including magnet wire	(X)	36.7
Paper and paperboard containers, including shipping sacks and other paper packaging supplies	(X)	3.9
All other materials and components, parts, containers, and supplies	(X)	79.9
Materials, ingredients, containers, and supplies, nsk	(X)	202.6

Source: 1992 Economic Census. Explanation of symbols used: (D): Withheld to avoid disclosure of competitive data; na: Not available; (S): Withheld because statistical norms were not met; (X): Not applicable; (Z): Less than half the unit shown; nec: Not elsewhere classified; nsk: Not specified by kind; - : zero; * : 10-19 percent estimated; ** : 20-29 percent estimated.

PRODUCT SHARE DETAILS

Product or Product Class	% Share	Product or Product Class	% Share
Coils, transformers, reactors, and chokes for electronic		circuitry	100.00

Source: 1992 Economic Census. The values shown are percent of total shipments in an industry. Values of indented subcategories are summed in the main heading. The symbol (D) appears when data are withheld to prevent disclosure of competitive information. The abbreviation nsk stands for 'not specified by kind' and nec for 'not elsewhere classified'.

INPUTS AND OUTPUTS FOR ELECTRONIC COMPONENTS NEC

Economic Sector or Industry Providing Inputs	%	Sector	Economic Sector or Industry Buying Outputs	%	Sector
Miscellaneous plastics products	16.2	Manufg.	Radio & TV communication equipment	13.5	Manufg.
Imports	15.7	Foreign	Exports	9.8	Foreign
Electronic components nec	15.5	Manufg.	Telephone & telegraph apparatus	9.6	Manufg.
Wholesale trade	10.9	Trade	Electronic computing equipment	9.4	Manufg.
Plating & polishing	5.2	Manufg.	Electronic components nec	9.1	Manufg.
Primary nonferrous metals, nec	2.6	Manufg.	Radio & TV receiving sets	5.8	Manufg.
Electric services (utilities)	2.0	Util.	Guided missiles & space vehicles	3.2	Manufg.
Semiconductors & related devices	1.9	Manufg.	Personal consumption expenditures	2.9	
Advertising	1.5	Services	X-ray apparatus & tubes	2.6	Manufg.
Communications, except radio & TV	1.4	Util.	Aircraft	2.2	Manufg.
Metal stampings, nec	1.3	Manufg.	Instruments to measure electricity	2.2	Manufg.
Banking	1.2	Fin/R.E.	Aircraft & missile equipment, nec	2.0	Manufg.
Aluminum rolling & drawing	1.1	Manufg.	Semiconductors & related devices	2.0	Manufg.
Copper rolling & drawing	1.1	Manufg.	Industrial controls	1.9	Manufg.
Nonferrous wire drawing & insulating	1.1	Manufg.	Optical instruments & lenses	1.7	Manufg.
Air transportation	1.1	Util.	Computer & data processing services	1.7	Services
Metal coating & allied services	1.0	Manufg.	Federal Government purchases, national defense	1.7	Fed Govt
Blast furnaces & steel mills	0.9	Manufg.	Games, toys, & children's vehicles	1.6	Manufg.
Maintenance of nonfarm buildings nec	0.8	Constr.	Mechanical measuring devices	1.3	Manufg.
Plastics materials & resins	0.8	Manufg.	Calculating & accounting machines	1.2	Manufg.
Motor freight transportation & warehousing	0.8	Util.	Electron tubes	1.2	Manufg.
Real estate	0.8	Fin/R.E.	Photographic equipment & supplies	1.1	Manufg.
Chemical preparations, nec	0.7	Manufg.	Radio & TV broadcasting	1.0	Util.

Continued on next page.

INPUTS AND OUTPUTS FOR ELECTRONIC COMPONENTS NEC - Continued

Economic Sector or Industry Providing Inputs	%	Sector	Economic Sector or Industry Buying Outputs	%	Sector
Petroleum refining	0.7	Manufg.	Change in business inventories	1.0	In House
Screw machine and related products	0.7	Manufg.	Communications, except radio & TV	0.8	Util.
Eating & drinking places	0.7	Trade	Federal Government purchases, nondefense	0.7	Fed Govt
Equipment rental & leasing services	0.6	Services	Engine electrical equipment	0.6	Manufg.
Hotels & lodging places	0.6	Services	Musical instruments	0.6	Manufg.
Cyclic crudes and organics	0.5	Manufg.	Surgical appliances & supplies	0.6	Manufg.
Electronic computing equipment	0.5	Manufg.	Switchgear & switchboard apparatus	0.6	Manufg.
Paperboard containers & boxes	0.5	Manufg.	Phonograph records & tapes	0.5	Manufg.
Sheet metal work	0.5	Manufg.	Management & consulting services & labs	0.5	Services
Adhesives & sealants	0.4	Manufg.	Engineering & scientific instruments	0.4	Manufg.
Machinery, except electrical, nec	0.4	Manufg.	Surgical & medical instruments	0.4	Manufg.
Nonferrous rolling & drawing, nec	0.4	Manufg.	Banking	0.4	Fin/R.E.
Gas production & distribution (utilities)	0.4	Util.	Environmental controls	0.3	Manufg.
Computer & data processing services	0.4	Services	Typewriters & office machines, nec	0.3	Manufg.
Legal services	0.4	Services	Libraries, vocation education	0.3	Services
Carbon & graphite products	0.3	Manufg.	Aircraft & missile engines & engine parts	0.2	Manufg.
Fabricated rubber products, nec	0.3	Manufg.	Ammunition, except for small arms, nec	0.2	Manufg.
Primary metal products, nec	0.3	Manufg.	Electrical equipment & supplies, nec	0.2	Manufg.
Special dies & tools & machine tool accessories	0.3	Manufg.	Detective & protective services	0.2	Services
Wiring devices	0.3	Manufg.	Elementary & secondary schools	0.2	Services
Management & consulting services & labs	0.3	Services	Gross private fixed investment	0.2	Cap Inv
Noncomparable imports	0.3	Foreign	Electrical industrial apparatus, nec	0.1	Manufg.
Aluminum castings	0.2	Manufg.	Motor vehicle parts & accessories	0.1	Manufg.
Instruments to measure electricity	0.2	Manufg.	Motor vehicles & car bodies	0.1	Manufg.
Miscellaneous fabricated wire products	0.2	Manufg.	Scales & balances	0.1	Manufg.
Sanitary services, steam supply, irrigation	0.2	Util.	S/L Govt. purch., higher education	0.1	S/L Govt
Accounting, auditing & bookkeeping	0.2	Services			
U.S. Postal Service	0.2	Gov't			
Abrasive products	0.1	Manufg.			
Glass & glass products, except containers	0.1	Manufg.			
Industrial inorganic chemicals, nec	0.1	Manufg.			
Lubricating oils & greases	0.1	Manufg.			
Manifold business forms	0.1	Manufg.			
Metal foil & leaf	0.1	Manufg.			
Paper coating & glazing	0.1	Manufg.			
Railroads & related services	0.1	Util.			
Insurance carriers	0.1	Fin/R.E.			
Automotive rental & leasing, without drivers	0.1	Services			
Colleges, universities, & professional schools	0.1	Services			

Source: Benchmark Input-Output Accounts for the U.S. Economy, 1982, U.S. Department of Commerce, Washington, D.C., July 1991. Data, as reported in the source, are organized by the 1977 SIC structure in use in 1982 but have been matched, as closely as is possible, to the 1987 SIC structure used in this book.

OCCUPATIONS EMPLOYED BY SIC 367 - ELECTRONIC COMPONENTS AND ACCESSORIES

Occupation	% of Total 1994	Change to 2005	Occupation	% of Total 1994	Change to 2005
Electrical & electronic assemblers	9.8	-8.4	Secretaries, ex legal & medical	1.6	-7.4
Electrical & electronic equipment assemblers	5.7	1.8	Industrial production managers	1.5	-28.8
Electrical & electronic technicians,technologists	5.4	22.1	Electromechanical equipment assemblers	1.4	11.9
Inspectors, testers, & graders, precision	5.4	-28.8	Coil winders, tapers, & finishers	1.3	-18.6
Electronic semiconductor processors	5.3	-1.4	Production, planning, & expediting clerks	1.3	1.8
Electrical & electronics engineers	4.4	29.8	Industrial engineers, ex safety engineers	1.3	12.0
Assemblers, fabricators, & hand workers nec	4.0	1.8	Electrolytic plating machine workers	1.2	11.9
Blue collar worker supervisors	3.4	-14.1	Traffic, shipping, & receiving clerks	1.2	-2.1
Sales & related workers nec	2.2	1.8	Freight, stock, & material movers, hand	1.2	-18.6
Computer engineers	2.2	50.7	Precision assemblers nec	1.1	62.9
Engineering, mathematical, & science managers	1.8	15.6	Stock clerks	1.0	-17.3
Engineering technicians & technologists nec	1.8	1.8	Engineers nec	1.0	52.6
General managers & top executives	1.6	-3.4	General office clerks	1.0	-13.2

Source: Industry-Occupation Matrix, Bureau of Labor Statistics. These data relate to one or more 3-digit SIC industry groups rather than to a single 4-digit SIC. The change reported for each occupation to the year 2005 is a percent of growth or decline as estimated by the Bureau of Labor Statistics. The abbreviation nec stands for 'not elsewhere classified'.

LOCATION BY STATE AND REGIONAL CONCENTRATION

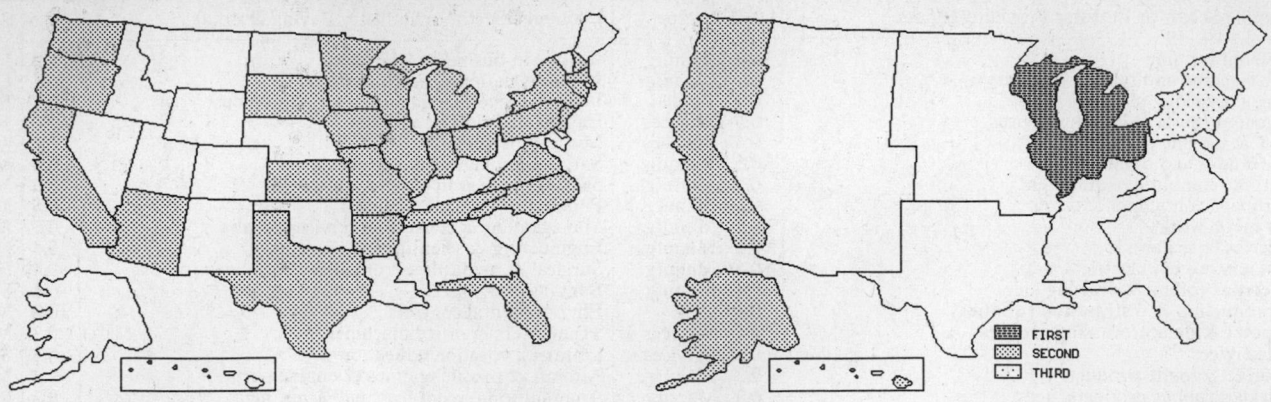

INDUSTRY DATA BY STATE

| State | Establish-ments | Shipments | | | Employment | | | | Cost as % of Shipments | Investment per Employee ($) |
		Total ($ mil)	% of U.S.	Per Establ.	Total Number	% of U.S.	Per Establ.	Wages ($/hour)		
California	84	191.7	16.9	2.3	2,500	13.0	30	8.71	45.4	1,120
Illinois	52	144.4	12.7	2.8	2,900	15.1	56	6.86	39.4	1,000
New York	37	104.2	9.2	2.8	2,300	12.0	62	7.62	34.5	-
Indiana	15	62.1	5.5	4.1	1,000	5.2	67	7.11	36.6	1,000
Ohio	15	55.3	4.9	3.7	700	3.6	47	9.30	37.8	-
Massachusetts	22	53.4	4.7	2.4	700	3.6	32	9.00	29.8	857
New Jersey	14	51.5	4.5	3.7	800	4.2	57	9.64	41.6	-
Minnesota	11	39.3	3.5	3.6	800	4.2	73	5.92	42.2	750
Michigan	14	36.2	3.2	2.6	500	2.6	36	9.00	42.8	800
Florida	17	34.3	3.0	2.0	800	4.2	47	7.27	37.9	-
Connecticut	13	28.8	2.5	2.2	600	3.1	46	7.50	36.8	-
Texas	19	28.6	2.5	1.5	400	2.1	21	7.67	37.4	500
Wisconsin	12	25.4	2.2	2.1	600	3.1	50	7.63	38.2	-
Arizona	12	18.3	1.6	1.5	400	2.1	33	6.20	35.5	1,500
Oregon	11	10.5	0.9	1.0	200	1.0	18	10.33	32.4	500
Pennsylvania	14	(D)	-	-	375 *	2.0	27	-	-	267
New Hampshire	8	(D)	-	-	375 *	2.0	47	-	-	-
North Carolina	7	(D)	-	-	175 *	0.9	25	-	-	-
Missouri	5	(D)	-	-	375 *	2.0	75	-	-	-
Washington	5	(D)	-	-	175 *	0.9	35	-	-	-
Iowa	3	(D)	-	-	375 *	2.0	125	-	-	-
South Dakota	3	(D)	-	-	750 *	3.9	250	-	-	-
Virginia	3	(D)	-	-	175 *	0.9	58	-	-	-
Arkansas	2	(D)	-	-	375 *	2.0	188	-	-	-
Tennessee	2	(D)	-	-	375 *	2.0	188	-	-	-
Kansas	1	(D)	-	-	175 *	0.9	175	-	-	-

Source: 1992 *Economic Census*. The states are in descending order of shipments or establishments (if shipment data are missing for the majority). The symbol (D) appears when data are withheld to prevent disclosure of competitive information. States marked with (D) are sorted by number of establishments. A dash (-) indicates that the data element cannot be calculated; * indicates the midpoint of a range.

3678 - ELECTRONIC CONNECTORS

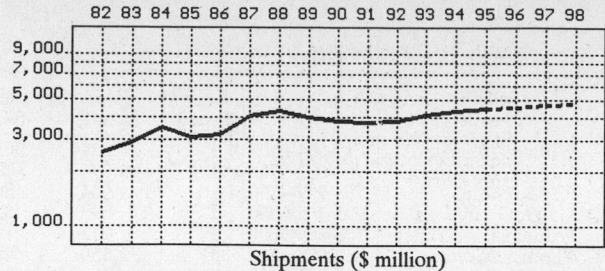

Shipments ($ million)

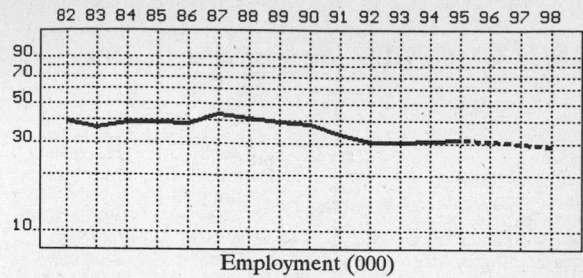

Employment (000)

GENERAL STATISTICS

Year	Com-panies	Establishments		Employment			Compensation		Production ($ million)			
		Total	with 20 or more employees	Total (000)	Production Workers (000)	Hours (Mil)	Payroll ($ mil)	Wages ($/hr)	Cost of Materials	Value Added by Manufacture	Value of Shipments	Capital Invest.
1982	148	198	159	39.7	28.0	54.2	687.5	7.72	974.9	1,633.5	2,565.0	168.7
1983		201	161	36.7	26.7	52.6	697.9	8.15	1,042.5	1,880.0	2,899.0	146.7
1984		204	163	39.2	29.2	58.1	828.4	8.65	1,299.5	2,232.3	3,491.6	184.8
1985		206	164	39.2	28.2	52.8	818.3	9.55	1,086.4	1,968.6	3,097.8	207.3
1986		209	161	38.6	27.6	53.6	872.1	9.66	1,094.0	2,137.0	3,211.8	125.4
1987	220	271	190	43.8	30.4	61.5	1,013.2	9.56	1,516.9	2,564.8	4,065.0	153.2
1988		275	197	41.5	28.9	57.4	996.0	10.22	1,704.0	2,626.6	4,333.2	132.5
1989		255	186	39.7	26.4	55.0	938.0	10.18	1,482.4	2,538.5	4,037.2	146.1
1990		249	183	38.1	26.2	54.1	958.0	10.32	1,427.5	2,389.3	3,820.9	173.5
1991		251	183	33.6	23.7	48.3	887.5	10.94	1,321.3	2,334.7	3,751.2	142.2
1992	240	285	188	30.7	21.1	42.4	909.2	11.92	1,390.2	2,385.9	3,773.5	144.3
1993		283	188	30.7	20.9	42.2	916.3	12.22	1,514.5	2,600.0	4,112.3	210.7
1994		295P	197P	30.8	21.7	45.8	954.3	12.29	1,686.2	2,637.0	4,304.3	200.4
1995		303P	200P	31.7P	21.5P	45.1P	1,010.5P	12.68P	1,743.4P	2,726.4P	4,450.2P	173.6P
1996		311P	203P	31.0P	20.8P	44.1P	1,028.8P	13.05P	1,788.1P	2,796.4P	4,564.4P	174.9P
1997		320P	206P	30.2P	20.2P	43.0P	1,047.0P	13.42P	1,832.8P	2,866.3P	4,678.6P	176.3P
1998		328P	209P	29.4P	19.5P	42.0P	1,065.3P	13.79P	1,877.6P	2,936.3P	4,792.8P	177.6P

Sources: 1982, 1987, 1992 *Economic Census*; *Annual Survey of Manufactures*, 83-86, 88-91, 93-94. Establishment counts for non-Census years are from *County Business Patterns*; establishment values for 83-84 are extrapolations. 'P's show projections by the editors. Industries reclassified in 87 will not have data for prior years.

INDICES OF CHANGE

Year	Com-panies	Establishments		Employment			Compensation		Production ($ million)			
		Total	with 20 or more employees	Total (000)	Production Workers (000)	Hours (Mil)	Payroll ($ mil)	Wages ($/hr)	Cost of Materials	Value Added by Manufacture	Value of Shipments	Capital Invest.
1982	62	69	85	129	133	128	76	65	70	68	68	117
1983		71	86	120	127	124	77	68	75	79	77	102
1984		72	87	128	138	137	91	73	93	94	93	128
1985		72	87	128	134	125	90	80	78	83	82	144
1986		73	86	126	131	126	96	81	79	90	85	87
1987	92	95	101	143	144	145	111	80	109	107	108	106
1988		96	105	135	137	135	110	86	123	110	115	92
1989		89	99	129	125	130	103	85	107	106	107	101
1990		87	97	124	124	128	105	87	103	100	101	120
1991		88	97	109	112	114	98	92	95	98	99	99
1992	100	100	100	100	100	100	100	100	100	100	100	100
1993		99	100	100	99	100	101	103	109	109	109	146
1994		103P	105P	100	103	108	105	103	121	111	114	139
1995		106P	106P	103P	102P	106P	111P	106P	125P	114P	118P	120P
1996		109P	108P	101P	99P	104P	113P	109P	129P	117P	121P	121P
1997		112P	110P	98P	96P	102P	115P	113P	132P	120P	124P	122P
1998		115P	111P	96P	92P	99P	117P	116P	135P	123P	127P	123P

Sources: Same as General Statistics. Values reflect change from the base year, 1992. Values above 100 mean greater than 92, values below 100 mean less than 92, and a value of 100 in the 82-91 or 93-98 period means same as 92. 'P's mark projections by the editors.

SELECTED RATIOS

For 1994	Avg. of All Manufact.	Analyzed Industry	Index	For 1994	Avg. of All Manufact.	Analyzed Industry	Index
Employees per Establishment	49	105	213	Value Added per Production Worker	134,084	121,521	91
Payroll per Establishment	1,500,273	3,239,241	216	Cost per Establishment	5,045,178	5,723,575	113
Payroll per Employee	30,620	30,984	101	Cost per Employee	102,970	54,747	53
Production Workers per Establishment	34	74	215	Cost per Production Worker	146,988	77,705	53
Wages per Establishment	853,319	1,910,626	224	Shipments per Establishment	9,576,895	14,610,358	153
Wages per Production Worker	24,861	25,939	104	Shipments per Employee	195,460	139,750	71
Hours per Production Worker	2,056	2,111	103	Shipments per Production Worker	279,017	198,355	71
Wages per Hour	12.09	12.29	102	Investment per Establishment	321,011	680,230	212
Value Added per Establishment	4,602,255	8,950,936	194	Investment per Employee	6,552	6,506	99
Value Added per Employee	93,930	85,617	91	Investment per Production Worker	9,352	9,235	99

Sources: Same as General Statistics. The 'Average of All Manufacturing' column represents the average of all manufacturing industries reported for the most recent complete year available. The Index shows the relationship between the Average and the Analyzed Industry. For example, 100 means that they are equal; 500 that the Analyzed Industry is five times the average; 50 means that the Analyzed Industry is half the national average. The abbreviation 'na' is used to show that data are 'not available'.

LEADING COMPANIES   Number shown: **42**   Total sales ($ mil): **7,715**   Total employment (000): **60.9**

Company Name	Address				CEO Name	Phone	Co. Type	Sales ($ mil)	Empl. (000)
AMP Inc	PO Box 3608	Harrisburg	PA	17105	William J Hudson	717-564-0100	P	4,027	30.4
Thomas and Betts Corp	1555 Lynnfield Rd	Memphis	TN	38119	T Kevin Dunnigan	901-682-7766	P	1,076	7.4
Molex Inc	2222 Wellington Ct	Lisle	IL	60532	F A Krehbiel	708-969-4550	P	964	8.2
Amphenol Corp	PO Box 5030	Wallingford	CT	06492	L J DeGeorge	203-265-8900	P	693	5.3
Methode Electronics Inc	7444 W Wilson Av	Chicago	IL	60656	William J McGinley	708-867-9600	P	213	1.8
Minnesota Mining & Mfg	PO Box 2963	Austin	TX	78769	Thomas Niccum	512-984-1800	D	150•	1.4
ITT Cannon	1851 E Deere Av	Santa Ana	CA	92705	Iain Duffin	714-261-5300	D	100	1.0
Glenair Inc	1211 Airway St	Glendale	CA	91201	Peter Kaufman	818-247-6000	R	56•	0.5
Mill-Max Manufacturing Corp	190 Pine Hollow Rd	Oyster Bay	NY	11771	RL Bahnik	516-922-6000	R	39	0.2
Homac MFG Co	12 Southland Rd	Ormond Beach	FL	32174	Eugene McGrane Jr	904-677-9110	R	30	0.5
Kings Electronics Company Inc	40 Marbledale Rd	Tuckahoe	NY	10707	Robert A Dock	914-793-5000	R	30	0.3
Molex-ETC Inc	4650 62nd Av N	Pinellas Park	FL	34665	Edwin Parkinson	813-521-2700	S	28	0.4
Positronic Industries	PO Box 8247	Springfield	MO	65801	Jack T Gentry	417-866-2322	R	28	0.5
Malco Interconnector Syst	201 Progress Dr	Montgomeryv	PA	18936	Charles Becker	215-699-5373	D	25	0.3
Trompeter Electronics Inc	31186 La Baya Dr	Westlake Vil	CA	91362	Jack Kantola	818-707-2020	S	25	0.1
Weidmuller Inc	821 Southlake Blv	Richmond	VA	23236	R Douglas White	804-794-2877	R	25	0.1
Airborn Inc	4321 Airborn Dr	Addison	TX	75001	Jay G McKie	214-931-3200	R	23	0.3
ITT Sealectro	585 E Main St	New Britain	CT	06051	Dennis Reed	203-223-2700	D	20	0.3
Aries Electronics Inc	PO Box 130	Frenchtown	NJ	08825	William Y Sinclair	908-996-6841	R	17	0.2
Zero Defects Inc	615 E 43rd St	Boise	ID	83714	Larry Hyatt	208-322-0500	R	16•	0.2
Circuit Assembly Corp	18 Thomas St	Irvine	CA	92718	R Lang	714-855-7887	R	15	0.3
Applied Engineering Products	PO Box 510	New Haven	CT	06513	Benjamin P Trivelli	203-776-2813	R	14•	0.1
Richards Manufacturing Ltd	517 Lyons Av	Irvington	NJ	07111	Horace Bier	201-371-1771	R	13	0.1
Hermetic Seal Corp	4232 Temple City	Rosemead	CA	91770	A Goldfarb	818-443-8931	S	11	0.2
Special Mine Services Inc	PO Box 188	West Frankfort	IL	62896	Les Huntsman	618-932-2151	R	9•	<0.1
Leviton Telcom	2222 222nd St SE	Bothell	WA	98021	Jay Garthwaite	206-486-2222	D	8•	0.1
WECO Electrical Connectors	Trimex Bldg	Mooers	NY	12958	Heiner Kammann	518-298-4810	R	8•	<0.1
George Diamond and Company	108 Wilmot Rd	Deerfield	IL	60015	George Diamond	708-940-8100	R	7	<0.1
Kemlon Prod & Development	PO Box 14666	Houston	TX	77221	S Ring	713-747-5020	R	6	0.2
Dossert Corp	500 Captain Neville	Waterbury	CT	06705	William Pastor	203-573-1616	R	5	<0.1
Modular Devices Inc	1 Roned Rd	Shirley	NY	11967	SE Summer	516-345-3100	R	5•	<0.1
Eby Co	4300 H St	Philadelphia	PA	19124	J Albrecht	215-537-4700	R	5	<0.1
IEH Corp	140 58th St	Brooklyn	NY	11220	Michael Offerman	718-492-4440	P	5	0.1
Crane Electronics Inc	4700 Smith Rd	Cincinnati	OH	45212	John Habbert	513-631-4700	R	4•	<0.1
Hobson Brothers Inc	4940 W Lawrence	Chicago	IL	60630	EJ Hobson	312-283-3600	R	3	<0.1
Wieland Inc	49 International Rd	Burgaw	NC	28425	Tony Chiarello	919-259-5050	S	3•	<0.1
Automatic Connector	400 Moreland Rd	Commack	NY	11725	Pierre D Lax	516-543-5000	R	2•	<0.1
E-Z-Hook	PO Box 660729	Arcadia	CA	91066	Phelps M Wood	818-446-6175	D	2	<0.1
Harting Electronik Inc	2155 Stonington	Hoffman Est	IL	60195	Peter Tillmann	708-519-7700	R	2•	<0.1
AirMouse Remote Controls Inc	PO Box 100	Williston	VT	05495	James D Richards	802-878-9600	R	1	<0.1
Electronic Contract Svcs Corp	14711 Sinclair Cir	Tustin	CA	92680	Barrett N Brown	714-730-5225	R	1•	<0.1
On-Shore Technology Inc	1917 W 3rd St	Tempe	AZ	85281	Serge Mandell	602-921-3000	R	1•	<0.1

Source: Ward's Business Directory of U.S. Private and Public Companies, Volumes 1 and 2, 1996. The company type code used is as follows: P - Public, R - Private, S - Subsidiary, D - Division, J - Joint Venture, A - Affiliate, G - Group. Sales are in millions of dollars, employees are in thousands. An asterisk (*) indicates an estimated sales volume. The symbol < stands for 'less than'. Company names and addresses are truncated, in some cases, to fit into the available space.

MATERIALS CONSUMED

Material	Quantity	Delivered Cost ($ million)
Materials, ingredients, containers, and supplies	(X)	1,206.0
Tube blanks	(X)	(D)
Printed circuit boards (without inserted components) for electronic circuitry	(X)	4.5
Printed circuit assemblies, loaded boards or modules	(X)	6.3
Semiconductors, including transistors, diodes, rectifiers, and integrated circuits for electronic circuitry	(X)	6.3
Capacitors for electronic circuitry	(X)	4.9
Resistors for electronic circuitry	(X)	0.2
Other components and accessories for electronic circuitry, nec, except tubes	(X)	17.1
Silicon, hyperpure	(X)	(D)
Gold and other precious metals, all forms (including ingot, sheet, strip, solder, plating, electrodes, etc.)	(X)	135.2
Doped chemicals, and other doped materials for electronic use	(X)	3.8
Ferrites (powder and paste)	(X)	(D)
Metal powders	(X)	(D)
Electronic computing equipment	(X)	(D)
Current-carrying wiring devices	(X)	19.1
Electronic communication equipment	(X)	(D)
Electrical instrument mechanisms and meter movements (including instrument relays)	(X)	0.4
Optical instruments and lenses (except sighting, tracking, and fire control)	(X)	(D)
Plastics resins consumed in the form of granules, pellets, powders, liquids, etc.	(X)	65.6
Fabricated plastics products (except gaskets, hoses, and belting)	(X)	120.1
Sheet metal products, except stampings	(X)	5.7
Metal stampings	(X)	140.2
Metal bolts, nuts, screws, washers, rivets, and other screw machine products	(X)	83.5
Other fabricated metal products (except forgings)	(X)	46.7
Forgings	(X)	0.1
Castings (rough and semifinished)	(X)	15.6
Steel shapes and forms	(X)	15.7
Copper and copper-base alloy shapes and forms	(X)	40.8
Aluminum and aluminum-base alloy shapes and forms	(X)	21.4
Other nonferrous shapes and forms	(X)	19.0
Insulated wire and cable, including magnet wire	(X)	30.7
Paper and paperboard containers, including shipping sacks and other paper packaging supplies	(X)	14.0
All other materials and components, parts, containers, and supplies	(X)	217.0
Materials, ingredients, containers, and supplies, nsk	(X)	168.0

Source: 1992 Economic Census. Explanation of symbols used: (D): Withheld to avoid disclosure of competitive data; na: Not available; (S): Withheld because statistical norms were not met; (X): Not applicable; (Z): Less than half the unit shown; nec: Not elsewhere classified; nsk: Not specified by kind; - : zero; * : 10-19 percent estimated; ** : 20-29 percent estimated.

PRODUCT SHARE DETAILS

Product or Product Class	% Share	Product or Product Class	% Share
Connectors for electronic circuitry	100.00	circuitry	13.97
Coaxial (RF) connectors for electronic circuitry	11.40	Printed circuit connectors for electronic circuitry	22.08
Cylindrical connectors for electronic circuitry	14.67	Other connectors for electronic circuitry, including parts	31.62
Rack and panel (rectangular) connectors for electronic		Connectors for electronic circuitry, nsk	6.26

Source: 1992 Economic Census. The values shown are percent of total shipments in an industry. Values of indented subcategories are summed in the main heading. The symbol (D) appears when data are withheld to prevent disclosure of competitive information. The abbreviation nsk stands for 'not specified by kind' and nec for 'not elsewhere classified'.

INPUTS AND OUTPUTS FOR ELECTRONIC COMPONENTS NEC

Economic Sector or Industry Providing Inputs	%	Sector	Economic Sector or Industry Buying Outputs	%	Sector
Miscellaneous plastics products	16.2	Manufg.	Radio & TV communication equipment	13.5	Manufg.
Imports	15.7	Foreign	Exports	9.8	Foreign
Electronic components nec	15.5	Manufg.	Telephone & telegraph apparatus	9.6	Manufg.
Wholesale trade	10.9	Trade	Electronic computing equipment	9.4	Manufg.
Plating & polishing	5.2	Manufg.	Electronic components nec	9.1	Manufg.
Primary nonferrous metals, nec	2.6	Manufg.	Radio & TV receiving sets	5.8	Manufg.
Electric services (utilities)	2.0	Util.	Guided missiles & space vehicles	3.2	Manufg.
Semiconductors & related devices	1.9	Manufg.	Personal consumption expenditures	2.9	
Advertising	1.5	Services	X-ray apparatus & tubes	2.6	Manufg.
Communications, except radio & TV	1.4	Util.	Aircraft	2.2	Manufg.
Metal stampings, nec	1.3	Manufg.	Instruments to measure electricity	2.2	Manufg.
Banking	1.2	Fin/R.E.	Aircraft & missile equipment, nec	2.0	Manufg.
Aluminum rolling & drawing	1.1	Manufg.	Semiconductors & related devices	2.0	Manufg.
Copper rolling & drawing	1.1	Manufg.	Industrial controls	1.9	Manufg.
Nonferrous wire drawing & insulating	1.1	Manufg.	Optical instruments & lenses	1.7	Manufg.
Air transportation	1.1	Util.	Computer & data processing services	1.7	Services
Metal coating & allied services	1.0	Manufg.	Federal Government purchases, national defense	1.7	Fed Govt
Blast furnaces & steel mills	0.9	Manufg.	Games, toys, & children's vehicles	1.6	Manufg.
Maintenance of nonfarm buildings nec	0.8	Constr.	Mechanical measuring devices	1.3	Manufg.
Plastics materials & resins	0.8	Manufg.	Calculating & accounting machines	1.2	Manufg.

Continued on next page.

INPUTS AND OUTPUTS FOR ELECTRONIC COMPONENTS NEC - Continued

Economic Sector or Industry Providing Inputs	%	Sector	Economic Sector or Industry Buying Outputs	%	Sector
Motor freight transportation & warehousing	0.8	Util.	Electron tubes	1.2	Manufg.
Real estate	0.8	Fin/R.E.	Photographic equipment & supplies	1.1	Manufg.
Chemical preparations, nec	0.7	Manufg.	Radio & TV broadcasting	1.0	Util.
Petroleum refining	0.7	Manufg.	Change in business inventories	1.0	In House
Screw machine and related products	0.7	Manufg.	Communications, except radio & TV	0.8	Util.
Eating & drinking places	0.7	Trade	Federal Government purchases, nondefense	0.7	Fed Govt
Equipment rental & leasing services	0.6	Services	Engine electrical equipment	0.6	Manufg.
Hotels & lodging places	0.6	Services	Musical instruments	0.6	Manufg.
Cyclic crudes and organics	0.5	Manufg.	Surgical appliances & supplies	0.6	Manufg.
Electronic computing equipment	0.5	Manufg.	Switchgear & switchboard apparatus	0.6	Manufg.
Paperboard containers & boxes	0.5	Manufg.	Phonograph records & tapes	0.5	Manufg.
Sheet metal work	0.5	Manufg.	Management & consulting services & labs	0.5	Services
Adhesives & sealants	0.4	Manufg.	Engineering & scientific instruments	0.4	Manufg.
Machinery, except electrical, nec	0.4	Manufg.	Surgical & medical instruments	0.4	Manufg.
Nonferrous rolling & drawing, nec	0.4	Manufg.	Banking	0.4	Fin/R.E.
Gas production & distribution (utilities)	0.4	Util.	Environmental controls	0.3	Manufg.
Computer & data processing services	0.4	Services	Typewriters & office machines, nec	0.3	Manufg.
Legal services	0.4	Services	Libraries, vocation education	0.3	Services
Carbon & graphite products	0.3	Manufg.	Aircraft & missile engines & engine parts	0.2	Manufg.
Fabricated rubber products, nec	0.3	Manufg.	Ammunition, except for small arms, nec	0.2	Manufg.
Primary metal products, nec	0.3	Manufg.	Electrical equipment & supplies, nec	0.2	Manufg.
Special dies & tools & machine tool accessories	0.3	Manufg.	Detective & protective services	0.2	Services
Wiring devices	0.3	Manufg.	Elementary & secondary schools	0.2	Services
Management & consulting services & labs	0.3	Services	Gross private fixed investment	0.2	Cap Inv
Noncomparable imports	0.3	Foreign	Electrical industrial apparatus, nec	0.1	Manufg.
Aluminum castings	0.2	Manufg.	Motor vehicle parts & accessories	0.1	Manufg.
Instruments to measure electricity	0.2	Manufg.	Motor vehicles & car bodies	0.1	Manufg.
Miscellaneous fabricated wire products	0.2	Manufg.	Scales & balances	0.1	Manufg.
Sanitary services, steam supply, irrigation	0.2	Util.	S/L Govt. purch., higher education	0.1	S/L Govt
Accounting, auditing & bookkeeping	0.2	Services			
U.S. Postal Service	0.2	Gov't			
Abrasive products	0.1	Manufg.			
Glass & glass products, except containers	0.1	Manufg.			
Industrial inorganic chemicals, nec	0.1	Manufg.			
Lubricating oils & greases	0.1	Manufg.			
Manifold business forms	0.1	Manufg.			
Metal foil & leaf	0.1	Manufg.			
Paper coating & glazing	0.1	Manufg.			
Railroads & related services	0.1	Util.			
Insurance carriers	0.1	Fin/R.E.			
Automotive rental & leasing, without drivers	0.1	Services			
Colleges, universities, & professional schools	0.1	Services			

Source: *Benchmark Input-Output Accounts for the U.S. Economy, 1982*, U.S. Department of Commerce, Washington, D.C., July 1991. Data, as reported in the source, are organized by the 1977 SIC structure in use in 1982 but have been matched, as closely as is possible, to the 1987 SIC structure used in this book.

OCCUPATIONS EMPLOYED BY SIC 367 - ELECTRONIC COMPONENTS AND ACCESSORIES

Occupation	% of Total 1994	Change to 2005	Occupation	% of Total 1994	Change to 2005
Electrical & electronic assemblers	9.8	-8.4	Secretaries, ex legal & medical	1.6	-7.4
Electrical & electronic equipment assemblers	5.7	1.8	Industrial production managers	1.5	-28.8
Electrical & electronic technicians,technologists	5.4	22.1	Electromechanical equipment assemblers	1.4	11.9
Inspectors, testers, & graders, precision	5.4	-28.8	Coil winders, tapers, & finishers	1.3	-18.6
Electronic semiconductor processors	5.3	-1.4	Production, planning, & expediting clerks	1.3	1.8
Electrical & electronics engineers	4.4	29.8	Industrial engineers, ex safety engineers	1.3	12.0
Assemblers, fabricators, & hand workers nec	4.0	1.8	Electrolytic plating machine workers	1.2	11.9
Blue collar worker supervisors	3.4	-14.1	Traffic, shipping, & receiving clerks	1.2	-2.1
Sales & related workers nec	2.2	1.8	Freight, stock, & material movers, hand	1.2	-18.6
Computer engineers	2.2	50.7	Precision assemblers nec	1.1	62.9
Engineering, mathematical, & science managers	1.8	15.6	Stock clerks	1.0	-17.3
Engineering technicians & technologists nec	1.8	1.8	Engineers nec	1.0	52.6
General managers & top executives	1.6	-3.4	General office clerks	1.0	-13.2

Source: *Industry-Occupation Matrix*, Bureau of Labor Statistics. These data relate to one or more 3-digit SIC industry groups rather than to a single 4-digit SIC. The change reported for each occupation to the year 2005 is a percent of growth or decline as estimated by the Bureau of Labor Statistics. The abbreviation nec stands for 'not elsewhere classified'.

LOCATION BY STATE AND REGIONAL CONCENTRATION

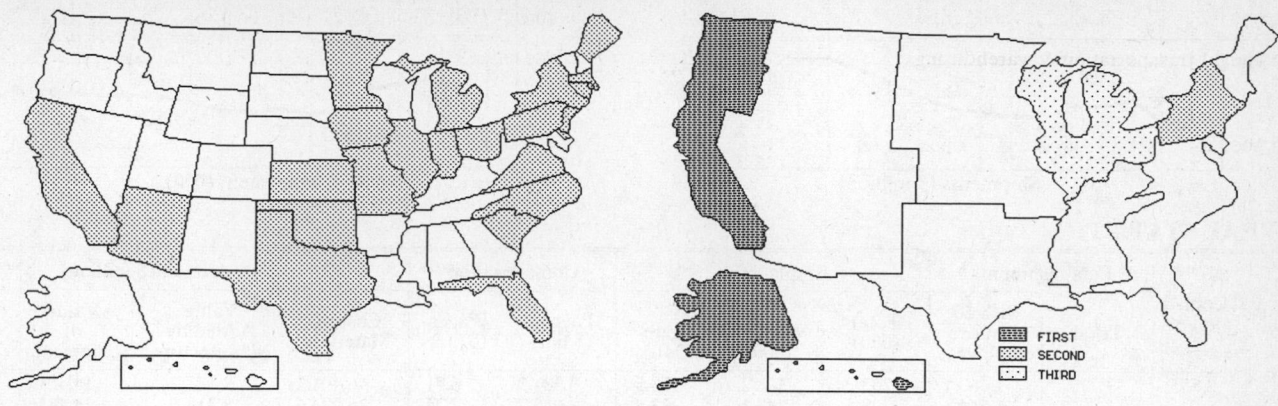

FIRST
SECOND
THIRD

INDUSTRY DATA BY STATE

| State | Establish-ments | Shipments | | | Employment | | | | Cost as % of Shipments | Investment per Employee ($) |
		Total ($ mil)	% of U.S.	Per Establ.	Total Number	% of U.S.	Per Establ.	Wages ($/hour)		
Pennsylvania	32	727.7	19.3	22.7	4,100	13.4	128	11.47	43.7	7,512
California	71	715.3	19.0	10.1	8,100	26.4	114	11.56	34.2	3,568
New York	20	303.2	8.0	15.2	3,200	10.4	160	13.24	30.6	-
Missouri	5	269.4	7.1	53.9	2,200	7.2	440	15.31	41.5	-
Connecticut	13	153.4	4.1	11.8	1,500	4.9	115	11.92	38.6	2,267
Illinois	15	124.6	3.3	8.3	1,400	4.6	93	10.53	30.6	4,786
Massachusetts	15	91.2	2.4	6.1	1,100	3.6	73	10.35	31.9	2,636
New Jersey	13	72.2	1.9	5.6	800	2.6	62	10.00	42.8	2,500
Ohio	8	64.1	1.7	8.0	500	1.6	63	9.20	20.7	2,800
Minnesota	8	52.4	1.4	6.6	700	2.3	88	10.00	35.9	-
Rhode Island	6	19.1	0.5	3.2	300	1.0	50	7.75	38.7	-
Texas	10	(D)	-	-	750 *	2.4	75	-	-	-
Arizona	8	(D)	-	-	750 *	2.4	94	-	-	3,467
Florida	8	(D)	-	-	375 *	1.2	47	-	-	1,067
Indiana	8	(D)	-	-	750 *	2.4	94	-	-	-
North Carolina	6	(D)	-	-	1,750 *	5.7	292	-	-	-
Virginia	5	(D)	-	-	750 *	2.4	150	-	-	-
Michigan	4	(D)	-	-	375 *	1.2	94	-	-	-
Kansas	2	(D)	-	-	175 *	0.6	88	-	-	-
Oklahoma	2	(D)	-	-	375 *	1.2	188	-	-	-
South Carolina	2	(D)	-	-	375 *	1.2	188	-	-	-
Delaware	1	(D)	-	-	175 *	0.6	175	-	-	-
Iowa	1	(D)	-	-	175 *	0.6	175	-	-	-
Maine	1	(D)	-	-	175 *	0.6	175	-	-	-

Source: 1992 *Economic Census*. The states are in descending order of shipments or establishments (if shipment data are missing for the majority). The symbol (D) appears when data are withheld to prevent disclosure of competitive information. States marked with (D) are sorted by number of establishments. A dash (-) indicates that the data element cannot be calculated; * indicates the midpoint of a range.

3679 - ELECTRONIC COMPONENTS, NEC

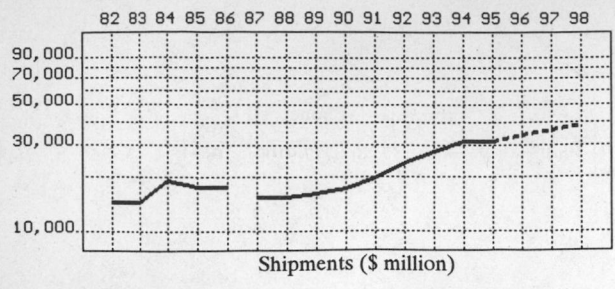

Shipments ($ million)

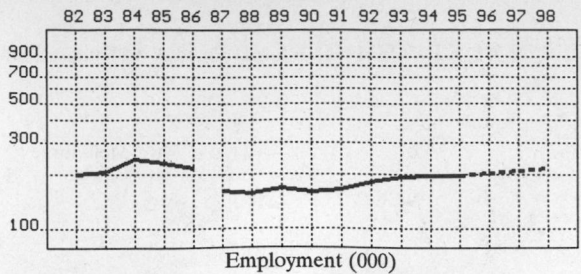

Employment (000)

GENERAL STATISTICS

Year	Com- panies	Establishments		Employment			Compensation		Production ($ million)			
		Total	with 20 or more employees	Total (000)	Production Workers (000)	Hours (Mil)	Payroll ($ mil)	Wages ($/hr)	Cost of Materials	Value Added by Manufacture	Value of Shipments	Capital Invest.
1982	3,575	3,770	1,463	203.2	138.2	270.5	3,584.9	6.91	6,401.1	8,127.2	14,401.6	893.5
1983		3,706	1,564	208.5	143.1	283.8	3,920.7	7.24	6,094.6	8,318.6	14,415.9	691.3
1984		3,642	1,665	243.3	167.7	327.6	4,722.2	7.48	8,433.4	10,833.0	19,119.6	1,090.3
1985		3,577	1,767	232.1	152.8	297.7	4,748.8	7.89	7,652.1	9,996.3	17,523.7	1,062.8
1986		3,483	1,669	217.7	140.4	277.6	4,767.2	8.24	7,776.4	9,506.3	17,514.6	803.8
1987*	2,561	2,900	1,185	162.6	97.3	187.9	3,890.7	8.93	7,285.2	8,383.2	15,438.5	533.3
1988		2,846	1,326	157.6	99.9	196.9	3,738.1	8.98	7,139.2	8,254.0	15,299.4	448.1
1989		2,499	1,207	166.8	103.3	202.9	4,013.8	8.94	8,403.6	8,168.2	16,122.5	539.4
1990		2,441	1,184	161.0	101.1	203.2	4,107.6	9.06	8,380.0	8,727.1	17,222.4	485.0
1991		2,503	1,187	165.4	101.6	219.5	4,530.5	8.99	9,875.8	9,513.0	19,450.3	540.3
1992	3,108	3,295	1,347	182.4	109.2	238.8	5,180.5	9.32	11,842.4	11,925.9	23,869.9	740.8
1993		2,995	1,290	195.2	115.5	257.2	5,857.6	9.27	16,051.6	11,148.9	27,687.1	991.9
1994		2,952P	1,295P	196.5	118.9	261.6	5,964.7	9.33	16,872.5	14,921.5	31,609.5	990.6
1995		2,995P	1,307P	199.0P	119.1P	271.4P	6,216.1P	9.39P	16,859.0P	14,909.5P	31,584.2P	1,011.2P
1996		3,037P	1,319P	204.6P	122.0P	282.6P	6,561.8P	9.45P	18,133.7P	16,036.9P	33,972.3P	1,089.5P
1997		3,079P	1,331P	210.3P	125.0P	293.8P	6,907.5P	9.51P	19,408.5P	17,164.2P	36,360.5P	1,167.8P
1998		3,122P	1,343P	216.0P	127.9P	305.1P	7,253.2P	9.58P	20,683.2P	18,291.6P	38,748.7P	1,246.1P

Sources: 1982, 1987, 1992 *Economic Census*; *Annual Survey of Manufactures*, 83-86, 88-91, 93-94. Establishment counts are from *County Business Patterns* for non-Census years; establishment counts for 83-84 are extrapolations. * indicates that industry content changed in 87; earlier years use 77 SICs. 'P's mark projections.

INDICES OF CHANGE

Year	Com- panies	Establishments		Employment			Compensation		Production ($ million)			
		Total	with 20 or more employees	Total (000)	Production Workers (000)	Hours (Mil)	Payroll ($ mil)	Wages ($/hr)	Cost of Materials	Value Added by Manufacture	Value of Shipments	Capital Invest.
1982	115	114	109	111	127	113	69	74	54	68	60	121
1983		112	116	114	131	119	76	78	51	70	60	93
1984		111	124	133	154	137	91	80	71	91	80	147
1985		109	131	127	140	125	92	85	65	84	73	143
1986		106	124	119	129	116	92	88	66	80	73	109
1987*	82	88	88	89	89	79	75	96	62	70	65	72
1988		86	98	86	91	82	72	96	60	69	64	60
1989		76	90	91	95	85	77	96	71	68	68	73
1990		74	88	88	93	85	79	97	71	73	72	65
1991		76	88	91	93	92	87	96	83	80	81	73
1992	100	100	100	100	100	100	100	100	100	100	100	100
1993		91	96	107	106	108	113	99	136	93	116	134
1994		90P	96P	108	109	110	115	100	142	125	132	134
1995		91P	97P	109P	109P	114P	120P	101P	142P	125P	132P	136P
1996		92P	98P	112P	112P	118P	127P	101P	153P	134P	142P	147P
1997		93P	99P	115P	114P	123P	133P	102P	164P	144P	152P	158P
1998		95P	100P	118P	117P	128P	140P	103P	175P	153P	162P	168P

Sources: Same as General Statistics. Values reflect change from the base year, 1992. Values above 100 mean greater than 92, values below 100 mean less than 92, and a value of 100 in the 82-91 or 93-98 period means same as 92. * indicates that industry content changed in 87. Data for earlier years are in 77 SIC format.

SELECTED RATIOS

For 1994	Avg. of All Manufact.	Analyzed Industry	Index	For 1994	Avg. of All Manufact.	Analyzed Industry	Index
Employees per Establishment	49	67	136	Value Added per Production Worker	134,084	125,496	94
Payroll per Establishment	1,500,273	2,020,367	135	Cost per Establishment	5,045,178	5,715,063	113
Payroll per Employee	30,620	30,355	99	Cost per Employee	102,970	85,865	83
Production Workers per Establishment	34	40	117	Cost per Production Worker	146,988	141,905	97
Wages per Establishment	853,319	826,725	97	Shipments per Establishment	9,576,895	10,706,789	112
Wages per Production Worker	24,861	20,528	83	Shipments per Employee	195,460	160,863	82
Hours per Production Worker	2,056	2,200	107	Shipments per Production Worker	279,017	265,849	95
Wages per Hour	12.09	9.33	77	Investment per Establishment	321,011	335,537	105
Value Added per Establishment	4,602,255	5,054,219	110	Investment per Employee	6,552	5,041	77
Value Added per Employee	93,930	75,936	81	Investment per Production Worker	9,352	8,331	89

Sources: Same as General Statistics. The 'Average of All Manufacturing' column represents the average of all manufacturing industries reported for the most complete year available. The Index shows the relationship between the Average and the Analyzed Industry. For example, 100 means that they are equal; 500 that the Analyzed Industry is five times the average; 50 means that the Analyzed Industry is half the national average. The abbreviation 'na' is used to show that data are 'not available'.

LEADING COMPANIES   Number shown: **75**   Total sales ($ mil): **19,211**   Total employment (000): **165.9**

Company Name	Address				CEO Name	Phone	Co. Type	Sales ($ mil)	Empl. (000)
Harris Corp	1025 W NASA Blv	Melbourne	FL	32919	Phillip W Farmer	407-727-9100	P	3,336	28.2
General Instrument Corp	181 W Madison St	Chicago	IL	60602	Daniel F Akerson	312-541-5000	P	2,036	11.0
SCI Systems Inc	PO Box 1000	Huntsville	AL	35807	Olin B King	205-882-4601	P	1,853	12.0
ITT Defense and Electronics Inc	1650 Tysons Blv	McLean	VA	22102	Louis J Giuliano	703-790-6300	S	1,710	15.0
Read-Rite Corp	345 Los Coches St	Milpitas	CA	95035	Cyril J Yansouni	408-262-6700	P	639	18.5
Andrew Corp	10500 W 153rd St	Orland Park	IL	60462	Floyd L English	708-349-3300	P	558	3.1
Augat Inc	PO Box 448	Mansfield	MA	02048	William R Fenoglio	508-543-4300	P	530	4.4
ESCO Electronics Corp	8100 W Florissant	St Louis	MO	63136	Dennis J Moore	314-553-7777	P	474	3.7
Philips Components	2001 W Blue Heron	Riviera Beach	FL	33404	Robert Chlebek	407-881-3200	D	450•	5.5
Magnavox Electronics Syst Co	1313 Production Rd	Fort Wayne	IN	46808	David Molfenter	219-429-6000	R	440	3.5
Sanders	PO Box 868	Nashua	NH	03061	John R Kreick	603-885-4321	S	410•	4.5
Mitsumi Electronics Corp	6210 N Beltline Rd	Irving	TX	75063	Tak Kumagai	214-550-7300	S	400	<0.1
Northrop Grumman	600 Hicks Rd	Rolling Mdws	IL	60008	Robert W Mendell	708-259-9600	D	350	2.5
Phillips Components	2001 W Blue Heron	Riviera Beach	FL	33404	Dennis Horowitz	407-881-3200	S	320•	3.5
CTS Corp	905 West Blv N	Elkhart	IN	46514	Joseph P Walker	219-293-7511	P	269	4.1
Park Electrochemical Corp	5 Dakota Dr	Lake Success	NY	11042	Jerry Shore	516-354-4100	P	253	1.8
Amer Power Conversion Corp	PO Box 278	West Kingston	RI	02892	R B Dowdell Jr	401-789-5735	P	250	0.9
Oak Industries Inc	1000 Winter St	Waltham	MA	02154	Williams S Antle III	617-890-0400	P	249	2.8
BMC Industries Inc	2 Appletree Sq	Minneapolis	MN	55425	Paul B Burke	612-851-6000	P	220	1.8
GEC-Marconi	PO Box 975	Wayne	NJ	07474	Mark H Ronald	201-633-6000	S	200	1.2
Kimball Electronics Group	1038 E 15th St	Jasper	IN	47549	Larry Kuntz	812-482-1600	S	200	1.5
Tech-Sym Corp	10500 Westoffice Dr	Houston	TX	77042	Wendell W Gamel	713-785-7790	P	198	2.0
Kent Electronics Corp	7433 Harwin Dr	Houston	TX	77036	Morrie K Abramson	713-780-7770	P	193	0.9
EMD Associates Inc	4065 Theurer Blv	Winona	MN	55987	David W Fradin	507-452-8932	R	160	0.9
Sparton Corp	2400 E Ganson St	Jackson	MI	49202	Richard Nichols	517-787-8600	D	153	1.0
Technitrol Inc	1210 Northbrook	Trevose	PA	19053	Roy E Hock	215-355-2900	P	146	3.1
Proxima Corp	9440 Carroll Pk Dr	San Diego	CA	92121	Kenneth E Olson	619-457-5500	P	136	0.6
Zytec Corp	7575 Mkt Place Dr	Eden Prairie	MN	55344	Ronald D Schmidt	612-941-1100	P	128	1.0
In Focus Systems Inc	27700B SW Parkway	Wilsonville	OR	97070	John V Harker	503-658-8888	P	123	0.4
Satellite Transmission Systems	125 Kennedy Dr	Hauppauge	NY	11788	Brian Maloney	516-231-1919	S	120•	0.5
Kavlico Corp	14501 W LA	Moorpark	CA	93021	M Gibson	805-523-2000	R	120•	1.3
Electromagnetic Sciences Inc	PO Box 7700	Norcross	GA	30091	Thomas E Sharon	404-263-9200	P	118	1.0
Sunward Technologies Inc	5828 Pacific Ctr	San Diego	CA	92121	Gregorio Reyes	619-587-9140	P	117	2.6
Stanford Telecommunications	PO Box 3733	Sunnyvale	CA	94088	James J Spilker Jr	408-745-0818	P	114	1.0
Electrode Corp	100 7th Av	Chardon	OH	44024	D Baxendale	216-285-0300	D	100	0.3
Optrex America Inc	44160ymouth Oaks	Plymouth	MI	48170	Makoto Yamamoto	313-416-8500	S	100	<0.1
Dynamics Corporation	475 Steamboat Rd	Greenwich	CT	06830	Andrew Lozyniak	203-869-3211	P	97	1.1
Sheldahl Inc	PO Box 170	Northfield	MN	55057	James E Donaghy	507-663-8000	P	88	1.0
Shugart Corp	9292 Jeronimo Rd	Irvine	CA	92718	Dennis Narlinger	714-770-1100	R	85•	1.0
Burndy Corp	51 Richards Av	Norwalk	CT	06856	John Mayo	203-838-4444	S	76•	1.0
Alpine Group Inc	1790 Broadway	New York	NY	10019	Steven S Elbaum	212-757-3333	P	74	0.6
General Automotive Specialty	US Hwy 1 & 130	N Brunswick	NJ	08902	Jesse Herman	908-545-7000	R	69•	0.8
Electro Scientific Industries Inc	13900 NW Science	Portland	OR	97229	D R VanLuvanee	503-641-4141	P	68	0.5
Pulse Engineering Inc	PO Box 12235	San Diego	CA	92112	David R Flowers	619-674-8100	P	65	0.4
Siemens Components Inc	10950 N Tantau Av	Cupertino	CA	95014	Alex Luepp	408-777-4500	D	65	0.2
United Technologies	PO Box 1049	Traverse City	MI	49685	Richard Green	616-947-3000	D	65•	0.9
Amtech Corp	17304 Preston Rd	Dallas	TX	75252	GR Mortenson	214-733-6600	P	62	0.3
Escod Industries	2411 N Oak St	Myrtle Beach	SC	29577	J D Oberlender	803-946-6200	D	60	0.5
RLC Electronics Inc	83 Radio Cir	Mount Kisco	NY	10549	CA Borck	914-241-1334	R	60	<0.1
REMEC Inc	9404 Chesapeake	San Diego	CA	92123	Ronald R Ragland	619-560-1301	R	59	0.7
Teledyne Electronic	PO Box 326	Lewisburg	TN	37091	Don Woods	615-359-4531	D	54	0.4
Periphonics Corp	4000 Veterans Mem	Bohemia	NY	11716	Peter J Cohen	516-467-0500	P	52	0.4
Lectron Products Inc	1400 S Livernois	Rochester Hills	MI	48307	Walter McPhail	313-656-0880	R	50•	0.8
Quality Technologies Corp	610 N Mary Av	Sunnyvale	CA	94086	Ralph Simon	408-720-1440	R	50	1.4
Bergquist Company Inc	5300 Edina Ind'l Bl	Edina	MN	55439	C Bergquist Jr	612-835-2322	R	46	0.3
Celwave	2 Ryan Rd	Marlboro	NJ	07746	Richard Tallon	908-462-1880	D	46•	0.5
Lambda Electronics Inc	515 Broad Hollow	Melville	NY	11747	Joshua Hauser	516-694-4200	S	46•	0.5
Manu-Tronics Inc	8701 100th St	Kenosha	WI	53142	Roger Mayer	414-947-7700	R	45	0.4
Accudyne Corp	PO Box 1429	Janesville	WI	53547	Edmund Durkee	608-752-4053	D	41•	0.5
Circle S Industries Inc	PO Box 321429	Birmingham	AL	35232	LD Striplin Jr	205-875-4040	R	40	0.1
Collmer Semiconductor Inc	14368 Proton Dr	Dallas	TX	75244	Jan Collmer	214-233-1589	R	40	0.2
Datel Inc	11 Cabot Blv	Mansfield	MA	02048	NG Tagaris	508-339-3000	R	40	0.4
Dolby Laboratories Inc	100 Potrero Av	San Francisco	CA	94103	Ray Dolby	415-558-0200	R	40	0.3
GW Lisk Company Inc	2 South St	Clifton Springs	NY	14432	IA Morris Jr	315-462-2611	R	40	0.4
M/A-COM Omni Spectra Inc	100 Chelmsford St	Lowell	MA	01851	Thomas Vanderslice	508-442-5000	S	40•	0.6
Teledyne Electronic	1274 Terra Bella	Mountain View	CA	94043	Charlie Kelley	415-962-6944	D	40	0.5
Varian Associates	150 Sohier Rd	Beverly	MA	01915	Dennis Gleason	508-922-6000	D	40	0.3
Johnson Matthey Electronics Inc	15128 E Euclid Av	Spokane	WA	99216	Geoff Wilde	509-924-2200	S	39	0.4
Alpha Wire Corp	711 Lidgerwood Av	Elizabeth	NJ	07207	Philip R Cowen	908-925-8000	R	38	0.4
Mid-South Electrics	711 N 8th St	Gadsden	AL	35901	David Smith	205-543-9683	S	38	0.4
Pure Carbon Co	441 Hall Av	St Marys	PA	15857	Scott Brown	814-781-1573	D	38•	0.3
Ark-Les Corp	51 Water St	Watertown	MA	02172	Bruce Mac Neil	617-924-2330	R	37	0.4
Centurion International Inc	3425 N 44th St	Lincoln	NE	68504	Gary Kuck	402-467-4491	R	36•	0.4
First Inertia Switch Ltd	PO Box 704	Grand Blanc	MI	48439	Alan Adams	810-695-8333	S	36	0.2
Aavid Thermal Technologies Inc	PO Box 400	Laconia	NH	03247	Alan Beane	603-528-3400	R	35	0.7

Source: Ward's Business Directory of U.S. Private and Public Companies, Volumes 1 and 2, 1996. The company type code used is as follows: P - Public, R - Private, S - Subsidiary, D - Division, J - Joint Venture, A - Affiliate, G - Group. Sales are in millions of dollars, employees are in thousands. An asterisk (*) indicates an estimated sales volume. The symbol < stands for 'less than'. Company names and addresses are truncated, in some cases, to fit into the available space.

MATERIALS CONSUMED

Material	Quantity	Delivered Cost ($ million)
Materials, ingredients, containers, and supplies	(X)	10,892.3
Tube blanks	(X)	7.9
Printed circuit boards (without inserted components) for electronic circuitry	(X)	445.0
Semiconductors, including transistors, diodes, rectifiers, and integrated circuits for electronic circuitry	(X)	1,238.1
Capacitors for electronic circuitry	(X)	177.1
Resistors for electronic circuitry	(X)	99.8
Other components and accessories for electronic circuitry, nec, except tubes	(X)	2,993.4
Silicon, hyperpure	(X)	3.4
Gold and other precious metals, all forms (including ingot, sheet, strip, solder, plating, electrodes, etc.)	(X)	43.8
Doped chemicals, and other doped materials for electronic use	(X)	3.1
Ferrites (powder and paste)	(X)	15.3
Metal powders	(X)	22.0
Electronic computing equipment	(X)	(D)
Current-carrying wiring devices	(X)	(D)
Electronic communication equipment	(X)	53.6
Electrical instrument mechanisms and meter movements (including instrument relays)	(X)	31.5
Optical instruments and lenses (except sighting, tracking, and fire control)	(X)	5.9
Plastics resins consumed in the form of granules, pellets, powders, liquids, etc.	(X)	42.3
Fabricated plastics products (except gaskets, hoses, and belting)	(X)	55.2
Sheet metal products, except stampings	(X)	76.2
Metal stampings	(X)	68.6
Metal bolts, nuts, screws, washers, rivets, and other screw machine products	(X)	44.9
Other fabricated metal products (except forgings)	(X)	80.5
Forgings	(X)	1.4
Castings (rough and semifinished)	(X)	33.3
Steel shapes and forms	(X)	25.1
Copper and copper-base alloy shapes and forms	(X)	21.2
Aluminum and aluminum-base alloy shapes and forms	(X)	30.3
Other nonferrous shapes and forms	(X)	12.5
Insulated wire and cable, including magnet wire	(X)	158.5
Paper and paperboard containers, including shipping sacks and other paper packaging supplies	(X)	73.7
All other materials and components, parts, containers, and supplies	(X)	1,047.8
Materials, ingredients, containers, and supplies, nsk	(X)	3,358.0

Source: 1992 Economic Census. Explanation of symbols used: (D): Withheld to avoid disclosure of competitive data; na: Not available; (S): Withheld because statistical norms were not met; (X): Not applicable; (Z): Less than half the unit shown; nec: Not elsewhere classified; nsk: Not specified by kind; - : zero; * : 10-19 percent estimated; ** : 20-29 percent estimated.

PRODUCT SHARE DETAILS

Product or Product Class	% Share	Product or Product Class	% Share
Electronic components, not elsewhere classified	100.00	components)	56.21
Crystals, filters, piezoelectric, and other related electronic devices, except microwave filters	2.03	Electronic components, nec	19.81
Microwave components and devices, except antennae, tubes, and semiconductors	5.20	Electronic components, earphones and headsets (except telephone), phonograph cartridges and pickups, and phonograph needle and styli	0.78
Transducers, electrical/electronic input or output, n.e.c.	3.11	Other electronic components, nec (including static power supply converters, and cable harness assemblies)	97.68
Switches, mechanical, for electronic circuitry	1.84	Electronic components, nec, nsk	1.54
Printed circuit assemblies, loaded boards or modules (printed circuit boards with inserted electronic		Electronic components, not elsewhere classified, nsk	11.79

Source: 1992 Economic Census. The values shown are percent of total shipments in an industry. Values of indented subcategories are summed in the main heading. The symbol (D) appears when data are withheld to prevent disclosure of competitive information. The abbreviation nsk stands for 'not specified by kind' and nec for 'not elsewhere classified'.

INPUTS AND OUTPUTS FOR ELECTRONIC COMPONENTS NEC

Economic Sector or Industry Providing Inputs	%	Sector	Economic Sector or Industry Buying Outputs	%	Sector
Miscellaneous plastics products	16.2	Manufg.	Radio & TV communication equipment	13.5	Manufg.
Imports	15.7	Foreign	Exports	9.8	Foreign
Electronic components nec	15.5	Manufg.	Telephone & telegraph apparatus	9.6	Manufg.
Wholesale trade	10.9	Trade	Electronic computing equipment	9.4	Manufg.
Plating & polishing	5.2	Manufg.	Electronic components nec	9.1	Manufg.
Primary nonferrous metals, nec	2.6	Manufg.	Radio & TV receiving sets	5.8	Manufg.
Electric services (utilities)	2.0	Util.	Guided missiles & space vehicles	3.2	Manufg.
Semiconductors & related devices	1.9	Manufg.	Personal consumption expenditures	2.9	
Advertising	1.5	Services	X-ray apparatus & tubes	2.6	Manufg.
Communications, except radio & TV	1.4	Util.	Aircraft	2.2	Manufg.
Metal stampings, nec	1.3	Manufg.	Instruments to measure electricity	2.2	Manufg.
Banking	1.2	Fin/R.E.	Aircraft & missile equipment, nec	2.0	Manufg.
Aluminum rolling & drawing	1.1	Manufg.	Semiconductors & related devices	2.0	Manufg.
Copper rolling & drawing	1.1	Manufg.	Industrial controls	1.9	Manufg.
Nonferrous wire drawing & insulating	1.1	Manufg.	Optical instruments & lenses	1.7	Manufg.
Air transportation	1.1	Util.	Computer & data processing services	1.7	Services

Continued on next page.

INPUTS AND OUTPUTS FOR ELECTRONIC COMPONENTS NEC - Continued

Economic Sector or Industry Providing Inputs	%	Sector	Economic Sector or Industry Buying Outputs	%	Sector
Metal coating & allied services	1.0	Manufg.	Federal Government purchases, national defense	1.7	Fed Govt
Blast furnaces & steel mills	0.9	Manufg.	Games, toys, & children's vehicles	1.6	Manufg.
Maintenance of nonfarm buildings nec	0.8	Constr.	Mechanical measuring devices	1.3	Manufg.
Plastics materials & resins	0.8	Manufg.	Calculating & accounting machines	1.2	Manufg.
Motor freight transportation & warehousing	0.8	Util.	Electron tubes	1.2	Manufg.
Real estate	0.8	Fin/R.E.	Photographic equipment & supplies	1.1	Manufg.
Chemical preparations, nec	0.7	Manufg.	Radio & TV broadcasting	1.0	Util.
Petroleum refining	0.7	Manufg.	Change in business inventories	1.0	In House
Screw machine and related products	0.7	Manufg.	Communications, except radio & TV	0.8	Util.
Eating & drinking places	0.7	Trade	Federal Government purchases, nondefense	0.7	Fed Govt
Equipment rental & leasing services	0.6	Services	Engine electrical equipment	0.6	Manufg.
Hotels & lodging places	0.6	Services	Musical instruments	0.6	Manufg.
Cyclic crudes and organics	0.5	Manufg.	Surgical appliances & supplies	0.6	Manufg.
Electronic computing equipment	0.5	Manufg.	Switchgear & switchboard apparatus	0.6	Manufg.
Paperboard containers & boxes	0.5	Manufg.	Phonograph records & tapes	0.5	Manufg.
Sheet metal work	0.5	Manufg.	Management & consulting services & labs	0.5	Services
Adhesives & sealants	0.4	Manufg.	Engineering & scientific instruments	0.4	Manufg.
Machinery, except electrical, nec	0.4	Manufg.	Surgical & medical instruments	0.4	Manufg.
Nonferrous rolling & drawing, nec	0.4	Manufg.	Banking	0.4	Fin/R.E.
Gas production & distribution (utilities)	0.4	Util.	Environmental controls	0.3	Manufg.
Computer & data processing services	0.4	Services	Typewriters & office machines, nec	0.3	Manufg.
Legal services	0.4	Services	Libraries, vocation education	0.3	Services
Carbon & graphite products	0.3	Manufg.	Aircraft & missile engines & engine parts	0.2	Manufg.
Fabricated rubber products, nec	0.3	Manufg.	Ammunition, except for small arms, nec	0.2	Manufg.
Primary metal products, nec	0.3	Manufg.	Electrical equipment & supplies, nec	0.2	Manufg.
Special dies & tools & machine tool accessories	0.3	Manufg.	Detective & protective services	0.2	Services
Wiring devices	0.3	Manufg.	Elementary & secondary schools	0.2	Services
Management & consulting services & labs	0.3	Services	Gross private fixed investment	0.2	Cap Inv
Noncomparable imports	0.3	Foreign	Electrical industrial apparatus, nec	0.1	Manufg.
Aluminum castings	0.2	Manufg.	Motor vehicle parts & accessories	0.1	Manufg.
Instruments to measure electricity	0.2	Manufg.	Motor vehicles & car bodies	0.1	Manufg.
Miscellaneous fabricated wire products	0.2	Manufg.	Scales & balances	0.1	Manufg.
Sanitary services, steam supply, irrigation	0.2	Util.	S/L Govt. purch., higher education	0.1	S/L Govt
Accounting, auditing & bookkeeping	0.2	Services			
U.S. Postal Service	0.2	Gov't			
Abrasive products	0.1	Manufg.			
Glass & glass products, except containers	0.1	Manufg.			
Industrial inorganic chemicals, nec	0.1	Manufg.			
Lubricating oils & greases	0.1	Manufg.			
Manifold business forms	0.1	Manufg.			
Metal foil & leaf	0.1	Manufg.			
Paper coating & glazing	0.1	Manufg.			
Railroads & related services	0.1	Util.			
Insurance carriers	0.1	Fin/R.E.			
Automotive rental & leasing, without drivers	0.1	Services			
Colleges, universities, & professional schools	0.1	Services			

Source: Benchmark Input-Output Accounts for the U.S. Economy, 1982, U.S. Department of Commerce, Washington, D.C., July 1991. Data, as reported in the source, are organized by the 1977 SIC structure in use in 1982 but have been matched, as closely as is possible, to the 1987 SIC structure used in this book.

OCCUPATIONS EMPLOYED BY SIC 367 - ELECTRONIC COMPONENTS AND ACCESSORIES

Occupation	% of Total 1994	Change to 2005	Occupation	% of Total 1994	Change to 2005
Electrical & electronic assemblers	9.8	-8.4	Secretaries, ex legal & medical	1.6	-7.4
Electrical & electronic equipment assemblers	5.7	1.8	Industrial production managers	1.5	-28.8
Electrical & electronic technicians,technologists	5.4	22.1	Electromechanical equipment assemblers	1.4	11.9
Inspectors, testers, & graders, precision	5.4	-28.8	Coil winders, tapers, & finishers	1.3	-18.6
Electronic semiconductor processors	5.3	-1.4	Production, planning, & expediting clerks	1.3	1.8
Electrical & electronics engineers	4.4	29.8	Industrial engineers, ex safety engineers	1.3	12.0
Assemblers, fabricators, & hand workers nec	4.0	1.8	Electrolytic plating machine workers	1.2	11.9
Blue collar worker supervisors	3.4	-14.1	Traffic, shipping, & receiving clerks	1.2	-2.1
Sales & related workers nec	2.2	1.8	Freight, stock, & material movers, hand	1.2	-18.6
Computer engineers	2.2	50.7	Precision assemblers nec	1.1	62.9
Engineering, mathematical, & science managers	1.8	15.6	Stock clerks	1.0	-17.3
Engineering technicians & technologists nec	1.8	1.8	Engineers nec	1.0	52.6
General managers & top executives	1.6	-3.4	General office clerks	1.0	-13.2

Source: Industry-Occupation Matrix, Bureau of Labor Statistics. These data relate to one or more 3-digit SIC industry groups rather than to a single 4-digit SIC. The change reported for each occupation to the year 2005 is a percent of growth or decline as estimated by the Bureau of Labor Statistics. The abbreviation nec stands for 'not elsewhere classified'.

LOCATION BY STATE AND REGIONAL CONCENTRATION

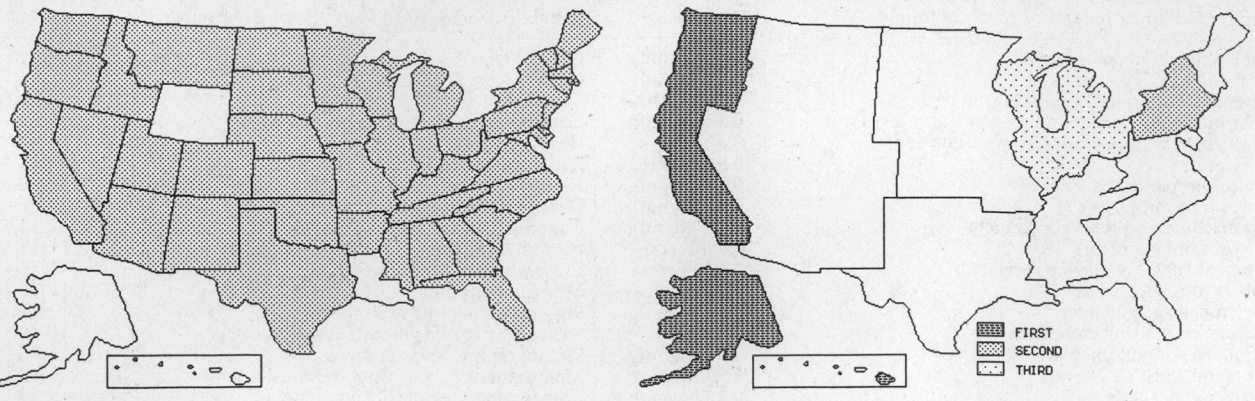

FIRST
SECOND
THIRD

INDUSTRY DATA BY STATE

| State | Establish- ments | Shipments | | | Employment | | | | Cost as % of Shipments | Investment per Employee ($) |
		Total ($ mil)	% of U.S.	Per Establ.	Total Number	% of U.S.	Per Establ.	Wages ($/hour)		
California	815	5,859.8	24.5	7.2	39,900	21.9	49	7.90	43.5	5,135
Texas	189	3,943.4	16.5	20.9	20,100	11.0	106	13.11	59.7	5,766
North Carolina	50	2,401.3	10.1	48.0	7,300	4.0	146	10.94	60.1	3,986
Massachusetts	202	1,406.0	5.9	7.0	12,600	6.9	62	11.58	40.9	3,762
New York	213	918.7	3.8	4.3	11,500	6.3	54	10.25	42.2	-
Minnesota	72	803.6	3.4	11.2	6,200	3.4	86	11.86	45.9	5,290
Florida	141	764.5	3.2	5.4	7,400	4.1	52	9.06	49.3	4,149
New Jersey	160	681.2	2.9	4.3	6,200	3.4	39	8.16	41.1	2,597
Pennsylvania	157	604.3	2.5	3.8	7,500	4.1	48	8.58	41.2	2,507
Arizona	66	580.5	2.4	8.8	4,300	2.4	65	10.36	41.0	-
Illinois	131	509.2	2.1	3.9	5,800	3.2	44	9.35	41.1	2,155
Colorado	57	412.6	1.7	7.2	4,400	2.4	77	8.22	67.0	3,136
Wisconsin	53	383.6	1.6	7.2	3,900	2.1	74	7.41	50.4	3,436
Ohio	96	374.4	1.6	3.9	3,400	1.9	35	9.40	45.4	2,235
Connecticut	82	368.0	1.5	4.5	3,400	1.9	41	10.16	39.3	4,588
Indiana	58	341.4	1.4	5.9	3,800	2.1	66	9.02	49.3	2,526
Michigan	79	265.1	1.1	3.4	2,500	1.4	32	9.38	53.8	1,840
Virginia	44	264.3	1.1	6.0	1,800	1.0	41	7.52	45.8	3,889
New Hampshire	55	231.2	1.0	4.2	2,600	1.4	47	10.19	42.5	4,077
Washington	55	204.3	0.9	3.7	1,900	1.0	35	9.05	46.7	3,842
Oklahoma	34	196.6	0.8	5.8	2,300	1.3	68	9.79	43.4	-
Oregon	51	186.3	0.8	3.7	2,300	1.3	45	9.38	46.8	2,652
Maryland	49	124.8	0.5	2.5	1,400	0.8	29	10.86	33.4	1,286
New Mexico	14	103.9	0.4	7.4	1,400	0.8	100	6.71	27.9	-
Missouri	31	92.7	0.4	3.0	1,100	0.6	35	7.29	57.4	1,364
Arkansas	20	90.1	0.4	4.5	1,200	0.7	60	7.07	52.5	1,250
South Carolina	18	87.1	0.4	4.8	1,200	0.7	67	8.53	36.2	1,000
Tennessee	34	85.0	0.4	2.5	1,000	0.5	29	7.56	48.6	1,100
Utah	25	64.4	0.3	2.6	700	0.4	28	8.13	55.7	-
Kentucky	17	56.2	0.2	3.3	1,000	0.5	59	8.14	33.5	-
Iowa	12	53.5	0.2	4.5	700	0.4	58	7.43	41.1	-
Kansas	43	53.3	0.2	1.2	800	0.4	19	7.60	39.4	1,375
Rhode Island	11	36.2	0.2	3.3	400	0.2	36	8.80	47.0	2,500
Maine	13	22.5	0.1	1.7	400	0.2	31	7.40	35.6	-
Georgia	29	(D)	-	-	1,750 *	1.0	60	-	-	-
Alabama	28	(D)	-	-	7,500 *	4.1	268	-	-	-
Vermont	16	(D)	-	-	750 *	0.4	47	-	-	-
Nevada	11	(D)	-	-	175 *	0.1	16	-	-	-
Mississippi	10	(D)	-	-	375 *	0.2	38	-	-	-
Idaho	8	(D)	-	-	750 *	0.4	94	-	-	3,333
South Dakota	7	(D)	-	-	1,750 *	1.0	250	-	-	1,600
Nebraska	6	(D)	-	-	175 *	0.1	29	-	-	-
West Virginia	6	(D)	-	-	175 *	0.1	29	-	-	-
Montana	5	(D)	-	-	175 *	0.1	35	-	-	-
North Dakota	5	(D)	-	-	175 *	0.1	35	-	-	-
Delaware	3	(D)	-	-	175 *	0.1	58	-	-	-

Source: 1992 *Economic Census*. The states are in descending order of shipments or establishments (if shipment data are missing for the majority). The symbol (D) appears when data are withheld to prevent disclosure of competitive information. States marked with (D) are sorted by number of establishments. A dash (-) indicates that the data element cannot be calculated; * indicates the midpoint of a range.

3691 - STORAGE BATTERIES

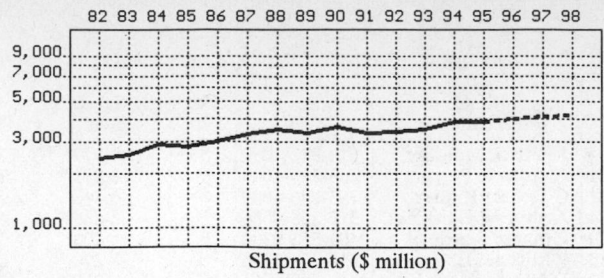
Shipments ($ million)

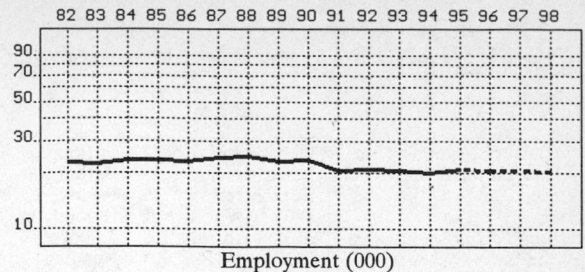
Employment (000)

GENERAL STATISTICS

Year	Com-panies	Establishments		Employment			Compensation		Production ($ million)			
		Total	with 20 or more employees	Total (000)	Production Workers (000)	Hours (Mil)	Payroll ($ mil)	Wages ($/hr)	Cost of Materials	Value Added by Manufacture	Value of Shipments	Capital Invest.
1982	129	201	123	22.9	18.0	34.4	473.6	10.10	1,196.2	1,203.0	2,431.3	109.7
1983		194	117	22.4	18.1	34.9	486.8	10.48	1,217.1	1,331.8	2,542.6	98.1
1984		187	111	23.7	19.0	38.5	563.8	11.07	1,386.1	1,550.0	2,916.3	132.1
1985		179	105	23.5	18.6	36.1	567.2	11.78	1,263.0	1,535.3	2,797.1	118.0
1986		178	111	23.2	18.3	37.1	594.6	11.98	1,348.3	1,639.0	2,991.8	109.8
1987	125	190	117	24.2	19.4	38.4	626.4	12.33	1,623.9	1,705.1	3,303.1	111.7
1988		185	114	24.9	19.8	39.8	632.6	11.86	1,775.4	1,743.2	3,484.4	140.8
1989		172	105	23.2	19.5	37.8	622.1	12.24	1,767.4	1,580.7	3,358.5	119.4
1990		153	102	23.5	18.4	36.7	623.9	12.51	1,947.5	1,718.8	3,625.8	126.5
1991		149	92	21.0	16.6	33.8	579.9	12.56	1,740.6	1,577.1	3,313.3	110.1
1992	110	154	91	21.1	16.9	33.5	596.7	12.77	1,633.3	1,774.0	3,409.5	120.0
1993		146	91	20.9	16.9	34.2	613.1	13.16	1,711.1	1,805.1	3,520.3	111.9
1994		143P	89P	20.3	16.5	35.5	663.6	14.04	1,891.2	2,023.6	3,877.2	158.5
1995		138P	87P	21.1P	17.0P	35.2P	664.8P	13.88P	1,905.2P	2,038.5P	3,905.8P	133.2P
1996		133P	84P	20.9P	16.9P	35.1P	675.7P	14.14P	1,954.5P	2,091.3P	4,007.0P	135.0P
1997		128P	82P	20.6P	16.7P	35.0P	686.7P	14.40P	2,003.8P	2,144.1P	4,108.1P	136.9P
1998		123P	79P	20.4P	16.6P	34.8P	697.7P	14.66P	2,053.2P	2,196.9P	4,209.3P	138.7P

Sources: 1982, 1987, 1992 *Economic Census*; *Annual Survey of Manufactures*, 83-86, 88-91, 93-94. Establishment counts for non-Census years are from *County Business Patterns*; establishment values for 83-84 are extrapolations. 'P's show projections by the editors. Industries reclassified in 87 will not have data for prior years.

INDICES OF CHANGE

Year	Com-panies	Establishments		Employment			Compensation		Production ($ million)			
		Total	with 20 or more employees	Total (000)	Production Workers (000)	Hours (Mil)	Payroll ($ mil)	Wages ($/hr)	Cost of Materials	Value Added by Manufacture	Value of Shipments	Capital Invest.
1982	117	131	135	109	107	103	79	79	73	68	71	91
1983		126	129	106	107	104	82	82	75	75	75	82
1984		121	122	112	112	115	94	87	85	87	86	110
1985		116	115	111	110	108	95	92	77	87	82	98
1986		116	122	110	108	111	100	94	83	92	88	91
1987	114	123	129	115	115	115	105	97	99	96	97	93
1988		120	125	118	117	119	106	93	109	98	102	117
1989		112	115	110	115	113	104	96	108	89	99	99
1990		99	112	111	109	110	105	98	119	97	106	105
1991		97	101	100	98	101	97	98	107	89	97	92
1992	100	100	100	100	100	100	100	100	100	100	100	100
1993		95	100	99	100	102	103	103	105	102	103	93
1994		93P	98P	96	98	106	111	110	116	114	114	132
1995		89P	95P	100P	101P	105P	111P	109P	117P	115P	115P	111P
1996		86P	92P	99P	100P	105P	113P	111P	120P	118P	118P	113P
1997		83P	90P	98P	99P	104P	115P	113P	123P	121P	120P	114P
1998		80P	87P	97P	98P	104P	117P	115P	126P	124P	123P	116P

Sources: Same as General Statistics. Values reflect change from the base year, 1992. Values above 100 mean greater than 92, values below 100 mean less than 92, and a value of 100 in the 82-91 or 93-98 period means same as 92. 'P's mark projections by the editors.

SELECTED RATIOS

For 1994	Avg. of All Manufact.	Analyzed Industry	Index	For 1994	Avg. of All Manufact.	Analyzed Industry	Index
Employees per Establishment	49	142	291	Value Added per Production Worker	134,084	122,642	91
Payroll per Establishment	1,500,273	4,655,357	310	Cost per Establishment	5,045,178	13,267,347	263
Payroll per Employee	30,620	32,690	107	Cost per Employee	102,970	93,163	90
Production Workers per Establishment	34	116	337	Cost per Production Worker	146,988	114,618	78
Wages per Establishment	853,319	3,496,569	410	Shipments per Establishment	9,576,895	27,199,745	284
Wages per Production Worker	24,861	30,207	122	Shipments per Employee	195,460	190,995	98
Hours per Production Worker	2,056	2,152	105	Shipments per Production Worker	279,017	234,982	84
Wages per Hour	12.09	14.04	116	Investment per Establishment	321,011	1,111,926	346
Value Added per Establishment	4,602,255	14,196,173	308	Investment per Employee	6,552	7,808	119
Value Added per Employee	93,930	99,685	106	Investment per Production Worker	9,352	9,606	103

Sources: Same as General Statistics. The 'Average of All Manufacturing' column represents the average of all manufacturing industries reported for the most recent complete year available. The Index shows the relationship between the Average and the Analyzed Industry. For example, 100 means that they are equal; 500 that the Analyzed Industry is five times the average; 50 means that the Analyzed Industry is half the national average. The abbreviation 'na' is used to show that data are 'not available'.

LEADING COMPANIES   Number shown: 47   Total sales ($ mil): 8,124   Total employment (000): 55.2

Company Name	Address				CEO Name	Phone	Co. Type	Sales ($ mil)	Empl. (000)
Eveready Battery Company Inc	Checkerboard Sq	St Louis	MO	63164	J Patrick Mulcahy	314-982-1000	S	1,979	17.5
Duracell International Inc	Berkshire Corp	Bethel	CT	06801	Charles R Perrin	203-796-4000	P	1,871	7.7
Duracell Inc	Berkshire Corp	Bethel	CT	06801	C Robert Kidder	203-796-4000	S	1,742	7.7
Exide Corp	1400 N Woodward	Bloomfield Hls	MI	48034	Arthur M Hawkins	313-258-0080	P	680	4.1
GNB Battery Technologies	375 Northridge Rd	Atlanta	GA	30350	Graham Spurling	404-551-0300	S	400	3.5
Yuasa-Exide Inc	PO Box 14145	Reading	PA	19612	P Michael Ehlerman	610-208-1908	S	290	2.0
East Penn Manufacturing	PO Box 147	Lyon Station	PA	19536	D Breidegam Jr	610-682-6361	R	260	2.6
Energizer Power Systems	PO Box 147114	Gainesville	FL	32614	Joe McClandthan	904-462-3911	D	190	3.6
GNB Industrial Battery Co	829 Parkview Blv	Lombard	IL	60148	Jon Adamson	708-629-5200	S	120	0.9
Charter Power Systems Inc	PO Box 239	Plym Meeting	PA	19462	Alfred Weber	610-828-9000	P	112	1.4
Douglas Battery Mfg Co	PO Box 12159	Winston-Salem	NC	27107	T S Douglas III	910-650-7000	R	53	0.9
Hawker Energy Products Inc	4318 Rainbow Blv	Kansas City	KS	66103	William D Vlasich	913-362-4898	S	48•	0.4
Power-Sonic Corp	PO Box 5242	Redwood City	CA	94063	Guy Clum	415-364-5001	R	45	<0.1
Palos Verdes Building Corp	1675 Sampson St	Corona	CA	91719	John Anderson	909-371-8090	R	32	0.2
Eagle-Picher Industries Inc	PO Box 130	Seneca	MO	64865	Mark Jost	417-776-2256	D	30	0.3
Ramcar Battery Co	2700 Carrier Av	City of Com	CA	90040	C Crowe	213-726-1212	R	30	0.4
Marathon Power Techn Co	PO Box 8233	Waco	TX	76714	Al Samuelsen	817-776-0650	S	24•	0.2
Crown Battery Manufacturing	PO Box 990	Fremont	OH	43420	Lee Koenig	419-334-7181	R	22•	0.2
Technocell	9163 Siempre Viva	San Diego	CA	92173	Javad Aliavadi	619-661-2020	D	21•	<0.1
Standard Industries Inc	PO Box 27500	San Antonio	TX	78227	G Z Dubinski Sr	210-623-3131	R	20	0.3
New Castle Battery Mfg	PO Box 5040	New Castle	PA	16105	Steven F Hoye	412-658-5501	R	16	0.1
Yardney Technical Products Inc	82 Mechanic St	Pawcatuck	CT	06379	Richard Scibelli	203-599-1100	R	15	0.2
Voltmaster Company Inc	PO Box 288	Corydon	IA	50060	RA Winslow	515-872-2044	R	13	0.1
Power Battery Co	543 E 42nd St	Paterson	NJ	07513	W Rasmussen	201-523-8630	R	13	0.3
Portable Energy Products Inc	940 Disc Dr	Scotts Valley	CA	95066	Robert Teal	408-439-5100	R	12	0.1
Quick Cable Corp	2501 Eaton Ln	Racine	WI	53404	John Shannon Jr	414-637-8363	R	12•	0.1
Saft Nife Inc	711 Industrial Blv	Valdosta	GA	31601	Frank D Westfall Jr	912-247-2331	S	12	<0.1
Acme Battery Manufacturing Co	3340 Morganford	St Louis	MO	63116	H S Gershenson	314-776-2980	R	10	0.1
Ovonic Battery Company Inc	1826 Northwood Dr	Troy	MI	48084	Shubhash K Dhar	810-362-1750	S	7•	0.1
Acme Electric Corp	528 W 21st St	Tempe	AZ	85282	M Anderman	602-894-6864	D	6•	<0.1
Mixon Inc	2286 Capp Rd	St Paul	MN	55114	Gordon W Mixon	612-646-2707	R	6	<0.1
Pro Battery Inc	3941 Oakcliff	Atlanta	GA	30340	EF Sherry	404-449-5900	R	5	<0.1
Yuasa Exide Industrial Battery	3470 Depot Rd	Hayward	CA	94545	John Shea	510-887-8080	D	5•	<0.1
Battery Engineering Inc	1636 Hyde Park Av	Hyde Park	MA	02136	James Estein	617-361-7555	S	5	<0.1
VDO-PAK Inc	PO Box 290969	Port Orange	FL	32129	Phil McReynolds	904-756-9770	S	4	<0.1
Battery Technology Inc	5700 Bandini Blv	Commerce	CA	90040	Tommy Tong	213-728-7874	R	3	<0.1
Ace Radio Control Inc	PO Box 472	Higginsville	MO	64037	Joseph Kessinger	816-584-7121	R	3	<0.1
RCI Inc	11800 E Gr Rvr	Brighton	MI	48116	Richard Chysler	810-229-0122	R	2•	<0.1
Plainview Batteries Inc	23 Newtown Rd	Plainview	NY	11803	Lynn Erde	516-249-2873	R	1	<0.1
Aristo-Craft Inc	346 Bergen Av	Jersey City	NJ	07304	Irwin S Polk	201-332-8100	R	1•	<0.1
Electruk Battery Co	4922 I D A Park Dr	Lockport	NY	14094	Don DiBacco	716-439-1033	R	1•	<0.1
Keystone Battery	35 Holton St	Winchester	MA	01890	Edward I Gray	617-729-8333	R	1	<0.1
Matsi Inc	430 10th St NW	Atlanta	GA	30318	John Russell	404-876-8009	R	1•	<0.1
VST Power Systems Inc	1620 Sudbury Rd	Concord	MA	01742	Vince Fedele	508-287-4600	R	1•	<0.1
Green Manufacturing Company	PO Box 26	Terrell	TX	75160	JR Green	214-524-1919	R	1•	<0.1
Norton Battery Mfg Co	PO Box 2440	Rialto	CA	92377	Carol Johnson	909-877-2161	R	1	<0.1
Lynwood Battery Mfg Co	4504 E Washington	Los Angeles	CA	90040	TW Kirk	213-263-8866	R	0	<0.1

Source: *Ward's Business Directory of U.S. Private and Public Companies*, Volumes 1 and 2, 1996. The company type code used is as follows: P - Public, R - Private, S - Subsidiary, D - Division, J - Joint Venture, A - Affiliate, G - Group. Sales are in millions of dollars, employees are in thousands. An asterisk (•) indicates an estimated sales volume. The symbol < stands for 'less than'. Company names and addresses are truncated, in some cases, to fit into the available space.

MATERIALS CONSUMED

Material	Quantity	Delivered Cost ($ million)
Materials, ingredients, containers, and supplies	(X)	1,513.4
Metal stampings	(X)	55.0
Iron and steel castings (rough and semifinished)	(X)	(D)
Nonferrous (aluminum, copper, etc.) castings (rough and semifinished)	(X)	(D)
Refined unalloyed lead shapes and forms	(X)	277.2
Antimonial lead	(X)	177.3
Lead-calcium alloyed	(X)	86.3
Steel sheet and strip, including tin plate	(X)	(D)
All other steel shapes and forms	(X)	(D)
Zinc and zinc-base alloy shapes and forms	(X)	(D)
Other nonferrous shapes and forms	(X)	(D)
Litharge	(X)	43.1
Sulfuric acid (new and spent) (100 percent H2SO4)	(X)	24.8
Other industrial inorganic chemicals (including mercury oxide and silver oxide)	(X)	22.2
Plastics resins consumed in the form of granules, pellets, powders, liquids, etc.	(X)	(D)
Plastics products consumed in the form of sheets, rods, tubes, film, and other shapes	(X)	38.6
Fabricated plastics products (except gaskets, hoses, and belting)	(X)	235.0
Paperboard containers, boxes, and corrugated paperboard	(X)	20.4
Carbon and graphite electrodes, and other carbon and graphite products for electrical use	(X)	(D)
All other materials and components, parts, containers, and supplies	(X)	289.6
Materials, ingredients, containers, and supplies, nsk	(X)	166.1

Source: 1992 Economic Census. Explanation of symbols used: (D): Withheld to avoid disclosure of competitive data; na: Not available; (S): Withheld because statistical norms were not met; (X): Not applicable; (Z): Less than half the unit shown; nec: Not elsewhere classified; nsk: Not specified by kind; - : zero; * : 10-19 percent estimated; ** : 20-29 percent estimated.

PRODUCT SHARE DETAILS

Product or Product Class	% Share	Product or Product Class	% Share
Storage batteries	100.00	Communication lead acid storage batteries (for central office telephone supervisory equipment, telemetering, and microwave), larger than BCI dimensional size group 8D (1.5 cu ft or .042 cu m)	16.06
Storage batteries, lead acid type, BCI dimensional size group 8D (1.5 cu ft or .042 cu m and smaller)	65.09		
Starting, lighting, and ignition (SLI) type lead acid storage batteries for original equipment, BCI dimensional size group 8D (1.5 cu ft or .042 cu m and smaller)	18.26	Standby emergency power lead acid storage batteries, larger than BCI dimensional size group 8D (1.5 cu ft or .042 cu m)	33.78
Starting, lighting, and ignition (SLI) type lead acid storage batteries for replacement, BCI dimensional size group 8D (1.5 cu ft or .042 cu m and smaller)	80.69	Other lead acid storage batteries, larger than BCI dimensional size group 8D (1.5 cu ft) (incl. starting, lighting, and ignition (SLI type) for aircraft, marine, fire alarms, and railroad signaling)	7.83
Lead acid storage batteries other than (SLI) type, BCI dimensional size group 8D (1.5 cu ft or .042 cu m and smaller)	0.43	Storage batteries, lead acid type, larger than bci dimensional size group 8d (1.5 cu ft or .042 cu m), nsk	2.25
Storage batteries, lead acid type, bci dimensional size group 8d (1.5 cu ft or .042 cu m and smaller), nsk	0.61	Storage batteries, except lead acid, including parts for all storage batteries	14.19
Storage batteries, lead acid type, larger than BCI dimensional size group 8D (1.5 cu ft or .042 cu m)	17.26	Nickel cadmium storage batteries (sealed or vented)	89.91
		Storage batteries other than nickel cadmium or lead acid	8.09
Industrial truck motive power type lead acid storage batteries, larger than BCI dimensional size group 8D (1.5 cu ft or .042 cu m)	30.30	Parts for all storage batteries (excluding cases and containers)	1.72
Other motive power type lead acid storage batteries (including mining and industrial locomotive), larger than BCI dimensional size group 8D (1.5 cu ft or .042 cu m)	9.78	Storage batteries, except lead acid, including parts for all storage batteries, nsk	0.29
		Storage batteries, nsk	3.46

Source: 1992 Economic Census. The values shown are percent of total shipments in an industry. Values of indented subcategories are summed in the main heading. The symbol (D) appears when data are withheld to prevent disclosure of competitive information. The abbreviation nsk stands for 'not specified by kind' and nec for 'not elsewhere classified'.

INPUTS AND OUTPUTS FOR STORAGE BATTERIES

Economic Sector or Industry Providing Inputs	%	Sector	Economic Sector or Industry Buying Outputs	%	Sector
Storage batteries	16.6	Manufg.	Personal consumption expenditures	49.2	
Primary lead	16.4	Manufg.	Storage batteries	15.6	Manufg.
Cyclic crudes and organics	13.9	Manufg.	Motor vehicles & car bodies	6.2	Manufg.
Wholesale trade	9.8	Trade	Exports	5.3	Foreign
Inorganic pigments	6.0	Manufg.	Gross private fixed investment	5.3	Cap Inv
Miscellaneous plastics products	5.9	Manufg.	State & local government enterprises, nec	1.8	Gov't
Imports	5.5	Foreign	Retail trade, except eating & drinking	1.2	Trade
Advertising	3.0	Services	Accounting, auditing & bookkeeping	1.0	Services
Electric services (utilities)	2.4	Util.	Radio & TV communication equipment	0.9	Manufg.
Motor freight transportation & warehousing	1.8	Util.	Wholesale trade	0.8	Trade
Petroleum refining	1.7	Manufg.	Boat building & repairing	0.5	Manufg.
Banking	1.6	Fin/R.E.	Elevators & moving stairways	0.5	Manufg.
Fabricated rubber products, nec	1.2	Manufg.	Oil field machinery	0.5	Manufg.
Air transportation	1.2	Util.	Electric services (utilities)	0.5	Util.
Nonferrous rolling & drawing, nec	1.1	Manufg.	Child day care services	0.5	Services

Continued on next page.

INPUTS AND OUTPUTS FOR STORAGE BATTERIES - Continued

Economic Sector or Industry Providing Inputs	%	Sector	Economic Sector or Industry Buying Outputs	%	Sector
Industrial inorganic chemicals, nec	0.9	Manufg.	Feed grains	0.4	Agric.
Paperboard containers & boxes	0.9	Manufg.	Industrial trucks & tractors	0.4	Manufg.
Gas production & distribution (utilities)	0.7	Util.	Motor freight transportation & warehousing	0.3	Util.
Railroads & related services	0.7	Util.	Railroads & related services	0.3	Util.
Hotels & lodging places	0.7	Services	Automotive rental & leasing, without drivers	0.3	Services
Synthetic rubber	0.6	Manufg.	State & local electric utilities	0.3	Gov't
Plastics materials & resins	0.5	Manufg.	Meat animals	0.2	Agric.
Communications, except radio & TV	0.4	Util.	Office buildings	0.2	Constr.
Eating & drinking places	0.4	Trade	Farm machinery & equipment	0.2	Manufg.
Maintenance of nonfarm buildings nec	0.3	Constr.	Gas production & distribution (utilities)	0.2	Util.
Alkalies & chlorine	0.3	Manufg.	Transit & bus transportation	0.2	Util.
Metal stampings, nec	0.3	Manufg.	Eating & drinking places	0.2	Trade
Real estate	0.3	Fin/R.E.	U.S. Postal Service	0.2	Gov't
Automotive rental & leasing, without drivers	0.3	Services	S/L Govt. purch., elem. & secondary education	0.2	S/L Govt
Computer & data processing services	0.3	Services	S/L Govt. purch., natural resource & recreation.	0.2	S/L Govt
Equipment rental & leasing services	0.3	Services	Agricultural, forestry, & fishery services	0.1	Agric.
Noncomparable imports	0.3	Foreign	Oil bearing crops	0.1	Agric.
Scrap	0.3	Scrap	Industrial buildings	0.1	Constr.
Blast furnaces & steel mills	0.2	Manufg.	Maintenance of nonfarm buildings nec	0.1	Constr.
Machinery, except electrical, nec	0.2	Manufg.	Maintenance of petroleum & natural gas wells	0.1	Constr.
Nonferrous castings, nec	0.2	Manufg.	Nonfarm residential structure maintenance	0.1	Constr.
Paints & allied products	0.2	Manufg.	Residential 1-unit structures, nonfarm	0.1	Constr.
Insurance carriers	0.2	Fin/R.E.	Rolling mill machinery	0.1	Manufg.
Automotive repair shops & services	0.2	Services	Semiconductors & related devices	0.1	Manufg.
Legal services	0.2	Services	Truck & bus bodies	0.1	Manufg.
Management & consulting services & labs	0.2	Services	Water transportation	0.1	Util.
Carbon & graphite products	0.1	Manufg.	Real estate	0.1	Fin/R.E.
Special dies & tools & machine tool accessories	0.1	Manufg.	Doctors & dentists	0.1	Services
Water transportation	0.1	Util.	Hospitals	0.1	Services
Royalties	0.1	Fin/R.E.	S/L Govt. purch., other general government	0.1	S/L Govt
Accounting, auditing & bookkeeping	0.1	Services	S/L Govt. purch., police	0.1	S/L Govt
Theatrical producers, bands, entertainers	0.1	Services			

Source: Benchmark Input-Output Accounts for the U.S. Economy, 1982, U.S. Department of Commerce, Washington, D.C., July 1991. Data, as reported in the source, are organized by the 1977 SIC structure in use in 1982 but have been matched, as closely as is possible, to the 1987 SIC structure used in this book.

OCCUPATIONS EMPLOYED BY SIC 369 - MISC ELECTRICAL EQUIPMENT AND SUPPLIES

Occupation	% of Total 1994	Change to 2005	Occupation	% of Total 1994	Change to 2005
Assemblers, fabricators, & hand workers nec	14.5	1.0	Secretaries, ex legal & medical	1.3	-8.1
Electrical & electronic assemblers	10.7	-9.1	Electrical & electronic technicians,technologists	1.3	1.0
Machine operators nec	4.6	-28.8	Tool & die makers	1.3	-18.4
Electrical & electronic equipment assemblers	4.0	1.0	Hand packers & packagers	1.3	-13.4
Blue collar worker supervisors	3.4	-11.6	Engineering technicians & technologists nec	1.2	1.0
Inspectors, testers, & graders, precision	3.3	-29.3	Freight, stock, & material movers, hand	1.2	-19.2
Industrial machinery mechanics	2.4	36.3	Mechanical engineers	1.2	22.2
Electrical & electronics engineers	1.9	7.5	Managers & administrators nec	1.1	1.0
Engineers nec	1.7	51.5	Helpers, laborers, & material movers nec	1.1	1.0
Sales & related workers nec	1.5	1.0	Traffic, shipping, & receiving clerks	1.0	-2.8
Plastic molding machine workers	1.4	21.2	Industrial production managers	1.0	1.0
General managers & top executives	1.4	-4.2	Electromechanical equipment assemblers	1.0	11.1

Source: Industry-Occupation Matrix, Bureau of Labor Statistics. These data relate to one or more 3-digit SIC industry groups rather than to a single 4-digit SIC. The change reported for each occupation to the year 2005 is a percent of growth or decline as estimated by the Bureau of Labor Statistics. The abbreviation nec stands for 'not elsewhere classified'.

LOCATION BY STATE AND REGIONAL CONCENTRATION

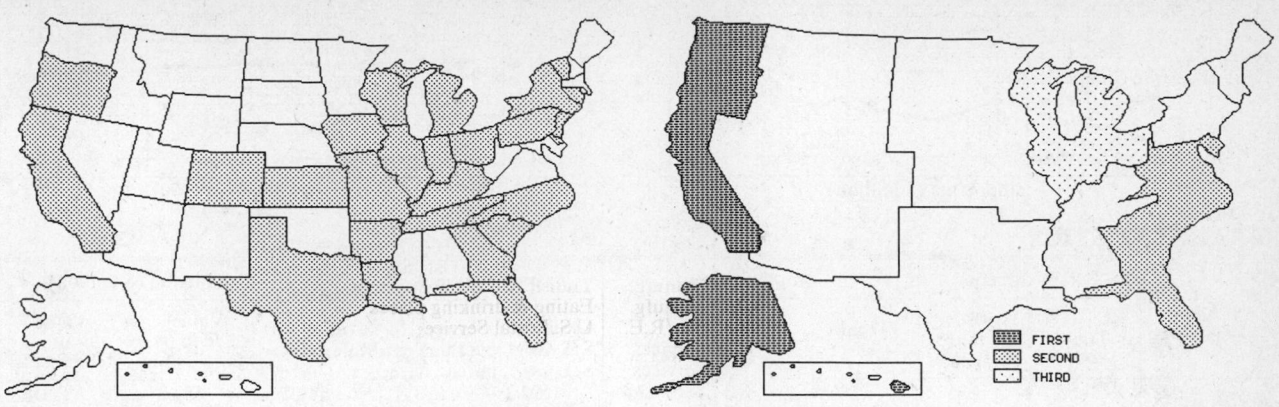

FIRST
SECOND
THIRD

INDUSTRY DATA BY STATE

| State | Establish-ments | Shipments | | | Employment | | | | Cost as % of Shipments | Investment per Employee ($) |
		Total ($ mil)	% of U.S.	Per Establ.	Total Number	% of U.S.	Per Establ.	Wages ($/hour)		
Pennsylvania	6	354.3	10.4	59.0	3,000	14.2	500	11.47	45.0	7,100
California	26	352.2	10.3	13.5	2,200	10.4	85	12.94	51.5	4,591
Georgia	8	301.6	8.8	37.7	1,400	6.6	175	15.00	55.6	7,929
Indiana	5	217.1	6.4	43.4	900	4.3	180	17.88	40.9	3,333
North Carolina	6	189.5	5.6	31.6	1,100	5.2	183	13.56	49.6	-
Illinois	10	186.4	5.5	18.6	1,000	4.7	100	11.56	67.4	2,900
Texas	11	154.8	4.5	14.1	1,300	6.2	118	10.52	54.8	10,077
Missouri	5	133.6	3.9	26.7	1,300	6.2	260	10.90	48.6	-
New York	9	71.5	2.1	7.9	400	1.9	44	13.33	40.8	-
Tennessee	7	54.0	1.6	7.7	500	2.4	71	9.00	70.7	3,600
Connecticut	6	36.0	1.1	6.0	400	1.9	67	8.67	36.1	-
Florida	8	(D)	-	-	1,750 *	8.3	219	-	-	-
Iowa	5	(D)	-	-	750 *	3.6	150	-	-	-
New Jersey	4	(D)	-	-	750 *	3.6	188	-	-	-
Oregon	4	(D)	-	-	375 *	1.8	94	-	-	-
Colorado	3	(D)	-	-	175 *	0.8	58	-	-	-
Kansas	3	(D)	-	-	750 *	3.6	250	-	-	-
Michigan	3	(D)	-	-	375 *	1.8	125	-	-	-
Kentucky	2	(D)	-	-	375 *	1.8	188	-	-	-
Louisiana	2	(D)	-	-	375 *	1.8	188	-	-	-
Ohio	2	(D)	-	-	750 *	3.6	375	-	-	-
South Carolina	2	(D)	-	-	375 *	1.8	188	-	-	-
Arkansas	1	(D)	-	-	175 *	0.8	175	-	-	-
Delaware	1	(D)	-	-	375 *	1.8	375	-	-	-
Vermont	1	(D)	-	-	375 *	1.8	375	-	-	-
Wisconsin	1	(D)	-	-	375 *	1.8	375	-	-	-

Source: 1992 *Economic Census*. The states are in descending order of shipments or establishments (if shipment data are missing for the majority). The symbol (D) appears when data are withheld to prevent disclosure of competitive information. States marked with (D) are sorted by number of establishments. A dash (-) indicates that the data element cannot be calculated; * indicates the midpoint of a range.

3692 - PRIMARY BATTERIES, DRY & WET

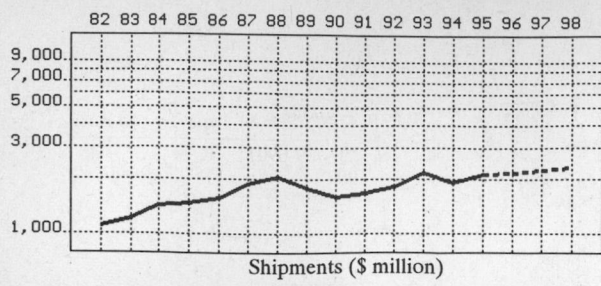

Shipments ($ million)

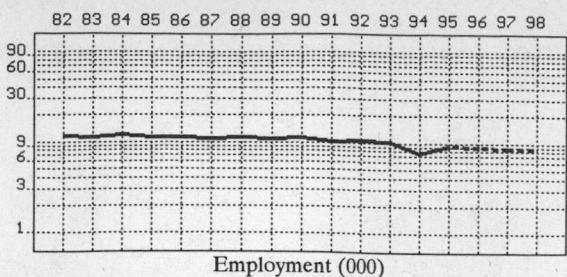

Employment (000)

GENERAL STATISTICS

Year	Com-panies	Establishments		Employment			Compensation		Production ($ million)			
		Total	with 20 or more employees	Total (000)	Production Workers (000)	Hours (Mil)	Payroll ($ mil)	Wages ($/hr)	Cost of Materials	Value Added by Manufacture	Value of Shipments	Capital Invest.
1982	36	55	35	11.7	9.1	17.6	192.3	7.78	531.7	559.2	1,101.8	42.9
1983		53	35	11.4	8.7	16.5	194.2	8.11	620.8	585.1	1,206.9	23.9
1984		51	35	11.9	9.0	16.9	215.0	8.20	731.5	695.9	1,426.1	37.7
1985		50	35	11.4	8.6	16.2	210.7	8.63	712.0	778.3	1,479.1	65.8
1986		61	35	11.0	8.3	15.7	218.3	8.85	759.7	803.3	1,570.4	47.0
1987	59	72	36	10.7	8.2	16.0	231.6	9.72	753.0	1,132.0	1,877.8	60.8
1988		76	41	11.1	8.6	17.2	249.0	9.86	808.8	1,276.5	2,079.8	46.0
1989		71	42	10.9	8.2	15.9	243.0	10.04	865.6	912.8	1,791.3	31.2
1990		80	41	11.1	8.0	15.8	247.6	10.58	817.6	761.9	1,580.3	34.9
1991		71	38	10.3	7.9	16.6	250.3	10.11	863.7	787.3	1,672.3	38.9
1992	53	68	32	10.4	7.9	16.0	271.9	11.23	933.9	871.0	1,823.7	52.5
1993		60	31	10.1	7.7	15.8	275.0	11.91	1,159.3	1,006.2	2,158.2	76.2
1994		76P	36P	7.7	6.4	11.9	211.4	12.82	1,041.3	891.0	1,925.9	80.6
1995		77P	36P	9.2P	7.1P	14.5P	267.5P	12.51P	1,147.7P	982.1P	2,122.8P	65.0P
1996		79P	37P	9.0P	7.0P	14.2P	272.7P	12.89P	1,182.8P	1,012.1P	2,187.6P	67.2P
1997		81P	37P	8.8P	6.8P	14.0P	277.8P	13.27P	1,217.9P	1,042.1P	2,252.5P	69.5P
1998		83P	37P	8.6P	6.6P	13.8P	282.9P	13.66P	1,252.9P	1,072.1P	2,317.3P	71.8P

Sources: 1982, 1987, 1992 *Economic Census; Annual Survey of Manufactures,* 83-86, 88-91, 93-94. Establishment counts for non-Census years are from *County Business Patterns*; establishment values for 83-84 are extrapolations. 'P's show projections by the editors. Industries reclassified in 87 will not have data for prior years.

INDICES OF CHANGE

Year	Com-panies	Establishments		Employment			Compensation		Production ($ million)			
		Total	with 20 or more employees	Total (000)	Production Workers (000)	Hours (Mil)	Payroll ($ mil)	Wages ($/hr)	Cost of Materials	Value Added by Manufacture	Value of Shipments	Capital Invest.
1982	68	81	109	113	115	110	71	69	57	64	60	82
1983		78	109	110	110	103	71	72	66	67	66	46
1984		75	109	114	114	106	79	73	78	80	78	72
1985		74	109	110	109	101	77	77	76	89	81	125
1986		90	109	106	105	98	80	79	81	92	86	90
1987	111	106	113	103	104	100	85	87	81	130	103	116
1988		112	128	107	109	108	92	88	87	147	114	88
1989		104	131	105	104	99	89	89	93	105	98	59
1990		118	128	107	101	99	91	94	88	87	87	66
1991		104	119	99	100	104	92	90	92	90	92	74
1992	100	100	100	100	100	100	100	100	100	100	100	100
1993		88	97	97	97	99	101	106	124	116	118	145
1994		111P	114P	74	81	74	78	114	112	102	106	154
1995		114P	114P	89P	90P	90P	98P	111P	123P	113P	116P	124P
1996		117P	114P	87P	88P	89P	100P	115P	127P	116P	120P	128P
1997		119P	114P	85P	86P	88P	102P	118P	130P	120P	124P	132P
1998		122P	114P	83P	84P	86P	104P	122P	134P	123P	127P	137P

Sources: Same as General Statistics. Values reflect change from the base year, 1992. Values above 100 mean greater than 92, values below 100 mean less than 92, and a value of 100 in the 82-91 or 93-98 period means same as 92. 'P's mark projections by the editors.

SELECTED RATIOS

For 1994	Avg. of All Manufact.	Analyzed Industry	Index	For 1994	Avg. of All Manufact.	Analyzed Industry	Index
Employees per Establishment	49	102	208	Value Added per Production Worker	134,084	139,219	104
Payroll per Establishment	1,500,273	2,793,273	186	Cost per Establishment	5,045,178	13,758,919	273
Payroll per Employee	30,620	27,455	90	Cost per Employee	102,970	135,234	131
Production Workers per Establishment	34	85	246	Cost per Production Worker	146,988	162,703	111
Wages per Establishment	853,319	2,015,781	236	Shipments per Establishment	9,576,895	25,447,327	266
Wages per Production Worker	24,861	23,837	96	Shipments per Employee	195,460	250,117	128
Hours per Production Worker	2,056	1,859	90	Shipments per Production Worker	279,017	300,922	108
Wages per Hour	12.09	12.82	106	Investment per Establishment	321,011	1,064,985	332
Value Added per Establishment	4,602,255	11,772,973	256	Investment per Employee	6,552	10,468	160
Value Added per Employee	93,930	115,714	123	Investment per Production Worker	9,352	12,594	135

Sources: Same as General Statistics. The 'Average of All Manufacturing' column represents the average of all manufacturing industries reported for the most recent complete year available. The Index shows the relationship between the Average and the Analyzed Industry. For example, 100 means that they are equal; 500 that the Analyzed Industry is five times the average; 50 means that the Analyzed Industry is half the national average. The abbreviation 'na' is used to show that data are 'not available'.

LEADING COMPANIES   Number shown: 8   Total sales ($ mil): 89   Total employment (000): 1.0

Company Name	Address				CEO Name	Phone	Co. Type	Sales ($ mil)	Empl. (000)
Wilson Greatbatch Ltd	10000 Wehrle Dr	Clarence	NY	14031	Edward Voboril	716-759-6901	R	48	0.5
US Battery Manufacturing Co	1675 Sampson St	Corona	CA	91719	John Anderson	909-371-8090	D	17*	0.2
Bren-Tronics Inc	10 Brayton Ct	Commack	NY	11725	LA Brenna	516-499-5155	R	7	0.1
Ultralife Batteries Inc	PO Box 622	Newark	NY	14513	Bruce Jagid	315-332-7100	P	5	0.2
Whittaker Corp	3850 Olive St	Denver	CO	80207	Mike Erixon	303-388-4836	D	5*	<0.1
Industrial Battery Engineering	9121 De Garmo Av	Sun Valley	CA	91352	Birger Holmquist	818-767-7067	R	3*	<0.1
Reaco Battery Service Corp	Rte 1	Johnston City	IL	62951	SK Gamster	618-983-5441	R	3	<0.1
Accurate Fabricating Co	PO Drawer 5159	Statesville	NC	28687	Barbara W Lee	704-872-7411	R	1	<0.1

Source: *Ward's Business Directory of U.S. Private and Public Companies*, Volumes 1 and 2, 1996. The company type code used is as follows: P - Public, R - Private, S - Subsidiary, D - Division, J - Joint Venture, A - Affiliate, G - Group. Sales are in millions of dollars, employees are in thousands. An asterisk (*) indicates an estimated sales volume. The symbol < stands for 'less than'. Company names and addresses are truncated, in some cases, to fit into the available space.

MATERIALS CONSUMED

Material	Quantity	Delivered Cost ($ million)
Materials, ingredients, containers, and supplies	(X)	899.4
Metal stampings	(X)	69.9
All other fabricated metal products (except forgings)	(X)	(D)
Forgings	(X)	(D)
Nonferrous (aluminum, copper, etc.) castings (rough and semifinished)	(X)	(D)
Refined unalloyed lead shapes and forms	(X)	(D)
Steel sheet and strip, including tin plate	(X)	38.5
All other steel shapes and forms	(X)	34.4
Zinc and zinc-base alloy shapes and forms	(X)	56.0
Other nonferrous shapes and forms	(X)	(D)
Sulfuric acid (new and spent) (100 percent H2SO4)	(X)	(D)
Other industrial inorganic chemicals (including mercury oxide and silver oxide)	(X)	70.1
Plastics resins consumed in the form of granules, pellets, powders, liquids, etc.	(X)	10.2
Plastics products consumed in the form of sheets, rods, tubes, film, and other shapes	(X)	30.3
Fabricated plastics products (except gaskets, hoses, and belting)	(X)	6.1
Paperboard containers, boxes, and corrugated paperboard	(X)	43.5
Carbon and graphite electrodes, and other carbon and graphite products for electrical use	(X)	(D)
All other materials and components, parts, containers, and supplies	(X)	455.5
Materials, ingredients, containers, and supplies, nsk	(X)	39.9

Source: 1992 *Economic Census*. Explanation of symbols used: (D): Withheld to avoid disclosure of competitive data; na: Not available; (S): Withheld because statistical norms were not met; (X): Not applicable; (Z): Less than half the unit shown; nec: Not elsewhere classified; nsk: Not specified by kind; - : zero; * : 10-19 percent estimated; ** : 20-29 percent estimated.

PRODUCT SHARE DETAILS

Product or Product Class	% Share	Product or Product Class	% Share
Primary batteries, dry and wet	100.00	Other primary cells and batteries, dry and wet, having an external volume not exceeding 300 cc	15.50
Manganese dioxide primary cells and batteries, dry and wet, having an external volume not exceeding 300 cc	47.05	Primary cells and batteries, dry and wet, having an external volume exceeding 300 cc (18.3 cu in.)	23.54
Silver oxide primary cells and batteries, dry and wet, having an external volume not exceeding 300 cc	2.95	Parts for primary batteries, dry and wet (excluding cases and containers)	5.17
Mercuric oxide primary cells and batteries, dry and wet, having an external volume not exceeding 300 cc	0.51		

Source: 1992 *Economic Census*. The values shown are percent of total shipments in an industry. Values of indented subcategories are summed in the main heading. The symbol (D) appears when data are withheld to prevent disclosure of competitive information. The abbreviation nsk stands for 'not specified by kind' and nec for 'not elsewhere classified'.

INPUTS AND OUTPUTS FOR PRIMARY BATTERIES, DRY & WET

Economic Sector or Industry Providing Inputs	%	Sector	Economic Sector or Industry Buying Outputs	%	Sector
Nonferrous metal ores, except copper	12.9	Mining	Personal consumption expenditures	61.7	
Fabricated metal products, nec	12.9	Manufg.	Exports	10.1	Foreign
Imports	9.9	Foreign	Management & consulting services & labs	5.9	Services
Cyclic crudes and organics	9.2	Manufg.	Calculating & accounting machines	4.3	Manufg.
Wholesale trade	7.1	Trade	Primary batteries, dry & wet	4.1	Manufg.
Metal stampings, nec	6.4	Manufg.	Federal Government purchases, national defense	3.2	Fed Govt
Primary batteries, dry & wet	4.8	Manufg.	Lawn & garden equipment	1.6	Manufg.
Industrial inorganic chemicals, nec	4.7	Manufg.	S/L Govt. purch., other general government	1.6	S/L Govt
Advertising	4.0	Services	Medical & health services, nec	1.2	Services
Nonferrous rolling & drawing, nec	2.7	Manufg.	Communications, except radio & TV	1.1	Util.

Continued on next page.

INPUTS AND OUTPUTS FOR PRIMARY BATTERIES, DRY & WET - Continued

Economic Sector or Industry Providing Inputs	%	Sector	Economic Sector or Industry Buying Outputs	%	Sector
Carbon & graphite products	2.5	Manufg.	Photographic equipment & supplies	0.9	Manufg.
Fabricated rubber products, nec	2.3	Manufg.	X-ray apparatus & tubes	0.9	Manufg.
Motor freight transportation & warehousing	2.1	Util.	Federal Government purchases, nondefense	0.7	Fed Govt
Banking	1.9	Fin/R.E.	Watches, clocks, & parts	0.6	Manufg.
Primary lead	1.8	Manufg.	Mechanical measuring devices	0.4	Manufg.
Miscellaneous plastics products	1.7	Manufg.	Dental equipment & supplies	0.3	Manufg.
Paperboard containers & boxes	1.6	Manufg.	Blast furnaces & steel mills	0.2	Manufg.
Railroads & related services	1.6	Util.	Electric housewares & fans	0.2	Manufg.
Electric services (utilities)	1.3	Util.	Farm machinery & equipment	0.2	Manufg.
Nonferrous castings, nec	1.0	Manufg.	Local government passenger transit	0.2	Gov't
Petroleum refining	0.7	Manufg.	Air transportation	0.1	Util.
Communications, except radio & TV	0.5	Util.	S/L Govt. purch., health & hospitals	0.1	S/L Govt
Eating & drinking places	0.5	Trade			
Maintenance of nonfarm buildings nec	0.4	Constr.			
Gas production & distribution (utilities)	0.4	Util.			
Computer & data processing services	0.4	Services			
Theatrical producers, bands, entertainers	0.4	Services			
Machinery, except electrical, nec	0.3	Manufg.			
Plastics materials & resins	0.3	Manufg.			
Engineering, architectural, & surveying services	0.3	Services			
Alkalies & chlorine	0.2	Manufg.			
Special dies & tools & machine tool accessories	0.2	Manufg.			
Water transportation	0.2	Util.			
Royalties	0.2	Fin/R.E.			
Automotive rental & leasing, without drivers	0.2	Services			
Equipment rental & leasing services	0.2	Services			
Legal services	0.2	Services			
Management & consulting services & labs	0.2	Services			
Noncomparable imports	0.2	Foreign			
Insurance carriers	0.1	Fin/R.E.			
Security & commodity brokers	0.1	Fin/R.E.			
Accounting, auditing & bookkeeping	0.1	Services			
Automotive repair shops & services	0.1	Services			
Electrical repair shops	0.1	Services			
U.S. Postal Service	0.1	Gov't			

Source: Benchmark Input-Output Accounts for the U.S. Economy, 1982, U.S. Department of Commerce, Washington, D.C., July 1991. Data, as reported in the source, are organized by the 1977 SIC structure in use in 1982 but have been matched, as closely as is possible, to the 1987 SIC structure used in this book.

OCCUPATIONS EMPLOYED BY SIC 369 - MISC ELECTRICAL EQUIPMENT AND SUPPLIES

Occupation	% of Total 1994	Change to 2005	Occupation	% of Total 1994	Change to 2005
Assemblers, fabricators, & hand workers nec	14.5	1.0	Secretaries, ex legal & medical	1.3	-8.1
Electrical & electronic assemblers	10.7	-9.1	Electrical & electronic technicians,technologists	1.3	1.0
Machine operators nec	4.6	-28.8	Tool & die makers	1.3	-18.4
Electrical & electronic equipment assemblers	4.0	1.0	Hand packers & packagers	1.3	-13.4
Blue collar worker supervisors	3.4	-11.6	Engineering technicians & technologists nec	1.2	1.0
Inspectors, testers, & graders, precision	3.3	-29.3	Freight, stock, & material movers, hand	1.2	-19.2
Industrial machinery mechanics	2.4	36.3	Mechanical engineers	1.2	22.2
Electrical & electronics engineers	1.9	7.5	Managers & administrators nec	1.1	1.0
Engineers nec	1.7	51.5	Helpers, laborers, & material movers nec	1.1	1.0
Sales & related workers nec	1.5	1.0	Traffic, shipping, & receiving clerks	1.0	-2.8
Plastic molding machine workers	1.4	21.2	Industrial production managers	1.0	1.0
General managers & top executives	1.4	-4.2	Electromechanical equipment assemblers	1.0	11.1

Source: Industry-Occupation Matrix, Bureau of Labor Statistics. These data relate to one or more 3-digit SIC industry groups rather than to a single 4-digit SIC. The change reported for each occupation to the year 2005 is a percent of growth or decline as estimated by the Bureau of Labor Statistics. The abbreviation nec stands for 'not elsewhere classified'.

LOCATION BY STATE AND REGIONAL CONCENTRATION

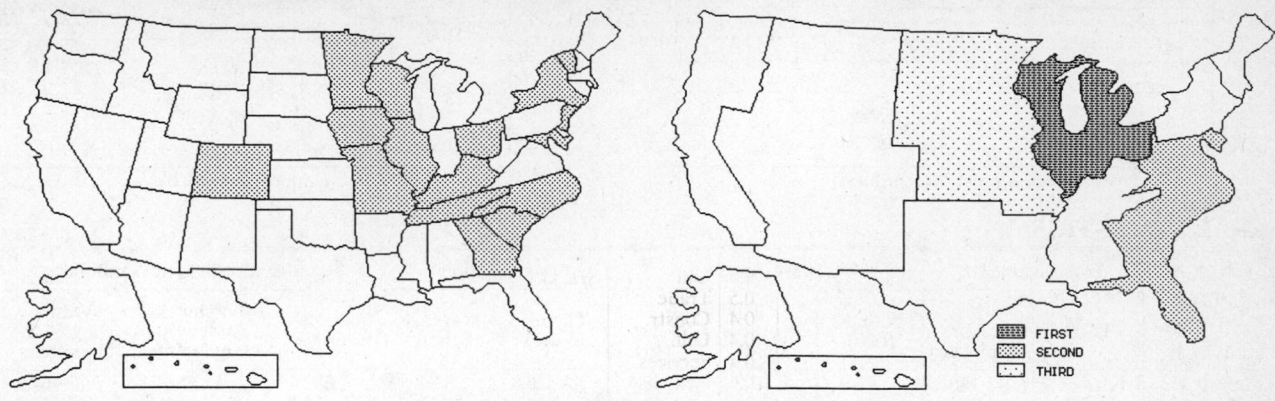

FIRST
SECOND
THIRD

INDUSTRY DATA BY STATE

State	Establish-ments	Shipments			Employment				Cost as % of Shipments	Investment per Employee ($)
		Total ($ mil)	% of U.S.	Per Establ.	Total Number	% of U.S.	Per Establ.	Wages ($/hour)		
Minnesota	5	33.1	1.8	6.6	400	3.8	80	10.67	30.5	-
Wisconsin	8	(D)	-	-	1,750 *	16.8	219	-	-	-
New York	6	(D)	-	-	750 *	7.2	125	-	-	-
North Carolina	6	(D)	-	-	1,750 *	16.8	292	-	-	-
Illinois	3	(D)	-	-	175 *	1.7	58	-	-	-
Missouri	3	(D)	-	-	1,750 *	16.8	583	-	-	-
New Jersey	3	(D)	-	-	750 *	7.2	250	-	-	-
Colorado	2	(D)	-	-	175 *	1.7	88	-	-	-
Maryland	2	(D)	-	-	175 *	1.7	88	-	-	-
Tennessee	2	(D)	-	-	750 *	7.2	375	-	-	-
Georgia	1	(D)	-	-	375 *	3.6	375	-	-	-
Iowa	1	(D)	-	-	375 *	3.6	375	-	-	-
Kentucky	1	(D)	-	-	375 *	3.6	375	-	-	-
Ohio	1	(D)	-	-	175 *	1.7	175	-	-	-
South Carolina	1	(D)	-	-	750 *	7.2	750	-	-	-
Vermont	1	(D)	-	-	375 *	3.6	375	-	-	-

Source: 1992 *Economic Census*. The states are in descending order of shipments or establishments (if shipment data are missing for the majority). The symbol (D) appears when data are withheld to prevent disclosure of competitive information. States marked with (D) are sorted by number of establishments. A dash (-) indicates that the data element cannot be calculated; * indicates the midpoint of a range.

3694 - ENGINE ELECTRICAL EQUIPMENT

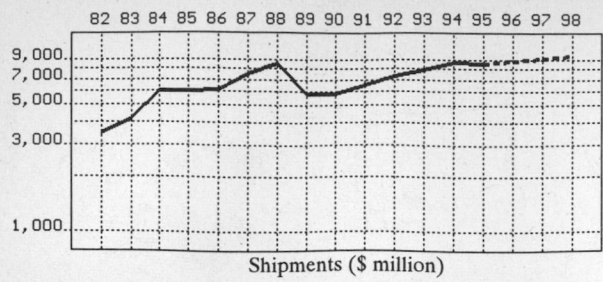

Shipments ($ million)

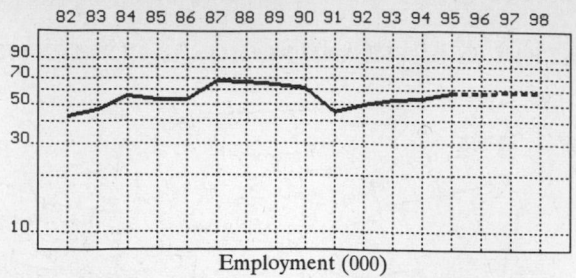

Employment (000)

GENERAL STATISTICS

Year	Companies	Establishments Total	Establishments with 20 or more employees	Employment Total (000)	Production Workers (000)	Hours (Mil)	Payroll ($ mil)	Wages ($/hr)	Cost of Materials	Value Added by Manufacture	Value of Shipments	Capital Invest.
1982	393	433	196	42.9	32.3	60.5	820.5	8.84	1,582.1	1,851.5	3,464.3	78.4
1983		433	206	46.6	36.4	68.8	939.8	9.31	1,873.7	2,364.3	4,212.4	112.7
1984		433	216	55.3	43.4	85.1	1,202.8	10.07	2,732.7	3,280.3	5,971.3	166.0
1985		434	226	52.9	40.5	78.8	1,224.7	10.96	2,820.9	3,121.0	5,925.3	222.2
1986		437	227	53.2	41.1	80.8	1,248.8	11.03	3,027.8	3,004.0	6,071.1	186.2
1987	431	486	235	67.1	51.8	102.5	1,713.2	11.56	4,040.3	3,432.4	7,472.6	237.7
1988		476	244	66.6	52.0	104.0	1,775.2	11.94	4,840.3	3,627.5	8,490.6	172.2
1989		450	228	64.3	45.6	91.4	1,704.0	12.54	4,601.3	3,920.3	5,817.0	271.0
1990		456	227	61.7	36.2	72.7	1,258.3	11.75	2,975.7	2,846.7	5,810.5	197.2
1991		461	214	44.9	34.7	69.2	1,312.9	12.97	3,177.9	3,201.5	6,407.8	158.6
1992	475	522	237	50.0	39.4	77.2	1,407.7	12.87	3,441.4	3,752.7	7,238.7	167.1
1993		512	225	52.8	41.8	84.4	1,481.7	12.59	3,918.5	3,957.7	7,832.7	201.8
1994		507p	237p	53.8	42.8	86.7	1,612.8	13.22	4,383.5	4,384.5	8,683.2	209.2
1995		514p	239p	57.3p	42.9p	87.4p	1,690.9p	13.91p	4,282.3p	4,283.3p	8,482.7p	225.4p
1996		521p	241p	57.7p	43.2p	88.2p	1,737.9p	14.25p	4,431.4p	4,432.4p	8,778.1p	231.4p
1997		528p	243p	58.0p	43.4p	89.0p	1,784.9p	14.59p	4,580.5p	4,581.5p	9,073.4p	237.5p
1998		535p	245p	58.4p	43.6p	89.8p	1,831.9p	14.94p	4,729.6p	4,730.7p	9,368.8p	243.5p

Sources: 1982, 1987, 1992 *Economic Census*; *Annual Survey of Manufactures*, 83-86, 88-91, 93-94. Establishment counts for non-Census years are from *County Business Patterns*; establishment values for 83-84 are extrapolations. 'P's show projections by the editors. Industries reclassified in 87 will not have data for prior years.

INDICES OF CHANGE

Year	Companies	Establishments Total	Establishments with 20 or more employees	Employment Total (000)	Production Workers (000)	Hours (Mil)	Payroll ($ mil)	Wages ($/hr)	Cost of Materials	Value Added by Manufacture	Value of Shipments	Capital Invest.
1982	83	83	83	86	82	78	58	69	46	49	48	47
1983		83	87	93	92	89	67	72	54	63	58	67
1984		83	91	111	110	110	85	78	79	87	82	99
1985		83	95	106	103	102	87	85	82	83	82	133
1986		84	96	106	104	105	89	86	88	80	84	111
1987	91	93	99	134	131	133	122	90	117	91	103	142
1988		91	103	133	132	135	126	93	141	97	117	103
1989		86	96	129	116	118	121	97	134	104	80	162
1990		87	96	123	92	94	89	91	86	76	80	118
1991		88	90	90	88	90	93	101	92	85	89	95
1992	100	100	100	100	100	100	100	100	100	100	100	100
1993		98	95	106	106	109	105	98	114	105	108	121
1994		97p	100p	108	109	112	115	103	127	117	120	125
1995		98p	101p	115p	109p	113p	120p	108p	124p	114p	117p	135p
1996		100p	102p	115p	110p	114p	123p	111p	129p	118p	121p	138p
1997		101p	103p	116p	110p	115p	127p	113p	133p	122p	125p	142p
1998		102p	104p	117p	111p	116p	130p	116p	137p	126p	129p	146p

Sources: Same as General Statistics. Values reflect change from the base year, 1992. Values above 100 mean greater than 92, values below 100 mean less than 92, and a value of 100 in the 82-91 or 93-98 period means same as 92. 'P's mark projections by the editors.

SELECTED RATIOS

For 1994	Avg. of All Manufact.	Analyzed Industry	Index	For 1994	Avg. of All Manufact.	Analyzed Industry	Index
Employees per Establishment	49	106	217	Value Added per Production Worker	134,084	102,442	76
Payroll per Establishment	1,500,273	3,183,253	212	Cost per Establishment	5,045,178	8,651,903	171
Payroll per Employee	30,620	29,978	98	Cost per Employee	102,970	81,478	79
Production Workers per Establishment	34	84	246	Cost per Production Worker	146,988	102,418	70
Wages per Establishment	853,319	2,262,253	265	Shipments per Establishment	9,576,895	17,138,407	179
Wages per Production Worker	24,861	26,780	108	Shipments per Employee	195,460	161,398	83
Hours per Production Worker	2,056	2,026	99	Shipments per Production Worker	279,017	202,879	73
Wages per Hour	12.09	13.22	109	Investment per Establishment	321,011	412,907	129
Value Added per Establishment	4,602,255	8,653,877	188	Investment per Employee	6,552	3,888	59
Value Added per Employee	93,930	81,496	87	Investment per Production Worker	9,352	4,888	52

Sources: Same as General Statistics. The 'Average of All Manufacturing' column represents the average of all manufacturing industries reported for the most recent complete year available. The Index shows the relationship between the Average and the Analyzed Industry. For example, 100 means that they are equal; 500 that the Analyzed Industry is five times the average; 50 means that the Analyzed Industry is half the national average. The abbreviation 'na' is used to show that data are 'not available'.

LEADING COMPANIES   Number shown: 42   Total sales ($ mil): 8,727   Total employment (000): 69.5

Company Name	Address				CEO Name	Phone	Co. Type	Sales ($ mil)	Empl. (000)
Delco Electronics Corp	1 Corporate Ctr	Kokomo	IN	46904	Gary W Dickinson	317-457-8461	S	4,300	30.0
Echlin Inc	100 Double Beach	Branford	CT	06405	F J Mancheski	203-481-5751	P	2,230	20.6
Champion Spark Plug	PO Box 910	Toledo	OH	43661	David Cartwright	419-535-2567	D	490	5.0
NGK Spark Plugs	8 Whatney	Irvine	CA	92718	Noboru Torii	714-855-8278	S	330	0.3
Regal-Beloit Corp	200 State St	Beloit	WI	53511	James L Packard	608-364-8800	P	243	2.4
Echlin Automotive	Echlin Rd & US 1	Branford	CT	06405	Scott Greer	203-481-5771	D	180	0.8
Prestolite Wire Corp	32871 Middlebelt	Farmington Hls	MI	48334	George L Joeckel	313-626-1336	R	180	1.2
Prestolite Electric Inc	2100 Commonw	Ann Arbor	MI	48105	P Kim Packard	313-930-6690	R	125	1.7
Motorola AIEG	3740 N Austin Rd	Seguin	TX	78155	Tom Marecek	210-379-8850	D	99*	1.0
BG Automotive Motors Inc	250 E Main St	Hendersonville	TN	37075	Dave Robinson	615-822-2800	J	71	0.3
Wells Mfg Corp	PO Box 70	Fond du Lac	WI	54936	WA Allen	414-922-5900	S	70	1.0
Wabash Magnetics	PO Box 829	Huntington	IN	46750	CJ Gallagher	219-356-8300	S	58	1.0
Hoskins Manufacturing Co	PO Box 218	Hamburg	MI	48139	Jerome Reinke	810-231-1900	S	53*	0.4
RE Phelon Company Inc	895 University Pkwy	Aiken	SC	29801	Russell D Phelon	803-643-7820	R	38*	0.4
Philips Technologies	813 S Grandstaff St	Auburn	IN	46706	Jerry Gabbard	219-925-8700	D	35	0.4
Autotronic Controls Corp	1490 Brennan	El Paso	TX	79936	Jack Priegel	915-857-5200	R	23	0.4
Motorcar Parts and Accessories	2727 Maricopa St	Torrance	CA	90503	Mel Marks	310-212-7910	P	21	0.2
Transpo Electronics Inc	2150 Brengle Av	Orlando	FL	32808	FO Oropeza	407-298-4563	R	20	0.3
Syncro Corp	PO Box 427	Arab	AL	35016	Blair E Stentz	205-586-6045	R	18	0.2
CE Niehoff and Co	2021 Lee St	Evanston	IL	60202	George Buhrfeind	708-866-6030	R	16	0.2
Minowitz Manufacturing Co	27941 Groesbeck	Roseville	MI	48066	Paul Pereira	313-779-5940	R	14*	0.1
TM Morris Manufacturing Co	PO Box 658	Logansport	IN	46947	T M Morris Sr	219-722-4040	R	14	0.4
Mallory Inc	550 Mallory Way	Carson City	NV	89701	Gregg Koechcien	702-882-6600	R	13	0.3
Del City Wire Company Inc	PO Box 95609	Oklahoma City	OK	73143	J Swartzendruber	405-672-4515	R	12*	0.1
Ampere Automotive Corp	3500 N Kostner	Chicago	IL	60641	Mike Cohen	312-685-4745	R	11	0.1
Mid-State Automotive	1420 Gover Ln	Ferguson	KY	42533	JL Thacker	606-679-4339	R	10*	0.3
Zenith Ignitions Inc	18 Furler St	Totowa	NJ	07512	Moses Goldstein	201-785-1500	R	8*	0.1
Burden Automotive Electric Inc	4914 S Paulina St	Chicago	IL	60609	Walter Burden	312-737-6300	R	6*	<0.1
Fargo Assembly of Indiana Co	800 S Cleveland	Mishawaka	IN	46544	Dale Cunningham	219-259-3728	R	5	<0.1
Willamette Electric Products Co	810 N Graham St	Portland	OR	97227	Elliott Quinn	503-288-7361	R	5	<0.1
PerTronix Inc	1268 E Edna Pl	Covina	CA	91724	Tom Reh	818-331-4801	R	4	<0.1
Gauss Corp	PO Box 660	Scarborough	ME	04070	R D Sampson Sr	207-883-4121	R	4	<0.1
Tennessee Armature	PO Box 27	Knoxville	TN	37901	RF Thomas	615-524-3681	S	4	<0.1
Nutek Inc	PO Drawer 17547	Pensacola	FL	32522	JL Taylor	904-478-1793	R	4	<0.1
Electrodyne	PO Box 660	Scarborough	ME	04070	R D Sampson Sr	207-883-4121	D	3	<0.1
Jacobs Electronics	500 N Baird St	Midland	TX	79701	Christopher Jacobs	915-685-3345	R	3	<0.1
Keystone Cable Corp	1801 W Courtland	Philadelphia	PA	19140	Thomas A Scott	215-457-3123	R	3	<0.1
Laketronics Inc	11153 NE Wawasee	Syracuse	IN	46567	RC McNary	219-856-4588	R	2	<0.1
United States Energy Corp	1600 N Missile Way	Anaheim	CA	92801	Pamela Higgins	714-871-8185	R	1	<0.1
Arrowhead Stator Rotor Inc	3829 Jefferson NE	Minneapolis	MN	55421	John Berger	612-788-9631	R	1	<0.1
Trio Aviation Inc	PO Box 29178	Dallas	TX	75229	Jack M Sherman III	214-241-2101	R	1	<0.1
Wrangler Power Products Inc	PO Box 12109	Prescott	AZ	86304	Marilyn J Jones	602-945-1514	R	1	<0.1

Source: *Ward's Business Directory of U.S. Private and Public Companies*, Volumes 1 and 2, 1996. The company type code used is as follows: P - Public, R - Private, S - Subsidiary, D - Division, J - Joint Venture, A - Affiliate, G - Group. Sales are in millions of dollars, employees are in thousands. An asterisk (*) indicates an estimated sales volume. The symbol < stands for 'less than'. Company names and addresses are truncated, in some cases, to fit into the available space.

MATERIALS CONSUMED

Material	Quantity	Delivered Cost ($ million)
Materials, ingredients, containers, and supplies	(X)	3,261.0
Metal bolts, nuts, screws, washers, rivets, and other screw machine products	(X)	123.1
Iron and steel castings (rough and semifinished)	(X)	13.9
Aluminum and aluminum-base alloy castings (rough and semifinished)	(X)	37.3
Other nonferrous castings (rough and semifinished)	(X)	1.3
Steel shapes and forms	(X)	156.9
Copper and copper-base alloy shapes and forms	(X)	18.1
All other nonferrous shapes and forms	(X)	30.4
Insulated wire and cable (except magnet wire)	(X)	108.7
Magnet wire	(X)	48.4
Ball and roller bearings (mounted or unmounted)	(X)	37.2
Mechanical speed changers, gears, and industrial high-speed drives	(X)	34.1
Paperboard containers, boxes, and corrugated paperboard	(X)	35.2
Plastics resins consumed in the form of granules, pellets, powders, liquids, etc.	(X)	29.8
Plastics products consumed in the form of sheets, rods, tubes, film, and other shapes	(X)	15.6
Fabricated plastics products (except gaskets, hoses, and belting)	(X)	60.7
Fabricated rubber products, except tires, tubes, hose, belting, and gaskets	(X)	46.4
Current-carrying wiring devices	(X)	35.8
Printed circuit boards (without inserted components) for electronic circuitry	(X)	592.6
Other components and accessories for electronic circuitry, nec, except tubes	(X)	327.9
Used engine electrical equipment	(X)	79.0
All other materials and components, parts, containers, and supplies	(X)	1,038.3
Materials, ingredients, containers, and supplies, nsk	(X)	390.2

Source: 1992 *Economic Census*. Explanation of symbols used: (D): Withheld to avoid disclosure of competitive data; na: Not available; (S): Withheld because statistical norms were not met; (X): Not applicable; (Z): Less than half the unit shown; nec: Not elsewhere classified; nsk: Not specified by kind; - : zero; * : 10-19 percent estimated; ** : 20-29 percent estimated.

PRODUCT SHARE DETAILS

Product or Product Class	% Share	Product or Product Class	% Share
Engine electrical equipment	100.00	Cranking motors (starters), nsk	1.93
Ignition harness and cable sets	14.20	Spark plugs (all types)	6.25
Automotive type ignition harness sets	68.27	Complete engine electrical equipment, nec	38.18
Other ignition harness sets (including tractor, stationary		Complete engine electrical ignition coils (all types)	13.34
engine, and aircraft)	7.87	Complete engine electrical distributors (all types)	10.34
Automotive type engine electrical cable sets	16.65	Complete engine magnetos, magneto-dynamos, and	
Aircraft and other type engine electrical cable sets	2.50	magnetic flywheels	2.41
Ignition harness and cable sets, nsk	4.70	Other complete ignition equipment (including electronic	
Battery charging alternators, generators, and regulators	17.43	ignitions)	10.02
Automotive battery charging alternators and generators,		Electronic systems for complete engine control, using	
new	54.16	computers or microprocessors	63.20
Other new battery charging alternators and generators for		Complete engine electrical equipment, nec, nsk	0.69
internal combustion engines	3.86	Parts for engine electrical and/or electronic equipment	6.53
Rebuilt battery charging alternators and generators for		Armatures, field coils, and drive-end housings for engine	
internal combustion engines, all types	35.30	electrical cranking motors	9.60
Regulators for battery charging alternators and generators		Armatures and field coils for alternators and generators	7.32
(new and rebuilt)	1.37	Ignition distributor heads and rotors	27.17
Battery charging alternators, generators, and regulators, nsk	5.31	Ignition distributor breaker point sets	6.25
Cranking motors (starters)	13.69	Other parts for engine electrical and/or electronic	
Automotive starting (engine cranking) motors, new	53.69	equipment	49.54
Other new starting (engine cranking) motors	8.40	Parts for engine electrical and/or electronic equipment, nsk	0.12
Rebuilt starting (engine cranking) motors, all types	35.99	Engine electrical equipment, nsk	3.73

Source: 1992 *Economic Census*. The values shown are percent of total shipments in an industry. Values of indented subcategories are summed in the main heading. The symbol (D) appears when data are withheld to prevent disclosure of competitive information. The abbreviation nsk stands for 'not specified by kind' and nec for 'not elsewhere classified'.

INPUTS AND OUTPUTS FOR ENGINE ELECTRICAL EQUIPMENT

Economic Sector or Industry Providing Inputs	%	Sector	Economic Sector or Industry Buying Outputs	%	Sector
Imports	18.0	Foreign	Motor vehicles & car bodies	25.5	Manufg.
Engine electrical equipment	15.9	Manufg.	Engine electrical equipment	11.3	Manufg.
Wholesale trade	13.0	Trade	Personal consumption expenditures	10.8	
Blast furnaces & steel mills	5.1	Manufg.	Exports	10.2	Foreign
Electronic components nec	4.9	Manufg.	Automotive repair shops & services	9.5	Services
Advertising	3.6	Services	Motor vehicle parts & accessories	6.7	Manufg.
Nonferrous wire drawing & insulating	2.9	Manufg.	Internal combustion engines, nec	4.0	Manufg.
Used & secondhand goods	2.6	Scrap	Federal Government purchases, national defense	3.1	Fed Govt
Air transportation	2.1	Util.	Real estate	2.7	Fin/R.E.
Semiconductors & related devices	1.8	Manufg.	Farm machinery & equipment	1.7	Manufg.
Miscellaneous plastics products	1.7	Manufg.	Aircraft	1.5	Manufg.
Banking	1.7	Fin/R.E.	Aircraft & missile engines & engine parts	1.4	Manufg.
Screw machine and related products	1.6	Manufg.	Ship building & repairing	1.0	Manufg.
Wiring devices	1.4	Manufg.	Feed grains	0.8	Agric.
Electric services (utilities)	1.4	Util.	Motor freight transportation & warehousing	0.7	Util.
Hotels & lodging places	1.3	Services	Retail trade, except eating & drinking	0.7	Trade
Fabricated metal products, nec	1.2	Manufg.	Wholesale trade	0.6	Trade
Petroleum refining	1.1	Manufg.	Meat animals	0.5	Agric.
Copper rolling & drawing	1.0	Manufg.	Local government passenger transit	0.5	Gov't
Plastics materials & resins	1.0	Manufg.	Oil bearing crops	0.3	Agric.
Motor freight transportation & warehousing	1.0	Util.	Boat building & repairing	0.3	Manufg.
Paperboard containers & boxes	0.9	Manufg.	Food grains	0.2	Agric.
Primary aluminum	0.8	Manufg.	Lawn & garden equipment	0.2	Manufg.
Communications, except radio & TV	0.8	Util.	Miscellaneous plastics products	0.2	Manufg.
Carbon & graphite products	0.7	Manufg.	Motor homes (made on purchased chassis)	0.2	Manufg.
Metal stampings, nec	0.7	Manufg.	Truck & bus bodies	0.2	Manufg.
Eating & drinking places	0.7	Trade	Transit & bus transportation	0.2	Util.
Aluminum castings	0.6	Manufg.	Water transportation	0.2	Util.
Metal coating & allied services	0.6	Manufg.	Automotive rental & leasing, without drivers	0.2	Services
Maintenance of nonfarm buildings nec	0.5	Constr.	S/L Govt. purch., elem. & secondary education	0.2	S/L Govt
Ball & roller bearings	0.5	Manufg.	Tobacco	0.1	Agric.
Equipment rental & leasing services	0.5	Services	Vegetables	0.1	Agric.
Machinery, except electrical, nec	0.4	Manufg.	Aircraft & missile equipment, nec	0.1	Manufg.
Power transmission equipment	0.4	Manufg.	Legal services	0.1	Services
Gas production & distribution (utilities)	0.4	Util.	Federal Government purchases, nondefense	0.1	Fed Govt
Computer & data processing services	0.4	Services	S/L Govt. purch., highways	0.1	S/L Govt
Legal services	0.4	Services	S/L Govt. purch., other general government	0.1	S/L Govt
Fabricated rubber products, nec	0.3	Manufg.	S/L Govt. purch., police	0.1	S/L Govt
Iron & steel foundries	0.3	Manufg.			
Special dies & tools & machine tool accessories	0.3	Manufg.			
Retail trade, except eating & drinking	0.3	Trade			
Management & consulting services & labs	0.3	Services			
Scrap	0.3	Scrap			
Aluminum rolling & drawing	0.2	Manufg.			
Brass, bronze, & copper castings	0.2	Manufg.			
Miscellaneous fabricated wire products	0.2	Manufg.			
Nonferrous castings, nec	0.2	Manufg.			
Nonferrous rolling & drawing, nec	0.2	Manufg.			

Continued on next page.

INPUTS AND OUTPUTS FOR ENGINE ELECTRICAL EQUIPMENT - Continued

Economic Sector or Industry Providing Inputs	%	Sector	Economic Sector or Industry Buying Outputs	%	Sector
Railroads & related services	0.2	Util.			
Insurance carriers	0.2	Fin/R.E.			
Accounting, auditing & bookkeeping	0.2	Services			
Automotive rental & leasing, without drivers	0.2	Services			
Automotive repair shops & services	0.2	Services			
U.S. Postal Service	0.2	Gov't			
Noncomparable imports	0.2	Foreign			
Iron & ferroalloy ores	0.1	Mining			
Abrasive products	0.1	Manufg.			
Felt goods, nec	0.1	Manufg.			
Lubricating oils & greases	0.1	Manufg.			
Manifold business forms	0.1	Manufg.			
Real estate	0.1	Fin/R.E.			
Royalties	0.1	Fin/R.E.			

Source: Benchmark Input-Output Accounts for the U.S. Economy, 1982, U.S. Department of Commerce, Washington, D.C., July 1991. Data, as reported in the source, are organized by the 1977 SIC structure in use in 1982 but have been matched, as closely as is possible, to the 1987 SIC structure used in this book.

OCCUPATIONS EMPLOYED BY SIC 369 - MISC ELECTRICAL EQUIPMENT AND SUPPLIES

Occupation	% of Total 1994	Change to 2005	Occupation	% of Total 1994	Change to 2005
Assemblers, fabricators, & hand workers nec	14.5	1.0	Secretaries, ex legal & medical	1.3	-8.1
Electrical & electronic assemblers	10.7	-9.1	Electrical & electronic technicians,technologists	1.3	1.0
Machine operators nec	4.6	-28.8	Tool & die makers	1.3	-18.4
Electrical & electronic equipment assemblers	4.0	1.0	Hand packers & packagers	1.3	-13.4
Blue collar worker supervisors	3.4	-11.6	Engineering technicians & technologists nec	1.2	1.0
Inspectors, testers, & graders, precision	3.3	-29.3	Freight, stock, & material movers, hand	1.2	-19.2
Industrial machinery mechanics	2.4	36.3	Mechanical engineers	1.2	22.2
Electrical & electronics engineers	1.9	7.5	Managers & administrators nec	1.1	1.0
Engineers nec	1.7	51.5	Helpers, laborers, & material movers nec	1.1	1.0
Sales & related workers nec	1.5	1.0	Traffic, shipping, & receiving clerks	1.0	-2.8
Plastic molding machine workers	1.4	21.2	Industrial production managers	1.0	1.0
General managers & top executives	1.4	-4.2	Electromechanical equipment assemblers	1.0	11.1

Source: Industry-Occupation Matrix, Bureau of Labor Statistics. These data relate to one or more 3-digit SIC industry groups rather than to a single 4-digit SIC. The change reported for each occupation to the year 2005 is a percent of growth or decline as estimated by the Bureau of Labor Statistics. The abbreviation nec stands for 'not elsewhere classified'.

LOCATION BY STATE AND REGIONAL CONCENTRATION

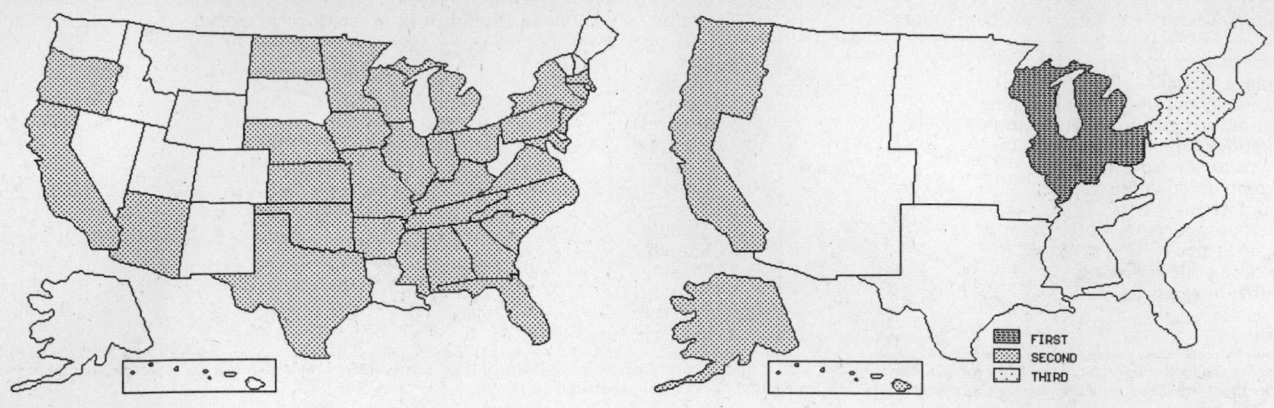

INDUSTRY DATA BY STATE

| State | Establish-ments | Shipments | | | Employment | | | | Cost as % of Shipments | Investment per Employee ($) |
		Total ($ mil)	% of U.S.	Per Establ.	Total Number	% of U.S.	Per Establ.	Wages ($/hour)		
Texas	46	641.8	8.9	14.0	3,100	6.2	67	8.07	50.0	8,548
Michigan	34	412.2	5.7	12.1	2,200	4.4	65	17.26	50.2	5,500
New York	30	394.2	5.4	13.1	2,600	5.2	87	11.10	29.2	2,577
Ohio	21	333.1	4.6	15.9	2,900	5.8	138	10.92	38.9	-
Mississippi	16	327.4	4.5	20.5	2,600	5.2	163	11.59	58.3	-
Georgia	19	276.5	3.8	14.6	1,600	3.2	84	13.00	47.3	1,438
Illinois	34	173.9	2.4	5.1	2,300	4.6	68	8.33	49.1	1,478
South Carolina	4	83.9	1.2	21.0	900	1.8	225	7.32	54.8	-
Tennessee	11	45.9	0.6	4.2	500	1.0	45	7.57	67.5	-
Kentucky	8	36.2	0.5	4.5	600	1.2	75	8.11	48.6	1,000
Oregon	9	35.9	0.5	4.0	400	0.8	44	10.67	32.6	1,750
Missouri	17	25.7	0.4	1.5	400	0.8	24	8.00	49.8	3,000
New Jersey	12	23.4	0.3	2.0	300	0.6	25	11.75	51.3	1,333
North Carolina	4	14.2	0.2	3.5	200	0.4	50	9.00	51.4	-
California	58	(D)	-	-	3,750 *	7.5	65	-	-	-
Florida	24	(D)	-	-	750 *	1.5	31	-	-	-
Indiana	24	(D)	-	-	7,500 *	15.0	313	-	-	-
Pennsylvania	21	(D)	-	-	3,750 *	7.5	179	-	-	-
Massachusetts	14	(D)	-	-	750 *	1.5	54	-	-	-
Wisconsin	12	(D)	-	-	3,750 *	7.5	313	-	-	-
Iowa	10	(D)	-	-	1,750 *	3.5	175	-	-	-
Alabama	9	(D)	-	-	3,750 *	7.5	417	-	-	-
Arizona	8	(D)	-	-	750 *	1.5	94	-	-	667
Kansas	8	(D)	-	-	1,750 *	3.5	219	-	-	-
Minnesota	8	(D)	-	-	175 *	0.4	22	-	-	1,143
Oklahoma	8	(D)	-	-	750 *	1.5	94	-	-	-
Connecticut	6	(D)	-	-	1,750 *	3.5	292	-	-	-
Virginia	6	(D)	-	-	750 *	1.5	125	-	-	-
Arkansas	4	(D)	-	-	750 *	1.5	188	-	-	-
Nebraska	3	(D)	-	-	175 *	0.4	58	-	-	-
North Dakota	3	(D)	-	-	175 *	0.4	58	-	-	-
Maryland	1	(D)	-	-	375 *	0.8	375	-	-	-

Source: 1992 *Economic Census*. The states are in descending order of shipments or establishments (if shipment data are missing for the majority). The symbol (D) appears when data are withheld to prevent disclosure of competitive information. States marked with (D) are sorted by number of establishments. A dash (-) indicates that the data element cannot be calculated; * indicates the midpoint of a range.

3695 - MAGNETIC & OPTICAL RECORDING MEDIA

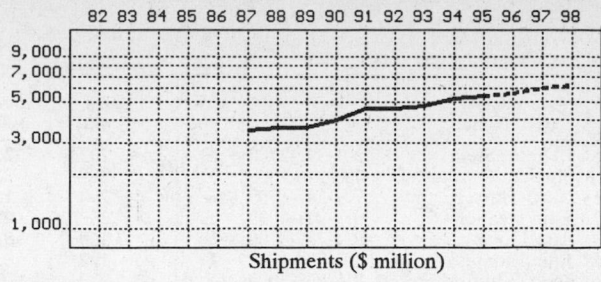

Shipments ($ million)

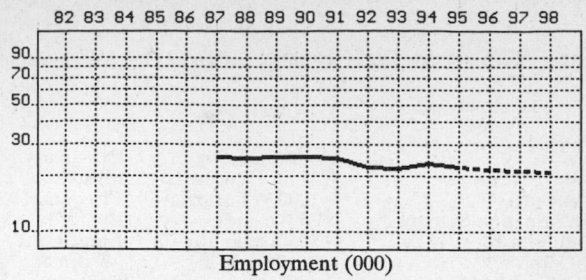

Employment (000)

GENERAL STATISTICS

Year	Companies	Establishments		Employment			Compensation		Production ($ million)			
		Total	with 20 or more employees	Total (000)	Production Workers (000)	Hours (Mil)	Payroll ($ mil)	Wages ($/hr)	Cost of Materials	Value Added by Manufacture	Value of Shipments	Capital Invest.
1982												
1983												
1984												
1985												
1986												
1987	181	200	92	25.6	16.3	33.0	638.6	9.82	1,836.6	1,687.2	3,504.0	225.6
1988		191	82	25.1	16.2	32.6	662.1	10.32	2,067.0	1,593.5	3,630.7	269.1
1989		214	96	25.5	16.9	32.6	651.1	10.78	2,080.0	1,542.8	3,644.2	219.8
1990		215	95	25.5	16.1	31.4	651.9	11.29	2,365.9	1,675.6	4,032.1	286.8
1991		243	103	25.4	16.1	33.2	734.7	11.24	2,734.8	1,959.9	4,615.9	305.2
1992	239	261	97	22.6	15.1	32.2	695.0	11.30	2,513.6	2,091.5	4,641.3	394.2
1993		261	93	22.2	15.0	31.2	689.6	12.18	2,406.1	2,388.0	4,765.5	337.9
1994		277P	100P	23.7	15.5	30.3	778.5	14.16	2,637.5	2,615.2	5,256.3	325.7
1995		289P	101P	22.5P	15.0P	30.7P	759.0P	13.59P	2,716.6P	2,693.6P	5,413.8P	380.5P
1996		302P	103P	22.1P	14.8P	30.4P	774.9P	14.08P	2,845.1P	2,821.0P	5,670.0P	399.4P
1997		314P	104P	21.6P	14.6P	30.1P	790.7P	14.57P	2,973.6P	2,948.5P	5,926.1P	418.3P
1998		327P	105P	21.2P	14.4P	29.8P	806.6P	15.06P	3,102.1P	3,075.9P	6,182.2P	437.2P

Sources: 1982, 1987, 1992 *Economic Census*; *Annual Survey of Manufactures*, 83-86, 88-91, 93-94. Establishment counts for non-Census years are from *County Business Patterns*; establishment values for 83-84 are extrapolations. 'P's show projections by the editors. Industries reclassified in 87 will not have data for prior years.

INDICES OF CHANGE

Year	Companies	Establishments		Employment			Compensation		Production ($ million)			
		Total	with 20 or more employees	Total (000)	Production Workers (000)	Hours (Mil)	Payroll ($ mil)	Wages ($/hr)	Cost of Materials	Value Added by Manufacture	Value of Shipments	Capital Invest.
1982												
1983												
1984												
1985												
1986												
1987	76	77	95	113	108	102	92	87	73	81	75	57
1988		73	85	111	107	101	95	91	82	76	78	68
1989		82	99	113	112	101	94	95	83	74	79	56
1990		82	98	113	107	98	94	100	94	80	87	73
1991		93	106	112	107	103	106	99	109	94	99	77
1992	100	100	100	100	100	100	100	100	100	100	100	100
1993		100	96	98	99	97	99	108	96	114	103	86
1994		106P	103P	105	103	94	112	125	105	125	113	83
1995		111P	104P	100P	99P	95P	109P	120P	108P	129P	117P	97P
1996		116P	106P	98P	98P	94P	111P	125P	113P	135P	122P	101P
1997		120P	107P	96P	97P	93P	114P	129P	118P	141P	128P	106P
1998		125P	109P	94P	95P	93P	116P	133P	123P	147P	133P	111P

Sources: Same as General Statistics. Values reflect change from the base year, 1992. Values above 100 mean greater than 92, values below 100 mean less than 92, and a value of 100 in the 82-91 or 93-98 period means same as 92. 'P's mark projections by the editors.

SELECTED RATIOS

For 1994	Avg. of All Manufact.	Analyzed Industry	Index	For 1994	Avg. of All Manufact.	Analyzed Industry	Index
Employees per Establishment	49	86	175	Value Added per Production Worker	134,084	168,723	126
Payroll per Establishment	1,500,273	2,813,371	188	Cost per Establishment	5,045,178	9,531,492	189
Payroll per Employee	30,620	32,848	107	Cost per Employee	102,970	111,287	108
Production Workers per Establishment	34	56	163	Cost per Production Worker	146,988	170,161	116
Wages per Establishment	853,319	1,550,509	182	Shipments per Establishment	9,576,895	18,995,405	198
Wages per Production Worker	24,861	27,681	111	Shipments per Employee	195,460	221,785	113
Hours per Production Worker	2,056	1,955	95	Shipments per Production Worker	279,017	339,116	122
Wages per Hour	12.09	14.16	117	Investment per Establishment	321,011	1,177,026	367
Value Added per Establishment	4,602,255	9,450,903	205	Investment per Employee	6,552	13,743	210
Value Added per Employee	93,930	110,346	117	Investment per Production Worker	9,352	21,013	225

Sources: Same as General Statistics. The 'Average of All Manufacturing' column represents the average of all manufacturing industries reported for the most recent complete year available. The Index shows the relationship between the Average and the Analyzed Industry. For example, 100 means that they are equal; 500 that the Analyzed Industry is five times the average; 50 means that the Analyzed Industry is half the national average. The abbreviation 'na' is used to show that data are 'not available'.

LEADING COMPANIES   Number shown: **35**   Total sales ($ mil): **6,732**   Total employment (000): **31.1**

Company Name	Address				CEO Name	Phone	Co. Type	Sales ($ mil)	Empl. (000)
BASF Corp	8 Campus Dr	Parsippany	NJ	07054	J Dieter Stein	201-397-2700	S	5,202	22.0
Komag Inc	275 S Hillview Dr	Milpitas	CA	95035	Stephen C Johnson	408-946-2300	P	385	3.5
Verbatim Corp	1200 WT Harris	Charlotte	NC	28262	Nicky Hartery	704-547-6500	S	290*	1.3
BASF Information Systems	35 Crosby Dr	Bedford	MA	01730	John Healion	617-271-4000	S	250	0.2
Fuji Photo Film USA Inc	555 Taxter Rd	Elmsford	NY	10523	Sam Inoue	914-789-8100	S	160	0.9
Nashua Corp	PO Box 3000	Merrimack	NH	03054	John Montesi	603-880-1234	D	112	0.8
Graham Magnetics Inc	4001 Airport	Bedford	TX	76021	Scott Whittenburg	817-868-5000	S	80	0.5
Micrographic Technology Corp	520 Logue Av	Mountain View	CA	94043	A Oosthuizen	415-965-3700	R	39	0.2
Plasmon Data Inc	1654 Centre Pte Dr	Milpitas	CA	95135	Peter Helfet	408-956-9400	R	36	0.2
Meridian Data Inc	5615 Scotts Val Dr	Scotts Valley	CA	95066	B Allen Lay	408-438-3100	P	20	<0.1
Syncom Technologies Inc	1000 Syncom Dr	Mitchell	SD	57301	Mark Anderson	605-996-8200	R	20	0.2
Gigatek Memory Systems Inc	1989 Palomar	Carlsbad	CA	92009	Andrew Wrobel	619-438-9010	R	15*	<0.1
TDK Magnetic Tape Corp	611 Hwy 74 S	Peachtree City	GA	30269	N Nakamora	404-487-5200	S	12	0.3
AM Holding Inc	PO Box 470848	Tulsa	OK	74147	Philip Odem	918-669-5181	R	10	<0.1
Innovative Data Technology	5340 Eastgate Mall	San Diego	CA	92121	Dale Spencer	619-587-0555	R	10	<0.1
Sigma Designs Imaging Systems	2701 Bayview Dr	Fremont	CA	93438	Bryan J Foertsch	510-770-1189	S	10	<0.1
Wabash Computer Products Inc	PO Box 470848	Tulsa	OK	74147	Philip Odom	918-669-5181	S	10	<0.1
InterMag Inc	4910 Raley Blv	Sacramento	CA	95838	Gary Steinberg	916-568-6744	S	8*	<0.1
US Optical Disc Inc	1 Eagle Dr	Sanford	ME	04073	Roland Demers	207-324-1124	S	8*	0.1
Kinesix	9800 Richmond	Houston	TX	77042	C Steve Pringle	713-953-8300	D	8	<0.1
National Micronetics Inc	71 Smith Av	Kingston	NY	12401	Yoon H Choo	914-338-0333	P	7	0.1
Electro Magnetic Products Inc	PO Box 87	Moorestown	NJ	08057	Gordon C Mason	609-235-3011	R	5	<0.1
Memory Control Techn Corp	2430 S 156 Cir	Omaha	NE	68130	Gerald C Korth	402-333-3100	R	5	<0.1
Factory Direct Disks Inc	6230 SW Patton Rd	Portland	OR	97221	Michael Bloom	503-291-6195	S	4	0.2
Iomega Corp	15110 Av of Science	San Diego	CA	92128	Farouk Al-Nasser	619-673-5500	S	4*	<0.1
DataDisc Inc	Route 3	Gainesville	VA	22065	Royce White	703-347-2111	R	3	<0.1
Brush Industries Inc	PO Box 638	Sunbury	PA	17801	Ed Bedell	717-286-5611	R	3	<0.1
Records Reserve Corp	56 Harvester Av	Batavia	NY	14020	John W Tompkins	716-344-2600	R	3	<0.1
Brown Disc Products Company	1120-B Elkton Dr	Co Springs	CO	80907	RE Rider	719-593-1015	P	3	<0.1
Magnetic Products Corp	15444 Cabrito Rd	Van Nuys	CA	91406	Gary R Hardt	818-780-2981	R	3	<0.1
American Digital Systems Inc	490 Boston Post Rd	Sudbury	MA	01776	Alan R Kivnik	508-443-7711	R	3	<0.1
Carisys Inc	6447 Warren Dr	Norcross	GA	30093	Michael W Smith	404-409-0098	R	2*	<0.1
Carpel Video	429 E Patrick St	Frederick	MD	21701	Andy Carpel	301-694-3500	R	1	<0.1
Santa Barbara Engineering Inc	14525 N 79th St	Scottsdale	AZ	85260	Edward Podwojski	602-948-8651	R	1	<0.1
DIC Digital Supply Corp	500 Frank W Burr	Teaneck	NJ	07666	Joseph Martinez	201-692-7700	S	1	<0.1

Source: *Ward's Business Directory of U.S. Private and Public Companies*, Volumes 1 and 2, 1996. The company type code used is as follows: P - Public, R - Private, S - Subsidiary, D - Division, J - Joint Venture, A - Affiliate, G - Group. Sales are in millions of dollars, employees are in thousands. An asterisk (*) indicates an estimated sales volume. The symbol < stands for 'less than'. Company names and addresses are truncated, in some cases, to fit into the available space.

MATERIALS CONSUMED

Material	Quantity	Delivered Cost ($ million)
Materials, ingredients, containers, and supplies	(X)	1,960.0
Paperboard containers, boxes, and corrugated paperboard	(X)	113.7
Plastics products consumed in the form of sheets, rods, tubes, film, and other shapes . . .	(X)	346.3
Fabricated plastics products (except gaskets, hoses, and belting)	(X)	151.0
Plastics resins consumed in the form of granules, pellets, powders, liquids, etc. . . .	(X)	200.1
Metal bolts, nuts, screws, washers, rivets, and other screw machine products	(X)	111.2
Castings (rough and semifinished)	(X)	(D)
Steel shapes and forms	(X)	34.8
Aluminum and aluminum-base alloy shapes and forms	(X)	91.3
Other nonferrous shapes and forms	(X)	(D)
Ferrites (powder and paste)	(X)	41.1
Other metal powders, including chromium	(X)	(D)
Current-carrying wiring devices	(X)	(D)
All other materials and components, parts, containers, and supplies	(X)	693.9
Materials, ingredients, containers, and supplies, nsk	(X)	176.1

Source: 1992 *Economic Census*. Explanation of symbols used: (D): Withheld to avoid disclosure of competitive data; na: Not available; (S): Withheld because statistical norms were not met; (X): Not applicable; (Z): Less than half the unit shown; nec: Not elsewhere classified; nsk: Not specified by kind; - : zero; * : 10-19 percent estimated; ** : 20-29 percent estimated.

PRODUCT SHARE DETAILS

Product or Product Class	% Share	Product or Product Class	% Share
Magnetic recording media	100.00		

Source: 1992 *Economic Census*. The values shown are percent of total shipments in an industry. Values of indented subcategories are summed in the main heading. The symbol (D) appears when data are withheld to prevent disclosure of competitive information. The abbreviation nsk stands for 'not specified by kind' and nec for 'not elsewhere classified'.

INPUTS AND OUTPUTS FOR ELECTRONIC COMPUTING EQUIPMENT

Economic Sector or Industry Providing Inputs	%	Sector	Economic Sector or Industry Buying Outputs	%	Sector
Electronic computing equipment	29.4	Manufg.	Gross private fixed investment	45.6	Cap Inv
Wholesale trade	11.9	Trade	Exports	21.6	Foreign
Imports	10.2	Foreign	Electronic computing equipment	17.8	Manufg.
Electronic components nec	9.0	Manufg.	Federal Government purchases, national defense	4.6	Fed Govt
Semiconductors & related devices	5.0	Manufg.	Federal Government purchases, nondefense	1.6	Fed Govt
Industrial controls	2.8	Manufg.	Computer & data processing services	1.4	Services
Miscellaneous plastics products	2.7	Manufg.	Change in business inventories	1.3	In House
Air transportation	2.1	Util.	Personal consumption expenditures	1.0	
Banking	1.6	Fin/R.E.	Radio & TV communication equipment	0.8	Manufg.
Hotels & lodging places	1.5	Services	Typewriters & office machines, nec	0.5	Manufg.
Communications, except radio & TV	1.4	Util.	S/L Govt. purch., elem. & secondary education	0.4	S/L Govt
Electric services (utilities)	1.3	Util.	S/L Govt. purch., higher education	0.4	S/L Govt
Sheet metal work	1.2	Manufg.	S/L Govt. purch., other general government	0.4	S/L Govt
Eating & drinking places	1.2	Trade	Calculating & accounting machines	0.3	Manufg.
Real estate	1.2	Fin/R.E.	Banking	0.3	Fin/R.E.
Noncomparable imports	1.1	Foreign	Accounting, auditing & bookkeeping	0.3	Services
Petroleum refining	1.0	Manufg.	Electronic components nec	0.2	Manufg.
Paperboard containers & boxes	0.9	Manufg.	Instruments to measure electricity	0.2	Manufg.
Motors & generators	0.8	Manufg.	Insurance carriers	0.2	Fin/R.E.
Switchgear & switchboard apparatus	0.8	Manufg.	Mechanical measuring devices	0.1	Manufg.
Legal services	0.7	Services	Retail trade, except eating & drinking	0.1	Trade
Maintenance of nonfarm buildings nec	0.5	Constr.	Wholesale trade	0.1	Trade
Gaskets, packing & sealing devices	0.5	Manufg.			
Metal stampings, nec	0.5	Manufg.			
Wiring devices	0.5	Manufg.			
Management & consulting services & labs	0.5	Services			
Aluminum castings	0.4	Manufg.			
Electron tubes	0.4	Manufg.			
Screw machine and related products	0.4	Manufg.			
Advertising	0.4	Services			
Equipment rental & leasing services	0.4	Services			
Aluminum rolling & drawing	0.3	Manufg.			
Blast furnaces & steel mills	0.3	Manufg.			
Fabricated metal products, nec	0.3	Manufg.			
Iron & steel foundries	0.3	Manufg.			
Miscellaneous fabricated wire products	0.3	Manufg.			
Nonferrous wire drawing & insulating	0.3	Manufg.			
Plating & polishing	0.3	Manufg.			
Motor freight transportation & warehousing	0.3	Util.			
Accounting, auditing & bookkeeping	0.3	Services			
Computer & data processing services	0.3	Services			
Electrical industrial apparatus, nec	0.2	Manufg.			
Machinery, except electrical, nec	0.2	Manufg.			
Manifold business forms	0.2	Manufg.			
Transformers	0.2	Manufg.			
Insurance carriers	0.2	Fin/R.E.			
Royalties	0.2	Fin/R.E.			
Security & commodity brokers	0.2	Fin/R.E.			
Automotive rental & leasing, without drivers	0.2	Services			
Ball & roller bearings	0.1	Manufg.			
Die-cut paper & board	0.1	Manufg.			
Nonferrous rolling & drawing, nec	0.1	Manufg.			
Gas production & distribution (utilities)	0.1	Util.			
Automotive repair shops & services	0.1	Services			
Photofinishing labs, commercial photography	0.1	Services			
U.S. Postal Service	0.1	Gov't			

Source: Benchmark Input-Output Accounts for the U.S. Economy, 1982, U.S. Department of Commerce, Washington, D.C., July 1991. Data, as reported in the source, are organized by the 1977 SIC structure in use in 1982 but have been matched, as closely as is possible, to the 1987 SIC structure used in this book.

OCCUPATIONS EMPLOYED BY SIC 369 - MISC ELECTRICAL EQUIPMENT AND SUPPLIES

Occupation	% of Total 1994	Change to 2005	Occupation	% of Total 1994	Change to 2005
Assemblers, fabricators, & hand workers nec	14.5	1.0	Secretaries, ex legal & medical	1.3	-8.1
Electrical & electronic assemblers	10.7	-9.1	Electrical & electronic technicians,technologists	1.3	1.0
Machine operators nec	4.6	-28.8	Tool & die makers	1.3	-18.4
Electrical & electronic equipment assemblers	4.0	1.0	Hand packers & packagers	1.3	-13.4
Blue collar worker supervisors	3.4	-11.6	Engineering technicians & technologists nec	1.2	1.0
Inspectors, testers, & graders, precision	3.3	-29.3	Freight, stock, & material movers, hand	1.2	-19.2
Industrial machinery mechanics	2.4	36.3	Mechanical engineers	1.2	22.2
Electrical & electronics engineers	1.9	7.5	Managers & administrators nec	1.1	1.0
Engineers nec	1.7	51.5	Helpers, laborers, & material movers nec	1.1	1.0
Sales & related workers nec	1.5	1.0	Traffic, shipping, & receiving clerks	1.0	-2.8
Plastic molding machine workers	1.4	21.2	Industrial production managers	1.0	1.0
General managers & top executives	1.4	-4.2	Electromechanical equipment assemblers	1.0	11.1

Source: Industry-Occupation Matrix, Bureau of Labor Statistics. These data relate to one or more 3-digit SIC industry groups rather than to a single 4-digit SIC. The change reported for each occupation to the year 2005 is a percent of growth or decline as estimated by the Bureau of Labor Statistics. The abbreviation nec stands for 'not elsewhere classified'.

LOCATION BY STATE AND REGIONAL CONCENTRATION

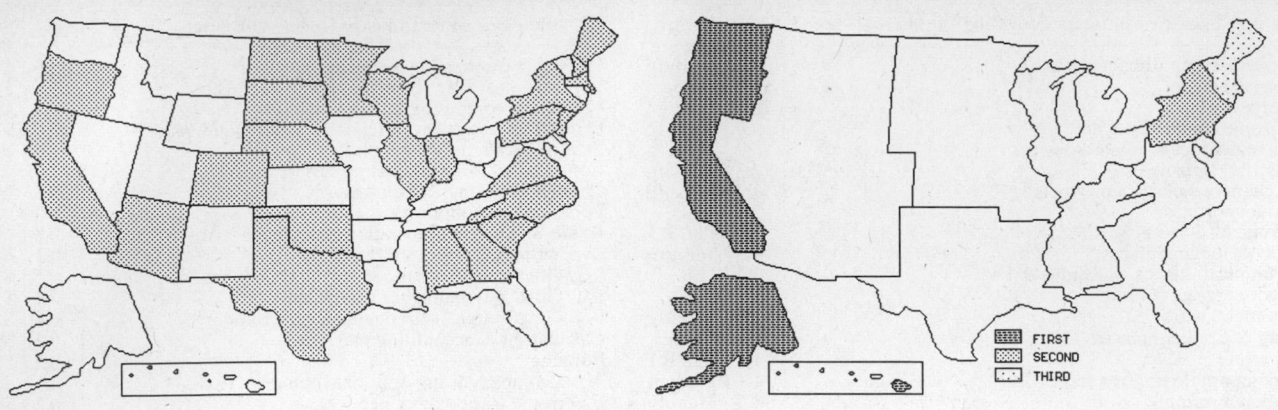

FIRST
SECOND
THIRD

INDUSTRY DATA BY STATE

| State | Establish-ments | Shipments | | | Employment | | | | Cost as % of Shipments | Investment per Employee ($) |
		Total ($ mil)	% of U.S.	Per Establ.	Total Number	% of U.S.	Per Establ.	Wages ($/hour)		
California	88	1,607.8	34.6	18.3	8,700	38.5	99	12.47	49.5	14,655
Alabama	3	560.5	12.1	186.8	3,100	13.7	1,033	11.46	65.4	9,968
Colorado	8	23.2	0.5	2.9	200	0.9	25	20.00	59.5	4,500
New Jersey	10	21.9	0.5	2.2	100	0.4	10	9.00	39.3	6,000
New York	16	(D)	-	-	375 *	1.7	23	-	-	3,733
Massachusetts	13	(D)	-	-	1,750 *	7.7	135	-	-	-
Pennsylvania	11	(D)	-	-	175 *	0.8	16	-	-	5,143
Texas	8	(D)	-	-	750 *	3.3	94	-	-	-
New Hampshire	7	(D)	-	-	375 *	1.7	54	-	-	-
Illinois	6	(D)	-	-	375 *	1.7	63	-	-	-
North Carolina	6	(D)	-	-	375 *	1.7	63	-	-	-
Oklahoma	5	(D)	-	-	750 *	3.3	150	-	-	-
Oregon	5	(D)	-	-	175 *	0.8	35	-	-	5,143
Arizona	4	(D)	-	-	375 *	1.7	94	-	-	-
Georgia	4	(D)	-	-	750 *	3.3	188	-	-	-
Virginia	4	(D)	-	-	375 *	1.7	94	-	-	-
Wisconsin	4	(D)	-	-	750 *	3.3	188	-	-	-
Maine	3	(D)	-	-	750 *	3.3	250	-	-	-
Minnesota	3	(D)	-	-	175 *	0.8	58	-	-	-
Nebraska	2	(D)	-	-	375 *	1.7	188	-	-	-
Indiana	1	(D)	-	-	750 *	3.3	750	-	-	-
North Dakota	1	(D)	-	-	750 *	3.3	750	-	-	-
South Carolina	1	(D)	-	-	375 *	1.7	375	-	-	-
South Dakota	1	(D)	-	-	375 *	1.7	375	-	-	-

Source: 1992 *Economic Census*. The states are in descending order of shipments or establishments (if shipment data are missing for the majority). The symbol (D) appears when data are withheld to prevent disclosure of competitive information. States marked with (D) are sorted by number of establishments. A dash (-) indicates that the data element cannot be calculated; * indicates the midpoint of a range.

3699 - ELECTRICAL EQUIPMENT & SUPPLIES, NEC

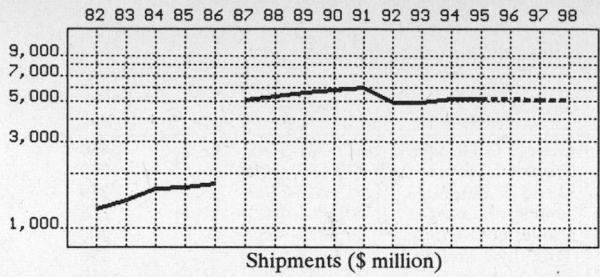

Shipments ($ million)

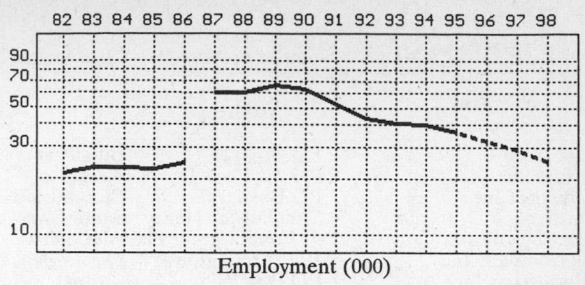

Employment (000)

GENERAL STATISTICS

Year	Com-panies	Establishments		Employment			Compensation		Production ($ million)			
		Total	with 20 or more employees	Total (000)	Production Workers (000)	Hours (Mil)	Payroll ($ mil)	Wages ($/hr)	Cost of Materials	Value Added by Manufacture	Value of Shipments	Capital Invest.
1982	713	748	235	21.6	15.9	29.2	315.1	6.33	615.9	645.5	1,271.8	43.1
1983		719	247	23.3	16.4	31.0	347.1	6.28	647.2	744.2	1,419.1	37.8
1984		690	259	23.4	17.7	34.1	377.5	6.80	807.0	865.8	1,660.4	39.4
1985		662	270	22.9	17.3	32.6	380.1	7.28	831.2	836.9	1,666.0	46.5
1986		656	280	24.7	18.4	35.2	425.1	7.49	862.1	913.9	1,766.9	39.9
1987*	1,324	1,379	473	60.3	32.4	63.0	1,466.4	8.70	2,198.7	2,901.9	5,056.1	180.6
1988		1,342	519	60.6	31.9	60.7	1,557.7	9.36	2,227.0	3,208.2	5,328.2	166.6
1989		1,249	509	66.4	29.0	56.4	1,633.1	10.50	2,259.4	3,359.3	5,666.5	170.5
1990		1,185	511	63.4	28.6	54.9	1,683.4	10.11	2,558.9	3,273.3	5,848.3	178.4
1991		1,192	504	52.7	25.9	51.2	1,622.5	10.98	2,468.7	3,500.6	6,015.7	196.8
1992	831	892	419	43.4	20.9	39.9	1,375.6	10.42	2,035.2	2,810.1	4,934.0	118.7
1993		912	427	40.9	19.7	37.8	1,287.5	10.26	2,198.5	2,823.3	4,963.0	112.0
1994		828P	431P	40.6	20.3	39.3	1,350.9	10.37	2,278.5	2,861.1	5,148.1	126.8
1995		743P	419P	36.6P	16.8P	32.5P	1,336.8P	10.99P	2,300.5P	2,888.8P	5,197.9P	114.2P
1996		659P	407P	32.8P	14.8P	28.6P	1,301.2P	11.19P	2,283.6P	2,867.5P	5,159.7P	104.8P
1997		575P	395P	29.1P	12.7P	24.6P	1,265.5P	11.39P	2,266.7P	2,846.3P	5,121.4P	95.4P
1998		491P	382P	25.3P	10.7P	20.6P	1,229.9P	11.59P	2,249.8P	2,825.0P	5,083.2P	86.1P

Sources: 1982, 1987, 1992 *Economic Census; Annual Survey of Manufactures*, 83-86, 88-91, 93-94. Establishment counts are from *County Business Patterns* for non-Census years; establishment counts for 83-84 are extrapolations. * indicates that industry content changed in 87; earlier years use 77 SICs. 'P's mark projections.

INDICES OF CHANGE

Year	Com-panies	Establishments		Employment			Compensation		Production ($ million)			
		Total	with 20 or more employees	Total (000)	Production Workers (000)	Hours (Mil)	Payroll ($ mil)	Wages ($/hr)	Cost of Materials	Value Added by Manufacture	Value of Shipments	Capital Invest.
1982	86	84	56	50	76	73	23	61	30	23	26	36
1983		81	59	54	78	78	25	60	32	26	29	32
1984		77	62	54	85	85	27	65	40	31	34	33
1985		74	64	53	83	82	28	70	41	30	34	39
1986		74	67	57	88	88	31	72	42	33	36	34
1987*	159	155	113	139	155	158	107	83	108	103	102	152
1988		150	124	140	153	152	113	90	109	114	108	140
1989		140	121	153	139	141	119	101	111	120	115	144
1990		133	122	146	137	138	122	97	126	116	119	150
1991		134	120	121	124	128	118	105	121	125	122	166
1992	100	100	100	100	100	100	100	100	100	100	100	100
1993		102	102	94	94	95	94	98	108	100	101	94
1994		93P	103P	94	97	98	98	100	112	102	104	107
1995		83P	100P	84P	81P	82P	97P	105P	113P	103P	105P	96P
1996		74P	97P	76P	71P	72P	95P	107P	112P	102P	105P	88P
1997		64P	94P	67P	61P	62P	92P	109P	111P	101P	104P	80P
1998		55P	91P	58P	51P	52P	89P	111P	111P	101P	103P	73P

Sources: Same as General Statistics. Values reflect change from the base year, 1992. Values above 100 mean greater than 92, values below 100 mean less than 92, and a value of 100 in the 82-91 or 93-98 period means same as 92. * indicates that industry content changed in 87. Data for earlier years are in 77 SIC format.

SELECTED RATIOS

For 1994	Avg. of All Manufact.	Analyzed Industry	Index	For 1994	Avg. of All Manufact.	Analyzed Industry	Index
Employees per Establishment	49	49	100	Value Added per Production Worker	134,084	140,941	105
Payroll per Establishment	1,500,273	1,632,367	109	Cost per Establishment	5,045,178	2,753,237	55
Payroll per Employee	30,620	33,273	109	Cost per Employee	102,970	56,121	55
Production Workers per Establishment	34	25	71	Cost per Production Worker	146,988	112,241	76
Wages per Establishment	853,319	492,454	58	Shipments per Establishment	9,576,895	6,220,732	65
Wages per Production Worker	24,861	20,076	81	Shipments per Employee	195,460	126,800	65
Hours per Production Worker	2,056	1,936	94	Shipments per Production Worker	279,017	253,601	91
Wages per Hour	12.09	10.37	86	Investment per Establishment	321,011	153,219	48
Value Added per Establishment	4,602,255	3,457,224	75	Investment per Employee	6,552	3,123	48
Value Added per Employee	93,930	70,470	75	Investment per Production Worker	9,352	6,246	67

Sources: Same as General Statistics. The 'Average of All Manufacturing' column represents the average of all manufacturing industries reported for the most recent complete year available. The Index shows the relationship between the Average and the Analyzed Industry. For example, 100 means that they are equal; 500 that the Analyzed Industry is five times the average; 50 means that the Analyzed Industry is half the national average. The abbreviation 'na' is used to show that data are 'not available'.

LEADING COMPANIES   Number shown: 75   Total sales ($ mil): 2,792   Total employment (000): 22.4

Company Name	Address				CEO Name	Phone	Co. Type	Sales ($ mil)	Empl. (000)
CAE-Link Corp	PO Box 1237	Binghamton	NY	13902	George G Houser	607-721-5465	S	460	3.6
Hughes Training Inc	621 Six Flags Dr	Arlington	TX	76011	Stewart I Moore	817-695-3000	S	250	1.1
Core Industries Inc	PO Box 2000	Bloomfield Hls	MI	48304	David R Zimmer	313-642-3400	P	219	2.3
Coherent Inc	5100 Patrick Henry	Santa Clara	CA	95054	James L Hobart	408-764-4000	P	215	1.3
Thomson Corporation	99 Canal Center Plz	Alexandria	VA	22314	Daniel H O'Brien	703-838-9685	S	200	1.1
Edwards Company Inc	195 Farmington Av	Farmington	CT	06034	Timothy J Mellen	203-678-0410	S	190	2.0
Woods Industries Inc	PO Box 2675	Carmel	IN	46032	Fred Stauder	317-844-7261	R	110•	0.8
Simulation Syst & Services Techn	8930 Stanford Blv	Columbia	MD	21045	William Kuhlmann	410-312-3500	S	61•	0.6
FlightSafety International Inc	2700 N Hemlock Cir	Broken Arrow	OK	74012	Dennis Gulasy	918-251-0500	D	57•	0.4
Reflectone Inc	4908 Tampa W Blvd	Tampa	FL	33634	Richard G Snyder	813-885-7481	P	54	0.7
Branson Ultrasonics Corp	41 Eagle Rd	Danbury	CT	06813	Charles Nims	203-796-0400	S	50	0.5
ENI	100 Highpower Rd	Rochester	NY	14623	John Stratakos	716-427-8300	D	48	0.4
Linear Corp	2055 Nogal	Carlsbad	CA	92009	Grant Rummell	619-438-7000	S	40	0.2
Keyence Corporation of America	50 Tice Blv	Woodcliff Lake	NJ	07675	Yoshi Akiyama	201-930-1400	S	30	<0.1
New PIG Corp	PO Box 304	Tipton	PA	16684	Nino Vella	814-684-0101	R	30•	0.3
SBS Engineering Inc	5550 Midway Pk N	Albuquerque	NM	87109	Andrew C Cruce	505-345-5353	P	29	0.2
DeJay Corp	3300 PGA Blv	P Bch Gardens	FL	33410	David Blotnick	407-626-0600	R	27•	0.3
Republic Industries Inc	7350 W Wilson Av	Harwood H	IL	60656	Matthew Kass	708-867-7400	R	27	0.3
American Magnetics Corp	740 Watson Ctr Rd	Carson	CA	90745	Robert Sawyer	213-775-8651	R	25	0.3
TSC	2950 31st St	Santa Monica	CA	90405	Robert Graziano	310-450-9755	S	24	0.2
Brasch Manufacturing Company	11880 Dorsett Rd	Maryland H	MO	63043	JF Brasch	314-291-0440	R	22	0.2
L and R Manufacturing Co	577 Elm St	Kearny	NJ	07032	JJ Lazarus	201-991-5330	R	22	0.2
Thomson Training & Simulation	7041 E 15th St	Tulsa	OK	74112	Patrick Oszczeda	918-836-4621	S	22	0.2
Whitney Blake of Vermont	PO Box 579	Bellows Falls	VT	05101	Roland Scott	802-463-9558	R	20	0.2
Identix Inc	510 N Pastoria Av	Sunnyvale	CA	94086	Randall C Fowler	408-739-2000	P	20	0.2
Chemcut Equipment Group	500 Science Park Rd	State College	PA	16803	Kenneth A Slocumb	814-238-0514	D	20•	0.2
Herley-Vega Systems	10 Industry Dr	Lancaster	PA	17603	Myron Levy	717-397-2777	D	20	0.2
Delex Systems Inc	1953 Gallows Rd	Vienna	VA	22182	HL Piper	703-734-8300	R	18	0.2
Frasca International Inc	906 E Airport Rd	Urbana	IL	61801	Rudy Frasca	217-344-9200	R	18	0.1
Environmental Tectonics Corp	County Line Ind Pk	Southampton	PA	18966	William F Mitchell	215-355-9100	P	17	0.2
Griffin Technology Inc	1133 Corporate Dr	Farmington	NY	14425	Robert S Urland	716-924-7121	P	17	0.2
ASI Robotic Systems Inc	1250 Crooks Rd	Clawson	MI	48017	John S Cargill	810-288-5070	S	16	<0.1
Allister Manufacturing Company	315 Willowbrook Ln	West Chester	PA	19382	Robert Holland	610-436-6190	S	15	<0.1
Bronner Display	PO Box 176	Frankenmuth	MI	48734	Wallace J Bronner	517-652-9931	R	15	0.4
Cartwright Electronics Inc	655 W Valencia Dr	Fullerton	CA	92632	BL Leathers	714-525-2300	R	15	0.1
Dor-O-Matic Products Corp	6800 Indrial Loop	Greendale	WI	53129	Tom Shea	414-421-1000	S	14	<0.1
Moore-O-Matic Inc	2055 Nogal	Carlsbad	CA	92009	Grant Rummell	619-431-1535	S	14	0.1
Air Lock Inc	108 Gulf St	Milford	CT	06460	JC Edwards	203-878-4691	R	13•	0.1
Copley Controls Corp	410 University Av	Westwood	MA	02090	Matthew Lorber	617-329-8200	R	13	0.1
Electrovert USA Corp	111 Carrier	Grand Prairie	TX	75050	G Anderson	214-606-1900	S	13	0.3
Fi-Shock Inc	5360 S National Dr	Knoxville	TN	37914	Tom Boyd	615-524-7380	R	13	0.1
MR Christmas Inc	41 Madison Av	New York	NY	10010	Merrill Hermanson	212-889-7220	R	13	<0.1
Detex Corp	302 Detex Dr	New Braunfels	TX	78130	Philip N Haselton	210-629-2900	R	12	0.1
Doron Precision System Inc	PO Box 400	Binghamton	NY	13902	Carl J Wenzinger	607-772-1610	R	12	0.1
Electri-Wire Corp	N 26 W 23315 Paul	Pewaukee	WI	53072	James Cerny	414-548-3700	R	12	0.2
Convergent Energy	1 Picker Rd	Sturbridge	MA	01566	Nathan Monty	508-347-2681	S	11•	0.1
Parker McCrory Mfg Co	2000 Forest Av	Kansas City	MO	64108	Ken Turner	816-221-2000	R	11•	0.1
Alabama Specialty Products Inc	PO Box 8	Munford	AL	36268	Don Johnson	205-358-4202	R	10	<0.1
CHA Industries Inc	4201 Business Ctr	Fremont	CA	94538	Richard Herrmann	510-683-8554	R	10	0.1
Ebtec Corp	PO Box 465	Agawam	MA	01001	David Maggs	413-786-0393	S	10	<0.1
Electri-Cord Manufacturing	312 Main St	Westfield	PA	16950	Mitch Samuels	814-367-2265	R	10	0.2
Seattle Silicon Corp	4122 128th Av SE	Bellevue	WA	98006	John V Celms	206-957-4422	R	10	<0.1
A-B Lasers Inc	4 Craig Rd	Acton	MA	01720	James L Shephard	508-635-9100	S	9•	<0.1
Visual Simulation Systems	5695 Campus Pkwy	St Louis	MO	63042	Albert L Ueltschi	314-551-8400	D	9•	<0.1
Technical Devices Co	560 Alaska Av	Torrance	CA	90503	DN Winther	310-618-8437	R	9	<0.1
Binghamton Simulator Company	4 Chenango St	Binghamton	NY	13901	John Morelli	607-722-6177	R	8	<0.1
Centerex Inc	19 Evergreen Dr	Portland	ME	04103	Jim Mitchell	207-797-4777	R	8	<0.1
Erbtec Engineering Inc	2760 29th St	Boulder	CO	80301	Lee Erb	303-447-8750	R	8	<0.1
Invisible Fence Company Inc	355 Phoenixville Pk	Malvern	PA	19355	Helen S Weary	215-651-0999	R	8•	<0.1
Lewis Corp	102 Willenbrock Rd	Oxford	CT	06478	Bob Hunter	203-264-3100	R	8	<0.1
Maverick Industries Inc	94 Mayfield Av	Edison	NJ	08837	EH Mackin	908-417-9666	R	8	<0.1
Novatron Corp	6000 Rinke Av	Warren	MI	48091	Robert J Phillips	810-755-1300	R	8	<0.1
Ney Ultrasonics Inc	1280 Blue Hills Av	Bloomfield	CT	06002	Ronald N Cerny	203-286-6149	S	8	<0.1
Appalachian Electr Instruments	PO Box 518	Ronceverte	WV	24970	Creigh Nickell	304-647-5855	R	7	0.1
Dare Products Inc	860 Betterly Rd	Battle Creek	MI	49016	Robert Wilson Jr	616-965-2307	R	7	0.1
Ehrhorn	4975 N 30th St	Co Springs	CO	80919	Richard W Ehrhorn	719-260-1191	R	7	<0.1
Guest Company Inc	48 Elm St	Meriden	CT	06450		203-238-0550	R	7	<0.1
Hobart Lasers	1191 Trade Rd E	Troy	OH	45373	James A Anderson	513-332-5222	D	7	<0.1
PTR-Precision Technologies Inc	120 Post Rd	Enfield	CT	06082	Gottfried Kuesters	203-741-2281	R	7	<0.1
Radiation Dynamics Inc	151 Heartland Blv	Edgewood	NY	11717	Bernie Kestler	516-254-6800	S	7	<0.1
Carco Electronics	195 Constitution Dr	Menlo Park	CA	94025	John M Carter	415-321-8174	P	7	<0.1
Three Sixty Services Inc	12623 Newburgh Rd	Livonia	MI	48150	Kenneth Pickl	313-591-9360	R	6	<0.1
Laser Technology Inc	7070 S Tucson Way	Englewood	CO	80112	David Williams	303-649-1000	P	5	<0.1
Bird-X Inc	730 W Lake St	Chicago	IL	60661	Richard Seid	312-648-2191	R	5	<0.1
AETEK International Inc	1750 N Van Dyke	Plainfield	IL	60544	David Harbourne	815-436-2304	S	5	<0.1

Source: Ward's Business Directory of U.S. Private and Public Companies, Volumes 1 and 2, 1996. The company type code used is as follows: P - Public, R - Private, S - Subsidiary, D - Division, J - Joint Venture, A - Affiliate, G - Group. Sales are in millions of dollars, employees are in thousands. An asterisk (*) indicates an estimated sales volume. The symbol < stands for 'less than'. Company names and addresses are truncated, in some cases, to fit into the available space.

MATERIALS CONSUMED

Material	Quantity	Delivered Cost ($ million)
Materials, ingredients, containers, and supplies	(X)	1,774.6
Sheet metal products, except stampings	(X)	41.5
Metal bolts, nuts, screws, washers, rivets, and other screw machine products	(X)	21.5
Metal stampings	(X)	18.1
Other fabricated metal products (except forgings)	(X)	74.4
Forgings	(X)	0.1
Castings (rough and semifinished)	(X)	15.5
Steel shapes and forms	(X)	6.9
Copper and copper-base alloy shapes and forms	(X)	32.4
Aluminum and aluminum-base alloy shapes and forms	(X)	23.7
Other nonferrous shapes and forms	(X)	7.3
Printed circuit boards (without inserted components) for electronic circuitry	(X)	52.7
Printed circuit assemblies, loaded boards or modules	(X)	56.8
Semiconductors, including transistors, diodes, rectifiers, and integrated circuits for electronic circuitry	(X)	64.3
Capacitors for electronic circuitry	(X)	16.4
Resistors for electronic circuitry	(X)	10.8
Other components and accessories for electronic circuitry, nec, except tubes	(X)	71.5
Insulated wire and cable, including magnet wire	(X)	101.3
Current-carrying wiring devices	(X)	24.2
Fractional horsepower electric motors (less than 1 hp)	(X)	(D)
Electronic communication equipment	(X)	18.1
Electrical instrument mechanisms and meter movements (including instrument relays)	(X)	14.7
Electronic computing equipment	(X)	64.5
Optical instruments and lenses (except sighting, tracking, and fire control)	(X)	25.7
Automatic garage door controllers	(X)	(D)
Fabricated plastics products (except gaskets, hoses, and belting)	(X)	34.2
Paper and paperboard containers, including shipping sacks and other paper packaging supplies	(X)	19.6
All other materials and components, parts, containers, and supplies	(X)	389.3
Materials, ingredients, containers, and supplies, nsk	(X)	530.4

Source: 1992 *Economic Census*. Explanation of symbols used: (D): Withheld to avoid disclosure of competitive data; na: Not available; (S): Withheld because statistical norms were not met; (X): Not applicable; (Z): Less than half the unit shown; nec: Not elsewhere classified; nsk: Not specified by kind; - : zero; * : 10-19 percent estimated; ** : 20-29 percent estimated.

PRODUCT SHARE DETAILS

Product or Product Class	% Share	Product or Product Class	% Share
Electrical machinery, equipment, and supplies, nec	100.00	Electrical door openers, except garage door openers	13.05
Electrical teaching machines, teaching aids, trainers, and simulators, including kits	26.44	Electric gongs, chimes, bells, etc.	9.32
		Electrical insect killers	3.28
Laser systems and equipment, except communication	16.24	Electrical outboard motors for boats (complete propulsion unit)	74.08
Ultrasonic equipment (except medical and dental)	2.80		
Apparatus wire and cordage manufactured from purchased insulated wire	6.24	Electrical products, nec (excluding garage door openers), nsk	0.29
Electronic systems and equipment, nec (including automatic garage door openers, and amplifiers)	23.91	Electrical machinery, equipment, and supplies, not elsewhere classified, nsk	13.54
Electrical products, nec (excluding garage door openers)	10.83		

Source: 1992 *Economic Census*. The values shown are percent of total shipments in an industry. Values of indented subcategories are summed in the main heading. The symbol (D) appears when data are withheld to prevent disclosure of competitive information. The abbreviation nsk stands for 'not specified by kind' and nec for 'not elsewhere classified'.

INPUTS AND OUTPUTS FOR ELECTRICAL EQUIPMENT & SUPPLIES, NEC

Economic Sector or Industry Providing Inputs	%	Sector	Economic Sector or Industry Buying Outputs	%	Sector
Imports	31.4	Foreign	Exports	40.2	Foreign
Semiconductors & related devices	6.0	Manufg.	Personal consumption expenditures	22.1	
Wholesale trade	5.9	Trade	Industrial buildings	7.2	Constr.
Eating & drinking places	5.8	Trade	Electric lamps	3.7	Manufg.
Nonferrous wire drawing & insulating	3.5	Manufg.	Office buildings	3.5	Constr.
Legal services	2.5	Services	Machinery, except electrical, nec	2.2	Manufg.
Primary copper	2.4	Manufg.	Maintenance of nonfarm buildings nec	1.7	Constr.
Electronic components nec	2.2	Manufg.	Nonfarm residential structure maintenance	1.7	Constr.
Miscellaneous plastics products	2.2	Manufg.	Lighting fixtures & equipment	1.4	Manufg.
Management & consulting services & labs	2.2	Services	Residential 1-unit structures, nonfarm	1.3	Constr.
Advertising	2.1	Services	Gross private fixed investment	1.3	Cap Inv
Maintenance of nonfarm buildings nec	1.8	Constr.	Construction of educational buildings	1.2	Constr.
Wiring devices	1.8	Manufg.	Residential additions/alterations, nonfarm	1.2	Constr.
Cyclic crudes and organics	1.7	Manufg.	Electrical equipment & supplies, nec	1.0	Manufg.
Water supply & sewage systems	1.6	Util.	Residential garden apartments	0.8	Constr.
Machinery, except electrical, nec	1.5	Manufg.	Federal Government purchases, national defense	0.8	Fed Govt
Metal stampings, nec	1.3	Manufg.	Federal Government purchases, nondefense	0.8	Fed Govt
Banking	1.3	Fin/R.E.	Construction of hospitals	0.7	Constr.
Aluminum rolling & drawing	1.2	Manufg.	Telephone & telegraph apparatus	0.6	Manufg.

Continued on next page.

INPUTS AND OUTPUTS FOR ELECTRICAL EQUIPMENT & SUPPLIES, NEC - Continued

Economic Sector or Industry Providing Inputs	%	Sector	Economic Sector or Industry Buying Outputs	%	Sector
Electric services (utilities)	1.2	Util.	S/L Govt. purch., public assistance & relief	0.6	S/L Govt
Electrical equipment & supplies, nec	1.1	Manufg.	Construction of stores & restaurants	0.5	Constr.
Fabricated metal products, nec	1.1	Manufg.	Warehouses	0.4	Constr.
Paperboard containers & boxes	1.0	Manufg.	Residential high-rise apartments	0.3	Constr.
Plastics materials & resins	1.0	Manufg.	Games, toys, & children's vehicles	0.3	Manufg.
Motor freight transportation & warehousing	1.0	Util.	Logging camps & logging contractors	0.3	Manufg.
Accounting, auditing & bookkeeping	1.0	Services	Turbines & turbine generator sets	0.3	Manufg.
Manifold business forms	0.8	Manufg.	S/L Govt. purch., highways	0.3	S/L Govt
Special dies & tools & machine tool accessories	0.8	Manufg.	Agricultural, forestry, & fishery services	0.2	Agric.
Air transportation	0.8	Util.	Coal	0.2	Mining
Communications, except radio & TV	0.8	Util.	Crude petroleum & natural gas	0.2	Mining
Electric lamps	0.6	Manufg.	Construction of nonfarm buildings nec	0.2	Constr.
Petroleum refining	0.6	Manufg.	Hotels & motels	0.2	Constr.
Photofinishing labs, commercial photography	0.6	Services	Resid. & other health facility construction	0.2	Constr.
Blast furnaces & steel mills	0.5	Manufg.	Residential 2-4 unit structures, nonfarm	0.2	Constr.
Nonferrous rolling & drawing, nec	0.5	Manufg.	Boat building & repairing	0.2	Manufg.
Real estate	0.5	Fin/R.E.	Miscellaneous plastics products	0.2	Manufg.
Copper rolling & drawing	0.4	Manufg.	X-ray apparatus & tubes	0.2	Manufg.
Gas production & distribution (utilities)	0.4	Util.	Nonferrous metal ores, except copper	0.1	Mining
Transit & bus transportation	0.4	Util.	Amusement & recreation building construction	0.1	Constr.
Equipment rental & leasing services	0.4	Services	Farm service facilities	0.1	Constr.
Hotels & lodging places	0.4	Services	Dental equipment & supplies	0.1	Manufg.
Noncomparable imports	0.4	Foreign	Guided missiles & space vehicles	0.1	Manufg.
Abrasive products	0.3	Manufg.	Optical instruments & lenses	0.1	Manufg.
Lubricating oils & greases	0.3	Manufg.	Pipe, valves, & pipe fittings	0.1	Manufg.
Photographic equipment & supplies	0.3	Manufg.			
Wood products, nec	0.3	Manufg.			
Computer & data processing services	0.3	Services			
Personnel supply services	0.3	Services			
Industrial inorganic chemicals, nec	0.2	Manufg.			
Miscellaneous fabricated wire products	0.2	Manufg.			
Periodicals	0.2	Manufg.			
Business services nec	0.2	Services			
Engineering, architectural, & surveying services	0.2	Services			
Theatrical producers, bands, entertainers	0.2	Services			
Envelopes	0.1	Manufg.			
Mechanical measuring devices	0.1	Manufg.			
Metal heat treating	0.1	Manufg.			
Railroads & related services	0.1	Util.			
Business/professional associations	0.1	Services			
Detective & protective services	0.1	Services			
Electrical repair shops	0.1	Services			

Source: Benchmark Input-Output Accounts for the U.S. Economy, 1982, U.S. Department of Commerce, Washington, D.C., July 1991. Data, as reported in the source, are organized by the 1977 SIC structure in use in 1982 but have been matched, as closely as is possible, to the 1987 SIC structure used in this book.

OCCUPATIONS EMPLOYED BY SIC 369 - MISC ELECTRICAL EQUIPMENT AND SUPPLIES

Occupation	% of Total 1994	Change to 2005	Occupation	% of Total 1994	Change to 2005
Assemblers, fabricators, & hand workers nec	14.5	1.0	Secretaries, ex legal & medical	1.3	-8.1
Electrical & electronic assemblers	10.7	-9.1	Electrical & electronic technicians,technologists	1.3	1.0
Machine operators nec	4.6	-28.8	Tool & die makers	1.3	-18.4
Electrical & electronic equipment assemblers	4.0	1.0	Hand packers & packagers	1.3	-13.4
Blue collar worker supervisors	3.4	-11.6	Engineering technicians & technologists nec	1.2	1.0
Inspectors, testers, & graders, precision	3.3	-29.3	Freight, stock, & material movers, hand	1.2	-19.2
Industrial machinery mechanics	2.4	36.3	Mechanical engineers	1.2	22.2
Electrical & electronics engineers	1.9	7.5	Managers & administrators nec	1.1	1.0
Engineers nec	1.7	51.5	Helpers, laborers, & material movers nec	1.1	1.0
Sales & related workers nec	1.5	1.0	Traffic, shipping, & receiving clerks	1.0	-2.8
Plastic molding machine workers	1.4	21.2	Industrial production managers	1.0	1.0
General managers & top executives	1.4	-4.2	Electromechanical equipment assemblers	1.0	11.1

Source: Industry-Occupation Matrix, Bureau of Labor Statistics. These data relate to one or more 3-digit SIC industry groups rather than to a single 4-digit SIC. The change reported for each occupation to the year 2005 is a percent of growth or decline as estimated by the Bureau of Labor Statistics. The abbreviation nec stands for 'not elsewhere classified'.

LOCATION BY STATE AND REGIONAL CONCENTRATION

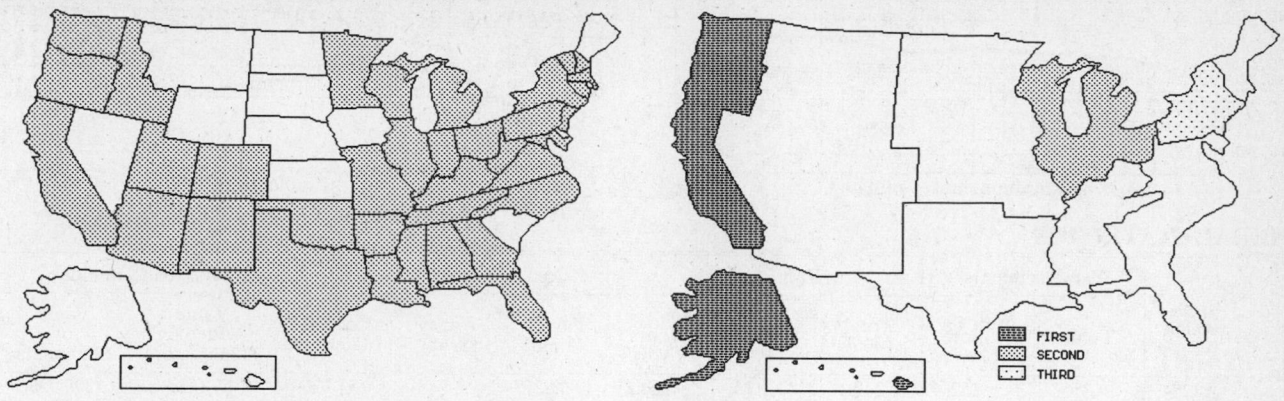

FIRST
SECOND
THIRD

INDUSTRY DATA BY STATE

| State | Establish- ments | Shipments | | | Employment | | | | Cost as % of Shipments | Investment per Employee ($) |
		Total ($ mil)	% of U.S.	Per Establ.	Total Number	% of U.S.	Per Establ.	Wages ($/hour)		
California	158	1,064.5	21.6	6.7	8,300	19.1	53	13.46	38.3	4,253
Florida	62	622.9	12.6	10.0	5,100	11.8	82	9.61	34.8	3,039
New York	59	535.8	10.9	9.1	4,300	9.9	73	10.73	33.0	2,047
Ohio	39	460.7	9.3	11.8	3,500	8.1	90	10.47	44.0	3,829
Illinois	49	266.1	5.4	5.4	3,300	7.6	67	9.00	40.8	2,091
Indiana	37	198.3	4.0	5.4	2,200	5.1	59	8.29	49.8	-
Massachusetts	37	162.2	3.3	4.4	1,200	2.8	32	11.45	54.9	3,000
Connecticut	35	156.2	3.2	4.5	1,500	3.5	43	10.92	33.9	3,067
Michigan	27	129.9	2.6	4.8	1,100	2.5	41	10.40	44.1	3,091
Minnesota	26	121.7	2.5	4.7	1,000	2.3	38	9.08	55.6	3,800
New Jersey	30	121.7	2.5	4.1	1,000	2.3	33	11.36	40.6	1,800
Pennsylvania	40	103.8	2.1	2.6	1,200	2.8	30	9.91	46.1	-
Virginia	12	98.4	2.0	8.2	700	1.6	58	9.14	39.5	1,000
Wisconsin	13	58.5	1.2	4.5	500	1.2	38	9.17	46.3	1,800
Oregon	14	58.3	1.2	4.2	600	1.4	43	10.20	36.9	-
Maryland	11	56.3	1.1	5.1	500	1.2	45	10.00	40.3	-
Colorado	15	46.4	0.9	3.1	400	0.9	27	9.00	37.1	4,750
Utah	10	37.0	0.7	3.7	400	0.9	40	7.50	40.0	1,500
Georgia	12	28.0	0.6	2.3	300	0.7	25	7.50	38.9	1,000
Washington	12	26.8	0.5	2.2	200	0.5	17	14.00	62.7	3,500
North Carolina	16	26.0	0.5	1.6	300	0.7	19	7.20	55.0	1,000
Arkansas	9	25.8	0.5	2.9	500	1.2	56	8.00	53.5	1,400
Arizona	12	21.3	0.4	1.8	300	0.7	25	9.50	30.5	1,333
Texas	46	(D)	-	-	1,750 *	4.0	38	-	-	-
Oklahoma	16	(D)	-	-	1,750 *	4.0	109	-	-	-
Tennessee	16	(D)	-	-	375 *	0.9	23	-	-	800
Alabama	12	(D)	-	-	750 *	1.7	63	-	-	-
New Hampshire	8	(D)	-	-	175 *	0.4	22	-	-	-
Missouri	7	(D)	-	-	375 *	0.9	54	-	-	1,867
New Mexico	7	(D)	-	-	175 *	0.4	25	-	-	-
Louisiana	5	(D)	-	-	175 *	0.4	35	-	-	1,143
Vermont	4	(D)	-	-	175 *	0.4	44	-	-	-
Mississippi	3	(D)	-	-	175 *	0.4	58	-	-	-
Idaho	2	(D)	-	-	175 *	0.4	88	-	-	-
West Virginia	2	(D)	-	-	175 *	0.4	88	-	-	-
Kentucky	1	(D)	-	-	175 *	0.4	175	-	-	-

Source: 1992 *Economic Census.* The states are in descending order of shipments or establishments (if shipment data are missing for the majority). The symbol (D) appears when data are withheld to prevent disclosure of competitive information. States marked with (D) are sorted by number of establishments. A dash (-) indicates that the data element cannot be calculated; * indicates the midpoint of a range.

3711 - MOTOR VEHICLES & CAR BODIES

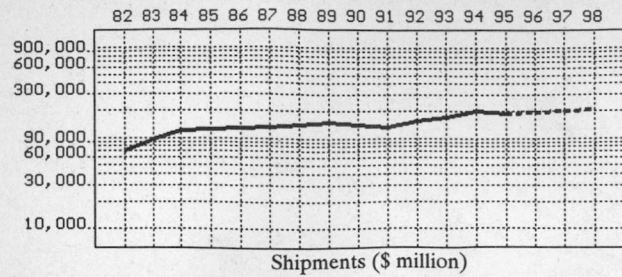

Shipments ($ million)

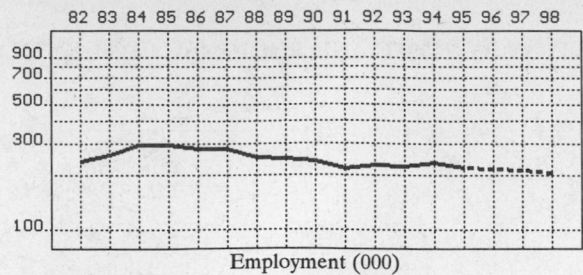

Employment (000)

GENERAL STATISTICS

Year	Companies	Establishments		Employment			Compensation		Production ($ million)			
		Total	with 20 or more employees	Total (000)	Production Workers (000)	Hours (Mil)	Payroll ($ mil)	Wages ($/hr)	Cost of Materials	Value Added by Manufacture	Value of Shipments	Capital Invest.
1982	284	355	152	240.1	193.5	364.2	6,821.9	14.45	55,520.0	15,455.8	70,739.7	2,368.3
1983		349	157	260.7	216.5	446.2	8,266.6	14.75	73,818.2	22,608.0	95,930.8	1,106.4
1984		343	162	296.2	247.6	564.4	10,192.2	14.46	90,435.0	27,668.1	118,066.0	2,420.9
1985		338	166	295.8	249.7	522.4	10,670.3	16.70	94,220.6	28,061.1	122,327.4	2,904.7
1986		348	167	280.5	233.8	479.5	10,261.7	17.22	93,965.0	31,846.0	125,869.6	3,912.8
1987	351	413	174	281.3	235.5	472.7	10,376.4	17.33	97,520.4	36,117.7	133,345.6	4,121.4
1988		391	183	250.3	213.6	448.2	10,121.2	18.68	102,364.8	39,762.2	142,059.6	1,136.6
1989		393	187	246.6	212.5	440.2	10,390.9	19.40	102,345.2	46,873.4	149,315.2	2,373.9
1990		406	183	239.8	200.0	399.7	10,060.0	20.31	101,130.8	39,504.4	140,417.0	3,004.4
1991		408	171	218.1	178.5	367.8	9,802.5	21.32	88,403.2	45,146.9	133,861.2	3,261.9
1992	400	456	161	228.4	193.3	397.3	10,438.8	21.66	107,636.6	45,262.2	152,948.5	2,989.5
1993		466	160	224.2	191.0	409.6	11,154.2	22.61	120,458.8	47,272.0	167,825.8	4,033.9
1994		459P	176P	234.0	202.5	447.5	12,437.7	23.35	144,809.9	52,917.7	197,553.7	4,245.7
1995		470P	178P	221.3P	190.0P	404.3P	11,850.6P	24.16P	136,455.7P	49,864.9P	186,156.7P	3,901.7P
1996		481P	179P	216.6P	186.8P	398.8P	12,104.0P	24.95P	141,850.9P	51,836.4P	193,516.9P	4,042.8P
1997		492P	180P	212.0P	183.5P	393.2P	12,357.5P	25.74P	147,246.0P	53,807.9P	200,877.1P	4,183.9P
1998		502P	181P	207.4P	180.2P	387.7P	12,610.9P	26.52P	152,641.1P	55,779.5P	208,237.3P	4,325.1P

Sources: 1982, 1987, 1992 *Economic Census*; *Annual Survey of Manufactures*, 83-86, 88-91, 93-94. Establishment counts for non-Census years are from *County Business Patterns*; establishment values for 83-84 are extrapolations. 'P's show projections by the editors. Industries reclassified in 87 will not have data for prior years.

INDICES OF CHANGE

Year	Companies	Establishments		Employment			Compensation		Production ($ million)			
		Total	with 20 or more employees	Total (000)	Production Workers (000)	Hours (Mil)	Payroll ($ mil)	Wages ($/hr)	Cost of Materials	Value Added by Manufacture	Value of Shipments	Capital Invest.
1982	71	78	94	105	100	92	65	67	52	34	46	79
1983		77	98	114	112	112	79	68	69	50	63	37
1984		75	101	130	128	142	98	67	84	61	77	81
1985		74	103	130	129	131	102	77	88	62	80	97
1986		76	104	123	121	121	98	80	87	70	82	131
1987	88	91	108	123	122	119	99	80	91	80	87	138
1988		86	114	110	111	113	97	86	95	88	93	38
1989		86	116	108	110	111	100	90	95	104	98	79
1990		89	114	105	103	101	96	94	94	87	92	100
1991		89	106	95	92	93	94	98	82	100	88	109
1992	100	100	100	100	100	100	100	100	100	100	100	100
1993		102	99	98	99	103	107	104	112	104	110	135
1994		101P	110P	102	105	113	119	108	135	117P	129	142
1995		103P	110P	97P	98P	102P	114P	112P	127P	110P	122P	131P
1996		105P	111P	95P	97P	100P	116P	115P	132P	115P	127P	135P
1997		108P	112P	93P	95P	99P	118P	119P	137P	119P	131P	140P
1998		110P	112P	91P	93P	98P	121P	122P	142P	123P	136P	145P

Sources: Same as General Statistics. Values reflect change from the base year, 1992. Values above 100 mean greater than 92, values below 100 mean less than 92, and a value of 100 in the 82-91 or 93-98 period means same as 92. 'P's mark projections by the editors.

SELECTED RATIOS

For 1994	Avg. of All Manufact.	Analyzed Industry	Index	For 1994	Avg. of All Manufact.	Analyzed Industry	Index
Employees per Establishment	49	510	1,040	Value Added per Production Worker	134,084	261,322	195
Payroll per Establishment	1,500,273	27,091,126	1,806	Cost per Establishment	5,045,178	315,417,095	6,252
Payroll per Employee	30,620	53,153	174	Cost per Employee	102,970	618,846	601
Production Workers per Establishment	34	441	1,285	Cost per Production Worker	146,988	715,111	487
Wages per Establishment	853,319	22,759,719	2,667	Shipments per Establishment	9,576,895	430,300,789	4,493
Wages per Production Worker	24,861	51,601	208	Shipments per Employee	195,460	844,247	432
Hours per Production Worker	2,056	2,210	107	Shipments per Production Worker	279,017	975,574	350
Wages per Hour	12.09	23.35	193	Investment per Establishment	321,011	9,247,754	2,881
Value Added per Establishment	4,602,255	115,262,473	2,504	Investment per Employee	6,552	18,144	277
Value Added per Employee	93,930	226,144	241	Investment per Production Worker	9,352	20,966	224

Sources: Same as General Statistics. The 'Average of All Manufacturing' column represents the average of all manufacturing industries reported for the most recent complete year available. The Index shows the relationship between the Average and the Analyzed Industry. For example, 100 means that they are equal; 500 that the Analyzed Industry is five times the average; 50 means that the Analyzed Industry is half the national average. The abbreviation 'na' is used to show that data are 'not available'.

LEADING COMPANIES   Number shown: 72   Total sales ($ mil): 384,671   Total employment (000): 1,333.9

Company Name	Address				CEO Name	Phone	Co. Type	Sales ($ mil)	Empl. (000)
General Motors Corp	3044 W Grand Blv	Detroit	MI	48202	John F Smith Jr	313-556-5000	P	154,951	692.8
Ford Motor Co	PO Box 1899	Dearborn	MI	48121	Alex Trotman	313-322-3000	P	128,439	337.8
Chrysler Corp	12000 Chrysler Dr	Highland Park	MI	48288	Robert J Eaton	313-956-5741	P	52,224	121.0
Ford Motor Co	17000 Oakwood	Dearborn	MI	48121	W Dale McKeehan	313-322-7715	D	14,970	67.8
Delco Chassis	PO Box 1042	Dayton	OH	45401	George G Johnston	513-455-9204	D	4,700	27.9
Navistar Intern Trans Corp	455 Cityfront Plz Dr	Chicago	IL	60611	James C Cotting	312-836-2000	S	4,694	13.6
PACCAR Inc	PO Box 1518	Bellevue	WA	98009	Charles M Pigott	206-455-7400	P	4,285	14.6
Freightliner Corp	4747 N Channel Av	Portland	OR	97217	James Hebe	503-735-8000	S	3,900	9.1
Saturn Corp	1420ephenson Hwy	Troy	MI	48007	Richard LeFauve	313-524-5000	S	3,560*	8.8
New United Motor Mfg	45500 Fremont Blv	Fremont	CA	94538	Iwao Itoh	510-498-5500	J	3,000	4.4
Nissan Motor Mfg Corp USA	Nissan Dr	Smyrna	TN	37167	Jerry L Benefield	615-459-1400	S	2,900	5.9
Diamond-Star Motors Inc	100 N Diamondar	Normal	IL	61761	Tsuneo Ohinouye	309-888-8000	S	2,287	3.7
Volvo GM Heavy Truck Corp	PO Box 26115	Greensboro	NC	27402	Karl-Erling Trogen	910-279-2000	S	1,000	4.2
Toyota Motor Mfg USA	1001 Ch Bloss	Georgetown	KY	40324	Mikio Kitano	502-868-2000	S	910	5.5
AutoAlliance International Inc	1 International Dr	Flat Rock	MI	48134	James Solberg	313-782-7800	R	480	3.8
PACCAR Peterbilt Motors Co	1700 Woodbrook St	Denton	TX	76205	Thomas Plimpton	817-591-4000	D	440*	1.9
Blue Bird Corp	PO Box 7839	Macon	GA	31210	Paul Glaske	912-757-7100	R	290*	1.5
Cobra Industries Inc	PO Box 124	Goshen	IN	46526	Dale R Glon	219-534-1418	P	239	1.2
Spartan Motors Inc	PO Box 440	Charlotte	MI	48813	George W Sztykiel	517-543-6400	P	192	0.5
Transportation Mfg Corp	PO Box 5670	Roswell	NM	88202	Harold Zuschlag	505-347-2011	S	190	1.0
Gillig Corp	PO Box 3008	Hayward	CA	94540	Dennis L Howard	510-785-1500	S	150	0.5
Marmon Motor Co	PO Box 462009	Garland	TX	75046	Stratton Georgoulis	214-276-5121	D	70	0.3
Andover Industries Inc	PO Box 459	Andover	OH	44003	John O'Neill	216-293-5900	R	66*	0.5
Elgin Sweeper Co	1300 W Bartlett Rd	Elgin	IL	60120	RB Parsons	708-741-5370	S	60	0.3
Wheeled Coach Industries Inc	PO Box 677339	Orlando	FL	32867	Robert L Collins	407-677-7777	S	56	0.4
Collins Bus Corp	PO Box 2946	Hutchinson	KS	67504	Ron Loomis	316-662-9000	S	45	0.3
Athey Products Corp	PO Box 669	Raleigh	NC	27602	James D Cloonan	919-556-5171	P	40	0.3
FWD Corp	105 E 12th St	Clintonville	WI	54929	Jim Green	715-823-2141	R	37*	0.5
Simon Ladder Towers Inc	64 Cocalico Creek	Ephrata	PA	17522	Steve Gerber	717-859-1176	S	35	0.3
Guzzler Manufacturing Inc	PO Box 66	Birmingham	AL	35201	Wilbur A La Ganke	205-591-2477	S	34*	0.2
De Tomaso Industries Inc	PO Box 856	Red Bank	NJ	07701	Santiago De Tomaso	908-842-7200	P	32	0.3
Eagle Coach Corp	2045 Les Mauldin	Brownsville	TX	78521	J Jimenez-Guzman	210-541-3111	R	32*	0.2
Johnston Sweeper Co	4651b Shaefer Av	Chino	CA	91710	Hugh Lamond	909-613-5600	S	31*	0.2
Chubb National Foam Inc	PO Box 270	Exton	PA	19341	John B Dowling	215-363-1400	S	30	0.2
Rockwood	3010 College Av	Goshen	IN	46526	Dale Glon	219-534-3645	D	30	0.2
WS Darley and Co	2000 Anson Dr	Melrose Park	IL	60160	William J Darley	708-345-8050	R	30	0.2
O'Gara, Hess	9113 Le Saint Dr	Fairfield	OH	45014	Tom O'Gara	513-874-2112	R	26	0.1
B and B Homes Corp	PO Box 2349	Mills	WY	82644	Anthony Ingram	307-235-1525	R	24	0.1
Fontaine Modification Co	9827 Mount Holly	Charlotte	NC	28214	Jim Van Winkle	704-391-1355	S	20	<0.1
All American Racers Inc	2334 S Broadway	Santa Ana	CA	92707	Daniel S Gurney	714-540-1771	R	19*	0.1
Road Rescue Inc	1133 Rankin St	St Paul	MN	55116	N J Conzemius	612-699-5588	R	18	0.1
Chance Coach Inc	4219 Irving St	Wichita	KS	67209	Scott Culbertson	316-942-7411	S	17	0.4
Tee Jay Industries Inc	34272 Doreka Av	Fraser	MI	48026	Thomas J Hogan	810-296-5160	D	17	0.2
HME Inc	1950 Byron Ctr Av	Wyoming	MI	49509		616-534-1463	R	16*	<0.1
Panda Motors Corp	8000 Twrs Crescent	Vienna	VA	22182	Bo Hi-Pak	703-761-7000	R	12	0.2
Frink America Inc	205 Webb St	Clayton	NY	13624	David Lowery	315-686-5531	R	10	<0.1
Dodgen Industries Inc	PO Box 39	Humboldt	IA	50548	JN Dodgen	515-332-3755	R	8	<0.1
JMX Inc	2790 Ranchview Ln	Minneapolis	MN	55447	Gary Henriksen	612-559-3300	R	6	<0.1
General Safety Equipment Inc	PO Box 219	Wyoming	MN	55092	Kevin Kirvida	612-462-1000	R	5	<0.1
Stratus Specialty Vehicles Inc	PO Box 10649	Kansas City	MO	64188	Gene Knisley	816-734-5000	R	4	<0.1
Bowie Manufacturing Inc	313 S Hancock St	Lake City	IA	51449	Marion E Peterson	712-464 3191	R	4	<0.1
Quarter Master Industries Inc	510 Telser Rd	Lake Zurich	IL	60047	Edgar Stoffels	708-540-8999	R	4	<0.1
Swab Wagon Company Inc	1 Chestnut Av	Elizabethville	PA	17023	Jonas B Margerum	717-362-8151	R	4	0.1
Astoria Industries Inc	201 Industrial Park	Jackson	MN	56143		507-847-5300	R	3	<0.1
Excalibur Automobile Corp	1800 S 108th St	Milwaukee	WI	53214	Udo Geitlinger	414-771-7171	R	3*	<0.1
Little Falls Machine Inc	305 3rd St SW	Little Falls	MN	56345	Raymond Schulte	612-632-9266	R	3	<0.1
Schwartz Industries Inc	5010 Beechmont Dr	Huntsville	AL	35811	Mark Schwartz	205-851-1200	R	3*	<0.1
Riley and Scott Inc	310 Gasoline Alley	Indianapolis	IN	46222	Bob Riley	317-248-9470	R	3	<0.1
Brubaker Group	10560 Dolcedo Way	Los Angeles	CA	90077	Curtis M Brubaker	310-472-4766	R	2	<0.1
S & W Race Cars	11 Mennonite	Spring City	PA	19475	Walter Weney	610-948-7303	R	2	<0.1
SCCA Enterprises Inc	7476 S Eagle St	Englewood	CO	80112	Martyn Thake	303-693-2111	R	2	<0.1
EJ Murphy Co	17 Winter St	Woodville	MA	01784	Edward Murphy	508-435-3431	R	2	<0.1
Total Performance Inc	400 S Orchard St	Wallingford	CT	06492	Michael V Lauria	203-265-7107	R	2	<0.1
Gowans-Knight Company Inc	49 Knight St	Watertown	CT	06795	C Palmer	203-274-8801	R	1*	<0.1
Kimball Products Inc	PO Box 792	Benton Harbor	MI	49022	Ron Kimball	616-925-9535	R	1	<0.1
Solar Car Corp	1300 Lk Wash	Melbourne	FL	32935	Douglas Cobb	407-254-2997	R	1	<0.1
Kugel Komponents Inc	451 Pk Industrial Dr	La Habra	CA	90631	Jerry Kugel	310-691-7006	R	1	<0.1
R Straman Co	846 Production Pl	Newport Beach	CA	92663	Richard Straman	714-548-8515	R	1	<0.1
Monster Motorsports	PO Box 461077	Escondido	CA	92046	Dave Hopps	619-738-7592	R	1	<0.1
Besasie Automobile Company	2809 S 5th Ct	Milwaukee	WI	53207	Raymond Besasie	414-483-7004	R	1	<0.1
AAC Inc	1250 Albert St	Youngstown	OH	44505	John J Cafaro	216-744-1523	R	0	<0.1
Vector Aeromotive Corp	7601 Centurion	Jacksonville	FL	32256	D Peter Rose	904-645-0505	P	0	<0.1

Source: Ward's Business Directory of U.S. Private and Public Companies, Volumes 1 and 2, 1996. The company type code used is as follows: P - Public, R - Private, S - Subsidiary, D - Division, J - Joint Venture, A - Affiliate, G - Group. Sales are in millions of dollars, employees are in thousands. An asterisk (*) indicates an estimated sales volume. The symbol < stands for 'less than'. Company names and addresses are truncated, in some cases, to fit into the available space.

MATERIALS CONSUMED

Material	Quantity	Delivered Cost ($ million)
Materials, ingredients, containers, and supplies	(X)	105,521.7
Gasoline engines and parts specially designed for gasoline engines	(X)	14,661.0
Diesel engines and parts specially designed for diesel engines	(X)	1,707.0
Axles and axle parts	(X)	3,402.2
Other drive train components and parts	(X)	13,421.6
Car bodies	(X)	682.9
Refrigeration compressors, compressor units, condensing units, and other heat transfer equipment	(X)	2,606.3
Shocks, struts, and other suspension equipment and parts	(X)	1,809.4
Exhaust systems and parts	(X)	1,376.8
Machine tool accessories, including cutting tools	(X)	(D)
Fluid power pumps, motors, and hydrostatic transmissions (hydraulic and pneumatic)	(X)	353.2
Fluid power valves (hydraulic and pneumatic)	(X)	203.6
Fluid power hose or tube fittings and assemblies (hydraulic and pneumatic)	(X)	123.9
Fluid power filters (hydraulic and pneumatic)	(X)	16.9
Other fluid power products (hydraulic and pneumatic)	(X)	255.3
Automotive stampings (including body parts, hubcaps, fenders, etc.)	(X)	11,981.4
Steel springs, except wire	(X)	676.8
Motor vehicle metal hardware (lock units, door and window handles, hinges, etc.), except forgings	(X)	573.7
Metal bolts, nuts, screws, washers, rivets, and other screw machine products	(X)	1,071.4
Other fabricated metal products, except forgings	(X)	972.8
Forgings	(X)	2.6
Castings (rough and semifinished)	(X)	(D)
Steel sheet and strip, including tin plate	(X)	(D)
All other steel shapes and forms	(X)	(D)
Nonferrous shapes and forms	(X)	28.2
Ball and roller bearings (mounted or unmounted)	(X)	13.5
Pneumatic tires and inner tubes	(X)	2,461.2
Rubber and plastics hose and belting	(X)	209.0
Fabricated rubber products, except tires, tubes, hose, belting, and gaskets	(X)	243.8
Gaskets (all types), and packing and sealing devices	(X)	(D)
Fabricated plastics products, including components, housings, accessories, etc.	(X)	1,129.9
Glass and glass products including windows and mirrors	(X)	1,652.7
Seats (purchased separately) for automobiles, trucks, and buses	(X)	2,616.6
Seat covers, seat belts, and shoulder harnesses	(X)	954.6
Automotive air bag assemblies and parts thereof	(X)	(D)
Automotive trimmings, textile (panels, headliners, etc.)	(X)	3,364.6
Carpeting	(X)	505.4
Ceramic and ceramic composite parts, components, and accessories	(X)	(D)
Glues and adhesives	(X)	841.7
Paints, varnishes, lacquers, stains, shellacs, japans, enamels, and allied products	(X)	1,280.0
Engine electrical equipment, including spark plugs, magnetos, generators, starters, etc.	(X)	1,745.5
Motor vehicle lighting fixtures	(X)	569.2
Automotive lamps (bulbs and sealed beams)	(X)	401.2
Storage batteries, automotive	(X)	270.4
Automotive radios and loudspeakers	(X)	2,046.0
Motor vehicle clusters, meters, and gauges, except electrical (including speedometers, fuel level)	(X)	750.0
All other materials and components, parts, containers, and supplies	(X)	25,216.5
Materials, ingredients, containers, and supplies, nsk	(X)	1,987.1

Source: 1992 Economic Census. Explanation of symbols used: (D): Withheld to avoid disclosure of competitive data; na: Not available; (S): Withheld because statistical norms were not met; (X): Not applicable; (Z): Less than half the unit shown; nec: Not elsewhere classified; nsk: Not specified by kind; - : zero; * : 10-19 percent estimated; ** : 20-29 percent estimated.

PRODUCT SHARE DETAILS

Product or Product Class	% Share	Product or Product Class	% Share
Motor vehicles and car bodies	100.00	Buses, including military and firefighting vehicles (chassis of own manufacture), nsk	3.52
Passenger cars, knockdown or assembled, chassis for sale separately and passenger car bodies	56.99	Military vehicles, wheeled tactical vehicles or carriers, excluding tanks and self-propelled weapons	(D)
Complete passenger vehicles, knockdown or assembled	99.95	Trucks, truck tractors, and bus chassis (chassis of own manufacture) 10,000 pounds or less	33.38
Passenger cars, knockdown or assembled, chassis for sale separately and passenger car bodies, nsk	0.05	Trucks, truck tractors, and bus chassis (chassis of own manufacture) 19,501 to 33,000 pounds	3.18
Buses, including military and firefighting vehicles (chassis of own manufacture)	0.73	Trucks, truck tractors, and bus chassis (chassis of own manufacture) 33,001 pounds or more	4.29
Buses, including military (except trolley buses) (chassis of own manufacture)	85.73	Motor vehicles and car bodies, nsk, total	(D)
Firefighting vehicles (chassis of own manufacture)	10.75		

Source: 1992 Economic Census. The values shown are percent of total shipments in an industry. Values of indented subcategories are summed in the main heading. The symbol (D) appears when data are withheld to prevent disclosure of competitive information. The abbreviation nsk stands for 'not specified by kind' and nec for 'not elsewhere classified'.

INPUTS AND OUTPUTS FOR MOTOR VEHICLES & CAR BODIES

Economic Sector or Industry Providing Inputs	%	Sector	Economic Sector or Industry Buying Outputs	%	Sector
Imports	33.0	Foreign	Personal consumption expenditures	51.5	
Motor vehicle parts & accessories	20.3	Manufg.	Gross private fixed investment	35.0	Cap Inv
Wholesale trade	9.0	Trade	Exports	5.4	Foreign
Automotive stampings	6.4	Manufg.	Motor vehicles & car bodies	3.5	Manufg.
Motor vehicles & car bodies	4.2	Manufg.	S/L Govt. purch., elem. & secondary education	0.8	S/L Govt
Internal combustion engines, nec	2.1	Manufg.	S/L Govt. purch., transit utilities	0.8	S/L Govt
Refrigeration & heating equipment	2.1	Manufg.	Federal Government purchases, national defense	0.6	Fed Govt
Tires & inner tubes	1.8	Manufg.	S/L Govt. purch., highways	0.5	S/L Govt
Automotive & apparel trimmings	1.7	Manufg.	S/L Govt. purch., other general government	0.4	S/L Govt
Miscellaneous plastics products	1.4	Manufg.	S/L Govt. purch., police	0.3	S/L Govt
Motor freight transportation & warehousing	1.3	Util.	Motor homes (made on purchased chassis)	0.2	Manufg.
Engine electrical equipment	1.1	Manufg.	Truck & bus bodies	0.1	Manufg.
Fabricated rubber products, nec	1.1	Manufg.	Federal Government purchases, nondefense	0.1	Fed Govt
Hardware, nec	1.1	Manufg.	S/L Govt. purch., fire	0.1	S/L Govt
Advertising	0.9	Services	S/L Govt. purch., natural resource & recreation.	0.1	S/L Govt
Railroads & related services	0.8	Util.	S/L Govt. purch., water	0.1	S/L Govt
Glass & glass products, except containers	0.6	Manufg.			
Special dies & tools & machine tool accessories	0.6	Manufg.			
Iron & steel foundries	0.5	Manufg.			
Paints & allied products	0.5	Manufg.			
Radio & TV receiving sets	0.5	Manufg.			
Electric services (utilities)	0.5	Util.			
Blast furnaces & steel mills	0.4	Manufg.			
Carburetors, pistons, rings, & valves	0.4	Manufg.			
Lighting fixtures & equipment	0.4	Manufg.			
Mechanical measuring devices	0.4	Manufg.			
Screw machine and related products	0.4	Manufg.			
Air transportation	0.4	Util.			
Maintenance of nonfarm buildings nec	0.3	Constr.			
Fabricated textile products, nec	0.3	Manufg.			
Petroleum refining	0.3	Manufg.			
Steel springs, except wire	0.3	Manufg.			
Floor coverings	0.2	Manufg.			
Semiconductors & related devices	0.2	Manufg.			
Storage batteries	0.2	Manufg.			
Truck & bus bodies	0.2	Manufg.			
Gas production & distribution (utilities)	0.2	Util.			
Banking	0.2	Fin/R.E.			
Adhesives & sealants	0.1	Manufg.			
Aluminum rolling & drawing	0.1	Manufg.			
Gaskets, packing & sealing devices	0.1	Manufg.			
Machinery, except electrical, nec	0.1	Manufg.			
Metal coating & allied services	0.1	Manufg.			
Miscellaneous fabricated wire products	0.1	Manufg.			
Plastics materials & resins	0.1	Manufg.			
Public building furniture	0.1	Manufg.			
Communications, except radio & TV	0.1	Util.			
Eating & drinking places	0.1	Trade			
Equipment rental & leasing services	0.1	Services			
U.S. Postal Service	0.1	Gov't			
Noncomparable imports	0.1	Foreign			

Source: Benchmark Input-Output Accounts for the U.S. Economy, 1982, U.S. Department of Commerce, Washington, D.C., July 1991. Data, as reported in the source, are organized by the 1977 SIC structure in use in 1982 but have been matched, as closely as is possible, to the 1987 SIC structure used in this book.

OCCUPATIONS EMPLOYED BY SIC 371 - MOTOR VEHICLES AND EQUIPMENT

Occupation	% of Total 1994	Change to 2005	Occupation	% of Total 1994	Change to 2005
Assemblers, fabricators, & hand workers nec	21.4	-12.2	Tool & die makers	1.6	-8.8
Inspectors, testers, & graders, precision	3.4	-51.1	Mechanics, installers, & repairers nec	1.6	-29.8
Engineers nec	3.4	5.4	Machine tool cutting & forming etc. nec	1.6	5.4
Blue collar worker supervisors	3.4	-11.4	Industrial truck & tractor operators	1.6	-29.8
Machine operators nec	2.9	-30.3	Electricians	1.6	-17.6
Machine tool cutting operators, metal & plastic	2.6	-55.0	Combination machine tool operators	1.5	40.5
Welders & cutters	2.4	-29.8	Machine forming operators, metal & plastic	1.5	-56.1
Managers & administrators nec	2.1	-12.2	Industrial machinery mechanics	1.4	-42.9
Engineering technicians & technologists nec	1.8	-12.2	Sales & related workers nec	1.2	-12.2
Helpers, laborers, & material movers nec	1.8	-12.2	Machine builders	1.1	31.7
Metal & plastic machine workers nec	1.7	55.6	Sheet metal workers & duct installers	1.0	-56.1
Welding machine setters, operators	1.7	-12.2			

Source: Industry-Occupation Matrix, Bureau of Labor Statistics. These data relate to one or more 3-digit SIC industry groups rather than to a single 4-digit SIC. The change reported for each occupation to the year 2005 is a percent of growth or decline as estimated by the Bureau of Labor Statistics. The abbreviation nec stands for 'not elsewhere classified'.

LOCATION BY STATE AND REGIONAL CONCENTRATION

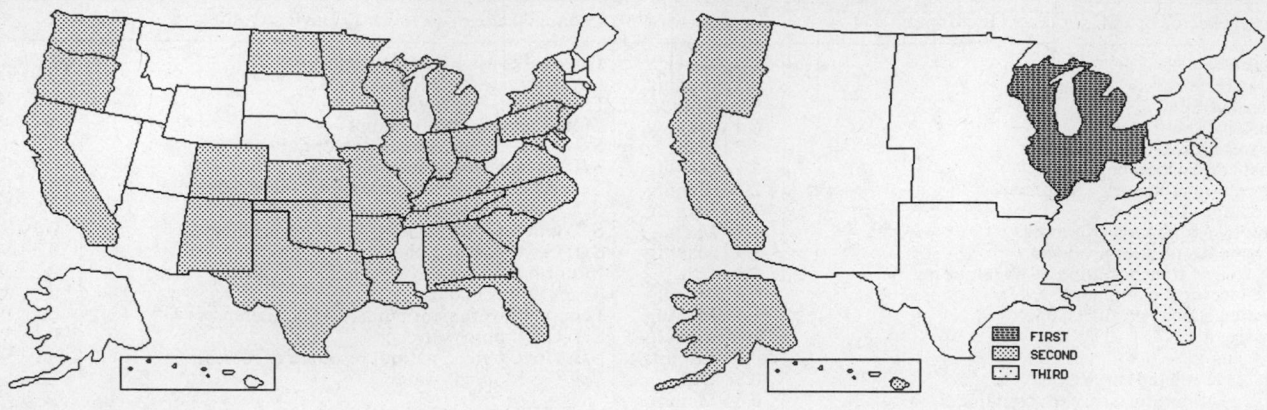

FIRST
SECOND
THIRD

INDUSTRY DATA BY STATE

State	Establish-ments	Shipments			Employment				Cost as % of Shipments	Investment per Employee ($)
		Total ($ mil)	% of U.S.	Per Establ.	Total Number	% of U.S.	Per Establ.	Wages ($/hour)		
Michigan	49	42,091.3	27.5	859.0	76,900	33.7	1,569	22.73	69.5	6,164
Ohio	29	23,032.0	15.1	794.2	35,700	15.6	1,231	21.96	66.3	8,975
Missouri	14	16,070.7	10.5	1,147.9	15,200	6.7	1,086	23.19	71.3	-
Illinois	19	7,766.8	5.1	408.8	8,500	3.7	447	20.67	75.8	-
Indiana	34	4,983.1	3.3	146.6	5,800	2.5	171	22.10	64.0	3,966
California	57	2,980.1	1.9	52.3	8,000	3.5	140	20.39	78.4	-
Arkansas	6	39.6	0.0	6.6	100	0.0	17	11.00	71.7	-
Florida	22	(D)	-	-	375 *	0.2	17	-	-	-
Texas	20	(D)	-	-	3,750 *	1.6	188	-	-	-
Pennsylvania	16	(D)	-	-	1,750 *	0.8	109	-	-	-
Wisconsin	15	(D)	-	-	7,500 *	3.3	500	-	-	-
Georgia	14	(D)	-	-	7,500 *	3.3	536	-	-	-
New York	12	(D)	-	-	3,750 *	1.6	313	-	-	-
North Carolina	12	(D)	-	-	1,750 *	0.8	146	-	-	-
New Jersey	10	(D)	-	-	1,750 *	0.8	175	-	-	-
Kentucky	9	(D)	-	-	17,500 *	7.7	1,944	-	-	-
Kansas	8	(D)	-	-	3,750 *	1.6	469	-	-	-
Oregon	8	(D)	-	-	1,750 *	0.8	219	-	-	-
Washington	8	(D)	-	-	750 *	0.3	94	-	-	-
Oklahoma	7	(D)	-	-	3,750 *	1.6	536	-	-	-
Alabama	6	(D)	-	-	375 *	0.2	63	-	-	-
Colorado	5	(D)	-	-	375 *	0.2	75	-	-	-
Tennessee	5	(D)	-	-	17,500 *	7.7	3,500	-	-	-
Virginia	5	(D)	-	-	1,750 *	0.8	350	-	-	-
Louisiana	4	(D)	-	-	1,750 *	0.8	438	-	-	-
Maryland	4	(D)	-	-	3,750 *	1.6	938	-	-	-
Minnesota	4	(D)	-	-	1,750 *	0.8	438	-	-	-
North Dakota	4	(D)	-	-	375 *	0.2	94	-	-	-
South Carolina	4	(D)	-	-	750 *	0.3	188	-	-	-
New Mexico	3	(D)	-	-	750 *	0.3	250	-	-	-
Delaware	2	(D)	-	-	7,500 *	3.3	3,750	-	-	-

Source: 1992 *Economic Census*. The states are in descending order of shipments or establishments (if shipment data are missing for the majority). The symbol (D) appears when data are withheld to prevent disclosure of competitive information. States marked with (D) are sorted by number of establishments. A dash (-) indicates that the data element cannot be calculated; * indicates the midpoint of a range.

3713 - TRUCK & BUS BODIES

82 83 84 85 86 87 88 89 90 91 92 93 94 95 96 97 98

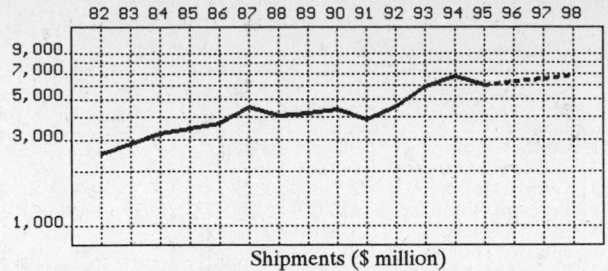

Shipments ($ million)

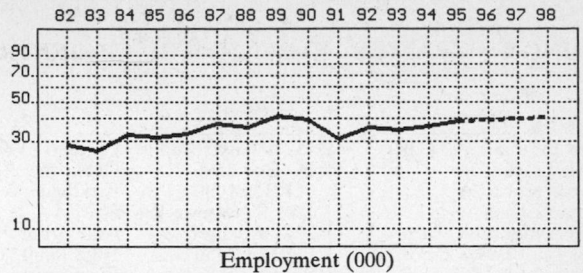

Employment (000)

GENERAL STATISTICS

Year	Companies	Establishments		Employment			Compensation		Production ($ million)			
		Total	with 20 or more employees	Total (000)	Production Workers (000)	Hours (Mil)	Payroll ($ mil)	Wages ($/hr)	Cost of Materials	Value Added by Manufacture	Value of Shipments	Capital Invest.
1982	637	681	278	28.5	21.5	40.6	502.4	8.59	1,331.1	1,195.5	2,510.9	51.7
1983		669	286	26.6	20.1	39.8	490.3	8.68	1,327.2	1,515.7	2,833.6	49.8
1984		657	294	32.8	25.1	49.9	624.6	8.86	1,807.7	1,498.5	3,255.7	71.1
1985		646	303	31.7	24.0	46.5	642.3	9.55	1,904.2	1,597.4	3,473.2	63.0
1986		638	315	33.2	24.8	50.0	709.7	9.70	2,031.8	1,715.0	3,686.3	82.3
1987	658	716	325	37.8	28.9	58.3	851.3	10.02	2,672.2	1,978.4	4,588.7	88.5
1988		718	345	36.2	27.9	55.3	809.4	9.93	2,333.9	1,767.6	4,076.1	68.9
1989		680	337	41.7	28.6	56.0	812.1	9.87	2,426.3	1,760.5	4,236.4	117.1
1990		657	318	39.6	28.3	56.1	839.2	9.99	2,525.2	1,809.3	4,382.2	62.8
1991		662	289	31.1	23.5	47.7	719.1	9.91	2,214.4	1,598.9	3,867.7	58.1
1992	622	676	301	35.9	26.2	52.4	868.6	10.77	2,817.5	1,791.9	4,594.7	71.2
1993		649	292	34.9	25.9	52.9	883.7	10.99	3,503.1	2,366.4	5,876.0	78.7
1994		669P	316P	36.5	27.3	56.7	961.0	11.30	3,735.6	2,969.0	6,674.5	171.6
1995		669P	318P	38.8P	28.4P	58.1P	983.5P	11.28P	3,355.6P	2,667.0P	5,995.6P	111.9P
1996		669P	319P	39.5P	28.8P	59.1P	1,017.3P	11.48P	3,502.5P	2,783.8P	6,258.0P	116.5P
1997		668P	321P	40.1P	29.2P	60.1P	1,051.1P	11.69P	3,649.4P	2,900.5P	6,520.5P	121.1P
1998		668P	322P	40.8P	29.6P	61.1P	1,084.8P	11.89P	3,796.3P	3,017.3P	6,783.0P	125.7P

Sources: 1982, 1987, 1992 *Economic Census*; *Annual Survey of Manufactures*, 83-86, 88-91, 93-94. Establishment counts for non-Census years are from *County Business Patterns*; establishment values for 83-84 are extrapolations. 'P's show projections by the editors. Industries reclassified in 87 will not have data for prior years.

INDICES OF CHANGE

Year	Companies	Establishments		Employment			Compensation		Production ($ million)			
		Total	with 20 or more employees	Total (000)	Production Workers (000)	Hours (Mil)	Payroll ($ mil)	Wages ($/hr)	Cost of Materials	Value Added by Manufacture	Value of Shipments	Capital Invest.
1982	102	101	92	79	82	77	58	80	47	67	55	73
1983		99	95	74	77	76	56	81	47	85	62	70
1984		97	98	91	96	95	72	82	64	84	71	100
1985		96	101	88	92	89	74	89	68	89	76	88
1986		94	105	92	95	95	82	90	72	96	80	116
1987	106	106	108	105	110	111	98	93	95	110	100	124
1988		106	115	101	106	106	93	92	83	99	89	97
1989		101	112	116	109	107	93	92	86	98	92	164
1990		97	106	110	108	107	97	93	90	101	95	88
1991		98	96	87	90	91	83	92	79	89	84	82
1992	100	100	100	100	100	100	100	100	100	100	100	100
1993		96	97	97	99	101	102	102	124	132	128	111
1994		99P	105P	102	104	108	111	105	133	166	145	241
1995		99P	106P	108P	108P	111P	113P	105P	119P	149P	130P	157P
1996		99P	106P	110P	110P	113P	117P	107P	124P	155P	136P	164P
1997		99P	107P	112P	111P	115P	121P	109P	130P	162P	142P	170P
1998		99P	107P	114P	113P	117P	125P	110P	135P	168P	148P	177P

Sources: Same as General Statistics. Values reflect change from the base year, 1992. Values above 100 mean greater than 92, values below 100 mean less than 92, and a value of 100 in the 82-91 or 93-98 period means same as 92. 'P's mark projections by the editors.

SELECTED RATIOS

For 1994	Avg. of All Manufact.	Analyzed Industry	Index	For 1994	Avg. of All Manufact.	Analyzed Industry	Index
Employees per Establishment	49	55	111	Value Added per Production Worker	134,084	108,755	81
Payroll per Establishment	1,500,273	1,436,180	96	Cost per Establishment	5,045,178	5,582,719	111
Payroll per Employee	30,620	26,329	86	Cost per Employee	102,970	102,345	99
Production Workers per Establishment	34	41	119	Cost per Production Worker	146,988	136,835	93
Wages per Establishment	853,319	957,518	112	Shipments per Establishment	9,576,895	9,974,798	104
Wages per Production Worker	24,861	23,469	94	Shipments per Employee	195,460	182,863	94
Hours per Production Worker	2,056	2,077	101	Shipments per Production Worker	279,017	244,487	88
Wages per Hour	12.09	11.30	93	Investment per Establishment	321,011	256,450	80
Value Added per Establishment	4,602,255	4,437,063	96	Investment per Employee	6,552	4,701	72
Value Added per Employee	93,930	81,342	87	Investment per Production Worker	9,352	6,286	67

Sources: Same as General Statistics. The 'Average of All Manufacturing' column represents the average of all manufacturing industries reported for the most recent complete year available. The Index shows the relationship between the Average and the Analyzed Industry. For example, 100 means that they are equal; 500 that the Analyzed Industry is five times the average; 50 means that the Analyzed Industry is half the national average. The abbreviation 'na' is used to show that data are 'not available'.

LEADING COMPANIES   Number shown: **75**   Total sales ($ mil): **10,073**   Total employment (000): **41.0**

Company Name	Address				CEO Name	Phone	Co. Type	Sales ($ mil)	Empl. (000)
Navistar International Corp	455 N Cityfront Plz	Chicago	IL	60611	James C Cotting	312-836-2000	P	5,337	14.9
Oshkosh Truck Corp	PO Box 2566	Oshkosh	WI	54903	R Eugene Goodson	414-235-9151	P	635	1.5
Kenworth Truck Co	PO Box 1000	Kirkland	WA	98083	Gary S Moore	206-828-5000	D	460•	2.1
Grumman Allied	1801 S Nottawa Rd	Sturgis	MI	49091	Leonard Rothenberg	616-659-0200	D	400	2.2
General Automotive Corp	2015 Washtenaw	Ann Arbor	MI	48104	Mark J Obert	313-994-8000	R	320	1.3
Thomas Built Buses Inc	PO Box 2450	High Point	NC	27261	John W Thomas Jr	919-889-4871	R	240	1.4
Altec Industries Inc	PO Box 10264	Birmingham	AL	35202	Thomas L Merrill	205-991-7733	R	220	1.5
TABC Inc	PO Box 2140	Long Beach	CA	90801	Y Onishi	310-428-3604	S	215	0.4
Bus Industries of America Inc	PO Box 449	Oriskany	NY	13424	James T Roddy	315-768-8101	S	200	0.6
Heil Co	205 Bishops Way	Brookfield	WI	53005	Lawrence Gray	414-789-5500	R	190•	1.1
Supreme Industries Inc	PO Box 237	Goshen	IN	46526	Herbert M Gardner	219-642-3070	P	137	1.4
AmTran Corp	PO Box 6000	Conway	AR	72033	William S Bankston	501-327-7761	R	110	1.0
Grumman Olson	1801 S Nottawa Rd	Sturgis	MI	49091	James A McConnell	616-659-0200	D	100	0.9
Utilimaster	PO Box 585	Wakarusa	IN	46573	Martin Snoey	219-862-4561	D	100	0.4
Waste-Quip Inc	25800 Science Pk Dr	Beachwood	OH	44122	George L Schneider	216-292-2554	R	100	0.3
Carpenter Manufacturing Inc	1500 W Main St	Mitchell	IN	47446	Tim Durham	812-849-3131	R	90	0.5
Holan Manufacturing Inc	1104 Everee Rd	Griffin	GA	30223	Van Walbridge	404-227-9423	S	71•	0.4
Supreme Corp	PO Box 463	Goshen	IN	46526	Omer G Kropf	219-533-0331	S	70	0.9
CCC Industries Inc	PO Box 582891	Tulsa	OK	74158	A E Mascarin	918-836-1651	R	53	0.3
Stahl/Scott Fetzer Company Inc	3201 Lincoln	Wooster	OH	44691	Bob McBride	216-264-7441	S	53•	0.3
Hackney and Sons Inc	PO Box 880	Washington	NC	27889	Jay Troger	919-946-6521	R	51	<0.1
McLaughlin Body Company Inc	2430 River Dr	Moline	IL	61265	Bud Pierce	309-762-7755	R	50	0.3
Metrotrans Corp	777 Greenbelt Pky	Griffin	GA	30223	Michael Walden	404-229-5995	R	50	0.3
Miller Industries Inc	PO Box 828	Ooltewah	TN	37363	William G Miller	615-238-4171	P	46	0.3
Leach Company Inc	PO Box 2608	Oshkosh	WI	54903	Frederick E Leach	414-231-2770	R	40	0.4
Union City Body Company Inc	PO Box 190	Union City	IN	47390	Andrew Taitz	317-964-3121	R	40	0.6
Champion Motor Coach Inc	PO Box 158	Imlay City	MI	48444	Tom Ensch	810-724-6474	S	38•	0.2
Crane Carrier Co	PO Box 582891	Tulsa	OK	74158	Edward Mascarin	918-836-1651	S	35•	0.2
Reading Body Works Inc	PO Box 650	Shillington	PA	19607	I Suknow	610-775-3301	R	32	0.3
Sutphen Corp	7000 Columbus	Amlin	OH	43002	Thomas Sutphen	614-889-1005	R	32•	0.1
Monroe Truck Equipment Inc	1051 W 7th St	Monroe	WI	53566	Richard L Feller	608-328-8127	R	30	0.2
Kidron Inc	PO Box 17	Kidron	OH	44636	Robert Sommer	216-857-3011	R	27	0.3
Blitz Corp	4525 W 26th St	Chicago	IL	60623	Carmont Blitz	312-762-7600	R	26	0.2
Hesse Corp	6700 St John Av	Kansas City	MO	64125	Michael Ireland	816-483-7808	R	26•	0.2
Pak-Mor Manufacturing Co	PO Box 14147	San Antonio	TX	78214	JV Thurmond Jr	210-923-4317	R	25	0.3
Murphy Body Co	PO Box 2009	Wilson	NC	27893	CW Layman	919-291-2191	R	24•	0.1
R-S Truck Body Company Inc	State Rte 1428	Allen	KY	41601	WL Smith	606-874-2151	R	20	0.2
Sherrod Vans Inc	6464 Greenland Rd	Jacksonville	FL	32258	Jack C Sherrod	904-268-3321	R	20	0.1
Benson Truck Bodies Inc	PO Box 49	Mineralwells	WV	26150	Tom Jones	304-489-9020	S	19	0.2
Rawson-Koenig Inc	2301 Central Pkwy	Houston	TX	77092	Thomas C Rawson	713-688-4414	P	17	0.3
Smeal Fire Apparatus Co	Hwy 91 N	Snyder	NE	68664	Delwin Smeal	402-568-2221	R	17•	0.1
Johnson Welding & Mfg	PO Box 480	Rice Lake	WI	54868	John C Peterson	715-234-7071	R	16	0.2
Medical Coaches Inc	PO Box 129	Oneonta	NY	13820	Geoffrey A Smith	607-432-1333	R	16	<0.1
Blue Bird Midwest	PO Box 180	Mount Pleasant	IA	52641	Richard Benedict	319-385-2231	D	13•	0.2
Gooseneck Trailer Mfg	PO Box 832	Bryan	TX	77806	David S Carrabba	409-778-0034	R	13	0.2
Phoenix Manufacturing Inc	PO Box 97	Nanticoke	PA	18634	SB Mermelstein	717-735-1800	R	13•	0.1
Centennial Body	PO Box 708	Columbus	GA	31993	David Parker	706-323-6446	D	12	<0.1
Fleet Equipment Corp	567 Commerce St	Franklin Lakes	NJ	07417	Richard D Pearson	201-337-7332	R	12	<0.1
Osterlund Inc	PO Box 1951	Harrisburg	PA	17105	Jan Osterlund	717-564-8070	R	12	<0.1
Somerset Welding and Steel	RD #2	Somerset	PA	15501	S William Riggs	814-443-2671	D	12	0.1
Custom Coach Corp	1400 Dublin Rd	Columbus	OH	43215	Buford Chester	614-481-8881	S	12	<0.1
Dailey Body Co	440 High St	Oakland	CA	94601	PB Kelly	510-534-1423	R	10	0.1
Hackney Brothers Inc	PO Box 2728	Wilson	NC	27894	Robert H Hackney	919-237-8171	R	10	0.2
Henderson Manufacturing	PO Box 40	Manchester	IA	52057	Steve Hedtke	319-927-2828	D	10•	0.1
Hobbs International Inc	PO Box 59	Norwalk	CT	06856	William H Bayles Jr	203-838-4151	R	10	<0.1
Truck Cab Manufacturers Inc	PO Box 58400	Cincinnati	OH	45258	James D Weber	513-922-1300	R	10	0.1
Arkansas Trailer Manufacturing	PO Box 4080	Little Rock	AR	72214	Guy Campbell	501-666-5417	R	9	<0.1
Attbar Inc	1721 NW 262nd St	Richfield	WA	98642	Jack J Barchek	360-887-3580	R	8•	0.2
Auto Truck Inc	1160 N Ellis St	Bensenville	IL	60106	James E Dondlinger	708-860-5600	R	8•	<0.1
Diamond B Body & Trailer	8405 Loch Lomond	Pico Rivera	CA	90660	DL Boicr	310-948-3497	R	8	<0.1
Equipment Innovators Inc	800 Industrial Pk Dr	Marietta	GA	30062	Ande Evers	404-427-9467	R	8	<0.1
MCT Custom Truck Bodies Inc	3155 Industrial Dr	Hernando	MS	38632	Rick Moore	601-429-0054	R	8	<0.1
Morse Manufacturing Inc	44 Chocksett Rd	Sterling	MA	01564	Steven F Schneider	508-422-8203	R	8•	<0.1
Weld Built Body Company Inc	276 Long Island Av	Wyandanch	NY	11798	Joseph Milan	516-643-9700	R	8•	<0.1
Tafco Equipment Co	PO Box 339	Blue Earth	MN	56013	Terry Ankeny	507-526-3247	R	7	<0.1
Atlas Truck Bodies Inc	PO Box 3370	Ontario	CA	91761	Canuto Espinoza	909-983-5669	R	6	<0.1
Cleveland Hardware	3270 E 79th St	Cleveland	OH	44104	William E Hoban	216-641-5200	R	6	<0.1
Pacific Utility Body Co	PO Box 307	San Lorenzo	CA	94580	Barbara Goecks	510-278-7400	R	6	<0.1
Parkhurst Manufacturing	PO Box 1323	Sedalia	MO	65302	W R Parkhurst II	816-826-8685	R	6	<0.1
Steelweld Equipment Company	PO Box 440	St Clair	MO	63077	RS Anderson	314-629-3704	R	6	<0.1
Arrow Truck Bodies & Equip	1639 S Campus Av	Ontario	CA	91761	Raymond A Glaze	909-947-3991	R	5•	<0.1
Northwest Bodies Inc	1 Industrial Park	Manson	IA	50563	RJ Wolf	712-469-3341	R	5•	<0.1
Simpson Equipment Corp	PO Box 2229	Wilson	NC	27894	JR Simpson	919-291-4105	R	5	<0.1
Thiele Industries Inc	111 Spruce St	Windber	PA	15963	David Romano	814-467-4504	R	5	<0.1
Unicell Body Co	571 Howard St	Buffalo	NY	14206	Roger Martin	716-853-8628	S	5	<0.1

Source: Ward's Business Directory of U.S. Private and Public Companies, Volumes 1 and 2, 1996. The company type code used is as follows: P - Public, R - Private, S - Subsidiary, D - Division, J - Joint Venture, A - Affiliate, G - Group. Sales are in millions of dollars, employees are in thousands. An asterisk (*) indicates an estimated sales volume. The symbol < stands for 'less than'. Company names and addresses are truncated, in some cases, to fit into the available space.

MATERIALS CONSUMED

Material	Quantity	Delivered Cost ($ million)
Materials, ingredients, containers, and supplies	(X)	2,587.4
Motor vehicle metal hardware (lock units, door and window handles, hinges, etc.), except forgings	(X)	71.6
Metal bolts, nuts, screws, washers, rivets, and other screw machine products	(X)	24.5
Other fabricated metal products, except forgings	(X)	72.6
Forgings	(X)	3.9
Castings (rough and semifinished)	(X)	5.5
Steel bars, bar shapes, and plates	(X)	40.6
Steel sheet and strip, including tin plate	(X)	129.0
Steel structural shapes and sheet piling	(X)	29.9
All other steel shapes and forms	(X)	30.9
Aluminum and aluminum-base alloy sheet, plate, foil, and welded tubing	(X)	82.6
Aluminum and aluminum-base alloy extruded shapes, including extruded rod, bar, pipe, tube, etc.	(X)	54.3
Other aluminum and aluminum-base alloy shapes and forms	(X)	14.5
Other nonferrous shapes and forms	(X)	2.8
Purchased chassis for vehicles (excluding passenger cars)	(X)	333.4
Truck bodies	(X)	38.8
Transmissions and parts	(X)	35.1
Shocks, struts, and other suspension equipment and parts	(X)	11.7
Axles, brakes, drums, rims, wheels, and other metal motor wheel parts	(X)	42.3
Pneumatic tires and inner tubes	(X)	13.4
Paints, varnishes, lacquers, stains, shellacs, japans, enamels, and allied products	(X)	43.7
Glass and glass products including windows and mirrors	(X)	19.5
Rough and dressed lumber	(X)	29.9
Fluid power products	(X)	72.8
Motor vehicle lighting fixtures	(X)	19.3
Automotive lamps (bulbs and sealed beams)	(X)	1.4
Seats (purchased separately) for automobiles, trucks, and buses	(X)	20.5
Fabricated plastics products, including components, housings, accessories, etc.	(X)	20.6
All other materials and components, parts, containers, and supplies	(X)	677.0
Materials, ingredients, containers, and supplies, nsk	(X)	645.7

Source: 1992 *Economic Census*. Explanation of symbols used: (D): Withheld to avoid disclosure of competitive data; na: Not available; (S): Withheld because statistical norms were not met; (X): Not applicable; (Z): Less than half the unit shown; nec: Not elsewhere classified; nsk: Not specified by kind; - : zero; * : 10-19 percent estimated; ** : 20-29 percent estimated.

PRODUCT SHARE DETAILS

Product or Product Class	% Share	Product or Product Class	% Share
Truck and bus bodies	100.00	Wrecker truck bodies for sale separately	3.27
Truck, bus, and other vehicle bodies (except passenger car bodies) for sale separatly	51.78	Other truck and vehicle bodies, except passenger cars, for sale separately	12.64
Bus bodies for sale separately	16.09	Truck, bus, and other vehicle bodies (except passenger car bodies) for sale separatly, nsk	5.78
Truck cabs for sale separately	24.35	Complete vehicles produced on purchased chassis	36.62
Vans with unit body-cab for sale separately	5.40	Buses, complete, produced on purchased chassis	20.71
Refrigerated (except food service) van bodies with separate cab for sale separately	1.12	Ambulance and rescue vehicles complete, produced on purchased chassis	9.56
Food service van bodies with separate cab for sale separately	2.26	Vans complete, produced on purchased chassis	0.98
Other van bodies with separate cab for sale separately	6.18	Tank trucks complete, produced on purchased chassis	3.35
Tank bodies cab for sale separately	0.84	Beverage trucks complete, produced on purchased chassis	(D)
Rear loading garbage and refuse truck bodies (packer-types) for sale separately	2.17	Dump trucks complete, produced on purchased chassis	0.76
Side loading garbage and refuse truck bodies (packer-types) for sale separately	1.26	Firefighting vehicles, complete, produced on purchased chassis	21.76
Other garbage and refuse truck bodies (packer-types) for sale separately	3.93	Utility line service trucks complete, produced on purchased chassis	(D)
Beverage truck bodies for sale separately	1.85	Other mobile service type trucks complete, produced on purchased chassis	4.85
Dump truck bodies for sale separately	5.22	Other trucks and complete vehicles produced on purchased chassis	25.98
Stake and platform truck bodies for sale separately	2.00	Complete vehicles produced on purchased chassis, nsk	11.26
Utility line service truck bodies for sale separately	4.78	Truck and bus bodies, nsk	11.60
Other mobile service type truck bodies for sale separately	0.85		

Source: 1992 *Economic Census*. The values shown are percent of total shipments in an industry. Values of indented subcategories are summed in the main heading. The symbol (D) appears when data are withheld to prevent disclosure of competitive information. The abbreviation nsk stands for 'not specified by kind' and nec for 'not elsewhere classified'.

INPUTS AND OUTPUTS FOR TRUCK & BUS BODIES

Economic Sector or Industry Providing Inputs	%	Sector	Economic Sector or Industry Buying Outputs	%	Sector
Imports	11.6	Foreign	Gross private fixed investment	81.0	Cap Inv
Blast furnaces & steel mills	11.5	Manufg.	Motor vehicles & car bodies	7.5	Manufg.
Motor vehicles & car bodies	9.6	Manufg.	Federal Government purchases, national defense	6.9	Fed Govt
Wholesale trade	7.1	Trade	Exports	2.4	Foreign
Aluminum rolling & drawing	5.8	Manufg.	Change in business inventories	0.7	In House
Advertising	5.7	Services	Truck & bus bodies	0.6	Manufg.
Pumps & compressors	3.2	Manufg.	Farm machinery & equipment	0.5	Manufg.
Carburetors, pistons, rings, & valves	2.9	Manufg.	Construction machinery & equipment	0.3	Manufg.
Automotive stampings	2.6	Manufg.			
Internal combustion engines, nec	2.3	Manufg.			
Maintenance of nonfarm buildings nec	1.5	Constr.			
Paints & allied products	1.5	Manufg.			
Electric services (utilities)	1.5	Util.			
Hardware, nec	1.3	Manufg.			
Refrigeration & heating equipment	1.3	Manufg.			
Petroleum refining	1.2	Manufg.			
Motor freight transportation & warehousing	1.2	Util.			
Motor vehicle parts & accessories	1.1	Manufg.			
Truck & bus bodies	0.9	Manufg.			
Communications, except radio & TV	0.9	Util.			
Fabricated textile products, nec	0.8	Manufg.			
Metal stampings, nec	0.8	Manufg.			
Primary lead	0.8	Manufg.			
Primary zinc	0.8	Manufg.			
Screw machine and related products	0.8	Manufg.			
Glass & glass products, except containers	0.7	Manufg.			
Pipe, valves, & pipe fittings	0.7	Manufg.			
Power transmission equipment	0.7	Manufg.			
Gas production & distribution (utilities)	0.7	Util.			
Eating & drinking places	0.7	Trade			
Engine electrical equipment	0.6	Manufg.			
Machinery, except electrical, nec	0.6	Manufg.			
Metal foil & leaf	0.6	Manufg.			
Special dies & tools & machine tool accessories	0.6	Manufg.			
Truck trailers	0.6	Manufg.			
Real estate	0.6	Fin/R.E.			
Equipment rental & leasing services	0.6	Services			
Abrasive products	0.5	Manufg.			
Fabricated metal products, nec	0.5	Manufg.			
Metal coating & allied services	0.5	Manufg.			
Miscellaneous plastics products	0.5	Manufg.			
Railroads & related services	0.5	Util.			
Banking	0.5	Fin/R.E.			
Coated fabrics, not rubberized	0.4	Manufg.			
Veneer & plywood	0.4	Manufg.			
Die-cut paper & board	0.3	Manufg.			
Fabricated rubber products, nec	0.3	Manufg.			
Metal heat treating	0.3	Manufg.			
Padding & upholstery filling	0.3	Manufg.			
Sawmills & planning mills, general	0.3	Manufg.			
Tires & inner tubes	0.3	Manufg.			
Transportation equipment, nec	0.3	Manufg.			
Engineering, architectural, & surveying services	0.3	Services			
Legal services	0.3	Services			
Management & consulting services & labs	0.3	Services			
U.S. Postal Service	0.3	Gov't			
Adhesives & sealants	0.2	Manufg.			
Automotive & apparel trimmings	0.2	Manufg.			
Iron & steel foundries	0.2	Manufg.			
Lighting fixtures & equipment	0.2	Manufg.			
Lubricating oils & greases	0.2	Manufg.			
Mineral wool	0.2	Manufg.			
Nonferrous rolling & drawing, nec	0.2	Manufg.			
Nonferrous wire drawing & insulating	0.2	Manufg.			
Plastics materials & resins	0.2	Manufg.			
Public building furniture	0.2	Manufg.			
Storage batteries	0.2	Manufg.			
Air transportation	0.2	Util.			
Insurance carriers	0.2	Fin/R.E.			
Accounting, auditing & bookkeeping	0.2	Services			
Automotive rental & leasing, without drivers	0.2	Services			
Automotive repair shops & services	0.2	Services			
Computer & data processing services	0.2	Services			
Laundry, dry cleaning, shoe repair	0.2	Services			
Aluminum castings	0.1	Manufg.			
Industrial gases	0.1	Manufg.			
Manifold business forms	0.1	Manufg.			
Metal doors, sash, & trim	0.1	Manufg.			

Continued on next page.

INPUTS AND OUTPUTS FOR TRUCK & BUS BODIES - Continued

Economic Sector or Industry Providing Inputs	%	Sector	Economic Sector or Industry Buying Outputs	%	Sector
Motors & generators	0.1	Manufg.			
Sanitary services, steam supply, irrigation	0.1	Util.			
Retail trade, except eating & drinking	0.1	Trade			
Colleges, universities, & professional schools	0.1	Services			

Source: *Benchmark Input-Output Accounts for the U.S. Economy, 1982*, U.S. Department of Commerce, Washington, D.C., July 1991. Data, as reported in the source, are organized by the 1977 SIC structure in use in 1982 but have been matched, as closely as is possible, to the 1987 SIC structure used in this book.

OCCUPATIONS EMPLOYED BY SIC 371 - MOTOR VEHICLES AND EQUIPMENT

Occupation	% of Total 1994	Change to 2005	Occupation	% of Total 1994	Change to 2005
Assemblers, fabricators, & hand workers nec	21.4	-12.2	Tool & die makers	1.6	-8.8
Inspectors, testers, & graders, precision	3.4	-51.1	Mechanics, installers, & repairers nec	1.6	-29.8
Engineers nec	3.4	5.4	Machine tool cutting & forming etc. nec	1.6	5.4
Blue collar worker supervisors	3.4	-11.4	Industrial truck & tractor operators	1.6	-29.8
Machine operators nec	2.9	-30.3	Electricians	1.6	-17.6
Machine tool cutting operators, metal & plastic	2.6	-55.0	Combination machine tool operators	1.5	40.5
Welders & cutters	2.4	-29.8	Machine forming operators, metal & plastic	1.5	-56.1
Managers & administrators nec	2.1	-12.2	Industrial machinery mechanics	1.4	-42.9
Engineering technicians & technologists nec	1.8	-12.2	Sales & related workers nec	1.2	-12.2
Helpers, laborers, & material movers nec	1.8	-12.2	Machine builders	1.1	31.7
Metal & plastic machine workers nec	1.7	55.6	Sheet metal workers & duct installers	1.0	-56.1
Welding machine setters, operators	1.7	-12.2			

Source: *Industry-Occupation Matrix*, Bureau of Labor Statistics. These data relate to one or more 3-digit SIC industry groups rather than to a single 4-digit SIC. The change reported for each occupation to the year 2005 is a percent of growth or decline as estimated by the Bureau of Labor Statistics. The abbreviation nec stands for 'not elsewhere classified'.

LOCATION BY STATE AND REGIONAL CONCENTRATION

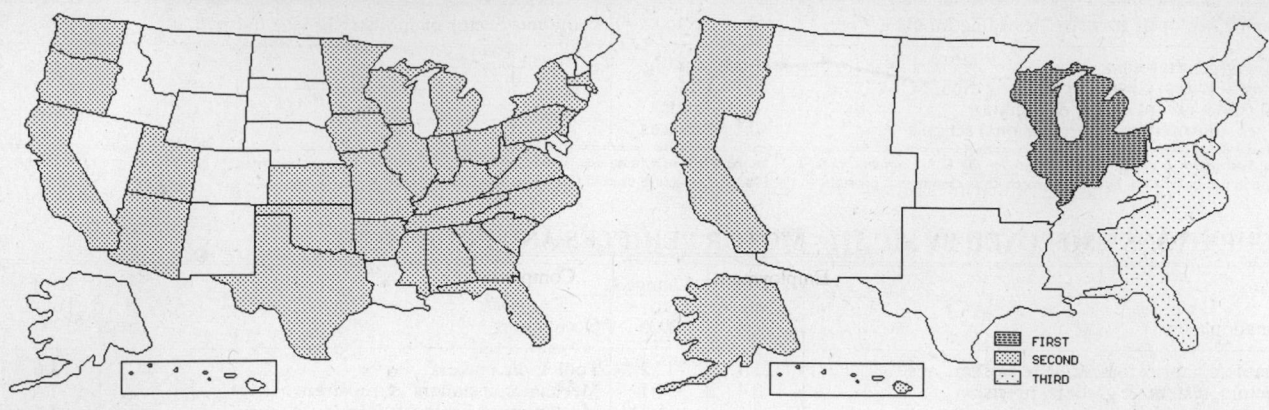

FIRST
SECOND
THIRD

INDUSTRY DATA BY STATE

State	Establish-ments	Shipments			Employment				Cost as % of Shipments	Investment per Employee ($)
		Total ($ mil)	% of U.S.	Per Establ.	Total Number	% of U.S.	Per Establ.	Wages ($/hour)		
Pennsylvania	50	582.3	12.7	11.6	3,500	9.7	70	10.96	61.6	1,629
California	95	578.8	12.6	6.1	2,800	7.8	29	10.42	55.6	-
Indiana	41	469.5	10.2	11.5	4,900	13.6	120	10.84	62.5	898
Wisconsin	19	300.7	6.5	15.8	2,400	6.7	126	13.64	62.3	-
Florida	25	283.3	6.2	11.3	2,300	6.4	92	9.44	59.8	1,087
Ohio	31	249.6	5.4	8.1	2,400	6.7	77	12.00	54.8	1,042
Texas	37	171.9	3.7	4.6	1,600	4.5	43	9.13	63.5	-
Iowa	14	135.4	2.9	9.7	900	2.5	64	11.86	68.5	1,333
Alabama	18	107.5	2.3	6.0	600	1.7	33	8.08	61.0	500
Michigan	19	90.4	2.0	4.8	800	2.2	42	12.19	50.2	1,875
Arkansas	5	82.6	1.8	16.5	1,000	2.8	200	11.64	90.1	-
Arizona	11	66.1	1.4	6.0	400	1.1	36	8.43	60.4	-
Tennessee	14	50.2	1.1	3.6	400	1.1	29	10.50	54.4	-
Minnesota	17	43.4	0.9	2.6	400	1.1	24	11.80	59.4	750
New Jersey	11	18.6	0.4	1.7	200	0.6	18	10.50	52.7	-
North Carolina	30	(D)	-	-	1,750 *	4.9	58	-	-	-
New York	28	(D)	-	-	1,750 *	4.9	63	-	-	400
Georgia	25	(D)	-	-	1,750 *	4.9	70	-	-	1,086
Illinois	19	(D)	-	-	750 *	2.1	39	-	-	-
Kansas	16	(D)	-	-	1,750 *	4.9	109	-	-	-
Missouri	15	(D)	-	-	375 *	1.0	25	-	-	800
Kentucky	14	(D)	-	-	375 *	1.0	27	-	-	1,333
Virginia	14	(D)	-	-	750 *	2.1	54	-	-	-
Oklahoma	13	(D)	-	-	375 *	1.0	29	-	-	-
Oregon	10	(D)	-	-	175 *	0.5	18	-	-	-
Massachusetts	9	(D)	-	-	375 *	1.0	42	-	-	-
Mississippi	9	(D)	-	-	750 *	2.1	83	-	-	-
Washington	7	(D)	-	-	175 *	0.5	25	-	-	-
Utah	6	(D)	-	-	175 *	0.5	29	-	-	-
West Virginia	6	(D)	-	-	175 *	0.5	29	-	-	1,143
South Carolina	4	(D)	-	-	175 *	0.5	44	-	-	-

Source: 1992 *Economic Census*. The states are in descending order of shipments or establishments (if shipment data are missing for the majority). The symbol (D) appears when data are withheld to prevent disclosure of competitive information. States marked with (D) are sorted by number of establishments. A dash (-) indicates that the data element cannot be calculated; * indicates the midpoint of a range.

3714 - MOTOR VEHICLE PARTS & ACCESSORIES

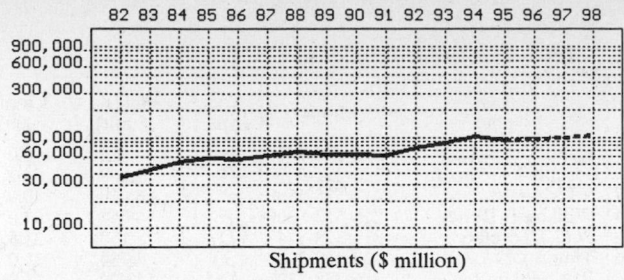

Shipments ($ million)

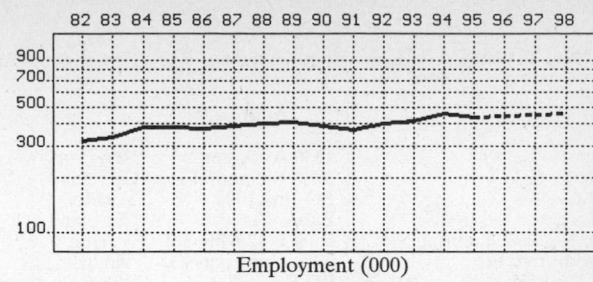

Employment (000)

GENERAL STATISTICS

Year	Com-panies	Establishments		Employment			Compensation		Production ($ million)			
		Total	with 20 or more employees	Total (000)	Production Workers (000)	Hours (Mil)	Payroll ($ mil)	Wages ($/hr)	Cost of Materials	Value Added by Manufacture	Value of Shipments	Capital Invest.
1982	2,000	2,420	1,111	321.4	251.1	481.8	7,614.0	11.65	19,007.0	16,764.6	36,293.1	1,791.7
1983		2,372	1,124	338.0	267.3	549.0	8,876.2	12.26	23,092.6	21,592.6	44,415.4	1,056.2
1984		2,324	1,137	381.6	306.3	645.2	10,649.9	12.69	28,986.1	23,887.7	52,583.1	1,725.0
1985		2,277	1,150	385.4	308.7	645.2	11,649.6	13.85	31,551.5	26,093.7	57,931.0	2,962.0
1986		2,276	1,158	376.6	298.7	622.6	11,537.3	14.06	33,019.5	24,373.5	57,393.9	2,644.0
1987	2,291	2,807	1,323	389.1	309.0	636.9	11,946.8	14.17	35,570.0	26,425.7	62,007.3	2,300.6
1988		2,737	1,351	400.8	321.3	673.3	13,228.6	15.01	40,664.7	28,731.4	69,048.5	1,932.0
1989		2,689	1,399	407.8	315.7	648.8	12,856.5	15.19	39,177.0	26,458.4	65,682.8	2,964.0
1990		2,725	1,394	386.0	308.4	624.1	13,037.0	15.69	38,354.5	26,871.4	64,875.4	3,446.0
1991		2,797	1,395	369.9	290.5	586.5	12,586.6	15.82	38,284.2	25,212.9	63,604.2	3,403.1
1992	2,713	3,246	1,454	400.2	313.4	647.0	13,955.6	15.85	43,951.9	30,927.9	75,070.7	3,647.6
1993		3,216	1,465	417.1	329.5	700.2	15,308.8	16.44	49,857.6	36,164.5	85,696.2	3,967.7
1994		3,188P	1,530P	454.1	364.1	789.4	17,229.1	16.64	57,871.9	43,597.5	100,983.1	4,211.9
1995		3,269P	1,567P	435.1P	344.5P	728.8P	16,567.3P	17.40P	52,708.9P	39,708.0P	91,973.9P	4,325.5P
1996		3,351P	1,605P	442.0P	349.9P	742.2P	17,170.5P	17.80P	54,976.5P	41,416.3P	95,930.8P	4,547.3P
1997		3,433P	1,642P	448.9P	355.4P	755.7P	17,773.8P	18.21P	57,244.1P	43,124.6P	99,887.7P	4,769.0P
1998		3,514P	1,679P	455.8P	360.8P	769.1P	18,377.1P	18.61P	59,511.8P	44,832.9P	103,844.6P	4,990.8P

Sources: 1982, 1987, 1992 *Economic Census*; *Annual Survey of Manufactures*, 83-86, 88-91, 93-94. Establishment counts for non-Census years are from *County Business Patterns*; establishment values for 83-84 are extrapolations. 'P's show projections by the editors. Industries reclassified in 87 will not have data for prior years.

INDICES OF CHANGE

Year	Com-panies	Establishments		Employment			Compensation		Production ($ million)			
		Total	with 20 or more employees	Total (000)	Production Workers (000)	Hours (Mil)	Payroll ($ mil)	Wages ($/hr)	Cost of Materials	Value Added by Manufacture	Value of Shipments	Capital Invest.
1982	74	75	76	80	80	74	55	74	43	54	48	49
1983		73	77	84	85	85	64	77	53	70	59	29
1984		72	78	95	98	100	76	80	66	77	70	47
1985		70	79	96	99	100	83	87	72	84	77	81
1986		70	80	94	95	96	83	89	75	79	76	72
1987	84	86	91	97	99	98	86	89	81	85	83	63
1988		84	93	100	103	104	95	95	93	93	92	53
1989		83	96	102	101	100	92	96	89	86	87	81
1990		84	96	96	98	96	93	99	87	87	86	94
1991		86	96	92	93	91	90	100	87	82	85	93
1992	100	100	100	100	100	100	100	100	100	100	100	100
1993		99	101	104	105	108	110	104	113	117	114	109
1994		98P	105P	113	116	122	123	105	132	141	135	115
1995		101P	108P	109P	110P	113P	119P	110P	120P	128P	123P	119P
1996		103P	110P	110P	112P	115P	123P	112P	125P	134P	128P	125P
1997		106P	113P	112P	113P	117P	127P	115P	130P	139P	133P	131P
1998		108P	115P	114P	115P	119P	132P	117P	135P	145P	138P	137P

Sources: Same as General Statistics. Values reflect change from the base year, 1992. Values above 100 mean greater than 92, values below 100 mean less than 92, and a value of 100 in the 82-91 or 93-98 period means same as 92. 'P's mark projections by the editors.

SELECTED RATIOS

For 1994	Avg. of All Manufact.	Analyzed Industry	Index	For 1994	Avg. of All Manufact.	Analyzed Industry	Index
Employees per Establishment	49	142	291	Value Added per Production Worker	134,084	119,740	89
Payroll per Establishment	1,500,273	5,404,925	360	Cost per Establishment	5,045,178	18,154,941	360
Payroll per Employee	30,620	37,941	124	Cost per Employee	102,970	127,443	124
Production Workers per Establishment	34	114	333	Cost per Production Worker	146,988	158,945	108
Wages per Establishment	853,319	4,120,762	483	Shipments per Establishment	9,576,895	31,679,316	331
Wages per Production Worker	24,861	36,077	145	Shipments per Employee	195,460	222,381	114
Hours per Production Worker	2,056	2,168	105	Shipments per Production Worker	279,017	277,350	99
Wages per Hour	12.09	16.64	138	Investment per Establishment	321,011	1,321,311	412
Value Added per Establishment	4,602,255	13,676,932	297	Investment per Employee	6,552	9,275	142
Value Added per Employee	93,930	96,009	102	Investment per Production Worker	9,352	11,568	124

Sources: Same as General Statistics. The 'Average of All Manufacturing' column represents the average of all manufacturing industries reported for the most recent complete year available. The Index shows the relationship between the Average and the Analyzed Industry. For example, 100 means that they are equal; 500 that the Analyzed Industry is five times the average; 50 means that the Analyzed Industry is half the national average. The abbreviation 'na' is used to show that data are 'not available'.

LEADING COMPANIES Number shown: **75** Total sales ($ mil): **110,812** Total employment (000): **761.3**

Company Name	Address				CEO Name	Phone	Co. Type	Sales ($ mil)	Empl. (000)
ITT Corp	1330 Av Amer	New York	NY	10019	Rand V Araskog	212-258-1000	P	23,620	110.0
TRW Inc	1900 Richmond Rd	Cleveland	OH	44124	Joseph T Gorman	216-291-7000	P	9,087	64.2
Dana Corp	PO Box 1000	Toledo	OH	43697	S J Morcott	419-535-4500	P	6,610	40.0
Eaton Corp	Eaton Ctr	Cleveland	OH	44114	William E Butler	216-523-5000	P	6,052	51.0
Delphi Interior & Lighting Syst	6600 E 12 Mile Rd	Warren	MI	48092	Paul J Tosch	810-578-3247	D	5,150	31.5
ITT Automotive Inc	3000 University Dr	Auburn Hills	MI	48326	Timothy D Leuliette	810-340-3000	S	4,784	35.0
Packard Electric	PO Box 431	Warren	OH	44486	R A Schlais Jr	216-373-2121	D	4,000	65.0
Rockwell International Corp	2135 W Maple Rd	Troy	MI	48084	R L Roudebush	810-435-1000	D	3,991	17.3
Rockwell Automotive	2135 W Maple Rd	Troy	MI	48084	Larry D Yost	810-435-1000	D	2,536	17.0
United Technologies Automotive	5200 Auto Club Dr	Dearborn	MI	48126	Norman R Bodine	313-593-9600	S	2,400	32.0
Varity Corp	672 Delaware Av	Buffalo	NY	14209	Victor A Rice	716-888-8000	P	2,268	10.5
Arvin Industries Inc	PO Box 3000	Columbus	IN	47202	Byron O Pond	812-379-3000	P	2,040	13.3
Ford Motor Co	14425 Sheldon Rd	Plymouth	MI	48170	Frank J Croskey	313-451-8750	D	2,000	11.8
Mark IV Industries Inc	PO Box 810	Amherst	NY	14226	Sal H Alfiero	716-689-4972	P	1,913	16.2
Federal-Mogul Corp	PO Box 1966	Detroit	MI	48235	Dennis J Gormley	313-354-7700	P	1,896	16.2
Tenneco Automotive	100 Tri-State	Lincolnshire	IL	60069	Richard A Snell	708-948-0900	D	1,808	14.8
Nippondenso America Inc	24777 Denso Dr	Southfield	MI	48086	Akira Kataoka	313-350-7500	S	1,800	6.0
Robert Bosch Corp	2800 S 25th Av	Broadview	IL	60153	Ranier Hahn	708-865-5200	S	1,800	5.5
MascoTech Inc	21001 Van Born Rd	Taylor	MI	48180	R A Manoogian	313-274-7405	P	1,702	12.7
AO Smith Corp	PO Box 23972	Milwaukee	WI	53223	Robert J O'Toole	414-359-4000	P	1,373	12.1
Collins & Aikman	8320 University	Charlotte	NC	28262	S A Schwarzman	704-548-2350	P	1,306	12.0
New Venture Gear Inc	1650 Research Dr	Troy	MI	48083	EA Reickert	313-680-4900	R	1,033	4.0
TRW Steering & Suspension Syst	PO Box 8008	Sterling Hts	MI	48311	James Remick	313-977-1000	D	1,000	2.0
Borg-Warner Automotive Inc	200 S Michigan Av	Chicago	IL	60604	J Gordon Amedee	312-322-8500	P	985	6.6
Modine Manufacturing Co	1500 DeKoven Av	Racine	WI	53403	Richard T Savage	414-636-1200	P	913	7.6
Standard Products Co	2130 W 110th St	Cleveland	OH	44102	James S Reid Jr	216-281-8300	P	872	9.5
Allison Transmission	PO Box 894	Indianapolis	IN	46206	John F Smith	317-242-5000	D	850	4.2
Alco Fujikura Ltd	105 Westpark Dr	Brentwood	TN	37027	Robert H Barton	615-370-2100	S	825	12.0
TBG Industries Inc	565 5th Av	New York	NY	10017	William Blue	212-850-8500	S	770•	7.0
Moog Automotive Inc	PO Box 7224	St Louis	MO	63177	John Collins	314-385-3400	S	665	3.5
Standard Motor Products Inc	37-18 Northern Blv	Long Island Ct	NY	11101	Lawrence I Sills	718-392-0200	P	641	3.3
AO Smith Automotive Prod Co	PO Box 584	Milwaukee	WI	53201	Robert J O'Toole	414-447-4000	D	606	3.0
Hayes Wheels International Inc	38481 Huron River	Romulus	MI	48174	Ranko Cucuz	313-941-2000	P	538	3.0
Rexnord Corp	4701 W Greenfield	Milwaukee	WI	53214	James R Swenson	414-643-3000	P	533	4.8
Morton International	3350 Airport Rd	Ogden	UT	84405	K D Holmgren	801-625-4800	D	524	2.5
Excel Industries Inc	PO Box 3118	Elkhart	IN	46515	James J Lohman	219-264-2131	P	516	3.4
Automotive Industries Holding	4508 IDS Ctr	Minneapolis	MN	55402	Rick Sommer	612-332-6828	P	513	4.1
Superior Industries International	7800 Woodley Av	Van Nuys	CA	91406	Louis L Borick	818-781-4973	P	457	4.5
T and N Industries Inc	777 E Eisenhower	Ann Arbor	MI	48108	Rita Grisham	313-663-6749	S	450	5.2
Brake Parts Inc	4400 Prime Pkwy	McHenry	IL	60050	Larry Pavey	815-363-9000	D	440	3.0
Purolator Products Co	6120 S Yale Av	Tulsa	OK	74136	Roman E Boruta	918-481-2500	P	436	3.6
Powertrain Systems	5401 Kilgore Av	Muncie	IN	47304	Alan Straub	317-286-6100	D	400	2.1
Rockwell International Corp	747 Advance St	Brighton	MI	48116	Wolpram Ruppel	313-227-2001	D	400	<0.1
Woodbridge Holdings Inc	105 Terry Dr	Newtown	PA	18940	TR Beamish	215-968-1400	S	380	2.0
First Brands Corp	PO Box 1911	Danbury	CT	06813	Ray O Pinion	203-731-2300	D	379	0.2
Spicer Driveshaft	PO Box 955	Toledo	OH	43697	Robert Fesenmyer	419-866-2600	D	350•	2.7
Spicer Heavy Axle	PO Box 2229	Fort Wayne	IN	46801	William Hoffmann	219-483-7174	D	350	0.9
Automotive Industries Inc	2998 Waterview Dr	Rochester Hills	MI	48309	F F Sommer	810-853-3040	S	349	4.1
Parish Light Vehicle	PO Box 13459	Reading	PA	19612	Michael Greene	215-371-7000	D	340	2.8
Truck Components Inc	302 Peoples Av	Rockford	IL	61104	Thomas W Cook	815-964-8725	P	313	2.0
Purolator Prod Co	6120 S Yale Av	Tulsa	OK	74136	Steve Thies	918-481-2300	D	301	2.0
Wynn's International Inc	PO Box 14143	Orange	CA	92668	James Carroll	714-938-3700	P	293	2.1
Stant Corp	425 Commerce Dr	Richmond	IN	47374	David R Paridy	317-962-6655	P	288	8.1
Breed Technologies Inc	PO Box 95023	Lakeland	FL	33804	Allen K Breed	813-683-2412	P	278	3.2
Clarcor Inc	PO Box 7007	Rockford	IL	61125	Lawrence E Gloyd	815-962-8867	P	270	2.2
Alphabet	8700 E Market St	Warren	OH	44484	D Draime	216-856-3366	S	270	2.5
Stoneridge Inc	9400 E Market St	Warren	OH	44484	David M Draime	216-856-2443	S	270•	2.5
ASC Inc	1 Sunroof Ctr	Southgate	MI	48195	Heinz C Prechter	313-285-4911	R	260•	3.0
Motor Wheel Corp	2501 Woodlake Cir	Okemos	MI	48864	Joseph C Overbeck	517-337-5700	R	260	2.0
American Racing Equipment Inc	19067 S Reyes Av	R Dominguez	CA	90221	John Onder	310-635-7806	S	250	2.0
Walbro Automotive Corp	PO Box 215257	Auburn Hills	MI	48231	Gary L Vollmar	313-377-1800	S	250	1.5
Champion Laboratories Inc	4th & Walnut Sts	Albion	IL	62806	Thomas A Mowatt	618-445-6011	S	240•	2.2
Gabriel Ride Control Prod	100 Westwood Pl	Brentwood	TN	37027	E Leon Viars	615-221-7433	D	240	2.4
Atwood Industries Inc	1400 Eddy Av	Rockford	IL	61103	Kraig D Pierceson	815-877-5771	S	220	2.4
Cardone Industries Inc	5670 Rising Sun Av	Philadelphia	PA	19120	Michael Cardone	215-722-9700	R	220•	2.0
EIS Brake Parts	129 Worth	Berlin	CT	06037	Dan Carboni	203-828-8290	D	200	0.5
Federal-Mogul Corp	2841 N Spring Av	St Louis	MO	63107	WW Bilkey Jr	314-289-7571	D	200	0.6
Lacks Industries Inc	5460 Cascade SE	Grand Rapids	MI	49546	Richard Lacks Sr	616-949-6570	S	200	1.4
Midland Brake Inc	10930 N Pomona	Kansas City	MO	64153	Ronald E Seufert	816-891-2470	S	200	1.5
Tokico	17225 Federal Dr	Allen Park	MI	48101	K Murofushi	313-336-5310	S	200	0.5
Holley Automotive	11955 E 9 Mile Rd	Warren	MI	48090	Philip M Peters	810-497-4000	D	195	0.9
Harman Automotive Inc	30665 Northwestern	Farmington Hls	MI	48334	Brian D Benninger	313-626-4300	S	190•	1.8
MascoTech Automotive Syst	275 Rex Blv	Auburn Hills	MI	48326	Alfred Grava	313-852-5700	S	190•	2.3
Clark Automotive Products Corp	PO Box 7008	South Bend	IN	46634	F E Bertolaccini	219-239-0155	P	184	2.4
AP Parts Marketing Co	1800 Indian	Maumee	OH	43537	James Ashford	419-891-8400	R	180	1.7

Source: *Ward's Business Directory of U.S. Private and Public Companies*, Volumes 1 and 2, 1996. The company type code used is as follows: P - Public, R - Private, S - Subsidiary, D - Division, J - Joint Venture, A - Affiliate, G - Group. Sales are in millions of dollars, employees are in thousands. An asterisk (•) indicates an estimated sales volume. The symbol < stands for 'less than'. Company names and addresses are truncated, in some cases, to fit into the available space.

MATERIALS CONSUMED

Material	Quantity	Delivered Cost ($ million)
Materials, ingredients, containers, and supplies	(X)	40,220.9
Fluid power pumps, motors, and hydrostatic transmissions (hydraulic and pneumatic)	(X)	226.8
Fluid power valves (hydraulic and pneumatic)	(X)	93.9
Fluid power cylinders and rotary actuators (hydraulic and pneumatic)	(X)	65.8
Fluid power hose or tube fittings and assemblies (hydraulic and pneumatic)	(X)	165.0
Fluid power filters (hydraulic and pneumatic)	(X)	15.3
Other fluid power products (hydraulic and pneumatic)	(X)	97.5
Automotive stampings (including body parts, hubcaps, fenders, etc.)	(X)	2,159.2
Metal bolts, nuts, screws, washers, rivets, and other screw machine products	(X)	983.2
Other fabricated metal products, except fluid power and forgings	(X)	4,510.8
Forgings	(X)	1,550.2
Iron and steel castings (rough and semifinished)	(X)	4,266.4
Aluminum and aluminum-base alloy castings (rough and semifinished)	(X)	2,451.0
Other nonferrous castings (rough and semifinished)	(X)	246.7
Steel bars, bar shapes, and plates	(X)	1,133.9
Steel sheet and strip, including tin plate	(X)	1,948.2
All other steel shapes and forms	(X)	788.3
Copper and copper-base alloy shapes and forms	(X)	251.2
Aluminum and aluminum-base alloy shapes and forms	(X)	541.8
Other nonferrous shapes and forms	(X)	376.1
Ball bearings (mounted or unmounted)	(X)	285.9
Roller bearings (mounted or unmounted)	(X)	615.3
Fabricated plastics products (except gaskets, hoses, and belting)	(X)	706.6
Plastics products consumed in the form of sheets, rods, tubes, film, and other shapes	(X)	147.7
Plastics resins consumed in the form of granules, pellets, powders, liquids, etc.	(X)	481.1
Fabricated rubber products, except tires, tubes, hose, belting, and gaskets	(X)	430.1
Rubber and plastics hose and belting	(X)	197.1
Ceramic raw materials, including powders, chemicals, and fibers (excluding refractory uses)	(X)	332.3
Ceramic and ceramic composite parts, components, and accessories	(X)	124.0
Gaskets (all types), and packing and sealing devices	(X)	232.0
Paints, varnishes, lacquers, stains, shellacs, japans, enamels, and allied products	(X)	403.5
Glues and adhesives	(X)	41.7
Flexible packaging materials	(X)	59.1
Paper and paperboard containers	(X)	347.8
Engine electrical equipment, including spark plugs, magnetos, generators, starters, etc.	(X)	1,516.8
Resistors, capacitors, transformers, electron tubes, semiconductors, and other electronic components	(X)	1,285.9
All other materials and components, parts, containers, and supplies	(X)	9,967.7
Materials, ingredients, containers, and supplies, nsk	(X)	1,174.9

Source: 1992 *Economic Census*. Explanation of symbols used: (D): Withheld to avoid disclosure of competitive data; na: Not available; (S): Withheld because statistical norms were not met; (X): Not applicable; (Z): Less than half the unit shown; nec: Not elsewhere classified; nsk: Not specified by kind; - : zero; * : 10-19 percent estimated; ** : 20-29 percent estimated.

PRODUCT SHARE DETAILS

Product or Product Class	% Share	Product or Product Class	% Share
Motor vehicle parts and accessories	100.00	Gasoline engine cooling fans (including hubs and clutches), new, for motor vehicles	0.82
Gasoline engines and gasoline engine parts for motor vehicles, new	23.00	Gasoline engine radiators, complete, new, for motor vehicles	4.55
Gasoline engines, new (with or without cylinder heads, fuel pumps, water pumps, and other standard accessories), for motor vehicles	61.21	Gasoline engine radiator shells and cores, new, for motor vehicles	0.41
Intake manifolds complete, produced on purchased chassis	0.87	Gasoline engine thermostats (engine cooling system), new, for motor vehicles	(D)
Gasoline engine exhaust manifolds, new, for motor vehicles	0.86	Gasoline engine PCV (positive crankcase ventilation) valves, new, for motor vehicles	0.10
Gasoline engine crankshafts, new, for motor vehicles	0.35	All other parts and accessories for gasoline engines, new, for motor vehicles	10.48
Gasoline engine camshafts, new, for motor vehicles	0.77	Gasoline engines and gasoline engine parts for motor vehicles, new, nsk	0.70
Gasoline engine rocker arms and parts, new, for motor vehicles	0.32	Filters for internal combustion engines and motor vehicles, new	2.89
Gasoline engine valve guides, seats, and tappets, new for motor vehicles	1.28	Oil filters for internal combustion engines and motor vehicles, new, light-duty (car and light truck)	30.86
Gasoline engine fuel injection systems, new, for motor vehicles	6.15	Oil filters for internal combustion engines and motor vehicles, new, heavy-duty	13.42
Gasoline engine flywheels and flexplates, new, for motor vehicles	0.21	Fuel filters for internal combustion engines and motor vehicles, new, light-duty (car and light truck)	9.56
Gasoline engine timing gears, sprockets, and chains, new, for motor vehicles	0.51	Fuel filters for internal combustion engines and motor vehicles, new, heavy-duty	7.30
Gasoline engine main engine bearings (halves), new, for motor vehicles	0.56	Air filters for internal combustion engines and motor vehicles, new, light-duty (car and light truck)	12.42
Gasoline engine connecting rod, engine bearings (halves), new, for motor vehicles	0.53	Air filters for internal combustion engines and motor vehicles, new, heavy-duty	15.32
Other gasoline engine bearings (halves) (balance shaft, camshaft, etc.), new, for motor vehicles	(D)	Other filters for internal combustion engines and motor vehicles, new, including coolant and hydraulic	10.90
Gasoline engine oil pumps, new, for motor vehicles	0.49	Filters for internal combustion engines and motor vehicles, new, nsk	0.21
Gasoline engine power steering pumps, new, for motor vehicles	(D)	Exhaust system parts, new	4.27
Gasoline engine fuel pump assemblies (excluding kits), new, for motor vehicles	3.73		
Gasoline engine water pump assemblies (excluding kits), new, for motor vehicles	0.80		

Continued on next page.

PRODUCT SHARE DETAILS - Continued

Product or Product Class	% Share	Product or Product Class	% Share
Exhaust system mufflers, including standard, sports or glass pack, and resonators, new, for motor vehicles	28.99	Motor vehicle brake rotors/discs (with or without hub), sold separately, new	5.78
Exhaust system pipes, including exhaust, intermediate, connecting, crossover, tail, and side pipes, new, for motor vehicles	19.69	Motor vehicle brake shoes (with or without lining), sold separately, new	2.68
Exhaust system catalytic converters, new, for motor vehicles	50.90	Motor vehicle metallic or semimetallic brake linings, except asbestos, new	6.94
Exhaust system parts, new, nsk	0.42	Motor vehicle air brake power actuation units, new	1.71
Drive train components, new, except wheels and brakes	26.32	Motor vehicle hydraulic brake power actuation units, new	(D)
Car and light truck drive train components, manual transmissions (except auxiliary and parts), new, except wheels and brakes	1.68	Motor vehicle vacuum brake power actuation units, new	(D)
		Other motor vehicle brake parts, new	23.12
Car and light truck drive train components, automatic transmissions (except auxiliary and parts), new, except wheels and brakes	22.59	Brake parts and assemblies, new, nsk	4.56
		All other motor vehicle parts and accessories, new, nec	27.49
		Motor vehicle bumper assemblies, bumpers, and parts, new	4.45
Heavy truck and bus drive train components, manual transmissions (except auxiliary and parts), new, except wheels and brakes	2.85	Automotive frames, new	4.33
		Motor vehicle fuel tanks, new	1.99
Heavy truck and bus drive train components, automatic transmissions (except auxiliary and parts), new, except wheels and brakes	(D)	Motor vehicle heaters, heater cores, and other heater parts, new	1.56
		Motor vehicle shock absorbers, new	3.67
Heavy truck and bus drive train components, parts for manual transmissions (except auxiliary and parts), new, except wheels and brakes	0.99	Motor vehicle tie rod ends, new	0.90
		Motor vehicle steering idler arms, drag links, and control arms, new	1.99
Heavy truck and bus drive train components, parts for automatic transmissions (except auxiliary and parts), new, except wheels and brakes	9.21	Motor vehicle ball joints, new	0.88
		Automotive air-conditioning hose assemblies, new	2.13
		Automotive power steering hose assemblies, new	0.45
Motor vehicle drive train components, transaxles, new, except wheels and brakes	(D)	Automotive brake hose assemblies, new	0.36
		Motor vehicle cruise control units, new	1.12
Motor vehicle drive train components, clutch disc and facing assemblies, new, except wheels and brakes	1.98	Motor vehicle windshield wiper blades, new	0.96
		Motor vehicle windshield washer pumps, new	0.20
Motor vehicle drive train components, gear shifters, new, except wheels and brakes	1.37	Motor vehicle steering wheels, columns, and gearboxes, new	6.49
		Motor vehicle convertible tops, new	0.93
Motor vehicle drive train components, drive shafts, new, except wheels and brakes	4.34	Motor vehicle sunroofs and parts, new	0.19
		Motor vehicle fifth wheels, new	0.53
Motor vehicle drive train components, universal joints, new, except wheels and brakes	(D)	Automotive air bag assemblies and parts thereof, new	7.71
Motor vehicle drive train components, axles and axle parts, new, except wheels and brakes	29.69	All other motor vehicle parts and accessories for cars, trucks, and buses, new	55.00
Motor vehicle drive train components, wheel hubs, sold separately, new, except wheels and brakes	0.65	All other motor vehicle parts and accessories, new, nec, nsk	4.16
		Rebuilt parts for motor vehicles, excluding carburetors and engine electrical equipment	1.85
Other motor vehicle drive train components, new, except wheels and brakes	9.87	Rebuilt motor vehicle fuel pumps	0.14
		Rebuilt motor vehicle water pumps	7.67
Drive train components, new, except wheels and brakes, nsk	0.55	Rebuilt motor vehicle clutch discs and pressure plates	8.67
Motor vehicle wheels, new	2.61	Car and light truck rebuilt gasoline engines	15.67
Car and light truck wheels, aluminum, new	43.63	Heavy truck and bus rebuilt gasoline engines	2.01
Other car and light truck wheels, including combination, new	37.64	Car and light truck rebuilt automatic transmissions, including drive lines and axles	4.73
Heavy truck and bus type wheels, including those for truck trailers and trailer coaches	17.04	Car and light truck rebuilt manual (standard) transmissions, including drive lines and axles	0.36
Motor vehicle wheels, new, nsk	1.69	Heavy truck and bus rebuilt transmissions, including drive lines and axles	0.28
Motor vehicle brake parts and assemblies, new	8.06	Rebuilt motor vehicle brake shoe assemblies (drum brake)	7.81
Motor vehicle wheel brake cylinders, sold separately, new	1.58	Rebuilt motor vehicle brake caliper assemblies (disc brake)	7.38
Motor vehicle master brake cylinders, sold separately, new	5.25	Rebuilt motor vehicle brake master cylinders	3.22
Motor vehicle brake valves, new	(D)	Rebuilt motor vehicle air brake power actuation units	(D)
Motor vehicle brake assemblies (drum), including backing plates, shoes, linings (except asbestos), cylinders, etc., if sold together, new	7.15	Rebuilt motor vehicle vacuum brake power actuation units	(D)
		Rebuilt motor vehicle hydraulic brake power actuation units	0.20
Motor vehicle brake assemblies (disc/caliper), including rotors, calipers, pads (except asbestos), cylinders, etc., if sold together, new	22.56	Rebuilt motor vehicle power steering pumps	3.79
		Rebuilt motor vehicle rack and pinion steering assemblies	4.04
		Other rebuilt motor vehicle parts, excluding carburetors and engine electrical equipment	22.21
Motor vehicle brake drums (with or without hub), sold separately, new	5.96	Rebuilt parts for motor vehicles, excluding carburetors and engine electrical equipment, nsk	7.95
		Motor vehicle parts and accessories, nsk	3.51

Source: 1992 *Economic Census*. The values shown are percent of total shipments in an industry. Values of indented subcategories are summed in the main heading. The symbol (D) appears when data are withheld to prevent disclosure of competitive information. The abbreviation nsk stands for 'not specified by kind' and nec for 'not elsewhere classified'.

INPUTS AND OUTPUTS FOR MOTOR VEHICLE PARTS & ACCESSORIES

Economic Sector or Industry Providing Inputs	%	Sector	Economic Sector or Industry Buying Outputs	%	Sector
Imports	19.3	Foreign	Motor vehicles & car bodies	36.9	Manufg.
Motor vehicle parts & accessories	18.4	Manufg.	Automotive repair shops & services	19.8	Services
Blast furnaces & steel mills	8.7	Manufg.	Exports	15.9	Foreign
Iron & steel foundries	7.8	Manufg.	Motor vehicle parts & accessories	11.8	Manufg.
Automotive repair shops & services	3.7	Services	Personal consumption expenditures	4.4	
Advertising	2.6	Services	Change in business inventories	2.0	In House
Wholesale trade	2.5	Trade	Sanitary services, steam supply, irrigation	0.8	Util.
Aluminum castings	2.4	Manufg.	Truck trailers	0.5	Manufg.
Carburetors, pistons, rings, & valves	2.4	Manufg.	Motor freight transportation & warehousing	0.5	Util.
Machinery, except electrical, nec	2.2	Manufg.	Retail trade, except eating & drinking	0.5	Trade
Iron & steel forgings	1.9	Manufg.	Wholesale trade	0.5	Trade
Electric services (utilities)	1.9	Util.	Mobile homes	0.3	Manufg.
Miscellaneous plastics products	1.8	Manufg.	Local government passenger transit	0.3	Gov't
Power transmission equipment	1.4	Manufg.	U.S. Postal Service	0.3	Gov't
Noncomparable imports	1.4	Foreign	S/L Govt. purch., elem. & secondary education	0.3	S/L Govt
Ball & roller bearings	1.0	Manufg.	Feed grains	0.2	Agric.
Screw machine and related products	1.0	Manufg.	Transit & bus transportation	0.2	Util.
Automotive stampings	0.9	Manufg.	Automotive rental & leasing, without drivers	0.2	Services
Maintenance of nonfarm buildings nec	0.8	Constr.	Federal Government purchases, national defense	0.2	Fed Govt
Engine electrical equipment	0.8	Manufg.	S/L Govt. purch., other general government	0.2	S/L Govt
Petroleum refining	0.8	Manufg.	S/L Govt. purch., police	0.2	S/L Govt
Plastics materials & resins	0.8	Manufg.	Meat animals	0.1	Agric.
Gas production & distribution (utilities)	0.7	Util.	Maintenance of nonfarm buildings nec	0.1	Constr.
Used & secondhand goods	0.7	Scrap	Farm machinery & equipment	0.1	Manufg.
Asbestos products	0.6	Manufg.	Motorcycles, bicycles, & parts	0.1	Manufg.
Copper rolling & drawing	0.6	Manufg.	Transportation equipment, nec	0.1	Manufg.
Fabricated rubber products, nec	0.6	Manufg.	Travel trailers & campers	0.1	Manufg.
Plating & polishing	0.6	Manufg.			
Motor freight transportation & warehousing	0.6	Util.			
Primary copper	0.5	Manufg.			
Refrigeration & heating equipment	0.5	Manufg.			
Sheet metal work	0.5	Manufg.			
U.S. Postal Service	0.5	Gov't			
Communications, except radio & TV	0.4	Util.			
Eating & drinking places	0.4	Trade			
Equipment rental & leasing services	0.4	Services			
Abrasive products	0.3	Manufg.			
Aluminum rolling & drawing	0.3	Manufg.			
Gaskets, packing & sealing devices	0.3	Manufg.			
Special dies & tools & machine tool accessories	0.3	Manufg.			
Retail trade, except eating & drinking	0.3	Trade			
Banking	0.3	Fin/R.E.			
Computer & data processing services	0.3	Services			
Chemical preparations, nec	0.2	Manufg.			
Miscellaneous fabricated wire products	0.2	Manufg.			
Motors & generators	0.2	Manufg.			
Nonferrous wire drawing & insulating	0.2	Manufg.			
Paints & allied products	0.2	Manufg.			
Paperboard containers & boxes	0.2	Manufg.			
Primary aluminum	0.2	Manufg.			
Primary metal products, nec	0.2	Manufg.			
Semiconductors & related devices	0.2	Manufg.			
Legal services	0.2	Services			
Miscellaneous repair shops	0.2	Services			
Electronic components nec	0.1	Manufg.			
Lubricating oils & greases	0.1	Manufg.			
Metal coating & allied services	0.1	Manufg.			
Metal heat treating	0.1	Manufg.			
Pipe, valves, & pipe fittings	0.1	Manufg.			
Rubber & plastics hose & belting	0.1	Manufg.			
Insurance carriers	0.1	Fin/R.E.			
Real estate	0.1	Fin/R.E.			
Accounting, auditing & bookkeeping	0.1	Services			
Automotive rental & leasing, without drivers	0.1	Services			
Management & consulting services & labs	0.1	Services			

Source: Benchmark Input-Output Accounts for the U.S. Economy, 1982, U.S. Department of Commerce, Washington, D.C., July 1991. Data, as reported in the source, are organized by the 1977 SIC structure in use in 1982 but have been matched, as closely as is possible, to the 1987 SIC structure used in this book.

OCCUPATIONS EMPLOYED BY SIC 371 - MOTOR VEHICLES AND EQUIPMENT

Occupation	% of Total 1994	Change to 2005	Occupation	% of Total 1994	Change to 2005
Assemblers, fabricators, & hand workers nec	21.4	-12.2	Tool & die makers	1.6	-8.8
Inspectors, testers, & graders, precision	3.4	-51.1	Mechanics, installers, & repairers nec	1.6	-29.8
Engineers nec	3.4	5.4	Machine tool cutting & forming etc. nec	1.6	5.4
Blue collar worker supervisors	3.4	-11.4	Industrial truck & tractor operators	1.6	-29.8
Machine operators nec	2.9	-30.3	Electricians	1.6	-17.6
Machine tool cutting operators, metal & plastic	2.6	-55.0	Combination machine tool operators	1.5	40.5
Welders & cutters	2.4	-29.8	Machine forming operators, metal & plastic	1.5	-56.1
Managers & administrators nec	2.1	-12.2	Industrial machinery mechanics	1.4	-42.9
Engineering technicians & technologists nec	1.8	-12.2	Sales & related workers nec	1.2	-12.2
Helpers, laborers, & material movers nec	1.8	-12.2	Machine builders	1.1	31.7
Metal & plastic machine workers nec	1.7	55.6	Sheet metal workers & duct installers	1.0	-56.1
Welding machine setters, operators	1.7	-12.2			

Source: Industry-Occupation Matrix, Bureau of Labor Statistics. These data relate to one or more 3-digit SIC industry groups rather than to a single 4-digit SIC. The change reported for each occupation to the year 2005 is a percent of growth or decline as estimated by the Bureau of Labor Statistics. The abbreviation nec stands for 'not elsewhere classified'.

LOCATION BY STATE AND REGIONAL CONCENTRATION

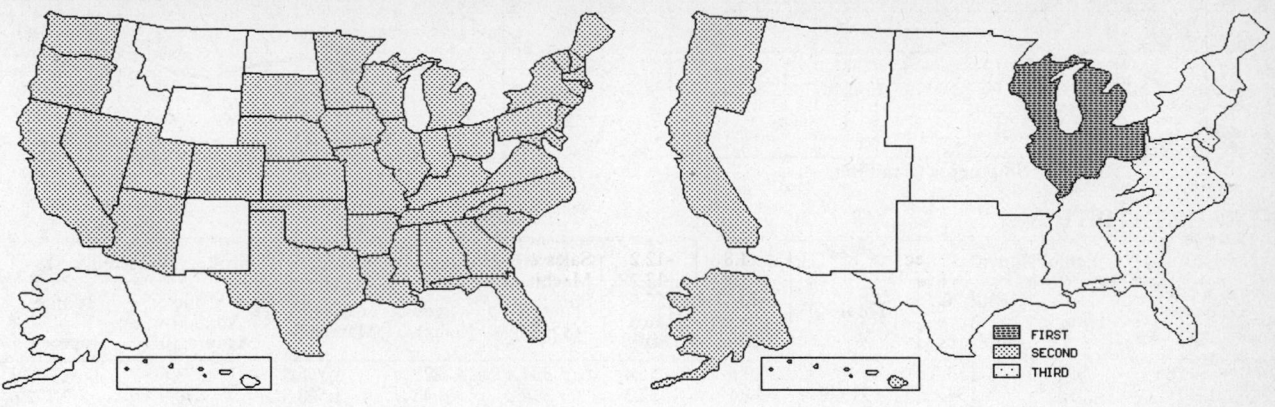

FIRST
SECOND
THIRD

INDUSTRY DATA BY STATE

State	Establish-ments	Shipments			Employment				Cost as % of Shipments	Investment per Employee ($)
		Total ($ mil)	% of U.S.	Per Establ.	Total Number	% of U.S.	Per Establ.	Wages ($/hour)		
Michigan	434	21,907.3	29.2	50.5	104,900	26.2	242	19.65	61.4	8,567
Ohio	247	13,591.2	18.1	55.0	55,400	13.8	224	18.51	61.5	21,755
Indiana	194	8,416.9	11.2	43.4	50,300	12.6	259	16.64	52.3	9,062
North Carolina	83	2,553.8	3.4	30.8	12,600	3.1	152	12.25	58.8	7,341
Illinois	133	1,998.6	2.7	15.0	14,100	3.5	106	10.66	50.9	-
Wisconsin	66	1,927.4	2.6	29.2	9,700	2.4	147	16.64	59.2	5,082
California	451	1,680.8	2.2	3.7	15,900	4.0	35	9.91	51.5	-
Pennsylvania	88	1,245.6	1.7	14.2	8,500	2.1	97	17.23	44.7	3,753
Mississippi	40	940.0	1.3	23.5	5,700	1.4	142	10.64	59.4	2,404
Alabama	41	937.6	1.2	22.9	5,800	1.4	141	17.92	51.2	2,655
Virginia	34	930.6	1.2	27.4	5,900	1.5	174	12.36	58.1	4,695
Iowa	49	757.4	1.0	15.5	4,700	1.2	96	12.18	44.7	3,596
Arkansas	38	688.7	0.9	18.1	4,900	1.2	129	8.92	51.7	3,490
Texas	173	652.4	0.9	3.8	5,600	1.4	32	9.98	65.3	5,125
Georgia	73	626.0	0.8	8.6	4,700	1.2	64	10.21	50.9	4,553
Nebraska	15	399.8	0.5	26.7	2,400	0.6	160	10.20	39.9	-
Arizona	50	329.1	0.4	6.6	2,000	0.5	40	8.38	57.6	6,750
Florida	103	289.2	0.4	2.8	2,500	0.6	24	8.14	49.7	-
Kansas	23	259.0	0.3	11.3	2,000	0.5	87	10.03	52.4	-
Connecticut	28	251.3	0.3	9.0	2,400	0.6	86	12.38	50.8	-
Oregon	48	212.4	0.3	4.4	1,600	0.4	33	14.81	61.1	-
Tennessee	111	(D)	-	-	17,500 *	4.4	158	-	-	-
New York	109	(D)	-	-	17,500 *	4.4	161	-	-	4,423
Missouri	97	(D)	-	-	17,500 *	4.4	180	-	-	2,989
New Jersey	66	(D)	-	-	1,750 *	0.4	27	-	-	-
Washington	56	(D)	-	-	750 *	0.2	13	-	-	-
Kentucky	55	(D)	-	-	7,500 *	1.9	136	-	-	-
Oklahoma	52	(D)	-	-	3,750 *	0.9	72	-	-	3,573
Colorado	46	(D)	-	-	1,750 *	0.4	38	-	-	-
Minnesota	43	(D)	-	-	1,750 *	0.4	41	-	-	-
Massachusetts	34	(D)	-	-	750 *	0.2	22	-	-	-
South Carolina	32	(D)	-	-	7,500 *	1.9	234	-	-	-
Maryland	26	(D)	-	-	750 *	0.2	29	-	-	1,467
Louisiana	19	(D)	-	-	375 *	0.1	20	-	-	-
Utah	18	(D)	-	-	3,750 *	0.9	208	-	-	-
Nevada	12	(D)	-	-	375 *	0.1	31	-	-	-
Maine	11	(D)	-	-	175 *	0.0	16	-	-	-
Rhode Island	8	(D)	-	-	375 *	0.1	47	-	-	-
South Dakota	6	(D)	-	-	175 *	0.0	29	-	-	-
New Hampshire	4	(D)	-	-	750 *	0.2	188	-	-	-
Vermont	3	(D)	-	-	375 *	0.1	125	-	-	-

Source: 1992 Economic Census. The states are in descending order of shipments or establishments (if shipment data are missing for the majority). The symbol (D) appears when data are withheld to prevent disclosure of competitive information. States marked with (D) are sorted by number of establishments. A dash (-) indicates that the data element cannot be calculated; * indicates the midpoint of a range.

3715 - TRUCK TRAILERS

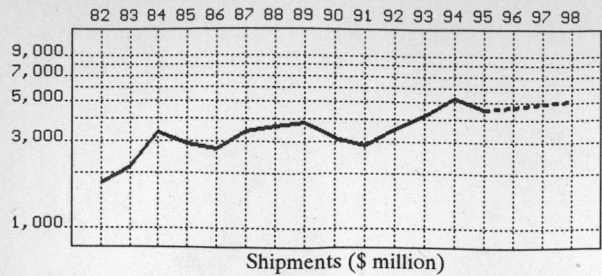

Shipments ($ million)

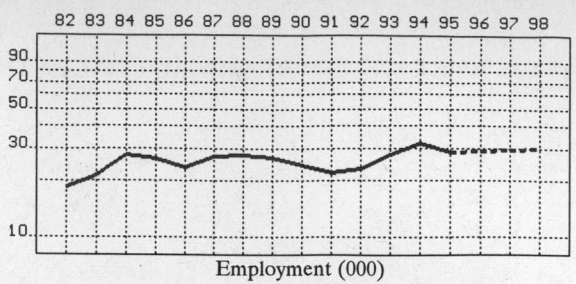

Employment (000)

GENERAL STATISTICS

Year	Com-panies	Establishments		Employment			Compensation		Production ($ million)			
		Total	with 20 or more employees	Total (000)	Production Workers (000)	Hours (Mil)	Payroll ($ mil)	Wages ($/hr)	Cost of Materials	Value Added by Manufacture	Value of Shipments	Capital Invest.
1982	285	324	153	18.5	13.6	26.4	334.4	8.29	1,166.8	573.6	1,773.3	37.3
1983		315	154	21.6	16.4	32.3	394.0	8.42	1,508.3	670.9	2,171.6	38.8
1984		306	155	27.9	22.1	43.5	524.0	8.70	2,302.6	1,036.5	3,311.1	35.8
1985		298	156	26.4	20.4	39.6	533.1	9.49	2,116.8	817.1	2,931.8	66.1
1986		304	156	23.6	18.1	36.0	492.4	9.66	1,898.8	786.8	2,714.3	31.6
1987	308	337	182	27.5	21.7	43.1	570.2	9.43	2,314.3	1,130.8	3,433.5	45.7
1988		332	183	27.7	21.8	42.5	573.0	9.48	2,566.1	1,069.7	3,634.8	57.2
1989		311	177	27.2	22.0	43.1	594.1	9.72	2,747.6	1,070.6	3,828.8	45.2
1990		317	179	24.4	19.4	38.7	548.2	9.80	2,248.7	869.0	3,122.0	37.7
1991		327	169	22.1	16.9	34.5	515.2	9.97	2,066.4	742.0	2,832.4	45.6
1992	310	339	157	23.4	18.7	38.4	565.9	10.34	2,459.0	1,093.5	3,545.5	30.9
1993		351	160	27.9	23.0	47.8	692.1	10.58	3,105.6	1,070.5	4,172.2	66.7
1994		339P	174P	32.1	26.6	56.0	812.1	10.71	3,821.6	1,487.1	5,273.7	116.1
1995		342P	175P	28.6P	23.5P	48.8P	727.1P	10.89P	3,298.4P	1,283.5P	4,551.7P	71.2P
1996		345P	176P	29.1P	24.0P	50.0P	752.4P	11.07P	3,429.3P	1,334.4P	4,732.3P	74.2P
1997		347P	178P	29.5P	24.5P	51.3P	777.7P	11.26P	3,560.1P	1,385.3P	4,912.8P	77.2P
1998		350P	179P	30.0P	25.0P	52.5P	803.0P	11.45P	3,690.9P	1,436.2P	5,093.3P	80.2P

Sources: 1982, 1987, 1992 *Economic Census*; *Annual Survey of Manufactures*, 83-86, 88-91, 93-94. Establishment counts for non-Census years are from *County Business Patterns*; establishment values for 83-84 are extrapolations. 'P's show projections by the editors. Industries reclassified in 87 will not have data for prior years.

INDICES OF CHANGE

Year	Com-panies	Establishments		Employment			Compensation		Production ($ million)			
		Total	with 20 or more employees	Total (000)	Production Workers (000)	Hours (Mil)	Payroll ($ mil)	Wages ($/hr)	Cost of Materials	Value Added by Manufacture	Value of Shipments	Capital Invest.
1982	92	96	97	79	73	69	59	80	47	52	50	121
1983		93	98	92	88	84	70	81	61	61	61	126
1984		90	99	119	118	113	93	84	94	95	93	116
1985		88	99	113	109	103	94	92	86	75	83	214
1986		90	99	101	97	94	87	93	77	72	77	102
1987	99	99	116	118	116	112	101	91	94	103	97	148
1988		98	117	118	117	111	101	92	104	98	103	185
1989		92	113	116	118	112	105	94	112	98	108	146
1990		94	114	104	104	101	97	95	91	79	88	122
1991		96	108	94	90	90	91	96	84	68	80	148
1992	100	100	100	100	100	100	100	100	100	100	100	100
1993		104	102	119	123	124	122	102	126	98	118	216
1994		100P	111P	137	142	146	144	104	155	136	149	376
1995		101P	112P	122P	126P	127P	128P	105P	134P	117P	128P	231P
1996		102P	112P	124P	128P	130P	133P	107P	139P	122P	133P	240P
1997		102P	113P	126P	131P	133P	137P	109P	145P	127P	139P	250P
1998		103P	114P	128P	134P	137P	142P	111P	150P	131P	144P	260P

Sources: Same as General Statistics. Values reflect change from the base year, 1992. Values above 100 mean greater than 92, values below 100 mean less than 92, and a value of 100 in the 82-91 or 93-98 period means same as 92. 'P's mark projections by the editors.

SELECTED RATIOS

For 1994	Avg. of All Manufact.	Analyzed Industry	Index	For 1994	Avg. of All Manufact.	Analyzed Industry	Index
Employees per Establishment	49	95	193	Value Added per Production Worker	134,084	55,906	42
Payroll per Establishment	1,500,273	2,393,650	160	Cost per Establishment	5,045,178	11,264,094	223
Payroll per Employee	30,620	25,299	83	Cost per Employee	102,970	119,053	116
Production Workers per Establishment	34	78	228	Cost per Production Worker	146,988	143,669	98
Wages per Establishment	853,319	1,767,781	207	Shipments per Establishment	9,576,895	15,544,132	162
Wages per Production Worker	24,861	22,547	91	Shipments per Employee	195,460	164,290	84
Hours per Production Worker	2,056	2,105	102	Shipments per Production Worker	279,017	198,259	71
Wages per Hour	12.09	10.71	89	Investment per Establishment	321,011	342,203	107
Value Added per Establishment	4,602,255	4,383,199	95	Investment per Employee	6,552	3,617	55
Value Added per Employee	93,930	46,327	49	Investment per Production Worker	9,352	4,365	47

Sources: Same as General Statistics. The 'Average of All Manufacturing' column represents the average of all manufacturing industries reported for the most recent complete year available. The Index shows the relationship between the Average and the Analyzed Industry. For example, 100 means that they are equal; 500 that the Analyzed Industry is five times the average; 50 means that the Analyzed Industry is half the national average. The abbreviation 'na' is used to show that data are 'not available'.

LEADING COMPANIES   Number shown: 51   Total sales ($ mil): 3,225   Total employment (000): 22.1

Company Name	Address				CEO Name	Phone	Co. Type	Sales ($ mil)	Empl. (000)
Great Dane Holdings Inc	2016 N Pitcher St	Kalamazoo	MI	49007	David R Markin	616-343-6121	R	1,096	5.8
Monon Corp	PO Box 655	Monon	IN	47959	TJ Rosby	219-253-6621	R	300	2.7
Stoughton Trailers Inc	PO Box 606	Stoughton	WI	53589	Donald D Wahlin	608-873-2500	R	270	1.6
Strick Corp	PO Box 9	Fairless Hills	PA	19030	Frank Katz	215-949-3600	R	250	1.3
Trailmobile Inc	200 E Randolph Dr	Chicago	IL	60601	Samuel T Rozzi	312-861-1190	S	200*	1.7
Dorsey Trailers Inc	2727 Paces Ferry Rd	Atlanta	GA	30339	Marilyn R Marks	404-438-9595	R	172	1.3
Lufkin Industries Inc	PO Box 849	Lufkin	TX	75902	Jim Barber	409-637-5534	D	87	0.6
Kentucky Manufacturing Co	PO Box 17185	Louisville	KY	40217	Robert C Tway III	502-637-2551	R	65*	0.4
Oshkosh Trailers	1512 38th Av E	Bradenton	FL	34208	John Randjelovic	813-748-3900	D	60	0.5
Pines Trailer LP Kewanee	324 Main St	Kewanee	IL	61443	Phillip Pines	309-853-3566	D	59*	0.4
Wells Cargo Inc	PO Box 728-995	Elkhart	IN	46515	JM Wells	219-264-9661	R	58	0.6
Wilson Trailer Co	4400 S Lewis Blv	Sioux City	IA	51106	Wilson G Persinger	712-943-5591	R	53*	0.5
Alloy Trailers Inc	PO Box 19208	Spokane	WA	99219	Frank E Pignanelli	509-455-8650	R	50	0.4
Fruehauf International Ltd	PO Box 44913	Indianapolis	IN	46244	Thomas B Roller	317-630-3000	S	50	0.7
Transcraft Corp	PO Drawer 500	Anna	IL	62906	M R Cunningham	618-833-5151	R	50	0.3
Mark Line Industries Inc	PO Box 277	Bristol	IN	46507	LM Arnold	219-825-5851	R	40	0.4
Polar Tank of Delaware Co	12810 County Rd 17	Holdingford	MN	56340	James Jungles	612-746-2255	R	36*	0.3
General Engines Company Inc	4425 Hwy 27 S	Lake Wales	FL	33853	Frank W Flowers Jr	813-638-1421	R	25	0.1
WW Trailer Company Inc	PO Box 807	Madill	OK	73446	Harold G Watkins	405-795-5571	R	23*	0.2
TI-Brook Inc	PO Box 300	Brookville	PA	15825	Larry C Sessions	814-849-2342	R	20	0.1
Kinedyne Corp	PO Box 5207	North Branch	NJ	08876	James Klausmann	201-231-1800	R	19	0.2
Peerless Corp	PO Box 760	Paragould	AR	72450	Fred E Workman	501-236-7753	S	17	0.1
Clement Industries Inc	PO Box 914	Minden	LA	71055	W Glenn Hicks	318-377-2776	R	16	0.2
Reliance Trailer Manufacturing	7911 Redwood Dr	Cotati	CA	94931	Brian Ling	707-795-0081	R	16*	0.1
Talbert Manufacturing Inc	1628 W State Rd	Rensselaer	IN	47978	Davies Wakefield	219-866-7141	R	15	0.2
Road Systems Inc	8432 Almeria Av	Fontana	CA	92335	Lynn Reinbolt	714-350-2400	S	14	0.1
West-Mark	PO Box 100	Ceres	CA	95307	Grant Smith	209-537-4747	R	14	<0.1
Modern Inc	PO Box 790	Beaumont	TX	77704	Will Crenshaw	409-833-2665	R	13	0.2
Kiefer Built Inc	PO Box 88	Kanawha	IA	50447	David Bennett	515-762-3201	R	12*	0.1
Utility Tool and Body Company	PO Box 360	Clintonville	WI	54929	Glenn Giersbach	715-823-3167	R	12	<0.1
Trailmaster Tanks Inc	PO Box 161759	Fort Worth	TX	76161	Richard L Frank	817-232-0900	R	11	0.1
Western World Inc	200 N Kit Av	Caldwell	ID	83605	Robert Bushnell Jr	208-459-0842	R	11*	0.1
Truck Equipment Service	800 Oak St	Lincoln	NE	68521	E Churda	402-476-3225	R	10	<0.1
Parker Inc	PO Box 3305	Allentown	PA	18106	Barry Hale	215-395-0371	R	8	<0.1
American Carrier Equipment Inc	PO Box 2615	Fresno	CA	93745	P Sweet	209-442-1500	R	8	<0.1
Dakota Manufacturing Co	1909 S Rowley	Mitchell	SD	57301	AD Oehlerking	605-996-5571	R	7	0.2
General Trailer Company Inc	PO Box G	Springfield	OR	97477	Kenneth J Schmidt	503-746-2506	R	7	<0.1
Steco Inc	PO Box 3127	Enid	OK	73702	Pam Smith	918-000-0000	S	7*	<0.1
Custom Trailers Inc	PO Box 310	Springfield	MO	65801	James Jungels	417-862-5526	S	6	<0.1
Rogers Brothers Corp	100 Orchard St	Albion	PA	16401	Mark Kulyk	814-756-4121	R	6	<0.1
Klein Products Inc	PO Box 3700	Ontario	CA	91761	Richard F Klein	909-460-4546	R	5	<0.1
PDI Ground Support Systems	7750 Hub Pkwy	Valley View	OH	44125	Greg Widell	216-524-7310	R	5	<0.1
Redi Haul Trailers Inc	PO Box 803	Fairmont	MN	56031	Clayton R Leach	507-238-4231	R	3	<0.1
Alabama Trailer Company Inc	3100 10th Av N	Birmingham	AL	35203	Glen Miller	205-328-4210	R	3	<0.1
Guthrie Trailer Sales	PO Box 1026	Great Bend	KS	67530	Kenneth R Guthrie	316-793-5418	R	3*	<0.1
Leland Trailer and Equipment	PO Box 11217	Spokane	WA	99211	Richard C DeSmet	509-535-0291	R	3	<0.1
MOEX Corp	10466 Sunnyside S	Jefferson	OR	97352	Robert A Christman	503-391-7590	D	3	<0.1
Hill Manufacturing Co	PO Box 108	Hamburg	PA	19526	William T Hill	610-562-2207	R	3	<0.1
Cronkhite Industries Inc	PO Box 877	Danville	IL	61834	Cynthia S Cronkhite	217-443-3700	R	2*	<0.1
Star Transport Trailers	PO Box 403	Sunnyside	WA	98944	Homer Waller	509-837-3136	R	2	<0.1
Boston Trailer Mfg Co	1 Production Rd	Walpole	MA	02081	Mike MacFarland	508-668-2242	S	1	<0.1

Source: *Ward's Business Directory of U.S. Private and Public Companies*, Volumes 1 and 2, 1996. The company type code used is as follows: P - Public, R - Private, S - Subsidiary, D - Division, J - Joint Venture, A - Affiliate, G - Group. Sales are in millions of dollars, employees are in thousands. An asterisk (*) indicates an estimated sales volume. The symbol < stands for 'less than'. Company names and addresses are truncated, in some cases, to fit into the available space.

MATERIALS CONSUMED

Material	Quantity	Delivered Cost ($ million)
Materials, ingredients, containers, and supplies	(X)	2,287.3
Motor vehicle metal hardware (lock units, door and window handles, hinges, etc.), except forgings	(X)	10.8
Metal bolts, nuts, screws, washers, and other screw machine products	(X)	43.0
Other fabricated metal products, except forgings	(X)	115.9
Forgings	(X)	2.2
Castings (rough and semifinished)	(X)	5.0
Steel bars, bar shapes, and plates	(X)	58.4
Steel sheet and strip, including tin plate	(X)	86.6
Steel structural shapes and sheet piling	(X)	34.2
All other steel shapes and forms	(X)	21.3
Aluminum and aluminum-base alloy sheet, plate, foil, and welded tubing	(X)	183.5
Aluminum and aluminum-base alloy extruded shapes, including extruded rod, bar, pipe, tube, etc.	(X)	163.1
Other aluminum and aluminum-base alloy shapes and forms	(X)	7.1
Other nonferrous shapes and forms	(X)	0.3
Purchased chassis for vehicles (excluding passenger cars)	(X)	16.6
Truck bodies	(X)	(D)
Transmissions and parts	(X)	(D)

Continued on next page.

MATERIALS CONSUMED - Continued

Material	Quantity	Delivered Cost ($ million)
Shocks, struts, and other suspension equipment and parts	(X)	86.8
Axles, brakes, drums, rims, wheels, and other metal motor wheel parts	(X)	264.9
Pneumatic tires and inner tubes	(X)	223.5
Paints, varnishes, lacquers, stains, shellacs, japans, enamels, and allied products	(X)	34.0
Glass and glass products including windows and mirrors	(X)	(D)
Rough and dressed lumber	(X)	95.7
Fluid power products	(X)	19.2
Motor vehicle lighting fixtures	(X)	20.1
Automotive lamps (bulbs and sealed beams)	(X)	(D)
Seats (purchased separately) for automobiles, trucks, and buses	(X)	(D)
Fabricated plastics products, including components, housings, accessories, etc.	(X)	12.5
All other materials and components, parts, containers, and supplies	(X)	492.8
Materials, ingredients, containers, and supplies, nsk	(X)	276.7

Source: 1992 Economic Census. Explanation of symbols used: (D): Withheld to avoid disclosure of competitive data; na: Not available; (S): Withheld because statistical norms were not met; (X): Not applicable; (Z): Less than half the unit shown; nec: Not elsewhere classified; nsk: Not specified by kind; - : zero; * : 10-19 percent estimated; ** : 20-29 percent estimated.

PRODUCT SHARE DETAILS

Product or Product Class	% Share	Product or Product Class	% Share
Truck trailers	100.00	Truck trailers and chassis, less than 10,000 lb (4.54 metric tons) maximum load rating per axle.	7.83
Truck trailers and chassis, 10,000 lb (4.54 metric tons) maximum load rating per axle or more.	85.99	Truck trailers, nsk	6.18

Source: 1992 Economic Census. The values shown are percent of total shipments in an industry. Values of indented subcategories are summed in the main heading. The symbol (D) appears when data are withheld to prevent disclosure of competitive information. The abbreviation nsk stands for 'not specified by kind' and nec for 'not elsewhere classified'.

INPUTS AND OUTPUTS FOR TRUCK TRAILERS

Economic Sector or Industry Providing Inputs	%	Sector	Economic Sector or Industry Buying Outputs	%	Sector
Motor vehicle parts & accessories	14.0	Manufg.	Gross private fixed investment	91.6	Cap Inv
Wholesale trade	12.0	Trade	Exports	4.2	Foreign
Aluminum rolling & drawing	10.7	Manufg.	Federal Government purchases, national defense	3.4	Fed Govt
Blast furnaces & steel mills	10.3	Manufg.	Truck & bus bodies	0.5	Manufg.
Tires & inner tubes	9.5	Manufg.	Logging camps & logging contractors	0.1	Manufg.
Refrigeration & heating equipment	7.7	Manufg.			
Pipe, valves, & pipe fittings	3.5	Manufg.			
Hardwood dimension & flooring mills	2.7	Manufg.			
Motor freight transportation & warehousing	1.7	Util.			
Advertising	1.4	Services			
Metal stampings, nec	1.3	Manufg.			
Miscellaneous plastics products	1.3	Manufg.			
Electric services (utilities)	1.3	Util.			
Sawmills & planning mills, general	1.2	Manufg.			
Fabricated rubber products, nec	1.1	Manufg.			
Maintenance of nonfarm buildings nec	1.0	Constr.			
Petroleum refining	0.9	Manufg.			
Veneer & plywood	0.9	Manufg.			
Railroads & related services	0.9	Util.			
Screw machine and related products	0.8	Manufg.			
Paints & allied products	0.7	Manufg.			
Plastics materials & resins	0.6	Manufg.			
Power transmission equipment	0.6	Manufg.			
Coated fabrics, not rubberized	0.5	Manufg.			
Metal coating & allied services	0.5	Manufg.			
Special dies & tools & machine tool accessories	0.5	Manufg.			
Steel springs, except wire	0.5	Manufg.			
Communications, except radio & TV	0.5	Util.			
Gas production & distribution (utilities)	0.5	Util.			
Eating & drinking places	0.5	Trade			
Adhesives & sealants	0.4	Manufg.			
Fabricated metal products, nec	0.4	Manufg.			
Hardware, nec	0.4	Manufg.			
Industrial trucks & tractors	0.4	Manufg.			
Machinery, except electrical, nec	0.4	Manufg.			
Mineral wool	0.4	Manufg.			
Public building furniture	0.4	Manufg.			
Wood products, nec	0.4	Manufg.			
Abrasive products	0.3	Manufg.			
Padding & upholstery filling	0.3	Manufg.			
Banking	0.3	Fin/R.E.			
Real estate	0.3	Fin/R.E.			

Continued on next page.

INPUTS AND OUTPUTS FOR TRUCK TRAILERS - Continued

Economic Sector or Industry Providing Inputs	%	Sector	Economic Sector or Industry Buying Outputs	%	Sector
Equipment rental & leasing services	0.3	Services			
U.S. Postal Service	0.3	Gov't			
Automotive stampings	0.2	Manufg.			
Ball & roller bearings	0.2	Manufg.			
Broadwoven fabric mills	0.2	Manufg.			
Iron & steel foundries	0.2	Manufg.			
Lubricating oils & greases	0.2	Manufg.			
Metal heat treating	0.2	Manufg.			
Motor vehicles & car bodies	0.2	Manufg.			
Transportation equipment, nec	0.2	Manufg.			
Air transportation	0.2	Util.			
Automotive rental & leasing, without drivers	0.2	Services			
Automotive repair shops & services	0.2	Services			
Legal services	0.2	Services			
Management & consulting services & labs	0.2	Services			
Lighting fixtures & equipment	0.1	Manufg.			
Primary aluminum	0.1	Manufg.			
Truck trailers	0.1	Manufg.			
Credit agencies other than banks	0.1	Fin/R.E.			
Insurance carriers	0.1	Fin/R.E.			
Accounting, auditing & bookkeeping	0.1	Services			
Colleges, universities, & professional schools	0.1	Services			
Computer & data processing services	0.1	Services			
Laundry, dry cleaning, shoe repair	0.1	Services			

Source: Benchmark Input-Output Accounts for the U.S. Economy, 1982, U.S. Department of Commerce, Washington, D.C., July 1991. Data, as reported in the source, are organized by the 1977 SIC structure in use in 1982 but have been matched, as closely as is possible, to the 1987 SIC structure used in this book.

OCCUPATIONS EMPLOYED BY SIC 371 - MOTOR VEHICLES AND EQUIPMENT

Occupation	% of Total 1994	Change to 2005	Occupation	% of Total 1994	Change to 2005
Assemblers, fabricators, & hand workers nec	21.4	-12.2	Tool & die makers	1.6	-8.8
Inspectors, testers, & graders, precision	3.4	-51.1	Mechanics, installers, & repairers nec	1.6	-29.8
Engineers nec	3.4	5.4	Machine tool cutting & forming etc. nec	1.6	5.4
Blue collar worker supervisors	3.4	-11.4	Industrial truck & tractor operators	1.6	-29.8
Machine operators nec	2.9	-30.3	Electricians	1.6	-17.6
Machine tool cutting operators, metal & plastic	2.6	-55.0	Combination machine tool operators	1.5	40.5
Welders & cutters	2.4	-29.8	Machine forming operators, metal & plastic	1.5	-56.1
Managers & administrators nec	2.1	-12.2	Industrial machinery mechanics	1.4	-42.9
Engineering technicians & technologists nec	1.8	-12.2	Sales & related workers nec	1.2	-12.2
Helpers, laborers, & material movers nec	1.8	-12.2	Machine builders	1.1	31.7
Metal & plastic machine workers nec	1.7	55.6	Sheet metal workers & duct installers	1.0	-56.1
Welding machine setters, operators	1.7	-12.2			

Source: Industry-Occupation Matrix, Bureau of Labor Statistics. These data relate to one or more 3-digit SIC industry groups rather than to a single 4-digit SIC. The change reported for each occupation to the year 2005 is a percent of growth or decline as estimated by the Bureau of Labor Statistics. The abbreviation nec stands for 'not elsewhere classified'.

LOCATION BY STATE AND REGIONAL CONCENTRATION

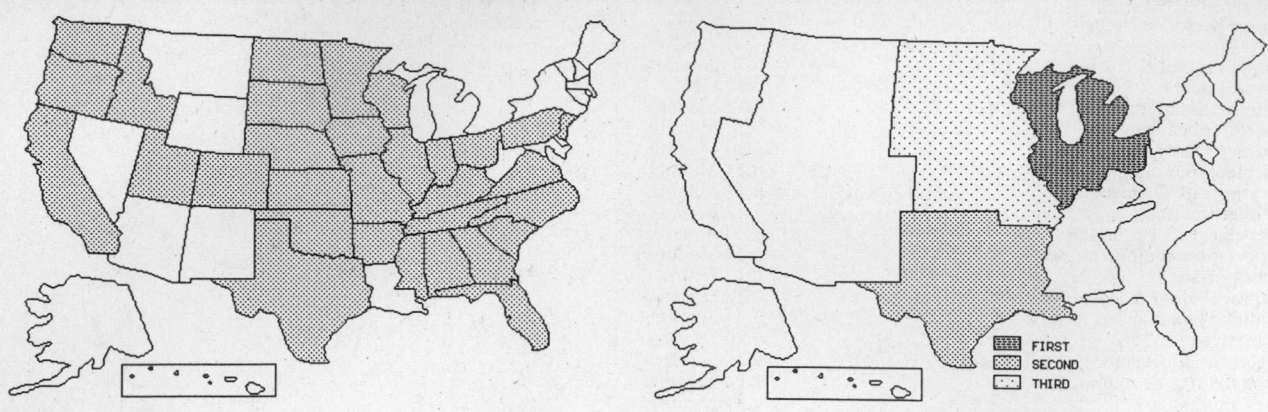

FIRST
SECOND
THIRD

INDUSTRY DATA BY STATE

State	Establish-ments	Shipments			Employment				Cost as % of Shipments	Investment per Employee ($)
		Total ($ mil)	% of U.S.	Per Establ.	Total Number	% of U.S.	Per Establ.	Wages ($/hour)		
Indiana	19	793.6	22.4	41.8	4,100	17.5	216	11.82	72.9	2,415
Illinois	10	397.5	11.2	39.7	2,300	9.8	230	9.79	56.4	522
Pennsylvania	19	260.5	7.3	13.7	1,200	5.1	63	11.22	81.1	-
Alabama	13	210.9	5.9	16.2	1,400	6.0	108	9.72	68.9	429
Nebraska	7	132.9	3.7	19.0	900	3.8	129	9.64	67.0	-
California	28	103.2	2.9	3.7	900	3.8	32	13.20	77.8	1,111
Tennessee	7	72.6	2.0	10.4	600	2.6	86	10.33	63.8	-
Florida	16	55.7	1.6	3.5	600	2.6	38	9.11	72.0	-
Washington	9	51.6	1.5	5.7	400	1.7	44	9.14	61.0	-
North Carolina	9	40.8	1.2	4.5	300	1.3	33	10.00	60.3	1,667
Colorado	4	33.2	0.9	8.3	300	1.3	75	9.50	46.1	-
Oklahoma	11	30.3	0.9	2.8	400	1.7	36	9.17	65.3	500
Kansas	8	21.1	0.6	2.6	200	0.9	25	9.50	50.2	-
Idaho	4	18.1	0.5	4.5	200	0.9	50	9.50	59.7	500
South Carolina	6	9.3	0.3	1.5	100	0.4	17	8.00	57.0	1,000
Texas	33	(D)	-	-	750 *	3.2	23	-	-	667
Wisconsin	16	(D)	-	-	1,750 *	7.5	109	-	-	1,600
Ohio	12	(D)	-	-	750 *	3.2	63	-	-	-
Iowa	11	(D)	-	-	1,750 *	7.5	159	-	-	-
Missouri	11	(D)	-	-	750 *	3.2	68	-	-	-
Georgia	10	(D)	-	-	750 *	3.2	75	-	-	-
Oregon	10	(D)	-	-	375 *	1.6	38	-	-	-
Arkansas	7	(D)	-	-	175 *	0.7	25	-	-	1,143
New Jersey	6	(D)	-	-	375 *	1.6	63	-	-	-
South Dakota	6	(D)	-	-	750 *	3.2	125	-	-	-
Minnesota	5	(D)	-	-	375 *	1.6	75	-	-	-
Utah	4	(D)	-	-	375 *	1.6	94	-	-	-
Virginia	4	(D)	-	-	375 *	1.6	94	-	-	-
Kentucky	2	(D)	-	-	375 *	1.6	188	-	-	-
Mississippi	2	(D)	-	-	375 *	1.6	188	-	-	-
North Dakota	2	(D)	-	-	175 *	0.7	88	-	-	-

Source: 1992 *Economic Census*. The states are in descending order of shipments or establishments (if shipment data are missing for the majority). The symbol (D) appears when data are withheld to prevent disclosure of competitive information. States marked with (D) are sorted by number of establishments. A dash (-) indicates that the data element cannot be calculated; * indicates the midpoint of a range.

3716 - MOTOR HOMES

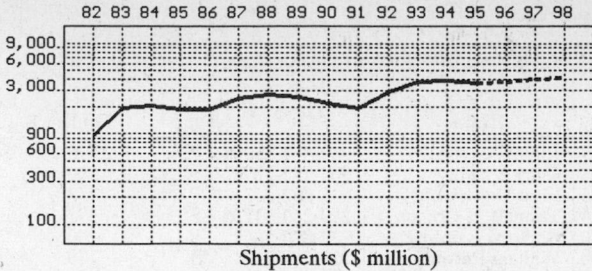

Shipments ($ million)

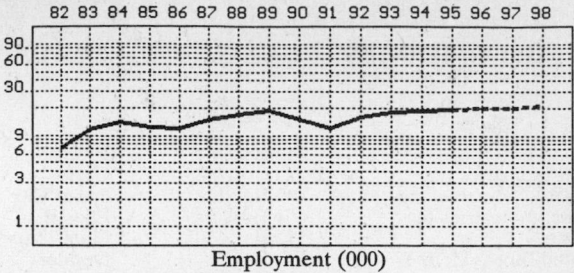

Employment (000)

GENERAL STATISTICS

Year	Companies	Establishments Total	Establishments with 20 or more employees	Employment Total (000)	Employment Production Workers (000)	Employment Hours (Mil)	Compensation Payroll ($ mil)	Compensation Wages ($/hr)	Production Cost of Materials	Production Value Added by Manufacture	Production Value of Shipments	Capital Invest.
1982	79	87	55	7.2	5.7	10.2	120.6	7.77	653.1	304.0	952.7	15.5
1983		97	59	11.7	9.2	19.1	246.8	7.32	1,388.4	582.2	1,934.3	13.9
1984		107	63	14.2	11.7	22.4	244.4	7.25	1,446.7	613.9	2,080.1	28.3
1985		118	66	12.2	9.7	18.9	223.2	7.93	1,337.1	558.1	1,872.2	41.8
1986		109	65	12.0	9.5	18.2	232.4	8.46	1,382.3	492.0	1,887.2	22.3
1987	144	165	81	15.1	11.8	23.4	316.9	9.06	1,811.4	688.6	2,486.8	18.7
1988		174	99	17.0	13.5	25.9	357.1	9.15	2,032.1	755.1	2,756.4	23.6
1989		163	103	18.8	12.3	24.2	336.0	9.30	1,853.5	699.4	2,562.7	24.2
1990		152	87	14.9	11.0	21.0	308.0	9.60	1,571.8	594.5	2,167.2	24.9
1991		144	70	11.8	9.5	18.4	259.9	9.30	1,334.3	631.9	1,935.4	17.2
1992	122	145	83	16.1	13.1	24.9	367.3	9.89	2,063.3	842.9	2,960.9	19.0
1993		124	69	18.2	15.1	27.7	438.8	10.75	2,810.2	957.6	3,795.0	23.4
1994		165P	90P	19.0	15.9	30.0	494.9	11.09	2,893.3	1,077.6	3,954.9	40.5
1995		170P	92P	19.1P	15.2P	28.7P	456.6P	11.08P	2,655.0P	988.8P	3,629.1P	27.8P
1996		175P	94P	19.7P	15.7P	29.6P	478.4P	11.38P	2,782.2P	1,036.2P	3,803.1P	28.4P
1997		180P	97P	20.4P	16.3P	30.6P	500.3P	11.67P	2,909.5P	1,083.6P	3,977.1P	28.9P
1998		185P	99P	21.0P	16.8P	31.6P	522.1P	11.97P	3,036.8P	1,131.1P	4,151.1P	29.4P

Sources: 1982, 1987, 1992 *Economic Census; Annual Survey of Manufactures*, 83-86, 88-91, 93-94. Establishment counts for non-Census years are from *County Business Patterns*; establishment values for 83-84 are extrapolations. 'P's show projections by the editors. Industries reclassified in 87 will not have data for prior years.

INDICES OF CHANGE

Year	Companies	Establishments Total	Establishments with 20 or more employees	Employment Total (000)	Employment Production Workers (000)	Employment Hours (Mil)	Compensation Payroll ($ mil)	Compensation Wages ($/hr)	Production Cost of Materials	Production Value Added by Manufacture	Production Value of Shipments	Capital Invest.
1982	65	60	66	45	44	41	33	79	32	36	32	82
1983		67	71	73	70	77	67	74	67	69	65	73
1984		74	76	88	89	90	67	73	70	73	70	149
1985		81	80	76	74	76	61	80	65	66	63	220
1986		75	78	75	73	73	63	86	67	58	64	117
1987	118	114	98	94	90	94	86	92	88	82	84	98
1988		120	119	106	103	104	97	93	98	90	93	124
1989		112	124	117	94	97	91	94	90	83	87	127
1990		105	105	93	84	84	84	97	76	71	73	131
1991		99	84	73	73	74	71	94	65	75	65	91
1992	100	100	100	100	100	100	100	100	100	100	100	100
1993		86	83	113	115	111	119	109	136	114	128	123
1994		114P	108P	118	121	120	135	112	140	128	134	213
1995		117P	111P	118P	116P	115P	124P	112P	129P	117P	123P	147P
1996		121P	114P	122P	120P	119P	130P	115P	135P	123P	128P	149P
1997		124P	117P	127P	124P	123P	136P	118P	141P	129P	134P	152P
1998		128P	119P	131P	129P	127P	142P	121P	147P	134P	140P	155P

Sources: Same as General Statistics. Values reflect change from the base year, 1992. Values above 100 mean greater than 92, values below 100 mean less than 92, and a value of 100 in the 82-91 or 93-98 period means same as 92. 'P's mark projections by the editors.

SELECTED RATIOS

For 1994	Avg. of All Manufact.	Analyzed Industry	Index	For 1994	Avg. of All Manufact.	Analyzed Industry	Index
Employees per Establishment	49	115	235	Value Added per Production Worker	134,084	67,774	51
Payroll per Establishment	1,500,273	3,003,255	200	Cost per Establishment	5,045,178	17,557,723	348
Payroll per Employee	30,620	26,047	85	Cost per Employee	102,970	152,279	148
Production Workers per Establishment	34	96	281	Cost per Production Worker	146,988	181,969	124
Wages per Establishment	853,319	2,018,959	237	Shipments per Establishment	9,576,895	23,999,945	251
Wages per Production Worker	24,861	20,925	84	Shipments per Employee	195,460	208,153	106
Hours per Production Worker	2,056	1,887	92	Shipments per Production Worker	279,017	248,736	89
Wages per Hour	12.09	11.09	92	Investment per Establishment	321,011	245,771	77
Value Added per Establishment	4,602,255	6,539,316	142	Investment per Employee	6,552	2,132	33
Value Added per Employee	93,930	56,716	60	Investment per Production Worker	9,352	2,547	27

Sources: Same as General Statistics. The 'Average of All Manufacturing' column represents the average of all manufacturing industries reported for the most recent complete year available. The Index shows the relationship between the Average and the Analyzed Industry. For example, 100 means that they are equal; 500 that the Analyzed Industry is five times the average; 50 means that the Analyzed Industry is half the national average. The abbreviation 'na' is used to show that data are 'not available'.

LEADING COMPANIES   Number shown: **18**   Total sales ($ mil): **5,252**   Total employment (000): **27.1**

Company Name	Address				CEO Name	Phone	Co. Type	Sales ($ mil)	Empl. (000)
Fleetwood Enterprises Inc	PO Box 7638	Riverside	CA	92513	John C Crean	909-351-3500	P	2,370	14.0
Mark III Industries Inc	PO Box 2525	Ocala	FL	34478	Clark Vitulli	904-732-5878	R	900	1.2
Winnebago Industries Inc	PO Box 152	Forest City	IA	50436	Fred Dohrmann	515-582-3535	P	452	2.7
Coachmen Industries Inc	601 E Bearsley Av	Elkhart	IN	46514	Thomas H Corson	219-262-0123	P	394	2.8
Jayco Inc	PO Box 460	Middlebury	IN	46540	Bernard Lambright	219-825-5861	R	200	1.1
Fleetwood Motor Homes	PO Box 31	Decatur	IN	46733	William Pettine	219-728-2121	S	178	0.8
SMC Corp	30725 Diamond Hill	Harrisburg	OR	97446	Mathew M Perlot	503-995-8214	P	120	0.9
Monaco Coach Corp	325 E 1st St	Junction City	OR	97448	Kay L Toolson	503-998-1068	P	107	0.8
Fleetwood Motor Homes	PO Drawer 5	Paxinos	PA	17860	John Weiss	717-644-0817	S	100	0.4
Newmar Corp	PO Box 30	Nappanee	IN	46550	Virgil Miller	219-773-7791	R	100	0.5
Four Winds International Corp	PO Box 1486	Elkhart	IN	46515	Steve Hicks	219-266-1111	S	95	0.3
National RV Holdings Inc	3411 N Perris Blv	Perris	CA	92571	Wayne M Mertes	909-943-6007	P	72	0.4
National RV Inc	3411 N Perris Blv	Perris	CA	92571	Wayne M Mertes	909-943-6007	S	72	0.4
Blue Bird Wanderlodge	PO Box 1259	Fort Valley	GA	31030	Paul Glaske	912-825-2021	D	35	0.2
Foretravel Inc	1221 NW Stallings	Nacogdoches	TX	75964	Clarence M Fore	409-564-8367	R	33	0.3
Rexhall Industries Inc	PO Box 905	Santa Clarita	CA	91380	William J Rex	805-253-1295	P	12	0.3
Barth Inc	PO Box 768	Milford	IN	46542	MD Umbaugh	219-658-9401	R	10	0.1
Custom Camp Vans and Service	7575 Jurupa Av	Riverside	CA	92504	Darrell Jones	909-359-3443	R	2	<0.1

Source: Ward's Business Directory of U.S. Private and Public Companies, Volumes 1 and 2, 1996. The company type code used is as follows: P - Public, R - Private, S - Subsidiary, D - Division, J - Joint Venture, A - Affiliate, G - Group. Sales are in millions of dollars, employees are in thousands. An asterisk (*) indicates an estimated sales volume. The symbol < stands for 'less than'. Company names and addresses are truncated, in some cases, to fit into the available space.

MATERIALS CONSUMED

Material	Quantity	Delivered Cost ($ million)
Materials, ingredients, containers, and supplies	(X)	2,033.6
Trailer axles, wheels, brakes, undercarriages, and other metal vehicular parts	(X)	95.9
Pneumatic tires and inner tubes	(X)	(D)
Purchased chassis for motor homes	(X)	518.3
Household appliances, including refrigerators, cooking equipment, and other household appliances	(X)	71.9
Air-conditioning equipment	(X)	23.6
Metal heating equipment (except electric)	(X)	9.4
Metal doors and door units, windows and window units	(X)	24.3
Metal plumbing fixtures, fittings, and trim (including enameled) (except forgings)	(X)	8.8
Sheet metal products, except stampings	(X)	15.5
Metal bolts, nuts, screws, washers, rivets, and other screw machine products	(X)	9.7
Castings (rough and semifinished)	(X)	(D)
Steel shapes and forms	(X)	14.8
Aluminum and aluminum-base alloy sheet, plate, foil, and welded tubing	(X)	32.5
Current-carrying wiring devices	(X)	47.6
Plywood	(X)	40.4
Dressed lumber	(X)	37.9
Millwork, wood (including wood doors, window sash, moldings, and cabinets)	(X)	30.6
Glass and glass products including windows and mirrors	(X)	15.0
Fabricated plastics products (except gaskets, hoses, and belting)	(X)	22.8
Plastics products consumed in the form of sheets, rods, tubes, film, and other shapes	(X)	23.4
Plastics resins consumed in the form of granules, pellets, powders, liquids, etc.	(X)	8.7
Paints, varnishes, lacquers, stains, shellacs, japans, enamels, and allied products	(X)	8.9
Carpeting	(X)	21.0
Curtains and draperies	(X)	17.2
All other materials and components, parts, containers, and supplies	(X)	460.2
Materials, ingredients, containers, and supplies, nsk	(X)	469.5

Source: 1992 *Economic Census*. Explanation of symbols used: (D): Withheld to avoid disclosure of competitive data; na: Not available; (S): Withheld because statistical norms were not met; (X): Not applicable; (Z): Less than half the unit shown; nec: Not elsewhere classified; nsk: Not specified by kind; - : zero; * : 10-19 percent estimated; ** : 20-29 percent estimated.

PRODUCT SHARE DETAILS

Product or Product Class	% Share	Product or Product Class	% Share
Motor homes produced on purchased chassis	100.00	Motor homes built on purchased chassis, van camper (Type B)	0.28
Motor homes built on purchased chassis, conventional (Type A)	50.14	Motor homes built on purchased chassis, converted vans not qualifying as van campers (Type B)	29.84
Motor homes built on purchased chassis, chopped van (Type C)	14.10		

Source: 1992 *Economic Census*. The values shown are percent of total shipments in an industry. Values of indented subcategories are summed in the main heading. The symbol (D) appears when data are withheld to prevent disclosure of competitive information. The abbreviation nsk stands for 'not specified by kind' and nec for 'not elsewhere classified'.

INPUTS AND OUTPUTS FOR MOTOR HOMES (MADE ON PURCHASED CHASSIS)

Economic Sector or Industry Providing Inputs	%	Sector	Economic Sector or Industry Buying Outputs	%	Sector
Motor vehicles & car bodies	33.9	Manufg.	Personal consumption expenditures	95.1	
Wholesale trade	8.1	Trade	Change in business inventories	1.9	In House
Advertising	3.6	Services	Exports	1.6	Foreign
Internal combustion engines, nec	2.7	Manufg.	Motor homes (made on purchased chassis)	1.4	Manufg.
Veneer & plywood	2.7	Manufg.			
Public building furniture	2.6	Manufg.			
Hardware, nec	2.5	Manufg.			
Glass & glass products, except containers	2.2	Manufg.			
Motor homes (made on purchased chassis)	2.2	Manufg.			
Management & consulting services & labs	2.1	Services			
Automotive & apparel trimmings	2.0	Manufg.			
Metal stampings, nec	1.6	Manufg.			
Coated fabrics, not rubberized	1.4	Manufg.			
Refrigeration & heating equipment	1.3	Manufg.			
Motor freight transportation & warehousing	1.3	Util.			
Aluminum rolling & drawing	1.2	Manufg.			
Engine electrical equipment	1.2	Manufg.			
Miscellaneous plastics products	1.2	Manufg.			
Air transportation	1.1	Util.			
Pipe, valves, & pipe fittings	1.0	Manufg.			
Nonferrous wire drawing & insulating	0.9	Manufg.			
Padding & upholstery filling	0.9	Manufg.			
Communications, except radio & TV	0.9	Util.			
Maintenance of nonfarm buildings nec	0.8	Constr.			
Adhesives & sealants	0.8	Manufg.			
Nonferrous rolling & drawing, nec	0.8	Manufg.			
Power transmission equipment	0.8	Manufg.			
Railroads & related services	0.8	Util.			
Metal doors, sash, & trim	0.7	Manufg.			
Petroleum refining	0.7	Manufg.			
Sawmills & planning mills, general	0.7	Manufg.			
Screw machine and related products	0.7	Manufg.			
Tires & inner tubes	0.7	Manufg.			
Metal coating & allied services	0.6	Manufg.			
Paints & allied products	0.6	Manufg.			
Fabricated metal products, nec	0.5	Manufg.			
Wood products, nec	0.5	Manufg.			
Electric services (utilities)	0.5	Util.			
Automotive stampings	0.4	Manufg.			
Machinery, except electrical, nec	0.4	Manufg.			
Steel springs, except wire	0.4	Manufg.			
Storage batteries	0.4	Manufg.			
Banking	0.4	Fin/R.E.			
Broadwoven fabric mills	0.3	Manufg.			
Cyclic crudes and organics	0.3	Manufg.			
Lighting fixtures & equipment	0.3	Manufg.			
Plumbing fixture fittings & trim	0.3	Manufg.			
Special dies & tools & machine tool accessories	0.3	Manufg.			
Transportation equipment, nec	0.3	Manufg.			
Truck & bus bodies	0.3	Manufg.			
Wiring devices	0.3	Manufg.			
Eating & drinking places	0.3	Trade			
Abrasive products	0.2	Manufg.			
Blast furnaces & steel mills	0.2	Manufg.			
Fabricated rubber products, nec	0.2	Manufg.			
Household refrigerators & freezers	0.2	Manufg.			
Iron & steel foundries	0.2	Manufg.			
Mineral wool	0.2	Manufg.			
Motor vehicle parts & accessories	0.2	Manufg.			
Motors & generators	0.2	Manufg.			
Plastics materials & resins	0.2	Manufg.			
Primary aluminum	0.2	Manufg.			
Real estate	0.2	Fin/R.E.			
Equipment rental & leasing services	0.2	Services			
Theatrical producers, bands, entertainers	0.2	Services			
Gaskets, packing & sealing devices	0.1	Manufg.			
Heating equipment, except electric	0.1	Manufg.			
Lubricating oils & greases	0.1	Manufg.			
Machine tools, metal cutting types	0.1	Manufg.			
Metal heat treating	0.1	Manufg.			
Truck trailers	0.1	Manufg.			
Gas production & distribution (utilities)	0.1	Util.			
Sanitary services, steam supply, irrigation	0.1	Util.			
Insurance carriers	0.1	Fin/R.E.			
Accounting, auditing & bookkeeping	0.1	Services			

Continued on next page.

INPUTS AND OUTPUTS FOR MOTOR HOMES (MADE ON PURCHASED CHASSIS) - Continued

Economic Sector or Industry Providing Inputs	%	Sector	Economic Sector or Industry Buying Outputs	%	Sector
Automotive rental & leasing, without drivers	0.1	Services			
Automotive repair shops & services	0.1	Services			
Business services nec	0.1	Services			
Colleges, universities, & professional schools	0.1	Services			
Computer & data processing services	0.1	Services			
Legal services	0.1	Services			

Source: Benchmark Input-Output Accounts for the U.S. Economy, 1982, U.S. Department of Commerce, Washington, D.C., July 1991. Data, as reported in the source, are organized by the 1977 SIC structure in use in 1982 but have been matched, as closely as is possible, to the 1987 SIC structure used in this book.

OCCUPATIONS EMPLOYED BY SIC 371 - MOTOR VEHICLES AND EQUIPMENT

Occupation	% of Total 1994	Change to 2005	Occupation	% of Total 1994	Change to 2005
Assemblers, fabricators, & hand workers nec	21.4	-12.2	Tool & die makers	1.6	-8.8
Inspectors, testers, & graders, precision	3.4	-51.1	Mechanics, installers, & repairers nec	1.6	-29.8
Engineers nec	3.4	5.4	Machine tool cutting & forming etc. nec	1.6	5.4
Blue collar worker supervisors	3.4	-11.4	Industrial truck & tractor operators	1.6	-29.8
Machine operators nec	2.9	-30.3	Electricians	1.6	-17.6
Machine tool cutting operators, metal & plastic	2.6	-55.0	Combination machine tool operators	1.5	40.5
Welders & cutters	2.4	-29.8	Machine forming operators, metal & plastic	1.5	-56.1
Managers & administrators nec	2.1	-12.2	Industrial machinery mechanics	1.4	-42.9
Engineering technicians & technologists nec	1.8	-12.2	Sales & related workers nec	1.2	-12.2
Helpers, laborers, & material movers nec	1.8	-12.2	Machine builders	1.1	31.7
Metal & plastic machine workers nec	1.7	55.6	Sheet metal workers & duct installers	1.0	-56.1
Welding machine setters, operators	1.7	-12.2			

Source: Industry-Occupation Matrix, Bureau of Labor Statistics. These data relate to one or more 3-digit SIC industry groups rather than to a single 4-digit SIC. The change reported for each occupation to the year 2005 is a percent of growth or decline as estimated by the Bureau of Labor Statistics. The abbreviation nec stands for 'not elsewhere classified'.

LOCATION BY STATE AND REGIONAL CONCENTRATION

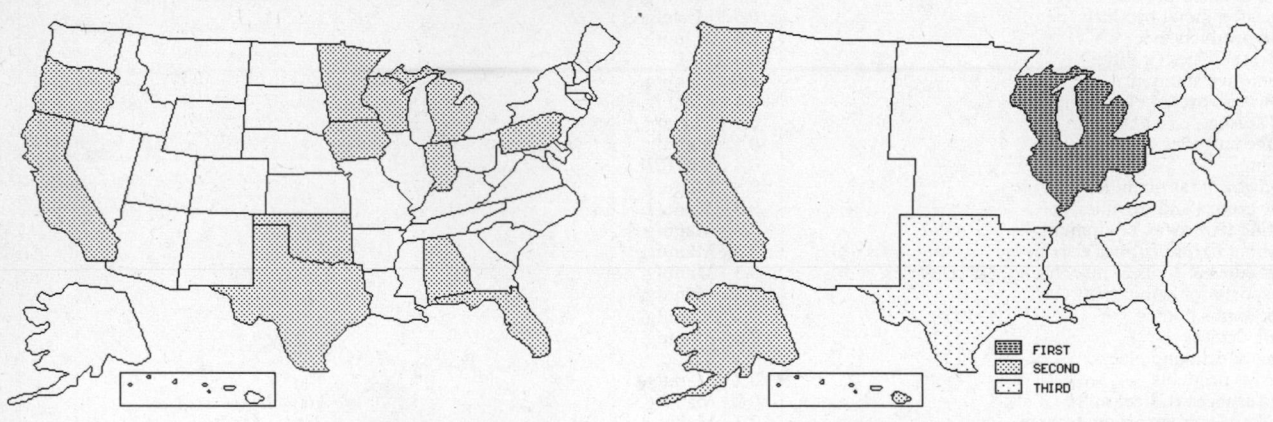

FIRST
SECOND
THIRD

INDUSTRY DATA BY STATE

State	Establish-ments	Shipments			Employment				Cost as % of Shipments	Investment per Employee ($)
		Total ($ mil)	% of U.S.	Per Establ.	Total Number	% of U.S.	Per Establ.	Wages ($/hour)		
Indiana	56	1,233.8	41.7	22.0	5,700	35.4	102	10.93	75.4	1,000
California	17	440.5	14.9	25.9	2,300	14.3	135	9.23	71.7	1,261
Oregon	4	171.1	5.8	42.8	1,100	6.8	275	8.32	69.7	-
Minnesota	5	21.3	0.7	4.3	200	1.2	40	7.67	66.7	-
Texas	9	(D)	-	-	375 *	2.3	42	-	-	800
Michigan	6	(D)	-	-	750 *	4.7	125	-	-	-
Florida	4	(D)	-	-	1,750 *	10.9	438	-	-	-
Alabama	3	(D)	-	-	750 *	4.7	250	-	-	-
Iowa	3	(D)	-	-	3,750 *	23.3	1,250	-	-	-
Pennsylvania	3	(D)	-	-	375 *	2.3	125	-	-	-
Wisconsin	2	(D)	-	-	175 *	1.1	88	-	-	-
Oklahoma	1	(D)	-	-	175 *	1.1	175	-	-	-

Source: 1992 Economic Census. The states are in descending order of shipments or establishments (if shipment data are missing for the majority). The symbol (D) appears when data are withheld to prevent disclosure of competitive information. States marked with (D) are sorted by number of establishments. A dash (-) indicates that the data element cannot be calculated; * indicates the midpoint of a range.

3721 - AIRCRAFT

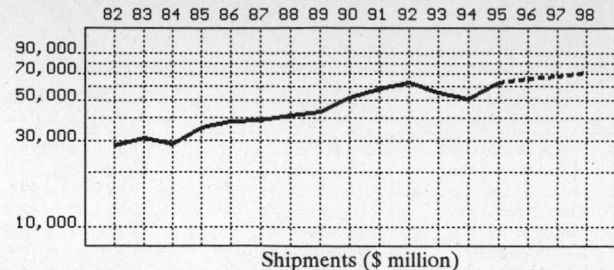

Shipments ($ million)

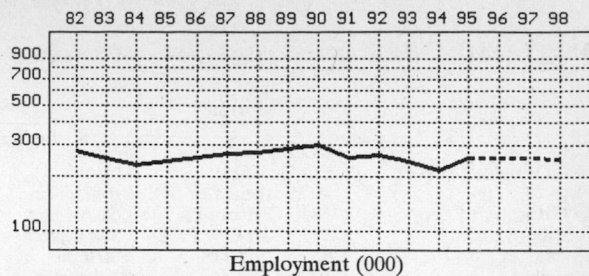

Employment (000)

GENERAL STATISTICS

| Year | Companies | Establishments | | Employment | | | Compensation | | Production ($ million) | | | |
		Total	with 20 or more employees	Total (000)	Production Workers (000)	Hours (Mil)	Payroll ($ mil)	Wages ($/hr)	Cost of Materials	Value Added by Manufacture	Value of Shipments	Capital Invest.
1982	137	166	87	275.4	138.8	272.7	7,750.1	12.91	15,716.6	15,717.4	28,047.4	840.3
1983		168	85	250.9	120.6	238.9	7,562.0	13.60	14,910.0	14,012.5	30,522.0	621.8
1984		170	83	232.5	115.5	228.2	7,456.9	14.05	1,577.2	15,498.2	28,453.2	860.8
1985		173	80	241.8	121.9	238.6	8,006.8	14.60	17,482.1	17,096.3	34,976.5	1,013.3
1986		151	77	256.7	135.3	269.5	8,983.9	14.71	22,167.9	15,160.7	38,184.3	1,108.2
1987	137	155	79	268.2	141.5	282.3	9,679.5	15.38	23,140.9	17,311.0	39,092.7	1,052.1
1988		163	85	274.2	140.0	274.3	10,015.2	16.16	26,140.8	18,218.6	41,493.7	1,029.6
1989		169	81	287.7	140.3	272.0	10,468.6	16.40	29,723.8	20,363.6	43,338.9	1,269.6
1990		181	90	300.0	139.7	268.1	11,224.7	17.02	33,171.2	20,235.4	51,369.6	1,020.9
1991		198	94	258.3	125.3	244.8	10,324.1	17.82	36,077.2	23,090.6	58,090.2	1,046.1
1992	151	182	103	264.9	122.1	227.0	11,498.9	19.98	36,133.3	25,157.1	62,980.8	1,661.3
1993		260	116	241.2	104.4	198.7	10,790.0	19.90	33,206.4	22,903.3	55,119.8	1,154.4
1994		211P	103P	217.9	92.8	172.7	10,312.1	20.77	25,778.3	23,606.4	50,944.0	872.2
1995		216P	105P	255.0P	114.0P	214.4P	11,848.0P	20.93P	31,599.8P	28,937.4P	62,448.6P	1,280.7P
1996		221P	107P	254.4P	112.3P	210.0P	12,177.1P	21.58P	32,985.6P	30,206.4P	65,187.2P	1,314.8P
1997		226P	109P	253.9P	110.5P	205.6P	12,506.3P	22.23P	34,371.4P	31,475.5P	67,925.9P	1,348.8P
1998		232P	111P	253.3P	108.8P	201.2P	12,835.4P	22.87P	35,757.2P	32,744.5P	70,664.6P	1,382.9P

Sources: 1982, 1987, 1992 *Economic Census*; *Annual Survey of Manufactures*, 83-86, 88-91, 93-94. Establishment counts for non-Census years are from *County Business Patterns*; establishment values for 83-84 are extrapolations. 'P's show projections by the editors. Industries reclassified in 87 will not have data for prior years.

INDICES OF CHANGE

| Year | Companies | Establishments | | Employment | | | Compensation | | Production ($ million) | | | |
		Total	with 20 or more employees	Total (000)	Production Workers (000)	Hours (Mil)	Payroll ($ mil)	Wages ($/hr)	Cost of Materials	Value Added by Manufacture	Value of Shipments	Capital Invest.
1982	91	91	84	104	114	120	67	65	43	62	45	51
1983		92	83	95	99	105	66	68	41	56	48	37
1984		93	81	88	95	101	65	70	4	62	45	52
1985		95	78	91	100	105	70	73	48	68	56	61
1986		83	75	97	111	119	78	74	61	60	61	67
1987	91	85	77	101	116	124	84	77	64	69	62	63
1988		90	83	104	115	121	87	81	72	72	66	62
1989		93	79	109	115	120	91	82	82	81	69	76
1990		99	87	113	114	118	98	85	92	80	82	61
1991		109	91	98	103	108	90	89	100	92	92	63
1992	100	100	100	100	100	100	100	100	100	100	100	100
1993		143	113	91	86	88	94	100	92	91	88	69
1994		116P	100P	82	76	76	90	104	71	94	81	53
1995		119P	102P	96P	93P	94P	103P	105P	87P	115P	99P	77P
1996		122P	104P	96P	92P	93P	106P	108P	91P	120P	104P	79P
1997		124P	106P	96P	91P	91P	109P	111P	95P	125P	108P	81P
1998		127P	108P	96P	89P	89P	112P	114P	99P	130P	112P	83P

Sources: Same as General Statistics. Values reflect change from the base year, 1992. Values above 100 mean greater than 92, values below 100 mean less than 92, and a value of 100 in the 82-91 or 93-98 period means same as 92. 'P's mark projections by the editors.

SELECTED RATIOS

For 1994	Avg. of All Manufact.	Analyzed Industry	Index	For 1994	Avg. of All Manufact.	Analyzed Industry	Index
Employees per Establishment	49	1,032	2,106	Value Added per Production Worker	134,084	254,379	190
Payroll per Establishment	1,500,273	48,840,947	3,255	Cost per Establishment	5,045,178	122,093,132	2,420
Payroll per Employee	30,620	47,325	155	Cost per Employee	102,970	118,303	115
Production Workers per Establishment	34	440	1,281	Cost per Production Worker	146,988	277,783	189
Wages per Establishment	853,319	16,988,921	1,991	Shipments per Establishment	9,576,895	241,284,822	2,519
Wages per Production Worker	24,861	38,653	155	Shipments per Employee	195,460	233,795	120
Hours per Production Worker	2,056	1,861	91	Shipments per Production Worker	279,017	548,966	197
Wages per Hour	12.09	20.77	172	Investment per Establishment	321,011	4,130,980	1,287
Value Added per Establishment	4,602,255	111,806,415	2,429	Investment per Employee	6,552	4,003	61
Value Added per Employee	93,930	108,336	115	Investment per Production Worker	9,352	9,399	100

Sources: Same as General Statistics. The 'Average of All Manufacturing' column represents the average of all manufacturing industries reported for the most recent complete year available. The Index shows the relationship between the Average and the Analyzed Industry. For example, 100 means that they are equal; 500 that the Analyzed Industry is five times the average; 50 means that the Analyzed Industry is half the national average. The abbreviation 'na' is used to show that data are 'not available'.

LEADING COMPANIES   Number shown: **64**   Total sales ($ mil): **65,661**   Total employment (000): **376.2**

Company Name	Address				CEO Name	Phone	Co. Type	Sales ($ mil)	Empl. (000)
Boeing Co	PO Box 3707	Seattle	WA	98124	Frank A Shrontz	206-655-2121	P	21,924	115.0
McDonnell Douglas Corp	PO Box 516	St Louis	MO	63166	H C Stonecipher	314-232-0232	P	13,176	65.8
Textron Inc	40 Westminster St	Providence	RI	02903	James F Hardymon	401-421-2800	P	9,683	53.0
Northrop Grumman Corp	1840 Century Pk E	Los Angeles	CA	90067	Kent Kresa	310-553-6262	P	6,711	42.4
General Dynamics Corp	3190 Fairview Pk Dr	Falls Church	VA	22042	James R Mellor	703-876-3000	P	3,187	30.5
Lockheed Martin	86 S Cobb Dr	Marietta	GA	30063	JA Blackwell	404-494-5432	D	2,329	11.0
Sikorsky Aircraft	6900 Main St	Bridgeport	CT	06601	Eugene Buckley	203-386-4000	D	2,300	13.0
Northrop Corp	1 Northrop Av	Hawthorne	CA	90250	Wallace C Solberg	310-332-1000	D	1,537	8.8
Kaman Corp	PO Box 1	Bloomfield	CT	06002	Charles H Kaman	203-243-8311	P	821	5.2
Cessna Aircraft Co	PO Box 7706	Wichita	KS	67277	Russell W Meyer Jr	316-946-6000	S	810•	5.4
Vought Aircraft Co	PO Box 655907	Dallas	TX	75265	Gordon L Williams	214-266-2011	S	800	5.2
Bell Boeing Corp	1235 J Davis	Arlington	VA	22202	Stuart Dodge	703-414-3350	J	650•	4.0
Lockheed Martin	1600 E Pioneer	Arlington	TX	76010	Robert E Tokerud	817-261-0295	S	400	6.0
Falcon Jet Corp	East 15 Midland	Paramus	NJ	07652	J Georges	201-262-0800	S	310•	1.0
Dee Howard Co	PO Box 469001	San Antonio	TX	78246	Matthew Donohue	210-828-1341	S	210•	1.3
American Eurocopter Corp	2701 Forum Dr	Grand Prairie	TX	75052	Dave Smith	214-641-0000	S	165	0.3
K-C Aviation Inc	PO Box 7145	Dallas	TX	75209	RW Emery	214-902-7500	S	76	1.8
Fairchild Aircraft Inc	PO Box 790490	San Antonio	TX	78279	Carl Albert	210-824-9421	R	58	1.0
Mobile Aerospace Engineering	2100 9th St	Mobile	AL	36615	Bob Tan	334-438-8888	S	54	0.5
Piper Aircraft Corp	2926 Piper Dr	Vero Beach	FL	32960	Chuck Suma	407-567-4361	R	50	0.4
Robinson Helicopter Co	2901 Airport Dr	Torrance	CA	90505	Frank Robinson	310-539-0508	R	50	0.5
Schweizer Aircraft Corp	PO Box 147	Elmira	NY	14902	LE Schweizer	607-739-3821	R	42•	0.4
Mooney Aircraft Corp	L Schreiner Field	Kerrville	TX	78029	Jacques Esculier	210-896-6000	S	28	0.5
Mooney Holding Corp	L Schreiner Field	Kerrville	TX	78029	Jacques Esculier	210-896-6000	R	28	0.5
Aydin Vector	PO Box 328	Newtown	PA	18940	John Vanderslice	215-968-4271	D	26	0.4
Bizjet Intern Sales and Support	3515 N Sheridan Rd	Tulsa	OK	74115	WL Butch Walker	918-832-7733	R	19•	0.1
General Mechatronics Corp	60 Milbar Blv	Farmingdale	NY	11735	Daniel D'Addario	516-249-7900	R	16•	0.1
Snow Aviation International Inc	7201 Paul Tibbets St	Columbus	OH	43217	Harry T Snow Jr	614-492-7669	R	16•	0.1
UNC Helicopter Inc	PO Box 1088	Ozark	AL	36361	Dick Joyce	334-774-2529	R	16•	0.1
Space	PO Box 462009	Garland	TX	75046	S J Georgoulis	214-494-2441	D	14	0.3
Aurora Flight Sciences Corp	9950 Wakeman Dr	Manassas	VA	22111	John Langford	703-369-3633	R	12•	<0.1
Scaled Composites Inc	Airpt 1624	Mojave	CA	93501	Burt Rutan	805-824-4541	S	12	<0.1
Enstrom Helicopter Corp	PO Box 490	Menominee	MI	49858	Robert M Tuttle	906-863-1200	R	11•	<0.1
Advanced Techn & Research	14201 Myerlake Cir	Clearwater	FL	34620	William Higgins	813-539-8585	R	9	0.1
Mattituck Aviation Corp	PO Box 1432	Mattituck	NY	11952	Jay Wickham	516-298-8330	R	9•	<0.1
Lancair International Inc	2244 Airport Way	Redmond	OR	97756	Lance Neibauer	503-923-2233	R	8•	<0.1
Rotorway International Inc	4141 W Chandler	Chandler	AZ	85226	John Netherwood	602-278-8899	R	8•	<0.1
Commander Aircraft Co	7200 NW 63rd St	Bethany	OK	73008	Wirt D Walker III	405-495-8080	P	8	0.1
Skystar Aircraft Corp	100 N Kings Rd	Nampa	ID	83687	Phil Reed	208-466-1711	R	7	<0.1
Stoddard-Hamilton Aircraft Inc	18701 58th Av NE	Arlington	WA	98223	Kelly Lee	206-435-8533	R	6	<0.1
Worldwide Aeros Corp	485 Aviator Dr	Atwater	CA	95301	Igor Pasternak	209-357-7000	R	6	0.3
Balloon Works Inc	PO Box 6237	Statesville	NC	28687	E Conn	704-878-9501	R	5	<0.1
Helicomb International Inc	1402 S 69th Av E	Tulsa	OK	74112	Bob Austin	918-835-3999	R	5	<0.1
Peregrine Aircraft Co	2207 Bellanca Av	Minden	NV	89423	J M Henderson Jr	702-782-1000	R	5	<0.1
Northrop Grumman Intern	1000 Wilson Blv	Arlington	VA	22209	William James	703-875-8460	S	4•	<0.1
Orion Aviation	27520 Hawthorne	Rolling Hls Est	CA	90274	Thomas Whinfrey	310-544-4844	R	4	<0.1
American General Aircraft Corp	RR 1	Greenville	MS	38703	Robert E Crowley	601-332-2422	R	4	<0.1
Bede Jet Corp	18421 Edison Av	Chesterfield	MO	63005	James R Bede	314-537-2333	R	4	<0.1
Apache Enterprises Inc	2985 Red Hawk Dr	Grand Prairie	TX	75051	Doug Gadberry	214-647-8500	R	3	<0.1
BAI Aerosystems Inc	PO Box 1600	Easton	MD	21601	Richard Bernstein	410-820-7500	R	3	<0.1
California Helicopter Intern	2935 Golf Course	Ventura	CA	93003	Gary Podolny	805-644-5800	R	3•	<0.1
Maule Air Inc	2099 GA Hwy 133 S	Moultrie	GA	31768	BD Maule	912-985-2045	R	3	<0.1
Moller International Inc	1222 Research Pk	Davis	CA	95616	Paul S Moller	916-756-5086	R	3	<0.1
Star of Phoenix Aircraft Corp	4710 E Falcon Dr	Mesa	AZ	85205	Charles Kallmann	602-832-5500	R	3	<0.1
Aerostar Aircraft Corp	S 3608 Davison Blvd	Spokane	WA	99204	Steven Speer	509-455-8872	R	2	<0.1
AJ Aerospace Inc	1721 Wilshire Blv	Austin	TX	78722	Matt Gordon	512-499-1618	S	2	<0.1
American Blimp Corp	1900 NE 25th Av	Hillsboro	OR	97124	James R Thiele	503-693-1611	R	2•	<0.1
Conrad Co	1304 Farmville Rd	Memphis	TN	38122	Brenda Dabbs	901-323-5926	R	2	<0.1
Freewing Aerial Robotics Corp	UMD/TAP	College Park	MD	20742	Hugh J Schmittle	301-314-7794	R	1•	<0.1
Questair Inc	3800 N Mcaree Rd	Waukegan	IL	60087	Robert McCallen	708-244-0005	R	1	<0.1
Aerofab Inc	PO Box 312	Sanford	ME	04073	Armand E Rivard	207-324-3916	R	1•	<0.1
Bellanca Inc	PO Box 964	Alexandria	MN	56308	Charles F Holm	612-762-1501	R	1	<0.1
Sadler Aircraft Corp	8225 E Montebello	Scottsdale	AZ	85250	William G Sadler	602-483-8661	R	1	<0.1
Aerolites Inc	12104 David Rd	Welsh	LA	70591	Daniel Roche	318-734-3865	R	0	<0.1

Source: Ward's Business Directory of U.S. Private and Public Companies, Volumes 1 and 2, 1996. The company type code used is as follows: P - Public, R - Private, S - Subsidiary, D - Division, J - Joint Venture, A - Affiliate, G - Group. Sales are in millions of dollars, employees are in thousands. An asterisk (•) indicates an estimated sales volume. The symbol < stands for 'less than'. Company names and addresses are truncated, in some cases, to fit into the available space.

MATERIALS CONSUMED

Material	Quantity	Delivered Cost ($ million)
Materials, ingredients, containers, and supplies	(X)	34,477.0
Aircraft engines	(X)	15,637.1
Aircraft propellers and parts thereof	(X)	45.0
Aircraft seats	(X)	8.7
Radio communication systems and equipment	(X)	655.8
Navigational systems and equipment (NAV AIDS)	(X)	254.1
Search, detection, tracking systems (RADAR, SONAR, etc.)	(X)	3,356.1
Resistors, capacitors, transformers, electron tubes, semiconductors, and other electronic components	(X)	285.3
Resin matrix composits	(X)	(D)
Other matrix composits, including ceramic, carbon, metal, etc.	(X)	41.6
Complete mechanical, hydraulic and pneumatic subassemblies	(X)	163.9
Fluid power pumps, motors, and hydrostatic transmissions (hydraulic and pneumatic)	(X)	149.6
Fluid power valves (except complete assemblies)	(X)	147.3
Fluid power hose or tube fittings and assemblies (hydraulic and pneumatic)	(X)	101.4
Fluid power cylinders and rotary actuators (except complete assemblies)	(X)	154.7
Fluid power filters (hydraulic and pneumatic)	(X)	158.1
Ball and roller bearings (mounted or unmounted)	(X)	(D)
Cutting tools for machine tools	(X)	(D)
Aircraft metal hardware (except forgings)	(X)	215.7
Metal bolts, nuts, screws, washers, rivets, and other screw machine products	(X)	447.6
Iron and steel forgings	(X)	(D)
Aluminum and aluminum-base alloy forgings	(X)	(D)
Titanium and titanium-base alloy forgings	(X)	(D)
Other forgings	(X)	(D)
Iron and steel castings (rough and semifinished)	(X)	(D)
Aluminum and aluminum-base alloy castings (rough and semifinished)	(X)	(D)
Other nonferrous castings (rough and semifinished)	(X)	21.2
Steel bars, bar shapes, and plates	(X)	115.1
Aluminum and aluminum-base alloy sheet, plate, foil, and welded tubing	(X)	268.1
All other aluminum and aluminum-base alloy shapes and forms	(X)	54.4
Copper and copper-base alloy shapes and forms	(X)	(D)
Titanium and titanium-base alloy shapes and forms	(X)	(D)
Other nonferrous shapes and forms	(X)	(D)
Paints, varnishes, lacquers, stains, shellacs, japans, enamels, and allied products	(X)	32.1
All other materials and components, parts, containers, and supplies	(X)	8,859.9
Materials, ingredients, containers, and supplies, nsk	(X)	2,310.8

Source: 1992 *Economic Census*. Explanation of symbols used: (D): Withheld to avoid disclosure of competitive data; na: Not available; (S): Withheld because statistical norms were not met; (X): Not applicable; (Z): Less than half the unit shown; nec: Not elsewhere classified; nsk: Not specified by kind; - : zero; * : 10-19 percent estimated; ** : 20-29 percent estimated.

PRODUCT SHARE DETAILS

Product or Product Class	% Share	Product or Product Class	% Share
Aircraft	100.00	Modification, conversion, and overhaul of previously accepted aircraft, nsk	0.11
Military aircraft (including all aircraft for U.S. military and any other aircraft built to military specifications)	29.12	Other aeronautical services on complete aircraft, nec	7.85
Civilian aircraft	55.85	Research and development on complete aircraft for military customers	35.20
Modification, conversion, and overhaul of previously accepted aircraft	6.59	All other aeronautical services on complete aircraft for military customers	50.09
Modification, conversion, and overhaul of U.S. military aircraft and all other aircraft built to military specifications	67.65	All other aeronautical services on complete aircraft for civilian customers	14.65
Modification, conversion, and overhaul of previously accepted aircradt for civilian customers	32.24	Other aeronautical services on complete aircraft, nec, nsk	0.06
		Aircraft, nsk	0.59

Source: 1992 *Economic Census*. The values shown are percent of total shipments in an industry. Values of indented subcategories are summed in the main heading. The symbol (D) appears when data are withheld to prevent disclosure of competitive information. The abbreviation nsk stands for 'not specified by kind' and nec for 'not elsewhere classified'.

INPUTS AND OUTPUTS FOR AIRCRAFT

Economic Sector or Industry Providing Inputs	%	Sector	Economic Sector or Industry Buying Outputs	%	Sector
Aircraft & missile engines & engine parts	25.0	Manufg.	Federal Government purchases, national defense	37.6	Fed Govt
Aircraft & missile equipment, nec	11.4	Manufg.	Exports	26.2	Foreign
Mechanical measuring devices	6.0	Manufg.	Gross private fixed investment	21.2	Cap Inv
Imports	5.9	Foreign	Change in business inventories	11.6	In House
Wholesale trade	4.5	Trade	Aircraft & missile equipment, nec	1.7	Manufg.
Radio & TV communication equipment	3.9	Manufg.	Federal Government purchases, nondefense	1.0	Fed Govt
Advertising	3.4	Services	Personal consumption expenditures	0.4	
Banking	3.1	Fin/R.E.			
Electronic components nec	2.7	Manufg.			
Special dies & tools & machine tool accessories	2.5	Manufg.			
Air transportation	2.5	Util.			

Continued on next page.

INPUTS AND OUTPUTS FOR AIRCRAFT - Continued

Economic Sector or Industry Providing Inputs	%	Sector	Economic Sector or Industry Buying Outputs	%	Sector
Miscellaneous plastics products	2.4	Manufg.			
Machinery, except electrical, nec	2.2	Manufg.			
Hotels & lodging places	1.7	Services			
Aluminum rolling & drawing	1.2	Manufg.			
Screw machine and related products	1.2	Manufg.			
Electric services (utilities)	1.2	Util.			
Semiconductors & related devices	1.1	Manufg.			
Communications, except radio & TV	0.9	Util.			
Engineering & scientific instruments	0.8	Manufg.			
Petroleum refining	0.8	Manufg.			
Public building furniture	0.8	Manufg.			
Engineering, architectural, & surveying services	0.8	Services			
Metal stampings, nec	0.7	Manufg.			
Nonferrous forgings	0.7	Manufg.			
Equipment rental & leasing services	0.7	Services			
Maintenance of nonfarm buildings nec	0.6	Constr.			
Fabricated textile products, nec	0.5	Manufg.			
Optical instruments & lenses	0.5	Manufg.			
Eating & drinking places	0.5	Trade			
Detective & protective services	0.5	Services			
Abrasive products	0.4	Manufg.			
Nonferrous wire drawing & insulating	0.4	Manufg.			
Real estate	0.4	Fin/R.E.			
Legal services	0.4	Services			
U.S. Postal Service	0.4	Gov't			
Automotive & apparel trimmings	0.3	Manufg.			
Broadwoven fabric mills	0.3	Manufg.			
Engine electrical equipment	0.3	Manufg.			
Hardware, nec	0.3	Manufg.			
Nonferrous rolling & drawing, nec	0.3	Manufg.			
Gas production & distribution (utilities)	0.3	Util.			
Motor freight transportation & warehousing	0.3	Util.			
Aluminum castings	0.2	Manufg.			
Blast furnaces & steel mills	0.2	Manufg.			
Fabricated structural metal	0.2	Manufg.			
Industrial controls	0.2	Manufg.			
Metal heat treating	0.2	Manufg.			
Paints & allied products	0.2	Manufg.			
Pipe, valves, & pipe fittings	0.2	Manufg.			
Plating & polishing	0.2	Manufg.			
Pumps & compressors	0.2	Manufg.			
Colleges, universities, & professional schools	0.2	Services			
Computer & data processing services	0.2	Services			
Management & consulting services & labs	0.2	Services			
Photofinishing labs, commercial photography	0.2	Services			
Ball & roller bearings	0.1	Manufg.			
Fabricated metal products, nec	0.1	Manufg.			
Fabricated rubber products, nec	0.1	Manufg.			
Iron & steel forgings	0.1	Manufg.			
Manifold business forms	0.1	Manufg.			
Insurance carriers	0.1	Fin/R.E.			
Accounting, auditing & bookkeeping	0.1	Services			
Miscellaneous repair shops	0.1	Services			

Source: Benchmark Input-Output Accounts for the U.S. Economy, 1982, U.S. Department of Commerce, Washington, D.C., July 1991. Data, as reported in the source, are organized by the 1977 SIC structure in use in 1982 but have been matched, as closely as is possible, to the 1987 SIC structure used in this book.

OCCUPATIONS EMPLOYED BY SIC 372 - AIRCRAFT AND PARTS

Occupation	% of Total 1994	Change to 2005	Occupation	% of Total 1994	Change to 2005
Aeronautical & astronautical engineers	5.6	-5.1	Engineering, mathematical, & science managers	1.8	18.0
Engineers nec	4.5	15.1	Secretaries, ex legal & medical	1.7	-12.7
Inspectors, testers, & graders, precision	4.5	-13.7	Tool & die makers	1.7	32.8
Management support workers nec	4.3	-4.1	NC machine tool operators, metal & plastic	1.5	-4.1
Aircraft assemblers, precision	4.1	-4.1	Stock clerks	1.4	-22.0
Blue collar worker supervisors	3.1	-4.6	Sheet metal workers & duct installers	1.4	-4.1
Machinists	3.1	-4.1	Purchasing agents, ex trade & farm products	1.4	-4.1
Assemblers, fabricators, & hand workers nec	3.0	-4.1	General office clerks	1.3	-18.2
Systems analysts	2.6	53.4	Industrial production managers	1.3	-37.7
Engineering technicians & technologists nec	2.4	-13.7	Professional workers nec	1.3	15.1
Production, planning, & expediting clerks	2.4	-4.1	Electrical & electronic technicians,technologists	1.2	-4.1
Industrial engineers, ex safety engineers	2.3	26.6	Machine tool cutting & forming etc. nec	1.1	-52.0
Mechanical engineers	2.2	-5.0	Managers & administrators nec	1.0	-4.1
Electrical & electronics engineers	2.0	2.1			

Source: Industry-Occupation Matrix, Bureau of Labor Statistics. These data relate to one or more 3-digit SIC industry groups rather than to a single 4-digit SIC. The change reported for each occupation to the year 2005 is a percent of growth or decline as estimated by the Bureau of Labor Statistics. The abbreviation nec stands for 'not elsewhere classified'.

LOCATION BY STATE AND REGIONAL CONCENTRATION

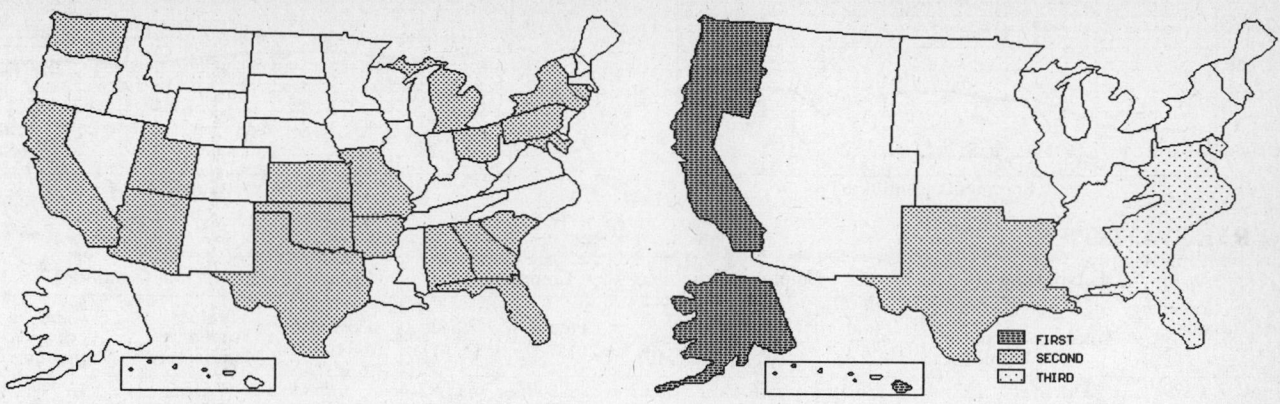

FIRST
SECOND
THIRD

INDUSTRY DATA BY STATE

State	Establish-ments	Shipments			Employment				Cost as % of Shipments	Investment per Employee ($)
		Total ($ mil)	% of U.S.	Per Establ.	Total Number	% of U.S.	Per Establ.	Wages ($/hour)		
California	36	13,015.6	20.7	361.5	58,900	22.2	1,636	23.53	58.5	3,024
Texas	26	4,835.6	7.7	186.0	29,600	11.2	1,138	19.77	35.7	1,213
Michigan	3	6.9	0.0	2.3	100	0.0	33	10.50	50.7	-
Florida	16	(D)	-	-	1,750 *	0.7	109	-	-	-
Arizona	8	(D)	-	-	7,500 *	2.8	938	-	-	-
Washington	8	(D)	-	-	75,000 *	28.3	9,375	-	-	-
Connecticut	7	(D)	-	-	17,500 *	6.6	2,500	-	-	-
Georgia	7	(D)	-	-	17,500 *	6.6	2,500	-	-	-
Kansas	7	(D)	-	-	17,500 *	6.6	2,500	-	-	-
Alabama	6	(D)	-	-	3,750 *	1.4	625	-	-	1,893
Arkansas	6	(D)	-	-	1,750 *	0.7	292	-	-	-
Ohio	6	(D)	-	-	1,750 *	0.7	292	-	-	-
Maryland	3	(D)	-	-	175 *	0.1	58	-	-	-
New York	3	(D)	-	-	17,500 *	6.6	5,833	-	-	-
Pennsylvania	3	(D)	-	-	7,500 *	2.8	2,500	-	-	-
Utah	3	(D)	-	-	750 *	0.3	250	-	-	-
Oklahoma	2	(D)	-	-	750 *	0.3	375	-	-	-
Missouri	1	(D)	-	-	17,500 *	6.6	17,500	-	-	-
South Carolina	1	(D)	-	-	1,750 *	0.7	1,750	-	-	-

Source: 1992 *Economic Census*. The states are in descending order of shipments or establishments (if shipment data are missing for the majority). The symbol (D) appears when data are withheld to prevent disclosure of competitive information. States marked with (D) are sorted by number of establishments. A dash (-) indicates that the data element cannot be calculated; * indicates the midpoint of a range.

3724 - AIRCRAFT ENGINES & ENGINE PARTS

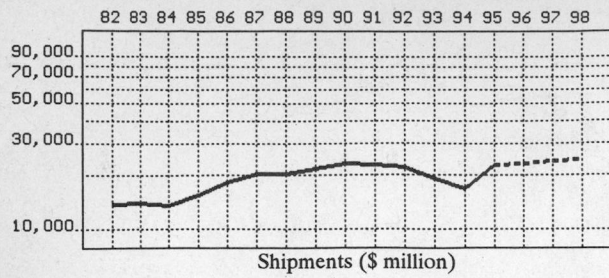

Shipments ($ million)

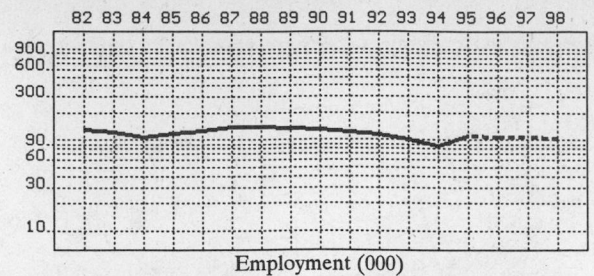

Employment (000)

GENERAL STATISTICS

| Year | Com-panies | Establishments | | Employment | | | Compensation | | Production ($ million) | | | |
		Total	with 20 or more employees	Total (000)	Production Workers (000)	Hours (Mil)	Payroll ($ mil)	Wages ($/hr)	Cost of Materials	Value Added by Manufacture	Value of Shipments	Capital Invest.
1982	279	338	223	130.5	76.4	153.7	3,540.3	11.82	6,258.9	7,565.2	13,799.1	440.5
1983		339	226	122.0	69.5	142.1	3,546.4	12.79	6,204.5	7,720.2	14,112.0	440.2
1984		340	229	109.1	64.0	134.5	3,388.3	12.41	6,205.2	7,824.7	13,659.2	632.7
1985		342	232	118.6	70.3	145.8	3,984.2	13.41	7,107.3	8,462.1	15,389.9	692.0
1986		348	239	127.5	74.7	156.9	4,388.2	13.86	7,960.9	10,791.1	18,214.2	857.6
1987	372	453	289	139.6	79.8	166.2	4,814.0	14.23	9,096.4	11,700.4	20,262.1	746.8
1988		441	277	141.4	76.8	165.3	4,988.8	14.29	9,247.4	11,157.4	20,338.7	692.6
1989		435	281	134.2	76.2	155.5	4,710.0	14.81	9,648.8	11,808.5	21,565.8	717.6
1990		442	278	133.6	72.6	147.8	4,829.6	15.68	10,310.5	12,059.1	22,812.8	784.7
1991		437	287	122.3	67.3	141.6	4,822.0	15.56	10,078.5	12,278.1	22,746.2	770.6
1992	338	442	281	116.7	64.2	130.8	4,851.7	16.97	9,019.2	11,445.4	21,968.5	590.2
1993		423	271	102.9	53.6	111.0	4,142.9	17.15	8,873.8	10,047.1	18,946.1	439.5
1994		473P	300P	86.9	46.4	94.5	3,786.0	18.02	8,455.3	7,956.5	16,663.7	437.7
1995		485P	306P	110.1P	58.0P	120.2P	4,814.9P	18.07P	11,477.6P	10,800.5P	22,620.1P	629.1P
1996		496P	312P	108.4P	56.5P	117.1P	4,889.6P	18.56P	11,776.4P	11,081.7P	23,208.9P	628.4P
1997		508P	318P	106.7P	55.0P	113.9P	4,964.4P	19.04P	12,075.2P	11,362.8P	23,797.7P	627.6P
1998		519P	325P	105.0P	53.5P	110.8P	5,039.1P	19.52P	12,373.9P	11,644.0P	24,386.5P	626.9P

Sources: 1982, 1987, 1992 *Economic Census*; *Annual Survey of Manufactures*, 83-86, 88-91, 93-94. Establishment counts for non-Census years are from *County Business Patterns*; establishment values for 83-84 are extrapolations. 'P's show projections by the editors. Industries reclassified in 87 will not have data for prior years.

INDICES OF CHANGE

| Year | Com-panies | Establishments | | Employment | | | Compensation | | Production ($ million) | | | |
		Total	with 20 or more employees	Total (000)	Production Workers (000)	Hours (Mil)	Payroll ($ mil)	Wages ($/hr)	Cost of Materials	Value Added by Manufacture	Value of Shipments	Capital Invest.
1982	83	76	79	112	119	118	73	70	69	66	63	75
1983		77	80	105	108	109	73	75	69	67	64	75
1984		77	81	93	100	103	70	73	69	68	62	107
1985		77	83	102	110	111	82	79	79	74	70	117
1986		79	85	109	116	120	90	82	88	94	83	145
1987	110	102	103	120	124	127	99	84	101	102	92	127
1988		100	99	121	120	126	103	84	103	97	93	117
1989		98	100	115	119	119	97	87	107	103	98	122
1990		100	99	114	113	113	100	92	114	105	104	133
1991		99	102	105	105	108	99	92	112	107	104	131
1992	100	100	100	100	100	100	100	100	100	100	100	100
1993		96	96	88	83	85	85	101	98	88	86	74
1994		107P	107P	74	72	72	78	106	94	70	76	74
1995		110P	109P	94P	90P	92P	99P	107P	127P	94P	103P	107P
1996		112P	111P	93P	88P	89P	101P	109P	131P	97P	106P	106P
1997		115P	113P	91P	86P	87P	102P	112P	134P	99P	108P	106P
1998		117P	115P	90P	83P	85P	104P	115P	137P	102P	111P	106P

Sources: Same as General Statistics. Values reflect change from the base year, 1992. Values above 100 mean greater than 92, values below 100 mean less than 92, and a value of 100 in the 82-91 or 93-98 period means same as 92. 'P's mark projections by the editors.

SELECTED RATIOS

For 1994	Avg. of All Manufact.	Analyzed Industry	Index	For 1994	Avg. of All Manufact.	Analyzed Industry	Index
Employees per Establishment	49	184	375	Value Added per Production Worker	134,084	171,476	128
Payroll per Establishment	1,500,273	8,002,434	533	Cost per Establishment	5,045,178	17,871,891	354
Payroll per Employee	30,620	43,567	142	Cost per Employee	102,970	97,299	94
Production Workers per Establishment	34	98	286	Cost per Production Worker	146,988	182,226	124
Wages per Establishment	853,319	3,599,383	422	Shipments per Establishment	9,576,895	35,221,912	368
Wages per Production Worker	24,861	36,700	148	Shipments per Employee	195,460	191,757	98
Hours per Production Worker	2,056	2,037	99	Shipments per Production Worker	279,017	359,131	129
Wages per Hour	12.09	18.02	149	Investment per Establishment	321,011	925,163	288
Value Added per Establishment	4,602,255	16,817,582	365	Investment per Employee	6,552	5,037	77
Value Added per Employee	93,930	91,559	97	Investment per Production Worker	9,352	9,433	101

Sources: Same as General Statistics. The 'Average of All Manufacturing' column represents the average of all manufacturing industries reported for the most recent complete year available. The Index shows the relationship between the Average and the Analyzed Industry. For example, 100 means that they are equal; 500 that the Analyzed Industry is five times the average; 50 means that the Analyzed Industry is half the national average. The abbreviation 'na' is used to show that data are 'not available'.

LEADING COMPANIES Number shown: 75   Total sales ($ mil): 28,536   Total employment (000): 192.5

Company Name	Address				CEO Name	Phone	Co. Type	Sales ($ mil)	Empl. (000)
GE Aircraft Engines	1 Neumann	Cincinnati	OH	45215	Eugene F Murphy	513-243-6136	D	6,580	29.0
AlliedSignal Aerospace	2525 W 190th St	Torrance	CA	90504	Daniel Burnham	213-321-5000	S	6,000	49.0
Pratt and Whitney	400 Main St	East Hartford	CT	06108	Karl J Krapek	203-565-4321	S	5,900	41.0
Teledyne Inc	1901 of the Stars	Los Angeles	CA	90067	William P Rutledge	310-277-3311	P	2,492	21.0
SCI Systems Inc	PO Box 1000	Huntsville	AL	35807	Olin B King	205-882-4800	D	1,853	12.0
Rohr Inc	PO Box 878	Chula Vista	CA	92012	Richard H Rau	619-691-4111	P	918	4.9
Chromalloy Gas Turbine Corp	PO Box 200150	San Antonio	TX	78220	Martin Weinstein	210-333-6010	S	713	7.5
Textron Lycoming	550 Main St	Stratford	CT	06497	David Assard	203-385-2000	D	670	2.4
Parker Bertea Aerospace Group	18321 Jamboree Rd	Irvine	CA	92715	Steve Hayes	714-833-3000	D	635	4.0
Allison Engine Co	PO Box 420	Indianapolis	IN	46206	F Blake Wallace	317-230-2000	S	580•	5.0
PCC Airfoils Inc	25201 Chagrin Blv	Beachwood	OH	44122	Peter Waite	216-831-3590	S	205	1.9
Greenwich Air Services Inc	PO Box 522187	Miami	FL	33152	E P Conese Sr	305-526-7000	P	105	0.9
Greenwich Company Ltd	116 Aragon Av	Coral Gables	FL	33134	Orlando Machado	305-526-7090	R	105	0.9
Teledyne Aircraft Products Inc	1901 of the Stars	Los Angeles	CA	90067	Hudson Drake	310-277-3311	S	100	0.7
Williams International Corp	PO Box 200	Walled Lake	MI	48390	Sam B Williams	313-624-5200	R	100•	0.7
Rolls-Royce Inc	11911 Freedom Dr	Reston	VA	22090	J Sandford	703-834-1700	S	93•	0.3
Napier Fields	PO Box 929	Dothan	AL	36302	Joseph J Walter	205-983-4571	D	90	0.6
Barnes Aerospace	169 Kennedy Rd	Windsor	CT	06095	David E Berges	203-298-7740	D	80	0.5
Chem-Tronics Inc	1150 W Bradley Av	El Cajon	CA	92020	James Legler	619-448-2320	S	77•	0.5
Parker Hannifin Corp	18321 Jamboree Rd	Irvine	CA	92715	Jim Sabin	714-833-3000	D	75	0.5
Lear Romec	PO Box 4014	Elyria	OH	44036	CB Reimer	216-323-3211	D	73	0.6
Pall Aeropower Corp	6301 49th St N	Pinellas Park	FL	34665	Robert Simkins	813-522-3111	S	72•	0.5
Litton Systems Inc	4545 S Western	Chicago	IL	60609	E Hill	312-847-4211	D	62	0.5
Fuel Systems Textron Inc	700 N Centennial St	Zeeland	MI	49464	Michael Boston	616-772-9171	S	60	0.5
Textron Specialty Materials Inc	2 Industrial Av	Lowell	MA	01851	Paul R Hoffman	508-452-8961	S	57	0.3
Aircraft Gear Corp	611 Beacon St	Rockford	IL	61111	James Olson	815-877-7473	S	44•	0.3
Turbomeca Engine Corp	2709 Forum Dr	Grand Prairie	TX	75052	Dennis Nichols	214-606-7600	S	40	0.1
Windsor Manufacturing Co	169 Kennedy Rd	Windsor	CT	06095	Doug Vaday	203-688-6411	D	38	0.2
Heico Corp	3000 Taft St	Hollywood	FL	33021	L A Mendelson	305-987-6101	P	32	0.3
Aeronca Inc	1712 Germantown	Middletown	OH	45042	James O Stine	513-422-2751	S	30	0.2
Meco Inc	PO Box 670	Paris	IL	61944	Russell L Magers	217-465-7575	R	28•	0.2
Jet Products Corp	9106 Balboa Av	San Diego	CA	92123	Ronald Blair	619-278-8400	R	26	0.2
Derlan Inc	2040 E Dyer Rd	Santa Ana	CA	92705	Terry Swain	714-250-3123	S	25•	0.2
Hitchcock Industries Inc	8701 Harriet Av S	Bloomington	MN	55420	T R Hitchcock	612-881-1000	R	25	0.3
New England Airfoil Prod	Spring Ln	Farmington	CT	06032	D Simm	203-677-1376	D	25	0.3
NORDAM Group	PO Box 1220	Springdale	AR	72765	Steve Pack	501-750-3600	D	25	0.2
West Star Aviation Inc	796 Heritage Way	Grand Junction	CO	81506	Bernard Buescher	303-243-7500	R	24	0.2
Exotic Metals Forming Co	5411 S 226th St	Kent	WA	98032	DR Lindsey	206-395-3710	R	23•	0.2
King Fifth Wheel Co	PO Box 68	Mountain Top	PA	18707	T Bellisario	717-474-6371	S	22	0.3
Airfoil Forging Textron Inc	23555 Euclid Av	Euclid	OH	44117	Michael Everhart	216-692-5200	S	20	0.1
Palmer Manufacturing Company	PO Box K	Malden	MA	02148	Frank Moda Jr	617-321-0480	R	20	0.1
World Aerospace Corp	8625 Monticello	Maple Grove	MN	55369	IE Phelps	612-424-8999	R	20	<0.1
Engine Components Inc	PO Box 17099	San Antonio	TX	78217	Gary Garvens	210-828-3131	S	19	0.2
HEICO Aerospace Corp	3000 Taft St	Hollywood	FL	33021	Eric A Mendelson	305-987-6101	S	19•	0.2
Aeroforge Corp	1200 W Jackson St	Muncie	IN	47303	Donald Wheeldon	317-747-7147	R	17	0.1
Jet Avion Corp	3000 Taft St	Hollywood	FL	33021	Eric Mendelson	305-987-6101	S	17•	0.2
Delta Industries	Bradley Park Rd	East Granby	CT	06026	William Evans	203-653-5041	R	16	0.1
Jarvis Airfoil Inc	528 Glastonbury Rd	Portland	CT	06480	Wal Jarvis	203-342-5000	R	16	0.1
NCI Inc	401 Sweeten Creek	Asheville	NC	28803	Michel Besson	704-274-4540	S	15•	0.1
CFAN	1000 Techn Way	San Marcos	TX	78666	Bob Baeumel	512-353-2832	J	14•	0.1
Corry Industries Inc	519 W Main St	Corry	PA	16407	Wilbur J Janszen	814-664-9611	R	14	0.2
Praxair Surface Technologies Inc	1234 Atlantic St	N Kansas City	MO	64116	Eric Wolber	816-556-4600	R	13	0.2
BH Aircraft Company Inc	441 E POawer H	Farmingdale	NY	11735	Vincent Kearns	516-249-5000	R	13	0.1
Electro-Jet Tool & Mfg	10400 Evendale Dr	Cincinnati	OH	45241	Paul Weber	513-563-0800	R	12	0.1
Gentz Industries Inc	23600 Schoenherr	Warren	MI	48089	Donald Duckett	810-772-2500	S	12•	<0.1
L and S Machine Company Inc	PO Box 12264	Wichita	KS	67277	Donald F Chmelka	316-942-0181	S	12	<0.1
SIFCO Custom Machining	2430 Winnetka N	Minneapolis	MN	55427	M Gonior	612-544-3511	D	12	0.1
Trilectron Industries Inc	12297 US Hwy 41 N	Palmetto	FL	34220	Charles Kott	813-723-1841	R	12	0.1
Selmet Inc	PO Box 689	Albany	OR	97321	Randy B Turner	503-926-7731	R	11•	0.1
Soloy Corp	450 Kennedy	Olympia	WA	98501	JI Soloy	206-754-7000	R	11	<0.1
Westfield Gage Company Inc	S Broad St	Westfield	MA	01085	Louis A Filios	413-568-3344	R	11	0.1
Basmat Inc	1531 W 240th St	Harbor City	CA	90710	John W Basso	310-325-2063	R	10•	<0.1
CEF Industries Inc	419 Interstate Dr	Addison	IL	60101	John Olson	708-628-2300	R	10	<0.1
PATS Inc	9570 Berger Rd	Columbia	MD	21046	Harvey O Patrick	410-381-1000	R	10	0.1
Lewis Engineering Co	238 Water St	Naugatuck	CT	06770	KJ Luczaj	203-597-6900	D	9•	0.1
Barridon Corp	PO Box 960	Hartford	CT	06143	Joseph Gantman	203-951-9736	R	8	0.1
Chromalloy Precision Products	PO Box 1337	Hurst	TX	76053	Shawn O'Keeffe	817-283-4681	D	8	<0.1
Whitcraft Corp	PO Box 128	Eastford	CT	06242	David Buchholz	203-974-0786	R	8	0.1
Aircraft Precision Products Inc	31000 Lahser Rd	Birmingham	MI	48025	W Henderson Jr	810-645-9100	R	8	<0.1
Moog Inc Engine Controls	Seneca & Jamisons	East Aurora	NY	14052	Paul Bement	716-652-2000	D	7•	<0.1
Projects Inc	65 Sequin Dr	Glastonbury	CT	06033	FM Kenyon	203-633-4615	R	7	0.1
AIW-Alton Inc	PO Box 0713	Windsor	CT	06095	William E Hoffberg	203-683-0733	R	7	<0.1
Symetrics Inc	3353 Old Conejo Rd	Newbury Park	CA	91320	John H Calvin	805-498-4586	R	6	0.1
Chromalloy Aircraft Structures	1234 Wellington Pl	Wichita	KS	67203	R Gilbert	316-262-1494	S	6	0.1
LPI Industries Corp	3000 Taft St	Hollywood	FL	33021	James Reum	305-989-3399	S	6	<0.1

Source: Ward's Business Directory of U.S. Private and Public Companies, Volumes 1 and 2, 1996. The company type code used is as follows: P - Public, R - Private, S - Subsidiary, D - Division, J - Joint Venture, A - Affiliate, G - Group. Sales are in millions of dollars, employees are in thousands. An asterisk (*) indicates an estimated sales volume. The symbol < stands for 'less than'. Company names and addresses are truncated, in some cases, to fit into the available space.

MATERIALS CONSUMED

Material	Quantity	Delivered Cost ($ million)
Materials, ingredients, containers, and supplies	(X)	7,712.7
Aircraft engines	(X)	(D)
Aircraft engine parts (except instruments)	(X)	4,350.2
Structural fuselage components, excluding instruments	(X)	(D)
Structural landing gear components	(X)	(D)
Other structural components (airframe), including engine mounts, excluding instruments	(X)	8.7
Aircraft propellers and parts thereof	(X)	(D)
Radio communication systems and equipment	(X)	(D)
Navigational systems and equipment (NAV AIDS)	(X)	(D)
Search, detection, tracking systems (RADAR, SONAR, etc.)	(X)	(D)
Resistors, capacitors, transformers, electron tubes, semiconductors, and other electronic components	(X)	8.7
Resin matrix composits	(X)	6.5
Other matrix composits, including ceramic, carbon, metal, etc.	(X)	24.1
Complete mechanical, hydraulic and pneumatic subassemblies	(X)	7.3
Fluid power pumps, motors, and hydrostatic transmissions (hydraulic and pneumatic)	(X)	3.0
Ball and roller bearings (mounted or unmounted)	(X)	(D)
Cutting tools for machine tools	(X)	(D)
Aircraft metal hardware (except forgings)	(X)	104.4
Metal bolts, nuts, screws, washers, rivets, and other screw machine products	(X)	57.6
Other fabricated metal products, except fluid power and forgings	(X)	69.7
Iron and steel forgings	(X)	127.9
Titanium and titanium-base alloy forgings	(X)	137.5
Other forgings	(X)	340.4
Iron and steel castings (rough and semifinished)	(X)	103.4
Aluminum and aluminum-base alloy castings (rough and semifinished)	(X)	71.4
Other nonferrous castings (rough and semifinished)	(X)	202.1
Steel bars, bar shapes, and plates	(X)	53.7
Steel sheet and strip, including tin plate	(X)	22.0
All other steel shapes and forms	(X)	18.2
Aluminum and aluminum-base alloy sheet, plate, foil, and welded tubing	(X)	5.0
All other aluminum and aluminum-base alloy shapes and forms	(X)	13.6
Titanium and titanium-base alloy shapes and forms	(X)	28.9
Other nonferrous shapes and forms	(X)	52.6
Paints, varnishes, lacquers, stains, shellacs, japans, enamels, and allied products	(X)	0.9
All other materials and components, parts, containers, and supplies	(X)	586.2
Materials, ingredients, containers, and supplies, nsk	(X)	855.8

Source: 1992 *Economic Census*. Explanation of symbols used: (D): Withheld to avoid disclosure of competitive data; na: Not available; (S): Withheld because statistical norms were not met; (X): Not applicable; (Z): Less than half the unit shown; nec: Not elsewhere classified; nsk: Not specified by kind; - : zero; * : 10-19 percent estimated; ** : 20-29 percent estimated.

PRODUCT SHARE DETAILS

Product or Product Class	% Share	Product or Product Class	% Share
Aircraft engines and engine parts	100.00	civilian aircraft	28.06
Military engines (for U.S. military aircraft and any other		Aircraft engine parts and accessories	37.54
aircraft built to military specifications)	15.33	Aircraft parts and accessories for military spark ignition	
Aircraft engines for civilian aircraft	32.51	reciprocating or rotary internal combustion engines	20.17
Aeronautical services on aircraft engines	12.77	Aircraft parts and accessories for other military engines	19.32
Research and development work on aircraft engines for		Aircraft parts and accessories for civilian spark ignition	
civilian aircraft engines	44.91	reciprocating or rotary internal combustion engines	24.95
All other aeronautical services on aircraft engines for U.S.		Aircraft parts and accessories for other civilian engines	33.71
military aircraft and all other engines built to military		Aircraft engine parts and accessories, nsk	1.86
specifications	27.03	Aircraft engines and engine parts, nsk	1.86
All other aeronautical services on aircraft engines for			

Source: 1992 *Economic Census*. The values shown are percent of total shipments in an industry. Values of indented subcategories are summed in the main heading. The symbol (D) appears when data is withheld to prevent disclosure of competitive information. The abbreviation nsk stands for 'not specified by kind' and nec for 'not elsewhere classified'.

INPUTS AND OUTPUTS FOR AIRCRAFT & MISSILE ENGINES & ENGINE PARTS

Economic Sector or Industry Providing Inputs	%	Sector	Economic Sector or Industry Buying Outputs	%	Sector
Aircraft & missile equipment, nec	19.6	Manufg.	Federal Government purchases, national defense	36.4	Fed Govt
Aircraft & missile engines & engine parts	11.5	Manufg.	Aircraft	29.5	Manufg.
Imports	8.0	Foreign	Exports	15.4	Foreign
Iron & steel forgings	6.7	Manufg.	Aircraft & missile engines & engine parts	6.6	Manufg.
Nonferrous forgings	4.9	Manufg.	Federal Government purchases, nondefense	4.8	Fed Govt
Advertising	4.3	Services	Air transportation	4.2	Util.
Wholesale trade	3.1	Trade	Gross private fixed investment	1.1	Cap Inv
Blast furnaces & steel mills	2.7	Manufg.	Aircraft & missile equipment, nec	0.7	Manufg.
Banking	2.5	Fin/R.E.	Change in business inventories	0.6	In House
Hotels & lodging places	2.5	Services	Electrical industrial apparatus, nec	0.2	Manufg.
Electric services (utilities)	2.0	Util.	Guided missiles & space vehicles	0.2	Manufg.

Continued on next page.

INPUTS AND OUTPUTS FOR AIRCRAFT & MISSILE ENGINES & ENGINE PARTS - Continued

Economic Sector or Industry Providing Inputs	%	Sector	Economic Sector or Industry Buying Outputs	%	Sector
Air transportation	1.7	Util.	Machinery, except electrical, nec	0.2	Manufg.
Machinery, except electrical, nec	1.6	Manufg.	Transportation equipment, nec	0.2	Manufg.
Nonferrous castings, nec	1.6	Manufg.	Motor vehicles & car bodies	0.1	Manufg.
Iron & steel foundries	1.4	Manufg.			
Equipment rental & leasing services	1.3	Services			
Aluminum castings	1.1	Manufg.			
Nonferrous rolling & drawing, nec	1.1	Manufg.			
Communications, except radio & TV	1.1	Util.			
Maintenance of nonfarm buildings nec	1.0	Constr.			
Noncomparable imports	1.0	Foreign			
Ball & roller bearings	0.9	Manufg.			
Special dies & tools & machine tool accessories	0.9	Manufg.			
Petroleum refining	0.8	Manufg.			
Screw machine and related products	0.8	Manufg.			
Computer & data processing services	0.8	Services			
Engineering, architectural, & surveying services	0.8	Services			
Metal stampings, nec	0.7	Manufg.			
Pipe, valves, & pipe fittings	0.7	Manufg.			
Plating & polishing	0.7	Manufg.			
Motor freight transportation & warehousing	0.7	Util.			
Engine electrical equipment	0.6	Manufg.			
Pumps & compressors	0.6	Manufg.			
Gas production & distribution (utilities)	0.6	Util.			
Eating & drinking places	0.6	Trade			
Aluminum rolling & drawing	0.5	Manufg.			
Legal services	0.5	Services			
Miscellaneous repair shops	0.5	Services			
U.S. Postal Service	0.5	Gov't			
Electronic components nec	0.4	Manufg.			
Fabricated rubber products, nec	0.4	Manufg.			
Abrasive products	0.3	Manufg.			
Broadwoven fabric mills	0.3	Manufg.			
Hand & edge tools, nec	0.3	Manufg.			
Machine tools, metal cutting types	0.3	Manufg.			
Real estate	0.3	Fin/R.E.			
Management & consulting services & labs	0.3	Services			
Manifold business forms	0.2	Manufg.			
Metal heat treating	0.2	Manufg.			
Metalworking machinery, nec	0.2	Manufg.			
Optical instruments & lenses	0.2	Manufg.			
Radio & TV communication equipment	0.2	Manufg.			
Royalties	0.2	Fin/R.E.			
Accounting, auditing & bookkeeping	0.2	Services			
Colleges, universities, & professional schools	0.2	Services			
Electrical repair shops	0.2	Services			
Photofinishing labs, commercial photography	0.2	Services			
Metal coating & allied services	0.1	Manufg.			
Paints & allied products	0.1	Manufg.			
Veneer & plywood	0.1	Manufg.			
Railroads & related services	0.1	Util.			
Sanitary services, steam supply, irrigation	0.1	Util.			
Water transportation	0.1	Util.			
Insurance carriers	0.1	Fin/R.E.			

Source: Benchmark Input-Output Accounts for the U.S. Economy, 1982, U.S. Department of Commerce, Washington, D.C., July 1991. Data, as reported in the source, are organized by the 1977 SIC structure in use in 1982 but have been matched, as closely as is possible, to the 1987 SIC structure used in this book.

OCCUPATIONS EMPLOYED BY SIC 372 - AIRCRAFT AND PARTS

Occupation	% of Total 1994	Change to 2005	Occupation	% of Total 1994	Change to 2005
Aeronautical & astronautical engineers	5.6	-5.1	Engineering, mathematical, & science managers	1.8	18.0
Engineers nec	4.5	15.1	Secretaries, ex legal & medical	1.7	-12.7
Inspectors, testers, & graders, precision	4.5	-13.7	Tool & die makers	1.7	32.8
Management support workers nec	4.3	-4.1	NC machine tool operators, metal & plastic	1.5	-4.1
Aircraft assemblers, precision	4.1	-4.1	Stock clerks	1.4	-22.0
Blue collar worker supervisors	3.1	-4.6	Sheet metal workers & duct installers	1.4	-4.1
Machinists	3.1	-4.1	Purchasing agents, ex trade & farm products	1.4	-4.1
Assemblers, fabricators, & hand workers nec	3.0	-4.1	General office clerks	1.3	-18.2
Systems analysts	2.6	53.4	Industrial production managers	1.3	-37.7
Engineering technicians & technologists nec	2.4	-13.7	Professional workers nec	1.3	15.1
Production, planning, & expediting clerks	2.4	-4.1	Electrical & electronic technicians,technologists	1.2	-4.1
Industrial engineers, ex safety engineers	2.3	26.6	Machine tool cutting & forming etc. nec	1.1	-52.0
Mechanical engineers	2.2	-5.0	Managers & administrators nec	1.0	-4.1
Electrical & electronics engineers	2.0	2.1			

Source: Industry-Occupation Matrix, Bureau of Labor Statistics. These data relate to one or more 3-digit SIC industry groups rather than to a single 4-digit SIC. The change reported for each occupation to the year 2005 is a percent of growth or decline as estimated by the Bureau of Labor Statistics. The abbreviation nec stands for 'not elsewhere classified'.

LOCATION BY STATE AND REGIONAL CONCENTRATION

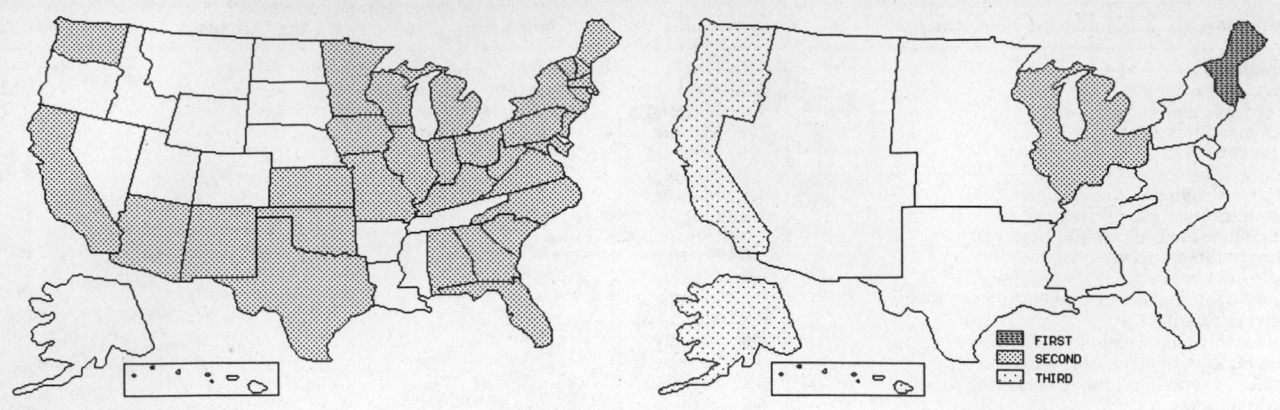

FIRST
SECOND
THIRD

INDUSTRY DATA BY STATE

| State | Establish-ments | Shipments | | | Employment | | | | Cost as % of Shipments | Investment per Employee ($) |
		Total ($ mil)	% of U.S.	Per Establ.	Total Number	% of U.S.	Per Establ.	Wages ($/hour)		
Connecticut	63	4,148.6	18.9	65.9	24,000	20.6	381	18.25	46.0	-
Florida	34	2,380.1	10.8	70.0	9,300	8.0	274	15.46	38.6	3,774
California	65	976.0	4.4	15.0	6,900	5.9	106	18.65	48.1	4,203
Indiana	12	946.2	4.3	78.8	7,900	6.8	658	16.42	47.7	-
Texas	31	549.2	2.5	17.7	3,600	3.1	116	13.46	58.5	5,306
Georgia	10	287.3	1.3	28.7	1,900	1.6	190	14.75	42.8	-
Michigan	26	258.5	1.2	9.9	2,400	2.1	92	13.67	37.1	2,625
Illinois	10	230.6	1.0	23.1	2,100	1.8	210	15.46	41.3	-
Washington	10	49.6	0.2	5.0	300	0.3	30	20.00	41.5	4,000
Ohio	29	(D)	-	-	17,500 *	15.0	603	-	-	-
Massachusetts	23	(D)	-	-	7,500 *	6.4	326	-	-	3,467
New York	21	(D)	-	-	3,750 *	3.2	179	-	-	3,440
Arizona	18	(D)	-	-	7,500 *	6.4	417	-	-	-
Oklahoma	14	(D)	-	-	1,750 *	1.5	125	-	-	-
Pennsylvania	13	(D)	-	-	3,750 *	3.2	288	-	-	-
North Carolina	7	(D)	-	-	1,750 *	1.5	250	-	-	4,971
New Jersey	5	(D)	-	-	1,750 *	1.5	350	-	-	-
Kansas	4	(D)	-	-	750 *	0.6	188	-	-	-
Minnesota	3	(D)	-	-	175 *	0.1	58	-	-	571
New Hampshire	3	(D)	-	-	750 *	0.6	250	-	-	-
Vermont	3	(D)	-	-	1,750 *	1.5	583	-	-	-
Virginia	3	(D)	-	-	375 *	0.3	125	-	-	-
Wisconsin	3	(D)	-	-	750 *	0.6	250	-	-	-
Iowa	2	(D)	-	-	375 *	0.3	188	-	-	-
Maryland	2	(D)	-	-	375 *	0.3	188	-	-	-
New Mexico	2	(D)	-	-	1,750 *	1.5	875	-	-	-
South Carolina	2	(D)	-	-	175 *	0.1	88	-	-	-
Alabama	1	(D)	-	-	750 *	0.6	750	-	-	-
Kentucky	1	(D)	-	-	750 *	0.6	750	-	-	-
Maine	1	(D)	-	-	1,750 *	1.5	1,750	-	-	-
Missouri	1	(D)	-	-	175 *	0.1	175	-	-	-
West Virginia	1	(D)	-	-	375 *	0.3	375	-	-	-

Source: 1992 *Economic Census*. The states are in descending order of shipments or establishments (if shipment data are missing for the majority). The symbol (D) appears when data are withheld to prevent disclosure of competitive information. States marked with (D) are sorted by number of establishments. A dash (-) indicates that the data element cannot be calculated; * indicates the midpoint of a range.

3728 - AIRCRAFT EQUIPMENT, NEC

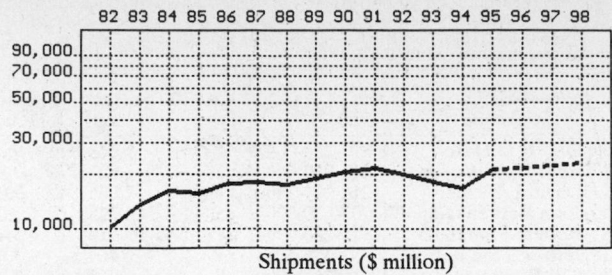

Shipments ($ million)

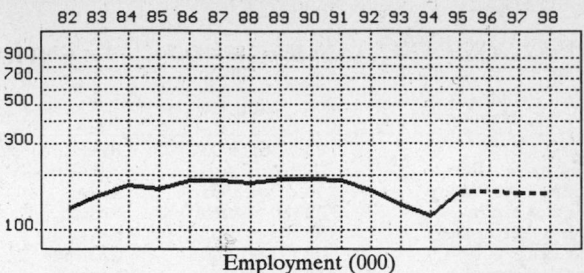

Employment (000)

GENERAL STATISTICS

| Year | Com-panies | Establishments | | Employment | | | Compensation | | Production ($ million) | | | |
		Total	with 20 or more employees	Total (000)	Production Workers (000)	Hours (Mil)	Payroll ($ mil)	Wages ($/hr)	Cost of Materials	Value Added by Manufacture	Value of Shipments	Capital Invest.
1982	912	966	419	132.8	73.5	146.6	3,429.4	11.26	3,989.0	6,188.1	10,193.1	402.9
1983		946	422	154.1	82.9	166.1	4,342.1	12.29	5,441.4	8,366.3	13,477.3	467.9
1984		926	425	175.2	93.9	195.2	5,244.8	12.68	6,528.5	10,257.1	16,217.7	597.1
1985		907	428	167.8	96.1	202.0	5,175.9	13.06	5,872.7	10,239.9	15,691.6	723.3
1986		910	434	186.4	105.2	218.9	5,972.3	13.50	6,312.6	11,804.2	17,904.6	851.9
1987	924	1,013	473	188.2	104.0	217.4	6,087.5	13.79	6,504.5	11,778.7	17,949.3	737.2
1988		986	481	181.0	98.3	209.0	5,964.9	13.85	6,803.3	11,116.7	17,720.1	639.5
1989		972	504	191.8	103.2	213.7	6,604.0	14.55	7,497.0	12,717.3	19,074.9	812.9
1990		1,016	520	190.1	111.3	225.3	6,851.9	14.89	8,091.0	12,608.7	20,457.9	815.4
1991		1,045	510	187.3	107.5	225.7	6,739.7	15.05	7,295.6	13,677.6	21,544.4	1,006.2
1992	1,030	1,121	461	165.3	93.6	194.3	6,162.1	16.44	6,108.4	12,636.6	19,834.6	1,132.1
1993		1,064	430	139.4	78.4	163.3	5,844.4	18.37	5,691.5	11,638.1	18,264.3	713.1
1994		1,085P	499P	119.5	65.4	135.5	5,348.2	19.12	5,016.2	12,047.3	16,679.4	655.7
1995		1,099P	505P	162.9P	92.3P	193.2P	6,815.2P	18.45P	6,374.9P	15,310.5P	21,197.3P	955.6P
1996		1,114P	511P	162.2P	92.2P	193.1P	6,978.2P	19.02P	6,542.0P	15,711.8P	21,752.8P	987.1P
1997		1,129P	517P	161.5P	92.0P	193.1P	7,141.1P	19.58P	6,709.1P	16,113.0P	22,308.4P	1,018.6P
1998		1,143P	523P	160.8P	91.9P	193.1P	7,304.1P	20.14P	6,876.2P	16,514.3P	22,863.9P	1,050.1P

Sources: 1982, 1987, 1992 *Economic Census*; *Annual Survey of Manufactures*, 83-86, 88-91, 93-94. Establishment counts for non-Census years are from *County Business Patterns*; establishment values for 83-84 are extrapolations. 'P's show projections by the editors. Industries reclassified in 87 will not have data for prior years.

INDICES OF CHANGE

| Year | Com-panies | Establishments | | Employment | | | Compensation | | Production ($ million) | | | |
		Total	with 20 or more employees	Total (000)	Production Workers (000)	Hours (Mil)	Payroll ($ mil)	Wages ($/hr)	Cost of Materials	Value Added by Manufacture	Value of Shipments	Capital Invest.
1982	89	86	91	80	79	75	56	68	65	49	51	36
1983		84	92	93	89	85	70	75	89	66	68	41
1984		83	92	106	100	100	85	77	107	81	82	53
1985		81	93	102	103	104	84	79	96	81	79	64
1986		81	94	113	112	113	97	82	103	93	90	75
1987	90	90	103	114	111	112	99	84	106	93	90	65
1988		88	104	109	105	108	97	84	111	88	89	56
1989		87	109	116	110	110	107	89	123	101	96	72
1990		91	113	115	119	116	111	91	132	100	103	72
1991		93	111	113	115	116	109	92	119	108	109	89
1992	100	100	100	100	100	100	100	100	100	100	100	100
1993		95	93	84	84	84	95	112	93	92	92	63
1994		97P	108P	72	70	70	87	116	82	95	84	58
1995		98P	109P	99P	99P	99P	111P	112P	104P	121P	107P	84P
1996		99P	111P	98P	98P	99P	113P	116P	107P	124P	110P	87P
1997		101P	112P	98P	98P	99P	116P	119P	110P	128P	112P	90P
1998		102P	113P	97P	98P	99P	119P	122P	113P	131P	115P	93P

Sources: Same as General Statistics. Values reflect change from the base year, 1992. Values above 100 mean greater than 92, values below 100 mean less than 92, and a value of 100 in the 82-91 or 93-98 period means same as 92. 'P's mark projections by the editors.

SELECTED RATIOS

For 1994	Avg. of All Manufact.	Analyzed Industry	Index	For 1994	Avg. of All Manufact.	Analyzed Industry	Index
Employees per Establishment	49	110	225	Value Added per Production Worker	134,084	184,209	137
Payroll per Establishment	1,500,273	4,931,214	329	Cost per Establishment	5,045,178	4,625,099	92
Payroll per Employee	30,620	44,755	146	Cost per Employee	102,970	41,977	41
Production Workers per Establishment	34	60	176	Cost per Production Worker	146,988	76,700	52
Wages per Establishment	853,319	2,388,765	280	Shipments per Establishment	9,576,895	15,378,947	161
Wages per Production Worker	24,861	39,614	159	Shipments per Employee	195,460	139,577	71
Hours per Production Worker	2,056	2,072	101	Shipments per Production Worker	279,017	255,037	91
Wages per Hour	12.09	19.12	158	Investment per Establishment	321,011	604,577	188
Value Added per Establishment	4,602,255	11,108,001	241	Investment per Employee	6,552	5,487	84
Value Added per Employee	93,930	100,814	107	Investment per Production Worker	9,352	10,026	107

Sources: Same as General Statistics. The 'Average of All Manufacturing' column represents the average of all manufacturing industries reported for the most recent complete year available. The Index shows the relationship between the Average and the Analyzed Industry. For example, 100 means that they are equal; 500 that the Analyzed Industry is five times the average; 50 means that the Analyzed Industry is half the national average. The abbreviation 'na' is used to show that data are 'not available'.

LEADING COMPANIES   Number shown: **75**   Total sales ($ mil): **10,001**   Total employment (000): **79.7**

Company Name	Address				CEO Name	Phone	Co. Type	Sales ($ mil)	Empl. (000)
Sundstrand Corp	PO Box 7003	Rockford	IL	61125	Don R O'Hare	815-226-6000	P	1,373	9.2
Coltec Industries Inc	430 Park Av	New York	NY	10022	John W Guffey Jr	212-940-0400	P	1,327	9.8
BF Goodrich Aerospace	PO Box 5501	Akron	OH	44334	David L Burner	216-374-2200	D	1,000	9.0
Lucas Aerospace Inc	11180 Sunrise Val	Reston	VA	22091	John Berkenkamp	703-620-8901	S	825	7.0
Sundstrand Aerospace	PO Box 7002	Rockford	IL	61125	Robert J Smuland	815-226-6000	D	710	5.4
Fairchild Corp	PO Box 10803	Chantilly	VA	22021	Jeffrey J Steiner	703-478-5800	P	464	3.6
Textron Aerostructures	PO Box 210	Nashville	TN	37202	CW Wells	615-361-2000	S	265	1.9
BE Aerospace Inc	1400 Corporate Ctr	Wellington	FL	33414	Amin J Khoury	407-791-5000	P	229	1.8
Convair	PO Box 85377	San Diego	CA	92186	Art Weitch	619-573-8000	D	190•	1.9
Kaman Aerospace Corp	PO Box 2	Bloomfield	CT	06002	Walter R Kozlow	203-243-7376	S	175	1.2
Loral Electro-Optical Systems	300 N Halstead St	Pasadena	CA	91107	Robert Mueller	818-351-5555	S	130	1.0
Menasco Aerosystems	4000 Hwy 157	Euless	TX	76040	RC Grill	817-283-4471	D	130•	1.0
Ketema Inc	501 S Cherry St	Denver	CO	80222	H H Williamson III	303-331-0940	P	127	1.1
Fairchild Fastener Group	3000 Lomita Blv	Torrance	CA	90505	Joseph Hood	310-784-0700	S	120•	2.0
Sundstrand Power Systems	4400 Ruffin Rd	San Diego	CA	92123	Omar Winter	619-569-4400	D	120	1.1
Hazeltine Corp	450 E Pulaski Rd	Greenlawn	NY	11740	Angelo Filosa	516-261-7000	S	110	1.0
OEA Inc	PO Box 100488	Denver	CO	80250	Ahmed D Kafadar	303-693-1248	P	110	0.9
Kaiser Electronics	2701 Orchard Pkwy	San Jose	CA	95134	Allen Gates	408-432-3000	D	100	0.7
Lockheed Martin Control Syst	600 Main St	Johnson City	NY	13790	JD Scanlon	607-770-2000	D	100•	1.0
AlliedSignal Aircraft	3520 Westmoor St	South Bend	IN	46628	Thomas A Johnson	219-231-2000	D	97•	0.8
Heath Tecna Aerospace Co	19819 84th Av S	Kent	WA	98032		206-872-7500	S	95	1.0
Fansteel Inc	1 Tantalum Pl	North Chicago	IL	60064	KR Garrity	708-689-4900	P	89	0.8
DeCrane Aircraft Holdings Inc	155 Montrose	Copley	OH	44321	R Jack DeCrane	216-668-3061	R	80•	0.4
Hydro-Aire	PO Box 7722	Burbank	CA	91510	RL Gruber	818-842-6121	D	70	0.4
IFE	PO Box 22008	Santa Ana	CA	92702	John Landstrom	714-698-1600	D	66•	0.4
Whittaker Controls Inc	12838 Saticoy St	N Hollywood	CA	91605	Jay Fernandez	818-765-8160	S	65	0.3
Ducommun Inc	23301 S Wilmington	Carson	CA	90745	Norman A Barkeley	310-513-7200	P	62	0.7
Ellanef Manufacturing Corp	97-11 50th Av	Corona	NY	11368	Mingo V Logothetis	718-699-4000	R	60	0.5
Sabreliner Corp	7733 Forsyth Blv	St Louis	MO	63105	F Holmes Lamoreux	314-863-6880	R	57•	0.5
Lucas Aerospace	PO Box 457	Utica	NY	13503	Frank Robilotto	315-793-1200	S	56	0.3
Jet Electronics and Technology	PO Box 873	Grand Rapids	MI	49588	Wes Perry	616-949-6600	S	55	0.5
Bird-Johnson Co	110 Norfolk St	Walpole	MA	02081	Peter J Gwyn	508-668-9610	S	50	0.3
Lucas Western Inc	PO Box 680910	Park City	UT	84068	Ron Strobl	801-649-1900	D	50	0.4
Sargent-Fletcher Co	9400 Flair Dr	El Monte	CA	91731	Gordon Smith	818-443-7171	R	50	0.3
Arrowhead Products	4411 Katella Av	Los Alamitos	CA	90720	D Shanahan	714-828-7770	D	49•	0.4
Delavan Gas Turbine Products	811 4th St	W Des Moines	IA	50265	James R Baker	515-274-1561	D	48	0.4
Landoll Corp	1700 May St	Marysville	KS	66508	Don Landoll	913-562-5381	R	48	0.5
Dynamic Controls Corp	PO Box 73	South Windsor	CT	06074	TP Farkas	203-528-9971	R	45	0.4
Korry Electronics Co	901 Dexter Av N	Seattle	WA	98109	David Elkins	206-281-1300	S	45•	0.4
Mamco Manufacturing Inc	PO Box 70645	Seattle	WA	98107	August Riehl	206-789-1111	R	45•	0.4
GEC-Marconi Aerospace Inc	110 Algonquin Pkwy	Whippany	NJ	07981	David A Sapio	201-428-9898	S	42	0.4
Hydro-Mill Co	9301 Mason Av	Chatsworth	CA	91311	Gloria Coppin	818-341-1314	R	41	0.3
Monitor Aerospace Corp	1000 New Horizons	Amityville	NY	11701	Doug Monitto	516-957-2300	R	40	0.4
PL Porter Co	6355 De Soto Av	Woodland Hills	CA	91367	Gregory Lennox	818-884-7260	R	40	0.5
Schneller Inc	PO Box 670	Kent	OH	44240	Donald R Cardis	216-673-1400	R	40	0.2
Aeroquip Corp	300 S East Av	Jackson	MI	49203	Roger Kremer	517-787-8121	D	39•	0.4
Kaiser Electroprecision Inc	17000 S Red Hill	Irvine	CA	92714	Dennis L Weaver	714-250-1015	S	39	0.3
Micro Craft Inc	PO Box 370	Tullahoma	TN	37388	Dan J Marcum	615-455-2664	R	39	0.4
Smith Industries Inc	PO Box 5389	Clearwater	FL	34618	William F Talley	813-531-7781	S	39	0.4
JC Carter Company Inc	671 W 17th St	Costa Mesa	CA	92627	R Veloz	714-548-3421	R	37	0.2
Rohr Inc	18238 Showalter Rd	Hagerstown	MD	21742	Phil Vacca	301-790-9500	D	36•	0.2
American Fuel Cell	PO Box 887	Magnolia	AR	71753	John G Ball	501-234-3381	R	35	0.6
Hartzell Propeller Inc	1 Propeller Pl	Piqua	OH	45356	Art Disbrow	513-778-4200	R	34	0.3
AHF Ducommun Inc	PO Box 2310	Gardena	CA	90247	Robert Hansen	310-380-5390	S	33	0.2
Dowty Aerospace Aviation Svcs	PO Box 5000	Sterling	VA	20167	AT White	703-450-8200	S	32	0.2
Litton Industries Inc	PO Box 4508	Davenport	IA	52808	John Heffernan	319-383-6000	D	32	0.3
Universal Propulsion Company	25401 N Central	Phoenix	AZ	85027	Harold G Watson	602-869-8067	S	32	0.2
Arrow Gear Co	2301 Curtiss St	Downers Grove	IL	60515	Joseph L Arvin	708-969-7640	R	30•	0.2
Assurance Technology Corp	84 South St	Carlisle	MA	01741	H Learue Renfro	508-369-8848	R	30	0.3
AVICOM International Inc	2100 E Alosta Av	Glendora	CA	91740	RA Bertagna	818-857-0061	S	30	0.4
Welco	1515 N A St	Wellington	KS	67152	Duane Creveling	316-326-5921	D	30	0.3
Ozone Industries Inc	101-32 101st St	Ozone Park	NY	11416	Robert Terenzi	718-845-5200	S	29	0.3
Arkwin Industries Inc	686 Main St	Westbury	NY	11590	Daniel Berlin	516-333-2640	R	28	0.2
Contraves Simulation	5902 Breckenridge	Tampa	FL	33610	Roger Charbonneau	813-628-6100	D	28	0.1
Sargent Controls and Aerospace	5675 W Burlingame	Tucson	AZ	85743	Donald Tarquin	602-744-1000	S	28•	0.3
AGC Inc	106 Evansville Av	Meriden	CT	06450	William Winaker	203-235-3361	R	26	0.3
Bruce Industries Inc	PO Box 1700	Dayton	NV	89403	Frank B Bruce	702-246-0101	R	26•	0.2
ACR Industries Inc	15375 23 Mile Rd	Macomb	MI	48042	Roger Blanchard	810-781-2800	R	25	0.1
McCauley Accessory	3535 McCauley Dr	Vandalia	OH	45377	William Buckles	513-890-5246	D	25	0.2
Meyer Tool Inc	3064 Colerain Av	Cincinnati	OH	45225	Arlyn T Easton	513-681-7362	R	25	0.3
MC Gill Corp	4056 Easy St	El Monte	CA	91731	Stephen E Gill	818-443-4022	R	25	0.1
Parker Hannifin Corp	PO Box 4032	Elyria	OH	44036	John R Hruska	216-284-6300	D	25	0.2
Purdy Corp	PO Box 1898	Manchester	CT	06045	JM Purdy Jr	203-649-0000	R	25	0.1
Rogerson Aircraft Corp	2485 DaVinchi	Irvine	CA	92714	Michael Rogerson	714-660-0666	R	25	0.2
Utica Corp	2 Halsey Rd	Whitesboro	NY	13492	Gordon Johnston	315-768-2008	S	25	0.2

Source: Ward's Business Directory of U.S. Private and Public Companies, Volumes 1 and 2, 1996. The company type code used is as follows: P - Public, R - Private, S - Subsidiary, D - Division, J - Joint Venture, A - Affiliate, G - Group. Sales are in millions of dollars, employees are in thousands. An asterisk (•) indicates an estimated sales volume. The symbol < stands for 'less than'. Company names and addresses are truncated, in some cases, to fit into the available space.

MATERIALS CONSUMED

Material	Quantity	Delivered Cost ($ million)
Materials, ingredients, containers, and supplies	(X)	4,988.8
Aircraft engines	(X)	80.0
Structural fuselage components, excluding instruments	(X)	511.1
Aircraft propellers and parts thereof	(X)	7.7
Aircraft seats	(X)	(D)
Radio communication systems and equipment	(X)	24.8
Navigational systems and equipment (NAV AIDS)	(X)	17.0
Search, detection, tracking systems (RADAR, SONAR, etc.)	(X)	46.5
Resistors, capacitors, transformers, electron tubes, semiconductors, and other electronic components	(X)	162.9
Resin matrix composits	(X)	78.5
Other matrix composits, including ceramic, carbon, metal, etc.	(X)	84.7
Complete mechanical, hydraulic and pneumatic subassemblies	(X)	94.5
Fluid power pumps, motors, and hydrostatic transmissions (hydraulic and pneumatic)	(X)	15.3
Fluid power valves (except complete assemblies)	(X)	57.6
Fluid power hose or tube fittings and assemblies (hydraulic and pneumatic)	(X)	5.4
Fluid power cylinders and rotary actuators (except complete assemblies)	(X)	8.5
Fluid power filters (hydraulic and pneumatic)	(X)	4.6
Other fluid power products (hydraulic and pneumatic)	(X)	47.1
Ball and roller bearings (mounted or unmounted)	(X)	48.9
Cutting tools for machine tools	(X)	41.8
Aircraft metal hardware (except forgings)	(X)	285.8
Metal bolts, nuts, screws, washers, rivets, and other screw machine products	(X)	152.5
Other fabricated metal products, except fluid power and forgings	(X)	251.2
Iron and steel forgings	(X)	46.6
Aluminum and aluminum-base alloy forgings	(X)	79.8
Titanium and titanium-base alloy forgings	(X)	44.5
Other forgings	(X)	37.0
Iron and steel castings (rough and semifinished)	(X)	45.8
Aluminum and aluminum-base alloy castings (rough and semifinished)	(X)	51.9
Other nonferrous castings (rough and semifinished)	(X)	54.7
Steel bars, bar shapes, and plates	(X)	38.1
Steel sheet and strip, including tin plate	(X)	22.6
All other steel shapes and forms	(X)	10.4
Aluminum and aluminum-base alloy sheet, plate, foil, and welded tubing	(X)	138.7
All other aluminum and aluminum-base alloy shapes and forms	(X)	75.3
Copper and copper-base alloy shapes and forms	(X)	1.5
Titanium and titanium-base alloy shapes and forms	(X)	151.5
Paints, varnishes, lacquers, stains, shellacs, japans, enamels, and allied products	(X)	(D)
All other materials and components, parts, containers, and supplies	(X)	1,309.5
Materials, ingredients, containers, and supplies, nsk	(X)	817.5

Source: 1992 Economic Census. Explanation of symbols used: (D): Withheld to avoid disclosure of competitive data; na: Not available; (S): Withheld because statistical norms were not met; (X): Not applicable; (Z): Less than half the unit shown; nec: Not elsewhere classified; nsk: Not specified by kind; - : zero; * : 10-19 percent estimated; ** : 20-29 percent estimated.

PRODUCT SHARE DETAILS

Product or Product Class	% Share	Product or Product Class	% Share
Aircraft parts and auxiliary equipment, nec	100.00	Aircraft pneumatic subassemblies for civilian aircraft	25.30
Aircraft propellers and helicopter rotors	2.34	Aircraft hydraulic and pneumatic subassemblies, nsk	0.16
Complete aircraft propellers, excluding helicopter rotors	16.73	Aircraft parts and auxiliary equipment, excluding hydraulic	
Aircraft propeller blades	12.48	and pneumatic subassemblies and engines	80.93
Aircraft propeller parts, except propeller blades	15.88	Aircraft mechanical power transmission equipment for U.S.	
Helicopter rotors and parts	51.72	military aircraft and all other aircraft built to military	
Aircraft propellers and helicopter rotors, nsk	3.21	specifications	3.29
Research and development on aircraft parts (except		Aircraft mechanical power transmission equipment for	
engines)	5.92	civilian aircraft	4.19
Research and development on aircraft parts (except		Aircraft landing gear for U.S. military aircraft and all other	
engines) for civilian aircraft	99.96	aircraft built to military specifications	4.87
Research and development on aircraft parts (except		Aircraft landing gear for civilian aircraft	3.52
engines), nsk	0.04	Other aircraft subassemblies and parts for U.S. military	
Aircraft hydraulic and pneumatic subassemblies	6.66	aircraft and all other aircraft built to military	
Aircraft hydraulic subassemblies for U.S. military aircraft		specifications	21.98
and all other aircraft built to military specifications	13.13	Other aircraft subassemblies and parts for civilian aircraft	61.86
Aircraft hydraulic subassemblies for civilian aircraft	52.52	Aircraft parts and auxiliary equipment, excluding hydraulic	
Aircraft pneumatic subassemblies for U.S. military aircraft		and pneumatic subassemblies and engines, nsk	0.29
and all other aircraft built to military specifications	8.88	Aircraft parts and auxiliary equipment, nec, nsk	4.15

Source: 1992 Economic Census. The values shown are percent of total shipments in an industry. Values of indented subcategories are summed in the main heading. The symbol (D) appears when data are withheld to prevent disclosure of competitive information. The abbreviation nsk stands for 'not specified by kind' and nec for 'not elsewhere classified'.

INPUTS AND OUTPUTS FOR AIRCRAFT & MISSILE EQUIPMENT, NEC

Economic Sector or Industry Providing Inputs	%	Sector	Economic Sector or Industry Buying Outputs	%	Sector
Imports	18.9	Foreign	Exports	31.4	Foreign
Aircraft & missile equipment, nec	9.0	Manufg.	Federal Government purchases, national defense	29.2	Fed Govt
Aircraft	7.3	Manufg.	Aircraft	13.4	Manufg.
Electronic components nec	6.3	Manufg.	Aircraft & missile engines & engine parts	11.3	Manufg.
Semiconductors & related devices	5.8	Manufg.	Guided missiles & space vehicles	5.8	Manufg.
Wholesale trade	4.2	Trade	Aircraft & missile equipment, nec	4.2	Manufg.
Advertising	3.7	Services	Air transportation	2.3	Util.
Banking	2.4	Fin/R.E.	Federal Government purchases, nondefense	1.7	Fed Govt
Aluminum rolling & drawing	2.0	Manufg.	Change in business inventories	0.5	In House
Electric services (utilities)	2.0	Util.			
Optical instruments & lenses	1.7	Manufg.			
Hotels & lodging places	1.7	Services			
Iron & steel forgings	1.5	Manufg.			
Nonferrous forgings	1.5	Manufg.			
Aircraft & missile engines & engine parts	1.4	Manufg.			
Communications, except radio & TV	1.4	Util.			
Equipment rental & leasing services	1.3	Services			
Air transportation	1.2	Util.			
Blast furnaces & steel mills	1.1	Manufg.			
Machinery, except electrical, nec	1.1	Manufg.			
Nonferrous rolling & drawing, nec	1.1	Manufg.			
Screw machine and related products	1.1	Manufg.			
Public building furniture	1.0	Manufg.			
Business services nec	1.0	Services			
Engineering, architectural, & surveying services	0.9	Services			
Maintenance of nonfarm buildings nec	0.8	Constr.			
Metal stampings, nec	0.8	Manufg.			
Nonferrous castings, nec	0.7	Manufg.			
Detective & protective services	0.7	Services			
Miscellaneous plastics products	0.6	Manufg.			
Special dies & tools & machine tool accessories	0.6	Manufg.			
Gas production & distribution (utilities)	0.6	Util.			
Eating & drinking places	0.6	Trade			
Real estate	0.6	Fin/R.E.			
Ball & roller bearings	0.5	Manufg.			
Fabricated rubber products, nec	0.5	Manufg.			
Petroleum refining	0.5	Manufg.			
Plating & polishing	0.5	Manufg.			
Power transmission equipment	0.5	Manufg.			
Legal services	0.5	Services			
U.S. Postal Service	0.5	Gov't			
Abrasive products	0.4	Manufg.			
Aluminum castings	0.4	Manufg.			
Machine tools, metal cutting types	0.4	Manufg.			
Motor freight transportation & warehousing	0.4	Util.			
Cyclic crudes and organics	0.3	Manufg.			
Fabricated metal products, nec	0.3	Manufg.			
Fabricated structural metal	0.3	Manufg.			
Hardware, nec	0.3	Manufg.			
Pipe, valves, & pipe fittings	0.3	Manufg.			
Plastics materials & resins	0.3	Manufg.			
Computer & data processing services	0.3	Services			
Management & consulting services & labs	0.3	Services			
Asbestos products	0.2	Manufg.			
Chemical preparations, nec	0.2	Manufg.			
Iron & steel foundries	0.2	Manufg.			
Manifold business forms	0.2	Manufg.			
Metal heat treating	0.2	Manufg.			
Miscellaneous fabricated wire products	0.2	Manufg.			
Motors & generators	0.2	Manufg.			
Nonferrous wire drawing & insulating	0.2	Manufg.			
Paints & allied products	0.2	Manufg.			
Primary metal products, nec	0.2	Manufg.			
Radio & TV communication equipment	0.2	Manufg.			
Wood pallets & skids	0.2	Manufg.			
Accounting, auditing & bookkeeping	0.2	Services			
Business/professional associations	0.2	Services			
Miscellaneous repair shops	0.2	Services			
Photofinishing labs, commercial photography	0.2	Services			
Broadwoven fabric mills	0.1	Manufg.			
Pumps & compressors	0.1	Manufg.			
Insurance carriers	0.1	Fin/R.E.			
Royalties	0.1	Fin/R.E.			
Colleges, universities, & professional schools	0.1	Services			
Noncomparable imports	0.1	Foreign			

Source: Benchmark Input-Output Accounts for the U.S. Economy, 1982, U.S. Department of Commerce, Washington, D.C., July 1991. Data, as reported in the source, are organized by the 1977 SIC structure in use in 1982 but have been matched, as closely as is possible, to the 1987 SIC structure used in this book.

OCCUPATIONS EMPLOYED BY SIC 372 - AIRCRAFT AND PARTS

Occupation	% of Total 1994	Change to 2005	Occupation	% of Total 1994	Change to 2005
Aeronautical & astronautical engineers	5.6	-5.1	Engineering, mathematical, & science managers	1.8	18.0
Engineers nec	4.5	15.1	Secretaries, ex legal & medical	1.7	-12.7
Inspectors, testers, & graders, precision	4.5	-13.7	Tool & die makers	1.7	32.8
Management support workers nec	4.3	-4.1	NC machine tool operators, metal & plastic	1.5	-4.1
Aircraft assemblers, precision	4.1	-4.1	Stock clerks	1.4	-22.0
Blue collar worker supervisors	3.1	-4.6	Sheet metal workers & duct installers	1.4	-4.1
Machinists	3.1	-4.1	Purchasing agents, ex trade & farm products	1.4	-4.1
Assemblers, fabricators, & hand workers nec	3.0	-4.1	General office clerks	1.3	-18.2
Systems analysts	2.6	53.4	Industrial production managers	1.3	-37.7
Engineering technicians & technologists nec	2.4	-13.7	Professional workers nec	1.3	15.1
Production, planning, & expediting clerks	2.4	-4.1	Electrical & electronic technicians,technologists	1.2	-4.1
Industrial engineers, ex safety engineers	2.3	26.6	Machine tool cutting & forming etc. nec	1.1	-52.0
Mechanical engineers	2.2	-5.0	Managers & administrators nec	1.0	-4.1
Electrical & electronics engineers	2.0	2.1			

Source: Industry-Occupation Matrix, Bureau of Labor Statistics. These data relate to one or more 3-digit SIC industry groups rather than to a single 4-digit SIC. The change reported for each occupation to the year 2005 is a percent of growth or decline as estimated by the Bureau of Labor Statistics. The abbreviation nec stands for 'not elsewhere classified'.

LOCATION BY STATE AND REGIONAL CONCENTRATION

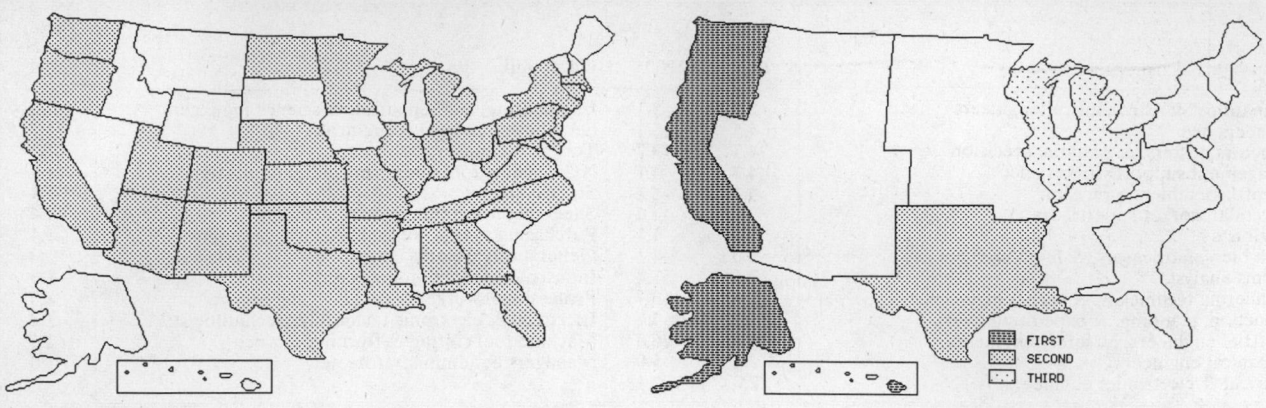

FIRST
SECOND
THIRD

INDUSTRY DATA BY STATE

| State | Establish-ments | Shipments | | | Employment | | | | Cost as % of Shipments | Investment per Employee ($) |
		Total ($ mil)	% of U.S.	Per Establ.	Total Number	% of U.S.	Per Establ.	Wages ($/hour)		
California	292	5,710.0	28.8	19.6	48,200	29.2	165	15.92	33.8	3,330
Washington	118	2,054.2	10.4	17.4	18,000	10.9	153	16.79	20.0	-
Texas	83	1,626.1	8.2	19.6	18,900	11.4	228	18.38	38.9	2,582
Connecticut	56	1,385.3	7.0	24.7	11,100	6.7	198	18.19	19.4	6,649
New York	54	589.7	3.0	10.9	3,200	1.9	59	12.86	56.6	-
Oklahoma	30	456.0	2.3	15.2	4,800	2.9	160	18.60	28.7	2,125
Michigan	34	232.4	1.2	6.8	1,800	1.1	53	15.67	44.3	-
New Jersey	20	229.0	1.2	11.5	1,900	1.1	95	13.43	38.4	-
Colorado	15	91.7	0.5	6.1	900	0.5	60	14.50	29.6	-
North Carolina	6	27.6	0.1	4.6	300	0.2	50	10.67	35.9	-
Massachusetts	5	10.4	0.1	2.1	100	0.1	20	17.00	36.5	-
Kansas	70	(D)	-	-	17,500 *	10.6	250	-	-	-
Florida	51	(D)	-	-	1,750 *	1.1	34	-	-	-
Ohio	42	(D)	-	-	7,500 *	4.5	179	-	-	-
Arizona	26	(D)	-	-	3,750 *	2.3	144	-	-	4,107
Missouri	25	(D)	-	-	750 *	0.5	30	-	-	-
Indiana	20	(D)	-	-	1,750 *	1.1	88	-	-	-
Oregon	18	(D)	-	-	1,750 *	1.1	97	-	-	-
Pennsylvania	16	(D)	-	-	750 *	0.5	47	-	-	1,467
Arkansas	14	(D)	-	-	1,750 *	1.1	125	-	-	857
Illinois	14	(D)	-	-	3,750 *	2.3	268	-	-	-
Utah	14	(D)	-	-	750 *	0.5	54	-	-	-
Minnesota	11	(D)	-	-	175 *	0.1	16	-	-	2,286
Tennessee	10	(D)	-	-	3,750 *	2.3	375	-	-	2,400
Alabama	9	(D)	-	-	750 *	0.5	83	-	-	-
Georgia	9	(D)	-	-	1,750 *	1.1	194	-	-	-
Maryland	6	(D)	-	-	750 *	0.5	125	-	-	-
New Mexico	5	(D)	-	-	175 *	0.1	35	-	-	-
Louisiana	3	(D)	-	-	175 *	0.1	58	-	-	-
Virginia	3	(D)	-	-	375 *	0.2	125	-	-	-
Nebraska	2	(D)	-	-	375 *	0.2	188	-	-	-
West Virginia	2	(D)	-	-	375 *	0.2	188	-	-	-
North Dakota	1	(D)	-	-	375 *	0.2	375	-	-	-

Source: 1992 *Economic Census*. The states are in descending order of shipments or establishments (if shipment data are missing for the majority). The symbol (D) appears when data are withheld to prevent disclosure of competitive information. States marked with (D) are sorted by number of establishments. A dash (-) indicates that the data element cannot be calculated; * indicates the midpoint of a range.

3731 - SHIP BUILDING & REPAIRING

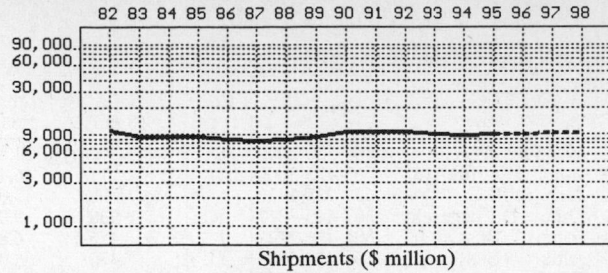

Shipments ($ million)

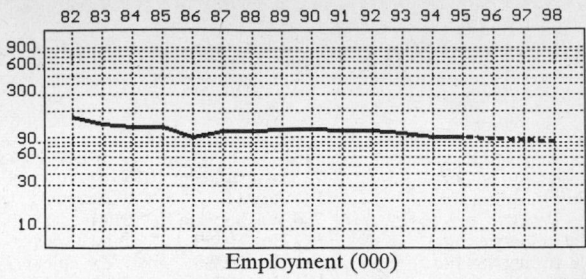

Employment (000)

GENERAL STATISTICS

Year	Companies	Establishments		Employment			Compensation		Production ($ million)			
		Total	with 20 or more employees	Total (000)	Production Workers (000)	Hours (Mil)	Payroll ($ mil)	Wages ($/hr)	Cost of Materials	Value Added by Manufacture	Value of Shipments	Capital Invest.
1982	616	690	379	166.9	130.8	259.6	3,740.7	10.68	4,593.6	6,385.6	10,979.2	438.6
1983		648	349	141.0	107.0	208.4	3,294.8	11.11	3,919.0	5,572.0	9,487.1	304.8
1984		606	319	132.7	102.5	203.4	3,332.0	11.45	3,778.1	5,867.6	9,643.6	342.8
1985		564	289	130.3	99.0	194.0	3,315.9	11.74	3,618.0	5,739.8	9,357.7	296.0
1986		537	272	100.6	90.3	179.3	3,181.3	11.98	3,413.8	5,426.1	8,839.9	214.2
1987	547	590	287	120.2	90.5	178.1	3,217.9	12.16	3,291.7	5,212.7	8,504.4	273.0
1988		566	295	120.1	89.5	177.5	3,254.3	12.30	3,570.9	5,222.1	8,793.0	237.9
1989		540	296	121.7	87.9	177.4	3,386.2	12.31	3,986.6	5,653.7	9,640.2	193.0
1990		532	281	122.0	91.0	184.6	3,605.8	12.59	4,492.8	6,362.8	10,855.7	226.8
1991		551	269	120.8	90.1	183.9	3,679.8	12.97	4,494.7	6,354.2	10,848.8	172.2
1992	562	598	267	118.3	87.1	176.8	3,624.1	12.98	4,065.2	6,543.2	10,608.3	128.4
1993		613	269	111.5	81.9	159.0	3,362.6	13.45	4,006.8	5,957.2	9,964.0	155.8
1994		544P	246P	101.5	75.6	151.7	3,311.1	13.98	4,103.7	5,760.3	9,864.0	140.7
1995		538P	238P	101.3P	73.1P	147.9P	3,448.1P	13.93P	4,223.7P	5,928.8P	10,152.5P	93.5P
1996		531P	231P	98.1P	70.1P	142.3P	3,453.9P	14.16P	4,244.7P	5,958.3P	10,203.0P	72.6P
1997		525P	223P	94.9P	67.1P	136.7P	3,459.6P	14.40P	4,265.7P	5,987.8P	10,253.5P	51.6P
1998		518P	215P	91.7P	64.1P	131.1P	3,465.3P	14.63P	4,286.8P	6,017.3P	10,304.0P	30.6P

Sources: 1982, 1987, 1992 *Economic Census*; *Annual Survey of Manufactures*, 83-86, 88-91, 93-94. Establishment counts for non-Census years are from *County Business Patterns*; establishment values for 83-84 are extrapolations. 'P's show projections by the editors. Industries reclassified in 87 will not have data for prior years.

INDICES OF CHANGE

Year	Companies	Establishments		Employment			Compensation		Production ($ million)			
		Total	with 20 or more employees	Total (000)	Production Workers (000)	Hours (Mil)	Payroll ($ mil)	Wages ($/hr)	Cost of Materials	Value Added by Manufacture	Value of Shipments	Capital Invest.
1982	110	115	142	141	150	147	103	82	113	98	103	342
1983		108	131	119	123	118	91	86	96	85	89	237
1984		101	119	112	118	115	92	88	93	90	91	267
1985		94	108	110	114	110	91	90	89	88	88	231
1986		90	102	85	104	101	88	92	84	83	83	167
1987	97	99	107	102	104	101	89	94	81	80	80	213
1988		95	110	102	103	100	90	95	88	80	83	185
1989		90	111	103	101	100	93	95	98	86	91	150
1990		89	105	103	104	104	99	97	111	97	102	177
1991		92	101	102	103	104	102	100	111	97	102	134
1992	100	100	100	100	100	100	100	100	100	100	100	100
1993		103	101	94	94	90	93	104	99	91	94	121
1994		91P	92P	86	87	86	91	108	101	88	93	110
1995		90P	89P	86P	84P	84P	95P	107P	104P	91P	96P	73P
1996		89P	86P	83P	80P	80P	95P	109P	104P	91P	96P	57P
1997		88P	83P	80P	77P	77P	95P	111P	105P	92P	97P	40P
1998		87P	80P	78P	74P	74P	96P	113P	105P	92P	97P	24P

Sources: Same as General Statistics. Values reflect change from the base year, 1992. Values above 100 mean greater than 92, values below 100 mean less than 92, and a value of 100 in the 82-91 or 93-98 period means same as 92. 'P's mark projections by the editors.

SELECTED RATIOS

For 1994	Avg. of All Manufact.	Analyzed Industry	Index	For 1994	Avg. of All Manufact.	Analyzed Industry	Index
Employees per Establishment	49	187	381	Value Added per Production Worker	134,084	76,194	57
Payroll per Establishment	1,500,273	6,086,072	406	Cost per Establishment	5,045,178	7,542,936	150
Payroll per Employee	30,620	32,622	107	Cost per Employee	102,970	40,431	39
Production Workers per Establishment	34	139	405	Cost per Production Worker	146,988	54,282	37
Wages per Establishment	853,319	3,898,141	457	Shipments per Establishment	9,576,895	18,130,838	189
Wages per Production Worker	24,861	28,052	113	Shipments per Employee	195,460	97,182	50
Hours per Production Worker	2,056	2,007	98	Shipments per Production Worker	279,017	130,476	47
Wages per Hour	12.09	13.98	116	Investment per Establishment	321,011	258,618	81
Value Added per Establishment	4,602,255	10,587,902	230	Investment per Employee	6,552	1,386	21
Value Added per Employee	93,930	56,752	60	Investment per Production Worker	9,352	1,861	20

Sources: Same as General Statistics. The 'Average of All Manufacturing' column represents the average of all manufacturing industries reported for the most recent complete year available. The Index shows the relationship between the Average and the Analyzed Industry. For example, 100 means that they are equal; 500 that the Analyzed Industry is five times the average; 50 means that the Analyzed Industry is half the national average. The abbreviation 'na' is used to show that data are 'not available'.

LEADING COMPANIES Number shown: **75** Total sales ($ mil): **8,989** Total employment (000): **101.8**

Company Name	Address				CEO Name	Phone	Co. Type	Sales ($ mil)	Empl. (000)
Newport News Shipbuilding	4101 Washington	Newport News	VA	23607	WR Phillips Jr	804-380-2000	S	1,900	21.8
General Dynamics Corp	75 Eastern Point Rd	Groton	CT	06340	James E Turner Jr	203-433-3000	D	1,711	16.4
Ingalls Shipbuilding Inc	PO Box 149	Pascagoula	MS	39568	Jerry St Pe	601-935-1122	S	1,000	15.0
Bath Iron Works Corp	700 Washington St	Bath	ME	04530	Duane D Fitzgerald	207-443-3311	R	930	9.0
Avondale Industries Inc	PO Box 50280	New Orleans	LA	70150	Albert L Bossier Jr	504-436-2121	P	476	6.2
Avondale Industries Inc	PO Box 50280	New Orleans	LA	70150	Albert L Bossier	504-436-5393	D	470*	6.0
J Ray McDermott Inc	PO Box 188	Morgan City	LA	70381	RH Rawle	504-631-2561	D	295	2.6
NASSCO Holdings Inc	PO Box 85278	San Diego	CA	92138	Richard Vortmann	619-544-3400	R	260	3.2
Trinity Marine Group	13085 Indrial Seawy	Gulfport	MS	39505	John Dane III	601-896-0029	D	200	2.0
Bethship	Sparrows Pt Shpyd	Sparrows Point	MD	21219	Dave Watson	410-388-7702	D	80	0.8
Metro Machine Corp	PO Box 1860	Norfolk	VA	23501	R A Goldbach	804-543-6801	R	80	0.7
MARCO Seattle	2300 W Commod	Seattle	WA	98199	Peter G Schmidt	206-285-3200	R	75*	0.8
West State Inc	PO Box 4768	Portland	OR	97208	Tore Steen	503-285-9706	R	75	0.8
Bollinger Machine Shop	PO Box 250	Lockport	LA	70374	Donald T Bollinger	504-532-2554	R	70	0.8
Service Marine Industries Inc	PO Box 3606	Morgan City	LA	70381	Mike Clute	504-631-0511	R	67	0.3
Todd Shipyards Corp	PO Box 3806	Seattle	WA	98124	P WE Hodgson	206-223-1560	P	54	0.6
Atlantic Marine Holding Co	8500 Heckscher Dr	Jacksonville	FL	32226	George W Gibbs	904-251-3111	R	53*	0.7
Bender Shipbuilding and Repair	PO Box 42	Mobile	AL	36601	TB Bender Jr	205-431-8000	R	51*	0.6
Jeffboat	PO Box 610	Jeffersonville	IN	47130	RW Greene	812-288-0130	D	50	0.6
Viking Yacht Co	Rte 9 & Bass River	New Gretna	NJ	08224	William J Healey	609-296-6000	R	48	0.6
Hatteras Yachts	2100 Kivett Dr	High Point	NC	27260	D Alton Herndon	919-889-6621	S	46	1.0
North Florida Shipyards Inc	PO Box 3255	Jacksonville	FL	32206	Joseph B Shiffert	904-354-3278	R	45*	0.5
Production Management Cos	PO Box 44	Harvey	LA	70059	Michael C Sport	504-366-3594	R	45	0.7
Service Engineering Industries	PO Box 7714	San Francisco	CA	94120	Orlindo Barsetti	415-957-1777	R	42*	0.5
Trinity Indust Moss Point Marine	PO Box 1310	Escatawpa	MS	39552	Dan Shrahan	601-475-6885	D	41	0.4
Service Engineering Co	PO Box 7714	San Francisco	CA	94120	Orlindo Barsetti	415-957-1777	S	40*	0.5
Quality Shipyards Inc	PO Box 1817	Houma	LA	70361	Bobby Barthel	504-876-4846	S	37*	0.4
Marine Hydraulics International	543 E Indian River	Norfolk	VA	23523	James S Hong	804-545-6400	R	35	0.4
Detyens Shipyards Inc	Rte 2	Mount Pleasant	SC	29464	D Loy Stewart	803-884-2811	R	33*	0.4
Hopeman Brothers Inc	PO Box 820	Waynesboro	VA	22980		703-949-9200	S	30	0.5
Leevac Shipyards Inc	PO Box 1190	Jennings	LA	70546	Fred Stokes	318-824-2210	S	30	0.3
Texas Drydock Inc	PO Box 968	Orange	TX	77631	Don Covington	409-883-0954	R	29*	0.4
Nashville Bridge Co	PO Box 239	Nashville	TN	37202	A Zang	615-244-2050	R	28	0.3
Colonna's Shipyard Inc	400 E Indian River	Norfolk	VA	23523	Thomas Godfrey	804-545-2414	R	27	0.3
Pacific Ship Repair & Fabrication	PO Box 13428	San Diego	CA	92170	Dennis R Shaw	619-232-3200	R	27	0.3
Craig Systems Corp	10 Industrial Way	Amesbury	MA	01913	Harley W Waite Jr	508-388-5662	S	26	0.1
Campbell Industries Inc	PO Box 1870	San Diego	CA	92112	Robert F Allen	619-233-7115	S	25	0.4
General Ship Corp	300 Northern Av	South Boston	MA	02210	Arnold Mende	617-261-4200	R	25	0.2
Marinette Marine Corp	1600 Ely St	Marinette	WI	54143	Daniel L Gulling	715-735-9341	R	25	0.4
HBC Barge Inc	PO Box 510	Brownsville	PA	15417	W Ray Wallace	412-785-6100	S	20	0.3
Tollycraft Corp	2200 Clinton Av	Kelso	WA	98626	DR Cooley	206-423-5160	R	20	0.2
Platzer Shipyard Inc	PO Box 24399	Houston	TX	77229	Neal S Platzer	713-453-7251	R	17*	0.2
Production Management Indust	PO Box 44	Harvey	LA	70059	Michael C Sport	504-363-5900	D	17	0.3
Donco Industries Inc	2401 Union St	Oakland	CA	94607	Donald Manning	510-272-9922	R	16	<0.1
Global Movable Offshore Inc	PO Box 67	Amelia	LA	70340	William Dore	504-631-2124	S	16*	0.2
Houma Fabricators Inc	1100 Oak St	Houma	LA	70363	OE Monnier Jr	504-879-3346	S	15	0.1
Metal Trades Inc	PO Box 129	Hollywood	SC	29449	RB Corbin	803-889-6441	R	15*	0.2
Newpark Shipbuilding&Repair	PO Box 5426	Houston	TX	77262	Joseph P O'Toole	713-928-5051	S	15	0.2
Lake Union Drydock Co	1515 Fairview Av E	Seattle	WA	98102	George N Neilson	206-323-6400	R	14	0.1
Buck Kreihs Company Inc	PO Box 53305	New Orleans	LA	70153	Albert N Kreihs	504-524-7681	R	13	0.2
Caddell Dry Dock and Repair	PO Box 327	Staten Island	NY	10310	Steven Kalil	718-442-2112	R	13	0.2
Main Iron Works Inc	PO Box 1918	Houma	LA	70361	Leroy Molaison	504-876-6302	R	13	0.1
Park-Ohio Industries Inc	PO Box 05935	Cleveland	OH	44105	Edward Crawford	216-341-2300	D	13	0.1
Al Larson Boat Shop Inc	PO Box 3098	Terminal Island	CA	90731	Gloria Wall	310-514-4100	R	13	0.2
Bay Shipbuilding Co	PO Box 830	Sturgeon Bay	WI	54235	Bruce Shaw	414-743-5524	D	12	0.2
Christensen Shipyards Ltd	4400 Columbia	Vancouver	WA	98661	Dave Christensen	206-695-7671	R	12	0.2
Gulf Marine Repair Corp	1200 Sertoma Dr	Tampa	FL	33605	Aaron W Hendry	813-247-3153	R	11	0.1
American Marine Corp	PO Box 8126	New Orleans	LA	70182	Peter Durant	504-242-5200	R	10*	0.1
Bay Ship and Yacht Co	PO Box 4014	Alameda	CA	94501	William Elliott	510-237-0140	R	10*	0.2
Conrad Industries Inc	PO Box 790	Morgan City	LA	70381	JP Conrad	504-384-3060	R	10	0.1
Dixie Machine Welding	PO Box 53355	New Orleans	LA	70153	Carl M Roussel	504-581-3088	S	10	0.1
Duwamish Shipyard Inc	5658 W Marginal	Seattle	WA	98106	DM Larsen	206-767-4880	R	10	0.1
Louisiana Dock Co	5004 River Rd	Harahan	LA	70123	CW Kinzeler	504-733-4190	S	10*	0.1
PacOrd Inc	240 W 30th St	National City	CA	91950	Bob Hartsock	619-336-2000	S	10	0.1
Golten's NY Corp	160 Van Brunt St	Brooklyn	NY	11231	Norman S Golten	718-855-7200	R	8	<0.1
Marine Industries Northwest Inc	PO Box 1275	Tacoma	WA	98401	DA Slater	206-627-9136	R	8*	<0.1
Pacific Fisherman Inc	5351 24th Av NW	Seattle	WA	98107	P Ballinger	206-784-2562	R	8	<0.1
American Shipyard Corp	1 Washington St	Newport	RI	02840	Kreso Bezmalinovic	401-846-6700	R	7*	<0.1
Dakota Creek Industries Inc	PO Box 218	Anacortes	WA	98221	Richard Nelson	206-293-2931	R	7*	<0.1
Gladding-Hearn Shipbuilding	PO Box 300	Somerset	MA	02726	George Duclos	508-676-8596	R	7	<0.1
Fraser Shipyards Inc	PO Box 997	Superior	WI	54880	Trevor White	715-394-7787	S	6	0.1
JBF Scientific Company Inc	PO Box 139	SW Harbor	ME	04679	Ted Kleinman	207-244-9611	R	6*	<0.1
Orange Shipbuilding Co	PO Box 1670	Orange	TX	77630	Thomas E Clary	409-883-6666	R	6	<0.1
Allied Shipyard Inc	PO Box 1240	Larose	LA	70373	Ronald Callais	504-693-3323	R	6	0.1
Deep Ocean Engineering Inc	1431 Doolittle Dr	San Leandro	CA	94577	Philip J Ballou	510-562-9300	R	5	<0.1

Source: Ward's Business Directory of U.S. Private and Public Companies, Volumes 1 and 2, 1996. The company type code used is as follows: P - Public, R - Private, S - Subsidiary, D - Division, J - Joint Venture, A - Affiliate, G - Group. Sales are in millions of dollars, employees are in thousands. An asterisk (*) indicates an estimated sales volume. The symbol < stands for 'less than'. Company names and addresses are truncated, in some cases, to fit into the available space.

MATERIALS CONSUMED

Material	Quantity	Delivered Cost ($ million)
Materials, ingredients, containers, and supplies	(X)	3,525.6
Diesel and semidiesel engines	(X)	97.0
Integral horsepower electric motors and generators (1 hp or more)	(X)	76.5
Engine electrical equipment, including spark plugs, magnetos, generators, starters, etc.	(X)	68.0
Mechanical speed changers, gears, and industrial high-speed drives	(X)	88.6
Numerical controls for metalworking machinery (except programmable)	(X)	11.2
Fluid power pumps, motors, and hydrostatic transmissions (hydraulic and pneumatic)	(X)	27.5
Fluid power valves (hydraulic and pneumatic)	(X)	54.6
Fluid power cylinders and rotary actuators (hydraulic and pneumatic)	(X)	1.8
Fluid power hose or tube fittings and assemblies (hydraulic and pneumatic)	(X)	35.2
Fluid power filters (hydraulic and pneumatic)	(X)	11.8
Other fluid power products (hydraulic and pneumatic)	(X)	18.7
Fabricated structural metal for ships and barges (except forgings)	(X)	113.5
Metal boilers, condensers, and parts thereof (except forgings)	(X)	11.4
Metal bolts, nuts, screws, washers, rivets, and other screw machine products	(X)	41.8
Other fabricated metal products (except fluid power products and forgings)	(X)	23.7
Forgings	(X)	4.1
Castings (rough and semifinished)	(X)	20.7
Steel bars, bar shapes, and plates	(X)	281.5
Steel sheet and strip, including tin plate	(X)	26.2
Steel structural shapes and sheet piling	(X)	64.6
All other steel shapes and forms	(X)	24.9
Aluminum and aluminum-base alloy sheet, plate, foil, and welded tubing	(X)	15.9
Other aluminum and aluminum-base alloy shapes and forms	(X)	3.0
Other nonferrous shapes and forms	(X)	26.9
Construction machinery and parts thereof, including shipwinches, cranes, derricks, and capstans	(X)	42.9
Paints, varnishes, lacquers, stains, shellacs, japans, enamels, and allied products	(X)	61.3
Dressed lumber	(X)	12.6
All other materials and components, parts, containers, and supplies	(X)	1,555.7
Materials, ingredients, containers, and supplies, nsk	(X)	704.1

Source: 1992 *Economic Census*. Explanation of symbols used: (D): Withheld to avoid disclosure of competitive data; na: Not available; (S): Withheld because statistical norms were not met; (X): Not applicable; (Z): Less than half the unit shown; nec: Not elsewhere classified; nsk: Not specified by kind; - : zero; * : 10-19 percent estimated; ** : 20-29 percent estimated.

PRODUCT SHARE DETAILS

Product or Product Class	% Share	Product or Product Class	% Share
Ship building and repairing	100.00	Self-propelled commercial fishing trawlers, nonmilitary, new construction	8.07
Nonpropelled ships, new construction	4.32	Other self-propelled commercial fishing vessels (including seiners), nonmilitary, new construction	3.76
Military and nonmilitary nonpropelled barges, all types, new construction	83.33	Self-propelled tugboats and towboats, including integrated tug/barge combination, nonmilitary, new construction	1.65
Military and nonmilitary nonpropelled drilling/production platforms, new construction	13.38	Self-propelled ferryboats, nonmilitary, new construction	2.36
Nonpropelled ships, new construction, nsk	3.29	Other self-propelled ships (including container and trailer ships), nonmilitary, new construction	63.55
Military, self-propelled ships, including combat ships, troop transport vessels, fleet auxiliaries, service craft, etc., new construction	58.07	Self-propelled ships, nonmilitary, new construction, nsk	1.80
		Ship repair, military	18.95
Self-propelled ships, nonmilitary, new construction	6.24	Ship conversions and reconversions, military	19.52
Self-propelled yachts, nonmilitary, 65 ft or more in length (requires a professional crew as specified by the Coast Guard), new construction	16.63	All other ship repairs, military	80.01
		Ship repair, military, nsk	0.47
Self-propelled dry bulk carriers, nonmilitary, new construction	0.32	Ship repair, nonmilitary	8.39
		Ship conversions and reconversions, nonmilitary	13.44
Military and nonmilitary nonpropelled tankers, new construction	0.09	All other ship repairs, nonmilitary	83.97
		Ship repair, nonmilitary, nsk	2.58
Self-propelled support vessels for offshore drilling and mining, nonmilitary, new construction	1.74	Ship building and repairing, nsk	4.03

Source: 1992 *Economic Census*. The values shown are percent of total shipments in an industry. Values of indented subcategories are summed in the main heading. The symbol (D) appears when data are withheld to prevent disclosure of competitive information. The abbreviation nsk stands for 'not specified by kind' and nec for 'not elsewhere classified'.

INPUTS AND OUTPUTS FOR SHIP BUILDING & REPAIRING

Economic Sector or Industry Providing Inputs	%	Sector	Economic Sector or Industry Buying Outputs	%	Sector
Blast furnaces & steel mills	11.2	Manufg.	Federal Government purchases, national defense	49.6	Fed Govt
Wholesale trade	8.0	Trade	Gross private fixed investment	24.0	Cap Inv
Maintenance of nonfarm buildings nec	7.9	Constr.	Exports	10.8	Foreign
Royalties	5.2	Fin/R.E.	Water transportation	9.2	Util.
Construction machinery & equipment	4.8	Manufg.	Federal Government purchases, nondefense	3.4	Fed Govt
Imports	3.8	Foreign	Commercial fishing	2.4	Agric.
Internal combustion engines, nec	3.0	Manufg.	Ship building & repairing	0.3	Manufg.
Fabricated structural metal	2.8	Manufg.	S/L Govt. purch., transit utilities	0.1	S/L Govt
Electric services (utilities)	2.5	Util.			
Petroleum refining	2.3	Manufg.			
Power transmission equipment	2.0	Manufg.			
Blowers & fans	1.9	Manufg.			
Fabricated plate work (boiler shops)	1.8	Manufg.			
Pipe, valves, & pipe fittings	1.8	Manufg.			
Engineering & scientific instruments	1.6	Manufg.			
Machinery, except electrical, nec	1.6	Manufg.			
Paints & allied products	1.6	Manufg.			
Radio & TV communication equipment	1.6	Manufg.			
Advertising	1.4	Services			
Engineering, architectural, & surveying services	1.4	Services			
Equipment rental & leasing services	1.4	Services			
Industrial controls	1.2	Manufg.			
Pumps & compressors	1.1	Manufg.			
Screw machine and related products	1.0	Manufg.			
Communications, except radio & TV	1.0	Util.			
Motor freight transportation & warehousing	1.0	Util.			
Banking	1.0	Fin/R.E.			
Mineral wool	0.9	Manufg.			
Furniture & fixtures, nec	0.8	Manufg.			
Nonferrous wire drawing & insulating	0.8	Manufg.			
Eating & drinking places	0.8	Trade			
Aluminum rolling & drawing	0.7	Manufg.			
Brass, bronze, & copper castings	0.7	Manufg.			
Engine electrical equipment	0.7	Manufg.			
Copper rolling & drawing	0.6	Manufg.			
Motors & generators	0.6	Manufg.			
Ship building & repairing	0.6	Manufg.			
Wood products, nec	0.6	Manufg.			
Real estate	0.6	Fin/R.E.			
Computer & data processing services	0.6	Services			
Automotive & apparel trimmings	0.5	Manufg.			
Boat building & repairing	0.5	Manufg.			
Fabricated metal products, nec	0.5	Manufg.			
Particleboard	0.5	Manufg.			
Refrigeration & heating equipment	0.5	Manufg.			
Industrial gases	0.4	Manufg.			
Lubricating oils & greases	0.4	Manufg.			
Special dies & tools & machine tool accessories	0.4	Manufg.			
Welding apparatus, electric	0.4	Manufg.			
Automotive rental & leasing, without drivers	0.4	Services			
Canvas & related products	0.3	Manufg.			
Iron & steel foundries	0.3	Manufg.			
Sawmills & planning mills, general	0.3	Manufg.			
Service industry machines, nec	0.3	Manufg.			
Gas production & distribution (utilities)	0.3	Util.			
Railroads & related services	0.3	Util.			
Sanitary services, steam supply, irrigation	0.3	Util.			
Automotive repair shops & services	0.3	Services			
Legal services	0.3	Services			
Management & consulting services & labs	0.3	Services			
Miscellaneous repair shops	0.3	Services			
U.S. Postal Service	0.3	Gov't			
Abrasive products	0.2	Manufg.			
Ball & roller bearings	0.2	Manufg.			
Hardware, nec	0.2	Manufg.			
Miscellaneous plastics products	0.2	Manufg.			
Nonferrous rolling & drawing, nec	0.2	Manufg.			
Power driven hand tools	0.2	Manufg.			
Insurance carriers	0.2	Fin/R.E.			
Accounting, auditing & bookkeeping	0.2	Services			
Hotels & lodging places	0.2	Services			
Fabricated rubber products, nec	0.1	Manufg.			
General industrial machinery, nec	0.1	Manufg.			
Hand saws & saw blades	0.1	Manufg.			
Hardwood dimension & flooring mills	0.1	Manufg.			
Machine tools, metal cutting types	0.1	Manufg.			
Manifold business forms	0.1	Manufg.			
Narrow fabric mills	0.1	Manufg.			

Continued on next page.

INPUTS AND OUTPUTS FOR SHIP BUILDING & REPAIRING - Continued

Economic Sector or Industry Providing Inputs	%	Sector	Economic Sector or Industry Buying Outputs	%	Sector
Semiconductors & related devices	0.1	Manufg.			
Transformers	0.1	Manufg.			
Turbines & turbine generator sets	0.1	Manufg.			
Veneer & plywood	0.1	Manufg.			
Water transportation	0.1	Util.			
Retail trade, except eating & drinking	0.1	Trade			
Security & commodity brokers	0.1	Fin/R.E.			
Detective & protective services	0.1	Services			
Electrical repair shops	0.1	Services			
Laundry, dry cleaning, shoe repair	0.1	Services			

Source: Benchmark Input-Output Accounts for the U.S. Economy, 1982, U.S. Department of Commerce, Washington, D.C., July 1991. Data, as reported in the source, are organized by the 1977 SIC structure in use in 1982 but have been matched, as closely as is possible, to the 1987 SIC structure used in this book.

OCCUPATIONS EMPLOYED BY SIC 373 - SHIP AND BOAT BUILDING AND REPAIRING

Occupation	% of Total 1994	Change to 2005	Occupation	% of Total 1994	Change to 2005
Welders & cutters	10.7	-17.5	Drafters	1.9	-35.7
Assemblers, fabricators, & hand workers nec	8.1	-17.4	Painters & paperhangers	1.9	-17.4
Shipfitters	6.3	-0.9	Sheet metal workers & duct installers	1.9	-17.5
Blue collar worker supervisors	5.6	-18.2	General managers & top executives	1.8	-21.7
Plumbers, pipefitters, & steamfitters	5.0	-17.5	Inspectors, testers, & graders, precision	1.7	-17.4
Carpenters	3.6	-34.0	Stock clerks	1.3	-32.9
Electricians	3.3	-22.5	Production, planning, & expediting clerks	1.3	-17.5
Machinists	3.2	-17.4	Helpers, construction trades	1.3	-17.4
Helpers, laborers, & material movers nec	2.9	-17.5	Grinders & polishers, hand	1.2	-17.5
Painters, transportation equipment	2.7	-0.9	Industrial production managers	1.1	-17.5
Riggers	2.2	-13.1	Welding machine setters, operators	1.0	-25.7
General office clerks	2.0	-29.6			

Source: Industry-Occupation Matrix, Bureau of Labor Statistics. These data relate to one or more 3-digit SIC industry groups rather than to a single 4-digit SIC. The change reported for each occupation to the year 2005 is a percent of growth or decline as estimated by the Bureau of Labor Statistics. The abbreviation nec stands for 'not elsewhere classified'.

LOCATION BY STATE AND REGIONAL CONCENTRATION

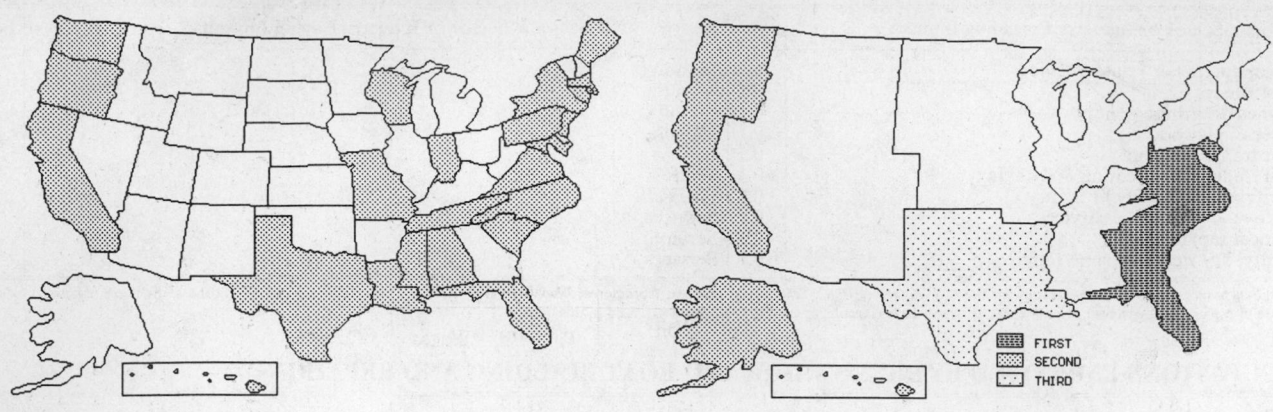

FIRST
SECOND
THIRD

INDUSTRY DATA BY STATE

| State | Establish-ments | Shipments | | | Employment | | | | Cost as % of Shipments | Investment per Employee ($) |
		Total ($ mil)	% of U.S.	Per Establ.	Total Number	% of U.S.	Per Establ.	Wages ($/hour)		
Virginia	41	2,713.6	25.6	66.2	28,600	24.2	698	14.24	34.4	1,080
Louisiana	66	1,240.6	11.7	18.8	13,200	11.2	200	10.09	50.0	992
California	67	851.9	8.0	12.7	9,800	8.3	146	13.24	42.1	878
Florida	69	280.4	2.6	4.1	3,600	3.0	52	12.28	51.4	1,861
Washington	46	269.5	2.5	5.9	3,000	2.5	65	15.33	36.1	1,333
Wisconsin	9	208.5	2.0	23.2	1,600	1.4	178	11.65	60.9	1,750
Alabama	23	181.4	1.7	7.9	1,700	1.4	74	11.36	52.9	1,882
Oregon	19	180.9	1.7	9.5	2,100	1.8	111	14.58	39.7	1,095
Texas	48	178.0	1.7	3.7	2,300	1.9	48	11.75	40.3	1,957
South Carolina	9	74.1	0.7	8.2	900	0.8	100	11.21	39.9	-
New York	28	72.7	0.7	2.6	800	0.7	29	14.92	37.1	1,500
Massachusetts	15	49.7	0.5	3.3	400	0.3	27	13.00	38.6	-
New Jersey	20	37.2	0.4	1.9	500	0.4	25	10.00	40.1	1,800
Rhode Island	7	21.7	0.2	3.1	300	0.3	43	12.20	39.6	-
Mississippi	17	(D)	-	-	17,500 *	14.8	1,029	-	-	-
Maine	15	(D)	-	-	7,500 *	6.3	500	-	-	-
Pennsylvania	13	(D)	-	-	750 *	0.6	58	-	-	-
Maryland	12	(D)	-	-	1,750 *	1.5	146	-	-	-
Connecticut	9	(D)	-	-	17,500 *	14.8	1,944	-	-	-
North Carolina	8	(D)	-	-	1,750 *	1.5	219	-	-	-
Tennessee	6	(D)	-	-	375 *	0.3	63	-	-	-
Indiana	5	(D)	-	-	750 *	0.6	150	-	-	-
Missouri	5	(D)	-	-	375 *	0.3	75	-	-	-
Hawaii	4	(D)	-	-	175 *	0.1	44	-	-	-

Source: 1992 *Economic Census*. The states are in descending order of shipments or establishments (if shipment data are missing for the majority). The symbol (D) appears when data are withheld to prevent disclosure of competitive information. States marked with (D) are sorted by number of establishments. A dash (-) indicates that the data element cannot be calculated; * indicates the midpoint of a range.

3732 - BOAT BUILDING & REPAIRING

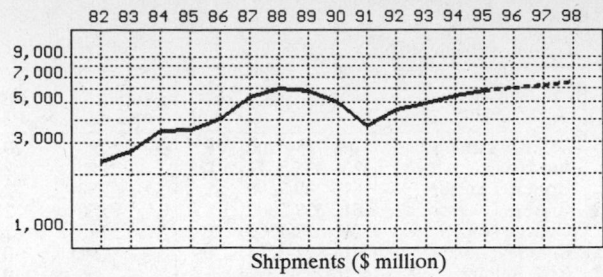

Shipments ($ million)

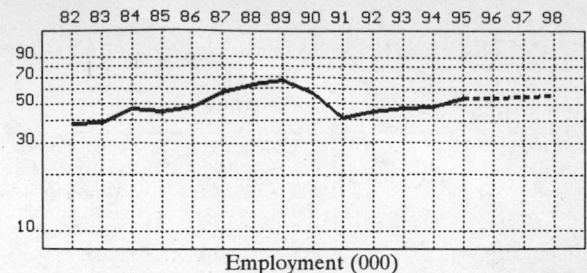

Employment (000)

GENERAL STATISTICS

Year	Companies	Establishments		Employment			Compensation		Production ($ million)			
		Total	with 20 or more employees	Total (000)	Production Workers (000)	Hours (Mil)	Payroll ($ mil)	Wages ($/hr)	Cost of Materials	Value Added by Manufacture	Value of Shipments	Capital Invest.
1982	1,833	1,876	384	38.2	30.7	59.7	585.2	7.04	1,229.1	1,123.0	2,347.2	57.5
1983		1,858	393	38.9	32.0	60.4	634.1	7.54	1,451.5	1,252.3	2,690.1	40.1
1984		1,840	402	46.3	38.5	73.8	770.4	7.52	1,962.9	1,492.8	3,452.4	90.0
1985		1,822	410	45.1	35.7	69.3	786.8	7.91	2,028.1	1,466.7	3,496.4	79.9
1986		1,787	412	48.1	39.0	78.8	892.4	8.17	2,239.7	1,804.2	4,048.7	87.8
1987	2,108	2,176	433	57.2	47.1	94.3	1,047.7	8.13	2,947.3	2,444.1	5,352.5	133.8
1988		2,077	473	62.8	51.9	103.0	1,161.2	8.22	3,324.5	2,679.4	5,935.1	140.6
1989		1,975	471	66.9	49.2	99.3	1,183.6	8.44	3,242.6	2,522.3	5,739.3	120.3
1990		2,032	451	57.0	42.7	84.8	1,060.8	8.68	2,792.1	2,191.8	4,998.0	83.4
1991		2,042	387	40.8	31.7	62.2	824.2	9.17	2,058.1	1,605.0	3,675.6	38.7
1992	2,376	2,455	413	44.5	34.4	68.7	1,005.8	9.62	2,609.4	1,991.4	4,599.3	63.1
1993		2,435	411	47.0	36.8	71.9	1,033.4	9.57	2,919.4	2,066.9	4,975.3	83.2
1994		2,360P	438P	48.1	38.9	78.8	1,096.8	9.77	3,144.0	2,288.6	5,425.1	99.2
1995		2,410P	441P	53.4P	41.2P	83.0P	1,183.0P	9.98P	3,360.7P	2,446.3P	5,799.0P	94.1P
1996		2,461P	444P	54.0P	41.5P	83.8P	1,219.2P	10.20P	3,479.5P	2,532.8P	6,004.0P	95.3P
1997		2,511P	446P	54.6P	41.8P	84.6P	1,255.4P	10.42P	3,598.3P	2,619.3P	6,208.9P	96.5P
1998		2,562P	449P	55.2P	42.1P	85.4P	1,291.6P	10.64P	3,717.0P	2,705.7P	6,413.9P	97.6P

Sources: 1982, 1987, 1992 *Economic Census*; *Annual Survey of Manufactures*, 83-86, 88-91, 93-94. Establishment counts for non-Census years are from *County Business Patterns*; establishment values for 83-84 are extrapolations. 'P's show projections by the editors. Industries reclassified in 87 will not have data for prior years.

INDICES OF CHANGE

Year	Companies	Establishments		Employment			Compensation		Production ($ million)			
		Total	with 20 or more employees	Total (000)	Production Workers (000)	Hours (Mil)	Payroll ($ mil)	Wages ($/hr)	Cost of Materials	Value Added by Manufacture	Value of Shipments	Capital Invest.
1982	77	76	93	86	89	87	58	73	47	56	51	91
1983		76	95	87	93	88	63	78	56	63	58	64
1984		75	97	104	112	107	77	78	75	75	75	143
1985		74	99	101	104	101	78	82	78	74	76	127
1986		73	100	108	113	115	89	85	86	91	88	139
1987	89	89	105	129	137	137	104	85	113	123	116	212
1988		85	115	141	151	150	115	85	127	135	129	223
1989		80	114	150	143	145	118	88	124	127	125	191
1990		83	109	128	124	123	105	90	107	110	109	132
1991		83	94	92	92	91	82	95	79	81	80	61
1992	100	100	100	100	100	100	100	100	100	100	100	100
1993		99	100	106	107	105	103	99	112	104	108	132
1994		96P	106P	108	113	115	109	102	120	115	118	157
1995		98P	107P	120P	120P	121P	118P	104P	129P	123P	126P	149P
1996		100P	107P	121P	121P	122P	121P	106P	133P	127P	131P	151P
1997		102P	108P	123P	122P	123P	125P	108P	138P	132P	135P	153P
1998		104P	109P	124P	122P	124P	128P	111P	142P	136P	139P	155P

Sources: Same as General Statistics. Values reflect change from the base year, 1992. Values above 100 mean greater than 92, values below 100 mean less than 92, and a value of 100 in the 82-91 or 93-98 period means same as 92. 'P's mark projections by the editors.

SELECTED RATIOS

For 1994	Avg. of All Manufact.	Analyzed Industry	Index	For 1994	Avg. of All Manufact.	Analyzed Industry	Index
Employees per Establishment	49	20	42	Value Added per Production Worker	134,084	58,833	44
Payroll per Establishment	1,500,273	464,808	31	Cost per Establishment	5,045,178	1,332,383	26
Payroll per Employee	30,620	22,802	74	Cost per Employee	102,970	65,364	63
Production Workers per Establishment	34	16	48	Cost per Production Worker	146,988	80,823	55
Wages per Establishment	853,319	326,263	38	Shipments per Establishment	9,576,895	2,299,081	24
Wages per Production Worker	24,861	19,791	80	Shipments per Employee	195,460	112,788	58
Hours per Production Worker	2,056	2,026	99	Shipments per Production Worker	279,017	139,463	50
Wages per Hour	12.09	9.77	81	Investment per Establishment	321,011	42,040	13
Value Added per Establishment	4,602,255	969,877	21	Investment per Employee	6,552	2,062	31
Value Added per Employee	93,930	47,580	51	Investment per Production Worker	9,352	2,550	27

Sources: Same as General Statistics. The 'Average of All Manufacturing' column represents the average of all manufacturing industries reported for the most recent complete year available. The Index shows the relationship between the Average and the Analyzed Industry. For example, 100 means that they are equal; 500 that the Analyzed Industry is five times the average; 50 means that the Analyzed Industry is half the national average. The abbreviation 'na' is used to show that data are 'not available'.

LEADING COMPANIES   Number shown: 75   Total sales ($ mil): 5,466   Total employment (000): 46.8

Company Name	Address				CEO Name	Phone	Co. Type	Sales ($ mil)	Empl. (000)
Brunswick Corp	1 N Field Ct	Lake Forest	IL	60045	Peter N Larson	708-735-4700	P	2,700	20.8
Brunswick Corp	PO Box 9029	Everett	WA	98206	Jim Hoag	206-435-5571	D	320*	2.7
Genmar Industries Inc	100 S 5th St	Minneapolis	MN	55402	Robert J Sutter	612-339-7600	S	250*	2.5
Minstar Inc	100 S 5th St	Minneapolis	MN	55402	Robert J Sutter	612-339-7900	S	250*	2.5
Meridian Sports Inc	625 Madison Av	New York	NY	10022	George Napier	212-527-6300	P	188	1.5
Carver Boat Corp	PO Box 1010	Pulaski	WI	54162	Ken Severinson	414-822-3214	S	92*	0.9
Four Winns Inc	4 Winn Way	Cadillac	MI	49601	Rick Fulmer	616-775-1351	S	88*	0.8
Larson Boat	Paul Larson Mem	Little Falls	MN	56345	Clint Moore	612-632-5481	D	79	0.5
Wood Manufacturing Company	PO Box 179	Flippin	AR	72634	Randy Hopper	501-453-2222	R	65	0.8
Godfrey Conveyor Company Inc	PO Box 1088	Elkhart	IN	46515	Robert J Deputy	219-522-8381	R	60	0.5
Regal Marine Industries Inc	2300 Jetport Dr	Orlando	FL	32809	Paul Kuck	407-851-4360	R	60	0.5
S2 Yachts Inc	725 E 40th St	Holland	MI	49423	Leon R Slikkers	616-392-7163	R	56	0.5
Mariah Boats Inc	PO Box 1300	Benton	IL	62812	Jimmy J Fulks	618-435-5300	R	50	0.4
Tracker Marine Corp	1915C S Campbell	Springfield	MO	65807	Ham Hamberger	417-882-4444	R	50	2.1
Boston Whaler Inc	4121 US Hwy 1	Edgewater	FL	32141	F Douglas Fonte	904-428-0057	S	50	0.5
Skeeter Products Inc	PO Box 230	Kilgore	TX	75663	Ken Burroughs	903-984-0541	S	50	0.1
Hood Enterprises Inc	1 Little Harbor	Portsmouth	RI	02871	Rick Hood	401-683-7000	R	47*	0.2
Catalina Yachts Inc	21200 Victory Blv	Woodland Hills	CA	91367	FW Butler	818-884-7700	R	42	0.3
Peterson Builders Inc	PO Box 650	Sturgeon Bay	WI	54235	EL Peterson	414-743-5574	R	40	0.4
Grady White Boats Inc	PO Box 1527	Greenville	NC	27835	Eddie Smith Jr	919-752-2111	R	35	0.4
Porter Inc	PO Box 1003	Decatur	IN	46733	Scott Porter	219-724-9111	R	35	0.3
Sunbird Boat Company Inc	2348 Shop Rd	Columbia	SC	29201	Blaine Timmer	803-799-1125	S	35	0.3
Thunderbird Products	2200 W Monroe St	Decatur	IN	46733	Victor B Porter	219-724-9111	S	35	0.3
Correct Craft Inc	6100 S Orange	Orlando	FL	32809	WN Meloon	407-855-4141	R	33	0.2
Lund Boat Co	PO Box 248	NY Mills	MN	56567	Larry Lovold	218-385-2235	D	33*	0.3
Teleflex Inc	640 N Lewis Rd	Limerick	PA	19468	Robert M DiPietro	610-495-7011	D	32	<0.1
Sea Nymph Inc	PO Box 337	Syracuse	IN	46567	Bill Ek	219-457-3131	S	31	0.4
Hunter Marine Corp	Rte 441	Alachua	FL	32615	Warren R Luhrs	904-462-3077	R	30*	0.3
KCS International Inc	804 Pecor St	Oconto	WI	54153	KC Stock	414-834-2211	R	30	0.3
Seaswirl Boats Inc	PO Box 167	Culver	OR	97734	Curt Olson	503-546-5011	S	28*	0.2
Beneteau USA Inc	8720 Red Oak Blv	Charlotte	NC	28217	JF de Premorel	704-527-8244	S	25	0.2
Crestliner Boats Inc	609 13th Av NE	Little Falls	MN	56345	Al Kuebeleck	612-632-6686	S	25*	0.3
Fiberglass Engineering Inc	PO Box 29	Neodesha	KS	66757	Pack St Clair	316-325-2653	R	25	0.3
Ocean Yachts Inc	PO Box 312	Egg Harbor Ct	NJ	08215	John E Leek III	609-965-4616	R	25	0.1
Sport-Craft Inc	Rte 1	Perry	FL	32347	Joe R Roberts	904-584-5679	R	24	0.3
Fountain Powerboat Industries	PO Drawer 457	Washington	NC	27889	R M Fountain Jr	919-975-2000	P	22	0.3
Magnum Marine Corp	2900 NE 188th St	N Miami Beach	FL	33180	Katrin Theodoli	305-931-4292	R	20	0.1
Mako Marine Inc	4355 NW 128th St	Opa Locka	FL	33054	Robert C Schwebke	305-685-6591	R	19	0.2
Alumacraft Boat Co	315 W Saint Julian	St Peter	MN	56082	Vern Berglin	507-931-1050	R	17*	0.2
General Marine Industries LP	6725 Bayline Dr	Panama City	FL	32404	Richard Genth	904-769-0311	R	17	0.2
Mastercrafters Corp	3200 Industrial Blv	Winnsboro	LA	71295	Steve Berry	318-435-9431	D	17	0.2
Nichols Brothers Boat Builders	PO Box 580	Freeland	WA	98249	Matthew J Nichols	360-321-5500	R	17	0.2
Cobia Boat Co	2000 Cobia Dr	Vonore	TN	37885	Edward Atchley	615-884-6881	R	16*	<0.1
Vivian Industries Inc	PO Box 232	Vivian	LA	71082	Ernie Avra	318-375-3241	R	16	0.1
Hinckley Co	PO Box 699	SW Harbor	ME	04679	Robert Hinckley	207-244-5531	R	16	0.2
Delta Marine Industries Inc	1608 S 96th St	Seattle	WA	98108	Ivor A Jones	206-763-2383	R	15	0.2
Spencer Boat Company Inc	4200 Poinsettia Av	W Palm Beach	FL	33407	Edward Bronstein	407-844-1800	R	14	0.2
Invader Marine Inc	PO Box 420	Giddings	TX	78942	Earl Mueller	409-542-3101	R	13*	0.1
Morgan	7200 Bryan Dairy	Largo	FL	34647	Frank Butler	813-544-6681	D	13	0.2
Somerset Marine Inc	5000 S US Hwy 27	Somerset	KY	42501	James E Sharpe	606-679-9393	R	13*	0.1
Survival Systems International	PO Box 1855	Valley Center	CA	92082	George L Beatty	619-749-6800	R	13	0.1
Willard Marine Inc	1250 N Grove St	Anaheim	CA	92806	GL Angle	714-666-2150	R	12	<0.1
American Marine Ltd	811 E Maple	Mora	MN	55051	Robert Clarin	612-679-3811	R	10	0.1
Bradford Marine Inc	3051 State Rd	Ft Lauderdale	FL	33312	J C Smallwood	305-791-3800	R	10*	0.1
Gibson Fiberglass Products Inc	308 Church St	Goodlettsville	TN	37072	W C Brummett Jr	615-859-1351	R	10	<0.1
Marisco Ltd	91-607 Malakole Rd	Kapolei	HI	96707	Fred Ananwati	808-682-1333	R	10*	0.1
Phoenix Marine Enterprises Inc	1775 W Okeechobee	Hialeah	FL	33010	Frank Piedra	305-887-5625	R	10	0.1
Skipperliner Industries Inc	621 Park Plaza Dr	La Crosse	WI	54601	Noel C Jordan	608-784-5110	R	10	0.1
UniGrace Inc	PO Box 210	Monticello	AR	71655	Zach McClendon Jr	501-367-9755	R	10	0.2
Laudau Boats Inc	PO Box 750	Lebanon	MO	65536	Charles Clinard	417-532-9126	R	9*	<0.1
Merrill Stevens Drydock Co	PO Box 011980	Miami	FL	33101	Fred Kirtland	305-324-5211	R	8	<0.1
Pacific Seacraft Corp	1301 Orangeth	Fullerton	CA	92631	TaiKwee Lee	714-879-1610	S	8	<0.1
Robert E Derecktor Inc	311 E Post Rd	Mamaroneck	NY	10543	E Paul Derecktor	914-698-5020	R	8	0.1
Sovereign Yachts Corp	7814 8th Av S	Seattle	WA	98108	Bruce Reagan	206-767-4497	R	8*	<0.1
Larsen Marine Service	625 Sea Horse Dr	Waukegan	IL	60085	Gerald N Larsen	708-336-5456	R	8	<0.1
Sabre Corp	PO Box 134	South Casco	ME	04077	Edward Miller	207-655-3831	R	8	<0.1
Forester Boats	180 Industrial Pk E	Sauk Rapids	MN	56379	Thomas Benkoske	612-252-4304	R	7	<0.1
Merritt's Boat and Engine Works	2931 NE 16th St	Pompano Bch	FL	33062	Allen Merritt	305-941-0118	R	7	<0.1
Hobie Cat Co	PO Box 1008	Oceanside	CA	92054	Scott Foresman	619-758-9100	R	6	<0.1
Little Harbor Custom Yachts	1 Little Harbor	Portsmouth	RI	02871	FG Hood	401-683-5600	D	6*	<0.1
Polar Kraft Manufacturing Co	PO Box 708	Olive Branch	MS	38654	Larry Deputy	601-895-5576	D	6	<0.1
SeaArk Marine Inc	PO Box 210	Monticello	AR	71655	Zach McClendon	501-367-9755	S	6	0.1
Albin Marine Inc	PO Box 228	Cos Cob	CT	06807	Frederick Peters	203-661-4341	R	5	<0.1
Black Watch	1 Little Harbor	Portsmouth	RI	02871	Rick Hood	401-683-2797	D	5*	<0.1
Corsair Marine Inc	150 Reed Ct	Chula Vista	CA	91911		619-585-3005	R	5	<0.1

Source: Ward's Business Directory of U.S. Private and Public Companies, Volumes 1 and 2, 1996. The company type code used is as follows: P - Public, R - Private, S - Subsidiary, D - Division, J - Joint Venture, A - Affiliate, G - Group. Sales are in millions of dollars, employees are in thousands. An asterisk (*) indicates an estimated sales volume. The symbol < stands for 'less than'. Company names and addresses are truncated, in some cases, to fit into the available space.

MATERIALS CONSUMED

Material	Quantity	Delivered Cost ($ million)
Materials, ingredients, containers, and supplies	(X)	2,383.5
Diesel and semidiesel engines	(X)	70.6
Gasoline and other internal combustion engines	(X)	411.4
Integral horsepower electric motors and generators (1 hp or more)	(X)	23.0
Boat propellers	(X)	28.4
Marine metal hardware	(X)	66.4
Metal bolts, nuts, screws, washers, rivets, and other screw machine products	(X)	26.2
Other fabricated metal products (except forgings)	(X)	25.8
Forgings	(X)	4.9
Castings (rough and semifinished)	(X)	6.2
Steel shapes and forms	(X)	17.8
Aluminum and aluminum-base alloy sheet, plate, foil, and welded tubing	(X)	47.4
Aluminum and aluminum-base alloy extruded shapes, including extruded rod, bar, pipe, tube, etc.	(X)	23.4
Other aluminum and aluminum-base alloy shapes and forms	(X)	2.2
Other nonferrous shapes and forms	(X)	3.7
Plastics resins consumed in the form of granules, pellets, powders, liquids, etc.	(X)	74.4
Plastics products consumed in the form of sheets, rods, tubes, film, and other shapes	(X)	24.8
Glass fiber, textile type, bonded mat type, etc.	(X)	66.9
Dressed lumber	(X)	20.0
Plywood	(X)	46.9
Carpeting	(X)	24.5
Canvas products	(X)	22.3
Paints, varnishes, lacquers, stains, shellacs, japans, enamels, and allied products	(X)	26.4
Marine nautical and navigation equipment operating by radio signal	(X)	38.7
Bilge pumps	(X)	4.5
All other materials and components, parts, containers, and supplies	(X)	400.7
Materials, ingredients, containers, and supplies, nsk	(X)	876.2

Source: 1992 *Economic Census.* Explanation of symbols used: (D): Withheld to avoid disclosure of competitive data; na: Not available; (S): Withheld because statistical norms were not met; (X): Not applicable; (Z): Less than half the unit shown; nec: Not elsewhere classified; nsk: Not specified by kind; - : zero; * : 10-19 percent estimated; ** : 20-29 percent estimated.

PRODUCT SHARE DETAILS

Product or Product Class	% Share	Product or Product Class	% Share
Boat building and repairing.	100.00	auxiliary power and lifeboats)	36.06
Outboard motorboats, including commercial and military (except sailboats with auxiliary power and lifeboats)	26.22	Inboard cabin cruisers, 40 ft (12.19 m) or more in length (professional crew not required by Coast Guard), including commercial and military (except sailboats with auxiliary power lifeboats)	25.57
Outboard motor runabouts, wood or metal, runabouts, including commerical and military (except sailboats with auxiliary power and lifeboats)	6.67	Other inboard motorboats (including houseboats), including commercial and military (except sailboats with auxiliary power lifeboats)	8.23
Outboard motor utility boats, wood or metal, including commercial and military (except sailboats with auxiliary power and lifeboats)	3.40	Inboard motorboats, nsk	10.31
Outboard bass boats, wood or metal	7.74	Inboard-outdrive boats	24.34
Outboard motor pontoon boats, wood or metal, including commercial and military (except sailboats with auxiliary power and lifeboats)	14.51	Inboard-outdrive houseboats, including commercial and military (except sailboats with auxiliary power lifeboats)	3.13
Other outboard motor boats, wood or metal (including cabin cruisers and center consoles), including commerical and military (except sailboats with auxiliary power and lifeboats)	2.45	Inboard-outdrive runabouts, including commercial and military (except sailboats with auxiliary power lifeboats)	58.25
Outboard motor runabouts, plastics (reinforced), fiberglass, including commercial and military (except sailboats with auxiliary power and lifeboats)	10.62	Inboard-outdrive cabin cruisers, including commercial and military (except sailboats with auxiliary power lifeboats)	27.68
Outboard motor utility boats, plastics (reinforced), fiberglass, including commercial and military (except sailboats with auxiliary power and lifeboats)	2.86	Inboard-outdrive center consoles, including commercial and military (except sailboats with auxiliary power lifeboats)	0.39
Outboard cabin cruisers, plastics (reinforced), fiberglass, including commercial and military (except sailboats with auxiliary power and lifeboats)	1.99	Other inboard-outdrive motorboats, including commercial and military (except sailboats with auxiliary power lifeboats)	7.07
Center console outboard motorboats, plastics (reinforced), fiberglass, including commercial and military (except sailboats with auxiliary power and lifeboats)	10.61	Inboard-outdrive boats, nsk	3.49
Outboard bass boats, plastics (reinforced), fiberglass, including commercial and military (except sailboats with auxiliary power and lifeboats)	20.34	All other boats (excluding military and commercial)	5.90
		Sailboats without auxiliary motor, all sizes (excluding military and commercial)	6.55
Other outboard motorboats, plastics (reinforced), fiberglass, including commercial and military (except sailboats with auxiliary power and lifeboats)	16.06	Sailboats with auxiliary motor, not more than 6.5 m (21.33 ft) in length (excluding military and commercial)	4.01
Outboard motorboats, including commercial and military (except sailboats with auxiliary power and lifeboats), nsk	2.74	Sailboats with auxiliary motor, more than 6.5 m (21.33 ft) but not more than 9.0 m (29.53 ft) in length (excluding military and commercial)	9.64
Inboard motorboats	16.27	Sailboats with auxiliary motor, more than 9.0 m (29.53 ft) but not more than 12.0 m (39.03 ft) in length (excluding military and commercial)	18.45
Inboard runabouts, including commercial and military (except sailboats with auxiliary power and lifeboats)	19.83	Sailboats with auxiliary motor, more than 12.0 m (39.03 ft) in length (excluding military and commercial)	20.79
Inboard cabin cruisers, less than 40 ft (12.19 m) in length, including commercial and military (except sailboats with		Canoes (made from all types of materials) (excluding military and commercial)	9.72
		All other boats, nec (excluding military and commercial)	30.62
		All other boats (excluding military and commercial), nsk	0.28
		Boat repair, military and nonmilitary	7.23
		Boat building and repairing, nsk	20.03

Source: 1992 *Economic Census.* The values shown are percent of total shipments in an industry. Values of indented subcategories are summed in the main heading. The symbol (D) appears when data are withheld to prevent disclosure of competitive information. The abbreviation nsk stands for 'not specified by kind' and nec for 'not elsewhere classified'.

INPUTS AND OUTPUTS FOR BOAT BUILDING & REPAIRING

Economic Sector or Industry Providing Inputs	%	Sector	Economic Sector or Industry Buying Outputs	%	Sector
Internal combustion engines, nec	12.7	Manufg.	Personal consumption expenditures	79.4	
Imports	10.7	Foreign	Gross private fixed investment	6.9	Cap Inv
Wholesale trade	8.9	Trade	Exports	2.4	Foreign
Fabricated structural metal	7.0	Manufg.	Amusement & recreation services nec	2.2	Services
Royalties	4.5	Fin/R.E.	Commercial fishing	2.0	Agric.
Maintenance of nonfarm buildings nec	4.3	Constr.	Water transportation	1.5	Util.
Boat building & repairing	3.3	Manufg.	Federal Government purchases, national defense	1.5	Fed Govt
Glass & glass products, except containers	3.1	Manufg.	Boat building & repairing	1.4	Manufg.
Plastics materials & resins	3.0	Manufg.	Change in business inventories	0.8	In House
Hardware, nec	2.8	Manufg.	Ship building & repairing	0.6	Manufg.
Aluminum rolling & drawing	2.3	Manufg.	Federal Government purchases, nondefense	0.5	Fed Govt
Veneer & plywood	1.5	Manufg.	State & local government enterprises, nec	0.4	Gov't
Blast furnaces & steel mills	1.4	Manufg.	S/L Govt. purch., elem. & secondary education	0.2	S/L Govt
Coated fabrics, not rubberized	1.3	Manufg.	S/L Govt. purch., higher education	0.2	S/L Govt
Paints & allied products	1.3	Manufg.	S/L Govt. purch., natural resource & recreation.	0.2	S/L Govt
Petroleum refining	1.3	Manufg.			
Advertising	1.2	Services			
Sawmills & planning mills, general	1.1	Manufg.			
Automotive & apparel trimmings	1.0	Manufg.			
Wood products, nec	1.0	Manufg.			
Electric services (utilities)	1.0	Util.			
Motor freight transportation & warehousing	1.0	Util.			
Engineering, architectural, & surveying services	1.0	Services			
Particleboard	0.9	Manufg.			
Screw machine and related products	0.8	Manufg.			
Real estate	0.8	Fin/R.E.			
Canvas & related products	0.7	Manufg.			
Engineering & scientific instruments	0.7	Manufg.			
Power transmission equipment	0.7	Manufg.			
Storage batteries	0.7	Manufg.			
Adhesives & sealants	0.6	Manufg.			
Fabricated metal products, nec	0.6	Manufg.			
Floor coverings	0.6	Manufg.			
Communications, except radio & TV	0.6	Util.			
Brass, bronze, & copper castings	0.5	Manufg.			
Engine electrical equipment	0.5	Manufg.			
Machinery, except electrical, nec	0.5	Manufg.			
Miscellaneous plastics products	0.5	Manufg.			
Motors & generators	0.5	Manufg.			
Eating & drinking places	0.5	Trade			
Equipment rental & leasing services	0.5	Services			
Gaskets, packing & sealing devices	0.4	Manufg.			
Nonferrous castings, nec	0.4	Manufg.			
Nonferrous rolling & drawing, nec	0.4	Manufg.			
Pipe, valves, & pipe fittings	0.4	Manufg.			
Gas production & distribution (utilities)	0.4	Util.			
Railroads & related services	0.4	Util.			
Ball & roller bearings	0.3	Manufg.			
Blowers & fans	0.3	Manufg.			
Fabricated plate work (boiler shops)	0.3	Manufg.			
Fabricated rubber products, nec	0.3	Manufg.			
Industrial gases	0.3	Manufg.			
Lighting fixtures & equipment	0.3	Manufg.			
Nonferrous forgings	0.3	Manufg.			
Radio & TV communication equipment	0.3	Manufg.			
Refrigeration & heating equipment	0.3	Manufg.			
Banking	0.3	Fin/R.E.			
Automotive rental & leasing, without drivers	0.3	Services			
U.S. Postal Service	0.3	Gov't			
Aluminum castings	0.2	Manufg.			
Cordage & twine	0.2	Manufg.			
Lubricating oils & greases	0.2	Manufg.			
Metal stampings, nec	0.2	Manufg.			
Miscellaneous fabricated wire products	0.2	Manufg.			
Nonferrous wire drawing & insulating	0.2	Manufg.			
Padding & upholstery filling	0.2	Manufg.			
Special dies & tools & machine tool accessories	0.2	Manufg.			
Automotive repair shops & services	0.2	Services			
Computer & data processing services	0.2	Services			
Legal services	0.2	Services			
Management & consulting services & labs	0.2	Services			
Abrasive products	0.1	Manufg.			
Electrical equipment & supplies, nec	0.1	Manufg.			
Industrial controls	0.1	Manufg.			
Metal coating & allied services	0.1	Manufg.			
Pumps & compressors	0.1	Manufg.			
Rubber & plastics hose & belting	0.1	Manufg.			
Sanitary services, steam supply, irrigation	0.1	Util.			

Continued on next page.

INPUTS AND OUTPUTS FOR BOAT BUILDING & REPAIRING - Continued

Economic Sector or Industry Providing Inputs	%	Sector	Economic Sector or Industry Buying Outputs	%	Sector
Water transportation	0.1	Util.			
Insurance carriers	0.1	Fin/R.E.			
Accounting, auditing & bookkeeping	0.1	Services			
Business/professional associations	0.1	Services			

Source: *Benchmark Input-Output Accounts for the U.S. Economy, 1982*, U.S. Department of Commerce, Washington, D.C., July 1991. Data, as reported in the source, are organized by the 1977 SIC structure in use in 1982 but have been matched, as closely as is possible, to the 1987 SIC structure used in this book.

OCCUPATIONS EMPLOYED BY SIC 373 - SHIP AND BOAT BUILDING AND REPAIRING

Occupation	% of Total 1994	Change to 2005	Occupation	% of Total 1994	Change to 2005
Welders & cutters	10.7	-17.5	Drafters	1.9	-35.7
Assemblers, fabricators, & hand workers nec	8.1	-17.4	Painters & paperhangers	1.9	-17.4
Shipfitters	6.3	-0.9	Sheet metal workers & duct installers	1.9	-17.5
Blue collar worker supervisors	5.6	-18.2	General managers & top executives	1.8	-21.7
Plumbers, pipefitters, & steamfitters	5.0	-17.5	Inspectors, testers, & graders, precision	1.7	-17.4
Carpenters	3.6	-34.0	Stock clerks	1.3	-32.9
Electricians	3.3	-22.5	Production, planning, & expediting clerks	1.3	-17.5
Machinists	3.2	-17.4	Helpers, construction trades	1.3	-17.4
Helpers, laborers, & material movers nec	2.9	-17.5	Grinders & polishers, hand	1.2	-17.5
Painters, transportation equipment	2.7	-0.9	Industrial production managers	1.1	-17.5
Riggers	2.2	-13.1	Welding machine setters, operators	1.0	-25.7
General office clerks	2.0	-29.6			

Source: *Industry-Occupation Matrix*, Bureau of Labor Statistics. These data relate to one or more 3-digit SIC industry groups rather than to a single 4-digit SIC. The change reported for each occupation to the year 2005 is a percent of growth or decline as estimated by the Bureau of Labor Statistics. The abbreviation nec stands for 'not elsewhere classified'.

LOCATION BY STATE AND REGIONAL CONCENTRATION

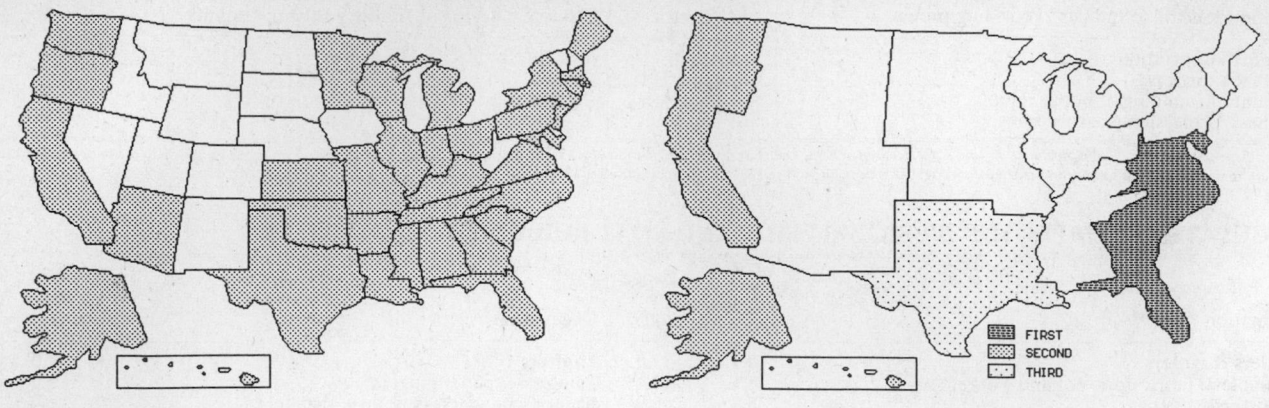

INDUSTRY DATA BY STATE

| State | Establish-ments | Shipments | | | Employment | | | | Cost as % of Shipments | Investment per Employee ($) |
		Total ($ mil)	% of U.S.	Per Establ.	Total Number	% of U.S.	Per Establ.	Wages ($/hour)		
Florida	465	878.8	19.1	1.9	9,000	20.2	19	9.84	55.8	1,233
California	231	255.6	5.6	1.1	2,900	6.5	13	10.40	53.7	1,000
Washington	186	252.0	5.5	1.4	3,200	7.2	17	10.82	47.4	1,250
Georgia	41	218.4	4.7	5.3	1,700	3.8	41	9.15	64.1	-
Michigan	84	209.3	4.6	2.5	1,700	3.8	20	9.26	46.5	-
Texas	116	118.0	2.6	1.0	1,200	2.7	10	8.28	59.8	917
Arkansas	38	111.1	2.4	2.9	1,200	2.7	32	7.72	51.4	750
South Carolina	37	100.3	2.2	2.7	800	1.8	22	7.62	55.5	-
Wisconsin	34	95.5	2.1	2.8	900	2.0	26	9.53	55.4	1,333
Oregon	52	81.2	1.8	1.6	700	1.6	13	9.91	63.8	-
New Jersey	52	70.6	1.5	1.4	700	1.6	13	10.92	46.0	2,571
Massachusetts	74	65.9	1.4	0.9	700	1.6	9	12.33	53.3	2,571
Rhode Island	33	63.7	1.4	1.9	700	1.6	21	10.80	55.1	1,571
Virginia	65	56.2	1.2	0.9	800	1.8	12	10.91	48.4	875
New York	82	46.2	1.0	0.6	600	1.3	7	10.00	53.7	667
Hawaii	15	29.9	0.7	2.0	200	0.4	13	10.25	45.8	-
Alabama	49	29.3	0.6	0.6	400	0.9	8	9.33	52.6	750
Connecticut	31	15.2	0.3	0.5	200	0.4	6	12.67	52.0	2,000
Pennsylvania	19	6.0	0.1	0.3	100	0.2	5	14.00	56.7	-
Louisiana	86	(D)	-	-	1,750 *	3.9	20	-	-	-
North Carolina	85	(D)	-	-	1,750 *	3.9	21	-	-	-
Maine	82	(D)	-	-	750 *	1.7	9	-	-	2,800
Maryland	71	(D)	-	-	750 *	1.7	11	-	-	-
Tennessee	54	(D)	-	-	3,750 *	8.4	69	-	-	-
Ohio	41	(D)	-	-	750 *	1.7	18	-	-	-
Minnesota	36	(D)	-	-	1,750 *	3.9	49	-	-	-
Missouri	36	(D)	-	-	1,750 *	3.9	49	-	-	-
Illinois	30	(D)	-	-	750 *	1.7	25	-	-	-
Indiana	27	(D)	-	-	1,750 *	3.9	65	-	-	-
Arizona	24	(D)	-	-	375 *	0.8	16	-	-	2,133
Mississippi	24	(D)	-	-	375 *	0.8	16	-	-	-
Alaska	20	(D)	-	-	175 *	0.4	9	-	-	571
Kentucky	20	(D)	-	-	750 *	1.7	38	-	-	533
Oklahoma	18	(D)	-	-	375 *	0.8	21	-	-	-
Kansas	5	(D)	-	-	175 *	0.4	35	-	-	-

Source: 1992 *Economic Census*. The states are in descending order of shipments or establishments (if shipment data are missing for the majority). The symbol (D) appears when data are withheld to prevent disclosure of competitive information. States marked with (D) are sorted by number of establishments. A dash (-) indicates that the data element cannot be calculated; * indicates the midpoint of a range.

3743 - RAILROAD EQUIPMENT

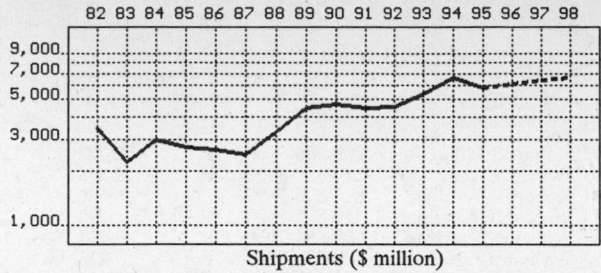

82 83 84 85 86 87 88 89 90 91 92 93 94 95 96 97 98

Shipments ($ million)

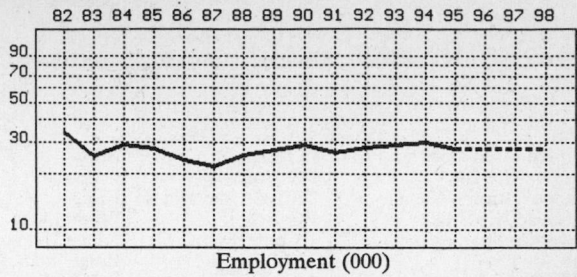

82 83 84 85 86 87 88 89 90 91 92 93 94 95 96 97 98

Employment (000)

GENERAL STATISTICS

| Year | Com-panies | Establishments | | Employment | | | Compensation | | Production ($ million) | | | |
		Total	with 20 or more employees	Total (000)	Production Workers (000)	Hours (Mil)	Payroll ($ mil)	Wages ($/hr)	Cost of Materials	Value Added by Manufacture	Value of Shipments	Capital Invest.
1982	158	200	107	34.5	22.8	38.9	790.4	12.31	1,770.5	1,491.8	3,456.6	142.3
1983				25.0	16.2	29.6	634.8	12.42	1,176.1	1,036.5	2,248.1	140.4
1984				29.5	20.5	38.7	789.7	12.73	1,595.0	1,485.9	3,021.2	123.4
1985				27.6	18.9	35.9	762.2	13.26	1,391.4	1,286.1	2,735.5	79.2
1986				24.0	16.1	30.6	687.4	13.76	1,256.3	1,198.1	2,613.0	58.4
1987	150	174	93	22.1	14.3	29.1	630.6	12.02	1,237.1	1,295.4	2,470.9	58.9
1988				25.9	17.9	36.0	742.1	12.40	1,807.3	1,482.8	3,285.9	61.3
1989		174	98	27.4	19.6	40.4	840.8	13.25	2,562.1	1,940.5	4,428.6	67.4
1990				29.5	21.0	41.0	870.1	12.99	2,763.1	1,839.2	4,693.6	94.5
1991				26.4	19.4	37.7	819.8	13.84	2,645.7	1,689.6	4,449.7	96.1
1992	154	206	123	28.2	20.6	40.3	897.5	13.80	2,743.3	1,961.6	4,588.8	94.9
1993				29.0	21.1	42.5	955.8	14.07	3,160.3	2,190.9	5,309.5	101.6
1994				30.2	22.0	45.7	1,076.3	14.98	4,074.4	2,534.8	6,569.3	112.4
1995				27.7P	20.7P	43.2P	980.6P	14.37P	3,608.2P	2,244.7P	5,817.6P	81.0P
1996				27.7P	20.9P	44.0P	1,005.3P	14.54P	3,783.7P	2,354.0P	6,100.6P	79.0P
1997				27.7P	21.1P	44.8P	1,030.0P	14.70P	3,959.3P	2,463.2P	6,383.7P	77.1P
1998				27.7P	21.3P	45.6P	1,054.8P	14.86P	4,134.8P	2,572.4P	6,666.7P	75.1P

Sources: 1982, 1987, 1992 *Economic Census*; *Annual Survey of Manufactures*, 83-86, 88-91, 93-94. Establishment counts for non-Census years are from *County Business Patterns*; establishment values for 83-84 are extrapolations. 'P's show projections by the editors. Industries reclassified in 87 will not have data for prior years.

INDICES OF CHANGE

| Year | Com-panies | Establishments | | Employment | | | Compensation | | Production ($ million) | | | |
		Total	with 20 or more employees	Total (000)	Production Workers (000)	Hours (Mil)	Payroll ($ mil)	Wages ($/hr)	Cost of Materials	Value Added by Manufacture	Value of Shipments	Capital Invest.
1982	103	97	87	122	111	97	88	89	65	76	75	150
1983				89	79	73	71	90	43	53	49	148
1984				105	100	96	88	92	58	76	66	130
1985				98	92	89	85	96	51	66	60	83
1986				85	78	76	77	100	46	61	57	62
1987	97	84	76	78	69	72	70	87	45	66	54	62
1988				92	87	89	83	90	66	76	72	65
1989		84	80	97	95	100	94	96	93	99	97	71
1990				105	102	102	97	94	101	94	102	100
1991				94	94	94	91	100	96	86	97	101
1992	100	100	100	100	100	100	100	100	100	100	100	100
1993				103	102	105	106	102	115	112	116	107
1994				107	107	113	120	109	149	129	143	118
1995				98P	100P	107P	109P	104P	132P	114P	127P	85P
1996				98P	101P	109P	112P	105P	138P	120P	133P	83P
1997				98P	102P	111P	115P	107P	144P	126P	139P	81P
1998				98P	103P	113P	118P	108P	151P	131P	145P	79P

Sources: Same as General Statistics. Values reflect change from the base year, 1992. Values above 100 mean greater than 92, values below 100 mean less than 92, and a value of 100 in the 82-91 or 93-98 period means same as 92. 'P's mark projections by the editors.

SELECTED RATIOS

For 1992	Avg. of All Manufact.	Analyzed Industry	Index	For 1992	Avg. of All Manufact.	Analyzed Industry	Index
Employees per Establishment	46	137	300	Value Added per Production Worker	122,353	95,223	78
Payroll per Establishment	1,332,320	4,356,796	327	Cost per Establishment	4,239,462	13,316,990	314
Payroll per Employee	29,181	31,826	109	Cost per Employee	92,853	97,280	105
Production Workers per Establishment	31	100	318	Cost per Production Worker	135,003	133,170	99
Wages per Establishment	734,496	2,699,709	368	Shipments per Establishment	8,100,800	22,275,728	275
Wages per Production Worker	23,390	26,997	115	Shipments per Employee	177,425	162,723	92
Hours per Production Worker	2,025	1,956	97	Shipments per Production Worker	257,966	222,757	86
Wages per Hour	11.55	13.80	119	Investment per Establishment	278,244	460,680	166
Value Added per Establishment	3,842,210	9,522,330	248	Investment per Employee	6,094	3,365	55
Value Added per Employee	84,153	69,560	83	Investment per Production Worker	8,861	4,607	52

Sources: Same as General Statistics. The 'Average of All Manufacturing' column represents the average of all manufacturing industries reported for the most recent complete year available. The Index shows the relationship between the Average and the Analyzed Industry. For example, 100 means that they are equal; 500 that the Analyzed Industry is five times the average; 50 means that the Analyzed Industry is half the national average. The abbreviation 'na' is used to show that data are 'not available'.

LEADING COMPANIES   Number shown: 37   Total sales ($ mil): 7,487   Total employment (000): 46.3

Company Name	Address				CEO Name	Phone	Co. Type	Sales ($ mil)	Empl. (000)
Morrison Knudsen Corp	PO Box 73	Boise	ID	83729	Robert S Miller Jr	208-386-5000	P	2,722	12.4
Trinity Industries Inc	PO Box 568887	Dallas	TX	75356	W Ray Wallace	214-631-4420	P	2,315	14.7
GE Transportation Systems	2901 E Lake Rd	Erie	PA	16531	Robert L Nardelli	814-875-5677	D	560	6.0
Westinghouse Air Brake Co	PO Box 1	Wilmerding	PA	15148	William E Kassling	412-825-1000	R	390	1.8
Greenbrier Companies Inc	1 Centerpointe Dr	Lake Oswego	OR	97035	William A Furman	503-684-7000	P	234	1.6
MK Rail Corp	720 Park Blv	Boise	ID	83729	William J Agee	208-386-5209	P	218	1.9
Gunderson Inc	4350 NW Front Av	Portland	OR	97210	L Clark Wood	503-228-9281	S	180*	1.4
Firmont Tamper	PO Box 415	Fairmont	MN	56031	G Robert Newman	507-235-3361	D	110*	0.9
Miner Enterprises Inc	1200 E State St	Geneva	IL	60134	David W Withall	708-232-3000	R	100	0.5
Safetran Systems Corp	4650 Main St NE	Minneapolis	MN	55421	GL Kline	612-572-1400	S	100	0.7
Portec Inc	100 Field Dr	Lake Forest	IL	60045	Michael T Yonker	708-735-2800	P	96	0.8
C Jim Stewart and Stevenson	PO Box 1637	Houston	TX	77251	Bob H O'Neal	713-671-6300	S	66	0.3
ABB Traction Inc	E 18th St	Elmira Heights	NY	14903	L R Lansford	607-732-5251	S	63	0.7
Siemens Duewag Corp	3035 Prospect Pk	R Cordova	CA	95670	Gunter Ernst	916-852-0180	S	60	0.2
Kershaw Manufacturing	PO Box 244100	Montgomery	AL	36124	Royce Kershaw Jr	334-215-1000	R	35	0.3
Portec Inc	PO Box 38250	Pittsburgh	PA	15238	John S Cooper	412-782-6000	D	34	<0.1
TTX Co	1 Hamburg Rd	North Augusta	SC	29841	Robert G Lang	803-279-1922	D	32*	0.3
Trackmobile Inc	1602 Executive Dr	La Grange	GA	30240	Richard L Lich	706-884-6651	S	25	<0.1
Graham-White Mfg Co	PO Box 1099	Salem	VA	24153	James S Frantz	703-387-5620	R	17	0.3
National Railway Equipment Co	PO Box 2270	Dixmoor	IL	60426	Lawrence J Beal	708-388-6002	R	17	0.2
Livingston Rebuild Center Inc	704 E Gallatin	Livingston	MT	59047	Randy Peterson	406-222-1200	R	14	0.2
Plymouth Locomotive	607 Bell St	Plymouth	OH	44865	Richard Gullett	419-687-4641	R	13	0.1
Buffalo Brake Beam Co	400 Ingham Av	Lackawanna	NY	14218	RG Adams	716-823-4200	R	10	<0.1
Prime Corp	7730 S 6th St	Oak Creek	WI	53154	Morris R Liles	414-764-1400	R	10	0.2
Brammall Inc	PO Box 208	Angola	IN	46703	JE Bledsoe	219-665-3176	S	9	0.1
United Rail Car Manufacturing	44 Wells Av	Yonkers	NY	10701	K Takai	914-376-4714	R	9	0.1
LB Foster Co	415 Holiday Dr	Pittsburgh	PA	15220	Hank Ortwein Jr	412-928-3500	S	8*	<0.1
Texana Tank Car & Mfg	PO Box 550	Nash	TX	75569	SM Brooks	903-838-5564	R	8	<0.1
Touchstone Inc	PO Box 7568	Jackson	TN	38308	Ted E Nelson	901-424-5060	S	8	0.1
Triax-Davis Inc	PO Box 597	Benton Harbor	MI	49023	Robert G Jackson	616-925-2600	R	6	<0.1
Fleet Body Equipment	200 NW Harlem Rd	Kansas City	MO	64116	Russel Wattson	816-842-4065	R	4*	<0.1
Mid-America Car Inc	PO Box 33543	Kansas City	MO	64120	Curtis Blanc	816-483-5303	R	4	<0.1
APS Systems	3535 W 5th St	Oxnard	CA	93030	Nick Advani	805-984-0300	R	3	<0.1
Teleweld Inc	416 N Park St	Streator	IL	61364	JM Rithmiller	815-672-4561	R	3	<0.1
Alabama Railcar Service	PO Box 968	Ozark	AL	36361	Carey D Harper III	205-774-2621	R	2	<0.1
Nolan Co	1016 9th St SW	Canton	OH	44707	James L Anderson	216-453-7922	R	1	<0.1
Briggs and Turivas Inc	PO Box 517	River Forest	IL	60305	Peter Briggs	708-771-5584	R	0*	<0.1

Source: Ward's Business Directory of U.S. Private and Public Companies, Volumes 1 and 2, 1996. The company type code used is as follows: P - Public, R - Private, S - Subsidiary, D - Division, J - Joint Venture, A - Affiliate, G - Group. Sales are in millions of dollars, employees are in thousands. An asterisk (*) indicates an estimated sales volume. The symbol < stands for 'less than'. Company names and addresses are truncated, in some cases, to fit into the available space.

MATERIALS CONSUMED

Material	Quantity	Delivered Cost ($ million)
Materials, ingredients, containers, and supplies	(X)	2,591.0
Railway electrical control equipment	(X)	161.0
Brake equipment, truck assemblies, hooks and other coupling devices, buffers, and parts	(X)	312.7
Roller bearings (mounted or unmounted)	(X)	87.8
Mechanical speed changers, gears, and industrial high-speed drives	(X)	32.1
Fluid power products	(X)	38.3
Metal bolts, nuts, screws, washers, rivets, and other screw machine products	(X)	69.6
Other fabricated metal products (except fluid power products and forgings)	(X)	210.2
Forged iron and steel wheels and axles	(X)	194.9
Other iron and steel forgings, except wheels and axles	(X)	46.0
Iron and steel castings (rough and semifinished)	(X)	227.0
Nonferrous (aluminum, copper, etc.) castings (rough and semifinished)	(X)	5.1
Steel bars, bar shapes, and plates	(X)	275.1
Steel sheet and strip, including tin plate	(X)	87.9
Steel structural shapes and sheet piling	(X)	30.8
All other steel shapes and forms	(X)	66.9
Copper and copper-base alloy shapes and forms	(X)	54.4
All other materials and components, parts, containers, and supplies	(X)	481.8
Materials, ingredients, containers, and supplies, nsk	(X)	209.5

Source: 1992 *Economic Census*. Explanation of symbols used: (D): Withheld to avoid disclosure of competitive data; na: Not available; (S): Withheld because statistical norms were not met; (X): Not applicable; (Z): Less than half the unit shown; nec: Not elsewhere classified; nsk: Not specified by kind; - : zero; * : 10-19 percent estimated; ** : 20-29 percent estimated.

PRODUCT SHARE DETAILS

Product or Product Class	% Share	Product or Product Class	% Share
Railroad equipment	100.00	Rebuilt passenger and freight train cars	9.24
Locomotives, both new and rebuilt, and parts	36.96	Self-propelled and nonself-propelled streetcars, subway	
Locomotives, both new and rebuilt	53.00	cars, rapid transit cars, trolley buses, etc..	30.17
Parts for locomotives	46.98	Other work and service railroad vehicles (excluding	
Locomotives, both new and rebuilt, and parts, nsk	0.02	locomotive cranes).	3.45
Freight train and passenger train cars, new, excluding parts .	33.98	Airbrake equipment for railroad and streetcars	13.68
Freight train and passenger train box cars (AAR types A,		Other brake equipment for railroad and streetcars	1.65
B), new, excluding parts	60.62	Hooks and other coupling devices, buffers, and parts	
Other freight train and passenger train cars, new, excluding		thereof for railroad and streetcars	3.16
parts	39.22	Other railroad and streetcar parts and accessories, including	
Freight train and passenger train cars, new, excluding parts,		truck assemblies	38.38
nsk	0.16	Street, subway, trolley, and rapid transit cars, all rebuilt	
Street, subway, trolley, and rapid transit cars, all rebuilt		railcars, and parts for all railcars, nsk	0.28
railcars, and parts for all railcars	26.94	Railroad equipment, nsk	2.12

Source: 1992 *Economic Census*. The values shown are percent of total shipments in an industry. Values of indented subcategories are summed in the main heading. The symbol (D) appears when data are withheld to prevent disclosure of competitive information. The abbreviation nsk stands for 'not specified by kind' and nec for 'not elsewhere classified'.

INPUTS AND OUTPUTS FOR RAILROAD EQUIPMENT

Economic Sector or Industry Providing Inputs	%	Sector	Economic Sector or Industry Buying Outputs	%	Sector
Railroad equipment	15.8	Manufg.	Gross private fixed investment	58.3	Cap Inv
Blast furnaces & steel mills	9.4	Manufg.	Exports	12.4	Foreign
Iron & steel foundries	6.6	Manufg.	Railroads & related services	10.7	Util.
Wholesale trade	6.3	Trade	Railroad equipment	10.5	Manufg.
Iron & steel forgings	5.0	Manufg.	S/L Govt. purch., transit utilities	4.4	S/L Govt
Imports	5.0	Foreign	Local government passenger transit	2.2	Gov't
Blowers & fans	3.4	Manufg.	S/L Govt. purch., natural resource & recreation.	0.8	S/L Govt
Power transmission equipment	3.3	Manufg.	Federal Government enterprises nec	0.2	Gov't
Internal combustion engines, nec	3.0	Manufg.	Federal Government purchases, national defense	0.2	Fed Govt
Petroleum refining	2.9	Manufg.	Copper ore	0.1	Mining
Motors & generators	2.8	Manufg.			
Maintenance of nonfarm buildings nec	2.6	Constr.			
Switchgear & switchboard apparatus	2.4	Manufg.			
Nonferrous forgings	2.3	Manufg.			
Steel springs, except wire	1.6	Manufg.			
Electric services (utilities)	1.6	Util.			
Ball & roller bearings	1.3	Manufg.			
Fabricated metal products, nec	1.1	Manufg.			
Industrial controls	1.0	Manufg.			
Motor freight transportation & warehousing	1.0	Util.			
Automotive & apparel trimmings	0.9	Manufg.			
Cyclic crudes and organics	0.8	Manufg.			
Pumps & compressors	0.8	Manufg.			
Screw machine and related products	0.8	Manufg.			
Communications, except radio & TV	0.7	Util.			
Gas production & distribution (utilities)	0.7	Util.			
Banking	0.7	Fin/R.E.			
Machinery, except electrical, nec	0.6	Manufg.			
Equipment rental & leasing services	0.6	Services			
U.S. Postal Service	0.6	Gov't			
Copper rolling & drawing	0.5	Manufg.			
Glass & glass products, except containers	0.5	Manufg.			
Hardware, nec	0.5	Manufg.			
Metal coating & allied services	0.5	Manufg.			
Nonferrous wire drawing & insulating	0.5	Manufg.			
Refrigeration & heating equipment	0.5	Manufg.			
Sawmills & planning mills, general	0.5	Manufg.			
Railroads & related services	0.5	Util.			
Automotive rental & leasing, without drivers	0.5	Services			
Fabricated rubber products, nec	0.4	Manufg.			
Paints & allied products	0.4	Manufg.			
Automotive repair shops & services	0.4	Services			
Abrasive products	0.3	Manufg.			
Primary metal products, nec	0.3	Manufg.			
Rubber & plastics hose & belting	0.3	Manufg.			
Special dies & tools & machine tool accessories	0.3	Manufg.			
Eating & drinking places	0.3	Trade			
Real estate	0.3	Fin/R.E.			
Advertising	0.3	Services			
Computer & data processing services	0.3	Services			
Engineering, architectural, & surveying services	0.3	Services			
Aluminum castings	0.2	Manufg.			
Aluminum rolling & drawing	0.2	Manufg.			
Hardwood dimension & flooring mills	0.2	Manufg.			
Industrial gases	0.2	Manufg.			
Industrial patterns	0.2	Manufg.			

Continued on next page.

INPUTS AND OUTPUTS FOR RAILROAD EQUIPMENT - Continued

Economic Sector or Industry Providing Inputs	%	Sector	Economic Sector or Industry Buying Outputs	%	Sector
Machine tools, metal cutting types	0.2	Manufg.			
Metal heat treating	0.2	Manufg.			
Miscellaneous fabricated wire products	0.2	Manufg.			
Miscellaneous plastics products	0.2	Manufg.			
Nonferrous rolling & drawing, nec	0.2	Manufg.			
Transportation equipment, nec	0.2	Manufg.			
Air transportation	0.2	Util.			
Water transportation	0.2	Util.			
Retail trade, except eating & drinking	0.2	Trade			
Insurance carriers	0.2	Fin/R.E.			
Business services nec	0.2	Services			
Legal services	0.2	Services			
Miscellaneous repair shops	0.2	Services			
Coated fabrics, not rubberized	0.1	Manufg.			
Metal stampings, nec	0.1	Manufg.			
Motor vehicle parts & accessories	0.1	Manufg.			
Tires & inner tubes	0.1	Manufg.			
Veneer & plywood	0.1	Manufg.			
Sanitary services, steam supply, irrigation	0.1	Util.			
Accounting, auditing & bookkeeping	0.1	Services			
Hotels & lodging places	0.1	Services			
Management & consulting services & labs	0.1	Services			

Source: Benchmark Input-Output Accounts for the U.S. Economy, 1982, U.S. Department of Commerce, Washington, D.C., July 1991. Data, as reported in the source, are organized by the 1977 SIC structure in use in 1982 but have been matched, as closely as is possible, to the 1987 SIC structure used in this book.

OCCUPATIONS EMPLOYED BY SIC 374 - TRANSPORTATION EQUIPMENT NEC

Occupation	% of Total 1994	Change to 2005	Occupation	% of Total 1994	Change to 2005
Assemblers, fabricators, & hand workers nec	21.4	11.7	Painters, transportation equipment	1.5	8.6
Welders & cutters	9.3	8.0	Helpers, laborers, & material movers nec	1.3	7.5
Blue collar worker supervisors	4.3	3.0	NC machine tool operators, metal & plastic	1.3	29.8
Mechanics, installers, & repairers nec	3.8	-13.7	Combination machine tool operators	1.2	22.0
Welding machine setters, operators	3.0	-0.3	Industrial production managers	1.1	10.1
Sales & related workers nec	2.7	10.0	Coating, painting, & spraying machine workers	1.1	11.6
Inspectors, testers, & graders, precision	2.5	7.9	Maintenance repairers, general utility	1.1	-2.2
General managers & top executives	2.2	4.8	General office clerks	1.1	-5.7
Machinists	2.1	6.8	Grinders & polishers, hand	1.1	11.7
Stock clerks	1.7	-13.1	Lathe & turning machine tool operators	1.1	-14.2
Freight, stock, & material movers, hand	1.6	-12.9	Bookkeeping, accounting, & auditing clerks	1.0	-17.5
Traffic, shipping, & receiving clerks	1.5	7.5	Extraction & related workers nec	1.0	9.1
Secretaries, ex legal & medical	1.5	0.1	Truck drivers light & heavy	1.0	15.1
Machine tool cutting operators, metal & plastic	1.5	-8.1			

Source: Industry-Occupation Matrix, Bureau of Labor Statistics. These data relate to one or more 3-digit SIC industry groups rather than to a single 4-digit SIC. The change reported for each occupation to the year 2005 is a percent of growth or decline as estimated by the Bureau of Labor Statistics. The abbreviation nec stands for 'not elsewhere classified'.

LOCATION BY STATE AND REGIONAL CONCENTRATION

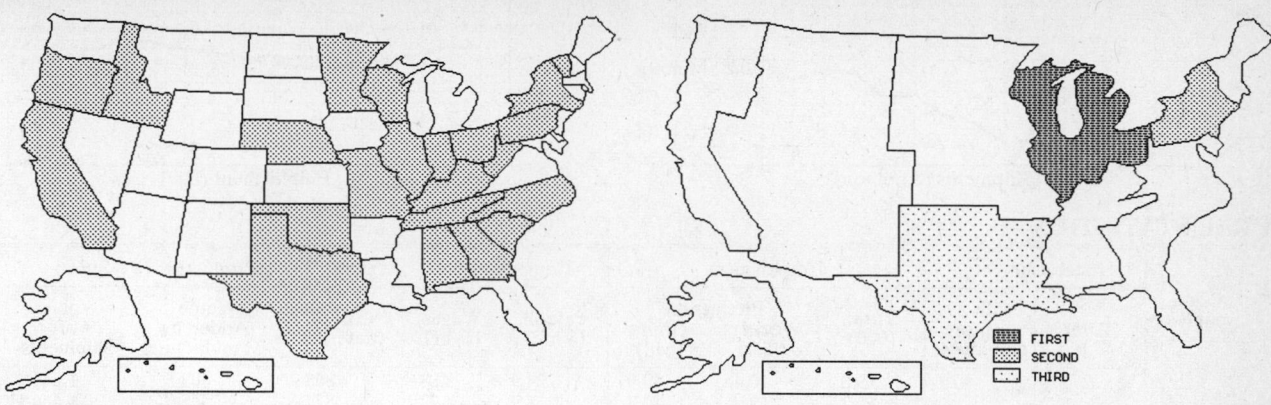

INDUSTRY DATA BY STATE

| State | Establish-ments | Shipments | | | Employment | | | | Cost as % of Shipments | Investment per Employee ($) |
		Total ($ mil)	% of U.S.	Per Establ.	Total Number	% of U.S.	Per Establ.	Wages ($/hour)		
Pennsylvania	28	1,609.5	35.1	57.5	9,700	34.4	346	15.44	57.0	4,598
Illinois	38	1,100.7	24.0	29.0	5,400	19.1	142	18.01	57.2	704
New York	11	267.2	5.8	24.3	1,800	6.4	164	12.63	66.9	-
Ohio	11	124.3	2.7	11.3	900	3.2	82	10.00	55.5	889
Wisconsin	5	23.5	0.5	4.7	200	0.7	40	12.50	64.7	-
Minnesota	4	8.4	0.2	2.1	100	0.4	25	6.50	47.6	-
Texas	19	(D)	-	-	1,750 *	6.2	92	-	-	-
Georgia	7	(D)	-	-	750 *	2.7	107	-	-	-
California	6	(D)	-	-	175 *	0.6	29	-	-	-
Missouri	6	(D)	-	-	375 *	1.3	63	-	-	-
South Carolina	6	(D)	-	-	750 *	2.7	125	-	-	-
Kentucky	5	(D)	-	-	375 *	1.3	75	-	-	-
North Carolina	5	(D)	-	-	175 *	0.6	35	-	-	-
Alabama	4	(D)	-	-	750 *	2.7	188	-	-	-
Indiana	4	(D)	-	-	1,750 *	6.2	438	-	-	-
Nebraska	3	(D)	-	-	175 *	0.6	58	-	-	-
Oregon	3	(D)	-	-	1,750 *	6.2	583	-	-	-
Tennessee	3	(D)	-	-	175 *	0.6	58	-	-	-
West Virginia	3	(D)	-	-	375 *	1.3	125	-	-	-
Oklahoma	2	(D)	-	-	175 *	0.6	88	-	-	-
Idaho	1	(D)	-	-	750 *	2.7	750	-	-	-
Vermont	1	(D)	-	-	375 *	1.3	375	-	-	-

Source: 1992 *Economic Census*. The states are in descending order of shipments or establishments (if shipment data are missing for the majority). The symbol (D) appears when data are withheld to prevent disclosure of competitive information. States marked with (D) are sorted by number of establishments. A dash (-) indicates that the data element cannot be calculated; * indicates the midpoint of a range.

3751 - MOTORCYCLES, BICYCLES & PARTS

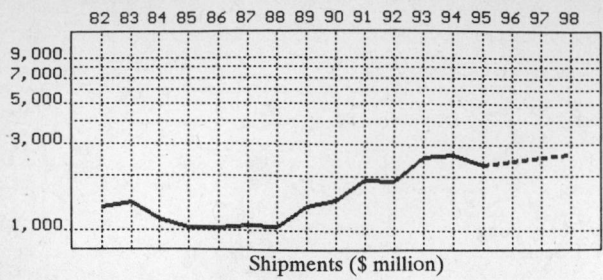

Shipments ($ million)

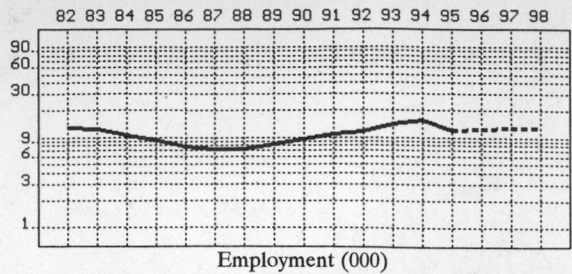

Employment (000)

GENERAL STATISTICS

| Year | Com-panies | Establishments | | Employment | | | Compensation | | Production ($ million) | | | |
		Total	with 20 or more employees	Total (000)	Production Workers (000)	Hours (Mil)	Payroll ($ mil)	Wages ($/hr)	Cost of Materials	Value Added by Manufacture	Value of Shipments	Capital Invest.
1982	269	273	67	13.0	10.4	19.0	224.9	8.88	888.9	402.5	1,341.1	23.1
1983				12.2	9.9	20.3	238.9	8.77	874.6	545.7	1,436.1	26.5
1984				10.4	8.5	16.7	187.6	8.05	697.8	438.7	1,152.6	20.3
1985				9.2	7.4	15.3	168.7	7.94	644.6	402.3	1,044.0	23.4
1986				7.9	6.3	13.7	169.4	9.01	617.5	430.1	1,032.0	21.4
1987	242	246	56	7.4	5.8	12.1	158.3	9.17	679.6	367.8	1,062.6	30.7
1988				7.5	5.8	12.0	164.0	9.62	674.4	385.9	1,056.8	16.7
1989		245	57	8.4	6.7	13.5	190.6	10.13	884.6	501.7	1,369.6	25.9
1990				9.4	7.6	14.8	213.5	10.26	928.6	570.8	1,475.8	24.0
1991				10.8	8.5	16.7	276.6	11.69	1,177.6	741.0	1,913.9	61.5
1992	244	248	51	11.7	9.0	17.1	301.7	11.58	1,146.2	744.5	1,878.3	45.6
1993				14.1	10.5	19.5	409.3	13.53	1,362.0	1,241.0	2,565.6	50.6
1994				15.7	11.7	20.9	482.6	14.71	1,488.6	1,148.8	2,632.1	92.3
1995				12.1P	9.1P	16.9P	372.0P	13.63P	1,305.6P	1,007.6P	2,308.5P	64.4P
1996				12.3P	9.2P	17.0P	390.1P	14.11P	1,368.1P	1,055.8P	2,419.0P	68.6P
1997				12.6P	9.3P	17.1P	408.2P	14.59P	1,430.5P	1,104.0P	2,529.4P	72.7P
1998				12.8P	9.4P	17.2P	426.3P	15.07P	1,493.0P	1,152.2P	2,639.9P	76.8P

Sources: 1982, 1987, 1992 *Economic Census*; *Annual Survey of Manufactures*, 83-86, 88-91, 93-94. Establishment counts for non-Census years are from *County Business Patterns*; establishment values for 83-84 are extrapolations. 'P's show projections by the editors. Industries reclassified in 87 will not have data for prior years.

INDICES OF CHANGE

| Year | Com-panies | Establishments | | Employment | | | Compensation | | Production ($ million) | | | |
		Total	with 20 or more employees	Total (000)	Production Workers (000)	Hours (Mil)	Payroll ($ mil)	Wages ($/hr)	Cost of Materials	Value Added by Manufacture	Value of Shipments	Capital Invest.
1982	110	110	131	111	116	111	75	77	78	54	71	51
1983				104	110	119	79	76	76	73	76	58
1984				89	94	98	62	70	61	59	61	45
1985				79	82	89	56	69	56	54	56	51
1986				68	70	80	56	78	54	58	55	47
1987	99	99	110	63	64	71	52	79	59	49	57	67
1988				64	64	70	54	83	59	52	56	37
1989		99	112	72	74	79	63	87	77	67	73	57
1990				80	84	87	71	89	81	77	79	53
1991				92	94	98	92	101	103	100	102	135
1992	100	100	100	100	100	100	100	100	100	100	100	100
1993				121	117	114	136	117	119	167	137	111
1994				134	130	122	160	127	130	154	140	202
1995				104P	101P	99P	123P	118P	114P	135P	123P	141P
1996				105P	102P	100P	129P	122P	119P	142P	129P	150P
1997				107P	103P	100P	135P	126P	125P	148P	135P	159P
1998				109P	104P	101P	141P	130P	130P	155P	141P	168P

Sources: Same as General Statistics. Values reflect change from the base year, 1992. Values above 100 mean greater than 92, values below 100 mean less than 92, and a value of 100 in the 82-91 or 93-98 period means same as 92. 'P's mark projections by the editors.

SELECTED RATIOS

For 1992	Avg. of All Manufact.	Analyzed Industry	Index	For 1992	Avg. of All Manufact.	Analyzed Industry	Index
Employees per Establishment	46	47	103	Value Added per Production Worker	122,353	82,722	68
Payroll per Establishment	1,332,320	1,216,532	91	Cost per Establishment	4,239,462	4,621,774	109
Payroll per Employee	29,181	25,786	88	Cost per Employee	92,853	97,966	106
Production Workers per Establishment	31	36	116	Cost per Production Worker	135,003	127,356	94
Wages per Establishment	734,496	798,460	109	Shipments per Establishment	8,100,800	7,573,790	93
Wages per Production Worker	23,390	22,002	94	Shipments per Employee	177,425	160,538	90
Hours per Production Worker	2,025	1,900	94	Shipments per Production Worker	257,966	208,700	81
Wages per Hour	11.55	11.58	100	Investment per Establishment	278,244	183,871	66
Value Added per Establishment	3,842,210	3,002,016	78	Investment per Employee	6,094	3,897	64
Value Added per Employee	84,153	63,632	76	Investment per Production Worker	8,861	5,067	57

Sources: Same as General Statistics. The 'Average of All Manufacturing' column represents the average of all manufacturing industries reported for the most recent complete year available. The Index shows the relationship between the Average and the Analyzed Industry. For example, 100 means that they are equal; 500 that the Analyzed Industry is five times the average; 50 means that the Analyzed Industry is half the national average. The abbreviation 'na' is used to show that data are 'not available'.

LEADING COMPANIES Number shown: **40** Total sales ($ mil): **4,893** Total employment (000): **31.8**

Company Name	Address				CEO Name	Phone	Co. Type	Sales ($ mil)	Empl. (000)
Harley-Davidson Inc	PO Box 653	Milwaukee	WI	53201	Richard F Teerlink	414-342-4680	P	1,217	6.0
Harley-Davidson Inc	PO Box 653	Milwaukee	WI	53201	Jeffrey L Bleustein	414-342-4680	D	933	4.1
Huffy Corp	PO Box 1204	Dayton	OH	45401	Harry A Shaw lll	513-866-6251	P	719	7.0
Roadmaster Industries Inc	250 Spring St NW	Atlanta	GA	30303	Henry Fong	404-586-9000	P	456	5.8
Kawasaki Motors	PO Box 81469	Lincoln	NE	68501	T Saeki	402-476-6600	S	436	0.7
Roadmaster Corp	PO Box 344	Olney	IL	62450	Ed Shake	618-393-2991	S	220*	2.7
Trek Bicycle Corp	PO Box 183	Waterloo	WI	53594	Richard A Burke	414-478-2191	S	200*	0.8
Huffy Bicycle	PO Box 1204	Dayton	OH	45401	Barry Ryan	513-866-6251	D	160	2.2
Specialized Bicycle Components	15130 Concord Cir	Morgan Hill	CA	95037	Michael Sinyard	408-779-6229	R	160	0.3
Cannondale Corp	9 Brookside Pl	Georgetown	CT	06829	J S Montgomery	203-544-9800	P	102	0.7
Schwinn Bicycle and Fitness/LP	1690 38th St	Boulder	CO	80301	Ralph Murray	303-939-0100	R	71	0.3
Ross Bicycles USA Ltd	51 Executive Blv	Farmingdale	NY	11735	Larry Goldmeier	516-249-6000	S	65	<0.1
Persons-Majestic Mfg Co	21-31 Hamilton St	Monroeville	OH	44847	R D Sanderson	419-465-2504	R	28	0.2
Faulhaber Co	21-31 Hamilton St	Monroeville	OH	44847	Richard Sanderson	419-465-2504	S	25*	0.2
Answer Products Inc	27460 Scott Av	Valencia	CA	91355	Eddie Cole	805-257-4411	S	17*	0.1
Columbia Manufacturing Inc	1 Cycle St	Westfield	MA	01085	Kenneth Howard	413-562-3664	R	11	<0.1
Burley Design Cooperative	4080 Stewart Rd	Eugene	OR	97402	Bruce Creps	503-687-1644	R	8*	<0.1
Excel International Group Inc	314 E Main St	Barrington	IL	60010	Thomas Nestrud	708-428-1350	R	8	<0.1
AC International	9115-1 Dice Rd	Santa Fe Sprgs	CA	90670	Randy Kirk	310-946-1616	R	6	<0.1
Buell Motorcycle Co	2815 Meyer Ln	East Troy	WI	53120	Erik Buell	414-642-2020	R	6	<0.1
Worksman Trading Corp	94-15 100th St	Ozone Park	NY	11416	Jeff Mishkin	718-322-2000	R	6	<0.1
Custom Chrome Manufacturing	12950 San Fernando	Sylmar	CA	91342	Arthur LoGreco	818-367-5747	S	5*	<0.1
Sandpoint Design Inc	120A Lee Rd	Watsonville	CA	95076	Preston Sandusky	408-724-9079	R	5	<0.1
Emory Manufacturing Co	PO Box 3812	Jacksonville	FL	32206	Clayton Smith	904-354-3339	R	4*	<0.1
Fat City Cycles Inc	PO Box 1439	S Glens Falls	NY	12803	Chris Chance	518-747-8020	R	4	<0.1
American Cycle Systems Inc	PO Box 2597	City of Industry	CA	91746	CW Stephens	818-961-3942	R	3	<0.1
Arthur Fulmer Inc	122 Gayoso Av	Memphis	TN	38103	Arthur Fulmer Jr	901-525-5711	R	3*	<0.1
Merlin Metalworks Inc	40 Smith Pl	Cambridge	MA	02138	Gwyndaf Jones	617-661-6688	R	3	<0.1
Santana Cycles Inc	1324 E Arrow Hwy	La Verne	CA	91750	William McCready	909-621-6943	R	3	<0.1
Cosmopolitan Motors Inc	301 Jacksonville Rd	Hatboro	PA	19040	Lawrence Wise	215-672-9100	R	2	<0.1
Dahon California Inc	5741 Buckingham	Culver City	CA	90230	David Hon	310-417-3456	R	2	<0.1
Cycle Shack Inc	1104 San Mateo Av	S San Francisco	CA	94080	SA Reedy	415-583-7014	R	1*	<0.1
Sifton Motorcycle Products Inc	943 Bransten Rd	San Carlos	CA	94070	Edward A Poleselli	415-592-2203	R	1*	<0.1
Sore Saddle Cyclery Inc	PO Box 2480	Steamboat Sp	CO	80477	Mike Sanders	303-879-1675	R	1	<0.1
IMS Products Inc	6240 Box Spgs Blvd	Riverside	CA	92507	CH Wheat	909-653-7720	R	0*	<0.1
Moot's Mountain Bikes Inc	PO Box 2480	Steamboat Sp	CO	80477	Mike Sanders	303-879-1675	S	0	<0.1
Greendale Bicycle Co	5610 S Division Av	Grand Rapids	MI	49548	Mark Groendal	616-530-5556	R	0	<0.1
Centerline Sports Inc	370 S Crenshaw	Torrance	CA	90503	Michael Holman	310-787-7094	R	0	<0.1
Classified Motorcycle Co	PO Box 565	Carmel Valley	CA	93924	Dan Wilson	408-659-0329	R	0*	<0.1
CV Posi-Drive Corp	3832 148th Av NE	Redmond	WA	98052	Bill Terry	206-867-5477	R	0*	<0.1

Source: Ward's Business Directory of U.S. Private and Public Companies, Volumes 1 and 2, 1996. The company type code used is as follows: P - Public, R - Private, S - Subsidiary, D - Division, J - Joint Venture, A - Affiliate, G - Group. Sales are in millions of dollars, employees are in thousands. An asterisk (*) indicates an estimated sales volume. The symbol < stands for 'less than'. Company names and addresses are truncated, in some cases, to fit into the available space.

MATERIALS CONSUMED

Material	Quantity	Delivered Cost ($ million)
Materials, ingredients, containers, and supplies	(X)	977.0
Frames, forks, and parts thereof, bicycle	(X)	42.0
Wheel rims and spokes, bicycle	(X)	11.0
Bicycle hubs (other than coaster braking hubs and hub brakes), and freewheel sprocket wheels	(X)	15.1
Seats (saddles), bicycle	(X)	28.7
Fabricated metal products (except castings and forgings)	(X)	33.2
Iron and steel castings (rough and semifinished)	(X)	3.5
Aluminum and aluminum-base alloy castings (rough and semifinished)	(X)	6.7
Steel sheet and strip, including tin plate	(X)	24.3
Steel tubing	(X)	19.3
All other steel shapes and forms	(X)	7.0
All other materials and components, parts, containers, and supplies	(X)	653.3
Materials, ingredients, containers, and supplies, nsk	(X)	132.8

Source: 1992 *Economic Census*. Explanation of symbols used: (D): Withheld to avoid disclosure of competitive data; na: Not available; (S): Withheld because statistical norms were not met; (X): Not applicable; (Z): Less than half the unit shown; nec: Not elsewhere classified; nsk: Not specified by kind; - : zero; * : 10-19 percent estimated; ** : 20-29 percent estimated.

PRODUCT SHARE DETAILS

Product or Product Class	% Share	Product or Product Class	% Share
Motorcycles, bicycles, and parts	100.00	Wheel rims and spokes for bicycles, unicycles, and adult tricycles	1.79
Bicycles and parts (excluding children's two-wheel sidewalk cycles with solid or semipneumatic tires)	50.33	Brakes, including coaster braking hubs and hub brakes, and parts thereof for bicycles, unicycles, and adult tricycles	0.05
Bicycles, complete with both wheels 20 inches in diameter, all speeds (excluding children's two-wheel sidewalk cycles with solid or semipneumatic tires)	10.80	Seats (saddles) for bicycles, unicycles, and adult tricycles	1.41
		Other parts for bicycles, unicycles, and adult tricycles	6.20
Lightweight, road bicycles with both wheels 26 inches or more in diameter, all speeds (excluding children's two-wheel sidewalk cycles with solid or semipneumatic tires)	3.72	Bicycles and parts (excluding children's two-wheel sidewalk cycles with solid or semipneumatic tires), nsk	4.86
Other bicycles, incl. mountain, all terrain, and cruisers with both wheels 26 inches or more in diameter, all speeds (excluding children's two-wheel sidewalk cycles with solid or semipneumatic tires)	29.42	Motorcycles, including three-wheel (excluding sidecars), motorbikes, motor scooters, mopeds, and parts	41.18
		Motorcycles, including three-wheel (excluding sidecars), motor scooters, motorbikes, and mopeds	(D)
Other cycles, including unicycles and adult tricycles (excluding children's two-wheel sidewalk cycles with solid or semipneumatic tires)	37.42	Parts for motorcycles (including sidecars), motorbikes, motor scooters, and mopeds (including bicycle engines)	(D)
		Motorcycles, motorbikes, motor scooters, mopeds, and parts, nsk	0.25
Frames, forks, and parts thereof for bicycles, unicycles, and adult tricycles	4.35	Motorcycles, bicycles, and parts, nsk	8.49

Source: 1992 *Economic Census.* The values shown are percent of total shipments in an industry. Values of indented subcategories are summed in the main heading. The symbol (D) appears when data are withheld to prevent disclosure of competitive information. The abbreviation nsk stands for 'not specified by kind' and nec for 'not elsewhere classified'.

INPUTS AND OUTPUTS FOR MOTORCYCLES, BICYCLES, & PARTS

Economic Sector or Industry Providing Inputs	%	Sector	Economic Sector or Industry Buying Outputs	%	Sector
Imports	56.1	Foreign	Personal consumption expenditures	80.3	
Wholesale trade	6.8	Trade	Miscellaneous repair shops	10.8	Services
Internal combustion engines, nec	6.7	Manufg.	Exports	3.6	Foreign
Power transmission equipment	5.7	Manufg.	Gross private fixed investment	2.1	Cap Inv
Blast furnaces & steel mills	4.8	Manufg.	S/L Govt. purch., police	1.5	S/L Govt
Tires & inner tubes	2.8	Manufg.	Motorcycles, bicycles, & parts	1.0	Manufg.
Screw machine and related products	2.0	Manufg.	Change in business inventories	0.4	In House
Motor vehicle parts & accessories	1.8	Manufg.	S/L Govt. purch., correction	0.3	S/L Govt
Automotive & apparel trimmings	1.3	Manufg.			
Plating & polishing	1.2	Manufg.			
Iron & steel foundries	1.1	Manufg.			
Metal stampings, nec	1.1	Manufg.			
Motorcycles, bicycles, & parts	1.0	Manufg.			
Fabricated metal products, nec	0.7	Manufg.			
Motor freight transportation & warehousing	0.6	Util.			
Maintenance of nonfarm buildings nec	0.4	Constr.			
Rubber & plastics hose & belting	0.4	Manufg.			
Electric services (utilities)	0.4	Util.			
Lighting fixtures & equipment	0.3	Manufg.			
Paints & allied products	0.3	Manufg.			
Paperboard containers & boxes	0.3	Manufg.			
Gas production & distribution (utilities)	0.3	Util.			
Aluminum castings	0.2	Manufg.			
Metal coating & allied services	0.2	Manufg.			
Miscellaneous plastics products	0.2	Manufg.			
Plastics materials & resins	0.2	Manufg.			
Communications, except radio & TV	0.2	Util.			
Railroads & related services	0.2	Util.			
Eating & drinking places	0.2	Trade			
Banking	0.2	Fin/R.E.			
Aluminum rolling & drawing	0.1	Manufg.			
Gaskets, packing & sealing devices	0.1	Manufg.			
Iron & steel forgings	0.1	Manufg.			
Machinery, except electrical, nec	0.1	Manufg.			
Air transportation	0.1	Util.			
Real estate	0.1	Fin/R.E.			
Equipment rental & leasing services	0.1	Services			
Theatrical producers, bands, entertainers	0.1	Services			
Noncomparable imports	0.1	Foreign			

Source: Benchmark Input-Output Accounts for the U.S. Economy, 1982, U.S. Department of Commerce, Washington, D.C., July 1991. Data, as reported in the source, are organized by the 1977 SIC structure in use in 1982 but have been matched, as closely as is possible, to the 1987 SIC structure used in this book.

OCCUPATIONS EMPLOYED BY SIC 375 - TRANSPORTATION EQUIPMENT NEC

Occupation	% of Total 1994	Change to 2005	Occupation	% of Total 1994	Change to 2005
Assemblers, fabricators, & hand workers nec	21.4	11.7	Painters, transportation equipment	1.5	8.6
Welders & cutters	9.3	8.0	Helpers, laborers, & material movers nec	1.3	7.5
Blue collar worker supervisors	4.3	3.0	NC machine tool operators, metal & plastic	1.3	29.8
Mechanics, installers, & repairers nec	3.8	-13.7	Combination machine tool operators	1.2	22.0
Welding machine setters, operators	3.0	-0.3	Industrial production managers	1.1	10.1
Sales & related workers nec	2.7	10.0	Coating, painting, & spraying machine workers	1.1	11.6
Inspectors, testers, & graders, precision	2.5	7.9	Maintenance repairers, general utility	1.1	-2.2
General managers & top executives	2.2	4.8	General office clerks	1.1	-5.7
Machinists	2.1	6.8	Grinders & polishers, hand	1.1	11.7
Stock clerks	1.7	-13.1	Lathe & turning machine tool operators	1.1	-14.2
Freight, stock, & material movers, hand	1.6	-12.9	Bookkeeping, accounting, & auditing clerks	1.0	-17.5
Traffic, shipping, & receiving clerks	1.5	7.5	Extraction & related workers nec	1.0	9.1
Secretaries, ex legal & medical	1.5	0.1	Truck drivers light & heavy	1.0	15.1
Machine tool cutting operators, metal & plastic	1.5	-8.1			

Source: *Industry-Occupation Matrix*, Bureau of Labor Statistics. These data relate to one or more 3-digit SIC industry groups rather than to a single 4-digit SIC. The change reported for each occupation to the year 2005 is a percent of growth or decline as estimated by the Bureau of Labor Statistics. The abbreviation nec stands for 'not elsewhere classified'.

LOCATION BY STATE AND REGIONAL CONCENTRATION

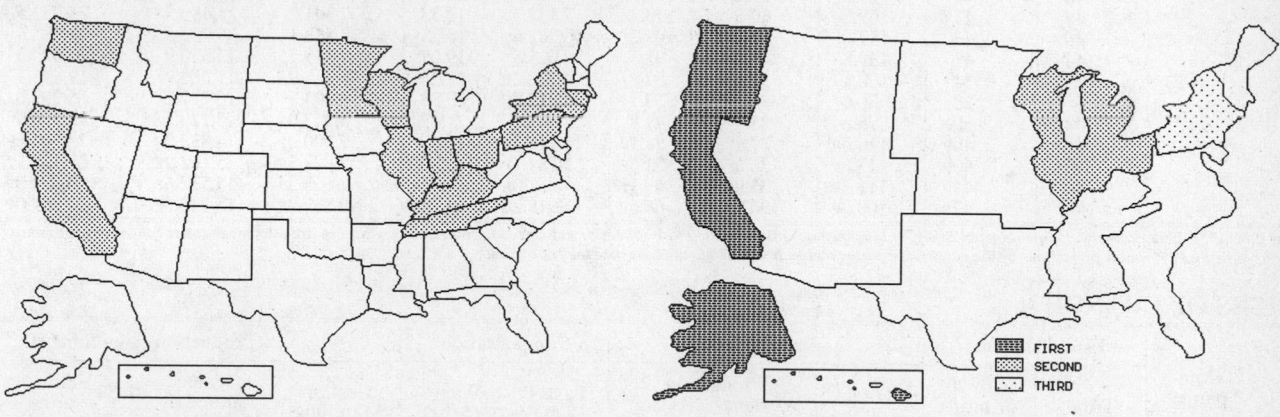

FIRST
SECOND
THIRD

INDUSTRY DATA BY STATE

State	Establish-ments	Shipments			Employment				Cost as % of Shipments	Investment per Employee ($)
		Total ($ mil)	% of U.S.	Per Establ.	Total Number	% of U.S.	Per Establ.	Wages ($/hour)		
Wisconsin	9	272.2	14.5	30.2	1,500	12.8	167	9.73	65.4	4,067
Washington	12	87.3	4.6	7.3	500	4.3	42	10.00	76.9	-
Minnesota	7	12.4	0.7	1.8	100	0.9	14	11.00	46.0	-
California	107	(D)	-	-	1,750 *	15.0	16	-	-	3,371
Illinois	10	(D)	-	-	1,750 *	15.0	175	-	-	-
Ohio	8	(D)	-	-	1,750 *	15.0	219	-	-	-
New York	7	(D)	-	-	175 *	1.5	25	-	-	-
Tennessee	6	(D)	-	-	175 *	1.5	29	-	-	-
Indiana	5	(D)	-	-	175 *	1.5	35	-	-	-
Pennsylvania	4	(D)	-	-	1,750 *	15.0	438	-	-	-
Connecticut	3	(D)	-	-	750 *	6.4	250	-	-	-
Kentucky	1	(D)	-	-	175 *	1.5	175	-	-	-

Source: 1992 *Economic Census*. The states are in descending order of shipments or establishments (if shipment data are missing for the majority). The symbol (D) appears when data are withheld to prevent disclosure of competitive information. States marked with (D) are sorted by number of establishments. A dash (-) indicates that the data element cannot be calculated; * indicates the midpoint of a range.

3761 - GUIDED MISSILES & SPACE VEHICLES

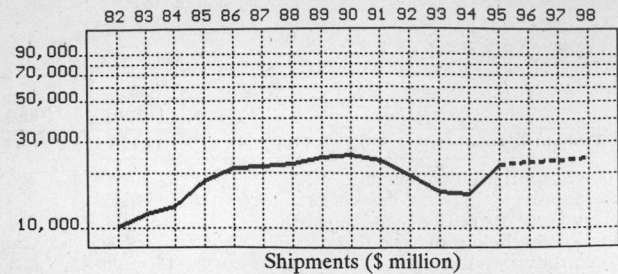

Shipments ($ million)

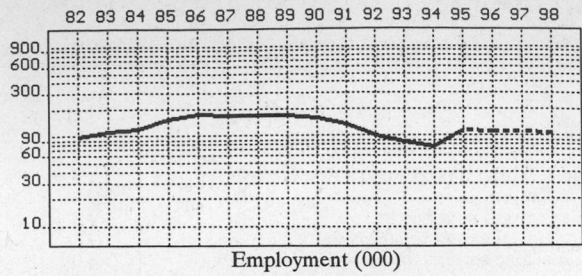

Employment (000)

GENERAL STATISTICS

Year	Com-panies	Establishments		Employment			Compensation		Production ($ million)			
		Total	with 20 or more employees	Total (000)	Production Workers (000)	Hours (Mil)	Payroll ($ mil)	Wages ($/hr)	Cost of Materials	Value Added by Manufacture	Value of Shipments	Capital Invest.
1982	16	29	28	99.6	35.9	70.5	3,159.4	12.99	3,652.1	7,025.5	10,218.6	293.3
1983		30	28	110.7	42.7	80.5	3,469.3	13.86	4,068.5	8,260.8	11,870.6	437.6
1984		31	28	120.9	43.7	86.6	4,118.8	13.89	4,205.6	10,047.0	13,191.5	686.3
1985		32	29	154.3	53.4	105.8	5,473.3	14.10	6,098.7	12,576.7	18,087.1	968.8
1986		33	31	174.2	63.5	122.9	6,301.1	15.72	7,226.8	14,120.6	21,401.3	977.7
1987	19	40	38	166.7	62.7	121.4	6,414.8	16.64	6,791.3	15,072.7	21,565.8	818.7
1988		40	40	169.0	61.3	120.8	6,786.9	17.19	8,261.1	15,099.2	22,512.8	757.5
1989		44	41	169.5	60.1	118.3	7,131.3	17.53	9,301.2	15,480.5	23,982.5	794.8
1990		46	43	157.0	54.2	109.2	6,742.6	18.25	9,588.1	15,782.5	25,082.6	659.0
1991		44	41	135.8	45.0	91.3	6,025.5	19.07	8,219.7	13,550.9	23,399.3	449.9
1992	23	38	31	100.1	31.5	57.8	4,722.2	22.23	6,788.9	11,080.9	19,675.1	306.7
1993		35	28	86.6	27.6	53.1	4,017.6	20.88	5,804.5	10,090.2	15,799.6	307.6
1994		44P	39P	76.8	25.0	49.8	3,806.2	21.02	6,277.0	10,120.5	15,396.9	329.0
1995		46P	40P	116.0P	37.6P	74.1P	5,716.5P	22.47P	9,200.0P	14,833.2P	22,566.6P	438.5P
1996		47P	40P	113.6P	36.3P	71.6P	5,784.1P	23.23P	9,429.3P	15,202.9P	23,129.1P	415.6P
1997		48P	41P	111.3P	35.0P	69.1P	5,851.6P	23.98P	9,658.5P	15,572.6P	23,691.5P	392.7P
1998		49P	42P	108.9P	33.7P	66.6P	5,919.2P	24.74P	9,887.8P	15,942.3P	24,254.0P	369.8P

Sources: 1982, 1987, 1992 *Economic Census*; *Annual Survey of Manufactures*, 83-86, 88-91, 93-94. Establishment counts for non-Census years are from *County Business Patterns*; establishment values for 83-84 are extrapolations. 'P's show projections by the editors. Industries reclassified in 87 will not have data for prior years.

INDICES OF CHANGE

Year	Com-panies	Establishments		Employment			Compensation		Production ($ million)			
		Total	with 20 or more employees	Total (000)	Production Workers (000)	Hours (Mil)	Payroll ($ mil)	Wages ($/hr)	Cost of Materials	Value Added by Manufacture	Value of Shipments	Capital Invest.
1982	70	76	90	100	114	122	67	58	54	63	52	96
1983		79	90	111	136	139	73	62	60	75	60	143
1984		82	90	121	139	150	87	62	62	91	67	224
1985		84	94	154	170	183	116	63	90	113	92	316
1986		87	100	174	202	213	133	71	106	127	109	319
1987	83	105	123	167	199	210	136	75	100	136	110	267
1988		105	129	169	195	209	144	77	122	136	114	247
1989		116	132	169	191	205	151	79	137	140	122	259
1990		121	139	157	172	189	143	82	141	142	127	215
1991		116	132	136	143	158	128	86	121	122	119	147
1992	100	100	100	100	100	100	100	100	100	100	100	100
1993		92	90	87	88	92	85	94	85	91	80	100
1994		117P	125P	77	79	86	81	95	92	91	78	107
1995		120P	128P	116P	119P	128P	121P	101P	136P	134P	115P	143P
1996		123P	130P	113P	115P	124P	122P	104P	139P	137P	118P	136P
1997		126P	133P	111P	111P	120P	124P	108P	142P	141P	120P	128P
1998		129P	135P	109P	107P	115P	125P	111P	146P	144P	123P	121P

Sources: Same as General Statistics. Values reflect change from the base year, 1992. Values above 100 mean greater than 92, values below 100 mean less than 92, and a value of 100 in the 82-91 or 93-98 period means same as 92. 'P's mark projections by the editors.

SELECTED RATIOS

For 1994	Avg. of All Manufact.	Analyzed Industry	Index	For 1994	Avg. of All Manufact.	Analyzed Industry	Index
Employees per Establishment	49	1,731	3,532	Value Added per Production Worker	134,084	404,820	302
Payroll per Establishment	1,500,273	85,766,200	5,717	Cost per Establishment	5,045,178	141,441,448	2,803
Payroll per Employee	30,620	49,560	162	Cost per Employee	102,970	81,732	79
Production Workers per Establishment	34	563	1,641	Cost per Production Worker	146,988	251,080	171
Wages per Establishment	853,319	23,587,756	2,764	Shipments per Establishment	9,576,895	346,942,779	3,623
Wages per Production Worker	24,861	41,872	168	Shipments per Employee	195,460	200,480	103
Hours per Production Worker	2,056	1,992	97	Shipments per Production Worker	279,017	615,876	221
Wages per Hour	12.09	21.02	174	Investment per Establishment	321,011	7,413,452	2,309
Value Added per Establishment	4,602,255	228,048,139	4,955	Investment per Employee	6,552	4,284	65
Value Added per Employee	93,930	131,777	140	Investment per Production Worker	9,352	13,160	141

Sources: Same as General Statistics. The 'Average of All Manufacturing' column represents the average of all manufacturing industries reported for the most recent complete year available. The Index shows the relationship between the Average and the Analyzed Industry. For example, 100 means that they are equal; 500 that the Analyzed Industry is five times the average; 50 means that the Analyzed Industry is half the national average. The abbreviation 'na' is used to show that data are 'not available'.

LEADING COMPANIES   Number shown: 14   Total sales ($ mil): 12,990   Total employment (000): 83.4

Company Name	Address				CEO Name	Phone	Co. Type	Sales ($ mil)	Empl. (000)
Rockwell International Corp	2201 Seal Beach	Seal Beach	CA	90740	Donald R Beall	310-797-3311	P	11,123	71.9
Aerojet General Corp	PO Box 13222	Sacramento	CA	95813	Roger J Ramseier	916-355-1000	S	594	3.4
Lockheed Martin	103 Chesapeake Pk	Baltimore	MD	21220	Robert B Coutts	410-682-1000	S	350	1.5
Orbital Sciences Corp	21700 Atlantic Blv	Dulles	VA	20166	David W Thompson	703-406-5000	P	222	1.8
Loral Aeronutronic	PO Box 7004	R S Margari	CA	92688	WS Buttrill	714-459-3000	D	190*	1.4
Lockheed Martin	PO Box 17100	Austin	TX	78760	Dave Penrose	512-386-0000	D	170	0.8
BEI Electronics Inc	1 Post St	San Francisco	CA	94104	Peter G Paraskos	415-956-4477	P	139	1.3
Orbital Sciences Corp	3380 S Price Rd	Chandler	AZ	85248	Don Tutwiler	602-899-6000	D	75	0.5
BEI Defense Systems Company	PO Box 155429	Fort Worth	TX	76155	Edward Smith	817-685-7066	S	50	0.4
Brunswick Corp	1 N Field Ct	Lake Forest	IL	60045	Robert Sigrist	708-735-4821	D	50*	0.3
Spectrum Astro Inc	3601 Aviation Blv	Manhattan Bch	CA	90266	W David Thompson	310-643-9303	R	18	<0.1
Entron Industries LP	70-31 84th St	Glendale	NY	11385	Joseph Ross	718-894-8100	R	4*	<0.1
L'Garde Inc	15181 Woodlawn	Tustin	CA	92680	Mitch Thomas	714-259-0771	R	4	<0.1
PacAstro	520 Huntmar Pk Dr	Herndon	VA	22070	Richard Fleeter	703-709-2240	R	1*	<0.1

Source: Ward's Business Directory of U.S. Private and Public Companies, Volumes 1 and 2, 1996. The company type code used is as follows: P - Public, R - Private, S - Subsidiary, D - Division, J - Joint Venture, A - Affiliate, G - Group. Sales are in millions of dollars, employees are in thousands. An asterisk (*) indicates an estimated sales volume. The symbol < stands for 'less than'. Company names and addresses are truncated, in some cases, to fit into the available space.

MATERIALS CONSUMED

Material	Quantity	Delivered Cost ($ million)
Materials, ingredients, containers, and supplies	(X)	5,240.9
Guided missile and space vehicle engines and parts	(X)	624.1
Guided missile and space vehicle propulsion units and parts	(X)	(D)
Guided missile and space vehicle airframe parts	(X)	(D)
Radio communication systems and equipment	(X)	778.4
Resistors, capacitors, transformers, electron tubes, semiconductors, and other electronic components	(X)	(D)
Other matrix composits, including ceramic, carbon, metal, etc.	(X)	8.4
Complete mechanical, hydraulic and pneumatic subassemblies	(X)	77.5
Fluid power products	(X)	28.7
Metal bolts, nuts, screws, washers, rivets, and other screw machine products	(X)	13.7
Other fabricated metal products (except fluid power products and forgings)	(X)	(D)
Iron and steel forgings	(X)	(D)
Nonferrous forgings	(X)	(D)
Iron and steel castings (rough and semifinished)	(X)	13.6
Metal shapes and forms, except castings, forgings, and fabricated metal products	(X)	12.7
Chemicals, all types (including propellants)	(X)	(D)
Special dies, tools, die sets, jigs, and fixtures, except cutting tools for machine tools	(X)	(D)
All other materials and components, parts, containers, and supplies	(X)	1,682.8
Materials, ingredients, containers, and supplies, nsk	(X)	816.4

Source: 1992 *Economic Census.* Explanation of symbols used: (D): Withheld to avoid disclosure of competitive data; na: Not available; (S): Withheld because statistical norms were not met; (X): Not applicable; (Z): Less than half the unit shown; nec: Not elsewhere classified; nsk: Not specified by kind; - : zero; * : 10-19 percent estimated; ** : 20-29 percent estimated.

PRODUCT SHARE DETAILS

Product or Product Class	% Share	Product or Product Class	% Share
Guided missiles and space vehicles	100.00	Research and development on complete space vehicles for U.S. Government military customers.	37.25
Complete guided missiles	28.25		
Research and development on complete guided missiles	9.77	Research and development on complete space vehicles for other customers	62.75
Other services on complete guided missiles	10.13		
Complete space vehicles (excluding propulsion systems)	41.69	All other services on complete space vehicles	5.89
Complete space vehicles (escluding propulsion systems) for U.S. Government military customers.	75.93	All other services on complete space vehicles for U.S. Government military customers	86.67
Complete space vehicles (excluding propulsion systems) for other customers	24.07	All other services on complete space vehicles for other customers	13.33
Research and development on complete space vehicles	4.25	Guided missiles and space vehicles, nsk	0.01

Source: 1992 *Economic Census.* The values shown are percent of total shipments in an industry. Values of indented subcategories are summed in the main heading. The symbol (D) appears when data are withheld to prevent disclosure of competitive information. The abbreviation nsk stands for 'not specified by kind' and nec for 'not elsewhere classified'.

INPUTS AND OUTPUTS FOR AIRCRAFT

Economic Sector or Industry Providing Inputs	%	Sector	Economic Sector or Industry Buying Outputs	%	Sector
Aircraft & missile engines & engine parts	25.0	Manufg.	Federal Government purchases, national defense	37.6	Fed Govt
Aircraft & missile equipment, nec	11.4	Manufg.	Exports	26.2	Foreign
Mechanical measuring devices	6.0	Manufg.	Gross private fixed investment	21.2	Cap Inv
Imports	5.9	Foreign	Change in business inventories	11.6	In House
Wholesale trade	4.5	Trade	Aircraft & missile equipment, nec	1.7	Manufg.
Radio & TV communication equipment	3.9	Manufg.	Federal Government purchases, nondefense	1.0	Fed Govt
Advertising	3.4	Services	Personal consumption expenditures	0.4	
Banking	3.1	Fin/R.E.			
Electronic components nec	2.7	Manufg.			
Special dies & tools & machine tool accessories	2.5	Manufg.			
Air transportation	2.5	Util.			
Miscellaneous plastics products	2.4	Manufg.			
Machinery, except electrical, nec	2.2	Manufg.			
Hotels & lodging places	1.7	Services			
Aluminum rolling & drawing	1.2	Manufg.			
Screw machine and related products	1.2	Manufg.			
Electric services (utilities)	1.2	Util.			
Semiconductors & related devices	1.1	Manufg.			
Communications, except radio & TV	0.9	Util.			
Engineering & scientific instruments	0.8	Manufg.			
Petroleum refining	0.8	Manufg.			
Public building furniture	0.8	Manufg.			
Engineering, architectural, & surveying services	0.8	Services			
Metal stampings, nec	0.7	Manufg.			
Nonferrous forgings	0.7	Manufg.			
Equipment rental & leasing services	0.7	Services			
Maintenance of nonfarm buildings nec	0.6	Constr.			
Fabricated textile products, nec	0.5	Manufg.			
Optical instruments & lenses	0.5	Manufg.			
Eating & drinking places	0.5	Trade			
Detective & protective services	0.5	Services			
Abrasive products	0.4	Manufg.			
Nonferrous wire drawing & insulating	0.4	Manufg.			
Real estate	0.4	Fin/R.E.			
Legal services	0.4	Services			
U.S. Postal Service	0.4	Gov't			
Automotive & apparel trimmings	0.3	Manufg.			
Broadwoven fabric mills	0.3	Manufg.			
Engine electrical equipment	0.3	Manufg.			
Hardware, nec	0.3	Manufg.			
Nonferrous rolling & drawing, nec	0.3	Manufg.			
Gas production & distribution (utilities)	0.3	Util.			
Motor freight transportation & warehousing	0.3	Util.			
Aluminum castings	0.2	Manufg.			
Blast furnaces & steel mills	0.2	Manufg.			
Fabricated structural metal	0.2	Manufg.			
Industrial controls	0.2	Manufg.			
Metal heat treating	0.2	Manufg.			
Paints & allied products	0.2	Manufg.			
Pipe, valves, & pipe fittings	0.2	Manufg.			
Plating & polishing	0.2	Manufg.			
Pumps & compressors	0.2	Manufg.			
Colleges, universities, & professional schools	0.2	Services			
Computer & data processing services	0.2	Services			
Management & consulting services & labs	0.2	Services			
Photofinishing labs, commercial photography	0.2	Services			
Ball & roller bearings	0.1	Manufg.			
Fabricated metal products, nec	0.1	Manufg.			
Fabricated rubber products, nec	0.1	Manufg.			
Iron & steel forgings	0.1	Manufg.			
Manifold business forms	0.1	Manufg.			
Insurance carriers	0.1	Fin/R.E.			
Accounting, auditing & bookkeeping	0.1	Services			
Miscellaneous repair shops	0.1	Services			

Source: Benchmark Input-Output Accounts for the U.S. Economy, 1982, U.S. Department of Commerce, Washington, D.C., July 1991. Data, as reported in the source, are organized by the 1977 SIC structure in use in 1982 but have been matched, as closely as is possible, to the 1987 SIC structure used in this book.

OCCUPATIONS EMPLOYED BY SIC 376 - GUIDED MISSILES, SPACE VEHICLES, AND PARTS

Occupation	% of Total 1994	Change to 2005	Occupation	% of Total 1994	Change to 2005
Engineers nec	13.9	1.6	Professional workers nec	2.1	1.6
Electrical & electronics engineers	5.5	-9.9	Electrical & electronic technicians,technologists	1.8	-32.3
Management support workers nec	3.7	-15.3	Industrial engineers, ex safety engineers	1.8	-6.9
Mechanical engineers	3.6	-6.8	Managers & administrators nec	1.6	-15.4
Inspectors, testers, & graders, precision	3.4	-15.3	Machinists	1.5	-15.3
Engineering technicians & technologists nec	3.2	-23.8	General office clerks	1.5	-27.8
Engineering, mathematical, & science managers	3.1	20.2	Industrial production managers	1.3	-15.4
Secretaries, ex legal & medical	3.1	-22.9	Purchasing agents, ex trade & farm products	1.3	-15.3
Electrical & electronic equipment assemblers	2.9	-6.9	Clerical support workers nec	1.3	-32.3
Blue collar worker supervisors	2.5	-18.0	Aircraft assemblers, precision	1.2	-66.1
Aeronautical & astronautical engineers	2.4	-15.4	Production, planning, & expediting clerks	1.2	-15.3
Computer programmers	2.3	-31.4	Computer engineers	1.1	25.4
Systems analysts	2.2	35.5	Stock clerks	1.0	-31.3
Assemblers, fabricators, & hand workers nec	2.1	-15.4			

Source: *Industry-Occupation Matrix*, Bureau of Labor Statistics. These data relate to one or more 3-digit SIC industry groups rather than to a single 4-digit SIC. The change reported for each occupation to the year 2005 is a percent of growth or decline as estimated by the Bureau of Labor Statistics. The abbreviation nec stands for 'not elsewhere classified'.

LOCATION BY STATE AND REGIONAL CONCENTRATION

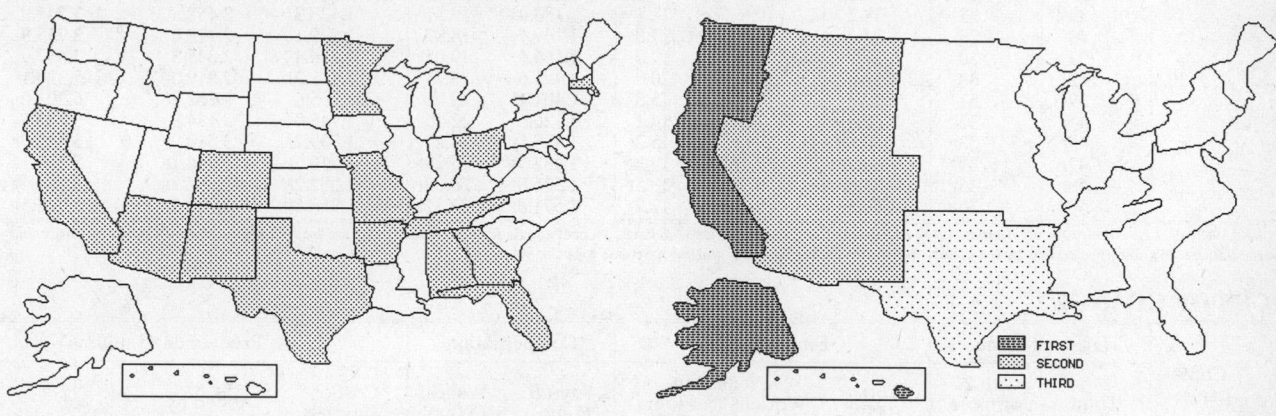

FIRST
SECOND
THIRD

INDUSTRY DATA BY STATE

State	Establish-ments	Shipments Total ($ mil)	Shipments % of U.S.	Shipments Per Establ.	Employment Total Number	Employment % of U.S.	Employment Per Establ.	Wages ($/hour)	Cost as % of Shipments	Investment per Employee ($)
California	15	11,587.8	58.9	772.5	63,500	63.4	4,233	19.16	32.0	3,772
Arizona	3	(D)	-	-	3,750 *	3.7	1,250	-	-	-
Massachusetts	3	(D)	-	-	3,750 *	3.7	1,250	-	-	-
Texas	3	(D)	-	-	7,500 *	7.5	2,500	-	-	-
Alabama	2	(D)	-	-	3,750 *	3.7	1,875	-	-	-
Georgia	2	(D)	-	-	1,750 *	1.7	875	-	-	-
New Mexico	2	(D)	-	-	175 *	0.2	88	-	-	-
Arkansas	1	(D)	-	-	375 *	0.4	375	-	-	-
Colorado	1	(D)	-	-	7,500 *	7.5	7,500	-	-	-
Florida	1	(D)	-	-	1,750 *	1.7	1,750	-	-	-
Minnesota	1	(D)	-	-	175 *	0.2	175	-	-	-
Missouri	1	(D)	-	-	3,750 *	3.7	3,750	-	-	-
Ohio	1	(D)	-	-	1,750 *	1.7	1,750	-	-	-
Tennessee	1	(D)	-	-	1,750 *	1.7	1,750	-	-	-

Source: 1992 *Economic Census*. The states are in descending order of shipments or establishments (if shipment data are missing for the majority). The symbol (D) appears when data are withheld to prevent disclosure of competitive information. States marked with (D) are sorted by number of establishments. A dash (-) indicates that the data element cannot be calculated; * indicates the midpoint of a range.

3764 - SPACE PROPULSION UNITS & PARTS

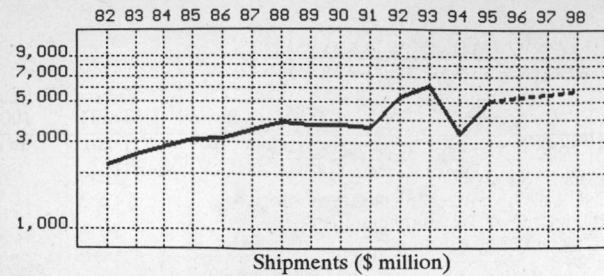

82 83 84 85 86 87 88 89 90 91 92 93 94 95 96 97 98

Shipments ($ million)

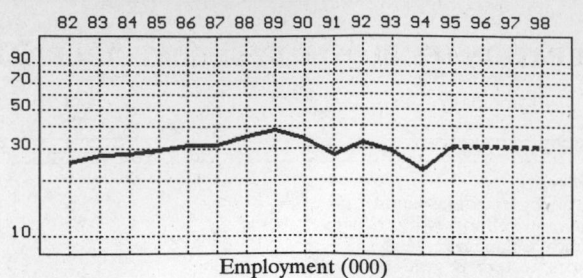

82 83 84 85 86 87 88 89 90 91 92 93 94 95 96 97 98

Employment (000)

GENERAL STATISTICS

| Year | Companies | Establishments | | Employment | | | Compensation | | Production ($ million) | | | |
		Total	with 20 or more employees	Total (000)	Production Workers (000)	Hours (Mil)	Payroll ($ mil)	Wages ($/hr)	Cost of Materials	Value Added by Manufacture	Value of Shipments	Capital Invest.
1982	20	27	25	25.3	10.8	23.4	737.1	11.24	737.2	1,534.0	2,221.2	95.8
1983		27	25	27.6	11.7	23.8	848.8	12.86	887.8	1,694.1	2,577.3	106.1
1984		27	25	28.2	11.7	25.5	902.3	12.05	965.6	1,853.9	2,802.7	148.2
1985		26	24	29.8	11.3	24.5	1,034.2	13.87	1,205.6	1,879.7	3,110.0	233.5
1986		29	25	31.4	11.2	22.3	1,089.6	15.25	1,200.2	1,974.1	3,125.1	280.4
1987	27	35	28	31.8	11.2	22.2	1,174.7	15.77	1,286.1	2,314.2	3,537.1	194.4
1988		35	28	35.3	12.6	24.8	1,351.8	17.08	1,424.1	2,564.2	3,881.3	208.9
1989		39	31	38.2	10.6	18.1	1,183.4	19.51	1,423.3	2,452.5	3,746.9	262.0
1990		40	29	34.0	10.6	17.8	1,169.7	18.87	1,339.1	2,412.0	3,755.8	181.9
1991		37	30	27.7	9.5	17.7	1,166.6	19.05	1,230.4	2,345.8	3,657.9	102.3
1992	30	42	33	32.3	13.4	20.9	1,495.6	23.74	2,181.3	2,819.0	5,328.1	128.2
1993		39	31	29.2	8.9	15.8	1,401.1	22.03	1,255.6	4,282.2	6,201.0	85.4
1994		44P	33P	22.8	7.8	14.1	1,126.6	22.10	1,148.0	2,184.8	3,373.6	68.9
1995		45P	33P	30.8P	9.6P	15.2P	1,438.5P	24.27P	1,739.2P	3,310.0P	5,111.0P	127.8P
1996		47P	34P	30.9P	9.4P	14.3P	1,482.7P	25.29P	1,810.7P	3,446.1P	5,321.1P	123.1P
1997		48P	35P	31.0P	9.3P	13.5P	1,526.8P	26.30P	1,882.2P	3,582.2P	5,531.3P	118.3P
1998		50P	36P	31.1P	9.1P	12.7P	1,571.0P	27.31P	1,953.8P	3,718.3P	5,741.5P	113.5P

Sources: 1982, 1987, 1992 *Economic Census*; *Annual Survey of Manufactures*, 83-86, 88-91, 93-94. Establishment counts for non-Census years are from *County Business Patterns*; establishment values for 83-84 are extrapolations. 'P's show projections by the editors. Industries reclassified in 87 will not have data for prior years.

INDICES OF CHANGE

| Year | Companies | Establishments | | Employment | | | Compensation | | Production ($ million) | | | |
		Total	with 20 or more employees	Total (000)	Production Workers (000)	Hours (Mil)	Payroll ($ mil)	Wages ($/hr)	Cost of Materials	Value Added by Manufacture	Value of Shipments	Capital Invest.
1982	67	64	76	78	81	112	49	47	34	54	42	75
1983		64	76	85	87	114	57	54	41	60	48	83
1984		64	76	87	87	122	60	51	44	66	53	116
1985		62	73	92	84	117	69	58	55	67	58	182
1986		69	76	97	84	107	73	64	55	70	59	219
1987	90	83	85	98	84	106	79	66	59	82	66	152
1988		83	85	109	94	119	90	72	65	91	73	163
1989		93	94	118	79	87	79	82	65	87	70	204
1990		95	88	105	79	85	78	79	61	86	70	142
1991		88	91	86	71	85	78	80	56	83	69	80
1992	100	100	100	100	100	100	100	100	100	100	100	100
1993		93	94	90	66	76	94	93	58	152	116	67
1994		104P	99P	71	58	67	75	93	53	78	63	54
1995		107P	102P	95P	72P	73P	96P	102P	80P	117P	96P	100P
1996		111P	104P	96P	70P	69P	99P	107P	83P	122P	100P	96P
1997		115P	106P	96P	69P	65P	102P	111P	86P	127P	104P	92P
1998		118P	108P	96P	68P	61P	105P	115P	90P	132P	108P	89P

Sources: Same as General Statistics. Values reflect change from the base year, 1992. Values above 100 mean greater than 92, values below 100 mean less than 92, and a value of 100 in the 82-91 or 93-98 period means same as 92. 'P's mark projections by the editors.

SELECTED RATIOS

For 1994	Avg. of All Manufact.	Analyzed Industry	Index	For 1994	Avg. of All Manufact.	Analyzed Industry	Index
Employees per Establishment	49	524	1,069	Value Added per Production Worker	134,084	280,103	209
Payroll per Establishment	1,500,273	25,889,833	1,726	Cost per Establishment	5,045,178	26,381,616	523
Payroll per Employee	30,620	49,412	161	Cost per Employee	102,970	50,351	49
Production Workers per Establishment	34	179	522	Cost per Production Worker	146,988	147,179	100
Wages per Establishment	853,319	7,160,954	839	Shipments per Establishment	9,576,895	77,527,020	810
Wages per Production Worker	24,861	39,950	161	Shipments per Employee	195,460	147,965	76
Hours per Production Worker	2,056	1,808	88	Shipments per Production Worker	279,017	432,513	155
Wages per Hour	12.09	22.10	183	Investment per Establishment	321,011	1,583,357	493
Value Added per Establishment	4,602,255	50,207,799	1,091	Investment per Employee	6,552	3,022	46
Value Added per Employee	93,930	95,825	102	Investment per Production Worker	9,352	8,833	94

Sources: Same as General Statistics. The 'Average of All Manufacturing' column represents the average of all manufacturing industries reported for the most recent complete year available. The Index shows the relationship between the Average and the Analyzed Industry. For example, 100 means that they are equal; 500 that the Analyzed Industry is five times the average; 50 means that the Analyzed Industry is half the national average. The abbreviation 'na' is used to show that data are 'not available'.

LEADING COMPANIES
Number shown: **19**   Total sales ($ mil): **5,237**   Total employment (000): **37.2**

Company Name	Address				CEO Name	Phone	Co. Type	Sales ($ mil)	Empl. (000)
GenCorp Inc	175 Ghent Rd	Fairlawn	OH	44333	John B Yasinsky	216-869-4200	P	1,740	13.0
Thiokol Corp	2475 Washington	Ogden	UT	84401	James R Wilson	801-629-2270	P	1,058	8.0
Rocketdyne	PO Box 7922	Canoga Park	CA	91309	Robert Paster	818-586-1000	D	920*	6.6
Lockheed Martin	PO Box 29304	New Orleans	LA	70189	G Thomas Marsh	504-257-3311	D	320	2.7
Thiokol Corp	PO Box 689	Brigham City	UT	84302	WW Brant	801-863-3679	D	312	1.1
USBI Co	PO Box 1900	Huntsville	AL	35807	Joseph Zimonis	205-721-2770	S	200	1.5
United Technologies Corp	PO Box 49028	San Jose	CA	95161	Douglas A North	408-779-9121	D	180	1.0
Allegany Ballistics Laboratory	PO Box 210	Rocket Center	WV	26726	John Hixon	304-726-5000	D	160	0.7
Atlantic Research Corp	1577 Spring Hill Rd	Vienna	VA	22182	Antonio L Savoca	703-448-2900	S	120*	0.9
Olin Aerospace Co	PO Box 97009	Redmond	WA	98073	WW Smith	206-885-5000	S	80	0.6
Thiokol Corp	PO Box 400006	Huntsville	AL	35815	James R Wilson	205-882-8000	D	39*	0.3
Orbital Sciences Corp	21700 Atlantic Blv	Sterling	VA	20766	Dave Thomson	703-802-8000	D	29*	0.2
Kaiser Aerotech	880 Doolittle Dr	San Leandro	CA	94577	Mel Fisher	510-562-2456	D	25	0.3
Kaiser Marquardt	16555 Saticoy St	Van Nuys	CA	91406	Harry Halamandaris	818-989-6400	S	20	0.2
Space Vector Corp	9223 Deering Av	Chatsworth	CA	91311	JH Jerger	818-886-6500	S	13*	0.1
Atlantic Research Corp	PO Box 300	Niagara Falls	NY	14304	GL Greene	716-731-6000	D	12	<0.1
Imperial Defense Systems Inc	313 Cir of Progress	Pottstown	PA	19464	John P Koser	610-326-3255	R	5	<0.1
American Rocket Co	5126 Ralston St	Ventura	CA	93003	Briam Hughes	805-987-8970	R	4	<0.1
ABB Vetco Gray Inc	PO Box 2291	Houston	TX	77252	Frank Adamek	713-688-2526	D	1	<0.1

Source: Ward's Business Directory of U.S. Private and Public Companies, Volumes 1 and 2, 1996. The company type code used is as follows: P - Public, R - Private, S - Subsidiary, D - Division, J - Joint Venture, A - Affiliate, G - Group. Sales are in millions of dollars, employees are in thousands. An asterisk (*) indicates an estimated sales volume. The symbol < stands for 'less than'. Company names and addresses are truncated, in some cases, to fit into the available space.

MATERIALS CONSUMED

Material	Quantity	Delivered Cost ($ million)
Materials, ingredients, containers, and supplies	(X)	2,020.6
Guided missile and space vehicle engines and parts	(X)	393.5
Navigational systems and equipment (NAV AIDS)	(X)	(D)
Flight, navigational, airframe, and engine indicators, instruments, etc.	(X)	(D)
Resistors, capacitors, transformers, electron tubes, semiconductors, and other electronic components	(X)	5.4
Resin matrix composits	(X)	(D)
Other matrix composits, including ceramic, carbon, metal, etc.	(X)	(D)
Complete mechanical, hydraulic and pneumatic subassemblies	(X)	(D)
Fluid power products	(X)	(D)
Metal bolts, nuts, screws, washers, rivets, and other screw machine products	(X)	5.7
Other fabricated metal products (except fluid power products and forgings)	(X)	59.5
Iron and steel forgings	(X)	26.7
Iron and steel castings (rough and semifinished)	(X)	6.0
Aluminum and aluminum-base alloy castings (rough and semifinished)	(X)	2.0
Other nonferrous castings (rough and semifinished)	(X)	2.4
Metal shapes and forms, except castings, forgings, and fabricated metal products	(X)	9.5
Chemicals, all types (including propellants)	(X)	(D)
Special dies, tools, die sets, jigs, and fixtures, except cutting tools for machine tools	(X)	7.2
All other materials and components, parts, containers, and supplies	(X)	372.8
Materials, ingredients, containers, and supplies, nsk	(X)	813.3

Source: 1992 *Economic Census*. Explanation of symbols used: (D): Withheld to avoid disclosure of competitive data; na: Not available; (S): Withheld because statistical norms were not met; (X): Not applicable; (Z): Less than half the unit shown; nec: Not elsewhere classified; nsk: Not specified by kind; - : zero; * : 10-19 percent estimated; ** : 20-29 percent estimated.

PRODUCT SHARE DETAILS

Product or Product Class	% Share	Product or Product Class	% Share
Space propulsion units and parts	100.00	Other services on complete missile or space vehicle engines and/or propulsion units	24.10
Complete missile or space vehicle engines and/or propulsion units	51.85	Other services on complete missile or space vehicle engines and/or propulsion units for U.S. Government military customers	(D)
Complete missile or space vehicle engines and/or propulsion units for U.S. Government military customers	51.00	Other services on complete missile or space vehicle engines and/or propulsion units for U.S. Government nonmilitary customers	(D)
Complete missile or space vehicle engines and/or propulsion units for U.S. Government nonmilitary customers	38.42	Other services on complete missile or space vehicle engines and/or propulsion units for other customers	(D)
Complete missile or space vehicle engines and/or propulsion units for other customers	10.59	Missile and space vehicle engine and/or propulsion parts and accessories	11.57
Research and development on complete missile or space vehicle engines and/or propulsion units	11.46	Missile and space vehicle engine and/or propulsion parts and accessories for U.S. Government military customers	79.88
Research and development on complete missile or space vehicle engines and/or propulsion units for U.S. Government military customers	69.84	Missile and space vehicle engine and/or propulsion parts and accessories for U.S. Government nonmilitary customers	9.04
Research and development on complete missile or space vehicle engines and/or propulsion units for U.S. Government nonmilitary customers	19.84	Missile and space vehicle engine and/or propulsion parts and accessories for other customers	10.48
Research and development on complete missile or space vehicle engines and/or propulsion units for other customers	10.17	Missile and space vehicle engine and/or propulsion parts and accessories, nsk	0.62
Research and development on complete missile or space vehicle engines and/or propulsion units, nsk	0.16	Space propulsion units and parts, nsk	1.02

Source: 1992 *Economic Census*. The values shown are percent of total shipments in an industry. Values of indented subcategories are summed in the main heading. The symbol (D) appears when data are withheld to prevent disclosure of competitive information. The abbreviation nsk stands for 'not specified by kind' and nec for 'not elsewhere classified'.

INPUTS AND OUTPUTS FOR AIRCRAFT & MISSILE ENGINES & ENGINE PARTS

Economic Sector or Industry Providing Inputs	%	Sector	Economic Sector or Industry Buying Outputs	%	Sector
Aircraft & missile equipment, nec	19.6	Manufg.	Federal Government purchases, national defense	36.4	Fed Govt
Aircraft & missile engines & engine parts	11.5	Manufg.	Aircraft	29.5	Manufg.
Imports	8.0	Foreign	Exports	15.4	Foreign
Iron & steel forgings	6.7	Manufg.	Aircraft & missile engines & engine parts	6.6	Manufg.
Nonferrous forgings	4.9	Manufg.	Federal Government purchases, nondefense	4.8	Fed Govt
Advertising	4.3	Services	Air transportation	4.2	Util.
Wholesale trade	3.1	Trade	Gross private fixed investment	1.1	Cap Inv
Blast furnaces & steel mills	2.7	Manufg.	Aircraft & missile equipment, nec	0.7	Manufg.
Banking	2.5	Fin/R.E.	Change in business inventories	0.6	In House
Hotels & lodging places	2.5	Services	Electrical industrial apparatus, nec	0.2	Manufg.
Electric services (utilities)	2.0	Util.	Guided missiles & space vehicles	0.2	Manufg.
Air transportation	1.7	Util.	Machinery, except electrical, nec	0.2	Manufg.
Machinery, except electrical, nec	1.6	Manufg.	Transportation equipment, nec	0.2	Manufg.
Nonferrous castings, nec	1.6	Manufg.	Motor vehicles & car bodies	0.1	Manufg.
Iron & steel foundries	1.4	Manufg.			
Equipment rental & leasing services	1.3	Services			
Aluminum castings	1.1	Manufg.			
Nonferrous rolling & drawing, nec	1.1	Manufg.			
Communications, except radio & TV	1.1	Util.			
Maintenance of nonfarm buildings nec	1.0	Constr.			
Noncomparable imports	1.0	Foreign			
Ball & roller bearings	0.9	Manufg.			
Special dies & tools & machine tool accessories	0.9	Manufg.			
Petroleum refining	0.8	Manufg.			
Screw machine and related products	0.8	Manufg.			
Computer & data processing services	0.8	Services			
Engineering, architectural, & surveying services	0.8	Services			
Metal stampings, nec	0.7	Manufg.			
Pipe, valves, & pipe fittings	0.7	Manufg.			
Plating & polishing	0.7	Manufg.			
Motor freight transportation & warehousing	0.7	Util.			
Engine electrical equipment	0.6	Manufg.			
Pumps & compressors	0.6	Manufg.			
Gas production & distribution (utilities)	0.6	Util.			
Eating & drinking places	0.6	Trade			
Aluminum rolling & drawing	0.5	Manufg.			
Legal services	0.5	Services			
Miscellaneous repair shops	0.5	Services			
U.S. Postal Service	0.5	Gov't			
Electronic components nec	0.4	Manufg.			
Fabricated rubber products, nec	0.4	Manufg.			
Abrasive products	0.3	Manufg.			
Broadwoven fabric mills	0.3	Manufg.			
Hand & edge tools, nec	0.3	Manufg.			
Machine tools, metal cutting types	0.3	Manufg.			
Real estate	0.3	Fin/R.E.			
Management & consulting services & labs	0.3	Services			

Continued on next page.

INPUTS AND OUTPUTS FOR AIRCRAFT & MISSILE ENGINES & ENGINE PARTS - Continued

Economic Sector or Industry Providing Inputs	%	Sector	Economic Sector or Industry Buying Outputs	%	Sector
Manifold business forms	0.2	Manufg.			
Metal heat treating	0.2	Manufg.			
Metalworking machinery, nec	0.2	Manufg.			
Optical instruments & lenses	0.2	Manufg.			
Radio & TV communication equipment	0.2	Manufg.			
Royalties	0.2	Fin/R.E.			
Accounting, auditing & bookkeeping	0.2	Services			
Colleges, universities, & professional schools	0.2	Services			
Electrical repair shops	0.2	Services			
Photofinishing labs, commercial photography	0.2	Services			
Metal coating & allied services	0.1	Manufg.			
Paints & allied products	0.1	Manufg.			
Veneer & plywood	0.1	Manufg.			
Railroads & related services	0.1	Util.			
Sanitary services, steam supply, irrigation	0.1	Util.			
Water transportation	0.1	Util.			
Insurance carriers	0.1	Fin/R.E.			

Source: Benchmark Input-Output Accounts for the U.S. Economy, 1982, U.S. Department of Commerce, Washington, D.C., July 1991. Data, as reported in the source, are organized by the 1977 SIC structure in use in 1982 but have been matched, as closely as is possible, to the 1987 SIC structure used in this book.

OCCUPATIONS EMPLOYED BY SIC 376 - GUIDED MISSILES, SPACE VEHICLES, AND PARTS

Occupation	% of Total 1994	Change to 2005	Occupation	% of Total 1994	Change to 2005
Engineers nec	13.9	1.6	Professional workers nec	2.1	1.6
Electrical & electronics engineers	5.5	-9.9	Electrical & electronic technicians,technologists	1.8	-32.3
Management support workers nec	3.7	-15.3	Industrial engineers, ex safety engineers	1.8	-6.9
Mechanical engineers	3.6	-6.8	Managers & administrators nec	1.6	-15.4
Inspectors, testers, & graders, precision	3.4	-15.3	Machinists	1.5	-15.3
Engineering technicians & technologists nec	3.2	-23.8	General office clerks	1.5	-27.8
Engineering, mathematical, & science managers	3.1	20.2	Industrial production managers	1.3	-15.4
Secretaries, ex legal & medical	3.1	-22.9	Purchasing agents, ex trade & farm products	1.3	-15.3
Electrical & electronic equipment assemblers	2.9	-6.9	Clerical support workers nec	1.3	-32.3
Blue collar worker supervisors	2.5	-18.0	Aircraft assemblers, precision	1.2	-66.1
Aeronautical & astronautical engineers	2.4	-15.4	Production, planning, & expediting clerks	1.2	-15.3
Computer programmers	2.3	-31.4	Computer engineers	1.1	25.4
Systems analysts	2.2	35.5	Stock clerks	1.0	-31.3
Assemblers, fabricators, & hand workers nec	2.1	-15.4			

Source: Industry-Occupation Matrix, Bureau of Labor Statistics. These data relate to one or more 3-digit SIC industry groups rather than to a single 4-digit SIC. The change reported for each occupation to the year 2005 is a percent of growth or decline as estimated by the Bureau of Labor Statistics. The abbreviation nec stands for 'not elsewhere classified'.

LOCATION BY STATE AND REGIONAL CONCENTRATION

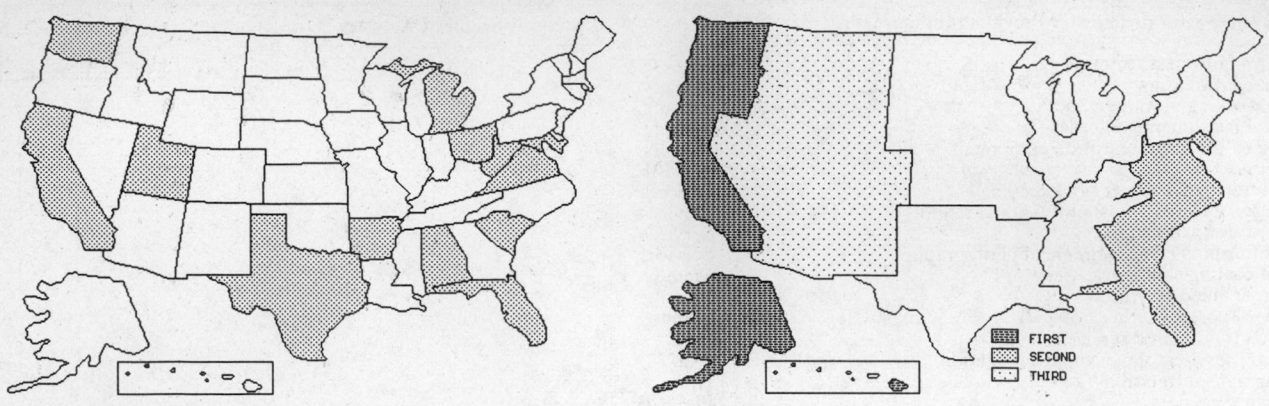

INDUSTRY DATA BY STATE

| State | Establish-ments | Shipments | | | Employment | | | | Cost as % of Shipments | Investment per Employee ($) |
		Total ($ mil)	% of U.S.	Per Establ.	Total Number	% of U.S.	Per Establ.	Wages ($/hour)		
California	13	2,932.0	55.0	225.5	16,200	50.2	1,246	33.74	41.3	3,802
Utah	4	(D)	-	-	7,500 *	23.2	1,875	-	-	-
Virginia	3	(D)	-	-	750 *	2.3	250	-	-	-
Alabama	2	(D)	-	-	750 *	2.3	375	-	-	-
Florida	2	(D)	-	-	750 *	2.3	375	-	-	-
Texas	2	(D)	-	-	750 *	2.3	375	-	-	-
Arkansas	1	(D)	-	-	375 *	1.2	375	-	-	-
Maryland	1	(D)	-	-	375 *	1.2	375	-	-	-
Michigan	1	(D)	-	-	750 *	2.3	750	-	-	-
Ohio	1	(D)	-	-	375 *	1.2	375	-	-	-
South Carolina	1	(D)	-	-	175 *	0.5	175	-	-	-
Washington	1	(D)	-	-	750 *	2.3	750	-	-	-
West Virginia	1	(D)	-	-	750 *	2.3	750	-	-	-

Source: 1992 *Economic Census*. The states are in descending order of shipments or establishments (if shipment data are missing for the majority). The symbol (D) appears when data are withheld to prevent disclosure of competitive information. States marked with (D) are sorted by number of establishments. A dash (-) indicates that the data element cannot be calculated; * indicates the midpoint of a range.

3769 - SPACE VEHICLE EQUIPMENT, NEC

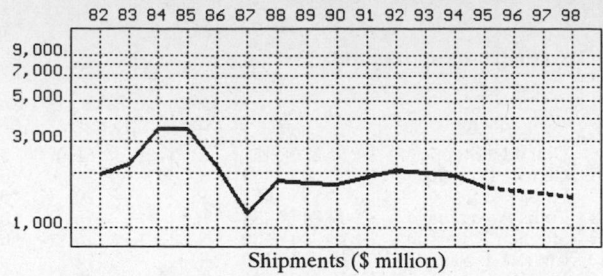

Shipments ($ million)

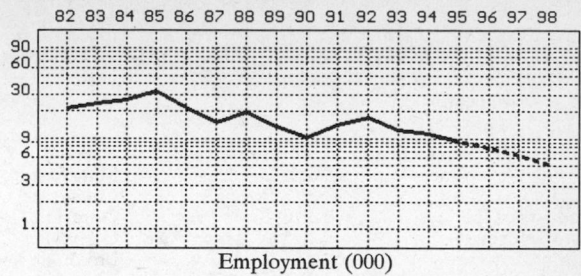

Employment (000)

GENERAL STATISTICS

| Year | Com-panies | Establishments | | Employment | | | Compensation | | Production ($ million) | | | |
		Total	with 20 or more employees	Total (000)	Production Workers (000)	Hours (Mil)	Payroll ($ mil)	Wages ($/hr)	Cost of Materials	Value Added by Manufacture	Value of Shipments	Capital Invest.
1982	45	49	33	21.4	13.0	26.2	584.6	11.62	645.1	1,297.1	1,958.3	72.4
1983		50	34	24.4	13.5	27.7	735.4	12.75	579.1	1,728.3	2,264.0	85.0
1984		51	35	26.7	15.5	34.1	905.6	13.58	836.2	2,737.5	3,502.0	125.0
1985		51	36	33.7	19.3	39.1	1,141.8	14.99	936.3	2,652.0	3,539.8	153.3
1986		54	35	22.1	10.4	20.6	763.5	14.08	519.3	1,686.6	2,169.7	68.7
1987	61	66	43	15.1	7.9	15.6	524.8	15.24	313.7	869.9	1,182.2	62.3
1988		77	48	19.4	9.5	18.8	686.5	15.47	540.0	1,267.7	1,799.6	59.8
1989		58	39	13.8	9.1	17.8	675.4	16.56	560.6	1,209.6	1,768.3	64.2
1990		57	34	10.1	8.0	15.3	525.5	16.61	612.4	1,089.7	1,715.6	28.0
1991		53	34	14.2	7.7	14.5	558.2	18.36	752.5	1,206.4	1,907.3	31.4
1992	54	60	37	17.2	7.7	15.0	694.5	18.09	643.8	1,456.4	2,070.7	41.0
1993		63	40	12.3	5.6	11.1	532.6	19.95	778.3	1,289.2	2,014.9	25.3
1994		64P	40P	11.1	5.5	11.6	520.3	20.48	813.5	1,045.8	1,944.4	38.3
1995		66P	40P	9.2P	4.3P	7.9P	514.5P	20.74P	698.4P	897.8P	1,669.2P	16.3P
1996		67P	40P	7.8P	3.4P	6.1P	490.8P	21.42P	670.2P	861.5P	1,601.8P	9.3P
1997		68P	41P	6.5P	2.6P	4.3P	467.0P	22.10P	641.9P	825.3P	1,534.4P	2.2P
1998		69P	41P	5.2P	1.7P	2.5P	443.3P	22.78P	613.7P	789.0P	1,466.9P	

Sources: 1982, 1987, 1992 *Economic Census*; *Annual Survey of Manufactures*, 83-86, 88-91, 93-94. Establishment counts for non-Census years are from *County Business Patterns*; establishment values for 83-84 are extrapolations. 'P's show projections by the editors. Industries reclassified in 87 will not have data for prior years.

INDICES OF CHANGE

| Year | Com-panies | Establishments | | Employment | | | Compensation | | Production ($ million) | | | |
		Total	with 20 or more employees	Total (000)	Production Workers (000)	Hours (Mil)	Payroll ($ mil)	Wages ($/hr)	Cost of Materials	Value Added by Manufacture	Value of Shipments	Capital Invest.
1982	83	82	89	124	169	175	84	64	100	89	95	177
1983		83	92	142	175	185	106	70	90	119	109	207
1984		85	95	155	201	227	130	75	130	188	169	305
1985		85	97	196	251	261	164	83	145	182	171	374
1986		90	95	128	135	137	110	78	81	116	105	168
1987	113	110	116	88	103	104	76	84	49	60	57	152
1988		128	130	113	123	125	99	86	84	87	87	146
1989		97	105	80	118	119	97	92	87	83	85	157
1990		95	92	59	104	102	76	92	95	75	83	68
1991		88	92	83	100	97	80	101	117	83	92	77
1992	100	100	100	100	100	100	100	100	100	100	100	100
1993		105	108	72	73	74	77	110	121	89	97	62
1994		107P	107P	65	71	77	75	113	126	72	94	93
1995		109P	108P	53P	56P	53P	74P	115P	108P	62P	81P	40P
1996		111P	109P	46P	45P	41P	71P	118P	104P	59P	77P	23P
1997		113P	110P	38P	34P	29P	67P	122P	100P	57P	74P	5P
1998		115P	111P	30P	23P	17P	64P	126P	95P	54P	71P	

Sources: Same as General Statistics. Values reflect change from the base year, 1992. Values above 100 mean greater than 92, values below 100 mean less than 92, and a value of 100 in the 82-91 or 93-98 period means same as 92. 'P's mark projections by the editors.

SELECTED RATIOS

For 1994	Avg. of All Manufact.	Analyzed Industry	Index	For 1994	Avg. of All Manufact.	Analyzed Industry	Index
Employees per Establishment	49	172	351	Value Added per Production Worker	134,084	190,145	142
Payroll per Establishment	1,500,273	8,068,562	538	Cost per Establishment	5,045,178	12,615,367	250
Payroll per Employee	30,620	46,874	153	Cost per Employee	102,970	73,288	71
Production Workers per Establishment	34	85	248	Cost per Production Worker	146,988	147,909	101
Wages per Establishment	853,319	3,684,090	432	Shipments per Establishment	9,576,895	30,152,820	315
Wages per Production Worker	24,861	43,194	174	Shipments per Employee	195,460	175,171	90
Hours per Production Worker	2,056	2,109	103	Shipments per Production Worker	279,017	353,527	127
Wages per Hour	12.09	20.48	169	Investment per Establishment	321,011	593,938	185
Value Added per Establishment	4,602,255	16,217,763	352	Investment per Employee	6,552	3,450	53
Value Added per Employee	93,930	94,216	100	Investment per Production Worker	9,352	6,964	74

Sources: Same as General Statistics. The 'Average of All Manufacturing' column represents the average of all manufacturing industries reported for the most recent complete year available. The Index shows the relationship between the Average and the Analyzed Industry. For example, 100 means that they are equal; 500 that the Analyzed Industry is five times the average; 50 means that the Analyzed Industry is half the national average. The abbreviation 'na' is used to show that data are 'not available'.

LEADING COMPANIES   Number shown: 11   Total sales ($ mil): 201   Total employment (000): 1.8

Company Name	Address				CEO Name	Phone	Co. Type	Sales ($ mil)	Empl. (000)
Intercontinental Manufacturing	PO Box 461148	Garland	TX	75046	Donald J Steppe	214-276-5131	S	58	0.6
Engineering Group Inc	PO Box 250	Manchester	CT	06045	George Longtin	203-643-2473	S	54*	0.4
Alpha Q Inc	PO Box 536	Colchester	CT	06415	Steve Prout	203-537-4681	R	20*	0.1
Arral Industries Inc	2101 Carr Priv	Ontario	CA	91761	Louis A Arranaga	909-947-6585	R	18*	0.1
TA Manufacturing Co	375 W Arden Av	Glendale	CA	91203	Stephen R Larson	818-240-4600	S	16	0.1
Aerotherm Corp	580 Clyde Av	Mountain View	CA	94043	Charles T Nardo	415-961-6100	S	14*	0.2
AC Inc	PO Box 17069	Huntsville	AL	35810	John R Riche	205-851-9020	R	9	0.1
Programmed Composites Inc	380 Cliffwood Park	Brea	CA	92621	John Van Doren	714-671-3890	S	6	<0.1
Precision Machine & Engineering	2746 W Palm Ln	Phoenix	AZ	85009	Mark W Stern	602-269-6143	R	4	<0.1
Superior Thread Rolling	6926 Farmdale Av	N Hollywood	CA	91605	Tom Mundy	818-982-3807	R	2*	<0.1
Truax Engineering Inc	920 B Rancheros Dr	San Marcos	CA	92069	Robert C Truax	619-738-0534	R	0*	<0.1

Source: Ward's Business Directory of U.S. Private and Public Companies, Volumes 1 and 2, 1996. The company type code used is as follows: P - Public, R - Private, S - Subsidiary, D - Division, J - Joint Venture, A - Affiliate, G - Group. Sales are in millions of dollars, employees are in thousands. An asterisk (*) indicates an estimated sales volume. The symbol < stands for 'less than'. Company names and addresses are truncated, in some cases, to fit into the available space.

MATERIALS CONSUMED

Material	Quantity	Delivered Cost ($ million)
Materials, ingredients, containers, and supplies	(X)	536.8
Guided missile and space vehicle propulsion units and parts	(X)	(D)
Guided missile and space vehicle airframe parts	(X)	(D)
Resistors, capacitors, transformers, electron tubes, semiconductors, and other electronic components	(X)	(D)
Resin matrix composits	(X)	8.4
Complete mechanical, hydraulic and pneumatic subassemblies	(X)	(D)
Fluid power products	(X)	(D)
Metal bolts, nuts, screws, washers, rivets, and other screw machine products	(X)	3.8
Other fabricated metal products (except fluid power products and forgings)	(X)	9.8
Iron and steel forgings	(X)	5.1
Iron and steel castings (rough and semifinished)	(X)	(D)
Aluminum and aluminum-base alloy castings (rough and semifinished)	(X)	(D)
Metal shapes and forms, except castings, forgings, and fabricated metal products	(X)	(D)
Chemicals, all types (including propellants)	(X)	1.1
Special dies, tools, die sets, jigs, and fixtures, except cutting tools for machine tools	(X)	(D)
All other materials and components, parts, containers, and supplies	(X)	238.0
Materials, ingredients, containers, and supplies, nsk	(X)	217.1

*Source: 1992 Economic Census. Explanation of symbols used: (D): Withheld to avoid disclosure of competitive data; na: Not available; (S): Withheld because statistical norms were not met; (X): Not applicable; (Z): Less than half the unit shown; nec: Not elsewhere classified; nsk: Not specified by kind; - : zero; * : 10-19 percent estimated; ** : 20-29 percent estimated.*

PRODUCT SHARE DETAILS

Product or Product Class	% Share	Product or Product Class	% Share
Space vehicle equipment, nec	100.00	subassemblies, nsk	0.61
Missile and space vehicle components, parts, and subassemblies	83.52	Research and development on missile and space vehicle parts and components, nec	15.63
Missile and space vehicle components, parts, and subassemblies, airframes for U.S. Government military customers	3.28	Research and development on missile and space vehicle parts and components, nec, airframes and space capsules for U.S. Government military customers	12.44
Missile and space vehicle components, parts, and subassemblies, space capsules for U.S. Government military customers	4.20	All other research and development on missile and space vehicle parts and components, nec	59.89
All other missile and space vehicle components, parts, and subassemblies for U.S. Government military customers	62.94	Research and development on missile and space vehicle parts and components, nec for U.S. Government nonmilitary customers	19.73
Missile and space vehicle components, parts, and subassemblies for U.S. Government nonmilitary customers	24.38	Research and development on missile and space vehicle parts and components, nec for other customers	5.46
Missile and space vehicle components, parts, and subassemblies for other customers	4.60	Research and development on missile and space vehicle parts and components, nec, nsk	2.48
Missile and space vehicle components, parts, and		Space vehicle equipment, nec, nsk	0.86

Source: 1992 Economic Census. The values shown are percent of total shipments in an industry. Values of indented subcategories are summed in the main heading. The symbol (D) appears when data are withheld to prevent disclosure of competitive information. The abbreviation nsk stands for 'not specified by kind' and nec for 'not elsewhere classified'.

INPUTS AND OUTPUTS FOR AIRCRAFT & MISSILE EQUIPMENT, NEC

Economic Sector or Industry Providing Inputs	%	Sector	Economic Sector or Industry Buying Outputs	%	Sector
Imports	18.9	Foreign	Exports	31.4	Foreign
Aircraft & missile equipment, nec	9.0	Manufg.	Federal Government purchases, national defense	29.2	Fed Govt
Aircraft	7.3	Manufg.	Aircraft	13.4	Manufg.
Electronic components nec	6.3	Manufg.	Aircraft & missile engines & engine parts	11.3	Manufg.
Semiconductors & related devices	5.8	Manufg.	Guided missiles & space vehicles	5.8	Manufg.
Wholesale trade	4.2	Trade	Aircraft & missile equipment, nec	4.2	Manufg.
Advertising	3.7	Services	Air transportation	2.3	Util.
Banking	2.4	Fin/R.E.	Federal Government purchases, nondefense	1.7	Fed Govt
Aluminum rolling & drawing	2.0	Manufg.	Change in business inventories	0.5	In House
Electric services (utilities)	2.0	Util.			
Optical instruments & lenses	1.7	Manufg.			
Hotels & lodging places	1.7	Services			
Iron & steel forgings	1.5	Manufg.			
Nonferrous forgings	1.5	Manufg.			
Aircraft & missile engines & engine parts	1.4	Manufg.			
Communications, except radio & TV	1.4	Util.			
Equipment rental & leasing services	1.3	Services			
Air transportation	1.2	Util.			
Blast furnaces & steel mills	1.1	Manufg.			
Machinery, except electrical, nec	1.1	Manufg.			
Nonferrous rolling & drawing, nec	1.1	Manufg.			
Screw machine and related products	1.1	Manufg.			
Public building furniture	1.0	Manufg.			
Business services nec	1.0	Services			
Engineering, architectural, & surveying services	0.9	Services			
Maintenance of nonfarm buildings nec	0.8	Constr.			
Metal stampings, nec	0.8	Manufg.			
Nonferrous castings, nec	0.7	Manufg.			
Detective & protective services	0.7	Services			
Miscellaneous plastics products	0.6	Manufg.			
Special dies & tools & machine tool accessories	0.6	Manufg.			
Gas production & distribution (utilities)	0.6	Util.			
Eating & drinking places	0.6	Trade			
Real estate	0.6	Fin/R.E.			
Ball & roller bearings	0.5	Manufg.			
Fabricated rubber products, nec	0.5	Manufg.			
Petroleum refining	0.5	Manufg.			
Plating & polishing	0.5	Manufg.			
Power transmission equipment	0.5	Manufg.			
Legal services	0.5	Services			
U.S. Postal Service	0.5	Gov't			
Abrasive products	0.4	Manufg.			
Aluminum castings	0.4	Manufg.			
Machine tools, metal cutting types	0.4	Manufg.			
Motor freight transportation & warehousing	0.4	Util.			
Cyclic crudes and organics	0.3	Manufg.			
Fabricated metal products, nec	0.3	Manufg.			
Fabricated structural metal	0.3	Manufg.			
Hardware, nec	0.3	Manufg.			
Pipe, valves, & pipe fittings	0.3	Manufg.			
Plastics materials & resins	0.3	Manufg.			
Computer & data processing services	0.3	Services			
Management & consulting services & labs	0.3	Services			
Asbestos products	0.2	Manufg.			
Chemical preparations, nec	0.2	Manufg.			
Iron & steel foundries	0.2	Manufg.			
Manifold business forms	0.2	Manufg.			
Metal heat treating	0.2	Manufg.			
Miscellaneous fabricated wire products	0.2	Manufg.			
Motors & generators	0.2	Manufg.			
Nonferrous wire drawing & insulating	0.2	Manufg.			
Paints & allied products	0.2	Manufg.			
Primary metal products, nec	0.2	Manufg.			
Radio & TV communication equipment	0.2	Manufg.			
Wood pallets & skids	0.2	Manufg.			
Accounting, auditing & bookkeeping	0.2	Services			
Business/professional associations	0.2	Services			
Miscellaneous repair shops	0.2	Services			
Photofinishing labs, commercial photography	0.2	Services			
Broadwoven fabric mills	0.1	Manufg.			
Pumps & compressors	0.1	Manufg.			
Insurance carriers	0.1	Fin/R.E.			
Royalties	0.1	Fin/R.E.			
Colleges, universities, & professional schools	0.1	Services			
Noncomparable imports	0.1	Foreign			

Source: Benchmark Input-Output Accounts for the U.S. Economy, 1982, U.S. Department of Commerce, Washington, D.C., July 1991. Data, as reported in the source, are organized by the 1977 SIC structure in use in 1982 but have been matched, as closely as is possible, to the 1987 SIC structure used in this book.

OCCUPATIONS EMPLOYED BY SIC 376 - GUIDED MISSILES, SPACE VEHICLES, AND PARTS

Occupation	% of Total 1994	Change to 2005	Occupation	% of Total 1994	Change to 2005
Engineers nec	13.9	1.6	Professional workers nec	2.1	1.6
Electrical & electronics engineers	5.5	-9.9	Electrical & electronic technicians,technologists	1.8	-32.3
Management support workers nec	3.7	-15.3	Industrial engineers, ex safety engineers	1.8	-6.9
Mechanical engineers	3.6	-6.8	Managers & administrators nec	1.6	-15.4
Inspectors, testers, & graders, precision	3.4	-15.3	Machinists	1.5	-15.3
Engineering technicians & technologists nec	3.2	-23.8	General office clerks	1.5	-27.8
Engineering, mathematical, & science managers	3.1	20.2	Industrial production managers	1.3	-15.4
Secretaries, ex legal & medical	3.1	-22.9	Purchasing agents, ex trade & farm products	1.3	-15.3
Electrical & electronic equipment assemblers	2.9	-6.9	Clerical support workers nec	1.3	-32.3
Blue collar worker supervisors	2.5	-18.0	Aircraft assemblers, precision	1.2	-66.1
Aeronautical & astronautical engineers	2.4	-15.4	Production, planning, & expediting clerks	1.2	-15.3
Computer programmers	2.3	-31.4	Computer engineers	1.1	25.4
Systems analysts	2.2	35.5	Stock clerks	1.0	-31.3
Assemblers, fabricators, & hand workers nec	2.1	-15.4			

Source: Industry-Occupation Matrix, Bureau of Labor Statistics. These data relate to one or more 3-digit SIC industry groups rather than to a single 4-digit SIC. The change reported for each occupation to the year 2005 is a percent of growth or decline as estimated by the Bureau of Labor Statistics. The abbreviation nec stands for 'not elsewhere classified'.

LOCATION BY STATE AND REGIONAL CONCENTRATION

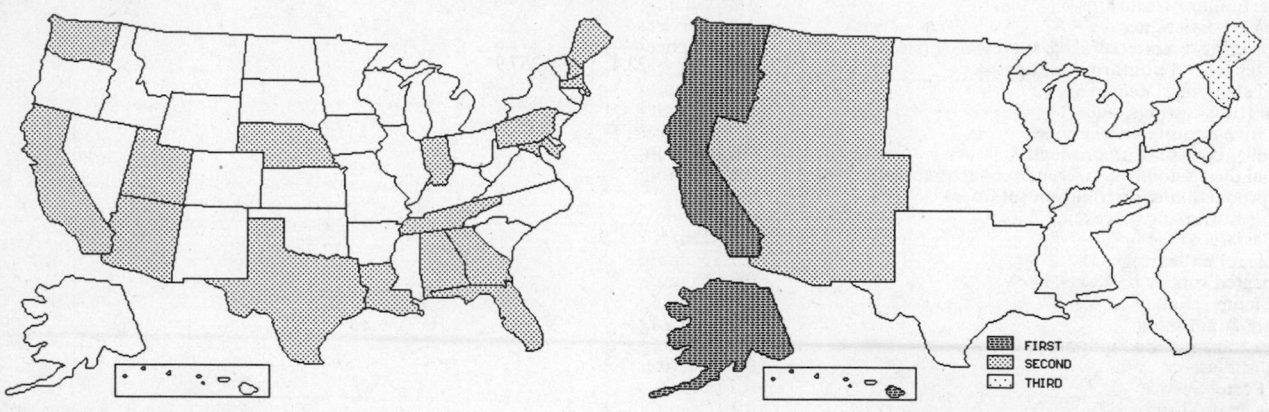

FIRST
SECOND
THIRD

INDUSTRY DATA BY STATE

State	Establish-ments	Shipments Total ($ mil)	Shipments % of U.S.	Shipments Per Establ.	Employment Total Number	Employment % of U.S.	Employment Per Establ.	Wages ($/hour)	Cost as % of Shipments	Investment per Employee ($)
California	21	176.6	8.5	8.4	1,800	10.5	86	12.09	27.2	3,000
Texas	3	14.2	0.7	4.7	200 *	1.2	67	-	19.7	-
Arizona	5	(D)	-	-	1,750 *	10.2	350	-	-	-
Alabama	3	(D)	-	-	375 *	2.2	125	-	-	-
Massachusetts	3	(D)	-	-	175 *	1.0	58	-	-	-
Florida	2	(D)	-	-	175 *	1.0	88	-	-	-
Indiana	2	(D)	-	-	175 *	1.0	88	-	-	-
Maryland	2	(D)	-	-	3,750 *	21.8	1,875	-	-	-
New Hampshire	2	(D)	-	-	375 *	2.2	188	-	-	-
Pennsylvania	2	(D)	-	-	3,750 *	21.8	1,875	-	-	-
Georgia	1	(D)	-	-	175 *	1.0	175	-	-	-
Louisiana	1	(D)	-	-	3,750 *	21.8	3,750	-	-	-
Maine	1	(D)	-	-	375 *	2.2	375	-	-	-
Nebraska	1	(D)	-	-	175 *	1.0	175	-	-	-
Tennessee	1	(D)	-	-	175 *	1.0	175	-	-	-
Utah	1	(D)	-	-	1,750 *	10.2	1,750	-	-	-
Washington	1	(D)	-	-	175 *	1.0	175	-	-	-

Source: 1992 Economic Census. The states are in descending order of shipments or establishments (if shipment data are missing for the majority). The symbol (D) appears when data are withheld to prevent disclosure of competitive information. States marked with (D) are sorted by number of establishments. A dash (-) indicates that the data element cannot be calculated; * indicates the midpoint of a range.

3792 - TRAVEL TRAILERS & CAMPERS

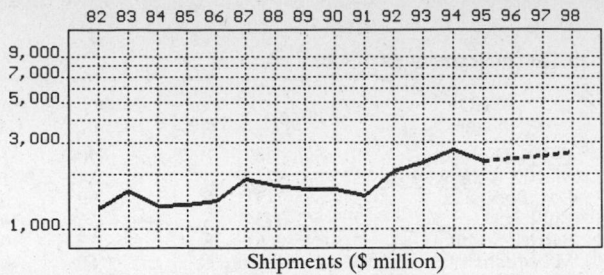

Shipments ($ million)

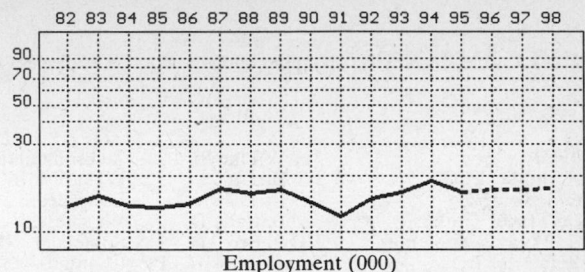

Employment (000)

GENERAL STATISTICS

Year	Companies	Establishments		Employment			Compensation		Production ($ million)			
		Total	with 20 or more employees	Total (000)	Production Workers (000)	Hours (Mil)	Payroll ($ mil)	Wages ($/hr)	Cost of Materials	Value Added by Manufacture	Value of Shipments	Capital Invest.
1982	446	492	142	13.8	11.2	20.4	221.6	7.34	837.5	441.8	1,280.8	16.9
1983		484	155	15.6	12.7	24.2	274.8	7.68	998.6	649.6	1,617.7	20.7
1984		476	168	13.7	11.3	21.1	226.9	7.34	881.2	460.6	1,338.0	16.4
1985		468	180	13.5	10.9	20.5	229.2	7.60	874.2	478.8	1,354.2	14.5
1986		458	173	14.2	11.0	20.4	257.7	8.15	916.5	506.7	1,424.8	18.5
1987	384	427	151	17.2	13.7	25.2	316.2	8.49	1,183.1	696.1	1,868.8	16.0
1988		388	146	15.9	12.7	23.7	306.4	8.66	1,091.9	637.8	1,730.2	10.9
1989		359	149	16.8	11.9	21.6	283.2	8.81	1,053.3	604.9	1,657.3	15.3
1990		338	133	14.3	11.0	21.5	284.8	8.99	1,050.2	622.8	1,657.5	14.1
1991		317	117	11.9	9.2	18.0	271.8	9.66	961.9	556.5	1,523.1	7.7
1992	269	302	126	15.2	12.0	23.4	357.9	10.24	1,326.9	774.9	2,082.6	18.0
1993		302	121	16.3	13.1	24.8	367.7	9.77	1,512.3	806.1	2,319.0	26.4
1994		268P	120P	18.9	14.9	28.9	435.1	9.94	1,814.8	898.9	2,714.4	21.1
1995		248P	116P	16.5P	12.7P	24.7P	387.9P	10.43P	1,568.2P	776.7P	2,345.5P	17.8P
1996		228P	112P	16.7P	12.9P	25.0P	401.2P	10.68P	1,626.4P	805.6P	2,432.6P	18.0P
1997		207P	108P	16.9P	13.0P	25.3P	414.5P	10.93P	1,684.6P	834.4P	2,519.7P	18.1P
1998		187P	104P	17.1P	13.1P	25.6P	427.8P	11.19P	1,742.8P	863.3P	2,606.8P	18.3P

Sources: 1982, 1987, 1992 *Economic Census*; *Annual Survey of Manufactures*, 83-86, 88-91, 93-94. Establishment counts for non-Census years are from *County Business Patterns*; establishment values for 83-84 are extrapolations. 'P's show projections by the editors. Industries reclassified in 87 will not have data for prior years.

INDICES OF CHANGE

Year	Companies	Establishments		Employment			Compensation		Production ($ million)			
		Total	with 20 or more employees	Total (000)	Production Workers (000)	Hours (Mil)	Payroll ($ mil)	Wages ($/hr)	Cost of Materials	Value Added by Manufacture	Value of Shipments	Capital Invest.
1982	166	163	113	91	93	87	62	72	63	57	62	94
1983		160	123	103	106	103	77	75	75	84	78	115
1984		158	133	90	94	90	63	72	66	59	64	91
1985		155	143	89	91	88	64	74	66	62	65	81
1986		152	137	93	92	87	72	80	69	65	68	103
1987	143	141	120	113	114	108	88	83	89	90	90	89
1988		128	116	105	106	101	86	85	82	82	83	61
1989		119	118	111	99	92	79	86	79	78	80	85
1990		112	106	94	92	92	80	88	79	80	80	78
1991		105	93	78	77	77	76	94	72	72	73	43
1992	100	100	100	100	100	100	100	100	100	100	100	100
1993		100	96	107	109	106	103	95	114	104	111	147
1994		89P	96P	124	124	124	122	97	137	116	130	117
1995		82P	92P	109P	106P	105P	108P	102P	118P	100P	113P	99P
1996		75P	89P	110P	107P	107P	112P	104P	123P	104P	117P	100P
1997		69P	86P	111P	108P	108P	116P	107P	127P	108P	121P	101P
1998		62P	83P	113P	109P	109P	120P	109P	131P	111P	125P	102P

Sources: Same as General Statistics. Values reflect change from the base year, 1992. Values above 100 mean greater than 92, values below 100 mean less than 92, and a value of 100 in the 82-91 or 93-98 period means same as 92. 'P's mark projections by the editors.

SELECTED RATIOS

For 1994	Avg. of All Manufact.	Analyzed Industry	Index	For 1994	Avg. of All Manufact.	Analyzed Industry	Index
Employees per Establishment	49	70	144	Value Added per Production Worker	134,084	60,329	45
Payroll per Establishment	1,500,273	1,620,576	108	Cost per Establishment	5,045,178	6,759,413	134
Payroll per Employee	30,620	23,021	75	Cost per Employee	102,970	96,021	93
Production Workers per Establishment	34	55	162	Cost per Production Worker	146,988	121,799	83
Wages per Establishment	853,319	1,069,952	125	Shipments per Establishment	9,576,895	10,110,068	106
Wages per Production Worker	24,861	19,280	78	Shipments per Employee	195,460	143,619	73
Hours per Production Worker	2,056	1,940	94	Shipments per Production Worker	279,017	182,174	65
Wages per Hour	12.09	9.94	82	Investment per Establishment	321,011	78,589	24
Value Added per Establishment	4,602,255	3,348,047	73	Investment per Employee	6,552	1,116	17
Value Added per Employee	93,930	47,561	51	Investment per Production Worker	9,352	1,416	15

Sources: Same as General Statistics. The 'Average of All Manufacturing' column represents the average of all manufacturing industries reported for the most recent complete year available. The Index shows the relationship between the Average and the Analyzed Industry. For example, 100 means that they are equal; 500 that the Analyzed Industry is five times the average; 50 means that the Analyzed Industry is half the national average. The abbreviation 'na' is used to show that data are 'not available'.

LEADING COMPANIES   Number shown: 39   Total sales ($ mil): 1,486   Total employment (000): 9.3

Company Name	Address				CEO Name	Phone	Co. Type	Sales ($ mil)	Empl. (000)
Thor Industries Inc	419 W Pike St	Jackson Center	OH	45334	W FB Thompson	513-596-6849	P	490	1.9
Holiday Rambler LLC	PO Box 465	Wakarusa	IN	46573	Martin R Snowey	219-862-7211	S	284	1.9
Kit Manufacturing Co	PO Box 848	Long Beach	CA	90801	Dan Pocapalia	310-595-7451	P	90	0.7
Fleetwood Travel Trailers	380 Battaile Dr	Winchester	VA	22601	Bill Look	703-662-3436	S	76*	0.2
Fleetwood Folding Trailers Inc	RD 2, Box 111	Somerset	PA	15501	Patrick Scanlon	814-445-9661	S	72	0.6
Nu-Wa Industries Inc	Safari Industrial Pk	Chanute	KS	66720	MS Mitchell	316-431-2088	R	60*	0.5
Fleetwood Travel Trailers	PO Box 1247	Pendleton	OR	97801	Ward King Jr	503-276-1244	S	53*	0.4
Airstream Inc	419 W Pike St	Jackson Center	OH	45334	Lawrence J Huttle	513-596-6111	S	50	0.4
Shasta Industries Inc	PO Box 631	Middlebury	IN	46540	William Snook	219-825-8555	S	35	0.2
Coachman Industries of Georgia	PO Box 948	Fitzgerald	GA	31750	Lawton Tinley III	912-423-5471	S	30	0.1
Fleetwood Travel Trailers	PO Box 810	Rialto	CA	92376	Glenn F Kummer	909-874-2223	S	30*	0.2
Western Recreational Vehicles	PO Box 9547	Yakima	WA	98909	Ronald Doyle	509-457-4133	R	25	0.3
Glasstite Inc	RR1	Dunnell	MN	56127	Ron Moquist	507-695-2378	S	20*	0.2
Sunline Coach Co	245 S Muddy Creek	Denver	PA	17517	Larry Lawrence	717-336-2858	R	20	0.2
Vanguard Industries of Michigan	PO Box 802	Colon	MI	49040	LO Landey	616-432-3271	R	17	0.2
Hi-Lo Trailer Company Inc	145 Elm St	Butler	OH	44822	James Snyder	419-883-3000	R	15*	0.1
Skamper Corp	PO Box 338	Bristol	IN	46507	Steven T Decker	219-848-7411	R	14	0.1
Tri-Glas Inc	PO Box 1088	Daleville	AL	36322	Robert E Ostendorf	205-598-2422	R	14	0.3
Viking Recreational Vehicles Inc	PO Box 549	Centreville	MI	49032	Patrick M Murphy	616-467-6321	S	14	0.1
Casa Villa Inc	PO Box 567	Wakarusa	IN	46573	Bob Bozzo	219-862-4531	R	8*	<0.1
Serro Travel Trailer Company	450 Arona Rd	Irwin	PA	15642	Gary Pirschl	412-863-3407	R	8	<0.1
Born Free Motorcoach	PO Box 39	Humboldt	IA	50548	John N Dodgen	515-332-3755	D	7	<0.1
Franklin Coach Co	PO Box 152	Nappanee	IN	46550	Paul Able	219-773-4106	R	7	<0.1
Six-Pac Industries Inc	1428 E 6th St	Corona	CA	91719	Cleon E Benson III	909-737-8232	R	7*	<0.1
Beck Industries LP	PO Box 8	Elkhart	IN	46515	Roy Beck Jr	219-294-5621	R	6	<0.1
Sellers Manufacturing Inc	PO Box 398	Milford	IN	46542	Howard E Sellers	219-658-9461	R	6	<0.1
Galaxie Corp	PO Box 820	Bristol	IN	46507	Eugene Potterbaum	219-848-4441	R	4	<0.1
Hop-Cap Inc	1345 W North St	Bremen	IN	46506	JL Bope	219-546-4939	R	4	<0.1
Allen Camper Manufacturing	Rte 1	Allen	OK	74825	Jerry Peay	405-857-2413	R	3	<0.1
Excel Trailer Company Inc	5111 Grumman Dr	Carson City	NV	89706	Charles Wahl	702-885-0808	R	3	<0.1
S Winchester's Originals Inc	PO Box 3480	La Habra	CA	90631	Richard C Everett	714-738-4971	R	3	<0.1
USA Venturcraft Corp	PO Box 1039	Abilene	TX	79604	Stuart Hall	915-673-5267	R	3	<0.1
Evelands Inc	PO Box 2	Backus	MN	56435	Kent Eveland	218-947-4932	R	2	<0.1
Floyd Manufacturing	2711 Nevada Av	Norfolk	VA	23509	Jeff L Floyd	804-855-0244	R	2	<0.1
Viking Camper Supply Inc	9025 Zachary Ln N	Maple Grove	MN	55369	Charles Achtelik	612-424-6960	R	1	<0.1
Silver Star RV	4741 Murietta St	Chino	CA	91710	Rolf Zuschlag	909-591-0416	R	1	<0.1
TPD California Trailers Inc	8530 Fruitridge Rd	Sacramento	CA	95826	Victor Takehara	916-381-0532	R	1	<0.1
Bentley Welding Inc	1510-B SW Austin	Roseburg	OR	97470	Doug Bentley	503-679-7849	R	1*	<0.1
Safari Motor Coaches Inc	PO Box 740	Harrisburg	OR	97446	Matthew Perlot	503-995-8214	S	1*	<0.1

Source: *Ward's Business Directory of U.S. Private and Public Companies*, Volumes 1 and 2, 1996. The company type code used is as follows: P - Public, R - Private, S - Subsidiary, D - Division, J - Joint Venture, A - Affiliate, G - Group. Sales are in millions of dollars, employees are in thousands. An asterisk (*) indicates an estimated sales volume. The symbol < stands for 'less than'. Company names and addresses are truncated, in some cases, to fit into the available space.

MATERIALS CONSUMED

Material	Quantity	Delivered Cost ($ million)
Materials, ingredients, containers, and supplies	(X)	1,227.8
Trailer axles, wheels, brakes, undercarriages, and other metal vehicular parts	(X)	74.5
Pneumatic tires and inner tubes	(X)	18.5
Purchased chassis for motor homes	(X)	(D)
Household appliances, including refrigerators, cooking equipment, and other household appliances	(X)	87.5
Air-conditioning equipment	(X)	32.8
Metal heating equipment (except electric)	(X)	21.1
Metal doors and door units, windows and window units	(X)	51.8
Metal plumbing fixtures, fittings, and trim (including enameled) (except forgings)	(X)	14.7
Sheet metal products, except stampings	(X)	22.4
Metal bolts, nuts, screws, washers, rivets, and other screw machine products	(X)	20.3
Other fabricated metal products, except forgings	(X)	(D)
Forgings	(X)	(D)
Castings (rough and semifinished)	(X)	14.2
Steel shapes and forms	(X)	31.6
Aluminum and aluminum-base alloy sheet, plate, foil, and welded tubing	(X)	31.6
All other aluminum and aluminum-base alloy shapes and forms	(X)	10.7
Other nonferrous shapes and forms	(X)	4.1
Current-carrying wiring devices	(X)	27.6
Plywood	(X)	70.3
Dressed lumber	(X)	57.0
Millwork, wood (including wood doors, window sash, moldings, and cabinets)	(X)	27.7
Glass and glass products including windows and mirrors	(X)	50.4
Fabricated plastics products (except gaskets, hoses, and belting)	(X)	18.8
Plastics products consumed in the form of sheets, rods, tubes, film, and other shapes	(X)	16.7
Plastics resins consumed in the form of granules, pellets, powders, liquids, etc.	(X)	11.8
Paints, varnishes, lacquers, stains, shellacs, japans, enamels, and allied products	(X)	11.0
Carpeting	(X)	15.7
Curtains and draperies	(X)	22.0
All other materials and components, parts, containers, and supplies	(X)	251.3
Materials, ingredients, containers, and supplies, nsk	(X)	180.7

Source: 1992 *Economic Census.* Explanation of symbols used: (D): Withheld to avoid disclosure of competitive data; na: Not available; (S): Withheld because statistical norms were not met; (X): Not applicable; (Z): Less than half the unit shown; nec: Not elsewhere classified; nsk: Not specified by kind; - : zero; * : 10-19 percent estimated; ** : 20-29 percent estimated.

PRODUCT SHARE DETAILS

Product or Product Class	% Share	Product or Product Class	% Share
Travel trailers and campers	100.00	Fifth wheel travel trailers, less than 30 ft (9.144 m) in length	18.54
Travel trailers	65.62	Fifth wheel travel trailers, 30 ft (9.144 m) or more in length	25.04
Conventional travel trailers, less than 20 ft (6.096 m) in length	4.66	Travel trailers, nsk	3.00
		Camping trailers, campers, pickup covers, and parts	26.56
Conventional travel trailers, 20 ft (6.096 m) to 24 ft 11 in. (7.595 m) in length	18.90	Folddown camping trailers	29.30
Conventional travel trailers, 25 ft (7.620 m) to 29 ft 11 in. (9.118 m) in length	13.61	Truck (pickup) campers (for sliding on and off trucks)	9.41
		Truck (pickup) caps or box covers	31.53
Conventional travel trailers, 30 ft (9.144 m) or more, including park models	16.25	Bodies for travel and camping trailers	27.48
		Camping trailers, campers, pickup covers, and parts, nsk	2.27
		Travel trailers and campers, nsk	7.82

Source: 1992 *Economic Census.* The values shown are percent of total shipments in an industry. Values of indented subcategories are summed in the main heading. The symbol (D) appears when data are withheld to prevent disclosure of competitive information. The abbreviation nsk stands for 'not specified by kind' and nec for 'not elsewhere classified'.

INPUTS AND OUTPUTS FOR TRAVEL TRAILERS & CAMPERS

Economic Sector or Industry Providing Inputs	%	Sector	Economic Sector or Industry Buying Outputs	%	Sector
Travel trailers & campers	11.3	Manufg.	Personal consumption expenditures	85.8	
Wholesale trade	9.1	Trade	Travel trailers & campers	11.0	Manufg.
Motor vehicle parts & accessories	5.6	Manufg.	Gross private fixed investment	1.2	Cap Inv
Aluminum rolling & drawing	5.5	Manufg.	Change in business inventories	1.1	In House
Motor vehicles & car bodies	4.3	Manufg.	Exports	0.7	Foreign
Veneer & plywood	3.8	Manufg.			
Building paper & board mills	3.5	Manufg.			
Metal doors, sash, & trim	3.0	Manufg.			
Glass & glass products, except containers	2.6	Manufg.			
Automotive & apparel trimmings	2.4	Manufg.			
Motor freight transportation & warehousing	2.1	Util.			
Household cooking equipment	2.0	Manufg.			
Household refrigerators & freezers	2.0	Manufg.			
Sawmills & planning mills, general	1.9	Manufg.			
Refrigeration & heating equipment	1.8	Manufg.			
Heating equipment, except electric	1.7	Manufg.			
Millwork	1.7	Manufg.			

Continued on next page.

INPUTS AND OUTPUTS FOR TRAVEL TRAILERS & CAMPERS - Continued

Economic Sector or Industry Providing Inputs	%	Sector	Economic Sector or Industry Buying Outputs	%	Sector
Miscellaneous plastics products	1.7	Manufg.			
Wood household furniture	1.7	Manufg.			
Tires & inner tubes	1.6	Manufg.			
Blast furnaces & steel mills	1.4	Manufg.			
Machinery, except electrical, nec	1.3	Manufg.			
Maintenance of nonfarm buildings nec	1.2	Constr.			
Curtains & draperies	1.2	Manufg.			
Screw machine and related products	1.1	Manufg.			
Railroads & related services	1.1	Util.			
Coated fabrics, not rubberized	1.0	Manufg.			
Metal coating & allied services	1.0	Manufg.			
Sheet metal work	1.0	Manufg.			
Banking	1.0	Fin/R.E.			
Floor coverings	0.9	Manufg.			
Wiring devices	0.9	Manufg.			
Hardware, nec	0.8	Manufg.			
Household appliances, nec	0.8	Manufg.			
Plastics materials & resins	0.8	Manufg.			
Wood products, nec	0.8	Manufg.			
Lighting fixtures & equipment	0.7	Manufg.			
Padding & upholstery filling	0.7	Manufg.			
Plumbing fixture fittings & trim	0.7	Manufg.			
Communications, except radio & TV	0.7	Util.			
Electric services (utilities)	0.7	Util.			
Paints & allied products	0.6	Manufg.			
Eating & drinking places	0.6	Trade			
Hard surface floor coverings	0.5	Manufg.			
Nonferrous wire drawing & insulating	0.5	Manufg.			
Automotive stampings	0.4	Manufg.			
Petroleum refining	0.4	Manufg.			
Wood kitchen cabinets	0.4	Manufg.			
Real estate	0.4	Fin/R.E.			
Canvas & related products	0.3	Manufg.			
Gaskets, packing & sealing devices	0.3	Manufg.			
Gas production & distribution (utilities)	0.3	Util.			
Advertising	0.3	Services			
Engineering, architectural, & surveying services	0.3	Services			
Abrasive products	0.2	Manufg.			
Fabricated metal products, nec	0.2	Manufg.			
Fabricated plate work (boiler shops)	0.2	Manufg.			
Fabricated rubber products, nec	0.2	Manufg.			
Mineral wool	0.2	Manufg.			
Signs & advertising displays	0.2	Manufg.			
Special dies & tools & machine tool accessories	0.2	Manufg.			
Water transportation	0.2	Util.			
Equipment rental & leasing services	0.2	Services			
Legal services	0.2	Services			
Management & consulting services & labs	0.2	Services			
Theatrical producers, bands, entertainers	0.2	Services			
Copper rolling & drawing	0.1	Manufg.			
Industrial gases	0.1	Manufg.			
Metal heat treating	0.1	Manufg.			
Miscellaneous fabricated wire products	0.1	Manufg.			
Radio & TV receiving sets	0.1	Manufg.			
Upholstered household furniture	0.1	Manufg.			
Sanitary services, steam supply, irrigation	0.1	Util.			
Insurance carriers	0.1	Fin/R.E.			
Accounting, auditing & bookkeeping	0.1	Services			
Computer & data processing services	0.1	Services			

Source: Benchmark Input-Output Accounts for the U.S. Economy, 1982, U.S. Department of Commerce, Washington, D.C., July 1991. Data, as reported in the source, are organized by the 1977 SIC structure in use in 1982 but have been matched, as closely as is possible, to the 1987 SIC structure used in this book.

OCCUPATIONS EMPLOYED BY SIC 379 - TRANSPORTATION EQUIPMENT NEC

Occupation	% of Total 1994	Change to 2005	Occupation	% of Total 1994	Change to 2005
Assemblers, fabricators, & hand workers nec	21.4	11.7	Painters, transportation equipment	1.5	8.6
Welders & cutters	9.3	8.0	Helpers, laborers, & material movers nec	1.3	7.5
Blue collar worker supervisors	4.3	3.0	NC machine tool operators, metal & plastic	1.3	29.8
Mechanics, installers, & repairers nec	3.8	-13.7	Combination machine tool operators	1.2	22.0
Welding machine setters, operators	3.0	-0.3	Industrial production managers	1.1	10.1
Sales & related workers nec	2.7	10.0	Coating, painting, & spraying machine workers	1.1	11.6
Inspectors, testers, & graders, precision	2.5	7.9	Maintenance repairers, general utility	1.1	-2.2
General managers & top executives	2.2	4.8	General office clerks	1.1	-5.7
Machinists	2.1	6.8	Grinders & polishers, hand	1.1	11.7
Stock clerks	1.7	-13.1	Lathe & turning machine tool operators	1.1	-14.2
Freight, stock, & material movers, hand	1.6	-12.9	Bookkeeping, accounting, & auditing clerks	1.0	-17.5
Traffic, shipping, & receiving clerks	1.5	7.5	Extraction & related workers nec	1.0	9.1
Secretaries, ex legal & medical	1.5	0.1	Truck drivers light & heavy	1.0	15.1
Machine tool cutting operators, metal & plastic	1.5	-8.1			

Source: Industry-Occupation Matrix, Bureau of Labor Statistics. These data relate to one or more 3-digit SIC industry groups rather than to a single 4-digit SIC. The change reported for each occupation to the year 2005 is a percent of growth or decline as estimated by the Bureau of Labor Statistics. The abbreviation nec stands for 'not elsewhere classified'.

LOCATION BY STATE AND REGIONAL CONCENTRATION

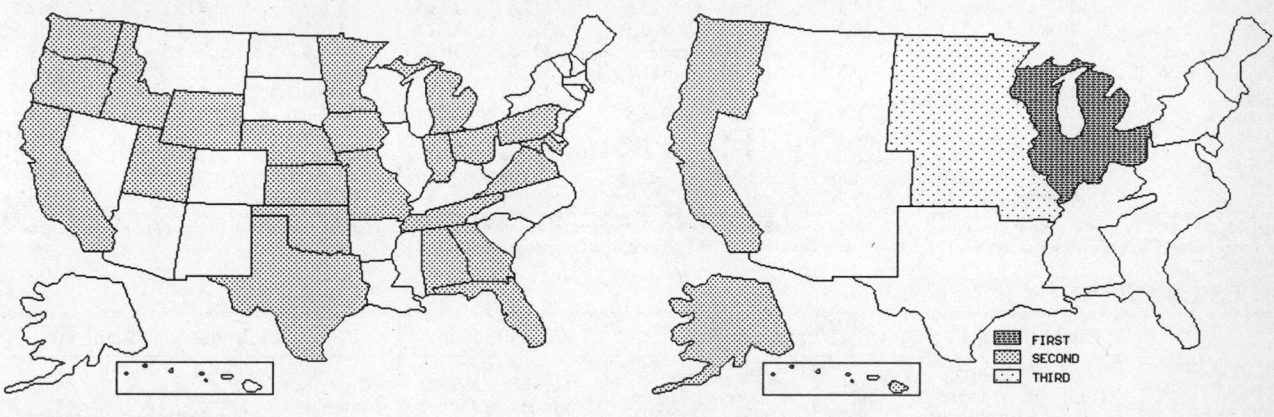

FIRST
SECOND
THIRD

INDUSTRY DATA BY STATE

State	Establish-ments	Shipments Total ($ mil)	Shipments % of U.S.	Shipments Per Establ.	Employment Total Number	Employment % of U.S.	Employment Per Establ.	Wages ($/hour)	Cost as % of Shipments	Investment per Employee ($)
Indiana	53	833.7	40.0	15.7	5,100	33.6	96	11.56	65.6	1,569
California	44	179.3	8.6	4.1	1,500	9.9	34	9.35	66.5	867
Pennsylvania	16	132.5	6.4	8.3	1,100	7.2	69	11.14	52.3	545
Oregon	13	129.4	6.2	10.0	900	5.9	69	10.21	64.9	-
Ohio	6	110.4	5.3	18.4	800	5.3	133	11.64	62.0	-
Kansas	10	108.8	5.2	10.9	900	5.9	90	8.57	55.4	-
Texas	18	91.8	4.4	5.1	700	4.6	39	8.83	69.7	-
Michigan	10	50.9	2.4	5.1	500	3.3	50	11.86	60.1	400
Idaho	6	37.9	1.8	6.3	400	2.6	67	7.57	64.9	-
Iowa	7	31.5	1.5	4.5	200	1.3	29	8.00	47.6	-
Florida	9	24.6	1.2	2.7	200	1.3	22	9.67	61.8	-
Alabama	7	17.3	0.8	2.5	200	1.3	29	7.00	42.2	-
Tennessee	6	13.5	0.6	2.3	200	1.3	33	7.50	57.0	-
Minnesota	8	12.9	0.6	1.6	100	0.7	13	9.50	51.2	-
Missouri	9	(D)	-	-	175 *	1.2	19	-	-	-
Utah	7	(D)	-	-	175 *	1.2	25	-	-	-
Washington	5	(D)	-	-	175 *	1.2	35	-	-	-
Georgia	4	(D)	-	-	175 *	1.2	44	-	-	-
Oklahoma	4	(D)	-	-	175 *	1.2	44	-	-	-
Virginia	4	(D)	-	-	375 *	2.5	94	-	-	-
Maryland	1	(D)	-	-	375 *	2.5	375	-	-	-
Nebraska	1	(D)	-	-	175 *	1.2	175	-	-	-
Wyoming	1	(D)	-	-	175 *	1.2	175	-	-	-

Source: 1992 Economic Census. The states are in descending order of shipments or establishments (if shipment data are missing for the majority). The symbol (D) appears when data are withheld to prevent disclosure of competitive information. States marked with (D) are sorted by number of establishments. A dash (-) indicates that the data element cannot be calculated; * indicates the midpoint of a range.

3795 - TANKS & TANK COMPONENTS

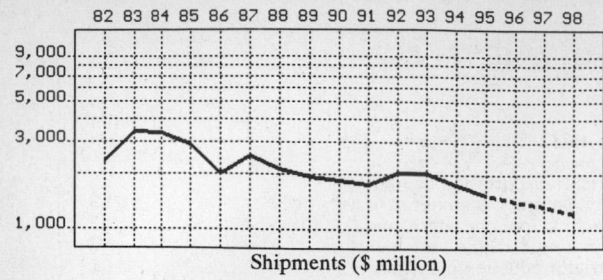

Shipments ($ million)

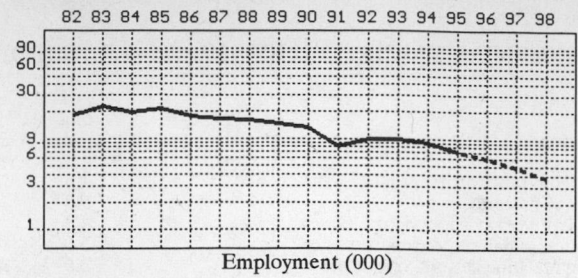

Employment (000)

GENERAL STATISTICS

Year	Companies	Establishments		Employment			Compensation		Production ($ million)			
		Total	with 20 or more employees	Total (000)	Production Workers (000)	Hours (Mil)	Payroll ($ mil)	Wages ($/hr)	Cost of Materials	Value Added by Manufacture	Value of Shipments	Capital Invest.
1982	37	43	36	18.1	12.5	23.1	473.9	12.54	1,503.3	1,153.6	2,343.7	114.5
1983		42	36	22.4	14.7	26.6	579.6	12.96	2,133.1	1,397.2	3,418.5	88.3
1984		41	36	19.2	12.5	24.3	540.6	13.18	1,945.9	1,291.4	3,314.0	81.6
1985		41	35	21.2	14.1	26.6	578.8	12.94	1,761.2	1,055.2	2,887.5	32.4
1986		43	34	17.8	11.6	23.6	548.1	14.50	1,124.5	1,073.7	2,010.1	28.5
1987	50	56	39	16.4	11.0	22.7	498.7	13.66	1,454.0	969.0	2,521.8	39.9
1988		54	37	16.1	10.9	20.8	498.2	14.97	1,308.6	899.5	2,146.0	47.2
1989		49	35	14.4	7.0	12.8	417.4	17.95	1,122.1	875.3	1,947.2	42.5
1990		46	31	13.1	5.5	10.7	353.6	18.22	1,015.5	694.5	1,846.5	27.4
1991		45	29	8.1	4.9	9.8	327.5	18.02	864.7	761.9	1,749.2	22.0
1992	36	42	31	9.8	5.7	11.3	362.9	16.77	931.3	1,157.8	2,034.7	15.6
1993		42	28	9.9	5.6	10.2	378.0	19.42	908.2	1,147.9	2,043.1	17.2
1994		47P	29P	8.9	4.9	8.5	356.0	20.44	730.0	937.5	1,768.8	16.6
1995		47P	29P	7.1P	3.1P	5.9P	314.5P	20.47P	639.7P	821.6P	1,550.1P	
1996		47P	28P	6.0P	2.2P	4.2P	294.4P	21.13P	594.9P	764.0P	1,441.5P	
1997		47P	27P	4.8P	1.3P	2.6P	274.4P	21.80P	550.1P	706.5P	1,332.9P	
1998		47P	27P	3.7P	0.4P	0.9P	254.3P	22.46P	505.3P	648.9P	1,224.4P	

Sources: 1982, 1987, 1992 *Economic Census*; *Annual Survey of Manufactures*, 83-86, 88-91, 93-94. Establishment counts for non-Census years are from *County Business Patterns*; establishment values for 83-84 are extrapolations. 'P's show projections by the editors. Industries reclassified in 87 will not have data for prior years.

INDICES OF CHANGE

Year	Companies	Establishments		Employment			Compensation		Production ($ million)			
		Total	with 20 or more employees	Total (000)	Production Workers (000)	Hours (Mil)	Payroll ($ mil)	Wages ($/hr)	Cost of Materials	Value Added by Manufacture	Value of Shipments	Capital Invest.
1982	103	102	116	185	219	204	131	75	161	100	115	734
1983		100	116	229	258	235	160	77	229	121	168	566
1984		98	116	196	219	215	149	79	209	112	163	523
1985		98	113	216	247	235	159	77	189	91	142	208
1986		102	110	182	204	209	151	86	121	93	99	183
1987	139	133	126	167	193	201	137	81	156	84	124	256
1988		129	119	164	191	184	137	89	141	78	105	303
1989		117	113	147	123	113	115	107	120	76	96	272
1990		110	100	134	96	95	97	109	109	60	91	176
1991		107	94	83	86	87	90	107	93	66	86	141
1992	100	100	100	100	100	100	100	100	100	100	100	100
1993		100	90	101	98	90	104	116	98	99	100	110
1994		111P	95P	91	86	75	98	122	78	81	87	106
1995		112P	92P	73P	54P	53P	87P	122P	69P	71P	76P	
1996		112P	90P	61P	38P	38P	81P	126P	64P	66P	71P	
1997		113P	88P	49P	23P	23P	76P	130P	59P	61P	66P	
1998		113P	86P	38P	7P	8P	70P	134P	54P	56P	60P	

Sources: Same as General Statistics. Values reflect change from the base year, 1992. Values above 100 mean greater than 92, values below 100 mean less than 92, and a value of 100 in the 82-91 or 93-98 period means same as 92. 'P's mark projections by the editors.

SELECTED RATIOS

For 1994	Avg. of All Manufact.	Analyzed Industry	Index	For 1994	Avg. of All Manufact.	Analyzed Industry	Index
Employees per Establishment	49	191	389	Value Added per Production Worker	134,084	191,327	143
Payroll per Establishment	1,500,273	7,631,049	509	Cost per Establishment	5,045,178	15,647,938	310
Payroll per Employee	30,620	40,000	131	Cost per Employee	102,970	82,022	80
Production Workers per Establishment	34	105	306	Cost per Production Worker	146,988	148,980	101
Wages per Establishment	853,319	3,724,209	436	Shipments per Establishment	9,576,895	37,915,167	396
Wages per Production Worker	24,861	35,457	143	Shipments per Employee	195,460	198,742	102
Hours per Production Worker	2,056	1,735	84	Shipments per Production Worker	279,017	360,980	129
Wages per Hour	12.09	20.44	169	Investment per Establishment	321,011	355,830	111
Value Added per Establishment	4,602,255	20,095,810	437	Investment per Employee	6,552	1,865	28
Value Added per Employee	93,930	105,337	112	Investment per Production Worker	9,352	3,388	36

Sources: Same as General Statistics. The 'Average of All Manufacturing' column represents the average of all manufacturing industries reported for the most recent complete year available. The Index shows the relationship between the Average and the Analyzed Industry. For example, 100 means that they are equal; 500 that the Analyzed Industry is five times the average; 50 means that the Analyzed Industry is half the national average. The abbreviation 'na' is used to show that data are 'not available'.

LEADING COMPANIES Number shown: 6 Total sales ($ mil): 2,251 Total employment (000): 16.1

Company Name	Address				CEO Name	Phone	Co. Type	Sales ($ mil)	Empl. (000)
General Dynamics Corp	38500 Mound Rd	Sterling Hts	MI	48310	Roger E Tetrault	810-825-4000	D	1,286	12.3
FMC Corp	PO Box 58123	Santa Clara	CA	95052	Peter C Woglom	408-289-0111	D	630	2.3
Cadillac Gage Textron Inc	6600 Plaza Dr	New Orleans	LA	70127	John J Kelly	504-245-6978	S	250	0.9
General Dynamics Services Co	PO Box 760	Troy	MI	48099	AW Carion	313-244-7000	S	42*	0.3
Michigan Fleet	1988 Alpine Av NW	Grand Rapids	MI	49504	W Venema	616-364-8555	D	38*	0.3
Tamco Industries Inc	1390 Neubrecht Rd	Lima	OH	45801	Wesley A Lytle	419-227-6030	R	5	<0.1

Source: Ward's Business Directory of U.S. Private and Public Companies, Volumes 1 and 2, 1996. The company type code used is as follows: P - Public, R - Private, S - Subsidiary, D - Division, J - Joint Venture, A - Affiliate, G - Group. Sales are in millions of dollars, employees are in thousands. An asterisk (*) indicates an estimated sales volume. The symbol < stands for 'less than'. Company names and addresses are truncated, in some cases, to fit into the available space.

MATERIALS CONSUMED

Material	Quantity	Delivered Cost ($ million)
Materials, ingredients, containers, and supplies	(X)	898.9
Iron and steel castings (rough and semifinished)	(X)	19.2
Aluminum and aluminum-base alloy castings (rough and semifinished)	(X)	21.4
Iron and steel forgings	(X)	77.3
Steel shapes and forms	(X)	63.9
Machine tool accessories, including cutting tools	(X)	6.0
All other materials and components, parts, containers, and supplies	(X)	694.9
Materials, ingredients, containers, and supplies, nsk	(X)	16.2

Source: 1992 Economic Census. Explanation of symbols used: (D): Withheld to avoid disclosure of competitive data; na: Not available; (S): Withheld because statistical norms were not met; (X): Not applicable; (Z): Less than half the unit shown; nec: Not elsewhere classified; nsk: Not specified by kind; - : zero; * : 10-19 percent estimated; ** : 20-29 percent estimated.

PRODUCT SHARE DETAILS

Product or Product Class	% Share	Product or Product Class	% Share
Tanks and tank components	100.00	Tank self-propelled weapons and parts	42.30
Tanks and parts	56.83		

Source: 1992 Economic Census. The values shown are percent of total shipments in an industry. Values of indented subcategories are summed in the main heading. The symbol (D) appears when data are withheld to prevent disclosure of competitive information. The abbreviation nsk stands for 'not specified by kind' and nec for 'not elsewhere classified'.

INPUTS AND OUTPUTS FOR TANKS & TANK COMPONENTS

Economic Sector or Industry Providing Inputs	%	Sector	Economic Sector or Industry Buying Outputs	%	Sector
Tanks & tank components	33.6	Manufg.	Exports	41.6	Foreign
Aluminum rolling & drawing	11.9	Manufg.	Federal Government purchases, national defense	29.9	Fed Govt
Blast furnaces & steel mills	11.1	Manufg.	Tanks & tank components	18.1	Manufg.
Power transmission equipment	5.7	Manufg.	Change in business inventories	10.3	In House
Wholesale trade	5.6	Trade			
Iron & steel forgings	3.1	Manufg.			
Iron & steel foundries	3.0	Manufg.			
Nonferrous forgings	2.9	Manufg.			
Nonferrous wire drawing & insulating	2.7	Manufg.			
Fabricated rubber products, nec	1.7	Manufg.			
Internal combustion engines, nec	1.7	Manufg.			
Imports	1.5	Foreign			
Air transportation	1.0	Util.			
Maintenance of nonfarm buildings nec	0.9	Constr.			
Motors & generators	0.9	Manufg.			
Plastics materials & resins	0.9	Manufg.			
Motor freight transportation & warehousing	0.9	Util.			
Aluminum castings	0.8	Manufg.			
Electric services (utilities)	0.8	Util.			
Railroads & related services	0.7	Util.			
Advertising	0.7	Services			
Business services nec	0.7	Services			
Metal stampings, nec	0.6	Manufg.			
Screw machine and related products	0.5	Manufg.			
Nonferrous rolling & drawing, nec	0.4	Manufg.			
Gas production & distribution (utilities)	0.4	Util.			
Eating & drinking places	0.4	Trade			
Fabricated metal products, nec	0.3	Manufg.			

Continued on next page.

INPUTS AND OUTPUTS FOR TANKS & TANK COMPONENTS - Continued

Economic Sector or Industry Providing Inputs	%	Sector	Economic Sector or Industry Buying Outputs	%	Sector
Communications, except radio & TV	0.3	Util.			
Banking	0.3	Fin/R.E.			
Real estate	0.3	Fin/R.E.			
Equipment rental & leasing services	0.3	Services			
Coal	0.2	Mining			
Machinery, except electrical, nec	0.2	Manufg.			
Sanitary services, steam supply, irrigation	0.2	Util.			
Computer & data processing services	0.2	Services			
Hotels & lodging places	0.2	Services			
Scrap	0.2	Scrap			
Abrasive products	0.1	Manufg.			
Machine tools, metal cutting types	0.1	Manufg.			
Special dies & tools & machine tool accessories	0.1	Manufg.			
Royalties	0.1	Fin/R.E.			
Legal services	0.1	Services			
Management & consulting services & labs	0.1	Services			

Source: Benchmark Input-Output Accounts for the U.S. Economy, 1982, U.S. Department of Commerce, Washington, D.C., July 1991. Data, as reported in the source, are organized by the 1977 SIC structure in use in 1982 but have been matched, as closely as is possible, to the 1987 SIC structure used in this book.

OCCUPATIONS EMPLOYED BY SIC 379 - TRANSPORTATION EQUIPMENT NEC

Occupation	% of Total 1994	Change to 2005	Occupation	% of Total 1994	Change to 2005
Assemblers, fabricators, & hand workers nec	21.4	11.7	Painters, transportation equipment	1.5	8.6
Welders & cutters	9.3	8.0	Helpers, laborers, & material movers nec	1.3	7.5
Blue collar worker supervisors	4.3	3.0	NC machine tool operators, metal & plastic	1.3	29.8
Mechanics, installers, & repairers nec	3.8	-13.7	Combination machine tool operators	1.2	22.0
Welding machine setters, operators	3.0	-0.3	Industrial production managers	1.1	10.1
Sales & related workers nec	2.7	10.0	Coating, painting, & spraying machine workers	1.1	11.6
Inspectors, testers, & graders, precision	2.5	7.9	Maintenance repairers, general utility	1.1	-2.2
General managers & top executives	2.2	4.8	General office clerks	1.1	-5.7
Machinists	2.1	6.8	Grinders & polishers, hand	1.1	11.7
Stock clerks	1.7	-13.1	Lathe & turning machine tool operators	1.1	-14.2
Freight, stock, & material movers, hand	1.6	-12.9	Bookkeeping, accounting, & auditing clerks	1.0	-17.5
Traffic, shipping, & receiving clerks	1.5	7.5	Extraction & related workers nec	1.0	9.1
Secretaries, ex legal & medical	1.5	0.1	Truck drivers light & heavy	1.0	15.1
Machine tool cutting operators, metal & plastic	1.5	-8.1			

Source: Industry-Occupation Matrix, Bureau of Labor Statistics. These data relate to one or more 3-digit SIC industry groups rather than to a single 4-digit SIC. The change reported for each occupation to the year 2005 is a percent of growth or decline as estimated by the Bureau of Labor Statistics. The abbreviation nec stands for 'not elsewhere classified'.

LOCATION BY STATE AND REGIONAL CONCENTRATION

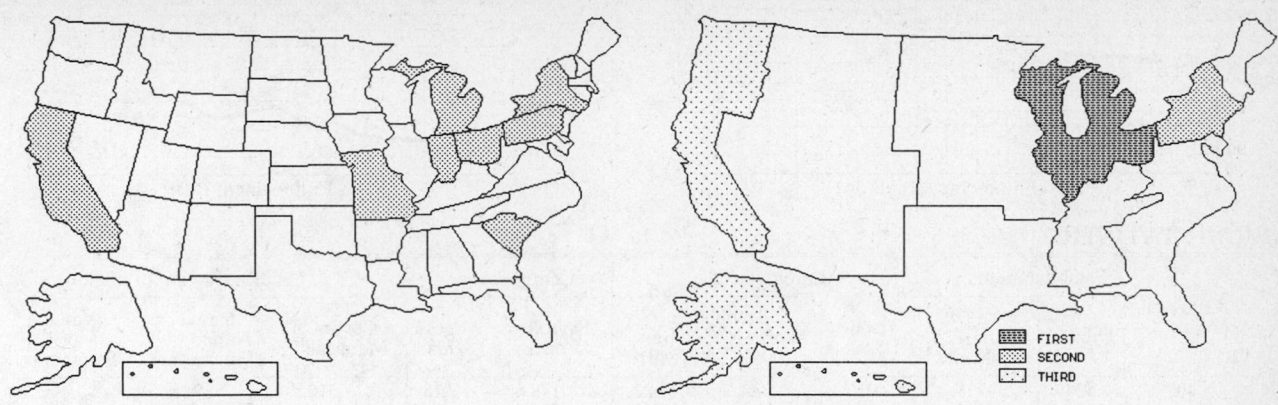

FIRST
SECOND
THIRD

INDUSTRY DATA BY STATE

State	Establish-ments	Shipments			Employment				Cost as % of Shipments	Investment per Employee ($)
		Total ($ mil)	% of U.S.	Per Establ.	Total Number	% of U.S.	Per Establ.	Wages ($/hour)		
Michigan	14	146.2	7.2	10.4	1,300	13.3	93	15.35	59.4	3,000
New York	4	14.5	0.7	3.6	100	1.0	* 25	17.00	37.2	1,000
Ohio	7	(D)	-	-	1,750 *	17.9	250	-	-	-
California	4	(D)	-	-	3,750 *	38.3	938	-	-	-
Pennsylvania	3	(D)	-	-	3,750 *	38.3	1,250	-	-	-
Indiana	2	(D)	-	-	175 *	1.8	88	-	-	-
Missouri	1	(D)	-	-	175 *	1.8	175	-	-	-
South Carolina	1	(D)	-	-	375 *	3.8	375	-	-	-

Source: 1992 *Economic Census*. The states are in descending order of shipments or establishments (if shipment data are missing for the majority). The symbol (D) appears when data are withheld to prevent disclosure of competitive information. States marked with (D) are sorted by number of establishments. A dash (-) indicates that the data element cannot be calculated; * indicates the midpoint of a range.

3799 - TRANSPORTATION EQUIPMENT, NEC

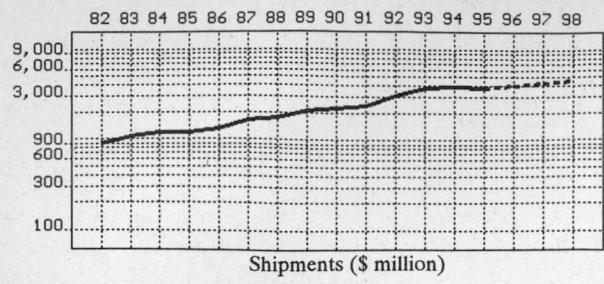

Shipments ($ million)

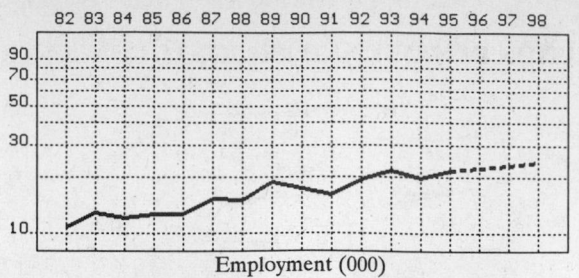

Employment (000)

GENERAL STATISTICS

Year	Companies	Establishments Total	Establishments with 20 or more employees	Employment Total (000)	Employment Production Workers (000)	Employment Hours (Mil)	Compensation Payroll ($ mil)	Compensation Wages ($/hr)	Production Cost of Materials	Production Value Added by Manufacture	Production Value of Shipments	Production Capital Invest.
1982	407	424	119	10.7	7.4	14.2	178.6	7.18	510.9	372.4	886.2	14.8
1983		416	122	12.8	9.2	17.2	205.4	6.88	643.8	438.1	1,073.0	16.5
1984		408	125	12.1	8.7	17.1	211.5	7.61	678.8	538.7	1,196.9	20.8
1985		401	129	12.6	8.8	16.6	215.5	7.56	705.8	490.2	1,180.2	30.2
1986		405	127	12.7	9.1	17.3	239.2	7.33	784.0	516.1	1,306.2	29.0
1987	617	635	174	15.4	11.1	21.4	306.6	7.73	989.6	664.9	1,642.1	37.6
1988		577	183	15.2	10.8	20.9	306.3	7.89	1,121.0	682.8	1,779.1	26.5
1989		564	189	18.8	11.2	20.6	314.0	8.28	1,385.3	753.9	2,095.4	57.1
1990		568	179	17.3	11.4	21.4	329.7	8.70	1,479.5	798.3	2,241.5	43.0
1991		568	154	16.1	11.1	21.0	359.8	9.12	1,577.9	788.4	2,425.5	56.4
1992	654	685	190	19.5	13.8	26.7	449.2	9.30	1,969.3	1,112.1	3,087.0	61.6
1993		691	198	21.9	15.6	30.5	514.3	9.64	2,422.0	1,407.9	3,806.9	86.0
1994		704P	206P	19.9	14.8	28.5	469.4	9.76	2,610.6	1,360.4	3,947.2	97.5
1995		731P	213P	21.7P	15.2P	29.2P	502.3P	9.92P	2,518.1P	1,312.2P	3,807.3P	88.0P
1996		758P	221P	22.5P	15.8P	30.3P	529.0P	10.16P	2,684.0P	1,398.6P	4,058.2P	94.2P
1997		785P	228P	23.4P	16.4P	31.5P	555.7P	10.41P	2,849.9P	1,485.1P	4,309.0P	100.4P
1998		812P	236P	24.2P	17.0P	32.6P	582.5P	10.65P	3,015.8P	1,571.6P	4,559.9P	106.6P

Sources: 1982, 1987, 1992 *Economic Census*; *Annual Survey of Manufactures*, 83-86, 88-91, 93-94. Establishment counts for non-Census years are from *County Business Patterns*; establishment values for 83-84 are extrapolations. 'P's show projections by the editors. Industries reclassified in 87 will not have data for prior years.

INDICES OF CHANGE

Year	Companies	Establishments Total	Establishments with 20 or more employees	Employment Total (000)	Employment Production Workers (000)	Employment Hours (Mil)	Compensation Payroll ($ mil)	Compensation Wages ($/hr)	Production Cost of Materials	Production Value Added by Manufacture	Production Value of Shipments	Production Capital Invest.
1982	62	62	63	55	54	53	40	77	26	33	29	24
1983		61	64	66	67	64	46	74	33	39	35	27
1984		60	66	62	63	64	47	82	34	48	39	34
1985		59	68	65	64	62	48	81	36	44	38	49
1986		59	67	65	66	65	53	79	40	46	42	47
1987	94	93	92	79	80	80	68	83	50	60	53	61
1988		84	96	78	78	78	68	85	57	61	58	43
1989		82	99	96	81	77	70	89	70	68	68	93
1990		83	94	89	83	80	73	94	75	72	73	70
1991		83	81	83	80	79	80	98	80	71	79	92
1992	100	100	100	100	100	100	100	100	100	100	100	100
1993		101	104	112	113	114	114	104	123	127	123	140
1994		103P	108P	102	107	107	104	105	133	122	128	158
1995		107P	112P	111P	110P	109P	112P	107P	128P	118P	123P	143P
1996		111P	116P	115P	114P	114P	118P	109P	136P	126P	131P	153P
1997		115P	120P	120P	119P	118P	124P	112P	145P	134P	140P	163P
1998		119P	124P	124P	123P	122P	130P	114P	153P	141P	148P	173P

Sources: Same as General Statistics. Values reflect change from the base year, 1992. Values above 100 mean greater than 92, values below 100 mean less than 92, and a value of 100 in the 82-91 or 93-98 period means same as 92. 'P's mark projections by the editors.

SELECTED RATIOS

For 1994	Avg. of All Manufact.	Analyzed Industry	Index	For 1994	Avg. of All Manufact.	Analyzed Industry	Index
Employees per Establishment	49	28	58	Value Added per Production Worker	134,084	91,919	69
Payroll per Establishment	1,500,273	666,546	44	Cost per Establishment	5,045,178	3,707,042	73
Payroll per Employee	30,620	23,588	77	Cost per Employee	102,970	131,186	127
Production Workers per Establishment	34	21	61	Cost per Production Worker	146,988	176,392	120
Wages per Establishment	853,319	394,986	46	Shipments per Establishment	9,576,895	5,605,009	59
Wages per Production Worker	24,861	18,795	76	Shipments per Employee	195,460	198,352	101
Hours per Production Worker	2,056	1,926	94	Shipments per Production Worker	279,017	266,703	96
Wages per Hour	12.09	9.76	81	Investment per Establishment	321,011	138,450	43
Value Added per Establishment	4,602,255	1,931,763	42	Investment per Employee	6,552	4,899	75
Value Added per Employee	93,930	68,362	73	Investment per Production Worker	9,352	6,588	70

Sources: Same as General Statistics. The 'Average of All Manufacturing' column represents the average of all manufacturing industries reported for the most recent complete year available. The Index shows the relationship between the Average and the Analyzed Industry. For example, 100 means that they are equal; 500 that the Analyzed Industry is five times the average; 50 means that the Analyzed Industry is half the national average. The abbreviation 'na' is used to show that data are 'not available'.

LEADING COMPANIES Number shown: 39   Total sales ($ mil): 3,125   Total employment (000): 16.5

Company Name	Address				CEO Name	Phone	Co. Type	Sales ($ mil)	Empl. (000)
Polaris Industries LP	1225 N Hwy 169	Minneapolis	MN	55441	W Hall Wendel Jr	612-542-0500	P	826	2.8
TriMas Corp	315 E Eisenhower	Ann Arbor	MI	48108	Brian P Campbell	313-747-7025	P	536	3.4
Polaris Industries Partners LP	33 Flying Point Rd	Southampton	NY	11968	W Hall Wendel Jr	516-283-1915	S	528	2.5
E-Z-Go Textron	PO Box 388	Augusta	GA	30913	LT Walden Jr	706-798-4311	D	230	1.1
Club Car Inc	PO Box 204658	Augusta	GA	30917	George H Inman	706-863-3000	P	186	0.7
AEG Westinghouse	1501 Lebanon	Pittsburgh	PA	15236	Ray Betler	412-655-5700	J	170	1.0
Cushman Inc	PO Box 82409	Lincoln	NE	68501	Jerry Ogren	402-475-9581	S	92	0.6
Ransomes America Corp	PO Box 82409	Lincoln	NE	68501	Irv E Aal	402-475-9581	S	92*	0.6
HammerBlow Corp	PO Box 419	Wausau	WI	54402	Tom Reinhart	715-842-0561	R	51*	0.3
EZ Loader Boat Trailers Inc	PO Box 3263	Spokane	WA	99220	Randy D Johnson	509-489-0181	R	50	0.4
Midwest Industries Inc	Hwy 59 & Hwy 175	Ida Grove	IA	51445	Byron Godbersen	712-364-3365	R	39*	0.3
Valley Industries Inc	32501 Dequiandre	Madison H	MI	48071	Roger Morgan	810-588-6900	R	33	0.2
Ryder Automotive Operations	PO Box 104	Buffalo	NY	14240	Bruce Lamar	716-894-3120	S	32*	0.2
Perry Tritech Inc	821 Jupiter Park Dr	Jupiter	FL	33458	Kevin Peterson	407-743-7000	R	30	0.1
ARDCO Industries	PO Box 451960	Houston	TX	77245	Paul Manchina	713-433-6751	D	28	0.2
LMC Operating Corp	PO Box 407	Logan	UT	84321		801-753-0220	R	25	0.2
Quality 'S' Mfg Inc	3801 N 43rd Av	Phoenix	AZ	85019	Jim D Weir	602-233-3499	R	23	0.2
Draw-Tite Inc	40500 Van Born Rd	Canton	MI	48188	JL Mellow	313-722-7800	R	22	0.4
Reese Products	PO Box 1706	Elkhart	IN	46515	Bob Mater	219-264-7564	D	21	0.2
Spartan Products Inc	275 E Marie Av	West St Paul	MN	55118	Arthur Freeman	612-451-1751	R	16*	<0.1
Electrical Rebuilders Inc	8600 Rheem Av	South Gate	CA	90280	MA Klapper	310-217-9730	R	14	0.4
Columbia ParCar Corp	PO Box 30	Reedsburg	WI	53959	Todd L Sauey	608-524-8888	R	12	<0.1
Rigid Hitch Inc	9216 Grand Av	Minneapolis	MN	55420	JD Skahen Jr	612-888-6799	R	12	<0.1
K and P Manufacturing	950 W Foothill Blv	Azusa	CA	91702	TJ Pierson II	818-334-0334	R	9	<0.1
CCI	2666 S C Club	Warsaw	IN	46580	Mark Randall	219-267-2222	D	7*	<0.1
Colet Special Vehicles Inc	575 W San Carlos St	San Jose	CA	95126	Ralph Colet	408-294-3643	R	6	<0.1
Riker Products Inc	PO Box 6976	Toledo	OH	43612	Phillip Meuser	419-729-1626	R	6	<0.1
Trail-Rite Inc	3100 W Central Av	Santa Ana	CA	92704	Donald R Williams	714-556-4540	R	6	<0.1
Unique Functional Products	135 Sunshine Ln	San Marcos	CA	92069	RH Troester	619-744-1610	R	6*	<0.1
Recreatives Industries Inc	60 Depot St	Buffalo	NY	14206	JH Wallach	716-855-2226	R	4	<0.1
Rolligon Corp	10635 Brighton Ln	Stafford	TX	77477	John G Holland	713-495-1140	R	4	0.1
Riblet Tramway Co	PO Box 3523	Spokane	WA	99220	TR Sowder	509-483-8555	R	3	<0.1
Spreuer and Son Inc	115 E Spring St	Lagrange	IN	46761	Phil M Spreuer	219-463-3513	R	2*	<0.1
Steelco Inc	PO Drawer 670	Santa Fe	TX	77510	Joyce Botman	409-925-2526	R	2*	<0.1
Walnut Industries Inc	PO Box 624	Bensalem	PA	19020	David Blatt	215-638-7847	R	1*	<0.1
American Performance Products	5839 Hamilton	Middletown	OH	45044	AR Gardner	513-539-9900	R	1*	<0.1
Lite Equipment Leasing Corp	PO Box 172	Annapolis Jnc	MD	20701	Tim Hazard	301-621-1950	R	1	<0.1
EZ Steer Inc	3106 Hill Av	Everett	WA	98201	Ronald W Wolcott	206-259-5476	R	1	<0.1
Stidham Horse Trailers Inc	PO Box 768	Chickasha	OK	73018	Ronnie Stidham	405-224-1302	R	1	<0.1

Source: *Ward's Business Directory of U.S. Private and Public Companies*, Volumes 1 and 2, 1996. The company type code used is as follows: P - Public, R - Private, S - Subsidiary, D - Division, J - Joint Venture, A - Affiliate, G - Group. Sales are in millions of dollars, employees are in thousands. An asterisk (*) indicates an estimated sales volume. The symbol < stands for 'less than'. Company names and addresses are truncated, in some cases, to fit into the available space.

MATERIALS CONSUMED

Material	Quantity	Delivered Cost ($ million)
Materials, ingredients, containers, and supplies	(X)	1,823.8
Trailer axles, wheels, brakes, undercarriages, and other metal vehicular parts	(X)	329.7
Internal combustion engines, gasoline	(X)	270.2
Pneumatic tires and inner tubes	(X)	41.7
Plastics products consumed in the form of sheets, rods, tubes, film, and other shapes	(X)	163.6
Paints, varnishes, lacquers, stains, shellacs, japans, enamels, and allied products	(X)	29.8
Fabricated metal products (except castings and forgings)	(X)	140.5
Forgings	(X)	(D)
Iron and steel castings (rough and semifinished)	(X)	8.5
Aluminum and aluminum-base alloy castings (rough and semifinished)	(X)	9.3
Other nonferrous castings (rough and semifinished)	(X)	(D)
Steel bars, bar shapes, and plates	(X)	42.2
Steel sheet and strip, including tin plate	(X)	25.4
Steel structural shapes and sheet piling	(X)	25.9
All other steel shapes and forms	(X)	19.8
Aluminum and aluminum-base alloy shapes and forms	(X)	24.3
Other nonferrous shapes and forms	(X)	(D)
All other materials and components, parts, containers, and supplies	(X)	368.2
Materials, ingredients, containers, and supplies, nsk	(X)	316.9

Source: 1992 *Economic Census*. Explanation of symbols used: (D): Withheld to avoid disclosure of competitive data; na: Not available; (S): Withheld because statistical norms were not met; (X): Not applicable; (Z): Less than half the unit shown; nec: Not elsewhere classified; nsk: Not specified by kind; - : zero; * : 10-19 percent estimated; ** : 20-29 percent estimated.

PRODUCT SHARE DETAILS

Product or Product Class	% Share	Product or Product Class	% Share
Transportation equipment, nec	100.00	Automobile and light truck trailers, nsk	2.82
Golf carts and industrial in-plant personnel carriers, self-propelled, and parts	13.44	Transportation equipment, nec, including all-terrain vehicles	50.82
Self-propelled golf carts (electric and gasoline) for carrying passengers and/or industrial in-plant person carriers	89.87	All-terrain vehicles, gasoline or electric, for transport of people or goods designed to traverse all types of terrain	24.32
Parts for self-propelled golf carts and/or industrial in-plant personnel carriers	10.11	Parts for transportation equipment, all-terrain vehicles	0.98
Automobile and light truck trailers	22.99	Trailer hitches (for travel trailers, automobile trailers, and light duty truck trailers)	9.64
Automobile and light truck horse trailers, excluding those pulled by truck tractors	22.47	Other miscellaneous transportation equipment, including snowmobiles and wheelbarrows	50.52
Automobile and light truck boat trailers	34.24	Parts for automobile and light truck trailers and other transportation equipment	13.89
Automobile and light truck mobile equipment trailers	10.17	Transportation equipment, nec, including all-terrain vehicles, nsk	0.65
Other automobile and light truck trailers, including general utility, commercial display, etc., for transport of goods	18.28	Transportation equipment, nec, nsk	12.74
Other automobile and light truck trailers, including general utility, commercial display, etc., for other uses	12.00		

Source: 1992 *Economic Census*. The values shown are percent of total shipments in an industry. Values of indented subcategories are summed in the main heading. The symbol (D) appears when data are withheld to prevent disclosure of competitive information. The abbreviation nsk stands for 'not specified by kind' and nec for 'not elsewhere classified'.

INPUTS AND OUTPUTS FOR TRANSPORTATION EQUIPMENT, NEC

Economic Sector or Industry Providing Inputs	%	Sector	Economic Sector or Industry Buying Outputs	%	Sector
Transportation equipment, nec	14.6	Manufg.	Gross private fixed investment	33.3	Cap Inv
Wholesale trade	14.4	Trade	Personal consumption expenditures	25.7	
Imports	14.1	Foreign	Transportation equipment, nec	10.4	Manufg.
Motor vehicle parts & accessories	7.6	Manufg.	Exports	7.6	Foreign
Blast furnaces & steel mills	7.5	Manufg.	Water transportation	4.4	Util.
Aircraft & missile engines & engine parts	4.0	Manufg.	S/L Govt. purch., elem. & secondary education	1.9	S/L Govt
Tires & inner tubes	3.1	Manufg.	Banking	1.1	Fin/R.E.
Aluminum rolling & drawing	2.9	Manufg.	Change in business inventories	1.1	In House
Plastics materials & resins	2.6	Manufg.	S/L Govt. purch., higher education	1.1	S/L Govt
Motor freight transportation & warehousing	2.1	Util.	Amusement & recreation services nec	1.0	Services
Internal combustion engines, nec	2.0	Manufg.	S/L Govt. purch., other education & libraries	0.9	S/L Govt
Aluminum castings	1.7	Manufg.	Motor vehicles & car bodies	0.7	Manufg.
Nonferrous wire drawing & insulating	1.1	Manufg.	Credit agencies other than banks	0.7	Fin/R.E.
Aircraft	1.0	Manufg.	Security & commodity brokers	0.7	Fin/R.E.
Aircraft & missile equipment, nec	1.0	Manufg.	Membership sports & recreation clubs	0.7	Services
Paints & allied products	1.0	Manufg.	Agricultural, forestry, & fishery services	0.6	Agric.
Maintenance of nonfarm buildings nec	0.9	Constr.	Communications, except radio & TV	0.6	Util.
Electric services (utilities)	0.9	Util.	Games, toys, & children's vehicles	0.5	Manufg.
Veneer & plywood	0.8	Manufg.	Railroad equipment	0.5	Manufg.
Ball & roller bearings	0.7	Manufg.	Electric services (utilities)	0.5	Util.
Sawmills & planning mills, general	0.7	Manufg.	Freight forwarders	0.4	Util.
Screw machine and related products	0.7	Manufg.	Automotive rental & leasing, without drivers	0.4	Services
Real estate	0.7	Fin/R.E.	State & local government enterprises, nec	0.4	Gov't
Building paper & board mills	0.6	Manufg.	Mobile homes	0.3	Manufg.
Fabricated rubber products, nec	0.6	Manufg.	Ship building & repairing	0.3	Manufg.
Iron & steel forgings	0.6	Manufg.	Truck & bus bodies	0.3	Manufg.
Iron & steel foundries	0.6	Manufg.	Truck trailers	0.3	Manufg.
Miscellaneous plastics products	0.6	Manufg.	Gas production & distribution (utilities)	0.3	Util.
Railroads & related services	0.6	Util.	Motor freight transportation & warehousing	0.3	Util.
Communications, except radio & TV	0.5	Util.	Colleges, universities, & professional schools	0.3	Services
Equipment rental & leasing services	0.5	Services	Hotels & lodging places	0.3	Services
Gaskets, packing & sealing devices	0.4	Manufg.	Motion pictures	0.3	Services
Metal stampings, nec	0.4	Manufg.	S/L Govt. purch., water	0.3	S/L Govt
Eating & drinking places	0.4	Trade	Logging camps & logging contractors	0.2	Manufg.
Banking	0.4	Fin/R.E.	Motor homes (made on purchased chassis)	0.2	Manufg.
Advertising	0.4	Services	Transit & bus transportation	0.2	Util.
Personnel supply services	0.4	Services	Hospitals	0.2	Services
Electronic components nec	0.3	Manufg.	Boat building & repairing	0.1	Manufg.
Housefurnishings, nec	0.3	Manufg.	Automotive repair shops & services	0.1	Services
Paperboard containers & boxes	0.3	Manufg.			
Gas production & distribution (utilities)	0.3	Util.			
Sanitary services, steam supply, irrigation	0.3	Util.			
Water transportation	0.3	Util.			
Theatrical producers, bands, entertainers	0.3	Services			
Abrasive products	0.2	Manufg.			
Lighting fixtures & equipment	0.2	Manufg.			
Machinery, except electrical, nec	0.2	Manufg.			
Metal doors, sash, & trim	0.2	Manufg.			
Petroleum refining	0.2	Manufg.			
Sheet metal work	0.2	Manufg.			
Detective & protective services	0.2	Services			
Engineering, architectural, & surveying services	0.2	Services			
Miscellaneous fabricated wire products	0.1	Manufg.			
Plumbing fixture fittings & trim	0.1	Manufg.			
Special dies & tools & machine tool accessories	0.1	Manufg.			
Wiring devices	0.1	Manufg.			
Accounting, auditing & bookkeeping	0.1	Services			
Computer & data processing services	0.1	Services			
Laundry, dry cleaning, shoe repair	0.1	Services			
Legal services	0.1	Services			
Management & consulting services & labs	0.1	Services			
U.S. Postal Service	0.1	Gov't			

Source: Benchmark Input-Output Accounts for the U.S. Economy, 1982, U.S. Department of Commerce, Washington, D.C., July 1991. Data, as reported in the source, are organized by the 1977 SIC structure in use in 1982 but have been matched, as closely as is possible, to the 1987 SIC structure used in this book.

OCCUPATIONS EMPLOYED BY SIC 379 - TRANSPORTATION EQUIPMENT NEC

Occupation	% of Total 1994	Change to 2005	Occupation	% of Total 1994	Change to 2005
Assemblers, fabricators, & hand workers nec	21.4	11.7	Painters, transportation equipment	1.5	8.6
Welders & cutters	9.3	8.0	Helpers, laborers, & material movers nec	1.3	7.5
Blue collar worker supervisors	4.3	3.0	NC machine tool operators, metal & plastic	1.3	29.8
Mechanics, installers, & repairers nec	3.8	-13.7	Combination machine tool operators	1.2	22.0
Welding machine setters, operators	3.0	-0.3	Industrial production managers	1.1	10.1
Sales & related workers nec	2.7	10.0	Coating, painting, & spraying machine workers	1.1	11.6
Inspectors, testers, & graders, precision	2.5	7.9	Maintenance repairers, general utility	1.1	-2.2
General managers & top executives	2.2	4.8	General office clerks	1.1	-5.7
Machinists	2.1	6.8	Grinders & polishers, hand	1.1	11.7
Stock clerks	1.7	-13.1	Lathe & turning machine tool operators	1.1	-14.2
Freight, stock, & material movers, hand	1.6	-12.9	Bookkeeping, accounting, & auditing clerks	1.0	-17.5
Traffic, shipping, & receiving clerks	1.5	7.5	Extraction & related workers nec	1.0	9.1
Secretaries, ex legal & medical	1.5	0.1	Truck drivers light & heavy	1.0	15.1
Machine tool cutting operators, metal & plastic	1.5	-8.1			

Source: *Industry-Occupation Matrix*, Bureau of Labor Statistics. These data relate to one or more 3-digit SIC industry groups rather than to a single 4-digit SIC. The change reported for each occupation to the year 2005 is a percent of growth or decline as estimated by the Bureau of Labor Statistics. The abbreviation nec stands for 'not elsewhere classified'.

LOCATION BY STATE AND REGIONAL CONCENTRATION

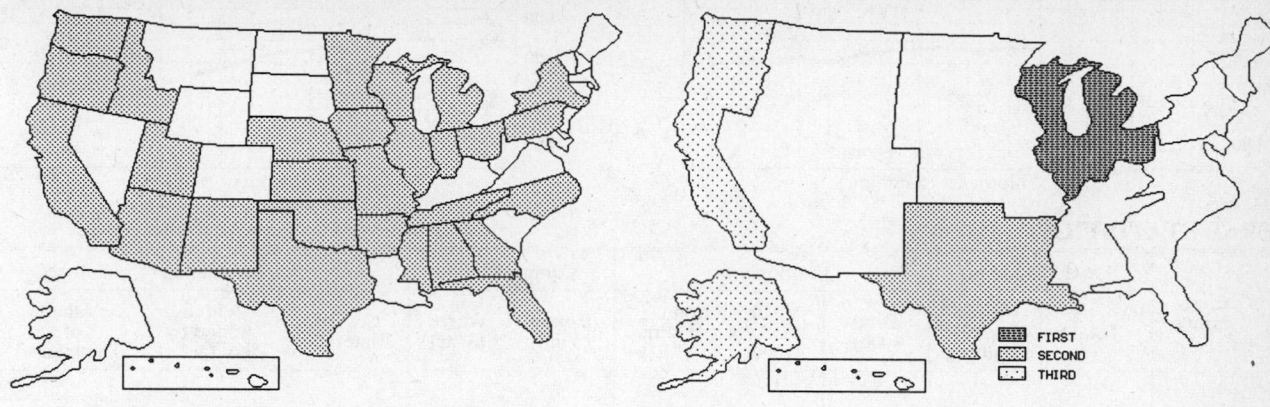

FIRST
SECOND
THIRD

INDUSTRY DATA BY STATE

State	Establish-ments	Shipments			Employment				Cost as % of Shipments	Investment per Employee ($)
		Total ($ mil)	% of U.S.	Per Establ.	Total Number	% of U.S.	Per Establ.	Wages ($/hour)		
Michigan	26	164.1	5.3	6.3	1,400	7.2	54	9.16	49.0	6,500
Texas	63	109.1	3.5	1.7	1,100	5.6	17	7.44	58.6	1,273
Oklahoma	26	107.3	3.5	4.1	1,300	6.7	50	7.55	60.4	-
Iowa	16	103.8	3.4	6.5	1,100	5.6	69	8.53	72.9	-
New York	15	62.2	2.0	4.1	400	2.1	27	10.17	50.8	2,250
Tennessee	19	57.8	1.9	3.0	600	3.1	32	10.14	53.6	-
Florida	42	41.0	1.3	1.0	400	2.1	10	8.60	58.3	-
Arizona	14	34.4	1.1	2.5	200	1.0	14	9.00	64.2	-
Oregon	22	27.9	0.9	1.3	300	1.5	14	8.75	47.0	-
Idaho	9	23.0	0.7	2.6	300	1.5	33	9.00	72.2	-
Mississippi	8	21.2	0.7	2.7	200	1.0	25	6.00	66.0	-
Illinois	22	19.7	0.6	0.9	100	0.5	5	9.00	57.9	10,000
Alabama	14	17.2	0.6	1.2	200	1.0	14	6.67	59.9	-
Kansas	10	15.5	0.5	1.5	200	1.0	20	9.33	71.0	-
North Carolina	14	13.7	0.4	1.0	100	0.5	7	12.50	66.4	1,000
California	66	(D)	-	-	750 *	3.8	11	-	-	-
Missouri	40	(D)	-	-	375 *	1.9	9	-	-	-
Indiana	33	(D)	-	-	1,750 *	9.0	53	-	-	-
Pennsylvania	33	(D)	-	-	375 *	1.9	11	-	-	-
Wisconsin	21	(D)	-	-	750 *	3.8	36	-	-	-
Arkansas	20	(D)	-	-	375 *	1.9	19	-	-	-
Ohio	20	(D)	-	-	750 *	3.8	38	-	-	-
Georgia	18	(D)	-	-	3,750 *	19.2	208	-	-	-
Minnesota	15	(D)	-	-	1,750 *	9.0	117	-	-	-
Washington	14	(D)	-	-	375 *	1.9	27	-	-	-
Utah	11	(D)	-	-	375 *	1.9	34	-	-	-
Nebraska	8	(D)	-	-	1,750 *	9.0	219	-	-	-
New Mexico	3	(D)	-	-	175 *	0.9	58	-	-	-

Source: 1992 *Economic Census*. The states are in descending order of shipments or establishments (if shipment data are missing for the majority). The symbol (D) appears when data are withheld to prevent disclosure of competitive information. States marked with (D) are sorted by number of establishments. A dash (-) indicates that the data element cannot be calculated; * indicates the midpoint of a range.

3812 - SEARCH AND NAVIGATION EQUIPMENT

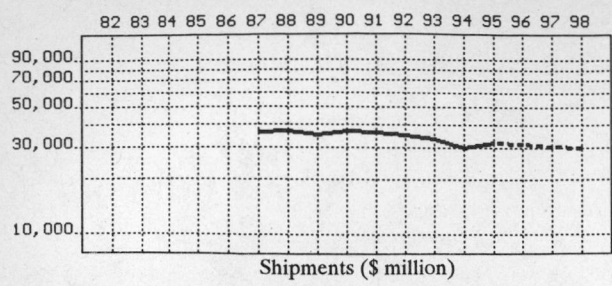

Shipments ($ million)

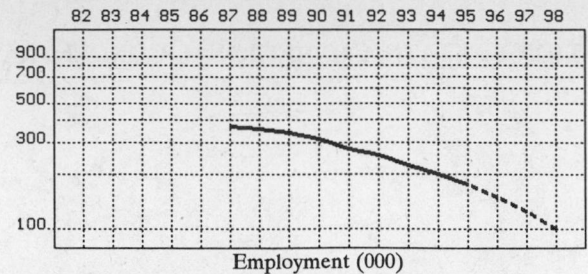

Employment (000)

GENERAL STATISTICS

Year	Companies	Establishments		Employment			Compensation		Production ($ million)			
		Total	with 20 or more employees	Total (000)	Production Workers (000)	Hours (Mil)	Payroll ($ mil)	Wages ($/hr)	Cost of Materials	Value Added by Manufacture	Value of Shipments	Capital Invest.
1982												
1983												
1984												
1985												
1986												
1987	920	1,084	508	369.4	158.8	314.3	12,373.0	14.21	12,208.3	24,738.7	36,266.8	1,439.0
1988				361.3	155.3	297.1	12,547.3	14.99	11,510.2	24,666.7	36,596.5	1,368.6
1989	895		502	339.5	140.8	276.7	12,445.3	14.99	10,874.7	23,924.5	35,295.4	1,366.9
1990				313.6	130.3	281.7	12,257.9	14.49	11,275.3	24,931.9	36,733.5	1,124.5
1991				279.8	112.3	258.0	11,630.7	14.08	11,401.5	23,672.3	36,213.4	829.9
1992	634	769	409	255.0	103.6	203.1	11,056.2	17.29	10,115.8	24,411.1	35,266.1	859.1
1993				225.1	88.5	170.8	10,123.0	18.79	8,844.2	24,031.3	33,546.2	706.1
1994				199.4	79.7	156.7	9,395.5	18.86	7,928.2	21,900.4	30,103.3	648.9
1995				177.3P	66.7P	138.8P	9,455.8P	19.07P	8,386.0P	23,165.1P	31,841.7P	471.7P
1996				151.6P	54.6P	115.2P	9,006.3P	19.76P	8,201.0P	22,654.1P	31,139.3P	344.8P
1997				125.9P	42.4P	91.6P	8,556.8P	20.45P	8,016.0P	22,143.1P	30,436.9P	217.9P
1998				100.2P	30.3P	68.1P	8,107.3P	21.14P	7,831.1P	21,632.0P	29,734.4P	91.0P

Sources: 1982, 1987, 1992 *Economic Census*; *Annual Survey of Manufactures*, 83-86, 88-91, 93-94. Establishment counts for non-Census years are from *County Business Patterns*; establishment values for 83-84 are extrapolations. 'P's show projections by the editors. Industries reclassified in 87 will not have data for prior years.

INDICES OF CHANGE

Year	Companies	Establishments		Employment			Compensation		Production ($ million)			
		Total	with 20 or more employees	Total (000)	Production Workers (000)	Hours (Mil)	Payroll ($ mil)	Wages ($/hr)	Cost of Materials	Value Added by Manufacture	Value of Shipments	Capital Invest.
1982												
1983												
1984												
1985												
1986												
1987	145	141	124	145	153	155	112	82	121	101	103	168
1988				142	150	146	113	87	114	101	104	159
1989		116	123	133	136	136	113	87	108	98	100	159
1990				123	126	139	111	84	111	102	104	131
1991				110	108	127	105	81	113	97	103	97
1992	100	100	100	100	100	100	100	100	100	100	100	100
1993				88	85	84	92	109	87	98	95	82
1994				78	77	77	85	109	78	90	85	76
1995				70P	64P	68P	86P	110P	83P	95P	90P	55P
1996				59P	53P	57P	81P	114P	81P	93P	88P	40P
1997				49P	41P	45P	77P	118P	79P	91P	86P	25P
1998				39P	29P	34P	73P	122P	77P	89P	84P	11P

Sources: Same as General Statistics. Values reflect change from the base year, 1992. Values above 100 mean greater than 92, values below 100 mean less than 92, and a value of 100 in the 82-91 or 93-98 period means same as 92. 'P's mark projections by the editors.

SELECTED RATIOS

For 1992	Avg. of All Manufact.	Analyzed Industry	Index	For 1992	Avg. of All Manufact.	Analyzed Industry	Index
Employees per Establishment	46	332	726	Value Added per Production Worker	122,353	235,628	193
Payroll per Establishment	1,332,320	14,377,373	1,079	Cost per Establishment	4,239,462	13,154,486	310
Payroll per Employee	29,181	43,358	149	Cost per Employee	92,853	39,670	43
Production Workers per Establishment	31	135	429	Cost per Production Worker	135,003	97,643	72
Wages per Establishment	734,496	4,566,449	622	Shipments per Establishment	8,100,800	45,859,688	566
Wages per Production Worker	23,390	33,896	145	Shipments per Employee	177,425	138,298	78
Hours per Production Worker	2,025	1,960	97	Shipments per Production Worker	257,966	340,406	132
Wages per Hour	11.55	17.29	150	Investment per Establishment	278,244	1,117,165	402
Value Added per Establishment	3,842,210	31,743,953	826	Investment per Employee	6,094	3,369	55
Value Added per Employee	84,153	95,730	114	Investment per Production Worker	8,861	8,292	94

Sources: Same as General Statistics. The 'Average of All Manufacturing' column represents the average of all manufacturing industries reported for the most recent complete year available. The Index shows the relationship between the Average and the Analyzed Industry. For example, 100 means that they are equal; 500 that the Analyzed Industry is five times the average; 50 means that the Analyzed Industry is half the national average. The abbreviation 'na' is used to show that data are 'not available'.

LEADING COMPANIES Number shown: **74**   Total sales ($ mil): **80,468**   Total employment (000): **552.8**

Company Name	Address				CEO Name	Phone	Co. Type	Sales ($ mil)	Empl. (000)
Lockheed Martin Corp	6801 Rockledge Dr	Bethesda	MD	20817	Daniel M Tellep	301-897-6000	P	22,915	173.0
GM Hughes Electronics Corp	PO Box 80028	Los Angeles	CA	90080	CM Armstrong	310-568-7200	P	14,099	79.0
AlliedSignal Inc	PO Box 2245	Morristown	NJ	07962	Lawrence A Bossidy	201-455-2000	P	12,817	87.5
Raytheon Co	141 Spring St	Lexington	MA	02173	Dennis J Picard	617-862-6600	P	10,013	60.2
Loral Corp	600 3rd Av	New York	NY	10016	Bernard L Schwartz	212-697-1105	P	5,484	28.9
E-Systems Inc	PO Box 660248	Dallas	TX	75266	A Lowell Lawson	214-661-1000	P	2,028	15.8
Collins Air Transport	400 Collins Rd NE	Cedar Rapids	IA	52498	Robert J Tibor	319-395-1000	D	1,100*	6.5
Collins Commercial Avionics	400 Collins Rd NE	Cedar Rapids	IA	52498	John R Girotto	319-395-1000	D	1,100	6.5
Hughes Aircraft Co	1901 W Malvern	Fullerton	CA	92634	J Weaver	714-732-3232	D	850*	7.0
Tracor Inc	6500 Tracor Ln	Austin	TX	78725	James B Skaggs	512-926-2800	P	694	9.7
Sensormatic Electronics Corp	500 NW 12th Av	Deerfield Bch	FL	33442	Ronald G Assaf	305-427-9700	P	656	5.5
E-Systems Inc	PO Box 660023	Dallas	TX	75266	MD Williamson	214-272-0515	D	620*	5.0
E-Systems Inc	PO Box 6056	Greenville	TX	75403	BD Cullin	903-455-3450	D	560*	4.5
Astronautics Corporation	PO Box 523	Milwaukee	WI	53201	Nathaniel K Zelazo	414-447-8200	R	380	3.5
Talley Industries Inc	2702 N 44th St	Phoenix	AZ	85008	W H Mallender	602-957-7711	P	328	2.4
Moog Inc	Seneca & Jamisons	East Aurora	NY	14052	Robert T Brady	716-652-2000	P	307	3.1
GDE Systems Inc	PO Box 509009	San Diego	CA	92150	Terry A Straeter	619-675-2600	D	282	1.8
Teledyne Brown Engineering Co	300 Sparkman Dr	Huntsville	AL	35805	Jim McGovern	205-726-1000	S	280	3.0
Northrop Corp	PO Box 5032	Hawthorne	CA	90251	Robert L Silverstein	213-600-3000	D	270	1.2
Sparton Corp	2400 E Ganson St	Jackson	MI	49202	John J Smith	517-787-8600	P	233	2.3
Melpar	7700 Arlington Blv	Falls Church	VA	22046	LF Judd	703-560-5000	D	230*	2.2
Kearfott Guidance	150 Totowa Rd	Wayne	NJ	07474	Ronald Zelazo	201-785-6993	S	225	3.0
United Industrial Corp	18 E 48th St	New York	NY	10017	P David Bocksch	212-752-8787	P	210	1.9
Honeywell Inc	13350 US Hwy 19 N	Clearwater	FL	34624	Bill Poe	813-531-4611	D	200*	1.7
Raytheon Co	6380 Hollister Av	Goleta	CA	93117	Jack L Gressingh	805-967-5511	D	200	1.2
Raytheon Marine Co	676 Island Pond Rd	Manchester	NH	03109	Robert Unger	603-647-7530	D	200*	0.2
Westinghouse Norden Systems	PO Box 5300	Norwalk	CT	06856	Jack Wohler	203-852-5000	S	200	1.2
AAI Corp	PO Box 126	Hunt Valley	MD	21030	R R Erkenneff	410-666-1400	S	180	1.6
Trimble Navigation Ltd	PO Box 3642	Sunnyvale	CA	94088	Charles R Trimble	408-481-8000	P	176	0.9
Smiths Industies	4141 Eastern Av SE	Grand Rapids	MI	49518	Richard Tierney	616-241-7000	D	160	1.7
Electrospace Systems Inc	PO Box 831359	Richardson	TX	75083	M D McAnally	214-470-2000	S	153	1.2
Meggitt-USA Inc	540 N Commercial	Manchester	NH	03101	Bennett F Moore	603-669-9971	S	150	1.5
Precision Standard Inc	1943 50th St	Birmingham	AL	35212	Matthew L Gold	205-591-3009	P	149	2.4
Sperry Marine Inc	1070 Seminole Trail	Charlottesville	VA	22901	George A Sawyer	804-974-2000	R	140	0.9
Loral Infrared & Imaging Syst	2 Forbes Rd	Lexington	MA	02173	John S Dehne	617-862-6222	S	130	0.8
Whittaker Corp	10880 Wilshire Blv	Los Angeles	CA	90024	Thomas Brancati	310-475-9411	P	126	0.8
AEL Industries Inc	305 Richardson Rd	Lansdale	PA	19446	Leon Riebman	215-822-3238	P	125	1.1
Sparton Corp	PO Box 788	De Leon Sp	FL	32130	DE Johnson	904-985-4631	D	120*	0.6
Interstate Electronics Corp	1001 E Ball Rd	Anaheim	CA	92803	LA LaCotti	714-758-0500	S	113	1.1
HR Textron Inc	25200 W Rye	Valencia	CA	91355	Bradley W Spahr	805-294-6000	S	105	0.9
AlliedSignal Ocean Systems	15825 Roxford St	Sylmar	CA	91342	Robert Scrofano	818-367-0111	S	100*	0.9
BF Goodrich Co	Panton Rd	Vergennes	VT	05491	Ron Hodges	802-877-2911	D	100	0.8
ITT Gilfillan	7821 Orion Av	Van Nuys	CA	91406	Victor N Rios	818-988-2600	D	100	0.7
Westinghouse Electric Corp	PO Box 1488	Annapolis	MD	21404	WL Dunkle	410-260-5000	D	100*	0.8
Datron Inc	8 Griffin Rd N	Windsor	CT	06095	T Stephen Melvin	203-688-6855	R	98*	1.0
Nicolet Instrument Corp	5225 Verona Rd	Madison	WI	53744	Robert Rosenthal	608-276-6100	S	97*	0.8
Litton Industries Inc	PO Box 7012	San Jose	CA	95150	Clayton Williams	408-365-4747	D	94	0.8
EDO Corp	14-04 111th St	College Point	NY	11356	Frank A Fariello	718-321-4000	P	91	0.8
Lowrance Electronics Inc	12000 E Skelly Dr	Tulsa	OK	74128	Darrell J Lowrance	918-437-6881	P	81	0.7
Lear Astronics Corp	3400 Airport Av	Santa Monica	CA	90406	Dave Dallob	310-915-6881	S	80	0.8
Wilcox Electric Inc	2001 NE 46th	Kansas City	MO	64116	D Welde	816-453-2600	S	80	0.6
LaBarge Inc	PO Box 14499	St Louis	MO	63178	Craig E LaBarge	314-231-5960	P	73	1.0
Montek	2268 S 3270 W	Salt Lake City	UT	84119	DA Williams	801-973-4300	D	70	0.6
Universal Avionic Systems Corp	3260 E Lerdo Rd	Tucson	AZ	85706	Hubert L Naimer	602-295-2300	R	70	0.2
EDO Corp	14-04 111th St	College Point	NY	11356	Ira Kaplan	718-321-4000	D	67	0.7
ECC International Corp	175 Strafford Av	Wayne	PA	19087	George W Murphy	215-687-2600	P	63	0.9
Magnavox Electronics Syst Co	2829 Maricopa St	Torrance	CA	90503	Vito F Brenna	310-618-1200	D	63	0.9
Metric Systems Corp	645 Anchors St	Ft Walton Bch	FL	32548	Coy J Scribner	904-244-9600	S	62*	0.7
Marvin Engineering Co	260 W Beach Av	Inglewood	CA	90302	Marvin Gussman	213-678-1281	R	60	0.4
Watkins-Johnson Co	700 Quince Orchard	Gaithersburg	MD	20878	William T Bruff	301-948-7550	D	58	0.6
Diagnostic/Retrieval Systems	16 Thorton Rd	Oakland	NJ	07436	Mark S Newman	201-337-3800	P	58	0.5
Teledyne Controls Inc	12333 W Olympic	Los Angeles	CA	90064	Jeff Amacker	310-820-4616	R	49	0.4
Kollsman	220 Daniel Webster	Merrimack	NH	03054	Ron Wright	603-889-2500	D	45*	0.5
TCOM LP	7115 T Edison Dr	Columbia	MD	21046	J Bitonti	410-312-2300	R	45	0.2
NavCom Defense Electronics Inc	4323 Arden Dr	El Monte	CA	91731	Cifford Christ	818-442-0123	R	44*	0.2
GEC-Marconi Avionics Inc	PO Box 81999	Atlanta	GA	30366	William Broyles	404-448-1947	S	43*	0.4
Buffton Corp	226 Bailey Av	Fort Worth	TX	76107	Robert H McLean	817-332-4761	P	41	0.6
Opto Mechanik Inc	PO Box 361907	Melbourne	FL	32936	Ottmar Dippold	407-254-1212	P	35	0.2
Sippican Inc	PO Box 861	Marion	MA	02738	William E Walsh	508-748-1160	R	35	0.2
Techsonic Industries Inc	5 Hummingbird Ln	Eufaula	AL	36027	J R Balkcom Jr	205-687-6613	S	35*	0.3
Whitehall Corp	PO Box 29709	Dallas	TX	75229	George F Baker	214-247-8747	P	32	0.5
Pacer Systems Inc	900 Techn Pk Dr	Billerica	MA	01821	John C Rennie	508-667-8800	P	30	0.3
Loral Conic Corp	9020 Balboa Av	San Diego	CA	92123	Gus Schneidau	619-279-0411	S	30	0.4
Systron Donner Inertial	2700 Systron Dr	Concord	CA	94518	David Pike	510-682-6161	D	30	0.3

Source: *Ward's Business Directory of U.S. Private and Public Companies*, Volumes 1 and 2, 1996. The company type code used is as follows: P - Public, R - Private, S - Subsidiary, D - Division, J - Joint Venture, A - Affiliate, G - Group. Sales are in millions of dollars, employees are in thousands. An asterisk (*) indicates an estimated sales volume. The symbol < stands for 'less than'. Company names and addresses are truncated, in some cases, to fit into the available space.

MATERIALS CONSUMED

Material	Quantity	Delivered Cost ($ million)
Materials, ingredients, containers, and supplies	(X)	8,758.8
Printed circuit boards (without inserted components) for electronic circuitry	(X)	315.5
Printed circuit assemblies, loaded boards or modules	(X)	219.3
Semiconductors, including transistors, diodes, rectifiers, and integrated circuits for electronic circuitry	(X)	551.9
Capacitors for electronic circuitry	(X)	93.5
Resistors for electronic circuitry	(X)	82.9
Other components and accessories for electronic circuitry, nec, except tubes	(X)	948.6
Electronic communication equipment	(X)	664.4
Electrical instrument mechanisms and meter movements (including instrument relays)	(X)	228.0
Electronic computing equipment	(X)	179.0
Current-carrying wiring devices	(X)	76.5
Insulated wire and cable, including magnet wire	(X)	91.9
Loudspeakers, microphones, and tuners (all types)	(X)	9.6
Fractional horsepower electric motors (less than 1 hp)	(X)	31.2
Silicon, hyperpure	(X)	1.4
Plastics resins consumed in the form of granules, pellets, powders, liquids, etc.	(X)	12.8
Fabricated plastics products (except gaskets, hoses, and belting)	(X)	36.7
Sheet metal products, except stampings	(X)	188.2
Metal stampings	(X)	27.0
Metal bolts, nuts, screws, washers, rivets, and other screw machine products	(X)	105.3
Other fabricated metal products (except forgings)	(X)	252.2
Forgings	(X)	33.0
Castings (rough and semifinished)	(X)	145.8
Steel shapes and forms	(X)	80.5
Aluminum and aluminum-base alloy shapes and forms	(X)	75.8
Other nonferrous shapes and forms	(X)	36.7
Paper and paperboard containers, including shipping sacks and other paper packaging supplies	(X)	19.8
All other materials and components, parts, containers, and supplies	(X)	2,438.9
Materials, ingredients, containers, and supplies, nsk	(X)	1,812.1

*Source: 1992 Economic Census. Explanation of symbols used: (D): Withheld to avoid disclosure of competitive data; na: Not available; (S): Withheld because statistical norms were not met; (X): Not applicable; (Z): Less than half the unit shown; nec: Not elsewhere classified; nsk: Not specified by kind; - : zero; * : 10-19 percent estimated; ** : 20-29 percent estimated.*

PRODUCT SHARE DETAILS

Product or Product Class	% Share	Product or Product Class	% Share
Search, detection, nagivation, guidance, aeronautical and nautical systems, instruments, and equipment	100.00	Search, detection, navigation, and guidance systems and equipment	90.79
Aeronautical, nautical, and navigational instruments, not sending or receiving radio signals, except aircraft engine instruments	7.41	Search, detection, nagivation, guidance, aeronautical, and nautical systems, instruments, and equipment, nsk	1.80

Source: 1992 Economic Census. The values shown are percent of total shipments in an industry. Values of indented subcategories are summed in the main heading. The symbol (D) appears when data are withheld to prevent disclosure of competitive information. The abbreviation nsk stands for 'not specified by kind' and nec for 'not elsewhere classified'.

INPUTS AND OUTPUTS FOR ENGINEERING & SCIENTIFIC INSTRUMENTS

Economic Sector or Industry Providing Inputs	%	Sector	Economic Sector or Industry Buying Outputs	%	Sector
Imports	14.6	Foreign	Gross private fixed investment	41.5	Cap Inv
Engineering & scientific instruments	12.1	Manufg.	Exports	26.4	Foreign
Wholesale trade	8.7	Trade	Federal Government purchases, national defense	7.5	Fed Govt
Electronic components nec	6.8	Manufg.	Engineering & scientific instruments	5.2	Manufg.
Veneer & plywood	4.7	Manufg.	Aircraft	4.2	Manufg.
Blast furnaces & steel mills	3.6	Manufg.	Ship building & repairing	2.4	Manufg.
Advertising	3.6	Services	Radio & TV communication equipment	2.3	Manufg.
Instruments to measure electricity	2.7	Manufg.	S/L Govt. purch., health & hospitals	2.1	S/L Govt
Semiconductors & related devices	2.4	Manufg.	Mechanical measuring devices	1.4	Manufg.
Communications, except radio & TV	2.1	Util.	Federal Government purchases, nondefense	1.3	Fed Govt
Electric services (utilities)	2.0	Util.	Change in business inventories	0.8	In House
Petroleum refining	1.9	Manufg.	S/L Govt. purch., higher education	0.6	S/L Govt
Eating & drinking places	1.8	Trade	S/L Govt. purch., police	0.6	S/L Govt
Motors & generators	1.7	Manufg.	S/L Govt. purch., sewerage	0.6	S/L Govt
Industrial patterns	1.4	Manufg.	S/L Govt. purch., sanitation	0.5	S/L Govt
Fabricated metal products, nec	1.3	Manufg.	Boat building & repairing	0.4	Manufg.
Machinery, except electrical, nec	1.2	Manufg.	S/L Govt. purch., elem. & secondary education	0.4	S/L Govt
Fabricated rubber products, nec	1.1	Manufg.	Blast furnaces & steel mills	0.3	Manufg.
Sheet metal work	1.1	Manufg.	Miscellaneous plastics products	0.3	Manufg.
Aluminum rolling & drawing	1.0	Manufg.	Watches, clocks, & parts	0.3	Manufg.
Miscellaneous plastics products	0.9	Manufg.	S/L Govt. purch., highways	0.3	S/L Govt
Screw machine and related products	0.9	Manufg.	S/L Govt. purch., natural resource & recreation.	0.3	S/L Govt
Banking	0.9	Fin/R.E.			
Legal services	0.9	Services			

Continued on next page.

INPUTS AND OUTPUTS FOR ENGINEERING & SCIENTIFIC INSTRUMENTS - Continued

Economic Sector or Industry Providing Inputs	%	Sector	Economic Sector or Industry Buying Outputs	%	Sector
Maintenance of nonfarm buildings nec	0.8	Constr.			
Electronic computing equipment	0.8	Manufg.			
Motor freight transportation & warehousing	0.8	Util.			
Equipment rental & leasing services	0.8	Services			
Glass & glass products, except containers	0.7	Manufg.			
Paperboard containers & boxes	0.7	Manufg.			
Real estate	0.7	Fin/R.E.			
Management & consulting services & labs	0.7	Services			
U.S. Postal Service	0.7	Gov't			
Aluminum castings	0.6	Manufg.			
Gaskets, packing & sealing devices	0.6	Manufg.			
Special dies & tools & machine tool accessories	0.6	Manufg.			
Ball & roller bearings	0.5	Manufg.			
Paints & allied products	0.5	Manufg.			
Manifold business forms	0.4	Manufg.			
Metal stampings, nec	0.4	Manufg.			
Gas production & distribution (utilities)	0.4	Util.			
Accounting, auditing & bookkeeping	0.4	Services			
Converted paper products, nec	0.3	Manufg.			
Lighting fixtures & equipment	0.3	Manufg.			
Railroads & related services	0.3	Util.			
Automotive rental & leasing, without drivers	0.3	Services			
Business/professional associations	0.3	Services			
Computer & data processing services	0.3	Services			
Engineering, architectural, & surveying services	0.3	Services			
Abrasive products	0.2	Manufg.			
Iron & steel foundries	0.2	Manufg.			
Lubricating oils & greases	0.2	Manufg.			
Machine tools, metal cutting types	0.2	Manufg.			
Nonferrous castings, nec	0.2	Manufg.			
Photographic equipment & supplies	0.2	Manufg.			
Plastics materials & resins	0.2	Manufg.			
Plating & polishing	0.2	Manufg.			
Sawmills & planning mills, general	0.2	Manufg.			
Transformers	0.2	Manufg.			
Wiring devices	0.2	Manufg.			
Air transportation	0.2	Util.			
Credit agencies other than banks	0.2	Fin/R.E.			
Insurance carriers	0.2	Fin/R.E.			
Royalties	0.2	Fin/R.E.			
Automotive repair shops & services	0.2	Services			
Photofinishing labs, commercial photography	0.2	Services			
Theatrical producers, bands, entertainers	0.2	Services			
Electron tubes	0.1	Manufg.			
Miscellaneous fabricated wire products	0.1	Manufg.			
Rubber & plastics hose & belting	0.1	Manufg.			
Transit & bus transportation	0.1	Util.			
Water supply & sewage systems	0.1	Util.			
Water transportation	0.1	Util.			
Motion pictures	0.1	Services			
Personnel supply services	0.1	Services			
Noncomparable imports	0.1	Foreign			

Source: *Benchmark Input Output Accounts for the U.S. Economy, 1982,* U.S. Department of Commerce, Washington, D.C., July 1991. Data, as reported in the source, are organized by the 1977 SIC structure in use in 1982 but have been matched, as closely as is possible, to the 1987 SIC structure used in this book.

OCCUPATIONS EMPLOYED BY SIC 381 - SEARCH AND NAVIGATION EQUIPMENT

Occupation	% of Total 1994	Change to 2005	Occupation	% of Total 1994	Change to 2005
Electrical & electronics engineers	13.1	-9.2	Mechanical engineers	2.2	-21.8
Electrical & electronic equipment assemblers	5.8	-29.0	Professional workers nec	2.1	-14.7
Electrical & electronic technicians,technologists	5.6	-29.0	Managers & administrators nec	1.9	-29.0
Inspectors, testers, & graders, precision	4.7	-50.3	Industrial production managers	1.5	-28.9
Engineering, mathematical, & science managers	4.1	-19.3	Administrative services managers	1.4	-43.2
Industrial engineers, ex safety engineers	3.4	-21.8	Stock clerks	1.4	-42.3
Engineers nec	3.3	-14.8	General office clerks	1.3	-39.4
Secretaries, ex legal & medical	3.3	-35.3	Assemblers, fabricators, & hand workers nec	1.2	-29.0
Production, planning, & expediting clerks	2.9	-29.0	General managers & top executives	1.2	-32.6
Management support workers nec	2.7	-29.0	Purchasing agents, ex trade & farm products	1.1	-28.9
Electrical & electronic assemblers	2.7	-36.1	Electromechanical equipment assemblers	1.1	-21.8
Blue collar worker supervisors	2.5	-38.8	Drafters	1.1	-44.7
Engineering technicians & technologists nec	2.5	-29.0	Accountants & auditors	1.1	-29.0
Systems analysts	2.4	13.7	Clerical support workers nec	1.0	-43.2

Source: *Industry-Occupation Matrix,* Bureau of Labor Statistics. These data relate to one or more 3-digit SIC industry groups rather than to a single 4-digit SIC. The change reported for each occupation to the year 2005 is a percent of growth or decline as estimated by the Bureau of Labor Statistics. The abbreviation nec stands for 'not elsewhere classified'.

LOCATION BY STATE AND REGIONAL CONCENTRATION

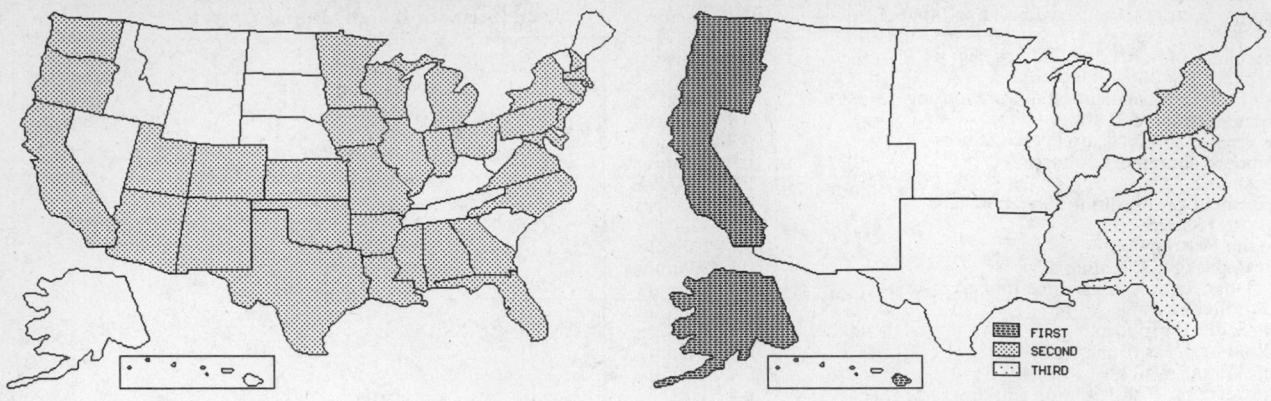

FIRST
SECOND
THIRD

INDUSTRY DATA BY STATE

State	Establish-ments	Shipments			Employment				Cost as % of Shipments	Investment per Employee ($)
		Total ($ mil)	% of U.S.	Per Establ.	Total Number	% of U.S.	Per Establ.	Wages ($/hour)		
California	164	9,079.2	25.7	55.4	65,400	25.6	399	18.32	25.9	2,644
New York	72	3,873.1	11.0	53.8	22,300	8.7	310	21.32	25.0	4,184
Texas	50	2,864.5	8.1	57.3	23,000	9.0	460	14.31	25.1	3,643
Florida	55	2,667.3	7.6	48.5	18,700	7.3	340	12.18	28.4	4,540
Maryland	19	2,572.4	7.3	135.4	17,700	6.9	932	26.02	27.4	3,441
Massachusetts	49	2,490.8	7.1	50.8	16,300	6.4	333	18.81	37.3	2,472
New Jersey	46	2,281.2	6.5	49.6	15,100	5.9	328	21.88	37.6	3,497
Arizona	12	1,545.6	4.4	128.8	9,000	3.5	750	14.05	31.3	5,522
Virginia	13	1,371.7	3.9	105.5	8,800	3.5	677	12.28	22.2	6,193
Illinois	15	494.5	1.4	33.0	4,200	1.6	280	13.27	29.3	-
Missouri	7	392.1	1.1	56.0	4,900	1.9	700	15.04	43.6	-
Washington	27	373.2	1.1	13.8	3,500	1.4	130	12.38	18.3	3,314
Connecticut	23	370.9	1.1	16.1	4,000	1.6	174	13.73	34.2	2,375
Pennsylvania	24	353.8	1.0	14.7	2,700	1.1	113	13.93	40.0	4,185
Michigan	13	262.8	0.7	20.2	2,900	1.1	223	15.07	23.7	2,586
Oregon	15	202.9	0.6	13.5	1,700	0.7	113	11.27	29.0	-
North Carolina	8	140.2	0.4	17.5	1,100	0.4	138	11.80	59.1	-
Ohio	20	122.9	0.3	6.1	1,100	0.4	55	14.07	26.0	2,000
Alabama	8	122.2	0.3	15.3	900	0.4	113	9.23	62.4	2,333
Georgia	12	53.9	0.2	4.5	600	0.2	50	15.00	34.0	-
Indiana	4	16.9	0.0	4.2	200	0.1	50	11.00	24.3	-
Kansas	15	(D)	-	-	3,750 *	1.5	250	-	-	1,413
Colorado	13	(D)	-	-	3,750 *	1.5	288	-	-	-
New Hampshire	10	(D)	-	-	7,500 *	2.9	750	-	-	-
Wisconsin	8	(D)	-	-	375 *	0.1	47	-	-	-
Minnesota	7	(D)	-	-	3,750 *	1.5	536	-	-	-
Utah	6	(D)	-	-	1,750 *	0.7	292	-	-	3,200
Louisiana	5	(D)	-	-	175 *	0.1	35	-	-	-
New Mexico	5	(D)	-	-	1,750 *	0.7	350	-	-	-
Rhode Island	5	(D)	-	-	1,750 *	0.7	350	-	-	-
Iowa	4	(D)	-	-	7,500 *	2.9	1,875	-	-	-
Oklahoma	4	(D)	-	-	750 *	0.3	188	-	-	-
Arkansas	3	(D)	-	-	375 *	0.1	125	-	-	-
Mississippi	3	(D)	-	-	750 *	0.3	250	-	-	-

Source: 1992 *Economic Census*. The states are in descending order of shipments or establishments (if shipment data are missing for the majority). The symbol (D) appears when data are withheld to prevent disclosure of competitive information. States marked with (D) are sorted by number of establishments. A dash (-) indicates that the data element cannot be calculated; * indicates the midpoint of a range.

3821 - LABORATORY APPARATUS & FURNITURE

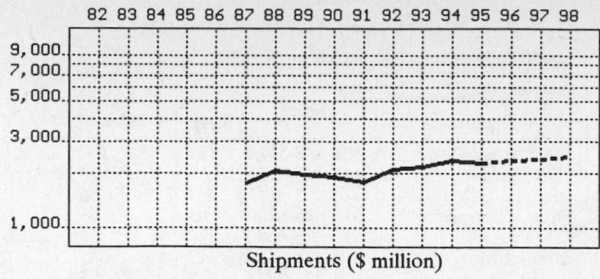

Shipments ($ million)

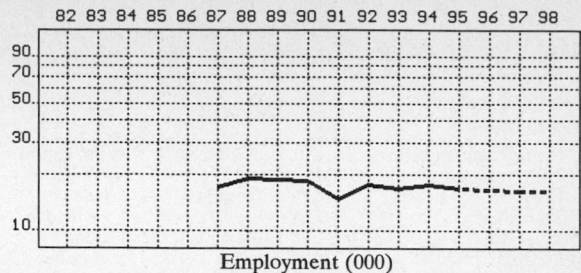

Employment (000)

GENERAL STATISTICS

| Year | Com-panies | Establishments | | Employment | | | Compensation | | Production ($ million) | | | |
		Total	with 20 or more employees	Total (000)	Production Workers (000)	Hours (Mil)	Payroll ($ mil)	Wages ($/hr)	Cost of Materials	Value Added by Manufacture	Value of Shipments	Capital Invest.
1982												
1983												
1984												
1985												
1986												
1987	246	260	124	17.1	9.6	19.2	440.9	10.17	639.8	1,142.4	1,769.3	52.3
1988		295	157	19.3	11.2	23.2	531.5	10.14	777.3	1,301.9	2,068.8	66.1
1989		262	131	19.0	9.8	20.3	519.0	10.57	728.0	1,238.7	1,969.8	58.1
1990		250	130	18.6	9.1	19.1	529.4	10.80	682.2	1,209.7	1,916.7	59.5
1991	246	246	123	14.8	6.9	14.6	485.7	11.41	650.5	1,100.1	1,782.5	52.7
1992	330	342	143	17.7	9.0	18.4	571.6	11.63	817.1	1,314.9	2,106.0	55.4
1993		346	143	16.9	8.2	16.7	567.1	11.86	818.3	1,364.3	2,172.7	42.9
1994		334P	139P	17.8	8.8	17.9	603.2	11.89	913.3	1,399.0	2,360.5	39.7
1995		346P	140P	16.9P	7.7P	15.9P	607.6P	12.37P	883.1P	1,352.8P	2,282.5P	41.6P
1996		358P	140P	16.7P	7.4P	15.3P	624.6P	12.66P	905.8P	1,387.6P	2,341.2P	39.0P
1997		370P	141P	16.5P	7.1P	14.7P	641.6P	12.95P	928.6P	1,422.4P	2,400.0P	36.4P
1998		382P	142P	16.3P	6.8P	14.1P	658.6P	13.24P	951.3P	1,457.2P	2,458.7P	33.8P

Sources: 1982, 1987, 1992 *Economic Census*; *Annual Survey of Manufactures*, 83-86, 88-91, 93-94. Establishment counts for non-Census years are from *County Business Patterns*; establishment values for 83-84 are extrapolations. 'P's show projections by the editors. Industries reclassified in 87 will not have data for prior years.

INDICES OF CHANGE

| Year | Com-panies | Establishments | | Employment | | | Compensation | | Production ($ million) | | | |
		Total	with 20 or more employees	Total (000)	Production Workers (000)	Hours (Mil)	Payroll ($ mil)	Wages ($/hr)	Cost of Materials	Value Added by Manufacture	Value of Shipments	Capital Invest.
1982												
1983												
1984												
1985												
1986												
1987	75	76	87	97	107	104	77	87	78	87	84	94
1988		86	110	109	124	126	93	87	95	99	98	119
1989		77	92	107	109	110	91	91	89	94	94	105
1990		73	91	105	101	104	93	93	83	92	91	107
1991		72	86	84	77	79	85	98	80	84	85	95
1992	100	100	100	100	100	100	100	100	100	100	100	100
1993		101	100	95	91	91	99	102	100	104	103	77
1994		98P	97P	101	98	97	106	102	112	106	112	72
1995		101P	98P	95P	86P	86P	106P	106P	108P	103P	108P	75P
1996		105P	98P	94P	83P	83P	109P	109P	111P	106P	111P	70P
1997		108P	99P	93P	79P	80P	112P	111P	114P	108P	114P	66P
1998		112P	99P	92P	76P	76P	115P	114P	116P	111P	117P	61P

Sources: Same as General Statistics. Values reflect change from the base year, 1992. Values above 100 mean greater than 92, values below 100 mean less than 92, and a value of 100 in the 82-91 or 93-98 period means same as 92. 'P's mark projections by the editors.

SELECTED RATIOS

For 1994	Avg. of All Manufact.	Analyzed Industry	Index	For 1994	Avg. of All Manufact.	Analyzed Industry	Index
Employees per Establishment	49	53	109	Value Added per Production Worker	134,084	158,977	119
Payroll per Establishment	1,500,273	1,806,761	120	Cost per Establishment	5,045,178	2,735,601	54
Payroll per Employee	30,620	33,888	111	Cost per Employee	102,970	51,309	50
Production Workers per Establishment	34	26	77	Cost per Production Worker	146,988	103,784	71
Wages per Establishment	853,319	637,491	75	Shipments per Establishment	9,576,895	7,070,389	74
Wages per Production Worker	24,861	24,185	97	Shipments per Employee	195,460	132,612	68
Hours per Production Worker	2,056	2,034	99	Shipments per Production Worker	279,017	268,239	96
Wages per Hour	12.09	11.89	98	Investment per Establishment	321,011	118,913	37
Value Added per Establishment	4,602,255	4,190,415	91	Investment per Employee	6,552	2,230	34
Value Added per Employee	93,930	78,596	84	Investment per Production Worker	9,352	4,511	48

Sources: Same as General Statistics. The 'Average of All Manufacturing' column represents the average of all manufacturing industries reported for the most recent complete year available. The Index shows the relationship between the Average and the Analyzed Industry. For example, 100 means that they are equal; 500 that the Analyzed Industry is five times the average; 50 means that the Analyzed Industry is half the national average. The abbreviation 'na' is used to show that data are 'not available'.

LEADING COMPANIES   Number shown: 47   Total sales ($ mil): 1,132   Total employment (000): 8.8

Company Name	Address				CEO Name	Phone	Co. Type	Sales ($ mil)	Empl. (000)
Edwards High Vacuum Intern	301 Ballardvale St	Wilmington	MA	01887	Mark Rosenzweig	508-658-5410	D	130	1.0
Fisher Hamilton Scientific Inc	PO Box 137	Two Rivers	WI	54241	Carl Bretko	414-793-1121	S	120	0.9
Newport Corp	1791 Deere Av	Irvine	CA	92714	Richard E Schmidt	714-863-3144	P	86	0.6
Corning Costar Corp	1 Alewife Ctr	Cambridge	MA	02140	Paul A Looney	617-868-6200	S	75	0.6
Thermotron Industries	291 Kollen Park Dr	Holland	MI	49423	DJ O'Keefe	616-393-4580	D	66•	0.5
Midmark Corp	60 Vista Dr	Versailles	OH	45380	James Eiting	513-526-3662	R	65	0.5
Forma Scientific Inc	PO Box 649	Marietta	OH	45750	Lewis Rosenblum	614-373-4763	S	55	0.4
Nova Biomedical Corp	200 Prospect St	Waltham	MA	02254	Francis Manganero	617-894-0800	R	55	0.5
Buehler Ltd	PO Box 1	Lake Bluff	IL	60044	Peter R Strong	708-295-6500	D	50	0.3
Barnstead Thermolyne Corp	2555 Kerper Blv	Dubuque	IA	52004	Ken Townsend	319-556-2241	S	46•	0.3
Zymark Corp	Zymark Ctr	Hopkinton	MA	01748	Frank Zenie	508-435-9500	R	40	0.3
Elkay Products Inc	3300 Hunting	Fort Wayne	IN	46809	Norman Loney	219-747-4102	R	35•	0.3
Labconco Corp	8811 Prospect Av	Kansas City	MO	64132	JN McConnell	816-333-8811	R	31•	0.2
Baker Company Inc	Sanford Airport	Sanford	ME	04073	D Eagleson	207-324-8773	R	29	0.2
CEM Corp	PO Box 200	Matthews	NC	28106	Michael J Collins	704-821-7015	P	29	0.2
Precision Scientific Inc	3737 W Cortland St	Chicago	IL	60647	Michael Bayles	312-227-2660	D	25	0.3
Savant Instruments Inc	100 Colin Dr	Holbrook	NY	11741	Ronald Robertson	516-244-2929	S	18	<0.1
Pasco Scientific Corp	10101 Foothills Blv	Roseville	CA	95678	Paul Stokstad	916-786-3800	R	15•	0.1
Hotpack Corp	10940 Dutton Rd	Philadelphia	PA	19154	Jay Silverman	215-824-1700	S	13	0.1
Jamestown Metal Products Inc	178 Blackstone Av	Jamestown	NY	14701	Charles Heinzelman	716-665-5313	R	13	0.1
Collegedale Casework Inc	PO Box 810	Collegedale	TN	37315	William C Cook	615-238-4131	R	12•	0.1
Virtis Company Inc	815 Rte 208	Gardiner	NY	12525	SG Bart Jr	914-255-5000	S	11	0.1
Rheodyne Inc	PO Box 996	Cotati	CA	94931	Lowell Gordon	707-664-9050	R	11	<0.1
General Equip Manufacturers	PO Box 836	Crystal Springs	MS	39059	Victor L Smith	601-892-2731	S	11	<0.1
Bellco Glass Inc	340 Edrudo Rd	Vineland	NJ	08360	Steven J Harker	609-691-1075	R	10•	0.2
United Metal Fabricators Inc	409 Eisenhower	Johnstown	PA	15904	Mark J Romano	814-266-8726	R	8	0.1
Burleigh Instruments Inc	Burleigh Park	Fishers	NY	14453	DJ Farrell	716-924-9355	R	7	<0.1
Glas-Col Apparatus Co	PO Box 2128	Terre Haute	IN	47802	RE Rickert	812-235-6167	S	7	<0.1
Wilkens-Anderson Co	4525 W Division St	Chicago	IL	60651	BE Wilkens	312-384-4433	R	7	<0.1
Misonix Inc	1938 New Hwy	Farmingdale	NY	11735	Michael Juliano	516-694-9555	P	7	<0.1
United Hospital Supply Corp	PO Box 1238	Burlington	NJ	08016		609-387-7580	R	6•	<0.1
Hyperion Inc	14100 SW 136th St	Miami	FL	33186	Eduard Botz	305-238-3020	R	5•	<0.1
Microgon Inc	23152 Verdugo Dr	Laguna Hills	CA	92653	James P Budimlya	714-581-3880	R	4•	<0.1
Pedigo Products Inc	4000 Columbia	Vancouver	WA	98661	Richard R Pedigo	206-695-3500	R	4•	<0.1
Perma Pure Inc	PO Box 2105	Toms River	NJ	08754	David A Leighty	908-244-0010	S	4	<0.1
Intelligent Enclosures Corp	4351 Shackleford	Norcross	GA	30093	Robert M Genco	404-564-5640	S	4	<0.1
B-J Enteprises Inc	173 Queen Av SE	Albany	OR	97321	James D Nydigger	503-926-9968	R	3	<0.1
CSC Scientific Company Inc	8315 Lee Hwy	Fairfax	VA	22031	A Gatenby	703-876-4030	R	3	<0.1
ThermoGenesis Corp	11431 Sunrise	R Cordova	CA	95742	Philip H Coelho	916-638-8357	P	3	<0.1
Marvac Scientific Manufacturing	2931 Cloverdale Av	Concord	CA	94518	George Marin	510-825-4636	R	2	<0.1
Scientech Inc	5649 Arapahoe Av	Boulder	CO	80303	TW Orourke	303-444-1361	R	2	<0.1
BioScan Inc	4590 MacArthur N	Washington	DC	20007	Seth D Shulman	202-338-0974	R	2	<0.1
Biomedical Devices Co	40 W Howard St	Pontiac	MI	48342	Denis A Ferkany	810-335-3535	R	1	<0.1
BioQuip Products Inc	17803 LaSalle Av	Gardena	CA	90248	Richard P Fall	310-324-0620	R	1•	<0.1
Science First Inc	95 Botsford Pl	Buffalo	NY	14216	Nancy Bell	716-874-0133	R	1	<0.1
Systec Inc	3816 Chandler Dr	Minneapolis	MN	55421	Thomas Thielen	612-788-9701	R	1•	<0.1
S and G Enterprises Inc	N115 Edison	Germantown	WI	53022	LC Griffith	414-251-8300	R	1	<0.1

Source: *Ward's Business Directory of U.S. Private and Public Companies*, Volumes 1 and 2, 1996. The company type code used is as follows: P - Public, R - Private, S - Subsidiary, D - Division, J - Joint Venture, A - Affiliate, G - Group. Sales are in millions of dollars, employees are in thousands. An asterisk (•) indicates an estimated sales volume. The symbol < stands for 'less than'. Company names and addresses are truncated, in some cases, to fit into the available space.

MATERIALS CONSUMED

Material	Quantity	Delivered Cost ($ million)
Materials, ingredients, containers, and supplies	(X)	685.1
Printed circuit boards (without inserted components) for electronic circuitry	(X)	14.8
Printed circuit assemblies, loaded boards or modules	(X)	29.8
Semiconductors, including transistors, diodes, rectifiers, and integrated circuits for electronic circuitry	(X)	7.5
Capacitors for electronic circuitry	(X)	1.9
Resistors for electronic circuitry	(X)	1.6
Other components and accessories for electronic circuitry, nec, except tubes	(X)	8.8
Current-carrying wiring devices	(X)	8.8
Electrical transmission, distribution, and control equipment	(X)	1.6
Electronic computing equipment	(X)	4.3
Electrical instrument mechanisms and meter movements (including instrument relays)	(X)	3.6
Electrical measuring instruments and parts, not listed elsewhere	(X)	1.7
Fractional horsepower electric motors and generators (less than 1 hp) including timing motors	(X)	7.2
Plastics resins consumed in the form of granules, pellets, powders, liquids, etc.	(X)	9.2
Fabricated plastics products (except gaskets, hoses, and belting)	(X)	11.8
Sheet metal products, except stampings	(X)	49.1
Metal stampings	(X)	1.3
Fabricated metal wire products (including wire rope, cable, springs, etc.)	(X)	1.2
Metal bolts, nuts, screws, washers, rivets, and other screw machine products	(X)	5.4
Other fabricated metal products (except forgings)	(X)	25.4
Forgings	(X)	(D)

Continued on next page.

MATERIALS CONSUMED - Continued

Material	Quantity	Delivered Cost ($ million)
Iron and steel castings (rough and semifinished)	(X)	1.6
Aluminum and aluminum-base alloy castings (rough and semifinished)	(X)	2.6
Other nonferrous castings (rough and semifinished)	(X)	(D)
Steel shapes and forms	(X)	15.6
Copper and copper-base alloy shapes and forms	(X)	0.8
Aluminum and aluminum-base alloy shapes and forms	(X)	3.0
Other nonferrous shapes and forms	(X)	2.1
Glass and glass products (excluding windows and mirrors)	(X)	8.7
Paper and paperboard products except paperboard boxes, containers, and corrugated paperboard	(X)	0.8
Paper and paperboard containers, including shipping sacks and other paper packaging supplies	(X)	6.9
All other materials and components, parts, containers, and supplies	(X)	288.2
Materials, ingredients, containers, and supplies, nsk	(X)	146.0

Source: 1992 Economic Census. Explanation of symbols used: (D): Withheld to avoid disclosure of competitive data; na: Not available; (S): Withheld because statistical norms were not met; (X): Not applicable; (Z): Less than half the unit shown; nec: Not elsewhere classified; nsk: Not specified by kind; - : zero; * : 10-19 percent estimated; ** : 20-29 percent estimated.

PRODUCT SHARE DETAILS

Product or Product Class	% Share	Product or Product Class	% Share
Laboratory apparatus and furniture	100.00	Laboratory furniture and parts sold separately.	15.67
Laboratory and scientific apparatus	76.84		

Source: 1992 Economic Census. The values shown are percent of total shipments in an industry. Values of indented subcategories are summed in the main heading. The symbol (D) appears when data are withheld to prevent disclosure of competitive information. The abbreviation nsk stands for 'not specified by kind' and nec for 'not elsewhere classified'.

INPUTS AND OUTPUTS FOR ENGINEERING & SCIENTIFIC INSTRUMENTS

Economic Sector or Industry Providing Inputs	%	Sector	Economic Sector or Industry Buying Outputs	%	Sector
Imports	14.6	Foreign	Gross private fixed investment	41.5	Cap Inv
Engineering & scientific instruments	12.1	Manufg.	Exports	26.4	Foreign
Wholesale trade	8.7	Trade	Federal Government purchases, national defense	7.5	Fed Govt
Electronic components nec	6.8	Manufg.	Engineering & scientific instruments	5.2	Manufg.
Veneer & plywood	4.7	Manufg.	Aircraft	4.2	Manufg.
Blast furnaces & steel mills	3.6	Manufg.	Ship building & repairing	2.4	Manufg.
Advertising	3.6	Services	Radio & TV communication equipment	2.3	Manufg.
Instruments to measure electricity	2.7	Manufg.	S/L Govt. purch., health & hospitals	2.1	S/L Govt
Semiconductors & related devices	2.4	Manufg.	Mechanical measuring devices	1.4	Manufg.
Communications, except radio & TV	2.1	Util.	Federal Government purchases, nondefense	1.3	Fed Govt
Electric services (utilities)	2.0	Util.	Change in business inventories	0.8	In House
Petroleum refining	1.9	Manufg.	S/L Govt. purch., higher education	0.6	S/L Govt
Eating & drinking places	1.8	Trade	S/L Govt. purch., police	0.6	S/L Govt
Motors & generators	1.7	Manufg.	S/L Govt. purch., sewerage	0.6	S/L Govt
Industrial patterns	1.4	Manufg.	S/L Govt. purch., sanitation	0.5	S/L Govt
Fabricated metal products, nec	1.3	Manufg.	Boat building & repairing	0.4	Manufg.
Machinery, except electrical, nec	1.2	Manufg.	S/L Govt. purch., elem. & secondary education	0.4	S/L Govt
Fabricated rubber products, nec	1.1	Manufg.	Blast furnaces & steel mills	0.3	Manufg.
Sheet metal work	1.1	Manufg.	Miscellaneous plastics products	0.3	Manufg.
Aluminum rolling & drawing	1.0	Manufg.	Watches, clocks, & parts	0.3	Manufg.
Miscellaneous plastics products	0.9	Manufg.	S/L Govt. purch., highways	0.3	S/L Govt
Screw machine and related products	0.9	Manufg.	S/L Govt. purch., natural resource & recreation.	0.3	S/L Govt
Banking	0.9	Fin/R.E.			
Legal services	0.9	Services			
Maintenance of nonfarm buildings nec	0.8	Constr.			
Electronic computing equipment	0.8	Manufg.			
Motor freight transportation & warehousing	0.8	Util.			
Equipment rental & leasing services	0.8	Services			
Glass & glass products, except containers	0.7	Manufg.			
Paperboard containers & boxes	0.7	Manufg.			
Real estate	0.7	Fin/R.E.			
Management & consulting services & labs	0.7	Services			
U.S. Postal Service	0.7	Gov't			
Aluminum castings	0.6	Manufg.			
Gaskets, packing & sealing devices	0.6	Manufg.			
Special dies & tools & machine tool accessories	0.6	Manufg.			
Ball & roller bearings	0.5	Manufg.			
Paints & allied products	0.5	Manufg.			
Manifold business forms	0.4	Manufg.			
Metal stampings, nec	0.4	Manufg.			
Gas production & distribution (utilities)	0.4	Util.			
Accounting, auditing & bookkeeping	0.4	Services			
Converted paper products, nec	0.3	Manufg.			
Lighting fixtures & equipment	0.3	Manufg.			

Continued on next page.

INPUTS AND OUTPUTS FOR ENGINEERING & SCIENTIFIC INSTRUMENTS - Continued

Economic Sector or Industry Providing Inputs	%	Sector	Economic Sector or Industry Buying Outputs	%	Sector
Railroads & related services	0.3	Util.			
Automotive rental & leasing, without drivers	0.3	Services			
Business/professional associations	0.3	Services			
Computer & data processing services	0.3	Services			
Engineering, architectural, & surveying services	0.3	Services			
Abrasive products	0.2	Manufg.			
Iron & steel foundries	0.2	Manufg.			
Lubricating oils & greases	0.2	Manufg.			
Machine tools, metal cutting types	0.2	Manufg.			
Nonferrous castings, nec	0.2	Manufg.			
Photographic equipment & supplies	0.2	Manufg.			
Plastics materials & resins	0.2	Manufg.			
Plating & polishing	0.2	Manufg.			
Sawmills & planning mills, general	0.2	Manufg.			
Transformers	0.2	Manufg.			
Wiring devices	0.2	Manufg.			
Air transportation	0.2	Util.			
Credit agencies other than banks	0.2	Fin/R.E.			
Insurance carriers	0.2	Fin/R.E.			
Royalties	0.2	Fin/R.E.			
Automotive repair shops & services	0.2	Services			
Photofinishing labs, commercial photography	0.2	Services			
Theatrical producers, bands, entertainers	0.2	Services			
Electron tubes	0.1	Manufg.			
Miscellaneous fabricated wire products	0.1	Manufg.			
Rubber & plastics hose & belting	0.1	Manufg.			
Transit & bus transportation	0.1	Util.			
Water supply & sewage systems	0.1	Util.			
Water transportation	0.1	Util.			
Motion pictures	0.1	Services			
Personnel supply services	0.1	Services			
Noncomparable imports	0.1	Foreign			

Source: Benchmark Input-Output Accounts for the U.S. Economy, 1982, U.S. Department of Commerce, Washington, D.C., July 1991. Data, as reported in the source, are organized by the 1977 SIC structure in use in 1982 but have been matched, as closely as is possible, to the 1987 SIC structure used in this book.

OCCUPATIONS EMPLOYED BY SIC 382 - MEASURING AND CONTROLLING DEVICES

Occupation	% of Total 1994	Change to 2005	Occupation	% of Total 1994	Change to 2005
Assemblers, fabricators, & hand workers nec	7.4	-11.4	Engineering, mathematical, & science managers	1.7	0.6
Electrical & electronics engineers	5.6	13.4	Industrial production managers	1.5	-11.4
Electrical & electronic equipment assemblers	5.4	-11.4	General office clerks	1.4	-24.4
Sales & related workers nec	4.3	-11.4	Bookkeeping, accounting, & auditing clerks	1.4	-33.6
Electrical & electronic technicians,technologists	3.7	-11.4	Purchasing agents, ex trade & farm products	1.3	-11.4
Electrical & electronic assemblers	3.5	-20.3	Production, planning, & expediting clerks	1.3	-11.4
Inspectors, testers, & graders, precision	3.3	-38.0	Traffic, shipping, & receiving clerks	1.3	-14.8
Electromechanical equipment assemblers	3.2	-2.5	Drafters	1.3	-31.0
Blue collar worker supervisors	3.2	-24.1	Computer engineers	1.3	31.2
General managers & top executives	2.4	-15.9	Stock clerks	1.2	-28.0
Secretaries, ex legal & medical	2.4	-19.3	Engineers nec	1.2	6.3
Machinists	2.0	-33.6	Management support workers nec	1.1	-11.4
Mechanical engineers	1.8	-2.5	Managers & administrators nec	1.1	-11.4
Engineering technicians & technologists nec	1.7	-11.4	Computer programmers	1.1	-28.2
Marketing, advertising, & PR managers	1.7	-11.4			

Source: Industry-Occupation Matrix, Bureau of Labor Statistics. These data relate to one or more 3-digit SIC industry groups rather than to a single 4-digit SIC. The change reported for each occupation to the year 2005 is a percent of growth or decline as estimated by the Bureau of Labor Statistics. The abbreviation nec stands for 'not elsewhere classified'.

LOCATION BY STATE AND REGIONAL CONCENTRATION

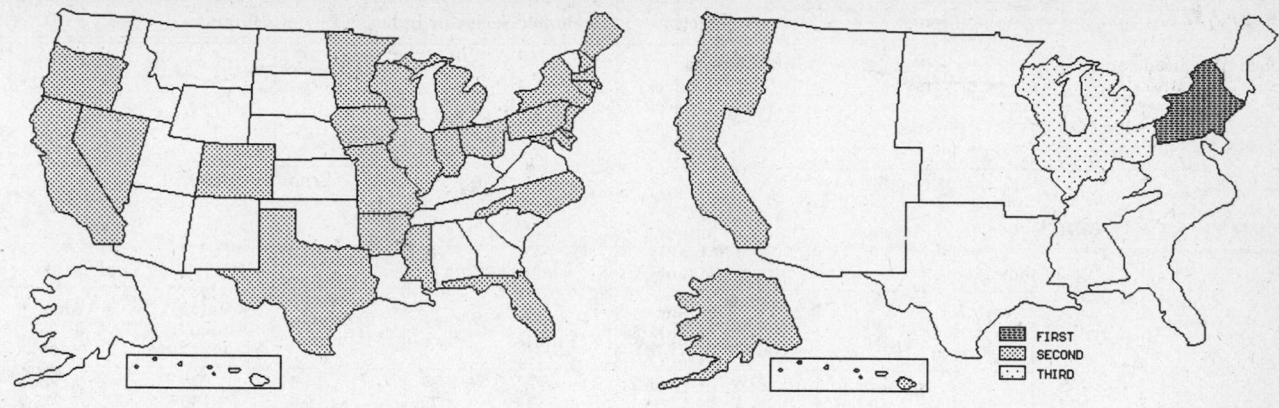

FIRST
SECOND
THIRD

INDUSTRY DATA BY STATE

| State | Establish-ments | Shipments | | | Employment | | | | Cost as % of Shipments | Investment per Employee ($) |
		Total ($ mil)	% of U.S.	Per Establ.	Total Number	% of U.S.	Per Establ.	Wages ($/hour)		
California	64	328.2	15.6	5.1	2,600	14.7	41	12.08	31.4	7,423
Wisconsin	9	150.4	7.1	16.7	1,400	7.9	156	11.60	40.5	3,357
Pennsylvania	28	149.0	7.1	5.3	1,600	9.0	57	13.28	42.1	2,063
New Jersey	23	141.9	6.7	6.2	1,500	8.5	65	13.33	38.4	1,067
Illinois	18	136.2	6.5	7.6	1,200	6.8	67	12.07	43.5	1,667
New York	24	133.6	6.3	5.6	1,000	5.6	42	9.55	36.5	3,000
Ohio	14	93.1	4.4	6.7	800	4.5	57	14.33	39.5	-
Massachusetts	25	93.0	4.4	3.7	700	4.0	28	13.83	34.4	2,143
Minnesota	7	57.5	2.7	8.2	500	2.8	71	12.40	28.5	3,200
Texas	12	52.1	2.5	4.3	500	2.8	42	9.57	40.9	2,000
Florida	13	25.6	1.2	2.0	300	1.7	23	8.67	30.1	-
Maryland	9	21.7	1.0	2.4	200	1.1	22	11.00	31.3	1,500
Indiana	9	9.8	0.5	1.1	100	0.6	11	11.00	34.7	2,000
Oregon	6	8.7	0.4	1.5	100	0.6	17	9.00	48.3	-
Michigan	13	(D)	-	-	750 *	4.2	58	-	-	800
Connecticut	7	(D)	-	-	375 *	2.1	54	-	-	-
North Carolina	7	(D)	-	-	750 *	4.2	107	-	-	-
Colorado	5	(D)	-	-	175 *	1.0	35	-	-	-
Delaware	5	(D)	-	-	1,750 *	9.9	350	-	-	-
Missouri	5	(D)	-	-	175 *	1.0	35	-	-	-
New Hampshire	5	(D)	-	-	375 *	2.1	75	-	-	-
Mississippi	4	(D)	-	-	175 *	1.0	44	-	-	-
Iowa	3	(D)	-	-	375 *	2.1	125	-	-	-
Nevada	2	(D)	-	-	375 *	2.1	188	-	-	-
Arkansas	1	(D)	-	-	175 *	1.0	175	-	-	-
Maine	1	(D)	-	-	175 *	1.0	175	-	-	-

Source: 1992 *Economic Census*. The states are in descending order of shipments or establishments (if shipment data are missing for the majority). The symbol (D) appears when data are withheld to prevent disclosure of competitive information. States marked with (D) are sorted by number of establishments. A dash (-) indicates that the data element cannot be calculated; * indicates the midpoint of a range.

3822 - ENVIRONMENTAL CONTROLS

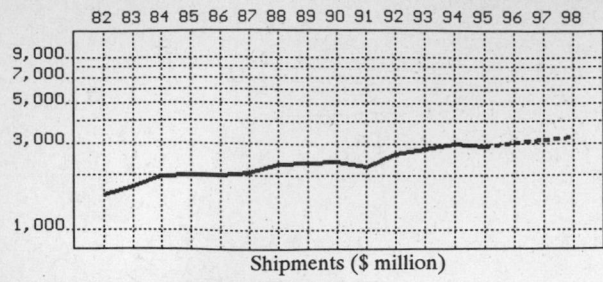

Shipments ($ million)

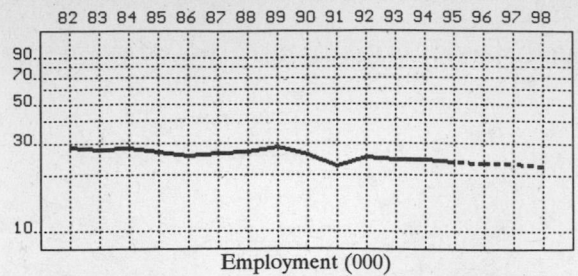

Employment (000)

GENERAL STATISTICS

Year	Companies	Establishments		Employment			Compensation		Production ($ million)			
		Total	with 20 or more employees	Total (000)	Production Workers (000)	Hours (Mil)	Payroll ($ mil)	Wages ($/hr)	Cost of Materials	Value Added by Manufacture	Value of Shipments	Capital Invest.
1982	221	245	89	28.8	20.6	36.2	497.5	8.34	514.3	1,025.7	1,549.1	66.8
1983		237	93	27.9	20.4	38.1	539.5	8.72	616.6	1,130.5	1,745.2	67.7
1984		229	97	28.2	20.9	38.5	574.2	9.32	684.2	1,303.6	1,966.1	57.9
1985		222	100	27.1	19.5	36.9	580.9	9.64	669.8	1,318.2	1,989.3	63.6
1986		229	100	25.8	18.5	35.7	575.6	9.81	687.3	1,278.2	1,990.4	49.8
1987	233	254	106	26.5	18.6	36.2	602.4	9.87	759.9	1,302.7	2,068.8	66.3
1988		259	122	27.1	19.5	38.0	643.8	10.09	861.4	1,444.3	2,291.1	57.0
1989		242	115	28.5	18.2	35.4	613.7	10.21	889.0	1,471.5	2,336.3	66.3
1990		243	116	26.0	18.2	35.2	664.8	10.42	934.2	1,461.6	2,396.0	61.2
1991		258	118	22.5	14.9	27.7	615.2	11.29	892.4	1,297.7	2,243.7	56.0
1992	294	318	130	25.0	16.8	32.1	685.4	11.09	997.1	1,633.0	2,607.1	81.3
1993		329	144	24.4	17.2	33.2	697.0	11.29	1,069.2	1,732.9	2,812.9	87.6
1994		301P	139P	24.5	17.2	33.7	691.7	11.30	1,122.1	1,840.1	2,967.9	75.9
1995		308P	143P	23.7P	15.9P	31.5P	717.4P	11.81P	1,107.8P	1,816.7P	2,930.2P	75.5P
1996		315P	147P	23.4P	15.5P	31.0P	732.2P	12.05P	1,145.8P	1,878.9P	3,030.5P	76.8P
1997		322P	152P	23.0P	15.2P	30.5P	747.0P	12.29P	1,183.7P	1,941.1P	3,130.8P	78.2P
1998		329P	156P	22.6P	14.8P	30.0P	761.8P	12.54P	1,221.6P	2,003.3P	3,231.1P	79.6P

Sources: 1982, 1987, 1992 *Economic Census*; *Annual Survey of Manufactures*, 83-86, 88-91, 93-94. Establishment counts for non-Census years are from *County Business Patterns*; establishment values for 83-84 are extrapolations. 'P's show projections by the editors. Industries reclassified in 87 will not have data for prior years.

INDICES OF CHANGE

Year	Companies	Establishments		Employment			Compensation		Production ($ million)			
		Total	with 20 or more employees	Total (000)	Production Workers (000)	Hours (Mil)	Payroll ($ mil)	Wages ($/hr)	Cost of Materials	Value Added by Manufacture	Value of Shipments	Capital Invest.
1982	75	77	68	115	123	113	73	75	52	63	59	82
1983		75	72	112	121	119	79	79	62	69	67	83
1984		72	75	113	124	120	84	84	69	80	75	71
1985		70	77	108	116	115	85	87	67	81	76	78
1986		72	77	103	110	111	84	88	69	78	76	61
1987	79	80	82	106	111	113	88	89	76	80	79	82
1988		81	94	108	116	118	94	91	86	88	88	70
1989		76	88	114	108	110	90	92	89	90	90	82
1990		76	89	104	108	110	97	94	94	90	92	75
1991		81	91	90	89	86	90	102	89	79	86	69
1992	100	100	100	100	100	100	100	100	100	100	100	100
1993		103	111	98	102	103	102	102	107	106	108	108
1994		95P	107P	98	102	105	101	102	113	113	114	93
1995		97P	110P	95P	95P	98P	105P	106P	111P	111P	112P	93P
1996		99P	113P	93P	92P	97P	107P	109P	115P	115P	116P	95P
1997		101P	117P	92P	90P	95P	109P	111P	119P	119P	120P	96P
1998		103P	120P	90P	88P	93P	111P	113P	123P	123P	124P	98P

Sources: Same as General Statistics. Values reflect change from the base year, 1992. Values above 100 mean greater than 92, values below 100 mean less than 92, and a value of 100 in the 82-91 or 93-98 period means same as 92. 'P's mark projections by the editors.

SELECTED RATIOS

For 1994	Avg. of All Manufact.	Analyzed Industry	Index	For 1994	Avg. of All Manufact.	Analyzed Industry	Index
Employees per Establishment	49	81	166	Value Added per Production Worker	134,084	106,983	80
Payroll per Establishment	1,500,273	2,298,122	153	Cost per Establishment	5,045,178	3,728,095	74
Payroll per Employee	30,620	28,233	92	Cost per Employee	102,970	45,800	44
Production Workers per Establishment	34	57	166	Cost per Production Worker	146,988	65,238	44
Wages per Establishment	853,319	1,265,213	148	Shipments per Establishment	9,576,895	9,860,629	103
Wages per Production Worker	24,861	22,140	89	Shipments per Employee	195,460	121,139	62
Hours per Production Worker	2,056	1,959	95	Shipments per Production Worker	279,017	172,552	62
Wages per Hour	12.09	11.30	93	Investment per Establishment	321,011	252,172	79
Value Added per Establishment	4,602,255	6,113,597	133	Investment per Employee	6,552	3,098	47
Value Added per Employee	93,930	75,106	80	Investment per Production Worker	9,352	4,413	47

Sources: Same as General Statistics. The 'Average of All Manufacturing' column represents the average of all manufacturing industries reported for the most recent complete year available. The Index shows the relationship between the Average and the Analyzed Industry. For example, 100 means that they are equal; 500 that the Analyzed Industry is five times the average; 50 means that the Analyzed Industry is half the national average. The abbreviation 'na' is used to show that data are 'not available'.

LEADING COMPANIES   Number shown: **75**   Total sales ($ mil): **10,653**   Total employment (000): **90.5**

Company Name	Address				CEO Name	Phone	Co. Type	Sales ($ mil)	Empl. (000)
Honeywell Inc	PO Box 524	Minneapolis	MN	55440	M R Bonsignore	612-951-1000	P	6,057	50.8
Hamilton Standard Electr Syst	1 Hamilton Rd	Windsor Locks	CT	06096	Robert N Kuhn	203-654-6000	S	1,000	11.0
Robertshaw Controls Co	2809 Emerywood	Richmond	VA	23294	Leslie J Jezuit	804-756-6500	S	650•	6.5
Landis and Gyr Inc	1000 Deerfield	Buffalo Grove	IL	60089	John Grad	708-215-1022	S	527	4.0
Therm-O-Disc Inc	1320 S Main St	Mansfield	OH	44907	JA Kight	419-525-8500	S	310	1.1
Watsco Inc	2665 S Bayshore Dr	Coconut Grove	FL	33133	Albert H Nahmad	305-858-0828	P	284	0.9
Research-Cottrell Companies	PO Box 1500	Somerville	NJ	08876	George Mammola	908-685-4000	S	277	0.8
White-Rodgers	9797 Reavis Rd	St Louis	MO	63123	John Cichy	314-577-1300	D	250	3.0
Sterling Inc	5200 W Clinton Av	Milwaukee	WI	53223	Roger Lang	414-354-0970	R	150	0.8
Robertshaw Controls Co	1 Robertshaw Dr	New Stanton	PA	15672	David F Rochette	412-925-7211	D	87	1.5
Grayson Controls	100 W Victoria St	Long Beach	CA	90805	John Hartsworm	310-638-6111	D	81	0.8
Simicon	11768 James St	Holland	MI	49424	Don Naab	616-396-1467	D	70	0.4
Andover Controls Corp	300 Brickstone Sq	Andover	MA	01810	William J Lapointe	508-470-0555	S	65	0.2
CSI Control Systems Intern	PO Box 59469	Dallas	TX	75229	Wayne Stevenson	214-323-1111	R	60	0.4
Dwyer Instruments Inc	PO Box 373	Michigan City	IN	46361	Steve Clark	219-879-8000	R	60	0.8
Indak Manufacturing Corp	1915 Techny Rd	Northbrook	IL	60062	Martin R Cobb	708-272-0343	R	54	0.7
Channel Products Inc	7100 Wilson Mills	Chesterland	OH	44026	Ken Pim	216-423-0113	S	40	0.4
Omega Engineering Inc	PO Box 407	Stamford	CT	06907	BR Hollander	203-359-1660	R	40•	0.4
Jameson Home Products Inc	2820 Thatcher Rd	Downers Grove	IL	60515	J McCrink	708-963-2850	R	38	0.4
Yellow Springs Instrument Inc	PO Box 279	Yellow Springs	OH	45387	M Von Matthiessen	513-767-7241	R	30	0.3
Peco Manufacturing Company	PO Box 82189	Portland	OR	97282	MJ McChord	503-233-6401	R	25	0.3
Stewart Warner Instrument	580 Slawin Ct	Mt Prospect	IL	60056	Mike Benner	708-803-0200	S	25	0.4
Novar Controls Corp	24 Brown St	Barberton	OH	44203	Dominic Federico	216-745-0074	S	23•	0.2
Marshalltown Instruments Inc	108 S Colorado Av	Hastings	NE	68901	Bill Cunningham	402-463-6851	S	20	0.3
Partlow Corp	2 Campion Rd	New Hartford	NY	13413	Lawrence C Curtis	315-797-2222	S	20	0.2
Magnum Corp	32400 Telegraph Rd	Bingham Farms	MI	48025	Martin Abel	810-433-1170	R	19	0.2
Powers Process Controls	3400 Oakton St	Skokie	IL	60076	HL Singer	708-673-6700	S	18	0.1
Anarad Inc	534 E Ortega St	Santa Barbara	CA	93103	Don E Burrows	805-963-6583	S	15	<0.1
Columbus Electric Mfg Co	485 Indrial Pk Rd	Piney Flats	TN	37686	Leonard Runyan	615-538-8191	S	15•	0.2
Cox and Company Inc	200 Varick St	New York	NY	10014	W Achenbaum	212-366-0200	R	15•	0.2
Scully Signal Co	70 Industrial Way	Wilmington	MA	01887	RG Scully	617-729-7510	R	15	0.1
Phoenix Controls Corp	55 Chapel St	Newton	MA	02158	Gordon Sharp	617-964-6670	R	14	0.1
J and J Register	12504 Weaver Rd	El Paso	TX	79927	C Smith	915-852-9111	D	12•	0.3
Kreuter Manufacturing Company	PO Box 497	New Paris	IN	46553	Wayne Kehler	219-831-5250	R	12	0.1
Sensors and Switches Inc	PO Box 3297	Lexington	OH	44904	Ed Costin	419-884-1311	S	12•	0.3
Athena Controls Inc	5145 Campus Dr	Plym Meeting	PA	19462	Thomas McCool	610-828-2490	S	11	<0.1
Automated Logic Corp	1283 Kennest	Marietta	GA	30066	Gerry Hull	404-423-7474	R	11•	<0.1
Pyromation Inc	PO Box 5601	Fort Wayne	IN	46895	Pete Wilson	219-484-2580	R	11	<0.1
Ryan Instruments LP	PO Box 599	Redmond	WA	98073	John P Vache	206-883-7926	R	11	<0.1
Mercury Instruments Inc	3940 Virginia Av	Cincinnati	OH	45227	Richard Hannon	513-272-1111	R	10•	0.1
Pneu Devices Inc	72 Santa Felicia Dr	Santa Barbara	CA	93117	Merlin E Rossow	805-968-0702	S	10	<0.1
Syscon International Inc	1108 S High St	South Bend	IN	46601	Steven J Thomas	219-232-3900	R	10	<0.1
Syscon-RKC	1108 S High St	South Bend	IN	46601	Steven J Thomas	219-232-4836	D	10	<0.1
Temptronic Corp	55 Chapel St	Newton	MA	02158	TG Gerendas	617-969-2501	R	10	<0.1
FasTest Inc	2315 Hampden Av	St Paul	MN	55114	Stanlee Meisinger	612-645-6266	R	9•	<0.1
Nexus Custom Electronics Inc	PO Box 250	Brandon	VT	05733	Victor Giglio	802-247-6811	S	9	0.1
Amtech Systems Inc	131 S Clark Dr	Tempe	AZ	85281	Jong S Whang	602-967-5146	P	8	0.2
Meriam Instrument	10920 Madison Av	Cleveland	OH	44102	Charles O'Brien	216-281-1100	D	8	<0.1
Flair International Corp	600 O Willets	Hauppauge	NY	11788	Jamie Frank	516-234-3600	R	8	0.1
Hartel Corp	PO Box 41	Fort Atkinson	WI	53538	Douglas E Hartel	414-563-8461	R	8	<0.1
Tenney Engineering Inc	1090 Springfield Rd	Union	NJ	07083	Robert S Schiffman	908-686-7870	P	7	<0.1
Airguide Instrument Co	2210 W Wabansia	Chicago	IL	60647	John Alstadt	312-486-3000	S	7•	<0.1
Bimet Corp	PO Box 518	Morris	IL	60450	JA Peterson	815-942-2600	R	7	0.1
PSG Industries Inc	PO Box 157	Perkasie	PA	18944	HN Bender	215-257-3621	R	6	0.1
Russells Technical Products Inc	1145 S Washington	Holland	MI	49423	Gary Molenaar	616-392-3161	R	6	<0.1
Tempset Inc	4204 Miami St	St Louis	MO	63116	VG Gaskell Sr	314-772-8855	R	6	0.1
Trig Inc	2247 US Hwy 127 S	Frankfort	KY	40601	Harold O Henning	502-227-4531	R	6	0.2
Sutron Corp	21300 Ridgetop Cir	Sterling	VA	20166	Raul S McQuivey	703-406-2800	P	5	<0.1
ACDC Inc	PO Box 230	Milford	OH	45150	David S Parker	513-248-1820	R	5•	<0.1
Adams Manufacturing Co	9790 Midwest Av	Cleveland	OH	44125	M L Schonberger	216-587-6801	R	5	<0.1
Brule CE and E Inc	13920 Western Av	Blue Island	IL	60406	Charles W Friedrich	708-388-7900	R	5	<0.1
Cormetech Inc	5000 International	Durham	NC	27712	Gerhard F Koenig	919-620-3000	J	5	<0.1
Inncom International Inc	PO Box 966	Old Lyme	CT	06371	D W Buckingham	203-434-7777	R	5	<0.1
Pressure Cool Co	83-801 Av 45	Indio	CA	92202	DA York	619-347-2366	D	5	0.1
Voltec Inc	3075 Washington	McMurray	PA	15317	Larry J Kucera	412-941-5360	S	5	<0.1
Walter G Legge Company Inc	444 Central Av	Peekskill	NY	10566	Walter Wowtschuk	914-737-5040	R	5	<0.1
Anello Corp	2641 Walnut Av	Tustin	CA	92680	Peter J Anello	714-669-9940	D	4	<0.1
Budzar Industries Inc	26953 Tungsten Rd	Euclid	OH	44132	David Young	216-261-9191	R	4	<0.1
LIGHTSTAT Inc	PO Box 326	Canton	CT	06019	R H Eigenbrod	203-693-2444	R	4	<0.1
Teletrol Systems Inc	324 Commercial St	Manchester	NH	03101	John Petze	603-645-6061	R	4	<0.1
Allteq Industries Inc	930 Auburn Ct	Fremont	CA	94538	Larry Chiponis	510-770-8181	R	3	<0.1
Crest Manufacturing Co	5 Hood Dr	Lincoln	RI	02865	Gary Hood	401-333-1350	R	3•	<0.1
PPC Industries Inc	3000 E Marshall	Longview	TX	75601	William Fisher	903-758-3395	S	3•	<0.1
Swanson Eng & Mfg Co	1133 E Redondo	Inglewood	CA	90302	Nels E Swanson	213-678-5100	R	3	<0.1
AET Systems Inc	77 Accord Park Dr	Norwell	MA	02061	James R Andrew	617-871-4801	R	2	<0.1

Source: Ward's Business Directory of U.S. Private and Public Companies, Volumes 1 and 2, 1996. The company type code used is as follows: P - Public, R - Private, S - Subsidiary, D - Division, J - Joint Venture, A - Affiliate, G - Group. Sales are in millions of dollars, employees are in thousands. An asterisk (*) indicates an estimated sales volume. The symbol < stands for 'less than'. Company names and addresses are truncated, in some cases, to fit into the available space.

MATERIALS CONSUMED

Material	Quantity	Delivered Cost ($ million)
Materials, ingredients, containers, and supplies	(X)	805.7
Printed circuit boards (without inserted components) for electronic circuitry	(X)	30.4
Printed circuit assemblies, loaded boards or modules	(X)	8.7
Semiconductors, including transistors, diodes, rectifiers, and integrated circuits for electronic circuitry	(X)	41.6
Capacitors for electronic circuitry	(X)	9.0
Resistors for electronic circuitry	(X)	7.6
Other components and accessories for electronic circuitry, nec, except tubes	(X)	47.5
Current-carrying wiring devices	(X)	12.3
Electrical transmission, distribution, and control equipment	(X)	7.4
Electronic computing equipment	(X)	7.2
Electrical instrument mechanisms and meter movements (including instrument relays)	(X)	2.0
Fractional horsepower electric motors and generators (less than 1 hp) including timing motors	(X)	8.4
Plastics resins consumed in the form of granules, pellets, powders, liquids, etc.	(X)	19.4
Fabricated plastics products (except gaskets, hoses, and belting)	(X)	24.2
Sheet metal products, except stampings	(X)	21.4
Metal stampings	(X)	33.5
Fabricated metal wire products (including wire rope, cable, springs, etc.)	(X)	11.2
Metal bolts, nuts, screws, washers, rivets, and other screw machine products	(X)	38.2
Other fabricated metal products (except forgings)	(X)	27.1
Forgings	(X)	0.1
Iron and steel castings (rough and semifinished)	(X)	4.7
Aluminum and aluminum-base alloy castings (rough and semifinished)	(X)	34.0
Other nonferrous castings (rough and semifinished)	(X)	5.2
Steel shapes and forms	(X)	14.8
Copper and copper-base alloy shapes and forms	(X)	18.3
Aluminum and aluminum-base alloy shapes and forms	(X)	9.5
Other nonferrous shapes and forms	(X)	7.2
Glass and glass products (excluding windows and mirrors)	(X)	2.3
Paper and paperboard products except paperboard boxes, containers, and corrugated paperboard	(X)	3.1
Paper and paperboard containers, including shipping sacks and other paper packaging supplies	(X)	7.7
All other materials and components, parts, containers, and supplies	(X)	141.3
Materials, ingredients, containers, and supplies, nsk	(X)	200.5

Source: 1992 *Economic Census*. Explanation of symbols used: (D): Withheld to avoid disclosure of competitive data; na: Not available; (S): Withheld because statistical norms were not met; (X): Not applicable; (Z): Less than half the unit shown; nec: Not elsewhere classified; nsk: Not specified by kind; - : zero; * : 10-19 percent estimated; ** : 20-29 percent estimated.

PRODUCT SHARE DETAILS

Product or Product Class	% Share	Product or Product Class	% Share
Environmental controls	100.00		

Source: 1992 *Economic Census*. The values shown are percent of total shipments in an industry. Values of indented subcategories are summed in the main heading. The symbol (D) appears when data are withheld to prevent disclosure of competitive information. The abbreviation nsk stands for 'not specified by kind' and nec for 'not elsewhere classified'.

INPUTS AND OUTPUTS FOR ENVIRONMENTAL CONTROLS

Economic Sector or Industry Providing Inputs	%	Sector	Economic Sector or Industry Buying Outputs	%	Sector
Cyclic crudes and organics	11.1	Manufg.	Nonfarm residential structure maintenance	9.7	Constr.
Wholesale trade	10.4	Trade	Office buildings	9.7	Constr.
Electronic components nec	8.6	Manufg.	Maintenance of nonfarm buildings nec	9.4	Constr.
Blast furnaces & steel mills	4.7	Manufg.	Refrigeration & heating equipment	9.3	Manufg.
Metal stampings, nec	3.4	Manufg.	Exports	7.8	Foreign
Petroleum refining	3.0	Manufg.	Residential 1-unit structures, nonfarm	6.3	Constr.
Plating & polishing	2.9	Manufg.	Industrial buildings	5.3	Constr.
Screw machine and related products	2.8	Manufg.	Household cooking equipment	4.8	Manufg.
Advertising	2.7	Services	Construction of hospitals	4.5	Constr.
Electric services (utilities)	2.6	Util.	Household appliances, nec	3.8	Manufg.
Copper rolling & drawing	2.0	Manufg.	Construction of stores & restaurants	3.3	Constr.
Eating & drinking places	1.9	Trade	Construction of educational buildings	2.8	Constr.
Imports	1.8	Foreign	Residential additions/alterations, nonfarm	2.6	Constr.
Communications, except radio & TV	1.6	Util.	Heating equipment, except electric	2.3	Manufg.
Fabricated rubber products, nec	1.5	Manufg.	Household laundry equipment	2.0	Manufg.
Miscellaneous plastics products	1.4	Manufg.	Household refrigerators & freezers	2.0	Manufg.
Industrial inorganic chemicals, nec	1.3	Manufg.	Residential garden apartments	1.9	Constr.
Instruments to measure electricity	1.3	Manufg.	Electric housewares & fans	1.9	Manufg.
Machinery, except electrical, nec	1.3	Manufg.	Hotels & motels	1.6	Constr.
U.S. Postal Service	1.2	Gov't	Construction of nonfarm buildings nec	1.2	Constr.
Aluminum castings	1.1	Manufg.	Blowers & fans	1.0	Manufg.
Electronic computing equipment	1.1	Manufg.	Federal Government purchases, national defense	1.0	Fed Govt
Aluminum rolling & drawing	1.0	Manufg.	Warehouses	0.8	Constr.
Nonferrous wire drawing & insulating	1.0	Manufg.	Amusement & recreation building construction	0.7	Constr.
Motors & generators	0.9	Manufg.	Resid. & other health facility construction	0.7	Constr.

Continued on next page.

INPUTS AND OUTPUTS FOR ENVIRONMENTAL CONTROLS - Continued

Economic Sector or Industry Providing Inputs	%	Sector	Economic Sector or Industry Buying Outputs	%	Sector
Semiconductors & related devices	0.9	Manufg.	Construction of religious buildings	0.5	Constr.
Gas production & distribution (utilities)	0.9	Util.	Residential 2-4 unit structures, nonfarm	0.5	Constr.
Motor freight transportation & warehousing	0.9	Util.	Electric utility facility construction	0.4	Constr.
Banking	0.9	Fin/R.E.	Maintenance of farm service facilities	0.4	Constr.
Legal services	0.9	Services	Residential high-rise apartments	0.4	Constr.
Maintenance of nonfarm buildings nec	0.8	Constr.	Household vacuum cleaners	0.4	Manufg.
Ball & roller bearings	0.8	Manufg.	Telephone & telegraph facility construction	0.3	Constr.
Miscellaneous fabricated wire products	0.8	Manufg.	Dormitories & other group housing	0.2	Constr.
Plastics materials & resins	0.8	Manufg.	Federal Government purchases, nondefense	0.2	Fed Govt
Wiring devices	0.8	Manufg.	Local transit facility construction	0.1	Constr.
Real estate	0.8	Fin/R.E.			
Gaskets, packing & sealing devices	0.7	Manufg.			
Paperboard containers & boxes	0.7	Manufg.			
Signs & advertising displays	0.7	Manufg.			
Special dies & tools & machine tool accessories	0.7	Manufg.			
Transformers	0.7	Manufg.			
Equipment rental & leasing services	0.7	Services			
Management & consulting services & labs	0.7	Services			
Commercial printing	0.6	Manufg.			
Metal coating & allied services	0.6	Manufg.			
Automotive rental & leasing, without drivers	0.6	Services			
Industrial controls	0.5	Manufg.			
Sheet metal work	0.5	Manufg.			
Accounting, auditing & bookkeeping	0.5	Services			
Fabricated metal products, nec	0.4	Manufg.			
Manifold business forms	0.4	Manufg.			
Nonferrous castings, nec	0.4	Manufg.			
Nonferrous rolling & drawing, nec	0.4	Manufg.			
Railroads & related services	0.4	Util.			
Automotive repair shops & services	0.4	Services			
Computer & data processing services	0.4	Services			
Iron & steel foundries	0.3	Manufg.			
Machine tools, metal cutting types	0.3	Manufg.			
Manufacturing industries, nec	0.3	Manufg.			
Sawmills & planning mills, general	0.3	Manufg.			
Air transportation	0.3	Util.			
Water transportation	0.3	Util.			
Credit agencies other than banks	0.3	Fin/R.E.			
Insurance carriers	0.3	Fin/R.E.			
Business/professional associations	0.3	Services			
Engineering, architectural, & surveying services	0.3	Services			
Abrasive products	0.2	Manufg.			
Alkalies & chlorine	0.2	Manufg.			
Brass, bronze, & copper castings	0.2	Manufg.			
Lubricating oils & greases	0.2	Manufg.			
Machine tools, metal forming types	0.2	Manufg.			
Photographic equipment & supplies	0.2	Manufg.			
Security & commodity brokers	0.2	Fin/R.E.			
Personnel supply services	0.2	Services			
Photofinishing labs, commercial photography	0.2	Services			
Motor vehicle parts & accessories	0.1	Manufg.			
Tires & inner tubes	0.1	Manufg.			
Sanitary services, steam supply, irrigation	0.1	Util.			
Transit & bus transportation	0.1	Util.			
Retail trade, except eating & drinking	0.1	Trade			
Royalties	0.1	Fin/R.E.			
Laundry, dry cleaning, shoe repair	0.1	Services			
State & local government enterprises, nec	0.1	Gov't			

Source: Benchmark Input-Output Accounts for the U.S. Economy, 1982, U.S. Department of Commerce, Washington, D.C., July 1991. Data, as reported in the source, are organized by the 1977 SIC structure in use in 1982 but have been matched, as closely as is possible, to the 1987 SIC structure used in this book.

OCCUPATIONS EMPLOYED BY SIC 382 - MEASURING AND CONTROLLING DEVICES

Occupation	% of Total 1994	Change to 2005	Occupation	% of Total 1994	Change to 2005
Assemblers, fabricators, & hand workers nec	7.4	-11.4	Engineering, mathematical, & science managers	1.7	0.6
Electrical & electronics engineers	5.6	13.4	Industrial production managers	1.5	-11.4
Electrical & electronic equipment assemblers	5.4	-11.4	General office clerks	1.4	-24.4
Sales & related workers nec	4.3	-11.4	Bookkeeping, accounting, & auditing clerks	1.4	-33.6
Electrical & electronic technicians,technologists	3.7	-11.4	Purchasing agents, ex trade & farm products	1.3	-11.4
Electrical & electronic assemblers	3.5	-20.3	Production, planning, & expediting clerks	1.3	-11.4
Inspectors, testers, & graders, precision	3.3	-38.0	Traffic, shipping, & receiving clerks	1.3	-14.8
Electromechanical equipment assemblers	3.2	-2.5	Drafters	1.3	-31.0
Blue collar worker supervisors	3.2	-24.1	Computer engineers	1.3	31.2
General managers & top executives	2.4	-15.9	Stock clerks	1.2	-28.0
Secretaries, ex legal & medical	2.4	-19.3	Engineers nec	1.2	6.3
Machinists	2.0	-33.6	Management support workers nec	1.1	-11.4
Mechanical engineers	1.8	-2.5	Managers & administrators nec	1.1	-11.4
Engineering technicians & technologists nec	1.7	-11.4	Computer programmers	1.1	-28.2
Marketing, advertising, & PR managers	1.7	-11.4			

Source: Industry-Occupation Matrix, Bureau of Labor Statistics. These data relate to one or more 3-digit SIC industry groups rather than to a single 4-digit SIC. The change reported for each occupation to the year 2005 is a percent of growth or decline as estimated by the Bureau of Labor Statistics. The abbreviation nec stands for 'not elsewhere classified'.

LOCATION BY STATE AND REGIONAL CONCENTRATION

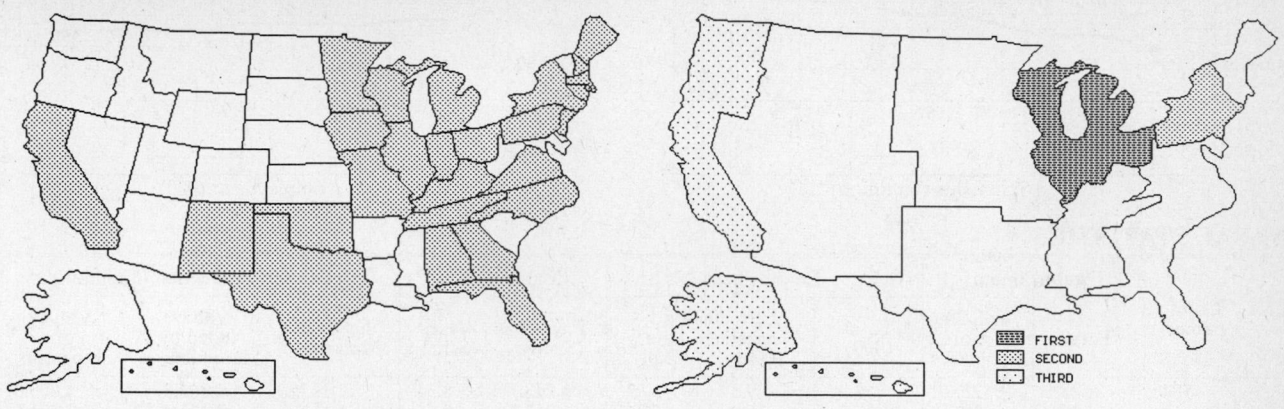

INDUSTRY DATA BY STATE

State	Establish-ments	Shipments			Employment				Cost as % of Shipments	Investment per Employee ($)
		Total ($ mil)	% of U.S.	Per Establ.	Total Number	% of U.S.	Per Establ.	Wages ($/hour)		
Illinois	18	492.8	18.9	27.4	4,100	16.4	228	13.37	45.4	4,683
California	42	214.1	8.2	5.1	1,700	6.8	40	11.52	39.8	-
Indiana	9	193.4	7.4	21.5	1,800	7.2	200	11.08	40.4	3,056
Ohio	21	170.6	6.5	8.1	2,600	10.4	124	8.69	27.6	2,231
Kentucky	4	111.0	4.3	27.8	1,300	5.2	325	7.89	34.7	1,769
Tennessee	8	100.2	3.8	12.5	1,200	4.8	150	11.26	39.1	-
Texas	13	91.3	3.5	7.0	700	2.8	54	9.42	35.3	2,286
Michigan	18	83.7	3.2	4.7	600	2.4	33	10.00	42.8	-
Massachusetts	12	60.4	2.3	5.0	700	2.8	58	9.86	35.3	1,429
Pennsylvania	18	57.0	2.2	3.2	800	3.2	44	11.30	51.2	6,500
Wisconsin	14	54.6	2.1	3.9	700	2.8	50	10.56	41.6	-
Florida	19	53.6	2.1	2.8	500	2.0	26	8.86	41.0	3,600
Rhode Island	5	49.6	1.9	9.9	700	2.8	140	7.50	23.2	1,143
New York	17	44.5	1.7	2.6	500	2.0	29	9.67	42.2	-
Georgia	7	26.0	1.0	3.7	300	1.2	43	8.75	28.1	333
New Jersey	9	22.6	0.9	2.5	200	0.8	22	15.00	35.8	2,500
Connecticut	8	17.2	0.7	2.2	200	0.8	25	13.50	24.4	1,500
Maine	4	15.2	0.6	3.8	200	0.8	50	8.00	58.6	1,000
North Carolina	8	14.6	0.6	1.8	200	0.8	25	10.00	29.5	1,500
Minnesota	13	(D)	-	-	3,750 *	15.0	288	-	-	-
Missouri	5	(D)	-	-	1,750 *	7.0	350	-	-	-
Oklahoma	5	(D)	-	-	175 *	0.7	35	-	-	-
New Hampshire	4	(D)	-	-	175 *	0.7	44	-	-	-
Alabama	3	(D)	-	-	375 *	1.5	125	-	-	-
Iowa	3	(D)	-	-	175 *	0.7	58	-	-	-
Virginia	3	(D)	-	-	175 *	0.7	58	-	-	-
New Mexico	1	(D)	-	-	175 *	0.7	175	-	-	-

Source: 1992 *Economic Census*. The states are in descending order of shipments or establishments (if shipment data are missing for the majority). The symbol (D) appears when data are withheld to prevent disclosure of competitive information. States marked with (D) are sorted by number of establishments. A dash (-) indicates that the data element cannot be calculated; * indicates the midpoint of a range.

3823 - PROCESS CONTROL INSTRUMENTS

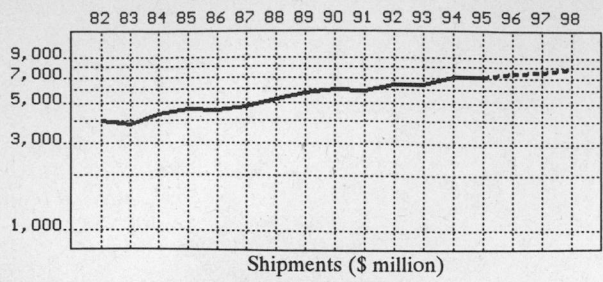

Shipments ($ million)

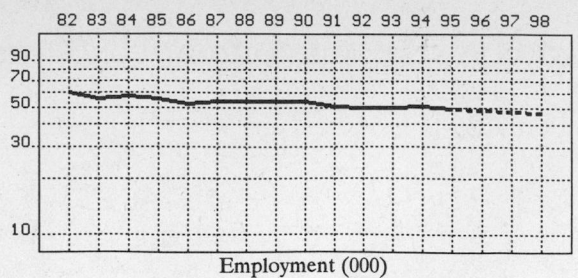

Employment (000)

GENERAL STATISTICS

Year	Companies	Establishments		Employment			Compensation		Production ($ million)			
		Total	with 20 or more employees	Total (000)	Production Workers (000)	Hours (Mil)	Payroll ($ mil)	Wages ($/hr)	Cost of Materials	Value Added by Manufacture	Value of Shipments	Capital Invest.
1982	584	625	288	59.9	29.8	57.3	1,243.7	8.31	1,167.8	2,804.3	4,005.8	126.9
1983		628	301	55.5	26.9	51.6	1,213.3	9.16	1,113.3	2,611.4	3,781.5	101.6
1984		631	314	57.4	30.3	59.7	1,353.0	9.30	1,352.3	3,017.6	4,307.9	131.3
1985		635	326	55.3	28.3	55.3	1,383.4	10.03	1,590.0	3,046.2	4,609.6	149.9
1986		630	316	52.2	26.2	51.5	1,350.8	10.31	1,575.4	2,924.4	4,535.4	148.0
1987	708	785	343	53.3	26.7	53.3	1,476.5	10.51	1,601.4	3,205.2	4,788.2	129.3
1988		757	355	53.7	26.8	52.9	1,552.0	10.49	1,918.6	3,328.2	5,248.9	129.9
1989		721	365	53.8	28.0	56.0	1,672.2	11.18	2,060.8	3,700.3	5,693.1	135.8
1990		748	367	53.4	26.1	52.8	1,730.5	11.38	2,169.7	3,764.7	5,924.0	150.6
1991		789	360	50.4	23.7	47.2	1,654.6	11.39	2,078.8	3,765.7	5,903.5	346.8
1992	817	885	358	50.1	24.0	47.3	1,764.8	12.32	2,137.7	4,182.9	6,360.4	158.1
1993		906	370	50.0	23.6	47.1	1,790.4	12.32	2,118.8	4,238.0	6,356.1	133.3
1994		895P	386P	50.4	24.3	50.5	1,876.8	12.31	2,446.7	4,568.6	7,012.1	226.9
1995		920P	394P	48.7P	23.2P	47.4P	1,931.7P	12.95P	2,465.8P	4,604.3P	7,066.8P	215.6P
1996		946P	401P	48.0P	22.7P	46.7P	1,987.2P	13.28P	2,555.3P	4,771.4P	7,323.3P	223.6P
1997		971P	408P	47.3P	22.2P	46.0P	2,042.7P	13.60P	2,644.8P	4,938.5P	7,579.8P	231.7P
1998		997P	416P	46.6P	21.7P	45.3P	2,098.2P	13.92P	2,734.3P	5,105.6P	7,836.4P	239.8P

Sources: 1982, 1987, 1992 *Economic Census*; *Annual Survey of Manufactures*, 83-86, 88-91, 93-94. Establishment counts for non-Census years are from *County Business Patterns*; establishment values for 83-84 are extrapolations. 'P's show projections by the editors. Industries reclassified in 87 will not have data for prior years.

INDICES OF CHANGE

Year	Companies	Establishments		Employment			Compensation		Production ($ million)			
		Total	with 20 or more employees	Total (000)	Production Workers (000)	Hours (Mil)	Payroll ($ mil)	Wages ($/hr)	Cost of Materials	Value Added by Manufacture	Value of Shipments	Capital Invest.
1982	71	71	80	120	124	121	70	67	55	67	63	80
1983		71	84	111	112	109	69	74	52	62	59	64
1984		71	88	115	126	126	77	75	63	72	68	83
1985		72	91	110	118	117	78	81	74	73	72	95
1986		71	88	104	109	109	77	84	74	70	71	94
1987	87	89	96	106	111	113	84	85	75	77	75	82
1988		86	99	107	112	112	88	85	90	80	83	82
1989		81	102	107	117	118	95	91	96	88	90	86
1990		85	103	107	109	112	98	92	101	90	93	95
1991		89	101	101	99	100	94	92	97	90	93	219
1992	100	100	100	100	100	100	100	100	100	100	100	100
1993		102	103	100	98	100	101	100	99	101	100	84
1994		101P	108P	101	101	107	106	100	114	109	110	144
1995		104P	110P	97P	96P	100P	109P	105P	115P	110P	111P	136P
1996		107P	112P	96P	94P	99P	113P	108P	120P	114P	115P	141P
1997		110P	114P	94P	92P	97P	116P	110P	124P	118P	119P	147P
1998		113P	116P	93P	90P	96P	119P	113P	128P	122P	123P	152P

Sources: Same as General Statistics. Values reflect change from the base year, 1992. Values above 100 mean greater than 92, values below 100 mean less than 92, and a value of 100 in the 82-91 or 93-98 period means same as 92. 'P's mark projections by the editors.

SELECTED RATIOS

For 1994	Avg. of All Manufact.	Analyzed Industry	Index	For 1994	Avg. of All Manufact.	Analyzed Industry	Index
Employees per Establishment	49	56	115	Value Added per Production Worker	134,084	188,008	140
Payroll per Establishment	1,500,273	2,097,693	140	Cost per Establishment	5,045,178	2,734,669	54
Payroll per Employee	30,620	37,238	122	Cost per Employee	102,970	48,546	47
Production Workers per Establishment	34	27	79	Cost per Production Worker	146,988	100,687	69
Wages per Establishment	853,319	694,822	81	Shipments per Establishment	9,576,895	7,837,402	82
Wages per Production Worker	24,861	25,583	103	Shipments per Employee	195,460	139,129	71
Hours per Production Worker	2,056	2,078	101	Shipments per Production Worker	279,017	288,564	103
Wages per Hour	12.09	12.31	102	Investment per Establishment	321,011	253,605	79
Value Added per Establishment	4,602,255	5,106,310	111	Investment per Employee	6,552	4,502	69
Value Added per Employee	93,930	90,647	97	Investment per Production Worker	9,352	9,337	100

Sources: Same as General Statistics. The 'Average of All Manufacturing' column represents the average of all manufacturing industries reported for the most recent complete year available. The Index shows the relationship between the Average and the Analyzed Industry. For example, 100 means that they are equal; 500 that the Analyzed Industry is five times the average; 50 means that the Analyzed Industry is half the national average. The abbreviation 'na' is used to show that data are 'not available'.

LEADING COMPANIES   Number shown: **75**   Total sales ($ mil): **10,348**   Total employment (000): **82.6**

Company Name	Address				CEO Name	Phone	Co. Type	Sales ($ mil)	Empl. (000)
Thermo Electron Corp	PO Box 9046	Waltham	MA	02254	G N Hatsopoulos	617-622-1000	P	1,585	10.2
EG and G Inc	45 William St	Wellesley	MA	02181	John M Kucharski	617-237-5100	P	1,333	14.0
Rosemount Inc	12001 Techn Dr	Eden Prairie	MN	55344	Robert Bateman	612-941-5560	S	1,070*	10.0
Elsag Bailey Process Automation	29801 Euclid Av	Wickliffe	OH	44092	Vincenzo Canatelli	216-585-8500	S	680	6.0
Foxboro Co	33 Commercial St	Foxboro	MA	02035	Alan Yurko	508-543-8750	S	540	5.0
IDEA Inc	29 Dunham Rd	Billerica	MA	01821	Gautam Gupta	508-663-6878	S	537	1.3
Simpson Industries Inc	32100 Telegraph Rd	Bingham Farms	MI	48025	Roy E Parrott	810-540-6200	P	357	2.1
Modicon Inc	1 High St	North Andover	MA	01845	Paul White	508-794-0800	S	300	2.0
Barber-Colman Co	PO Box 2940	Loves Park	IL	61132	WA Cashin	815-397-7400	S	280	2.5
Fisher Controls International Inc	8301 Cameron Rd	Austin	TX	78753	John Berra	512-535-2190	D	210*	0.6
Analogic Corp	8 Centennial Dr	Peabody	MA	01960	Bruce R Rusch	508-977-3000	P	194	1.4
Serv-Tech Inc	5200 Cedar Crest	Houston	TX	77087	Richard L Daerr	713-644-9974	P	181	2.1
High Voltage Engineering Corp	401 Edgewater Pl	Wakefield	MA	01880	Clifford Press	617-224-1001	R	150	1.4
Raven Industries Inc	PO Box 5107	Sioux Falls	SD	57117	D A Christensen	605-336-2750	P	122	1.4
United States Gauge	900 Clymer Av	Sellersville	PA	18960	Vito Parato	215-257-6531	D	120	0.8
K-Tron International Inc	1810 Chapel Av W	Cherry Hill	NJ	08002	Marcel O Rohr	609-661-6240	P	105	0.7
Cohu Inc	5755 Kearny Villa	San Diego	CA	92123	James W Barnes	619-277-6700	P	103	0.6
Moore Products Co	Sumneytown Pike	Spring House	PA	19477	William B Moore	215-646-7400	P	101	1.2
Spectra-Physics Scanning Systems	959 Terry St	Eugene	OR	97402	John O'Brien	503-683-5700	S	100	0.6
NEWFLO Corp	301 Camp Craft Rd	Austin	TX	78746	J Jack Watson	512-314-8500	R	95	0.9
Panametrics Inc	221 Crescent St	Waltham	MA	02154	Francis B Sellers	617-899-2719	R	87	0.7
Chart Industries Inc	35555 Curtis Blv	Eastlake	OH	44095	Arthur S Holmes	216-946-2525	P	84	0.7
Autoclave Engineers Inc	PO Box 5051	Erie	PA	16512	William F Schilling	814-838-5700	P	77	0.7
Compressor Controls Corp	11359 Aurora Av	Des Moines	IA	50332	Naum Staroselsky	515-270-0857	S	70	0.3
Fairchild Industrial Products Co	3920 West Pt Blvd	Winston-Salem	NC	27103	Hugh Steele	910-659-3400	R	67*	0.7
Eurotherm International Inc	11513 Sunset Hills	Reston	VA	22090	TR Brewer	703-471-4564	S	65*	0.7
Ontario Corp	123 E Adams St	Muncie	IN	47305	Van P Smith	317-747-0747	R	64*	0.6
Coherent Inc	5100 Patrick Henry	Santa Clara	CA	95054	Bernard Couillaud	415-493-2111	D	60	0.3
Daniel Flow Products Inc	PO Box 19097	Houston	TX	77224	Larry Irving Sr	713-467-6000	D	60	0.7
IRD Mechanalysis Inc	6150 Huntley Rd	Columbus	OH	43229	Steve Young	614-885-5376	S	58	0.3
Tencor Instruments Inc	2400 Charleston Rd	Mountain View	CA	94043	Jon D Tompkins	415-969-6784	P	58	0.8
Olympus America Inc	2 Corporate Ctr Dr	Melville	NY	11747	Pete Lorenz	516-844-5888	D	54	<0.1
Ronan Engineering Co	PO Box 1275	Woodland Hills	CA	91365	JA Hewitson	818-883-5211	R	50	0.5
TSI Inc	PO Box 64394	St Paul	MN	55164	Leroy M Fingerson	612-483-0900	P	49	0.4
Tylan General Inc	9577 Chesapeake	San Diego	CA	92123	David J Ferran	619-571-1222	P	48	0.3
Emerson Electric Co	4550 Old Whitley	London	KY	40741	Terry Bernhold	606-864-2241	D	46	0.4
ABB Kent-Taylor Inc	PO Box 20550	Rochester	NY	14602	Rick Keane	716-292-6050	S	43*	0.4
Forney Corp	PO Box 189	Addison	TX	75001	Brent Ehmke	214-458-6100	S	43*	0.4
Fisher-Rosemount Systems Inc	2400 Barranca Pkwy	Irvine	CA	92714	J Coletti	714-863-1181	D	42	0.2
Unit Instruments Inc	22600 Savi Ranch	Yorba Linda	CA	92687	Mike Doyle	714-921-2640	S	42	0.4
Amicon Inc	72 Cherry Hill Dr	Beverly	MA	01915	JD Palm	508-777-3622	S	40*	0.4
Digilab	237 Putnam Av	Cambridge	MA	02139	Jerry Keahl	617-868-4330	D	40	0.2
Industrial Dynamics Ltd	2729 Lomita Blv	Torrance	CA	90505	FL Calhoun	310-325-5633	R	40	0.4
Isco Inc	PO Box 5347	Lincoln	NE	68505	Robert W Allington	402-464-0231	P	39	0.4
Medar Inc	38700 Grand River	Farmington Hls	MI	48335	Charles J Drake	313-477-3900	P	37	0.2
Bacharach Inc	625 Alpha Dr	Pittsburgh	PA	15238	P Zito	412-963-2000	R	35	0.4
Eurotherm Controls Inc	11485 Sunset Hills	Reston	VA	22090	W V Perry	703-471-4870	S	35	0.2
Magnetrol International Inc	5300 Belmont Rd	Downers Grove	IL	60515	JG Stevenson	708-969-4000	R	35	0.3
SH Leggitt Co	PO Box 946	San Marcos	TX	78667	DC Leggitt Sr	512-396-0707	R	35*	0.3
Tescom Corp	12616 Indrial Blvd	Elk River	MN	55330	Donald E Glesmann	612-441-6330	R	35	0.4
Uniphase Corp	163 Baypointe Pkwy	San Jose	CA	95134	Kevin N Kalkhoven	408-434-1800	P	33	0.2
Frank W Murphy Manufacturing	PO Box 470248	Tulsa	OK	74147	Frank W Murphy Jr	918-627-3550	R	32	0.5
AMETEK Inc	150 Freeport Rd	Pittsburgh	PA	15238	Thomas F Mangold	412-828-9040	D	30	0.2
Gables Engineering Inc	PO Box 140880	Coral Gables	FL	33114	VE Clarke	305-442-2578	R	30	0.2
Moore Industries-International	16650 Schoenborn	Sepulveda	CA	91343	Robert E Maroney	818-894-7111	R	30	0.3
Pro-Log Corp	12 Upper Ragsdale	Monterey	CA	93940	R P McClellan	408-372-4593	R	30	0.2
SOR Inc	14685 W 105th St	Lenexa	KS	66215	Lew Goetz	913-888-2630	R	30	0.2
Watlow Gordon Co	PO Box 500	Richmond	IL	60071	Gary Neal	815-678-2211	S	30*	0.3
K-Tron North America Corp	Rte 55 & Rte 553	Pitman	NJ	08071	Kevin Bowen	609-589-0500	S	29*	0.2
WIKA Instrument Corp	1000 Wiegand Blv	Lawrenceville	GA	30243	W Dreftmeier	404-513-8200	S	29	0.2
Liberty Technologies Inc	555 North Ln	Conshohocken	PA	19428	R Nim Evatt	215-834-0330	P	28	0.3
Isco Inc	PO Box 82531	Lincoln	NE	68501	Douglas M Grant	402-474-2233	D	26	0.3
Raytek Inc	1201 Shaffer Rd	Santa Cruz	CA	95061	Cliff Warren	408-458-1110	R	26	0.2
Amot Controls Corp	401 First St	Richmond	CA	94801	Larry Christensen	510-236-8300	S	25	0.4
Barnes Engineering	PO Box 867	Shelton	CT	06484	Harry M Elmendorf	203-926-1777	D	25*	0.2
Eberline Instrument Corp	PO Box 2108	Santa Fe	NM	87504	Don Hannah	505-471-3232	S	25*	0.2
Rochester Gauges Incorporated	PO Box 29242	Dallas	TX	75229	JW La Due	214-241-2161	R	24	0.4
Maxitrol Co	PO Box 2230	Southfield	MI	48037	Frank Kern III	810-356-1400	R	23	0.2
HO Trerice Co	12950 W 8 Mile Rd	Oak Park	MI	48237	Allan D Feys	810-399-8000	R	22	0.2
Great Lakes Instruments Inc	8855 N 55th St	Milwaukee	WI	53223	Chris Dreher	414-355-3601	S	21	0.2
Daedel	PO Box 500	Harrison City	PA	15636	Robert Rebich	412-744-4451	D	21*	0.2
American Technical Service	PO Box 508	Norcross	GA	30091	Robert S Russo	404-447-9444	R	20	0.3
BLH Electronics	75 Shawmut Rd	Canton	MA	02021	T Selig	617-821-2000	S	20	0.2
Brookfield Eng Laboratories	240 Cushing St	Stoughton	MA	02072	Louis Dicorpo	617-344-4310	R	20	0.2
Columbia Scientific Indust Corp	PO Box 203190	Austin	TX	78720	John E Sullivan	512-258-5191	S	20	0.1

Source: Ward's Business Directory of U.S. Private and Public Companies, Volumes 1 and 2, 1996. The company type code used is as follows: P - Public, R - Private, S - Subsidiary, D - Division, J - Joint Venture, A - Affiliate, G - Group. Sales are in millions of dollars, employees are in thousands. An asterisk (*) indicates an estimated sales volume. The symbol < stands for 'less than'. Company names and addresses are truncated, in some cases, to fit into the available space.

MATERIALS CONSUMED

Material	Quantity	Delivered Cost ($ million)
Materials, ingredients, containers, and supplies	(X)	1,918.5
Printed circuit boards (without inserted components) for electronic circuitry	(X)	46.9
Printed circuit assemblies, loaded boards or modules	(X)	70.5
Semiconductors, including transistors, diodes, rectifiers, and integrated circuits for electronic circuitry	(X)	68.9
Capacitors for electronic circuitry	(X)	13.9
Resistors for electronic circuitry	(X)	60.9
Other components and accessories for electronic circuitry, nec, except tubes	(X)	43.7
Current-carrying wiring devices	(X)	15.5
Electrical transmission, distribution, and control equipment	(X)	14.9
Electronic computing equipment	(X)	84.8
Electrical instrument mechanisms and meter movements (including instrument relays)	(X)	24.5
Electrical measuring instruments and parts, not listed elsewhere	(X)	66.8
Fractional horsepower electric motors and generators (less than 1 hp) including timing motors	(X)	6.0
Plastics resins consumed in the form of granules, pellets, powders, liquids, etc.	(X)	2.6
Fabricated plastics products (except gaskets, hoses, and belting)	(X)	26.0
Sheet metal products, except stampings	(X)	62.0
Metal stampings	(X)	22.9
Fabricated metal wire products (including wire rope, cable, springs, etc.)	(X)	13.4
Metal bolts, nuts, screws, washers, rivets, and other screw machine products	(X)	26.8
Other fabricated metal products (except forgings)	(X)	35.9
Forgings	(X)	4.1
Iron and steel castings (rough and semifinished)	(X)	32.6
Aluminum and aluminum-base alloy castings (rough and semifinished)	(X)	19.1
Other nonferrous castings (rough and semifinished)	(X)	5.1
Steel shapes and forms	(X)	22.4
Copper and copper-base alloy shapes and forms	(X)	7.1
Aluminum and aluminum-base alloy shapes and forms	(X)	8.6
Other nonferrous shapes and forms	(X)	8.6
Glass and glass products (excluding windows and mirrors)	(X)	7.5
Paper and paperboard products except paperboard boxes, containers, and corrugated paperboard	(X)	9.5
Paper and paperboard containers, including shipping sacks and other paper packaging supplies	(X)	10.0
All other materials and components, parts, containers, and supplies	(X)	373.7
Materials, ingredients, containers, and supplies, nsk	(X)	703.4

Source: 1992 Economic Census. Explanation of symbols used: (D): Withheld to avoid disclosure of competitive data; na: Not available; (S): Withheld because statistical norms were not met; (X): Not applicable; (Z): Less than half the unit shown; nec: Not elsewhere classified; nsk: Not specified by kind; - : zero; * : 10-19 percent estimated; ** : 20-29 percent estimated.

PRODUCT SHARE DETAILS

Product or Product Class	% Share	Product or Product Class	% Share
Process control instruments	100.00		

Source: 1992 Economic Census. The values shown are percent of total shipments in an industry. Values of indented subcategories are summed in the main heading. The symbol (D) appears when data are withheld to prevent disclosure of competitive information. The abbreviation nsk stands for 'not specified by kind' and nec for 'not elsewhere classified'.

INPUTS AND OUTPUTS FOR ENGINEERING & SCIENTIFIC INSTRUMENTS

Economic Sector or Industry Providing Inputs	%	Sector	Economic Sector or Industry Buying Outputs	%	Sector
Imports	14.6	Foreign	Gross private fixed investment	41.5	Cap Inv
Engineering & scientific instruments	12.1	Manufg.	Exports	26.4	Foreign
Wholesale trade	8.7	Trade	Federal Government purchases, national defense	7.5	Fed Govt
Electronic components nec	6.8	Manufg.	Engineering & scientific instruments	5.2	Manufg.
Veneer & plywood	4.7	Manufg.	Aircraft	4.2	Manufg.
Blast furnaces & steel mills	3.6	Manufg.	Ship building & repairing	2.4	Manufg.
Advertising	3.6	Services	Radio & TV communication equipment	2.3	Manufg.
Instruments to measure electricity	2.7	Manufg.	S/L Govt. purch., health & hospitals	2.1	S/L Govt
Semiconductors & related devices	2.4	Manufg.	Mechanical measuring devices	1.4	Manufg.
Communications, except radio & TV	2.1	Util.	Federal Government purchases, nondefense	1.3	Fed Govt
Electric services (utilities)	2.0	Util.	Change in business inventories	0.8	In House
Petroleum refining	1.9	Manufg.	S/L Govt. purch., higher education	0.6	S/L Govt
Eating & drinking places	1.8	Trade	S/L Govt. purch., police	0.6	S/L Govt
Motors & generators	1.7	Manufg.	S/L Govt. purch., sewerage	0.6	S/L Govt
Industrial patterns	1.4	Manufg.	S/L Govt. purch., sanitation	0.5	S/L Govt
Fabricated metal products, nec	1.3	Manufg.	Boat building & repairing	0.4	Manufg.
Machinery, except electrical, nec	1.2	Manufg.	S/L Govt. purch., elem. & secondary education	0.4	S/L Govt
Fabricated rubber products, nec	1.1	Manufg.	Blast furnaces & steel mills	0.3	Manufg.
Sheet metal work	1.1	Manufg.	Miscellaneous plastics products	0.3	Manufg.
Aluminum rolling & drawing	1.0	Manufg.	Watches, clocks, & parts	0.3	Manufg.
Miscellaneous plastics products	0.9	Manufg.	S/L Govt. purch., highways	0.3	S/L Govt
Screw machine and related products	0.9	Manufg.	S/L Govt. purch., natural resource & recreation.	0.3	S/L Govt
Banking	0.9	Fin/R.E.			
Legal services	0.9	Services			

Continued on next page.

INPUTS AND OUTPUTS FOR ENGINEERING & SCIENTIFIC INSTRUMENTS - Continued

Economic Sector or Industry Providing Inputs	%	Sector	Economic Sector or Industry Buying Outputs	%	Sector
Maintenance of nonfarm buildings nec	0.8	Constr.			
Electronic computing equipment	0.8	Manufg.			
Motor freight transportation & warehousing	0.8	Util.			
Equipment rental & leasing services	0.8	Services			
Glass & glass products, except containers	0.7	Manufg.			
Paperboard containers & boxes	0.7	Manufg.			
Real estate	0.7	Fin/R.E.			
Management & consulting services & labs	0.7	Services			
U.S. Postal Service	0.7	Gov't			
Aluminum castings	0.6	Manufg.			
Gaskets, packing & sealing devices	0.6	Manufg.			
Special dies & tools & machine tool accessories	0.6	Manufg.			
Ball & roller bearings	0.5	Manufg.			
Paints & allied products	0.5	Manufg.			
Manifold business forms	0.4	Manufg.			
Metal stampings, nec	0.4	Manufg.			
Gas production & distribution (utilities)	0.4	Util.			
Accounting, auditing & bookkeeping	0.4	Services			
Converted paper products, nec	0.3	Manufg.			
Lighting fixtures & equipment	0.3	Manufg.			
Railroads & related services	0.3	Util.			
Automotive rental & leasing, without drivers	0.3	Services			
Business/professional associations	0.3	Services			
Computer & data processing services	0.3	Services			
Engineering, architectural, & surveying services	0.3	Services			
Abrasive products	0.2	Manufg.			
Iron & steel foundries	0.2	Manufg.			
Lubricating oils & greases	0.2	Manufg.			
Machine tools, metal cutting types	0.2	Manufg.			
Nonferrous castings, nec	0.2	Manufg.			
Photographic equipment & supplies	0.2	Manufg.			
Plastics materials & resins	0.2	Manufg.			
Plating & polishing	0.2	Manufg.			
Sawmills & planning mills, general	0.2	Manufg.			
Transformers	0.2	Manufg.			
Wiring devices	0.2	Manufg.			
Air transportation	0.2	Util.			
Credit agencies other than banks	0.2	Fin/R.E.			
Insurance carriers	0.2	Fin/R.E.			
Royalties	0.2	Fin/R.E.			
Automotive repair shops & services	0.2	Services			
Photofinishing labs, commercial photography	0.2	Services			
Theatrical producers, bands, entertainers	0.2	Services			
Electron tubes	0.1	Manufg.			
Miscellaneous fabricated wire products	0.1	Manufg.			
Rubber & plastics hose & belting	0.1	Manufg.			
Transit & bus transportation	0.1	Util.			
Water supply & sewage systems	0.1	Util.			
Water transportation	0.1	Util.			
Motion pictures	0.1	Services			
Personnel supply services	0.1	Services			
Noncomparable imports	0.1	Foreign			

Source: Benchmark Input-Output Accounts for the U.S. Economy, 1982, U.S. Department of Commerce, Washington, D.C., July 1991. Data, as reported in the source, are organized by the 1977 SIC structure in use in 1982 but have been matched, as closely as is possible, to the 1987 SIC structure used in this book.

OCCUPATIONS EMPLOYED BY SIC 382 - MEASURING AND CONTROLLING DEVICES

Occupation	% of Total 1994	Change to 2005	Occupation	% of Total 1994	Change to 2005
Assemblers, fabricators, & hand workers nec	7.4	-11.4	Engineering, mathematical, & science managers	1.7	0.6
Electrical & electronics engineers	5.6	13.4	Industrial production managers	1.5	-11.4
Electrical & electronic equipment assemblers	5.4	-11.4	General office clerks	1.4	-24.4
Sales & related workers nec	4.3	-11.4	Bookkeeping, accounting, & auditing clerks	1.4	-33.6
Electrical & electronic technicians,technologists	3.7	-11.4	Purchasing agents, ex trade & farm products	1.3	-11.4
Electrical & electronic assemblers	3.5	-20.3	Production, planning, & expediting clerks	1.3	-11.4
Inspectors, testers, & graders, precision	3.3	-38.0	Traffic, shipping, & receiving clerks	1.3	-14.8
Electromechanical equipment assemblers	3.2	-2.5	Drafters	1.3	-31.0
Blue collar worker supervisors	3.2	-24.1	Computer engineers	1.3	31.2
General managers & top executives	2.4	-15.9	Stock clerks	1.2	-28.0
Secretaries, ex legal & medical	2.4	-19.3	Engineers nec	1.2	6.3
Machinists	2.0	-33.6	Management support workers nec	1.1	-11.4
Mechanical engineers	1.8	-2.5	Managers & administrators nec	1.1	-11.4
Engineering technicians & technologists nec	1.7	-11.4	Computer programmers	1.1	-28.2
Marketing, advertising, & PR managers	1.7	-11.4			

Source: Industry-Occupation Matrix, Bureau of Labor Statistics. These data relate to one or more 3-digit SIC industry groups rather than to a single 4-digit SIC. The change reported for each occupation to the year 2005 is a percent of growth or decline as estimated by the Bureau of Labor Statistics. The abbreviation nec stands for 'not elsewhere classified'.

LOCATION BY STATE AND REGIONAL CONCENTRATION

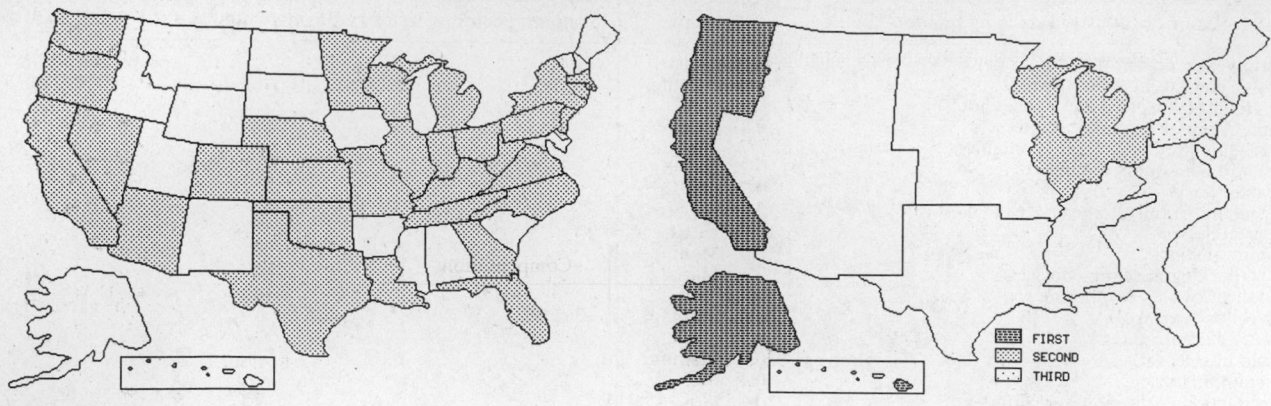

FIRST
SECOND
THIRD

INDUSTRY DATA BY STATE

| State | Establish-ments | Shipments | | | Employment | | | | Cost as % of Shipments | Investment per Employee ($) |
		Total ($ mil)	% of U.S.	Per Establ.	Total Number	% of U.S.	Per Establ.	Wages ($/hour)		
California	173	946.6	14.9	5.5	7,700	15.4	45	12.71	37.3	3,506
Pennsylvania	66	941.4	14.8	14.3	6,800	13.6	103	13.08	32.7	2,441
Ohio	41	626.6	9.9	15.3	4,700	9.4	115	15.77	36.4	3,234
Massachusetts	54	484.7	7.6	9.0	4,100	8.2	76	11.98	31.6	1,902
Texas	64	343.1	5.4	5.4	2,900	5.8	45	12.33	39.8	2,759
Connecticut	29	333.7	5.2	11.5	2,700	5.4	93	13.03	35.2	2,111
Illinois	49	311.2	4.9	6.4	2,500	5.0	51	10.41	33.6	4,400
Minnesota	20	284.6	4.5	14.2	3,200	6.4	160	11.13	46.0	-
New York	38	272.4	4.3	7.2	2,300	4.6	61	10.36	38.3	6,609
Colorado	24	168.1	2.6	7.0	1,100	2.2	46	11.91	23.5	6,545
Oklahoma	11	159.8	2.5	14.5	1,200	2.4	109	16.82	43.1	2,083
Michigan	38	106.4	1.7	2.8	1,000	2.0	26	11.90	36.3	2,200
Indiana	18	98.1	1.5	5.4	1,000	2.0	56	10.73	33.7	1,900
Wisconsin	21	88.9	1.4	4.2	900	1.8	43	12.45	34.9	1,667
New Jersey	39	84.6	1.3	2.2	800	1.6	21	11.57	37.1	1,750
Louisiana	15	70.2	1.1	4.7	900	1.8	60	15.00	37.5	1,000
Florida	23	48.2	0.8	2.1	400	0.8	17	13.33	32.8	-
Washington	13	33.7	0.5	2.6	300	0.6	23	10.67	42.1	2,000
Virginia	11	31.9	0.5	2.9	200	0.4	18	9.50	43.3	2,500
North Carolina	17	31.6	0.5	1.9	300	0.6	18	9.33	35.8	1,667
Georgia	12	30.0	0.5	2.5	200	0.4	17	11.00	37.3	-
Rhode Island	6	18.1	0.3	3.0	200	0.4	33	7.00	18.2	-
Oregon	13	17.3	0.3	1.3	200	0.4	15	15.00	41.6	1,000
Nevada	4	17.0	0.3	4.3	100	0.2	25	12.00	28.8	-
Missouri	11	16.5	0.3	1.5	200	0.4	18	8.00	36.4	-
Tennessee	8	10.7	0.2	1.3	500	1.0	63	6.86	32.7	-
Arizona	14	(D)	-	-	1,750 *	3.5	125	-	-	-
Nebraska	4	(D)	-	-	750 *	1.5	188	-	-	-
West Virginia	3	(D)	-	-	375 *	0.7	125	-	-	-
Kansas	2	(D)	-	-	175 *	0.3	88	-	-	-
Kentucky	1	(D)	-	-	375 *	0.7	375	-	-	-

Source: 1992 Economic Census. The states are in descending order of shipments or establishments (if shipment data are missing for the majority). The symbol (D) appears when data are withheld to prevent disclosure of competitive information. States marked with (D) are sorted by number of establishments. A dash (-) indicates that the data element cannot be calculated; * indicates the midpoint of a range.

3824 - FLUID METERS & COUNTING DEVICES

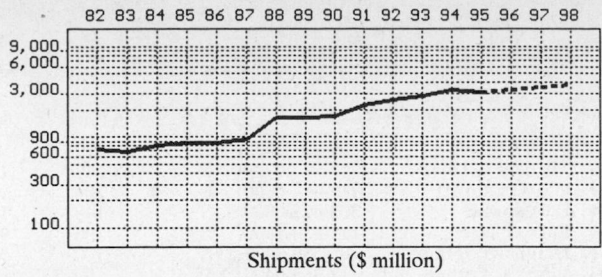

Shipments ($ million)

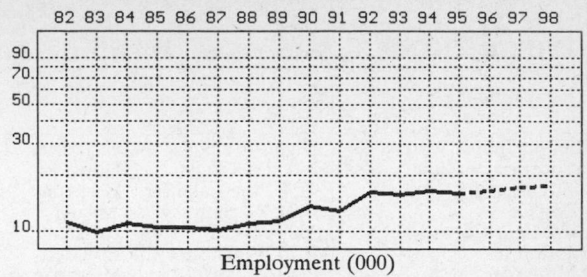

Employment (000)

GENERAL STATISTICS

Year	Companies	Establishments		Employment			Compensation		Production ($ million)			
		Total	with 20 or more employees	Total (000)	Production Workers (000)	Hours (Mil)	Payroll ($ mil)	Wages ($/hr)	Cost of Materials	Value Added by Manufacture	Value of Shipments	Capital Invest.
1982	132	144	69	11.0	6.9	13.0	197.3	8.39	266.1	459.2	726.7	27.5
1983		140	68	9.8	6.2	11.8	189.7	8.81	259.3	422.9	692.4	18.0
1984		136	67	10.9	6.5	12.2	230.9	9.80	298.2	529.5	801.9	26.3
1985		131	65	10.4	6.2	11.8	236.6	9.95	310.2	554.7	865.3	41.0
1986		125	66	10.4	6.2	12.7	237.3	8.96	298.1	548.3	858.6	29.0
1987	138	158	61	10.1	6.5	12.7	237.1	9.44	381.2	566.5	938.6	34.9
1988		162	64	10.9	7.3	14.0	267.5	10.41	659.8	976.5	1,659.0	40.4
1989		159	62	11.3	6.8	12.9	269.4	10.43	672.6	987.6	1,656.9	38.2
1990		162	64	13.7	6.6	12.9	278.6	10.74	683.3	976.7	1,665.9	57.0
1991		164	63	12.8	8.3	16.3	388.2	13.13	986.9	1,260.1	2,246.8	81.8
1992	181	193	74	16.2	11.3	21.6	533.7	15.36	1,117.5	1,469.2	2,601.5	74.1
1993		197	78	15.8	10.7	22.0	554.1	15.55	1,333.3	1,486.3	2,823.8	76.1
1994		190P	69P	16.6	11.8	25.0	607.4	16.70	1,466.8	1,861.7	3,299.8	92.3
1995		196P	70P	16.1P	10.8P	22.0P	558.4P	15.97P	1,392.4P	1,767.3P	3,132.5P	89.4P
1996		201P	70P	16.7P	11.2P	23.0P	591.7P	16.63P	1,489.6P	1,890.6P	3,351.1P	95.2P
1997		206P	70P	17.2P	11.7P	23.9P	625.1P	17.29P	1,586.7P	2,013.9P	3,569.6P	101.0P
1998		212P	71P	17.8P	12.1P	24.9P	658.4P	17.95P	1,683.9P	2,137.2P	3,788.1P	106.8P

Sources: 1982, 1987, 1992 *Economic Census*; *Annual Survey of Manufactures*, 83-86, 88-91, 93-94. Establishment counts for non-Census years are from *County Business Patterns*; establishment values for 83-84 are extrapolations. 'P's show projections by the editors. Industries reclassified in 87 will not have data for prior years.

INDICES OF CHANGE

Year	Companies	Establishments		Employment			Compensation		Production ($ million)			
		Total	with 20 or more employees	Total (000)	Production Workers (000)	Hours (Mil)	Payroll ($ mil)	Wages ($/hr)	Cost of Materials	Value Added by Manufacture	Value of Shipments	Capital Invest.
1982	73	75	93	68	61	60	37	55	24	31	28	37
1983		73	92	60	55	55	36	57	23	29	27	24
1984		70	91	67	58	56	43	64	27	36	31	35
1985		68	88	64	55	55	44	65	28	38	33	55
1986		65	89	64	55	59	44	58	27	37	33	39
1987	76	82	82	62	58	59	44	61	34	39	36	47
1988		84	86	67	65	65	50	68	59	66	64	55
1989		82	84	70	60	60	50	68	60	67	64	52
1990		84	86	85	58	60	52	70	61	66	64	77
1991		85	85	79	73	75	73	85	88	86	86	110
1992	100	100	100	100	100	100	100	100	100	100	100	100
1993		102	105	98	95	102	104	101	119	101	109	103
1994		99P	94P	102	104	116	114	109	131	127	127	125
1995		101P	94P	100P	96P	102P	105P	104P	125P	120P	120P	121P
1996		104P	95P	103P	99P	106P	111P	108P	133P	129P	129P	129P
1997		107P	95P	106P	103P	111P	117P	113P	142P	137P	137P	136P
1998		110P	96P	110P	107P	115P	123P	117P	151P	145P	146P	144P

Sources: Same as General Statistics. Values reflect change from the base year, 1992. Values above 100 mean greater than 92, values below 100 mean less than 92, and a value of 100 in the 82-91 or 93-98 period means same as 92. 'P's mark projections by the editors.

SELECTED RATIOS

For 1994	Avg. of All Manufact.	Analyzed Industry	Index	For 1994	Avg. of All Manufact.	Analyzed Industry	Index
Employees per Establishment	49	87	178	Value Added per Production Worker	134,084	157,771	118
Payroll per Establishment	1,500,273	3,190,228	213	Cost per Establishment	5,045,178	7,704,027	153
Payroll per Employee	30,620	36,590	119	Cost per Employee	102,970	88,361	86
Production Workers per Establishment	34	62	181	Cost per Production Worker	146,988	124,305	85
Wages per Establishment	853,319	2,192,822	257	Shipments per Establishment	9,576,895	17,331,434	181
Wages per Production Worker	24,861	35,381	142	Shipments per Employee	195,460	198,783	102
Hours per Production Worker	2,056	2,119	103	Shipments per Production Worker	279,017	279,644	100
Wages per Hour	12.09	16.70	138	Investment per Establishment	321,011	484,784	151
Value Added per Establishment	4,602,255	9,778,147	212	Investment per Employee	6,552	5,560	85
Value Added per Employee	93,930	112,151	119	Investment per Production Worker	9,352	7,822	84

Sources: Same as General Statistics. The 'Average of All Manufacturing' column represents the average of all manufacturing industries reported for the most recent complete year available. The Index shows the relationship between the Average and the Analyzed Industry. For example, 100 means that they are equal; 500 that the Analyzed Industry is five times the average; 50 means that the Analyzed Industry is half the national average. The abbreviation 'na' is used to show that data are 'not available'.

LEADING COMPANIES   Number shown: 39   Total sales ($ mil): 1,529   Total employment (000): 11.4

Company Name	Address				CEO Name	Phone	Co. Type	Sales ($ mil)	Empl. (000)
Schlumberger Indust Water	Highway 229 S	Tallassee	AL	36078	Pat Mahoney	205-283-6555	D	250	0.7
Moorco International Inc	2800 Post Oak	Houston	TX	77056	Michael L Tiner	713-993-0999	P	211	1.4
Daniel Industries Inc	9753 Pine Lake Dr	Houston	TX	77055	WA Griffin III	713-467-6000	P	204	1.5
Milton Roy Co	14845 W 64th Av	Arvada	CO	80004	Ed Laprade	303-425-0800	S	140	1.5
American Meter Co	300 Welsh Rd	Horsham	PA	19044	Larry F Neely	215-830-1800	S	100	0.8
Micro Motion Inc	7070 Winchester Cir	Boulder	CO	80301	Dennis G Perkins	303-530-8400	S	100	0.5
Badger Meter Inc	PO Box 23099	Milwaukee	WI	53223	James L Forbes	414-355-0400	P	99	0.9
Harper-Wyman Co	3600 Thayer Ct	Aurora	IL	60504	Ray Mehra	708-978-8000	S	75	0.9
Smith Blair Inc	PO Box 5337	Texarkana	TX	75505	Al Jurkonis	501-773-5127	S	35*	0.3
Marshall Gas Controls Inc	1000 Civic Ctr Dr	San Marcos	TX	78666	Don C Leggitt Jr	512-396-2257	D	33*	0.2
Liquid Controls Corp	105 Albrecht Dr	Lake Bluff	IL	60044	Frederick Wacker III	708-295-1050	R	30	0.2
Vickers Electromechanical	PO Box 872	Grand Rapids	MI	49588	Ronald Modreski	616-949-1090	D	30	0.2
Schlumberger Industries Inc	1310 Emerald Rd	Greenwood	SC	29646	Skip Tierno	803-223-1212	D	26	0.3
Dresser Measurement	PO Box 42176	Houston	TX	77242	EA Svinky	713-972-5000	D	22	0.2
L and J Engineering Inc	5911 Butterfield Rd	Hillside	IL	60162	Louis Janotta	708-236-6000	S	21	0.2
Capital Controls Co	3000 Advance Lane	Colmar	PA	18915	S Fierce	215-997-4000	R	20	0.1
POM Inc	PO Box 430	Russellville	AR	72801	S Ward	501-968-2880	R	18	0.2
GH Flow Automation	9303 W S Houston	Houston	TX	77088	CR Duarte	713-272-0404	S	17	0.2
Spirit of America Corp	10206 Lima Rd	Fort Wayne	IN	46818	JC Brown	219-489-2511	R	12	0.4
Hedland Flow Meters	PO Box 1405	Racine	WI	53401	John Erskine	414-639-6770	D	10	<0.1
Teleflex Inc	1816 57th St	Sarasota	FL	34243	William Rogers	813-355-7721	D	10	0.2
Gammon Technical Products Inc	PO Box 400	Manasquan	NJ	08736	HM Gammon	908-223-4600	R	8	<0.1
Key Instruments	250 Andrews Rd	Trevose	PA	19053		215-357-0893	R	8*	<0.1
Max Machinery Inc	1420 Healdsburg	Healdsburg	CA	95448	J Max	707-433-7281	R	8	<0.1
Electro-Sensors Inc	6111 Blue Circle Dr	Minnetonka	MN	55343	James P Slattery	612-941-8171	P	6	<0.1
Redington Counters Inc	130 Addison Rd	Windsor	CT	06095	C Lombardi	203-688-6205	R	6	0.1
TM Analytic Inc	381 Beinoris Dr	Wood Dale	IL	60191	Donald Winkelmann	708-860-9122	R	5	<0.1
Auto Meter Products Inc	413 W Elm St	Sycamore	IL	60178	JV Westberg	815-895-8141	R	3	0.2
Eugene Ernst Products Co	PO Box 427	Farmingdale	NJ	07727	Roger Ernst	908-938-5641	R	3	<0.1
JY Taylor Manufacturing Co	PO Box 461585	Garland	TX	75046	Leta S Taylor	214-276-1148	R	3	<0.1
Quantum Dynamics Inc	6414 Independence	Woodland Hills	CA	91367	Frederick F Liu	818-719-0142	R	3	<0.1
Anadex Inc	2260 Townsgate Rd	Westlake Vil	CA	91361	JR Weaver	805-987-9660	R	2	<0.1
Carlon Meter Company Inc	1710 Eaton Dr	Grand Haven	MI	49417	William L McAlister	616-842-0420	R	2	<0.1
Olympic Controls Inc	12818 Century Dr	Stafford	TX	77477	John Peterson	713-240-1540	R	2	<0.1
Scientific Instruments Inc	518 W Cherry St	Milwaukee	WI	53212	James Buraczewski	414-263-1600	R	2	<0.1
Proel Systems USA Inc	12320 Globe Rd	Livonia	MI	48150	Dominic Persichini	313-591-2424	S	2	<0.1
FS Brainard and Co	PO Box 366	Burlington	NJ	08016	FS Brainard Jr	609-387-4300	R	1	<0.1
MCO/EASTECH Inc	26 W Highland Av	Atlantic Hghlds	NJ	07716	P McCamely	908-291-3500	R	1	<0.1
Spangler Valve Co	505 S Vermont Av	Glendora	CA	91740	WE Jorgensen	818-335-4028	R	1*	<0.1

Source: Ward's Business Directory of U.S. Private and Public Companies, Volumes 1 and 2, 1996. The company type code used is as follows: P - Public, R - Private, S - Subsidiary, D - Division, J - Joint Venture, A - Affiliate, G - Group. Sales are in millions of dollars, employees are in thousands. An asterisk (*) indicates an estimated sales volume. The symbol < stands for 'less than'. Company names and addresses are truncated, in some cases, to fit into the available space.

MATERIALS CONSUMED

Material	Quantity	Delivered Cost ($ million)
Materials, ingredients, containers, and supplies	(X)	1,039.3
Printed circuit boards (without inserted components) for electronic circuitry	(X)	42.5
Printed circuit assemblies, loaded boards or modules	(X)	19.9
Semiconductors, including transistors, diodes, rectifiers, and integrated circuits for electronic circuitry	(X)	79.8
Capacitors for electronic circuitry	(X)	22.6
Resistors for electronic circuitry	(X)	7.3
Other components and accessories for electronic circuitry, nec, except tubes	(X)	3.9
Current-carrying wiring devices	(X)	12.1
Electrical transmission, distribution, and control equipment	(X)	5.6
Electronic computing equipment	(X)	11.4
Electrical instrument mechanisms and meter movements (including instrument relays)	(X)	26.5
Electrical measuring instruments and parts, not listed elsewhere	(X)	(D)
Fractional horsepower electric motors and generators (less than 1 hp) including timing motors	(X)	7.3
Plastics resins consumed in the form of granules, pellets, powders, liquids, etc.	(X)	46.5
Fabricated plastics products (except gaskets, hoses, and belting)	(X)	27.7
Sheet metal products, except stampings	(X)	10.4
Metal stampings	(X)	16.5
Fabricated metal wire products (including wire rope, cable, springs, etc.)	(X)	7.8
Metal bolts, nuts, screws, washers, rivets, and other screw machine products	(X)	17.9
Other fabricated metal products (except forgings)	(X)	8.4
Forgings	(X)	(D)
Iron and steel castings (rough and semifinished)	(X)	23.9
Aluminum and aluminum-base alloy castings (rough and semifinished)	(X)	34.4
Other nonferrous castings (rough and semifinished)	(X)	17.4
Steel shapes and forms	(X)	19.9
Copper and copper-base alloy shapes and forms	(X)	1.2
Aluminum and aluminum-base alloy shapes and forms	(X)	9.0
Other nonferrous shapes and forms	(X)	(D)
Glass and glass products (excluding windows and mirrors)	(X)	2.4
Paper and paperboard products except paperboard boxes, containers, and corrugated paperboard	(X)	0.5
Paper and paperboard containers, including shipping sacks and other paper packaging supplies	(X)	9.2
All other materials and components, parts, containers, and supplies	(X)	429.5
Materials, ingredients, containers, and supplies, nsk	(X)	94.7

Source: 1992 *Economic Census*. Explanation of symbols used: (D): Withheld to avoid disclosure of competitive data; na: Not available; (S): Withheld because statistical norms were not met; (X): Not applicable; (Z): Less than half the unit shown; nec: Not elsewhere classified; nsk: Not specified by kind; - : zero; * : 10-19 percent estimated; ** : 20-29 percent estimated.

PRODUCT SHARE DETAILS

Product or Product Class	% Share	Product or Product Class	% Share
Fluid meters and counting devices	100.00	Motor vehicle instruments	58.27
Integrating and totalizing meters for gas and liquids	28.35	Fluid meters and counting devices, nsk	3.44
Counting devices, excluding motor vehicle instruments	9.95		

Source: 1992 *Economic Census*. The values shown are percent of total shipments in an industry. Values of indented subcategories are summed in the main heading. The symbol (D) appears when data are withheld to prevent disclosure of competitive information. The abbreviation nsk stands for 'not specified by kind' and nec for 'not elsewhere classified'.

INPUTS AND OUTPUTS FOR MECHANICAL MEASURING DEVICES

Economic Sector or Industry Providing Inputs	%	Sector	Economic Sector or Industry Buying Outputs	%	Sector
Imports	24.6	Foreign	Gross private fixed investment	36.4	Cap Inv
Electronic components nec	8.6	Manufg.	Exports	21.8	Foreign
Wholesale trade	7.8	Trade	Aircraft	13.3	Manufg.
Instruments to measure electricity	3.9	Manufg.	Motor vehicles & car bodies	4.5	Manufg.
Automotive repair shops & services	3.7	Services	Sanitary services, steam supply, irrigation	3.0	Util.
Advertising	2.7	Services	Federal Government purchases, national defense	1.7	Fed Govt
Blast furnaces & steel mills	2.5	Manufg.	Commercial printing	1.4	Manufg.
Semiconductors & related devices	2.4	Manufg.	Federal Government purchases, nondefense	1.1	Fed Govt
Mechanical measuring devices	2.2	Manufg.	Mechanical measuring devices	1.0	Manufg.
Communications, except radio & TV	2.2	Util.	Electric services (utilities)	0.8	Util.
Electric services (utilities)	2.1	Util.	Cyclic crudes and organics	0.7	Manufg.
Petroleum refining	2.0	Manufg.	Newspapers	0.7	Manufg.
Electronic computing equipment	1.6	Manufg.	Personal consumption expenditures	0.6	
Screw machine and related products	1.4	Manufg.	Measuring & dispensing pumps	0.6	Manufg.
Engineering & scientific instruments	1.3	Manufg.	Electric utility facility construction	0.5	Constr.
Equipment rental & leasing services	1.3	Services	Wholesale trade	0.5	Trade
Aluminum castings	1.2	Manufg.	Retail trade, except eating & drinking	0.4	Trade
Sheet metal work	1.2	Manufg.	Automotive repair shops & services	0.4	Services
Aluminum rolling & drawing	1.1	Manufg.	Business services nec	0.4	Services
Fabricated metal products, nec	1.1	Manufg.	S/L Govt. purch., health & hospitals	0.4	S/L Govt
Iron & steel foundries	1.1	Manufg.	Periodicals	0.3	Manufg.
Metal stampings, nec	1.0	Manufg.	Computer & data processing services	0.3	Services

Continued on next page.

INPUTS AND OUTPUTS FOR MECHANICAL MEASURING DEVICES - Continued

Economic Sector or Industry Providing Inputs	%	Sector	Economic Sector or Industry Buying Outputs	%	Sector
Paperboard containers & boxes	1.0	Manufg.	S/L Govt. purch., higher education	0.3	S/L Govt
Real estate	1.0	Fin/R.E.	Industrial buildings	0.2	Constr.
Nonferrous rolling & drawing, nec	0.9	Manufg.	Office buildings	0.2	Constr.
Eating & drinking places	0.9	Trade	Book publishing	0.2	Manufg.
Commercial printing	0.8	Manufg.	Electronic computing equipment	0.2	Manufg.
Fabricated rubber products, nec	0.8	Manufg.	Manifold business forms	0.2	Manufg.
Wiring devices	0.8	Manufg.	Miscellaneous plastics products	0.2	Manufg.
Maintenance of nonfarm buildings nec	0.7	Constr.	Miscellaneous publishing	0.2	Manufg.
Brass, bronze, & copper castings	0.7	Manufg.	Motor vehicle parts & accessories	0.2	Manufg.
Motors & generators	0.7	Manufg.	Gas production & distribution (utilities)	0.2	Util.
U.S. Postal Service	0.7	Gov't	Hospitals	0.2	Services
Nonferrous metal ores, except copper	0.6	Mining	Blankbooks & looseleaf binders	0.1	Manufg.
Copper rolling & drawing	0.6	Manufg.	Glass & glass products, except containers	0.1	Manufg.
Industrial controls	0.6	Manufg.	Nitrogenous & phosphatic fertilizers	0.1	Manufg.
Machinery, except electrical, nec	0.6	Manufg.	Paper mills, except building paper	0.1	Manufg.
Miscellaneous plastics products	0.6	Manufg.	Paperboard containers & boxes	0.1	Manufg.
Transformers	0.6	Manufg.	Petroleum refining	0.1	Manufg.
Glass & glass products, except containers	0.5	Manufg.	Water supply & sewage systems	0.1	Util.
Nonferrous wire drawing & insulating	0.5	Manufg.	Colleges, universities, & professional schools	0.1	Services
Plastics materials & resins	0.5	Manufg.	Photofinishing labs, commercial photography	0.1	Services
Motor freight transportation & warehousing	0.5	Util.	S/L Govt. purch., elem. & secondary education	0.1	S/L Govt
Banking	0.5	Fin/R.E.	S/L Govt. purch., gas & electric utilities	0.1	S/L Govt
Legal services	0.5	Services			
Gas production & distribution (utilities)	0.4	Util.			
Automotive rental & leasing, without drivers	0.4	Services			
Computer & data processing services	0.4	Services			
Management & consulting services & labs	0.4	Services			
Miscellaneous fabricated wire products	0.3	Manufg.			
Plating & polishing	0.3	Manufg.			
Special dies & tools & machine tool accessories	0.3	Manufg.			
Noncomparable imports	0.3	Foreign			
Ball & roller bearings	0.2	Manufg.			
Converted paper products, nec	0.2	Manufg.			
Electron tubes	0.2	Manufg.			
Gaskets, packing & sealing devices	0.2	Manufg.			
Machine tools, metal cutting types	0.2	Manufg.			
Manifold business forms	0.2	Manufg.			
Nonferrous castings, nec	0.2	Manufg.			
Air transportation	0.2	Util.			
Insurance carriers	0.2	Fin/R.E.			
Accounting, auditing & bookkeeping	0.2	Services			
Abrasive products	0.1	Manufg.			
Lubricating oils & greases	0.1	Manufg.			
Primary batteries, dry & wet	0.1	Manufg.			
Sawmills & planning mills, general	0.1	Manufg.			
Switchgear & switchboard apparatus	0.1	Manufg.			
X-ray apparatus & tubes	0.1	Manufg.			
Royalties	0.1	Fin/R.E.			
Engineering, architectural, & surveying services	0.1	Services			
Theatrical producers, bands, entertainers	0.1	Services			

Source: Benchmark Input-Output Accounts for the U.S. Economy, 1982, U.S. Department of Commerce, Washington, D.C., July 1991. Data, as reported in the source, are organized by the 1977 SIC structure in use in 1982 but have been matched, as closely as is possible, to the 1987 SIC structure used in this book.

OCCUPATIONS EMPLOYED BY SIC 382 - MEASURING AND CONTROLLING DEVICES

Occupation	% of Total 1994	Change to 2005	Occupation	% of Total 1994	Change to 2005
Assemblers, fabricators, & hand workers nec	7.4	-11.4	Engineering, mathematical, & science managers	1.7	0.6
Electrical & electronics engineers	5.6	13.4	Industrial production managers	1.5	-11.4
Electrical & electronic equipment assemblers	5.4	-11.4	General office clerks	1.4	-24.4
Sales & related workers nec	4.3	-11.4	Bookkeeping, accounting, & auditing clerks	1.4	-33.6
Electrical & electronic technicians,technologists	3.7	-11.4	Purchasing agents, ex trade & farm products	1.3	-11.4
Electrical & electronic assemblers	3.5	-20.3	Production, planning, & expediting clerks	1.3	-11.4
Inspectors, testers, & graders, precision	3.3	-38.0	Traffic, shipping, & receiving clerks	1.3	-14.8
Electromechanical equipment assemblers	3.2	-2.5	Drafters	1.3	-31.0
Blue collar worker supervisors	3.2	-24.1	Computer engineers	1.3	31.2
General managers & top executives	2.4	-15.9	Stock clerks	1.2	-28.0
Secretaries, ex legal & medical	2.4	-19.3	Engineers nec	1.2	6.3
Machinists	2.0	-33.6	Management support workers nec	1.1	-11.4
Mechanical engineers	1.8	-2.5	Managers & administrators nec	1.1	-11.4
Engineering technicians & technologists nec	1.7	-11.4	Computer programmers	1.1	-28.2
Marketing, advertising, & PR managers	1.7	-11.4			

Source: Industry-Occupation Matrix, Bureau of Labor Statistics. These data relate to one or more 3-digit SIC industry groups rather than to a single 4-digit SIC. The change reported for each occupation to the year 2005 is a percent of growth or decline as estimated by the Bureau of Labor Statistics. The abbreviation nec stands for 'not elsewhere classified'.

LOCATION BY STATE AND REGIONAL CONCENTRATION

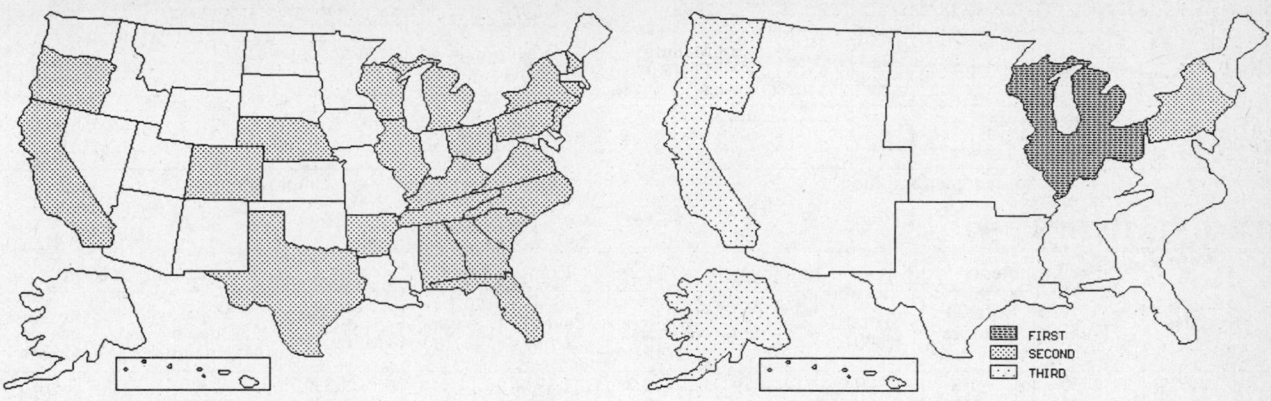

FIRST
SECOND
THIRD

INDUSTRY DATA BY STATE

State	Establish-ments	Shipments			Employment				Cost as % of Shipments	Investment per Employee ($)
		Total ($ mil)	% of U.S.	Per Establ.	Total Number	% of U.S.	Per Establ.	Wages ($/hour)		
Pennsylvania	11	328.3	12.6	29.8	2,100	13.0	191	14.80	34.4	6,286
Illinois	12	100.2	3.9	8.4	900	5.6	75	9.40	35.3	3,333
California	23	95.6	3.7	4.2	700	4.3	30	15.25	50.2	-
Ohio	9	75.7	2.9	8.4	400	2.5	44	8.63	64.7	-
Texas	17	60.0	2.3	3.5	500	3.1	29	14.00	30.5	2,000
New Jersey	10	32.1	1.2	3.2	300	1.9	30	4.33	43.0	1,000
Connecticut	9	17.7	0.7	2.0	300	1.9	33	8.50	39.5	667
Florida	7	13.8	0.5	2.0	100	0.6	14	5.00	50.7	2,000
New York	11	9.1	0.3	0.8	100	0.6	9	9.00	30.8	-
Michigan	10	(D)	-	-	7,500 *	46.3	750	-	-	-
North Carolina	7	(D)	-	-	375 *	2.3	54	-	-	-
Oregon	7	(D)	-	-	175 *	1.1	25	-	-	1,714
Wisconsin	7	(D)	-	-	750 *	4.6	107	-	-	-
South Carolina	4	(D)	-	-	375 *	2.3	94	-	-	-
New Hampshire	3	(D)	-	-	175 *	1.1	58	-	-	-
Alabama	2	(D)	-	-	750 *	4.6	375	-	-	-
Arkansas	2	(D)	-	-	375 *	2.3	188	-	-	-
Colorado	2	(D)	-	-	175 *	1.1	88	-	-	-
Kentucky	2	(D)	-	-	375 *	2.3	188	-	-	-
Nebraska	2	(D)	-	-	750 *	4.6	375	-	-	-
Tennessee	2	(D)	-	-	175 *	1.1	88	-	-	-
Georgia	1	(D)	-	-	375 *	2.3	375	-	-	-
Virginia	1	(D)	-	-	750 *	4.6	750	-	-	-

Source: 1992 *Economic Census*. The states are in descending order of shipments or establishments (if shipment data are missing for the majority). The symbol (D) appears when data are withheld to prevent disclosure of competitive information. States marked with (D) are sorted by number of establishments. A dash (-) indicates that the data element cannot be calculated; * indicates the midpoint of a range.

3825 - INSTRUMENTS TO MEASURE ELECTRICITY

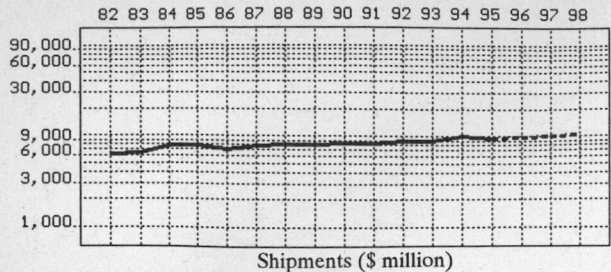

Shipments ($ million)

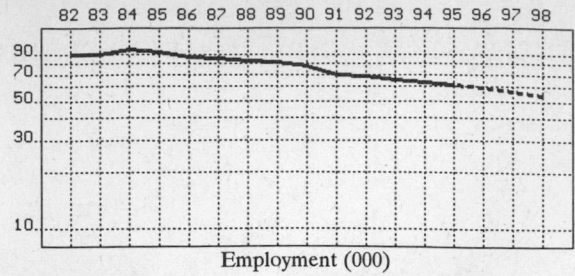

Employment (000)

GENERAL STATISTICS

Year	Companies	Establishments		Employment			Compensation		Production ($ million)			
		Total	with 20 or more employees	Total (000)	Production Workers (000)	Hours (Mil)	Payroll ($ mil)	Wages ($/hr)	Cost of Materials	Value Added by Manufacture	Value of Shipments	Capital Invest.
1982	675	750	353	89.9	48.9	92.7	1,893.6	8.18	1,850.7	4,305.4	6,120.1	309.9
1983		753	367	89.7	49.5	94.1	2,052.3	9.14	2,129.0	4,413.3	6,484.4	272.6
1984		756	381	95.8	53.0	101.7	2,356.1	9.71	2,589.5	5,371.3	7,810.5	418.0
1985		759	395	92.4	48.3	93.1	2,293.6	10.63	2,474.0	5,169.8	7,705.2	343.1
1986		754	379	86.2	44.3	89.2	2,356.1	11.05	2,407.1	4,535.2	6,940.5	290.4
1987	862	930	412	85.2	43.9	91.4	2,476.7	11.00	2,662.4	5,090.9	7,703.3	307.5
1988		905	395	82.9	42.1	84.7	2,511.1	11.70	2,801.4	5,198.0	7,984.4	214.8
1989		881	410	81.3	39.6	78.7	2,485.5	12.24	2,798.9	5,206.1	7,919.9	304.7
1990		873	392	77.1	38.7	77.3	2,603.9	12.88	3,041.0	5,352.4	8,389.7	292.7
1991		903	390	69.3	34.0	73.9	2,496.8	12.47	2,800.3	5,455.6	8,239.7	257.2
1992	900	964	388	68.7	32.3	63.4	2,549.0	14.15	3,091.2	5,721.1	8,873.3	324.7
1993		951	351	64.9	30.5	59.9	2,462.7	14.56	3,102.8	5,594.2	8,746.1	401.6
1994		986P	391P	62.6	31.8	64.3	2,547.3	14.85	3,600.5	6,660.7	10,097.8	333.3
1995		1,007P	392P	61.7P	28.3P	59.2P	2,693.4P	15.40P	3,430.9P	6,346.9P	9,622.1P	319.1P
1996		1,029P	393P	59.0P	26.4P	56.0P	2,736.6P	15.92P	3,517.4P	6,506.9P	9,864.6P	320.0P
1997		1,050P	394P	56.3P	24.5P	52.8P	2,779.8P	16.45P	3,603.8P	6,666.9P	10,107.2P	320.8P
1998		1,071P	395P	53.7P	22.7P	49.5P	2,823.0P	16.97P	3,690.3P	6,826.9P	10,349.7P	321.7P

Sources: 1982, 1987, 1992 Economic Census; Annual Survey of Manufactures, 83-86, 88-91, 93-94. Establishment counts for non-Census years are from County Business Patterns; establishment values for 83-84 are extrapolations. 'P's show projections by the editors. Industries reclassified in 87 will not have data for prior years.

INDICES OF CHANGE

Year	Companies	Establishments		Employment			Compensation		Production ($ million)			
		Total	with 20 or more employees	Total (000)	Production Workers (000)	Hours (Mil)	Payroll ($ mil)	Wages ($/hr)	Cost of Materials	Value Added by Manufacture	Value of Shipments	Capital Invest.
1982	75	78	91	131	151	146	74	58	60	75	69	95
1983		78	95	131	153	148	81	65	69	77	73	84
1984		78	98	139	164	160	92	69	84	94	88	129
1985		79	102	134	150	147	90	75	80	90	87	106
1986		78	98	125	137	141	92	78	78	79	78	89
1987	96	96	106	124	136	144	97	78	86	89	87	95
1988		94	102	121	130	134	99	83	91	91	90	66
1989		91	106	118	123	124	98	87	91	91	89	94
1990		91	101	112	120	122	102	91	98	94	95	90
1991		94	101	101	105	117	98	88	91	95	93	79
1992	100	100	100	100	100	100	100	100	100	100	100	100
1993		99	90	94	94	94	97	103	100	98	99	124
1994		102P	101P	91	98	101	100	105	116	116	114	103
1995		104P	101P	90P	88P	93P	106P	109P	111P	111P	108P	98P
1996		107P	101P	86P	82P	88P	107P	113P	114P	114P	111P	99P
1997		109P	102P	82P	76P	83P	109P	116P	117P	117P	114P	99P
1998		111P	102P	78P	70P	78P	111P	120P	119P	119P	117P	99P

Sources: Same as General Statistics. Values reflect change from the base year, 1992. Values above 100 mean greater than 92, values below 100 mean less than 92, and a value of 100 in the 82-91 or 93-98 period means same as 92. 'P's mark projections by the editors.

SELECTED RATIOS

For 1994	Avg. of All Manufact.	Analyzed Industry	Index	For 1994	Avg. of All Manufact.	Analyzed Industry	Index
Employees per Establishment	49	63	130	Value Added per Production Worker	134,084	209,456	156
Payroll per Establishment	1,500,273	2,583,230	172	Cost per Establishment	5,045,178	3,651,286	72
Payroll per Employee	30,620	40,692	133	Cost per Employee	102,970	57,516	56
Production Workers per Establishment	34	32	94	Cost per Production Worker	146,988	113,223	77
Wages per Establishment	853,319	968,323	113	Shipments per Establishment	9,576,895	10,240,232	107
Wages per Production Worker	24,861	30,027	121	Shipments per Employee	195,460	161,307	83
Hours per Production Worker	2,056	2,022	98	Shipments per Production Worker	279,017	317,541	114
Wages per Hour	12.09	14.85	123	Investment per Establishment	321,011	338,001	105
Value Added per Establishment	4,602,255	6,754,651	147	Investment per Employee	6,552	5,324	81
Value Added per Employee	93,930	106,401	113	Investment per Production Worker	9,352	10,481	112

Sources: Same as General Statistics. The 'Average of All Manufacturing' column represents the average of all manufacturing industries reported for the most recent complete year available. The Index shows the relationship between the Average and the Analyzed Industry. For example, 100 means that they are equal; 500 that the Analyzed Industry is five times the average; 50 means that the Analyzed Industry is half the national average. The abbreviation 'na' is used to show that data are 'not available'.

LEADING COMPANIES   Number shown: **75**   Total sales ($ mil): **6,539**   Total employment (000): **43.9**

Company Name	Address				CEO Name	Phone	Co. Type	Sales ($ mil)	Empl. (000)
Tektronix Inc	PO Box 1000	Wilsonville	OR	97070	Jerome J Meyer	503-627-7111	P	1,318	8.5
Teradyne Inc	321 Harrison Av	Boston	MA	02118	A V d'Arbeloff	617-482-2700	P	667	4.0
TRW Technar Inc	5462 N Irwindale	Irwindale	CA	91706	Tom Doyle	818-334-0250	S	500*	1.3
Fluke Corp	PO Box 9090	Everett	WA	98206	W G Parzybok Jr	206-347-6100	P	358	2.5
Allen Group Inc	25101 Chagrin Blv	Beachwood	OH	44122	Robert G Paul	216-765-5800	P	331	2.7
ATE	1601 Technology Dr	San Jose	CA	95110	Irwin Pfister	408-453-0123	D	300	1.2
LTX Corp	LTX Pk	Westwood	MA	02090	Graham C Miller	617-461-1000	P	168	0.9
KLA Instruments Corp	160 Rio Robles	San Jose	CA	95134	Kenneth Levy	408-434-4200	P	167	1.0
Automotive Diagnostics	8001 Angling Rd	Kalamazoo	MI	49002	Ron Ortiz	616-329-7600	S	160	1.6
Credence Systems Corp	3500 W Warren Av	Fremont	CA	94538	Elwood H Spedden	510-657-7400	P	104	0.4
Megatest Corp	1321 Ridder Pk Dr	San Jose	CA	95131	John E Halter	408-437-9400	P	100	0.5
Hekimian Laboratories Inc	15200 Omega Dr	Rockville	MD	20850	Robert M Ginnings	301-590-3600	S	100	0.3
Lucas Assembly	12841 Stark Rd	Livonia	MI	48150	Ralph Dautton	313-522-9680	S	100	0.7
Signal Technology Corp	955 Benecia Av	Sunnyvale	CA	94086	Dale L Peterson	408-730-6318	P	97	1.0
Lucas Assembly	12841 Stark Rd	Livonia	MI	48150	David Watkins	313-522-9680	D	90	0.6
Keithley Instruments Inc	28775 Aurora Rd	Cleveland	OH	44139	Joseph P Keithley	216-248-0400	P	89	0.6
Everett Charles Technologies	700 E Harrison Av	Pomona	CA	91767	Dave Van Loan	909-625-5551	R	75	0.4
Torrey Investments Inc	11995 El Cam Real	San Diego	CA	92130	Terrence Gooding	619-793-2300	R	69*	0.5
Wavetek Corp	11995 El Cam Real	San Diego	CA	92130	Terence Gooding	619-793-2300	S	69	0.3
IFR Systems Inc	10200 W York St	Wichita	KS	67215	Alfred H Hunt III	316-522-4981	P	65	0.7
Broadband Commun Prod	17 E Hibiscus Blv	Melbourne	FL	32901		407-984-3671	R	63*	0.5
Hitech Instruments	Robinair Way	Montpelier	OH	43543	William Clogg	614-884-7250	R	63*	0.5
LeCroy Corp	700 Chestnut Ridge	Chestnut Ridge	NY	10977	Lutz Henckels	914-425-2000	R	62	0.5
Tekelec Inc	26580 W Agoura Rd	Calabasas	CA	91302	Philip J Alford	818-880-5656	P	61	0.3
Therma-Wave Inc	47320 Msn Fls Ct	Fremont	CA	94539	Allan Rosencwaig	510-490-3663	J	51	0.2
Bourns Integrated Technologies	2533 N 1500 W	Ogden	UT	84404	Richard Box	801-786-6200	D	50	1.0
Clarostat Sensors & Controls	1500 International	Richardson	TX	75081	Dana Skaddon	214-479-1122	D	50	0.6
Whistler Corp	5 Liberty Way	Westford	MA	01886	Jack Shirman	508-692-3000	S	50	0.2
KDI Precision Products Inc	3975 McMann Rd	Cincinnati	OH	45245	Al Dilz	513-943-2000	S	45*	0.4
Micro Component Technology	3850 N Victoria St	St Paul	MN	55126	Daniel J Hill	612-482-5100	P	44	0.3
ADE Corp	77 Rowe St	Newton	MA	02166	Robert Abbe	617-969-0600	R	40	0.2
Hobbs Corp	Yale Blv & Ash St	Springfield	IL	62794	Willie P Ozbirn	217-753-7600	S	40	0.4
Dracon	809 Calle Plano	Camarillo	CA	93012	Thomas Erdmann	805-987-9511	D	37*	0.4
Integrated Measurement Systems	9525 SW Gemini Dr	Beaverton	OR	97005	Keith Barnes	503-626-7117	S	37*	0.2
Actron Manufacturing Company	9999 Walford Av	Cleveland	OH	44102	TF Slater	216-651-9200	R	35	0.2
Teledyne Relay	12525 Daphine Av	Hawthorne	CA	90250	Walter Strack	213-777-0077	D	35*	0.3
DIT-MCO International Corp	5612 Brighton Ter	Kansas City	MO	64130	RK Thompson	816-444-9700	S	33	0.3
RADA Electronics Indust Ltd	80 Express St	Plainview	NY	11803	Haim Nissenson	212-734-8340	P	33	0.2
Yokogawa Corporation	2 Dart Rd	Newnan	GA	30265	Ren Shabata	404-253-7000	S	33	0.3
Coils Inc	11716 Algonquin Rd	Huntley	IL	60142	J Plunkett	708-669-5115	R	28*	0.4
Bowmar Instrument Corp	5080 N 40th St	Phoenix	AZ	85018	Gardiner S Dutton	602-957-0271	P	28	0.2
Aetrium Inc	2350 Helen St	North St Paul	MN	55109		612-770-2000	P	26	0.3
AR Test Systems	7401 Boston Blv	Springfield	VA	22153	Robert Coackley	703-644-9000	D	26*	0.1
Telecom Solutions	85 W Tasman Dr	San Jose	CA	95134	D Ronald Duren	408-433-1900	D	26*	0.2
Curtis Instruments Inc	200 Kisco Av	Mount Kisco	NY	10549	EM Marwell	914-666-2971	R	25	0.4
Dranetz Technologies Inc	PO Box 4019	Edison	NJ	08818	Colin Baxter	908-287-3680	S	25	0.3
ITT Pomona Electronics	1500 E 9th St	Pomona	CA	91766	Tom Barber	909-469-2900	D	25	0.2
Rochester Instrument Systems	255 N Union St	Rochester	NY	14605	Gerald Schaefer	716-263-7700	S	25*	0.2
Datum Inc	1363 S State College	Anaheim	CA	92806	Louis B Horwitz	714-533-6333	P	25	0.3
Laser Precision Corp	32242 Adelanto	S J Capistrano	CA	92675	C Frederick Sehnert	714-489-2991	P	24	0.2
Thermo Voltek Corp	PO Box 2878	Woburn	MA	01888	John W Wood Jr	617-622-1000	P	24	0.1
Reliability Inc	PO Box 218370	Houston	TX	77218	Larry Edwards	713-492-0550	P	23	0.4
Zygo Corp	Laurel Brook Rd	Middlefield	CT	06455	Gary K Willis	203-347-8506	P	23	0.2
Hickok Inc	10514 DuPont Av	Cleveland	OH	44108	Robert L Bauman	216-541-8060	P	23	0.3
Daymarc Corp	301 Second Av	Waltham	MA	02154	MW Bosch	617-890-2345	S	22	0.1
Simpson Electric Co	853 Dundee Av	Elgin	IL	60120	Edward M Herter	708-697-2260	R	22	0.3
CXR Telcom Corp	2040 Fortune Dr	San Jose	CA	95131	Henry Mourad	408-435-8520	S	22	0.1
Technology Research Corp	5250 140th Av N	Clearwater	FL	34620	Robert S Wiggins	813-535-0572	P	20	0.3
Aseco Corp	500 Donald Lynch	Marlboro	MA	01752	Carl S Archer Jr	508-481-8896	P	20	0.1
Commonwealth Scientific Corp	500 Pendleton St	Alexandria	VA	22314	G Thompson	703-548-0800	R	20	0.1
Numerx Corp	1400 N Providence	Media	PA	19063	Eugene J White	610-892-0316	R	20	0.3
CNA Cable Network Analysis	625 SE Salmon Av	Redmond	OR	97756	Gerhard Beenen	503-923-0333	D	19*	0.2
Bird Electronic Corp	30303 Aurora Rd	Cleveland	OH	44139	B Bird	216-248-1200	R	19	0.2
Datcon Instrument Co	PO Box 128	E Petersburg	PA	17520	Steve Wener	717-569-5713	S	19	0.3
Cascade Microtech Inc	14255 Brigad	Beaverton	OR	97005	Eric Strid	503-626-8245	R	17*	0.1
AVO Biddle Instruments	510 Township Line	Blue Bell	PA	19422	Paul Ochadlick	215-646-9200	D	16*	0.2
Comlinear Corp	4800 Wheaton Dr	Fort Collins	CO	80525	Willem Andersen	303-226-0500	R	16	0.1
Media Logic Inc	PO Box 2258	Plainville	MA	02762	David R Lennox	508-695-2006	P	15	<0.1
Sonic Solutions	1891 E Francisco	San Rafael	CA	94901	Robert J Doris	415-485-4800	P	15	<0.1
Trio-Tech International	355 Parkside Dr	San Fernando	CA	91340	SW Yong	818-365-9200	P	15	0.6
JcAIR Inc	400 Industrial Pkwy	Indust Apt	KS	66031	LeWayne Rothers	913-764-2452	S	15	0.2
Racal Instruments Inc	4 Goodyear St	Irvine	CA	92718	Gordon Taylor	714-859-8999	S	15	0.1
SenSym Inc	1804 McCarthy Blv	Milpitas	CA	95035	Chris Cartsonas	408-954-1100	S	15	0.1
Technical Instrument Co	348 6th St	San Francisco	CA	94103	Francis E Lundy	415-431-8231	R	15	<0.1
Technicorp Inc	840 Church Rd	Mechanicsburg	PA	17055	Fred Johnston	717-766-0223	R	15	0.2

Source: Ward's Business Directory of U.S. Private and Public Companies, Volumes 1 and 2, 1996. The company type code used is as follows: P - Public, R - Private, S - Subsidiary, D - Division, J - Joint Venture, A - Affiliate, G - Group. Sales are in millions of dollars, employees are in thousands. An asterisk (*) indicates an estimated sales volume. The symbol < stands for 'less than'. Company names and addresses are truncated, in some cases, to fit into the available space.

MATERIALS CONSUMED

Material	Quantity	Delivered Cost ($ million)
Materials, ingredients, containers, and supplies	(X)	2,578.7
Printed circuit boards (without inserted components) for electronic circuitry	(X)	134.3
Printed circuit assemblies, loaded boards or modules	(X)	187.0
Semiconductors, including transistors, diodes, rectifiers, and integrated circuits for electronic circuitry	(X)	260.9
Capacitors for electronic circuitry	(X)	58.2
Resistors for electronic circuitry	(X)	45.4
Other components and accessories for electronic circuitry, nec, except tubes	(X)	124.0
Current-carrying wiring devices	(X)	60.1
Electrical transmission, distribution, and control equipment	(X)	95.3
Electronic computing equipment	(X)	49.8
Electrical instrument mechanisms and meter movements (including instrument relays)	(X)	23.5
Electrical measuring instruments and parts, not listed elsewhere	(X)	107.8
Fractional horsepower electric motors and generators (less than 1 hp) including timing motors	(X)	3.8
Plastics resins consumed in the form of granules, pellets, powders, liquids, etc.	(X)	23.5
Fabricated plastics products (except gaskets, hoses, and belting)	(X)	47.0
Sheet metal products, except stampings	(X)	83.6
Metal stampings	(X)	13.6
Fabricated metal wire products (including wire rope, cable, springs, etc.)	(X)	20.8
Metal bolts, nuts, screws, washers, rivets, and other screw machine products	(X)	26.9
Other fabricated metal products (except forgings)	(X)	32.6
Forgings	(X)	0.2
Iron and steel castings (rough and semifinished)	(X)	3.9
Aluminum and aluminum-base alloy castings (rough and semifinished)	(X)	8.5
Other nonferrous castings (rough and semifinished)	(X)	0.4
Steel shapes and forms	(X)	17.0
Copper and copper-base alloy shapes and forms	(X)	6.8
Aluminum and aluminum-base alloy shapes and forms	(X)	16.7
Other nonferrous shapes and forms	(X)	3.5
Glass and glass products (excluding windows and mirrors)	(X)	18.7
Paper and paperboard products except paperboard boxes, containers, and corrugated paperboard	(X)	19.8
Paper and paperboard containers, including shipping sacks and other paper packaging supplies	(X)	18.0
All other materials and components, parts, containers, and supplies	(X)	408.2
Materials, ingredients, containers, and supplies, nsk	(X)	659.0

Source: 1992 Economic Census. Explanation of symbols used: (D): Withheld to avoid disclosure of competitive data; na: Not available; (S): Withheld because statistical norms were not met; (X): Not applicable; (Z): Less than half the unit shown; nec: Not elsewhere classified; nsk: Not specified by kind; - : zero; * : 10-19 percent estimated; ** : 20-29 percent estimated.

PRODUCT SHARE DETAILS

Product or Product Class	% Share	Product or Product Class	% Share
Instruments to measure electricity	100.00	communication circuits, and motors	81.94
Integrating instruments, electrical	5.52	Other instruments to measure electricity	6.94
Test equipment for testing electrical, radio and		Instruments to measure electricity, nsk	5.60

Source: 1992 Economic Census. The values shown are percent of total shipments in an industry. Values of indented subcategories are summed in the main heading. The symbol (D) appears when data are withheld to prevent disclosure of competitive information. The abbreviation nsk stands for 'not specified by kind' and nec for 'not elsewhere classified'.

INPUTS AND OUTPUTS FOR INSTRUMENTS TO MEASURE ELECTRICITY

Economic Sector or Industry Providing Inputs	%	Sector	Economic Sector or Industry Buying Outputs	%	Sector
Electronic components nec	17.5	Manufg.	Gross private fixed investment	54.4	Cap Inv
Instruments to measure electricity	13.0	Manufg.	Exports	22.7	Foreign
Industrial controls	10.4	Manufg.	Federal Government purchases, national defense	8.5	Fed Govt
Wholesale trade	10.4	Trade	Instruments to measure electricity	5.8	Manufg.
Semiconductors & related devices	6.0	Manufg.	Mechanical measuring devices	2.2	Manufg.
Imports	5.9	Foreign	Federal Government purchases, nondefense	1.7	Fed Govt
Advertising	3.6	Services	Radio & TV communication equipment	1.2	Manufg.
Electronic computing equipment	2.5	Manufg.	Engineering & scientific instruments	0.6	Manufg.
Communications, except radio & TV	2.2	Util.	Electronic components nec	0.4	Manufg.
Electric services (utilities)	2.1	Util.	Change in business inventories	0.4	In House
Plating & polishing	1.8	Manufg.	Crude petroleum & natural gas	0.3	Mining
Sheet metal work	1.4	Manufg.	Aircraft	0.3	Manufg.
Real estate	1.3	Fin/R.E.	Industrial controls	0.3	Manufg.
Eating & drinking places	1.2	Trade	S/L Govt. purch., gas & electric utilities	0.2	S/L Govt
Equipment rental & leasing services	1.1	Services	Environmental controls	0.1	Manufg.
Screw machine and related products	0.9	Manufg.	Semiconductors & related devices	0.1	Manufg.
Wiring devices	0.9	Manufg.	Switchgear & switchboard apparatus	0.1	Manufg.
Blast furnaces & steel mills	0.8	Manufg.	Electric services (utilities)	0.1	Util.
Machinery, except electrical, nec	0.8	Manufg.	S/L Govt. purch., higher education	0.1	S/L Govt
Maintenance of nonfarm buildings nec	0.7	Constr.			
Aluminum rolling & drawing	0.7	Manufg.			
Gaskets, packing & sealing devices	0.7	Manufg.			

Continued on next page.

INPUTS AND OUTPUTS FOR INSTRUMENTS TO MEASURE ELECTRICITY - Continued

Economic Sector or Industry Providing Inputs	%	Sector	Economic Sector or Industry Buying Outputs	%	Sector
Banking	0.7	Fin/R.E.			
Nonferrous wire drawing & insulating	0.6	Manufg.			
Legal services	0.6	Services			
Glass & glass products, except containers	0.5	Manufg.			
Metal stampings, nec	0.5	Manufg.			
Miscellaneous plastics products	0.5	Manufg.			
Paperboard containers & boxes	0.5	Manufg.			
Gas production & distribution (utilities)	0.5	Util.			
Management & consulting services & labs	0.5	Services			
Electron tubes	0.4	Manufg.			
Metal coating & allied services	0.4	Manufg.			
Motors & generators	0.4	Manufg.			
Primary nonferrous metals, nec	0.4	Manufg.			
Special dies & tools & machine tool accessories	0.4	Manufg.			
Computer & data processing services	0.4	Services			
Aluminum castings	0.3	Manufg.			
Copper rolling & drawing	0.3	Manufg.			
Plastics materials & resins	0.3	Manufg.			
Air transportation	0.3	Util.			
Motor freight transportation & warehousing	0.3	Util.			
Accounting, auditing & bookkeeping	0.3	Services			
U.S. Postal Service	0.3	Gov't			
Abrasive products	0.2	Manufg.			
Commercial printing	0.2	Manufg.			
Converted paper products, nec	0.2	Manufg.			
Fabricated metal products, nec	0.2	Manufg.			
Manifold business forms	0.2	Manufg.			
Miscellaneous fabricated wire products	0.2	Manufg.			
Switchgear & switchboard apparatus	0.2	Manufg.			
Royalties	0.2	Fin/R.E.			
Engineering, architectural, & surveying services	0.2	Services			
Ball & roller bearings	0.1	Manufg.			
Iron & steel foundries	0.1	Manufg.			
Lubricating oils & greases	0.1	Manufg.			
Nonferrous rolling & drawing, nec	0.1	Manufg.			
Petroleum refining	0.1	Manufg.			
Photographic equipment & supplies	0.1	Manufg.			
Transformers	0.1	Manufg.			
Sanitary services, steam supply, irrigation	0.1	Util.			
Automotive repair shops & services	0.1	Services			
Photofinishing labs, commercial photography	0.1	Services			

Source: *Benchmark Input-Output Accounts for the U.S. Economy, 1982*, U.S. Department of Commerce, Washington, D.C., July 1991. Data, as reported in the source, are organized by the 1977 SIC structure in use in 1982 but have been matched, as closely as is possible, to the 1987 SIC structure used in this book.

OCCUPATIONS EMPLOYED BY SIC 382 - MEASURING AND CONTROLLING DEVICES

Occupation	% of Total 1994	Change to 2005	Occupation	% of Total 1994	Change to 2005
Assemblers, fabricators, & hand workers nec	7.4	-11.4	Engineering, mathematical, & science managers	1.7	0.6
Electrical & electronics engineers	5.6	13.4	Industrial production managers	1.5	-11.4
Electrical & electronic equipment assemblers	5.4	-11.4	General office clerks	1.4	-24.4
Sales & related workers nec	4.3	-11.4	Bookkeeping, accounting, & auditing clerks	1.4	-33.6
Electrical & electronic technicians,technologists	3.7	-11.4	Purchasing agents, ex trade & farm products	1.3	-11.4
Electrical & electronic assemblers	3.5	-20.3	Production, planning, & expediting clerks	1.3	-11.4
Inspectors, testers, & graders, precision	3.3	-38.0	Traffic, shipping, & receiving clerks	1.3	-14.8
Electromechanical equipment assemblers	3.2	-2.5	Drafters	1.3	-31.0
Blue collar worker supervisors	3.2	-24.1	Computer engineers	1.3	31.2
General managers & top executives	2.4	-15.9	Stock clerks	1.2	-28.0
Secretaries, ex legal & medical	2.4	-19.3	Engineers nec	1.2	6.3
Machinists	2.0	-33.6	Management support workers nec	1.1	-11.4
Mechanical engineers	1.8	-2.5	Managers & administrators nec	1.1	-11.4
Engineering technicians & technologists nec	1.7	-11.4	Computer programmers	1.1	-28.2
Marketing, advertising, & PR managers	1.7	-11.4			

Source: *Industry-Occupation Matrix*, Bureau of Labor Statistics. These data relate to one or more 3-digit SIC industry groups rather than to a single 4-digit SIC. The change reported for each occupation to the year 2005 is a percent of growth or decline as estimated by the Bureau of Labor Statistics. The abbreviation nec stands for 'not elsewhere classified'.

LOCATION BY STATE AND REGIONAL CONCENTRATION

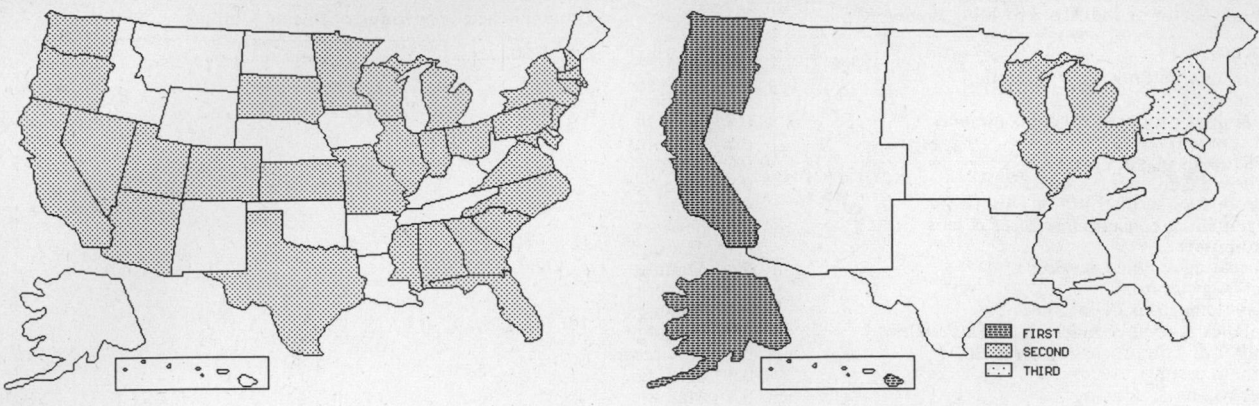

FIRST
SECOND
THIRD

INDUSTRY DATA BY STATE

State	Establish-ments	Shipments			Employment				Cost as % of Shipments	Investment per Employee ($)
		Total ($ mil)	% of U.S.	Per Establ.	Total Number	% of U.S.	Per Establ.	Wages ($/hour)		
California	268	2,438.0	27.5	9.1	17,600	25.6	66	17.02	31.9	4,977
Illinois	41	659.5	7.4	16.1	2,900	4.2	71	11.54	40.4	22,138
Massachusetts	55	651.5	7.3	11.8	5,200	7.6	95	17.64	23.1	5,000
Washington	35	607.9	6.9	17.4	4,200	6.1	120	18.84	30.6	5,167
New York	56	579.6	6.5	10.3	4,600	6.7	82	14.67	34.5	3,174
Colorado	34	577.2	6.5	17.0	4,200	6.1	124	13.69	38.0	-
New Jersey	36	258.8	2.9	7.2	2,000	2.9	56	15.06	28.6	3,300
Ohio	35	201.2	2.3	5.7	2,000	2.9	57	10.11	40.3	2,350
Florida	28	194.5	2.2	6.9	1,800	2.6	64	11.71	42.4	4,167
New Hampshire	19	192.7	2.2	10.1	2,300	3.3	121	13.77	41.0	2,304
Texas	50	167.5	1.9	3.3	1,500	2.2	30	9.25	38.0	1,933
Pennsylvania	34	162.3	1.8	4.8	1,400	2.0	41	11.29	37.0	1,643
Georgia	11	134.2	1.5	12.2	700	1.0	64	8.83	45.0	3,000
North Carolina	8	121.3	1.4	15.2	1,200	1.7	150	14.36	37.7	2,250
Indiana	13	111.0	1.3	8.5	1,300	1.9	100	12.41	36.8	2,462
Wisconsin	11	86.5	1.0	7.9	900	1.3	82	10.70	41.7	3,111
Minnesota	21	83.4	0.9	4.0	700	1.0	33	13.50	30.6	1,714
Connecticut	26	74.7	0.8	2.9	700	1.0	27	10.00	32.5	1,714
Virginia	11	39.0	0.4	3.5	400	0.6	36	14.00	22.8	1,500
Arizona	18	28.5	0.3	1.6	300	0.4	17	12.50	40.7	1,667
Missouri	10	25.5	0.3	2.5	300	0.4	30	11.00	33.7	2,333
Rhode Island	7	17.2	0.2	2.5	200	0.3	29	11.00	26.2	1,000
Michigan	30	(D)	-	-	750 *	1.1	25	-	-	1,867
Oregon	27	(D)	-	-	7,500 *	10.9	278	-	-	-
Maryland	14	(D)	-	-	750 *	1.1	54	-	-	-
Alabama	7	(D)	-	-	750 *	1.1	107	-	-	-
Kansas	5	(D)	-	-	750 *	1.1	150	-	-	-
Nevada	4	(D)	-	-	1,750 *	2.5	438	-	-	-
Utah	4	(D)	-	-	175 *	0.3	44	-	-	-
South Carolina	3	(D)	-	-	750 *	1.1	250	-	-	-
Mississippi	1	(D)	-	-	175 *	0.3	175	-	-	-
South Dakota	1	(D)	-	-	175 *	0.3	175	-	-	-

Source: 1992 *Economic Census*. The states are in descending order of shipments or establishments (if shipment data are missing for the majority). The symbol (D) appears when data are withheld to prevent disclosure of competitive information. States marked with (D) are sorted by number of establishments. A dash (-) indicates that the data element cannot be calculated; * indicates the midpoint of a range.

3826 - LABORATORY ANALYTICAL INSTRUMENTS

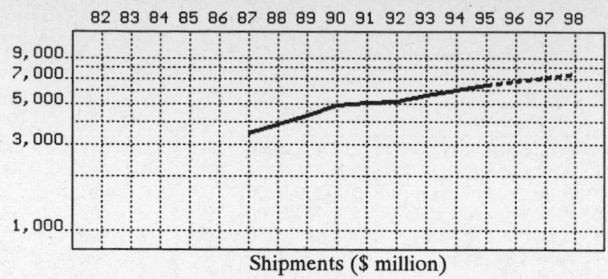

Shipments ($ million)

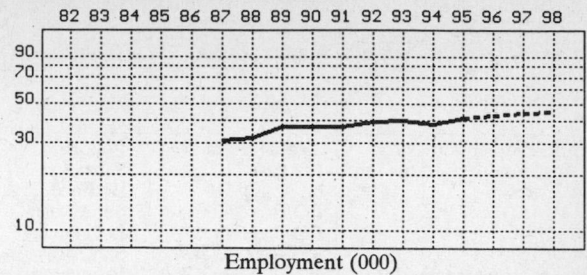

Employment (000)

GENERAL STATISTICS

| Year | Com-panies | Establishments | | Employment | | | Compensation | | Production ($ million) | | | |
		Total	with 20 or more employees	Total (000)	Production Workers (000)	Hours (Mil)	Payroll ($ mil)	Wages ($/hr)	Cost of Materials	Value Added by Manufacture	Value of Shipments	Capital Invest.
1982												
1983												
1984												
1985												
1986												
1987	528	562	207	31.2	13.5	26.7	892.9	10.76	1,363.2	2,107.1	3,468.2	125.5
1988		578	224	32.2	13.6	27.6	1,012.6	11.53	1,447.9	2,458.9	3,863.0	167.0
1989		554	217	37.5	15.3	30.8	1,142.5	11.51	1,598.1	2,776.4	4,306.1	163.4
1990		530	221	37.0	15.1	30.6	1,285.3	11.79	1,875.6	3,018.7	4,906.1	151.9
1991		538	215	37.0	14.7	29.8	1,343.5	12.30	1,965.0	3,134.0	5,070.6	195.3
1992	551	593	227	39.7	15.2	29.6	1,478.1	13.32	2,205.5	3,004.8	5,191.3	227.8
1993		581	220	40.3	15.5	31.1	1,558.7	13.51	2,286.6	3,356.8	5,657.3	212.6
1994		572P	225P	38.4	15.1	31.2	1,482.2	14.09	2,369.6	3,603.5	5,973.4	176.1
1995		575P	226P	41.9P	15.8P	32.1P	1,698.8P	14.45P	2,529.2P	3,846.2P	6,375.6P	221.3P
1996		578P	228P	43.0P	16.1P	32.6P	1,793.1P	14.91P	2,667.7P	4,056.8P	6,724.8P	231.1P
1997		580P	229P	44.2P	16.3P	33.1P	1,887.4P	15.38P	2,806.2P	4,267.4P	7,073.9P	240.8P
1998		583P	231P	45.4P	16.5P	33.7P	1,981.7P	15.85P	2,944.7P	4,478.0P	7,423.1P	250.6P

Sources: 1982, 1987, 1992 *Economic Census*; *Annual Survey of Manufactures*, 83-86, 88-91, 93-94. Establishment counts for non-Census years are from *County Business Patterns*; establishment values for 83-84 are extrapolations. 'P's show projections by the editors. Industries reclassified in 87 will not have data for prior years.

INDICES OF CHANGE

| Year | Com-panies | Establishments | | Employment | | | Compensation | | Production ($ million) | | | |
		Total	with 20 or more employees	Total (000)	Production Workers (000)	Hours (Mil)	Payroll ($ mil)	Wages ($/hr)	Cost of Materials	Value Added by Manufacture	Value of Shipments	Capital Invest.
1982												
1983												
1984												
1985												
1986												
1987	96	95	91	79	89	90	60	81	62	70	67	55
1988		97	99	81	89	93	69	87	66	82	74	73
1989		93	96	94	101	104	77	86	72	92	83	72
1990		89	97	93	99	103	87	89	85	100	95	67
1991		91	95	93	97	101	91	92	89	104	98	86
1992	100	100	100	100	100	100	100	100	100	100	100	100
1993		98	97	102	102	105	105	101	104	112	109	93
1994		97P	99P	97	99	105	100	106	107	120	115	77
1995		97P	100P	106P	104P	108P	115P	108P	115P	128P	123P	97P
1996		97P	100P	108P	106P	110P	121P	112P	121P	135P	130P	101P
1997		98P	101P	111P	107P	112P	128P	115P	127P	142P	136P	106P
1998		98P	102P	114P	109P	114P	134P	119P	134P	149P	143P	110P

Sources: Same as General Statistics. Values reflect change from the base year, 1992. Values above 100 mean greater than 92, values below 100 mean less than 92, and a value of 100 in the 82-91 or 93-98 period means same as 92. 'P's mark projections by the editors.

SELECTED RATIOS

For 1994	Avg. of All Manufact.	Analyzed Industry	Index	For 1994	Avg. of All Manufact.	Analyzed Industry	Index
Employees per Establishment	49	67	137	Value Added per Production Worker	134,084	238,642	178
Payroll per Establishment	1,500,273	2,589,319	173	Cost per Establishment	5,045,178	4,139,556	82
Payroll per Employee	30,620	38,599	126	Cost per Employee	102,970	61,708	60
Production Workers per Establishment	34	26	77	Cost per Production Worker	146,988	156,927	107
Wages per Establishment	853,319	767,970	90	Shipments per Establishment	9,576,895	10,435,188	109
Wages per Production Worker	24,861	29,113	117	Shipments per Employee	195,460	155,557	80
Hours per Production Worker	2,056	2,066	100	Shipments per Production Worker	279,017	395,589	142
Wages per Hour	12.09	14.09	117	Investment per Establishment	321,011	307,637	96
Value Added per Establishment	4,602,255	6,295,109	137	Investment per Employee	6,552	4,586	70
Value Added per Employee	93,930	93,841	100	Investment per Production Worker	9,352	11,662	125

Sources: Same as General Statistics. The 'Average of All Manufacturing' column represents the average of all manufacturing industries reported for the most recent complete year available. The Index shows the relationship between the Average and the Analyzed Industry. For example, 100 means that they are equal; 500 that the Analyzed Industry is five times the average; 50 means that the Analyzed Industry is half the national average. The abbreviation 'na' is used to show that data are 'not available'.

LEADING COMPANIES   Number shown: 75   Total sales ($ mil): 3,548   Total employment (000): 24.1

Company Name	Address				CEO Name	Phone	Co. Type	Sales ($ mil)	Empl. (000)
Beckman Instruments Inc	2500 Harbor Blv	Fullerton	CA	92634	Louis T Rosso	714-871-4848	P	909	6.9
Thermo Instrument Systems Inc	PO Box 2108	Santa Fe	NM	87504	Arvin H Smith	505-471-3232	P	662	4.0
Millipore Corp	80 Ashby Rd	Bedford	MA	01730	John A Gilmartin	617-275-9200	P	497	3.1
Dionex Corp	PO Box 3603	Sunnyvale	CA	94088	A Blaine Bowman	408-737-0700	P	109	0.6
Hach Co	PO Box 389	Loveland	CO	80539	Kathryn C Hach	303-669-3050	P	100	0.9
Finnigan Corp	355 River Oaks	San Jose	CA	95134	R W Chapman	408-433-4800	S	92	0.8
Roche Diagnostic Systems Inc	1080 US Hwy 202	Branchburg	NJ	08876	Carlo Medici	908-253-7200	S	84*	0.6
Baird Corp	125 Middlesex Tpk	Bedford	MA	01730	Michael W Routh	617-276-6000	S	70	0.4
Horiba Instruments Inc	17671 Armstrong	Irvine	CA	92714	Yuji Hayashi	714-250-4811	S	65	0.2
Abbott Diagnostics	5440 Patrick Henry	Santa Clara	CA	95054	C Monahan	408-982-4850	R	63*	0.4
Thermo Jarrell Ash Corp	27 Forge Pkwy	Franklin	MA	02038	Earl Lewis	508-520-1880	D	60	0.3
Andros Inc	2332 4th St	Berkeley	CA	94710	Dane Nelson	510-849-5700	P	58	0.2
Perkin-Elmer Corp	6509 Flying Cloud	Eden Prairie	MN	55344	Paul W Palmberg	612-828-6100	D	52	0.4
Ionics Inc	65 Grove St	Watertown	MA	02172	A Goldstein	617-926-2500	D	40	0.4
New Brunswick Scientific	PO Box 4005	Edison	NJ	08818	Ezra Weisman	908-287-1200	P	36	0.4
Rheometrics Inc	1 Possumtown Rd	Piscataway	NJ	08854	Robert E Davis	908-560-8550	P	35	0.2
Industrial Scientific Corp	1001 Oakdale Rd	Oakdale	PA	15071	Kent D McElhattan	412-788-4353	P	31	0.2
Micromeritics Instrument Corp	1 Micromeritics Dr	Norcross	GA	30093	W Hendricks	404-662-3633	R	25	0.2
Molecular Devices Corp	1311 Orleans Dr	Sunnyvale	CA	94025	James p Iuliano	408-747-1700	R	25	0.1
Spex Group Inc	3880 Park Av	Edison	NJ	08820	N Stein	908-549-7144	P	23*	0.2
NORAN Instruments Inc	2551 W Beltline	Middleton	WI	53562	Stephen M Dillard	608-831-6511	S	22*	0.2
Fisons Instruments	55 Cherry Hill Dr	Beverly	MA	01915	Chuck Eichten	508-524-1000	D	21*	0.2
Shimadzu Scientific Instruments	7102 Riverwood Rd	Columbia	MD	21046	Shigehiko Hattori	410-381-1227	S	21*	0.2
Apothecary Products Inc	11531 Rupp Dr	Burnsville	MN	55337	T Noble	612-890-1940	R	20	0.2
Foxboro Co	PO Box 500	E Bridgewater	MA	02333	Robert R Hicks	508-378-5556	D	20	0.1
Lee Scientific	1515 W 2200 S	Salt Lake City	UT	84119	Brent Middleton	801-972-9292	D	20	<0.1
Stanford Research Systems Inc	1290 Reamwood	Sunnyvale	CA	94089	William R Green	408-744-9040	R	20	<0.1
Spectrum Systems Inc	3410 W 9 Mile Rd	Pensacola	FL	32526	HG Jones	904-944-3392	R	20	<0.1
Dynatech Laboratories Inc	14340 Sullyfield Cir	Chantilly	VA	22021	Tim Ellis	703-631-7800	S	18	0.1
Fisons Instruments Mfg	24911nue Stanford	Valencia	CA	91355	Keith Paul	805-295-0019	S	18*	0.2
Bioanalytical Systems Inc	2701 Kent Av	West Lafayette	IN	47906	Peter Kissinger	317-463-4527	R	16	<0.1
Ceramatec Inc	2425 S 900 W	Salt Lake City	UT	84119	A Bjorseth	801-972-2455	S	16	0.1
Meridian Instruments Inc	2310 Science Pkwy	Okemos	MI	48864	John Duffendack	517-349-7200	R	16	<0.1
TopoMetrix Corp	5403 Betsy Ross Dr	Santa Clara	CA	95054	Gary Aden	408-982-9700	R	15	<0.1
Advanced Instruments Inc	2 Technology Way	Norwood	MA	02062	Romano Micciche	617-449-3000	R	13	0.1
Gilson Inc	PO Box 620027	Middleton	WI	53562	Robert E Gilson	608-836-1551	R	13	<0.1
United Sciences Inc	5310 N Pioneer Rd	Gibsonia	PA	15044	John E Traina	412-443-8610	S	13	<0.1
Isco Inc	PO Box 5347	Lincoln	NE	68505	Robert W Allington	402-464-0231	D	12	0.2
ESA Inc	22 Alpha Rd	Chelmsford	MA	01824	Alvin V Block	508-250-7000	R	12	0.1
NIRSystems Inc	12101 Tech Rd	Silver Spring	MD	20904	Donald Webster	301-680-9600	S	12	<0.1
Dynatech Nevada Inc	2000 Arrowhead Dr	Carson City	NV	89706	T Hendricks	702-883-3400	R	11	<0.1
Energy Concepts Inc	7440 N Long Av	Skokie	IL	60077	Richard E Gibbons	708-676-4300	R	10	<0.1
Humboldt Manufacturing Co	7300 W Agatite Av	Norridge	IL	60656	DE Burgess	708-456-6300	R	10	<0.1
Spex Chemical and Sample Prep	203 Norcross Av	Metuchen	NJ	08840	Gilbert Hayat	908-549-7144	S	10	<0.1
UTI Instruments Co	2030 Fortune Dr	San Jose	CA	95131	Philip Merritt	408-428-9400	R	10	<0.1
CELLEX BIOSCIENCES Inc	8500 Evergreen	Coon Rapids	MN	55433	Eugene I Schuster	612-786-0302	P	8	<0.1
Digital Instruments Inc	520 E Montecito St	Santa Barbara	CA	93103	Virgil Elings	805-899-3380	R	8	<0.1
Genus Inc	4 Mulliken Way	Newburyport	MA	01950	John Aldeborgh	508-463-1500	D	8*	<0.1
Alltech Associates Inc	2051 Waukegan Rd	Deerfield	IL	60015	Richard Dolan	708-948-8600	R	7	0.2
Interactive Video Systems Inc	45 Winthrop St	Concord	MA	01742	Hans Hoyer	508-371-2600	R	7*	<0.1
SLM Instruments Inc	820 Linden Av	Rochester	NY	14625	Stuart Karon	716-248-4110	S	7	<0.1
World Precision Instruments Inc	175 Sarasota Center	Sarasota	FL	34240	H Fein	813-371-1003	R	6	<0.1
Gow-Mac Instrument Co	PO Box 25444	Lehigh Valley	PA	18002	R Mathieu	610-954-9000	R	6	<0.1
Baseline Industries Inc	PO Box 649	Lyons	CO	80540	Robert Forsberg	303-823-6661	S	5*	<0.1
Columbus Instruments	950 N Hague Av	Columbus	OH	43204	Jan Czekajewski	614-488-6176	R	5	<0.1
Paar Physica USA Inc	1090 K Georges	Edison	NJ	08837	Sean W Race	908-738-6640	R	5	<0.1
Innovative Sensors Inc	4745 E Bryson St	Anaheim	CA	92807	GL Bukamier	714-779-8781	R	4	<0.1
Regis Technologies Inc	8210 Austin Av	Morton Grove	IL	60053	LJ Glunz	708-967-6000	R	4	<0.1
Techne Inc	3700 Brunswick Pk	Princeton	NJ	08540	Neil R Pope	609-452-9275	R	4	<0.1
Teletrac Inc	137 Aero Camino	Santa Barbara	CA	93117	Richard Howitt	805-968-4333	R	4	<0.1
Biomolecular Separation Inc	2325 Robb Dr	Reno	NV	89523	ER Tarantino	702-746-2200	R	3	<0.1
GBC Scientific Equipment Inc	3930 Ventura Dr	Arlington H	IL	60004	Ron Grey	708-506-1900	R	3*	<0.1
Inter-Continental Microwave Inc	1515 Wyatt Dr	Santa Clara	CA	95054	Werner Schuerch	408-727-1596	R	3	<0.1
Wescor Inc	459 S Main St	Logan	UT	84321	Wayne K Barlow	801-752-6011	R	3*	<0.1
Alcott Chromatography Inc	1770 Corporate Dr	Norcross	GA	30093	Thomas Alcott	404-279-2521	S	3	<0.1
Holometrix Inc	25 Wiggins Av	Bedford	MA	01730	Edward Estey	617-275-3300	P	3	<0.1
Unimetrics Corp	501 Earl Rd	Shorewood	IL	60436	WE Armstrong	815-741-1370	R	3	<0.1
Ophir Corp	10184 W Belleview	Littleton	CO	80127	Donald Rottner	303-933-2200	R	2	<0.1
Genetronics Inc	11199 Sorrento Val	San Diego	CA	92121	Lois J Crandell	619-597-6006	R	2	<0.1
RP Cargille Laboratories Inc	55 Commerce Rd	Cedar Grove	NJ	07009	John J Cargille	201-239-6633	R	2	<0.1
Angstrom Inc	PO Box 248	Belleville	MI	48112	John R Schneider	313-697-8058	R	2	<0.1
Biolog Inc	3938 Trust Way	Hayward	CA	94545	Barry R Bochner	510-785-2564	R	2*	<0.1
Biotronics Technologies Inc	W 226 N 555	Waukesha	WI	53186	Charles G Terrizzi	414-475-7653	R	2	<0.1
Imaging Products International	4425 Brookfield	Chantilly	VA	22021	Howard Neal	703-803-0211	R	2	<0.1
North Carolina SRT Inc	1018 Morrisville	Morrisville	NC	27560	Henry Kopf	919-469-5848	R	2	<0.1

Source: Ward's Business Directory of U.S. Private and Public Companies, Volumes 1 and 2, 1996. The company type code used is as follows: P - Public, R - Private, S - Subsidiary, D - Division, J - Joint Venture, A - Affiliate, G - Group. Sales are in millions of dollars, employees are in thousands. An asterisk (*) indicates an estimated sales volume. The symbol < stands for 'less than'. Company names and addresses are truncated, in some cases, to fit into the available space.

MATERIALS CONSUMED

Material	Quantity	Delivered Cost ($ million)
Materials, ingredients, containers, and supplies	(X)	1,899.2
Printed circuit boards (without inserted components) for electronic circuitry	(X)	23.8
Printed circuit assemblies, loaded boards or modules	(X)	71.0
Semiconductors, including transistors, diodes, rectifiers, and integrated circuits for electronic circuitry	(X)	38.8
Capacitors for electronic circuitry	(X)	7.5
Resistors for electronic circuitry	(X)	9.3
Other components and accessories for electronic circuitry, nec, except tubes	(X)	50.6
Current-carrying wiring devices	(X)	21.7
Electrical transmission, distribution, and control equipment	(X)	26.2
Electronic computing equipment	(X)	65.6
Electrical instrument mechanisms and meter movements (including instrument relays)	(X)	18.5
Electrical measuring instruments and parts, not listed elsewhere	(X)	75.1
Fractional horsepower electric motors and generators (less than 1 hp) including timing motors	(X)	11.5
Plastics resins consumed in the form of granules, pellets, powders, liquids, etc.	(X)	1.9
Fabricated plastics products (except gaskets, hoses, and belting)	(X)	43.1
Sheet metal products, except stampings	(X)	74.3
Metal stampings	(X)	6.4
Fabricated metal wire products (including wire rope, cable, springs, etc.)	(X)	7.3
Metal bolts, nuts, screws, washers, rivets, and other screw machine products	(X)	37.3
Other fabricated metal products (except forgings)	(X)	47.9
Iron and steel castings (rough and semifinished)	(X)	1.6
Aluminum and aluminum-base alloy castings (rough and semifinished)	(X)	3.3
Other nonferrous castings (rough and semifinished)	(X)	1.1
Steel shapes and forms	(X)	7.9
Aluminum and aluminum-base alloy shapes and forms	(X)	13.2
Other nonferrous shapes and forms	(X)	2.4
Glass and glass products (excluding windows and mirrors)	(X)	47.0
Paper and paperboard products except paperboard boxes, containers, and corrugated paperboard	(X)	6.5
Paper and paperboard containers, including shipping sacks and other paper packaging supplies	(X)	19.3
All other materials and components, parts, containers, and supplies	(X)	417.4
Materials, ingredients, containers, and supplies, nsk	(X)	741.7

Source: 1992 *Economic Census*. Explanation of symbols used: (D): Withheld to avoid disclosure of competitive data; na: Not available; (S): Withheld because statistical norms were not met; (X): Not applicable; (Z): Less than half the unit shown; nec: Not elsewhere classified; nsk: Not specified by kind; - : zero; * : 10-19 percent estimated; ** : 20-29 percent estimated.

PRODUCT SHARE DETAILS

Product or Product Class	% Share	Product or Product Class	% Share
Analytical instruments	100.00		

Source: 1992 *Economic Census*. The values shown are percent of total shipments in an industry. Values of indented subcategories are summed in the main heading. The symbol (D) appears when data are withheld to prevent disclosure of competitive information. The abbreviation nsk stands for 'not specified by kind' and nec for 'not elsewhere classified'.

INPUTS AND OUTPUTS FOR ENGINEERING & SCIENTIFIC INSTRUMENTS

Economic Sector or Industry Providing Inputs	%	Sector	Economic Sector or Industry Buying Outputs	%	Sector
Imports	14.6	Foreign	Gross private fixed investment	41.5	Cap Inv
Engineering & scientific instruments	12.1	Manufg.	Exports	26.4	Foreign
Wholesale trade	8.7	Trade	Federal Government purchases, national defense	7.5	Fed Govt
Electronic components nec	6.8	Manufg.	Engineering & scientific instruments	5.2	Manufg.
Veneer & plywood	4.7	Manufg.	Aircraft	4.2	Manufg.
Blast furnaces & steel mills	3.6	Manufg.	Ship building & repairing	2.4	Manufg.
Advertising	3.6	Services	Radio & TV communication equipment	2.3	Manufg.
Instruments to measure electricity	2.7	Manufg.	S/L Govt. purch., health & hospitals	2.1	S/L Govt
Semiconductors & related devices	2.4	Manufg.	Mechanical measuring devices	1.4	Manufg.
Communications, except radio & TV	2.1	Util.	Federal Government purchases, nondefense	1.3	Fed Govt
Electric services (utilities)	2.0	Util.	Change in business inventories	0.8	In House
Petroleum refining	1.9	Manufg.	S/L Govt. purch., higher education	0.6	S/L Govt
Eating & drinking places	1.8	Trade	S/L Govt. purch., police	0.6	S/L Govt
Motors & generators	1.7	Manufg.	S/L Govt. purch., sewerage	0.6	S/L Govt
Industrial patterns	1.4	Manufg.	S/L Govt. purch., sanitation	0.5	S/L Govt
Fabricated metal products, nec	1.3	Manufg.	Boat building & repairing	0.4	Manufg.
Machinery, except electrical, nec	1.2	Manufg.	S/L Govt. purch., elem. & secondary education	0.4	S/L Govt
Fabricated rubber products, nec	1.1	Manufg.	Blast furnaces & steel mills	0.3	Manufg.
Sheet metal work	1.1	Manufg.	Miscellaneous plastics products	0.3	Manufg.
Aluminum rolling & drawing	1.0	Manufg.	Watches, clocks, & parts	0.3	Manufg.
Miscellaneous plastics products	0.9	Manufg.	S/L Govt. purch., highways	0.3	S/L Govt
Screw machine and related products	0.9	Manufg.	S/L Govt. purch., natural resource & recreation.	0.3	S/L Govt
Banking	0.9	Fin/R.E.			
Legal services	0.9	Services			
Maintenance of nonfarm buildings nec	0.8	Constr.			
Electronic computing equipment	0.8	Manufg.			

Continued on next page.

INPUTS AND OUTPUTS FOR ENGINEERING & SCIENTIFIC INSTRUMENTS - Continued

Economic Sector or Industry Providing Inputs	%	Sector	Economic Sector or Industry Buying Outputs	%	Sector
Motor freight transportation & warehousing	0.8	Util.			
Equipment rental & leasing services	0.8	Services			
Glass & glass products, except containers	0.7	Manufg.			
Paperboard containers & boxes	0.7	Manufg.			
Real estate	0.7	Fin/R.E.			
Management & consulting services & labs	0.7	Services			
U.S. Postal Service	0.7	Gov't			
Aluminum castings	0.6	Manufg.			
Gaskets, packing & sealing devices	0.6	Manufg.			
Special dies & tools & machine tool accessories	0.6	Manufg.			
Ball & roller bearings	0.5	Manufg.			
Paints & allied products	0.5	Manufg.			
Manifold business forms	0.4	Manufg.			
Metal stampings, nec	0.4	Manufg.			
Gas production & distribution (utilities)	0.4	Util.			
Accounting, auditing & bookkeeping	0.4	Services			
Converted paper products, nec	0.3	Manufg.			
Lighting fixtures & equipment	0.3	Manufg.			
Railroads & related services	0.3	Util.			
Automotive rental & leasing, without drivers	0.3	Services			
Business/professional associations	0.3	Services			
Computer & data processing services	0.3	Services			
Engineering, architectural, & surveying services	0.3	Services			
Abrasive products	0.2	Manufg.			
Iron & steel foundries	0.2	Manufg.			
Lubricating oils & greases	0.2	Manufg.			
Machine tools, metal cutting types	0.2	Manufg.			
Nonferrous castings, nec	0.2	Manufg.			
Photographic equipment & supplies	0.2	Manufg.			
Plastics materials & resins	0.2	Manufg.			
Plating & polishing	0.2	Manufg.			
Sawmills & planning mills, general	0.2	Manufg.			
Transformers	0.2	Manufg.			
Wiring devices	0.2	Manufg.			
Air transportation	0.2	Util.			
Credit agencies other than banks	0.2	Fin/R.E.			
Insurance carriers	0.2	Fin/R.E.			
Royalties	0.2	Fin/R.E.			
Automotive repair shops & services	0.2	Services			
Photofinishing labs, commercial photography	0.2	Services			
Theatrical producers, bands, entertainers	0.2	Services			
Electron tubes	0.1	Manufg.			
Miscellaneous fabricated wire products	0.1	Manufg.			
Rubber & plastics hose & belting	0.1	Manufg.			
Transit & bus transportation	0.1	Util.			
Water supply & sewage systems	0.1	Util.			
Water transportation	0.1	Util.			
Motion pictures	0.1	Services			
Personnel supply services	0.1	Services			
Noncomparable imports	0.1	Foreign			

Source: Benchmark Input-Output Accounts for the U.S. Economy, 1982, U.S. Department of Commerce, Washington, D.C., July 1991. Data, as reported in the source, are organized by the 1977 SIC structure in use in 1982 but have been matched, as closely as is possible, to the 1987 SIC structure used in this book.

OCCUPATIONS EMPLOYED BY SIC 382 - MEASURING AND CONTROLLING DEVICES

Occupation	% of Total 1994	Change to 2005	Occupation	% of Total 1994	Change to 2005
Assemblers, fabricators, & hand workers nec	7.4	-11.4	Engineering, mathematical, & science managers	1.7	0.6
Electrical & electronics engineers	5.6	13.4	Industrial production managers	1.5	-11.4
Electrical & electronic equipment assemblers	5.4	-11.4	General office clerks	1.4	-24.4
Sales & related workers nec	4.3	-11.4	Bookkeeping, accounting, & auditing clerks	1.4	-33.6
Electrical & electronic technicians,technologists	3.7	-11.4	Purchasing agents, ex trade & farm products	1.3	-11.4
Electrical & electronic assemblers	3.5	-20.3	Production, planning, & expediting clerks	1.3	-11.4
Inspectors, testers, & graders, precision	3.3	-38.0	Traffic, shipping, & receiving clerks	1.3	-14.8
Electromechanical equipment assemblers	3.2	-2.5	Drafters	1.3	-31.0
Blue collar worker supervisors	3.2	-24.1	Computer engineers	1.3	31.2
General managers & top executives	2.4	-15.9	Stock clerks	1.2	-28.0
Secretaries, ex legal & medical	2.4	-19.3	Engineers nec	1.2	6.3
Machinists	2.0	-33.6	Management support workers nec	1.1	-11.4
Mechanical engineers	1.8	-2.5	Managers & administrators nec	1.1	-11.4
Engineering technicians & technologists nec	1.7	-11.4	Computer programmers	1.1	-28.2
Marketing, advertising, & PR managers	1.7	-11.4			

Source: Industry-Occupation Matrix, Bureau of Labor Statistics. These data relate to one or more 3-digit SIC industry groups rather than to a single 4-digit SIC. The change reported for each occupation to the year 2005 is a percent of growth or decline as estimated by the Bureau of Labor Statistics. The abbreviation nec stands for 'not elsewhere classified'.

LOCATION BY STATE AND REGIONAL CONCENTRATION

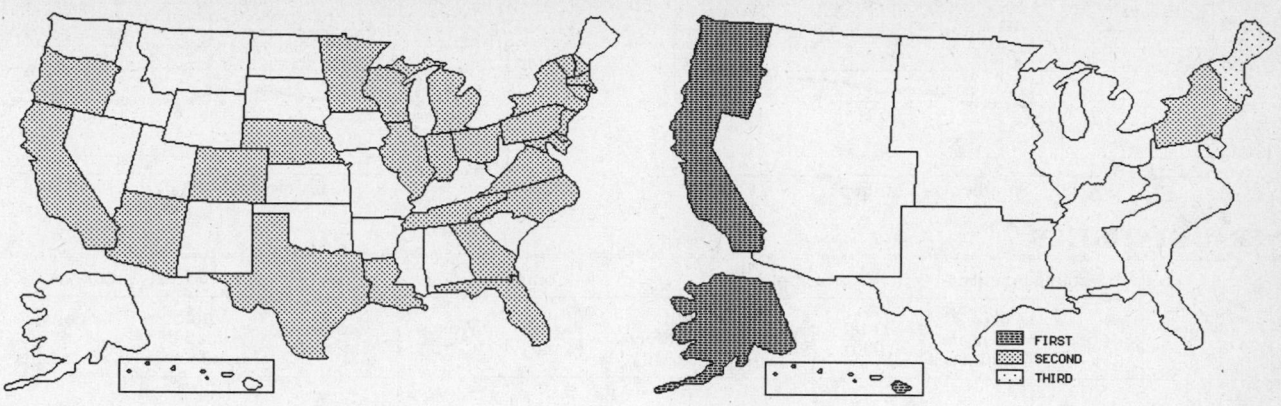

FIRST
SECOND
THIRD

INDUSTRY DATA BY STATE

State	Establish-ments	Shipments			Employment				Cost as % of Shipments	Investment per Employee ($)
		Total ($ mil)	% of U.S.	Per Establ.	Total Number	% of U.S.	Per Establ.	Wages ($/hour)		
California	133	1,488.3	28.7	11.2	10,200	25.7	77	16.56	42.4	4,402
Massachusetts	62	894.9	17.2	14.4	5,600	14.1	90	12.72	35.3	2,750
Florida	19	424.7	8.2	22.4	4,100	10.3	216	14.59	57.4	-
Texas	40	258.1	5.0	6.5	3,500	8.8	88	9.04	52.5	-
Pennsylvania	36	253.5	4.9	7.0	2,400	6.0	67	16.22	43.2	3,167
New Jersey	28	176.3	3.4	6.3	1,200	3.0	43	15.33	45.7	2,917
Wisconsin	12	161.0	3.1	13.4	1,200	3.0	100	10.78	37.0	3,667
Ohio	24	139.8	2.7	5.8	1,400	3.5	58	9.55	43.1	2,786
New York	27	120.0	2.3	4.4	1,000	2.5	37	13.43	42.5	4,100
Michigan	16	114.7	2.2	7.2	1,100	2.8	69	9.62	47.2	2,364
Colorado	16	89.1	1.7	5.6	1,100	2.8	69	11.78	44.3	5,364
Minnesota	10	69.1	1.3	6.9	500	1.3	50	16.25	35.7	-
Illinois	19	61.9	1.2	3.3	400	1.0	21	12.00	44.9	2,500
Oregon	11	51.6	1.0	4.7	400	1.0	36	12.00	43.0	2,500
New Hampshire	7	46.1	0.9	6.6	300	0.8	43	10.50	43.2	2,333
Virginia	9	39.9	0.8	4.4	300	0.8	33	10.50	34.3	4,000
Maryland	18	35.0	0.7	1.9	200	0.5	11	12.50	33.4	5,500
Arizona	11	30.6	0.6	2.8	300	0.8	27	14.00	31.7	2,667
Nebraska	5	29.2	0.6	5.8	400	1.0	80	13.00	23.6	-
Indiana	5	16.4	0.3	3.3	200	0.5	40	7.00	27.4	-
Louisiana	3	7.4	0.1	2.5	100	0.3	33	6.00	32.4	2,000
Connecticut	12	(D)	-	-	1,750 *	4.4	146	-	-	-
North Carolina	6	(D)	-	-	175 *	0.4	29	-	-	-
Georgia	5	(D)	-	-	375 *	0.9	75	-	-	1,333
Tennessee	5	(D)	-	-	375 *	0.9	75	-	-	-
Delaware	4	(D)	-	-	1,750 *	4.4	438	-	-	-
Rhode Island	4	(D)	-	-	375 *	0.9	94	-	-	-
Vermont	2	(D)	-	-	175 *	0.4	88	-	-	-

Source: 1992 *Economic Census*. The states are in descending order of shipments or establishments (if shipment data are missing for the majority). The symbol (D) appears when data are withheld to prevent disclosure of competitive information. States marked with (D) are sorted by number of establishments. A dash (-) indicates that the data element cannot be calculated; * indicates the midpoint of a range.

3827 - OPTICAL INSTRUMENTS & LENSES

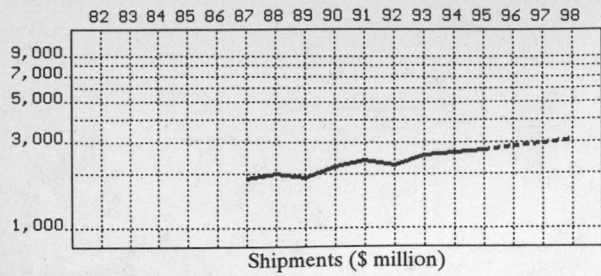

Shipments ($ million)

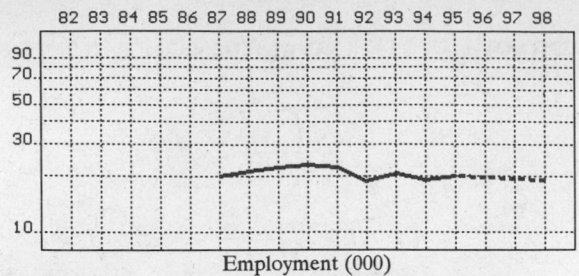

Employment (000)

GENERAL STATISTICS

Year	Companies	Establishments		Employment			Compensation		Production ($ million)			
		Total	with 20 or more employees	Total (000)	Production Workers (000)	Hours (Mil)	Payroll ($ mil)	Wages ($/hr)	Cost of Materials	Value Added by Manufacture	Value of Shipments	Capital Invest.
1982												
1983												
1984												
1985												
1986												
1987	236	250	127	20.1	11.3	21.9	581.6	11.91	694.7	1,167.8	1,863.6	83.3
1988		266	131	21.3	11.7	22.9	630.1	12.04	786.8	1,251.7	2,001.4	83.3
1989		267	130	22.5	11.9	23.3	627.4	12.34	749.2	1,186.3	1,917.5	72.4
1990		260	136	23.4	12.6	26.2	702.5	12.71	874.2	1,326.7	2,217.7	77.2
1991		285	138	22.4	11.1	23.0	829.6	13.10	879.7	1,342.0	2,380.4	77.6
1992	415	425	167	18.9	9.4	19.8	679.9	12.96	836.0	1,435.0	2,262.9	65.0
1993		430	164	20.8	10.3	21.7	756.1	13.35	962.2	1,592.9	2,553.9	74.5
1994		437P	169P	19.3	10.8	21.5	736.3	13.52	946.1	1,704.0	2,632.6	105.0
1995		468P	176P	20.0P	10.1P	21.3P	799.9P	13.82P	980.9P	1,766.6P	2,729.3P	84.4P
1996		500P	183P	19.8P	9.9P	21.1P	823.7P	14.06P	1,020.8P	1,838.6P	2,840.6P	85.4P
1997		531P	190P	19.5P	9.6P	20.8P	847.5P	14.29P	1,060.8P	1,910.6P	2,951.8P	86.4P
1998		562P	196P	19.3P	9.4P	20.5P	871.3P	14.53P	1,100.8P	1,982.6P	3,063.1P	87.5P

Sources: 1982, 1987, 1992 *Economic Census*; *Annual Survey of Manufactures*, 83-86, 88-91, 93-94. Establishment counts for non-Census years are from *County Business Patterns*; establishment values for 83-84 are extrapolations. 'P's show projections by the editors. Industries reclassified in 87 will not have data for prior years.

INDICES OF CHANGE

Year	Companies	Establishments		Employment			Compensation		Production ($ million)			
		Total	with 20 or more employees	Total (000)	Production Workers (000)	Hours (Mil)	Payroll ($ mil)	Wages ($/hr)	Cost of Materials	Value Added by Manufacture	Value of Shipments	Capital Invest.
1982												
1983												
1984												
1985												
1986												
1987	57	59	76	106	120	111	86	92	83	81	82	128
1988		63	78	113	124	116	93	93	94	87	88	128
1989		63	78	119	127	118	92	95	90	83	85	111
1990		61	81	124	134	132	103	98	105	92	98	119
1991		67	83	119	118	116	122	101	105	94	105	119
1992	100	100	100	100	100	100	100	100	100	100	100	100
1993		101	98	110	110	110	111	103	115	111	113	115
1994		103P	101P	102	115	109	108	104	113	119	116	162
1995		110P	105P	106P	107P	108P	118P	107P	117P	123P	121P	130P
1996		118P	109P	105P	105P	106P	121P	108P	122P	128P	126P	131P
1997		125P	114P	103P	102P	105P	125P	110P	127P	133P	130P	133P
1998		132P	118P	102P	100P	104P	128P	112P	132P	138P	135P	135P

Sources: Same as General Statistics. Values reflect change from the base year, 1992. Values above 100 mean greater than 92, values below 100 mean less than 92, and a value of 100 in the 82-91 or 93-98 period means same as 92. 'P's mark projections by the editors.

SELECTED RATIOS

For 1994	Avg. of All Manufact.	Analyzed Industry	Index	For 1994	Avg. of All Manufact.	Analyzed Industry	Index
Employees per Establishment	49	44	90	Value Added per Production Worker	134,084	157,778	118
Payroll per Establishment	1,500,273	1,684,897	112	Cost per Establishment	5,045,178	2,164,989	43
Payroll per Employee	30,620	38,150	125	Cost per Employee	102,970	49,021	48
Production Workers per Establishment	34	25	72	Cost per Production Worker	146,988	87,602	60
Wages per Establishment	853,319	665,172	78	Shipments per Establishment	9,576,895	6,024,256	63
Wages per Production Worker	24,861	26,915	108	Shipments per Employee	195,460	136,404	70
Hours per Production Worker	2,056	1,991	97	Shipments per Production Worker	279,017	243,759	87
Wages per Hour	12.09	13.52	112	Investment per Establishment	321,011	240,275	75
Value Added per Establishment	4,602,255	3,899,314	85	Investment per Employee	6,552	5,440	83
Value Added per Employee	93,930	88,290	94	Investment per Production Worker	9,352	9,722	104

Sources: Same as General Statistics. The 'Average of All Manufacturing' column represents the average of all manufacturing industries reported for the most recent complete year available. The Index shows the relationship between the Average and the Analyzed Industry. For example, 100 means that they are equal; 500 that the Analyzed Industry is five times the average; 50 means that the Analyzed Industry is half the national average. The abbreviation 'na' is used to show that data are 'not available'.

LEADING COMPANIES   Number shown: 75   Total sales ($ mil): 1,863   Total employment (000): 15.8

Company Name	Address				CEO Name	Phone	Co. Type	Sales ($ mil)	Empl. (000)
Electro-Optical Systems	2000 E El Segundo	El Segundo	CA	90245	C Richard Jones	310-616-4636	D	550*	5.0
Optical Coating Laboratory Inc	2789 Northpoint	Santa Rosa	CA	95407	H M Dwight Jr	707-545-6440	P	132	1.2
Hughes Danbury Optical Systems	100 Wooster Hght	Danbury	CT	06810	John C Rich	203-797-5000	S	100	0.8
ITEK Optical Systems	10 Maguire Rd	Lexington	MA	02173	James H Frey	617-276-2000	D	80	0.4
Recon-Optical Inc	550 W Northwest	Barrington	IL	60010	Gordon Bourns	708-381-2400	S	70	0.5
Melles Griot Inc	1770 Kettering St	Irvine	CA	92714	Paul Kenrick	714-556-8200	S	57*	0.6
Intevac	3550 Bassett St	Santa Clara	CA	95054	Norman Pond	408-986-9888	S	55	0.3
ILC Technology Inc	399 Java Dr	Sunnyvale	CA	94089	H C Baumgartner	408-745-7900	P	52	0.5
Leupold and Stevens Inc	PO Box 688	Beaverton	OR	97075	WK Wildauer	503-646-9171	R	50	0.4
Schott Fiber Optics Inc	122 Charlton St	Southbridge	MA	01550	Brian A Edney	508-765-1680	S	45	0.3
Precision Aerotech Inc	7777 Fay Av	La Jolla	CA	92037	R W Detweiler	619-456-2992	P	42	0.3
Kollmorgen Corp	347 King St	Northampton	MA	01060	Daniel F Desmond	413-586-2330	D	38	0.3
US Precision Lens Inc	4000 McMann Rd	Cincinnati	OH	45245	David Hinchman	513-752-7000	S	34	0.6
Molecular Dynamics Inc	928 E Arques Av	Sunnyvale	CA	94086	Jay T Flatley	408-773-1222	P	34	0.2
National Crane Corp	11200 N 148th St	Waverly	NE	68462	TJ Urbanek	402-786-6300	S	30*	0.4
Perceptron Inc	23855 Research Dr	Farmington Hls	MI	48335	Dwight D Carlson	313-478-7710	P	28	0.1
Optical Corporation of America	7421 Orangewood	Garden Grove	CA	92641	Don Johnson	714-895-1667	R	26*	0.2
Meade Instruments Corp	16542 Millikan Av	Irvine	CA	92714	Steve Murdock	714-756-2291	R	25	0.2
Germanow-Simon Corp	408 St Paul St	Rochester	NY	14605	Lew Germanow	716-232-1440	R	20	0.2
II-VI Inc	375 Saxonburg Blv	Saxonburg	PA	16056	Carl J Johnson	412-352-4455	P	19	0.2
BEI Sensors & Motion Syst Co	PO Box 3838	Little Rock	AR	72203	Ronald R Roberts	501-851-4000	D	17	0.2
David White Inc	PO Box 1007	Germantown	WI	53022	Tony L Mihalovich	414-251-8100	P	16	0.2
LOH Optical Machinery Inc	PO Box 664	Germantown	WI	53022	M S-Wetekam	414-255-6001	S	15	<0.1
Pennsylvania Optical	234 S 8th St	Reading	PA	19603	Anthony Fabrizio	610-376-5701	S	15	0.2
Van Cort Instruments Inc	29 E Industrial Dr	Northampton	MA	01060	L Erik Van Cort	413-586-9800	R	13	0.1
Tinsley Laboratories Inc	3900 Lakeside Dr	Richmond	CA	94806	Robert J Aronno	510-222-8110	P	13	<0.1
Park Scientific Instruments	1171 Borregas Av	Sunnyvale	CA	94089	Sang-il Park	408-747-1600	R	12	<0.1
OCA MicroCoatings	170 Locke Dr	Marlborough	MA	01752	John Viggiano	508-481-9860	D	11	0.1
Oriel Corp	PO Box 872	Stratford	CT	06497	Eugene Arthurs	203-377-8282	R	11*	0.1
JML Optical Industries Inc	690 Portland Av	Rochester	NY	14621	J M Labozzo II	716-342-8900	R	10	<0.1
Laser Power Corp	12777 High Bluff Dr	San Diego	CA	92130	Glenn H Sherman	619-755-0700	R	10	0.1
Miller-Holzwarth	PO Box 270	Salem	OH	44460	Tom Lymenstull	216-337-8736	D	10*	<0.1
NAVITAR Inc	200 Commerce Dr	Rochester	NY	14623	Jeremy Goldstein	716-359-4000	R	10*	<0.1
OFC Corp	2 Mercer Rd	Natick	MA	01760	John F Blais	508-655-1650	R	10	0.1
Precision Optics Corporation Inc	612 Indrial Way W	Eatontown	NJ	07724	Walter A Merkl	908-542-4801	P	9	<0.1
Plummer Precision Optics	601 Montgomery	Pennsburg	PA	18073	JL Plummer	215-679-6272	R	9	<0.1
Automatic Inspection Devices	1700 N Westwood	Toledo	OH	43607	Charles J Drake	419-536-1983	S	8	<0.1
Exotic Materials Inc	36570 Briggs Rd	Murrieta	CA	92563	Dick Sharman	909-926-2994	R	8	<0.1
Ferson Optics Inc	2006 Government St	Ocean Springs	MS	39564	Louis S Peters	601-875-8146	R	8*	<0.1
Industrial Technologies Inc	1 Trefoil Dr	Trumbull	CT	06611	Gerald W Stewart	203-268-8000	P	8	<0.1
Burris Company Inc	PO Box 1747	Greeley	CO	80632	JP McCarty	303-356-1670	S	8	<0.1
American Polarizers Inc	141 S 7th St	Reading	PA	19602	WH Bentley	610-373-5177	R	7	<0.1
Rosin Optical Co	6233 W Cermak Rd	Berwyn	IL	60402	Sorrel Rosin	708-749-2020	R	7*	0.1
Deltronic Corp	3900 W Segerstrom	Santa Ana	CA	92704	Richard Larzelere	714-545-0401	R	6*	<0.1
Ilex	55 Science Pkwy	Rochester	NY	14620	Jo Bunkenburg	716-244-7220	D	6*	<0.1
Illmo RX Services Inc	52 Progress Pkwy	Maryland H	MO	63043	John Gerber	314-434-6858	R	6*	<0.1
Minuteman Laboratories Inc	530 Main St	Acton	MA	01720	JM Nihen	508-263-2632	R	6	<0.1
Olympus America Inc	4 Nevada Dr	Lake Success	NY	11042	Daniel D Biondi	516-488-3880	D	6	<0.1
Renco Encoders Inc	26 Coromar Dr	Goleta	CA	93117	Ludwig Wagatha	805-968-1525	S	6	<0.1
Solar Kinetics Inc	10635 King William	Dallas	TX	75220	Gus Hutchinson	214-556-2376	R	6	<0.1
Photon Technology International	1 Deerpark Dr	S Brunswick	NJ	08852	Charles G Marianik	908-329-0910	P	5	<0.1
Bond Optics Inc	PO Box 422	Lebanon	NH	03766	L A Guaraldi Jr	603-448-2300	R	5	<0.1
Diversified Optical Products Inc	282 Main St	Salem	NH	03079	Larry Kessler	603-898-1880	R	5	<0.1
Liberty Mirror Co	851 3rd Av	Brackenridge	PA	15014	Stephen W Williams	412-224-1800	D	5	<0.1
Micro-Vu Corp	7909 Conde Ln	Windsor	CA	95492	EP Amormino	707-838-6272	R	5*	<0.1
Spectrex Inc	218 Little Falls Rd	Cedar Grove	NJ	07009	Eric J Zinn	201-239-8398	R	5*	<0.1
French Reflection Inc	820 S Robertson	Los Angeles	CA	90035	Richard P Myers Jr	310-659-3800	R	4	<0.1
Metavac Inc	4000 Point St	Holtsville	NY	11742	Jeff Witherwax	516-447-7700	S	4	<0.1
OPTEM International Inc	78 Schuyler Baldwin	Fairport	NY	14450	John Amarel	716-223-2370	R	4*	<0.1
Vogelin Optical Company Inc	PO Box 95	Norwood	MN	55368	Scott Pollock	612-466-5517	R	4	<0.1
Bernell Corp	PO Box 4637	South Bend	IN	46601	JS Beardsley	219-234-3200	R	4	<0.1
Century Precision Optics	10713 Burbank Blv	N Hollywood	CA	91601	SE Manios	818-766-3715	R	4	<0.1
Questar Corp	Rte 202	New Hope	PA	18938	Douglas M Knight	215-862-5277	R	4	<0.1
Buhl Optical Co	1009 Beech Av	Pittsburgh	PA	15233	Irving S Stapsy	412-321-0076	R	3	<0.1
Electro Optical Instruments Inc	2 McLaren	Irvine	CA	92718	Joseph F Hall Jr	714-581-7430	R	3	<0.1
Fotec Inc	529 Main St	Boston	MA	02129	James E Hayes	617-241-7810	R	3	<0.1
JL Wood Optical Systems	1361 E Edinger Av	Santa Ana	CA	92705	James L Wood	714-835-1888	R	3*	<0.1
Lenox Instrument Company Inc	265 Andrews Rd	Trevose	PA	19053	John Lang	215-322-9990	R	3	<0.1
Opkor Inc	740 Driving Pk Av	Rochester	NY	14613	Anthony J Lapaglia	716-458-5390	R	3	<0.1
PLX Inc	40 W Jefryn Blv	Deer Park	NY	11729	Morton Lipkins	516-586-4190	R	3	<0.1
Pyramid Optical Corp	10871 Forbes Av	Garden Grove	CA	92643	Shingo Enomoto	714-265-1100	S	3	<0.1
Synthetic Vision Systems	PO Box 1910	Ann Arbor	MI	48106	Thomas Swain	313-665-1140	D	3*	<0.1
Harold Johnson Optical Labs	1826 W 169th St	Gardena	CA	90247	H F Johnson Jr	310-327-3051	R	3	<0.1
General Scientific Corp	77 Enterprise Dr	Ann Arbor	MI	48103	Harvey Bauss	313-996-9200	R	2	<0.1
Brunton Co	620 E Monroe Av	Riverton	WY	82501	JL Larsen	307-856-6559	R	2	<0.1

Source: Ward's Business Directory of U.S. Private and Public Companies, Volumes 1 and 2, 1996. The company type code used is as follows: P - Public, R - Private, S - Subsidiary, D - Division, J - Joint Venture, A - Affiliate, G - Group. Sales are in millions of dollars, employees are in thousands. An asterisk (*) indicates an estimated sales volume. The symbol < stands for 'less than'. Company names and addresses are truncated, in some cases, to fit into the available space.

MATERIALS CONSUMED

Material	Quantity	Delivered Cost ($ million)
Materials, ingredients, containers, and supplies	(X)	729.7
Printed circuit boards (without inserted components) for electronic circuitry	(X)	12.1
Printed circuit assemblies, loaded boards or modules	(X)	6.0
Semiconductors, including transistors, diodes, rectifiers, and integrated circuits for electronic circuitry	(X)	12.2
Capacitors for electronic circuitry	(X)	3.9
Resistors for electronic circuitry	(X)	2.9
Other components and accessories for electronic circuitry, nec, except tubes	(X)	(D)
Current-carrying wiring devices	(X)	1.2
Electrical transmission, distribution, and control equipment	(X)	5.2
Electronic computing equipment	(X)	12.0
Electrical instrument mechanisms and meter movements (including instrument relays)	(X)	4.7
Electrical measuring instruments and parts, not listed elsewhere	(X)	1.8
Fractional horsepower electric motors and generators (less than 1 hp) including timing motors	(X)	0.9
Plastics resins consumed in the form of granules, pellets, powders, liquids, etc.	(X)	5.5
Fabricated plastics products (except gaskets, hoses, and belting)	(X)	5.6
Sheet metal products, except stampings	(X)	11.3
Metal stampings	(X)	3.9
Fabricated metal wire products (including wire rope, cable, springs, etc.)	(X)	0.8
Metal bolts, nuts, screws, washers, rivets, and other screw machine products	(X)	3.4
Other fabricated metal products (except forgings)	(X)	30.0
Forgings	(X)	(D)
Iron and steel castings (rough and semifinished)	(X)	6.1
Aluminum and aluminum-base alloy castings (rough and semifinished)	(X)	12.0
Other nonferrous castings (rough and semifinished)	(X)	1.0
Steel shapes and forms	(X)	6.1
Copper and copper-base alloy shapes and forms	(X)	2.2
Aluminum and aluminum-base alloy shapes and forms	(X)	7.7
Other nonferrous shapes and forms	(X)	2.2
Glass and glass products (excluding windows and mirrors)	(X)	88.0
Paper and paperboard products except paperboard boxes, containers, and corrugated paperboard	(X)	0.5
Paper and paperboard containers, including shipping sacks and other paper packaging supplies	(X)	7.8
All other materials and components, parts, containers, and supplies	(X)	179.4
Materials, ingredients, containers, and supplies, nsk	(X)	281.3

Source: 1992 Economic Census. Explanation of symbols used: (D): Withheld to avoid disclosure of competitive data; na: Not available; (S): Withheld because statistical norms were not met; (X): Not applicable; (Z): Less than half the unit shown; nec: Not elsewhere classified; nsk: Not specified by kind; - : zero; * : 10-19 percent estimated; ** : 20-29 percent estimated.

PRODUCT SHARE DETAILS

Product or Product Class	% Share	Product or Product Class	% Share
Optical instruments and lenses	100.00	Other optical instruments and lenses (except sighting, tracking, and fire-control)	88.77
Sighting, tracking, and fire-control equipment, optical-type	34.38	Optical instruments and lenses, nec, nsk	6.44
Optical instruments and lenses, nec	62.02	Optical instruments and lenses, nsk	3.60
Binoculars and astronomical instruments	4.79		

Source: 1992 Economic Census. The values shown are percent of total shipments in an industry. Values of indented subcategories are summed in the main heading. The symbol (D) appears when data are withheld to prevent disclosure of competitive information. The abbreviation nsk stands for 'not specified by kind' and nec for 'not elsewhere classified'.

INPUTS AND OUTPUTS FOR OPTICAL INSTRUMENTS & LENSES

Economic Sector or Industry Providing Inputs	%	Sector	Economic Sector or Industry Buying Outputs	%	Sector
Imports	22.2	Foreign	Gross private fixed investment	60.8	Cap Inv
Electronic components nec	13.8	Manufg.	Exports	14.3	Foreign
Miscellaneous plastics products	12.9	Manufg.	Federal Government purchases, national defense	8.7	Fed Govt
Accounting, auditing & bookkeeping	9.1	Services	Federal Government purchases, nondefense	3.6	Fed Govt
Wholesale trade	8.1	Trade	Personal consumption expenditures	2.8	
Metal stampings, nec	4.8	Manufg.	Aircraft & missile equipment, nec	2.7	Manufg.
Glass & glass products, except containers	2.2	Manufg.	Aircraft	2.0	Manufg.
Optical instruments & lenses	2.1	Manufg.	Optical instruments & lenses	1.3	Manufg.
Petroleum refining	2.1	Manufg.	Radio & TV communication equipment	0.8	Manufg.
Semiconductors & related devices	2.1	Manufg.	S/L Govt. purch., elem. & secondary education	0.8	S/L Govt
Communications, except radio & TV	1.7	Util.	Change in business inventories	0.7	In House
Business services nec	1.6	Services	Ophthalmic goods	0.5	Manufg.
Electric services (utilities)	1.4	Util.	Aircraft & missile engines & engine parts	0.4	Manufg.
Banking	1.3	Fin/R.E.	Photographic equipment & supplies	0.4	Manufg.
Air transportation	0.8	Util.	S/L Govt. purch., higher education	0.2	S/L Govt
Advertising	0.8	Services			
Eating & drinking places	0.7	Trade			
Real estate	0.7	Fin/R.E.			
Equipment rental & leasing services	0.7	Services			
Screw machine and related products	0.6	Manufg.			
Motor freight transportation & warehousing	0.6	Util.			

Continued on next page.

INPUTS AND OUTPUTS FOR OPTICAL INSTRUMENTS & LENSES - Continued

Economic Sector or Industry Providing Inputs	%	Sector	Economic Sector or Industry Buying Outputs	%	Sector
Noncomparable imports	0.6	Foreign			
Adhesives & sealants	0.5	Manufg.			
Hotels & lodging places	0.5	Services			
Maintenance of nonfarm buildings nec	0.4	Constr.			
Gaskets, packing & sealing devices	0.4	Manufg.			
Automotive rental & leasing, without drivers	0.4	Services			
Ball & roller bearings	0.3	Manufg.			
Chemical preparations, nec	0.3	Manufg.			
Fabricated metal products, nec	0.3	Manufg.			
Machinery, except electrical, nec	0.3	Manufg.			
Plastics materials & resins	0.3	Manufg.			
Gas production & distribution (utilities)	0.3	Util.			
Credit agencies other than banks	0.3	Fin/R.E.			
Automotive repair shops & services	0.3	Services			
Business/professional associations	0.3	Services			
Legal services	0.3	Services			
Industrial controls	0.2	Manufg.			
Nonferrous rolling & drawing, nec	0.2	Manufg.			
Paperboard containers & boxes	0.2	Manufg.			
Insurance carriers	0.2	Fin/R.E.			
Computer & data processing services	0.2	Services			
Management & consulting services & labs	0.2	Services			
Abrasive products	0.1	Manufg.			
Manifold business forms	0.1	Manufg.			
Special dies & tools & machine tool accessories	0.1	Manufg.			
Sanitary services, steam supply, irrigation	0.1	Util.			
Retail trade, except eating & drinking	0.1	Trade			
Royalties	0.1	Fin/R.E.			
Security & commodity brokers	0.1	Fin/R.E.			
U.S. Postal Service	0.1	Gov't			

Source: Benchmark Input-Output Accounts for the U.S. Economy, 1982, U.S. Department of Commerce, Washington, D.C., July 1991. Data, as reported in the source, are organized by the 1977 SIC structure in use in 1982 but have been matched, as closely as is possible, to the 1987 SIC structure used in this book.

OCCUPATIONS EMPLOYED BY SIC 382 - MEASURING AND CONTROLLING DEVICES

Occupation	% of Total 1994	Change to 2005	Occupation	% of Total 1994	Change to 2005
Assemblers, fabricators, & hand workers nec	7.4	-11.4	Engineering, mathematical, & science managers	1.7	0.6
Electrical & electronics engineers	5.6	13.4	Industrial production managers	1.5	-11.4
Electrical & electronic equipment assemblers	5.4	-11.4	General office clerks	1.4	-24.4
Sales & related workers nec	4.3	-11.4	Bookkeeping, accounting, & auditing clerks	1.4	-33.6
Electrical & electronic technicians,technologists	3.7	-11.4	Purchasing agents, ex trade & farm products	1.3	-11.4
Electrical & electronic assemblers	3.5	-20.3	Production, planning, & expediting clerks	1.3	-11.4
Inspectors, testers, & graders, precision	3.3	-38.0	Traffic, shipping, & receiving clerks	1.3	-14.8
Electromechanical equipment assemblers	3.2	-2.5	Drafters	1.3	-31.0
Blue collar worker supervisors	3.2	-24.1	Computer engineers	1.3	31.2
General managers & top executives	2.4	-15.9	Stock clerks	1.2	-28.0
Secretaries, ex legal & medical	2.4	-19.3	Engineers nec	1.2	6.3
Machinists	2.0	-33.6	Management support workers nec	1.1	-11.4
Mechanical engineers	1.8	-2.5	Managers & administrators nec	1.1	-11.4
Engineering technicians & technologists nec	1.7	-11.4	Computer programmers	1.1	-28.2
Marketing, advertising, & PR managers	1.7	-11.4			

Source: Industry-Occupation Matrix, Bureau of Labor Statistics. These data relate to one or more 3-digit SIC industry groups rather than to a single 4-digit SIC. The change reported for each occupation to the year 2005 is a percent of growth or decline as estimated by the Bureau of Labor Statistics. The abbreviation nec stands for 'not elsewhere classified'.

LOCATION BY STATE AND REGIONAL CONCENTRATION

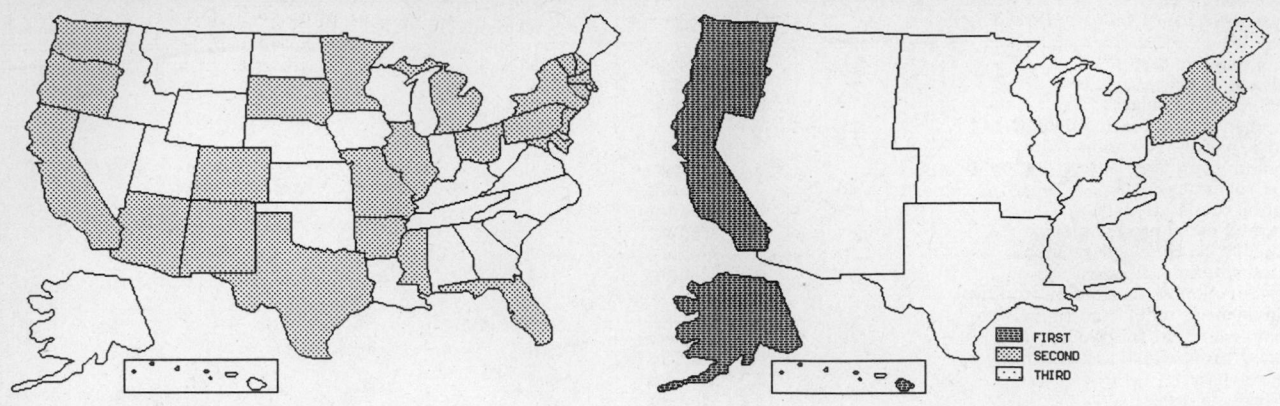

INDUSTRY DATA BY STATE

| State | Establish-ments | Shipments | | | Employment | | | | Cost as % of Shipments | Investment per Employee ($) |
		Total ($ mil)	% of U.S.	Per Establ.	Total Number	% of U.S.	Per Establ.	Wages ($/hour)		
California	103	933.3	41.2	9.1	6,500	34.4	63	12.97	39.5	3,138
Massachusetts	39	274.8	12.1	7.0	2,300	12.2	59	14.73	43.2	3,174
Connecticut	9	212.5	9.4	23.6	1,700	9.0	189	21.58	23.7	-
New York	40	137.0	6.1	3.4	1,300	6.9	33	12.12	36.5	2,385
Oregon	13	99.0	4.4	7.6	900	4.8	69	14.18	35.1	-
Ohio	9	95.8	4.2	10.6	800	4.2	89	8.25	30.1	5,375
Florida	20	86.7	3.8	4.3	800	4.2	40	10.78	31.5	6,750
New Jersey	20	78.6	3.5	3.9	500	2.6	25	14.00	48.5	2,600
Pennsylvania	24	45.5	2.0	1.9	600	3.2	25	12.57	37.8	3,333
Colorado	16	34.6	1.5	2.2	400	2.1	25	8.67	39.0	2,500
Maryland	7	34.0	1.5	4.9	500	2.6	71	13.17	29.4	-
New Hampshire	10	19.6	0.9	2.0	300	1.6	30	13.33	34.7	1,000
Vermont	6	15.8	0.7	2.6	200	1.1	33	12.50	33.5	1,500
Illinois	17	15.3	0.7	0.9	200	1.1	12	13.00	26.1	1,500
Texas	9	11.9	0.5	1.3	100	0.5	11	6.50	43.7	-
Arizona	9	11.2	0.5	1.2	100	0.5	11	9.00	37.5	-
Washington	5	10.4	0.5	2.1	100	0.5	20	13.00	29.8	-
Mississippi	4	10.0	0.4	2.5	200	1.1	50	13.50	29.0	1,000
Michigan	12	(D)	-	-	175 *	0.9	15	-	-	-
Minnesota	10	(D)	-	-	175 *	0.9	18	-	-	5,143
Missouri	4	(D)	-	-	175 *	0.9	44	-	-	-
New Mexico	3	(D)	-	-	175 *	0.9	58	-	-	-
Arkansas	2	(D)	-	-	175 *	0.9	88	-	-	-
South Dakota	1	(D)	-	-	175 *	0.9	175	-	-	-

Source: 1992 *Economic Census*. The states are in descending order of shipments or establishments (if shipment data are missing for the majority). The symbol (D) appears when data are withheld to prevent disclosure of competitive information. States marked with (D) are sorted by number of establishments. A dash (-) indicates that the data element cannot be calculated; * indicates the midpoint of a range.

3829 - MEASURING & CONTROLLING DEVICES, NEC

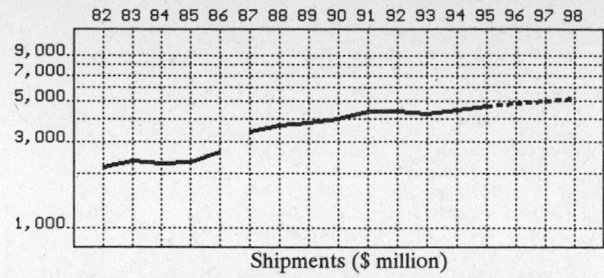

82 83 84 85 86 87 88 89 90 91 92 93 94 95 96 97 98

Shipments ($ million)

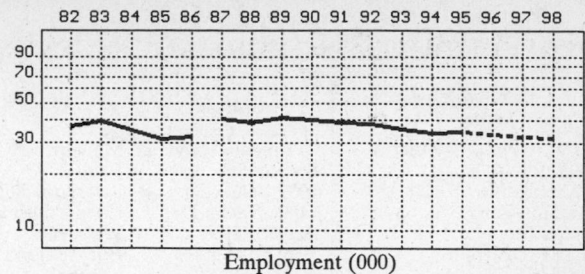

82 83 84 85 86 87 88 89 90 91 92 93 94 95 96 97 98

Employment (000)

GENERAL STATISTICS

| Year | Companies | Establishments | | Employment | | | Compensation | | Production ($ million) | | | |
		Total	with 20 or more employees	Total (000)	Production Workers (000)	Hours (Mil)	Payroll ($ mil)	Wages ($/hr)	Cost of Materials	Value Added by Manufacture	Value of Shipments	Capital Invest.
1982	692	716	258	37.4	17.6	34.6	797.4	8.45	785.3	1,381.3	2,194.6	70.1
1983		700	262	39.7	17.4	33.6	938.8	9.69	854.0	1,465.9	2,358.3	54.6
1984		684	266	35.0	18.1	35.5	849.3	9.26	898.5	1,396.9	2,282.5	93.5
1985		667	269	31.9	17.9	34.7	770.1	9.59	831.0	1,532.7	2,310.5	85.8
1986		651	274	32.6	17.3	35.2	845.5	9.63	965.7	1,640.4	2,618.0	78.7
1987*	939	971	304	41.0	20.2	39.8	1,098.8	10.40	1,228.1	2,259.0	3,442.0	104.0
1988		928	333	38.8	20.7	40.8	1,102.3	10.59	1,349.1	2,368.5	3,698.6	116.8
1989		868	327	41.8	20.4	41.5	1,117.8	10.78	1,394.7	2,404.2	3,828.8	147.2
1990		843	327	40.4	18.1	36.2	1,155.4	11.86	1,443.1	2,518.7	4,039.7	126.8
1991		829	309	38.7	19.5	40.1	1,256.7	11.95	1,620.0	2,741.0	4,395.3	131.5
1992	977	1,006	318	38.1	19.3	38.1	1,305.4	12.69	1,584.2	2,809.5	4,400.1	180.1
1993		928	307	35.7	17.9	35.6	1,269.9	13.56	1,541.9	2,748.7	4,286.6	123.0
1994		909P	312P	34.5	17.9	36.3	1,237.2	12.92	1,685.7	2,789.7	4,455.5	102.1
1995		908P	311P	34.7P	17.5P	35.5P	1,325.3P	13.90P	1,784.5P	2,953.3P	4,716.8P	135.4P
1996		908P	310P	33.8P	17.2P	34.8P	1,354.7P	14.35P	1,839.1P	3,043.5P	4,860.8P	136.9P
1997		907P	308P	32.9P	16.8P	34.2P	1,384.1P	14.81P	1,893.6P	3,133.7P	5,004.9P	138.3P
1998		907P	307P	32.0P	16.4P	33.5P	1,413.6P	15.26P	1,948.1P	3,223.9P	5,149.0P	139.7P

Sources: 1982, 1987, 1992 *Economic Census*; *Annual Survey of Manufactures*, 83-86, 88-91, 93-94. Establishment counts are from *County Business Patterns* for non-Census years; establishment counts for 83-84 are extrapolations. * indicates that industry content changed in 87; earlier years use 77 SICs. 'P's mark projections.

INDICES OF CHANGE

| Year | Companies | Establishments | | Employment | | | Compensation | | Production ($ million) | | | |
		Total	with 20 or more employees	Total (000)	Production Workers (000)	Hours (Mil)	Payroll ($ mil)	Wages ($/hr)	Cost of Materials	Value Added by Manufacture	Value of Shipments	Capital Invest.
1982	71	71	81	98	91	91	61	67	50	49	50	39
1983		70	82	104	90	88	72	76	54	52	54	30
1984		68	84	92	94	93	65	73	57	50	52	52
1985		66	85	84	93	91	59	76	52	55	53	48
1986		65	86	86	90	92	65	76	61	58	59	44
1987*	96	97	96	108	105	104	84	82	78	80	78	58
1988		92	105	102	107	107	84	83	85	84	84	65
1989		86	103	110	106	109	86	85	88	86	87	82
1990		84	103	106	94	95	89	93	91	90	92	70
1991		82	97	102	101	105	96	94	102	98	100	73
1992	100	100	100	100	100	100	100	100	100	100	100	100
1993		92	97	94	93	93	97	107	97	98	97	68
1994		90P	98P	91	93	95	95	102	106	99	101	57
1995		90P	98P	91P	91P	93P	102P	110P	113P	105P	107P	75P
1996		90P	97P	89P	89P	91P	104P	113P	116P	108P	110P	76P
1997		90P	97P	86P	87P	90P	106P	117P	120P	112P	114P	77P
1998		90P	96P	84P	85P	88P	108P	120P	123P	115P	117P	78P

Sources: Same as General Statistics. Values reflect change from the base year, 1992. Values above 100 mean greater than 92, values below 100 mean less than 92, and a value of 100 in the 82-91 or 93-98 period means same as 92. * indicates that industry content changed in 87. Data for earlier years are in 77 SIC format.

SELECTED RATIOS

For 1994	Avg. of All Manufact.	Analyzed Industry	Index	For 1994	Avg. of All Manufact.	Analyzed Industry	Index
Employees per Establishment	49	38	77	Value Added per Production Worker	134,084	155,849	116
Payroll per Establishment	1,500,273	1,361,484	91	Cost per Establishment	5,045,178	1,855,039	37
Payroll per Employee	30,620	35,861	117	Cost per Employee	102,970	48,861	47
Production Workers per Establishment	34	20	57	Cost per Production Worker	146,988	94,173	64
Wages per Establishment	853,319	516,109	60	Shipments per Establishment	9,576,895	4,903,081	51
Wages per Production Worker	24,861	26,201	105	Shipments per Employee	195,460	129,145	66
Hours per Production Worker	2,056	2,028	99	Shipments per Production Worker	279,017	248,911	89
Wages per Hour	12.09	12.92	107	Investment per Establishment	321,011	112,357	35
Value Added per Establishment	4,602,255	3,069,942	67	Investment per Employee	6,552	2,959	45
Value Added per Employee	93,930	80,861	86	Investment per Production Worker	9,352	5,704	61

Sources: Same as General Statistics. The 'Average of All Manufacturing' column represents the average of all manufacturing industries reported for the most recent complete year available. The Index shows the relationship between the Average and the Analyzed Industry. For example, 100 means that they are equal; 500 that the Analyzed Industry is five times the average; 50 means that the Analyzed Industry is half the national average. The abbreviation 'na' is used to show that data are 'not available'.

LEADING COMPANIES Number shown: **75**   Total sales ($ mil): **5,631**   Total employment (000): **56.2**

Company Name	Address				CEO Name	Phone	Co. Type	Sales ($ mil)	Empl. (000)
Vishay Intertechnology Inc	63 Lincoln Hwy	Malvern	PA	19355	Felix Zandman	215-644-1300	P	988	16.8
AMETEK Inc	Station Sq	Paoli	PA	19301	Walter E Blankley	215-647-2121	P	808	6.2
Imo Industries Inc	1009 Lenox Dr	Lawrenceville	NJ	08648	Donald K Farrar	609-896-7600	P	464	6.2
MTS Systems Corp	14000 Techn Dr	Eden Prairie	MN	55344	Donald M Sullivan	612-937-4000	P	201	1.6
Thermedics Inc	PO Box 2999	Woburn	MA	01888	John W Wood Jr	617-622-1000	P	155	1.0
Computer Products Inc	7900 Glades Rd	Boca Raton	FL	33434	J M O'Donnell	407-451-1000	P	155	1.6
Instron Corp	100 Royall St	Canton	MA	02021	James M McConnell	617-828-2500	P	136	1.1
Electroglas Inc	2901 Coronado Dr	Santa Clara	CA	95054	Neil R Bonke	408-727-6500	P	112	0.5
Analytical Technology Inc	529 Main St	Boston	MA	02129	William J Kennedy	617-242-3900	R	110*	0.9
Brooks Instrument	407 W Vine St	Hatfield	PA	19440	Robert Heffernan	215-362-3500	D	110	1.5
EG and G Energy Measurements	PO Box 1912	Las Vegas	NV	89125	Peter Zavattaro	702-295-5511	S	110*	1.3
Topcon America Corp	65 W Century Rd	Paramus	NJ	07652	Bob Iguchi	201-261-9450	S	110	0.3
MKS Instruments Inc	6 Shattuck Rd	Andover	MA	01810	John R Bertucci	508-975-2350	R	100	0.8
Viggo-Spectramed	1911 Williams Dr	Oxnard	CA	93030	Dick Grant	805-983-1300	D	100	0.8
Input/Output Inc	12300 Parc Crest Dr	Stafford	TX	77477	Gary D Owens	713-933-3339	P	96	0.4
MFS Network Technologies Inc	200 Kiewit Plz	Omaha	NE	68131	Royce Holland	402-342-2052	S	92	0.3
Hy-Cal Engineering	9650 Telstar Av	El Monte	CA	91731		818-444-4000	S	88	0.1
Applied Automation Inc	PO Box 9999	Bartlesville	OK	74005	David Snow	918-662-7000	S	70	0.6
Mechanical Technology Inc	968 Albany-Shaker	Latham	NY	12110	R Wayne Diesel	518-785-2211	P	68	0.5
VDO Yazaki Corp	188 Brooke Rd	Winchester	VA	22603	Uli Baur	703-665-0100	J	67	0.7
AVO International Inc	4651 Westmld	Dallas	TX	75237	Andrew R Bach	214-333-3201	S	56*	0.7
Delta Design Inc	PO Box 421	San Diego	CA	92112	James A Donahue	619-292-5000	S	53	0.4
Tel Tru Manufacturing Co	408 St Paul St	Rochester	NY	14605	L Germanow	716-232-1440	R	52	0.3
Delavan Inc	811 4th St	W Des Moines	IA	50265	James R Baker	515-274-1561	S	48	0.4
Fife Corp	PO Box 26508	Oklahoma City	OK	73126	J Richard Webb	405-755-1600	R	48*	0.4
Minco Products Inc	7300 Commerce Ln	Minneapolis	MN	55432	K Schurr	612-571-3121	R	44	0.6
OYO Geospace Corp	7334 N Gessner Rd	Houston	TX	77040	Satoru Ohya	713-939-9700	S	44	0.4
Environmental Technologies	PO Box 9840	Towson	MD	21284	Timothy Karpetsky	410-321-5200	R	40	0.4
Inex Vision Systems	13327 US Hwy 19 N	Clearwater	FL	34624	Jim Schemenaur	813-535-5502	S	40	0.2
Microtest Inc	4747 N 22nd St	Phoenix	AZ	85016	Richard G Meise	602-957-6400	P	38	0.2
Metrologic Instruments Inc	Coles Rd	Blackwood	NJ	08012	C Harry Knowles	609-228-8100	P	36	0.3
GFI-GENFARE	751 Pratt Blv	Elk Grove Vill	IL	60007	James A Pacelli	708-593-8855	D	35	0.2
Telkee	PO Box 19903-7054	Dover	DE	19903	A A Salamone	302-678-7800	D	35*	0.3
CPG International Inc	535 Madison Av	New York	NY	10022	Richard Henshaw	212-421-3125	R	34*	0.4
United Electric Controls Co	PO Box 9143	Watertown	MA	02272	R Reis	617-926-1000	R	32	0.4
Washington Scientific Industries	2605 W Wayzata	Long Lake	MN	55356	Michael J Pudil	612-473-1271	P	31	0.3
ITI MOVATS Inc	2825 Cobb	Kennesaw	GA	30144	Donald L Janecek	404-424-6343	S	30	0.1
Measurements Group Inc	PO Box 27777	Raleigh	NC	27611	Henry Landau	919-365-3800	S	29	0.5
AVL Powertrain Engineering Inc	41169 Vincenti Ct	Novi	MI	48375	J C McCandless	313-477-3399	R	28	0.3
SafetyTek Corp	49050 Milmont Dr	Fremont	CA	94538	James B Hawkins	510-226-9600	P	28	0.2
Air Gage Co	12170 Globe St	Livonia	MI	48150	D Persichini	313-591-9220	R	28	0.2
Victoreen Inc	6000 Cochran Rd	Cleveland	OH	44139	Charles Francisco	216-248-9300	S	27	0.3
Dynisco Inc	4 Commercial St	Sharon	MA	02067	Roger Brooks	617-784-8400	S	27	0.2
Monitor Labs Inc	74 Inverness Dr E	Englewood	CO	80112	Allan Budd	303-792-3300	S	26	0.2
Particle Measuring Systems Inc	5475 Airport Blv	Boulder	CO	80301	Robert Knollenberg	303-443-7100	R	26	0.2
Opal Inc	2903 Bunker Hill Ln	Santa Clara	CA	95054	Rafi Yizhar	408-727-6060	P	25	0.2
Gas Tech Inc	8407 Central Av	Newark	CA	94560	Carl Mazzuca	510-794-6200	S	24*	0.2
PCB Piezotronics Inc	3425 Walden Av	Depew	NY	14043		716-684-0001	R	24	0.3
Fairfield Industries Inc	PO Box 42154	Houston	TX	77042	Walter Pharris	713-981-8181	S	23	0.3
Veeco Instruments Inc	Terminal Dr	Plainview	NY	11803	Ed Braun	516-349-8300	D	23*	0.2
Dieterich Standard Corp	PO Box 9000	Boulder	CO	80301	Gene Shanahan	303-530-9600	S	23	<0.1
Reuter-Stokes Inc	8499 Darrow Rd	Twinsburg	OH	44087	David R McLemore	216-425-3755	S	22	<0.1
X-Ray Industries Inc	1961 Thunderbird	Troy	MI	48084	Scott Thams	313-362-2242	R	22	0.2
American Sigma Inc	11601 Maple Ridge	Medina	NY	14103	W Hungerford	716-798-5580	R	20	0.1
Action Instruments Company	8601 Aero Dr	San Diego	CA	92123	Jim Pinto	619-279-5726	R	20	0.2
DICKEY-john Corp	PO Box 10	Auburn	IL	62615	CS Davis III	217-438-3371	S	20	0.4
Figgie International	PO Box 1349	Fletcher	NC	28732	Robert F Fitch	704-684-5178	D	20	0.1
General Monitors Inc	26776 Simpatica Cir	Lake Forest	CA	92630	Don S Edwards	714-581-4464	R	20*	0.1
Hunter Associates Laboratory	11491 Sunset Hills	Reston	VA	22090	Philip S Hunter	703-471-6870	R	20	0.2
Lucas NovaSensor	1055 Mission Ct	Fremont	CA	94539	John Pendergrass	510-490-9100	S	20	0.1
SKF Condition Monitoring Inc	4141 Ruffin Rd	San Diego	CA	92123	Marianne Ericsson	619-496-3400	S	20	0.1
Structural Test Products	510 Cottonwood Dr	Milpitas	CA	95035	Ashwin Vora	408-432-1000	D	20	0.1
Sensotec Inc	1200 Chesapeake	Columbus	OH	43212	John Easton	614-486-7723	R	19	0.2
Wright-K Technology Inc	2025 E Genesee St	Saginaw	MI	48601	RP Floeter	517-752-3103	R	19	0.3
AGR International Inc	615 Whitestown Rd	Butler	PA	16001	HM Dimmick	412-482-2163	R	18	0.2
Detector Electronics Corp	6901 W 110th St	Minneapolis	MN	55438	Ross Colwell	612-941-5665	S	18	0.3
Princeton Gamma-Tech Inc	1200 State Rd	Princeton	NJ	08540	John Patterson	609-924-7310	S	18*	0.1
Faria Corp	385 Norwich	Uncasville	CT	06382	David A Blackburn	203-848-9271	R	17	0.3
Photon Kinetics Inc	9405 SW Gemini Dr	Beaverton	OR	97005	Gene White	503-644-1960	S	16*	0.1
Atlas Electric Devices Co	4114 N Ravenswood	Chicago	IL	60613	WW Lane	312-327-4520	R	15	0.2
Hallcrest Inc	1820 Pickwick Ln	Glenview	IL	60025	Rich Brett	708-998-8580	R	15	<0.1
Intoximeters Inc	1901 Locust St	St Louis	MO	63103	MR Forrester	314-241-1158	R	15	<0.1
Conax Buffalo Corp	2300 Walden Av	Buffalo	NY	14225	RA Fox	716-684-4500	S	14*	0.2
Cymbolic Sciences International	26072 Merit Cir	Laguna Hills	CA	92653	Kenneth Smith	714-582-3515	R	14	0.2
HNU Systems Inc	160 Charlemont St	Newton Hghlds	MA	02161	John D Driscoll	617-964-6690	R	14	<0.1

Source: Ward's Business Directory of U.S. Private and Public Companies, Volumes 1 and 2, 1996. The company type code used is as follows: P - Public, R - Private, S - Subsidiary, D - Division, J - Joint Venture, A - Affiliate, G - Group. Sales are in millions of dollars, employees are in thousands. An asterisk (*) indicates an estimated sales volume. The symbol < stands for 'less than'. Company names and addresses are truncated, in some cases, to fit into the available space.

MATERIALS CONSUMED

Material	Quantity	Delivered Cost ($ million)
Materials, ingredients, containers, and supplies	(X)	1,348.7
Printed circuit boards (without inserted components) for electronic circuitry	(X)	35.8
Printed circuit assemblies, loaded boards or modules	(X)	79.6
Semiconductors, including transistors, diodes, rectifiers, and integrated circuits for electronic circuitry	(X)	38.8
Capacitors for electronic circuitry	(X)	10.7
Resistors for electronic circuitry	(X)	8.9
Other components and accessories for electronic circuitry, nec, except tubes	(X)	38.9
Current-carrying wiring devices	(X)	15.8
Electrical transmission, distribution, and control equipment	(X)	17.6
Electronic computing equipment	(X)	51.1
Electrical instrument mechanisms and meter movements (including instrument relays)	(X)	27.7
Electrical measuring instruments and parts, not listed elsewhere	(X)	12.3
Fractional horsepower electric motors and generators (less than 1 hp) including timing motors	(X)	13.4
Plastics resins consumed in the form of granules, pellets, powders, liquids, etc.	(X)	8.3
Fabricated plastics products (except gaskets, hoses, and belting)	(X)	23.4
Sheet metal products, except stampings	(X)	34.2
Metal stampings	(X)	10.1
Fabricated metal wire products (including wire rope, cable, springs, etc.)	(X)	8.5
Metal bolts, nuts, screws, washers, rivets, and other screw machine products	(X)	14.3
Other fabricated metal products (except forgings)	(X)	63.8
Forgings	(X)	1.7
Iron and steel castings (rough and semifinished)	(X)	23.1
Aluminum and aluminum-base alloy castings (rough and semifinished)	(X)	15.8
Other nonferrous castings (rough and semifinished)	(X)	2.8
Steel shapes and forms	(X)	20.2
Copper and copper-base alloy shapes and forms	(X)	7.5
Aluminum and aluminum-base alloy shapes and forms	(X)	13.3
Other nonferrous shapes and forms	(X)	14.7
Glass and glass products (excluding windows and mirrors)	(X)	8.8
Paper and paperboard products except paperboard boxes, containers, and corrugated paperboard	(X)	3.8
Paper and paperboard containers, including shipping sacks and other paper packaging supplies	(X)	8.5
All other materials and components, parts, containers, and supplies	(X)	321.1
Materials, ingredients, containers, and supplies, nsk	(X)	394.1

Source: 1992 *Economic Census*. Explanation of symbols used: (D): Withheld to avoid disclosure of competitive data; na: Not available; (S): Withheld because statistical norms were not met; (X): Not applicable; (Z): Less than half the unit shown; nec: Not elsewhere classified; nsk: Not specified by kind; - : zero; * : 10-19 percent estimated; ** : 20-29 percent estimated.

PRODUCT SHARE DETAILS

Product or Product Class	% Share	Product or Product Class	% Share
Measuring and controlling devices, nec	100.00	Commercial, geophysical, meteorological, and general-purpose instruments	32.50
Aircraft engine instruments, except flight	14.35	Survey and drafting instruments and apparatus, including photogrammetric equipment	5.61
Physical properties testing and inspection equipment and kinematic testing and measuring equipment	26.99		
Nuclear radiation detection and monitoring instruments	13.36	Measuring and controlling devices, nec, nsk	7.18

Source: 1992 *Economic Census*. The values shown are percent of total shipments in an industry. Values of indented subcategories are summed in the main heading. The symbol (D) appears when data are withheld to prevent disclosure of competitive information. The abbreviation nsk stands for 'not specified by kind' and nec for 'not elsewhere classified'.

INPUTS AND OUTPUTS FOR ENGINEERING & SCIENTIFIC INSTRUMENTS

Economic Sector or Industry Providing Inputs	%	Sector	Economic Sector or Industry Buying Outputs	%	Sector
Imports	14.6	Foreign	Gross private fixed investment	41.5	Cap Inv
Engineering & scientific instruments	12.1	Manufg.	Exports	26.4	Foreign
Wholesale trade	8.7	Trade	Federal Government purchases, national defense	7.5	Fed Govt
Electronic components nec	6.8	Manufg.	Engineering & scientific instruments	5.2	Manufg.
Veneer & plywood	4.7	Manufg.	Aircraft	4.2	Manufg.
Blast furnaces & steel mills	3.6	Manufg.	Ship building & repairing	2.4	Manufg.
Advertising	3.6	Services	Radio & TV communication equipment	2.3	Manufg.
Instruments to measure electricity	2.7	Manufg.	S/L Govt. purch., health & hospitals	2.1	S/L Govt
Semiconductors & related devices	2.4	Manufg.	Mechanical measuring devices	1.4	Manufg.
Communications, except radio & TV	2.1	Util.	Federal Government purchases, nondefense	1.3	Fed Govt
Electric services (utilities)	2.0	Util.	Change in business inventories	0.8	In House
Petroleum refining	1.9	Manufg.	S/L Govt. purch., higher education	0.6	S/L Govt
Eating & drinking places	1.8	Trade	S/L Govt. purch., police	0.6	S/L Govt
Motors & generators	1.7	Manufg.	S/L Govt. purch., sewerage	0.6	S/L Govt
Industrial patterns	1.4	Manufg.	S/L Govt. purch., sanitation	0.5	S/L Govt
Fabricated metal products, nec	1.3	Manufg.	Boat building & repairing	0.4	Manufg.
Machinery, except electrical, nec	1.2	Manufg.	S/L Govt. purch., elem. & secondary education	0.4	S/L Govt
Fabricated rubber products, nec	1.1	Manufg.	Blast furnaces & steel mills	0.3	Manufg.
Sheet metal work	1.1	Manufg.	Miscellaneous plastics products	0.3	Manufg.
Aluminum rolling & drawing	1.0	Manufg.	Watches, clocks, & parts	0.3	Manufg.

Continued on next page.

INPUTS AND OUTPUTS FOR ENGINEERING & SCIENTIFIC INSTRUMENTS - Continued

Economic Sector or Industry Providing Inputs	%	Sector	Economic Sector or Industry Buying Outputs	%	Sector
Miscellaneous plastics products	0.9	Manufg.	S/L Govt. purch., highways	0.3	S/L Govt
Screw machine and related products	0.9	Manufg.	S/L Govt. purch., natural resource & recreation.	0.3	S/L Govt
Banking	0.9	Fin/R.E.			
Legal services	0.9	Services			
Maintenance of nonfarm buildings nec	0.8	Constr.			
Electronic computing equipment	0.8	Manufg.			
Motor freight transportation & warehousing	0.8	Util.			
Equipment rental & leasing services	0.8	Services			
Glass & glass products, except containers	0.7	Manufg.			
Paperboard containers & boxes	0.7	Manufg.			
Real estate	0.7	Fin/R.E.			
Management & consulting services & labs	0.7	Services			
U.S. Postal Service	0.7	Gov't			
Aluminum castings	0.6	Manufg.			
Gaskets, packing & sealing devices	0.6	Manufg.			
Special dies & tools & machine tool accessories	0.6	Manufg.			
Ball & roller bearings	0.5	Manufg.			
Paints & allied products	0.5	Manufg.			
Manifold business forms	0.4	Manufg.			
Metal stampings, nec	0.4	Manufg.			
Gas production & distribution (utilities)	0.4	Util.			
Accounting, auditing & bookkeeping	0.4	Services			
Converted paper products, nec	0.3	Manufg.			
Lighting fixtures & equipment	0.3	Manufg.			
Railroads & related services	0.3	Util.			
Automotive rental & leasing, without drivers	0.3	Services			
Business/professional associations	0.3	Services			
Computer & data processing services	0.3	Services			
Engineering, architectural, & surveying services	0.3	Services			
Abrasive products	0.2	Manufg.			
Iron & steel foundries	0.2	Manufg.			
Lubricating oils & greases	0.2	Manufg.			
Machine tools, metal cutting types	0.2	Manufg.			
Nonferrous castings, nec	0.2	Manufg.			
Photographic equipment & supplies	0.2	Manufg.			
Plastics materials & resins	0.2	Manufg.			
Plating & polishing	0.2	Manufg.			
Sawmills & planning mills, general	0.2	Manufg.			
Transformers	0.2	Manufg.			
Wiring devices	0.2	Manufg.			
Air transportation	0.2	Util.			
Credit agencies other than banks	0.2	Fin/R.E.			
Insurance carriers	0.2	Fin/R.E.			
Royalties	0.2	Fin/R.E.			
Automotive repair shops & services	0.2	Services			
Photofinishing labs, commercial photography	0.2	Services			
Theatrical producers, bands, entertainers	0.2	Services			
Electron tubes	0.1	Manufg.			
Miscellaneous fabricated wire products	0.1	Manufg.			
Rubber & plastics hose & belting	0.1	Manufg.			
Transit & bus transportation	0.1	Util.			
Water supply & sewage systems	0.1	Util.			
Water transportation	0.1	Util.			
Motion pictures	0.1	Services			
Personnel supply services	0.1	Services			
Noncomparable imports	0.1	Foreign			

Source: Benchmark Input-Output Accounts for the U.S. Economy, 1982, U.S. Department of Commerce, Washington, D.C., July 1991. Data, as reported in the source, are organized by the 1977 SIC structure in use in 1982 but have been matched, as closely as is possible, to the 1987 SIC structure used in this book.

OCCUPATIONS EMPLOYED BY SIC 382 - MEASURING AND CONTROLLING DEVICES

Occupation	% of Total 1994	Change to 2005	Occupation	% of Total 1994	Change to 2005
Assemblers, fabricators, & hand workers nec	7.4	-11.4	Engineering, mathematical, & science managers	1.7	0.6
Electrical & electronics engineers	5.6	13.4	Industrial production managers	1.5	-11.4
Electrical & electronic equipment assemblers	5.4	-11.4	General office clerks	1.4	-24.4
Sales & related workers nec	4.3	-11.4	Bookkeeping, accounting, & auditing clerks	1.4	-33.6
Electrical & electronic technicians,technologists	3.7	-11.4	Purchasing agents, ex trade & farm products	1.3	-11.4
Electrical & electronic assemblers	3.5	-20.3	Production, planning, & expediting clerks	1.3	-11.4
Inspectors, testers, & graders, precision	3.3	-38.0	Traffic, shipping, & receiving clerks	1.3	-14.8
Electromechanical equipment assemblers	3.2	-2.5	Drafters	1.3	-31.0
Blue collar worker supervisors	3.2	-24.1	Computer engineers	1.3	31.2
General managers & top executives	2.4	-15.9	Stock clerks	1.2	-28.0
Secretaries, ex legal & medical	2.4	-19.3	Engineers nec	1.2	6.3
Machinists	2.0	-33.6	Management support workers nec	1.1	-11.4
Mechanical engineers	1.8	-2.5	Managers & administrators nec	1.1	-11.4
Engineering technicians & technologists nec	1.7	-11.4	Computer programmers	1.1	-28.2
Marketing, advertising, & PR managers	1.7	-11.4			

Source: *Industry-Occupation Matrix*, Bureau of Labor Statistics. These data relate to one or more 3-digit SIC industry groups rather than to a single 4-digit SIC. The change reported for each occupation to the year 2005 is a percent of growth or decline as estimated by the Bureau of Labor Statistics. The abbreviation nec stands for 'not elsewhere classified'.

LOCATION BY STATE AND REGIONAL CONCENTRATION

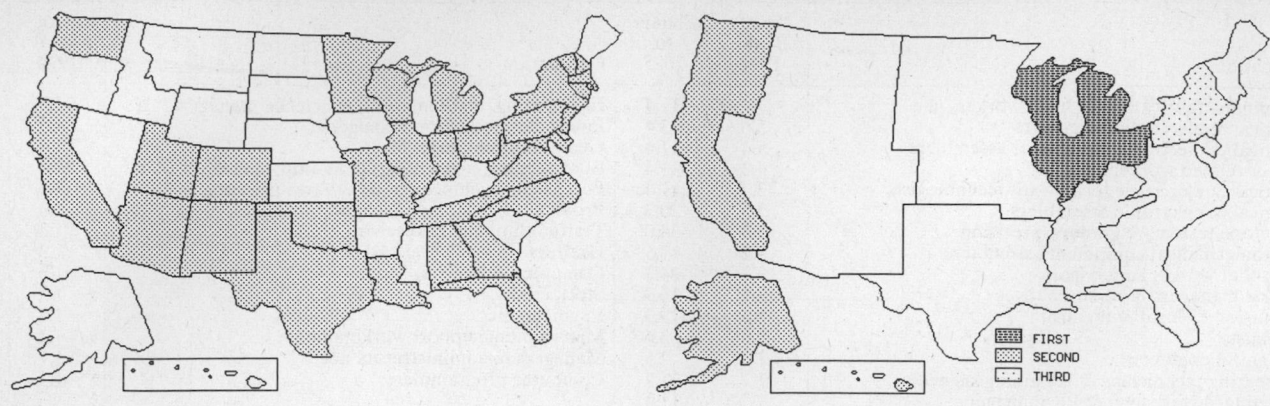

FIRST
SECOND
THIRD

INDUSTRY DATA BY STATE

| State | Establish-ments | Shipments | | | Employment | | | | Cost as % of Shipments | Investment per Employee ($) |
		Total ($ mil)	% of U.S.	Per Establ.	Total Number	% of U.S.	Per Establ.	Wages ($/hour)		
California	163	596.7	13.6	3.7	6,000	15.7	37	13.83	25.0	3,250
Texas	73	594.8	13.5	8.1	3,700	9.7	51	11.50	48.6	18,649
Ohio	67	431.6	9.8	6.4	3,200	8.4	48	11.94	32.8	3,188
Massachusetts	58	308.6	7.0	5.3	2,500	6.6	43	18.26	32.3	3,040
Pennsylvania	50	245.2	5.6	4.9	3,000	7.9	60	12.19	39.1	2,400
Illinois	44	228.4	5.2	5.2	1,400	3.7	32	9.92	34.6	3,143
New Jersey	43	226.2	5.1	5.3	1,900	5.0	44	11.28	43.6	8,684
Minnesota	26	206.2	4.7	7.9	1,600	4.2	62	19.59	29.1	4,875
New York	51	201.4	4.6	3.9	2,200	5.8	43	12.61	45.2	3,773
Connecticut	38	175.9	4.0	4.6	1,800	4.7	47	12.69	23.7	4,167
Florida	44	135.7	3.1	3.1	1,200	3.1	27	9.19	42.4	1,250
North Carolina	23	111.3	2.5	4.8	1,000	2.6	43	10.56	31.6	3,600
Michigan	52	111.1	2.5	2.1	1,100	2.9	21	11.82	38.1	1,455
Maryland	19	99.6	2.3	5.2	800	2.1	42	11.38	39.5	2,000
Tennessee	16	95.9	2.2	6.0	700	1.8	44	13.75	39.1	2,714
Washington	29	68.5	1.6	2.4	900	2.4	31	12.67	26.3	2,111
Wisconsin	17	48.8	1.1	2.9	400	1.0	24	8.40	44.5	3,000
Georgia	10	36.2	0.8	3.6	300	0.8	30	9.67	36.5	2,000
New Hampshire	8	33.7	0.8	4.2	300	0.8	38	14.50	32.6	2,333
Colorado	29	29.9	0.7	1.0	300	0.8	10	9.67	45.2	2,333
Indiana	19	28.1	0.6	1.5	300	0.8	16	13.50	38.1	1,000
Utah	6	23.6	0.5	3.9	200	0.5	33	9.00	31.8	2,500
Missouri	9	22.1	0.5	2.5	200	0.5	22	9.00	22.6	3,500
Oklahoma	10	19.5	0.4	2.0	200	0.5	20	11.00	43.1	-
Arizona	15	(D)	-	-	175 *	0.5	12	-	-	1,714
Alabama	7	(D)	-	-	175 *	0.5	25	-	-	571
Rhode Island	7	(D)	-	-	750 *	2.0	107	-	-	-
Virginia	7	(D)	-	-	175 *	0.5	25	-	-	-
New Mexico	6	(D)	-	-	175 *	0.5	29	-	-	1,143
Louisiana	4	(D)	-	-	175 *	0.5	44	-	-	-
Vermont	3	(D)	-	-	750 *	2.0	250	-	-	-
West Virginia	1	(D)	-	-	175 *	0.5	175	-	-	-

Source: 1992 *Economic Census*. The states are in descending order of shipments or establishments (if shipment data are missing for the majority). The symbol (D) appears when data are withheld to prevent disclosure of competitive information. States marked with (D) are sorted by number of establishments. A dash (-) indicates that the data element cannot be calculated; * indicates the midpoint of a range.

3841 - SURGICAL & MEDICAL INSTRUMENTS

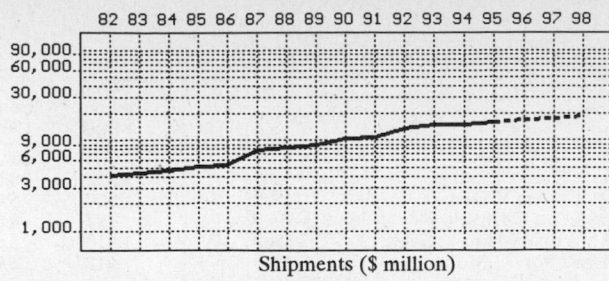

Shipments ($ million)

82 83 84 85 86 87 88 89 90 91 92 93 94 95 96 97 98

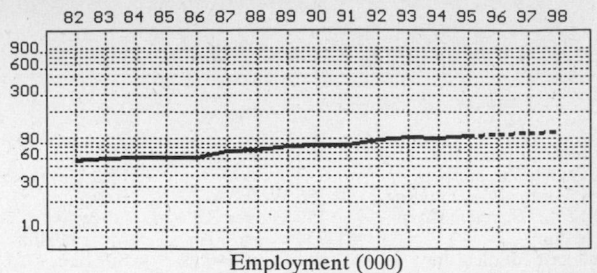

Employment (000)

82 83 84 85 86 87 88 89 90 91 92 93 94 95 96 97 98

GENERAL STATISTICS

| Year | Com-panies | Establishments | | Employment | | | Compensation | | Production ($ million) | | | |
		Total	with 20 or more employees	Total (000)	Production Workers (000)	Hours (Mil)	Payroll ($ mil)	Wages ($/hr)	Cost of Materials	Value Added by Manufacture	Value of Shipments	Capital Invest.
1982	760	859	312	56.9	38.5	74.3	999.5	6.97	1,252.5	2,884.8	4,084.5	230.2
1983		876	337	60.2	40.1	76.6	1,130.2	7.44	1,362.6	2,990.1	4,343.2	191.2
1984		893	362	61.3	39.7	77.3	1,215.5	7.78	1,529.4	3,155.6	4,629.6	219.1
1985		910	387	61.4	39.3	77.8	1,290.7	8.12	1,603.4	3,528.1	5,081.6	202.9
1986		950	390	62.6	39.4	79.3	1,386.2	8.66	1,814.7	3,575.2	5,346.8	217.2
1987	1,031	1,136	442	73.1	45.4	93.6	1,785.8	8.91	2,598.9	5,202.2	7,779.5	354.6
1988		1,137	460	75.7	46.7	93.1	1,918.5	9.57	2,668.3	5,683.2	8,258.6	384.8
1989		1,144	484	83.1	50.7	99.1	2,187.2	9.58	2,958.4	6,059.8	8,971.6	403.3
1990		1,127	498	86.0	53.8	104.2	2,433.8	9.81	3,219.0	7,077.5	10,261.6	468.7
1991		1,184	508	87.7	53.1	103.6	2,591.4	10.59	3,352.6	7,431.7	10,710.3	535.6
1992	1,216	1,340	551	98.2	58.5	113.1	3,095.3	10.86	4,063.5	9,397.8	13,384.9	688.8
1993		1,379	554	103.1	61.4	119.7	3,305.2	11.32	4,534.6	10,664.2	15,113.0	809.0
1994		1,387P	587P	100.5	60.0	119.1	3,403.6	11.52	4,480.7	10,272.2	14,811.6	624.4
1995		1,435P	610P	106.9P	63.1P	123.9P	3,565.5P	11.99P	4,731.9P	10,848.2P	15,642.1P	751.6P
1996		1,482P	633P	111.1P	65.2P	128.1P	3,781.0P	12.37P	5,033.0P	11,538.4P	16,637.4P	800.4P
1997		1,530P	655P	115.2P	67.3P	132.3P	3,996.5P	12.75P	5,334.1P	12,228.7P	17,632.6P	849.2P
1998		1,577P	678P	119.4P	69.4P	136.4P	4,212.0P	13.13P	5,635.2P	12,918.9P	18,627.9P	898.1P

Sources: 1982, 1987, 1992 *Economic Census*; *Annual Survey of Manufactures*, 83-86, 88-91, 93-94. Establishment counts for non-Census years are from *County Business Patterns*; establishment values for 83-84 are extrapolations. 'P's show projections by the editors. Industries reclassified in 87 will not have data for prior years.

INDICES OF CHANGE

| Year | Com-panies | Establishments | | Employment | | | Compensation | | Production ($ million) | | | |
		Total	with 20 or more employees	Total (000)	Production Workers (000)	Hours (Mil)	Payroll ($ mil)	Wages ($/hr)	Cost of Materials	Value Added by Manufacture	Value of Shipments	Capital Invest.
1982	63	64	57	58	66	66	32	64	31	31	31	33
1983		65	61	61	69	68	37	69	34	32	32	28
1984		67	66	62	68	68	39	72	38	34	35	32
1985		68	70	63	67	69	42	75	39	38	38	29
1986		71	71	64	67	70	45	80	45	38	40	32
1987	85	85	80	74	78	83	58	82	64	55	58	51
1988		85	83	77	80	82	62	88	66	60	62	56
1989		85	88	85	87	88	71	88	73	64	67	59
1990		84	90	88	92	92	79	90	79	75	77	68
1991		88	92	89	91	92	84	98	83	79	80	78
1992	100	100	100	100	100	100	100	100	100	100	100	100
1993		103	101	105	105	106	107	104	112	113	113	117
1994		104P	107P	102	103	105	110	106	110	109	111	91
1995		107P	111P	109P	108P	110P	115P	110P	116P	115P	117P	109P
1996		111P	115P	113P	111P	113P	122P	114P	124P	123P	124P	116P
1997		114P	119P	117P	115P	117P	129P	117P	131P	130P	132P	123P
1998		118P	123P	122P	119P	121P	136P	121P	139P	137P	139P	130P

Sources: Same as General Statistics. Values reflect change from the base year, 1992. Values above 100 mean greater than 92, values below 100 mean less than 92, and a value of 100 in the 82-91 or 93-98 period means same as 92. 'P's mark projections by the editors.

SELECTED RATIOS

For 1994	Avg. of All Manufact.	Analyzed Industry	Index	For 1994	Avg. of All Manufact.	Analyzed Industry	Index
Employees per Establishment	49	72	148	Value Added per Production Worker	134,084	171,203	128
Payroll per Establishment	1,500,273	2,453,876	164	Cost per Establishment	5,045,178	3,230,427	64
Payroll per Employee	30,620	33,867	111	Cost per Employee	102,970	44,584	43
Production Workers per Establishment	34	43	126	Cost per Production Worker	146,988	74,678	51
Wages per Establishment	853,319	989,187	116	Shipments per Establishment	9,576,895	10,678,642	112
Wages per Production Worker	24,861	22,867	92	Shipments per Employee	195,460	147,379	75
Hours per Production Worker	2,056	1,985	97	Shipments per Production Worker	279,017	246,860	88
Wages per Hour	12.09	11.52	95	Investment per Establishment	321,011	450,170	140
Value Added per Establishment	4,602,255	7,405,894	161	Investment per Employee	6,552	6,213	95
Value Added per Employee	93,930	102,211	109	Investment per Production Worker	9,352	10,407	111

Sources: Same as General Statistics. The 'Average of All Manufacturing' column represents the average of all manufacturing industries reported for the most recent complete year available. The Index shows the relationship between the Average and the Analyzed Industry. For example, 100 means that they are equal; 500 that the Analyzed Industry is five times the average; 50 means that the Analyzed Industry is half the national average. The abbreviation 'na' is used to show that data are 'not available'.

LEADING COMPANIES Number shown: **75**   Total sales ($ mil): **25,754**   Total employment (000): **170.7**

Company Name	Address				CEO Name	Phone	Co. Type	Sales ($ mil)	Empl. (000)
Baxter International Inc	1 Baxter Pkwy	Deerfield	IL	60015	V R Loucks Jr	708-948-2000	P	9,324	53.5
Becton Dickinson and Co	1 Becton Dr	Franklin Lakes	NJ	07417	Clateo Castellini	201-847-6800	P	2,559	18.6
Siemens Medical Systems Inc	186 Wood Av S	Iselin	NJ	08830	Robert V Dumke	908-321-4500	S	1,500	5.0
United States Surgical Corp	150 Glover Av	Norwalk	CT	06856	Leon C Hirsch	203-845-1000	P	1,197	8.0
CR Bard Inc	730 Central Av	Murray Hill	NJ	07974	William H Longfield	908-277-8000	P	1,018	8.6
Mallinckrodt Medical Inc	PO Box 5840	St Louis	MO	63134	Robert G Moussa	314-895-2000	S	912	4.9
Stryker Corp	PO Box 4085	Kalamazoo	MI	49003	John W Brown	616-385-2600	P	682	4.2
Coulter Electronics Inc	600 W 20th St	Hialeah	FL	33010	Joseph R Coulter Jr	305-885-0131	R	640*	5.0
Boehringer Mannheim Corp	PO Box 50457	Indianapolis	IN	46250	Andre de Bruin	317-845-2000	S	510	3.3
Boston Scientific Corp	1 Boston Scientific	Natick	MA	01760	Peter M Nicholas	508-650-8000	P	449	2.8
Kendall Healthcare Products Co	15 Hampshire Rd	Mansfield	MA	02048	Richard J Meelia	508-261-8000	D	392	4.5
Cordis Corp	PO Box 025700	Miami	FL	33102	Robert C Strauss	305-824-2000	P	337	3.4
Puritan-Bennett Corp	9401 Indian Creek	Overland Park	KS	66210	Burton A Dole Jr	913-661-0444	P	336	2.0
COBE Laboratories Inc	1185 Oak St	Lakewood	CO	80215	Mats Wahlstrom	303-232-6800	S	313	2.0
SCIMED Life Systems Inc	1 Scimed Pl	Maple Grove	MN	55311	Dale A Spencer	612-494-1700	S	265	1.8
Bard Urological	8195 Industrial Blv	Covington	GA	30209	PJ Ehret	404-786-9051	D	260	3.0
Hollister Inc	2000 Hollister Dr	Libertyville	IL	60048	Michael C Winn	708-680-1000	R	250*	2.0
Haemonetics Corp	400 Wood Rd	Braintree	MA	02184	John F White	617-848-7100	P	248	1.1
Pharmacia Biotech Inc	800 Centennial Av	Piscataway	NJ	08855	Michael Wohler	908-457-8000	S	240*	1.9
Valleylab Inc	5920 Longbow Dr	Boulder	CO	80301	George A Stewart	303-530-2300	S	200*	1.1
Diasonics Ultrasound Inc	1565 Barber Ln	Milpitas	CA	95035	Bruce N Moore	408-432-9000	P	195	1.0
Arrow International Inc	PO Box 12888	Reading	PA	19612	Marlin J Miller Jr	215-378-0131	P	179	1.5
Vascular Access	9450 S State St	Sandy	UT	84070	Robert Adrion	801-255-6851	D	170	1.3
Welch Allyn Inc	4341 State Street	Skaneateles Fls	NY	13153	William F Allyn	315-685-4100	R	160	1.7
Schneider	5905 Nathan Ln	Minneapolis	MN	55442	Joseph Laptewicz	612-550-5500	D	159	0.8
Sterile Concepts Holdings Inc	5100 Commerce Rd	Richmond	VA	23234	Paul J Woo Jr	804-275-0200	P	132	0.6
Sterile Concepts Inc	5100 Commerce Rd	Richmond	VA	23234	Paul J Woo Jr	804-275-0200	S	132	0.6
Smith and Nephew Dyonics Inc	160 Dascomb Rd	Andover	MA	01810	Charles Federico	508-470-2800	S	130*	0.5
Maxxim Medical Inc	104 Industrial Blv	Sugar Land	TX	77478	K W Davidson	713-240-5588	P	130	2.3
Bennett Group	2200 Faraday Av	Carlsbad	CA	92008	Ron Rakin	619-929-4000	D	112	0.6
Davol Inc	PO Box 8500	Cranston	RI	02920	Edward Kelly	401-463-7000	S	110	1.0
Baxter Healthcare Corp IV	Rte 120 and Wilson	Round Lake	IL	60073	Jack McGinley	708-270-4316	D	100	0.8
Symbiosis Corp	8600 NW 41 St	Miami	FL	33166	Kevin Smith	305-597-4000	S	97*	0.8
Medex Inc	3637 Lacon Rd	Hilliard	OH	43026	Phillip D Messinger	614-876-2413	P	96	1.3
Graham-Field Health Products	400 Rabro Dr E	Hauppauge	NY	11788	Irwin Selinger	516-582-5900	P	94	0.5
Healthdyne Technologies Inc	1255 Kennest	Marietta	GA	30066	Craig Reynolds	404-499-1212	P	89	0.4
Cell Robotics International Inc	2715 Broadbent N	Albuquerque	NM	87107	Craig T Rogers	505-343-1131	P	81*	<0.1
Linvatec Corp	11311 Concept Blv	Largo	FL	34643	George Kempsell	813-392-6464	S	80	0.7
Vital Signs Inc	20 Campus Rd	Totowa	NJ	07512	Terence D Wall	201-790-1330	P	80	0.7
Medrad Inc	271 Kappa Dr	Pittsburgh	PA	15238	Thomas H Witmer	412-967-9700	P	78	0.7
Respironics Inc	1001 Murry Ridge	Murrysville	PA	15668	Gerald E McGinnis	412-733-0200	P	78	1.1
Corometrics Medical Systems	61 Barnes Pk Rd N	Wallingford	CT	06492	Ann Arneson	203-265-5631	D	75	0.7
Diagnostic Instrument Systems	7 Loveton Cir	Sparks	MD	21152	Vince Forlenza	410-316-4000	D	75	0.8
Allied Healthcare Products Inc	1720 Sublette Av	St Louis	MO	63110	David V LaRusso	314-771-2400	P	74	0.5
Hudson RCI Inc	PO Box 9020	Temecula	CA	92389	Richard Johansen	909-676-5611	R	70*	0.5
Cabot Medical Corp	2021 Cabot Blv W	Langhorne	PA	19047	Warren G Wood	215-752-8300	P	68	0.4
IMED Corp	9775 Businesspark	San Diego	CA	92131	Joseph W Cuhn	619-566-9000	S	66*	0.6
Infusion Technology Inc	35 Cherry Hill Dr	Danvers	MA	01923	Robert L Miller	508-774-7277	R	65	<0.1
Metronic Electromedics Inc	18501 E Plaza Dr	Parker	CO	80134	Alfonzo Sorrato	303-840-4000	S	65*	0.5
Smith and Nephew Perry	1875 Harsh SE	Massillon	OH	44646	Don Urbanowicz	216-833-2811	D	65	0.9
Sarns Inc	PO Box 1247	Ann Arbor	MI	48106	Tom Engels	313-663-4145	S	64*	0.5
Empi Inc	5255 E River Rd	Minneapolis	MN	55421	J E Laptewicz Jr	612-586-7300	P	61	0.5
LUMEX	81 Spence St	Bay Shore	NY	11706	James E George	516-273-2200	D	61	0.5
Mallinckrodt Anesthesiology	675 McDonnell Blv	St Louis	MO	63134	Charles R Clark	314-895-2000	D	58	0.8
Acme United Corp	129 Marconi Av	Fairfield	CT	06606	Dwight C Wheeler	203-372-4458	P	53	0.6
North American Drager Inc	148-B Quarry Rd	Telford	PA	18969	Peter J Schreiber	215-723-9824	R	51*	0.4
Nedaes Inc	2850 Colonnades Ct	Norcross	GA	30071	Robert P Dutlinger	404-448-6684	R	45	0.3
CONMED Corp	310 Broad St	Utica	NY	13501	Eugene R Corasanti	315-797-8375	P	43	0.8
INCSTAR Corp	PO Box 285	Stillwater	MN	55082	John J Booth	612-439-9710	P	43	0.3
Techne Corp	614 McKinley Pl NE	Minneapolis	MN	55413	Thomas E Oland	612-379-8854	P	40	0.3
Utah Medical Products Inc	7043 S 300 W	Midvale	UT	84047	Kevin L Cornwell	801-566-1200	P	40	0.4
InnoServ Technologies Inc	1611 Pomona Rd	Corona	CA	91720	Alan D Margulis	909-736-3700	S	38	0.2
Target Therapeutics Inc	PO Box 5120	Fremont	CA	94537	Gary R Bang	510-440-7700	P	35	0.2
Palomar Medical Technologies	66 Cherry Hill Dr	Beverly	MA	01915	Steven Georgiev	508-921-9300	R	35	0.3
Bird Products Corp	1100 Bird Ctr Dr	Palm Springs	CA	92622	Phil Troilo	619-778-7200	S	34	0.2
Merit Medical Systems Inc	1600 W Merit Pkwy	South Jordan	UT	84095	F P Lampropoulos	801-253-1600	P	33	0.5
Gettig Technologies Inc	1 Streamside Pl	Spring Mills	PA	16875	WA Gettig	814-422-8892	P	33	0.6
Virginia Industries Inc	1022 Elm St	Rocky Hill	CT	06067	L Thompson Jr	203-563-0111	R	33*	0.5
Block Medical Inc	5957 Landau Ct	Carlsbad	CA	92008	Emil Soika	619-431-1501	S	32*	0.3
MiniMed Technologies	12744 San Fernando	Sylmar	CA	91342	Al Mann	818-362-5958	R	32	0.3
West Company Inc	PO Box 645	Lionville	PA	19341	William G Little	610-594-2900	D	32	0.3
LIFECARE International Inc	655 Aspen Ridge Dr	Lafayette	CO	80026	James C Campbell	303-666-9234	R	31*	0.3
Criticare Systems Inc	20925 Crossroads	Waukesha	WI	53186	G J Von der Ruhr	414-798-8282	P	30	0.1
Bard Access Systems	5425 A Earhart	Salt Lake City	UT	84116	Guy Jordan	801-975-1700	D	30	0.3
LUNAR Corp	313 W Beltline Hwy	Madison	WI	53713	Richard B Mazess	608-274-2663	P	30	0.2

Source: Ward's Business Directory of U.S. Private and Public Companies, Volumes 1 and 2, 1996. The company type code used is as follows: P - Public, R - Private, S - Subsidiary, D - Division, J - Joint Venture, A - Affiliate, G - Group. Sales are in millions of dollars, employees are in thousands. An asterisk (*) indicates an estimated sales volume. The symbol < stands for 'less than'. Company names and addresses are truncated, in some cases, to fit into the available space.

MATERIALS CONSUMED

Material	Quantity	Delivered Cost ($ million)
Materials, ingredients, containers, and supplies	(X)	3,558.8
Surgical and orthopedic supplies	(X)	474.9
Resistors, capacitors, transformers, electron tubes, semiconductors, and other electronic components	(X)	178.8
Metal bolts, nuts, screws, washers, rivets, and other screw machine products	(X)	64.9
Other fabricated metal products (except castings and forgings)	(X)	136.3
Iron and steel forgings	(X)	16.6
Nonferrous forgings	(X)	2.1
Iron and steel castings (rough and semifinished)	(X)	18.4
Nonferrous (aluminum, copper, etc.) castings (rough and semifinished)	(X)	7.3
Steel shapes and forms	(X)	48.8
Nonferrous shapes and forms	(X)	23.1
Nonwoven fabrics	(X)	5.5
Broadwoven fabrics	(X)	13.3
Plastics resins consumed in the form of granules, pellets, powders, liquids, etc.	(X)	140.6
Plastics products consumed in the form of sheets, rods, tubes, film, and other shapes	(X)	146.5
Fabricated plastics products	(X)	262.0
Fabricated rubber products, except tires, tubes, hose, belting, and gaskets	(X)	56.3
Adhesives and sealants	(X)	5.6
Glass and glass products, except photographic and projection lenses and prisms	(X)	17.6
Paperboard containers, boxes, and corrugated paperboard	(X)	79.4
Paper and paperboard products except paperboard boxes, containers, and corrugated paperboard	(X)	48.3
All other materials and components, parts, containers, and supplies	(X)	845.3
Materials, ingredients, containers, and supplies, nsk	(X)	967.2

Source: 1992 *Economic Census*. Explanation of symbols used: (D): Withheld to avoid disclosure of competitive data; na: Not available; (S): Withheld because statistical norms were not met; (X): Not applicable; (Z): Less than half the unit shown; nec: Not elsewhere classified; nsk: Not specified by kind; - : zero; * : 10-19 percent estimated; ** : 20-29 percent estimated.

PRODUCT SHARE DETAILS

Product or Product Class	% Share	Product or Product Class	% Share
Surgical and medical instruments	100.00	donor kits	7.11
Surgical and medical instruments and apparatus	88.33	Surgical and medical catheters	17.07
Surgical and medical instruments, including suture needles, eye, ear, nose, and throat instruments	18.41	Surgical and medical mechanical therapy appliances	1.89
		Other surgical and medical instruments	19.21
Orthopedic instruments, excluding eye, ear, nose, and throat instruments	2.62	Parts for surgical and medical instruments and apparatus	4.28
		Surgical and medical instruments and apparatus, nsk	2.70
Metabolism and blood-pressure diagnostic apparatus	2.10	Hospital furniture	3.13
Other diagnostic apparatus, including optical diagnostic apparatus	9.58	Operating room furniture, including tables, cases, cabinets, etc.	35.43
Surgical and medical syringes	5.95	Patient room furniture, including cabinets, overbed tables, desks, dressers, etc., but excluding beds and chairs	33.00
Surgical and medical hypodermic needles	1.84	Other hospital furniture excluding operating and patient room furniture, beds, and instruments	29.15
Surgical and medical anesthesia apparatus and instruments	3.81	Hospital furniture, nsk	2.45
Surgical and medical bone plates, screws, and nails, and other internal fixation devices or appliances	3.41	Surgical and medical instruments, nsk	8.53
Surgical and medical blood transfusion, I.V. equipment, and			

Source: 1992 *Economic Census*. The values shown are percent of total shipments in an industry. Values of indented subcategories are summed in the main heading. The symbol (D) appears when data are withheld to prevent disclosure of competitive information. The abbreviation nsk stands for 'not specified by kind' and nec for 'not elsewhere classified'.

INPUTS AND OUTPUTS FOR SURGICAL & MEDICAL INSTRUMENTS

Economic Sector or Industry Providing Inputs	%	Sector	Economic Sector or Industry Buying Outputs	%	Sector
Imports	13.3	Foreign	Gross private fixed investment	62.9	Cap Inv
Primary copper	11.7	Manufg.	Exports	10.1	Foreign
Wholesale trade	7.0	Trade	S/L Govt. purch., health & hospitals	7.6	S/L Govt
Surgical appliances & supplies	6.3	Manufg.	Medical & health services, nec	4.8	Services
Miscellaneous plastics products	5.5	Manufg.	Doctors & dentists	3.5	Services
Metal stampings, nec	4.4	Manufg.	Hospitals	3.5	Services
Advertising	4.0	Services	Federal Government purchases, nondefense	1.6	Fed Govt
Electronic components nec	3.7	Manufg.	Federal Government purchases, national defense	1.5	Fed Govt
Plastics materials & resins	3.2	Manufg.	Surgical appliances & supplies	1.4	Manufg.
Fabricated rubber products, nec	2.7	Manufg.	Change in business inventories	1.3	In House
Paperboard containers & boxes	2.0	Manufg.	Personal consumption expenditures	0.7	
Blast furnaces & steel mills	1.9	Manufg.	Drugs	0.4	Manufg.
Electric services (utilities)	1.8	Util.	S/L Govt. purch., higher education	0.3	S/L Govt
Eating & drinking places	1.7	Trade	S/L Govt. purch., public assistance & relief	0.2	S/L Govt
Paper mills, except building paper	1.6	Manufg.			
Petroleum refining	1.6	Manufg.			
Metal coating & allied services	1.4	Manufg.			
Primary nonferrous metals, nec	1.3	Manufg.			
Communications, except radio & TV	1.2	Util.			
Cyclic crudes and organics	0.9	Manufg.			

Continued on next page.

INPUTS AND OUTPUTS FOR SURGICAL & MEDICAL INSTRUMENTS - Continued

Economic Sector or Industry Providing Inputs	%	Sector	Economic Sector or Industry Buying Outputs	%	Sector
Motor freight transportation & warehousing	0.9	Util.			
Metal foil & leaf	0.8	Manufg.			
Paints & allied products	0.8	Manufg.			
Paperboard mills	0.8	Manufg.			
Sanitary services, steam supply, irrigation	0.8	Util.			
Royalties	0.8	Fin/R.E.			
Copper rolling & drawing	0.7	Manufg.			
Machinery, except electrical, nec	0.7	Manufg.			
Rubber & plastics hose & belting	0.7	Manufg.			
Screw machine and related products	0.7	Manufg.			
Banking	0.7	Fin/R.E.			
Legal services	0.7	Services			
Fabricated metal products, nec	0.6	Manufg.			
Veneer & plywood	0.6	Manufg.			
Real estate	0.6	Fin/R.E.			
Equipment rental & leasing services	0.6	Services			
Management & consulting services & labs	0.6	Services			
Maintenance of nonfarm buildings nec	0.5	Constr.			
Gaskets, packing & sealing devices	0.5	Manufg.			
Glass & glass products, except containers	0.5	Manufg.			
Broadwoven fabric mills	0.4	Manufg.			
Iron & steel foundries	0.4	Manufg.			
Special dies & tools & machine tool accessories	0.4	Manufg.			
Railroads & related services	0.4	Util.			
Accounting, auditing & bookkeeping	0.4	Services			
Abrasive products	0.3	Manufg.			
Hardware, nec	0.3	Manufg.			
Manifold business forms	0.3	Manufg.			
Primary metal products, nec	0.3	Manufg.			
Semiconductors & related devices	0.3	Manufg.			
Gas production & distribution (utilities)	0.3	Util.			
Automotive rental & leasing, without drivers	0.3	Services			
Computer & data processing services	0.3	Services			
Noncomparable imports	0.3	Foreign			
Iron & steel forgings	0.2	Manufg.			
Lubricating oils & greases	0.2	Manufg.			
Nonferrous rolling & drawing, nec	0.2	Manufg.			
Nonwoven fabrics	0.2	Manufg.			
Paper coating & glazing	0.2	Manufg.			
Sawmills & planning mills, general	0.2	Manufg.			
Insurance carriers	0.2	Fin/R.E.			
Automotive repair shops & services	0.2	Services			
Hotels & lodging places	0.2	Services			
Adhesives & sealants	0.1	Manufg.			
Apparel made from purchased materials	0.1	Manufg.			
Household furniture, nec	0.1	Manufg.			
Industrial inorganic chemicals, nec	0.1	Manufg.			
Photographic equipment & supplies	0.1	Manufg.			
Surgical & medical instruments	0.1	Manufg.			
Transit & bus transportation	0.1	Util.			
Personnel supply services	0.1	Services			
U.S. Postal Service	0.1	Gov't			

Source: Benchmark Input-Output Accounts for the U.S. Economy, 1982, U.S. Department of Commerce, Washington, D.C., July 1991. Data, as reported in the source, are organized by the 1977 SIC structure in use in 1982 but have been matched, as closely as is possible, to the 1987 SIC structure used in this book.

OCCUPATIONS EMPLOYED BY SIC 384 - MEDICAL INSTRUMENTS AND SUPPLIES

Occupation	% of Total 1994	Change to 2005	Occupation	% of Total 1994	Change to 2005
Assemblers, fabricators, & hand workers nec	16.5	17.4	Mechanical engineers	1.5	29.3
Sales & related workers nec	5.3	17.4	Electrical & electronic equipment assemblers	1.5	17.4
Inspectors, testers, & graders, precision	4.3	-17.8	Precision workers nec	1.5	5.7
Blue collar worker supervisors	3.3	-0.0	Electrical & electronics engineers	1.4	49.1
Precision assemblers nec	2.4	88.0	Industrial production managers	1.4	17.4
Secretaries, ex legal & medical	2.3	6.9	Packaging & filling machine operators	1.3	17.4
General managers & top executives	1.8	11.4	Electrical & electronic technicians, technologists	1.3	17.4
Traffic, shipping, & receiving clerks	1.8	13.0	Engineering, mathematical, & science managers	1.3	33.4
Machine operators nec	1.7	-17.2	Bookkeeping, accounting, & auditing clerks	1.2	-11.9
Electrical & electronic assemblers	1.7	5.7	Production, planning, & expediting clerks	1.2	17.4
Engineering technicians & technologists nec	1.6	17.4	General office clerks	1.2	0.2
Marketing, advertising, & PR managers	1.5	17.4	Adjustment clerks	1.1	40.9
Hand packers & packagers	1.5	0.6	Machinists	1.1	-11.9
Sewing machine operators, non-garment	1.5	76.2	Managers & administrators nec	1.1	17.4

Source: Industry-Occupation Matrix, Bureau of Labor Statistics. These data relate to one or more 3-digit SIC industry groups rather than to a single 4-digit SIC. The change reported for each occupation to the year 2005 is a percent of growth or decline as estimated by the Bureau of Labor Statistics. The abbreviation nec stands for 'not elsewhere classified'.

LOCATION BY STATE AND REGIONAL CONCENTRATION

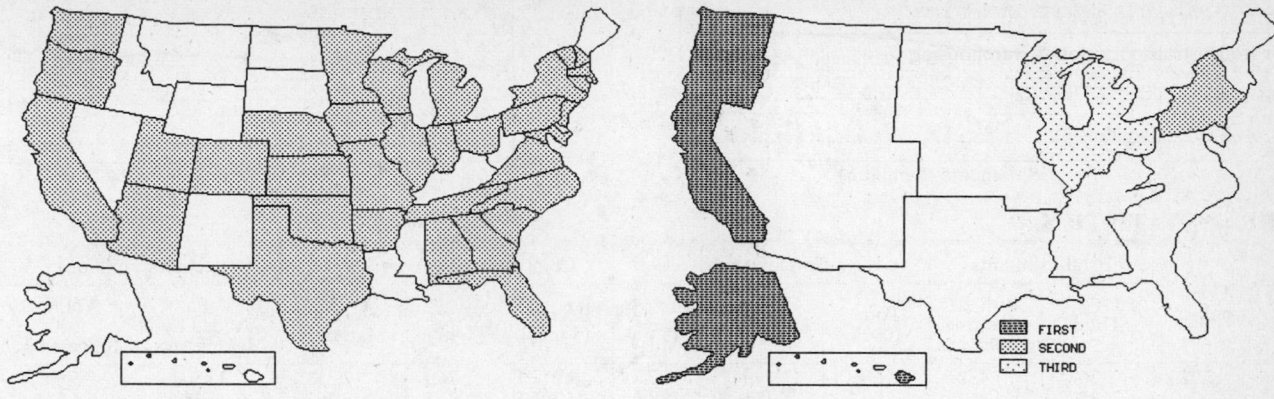

FIRST
SECOND
THIRD

INDUSTRY DATA BY STATE

| State | Establish- ments | Shipments | | | Employment | | | | Cost as % of Shipments | Investment per Employee ($) |
		Total ($ mil)	% of U.S.	Per Establ.	Total Number	% of U.S.	Per Establ.	Wages ($/hour)		
California	265	2,859.1	21.4	10.8	18,000	18.3	68	11.48	31.9	8,700
Connecticut	44	1,596.7	11.9	36.3	6,200	6.3	141	13.66	21.3	15,952
Massachusetts	91	1,331.9	10.0	14.6	9,300	9.5	102	11.36	28.0	5,452
Minnesota	65	676.7	5.1	10.4	5,300	5.4	82	11.09	21.8	8,453
New York	73	663.3	5.0	9.1	7,800	7.9	107	10.94	32.9	3,423
Pennsylvania	69	589.7	4.4	8.5	4,700	4.8	68	12.29	35.0	10,255
Florida	67	510.1	3.8	7.6	5,100	5.2	76	8.75	22.8	4,686
Texas	56	431.6	3.2	7.7	4,500	4.6	80	10.62	43.9	6,289
Missouri	35	408.3	3.1	11.7	3,600	3.7	103	10.73	33.7	2,944
Illinois	61	367.8	2.7	6.0	3,100	3.2	51	9.31	33.1	4,258
Colorado	48	312.7	2.3	6.5	2,200	2.2	46	11.17	44.5	5,636
Ohio	42	286.4	2.1	6.8	1,500	1.5	36	11.37	38.3	3,733
Utah	23	280.9	2.1	12.2	2,400	2.4	104	7.61	26.4	7,250
Michigan	31	270.4	2.0	8.7	2,200	2.2	71	9.90	32.9	8,909
Indiana	32	259.6	1.9	8.1	2,500	2.5	78	10.58	26.5	4,640
Georgia	18	252.4	1.9	14.0	1,600	1.6	89	11.26	19.6	10,125
North Carolina	21	232.1	1.7	11.1	2,400	2.4	114	9.63	46.1	4,292
New Jersey	61	202.5	1.5	3.3	2,200	2.2	36	10.72	31.0	4,227
New Hampshire	20	133.7	1.0	6.7	1,300	1.3	65	13.75	35.8	2,154
Maryland	23	91.7	0.7	4.0	700	0.7	30	13.44	63.2	-
Washington	30	69.8	0.5	2.3	800	0.8	27	10.71	29.2	14,875
Arizona	16	60.7	0.5	3.8	600	0.6	38	10.50	36.7	1,667
Oregon	20	40.1	0.3	2.0	300	0.3	15	9.25	31.4	5,000
Tennessee	20	16.9	0.1	0.8	300	0.3	15	9.67	37.9	2,000
Wisconsin	18	(D)	-	-	1,750 *	1.8	97	-	-	1,943
Iowa	9	(D)	-	-	175 *	0.2	19	-	-	1,714
Nebraska	9	(D)	-	-	1,750 *	1.8	194	-	-	-
Virginia	9	(D)	-	-	750 *	0.8	83	-	-	-
South Carolina	7	(D)	-	-	1,750 *	1.8	250	-	-	-
Alabama	6	(D)	-	-	375 *	0.4	63	-	-	-
Kansas	5	(D)	-	-	750 *	0.8	150	-	-	-
Oklahoma	5	(D)	-	-	375 *	0.4	75	-	-	-
Arkansas	2	(D)	-	-	1,750 *	1.8	875	-	-	-
Vermont	2	(D)	-	-	175 *	0.2	88	-	-	-

Source: 1992 *Economic Census*. The states are in descending order of shipments or establishments (if shipment data are missing for the majority). The symbol (D) appears when data are withheld to prevent disclosure of competitive information. States marked with (D) are sorted by number of establishments. A dash (-) indicates that the data element cannot be calculated; * indicates the midpoint of a range.

3842 - SURGICAL APPLIANCES & SUPPLIES

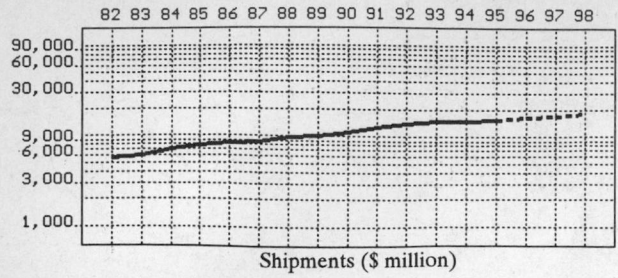

Shipments ($ million)

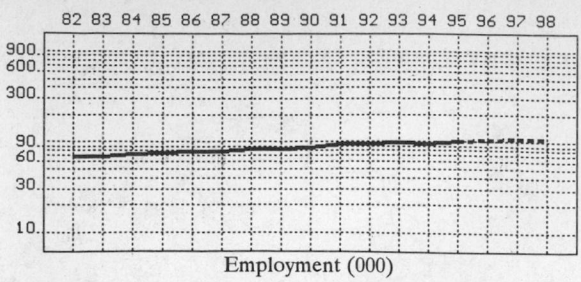

Employment (000)

GENERAL STATISTICS

| Year | Com-panies | Establishments | | Employment | | | Compensation | | Production ($ million) | | | |
		Total	with 20 or more employees	Total (000)	Production Workers (000)	Hours (Mil)	Payroll ($ mil)	Wages ($/hr)	Cost of Materials	Value Added by Manufacture	Value of Shipments	Capital Invest.
1982	1,224	1,367	452	68.8	46.1	87.8	1,211.1	7.21	2,233.1	3,450.8	5,667.1	187.7
1983		1,357	462	69.3	45.6	88.0	1,301.5	7.69	2,302.8	3,738.3	6,044.1	180.4
1984		1,347	472	73.8	49.5	95.4	1,441.4	7.79	2,760.3	4,422.9	7,116.9	260.9
1985		1,337	483	76.3	49.2	95.3	1,606.6	8.34	2,872.7	4,975.1	7,864.7	348.3
1986		1,333	473	77.7	48.6	93.6	1,722.7	8.59	2,971.6	5,379.6	8,290.8	254.5
1987	1,326	1,500	550	78.5	51.1	99.5	1,785.8	8.79	3,145.7	5,443.7	8,534.1	215.3
1988		1,464	564	82.1	52.8	105.0	1,910.0	8.98	3,465.3	6,539.5	9,827.9	219.7
1989		1,429	565	84.7	54.3	104.0	2,011.4	9.35	3,725.7	6,504.4	10,187.1	239.9
1990		1,446	560	85.9	55.8	109.8	2,194.9	9.60	4,058.3	7,163.1	11,127.6	265.5
1991		1,508	568	93.3	60.6	118.3	2,509.8	10.12	4,622.9	8,044.3	12,554.7	425.4
1992	1,572	1,764	667	96.8	61.6	119.7	2,865.2	10.83	4,963.7	8,939.3	13,842.8	504.3
1993		1,767	656	98.0	61.9	118.9	2,955.7	10.93	5,238.2	9,276.1	14,553.0	433.3
1994		1,695P	663P	96.1	59.4	114.4	3,000.3	11.21	5,170.8	9,366.4	14,572.6	374.1
1995		1,730P	682P	101.4P	63.6P	123.7P	3,139.0P	11.51P	5,522.3P	10,003.1P	15,563.2P	440.5P
1996		1,765P	701P	104.0P	65.1P	126.6P	3,296.0P	11.84P	5,803.6P	10,512.7P	16,356.0P	460.5P
1997		1,800P	720P	106.6P	66.5P	129.4P	3,453.1P	12.17P	6,084.9P	11,022.2P	17,148.7P	480.4P
1998		1,835P	739P	109.2P	67.9P	132.2P	3,610.1P	12.50P	6,366.2P	11,531.7P	17,941.5P	500.4P

Sources: 1982, 1987, 1992 *Economic Census*; *Annual Survey of Manufactures*, 83-86, 88-91, 93-94. Establishment counts for non-Census years are from *County Business Patterns*; establishment values for 83-84 are extrapolations. 'P's show projections by the editors. Industries reclassified in 87 will not have data for prior years.

INDICES OF CHANGE

| Year | Com-panies | Establishments | | Employment | | | Compensation | | Production ($ million) | | | |
		Total	with 20 or more employees	Total (000)	Production Workers (000)	Hours (Mil)	Payroll ($ mil)	Wages ($/hr)	Cost of Materials	Value Added by Manufacture	Value of Shipments	Capital Invest.
1982	78	77	68	71	75	73	42	67	45	39	41	37
1983		77	69	72	74	74	45	71	46	42	44	36
1984		76	71	76	80	80	50	72	56	49	51	52
1985		76	72	79	80	80	56	77	58	56	57	69
1986		76	71	80	79	78	60	79	60	60	60	50
1987	84	85	82	81	83	83	62	81	63	61	62	43
1988		83	85	85	86	88	67	83	70	73	71	44
1989		81	85	87	88	87	70	86	75	73	74	48
1990		82	84	89	91	92	77	89	82	80	80	53
1991		85	85	96	98	99	88	93	93	90	91	84
1992	100	100	100	100	100	100	100	100	100	100	100	100
1993		100	98	101	100	99	103	101	106	104	105	86
1994		96P	99P	99	96	96	105	104	104	105	105	74
1995		98P	102P	105P	103P	103P	110P	106P	111P	112P	112P	87P
1996		100P	105P	107P	106P	106P	115P	109P	117P	118P	118P	91P
1997		102P	108P	110P	108P	108P	121P	112P	123P	123P	124P	95P
1998		104P	111P	113P	110P	110P	126P	115P	128P	129P	130P	99P

Sources: Same as General Statistics. Values reflect change from the base year, 1992. Values above 100 mean greater than 92, values below 100 mean less than 92, and a value of 100 in the 82-91 or 93-98 period means same as 92. 'P's mark projections by the editors.

SELECTED RATIOS

For 1994	Avg. of All Manufact.	Analyzed Industry	Index	For 1994	Avg. of All Manufact.	Analyzed Industry	Index
Employees per Establishment	49	57	116	Value Added per Production Worker	134,084	157,684	118
Payroll per Establishment	1,500,273	1,769,851	118	Cost per Establishment	5,045,178	3,050,210	60
Payroll per Employee	30,620	31,221	102	Cost per Employee	102,970	53,806	52
Production Workers per Establishment	34	35	102	Cost per Production Worker	146,988	87,051	59
Wages per Establishment	853,319	756,491	89	Shipments per Establishment	9,576,895	8,596,252	90
Wages per Production Worker	24,861	21,590	87	Shipments per Employee	195,460	151,640	78
Hours per Production Worker	2,056	1,926	94	Shipments per Production Worker	279,017	245,330	88
Wages per Hour	12.09	11.21	93	Investment per Establishment	321,011	220,678	69
Value Added per Establishment	4,602,255	5,525,159	120	Investment per Employee	6,552	3,893	59
Value Added per Employee	93,930	97,465	104	Investment per Production Worker	9,352	6,298	67

Sources: Same as General Statistics. The 'Average of All Manufacturing' column represents the average of all manufacturing industries reported for the most recent complete year available. The Index shows the relationship between the Average and the Analyzed Industry. For example, 100 means that they are equal; 500 that the Analyzed Industry is five times the average; 50 means that the Analyzed Industry is half the national average. The abbreviation 'na' is used to show that data are 'not available'.

LEADING COMPANIES Number shown: 75   Total sales ($ mil): 12,614   Total employment (000): 99.2

Company Name	Address				CEO Name	Phone	Co. Type	Sales ($ mil)	Empl. (000)
Pfizer Hospital Products Group	235 E 42nd St	New York	NY	10017	David M Fitzgerald	212-573-2323	S	1,082	5.9
Sulzer Inc	200 Park Av	New York	NY	10166	Michel Bally	212-949-0999	S	1,000	4.5
Kendall Co	15 Hampshire St	Mansfield	MA	02048	Richard A Gilleland	508-261-8000	S	816	8.5
Kendall International Inc	15 Hampshire St	Mansfield	MA	02048	Richard A Gilleland	508-261-8000	S	816	8.5
Howmedica Inc	359 Veterans Blv	Rutherford	NJ	07070	James Barbarito	201-507-7300	S	657	2.0
Chemed Corp	255 E 5th St	Cincinnati	OH	45202	Edward L Hutton	513-762-6900	P	645	6.6
Ethicon Inc	PO Box 151	Somerville	NJ	08876	Frank J Ryan	908-218-0707	S	500	5.0
AMSCO International Inc	112 Washington Pl	Pittsburgh	PA	15219	Daniel R Barry	412-338-6500	P	483	3.1
American Sterilizer Co	500 Grant St	Pittsburgh	PA	15219	David A Nelson	412-338-6500	S	480*	3.0
Mine Safety Appliances Co	PO Box 426	Pittsburgh	PA	15230	John T Ryan III	412-967-3000	P	460	4.4
Invacare Corp	PO Box 4028	Elyria	OH	44036	A Malachi Mixon III	216-329-6000	P	411	3.3
Medline Industries Inc	1 Medline Pl	Mundelein	IL	60060	James Mills	708-949-5500	R	380	2.0
Biomet Inc	PO Box 587	Warsaw	IN	46581	Dane A Miller	219-267-6639	P	373	1.9
St Jude Medical Inc	1 Lillehei Plz	St Paul	MN	55117	R A Matricaria	612-483-2000	P	360	2.2
Figgie International Inc	4420 Sherwin Rd	Willoughby	OH	44094	John P Reilly	216-953-2700	P	319	6.9
Sunrise Medical Inc	2382 Faraday Av	Carlsbad	CA	92008	Richard H Chandler	310-328-8018	P	319	2.6
Davis + Geck	1 Casper St	Danbury	CT	06810	Jeffrey P Ashpitz	203-743-4451	D	300	2.0
Starkey Labs Inc	6700 Washington S	Eden Prairie	MN	55344	William F Austin	612-941-6401	R	210*	1.9
Sofamor Danek Group Inc	3092 Directors Row	Memphis	TN	38131	ER Pickard	901-396-2695	P	162	0.6
Lumex Inc	81 Spence St	Bay Shore	NY	11706	James E George	516-273-2200	P	131	1.2
Professional Medical Products	PO Box 3288	Greenwood	SC	29648	C Birge Sigety	803-223-4281	R	130	1.2
Mentor Corp	5425 Hollister Av	Santa Barbara	CA	93111	C J Conway	805-681-6000	P	124	0.8
Tecnol Medical Products Inc	7201 Industrial Park	Fort Worth	TX	76180	Van Hubbard	817-581-6424	P	121	1.8
Scott Aviation	225 Erie St	Lancaster	NY	14086	Glen Lindeman	716-683-5100	D	110	0.8
Wright Medical Technology Inc	PO Box 100	Arlington	TN	38002	Herbert W Korthoff	901-867-9971	R	100	0.5
Graham-Field Inc	400 Rabro Dr E	Hauppauge	NY	11788	Harvey Diamond	516-582-5900	S	93	0.6
American White Cross Inc	349 Lake Rd	Dayville	CT	06241	Howard Koenig	203-774-8541	P	90	0.9
Beltone Electronics Corp	4201 W Victoria St	Chicago	IL	60646	Lawrence M Posen	312-583-3600	R	90	1.0
Meadox Medicals Inc	112 Bauer Dr	Oakland	NJ	07436	Eleanor Gackstatter	201-337-6126	R	83*	0.8
American Medical Systems Inc	11001 Bren Rd E	Minnetonka	MN	55343	Lloyd R Armstrong	612-933-4666	S	81	0.4
Everest and Jennings Intern Ltd	1100 Corp Square	St Louis	MO	63132	Bevil J Hogg	314-995-7000	P	79	0.7
STERIS Corp	5960 Heusley Rd	Mentor	OH	44060	William R Sanford	216-354-2600	P	64	0.3
ReSound Corp	220 Saginaw Dr	Redwood City	CA	94063	Peter Riepenhausen	415-780-7800	P	62	0.2
CarboMedics Inc	1300 Anders	Austin	TX	78752	Terry Marlatt	512-873-3200	S	62*	0.6
Kappler USA Inc	PO Box 218	Guntersville	AL	35976	Henry B Swoope	205-582-2195	S	60	0.5
Minnesota Mining & Mfg	610 N County Rd 19	Aberdeen	SD	57401	Lowell Christensen	605-229-5002	D	60*	0.6
NAMIC USA Corp	Pruyn's Island	Glens Falls	NY	12801	Cynthia L Morris	518-748-0067	P	57	0.7
Barnhardt Manufacturing	PO Box 34276	Charlotte	NC	28234	Tom Barnhardt III	704-376-0380	R	55*	0.5
Clinipad Corp	66 High St	Guilford	CT	06437	David A Greenberg	203-453-6543	R	55*	0.5
Bird Medical Technologies Inc	1100 Bird Ctr Dr	Palm Springs	CA	92262	Felix T Troilo	619-778-7200	P	54	0.4
Kellogg Industries	PO Box 320	Jackson	MI	49204	Richard Stockman	517-782-0579	D	53	0.1
JE Hanger Inc	7700 Old Gtwn	Bethesda	MD	20814	Ivan Sabel	301-986-0701	S	50*	0.5
JE Hanger Inc	PO Box 406	Alpharetta	GA	30239	HE Thranhardt	404-442-9870	R	49*	0.5
Minntech Corp	14905 28th Av N	Minneapolis	MN	55447	Louis C Cosentino	612-553-3300	P	48	0.3
ILC Dover Inc	PO Box 266	Frederica	DE	19946	Homer Reihm	302-335-3911	S	40	0.4
American Medical Electronics	250 E Arapaho Rd	Richardson	TX	75081	John F Clifford	214-918-8300	P	38	0.3
SBS Enterprises Inc	PO Box 2501	Waco	TX	76702	Steven B Smith	817-772-6000	R	38*	0.4
Spenco Medical Corp	PO Box 2501	Waco	TX	76702	Steven B Smith	817-772-6000	S	38*	0.4
Chattanooga Group Inc	PO Box 489	Hixson	TN	37343	Paul Chapman	615-870-2281	R	34*	0.3
Span America Medical Systems	PO Box 5231	Greenville	SC	29606	Charles B Mitchell	803-288-8877	P	33	0.2
Smith and Nephew Rolyan Inc	PO Box 1005	Germantown	WI	53022	John Clark	414-251-7841	S	33*	0.3
NDM Acquisition Corp	PO Box 1408	Dayton	OH	45401	William Shea	513-456-5585	S	32	0.3
Wright and Filippis Inc	2845 Crooks Rd	Rochester Hills	MI	48309	Gene Fillippis	313-853-1888	R	32	0.3
Resistance Technology Inc	1260 Red Fox Rd	Arden Hills	MN	55112	Mark Gorder	612-636-9770	S	31*	0.3
Boss Manufacturing Co	221 W 1st St	Kewanee	IL	61443	Richard Bern	309-852-2131	S	30	0.6
Electric Mobility Corp	591 Mantua Blv	Sewell	NJ	08080	Mike Flowers	609-468-0270	R	30	0.2
Ling Products Inc	570 Enterprise Dr	Neenah	WI	54957	Franz Stadtnueller	414-725-8491	S	30	0.1
Point Blank Body Armor LP	185 Dixon Av	Amityville	NY	11701	Lawrence Kaucheck	516-842-3900	R	30	0.2
Sutter Corp	9425 Chesapeake	San Diego	CA	92123	Vince Estrada	619-569-8148	S	30	0.3
Eastco Industrial Safety Corp	130 W 10th St	Huntington St	NY	11746	Alan E Densen	516-427-1802	P	29*	0.3
Conco Medical Co	481 Lakeshore Pkwy	Rock Hill	SC	29730	Bruce Griffiths	803-325-7600	R	27	0.3
Synthes USA	PO Box 366	Monument	CO	80132	Hansjorg Wyss	719-481-3021	P	27	0.4
Microtek Medical Inc	PO Box 2487	Columbus	MS	39704	Kimber L Vought	601-327-1863	P	27	0.5
Ortho-Kinetics Inc	PO Box 1647	Waukesha	WI	53187	Edward J Gaffney	414-542-6060	R	26	0.3
Augustine Medical Inc	10393 W 70th St	Eden Prairie	MN	55344	John Thomas	612-947-1200	P	25	0.2
Orthopedic Systems Inc	30031 Ahern Av	Union City	CA	94587	Allan Epstein	510-632-3824	R	24*	0.2
B and F Medical Products Inc	1421 Expressway N	Toledo	OH	43608	Robert L Weaver	419-729-0606	S	23	0.1
Dyna Corp	6300 Yarrow Dr	Carlsbad	CA	92009	Robert DeBussy	619-438-2511	R	23*	0.2
Research Medical Inc	6864 S 300 W	Midvale	UT	84047	Gary L Crocker	801-972-5500	P	22	0.2
Orthomet Inc	6301 Cecilia Cir	Minneapolis	MN	55439	RD Nikolaev	612-944-6112	P	22	0.2
OK-1 Manufacturing Co	PO Box 736	Altus	OK	73522	Joe Courtney	405-482-9066	R	22	0.2
PyMaH Corp	500 Rte 202 N	Flemington	NJ	08822	Bernard M Hanafin	908-788-4000	R	22*	0.2
Cygnus Therapeutic Systems	400 Penobscot Dr	Redwood City	CA	94063	Gregory B Lawless	415-369-4300	P	21	0.2
Coloplast Corp	1955 W Oak Cir	Marietta	GA	30062	Joachim Rechenberg	404-426-6362	S	20	0.2
Ferno Ille	70 Weil Way	Wilmington	OH	45177	Wayne Smith	513-382-1451	D	20	0.3

Source: Ward's Business Directory of U.S. Private and Public Companies, Volumes 1 and 2, 1996. The company type code used is as follows: P - Public, R - Private, S - Subsidiary, D - Division, J - Joint Venture, A - Affiliate, G - Group. Sales are in millions of dollars, employees are in thousands. An asterisk (*) indicates an estimated sales volume. The symbol < stands for 'less than'. Company names and addresses are truncated, in some cases, to fit into the available space.

MATERIALS CONSUMED

Material	Quantity	Delivered Cost ($ million)
Materials, ingredients, containers, and supplies	(X)	4,150.9
Surgical and orthopedic supplies	(X)	319.3
Resistors, capacitors, transformers, electron tubes, semiconductors, and other electronic components	(X)	123.8
Metal bolts, nuts, screws, washers, rivets, and other screw machine products	(X)	35.0
Other fabricated metal products (except castings and forgings)	(X)	120.2
Iron and steel forgings	(X)	40.9
Nonferrous forgings	(X)	12.6
Iron and steel castings (rough and semifinished)	(X)	91.2
Nonferrous (aluminum, copper, etc.) castings (rough and semifinished)	(X)	25.2
Steel shapes and forms	(X)	101.2
Nonferrous shapes and forms	(X)	35.0
Nonwoven fabrics	(X)	466.0
Broadwoven fabrics	(X)	252.3
Plastics resins consumed in the form of granules, pellets, powders, liquids, etc.	(X)	110.4
Plastics products consumed in the form of sheets, rods, tubes, film, and other shapes	(X)	154.6
Fabricated plastics products	(X)	149.8
Fabricated rubber products, except tires, tubes, hose, belting, and gaskets	(X)	49.6
Adhesives and sealants	(X)	41.1
Glass and glass products, except photographic and projection lenses and prisms	(X)	6.1
Paperboard containers, boxes, and corrugated paperboard	(X)	137.1
Paper and paperboard products except paperboard boxes, containers, and corrugated paperboard	(X)	81.7
All other materials and components, parts, containers, and supplies	(X)	1,065.6
Materials, ingredients, containers, and supplies, nsk	(X)	732.1

Source: 1992 *Economic Census.* Explanation of symbols used: (D): Withheld to avoid disclosure of competitive data; na: Not available; (S): Withheld because statistical norms were not met; (X): Not applicable; (Z): Less than half the unit shown; nec: Not elsewhere classified; nsk: Not specified by kind; - : zero; * : 10-19 percent estimated; ** : 20-29 percent estimated.

PRODUCT SHARE DETAILS

Product or Product Class	% Share	Product or Product Class	% Share
Surgical appliances and supplies	100.00	Patient transport devices, wheel chairs	2.90
Surgical, orthopedic or fracture, prosthetic, and therapeutic appliances and supplies	77.41	Other patient transport devices, including stretchers, tables, etc., except wheel chairs	1.46
Artificial joints, orthopedic and prosthetic appliances	15.49	Therapeutic appliances and supplies, hydrotherapy equipment, including full body and limb tanks (portable and stationary)	0.37
Artificial limbs, orthopedic and prosthetic appliances	0.91	Other therapeutic appliances and supplies, excluding electromedical	1.63
Braces, mechanical	1.42	Surgical kits	5.79
Elastic braces, suspensories, and other elastic supports	2.24	Other surgical and orthopedic items not included in above categories	18.60
Elastic stockings	0.47	Parts for surgical, orthopedic, prosthetic, and therapeutic appliances and supplies	1.62
Surgical corsets	0.21	Surgical, orthopedic, prosthetic, and therapeutic appliances and supplies, nsk	1.47
Splints and trusses	0.69	Personal industrial safety devices	10.67
Crutches, canes (orthopedic), and other walking assistance devices	0.86	Industrial safety devices, respiratory protection equipment, including gas masks, abrasive masks, canister masks, etc.	36.26
Arch supports and other foot appliances	1.87	Industrial safety devices, helmets (hardhats)	4.91
Intraocular lenses, orthopedic and prosthetic appliances	3.02	Industrial safety devices, eye and face protection devices (face shields, welding helmets, masks), excluding industrial goggles and eye protectors	7.58
Other orthopedic and prosthetic appliances	5.35	Industrial safety devices, protective clothing, except shoes	27.14
Surgical dressings, bandages, elastic	0.42	First aid, snake bite, and burn kits, both household and industrial types	3.00
Surgical dressings, bandages, other, including muslin, plaster of paris, etc., excluding self-adhering bandages	1.11	Other personal safety devices, including motorcycle and auto racing helmets	17.80
Surgical dressings, adhesive plaster, medicated and nonmedicated, including self-adhering bandages	2.83	Personal industrial safety devices, nsk	3.31
Surgical dressings, gauze (absorbent and packing)	0.59	Electronic hearing aids, complete units	2.98
Surgical dressings, cotton, including cotton balls (sterile and nonsterile)	0.79	Surgical appliances and supplies, nsk	8.93
Other surgical dressings, including sponges, compresses, pads, etc.	4.40		
Disposable surgical drapes, including O/B and O/R packs	6.54		
Disposable incontinent pads, bedpads, and adult diapers	7.81		
Sterile surgical sutures	5.47		
Breathing devices, excluding anesthetic apparatus but including incubators, respirators, resuscitators, inhalators, etc.	3.65		

Source: 1992 *Economic Census.* The values shown are percent of total shipments in an industry. Values of indented subcategories are summed in the main heading. The symbol (D) appears when data are withheld to prevent disclosure of competitive information. The abbreviation nsk stands for 'not specified by kind' and nec for 'not elsewhere classified'.

INPUTS AND OUTPUTS FOR SURGICAL APPLIANCES & SUPPLIES

Economic Sector or Industry Providing Inputs	%	Sector	Economic Sector or Industry Buying Outputs	%	Sector
Imports	10.4	Foreign	Personal consumption expenditures	22.0	
Broadwoven fabric mills	8.9	Manufg.	S/L Govt. purch., health & hospitals	14.9	S/L Govt
Surgical appliances & supplies	8.5	Manufg.	Hospitals	14.0	Services
Wholesale trade	7.6	Trade	Gross private fixed investment	12.0	Cap Inv
Electronic components nec	4.2	Manufg.	Exports	7.6	Foreign
Metal stampings, nec	3.8	Manufg.	Surgical appliances & supplies	4.7	Manufg.
Advertising	3.8	Services	Doctors & dentists	3.9	Services
Nonwoven fabrics	3.7	Manufg.	Surgical & medical instruments	2.8	Manufg.
Miscellaneous plastics products	3.0	Manufg.	S/L Govt. purch., public assistance & relief	2.6	S/L Govt
Die-cut paper & board	2.4	Manufg.	S/L Govt. purch., fire	2.2	S/L Govt
Surgical & medical instruments	2.1	Manufg.	Federal Government purchases, national defense	1.8	Fed Govt
Paperboard containers & boxes	2.0	Manufg.	Federal Government purchases, nondefense	1.8	Fed Govt
Plastics materials & resins	1.8	Manufg.	Nursing & personal care facilities	1.4	Services
Electric services (utilities)	1.6	Util.	Change in business inventories	0.9	In House
Banking	1.4	Fin/R.E.	S/L Govt. purch., police	0.9	S/L Govt
Blast furnaces & steel mills	1.2	Manufg.	Industrial buildings	0.4	Constr.
Metal coating & allied services	1.2	Manufg.	Electric utility facility construction	0.3	Constr.
Narrow fabric mills	1.2	Manufg.	Maintenance of nonfarm buildings nec	0.3	Constr.
Eating & drinking places	1.2	Trade	S/L Govt. purch., natural resource & recreation.	0.3	S/L Govt
Business services nec	1.2	Services	Wholesale trade	0.2	Trade
Adhesives & sealants	1.1	Manufg.	S/L Govt. purch., correction	0.2	S/L Govt
Communications, except radio & TV	1.1	Util.	S/L Govt. purch., elem. & secondary education	0.2	S/L Govt
Motor freight transportation & warehousing	1.1	Util.	S/L Govt. purch., other education & libraries	0.2	S/L Govt
Petroleum refining	1.0	Manufg.	Construction of hospitals	0.1	Constr.
Fabricated rubber products, nec	0.9	Manufg.	Construction of stores & restaurants	0.1	Constr.
Meat packing plants	0.9	Manufg.	Highway & street construction	0.1	Constr.
Paper mills, except building paper	0.9	Manufg.	Maintenance of highways & streets	0.1	Constr.
Job training & related services	0.9	Services	Sewer system facility construction	0.1	Constr.
Cyclic crudes and organics	0.8	Manufg.	Child day care services	0.1	Services
Royalties	0.8	Fin/R.E.			
Aluminum rolling & drawing	0.7	Manufg.			
Cottonseed oil mills	0.7	Manufg.			
Real estate	0.7	Fin/R.E.			
Equipment rental & leasing services	0.7	Services			
Maintenance of nonfarm buildings nec	0.6	Constr.			
Felt goods, nec	0.6	Manufg.			
Gas production & distribution (utilities)	0.6	Util.			
Engineering, architectural, & surveying services	0.6	Services			
Chemical preparations, nec	0.5	Manufg.			
Fabricated metal products, nec	0.5	Manufg.			
Gaskets, packing & sealing devices	0.5	Manufg.			
Iron & steel foundries	0.5	Manufg.			
Machinery, except electrical, nec	0.5	Manufg.			
Metal cans	0.5	Manufg.			
Legal services	0.5	Services			
Iron & steel forgings	0.4	Manufg.			
Nonferrous wire drawing & insulating	0.4	Manufg.			
Processed textile waste	0.4	Manufg.			
Management & consulting services & labs	0.4	Services			
Noncomparable imports	0.4	Foreign			
Coated fabrics, not rubberized	0.3	Manufg.			
Glass & glass products, except containers	0.3	Manufg.			
Screw machine and related products	0.3	Manufg.			
Special dies & tools & machine tool accessories	0.3	Manufg.			
Thread mills	0.3	Manufg.			
Railroads & related services	0.3	Util.			
Sanitary services, steam supply, irrigation	0.3	Util.			
Accounting, auditing & bookkeeping	0.3	Services			
Business/professional associations	0.3	Services			
Computer & data processing services	0.3	Services			
Hotels & lodging places	0.3	Services			
U.S. Postal Service	0.3	Gov't			
Abrasive products	0.2	Manufg.			
Asbestos products	0.2	Manufg.			
Copper rolling & drawing	0.2	Manufg.			
Hardware, nec	0.2	Manufg.			
Manifold business forms	0.2	Manufg.			
Miscellaneous fabricated wire products	0.2	Manufg.			
Nonferrous rolling & drawing, nec	0.2	Manufg.			
Paper coating & glazing	0.2	Manufg.			
Primary metal products, nec	0.2	Manufg.			
Sawmills & planning mills, general	0.2	Manufg.			
Air transportation	0.2	Util.			
Water transportation	0.2	Util.			
Credit agencies other than banks	0.2	Fin/R.E.			
Insurance carriers	0.2	Fin/R.E.			
Automotive rental & leasing, without drivers	0.2	Services			
Coal	0.1	Mining			

Continued on next page.

INPUTS AND OUTPUTS FOR SURGICAL APPLIANCES & SUPPLIES - Continued

Economic Sector or Industry Providing Inputs	%	Sector	Economic Sector or Industry Buying Outputs	%	Sector
Industrial inorganic chemicals, nec	0.1	Manufg.			
Lubricating oils & greases	0.1	Manufg.			
Automotive repair shops & services	0.1	Services			
Personnel supply services	0.1	Services			

Source: Benchmark Input-Output Accounts for the U.S. Economy, 1982, U.S. Department of Commerce, Washington, D.C., July 1991. Data, as reported in the source, are organized by the 1977 SIC structure in use in 1982 but have been matched, as closely as is possible, to the 1987 SIC structure used in this book.

OCCUPATIONS EMPLOYED BY SIC 384 - MEDICAL INSTRUMENTS AND SUPPLIES

Occupation	% of Total 1994	Change to 2005	Occupation	% of Total 1994	Change to 2005
Assemblers, fabricators, & hand workers nec	16.5	17.4	Mechanical engineers	1.5	29.3
Sales & related workers nec	5.3	17.4	Electrical & electronic equipment assemblers	1.5	17.4
Inspectors, testers, & graders, precision	4.3	-17.8	Precision workers nec	1.5	5.7
Blue collar worker supervisors	3.3	-0.0	Electrical & electronics engineers	1.4	49.1
Precision assemblers nec	2.4	88.0	Industrial production managers	1.4	17.4
Secretaries, ex legal & medical	2.3	6.9	Packaging & filling machine operators	1.3	17.4
General managers & top executives	1.8	11.4	Electrical & electronic technicians,technologists	1.3	17.4
Traffic, shipping, & receiving clerks	1.8	13.0	Engineering, mathematical, & science managers	1.3	33.4
Machine operators nec	1.7	-17.2	Bookkeeping, accounting, & auditing clerks	1.2	-11.9
Electrical & electronic assemblers	1.7	5.7	Production, planning, & expediting clerks	1.2	17.4
Engineering technicians & technologists nec	1.6	17.4	General office clerks	1.2	0.2
Marketing, advertising, & PR managers	1.5	17.4	Adjustment clerks	1.1	40.9
Hand packers & packagers	1.5	0.6	Machinists	1.1	-11.9
Sewing machine operators, non-garment	1.5	76.2	Managers & administrators nec	1.1	17.4

Source: Industry-Occupation Matrix, Bureau of Labor Statistics. These data relate to one or more 3-digit SIC industry groups rather than to a single 4-digit SIC. The change reported for each occupation to the year 2005 is a percent of growth or decline as estimated by the Bureau of Labor Statistics. The abbreviation nec stands for 'not elsewhere classified'.

LOCATION BY STATE AND REGIONAL CONCENTRATION

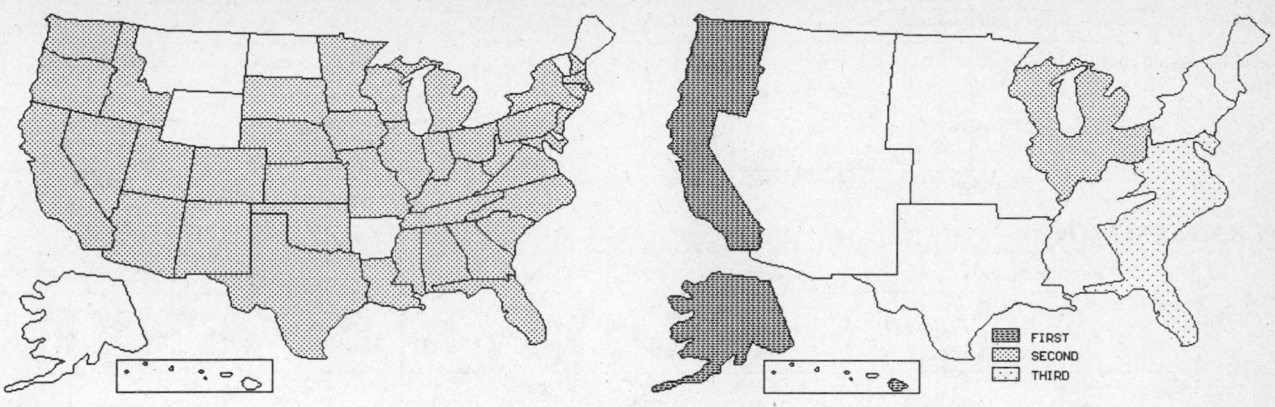

FIRST
SECOND
THIRD

INDUSTRY DATA BY STATE

State	Establish-ments	Shipments			Employment				Cost as % of Shipments	Investment per Employee ($)
		Total ($ mil)	% of U.S.	Per Establ.	Total Number	% of U.S.	Per Establ.	Wages ($/hour)		
California	259	1,525.4	11.0	5.9	12,500	12.9	48	9.42	36.3	4,200
Texas	102	1,194.6	8.6	11.7	6,600	6.8	65	9.95	51.4	5,030
New Jersey	57	1,114.9	8.1	19.6	7,800	8.1	137	14.71	33.5	4,038
Indiana	37	1,090.8	7.9	29.5	4,800	5.0	130	18.02	15.0	10,750
Ohio	88	899.0	6.5	10.2	7,200	7.4	82	11.34	27.7	-
North Carolina	50	783.9	5.7	15.7	3,300	3.4	66	11.94	37.0	11,727
Tennessee	38	758.2	5.5	20.0	4,300	4.4	113	12.46	29.0	6,465
Pennsylvania	90	680.2	4.9	7.6	5,000	5.2	56	12.97	33.1	4,560
Minnesota	66	526.3	3.8	8.0	5,400	5.6	82	10.05	31.2	2,500
Florida	124	411.8	3.0	3.3	3,400	3.5	27	9.95	29.3	4,500
New York	118	399.7	2.9	3.4	4,000	4.1	34	9.78	37.8	1,850
Georgia	35	345.1	2.5	9.9	2,700	2.8	77	9.71	57.4	1,852
Wisconsin	40	342.2	2.5	8.6	2,200	2.3	55	11.25	40.8	3,000
Illinois	57	326.4	2.4	5.7	2,900	3.0	51	9.13	55.8	1,759
Connecticut	26	322.8	2.3	12.4	1,900	2.0	73	10.67	42.0	1,263
Massachusetts	60	295.2	2.1	4.9	2,300	2.4	38	10.50	34.6	3,696
Michigan	61	295.0	2.1	4.8	2,300	2.4	38	10.84	41.6	-
Arizona	31	278.3	2.0	9.0	1,500	1.5	48	9.44	34.7	2,067
Rhode Island	16	232.4	1.7	14.5	2,100	2.2	131	9.54	43.6	3,333
Virginia	26	211.2	1.5	8.1	1,300	1.3	50	8.40	55.6	1,231
South Carolina	13	163.2	1.2	12.6	1,200	1.2	92	9.04	49.3	3,083
Missouri	28	148.4	1.1	5.3	1,200	1.2	43	7.29	42.9	7,250
Mississippi	12	95.8	0.7	8.0	1,000	1.0	83	7.69	44.5	-
Colorado	36	90.5	0.7	2.5	800	0.8	22	8.78	32.7	2,250
Utah	19	82.1	0.6	4.3	800	0.8	42	8.36	35.9	2,500
Alabama	27	61.4	0.4	2.3	1,000	1.0	37	5.73	37.3	-
Maryland	23	60.2	0.4	2.6	500	0.5	22	9.60	54.2	1,600
Washington	36	59.3	0.4	1.6	600	0.6	17	9.00	36.3	3,167
Kentucky	18	56.2	0.4	3.1	900	0.9	50	6.75	53.2	1,778
Oklahoma	16	46.8	0.3	2.9	400	0.4	25	8.40	26.5	1,500
New Hampshire	11	40.6	0.3	3.7	400	0.4	36	8.80	46.6	-
Louisiana	11	13.7	0.1	1.2	100	0.1	9	15.00	31.4	2,000
Nevada	8	13.2	0.1	1.6	100	0.1	13	7.00	36.4	2,000
Idaho	11	6.6	0.0	0.6	100	0.1	9	11.00	33.3	1,000
Oregon	22	(D)	-	-	375 *	0.4	17	-	-	-
Iowa	16	(D)	-	-	175 *	0.2	11	-	-	-
Kansas	16	(D)	-	-	750 *	0.8	47	-	-	3,600
New Mexico	9	(D)	-	-	1,750 *	1.8	194	-	-	1,486
Nebraska	5	(D)	-	-	375 *	0.4	75	-	-	-
South Dakota	5	(D)	-	-	750 *	0.8	150	-	-	-
West Virginia	5	(D)	-	-	175 *	0.2	35	-	-	-

Source: 1992 *Economic Census*. The states are in descending order of shipments or establishments (if shipment data are missing for the majority). The symbol (D) appears when data are withheld to prevent disclosure of competitive information. States marked with (D) are sorted by number of establishments. A dash (-) indicates that the data element cannot be calculated; * indicates the midpoint of a range.

3843 - DENTAL EQUIPMENT & SUPPLIES

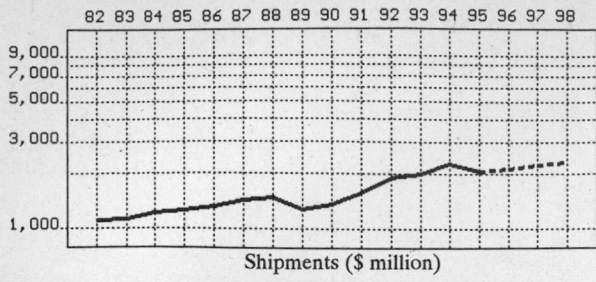

Shipments ($ million)

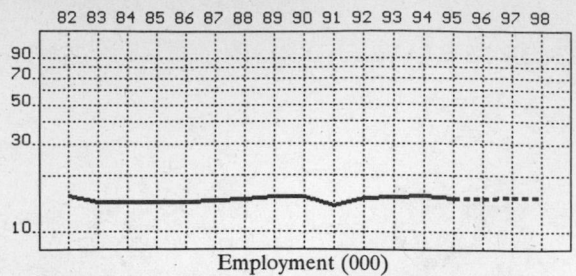

Employment (000)

GENERAL STATISTICS

| Year | Com-panies | Establishments | | Employment | | | Compensation | | Production ($ million) | | | |
		Total	with 20 or more employees	Total (000)	Production Workers (000)	Hours (Mil)	Payroll ($ mil)	Wages ($/hr)	Cost of Materials	Value Added by Manufacture	Value of Shipments	Capital Invest.
1982	438	485	131	15.5	9.8	18.8	281.3	7.02	473.3	642.8	1,111.7	21.6
1983		479	130	14.4	9.0	17.5	271.7	7.11	481.2	612.9	1,117.2	15.9
1984		473	129	14.4	9.3	18.2	290.4	7.96	503.0	720.3	1,213.2	29.6
1985		468	128	14.4	9.2	17.4	320.7	9.11	508.7	740.6	1,256.5	24.4
1986		453	121	14.5	8.7	16.7	336.3	9.87	531.1	798.1	1,317.6	33.5
1987	472	504	129	14.6	8.7	17.2	342.4	9.35	525.3	880.7	1,420.7	31.5
1988		503	126	14.9	8.9	17.3	360.2	9.45	528.3	965.3	1,473.4	30.0
1989		500	126	15.3	8.4	15.3	315.5	10.31	428.2	851.9	1,277.1	27.3
1990		505	130	15.4	8.4	15.1	345.0	11.34	503.4	890.1	1,364.7	24.4
1991		537	133	13.8	8.6	16.0	383.1	11.23	585.5	1,000.5	1,576.2	32.2
1992	589	611	129	15.1	8.9	16.9	458.1	11.50	731.7	1,202.9	1,910.0	48.6
1993		645	143	15.4	9.2	17.0	470.7	11.89	752.4	1,264.6	2,011.9	58.5
1994		598P	134P	15.5	9.8	18.9	477.5	11.43	845.8	1,440.8	2,273.6	52.0
1995		611P	134P	15.2P	8.9P	16.5P	474.1P	12.69P	769.6P	1,311.0P	2,068.8P	51.2P
1996		624P	135P	15.2P	8.8P	16.4P	490.7P	13.10P	800.6P	1,363.8P	2,152.0P	53.8P
1997		637P	135P	15.3P	8.8P	16.3P	507.3P	13.51P	831.5P	1,416.5P	2,235.2P	56.4P
1998		650P	136P	15.3P	8.8P	16.2P	523.9P	13.92P	862.5P	1,469.2P	2,318.4P	59.0P

Sources: 1982, 1987, 1992 *Economic Census*; *Annual Survey of Manufactures*, 83-86, 88-91, 93-94. Establishment counts for non-Census years are from *County Business Patterns*; establishment values for 83-84 are extrapolations. 'P's show projections by the editors. Industries reclassified in 87 will not have data for prior years.

INDICES OF CHANGE

| Year | Com-panies | Establishments | | Employment | | | Compensation | | Production ($ million) | | | |
		Total	with 20 or more employees	Total (000)	Production Workers (000)	Hours (Mil)	Payroll ($ mil)	Wages ($/hr)	Cost of Materials	Value Added by Manufacture	Value of Shipments	Capital Invest.
1982	74	79	102	103	110	111	61	61	65	53	58	44
1983		78	101	95	101	104	59	62	66	51	58	33
1984		77	100	95	104	108	63	69	69	60	64	61
1985		77	99	95	103	103	70	79	70	62	66	50
1986		74	94	96	98	99	73	86	73	66	69	69
1987	80	82	100	97	98	102	75	81	72	73	74	65
1988		82	98	99	100	102	79	82	72	80	77	62
1989		82	98	101	94	91	69	90	59	71	67	56
1990		83	101	102	94	89	75	99	69	74	71	50
1991		88	103	91	97	95	84	98	80	83	83	66
1992	100	100	100	100	100	100	100	100	100	100	100	100
1993		106	111	102	103	101	103	103	103	105	105	120
1994		98P	104P	103	110	112	104	99	116	120	119	107
1995		100P	104P	101P	100P	97P	103P	110P	105P	109P	108P	105P
1996		102P	104P	101P	99P	97P	107P	114P	109P	113P	113P	111P
1997		104P	105P	101P	99P	96P	111P	117P	114P	118P	117P	116P
1998		106P	105P	102P	99P	96P	114P	121P	118P	122P	121P	121P

Sources: Same as General Statistics. Values reflect change from the base year, 1992. Values above 100 mean greater than 92, values below 100 mean less than 92, and a value of 100 in the 82-91 or 93-98 period means same as 92. 'P's mark projections by the editors.

SELECTED RATIOS

For 1994	Avg. of All Manufact.	Analyzed Industry	Index	For 1994	Avg. of All Manufact.	Analyzed Industry	Index
Employees per Establishment	49	26	53	Value Added per Production Worker	134,084	147,020	110
Payroll per Establishment	1,500,273	798,293	53	Cost per Establishment	5,045,178	1,414,023	28
Payroll per Employee	30,620	30,806	101	Cost per Employee	102,970	54,568	53
Production Workers per Establishment	34	16	48	Cost per Production Worker	146,988	86,306	59
Wages per Establishment	853,319	361,158	42	Shipments per Establishment	9,576,895	3,801,044	40
Wages per Production Worker	24,861	22,044	89	Shipments per Employee	195,460	146,684	75
Hours per Production Worker	2,056	1,929	94	Shipments per Production Worker	279,017	232,000	83
Wages per Hour	12.09	11.43	95	Investment per Establishment	321,011	86,934	27
Value Added per Establishment	4,602,255	2,408,754	52	Investment per Employee	6,552	3,355	51
Value Added per Employee	93,930	92,955	99	Investment per Production Worker	9,352	5,306	57

Sources: Same as General Statistics. The 'Average of All Manufacturing' column represents the average of all manufacturing industries reported for the most recent complete year available. The Index shows the relationship between the Average and the Analyzed Industry. For example, 100 means that they are equal; 500 that the Analyzed Industry is five times the average; 50 means that the Analyzed Industry is half the national average. The abbreviation 'na' is used to show that data are 'not available'.

LEADING COMPANIES   Number shown: **57**   Total sales ($ mil): **1,679**   Total employment (000): **14.4**

Company Name	Address				CEO Name	Phone	Co. Type	Sales ($ mil)	Empl. (000)
Sybron International Corp	411 E Wisconsin	Milwaukee	WI	53202	Kenneth F Yontz	414-274-6600	P	383	3.4
Dentsply International Inc	717 Forest Av	Lake Forest	IL	60045	John J McDonough	708-367-9729	R	260*	3.4
Kerr Manufacturing Co	2210 E Alosta Av	Glendora	CA	91740	Steve Semmelmayer	818-852-0921	D	125	0.5
A-Dec Inc	PO Box 111	Newberg	OR	97132	GK Austin Jr	503-538-9471	R	104	0.7
Veratex Group	PO Box 4031	Troy	MI	48007	James H Devlin	313-588-2970	S	97*	0.5
Ormco Corp	1332 S Lone Hill	Glendora	CA	91740	Dan Even	909-596-0100	S	70	1.0
Andersen Group Inc	2 Douglas St	Bloomfield	CT	06002	Francis E Baker	203-242-0761	P	67	0.4
Den-Tal-Ez Inc	PO Box 896	Valley Forge	PA	19482	Richard V Trefz	215-666-9050	R	50*	0.5
JF Jelenko and Co	99 Business Park Dr	Armonk	NY	10504	Norman F Strate	914-273-8600	R	50	0.2
JM Ney Co	2 Douglas St	Bloomfield	CT	06002	Ronald N Cerny	203-242-2281	S	47*	0.4
Nobelpharma USA Inc	777 Oakmont Ln	Westmont	IL	60559	Thomas Nortoft	708-654-9100	S	45	0.2
Takara Belmont USA Inc	1 Belmont Dr	Somerset	NJ	08873	T Yanagihara	908-469-5000	S	42	<0.1
Coltene/Whaledent Inc	750 Corporate Dr	Mahwah	NJ	07430	Jerry Sullivan	201-512-8000	S	30	0.2
Hu-Friedy Manufacturing	3232 N Rockwell St	Chicago	IL	60618	Richard E Saslow	312-975-6100	R	30*	0.3
Hygenic Corp	1245 Home Av	Akron	OH	44310	Stewart Lorenzen	216-633-8460	R	30	0.3
Air Techniques Inc	70 Cantiague Rock	Hicksville	NY	11801	F Bader	516-433-7676	R	26	0.3
Implant Innovations Inc	3071 Continental Dr	W Palm Beach	FL	33407	Richard Lazzara	407-683-9028	R	20	0.2
A Company Inc	11436 Sorrento Val	San Diego	CA	92121	Don O'Connel	619-453-9010	S	17	0.3
Fuji Optical Systems Inc	1 Frassetto Way	Lincoln	NJ	07035	S Takada	201-305-3674	R	15*	<0.1
American Dental Laser Inc	2600 W Big Beaver	Troy	MI	48084	Anthony D Fiorillo	313-649-0000	P	11	<0.1
Buffalo Dental Manufacturing	99 Lafayette Dr	Syosset	NY	11791	Donald M Nevin	516-496-7200	R	10	<0.1
MPL Technologies	9400 King St	Franklin Park	IL	60131	Gustav A Scheuble	708-678-7555	D	10	<0.1
Vacudent	2601 S 2700 W	Salt Lake City	UT	84119	John Ward	801-972-3165	R	10	<0.1
Whip-Mix Corp	PO Box 17183	Louisville	KY	40217	Allen F Steinbock	502-637-1451	R	10	0.2
CMP Industries Inc	PO Box 350	Albany	NY	12201	William Regan	518-434-3147	P	8	<0.1
Jeneric-Pentron Inc	53 N Plains Indus	Wallingford	CT	06492	Gordon Cohen	203-265-7397	S	8	0.1
Keller Laboratories Inc	PO Box 22037	St Louis	MO	63126	William Keller	314-842-4320	R	8*	<0.1
Rinn Corp	1212 Abbott Dr	Elgin	IL	60123	A Leigh Harrington	708-742-1115	D	8	<0.1
Proma Inc	751 E Kingshill Pl	Carson	CA	90746	Cecil J Casillas	310-327-0035	R	7*	<0.1
Lancer Orthodontics Inc	253 Pawnee St	San Marcos	CA	92069	Douglas D Miller	619-744-5585	P	7	<0.1
ETM Corp	144 W Chestnut	Monrovia	CA	91016	Dan Even	818-359-8102	S	6	0.1
Aseptico Inc	PO Box 3209	Kirkland	WA	98083	Doug H Kazen	206-487-3157	R	5	<0.1
Crescent Dental Mfg Co	7750 W 47th St	Lyons	IL	60534	EL Chott	708-447-8050	R	5	<0.1
Harry J Bosworth Co	7227 N Hamlin Av	Skokie	IL	60076	M M Goldstein	708-679-3400	R	5	<0.1
Parkell Products Inc	155 Schmitt Blv	E Farmingdale	NY	11735	Dan Braet	516-249-1134	R	5*	<0.1
E and D Dental Products	71 Veronica Av	Somerset	NJ	08873	Ed Spehar	908-249-6000	R	5	<0.1
Denar Corp	901 E Cerritos Av	Anaheim	CA	92805	R Stern	714-776-9000	R	4*	<0.1
Midwest Orthodontic Mfg Co	4570 Progress Dr	Columbus	IN	47201	Jeffery L Fasnacht	812-376-0544	R	4*	<0.1
Wykle Research Inc	2222 Hot Spgs Rd	Carson City	NV	89706	Paul R Dempsey	702-887-7500	R	4	<0.1
American Tooth Industries	1200 Stellar Dr	Oxnard	CA	93033	B Pozzi	805-487-9868	R	3	<0.1
Biotec Inc	652 E Main St	Zeeland	MI	49464	Chuck De Pree	616-772-2133	R	3	<0.1
Ortho Arch Co	715 E Golf Rd	Schaumburg	IL	60173	Mike Zerafa	708-885-7805	R	3	<0.1
Orthopli Corp	10061 Sandm	Philadelphia	PA	19116	William P Tippy	215-671-1000	R	3*	<0.1
Brandt Industries Inc	4461 Bronx Blv	Bronx	NY	10470	Neil Brandt	718-994-0800	R	2	<0.1
Dolphin Imaging Systems Inc	28310 Av Crocker	Valencia	CA	91355	Gary Engel	805-295-8500	R	2	<0.1
ERC Co	2970 E Maria St	R Dominguez	CA	90221	MF Coy	310-603-2970	D	2	<0.1
Faro USA Corp	1320 Marsten Rd	Burlingame	CA	94010	Osvaldo Favonio	415-348-3763	S	2*	<0.1
Krohn Industries Inc	303 Veterans Blv	Carlstadt	NJ	07072	John M Krohn	201-933-9696	R	2*	<0.1
Myo-Tronics Inc	720 Olive Way	Seattle	WA	98101	WE Trimingham	206-622-2121	R	2	<0.1
Ohio Health Care Products Inc	7020 Kennedy Blv	North Bergen	NJ	07047	Tom Lake	201-869-2355	R	2*	<0.1
Minimatic Implant Technology	1225 Broken Sound	Boca Raton	FL	33487	Leon Shaw	407-997-7777	P	2	<0.1
Challenge Products Inc	PO Box 468	Osage Beach	MO	65065	Charles Bull	314-348-2227	S	1*	<0.1
Dental Concepts Inc	100 Clearbrook Rd	Elmsford	NY	10523	Peter Strauss	914-592-1860	R	1*	<0.1
Optiva Corp	13222 SE 30th St	Bellevue	WA	98005	David Giuliani	206-957-0970	R	1	<0.1
Wells Dental Inc	PO Box 106	Comptche	CA	95427	RB Wells	707-937-0521	R	1*	<0.1
Biological Rescue Products Inc	1100 E Hector St	Conshohocken	PA	19428	Gioia Davidovits	610-834-0905	R	0*	<0.1
Sid Jones Dental Studios	PO Box 247	Cass Lake	MN	56633	Sid Jones	218-335-6960	R	0	<0.1

Source: Ward's Business Directory of U.S. Private and Public Companies, Volumes 1 and 2, 1996. The company type code used is as follows: P - Public, R - Private, S - Subsidiary, D - Division, J - Joint Venture, A - Affiliate, G - Group. Sales are in millions of dollars, employees are in thousands. An asterisk (*) indicates an estimated sales volume. The symbol < stands for 'less than'. Company names and addresses are truncated, in some cases, to fit into the available space.

MATERIALS CONSUMED

Material	Quantity	Delivered Cost ($ million)
Materials, ingredients, containers, and supplies	(X)	572.9
Metal bolts, nuts, screws, washers, rivets, and other screw machine products	(X)	7.3
Other fabricated metal products (except castings and forgings)	(X)	54.2
Forgings	(X)	1.2
Castings (rough and semifinished)	(X)	5.6
Steel shapes and forms	(X)	12.7
Nonferrous shapes and forms	(X)	9.3
Precious metals (gold, platinum, etc.), all forms	(X)	136.5
Chemicals, all types, except resins	(X)	31.6
Resistors, capacitors, transformers, electron tubes, semiconductors, and other electronic components	(X)	8.8
Plastics resins consumed in the form of granules, pellets, powders, liquids, etc.	(X)	11.0
Plastics products consumed in the form of sheets, rods, tubes, film, and other shapes	(X)	10.9
Fabricated plastics products (except gaskets, hoses, and belting)	(X)	21.2
Glass and glass products, except photographic and projection lenses and prisms	(X)	15.1
Paper and paperboard products except paperboard boxes, containers, and corrugated paperboard	(X)	5.8
Paperboard containers, boxes, and corrugated paperboard	(X)	14.2
All other materials and components, parts, containers, and supplies	(X)	112.6
Materials, ingredients, containers, and supplies, nsk	(X)	114.9

Source: 1992 Economic Census. Explanation of symbols used: (D): Withheld to avoid disclosure of competitive data; na: Not available; (S): Withheld because statistical norms were not met; (X): Not applicable; (Z): Less than half the unit shown; nec: Not elsewhere classified; nsk: Not specified by kind; - : zero; * : 10-19 percent estimated; ** : 20-29 percent estimated.

PRODUCT SHARE DETAILS

Product or Product Class	% Share	Product or Product Class	% Share
Dental equipment and supplies	100.00	Professional supplies, dental cements and other nonmetallic filling materials	8.08
Dental equipment and supplies	59.28	Other dental professional supplies	31.33
Professional dental chairs	6.05	Dental professional equipment and supplies, nsk	2.02
Professional dental instrument delivery systems units	6.79	Dental laboratory equipment and supplies	25.70
Professional dental hand pieces	5.04	Dental laboratory equipment (furnaces, casting machines, lathes, benches, polishing units, flasks, blow pipes, etc.)	23.30
Other professional dental equipment, excluding x-ray	15.87	Dental laboratory supplies, precious metals	39.94
Professional supplies, dental hand instruments (forceps and pliers, broaches, cutting instruments, etc.)	9.48	Dental laboratory supplies, nonprecious metals	13.63
Professional supplies, burs, diamond points, abrasive points, wheels, disks, and similar tools for use with dental hand pieces	4.40	Dental laboratory supplies, teeth (excluding dentures)	9.15
		Other dental laboratory supplies (waxes, gypsums, etc.)	12.47
Professional supplies, dental alloys for amalgams	5.57	Dental laboratory equipment and supplies, nsk	1.52
Professional supplies, dental impression materials (alginates, silicones, etc.)	5.37	Dental equipment and supplies, nsk	15.02

Source: 1992 Economic Census. The values shown are percent of total shipments in an industry. Values of indented subcategories are summed in the main heading. The symbol (D) appears when data are withheld to prevent disclosure of competitive information. The abbreviation nsk stands for 'not specified by kind' and nec for 'not elsewhere classified'.

INPUTS AND OUTPUTS FOR DENTAL EQUIPMENT & SUPPLIES

Economic Sector or Industry Providing Inputs	%	Sector	Economic Sector or Industry Buying Outputs	%	Sector
Primary nonferrous metals, nec	25.7	Manufg.	Gross private fixed investment	34.3	Cap Inv
Nonferrous rolling & drawing, nec	10.6	Manufg.	Medical & health services, nec	32.9	Services
Wholesale trade	7.4	Trade	Doctors & dentists	18.8	Services
Imports	7.2	Foreign	Exports	10.0	Foreign
Blast furnaces & steel mills	3.5	Manufg.	S/L Govt. purch., health & hospitals	1.2	S/L Govt
Electronic components nec	3.4	Manufg.	Change in business inventories	0.9	In House
Advertising	2.9	Services	Federal Government purchases, national defense	0.8	Fed Govt
Noncomparable imports	2.7	Foreign	Federal Government purchases, nondefense	0.8	Fed Govt
Communications, except radio & TV	1.6	Util.	Dental equipment & supplies	0.4	Manufg.
Electric services (utilities)	1.4	Util.	S/L Govt. purch., correction	0.1	S/L Govt
Eating & drinking places	1.4	Trade			
Business services nec	1.4	Services			
Paperboard containers & boxes	1.3	Manufg.			
Petroleum refining	1.3	Manufg.			
Job training & related services	1.3	Services			
Miscellaneous plastics products	1.2	Manufg.			
Nonferrous wire drawing & insulating	1.0	Manufg.			
Motor freight transportation & warehousing	1.0	Util.			
Cyclic crudes and organics	0.9	Manufg.			
Metal stampings, nec	0.9	Manufg.			
Real estate	0.9	Fin/R.E.			
Coated fabrics, not rubberized	0.8	Manufg.			
Nonmetallic mineral products, nec	0.8	Manufg.			
Banking	0.8	Fin/R.E.			
Equipment rental & leasing services	0.8	Services			
Nonferrous metal ores, except copper	0.7	Mining			

Continued on next page.

INPUTS AND OUTPUTS FOR DENTAL EQUIPMENT & SUPPLIES - Continued

Economic Sector or Industry Providing Inputs	%	Sector	Economic Sector or Industry Buying Outputs	%	Sector
Gypsum products	0.7	Manufg.			
Metal coating & allied services	0.7	Manufg.			
Paints & allied products	0.7	Manufg.			
Maintenance of nonfarm buildings nec	0.6	Constr.			
Aluminum rolling & drawing	0.6	Manufg.			
Dental equipment & supplies	0.6	Manufg.			
Glass & glass products, except containers	0.6	Manufg.			
Plastics materials & resins	0.6	Manufg.			
Screw machine and related products	0.6	Manufg.			
Royalties	0.6	Fin/R.E.			
Fabricated metal products, nec	0.5	Manufg.			
Gaskets, packing & sealing devices	0.5	Manufg.			
Machinery, except electrical, nec	0.5	Manufg.			
Primary batteries, dry & wet	0.5	Manufg.			
Legal services	0.5	Services			
Management & consulting services & labs	0.5	Services			
Electrical equipment & supplies, nec	0.4	Manufg.			
Fabricated rubber products, nec	0.4	Manufg.			
Paper mills, except building paper	0.4	Manufg.			
Theatrical producers, bands, entertainers	0.4	Services			
Manifold business forms	0.3	Manufg.			
Special dies & tools & machine tool accessories	0.3	Manufg.			
Veneer & plywood	0.3	Manufg.			
Gas production & distribution (utilities)	0.3	Util.			
Accounting, auditing & bookkeeping	0.3	Services			
Computer & data processing services	0.3	Services			
U.S. Postal Service	0.3	Gov't			
Abrasive products	0.2	Manufg.			
Copper rolling & drawing	0.2	Manufg.			
Railroads & related services	0.2	Util.			
Automotive rental & leasing, without drivers	0.2	Services			
Automotive repair shops & services	0.2	Services			
Business/professional associations	0.2	Services			
Hotels & lodging places	0.2	Services			
Lubricating oils & greases	0.1	Manufg.			
Photographic equipment & supplies	0.1	Manufg.			
Rubber & plastics hose & belting	0.1	Manufg.			
Signs & advertising displays	0.1	Manufg.			
Air transportation	0.1	Util.			
Sanitary services, steam supply, irrigation	0.1	Util.			
Insurance carriers	0.1	Fin/R.E.			
Personnel supply services	0.1	Services			

Source: Benchmark Input-Output Accounts for the U.S. Economy, 1982, U.S. Department of Commerce, Washington, D.C., July 1991. Data, as reported in the source, are organized by the 1977 SIC structure in use in 1982 but have been matched, as closely as is possible, to the 1987 SIC structure used in this book.

OCCUPATIONS EMPLOYED BY SIC 384 - MEDICAL INSTRUMENTS AND SUPPLIES

Occupation	% of Total 1994	Change to 2005	Occupation	% of Total 1994	Change to 2005
Assemblers, fabricators, & hand workers nec	16.5	17.4	Mechanical engineers	1.5	29.3
Sales & related workers nec	5.3	17.4	Electrical & electronic equipment assemblers	1.5	17.4
Inspectors, testers, & graders, precision	4.3	-17.8	Precision workers nec	1.5	5.7
Blue collar worker supervisors	3.3	-0.0	Electrical & electronics engineers	1.4	49.1
Precision assemblers nec	2.4	88.0	Industrial production managers	1.4	17.4
Secretaries, ex legal & medical	2.3	6.9	Packaging & filling machine operators	1.3	17.4
General managers & top executives	1.8	11.4	Electrical & electronic technicians,technologists	1.3	17.4
Traffic, shipping, & receiving clerks	1.8	13.0	Engineering, mathematical, & science managers	1.3	33.4
Machine operators nec	1.7	-17.2	Bookkeeping, accounting, & auditing clerks	1.2	-11.9
Electrical & electronic assemblers	1.7	5.7	Production, planning, & expediting clerks	1.2	17.4
Engineering technicians & technologists nec	1.6	17.4	General office clerks	1.2	0.2
Marketing, advertising, & PR managers	1.5	17.4	Adjustment clerks	1.1	40.9
Hand packers & packagers	1.5	0.6	Machinists	1.1	-11.9
Sewing machine operators, non-garment	1.5	76.2	Managers & administrators nec	1.1	17.4

Source: Industry-Occupation Matrix, Bureau of Labor Statistics. These data relate to one or more 3-digit SIC industry groups rather than to a single 4-digit SIC. The change reported for each occupation to the year 2005 is a percent of growth or decline as estimated by the Bureau of Labor Statistics. The abbreviation nec stands for 'not elsewhere classified'.

LOCATION BY STATE AND REGIONAL CONCENTRATION

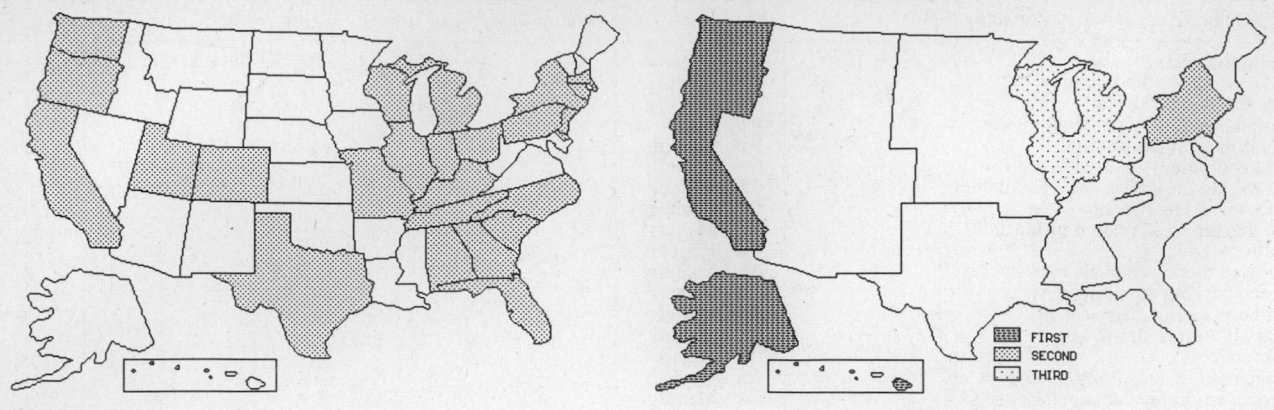

FIRST
SECOND
THIRD

INDUSTRY DATA BY STATE

State	Establish-ments	Shipments			Employment				Cost as % of Shipments	Investment per Employee ($)
		Total ($ mil)	% of U.S.	Per Establ.	Total Number	% of U.S.	Per Establ.	Wages ($/hour)		
California	124	536.5	28.1	4.3	3,800	25.2	31	13.95	29.0	3,684
Illinois	39	221.1	11.6	5.7	1,400	9.3	36	12.44	37.6	3,786
Oregon	19	91.6	4.8	4.8	800	5.3	42	12.30	29.4	4,875
Indiana	13	43.2	2.3	3.3	500	3.3	38	9.67	33.6	1,200
Florida	28	34.7	1.8	1.2	300	2.0	11	9.00	50.7	2,667
New Jersey	23	23.3	1.2	1.0	300	2.0	13	8.00	32.6	1,333
Massachusetts	14	20.2	1.1	1.4	100	0.7	7	10.00	32.7	-
Ohio	19	15.7	0.8	0.8	200	1.3	11	14.00	68.2	1,000
Tennessee	8	6.4	0.3	0.8	100	0.7	13	9.00	31.3	1,000
New York	48	(D)	-	-	1,750 *	11.6	36	-	-	-
Pennsylvania	35	(D)	-	-	1,750 *	11.6	50	-	-	-
Texas	24	(D)	-	-	375 *	2.5	16	-	-	-
Colorado	18	(D)	-	-	375 *	2.5	21	-	-	-
Michigan	14	(D)	-	-	750 *	5.0	54	-	-	-
North Carolina	14	(D)	-	-	375 *	2.5	27	-	-	-
Washington	14	(D)	-	-	175 *	1.2	13	-	-	-
Connecticut	13	(D)	-	-	750 *	5.0	58	-	-	-
Georgia	12	(D)	-	-	375 *	2.5	31	-	-	-
Missouri	10	(D)	-	-	175 *	1.2	18	-	-	-
South Carolina	8	(D)	-	-	175 *	1.2	22	-	-	-
Utah	7	(D)	-	-	375 *	2.5	54	-	-	-
Wisconsin	7	(D)	-	-	175 *	1.2	25	-	-	-
Alabama	4	(D)	-	-	175 *	1.2	44	-	-	-
Kentucky	4	(D)	-	-	175 *	1.2	44	-	-	-
Delaware	1	(D)	-	-	375 *	2.5	375	-	-	-

Source: 1992 *Economic Census*. The states are in descending order of shipments or establishments (if shipment data are missing for the majority). The symbol (D) appears when data are withheld to prevent disclosure of competitive information. States marked with (D) are sorted by number of establishments. A dash (-) indicates that the data element cannot be calculated; * indicates the midpoint of a range.

3844 - X-RAY APPARATUS & TUBES

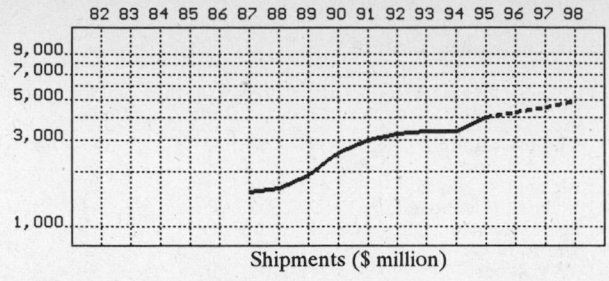

Shipments ($ million)

82 83 84 85 86 87 88 89 90 91 92 93 94 95 96 97 98

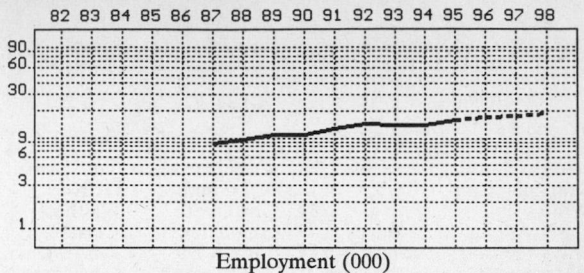

Employment (000)

82 83 84 85 86 87 88 89 90 91 92 93 94 95 96 97 98

GENERAL STATISTICS

| Year | Companies | Establishments | | Employment | | | Compensation | | Production ($ million) | | | |
		Total	with 20 or more employees	Total (000)	Production Workers (000)	Hours (Mil)	Payroll ($ mil)	Wages ($/hr)	Cost of Materials	Value Added by Manufacture	Value of Shipments	Capital Invest.
1982												
1983												
1984												
1985												
1986												
1987	69	75	43	8.7	5.5	11.2	257.1	12.21	727.5	834.5	1,554.3	32.0
1988		87	51	9.4	5.4	11.4	284.6	12.47	658.9	995.9	1,614.8	41.8
1989		80	49	10.7	5.5	11.8	308.6	12.93	853.1	1,113.6	1,925.8	56.9
1990		82	46	10.7	6.8	14.4	454.1	13.79	1,099.2	1,495.8	2,576.5	91.5
1991		93	48	13.0	6.6	13.1	483.8	14.37	1,321.2	1,683.9	3,011.4	65.3
1992	110	128	65	14.3	7.1	14.3	562.7	14.79	1,302.0	1,871.4	3,235.0	63.6
1993		132	60	14.2	6.9	13.7	572.9	15.44	1,586.2	1,770.8	3,372.3	85.6
1994		135P	63P	14.0	6.6	13.6	598.7	15.28	1,638.0	1,704.9	3,373.1	84.8
1995		144P	66P	15.8P	7.4P	14.8P	688.1P	16.19P	1,927.6P	2,006.3P	3,969.4P	96.4P
1996		154P	68P	16.7P	7.6P	15.2P	743.1P	16.69P	2,077.2P	2,162.0P	4,277.5P	103.3P
1997		163P	71P	17.6P	7.8P	15.6P	798.2P	17.20P	2,226.8P	2,317.8P	4,585.6P	110.3P
1998		173P	74P	18.5P	8.1P	16.0P	853.2P	17.70P	2,376.4P	2,473.5P	4,893.8P	117.2P

Sources: 1982, 1987, 1992 *Economic Census*; *Annual Survey of Manufactures*, 83-86, 88-91, 93-94. Establishment counts for non-Census years are from *County Business Patterns*; establishment values for 83-84 are extrapolations. 'P's show projections by the editors. Industries reclassified in 87 will not have data for prior years.

INDICES OF CHANGE

| Year | Companies | Establishments | | Employment | | | Compensation | | Production ($ million) | | | |
		Total	with 20 or more employees	Total (000)	Production Workers (000)	Hours (Mil)	Payroll ($ mil)	Wages ($/hr)	Cost of Materials	Value Added by Manufacture	Value of Shipments	Capital Invest.
1982												
1983												
1984												
1985												
1986												
1987	63	59	66	61	77	78	46	83	56	45	48	50
1988		68	78	66	76	80	51	84	51	53	50	66
1989		63	75	75	77	83	55	87	66	60	60	89
1990		64	71	75	96	101	81	93	84	80	80	144
1991		73	74	91	93	92	86	97	101	90	93	103
1992	100	100	100	100	100	100	100	100	100	100	100	100
1993		103	92	99	97	96	102	104	122	95	104	135
1994		105P	97P	98	93	95	106	103	126	91	104	133
1995		113P	101P	111P	104P	103P	122P	109P	148P	107P	123P	152P
1996		120P	105P	117P	107P	106P	132P	113P	160P	116P	132P	162P
1997		128P	110P	123P	110P	109P	142P	116P	171P	124P	142P	173P
1998		135P	114P	129P	114P	112P	152P	120P	183P	132P	151P	184P

Sources: Same as General Statistics. Values reflect change from the base year, 1992. Values above 100 mean greater than 92, values below 100 mean less than 92, and a value of 100 in the 82-91 or 93-98 period means same as 92. 'P's mark projections by the editors.

SELECTED RATIOS

For 1994	Avg. of All Manufact.	Analyzed Industry	Index	For 1994	Avg. of All Manufact.	Analyzed Industry	Index
Employees per Establishment	49	104	212	Value Added per Production Worker	134,084	258,318	193
Payroll per Establishment	1,500,273	4,444,221	296	Cost per Establishment	5,045,178	12,159,067	241
Payroll per Employee	30,620	42,764	140	Cost per Employee	102,970	117,000	114
Production Workers per Establishment	34	49	143	Cost per Production Worker	146,988	248,182	169
Wages per Establishment	853,319	1,542,583	181	Shipments per Establishment	9,576,895	25,038,918	261
Wages per Production Worker	24,861	31,486	127	Shipments per Employee	195,460	240,936	123
Hours per Production Worker	2,056	2,061	100	Shipments per Production Worker	279,017	511,076	183
Wages per Hour	12.09	15.28	126	Investment per Establishment	321,011	629,480	196
Value Added per Establishment	4,602,255	12,655,673	275	Investment per Employee	6,552	6,057	92
Value Added per Employee	93,930	121,779	130	Investment per Production Worker	9,352	12,848	137

Sources: Same as General Statistics. The 'Average of All Manufacturing' column represents the average of all manufacturing industries reported for the most recent complete year available. The Index shows the relationship between the Average and the Analyzed Industry. For example, 100 means that they are equal; 500 that the Analyzed Industry is five times the average; 50 means that the Analyzed Industry is half the national average. The abbreviation 'na' is used to show that data are 'not available'.

LEADING COMPANIES   Number shown: 30   Total sales ($ mil): 617   Total employment (000): 3.9

Company Name	Address				CEO Name	Phone	Co. Type	Sales ($ mil)	Empl. (000)
Siemens Medical Systems	2501 N Barrington	Hoffman Est	IL	60195	Thomas Cafarella	708-304-7700	D	170	0.8
OEC Medical Systems Inc	384 Wright Brothers	Salt Lake City	UT	84116	R N-Etienne	801-328-9300	P	98	0.5
Fischer Imaging Corp	12300 N Grant St	Denver	CO	80241	Morgan W Nields	303-452-6800	P	73	0.6
Hologic Inc	590 Lincoln St	Waltham	MA	02154	S David Ellenbogen	617-890-2300	P	38	0.2
Picker International Inc Dunlee	555 Commerce St	Aurora	IL	60504	John Newman	708-547-9535	D	35*	0.2
E-Systems Medical Electronics	PO Box 690390	San Antonio	TX	78269	Jim Karlak	210-641-8340	S	31	0.2
Liebel-Flarsheim Co	2111 E Galbraith	Cincinnati	OH	45215	JP Vollmer	513-761-2700	R	30*	0.3
Biodex Medical Systems Inc	20 Ramsey Rd	Shirley	NY	11967	James Reiss	516-924-9000	R	16	0.1
LumenX Co	3400 Gilchrist Rd	Mogadore	OH	44260	Charles Gilmore	216-784-4456	D	14	0.1
Siemens Industrial Automation	6300 Enterprise Ln	Madison	WI	53719	Martin Haase	608-276-3000	D	14	<0.1
Sierra Scientific	605 W California	Sunnyvale	CA	94086	D Gilblom	408-773-5600	D	12	<0.1
ASOMA Instruments Inc	11675 Jollyville Rd	Austin	TX	78759	EB King	512-258-6608	R	11*	<0.1
Bio-Imaging Research Inc	425 Barclay Blv	Lincolnshire	IL	60069	John F Moore	708-634-6425	R	7	<0.1
Edax International	91 McKee Dr	Mahwah	NJ	07430	Alan Devenish	201-529-4880	S	7	0.1
Medx Inc	925 Aec Dr	Wood Dale	IL	60191	Bob Moss	708-595-4400	R	7	<0.1
Modern Metal Products	PO Box 20388	Greensboro	NC	27420	H Shoenfeld	910-275-0421	D	7	<0.1
Norland Corp	6340 W Hackbarth	Fort Atkinson	WI	53538	Reynald Bonmati	414-563-8456	S	7	<0.1
Fischer Industries Inc	2630 Kaneville Ct	Geneva	IL	60134	Thomas Muchisky	708-232-2803	R	6	<0.1
MDH Industries Inc	426 W Duarte Rd	Monrovia	CA	91016	Paul Sunde	818-357-7921	R	5*	<0.1
Radcal Corp	426 W Duarte Rd	Monrovia	CA	91016	Paul Sunde	818-357-7921	S	5	<0.1
Trophy Radiology Inc	2252 NW Pkwy SE	Marietta	GA	30067	James G Dunn	404-859-9033	R	5	<0.1
Panoramic Corp	4321 Goshen Rd	Fort Wayne	IN	46818	Eric Stetzel	219-489-2291	R	4*	<0.1
Bar-Ray Products Inc	95 Monarch St	Littlestown	PA	17340	RC Lapof	717-359-9100	R	3	<0.1
Diano Corp	30 Commerce Way	Woburn	MA	01801	E Stelzner	617-935-4310	R	2	<0.1
Digivision Inc	5626 Oberlin Dr	San Diego	CA	92121	Sherman DeForest	619-458-1111	R	2	<0.1
Schonberg Research Corp	3300 Keller St	Santa Clara	CA	95054	Russell Schonberg	408-980-9729	R	2*	<0.1
Scientific Measurement Systems	2209 Donley Dr	Austin	TX	78758	Larry Secrest	512-837-4712	P	2	<0.1
Alvord Systems Inc	PO Box 489	Clairton	PA	15025	Thomas S Alvord	412-233-3910	R	1	<0.1
Carr Corp	1547 11th St	Santa Monica	CA	90401	John Carr	213-870-9237	R	1	<0.1
Connecticut Coining Inc	81 Beaver Brook Rd	Danbury	CT	06810	G Marciano	203-743-3861	R	1*	<0.1

Source: Ward's Business Directory of U.S. Private and Public Companies, Volumes 1 and 2, 1996. The company type code used is as follows: P - Public, R - Private, S - Subsidiary, D - Division, J - Joint Venture, A - Affiliate, G - Group. Sales are in millions of dollars, employees are in thousands. An asterisk (*) indicates an estimated sales volume. The symbol < stands for 'less than'. Company names and addresses are truncated, in some cases, to fit into the available space.

MATERIALS CONSUMED

Material	Quantity	Delivered Cost ($ million)
Materials, ingredients, containers, and supplies	(X)	1,079.9
Metal bolts, nuts, screws, washers, rivets, and other screw machine products	(X)	11.1
Metal stampings	(X)	(D)
All other fabricated metal products (except forgings)	(X)	99.1
Forgings	(X)	(D)
Castings (rough and semifinished)	(X)	1.0
Stainless steel shapes and forms	(X)	4.2
Other steel shapes and forms	(X)	5.2
Nonferrous shapes and forms	(X)	8.9
Purchased electronic computing and peripheral equipment	(X)	(D)
Transmittal, industrial, and special-purpose electron tubes, except x-ray	(X)	(D)
Semiconductors, including transistors, diodes, rectifiers, and integrated circuits for electronic circuitry	(X)	8.1
Capacitors for electronic circuitry	(X)	3.8
Resistors for electronic circuitry	(X)	3.7
Connectors for electronic circuitry	(X)	3.2
Other electronic components and accessories	(X)	169.3
Insulated wire and cable, except magnet wire	(X)	14.6
Plastics products consumed in the form of sheets, rods, tubes, film, and other shapes	(X)	5.7
Paperboard containers, boxes, and corrugated paperboard	(X)	1.6
All other materials and components, parts, containers, and supplies	(X)	350.6
Materials, ingredients, containers, and supplies, nsk	(X)	248.7

Source: 1992 *Economic Census*. Explanation of symbols used: (D): Withheld to avoid disclosure of competitive data; na: Not available; (S): Withheld because statistical norms were not met; (X): Not applicable; (Z): Less than half the unit shown; nec: Not elsewhere classified; nsk: Not specified by kind; - : zero; * : 10-19 percent estimated; ** : 20-29 percent estimated.

PRODUCT SHARE DETAILS

Product or Product Class	% Share	Product or Product Class	% Share
X-ray apparatus and tubes	100.00		

Source: 1992 *Economic Census*. The values shown are percent of total shipments in an industry. Values of indented subcategories are summed in the main heading. The symbol (D) appears when data is withheld to prevent disclosure of competitive information. The abbreviation nsk stands for 'not specified by kind' and nec for 'not elsewhere classified'.

INPUTS AND OUTPUTS FOR X-RAY APPARATUS & TUBES

Economic Sector or Industry Providing Inputs	%	Sector	Economic Sector or Industry Buying Outputs	%	Sector
Imports	22.9	Foreign	Gross private fixed investment	62.8	Cap Inv
Electronic components nec	22.4	Manufg.	Exports	21.1	Foreign
Advertising	7.2	Services	S/L Govt. purch., health & hospitals	4.2	S/L Govt
Wholesale trade	6.4	Trade	Hospitals	3.9	Services
Carbon & graphite products	6.0	Manufg.	Change in business inventories	2.5	In House
Metal coating & allied services	3.7	Manufg.	Doctors & dentists	2.0	Services
Semiconductors & related devices	2.9	Manufg.	Federal Government purchases, nondefense	1.4	Fed Govt
Electron tubes	1.9	Manufg.	Federal Government purchases, national defense	0.8	Fed Govt
Communications, except radio & TV	1.9	Util.	X-ray apparatus & tubes	0.5	Manufg.
Banking	1.8	Fin/R.E.	S/L Govt. purch., higher education	0.2	S/L Govt
Electric services (utilities)	1.3	Util.	Drugs	0.1	Manufg.
Electrical repair shops	1.3	Services	Photographic equipment & supplies	0.1	Manufg.
Air transportation	1.2	Util.	State & local electric utilities	0.1	Gov't
Miscellaneous plastics products	1.1	Manufg.			
Nonferrous rolling & drawing, nec	0.9	Manufg.			
Nonferrous wire drawing & insulating	0.9	Manufg.			
Paints & allied products	0.9	Manufg.			
Brass, bronze, & copper castings	0.8	Manufg.			
Plastics materials & resins	0.8	Manufg.			
X-ray apparatus & tubes	0.8	Manufg.			
Real estate	0.8	Fin/R.E.			
Metal stampings, nec	0.7	Manufg.			
Petroleum refining	0.7	Manufg.			
Screw machine and related products	0.7	Manufg.			
Equipment rental & leasing services	0.7	Services			
Photofinishing labs, commercial photography	0.7	Services			
Maintenance of nonfarm buildings nec	0.6	Constr.			
Ball & roller bearings	0.6	Manufg.			
Hotels & lodging places	0.6	Services			
Industrial controls	0.5	Manufg.			
Paperboard containers & boxes	0.4	Manufg.			
Primary batteries, dry & wet	0.4	Manufg.			
Eating & drinking places	0.4	Trade			
Noncomparable imports	0.4	Foreign			
Motor freight transportation & warehousing	0.3	Util.			
Royalties	0.3	Fin/R.E.			
Electrical equipment & supplies, nec	0.2	Manufg.			
Machinery, except electrical, nec	0.2	Manufg.			
Miscellaneous fabricated wire products	0.2	Manufg.			
Motors & generators	0.2	Manufg.			
Gas production & distribution (utilities)	0.2	Util.			
Credit agencies other than banks	0.2	Fin/R.E.			
Computer & data processing services	0.2	Services			
Legal services	0.2	Services			
Management & consulting services & labs	0.2	Services			
Fabricated rubber products, nec	0.1	Manufg.			
Glass & glass products, except containers	0.1	Manufg.			
Primary nonferrous metals, nec	0.1	Manufg.			
Sheet metal work	0.1	Manufg.			
Special dies & tools & machine tool accessories	0.1	Manufg.			
Wiring devices	0.1	Manufg.			
Insurance carriers	0.1	Fin/R.E.			
Accounting, auditing & bookkeeping	0.1	Services			
Automotive rental & leasing, without drivers	0.1	Services			

Source: Benchmark Input-Output Accounts for the U.S. Economy, 1982, U.S. Department of Commerce, Washington, D.C., July 1991. Data, as reported in the source, are organized by the 1977 SIC structure in use in 1982 but have been matched, as closely as is possible, to the 1987 SIC structure used in this book.

OCCUPATIONS EMPLOYED BY SIC 384 - MEDICAL INSTRUMENTS AND SUPPLIES

Occupation	% of Total 1994	Change to 2005	Occupation	% of Total 1994	Change to 2005
Assemblers, fabricators, & hand workers nec	16.5	17.4	Mechanical engineers	1.5	29.3
Sales & related workers nec	5.3	17.4	Electrical & electronic equipment assemblers	1.5	17.4
Inspectors, testers, & graders, precision	4.3	-17.8	Precision workers nec	1.5	5.7
Blue collar worker supervisors	3.3	-0.0	Electrical & electronics engineers	1.4	49.1
Precision assemblers nec	2.4	88.0	Industrial production managers	1.4	17.4
Secretaries, ex legal & medical	2.3	6.9	Packaging & filling machine operators	1.3	17.4
General managers & top executives	1.8	11.4	Electrical & electronic technicians,technologists	1.3	17.4
Traffic, shipping, & receiving clerks	1.8	13.0	Engineering, mathematical, & science managers	1.3	33.4
Machine operators nec	1.7	-17.2	Bookkeeping, accounting, & auditing clerks	1.2	-11.9
Electrical & electronic assemblers	1.7	5.7	Production, planning, & expediting clerks	1.2	17.4
Engineering technicians & technologists nec	1.6	17.4	General office clerks	1.2	0.2
Marketing, advertising, & PR managers	1.5	17.4	Adjustment clerks	1.1	40.9
Hand packers & packagers	1.5	0.6	Machinists	1.1	-11.9
Sewing machine operators, non-garment	1.5	76.2	Managers & administrators nec	1.1	17.4

Source: Industry-Occupation Matrix, Bureau of Labor Statistics. These data relate to one or more 3-digit SIC industry groups rather than to a single 4-digit SIC. The change reported for each occupation to the year 2005 is a percent of growth or decline as estimated by the Bureau of Labor Statistics. The abbreviation nec stands for 'not elsewhere classified'.

LOCATION BY STATE AND REGIONAL CONCENTRATION

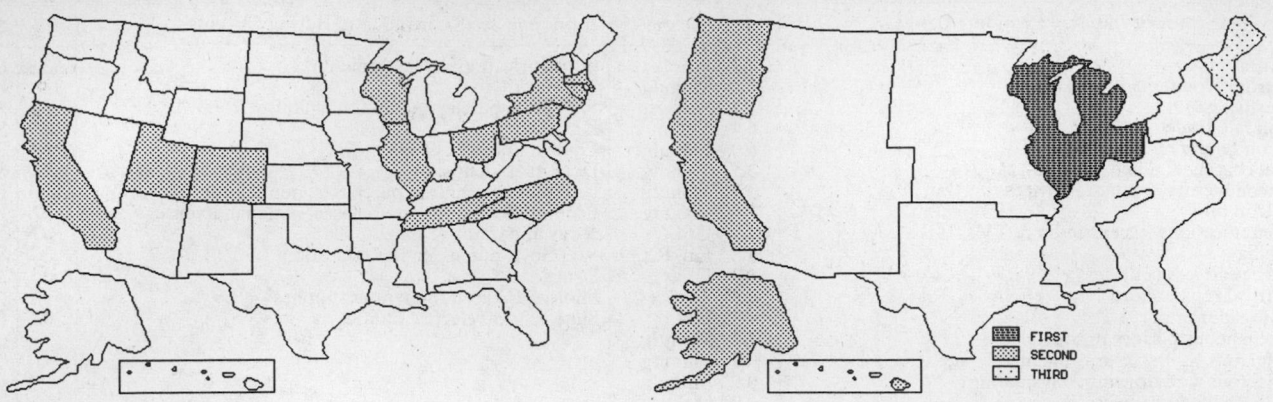

FIRST
SECOND
THIRD

INDUSTRY DATA BY STATE

State	Establish-ments	Shipments			Employment				Cost as % of Shipments	Investment per Employee ($)
		Total ($ mil)	% of U.S.	Per Establ.	Total Number	% of U.S.	Per Establ.	Wages ($/hour)		
California	30	835.9	25.8	27.9	4,700	32.9	157	17.72	43.3	4,681
Illinois	18	342.5	10.6	19.0	2,000	14.0	111	12.00	43.6	4,450
Ohio	10	318.4	9.8	31.8	900	6.3	90	12.00	49.1	-
Massachusetts	7	144.7	4.5	20.7	1,500	10.5	214	10.44	34.6	-
Wisconsin	6	(D)	-	-	1,750 *	12.2	292	-	-	-
Connecticut	5	(D)	-	-	175 *	1.2	35	-	-	-
New York	5	(D)	-	-	175 *	1.2	35	-	-	-
Colorado	4	(D)	-	-	375 *	2.6	94	-	-	-
North Carolina	3	(D)	-	-	375 *	2.6	125	-	-	-
Tennessee	3	(D)	-	-	175 *	1.2	58	-	-	-
Utah	3	(D)	-	-	1,750 *	12.2	583	-	-	-
Pennsylvania	2	(D)	-	-	750 *	5.2	375	-	-	-

Source: 1992 *Economic Census*. The states are in descending order of shipments or establishments (if shipment data are missing for the majority). The symbol (D) appears when data are withheld to prevent disclosure of competitive information. States marked with (D) are sorted by number of establishments. A dash (-) indicates that the data element cannot be calculated; * indicates the midpoint of a range.

3845 - ELECTROMEDICAL APPARATUS

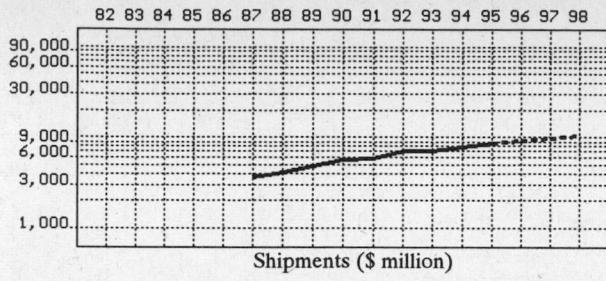

Shipments ($ million)

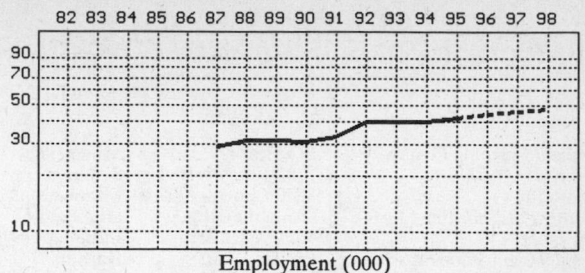

Employment (000)

GENERAL STATISTICS

Year	Com-panies	Establishments		Employment			Compensation		Production ($ million)			
		Total	with 20 or more employees	Total (000)	Production Workers (000)	Hours (Mil)	Payroll ($ mil)	Wages ($/hr)	Cost of Materials	Value Added by Manufacture	Value of Shipments	Capital Invest.
1982												
1983												
1984												
1985												
1986												
1987	195	224	144	29.2	13.2	25.6	850.5	10.21	1,251.3	2,368.8	3,576.7	114.7
1988		214	137	31.4	14.4	29.1	970.8	10.21	1,482.6	2,702.8	4,155.1	125.9
1989		232	149	31.8	14.1	29.2	1,062.8	10.91	1,764.2	3,093.2	4,821.4	174.0
1990		241	146	31.3	13.8	28.1	1,175.6	11.49	1,955.3	3,659.8	5,603.8	174.3
1991		257	143	33.2	14.5	29.4	1,249.6	11.74	2,018.6	3,712.9	5,742.9	181.3
1992	331	357	193	40.0	18.0	37.6	1,553.0	11.95	2,493.1	4,692.3	7,175.4	253.6
1993		363	186	39.9	17.6	36.3	1,580.3	12.12	2,318.8	4,863.5	7,186.3	240.9
1994		374P	190P	40.0	17.8	40.2	1,663.2	11.83	2,523.2	5,160.2	7,745.5	308.8
1995		400P	198P	42.3P	18.7P	40.8P	1,814.0P	12.61P	2,772.9P	5,670.8P	8,511.9P	313.4P
1996		426P	207P	44.1P	19.4P	42.7P	1,936.4P	12.90P	2,972.7P	6,079.5P	9,125.4P	339.4P
1997		452P	215P	45.8P	20.1P	44.7P	2,058.8P	13.18P	3,172.6P	6,488.3P	9,739.0P	365.3P
1998		478P	223P	47.5P	20.8P	46.6P	2,181.2P	13.47P	3,372.5P	6,897.1P	10,352.6P	391.3P

Sources: 1982, 1987, 1992 *Economic Census*; *Annual Survey of Manufactures*, 83-86, 88-91, 93-94. Establishment counts for non-Census years are from *County Business Patterns*; establishment values for 83-84 are extrapolations. 'P's show projections by the editors. Industries reclassified in 87 will not have data for prior years.

INDICES OF CHANGE

Year	Com-panies	Establishments		Employment			Compensation		Production ($ million)			
		Total	with 20 or more employees	Total (000)	Production Workers (000)	Hours (Mil)	Payroll ($ mil)	Wages ($/hr)	Cost of Materials	Value Added by Manufacture	Value of Shipments	Capital Invest.
1982												
1983												
1984												
1985												
1986												
1987	59	63	75	73	73	68	55	85	50	50	50	45
1988		60	71	78	80	77	63	85	59	58	58	50
1989		65	77	80	78	78	68	91	71	66	67	69
1990		68	76	78	77	75	76	96	78	78	78	69
1991		72	74	83	81	78	80	98	81	79	80	71
1992	100	100	100	100	100	100	100	100	100	100	100	100
1993		102	96	100	98	97	102	101	93	104	100	95
1994		105P	98P	100	99	107	107	99	101	110	108	122
1995		112P	103P	106P	104P	108P	117P	105P	111P	121P	119P	124P
1996		119P	107P	110P	108P	114P	125P	108P	119P	130P	127P	134P
1997		127P	111P	114P	112P	119P	133P	110P	127P	138P	136P	144P
1998		134P	116P	119P	116P	124P	140P	113P	135P	147P	144P	154P

Sources: Same as General Statistics. Values reflect change from the base year, 1992. Values above 100 mean greater than 92, values below 100 mean less than 92, and a value of 100 in the 82-91 or 93-98 period means same as 92. 'P's mark projections by the editors.

SELECTED RATIOS

For 1994	Avg. of All Manufact.	Analyzed Industry	Index	For 1994	Avg. of All Manufact.	Analyzed Industry	Index
Employees per Establishment	49	107	218	Value Added per Production Worker	134,084	289,899	216
Payroll per Establishment	1,500,273	4,450,459	297	Cost per Establishment	5,045,178	6,751,682	134
Payroll per Employee	30,620	41,580	136	Cost per Employee	102,970	63,080	61
Production Workers per Establishment	34	48	139	Cost per Production Worker	146,988	141,753	96
Wages per Establishment	853,319	1,272,539	149	Shipments per Establishment	9,576,895	20,725,726	216
Wages per Production Worker	24,861	26,717	107	Shipments per Employee	195,460	193,637	99
Hours per Production Worker	2,056	2,258	110	Shipments per Production Worker	279,017	435,140	156
Wages per Hour	12.09	11.83	98	Investment per Establishment	321,011	826,300	257
Value Added per Establishment	4,602,255	13,807,875	300	Investment per Employee	6,552	7,720	118
Value Added per Employee	93,930	129,005	137	Investment per Production Worker	9,352	17,348	185

Sources: Same as General Statistics. The 'Average of All Manufacturing' column represents the average of all manufacturing industries reported for the most recent complete year available. The Index shows the relationship between the Average and the Analyzed Industry. For example, 100 means that they are equal; 500 that the Analyzed Industry is five times the average; 50 means that the Analyzed Industry is half the national average. The abbreviation 'na' is used to show that data are 'not available'.

LEADING COMPANIES   Number shown: **75**   Total sales ($ mil): **13,631**   Total employment (000): **69.2**

Company Name	Address				CEO Name	Phone	Co. Type	Sales ($ mil)	Empl. (000)
GE Medical Systems Group	PO Box 414	Milwaukee	WI	53201	John M Trani	414-544-3011	D	4,174	15.0
Mallinckrodt Group Inc	7733 Forsyth Blv	St Louis	MO	63105	C Ray Holman	314-854-5200	P	1,940	10.2
Medtronic Inc	7000 Central NE	Minneapolis	MN	55432	William W George	612-574-4000	P	1,395	8.7
Picker International Inc	595 Miner Rd	Highland H	OH	44143	Cary J Nolan	216-473-3000	S	1,180	6.0
Guidant Corp	307 E McCarthy St	Indianapolis	IN	46225	Ronald W Dollens	317-971-2000	P	795	4.6
Advanced Techn Laboratories	PO Box 3003	Bothell	WA	98041	Dennis C Fill	206-487-7000	P	366	2.6
Acuson Corp	PO Box 7393	Mountain View	CA	94039	Samuel H Maslak	415-969-9112	P	351	1.6
Marquette Electronics Inc	8200 W Tower Av	Milwaukee	WI	53223	Michael J Cudahy	414-355-5000	P	254	1.5
SpaceLabs Medical Inc	PO Box 97013	Redmond	WA	98073	Carl A Lombardi	206-882-3700	P	247	1.5
Nellcor Inc	4280 Hacienda Dr	Pleasanton	CA	94588	CR Larkin	510-463-4000	P	235	1.7
Siemens Pacesetter Systems Inc	15900 Val View Ct	Fillmore	CA	91342	Guenther Jayensch	818-362-6822	S	200	0.9
Datascope Corp	14 Philips Pkwy	Montvale	NJ	07645	Lawrence Saper	201-265-8800	P	183	1.0
ADAC Laboratories	540 Alder Dr	Milpitas	CA	95035	Stanley D Officers	408-945-2990	P	157	0.6
DeVilbiss Health Care	PO Box 635	Somerset	PA	15501	Ray Dyer	814-443-4881	D	130	0.4
Siemens Medical Systems	PO Box 7002	Issaquah	WA	98027	Lothar Koob	206-392-9180	D	130*	0.5
Quinton Instrument Co	2121 Terry Av	Seattle	WA	98121	Anthony G Perri	206-223-7373	S	107	0.8
Ventritex Inc	709 E Evelyn Av	Sunnyvale	CA	94086	Frank M Fischer	408-738-4883	P	106	0.5
Physio-Control Corp	PO Box 97006	Redmond	WA	98073	Richard O Martin	206-867-4000	S	98*	0.9
Spectra-Physics Lasers Inc	PO Box 7013	Mountain View	CA	94039	Patrick L Edsell	415-961-2550	S	90	0.5
Circon Corp	460 Ward Dr	Santa Barbara	CA	93111	Richard A Auhll	805-967-0404	P	84	0.8
Siemens Medical Systems Inc	4040 Nelson Av	Concord	CA	94520	Dale Erickson	510-246-8200	D	80	0.5
Sony Electronics Inc	3 Paragon Dr	Montvale	NJ	07645	James T Sandy	201-930-7098	D	75	<0.1
Dornier Medical Systems Inc	1155 Roberts Blv	Kennesaw	GA	30144	David N Gill	404-426-1315	S	70	0.5
Ballard Medical Products	12050 Lone Peak	Draper	UT	84020	Dale H Ballard	801-572-6800	P	65	0.6
ICN Biomedicals Inc	3300 Hyland Av	Costa Mesa	CA	92626	Milan Panic	714-545-0113	S	59	0.5
Philips Ultrasound	29 Parker	Irvine	CA	92718	William Joyce	714-470-1300	S	51	0.3
Elscint Inc	505 Main St	Hackensack	NJ	07601	Thomas Spackman	201-342-2020	R	47*	0.2
SensorMedics Corp	22705 Savi Ranch	Yorba Linda	CA	92687	George Holmes	714-283-2228	R	46	0.3
Laserscope Surgical Systems	3052 Orchard Dr	San Jose	CA	95134	R V McCormick	408-943-0636	P	36	0.2
Acoustic Imaging Techn Corp	10027 S 51 St	Phoenix	AZ	85044	John T Kingsley	602-496-6681	S	36	0.3
Birtcher Medical Systems Inc	50 Technology Dr	Irvine	CA	92718	K C Cleveland	714-753-9400	P	34	0.1
Daig Corp	14901 DeVeau Pl	Minnetonka	MN	55345	J J Fleischhacker	612-933-4700	P	32	0.2
Camtronics Limited Medical Syst	PO Box 950	Hartland	WI	53029	Thomas W Lambert	414-367-0700	S	30	0.1
Shandon Lipshaw	171 Industry Dr	Pittsburgh	PA	15175	Colin Maddix	412-788-1133	S	30	0.1
Candela Laser Corp	530 Boston Post Rd	Wayland	MA	01778	Gerard E Puorro	508-358-7400	P	30	0.2
Endeavor Surgical Products Inc	Woodlnds Res	The Woodlands	TX	77381	Tom Early	713-363-4949	S	27	0.2
Aequitron Medical Inc	14800 28th Av N	Minneapolis	MN	55447	James B Hickey Jr	612-557-9200	P	26	0.2
Imatron Inc	389 Oyster Pt Blvd	S San Francisco	CA	94080	S Lewis Meyer	415-583-9964	P	25	0.1
Work Recovery Inc	2341 S Friebus Av	Tucson	AZ	85713	Thomas L Brandon	602-322-6634	P	25	0.2
Summit Technology Inc	21 Hickory Dr	Waltham	MA	02154	David F Muller	617-890-1234	P	24	0.2
Timeter Instrument Corp	1720 Sublette Av	St Louis	MO	63110	David V LaRusso	314-771-2400	D	24*	0.2
Plasma-Therm Inc	9509 Int Ct	St Petersburg	FL	33716	Ronald Deferrari	813-577-4999	P	23	<0.1
Medical Graphics Corp	350 Oak Grove	St Paul	MN	55127	C A Anderson	612-484-4874	P	23	0.2
Cybermedic Inc	6175 Longbow Dr	Boulder	CO	80301	Lockette E Wood	303-530-8280	S	22*	0.2
RELA Inc	6175 Longbow Dr	Boulder	CO	80301	John V Atanasoff	303-530-2626	P	22	0.2
Surgical Laser Technologies Inc	PO Box 880	Oaks	PA	19456	James R Appleby Jr	215-650-0700	P	22	0.2
Gish Biomedical Inc	2681 Kelvin Av	Irvine	CA	92714	Jack W Brown	714-756-5485	P	21	0.2
Colorado MEDTech Inc	6175 Longbow Dr	Boulder	CO	80301	John V Atanasoff	303-530-2660	P	21	0.2
Instromedix Inc	7431 NE Evergreen	Hillsboro	OR	97124	Greg T Semler	503-681-9000	R	18*	0.1
Biochem International Inc	238 Rockwood	Waukesha	WI	53188	David H Sanders	414-542-3100	P	18	<0.1
VISX Inc	3400 Central Expwy	Santa Clara	CA	95051	Mark B Logan	408-733-2020	P	18	0.1
Arrhythmia Research Technology	5910 Courtyard Dr	Austin	TX	78731	Anthony A Cetrone	512-343-6912	P	18	<0.1
Strato Medical Corp	123 Brimbal Av	Beverly	MA	01915	John Brooks	508-927-9419	S	17	<0.1
Mortara Instrument Inc	7865 N 86th St	Milwaukee	WI	53224	David Mortara	414-354-1600	R	17*	<0.1
TheraTek Group	5640 Airline Rd	Arlington	TN	38002	Erin McGurk	901-867-9971	D	17*	<0.1
Cytocare Inc	100 Columbia St	Aliso Viejo	CA	92656	Errol Payne	714-448-7700	P	16	<0.1
Warren E Collins Inc	220 Wood Rd	Braintree	MA	02184	W E Collins III	617-843-0610	R	16*	<0.1
Trimedyne Inc	PO Box 57001	Irvine	CA	92619	Marvin P Loeb	714-559-5300	P	16	0.1
CyberOptics Corp	2505 Kennedy NE	Minneapolis	MN	55413	Steven K Case	612-331-5702	P	15	0.1
Staodyn Inc	1225 Florida Av	Longmont	CO	80501	W Bayne Gibson	303-772-3631	P	15	0.2
Fonar Corp	110 Marcus Dr	Melville	NY	11747	Raymond Damadian	516-694-2929	P	15	0.2
Schwartz Electro-Optics Inc	3404 Orange	Orlando	FL	32804	William C Schwartz	407-298-1802	R	15	0.1
Vital Signs Minnesota Inc	12250 Nicollet Av S	Burnsville	MN	55337	Susan Brunsvold	612-894-7523	S	15	<0.1
PerSeptive Biosystems Inc	500 Old Connecticut	Framingham	MA	01701	Noubar B Afeyan	508-383-7700	P	14	0.2
Luxar Corp	19204 Northcreek	Bothell	WA	98011	Rose Hirsch	206-483-4142	R	13*	<0.1
Visiplex Instruments Ltd	250 Clearbrook Rd	Elmsford	NY	10523	Ed Costa	914-365-0190	D	13*	<0.1
Medstone International Inc	100 Columbia St	Aliso Viejo	CA	92656	David V Radlinski	714-448-7700	P	12	<0.1
Sunrise Technologies Inc	47257 Fremont Blv	Fremont	CA	94538	Arthur Vassiliadis	510-623-9001	P	12	<0.1
Spectranetics Corp	96 Talamine Ct	Co Springs	CO	80907	E Wyatt Cannady	719-633-8333	P	11	0.1
Hanovia Inc	100 Chestnut St	Newark	NJ	07105	Len Perre	201-589-4300	S	11	<0.1
JMAR Industries Inc	3956 Sorrento Val	San Diego	CA	92121	John S Martinez	619-535-1706	P	11	<0.1
LecTec Corp	10701 Red Circle Dr	Minnetonka	MN	55343	Thomas E Brunelle	612-933-2291	P	11	<0.1
Thermo Cardiosystems Inc	PO Box 2697	Woburn	MA	01888	Victor L Poirier	617-622-1000	P	10	<0.1
International Imaging Electronics	881 Remington Blv	Bolingbrook	IL	60440	Neal McGrath	708-378-4800	R	10	<0.1
Motion Analysis Corp	3617 Westwind Blv	Santa Rosa	CA	95403	Tom D Whitaker	707-579-6500	R	10	<0.1

Source: Ward's Business Directory of U.S. Private and Public Companies, Volumes 1 and 2, 1996. The company type code used is as follows: P - Public, R - Private, S - Subsidiary, D - Division, J - Joint Venture, A - Affiliate, G - Group. Sales are in millions of dollars, employees are in thousands. An asterisk (*) indicates an estimated sales volume. The symbol < stands for 'less than'. Company names and addresses are truncated, in some cases, to fit into the available space.

MATERIALS CONSUMED

Material	Quantity	Delivered Cost ($ million)
Materials, ingredients, containers, and supplies	(X)	2,109.6
Metal bolts, nuts, screws, washers, rivets, and other screw machine products	(X)	29.9
Metal stampings	(X)	25.1
All other fabricated metal products (except forgings)	(X)	118.4
Castings (rough and semifinished)	(X)	6.5
Stainless steel shapes and forms	(X)	12.0
Other steel shapes and forms	(X)	18.2
Nonferrous shapes and forms	(X)	4.2
Purchased electronic computing and peripheral equipment	(X)	283.1
Transmittal, industrial, and special-purpose electron tubes, except x-ray	(X)	8.4
Semiconductors, including transistors, diodes, rectifiers, and integrated circuits for electronic circuitry	(X)	132.2
Capacitors for electronic circuitry	(X)	18.4
Resistors for electronic circuitry	(X)	15.5
Connectors for electronic circuitry	(X)	20.1
Other electronic components and accessories	(X)	191.5
Insulated wire and cable, except magnet wire	(X)	68.8
Plastics products consumed in the form of sheets, rods, tubes, film, and other shapes	(X)	90.3
Paperboard containers, boxes, and corrugated paperboard	(X)	33.9
All other materials and components, parts, containers, and supplies	(X)	593.9
Materials, ingredients, containers, and supplies, nsk	(X)	439.2

Source: 1992 *Economic Census*. Explanation of symbols used: (D): Withheld to avoid disclosure of competitive data; na: Not available; (S): Withheld because statistical norms were not met; (X): Not applicable; (Z): Less than half the unit shown; nec: Not elsewhere classified; nsk: Not specified by kind; - : zero; * : 10-19 percent estimated; ** : 20-29 percent estimated.

PRODUCT SHARE DETAILS

Product or Product Class	% Share	Product or Product Class	% Share
Electromedical equipment	100.00		

Source: 1992 *Economic Census*. The values shown are percent of total shipments in an industry. Values of indented subcategories are summed in the main heading. The symbol (D) appears when data are withheld to prevent disclosure of competitive information. The abbreviation nsk stands for 'not specified by kind' and nec for 'not elsewhere classified'.

INPUTS AND OUTPUTS FOR SURGICAL & MEDICAL INSTRUMENTS

Economic Sector or Industry Providing Inputs	%	Sector	Economic Sector or Industry Buying Outputs	%	Sector
Imports	13.3	Foreign	Gross private fixed investment	62.9	Cap Inv
Primary copper	11.7	Manufg.	Exports	10.1	Foreign
Wholesale trade	7.0	Trade	S/L Govt. purch., health & hospitals	7.6	S/L Govt
Surgical appliances & supplies	6.3	Manufg.	Medical & health services, nec	4.8	Services
Miscellaneous plastics products	5.5	Manufg.	Doctors & dentists	3.5	Services
Metal stampings, nec	4.4	Manufg.	Hospitals	3.5	Services
Advertising	4.0	Services	Federal Government purchases, nondefense	1.6	Fed Govt
Electronic components nec	3.7	Manufg.	Federal Government purchases, national defense	1.5	Fed Govt
Plastics materials & resins	3.2	Manufg.	Surgical appliances & supplies	1.4	Manufg.
Fabricated rubber products, nec	2.7	Manufg.	Change in business inventories	1.3	In House
Paperboard containers & boxes	2.0	Manufg.	Personal consumption expenditures	0.7	
Blast furnaces & steel mills	1.9	Manufg.	Drugs	0.4	Manufg.
Electric services (utilities)	1.8	Util.	S/L Govt. purch., higher education	0.3	S/L Govt
Eating & drinking places	1.7	Trade	S/L Govt. purch., public assistance & relief	0.2	S/L Govt
Paper mills, except building paper	1.6	Manufg.			
Petroleum refining	1.6	Manufg.			
Metal coating & allied services	1.4	Manufg.			
Primary nonferrous metals, nec	1.3	Manufg.			
Communications, except radio & TV	1.2	Util.			
Cyclic crudes and organics	0.9	Manufg.			
Motor freight transportation & warehousing	0.9	Util.			
Metal foil & leaf	0.8	Manufg.			
Paints & allied products	0.8	Manufg.			
Paperboard mills	0.8	Manufg.			
Sanitary services, steam supply, irrigation	0.8	Util.			
Royalties	0.8	Fin/R.E.			
Copper rolling & drawing	0.7	Manufg.			
Machinery, except electrical, nec	0.7	Manufg.			
Rubber & plastics hose & belting	0.7	Manufg.			
Screw machine and related products	0.7	Manufg.			
Banking	0.7	Fin/R.E.			
Legal services	0.7	Services			
Fabricated metal products, nec	0.6	Manufg.			
Veneer & plywood	0.6	Manufg.			
Real estate	0.6	Fin/R.E.			
Equipment rental & leasing services	0.6	Services			
Management & consulting services & labs	0.6	Services			

Continued on next page.

INPUTS AND OUTPUTS FOR SURGICAL & MEDICAL INSTRUMENTS - Continued

Economic Sector or Industry Providing Inputs	%	Sector	Economic Sector or Industry Buying Outputs	%	Sector
Maintenance of nonfarm buildings nec	0.5	Constr.			
Gaskets, packing & sealing devices	0.5	Manufg.			
Glass & glass products, except containers	0.5	Manufg.			
Broadwoven fabric mills	0.4	Manufg.			
Iron & steel foundries	0.4	Manufg.			
Special dies & tools & machine tool accessories	0.4	Manufg.			
Railroads & related services	0.4	Util.			
Accounting, auditing & bookkeeping	0.4	Services			
Abrasive products	0.3	Manufg.			
Hardware, nec	0.3	Manufg.			
Manifold business forms	0.3	Manufg.			
Primary metal products, nec	0.3	Manufg.			
Semiconductors & related devices	0.3	Manufg.			
Gas production & distribution (utilities)	0.3	Util.			
Automotive rental & leasing, without drivers	0.3	Services			
Computer & data processing services	0.3	Services			
Noncomparable imports	0.3	Foreign			
Iron & steel forgings	0.2	Manufg.			
Lubricating oils & greases	0.2	Manufg.			
Nonferrous rolling & drawing, nec	0.2	Manufg.			
Nonwoven fabrics	0.2	Manufg.			
Paper coating & glazing	0.2	Manufg.			
Sawmills & planning mills, general	0.2	Manufg.			
Insurance carriers	0.2	Fin/R.E.			
Automotive repair shops & services	0.2	Services			
Hotels & lodging places	0.2	Services			
Adhesives & sealants	0.1	Manufg.			
Apparel made from purchased materials	0.1	Manufg.			
Household furniture, nec	0.1	Manufg.			
Industrial inorganic chemicals, nec	0.1	Manufg.			
Photographic equipment & supplies	0.1	Manufg.			
Surgical & medical instruments	0.1	Manufg.			
Transit & bus transportation	0.1	Util.			
Personnel supply services	0.1	Services			
U.S. Postal Service	0.1	Gov't			

Source: Benchmark Input-Output Accounts for the U.S. Economy, 1982, U.S. Department of Commerce, Washington, D.C., July 1991. Data, as reported in the source, are organized by the 1977 SIC structure in use in 1982 but have been matched, as closely as is possible, to the 1987 SIC structure used in this book.

OCCUPATIONS EMPLOYED BY SIC 384 - MEDICAL INSTRUMENTS AND SUPPLIES

Occupation	% of Total 1994	Change to 2005	Occupation	% of Total 1994	Change to 2005
Assemblers, fabricators, & hand workers nec	16.5	17.4	Mechanical engineers	1.5	29.3
Sales & related workers nec	5.3	17.4	Electrical & electronic equipment assemblers	1.5	17.4
Inspectors, testers, & graders, precision	4.3	-17.8	Precision workers nec	1.5	5.7
Blue collar worker supervisors	3.3	-0.0	Electrical & electronics engineers	1.4	49.1
Precision assemblers nec	2.4	88.0	Industrial production managers	1.4	17.4
Secretaries, ex legal & medical	2.3	6.9	Packaging & filling machine operators	1.3	17.4
General managers & top executives	1.8	11.4	Electrical & electronic technicians,technologists	1.3	17.4
Traffic, shipping, & receiving clerks	1.8	13.0	Engineering, mathematical, & science managers	1.3	33.4
Machine operators nec	1.7	-17.2	Bookkeeping, accounting, & auditing clerks	1.2	-11.9
Electrical & electronic assemblers	1.7	5.7	Production, planning, & expediting clerks	1.2	17.4
Engineering technicians & technologists nec	1.6	17.4	General office clerks	1.2	0.2
Marketing, advertising, & PR managers	1.5	17.4	Adjustment clerks	1.1	40.9
Hand packers & packagers	1.5	0.6	Machinists	1.1	-11.9
Sewing machine operators, non-garment	1.5	76.2	Managers & administrators nec	1.1	17.4

Source: Industry-Occupation Matrix, Bureau of Labor Statistics. These data relate to one or more 3-digit SIC industry groups rather than to a single 4-digit SIC. The change reported for each occupation to the year 2005 is a percent of growth or decline as estimated by the Bureau of Labor Statistics. The abbreviation nec stands for 'not elsewhere classified'.

LOCATION BY STATE AND REGIONAL CONCENTRATION

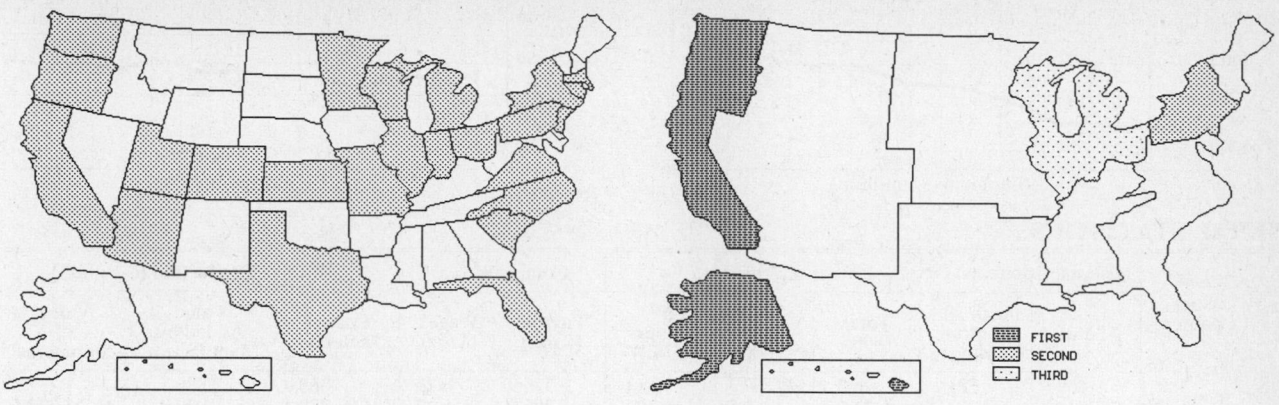

FIRST
SECOND
THIRD

INDUSTRY DATA BY STATE

State	Establish-ments	Shipments			Employment				Cost as % of Shipments	Investment per Employee ($)
		Total ($ mil)	% of U.S.	Per Establ.	Total Number	% of U.S.	Per Establ.	Wages ($/hour)		
California	75	1,786.2	24.9	23.8	8,300	20.8	111	16.07	27.6	6,337
Wisconsin	16	828.7	11.5	51.8	2,700	6.8	169	9.60	29.9	5,407
Massachusetts	21	710.0	9.9	33.8	3,200	8.0	152	16.93	44.5	4,313
Colorado	11	516.3	7.2	46.9	3,300	8.3	300	9.54	28.7	-
Minnesota	20	504.8	7.0	25.2	2,400	6.0	120	10.55	25.1	-
Washington	10	488.9	6.8	48.9	3,600	9.0	360	11.38	34.3	4,528
New York	28	254.3	3.5	9.1	2,000	5.0	71	11.55	35.3	3,150
Florida	13	246.9	3.4	19.0	1,800	4.5	138	10.67	50.6	4,111
Texas	16	238.2	3.3	14.9	1,800	4.5	113	12.06	23.4	-
New Jersey	16	215.5	3.0	13.5	1,100	2.8	69	13.40	45.5	7,364
Pennsylvania	18	197.2	2.7	11.0	1,500	3.8	83	10.38	60.0	2,600
Ohio	12	170.3	2.4	14.2	900	2.3	75	9.71	50.7	4,222
Utah	5	152.1	2.1	30.4	2,000	5.0	400	7.51	42.2	-
Connecticut	14	144.2	2.0	10.3	800	2.0	57	13.70	39.9	3,375
Oregon	9	128.0	1.8	14.2	800	2.0	89	11.20	49.5	2,625
Illinois	17	63.3	0.9	3.7	400	1.0	24	13.00	31.0	1,750
North Carolina	6	(D)	-	-	375 *	0.9	63	-	-	-
Indiana	5	(D)	-	-	375 *	0.9	75	-	-	-
Missouri	5	(D)	-	-	750 *	1.9	150	-	-	-
Arizona	4	(D)	-	-	175 *	0.4	44	-	-	-
Michigan	4	(D)	-	-	175 *	0.4	44	-	-	-
Kansas	3	(D)	-	-	375 *	0.9	125	-	-	-
South Carolina	2	(D)	-	-	375 *	0.9	188	-	-	-
Virginia	2	(D)	-	-	375 *	0.9	188	-	-	-

Source: 1992 *Economic Census*. The states are in descending order of shipments or establishments (if shipment data are missing for the majority). The symbol (D) appears when data are withheld to prevent disclosure of competitive information. States marked with (D) are sorted by number of establishments. A dash (-) indicates that the data element cannot be calculated; * indicates the midpoint of a range.

3851 - OPHTHALMIC GOODS

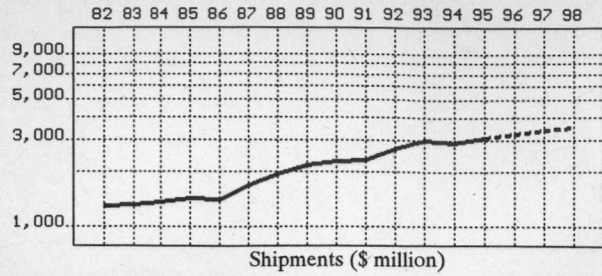

Shipments ($ million)

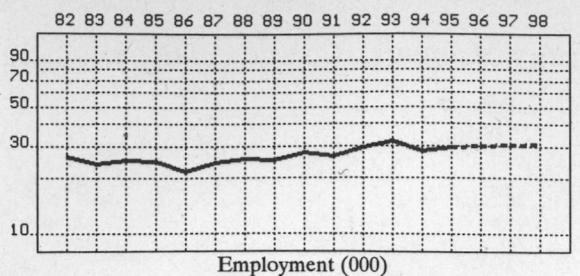

Employment (000)

GENERAL STATISTICS

Year	Com-panies	Establishments		Employment			Compensation		Production ($ million)			
		Total	with 20 or more employees	Total (000)	Production Workers (000)	Hours (Mil)	Payroll ($ mil)	Wages ($/hr)	Cost of Materials	Value Added by Manufacture	Value of Shipments	Capital Invest.
1982	389	409	123	26.3	17.3	35.1	416.9	6.13	388.0	886.5	1,287.2	41.6
1983				24.1	15.5	31.1	396.3	6.49	388.4	930.0	1,332.4	29.2
1984				25.1	16.6	33.1	433.0	6.64	398.3	1,009.1	1,350.7	63.5
1985				24.9	15.5	31.2	434.0	6.73	412.4	995.6	1,418.2	78.6
1986				21.7	14.3	28.6	414.6	7.26	472.2	933.5	1,411.2	92.6
1987	480	495	149	24.2	15.8	31.1	477.9	7.55	546.0	1,152.9	1,689.4	76.7
1988				25.5	17.6	33.8	499.9	8.02	621.3	1,345.1	1,945.8	88.6
1989		517	161	25.0	17.8	36.2	500.8	8.36	664.1	1,542.6	2,193.5	146.0
1990				28.0	19.8	40.3	605.0	8.62	672.5	1,625.6	2,274.7	137.2
1991				26.2	17.2	34.8	626.9	9.64	666.4	1,645.2	2,313.0	120.0
1992	526	569	150	29.6	19.9	40.2	716.3	9.59	748.0	1,950.6	2,692.1	202.4
1993				32.0	21.6	41.7	796.7	9.86	838.2	2,173.4	2,983.9	192.4
1994				28.2	18.8	36.4	719.1	10.09	865.9	2,095.6	2,928.5	200.3
1995				29.5P	20.2P	39.8P	769.4P	10.56P	910.1P	2,202.6P	3,078.0P	213.3P
1996				30.0P	20.6P	40.5P	802.0P	10.92P	956.2P	2,314.2P	3,234.0P	227.6P
1997				30.5P	21.0P	41.2P	834.6P	11.27P	1,002.3P	2,425.8P	3,389.9P	241.9P
1998				31.0P	21.4P	42.0P	867.2P	11.63P	1,048.5P	2,537.4P	3,545.9P	256.2P

Sources: 1982, 1987, 1992 *Economic Census*; *Annual Survey of Manufactures*, 83-86, 88-91, 93-94. Establishment counts for non-Census years are from *County Business Patterns*; establishment values for 83-84 are extrapolations. 'P's show projections by the editors. Industries reclassified in 87 will not have data for prior years.

INDICES OF CHANGE

Year	Com-panies	Establishments		Employment			Compensation		Production ($ million)			
		Total	with 20 or more employees	Total (000)	Production Workers (000)	Hours (Mil)	Payroll ($ mil)	Wages ($/hr)	Cost of Materials	Value Added by Manufacture	Value of Shipments	Capital Invest.
1982	74	72	82	89	87	87	58	64	52	45	48	21
1983				81	78	77	55	68	52	48	49	14
1984				85	83	82	60	69	53	52	50	31
1985				84	78	78	61	70	55	51	53	39
1986				73	72	71	58	76	63	48	52	46
1987	91	87	99	82	79	77	67	79	73	59	63	38
1988				86	88	84	70	84	83	69	72	44
1989		91	107	84	89	90	70	87	89	79	81	72
1990				95	99	100	84	90	90	83	84	68
1991				89	86	87	88	101	89	84	86	59
1992	100	100	100	100	100	100	100	100	100	100	100	100
1993				108	109	104	111	103	112	111	111	95
1994				95	94	91	100	105	116	107	109	99
1995				100P	102P	99P	107P	110P	122P	113P	114P	105P
1996				101P	104P	101P	112P	114P	128P	119P	120P	112P
1997				103P	106P	103P	117P	118P	134P	124P	126P	120P
1998				105P	108P	104P	121P	121P	140P	130P	132P	127P

Sources: Same as General Statistics. Values reflect change from the base year, 1992. Values above 100 mean greater than 92, values below 100 mean less than 92, and a value of 100 in the 82-91 or 93-98 period means same as 92. 'P's mark projections by the editors.

SELECTED RATIOS

For 1992	Avg. of All Manufact.	Analyzed Industry	Index	For 1992	Avg. of All Manufact.	Analyzed Industry	Index
Employees per Establishment	46	52	114	Value Added per Production Worker	122,353	98,020	80
Payroll per Establishment	1,332,320	1,258,875	94	Cost per Establishment	4,239,462	1,314,587	31
Payroll per Employee	29,181	24,199	83	Cost per Employee	92,853	25,270	27
Production Workers per Establishment	31	35	111	Cost per Production Worker	135,003	37,588	28
Wages per Establishment	734,496	677,536	92	Shipments per Establishment	8,100,800	4,731,283	58
Wages per Production Worker	23,390	19,373	83	Shipments per Employee	177,425	90,949	51
Hours per Production Worker	2,025	2,020	100	Shipments per Production Worker	257,966	135,281	52
Wages per Hour	11.55	9.59	83	Investment per Establishment	278,244	355,712	128
Value Added per Establishment	3,842,210	3,428,120	89	Investment per Employee	6,094	6,838	112
Value Added per Employee	84,153	65,899	78	Investment per Production Worker	8,861	10,171	115

Sources: Same as General Statistics. The 'Average of All Manufacturing' column opresents the average of all manufacturing industries reported for the most recent complete year available. The Index shows the relationship between the Average and the Analyzed Industry. For example, 100 means that they are equal; 500 that the Analyzed Industry is five times the average; 50 means that the Analyzed Industry is half the national average. The abbreviation 'na' is used to show that data are 'not available'.

LEADING COMPANIES Number shown: **75**   Total sales ($ mil): **4,792**   Total employment (000): **43.1**

Company Name	Address				CEO Name	Phone	Co. Type	Sales ($ mil)	Empl. (000)
Bausch and Lomb Inc	1 Lincoln First Sq	Rochester	NY	14601	Daniel E Gill	716-338-6000	P	1,872	15.9
Allergan Inc	PO Box 19534	Irvine	CA	92713	William C Shepherd	714-752-4500	P	947	4.9
Sola International Inc	2420 Sand Hill Rd	Menlo Park	CA	94025	John E Heine	415-324-6868	P	306*	5.4
Wesley-Jessen Co	400 W Superior St	Chicago	IL	60610	Charles Stroupe	312-751-6357	S	180*	1.8
Marchon and Marcolin Eyewear	35 Hub Dr	Melville	NY	11747	Larry Roth	516-755-2020	R	100	0.3
Omega Optical Co	13515 Nemmons	Dallas	TX	75234	Weldon B Lucas	214-241-4141	S	100	1.2
Cabot Safety Corp	90 Mechanic St	Southbridge	MA	01550	John D Curtin	508-764-5500	S	97*	1.2
Cooper Companies Inc	1 Bridge Plz	Fort Lee	NJ	07024	Allan E Rubenstein	201-585-5100	P	96	1.0
Vision Ease	PO Box 165117	Ft Lauderdale	FL	33316	Paul Burke	305-767-6222	D	70	1.0
Bacou USA Inc	10 Thurber Blv	Smithfield	RI	02917	Henri Bacou	401-232-1200	R	55	0.3
Uvex Safety Inc	10 Thurber Blv	Smithfield	RI	02917	Walter Stepan	401-232-1200	S	55	0.3
Uvex Safety LLC	10 Thurber Blv	Smithfield	RI	02917	Walter Stepan	401-232-1200	J	50	0.3
Parmelee Industries Inc	PO Box 15965	Lenexa	KS	66285	LA Sankpill	913-599-5555	R	47	0.5
US Safety	PO Box 15965	Shawnee Msn	KS	66285	L Alan Sankpill	913-599-5555	D	47	0.4
Corning Inc Optical Products	HP-CB-4-1	Corning	NY	14831	RG Ackerman	607-974-9000	D	45	1.0
Southern Optical Company Inc	PO Box 21328	Greensboro	NC	27420	Robert B Lawson	910-272-8146	R	43	0.4
CooperVision Inc	200 Willowb	Fairport	NY	14450	Thomas Bender	716-385-6810	S	41	0.3
New West Eyeworks	2109 W Southern	Tempe	AZ	85282	Barry J Feld	602-438-1330	P	40	0.6
Titmus Optical Inc	PO Box 191	Petersburg	VA	23804	EE Greene	804-732-6121	R	40	0.5
IOLAB Corp	500 Iolab Dr	Claremont	CA	91711	Bob Darretta	909-624-2020	S	37*	0.4
Cooper Development Co	455 E Middlefield	Mountain View	CA	94043	P G Montgomery	415-969-9030	P	36	0.2
Soderberg Inc	230 Eva St	St Paul	MN	55107	F A Soderberg	612-291-1400	R	35	0.5
Renaissance Eyewear Inc	50 S Avnue West	Cranford	NJ	07016	E Kauz	908-709-0110	R	26	0.3
Lantis Corp	461 5th Av	New York	NY	10017	Lon Moellentine	212-686-1500	R	25*	0.3
Wilshire Designs Inc	6120 Bristol Pkwy	Culver City	CA	90230	Richard Haft	310-417-8110	R	25	<0.1
Willson Safety Products	PO Box 622	Reading	PA	19603	Roger W Gehring	610-376-6161	D	22	0.2
STAAR Surgical Co	1911 Walker Av	Monrovia	CA	91016	John R Wolf	818-303-7902	P	20	<0.1
Sterling Vision Shoppes Inc	140 Macomb	Mount Clemens	MI	48043	R Dooley	313-468-7370	R	19*	0.2
Revo Inc	455 E Middlefield	Mountain View	CA	94043	Doug Lauer	415-962-0906	S	19	<0.1
Associated Optical Top Network	PO Box 655505	Dallas	TX	75265	Greg Solomon	214-285-8881	R	18*	0.2
Icare Industries Inc	PO Box 84000	St Petersburg	FL	33784	J Scott Payne	813-526-0501	R	15*	0.2
X-Cel Optical Co	806 S Benton Dr	Sauk Rapids	MN	56379	Joseph Doescher Sr	612-251-8404	R	15	0.3
Zylo Ware Corp	1136 46th Rd	Long Island City	NY	11101	RM Shyer	718-392-3900	R	15*	0.2
HL Bouton Company Inc	PO Box G	Buzzards Bay	MA	02532	David F Miller III	508-295-3300	S	13	0.1
Bell Optical Laboratory Inc	2510 Lance Dr	Dayton	OH	45409	Hank Zobrist	513-294-8022	R	13*	0.1
Dynoptic Inc	PO Box 84000	St Petersburg	FL	33784	J Scott Payne	813-522-3146	S	12	0.2
Rite-Style Optical Co	12240 Emmet St	Omaha	NE	68164	George Lee	402-492-8822	R	12	0.1
Dispensers Optical Service Corp	1815 Plantside Dr	Louisville	KY	40299	Charles S Arensberg	502-491-3440	R	10	0.1
Langley Optical Company Inc	8140 Marshall Dr	Lenexa	KS	66214	Mick Aslin	913-492-5379	R	10	0.2
Optical Services Inc	4905 S 97th St	Omaha	NE	68127	Sheldon I Rips	402-592-1212	R	10*	0.2
TKC Inc	401 Pearl St	Sioux City	IA	51102	Steve Conley	712-252-1519	R	10*	0.1
WOS Inc	PO Box 10387	Green Bay	WI	54307	Jim LaLuzerne	414-336-0690	R	10*	0.1
DAC Vision	380 E 167th St	Harvey	IL	60426	Dick Bullwinkle	708-333-4848	R	9*	<0.1
Art-Craft Optical Co	89 Allen St	Rochester	NY	14608	C Thomas Eagle	716-546-6640	R	8*	0.2
Con-Cise Lens Co	PO Box 2198	San Leandro	CA	94577	CF Moore	510-483-9400	R	8	<0.1
Fosta-Tek Optics Inc	320 Hamilton St	Leominster	MA	01453	John Morrison	508-534-6511	R	8	<0.1
Precision Optics Inc	PO Box 1288	St Cloud	MN	56302	Roy Hinkemeyer	612-251-8591	S	7	0.1
Sun Rams Products Inc	8736 Lion St	R Cucamonga	CA	91730	Johnny Yam	909-980-1160	R	7	0.1
Cumberland Optical Company	806 Olympic St	Nashville	TN	37203	Kenneth W Wyatt	615-254-5868	R	7	<0.1
Eye Kraft Optical Inc	PO Box 400	St Cloud	MN	56302	James T Negaard	612-251-0141	R	6	<0.1
Carskadden Optical Co	24 S 6th St	Zanesville	OH	43701	Steve Meyers	614-452-9306	R	6	<0.1
Williams Optical Laboratory Inc	PO Box 1246	Nashville	TN	37202	Gary Jones	615-256-6631	S	5	0.1
Lombart Lenses Ltd	PO Box 1693	Norfolk	VA	23501	John H Keenan	804-625-7866	S	4	0.1
American Consolidated Labs	6416 Parkland Dr	Sarasota	FL	34243	Wayne Smith	813-753-0383	P	4	<0.1
Halo Optical Corp	9 Phair St	Gloversville	NY	12078	Peter Leonardi	518-773-4256	R	4	<0.1
Hudson Universal Ltd	213 S Van Brunt St	Englewood	NJ	07631	H Strauss	201-567-7740	R	4	<0.1
OSD Envizion Co	1370 Willow Rd	Menlo Park	CA	94025	Jeffrey K Fergason	415-323-8262	J	4	<0.1
Unilens Corporation USA	10431 72nd St N	Largo	FL	34647	Al Vitale	813-544-2531	S	4	<0.1
Seoco Inc	6 Henson Pl	Champaign	IL	61820	Edward Schmidt	217-352-7865	R	4*	<0.1
Webster Lens Co	Tracy Ct	Webster	MA	01570	Nathaniel Chaffin	508-943-1550	R	4	<0.1
Epic Eyewear Group Inc	455 Pleasant St	Fall River	MA	02721	Robert Alexander	508-324-1985	S	3*	<0.1
Gateway Amsafe Inc	4722 Spring Rd	Cleveland	OH	44131	Kenneth E Love	216-749-1100	R	3	<0.1
Kontur Kontact Lens Co	200 S Garrard Blv	Richmond	CA	94801	DG Ewell	510-235-5225	R	3*	<0.1
Medical Developmental Research	2540 118th Av N	St Petersburg	FL	33716	Mark Robinson	813-572-6644	R	3	<0.1
Uvex Sports Inc	910 Douglas Pike	Smithfield	RI	02917	Walter Stepan	401-232-7670	S	3	<0.1
Stemaco Products Inc	5139 LaPeer Rd	Smiths Creek	MI	48074	Howard Stein	810-987-5151	R	3	<0.1
nanoFILM Co	10111 Sweet Val Dr	Valley View	OH	44125	Scott E Rickert	216-447-1199	R	2	<0.1
Prescription Optical Supply Inc	311 72nd Av S	St Cloud	MN	56301	Julius H Reimer	612-252-3113	R	2	<0.1
Co-Optics	413 River St Service	Oneonta	NY	13820	Edward V Krupp Jr	607-432-0557	R	2	<0.1
Sadler Brothers Inc	PO Box 3005	S Attleboro	MA	02703	Thomas M Sadler	508-761-8352	R	2	<0.1
Rochester Optical Mfg Co	38 Scio St	Rochester	NY	14604	Patrick C Ho	716-454-4250	R	2	<0.1
American Biocurve Inc	15970 Bernardo Ctr	San Diego	CA	92127	Reid Cole	619-487-8684	R	1*	<0.1
Breger Mueller Welt Corp	522 S Clinton St	Chicago	IL	60607	Joseph L Breger	312-922-9444	R	1	<0.1
Dietz Laboratories Inc	PO Box 1082	Fort Worth	TX	76101	Ed Dietz III	817-926-6611	R	1	<0.1
Moody Optical	623 15th Av	East Moline	IL	61244	Bob Kane	309-755-2114	R	1	<0.1

Source: Ward's Business Directory of U.S. Private and Public Companies, Volumes 1 and 2, 1996. The company type code used is as follows: P - Public, R - Private, S - Subsidiary, D - Division, J - Joint Venture, A - Affiliate, G - Group. Sales are in millions of dollars, employees are in thousands. An asterisk (*) indicates an estimated sales volume. The symbol < stands for 'less than'. Company names and addresses are truncated, in some cases, to fit into the available space.

MATERIALS CONSUMED

Material	Quantity	Delivered Cost ($ million)
Materials, ingredients, containers, and supplies	(X)	550.1
Lens blanks, optical and ophthalmic	(X)	144.7
Lenses and prisms for optical instruments, and sighting and fire control equipment	(X)	11.7
Plastics resins consumed in the form of granules, pellets, powders, liquids, etc.	(X)	55.0
Plastics products consumed in the form of sheets, rods, tubes, film, and other shapes	(X)	48.0
Paperboard containers, boxes, and corrugated paperboard	(X)	12.5
All other materials and components, parts, containers, and supplies	(X)	153.7
Materials, ingredients, containers, and supplies, nsk	(X)	124.4

Source: 1992 *Economic Census*. Explanation of symbols used: (D): Withheld to avoid disclosure of competitive data; na: Not available; (S): Withheld because statistical norms were not met; (X): Not applicable; (Z): Less than half the unit shown; nec: Not elsewhere classified; nsk: Not specified by kind; - : zero; * : 10-19 percent estimated; ** : 20-29 percent estimated.

PRODUCT SHARE DETAILS

Product or Product Class	% Share	Product or Product Class	% Share
Ophthalmic goods	100.00	Plastics ophthalmic multifocal lenses	54.48
Ophthalmic fronts and temples	6.97	Contact lenses	38.40
Ophthalmic plastics fronts, finished (with or without decoration)	43.68	Conventional (hard) contact lenses	10.24
Other Ophthalmic fronts, finished (with or without decoration)	22.26	Soft contact lenses	89.75
Ophthalmic plastic temples	22.02	Ophthalmic goods except fronts, temples, and lenses	16.71
Other Ophthalmic temples	11.99	Ophthalmic goods, industrial goggles, eye protectors, welding circles and plates, and mountings	24.55
Glass ophthalmic focus lenses	6.57	Ophthalmic goods, sun or glare glasses and sungoggles, ready-made	39.28
Glass ophthalmic single vision lenses (ground and polished and molded blanks)	39.24	Ophthalmic goods, parts for frames and mounting, except fronts and temples	10.55
Glass ophthalmic multifocal lenses (finished, semifinished, and molded blanks)	58.67	Other ophthalmic goods, plastics	12.11
Glass ophthalmic focus lenses, nsk	2.09	Other ophthalmic goods	12.61
Plastics ophthalmic focus lenses	17.62	Ophthalmic goods except fronts, temples, and lenses, nsk	0.90
Plastics ophthalmic single vision lenses	45.52	Ophthalmic goods, nsk	13.75

Source: 1992 *Economic Census*. The values shown are percent of total shipments in an industry. Values of indented subcategories are summed in the main heading. The symbol (D) appears when data are withheld to prevent disclosure of competitive information. The abbreviation nsk stands for 'not specified by kind' and nec for 'not elsewhere classified'.

INPUTS AND OUTPUTS FOR OPHTHALMIC GOODS

Economic Sector or Industry Providing Inputs	%	Sector	Economic Sector or Industry Buying Outputs	%	Sector
Imports	41.8	Foreign	Personal consumption expenditures	81.7	
Glass & glass products, except containers	9.4	Manufg.	Exports	6.6	Foreign
Wholesale trade	5.3	Trade	S/L Govt. purch., health & hospitals	4.9	S/L Govt
Petroleum refining	4.8	Manufg.	S/L Govt. purch., public assistance & relief	1.8	S/L Govt
Miscellaneous plastics products	3.4	Manufg.	Ophthalmic goods	0.5	Manufg.
Plastics materials & resins	2.8	Manufg.	Federal Government purchases, nondefense	0.4	Fed Govt
Nonferrous rolling & drawing, nec	2.4	Manufg.	Maintenance of electric utility facilities	0.2	Constr.
Optical instruments & lenses	2.3	Manufg.	Motor vehicle parts & accessories	0.1	Manufg.
Eating & drinking places	2.3	Trade	Motor vehicles & car bodies	0.1	Manufg.
Communications, except radio & TV	2.2	Util.	Miscellaneous repair shops	0.1	Services
Electric services (utilities)	2.2	Util.			
Air transportation	1.5	Util.			
Paperboard containers & boxes	1.0	Manufg.			
Equipment rental & leasing services	1.0	Services			
Hotels & lodging places	1.0	Services			
Machinery, except electrical, nec	0.9	Manufg.			
Automotive rental & leasing, without drivers	0.9	Services			
Legal services	0.9	Services			
Ophthalmic goods	0.8	Manufg.			
Advertising	0.8	Services			
Management & consulting services & labs	0.8	Services			
Banking	0.7	Fin/R.E.			
Insurance carriers	0.7	Fin/R.E.			
Real estate	0.7	Fin/R.E.			
Automotive repair shops & services	0.7	Services			
Gas production & distribution (utilities)	0.6	Util.			
Motor freight transportation & warehousing	0.6	Util.			
Maintenance of nonfarm buildings nec	0.5	Constr.			
Special dies & tools & machine tool accessories	0.5	Manufg.			
Accounting, auditing & bookkeeping	0.5	Services			
Abrasive products	0.4	Manufg.			
Gaskets, packing & sealing devices	0.4	Manufg.			
Manifold business forms	0.4	Manufg.			
U.S. Postal Service	0.4	Gov't			

Continued on next page.

INPUTS AND OUTPUTS FOR OPHTHALMIC GOODS - Continued

Economic Sector or Industry Providing Inputs	%	Sector	Economic Sector or Industry Buying Outputs	%	Sector
Lubricating oils & greases	0.3	Manufg.			
Computer & data processing services	0.3	Services			
Theatrical producers, bands, entertainers	0.3	Services			
Motor vehicle parts & accessories	0.2	Manufg.			
Photographic equipment & supplies	0.2	Manufg.			
Tires & inner tubes	0.2	Manufg.			
Retail trade, except eating & drinking	0.2	Trade			
Personnel supply services	0.2	Services			
Railroads & related services	0.1	Util.			
Transit & bus transportation	0.1	Util.			
Royalties	0.1	Fin/R.E.			
Engineering, architectural, & surveying services	0.1	Services			

Source: Benchmark Input-Output Accounts for the U.S. Economy, 1982, U.S. Department of Commerce, Washington, D.C., July 1991. Data, as reported in the source, are organized by the 1977 SIC structure in use in 1982 but have been matched, as closely as is possible, to the 1987 SIC structure used in this book.

OCCUPATIONS EMPLOYED BY SIC 385 - PROFESSIONAL AND SCIENTIFIC INSTRUMENTS NEC

Occupation	% of Total 1994	Change to 2005	Occupation	% of Total 1994	Change to 2005
Assemblers, fabricators, & hand workers nec	13.9	-8.1	Bookkeeping, accounting, & auditing clerks	1.9	-5.4
Inspectors, testers, & graders, precision	6.7	-31.9	General office clerks	1.8	3.6
Metal & plastic machine workers nec	4.8	6.4	Marketing, advertising, & PR managers	1.7	-6.0
Precision assemblers nec	4.6	-14.6	Freight, stock, & material movers, hand	1.7	-22.3
Sales & related workers nec	4.2	-3.3	Hand packers & packagers	1.5	-29.3
Blue collar worker supervisors	4.0	-13.8	Industrial production managers	1.5	-2.5
Order clerks, materials, merchandise, & service	3.4	-3.4	Engineering technicians & technologists nec	1.4	-2.7
Stock clerks	2.8	-22.9	Clerical supervisors & managers	1.3	-11.0
General managers & top executives	2.8	-11.6	Opticians, dispensing & measuring	1.3	2.1
Traffic, shipping, & receiving clerks	2.6	-9.4	Machine tool cutting operators, metal & plastic	1.2	-39.6
Secretaries, ex legal & medical	2.4	-4.2	Production, planning, & expediting clerks	1.2	-3.7
Machine operators nec	2.1	-22.4	Financial managers	1.1	-5.3
Adjustment clerks	2.1	-3.5	Maintenance repairers, general utility	1.0	-8.3

Source: Industry-Occupation Matrix, Bureau of Labor Statistics. These data relate to one or more 3-digit SIC industry groups rather than to a single 4-digit SIC. The change reported for each occupation to the year 2005 is a percent of growth or decline as estimated by the Bureau of Labor Statistics. The abbreviation nec stands for 'not elsewhere classified'.

LOCATION BY STATE AND REGIONAL CONCENTRATION

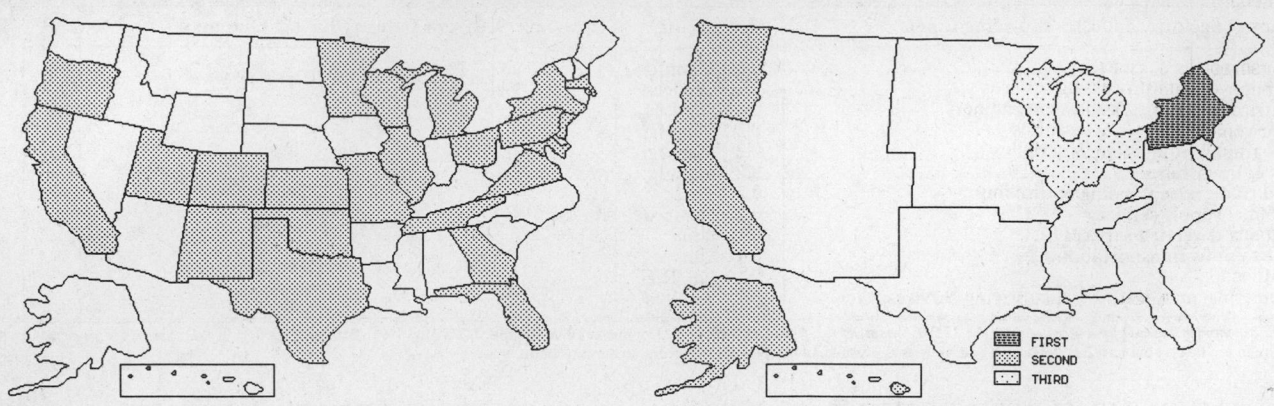

FIRST
SECOND
THIRD

INDUSTRY DATA BY STATE

| State | Establish-ments | Shipments | | | Employment | | | | Cost as % of Shipments | Investment per Employee ($) |
		Total ($ mil)	% of U.S.	Per Establ.	Total Number	% of U.S.	Per Establ.	Wages ($/hour)		
Florida	35	518.1	19.2	14.8	4,500	15.2	129	10.16	25.0	-
California	80	392.8	14.6	4.9	3,800	12.8	48	10.81	22.6	3,211
New York	51	375.1	13.9	7.4	4,600	15.5	90	10.95	22.4	4,761
Texas	36	208.8	7.8	5.8	2,200	7.4	61	6.71	56.2	2,591
Massachusetts	33	171.7	6.4	5.2	2,200	7.4	67	9.48	31.1	-
Minnesota	14	96.3	3.6	6.9	1,300	4.4	93	8.50	39.7	-
New Jersey	25	56.0	2.1	2.2	700	2.4	28	6.50	37.3	2,000
Pennsylvania	25	54.7	2.0	2.2	600	2.0	24	8.20	37.8	3,833
Maryland	10	52.0	1.9	5.2	1,100	3.7	110	8.28	32.3	-
Virginia	11	51.0	1.9	4.6	1,000	3.4	91	8.20	52.5	2,500
Ohio	18	30.0	1.1	1.7	400	1.4	22	5.67	48.7	2,250
Colorado	14	15.2	0.6	1.1	200	0.7	14	8.00	25.7	-
Tennessee	8	13.3	0.5	1.7	100	0.3	13	6.00	38.3	1,000
Michigan	21	7.8	0.3	0.4	100	0.3	5	7.50	29.5	3,000
Utah	9	5.9	0.2	0.7	100	0.3	11	8.00	18.6	1,000
Illinois	26	(D)	-	-	1,750 *	5.9	67	-	-	-
Georgia	11	(D)	-	-	1,750 *	5.9	159	-	-	-
Missouri	11	(D)	-	-	375 *	1.3	34	-	-	1,067
Oklahoma	10	(D)	-	-	175 *	0.6	18	-	-	1,714
Oregon	9	(D)	-	-	175 *	0.6	19	-	-	2,286
Kansas	7	(D)	-	-	375 *	1.3	54	-	-	-
New Mexico	6	(D)	-	-	175 *	0.6	29	-	-	-
Wisconsin	5	(D)	-	-	175 *	0.6	35	-	-	-
Rhode Island	3	(D)	-	-	375 *	1.3	125	-	-	-

Source: 1992 Economic Census. The states are in descending order of shipments or establishments (if shipment data are missing for the majority). The symbol (D) appears when data are withheld to prevent disclosure of competitive information. States marked with (D) are sorted by number of establishments. A dash (-) indicates that the data element cannot be calculated; * indicates the midpoint of a range.

3861 - PHOTOGRAPHIC EQUIPMENT & SUPPLIES

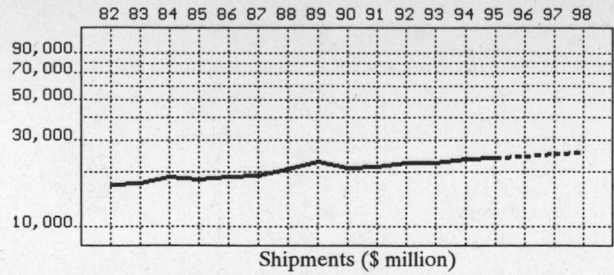

82 83 84 85 86 87 88 89 90 91 92 93 94 95 96 97 98

Shipments ($ million)

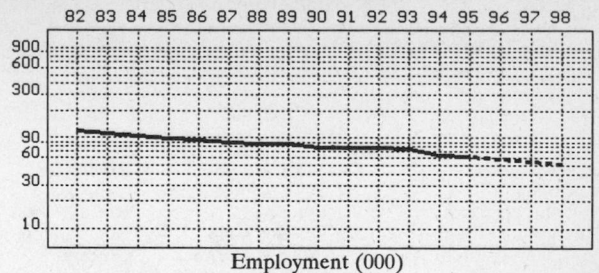

82 83 84 85 86 87 88 89 90 91 92 93 94 95 96 97 98

Employment (000)

GENERAL STATISTICS

| Year | Companies | Establishments | | Employment | | | Compensation | | Production ($ million) | | | |
		Total	with 20 or more employees	Total (000)	Production Workers (000)	Hours (Mil)	Payroll ($ mil)	Wages ($/hr)	Cost of Materials	Value Added by Manufacture	Value of Shipments	Capital Invest.
1982	723	795	289	119.3	64.1	123.7	3,193.1	11.78	5,859.7	10,859.5	17,037.5	752.6
1983				110.2	57.6	112.7	3,117.6	12.41	5,887.0	11,654.7	17,366.3	587.3
1984				104.0	53.7	108.4	3,137.8	12.92	5,682.4	12,960.9	18,701.9	665.4
1985				98.5	50.2	100.4	3,128.8	13.72	5,890.1	12,257.4	18,114.4	834.2
1986				94.6	47.6	95.0	2,870.2	13.00	6,110.5	12,355.9	18,580.4	697.0
1987	719	787	279	88.0	44.8	92.1	2,878.3	13.29	6,233.5	12,908.0	19,240.5	681.0
1988				87.5	43.9	92.8	2,963.4	13.11	6,638.0	14,223.2	20,545.8	809.7
1989		806	309	87.0	43.9	96.9	3,134.3	13.41	6,935.4	15,804.2	22,737.8	1,008.2
1990				79.3	41.2	91.8	2,937.4	13.69	6,439.2	14,527.2	21,018.2	1,008.6
1991				78.0	40.0	87.5	3,044.1	14.73	6,686.4	14,603.3	21,397.8	1,089.2
1992	831	904	264	77.5	39.4	90.7	3,069.3	14.65	7,058.7	14,885.4	22,149.8	808.1
1993				75.7	38.0	86.6	2,881.0	14.64	6,750.5	15,916.8	22,367.8	775.4
1994				64.8	34.8	77.7	2,716.6	15.80	7,011.4	16,057.0	23,367.6	755.3
1995				62.7P	31.7P	76.7P	2,844.8P	15.42P	7,157.9P	16,392.6P	23,855.9P	930.3P
1996				58.9P	29.6P	73.9P	2,821.8P	15.68P	7,314.5P	16,751.2P	24,377.9P	948.1P
1997				55.0P	27.5P	71.0P	2,798.8P	15.94P	7,471.2P	17,109.9P	24,899.9P	965.9P
1998				51.2P	25.5P	68.2P	2,775.9P	16.19P	7,627.8P	17,468.6P	25,421.9P	983.7P

Sources: 1982, 1987, 1992 *Economic Census*; *Annual Survey of Manufactures*, 83-86, 88-91, 93-94. Establishment counts for non-Census years are from *County Business Patterns*; establishment values for 83-84 are extrapolations. 'P's show projections by the editors. Industries reclassified in 87 will not have data for prior years.

INDICES OF CHANGE

| Year | Companies | Establishments | | Employment | | | Compensation | | Production ($ million) | | | |
		Total	with 20 or more employees	Total (000)	Production Workers (000)	Hours (Mil)	Payroll ($ mil)	Wages ($/hr)	Cost of Materials	Value Added by Manufacture	Value of Shipments	Capital Invest.
1982	87	88	109	154	163	136	104	80	83	73	77	93
1983				142	146	124	102	85	83	78	78	73
1984				134	136	120	102	88	81	87	84	82
1985				127	127	111	102	94	83	82	82	103
1986				122	121	105	94	89	87	83	84	86
1987	87	87	106	114	114	102	94	91	88	87	87	84
1988				113	111	102	97	89	94	96	93	100
1989		89	117	112	111	107	102	92	98	106	103	125
1990				102	105	101	96	93	91	98	95	125
1991				101	102	96	99	101	95	98	97	135
1992	100	100	100	100	100	100	100	100	100	100	100	100
1993				98	96	95	94	100	96	107	101	96
1994				84	88	86	89	108	99	108	105	93
1995				81P	80P	85P	93P	105P	101P	110P	108P	115P
1996				76P	75P	81P	92P	107P	104P	113P	110P	117P
1997				71P	70P	78P	91P	109P	106P	115P	112P	120P
1998				66P	65P	75P	90P	111P	108P	117P	115P	122P

Sources: Same as General Statistics. Values reflect change from the base year, 1992. Values above 100 mean greater than 92, values below 100 mean less than 92, and a value of 100 in the 82-91 or 93-98 period means same as 92. 'P's mark projections by the editors.

SELECTED RATIOS

For 1992	Avg. of All Manufact.	Analyzed Industry	Index	For 1992	Avg. of All Manufact.	Analyzed Industry	Index
Employees per Establishment	46	86	188	Value Added per Production Worker	122,353	377,802	309
Payroll per Establishment	1,332,320	3,395,243	255	Cost per Establishment	4,239,462	7,808,296	184
Payroll per Employee	29,181	39,604	136	Cost per Employee	92,853	91,080	98
Production Workers per Establishment	31	44	139	Cost per Production Worker	135,003	179,155	133
Wages per Establishment	734,496	1,469,862	200	Shipments per Establishment	8,100,800	24,501,991	302
Wages per Production Worker	23,390	33,725	144	Shipments per Employee	177,425	285,804	161
Hours per Production Worker	2,025	2,302	114	Shipments per Production Worker	257,966	562,178	218
Wages per Hour	11.55	14.65	127	Investment per Establishment	278,244	893,916	321
Value Added per Establishment	3,842,210	16,466,150	429	Investment per Employee	6,094	10,427	171
Value Added per Employee	84,153	192,070	228	Investment per Production Worker	8,861	20,510	231

Sources: Same as General Statistics. The 'Average of All Manufacturing' column represents the average of all manufacturing industries reported for the most recent complete year available. The Index shows the relationship between the Average and the Analyzed Industry. For example, 100 means that they are equal; 500 that the Analyzed Industry is five times the average; 50 means that the Analyzed Industry is half the national average. The abbreviation 'na' is used to show that data are 'not available'.

LEADING COMPANIES   Number shown: 75   Total sales ($ mil): 35,910   Total employment (000): 211.2

Company Name	Address				CEO Name	Phone	Co. Type	Sales ($ mil)	Empl. (000)
Xerox Corp	PO Box 1600	Stamford	CT	06904	Paul A Allaire	203-968-3000	P	17,837	87.6
Eastman Kodak Co	343 State St	Rochester	NY	14650	George MC Fisher	716-724-4000	P	13,557	96.3
Polaroid Corp	549 Technology Sq	Cambridge	MA	02139	I MacAllister Booth	617-386-2000	P	2,313	12.1
Graphic Controls Corp	PO Box 1271	Buffalo	NY	14240	Duane Hopper	716-853-7500	R	180	1.5
Anitec Imaging Products	PO Box 4444	Binghamton	NY	13902	Harry Lambroussis	607-774-3333	D	110	1.0
Document Management	6800 McCormick	Chicago	IL	60645	William J White	708-675-7600	D	100	1.2
YRJ Corp	5711 Hillcroft St	Houston	TX	77036	Yandall Rogers Jr	713-782-8580	R	100	0.8
Schoeller Technical Papers Inc	PO Box 250	Pulaski	NY	13142	David E Paulus	315-298-5133	S	95	0.3
Azon Corp	PO Box 290	Johnson City	NY	13790	William Bordages	607-797-2368	R	90	0.7
ILFORD Photo Corp	W 70 Century Rd	Paramus	NJ	07653	John Georges	201-265-6000	S	69•	0.1
LB Russell Chemicals Inc	14-33 31st Av	Long Island Ct	NY	11106	IM Ewig	718-721-8900	R	69	0.1
Arkwright Inc	538 Main St	Fiskeville	RI	02823	F Van Oudenhov	401-821-1000	S	66	0.3
PSC Inc	675 Basket Rd	Webster	NY	14580	L Michael Hone	716-265-1600	P	60	0.4
X-Rite Inc	3100 44th St SW	Grandville	MI	49418	Ted Thompson	616-534-7663	P	59	0.5
Copy Duplicating Products Inc	6636 Cedar Av S	Richfield	MN	55423	Jon Malinski	612-861-0555	R	56•	0.4
Concord Camera Corp	35 Mileed Way	Avenel	NJ	07001	Ira B Lambert	908-499-8280	P	55	0.2
CAI	550 W Northwest	Barrington	IL	60010	William L Owens	708-381-2400	D	50	0.4
CBM America Corp	10900 Wilshire Blv	Los Angeles	CA	90024	Z Nakagawa	310-209-1233	R	50•	<0.1
Da-Lite Screen Company Inc	PO Box 137	Warsaw	IN	46580	Rich Lundin	219-267-8101	R	50	0.4
Leaf Systems Inc	250 Turnpike Rd	Southborough	MA	01772	Robert Caspe	508-460-8300	R	50•	0.2
Minnesota Mining & Mfg	1999 Mount Read	Rochester	NY	14615	RE Soenen	716-458-2920	D	49•	0.4
CPAC Inc	2364 Leicester Rd	Leicester	NY	14481	T N Hendrickson	716-382-3223	P	44	0.2
Hughes-JVC Technology Corp	2310 Roble	Carlsbad	CA	92009	William Donaldson	619-929-5300	R	40	0.2
Panavision	18618 Oxnard St	Tarzana	CA	91356	John Farrand	818-881-1702	R	37•	0.3
Dietzgen Corp	250 Wille Rd	Des Plaines	IL	60018	Lawarence Kujovich	708-635-5200	R	32•	0.3
AFP Imaging Corp	250 Clearbrook Rd	Elmsford	NY	10523	Donald Rabinovitch	914-592-6100	P	30	0.2
Intern Communication Materials	PO Box 716	Connellsville	PA	15425	John Langley	412-628-1014	S	28	0.2
Kreonite Inc	PO Box 2099	Wichita	KS	67201	William K Oetting	316-263-1111	R	25	0.2
Omega Acquisition Corp	191 Shaeffer Av	Westminster	MD	21158	Charles Ezrine	410-857-6353	R	25	0.1
Graphic Enterprises of Ohio Inc	PO Box 3080	North Canton	OH	44720	Rich Jusseaume	216-494-9694	R	24	0.3
Quantel Inc	85 Old Kings N	Darien	CT	06820	Ken Ellis	203-656-3100	S	23	<0.1
AmPro Corp	525 J Rodes Blvd	Melbourne	FL	32934	David K Mutchler	407-254-3000	R	21•	0.2
Huey Co	10100 Franklin Av	Franklin Park	IL	60131	Huey C Shelton	708-671-6150	R	21	0.2
Lucht Inc	11201 Hampshire S	Bloomington	MN	55438	Stephen West	612-829-5444	R	21	0.2
Charles Beseler Co	1600 Lower Rd	Linden	NJ	07036	I Brightman	908-862-7999	R	20	0.3
Draper Shade and Screen	PO Box 425	Spiceland	IN	47385	L Pidgeon	317-987-7999	R	20	0.2
Photometrics Ltd	3440 E Britannia Dr	Tucson	AZ	85706	Robert Stevenson	602-889-9933	R	20•	0.1
Tamron Industries Inc	PO Box 388	Pt Washington	NY	11050	Hank Nagashima	516-484-8880	S	20	<0.1
Ballantyne of Omaha Inc	4350 McKinley St	Omaha	NE	68112	Ron H Echtenkamp	402-453-4444	S	18	0.1
Photo Control Corp	4800 Quebec Av N	Minneapolis	MN	55428	Leslie A Willig	612-537-3601	P	18	0.2
Optical Gaging Products Inc	850 Hudson Av	Rochester	NY	14621	Edward Polidor	716-544-0400	R	17	0.2
Tomoegawa	742 Glenn Av	Wheeling	IL	60090	Shigeru Yokoyama	708-541-3001	S	17•	0.1
Motion Analysis Systems	11633 Sorrento Val	San Diego	CA	92121	John Bagby	619-535-2908	D	16•	0.1
3M Visual Systems	130-5N 6801 River	Austin	TX	78726	C Agnew Meek	512-984-1800	D	15•	0.1
Leica Camera Inc	156 Ludlow Av	Northvale	NJ	07647	Roger Horn	201-767-7500	S	15	0.1
Minolta Corp	101 Williams Dr	Ramsey	NJ	07446	Hiro Fuji	201-825-4000	D	14	<0.1
NuArc Company Inc	6200 W Howard St	Niles	IL	60714	H Weisman	708-967-4400	R	14•	0.1
Rosco Laboratories Inc	36 Bush Av	Port Chester	NY	10573	Stan Miller	914-937-1300	R	14	0.1
Turner Bellows Inc	526 Child St	Rochester	NY	14606	Richard Pontarella	716-235-4456	R	14•	0.1
Kollmorgen Corp	9330 De Soto Av	Chatsworth	CA	91311	Ron Andersen	818-341-5151	D	12	<0.1
Nord Photo Engineering Inc	4800 Quebec Av N	Minneapolis	MN	55428	Leslie A Willig	612-537-7620	S	12	0.1
Coburn Corp	1650 Corporate W	Lakewood	NJ	08701	JW Coburn II	908-367-5511	R	11	0.1
Pinnacle Systems Inc	870 W Maude Av	Sunnyvale	CA	94086	Mark L Sanders	408-970-9787	P	10	<0.1
Equity Enterprises Inc	20 Squadron Blv	New City	NY	10936	Herbert Moelis	914-634-7676	R	10	<0.1
Kenro Corp	7001 Loisdale Rd	Springfield	VA	22150	Raymond Luca	703-971-1400	S	10	0.1
LogEtronics Corp	7001 Loisdale Rd	Springfield	VA	22150	Raymond J Luca	703-971-1400	S	10	0.1
Mole-Richardson Co	937 N Sycamore	Hollywood	CA	90038	WK Parker	213-851-0111	R	10	<0.1
PixelCraft Inc	PO Box 14467	Oakland	CA	94614	Albert E Sisto	510-562-2480	S	10•	<0.1
Stewart Filmscreen Corp	1161 W Sepulveda	Torrance	CA	90502	PH Stewart	310-326-1422	R	10•	<0.1
Microcircuit Engineering Corp	PO Box 570	Mount Holly	NJ	08060	Fred Cox	609-261-1400	R	9•	<0.1
Myou Video Corp	4783 Ruffner St	San Diego	CA	92111	David Stepp	619-268-1100	R	9	<0.1
Tiffen Manufacturing Corp	90 Oser Av	Hauppauge	NY	11788	Steven Tiffen	516-273-2500	R	9	0.1
Buhl Industries Inc	14-01 Maple Av	Fair Lawn	NJ	07410	Henry Kyhl	201-423-2800	R	9	0.1
Diazit Company Inc	US Rte 1	Youngsville	NC	27596	Robert Neeb	919-556-5188	R	8	<0.1
D/B Cameras	1600 E Valencia Dr	Fullerton	CA	92631	RF Holland	714-871-3020	D	8	<0.1
Enchanced Imaging Technologies	625 Alaska Av	Torrance	CA	90503	Drew Hofmman	805-987-8801	D	8•	<0.1
Allen Products Co	180 Wampus Ln	Milford	CT	06460	Ronald L Bailer	203-874-2563	R	8	<0.1
Mold-In Graphics Systems Inc	PO Box 1650	Clarkdale	AZ	86324	Michael Stevenson	602-634-8838	R	7•	<0.1
Chromaline Corp	4832 Grand Av	Duluth	MN	55807	Thomas L Erickson	218-628-2217	P	7	<0.1
Cinema Products Corp	3211 S La Cienega	Los Angeles	CA	90016	Ronald Lenney	310-836-7991	R	6•	<0.1
Leedal Inc	1918 S Prairie Av	Chicago	IL	60616	SL Levin	312-842-6588	R	6•	<0.1
Lowell-Light Manufacturing Inc	140 58th St	Brooklyn	NY	11220	Marvin Seligman	718-921-0600	R	6	<0.1
Neumade Products Corp	PO Box 5001	Norwalk	CT	06856	RN Jones	203-866-7600	R	6	<0.1
Photo-Sonics Inc	820 S Mariposa St	Burbank	CA	91506	John Kiel	818-842-2141	R	6	<0.1
National Research & Chemical	15600 New Century	Gardena	CA	90248	JF Atwill	310-515-1700	R	6	<0.1

Source: Ward's Business Directory of U.S. Private and Public Companies, Volumes 1 and 2, 1996. The company type code used is as follows: P - Public, R - Private, S - Subsidiary, D - Division, J - Joint Venture, A - Affiliate, G - Group. Sales are in millions of dollars, employees are in thousands. An asterisk (*) indicates an estimated sales volume. The symbol < stands for 'less than'. Company names and addresses are truncated, in some cases, to fit into the available space.

MATERIALS CONSUMED

Material	Quantity	Delivered Cost ($ million)
Materials, ingredients, containers, and supplies	(X)	6,119.4
Metal stampings	(X)	166.1
All other fabricated metal products (except forgings)	(X)	62.9
Iron and steel castings (rough and semifinished)	(X)	6.2
Aluminum and aluminum-base alloy castings (rough and semifinished)	(X)	87.0
Other nonferrous castings (rough and semifinished)	(X)	1.5
Steel shapes and forms	(X)	29.0
Nonferrous shapes and forms	(X)	12.7
Fractional horsepower electric motors (less than 1 hp)	(X)	107.6
Resistors, capacitors, transformers, electron tubes, semiconductors, and other electronic components	(X)	415.1
Paper and paperboard products (except packaging, photographic)	(X)	195.6
Photographic base papers	(X)	308.2
Paperboard containers, boxes, and corrugated paperboard	(X)	187.8
Inorganic chemicals, nec, except silver nitrate and prepared photographic chemicals	(X)	126.9
Silver nitrate	(X)	265.5
Synthetic organic chemicals, except prepared photographic chemicals	(X)	141.3
Prepared photographic chemicals	(X)	1,253.0
Plastics resins consumed in the form of granules, pellets, powders, liquids, etc.	(X)	220.2
Unsupported plastics film and sheet for photographic, mimeographic, X-ray, etc.	(X)	171.9
Fabricated plastics products (except gaskets, hoses, and belting)	(X)	176.6
Photographic and projection lenses and prisms	(X)	48.0
Light sensitive films and papers	(X)	281.4
All other materials and components, parts, containers, and supplies	(X)	1,155.2
Materials, ingredients, containers, and supplies, nsk	(X)	699.6

Source: 1992 *Economic Census.* Explanation of symbols used: (D): Withheld to avoid disclosure of competitive data; na: Not available; (S): Withheld because statistical norms were not met; (X): Not applicable; (Z): Less than half the unit shown; nec: Not elsewhere classified; nsk: Not specified by kind; - : zero; * : 10-19 percent estimated; ** : 20-29 percent estimated.

PRODUCT SHARE DETAILS

Product or Product Class	% Share	Product or Product Class	% Share
Photographic equipment and supplies	100.00	lengths of still picture roll film other than graphic arts film	(D)
Photographic equipment, still picture	4.86	Photographic presensitized printing plates, unexposed	6.55
Still process cameras for photoengraving and photolithography, including value of stands and attachments when shipped together	8.91	Phototypesetting and imagesetting film	2.10
Other still cameras, excluding photocopying, microfilming, blueprinting, and whiteprinting	2.10	Photographic sensitized graphic arts film	(D)
Projectors, except rear screen viewers	6.05	Other photographic sensitized film, plates, and slides, including microfilm and motion picture film	(D)
Still picture commercial-type processing equipment for film	13.86	Photographic sensitized film and plates, silver halide type (except x-ray), nsk	0.17
All other still picture commercial-type processing equipment (developing machines, motor-operated print washers and driers, etc.)	12.58	Sensitized photographic paper and cloth, silver halide type	(D)
Other still picture equipment, parts, attachments, and accessories, excluding projection screens	50.97	Sensitized photographic film, plates, paper, and cloth, other than silver halide type	7.06
Still picture equipment, nsk	5.53	Dry diazo print paper	21.60
Photocopying equipment (includes diffusion transfer, dye transfer, electrostatic, light and heat sensitive types, etc.)	(D)	Diazo paper reproducibles	(D)
Microfilming, blueprinting, and whiteprinting equipment	1.43	Diazo type film, except microfilm/microfiche	2.31
Microfilming equipment (including microfiche), cameras, including computer output	16.91	Diazo microfilm/microfiche	2.96
Microfilming equipment (including microfiche), microfilm reader-printers	10.23	All other diazo materials	(D)
Other microfilming equipment (including microfiche)	45.51	Photographic diffusion transfer materials (including both imager and receiver sheets)	(D)
Blueprinting and whiteprinting (direct process type) equipment	27.32	Photographic off-press color proofing media	(D)
Motion picture equipment	1.12	Photographic letterpress plates	(D)
8 mm and 16 mm motion picture equipment	52.96	Photographic flexographic plates	(D)
Projection screens (for motion picture and/or still projection)	24.18	Other photographic types, including sensitized film, plates, paper, and cloth for all other processes (including blueprint types)	(D)
Motion picture processing equipment, all types, excluding motion picture still type equipment and interchangeable types	21.45	Sensitized photographic film, plates, paper, and cloth, other than silver halide type, nsk	3.44
Motion picture equipment, nsk	1.41	Prepared photographic chemicals	9.80
Photographic sensitized film and plates, silver halide type (except x-ray)	29.62	Prepared chemicals, office copy toners	52.01
Photographic sensitized sheet film, pack film, and long		Prepared photographic chemicals	37.53
		Prepared plate chemicals	2.21
		Other prepared photographic chemicals	7.70
		Prepared photographic chemicals, nsk	0.55
		X-ray film and plates	(D)
		Photographic equipment and supplies, nsk	4.27

Source: 1992 *Economic Census.* The values shown are percent of total shipments in an industry. Values of indented subcategories are summed in the main heading. The symbol (D) appears when data are withheld to prevent disclosure of competitive information. The abbreviation nsk stands for 'not specified by kind' and nec for 'not elsewhere classified'.

INPUTS AND OUTPUTS FOR PHOTOGRAPHIC EQUIPMENT & SUPPLIES

Economic Sector or Industry Providing Inputs	%	Sector	Economic Sector or Industry Buying Outputs	%	Sector
Imports	25.4	Foreign	Gross private fixed investment	27.5	Cap Inv
Semiconductors & related devices	12.0	Manufg.	Personal consumption expenditures	12.7	
Wholesale trade	6.3	Trade	Exports	12.0	Foreign
Industrial inorganic chemicals, nec	5.6	Manufg.	S/L Govt. purch., health & hospitals	3.4	S/L Govt
Advertising	3.9	Services	Photofinishing labs, commercial photography	3.2	Services
Photographic equipment & supplies	3.8	Manufg.	Federal Government purchases, national defense	3.1	Fed Govt
Nonferrous rolling & drawing, nec	3.5	Manufg.	Banking	2.9	Fin/R.E.
Electronic components nec	2.8	Manufg.	S/L Govt. purch., higher education	2.9	S/L Govt
Miscellaneous plastics products	2.8	Manufg.	S/L Govt. purch., other general government	2.3	S/L Govt
Banking	2.3	Fin/R.E.	S/L Govt. purch., elem. & secondary education	2.1	S/L Govt
Noncomparable imports	2.2	Foreign	Photographic equipment & supplies	1.8	Manufg.
Plastics materials & resins	2.1	Manufg.	Hospitals	1.7	Services
Fabricated metal products, nec	1.6	Manufg.	Portrait, photographic studios	1.5	Services
Cyclic crudes and organics	1.2	Manufg.	Federal Government purchases, nondefense	1.5	Fed Govt
Paper coating & glazing	1.2	Manufg.	Commercial printing	1.2	Manufg.
Paper mills, except building paper	1.2	Manufg.	Insurance carriers	1.1	Fin/R.E.
Electric services (utilities)	1.1	Util.	Legal services	1.1	Services
Metal stampings, nec	1.0	Manufg.	Radio & TV broadcasting	0.8	Util.
Paperboard containers & boxes	0.9	Manufg.	Retail trade, except eating & drinking	0.8	Trade
Air transportation	0.8	Util.	Wholesale trade	0.8	Trade
Communications, except radio & TV	0.8	Util.	Colleges, universities, & professional schools	0.7	Services
Aluminum rolling & drawing	0.7	Manufg.	Doctors & dentists	0.7	Services
Motors & generators	0.7	Manufg.	Motion pictures	0.7	Services
Motor freight transportation & warehousing	0.7	Util.	Newspapers	0.6	Manufg.
Business services nec	0.7	Services	Lithographic platemaking & services	0.5	Manufg.
Maintenance of nonfarm buildings nec	0.6	Constr.	Religious organizations	0.5	Services
Railroads & related services	0.6	Util.	Typesetting	0.4	Manufg.
Eating & drinking places	0.6	Trade	Insurance agents, brokers, & services	0.4	Fin/R.E.
Business/professional associations	0.6	Services	Real estate	0.4	Fin/R.E.
Detective & protective services	0.6	Services	Detective & protective services	0.4	Services
Aluminum castings	0.5	Manufg.	Miscellaneous repair shops	0.4	Services
Gaskets, packing & sealing devices	0.5	Manufg.	Communications, except radio & TV	0.3	Util.
Petroleum refining	0.5	Manufg.	Credit agencies other than banks	0.3	Fin/R.E.
Gas production & distribution (utilities)	0.5	Util.	Business services nec	0.3	Services
Blast furnaces & steel mills	0.4	Manufg.	Medical & health services, nec	0.3	Services
Primary nonferrous metals, nec	0.4	Manufg.	S/L Govt. purch., other education & libraries	0.3	S/L Govt
Real estate	0.4	Fin/R.E.	S/L Govt. purch., public assistance & relief	0.3	S/L Govt
Equipment rental & leasing services	0.4	Services	Crude petroleum & natural gas	0.2	Mining
Job training & related services	0.4	Services	Motor freight transportation & warehousing	0.2	Util.
Legal services	0.4	Services	Security & commodity brokers	0.2	Fin/R.E.
Alkalies & chlorine	0.3	Manufg.	Accounting, auditing & bookkeeping	0.2	Services
Fabricated rubber products, nec	0.3	Manufg.	Elementary & secondary schools	0.2	Services
Machinery, except electrical, nec	0.3	Manufg.	Engineering, architectural, & surveying services	0.2	Services
Metal foil & leaf	0.3	Manufg.	Labor, civic, social, & fraternal associations	0.2	Services
Screw machine and related products	0.3	Manufg.	Management & consulting services & labs	0.2	Services
Water transportation	0.3	Util.	S/L Govt. purch., correction	0.2	S/L Govt
Credit agencies other than banks	0.3	Fin/R.E.	S/L Govt. purch., natural resource & recreation.	0.2	S/L Govt
Automotive repair shops & services	0.3	Services	S/L Govt. purch., police	0.2	S/L Govt
Computer & data processing services	0.3	Services	Book printing	0.1	Manufg.
Hotels & lodging places	0.3	Services	Electronic computing equipment	0.1	Manufg.
Management & consulting services & labs	0.3	Services	Electric services (utilities)	0.1	Util.
Coal	0.2	Mining	Personnel supply services	0.1	Services
Chemical preparations, nec	0.2	Manufg.	Social services, nec	0.1	Services
Felt goods, nec	0.2	Manufg.			
Lighting fixtures & equipment	0.2	Manufg.			
Miscellaneous fabricated wire products	0.2	Manufg.			
Optical instruments & lenses	0.2	Manufg.			
Special dies & tools & machine tool accessories	0.2	Manufg.			
Sanitary services, steam supply, irrigation	0.2	Util.			
Accounting, auditing & bookkeeping	0.2	Services			
Engineering, architectural, & surveying services	0.2	Services			
Services to dwellings & other buildings	0.2	Services			
U.S. Postal Service	0.2	Gov't			
Electric lamps	0.1	Manufg.			
Lubricating oils & greases	0.1	Manufg.			
Manifold business forms	0.1	Manufg.			
Primary batteries, dry & wet	0.1	Manufg.			
Insurance carriers	0.1	Fin/R.E.			
Royalties	0.1	Fin/R.E.			

Source: Benchmark Input-Output Accounts for the U.S. Economy, 1982, U.S. Department of Commerce, Washington, D.C., July 1991. Data, as reported in the source, are organized by the 1977 SIC structure in use in 1982 but have been matched, as closely as is possible, to the 1987 SIC structure used in this book.

OCCUPATIONS EMPLOYED BY SIC 386 - PHOTOGRAPHIC EQUIPMENT AND SUPPLIES

Occupation	% of Total 1994	Change to 2005	Occupation	% of Total 1994	Change to 2005
Assemblers, fabricators, & hand workers nec	5.3	-17.6	Engineering, mathematical, & science managers	1.6	-6.4
Engineering technicians & technologists nec	4.6	-17.6	Financial managers	1.4	-17.6
Blue collar worker supervisors	3.5	-30.8	Engineers nec	1.4	-1.1
Inspectors, testers, & graders, precision	3.2	-42.3	Industrial machinery mechanics	1.4	11.2
Chemical equipment controllers, operators	2.9	-25.9	Traffic, shipping, & receiving clerks	1.3	-20.8
Management support workers nec	2.8	-17.6	Bookkeeping, accounting, & auditing clerks	1.3	-38.2
Sales & related workers nec	2.7	-17.6	Industrial production managers	1.3	-17.6
Machine operators nec	2.6	-41.9	Production, planning, & expediting clerks	1.3	-17.7
Electrical & electronic equipment assemblers	2.5	-17.6	Science & mathematics technicians	1.2	-17.6
Secretaries, ex legal & medical	2.4	-25.0	Stock clerks	1.2	-33.0
Electrical & electronic technicians,technologists	2.4	-17.6	Chemical engineers	1.1	-17.7
Freight, stock, & material movers, hand	2.3	-34.1	Managers & administrators nec	1.1	-17.6
Electrical & electronics engineers	2.3	-12.3	Machinists	1.1	-38.2
Packaging & filling machine operators	2.1	-17.6	Precision workers nec	1.1	-25.9
General managers & top executives	2.1	-21.8	Adjustment clerks	1.1	-1.1
Electrical & electronic assemblers	1.9	-25.8	Systems analysts	1.1	31.9
Marketing, advertising, & PR managers	1.9	-17.6	Machine assemblers	1.0	-25.9
Coating, painting, & spraying machine workers	1.9	-17.6	Hand packers & packagers	1.0	-29.4
Mechanical engineers	1.8	-9.3			

Source: *Industry-Occupation Matrix*, Bureau of Labor Statistics. These data relate to one or more 3-digit SIC industry groups rather than to a single 4-digit SIC. The change reported for each occupation to the year 2005 is a percent of growth or decline as estimated by the Bureau of Labor Statistics. The abbreviation nec stands for 'not elsewhere classified'.

LOCATION BY STATE AND REGIONAL CONCENTRATION

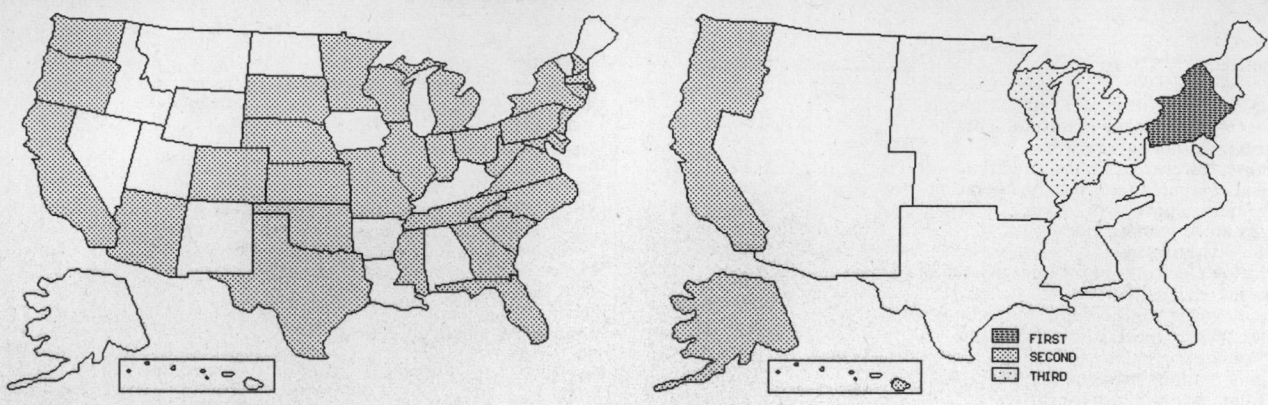

FIRST
SECOND
THIRD

INDUSTRY DATA BY STATE

State	Establish-ments	Shipments			Employment				Cost as % of Shipments	Investment per Employee ($)
		Total ($ mil)	% of U.S.	Per Establ.	Total Number	% of U.S.	Per Establ.	Wages ($/hour)		
Massachusetts	50	1,858.6	8.4	37.2	7,800	10.1	156	15.66	41.1	-
California	142	743.0	3.4	5.2	6,800	8.8	48	10.66	49.5	3,779
New Jersey	67	736.7	3.3	11.0	2,900	3.7	43	15.33	38.9	10,828
Illinois	76	464.9	2.1	6.1	3,000	3.9	39	13.00	46.3	2,067
Pennsylvania	40	380.3	1.7	9.5	1,400	1.8	35	13.79	33.3	6,571
Indiana	14	243.2	1.1	17.4	1,300	1.7	93	11.15	36.1	3,846
Texas	34	101.0	0.5	3.0	700	0.9	21	9.70	45.8	5,000
Missouri	20	99.3	0.4	5.0	700	0.9	35	10.86	33.2	5,429
Michigan	27	95.1	0.4	3.5	700	0.9	26	10.67	49.8	3,143
Ohio	23	88.0	0.4	3.8	600	0.8	26	11.17	34.3	4,833
Georgia	23	78.3	0.4	3.4	500	0.6	22	8.33	39.1	5,400
Virginia	23	78.2	0.4	3.4	400	0.5	17	9.20	56.6	4,250
Washington	17	77.3	0.3	4.5	400	0.5	24	11.00	38.8	5,250
Wisconsin	20	49.5	0.2	2.5	400	0.5	20	9.25	32.7	4,250
Florida	26	40.8	0.2	1.6	200	0.3	8	14.00	28.9	-
Maryland	9	36.3	0.2	4.0	200	0.3	22	12.00	47.9	-
South Carolina	9	28.1	0.1	3.1	300	0.4	33	8.00	54.1	1,000
Tennessee	10	25.2	0.1	2.5	200	0.3	20	7.00	40.1	2,500
Arizona	10	8.9	0.0	0.9	100	0.1	10	12.00	29.2	5,000
New York	112	(D)	-	-	37,500 *	48.4	335	-	-	-
Connecticut	18	(D)	-	-	375 *	0.5	21	-	-	7,200
Minnesota	18	(D)	-	-	375 *	0.5	21	-	-	5,067
North Carolina	16	(D)	-	-	1,750 *	2.3	109	-	-	-
Colorado	12	(D)	-	-	3,750 *	4.8	313	-	-	-
Oregon	11	(D)	-	-	375 *	0.5	34	-	-	-
Oklahoma	10	(D)	-	-	375 *	0.5	38	-	-	-
New Hampshire	8	(D)	-	-	375 *	0.5	47	-	-	-
Kansas	4	(D)	-	-	175 *	0.2	44	-	-	-
Mississippi	3	(D)	-	-	175 *	0.2	58	-	-	-
Rhode Island	3	(D)	-	-	175 *	0.2	58	-	-	-
West Virginia	3	(D)	-	-	375 *	0.5	125	-	-	-
Nebraska	1	(D)	-	-	175 *	0.2	175	-	-	-
South Dakota	1	(D)	-	-	175 *	0.2	175	-	-	-

Source: 1992 *Economic Census*. The states are in descending order of shipments or establishments (if shipment data are missing for the majority). The symbol (D) appears when data are withheld to prevent disclosure of competitive information. States marked with (D) are sorted by number of establishments. A dash (-) indicates that the data element cannot be calculated; * indicates the midpoint of a range.

3873 - WATCHES, CLOCKS, & WATCHCASES

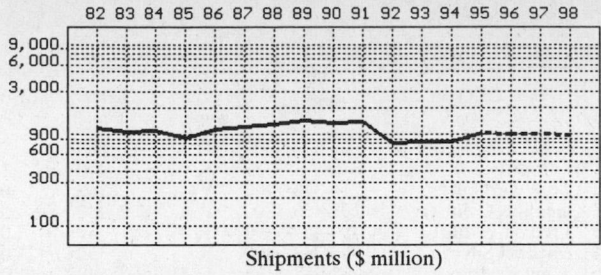

Shipments ($ million)

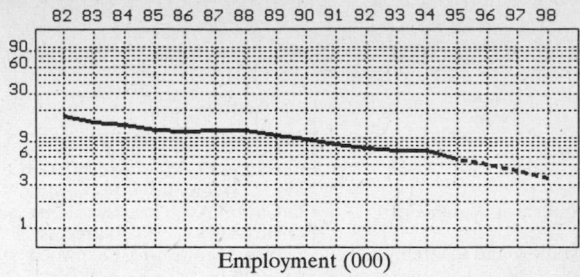

Employment (000)

GENERAL STATISTICS

| Year | Com-panies | Establishments | | Employment | | | Compensation | | Production ($ million) | | | |
		Total	with 20 or more employees	Total (000)	Production Workers (000)	Hours (Mil)	Payroll ($ mil)	Wages ($/hr)	Cost of Materials	Value Added by Manufacture	Value of Shipments	Capital Invest.
1982	227	237	87	16.8	12.3	22.4	248.1	6.71	688.2	483.5	1,187.6	14.6
1983				14.6	10.7	30.1	227.3	4.59	582.2	433.4	1,068.2	15.4
1984				13.4	10.3	19.7	223.7	7.07	573.9	531.5	1,094.6	12.1
1985				11.8	8.7	17.1	210.4	7.40	485.4	418.6	912.1	15.8
1986				11.6	8.5	16.8	222.7	7.61	635.8	509.7	1,147.6	28.6
1987	213	218	74	11.8	9.0	17.8	222.9	7.83	637.6	584.6	1,220.9	24.4
1988				12.1	9.0	17.8	230.9	7.84	676.6	644.7	1,295.1	16.6
1989		182	64	10.5	7.5	15.9	220.0	8.06	757.4	681.4	1,448.0	18.8
1990				9.4	6.9	14.7	218.3	8.69	693.7	665.4	1,360.2	20.4
1991				8.4	5.8	12.7	206.7	8.55	713.1	645.7	1,377.4	11.8
1992	179	180	52	7.6	5.6	11.0	171.9	8.35	382.0	423.5	811.6	22.3
1993				7.1	5.3	10.6	177.1	8.42	399.9	438.8	842.9	15.8
1994				7.1	5.0	11.3	184.6	8.38	375.9	455.6	822.6	16.7
1995				5.8P	4.1P	8.4P	179.5P	9.20P	470.6P	570.4P	1,029.9P	18.8P
1996				5.0P	3.5P	7.2P	174.8P	9.42P	464.6P	563.1P	1,016.7P	18.9P
1997				4.3P	3.0P	6.0P	170.0P	9.64P	458.6P	555.8P	1,003.5P	19.0P
1998				3.5P	2.4P	4.8P	165.3P	9.86P	452.5P	548.5P	990.3P	19.1P

Sources: 1982, 1987, 1992 *Economic Census*; *Annual Survey of Manufactures*, 83-86, 88-91, 93-94. Establishment counts for non-Census years are from *County Business Patterns*; establishment values for 83-84 are extrapolations. 'P's show projections by the editors. Industries reclassified in 87 will not have data for prior years.

INDICES OF CHANGE

| Year | Com-panies | Establishments | | Employment | | | Compensation | | Production ($ million) | | | |
		Total	with 20 or more employees	Total (000)	Production Workers (000)	Hours (Mil)	Payroll ($ mil)	Wages ($/hr)	Cost of Materials	Value Added by Manufacture	Value of Shipments	Capital Invest.
1982	127	132	167	221	220	204	144	80	180	114	146	65
1983				192	191	274	132	55	152	102	132	69
1984				176	184	179	130	85	150	126	135	54
1985				155	155	155	122	89	127	99	112	71
1986				153	152	153	130	91	166	120	141	128
1987	119	121	142	155	161	162	130	94	167	138	150	109
1988				159	161	162	134	94	177	152	160	74
1989		101	123	138	134	145	128	97	198	161	178	84
1990				124	123	134	127	104	182	157	168	91
1991				111	104	115	120	102	187	152	170	53
1992	100	100	100	100	100	100	100	100	100	100	100	100
1993				93	95	96	103	101	105	104	104	71
1994				93	89	103	107	100	98	108	101	75
1995				76P	73P	76P	104P	110P	123P	135P	127P	84P
1996				66P	63P	65P	102P	113P	122P	133P	125P	85P
1997				56P	53P	54P	99P	115P	120P	131P	124P	85P
1998				46P	43P	43P	96P	118P	118P	130P	122P	86P

Sources: Same as General Statistics. Values reflect change from the base year, 1992. Values above 100 mean greater than 92, values below 100 mean less than 92, and a value of 100 in the 82-91 or 93-98 period means same as 92. 'P's mark projections by the editors.

SELECTED RATIOS

For 1992	Avg. of All Manufact.	Analyzed Industry	Index	For 1992	Avg. of All Manufact.	Analyzed Industry	Index
Employees per Establishment	46	42	92	Value Added per Production Worker	122,353	75,625	62
Payroll per Establishment	1,332,320	955,000	72	Cost per Establishment	4,239,462	2,122,222	50
Payroll per Employee	29,181	22,618	78	Cost per Employee	92,853	50,263	54
Production Workers per Establishment	31	31	99	Cost per Production Worker	135,003	68,214	51
Wages per Establishment	734,496	510,278	69	Shipments per Establishment	8,100,800	4,508,889	56
Wages per Production Worker	23,390	16,402	70	Shipments per Employee	177,425	106,789	60
Hours per Production Worker	2,025	1,964	97	Shipments per Production Worker	257,966	144,929	56
Wages per Hour	11.55	8.35	72	Investment per Establishment	278,244	123,889	45
Value Added per Establishment	3,842,210	2,352,778	61	Investment per Employee	6,094	2,934	48
Value Added per Employee	84,153	55,724	66	Investment per Production Worker	8,861	3,982	45

Sources: Same as General Statistics. The 'Average of All Manufacturing' column represents the average of all manufacturing industries reported for the most recent complete year available. The Index shows the relationship between the Average and the Analyzed Industry. For example, 100 means that they are equal; 500 that the Analyzed Industry is five times the average; 50 means that the Analyzed Industry is half the national average. The abbreviation 'na' is used to show that data are 'not available'.

LEADING COMPANIES   Number shown: 32   Total sales ($ mil): 1,759   Total employment (000): 14.0

Company Name	Address				CEO Name	Phone	Co. Type	Sales ($ mil)	Empl. (000)
Timex Corp	PO Box 310	Waterbury	CT	06762	C Michael Jacobi	203-573-5000	R	850	7.5
Intermatic Inc	Intermatic Plz	Spring Grove	IL	60081	Lee Vinyard	815-675-2321	R	170	1.5
Fossil Inc	2280 N Greenville	Richardson	TX	75082	Tom Kartsotis	214-348-7400	P	162	0.4
North American Watch Corp	125 Chubb Av	Lyndhurst	NJ	07071	Gedalio Grinberg	201-460-4800	P	142	0.5
General Time Corp	PO Box 4125	Norcross	GA	30091	Fred M Pistilli	404-447-5300	R	59•	0.7
Harris Mallow and Contempra	651 New Hampshire	Lakewood	NJ	08701	Michael Silver	908-363-9400	R	50•	0.3
LINDEN	31 E 28th St	New York	NY	10016	Tony Rodriguez	212-255-5133	D	40•	0.4
Ingraham Time Products	PO Box 1609	Laurinburg	NC	28352	Ralph J Ronalter	919-276-3101	D	40	0.6
Benrus Watch Company Inc	1550 W Carroll Av	Chicago	IL	60607	IW Wein	312-243-3300	R	35	0.2
Sharp International Corp	484 Sunrise Hwy	Rockville Ct	NY	11570	Herbert Spitz	516-536-1600	R	30	0.1
Severin Montres Ltd	3 Mason St	Irvine	CA	92718	Severin Wunderman	714-472-0900	R	28•	0.3
Gruen Marketing Corp	150 Susquehanna	Exeter	PA	18643	Ed Mangiafico	717-655-2111	R	25	0.3
Jules Jurgensen and Rhapsody	101 W City Av	Bala Cynwyd	PA	19004	Morton Clayman	215-667-3500	R	20	0.1
Ridgeway Clock Co	PO Box 1371	Pulaski	VA	24301	Bernard C Wampler	703-980-8990	S	20	0.4
Emperor Clock Co	Emperor Industrial	Fairhope	AL	36532	Robert H Taupeka	334-928-2316	R	13	0.2
Helbros Watches Inc	101 W City Av	Bala Cynwyd	PA	19106	Alan Turin	215-667-3500	R	9•	<0.1
Belair Watch Corp	1995 Swarthmore	Lakewood	NJ	08701	Ernest Grunwald	908-905-0100	R	9•	<0.1
Lamontre Co	10-10 44th Av	Long Island Ct	NY	11101	Gary Latin	718-361-2520	R	8•	<0.1
Verdin Co	444 Reading Rd	Cincinnati	OH	45202	RJ Verdin Jr	513-241-4010	R	8	<0.1
Belove and Arienti Company Inc	PO Box 450417	Sunrise	FL	33345	Paul Ziegler	305-846-7540	R	7•	<0.1
Theo R Schwalm Inc	PO Box 4393	Lancaster	PA	17603	John R Gockley	717-397-3651	R	5	<0.1
Selco Custom Time Corp	8909 E 21st St	Tulsa	OK	74129	Larry J Abels	918-622-6100	R	5	<0.1
American Clock Maker	PO Box 326	Clintonville	WI	54929	Gregory H Smith	715-823-5101	R	4	<0.1
Cypress/IC Designs Inc	12020 113th Av NE	Kirkland	WA	98034	John Torod	206-821-9202	S	4•	<0.1
Franklin Instrument Company	PO Box 2949	Warminster	PA	18974	RM Fischer	215-355-7942	R	4	<0.1
Accusplit Inc	2290-A Ringwood	San Jose	CA	95131	WR Sutton	408-432-8228	R	3	<0.1
Chelsea Clock Company Inc	284 Everett Av	Chelsea	MA	02150	Richard F Leavitt	617-884-0250	R	3	<0.1
Pedre Company Inc	29 W 35th St	New York	NY	10016	P Gunshor	212-868-2935	R	2•	<0.1
California Clock Co	PO Box 9901	Fountain Val	CA	92708	W Young	714-545-4321	R	1	<0.1
Emdur Metal Products Inc	PO Box 1087	Camden	NJ	08101		609-541-1100	R	1	<0.1
Flagtime Inc	312 E 30th St	New York	NY	10016	Klaus Kurzina	212-679-1001	R	1	<0.1
Labs International Inc	225 Lafayette St	New York	NY	10012	Tibor Kalman	212-243-0082	D	1•	<0.1

Source: Ward's Business Directory of U.S. Private and Public Companies, Volumes 1 and 2, 1996. The company type code used is as follows: P - Public, R - Private, S - Subsidiary, D - Division, J - Joint Venture, A - Affiliate, G - Group. Sales are in millions of dollars, employees are in thousands. An asterisk (•) indicates an estimated sales volume. The symbol < stands for 'less than'. Company names and addresses are truncated, in some cases, to fit into the available space.

MATERIALS CONSUMED

Material	Quantity	Delivered Cost ($ million)
Materials, ingredients, containers, and supplies	(X)	310.0
Domestic watch movements or modules, without balance wheel and hairspring	(X)	16.6
Imported watch movements or modules, with balance wheel and hairspring	(X)	25.4
Domestic precious metal or precious metal clad watchcases	(X)	4.6
Imported (not made in the United States) precious metal or precious metal clad watchcases	(X)	6.2
Domestic watch parts	(X)	19.3
Imported watch parts	(X)	2.6
Watchbands	(X)	3.2
Face crystals	(X)	0.6
Precious metals (gold, platinum, etc.), all forms	(X)	11.5
Batteries, primary	(X)	0.8
Plastics resins consumed in the form of granules, pellets, powders, liquids, etc.	(X)	15.2
Metal bolts, nuts, screws, washers, rivets, and other screw machine products	(X)	2.9
Other fabricated metal products (except castings and forgings)	(X)	3.9
Paperboard containers, boxes, and corrugated paperboard	(X)	15.2
All other materials and components, parts, containers, and supplies	(X)	108.5
Materials, ingredients, containers, and supplies, nsk	(X)	73.6

Source: 1992 Economic Census. Explanation of symbols used: (D): Withheld to avoid disclosure of competitive data; na: Not available; (S): Withheld because statistical norms were not met; (X): Not applicable; (Z): Less than half the unit shown; nec: Not elsewhere classified; nsk: Not specified by kind; - : zero; • : 10-19 percent estimated; •• : 20-29 percent estimated.

PRODUCT SHARE DETAILS

Product or Product Class	% Share	Product or Product Class	% Share
Watches, clocks, watchcases, and parts	100.00	Other household clocks, complete, excluding alarm	6.51
Watches, watchcases, movements or modules, and watch parts	16.35	Household timing mechanisms, excluding time recording and time stamp machines	19.46
Watches	72.12	Commercial timing mechanisms, excluding time recording and time stamp machines	9.07
Watchcases, movements or modules, and watch parts	27.23	Other timing mechanisms, including military, excluding time recording and time stamp machines	0.42
Watches, watchcases, movements or modules, and watch parts, nsk	0.65	Timers and switch clocks with clock or watch movements or modules having dials or displays for telling time of day	1.95
Clocks, timing mechanisms, time switches, clock movements, clock cases, and parts	69.34	Clock movements and modules, complete	6.05
Clocks, alarm, excluding clock timers and timing mechanisms	9.31	Other clock parts (except timing motors)	1.64
Household clocks, wall, excluding alarm	16.78	Clocks, timing mechanisms, time switches, clock movements, clock cases, and parts, nsk	2.67
All other household clocks, including chime and strike, desk, mantel, etc., excluding alarm	26.14	Watches, clocks, watchcases, and parts, nsk	14.30

Source: 1992 *Economic Census.* The values shown are percent of total shipments in an industry. Values of indented subcategories are summed in the main heading. The symbol (D) appears when data are withheld to prevent disclosure of competitive information. The abbreviation nsk stands for 'not specified by kind' and nec for 'not elsewhere classified'.

INPUTS AND OUTPUTS FOR WATCHES, CLOCKS, & PARTS

Economic Sector or Industry Providing Inputs	%	Sector	Economic Sector or Industry Buying Outputs	%	Sector
Imports	49.9	Foreign	Personal consumption expenditures	66.9	
Watches, clocks, & parts	19.9	Manufg.	Watches, clocks, & parts	16.7	Manufg.
Wholesale trade	6.6	Trade	Exports	6.0	Foreign
Miscellaneous plastics products	3.2	Manufg.	Household laundry equipment	2.3	Manufg.
Semiconductors & related devices	2.5	Manufg.	Federal Government purchases, national defense	1.9	Fed Govt
Metal stampings, nec	1.7	Manufg.	Household cooking equipment	1.4	Manufg.
Nonferrous rolling & drawing, nec	1.3	Manufg.	Motor vehicles & car bodies	1.0	Manufg.
Advertising	1.2	Services	Water transportation	0.4	Util.
Plating & polishing	0.9	Manufg.	S/L Govt. purch., higher education	0.4	S/L Govt
Electric services (utilities)	0.7	Util.	S/L Govt. purch., public assistance & relief	0.4	S/L Govt
Engineering & scientific instruments	0.6	Manufg.	Electric housewares & fans	0.3	Manufg.
Primary nonferrous metals, nec	0.6	Manufg.	Household vacuum cleaners	0.3	Manufg.
Glass & glass products, except containers	0.5	Manufg.	Electronic computing equipment	0.2	Manufg.
Screw machine and related products	0.5	Manufg.	Photographic equipment & supplies	0.2	Manufg.
Fabricated metal products, nec	0.4	Manufg.	Refrigeration & heating equipment	0.2	Manufg.
Jewelers' materials & lapidary work	0.4	Manufg.	S/L Govt. purch., highways	0.2	S/L Govt
Motors & generators	0.4	Manufg.	Wholesale trade	0.1	Trade
Primary batteries, dry & wet	0.4	Manufg.	Hospitals	0.1	Services
Air transportation	0.4	Util.	Management & consulting services & labs	0.1	Services
Communications, except radio & TV	0.4	Util.			
Eating & drinking places	0.4	Trade			
Job training & related services	0.4	Services			
Jewelry, precious metal	0.3	Manufg.			
Machinery, except electrical, nec	0.3	Manufg.			
Miscellaneous fabricated wire products	0.3	Manufg.			
Paperboard containers & boxes	0.3	Manufg.			
Personal leather goods	0.3	Manufg.			
Plastics materials & resins	0.3	Manufg.			
Banking	0.3	Fin/R.E.			
Equipment rental & leasing services	0.3	Services			
U.S. Postal Service	0.3	Gov't			
Maintenance of nonfarm buildings nec	0.2	Constr.			
Manufacturing industries, nec	0.2	Manufg.			
Metal coating & allied services	0.2	Manufg.			
Nonferrous wire drawing & insulating	0.2	Manufg.			
Special dies & tools & machine tool accessories	0.2	Manufg.			
Gas production & distribution (utilities)	0.2	Util.			
Motor freight transportation & warehousing	0.2	Util.			
Real estate	0.2	Fin/R.E.			
Hotels & lodging places	0.2	Services			
Management & consulting services & labs	0.2	Services			
Insurance carriers	0.1	Fin/R.E.			
Accounting, auditing & bookkeeping	0.1	Services			
Computer & data processing services	0.1	Services			
Legal services	0.1	Services			
Theatrical producers, bands, entertainers	0.1	Services			

Source: Benchmark Input-Output Accounts for the U.S. Economy, 1982, U.S. Department of Commerce, Washington, D.C., July 1991. Data, as reported in the source, are organized by the 1977 SIC structure in use in 1982 but have been matched, as closely as is possible, to the 1987 SIC structure used in this book.

OCCUPATIONS EMPLOYED BY SIC 387 - PROFESSIONAL AND SCIENTIFIC INSTRUMENTS NEC

Occupation	% of Total 1994	Change to 2005	Occupation	% of Total 1994	Change to 2005
Assemblers, fabricators, & hand workers nec	13.9	-8.1	Bookkeeping, accounting, & auditing clerks	1.9	-5.4
Inspectors, testers, & graders, precision	6.7	-31.9	General office clerks	1.8	3.6
Metal & plastic machine workers nec	4.8	6.4	Marketing, advertising, & PR managers	1.7	-6.0
Precision assemblers nec	4.6	-14.6	Freight, stock, & material movers, hand	1.7	-22.3
Sales & related workers nec	4.2	-3.3	Hand packers & packagers	1.5	-29.3
Blue collar worker supervisors	4.0	-13.8	Industrial production managers	1.5	-2.5
Order clerks, materials, merchandise, & service	3.4	-3.4	Engineering technicians & technologists nec	1.4	-2.7
Stock clerks	2.8	-22.9	Clerical supervisors & managers	1.3	-11.0
General managers & top executives	2.8	-11.6	Opticians, dispensing & measuring	1.3	2.1
Traffic, shipping, & receiving clerks	2.6	-9.4	Machine tool cutting operators, metal & plastic	1.2	-39.6
Secretaries, ex legal & medical	2.4	-4.2	Production, planning, & expediting clerks	1.2	-3.7
Machine operators nec	2.1	-22.4	Financial managers	1.1	-5.3
Adjustment clerks	2.1	-3.5	Maintenance repairers, general utility	1.0	-8.3

Source: *Industry-Occupation Matrix*, Bureau of Labor Statistics. These data relate to one or more 3-digit SIC industry groups rather than to a single 4-digit SIC. The change reported for each occupation to the year 2005 is a percent of growth or decline as estimated by the Bureau of Labor Statistics. The abbreviation nec stands for 'not elsewhere classified'.

LOCATION BY STATE AND REGIONAL CONCENTRATION

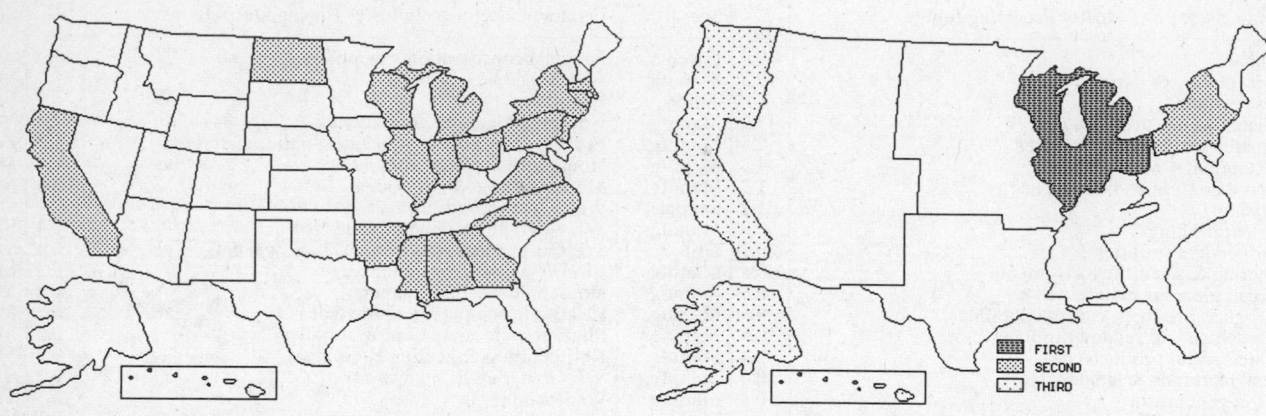

FIRST
SECOND
THIRD

INDUSTRY DATA BY STATE

State	Establish-ments	Shipments Total ($ mil)	Shipments % of U.S.	Shipments Per Establ.	Employment Total Number	Employment % of U.S.	Employment Per Establ.	Wages ($/hour)	Cost as % of Shipments	Investment per Employee ($)
Michigan	13	130.5	16.1	10.0	1,100	14.5	85	8.89	43.5	909
New Jersey	7	41.2	5.1	5.9	300	3.9	43	8.00	53.6	1,000
Virginia	3	39.9	4.9	13.3	400	5.3	133	7.43	49.9	3,000
California	24	28.0	3.4	1.2	300	3.9	13	7.00	52.9	667
Massachusetts	7	26.4	3.3	3.8	300	3.9	43	9.00	48.1	1,000
Pennsylvania	8	18.8	2.3	2.3	200	2.6	25	6.67	58.5	500
Ohio	9	8.2	1.0	0.9	100	1.3	11	9.00	41.5	1,000
New York	25	(D)	-	-	750 *	9.9	30	-	-	-
Illinois	14	(D)	-	-	1,750 *	23.0	125	-	-	-
Wisconsin	6	(D)	-	-	375 *	4.9	63	-	-	-
Connecticut	4	(D)	-	-	175 *	2.3	44	-	-	-
North Carolina	4	(D)	-	-	375 *	4.9	94	-	-	1,143
Alabama	3	(D)	-	-	175 *	2.3	58	-	-	-
Arkansas	3	(D)	-	-	375 *	4.9	125	-	-	-
Georgia	3	(D)	-	-	750 *	9.9	250	-	-	-
Indiana	3	(D)	-	-	375 *	4.9	125	-	-	-
Mississippi	1	(D)	-	-	175 *	2.3	175	-	-	-
North Dakota	1	(D)	-	-	175 *	2.3	175	-	-	-

Source: 1992 *Economic Census*. The states are in descending order of shipments or establishments (if shipment data are missing for the majority). The symbol (D) appears when data are withheld to prevent disclosure of competitive information. States marked with (D) are sorted by number of establishments. A dash (-) indicates that the data element cannot be calculated; * indicates the midpoint of a range.

3911 - JEWELRY, PRECIOUS METAL

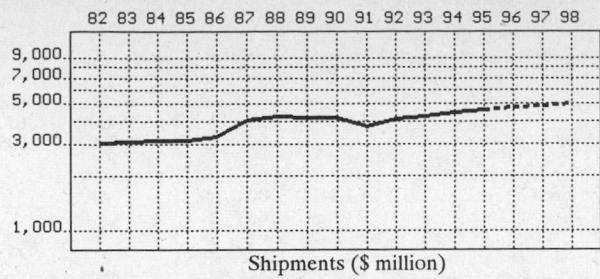

Shipments ($ million)

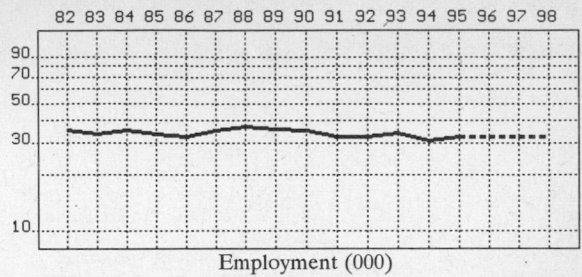

Employment (000)

GENERAL STATISTICS

| Year | Com-panies | Establishments | | Employment | | | Compensation | | Production ($ million) | | | |
		Total	with 20 or more employees	Total (000)	Production Workers (000)	Hours (Mil)	Payroll ($ mil)	Wages ($/hr)	Cost of Materials	Value Added by Manufacture	Value of Shipments	Capital Invest.
1982	2,159	2,193	367	35.3	23.7	43.6	535.8	6.94	1,779.3	1,278.8	3,039.1	33.4
1983		2,171	366	33.9	23.8	42.7	534.2	6.98	1,867.9	1,227.8	3,075.0	22.2
1984		2,149	365	35.3	24.5	43.1	576.6	7.57	1,893.6	1,287.8	3,179.8	29.1
1985		2,127	363	33.7	23.0	40.0	579.7	8.00	1,847.4	1,318.8	3,133.4	26.1
1986		2,152	353	32.6	22.2	41.7	564.9	7.59	1,886.4	1,415.3	3,302.5	27.3
1987	2,294	2,324	369	35.5	24.8	48.8	682.0	7.65	2,331.1	1,792.3	4,078.1	40.3
1988		2,238	364	37.1	25.6	51.4	727.7	7.54	2,375.7	1,870.4	4,273.3	26.4
1989		2,171	365	36.2	24.1	47.4	732.8	8.39	2,407.3	1,860.4	4,227.3	29.8
1990		2,147	365	35.3	24.6	46.4	749.2	8.80	2,329.2	1,869.4	4,180.1	43.2
1991		2,123	327	32.6	22.2	43.0	695.0	8.76	2,078.9	1,683.6	3,733.4	40.5
1992	2,180	2,201	322	32.3	22.3	44.3	753.5	9.41	2,378.2	1,757.0	4,163.8	36.5
1993		2,219	323	34.0	23.6	46.2	799.6	9.53	2,441.9	1,871.8	4,278.1	45.3
1994		2,195P	329P	31.1	20.1	39.6	756.4	9.78	2,501.7	1,978.1	4,459.2	49.2
1995		2,196P	325P	32.9P	22.3P	45.1P	826.9P	9.86P	2,598.4P	2,054.6P	4,631.6P	46.3P
1996		2,198P	321P	32.7P	22.1P	45.1P	849.5P	10.10P	2,666.8P	2,108.6P	4,753.5P	47.9P
1997		2,199P	317P	32.5P	22.0P	45.2P	872.2P	10.33P	2,735.2P	2,162.7P	4,875.3P	49.6P
1998		2,201P	314P	32.4P	21.8P	45.3P	894.8P	10.57P	2,803.5P	2,216.8P	4,997.2P	51.3P

Sources: 1982, 1987, 1992 *Economic Census*; *Annual Survey of Manufactures*, 83-86, 88-91, 93-94. Establishment counts for non-Census years are from *County Business Patterns*; establishment values for 83-84 are extrapolations. 'P's show projections by the editors. Industries reclassified in 87 will not have data for prior years.

INDICES OF CHANGE

| Year | Com-panies | Establishments | | Employment | | | Compensation | | Production ($ million) | | | |
		Total	with 20 or more employees	Total (000)	Production Workers (000)	Hours (Mil)	Payroll ($ mil)	Wages ($/hr)	Cost of Materials	Value Added by Manufacture	Value of Shipments	Capital Invest.
1982	99	100	114	109	106	98	71	74	75	73	73	92
1983		99	114	105	107	96	71	74	79	70	74	61
1984		98	113	109	110	97	77	80	80	73	76	80
1985		97	113	104	103	90	77	85	78	75	75	72
1986		98	110	101	100	94	75	81	79	81	79	75
1987	105	106	115	110	111	110	91	81	98	102	98	110
1988		102	113	115	115	116	97	80	100	106	103	72
1989		99	113	112	108	107	97	89	101	106	102	82
1990		98	113	109	110	105	99	94	98	106	100	118
1991		96	102	101	100	97	92	93	87	96	90	111
1992	100	100	100	100	100	100	100	100	100	100	100	100
1993		101	100	105	106	104	106	101	103	107	103	124
1994		100P	102P	96	90	89	100	104	105	113	107	135
1995		100P	101P	102P	100P	102P	110P	105P	109P	117P	111P	127P
1996		100P	100P	101P	99P	102P	113P	107P	112P	120P	114P	131P
1997		100P	99P	101P	98P	102P	116P	110P	115P	123P	117P	136P
1998		100P	97P	100P	98P	102P	119P	112P	118P	126P	120P	141P

Sources: Same as General Statistics. Values reflect change from the base year, 1992. Values above 100 mean greater than 92, values below 100 mean less than 92, and a value of 100 in the 82-91 or 93-98 period means same as 92. 'P's mark projections by the editors.

SELECTED RATIOS

For 1994	Avg. of All Manufact.	Analyzed Industry	Index	For 1994	Avg. of All Manufact.	Analyzed Industry	Index
Employees per Establishment	49	14	29	Value Added per Production Worker	134,084	98,413	73
Payroll per Establishment	1,500,273	344,649	23	Cost per Establishment	5,045,178	1,139,884	23
Payroll per Employee	30,620	24,322	79	Cost per Employee	102,970	80,441	78
Production Workers per Establishment	34	9	27	Cost per Production Worker	146,988	124,463	85
Wages per Establishment	853,319	176,465	21	Shipments per Establishment	9,576,895	2,031,807	21
Wages per Production Worker	24,861	19,268	78	Shipments per Employee	195,460	143,383	73
Hours per Production Worker	2,056	1,970	96	Shipments per Production Worker	279,017	221,851	80
Wages per Hour	12.09	9.78	81	Investment per Establishment	321,011	22,418	7
Value Added per Establishment	4,602,255	901,309	20	Investment per Employee	6,552	1,582	24
Value Added per Employee	93,930	63,605	68	Investment per Production Worker	9,352	2,448	26

Sources: Same as General Statistics. The 'Average of All Manufacturing' column represents the average of all manufacturing industries reported for the most recent complete year available. The Index shows the relationship between the Average and the Analyzed Industry. For example, 100 means that they are equal; 500 that the Analyzed Industry is five times the average; 50 means that the Analyzed Industry is half the national average. The abbreviation 'na' is used to show that data are 'not available'.

LEADING COMPANIES   Number shown: **75**   Total sales ($ mil): **4,264**   Total employment (000): **34.5**

Company Name	Address				CEO Name	Phone	Co. Type	Sales ($ mil)	Empl. (000)
Jostens Inc	5501 Norman Ctr	Minneapolis	MN	55497	R C Buhrmaster	612-830-3300	P	827	8.0
Franklin Mint Corp	The Franklin Mint	Franklin Ct	PA	19091	Thomas Durovsik	215-459-6000	R	800	4.7
Jostens Inc	5501 Norman Ctr	Minneapolis	MN	55437	Fred Bjork	612-830-3300	D	410•	6.0
Herff Jones Inc	4501 W 62nd St	Indianapolis	IN	46268	David Daly	317-297-3740	R	370•	3.0
Town and Country Corp	25 Union St	Chelsea	MA	02150	C William Carey	617-884-8500	P	288	2.4
Jan Bell Marketing Inc	13801 NW 14th St	Sunrise	FL	33323	Alan H Lipton	305-846-2705	P	275	1.5
OroAmerica Inc	443 N Varney St	Burbank	CA	91502	Guy Benhamou	818-848-5555	P	203	0.4
Michael Anthony Jewelers Inc	115 S MacQuesten	Mount Vernon	NY	10550	Michael Paolercio	914-699-0000	P	143	0.6
Swank Inc	6 Hazel St	Attleboro	MA	02703	Marshall Tulin	508-222-3400	P	125	1.5
Stuller Setting Inc	PO Box 52583	Lafayette	LA	70505	Matthew G Stuller	318-837-4100	R	77•	0.8
Andin International Inc	609 Greenwich St	New York	NY	10014	Ofer Azrielant	212-886-6000	R	61•	0.5
Beaucraft Inc	215 Georgia Av	Providence	RI	02905	Luigi Russo	401-461-2305	R	60	<0.1
Jewelmont Corp	800 Boone Av N	Minneapolis	MN	55427	Robert K Leeds	612-546-3800	R	43	0.4
Robbins Co	400 O'Neil Blv	Attleboro	MA	02703	Robert H Sweet	508-222-2900	R	40	0.3
Loren Industries Inc	2801 Greene St	Hollywood	FL	33020	Richard Goldstein	305-920-6622	R	33	0.2
EB Designs	389 5th Av	New York	NY	10016	Edgar Berebi	212-532-8888	R	30•	0.3
Edward D Sultan Company Ltd	PO Box 301	Honolulu	HI	96809	Edward D Sultan III	808-923-4971	R	25	0.3
T Sardelli and Sons Inc	195 DuPont Dr	Providence	RI	02907	Paul Sardelli	401-944-8510	R	25•	0.3
Howard H Sweet and Son Inc	60 Walton St	Attleboro	MA	02703	R W Crawford	508-222-9234	S	23	0.1
Empire Diamond Corp	350 5th Av	New York	NY	10001	IJ Brod	212-564-4777	R	20	<0.1
HR Terryberry Co	2033 Oak Indrial Dr	Grand Rapids	MI	49505	George Byam	616-458-1391	R	20	0.1
Ravel Inc	PO Box 10000	Pinellas Park	FL	34664	Kevin Tanaka	813-572-6360	R	19	<0.1
William Schneider Inc	16400 NW 15th Av	Miami	FL	33169	Francis J Dallahan	305-625-5171	R	17	0.1
Esposito Jewelry Inc	225 DuPont Dr	Providence	RI	02907	Joseph F Esposito	401-943-1900	R	16•	0.2
Karbra Co	62 W 47th St	New York	NY	10036	Joseph Roth	212-719-2500	R	16•	0.1
Foster Inc	PO Box 778	Attleboro	MA	02703	WJ Boots	508-222-1870	R	15•	<0.1
SH Clausin and Company Inc	41 N 12th St	Minneapolis	MN	55403	Thomas Wodarck	612-332-6565	R	15	<0.1
Jabel Inc	365 Coit St	Irvington	NJ	07111	AH Herman Jr	201-374-6000	R	14	0.1
Marathon Company Inc	90 O'Neil Blv	Attleboro	MA	02703	RL Forman	508-222-5544	R	14•	0.1
Bojar Company Inc	63 Baker St	Providence	RI	02905	William Bojar	401-785-1770	R	12•	0.1
Kinsley and Sons Inc	PO Box 7329	St Louis	MO	63177	Felix P Kinsley Jr	314-843-0400	R	12•	0.1
Fantasy Diamond Corp	1550 W Carroll Av	Chicago	IL	60607	Louis Price	312-243-3300	R	11	<0.1
Roman Research Inc	33 Riverside Dr	Pembroke	MA	02359	Sharon H Ryan	617-826-9700	R	10	0.1
Stamper Black Hills Gold Jewelry	PO Box 3210	Rapid City	SD	57709	RE Stamper	605-342-0751	R	10	0.1
Ullenberg Corp	420 Bell Av	Chattanooga	TN	37405	Ronald T Ullenberg	615-266-7758	R	10	<0.1
L and M Castings	1403 2nd Av	New Hyde Park	NY	11040	Steve Feld	516-354-1500	D	9•	<0.1
Goodman Jewelers Inc	30 W Washington St	Indianapolis	IN	46204	M Goodman	317-236-1000	R	8	0.1
Hammerman Brothers Inc	40 W 57th St	New York	NY	10019	B Hammerman	212-956-2800	R	8	<0.1
Leavens Awards Company Inc	41 Summer St	Attleboro	MA	02703	Dave Fleet	508-222-2930	R	8•	<0.1
Stanley Creations Inc	1414 Willow Av	Melrose Park	PA	19126	Stanley Needles	215-635-6200	R	8	<0.1
Valdawn Inc	600 Sylvan Av	Englewood Clfs	NJ	07632	Robert Reiss	201-871-1616	R	7	<0.1
Gem East Corp	2124 2nd Av	Seattle	WA	98121	Hal S Staehle	206-441-1700	R	7	<0.1
Ostbye and Anderson	10055 51st Av N	Minneapolis	MN	55442	John Macbean	612-553-1515	R	7•	<0.1
Strauss Jewelry Manufacturing	36 W 47th St	New York	NY	10036	Leo B Strauss	212-719-1516	R	7•	<0.1
Burr Patterson and Auld	PO Box 800	Elwood	IN	46036	Steve Short	317-552-7366	R	6•	<0.1
Byard F Brogan Inc	PO Box 369	Glenside	PA	19038	BF Brogan Jr	215-885-3550	R	6	<0.1
Erickson/Beamon Associates	498 7th Av	New York	NY	10018	Eric Beamon	212-643-4810	R	6•	<0.1
Leonore Doskow Inc	1 Doskow Rd	Montrose	NY	10548	David W Doskow	914-737-1335	R	6•	<0.1
Tiara Corp	2425 Oakton St	Evanston	IL	60202	John E Tagliaferro	708-570-4700	R	5	0.1
Scott Keating Designs	PO Box UU	Aspen	CO	81612	Scott Keating	303-927-4347	R	5•	<0.1
Charles Turi Jewelry Co	18 E 48th St	New York	NY	10017	Andrew Turi	212-355-5005	R	5	<0.1
Atkinson Trading Inc	1300 S 2nd St	Gallup	NM	87301	Joe Atkinson	505-722-4435	R	4	<0.1
Ben Silver Corp	149 King St	Charleston	SC	29401	Robert Prenner	803-577-4556	R	4	<0.1
Lestage Manufacturing Co	31 Larsen Way	N Attleboro	MA	02763	Donald LeStage III	508-695-7038	R	4•	<0.1
Regency Creations Inc	1 Plaza Rd	Greenvale	NY	11548	Solomon Gross	516-621-3220	R	4	<0.1
Simco Manufacturing Jewelers	62 W 47th St	New York	NY	10036	S Kopelowitz	212-575-8390	R	4	<0.1
Metal Arts Company Inc	1 American Ctr	Geneva	NY	14456	Stanley J Dahle	315-789-2200	P	3	<0.1
Art Deco Jewelry Inc	1 Weingeroff Blv	Cranston	RI	02910	Greg Weingeroff	401-467-2200	S	3•	0.1
David Webb Inc	445 Park Av	New York	NY	10022	Nina Silberstein	212-421-3030	R	3	<0.1
FL Thorpe and Company Inc	PO Box 547	Deadwood	SD	57732	Terry Sankey	605-578-2292	R	3•	<0.1
Garden Jewelry Mfg Corp	579 5th Av	New York	NY	10017	Mike Gottlieb	212-421-7813	R	3•	<0.1
Lo-Well Jewelry Corp	146 W 29th St	New York	NY	10001	J Hantman	212-594-5337	R	3	<0.1
LL Diamonds Inc	5878 Westheimer	Houston	TX	77057	Laura Levit	713-780-3336	R	3	<0.1
Samuel Platzer Company Inc	31 W 47th St	New York	NY	10036	David J Platzer	212-719-2000	R	3•	<0.1
Domenico Celi Manufacturing	412 W 6th St	Los Angeles	CA	90014	D Celi	213-623-3576	R	3	<0.1
International Cultured Pearls	71 W 47th St	New York	NY	10036	R Schwager	212-869-5141	R	3	<0.1
Oro-Cal Manufacturing	1720 Bird St	Oroville	CA	95965	David J Conner	916-533-5065	R	2	<0.1
Chicagoland Processing Corp	501 W Algonquin	Mt Prospect	IL	60056	John Obie	708-981-0310	R	2	<0.1
David Klein Manufacturing Inc	15 W 47th St	New York	NY	10036	D Klein	212-819-0555	R	2	<0.1
E Ringold Inc	245 W 29th St	New York	NY	10011	Erno Glauber	212-268-4907	R	2•	<0.1
J Jenkins Sons Company Inc	1801 Whitehead Rd	Baltimore	MD	21207	GP Sparagana	410-265-5200	S	2	<0.1
Kalibre	6900 E Indian	Scottsdale	AZ	85251	Harlene Korey	602-946-8055	R	2	<0.1
Ring Specialty Co	4900 Pearl E Cir	Boulder	CO	80301	Gerald E Rhodes	303-440-5507	R	2	<0.1
Robert S Fisher and Co	19 Liberty St	Newark	NJ	07102	Donald Roth	201-622-2658	R	2•	<0.1
W Ringold Inc	236 W 27th St	New York	NY	10001	William Glauber	212-206-7426	R	2	<0.1

Source: Ward's Business Directory of U.S. Private and Public Companies, Volumes 1 and 2, 1996. The company type code used is as follows: P - Public, R - Private, S - Subsidiary, D - Division, J - Joint Venture, A - Affiliate, G - Group. Sales are in millions of dollars, employees are in thousands. An asterisk (*) indicates an estimated sales volume. The symbol < stands for 'less than'. Company names and addresses are truncated, in some cases, to fit into the available space.

MATERIALS CONSUMED

Material	Quantity	Delivered Cost ($ million)
Materials, ingredients, containers, and supplies	(X)	2,096.0
Fabricated metal products, including forgings	(X)	91.3
Precious metals (gold, platinum, etc.), all forms	(X)	787.4
Other shapes and forms, including castings	(X)	42.4
Precious, semiprecious, and synthetic stones, and pearls; cut, polished, or drilled	(X)	257.1
Jewelers' findings, including joints, pins, clasps, chains, flat stock, etc.	(X)	111.1
Other jewlery, silverware, and plated ware	(X)	41.0
All other materials and components, parts, containers, and supplies	(X)	66.8
Materials, ingredients, containers, and supplies, nsk	(X)	699.0

Source: 1992 Economic Census. Explanation of symbols used: (D): Withheld to avoid disclosure of competitive data; na: Not available; (S): Withheld because statistical norms were not met; (X): Not applicable; (Z): Less than half the unit shown; nec: Not elsewhere classified; nsk: Not specified by kind; - : zero; * : 10-19 percent estimated; ** : 20-29 percent estimated.

PRODUCT SHARE DETAILS

Product or Product Class	% Share	Product or Product Class	% Share
Jewelry, precious metal	100.00	silver)	11.38
Jewelry, made of platinum metals and karat gold	70.45	Women's and children's jewelry (necklaces, bracelets, brooches, pins, clips, earrings, lockets, etc.) made of silver	
Fraternal, college, and school rings made of platinum metals and karat gold (complete)	10.01	(including platinum metals and karat gold clad to silver) .	51.83
Wedding rings made of platinum metals and karat gold (complete)	15.65	Watch attachments (bracelets for watches) made of silver (including platinum metals and karat gold clad to silver) .	0.17
Other rings made of platinum metals and karat gold	19.69	Other jewelry worn or carried about the person (rosaries, cigarette cases, lighters, compacts, vanity cases, etc.) made	
Ring mountings made of platinum metals and karat gold, for sale separately	5.07	of silver (including platinum metals and karat gold clad to silver)	11.38
Women's and children's jewelry (necklaces, bracelets, brooches, pins, clips, earrings, lockets, etc.) made of		Jewelry made of silver (including platinum metals and karat gold clad to silver), nsk	0.82
platinum metals and karat gold	34.29	Other jewelry, except costume	5.73
Watch attachments (bracelets for watches) made of platinum metals and karat gold	1.11	Other rings and ring mountings (except costume) made of base metal clad with precious metal	10.16
Organizational jewelry (fraternal, college, and school jewelry and emblems, and military insignia, excluding		Other men's jewelry (collar and cuff buttons, studs, watch chains, money clips, identification bracelets, scarf pins,	
rings) made of platinum metals and karat gold . . .	2.67	etc., except costume) made of base metal clad with precious metal	7.76
Other jewelry worn or carried about the person (watch chains, rosaries, cigarette cases, lighters, compacts, etc.)		Other women's and children's jewelry (necklaces, bracelets, brooches, pins, clips, earrings, lockets, etc., except	
made of platinum metals and karat gold	9.24	costume) made of base metal clad with precious metal .	38.22
Jewelry, made of platinum metals and karat gold, nsk . .	2.27	Other watch attachments (bracelets for watches, except costume) made of base metal clad with precious metal .	4.37
Jewelry made of silver (including platinum metals and karat gold clad to silver)	6.25	Jewelry of natural or cultured pearls	3.57
Rings and ring mountings made of silver (including platinum metals and karat gold clad to silver)	24.49	Jewelry of semiprecious or precious stones	33.90
Men's jewelry (collar and cuff buttons, studs, watch chains, money clips, identification bracelets, scarf pins, etc.) made		Other jewelry, except costume, nsk	2.02
of silver (including platinum metals and karat gold clad to		Jewelry, precious metal, nsk	17.57

Source: 1992 Economic Census. The values shown are percent of total shipments in an industry. Values of indented subcategories are summed in the main heading. The symbol (D) appears when data are withheld to prevent disclosure of competitive information. The abbreviation nsk stands for 'not specified by kind' and nec for 'not elsewhere classified'.

INPUTS AND OUTPUTS FOR JEWELRY, PRECIOUS METAL

Economic Sector or Industry Providing Inputs	%	Sector	Economic Sector or Industry Buying Outputs	%	Sector
Imports	32.9	Foreign	Personal consumption expenditures	92.2	
Primary nonferrous metals, nec	32.5	Manufg.	Exports	4.5	Foreign
Jewelers' materials & lapidary work	11.4	Manufg.	Jewelry, precious metal	0.8	Manufg.
Wholesale trade	8.8	Trade	Concrete products, nec	0.2	Manufg.
Advertising	2.0	Services	Legal services	0.2	Services
Jewelry, precious metal	1.1	Manufg.	Watches, clocks, & parts	0.1	Manufg.
Eating & drinking places	0.9	Trade	Retail trade, except eating & drinking	0.1	Trade
Converted paper products, nec	0.8	Manufg.	Wholesale trade	0.1	Trade
Personal leather goods	0.7	Manufg.	Banking	0.1	Fin/R.E.
Die-cut paper & board	0.6	Manufg.	Insurance carriers	0.1	Fin/R.E.
Petroleum refining	0.6	Manufg.	Hospitals	0.1	Services
Paperboard containers & boxes	0.4	Manufg.			
Electric services (utilities)	0.4	Util.			
Banking	0.4	Fin/R.E.			
Real estate	0.4	Fin/R.E.			
U.S. Postal Service	0.4	Gov't			
Maintenance of nonfarm buildings nec	0.3	Constr.			
Machinery, except electrical, nec	0.3	Manufg.			
Communications, except radio & TV	0.3	Util.			
Sanitary services, steam supply, irrigation	0.3	Util.			
Legal services	0.3	Services			

Continued on next page.

INPUTS AND OUTPUTS FOR JEWELRY, PRECIOUS METAL - Continued

Economic Sector or Industry Providing Inputs	%	Sector	Economic Sector or Industry Buying Outputs	%	Sector
Management & consulting services & labs	0.3	Services			
Photofinishing labs, commercial photography	0.3	Services			
Abrasive products	0.2	Manufg.			
Special dies & tools & machine tool accessories	0.2	Manufg.			
Air transportation	0.2	Util.			
Motor freight transportation & warehousing	0.2	Util.			
Insurance carriers	0.2	Fin/R.E.			
Royalties	0.2	Fin/R.E.			
Accounting, auditing & bookkeeping	0.2	Services			
Equipment rental & leasing services	0.2	Services			
Manifold business forms	0.1	Manufg.			
Automotive rental & leasing, without drivers	0.1	Services			
Business/professional associations	0.1	Services			
Electrical repair shops	0.1	Services			
Services to dwellings & other buildings	0.1	Services			

Source: Benchmark Input-Output Accounts for the U.S. Economy, 1982, U.S. Department of Commerce, Washington, D.C., July 1991. Data, as reported in the source, are organized by the 1977 SIC structure in use in 1982 but have been matched, as closely as is possible, to the 1987 SIC structure used in this book.

OCCUPATIONS EMPLOYED BY SIC 391 - JEWELRY, SILVERWARE, AND PLATED WARE

Occupation	% of Total 1994	Change to 2005	Occupation	% of Total 1994	Change to 2005
Precision metal workers nec	12.2	5.3	Secretaries, ex legal & medical	1.7	-20.1
Jewelers & silversmiths	10.4	-21.1	Order clerks, materials, merchandise, & service	1.6	-14.2
Assemblers, fabricators, & hand workers nec	7.6	-12.3	Tool & die makers	1.6	-29.2
Sales & related workers nec	4.7	-12.3	Machine forming operators, metal & plastic	1.6	-12.3
Grinders & polishers, hand	4.0	-12.3	Production, planning, & expediting clerks	1.5	-12.3
General managers & top executives	3.7	-16.8	Machine tool cutting operators, metal & plastic	1.5	-26.9
Precision workers nec	3.5	-21.0	Solderers & brazers	1.4	-12.2
Blue collar worker supervisors	3.4	-19.3	Clerical supervisors & managers	1.4	-10.3
General office clerks	2.7	-25.2	Industrial production managers	1.2	-12.1
Bookkeeping, accounting, & auditing clerks	2.6	-34.2	Metal molding machine workers	1.2	-3.5
Traffic, shipping, & receiving clerks	2.5	-15.5	Adjustment clerks	1.1	5.1
Inspectors, testers, & graders, precision	2.1	-12.3	Stock clerks	1.1	-28.7
Grinding machine operators, metal & plastic	2.1	-29.9	Designers, ex interior designers	1.0	-3.5
Hand packers & packagers	2.0	-24.7			

Source: Industry-Occupation Matrix, Bureau of Labor Statistics. These data relate to one or more 3-digit SIC industry groups rather than to a single 4-digit SIC. The change reported for each occupation to the year 2005 is a percent of growth or decline as estimated by the Bureau of Labor Statistics. The abbreviation nec stands for 'not elsewhere classified'.

LOCATION BY STATE AND REGIONAL CONCENTRATION

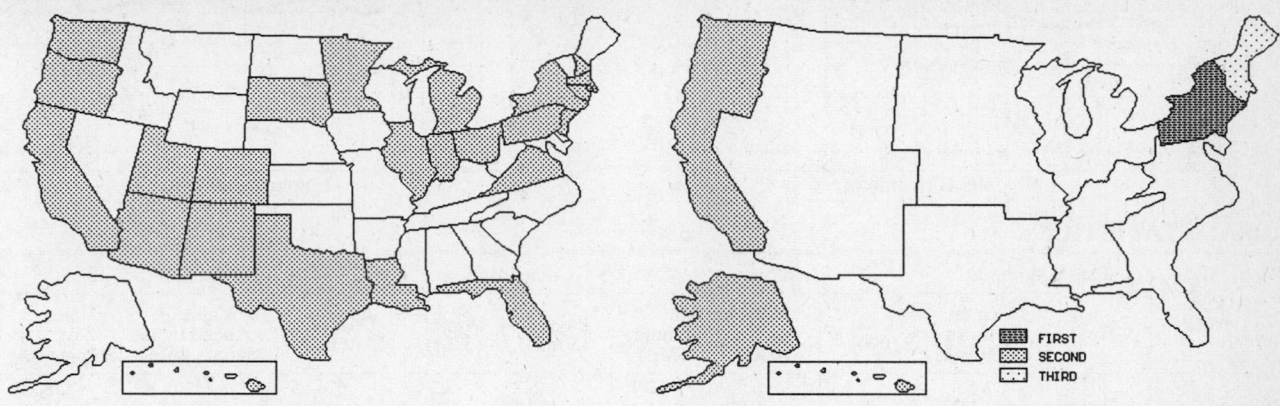

FIRST
SECOND
THIRD

INDUSTRY DATA BY STATE

State	Establish-ments	Shipments			Employment				Cost as % of Shipments	Investment per Employee ($)
		Total ($ mil)	% of U.S.	Per Establ.	Total Number	% of U.S.	Per Establ.	Wages ($/hour)		
New York	632	1,577.4	37.9	2.5	9,500	29.4	15	10.69	64.7	1,537
California	312	446.1	10.7	1.4	3,200	9.9	10	8.62	54.1	813
Rhode Island	186	333.9	8.0	1.8	3,300	10.2	18	9.19	51.2	939
Massachusetts	56	282.9	6.8	5.1	2,100	6.5	38	9.90	42.5	-
New Jersey	61	252.2	6.1	4.1	1,600	5.0	26	9.83	62.5	1,438
Texas	109	223.6	5.4	2.1	2,100	6.5	19	8.24	44.9	1,619
Florida	121	143.9	3.5	1.2	1,200	3.7	10	9.63	66.2	833
New Mexico	85	103.0	2.5	1.2	1,400	4.3	16	7.62	45.6	571
Illinois	50	99.4	2.4	2.0	800	2.5	16	10.00	52.2	1,250
South Dakota	14	90.9	2.2	6.5	1,200	3.7	86	8.00	36.0	1,333
Pennsylvania	46	86.4	2.1	1.9	500	1.5	11	10.14	58.3	1,200
Minnesota	19	46.5	1.1	2.4	400	1.2	21	7.83	63.7	750
Ohio	30	32.7	0.8	1.1	300	0.9	10	10.33	57.8	333
Connecticut	14	20.7	0.5	1.5	200	0.6	14	9.50	48.8	500
Oregon	23	13.6	0.3	0.6	200	0.6	9	8.50	40.4	500
Arizona	36	13.0	0.3	0.4	100	0.3	3	7.50	46.9	-
Washington	45	(D)	-	-	175 *	0.5	4	-	-	571
Hawaii	36	(D)	-	-	375 *	1.2	10	-	-	-
Colorado	34	(D)	-	-	175 *	0.5	5	-	-	571
Michigan	26	(D)	-	-	175 *	0.5	7	-	-	-
Indiana	19	(D)	-	-	750 *	2.3	39	-	-	-
Virginia	18	(D)	-	-	175 *	0.5	10	-	-	-
Utah	16	(D)	-	-	1,750 *	5.4	109	-	-	-
Louisiana	14	(D)	-	-	175 *	0.5	13	-	-	-
New Hampshire	10	(D)	-	-	175 *	0.5	18	-	-	-

Source: 1992 *Economic Census.* The states are in descending order of shipments or establishments (if shipment data are missing for the majority). The symbol (D) appears when data are withheld to prevent disclosure of competitive information. States marked with (D) are sorted by number of establishments. A dash (-) indicates that the data element cannot be calculated; * indicates the midpoint of a range.

3914 - SILVERWARE & PLATED WARE

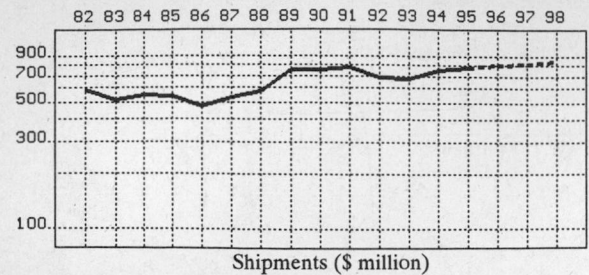

Shipments ($ million)

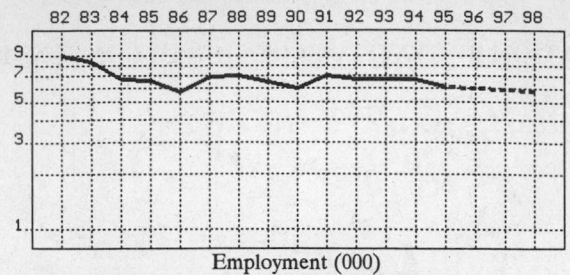

Employment (000)

GENERAL STATISTICS

Year	Companies	Establishments Total	Establishments with 20 or more employees	Employment Total (000)	Employment Production Workers (000)	Employment Hours (Mil)	Compensation Payroll ($ mil)	Compensation Wages ($/hr)	Production Cost of Materials	Production Value Added by Manufacture	Production Value of Shipments	Capital Invest.
1982	199	208	55	9.1	6.5	12.4	148.3	7.39	262.5	311.9	583.9	11.8
1983		202	55	8.4	5.9	11.1	137.4	7.90	232.0	290.7	510.8	9.6
1984		196	55	6.7	4.7	8.8	125.7	8.58	264.4	242.3	547.0	9.4
1985		191	54	6.6	4.3	8.6	126.8	8.59	268.6	273.4	541.4	10.9
1986		190	59	5.7	3.8	7.2	120.7	9.88	227.5	252.8	473.7	10.2
1987	205	209	52	6.9	4.9	8.9	132.7	9.21	254.5	276.1	528.7	9.9
1988		192	51	7.0	5.0	9.2	137.0	9.54	264.5	322.4	575.0	8.9
1989		183	50	6.5	5.9	9.6	167.6	11.12	292.7	470.6	749.5	15.6
1990		187	49	6.0	5.4	8.7	163.0	11.70	282.8	462.3	751.9	14.8
1991		202	45	7.0	5.3	8.7	169.5	12.56	300.8	501.9	774.3	22.8
1992	209	213	41	6.7	4.9	7.4	156.5	13.49	279.1	405.0	684.1	16.9
1993		211	42	6.7	4.8	7.3	163.1	14.37	288.4	392.3	674.9	14.4
1994		201P	42P	6.7	4.6	7.3	171.4	14.44	330.5	428.1	753.9	12.8
1995		202P	40P	6.1P	4.7P	6.9P	172.2P	14.97P	340.0P	440.4P	775.5P	17.2P
1996		202P	39P	6.0P	4.7P	6.6P	175.7P	15.59P	349.3P	452.4P	796.7P	17.8P
1997		203P	38P	5.9P	4.6P	6.3P	179.2P	16.20P	358.6P	464.5P	818.0P	18.4P
1998		203P	36P	5.7P	4.6P	6.0P	182.7P	16.81P	367.9P	476.5P	839.2P	19.0P

Sources: 1982, 1987, 1992 *Economic Census*; *Annual Survey of Manufactures*, 83-86, 88-91, 93-94. Establishment counts for non-Census years are from *County Business Patterns*; establishment values for 83-84 are extrapolations. 'P's show projections by the editors. Industries reclassified in 87 will not have data for prior years.

INDICES OF CHANGE

Year	Companies	Establishments Total	Establishments with 20 or more employees	Employment Total (000)	Employment Production Workers (000)	Employment Hours (Mil)	Compensation Payroll ($ mil)	Compensation Wages ($/hr)	Production Cost of Materials	Production Value Added by Manufacture	Production Value of Shipments	Capital Invest.
1982	95	98	134	136	133	168	95	55	94	77	85	70
1983		95	134	125	120	150	88	59	83	72	75	57
1984		92	134	100	96	119	80	64	95	60	80	56
1985		90	132	99	88	116	81	64	96	68	79	64
1986		89	144	85	78	97	77	73	82	62	69	60
1987	98	98	127	103	100	120	85	68	91	68	77	59
1988		90	124	104	102	124	88	71	95	80	84	53
1989		86	122	97	120	130	107	82	105	116	110	92
1990		88	120	90	110	118	104	87	101	114	110	88
1991		95	110	104	108	118	108	93	108	124	113	135
1992	100	100	100	100	100	100	100	100	100	100	100	100
1993		99	102	100	98	99	104	107	103	97	99	85
1994		95P	102P	100	94	99	110	107	118	106	110	76
1995		95P	99P	91P	97P	93P	110P	111P	122P	109P	113P	102P
1996		95P	95P	89P	96P	89P	112P	116P	125P	112P	116P	105P
1997		95P	92P	87P	95P	86P	115P	120P	128P	115P	120P	109P
1998		95P	88P	86P	94P	82P	117P	125P	132P	118P	123P	112P

Sources: Same as General Statistics. Values reflect change from the base year, 1992. Values above 100 mean greater than 92, values below 100 mean less than 92, and a value of 100 in the 82-91 or 93-98 period means same as 92. 'P's mark projections by the editors.

SELECTED RATIOS

For 1994	Avg. of All Manufact.	Analyzed Industry	Index	For 1994	Avg. of All Manufact.	Analyzed Industry	Index
Employees per Establishment	49	33	68	Value Added per Production Worker	134,084	93,065	69
Payroll per Establishment	1,500,273	851,453	57	Cost per Establishment	5,045,178	1,641,803	33
Payroll per Employee	30,620	25,582	84	Cost per Employee	102,970	49,328	48
Production Workers per Establishment	34	23	67	Cost per Production Worker	146,988	71,848	49
Wages per Establishment	853,319	523,648	61	Shipments per Establishment	9,576,895	3,745,100	39
Wages per Production Worker	24,861	22,916	92	Shipments per Employee	195,460	112,522	58
Hours per Production Worker	2,056	1,587	77	Shipments per Production Worker	279,017	163,891	59
Wages per Hour	12.09	14.44	119	Investment per Establishment	321,011	63,586	20
Value Added per Establishment	4,602,255	2,126,645	46	Investment per Employee	6,552	1,910	29
Value Added per Employee	93,930	63,896	68	Investment per Production Worker	9,352	2,783	30

Sources: Same as General Statistics. The 'Average of All Manufacturing' column represents the average of all manufacturing industries reported for the most recent complete year available. The Index shows the relationship between the Average and the Analyzed Industry. For example, 100 means that they are equal; 500 that the Analyzed Industry is five times the average; 50 means that the Analyzed Industry is half the national average. The abbreviation 'na' is used to show that data are 'not available'.

LEADING COMPANIES   Number shown: 27   Total sales ($ mil): 1,246   Total employment (000): 12.7

Company Name	Address				CEO Name	Phone	Co. Type	Sales ($ mil)	Empl. (000)
Oneida Ltd	Kenwood Av	Oneida	NY	13421	W D Matthews	315-361-3000	P	455	5.5
Oneida Silversmiths	Kenwood Av	Oneida	NY	13421	Gary L Moreau	315-361-3000	D	250	2.8
Syratech Corp	175 McClellan Hwy	East Boston	MA	02128	Leonard Florence	617-561-2200	P	242	1.0
WorldCrisa Corp	PO Box 5020	Wallingford	CT	06492	A Reed Hayes	203-265-8000	S	50	0.9
Towle Manufacturing Co	PO Box 9115	East Boston	MA	02128	Leonard Florence	617-568-1300	S	47	0.2
Reed and Barton Corp	144 W Britannia St	Taunton	MA	02780	Albert D Krebel	508-824-6611	R	43*	0.5
Wallace Silversmiths Inc	175 McClellan Hwy	East Boston	MA	02128	Leonard Florence	617-561-2200	S	24	0.3
Kirk Stieff Co	800 Wyman Park Dr	Baltimore	MD	21211	James Solomon	410-338-6000	S	21*	0.3
Rosemar Silver Company Inc	620 Spring St	North Dighton	MA	02764	Jessie Lorenco	508-823-7176	S	20*	<0.1
RS Owens and Co	5535 N Lynch Av	Chicago	IL	60630	S Siegel	312-282-6000	R	20	0.3
Utica Cutlery Co	PO Box 10527	Utica	NY	13503	A Edward Allen Jr	315-733-4663	R	17	0.1
Lance Corp	321 Central St	Hudson	MA	01749	Ronald K Larson	508-568-1401	R	10	0.1
Bruce Fox Inc	PO Box 89	New Albany	IN	47151	James Greer	812-945-3511	R	8	0.2
Rada Manufacturing Co	PO Box 838	Waverly	IA	50677	Gary Nelson	319-352-5454	R	8*	0.1
Web Silver Co	Butler & Glenwood	Philadelphia	PA	19124	JS Bass	215-744-9090	R	7	<0.1
Comstock Creations Inc	PO Box 2715	Durango	CO	81302	Don Muchow	303-247-3836	R	5	<0.1
Empire Silver Company Inc	6520 New Utrecht	Brooklyn	NY	11219	J Otranto	718-232-3389	R	4	<0.1
Christofle Silver Inc	373 Park Av S	New York	NY	10016	Thierry Chaunu	212-683-4616	S	3*	<0.1
Olde Country Reproductions Inc	PO Box 2617	York	PA	17405	W H Swartz Jr	717-848-1859	R	3	<0.1
Achievement Products Inc	PO Box 388	East Hanover	NJ	07936	Edward Van Rooyen	201-887-5090	R	2	<0.1
Gallo Pewter Sculptures Corp	PO Box 1996	Green Cv Spgs	FL	32043	Mel Yahre	904-284-3100	R	2*	<0.1
Carvel Hall Inc	PO Box 271	Crisfield	MD	21817	James A Hart	410-968-0500	R	1	<0.1
Randall Made Knives	PO Box 1988	Orlando	FL	32802	Gary Randall	407-855-8075	R	1	<0.1
RGT Enterprises Inc	8316 Jamison	Englewood	CO	80112	Martha M Thomas	303-793-3667	R	1	<0.1
Royal Silver Manufacturing	PO Box 10097	Norfolk	VA	23513	Lloyd M Gilbert Jr	804-855-6004	R	1	<0.1
Fellowship Foundry	2250 E 12th St	Oakland	CA	94601	Randell Moore	510-261-3292	R	1*	<0.1
House of Trophies	6102 N 16th St	Phoenix	AZ	85016	Jim L Rogers	602-285-0001	R	1	<0.1

Source: Ward's Business Directory of U.S. Private and Public Companies, Volumes 1 and 2, 1996. The company type code used is as follows: P - Public, R - Private, S - Subsidiary, D - Division, J - Joint Venture, A - Affiliate, G - Group. Sales are in millions of dollars, employees are in thousands. An asterisk (*) indicates an estimated sales volume. The symbol < stands for 'less than'. Company names and addresses are truncated, in some cases, to fit into the available space.

MATERIALS CONSUMED

Material	Quantity	Delivered Cost ($ million)
Materials, ingredients, containers, and supplies	(X)	195.1
Fabricated metal products (except castings and forgings)	(X)	22.4
Castings (rough and semifinished)	(X)	1.3
Forgings	(X)	0.2
Steel shapes and forms	(X)	45.3
Precious metals (gold, platinum, etc.), all forms	(X)	66.3
All other materials and components, parts, containers, and supplies	(X)	29.8
Materials, ingredients, containers, and supplies, nsk	(X)	29.9

Source: 1992 Economic Census. Explanation of symbols used: (D): Withheld to avoid disclosure of competitive data; na: Not available; (S): Withheld because statistical norms were not met; (X): Not applicable; (Z): Less than half the unit shown; nec: Not elsewhere classified; nsk: Not specified by kind; - : zero; * : 10-19 percent estimated; ** : 20-29 percent estimated.

PRODUCT SHARE DETAILS

Product or Product Class	% Share	Product or Product Class	% Share
Silverware and plated ware	100.00	toiletware, ecclesiastical ware, novelties, trophies, baby goods, and other plated ware)	2.86
Hollowware (including toiletware, ecclesiastical ware, novelties, trophies, baby goods, and other plated ware) . .	42.00	Pewter hollowware (including toiletware, ecclesiastical ware, novelties, trophies, baby goods, and other plated ware) .	34.99
Sterling silver hollowware (including toiletware, ecclesiastical ware, novelties, trophies, baby goods, and other plated ware)	7.43	Unplated hollowware of other metals (including stainless steel), and hollowware plated with other metals (including toiletware, ecclesiastical ware, novelties, trophies, baby goods, and others)	20.04
Electrosilverplated hollowware (including toiletware, ecclesiastical ware, novelties, trophies, baby goods, and other plated ware)	21.85	Hollowware (including toiletware, ecclesiastical ware, novelties, trophies, baby goods, and other plated ware), nsk	11.47
Precious metal hollowware, other than silver, whether or not clad with precious metal (including toiletware, ecclesiastical ware, novelties, trophies, baby goods, and other plated ware)	1.32	Flatware (including all knives, forks, spoons, and carving sets made wholly of metal)	51.08
Precious metal clad base metal hollowware (including		Silverware and plated ware, nsk	6.90

Source: 1992 Economic Census. The values shown are percent of total shipments in an industry. Values of indented subcategories are summed in the main heading. The symbol (D) appears when data are withheld to prevent disclosure of competitive information. The abbreviation nsk stands for 'not specified by kind' and nec for 'not elsewhere classified'.

INPUTS AND OUTPUTS FOR SILVERWARE & PLATED WARE

Economic Sector or Industry Providing Inputs	%	Sector	Economic Sector or Industry Buying Outputs	%	Sector
Imports	44.5	Foreign	Personal consumption expenditures	62.5	
Primary nonferrous metals, nec	15.8	Manufg.	Eating & drinking places	14.1	Trade
Copper rolling & drawing	5.1	Manufg.	Exports	7.7	Foreign
Blast furnaces & steel mills	4.4	Manufg.	S/L Govt. purch., higher education	4.7	S/L Govt
Wholesale trade	4.2	Trade	S/L Govt. purch., elem. & secondary education	3.9	S/L Govt
Advertising	2.5	Services	S/L Govt. purch., natural resource & recreation.	1.4	S/L Govt
Sanitary services, steam supply, irrigation	2.2	Util.	S/L Govt. purch., health & hospitals	1.0	S/L Govt
Paperboard containers & boxes	2.0	Manufg.	Silverware & plated ware	0.7	Manufg.
Fabricated metal products, nec	1.7	Manufg.	S/L Govt. purch., other general government	0.7	S/L Govt
Eating & drinking places	1.7	Trade	Insurance carriers	0.4	Fin/R.E.
U.S. Postal Service	1.3	Gov't	S/L Govt. purch., fire	0.4	S/L Govt
Electric services (utilities)	1.2	Util.	S/L Govt. purch., correction	0.3	S/L Govt
Silverware & plated ware	1.1	Manufg.	Federal Government purchases, national defense	0.2	Fed Govt
Maintenance of nonfarm buildings nec	0.9	Constr.	Federal Government purchases, nondefense	0.1	Fed Govt
Banking	0.7	Fin/R.E.			
Equipment rental & leasing services	0.7	Services			
Machinery, except electrical, nec	0.6	Manufg.			
Nonferrous rolling & drawing, nec	0.6	Manufg.			
Communications, except radio & TV	0.6	Util.			
Gas production & distribution (utilities)	0.6	Util.			
Motor freight transportation & warehousing	0.6	Util.			
Management & consulting services & labs	0.6	Services			
Abrasive products	0.5	Manufg.			
Real estate	0.5	Fin/R.E.			
Legal services	0.5	Services			
Petroleum refining	0.4	Manufg.			
Special dies & tools & machine tool accessories	0.4	Manufg.			
Water transportation	0.4	Util.			
Accounting, auditing & bookkeeping	0.4	Services			
Air transportation	0.3	Util.			
Royalties	0.3	Fin/R.E.			
Lubricating oils & greases	0.2	Manufg.			
Manifold business forms	0.2	Manufg.			
Insurance carriers	0.2	Fin/R.E.			
Hotels & lodging places	0.2	Services			
State & local government enterprises, nec	0.2	Gov't			
Photographic equipment & supplies	0.1	Manufg.			
Automotive repair shops & services	0.1	Services			
Business services nec	0.1	Services			
Laundry, dry cleaning, shoe repair	0.1	Services			
Personnel supply services	0.1	Services			

Source: Benchmark Input-Output Accounts for the U.S. Economy, 1982, U.S. Department of Commerce, Washington, D.C., July 1991. Data, as reported in the source, are organized by the 1977 SIC structure in use in 1982 but have been matched, as closely as is possible, to the 1987 SIC structure used in this book.

OCCUPATIONS EMPLOYED BY SIC 391 - JEWELRY, SILVERWARE, AND PLATED WARE

Occupation	% of Total 1994	Change to 2005	Occupation	% of Total 1994	Change to 2005
Precision metal workers nec	12.2	5.3	Secretaries, ex legal & medical	1.7	-20.1
Jewelers & silversmiths	10.4	-21.1	Order clerks, materials, merchandise, & service	1.6	-14.2
Assemblers, fabricators, & hand workers nec	7.6	-12.3	Tool & die makers	1.6	-29.2
Sales & related workers nec	4.7	-12.3	Machine forming operators, metal & plastic	1.6	-12.3
Grinders & polishers, hand	4.0	-12.3	Production, planning, & expediting clerks	1.5	-12.3
General managers & top executives	3.7	-16.8	Machine tool cutting operators, metal & plastic	1.5	-26.9
Precision workers nec	3.5	-21.0	Solderers & brazers	1.4	-12.2
Blue collar worker supervisors	3.4	-19.3	Clerical supervisors & managers	1.4	-10.3
General office clerks	2.7	-25.2	Industrial production managers	1.2	-12.1
Bookkeeping, accounting, & auditing clerks	2.6	-34.2	Metal molding machine workers	1.2	-3.5
Traffic, shipping, & receiving clerks	2.5	-15.5	Adjustment clerks	1.1	5.1
Inspectors, testers, & graders, precision	2.1	-12.3	Stock clerks	1.1	-28.7
Grinding machine operators, metal & plastic	2.1	-29.9	Designers, ex interior designers	1.0	-3.5
Hand packers & packagers	2.0	-24.7			

Source: Industry-Occupation Matrix, Bureau of Labor Statistics. These data relate to one or more 3-digit SIC industry groups rather than to a single 4-digit SIC. The change reported for each occupation to the year 2005 is a percent of growth or decline as estimated by the Bureau of Labor Statistics. The abbreviation nec stands for 'not elsewhere classified'.

LOCATION BY STATE AND REGIONAL CONCENTRATION

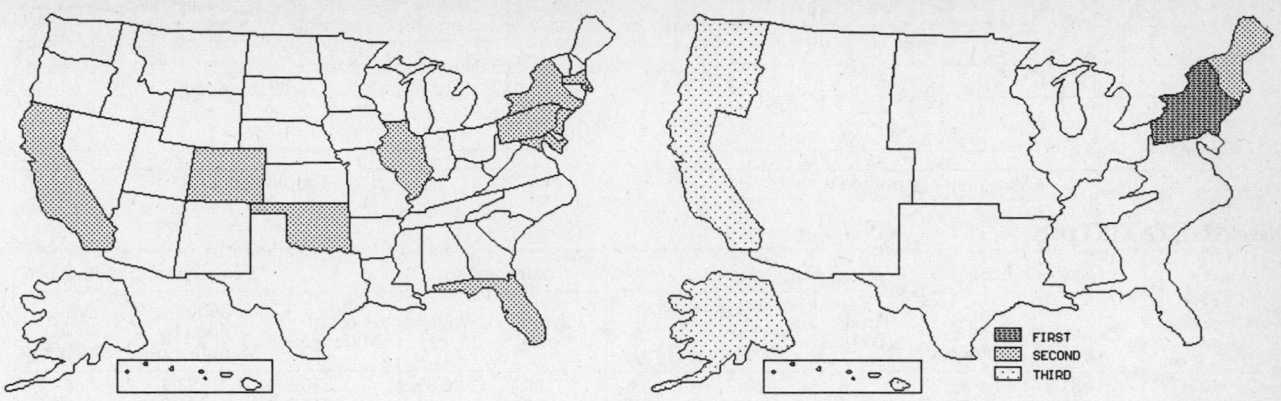

FIRST
SECOND
THIRD

INDUSTRY DATA BY STATE

State	Establish-ments	Shipments			Employment				Cost as % of Shipments	Investment per Employee ($)
		Total ($ mil)	% of U.S.	Per Establ.	Total Number	% of U.S.	Per Establ.	Wages ($/hour)		
Massachusetts	21	169.9	24.8	8.1	1,300	19.4	62	12.50	55.8	-
Rhode Island	6	37.3	5.5	6.2	400	6.0	67	9.00	29.2	1,000
Illinois	12	24.5	3.6	2.0	400	6.0	33	10.75	40.4	-
California	23	21.3	3.1	0.9	300	4.5	13	10.33	43.7	-
New Jersey	9	16.1	2.4	1.8	200	3.0	22	10.50	52.8	500
Pennsylvania	10	11.3	1.7	1.1	200	3.0	20	10.50	37.2	-
New York	28	(D)	-	-	1,750 *	26.1	63	-	-	-
Florida	10	(D)	-	-	175 *	2.6	18	-	-	-
Maryland	4	(D)	-	-	175 *	2.6	44	-	-	-
Colorado	2	(D)	-	-	175 *	2.6	88	-	-	-
Oklahoma	2	(D)	-	-	375 *	5.6	188	-	-	-

Source: 1992 *Economic Census*. The states are in descending order of shipments or establishments (if shipment data are missing for the majority). The symbol (D) appears when data are withheld to prevent disclosure of competitive information. States marked with (D) are sorted by number of establishments. A dash (-) indicates that the data element cannot be calculated; * indicates the midpoint of a range.

3915 - JEWELERS' MATERIALS & LAPIDARY WORK

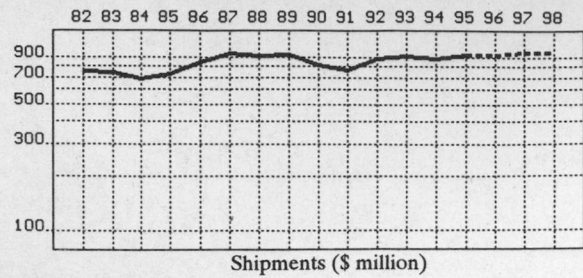

Shipments ($ million)

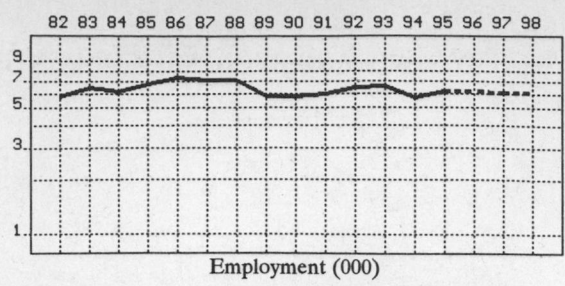

Employment (000)

GENERAL STATISTICS

Year	Companies	Establishments		Employment			Compensation		Production ($ million)			
		Total	with 20 or more employees	Total (000)	Production Workers (000)	Hours (Mil)	Payroll ($ mil)	Wages ($/hr)	Cost of Materials	Value Added by Manufacture	Value of Shipments	Capital Invest.
1982	477	481	82	5.8	4.5	8.8	86.1	6.09	536.8	193.7	755.5	7.1
1983		463	78	6.4	4.9	9.5	97.1	6.34	515.7	225.3	742.5	5.2
1984		445	74	6.2	4.7	9.2	102.0	6.75	456.4	234.8	685.5	8.0
1985		426	70	6.8	5.2	10.3	106.2	6.56	485.4	284.4	728.8	28.5
1986		428	75	7.4	5.7	11.4	126.5	7.29	565.6	281.9	846.2	5.8
1987	437	442	73	7.1	5.3	10.7	126.3	7.20	649.5	303.1	947.3	9.2
1988		405	64	7.2	5.0	10.2	132.4	7.46	651.8	270.3	920.5	6.0
1989		379	67	5.8	4.9	9.8	126.6	8.02	646.2	292.4	935.0	11.5
1990		384	66	5.8	4.5	9.1	128.0	8.89	564.1	259.0	822.2	7.1
1991		401	57	6.0	4.4	8.5	117.9	8.58	498.8	272.8	767.2	7.2
1992	419	421	60	6.4	4.2	8.4	156.0	9.70	530.8	335.3	877.8	12.9
1993		412	63	6.6	4.5	8.8	160.1	10.08	580.3	355.9	908.4	16.1
1994		382P	57P	5.7	3.9	8.3	142.2	9.39	599.7	316.7	881.9	7.9
1995		376P	55P	6.2P	4.3P	8.7P	158.5P	10.20P	629.0P	332.2P	925.0P	11.0P
1996		369P	53P	6.1P	4.2P	8.6P	163.5P	10.53P	638.0P	336.9P	938.2P	11.1P
1997		363P	51P	6.1P	4.1P	8.4P	168.4P	10.86P	647.0P	341.7P	951.5P	11.2P
1998		356P	49P	6.1P	4.0P	8.3P	173.4P	11.19P	656.0P	346.4P	964.7P	11.3P

Sources: 1982, 1987, 1992 *Economic Census*; *Annual Survey of Manufactures*, 83-86, 88-91, 93-94. Establishment counts for non-Census years are from *County Business Patterns*; establishment values for 83-84 are extrapolations. 'P's show projections by the editors. Industries reclassified in 87 will not have data for prior years.

INDICES OF CHANGE

Year	Companies	Establishments		Employment			Compensation		Production ($ million)			
		Total	with 20 or more employees	Total (000)	Production Workers (000)	Hours (Mil)	Payroll ($ mil)	Wages ($/hr)	Cost of Materials	Value Added by Manufacture	Value of Shipments	Capital Invest.
1982	114	114	137	91	107	105	55	63	101	58	86	55
1983		110	130	100	117	113	62	65	97	67	85	40
1984		106	123	97	112	110	65	70	86	70	78	62
1985		101	117	106	124	123	68	68	91	85	83	221
1986		102	125	116	136	136	81	75	107	84	96	45
1987	104	105	122	111	126	127	81	74	122	90	108	71
1988		96	107	113	119	121	85	77	123	81	105	47
1989		90	112	91	117	117	81	83	122	87	107	89
1990		91	110	91	107	108	82	92	106	77	94	55
1991		95	95	94	105	101	76	88	94	81	87	56
1992	100	100	100	100	100	100	100	100	100	100	100	100
1993		98	105	103	107	105	103	104	109	106	103	125
1994		91P	95P	89	93	99	91	97	113	94	100	61
1995		89P	91P	97P	101P	103P	102P	105P	118P	99P	105P	85P
1996		88P	88P	96P	100P	102P	105P	109P	120P	100P	107P	86P
1997		86P	85P	96P	98P	101P	108P	112P	122P	102P	108P	87P
1998		85P	82P	95P	96P	99P	111P	115P	124P	103P	110P	87P

Sources: Same as General Statistics. Values reflect change from the base year, 1992. Values above 100 mean greater than 92, values below 100 mean less than 92, and a value of 100 in the 82-91 or 93-98 period means same as 92. 'P's mark projections by the editors.

SELECTED RATIOS

For 1994	Avg. of All Manufact.	Analyzed Industry	Index	For 1994	Avg. of All Manufact.	Analyzed Industry	Index
Employees per Establishment	49	15	30	Value Added per Production Worker	134,084	81,205	61
Payroll per Establishment	1,500,273	372,133	25	Cost per Establishment	5,045,178	1,569,397	31
Payroll per Employee	30,620	24,947	81	Cost per Employee	102,970	105,211	102
Production Workers per Establishment	34	10	30	Cost per Production Worker	146,988	153,769	105
Wages per Establishment	853,319	203,959	24	Shipments per Establishment	9,576,895	2,307,906	24
Wages per Production Worker	24,861	19,984	80	Shipments per Employee	195,460	154,719	79
Hours per Production Worker	2,056	2,128	104	Shipments per Production Worker	279,017	226,128	81
Wages per Hour	12.09	9.39	78	Investment per Establishment	321,011	20,674	6
Value Added per Establishment	4,602,255	828,795	18	Investment per Employee	6,552	1,386	21
Value Added per Employee	93,930	55,561	59	Investment per Production Worker	9,352	2,026	22

Sources: Same as General Statistics. The 'Average of All Manufacturing' column represents the average of all manufacturing industries reported for the most recent complete year available. The Index shows the relationship between the Average and the Analyzed Industry. For example, 100 means that they are equal; 500 that the Analyzed Industry is five times the average; 50 means that the Analyzed Industry is half the national average. The abbreviation 'na' is used to show that data are 'not available'.

2130

LEADING COMPANIES   Number shown: **16**   Total sales ($ mil): **227**   Total employment (000): **0.9**

Company Name	Address				CEO Name	Phone	Co. Type	Sales ($ mil)	Empl. (000)
Lazare Kaplan International Inc	529 5th Av	New York	NY	10017	Leon Tempelsman	212-972-9700	P	158	0.3
Excell Manufacturing Co	200 Chestnut St	Providence	RI	02903	HM Kilguss	401-421-5957	R	18*	<0.1
Magic Novelty Company Inc	308 Dyckman St	New York	NY	10034	Alex Neuburger	212-304-2777	R	11	0.1
MS Co	61 School St	Attleboro	MA	02703	Kurt Schweinshaut	508-222-1700	R	11	<0.1
Ampex Casting Corp	23 W 47th St	New York	NY	10036	J Ipek	212-719-1318	R	4	<0.1
George H Fuller and Son Inc	151 Exchange St	Pawtucket	RI	02860	FP Mooney Jr	401-722-6530	R	4	<0.1
Lee's Manufacturing Co	1700 Smith St	N Providence	RI	02911	C Morvillo	401-353-1740	R	4*	<0.1
Newall Manufacturing Company	30 E Edams St	Chicago	IL	60603	TO Wright	312-236-2789	R	4	<0.1
James A Murphy and Son Inc	PO Box 3006	S Attleboro	MA	02703	James Murphy III	508-761-5060	R	3	<0.1
Bruce Diamond Corp	PO Box 420	Attleboro	MA	02703	RB Puleston	508-222-3755	R	2	<0.1
Eastern Reproduction Corp	1250 Main St	Waltham	MA	02154	Robert Maguire	617-893-0555	R	2	<0.1
I Kassoy Inc	16 Midland Av	Hicksville	NY	11801	I Teper	516-942-0560	R	2*	<0.1
Richard N Bird and Company	PO Box 569	Waltham	MA	02254	Carl J Cunningham	617-894-0160	R	2*	<0.1
Craftstones Inc	PO Box 847	Ramona	CA	92065	HE Walters	619-789-1620	R	1	<0.1
Contempo Lapidary Equip Mfg	12257 Foothill Blv	Sylmar	CA	91342	Ernie Wilson	818-899-1973	R	1*	<0.1
Max Duraffourg Gem Company	PO Box 568	Bedford	NY	10506	Pierre Zweigart	914-234-9784	R	1	<0.1

Source: *Ward's Business Directory of U.S. Private and Public Companies*, Volumes 1 and 2, 1996. The company type code used is as follows: P - Public, R - Private, S - Subsidiary, D - Division, J - Joint Venture, A - Affiliate, G - Group. Sales are in millions of dollars, employees are in thousands. An asterisk (*) indicates an estimated sales volume. The symbol < stands for 'less than'. Company names and addresses are truncated, in some cases, to fit into the available space.

MATERIALS CONSUMED

Material	Quantity	Delivered Cost ($ million)
Materials, ingredients, containers, and supplies	(X)	474.3
Fabricated metal products, including forgings	(X)	22.3
Precious metals (gold, platinum, etc.), all forms	(X)	125.9
Other shapes and forms, including castings	(X)	5.9
Precious, semiprecious, and synthetic stones, and pearls; cut, polished, or drilled	(X)	139.3
Jewelers' findings, including joints, pins, clasps, chains, flat stock, etc.	(X)	18.0
Other jewelry, silverware, and plated ware	(X)	0.2
All other materials and components, parts, containers, and supplies	(X)	10.4
Materials, ingredients, containers, and supplies, nsk	(X)	152.3

Source: 1992 *Economic Census*. Explanation of symbols used: (D): Withheld to avoid disclosure of competitive data; na: Not available; (S): Withheld because statistical norms were not met; (X): Not applicable; (Z): Less than half the unit shown; nec: Not elsewhere classified; nsk: Not specified by kind; - : zero; * : 10-19 percent estimated; ** : 20-29 percent estimated.

PRODUCT SHARE DETAILS

Product or Product Class	% Share	Product or Product Class	% Share
Jewelers' materials and lapidary work	100.00	except machine chain	51.25
Lapidary work and diamond cutting and polishing	19.46	Jewelers' machine chain of platinum and karat gold . . .	14.69
Diamonds cut or polished in the plant from own materials		Jewelers' findings and materials of silver	3.61
for jewelry purposes	89.72	Jewelers' findings and materials made of base metal clad	
Other cut or polished natural precious and semiprecious		with precious metal	30.00
stones including synthetic stones, industrial diamonds,		Jewelers' findings and materials of precious metal, nsk . .	0.47
and drilling of pearls from own materials	5.49	Jewelers' findings and shop stock products made of base	
Lapidary work and diamond cutting and polishing, nsk . .	4.85	metal not clad with precious metal	10.05
Jewelers' findings and materials of precious metal	61.68	Jewelers' materials and lapidary work, nsk	8.81
Jewelers' findings and materials of platinum and karat gold,			

Source: 1992 *Economic Census*. The values shown are percent of total shipments in an industry. Values of indented subcategories are summed in the main heading. The symbol (D) appears when data are withheld to prevent disclosure of competitive information. The abbreviation nsk stands for 'not specified by kind' and nec for 'not elsewhere classified'.

INPUTS AND OUTPUTS FOR JEWELERS' MATERIALS & LAPIDARY WORK

Economic Sector or Industry Providing Inputs	%	Sector	Economic Sector or Industry Buying Outputs	%	Sector
Imports	54.9	Foreign	Personal consumption expenditures	66.2	
Noncomparable imports	25.0	Foreign	Watch, clock, jewelry, & furniture repair	10.5	Services
Nonferrous rolling & drawing, nec	4.5	Manufg.	Jewelry, precious metal	10.4	Manufg.
Motor freight transportation & warehousing	4.4	Util.	Exports	7.8	Foreign
Jewelers' materials & lapidary work	2.5	Manufg.	Costume jewelry	2.0	Manufg.
Wholesale trade	2.0	Trade	Jewelers' materials & lapidary work	2.0	Manufg.
Cyclic crudes and organics	1.3	Manufg.	Colleges, universities, & professional schools	0.5	Services
Nonmetallic mineral services	1.2	Mining	Abrasive products	0.3	Manufg.
Nonmetallic mineral products, nec	0.6	Manufg.	Watches, clocks, & parts	0.2	Manufg.
Advertising	0.3	Services			
Industrial inorganic chemicals, nec	0.2	Manufg.			
Primary nonferrous metals, nec	0.2	Manufg.			
Railroads & related services	0.2	Util.			
Water transportation	0.2	Util.			
Eating & drinking places	0.2	Trade			
Banking	0.2	Fin/R.E.			
U.S. Postal Service	0.2	Gov't			
Jewelry, precious metal	0.1	Manufg.			
Paperboard containers & boxes	0.1	Manufg.			
Petroleum refining	0.1	Manufg.			
Air transportation	0.1	Util.			
Electric services (utilities)	0.1	Util.			

Source: Benchmark Input-Output Accounts for the U.S. Economy, 1982, U.S. Department of Commerce, Washington, D.C., July 1991. Data, as reported in the source, are organized by the 1977 SIC structure in use in 1982 but have been matched, as closely as is possible, to the 1987 SIC structure used in this book.

OCCUPATIONS EMPLOYED BY SIC 391 - JEWELRY, SILVERWARE, AND PLATED WARE

Occupation	% of Total 1994	Change to 2005	Occupation	% of Total 1994	Change to 2005
Precision metal workers nec	12.2	5.3	Secretaries, ex legal & medical	1.7	-20.1
Jewelers & silversmiths	10.4	-21.1	Order clerks, materials, merchandise, & service	1.6	-14.2
Assemblers, fabricators, & hand workers nec	7.6	-12.3	Tool & die makers	1.6	-29.2
Sales & related workers nec	4.7	-12.3	Machine forming operators, metal & plastic	1.6	-12.3
Grinders & polishers, hand	4.0	-12.3	Production, planning, & expediting clerks	1.5	-12.3
General managers & top executives	3.7	-16.8	Machine tool cutting operators, metal & plastic	1.5	-26.9
Precision workers nec	3.5	-21.0	Solderers & brazers	1.4	-12.2
Blue collar worker supervisors	3.4	-19.3	Clerical supervisors & managers	1.4	-10.3
General office clerks	2.7	-25.2	Industrial production managers	1.2	-12.1
Bookkeeping, accounting, & auditing clerks	2.6	-34.2	Metal molding machine workers	1.2	-3.5
Traffic, shipping, & receiving clerks	2.5	-15.5	Adjustment clerks	1.1	5.1
Inspectors, testers, & graders, precision	2.1	-12.3	Stock clerks	1.1	-28.7
Grinding machine operators, metal & plastic	2.1	-29.9	Designers, ex interior designers	1.0	-3.5
Hand packers & packagers	2.0	-24.7			

Source: Industry-Occupation Matrix, Bureau of Labor Statistics. These data relate to one or more 3-digit SIC industry groups rather than to a single 4-digit SIC. The change reported for each occupation to the year 2005 is a percent of growth or decline as estimated by the Bureau of Labor Statistics. The abbreviation nec stands for 'not elsewhere classified'.

LOCATION BY STATE AND REGIONAL CONCENTRATION

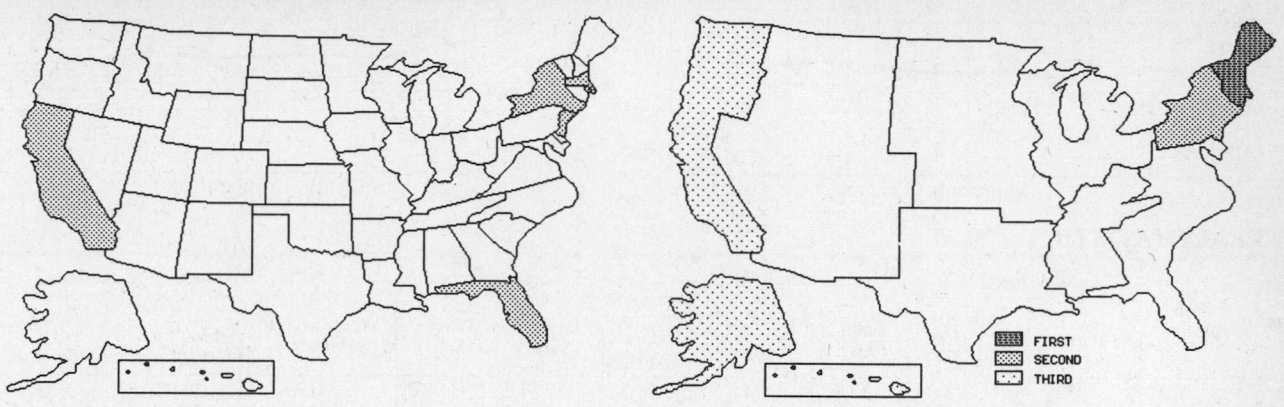

INDUSTRY DATA BY STATE

| State | Establish-ments | Shipments | | | Employment | | | | Cost as % of Shipments | Investment per Employee ($) |
		Total ($ mil)	% of U.S.	Per Establ.	Total Number	% of U.S.	Per Establ.	Wages ($/hour)		
Rhode Island	137	211.1	24.0	1.5	2,500	39.1	18	10.11	47.3	1,840
Massachusetts	20	127.1	14.5	6.4	1,000	15.6	50	8.54	58.6	2,400
New Jersey	13	47.7	5.4	3.7	300	4.7	23	8.00	65.0	1,667
California	28	11.2	1.3	0.4	200	3.1	7	7.00	32.1	-
New York	133	(D)	-	-	1,750 *	27.3	13	-	-	-
Florida	13	(D)	-	-	175 *	2.7	13	-	-	-

Source: 1992 *Economic Census*. The states are in descending order of shipments or establishments (if shipment data are missing for the majority). The symbol (D) appears when data are withheld to prevent disclosure of competitive information. States marked with (D) are sorted by number of establishments. A dash (-) indicates that the data element cannot be calculated; * indicates the midpoint of a range.

3931 - MUSICAL INSTRUMENTS

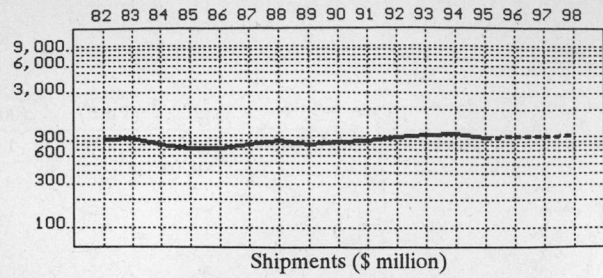

82 83 84 85 86 87 88 89 90 91 92 93 94 95 96 97 98

Shipments ($ million)

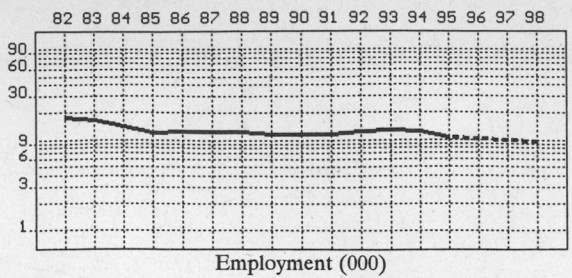

82 83 84 85 86 87 88 89 90 91 92 93 94 95 96 97 98

Employment (000)

GENERAL STATISTICS

| Year | Companies | Establishments | | Employment | | | Compensation | | Production ($ million) | | | |
		Total	with 20 or more employees	Total (000)	Production Workers (000)	Hours (Mil)	Payroll ($ mil)	Wages ($/hr)	Cost of Materials	Value Added by Manufacture	Value of Shipments	Capital Invest.
1982	403	452	137	17.8	14.0	25.7	257.8	6.87	415.5	494.8	915.9	20.1
1983				16.6	13.1	24.7	261.3	7.11	441.4	489.4	945.6	15.5
1984				14.0	10.8	20.8	228.6	7.40	384.5	421.8	811.4	19.3
1985				12.1	9.1	17.4	202.8	7.75	295.7	407.2	725.2	15.0
1986				12.2	9.4	18.8	211.8	7.46	301.3	443.5	742.2	11.0
1987	402	423	104	12.2	9.6	18.5	218.1	7.89	318.9	501.9	814.1	13.9
1988				12.3	9.4	18.4	235.2	8.46	334.3	538.8	875.8	24.0
1989		392	110	11.6	9.1	17.6	226.2	8.67	323.6	506.8	814.7	10.9
1990				11.7	9.1	17.7	233.8	8.72	335.0	547.7	872.9	14.0
1991				11.5	8.7	17.4	238.1	8.90	349.2	536.3	881.3	12.6
1992	437	461	104	12.2	9.4	17.9	272.7	9.87	405.7	588.4	981.3	13.8
1993				12.5	9.7	18.9	290.0	10.14	411.6	629.8	1,036.8	17.1
1994				12.3	9.8	19.3	280.9	9.63	420.3	637.6	1,061.6	11.4
1995				10.5P	8.2P	16.3P	266.6P	10.23P	391.3P	593.6P	988.4P	12.6P
1996				10.2P	7.9P	15.9P	270.0P	10.50P	397.3P	602.7P	1,003.4P	12.2P
1997				9.8P	7.6P	15.4P	273.4P	10.77P	403.2P	611.7P	1,018.5P	11.8P
1998				9.5P	7.3P	15.0P	276.7P	11.03P	409.2P	620.7P	1,033.5P	11.4P

Sources: 1982, 1987, 1992 *Economic Census*; *Annual Survey of Manufactures*, 83-86, 88-91, 93-94. Establishment counts for non-Census years are from *County Business Patterns*; establishment values for 83-84 are extrapolations. 'P's show projections by the editors. Industries reclassified in 87 will not have data for prior years.

INDICES OF CHANGE

| Year | Companies | Establishments | | Employment | | | Compensation | | Production ($ million) | | | |
		Total	with 20 or more employees	Total (000)	Production Workers (000)	Hours (Mil)	Payroll ($ mil)	Wages ($/hr)	Cost of Materials	Value Added by Manufacture	Value of Shipments	Capital Invest.
1982	92	98	132	146	149	144	95	70	102	84	93	146
1983				136	139	138	96	72	109	83	96	112
1984				115	115	116	84	75	95	72	83	140
1985				99	97	97	74	79	73	69	74	109
1986				100	100	105	78	76	74	75	76	80
1987	92	92	100	100	102	103	80	80	79	85	83	101
1988				101	100	103	86	86	82	92	89	174
1989		85	106	95	97	98	83	88	80	86	83	79
1990				96	97	99	86	88	83	93	89	101
1991				94	93	97	87	90	86	91	90	91
1992	100	100	100	100	100	100	100	100	100	100	100	100
1993				102	103	106	106	103	101	107	106	124
1994				101	104	108	103	98	104	108	108	83
1995				86P	87P	91P	98P	104P	96P	101P	101P	91P
1996				83P	84P	89P	99P	106P	98P	102P	102P	88P
1997				81P	81P	86P	100P	109P	99P	104P	104P	85P
1998				78P	78P	84P	101P	112P	101P	105P	105P	83P

Sources: Same as General Statistics. Values reflect change from the base year, 1992. Values above 100 mean greater than 92, values below 100 mean less than 92, and a value of 100 in the 82-91 or 93-98 period means same as 92. 'P's mark projections by the editors.

SELECTED RATIOS

For 1992	Avg. of All Manufact.	Analyzed Industry	Index	For 1992	Avg. of All Manufact.	Analyzed Industry	Index
Employees per Establishment	46	26	58	Value Added per Production Worker	122,353	62,596	51
Payroll per Establishment	1,332,320	591,540	44	Cost per Establishment	4,239,462	880,043	21
Payroll per Employee	29,181	22,352	77	Cost per Employee	92,853	33,254	36
Production Workers per Establishment	31	20	65	Cost per Production Worker	135,003	43,160	32
Wages per Establishment	734,496	383,239	52	Shipments per Establishment	8,100,800	2,128,633	26
Wages per Production Worker	23,390	18,795	80	Shipments per Employee	177,425	80,434	45
Hours per Production Worker	2,025	1,904	94	Shipments per Production Worker	257,966	104,394	40
Wages per Hour	11.55	9.87	85	Investment per Establishment	278,244	29,935	11
Value Added per Establishment	3,842,210	1,276,356	33	Investment per Employee	6,094	1,131	19
Value Added per Employee	84,153	48,230	57	Investment per Production Worker	8,861	1,468	17

Sources: Same as General Statistics. The 'Average of All Manufacturing' column represents the average of all manufacturing industries reported for the most recent complete year available. The Index shows the relationship between the Average and the Analyzed Industry. For example, 100 means that they are equal; 500 that the Analyzed Industry is five times the average; 50 means that the Analyzed Industry is half the national average. The abbreviation 'na' is used to show that data are 'not available'.

LEADING COMPANIES   Number shown: **75**   Total sales ($ mil): **818**   Total employment (000): **10.3**

Company Name	Address				CEO Name	Phone	Co. Type	Sales ($ mil)	Empl. (000)
Baldwin Piano and Organ Co	422 Wards Corner	Loveland	OH	45140	Karen Hendricks	513-576-4500	P	121	1.6
Selmer Company Inc	PO Box 310	Elkhart	IN	46515	Tom Burzycki	219-522-1675	R	85*	1.0
Vincent Bach-Selmer Co	600 Industrial Pkwy	Elkhart	IN	46516	Thomas T Burzycki	219-295-6730	S	51*	0.6
Kaman Music Corp	20 Old Windsor Rd	Bloomfield	CT	06002	William Kaman II	203-243-8353	S	40*	0.5
United Musical Instruments	1000 Indrial Pkwy	Elkhart	IN	46516	Robert Palmer	219-295-0079	R	40*	0.8
Wenger Corp	555 Park Dr	Owatonna	MN	55060	Timothy P Covey	507-455-4100	R	40	0.3
Gibson Guitar Corp	1818 Elm Hill Pike	Nashville	TN	37210	Henry Juszkiewicz	615-871-4500	R	29	0.5
Allen Organ Co	PO Box 36	Macungie	PA	18062	Steven A Markowitz	215-966-2200	P	29	0.5
Zildjian Co	22 Longwater Dr	Norwell	MA	02061	James Roberts	617-871-2200	R	26	<0.1
G Leblanc Corp	7001 Leblanc Blv	Kenosha	WI	53141	Vito Pascucci	414-658-1644	R	23*	0.4
Remo Inc	12804 Raymer St	N Hollywood	CA	91605	Remo D Belli	818-983-2600	R	22	0.3
United Musical Instruments	33999 Curtis Blv	East Lake	OH	44095	Robert Palmer	216-946-6100	R	20*	0.3
Ludwig Industries	2806 Mason St	Monroe	NC	28110	Tom Burzycki	704-289-6459	D	18	<0.1
Rico International	PO Box 661	Sun Valley	CA	91353	Marvin Snyder	818-767-7711	R	18*	0.2
Yamaha Music Manufacturing	PO Box 1237	Thomaston	GA	30286	Yoshi Ishiwaka	706-647-9601	S	18	0.2
Ernie Ball Inc	PO Box 4117	S L Obispo	CA	93403	Ernie Ball	805-544-7726	R	16*	0.2
PianoDisc	4111 N Freeway	Sacramento	CA	95834	Kirk Burgett	916-973-8710	D	16*	0.1
Burgett Inc	4111 N Freeway	Sacramento	CA	95834	Kirk Burgett	916-973-8710	R	12*	0.1
Rodgers Instrument Corp	1300 NE 25th Av	Hillsboro	OR	97124	Dennis Houlihan	503-648-4181	S	11*	0.1
Fred Gretsch Enterprises Ltd	PO Box 2468	Savannah	GA	31402	Fred Gretsch	912-748-1101	R	10	<0.1
Getzen Company Inc	PO Box 440	Elkhorn	WI	53121	Edward Getzen	414-723-4221	R	10*	0.1
GHS Corp	2813 Wilber Av	Battle Creek	MI	49015	Robert D McFee	616-968-3351	R	10	0.1
US Music Corp	2885 S James Dr	New Berlin	WI	53151	Darrold Orr	414-784-8388	R	10	0.1
Taylor Guitars	1940 Gillespie Way	El Cajon	CA	92020	Kurt Listug	619-258-1207	R	8	<0.1
Ovation Instruments	37 Greenwoods Rd	New Hartford	CT	06002	Chuck Bashaw	203-379-7575	D	8*	0.1
Frank Holton and Co	320 N Church St	Elkhorn	WI	53121	V Pascucci	414-723-2220	D	7*	0.1
Lowrey Organ Co	825 E 26th St	La Grange Park	IL	60525	Ippei Koga	708-352-3388	S	6	<0.1
Lyon and Healy Harps Inc	168 N Ogden Av	Chicago	IL	60607	Janet Harrell	312-786-1881	S	6*	<0.1
Mason and Hamlin Piano Co	35 Duncan St	Haverhill	MA	01830	Lloyd Meyer	508-372-8300	R	6	0.1
Schaff Piano Supply Co	451 Oakwood Rd	Lake Zurich	IL	60047	Dave Johnson	708-438-4556	R	6*	<0.1
Schantz Organ Co	626 S Walnut St	Orrville	OH	44667	John Schantz	216-682-6065	R	6	<0.1
Organ Supply Industries Inc	PO Box 8325	Erie	PA	16505	Dennis A Unks	814-835-2244	R	5*	<0.1
PRS Guitars Ltd	1812 Virginia Av	Annapolis	MD	21401	Clay J Evans	410-263-2701	R	5	<0.1
QRS Music Rolls Inc	1026 Niagara St	Buffalo	NY	14213	Richard A Dolan	716-885-4600	R	5	<0.1
Sound Enhancements Inc	185 Detroit Av	Cary	IL	60013	Randy Wright	708-639-4646	S	5	<0.1
Ultimate Support Systems Inc	PO Box 470	Fort Collins	CO	80522	James Dismore	970-493-4488	R	5	<0.1
Dean Markley Strings Inc	3350 Scott Blv	Santa Clara	CA	95054	Dean Markley	408-988-2456	R	4	<0.1
Drum Workshop Inc	101 Bernoulli Cir	Oxnard	CA	93030	Don Lombardi	805-485-6999	R	4*	<0.1
John Schadler and Sons Inc	PO Box 1068	Clifton	NJ	07014	EJ Schadler	201-777-3600	R	4	<0.1
Schulmerich Carillons Inc	Carillon Hill	Sellersville	PA	18960	Bern E Deichmann	215-257-2771	R	4	0.1
Wicks Organ Company Inc	PO Box 129	Highland	IL	62249	Martin M Wick	618-654-2191	R	4	<0.1
Carter Duncan Corp	5427 Hollister Ave	Santa Barbara	CA	93111	C Carter-Duncan	805-964-9610	R	4	<0.1
Austin Organs Inc	156 Woodland St	Hartford	CT	06105	Donald Austin	203-522-8293	R	3	<0.1
Pearl Corp	PO Box 1111240	Nashville	TN	37211	Ralph Miller	615-833-4477	S	3*	<0.1
Reuter Organ Co	PO Box 486	Lawrence	KS	66044	Albert Neutel	913-843-2622	R	3	<0.1
Posey Manufacturing Company	PO Box 418	Hoquiam	WA	98550	Frank Johnson	360-533-0565	R	3	<0.1
Alembic Inc	3005 Wiljan Ct	Santa Rosa	CA	95407	Susan Wickersham	707-523-2611	R	2	<0.1
BBE Sound Inc	5500 Bolsa Av	Huntington Bch	CA	92649	John C McLaren	714-897-6766	R	2	<0.1
Borisoff Engineering Co	8248 Rte	Interlaken	NY	14847	D Borisoff	607-532-9404	R	2*	<0.1
Kawai America Corp	2055 E University	Compton	CA	90224	Masao Yamamoto	310-631-1771	R	2	<0.1
Maas-Rowe Carillons Inc	2255 Meyers Av	Escondido	CA	92029	Paul H Rowe	619-743-1311	R	2	<0.1
Story and Clark Piano Co	269 Quaker Dr	Seneca	PA	16346	Edward E Keefer	814-676-6683	D	2*	<0.1
William Haynes Co	12 Piedmont St	Boston	MA	02116	Anna C Deveau	617-482-7456	R	2	<0.1
Oberheim	732 Kevin Ct	Oakland	CA	94621	Henry Juszkiewicz	510-635-9633	D	2	<0.1
Prestini Musical Instrument	PO Box 2296	Nogales	AZ	85628	Giuseppi Prestini	602-287-4931	R	1	<0.1
Kamaka Hawaii Inc	550 South St	Honolulu	HI	96813	Sam Kamaka Jr	808-531-3165	R	1*	<0.1
Kawai America Manufacturing	2001 Kawai Rd	Lincolnton	NC	28092	Ippei Koga	704-735-8766	S	1*	<0.1
Original Musical Instruments	18108 Redondo Cir	Huntington Bch	CA	92648	Chester Lizak	714-848-9823	R	1*	<0.1
Reisner Inc	PO Box 71	Hagerstown	MD	21740	David R Parker	301-733-2650	S	1*	<0.1
SIT Strings Co	815 S Broadway	Akron	OH	44311	Virgil L Lay	216-434-8010	R	1*	<0.1
Super-Sensitive Musical	6121 Porter Rd	Sarasota	FL	34240	JV Cavanaugh	813-371-0016	R	1*	<0.1
Tobias Guitars Inc	1050 Acorn Dr	Nashville	TN	37210	Ron Eubanks	615-872-8420	D	1*	<0.1
G and L Musical Products	5381 Production Dr	Huntington Bch	CA	92649	John C McLaren	714-897-6766	R	1	<0.1
Gulbransen Inc	2102 Hancock St	San Diego	CA	92110	Robert Hill	619-296-5760	R	1	<0.1
Music Maestro Inc	1639 11th St	Santa Monica	CA	90404	Larry Worchel	310-314-3888	R	1*	<0.1
Lewis and Hitchcock	8466-A Tyco Rd	Vienna	VA	22182	David L Selby	703-734-8585	R	1*	<0.1
Jaeckel Inc	1600 London Rd	Duluth	MN	55812	Daniel J Jaeckel	218-728-2394	R	1	<0.1
Lincoln Organ Co	4221 NW 37th St	Lincoln	NE	68524	Gene R Bedient	402-470-3675	R	1	<0.1
Wendin Inc	PO Box 3888	Spokane	WA	99220	Wendal S Jones	509-747-1224	R	1	<0.1
John Brombaugh and Associates	325 N Brooklyn Av	Eugene	OR	97403	John Brombaugh	503-726-9323	R	0	<0.1
Zuckerman Harpsichords Inc	PO Box 151	Stonington	CT	06378	Katherine Way	203-535-1715	R	0	<0.1
Kansas City Drumworks	1827 McGee St	Kansas City	MO	64108	Gary Boyle	816-471-3786	R	0	<0.1
Olivieri Reeds	7609 E Speedway	Tucson	AZ	85710	Charles Olivieri	602-290-8980	R	0	<0.1
American Drum Mfg Co	1150 S Kalamath St	Denver	CO	80223	Marshall Light	303-722-3844	R	0*	<0.1
Buscarino Guitars	9075 130th Av	Largo	FL	34643	John Buscarino	813-586-4992	R	0	<0.1

Source: Ward's Business Directory of U.S. Private and Public Companies, Volumes 1 and 2, 1996. The company type code used is as follows: P - Public, R - Private, S - Subsidiary, D - Division, J - Joint Venture, A - Affiliate, G - Group. Sales are in millions of dollars, employees are in thousands. An asterisk (*) indicates an estimated sales volume. The symbol < stands for 'less than'. Company names and addresses are truncated, in some cases, to fit into the available space.

MATERIALS CONSUMED

Material	Quantity	Delivered Cost ($ million)
Materials, ingredients, containers, and supplies	(X)	344.5
Rough and dressed lumber	(X)	46.7
Paperboard containers, boxes, and corrugated paperboard	(X)	5.4
Paints, varnishes, lacquers, stains, shellacs, japans, enamels, and allied products	(X)	4.8
Loudspeakers, microphones, and tuners (all types)	(X)	4.1
Broadcast, studio, and related electronic amplifiers	(X)	0.6
Electronic components and accessories, including circuit boards and recording heads	(X)	44.6
All other electronic and electrical equipment and components, except computer equipment	(X)	6.9
Parts specially designed for musical instruments, including actions, strings, mouthpieces, etc.	(X)	100.3
All other materials and components, parts, containers, and supplies	(X)	78.7
Materials, ingredients, containers, and supplies, nsk	(X)	52.3

Source: 1992 Economic Census. Explanation of symbols used: (D): Withheld to avoid disclosure of competitive data; na: Not available; (S): Withheld because statistical norms were not met; (X): Not applicable; (Z): Less than half the unit shown; nec: Not elsewhere classified; nsk: Not specified by kind; - : zero; * : 10-19 percent estimated; ** : 20-29 percent estimated.

PRODUCT SHARE DETAILS

Product or Product Class	% Share	Product or Product Class	% Share
Musical instruments	100.00	Brass wind musical instruments	12.93
Pianos	15.60	Nonelectronic fretted or string instruments (such as harps, harpsichords, guitars, banjos, etc.)	7.19
Vertical, upright, or console pianos, 37 in. or less in height	59.86		
Grand pianos	39.50	Electronic musical instruments, other than electronic organs and synthesizers	25.64
Pianos, nsk	0.64		
Organs	9.66	Percussion musical instruments (cymbals, drums, vibraphones (nonelectronic), etc.)	9.39
Pipe and reed organs	40.57		
Electronic organs	59.43	Other nonelectronic musical instruments, including accordions, harmonicas, bagpipes, etc.	2.62
Piano and organ parts	3.25		
Piano parts (actions, attachments, strings, tuning pins, etc.), except benches	53.24	Accessories and parts for other musical instruments, such as reed mouthpieces, strings (excluding piano strings), music stands, drummers' traps, etc.	27.31
Organ parts and materials, except benches	46.76		
Other musical instruments and parts	63.94	Other musical instruments and parts, nsk	0.56
Woodwind musical instruments	14.34	Musical instruments, nsk	7.55

Source: 1992 Economic Census. The values shown are percent of total shipments in an industry. Values of indented subcategories are summed in the main heading. The symbol (D) appears when data are withheld to prevent disclosure of competitive information. The abbreviation nsk stands for 'not specified by kind' and nec for 'not elsewhere classified'.

INPUTS AND OUTPUTS FOR MUSICAL INSTRUMENTS

Economic Sector or Industry Providing Inputs	%	Sector	Economic Sector or Industry Buying Outputs	%	Sector
Imports	32.3	Foreign	Personal consumption expenditures	49.0	
Electronic components nec	12.8	Manufg.	Gross private fixed investment	15.1	Cap Inv
Musical instruments	11.3	Manufg.	Exports	12.1	Foreign
Wholesale trade	6.7	Trade	Musical instruments	9.0	Manufg.
Advertising	6.0	Services	S/L Govt. purch., elem. & secondary education	5.8	S/L Govt
Veneer & plywood	3.1	Manufg.	Miscellaneous repair shops	5.3	Services
Sawmills & planning mills, general	3.0	Manufg.	S/L Govt. purch., higher education	1.4	S/L Govt
Furniture & fixtures, nec	2.3	Manufg.	Equipment rental & leasing services	1.2	Services
Brass, bronze, & copper castings	1.7	Manufg.	Change in business inventories	0.7	In House
Wood products, nec	1.5	Manufg.	S/L Govt. purch., other education & libraries	0.3	S/L Govt
Eating & drinking places	1.4	Trade	Federal Government purchases, national defense	0.2	Fed Govt
Electric services (utilities)	1.3	Util.			
Paints & allied products	1.0	Manufg.			
Semiconductors & related devices	1.0	Manufg.			
Luggage	0.8	Manufg.			
Radio & TV receiving sets	0.8	Manufg.			
U.S. Postal Service	0.8	Gov't			
Maintenance of nonfarm buildings nec	0.7	Constr.			
Plastics materials & resins	0.6	Manufg.			
Motor freight transportation & warehousing	0.6	Util.			
Banking	0.6	Fin/R.E.			
Real estate	0.6	Fin/R.E.			
Machinery, except electrical, nec	0.5	Manufg.			
Paperboard containers & boxes	0.5	Manufg.			
Petroleum refining	0.5	Manufg.			
Communications, except radio & TV	0.5	Util.			
Management & consulting services & labs	0.5	Services			
Abrasive products	0.4	Manufg.			
Railroads & related services	0.4	Util.			
Legal services	0.4	Services			
Special dies & tools & machine tool accessories	0.3	Manufg.			
Wood containers	0.3	Manufg.			
Air transportation	0.3	Util.			

Continued on next page.

INPUTS AND OUTPUTS FOR MUSICAL INSTRUMENTS - Continued

Economic Sector or Industry Providing Inputs	%	Sector	Economic Sector or Industry Buying Outputs	%	Sector
Gas production & distribution (utilities)	0.3	Util.			
Accounting, auditing & bookkeeping	0.3	Services			
Equipment rental & leasing services	0.3	Services			
Broadwoven fabric mills	0.2	Manufg.			
Felt goods, nec	0.2	Manufg.			
Manifold business forms	0.2	Manufg.			
Paperboard mills	0.2	Manufg.			
Royalties	0.2	Fin/R.E.			
Lubricating oils & greases	0.1	Manufg.			
Metal stampings, nec	0.1	Manufg.			
Miscellaneous plastics products	0.1	Manufg.			
Sanitary services, steam supply, irrigation	0.1	Util.			
Water transportation	0.1	Util.			
Automotive rental & leasing, without drivers	0.1	Services			
Computer & data processing services	0.1	Services			
Hotels & lodging places	0.1	Services			
Personnel supply services	0.1	Services			

Source: Benchmark Input-Output Accounts for the U.S. Economy, 1982, U.S. Department of Commerce, Washington, D.C., July 1991. Data, as reported in the source, are organized by the 1977 SIC structure in use in 1982 but have been matched, as closely as is possible, to the 1987 SIC structure used in this book.

OCCUPATIONS EMPLOYED BY SIC 393 - MANUFACTURED PRODUCTS, NEC

Occupation	% of Total 1994	Change to 2005	Occupation	% of Total 1994	Change to 2005
Assemblers, fabricators, & hand workers nec	17.0	4.8	Freight, stock, & material movers, hand	1.5	-16.1
Sales & related workers nec	4.3	4.8	Inspectors, testers, & graders, precision	1.5	4.9
Blue collar worker supervisors	4.0	-2.4	General office clerks	1.5	-10.6
General managers & top executives	3.8	-0.5	Designers, ex interior designers	1.5	15.3
Hand packers & packagers	3.7	-28.1	Electrical & electronic assemblers	1.4	4.8
Traffic, shipping, & receiving clerks	2.5	0.9	Industrial production managers	1.4	4.8
Machine operators nec	2.2	-7.6	Packaging & filling machine operators	1.4	4.9
Helpers, laborers, & material movers nec	2.2	4.8	Screen printing machine setters & set-up operators	1.2	4.8
Painting, coating, & decorating workers, hand	2.0	-5.7	Plastic molding machine workers	1.2	4.8
Precision workers nec	2.0	41.5	Machine feeders & offbearers	1.2	-5.7
Secretaries, ex legal & medical	2.0	-4.5	Cabinetmakers & bench carpenters	1.1	-31.9
Bookkeeping, accounting, & auditing clerks	1.8	-21.4	Order clerks, materials, merchandise, & service	1.1	2.6
Sheet metal workers & duct installers	1.8	4.8	Grinders & polishers, hand	1.1	4.8
Precision metal workers nec	1.8	36.3			

Source: Industry-Occupation Matrix, Bureau of Labor Statistics. These data relate to one or more 3-digit SIC industry groups rather than to a single 4-digit SIC. The change reported for each occupation to the year 2005 is a percent of growth or decline as estimated by the Bureau of Labor Statistics. The abbreviation nec stands for 'not elsewhere classified'.

LOCATION BY STATE AND REGIONAL CONCENTRATION

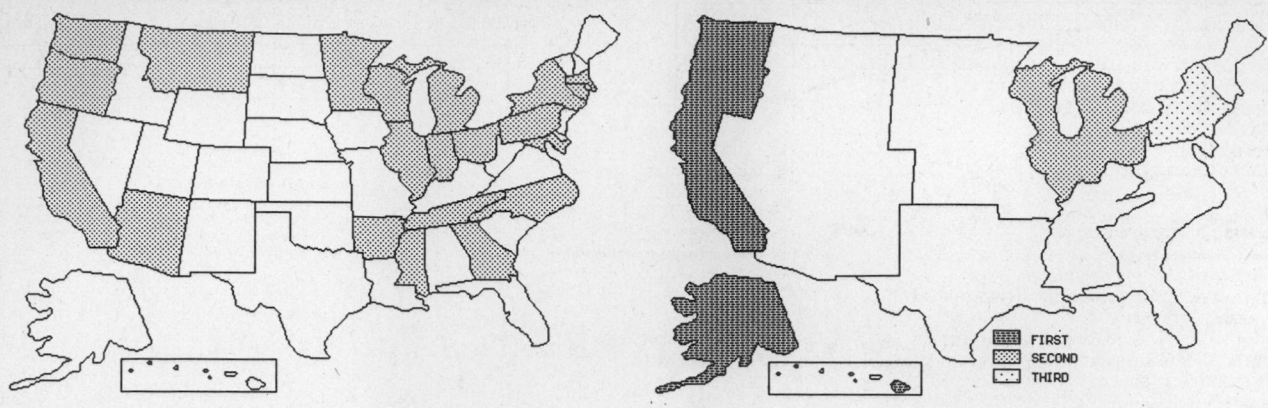

FIRST
SECOND
THIRD

INDUSTRY DATA BY STATE

| State | Establish-ments | Shipments | | | Employment | | | | Cost as % of Shipments | Investment per Employee ($) |
		Total ($ mil)	% of U.S.	Per Establ.	Total Number	% of U.S.	Per Establ.	Wages ($/hour)		
California	81	149.1	15.2	1.8	1,800	14.8	22	8.52	46.1	1,167
Indiana	17	128.1	13.1	7.5	1,700	13.9	100	12.38	38.4	1,118
Pennsylvania	23	108.1	11.0	4.7	1,300	10.7	57	10.60	41.3	1,231
New York	38	78.1	8.0	2.1	1,200	9.8	32	9.41	34.2	417
Illinois	32	56.2	5.7	1.8	600	4.9	19	9.78	42.0	667
Ohio	18	41.7	4.2	2.3	600	4.9	33	12.00	26.4	667
Michigan	11	37.8	3.9	3.4	400	3.3	36	11.67	48.9	2,000
Wisconsin	17	32.7	3.3	1.9	300	2.5	18	8.60	55.0	-
Massachusetts	23	31.2	3.2	1.4	400	3.3	17	12.00	24.0	2,750
North Carolina	12	27.8	2.8	2.3	300	2.5	25	7.25	58.6	333
Oregon	9	14.8	1.5	1.6	200	1.6	22	12.00	43.2	500
Maryland	9	10.3	1.0	1.1	200	1.6	22	7.33	36.9	-
Washington	18	(D)	-	-	175 *	1.4	10	-	-	-
Minnesota	12	(D)	-	-	375 *	3.1	31	-	-	-
Tennessee	9	(D)	-	-	375 *	3.1	42	-	-	-
Arkansas	8	(D)	-	-	375 *	3.1	47	-	-	-
Connecticut	8	(D)	-	-	175 *	1.4	22	-	-	-
Arizona	5	(D)	-	-	375 *	3.1	75	-	-	-
Mississippi	4	(D)	-	-	375 *	3.1	94	-	-	-
Georgia	3	(D)	-	-	175 *	1.4	58	-	-	-
Montana	2	(D)	-	-	175 *	1.4	88	-	-	-

Source: 1992 *Economic Census*. The states are in descending order of shipments or establishments (if shipment data are missing for the majority). The symbol (D) appears when data are withheld to prevent disclosure of competitive information. States marked with (D) are sorted by number of establishments. A dash (-) indicates that the data element cannot be calculated; * indicates the midpoint of a range.

3942 - DOLLS & STUFFED TOYS

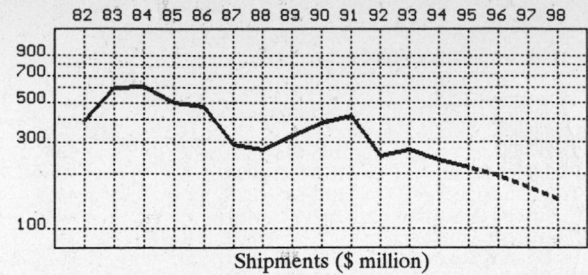

Shipments ($ million)

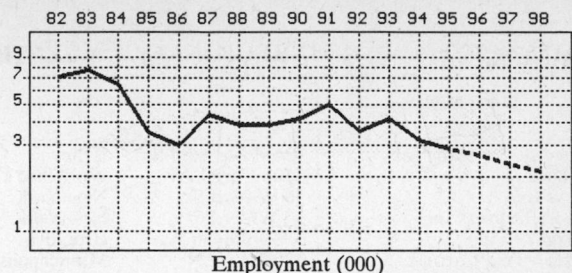

Employment (000)

GENERAL STATISTICS

Year	Companies	Establishments Total	with 20 or more employees	Total (000)	Production Workers (000)	Hours (Mil)	Payroll ($ mil)	Wages ($/hr)	Cost of Materials	Value Added by Manufacture	Value of Shipments	Capital Invest.
1982	232	236	70	7.1	5.7	10.2	84.7	4.55	189.2	208.1	396.3	7.1
1983		229	69	7.7	5.5	10.1	103.1	4.74	256.9	335.7	595.1	
1984		222	68	6.4	5.1	9.4	89.7	6.18	317.8	312.6	614.9	
1985		216	66	3.5	2.4	4.4	52.4	6.66	257.6	231.3	496.0	16.2
1986		188	64	3.0	1.8	3.3	48.4	7.00	235.6	235.0	470.6	
1987	191	197	53	4.4	3.3	5.7	57.9	5.44	154.7	150.3	294.0	4.4
1988		187	51	3.9	3.1	5.4	57.6	5.80	132.2	135.7	272.3	2.3
1989		176	48	3.9	3.4	6.0	62.5	5.73	146.0	184.7	322.5	10.4
1990		191	53	4.2	3.8	6.4	76.3	6.28	135.3	244.1	380.4	11.8
1991		188	52	5.0	3.8	5.9	87.4	8.71	152.1	274.0	419.9	11.4
1992	204	209	42	3.6	2.8	5.2	64.1	6.88	91.1	156.5	251.0	3.0
1993		221	43	4.2	3.1	6.0	73.0	6.73	90.1	183.2	272.6	2.7
1994		188P	39P	3.2	2.5	4.4	57.8	7.66	83.9	155.6	239.3	1.7
1995		185P	36P	2.9P	2.3P	4.0P	60.8P	7.74P	76.3P	141.5P	217.7P	
1996		183P	34P	2.6P	2.2P	3.7P	59.4P	7.94P	67.9P	125.9P	193.6P	
1997		180P	31P	2.4P	2.0P	3.3P	58.1P	8.14P	59.4P	110.2P	169.4P	
1998		177P	28P	2.1P	1.8P	3.0P	56.7P	8.34P	50.9P	94.5P	145.3P	

Sources: 1982, 1987, 1992 *Economic Census*; *Annual Survey of Manufactures*, 83-86, 88-91, 93-94. Establishment counts for non-Census years are from *County Business Patterns*; establishment values for 83-84 are extrapolations. 'P's show projections by the editors. Industries reclassified in 87 will not have data for prior years.

INDICES OF CHANGE

Year	Companies	Establishments Total	with 20 or more employees	Total (000)	Production Workers (000)	Hours (Mil)	Payroll ($ mil)	Wages ($/hr)	Cost of Materials	Value Added by Manufacture	Value of Shipments	Capital Invest.
1982	114	113	167	197	204	196	132	66	208	133	158	237
1983		110	164	214	196	194	161	69	282	215	237	
1984		106	162	178	182	181	140	90	349	200	245	
1985		103	157	97	86	85	82	97	283	148	198	540
1986		90	152	83	64	63	76	102	259	150	187	
1987	94	94	126	122	118	110	90	79	170	96	117	147
1988		89	121	108	111	104	90	84	145	87	108	77
1989		84	114	108	121	115	98	83	160	118	128	347
1990		91	126	117	136	123	119	91	149	156	152	393
1991		90	124	139	136	113	136	127	167	175	167	380
1992	100	100	100	100	100	100	100	100	100	100	100	100
1993		106	102	117	111	115	114	98	99	117	109	90
1994		90P	93P	89	89	85	90	111	92	99	95	57
1995		89P	87P	80P	83P	77P	95P	112P	84P	90P	87P	
1996		87P	80P	73P	77P	70P	93P	115P	74P	80P	77P	
1997		86P	74P	66P	71P	64P	91P	118P	65P	70P	68P	
1998		85P	68P	59P	64P	57P	88P	121P	56P	60P	58P	

Sources: Same as General Statistics. Values reflect change from the base year, 1992. Values above 100 mean greater than 92, values below 100 mean less than 92, and a value of 100 in the 82-91 or 93-98 period means same as 92. 'P's mark projections by the editors.

SELECTED RATIOS

For 1994	Avg. of All Manufact.	Analyzed Industry	Index	For 1994	Avg. of All Manufact.	Analyzed Industry	Index
Employees per Establishment	49	17	35	Value Added per Production Worker	134,084	62,240	46
Payroll per Establishment	1,500,273	307,670	21	Cost per Establishment	5,045,178	446,601	9
Payroll per Employee	30,620	18,062	59	Cost per Employee	102,970	26,219	25
Production Workers per Establishment	34	13	39	Cost per Production Worker	146,988	33,560	23
Wages per Establishment	853,319	179,407	21	Shipments per Establishment	9,576,895	1,273,796	13
Wages per Production Worker	24,861	13,482	54	Shipments per Employee	195,460	74,781	38
Hours per Production Worker	2,056	1,760	86	Shipments per Production Worker	279,017	95,720	34
Wages per Hour	12.09	7.66	63	Investment per Establishment	321,011	9,049	3
Value Added per Establishment	4,602,255	828,260	18	Investment per Employee	6,552	531	8
Value Added per Employee	93,930	48,625	52	Investment per Production Worker	9,352	680	7

Sources: Same as General Statistics. The 'Average of All Manufacturing' column represents the average of all manufacturing industries reported for the most recent complete year available. The Index shows the relationship between the Average and the Analyzed Industry. For example, 100 means that they are equal; 500 that the Analyzed Industry is five times the average; 50 means that the Analyzed Industry is half the national average. The abbreviation 'na' is used to show that data are 'not available'.

LEADING COMPANIES   Number shown: 21   Total sales ($ mil): 355   Total employment (000): 2.6

Company Name	Address				CEO Name	Phone	Co. Type	Sales ($ mil)	Empl. (000)
Ace Novelty Company Inc	13434 NE 16th St	Bellevue	WA	98005	B Mayers	206-644-1820	R	90*	1.0
Applause Inc	6101 Variel Av	Woodland Hills	CA	91365	James Klein	818-992-6000	R	78*	0.4
Just Toys Inc	50 W 23rd St	New York	NY	10010	Allan Rigberg	212-645-6335	P	43	<0.1
Happiness Express Inc	50 W 23rd St	New York	NY	10010	Joseph A Sutton	212-675-0461	P	40	<0.1
Goldberger Doll	538 Johnson Av	Brooklyn	NY	11237	Eugene Goldberger	718-366-5800	R	25	0.5
Manhattan Toy/Carousel	430 1st Av N	Minneapolis	MN	55401	Dean K Rizer Jr	612-337-9600	D	20	<0.1
K and B Brothers Inc	200 5th Av	New York	NY	10010	WA Levine	212-924-0673	R	10	<0.1
Original Appalachian Artworks	PO Box 714	Cleveland	GA	30528	X Roberts	706-865-2171	R	10	<0.1
Azrac Hanway International	1107 Broadway	New York	NY	10010	Ezra Hanway	212-675-3427	R	7	<0.1
Middleton Doll Co	1301 Washington	Belpre	OH	45714	Lee Middleton	614-423-1717	R	6*	<0.1
Douglas Company Inc	PO Drawer D	Keene	NH	03431	Scott T Clarke	603-352-3414	R	6	<0.1
Lovee Doll and Toy Company	303 Stanley Av	Brooklyn	NY	11207	IL Roth	718-257-5000	R	5*	0.1
Animal Fair Inc	PO Box 1326	Minneapolis	MN	55440	Larry Schoenecker	612-831-7200	R	4	<0.1
Fable Toy Company Inc	1710 Flushing Av	Ridgewood	NY	11385	Sam You	718-456-8500	R	3	<0.1
Commonwealth Toy and Novelty	45 W 25th St	New York	NY	10010	Steven Greenfield	212-242-4070	R	3*	<0.1
Well-Made Toy Mfg Co	184-10 Jamaica Av	Hollis	NY	11423	Fred F Catapano	718-454-1326	R	2*	<0.1
Cultural Exchange Corp	80 S 8th St	Minneapolis	MN	55402	Jacob R Miles III	612-339-1254	R	1	<0.1
Admiration Toy Company Inc	60 McLean Av	Yonkers	NY	10705	Sidney Newman	914-963-9400	R	1	<0.1
Geckostufs Inc	PO Box 27244	Honolulu	HI	96827	Susana Brown	808-676-7509	R	1	<0.1
Diversified Educational Services	20054 Greenfield	Detroit	MI	48235	Melvin Chapman	313-864-5522	R	1*	<0.1
Four B's Sales Co	7313 Ashcraft Av	Houston	TX	77081	Richard Brown	713-270-1144	R	0*	<0.1

Source: *Ward's Business Directory of U.S. Private and Public Companies*, Volumes 1 and 2, 1996. The company type code used is as follows: P - Public, R - Private, S - Subsidiary, D - Division, J - Joint Venture, A - Affiliate, G - Group. Sales are in millions of dollars, employees are in thousands. An asterisk (*) indicates an estimated sales volume. The symbol < stands for 'less than'. Company names and addresses are truncated, in some cases, to fit into the available space.

MATERIALS CONSUMED

Material	Quantity	Delivered Cost ($ million)
Materials, ingredients, containers, and supplies	(X)	76.6
Steel shapes and forms	(X)	(D)
All other nonferrous shapes and forms	(X)	(D)
Plastics resins consumed in the form of granules, pellets, powders, liquids, etc.	(X)	2.1
Plastics products consumed in the form of sheets, rods, tubes, film, and other shapes	(X)	0.8
Broadwoven fabrics (piece goods)	(X)	6.9
Paperboard (including news, chip, pasted, tablet, check, binders' board), except for shipping	(X)	0.9
Paperboard containers, boxes, and corrugated paperboard	(X)	0.9
Other paper products	(X)	0.2
Hardwood lumber, rough and dressed	(X)	(D)
Other wood products (except lumber)	(X)	(D)
Doll parts	(X)	15.7
All other materials and components, parts, containers, and supplies	(X)	16.7
Materials, ingredients, containers, and supplies, nsk	(X)	32.3

Source: 1992 *Economic Census*. Explanation of symbols used: (D): Withheld to avoid disclosure of competitive data; na: Not available; (S): Withheld because statistical norms were not met; (X): Not applicable; (Z): Less than half the unit shown; nec: Not elsewhere classified; nsk: Not specified by kind; - : zero; * : 10-19 percent estimated; ** : 20-29 percent estimated.

PRODUCT SHARE DETAILS

Product or Product Class	% Share	Product or Product Class	% Share
Dolls and stuffed toys	100.00	Doll parts (clothes, accessories, and playsets for dolls, including fashion dolls and action figures)	6.61
Stuffed dolls	11.50	Stuffed toy animals	36.73
Dolls, complete, more than 13 in., including mechanical/ electrical (except stuffed dolls)	3.99	Other stuffed toys	4.48
Dolls, complete, 13 in. or less, including fashion dolls, action figures, and collectors' miniatures (except stuffed)	6.44	Puppets, marionettes, and other animals and figures not stuffed	2.38

Source: 1992 *Economic Census*. The values shown are percent of total shipments in an industry. Values of indented subcategories are summed in the main heading. The symbol (D) appears when data are withheld to prevent disclosure of competitive information. The abbreviation nsk stands for 'not specified by kind' and nec for 'not elsewhere classified'.

INPUTS AND OUTPUTS FOR DOLLS

Economic Sector or Industry Providing Inputs	%	Sector	Economic Sector or Industry Buying Outputs	%	Sector
Imports	69.0	Foreign	Personal consumption expenditures	90.0	
Dolls	8.4	Manufg.	Dolls	6.0	Manufg.
Broadwoven fabric mills	4.7	Manufg.	Games, toys, & children's vehicles	1.7	Manufg.
Wholesale trade	3.4	Trade	Exports	1.3	Foreign
Paints & allied products	2.3	Manufg.	Change in business inventories	0.7	In House
Paperboard containers & boxes	1.2	Manufg.	Federal Government purchases, nondefense	0.3	Fed Govt
Plastics materials & resins	1.2	Manufg.			
Job training & related services	0.9	Services			
Paperboard mills	0.8	Manufg.			
U.S. Postal Service	0.8	Gov't			
Real estate	0.6	Fin/R.E.			
Advertising	0.6	Services			
Petroleum refining	0.4	Manufg.			
Electric services (utilities)	0.4	Util.			
Eating & drinking places	0.4	Trade			
Banking	0.4	Fin/R.E.			
Padding & upholstery filling	0.3	Manufg.			
Yarn mills & finishing of textiles, nec	0.3	Manufg.			
Communications, except radio & TV	0.3	Util.			
Motor freight transportation & warehousing	0.3	Util.			
Credit agencies other than banks	0.3	Fin/R.E.			
Maintenance of nonfarm buildings nec	0.2	Constr.			
Sanitary services, steam supply, irrigation	0.2	Util.			
Equipment rental & leasing services	0.2	Services			
Management & consulting services & labs	0.2	Services			
Machinery, except electrical, nec	0.1	Manufg.			
Air transportation	0.1	Util.			
Automotive rental & leasing, without drivers	0.1	Services			
Legal services	0.1	Services			

Source: Benchmark Input-Output Accounts for the U.S. Economy, 1982, U.S. Department of Commerce, Washington, D.C., July 1991. Data, as reported in the source, are organized by the 1977 SIC structure in use in 1982 but have been matched, as closely as is possible, to the 1987 SIC structure used in this book.

OCCUPATIONS EMPLOYED BY SIC 394 - TOYS AND SPORTING GOODS

Occupation	% of Total 1994	Change to 2005	Occupation	% of Total 1994	Change to 2005
Assemblers, fabricators, & hand workers nec	23.5	21.3	Coating, painting, & spraying machine workers	1.5	21.3
Plastic molding machine workers	4.3	21.3	Machine forming operators, metal & plastic	1.5	21.3
Blue collar worker supervisors	4.2	13.5	Painting, coating, & decorating workers, hand	1.4	21.3
Hand packers & packagers	3.6	4.0	Welders & cutters	1.4	21.3
Sales & related workers nec	2.9	21.3	General office clerks	1.4	3.4
General managers & top executives	2.8	15.1	Industrial production managers	1.3	21.2
Traffic, shipping, & receiving clerks	2.6	16.8	Machine tool cutting operators, metal & plastic	1.3	1.0
Inspectors, testers, & graders, precision	2.4	21.3	Electrical & electronic assemblers	1.2	21.4
Freight, stock, & material movers, hand	2.3	-2.9	Production, planning, & expediting clerks	1.1	21.3
Helpers, laborers, & material movers nec	2.0	21.3	Packaging & filling machine operators	1.1	21.3
Secretaries, ex legal & medical	1.9	10.4	Order clerks, materials, merchandise, & service	1.0	18.7
Machine operators nec	1.7	6.9	Maintenance repairers, general utility	1.0	9.2
Bookkeeping, accounting, & auditing clerks	1.6	-9.0			

Source: Industry-Occupation Matrix, Bureau of Labor Statistics. These data relate to one or more 3-digit SIC industry groups rather than to a single 4-digit SIC. The change reported for each occupation to the year 2005 is a percent of growth or decline as estimated by the Bureau of Labor Statistics. The abbreviation nec stands for 'not elsewhere classified'.

LOCATION BY STATE AND REGIONAL CONCENTRATION

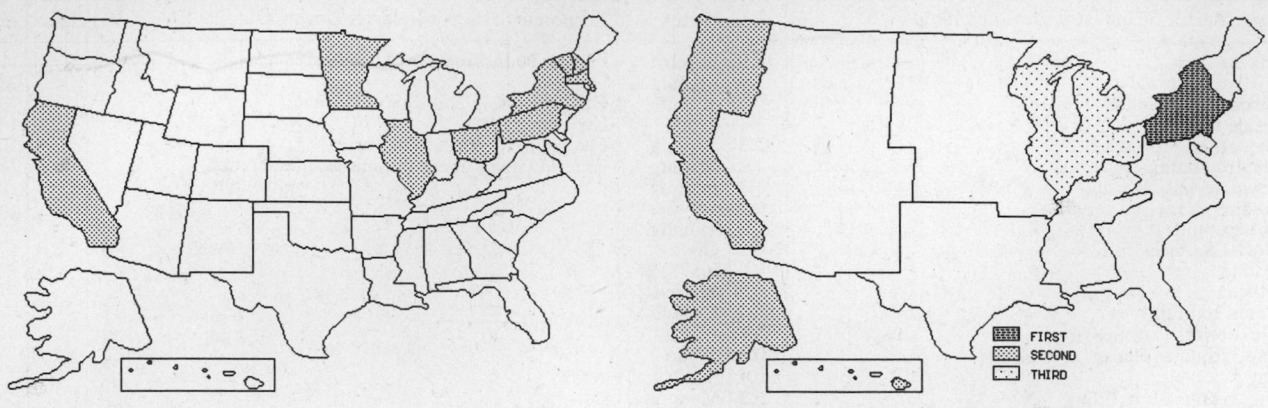

FIRST
SECOND
THIRD

INDUSTRY DATA BY STATE

State	Establish-ments	Shipments			Employment				Cost as % of Shipments	Investment per Employee ($)
		Total ($ mil)	% of U.S.	Per Establ.	Total Number	% of U.S.	Per Establ.	Wages ($/hour)		
New York	40	73.4	29.2	1.8	1,300	36.1	33	6.55	50.1	1,000
California	32	29.0	11.6	0.9	300	8.3	9	6.20	30.0	333
Pennsylvania	9	11.4	4.5	1.3	200	5.6	22	5.67	34.2	1,000
Ohio	5	9.7	3.9	1.9	100	2.8	20	10.00	45.4	-
Minnesota	7	7.2	2.9	1.0	100	2.8	14	6.00	37.5	-
Illinois	9	(D)	-	-	175 *	4.9	19	-	-	571
New Hampshire	3	(D)	-	-	375 *	10.4	125	-	-	-
Massachusetts	2	(D)	-	-	175 *	4.9	88	-	-	-
Vermont	1	(D)	-	-	175 *	4.9	175	-	-	-

Source: 1992 *Economic Census*. The states are in descending order of shipments or establishments (if shipment data are missing for the majority). The symbol (D) appears when data are withheld to prevent disclosure of competitive information. States marked with (D) are sorted by number of establishments. A dash (-) indicates that the data element cannot be calculated; * indicates the midpoint of a range.

3944 - GAMES, TOYS, & CHILDREN'S VEHICLES

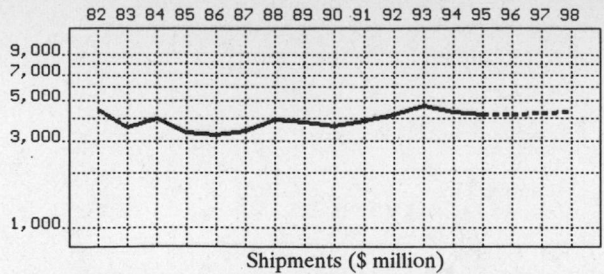

Shipments ($ million)

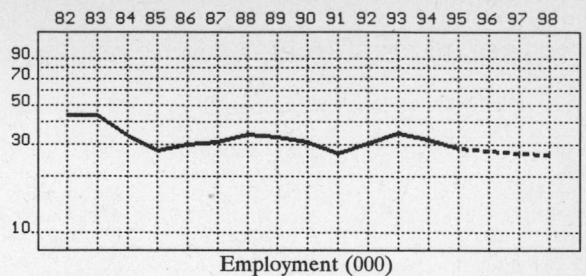

Employment (000)

GENERAL STATISTICS

Year	Companies	Establishments		Employment			Compensation		Production ($ million)			
		Total	with 20 or more employees	Total (000)	Production Workers (000)	Hours (Mil)	Payroll ($ mil)	Wages ($/hr)	Cost of Materials	Value Added by Manufacture	Value of Shipments	Capital Invest.
1982	732	780	253	43.8	32.2	59.4	625.0	6.12	1,854.5	2,622.6	4,476.0	111.5
1983		741	239	43.9	34.4	61.3	565.6	5.63	1,547.8	1,980.0	3,552.8	99.8
1984		702	225	33.5	26.1	48.7	530.9	6.72	1,965.9	2,120.6	4,020.2	66.6
1985		662	212	27.6	20.6	38.5	462.5	6.94	1,601.2	1,714.8	3,323.7	68.2
1986		629	187	30.0	22.3	41.7	504.3	7.33	1,603.8	1,663.1	3,264.7	112.4
1987	698	716	197	30.9	23.7	45.1	541.6	7.35	1,507.4	1,911.3	3,381.3	94.2
1988		717	211	34.1	26.5	49.7	602.8	7.43	1,791.4	2,201.8	3,950.8	96.0
1989		693	218	33.2	24.4	44.9	561.4	7.41	1,639.5	2,156.3	3,802.9	117.6
1990		719	208	31.0	19.9	39.7	571.6	7.94	1,672.7	1,911.8	3,622.9	78.7
1991		745	187	27.0	19.4	38.4	588.7	8.24	1,831.5	2,060.3	3,887.2	92.1
1992	895	917	213	30.6	23.0	46.1	673.6	8.50	1,818.0	2,389.8	4,186.2	137.3
1993		931	229	34.8	25.7	48.9	764.2	9.31	2,192.5	2,545.2	4,695.6	139.1
1994		837P	200P	32.1	25.1	50.0	761.1	9.45	1,965.5	2,319.6	4,313.3	138.7
1995		852P	197P	28.5P	20.8P	42.0P	708.5P	9.52P	1,913.6P	2,258.4P	4,199.4P	129.8P
1996		866P	195P	27.8P	20.2P	41.2P	724.5P	9.79P	1,934.2P	2,282.7P	4,244.7P	133.5P
1997		880P	192P	27.1P	19.6P	40.5P	740.5P	10.07P	1,954.8P	2,307.0P	4,289.9P	137.2P
1998		894P	190P	26.4P	19.0P	39.8P	756.5P	10.35P	1,975.4P	2,331.3P	4,335.1P	140.8P

Sources: 1982, 1987, 1992 *Economic Census*; *Annual Survey of Manufactures*, 83-86, 88-91, 93-94. Establishment counts for non-Census years are from *County Business Patterns*; establishment values for 83-84 are extrapolations. 'P's show projections by the editors. Industries reclassified in 87 will not have data for prior years.

INDICES OF CHANGE

Year	Companies	Establishments		Employment			Compensation		Production ($ million)			
		Total	with 20 or more employees	Total (000)	Production Workers (000)	Hours (Mil)	Payroll ($ mil)	Wages ($/hr)	Cost of Materials	Value Added by Manufacture	Value of Shipments	Capital Invest.
1982	82	85	119	143	140	129	93	72	102	110	107	81
1983		81	112	143	150	133	84	66	85	83	85	73
1984		77	106	109	113	106	79	79	108	89	96	49
1985		72	100	90	90	84	69	82	88	72	79	50
1986		69	88	98	97	90	75	86	88	70	78	82
1987	78	78	92	101	103	98	80	86	83	80	81	69
1988		78	99	111	115	108	89	87	99	92	94	70
1989		76	102	108	106	97	83	87	90	90	91	86
1990		78	98	101	87	86	85	93	92	80	87	57
1991		81	88	88	84	83	87	97	101	86	93	67
1992	100	100	100	100	100	100	100	100	100	100	100	100
1993		102	108	114	112	106	113	110	121	107	112	101
1994		91P	94P	105	109	108	113	111	108	97	103	101
1995		93P	93P	93P	90P	91P	105P	112P	105P	95P	100P	95P
1996		94P	91P	91P	88P	89P	108P	115P	106P	96P	101P	97P
1997		96P	90P	89P	85P	88P	110P	119P	108P	97P	102P	100P
1998		97P	89P	86P	83P	86P	112P	122P	109P	98P	104P	103P

Sources: Same as General Statistics. Values reflect change from the base year, 1992. Values above 100 mean greater than 92, values below 100 mean less than 92, and a value of 100 in the 82-91 or 93-98 period means same as 92. 'P's mark projections by the editors.

SELECTED RATIOS

For 1994	Avg. of All Manufact.	Analyzed Industry	Index	For 1994	Avg. of All Manufact.	Analyzed Industry	Index
Employees per Establishment	49	38	78	Value Added per Production Worker	134,084	92,414	69
Payroll per Establishment	1,500,273	908,825	61	Cost per Establishment	5,045,178	2,346,993	47
Payroll per Employee	30,620	23,710	77	Cost per Employee	102,970	61,231	59
Production Workers per Establishment	34	30	87	Cost per Production Worker	146,988	78,307	53
Wages per Establishment	853,319	564,210	66	Shipments per Establishment	9,576,895	5,150,488	54
Wages per Production Worker	24,861	18,825	76	Shipments per Employee	195,460	134,371	69
Hours per Production Worker	2,056	1,992	97	Shipments per Production Worker	279,017	171,845	62
Wages per Hour	12.09	9.45	78	Investment per Establishment	321,011	165,621	52
Value Added per Establishment	4,602,255	2,769,822	60	Investment per Employee	6,552	4,321	66
Value Added per Employee	93,930	72,262	77	Investment per Production Worker	9,352	5,526	59

Sources: Same as General Statistics. The 'Average of All Manufacturing' column represents the average of all manufacturing industries reported for the most recent complete year available. The Index shows the relationship between the Average and the Analyzed Industry. For example, 100 means that they are equal; 500 that the Analyzed Industry is five times the average; 50 means that the Analyzed Industry is half the national average. The abbreviation 'na' is used to show that data are 'not available'.

LEADING COMPANIES　　Number shown: 75　　Total sales ($ mil): 9,712　　Total employment (000): 61.4

Company Name	Address				CEO Name	Phone	Co. Type	Sales ($ mil)	Empl. (000)
Hasbro Inc	1027 Newport Av	Pawtucket	RI	02862	Alan G Hassenfeld	401-431-8697	P	2,670	12.5
Mattel Inc	333 Continental	El Segundo	CA	90245	John W Amerman	310-524-2000	P	2,046	21.0
Tyco Toys Inc	6000 Midlantic Dr	Mount Laurel	NJ	08054	Richard E Grey	609-234-7400	P	753	2.5
Fisher-Price Inc	636 Girard Av	East Aurora	NY	14052	Ronald J Jackson	716-687-3000	S	750	4.2
Milton Bradley Co	443 Shaker Rd	E Longmeadow	MA	01028	E David Wilson	413-525-6411	S	340•	2.0
SLM International Inc	200 5th Av	New York	NY	10010	Earl Takefman	212-675-0070	P	315	2.3
Ertl Company Inc	PO Box 500	Dyersville	IA	52040	George Volanakis	319-875-2000	S	200	1.0
LEGO Systems Inc	PO Box 1600	Enfield	CT	06083	Peter Eio	203-749-2291	R	190	1.3
Nintendo of America Inc	4820 150th Av NE	Redmond	WA	98052	Minoro Arkawa	206-882-2040	S	190•	1.3
Century Products Co	9600 Val View Rd	Macedonia	OH	44056	Frank Rumpeltin	216-468-2000	R	170	1.2
Toy Biz Inc	333 E 38th St	New York	NY	10016	Joseph M Ahearn	212-682-4700	P	156	<0.1
Lewis Galoob Toys Inc	500 Forbes Blv	S San Francisco	CA	94080	Mark Goldman	415-952-1678	P	134	0.2
Sega of America Inc	255 Shoreline Dr	Redwood City	CA	94065	Tom Kalinske	415-508-2800	S	130•	0.9
Bicycle Holding Inc	4590 Beech St	Cincinnati	OH	45212	RC Rule	513-396-5888	R	100	0.5
Rose Art Industries Inc	6 Regent St	Livingston	NJ	07039	Sidney Rosen	201-535-1313	R	100	1.0
Tyco Playtime Inc	1107 Broadway	New York	NY	10010	Martin Scheman	212-691-5898	S	100	<0.1
Revell-Monogram Inc	8601 Waukegan Rd	Morton Grove	IL	60053	Theodore J Eischeid	708-966-3500	R	96	0.6
Kenner Products	615 Elsinore Pl	Cincinnati	OH	45202	Bruce Stein	513-579-4000	D	76•	0.4
Catco Inc	529 W 42nd St	New York	NY	10036	Barbara E Carver	212-563-6363	R	70	<0.1
Pressman Toy Corp	200 5th Av	New York	NY	10010	James R Pressman	212-675-7910	R	56•	0.3
Lionel Trains Inc	50625 Richard W	Chesterfield	MI	48051	Richard Kughn	313-949-4100	R	55	0.6
Stuart Entertainment Inc	3211 Nebraska Av	Council Bluffs	IA	51501	Leonard A Stuart	712-323-1488	P	54	0.5
Ocean of America Inc	1870 Little Orchard	San Jose	CA	95125	Ray Musci	408-289-1411	R	50•	<0.1
Monogram Models Inc	8601 Waukegan Rd	Morton Grove	IL	60053	Theodore J Eischeid	708-966-3500	S	50	0.4
Educational Insights Inc	19560 S Rancho	Dominguez Hls	CA	90220	Jay A Cutler	310-884-1931	R	46	0.2
Cap Toys Inc	26201 Richmond Rd	Bedford Hts	OH	44146	John D Osher	216-292-6363	R	45	<0.1
Strombecker Corp	600 N Pulaski Rd	Chicago	IL	60624	Daniel B Shure	312-638-1000	R	45	0.4
Equity Marketing Inc	156 5th Av	New York	NY	10010	Stephen P Robeck	212-645-2333	P	44	<0.1
Processed Plastic Co	1001 Aucutt Rd	Montgomery	IL	60538	RS Bergman	708-892-7981	R	43	0.4
Douglas Press Inc	2810 Madison St	Bellwood	IL	60104	Frank Fienberg	708-547-8400	R	41•	0.3
Ohio Art Co	PO Box 111	Bryan	OH	43506	William C Killgallon	419-636-3141	P	41	0.3
Craft House Corp	328 N Westwood	Toledo	OH	43607	Sam Bushala	419-536-8351	S	35	0.2
Parker Brothers	50 Dunham Rd	Beverly	MA	01915	Robert Wann	508-927-7600	D	32•	0.2
Instant Ticket Factory	PO Box 544	Black Eagle	MT	59414	Pat Hoen	406-727-7812	R	29•	0.2
Smart Industries Corp	1626 Delaware Av	Des Moines	IA	50317	Gordon Smart	515-265-9900	R	29•	0.2
Spearhead Industries Inc	10 S 5th St	Minneapolis	MN	55402	Rich Barton	612-941-9171	S	26	<0.1
Paul-Son Gaming Corp	2121 Industrial Rd	Las Vegas	NV	89102	Paul S Endy	702-384-2425	P	21	0.6
Nylint Corp	1800 16th Av	Rockford	IL	61108	Theodore C Klint	815-397-2880	R	20	0.2
Dynamo Corp	2525 Handley	Fort Worth	TX	76118	William Rickett	817-284-0114	R	17	0.2
Action Performance Companies	2401 W 1st St	Tempe	AZ	85251	Fred W Wagenhals	602-894-0100	P	17	<0.1
Colorforms	133 Williams Dr	Ramsey	NJ	07446	Josh Kisleritz	201-327-2600	R	15•	0.1
Radio Flyer Inc	6515 W Grand Av	Chicago	IL	60635	MA Pasin	312-637-7100	R	15•	0.1
Atlas Model Railroad Company	378 Florence Av	Hillside	NJ	07205	T W Haedrich	908-687-0880	R	14•	0.1
Come Play Products Co	44 Suffolk St	Worcester	MA	01604	Mike Freelander	508-756-8353	R	14•	0.1
Peoria Plastics	9000 N University	Peoria	IL	61615	Gus Poulis	309-692-1700	D	14•	0.1
Universal Manufacturing Co	5450 Deramus St	Kansas City	MO	64120	Joe Wilner	816-231-2771	R	14•	0.3
Janex International Inc	21700 Oxnard St	Woodland Hills	CA	91367	Sheldon Morick	818-593-6777	P	13	<0.1
Century Products Co	3166 E Slauson Av	Los Angeles	CA	90058	Frank Rumpeltin	213-581-2299	D	12	<0.1
Fidelity Electronics International	2468 Harbor Cove	Fort Pierce	FL	34949	Manfred Egganer	305-597-1500	S	12•	0.1
RazorSoft International Inc	7321 N Broadway	Oklahoma City	OK	73116	Kyle Shelley	405-843-3505	R	12	<0.1
Creative Art Activities Inc	1802 Central Av	Cleveland	OH	44115	Phyllis Brody	216-589-4800	R	11•	<0.1
Trinity Products Inc	1901 E Linden Av	Linden	NJ	07036	Ernest N Provetti	908-862-1705	R	10	<0.1
American Laser Games Inc	4801 Lincoln Rd NE	Albuquerque	NM	87109	Robert Grebe	505-880-1718	R	10	<0.1
Amloid Corp	PO Box 557	Saddle Brook	NJ	07662	Michael Albarelli Jr	201-368-1000	R	10•	0.2
Brooklyn Products Inc	PO Box 218	Brooklyn	MI	49230	Robert Linenfelser	517-592-2185	R	10•	0.1
Cox Hobbies Inc	350 W Rincon St	Corona	CA	91720	William H Selzer	909-278-1282	R	10	<0.1
Konami	900 Deerfield Pkwy	Buffalo Grove	IL	60089	Ken Direnberger	708-215-5100	S	10•	0.1
Steven Manufacturing Co	104 Industrial Dr	Hermann	MO	65041	Tyler Bulkley	314-486-5494	R	10	0.1
University Games Corp	1633 Adrian Rd	Burlingame	CA	94010	Robert Moog	415-692-2500	R	10•	<0.1
James Industries Inc	PO Box 407	Hollidaysburg	PA	16648	Betty M James	814-695-5681	R	9	0.1
Micro-Trains Line Co	PO Box 1200	Talent	OR	97540	Keith Edwards	503-772-9890	R	8	<0.1
M Pressner and Company Inc	99 Gold St	Brooklyn	NY	11201	Jerry Pressner	718-858-1000	R	8	<0.1
Pacific Miniatures Inc	2710 Fly Av	City of Com	CA	90040	F Ouweleen Jr	213-721-4790	R	8•	<0.1
Esquire Novelty Corp	350 Forest Av	Amsterdam	NY	12010	Joe Gorecki	518-843-3000	S	7	0.1
Handi Craft Co	4433 Fyler Av	St Louis	MO	63116	C Rhodes	314-773-2979	R	7	0.1
Community Playthings	Rte 213	Rifton	NY	12471	Richard E Domer	914-658-8799	R	6	0.1
Go Fly A Kite Inc	PO Box AA	East Haddam	CT	06423	A Skwarek	203-873-8675	R	6•	<0.1
K and B Manufacturing Inc	PO Box 3000	Lk Havasu Ct	AZ	86405	John W Brodbeck	602-453-3030	R	6	<0.1
National Artcraft Co	23456 Mercantile	Cleveland	OH	44122	Jon Berrie	216-292-4944	R	6	<0.1
Topstone Industries Inc	81 Sand Pit Rd	Danbury	CT	06810	M Goldberg	203-792-2100	R	6	<0.1
Olmec Toys Inc	156 FIfth Av #1123	New York	NY	10010	Yla Eason	212-645-3660	R	6	<0.1
Sig Manufacturing Company Inc	401-7 S Front St	Montezuma	IA	50171	Hazel Sig-Hester	515-623-5154	R	6	<0.1
Bird Manufacturing	113 N Main St	Elkhorn	NE	68022	Fred Schweser	402-289-3779	R	5	<0.1
Cadaco	4300 W 47th St	Chicago	IL	60632	Mark E Hartelius	312-927-1500	D	5	<0.1
Comet-Montrose Ltd	1337 W 37th Pl	Chicago	IL	60609	Howard Williams	312-927-1900	R	5	0.1

Source: Ward's Business Directory of U.S. Private and Public Companies, Volumes 1 and 2, 1996. The company type code used is as follows: P - Public, R - Private, S - Subsidiary, D - Division, J - Joint Venture, A - Affiliate, G - Group. Sales are in millions of dollars, employees are in thousands. An asterisk (*) indicates an estimated sales volume. The symbol < stands for 'less than'. Company names and addresses are truncated, in some cases, to fit into the available space.

MATERIALS CONSUMED

Material	Quantity	Delivered Cost ($ million)
Materials, ingredients, containers, and supplies	(X)	1,390.9
Fabricated metal products, including forgings	(X)	36.6
Steel shapes and forms	(X)	18.4
All other nonferrous shapes and forms	(X)	25.1
Plastics resins consumed in the form of granules, pellets, powders, liquids, etc.	(X)	241.2
Plastics products consumed in the form of sheets, rods, tubes, film, and other shapes	(X)	226.5
Broadwoven fabrics (piece goods)	(X)	34.6
Paperboard (including news, chip, pasted, tablet, check, binders' board), except for shipping	(X)	63.2
Paperboard containers, boxes, and corrugated paperboard	(X)	126.7
Other paper products	(X)	21.3
Hardwood lumber, rough and dressed	(X)	15.3
Softwood lumber, rough and dressed	(X)	3.7
Other wood products (except lumber)	(X)	11.3
Electronic components and accessories, including circuit boards and recording heads	(X)	53.9
All other materials and components, parts, containers, and supplies	(X)	291.9
Materials, ingredients, containers, and supplies, nsk	(X)	221.5

Source: 1992 *Economic Census*. Explanation of symbols used: (D): Withheld to avoid disclosure of competitive data; na: Not available; (S): Withheld because statistical norms were not met; (X): Not applicable; (Z): Less than half the unit shown; nec: Not elsewhere classified; nsk: Not specified by kind; - : zero; * : 10-19 percent estimated; ** : 20-29 percent estimated.

PRODUCT SHARE DETAILS

Product or Product Class	% Share	Product or Product Class	% Share
Games, toys, and children's vehicles	100.00	Other toys, nec, except games, hobbies, and electronic toys	21.66
Baby carriages and children's vehicles, except bicycles with pneumatic tires	11.93	Toys; excluding games, hobbies, and electronic toys, nsk	1.06
Baby carriages and strollers	11.33	Hobbies: models (operating or static), craft, structural, and scientific equipment kits, sets, and individual units	12.44
Children's tricycles (including pedal and chain driven), plastics and metal tubular construction	13.16	Electrically operated model railroads (individual units, kits, sets, and accessories)	8.63
Other children's vehicles (automobiles, tractors, two-wheel sidewalk cycles, scooters, wagons, baby walkers, and sleds) (excluding bicycles with pneumatic tires)	75.51	Operating model cars, boats, planes, and other models (individual units, kits, and sets)	14.43
Toys, excluding games, hobbies, and electronic toys	47.87	Static models, other than plastics (all individual units, kits, sets, and structural kits, including railroad, car, boat, and plane)	2.00
Doll carriages, strollers, and doll carts	0.80	Plastics static models	17.94
Doll houses and furniture (excluding collectors' doll houses, miniatures, and accessories)	0.68	Components and accessories for all models (operating and static)	3.57
Toy trains and equipment (mechanical and electric)	3.86	Craft kits and supplies individually packaged or in bulk (decoupage, macrame, tiffany glass, beadery, etc.)	40.44
Road-racing sets (including accessories and parts), and mechanically powered toys excluding scale model operating type	(D)	Science: microscopes, chemistry sets, or any natural science kit or set (botany, minerology, electrical, etc.)	3.48
Plastics nonpowered transportation toys (nonriding, sold without accessories), except model kits, greater than 6 in. in length	5.29	Collectors' miniatures (doll houses, accessories, soldiers or historic figures, scale cars, aircraft, etc.) except dolls	5.72
Other nonpowered transportation toys (nonriding, sold without accessories), except model kits, greater than 6 in. in length	3.88	Hobbies: models ; craft, structural, and scientific equipment kits, sets, and individual units, nsk	3.77
Other nonpowered transportation toys (nonriding, sold without accessories), except model kits, 6 in. in length or less	0.82	Nonelectronic games	17.44
		Children's nonelectronic board games (under 12 yr)	(D)
Nonpowered transportation toy sets (nonriding, sold with accessories), except model kits	2.63	Family and other nonelectronic board games (chess, checkers, etc.) (12 yr and over)	27.08
Musical toys and toy musical instruments, except electronic	0.89	Sports-oriented nonelectronic action and skill games (football, baseball, etc.)	8.79
Infant toys, not elsewhere classified, except games, hobbies, and electronic toys	4.16	Nonsports-oriented nonelectronic action and skill games	17.62
		Puzzles	14.83
Construction sets and building toys	(D)	Other nonelectronic games	(D)
Preschool playsets and toys, not elsewhere classified (excluding infants' toys, building toys, and electronic toys)	13.75	Parts for games (excluding electronic parts)	2.58
		Nonelectronic games, nsk	1.55
Toy guns, gun sets, and rifles	1.98	Electronic games and toys (excluding disks, tapes, and cartridges)	1.77
Children's coloring books and picture-word books, except games	6.22	Home video games, for attachment to television receiver	(D)
Juvenile-scale sporting goods and inflatables (including sand, water, gardening toys, etc.)	5.33	Other home electronic games, not video	(D)
		Electronic toys, not elsewhere classified	(D)
Housekeeping and cooking toys (including tea sets and play tools)	10.80	Electronic games and toys (excluding disks, tapes, and cartridges), nsk	0.16
Parts for toys	3.41	Games, toys, and children's vehicles, nsk	8.56

Source: 1992 *Economic Census*. The values shown are percent of total shipments in an industry. Values of indented subcategories are summed in the main heading. The symbol (D) appears when data are withheld to prevent disclosure of competitive information. The abbreviation nsk stands for 'not specified by kind' and nec for 'not elsewhere classified'.

INPUTS AND OUTPUTS FOR GAMES, TOYS, & CHILDREN'S VEHICLES

Economic Sector or Industry Providing Inputs	%	Sector	Economic Sector or Industry Buying Outputs	%	Sector
Imports	32.5	Foreign	Personal consumption expenditures	88.0	
Electronic components nec	10.3	Manufg.	Exports	6.3	Foreign
Wholesale trade	7.0	Trade	S/L Govt. purch., other general government	1.9	S/L Govt
Plastics materials & resins	6.6	Manufg.	Games, toys, & children's vehicles	1.0	Manufg.
Paperboard containers & boxes	4.1	Manufg.	S/L Govt. purch., natural resource & recreation.	0.6	S/L Govt
Miscellaneous plastics products	3.8	Manufg.	S/L Govt. purch., elem. & secondary education	0.5	S/L Govt
Coated fabrics, not rubberized	3.0	Manufg.	S/L Govt. purch., higher education	0.5	S/L Govt
Advertising	2.3	Services	Residential care	0.4	Services
Wood products, nec	1.9	Manufg.	Child day care services	0.3	Services
Blast furnaces & steel mills	1.4	Manufg.	Federal Government purchases, national defense	0.2	Fed Govt
Games, toys, & children's vehicles	1.4	Manufg.			
Paperboard mills	1.4	Manufg.			
Book printing	1.3	Manufg.			
Petroleum refining	1.3	Manufg.			
Screw machine and related products	1.2	Manufg.			
Electric services (utilities)	1.1	Util.			
Fabricated textile products, nec	0.9	Manufg.			
Motor freight transportation & warehousing	0.9	Util.			
Veneer & plywood	0.8	Manufg.			
Banking	0.8	Fin/R.E.			
Adhesives & sealants	0.7	Manufg.			
Eating & drinking places	0.7	Trade			
Broadwoven fabric mills	0.6	Manufg.			
Canvas & related products	0.6	Manufg.			
Paints & allied products	0.6	Manufg.			
Sawmills & planning mills, general	0.6	Manufg.			
Aluminum rolling & drawing	0.5	Manufg.			
Dolls	0.5	Manufg.			
Metal stampings, nec	0.5	Manufg.			
Railroads & related services	0.5	Util.			
Real estate	0.5	Fin/R.E.			
Communications, except radio & TV	0.4	Util.			
U.S. Postal Service	0.4	Gov't			
Maintenance of nonfarm buildings nec	0.3	Constr.			
Chemical preparations, nec	0.3	Manufg.			
Commercial printing	0.3	Manufg.			
Internal combustion engines, nec	0.3	Manufg.			
Nonferrous rolling & drawing, nec	0.3	Manufg.			
Yarn mills & finishing of textiles, nec	0.3	Manufg.			
Air transportation	0.3	Util.			
Gas production & distribution (utilities)	0.3	Util.			
Automotive rental & leasing, without drivers	0.3	Services			
Equipment rental & leasing services	0.3	Services			
Job training & related services	0.3	Services			
Management & consulting services & labs	0.3	Services			
Electrical equipment & supplies, nec	0.2	Manufg.			
Hardwood dimension & flooring mills	0.2	Manufg.			
Leather tanning & finishing	0.2	Manufg.			
Machinery, except electrical, nec	0.2	Manufg.			
Miscellaneous fabricated wire products	0.2	Manufg.			
Paper coating & glazing	0.2	Manufg.			
Transportation equipment, nec	0.2	Manufg.			
Insurance carriers	0.2	Fin/R.E.			
Accounting, auditing & bookkeeping	0.2	Services			
Automotive repair shops & services	0.2	Services			
Legal services	0.2	Services			
Abrasive products	0.1	Manufg.			
Fabricated metal products, nec	0.1	Manufg.			
Hardware, nec	0.1	Manufg.			
Motors & generators	0.1	Manufg.			
Paper mills, except building paper	0.1	Manufg.			
Special dies & tools & machine tool accessories	0.1	Manufg.			
Sanitary services, steam supply, irrigation	0.1	Util.			
Credit agencies other than banks	0.1	Fin/R.E.			
Royalties	0.1	Fin/R.E.			
Engineering, architectural, & surveying services	0.1	Services			
Hotels & lodging places	0.1	Services			

Source: *Benchmark Input-Output Accounts for the U.S. Economy, 1982*, U.S. Department of Commerce, Washington, D.C., July 1991. Data, as reported in the source, are organized by the 1977 SIC structure in use in 1982 but have been matched, as closely as is possible, to the 1987 SIC structure used in this book.

OCCUPATIONS EMPLOYED BY SIC 394 - TOYS AND SPORTING GOODS

Occupation	% of Total 1994	Change to 2005	Occupation	% of Total 1994	Change to 2005
Assemblers, fabricators, & hand workers nec	23.5	21.3	Coating, painting, & spraying machine workers	1.5	21.3
Plastic molding machine workers	4.3	21.3	Machine forming operators, metal & plastic	1.5	21.3
Blue collar worker supervisors	4.2	13.5	Painting, coating, & decorating workers, hand	1.4	21.3
Hand packers & packagers	3.6	4.0	Welders & cutters	1.4	21.3
Sales & related workers nec	2.9	21.3	General office clerks	1.4	3.4
General managers & top executives	2.8	15.1	Industrial production managers	1.3	21.2
Traffic, shipping, & receiving clerks	2.6	16.8	Machine tool cutting operators, metal & plastic	1.3	1.0
Inspectors, testers, & graders, precision	2.4	21.3	Electrical & electronic assemblers	1.2	21.4
Freight, stock, & material movers, hand	2.3	-2.9	Production, planning, & expediting clerks	1.1	21.3
Helpers, laborers, & material movers nec	2.0	21.3	Packaging & filling machine operators	1.1	21.3
Secretaries, ex legal & medical	1.9	10.4	Order clerks, materials, merchandise, & service	1.0	18.7
Machine operators nec	1.7	6.9	Maintenance repairers, general utility	1.0	9.2
Bookkeeping, accounting, & auditing clerks	1.6	-9.0			

Source: *Industry-Occupation Matrix*, Bureau of Labor Statistics. These data relate to one or more 3-digit SIC industry groups rather than to a single 4-digit SIC. The change reported for each occupation to the year 2005 is a percent of growth or decline as estimated by the Bureau of Labor Statistics. The abbreviation nec stands for 'not elsewhere classified'.

LOCATION BY STATE AND REGIONAL CONCENTRATION

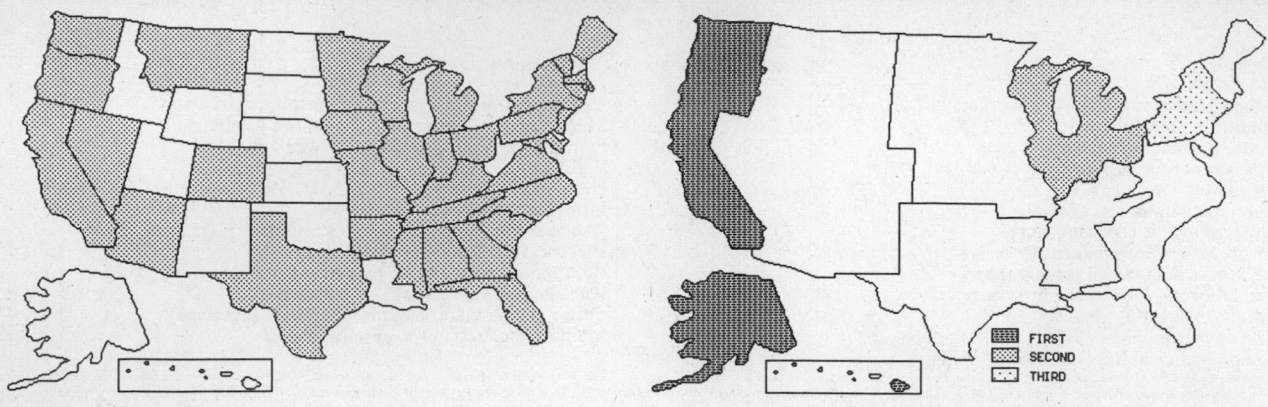

FIRST
SECOND
THIRD

INDUSTRY DATA BY STATE

| State | Establish-ments | Shipments | | | Employment | | | | Cost as % of Shipments | Investment per Employee ($) |
		Total ($ mil)	% of U.S.	Per Establ.	Total Number	% of U.S.	Per Establ.	Wages ($/hour)		
Ohio	44	544.4	13.0	12.4	3,600	11.8	82	9.17	44.5	6,444
New York	66	270.2	6.5	4.1	2,100	6.9	32	10.96	39.6	905
California	135	211.4	5.0	1.6	1,800	5.9	13	7.63	45.1	2,167
North Carolina	17	161.3	3.9	9.5	1,100	3.6	65	6.31	44.8	-
Texas	46	144.5	3.5	3.1	1,300	4.2	28	6.79	44.3	2,308
New Jersey	30	124.5	3.0	4.2	1,100	3.6	37	7.50	43.5	2,182
Pennsylvania	51	121.0	2.9	2.4	1,500	4.9	29	7.75	46.1	1,467
Michigan	22	115.6	2.8	5.3	1,200	3.9	55	8.20	33.6	5,917
Oregon	26	93.9	2.2	3.6	700	2.3	27	6.13	42.8	2,571
Maryland	12	37.3	0.9	3.1	400	1.3	33	7.17	52.8	-
Florida	27	32.2	0.8	1.2	300	1.0	11	8.67	48.1	2,000
Minnesota	19	22.4	0.5	1.2	100	0.3	5	9.50	54.5	-
Georgia	21	14.8	0.4	0.7	200	0.7	10	9.00	31.8	1,000
Washington	27	12.9	0.3	0.5	100	0.3	4	8.50	48.1	-
Nevada	12	12.8	0.3	1.1	100	0.3	8	7.00	53.1	3,000
Maine	17	10.3	0.2	0.6	100	0.3	6	6.00	44.7	2,000
Illinois	47	(D)	-	-	1,750 *	5.7	37	-	-	6,457
Massachusetts	29	(D)	-	-	3,750 *	12.3	129	-	-	-
Connecticut	24	(D)	-	-	1,750 *	5.7	73	-	-	-
Indiana	19	(D)	-	-	1,750 *	5.7	92	-	-	-
Tennessee	19	(D)	-	-	375 *	1.2	20	-	-	-
Virginia	18	(D)	-	-	175 *	0.6	10	-	-	1,143
Arkansas	17	(D)	-	-	375 *	1.2	22	-	-	-
Missouri	17	(D)	-	-	750 *	2.5	44	-	-	3,333
Wisconsin	17	(D)	-	-	175 *	0.6	10	-	-	1,714
Colorado	15	(D)	-	-	375 *	1.2	25	-	-	-
Vermont	14	(D)	-	-	175 *	0.6	13	-	-	-
Iowa	13	(D)	-	-	1,750 *	5.7	135	-	-	-
Arizona	9	(D)	-	-	175 *	0.6	19	-	-	-
Montana	9	(D)	-	-	175 *	0.6	19	-	-	-
Rhode Island	8	(D)	-	-	1,750 *	5.7	219	-	-	-
Kentucky	6	(D)	-	-	1,750 *	5.7	292	-	-	-
South Carolina	6	(D)	-	-	175 *	0.6	29	-	-	-
Alabama	5	(D)	-	-	175 *	0.6	35	-	-	-
Mississippi	5	(D)	-	-	375 *	1.2	75	-	-	-

Source: 1992 *Economic Census*. The states are in descending order of shipments or establishments (if shipment data are missing for the majority). The symbol (D) appears when data are withheld to prevent disclosure of competitive information. States marked with (D) are sorted by number of establishments. A dash (-) indicates that the data element cannot be calculated; * indicates the midpoint of a range.

3949 - SPORTING & ATHLETIC GOODS, NEC

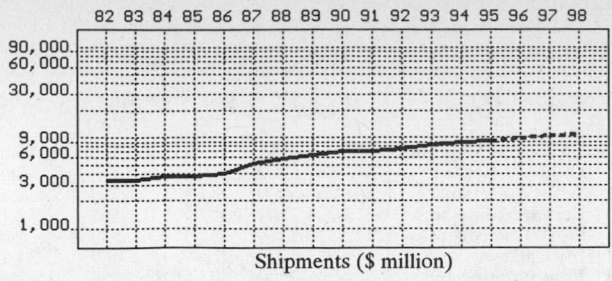

Shipments ($ million)

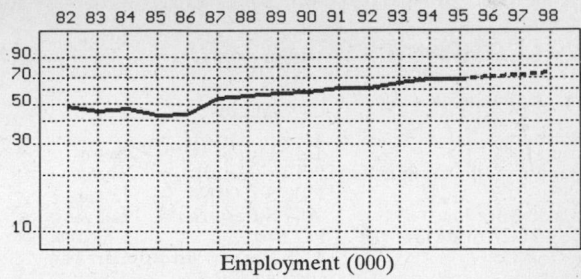

Employment (000)

GENERAL STATISTICS

| Year | Com-panies | Establishments | | Employment | | | Compensation | | Production ($ million) | | | |
		Total	with 20 or more employees	Total (000)	Production Workers (000)	Hours (Mil)	Payroll ($ mil)	Wages ($/hr)	Cost of Materials	Value Added by Manufacture	Value of Shipments	Capital Invest.
1982	1,452	1,553	460	47.8	36.0	67.7	677.7	6.29	1,591.2	1,778.0	3,376.8	76.5
1983		1,548	455	44.8	33.7	63.1	663.1	6.51	1,572.4	1,759.3	3,333.9	
1984		1,543	450	46.7	35.2	68.3	742.6	6.73	1,812.6	2,020.7	3,765.3	
1985		1,539	445	42.8	32.4	60.1	707.6	7.29	1,782.4	1,928.6	3,766.1	111.3
1986		1,523	428	43.2	31.8	61.5	729.1	7.31	1,844.9	2,131.2	3,960.6	
1987	1,710	1,800	490	53.6	39.9	75.0	911.2	7.40	2,344.3	2,798.3	5,123.0	133.2
1988		1,718	517	55.7	41.7	81.2	1,022.7	7.70	2,641.5	3,202.4	5,746.9	120.1
1989		1,743	515	57.5	45.4	84.5	1,128.4	8.26	3,069.9	3,547.1	6,509.6	166.2
1990		1,784	502	58.4	49.5	91.3	1,238.0	8.20	3,265.5	3,763.6	7,040.2	169.6
1991		1,881	499	61.3	45.4	90.3	1,247.3	8.20	3,287.4	3,753.6	7,035.7	138.2
1992	2,025	2,113	510	61.7	44.0	88.5	1,361.4	8.52	3,400.8	4,198.8	7,566.4	176.9
1993		2,204	556	64.4	46.8	93.7	1,438.4	8.74	3,834.1	4,626.0	8,459.4	217.7
1994		2,119P	542P	68.1	49.6	99.2	1,526.0	8.66	4,199.7	4,896.2	8,936.1	191.9
1995		2,176P	550P	68.5P	51.0P	101.2P	1,580.1P	9.14P	4,352.3P	5,074.1P	9,260.9P	
1996		2,234P	559P	70.6P	52.4P	104.4P	1,658.6P	9.34P	4,588.7P	5,349.7P	9,763.8P	
1997		2,291P	568P	72.6P	53.8P	107.6P	1,737.2P	9.55P	4,825.1P	5,625.3P	10,266.8P	
1998		2,348P	576P	74.6P	55.3P	110.8P	1,815.7P	9.76P	5,061.5P	5,900.9P	10,769.8P	

Sources: 1982, 1987, 1992 *Economic Census*; *Annual Survey of Manufactures*, 83-86, 88-91, 93-94. Establishment counts for non-Census years are from *County Business Patterns*; establishment values for 83-84 are extrapolations. 'P's show projections by the editors. Industries reclassified in 87 will not have data for prior years.

INDICES OF CHANGE

| Year | Com-panies | Establishments | | Employment | | | Compensation | | Production ($ million) | | | |
		Total	with 20 or more employees	Total (000)	Production Workers (000)	Hours (Mil)	Payroll ($ mil)	Wages ($/hr)	Cost of Materials	Value Added by Manufacture	Value of Shipments	Capital Invest.
1982	72	73	90	77	82	76	50	74	47	42	45	43
1983		73	89	73	77	71	49	76	46	42	44	
1984		73	88	76	80	77	55	79	53	48	50	
1985		73	87	69	74	68	52	86	52	46	50	63
1986		72	84	70	72	69	54	86	54	51	52	
1987	84	85	96	87	91	85	67	87	69	67	68	75
1988		81	101	90	95	92	75	90	78	76	76	68
1989		82	101	93	103	95	83	97	90	84	86	94
1990		84	98	95	113	103	91	96	96	90	93	96
1991		89	98	99	103	102	92	96	97	89	93	78
1992	100	100	100	100	100	100	100	100	100	100	100	100
1993		104	109	104	106	106	106	103	113	110	112	123
1994		100P	106P	110	113	112	112	102	123	117	118	108
1995		103P	108P	111P	116P	114P	116P	107P	128P	121P	122P	
1996		106P	110P	114P	119P	118P	122P	110P	135P	127P	129P	
1997		108P	111P	118P	122P	122P	128P	112P	142P	134P	136P	
1998		111P	113P	121P	126P	125P	133P	115P	149P	141P	142P	

Sources: Same as General Statistics. Values reflect change from the base year, 1992. Values above 100 mean greater than 92, values below 100 mean less than 92, and a value of 100 in the 82-91 or 93-98 period means same as 92. 'P's mark projections by the editors.

SELECTED RATIOS

For 1994	Avg. of All Manufact.	Analyzed Industry	Index	For 1994	Avg. of All Manufact.	Analyzed Industry	Index
Employees per Establishment	49	32	66	Value Added per Production Worker	134,084	98,714	74
Payroll per Establishment	1,500,273	720,213	48	Cost per Establishment	5,045,178	1,982,096	39
Payroll per Employee	30,620	22,408	73	Cost per Employee	102,970	61,670	60
Production Workers per Establishment	34	23	68	Cost per Production Worker	146,988	84,671	58
Wages per Establishment	853,319	405,449	48	Shipments per Establishment	9,576,895	4,217,493	44
Wages per Production Worker	24,861	17,320	70	Shipments per Employee	195,460	131,220	67
Hours per Production Worker	2,056	2,000	97	Shipments per Production Worker	279,017	180,163	65
Wages per Hour	12.09	8.66	72	Investment per Establishment	321,011	90,569	28
Value Added per Establishment	4,602,255	2,310,816	50	Investment per Employee	6,552	2,818	43
Value Added per Employee	93,930	71,897	77	Investment per Production Worker	9,352	3,869	41

Sources: Same as General Statistics. The 'Average of All Manufacturing' column represents the average of all manufacturing industries reported for the most recent complete year available. The Index shows the relationship between the Average and the Analyzed Industry. For example, 100 means that they are equal; 500 that the Analyzed Industry is five times the average; 50 means that the Analyzed Industry is half the national average. The abbreviation 'na' is used to show that data are 'not available'.

LEADING COMPANIES   Number shown: 75   Total sales ($ mil): 7,261   Total employment (000): 45.4

Company Name	Address				CEO Name	Phone	Co. Type	Sales ($ mil)	Empl. (000)
CML Group Inc	524 Main St	Acton	MA	01720	Charles M Leighton	508-264-4155	P	772	6.8
Spalding and Evenflo Companies	PO Box 30101	Tampa	FL	33630	Donald J Byrnes	813-887-5200	R	750*	2.0
Weslo Inc	1500 S 1000 W	Logan	UT	84321	Scott Watterson	801-750-5000	S	550	2.6
Anthony Industries Inc	4900 S Eastern Av	Los Angeles	CA	90040	Bernard I Forester	213-724-2800	P	502	3.7
Spalding Sports Worldwide	PO Box 901	Chicopee	MA	01021	G A Dickerman	413-536-1200	D	500	1.0
Callaway Golf Co	2285 Rutherford Rd	Carlsbad	CA	92008	Ely Callaway	619-931-1771	P	449	1.1
NordicTrack Inc	104 Peavy Rd	Chaska	MN	55318	Kent Flummerfelt	612-368-2500	S	378	2.4
Wilson Sporting Goods Co	8700 W Bryn Mawr	Chicago	IL	60631	John Riccitiello	312-714-6400	S	360*	3.2
Johnson Worldwide Associates	1326 Willow Rd	Sturtevant	WI	53177	John D Crabb	414-884-1500	P	284	1.3
Precision Shooting Equip Co	2727 N Fairview	Tucson	AZ	85705	Peter Shepley	602-884-9065	R	250*	0.4
Bell Sports Corp	10601 N Hayden Rd	Scottsdale	AZ	85260	Terry G Lee	602-951-0033	P	116	1.0
Bell Sports Inc	10601 N Hayden Rd	Scottsdale	AZ	85260	Terry G Lee	602-951-0033	S	116	1.0
Escalade Inc	817 Maxwell Av	Evansville	IN	47717	Robert E Griffin	812-467-1200	P	103	0.8
K2 Corp	19215 Vashon SW	Vashon	WA	98070	Rich Rodstien	206-463-3631	S	100*	0.8
Wellington Leisure Products Inc	1140 Monticello Rd	Madison	GA	30650	Frank H Carter	706-342-1916	S	100	1.5
First Team Sports Inc	2274 Woodale Dr	Mounds View	MN	55112	John Egart	612-780-4454	P	86	<0.1
Indian Industries Inc	PO Box 889	Evansville	IN	47706	CW Reed	812-426-2281	S	83	0.6
Variflex Inc	5152 N Com Av	Moorpark	CA	93021	Raymond H Losi	805-523-0322	P	80	0.1
Daiwa Corp	PO Box 3235	Garden Grove	CA	92641	Tad Suzuki	714-895-6645	S	75	0.2
CYBEX	2100 Smithtown Av	Ronkonkoma	NY	11779	James E George	516-585-9000	D	70	0.5
Sport Supply Group Inc	PO Box 7726	Dallas	TX	75209	M J Blumenfeld	214-484-9484	P	68	0.3
Life Fitness	10601 W Belmont	Franklin Park	IL	60131	Augie Nieto	708-288-3300	R	67*	0.6
Worth Inc	PO Box 88104	Tullahoma	TN	37388	Fred Bryan	615-455-0691	R	61	1.2
Riddell Sports Inc	900 3rd Av	New York	NY	10022	David Mauer	212-826-4300	P	55	0.7
Life Fitness	10601 W Belmont	Franklin Park	IL	60131	Augie Nieto	708-288-3300	S	54*	0.5
Ram Golf Corp	2020 Indian	Melrose Park	IL	60160	JR Hansberger	708-681-5800	R	50	0.5
True Fitness Technology Inc	865 Hoff Rd	O'Fallon	MO	63366	Larry Stallings	314-272-7100	R	50	0.2
Bollinger Industries Inc	222 W Airport Fwy	Irving	TX	75062	Glenn D Bollinger	214-445-0386	P	45	0.4
Daisy Manufacturing Company	PO Box 220	Rogers	AR	72757	Marvin W Griffin	501-636-1200	R	45*	0.5
Grandoe Corp	PO Box 713	Gloversville	NY	12078	R J Zuckerwar	518-725-8641	R	45	0.3
Muskin Leisure Products Inc	401 E Thomas St	Wilkes-Barre	PA	18705	Jean Pierre Parent	717-825-4501	S	45	0.4
Lifetime Products Inc	PO Box 1525	Clearfield	UT	84016	Barry Mower	801-776-1532	R	40	0.5
Iron Mountain Forge Corp	PO Box 897	Farmington	MO	63640	Terry Braxton	314-756-4591	D	39*	0.4
ProGroup Inc	6201 Mtn View	Ooltewah	TN	37363	WR Dooley	615-238-5890	P	38	0.8
Nash Manufacturing Inc	315 W Ripy St	Fort Worth	TX	76110	CE Nash III	817-926-5225	R	33*	0.3
Landscape Structures Inc	601 7th St S	Delano	MN	55328	Barbara King	612-972-3391	R	33	0.2
Riddell Inc	3670 N Milwaukee	Chicago	IL	60641	Can Cougill	312-794-1994	S	30	0.2
Shane Group Inc	PO Box 765	Hillsdale	MI	49242	DC Shaneour Jr	517-439-4316	R	30*	0.3
Valley Co	PO Box 656	Bay City	MI	48707	CP Milhem	517-892-4536	S	29*	0.3
Harvard Sports Inc	2640 E Del Amo	Compton	CA	90221	Bem Dadbeh	213-636-0691	S	28*	0.3
Columbia Industries Inc	5005 West Av	San Antonio	TX	78213	Michael T Albritton	210-344-9211	R	28*	0.3
Northwestern Golf Co	4701 N Ravenswood	Chicago	IL	60640	N Rosasco Jr	312-275-0500	R	28*	0.3
ParaBody Inc	14150 Sunfish Lake	Ramsey	MN	55303	Jerry Dettinger	612-422-0747	R	28	0.3
Unisen Inc	14352 Chambers Rd	Tustin	CA	92680	Robin Klaus	714-669-1660	R	27*	0.2
DBI/Sala	PO Box 46	Red Wing	MN	55066	Craig W Burow	612-388-8282	D	26*	0.2
Burton Manufacturing Company	PO Box 1669	Jasper	AL	35502	Don Ochseweiter	205-221-3630	R	25	0.2
Penguin Industries Inc	Airport Industrial	Coatesville	PA	19320	Robert D McNeil	610-384-6000	R	25	0.3
Shakespeare Fishing Tackle	3801 Westmore Dr	Columbia	SC	29223	Scott Hogsett	803-754-7000	D	25	<0.1
SR Smith Inc	PO Box 400	Canby	OR	97013	Robert B Sigler	503-266-2231	R	25	0.1
Tommy Armour Golf Co	8350 N Lehigh Av	Morton Grove	IL	60053	Robert F MacNally	708-966-6300	D	25	0.2
Troxel Co Cycling and Fitness	1333 30th St	San Diego	CA	92154	D Kenneth Mitchell	619-424-4880	D	25	0.3
Game Time Inc	PO Box 121	Fort Payne	AL	35967	RD Siragusa Jr	205-845-5610	D	20	0.3
Bear Archery Inc	4600 SW 41st Blv	Gainesville	FL	32608	Charles T Smith Jr	904-376-2327	S	19*	0.2
American Allsafe Co	99 Wales Av	Tonawanda	NY	14150	LM Kennedy	716-695-8300	S	18	0.2
Universal Gym Equipment Inc	PO Box 1270	Cedar Rapids	IA	52406	Mark Corey	319-365-7561	S	18*	0.2
Aerobics Inc	385 Main St	Little Falls	NJ	07424	Gerald J Staub	201-256-7805	R	16	<0.1
Ebonite International Inc	PO Box 746	Hopkinsville	KY	42241	William T Scheid	502-886-5261	R	16*	0.3
Paramount Fitness Corp	6450 E Bandini Blvd	Los Angeles	CA	90040	Jim Trissler	213-721-2121	R	16*	0.1
Sheldons' Inc	626 Center St	Antigo	WI	54409	JM Sheldon	715-623-2382	R	16	0.2
Soloflex Inc	570 NE 53rd St	Hillsboro	OR	97124	Jerry Wilson	503-640-8891	R	16	0.2
STEP Co	400 Interstate N	Atlanta	GA	30339	Richard P Boggs	404-859-9292	R	16	0.2
Allied Golf Co	4538 W Fullerton	Chicago	IL	60639	Edward Holda	312-772-7710	R	15	0.1
Brine Inc	47 Sumner St	Milford	MA	01757	Jim Devis	508-478-3250	R	15	<0.1
CurveMaster Inc	PO Box 7726	Dallas	TX	75209		214-484-9484	S	15	0.3
Fox Pool Corp	PO Box 549	York	PA	17405	Robert Seitz	717-764-8581	R	15	0.2
Huffy Corp	2021 MacArthur Rd	Waukesha	WI	53188	David Allen	414-548-0440	D	15	0.2
Poolsavers	1708 Gage Rd	Montebello	CA	90640	Terrance Fitch	213-726-8444	D	15	0.1
Jandy Industries Inc	PO Box 6101	Novato	CA	94949	Andrew Pansini Jr	415-382-8220	R	14	<0.1
Porter Athletic Equipment Co	PO Box 2500	Broadview	IL	60153	G Hege	708-338-2000	R	14	0.1
Eagle Creek	1740 La Costa	San Marcos	CA	92069	S Barker	619-471-7600	R	13*	0.1
Riedell Shoes Inc	PO Box 21	Red Wing	MN	55066	R Riegelman	612-388-8251	R	13	0.2
Roger Cleveland Golf Co	5630 Cerritos Av	Cypress	CA	90630	Patrice Hutin	714-821-4200	S	13*	0.1
Caswell International Corp	1221 Marshall NE	Minneapolis	MN	55413	Paul Faust	612-379-2000	R	12	<0.1
Dacor Corp	161 Northfield Rd	Northfield	IL	60093	A Richard Michka	708-446-9555	R	12	0.1
Luhr Jensen and Sons Inc	PO Box 297	Hood River	OR	97031	Phillip W Jensen	503-386-3811	R	12	0.3

Source: Ward's Business Directory of U.S. Private and Public Companies, Volumes 1 and 2, 1996. The company type code used is as follows: P - Public, R - Private, S - Subsidiary, D - Division, J - Joint Venture, A - Affiliate, G - Group. Sales are in millions of dollars, employees are in thousands. An asterisk (*) indicates an estimated sales volume. The symbol < stands for 'less than'. Company names and addresses are truncated, in some cases, to fit into the available space.

MATERIALS CONSUMED

Material	Quantity	Delivered Cost ($ million)
Materials, ingredients, containers, and supplies	(X)	2,854.7
Metal bolts, nuts, screws, washers, rivets, and other screw machine products	(X)	59.2
Other fabricated metal products (except castings and forgings)	(X)	149.8
Aluminum and aluminum-base alloy castings (rough and semifinished)	(X)	26.3
Other castings (rough and semifinished)	(X)	64.3
Forgings	(X)	2.3
Steel sheet and strip, including tin plate	(X)	54.7
All other steel shapes and forms	(X)	120.3
Aluminum and aluminum-base alloy sheet, plate, foil, and welded tubing	(X)	52.1
All other aluminum and aluminum-base alloy shapes and forms	(X)	14.7
Other nonferrous shapes and forms	(X)	25.9
Paints, varnishes, lacquers, stains, shellacs, japans, enamels, and allied products	(X)	33.4
Plastics resins consumed in the form of granules, pellets, powders, liquids, etc.	(X)	174.9
All other chemicals and allied products	(X)	15.5
Plastics products consumed in the form of sheets, rods, tubes, film, and other shapes	(X)	101.6
Cotton, wool, manmade fiber fabrics, etc.	(X)	76.5
Finished leather	(X)	17.4
Rough and dressed lumber	(X)	62.7
Parts specially designed for sporting goods	(X)	495.6
Paperboard containers, boxes, and corrugated paperboard	(X)	80.7
All other materials and components, parts, containers, and supplies	(X)	643.8
Materials, ingredients, containers, and supplies, nsk	(X)	583.0

Source: 1992 Economic Census. Explanation of symbols used: (D): Withheld to avoid disclosure of competitive data; na: Not available; (S): Withheld because statistical norms were not met; (X): Not applicable; (Z): Less than half the unit shown; nec: Not elsewhere classified; nsk: Not specified by kind; - : zero; * : 10-19 percent estimated; ** : 20-29 percent estimated.

PRODUCT SHARE DETAILS

Product or Product Class	% Share	Product or Product Class	% Share
Sporting and athletic goods, nec	100.00	Other bowling alley playing supplies (including pins, etc.)	6.12
Fishing tackle and equipment	7.06	Baseballs and softballs	2.04
Fishing rods, all types	8.74	Baseball mitts and gloves, including softball	(D)
Fishing reels, all types	(D)	Wood baseball bats, including softball bats	0.58
Fishing rod and reel combinations	(D)	Metal baseball bats, including softball bats	3.39
Fish hooks (including snelled hooks)	6.29	Footballs	1.47
Fishing casting plugs, spinners, spoons, flies, lures, and similar artificial baits	27.76	All inflatable athletic balls other than footballs (including basketballs, soccer balls, volleyballs, etc.)	0.27
Fishing tackle boxes	9.15	Tennis rackets, all types (strung and unstrung), and tennis balls	(D)
Other fishing equipment, including creels, fish and bait buckets, floats, furnished lines, sinkers, snap swivels, etc.	14.11	Other tennis equipment and accessories (excluding clothing, shoes, and nets)	0.37
Fishing tackle and equipment, nsk	2.72	Racquetball rackets and racquet balls	0.37
Golf equipment	25.33	Uninflatable athletic balls (handballs, table tennis balls, etc.), excluding golf, tennis, bowling, billard, and racquet	(D)
Golf balls	28.39	Archery equipment	8.68
Golf clubs, irons	28.68	Ice skates, shoe skates and roller rink skates, excluding clamp-on type roller skates	3.13
Golf clubs, woods	19.06	Wooden and plastics skateboards (including complete sets)	0.51
Golf bags	8.11	Winter sports equipment (bobsleds, toboggans, hockey goods, etc.), excluding clothing, protective equipment, and skates	4.13
Other golf equipment (carts for carrying golf bags, shafts sold as such, tees, etc.), excluding shoes and apparel	15.10	Snow skis and other snow-ski equipment (excluding clothing, body protective equipment, and shoes)	0.78
Golf equipment, nsk	0.66	Surfboards and sailboards	0.92
Playground equipment	5.51	Water skis	1.51
Home playground equipment, including swing sets, slides, seesaws, sandboxes, etc.	58.87	Underwater sports equipment (SCUBA) and skindiving equipment, excluding watches and cameras	4.78
Institutional and commercial playground equipment, heavy-duty (including swings, slides, etc.)	41.11	Football helmets	(D)
Playground equipment, nsk	0.03	Other sports helmets (including bicycle, excluding football, motorcycle, and auto racing)	3.75
Gymnasium and exercise equipment	19.72	Body protective equipment for all sports (masks, shoulder, chest, knee, and kidney pads, etc.), excluding helmets	3.10
Gymnasium and gymnastic apparatus and equipment (parallel and horizontal bars, balance beams, trampolines, mats, etc.)	6.24	Football, baseball, and soccer equipment, nec (including track, field, and miscellaneous athletic field equipment)	1.74
Weight lifting equipment (including belts, benches, and weights)	13.12	Wading pools and other above ground swimming pools less than 15 ft in diameter, not filtered	(D)
Multipurpose home gyms	10.49	Above ground swimming pools 15 ft. in diameter or more, filtered (completely manufactured)	4.90
Exercise cycles	7.30	Other sporting and athletic goods	22.99
Other health, physical fitness, and exercising equipment (treadmills, slant-boards, multistation training units, etc.)	61.85	Other sporting and athletic goods, nsk	2.31
Gymnasium and exercise equipment, nsk	1.02	Sporting and athletic goods, nec, nsk	8.50
Other sporting and athletic goods	33.89		
Billiard and pool tables	4.26		
Billiard and pool supplies (such as balls, cues, etc.) sold separately	1.12		
Bowling alleys and bowling pinsetters	5.72		
Bowling balls	3.26		

Source: 1992 Economic Census. The values shown are percent of total shipments in an industry. Values of indented subcategories are summed in the main heading. The symbol (D) appears when data are withheld to prevent disclosure of competitive information. The abbreviation nsk stands for 'not specified by kind' and nec for 'not elsewhere classified'.

INPUTS AND OUTPUTS FOR SPORTING & ATHLETIC GOODS, NEC

Economic Sector or Industry Providing Inputs	%	Sector	Economic Sector or Industry Buying Outputs	%	Sector
Imports	28.9	Foreign	Personal consumption expenditures	63.0	
Sporting & athletic goods, nec	10.2	Manufg.	Gross private fixed investment	9.9	Cap Inv
Wholesale trade	10.1	Trade	Exports	7.1	Foreign
Blast furnaces & steel mills	4.4	Manufg.	Sporting & athletic goods, nec	6.3	Manufg.
Plastics materials & resins	2.7	Manufg.	S/L Govt. purch., elem. & secondary education	5.5	S/L Govt
Miscellaneous plastics products	2.5	Manufg.	S/L Govt. purch., natural resource & recreation.	1.1	S/L Govt
Aluminum rolling & drawing	2.4	Manufg.	Change in business inventories	1.0	In House
Banking	2.2	Fin/R.E.	Amusement & recreation services nec	0.8	Services
Paperboard containers & boxes	1.9	Manufg.	S/L Govt. purch., higher education	0.8	S/L Govt
Chemical preparations, nec	1.6	Manufg.	Membership sports & recreation clubs	0.5	Services
Broadwoven fabric mills	1.5	Manufg.	S/L Govt. purch., public assistance & relief	0.4	S/L Govt
Fabricated textile products, nec	1.5	Manufg.	Elementary & secondary schools	0.3	Services
Advertising	1.5	Services	Hotels & lodging places	0.3	Services
Fabricated rubber products, nec	1.4	Manufg.	S/L Govt. purch., other general government	0.3	S/L Govt
Leather tanning & finishing	1.4	Manufg.	Wholesale trade	0.2	Trade
Pipe, valves, & pipe fittings	1.3	Manufg.	Bowling alleys, billiard & pool establishments	0.2	Services
Veneer & plywood	1.3	Manufg.	Residential care	0.2	Services
Electric services (utilities)	1.3	Util.	Federal Government purchases, national defense	0.2	Fed Govt
Synthetic rubber	1.2	Manufg.	Federal Government purchases, nondefense	0.2	Fed Govt
Petroleum refining	1.1	Manufg.	Agricultural, forestry, & fishery services	0.1	Agric.
Sawmills & planning mills, general	1.1	Manufg.	Colleges, universities, & professional schools	0.1	Services
Motor freight transportation & warehousing	1.0	Util.	S/L Govt. purch., correction	0.1	S/L Govt
Eating & drinking places	1.0	Trade			
Aluminum castings	0.9	Manufg.			
Paints & allied products	0.7	Manufg.			
Screw machine and related products	0.7	Manufg.			
Communications, except radio & TV	0.6	Util.			
Gas production & distribution (utilities)	0.6	Util.			
Real estate	0.6	Fin/R.E.			
U.S. Postal Service	0.6	Gov't			
Rubber & plastics hose & belting	0.5	Manufg.			
Job training & related services	0.5	Services			
Maintenance of nonfarm buildings nec	0.4	Constr.			
General industrial machinery, nec	0.4	Manufg.			
Semiconductors & related devices	0.4	Manufg.			
Signs & advertising displays	0.4	Manufg.			
Railroads & related services	0.4	Util.			
Management & consulting services & labs	0.4	Services			
Coated fabrics, not rubberized	0.3	Manufg.			
Cutstone & stone products	0.3	Manufg.			
Machinery, except electrical, nec	0.3	Manufg.			
Wood products, nec	0.3	Manufg.			
Air transportation	0.3	Util.			
Security & commodity brokers	0.3	Fin/R.E.			
Equipment rental & leasing services	0.3	Services			
Legal services	0.3	Services			
Abrasive products	0.2	Manufg.			
Apparel made from purchased materials	0.2	Manufg.			
Commercial printing	0.2	Manufg.			
Copper rolling & drawing	0.2	Manufg.			
Felt goods, nec	0.2	Manufg.			
Hardwood dimension & flooring mills	0.2	Manufg.			
Organic fibers, noncellulosic	0.2	Manufg.			
Special dies & tools & machine tool accessories	0.2	Manufg.			
Sanitary services, steam supply, irrigation	0.2	Util.			
Water transportation	0.2	Util.			
Accounting, auditing & bookkeeping	0.2	Services			
Automotive rental & leasing, without drivers	0.2	Services			
Automotive repair shops & services	0.2	Services			
Engineering, architectural, & surveying services	0.2	Services			
Adhesives & sealants	0.1	Manufg.			
Glass & glass products, except containers	0.1	Manufg.			
Lubricating oils & greases	0.1	Manufg.			
Manifold business forms	0.1	Manufg.			
Motors & generators	0.1	Manufg.			
Needles, pins, & fasteners	0.1	Manufg.			
Nonferrous rolling & drawing, nec	0.1	Manufg.			
Paper coating & glazing	0.1	Manufg.			
Particleboard	0.1	Manufg.			
Insurance carriers	0.1	Fin/R.E.			
Hotels & lodging places	0.1	Services			

Source: Benchmark Input-Output Accounts for the U.S. Economy, 1982, U.S. Department of Commerce, Washington, D.C., July 1991. Data, as reported in the source, are organized by the 1977 SIC structure in use in 1982 but have been matched, as closely as is possible, to the 1987 SIC structure used in this book.

OCCUPATIONS EMPLOYED BY SIC 394 - TOYS AND SPORTING GOODS

Occupation	% of Total 1994	Change to 2005	Occupation	% of Total 1994	Change to 2005
Assemblers, fabricators, & hand workers nec	23.5	21.3	Coating, painting, & spraying machine workers	1.5	21.3
Plastic molding machine workers	4.3	21.3	Machine forming operators, metal & plastic	1.5	21.3
Blue collar worker supervisors	4.2	13.5	Painting, coating, & decorating workers, hand	1.4	21.3
Hand packers & packagers	3.6	4.0	Welders & cutters	1.4	21.3
Sales & related workers nec	2.9	21.3	General office clerks	1.4	3.4
General managers & top executives	2.8	15.1	Industrial production managers	1.3	21.2
Traffic, shipping, & receiving clerks	2.6	16.8	Machine tool cutting operators, metal & plastic	1.3	1.0
Inspectors, testers, & graders, precision	2.4	21.3	Electrical & electronic assemblers	1.2	21.4
Freight, stock, & material movers, hand	2.3	-2.9	Production, planning, & expediting clerks	1.1	21.3
Helpers, laborers, & material movers nec	2.0	21.3	Packaging & filling machine operators	1.1	21.3
Secretaries, ex legal & medical	1.9	10.4	Order clerks, materials, merchandise, & service	1.0	18.7
Machine operators nec	1.7	6.9	Maintenance repairers, general utility	1.0	9.2
Bookkeeping, accounting, & auditing clerks	1.6	-9.0			

Source: *Industry-Occupation Matrix*, Bureau of Labor Statistics. These data relate to one or more 3-digit SIC industry groups rather than to a single 4-digit SIC. The change reported for each occupation to the year 2005 is a percent of growth or decline as estimated by the Bureau of Labor Statistics. The abbreviation nec stands for 'not elsewhere classified'.

LOCATION BY STATE AND REGIONAL CONCENTRATION

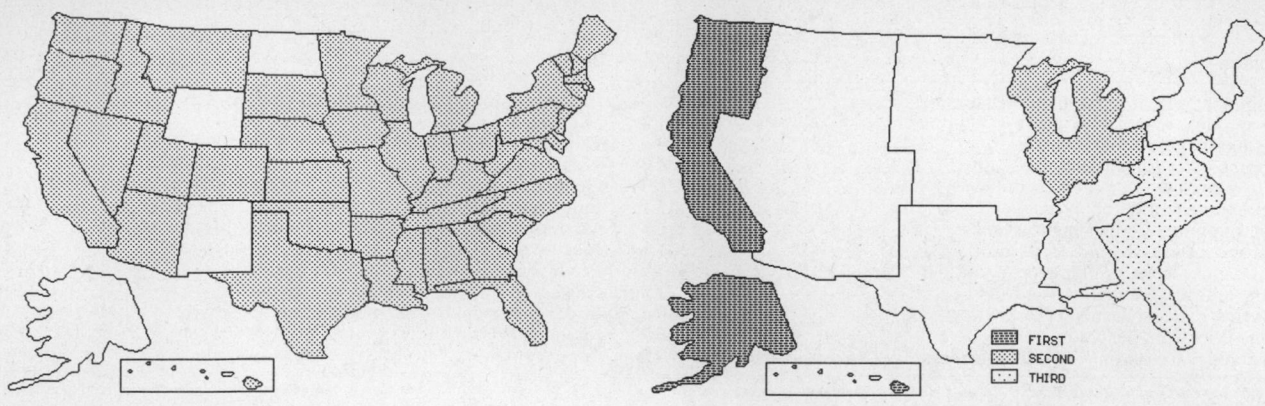

FIRST
SECOND
THIRD

INDUSTRY DATA BY STATE

State	Establish-ments	Shipments			Employment				Cost as % of Shipments	Investment per Employee ($)
		Total ($ mil)	% of U.S.	Per Establ.	Total Number	% of U.S.	Per Establ.	Wages ($/hour)		
California	345	1,099.4	14.5	3.2	8,300	13.5	24	8.14	45.7	3,602
Illinois	94	484.0	6.4	5.1	3,300	5.3	35	8.64	42.1	3,121
Minnesota	66	380.1	5.0	5.8	2,800	4.5	42	9.10	38.3	3,714
Texas	117	310.8	4.1	2.7	3,000	4.9	26	7.65	54.8	6,033
Pennsylvania	78	307.6	4.1	3.9	2,600	4.2	33	10.38	52.2	1,885
Arizona	23	305.9	4.0	13.3	2,200	3.6	96	7.20	44.8	2,273
Washington	88	274.0	3.6	3.1	2,600	4.2	30	9.81	43.0	1,577
Michigan	73	263.5	3.5	3.6	1,100	1.8	15	10.18	47.7	-
Wisconsin	75	245.3	3.2	3.3	2,000	3.2	27	7.19	47.5	3,100
Tennessee	33	219.1	2.9	6.6	1,900	3.1	58	9.00	45.2	3,368
Virginia	15	207.3	2.7	13.8	1,200	1.9	80	6.43	46.4	-
Oklahoma	40	206.4	2.7	5.2	1,300	2.1	33	7.74	47.0	2,846
New York	73	205.1	2.7	2.8	1,600	2.6	22	9.18	45.4	2,125
South Carolina	24	194.8	2.6	8.1	1,200	1.9	50	10.22	41.4	4,083
Missouri	62	160.3	2.1	2.6	1,700	2.8	27	7.15	45.0	1,647
Indiana	42	158.4	2.1	3.8	1,100	1.8	26	10.85	42.3	-
Oregon	75	133.1	1.8	1.8	1,300	2.1	17	9.73	30.7	1,154
Florida	130	119.2	1.6	0.9	1,400	2.3	11	8.25	46.2	1,357
Ohio	60	111.2	1.5	1.9	1,300	2.1	22	8.25	46.4	-
Georgia	48	80.7	1.1	1.7	800	1.3	17	8.09	45.7	5,375
Colorado	53	80.6	1.1	1.5	900	1.5	17	7.06	35.4	1,889
Vermont	10	71.2	0.9	7.1	400	0.6	40	10.80	56.2	1,250
Connecticut	20	40.6	0.5	2.0	500	0.8	25	11.17	40.6	-
Montana	19	34.2	0.5	1.8	600	1.0	32	7.44	37.4	667
Kansas	17	19.2	0.3	1.1	200	0.3	12	6.67	45.8	-
Nevada	13	14.6	0.2	1.1	200	0.3	15	8.50	39.0	1,000
New Hampshire	12	9.8	0.1	0.8	100	0.2	8	7.50	49.0	-
South Dakota	6	8.5	0.1	1.4	100	0.2	17	5.00	56.5	2,000
North Carolina	49	(D)	-	-	375 *	0.6	8	-	-	1,600
Alabama	43	(D)	-	-	3,750 *	6.1	87	-	-	-
Arkansas	34	(D)	-	-	1,750 *	2.8	51	-	-	-
Massachusetts	34	(D)	-	-	1,750 *	2.8	51	-	-	-
Utah	31	(D)	-	-	3,750 *	6.1	121	-	-	-
New Jersey	30	(D)	-	-	750 *	1.2	25	-	-	-
Kentucky	24	(D)	-	-	750 *	1.2	31	-	-	2,933
Mississippi	23	(D)	-	-	1,750 *	2.8	76	-	-	-
Idaho	22	(D)	-	-	175 *	0.3	8	-	-	-
Louisiana	22	(D)	-	-	750 *	1.2	34	-	-	-
Iowa	20	(D)	-	-	750 *	1.2	38	-	-	2,933
Nebraska	16	(D)	-	-	175 *	0.3	11	-	-	-
Maryland	15	(D)	-	-	175 *	0.3	12	-	-	-
Hawaii	11	(D)	-	-	175 *	0.3	16	-	-	-
Maine	6	(D)	-	-	175 *	0.3	29	-	-	1,143
West Virginia	3	(D)	-	-	175 *	0.3	58	-	-	-

Source: 1992 *Economic Census*. The states are in descending order of shipments or establishments (if shipment data are missing for the majority). The symbol (D) appears when data are withheld to prevent disclosure of competitive information. States marked with (D) are sorted by number of establishments. A dash (-) indicates that the data element cannot be calculated; * indicates the midpoint of a range.

3951 - PENS & MECHANICAL PENCILS

82 83 84 85 86 87 88 89 90 91 92 93 94 95 96 97 98

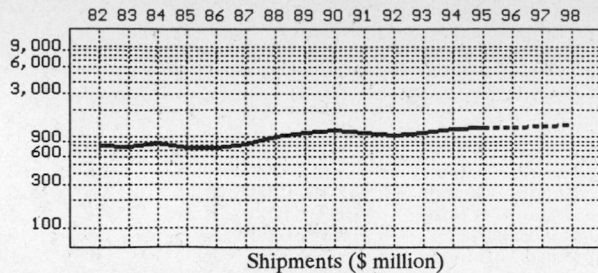

Shipments ($ million)

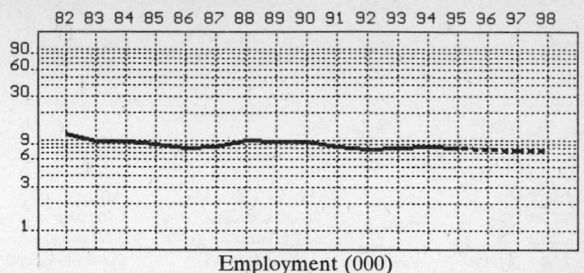

Employment (000)

GENERAL STATISTICS

| Year | Com-panies | Establishments | | Employment | | | Compensation | | Production ($ million) | | | |
		Total	with 20 or more employees	Total (000)	Production Workers (000)	Hours (Mil)	Payroll ($ mil)	Wages ($/hr)	Cost of Materials	Value Added by Manufacture	Value of Shipments	Capital Invest.
1982	129	136	69	11.6	8.8	16.8	174.8	6.68	346.6	472.6	809.4	27.8
1983		132	69	9.4	7.1	13.7	158.2	6.91	311.1	448.6	785.7	22.0
1984		128	69	9.6	7.3	13.3	172.8	8.41	356.7	519.6	860.9	18.4
1985		123	69	8.9	6.7	12.4	160.9	8.35	323.7	419.8	746.9	20.1
1986		118	65	8.2	6.2	12.0	151.9	8.33	312.6	444.3	757.9	19.6
1987	106	110	54	8.4	6.4	12.5	166.5	8.61	325.5	485.1	818.8	23.9
1988		104	56	9.9	7.3	14.5	197.3	8.57	413.5	603.2	1,009.1	37.2
1989		100	57	9.7	7.3	14.4	214.5	8.78	465.0	652.7	1,108.9	33.2
1990		99	55	9.7	6.9	12.9	215.6	9.64	540.9	682.4	1,205.8	45.8
1991		100	56	8.5	6.1	11.7	205.4	10.36	457.5	645.5	1,096.7	44.0
1992	103	106	52	7.9	5.8	11.2	197.0	10.50	383.4	657.7	1,048.5	34.1
1993		108	55	8.2	5.9	11.6	213.6	11.03	469.8	647.6	1,133.9	44.4
1994		93P	49P	8.5	5.9	12.0	231.8	10.45	515.2	690.2	1,230.0	32.4
1995		90P	48P	8.0P	5.6P	11.2P	228.7P	11.29P	523.2P	700.9P	1,249.1P	43.9P
1996		86P	46P	7.9P	5.5P	11.0P	234.3P	11.63P	539.9P	723.2P	1,288.9P	45.8P
1997		83P	44P	7.7P	5.3P	10.7P	239.9P	11.96P	556.6P	745.6P	1,328.7P	47.6P
1998		80P	43P	7.6P	5.2P	10.5P	245.6P	12.29P	573.2P	768.0P	1,368.6P	49.4P

Sources: 1982, 1987, 1992 *Economic Census*; *Annual Survey of Manufactures*, 83-86, 88-91, 93-94. Establishment counts for non-Census years are from *County Business Patterns*; establishment values for 83-84 are extrapolations. 'P's show projections by the editors. Industries reclassified in 87 will not have data for prior years.

INDICES OF CHANGE

| Year | Com-panies | Establishments | | Employment | | | Compensation | | Production ($ million) | | | |
		Total	with 20 or more employees	Total (000)	Production Workers (000)	Hours (Mil)	Payroll ($ mil)	Wages ($/hr)	Cost of Materials	Value Added by Manufacture	Value of Shipments	Capital Invest.
1982	125	128	133	147	152	150	89	64	90	72	77	82
1983		125	133	119	122	122	80	66	81	68	75	65
1984		121	133	122	126	119	88	80	93	79	82	54
1985		116	133	113	116	111	82	80	84	64	71	59
1986		111	125	104	107	107	77	79	82	68	72	57
1987	103	104	104	106	110	112	85	82	85	74	78	70
1988		98	108	125	126	129	100	82	108	92	96	109
1989		94	110	123	126	129	109	84	121	99	106	97
1990		93	106	123	119	115	109	92	141	104	115	134
1991		94	108	108	105	104	104	99	119	98	105	129
1992	100	100	100	100	100	100	100	100	100	100	100	100
1993		102	106	104	102	104	108	105	123	98	108	130
1994		88P	95P	108	102	107	118	100	134	105	117	95
1995		85P	92P	102P	97P	100P	116P	108P	136P	107P	119P	129P
1996		81P	88P	100P	94P	98P	119P	111P	141P	110P	123P	134P
1997		78P	85P	98P	92P	96P	122P	114P	145P	113P	127P	140P
1998		75P	82P	96P	89P	93P	125P	117P	150P	117P	131P	145P

Sources: Same as General Statistics. Values reflect change from the base year, 1992. Values above 100 mean greater than 92, values below 100 mean less than 92, and a value of 100 in the 82-91 or 93-98 period means same as 92. 'P's mark projections by the editors.

SELECTED RATIOS

For 1994	Avg. of All Manufact.	Analyzed Industry	Index	For 1994	Avg. of All Manufact.	Analyzed Industry	Index
Employees per Establishment	49	92	187	Value Added per Production Worker	134,084	116,983	87
Payroll per Establishment	1,500,273	2,497,763	166	Cost per Establishment	5,045,178	5,551,543	110
Payroll per Employee	30,620	27,271	89	Cost per Employee	102,970	60,612	59
Production Workers per Establishment	34	64	185	Cost per Production Worker	146,988	87,322	59
Wages per Establishment	853,319	1,351,249	158	Shipments per Establishment	9,576,895	13,253,878	138
Wages per Production Worker	24,861	21,254	85	Shipments per Employee	195,460	144,706	74
Hours per Production Worker	2,056	2,034	99	Shipments per Production Worker	279,017	208,475	75
Wages per Hour	12.09	10.45	86	Investment per Establishment	321,011	349,127	109
Value Added per Establishment	4,602,255	7,437,257	162	Investment per Employee	6,552	3,812	58
Value Added per Employee	93,930	81,200	86	Investment per Production Worker	9,352	5,492	59

Sources: Same as General Statistics. The 'Average of All Manufacturing' column represents the average of all manufacturing industries reported for the most recent complete year available. The Index shows the relationship between the Average and the Analyzed Industry. For example, 100 means that they are equal; 500 that the Analyzed Industry is five times the average; 50 means that the Analyzed Industry is half the national average. The abbreviation 'na' is used to show that data are 'not available'.

LEADING COMPANIES   Number shown: **30**   Total sales ($ mil): **1,643**   Total employment (000): **13.0**

Company Name	Address				CEO Name	Phone	Co. Type	Sales ($ mil)	Empl. (000)
Bic Corp	500 Bic Dr	Milford	CT	06460	Bruno Bich	203-783-2000	P	439	2.5
Hunt Manufacturing Co	2005 Market St	Philadelphia	PA	19103	Ronald J Naples	215-656-0300	P	288	2.2
Pen Holdings Inc	105 W Park Dr	Brentwood	TN	37027	Charles A Lieppe	615-370-9700	R	225	2.3
AT Cross Co	1 Albion Rd	Lincoln	RI	02865	Bradford R Boss	401-333-1200	P	177	1.7
Sanford Corp	2711 Washington	Bellwood	IL	60104	Robert Parker	708-547-6650	S	155	0.6
Sheaffer Inc	301 Avnue H	Fort Madison	IA	52627	David M Connors	319-372-3300	R	110*	1.0
Parker Pen USA	1400 N Parker Dr	Janesville	WI	53545	John Jacks	608-755-7000	S	52*	0.5
Pentel of America Ltd	2805 Columbia St	Torrance	CA	90509	Yukio Horie	310-320-3831	S	38*	0.3
Ritepoint Inc	969 Executive Pkwy	St Louis	MO	63141	G William Henry	314-434-9222	S	30	0.4
C and D Hit Inc	PO Box 10200	St Petersburg	FL	33733	A W Schmidt III	813-541-5561	S	19	0.3
Gold Bond Inc	5485 Hixson Pike	Hixson	TN	37343	DW Godsey	615-842-5844	R	16	0.2
Rotary Pen Corp	746 Colfax Av	Kenilworth	NJ	07033	W Shea	908-245-2437	R	12*	0.1
Accutec Inc	168 Main Av	Wallington	NJ	07057	Arnold Hendelman	201-471-3131	R	11*	0.1
Garland Industries Inc	1 S Main St	Coventry	RI	02816	Louise Lanoie	401-821-1450	R	11	0.1
Eversharp Pen Co	9240 W Belmont	Franklin Park	IL	60131	Bruce J Brizzolara	708-366-5030	R	10	0.1
Island Pen Manufacturing Corp	2004 McDonald Av	Brooklyn	NY	11223	Ida Cooper	718-376-5700	R	8*	0.1
Joseph Lipic Pen Co	2200 Gravois Av	St Louis	MO	63104	Leonard G Lipic	314-664-2111	R	7*	<0.1
Pencoa	117 State St	Westbury	NY	11590	R Perlmutter	516-997-2330	R	7	<0.1
Hartley Co	1987 Placentia Av	Costa Mesa	CA	92627	Nicholas W Smith	714-646-9641	R	6	<0.1
Fisher Space Pen Co	711 Yucca St	Boulder City	NV	89005	Paul Fisher	702-293-3011	R	4	<0.1
Mark-Tex Corp	161 Coolidge Av	Englewood	NJ	07631	George Pappageorge	201-567-4111	R	4*	<0.1
Arnold Pen Company Inc	PO Box 791	Petersburg	VA	23804	C Shepherd	804-733-6612	R	3	<0.1
Hub Pen Company Inc	26 Quincy Av	Braintree	MA	02184	Frank P Fleming	617-848-4555	R	3	<0.1
Anja Engineering Corp	PO Box 5555	Fontana	CA	92334	Glen Gibbs	909-360-2137	S	2*	<0.1
Yafa Pen Co	21306 Gault St	Canoga Park	CA	91303	Jerry Greenberg	818-704-8888	R	2	<0.1
Listo Pencil Corp	PO Drawer J	Alameda	CA	94501	Rick D Stuart	510-522-2910	R	2	<0.1
Blackfeet Writing Instruments	PO Box 729	Browning	MT	59417	Marty Meineke	406-338-2535	R	1	<0.1
Racon Manufacturing Company	240 Moonachie Av	Moonachie	NJ	07074	Ed Kuder	201-641-1212	R	1*	<0.1
Permacharge Corp	5930 Midway Pk	Albuquerque	NM	87109	Ioana McNamara	505-344-1415	R	0	<0.1
Lindy Pen	251 1/2 Grove Av	Verona	NJ	07044	Marty Meineke	201-239-6480	D	0	<0.1

Source: Ward's Business Directory of U.S. Private and Public Companies, Volumes 1 and 2, 1996. The company type code used is as follows: P - Public, R - Private, S - Subsidiary, D - Division, J - Joint Venture, A - Affiliate, G - Group. Sales are in millions of dollars, employees are in thousands. An asterisk (*) indicates an estimated sales volume. The symbol < stands for 'less than'. Company names and addresses are truncated, in some cases, to fit into the available space.

MATERIALS CONSUMED

Material	Quantity	Delivered Cost ($ million)
Materials, ingredients, containers, and supplies	(X)	330.5
Lumber and wood products, except furniture	(X)	(D)
Paperboard containers, boxes, and corrugated paperboard	(X)	33.8
Inorganic pigments	(X)	(D)
Organic color pigments, lakes, and toners	(X)	(D)
Paints, varnishes, lacquers, stains, shellacs, japans, enamels, and allied products	(X)	(D)
Other chemicals and allied products	(X)	16.3
Petroleum wax	(X)	(D)
Plastics products consumed in the form of sheets, rods, tubes, film, and other shapes	(X)	26.5
Natural graphite	(X)	0.7
Other stone, clay, glass, and concrete products	(X)	(D)
Fabricated metal products, including forgings	(X)	27.7
Parts for pens and mechanical pencils	(X)	155.4
All other materials and components, parts, containers, and supplies	(X)	52.2
Materials, ingredients, containers, and supplies, nsk	(X)	11.6

Source: 1992 Economic Census. Explanation of symbols used: (D): Withheld to avoid disclosure of competitive data; na: Not available; (S): Withheld because statistical norms were not met; (X): Not applicable; (Z): Less than half the unit shown; nec: Not elsewhere classified; nsk: Not specified by kind; - : zero; * : 10-19 percent estimated; ** : 20-29 percent estimated.

PRODUCT SHARE DETAILS

Product or Product Class	% Share	Product or Product Class	% Share
Pens and mechanical pencils	100.00	Markers, nsk	0.14
Pens	47.42	Other pens, mechanical pencils, and parts	21.03
Refillable ball point pens	50.53	Mechanical pencils, including clutch action and twist action	28.41
Nonrefillable ball point pens	37.22	Ball point pen refill cartridges	10.70
Roller pens	11.85	All other refill cartridges (for fountain and roller pens, fine point markers, etc.)	9.05
Pens, nsk	0.41		
Markers	29.53	All other pens and mechanical pencil parts (including pen points, renewal parts, fountain pens, desk sets, etc.)	50.81
Fine point markers (thin-line writing pens)	28.99	Other pens, mechanical pencils, and parts, nsk	1.04
Broad tipped markers (thick-line coloring pens and markers)	70.87	Pens and mechanical pencils, nsk	2.02

Source: 1992 Economic Census. The values shown are percent of total shipments in an industry. Values of indented subcategories are summed in the main heading. The symbol (D) appears when data are withheld to prevent disclosure of competitive information. The abbreviation nsk stands for 'not specified by kind' and nec for 'not elsewhere classified'.

INPUTS AND OUTPUTS FOR PENS & MECHANICAL PENCILS

Economic Sector or Industry Providing Inputs	%	Sector	Economic Sector or Industry Buying Outputs	%	Sector
Imports	16.4	Foreign	Personal consumption expenditures	52.9	
Fabricated rubber products, nec	15.3	Manufg.	Exports	11.6	Foreign
Advertising	14.6	Services	Legal services	3.8	Services
Wholesale trade	8.4	Trade	Pens & mechanical pencils	3.6	Manufg.
Pens & mechanical pencils	5.3	Manufg.	Insurance carriers	1.4	Fin/R.E.
Banking	4.7	Fin/R.E.	Change in business inventories	1.4	In House
Fabricated metal products, nec	4.2	Manufg.	S/L Govt. purch., other general government	1.4	S/L Govt
Paperboard containers & boxes	3.1	Manufg.	Wholesale trade	1.3	Trade
Petroleum refining	3.1	Manufg.	S/L Govt. purch., elem. & secondary education	1.3	S/L Govt
Miscellaneous plastics products	2.8	Manufg.	Retail trade, except eating & drinking	1.2	Trade
Felt goods, nec	2.4	Manufg.	Hospitals	1.2	Services
Nonmetallic mineral products, nec	2.1	Manufg.	Banking	1.0	Fin/R.E.
Job training & related services	1.9	Services	Federal Government purchases, nondefense	1.0	Fed Govt
Electric services (utilities)	1.5	Util.	Soap & other detergents	0.8	Manufg.
Eating & drinking places	1.2	Trade	Doctors & dentists	0.8	Services
Motor freight transportation & warehousing	1.1	Util.	Religious organizations	0.8	Services
Noncomparable imports	0.9	Foreign	S/L Govt. purch., higher education	0.8	S/L Govt
Communications, except radio & TV	0.7	Util.	Colleges, universities, & professional schools	0.7	Services
Automotive rental & leasing, without drivers	0.7	Services	Insurance agents, brokers, & services	0.6	Fin/R.E.
Electrical repair shops	0.6	Services	Real estate	0.6	Fin/R.E.
Maintenance of nonfarm buildings nec	0.5	Constr.	Federal Government purchases, national defense	0.6	Fed Govt
Miscellaneous fabricated wire products	0.5	Manufg.	S/L Govt. purch., health & hospitals	0.6	S/L Govt
Sanitary services, steam supply, irrigation	0.5	Util.	Credit agencies other than banks	0.4	Fin/R.E.
Royalties	0.5	Fin/R.E.	Accounting, auditing & bookkeeping	0.4	Services
Automotive repair shops & services	0.5	Services	Elementary & secondary schools	0.4	Services
U.S. Postal Service	0.5	Gov't	S/L Govt. purch., public assistance & relief	0.4	S/L Govt
Machinery, except electrical, nec	0.4	Manufg.	Industrial buildings	0.3	Constr.
Air transportation	0.4	Util.	Communications, except radio & TV	0.3	Util.
Real estate	0.4	Fin/R.E.	Business services nec	0.3	Services
Equipment rental & leasing services	0.4	Services	Engineering, architectural, & surveying services	0.3	Services
Legal services	0.4	Services	Management & consulting services & labs	0.3	Services
Management & consulting services & labs	0.4	Services	Crude petroleum & natural gas	0.2	Mining
Abrasive products	0.3	Manufg.	Residential high-rise apartments	0.2	Constr.
Special dies & tools & machine tool accessories	0.3	Manufg.	Commercial printing	0.2	Manufg.
Gas production & distribution (utilities)	0.3	Util.	Electric services (utilities)	0.2	Util.
Motor vehicle parts & accessories	0.2	Manufg.	Motor freight transportation & warehousing	0.2	Util.
Insurance carriers	0.2	Fin/R.E.	Security & commodity brokers	0.2	Fin/R.E.
Accounting, auditing & bookkeeping	0.2	Services	Labor, civic, social, & fraternal associations	0.2	Services
Hotels & lodging places	0.2	Services	Medical & health services, nec	0.2	Services
Lubricating oils & greases	0.1	Manufg.	Personnel supply services	0.2	Services
Manifold business forms	0.1	Manufg.	Social services, nec	0.2	Services
Tires & inner tubes	0.1	Manufg.	S/L Govt. purch., police	0.2	S/L Govt
Retail trade, except eating & drinking	0.1	Trade	Construction of hospitals	0.1	Constr.
			Construction of stores & restaurants	0.1	Constr.
			Dormitories & other group housing	0.1	Constr.
			Electronic computing equipment	0.1	Manufg.
			Newspapers	0.1	Manufg.
			Radio & TV communication equipment	0.1	Manufg.
			Eating & drinking places	0.1	Trade
			Advertising	0.1	Services
			Computer & data processing services	0.1	Services
			Hotels & lodging places	0.1	Services
			S/L Govt. purch., correction	0.1	S/L Govt
			S/L Govt. purch., natural resource & recreation.	0.1	S/L Govt

Source: *Benchmark Input-Output Accounts for the U.S. Economy, 1982*, U.S. Department of Commerce, Washington, D.C., July 1991. Data, as reported in the source, are organized by the 1977 SIC structure in use in 1982 but have been matched, as closely as is possible, to the 1987 SIC structure used in this book.

OCCUPATIONS EMPLOYED BY SIC 395 - MANUFACTURED PRODUCTS, NEC

Occupation	% of Total 1994	Change to 2005	Occupation	% of Total 1994	Change to 2005
Assemblers, fabricators, & hand workers nec	17.0	4.8	Freight, stock, & material movers, hand	1.5	-16.1
Sales & related workers nec	4.3	4.8	Inspectors, testers, & graders, precision	1.5	4.9
Blue collar worker supervisors	4.0	-2.4	General office clerks	1.5	-10.6
General managers & top executives	3.8	-0.5	Designers, ex interior designers	1.5	15.3
Hand packers & packagers	3.7	-28.1	Electrical & electronic assemblers	1.4	4.8
Traffic, shipping, & receiving clerks	2.5	0.9	Industrial production managers	1.4	4.8
Machine operators nec	2.2	-7.6	Packaging & filling machine operators	1.4	4.9
Helpers, laborers, & material movers nec	2.2	4.8	Screen printing machine setters & set-up operators	1.2	4.8
Painting, coating, & decorating workers, hand	2.0	-5.7	Plastic molding machine workers	1.2	4.8
Precision workers nec	2.0	41.5	Machine feeders & offbearers	1.2	-5.7
Secretaries, ex legal & medical	2.0	-4.5	Cabinetmakers & bench carpenters	1.1	-31.9
Bookkeeping, accounting, & auditing clerks	1.8	-21.4	Order clerks, materials, merchandise, & service	1.1	2.6
Sheet metal workers & duct installers	1.8	4.8	Grinders & polishers, hand	1.1	4.8
Precision metal workers nec	1.8	36.3			

Source: *Industry-Occupation Matrix*, Bureau of Labor Statistics. These data relate to one or more 3-digit SIC industry groups rather than to a single 4-digit SIC. The change reported for each occupation to the year 2005 is a percent of growth or decline as estimated by the Bureau of Labor Statistics. The abbreviation nec stands for 'not elsewhere classified'.

LOCATION BY STATE AND REGIONAL CONCENTRATION

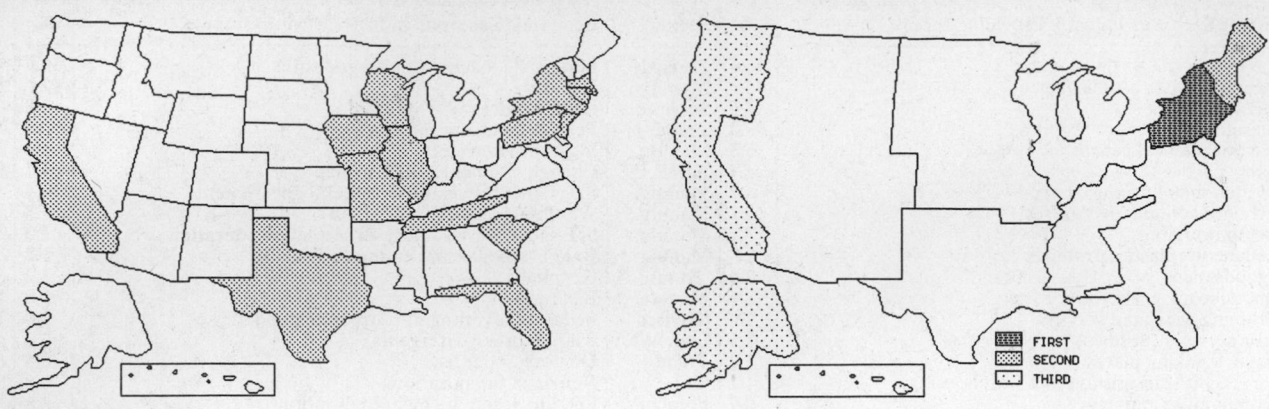

INDUSTRY DATA BY STATE

State	Establish-ments	Shipments			Employment				Cost as % of Shipments	Investment per Employee ($)
		Total ($ mil)	% of U.S.	Per Establ.	Total Number	% of U.S.	Per Establ.	Wages ($/hour)		
Rhode Island	6	159.2	15.2	26.5	1,400	17.7	233	13.33	23.6	-
New Jersey	17	57.3	5.5	3.4	600	7.6	35	9.67	46.1	2,833
New York	17	44.0	4.2	2.6	500	6.3	29	6.86	43.4	1,600
Pennsylvania	4	25.1	2.4	6.3	300	3.8	75	9.50	44.6	-
Connecticut	5	21.7	2.1	4.3	200	2.5	40	11.50	48.4	2,500
Missouri	3	16.9	1.6	5.6	200	2.5	67	7.00	42.0	2,000
Massachusetts	5	9.3	0.9	1.9	100	1.3	20	17.00	26.9	-
California	12	(D)	-	-	750 *	9.5	63	-	-	-
Florida	5	(D)	-	-	375 *	4.7	75	-	-	-
Illinois	5	(D)	-	-	750 *	9.5	150	-	-	-
Texas	4	(D)	-	-	375 *	4.7	94	-	-	-
Tennessee	3	(D)	-	-	750 *	9.5	250	-	-	-
Wisconsin	2	(D)	-	-	750 *	9.5	375	-	-	-
Iowa	1	(D)	-	-	750 *	9.5	750	-	-	-
South Carolina	1	(D)	-	-	175 *	2.2	175	-	-	-

Source: 1992 *Economic Census*. The states are in descending order of shipments or establishments (if shipment data are missing for the majority). The symbol (D) appears when data are withheld to prevent disclosure of competitive information. States marked with (D) are sorted by number of establishments. A dash (-) indicates that the data element cannot be calculated; * indicates the midpoint of a range.

3952 - LEAD PENCILS & ART GOODS

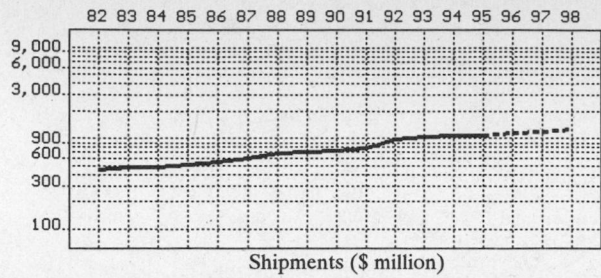

82 83 84 85 86 87 88 89 90 91 92 93 94 95 96 97 98

Shipments ($ million)

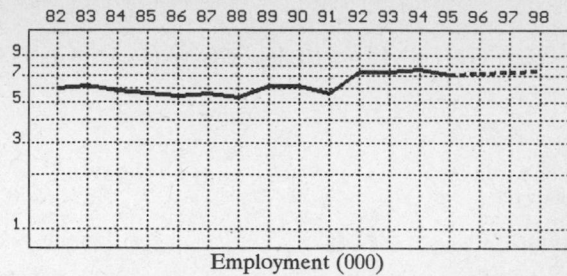

82 83 84 85 86 87 88 89 90 91 92 93 94 95 96 97 98

Employment (000)

GENERAL STATISTICS

Year	Com-panies	Establishments		Employment			Compensation		Production ($ million)			
		Total	with 20 or more employees	Total (000)	Production Workers (000)	Hours (Mil)	Payroll ($ mil)	Wages ($/hr)	Cost of Materials	Value Added by Manufacture	Value of Shipments	Capital Invest.
1982	144	148	43	6.0	4.3	8.0	87.5	6.07	199.4	244.8	445.9	16.2
1983		146	43	6.2	4.5	8.5	91.3	6.26	221.5	270.1	487.7	13.6
1984		144	43	5.8	4.2	8.2	94.5	6.61	213.8	270.7	484.2	15.6
1985		141	43	5.6	4.1	8.2	97.9	6.91	214.1	293.3	510.8	14.7
1986		136	40	5.4	4.0	7.9	101.1	7.85	239.8	315.4	547.0	12.9
1987	140	145	40	5.6	4.0	7.8	103.3	7.45	252.3	361.5	609.3	21.3
1988		139	41	5.3	4.0	7.6	97.1	8.14	297.0	394.4	676.0	14.1
1989		143	46	6.2	4.2	8.4	105.5	8.61	319.9	403.0	716.8	15.9
1990		145	43	6.2	4.0	8.2	103.1	8.13	349.6	407.2	745.9	19.0
1991		150	42	5.6	4.3	8.7	110.1	8.80	351.6	432.3	786.1	21.1
1992	162	164	45	7.3	5.4	10.4	149.3	8.44	408.9	576.8	975.9	23.0
1993		176	43	7.4	5.5	10.9	159.5	8.82	446.0	608.2	1,047.5	19.3
1994		161P	43P	7.6	5.8	11.3	165.1	8.75	458.7	639.7	1,089.2	22.2
1995		162P	43P	7.1P	5.2P	10.4P	153.8P	9.43P	456.8P	637.0P	1,084.7P	22.2P
1996		164P	44P	7.2P	5.3P	10.7P	159.7P	9.67P	479.8P	669.2P	1,139.4P	22.9P
1997		166P	44P	7.3P	5.5P	10.9P	165.6P	9.91P	502.9P	701.3P	1,194.1P	23.5P
1998		168P	44P	7.5P	5.6P	11.2P	171.4P	10.15P	525.9P	733.4P	1,248.8P	24.2P

Sources: 1982, 1987, 1992 *Economic Census*; *Annual Survey of Manufactures*, 83-86, 88-91, 93-94. Establishment counts for non-Census years are from *County Business Patterns*; establishment values for 83-84 are extrapolations. 'P's show projections by the editors. Industries reclassified in 87 will not have data for prior years.

INDICES OF CHANGE

Year	Com-panies	Establishments		Employment			Compensation		Production ($ million)			
		Total	with 20 or more employees	Total (000)	Production Workers (000)	Hours (Mil)	Payroll ($ mil)	Wages ($/hr)	Cost of Materials	Value Added by Manufacture	Value of Shipments	Capital Invest.
1982	89	90	96	82	80	77	59	72	49	42	46	70
1983		89	96	85	83	82	61	74	54	47	50	59
1984		88	96	79	78	79	63	78	52	47	50	68
1985		86	96	77	76	79	66	82	52	51	52	64
1986		83	89	74	74	76	68	93	59	55	56	56
1987	86	88	89	77	74	75	69	88	62	63	62	93
1988		85	91	73	74	73	65	96	73	68	69	61
1989		87	102	85	78	81	71	102	78	70	73	69
1990		88	96	85	74	79	69	96	85	71	76	83
1991		91	93	77	80	84	74	104	86	75	81	92
1992	100	100	100	100	100	100	100	100	100	100	100	100
1993		107	96	101	102	105	107	105	109	105	107	84
1994		98P	96P	104	107	109	111	104	112	111	112	97
1995		99P	97P	97P	97P	100P	103P	112P	112P	110P	111P	97P
1996		100P	97P	99P	99P	103P	107P	115P	117P	116P	117P	100P
1997		101P	97P	101P	101P	105P	111P	117P	123P	122P	122P	102P
1998		103P	97P	102P	102P	107P	115P	120P	129P	127P	128P	105P

Sources: Same as General Statistics. Values reflect change from the base year, 1992. Values above 100 mean greater than 92, values below 100 mean less than 92, and a value of 100 in the 82-91 or 93-98 period means same as 92. 'P's mark projections by the editors.

SELECTED RATIOS

For 1994	Avg. of All Manufact.	Analyzed Industry	Index	For 1994	Avg. of All Manufact.	Analyzed Industry	Index
Employees per Establishment	49	47	97	Value Added per Production Worker	134,084	110,293	82
Payroll per Establishment	1,500,273	1,028,563	69	Cost per Establishment	5,045,178	2,857,674	57
Payroll per Employee	30,620	21,724	71	Cost per Employee	102,970	60,355	59
Production Workers per Establishment	34	36	105	Cost per Production Worker	146,988	79,086	54
Wages per Establishment	853,319	615,985	72	Shipments per Establishment	9,576,895	6,785,652	71
Wages per Production Worker	24,861	17,047	69	Shipments per Employee	195,460	143,316	73
Hours per Production Worker	2,056	1,948	95	Shipments per Production Worker	279,017	187,793	67
Wages per Hour	12.09	8.75	72	Investment per Establishment	321,011	138,305	43
Value Added per Establishment	4,602,255	3,985,294	87	Investment per Employee	6,552	2,921	45
Value Added per Employee	93,930	84,171	90	Investment per Production Worker	9,352	3,828	41

Sources: Same as General Statistics. The 'Average of All Manufacturing' column represents the average of all manufacturing industries reported for the most recent complete year available. The Index shows the relationship between the Average and the Analyzed Industry. For example, 100 means that they are equal; 500 that the Analyzed Industry is five times the average; 50 means that the Analyzed Industry is half the national average. The abbreviation 'na' is used to show that data are 'not available'.

LEADING COMPANIES   Number shown: **19**   Total sales ($ mil): **637**   Total employment (000): **6.1**

Company Name	Address				CEO Name	Phone	Co. Type	Sales ($ mil)	Empl. (000)
Berol Corp	105 W Park Dr	Brentwood	TN	37027	Charles A Lieppe	615-370-9700	S	225	2.3
Binney and Smith Inc	1100 Church Ln	Easton	PA	18042	Tom Muller	215-253-6271	S	170	1.6
Koh-I-Noor Inc	PO Box 68	Bloomsbury	NJ	08804	Fred Clauser	908-479-4124	S	60	0.4
Plaid Enterprises	PO Box 7600	Norcross	GA	30091	Michael J McCooey	404-923-8200	D	46	0.4
American Art Clay Company Inc	4717 W 16th St	Indianapolis	IN	46222	LB Sandoe Jr	317-244-6871	R	20	0.2
Chartpak	1 River Rd	Leeds	MA	01053	Ralph Van Gilder	413-584-5446	S	15	0.2
Fantastic Graphics Inc	4710 Eisenhower	Tampa	FL	33634	Gilbert R Bailie	813-884-1332	R	15	<0.1
Badger Air Brush Co	9128 W Belmont	Franklin Park	IL	60131	WA Schlotfeldt	708-678-3104	R	14	<0.1
Paasche Airbrush Co	7440 W Lawrence	Harwood H	IL	60656	John Pettersen	708-867-9191	R	10	0.1
FM Brush Company Inc	70-02 72nd Pl	Glendale	NY	11385	Frederick Mink	718-821-5939	R	9*	0.1
Binney and Smith Inc	1100 Church Ln	Easton	PA	18042	Tom Muller	215-253-6271	D	8*	<0.1
Robert Simmons Inc	45 W 18th St	New York	NY	10011	George Disch	212-675-3136	R	8	0.1
Martin F Weber Co	2727 Southampton	Philadelphia	PA	19154	Mike Gorak	215-677-5600	S	8	<0.1
JR Moon Pencil Company Inc	PO Box 1309	Lewisburg	TN	37091	JR Moon	615-359-1501	R	6	0.1
Musgrave Pencil Company Inc	PO Box 290	Shelbyville	TN	37160	George H Hulan Jr	615-684-3611	R	6	<0.1
General Pencil Co	67 Fleet St	Jersey City	NJ	07306	OA Weissenborn	201-653-5351	R	6	<0.1
Duro Art Industries Inc	1832 W Juneway	Chicago	IL	60626	TC Rathslag	312-743-3430	R	5	0.1
Jensen's Inc	PO Box 320	Shelbyville	TN	37160		615-684-5021	R	4	<0.1
Crescent Bronze Powder	3400 N Avondale St	Chicago	IL	60618	M Lazarus	312-539-2441	R	3	<0.1

Source: Ward's Business Directory of U.S. Private and Public Companies, Volumes 1 and 2, 1996. The company type code used is as follows: P - Public, R - Private, S - Subsidiary, D - Division, J - Joint Venture, A - Affiliate, G - Group. Sales are in millions of dollars, employees are in thousands. An asterisk (*) indicates an estimated sales volume. The symbol < stands for 'less than'. Company names and addresses are truncated, in some cases, to fit into the available space.

MATERIALS CONSUMED

Material	Quantity	Delivered Cost ($ million)
Materials, ingredients, containers, and supplies	(X)	313.3
Lumber and wood products, except furniture	(X)	39.5
Paperboard containers, boxes, and corrugated paperboard	(X)	29.4
Inorganic pigments	(X)	9.1
Organic color pigments, lakes, and toners	(X)	8.6
Paints, varnishes, lacquers, stains, shellacs, japans, enamels, and allied products	(X)	23.2
Other chemicals and allied products	(X)	27.5
Petroleum wax	(X)	6.5
Plastics products consumed in the form of sheets, rods, tubes, film, and other shapes	(X)	16.2
Natural graphite	(X)	8.3
Other stone, clay, glass, and concrete products	(X)	1.6
Fabricated metal products, including forgings	(X)	21.2
Parts for pens and mechanical pencils	(X)	49.9
All other materials and components, parts, containers, and supplies	(X)	49.6
Materials, ingredients, containers, and supplies, nsk	(X)	22.5

Source: 1992 *Economic Census.* Explanation of symbols used: (D): Withheld to avoid disclosure of competitive data; na: Not available; (S): Withheld because statistical norms were not met; (X): Not applicable; (Z): Less than half the unit shown; nec: Not elsewhere classified; nsk: Not specified by kind; - : zero; * : 10-19 percent estimated; ** : 20-29 percent estimated.

PRODUCT SHARE DETAILS

Product or Product Class	% Share	Product or Product Class	% Share
Lead pencils and art goods	100.00	Artists' equipment (including children's school art equipment, airbrushes, drawing tables, boards, palettes, etc.)	42.43
Nonmechanical pencils, crayons, and chalk	53.53		
Nonmechanical (wood-cased) black graphite pencils	45.46		
Other nonmechanical (wood-cased) pencils (indelible, colored, etc.) and graphite and colored sticks	5.37	Other art materials (including modeling clay, other modeling material, chalk, water colors, tempera colors, etc.)	57.57
Crayons and chalk, except artists', including tailors' chalk	48.48		
Nonmechanical pencils, crayons, and chalk, nsk	0.69	Lead pencils and art goods, nsk	4.88
Artists' equipment	41.58		

Source: 1992 *Economic Census.* The values shown are percent of total shipments in an industry. Values of indented subcategories are summed in the main heading. The symbol (D) appears when data are withheld to prevent disclosure of competitive information. The abbreviation nsk stands for 'not specified by kind' and nec for 'not elsewhere classified'.

INPUTS AND OUTPUTS FOR LEAD PENCILS & ART GOODS

Economic Sector or Industry Providing Inputs	%	Sector	Economic Sector or Industry Buying Outputs	%	Sector
Wholesale trade	24.1	Trade	Personal consumption expenditures	42.8	
Wood products, nec	12.2	Manufg.	Business services nec	15.5	Services
Advertising	10.8	Services	Legal services	5.1	Services
Miscellaneous plastics products	7.3	Manufg.	Exports	3.4	Foreign
Imports	4.9	Foreign	Accounting, auditing & bookkeeping	2.8	Services
Petroleum refining	4.6	Manufg.	Advertising	2.8	Services
Gypsum products	2.8	Manufg.	S/L Govt. purch., other general government	1.7	S/L Govt
Motor freight transportation & warehousing	2.4	Util.	S/L Govt. purch., elem. & secondary education	1.5	S/L Govt
Metal stampings, nec	2.2	Manufg.	S/L Govt. purch., higher education	1.5	S/L Govt
Nonmetallic mineral products, nec	2.2	Manufg.	Insurance carriers	1.3	Fin/R.E.
Manufacturing industries, nec	1.7	Manufg.	Wholesale trade	1.2	Trade
Job training & related services	1.6	Services	Hospitals	1.2	Services
Fabricated metal products, nec	1.4	Manufg.	Federal Government purchases, nondefense	1.1	Fed Govt
Eating & drinking places	1.3	Trade	Retail trade, except eating & drinking	1.0	Trade
Felt goods, nec	1.2	Manufg.	Banking	1.0	Fin/R.E.
Electric services (utilities)	1.1	Util.	Religious organizations	0.9	Services
Inorganic pigments	1.0	Manufg.	Doctors & dentists	0.8	Services
Noncomparable imports	1.0	Foreign	Colleges, universities, & professional schools	0.7	Services
Clay, ceramic, & refractory minerals	0.8	Mining	Real estate	0.6	Fin/R.E.
Paints & allied products	0.8	Manufg.	Federal Government purchases, national defense	0.6	Fed Govt
Paperboard containers & boxes	0.8	Manufg.	S/L Govt. purch., health & hospitals	0.6	S/L Govt
Screw machine and related products	0.8	Manufg.	Insurance agents, brokers, & services	0.5	Fin/R.E.
Real estate	0.7	Fin/R.E.	S/L Govt. purch., natural resource & recreation.	0.5	S/L Govt
Fabricated rubber products, nec	0.6	Manufg.	S/L Govt. purch., public assistance & relief	0.5	S/L Govt
Synthetic rubber	0.6	Manufg.	Credit agencies other than banks	0.4	Fin/R.E.
Communications, except radio & TV	0.6	Util.	Communications, except radio & TV	0.3	Util.
Banking	0.6	Fin/R.E.	Business/professional associations	0.3	Services
Equipment rental & leasing services	0.6	Services	Elementary & secondary schools	0.3	Services
Maintenance of nonfarm buildings nec	0.5	Constr.	Engineering, architectural, & surveying services	0.3	Services
Chemical preparations, nec	0.5	Manufg.	Management & consulting services & labs	0.3	Services
Machinery, except electrical, nec	0.5	Manufg.	Federal electric utilities	0.3	Gov't
Sawmills & planning mills, general	0.5	Manufg.	S/L Govt. purch., police	0.3	S/L Govt
Railroads & related services	0.5	Util.	Crude petroleum & natural gas	0.2	Mining
Management & consulting services & labs	0.5	Services	Industrial buildings	0.2	Constr.
Abrasive products	0.4	Manufg.	Residential high-rise apartments	0.2	Constr.
Gas production & distribution (utilities)	0.4	Util.	Apparel made from purchased materials	0.2	Manufg.
Legal services	0.4	Services	Commercial printing	0.2	Manufg.
U.S. Postal Service	0.4	Gov't	Electronic computing equipment	0.2	Manufg.
Minerals, ground or treated	0.3	Manufg.	Lead pencils & art goods	0.2	Manufg.
Special dies & tools & machine tool accessories	0.3	Manufg.	Electric services (utilities)	0.2	Util.
Sanitary services, steam supply, irrigation	0.3	Util.	Motor freight transportation & warehousing	0.2	Util.
Water transportation	0.3	Util.	Security & commodity brokers	0.2	Fin/R.E.
Royalties	0.3	Fin/R.E.	Commercial sports, except racing	0.2	Services
Accounting, auditing & bookkeeping	0.3	Services	Labor, civic, social, & fraternal associations	0.2	Services
Lead pencils & art goods	0.2	Manufg.	Personnel supply services	0.2	Services
Lubricating oils & greases	0.2	Manufg.	Social services, nec	0.2	Services
Manifold business forms	0.2	Manufg.	S/L Govt. purch., correction	0.2	S/L Govt
Air transportation	0.2	Util.	S/L Govt. purch., highways	0.2	S/L Govt
Automotive rental & leasing, without drivers	0.2	Services	Construction of hospitals	0.1	Constr.
Automotive repair shops & services	0.2	Services	Construction of stores & restaurants	0.1	Constr.
Cyclic crudes and organics	0.1	Manufg.	Dormitories & other group housing	0.1	Constr.
Metal foil & leaf	0.1	Manufg.	Residential 1-unit structures, nonfarm	0.1	Constr.
Nonferrous rolling & drawing, nec	0.1	Manufg.	Newspapers	0.1	Manufg.
Insurance carriers	0.1	Fin/R.E.	Radio & TV communication equipment	0.1	Manufg.
			Eating & drinking places	0.1	Trade
			Computer & data processing services	0.1	Services
			Equipment rental & leasing services	0.1	Services
			Hotels & lodging places	0.1	Services
			Job training & related services	0.1	Services
			Medical & health services, nec	0.1	Services
			Nursing & personal care facilities	0.1	Services
			S/L Govt. purch., other education & libraries	0.1	S/L Govt

Source: Benchmark Input-Output Accounts for the U.S. Economy, 1982, U.S. Department of Commerce, Washington, D.C., July 1991. Data, as reported in the source, are organized by the 1977 SIC structure in use in 1982 but have been matched, as closely as is possible, to the 1987 SIC structure used in this book.

OCCUPATIONS EMPLOYED BY SIC 395 - MANUFACTURED PRODUCTS, NEC

Occupation	% of Total 1994	Change to 2005	Occupation	% of Total 1994	Change to 2005
Assemblers, fabricators, & hand workers nec	17.0	4.8	Freight, stock, & material movers, hand	1.5	-16.1
Sales & related workers nec	4.3	4.8	Inspectors, testers, & graders, precision	1.5	4.9
Blue collar worker supervisors	4.0	-2.4	General office clerks	1.5	-10.6
General managers & top executives	3.8	-0.5	Designers, ex interior designers	1.5	15.3
Hand packers & packagers	3.7	-28.1	Electrical & electronic assemblers	1.4	4.8
Traffic, shipping, & receiving clerks	2.5	0.9	Industrial production managers	1.4	4.8
Machine operators nec	2.2	-7.6	Packaging & filling machine operators	1.4	4.9
Helpers, laborers, & material movers nec	2.2	4.8	Screen printing machine setters & set-up operators	1.2	4.8
Painting, coating, & decorating workers, hand	2.0	-5.7	Plastic molding machine workers	1.2	4.8
Precision workers nec	2.0	41.5	Machine feeders & offbearers	1.2	-5.7
Secretaries, ex legal & medical	2.0	-4.5	Cabinetmakers & bench carpenters	1.1	-31.9
Bookkeeping, accounting, & auditing clerks	1.8	-21.4	Order clerks, materials, merchandise, & service	1.1	2.6
Sheet metal workers & duct installers	1.8	4.8	Grinders & polishers, hand	1.1	4.8
Precision metal workers nec	1.8	36.3			

Source: Industry-Occupation Matrix, Bureau of Labor Statistics. These data relate to one or more 3-digit SIC industry groups rather than to a single 4-digit SIC. The change reported for each occupation to the year 2005 is a percent of growth or decline as estimated by the Bureau of Labor Statistics. The abbreviation nec stands for 'not elsewhere classified'.

LOCATION BY STATE AND REGIONAL CONCENTRATION

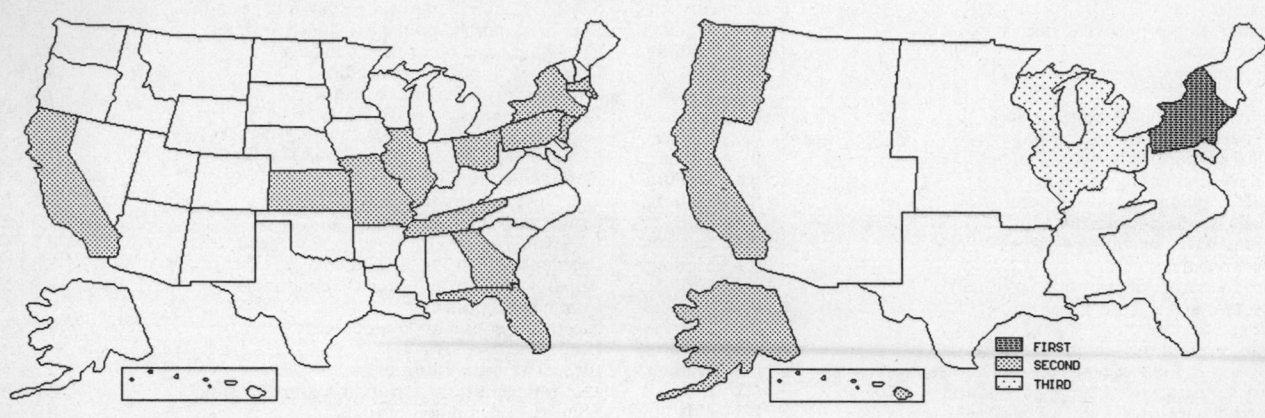

FIRST
SECOND
THIRD

INDUSTRY DATA BY STATE

State	Establish-ments	Shipments			Employment				Cost as % of Shipments	Investment per Employee ($)
		Total ($ mil)	% of U.S.	Per Establ.	Total Number	% of U.S.	Per Establ.	Wages ($/hour)		
Tennessee	8	324.9	33.3	40.6	2,200	30.1	275	6.94	46.5	-
Illinois	12	75.1	7.7	6.3	700	9.6	58	8.63	44.5	1,857
New Jersey	10	69.4	7.1	6.9	700	9.6	70	9.22	28.8	714
New York	21	19.9	2.0	0.9	400	5.5	19	6.33	37.2	2,250
Massachusetts	10	14.5	1.5	1.5	100	1.4	10	8.50	44.1	3,000
California	29	(D)	-	-	375 *	5.1	13	-	-	-
Florida	7	(D)	-	-	175 *	2.4	25	-	-	2,286
Pennsylvania	6	(D)	-	-	750 *	10.3	125	-	-	-
Georgia	5	(D)	-	-	375 *	5.1	75	-	-	-
Ohio	4	(D)	-	-	175 *	2.4	44	-	-	-
Missouri	2	(D)	-	-	375 *	5.1	188	-	-	-
Kansas	1	(D)	-	-	375 *	5.1	375	-	-	-

Source: 1992 Economic Census. The states are in descending order of shipments or establishments (if shipment data are missing for the majority). The symbol (D) appears when data are withheld to prevent disclosure of competitive information. States marked with (D) are sorted by number of establishments. A dash (-) indicates that the data element cannot be calculated; * indicates the midpoint of a range.

3953 - MARKING DEVICES

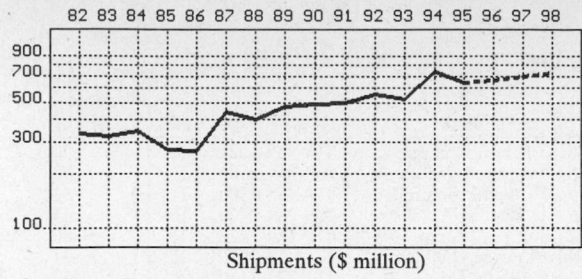

Shipments ($ million)

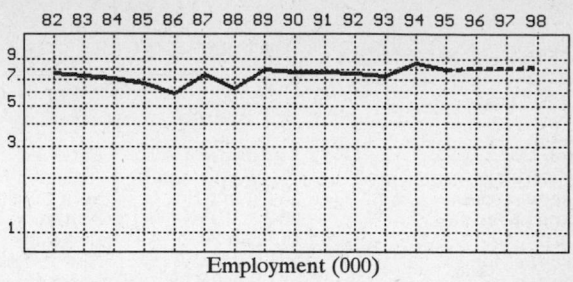

Employment (000)

GENERAL STATISTICS

| Year | Com-panies | Establishments | | Employment | | | Compensation | | Production ($ million) | | | |
		Total	with 20 or more employees	Total (000)	Production Workers (000)	Hours (Mil)	Payroll ($ mil)	Wages ($/hr)	Cost of Materials	Value Added by Manufacture	Value of Shipments	Capital Invest.
1982	593	627	88	7.6	5.5	10.2	107.1	6.44	119.0	221.4	337.6	7.6
1983		613	90	7.3	5.2	9.4	101.3	6.70	119.8	209.3	327.4	6.0
1984		599	92	7.1	4.8	9.3	111.2	7.85	115.8	228.3	348.8	15.8
1985		586	94	6.7	4.4	9.0	97.5	6.71	79.7	186.1	273.2	6.9
1986		571	97	5.9	3.7	7.6	92.1	6.51	84.0	186.8	271.7	16.8
1987	609	636	92	7.5	4.9	9.3	136.7	7.89	165.0	277.2	442.5	9.7
1988		596	94	6.3	4.3	9.2	127.2	7.51	133.3	265.3	399.1	1.1
1989		579	99	8.0	4.7	9.6	140.0	7.52	191.5	280.8	473.7	4.1
1990		591	96	7.8	4.8	9.9	143.3	8.04	185.9	295.2	485.6	4.0
1991		605	92	7.8	4.8	9.1	148.0	8.55	188.7	306.5	497.3	6.7
1992	628	649	94	7.6	4.3	8.5	160.8	8.51	202.4	355.4	553.2	9.3
1993		637	97	7.4	4.4	8.4	160.1	8.71	187.3	328.3	516.8	5.7
1994		618P	97P	8.6	5.4	9.9	183.3	8.80	268.7	486.0	732.6	17.5
1995		620P	98P	8.0P	4.6P	9.0P	177.8P	9.02P	234.0P	423.2P	637.9P	8.6P
1996		622P	98P	8.1P	4.6P	9.0P	184.5P	9.21P	244.6P	442.4P	666.8P	8.6P
1997		624P	99P	8.2P	4.5P	8.9P	191.1P	9.41P	255.2P	461.6P	695.7P	8.6P
1998		625P	99P	8.2P	4.5P	8.9P	197.7P	9.60P	265.8P	480.7P	724.7P	8.6P

Sources: 1982, 1987, 1992 *Economic Census*; *Annual Survey of Manufactures*, 83-86, 88-91, 93-94. Establishment counts for non-Census years are from *County Business Patterns*; establishment values for 83-84 are extrapolations. 'P's show projections by the editors. Industries reclassified in 87 will not have data for prior years.

INDICES OF CHANGE

| Year | Com-panies | Establishments | | Employment | | | Compensation | | Production ($ million) | | | |
		Total	with 20 or more employees	Total (000)	Production Workers (000)	Hours (Mil)	Payroll ($ mil)	Wages ($/hr)	Cost of Materials	Value Added by Manufacture	Value of Shipments	Capital Invest.
1982	94	97	94	100	128	120	67	76	59	62	61	82
1983		94	96	96	121	111	63	79	59	59	59	65
1984		92	98	93	112	109	69	92	57	64	63	170
1985		90	100	88	102	106	61	79	39	52	49	74
1986		88	103	78	86	89	57	76	42	53	49	181
1987	97	98	98	99	114	109	85	93	82	78	80	104
1988		92	100	83	100	108	79	88	66	75	72	12
1989		89	105	105	109	113	87	88	95	79	86	44
1990		91	102	103	112	116	89	94	92	83	88	43
1991		93	98	103	112	107	92	100	93	86	90	72
1992	100	100	100	100	100	100	100	100	100	100	100	100
1993		98	103	97	102	99	100	102	93	92	93	61
1994		95P	103P	113	126	116	114	103	133	137	132	188
1995		96P	104P	105P	106P	106P	111P	106P	116P	119P	115P	92P
1996		96P	105P	106P	106P	106P	115P	108P	121P	124P	121P	92P
1997		96P	105P	107P	106P	105P	119P	111P	126P	130P	126P	92P
1998		96P	106P	108P	105P	105P	123P	113P	131P	135P	131P	92P

Sources: Same as General Statistics. Values reflect change from the base year, 1992. Values above 100 mean greater than 92, values below 100 mean less than 92, and a value of 100 in the 82-91 or 93-98 period means same as 92. 'P's mark projections by the editors.

SELECTED RATIOS

For 1994	Avg. of All Manufact.	Analyzed Industry	Index	For 1994	Avg. of All Manufact.	Analyzed Industry	Index
Employees per Establishment	49	14	28	Value Added per Production Worker	134,084	90,000	67
Payroll per Establishment	1,500,273	296,391	20	Cost per Establishment	5,045,178	434,481	9
Payroll per Employee	30,620	21,314	70	Cost per Employee	102,970	31,244	30
Production Workers per Establishment	34	9	25	Cost per Production Worker	146,988	49,759	34
Wages per Establishment	853,319	140,871	17	Shipments per Establishment	9,576,895	1,184,595	12
Wages per Production Worker	24,861	16,133	65	Shipments per Employee	195,460	85,186	44
Hours per Production Worker	2,056	1,833	89	Shipments per Production Worker	279,017	135,667	49
Wages per Hour	12.09	8.80	73	Investment per Establishment	321,011	28,297	9
Value Added per Establishment	4,602,255	785,849	17	Investment per Employee	6,552	2,035	31
Value Added per Employee	93,930	56,512	60	Investment per Production Worker	9,352	3,241	35

Sources: Same as General Statistics. The 'Average of All Manufacturing' column represents the average of all manufacturing industries reported for the most recent complete year available. The Index shows the relationship between the Average and the Analyzed Industry. For example, 100 means that they are equal; 500 that the Analyzed Industry is five times the average; 50 means that the Analyzed Industry is half the national average. The abbreviation 'na' is used to show that data are 'not available'.

LEADING COMPANIES   Number shown: **35**   Total sales ($ mil): **539**   Total employment (000): **5.1**

Company Name	Address				CEO Name	Phone	Co. Type	Sales ($ mil)	Empl. (000)
Diagraph Corp	3401 Rider Trail S	Earth City	MO	63045	J R Brigham Jr	314-739-1221	R	70	0.4
Weber Marking Systems Inc	711 W Algonquin	Arlington H	IL	60005	Joseph A Weber Jr	708-364-8500	R	70	0.7
Cosco Industries Inc	56 Church St	Spring Valley	NY	10977	Robert F Schmidt	914-356-4300	R	52*	0.7
GM Nameplate Inc	2040 15th Av W	Seattle	WA	98119	D Root	206-284-2200	R	44	0.5
Marsh Co	PO Box 388	Belleville	IL	62222	John A Marsh	618-234-1122	R	40	0.4
Pamarco Inc	571 Central Av	New Providence	NJ	07974	Maurice Buckley	201-467-1070	R	31	0.2
Schwaab Inc	PO Box 26069	Milwaukee	WI	53226	William Zahn	414-771-4150	R	28*	0.4
Foilmark Inc	40 Melville Park Rd	Melville	NY	11747	Frank J Olsen Jr	516-694-7773	P	28	0.2
Consolidated Stamp of Chicago	7220 W Wilson Av	Harwood H	IL	60656	Robert Schmidt	708-867-5800	D	24*	0.3
Shachihata Incorporated USA	3305 Kashiwa St	Torrance	CA	90505	S Asano	310-530-4445	S	23	0.2
US Stamp and Sign	1480 Gould Dr	Cookeville	TN	38506	Donald Polak	615-432-4055	D	21*	0.1
M and R Marking Systems	100 Springfield Av	Piscataway	NJ	08855	L Sculler	908-562-9500	R	15	0.1
Helicoflex Co	PO Box 9889	Columbia	SC	29290	Didier Muller	803-783-1880	D	14	<0.1
Louis Melind Co	7631 N Austin Av	Skokie	IL	60076	David B Sterrett	708-581-2500	R	13	0.2
Volk Corp	23936 Industrial Pk	Farmington Hls	MI	48335	P Hankins	810-477-6700	R	7	<0.1
George T Schmidt Inc	6151 W Howard St	Niles	IL	60714	Ed Bottum	708-647-7117	R	6*	0.1
Norwood Marking Systems	2538 Wisconsin Av	Downers Grove	IL	60515	Jack Campbell	708-968-0646	D	6*	<0.1
CH Hanson Co	3630 N Wolf Rd	Franklin Park	IL	60131	Craig F Hanson	708-451-0500	R	5*	<0.1
Kiwi Coders Corp	265 E Messner Dr	Wheeling	IL	60090	AJ McKay	708-541-4511	R	5*	<0.1
Tacoma Rubber Stamp	919 Market St	Tacoma	WA	98402	Joseph R Lovely	206-383-5433	R	5	<0.1
Advance Corp	327 E York Av	St Paul	MN	55101	Glen Lorenz	612-771-9297	R	4*	<0.1
Menke Marking Devices Inc	PO Box 2986	Santa Fe Sprgs	CA	90670	Steve A Menke	714-994-0440	R	4	<0.1
Microfoam Inc	31 Faass Av	Utica	NY	13502	Donald Polak	315-797-1000	S	4*	<0.1
ME Cunningham International	PO Box 307	Ingomar	PA	15127	CW Kratz	412-369-9199	R	4	<0.1
Worcester Stamp Co	75 Webster St	Worcester	MA	01603	Fred Barry	508-756-7138	R	3	<0.1
Excelsior Marking Products	4524 Hudson Dr	Stow	OH	44224	Bruce D Horsfall	216-929-2802	R	3	<0.1
Huntington Park Rubber	2761 E Slauson Av	Huntington Pk	CA	90255	G Barlam	213-582-6461	R	2	<0.1
Marking Methods Inc	301 S Raymond Av	Alhambra	CA	91803	C Nichols	818-282-8823	R	2	<0.1
Mercury Marking Devices Inc	PO Box 86	Lake Elsinore	CA	92531	H Mosbacher	909-674-8717	R	2	<0.1
Hancock Wood Engineering Inc	PO Box 864	Iron Mountain	MI	49801	Jack Donnelly	906-774-3321	R	2	<0.1
Dyer Specialty Company Inc	1550 Corona Dr	Lk Havasu Ct	AZ	86403	E Dyer	602-453-8600	R	1	<0.1
Krengel Manufacturing Company	PO Box 673	New York	NY	10018	John A Collins III	212-564-8850	R	1	<0.1
Cleveland Rubber Stamp Works	3110 Payne Av	Cleveland	OH	44114	Theodore Cutts	216-241-6211	R	0	<0.1
Time To Market Associates Inc	PO Box 443	Verdi	NV	89439	R E Kmetovica	702-789-2924	R	0	<0.1
Stamptastics	3985 Harney St	San Diego	CA	92110	Dan LeRoy	619-296-6044	R	0	<0.1

Source: *Ward's Business Directory of U.S. Private and Public Companies*, Volumes 1 and 2, 1996. The company type code used is as follows: P - Public, R - Private, S - Subsidiary, D - Division, J - Joint Venture, A - Affiliate, G - Group. Sales are in millions of dollars, employees are in thousands. An asterisk (*) indicates an estimated sales volume. The symbol < stands for 'less than'. Company names and addresses are truncated, in some cases, to fit into the available space.

MATERIALS CONSUMED

Material	Quantity	Delivered Cost ($ million)
Materials, ingredients, containers, and supplies	(X)	161.2
Lumber and wood products, except furniture	(X)	2.1
Paperboard containers, boxes, and corrugated paperboard	(X)	1.1
Inorganic pigments	(X)	(D)
Organic color pigments, lakes, and toners	(X)	(D)
Paints, varnishes, lacquers, stains, shellacs, japans, enamels, and allied products	(X)	0.1
Other chemicals and allied products	(X)	1.5
Petroleum wax	(X)	(D)
Plastics products consumed in the form of sheets, rods, tubes, film, and other shapes	(X)	10.2
Other stone, clay, glass, and concrete products	(X)	(D)
Fabricated metal products, including forgings	(X)	9.8
Parts for pens and mechanical pencils	(X)	(D)
All other materials and components, parts, containers, and supplies	(X)	75.1
Materials, ingredients, containers, and supplies, nsk	(X)	58.6

Source: 1992 *Economic Census*. Explanation of symbols used: (D): Withheld to avoid disclosure of competitive data; na: Not available; (S): Withheld because statistical norms were not met; (X): Not applicable; (Z): Less than half the unit shown; nec: Not elsewhere classified; nsk: Not specified by kind; - : zero; * : 10-19 percent estimated; ** : 20-29 percent estimated.

PRODUCT SHARE DETAILS

Product or Product Class	% Share	Product or Product Class	% Share
Marking devices	100.00	goods.	22.80
Rubber and vinyl hand stamps, typeholder, and dies, custom and stock	16.81	Embossing seals, including notary, engineering, corporate, stationery, etc.	2.04
Rubber and vinyl permanently inked stamps, excluding print dies	5.46	Hand, letter and figure stamps, dies, types and type holders, and steel embossing and incising numbering heads.	9.27
Mechanical hand stamps, self inkers including daters, time and numbering stamps, and metal and rubber wheel band		Other marking devices, such as stencils, letters, figures, numerals, stamp pads, branding irons, etc.	22.93

Source: 1992 *Economic Census.* The values shown are percent of total shipments in an industry. Values of indented subcategories are summed in the main heading. The symbol (D) appears when data are withheld to prevent disclosure of competitive information. The abbreviation nsk stands for 'not specified by kind' and nec for 'not elsewhere classified'.

INPUTS AND OUTPUTS FOR MARKING DEVICES

Economic Sector or Industry Providing Inputs	%	Sector	Economic Sector or Industry Buying Outputs	%	Sector
Advertising	15.2	Services	Insurance carriers	5.5	Fin/R.E.
Blast furnaces & steel mills	14.8	Manufg.	Hospitals	4.9	Services
Wholesale trade	8.3	Trade	Retail trade, except eating & drinking	4.7	Trade
Metal stampings, nec	6.4	Manufg.	S/L Govt. purch., elem. & secondary education	4.5	S/L Govt
Wood products, nec	6.1	Manufg.	S/L Govt. purch., other general government	4.3	S/L Govt
Plastics materials & resins	4.3	Manufg.	Banking	3.8	Fin/R.E.
Miscellaneous plastics products	3.8	Manufg.	Wholesale trade	3.4	Trade
Fabricated rubber products, nec	3.7	Manufg.	Federal Government purchases, nondefense	3.0	Fed Govt
Eating & drinking places	3.4	Trade	S/L Govt. purch., higher education	3.0	S/L Govt
Security & commodity brokers	2.0	Fin/R.E.	Real estate	2.7	Fin/R.E.
Automotive repair shops & services	2.0	Services	Colleges, universities, & professional schools	2.6	Services
Imports	1.9	Foreign	Change in business inventories	2.3	In House
Real estate	1.8	Fin/R.E.	Insurance agents, brokers, & services	2.2	Fin/R.E.
Fabricated metal products, nec	1.6	Manufg.	Business/professional associations	2.1	Services
Electric services (utilities)	1.5	Util.	Religious organizations	1.9	Services
Maintenance of nonfarm buildings nec	1.2	Constr.	Credit agencies other than banks	1.7	Fin/R.E.
Machinery, except electrical, nec	1.2	Manufg.	Doctors & dentists	1.7	Services
Motor freight transportation & warehousing	1.2	Util.	S/L Govt. purch., health & hospitals	1.7	S/L Govt
Management & consulting services & labs	1.2	Services	Labor, civic, social, & fraternal associations	1.5	Services
Marking devices	1.1	Manufg.	Social services, nec	1.5	Services
Paperboard containers & boxes	1.1	Manufg.	Communications, except radio & TV	1.4	Util.
Legal services	1.1	Services	Accounting, auditing & bookkeeping	1.4	Services
Banking	1.0	Fin/R.E.	Elementary & secondary schools	1.3	Services
Abrasive products	0.9	Manufg.	S/L Govt. purch., public assistance & relief	1.2	S/L Govt
U.S. Postal Service	0.9	Gov't	Engineering, architectural, & surveying services	1.1	Services
Communications, except radio & TV	0.8	Util.	Legal services	1.0	Services
Gas production & distribution (utilities)	0.8	Util.	Personal consumption expenditures	0.9	
Special dies & tools & machine tool accessories	0.7	Manufg.	Crude petroleum & natural gas	0.9	Mining
Accounting, auditing & bookkeeping	0.7	Services	Motor freight transportation & warehousing	0.9	Util.
Electrical repair shops	0.7	Services	Industrial buildings	0.8	Constr.
Royalties	0.5	Fin/R.E.	Electric services (utilities)	0.8	Util.
Job training & related services	0.5	Services	Security & commodity brokers	0.8	Fin/R.E.
Lubricating oils & greases	0.4	Manufg.	Marking devices	0.7	Manufg.
Manifold business forms	0.4	Manufg.	Job training & related services	0.7	Services
Nonferrous rolling & drawing, nec	0.4	Manufg.	S/L Govt. purch., police	0.7	S/L Govt
Nonmetallic mineral products, nec	0.4	Manufg.	Commercial printing	0.6	Manufg.
Air transportation	0.4	Util.	Electronic computing equipment	0.6	Manufg.
Railroads & related services	0.4	Util.	Computer & data processing services	0.6	Services
Equipment rental & leasing services	0.4	Services	Management & consulting services & labs	0.6	Services
Sanitary services, steam supply, irrigation	0.3	Util.	Residential high-rise apartments	0.5	Constr.
Credit agencies other than banks	0.3	Fin/R.E.	Newspapers	0.5	Manufg.
Business/professional associations	0.3	Services	Eating & drinking places	0.5	Trade
Photographic equipment & supplies	0.2	Manufg.	Exports	0.5	Foreign
Transit & bus transportation	0.2	Util.	S/L Govt. purch., fire	0.5	S/L Govt
Business services nec	0.2	Services	Residential 1-unit structures, nonfarm	0.4	Constr.
Laundry, dry cleaning, shoe repair	0.2	Services	Drugs	0.4	Manufg.
Miscellaneous repair shops	0.2	Services	Radio & TV communication equipment	0.4	Manufg.
Personnel supply services	0.2	Services	Business services nec	0.4	Services
Distilled liquor, except brandy	0.1	Manufg.	Hotels & lodging places	0.4	Services
Gaskets, packing & sealing devices	0.1	Manufg.	Medical & health services, nec	0.4	Services
Hotels & lodging places	0.1	Services	Membership organizations nec	0.4	Services
			Federal Government purchases, national defense	0.4	Fed Govt
			S/L Govt. purch., correction	0.4	S/L Govt
			Construction of hospitals	0.3	Constr.
			Construction of stores & restaurants	0.3	Constr.
			Dormitories & other group housing	0.3	Constr.
			Aircraft	0.3	Manufg.
			Apparel made from purchased materials	0.3	Manufg.
			Electronic components nec	0.3	Manufg.
			Guided missiles & space vehicles	0.3	Manufg.
			Machinery, except electrical, nec	0.3	Manufg.
			Manufacturing industries, nec	0.3	Manufg.
			Miscellaneous plastics products	0.3	Manufg.

Continued on next page.

INPUTS AND OUTPUTS FOR MARKING DEVICES - Continued

Economic Sector or Industry Providing Inputs	%	Sector	Economic Sector or Industry Buying Outputs	%	Sector
			Motor vehicle parts & accessories	0.3	Manufg.
			Motor vehicles & car bodies	0.3	Manufg.
			Air transportation	0.3	Util.
			Gas production & distribution (utilities)	0.3	Util.
			Radio & TV broadcasting	0.3	Util.
			State & local government enterprises, nec	0.3	Gov't
			S/L Govt. purch., natural resource & recreation.	0.3	S/L Govt
			S/L Govt. purch., other education & libraries	0.3	S/L Govt
			Blast furnaces & steel mills	0.2	Manufg.
			Cyclic crudes and organics	0.2	Manufg.
			Petroleum refining	0.2	Manufg.
			Photographic equipment & supplies	0.2	Manufg.
			Semiconductors & related devices	0.2	Manufg.
			Railroads & related services	0.2	Util.
			Advertising	0.2	Services
			Libraries, vocation education	0.2	Services
			Federal Government enterprises nec	0.2	Gov't
			Agricultural, forestry, & fishery services	0.1	Agric.
			Landscape & horticultural services	0.1	Agric.
			Coal	0.1	Mining
			Construction of educational buildings	0.1	Constr.
			Aircraft & missile engines & engine parts	0.1	Manufg.
			Aircraft & missile equipment, nec	0.1	Manufg.
			Book publishing	0.1	Manufg.
			Bottled & canned soft drinks	0.1	Manufg.
			Broadwoven fabric mills	0.1	Manufg.
			Electrical equipment & supplies, nec	0.1	Manufg.
			Food preparations, nec	0.1	Manufg.
			Industrial inorganic chemicals, nec	0.1	Manufg.
			Oil field machinery	0.1	Manufg.
			Paper mills, except building paper	0.1	Manufg.
			Paperboard containers & boxes	0.1	Manufg.
			Periodicals	0.1	Manufg.
			Refrigeration & heating equipment	0.1	Manufg.
			Special dies & tools & machine tool accessories	0.1	Manufg.
			Surgical & medical instruments	0.1	Manufg.
			Telephone & telegraph apparatus	0.1	Manufg.
			Arrangement of passenger transportation	0.1	Util.
			Freight forwarders	0.1	Util.
			Transit & bus transportation	0.1	Util.
			Water transportation	0.1	Util.
			Automotive rental & leasing, without drivers	0.1	Services
			Automotive repair shops & services	0.1	Services
			Equipment rental & leasing services	0.1	Services
			Laundry, dry cleaning, shoe repair	0.1	Services
			Miscellaneous repair shops	0.1	Services
			Motion pictures	0.1	Services
			Nursing & personal care facilities	0.1	Services
			Personnel supply services	0.1	Services
			Portrait, photographic studios	0.1	Services
			Residential care	0.1	Services
			Services to dwellings & other buildings	0.1	Services
			U.S. Postal Service	0.1	Gov't

Source: Benchmark Input-Output Accounts for the U.S. Economy, 1982, U.S. Department of Commerce, Washington, D.C., July 1991. Data, as reported in the source, are organized by the 1977 SIC structure in use in 1982 but have been matched, as closely as is possible, to the 1987 SIC structure used in this book.

OCCUPATIONS EMPLOYED BY SIC 395 - MANUFACTURED PRODUCTS, NEC

Occupation	% of Total 1994	Change to 2005	Occupation	% of Total 1994	Change to 2005
Assemblers, fabricators, & hand workers nec	17.0	4.8	Freight, stock, & material movers, hand	1.5	-16.1
Sales & related workers nec	4.3	4.8	Inspectors, testers, & graders, precision	1.5	4.9
Blue collar worker supervisors	4.0	-2.4	General office clerks	1.5	-10.6
General managers & top executives	3.8	-0.5	Designers, ex interior designers	1.5	15.3
Hand packers & packagers	3.7	-28.1	Electrical & electronic assemblers	1.4	4.8
Traffic, shipping, & receiving clerks	2.5	0.9	Industrial production managers	1.4	4.8
Machine operators nec	2.2	-7.6	Packaging & filling machine operators	1.4	4.9
Helpers, laborers, & material movers nec	2.2	4.8	Screen printing machine setters & set-up operators	1.2	4.8
Painting, coating, & decorating workers, hand	2.0	-5.7	Plastic molding machine workers	1.2	4.8
Precision workers nec	2.0	41.5	Machine feeders & offbearers	1.2	-5.7
Secretaries, ex legal & medical	2.0	-4.5	Cabinetmakers & bench carpenters	1.1	-31.9
Bookkeeping, accounting, & auditing clerks	1.8	-21.4	Order clerks, materials, merchandise, & service	1.1	2.6
Sheet metal workers & duct installers	1.8	4.8	Grinders & polishers, hand	1.1	4.8
Precision metal workers nec	1.8	36.3			

Source: Industry-Occupation Matrix, Bureau of Labor Statistics. These data relate to one or more 3-digit SIC industry groups rather than to a single 4-digit SIC. The change reported for each occupation to the year 2005 is a percent of growth or decline as estimated by the Bureau of Labor Statistics. The abbreviation nec stands for 'not elsewhere classified'.

LOCATION BY STATE AND REGIONAL CONCENTRATION

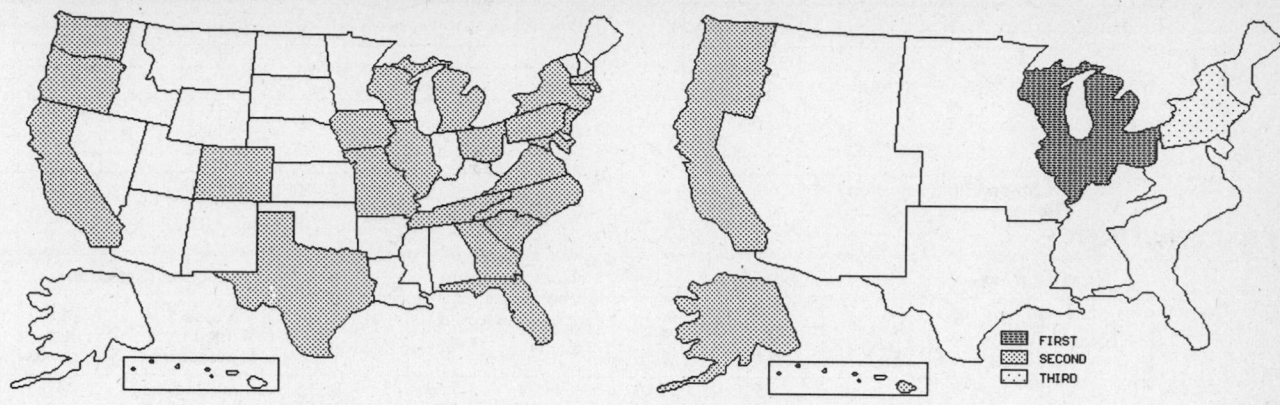

INDUSTRY DATA BY STATE

| State | Establish-ments | Shipments | | | Employment | | | | Cost as % of Shipments | Investment per Employee ($) |
		Total ($ mil)	% of U.S.	Per Establ.	Total Number	% of U.S.	Per Establ.	Wages ($/hour)		
California	82	81.4	14.7	1.0	1,100	14.5	13	8.92	34.3	1,455
Illinois	56	76.6	13.8	1.4	900	11.8	16	9.33	37.7	1,111
New Jersey	26	42.5	7.7	1.6	500	6.6	19	9.67	29.6	1,600
New York	39	26.9	4.9	0.7	400	5.3	10	7.83	42.4	1,000
Ohio	30	21.9	4.0	0.7	400	5.3	13	7.40	31.1	1,000
Wisconsin	15	19.3	3.5	1.3	400	5.3	27	11.00	16.6	500
Texas	45	16.1	2.9	0.4	300	3.9	7	9.33	34.2	1,333
Oregon	14	16.0	2.9	1.1	100	1.3	7	11.00	34.4	1,000
Connecticut	13	11.7	2.1	0.9	200	2.6	15	8.50	30.8	-
Tennessee	14	11.5	2.1	0.8	100	1.3	7	7.50	45.2	1,000
Maryland	8	10.0	1.8	1.3	100	1.3	13	7.00	27.0	1,000
South Carolina	7	9.4	1.7	1.3	100	1.3	14	11.00	37.2	2,000
Iowa	5	8.7	1.6	1.7	100	1.3	20	6.50	31.0	-
Florida	33	8.4	1.5	0.3	100	1.3	3	6.50	35.7	-
Colorado	17	8.3	1.5	0.5	200	2.6	12	7.50	33.7	-
North Carolina	12	7.5	1.4	0.6	200	2.6	17	6.00	29.3	-
Pennsylvania	28	(D)	-	-	375 *	4.9	13	-	-	1,067
Michigan	23	(D)	-	-	175 *	2.3	8	-	-	-
Washington	19	(D)	-	-	375 *	4.9	20	-	-	-
Massachusetts	18	(D)	-	-	175 *	2.3	10	-	-	571
Georgia	11	(D)	-	-	175 *	2.3	16	-	-	571
Missouri	11	(D)	-	-	375 *	4.9	34	-	-	-
Virginia	10	(D)	-	-	175 *	2.3	18	-	-	1,714

Source: 1992 *Economic Census*. The states are in descending order of shipments or establishments (if shipment data are missing for the majority). The symbol (D) appears when data are withheld to prevent disclosure of competitive information. States marked with (D) are sorted by number of establishments. A dash (-) indicates that the data element cannot be calculated; * indicates the midpoint of a range.

3955 - CARBON PAPER & INKED RIBBONS

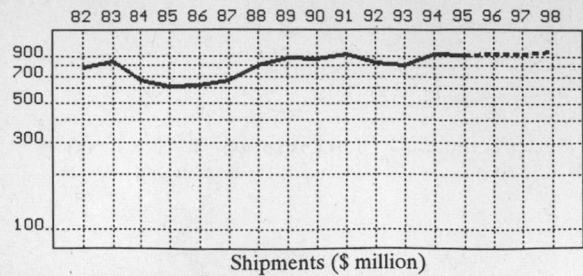

Shipments ($ million)

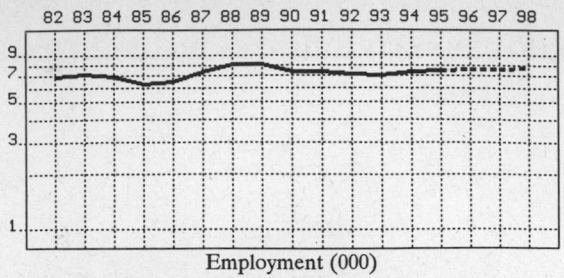

Employment (000)

GENERAL STATISTICS

| Year | Companies | Establishments | | Employment | | | Compensation | | Production ($ million) | | | |
		Total	with 20 or more employees	Total (000)	Production Workers (000)	Hours (Mil)	Payroll ($ mil)	Wages ($/hr)	Cost of Materials	Value Added by Manufacture	Value of Shipments	Capital Invest.
1982	93	115	67	6.9	5.2	10.3	115.4	7.53	463.5	324.1	778.8	23.1
1983		114	68	7.1	5.4	10.7	127.9	7.79	516.5	361.9	849.7	13.1
1984		113	69	6.9	5.2	10.3	118.5	7.27	361.1	308.6	661.6	22.2
1985		113	69	6.3	4.8	9.5	112.7	7.54	353.7	243.8	612.5	16.6
1986		110	64	6.4	5.0	10.3	118.9	7.46	349.6	278.1	621.8	17.5
1987	108	125	63	7.3	5.4	11.4	133.8	7.04	351.0	308.3	665.3	16.0
1988		121	66	8.1	6.2	13.3	153.8	6.98	454.3	342.4	796.3	16.0
1989		128	72	8.2	6.5	13.1	159.7	6.75	489.4	411.0	887.4	22.4
1990		127	67	7.4	5.6	12.1	149.3	7.13	488.6	395.2	872.7	20.1
1991		125	64	7.4	5.7	12.0	161.7	7.49	527.7	403.4	925.8	12.6
1992	117	133	62	7.2	5.4	10.9	159.4	8.32	432.2	399.0	831.1	11.3
1993		130	66	7.0	5.4	11.0	167.4	8.58	446.2	352.3	806.8	11.4
1994		133P	65P	7.3	5.5	11.1	168.7	8.77	471.0	466.2	933.6	6.3
1995		135P	64P	7.6P	5.8P	12.1P	177.3P	8.15P	456.8P	452.2P	905.5P	10.2P
1996		137P	64P	7.6P	5.8P	12.2P	182.3P	8.23P	465.3P	460.6P	922.3P	9.3P
1997		139P	64P	7.7P	5.9P	12.3P	187.3P	8.31P	473.8P	468.9P	939.1P	8.5P
1998		141P	63P	7.7P	5.9P	12.4P	192.3P	8.38P	482.2P	477.3P	955.9P	7.6P

Sources: 1982, 1987, 1992 *Economic Census*; *Annual Survey of Manufactures*, 83-86, 88-91, 93-94. Establishment counts for non-Census years are from *County Business Patterns*; establishment values for 83-84 are extrapolations. 'P's show projections by the editors. Industries reclassified in 87 will not have data for prior years.

INDICES OF CHANGE

| Year | Companies | Establishments | | Employment | | | Compensation | | Production ($ million) | | | |
		Total	with 20 or more employees	Total (000)	Production Workers (000)	Hours (Mil)	Payroll ($ mil)	Wages ($/hr)	Cost of Materials	Value Added by Manufacture	Value of Shipments	Capital Invest.
1982	79	86	108	96	96	94	72	91	107	81	94	204
1983		86	110	99	100	98	80	94	120	91	102	116
1984		85	111	96	96	94	74	87	84	77	80	196
1985		85	111	88	89	87	71	91	82	61	74	147
1986		83	103	89	93	94	75	90	81	70	75	155
1987	92	94	102	101	100	105	84	85	81	77	80	142
1988		91	106	113	115	122	96	84	105	86	96	142
1989		96	116	114	120	120	100	81	113	103	107	198
1990		95	108	103	104	111	94	86	113	99	105	178
1991		94	103	103	106	110	101	90	122	101	111	112
1992	100	100	100	100	100	100	100	100	100	100	100	100
1993		98	106	97	100	101	105	103	103	88	97	101
1994		100P	104P	101	102	102	106	105	109	117	112	56
1995		102P	104P	105P	107P	111P	111P	98P	106P	113P	109P	90P
1996		103P	103P	106P	108P	112P	114P	99P	108P	115P	111P	82P
1997		105P	103P	106P	109P	113P	118P	100P	110P	118P	113P	75P
1998		106P	102P	107P	109P	114P	121P	101P	112P	120P	115P	67P

Sources: Same as General Statistics. Values reflect change from the base year, 1992. Values above 100 mean greater than 92, values below 100 mean less than 92, and a value of 100 in the 82-91 or 93-98 period means same as 92. 'P's mark projections by the editors.

SELECTED RATIOS

For 1994	Avg. of All Manufact.	Analyzed Industry	Index	For 1994	Avg. of All Manufact.	Analyzed Industry	Index
Employees per Establishment	49	55	112	Value Added per Production Worker	134,084	84,764	63
Payroll per Establishment	1,500,273	1,264,244	84	Cost per Establishment	5,045,178	3,529,692	70
Payroll per Employee	30,620	23,110	75	Cost per Employee	102,970	64,521	63
Production Workers per Establishment	34	41	120	Cost per Production Worker	146,988	85,636	58
Wages per Establishment	853,319	729,522	85	Shipments per Establishment	9,576,895	6,996,435	73
Wages per Production Worker	24,861	17,699	71	Shipments per Employee	195,460	127,890	65
Hours per Production Worker	2,056	2,018	98	Shipments per Production Worker	279,017	169,745	61
Wages per Hour	12.09	8.77	73	Investment per Establishment	321,011	47,212	15
Value Added per Establishment	4,602,255	3,493,721	76	Investment per Employee	6,552	863	13
Value Added per Employee	93,930	63,863	68	Investment per Production Worker	9,352	1,145	12

Sources: Same as General Statistics. The 'Average of All Manufacturing' column represents the average of all manufacturing industries reported for the most recent complete year available. The Index shows the relationship between the Average and the Analyzed Industry. For example, 100 means that they are equal; 500 that the Analyzed Industry is five times the average; 50 means that the Analyzed Industry is half the national average. The abbreviation 'na' is used to show that data are 'not available'.

LEADING COMPANIES Number shown: 23 Total sales ($ mil): 767 Total employment (000): 6.5

Company Name	Address				CEO Name	Phone	Co. Type	Sales ($ mil)	Empl. (000)
Nu-kote Holding Inc	17950 Preston	Dallas	TX	75252	Larry H Holswade	214-250-2785	P	151	1.1
Bobbie Brooks Inc	3830 Kelley Av	Cleveland	OH	44114	RH Kanner	216-881-5300	S	96	1.0
Pubco Corp	3830 Kelley Av	Cleveland	OH	44114	Robert H Kanner	216-881-5300	P	96	1.0
NER Data Products Inc	307 S Delsea Dr	Glassboro	NJ	08028	Paul O'Neill	609-881-5524	R	81*	0.6
General Ribbon Corp	PO Box 3699	Chatsworth	CA	91313	Robert W Daggs	818-709-1234	R	74*	0.6
International Imaging Materials	310 Commerce Dr	Amherst	NY	14228	Jack O'Leary	716-691-6333	P	60	0.5
Barouh Eaton Allen Corp	67 Kent Av	Brooklyn	NY	11211	Vic Barouh	718-782-2601	R	50*	0.4
Buckeye Business Products Inc	3830 Kelley Av	Cleveland	OH	44114	Robert H Kanner	216-391-6300	D	23	<0.1
Curtis-Young Corp	1050 Taylors Ln	Cinnaminson	NJ	08077	Dean C Edwards	609-665-6650	R	20	0.2
Media Recovery Inc	PO Box 1407	Graham	TX	76046	Alan Myers	817-549-5462	R	18	0.1
Parana Supplies Corp	3625 Del Amo Blv	Torrance	CA	90503	Haryasu Murasawa	310-793-1325	R	17	0.1
Leedall Products Inc	351 W 35th St	New York	NY	10001	S Sharkey	212-563-5281	R	14	<0.1
Aspen Imaging International Inc	555 Aspen Ridge Dr	Lafayette	CO	80026	Peter C Williams	303-666-5750	P	14	0.2
Leedall Prod Manufacturing Co	130 Van Liew Av	Milltown	NJ	08850	Eugene B DiLuco	908-828-1045	R	13	0.1
Advanced Supplies Inc	137 4th St	Middletown	PA	17057	Luigi Rabbi	717-944-5551	R	10	<0.1
Golden Ribbon Corp	3075 75th St	Boulder	CO	80301	Bill Patterson	303-443-6966	R	7*	<0.1
California Ribbon & Carbon Co	4720 Eastern Av	City of Com	CA	90040	Robert T Picou	213-724-9100	R	5	<0.1
Indiana Carbon Company Inc	3164 N Shadeland	Indianapolis	IN	46226	Quinn Ray	317-547-9621	R	5*	<0.1
National Computer Ribbons	9566 Deereco Rd	Timonium	MD	21093	Bill Hardy	410-561-0200	R	5*	<0.1
American Ribbon and Tower	2895 W Prospect Rd	Ft Lauderdale	FL	33309	G Howard	305-733-4552	R	2	<0.1
Fair Publishing House Inc	PO Box 350	Norwalk	OH	44857	KF Doyle	419-668-3746	S	2	<0.1
Modern Ribbon and Carbon Inc	128-30 N 10th St	Philadelphia	PA	19107	Edward Sutherland	215-922-1545	R	1*	<0.1
SAS Industries Inc	3091 N Bay Dr	North Bend	OR	97459	John H Smith	503-756-2508	R	1	<0.1

Source: *Ward's Business Directory of U.S. Private and Public Companies*, Volumes 1 and 2, 1996. The company type code used is as follows: P - Public, R - Private, S - Subsidiary, D - Division, J - Joint Venture, A - Affiliate, G - Group. Sales are in millions of dollars, employees are in thousands. An asterisk (*) indicates an estimated sales volume. The symbol < stands for 'less than'. Company names and addresses are truncated, in some cases, to fit into the available space.

MATERIALS CONSUMED

Material	Quantity	Delivered Cost ($ million)
Materials, ingredients, containers, and supplies	(X)	376.5
Broadwoven fabrics (piece goods)	(X)	14.4
Narrow fabrics (12 inches or less in width)	(X)	56.0
Other textile mill products	(X)	(D)
Purchased (market) paper	(X)	42.6
Paperboard containers, boxes, and corrugated paperboard	(X)	20.4
Other paper and allied products	(X)	(D)
Carbon black	(X)	3.0
Printing ink	(X)	8.4
Other chemicals and allied products	(X)	29.5
All other materials and components, parts, containers, and supplies	(X)	112.9
Materials, ingredients, containers, and supplies, nsk	(X)	33.3

Source: 1992 *Economic Census*. Explanation of symbols used: (D): Withheld to avoid disclosure of competitive data; na: Not available; (S): Withheld because statistical norms were not met; (X): Not applicable; (Z): Less than half the unit shown; nec: Not elsewhere classified; nsk: Not specified by kind; - : zero; * : 10-19 percent estimated; ** : 20-29 percent estimated.

PRODUCT SHARE DETAILS

Product or Product Class	% Share	Product or Product Class	% Share
Carbon paper and inked ribbons	100.00	Other inked ribbons	5.00
Inked ribbons	78.90	Inked ribbons, nsk	2.17
Inked typewriter ribbons	17.87	Carbon paper, stencil paper, etc.	17.31
Inked computer (electronic data processing) ribbons	74.97	Carbon paper and inked ribbons, nsk	3.78

Source: 1992 *Economic Census*. The values shown are percent of total shipments in an industry. Values of indented subcategories are summed in the main heading. The symbol (D) appears when data are withheld to prevent disclosure of competitive information. The abbreviation nsk stands for 'not specified by kind' and nec for 'not elsewhere classified'.

INPUTS AND OUTPUTS FOR CARBON PAPER & INKED RIBBONS

Economic Sector or Industry Providing Inputs	%	Sector	Economic Sector or Industry Buying Outputs	%	Sector
Miscellaneous plastics products	24.7	Manufg.	Manifold business forms	15.5	Manufg.
Paper mills, except building paper	14.7	Manufg.	S/L Govt. purch., other general government	10.2	S/L Govt
Wholesale trade	12.7	Trade	Exports	8.6	Foreign
Advertising	10.6	Services	S/L Govt. purch., higher education	5.8	S/L Govt
Narrow fabric mills	6.5	Manufg.	Banking	4.2	Fin/R.E.
Imports	4.3	Foreign	S/L Govt. purch., elem. & secondary education	4.1	S/L Govt
Broadwoven fabric mills	3.4	Manufg.	Retail trade, except eating & drinking	3.1	Trade
Carbon black	2.4	Manufg.	Insurance carriers	3.1	Fin/R.E.
Paperboard containers & boxes	2.4	Manufg.	S/L Govt. purch., health & hospitals	3.0	S/L Govt
Fabricated metal products, nec	2.2	Manufg.	Personal consumption expenditures	2.9	
Motor freight transportation & warehousing	1.8	Util.	Wholesale trade	2.3	Trade
Printing ink	1.7	Manufg.	U.S. Postal Service	1.8	Gov't
Nonmetallic mineral products, nec	1.3	Manufg.	Commercial printing	1.6	Manufg.
Paper coating & glazing	1.1	Manufg.	Labor, civic, social, & fraternal associations	1.6	Services
Electric services (utilities)	1.1	Util.	Credit agencies other than banks	1.5	Fin/R.E.
Railroads & related services	0.9	Util.	Hospitals	1.5	Services
Chemical preparations, nec	0.8	Manufg.	Doctors & dentists	1.2	Services
Eating & drinking places	0.7	Trade	Electronic computing equipment	1.1	Manufg.
Banking	0.7	Fin/R.E.	Accounting, auditing & bookkeeping	1.1	Services
Security & commodity brokers	0.6	Fin/R.E.	Change in business inventories	1.1	In House
Gas production & distribution (utilities)	0.5	Util.	Real estate	1.0	Fin/R.E.
Real estate	0.5	Fin/R.E.	Air transportation	0.8	Util.
Communications, except radio & TV	0.4	Util.	Insurance agents, brokers, & services	0.8	Fin/R.E.
Maintenance of nonfarm buildings nec	0.3	Constr.	S/L Govt. purch., natural resource & recreation.	0.8	S/L Govt
Royalties	0.3	Fin/R.E.	Security & commodity brokers	0.7	Fin/R.E.
U.S. Postal Service	0.3	Gov't	Legal services	0.7	Services
Abrasive products	0.2	Manufg.	Newspapers	0.6	Manufg.
Machinery, except electrical, nec	0.2	Manufg.	Electric services (utilities)	0.6	Util.
Air transportation	0.2	Util.	Motor freight transportation & warehousing	0.6	Util.
Sanitary services, steam supply, irrigation	0.2	Util.	Business/professional associations	0.6	Services
Equipment rental & leasing services	0.2	Services	Elementary & secondary schools	0.6	Services
Legal services	0.2	Services	Federal Government purchases, national defense	0.6	Fed Govt
Management & consulting services & labs	0.2	Services	Radio & TV communication equipment	0.5	Manufg.
Special dies & tools & machine tool accessories	0.1	Manufg.	Communications, except radio & TV	0.5	Util.
Credit agencies other than banks	0.1	Fin/R.E.	Colleges, universities, & professional schools	0.5	Services
Accounting, auditing & bookkeeping	0.1	Services	Apparel made from purchased materials	0.4	Manufg.
			Railroads & related services	0.4	Util.
			Eating & drinking places	0.4	Trade
			Child day care services	0.4	Services
			Computer & data processing services	0.4	Services
			Management & consulting services & labs	0.4	Services
			Religious organizations	0.4	Services
			Federal Government purchases, nondefense	0.4	Fed Govt
			Freight forwarders	0.3	Util.
			Business services nec	0.3	Services
			Hotels & lodging places	0.3	Services
			Social services, nec	0.3	Services
			S/L Govt. purch., other education & libraries	0.3	S/L Govt
			Aircraft	0.2	Manufg.
			Drugs	0.2	Manufg.
			Electronic components nec	0.2	Manufg.
			Miscellaneous plastics products	0.2	Manufg.
			Semiconductors & related devices	0.2	Manufg.
			Telephone & telegraph apparatus	0.2	Manufg.
			Gas production & distribution (utilities)	0.2	Util.
			Engineering, architectural, & surveying services	0.2	Services
			Job training & related services	0.2	Services
			Medical & health services, nec	0.2	Services
			S/L Govt. purch., correction	0.2	S/L Govt
			S/L Govt. purch., police	0.2	S/L Govt
			Aircraft & missile engines & engine parts	0.1	Manufg.
			Aircraft & missile equipment, nec	0.1	Manufg.
			Blast furnaces & steel mills	0.1	Manufg.
			Bread, cake, & related products	0.1	Manufg.
			Machinery, except electrical, nec	0.1	Manufg.
			Motor vehicle parts & accessories	0.1	Manufg.
			Paperboard containers & boxes	0.1	Manufg.
			Arrangement of passenger transportation	0.1	Util.
			Radio & TV broadcasting	0.1	Util.
			Water transportation	0.1	Util.
			Advertising	0.1	Services
			Libraries, vocation education	0.1	Services
			Personnel supply services	0.1	Services
			State & local electric utilities	0.1	Gov't

Source: Benchmark Input-Output Accounts for the U.S. Economy, 1982, U.S. Department of Commerce, Washington, D.C., July 1991. Data, as reported in the source, are organized by the 1977 SIC structure in use in 1982 but have been matched, as closely as is possible, to the 1987 SIC structure used in this book.

OCCUPATIONS EMPLOYED BY SIC 395 - MANUFACTURED PRODUCTS, NEC

Occupation	% of Total 1994	Change to 2005	Occupation	% of Total 1994	Change to 2005
Assemblers, fabricators, & hand workers nec	17.0	4.8	Freight, stock, & material movers, hand	1.5	-16.1
Sales & related workers nec	4.3	4.8	Inspectors, testers, & graders, precision	1.5	4.9
Blue collar worker supervisors	4.0	-2.4	General office clerks	1.5	-10.6
General managers & top executives	3.8	-0.5	Designers, ex interior designers	1.5	15.3
Hand packers & packagers	3.7	-28.1	Electrical & electronic assemblers	1.4	4.8
Traffic, shipping, & receiving clerks	2.5	0.9	Industrial production managers	1.4	4.8
Machine operators nec	2.2	-7.6	Packaging & filling machine operators	1.4	4.9
Helpers, laborers, & material movers nec	2.2	4.8	Screen printing machine setters & set-up operators	1.2	4.8
Painting, coating, & decorating workers, hand	2.0	-5.7	Plastic molding machine workers	1.2	4.8
Precision workers nec	2.0	41.5	Machine feeders & offbearers	1.2	-5.7
Secretaries, ex legal & medical	2.0	-4.5	Cabinetmakers & bench carpenters	1.1	-31.9
Bookkeeping, accounting, & auditing clerks	1.8	-21.4	Order clerks, materials, merchandise, & service	1.1	2.6
Sheet metal workers & duct installers	1.8	4.8	Grinders & polishers, hand	1.1	4.8
Precision metal workers nec	1.8	36.3			

Source: Industry-Occupation Matrix, Bureau of Labor Statistics. These data relate to one or more 3-digit SIC industry groups rather than to a single 4-digit SIC. The change reported for each occupation to the year 2005 is a percent of growth or decline as estimated by the Bureau of Labor Statistics. The abbreviation nec stands for 'not elsewhere classified'.

LOCATION BY STATE AND REGIONAL CONCENTRATION

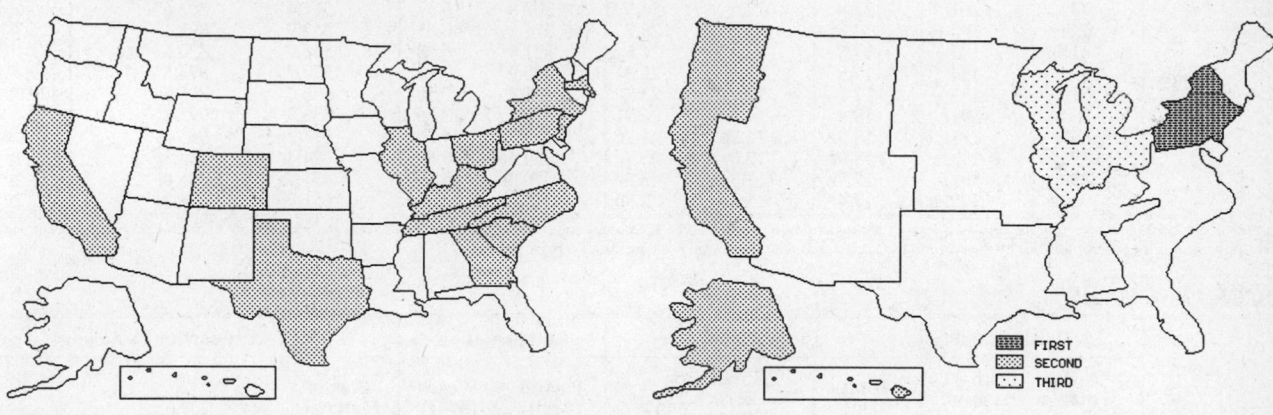

FIRST
SECOND
THIRD

INDUSTRY DATA BY STATE

State	Establish-ments	Shipments			Employment				Cost as % of Shipments	Investment per Employee ($)
		Total ($ mil)	% of U.S.	Per Establ.	Total Number	% of U.S.	Per Establ.	Wages ($/hour)		
California	19	155.1	18.7	8.2	1,200	16.7	63	9.88	50.0	2,167
New York	12	116.5	14.0	9.7	1,100	15.3	92	8.52	57.3	1,000
New Jersey	10	68.0	8.2	6.8	400	5.6	40	8.57	57.6	1,000
Texas	10	54.4	6.5	5.4	300	4.2	30	8.25	45.2	3,667
Colorado	9	50.4	6.1	5.6	700	9.7	78	8.30	50.2	714
Ohio	7	39.9	4.8	5.7	300	4.2	43	9.67	57.4	-
Georgia	4	36.2	4.4	9.1	400	5.6	100	8.20	66.6	-
Pennsylvania	8	33.1	4.0	4.1	400	5.6	50	8.40	57.4	500
North Carolina	4	22.1	2.7	5.5	200	2.8	50	7.67	48.4	-
Illinois	7	(D)	-	-	375 *	5.2	54	-	-	-
Tennessee	4	(D)	-	-	750 *	10.4	188	-	-	-
South Carolina	3	(D)	-	-	175 *	2.4	58	-	-	-
Kentucky	2	(D)	-	-	375 *	5.2	188	-	-	-
Massachusetts	1	(D)	-	-	175 *	2.4	175	-	-	-

Source: 1992 Economic Census. The states are in descending order of shipments or establishments (if shipment data are missing for the majority). The symbol (D) appears when data are withheld to prevent disclosure of competitive information. States marked with (D) are sorted by number of establishments. A dash (-) indicates that the data element cannot be calculated; * indicates the midpoint of a range.

3961 - COSTUME JEWELRY

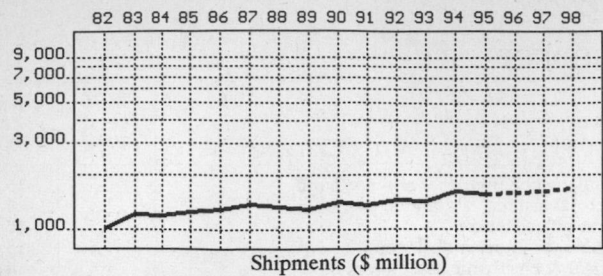

Shipments ($ million)

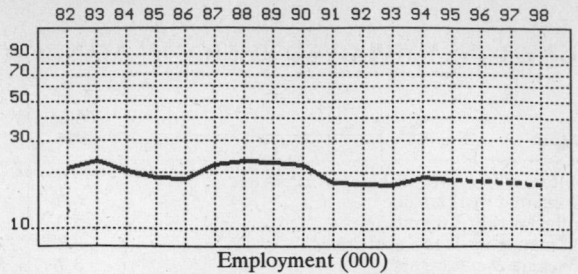

Employment (000)

GENERAL STATISTICS

| Year | Companies | Establishments | | Employment | | | Compensation | | Production ($ million) | | | |
		Total	with 20 or more employees	Total (000)	Production Workers (000)	Hours (Mil)	Payroll ($ mil)	Wages ($/hr)	Cost of Materials	Value Added by Manufacture	Value of Shipments	Capital Invest.
1982	780	785	186	21.1	15.7	29.4	263.3	5.37	414.5	588.9	1,027.1	16.6
1983		767	184	23.3	17.3	31.6	290.0	5.30	536.2	674.4	1,216.1	26.2
1984		749	182	20.5	14.6	29.0	292.8	6.10	487.4	712.8	1,185.8	13.1
1985		731	181	18.7	13.5	27.0	292.5	6.34	512.5	762.7	1,265.0	19.9
1986		711	185	18.5	12.7	26.7	303.1	6.30	509.3	782.4	1,292.3	19.0
1987	754	760	198	22.2	14.9	28.4	328.9	6.20	537.6	852.0	1,391.9	11.2
1988		722	197	23.3	15.6	29.7	343.2	5.95	557.9	822.5	1,333.1	13.8
1989		732	201	22.8	13.5	26.3	302.8	6.37	517.4	803.6	1,305.6	13.2
1990		768	192	22.3	13.1	25.7	315.8	6.65	542.9	892.4	1,415.7	11.5
1991		814	187	17.6	12.4	24.6	331.4	6.93	546.9	820.6	1,371.1	9.4
1992	888	895	184	17.5	12.6	25.3	330.9	7.25	570.2	874.5	1,449.4	12.8
1993		919	183	17.1	12.0	23.8	325.8	7.50	538.0	895.5	1,429.3	9.7
1994		854P	191P	18.8	14.3	26.9	376.4	8.06	633.5	998.8	1,627.4	13.6
1995		866P	191P	18.3P	12.2P	24.2P	358.5P	7.81P	611.2P	963.6P	1,570.0P	9.0P
1996		877P	191P	18.0P	12.0P	23.7P	364.6P	8.00P	624.4P	984.5P	1,604.1P	8.2P
1997		889P	192P	17.7P	11.7P	23.3P	370.8P	8.19P	637.7P	1,005.4P	1,638.2P	7.4P
1998		900P	192P	17.4P	11.4P	22.9P	377.0P	8.38P	651.0P	1,026.3P	1,672.2P	6.6P

Sources: 1982, 1987, 1992 *Economic Census*; *Annual Survey of Manufactures*, 83-86, 88-91, 93-94. Establishment counts for non-Census years are from *County Business Patterns*; establishment values for 83-84 are extrapolations. 'P's show projections by the editors. Industries reclassified in 87 will not have data for prior years.

INDICES OF CHANGE

| Year | Companies | Establishments | | Employment | | | Compensation | | Production ($ million) | | | |
		Total	with 20 or more employees	Total (000)	Production Workers (000)	Hours (Mil)	Payroll ($ mil)	Wages ($/hr)	Cost of Materials	Value Added by Manufacture	Value of Shipments	Capital Invest.
1982	88	88	101	121	125	116	80	74	73	67	71	130
1983		86	100	133	137	125	88	73	94	77	84	205
1984		84	99	117	116	115	88	84	85	82	82	102
1985		82	98	107	107	107	88	87	90	87	87	155
1986		79	101	106	101	106	92	87	89	89	89	148
1987	85	85	108	127	118	112	99	86	94	97	96	88
1988		81	107	133	124	117	104	82	98	94	92	108
1989		82	109	130	107	104	92	88	91	92	90	103
1990		86	104	127	104	102	95	92	95	102	98	90
1991		91	102	101	98	97	100	96	96	94	95	73
1992	100	100	100	100	100	100	100	100	100	100	100	100
1993		103	99	98	95	94	98	103	94	102	99	76
1994		95P	104P	107	113	106	114	111	111	114	112	106
1995		97P	104P	105P	97P	96P	108P	108P	107P	110P	108P	70P
1996		98P	104P	103P	95P	94P	110P	110P	110P	113P	111P	64P
1997		99P	104P	101P	93P	92P	112P	113P	112P	115P	113P	58P
1998		101P	104P	100P	91P	90P	114P	116P	114P	117P	115P	51P

Sources: Same as General Statistics. Values reflect change from the base year, 1992. Values above 100 mean greater than 92, values below 100 mean less than 92, and a value of 100 in the 82-91 or 93-98 period means same as 92. 'P's mark projections by the editors.

SELECTED RATIOS

For 1994	Avg. of All Manufact.	Analyzed Industry	Index	For 1994	Avg. of All Manufact.	Analyzed Industry	Index
Employees per Establishment	49	22	45	Value Added per Production Worker	134,084	69,846	52
Payroll per Establishment	1,500,273	440,640	29	Cost per Establishment	5,045,178	741,619	15
Payroll per Employee	30,620	20,021	65	Cost per Employee	102,970	33,697	33
Production Workers per Establishment	34	17	49	Cost per Production Worker	146,988	44,301	30
Wages per Establishment	853,319	253,818	30	Shipments per Establishment	9,576,895	1,905,147	20
Wages per Production Worker	24,861	15,162	61	Shipments per Employee	195,460	86,564	44
Hours per Production Worker	2,056	1,881	91	Shipments per Production Worker	279,017	113,804	41
Wages per Hour	12.09	8.06	67	Investment per Establishment	321,011	15,921	5
Value Added per Establishment	4,602,255	1,169,265	25	Investment per Employee	6,552	723	11
Value Added per Employee	93,930	53,128	57	Investment per Production Worker	9,352	951	10

Sources: Same as General Statistics. The 'Average of All Manufacturing' column represents the average of all manufacturing industries reported for the most recent complete year available. The Index shows the relationship between the Average and the Analyzed Industry. For example, 100 means that they are equal; 500 that the Analyzed Industry is five times the average; 50 means that the Analyzed Industry is half the national average. The abbreviation 'na' is used to show that data are 'not available'.

LEADING COMPANIES   Number shown: 49   Total sales ($ mil): 636   Total employment (000): 8.6

Company Name	Address				CEO Name	Phone	Co. Type	Sales ($ mil)	Empl. (000)
Artra Group Inc	500 Central Av	Northfield	IL	60093	John Harvey	708-441-6650	P	160	2.1
Napier Co	Napier Park	Meriden	CT	06450	Ronald J Meoni	203-237-5522	R	70	0.8
Accessories Associates Inc	4 Warren Av	N Providence	RI	02911	Gerald Cerce	401-231-3800	R	53*	0.8
Swarovski Jewelry US Ltd	1 Kenney Dr	Cranston	RI	02920	Tom Keller	401-463-6400	S	50	0.6
Lori Corp	500 Central Av	Northfield	IL	60093	Austin A Iodice	708-441-7300	P	46	0.8
Victoria Creations Inc	30 Jefferson Pk Rd	Warwick	RI	02888	Patricia Stensrud	401-467-7150	P	43	0.6
Weingeroff Enterprises Inc	1 Weingeroff Blv	Cranston	RI	02910	Gregg Weingeroff	401-467-2200	R	25	0.4
RJ Manufacturing Co	560 Metacom Av	Warren	RI	02885	Richard Andreolli	401-467-7150	D	24	0.3
Alice and Reed Group	185 Jefferson Blv	Warwick	RI	02888	Larry Cohen	401-739-8280	R	11	0.1
P and B Manufacturing Co	185 Jefferson Blv	Warwick	RI	02888	Larry Cohen	401-739-8280	S	11	0.1
Howard Eldon Ltd	1187 Coast Village	Santa Barbara	CA	93108	Howard Scar	818-989-4444	R	10	0.1
Jewelry Fashions Inc	213 W 35th St	New York	NY	10001	Ronald Rose	212-947-7700	R	10	0.1
Paramount Sales Company Inc	1718 N Central St	Knoxville	TN	37917	WT Starks	615-525-1165	R	10	0.1
Carolee Designs Inc	19 E Elm St	Greenwich	CT	06830	Carolee Friedlander	203-629-1139	R	9*	0.2
Darlene Jewelry Mfg Co	483 Main St	Pawtucket	RI	02860	BE Baccari	401-728-3300	R	9*	0.1
Rolo Manufacturing Company	274 Pine St	Providence	RI	02903	Ralph Posner	401-521-0800	R	9*	0.2
American Ring Co	19 Grosvenor Av	E Providence	RI	02914	R Calandrelli	401-438-9060	R	8*	0.1
Catherine Stein Designs Inc	8 W 38th St	New York	NY	10018	Catherine Stein	212-840-1188	R	8	<0.1
Donley Co	91 Geo Leven Dr	N Attleboro	MA	02761	Norine Zwolenski	508-699-7515	R	7*	<0.1
MJ Enterprises Inc	530 Wellington Av	Cranston	RI	02910	George Sadue	401-781-0832	R	7	<0.1
Don-Lin Jewelry Company Inc	39 Haskins St	Providence	RI	02903	D St Angelo Sr	401-274-0165	R	6	<0.1
Hobe Cie Ltd	138 S Columbus	Mount Vernon	NY	10553	Anthony Iati	914-664-2640	R	5*	0.1
RF Simmons Company Inc	PO Box 540	Attleboro	MA	02703	David W Cochran	508-222-6655	R	4	<0.1
Circle Jewelry Products Inc	148 W 24th St	New York	NY	10011	Michael Gartner	212-255-3608	R	4	<0.1
Bastian Co	122 N Genesee St	Geneva	NY	14456	Albert Cauwels	315-789-8000	S	3	<0.1
C and N Associates	70 Rosner Av	N Providence	RI	02904	Cal J Verduchi	401-353-4310	R	3	<0.1
J and K Sales Company Inc	PO Box 699	Pawtucket	RI	02862	Michael Jablecki	401-728-5320	R	3	0.1
Kirk's Folly Inc	389 5th Av	New York	NY	10016	Jennifer Kirk	212-683-9797	R	3*	<0.1
Rosecraft Inc	685 Social St	Woonsocket	RI	02895	Nancy Cass	401-331-1500	S	3	<0.1
Singh Corp	6 Saddle Rd	Cedar Knolls	NJ	07927	Ronjit Singh	201-267-8008	R	3*	<0.1
William Rand Inc	20 W 36th St	New York	NY	10018	William Rand	212-563-3252	R	3*	<0.1
Deltah Inc	789 Waterman Av	E Providence	RI	02914	Patricia Accardia	401-434-2250	R	2	<0.1
EL Tool and Die Company Inc	62 W 47th St	New York	NY	10036	Jack F Ricotta	212-719-4414	R	2	<0.1
Gauntlet Inc	2377 Market St	San Francisco	CA	94114	Jim Ward	415-431-3133	R	2*	<0.1
Museum Reproductions Inc	62 Harvard St	Brookline	MA	02145	Lars Messler	617-277-7707	R	2	<0.1
Plastic Craft Novelty Co	12 Dunham St	Attleboro	MA	02703	Peter Manickas	508-222-1486	R	2*	<0.1
Maeve Carr Designs Inc	15 W 36th St 14th Fl	New York	NY	10018	Maeve Carr	212-714-9140	R	2	<0.1
Ann Marie Inc	4919 Old Summer	Memphis	TN	38122	Mary Day	901-767-6783	R	2	<0.1
Arden Jewelry Mfg Co	10 Industrial Ln	Johnston	RI	02919	Steven Abrams	401-274-9800	R	1*	<0.1
Ramzor Inc	36 Colorado Av	Warwick	RI	02888	Ramon Zorabedian	401-738-8550	R	1*	<0.1
Joan Rivers Products Inc	8306 Wilshire Blv	Beverly Hills	CA	90211	Barry Greenburg	310-552-3131	R	1*	<0.1
Flattery Inc	1388 Chattahoochee	Atlanta	GA	30318	Susan L Richardson	404-352-0888	R	1	<0.1
Vogt Western Silversmiths Ltd	PO Box 2309	Turlock	CA	95381	Chet Vogt	209-667-2471	R	1*	<0.1
Facets by Spectrum	200 Fieldcrest Cir	Coppell	TX	75019	Jim Hanzlik	214-393-7281	R	0*	<0.1
CF Weston Inc	1707 Berkeley St	Santa Monica	CA	90404	Maria C Weston	310-828-8515	R	0*	<0.1
ELA Corp	225 Stevens Av	Solana Beach	CA	92075	Audrey L Eller	619-481-4161	R	0	<0.1
Law Design Specialties Inc	21424 N 7th Av	Phoenix	AZ	85027	Donna Law	602-258-2477	R	0	<0.1
Les Bernard Inc	417 5th Av	New York	NY	10016	Bernard Shapiro	212-679-8855	S	0	<0.1
SS/F Inc	1324 W Webster St	Chicago	IL	60614	Lynn Foster	312-549-0046	R	0	<0.1

Source: Ward's Business Directory of U.S. Private and Public Companies, Volumes 1 and 2, 1996. The company type code used is as follows: P - Public, R - Private, S - Subsidiary, D - Division, J - Joint Venture, A - Affiliate, G - Group. Sales are in millions of dollars, employees are in thousands. An asterisk (*) indicates an estimated sales volume. The symbol < stands for 'less than'. Company names and addresses are truncated, in some cases, to fit into the available space.

MATERIALS CONSUMED

Material	Quantity	Delivered Cost ($ million)
Materials, ingredients, containers, and supplies	(X)	432.2
Fabricated metal products, including forgings	(X)	24.0
Precious metals (gold, platinum, etc.), all forms	(X)	33.6
Other shapes and forms, including castings	(X)	11.2
Precious, semiprecious, and synthetic stones, and pearls; cut, polished, or drilled	(X)	56.2
Jewelers' findings, including joints, pins, clasps, chains, flat stock, etc.	(X)	80.9
Other jewelry, silverware, and plated ware	(X)	15.2
All other materials and components, parts, containers, and supplies	(X)	60.3
Materials, ingredients, containers, and supplies, nsk	(X)	150.8

Source: 1992 *Economic Census.* Explanation of symbols used: (D): Withheld to avoid disclosure of competitive data; na: Not available; (S): Withheld because statistical norms were not met; (X): Not applicable; (Z): Less than half the unit shown; nec: Not elsewhere classified; nsk: Not specified by kind; - : zero; * : 10-19 percent estimated; ** : 20-29 percent estimated.

PRODUCT SHARE DETAILS

Product or Product Class	% Share	Product or Product Class	% Share
Costume jewelry	100.00	Costume jewelry watch attachments (bracelets for watches) made of base metal, whether or not electroplated with gold, silver, chromium, etc.	6.37
Costume jewelry rings and ring mountings made of base metal, whether or not electroplated with gold, silver, chromium, etc..	6.42	Other costume jewelry worn or carried about the person (except compacts, vanity cases, cigar and cigarette cases, and lighters) made of base metal, whether or not electroplated with gold, silver, etc.	3.84
Men's costume jewelry (excluding watch attachments and rings) made of base metal, whether or not electroplated with gold, silver, chromium, etc.	2.42	Costume jewelry compacts and vanity cases, other than leather	1.14
Women's and children's costume jewelry and costume novelties (excluding watch attachments and rings) made of base metal, whether or not electroplated with gold, silver, chromium, etc.	60.91	Other costume jewelry, imitation pearls, and costume novelties made of plastics, wood, leather, etc.	6.11

Source: 1992 *Economic Census*. The values shown are percent of total shipments in an industry. Values of indented subcategories are summed in the main heading. The symbol (D) appears when data are withheld to prevent disclosure of competitive information. The abbreviation nsk stands for 'not specified by kind' and nec for 'not elsewhere classified'.

INPUTS AND OUTPUTS FOR COSTUME JEWELRY

Economic Sector or Industry Providing Inputs	%	Sector	Economic Sector or Industry Buying Outputs	%	Sector
Imports	43.8	Foreign	Personal consumption expenditures	93.7	
Advertising	9.9	Services	Exports	4.9	Foreign
Jewelers' materials & lapidary work	6.5	Manufg.	Costume jewelry	1.4	Manufg.
Wholesale trade	5.6	Trade			
Primary nonferrous metals, nec	5.0	Manufg.			
Personal leather goods	2.1	Manufg.			
Costume jewelry	2.0	Manufg.			
Petroleum refining	1.8	Manufg.			
Sanitary services, steam supply, irrigation	1.8	Util.			
Wood products, nec	1.7	Manufg.			
Nonferrous rolling & drawing, nec	1.5	Manufg.			
Eating & drinking places	1.5	Trade			
Miscellaneous plastics products	1.3	Manufg.			
Paperboard containers & boxes	1.0	Manufg.			
Job training & related services	1.0	Services			
Electric services (utilities)	0.9	Util.			
Banking	0.7	Fin/R.E.			
Real estate	0.7	Fin/R.E.			
U.S. Postal Service	0.7	Gov't			
Maintenance of nonfarm buildings nec	0.6	Constr.			
Communications, except radio & TV	0.6	Util.			
Glass & glass products, except containers	0.5	Manufg.			
Machinery, except electrical, nec	0.5	Manufg.			
Nonmetallic mineral products, nec	0.5	Manufg.			
Legal services	0.5	Services			
Management & consulting services & labs	0.5	Services			
Abrasive products	0.4	Manufg.			
Automotive rental & leasing, without drivers	0.4	Services			
Signs & advertising displays	0.3	Manufg.			
Special dies & tools & machine tool accessories	0.3	Manufg.			
Gas production & distribution (utilities)	0.3	Util.			
Motor freight transportation & warehousing	0.3	Util.			
Royalties	0.3	Fin/R.E.			
Accounting, auditing & bookkeeping	0.3	Services			
Automotive repair shops & services	0.3	Services			
Equipment rental & leasing services	0.3	Services			
Cyclic crudes and organics	0.2	Manufg.			
Industrial inorganic chemicals, nec	0.2	Manufg.			
Lubricating oils & greases	0.2	Manufg.			
Manifold business forms	0.2	Manufg.			
Paints & allied products	0.2	Manufg.			
Plastics materials & resins	0.2	Manufg.			
Air transportation	0.2	Util.			
Insurance carriers	0.2	Fin/R.E.			
Business/professional associations	0.1	Services			
Hotels & lodging places	0.1	Services			
Personnel supply services	0.1	Services			

Source: Benchmark Input-Output Accounts for the U.S. Economy, 1982, U.S. Department of Commerce, Washington, D.C., July 1991. Data, as reported in the source, are organized by the 1977 SIC structure in use in 1982 but have been matched, as closely as is possible, to the 1987 SIC structure used in this book.

OCCUPATIONS EMPLOYED BY SIC 396 - MANUFACTURED PRODUCTS, NEC

Occupation	% of Total 1994	Change to 2005	Occupation	% of Total 1994	Change to 2005
Assemblers, fabricators, & hand workers nec	17.0	4.8	Freight, stock, & material movers, hand	1.5	-16.1
Sales & related workers nec	4.3	4.8	Inspectors, testers, & graders, precision	1.5	4.9
Blue collar worker supervisors	4.0	-2.4	General office clerks	1.5	-10.6
General managers & top executives	3.8	-0.5	Designers, ex interior designers	1.5	15.3
Hand packers & packagers	3.7	-28.1	Electrical & electronic assemblers	1.4	4.8
Traffic, shipping, & receiving clerks	2.5	0.9	Industrial production managers	1.4	4.8
Machine operators nec	2.2	-7.6	Packaging & filling machine operators	1.4	4.9
Helpers, laborers, & material movers nec	2.2	4.8	Screen printing machine setters & set-up operators	1.2	4.8
Painting, coating, & decorating workers, hand	2.0	-5.7	Plastic molding machine workers	1.2	4.8
Precision workers nec	2.0	41.5	Machine feeders & offbearers	1.2	-5.7
Secretaries, ex legal & medical	2.0	-4.5	Cabinetmakers & bench carpenters	1.1	-31.9
Bookkeeping, accounting, & auditing clerks	1.8	-21.4	Order clerks, materials, merchandise, & service	1.1	2.6
Sheet metal workers & duct installers	1.8	4.8	Grinders & polishers, hand	1.1	4.8
Precision metal workers nec	1.8	36.3			

Source: *Industry-Occupation Matrix*, Bureau of Labor Statistics. These data relate to one or more 3-digit SIC industry groups rather than to a single 4-digit SIC. The change reported for each occupation to the year 2005 is a percent of growth or decline as estimated by the Bureau of Labor Statistics. The abbreviation nec stands for 'not elsewhere classified'.

LOCATION BY STATE AND REGIONAL CONCENTRATION

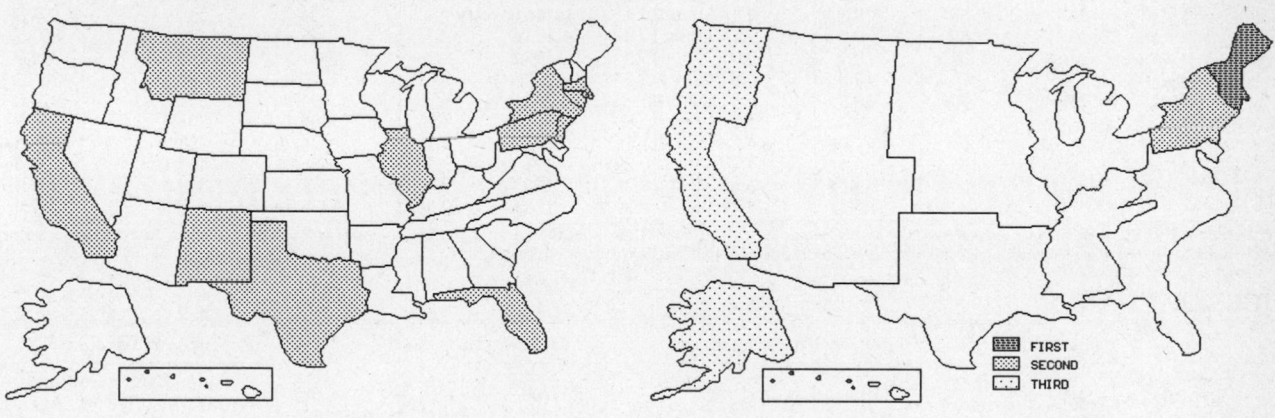

FIRST
SECOND
THIRD

INDUSTRY DATA BY STATE

State	Establish-ments	Shipments			Employment				Cost as % of Shipments	Investment per Employee ($)
		Total ($ mil)	% of U.S.	Per Establ.	Total Number	% of U.S.	Per Establ.	Wages ($/hour)		
Rhode Island	328	807.8	55.7	2.5	8,500	48.6	26	7.70	38.6	471
New York	153	159.0	11.0	1.0	2,500	14.3	16	7.35	42.1	680
California	97	133.9	9.2	1.4	1,900	10.9	20	6.32	27.4	1,368
Pennsylvania	12	25.7	1.8	2.1	200	1.1	17	6.33	52.1	-
Florida	31	13.2	0.9	0.4	200	1.1	6	7.67	40.2	-
Illinois	20	6.7	0.5	0.3	100	0.6	5	11.00	50.7	-
New Jersey	14	6.6	0.5	0.5	100	0.6	7	5.50	34.8	-
Massachusetts	35	(D)	-	-	1,750 *	10.0	50	-	-	-
Texas	25	(D)	-	-	175 *	1.0	7	-	-	-
New Mexico	20	(D)	-	-	375 *	2.1	19	-	-	-
Connecticut	10	(D)	-	-	750 *	4.3	75	-	-	-
Montana	1	(D)	-	-	175 *	1.0	175	-	-	-

Source: 1992 *Economic Census*. The states are in descending order of shipments or establishments (if shipment data are missing for the majority). The symbol (D) appears when data are withheld to prevent disclosure of competitive information. States marked with (D) are sorted by number of establishments. A dash (-) indicates that the data element cannot be calculated; * indicates the midpoint of a range.

3965 - FASTENERS BUTTONS NEEDLES & PINS

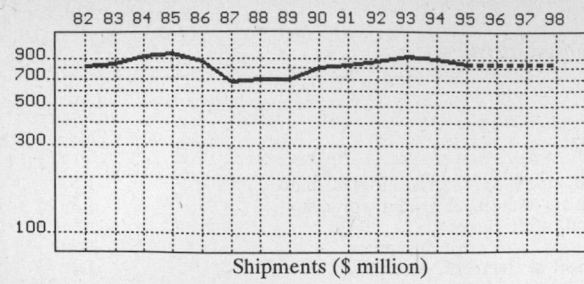

Shipments ($ million)

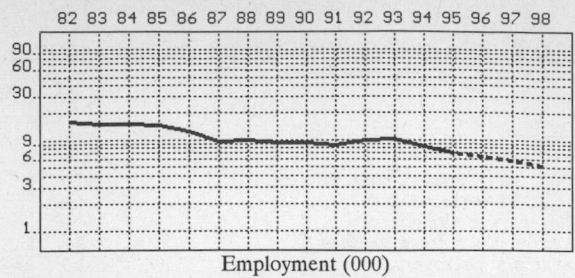

Employment (000)

GENERAL STATISTICS

| Year | Com-panies | Establishments | | Employment | | | Compensation | | Production ($ million) | | | |
		Total	with 20 or more employees	Total (000)	Production Workers (000)	Hours (Mil)	Payroll ($ mil)	Wages ($/hr)	Cost of Materials	Value Added by Manufacture	Value of Shipments	Capital Invest.
1982	333	356	132	16.1	12.7	25.7	212.5	5.60	370.5	446.6	823.0	41.6
1983				14.8	11.5	22.7	216.5	6.37	360.3	497.1	842.8	42.8
1984				15.0	11.8	23.9	233.0	6.36	402.8	526.7	921.5	77.6
1985				14.3	11.2	22.3	240.6	7.09	408.4	557.7	964.4	37.7
1986				12.4	9.8	18.5	215.6	7.64	388.6	480.4	869.7	30.9
1987	219	262	96	9.6	7.6	15.0	169.3	7.33	282.5	389.3	670.0	20.5
1988		260	91	9.8	7.4	14.1	177.5	7.98	291.8	423.0	699.0	39.7
1989		237	87	9.3	7.1	14.4	166.3	7.60	304.7	388.4	695.3	36.4
1990		237	87	9.2	7.1	14.8	174.8	7.91	345.1	471.1	807.2	42.2
1991		240	85	8.7	7.0	15.9	195.4	8.23	361.5	464.9	831.9	42.2
1992	221	238	91	10.0	6.8	15.5	225.9	8.69	379.6	493.8	868.2	119.1
1993		239	90	10.2	6.9	15.7	233.9	8.76	394.9	531.1	923.0	51.1
1994				8.4	5.9	12.8	192.9	8.90	402.1	503.0	898.8	45.7
1995				7.0P	4.7P	11.1P	193.4P	9.32P	374.7P	468.7P	837.5P	59.2P
1996				6.4P	4.2P	10.2P	191.9P	9.57P	375.0P	469.1P	838.3P	60.8P
1997				5.8P	3.6P	9.2P	190.4P	9.81P	375.4P	469.6P	839.1P	62.3P
1998				5.2P	3.1P	8.3P	188.8P	10.06P	375.7P	470.0P	839.9P	63.9P

Sources: 1982, 1987, 1992 *Economic Census*; *Annual Survey of Manufactures*, 83-86, 88-91, 93-94. Establishment counts for non-Census years are from *County Business Patterns*; establishment values for 83-84 are extrapolations. 'P's show projections by the editors. Industries reclassified in 87 will not have data for prior years.

INDICES OF CHANGE

| Year | Com-panies | Establishments | | Employment | | | Compensation | | Production ($ million) | | | |
		Total	with 20 or more employees	Total (000)	Production Workers (000)	Hours (Mil)	Payroll ($ mil)	Wages ($/hr)	Cost of Materials	Value Added by Manufacture	Value of Shipments	Capital Invest.
1982	151	150	145	161	187	166	94	64	98	90	95	35
1983				148	169	146	96	73	95	101	97	36
1984				150	174	154	103	73	106	107	106	65
1985				143	165	144	107	82	108	113	111	32
1986				124	144	119	95	88	102	97	100	26
1987	99	110	105	96	112	97	75	84	74	79	77	17
1988		109	100	98	109	91	79	92	77	86	81	33
1989		100	96	93	104	93	74	87	80	79	80	31
1990		100	96	92	104	95	77	91	91	95	93	35
1991		101	93	87	103	103	86	95	95	94	96	35
1992	100	100	100	100	100	100	100	100	100	100	100	100
1993		100	99	102	101	101	104	101	104	108	106	43
1994				84	87	83	85	102	106	102	104	38
1995				70P	70P	72P	86P	107P	99P	95P	96P	50P
1996				64P	61P	66P	85P	110P	99P	95P	97P	51P
1997				58P	53P	60P	84P	113P	99P	95P	97P	52P
1998				52P	45P	53P	84P	116P	99P	95P	97P	54P

Sources: Same as General Statistics. Values reflect change from the base year, 1992. Values above 100 mean greater than 92, values below 100 mean less than 92, and a value of 100 in the 82-91 or 93-98 period means same as 92. 'P's mark projections by the editors.

SELECTED RATIOS

For 1992	Avg. of All Manufact.	Analyzed Industry	Index	For 1992	Avg. of All Manufact.	Analyzed Industry	Index
Employees per Establishment	46	42	92	Value Added per Production Worker	122,353	72,618	59
Payroll per Establishment	1,332,320	949,160	71	Cost per Establishment	4,239,462	1,594,958	38
Payroll per Employee	29,181	22,590	77	Cost per Employee	92,853	37,960	41
Production Workers per Establishment	31	29	91	Cost per Production Worker	135,003	55,824	41
Wages per Establishment	734,496	565,945	77	Shipments per Establishment	8,100,800	3,647,899	45
Wages per Production Worker	23,390	19,808	85	Shipments per Employee	177,425	86,820	49
Hours per Production Worker	2,025	2,279	113	Shipments per Production Worker	257,966	127,676	49
Wages per Hour	11.55	8.69	75	Investment per Establishment	278,244	500,420	180
Value Added per Establishment	3,842,210	2,074,790	54	Investment per Employee	6,094	11,910	195
Value Added per Employee	84,153	49,380	59	Investment per Production Worker	8,861	17,515	198

Sources: Same as General Statistics. The 'Average of All Manufacturing' column represents the average of all manufacturing industries reported for the most recent complete year available. The Index shows the relationship between the Average and the Analyzed Industry. For example, 100 means that they are equal; 500 that the Analyzed Industry is five times the average; 50 means that the Analyzed Industry is half the national average. The abbreviation 'na' is used to show that data are 'not available'.

LEADING COMPANIES Number shown: **48** Total sales ($ mil): **836** Total employment (000): **8.9**

Company Name	Address				CEO Name	Phone	Co. Type	Sales ($ mil)	Empl. (000)
Coats Crafts North America	PO Box 24998	Greenville	SC	29616	Michael G Pratt	803-234-0331	D	200	2.0
MacLean-Fogg Co	1000 Allanson Rd	Mundelein	IL	60060	Barry Maclean	708-566-0010	R	110*	1.3
Scovill Fasteners Inc	PO Box 44	Clarkesville	GA	30523	Alan Masarek	706-754-4181	S	65	0.6
Universal Fasteners Inc	PO Box 240	Lawrenceburg	KY	40342	Mark Mizumoto	502-839-6971	S	35	0.3
QSN Manufacturing Inc	101 Frontier Way	Bensenville	IL	60106	Gary Mitchell	708-616-1500	R	34	0.2
Velcro USA Inc	406 Brown Av	Manchester	NH	03108	KT Krantz	603-669-4892	S	32	0.6
Zabin Industries Inc	3957 S Hill St	Los Angeles	CA	90037	Alan Faiola	213-749-1215	R	28	0.2
Montana Silversmiths Inc	PO Box 839	Columbus	MT	59019		406-322-4555	R	25	0.2
Decker Manufacturing Corp	703 N Clark St	Albion	MI	49224	Henry R Konkle	517-629-3955	P	21	0.1
C and C Metal Products Corp	456 Nordhoff Pl	Englewood	NJ	07631	Gerald Nathel	201-569-7300	R	20	0.2
Emsig Manufacturing Corp	253 W 35th St	New York	NY	10001	L Jacobs	212-563-5460	R	20	0.3
Ohio Rod Products Inc	PO Box 416	Versailles	IN	47042	Gary W Walston	812-689-6565	S	20	0.1
Dunlap Industries Inc	PO Box 459	Dunlap	TN	37327	Mike Kwasnik	615-267-5447	R	15*	0.2
PCI Group Inc	PO Box 900	New Bedford	MA	02741	John W Rachwalski	508-995-2641	R	15	0.2
Rau Fastener Inc	102 Westfield St	Providence	RI	02907	John Champagne	401-861-7100	R	13*	0.1
Ideal Fastener Corp	PO Box 548	Oxford	NC	27565	R Gut	919-693-3115	R	12	0.3
Cresthill Industries Inc	196 Ashburton Av	Yonkers	NY	10701	Chris Rie	914-965-9510	R	11	<0.1
Eagle Button Company Inc	415 14th St	Carlstadt	NJ	07072	E Simon	201-935-3990	S	10	0.2
ITW Waterbury Buckle Co	952 S Main St	Waterbury	CT	06721	Stan Reeve	203-753-1161	D	10*	0.1
Midwest Sintered Products Corp	13605 S Halsted St	Riverdale	IL	60627	Herbert Tews	708-849-5290	S	10	<0.1
Rocknel Fastener Inc	PO Box 7009	Rockford	IL	61125	Eugene J Werlich	815-397-5151	J	10	<0.1
Chandler Products Co	1491 Chardon Rd	Cleveland	OH	44117	James E Hanson	216-481-4400	R	10	0.1
Girard Fastener Company Inc	525 W 169th St	South Holland	IL	60473	GJ Brenneman	708-333-6133	R	9*	<0.1
Defiance Button Machine	50-05 47th Av	Woodside	NY	11377	Elliot Baritz	718-446-2234	R	8*	0.1
Howard Engineering Co	PO Box 1317	Naugatuck	CT	06770	Brian Howard	203-729-5213	R	8	<0.1
McKee Button Company Inc	PO Box 239	Muscatine	IA	52761	TF McKee	319-263-2421	R	8	0.1
Foster Needle Company Inc	PO Box 1027	Manitowoc	WI	54220	Edson P Foster Jr	414-682-6314	R	6	0.1
Monadnock Co	18301 E Arenth Av	City of Industry	CA	91748	Martin R Cohen	818-964-6581	R	6	<0.1
Southtec Inc	PO Box 1985	Albany	NY	12201	Steve Kelly	518-463-4234	S	6	0.2
Boye Needle	4343 N Ravenswood	Chicago	IL	60613	Donald Bodziak	312-472-0354	D	5*	<0.1
Horn Badge Co	5 S Easton Rd	Glenside	PA	19038	Walter W Nicholson	215-576-5700	R	5*	<0.1
Rensen Products Inc	Rte 1	Ahmeek	MI	49901	David L Sorensen	906-337-5113	R	5	<0.1
Rome Fastener Corp	PO Box 213	Milford	CT	06460	Stanley F Reiter	203-874-6719	R	5	0.1
Scolding Locks Corp	1520 W Rogers Av	Appleton	WI	54914	P Skaer	414-733-5561	R	5*	<0.1
Weber and Sons Button	PO Box 96	Muscatine	IA	52761	Ed W Weber Jr	319-263-9451	R	5	<0.1
American Robin	12800 NW 38th Av	Opa Locka	FL	33054	John Copeland	305-688-8100	D	4*	<0.1
Bellcraft Button Co	122 W 27th St	New York	NY	10001	Bernard Raider	212-929-3440	R	3*	<0.1
Hillwood Manufacturing Co	21700 St Clair Av	Cleveland	OH	44117	Charles K Hill II	216-531-0300	R	3	<0.1
Knobby Krafters Inc	PO Box 300	Attleboro	MA	02703	Nicholas Nerney	508-222-7272	R	3*	<0.1
Western Buckle Co	1757 N Paulina St	Chicago	IL	60622	Roy Mapes	312-384-6900	R	3	<0.1
Domar Buckle Mfg Corp	PO Box 1339	Elizabeth	NJ	07207	Robert Goldstein	908-289-8118	R	2*	<0.1
John M Dean Inc	PO Box 924	Putnam	CT	06260	Robert Main	203-928-7701	S	2	<0.1
North and Judd Inc	699 Middle St	Middletown	CT	06457	Al Garcia	203-632-2600	R	2*	<0.1
Texas Specialties Inc	PO Box 567783	Dallas	TX	75356	Julius Raden	214-741-3967	R	2*	<0.1
J and K Button Company Inc	319 W Mississippi	Muscatine	IA	52761	BH Hahn	319-263-1041	R	1	<0.1
Sta-Rite Ginnie Lou Inc	245 E South 1st St	Shelbyville	IL	62565	Robert N Bolinger	217-774-3921	R	1	<0.1
Wade Button Co	125 Jersey St	Harrison	NJ	07029	Ralph Osborne	201-483-3232	D	1	<0.1
Akim Company Inc	131 Ash St	Willimantic	CT	06226	Ben Schilberg	203-423-6378	R	1	<0.1

Source: *Ward's Business Directory of U.S. Private and Public Companies*, Volumes 1 and 2, 1996. The company type code used is as follows: P - Public, R - Private, S - Subsidiary, D - Division, J - Joint Venture, A - Affiliate, G - Group. Sales are in millions of dollars, employees are in thousands. An asterisk (*) indicates an estimated sales volume. The symbol < stands for 'less than'. Company names and addresses are truncated, in some cases, to fit into the available space.

MATERIALS CONSUMED

Material	Quantity	Delivered Cost ($ million)
Materials, ingredients, containers, and supplies	(X)	296.0
Fabricated metal products, including forgings	(X)	21.8
Steel shapes and forms	(X)	16.6
Aluminum and aluminum-base alloy shapes and forms	(X)	5.3
Copper and copper-base alloy shapes and forms	(X)	46.8
Other nonferrous shapes and forms	(X)	5.6
Cotton and manmade fiber fabrics, broadwoven and narrow woven	(X)	46.8
Plastics resins consumed in the form of granules, pellets, powders, liquids, etc.	(X)	17.7
Buttons, zippers, and slide fasteners	(X)	15.9
All other materials and components, parts, containers, and supplies	(X)	57.8
Materials, ingredients, containers, and supplies, nsk	(X)	61.5

Source: 1992 *Economic Census*. Explanation of symbols used: (D): Withheld to avoid disclosure of competitive data; na: Not available; (S): Withheld because statistical norms were not met; (X): Not applicable; (Z): Less than half the unit shown; nec: Not elsewhere classified; nsk: Not specified by kind; - : zero; * : 10-19 percent estimated; ** : 20-29 percent estimated.

PRODUCT SHARE DETAILS

Product or Product Class	% Share	Product or Product Class	% Share
Fasteners, buttons, needles, and pins	100.00	Zippers and slide fasteners, nsk	0.43
Buttons and parts (except precious or semiprecious metals and precious or semiprecious stones)	18.42	Needles, pins, fasteners (except slide), and similar notions	48.39
Metal sew-on type buttons (except precious or semiprecious metal)	11.91	Snap fasteners (all types)	22.80
Other metal buttons (including mechanically applied types, etc., except precious or semiprecious metals)	16.97	Buckles (including those covered with fabrics or other material, but excluding those used for costume jewelry and shoes)	26.78
Polyester buttons	38.68	Other fasteners (including tape fasteners, hook and eyes, rivet and burrs, trimmings, etc., except slide)	30.21
Other plastics buttons (including casein)	7.96	Hair curlers (except rubber and those designed for beauty parlor use)	(D)
Button blanks or molds, backs and parts for sale as such	0.66	Needles (except hypodermic, phonograph, and styli)	3.93
Other buttons (including all fabrics- covered, wood, bone, and hoof, vegetable ivory, leather, etc., except all precious and semiprecious metals and stones)	17.30	Common or toilet pins (except jewelry), including dressmakers' pins	(D)
Buttons and parts (except precious or semiprecious metals and precious or semiprecious stones), nsk	6.51	Other pins, hatpins, glasshead pins, safety pins, and plastics and metal hairpins (except jewelry)	12.17
Zippers and slide fasteners	28.46	Needles, pins, fasteners (except slide), and similar notions, nsk	0.15
Aluminum zippers and slide fasteners	4.60	Fasteners, buttons, needles, and pins, nsk	4.73
Brass zippers and slide fasteners	40.84		
Other metal zippers and slide fasteners	54.13		

Source: 1992 *Economic Census*. The values shown are percent of total shipments in an industry. Values of indented subcategories are summed in the main heading. The symbol (D) appears when data are withheld to prevent disclosure of competitive information. The abbreviation nsk stands for 'not specified by kind' and nec for 'not elsewhere classified'.

INPUTS AND OUTPUTS FOR NEEDLES, PINS, & FASTENERS

Economic Sector or Industry Providing Inputs	%	Sector	Economic Sector or Industry Buying Outputs	%	Sector
Imports	20.7	Foreign	Apparel made from purchased materials	40.7	Manufg.
Needles, pins, & fasteners	10.4	Manufg.	Personal consumption expenditures	11.7	
Advertising	9.5	Services	Shoes, except rubber	8.1	Manufg.
Wholesale trade	6.9	Trade	Needles, pins, & fasteners	7.6	Manufg.
Broadwoven fabric mills	6.5	Manufg.	Exports	5.5	Foreign
Copper rolling & drawing	4.8	Manufg.	Knit outerwear mills	4.7	Manufg.
Miscellaneous plastics products	4.8	Manufg.	Upholstered household furniture	3.1	Manufg.
Blast furnaces & steel mills	4.5	Manufg.	S/L Govt. purch., other general government	2.4	S/L Govt
Aluminum rolling & drawing	3.1	Manufg.	Women's handbags & purses	2.3	Manufg.
Plastics materials & resins	3.1	Manufg.	Leather goods, nec	1.9	Manufg.
Electric services (utilities)	2.7	Util.	Laundry, dry cleaning, shoe repair	1.7	Services
Narrow fabric mills	2.1	Manufg.	Canvas & related products	1.5	Manufg.
Petroleum refining	1.7	Manufg.	S/L Govt. purch., public assistance & relief	1.3	S/L Govt
Eating & drinking places	1.5	Trade	Fabricated textile products, nec	1.2	Manufg.
U.S. Postal Service	1.3	Gov't	Pleating & stitching	1.2	Manufg.
Paperboard containers & boxes	1.1	Manufg.	Luggage	1.0	Manufg.
Metal stampings, nec	0.9	Manufg.	Watch, clock, jewelry, & furniture repair	0.7	Services
Miscellaneous fabricated wire products	0.9	Manufg.	Automotive & apparel trimmings	0.5	Manufg.
Paints & allied products	0.9	Manufg.	Sporting & athletic goods, nec	0.5	Manufg.
Motor freight transportation & warehousing	0.9	Util.	Sewing machines	0.4	Manufg.
Banking	0.9	Fin/R.E.	Beauty & barber shops	0.3	Services
Maintenance of nonfarm buildings nec	0.7	Constr.	Knit underwear mills	0.2	Manufg.
Machinery, except electrical, nec	0.6	Manufg.	Knitting mills, nec	0.2	Manufg.
Communications, except radio & TV	0.6	Util.	Motion pictures	0.2	Services
Gas production & distribution (utilities)	0.6	Util.	Portrait, photographic studios	0.2	Services
Air transportation	0.5	Util.	S/L Govt. purch., health & hospitals	0.2	S/L Govt
Real estate	0.5	Fin/R.E.	Federal Government purchases, national defense	0.1	Fed Govt
Legal services	0.5	Services	S/L Govt. purch., correction	0.1	S/L Govt
Management & consulting services & labs	0.5	Services			
Abrasive products	0.4	Manufg.			
Special dies & tools & machine tool accessories	0.4	Manufg.			
Automotive rental & leasing, without drivers	0.4	Services			
Royalties	0.3	Fin/R.E.			
Accounting, auditing & bookkeeping	0.3	Services			
Automotive repair shops & services	0.3	Services			
Equipment rental & leasing services	0.3	Services			
Job training & related services	0.3	Services			
Lubricating oils & greases	0.2	Manufg.			
Manifold business forms	0.2	Manufg.			
Paper coating & glazing	0.2	Manufg.			
Railroads & related services	0.2	Util.			
Sanitary services, steam supply, irrigation	0.2	Util.			
Water transportation	0.2	Util.			
Insurance carriers	0.2	Fin/R.E.			
Hotels & lodging places	0.2	Services			
Security & commodity brokers	0.1	Fin/R.E.			
Computer & data processing services	0.1	Services			
Personnel supply services	0.1	Services			

Source: Benchmark Input-Output Accounts for the U.S. Economy, 1982, U.S. Department of Commerce, Washington, D.C., July 1991. Data, as reported in the source, are organized by the 1977 SIC structure in use in 1982 but have been matched, as closely as is possible, to the 1987 SIC structure used in this book.

OCCUPATIONS EMPLOYED BY SIC 396 - MANUFACTURED PRODUCTS, NEC

Occupation	% of Total 1994	Change to 2005	Occupation	% of Total 1994	Change to 2005
Assemblers, fabricators, & hand workers nec	17.0	4.8	Freight, stock, & material movers, hand	1.5	-16.1
Sales & related workers nec	4.3	4.8	Inspectors, testers, & graders, precision	1.5	4.9
Blue collar worker supervisors	4.0	-2.4	General office clerks	1.5	-10.6
General managers & top executives	3.8	-0.5	Designers, ex interior designers	1.5	15.3
Hand packers & packagers	3.7	-28.1	Electrical & electronic assemblers	1.4	4.8
Traffic, shipping, & receiving clerks	2.5	0.9	Industrial production managers	1.4	4.8
Machine operators nec	2.2	-7.6	Packaging & filling machine operators	1.4	4.9
Helpers, laborers, & material movers nec	2.2	4.8	Screen printing machine setters & set-up operators	1.2	4.8
Painting, coating, & decorating workers, hand	2.0	-5.7	Plastic molding machine workers	1.2	4.8
Precision workers nec	2.0	41.5	Machine feeders & offbearers	1.2	-5.7
Secretaries, ex legal & medical	2.0	-4.5	Cabinetmakers & bench carpenters	1.1	-31.9
Bookkeeping, accounting, & auditing clerks	1.8	-21.4	Order clerks, materials, merchandise, & service	1.1	2.6
Sheet metal workers & duct installers	1.8	4.8	Grinders & polishers, hand	1.1	4.8
Precision metal workers nec	1.8	36.3			

Source: Industry-Occupation Matrix, Bureau of Labor Statistics. These data relate to one or more 3-digit SIC industry groups rather than to a single 4-digit SIC. The change reported for each occupation to the year 2005 is a percent of growth or decline as estimated by the Bureau of Labor Statistics. The abbreviation nec stands for 'not elsewhere classified'.

LOCATION BY STATE AND REGIONAL CONCENTRATION

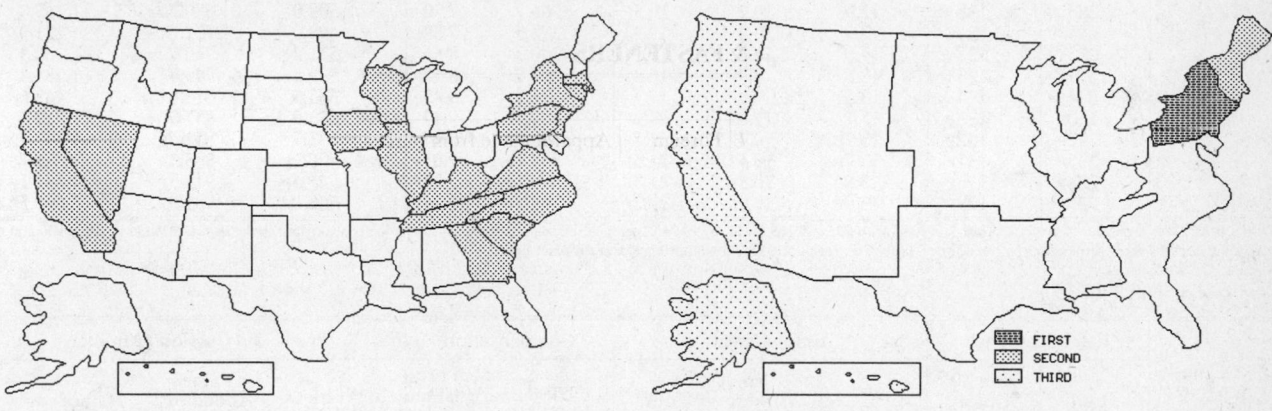

FIRST
SECOND
THIRD

INDUSTRY DATA BY STATE

State	Establish-ments	Shipments			Employment				Cost as % of Shipments	Investment per Employee ($)
		Total ($ mil)	% of U.S.	Per Establ.	Total Number	% of U.S.	Per Establ.	Wages ($/hour)		
Georgia	5	193.8	22.3	38.8	1,600	16.0	320	10.03	43.5	-
New York	64	157.0	18.1	2.5	1,800	18.0	28	8.70	43.8	2,778
New Jersey	11	51.4	5.9	4.7	1,100	11.0	100	7.64	41.2	-
North Carolina	7	48.8	5.6	7.0	700	7.0	100	6.57	48.8	-
Rhode Island	13	47.2	5.4	3.6	500	5.0	38	9.00	46.8	2,400
Pennsylvania	7	35.8	4.1	5.1	300	3.0	43	10.60	41.1	-
California	26	35.6	4.1	1.4	500	5.0	19	8.00	40.7	3,200
Illinois	11	26.2	3.0	2.4	400	4.0	36	7.60	45.0	-
Wisconsin	6	20.7	2.4	3.5	400	4.0	67	9.67	32.4	-
Iowa	7	17.7	2.0	2.5	300	3.0	43	7.25	45.2	-
Connecticut	16	(D)	-	-	750 *	7.5	47	-	-	-
Virginia	5	(D)	-	-	175 *	1.8	35	-	-	-
Massachusetts	4	(D)	-	-	175 *	1.8	44	-	-	-
Nevada	3	(D)	-	-	175 *	1.8	58	-	-	-
South Carolina	3	(D)	-	-	375 *	3.8	125	-	-	-
Tennessee	3	(D)	-	-	175 *	1.8	58	-	-	-
Kentucky	1	(D)	-	-	375 *	3.8	375	-	-	-

Source: 1992 *Economic Census.* The states are in descending order of shipments or establishments (if shipment data are missing for the majority). The symbol (D) appears when data are withheld to prevent disclosure of competitive information. States marked with (D) are sorted by number of establishments. A dash (-) indicates that the data element cannot be calculated; * indicates the midpoint of a range.

3991 - BROOMS & BRUSHES

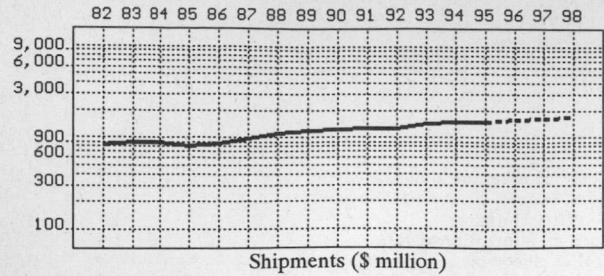

Shipments ($ million)

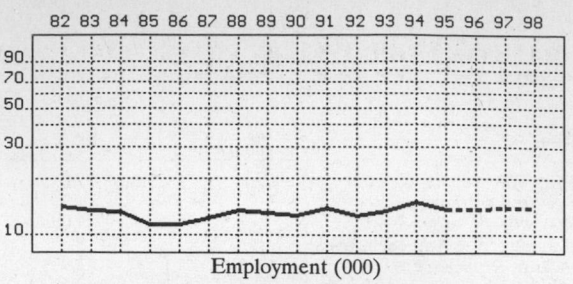

Employment (000)

GENERAL STATISTICS

Year	Companies	Establishments		Employment			Compensation		Production ($ million)			
		Total	with 20 or more employees	Total (000)	Production Workers (000)	Hours (Mil)	Payroll ($ mil)	Wages ($/hr)	Cost of Materials	Value Added by Manufacture	Value of Shipments	Capital Invest.
1982	320	341	139	14.1	10.5	19.4	197.1	6.34	376.9	432.5	818.1	17.2
1983		333	140	13.4	9.9	17.6	205.1	7.05	421.1	470.9	894.0	18.7
1984		325	141	13.2	9.9	18.0	202.6	6.74	402.9	482.4	878.2	17.7
1985		316	141	11.4	8.6	16.3	190.3	6.87	369.7	438.9	808.0	26.1
1986		309	148	11.4	8.7	16.9	202.3	7.04	386.2	479.3	859.3	27.6
1987	293	301	143	12.3	9.3	17.5	230.6	7.83	440.7	550.9	990.4	26.5
1988		281	143	13.4	9.9	18.8	267.8	8.26	500.3	616.5	1,105.1	37.9
1989		266	138	12.9	10.2	19.5	266.7	7.50	508.0	680.7	1,185.0	14.1
1990		264	134	12.7	10.4	19.3	278.8	7.88	498.0	731.0	1,221.8	17.4
1991		265	137	14.0	10.7	19.7	290.7	8.66	521.2	740.7	1,260.3	23.8
1992	260	279	128	12.6	8.9	18.8	291.7	8.59	539.5	745.4	1,281.4	35.8
1993		274	134	13.5	9.9	19.7	318.0	9.43	635.5	811.6	1,443.6	33.3
1994		250P	133P	15.1	11.5	24.2	357.4	9.13	673.9	851.9	1,512.8	56.1
1995		242P	132P	13.7P	10.4P	21.2P	345.0P	9.44P	673.7P	851.7P	1,512.4P	40.1P
1996		235P	131P	13.7P	10.4P	21.5P	358.1P	9.67P	700.2P	885.1P	1,571.7P	42.0P
1997		228P	130P	13.8P	10.5P	21.8P	371.1P	9.91P	726.6P	918.5P	1,631.1P	43.9P
1998		221P	129P	13.9P	10.6P	22.2P	384.2P	10.14P	753.0P	952.0P	1,690.5P	45.7P

Sources: 1982, 1987, 1992 *Economic Census*; *Annual Survey of Manufactures*, 83-86, 88-91, 93-94. Establishment counts for non-Census years are from *County Business Patterns*; establishment values for 83-84 are extrapolations. 'P's show projections by the editors. Industries reclassified in 87 will not have data for prior years.

INDICES OF CHANGE

Year	Companies	Establishments		Employment			Compensation		Production ($ million)			
		Total	with 20 or more employees	Total (000)	Production Workers (000)	Hours (Mil)	Payroll ($ mil)	Wages ($/hr)	Cost of Materials	Value Added by Manufacture	Value of Shipments	Capital Invest.
1982	123	122	109	112	118	103	68	74	70	58	64	48
1983		119	109	106	111	94	70	82	78	63	70	52
1984		116	110	105	111	96	69	78	75	65	69	49
1985		113	110	90	97	87	65	80	69	59	63	73
1986		111	116	90	98	90	69	82	72	64	67	77
1987	113	108	112	98	104	93	79	91	82	74	77	74
1988		101	112	106	111	100	92	96	93	83	86	106
1989		95	108	102	115	104	91	87	94	91	92	39
1990		95	105	101	117	103	96	92	92	98	95	49
1991		95	107	111	120	105	100	101	97	99	98	66
1992	100	100	100	100	100	100	100	100	100	100	100	100
1993		98	105	107	111	105	109	110	118	109	113	93
1994		89P	104P	120	129	129	123	106	125	114	118	157
1995		87P	103P	108P	116P	113P	118P	110P	125P	114P	118P	112P
1996		84P	103P	109P	117P	114P	123P	113P	130P	119P	123P	117P
1997		82P	102P	110P	118P	116P	127P	115P	135P	123P	127P	123P
1998		79P	101P	110P	119P	118P	132P	118P	140P	128P	132P	128P

Sources: Same as General Statistics. Values reflect change from the base year, 1992. Values above 100 mean greater than 92, values below 100 mean less than 92, and a value of 100 in the 82-91 or 93-98 period means same as 92. 'P's mark projections by the editors.

SELECTED RATIOS

For 1994	Avg. of All Manufact.	Analyzed Industry	Index	For 1994	Avg. of All Manufact.	Analyzed Industry	Index
Employees per Establishment	49	61	124	Value Added per Production Worker	134,084	74,078	55
Payroll per Establishment	1,500,273	1,432,291	95	Cost per Establishment	5,045,178	2,700,674	54
Payroll per Employee	30,620	23,669	77	Cost per Employee	102,970	44,629	43
Production Workers per Establishment	34	46	134	Cost per Production Worker	146,988	58,600	40
Wages per Establishment	853,319	885,448	104	Shipments per Establishment	9,576,895	6,062,590	63
Wages per Production Worker	24,861	19,213	77	Shipments per Employee	195,460	100,185	51
Hours per Production Worker	2,056	2,104	102	Shipments per Production Worker	279,017	131,548	47
Wages per Hour	12.09	9.13	76	Investment per Establishment	321,011	224,822	70
Value Added per Establishment	4,602,255	3,414,014	74	Investment per Employee	6,552	3,715	57
Value Added per Employee	93,930	56,417	60	Investment per Production Worker	9,352	4,878	52

Sources: Same as General Statistics. The 'Average of All Manufacturing' column represents the average of all manufacturing industries reported for the most recent complete year available. The Index shows the relationship between the Average and the Analyzed Industry. For example, 100 means that they are equal; 500 that the Analyzed Industry is five times the average; 50 means that the Analyzed Industry is half the national average. The abbreviation 'na' is used to show that data are 'not available'.

LEADING COMPANIES Number shown: **46** Total sales ($ mil): **902** Total employment (000): **6.1**

Company Name	Address				CEO Name	Phone	Co. Type	Sales ($ mil)	Empl. (000)
Oral B Laboratories	1 Lagoon Dr	Redwood City	CA	94065	Jorgen Wedel	415-598-5000	S	366	0.8
Wooster Brush Co	PO Box 6010	Wooster	OH	44691	Stan Welty Jr	216-264-4440	R	60	0.5
Osborn Manufacturing	5401 Hamilton Av	Cleveland	OH	44114	JB Tyler	216-361-1900	D	45	0.3
Linzer Products Corp	133-30 37th Av	Flushing	NY	11354	Alan Benson	718-961-0900	R	35	0.3
Kellogg Brush Manufacturing Co	PO Box 511	Easthampton	MA	01027	Robert Ryan	413-527-2450	S	30	0.3
Linzer Products Corp	PO Box 9608	N Hollywood	CA	91609	Alan Benson	818-763-6117	R	25	0.3
Rubbermaid	PO Box 1606	Greenville	NC	27835	Joe Gantz	919-758-4111	D	25	0.5
Sunshine Industries Inc	1111 E 200th St	Cleveland	OH	44117	Sheldon Leventhal	216-383-9000	R	25	0.1
Advance Milwaukee	PO Box 830	Menomonee Fls	WI	53051	Edward A Krautner	414-255-3200	R	23	0.2
Flo-Pac Corp	700 N Washington	Minneapolis	MN	55401	JA Bowen	612-332-6240	R	23	0.2
O-Cedar/Vining	PO Box 1606	Springfield	OH	45501	Godon Garrett	513-324-5596	R	22*	0.3
Adams Brush Manufacturing	94-02 104th St	Ozone Park	NY	11416	A Zurawin	718-846-8700	R	14	0.1
Flour City Brush Co	PO Box 9356	Minneapolis	MN	55440	Everett Madsen	612-332-8641	S	13	0.1
American Brush Co	PO Box 1490	Claremont	NH	03743	George Buzuvis	603-542-9951	D	12*	0.1
Bestt Rollr Inc	PO Box 550	Fond du Lac	WI	54935	KR Stoddart	414-922-6250	D	12	0.2
Libman Co	220 N Sheldon St	Arcola	IL	61910	Robert Libman	217-268-4151	R	12*	0.1
Milor Corp	511 Lancaster St	Leominster	MA	01453	Larry Gottsegen	508-534-4966	R	12	0.2
Harper Corp	PO Box 608	Fairfield	IA	52556	Barry Harper	515-472-5186	R	11	<0.1
Industrial Brush Corp	PO Box 2608	Pomona	CA	91769	J Cottam	909-591-9341	R	10	<0.1
Padco Inc	2220 Elm St SE	Minneapolis	MN	55414	Robert I Janssen	612-378-7270	R	10	<0.1
Magnolia Brush Manufacturing	1001 N Cedar St	Clarksville	TX	75426	Ken Bakus	903-427-2261	R	9	0.1
Lactona Corp	201 Commerce Dr	Montgomeryv	PA	18936	A Montemurro	215-368-2000	R	9	0.1
Sparta Brush Company Inc	402 S Black River St	Sparta	WI	54656	John A Larson	608-269-2151	R	8	<0.1
Brush Research Manufacturing	4642 E Floral Dr	Los Angeles	CA	90022	Tara Rands	213-261-2193	R	8	<0.1
France	PO Box 71	Paxton	IL	60957	Stanley Koshnick	217-379-2377	D	7*	<0.1
Helmac Products Corp	PO Box 73	Flint	MI	48501	Douglas R Taeckens	313-239-5000	R	7*	<0.1
Maugus Manufacturing Inc	PO Box 4096	Lancaster	PA	17604	Richard Seavey	717-299-5681	R	7*	<0.1
Pacific Coast Brush Company	11690 Pacific Av	Fontana	CA	92335	Theodore Alm	909-681-3747	S	7	<0.1
Superior Brush Co	3453 W 140th St	Cleveland	OH	44111	Theodore O Mertes	216-252-5144	R	7	<0.1
Oral Logic Inc	555 Renton Village	Renton	WA	98055	Bruce Merrell	206-227-9800	R	6*	<0.1
Prager Brush Company Inc	PO Box 93263	Atlanta	GA	30377	Leonard Butler	404-875-9292	R	5*	<0.1
Sanderson-MacLeod Inc	199 S Main St	Palmer	MA	01069	E Sanderson	413-283-3481	R	5	0.1
Creations by Alan Stuart Inc	49 W 38th St	New York	NY	10018	Stuart Kalinsky	212-719-5267	R	4	<0.1
Industrial Brush Company Inc	PO Box 869	Fairfield	NJ	07004	H Enchelmaier	201-575-0455	R	4	<0.1
Gordon Brush Manufacturing	2150 Sacramento St	Los Angeles	CA	90021	Kenneth L Rakusin	213-627-6889	R	3	<0.1
Milwaukee Dustless Brush Co	10930 W Lapham St	Milwaukee	WI	53214	George J Hunt	414-476-1147	R	3*	<0.1
Universal Brush Mfg Co	PO Box 1618	Harvey	IL	60426	CD Reid	708-331-1700	R	3	<0.1
Danline Inc	137 N Michigan Av	Kenilworth	NJ	07033	R Seth	908-245-5900	R	3	<0.1
Hamburg Broom Works Inc	PO Box 27	Hamburg	PA	19526	Richard E Stiller	215-562-3031	P	3	<0.1
A & B Brush Mfg Corp	1150 Three Ranch	Duarte	CA	91010	D Anawalt	818-303-8856	R	2	<0.1
Amro Manufacturing Company	3501 E 26th St	Vernon	CA	90023	Arnold Kline	213-264-7333	R	2	<0.1
Richards Brush Co	PO Box 3856	Seattle	WA	98124	WT Richards	206-623-3720	R	2	<0.1
A Steiert and Son Inc	PO Box 100	Hatfield	PA	19440	HA Steiert Jr	215-822-0567	R	2	<0.1
Ohio Brush Co	2680 Lisbon Rd	Cleveland	OH	44104	Ray A Gardner	216-791-3265	R	1	<0.1
Zimmerman Brush Co	900 W Lake St	Chicago	IL	60607	Yale Zimmerman	312-829-3262	R	1	<0.1
Brushtech Inc	PO Box 1130	Plattsburgh	NY	12901	AG Gunjian	518-563-8420	R	1	<0.1

Source: Ward's Business Directory of U.S. Private and Public Companies, Volumes 1 and 2, 1996. The company type code used is as follows: P - Public, R - Private, S - Subsidiary, D - Division, J - Joint Venture, A - Affiliate, G - Group. Sales are in millions of dollars, employees are in thousands. An asterisk (*) indicates an estimated sales volume. The symbol < stands for 'less than'. Company names and addresses are truncated, in some cases, to fit into the available space.

MATERIALS CONSUMED

Material	Quantity	Delivered Cost ($ million)
Materials, ingredients, containers, and supplies	(X)	446.3
Fabricated metal ferrules	(X)	7.9
Metal stampings	(X)	5.6
All other fabricated metal products (except castings and forgings)	(X)	7.2
Castings (rough and semifinished)	(X)	0.5
Steel brush wire	(X)	19.6
Other steel shapes and forms	(X)	11.3
Nonferrous shapes and forms	(X)	6.3
Broomcorn	(X)	9.1
Yarns and textiles made of cotton, wool, silk, and manmade fibers	(X)	35.8
Rough and dressed lumber	(X)	3.4
Wood brush handles and backs	(X)	30.0
Other lumber and wood products, except furniture	(X)	5.3
Paperboard containers, boxes, and corrugated paperboard	(X)	29.4
Plastics resins consumed in the form of granules, pellets, powders, liquids, etc.	(X)	47.7
Plastics products consumed in the form of sheets, rods, tubes, film, and other shapes	(X)	62.6
Dressed hair (including bristle and horsehair)	(X)	16.3
All other materials and components, parts, containers, and supplies	(X)	102.5
Materials, ingredients, containers, and supplies, nsk	(X)	45.9

Source: 1992 Economic Census. Explanation of symbols used: (D): Withheld to avoid disclosure of competitive data; na: Not available; (S): Withheld because statistical norms were not met; (X): Not applicable; (Z): Less than half the unit shown; nec: Not elsewhere classified; nsk: Not specified by kind; - : zero; * : 10-19 percent estimated; ** : 20-29 percent estimated.

PRODUCT SHARE DETAILS

Product or Product Class	% Share	Product or Product Class	% Share
Brooms and brushes	100.00	Replacement rollers	19.20
Brooms	12.97	Paint and varnish brushes, rollers, and pads, nsk	1.80
Household floor brooms	45.66	Other brushes	52.22
Other brooms (industrial brooms, whiskbrooms, toy brooms, hearth brooms, streetsweeping machine brooms, etc.)	52.55	Toothbrushes	32.95
		Hairbrushes	7.14
Brooms, nsk	1.72	Household maintenance brushes (floor, scrub, dusting, window, etc.), including any twisted-in-wire brushes	13.62
Paint and varnish brushes, rollers, and pads	29.82	Industrial maintenance brushes (floor, scrub, dusting, window, etc.), including any twisted-in-wire brushes	16.06
Whitewash, kalsomine, paperhanging, marking, and stenciling brushes made of pure bristle	12.09	Industrial brushes (except maintenance) (including power-driven, rotary, end, cup, jewelers' and dentists' brushes, etc.)	16.65
Whitewash, kalsomine, paperhanging, marking, and stenciling brushes made of synthetic bristle and other materials, including mixtures	35.23	Other brushes, including artists' brushes and hair pencils, except artists' airbrushes	13.48
Paint pads and holders	11.01	Other brushes, nsk	0.10
Complete paint roller, roller frame, and replacement roller units	15.92	Brooms and brushes, nsk	5.00
Roller frames	4.72		

Source: 1992 *Economic Census*. The values shown are percent of total shipments in an industry. Values of indented subcategories are summed in the main heading. The symbol (D) appears when data are withheld to prevent disclosure of competitive information. The abbreviation nsk stands for 'not specified by kind' and nec for 'not elsewhere classified'.

INPUTS AND OUTPUTS FOR BROOMS & BRUSHES

Economic Sector or Industry Providing Inputs	%	Sector	Economic Sector or Industry Buying Outputs	%	Sector
Imports	14.8	Foreign	Personal consumption expenditures	38.5	
Advertising	11.8	Services	S/L Govt. purch., elem. & secondary education	4.2	S/L Govt
Plastics materials & resins	11.0	Manufg.	Nonfarm residential structure maintenance	4.0	Constr.
Wholesale trade	7.6	Trade	Maintenance of nonfarm buildings nec	3.7	Constr.
Broadwoven fabric mills	6.2	Manufg.	Exports	3.1	Foreign
Manufacturing industries, nec	5.9	Manufg.	Residential additions/alterations, nonfarm	3.0	Constr.
Wood products, nec	5.7	Manufg.	Industrial buildings	2.8	Constr.
Blast furnaces & steel mills	3.5	Manufg.	Accounting, auditing & bookkeeping	2.6	Services
Miscellaneous crops	2.4	Agric.	Religious organizations	2.6	Services
Metal stampings, nec	2.3	Manufg.	Office buildings	2.1	Constr.
Motor freight transportation & warehousing	1.9	Util.	Residential 1-unit structures, nonfarm	1.9	Constr.
Eating & drinking places	1.9	Trade	Hotels & lodging places	1.7	Services
Electric services (utilities)	1.6	Util.	Services to dwellings & other buildings	1.6	Services
Veneer & plywood	1.4	Manufg.	Hospitals	1.4	Services
U.S. Postal Service	1.3	Gov't	Federal Government purchases, nondefense	1.3	Fed Govt
Noncomparable imports	1.2	Foreign	Beauty & barber shops	1.2	Services
Organic fibers, noncellulosic	1.0	Manufg.	Highway & street construction	0.9	Constr.
Paperboard containers & boxes	1.0	Manufg.	Retail trade, except eating & drinking	0.8	Trade
Sawmills & planning mills, general	1.0	Manufg.	S/L Govt. purch., health & hospitals	0.8	S/L Govt
Banking	1.0	Fin/R.E.	S/L Govt. purch., other general government	0.8	S/L Govt
Nonferrous wire drawing & insulating	0.9	Manufg.	Real estate	0.7	Fin/R.E.
Maintenance of nonfarm buildings nec	0.8	Constr.	Business services nec	0.7	Services
Petroleum refining	0.8	Manufg.	S/L Govt. purch., higher education	0.7	S/L Govt
Brooms & brushes	0.7	Manufg.	S/L Govt. purch., natural resource & recreation.	0.7	S/L Govt
Machinery, except electrical, nec	0.7	Manufg.	Construction of stores & restaurants	0.6	Constr.
Management & consulting services & labs	0.7	Services	Residential high-rise apartments	0.6	Constr.
Communications, except radio & TV	0.6	Util.	Colleges, universities, & professional schools	0.6	Services
Railroads & related services	0.6	Util.	Meat animals	0.5	Agric.
Real estate	0.6	Fin/R.E.	Construction of hospitals	0.5	Constr.
Legal services	0.6	Services	Hotels & motels	0.5	Constr.
Abrasive products	0.5	Manufg.	Eating & drinking places	0.5	Trade
Hardwood dimension & flooring mills	0.5	Manufg.	Advertising	0.5	Services
Radio & TV receiving sets	0.5	Manufg.	Business/professional associations	0.5	Services
Gas production & distribution (utilities)	0.5	Util.	Elementary & secondary schools	0.5	Services
Special dies & tools & machine tool accessories	0.4	Manufg.	Nursing & personal care facilities	0.5	Services
Air transportation	0.4	Util.	Feed grains	0.4	Agric.
Royalties	0.4	Fin/R.E.	Residential garden apartments	0.4	Constr.
Accounting, auditing & bookkeeping	0.4	Services	Brooms & brushes	0.4	Manufg.
Adhesives & sealants	0.3	Manufg.	Commercial printing	0.4	Manufg.
Aluminum rolling & drawing	0.3	Manufg.	Federal Government purchases, national defense	0.4	Fed Govt
Fabricated rubber products, nec	0.3	Manufg.	S/L Govt. purch., police	0.4	S/L Govt
Manifold business forms	0.3	Manufg.	S/L Govt. purch., public assistance & relief	0.4	S/L Govt
Miscellaneous plastics products	0.3	Manufg.	Construction of educational buildings	0.3	Constr.
Paints & allied products	0.3	Manufg.	Construction of nonfarm buildings nec	0.3	Constr.
Lubricating oils & greases	0.2	Manufg.	Maintenance of farm service facilities	0.3	Constr.
Automotive rental & leasing, without drivers	0.2	Services	Maintenance of water supply facilities	0.3	Constr.
Equipment rental & leasing services	0.2	Services	Residential 2-4 unit structures, nonfarm	0.3	Constr.
Hotels & lodging places	0.2	Services	Dairy farm products	0.2	Agric.
Personnel supply services	0.2	Services	Oil bearing crops	0.2	Agric.
Photographic equipment & supplies	0.1	Manufg.	Farm service facilities	0.2	Constr.
Sanitary services, steam supply, irrigation	0.1	Util.	Maintenance of highways & streets	0.2	Constr.
Transit & bus transportation	0.1	Util.	Maintenance of sewer facilities	0.2	Constr.
Credit agencies other than banks	0.1	Fin/R.E.	Newspapers	0.2	Manufg.

Continued on next page.

INPUTS AND OUTPUTS FOR BROOMS & BRUSHES - Continued

Economic Sector or Industry Providing Inputs	%	Sector	Economic Sector or Industry Buying Outputs	%	Sector
Insurance carriers	0.1	Fin/R.E.	Wholesale trade	0.2	Trade
Automotive repair shops & services	0.1	Services	Banking	0.2	Fin/R.E.
Business services nec	0.1	Services	Equipment rental & leasing services	0.2	Services
Business/professional associations	0.1	Services	Photofinishing labs, commercial photography	0.2	Services
Computer & data processing services	0.1	Services	Residential care	0.2	Services
Laundry, dry cleaning, shoe repair	0.1	Services	Social services, nec	0.2	Services
			Food grains	0.1	Agric.
			Poultry & eggs	0.1	Agric.
			Electric utility facility construction	0.1	Constr.
			Farm housing units & additions & alterations	0.1	Constr.
			Maintenance of electric utility facilities	0.1	Constr.
			Maintenance of farm residential buildings	0.1	Constr.
			Maintenance of nonbuilding facilities nec	0.1	Constr.
			Warehouses	0.1	Constr.
			Electric services (utilities)	0.1	Util.
			Amusement & recreation services nec	0.1	Services
			Job training & related services	0.1	Services
			State & local government enterprises, nec	0.1	Gov't
			U.S. Postal Service	0.1	Gov't

Source: Benchmark Input-Output Accounts for the U.S. Economy, 1982, U.S. Department of Commerce, Washington, D.C., July 1991. Data, as reported in the source, are organized by the 1977 SIC structure in use in 1982 but have been matched, as closely as is possible, to the 1987 SIC structure used in this book.

OCCUPATIONS EMPLOYED BY SIC 399 - MANUFACTURED PRODUCTS, NEC

Occupation	% of Total 1994	Change to 2005	Occupation	% of Total 1994	Change to 2005
Assemblers, fabricators, & hand workers nec	17.0	4.8	Freight, stock, & material movers, hand	1.5	-16.1
Sales & related workers nec	4.3	4.8	Inspectors, testers, & graders, precision	1.5	4.9
Blue collar worker supervisors	4.0	-2.4	General office clerks	1.5	-10.6
General managers & top executives	3.8	-0.5	Designers, ex interior designers	1.5	15.3
Hand packers & packagers	3.7	-28.1	Electrical & electronic assemblers	1.4	4.8
Traffic, shipping, & receiving clerks	2.5	0.9	Industrial production managers	1.4	4.8
Machine operators nec	2.2	-7.6	Packaging & filling machine operators	1.4	4.9
Helpers, laborers, & material movers nec	2.2	4.8	Screen printing machine setters & set-up operators	1.2	4.8
Painting, coating, & decorating workers, hand	2.0	-5.7	Plastic molding machine workers	1.2	4.8
Precision workers nec	2.0	41.5	Machine feeders & offbearers	1.2	-5.7
Secretaries, ex legal & medical	2.0	-4.5	Cabinetmakers & bench carpenters	1.1	-31.9
Bookkeeping, accounting, & auditing clerks	1.8	-21.4	Order clerks, materials, merchandise, & service	1.1	2.6
Sheet metal workers & duct installers	1.8	4.8	Grinders & polishers, hand	1.1	4.8
Precision metal workers nec	1.8	36.3			

Source: Industry-Occupation Matrix, Bureau of Labor Statistics. These data relate to one or more 3-digit SIC industry groups rather than to a single 4-digit SIC. The change reported for each occupation to the year 2005 is a percent of growth or decline as estimated by the Bureau of Labor Statistics. The abbreviation nec stands for 'not elsewhere classified'.

LOCATION BY STATE AND REGIONAL CONCENTRATION

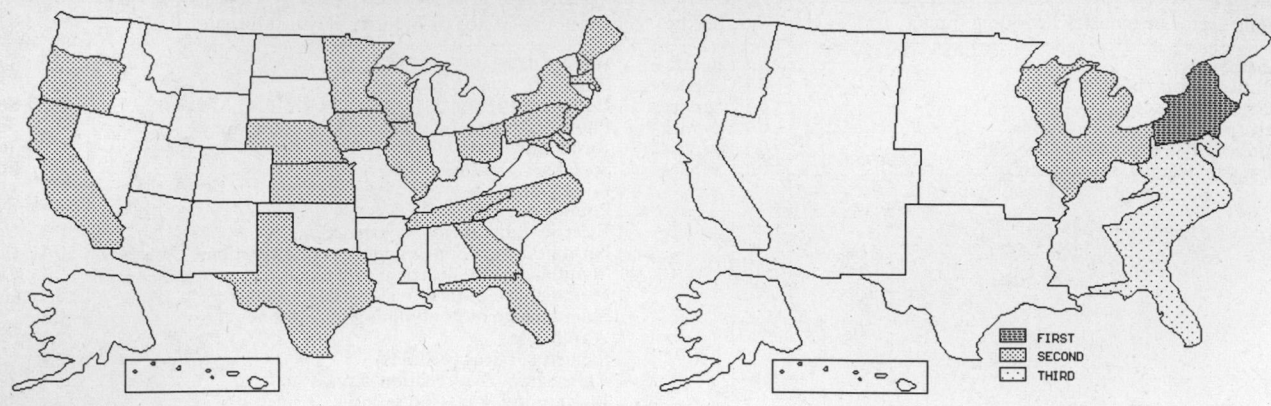

FIRST
SECOND
THIRD

INDUSTRY DATA BY STATE

| State | Establish-ments | Shipments | | | Employment | | | | Cost as % of Shipments | Investment per Employee ($) |
		Total ($ mil)	% of U.S.	Per Establ.	Total Number	% of U.S.	Per Establ.	Wages ($/hour)		
Ohio	27	191.7	15.0	7.1	1,700	13.5	63	10.79	44.8	2,118
Illinois	21	149.1	11.6	7.1	1,300	10.3	62	7.76	47.9	-
New York	35	100.4	7.8	2.9	1,000	7.9	29	7.80	45.2	-
Massachusetts	12	72.4	5.7	6.0	900	7.1	75	9.23	39.2	2,444
California	22	57.0	4.4	2.6	600	4.8	27	8.00	46.1	-
Pennsylvania	12	53.6	4.2	4.5	600	4.8	50	8.00	46.1	-
New Hampshire	3	35.4	2.8	11.8	300	2.4	100	7.80	41.0	-
Texas	19	20.8	1.6	1.1	200	1.6	11	6.67	45.7	1,500
Florida	10	13.5	1.1	1.4	100	0.8	10	7.50	42.2	1,000
Georgia	7	10.8	0.8	1.5	100	0.8	14	7.50	42.6	-
New Jersey	14	(D)	-	-	375 *	3.0	27	-	-	-
Wisconsin	11	(D)	-	-	1,750 *	13.9	159	-	-	-
North Carolina	8	(D)	-	-	750 *	6.0	94	-	-	-
Iowa	6	(D)	-	-	375 *	3.0	63	-	-	-
Tennessee	6	(D)	-	-	750 *	6.0	125	-	-	-
Kansas	3	(D)	-	-	750 *	6.0	250	-	-	-
Maryland	3	(D)	-	-	375 *	3.0	125	-	-	-
Oregon	3	(D)	-	-	375 *	3.0	125	-	-	-
Minnesota	2	(D)	-	-	175 *	1.4	88	-	-	-
Nebraska	2	(D)	-	-	175 *	1.4	88	-	-	-
Maine	1	(D)	-	-	175 *	1.4	175	-	-	-

Source: 1992 *Economic Census*. The states are in descending order of shipments or establishments (if shipment data are missing for the majority). The symbol (D) appears when data are withheld to prevent disclosure of competitive information. States marked with (D) are sorted by number of establishments. A dash (-) indicates that the data element cannot be calculated; * indicates the midpoint of a range.

3993 - SIGNS & ADVERTISING DISPLAYS

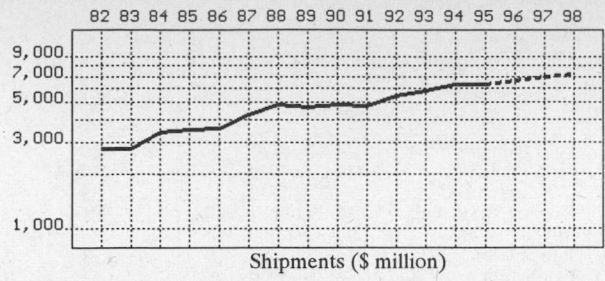

Shipments ($ million)

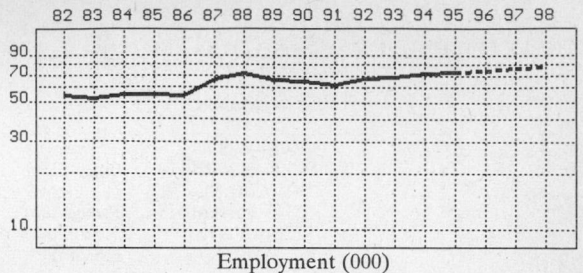

Employment (000)

GENERAL STATISTICS

Year	Com-panies	Establishments		Employment			Compensation		Production ($ million)			
		Total	with 20 or more employees	Total (000)	Production Workers (000)	Hours (Mil)	Payroll ($ mil)	Wages ($/hr)	Cost of Materials	Value Added by Manufacture	Value of Shipments	Capital Invest.
1982	3,156	3,248	667	53.6	37.1	67.2	884.8	7.56	1,141.3	1,603.3	2,739.2	63.3
1983		3,207	691	52.7	38.4	69.4	881.7	7.25	1,171.6	1,610.1	2,759.3	54.7
1984		3,166	715	54.6	38.9	73.3	1,013.2	7.86	1,380.8	2,049.1	3,418.1	68.1
1985		3,125	740	54.6	38.5	74.6	1,099.9	8.40	1,425.1	2,129.9	3,533.2	98.3
1986		3,155	772	53.7	36.7	74.3	1,144.2	8.54	1,431.2	2,143.3	3,557.5	75.2
1987	3,676	3,778	828	66.3	44.2	85.0	1,365.0	8.61	1,791.7	2,496.6	4,284.3	97.8
1988		3,634	830	72.3	48.8	93.0	1,457.6	8.67	2,033.7	2,772.4	4,794.4	99.8
1989		3,557	837	67.0	49.2	92.9	1,523.6	8.82	2,019.9	2,665.9	4,677.6	98.2
1990		3,670	801	65.1	44.4	85.3	1,525.8	9.20	2,214.4	2,613.8	4,826.5	107.8
1991		3,766	766	62.2	39.5	79.2	1,453.9	8.92	2,237.7	2,529.5	4,755.0	105.0
1992	4,467	4,577	762	67.5	42.5	86.1	1,679.9	9.69	2,279.2	3,164.8	5,420.1	96.9
1993		4,617	768	68.0	43.6	86.7	1,728.0	10.09	2,462.5	3,350.0	5,777.3	106.7
1994		4,429P	824P	72.0	46.2	93.9	1,845.9	9.79	2,580.5	3,718.9	6,279.2	144.3
1995		4,553P	833P	73.2P	46.7P	94.7P	1,917.5P	10.18P	2,601.3P	3,748.9P	6,329.9P	130.0P
1996		4,676P	842P	74.8P	47.4P	96.6P	1,998.0P	10.39P	2,716.4P	3,914.7P	6,609.8P	135.2P
1997		4,800P	851P	76.4P	48.0P	98.5P	2,078.5P	10.60P	2,831.4P	4,080.4P	6,889.6P	140.4P
1998		4,924P	861P	77.9P	48.7P	100.4P	2,159.0P	10.81P	2,946.4P	4,246.2P	7,169.5P	145.6P

Sources: 1982, 1987, 1992 *Economic Census*; *Annual Survey of Manufactures*, 83-86, 88-91, 93-94. Establishment counts for non-Census years are from *County Business Patterns*; establishment values for 83-84 are extrapolations. 'P's show projections by the editors. Industries reclassified in 87 will not have data for prior years.

INDICES OF CHANGE

Year	Com-panies	Establishments		Employment			Compensation		Production ($ million)			
		Total	with 20 or more employees	Total (000)	Production Workers (000)	Hours (Mil)	Payroll ($ mil)	Wages ($/hr)	Cost of Materials	Value Added by Manufacture	Value of Shipments	Capital Invest.
1982	71	71	88	79	87	78	53	78	50	51	51	65
1983		70	91	78	90	81	52	75	51	51	51	56
1984		69	94	81	92	85	60	81	61	65	63	70
1985		68	97	81	91	87	65	87	63	67	65	101
1986		69	101	80	86	86	68	88	63	68	66	78
1987	82	83	109	98	104	99	81	89	79	79	79	101
1988		79	109	107	115	108	87	89	89	88	88	103
1989		78	110	99	116	108	91	91	89	84	86	101
1990		80	105	96	104	99	91	95	97	83	89	111
1991		82	101	92	93	92	87	92	98	80	88	108
1992	100	100	100	100	100	100	100	100	100	100	100	100
1993		101	101	101	103	101	108	104	108	106	107	110
1994		97P	108P	107	109	109	110	101	113	118	116	149
1995		99P	109P	108P	110P	110P	114P	105P	114P	118P	117P	134P
1996		102P	111P	111P	111P	112P	119P	107P	119P	124P	122P	139P
1997		105P	112P	113P	113P	114P	124P	109P	124P	129P	127P	145P
1998		108P	113P	115P	114P	117P	129P	112P	129P	134P	132P	150P

Sources: Same as General Statistics. Values reflect change from the base year, 1992. Values above 100 mean greater than 92, values below 100 mean less than 92, and a value of 100 in the 82-91 or 93-98 period means same as 92. 'P's mark projections by the editors.

SELECTED RATIOS

For 1994	Avg. of All Manufact.	Analyzed Industry	Index	For 1994	Avg. of All Manufact.	Analyzed Industry	Index
Employees per Establishment	49	16	33	Value Added per Production Worker	134,084	80,496	60
Payroll per Establishment	1,500,273	416,776	28	Cost per Establishment	5,045,178	582,637	12
Payroll per Employee	30,620	25,638	84	Cost per Employee	102,970	35,840	35
Production Workers per Establishment	34	10	30	Cost per Production Worker	146,988	55,855	38
Wages per Establishment	853,319	207,559	24	Shipments per Establishment	9,576,895	1,417,747	15
Wages per Production Worker	24,861	19,898	80	Shipments per Employee	195,460	87,211	45
Hours per Production Worker	2,056	2,032	99	Shipments per Production Worker	279,017	135,913	49
Wages per Hour	12.09	9.79	81	Investment per Establishment	321,011	32,581	10
Value Added per Establishment	4,602,255	839,670	18	Investment per Employee	6,552	2,004	31
Value Added per Employee	93,930	51,651	55	Investment per Production Worker	9,352	3,123	33

Sources: Same as General Statistics. The 'Average of All Manufacturing' column represents the average of all manufacturing industries reported for the most recent complete year available. The Index shows the relationship between the Average and the Analyzed Industry. For example, 100 means that they are equal; 500 that the Analyzed Industry is five times the average; 50 means that the Analyzed Industry is half the national average. The abbreviation 'na' is used to show that data are 'not available'.

LEADING COMPANIES   Number shown: **75**   Total sales ($ mil): **1,970**   Total employment (000): **18.3**

Company Name	Address				CEO Name	Phone	Co. Type	Sales ($ mil)	Empl. (000)
Signmark	PO Box 2999	Milwaukee	WI	53201	Paul Gengler	414-228-1411	D	130•	1.0
Everbrite Inc	PO Box 20020	Greenfield	WI	53220	William J Fritz	414-529-3500	R	120•	1.4
Exhibitgroup Inc	2825 Carl Blv	Elk Grove Vill	IL	60007	C J Corsentino	708-350-5611	S	85	0.7
Skyline Displays Inc	12345 Portland Av	Burnsville	MN	55337	Gordon P Savoie	612-890-8392	R	81•	0.4
HMG Worldwide	475 10th Av	New York	NY	10018	Michael Wahl	212-736-2300	P	80	0.5
Plasti-Line Inc	PO Box 59043	Knoxville	TN	37950	James R Martin	615-938-1511	P	78	0.9
Giltspur Inc	500 Park Blv	Itasca	IL	60143	Paul Mullen	708-250-3930	S	63	0.8
Federal Sign Co	140 Tower Dr	Burr Ridge	IL	60521	Kevin S Stotmeister	708-887-6800	D	59	0.6
Gannett Transit	666 3rd Av	New York	NY	10017	Don Davidson	212-297-6400	D	50•	0.1
Thomas A Schutz Co	8710 Ferris Av	Morton Grove	IL	60053	Albert S Saia	708-965-7100	S	45	<0.1
Cummings Inc	200 12th Av S	Nashville	TN	37203	T L Cummings III	615-244-5555	S	42	0.6
KCS Industries Inc	5111 S 9th St	Milwaukee	WI	53221	Tim Battles	414-744-5111	S	42	0.6
RTC Industries Inc	3101 S Kedzie Av	Chicago	IL	60623	W Nathan	312-376-8200	R	41•	0.2
Gill Studios Inc	PO Box 2909	Shawnee Msn	KS	66201	Mark Gilman	913-888-4422	R	40	0.5
Daktronics Inc	331 32nd Av	Brookings	SD	57006	A J Kurtenbach	605-697-4000	P	35	0.6
CD Baird and Company Inc	5325 W Rogers St	West Allis	WI	53219	Fred Silloway	414-645-0340	R	34	0.4
Zimmerman Sign Co	8350 N Central	Dallas	TX	75206	Thomas Boner	214-691-8797	S	34	0.3
Trans-Lux Corp	110 Richards Av	Norwalk	CT	06856	Victor Liss	203-853-4321	P	34	0.5
Dualite Inc	1 Dualite Ln	Williamsburg	OH	45176	Frank W Schube	513-724-7100	R	32	0.3
Felbro Inc	3666 E Olympic	Los Angeles	CA	90023	Robert Feldner	213-263-8686	R	30•	0.3
Acme-Wiley Corp	2480 Greenleaf Av	Elk Grove Vill	IL	60007	Roger O'Neill	708-364-2250	R	26	0.3
Display Systems Inc	57-13 49th St	Maspeth	NY	11378	Ben Weshler	718-386-7700	R	26	0.4
MCA Sign Co	PO Box 555	Massillon	OH	44648	S Mollet IV	216-833-3165	R	25•	0.2
White Way Sign	1317 N Clybourn	Chicago	IL	60610	R B Flannery Jr	312-642-6580	R	25	0.2
Marlton Technologies Inc	111 Presidential	Bala Cynwyd	PA	19004	Robert B Ginsburg	215-664-6900	P	25	0.2
Yerger Brothers Inc	PO Box 1200	Lititz	PA	17543	John R Yerger	717-626-2145	R	23•	0.3
Ad Art Signs Inc	PO Box 8570	Stockton	CA	95208	Lou A Popais	209-931-0860	R	20	0.1
Chicago Display Co	1301 W Armitage	Melrose Park	IL	60160	Craig Binney	708-681-4340	R	20	0.3
Frank Mayer and Associates Inc	PO Box 105	Grafton	WI	53024	Frank W Mayer	414-377-4700	R	20	0.1
IDL Inc	535 Old Frankstown	Pittsburgh	PA	15239	Ralph Murray	412-798-2500	R	20	0.3
Trans World Marketing	360 Murray Hill	E Rutherford	NJ	07073	J Cavaluzzi	201-935-5565	R	20	0.1
Marketing Displays International	38271 W 12 Mile Rd	Farmington Hls	MI	48331	R Sarkisian	810-553-1900	R	18•	0.2
White Way Sign Co	1317 Clybourn Av	Chicago	IL	60610	R B Flannery Jr	312-642-6580	S	18	0.2
Central Sales Promotions	PO Box 53444	Oklahoma City	OK	73152	Gary M Watts	405-525-2335	S	18	0.2
Adaptive Micro Systems Inc	7840 N 86th St	Milwaukee	WI	53224	William Latz	414-357-2020	R	17•	<0.1
American Sign & Marketing Svcs	7430 Industrial Rd	Florence	KY	41042	Richard Jordan	606-371-2880	S	17	0.2
Insignia Systems Inc	10801 Red Circle Dr	Minnetonka	MN	55343	GL Hoffman	612-930-8200	P	16	0.2
Key Industries Inc	215 Taylor St	East Peoria	IL	61611	M Linderman	309-694-4241	P	16	0.2
Heath Northwest Inc	PO Box 9608	Yakima	WA	98909	Mike St Onge	509-248-2050	R	16	0.1
Work Area Protection Corp	PO Box 87	St Charles	IL	60174	Richard A Nelson	708-377-9100	S	15	<0.1
Design and Production Inc	7110 Rainwater Pl	Lorton	VA	22079	Jay F Barnwell	703-550-8640	S	15	0.1
General Clay Co	PO Box 999	Cape Girardeau	MO	63701	Jack Koutek	314-334-5041	R	15	0.1
KUX Manufacturing Co	12675 Burt Rd	Detroit	MI	48223	Norman Kuhar	313-255-6460	S	15	0.1
Rapid Mounting & Finishing Co	4300 W 47th St	Chicago	IL	60632	Earl Abramson	312-927-5000	R	15	0.3
Allen Displays Inc	6434 Burnt Poplar	Greensboro	NC	27409	Tom Allen	910-668-2791	R	14	0.1
Chicago Show Printing Co	8330 N Austin Av	Morton Grove	IL	60053	Robert Snediker Sr	708-470-5300	R	14	<0.1
Derse Inc	1234 N 62nd St	Milwaukee	WI	53213	William F Haney	414-257-2000	R	14•	0.2
New Dimensions Research Corp	260 Spagnoli Rd	Melville	NY	11747	JE Mason	516-694-1356	R	14	0.2
Structural Display Inc	12-12 33rd Av	Long Island Ct	NY	11106	William S Abbate	718-274-1136	R	14	0.1
Selecto Flash Inc	PO Box 879	West Orange	NJ	07052	James Peepas	201-677-3500	R	13•	0.1
Spanjer Brothers Inc	1160 N Howe St	Chicago	IL	60610	Clayton Spanjer	312-664-2900	R	13	0.2
United Wire Craft Inc	4935 W Le Moyne	Chicago	IL	60651	Sandy Carrigan	312-287-0800	R	13•	0.2
Allen Morrison Inc	PO Box 10245	Lynchburg	VA	24506	BL Reams	804-846-8461	R	12	0.1
Artkraft Strauss Sign Corp	830 12th Av	New York	NY	10019	Tama Starr	212-265-5155	R	12	0.2
E & T Plastic Manufacturing Co	45-33 37th St	Long Island Ct	NY	11101	Rudolf Thal	718-729-6226	R	12	0.1
Giltspur/Chicago	3225 S Western Av	Chicago	IL	60608	Paul Romer	312-376-3000	D	12	<0.1
Harmeson Manufacturing	PO Box 429	Frankfort	IN	46041	JW Kay	317-659-3388	R	12	0.2
Heath and Co	PO Box 22066	Tampa	FL	33622		813-855-4415	S	12	0.2
Jack Stone Company Inc	3131 Pennsy Dr	Landover	MD	20785	John Stone Jr	301-322-3323	R	12•	0.1
Sanders Manufacturing Co	1422 Lebanon Rd	Nashville	TN	37210	James J Sanders Jr	615-254-6611	R	12	0.1
Zumar Industries Inc	PO Box 11305	Tacoma	WA	98411	PJ Lemcke	206-472-4484	R	12•	<0.1
Process Displays Inc	PO Box 287	New Berlin	WI	53151	James Coffey	414-782-3600	R	11	<0.1
Sargent-Sowell Co	1185 108th St	Grand Prairie	TX	75050	Bill Montgomery	214-647-1525	R	11	0.1
Displaymasters Inc	4401 Quebec Av	Minneapolis	MN	55428	Steve D Gregerson	612-533-2271	R	10	<0.1
Niedermaier Inc	2828 N Paulina St	Chicago	IL	60657	J Niedermaier	312-528-8123	R	10	<0.1
Plastolite	6135 District Blv	Maywood	CA	90270	J Ledbetter	213-771-2098	R	10•	0.1
Stout Industries Inc	6425 W Florissant	St Louis	MO	63136	John R Woods	314-385-2280	R	10	0.1
Tusco Display Co	PO Box 175	Gnadenhutten	OH	44629	Michael R Lauber	614-254-4343	R	10	0.1
Vincent Printing Company Inc	PO Box 1057	Hixson	TN	37343	Tom Judge	615-875-0054	R	10	0.1
Visual Marketing Inc	154 W Erie St	Chicago	IL	60610	Marion D Cloud	312-664-9177	R	10	<0.1
Mulholland Harper Co	Rte 404	Denton	MD	21629	Patrick J Hanrahan	410-479-1300	R	10	0.1
Allied Plastics Inc	PO Box 9376	Minneapolis	MN	55440	David B Grover	612-553-7771	R	9	<0.1
Downing Displays Inc	550 Techne Ctr Dr	Milford	OH	45150	Michael J Scherer	513-621-7888	R	9•	<0.1
Field Manufacturing Corp	122 Eucalyptus Dr	El Segundo	CA	90245	Frank P Field	213-772-4161	R	9•	0.1
Milwaukee Sign Co	1964 Wisconsin Av	Grafton	WI	53024	Robert Aiken	414-375-5740	R	9•	0.1

Source: Ward's Business Directory of U.S. Private and Public Companies, Volumes 1 and 2, 1996. The company type code used is as follows: P - Public, R - Private, S - Subsidiary, D - Division, J - Joint Venture, A - Affiliate, G - Group. Sales are in millions of dollars, employees are in thousands. An asterisk (•) indicates an estimated sales volume. The symbol < stands for 'less than'. Company names and addresses are truncated, in some cases, to fit into the available space.

MATERIALS CONSUMED

Material	Quantity	Delivered Cost ($ million)
Materials, ingredients, containers, and supplies	(X)	1,796.1
Veneer and plywood	(X)	61.9
Paper and paperboard products including paperboard boxes, containers, and corrugated paperboard	(X)	154.6
Plastics resins consumed in the form of granules, pellets, powders, liquids, etc.	(X)	41.9
Paints, varnishes, lacquers, stains, shellacs, japans, enamels, and allied products	(X)	34.1
Plastics products consumed in the form of sheets, rods, tubes, film, and other shapes	(X)	178.3
Metal hardware, including hinges, handles, locks, casters, etc. (except castings and forgings)	(X)	33.7
All other fabricated metal products (except castings and forgings)	(X)	89.6
Castings (rough and semifinished)	(X)	2.9
Forgings	(X)	0.5
Steel shapes and forms	(X)	31.6
Nonferrous shapes and forms	(X)	38.6
Specialty transformers and fluorescent ballasts	(X)	51.3
All other materials and components, parts, containers, and supplies	(X)	420.4
Materials, ingredients, containers, and supplies, nsk	(X)	656.8

Source: 1992 *Economic Census*. Explanation of symbols used: (D): Withheld to avoid disclosure of competitive data; na: Not available; (S): Withheld because statistical norms were not met; (X): Not applicable; (Z): Less than half the unit shown; nec: Not elsewhere classified; nsk: Not specified by kind; - : zero; * : 10-19 percent estimated; ** : 20-29 percent estimated.

PRODUCT SHARE DETAILS

Product or Product Class	% Share	Product or Product Class	% Share
Signs and advertising specialties	100.00	Nonelectric, letterpress printed counter, floor display, point-of-purchase, and other display signs	2.74
Electric signs	24.31	Nonelectric, screen printed counter, floor display, point-of-purchase, and other display signs	19.58
Luminous tubing electric signs (neon, argon, hydrogen, etc.)	28.89		
Fluorescent lamp electric signs	34.47	Nonelectric, offset (lithographic) printed counter, floor display, point-of-purchase, and other display signs	4.84
Incandescent bulb, electronic variable message display signs	5.03		
Other incandescent bulb signs	6.00	Nonelectric, unprinted counter, floor display, point-of-purchase, and other display signs	12.06
All other electric signs (including combinations of luminous fluorescent and incandescent)	19.39	All other nonelectric screen printed signs (including sign letters)	3.39
Electric signs, nsk.	6.23	All other nonelectric signs (including sign letters)	25.73
Nonelectric signs	34.04	Nonelectric signs, nsk	5.12
Nonelectric screen printed metal signs	9.39	Advertising specialties	14.75
Other nonelectric metal signs	9.35	Signs and advertising specialties, nsk	26.90
Nonelectric screen printed wood signs	3.31		
Other nonelectric wood signs	4.50		

Source: 1992 *Economic Census*. The values shown are percent of total shipments in an industry. Values of indented subcategories are summed in the main heading. The symbol (D) appears when data are withheld to prevent disclosure of competitive information. The abbreviation nsk stands for 'not specified by kind' and nec for 'not elsewhere classified'.

INPUTS AND OUTPUTS FOR SIGNS & ADVERTISING DISPLAYS

Economic Sector or Industry Providing Inputs	%	Sector	Economic Sector or Industry Buying Outputs	%	Sector
Advertising	11.6	Services	Gross private fixed investment	68.5	Cap Inv
Wholesale trade	9.4	Trade	Highway & street construction	5.5	Constr.
Miscellaneous plastics products	8.8	Manufg.	Eating & drinking places	3.8	Trade
Nonferrous wire drawing & insulating	7.2	Manufg.	Wholesale trade	2.6	Trade
Veneer & plywood	5.6	Manufg.	Signs & advertising displays	2.1	Manufg.
Fabricated structural metal	5.3	Manufg.	Exports	1.9	Foreign
Aluminum rolling & drawing	4.1	Manufg.	Maintenance of highways & streets	1.4	Constr.
Wood products, nec	4.0	Manufg.	Refrigeration & heating equipment	0.9	Manufg.
Paperboard containers & boxes	2.7	Manufg.	Household vacuum cleaners	0.8	Manufg.
Die-cut paper & board	2.5	Manufg.	Motor vehicles & car bodies	0.7	Manufg.
Signs & advertising displays	2.5	Manufg.	Sporting & athletic goods, nec	0.6	Manufg.
Transformers	2.5	Manufg.	Change in business inventories	0.6	In House
Petroleum refining	2.4	Manufg.	Switchgear & switchboard apparatus	0.5	Manufg.
Hardware, nec	2.2	Manufg.	Social services, nec	0.5	Services
Eating & drinking places	2.2	Trade	Electric housewares & fans	0.4	Manufg.
Paints & allied products	1.6	Manufg.	Electronic computing equipment	0.4	Manufg.
Electric services (utilities)	1.6	Util.	Household laundry equipment	0.4	Manufg.
Real estate	1.6	Fin/R.E.	Retail trade, except eating & drinking	0.4	Trade
Motor freight transportation & warehousing	1.3	Util.	Banking	0.4	Fin/R.E.
Sawmills & planning mills, general	1.2	Manufg.	Hotels & lodging places	0.4	Services
Communications, except radio & TV	1.1	Util.	Job training & related services	0.4	Services
Banking	1.1	Fin/R.E.	S/L Govt. purch., natural resource & recreation.	0.4	S/L Govt
Maintenance of nonfarm buildings nec	1.0	Constr.	Nonbuilding facilities nec	0.3	Constr.
Fabricated metal products, nec	0.8	Manufg.	Calculating & accounting machines	0.3	Manufg.
Machinery, except electrical, nec	0.8	Manufg.	Environmental controls	0.3	Manufg.
Plastics materials & resins	0.8	Manufg.	Doctors & dentists	0.3	Services
Management & consulting services & labs	0.8	Services	Labor, civic, social, & fraternal associations	0.3	Services
Legal services	0.7	Services	Federal Government purchases, national defense	0.3	Fed Govt
U.S. Postal Service	0.7	Gov't	Federal Government purchases, nondefense	0.3	Fed Govt

Continued on next page.

INPUTS AND OUTPUTS FOR SIGNS & ADVERTISING DISPLAYS - Continued

Economic Sector or Industry Providing Inputs	%	Sector	Economic Sector or Industry Buying Outputs	%	Sector
Abrasive products	0.6	Manufg.	Construction of stores & restaurants	0.2	Constr.
Industrial controls	0.6	Manufg.	Costume jewelry	0.2	Manufg.
Special dies & tools & machine tool accessories	0.5	Manufg.	Hardware, nec	0.2	Manufg.
Wiring devices	0.5	Manufg.	Heating equipment, except electric	0.2	Manufg.
Gas production & distribution (utilities)	0.5	Util.	Household appliances, nec	0.2	Manufg.
Railroads & related services	0.5	Util.	Radio & TV receiving sets	0.2	Manufg.
Accounting, auditing & bookkeeping	0.5	Services	Typewriters & office machines, nec	0.2	Manufg.
Imports	0.5	Foreign	Automatic merchandising machines	0.1	Manufg.
Commercial printing	0.4	Manufg.	Construction machinery & equipment	0.1	Manufg.
Glass & glass products, except containers	0.4	Manufg.	Cutlery	0.1	Manufg.
Screw machine and related products	0.4	Manufg.	Fabricated plate work (boiler shops)	0.1	Manufg.
Air transportation	0.4	Util.	Household cooking equipment	0.1	Manufg.
Royalties	0.4	Fin/R.E.	Industrial trucks & tractors	0.1	Manufg.
Automotive rental & leasing, without drivers	0.4	Services	Mining machinery, except oil field	0.1	Manufg.
Electric lamps	0.3	Manufg.	Mobile homes	0.1	Manufg.
Manifold business forms	0.3	Manufg.	Motor vehicle parts & accessories	0.1	Manufg.
Automotive repair shops & services	0.3	Services	Small arms	0.1	Manufg.
Business/professional associations	0.3	Services	Travel trailers & campers	0.1	Manufg.
Equipment rental & leasing services	0.3	Services	Automotive repair shops & services	0.1	Services
Lubricating oils & greases	0.2	Manufg.			
Insurance carriers	0.2	Fin/R.E.			
Computer & data processing services	0.2	Services			
Personnel supply services	0.2	Services			
Apparel made from purchased materials	0.1	Manufg.			
Photographic equipment & supplies	0.1	Manufg.			
Switchgear & switchboard apparatus	0.1	Manufg.			
Transit & bus transportation	0.1	Util.			
Water transportation	0.1	Util.			
Retail trade, except eating & drinking	0.1	Trade			
Business services nec	0.1	Services			
Hotels & lodging places	0.1	Services			

Source: *Benchmark Input-Output Accounts for the U.S. Economy, 1982*, U.S. Department of Commerce, Washington, D.C., July 1991. Data, as reported in the source, are organized by the 1977 SIC structure in use in 1982 but have been matched, as closely as is possible, to the 1987 SIC structure used in this book.

OCCUPATIONS EMPLOYED BY SIC 399 - MANUFACTURED PRODUCTS, NEC

Occupation	% of Total 1994	Change to 2005	Occupation	% of Total 1994	Change to 2005
Assemblers, fabricators, & hand workers nec	17.0	4.8	Freight, stock, & material movers, hand	1.5	-16.1
Sales & related workers nec	4.3	4.8	Inspectors, testers, & graders, precision	1.5	4.9
Blue collar worker supervisors	4.0	-2.4	General office clerks	1.5	-10.6
General managers & top executives	3.8	-0.5	Designers, ex interior designers	1.5	15.3
Hand packers & packagers	3.7	-28.1	Electrical & electronic assemblers	1.4	4.8
Traffic, shipping, & receiving clerks	2.5	0.9	Industrial production managers	1.4	4.8
Machine operators nec	2.2	-7.6	Packaging & filling machine operators	1.4	4.9
Helpers, laborers, & material movers nec	2.2	4.8	Screen printing machine setters & set-up operators	1.2	4.8
Painting, coating, & decorating workers, hand	2.0	-5.7	Plastic molding machine workers	1.2	4.8
Precision workers nec	2.0	41.5	Machine feeders & offbearers	1.2	-5.7
Secretaries, ex legal & medical	2.0	-4.5	Cabinetmakers & bench carpenters	1.1	-31.9
Bookkeeping, accounting, & auditing clerks	1.8	-21.4	Order clerks, materials, merchandise, & service	1.1	2.6
Sheet metal workers & duct installers	1.8	4.8	Grinders & polishers, hand	1.1	4.8
Precision metal workers nec	1.8	36.3			

Source: *Industry-Occupation Matrix*, Bureau of Labor Statistics. These data relate to one or more 3-digit SIC industry groups rather than to a single 4-digit SIC. The change reported for each occupation to the year 2005 is a percent of growth or decline as estimated by the Bureau of Labor Statistics. The abbreviation nec stands for 'not elsewhere classified'.

LOCATION BY STATE AND REGIONAL CONCENTRATION

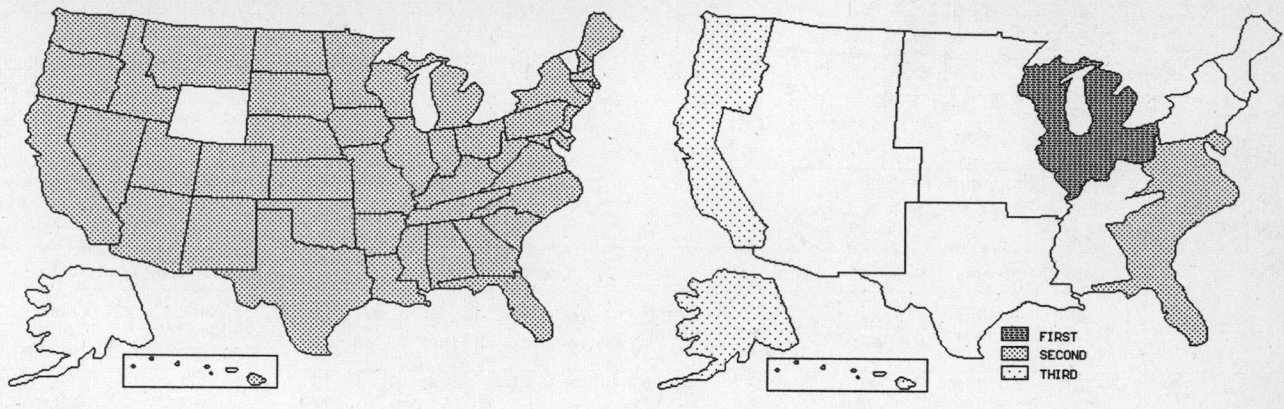

FIRST
SECOND
THIRD

INDUSTRY DATA BY STATE

State	Establish-ments	Shipments			Employment				Cost as % of Shipments	Investment per Employee ($)
		Total ($ mil)	% of U.S.	Per Establ.	Total Number	% of U.S.	Per Establ.	Wages ($/hour)		
Illinois	224	580.0	10.7	2.6	5,400	8.0	24	10.80	43.8	1,852
California	539	508.8	9.4	0.9	6,500	9.6	12	10.72	39.0	1,062
New York	341	502.5	9.3	1.5	6,500	9.6	19	9.47	39.5	1,108
Ohio	211	355.9	6.6	1.7	4,000	5.9	19	9.35	46.4	1,150
Wisconsin	115	349.5	6.4	3.0	3,400	5.0	30	9.41	46.8	2,235
New Jersey	166	279.9	5.2	1.7	3,300	4.9	20	9.95	45.4	1,242
Texas	346	257.2	4.7	0.7	3,600	5.3	10	9.38	39.0	1,000
Pennsylvania	185	251.9	4.6	1.4	2,800	4.1	15	9.94	39.8	2,571
Michigan	163	247.8	4.6	1.5	2,900	4.3	18	11.68	43.8	1,103
Florida	292	199.7	3.7	0.7	3,000	4.4	10	8.51	41.2	1,000
Missouri	88	161.0	3.0	1.8	2,900	4.3	33	9.45	41.6	1,517
Tennessee	112	148.6	2.7	1.3	2,100	3.1	19	8.70	44.3	1,952
Minnesota	103	139.9	2.6	1.4	1,900	2.8	18	9.20	36.0	1,316
Iowa	46	100.6	1.9	2.2	800	1.2	17	10.11	48.5	2,125
Washington	105	83.0	1.5	0.8	1,300	1.9	12	11.23	44.5	-
North Carolina	119	82.8	1.5	0.7	1,300	1.9	11	8.56	39.4	1,308
Alabama	54	78.3	1.4	1.5	800	1.2	15	8.25	50.7	4,125
Rhode Island	42	75.3	1.4	1.8	1,100	1.6	26	7.57	45.8	818
Maryland	87	72.4	1.3	0.8	900	1.3	10	11.00	44.6	1,556
Georgia	101	64.7	1.2	0.6	900	1.3	9	10.20	40.6	1,111
Massachusetts	89	62.5	1.2	0.7	800	1.2	9	9.75	38.4	750
Oregon	63	54.3	1.0	0.9	700	1.0	11	9.60	41.6	2,000
South Dakota	12	50.5	0.9	4.2	700	1.0	58	6.80	47.1	2,143
Connecticut	66	46.2	0.9	0.7	500	0.7	8	10.00	42.0	800
Arizona	79	44.1	0.8	0.6	600	0.9	8	10.25	37.9	1,333
Colorado	78	41.9	0.8	0.5	600	0.9	8	9.00	34.8	1,667
Kansas	42	39.7	0.7	0.9	700	1.0	17	8.33	35.5	1,143
Utah	42	37.6	0.7	0.9	500	0.7	12	8.88	33.0	1,400
Arkansas	38	26.0	0.5	0.7	300	0.4	8	9.50	41.9	-
Idaho	29	21.4	0.4	0.7	300	0.4	10	9.00	36.0	-
Oklahoma	51	20.1	0.4	0.4	400	0.6	8	7.20	40.3	1,000
South Carolina	42	18.3	0.3	0.4	300	0.4	7	9.25	41.5	667
Louisiana	32	17.3	0.3	0.5	200	0.3	6	10.33	41.6	1,500
West Virginia	16	17.3	0.3	1.1	200	0.3	13	10.00	34.7	-
Mississippi	22	15.9	0.3	0.7	200	0.3	9	9.50	59.7	1,500
Hawaii	26	11.1	0.2	0.4	100	0.1	4	8.50	40.5	-
Indiana	111	(D)	-	-	1,750 *	2.6	16	-	-	-
Virginia	77	(D)	-	-	1,750 *	2.6	23	-	-	743
Kentucky	37	(D)	-	-	750 *	1.1	20	-	-	-
Nebraska	31	(D)	-	-	175 *	0.3	6	-	-	2,857
Nevada	31	(D)	-	-	750 *	1.1	24	-	-	933
New Hampshire	20	(D)	-	-	175 *	0.3	9	-	-	571
Montana	19	(D)	-	-	175 *	0.3	9	-	-	571
New Mexico	17	(D)	-	-	175 *	0.3	10	-	-	-
Maine	15	(D)	-	-	175 *	0.3	12	-	-	-
North Dakota	13	(D)	-	-	175 *	0.3	13	-	-	-
D.C.	6	(D)	-	-	175 *	0.3	29	-	-	-

Source: 1992 *Economic Census*. The states are in descending order of shipments or establishments (if shipment data are missing for the majority). The symbol (D) appears when data are withheld to prevent disclosure of competitive information. States marked with (D) are sorted by number of establishments. A dash (-) indicates that the data element cannot be calculated; * indicates the midpoint of a range.

3995 - BURIAL CASKETS

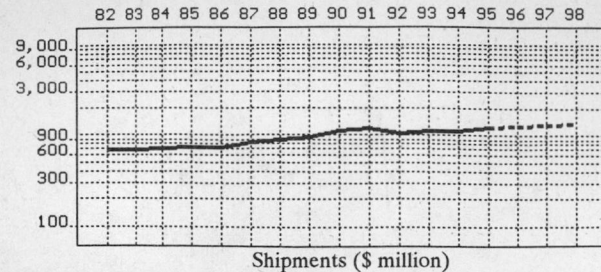

Shipments ($ million)

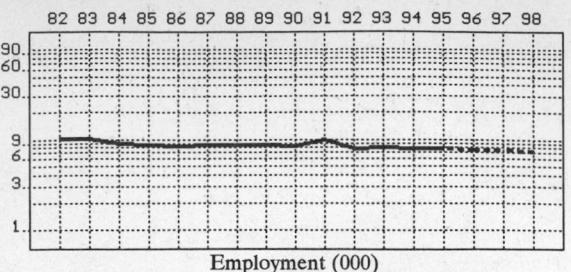

Employment (000)

GENERAL STATISTICS

| Year | Companies | Establishments | | Employment | | | Compensation | | Production ($ million) | | | |
		Total	with 20 or more employees	Total (000)	Production Workers (000)	Hours (Mil)	Payroll ($ mil)	Wages ($/hr)	Cost of Materials	Value Added by Manufacture	Value of Shipments	Capital Invest.
1982	269	301	110	10.3	8.1	15.3	145.1	6.41	309.7	374.6	682.1	36.7
1983		285	104	10.3	8.3	17.0	151.2	6.24	275.1	408.3	689.7	14.3
1984		269	98	9.0	6.9	13.6	147.4	7.16	332.6	373.5	704.9	27.2
1985		254	91	8.7	6.9	13.2	151.0	7.87	329.3	397.0	727.3	28.5
1986		234	89	8.4	6.6	13.2	155.0	7.95	338.5	403.9	738.3	29.7
1987	213	231	80	8.7	6.9	13.7	163.2	8.21	361.8	472.1	839.9	28.8
1988		215	86	8.6	6.6	13.3	169.9	8.76	390.9	497.5	884.6	31.1
1989		210	85	8.8	6.6	13.3	170.3	8.63	446.5	502.5	946.3	23.5
1990		205	75	8.5	8.0	17.8	228.6	8.90	539.6	579.8	1,093.5	26.9
1991		209	77	9.9	7.7	16.2	237.8	9.27	554.3	644.2	1,193.7	29.1
1992	195	211	67	7.8	6.0	13.4	196.2	9.78	390.6	660.6	1,053.3	28.5
1993		197	70	8.0	6.3	13.6	202.0	9.98	431.9	693.5	1,118.6	27.2
1994		177P	63P	7.9	6.2	13.4	199.6	10.44	411.7	722.5	1,128.0	18.7
1995		168P	60P	7.8P	6.2P	13.9P	224.1P	10.73P	449.6P	789.1P	1,231.9P	25.1P
1996		159P	56P	7.7P	6.1P	13.9P	230.6P	11.06P	466.5P	818.7P	1,278.2P	24.9P
1997		150P	53P	7.5P	6.0P	13.8P	237.1P	11.39P	483.4P	848.4P	1,324.6P	24.6P
1998		141P	49P	7.4P	5.9P	13.8P	243.7P	11.72P	500.3P	878.1P	1,370.9P	24.3P

Sources: 1982, 1987, 1992 *Economic Census*; *Annual Survey of Manufactures*, 83-86, 88-91, 93-94. Establishment counts for non-Census years are from *County Business Patterns*; establishment values for 83-84 are extrapolations. 'P's show projections by the editors. Industries reclassified in 87 will not have data for prior years.

INDICES OF CHANGE

| Year | Companies | Establishments | | Employment | | | Compensation | | Production ($ million) | | | |
		Total	with 20 or more employees	Total (000)	Production Workers (000)	Hours (Mil)	Payroll ($ mil)	Wages ($/hr)	Cost of Materials	Value Added by Manufacture	Value of Shipments	Capital Invest.
1982	138	143	164	132	135	114	74	66	79	57	65	129
1983		135	155	132	138	127	77	64	70	62	65	50
1984		127	146	115	115	101	75	73	85	57	67	95
1985		120	136	112	115	99	77	80	84	60	69	100
1986		111	133	108	110	99	79	81	87	61	70	104
1987	109	109	119	112	115	102	83	84	93	71	80	101
1988		102	128	110	110	99	87	90	100	75	84	109
1989		100	127	113	110	99	87	88	114	76	90	82
1990		97	112	109	133	133	117	91	138	88	104	94
1991		99	115	127	128	121	121	95	142	98	113	102
1992	100	100	100	100	100	100	100	100	100	100	100	100
1993		93	104	103	105	101	103	102	111	105	106	95
1994		84P	94P	101	103	100	102	107	105	109	107	66
1995		80P	89P	100P	104P	104P	114P	110P	115P	119P	117P	88P
1996		75P	84P	98P	102P	104P	118P	113P	119P	124P	121P	87P
1997		71P	78P	96P	100P	103P	121P	116P	124P	128P	126P	86P
1998		67P	73P	94P	98P	103P	124P	120P	128P	133P	130P	85P

Sources: Same as General Statistics. Values reflect change from the base year, 1992. Values above 100 mean greater than 92, values below 100 mean less than 92, and a value of 100 in the 82-91 or 93-98 period means same as 92. 'P's mark projections by the editors.

SELECTED RATIOS

For 1994	Avg. of All Manufact.	Analyzed Industry	Index	For 1994	Avg. of All Manufact.	Analyzed Industry	Index
Employees per Establishment	49	45	91	Value Added per Production Worker	134,084	116,532	87
Payroll per Establishment	1,500,273	1,128,746	75	Cost per Establishment	5,045,178	2,328,181	46
Payroll per Employee	30,620	25,266	83	Cost per Employee	102,970	52,114	51
Production Workers per Establishment	34	35	102	Cost per Production Worker	146,988	66,403	45
Wages per Establishment	853,319	791,118	93	Shipments per Establishment	9,576,895	6,378,888	67
Wages per Production Worker	24,861	22,564	91	Shipments per Employee	195,460	142,785	73
Hours per Production Worker	2,056	2,161	105	Shipments per Production Worker	279,017	181,935	65
Wages per Hour	12.09	10.44	86	Investment per Establishment	321,011	105,749	33
Value Added per Establishment	4,602,255	4,085,768	89	Investment per Employee	6,552	2,367	36
Value Added per Employee	93,930	91,456	97	Investment per Production Worker	9,352	3,016	32

Sources: Same as General Statistics. The 'Average of All Manufacturing' column represents the average of all manufacturing industries reported for the most recent complete year available. The Index shows the relationship between the Average and the Analyzed Industry. For example, 100 means that they are equal; 500 that the Analyzed Industry is five times the average; 50 means that the Analyzed Industry is half the national average. The abbreviation 'na' is used to show that data are 'not available'.

LEADING COMPANIES   Number shown: 17   Total sales ($ mil): 673   Total employment (000): 6.0

Company Name	Address				CEO Name	Phone	Co. Type	Sales ($ mil)	Empl. (000)
Batesville Casket Company Inc	1 Batesville Blv	Batesville	IN	47006	David J Hirt	812-934-7500	S	426	3.2
York Group Inc	9430 Old Katy Rd	Houston	TX	77055	Eldon Nuss	713-984-5500	R	115	1.1
Casket Shells Inc	1st St	Eynon	PA	18403	Joseph R Semon	717-876-2642	R	28*	0.3
York Metal Casket Co	197 George St	Marshfield	MO	65706	Dan Mills	417-468-6500	S	17*	0.2
Astral Industries Inc	PO Box 638	Lynn	IN	47355	Charles B Shaw	317-874-2525	R	16*	0.2
Clark Grave Vault Co	375 E 5th Av	Columbus	OH	43201	David A Beck	614-294-3761	R	15	0.1
Toccoa Casket Co	726 W Currahee St	Toccoa	GA	30577	James H Peak	706-886-3153	R	15	0.3
Clarksburg Casket Co	PO Box 66	Clarksburg	WV	26301	Marcus E Garrett	304-624-6471	R	13	0.1
Marsellus Casket Company Inc	PO Box 4968	Syracuse	NY	13204	Larry English	315-422-2306	R	8*	0.3
West Point Casket Co	PO Box 232	West Point	MS	39773	H Joe Trulove	601-494-4151	R	6	<0.1
Northwestern Casket Co	1707 Jefferson NE	Minneapolis	MN	55413	William Shields	612-789-4356	R	4	<0.1
Loretto Casket Company Inc	PO Box 66	Loretto	TN	38469	Ken Abercrombie	615-853-6921	R	4*	<0.1
Balanced Line Casket Co	PO Box 268	Cambridge City	IN	47327	James Peacock	317-478-3501	R	2*	<0.1
Franklin Casket Company Inc	PO Box 8	Brookville	IN	47012	George Musekamp	317-647-4124	R	2	<0.1
Sound Casket Co	PO Box 1023	Everett	WA	98206	Gordon Ropchan	206-259-6012	R	1*	<0.1
Hoegh Industries Inc	PO Box 311	Gladstone	MI	49837	Dennis J Hoegh	906-428-2151	R	1	<0.1
California Shell Company Inc	2458 Rosemead	S El Monte	CA	91733	Robin Yuan	818-444-1581	R	0*	<0.1

Source:. *Ward's Business Directory of U.S. Private and Public Companies*, Volumes 1 and 2, 1996. The company type code used is as follows: P - Public, R - Private, S - Subsidiary, D - Division, J - Joint Venture, A - Affiliate, G - Group. Sales are in millions of dollars, employees are in thousands. An asterisk (*) indicates an estimated sales volume. The symbol < stands for 'less than'. Company names and addresses are truncated, in some cases, to fit into the available space.

MATERIALS CONSUMED

Material	Quantity	Delivered Cost ($ million)
Materials, ingredients, containers, and supplies	(X)	357.1
Cotton, wool, manmade fiber fabrics, etc.	(X)	55.1
Rough and dressed lumber	(X)	40.0
Metal casket and casket shell hardware (except castings and forgings)	(X)	59.8
Other fabricated metal products (except castings and forgings)	(X)	11.8
Steel shapes and forms	(X)	69.0
Nonferrous shapes and forms	(X)	19.0
Paints, varnishes, lacquers, stains, shellacs, japans, enamels, and allied products	(X)	21.1
All other materials and components, parts, containers, and supplies	(X)	37.0
Materials, ingredients, containers, and supplies, nsk	(X)	44.4

Source: 1992 *Economic Census*. Explanation of symbols used: (D): Withheld to avoid disclosure of competitive data; na: Not available; (S): Withheld because statistical norms were not met; (X): Not applicable; (Z): Less than half the unit shown; nec: Not elsewhere classified; nsk: Not specified by kind; - : zero; * : 10-19 percent estimated; ** : 20-29 percent estimated.

PRODUCT SHARE DETAILS

Product or Product Class	% Share	Product or Product Class	% Share
Burial caskets	100.00	adult sizes only, nsk	0.05
Metal burial caskets and coffins completely lined and trimmed, adult sizes only	64.51	Other burial caskets and coffins and metal vaults	11.36
Steel burial caskets and coffins (excluding stainless steel)	71.83	Burial caskets and coffins other than metal or wood, completely lined and trimmed (including plastics, fiberglass, foam, etc.)	12.50
Other metal burial caskets and coffins (stainless steel, bronze, copper, etc.)	28.01	Metal burial casket shells, knocked-down and set-up, unlined or untrimmed	37.33
Metal caskets and coffins completely lined and trimmed, adult sizes only, nsk	0.15	Other burial casket shells, knocked-down and set-up, unlined or untrimmed (including wood, plastics, fiberglass, foam, masonite, cardboard, fiberboard, etc.)	9.83
Wood burial caskets and coffins, completely lined and trimmed, adult sizes only	21.33	Burial vaults (except concrete), burial boxes, and casket-shipping containers and cases	20.00
Cloth covered softwood burial caskets and coffins, completely lined and trimmed, adult sizes only	7.35	Other caskets and coffins and metal vaults, nsk	20.34
Hardwood burial caskets and coffins, completely lined and trimmed, adult sizes only	92.61	Burial caskets, nsk	2.80
Wood caskets and coffins, completely lined and trimmed,			

Source: 1992 *Economic Census*. The values shown are percent of total shipments in an industry. Values of indented subcategories are summed in the main heading. The symbol (D) appears when data are withheld to prevent disclosure of competitive information. The abbreviation nsk stands for 'not specified by kind' and nec for 'not elsewhere classified'.

INPUTS AND OUTPUTS FOR BURIAL CASKETS & VAULTS

Economic Sector or Industry Providing Inputs	%	Sector	Economic Sector or Industry Buying Outputs	%	Sector
Advertising	16.8	Services	Funeral service & crematories	99.1	Services
Blast furnaces & steel mills	16.3	Manufg.	Burial caskets & vaults	0.4	Manufg.
Wholesale trade	15.3	Trade	Change in business inventories	0.4	In House
Hardware, nec	12.0	Manufg.	Federal Government purchases, national defense	0.1	Fed Govt
Broadwoven fabric mills	10.3	Manufg.			
Sawmills & planning mills, general	4.8	Manufg.			
Paints & allied products	2.7	Manufg.			
Eating & drinking places	1.9	Trade			
U.S. Postal Service	1.7	Gov't			
Electric services (utilities)	1.4	Util.			
Motor freight transportation & warehousing	1.1	Util.			
Padding & upholstery filling	1.0	Manufg.			
Banking	1.0	Fin/R.E.			
Petroleum refining	0.9	Manufg.			
Gas production & distribution (utilities)	0.9	Util.			
Maintenance of nonfarm buildings nec	0.8	Constr.			
Machinery, except electrical, nec	0.7	Manufg.			
Real estate	0.7	Fin/R.E.			
Management & consulting services & labs	0.7	Services			
Burial caskets & vaults	0.6	Manufg.			
Legal services	0.6	Services			
Abrasive products	0.5	Manufg.			
Communications, except radio & TV	0.5	Util.			
Railroads & related services	0.5	Util.			
Sanitary services, steam supply, irrigation	0.5	Util.			
Royalties	0.5	Fin/R.E.			
Special dies & tools & machine tool accessories	0.4	Manufg.			
Air transportation	0.4	Util.			
Accounting, auditing & bookkeeping	0.4	Services			
Gaskets, packing & sealing devices	0.3	Manufg.			
Security & commodity brokers	0.3	Fin/R.E.			
Lubricating oils & greases	0.2	Manufg.			
Manifold business forms	0.2	Manufg.			
Automotive rental & leasing, without drivers	0.2	Services			
Automotive repair shops & services	0.2	Services			
Computer & data processing services	0.2	Services			
Equipment rental & leasing services	0.2	Services			
Hotels & lodging places	0.2	Services			
Particleboard	0.1	Manufg.			
Photographic equipment & supplies	0.1	Manufg.			
Insurance carriers	0.1	Fin/R.E.			
Business services nec	0.1	Services			
Business/professional associations	0.1	Services			
Laundry, dry cleaning, shoe repair	0.1	Services			
Personnel supply services	0.1	Services			

Source: Benchmark Input-Output Accounts for the U.S. Economy, 1982, U.S. Department of Commerce, Washington, D.C., July 1991. Data, as reported in the source, are organized by the 1977 SIC structure in use in 1982 but have been matched, as closely as is possible, to the 1987 SIC structure used in this book.

OCCUPATIONS EMPLOYED BY SIC 399 - MANUFACTURED PRODUCTS, NEC

Occupation	% of Total 1994	Change to 2005	Occupation	% of Total 1994	Change to 2005
Assemblers, fabricators, & hand workers nec	17.0	4.8	Freight, stock, & material movers, hand	1.5	-16.1
Sales & related workers nec	4.3	4.8	Inspectors, testers, & graders, precision	1.5	4.9
Blue collar worker supervisors	4.0	-2.4	General office clerks	1.5	-10.6
General managers & top executives	3.8	-0.5	Designers, ex interior designers	1.5	15.3
Hand packers & packagers	3.7	-28.1	Electrical & electronic assemblers	1.4	4.8
Traffic, shipping, & receiving clerks	2.5	0.9	Industrial production managers	1.4	4.8
Machine operators nec	2.2	-7.6	Packaging & filling machine operators	1.4	4.9
Helpers, laborers, & material movers nec	2.2	4.8	Screen printing machine setters & set-up operators	1.2	4.8
Painting, coating, & decorating workers, hand	2.0	-5.7	Plastic molding machine workers	1.2	4.8
Precision workers nec	2.0	41.5	Machine feeders & offbearers	1.2	-5.7
Secretaries, ex legal & medical	2.0	-4.5	Cabinetmakers & bench carpenters	1.1	-31.9
Bookkeeping, accounting, & auditing clerks	1.8	-21.4	Order clerks, materials, merchandise, & service	1.1	2.6
Sheet metal workers & duct installers	1.8	4.8	Grinders & polishers, hand	1.1	4.8
Precision metal workers nec	1.8	36.3			

Source: Industry-Occupation Matrix, Bureau of Labor Statistics. These data relate to one or more 3-digit SIC industry groups rather than to a single 4-digit SIC. The change reported for each occupation to the year 2005 is a percent of growth or decline as estimated by the Bureau of Labor Statistics. The abbreviation nec stands for 'not elsewhere classified'.

LOCATION BY STATE AND REGIONAL CONCENTRATION

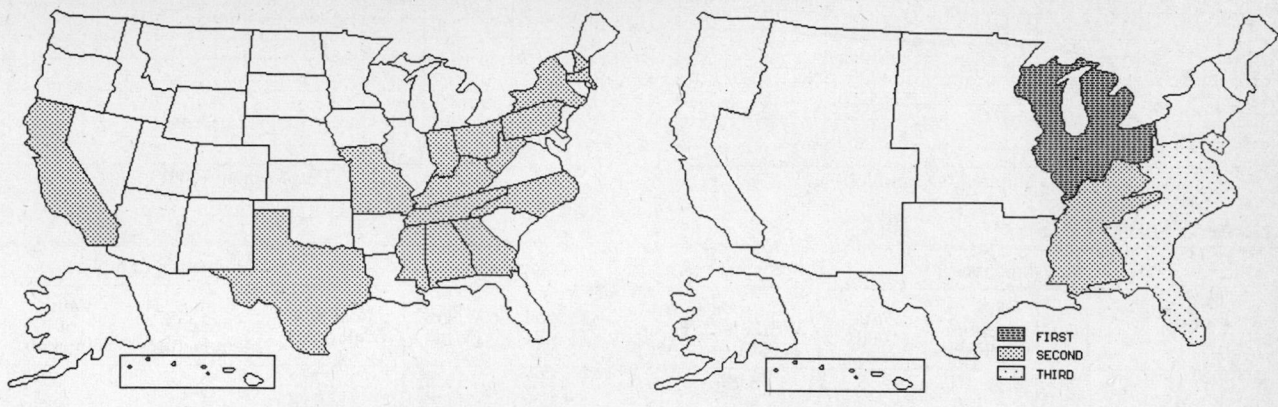

FIRST
SECOND
THIRD

INDUSTRY DATA BY STATE

State	Establish-ments	Shipments			Employment				Cost as % of Shipments	Investment per Employee ($)
		Total ($ mil)	% of U.S.	Per Establ.	Total Number	% of U.S.	Per Establ.	Wages ($/hour)		
Indiana	22	331.9	31.5	15.1	2,300	29.5	105	11.14	33.6	5,435
Pennsylvania	12	83.6	7.9	7.0	1,000	12.8	83	9.35	46.5	1,000
Georgia	10	18.8	1.8	1.9	300	3.8	30	6.60	45.2	667
Massachusetts	5	12.3	1.2	2.5	200	2.6	40	8.67	41.5	500
California	11	(D)	-	-	175 *	2.2	16	-	-	-
North Carolina	10	(D)	-	-	175 *	2.2	18	-	-	-
Alabama	9	(D)	-	-	175 *	2.2	19	-	-	-
Missouri	9	(D)	-	-	175 *	2.2	19	-	-	-
New York	9	(D)	-	-	750 *	9.6	83	-	-	1,333
Texas	8	(D)	-	-	175 *	2.2	22	-	-	-
Ohio	7	(D)	-	-	175 *	2.2	25	-	-	1,143
Kentucky	6	(D)	-	-	375 *	4.8	63	-	-	-
Mississippi	6	(D)	-	-	375 *	4.8	63	-	-	-
Tennessee	6	(D)	-	-	750 *	9.6	125	-	-	-
New Hampshire	3	(D)	-	-	375 *	4.8	125	-	-	-
West Virginia	3	(D)	-	-	175 *	2.2	58	-	-	-

Source: 1992 *Economic Census*. The states are in descending order of shipments or establishments (if shipment data are missing for the majority). The symbol (D) appears when data are withheld to prevent disclosure of competitive information. States marked with (D) are sorted by number of establishments. A dash (-) indicates that the data element cannot be calculated; * indicates the midpoint of a range.

3996 - HARD SURFACE FLOOR COVERINGS

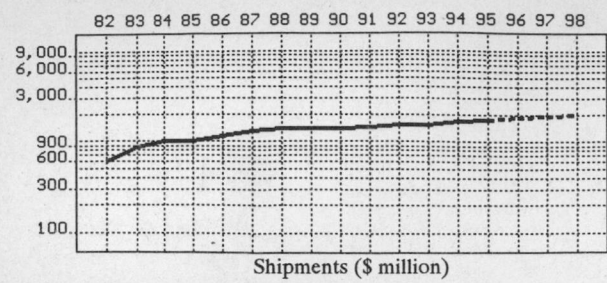

Shipments ($ million)

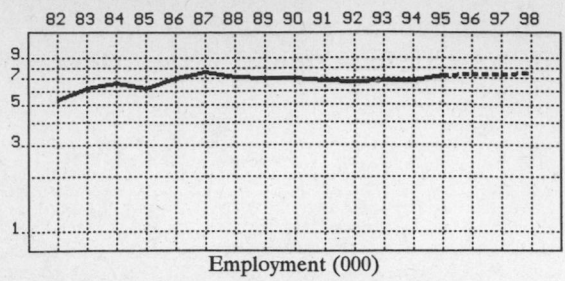

Employment (000)

GENERAL STATISTICS

Year	Com-panies	Establishments		Employment			Compensation		Production ($ million)			
		Total	with 20 or more employees	Total (000)	Production Workers (000)	Hours (Mil)	Payroll ($ mil)	Wages ($/hr)	Cost of Materials	Value Added by Manufacture	Value of Shipments	Capital Invest.
1982	12	17	10	5.3	3.9	7.3	112.1	10.73	270.3	323.4	604.1	18.2
1983		17	10	6.2	4.8	9.3	145.7	11.22	403.3	475.0	864.7	26.6
1984		17	10	6.6	5.2	10.1	165.9	12.14	440.3	579.6	1,010.2	54.3
1985		17	10	6.2	4.8	9.4	165.8	12.96	436.0	573.7	1,025.7	37.7
1986		17	11	7.0	5.5	11.3	197.1	12.86	461.9	693.4	1,155.4	52.4
1987	14	21	17	7.6	6.0	12.6	224.6	13.10	560.5	740.1	1,309.7	51.2
1988		21	17	7.2	5.6	11.9	222.6	13.50	590.8	795.1	1,380.0	68.1
1989		19	15	7.0	5.6	12.0	222.3	13.90	609.4	798.0	1,399.1	79.4
1990		19	15	7.0	5.5	11.2	225.6	14.78	594.3	793.2	1,377.3	62.9
1991		19	15	6.8	5.3	10.9	220.5	14.86	627.7	796.9	1,419.0	66.3
1992	21	28	18	6.7	5.2	11.2	240.2	16.02	616.5	902.7	1,521.1	65.7
1993		30	18	6.8	5.3	11.3	246.5	16.22	624.5	931.0	1,563.0	64.4
1994		26P	19P	6.8	5.4	11.3	251.6	16.68	664.2	1,011.6	1,674.9	46.7
1995		27P	20P	7.2P	5.7P	12.4P	274.5P	17.10P	705.9P	1,075.1P	1,780.1P	74.2P
1996		28P	21P	7.3P	5.8P	12.6P	284.7P	17.57P	735.7P	1,120.5P	1,855.2P	77.1P
1997		29P	22P	7.4P	5.9P	12.8P	294.9P	18.05P	765.5P	1,165.9P	1,930.3P	80.1P
1998		30P	22P	7.5P	5.9P	13.1P	305.1P	18.52P	795.3P	1,211.2P	2,005.4P	83.1P

Sources: 1982, 1987, 1992 *Economic Census*; *Annual Survey of Manufactures*, 83-86, 88-91, 93-94. Establishment counts for non-Census years are from *County Business Patterns*; establishment values for 83-84 are extrapolations. 'P's show projections by the editors. Industries reclassified in 87 will not have data for prior years.

INDICES OF CHANGE

Year	Com-panies	Establishments		Employment			Compensation		Production ($ million)			
		Total	with 20 or more employees	Total (000)	Production Workers (000)	Hours (Mil)	Payroll ($ mil)	Wages ($/hr)	Cost of Materials	Value Added by Manufacture	Value of Shipments	Capital Invest.
1982	57	61	56	79	75	65	47	67	44	36	40	28
1983		61	56	93	92	83	61	70	65	53	57	40
1984		61	56	99	100	90	69	76	71	64	66	83
1985		61	56	93	92	84	69	81	71	64	67	57
1986		61	61	104	106	101	82	80	75	77	76	80
1987	67	75	94	113	115	113	94	82	91	82	86	78
1988		75	94	107	108	106	93	84	96	88	91	104
1989		68	83	104	108	107	93	87	99	88	92	121
1990		68	83	104	106	100	94	92	96	88	91	96
1991		68	83	101	102	97	92	93	102	88	93	101
1992	100	100	100	100	100	100	100	100	100	100	100	100
1993		107	100	101	102	101	103	101	101	103	103	98
1994		94P	106P	101	104	101	105	104	108	112	110	71
1995		97P	111P	108P	110P	110P	114P	107P	115P	119P	117P	113P
1996		101P	115P	109P	111P	113P	119P	110P	119P	124P	122P	117P
1997		104P	120P	110P	113P	115P	123P	113P	124P	129P	127P	122P
1998		108P	124P	111P	114P	117P	127P	116P	129P	134P	132P	126P

Sources: Same as General Statistics. Values reflect change from the base year, 1992. Values above 100 mean greater than 92, values below 100 mean less than 92, and a value of 100 in the 82-91 or 93-98 period means same as 92. 'P's mark projections by the editors.

SELECTED RATIOS

For 1994	Avg. of All Manufact.	Analyzed Industry	Index	For 1994	Avg. of All Manufact.	Analyzed Industry	Index
Employees per Establishment	49	258	527	Value Added per Production Worker	134,084	187,333	140
Payroll per Establishment	1,500,273	9,548,936	636	Cost per Establishment	5,045,178	25,208,281	500
Payroll per Employee	30,620	37,000	121	Cost per Employee	102,970	97,676	95
Production Workers per Establishment	34	205	597	Cost per Production Worker	146,988	123,000	84
Wages per Establishment	853,319	7,153,504	838	Shipments per Establishment	9,576,895	63,567,223	664
Wages per Production Worker	24,861	34,904	140	Shipments per Employee	195,460	246,309	126
Hours per Production Worker	2,056	2,093	102	Shipments per Production Worker	279,017	310,167	111
Wages per Hour	12.09	16.68	138	Investment per Establishment	321,011	1,772,398	552
Value Added per Establishment	4,602,255	38,393,099	834	Investment per Employee	6,552	6,868	105
Value Added per Employee	93,930	148,765	158	Investment per Production Worker	9,352	8,648	92

Sources: Same as General Statistics. The 'Average of All Manufacturing' column supresents the average of all manufacturing industries reported for the most recent complete year available. The Index shows the relationship between the Average and the Analyzed Industry. For example, 100 means that they are equal; 500 that the Analyzed Industry is five times the average; 50 means that the Analyzed Industry is half the national average. The abbreviation 'na' is used to show that data are 'not available'.

LEADING COMPANIES Number shown: 2 Total sales ($ mil): 612 Total employment (000): 3.1

Company Name	Address				CEO Name	Phone	Co. Type	Sales ($ mil)	Empl. (000)
Mannington Mills Inc	PO Box 30	Salem	NJ	08079	Scott Smith	609-935-3000	R	600	3.0
Sport Court Inc	939 S 700 W	Salt Lake City	UT	84104	Dan Kotler	801-972-0260	R	12	<0.1

Source: Ward's Business Directory of U.S. Private and Public Companies, Volumes 1 and 2, 1996. The company type code used is as follows: P - Public, R - Private, S - Subsidiary, D - Division, J - Joint Venture, A - Affiliate, G - Group. Sales are in millions of dollars, employees are in thousands. An asterisk (*) indicates an estimated sales volume. The symbol < stands for 'less than'. Company names and addresses are truncated, in some cases, to fit into the available space.

MATERIALS CONSUMED

Material	Quantity	Delivered Cost ($ million)
Materials, ingredients, containers, and supplies	(X)	547.3
Pigments, organic and inorganic	(X)	29.0
Plastics resins consumed in the form of granules, pellets, powders, liquids, etc.	(X)	229.9
Other chemicals and allied products	(X)	41.6
All other materials and components, parts, containers, and supplies	(X)	239.7
Materials, ingredients, containers, and supplies, nsk	(X)	7.1

Source: 1992 Economic Census. Explanation of symbols used: (D): Withheld to avoid disclosure of competitive data; na: Not available; (S): Withheld because statistical norms were not met; (X): Not applicable; (Z): Less than half the unit shown; nec: Not elsewhere classified; nsk: Not specified by kind; - : zero; * : 10-19 percent estimated; ** : 20-29 percent estimated.

PRODUCT SHARE DETAILS

Product or Product Class	% Share	Product or Product Class	% Share
Hard surface floor coverings, nec	100.00		

Source: 1992 Economic Census. The values shown are percent of total shipments in an industry. Values of indented subcategories are summed in the main heading. The symbol (D) appears when data are withheld to prevent disclosure of competitive information. The abbreviation nsk stands for 'not specified by kind' and nec for 'not elsewhere classified'.

INPUTS AND OUTPUTS FOR HARD SURFACE FLOOR COVERINGS

Economic Sector or Industry Providing Inputs	%	Sector	Economic Sector or Industry Buying Outputs	%	Sector
Miscellaneous plastics products	14.2	Manufg.	Personal consumption expenditures	52.0	
Accounting, auditing & bookkeeping	11.3	Services	Residential 1-unit structures, nonfarm	10.1	Constr.
Plastics materials & resins	10.7	Manufg.	Residential additions/alterations, nonfarm	6.7	Constr.
Advertising	10.2	Services	Nonfarm residential structure maintenance	5.1	Constr.
Wholesale trade	7.6	Trade	Exports	4.6	Foreign
Imports	7.3	Foreign	Office buildings	3.7	Constr.
Building paper & board mills	6.4	Manufg.	Mobile homes	3.0	Manufg.
Banking	4.2	Fin/R.E.	Residential garden apartments	2.2	Constr.
Wood products, nec	3.3	Manufg.	Construction of hospitals	1.4	Constr.
Motor freight transportation & warehousing	3.1	Util.	Maintenance of nonfarm buildings nec	1.4	Constr.
Electric services (utilities)	1.9	Util.	Industrial buildings	1.3	Constr.
Sanitary services, steam supply, irrigation	1.8	Util.	Residential 2-4 unit structures, nonfarm	1.1	Constr.
Paperboard containers & boxes	1.4	Manufg.	Hotels & motels	1.0	Constr.
Railroads & related services	1.4	Util.	Sheet metal work	1.0	Manufg.
Industrial inorganic chemicals, nec	1.1	Manufg.	Travel trailers & campers	0.9	Manufg.
Business services nec	1.1	Services	Construction of stores & restaurants	0.8	Constr.
Petroleum refining	1.0	Manufg.	Construction of nonfarm buildings nec	0.6	Constr.
Eating & drinking places	1.0	Trade	Residential high-rise apartments	0.6	Constr.
Noncomparable imports	0.8	Foreign	Resid. & other health facility construction	0.5	Constr.
Real estate	0.7	Fin/R.E.	Construction of educational buildings	0.4	Constr.
Maintenance of nonfarm buildings nec	0.5	Constr.	Maintenance of farm residential buildings	0.3	Constr.
Nonmetallic mineral services	0.4	Mining	Prefabricated wood buildings	0.3	Manufg.
Adhesives & sealants	0.4	Manufg.	Amusement & recreation building construction	0.2	Constr.
Cyclic crudes and organics	0.4	Manufg.	Dormitories & other group housing	0.2	Constr.
Inorganic pigments	0.4	Manufg.	Farm housing units & additions & alterations	0.2	Constr.
Machinery, except electrical, nec	0.4	Manufg.	Maintenance of military facilities	0.1	Constr.
Gas production & distribution (utilities)	0.4	Util.	Warehouses	0.1	Constr.
Credit agencies other than banks	0.4	Fin/R.E.			
Business/professional associations	0.4	Services			
Computer & data processing services	0.4	Services			
Coal	0.3	Mining			
Abrasive products	0.3	Manufg.			
Air transportation	0.3	Util.			
Water transportation	0.3	Util.			
Royalties	0.3	Fin/R.E.			

Continued on next page.

INPUTS AND OUTPUTS FOR HARD SURFACE FLOOR COVERINGS - Continued

Economic Sector or Industry Providing Inputs	%	Sector	Economic Sector or Industry Buying Outputs	%	Sector
Security & commodity brokers	0.3	Fin/R.E.			
Legal services	0.3	Services			
Management & consulting services & labs	0.3	Services			
Clay, ceramic, & refractory minerals	0.2	Mining			
Alkalies & chlorine	0.2	Manufg.			
Hand & edge tools, nec	0.2	Manufg.			
Special dies & tools & machine tool accessories	0.2	Manufg.			
Communications, except radio & TV	0.2	Util.			
U.S. Postal Service	0.2	Gov't			
Lubricating oils & greases	0.1	Manufg.			
Manifold business forms	0.1	Manufg.			
Automotive rental & leasing, without drivers	0.1	Services			
Automotive repair shops & services	0.1	Services			
Electrical repair shops	0.1	Services			
Equipment rental & leasing services	0.1	Services			
Hotels & lodging places	0.1	Services			

Source: Benchmark Input-Output Accounts for the U.S. Economy, 1982, U.S. Department of Commerce, Washington, D.C., July 1991. Data, as reported in the source, are organized by the 1977 SIC structure in use in 1982 but have been matched, as closely as is possible, to the 1987 SIC structure used in this book.

OCCUPATIONS EMPLOYED BY SIC 399 - MANUFACTURED PRODUCTS, NEC

Occupation	% of Total 1994	Change to 2005	Occupation	% of Total 1994	Change to 2005
Assemblers, fabricators, & hand workers nec	17.0	4.8	Freight, stock, & material movers, hand	1.5	-16.1
Sales & related workers nec	4.3	4.8	Inspectors, testers, & graders, precision	1.5	4.9
Blue collar worker supervisors	4.0	-2.4	General office clerks	1.5	-10.6
General managers & top executives	3.8	-0.5	Designers, ex interior designers	1.5	15.3
Hand packers & packagers	3.7	-28.1	Electrical & electronic assemblers	1.4	4.8
Traffic, shipping, & receiving clerks	2.5	0.9	Industrial production managers	1.4	4.8
Machine operators nec	2.2	-7.6	Packaging & filling machine operators	1.4	4.9
Helpers, laborers, & material movers nec	2.2	4.8	Screen printing machine setters & set-up operators	1.2	4.8
Painting, coating, & decorating workers, hand	2.0	-5.7	Plastic molding machine workers	1.2	4.8
Precision workers nec	2.0	41.5	Machine feeders & offbearers	1.2	-5.7
Secretaries, ex legal & medical	2.0	-4.5	Cabinetmakers & bench carpenters	1.1	-31.9
Bookkeeping, accounting, & auditing clerks	1.8	-21.4	Order clerks, materials, merchandise, & service	1.1	2.6
Sheet metal workers & duct installers	1.8	4.8	Grinders & polishers, hand	1.1	4.8
Precision metal workers nec	1.8	36.3			

Source: Industry-Occupation Matrix, Bureau of Labor Statistics. These data relate to one or more 3-digit SIC industry groups rather than to a single 4-digit SIC. The change reported for each occupation to the year 2005 is a percent of growth or decline as estimated by the Bureau of Labor Statistics. The abbreviation nec stands for 'not elsewhere classified'.

LOCATION BY STATE AND REGIONAL CONCENTRATION

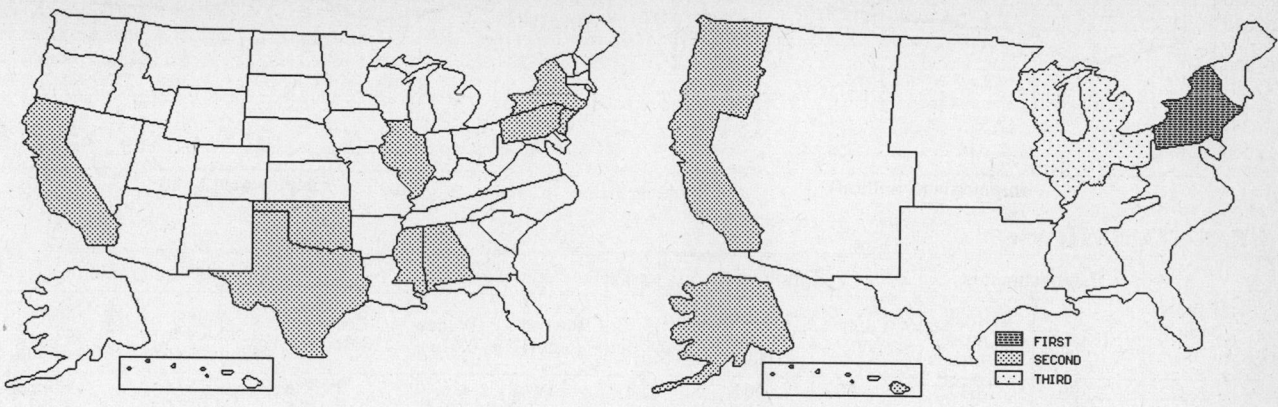

INDUSTRY DATA BY STATE

| State | Establish-ments | Shipments | | | Employment | | | | Cost as % of Shipments | Investment per Employee ($) |
		Total ($ mil)	% of U.S.	Per Establ.	Total Number	% of U.S.	Per Establ.	Wages ($/hour)		
Pennsylvania	3	703.7	46.3	234.6	2,900	43.3	967	18.27	36.7	-
New Jersey	5	302.8	19.9	60.6	1,600	23.9	320	15.14	43.5	-
California	5	(D)	-	-	175 *	2.6	35	-	-	-
Illinois	3	(D)	-	-	750 *	11.2	250	-	-	-
Mississippi	2	(D)	-	-	175 *	2.6	88	-	-	-
New York	2	(D)	-	-	375 *	5.6	188	-	-	-
Alabama	1	(D)	-	-	375 *	5.6	375	-	-	-
Oklahoma	1	(D)	-	-	175 *	2.6	175	-	-	-
Texas	1	(D)	-	-	175 *	2.6	175	-	-	-

Source: 1992 *Economic Census*. The states are in descending order of shipments or establishments (if shipment data are missing for the majority). The symbol (D) appears when data are withheld to prevent disclosure of competitive information. States marked with (D) are sorted by number of establishments. A dash (-) indicates that the data element cannot be calculated; * indicates the midpoint of a range.

3999 - MANUFACTURING INDUSTRIES, NEC

82 83 84 85 86 87 88 89 90 91 92 93 94 95 96 97 98

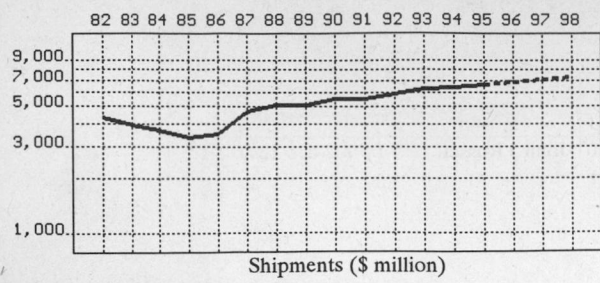

Shipments ($ million)

82 83 84 85 86 87 88 89 90 91 92 93 94 95 96 97 98

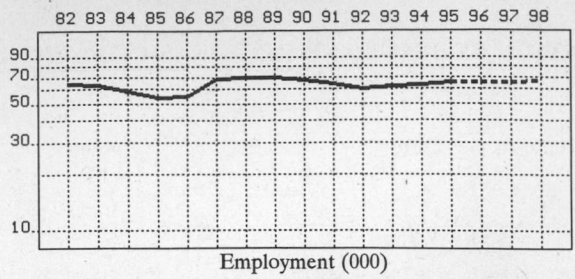

Employment (000)

GENERAL STATISTICS

Year	Com-panies	Establishments		Employment			Compensation		Production ($ million)			
		Total	with 20 or more employees	Total (000)	Production Workers (000)	Hours (Mil)	Payroll ($ mil)	Wages ($/hr)	Cost of Materials	Value Added by Manufacture	Value of Shipments	Capital Invest.
1982	3,842	3,898	637	64.0	46.8	85.8	944.0	6.40	2,173.8	2,157.0	4,322.9	93.2
1983		3,470	622	62.6	46.3	86.0	943.8	6.35	1,925.5	1,968.7	3,952.8	70.4
1984		3,042	607	58.2	43.3	80.1	913.5	6.51	1,631.4	2,048.8	3,663.3	87.1
1985		2,612	593	53.5	34.1	71.3	878.6	7.00	1,451.9	1,870.2	3,345.1	84.1
1986		2,395	591	54.7	40.4	75.0	905.3	7.06	1,485.5	1,991.9	3,488.0	72.4
1987	4,047	4,093	725	68.0	50.3	94.1	1,179.9	7.18	1,992.3	2,742.9	4,740.6	99.2
1988		3,569	754	70.3	51.3	95.6	1,255.0	7.39	2,182.4	2,954.1	5,122.8	71.3
1989		3,249	757	69.2	48.1	92.5	1,214.1	7.42	2,105.2	2,980.1	5,084.6	122.2
1990		3,217	732	67.1	50.8	96.5	1,284.1	7.68	2,326.0	3,176.2	5,482.4	108.9
1991		3,166	698	64.0	47.7	94.0	1,251.2	7.68	2,376.6	3,132.1	5,495.4	137.1
1992	3,208	3,252	689	60.0	41.3	82.4	1,362.8	8.75	2,538.0	3,373.1	5,881.3	213.8
1993		2,854	685	62.4	43.8	88.0	1,415.1	8.98	2,717.1	3,626.2	6,280.0	113.0
1994		3,064P	742P	62.9	44.2	88.7	1,400.9	8.88	2,791.1	3,645.7	6,404.2	175.5
1995		3,038P	753P	65.0P	46.2P	92.5P	1,488.5P	9.04P	2,854.7P	3,728.8P	6,550.2P	167.9P
1996		3,011P	763P	65.3P	46.3P	93.3P	1,536.9P	9.26P	2,959.6P	3,865.8P	6,790.8P	176.0P
1997		2,985P	773P	65.7P	46.4P	94.1P	1,585.3P	9.49P	3,064.4P	4,002.7P	7,031.3P	184.0P
1998		2,959P	784P	66.0P	46.5P	95.0P	1,633.6P	9.71P	3,169.3P	4,139.6P	7,271.9P	192.1P

Sources: 1982, 1987, 1992 *Economic Census*; *Annual Survey of Manufactures*, 83-86, 88-91, 93-94. Establishment counts for non-Census years are from *County Business Patterns*; establishment values for 83-84 are extrapolations. 'P's show projections by the editors. Industries reclassified in 87 will not have data for prior years.

INDICES OF CHANGE

Year	Com-panies	Establishments		Employment			Compensation		Production ($ million)			
		Total	with 20 or more employees	Total (000)	Production Workers (000)	Hours (Mil)	Payroll ($ mil)	Wages ($/hr)	Cost of Materials	Value Added by Manufacture	Value of Shipments	Capital Invest.
1982	120	120	92	107	113	104	69	73	86	64	74	44
1983		107	90	104	112	104	69	73	76	58	67	33
1984		94	88	97	105	97	67	74	64	61	62	41
1985		80	86	89	83	87	64	80	57	55	57	39
1986		74	86	91	98	91	66	81	59	59	59	34
1987	126	126	105	113	122	114	87	82	78	81	81	46
1988		110	109	117	124	116	92	84	86	88	87	33
1989		100	110	115	116	112	89	85	83	88	86	57
1990		99	106	112	123	117	94	88	92	94	93	51
1991		97	101	107	115	114	92	88	94	93	93	64
1992	100	100	100	100	100	100	100	100	100	100	100	100
1993		88	99	104	106	107	104	103	107	108	107	53
1994		94P	108P	105	107	108	103	101	110	108	109	82
1995		93P	109P	108P	112P	112P	109P	103P	112P	111P	111P	79P
1996		93P	111P	109P	112P	113P	113P	106P	117P	115P	115P	82P
1997		92P	112P	109P	112P	114P	116P	108P	121P	119P	120P	86P
1998		91P	114P	110P	113P	115P	120P	111P	125P	123P	124P	90P

Sources: Same as General Statistics. Values reflect change from the base year, 1992. Values above 100 mean greater than 92, values below 100 mean less than 92, and a value of 100 in the 82-91 or 93-98 period means same as 92. 'P's mark projections by the editors.

SELECTED RATIOS

For 1994	Avg. of All Manufact.	Analyzed Industry	Index	For 1994	Avg. of All Manufact.	Analyzed Industry	Index
Employees per Establishment	49	21	42	Value Added per Production Worker	134,084	82,482	62
Payroll per Establishment	1,500,273	457,220	30	Cost per Establishment	5,045,178	910,947	18
Payroll per Employee	30,620	22,272	73	Cost per Employee	102,970	44,374	43
Production Workers per Establishment	34	14	42	Cost per Production Worker	146,988	63,147	43
Wages per Establishment	853,319	257,072	30	Shipments per Establishment	9,576,895	2,090,175	22
Wages per Production Worker	24,861	17,820	72	Shipments per Employee	195,460	101,816	52
Hours per Production Worker	2,056	2,007	98	Shipments per Production Worker	279,017	144,891	52
Wages per Hour	12.09	8.88	73	Investment per Establishment	321,011	57,279	18
Value Added per Establishment	4,602,255	1,189,868	26	Investment per Employee	6,552	2,790	43
Value Added per Employee	93,930	57,960	62	Investment per Production Worker	9,352	3,971	42

Sources: Same as General Statistics. The 'Average of All Manufacturing' column represents the average of all manufacturing industries reported for the most recent complete year available. The Index shows the relationship between the Average and the Analyzed Industry. For example, 100 means that they are equal; 500 that the Analyzed Industry is five times the average; 50 means that the Analyzed Industry is half the national average. The abbreviation 'na' is used to show that data are 'not available'.

LEADING COMPANIES   Number shown: 75   Total sales ($ mil): 4,327   Total employment (000): 42.1

Company Name	Address				CEO Name	Phone	Co. Type	Sales ($ mil)	Empl. (000)
International Game Technology	5270 Neil Rd	Reno	NV	89510	John J Russell	702-688-1200	P	675	2.8
IGT-North America	520 S Rock Blv	Reno	NV	89502	John J Russell	702-688-0100	D	441	1.9
WMS Industries Inc	3401 N California	Chicago	IL	60618	Neil D Nicastro	312-728-2300	P	358	2.1
Enesco Corp	225 Windsor Dr	Itasca	IL	60143	Eugene Freedman	708-875-5300	S	300	0.6
Carbide/Graphite Group Inc	1 Gateway Ctr	Pittsburgh	PA	15222	NT Kaiser	412-562-3700	R	228	1.5
Blyth Industries Inc	2 Greenwich Plz	Greenwich	CT	06830	Robert B Goergen	203-661-1926	P	215	1.4
Mid-South Industries Inc	PO Box 322	Gadsden	AL	35999	David Gilchrist	205-442-3351	R	190	1.8
Windmere Corp	5980 Miami Lks Dr	Miami Lakes	FL	33014	David M Friedson	305-362-2611	P	181	12.4
Candle Corporation of America	999 E Touhy Av	Des Plaines	IL	60018	Thomas K Kreilick	708-294-1100	S	120	0.8
Day-Timers Inc	1 Willow Ln	East Texas	PA	18046	Loren Hulber	610-398-1151	S	120*	1.5
Stonhard Inc	PO Box 308	Maple Shade	NJ	08052	JM Stork	609-779-7500	S	120	0.7
Bally Gaming Inc	6601 S Bermuda Rd	Las Vegas	NV	89119	Hans Kloss	702-896-7700	S	59	0.5
Mikohn Gaming Corp	PO Box 98686	Las Vegas	NV	89119	David J Thompson	702-896-3890	P	58	0.8
Alcoa Composites Inc	801 Royal Oak Dr	Monrovia	CA	91016	David Rittichier	818-358-3211	S	51	0.4
Andis Co	1718 Layard Av	Racine	WI	53404	Matt Andis	414-634-3356	R	50	0.3
Rennoc Corp	3501 Southeast Blv	Vineland	NJ	08360	R Conner	609-327-5400	R	50*	0.6
Walter Kidde	1394 S Third St	Mebane	NC	27302	Guy Wannop	919-563-5911	D	50	0.3
Caffco International	PO Box 3508	Montgomery	AL	36109	Joseph D Patalono	205-272-2140	D	44	0.5
Colonial Candle of Cape Cod	232 Main St	Hyannis	MA	02601	Thomas Kreilick	508-775-2500	S	40	0.3
Sprigg Lane Investment Corp	PO Box 6668	Charlottesville	VA	22906	Mary Pollock	804-977-1402	R	37	0.9
Wahl Clipper Corp	2900 Locust St	Sterling	IL	61081	John F Wahl	815-625-6525	R	34	0.6
Santa's Best	770 N Frontage Rd	Northfield	IL	60093	WF Protz Jr	708-441-2034	R	33*	0.4
Galileo Electro-Optics Corp	PO Box 550	Sturbridge	MA	01566	William T Hanley	508-347-9191	P	30	0.2
DD Bean and Sons Co	PO Box 348	Jaffrey	NH	03452	DD Bean	603-532-8311	R	30	0.3
Metallized Products	2544 Terminal Dr S	St Petersburg	FL	33712	Joseph J O'Brien	813-327-2544	R	30	0.1
Smith Enterprises Inc	PO Box 12006	Rock Hill	SC	29730	Bob Loll	803-366-7101	R	30	0.5
Candle-Lite Co	PO Box 42364	Cincinnati	OH	45242	Robert Staab	513-563-1113	S	29*	0.3
CCL Custom Manufacturing Inc	6100 W Howard St	Niles	IL	60714	Dale Cook	708-967-8100	D	28	0.4
Buckeye Fire Equipment Co	PO Box 428	Kings Mt	NC	28086	Thomas J Bower	704-739-7415	R	26	0.3
Bon-Art International Inc	99 Evergreen Av	Newark	NJ	07114	Jan Wouters	201-623-6615	R	25	0.3
Heritage Display	550 Vandalia St	St Paul	MN	55114	Joel Turunen	612-646-7865	R	25	0.2
Lumi-Lite Candle Company Inc	PO Box 2	Norwich	OH	43767	William W Wilson	614-872-3248	R	25	0.3
Hughes Enterprises Inc	17291 Irvine Blv	Tustin	CA	92680	Roger L Hughes	714-665-2201	R	22	<0.1
Hughes Products Co	17291 Irvine Blv	Orange	CA	92680	Roger L Hughes	714-665-2201	D	22*	<0.1
LS	PO Box 10528	Charleston	SC	29411	JJ Evans	803-797-2500	D	22	0.2
American Educational Products	3101 Iris Av	Boulder	CO	80301	Paul D Whittle	303-443-0020	P	21	0.1
Bobrick Washroom Equipment	11611 Hart St	N Hollywood	CA	91605	Mark Louchheim	818-764-1000	R	21*	0.3
General Fire Extinguisher Corp	1685 Shermer Rd	Northbrook	IL	60062	Harry L Haulman	708-272-7500	R	20	0.2
Blue Ridge Mountain	PO Box 566	Ellijay	GA	30540	David Owen	706-276-2222	R	18	0.2
Paul-Son Dice and Card Inc	2121 Industrial Rd	Las Vegas	NV	89102	Eric Endy	702-384-2425	S	18	0.4
AI Root Co	529 S Flores St	San Antonio	TX	78204	John Root	210-223-2948	R	16*	0.3
Brady USA Inc	PO Box 2131	Milwaukee	WI	53201	David Schroeder	414-351-6600	D	16	0.2
Joplin Workshops Inc	PO Box 1609	Joplin	MO	64802	Ron Samson	417-781-2862	R	16*	0.2
J Kinderman and Sons Inc	22 Jackson St	Philadelphia	PA	19148	John Kinderman	215-271-7600	R	16	0.2
Maddak Inc	6 Industrial Rd	Pequannock	NJ	07440	Kirk Landberger	201-628-7600	R	16*	0.2
Replogle Globes Inc	2801 S 25th Av	Broadview	IL	60153	William Nickels	708-343-0900	R	16	0.2
Lazer-Tron Corp	4430 Willow Rd	Pleasanton	CA	94588	N B Petermeier	510-460-0873	P	15	<0.1
Lazer-Tron Ltd	4430 Willow Rd	Pleasanton	CA	94588	N B Petermeier	510-460-0873	S	15	<0.1
Peavey Corp	PO Box 14100	Lenexa	KS	66215	Zane Peavey	913-888-0600	R	15	<0.1
Poolmaster Inc	770 W Del Paso Rd	Sacramento	CA	95834	LH Tager	916-567-9800	R	15	0.1
W and F Products Inc	2299 Kenmore Av	Buffalo	NY	14207	Irwin Pastor	716-874-5850	R	15	0.4
Belvedere Co	One Belvedere Blv	Belvidere	IL	61008	M Holmes	815-544-3311	S	15	0.2
SKB Case Co	434 W Levers Pl	Orange	CA	92667	Steve Kottman	714-637-1252	R	13*	0.2
Stoughton Composites Inc	302 23rd St	Brodhead	WI	53520	Rob Sjostedt	608-897-8691	R	13*	0.2
Atlas Match Corp	1801 S Airport Cir	Euless	TX	76039	GD Walker	817-267-1500	R	12	0.2
Grenecker Wolf and Vine Inc	1345 S Herbert Av	Los Angeles	CA	90023	Jan Wouters	213-263-9000	R	12*	0.2
Ling Electronics Inc	4890 E La Palma	Anaheim	CA	92806	Stephen Sullivan	714-779-1900	S	12	<0.1
Muench-Kreuzer Candle Co	PO Box 4969	Syracuse	NY	13221	Fred Kiesinger	315-471-4515	S	12	0.1
Potter-Roemer Inc	16833 Edwards Rd	Cerritos	CA	90701	Lance McCabe	310-404-3753	S	12	0.1
Y-Tex Corp	PO Box 1450	Cody	WY	82414	Jerry V Payne	307-587-5515	R	12*	0.1
I-K-I Manufacturing Company	116 N Swift St	Edgerton	WI	53534	Stan Midtbo	608-884-3411	R	11*	0.1
Mace Security	PO Box 305	Saltsburg	PA	15681	Tom Breslin	412-639-3511	D	11	0.3
American Oak Preserving	PO Box 187	North Judson	IN	46366	Charles K Vorm	219-896-2171	R	10	0.1
Bob's Space Racers Inc	427 15th St	Daytona Beach	FL	32117	Jack Mendes	904-677-0761	R	10*	0.1
Hollander-Stapo Industries Inc	200 Syracuse Ct	Lakewood	NJ	08701	Stanley Pollinger	908-370-5050	R	10	0.1
New York Lighter Company Inc	1539 Schenectady	Brooklyn	NY	11234	John Nordstrom	718-338-4600	R	10	<0.1
Perrygraf	19365 Business Ctr	Northridge	CA	91324	Cathie Smith	818-993-1000	R	10	<0.1
Pharmasol Corp	1 Norfolk Av	South Easton	MA	02375	Scott McCaig	508-238-8501	S	10	<0.1
Seasons Inc	PO Box 190460	Little Rock	AR	72219	Terry Fry	501-562-6579	S	10	0.1
VFP Inc	PO Box 11927	Roanoke	VA	24022	Richard Poplstein	703-977-0500	R	10	0.1
Design Craftsmen Inc	PO Box 2126	Midland	MI	48641	Clark E Swayze	517-496-3220	R	9	<0.1
Hughes Identification Devices	14311 Chambers Rd	Tustin	CA	92680	Don Nelson	714-573-7270	S	9	<0.1
Silvestri Studio Inc	1733 Cordova St	Los Angeles	CA	90007	Terri Oltman	213-735-1481	R	9*	0.2
Texas Feathers Inc	PO Box 1118	Brownwood	TX	76804	Dale Parreck	915-646-1504	D	9	<0.1
Koken Manufacturing Company	1631 ML King	St Louis	MO	63106	T Yanagihara	314-231-7383	S	8	<0.1

Source: Ward's Business Directory of U.S. Private and Public Companies, Volumes 1 and 2, 1996. The company type code used is as follows: P - Public, R - Private, S - Subsidiary, D - Division, J - Joint Venture, A - Affiliate, G - Group. Sales are in millions of dollars, employees are in thousands. An asterisk (*) indicates an estimated sales volume. The symbol < stands for 'less than'. Company names and addresses are truncated, in some cases, to fit into the available space.

MATERIALS CONSUMED

Material	Quantity	Delivered Cost ($ million)
Materials, ingredients, containers, and supplies	(X)	2,104.1
Fabricated metal products (except castings and forgings)	(X)	163.5
Castings (rough and semifinished)	(X)	11.6
Forgings	(X)	4.7
Steel shapes and forms	(X)	63.9
Copper and copper-base alloy shapes and forms	(S) 1,000 s tons	15.6
Aluminum and aluminum-base alloy shapes and forms	(S) 1,000 s tons	19.2
Other nonferrous shapes and forms	(X)	5.0
Plastics products consumed in the form of sheets, rods, tubes, film, and other shapes	(X)	92.0
Plastics resins consumed in the form of granules, pellets, powders, liquids, etc.	(X)	42.1
Paperboard containers, boxes, and corrugated paperboard	(X)	81.6
Rough and dressed lumber	(X)	37.7
All other materials and components, parts, containers, and supplies	(X)	792.0
Materials, ingredients, containers, and supplies, nsk	(X)	775.4

Source: 1992 Economic Census. Explanation of symbols used: (D): Withheld to avoid disclosure of competitive data; na: Not available; (S): Withheld because statistical norms were not met; (X): Not applicable; (Z): Less than half the unit shown; nec: Not elsewhere classified; nsk: Not specified by kind; - : zero; * : 10-19 percent estimated; ** : 20-29 percent estimated.

PRODUCT SHARE DETAILS

Product or Product Class	% Share	Product or Product Class	% Share
Manufacturing industries, nec	100.00	materials owned by others	28.91
Chemical fire-extinguishing equipment and parts	8.15	Furs, dressed and dyed, nsk	12.89
Hand portable carbon dioxide fire extinguishers	3.13	Umbrellas and parasols (including parts)	0.99
Hand portable dry chemical fire extinguishers	35.18	Feathers, plumes, and artificial flowers	4.73
Other hand portable fire extinguishers (including foam, pressurized water, and halogenated agents)	21.72	Artificial Christmas trees, all types (metal, plastics, etc.)	41.65
		Plastics-type artificial flowers, fruits, and wreaths	3.06
Fixed fire-extinguishing systems, including inert gas, dry and wet chemical, and other chemical fire-extinguishing equipment	20.29	Other artificial flowers, fruits, and wreaths	20.21
		Feathers and plumes	15.84
Parts and attachments for chemical fire-extinguishing equipment	5.19	Feathers, plumes, and artificial flowers, nsk	19.27
		Miscellaneous fabricated products, nec	42.02
Chemical fire-extinguishing equipment and parts, nsk	14.51	Hair clippers, for human use, hand and electric	5.38
Coin-operated amusement machines	11.13	Barber and beauty shop furniture and equipment, including barber and beauty chairs	3.47
Coin-operated arcade and amusement center type electronic games	58.91	Beauty and barber shop accessories (including hair curlers, pads, and wraps)	1.14
Other coin-operated amusement machines, including nonelectronic arcade games and parts for all arcade games	40.94	Hair work, switches, toupees, and wigs	0.21
		Christmas tree ornaments and decorations (except glass and electrical)	5.93
Coin-operated amusement machines, nsk	0.17	Lamp shades, excluding plastics, metal, and glass	3.62
Matches	1.29	Potpourri (dried and chemically-preserved flowers, foilage, fruits, and vines)	4.78
Candles (including tapers)	6.95	Miscellaneous fabricated products, not elsewhere classified	72.11
Furs, dressed and dyed	0.49	Miscellaneous fabricated products, nec, nsk	3.35
Furs, dressed and dyed	58.59	Manufacturing industries, nec, nsk	24.24
Receipts for dressing and dyeing furs done in your plant on			

Source: 1992 Economic Census. The values shown are percent of total shipments in an industry. Values of indented subcategories are summed in the main heading. The symbol (D) appears when data are withheld to prevent disclosure of competitive information. The abbreviation nsk stands for 'not specified by kind' and nec for 'not elsewhere classified'.

INPUTS AND OUTPUTS FOR MANUFACTURING INDUSTRIES, NEC

Economic Sector or Industry Providing Inputs	%	Sector	Economic Sector or Industry Buying Outputs	%	Sector
Imports	29.5	Foreign	Gross private fixed investment	44.4	Cap Inv
Wholesale trade	7.8	Trade	Personal consumption expenditures	14.9	
Advertising	7.8	Services	Exports	4.0	Foreign
Cyclic crudes and organics	6.3	Manufg.	S/L Govt. purch., elem. & secondary education	3.5	S/L Govt
Eating & drinking places	3.5	Trade	Retail trade, except eating & drinking	3.4	Trade
Blast furnaces & steel mills	3.2	Manufg.	Wholesale trade	2.2	Trade
Manufacturing industries, nec	2.5	Manufg.	Manufacturing industries, nec	1.8	Manufg.
Paperboard containers & boxes	2.2	Manufg.	Religious organizations	1.8	Services
Machinery, except electrical, nec	2.1	Manufg.	Banking	1.5	Fin/R.E.
Fabricated metal products, nec	1.9	Manufg.	Communications, except radio & TV	1.2	Util.
Wood products, nec	1.8	Manufg.	Beauty & barber shops	1.1	Services
Maintenance of nonfarm buildings nec	1.7	Constr.	Eating & drinking places	1.0	Trade
Sawmills & planning mills, general	1.5	Manufg.	Water transportation	0.9	Util.
Motor freight transportation & warehousing	1.4	Util.	Insurance carriers	0.9	Fin/R.E.
Aluminum rolling & drawing	1.3	Manufg.	Federal Government purchases, national defense	0.8	Fed Govt
Paints & allied products	1.2	Manufg.	Credit agencies other than banks	0.7	Fin/R.E.
Petroleum refining	1.2	Manufg.	Hospitals	0.7	Services
Management & consulting services & labs	1.2	Services	Brooms & brushes	0.6	Manufg.
Chemical preparations, nec	1.1	Manufg.	Equipment rental & leasing services	0.6	Services
Legal services	1.1	Services	Miscellaneous repair shops	0.6	Services

Continued on next page.

INPUTS AND OUTPUTS FOR MANUFACTURING INDUSTRIES, NEC - Continued

Economic Sector or Industry Providing Inputs	%	Sector	Economic Sector or Industry Buying Outputs	%	Sector
Special dies & tools & machine tool accessories	1.0	Manufg.	Engineering, architectural, & surveying services	0.5	Services
Plastics materials & resins	0.9	Manufg.	Lighting fixtures & equipment	0.4	Manufg.
Yarn mills & finishing of textiles, nec	0.9	Manufg.	State & local government enterprises, nec	0.4	Gov't
Electric services (utilities)	0.9	Util.	S/L Govt. purch., higher education	0.4	S/L Govt
Die-cut paper & board	0.8	Manufg.	Motor freight transportation & warehousing	0.3	Util.
Miscellaneous plastics products	0.8	Manufg.	Real estate	0.3	Fin/R.E.
Automatic merchandising machines	0.7	Manufg.	Accounting, auditing & bookkeeping	0.3	Services
Industrial inorganic chemicals, nec	0.7	Manufg.	Management & consulting services & labs	0.3	Services
Accounting, auditing & bookkeeping	0.7	Services	S/L Govt. purch., natural resource & recreation.	0.3	S/L Govt
Abrasive products	0.6	Manufg.	S/L Govt. purch., public assistance & relief	0.3	S/L Govt
Fabricated textile products, nec	0.6	Manufg.	Commercial printing	0.2	Manufg.
Real estate	0.6	Fin/R.E.	Lithographic platemaking & services	0.2	Manufg.
Manifold business forms	0.5	Manufg.	Newspapers	0.2	Manufg.
Screw machine and related products	0.5	Manufg.	Air transportation	0.2	Util.
Veneer & plywood	0.5	Manufg.	Electric services (utilities)	0.2	Util.
Banking	0.5	Fin/R.E.	Insurance agents, brokers, & services	0.2	Fin/R.E.
Railroads & related services	0.4	Util.	Security & commodity brokers	0.2	Fin/R.E.
Fabricated rubber products, nec	0.3	Manufg.	Amusement & recreation services nec	0.2	Services
Lubricating oils & greases	0.3	Manufg.	Child day care services	0.2	Services
Metal stampings, nec	0.3	Manufg.	Hotels & lodging places	0.2	Services
Communications, except radio & TV	0.3	Util.	Job training & related services	0.2	Services
Gas production & distribution (utilities)	0.3	Util.	Labor, civic, social, & fraternal associations	0.2	Services
Water transportation	0.3	Util.	Legal services	0.2	Services
Security & commodity brokers	0.3	Fin/R.E.	Social services, nec	0.2	Services
Equipment rental & leasing services	0.3	Services	Crude petroleum & natural gas	0.1	Mining
Personnel supply services	0.3	Services	Lead pencils & art goods	0.1	Manufg.
U.S. Postal Service	0.3	Gov't	Miscellaneous plastics products	0.1	Manufg.
Broadwoven fabric mills	0.2	Manufg.	Radio & TV communication equipment	0.1	Manufg.
Copper rolling & drawing	0.2	Manufg.	Railroads & related services	0.1	Util.
Metal coating & allied services	0.2	Manufg.	Colleges, universities, & professional schools	0.1	Services
Miscellaneous fabricated wire products	0.2	Manufg.	Computer & data processing services	0.1	Services
Photographic equipment & supplies	0.2	Manufg.	Elementary & secondary schools	0.1	Services
Transit & bus transportation	0.2	Util.	Laundry, dry cleaning, shoe repair	0.1	Services
Credit agencies other than banks	0.2	Fin/R.E.	Medical & health services, nec	0.1	Services
Royalties	0.2	Fin/R.E.	Membership organizations nec	0.1	Services
Automotive rental & leasing, without drivers	0.2	Services	Personnel supply services	0.1	Services
Automotive repair shops & services	0.2	Services	U.S. Postal Service	0.1	Gov't
Business services nec	0.2	Services	Change in business inventories	0.1	In House
Noncomparable imports	0.2	Foreign	S/L Govt. purch., health & hospitals	0.1	S/L Govt
Alkalies & chlorine	0.1	Manufg.			
Industrial gases	0.1	Manufg.			
Meat packing plants	0.1	Manufg.			
Nonferrous wire drawing & insulating	0.1	Manufg.			
Insurance carriers	0.1	Fin/R.E.			
Laundry, dry cleaning, shoe repair	0.1	Services			

Source: Benchmark Input-Output Accounts for the U.S. Economy, 1982, U.S. Department of Commerce, Washington, D.C., July 1991. Data, as reported in the source, are organized by the 1977 SIC structure in use in 1982 but have been matched, as closely as is possible, to the 1987 SIC structure used in this book.

OCCUPATIONS EMPLOYED BY SIC 399 - MANUFACTURED PRODUCTS, NEC

Occupation	% of Total 1994	Change to 2005	Occupation	% of Total 1994	Change to 2005
Assemblers, fabricators, & hand workers nec	17.0	4.8	Freight, stock, & material movers, hand	1.5	-16.1
Sales & related workers nec	4.3	4.8	Inspectors, testers, & graders, precision	1.5	4.9
Blue collar worker supervisors	4.0	-2.4	General office clerks	1.5	-10.6
General managers & top executives	3.8	-0.5	Designers, ex interior designers	1.5	15.3
Hand packers & packagers	3.7	-28.1	Electrical & electronic assemblers	1.4	4.8
Traffic, shipping, & receiving clerks	2.5	0.9	Industrial production managers	1.4	4.8
Machine operators nec	2.2	-7.6	Packaging & filling machine operators	1.4	4.9
Helpers, laborers, & material movers nec	2.2	4.8	Screen printing machine setters & set-up operators	1.2	4.8
Painting, coating, & decorating workers, hand	2.0	-5.7	Plastic molding machine workers	1.2	4.8
Precision workers nec	2.0	41.5	Machine feeders & offbearers	1.2	-5.7
Secretaries, ex legal & medical	2.0	-4.5	Cabinetmakers & bench carpenters	1.1	-31.9
Bookkeeping, accounting, & auditing clerks	1.8	-21.4	Order clerks, materials, merchandise, & service	1.1	2.6
Sheet metal workers & duct installers	1.8	4.8	Grinders & polishers, hand	1.1	4.8
Precision metal workers nec	1.8	36.3			

Source: Industry-Occupation Matrix, Bureau of Labor Statistics. These data relate to one or more 3-digit SIC industry groups rather than to a single 4-digit SIC. The change reported for each occupation to the year 2005 is a percent of growth or decline as estimated by the Bureau of Labor Statistics. The abbreviation nec stands for 'not elsewhere classified'.

LOCATION BY STATE AND REGIONAL CONCENTRATION

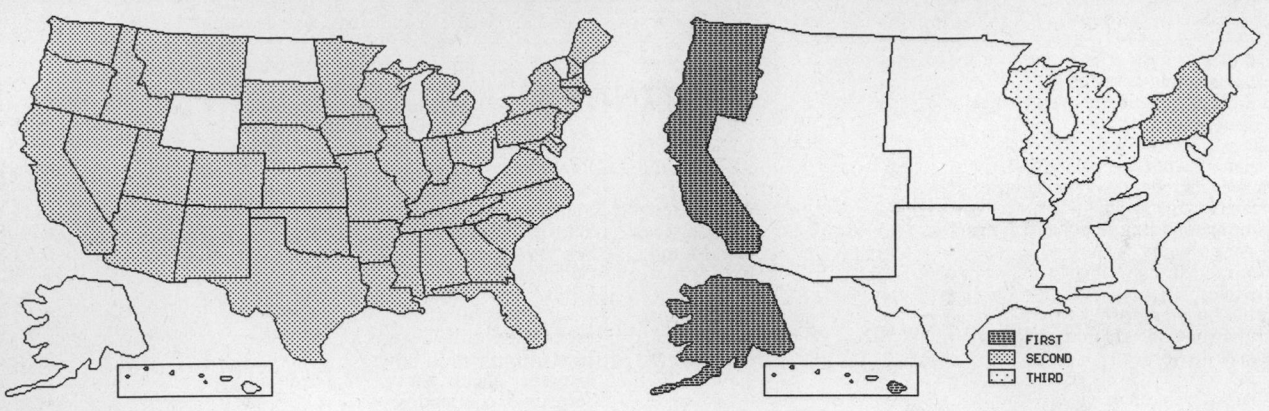

FIRST
SECOND
THIRD

INDUSTRY DATA BY STATE

State	Establish-ments	Shipments			Employment				Cost as % of Shipments	Investment per Employee ($)
		Total ($ mil)	% of U.S.	Per Establ.	Total Number	% of U.S.	Per Establ.	Wages ($/hour)		
California	479	818.7	13.9	1.7	7,800	13.0	16	9.15	41.3	4,667
Illinois	148	560.5	9.5	3.8	5,200	8.7	35	9.37	47.1	2,058
New York	320	428.9	7.3	1.3	5,200	8.7	16	8.72	44.9	1,288
Connecticut	51	348.0	5.9	6.8	2,000	3.3	39	15.08	34.1	-
Pennsylvania	149	307.1	5.2	2.1	3,200	5.3	21	9.56	35.3	2,031
North Carolina	86	265.2	4.5	3.1	2,500	4.2	29	8.86	44.1	3,040
New Jersey	137	252.9	4.3	1.8	2,900	4.8	21	8.43	41.5	1,103
Georgia	61	243.7	4.1	4.0	1,900	3.2	31	4.95	31.6	-
Ohio	126	242.2	4.1	1.9	2,400	4.0	19	8.09	41.3	2,875
Massachusetts	82	221.6	3.8	2.7	2,200	3.7	27	10.96	41.5	2,409
Texas	162	203.9	3.5	1.3	2,700	4.5	17	9.61	43.4	1,926
Nevada	26	199.5	3.4	7.7	1,700	2.8	65	9.63	68.3	-
Wisconsin	83	198.9	3.4	2.4	1,900	3.2	23	8.50	46.8	1,947
Tennessee	46	156.6	2.7	3.4	1,500	2.5	33	7.77	44.4	1,867
Michigan	149	131.4	2.2	0.9	1,600	2.7	11	10.78	39.4	1,688
Florida	185	128.8	2.2	0.7	1,800	3.0	10	7.22	39.8	1,333
Alabama	29	118.1	2.0	4.1	1,200	2.0	41	6.28	53.2	3,417
Virginia	40	107.4	1.8	2.7	900	1.5	23	10.86	50.0	2,667
Indiana	49	103.8	1.8	2.1	1,300	2.2	27	7.55	46.2	1,923
Minnesota	60	92.2	1.6	1.5	800	1.3	13	9.31	37.5	1,875
Missouri	57	90.4	1.5	1.6	1,200	2.0	21	6.38	40.6	1,167
Kentucky	31	63.1	1.1	2.0	800	1.3	26	7.60	35.0	1,875
Rhode Island	72	54.9	0.9	0.8	700	1.2	10	7.70	42.1	2,000
Arkansas	23	54.8	0.9	2.4	800	1.3	35	6.44	40.7	2,000
New Hampshire	13	20.9	0.4	1.6	300	0.5	23	11.67	37.8	667
Utah	27	20.8	0.4	0.8	300	0.5	11	7.50	39.9	1,000
Nebraska	17	19.9	0.3	1.2	300	0.5	18	6.80	35.7	-
Oklahoma	28	19.7	0.3	0.7	200	0.3	7	7.33	44.7	-
New Mexico	31	18.8	0.3	0.6	100	0.2	3	8.00	43.1	7,000
Maine	15	14.6	0.2	1.0	200	0.3	13	9.50	34.9	1,000
South Carolina	17	14.5	0.2	0.9	100	0.2	6	8.50	54.5	2,000
Mississippi	17	11.9	0.2	0.7	100	0.2	6	7.50	54.6	-
Louisiana	22	8.4	0.1	0.4	100	0.2	5	12.00	40.5	1,000
Washington	86	(D)	-	-	750 *	1.3	9	-	-	2,400
Colorado	58	(D)	-	-	750 *	1.3	13	-	-	-
Oregon	58	(D)	-	-	375 *	0.6	6	-	-	-
Arizona	40	(D)	-	-	375 *	0.6	9	-	-	-
Maryland	36	(D)	-	-	375 *	0.6	10	-	-	-
Iowa	23	(D)	-	-	375 *	0.6	16	-	-	-
Kansas	18	(D)	-	-	175 *	0.3	10	-	-	-
Montana	13	(D)	-	-	175 *	0.3	13	-	-	-
Idaho	11	(D)	-	-	175 *	0.3	16	-	-	571
South Dakota	8	(D)	-	-	175 *	0.3	22	-	-	-

Source: 1992 *Economic Census*. The states are in descending order of shipments or establishments (if shipment data are missing for the majority). The symbol (D) appears when data are withheld to prevent disclosure of competitive information. States marked with (D) are sorted by number of establishments. A dash (-) indicates that the data element cannot be calculated; * indicates the midpoint of a range.

SIC INDEX

The SIC Index shows all 4-digit SICs covered in *Manufacturing USA* in numerical order. A separate section, listing the industries in alphabetical order, follows. The Roman numerals I and II, which precede the page numbers, indicate the volume in which each SIC appears. In the alphabetical section, each industry name is followed by the SIC number (in parentheses) and then the page number, marked with a volume (I or II). This SIC structure is based on the 1987 definitions published in *Standard Industrial Classification Manual*, 1987, Office of Management and Budget. The abbreviation 'nec' stands for 'not elsewhere classified'. Please note that SIC 2067, Chewing gum, is missing from the listing (and from the body of the book). In recent years, the Bureau of the Census has stopped reporting details on this industry.

Roman numerals I and II indicate the volume in which pages appear. Volume I holds SICs 2011 through 3299; Volume II holds SICs 3312 through 3999.

Roman numerals I and II indicate the volume in which pages appear. Volume I holds SICs 2011 through 3299; Volume II holds SICs 3312 through 3999.

2205

Roman numerals I and II indicate the volume in which pages appear. Volume I holds SICs 2011 through 3299; Volume II holds SICs 3312 through 3999.

SIC Index

Roman numerals I and II indicate the volume in which pages appear. Volume I holds SICs 2011 through 3299; Volume II holds SICs 3312 through 3999.

PRODUCT INDEX

The Product Index holds the names of nearly 2,200 products, materials, and substances with references to all 4-digit SIC industries in which they are manufactured. Page references follow immediately behind the product and are marked with a Roman I or II to indicate the volume. If more than one page reference is provided, the phrases "Vol. I" or "Vol. II" show the beginning of each volume. After the page numbers is a listing, in brackets, of the SIC codes under which the product appears. The references are arranged sequentially; the first reference will be to the first occurrence of the product in *Manufacturing USA*.

2052]
Black plate, p. II-1171 [SIC 3312]
Black powder, p. I-884 [SIC 2892]
Blackstrap, p. I-115 [SIC 2061]
Blades, p. II-1317 [SIC 3425]
Blankbooks, p. I-744 [SIC 2782]
Blanketing, p. I-235 [SIC 2221]
Blankets, p. I-317 [SIC 2297]
Blasting accessories, p. I-884 [SIC 2892]
Blasting caps, p. I-884 [SIC 2892]
Blasting powder, p. I-884 [SIC 2892]
Bleaches, p. I-824 [SIC 2842]
Blended flour, p. I-82 [SIC 2045]
Blinds, p. I-598 [SIC 2591]
Blocks, concrete, p. I-1116 [SIC 3271]
Blocks, paving, p. I-908 [SIC 2951]
Blood derivatives, p. I-813 [SIC 2836]
Blouses, p. Vol. I - 257, 360, 389 [SICs 2253, 2331, 2361]
Blousettes, p. I-389 [SIC 2361]
Blow torches, p. II-1311 [SIC 3423]
Blowers, p. Vol. II - 1508, 1513, 1653 [SICs 3523, 3524, 3564]
Blown glass, p. I-1065 [SIC 3229]
Board games, p. I-687 [SIC 2679]
Board, building, p. I-531 [SIC 2493]
Board, die-cut, p. I-668 [SIC 2675]
Boats, p. Vol. I - 953, Vol. II - 1991 [SICs 3069, 3732]
Bobbins, p. I-535 [SIC 2499]
Bobsleds, p. II-2149 [SIC 3949]
Bodies, vehicle, p. Vol. II - 1942, 1947 [SICs 3711, 3713]
Boiler plate, p. II-1354 [SIC 3443]
Boilers, p. Vol. II - 1338, 1354, 1735 [SICs 3433, 3443, 3589]
Bolts, p. Vol. I - 459, Vol. II - 1387 [SICs 2411, 3452]
Bombs, p. II-1433 [SIC 3483]
Bone buttons, p. II-2176 [SIC 3965]
Book mailers, p. I-642 [SIC 2657]
Book printing, p. I-707 [SIC 2732]
Bookbinding, p. I-749 [SIC 2789]
Bookcases, p. Vol. I - 541, 570 [SICs 2511, 2521]
Books, p. Vol. I - 702, 707, 712 [SICs 2731, 2732, 2741]
Bookshelves, p. I-541 [SIC 2511]
Booster rockets, p. II-2006 [SIC 3761]
Boosters, p. II-1768 [SIC 3612]
Boot leather, p. I-1016 [SIC 3131]
Booths, p. I-602 [SIC 2599]
Boots, p. I-1024 [SIC 3143]
Boring machines, p. II-1557 [SIC 3541]
Botanicals, p. I-798 [SIC 2833]
Bottles, p. Vol. I - 984, 1061 [SICs 3085, 3221]
Bottoms, underwear, p. I-339 [SIC 2322]
Bowling balls, p. II-2149 [SIC 3949]
Bowling furniture, p. I-602 [SIC 2599]
Bowls, p. Vol. I - 1101, 1105 [SICs 3262, 3263]
Boxer shorts, p. I-339 [SIC 2322]
Boxes, p. Vol. I - 505, 514, 535, 622, 627, 642, 953 [SICs 2441, 2449, 2499, 2652, 2653, 2657, 3069]
Bracelets, p. Vol. II - 2121, 2172 [SICs 3911, 3961]
Braces, p. II-2086 [SIC 3842]
Braided fabrics, p. I-244 [SIC 2241]
Brake linings, p. I-1149 [SIC 3292]
Brakes, p. II-1953 [SIC 3714]
Bralettes, p. I-381 [SIC 2342]
Bran, p. I-78 [SIC 2044]
Brandy, p. I-167 [SIC 2084]
Brass, p. II-1279 [SIC 3366]
Brass winds, p. II-2134 [SIC 3931]
Brassieres, p. I-381 [SIC 2342]
Bread, p. Vol. I - 82, 100 [SICs 2045, 2051]
Breakfast foods, p. I-74 [SIC 2043]
Breakfronts, p. I-541 [SIC 2511]
Breeze, p. II-1171 [SIC 3312]
Brewers' rice, p. I-78 [SIC 2044]
Brewing malt, p. I-159 [SIC 2082]

Brick, p. Vol. I - 1080, 1116, 1162 [SICs 3251, 3271, 3297]
Bridge ties, p. I-527 [SIC 2491]
Bridges, p. II-1343 [SIC 3441]
Briefcases, p. I-1040 [SIC 3161]
Briefs, p. I-339 [SIC 2322]
Briquettes, p. I-923 [SIC 2999]
Bronze, p. II-1279 [SIC 3366]
Brooches, p. II-2121 [SIC 3911]
Brooders, p. II-1508 [SIC 3523]
Brooms, p. II-2180 [SIC 3991]
Brown sugar, p. I-118 [SIC 2062]
Brushes, p. II-2180 [SIC 3991]
Brushes, electric, p. II-1785 [SIC 3624]
Buckles, p. II-2176 [SIC 3965]
Buffets, p. I-541 [SIC 2511]
Buffing machines, p. II-1557 [SIC 3541]
Buffing wheels, p. I-1143 [SIC 3291]
Building paper, p. I-611 [SIC 2621]
Buildings, metal, p. II-1371 [SIC 3448]
Buildings, prefab, p. I-522 [SIC 2452]
Bulbs, p. Vol. II - 1825, 1937 [SICs 3641, 3699]
Bunk beds, p. I-541 [SIC 2511]
Burial garments, p. I-419 [SIC 2389]
Burial vaults, p. II-2190 [SIC 3995]
Burrs, p. II-2176 [SIC 3965]
Bus bodies, p. II-1947 [SIC 3713]
Bus engines, p. II-1502 [SIC 3519]
Bus. reference books, p. I-712 [SIC 2741]
Buses, p. II-1947 [SIC 3713]
Business forms, p. I-734 [SIC 2761]
Business periodicals, p. I-697 [SIC 2721]
Butcher knives, p. I-1307 [SIC 3421]
Butter, p. I-16 [SIC 2021]
Butterfat, p. I-34 [SIC 2026]
Buttermilk, p. I-34 [SIC 2026]
Buttons, p. II-2176 [SIC 3965]
Byproducts, p. Vol. I - 78, 159, 607 [SICs 2044, 2082, 2611]

Cabanas, p. II-1371 [SIC 3448]
Cabin cruisers, p. II-1991 [SIC 3732]
Cabinets, p. Vol. I - 485, 541, 552, 562 [SICs 2434, 2511, 2514, 2517]
Cable, p. Vol. I - 321, Vol. II - 1181, 1249 [SICs 2298, 3315, 3355]
Cable reels, p. I-535 [SIC 2499]
Cafeteria furniture, p. I-602 [SIC 2599]
Cake, p. I-100 [SIC 2051]
Cake mix, p. I-82 [SIC 2045]
Cake, cottonseed, p. I-139 [SIC 2074]
Cake, soybean, p. I-143 [SIC 2075]
Calcareous tufa, p. I-1139 [SIC 3281]
Calculating machines, p. II-1708 [SIC 3578]
Calendars, p. I-712 [SIC 2741]
Cameras, p. II-2111 [SIC 3861]
Campers, p. II-2019 [SIC 3792]
Can openers, p. II-1311 [SIC 3423]
Candles, p. II-2198 [SIC 3999]
Candy sticks, p. I-535 [SIC 2499]
Cane furniture, p. I-566 [SIC 2519]
Cane sugar, p. Vol. I - 115, 118 [SICs 2061, 2062]
Canes, p. Vol. II - 2086, 2198 [SICs 3842, 3999]
Canned food, p. Vol. I - 39, 44, 175, 185 [SICs 2032, 2033, 2086, 2091]
Canoes, p. II-1991 [SIC 3732]
Cans, p. Vol. I - 633, Vol. II - 1299 [SICs 2655, 3411]
Canvas products, p. I-436 [SIC 2394]
Capacitors, p. Vol. II - 1794, 1896 [SICs 3629, 3675]
Capes, fur, p. I-397 [SIC 2371]
Caps, p. Vol. I - 385, 419 [SICs 2353, 2389]
Caps, blasting, p. I-884 [SIC 2892]
Car bodies, p. II-1942 [SIC 3711]
Car seats (child), p. I-552 [SIC 2514]
Car washing equip., p. II-1735 [SIC 3589]
Carbon black, p. I-893 [SIC 2895]
Carbon dioxide, p. I-767 [SIC 2813]
Carbon paper, p. II-2168 [SIC 3955]

Carbon products, p. II-1785 [SIC 3624]
Carburetors, p. II-1741 [SIC 3592]
Card tables, p. Vol. I - 541, 552 [SICs 2511, 2514]
Carded cotton, p. I-287 [SIC 2269]
Cardigans, p. Vol. I - 257, 372 [SICs 2253, 2339]
Cards, greeting, p. Vol. I - 712, 740 [SICs 2741, 2771]
Cards, guide, p. I-668 [SIC 2675]
Cards, index, p. I-668 [SIC 2675]
Cards, picture, p. I-712 [SIC 2741]
Cards, playing, p. II-2143 [SIC 3944]
Cards, souvenir, p. I-712 [SIC 2741]
Cards, tabulating, p. I-668 [SIC 2675]
Carpet sweepers, p. II-1735 [SIC 3589]
Carpet tiles, p. I-454 [SIC 2399]
Carpet yarns, p. I-296 [SIC 2281]
Carpets, p. I-291 [SIC 2273]
Cars, p. II-1942 [SIC 3711]
Cars, railroad, p. II-1997 [SIC 3743]
Cartons, carrying, p. I-642 [SIC 2657]
Cartons, egg, p. I-687 [SIC 2679]
Cartons, liquids, p. I-638 [SIC 2656]
Cartridges, p. II-1429 [SIC 3482]
Carving sets, p. II-2126 [SIC 3914]
Cases, p. Vol. I - 535, 1040 [SICs 2499, 3161]
Cash registers, p. II-1708 [SIC 3578]
Casings, sausage, p. I-6 [SIC 2013]
Casings, tire, p. I-926 [SIC 3011]
Caskets, p. II-2190 [SIC 3995]
Cast iorn furniture, p. I-552 [SIC 2514]
Cast iron, p. II-1196 [SIC 3321]
Castings, p. Vol. II - 1196, 1202, 1265 [SICs 3321, 3322, 3363]
Casual suits, p. I-331 [SIC 2311]
Cat food, p. I-90 [SIC 2047]
Catalogs, p. Vol. I - 712, 717, 723, 728 [SICs 2741, 2752, 2754, 2759]
Catheters, p. II-2081 [SIC 3841]
Cathode ray tubes, p. II-1882 [SIC 3671]
Catsup, p. I-44 [SIC 2033]
Caulking compounds, p. I-878 [SIC 2891]
Caustic soda, p. I-763 [SIC 2812]
Cedar, p. I-459 [SIC 2411]
Cedar chests, p. I-541 [SIC 2511]
Cellos, p. II-2134 [SIC 3931]
Cellulosic fibers, p. I-790 [SIC 2823]
Cement, p. I-1076 [SIC 3241]
Cements, asphalt, p. I-913 [SIC 2952]
Ceramic tile, p. I-1085 [SIC 3253]
Ceramics, p. Vol. I - 1085, 1097, 1101, 1112 [SICs 3253, 3261, 3262, 3269]
Cereals, p. Vol. I - 39, 74 [SICs 2032, 2043]
Chain saws, p. Vol. II - 1317, 1583 [SICs 3425, 3546]
Chains, p. Vol. II - 2121, 2130 [SICs 3911, 3915]
Chairs, p. Vol. I - 541, 547, 552, 602 [SICs 2511, 2512, 2514, 2599]
Chairs, wheel, p. II-2086 [SIC 3842]
Chaise lounges, p. Vol. I - 541, 547, 552 [SICs 2511, 2512, 2514]
Chalk, p. II-2159 [SIC 3952]
Champagne, p. I-167 [SIC 2084]
Chandeliers, p. II-1839 [SIC 3645]
Charcoal briquettes, p. I-923 [SIC 2999]
Charts, hydrographic, p. I-712 [SIC 2741]
Chassis, p. Vol. II - 1942, 1947 [SICs 3711, 3713]
Checkbooks, p. I-744 [SIC 2782]
Cheese, p. I-20 [SIC 2022]
Chemical catalysts, p. I-776 [SIC 2819]
Chemicals, agricul., p. I-873 [SIC 2879]
Cherries, p. I-459 [SIC 2411]
Chests, p. Vol. I - 535, 541 [SICs 2499, 2511]
Chewing gum, p. I-131 [SIC 2066]
Chewing tobacco, p. I-223 [SIC 2131]
Chickens, p. I-11 [SIC 2015]
Chifforobes, p. I-541 [SIC 2511]
Chimes, p. II-1937 [SIC 3699]
China, p. I-1101 [SIC 3262]
China cabinets, p. I-541 [SIC 2511]

Roman numerals I and II indicate the volume in which pages appear. Volume I holds SICs 2011 through 3299; Volume II holds SICs 3312 through 3999.

Roman numerals I and II indicate the volume in which pages appear. Volume I holds SICs 2011 through 3299; Volume II holds SICs 3312 through 3999.

Roman numerals I and II indicate the volume in which pages appear. Volume I holds SICs 2011 through 3299; Volume II holds SICs 3312 through 3999.

Roman numerals I and II indicate the volume in which pages appear. Volume I holds SICs 2011 through 3299; Volume II holds SICs 3312 through 3999.

Product Index

Roman numerals I and II indicate the volume in which pages appear. Volume I holds SICs 2011 through 3299; Volume II holds SICs 3312 through 3999.

Mortars, p. Vol. I - 1162, Vol. II - 1442 [SICs 3297, 3489]
Mortisers, p. II-1609 [SIC 3553]
Mosaic tile, p. I-1085 [SIC 3253]
Mother's day cards, p. I-740 [SIC 2771]
Motor gasoline, p. I-902 [SIC 2911]
Motor homes, p. II-1965 [SIC 3716]
Motor vehicles, p. II-1953 [SIC 3714]
Motorboats, p. II-1991 [SIC 3732]
Motorcycles, p. II-2002 [SIC 3751]
Motors, p. II-1779 [SIC 3621]
Moulders, p. II-1609 [SIC 3553]
Mountings, p. II-2121 [SIC 3911]
Mouse feeds, p. I-90 [SIC 2047]
Mouthwashes, p. I-833 [SIC 2844]
Movable partitions, p. I-592 [SIC 2542]
Movements, clock, p. II-2117 [SIC 3873]
Movie cameras, p. II-2111 [SIC 3861]
Moving stairways, p. II-1535 [SIC 3534]
Mowers, lawn, p. II-1513 [SIC 3524]
Mowing machinery, p. II-1508 [SIC 3523]
Mufflers, automotive, p. II-1953 [SIC 3714]
Mufflers, clothing, p. I-343 [SIC 2323]
Multimedia kits, p. I-712 [SIC 2741]
Multiwall bags, p. I-664 [SIC 2674]
Mushrooms, p. I-44 [SIC 2033]
Music, sheet, p. I-712 [SIC 2741]
Musical instruments, p. II-2134 [SIC 3931]
Mutton, p. I-1 [SIC 2011]

Nails, p. Vol. II - 1181, 1475 [SICs 3315, 3496]
Naphtha, p. I-902 [SIC 2911]
Napkins, sanitary, p. I-673 [SIC 2676]
Narrow fabrics, p. I-244 [SIC 2241]
Nationality foods, p. Vol. I - 39, 64 [SICs 2032, 2038]
Natural cheese, p. I-20 [SIC 2022]
Necklaces, p. Vol. II - 2121, 2172 [SICs 3911, 3961]
Neckties, p. I-343 [SIC 2323]
Neckwear, p. Vol. I - 343, 372 [SICs 2323, 2339]
Nectars, p. I-44 [SIC 2033]
Needles, p. Vol. II - 2081, 2176 [SICs 3841, 3965]
Needlework, p. I-441 [SIC 2395]
Nets, fishing, p. I-321 [SIC 2298]
Netting, p. I-270 [SIC 2258]
Newsletters, p. I-712 [SIC 2741]
Newspapers, p. I-692 [SIC 2711]
Newsprint, p. I-611 [SIC 2621]
Nickel, p. II-1254 [SIC 3356]
Night tables, p. I-541 [SIC 2511]
Nightwear, p. Vol. I - 262, 266, 270, 335, 377 [SICs 2254, 2257, 2258, 2321, 2341]
Nitric acid, p. Vol. I - 776, 860 [SICs 2819, 2873]
Nitrile, p. I-786 [SIC 2822]
Nitrogen, p. I-767 [SIC 2813]
Nitroglycerine, p. I-884 [SIC 2892]
Nonwoven fabrics, p. I-317 [SIC 2297]
Noodle, egg, p. I-207 [SIC 2098]
Notary seals, p. II-2163 [SIC 3953]
Numbering stamps, p. II-2163 [SIC 3953]
Nutrients, p. I-803 [SIC 2834]
Nuts, p. Vol. I - 135, Vol. II - 1387 [SICs 2068, 3452]
Nylon, p. I-794 [SIC 2824]

Oak, p. I-459 [SIC 2411]
Oakum, p. I-326 [SIC 2299]
Oat, p. I-74 [SIC 2043]
Office furniture, p. Vol. I - 570, 575 [SICs 2521, 2522]
Office machines, p. II-1712 [SIC 3579]
Office supplies, p. I-668 [SIC 2675]
Offset printing, p. I-717 [SIC 2752]
Oil field machinery, p. II-1530 [SIC 3533]
Oil, cottonseed, p. I-139 [SIC 2074]
Oils, animal, p. I-151 [SIC 2077]
Oils, lubricating, p. Vol. I - 902, 917 [SICs 2911, 2992]
Oils, marine, p. I-151 [SIC 2077]

Ointments, toilet, p. I-833 [SIC 2844]
Ophthalmic goods, p. II-2106 [SIC 3851]
Optical instruments, p. II-2070 [SIC 3827]
Optical scanners, p. II-2070 [SIC 3827]
Ordnance, p. II-1442 [SIC 3489]
Orford, p. I-459 [SIC 2411]
Organic fibers, p. I-794 [SIC 2824]
Organosols, p. I-839 [SIC 2851]
Organs, p. II-2134 [SIC 3931]
Oriental soups, p. I-50 [SIC 2034]
Ornamental metal, p. II-1366 [SIC 3446]
Ornaments, p. II-2198 [SIC 3999]
Ottomans, p. I-547 [SIC 2512]
Outboard engines, p. II-1502 [SIC 3519]
Outboard motors, p. II-1937 [SIC 3699]
Outboards, p. II-1991 [SIC 3732]
Outdoor furniture, p. Vol. I - 541, 552 [SICs 2511, 2514]
Outdoor lighting, p. II-1853 [SIC 3648]
Outerwear, p. Vol. I - 257, 266, 270, 355, 372, 393 [SICs 2253, 2257, 2258, 2329, 2339, 2369]
Outlets, p. II-1829 [SIC 3643]
Ovens, p. Vol. II - 1670, 1799 [SICs 3567, 3631]
Overalls, p. I-351 [SIC 2326]
Overcoats, p. I-331 [SIC 2311]
Oxygen, p. I-767 [SIC 2813]
Oxygen tents, p. II-2081 [SIC 3841]

Pacifiers, p. I-953 [SIC 3069]
Packaging board, p. I-617 [SIC 2631]
Packaging machinery, p. II-1681 [SIC 3569]
Packaging paper, p. Vol. I - 611, 648 [SICs 2621, 2671]
Packaging, foil, p. II-1481 [SIC 3497]
Packing, p. I-941 [SIC 3053]
Padded envelopes, p. I-678 [SIC 2677]
Paddings, p. I-326 [SIC 2299]
Pads, p. Vol. I - 627, Vol. II - 2180 [SICs 2653, 3991]
Pads, automotive, p. I-326 [SIC 2299]
Pads, writing, p. I-683 [SIC 2678]
Pails, p. II-1303 [SIC 3412]
Paint brushes, p. II-2180 [SIC 3991]
Paints, p. I-839 [SIC 2851]
Pajamas, p. I-335 [SIC 2321]
Palings, p. I-535 [SIC 2499]
Pallets, p. Vol. I - 509, 627 [SICs 2448, 2653]
Pamphlets, p. Vol. I - 702, 707 [SICs 2731, 2732]
Pancake mix, p. I-82 [SIC 2045]
Panelboards, p. II-1774 [SIC 3613]
Panties, p. I-377 [SIC 2341]
Pantryware, p. II-1412 [SIC 3469]
Pants, p. Vol. I - 351, 372 [SICs 2326, 2339]
Pants, training, p. I-377 [SIC 2341]
Pantsuits, p. I-368 [SIC 2337]
Pantyhose, p. I-249 [SIC 2251]
Paper, p. Vol. I - 611, 648, 687 [SICs 2621, 2671, 2679]
Paperbacks, p. I-702 [SIC 2731]
Paperboard, p. I-617 [SIC 2631]
Papier-mache, p. I-687 [SIC 2679]
Parachutes, p. I-454 [SIC 2399]
Parasols, p. II-2198 [SIC 3999]
Particleboard, p. I-531 [SIC 2493]
Partitions, p. Vol. I - 586, 592, 627, Vol. II - 1366 [SICs 2541, 2542, 2653, 3446]
Passenger car bodies, p. II-1942 [SIC 3711]
Passenger trains, p. II-1997 [SIC 3743]
Pastes, tomato, p. I-44 [SIC 2033]
Pastries, p. I-100 [SIC 2051]
Patio furniture, p. Vol. I - 541, 552 [SICs 2511, 2514]
Patterns, clothing, p. I-712 [SIC 2741]
Patterns, industrial, p. II-1659 [SIC 3565]
Pavers, p. II-1518 [SIC 3531]
Paving asphalts, p. I-902 [SIC 2911]
Paving blocks, p. Vol. I - 908, 1139 [SICs 2951, 3281]
Paving mixtures, p. I-908 [SIC 2951]

PCs, p. II-1687 [SIC 3571]
Peanut butter, p. I-211 [SIC 2099]
Pearl buttons, p. II-2176 [SIC 3965]
Pectin, p. I-211 [SIC 2099]
Pedicure scissors, p. II-1307 [SIC 3421]
Pelts, p. I-1 [SIC 2011]
Pencils, p. Vol. II - 2155, 2159 [SICs 3951, 3952]
Pens, p. Vol. II - 1508, 2155 [SICs 3523, 3951]
Pepper, p. I-211 [SIC 2099]
Percussion instrum., p. II-2134 [SIC 3931]
Perfumes, p. Vol. I - 833, 854 [SICs 2844, 2869]
Periodicals, p. Vol. I - 697, 717, 723, 728 [SICs 2721, 2752, 2754, 2759]
Peripheral equipment, p. II-1703 [SIC 3577]
Personal computers, p. II-1687 [SIC 3571]
Pesticides, p. I-873 [SIC 2879]
Pet food, p. I-90 [SIC 2047]
Pettipants, p. I-377 [SIC 2341]
Pewter, p. II-2126 [SIC 3914]
Pharmaceuticals, p. I-803 [SIC 2834]
Phonograph needles, p. II-2176 [SIC 3965]
Phonograph records, p. II-1863 [SIC 3652]
Phonographs, p. II-1858 [SIC 3651]
Phosphatic fertiliz., p. I-864 [SIC 2874]
Phosphoric acid, p. Vol. I - 776, 864 [SICs 2819, 2874]
Photocopying, p. II-2111 [SIC 3861]
Photoengraving, p. I-758 [SIC 2796]
Photographic film, p. II-2111 [SIC 3861]
Pianos, p. II-2134 [SIC 3931]
Pickled fish, p. I-185 [SIC 2091]
Pickles, p. I-54 [SIC 2035]
Picnic tables, p. I-541 [SIC 2511]
Picture frames, p. I-535 [SIC 2499]
Picture postcards, p. I-712 [SIC 2741]
Pie crust, p. I-82 [SIC 2045]
Pies, p. Vol. I - 64, 100 [SICs 2038, 2051]
Pig iron, p. II-1171 [SIC 3312]
Pigeon feeds, p. I-90 [SIC 2047]
Pigments, p. Vol. I - 772, 849 [SICs 2816, 2865]
Piles, p. I-527 [SIC 2491]
Piles, wood, p. I-459 [SIC 2411]
Pillowcases, p. Vol. I - 230, 235, 427 [SICs 2211, 2221, 2392]
Pine oil, p. I-845 [SIC 2861]
Pins, p. Vol. I - 535, Vol. II - 2121, 2130, 2176 [SICs 2499, 3911, 3915, 3965]
Pins, bowling, p. II-2149 [SIC 3949]
Pipe, p. Vol. II - 1171, 1191, 1196, 1485 [SICs 3312, 3317, 3321, 3498]
Pipe cleaning, p. II-1735 [SIC 3589]
Pipe fittings, p. Vol. II - 1463, 1485 [SICs 3494, 3498]
Pipe, concrete, p. I-1121 [SIC 3272]
Pipes, tobacco, p. II-2198 [SIC 3999]
Piping, p. I-978 [SIC 3084]
Pistols, p. Vol. II - 1429, 1437 [SICs 3482, 3484]
Pistons, p. II-1741 [SIC 3592]
Pitches, p. Vol. I - 849, 913 [SICs 2865, 2952]
Planers, p. Vol. II - 1583, 1609 [SICs 3546, 3553]
Plaster, p. I-1135 [SIC 3275]
Plasticizers, p. I-854 [SIC 2869]
Plastics, p. I-781 [SIC 2821]
Plastisols, p. I-839 [SIC 2851]
Plate printing, p. I-728 [SIC 2759]
Plate, boiler, p. II-1354 [SIC 3443]
Plated ware, p. II-2126 [SIC 3914]
Plates, p. Vol. I - 638, 687, 1101, 1105 [SICs 2656, 2679, 3262, 3263]
Plates, printing, p. I-758 [SIC 2796]
Plating, p. II-1419 [SIC 3471]
Platinum, p. Vol. II - 1224, 1228, 2121 [SICs 3339, 3341, 3911]
Play suits, p. I-393 [SIC 2369]
Playing cards, p. II-2143 [SIC 3944]
Pleating, p. I-441 [SIC 2395]
Pliers, p. II-1311 [SIC 3423]
Plotters, p. II-1703 [SIC 3577]
Plows, p. II-1508 [SIC 3523]

Roman numerals I and II indicate the volume in which pages appear. Volume I holds SICs 2011 through 3299; Volume II holds SICs 3312 through 3999.

2215

Roman numerals I and II indicate the volume in which pages appear. Volume I holds SICs 2011 through 3299; Volume II holds SICs 3312 through 3999.

Product Index

Roman numerals I and II indicate the volume in which pages appear. Volume I holds SICs 2011 through 3299; Volume II holds SICs 3312 through 3999.

Roman numerals I and II indicate the volume in which pages appear. Volume I holds SICs 2011 through 3299; Volume II holds SICs 3312 through 3999.

2218

Product Index

Roman numerals I and II indicate the volume in which pages appear. Volume I holds SICs 2011 through 3299; Volume II holds SICs 3312 through 3999.

COMPANY INDEX

This index shows, in alphabetical order, more than 21,300 companies in *Manufacturing USA*. Organizations may be public or private companies, subsidiaries or divisions of companies, joint ventures or affiliates, or corporate groups. Each company entry is followed by one or more page numbers preceded by a volume number ("I-" or "II-") in the case of single appearances or the phrase "Vol. I" or "Vol. II" in the case of multiple appearances. After the page numbers, the SICs under which the company is listed follow in brackets preceded by "SIC" or "SICs". Some company names are abbreviated.

Roman numerals I and II indicate the volume in which pages appear. Volume I holds SICs 2011 through 3299; Volume II holds SICs 3312 through 3999.

Roman numerals I and II indicate the volume in which pages appear. Volume I holds SICs 2011 through 3299; Volume II holds SICs 3312 through 3999.

Algoma Hardwoods Inc, p. I-480 [SIC 2431]
Algoma Lumber Company Inc, p. I-465 [SIC 2421]
Algoma Net Co, p. I-553 [SIC 2514]
Algonquin Industries Inc, p. II-1234 [SIC 3351]
Algood Food Co, p. I-212 [SIC 2099]
Alice and Reed Group, p. II-2173 [SIC 3961]
Alice Manufacturing Company, p. I-795 [SIC 2824]
Alinabal Inc, p. II-1413 [SIC 3469]
Aline Components Inc, p. II-1764 [SIC 3599]
All Alaskan Seafoods Inc, p. I-190 [SIC 2092]
All-Amer Asphalt & Aggregates, p. I-909 [SIC 2951]
All-American Bottling Corp, p. I-176 [SIC 2086]
All-American Cooperative, p. I-869 [SIC 2875]
All American Emblem Corp, p. I-442 [SIC 2395]
All American Manufacturing Co, p. II-1334 [SIC 3432]
All American Metal Corp, p. II-1367 [SIC 3446]
All American Products Company, p. II-1578 [SIC 3545]
All American Racers Inc, p. II-1943 [SIC 3711]
All Felt Products Inc, p. I-327 [SIC 2299]
All Kind Quilting Corp, p. I-442 [SIC 2395]
All-Luminum Products Inc, p. I-553 [SIC 2514]
All Metals Service, p. II-1239 [SIC 3353]
All Systems Color Inc, p. I-759 [SIC 2796]
Allan Tool and Machine Co, p. II-1382 [SIC 3451]
Allco Chemical Corp, p. I-850 [SIC 2865]
Alle Processing Corp, p. I-7 [SIC 2013]
Allegan Metal Finishing, p. II-1420 [SIC 3471]
Allegany Ballistics Laboratory, p. II-2011 [SIC 3764]
Allegheny Ludlum Corp, p. II-1172 [SIC 3312]
Allegheny Paper Shredders Corp, p. II-1615 [SIC 3554]
Allen and Company Printers Inc, p. I-735 [SIC 2761]
Allen and Gibbons Logging Inc, p. I-460 [SIC 2411]
Allen and John Inc, p. I-741 [SIC 2771]
Allen-Bradley Company Inc, pp. Vol II - 1666, 1790, 1901 [SICs 3566, 3625, 3676]
Allen Brothers Milling Company, p. I-70 [SIC 2041]
Allen Camper Manufacturing, p. II-2020 [SIC 3792]
Allen Clark Inc, p. I-7 [SIC 2013]
Allen Co, p. I-1071 [SIC 3231]
Allen Displays Inc, p. II-2186 [SIC 3993]
Allen-Edmonds Shoe Corp, p. I-1025 [SIC 3143]
Allen Extruders Inc, p. I-961 [SIC 3081]
Allen Family Foods Inc, p. I-12 [SIC 2015]
Allen Group Inc, p. II-2061 [SIC 3825]
Allen Morrison Inc, p. II-2186 [SIC 3993]
Allen Organ Co, p. II-2135 [SIC 3931]
Allen Pattern of Michigan Inc, p. II-1568 [SIC 3543]
Allen Products Co, p. II-2112 [SIC 3861]
Allen Stevens, p. II-1409 [SIC 3466]
Allen Telecom Group Inc, p. II-1873 [SIC 3663]
Allen Wertz Candies, p. I-127 [SIC 2064]
Allenair Corp, p. II-1464 [SIC 3494]
Allens Manufacturing Company, p. I-1017 [SIC 3131]
Allergan Inc, p. II-2107 [SIC 3851]
Allerton Supply Company Inc, p. I-861 [SIC 2873]
Alliance Chemical Inc, p. I-850 [SIC 2865]
Alliance Machine Co, p. II-1547 [SIC 3536]
Alliance Peripheral Systems Inc, p. II-1694 [SIC 3572]
Alliance Semiconductor Corp, p. II-1892 [SIC 3674]
Alliant Techsystems Inc, p. II-1443 [SIC 3489]
Allied Asphalt Paving Co, p. I-914 [SIC 2952]
Allied Baltic Rubber Inc, p. I-948 [SIC 3061]
Allied Colloids Inc, p. I-819 [SIC 2841]
Allied Color Inc, p. I-850 [SIC 2865]
Allied Color Industries Inc, p. I-773 [SIC 2816]
Allied Construction Products, p. II-1519 [SIC 3531]
Allied Container Corp, p. I-506 [SIC 2441]
Allied Devices Corp, p. II-1666 [SIC 3566]
Allied Duralux Inc, p. I-433 [SIC 2393]
Allied Finishing Inc, p. II-1420 [SIC 3471]
Allied Foods Inc, p. I-91 [SIC 2047]
Allied Forest Products Inc, p. I-465 [SIC 2421]
Allied Gear and Machine Co, p. II-1660 [SIC 3565]
Allied Golf Co, p. II-2150 [SIC 3949]
Allied-Hastings Barrell&Drum, p. II-1304 [SIC 3412]
Allied Healthcare Products Inc, p. II-2082 [SIC 3841]
Allied Inc, p. I-909 [SIC 2951]

Allied Metal Co, p. II-1229 [SIC 3341]
Allied Mineral Products, p. I-1163 [SIC 3297]
Allied Plastics Company Inc, p. I-581 [SIC 2531]
Allied Plastics Inc, p. II-2186 [SIC 3993]
Allied Printing Service Inc, p. I-729 [SIC 2759]
Allied Products Corp, p. II-1509 [SIC 3523]
Allied Readymix Inc, p. I-1127 [SIC 3273]
Allied Record Co, p. II-1864 [SIC 3652]
Allied Research Corp, p. II-1434 [SIC 3483]
Allied Resinous Products Inc, p. I-967 [SIC 3082]
Allied Ring Corp, p. II-1742 [SIC 3592]
Allied Safe and Vault Company, p. II-1492 [SIC 3499]
Allied Shipyard Inc, p. II-1986 [SIC 3731]
Allied Sinterings Inc, p. II-1295 [SIC 3399]
Allied Telesis Inc, p. II-1704 [SIC 3577]
Allied Tube and Conduit Co, p. II-1835 [SIC 3644]
AlliedSignal, pp. Vol. I - 973, Vol. II - 1873 [SICs 3083, 3663]
AlliedSignal Aerospace, p. II-1975 [SIC 3724]
AlliedSignal Aircraft, p. II-1980 [SIC 3728]
AlliedSignal Inc, pp. Vol. I - 236, 795, Vol. II - 1295, 2035 [SICs 2221, 2824, 3399, 3812]
AlliedSignal Ocean Systems, p. II-2035 [SIC 3812]
Allis Mineral Systems, p. II-1519 [SIC 3531]
Allison Abrasives Inc, p. I-1144 [SIC 3291]
Allison Engine Co, p. II-1975 [SIC 3724]
Allison Manufacturing Co, p. I-390 [SIC 2361]
Allison Transmission, p. II-1954 [SIC 3714]
Allister Manufacturing Company, p. II-1938 [SIC 3699]
Allor Manufacturing Inc, p. II-1541 [SIC 3535]
Alloy Die Casting Company Inc, p. II-1266 [SIC 3363]
Alloy Metals Company Inc, p. II-1229 [SIC 3341]
Alloy Piping Products Company, p. II-1486 [SIC 3498]
Alloy Rods Corp, p. II-1594 [SIC 3548]
Alloy Stainless Products, p. II-1486 [SIC 3498]
Alloy Surfaces Company Inc, p. II-1425 [SIC 3479]
Alloy Trailers Inc, p. II-1961 [SIC 3715]
Alloyd Company Inc, p. II-1660 [SIC 3565]
Allsop Inc, p. II-1859 [SIC 3651]
Allstar Fasteners Inc, p. II-1382 [SIC 3451]
Alltech Associates Inc, p. II-2066 [SIC 3826]
Allteq Industries Inc, p. II-2045 [SIC 3822]
Alltrista Corp, p. II-1409 [SIC 3466]
Allwaste Glass Recycling Inc, p. I-1071 [SIC 3231]
Allwaste Recycling Inc, p. I-1071 [SIC 3231]
Allway Tools Inc, p. II-1318 [SIC 3425]
Allyn and Bacon, p. I-703 [SIC 2731]
ALM Surgical Equipment Inc, p. II-1854 [SIC 3648]
Almond Corp, p. II-1420 [SIC 3471]
ALOECORP, p. I-834 [SIC 2844]
Aloette Cosmetics Inc, p. I-834 [SIC 2844]
Alofs Manufacturing Company, p. II-1404 [SIC 3465]
Alon Processing Inc, p. II-1425 [SIC 3479]
Alorna Coat Corp, p. I-373 [SIC 2339]
ALP Industries Inc, p. II-1476 [SIC 3496]
Alpenrose Dairy Inc, p. I-35 [SIC 2026]
Alpha Associates Inc, p. I-310 [SIC 2295]
Alpha Baking Co, p. I-101 [SIC 2051]
Alpha/Beta Tube Corp, p. II-1172 [SIC 3312]
Alpha Bolt Co, p. II-1388 [SIC 3452]
Alpha Corporation of Tennessee, p. I-782 [SIC 2821]
Alpha Industries Inc, p. II-1892 [SIC 3674]
Alpha Metals Inc, p. II-1255 [SIC 3356]
Alpha Mills Corp, p. I-258 [SIC 2253]
Alpha/Owens-Corning, p. I-782 [SIC 2821]
Alpha Q Inc, p. II-2016 [SIC 3769]
Alpha Sintered Metals, p. II-1295 [SIC 3399]
Alpha Stamping Co, p. II-1413 [SIC 3469]
Alpha Technologies Inc, p. II-1873 [SIC 3663]
Alpha Technology Corp, p. II-1266 [SIC 3363]
Alpha Therapeutic Corp, p. I-814 [SIC 2836]
Alpha Wire Corp, p. II-1915 [SIC 3679]
Alphabet, p. II-1954 [SIC 3714]
Alphasonik Inc, p. II-1859 [SIC 3651]
Alphin Brothers Inc, p. I-7 [SIC 2013]
Alpine Alpa Cheese Factory Inc, p. I-21 [SIC 2022]
Alpine Engineered Products Inc, p. II-1355 [SIC 3443]
Alpine Group Inc, p. II-1915 [SIC 3679]

Alpine Packing Co, p. I-2 [SIC 2011]
ALPO Petfoods Inc, p. I-91 [SIC 2047]
Alps Sportswear Mfg Co, p. I-258 [SIC 2253]
Alsimag Technical Ceramics Inc, p. I-1167 [SIC 3299]
Alsons Corp, p. II-1334 [SIC 3432]
Alsy Lighting Inc, p. II-1826 [SIC 3641]
Alta Sales Inc, p. II-1447 [SIC 3491]
Altair Corp, p. I-95 [SIC 2048]
Altama Combat Boots, p. I-1025 [SIC 3143]
Altana Inc, p. I-804 [SIC 2834]
Altec Industries Inc, p. II-1948 [SIC 3713]
AlTech Specialty Steel Corp, p. II-1172 [SIC 3312]
Altera Corp, p. II-1892 [SIC 3674]
Alternative Pioneering Systems, p. II-1800 [SIC 3631]
Alto Dairy Cooperative, p. I-21 [SIC 2022]
Alto-Shaam Inc, p. II-1736 [SIC 3589]
Altron Inc, p. II-1887 [SIC 3672]
Altronic Inc, p. II-1769 [SIC 3612]
AluChem Inc, p. I-777 [SIC 2819]
Aluf Plastics, p. I-660 [SIC 2673]
Alufoil Products Company Inc, p. II-1239 [SIC 3353]
Alum-A-Therm Heat Treating, p. II-1290 [SIC 3398]
Alum-Alloy Company Inc, p. II-1275 [SIC 3365]
Aluma-Form Inc, p. II-1295 [SIC 3399]
Aluma-Glass Industries Inc, p. I-1071 [SIC 3231]
Aluma Shield Industries Inc, p. II-1372 [SIC 3448]
Aluma Systems USA Inc, p. II-1367 [SIC 3446]
Aluma Trim Inc, p. II-1244 [SIC 3354]
Alumacraft Boat Co, p. II-1992 [SIC 3732]
Alumax Extrusions Inc, p. II-1244 [SIC 3354]
Alumax Inc, p. II-1221 [SIC 3334]
Alumax Inc Building Products, p. II-1239 [SIC 3353]
Alumax Mill Products Inc, p. II-1239 [SIC 3353]
Aluminum Casting & Eng, p. II-1275 [SIC 3365]
Aluminum Coil Anodizing Corp, p. II-1420 [SIC 3471]
Aluminum Company of America, pp. Vol II - 1221, 1244 [SICs 3334, 3354]
Aluminum Dip Braze Co, p. II-1290 [SIC 3398]
Aluminum Finishing Corp, p. II-1420 [SIC 3471]
Aluminum Housewares Company, p. II-1275 [SIC 3365]
Aluminum Precision Products, p. II-1400 [SIC 3463]
Aluminum Products Company, p. II-1349 [SIC 3442]
Alva Allen Industries Inc, p. II-1563 [SIC 3542]
Alvah Bushnell Co, p. I-669 [SIC 2675]
Alvarado Dye House Inc, p. I-279 [SIC 2261]
Alvarado Manufacturing, p. II-1377 [SIC 3449]
Alves Precision Engineered Prod, p. II-1676 [SIC 3568]
Alvey Inc, p. II-1541 [SIC 3535]
Alvord-Polk Inc, p. II-1578 [SIC 3545]
Alvord Systems Inc, p. II-2098 [SIC 3844]
Alza Corp, p. I-804 [SIC 2834]
AM Best Company Inc, p. I-703 [SIC 2731]
Am Fab, p. I-581 [SIC 2531]
AM Graphics, p. II-1621 [SIC 3555]
AM Holding Inc, p. II-1934 [SIC 3695]
AM International Inc, p. II-1704 [SIC 3577]
AM Multigraphics, p. II-1713 [SIC 3579]
Am-Safe Inc, p. I-455 [SIC 2399]
Amac Enterprises Inc, p. II-1420 [SIC 3471]
Amacoil Inc, p. II-1600 [SIC 3549]
Amadas Industries Inc, p. II-1509 [SIC 3523]
Amalga Composites Inc, p. I-1066 [SIC 3229]
Amalgamated Sugar Co, p. I-123 [SIC 2063]
Amalloy Corp, p. II-1212 [SIC 3325]
Aman Brothers Inc, p. I-1117 [SIC 3271]
Amana Refrigeration Inc, p. II-1805 [SIC 3632]
Amanda Bent Bolt Co, p. II-1388 [SIC 3452]
Amarillo Gear Co, p. II-1666 [SIC 3566]
Amark Inc, p. I-1025 [SIC 3143]
Amatex Corp, p. I-283 [SIC 2262]
Amax Metals Recovery Inc, p. I-777 [SIC 2819]
Amax Plating Inc, p. II-1420 [SIC 3471]
Ambassador Industries, p. I-599 [SIC 2591]
Amber Milling Co, p. I-70 [SIC 2041]
Amcast Inc, p. II-1271 [SIC 3364]
Amcast Industrial Corp, pp. Vol II - 1275, 1492 [SICs 3365, 3499]

Roman numerals I and II indicate the volume in which pages appear. Volume I holds SICs 2011 through 3299; Volume II holds SICs 3312 through 3999.

Roman numerals I and II indicate the volume in which pages appear. Volume I holds SICs 2011 through 3299; Volume II holds SICs 3312 through 3999.

American Timber Co, p. I-465 [SIC 2421]
American Timber Homes Inc, p. I-523 [SIC 2452]
American Tool and Mold Inc, p. II-1266 [SIC 3363]
American Tool Companies Inc, p. II-1312 [SIC 3423]
American Tooth Industries, p. II-2093 [SIC 3843]
American Torch Tip Co, p. II-1594 [SIC 3548]
American Tourister Inc, p. I-1041 [SIC 3161]
American Trans-Coil Corp, p. II-1905 [SIC 3677]
American Trouser Inc, p. I-348 [SIC 2325]
American Ultraviolet Co, p. II-1854 [SIC 3648]
American Uniform Co, p. I-352 [SIC 2326]
American United Global Inc, p. I-942 [SIC 3053]
American United Products Inc, p. I-942 [SIC 3053]
American United Seal Inc, p. I-942 [SIC 3053]
American Vanguard Corp, p. I-874 [SIC 2879]
American Velvet Co, p. I-236 [SIC 2221]
American Walnut Company Inc, p. I-471 [SIC 2426]
American Water Heater Group, p. II-1822 [SIC 3639]
American Wax Inc, p. II-1605 [SIC 3552]
American Weavers LP, p. I-236 [SIC 2221]
American Western, p. I-1006 [SIC 3089]
American White Cross Inc, p. II-2087 [SIC 3842]
American Woodcraft Inc, p. I-587 [SIC 2541]
American Woodmark Corp, p. I-486 [SIC 2434]
American Woolen Co, p. I-241 [SIC 2231]
American Wyott Corp, p. II-1736 [SIC 3589]
Americana Art China Co, p. I-1102 [SIC 3262]
Americana Knitting Mills Inc, p. I-258 [SIC 2253]
Americo Manufacturing, p. I-1144 [SIC 3291]
Americraft Carton Inc, p. I-643 [SIC 2657]
AmeriMark Inc, p. II-1349 [SIC 3442]
Ameripol Synpol Co, p. I-787 [SIC 2822]
AmeriQuest Technologies Inc, p. II-1694 [SIC 3572]
AmeriSig Inc, p. I-729 [SIC 2759]
Ameritone Paint Corp, p. I-840 [SIC 2851]
Ameriwood Furniture, p. I-542 [SIC 2511]
Ameriwood Indust Intern Corp, p. I-542 [SIC 2511]
AmerLink Ltd, p. I-523 [SIC 2452]
Amerock Corp, p. II-1323 [SIC 3429]
Ameron Inc, pp. Vol. I - 979, 1122 [SICs 3084, 3272]
Ameron Pole Products, p. I-1122 [SIC 3272]
Amertec-Granada Inc, p. I-587 [SIC 2541]
Amerway Inc, p. II-1229 [SIC 3341]
Ames Company Inc, p. II-1464 [SIC 3494]
Ames Industrial Supply Co, p. I-787 [SIC 2822]
Ames Rubber Corp, p. I-954 [SIC 3069]
Ames Safety Envelope Co, p. I-679 [SIC 2677]
Amesbury Group Inc, p. I-942 [SIC 3053]
AMETEK Inc, pp. Vol II - 2051, 2076 [SICs 3823, 3829]
AMF Bakery Systems, p. II-1626 [SIC 3556]
AMF Bowling Inc, p. I-471 [SIC 2426]
AMG Resources Corp, p. II-1229 [SIC 3341]
Amgen Inc, p. I-804 [SIC 2834]
AMI Inc, p. II-1275 [SIC 3365]
Amicale Industries Inc, p. I-327 [SIC 2299]
Amicon Inc, p. II-2051 [SIC 3823]
Amida Industries Inc, p. II-1854 [SIC 3648]
Amigos Canning Company Inc, p. I-40 [SIC 2032]
Aminco Inc, p. I-834 [SIC 2844]
Amis Materials Co, p. I-1122 [SIC 3272]
Amistar Corp, p. II-1600 [SIC 3549]
Amital Spinning Corp, p. I-297 [SIC 2281]
Amity Leather Products Co, p. I-1049 [SIC 3172]
Amjems Inc, p. I-674 [SIC 2676]
Amkor Electronics Inc, p. II-1892 [SIC 3674]
Amloid Corp, p. II-2144 [SIC 3944]
Amoco Chemical Co, p. I-855 [SIC 2869]
Amoco Fabrics and Fibers Co, p. I-236 [SIC 2221]
Amoco Foam Products Co, p. I-991 [SIC 3086]
Amoco Performance Products, p. I-855 [SIC 2869]
Amoco Technology Co, p. II-1260 [SIC 3357]
Amorous Andi's, p. I-107 [SIC 2052]
Amos Eby Co, p. I-869 [SIC 2875]
Amos-Hill Associates Inc, p. I-491 [SIC 2435]
Amot Controls Corp, p. II-2051 [SIC 3823]
AMP-AKZO Co, p. II-1887 [SIC 3672]
AMP Inc, p. II-1910 [SIC 3678]
Ampacet Corp, p. I-773 [SIC 2816]

Ampad Corp, p. I-684 [SIC 2678]
Ampco Metal Inc, p. II-1280 [SIC 3366]
Ampco Products Inc, p. I-486 [SIC 2434]
Ampere Automotive Corp, p. II-1929 [SIC 3694]
Ampex Casting Corp, p. II-2131 [SIC 3915]
Ampex Corp, p. II-1694 [SIC 3572]
Ampex Systems Corp, p. II-1694 [SIC 3572]
Amphenol Commercial, p. II-1476 [SIC 3496]
Amphenol Corp, p. II-1910 [SIC 3678]
Amplaco Group Inc, p. I-352 [SIC 2326]
Amplex Corp, p. I-1144 [SIC 3291]
AmPro Corp, p. II-2112 [SIC 3861]
Amrep Inc, p. I-825 [SIC 2842]
Amro Manufacturing Company, p. II-2181 [SIC 3991]
Amron Corp, p. II-1443 [SIC 3489]
Amsat, p. II-1525 [SIC 3532]
AMSCO Healthcare, p. II-1854 [SIC 3648]
AMSCO International Inc, p. II-2087 [SIC 3842]
Amsco Products, p. II-1388 [SIC 3452]
Amspec Chemical Co, p. I-777 [SIC 2819]
AMSTED Industries Inc, p. II-1197 [SIC 3321]
Amster Novelty Co, p. I-446 [SIC 2396]
Amsterdam Printing & Litho, p. I-718 [SIC 2752]
AMT Industries Inc, p. II-1709 [SIC 3578]
Amtec Precision Products Inc, p. II-1382 [SIC 3451]
Amtech Corp, p. II-1915 [SIC 3679]
AMTECH Inc, p. II-1447 [SIC 3491]
Amtech Systems Inc, p. II-2045 [SIC 3822]
AmTran Corp, p. II-1948 [SIC 3713]
AMVAC Chemical Corp, p. I-874 [SIC 2879]
AMW Cuyuna Engine Co, p. II-1503 [SIC 3519]
Amway Corp, p. I-834 [SIC 2844]
AMX Corp, p. II-1859 [SIC 3651]
Amy Lynn of California, p. I-361 [SIC 2331]
Anadex Inc, p. II-2056 [SIC 3824]
Anaheim Custom Extruders Inc, p. I-967 [SIC 3082]
Anaheim Extrusion Company, p. II-1244 [SIC 3354]
Anaheim Foundry Co, p. II-1197 [SIC 3321]
Analog and Digital Systems Inc, p. II-1859 [SIC 3651]
Analog Devices Inc, p. II-1892 [SIC 3674]
Analogic Corp, p. II-2051 [SIC 3823]
Analytical Technology Inc, p. II-2076 [SIC 3829]
Anarad Inc, p. II-2045 [SIC 3822]
Anatomical Chart Co, p. I-973 [SIC 3083]
Anchor Audio Inc, p. II-1859 [SIC 3651]
Anchor Block and Concrete Co, p. I-1117 [SIC 3271]
Anchor Brewing Co, p. I-160 [SIC 2082]
Anchor Concrete Products, p. I-1117 [SIC 3271]
Anchor Continental Inc, p. I-654 [SIC 2672]
Anchor Crane and Hoist Service, p. II-1547 [SIC 3536]
Anchor-Darling Valve Co, p. II-1464 [SIC 3494]
Anchor Dyeing and Finishing, p. I-241 [SIC 2231]
Anchor Fabricators Inc, p. II-1377 [SIC 3449]
Anchor Gasoline Corp, p. I-903 [SIC 2911]
Anchor Glass Container Corp, p. I-1062 [SIC 3221]
Anchor Hocking Packaging Co, p. II-1409 [SIC 3466]
Anchor Packaging, p. I-782 [SIC 2821]
Anchor Post Products of Texas, p. II-1182 [SIC 3315]
Anchor Rubber Co, p. I-936 [SIC 3052]
Anchor Tool and Die Co, p. II-1572 [SIC 3544]
Anchor Wire Corp, p. II-1476 [SIC 3496]
Andersen 2000 Inc, p. II-1654 [SIC 3564]
Andersen Group Inc, p. II-2093 [SIC 3843]
Anderson Chemical Company, p. I-897 [SIC 2899]
Anderson-Cook Inc, p. II-1563 [SIC 3542]
Anderson Custom Processing Inc, p. I-87 [SIC 2046]
Anderson Development Co, p. I-782 [SIC 2821]
Anderson Erickson Dairy Co, p. I-35 [SIC 2026]
Anderson Fittings Inc, p. II-1334 [SIC 3432]
Anderson Forest Products Inc, p. I-510 [SIC 2448]
Anderson, Greenwood and Co, p. II-1447 [SIC 3491]
Anderson-Hickey Co, p. I-576 [SIC 2522]
Anderson International Corp, p. II-1626 [SIC 3556]
Anderson Iron Works Inc, p. II-1367 [SIC 3446]
Anderson Lithograph Co, p. I-718 [SIC 2752]
Anderson Pattern Inc, p. II-1568 [SIC 3543]
Anderson Power Products, p. II-1775 [SIC 3613]
Anderson Products Inc, p. II-1318 [SIC 3425]
Anderson-Snow Corp, p. II-1486 [SIC 3498]

Andes Candies Inc, p. I-127 [SIC 2064]
Andin International Inc, p. II-2122 [SIC 3911]
Andis Co, p. II-2199 [SIC 3999]
Andover Controls Corp, p. II-2045 [SIC 3822]
Andover Industries Inc, p. II-1943 [SIC 3711]
Andover Togs Inc, p. I-394 [SIC 2369]
Andover Wood Products Inc, p. I-480 [SIC 2431]
Andresen-Ryan Coffee Co, p. I-195 [SIC 2095]
Andrew Corp, p. II-1915 [SIC 3679]
Andrew Dutton Company Inc, p. I-599 [SIC 2591]
Andrew Jergens Co, p. I-834 [SIC 2844]
Andrew SciComm Inc, p. II-1878 [SIC 3669]
Andrews Apparel Co, p. I-356 [SIC 2329]
Andrews Glass Company Inc, p. I-1071 [SIC 3231]
Andrews Knitting Mills Inc, p. I-258 [SIC 2253]
Andrex Industries Corp, p. I-267 [SIC 2257]
Andritz Ruthner Inc, p. II-1736 [SIC 3589]
Andros Inc, p. II-2066 [SIC 3826]
Anello Corp, pp. Vol II - 1775, 2045 [SICs 3613, 3822]
Anemostat Products, p. II-1726 [SIC 3585]
Angel Echevarria Company Inc, p. I-558 [SIC 2515]
Angeles Metal Trim Co, p. II-1344 [SIC 3441]
Angelica Corp, p. I-356 [SIC 2329]
Angelina Hardwood Sales, p. I-471 [SIC 2426]
Angelique Imports Inc, p. I-361 [SIC 2331]
Angelus Sanitary Can, p. II-1660 [SIC 3565]
Anglo Fabrics Company Inc, p. I-241 [SIC 2231]
Angola Die Casting Inc, p. II-1266 [SIC 3363]
Angola Wire Products Inc, p. II-1476 [SIC 3496]
Angostura International Ltd, p. I-181 [SIC 2087]
Angstrohm Precision Inc, p. II-1901 [SIC 3676]
Angstrom Inc, p. II-2066 [SIC 3826]
ANGUS Chemical Co, p. I-855 [SIC 2869]
Angus-Palm Industries Inc, p. II-1492 [SIC 3499]
Anheuser-Busch Companies Inc, p. I-160 [SIC 2082]
Anheuser-Busch Inc, p. I-160 [SIC 2082]
Animal By-Products Corp, p. I-152 [SIC 2077]
Animal Fair Inc, p. II-2140 [SIC 3942]
Animated Advertising Inc, p. I-669 [SIC 2675]
Anitec Imaging Products, p. II-2112 [SIC 3861]
Anja Engineering Corp, p. II-2156 [SIC 3951]
Anko Products Inc, p. II-1753 [SIC 3594]
Ann Arbor Machine Co, p. II-1558 [SIC 3541]
Ann Marie Inc, p. II-2173 [SIC 3961]
Ann Sacks Tile and Stone, p. I-1086 [SIC 3253]
Annette Fashions Corp, p. I-369 [SIC 2337]
Annette Island Packing Co, p. I-186 [SIC 2091]
Annieglass Inc, p. I-1066 [SIC 3229]
Annin and Co, p. I-455 [SIC 2399]
Annona Manufacturing Co, p. I-480 [SIC 2431]
Anocoil Corp, p. I-759 [SIC 2796]
Anodizing Inc, p. II-1244 [SIC 3354]
Anorad Corp, p. II-1558 [SIC 3541]
Ansam Metals Corp, p. II-1229 [SIC 3341]
Anson Shirt Co, p. I-352 [SIC 2326]
Ansonia Copper and Brass Inc, p. II-1234 [SIC 3351]
Ansul Inc, p. II-1649 [SIC 3563]
Answer Products Inc, p. II-2003 [SIC 3751]
Antares Group Inc, p. II-1887 [SIC 3672]
Antenna Specialists Co, p. II-1873 [SIC 3663]
Anthem Technology Systems, p. II-1694 [SIC 3572]
Anthony Dally and Sons Inc, p. I-1140 [SIC 3281]
Anthony House Inc, p. I-424 [SIC 2391]
Anthony Industries Inc, p. II-2150 [SIC 3949]
Anthony Timberlands Inc, p. I-471 [SIC 2426]
Anthra Textile Company Inc, p. I-241 [SIC 2231]
Anthracite Industries Inc, p. I-924 [SIC 2999]
Anthro Corp, p. I-576 [SIC 2522]
Antibodies Inc, p. I-814 [SIC 2836]
Antigo Flour and Feed Co, p. I-70 [SIC 2041]
Antigua Mills Inc, p. I-292 [SIC 2273]
Antioch Publishing Co, p. I-684 [SIC 2678]
Antone's Records and Tapes, p. II-1864 [SIC 3652]
Anvil Cases Inc, p. I-634 [SIC 2655]
Anzon Inc, p. I-777 [SIC 2819]
AO Smith Automotive Prod Co, p. II-1954 [SIC 3714]
AO Smith Corp, p. II-1780 [SIC 3621]
AO Smith Harvestore Products, p. II-1355 [SIC 3443]
AP Green Industries Inc, p. I-1090 [SIC 3255]

Roman numerals I and II indicate the volume in which pages appear. Volume I holds SICs 2011 through 3299; Volume II holds SICs 3312 through 3999.

AP Green Refractories Inc, p. I-1163 [SIC 3297]
AP Parts Manufacturing Co, p. II-1192 [SIC 3317]
AP Parts Marketing Co, p. II-1954 [SIC 3714]
AP Techno Glass Co, p. I-1057 [SIC 3211]
APAC Carolina Inc, p. I-909 [SIC 2951]
Apache Enterprises Inc, p. II-1970 [SIC 3721]
Apache Nitrogen Products Inc, p. I-885 [SIC 2892]
Apache Plastics LP, p. I-979 [SIC 3084]
APAQ Technology Inc, p. II-1688 [SIC 3571]
Apex Alkali Products Co, p. I-918 [SIC 2992]
Apex Electronics Inc, p. II-1883 [SIC 3671]
Apex Mills Corp, p. I-271 [SIC 2258]
Apex Paper Box Corp, p. I-623 [SIC 2652]
Apex Wire and Cable Corp, p. II-1260 [SIC 3357]
APG, p. II-1709 [SIC 3578]
API Industries Inc, pp. Vol. I - 660, Vol. II - 1420 [SICs 2673, 3471]
APL Corp, p. I-674 [SIC 2676]
Apogee Enterprises Inc, p. I-1057 [SIC 3211]
Apogee Sound Inc, p. II-1859 [SIC 3651]
Apollo Colors Inc, p. I-850 [SIC 2865]
Apollo Metals Ltd, p. II-1420 [SIC 3471]
Apollo of the Ozarks Inc, p. I-897 [SIC 2899]
Apothecary Products Inc, p. II-2066 [SIC 3826]
Appalachian Electr Instruments, p. II-1938 [SIC 3699]
Appalachian Timber Service Inc, p. I-528 [SIC 2491]
Apparel America Inc, p. I-356 [SIC 2329]
Apparel Brands Inc, p. I-348 [SIC 2325]
Apparel Group, p. I-336 [SIC 2321]
Apparel Group Inc, p. I-369 [SIC 2337]
APPCO Process Equipment, p. II-1255 [SIC 3356]
Apperson Business Forms Inc, p. I-735 [SIC 2761]
Appert's Food Inc, p. I-7 [SIC 2013]
Applause Inc, p. II-2140 [SIC 3942]
Apple Computer Inc, p. II-1688 [SIC 3571]
Applegate Insulation Mfg, p. I-869 [SIC 2875]
Appleton Electric Co, p. II-1835 [SIC 3644]
Appleton Mills, p. I-236 [SIC 2221]
Appleton Papers Inc, p. I-654 [SIC 2672]
Appleton Supply Company Inc, p. II-1361 [SIC 3444]
Appliance Control Technology, p. II-1775 [SIC 3613]
Appliance Science Corp, p. II-1813 [SIC 3634]
Applied Automation Inc, p. II-2076 [SIC 3829]
Applied Coating Technology Inc, p. II-1420 [SIC 3471]
Applied Digital Data Systems, p. II-1699 [SIC 3575]
Applied Energy Company Inc, p. II-1753 [SIC 3594]
Applied Engineering Products, p. II-1910 [SIC 3678]
Applied Indust Materials, p. II-1178 [SIC 3313]
Applied Magnetics Corp, p. II-1694 [SIC 3572]
Applied Materials Inc, p. II-1632 [SIC 3559]
Applied Mechanical Energy, p. II-1676 [SIC 3568]
Applied Micro Circuits Corp, p. II-1892 [SIC 3674]
Applied Microbiology Inc, p. I-814 [SIC 2836]
Applied Power Inc, p. II-1454 [SIC 3492]
Applied Science Laboratories, p. I-799 [SIC 2833]
April Hill Inc, p. I-83 [SIC 2045]
APS Systems, p. II-1998 [SIC 3743]
AptarGroup Inc, p. II-1447 [SIC 3491]
APV Baker Inc, p. II-1626 [SIC 3556]
APV Crepaco Inc, p. II-1626 [SIC 3556]
APV Gaulin, p. II-1626 [SIC 3556]
AQUA-10 Laboratories, p. I-814 [SIC 2836]
Aqua-Aerobic Systems Inc, p. II-1682 [SIC 3569]
Aqua-Chem Inc, pp. Vol II - 1339, 1355, 1736 [SICs 3433, 3443, 3589]
Aqua-Dyne Inc, p. II-1753 [SIC 3594]
Aqua Glass Corp, p. I-1001 [SIC 3088]
Aqua-Mist Inc, p. II-1813 [SIC 3634]
Aqua-Trol Corp, p. II-1334 [SIC 3432]
Aqualon, p. I-897 [SIC 2899]
Aquapore Moisture Systems Inc, p. I-936 [SIC 3052]
Aquatic Industries Inc, p. I-1001 [SIC 3088]
Aqvila Inc, p. I-258 [SIC 2253]
AR Taylor Veneer Co, p. I-491 [SIC 2435]
AR Test Systems, p. II-2061 [SIC 3825]
Araban Coffee Company Inc, p. I-195 [SIC 2095]
Aramis Inc, p. I-834 [SIC 2844]
Arandell-Schmidt Corp, p. I-718 [SIC 2752]
Ararat Rock Products Inc, p. I-909 [SIC 2951]

Arban and Carosi Inc, p. I-1122 [SIC 3272]
Arbuckle Coffee Co, p. I-195 [SIC 2095]
Arc Machines Inc, p. II-1594 [SIC 3548]
ARC Rubber Inc, p. I-948 [SIC 3061]
Arcadian Corp, p. I-861 [SIC 2873]
Arcar Graphics Inc, p. I-889 [SIC 2893]
Arcata Redwood, p. I-465 [SIC 2421]
Arch Technology Corp, p. II-1372 [SIC 3448]
Archer Daniels Midland Co, pp. Vol. I - 70, 87 [SICs 2041, 2046]
Archer Wire International Corp, p. II-1476 [SIC 3496]
Archibald Candy Corp, p. I-127 [SIC 2064]
Architectural Reproductions Inc, p. I-1167 [SIC 3299]
Architectural Sculpture Ltd, p. I-1167 [SIC 3299]
Arco Alloys Corp, p. II-1229 [SIC 3341]
ARCO Chemical Co, p. I-777 [SIC 2819]
Arco Electronics Inc, p. II-1897 [SIC 3675]
Arcobasso Foods Inc, p. I-55 [SIC 2035]
Arcon Coating Mills Inc, p. I-649 [SIC 2671]
Arcos Alloys, p. II-1594 [SIC 3548]
Arctic Ice Cream Novelties LP, p. I-30 [SIC 2024]
Arctic Industries Inc, p. II-1805 [SIC 3632]
Arcy Plastic Laminates Inc, p. I-973 [SIC 3083]
Ardco Inc, p. I-1071 [SIC 3231]
ARDCO Industries, p. II-2029 [SIC 3799]
Ardell Industries, p. II-1308 [SIC 3421]
Arden/Benhar Mills, p. I-455 [SIC 2399]
Arden Corp, p. I-327 [SIC 2299]
Arden International Kitchens Inc, p. I-65 [SIC 2038]
Arden Jewelry Mfg Co, p. II-2173 [SIC 3961]
Ardney Ltd, p. I-413 [SIC 2386]
Ardrox Inc, p. I-855 [SIC 2869]
Area Electronics Systems Inc, p. II-1887 [SIC 3672]
ARES Inc, p. II-1443 [SIC 3489]
Argent Automotive Systems Inc, p. I-942 [SIC 3053]
Argentum USA Inc, p. I-365 [SIC 2335]
ARGO Industries Inc, p. II-1339 [SIC 3433]
ARI Industries Inc, p. II-1182 [SIC 3315]
Aridyne Corp, p. I-271 [SIC 2258]
Ariel Corp, p. II-1649 [SIC 3563]
Ariel Vineyards, p. I-168 [SIC 2084]
Ariens Co, p. II-1514 [SIC 3524]
Aries Electronics Inc, p. II-1910 [SIC 3678]
Aris Isotoner Inc, p. I-1037 [SIC 3151]
Aristech Chemical Corp, p. I-782 [SIC 2821]
Aristo-Craft Inc, p. II-1920 [SIC 3691]
Aristokraft Inc, p. I-486 [SIC 2434]
Aristotle Corp, p. I-382 [SIC 2342]
Arizona Catus Ranch, p. I-40 [SIC 2032]
Arizona Portland Cement Co, p. I-1077 [SIC 3241]
Arizona Sun Products Inc, p. I-799 [SIC 2833]
Ark-Ell Springs Inc, p. II-1470 [SIC 3495]
ARK Foundry & Mfg Co, p. II-1229 [SIC 3341]
Ark-Les Corp, p. II-1915 [SIC 3679]
Arkalite, p. I-1154 [SIC 3295]
Arkansas Face Veneer Company, p. I-491 [SIC 2435]
Arkansas Lighthouse, p. I-684 [SIC 2678]
Arkansas Lime Co, p. I-1132 [SIC 3274]
Arkansas Trailer Manufacturing, p. II-1948 [SIC 3713]
Arkay Packaging Corp, p. I-643 [SIC 2657]
Arkhola Sand and Gravel Co, p. I-909 [SIC 2951]
Arko Paper Products Company, p. I-643 [SIC 2657]
Arkwin Industries Inc, p. II-1980 [SIC 3728]
Arkwright Inc, p. II-2112 [SIC 3861]
Arkwright Mills, p. I-231 [SIC 2211]
Arland Tool and Manufacturing, p. II-1558 [SIC 3541]
Arlee Home Fashions Inc, p. I-424 [SIC 2391]
Arlington Hat Company Inc, p. I-275 [SIC 2259]
Arlington Industries Inc, p. II-1271 [SIC 3364]
Arm and Hammer, p. I-764 [SIC 2812]
Armbrust Corp, p. II-1476 [SIC 3496]
Armco Coshocton Operations, p. II-1172 [SIC 3312]
Armco Inc, p. II-1172 [SIC 3312]
Armor All Products Corp, p. I-879 [SIC 2891]
Armor Cote Corp, p. I-914 [SIC 2952]
Armotek Industries Inc, p. I-759 [SIC 2796]
Armour Food Ingredients, p. I-26 [SIC 2023]
Armour-Freeborn Foods, p. I-26 [SIC 2023]
Armour Swift-Eckrich Inc, p. I-7 [SIC 2013]

Armscorp USA Inc, p. II-1438 [SIC 3484]
Armstrong-Blum Mfg Co, p. II-1318 [SIC 3425]
Armstrong Brands Inc, p. I-997 [SIC 3087]
Armstrong Bros Tool Co, p. II-1312 [SIC 3423]
Armstrong International Inc, p. II-1464 [SIC 3494]
Armstrong Lumber Company, p. I-501 [SIC 2439]
Armstrong World Industries Inc, p. I-1086 [SIC 3253]
Arneson Foundry Inc, p. II-1212 [SIC 3325]
Arnold Engineering Co, p. II-1492 [SIC 3499]
Arnold Furniture Manufacturers, p. I-571 [SIC 2521]
Arnold Magnetics Corp, p. II-1795 [SIC 3629]
Arnold Pen Company Inc, p. II-2156 [SIC 3951]
Arntzen Corp, p. II-1486 [SIC 3498]
Aro Fluid Products, p. II-1584 [SIC 3546]
Aromat Corp, p. II-1790 [SIC 3625]
Aronsohn Tie Fabrics, p. I-236 [SIC 2221]
Arral Industries Inc, p. II-2016 [SIC 3769]
Array Technology Corp, p. II-1694 [SIC 3572]
Arrhythmia Research Technology, p. II-2102 [SIC 3845]
Arrow Ace Die Cutting Company, p. I-669 [SIC 2675]
Arrow Acme Co, p. II-1266 [SIC 3363]
Arrow Aluminum Castings Co, p. II-1275 [SIC 3365]
Arrow Business Forms & Labels, p. I-735 [SIC 2761]
Arrow Fastener Company Inc, p. II-1312 [SIC 3423]
Arrow Gear Co, p. II-1980 [SIC 3728]
Arrow Group Industries Inc, p. II-1372 [SIC 3448]
Arrow Inc, p. II-1344 [SIC 3441]
Arrow Industries Inc, p. I-649 [SIC 2671]
Arrow International Inc, p. II-2082 [SIC 3841]
Arrow Lock Manufacturing Co, p. II-1323 [SIC 3429]
Arrow Manufacturing Co, p. II-1460 [SIC 3493]
Arrow Paper Products Co, p. I-634 [SIC 2655]
Arrow Pattern and Foundry Co, p. I-536 [SIC 2499]
Arrow Pneumatics Inc, p. II-1312 [SIC 3423]
Arrow Road Construction Co, p. I-909 [SIC 2951]
Arrow Tru-Line Inc, p. II-1323 [SIC 3429]
Arrow Truck Bodies & Equip, p. II-1948 [SIC 3713]
Arrowhead Brass Products Inc, p. II-1334 [SIC 3432]
Arrowhead Mountain, p. I-176 [SIC 2086]
Arrowhead Products, p. II-1980 [SIC 3728]
Arrowhead Stator Rotor Inc, p. II-1929 [SIC 3694]
Arrowsmith Tool and Die Inc, p. II-1764 [SIC 3599]
Art-Craft Optical Co, p. II-2107 [SIC 3851]
Art Deco Jewelry Inc, p. II-2122 [SIC 3911]
Art Leather Manufacturing, p. I-745 [SIC 2782]
Art Line Inc, p. I-1113 [SIC 3269]
Art Needlecraft Inc, p. I-420 [SIC 2389]
Art-Phyl Creations, p. I-587 [SIC 2541]
Art's-Way Manufacturing, p. II-1509 [SIC 3523]
Artco-Bell Corp, p. I-581 [SIC 2531]
Artco Corp, p. I-593 [SIC 2542]
Artee Industries Inc, p. I-297 [SIC 2281]
Artex Manufacturing Company, p. I-356 [SIC 2329]
Arthur Court Designs Inc, p. II-1295 [SIC 3399]
Arthur Fulmer Inc, p. II-2003 [SIC 3751]
Arthur J Evers Corp, p. II-1621 [SIC 3555]
Arthur Matney Company Inc, p. I-834 [SIC 2844]
Artichoke Industries Inc, p. I-45 [SIC 2033]
Artisan House Inc, p. II-1367 [SIC 3446]
Artistic Cabinets, p. I-486 [SIC 2434]
Artistic Carton Co, p. I-643 [SIC 2657]
Artistic Creations Inc, p. I-373 [SIC 2339]
Artistic Finishes Inc, p. I-528 [SIC 2491]
Artistic Glass Products Co, p. I-1071 [SIC 3231]
Artistic Identification Systems, p. I-245 [SIC 2241]
Artistic Office Products, p. I-684 [SIC 2678]
Artistic Woven Labels, p. I-245 [SIC 2241]
Artkraft Strauss Sign Corp, p. II-2186 [SIC 3993]
ARTOS Engineering Co, p. II-1632 [SIC 3559]
Artra Group Inc, p. II-2173 [SIC 3961]
Artray Label Company Inc, p. I-245 [SIC 2241]
Artwork Reproduction Inc, p. I-446 [SIC 2396]
Arundel Corp, p. I-1127 [SIC 3273]
Arvco Container Corp, p. I-628 [SIC 2653]
Arvin Industries Inc, p. II-1954 [SIC 3714]
Arway Confections Inc, p. I-127 [SIC 2064]
ASA Corp, p. II-1859 [SIC 3651]
Asael Farr and Sons Co, p. I-30 [SIC 2024]

Roman numerals I and II indicate the volume in which pages appear. Volume I holds SICs 2011 through 3299; Volume II holds SICs 3312 through 3999.

Asahi Glass America Inc, p. I-1057 [SIC 3211]
Asante Technologies Inc, p. II-1704 [SIC 3577]
Asbury Carbons Inc, p. I-1154 [SIC 3295]
Asbury Graphite Mills Inc, p. I-1154 [SIC 3295]
Asbury Graphite of California, p. I-1154 [SIC 3295]
Asbury Park Press Inc, p. I-693 [SIC 2711]
ASC Inc, p. II-1954 [SIC 3714]
ASC Industries Inc, p. II-1897 [SIC 3675]
ASC Pacific Inc, p. II-1361 [SIC 3444]
Ascom Hasler Mailing Systems, p. II-1713 [SIC 3579]
Ascom Timeplex Inc, p. II-1873 [SIC 3663]
Ascom Warren Inc, p. II-1795 [SIC 3629]
Asea Brown Boveri, p. II-1355 [SIC 3443]
Asea Brown Boveri Inc, pp. Vol II - 1355, 1780 [SICs 3443, 3621]
Aseco Corp, p. II-2061 [SIC 3825]
Aseptico Inc, p. II-2093 [SIC 3843]
Ash Grove Cement Co, p. I-1077 [SIC 3241]
Ash-Lin Inc, p. I-515 [SIC 2449]
Ash Stevens Inc, p. I-799 [SIC 2833]
Ashaway Line & Twine Mfg Co, p. I-322 [SIC 2298]
Ashbach Construction Co, p. I-909 [SIC 2951]
Ashbrook Corp, p. II-1736 [SIC 3589]
Ashe Brick, p. I-1081 [SIC 3251]
Asheboro Elastics Corp, p. I-231 [SIC 2211]
Ashland Inc, p. I-903 [SIC 2911]
Ashley F Ward Inc, p. II-1382 [SIC 3451]
Ashley Furniture Industries Inc, p. I-542 [SIC 2511]
Ashta Chemicals Inc, p. I-764 [SIC 2812]
Ashtabula Rubber Co, p. I-954 [SIC 3069]
ASI of New York Inc, p. I-523 [SIC 2452]
ASI RCC Inc, p. I-1122 [SIC 3272]
ASI Robotic Systems Inc, p. II-1938 [SIC 3699]
Asko Inc, p. II-1312 [SIC 3423]
ASM Industries Inc, p. II-1638 [SIC 3561]
ASM Lithography, p. II-1632 [SIC 3559]
ASOMA Instruments Inc, p. II-2098 [SIC 3844]
Aspect Telecomm Corp, p. II-1868 [SIC 3661]
Aspects Inc, p. I-571 [SIC 2521]
Aspen Imaging International Inc, p. II-2169 [SIC 3955]
Aspen Publishers Inc, p. I-703 [SIC 2731]
Asphalt Cutbacks Inc, p. I-914 [SIC 2952]
Asphalt Paving Co, p. I-909 [SIC 2951]
ASR Recording Services, p. II-1864 [SIC 3652]
Assembly Fasteners Inc, p. II-1388 [SIC 3452]
Associated Bag Co, p. I-961 [SIC 3081]
Associated Ceramics, p. II-1795 [SIC 3629]
Associated Engineering Co, p. II-1769 [SIC 3612]
Associated Enterprises Inc, p. I-679 [SIC 2677]
Associated Equipment Corp, p. II-1795 [SIC 3629]
Associated Optical Top Network, p. II-2107 [SIC 3851]
Associated Pacific Machine Corp, p. II-1615 [SIC 3554]
Associated Printers Inc, p. I-735 [SIC 2761]
Associated Rubber Inc, p. I-948 [SIC 3061]
Associated Sand and Gravel, p. I-1127 [SIC 3273]
Associated Seafood Co, p. I-190 [SIC 2092]
Associated Spring, p. II-1470 [SIC 3495]
Associated Tagline Inc, p. I-861 [SIC 2873]
Associated Vending Co, p. II-1718 [SIC 3581]
Assurance Technology Corp, p. II-1980 [SIC 3728]
AST Research Inc, p. II-1688 [SIC 3571]
Astec Industries Inc, p. II-1519 [SIC 3531]
Astoria Industries Inc, p. II-1943 [SIC 3711]
Astral Industries Inc, p. II-2191 [SIC 3995]
Astro, p. II-1644 [SIC 3562]
Astro Air Inc, p. II-1726 [SIC 3585]
Astro Container Co, p. II-1304 [SIC 3412]
Astro Metallurgical Inc, p. II-1355 [SIC 3443]
Astro-Valcour Inc, p. I-1006 [SIC 3089]
Astronautics Corporation, p. II-2035 [SIC 3812]
Astronics Corp, p. I-643 [SIC 2657]
Astrosystems Inc, p. II-1795 [SIC 3629]
Astrotech International Corp, p. II-1355 [SIC 3443]
AT and G Co, p. II-1382 [SIC 3451]
AT & T Global Info Solutions, p. II-1892 [SIC 3674]
AT and T Nassau Metals Corp, p. II-1420 [SIC 3471]
AT and T Tridom, p. II-1873 [SIC 3663]
AT Cross Co, p. II-2156 [SIC 3951]

AT Foote Woodworking Co, p. I-571 [SIC 2521]
Atalla Corp, p. II-1713 [SIC 3579]
ATAS Aluminum Corp, p. II-1367 [SIC 3446]
Atchison Casting Corp, p. II-1212 [SIC 3325]
Atchison Products Inc, p. I-1053 [SIC 3199]
Atchley Controls, p. II-1454 [SIC 3492]
ATCO Rubber Products Inc, p. I-936 [SIC 3052]
Atcoflex Inc, p. I-936 [SIC 3052]
ATE, p. II-2061 [SIC 3825]
Ateeco Inc, p. I-65 [SIC 2038]
Athanor Group Inc, p. II-1382 [SIC 3451]
Athea Laboratories Inc, p. I-874 [SIC 2879]
Athena Controls Inc, p. II-2045 [SIC 3822]
Athena Industries Inc, p. I-593 [SIC 2542]
Athens Brick Co, p. I-1081 [SIC 3251]
Athens Corp, p. II-1632 [SIC 3559]
Atherton Foundry Products, p. II-1285 [SIC 3369]
Athey Products Corp, p. II-1943 [SIC 3711]
Athletic Bag Co, p. I-433 [SIC 2393]
Athol Manufacturing Corp, p. I-310 [SIC 2295]
ATI Group, p. I-318 [SIC 2297]
ATI Tools Inc, p. II-1312 [SIC 3423]
Atkins and Pearce Inc, p. I-245 [SIC 2241]
Atkins Inc, p. I-30 [SIC 2024]
Atkinson Industries Inc, p. II-1775 [SIC 3613]
Atkinson Trading Inc, p. II-2122 [SIC 3911]
Atkomatic Valve Company Inc, p. II-1447 [SIC 3491]
Atlanta Baking Company Inc, p. I-101 [SIC 2051]
Atlantic Coast Asphalt Co, p. I-909 [SIC 2951]
Atlantic Coast Carton, p. I-643 [SIC 2657]
Atlantic Cotton Mill, p. I-297 [SIC 2281]
Atlantic Fertilizer & Chemical, p. I-869 [SIC 2875]
Atlantic Magnetics Inc, p. II-1905 [SIC 3677]
Atlantic Marine Holding Co, p. II-1986 [SIC 3731]
Atlantic Meeco Inc, p. I-523 [SIC 2452]
Atlantic Metal Products Inc, p. II-1361 [SIC 3444]
Atlantic Paper Box Co, p. I-623 [SIC 2652]
Atlantic Research Corp, p. II-2011 [SIC 3764]
Atlantic Richfield Co, p. I-903 [SIC 2911]
Atlantic Sintered Metals Inc, p. II-1295 [SIC 3399]
Atlantic Spinners Inc, p. I-297 [SIC 2281]
Atlantic Spring Company Inc, p. II-1470 [SIC 3495]
Atlantic Steel Industries Inc, p. II-1172 [SIC 3312]
Atlantic Veneer Corp, p. I-491 [SIC 2435]
Atlantic Wire Co, p. II-1182 [SIC 3315]
Atlantic Wood, p. I-528 [SIC 2491]
Atlantic Wood Industries Inc, p. I-528 [SIC 2491]
Atlantic Zeiser Co, p. II-1605 [SIC 3552]
Atlantis Plastic Films Inc, p. I-961 [SIC 3081]
Atlantis Plastics Inc, p. I-961 [SIC 3081]
Atlas Bolt and Screw Co, p. II-1388 [SIC 3452]
Atlas Carpet Mills Inc, p. I-292 [SIC 2273]
Atlas Concrete Products Co, p. I-1117 [SIC 3271]
Atlas Container Corp, p. I-628 [SIC 2653]
Atlas Copco Compressors Inc, p. II-1649 [SIC 3563]
Atlas Copco Construction, p. II-1525 [SIC 3532]
Atlas Cylinder, p. II-1747 [SIC 3593]
Atlas Electric Devices Co, p. II-2076 [SIC 3829]
Atlas Foundry and Machine Co, p. II-1212 [SIC 3325]
Atlas Match Corp, p. II-2199 [SIC 3999]
Atlas Mill Supply, p. I-327 [SIC 2299]
Atlas Model Railroad Company, p. II-2144 [SIC 3944]
Atlas Spring Mfg Corp, p. I-558 [SIC 2515]
Atlas Technologies Inc, p. II-1600 [SIC 3549]
Atlas Truck Bodies Inc, p. II-1948 [SIC 3713]
Atlas Uniform Co, p. I-352 [SIC 2326]
Atlas Welding Accessories Inc, p. II-1594 [SIC 3548]
Atlas Wire Corp, p. II-1260 [SIC 3357]
Atmel Corp, p. II-1892 [SIC 3674]
Atmosphere Furnace Co, p. II-1671 [SIC 3567]
Atmosphere Processing Inc, p. II-1290 [SIC 3398]
Atols Tool and Mold Corp, p. II-1764 [SIC 3599]
Atotech USA Inc, p. I-777 [SIC 2819]
ATR Coil Company Inc, p. II-1905 [SIC 3677]
ATR Technologies Inc, p. II-1377 [SIC 3449]
Atrium Door and Window Co, p. I-480 [SIC 2431]
Atron Products and Services Inc, p. I-236 [SIC 2221]
Attbar Inc, p. II-1948 [SIC 3713]
Attleboro Refining, p. II-1225 [SIC 3339]

Attwood Corp, p. II-1323 [SIC 3429]
Attwoot Corp, p. I-437 [SIC 2394]
Atwater Inc, p. I-302 [SIC 2282]
Atwood Industries Inc, p. II-1954 [SIC 3714]
Auburn Ball Bearing, p. II-1644 [SIC 3562]
Auburn Consolidated Industries, p. II-1514 [SIC 3524]
Auburn Gear Inc, p. II-1666 [SIC 3566]
Auburn Hosiery Mills Inc, p. I-250 [SIC 2251]
Auburn Leather Co, p. I-1053 [SIC 3199]
Audio Computer Information, p. II-1864 [SIC 3652]
Audio Literature Inc, p. II-1864 [SIC 3652]
Audio Renaissance Tapes, p. II-1864 [SIC 3652]
Augat Inc, p. II-1915 [SIC 3679]
August Lotz Company Inc, p. I-536 [SIC 2499]
Augusta Bag Company Inc, p. I-433 [SIC 2393]
Augustine Medical Inc, p. II-2087 [SIC 3842]
Ault Inc, p. II-1795 [SIC 3629]
Aunt Jane's Foods, p. I-55 [SIC 2035]
Aurelius Manufacturing, p. II-1747 [SIC 3593]
Auric Corp, p. I-897 [SIC 2899]
Aurora Electronics Inc, p. II-1892 [SIC 3674]
Aurora Equipment Co, p. I-593 [SIC 2542]
Aurora Flight Sciences Corp, p. II-1970 [SIC 3721]
Aurora Industries Inc, p. II-1271 [SIC 3364]
Aurora Modular Industries, p. I-523 [SIC 2452]
Aurora Optics Inc, p. I-1066 [SIC 3229]
Aurora Paperboard, p. I-618 [SIC 2631]
Ausimont USA Inc, p. I-777 [SIC 2819]
Auspex Systems Inc, p. II-1704 [SIC 3577]
Austell Box Board Corp, p. I-669 [SIC 2675]
Austin Canvas Specialty Inc, p. I-437 [SIC 2394]
Austin Computer Systems Inc, p. II-1688 [SIC 3571]
Austin Innovations Inc, p. II-1840 [SIC 3645]
Austin Organs Inc, p. II-2135 [SIC 3931]
Austin Powder Co, p. I-885 [SIC 2892]
Austin Productions Inc, p. I-1113 [SIC 3269]
Austin Trading Company Inc, p. I-416 [SIC 2387]
Austin White Lime Co, p. I-1132 [SIC 3274]
Authentic Fitness Corp, p. I-356 [SIC 2329]
Auto Cast Inc, p. II-1271 [SIC 3364]
Auto Crane Co, p. II-1519 [SIC 3531]
Auto Meter Products Inc, p. II-2056 [SIC 3824]
Auto Truck Inc, p. II-1948 [SIC 3713]
AutoAlliance International Inc, p. II-1943 [SIC 3711]
Autoclave Engineers Inc, p. II-2051 [SIC 3823]
Autocon Technologies Inc, p. II-1790 [SIC 3625]
Autocrat Coffee Inc, p. I-195 [SIC 2095]
Autographic Services Inc, p. II-1621 [SIC 3555]
Autojectors Inc, p. II-1572 [SIC 3544]
Autologic Inc, p. II-1621 [SIC 3555]
Automata Inc, p. II-1887 [SIC 3672]
Automated Binding Company, p. I-750 [SIC 2789]
Automated Building Components, pp. Vol. I - 480, 501 [SICs 2431, 2439]
Automated Conveyor Systems, p. II-1541 [SIC 3535]
Automated Custom Food Svcs, p. II-1718 [SIC 3581]
Automated Logic Corp, p. II-2045 [SIC 3822]
Automated Machinery Systems, p. II-1626 [SIC 3556]
Automated Products Inc, p. I-501 [SIC 2439]
Automatic Bedding Corp, p. I-558 [SIC 2515]
Automatic Business Products, p. I-735 [SIC 2761]
Automatic Connector, p. II-1910 [SIC 3678]
Automatic Equipment Mfg Co, p. II-1509 [SIC 3523]
Automatic Feed Co, p. II-1632 [SIC 3559]
Automatic Handling Inc, p. II-1682 [SIC 3569]
Automatic Inspection Devices, p. II-2071 [SIC 3827]
Automatic Mach & Electronics, p. II-1626 [SIC 3556]
Automatic Machine Products Co, p. II-1464 [SIC 3494]
Automatic Power Inc, p. II-1854 [SIC 3648]
Automatic Screw Machine, p. II-1388 [SIC 3452]
Automatic Spring Products Co, p. II-1460 [SIC 3493]
Automatic Switch Co, p. II-1464 [SIC 3494]
Automatic Timing & Controls, p. II-1790 [SIC 3625]
Automatic Valve Corp, p. II-1464 [SIC 3494]
Automation International Inc, p. II-1594 [SIC 3548]
Automation Plating Corp, p. II-1420 [SIC 3471]
Automation Printing Co, p. I-708 [SIC 2732]
Automax Inc, p. II-1747 [SIC 3593]

Roman numerals I and II indicate the volume in which pages appear. Volume I holds SICs 2011 through 3299; Volume II holds SICs 3312 through 3999.

Roman numerals I and II indicate the volume in which pages appear. Volume I holds SICs 2011 through 3299; Volume II holds SICs 3312 through 3999.

Bard Urological, p. II-2082 [SIC 3841]
Bardane Manufacturing Co, p. II-1266 [SIC 3363]
Barden Corp, p. II-1644 [SIC 3562]
Bardes Corp, p. II-1830 [SIC 3643]
Bardex Corp, p. II-1547 [SIC 3536]
Bardon Trimount Inc, p. I-909 [SIC 2951]
Bardwell and McAlister Inc, p. II-1854 [SIC 3648]
Bardwil Industries Inc, p. I-428 [SIC 2392]
Bareman Dairy Inc, p. I-35 [SIC 2026]
Bareville Garment Corp, p. I-378 [SIC 2341]
Bargreen Coffee, p. I-195 [SIC 2095]
Bark River Culvert & Equip Co, p. II-1355 [SIC 3443]
Barker Brothers Inc, p. I-1144 [SIC 3291]
Barker Steel Company Inc, p. II-1344 [SIC 3441]
Barksdale Inc, p. II-1454 [SIC 3492]
Barmet Aluminum Corp, p. II-1244 [SIC 3354]
Barnes Aerospace, p. II-1975 [SIC 3724]
Barnes Engineering, p. II-2051 [SIC 3823]
Barnes Group Inc, p. II-1460 [SIC 3493]
Barnes Industries Inc, p. II-1644 [SIC 3562]
Barnes Pumps Inc, p. II-1638 [SIC 3561]
Barnett-Bates Corp, p. II-1367 [SIC 3446]
Barney Knitting Machinery, p. II-1605 [SIC 3552]
Barnhardt Manufacturing, p. II-2087 [SIC 3842]
Barnhill Trading Company Inc, p. I-486 [SIC 2434]
Barnstead Thermolyne Corp, p. II-2040 [SIC 3821]
Baroid Drilling Fluids Inc, p. I-918 [SIC 2992]
Baron-Abramson Inc, p. I-373 [SIC 2339]
Barouh Eaton Allen Corp, p. II-2169 [SIC 3955]
Barq's Inc, p. I-181 [SIC 2087]
Barr Enterprises Inc, p. I-91 [SIC 2047]
BARR Laboratories Inc, p. I-804 [SIC 2834]
Barrango Inc, p. I-1122 [SIC 3272]
Barrel O'Fun Snack Foods Co, p. I-199 [SIC 2096]
Barrett Carpet Mills Inc, p. I-292 [SIC 2273]
Barrett Refining Corp, p. I-903 [SIC 2911]
Barridon Corp, p. II-1975 [SIC 3724]
Barroncast Inc, p. II-1207 [SIC 3324]
Barrow Manufacturing Company, p. I-348 [SIC 2325]
Barry Blower, p. II-1654 [SIC 3564]
Barry Manufacturing Company, p. I-1033 [SIC 3149]
Barry Wehmiller Packaging Syst, p. II-1660 [SIC 3565]
Barry Wells, p. I-344 [SIC 2323]
Barth and Dreyfuss of California, p. I-327 [SIC 2299]
Barth Inc, p. II-1966 [SIC 3716]
Bartlett and Co, p. I-70 [SIC 2041]
Bartlett-Collins Co, p. I-1071 [SIC 3231]
Bartlett Milling Co, p. I-70 [SIC 2041]
Barton-Cotton Inc, p. I-741 [SIC 2771]
Barudan America Inc, p. II-1605 [SIC 3552]
Basalite, p. I-1117 [SIC 3271]
Basco, p. II-1355 [SIC 3443]
Basco Co, p. I-1071 [SIC 3231]
Base 10 Inc, p. I-156 [SIC 2079]
Base Inc, p. I-70 [SIC 2041]
Baseline Industries Inc, p. II-2066 [SIC 3826]
BASF Corp, p. II-1934 [SIC 3695]
BASF Information Systems, p. II-1934 [SIC 3695]
Bashlin Industries Inc, p. I-1053 [SIC 3199]
Basic Aluminum Castings Co, p. II-1266 [SIC 3363]
Basic American Foods, p. I-51 [SIC 2034]
Basic American Inc, p. I-51 [SIC 2034]
Basic Carbide Corp, p. II-1255 [SIC 3356]
Basic Comfort Inc, p. I-382 [SIC 2342]
Basic Vegetable Products LP, p. I-51 [SIC 2034]
Basketville Inc, p. I-536 [SIC 2499]
Basler Electric Co, p. II-1905 [SIC 3677]
Basmat Inc, p. II-1975 [SIC 3724]
BASS Inc, p. I-698 [SIC 2721]
Bassani Manufacturing Co, p. II-1486 [SIC 3498]
Bastian Co, p. II-2173 [SIC 3961]
Bata Shoe Company Inc, p. I-932 [SIC 3021]
Bates Abrasive Products Inc, p. I-1144 [SIC 3291]
Bates Container Inc, p. I-628 [SIC 2653]
Bates Manufacturing Co, p. II-1713 [SIC 3579]
Bates of Maine, p. I-231 [SIC 2211]
Bates Shoe Co, p. I-1025 [SIC 3143]
Batesville American Mfg Co, p. I-603 [SIC 2599]
Batesville Casket Company Inc, p. II-2191 [SIC 3995]

Batesville Products Inc, p. II-1271 [SIC 3364]
Bath Iron Works Corp, p. II-1986 [SIC 3731]
Bathroom Jewelry, p. II-1334 [SIC 3432]
Battenfeld-American Inc, p. I-918 [SIC 2992]
Battenfeld Gloucester Eng, p. II-1632 [SIC 3559]
Battenfeld Grease and Oil Corp, p. I-918 [SIC 2992]
Battery Engineering Inc, p. II-1920 [SIC 3691]
Battery Technology Inc, p. II-1920 [SIC 3691]
Batts Inc, p. I-1006 [SIC 3089]
Bauer Compressors Inc, p. II-1649 [SIC 3563]
Bauer Lamp Co, p. II-1840 [SIC 3645]
Bauer Welding, p. II-1486 [SIC 3498]
Baum Folder Corp, p. II-1615 [SIC 3554]
Bauman Carter Patterson Corp, p. I-332 [SIC 2311]
Baumer Foods Inc, p. I-55 [SIC 2035]
Baumfolder Corp, p. II-1615 [SIC 3554]
Bausch and Lomb Inc, p. II-2107 [SIC 3851]
Bautex Window Automation, p. I-424 [SIC 2391]
Bavarian Specialty Foods, p. I-112 [SIC 2053]
Bawden Printing Inc, p. I-718 [SIC 2752]
Baxter Corp, pp. Vol. I - 327, Vol. II - 1605 [SICs 2299, 3552]
Baxter Healthcare Corp IV, p. II-2082 [SIC 3841]
Baxter International Inc, p. II-2082 [SIC 3841]
Baxter Manufacturing Company, p. II-1626 [SIC 3556]
Baxter Tube Co, p. I-634 [SIC 2655]
Baxters Asphalt and Concrete, p. I-909 [SIC 2951]
Bay Cast Inc, p. II-1764 [SIC 3599]
Bay Mirror Inc, p. I-1071 [SIC 3231]
Bay Ship and Yacht Co, p. II-1986 [SIC 3731]
Bay Shipbuilding Co, p. II-1986 [SIC 3731]
Bay State Auto Spring Mfg, p. II-1460 [SIC 3493]
Bay State Bindery Inc, p. I-750 [SIC 2789]
Bay State Curtain Manufacturing, p. I-424 [SIC 2391]
Bay State Elevator Co, p. II-1536 [SIC 3534]
Bay State Milling Co, p. I-70 [SIC 2041]
Bay State/Sterling Inc, p. I-1144 [SIC 3291]
Bay Wood Homes Inc, p. I-523 [SIC 2452]
Bay Zinc Co, p. I-874 [SIC 2879]
Bayer Corp, p. I-855 [SIC 2869]
Bayless Bindery Inc, p. I-750 [SIC 2789]
Baylis Brothers Co, p. I-390 [SIC 2361]
Baylor Technology Inc, p. II-1531 [SIC 3533]
Bayou Land Seafood, p. I-190 [SIC 2092]
Bayou Pipe Coating Co, p. II-1425 [SIC 3479]
Bayou Steel Corp, p. II-1172 [SIC 3312]
BBE Sound Inc, p. II-2135 [SIC 3931]
BC Cook and Sons Enterprises, p. I-60 [SIC 2037]
BC Marketing Concepts Inc, p. I-160 [SIC 2082]
BC Rogers Poultry Inc, p. I-12 [SIC 2015]
BCO Industries Inc, p. II-1589 [SIC 3547]
BD Baggies, p. I-336 [SIC 2321]
BDK Holdings Inc, p. I-327 [SIC 2299]
BE Aerospace Inc, p. II-1980 [SIC 3728]
Bea Maurer Inc, p. I-506 [SIC 2441]
Beach Manufacturing Company, p. I-1071 [SIC 3231]
Beacon Container, p. I-628 [SIC 2653]
Beacon Converters Inc, p. I-991 [SIC 3086]
Beacon Light Products Inc, p. II-1826 [SIC 3641]
Beacon Manufacturing Co, pp. Vol. I - 236, 428 [SICs 2221, 2392]
Beacon Sweets Inc, p. I-127 [SIC 2064]
Beadex Manufacturing Company, p. I-879 [SIC 2891]
Beagle Manufacturing Company, p. I-593 [SIC 2542]
Beaird Industries Inc, p. II-1355 [SIC 3443]
Beam Corp, p. I-378 [SIC 2341]
Beam-Stream Inc, p. II-1883 [SIC 3671]
Bear Archery Inc, p. II-2150 [SIC 3949]
Bear Feet, p. I-1033 [SIC 3149]
Bear Graphics Inc, p. I-745 [SIC 2782]
Bear Paw Lumber Corp, p. I-471 [SIC 2426]
Beard Co, p. I-768 [SIC 2813]
Bearden Enterprises Inc, p. I-352 [SIC 2326]
Bearing Service of Pennsylvania, p. II-1644 [SIC 3562]
Bearium Metals Corp, p. II-1271 [SIC 3364]
Bearse Manufacturing Co, p. I-433 [SIC 2393]
Beasley Manufacturing Inc, p. II-1503 [SIC 3519]
Beatrice Cheese Inc, p. I-21 [SIC 2022]
Beau Interconnect Systems, p. II-1830 [SIC 3643]

Beaucraft Inc, p. II-2122 [SIC 3911]
Beaufort Fisheries Inc, p. I-152 [SIC 2077]
Beaulieu of America Inc, p. I-292 [SIC 2273]
Beaumont Birch Co, p. II-1541 [SIC 3535]
Beauti-Vue Products Corp, p. I-599 [SIC 2591]
BeautiControl Cosmetics Inc, p. I-834 [SIC 2844]
Beaver Shoe Co, p. I-1029 [SIC 3144]
Beavertown Block Company Inc, p. I-1117 [SIC 3271]
Beavertown Mills Inc, p. I-236 [SIC 2221]
Bechik Products Inc, p. I-428 [SIC 2392]
Beck and Beck Inc, p. I-1140 [SIC 3281]
Beck Flavors, p. I-181 [SIC 2087]
Beck Industries LP, p. II-2020 [SIC 3792]
Beck Manufacturing Inc, p. II-1464 [SIC 3494]
Becker Metal Works Inc, p. II-1207 [SIC 3324]
Beckett Bronze Co, p. II-1676 [SIC 3568]
Beckett Corp, p. II-1638 [SIC 3561]
Beckett Paper Co, p. I-612 [SIC 2621]
Beckman Instruments Inc, p. II-2066 [SIC 3826]
Beckwith Electric Co, p. II-1790 [SIC 3625]
Beckwith Elevator Co, p. II-1536 [SIC 3534]
Beco Helman Inc, p. I-378 [SIC 2341]
Becton Dickinson and Co, p. II-2082 [SIC 3841]
Bede Jet Corp, p. II-1970 [SIC 3721]
Bedford Chemical, p. I-855 [SIC 2869]
Bedford Gear, p. II-1666 [SIC 3566]
Bedford Industries Inc, p. II-1476 [SIC 3496]
Bee Bindery Inc, p. I-750 [SIC 2789]
Bee Gee Shrimp, p. I-190 [SIC 2092]
Bee Line Co, p. II-1600 [SIC 3549]
Beebe Rubber Co, p. I-948 [SIC 3061]
Beech-Nut Nutrition Corp, p. I-40 [SIC 2032]
Beef America Operating, p. I-2 [SIC 2011]
Beer Nuts Inc, p. I-136 [SIC 2068]
Beetle Plastics Inc, p. I-979 [SIC 3084]
Behm Quartz Industries Inc, p. I-1140 [SIC 3281]
Behnke Lubricants Inc, p. I-918 [SIC 2992]
Behr Process Corp, p. I-840 [SIC 2851]
Behr's Chocolates Inc, p. I-132 [SIC 2066]
BEI Defense Systems Company, p. II-2007 [SIC 3761]
BEI Electronics Inc, p. II-2007 [SIC 3761]
BEI Sensors & Motion Syst Co, p. II-2071 [SIC 3827]
Beiersdorf Inc, p. I-819 [SIC 2841]
Beistle Co, p. I-688 [SIC 2679]
Beitel Displays and Exhibits Inc, p. II-1372 [SIC 3448]
Bekaert Corp, p. II-1182 [SIC 3315]
Bel-Air Technologies Inc, p. II-1699 [SIC 3575]
Bel Aire Products Inc, p. I-1041 [SIC 3161]
Bel-Ami Knitwear Company Inc, p. I-258 [SIC 2253]
Bel Cheese, p. I-21 [SIC 2022]
Bel Fuse Inc, p. II-1775 [SIC 3613]
Belair Watch Corp, p. II-2118 [SIC 3873]
Belcam Inc, p. I-834 [SIC 2844]
Belcher, p. II-1203 [SIC 3322]
Belden Brick Co, p. I-1081 [SIC 3251]
Belden Inc, p. II-1260 [SIC 3357]
Belding Hausman Inc, p. I-236 [SIC 2221]
Belding Hausman Weldon Mill, p. I-236 [SIC 2221]
Belding Heminway Company Inc, p. I-306 [SIC 2284]
Belfab, p. II-1764 [SIC 3599]
Bell Boeing Corp, p. II-1970 [SIC 3721]
Bell Equipment USA Inc, p. II-1509 [SIC 3523]
Bell Flavors and Fragrances Inc, p. I-855 [SIC 2869]
Bell Lumber and Pole Co, p. I-528 [SIC 2491]
Bell National Corp, p. I-231 [SIC 2211]
Bell Optical Laboratory Inc, p. II-2107 [SIC 3851]
Bell Packaging Corp, p. I-628 [SIC 2653]
Bell Petroleum Services Inc, p. I-948 [SIC 3061]
Bell Sports Corp, p. II-2150 [SIC 3949]
Bell Sports Inc, p. II-2150 [SIC 3949]
Bellanca Inc, p. II-1970 [SIC 3721]
Bellco Glass Inc, p. II-2040 [SIC 3821]
Bellcraft Button Co, p. II-2177 [SIC 3965]
Bellefonte Lime Company Inc, p. I-1132 [SIC 3274]
Belmay Inc, p. I-834 [SIC 2844]
Belmont Corp, p. I-486 [SIC 2434]
Belmont Homes Inc, p. I-519 [SIC 2451]
Belmont Metals Inc, p. II-1225 [SIC 3339]
Belmont Textile Machinery, p. II-1605 [SIC 3552]

Roman numerals I and II indicate the volume in which pages appear. Volume I holds SICs 2011 through 3299; Volume II holds SICs 3312 through 3999.

Roman numerals I and II indicate the volume in which pages appear. Volume I holds SICs 2011 through 3299; Volume II holds SICs 3312 through 3999.

Biomune Co, p. I-814 [SIC 2836]
BioNebraska Inc, p. I-814 [SIC 2836]
Bioproducts Inc, p. I-95 [SIC 2048]
BioQuip Products Inc, p. II-2040 [SIC 3821]
BioScan Inc, p. II-2040 [SIC 3821]
Bioserv Corp, p. I-809 [SIC 2835]
BioSurface Technology Inc, p. I-814 [SIC 2836]
Biotec Inc, p. II-2093 [SIC 3843]
Biothane Corp, p. II-1736 [SIC 3589]
Biotronics Technologies Inc, p. II-2066 [SIC 3826]
BioWhittaker Inc, p. I-814 [SIC 2836]
BioZyme Inc, p. I-95 [SIC 2048]
Birch Machinery Co, p. II-1572 [SIC 3544]
Birchcraft Kitchens Inc, p. I-486 [SIC 2434]
Birchwood Lumber and Veneer, p. I-491 [SIC 2435]
Birchwood Manufacturing Co, p. I-491 [SIC 2435]
Bird Corp, p. I-914 [SIC 2952]
Bird Electronic Corp, p. II-2061 [SIC 3825]
Bird Island Soil Service Center, p. I-869 [SIC 2875]
Bird-Johnson Co, p. II-1980 [SIC 3728]
Bird Machine Co, p. II-1682 [SIC 3569]
Bird Manufacturing, p. II-2144 [SIC 3944]
Bird Medical Technologies Inc, p. II-2087 [SIC 3842]
Bird Products Corp, p. II-2082 [SIC 3841]
Bird-X Inc, p. II-1938 [SIC 3699]
Birdair Inc, p. I-437 [SIC 2394]
Birds Eye Veneer Co, p. I-491 [SIC 2435]
Birkett Mills, p. I-70 [SIC 2041]
Birmingham Steel Corp, p. II-1172 [SIC 3312]
Birmy Graphics Corp, p. II-1621 [SIC 3555]
Birt Inc, p. I-433 [SIC 2393]
Birtcher Medical Systems Inc, p. II-2102 [SIC 3845]
Birum Corp, p. II-1709 [SIC 3578]
Biscayne Holdings Inc, p. I-361 [SIC 2331]
Biscotti Inc, p. I-394 [SIC 2369]
Bishop Baking Co, p. I-101 [SIC 2051]
Bishop Freeman Co, p. II-1722 [SIC 3582]
Bishop-Wisecarver Corp, p. II-1676 [SIC 3568]
Bishopville Finishing, p. I-279 [SIC 2261]
Bison Canning Company Inc, p. I-45 [SIC 2033]
Bison Gear and Engineering Inc, p. II-1666 [SIC 3566]
Bitrek Corp, p. II-1464 [SIC 3494]
Bitrode Corp, p. II-1795 [SIC 3629]
Bituminous Roadways Inc, p. I-909 [SIC 2951]
BIW Cable Systems Inc, p. II-1476 [SIC 3496]
BIW Connector Systems Inc, p. II-1830 [SIC 3643]
Bizjet Intern Sales and Support, p. II-1970 [SIC 3721]
BJ Seaman and Co, p. I-433 [SIC 2393]
BL Curry and Sons Inc, p. I-491 [SIC 2435]
BL Downey Company Inc, p. II-1425 [SIC 3479]
Blachford Corp, p. I-918 [SIC 2992]
Black & Decker Corp, p. II-1578 [SIC 3545]
Black and Decker Corp, p. II-1584 [SIC 3546]
Black Bros Co, p. II-1610 [SIC 3553]
Black Clawson Co, p. II-1615 [SIC 3554]
Black Dot Group, p. I-754 [SIC 2791]
Black Magic Products Ltd, p. I-869 [SIC 2875]
Black Millwork Company Inc, p. I-480 [SIC 2431]
Black River Manufacturing Inc, p. II-1382 [SIC 3451]
Black Watch, p. II-1992 [SIC 3732]
Blackfeet Writing Instruments, p. II-2156 [SIC 3951]
Blackhawk Automotive Inc, p. II-1323 [SIC 3429]
Blackhawk Foundry and Machine, p. II-1197 [SIC 3321]
Blackhawk Leather Ltd, p. I-1013 [SIC 3111]
Blackhawk Manufacturing Inc, p. I-669 [SIC 2675]
Blackmer, p. II-1732 [SIC 3586]
Blacksmith Shop Co, p. I-542 [SIC 2511]
Blackstone Company Inc, p. I-480 [SIC 2431]
Blackstone Industries Inc, p. II-1584 [SIC 3546]
Blade Communications Inc, p. I-693 [SIC 2711]
Blair Mills LP, p. I-231 [SIC 2211]
Blakeslee Prestress Inc, p. I-1122 [SIC 3272]
Blakeway Metal Works Inc, p. II-1367 [SIC 3446]
Blako Industries Inc, p. I-961 [SIC 3081]
Blalock Manufacturing, p. I-471 [SIC 2426]
Blanchard Shirt Corp, p. I-336 [SIC 2321]
Blandin Paper Co, p. I-612 [SIC 2621]
Blanke Plastic Company Inc, p. I-985 [SIC 3085]

Blanks Color Imaging Inc, p. I-759 [SIC 2796]
Blasch Precision Ceramics Inc, p. I-1081 [SIC 3251]
Blaser Die Casting Co, p. II-1271 [SIC 3364]
Blauer Manufacturing Co, p. I-409 [SIC 2385]
Blaw-Knox Constr Equip Corp, p. II-1519 [SIC 3531]
Blenko Glass Company Inc, p. I-1066 [SIC 3229]
Blessings Corp, p. I-961 [SIC 3081]
Bleyer Industries Inc, p. I-688 [SIC 2679]
Bleyle Inc, p. I-361 [SIC 2331]
BLH Electronics, p. II-2051 [SIC 3823]
Blind Maker Inc, p. I-599 [SIC 2591]
Bliss and Laughlin Steel Co, p. II-1187 [SIC 3316]
Bliss Manufacturing Inc, pp. Vol. I - 378, Vol. II - 1404 [SICs 2341, 3465]
Blissfield Manufacturing Co, p. II-1192 [SIC 3317]
Blitz Corp, p. II-1948 [SIC 3713]
Blitz USA Inc, p. II-1300 [SIC 3411]
Bloch/New England Inc, p. I-428 [SIC 2392]
Block Industries Inc, p. I-336 [SIC 2321]
Block Medical Inc, p. II-2082 [SIC 3841]
Blocksom and Co, p. II-1654 [SIC 3564]
Blom Industries Inc, p. II-1404 [SIC 3465]
Bloom Brothers Co, p. I-1053 [SIC 3199]
Bloomsburg Mills Inc, pp. Vol. I - 231, 236 [SICs 2211, 2221]
Blount Inc, pp. Vol II - 1318, 1430, 1519 [SICs 3425, 3482, 3531]
Blue Bell Mattress Co, p. I-558 [SIC 2515]
Blue Bird Corp, p. II-1943 [SIC 3711]
Blue Bird Midwest, p. II-1948 [SIC 3713]
Blue Bird Wanderlodge, p. II-1966 [SIC 3716]
Blue Circle Cement, p. I-1077 [SIC 3241]
Blue Circle Raia Inc, p. I-1077 [SIC 3241]
Blue Coral Inc, p. I-825 [SIC 2842]
Blue Coral Systems, p. I-825 [SIC 2842]
Blue Cross Laboratories Inc, p. I-825 [SIC 2842]
Blue Diamond Growers, p. I-136 [SIC 2068]
Blue Grass Manufacturing, p. II-1795 [SIC 3629]
Blue Lustre Products Inc, p. II-1818 [SIC 3635]
Blue M Electric, p. II-1671 [SIC 3567]
Blue Magic Products Inc, p. I-897 [SIC 2899]
Blue Mountain Industries, p. I-322 [SIC 2298]
Blue Ox Industries Inc, p. I-501 [SIC 2439]
Blue Ribbon Label Corp, p. I-654 [SIC 2672]
Blue Ridge Carpet Mills, p. I-292 [SIC 2273]
Blue Ridge Farms Inc, p. I-212 [SIC 2099]
Blue Ridge Mountain, p. II-2199 [SIC 3999]
Blue Ridge Pressure Casting Inc, p. II-1266 [SIC 3363]
Blue Ridge Truss and Supply Inc, p. I-501 [SIC 2439]
Blue Ridge Veneer, p. I-496 [SIC 2436]
Blue Seal Feeds Inc, p. I-95 [SIC 2048]
Blue Sky Publishing, p. I-741 [SIC 2771]
Blue Star Leather Inc, p. I-745 [SIC 2782]
Blueside Company Inc, p. I-1013 [SIC 3111]
Blumcraft of Pittsburgh, p. II-1367 [SIC 3446]
Blumenthal Print Works Inc, p. I-231 [SIC 2211]
Blyth Industries Inc, p. II-2199 [SIC 3999]
BM Root Co, p. II-1610 [SIC 3553]
BMC Bil-Mac Corp, p. II-1382 [SIC 3451]
BMC Industries Inc, p. II-1915 [SIC 3679]
BNZ Materials Inc, p. I-1136 [SIC 3275]
Bo-Buck Mills Inc, p. I-245 [SIC 2241]
Bo-Witt Products Inc, p. II-1835 [SIC 3644]
Boart Longyear, p. II-1682 [SIC 3569]
Bob Allen Companies Inc, p. I-1041 [SIC 3161]
Bob's Candies Inc, p. I-127 [SIC 2064]
Bob's Space Racers Inc, p. II-2199 [SIC 3999]
Bob's Texas Style Potato Chips, p. I-199 [SIC 2096]
Bobber Products Company Inc, p. I-948 [SIC 3061]
Bobbie Brooks Inc, p. II-2169 [SIC 3955]
Bobrick Washroom Equipment, p. II-2199 [SIC 3999]
BOC Group Inc, p. I-768 [SIC 2813]
Boca Foods Inc, p. I-107 [SIC 2052]
Boca Research Inc, p. II-1704 [SIC 3577]
Bock Industries Inc, p. II-1192 [SIC 3317]
Bock Water Heaters Inc, p. II-1339 [SIC 3433]
Bockman Cos, p. I-729 [SIC 2759]
Bodie-Hoover Petroleum Corp, p. I-918 [SIC 2992]
Bodine Aluminum Inc, p. II-1275 [SIC 3365]

Bodine Corp, p. II-1600 [SIC 3549]
Bodine Electric Co, p. II-1780 [SIC 3621]
Body Slimmers Inc, p. I-378 [SIC 2341]
Boehme Filatex Inc, p. I-773 [SIC 2816]
Boehringer Mannheim, p. I-804 [SIC 2834]
Boehringer Mannheim Corp, p. II-2082 [SIC 3841]
Boeing Co, p. II-1970 [SIC 3721]
Boelter Industries Inc, p. I-643 [SIC 2657]
Bogner of America Inc, p. I-356 [SIC 2329]
Bohannon Brewing Co, p. I-160 [SIC 2082]
Bohle Machine Tools Inc, p. II-1558 [SIC 3541]
Boise Cascade Corp, p. I-612 [SIC 2621]
Bojar Company Inc, p. II-2122 [SIC 3911]
Bojud Knitting Mills Inc, p. I-271 [SIC 2258]
BOK Industries Inc, p. I-745 [SIC 2782]
Bollinger Industries Inc, p. II-2150 [SIC 3949]
Bollinger Machine Shop, p. II-1986 [SIC 3731]
Bollman Hat Co, p. I-386 [SIC 2353]
Bolt Beranek and Newman Inc, p. II-1688 [SIC 3571]
Bolt Technology Corp, p. II-1531 [SIC 3533]
Bolton-Emerson Americas Inc, p. II-1615 [SIC 3554]
Boltz Knitting Mill Inc, p. I-263 [SIC 2254]
Bomanite Corp, p. I-1117 [SIC 3271]
Bomark Inc, p. I-889 [SIC 2893]
Bomarko Inc, p. I-654 [SIC 2672]
Bommer Industries Inc, p. II-1323 [SIC 3429]
Bomont Mills Inc, p. I-236 [SIC 2221]
Bon-Art International Inc, p. II-2199 [SIC 3999]
Bon Secour Fisheries Inc, p. I-190 [SIC 2092]
Bon Ton Foods Inc, p. I-199 [SIC 2096]
Bonanza Materials Inc, p. I-1127 [SIC 3273]
Bonar Packaging Inc, p. I-649 [SIC 2671]
Bond Optics Inc, p. II-2071 [SIC 3827]
Bond Street, p. I-1049 [SIC 3172]
Bondhus Corp, p. II-1312 [SIC 3423]
Bongrain Cheese USA, p. I-21 [SIC 2022]
Bonhomme Shirtmakers Ltd, p. I-336 [SIC 2321]
Bonnavilla Homes, p. I-523 [SIC 2452]
Bonney Forge Corp, p. II-1464 [SIC 3494]
Bonnot Co, p. II-1563 [SIC 3542]
Bonny Products Inc, p. I-1106 [SIC 3263]
Boo-Boo-Baby Inc, p. I-394 [SIC 2369]
Book Covers Inc, p. I-669 [SIC 2675]
Bookbinders Co, p. I-750 [SIC 2789]
BookCrafters USA Inc, p. I-708 [SIC 2732]
Boose Aluminum Foundry, p. II-1275 [SIC 3365]
Booth Crystal Tips Inc, p. II-1718 [SIC 3581]
Booth Inc, p. I-510 [SIC 2448]
Booth Newspapers Inc, p. I-693 [SIC 2711]
Bootz Plumbingware Co, p. II-1330 [SIC 3431]
Boral Bricks Inc, p. I-1081 [SIC 3251]
Borden Chemicals & Plastics LP, p. I-855 [SIC 2869]
Borden Coatings, p. I-889 [SIC 2893]
Borden Inc, pp. Vol. I - 35, 496 [SICs 2026, 2436]
Borden Manufacturing Company, p. I-297 [SIC 2281]
Borden Pasta Inc, p. I-208 [SIC 2098]
Borden/Reaves Inc, p. II-1845 [SIC 3646]
Border Foods Inc, p. I-40 [SIC 2032]
Border Steel Mills Inc, p. II-1187 [SIC 3316]
Bordo Products Co, p. I-51 [SIC 2034]
Borg Textile Corp, p. I-267 [SIC 2257]
Borg-Warner Automotive Inc, p. II-1954 [SIC 3714]
Boride Products Inc, p. I-1167 [SIC 3299]
Borisoff Engineering Co, p. II-2135 [SIC 3931]
Borkholder Corp, p. I-542 [SIC 2511]
Born Free Motorcoach, p. II-2020 [SIC 3792]
Born Inc, p. II-1671 [SIC 3567]
Borroughs Corp, p. I-593 [SIC 2542]
Bosch Automation Products, p. II-1541 [SIC 3535]
Bose Corp, p. II-1859 [SIC 3651]
Boss Manufacturing Co, p. II-2087 [SIC 3842]
Bost Neckwear Company Inc, p. I-344 [SIC 2323]
Bostik Inc, p. I-879 [SIC 2891]
Boston Acoustics Inc, p. II-1859 [SIC 3651]
Boston Gear, p. II-1666 [SIC 3566]
Boston Metal Products Corp, p. I-593 [SIC 2542]
Boston Scientific Corp, p. II-2082 [SIC 3841]
Boston Technology Inc, p. II-1868 [SIC 3661]
Boston Trailer Mfg Co, p. II-1961 [SIC 3715]

Roman numerals I and II indicate the volume in which pages appear. Volume I holds SICs 2011 through 3299; Volume II holds SICs 3312 through 3999.

Roman numerals I and II indicate the volume in which pages appear. Volume I holds SICs 2011 through 3299; Volume II holds SICs 3312 through 3999.

2233

Company Index

Brown Paper Goods Co, p. I-660 [SIC 2673]
Brown Printing Co, p. I-698 [SIC 2721]
Brown Products Inc, p. I-688 [SIC 2679]
Brown Shoe Co, p. I-1029 [SIC 3144]
Brown Wood Preserving, p. I-528 [SIC 2491]
Brown Wood Products Co, p. I-536 [SIC 2499]
Brownell and Company Inc, p. I-322 [SIC 2298]
Browning Chemical Corp, p. I-897 [SIC 2899]
Broyhill Furniture Industries Inc, p. I-542 [SIC 2511]
Broyhill Investments Corp, p. I-548 [SIC 2512]
Brubaker Group, p. II-1943 [SIC 3711]
Brubaker Tool Corp, p. II-1578 [SIC 3545]
Bruce Diamond Corp, p. II-2131 [SIC 3915]
Bruce Foods Corp, p. I-45 [SIC 2033]
Bruce Fox Inc, p. II-2127 [SIC 3914]
Bruce Industries Inc, p. II-1980 [SIC 3728]
Bruderer Inc, p. II-1563 [SIC 3542]
Bruewer Woodwork Mfg Co, p. I-587 [SIC 2541]
Brule CE and E Inc, p. II-2045 [SIC 3822]
Brulin and Company Inc, p. I-825 [SIC 2842]
Brundidge Shirt Corp, p. I-258 [SIC 2253]
Bruner Corp, p. II-1736 [SIC 3589]
Bruning Paint Co, p. I-840 [SIC 2851]
Brunner-Hildebrand Lumber, p. II-1632 [SIC 3559]
Bruno Independent Living Aids, p. II-1552 [SIC 3537]
Brunswick Box Company Inc, p. I-510 [SIC 2448]
Brunswick Corp, p. II-1992 [SIC 3732]
Brunton Co, p. II-2071 [SIC 3827]
Brush Industries Inc, p. II-1934 [SIC 3695]
Brush Research Manufacturing, p. II-2181 [SIC 3991]
Brush Wellman Inc, p. II-1225 [SIC 3339]
Brushtech Inc, p. II-2181 [SIC 3991]
Bryan Foods Inc, p. I-2 [SIC 2011]
Bryan Industries Inc, p. I-390 [SIC 2361]
Bryant Grinder Corp, p. II-1558 [SIC 3541]
Bryant Manufacturing Company, p. I-548 [SIC 2512]
BryDet Development Corp, p. II-1525 [SIC 3532]
Bryson Capital Corp, p. I-669 [SIC 2675]
BS and CP, p. I-825 [SIC 2842]
BTG Inc, p. II-1615 [SIC 3554]
BTL Specialty Resins Corp, p. I-782 [SIC 2821]
BTM Corp, p. II-1563 [SIC 3542]
BTR Inc, p. I-942 [SIC 3053]
BTU International Inc, p. II-1632 [SIC 3559]
Bubbles Baking Co, p. I-112 [SIC 2053]
Buchan Industries Inc, p. I-745 [SIC 2782]
Buchanan Hardwood Inc, p. I-465 [SIC 2421]
Buchanan Industries Inc, p. II-1266 [SIC 3363]
Buck Company Inc, p. II-1197 [SIC 3321]
Buck Knives Inc, p. II-1308 [SIC 3421]
Buck Kreihs Company Inc, p. II-1986 [SIC 3731]
Buck Logansport Inc, p. II-1578 [SIC 3545]
Buckeye Business Products Inc, p. II-2169 [SIC 3955]
Buckeye Corrugated Inc, p. I-628 [SIC 2653]
Buckeye Fabric Finishing Co, p. I-310 [SIC 2295]
Buckeye Feed Mills Inc, p. I-95 [SIC 2048]
Buckeye Fire Equipment Co, p. II-2199 [SIC 3999]
Buckeye Rubber Products Inc, p. I-954 [SIC 3069]
Buckeye Steel Castings Co, p. II-1212 [SIC 3325]
Buckhorn Rubber Products Inc, p. I-948 [SIC 3061]
Buckingham-Virginia Slate Corp, p. I-1140 [SIC 3281]
Buckley Industries Inc, p. I-991 [SIC 3086]
Buckley Powder Co, p. I-885 [SIC 2892]
Buckman Laboratories Intern, p. I-855 [SIC 2869]
Buckstaff Company Inc, p. I-581 [SIC 2531]
Bucyrus Blades Inc, p. II-1519 [SIC 3531]
Bucyrus-Erie Co, p. II-1525 [SIC 3532]
Bud Industries Inc, p. II-1361 [SIC 3444]
Bud's Best Cookies Inc, p. I-107 [SIC 2052]
Budd Co, pp. Vol. I - 782, Vol. II - 1404 [SICs 2821, 3465]
Buddy Bar Casting Corp, p. II-1275 [SIC 3365]
Budge Industries Inc, p. I-437 [SIC 2394]
Budzar Industries Inc, p. II-2045 [SIC 3822]
Buehler Ltd, p. II-2040 [SIC 3821]
Buehler Products Inc, p. II-1780 [SIC 3621]
Buehner Corp, p. I-1122 [SIC 3272]
Buell Motorcycle Co, p. II-2003 [SIC 3751]
Buena Vista Winery Inc, p. I-168 [SIC 2084]

Buettner Brothers Lumber Co, p. I-501 [SIC 2439]
Buffalo Brake Beam Co, p. II-1998 [SIC 3743]
Buffalo Bullet Co, p. II-1430 [SIC 3482]
Buffalo Color Corp, p. I-855 [SIC 2869]
Buffalo Dental Manufacturing, p. II-2093 [SIC 3843]
Buffalo Inc, p. II-1704 [SIC 3577]
Buffalo Industries Inc, p. I-327 [SIC 2299]
Buffalo Maid Cabinets Inc, p. I-553 [SIC 2514]
Buffalo Metal Fabricating Corp, p. I-553 [SIC 2514]
Buffalo News, p. I-693 [SIC 2711]
Buffalo Pumps Inc, p. II-1638 [SIC 3561]
Buffalo Technologies Corp, p. II-1626 [SIC 3556]
Buffalo Tungsten Inc, p. II-1178 [SIC 3313]
Buffelen Woodworking Co, p. I-480 [SIC 2431]
Buffton Corp, p. II-2035 [SIC 3812]
Bugle Boy Industries Inc, p. I-336 [SIC 2321]
Buhl Industries Inc, p. II-2112 [SIC 3861]
Buhl Optical Co, p. II-2071 [SIC 3827]
Builders Brass Works Corp, p. II-1280 [SIC 3366]
Building Components Unlimited, p. I-501 [SIC 2439]
Bulk Lift International Inc, p. I-433 [SIC 2393]
Bullard Abrasives Inc, p. I-1144 [SIC 3291]
Bulldog Home Hardware, p. II-1388 [SIC 3452]
Bullen Companies Inc, p. I-825 [SIC 2842]
Bullen Midwest Inc, p. I-825 [SIC 2842]
Bullseye Glass Co, p. I-1071 [SIC 3231]
Bully Hill Vineyards Inc, p. I-168 [SIC 2084]
Bumkins International Inc, p. I-455 [SIC 2399]
Bundy Corp, p. II-1192 [SIC 3317]
Bunge Foods Corp, p. I-156 [SIC 2079]
Bunker Hill Foods, p. I-40 [SIC 2032]
Bunn-O-Matic Corp, p. II-1813 [SIC 3634]
Bunting Bearings Corp, p. II-1285 [SIC 3369]
Bunton Co, p. II-1514 [SIC 3524]
Burbank Steel Treating Company, p. II-1290 [SIC 3398]
Burch Manufacturing Company, p. I-506 [SIC 2441]
Burden Automotive Electric Inc, p. II-1929 [SIC 3694]
Burden China Company Inc, p. I-1102 [SIC 3262]
Burdick and Jackson Inc, p. I-855 [SIC 2869]
Bureau of Business Practice, p. I-698 [SIC 2721]
Bureau of Engraving Inc, p. I-718 [SIC 2752]
Burge Chemical Products Inc, p. I-819 [SIC 2841]
Burgess Industries Inc, p. II-1621 [SIC 3555]
Burgess-Manning Inc, p. II-1355 [SIC 3443]
Burgess-Norton Mfg Co, p. II-1492 [SIC 3499]
Burgess Products Inc, p. II-1514 [SIC 3524]
Burgett Inc, p. II-2135 [SIC 3931]
Burkart Foam Inc, p. I-991 [SIC 3086]
Burke E Porter Machinery Co, p. II-1682 [SIC 3569]
Burke Hosiery Mills Inc, p. I-254 [SIC 2252]
Burke Industries Inc, p. I-948 [SIC 3061]
Burke-Parsons-Bowlby Corp, p. I-528 [SIC 2491]
Burke Rubber Co, p. I-948 [SIC 3061]
Burkeen Manufacturing Co, p. II-1519 [SIC 3531]
Burkeville Veneer Co, p. I-491 [SIC 2435]
Burle Industries Inc, p. II-1873 [SIC 3663]
Burleigh Instruments Inc, p. II-2040 [SIC 3821]
Burlen Corp, p. I-378 [SIC 2341]
Burley Design Cooperative, p. II-2003 [SIC 3751]
Burlington Basket Company Inc, p. I-567 [SIC 2519]
Burlington House Upholstery, p. I-236 [SIC 2221]
Burlington Industries Inc, p. I-241 [SIC 2231]
Burlington Madison Yarn, p. I-297 [SIC 2281]
Burlington Menswear, p. I-297 [SIC 2281]
Burlington Scientific Corp, p. I-874 [SIC 2879]
Burndy Corp, p. II-1915 [SIC 3679]
Burner Systems International Inc, p. II-1339 [SIC 3433]
Burnett Manufacturing Corp, p. I-553 [SIC 2514]
Burnett Poultry Co, p. I-12 [SIC 2015]
Burnham and Brady Inc, p. I-132 [SIC 2066]
Burnham Corp, p. II-1197 [SIC 3321]
Burns Brick, p. I-1081 [SIC 3251]
Burns Construction Inc, p. I-501 [SIC 2439]
Burns-Philp Food Inc, p. I-212 [SIC 2099]
Burnside Manufacturing Co, p. II-1413 [SIC 3469]
Burr-Brown Corp, p. II-1892 [SIC 3674]
Burr Oak Tool and Gauge, p. II-1632 [SIC 3559]

Burr Patterson and Auld, p. II-2122 [SIC 3911]
Burrell Leder Beltech Inc, p. I-936 [SIC 3052]
Burris Company Inc, p. II-2071 [SIC 3827]
Burroughs Wellcome Co, p. I-804 [SIC 2834]
Burrows Paper Corp, pp. Vol. I - 612, 649 [SICs 2621, 2671]
Burtco Inc, p. II-1192 [SIC 3317]
Burtman Iron Works Inc, p. II-1552 [SIC 3537]
Burton Manufacturing Company, p. II-2150 [SIC 3949]
Bus Industries of America Inc, p. II-1948 [SIC 3713]
Buscarino Guitars, p. II-2135 [SIC 3931]
Busch Co, p. II-1654 [SIC 3564]
Buschman Co, p. II-1541 [SIC 3535]
Bush Hog Corp, p. II-1509 [SIC 3523]
Bush Industries Inc, p. I-542 [SIC 2511]
Bushman Equipment Inc, p. II-1547 [SIC 3536]
Bushman Press Inc, p. I-708 [SIC 2732]
Busse Hospital Disposables, p. I-674 [SIC 2676]
Busse Inc, p. II-1660 [SIC 3565]
Bussmann, p. II-1775 [SIC 3613]
Bustin Industrial Products Inc, p. II-1367 [SIC 3446]
Butler Automatic Inc, p. II-1615 [SIC 3554]
Butler Group Inc, p. I-593 [SIC 2542]
Butler Manufacturing Co, p. II-1509 [SIC 3523]
Butler Printing and Laminating, p. I-688 [SIC 2679]
Butter Krust Baking Co, p. I-101 [SIC 2051]
Butterball Farms Inc, p. I-212 [SIC 2099]
Butterball Turkey Co, p. I-12 [SIC 2015]
Butterick Company Inc, p. I-713 [SIC 2741]
Butternut Bread Bakeries, p. I-101 [SIC 2051]
Butwin Sportswear Co, p. I-356 [SIC 2329]
Buxton Co, p. I-1049 [SIC 3172]
BW Elliott Manufacturing, p. II-1676 [SIC 3568]
BW International Inc, p. II-1660 [SIC 3565]
BW/IP International Inc, p. II-1747 [SIC 3593]
BWI KartridgPak, p. II-1626 [SIC 3556]
BWIP Holding Inc, p. II-1638 [SIC 3561]
By Susan Inc, p. I-365 [SIC 2335]
Byard F Brogan Inc, p. II-2122 [SIC 3911]
Bybee Stone Company Inc, p. I-1140 [SIC 3281]
ByCobra Inc, p. I-455 [SIC 2399]
Byers Industries Inc, p. II-1682 [SIC 3569]
Byrd Cookie Co, p. I-107 [SIC 2052]
Byron Equipment Co, p. II-1509 [SIC 3523]
Byron Valve and Machine Co, p. II-1464 [SIC 3494]
Byron Weston, p. I-612 [SIC 2621]
Bystrom Brothers Inc, p. II-1382 [SIC 3451]

C-2 Office Gear Inc, p. I-576 [SIC 2522]
C and C Canvas Co, p. I-437 [SIC 2394]
C and C Manufacturing Inc, p. II-1747 [SIC 3593]
C and C Metal Products Corp, p. II-2177 [SIC 3965]
C and C Smith Lumber Company, p. I-471 [SIC 2426]
C and D Die Casting Company, p. II-1266 [SIC 3363]
C and D Hit Inc, p. II-2156 [SIC 3951]
C and F Packing Company Inc, p. I-7 [SIC 2013]
C and H Chemical Inc, p. I-819 [SIC 2841]
C and H Die Casting Inc, p. II-1266 [SIC 3363]
C and H Packaging Inc, p. I-649 [SIC 2671]
C and K Components Inc, p. II-1775 [SIC 3613]
C and M Corp, p. II-1260 [SIC 3357]
C and M Press Corp, p. I-708 [SIC 2732]
C and M Spring Engineering, p. II-1460 [SIC 3493]
C and N Associates, p. II-2173 [SIC 3961]
C and S Block Inc, p. I-1117 [SIC 3271]
C and S Valve Co, p. II-1447 [SIC 3491]
C-Case Corp, p. I-618 [SIC 2631]
C-Cor Electronics Inc, p. II-1873 [SIC 3663]
C-E Minerals Inc, p. I-1163 [SIC 3297]
C Hager & Sons Hinge Mfg, p. II-1323 [SIC 3429]
C Jim Stewart and Stevenson, p. II-1998 [SIC 3743]
C Mondavi and Sons, p. I-168 [SIC 2084]
C-Mor Co, p. I-599 [SIC 2591]
C Thorrez Industries Inc, p. II-1382 [SIC 3451]
CA Lawton Co, p. II-1615 [SIC 3554]
Cab-O-Sil, p. I-777 [SIC 2819]
Caberra Inc, p. I-361 [SIC 2331]

Roman numerals I and II indicate the volume in which pages appear. Volume I holds SICs 2011 through 3299; Volume II holds SICs 3312 through 3999.

Cabinet Maker Inc, p. I-486 [SIC 2434]
Cabinet Shop Machinery Inc, p. II-1610 [SIC 3553]
Cabinet Supply Inc, p. I-486 [SIC 2434]
Cable Design Technologies Inc, p. II-1172 [SIC 3312]
Cable Service Technologies Inc, p. II-1795 [SIC 3629]
Cablecraft, p. II-1476 [SIC 3496]
Cabo Rico Yachts Inc, p. I-437 [SIC 2394]
Cabot Corp, p. I-894 [SIC 2895]
Cabot Creamery Cooperative Inc, p. I-21 [SIC 2022]
Cabot Hosiery Mills Inc, p. I-250 [SIC 2251]
Cabot Medical Corp, p. II-2082 [SIC 3841]
Cabot Safety Corp, p. II-2107 [SIC 3851]
Caco Pacific Corp, p. II-1572 [SIC 3544]
Cadaco, p. II-2144 [SIC 3944]
Caddell Dry Dock and Repair, p. II-1986 [SIC 3731]
Cadet Manufacturing Co, p. II-1813 [SIC 3634]
Cadet Uniform Supply Co, p. I-352 [SIC 2326]
Cadillac Curtain Corp, p. I-424 [SIC 2391]
Cadillac Foods, p. I-91 [SIC 2047]
Cadillac Gage Textron Inc, p. II-2025 [SIC 3795]
Cadillac Oil Co, p. I-918 [SIC 2992]
Cadillac Products Inc, p. I-1006 [SIC 3089]
Cadillac Rubber and Plastics Inc, p. I-948 [SIC 3061]
Cadmus Communications Corp, p. I-729 [SIC 2759]
Cadmus Magazines, p. I-729 [SIC 2759]
Cady Bag Company Inc, p. I-433 [SIC 2393]
CAE-Link Corp, p. II-1938 [SIC 3699]
Cafe de Todd USA Inc, p. I-195 [SIC 2095]
Cafe Quick Enterprises Inc, p. II-1718 [SIC 3581]
Caffco International, p. II-2199 [SIC 3999]
Cagle's Inc, p. I-12 [SIC 2015]
Cahners Publishing Co, p. I-698 [SIC 2721]
CAI, p. II-2112 [SIC 3861]
Cain Cellars Inc, p. I-168 [SIC 2084]
Caire and Graugnard, p. I-116 [SIC 2061]
Cajun Bag and Supply Co, p. I-660 [SIC 2673]
Cajun Sugar Cooperative Inc, p. I-116 [SIC 2061]
Cal-Litho Color, p. I-759 [SIC 2796]
Cal Snap and Tab Corp, p. I-735 [SIC 2761]
Cal Themes Inc, p. I-420 [SIC 2389]
Cal-Van Tool, p. II-1312 [SIC 3423]
Calabro Cheese Corp, p. I-21 [SIC 2022]
Calcasieu Refining Co, p. I-903 [SIC 2911]
CalComp Inc, p. II-1704 [SIC 3577]
Calder Industries Inc, p. I-567 [SIC 2519]
Caldwell Culvert Co, p. II-1187 [SIC 3316]
Caldwell-Moser Leather, p. I-1013 [SIC 3111]
Caldwell Sugar Cooperative Inc, p. I-119 [SIC 2062]
Caldwell Tanks Inc, p. II-1355 [SIC 3443]
Calgon Carbon Corp, p. I-777 [SIC 2819]
Calgon Corp, p. I-897 [SIC 2899]
Caliber Computer Corp, p. II-1688 [SIC 3571]
Calidad Electronics Inc, p. II-1887 [SIC 3672]
California Acrylic Industries Inc, p. I-1001 [SIC 3088]
California Amplifier Inc, p. II-1873 [SIC 3663]
California Blended Products, p. I-1167 [SIC 3299]
California Cedar Products Co, p. I-536 [SIC 2499]
California Clock Co, p. II-2118 [SIC 3873]
California Combining Corp, p. I-973 [SIC 3083]
California Cooperative Creamery, p. I-35 [SIC 2026]
California Farm Products, p. I-60 [SIC 2037]
California Feather, p. I-428 [SIC 2392]
California Finished Metals Inc, p. II-1425 [SIC 3479]
California Flag and Sign Co, p. I-455 [SIC 2399]
California Flexrake Corp, p. II-1514 [SIC 3524]
California Gasket & Rubber, p. I-948 [SIC 3061]
California Graphite Machines, p. II-1525 [SIC 3532]
California Helicopter Intern, p. II-1970 [SIC 3721]
California Industrial Products, p. II-1413 [SIC 3469]
California Infanteen Togs, p. I-394 [SIC 2369]
California Kitchen, p. I-486 [SIC 2434]
California Magnetics, p. II-1864 [SIC 3652]
California Microwave, p. II-1873 [SIC 3663]
California Microwave Inc, p. II-1868 [SIC 3661]
California Offset Printers Inc, p. I-713 [SIC 2741]
California Optical Leather Inc, p. I-1049 [SIC 3172]
California Pellet Mill Co, p. I-1525 [SIC 3532]
California Products Corp, p. I-840 [SIC 2851]
California Protein Products Inc, p. I-152 [SIC 2077]

California Ranchwear Inc, p. I-336 [SIC 2321]
California Ribbon & Carbon Co, p. II-2169 [SIC 3955]
California Sample Co, p. I-750 [SIC 2789]
California Saw and Knife Works, p. II-1318 [SIC 3425]
California Shell Company Inc, p. II-2191 [SIC 3995]
California Soda Co, p. I-819 [SIC 2841]
California Stay Company Inc, p. I-1053 [SIC 3199]
California Steel and Tube, p. II-1187 [SIC 3316]
California Sun Dry Foods, p. I-51 [SIC 2034]
California Wine Co, p. I-168 [SIC 2084]
Calig Steel Drum Co, p. II-1304 [SIC 3412]
Calion Lumber Company Inc, p. I-471 [SIC 2426]
Calkins Manufacturing Co, p. II-1509 [SIC 3523]
Calkins Newspapers Inc, p. I-693 [SIC 2711]
Callaway Golf Co, p. II-2150 [SIC 3949]
Callidus Technologies Inc, p. II-1671 [SIC 3567]
Calmar Inc, p. II-1638 [SIC 3561]
Calmar Manufacturing Co, p. I-486 [SIC 2434]
CalMat Co, p. I-909 [SIC 2951]
Calnap Tanning Co, p. I-1013 [SIC 3111]
Calpine Containers Inc, p. I-515 [SIC 2449]
Calstone Company Inc, p. I-1117 [SIC 3271]
Caltex Petroleum Corp, p. I-903 [SIC 2911]
Calumet Brass Foundry Inc, p. II-1280 [SIC 3366]
Calvert Company Inc, p. I-501 [SIC 2439]
Calypso Panel Company Inc, p. I-491 [SIC 2435]
Cam Fran Tool Company Inc, p. II-1764 [SIC 3599]
Camac Corp, p. I-795 [SIC 2824]
Cambex Corp, p. II-1694 [SIC 3572]
Cambrex Corp, p. I-855 [SIC 2869]
Cambridge Dry Goods Company, p. I-373 [SIC 2339]
Cambridge Inc, p. II-1476 [SIC 3496]
Cambridge Lamps Inc, p. II-1840 [SIC 3645]
Cambridge Scientific Abstracts, p. I-713 [SIC 2741]
Cambridge SoundWorks Inc, p. II-1859 [SIC 3651]
Cambridge Tool & Mfg Co, p. II-1266 [SIC 3363]
Camcar Textron, p. II-1388 [SIC 3452]
Camco Chemical Co, p. I-825 [SIC 2842]
Camco International Inc, p. II-1531 [SIC 3533]
Camcraft Inc, p. II-1382 [SIC 3451]
Camden Industries Company Inc, p. I-587 [SIC 2541]
Camden Wire Company Inc, p. II-1260 [SIC 3357]
Camel Products, p. I-927 [SIC 3011]
Camelot Carpet Mills Inc, p. I-292 [SIC 2273]
Cameo Container Corp, p. I-628 [SIC 2653]
Cameron Coffee Company Inc, p. I-195 [SIC 2095]
Camillus Cutlery Co, p. II-1308 [SIC 3421]
Camirn Company Inc, p. I-528 [SIC 2491]
Camloc Products, p. II-1388 [SIC 3452]
Camp Cap Company Inc, p. I-386 [SIC 2353]
Camp Chef, p. II-1800 [SIC 3631]
Campbell Coffee Roasting Co, p. I-195 [SIC 2095]
Campbell Fittings Inc, p. II-1454 [SIC 3492]
Campbell Industries Inc, p. II-1986 [SIC 3731]
Campbell International Inc, p. II-1552 [SIC 3537]
Campbell Manufacturing, p. I-437 [SIC 2394]
Campbell Pattern Associates, p. II-1568 [SIC 3543]
Campbell Plastics, p. II-1404 [SIC 3465]
Campbell Soup Co, p. I-40 [SIC 2032]
Campbell Taggart Inc, p. I-101 [SIC 2051]
Campfire Charcoal Company Inc, p. I-846 [SIC 2861]
Campobello Foods, p. I-65 [SIC 2038]
Camptown Togs Inc, p. I-390 [SIC 2361]
Camtronics Limited Medical Syst, p. II-2102 [SIC 3845]
Can Lines Inc, p. II-1660 [SIC 3565]
Canada Dry of Delaware Valley, p. I-176 [SIC 2086]
Canamer International Inc, p. I-437 [SIC 2394]
Canandaigua Wine Company Inc, p. I-168 [SIC 2084]
Candela Laser Corp, p. II-2102 [SIC 3845]
Candes Systems Inc, p. II-1699 [SIC 3575]
Candle Corporation of America, p. II-2199 [SIC 3999]
Candle-Lite Co, p. II-2199 [SIC 3999]
Candlewick, p. I-297 [SIC 2281]
Candy and Co, p. I-819 [SIC 2841]
Cane Machine and Engineering, p. II-1509 [SIC 3523]
Cannon Bronze Corp, p. II-1676 [SIC 3568]
Cannon Conveyor Systems Inc, p. II-1660 [SIC 3565]
Cannon Equipment Southeast, p. II-1492 [SIC 3499]

Cannon Equipment West, p. I-593 [SIC 2542]
Cannon-Muskegon Corp, p. II-1229 [SIC 3341]
Cannondale Corp, p. II-2003 [SIC 3751]
Canon USA Inc, p. II-1704 [SIC 3577]
Cantek Metatron Corp, p. II-1854 [SIC 3648]
Cantex Inc, p. I-979 [SIC 3084]
Canton Mills Inc, p. I-869 [SIC 2875]
Canvas Products Co, p. I-437 [SIC 2394]
Canvas Specialty, p. I-437 [SIC 2394]
Canvasbacks Inc, p. I-373 [SIC 2339]
Canyon Materials Inc, p. I-1066 [SIC 3229]
Cap Toys Inc, p. II-2144 [SIC 3944]
Capacity of Texas Inc, p. II-1552 [SIC 3537]
Capaul Corp, p. I-1136 [SIC 3275]
Capco Inc, p. II-1775 [SIC 3613]
Cape Cod-Cricket Lane, p. I-365 [SIC 2335]
Cape Fear Feed Products Co, p. I-152 [SIC 2077]
Cape Industries, p. I-855 [SIC 2869]
Cape Industries Inc, p. II-1382 [SIC 3451]
Cape May Canners Inc, p. I-186 [SIC 2091]
Capel Inc, p. I-292 [SIC 2273]
Capital Bakers, p. I-101 [SIC 2051]
Capital Binding Co, p. I-750 [SIC 2789]
Capital City Press Inc, p. I-708 [SIC 2732]
Capital City Products, p. I-148 [SIC 2076]
Capital Controls Co, p. II-2056 [SIC 3824]
Capital Graphics Inc, p. I-735 [SIC 2761]
Capital Industries Inc, p. II-1355 [SIC 3443]
Capital Material Handling Inc, p. II-1547 [SIC 3536]
Capital Poly Bag Inc, p. I-660 [SIC 2673]
Capital Veneer Works Inc, p. I-491 [SIC 2435]
Capitol Cement Corp, p. I-1077 [SIC 3241]
Capitol Circuits Corp, p. II-1887 [SIC 3672]
Capitol City Container Corp, p. I-665 [SIC 2674]
Capitol Concrete Products Co, p. I-1117 [SIC 3271]
Capitol Engraving Co, p. I-759 [SIC 2796]
Capitol Glass & Aluminum Corp, p. I-1071 [SIC 3231]
Capitol Interior Inc, p. I-536 [SIC 2499]
Capitol Milling Co, p. I-70 [SIC 2041]
Capitol Products Corp, p. II-1244 [SIC 3354]
Capitol Records Inc, p. II-1864 [SIC 3652]
Capitol Saddlery, p. I-1053 [SIC 3199]
Capitol Stampings Corp, p. II-1676 [SIC 3568]
Capitol Technologies Inc, p. II-1572 [SIC 3544]
Capri Industries Inc, p. I-1071 [SIC 3231]
Capri Lighting Inc, p. II-1845 [SIC 3646]
Capsco Inc, p. II-1420 [SIC 3471]
Capstan Industries Inc, p. II-1323 [SIC 3429]
Captive-Aire Systems Inc, p. II-1813 [SIC 3634]
Captive Fasteners Corp, p. II-1388 [SIC 3452]
Capucci Creations Internationale, p. I-361 [SIC 2331]
Car Brite Inc, p. I-825 [SIC 2842]
Caradon Better-Bilt Inc, p. II-1349 [SIC 3442]
Carando, p. I-7 [SIC 2013]
Caraustar Industries Inc, p. I-623 [SIC 2652]
Caravali Coffees Inc, p. I-195 [SIC 2095]
Caravan Products Inc, p. I-83 [SIC 2045]
Carbide/Graphite Group Inc, p. II-2199 [SIC 3999]
Carbide International Inc, p. II-1558 [SIC 3541]
CarboMedics Inc, p. II-2087 [SIC 3842]
Carbon Products Operation Inc, p. II-1786 [SIC 3624]
Carbone of America, p. II-1786 [SIC 3624]
Carbonic Industries Corp, p. I-768 [SIC 2813]
Carbonic Reserves Inc, p. I-768 [SIC 2813]
Carbonite Filter Corp, p. I-1154 [SIC 3295]
Carborundum, p. I-1144 [SIC 3291]
Carborundum Co, p. I-1167 [SIC 3299]
Carbospheres Inc, p. II-1786 [SIC 3624]
Carburetion-J and S Inc, p. II-1742 [SIC 3592]
Carco Electronics, p. II-1938 [SIC 3699]
Cardell Cabinets Inc, p. I-486 [SIC 2434]
Cardinal Aluminum Co, p. II-1244 [SIC 3354]
Cardinal American Corp, p. II-1563 [SIC 3542]
Cardinal Glass Co, p. I-1057 [SIC 3211]
Cardinal IG Co, p. I-1071 [SIC 3231]
Cardinal Industrial Finish Inc, p. I-840 [SIC 2851]
Cardinal Industries Inc, p. II-1318 [SIC 3425]
Cardinal Scale Mfg Co, p. II-1759 [SIC 3596]
Cardinal Shoe Corp, p. I-1029 [SIC 3144]

Roman numerals I and II indicate the volume in which pages appear. Volume I holds SICs 2011 through 3299; Volume II holds SICs 3312 through 3999.

2235

Cardinal Wood Products Inc, p. I-510 [SIC 2448]
Cardkey Systems Inc, p. II-1878 [SIC 3669]
Cardone Industries Inc, p. II-1954 [SIC 3714]
Cardthartic Inc, p. I-741 [SIC 2771]
Cardwell Containers Inc, p. I-985 [SIC 3085]
Cardwell International Ltd, p. I-948 [SIC 3061]
Cardwell Machine Co, p. II-1626 [SIC 3556]
Carefree Aluminum Products, p. II-1349 [SIC 3442]
Carey Industries Inc, p. I-850 [SIC 2865]
Cargill Detroit Corp, p. II-1558 [SIC 3541]
Cargill Fertilizer Inc, p. I-865 [SIC 2874]
Cargill Inc, pp. Vol. I - 144, 869 [SICs 2075, 2875]
Cargill Soybean, p. I-144 [SIC 2075]
Carhartt Inc, p. I-352 [SIC 2326]
Caribou Mountaineering Inc, p. I-455 [SIC 2399]
Carisbrook Industries Inc, p. I-297 [SIC 2281]
Carisys Inc, p. II-1934 [SIC 3695]
Carl Buddig and Company Inc, p. I-7 [SIC 2013]
Carl Fischer Inc, p. I-713 [SIC 2741]
Carl G Wiklander Co, p. II-1621 [SIC 3555]
Carley Foundry Inc, p. II-1275 [SIC 3365]
Carley Inc, p. II-1826 [SIC 3641]
Carlin Manufacturing Inc, p. I-603 [SIC 2599]
Carlingswitch Inc, p. II-1830 [SIC 3643]
Carlisle Engineered Metals, p. II-1361 [SIC 3444]
Carlisle Geauga Co, p. I-1006 [SIC 3089]
Carlisle Plastics Inc, p. I-1006 [SIC 3089]
Carlisle SynTec Systems, p. I-954 [SIC 3069]
Carlon Meter Company Inc, p. II-2056 [SIC 3824]
Carlton Co, p. II-1318 [SIC 3425]
Carlton Forge Works, p. II-1395 [SIC 3462]
Carlton Manufacturing Inc, p. I-548 [SIC 2512]
Carlyle Golf Inc, p. I-336 [SIC 2321]
Carmen Foundations Inc, p. I-382 [SIC 2342]
Carmet Co, p. II-1578 [SIC 3545]
Carmi Molded Rubber Products, p. I-948 [SIC 3061]
Carnes Company Inc, p. II-1361 [SIC 3444]
Carnes Corp, p. II-1367 [SIC 3446]
Carnival Creations, p. I-382 [SIC 2342]
Carol Service Co, p. I-869 [SIC 2875]
Carolace Embroidery Company, p. I-271 [SIC 2258]
Carole Fabrics Inc, p. I-424 [SIC 2391]
Carole Wren Inc, p. I-373 [SIC 2339]
Carolee Designs Inc, p. II-2173 [SIC 3961]
Carolina Amato Inc, p. I-1037 [SIC 3151]
Carolina&Southern Processing, p. I-152 [SIC 2077]
Carolina Binding, p. I-446 [SIC 2396]
Carolina Business Furniture, p. I-581 [SIC 2531]
Carolina Canners Inc, p. I-176 [SIC 2086]
Carolina Coca-Cola Bottling Co, p. I-176 [SIC 2086]
Carolina Coml Heat Treating, p. II-1290 [SIC 3398]
Carolina Dairy Corp, p. I-35 [SIC 2026]
Carolina Dress Corp, p. I-365 [SIC 2335]
Carolina Furniture Works Inc, p. I-542 [SIC 2511]
Carolina Hardwoods LLC, p. I-471 [SIC 2426]
Carolina Hosiery Mills Inc, p. I-250 [SIC 2251]
Carolina Industries Inc, p. I-390 [SIC 2361]
Carolina Maid Products Inc, p. I-365 [SIC 2335]
Carolina Manufacturing Inc, p. I-420 [SIC 2389]
Carolina Mills Inc, p. I-297 [SIC 2281]
Carolina Mirror Co, p. I-1071 [SIC 3231]
Carolina Paper Box Company, p. I-623 [SIC 2652]
Carolina Quality Block Co, p. I-1117 [SIC 3271]
Carolina Shoe Co, p. I-1025 [SIC 3143]
Carolina Wood Preserving, p. I-528 [SIC 2491]
Carolmet Inc, p. II-1225 [SIC 3339]
Carolyn Collins Caviar Co, p. I-186 [SIC 2091]
Carolyn of Virginia Inc, p. I-405 [SIC 2384]
Caron Inc, p. I-365 [SIC 2335]
Carousel Carpet Mills Inc, p. I-292 [SIC 2273]
Carpel Video, p. II-1934 [SIC 3695]
Carpenter and Paterson Inc, p. II-1464 [SIC 3494]
Carpenter Co, p. I-991 [SIC 3086]
Carpenter Manufacturing Inc, p. II-1948 [SIC 3713]
Carpenter Rigging, p. I-437 [SIC 2394]
Carpenter Technology Corp, pp. Vol II - 1172, 1290, 1486 [SICs 3312, 3398, 3498]
Carpet Cushion Company Inc, p. I-327 [SIC 2299]
Carpeting Concepts, p. I-292 [SIC 2273]

Carpostan Industries Inc, p. I-231 [SIC 2211]
Carr Corp, p. II-2098 [SIC 3844]
Carr Lowrey Glass Co, p. I-1062 [SIC 3221]
Carriage Industries Inc, p. I-292 [SIC 2273]
Carrick Turning Works Inc, p. I-471 [SIC 2426]
Carrier Corp, p. II-1726 [SIC 3585]
Carrier Vibrating Equipment Inc, p. II-1541 [SIC 3535]
Carroll Co, p. I-825 [SIC 2842]
Carrollton Manufacturing, p. I-486 [SIC 2434]
Carskadden Optical Co, p. II-2107 [SIC 3851]
Carson and Gebel Ribbon Co, p. I-245 [SIC 2241]
Carson Manufacturing Co, p. II-1769 [SIC 3612]
Carson Products Co, p. I-834 [SIC 2844]
Carter Duncan Corp, p. II-2135 [SIC 3931]
Carter Footwear Inc, p. I-932 [SIC 3021]
Carter-Hoffmann Corp, p. II-1736 [SIC 3589]
Carter Machine Company Inc, p. II-1747 [SIC 3593]
Carter Manufacturing Company, p. I-515 [SIC 2449]
Carter Moore and Company Inc, p. I-795 [SIC 2824]
Carter Products Company Inc, p. II-1610 [SIC 3553]
Carter Traveler Co, p. II-1605 [SIC 3552]
Carter-Wallace Inc, p. I-804 [SIC 2834]
Carthage Co, p. I-373 [SIC 2339]
Carton-Craft Corp, p. I-643 [SIC 2657]
Cartridge Actuated Devices Inc, p. I-885 [SIC 2892]
Carts of Colorado Inc, p. I-587 [SIC 2541]
Cartwright Electronics Inc, p. II-1938 [SIC 3699]
Carus Corp, p. I-777 [SIC 2819]
Carvel Hall Inc, p. II-2127 [SIC 3914]
Carver Boat Corp, p. II-1992 [SIC 3732]
Carver Corp, p. II-1859 [SIC 3651]
Carville-National Leather Corp, p. I-1013 [SIC 3111]
Carving Craft Inc, p. I-471 [SIC 2426]
Casa Villa Inc, p. II-2020 [SIC 3792]
Casablanca Fan Co, p. II-1813 [SIC 3634]
Cascade Continental Foods Inc, p. I-639 [SIC 2656]
Cascade Corp, p. II-1552 [SIC 3537]
Cascade Die Casting Group Inc, p. II-1266 [SIC 3363]
Cascade Hardwood Inc, p. I-471 [SIC 2426]
Cascade Microtech Inc, p. II-2061 [SIC 3825]
Cascade-Niagara Falls Inc, p. I-618 [SIC 2631]
Cascade School Supplies, p. I-684 [SIC 2678]
Cascade Steel Rolling Mills Inc, p. II-1172 [SIC 3312]
Cascade Timber Company Inc, p. I-460 [SIC 2411]
Cascade Wood Components Inc, p. I-536 [SIC 2499]
Cascade Wood Products Inc, p. I-480 [SIC 2431]
Cascade Woolen Mill Inc, p. I-241 [SIC 2231]
Cascades Diamond Inc, p. I-688 [SIC 2679]
CasChem Inc, p. I-148 [SIC 2076]
Case Equipment Corp, p. II-1509 [SIC 3523]
Case Farms Inc, p. I-12 [SIC 2015]
Cases Inc, p. I-515 [SIC 2449]
Caseworks Furniture Mfg, p. I-581 [SIC 2531]
Caseworks International Inc, p. I-1041 [SIC 3161]
Cashco Inc, p. II-1447 [SIC 3491]
CASI-RUSCO Inc, p. II-1878 [SIC 3669]
Casket Shells Inc, p. II-2191 [SIC 3995]
Cass-Clay Creamery Inc, p. I-35 [SIC 2026]
Cass Screw Machine Prod Co, p. II-1382 [SIC 3451]
Cassie Cotillion Inc, p. I-378 [SIC 2341]
Cast-All Corp, p. I-973 [SIC 3083]
Cast Aluminum and Brass Corp, p. II-1275 [SIC 3365]
Cast Crete Tampa, p. I-1122 [SIC 3272]
Cast-Fab Technologies Inc, p. II-1197 [SIC 3321]
Cast Masters, p. II-1212 [SIC 3325]
Cast-Matic Corp, p. II-1266 [SIC 3363]
Cast-Rite Corp, p. II-1266 [SIC 3363]
Cast Specialties Inc, p. II-1271 [SIC 3364]
Castaic Clay Manufacturing Co, p. I-1081 [SIC 3251]
Castalloy Corp, p. II-1197 [SIC 3321]
CasTech Aluminum Group Inc, p. II-1239 [SIC 3353]
Casting Technology Inc, p. II-1207 [SIC 3324]
Castite Systems Inc, p. II-1295 [SIC 3399]
Castle Neckwear Inc, p. I-344 [SIC 2323]
Castle Rock Container Co, p. I-628 [SIC 2653]
Castleberry Knits Ltd, p. I-258 [SIC 2253]
Castlewood Apparel Corp, p. I-348 [SIC 2325]
Castrol Inc, p. I-918 [SIC 2992]

Castrol Industrial North America, p. I-918 [SIC 2992]
Castrol Industrial West Inc, p. I-918 [SIC 2992]
Castwell Products, p. II-1197 [SIC 3321]
Caswell International Corp, p. II-2150 [SIC 3949]
Caswell-Massey Company Ltd, p. I-819 [SIC 2841]
Catalina Lighting Inc, p. II-1840 [SIC 3645]
Catalina Marketing Corp, p. II-1704 [SIC 3577]
Catalina Products Corp, p. I-428 [SIC 2392]
Catalina Yachts Inc, p. II-1992 [SIC 3732]
Catalyst Semiconductor Inc, p. II-1892 [SIC 3674]
Catamount Brewing Co, p. I-160 [SIC 2082]
Cataphote Inc, p. I-1071 [SIC 3231]
Catawba-Charlab Inc, p. I-830 [SIC 2843]
Catawissa Lumber and Specialty, p. I-471 [SIC 2426]
Catawissa Valve and Fittings Co, p. II-1454 [SIC 3492]
Catching Engineering Inc, p. II-1747 [SIC 3593]
Catching Fluidpower Inc, p. II-1454 [SIC 3492]
Catco Inc, p. II-2144 [SIC 3944]
Caterpillar Inc, p. II-1552 [SIC 3537]
Caterpillar Paving Products Inc, p. I-1122 [SIC 3272]
Catherine's Rare Paper Inc, p. I-684 [SIC 2678]
Catherine Stein Designs Inc, p. II-2173 [SIC 3961]
CatHouse Fashions Inc, p. I-420 [SIC 2389]
Cato Oil and Grease Co, p. I-918 [SIC 2992]
Catoosa Knitting Mills Inc, p. I-258 [SIC 2253]
Cattiva Inc, p. I-365 [SIC 2335]
Cattleman's Meat Co, p. I-2 [SIC 2011]
Catty Corp, p. II-1660 [SIC 3565]
Cavaler Wire Products Inc, p. II-1182 [SIC 3315]
Cavalier Corp, p. II-1718 [SIC 3581]
Cavalier Homes Inc, p. I-519 [SIC 2451]
Cavalier Manufacturing Inc, p. I-548 [SIC 2512]
Cavanaugh Machine Works Inc, p. II-1552 [SIC 3537]
Cavco Industries Inc, p. I-523 [SIC 2452]
Cavel, p. I-236 [SIC 2221]
Cavert/Ace Baling Wire Co, p. II-1182 [SIC 3315]
Caviness Woodworking Co, p. I-536 [SIC 2499]
Caymus Vineyards, p. I-168 [SIC 2084]
CB Cummings and Sons Co, p. I-536 [SIC 2499]
CB Forms LP, p. I-735 [SIC 2761]
CB Manufacturing and Sales, p. II-1578 [SIC 3545]
CB Technical Sources Inc, p. I-603 [SIC 2599]
CBF Industries Inc, p. I-437 [SIC 2394]
CBI Industries Inc, p. II-1355 [SIC 3443]
CBI Laboratories Inc, p. I-834 [SIC 2844]
CBM America Corp, p. II-2112 [SIC 3861]
CBR Cement Corp, p. I-1077 [SIC 3241]
CBS Builders Supply Inc, p. I-501 [SIC 2439]
CCA Industries Inc, p. I-834 [SIC 2844]
CCC Industries Inc, p. II-1948 [SIC 3713]
CCI, p. II-2029 [SIC 3799]
CCL Custom Manufacturing Inc, p. II-2199 [SIC 3999]
CCL Label, p. I-729 [SIC 2759]
CCL Label Inc, p. I-729 [SIC 2759]
CCN International Inc, p. I-571 [SIC 2521]
CCP Industries Inc, p. II-231 [SIC 2211]
CCX Inc, p. II-1255 [SIC 3356]
CD Baird and Company Inc, p. II-2186 [SIC 3993]
CDF Corp, p. II-1304 [SIC 3412]
CDG Holdings Inc, p. I-373 [SIC 2339]
CDR Pigment and Dispersion, p. I-773 [SIC 2816]
CDS Mestel Construction Corp, p. I-587 [SIC 2541]
CE Cox Co, p. II-1600 [SIC 3549]
Ce De Candy Inc, p. I-127 [SIC 2064]
CE Larson and Sons Inc, p. II-1395 [SIC 3462]
CE Niehoff and Co, p. II-1929 [SIC 3694]
CE Shepherd Company Inc, p. II-1260 [SIC 3357]
Cecil Saydah Co, p. I-279 [SIC 2261]
Cecilware Corp, p. II-1736 [SIC 3589]
CECORP, p. II-1709 [SIC 3578]
Cedar Box Company Inc, p. I-506 [SIC 2441]
Cedar Crest Specialties Inc, p. I-30 [SIC 2024]
Cedar Grove Composting Inc, p. I-869 [SIC 2875]
Cedarapids Inc, p. II-1519 [SIC 3531]
Cedarmark Home Corp, p. I-523 [SIC 2452]
Cedartown Paper Board Co, p. I-688 [SIC 2679]
CEF Industries Inc, p. II-1975 [SIC 3724]
Celentano Brothers Inc, p. I-65 [SIC 2038]
Celestial Seasonings Inc, p. I-212 [SIC 2099]

Roman numerals I and II indicate the volume in which pages appear. Volume I holds SICs 2011 through 3299; Volume II holds SICs 3312 through 3999.

Roman numerals I and II indicate the volume in which pages appear. Volume I holds SICs 2011 through 3299; Volume II holds SICs 3312 through 3999.

Chem Polymer Corp, p. I-782 [SIC 2821]
Chem-Tec Equipment Co, p. II-1454 [SIC 3492]
Chem-Tronics Inc, p. II-1975 [SIC 3724]
Chemcraft, p. I-486 [SIC 2434]
Chemcut Equipment Group, p. II-1938 [SIC 3699]
Chemdal Corp, p. I-782 [SIC 2821]
Chemdal International, p. I-782 [SIC 2821]
Chemed Corp, p. II-2087 [SIC 3842]
Chemetals Inc, p. I-777 [SIC 2819]
Chemetron Railway Products, p. II-1519 [SIC 3531]
Chemfab Corp, p. I-236 [SIC 2221]
Chemgrate Corp, p. II-1367 [SIC 3446]
Chemi-Trol Chemical Co, pp. Vol II - 1355, 1584 [SICs 3443, 3546]
Chemical Exchange Industries, p. I-855 [SIC 2869]
Chemical Lime Co, p. I-1132 [SIC 3274]
Chemical Packaging Corp, p. I-825 [SIC 2842]
Chemidyne Corp, p. I-825 [SIC 2842]
Chemineer Inc, p. II-1632 [SIC 3559]
Chemlime NJ Inc, p. I-1132 [SIC 3274]
Chemoil Corp, p. I-903 [SIC 2911]
Chemonics Industries Inc, p. I-897 [SIC 2899]
Chemprene Inc, p. I-409 [SIC 2385]
ChemRex Inc, p. I-777 [SIC 2819]
Chemring Group Inc, p. II-1425 [SIC 3479]
Chemrock Corp, p. I-1154 [SIC 3295]
Chemstar Inc, p. I-825 [SIC 2842]
Chemtronics Inc, p. I-825 [SIC 2842]
Cheney Pulp and Paper Co, p. I-608 [SIC 2611]
Cher-Make Sausage Co, p. I-7 [SIC 2013]
Cheraw Yarn Mills Inc, p. I-297 [SIC 2281]
Cherco Compressors Inc, p. II-1649 [SIC 3563]
Cherokee Brick and Tile Co, p. I-1081 [SIC 3251]
Cherokee Finishing Co, p. I-279 [SIC 2261]
Cherokee Hosiery Mill Inc, p. I-254 [SIC 2252]
Cherokee Inc, p. I-336 [SIC 2321]
Cherokee Products Co, p. I-45 [SIC 2033]
Cherokee Shoe Co, p. I-1029 [SIC 3144]
Cherry Corp, p. II-1830 [SIC 3643]
Cherry Growers Inc, p. I-45 [SIC 2033]
Cherry Hill, p. I-542 [SIC 2511]
Cherry Hill Textiles Inc, p. I-288 [SIC 2269]
Cherry Semiconductor Corp, p. II-1892 [SIC 3674]
Chesapeake Consumer Prod Co, p. I-688 [SIC 2679]
Chesapeake Corp, pp. Vol. I - 618, 628 [SICs 2631, 2653]
Chesapeake Display, p. I-669 [SIC 2675]
Chesapeake Fish Company Inc, p. I-190 [SIC 2092]
Chesapeake Hardwood Products, p. I-491 [SIC 2435]
Chesapeake Packaging Co, p. I-628 [SIC 2653]
Chester Inc, p. I-869 [SIC 2875]
Chesterfield Mfg Corp, p. I-258 [SIC 2253]
Chesterfield Yarn Mill, p. I-297 [SIC 2281]
Chesterhill Stone Co, p. I-1132 [SIC 3274]
Chevron Corp, p. I-903 [SIC 2911]
Chevron USA Products, p. I-903 [SIC 2911]
Chic By HIS Inc, p. I-348 [SIC 2325]
Chic Lingerie Co, p. I-378 [SIC 2341]
Chicago Adhesive Products Co, p. I-879 [SIC 2891]
Chicago-Allis Mfg Corp, p. I-942 [SIC 3053]
Chicago Almond Inc, p. I-136 [SIC 2068]
Chicago Blower Corp, p. II-1654 [SIC 3564]
Chicago Bridge and Iron Co, p. II-1355 [SIC 3443]
Chicago Display Co, p. II-2186 [SIC 3993]
Chicago Dowel Company Inc, p. I-536 [SIC 2499]
Chicago Dryer Co, p. II-1722 [SIC 3582]
Chicago Faucet Co, p. II-1334 [SIC 3432]
Chicago Fire Brick Co, p. I-1090 [SIC 3255]
Chicago Gasket Co, p. I-942 [SIC 3053]
Chicago Gear-DO James Corp, p. II-1666 [SIC 3566]
Chicago Gear Works Inc, p. II-1666 [SIC 3566]
Chicago Litho-Plate Cies Ltd, p. I-759 [SIC 2796]
Chicago Lock Co, p. II-1323 [SIC 3429]
Chicago Magnesium Casting Co, p. II-1285 [SIC 3369]
Chicago Mailing Tube Co, p. I-634 [SIC 2655]
Chicago Manifold Products Co, p. II-1621 [SIC 3555]
Chicago Metallic Products Inc, p. II-1626 [SIC 3556]
Chicago Miniature Lamp Inc, p. II-1826 [SIC 3641]
Chicago Paving, p. I-909 [SIC 2951]

Chicago Press Corp, p. I-708 [SIC 2732]
Chicago Rivet and Machine Co, p. II-1563 [SIC 3542]
Chicago Show Printing Co, p. II-2186 [SIC 3993]
Chicago Specialty Mfg Co, p. II-1464 [SIC 3494]
Chicago Steel, p. II-1187 [SIC 3316]
Chicago Steel and Wire, p. II-1182 [SIC 3315]
Chicago Steel Construction, p. II-1344 [SIC 3441]
Chicago Steel Container Corp, p. II-1304 [SIC 3412]
Chicago Steel Rule, p. I-669 [SIC 2675]
Chicago Transparent Products, p. I-660 [SIC 2673]
Chicago White Metal Casting Inc, p. II-1271 [SIC 3364]
Chicago Wilcox Mfg Co, p. I-942 [SIC 3053]
Chicagoland Processing Corp, p. II-2122 [SIC 3911]
Chick Master Incubator Co, p. II-1509 [SIC 3523]
Chick Master International Inc, p. II-1509 [SIC 3523]
Chickasha Cotton Oil Co, p. I-140 [SIC 2074]
Chicken Noodle Inc, p. I-390 [SIC 2361]
Chief Agri/Industrial, p. II-1509 [SIC 3523]
Chief Apparel Inc, p. I-409 [SIC 2385]
Chief Automotive Systems Inc, p. II-1682 [SIC 3569]
Chief Industries Inc, p. II-1372 [SIC 3448]
Chikato Brothers Ice Co, p. I-204 [SIC 2097]
Childers Products Co, p. II-1361 [SIC 3444]
Chilton Co, p. I-698 [SIC 2721]
Chilton Metal Products, p. II-1413 [SIC 3469]
China Grove Textiles Inc, p. I-297 [SIC 2281]
Chinook Packing Co, p. I-190 [SIC 2092]
Chiodo Candy Co, p. I-127 [SIC 2064]
Chipcom Corp, p. II-1704 [SIC 3577]
Chips and Technologies Inc, p. II-1892 [SIC 3674]
Chiyoda America Inc, p. I-612 [SIC 2621]
Chloe's Closet, p. I-390 [SIC 2361]
Chloride Systems, p. II-1845 [SIC 3646]
Chock Full O'Nuts Corp, p. I-195 [SIC 2095]
Chocolate House Inc, p. I-127 [SIC 2064]
Choctaw Inc, p. I-1122 [SIC 3272]
Choctaw Maid Farms Inc, p. I-12 [SIC 2015]
Choice USA Beverages, p. I-176 [SIC 2086]
Chokio Equity Exchange Inc, p. I-869 [SIC 2875]
Chomerics Inc, p. I-942 [SIC 3053]
Chooljian Brothers Packing Co, p. I-51 [SIC 2034]
Chowan Veneer Company Inc, p. I-491 [SIC 2435]
Chris-Craft Industrial Products, p. I-961 [SIC 3081]
Chris Stone and Associates, p. I-288 [SIC 2269]
Christensen Shipyards Ltd, p. II-1986 [SIC 3731]
Christiana Companies Inc, p. II-1192 [SIC 3317]
Christiana Machine Co, p. II-1666 [SIC 3566]
Christie Brown and Co, p. I-199 [SIC 2096]
Christofle Silver Inc, p. II-2127 [SIC 3914]
Christy Refractories Co, p. I-1163 [SIC 3297]
Chromaline Corp, p. II-2112 [SIC 3861]
Chromalloy, p. II-1498 [SIC 3511]
Chromalloy Aircraft Structures, p. II-1975 [SIC 3724]
Chromalloy Casting Miami Corp, p. II-1285 [SIC 3369]
Chromalloy Gas Turbine Corp, p. II-1975 [SIC 3724]
Chromalloy Precision Products, p. II-1975 [SIC 3724]
Chromatic Technologies Inc, p. II-1260 [SIC 3357]
Chromium Corp, p. II-1503 [SIC 3519]
Chromium Industries Inc, p. II-1420 [SIC 3471]
Chronicle Publishing Co, p. I-693 [SIC 2711]
Chronomite Laboratories Inc, p. II-1822 [SIC 3639]
Chronos Richardson Inc, p. II-1759 [SIC 3596]
Chrysler Corp, p. II-1943 [SIC 3711]
CHS Acquisition Corp, p. II-1172 [SIC 3312]
Chubb National Foam Inc, p. II-1943 [SIC 3711]
Chums Ltd, p. I-1049 [SIC 3172]
Church and Chapel Metal Arts, p. II-1367 [SIC 3446]
Church and Dwight Company, p. I-764 [SIC 2812]
CI Hayes Inc, p. II-1671 [SIC 3567]
Ciao Bella Gelato Co, p. I-30 [SIC 2024]
Cibro Petroleum Inc, p. I-903 [SIC 2911]
CIDCO Inc, p. II-1868 [SIC 3661]
CIM Industrial Machinery Inc, p. II-1552 [SIC 3537]
Cimpl Packing Co, p. I-2 [SIC 2011]
Cin-Made Corp, p. I-634 [SIC 2655]
Cin-Tran Inc, p. II-1769 [SIC 3612]
Cincinnati Butchers Supply Co, p. II-1626 [SIC 3556]

Cincinnati Enquirer, p. I-693 [SIC 2711]
Cincinnati Gasket Packing Mfg, p. I-942 [SIC 3053]
Cincinnati Gear Co, p. II-1666 [SIC 3566]
Cincinnati Inc, p. II-1563 [SIC 3542]
Cincinnati Milacron, p. I-1144 [SIC 3291]
Cincinnati Milacron Inc, pp. Vol. I - 897, Vol. II - 1558, 1790 [SICs 2899, 3541, 3625]
Cincinnati Paperboard Corp, p. I-618 [SIC 2631]
Cincinnati Plastics Machinery, p. II-1632 [SIC 3559]
Cinder and Concrete Block Corp, p. I-1117 [SIC 3271]
Cinema Products Corp, p. II-2112 [SIC 3861]
Cinnabar Traders Ltd, p. I-336 [SIC 2321]
Cinram Inc, p. II-1864 [SIC 3652]
Circa Pharmaceuticals Inc, p. I-804 [SIC 2834]
Circle AW Products Co, p. II-1775 [SIC 3613]
Circle Jewelry Products Inc, p. II-2173 [SIC 3961]
Circle S Industries Inc, p. II-1915 [SIC 3679]
Circle Seal Controls, p. II-1447 [SIC 3491]
Circle Y of Yoakum Inc, p. I-1049 [SIC 3172]
Circlemaster Inc, p. II-1187 [SIC 3316]
Circon Corp, p. II-2102 [SIC 3845]
Circuit Assembly Corp, p. II-1910 [SIC 3678]
Circuit Components Inc, p. II-1897 [SIC 3675]
Circuit Foil USA Inc, p. II-1482 [SIC 3497]
Circuit Systems Inc, p. II-1887 [SIC 3672]
Circuits and Systems Inc, p. II-1759 [SIC 3596]
Cirrus Logic Inc, p. II-1892 [SIC 3674]
Cisco Systems Inc, p. II-1704 [SIC 3577]
Cissell Manufacturing Co, p. II-1722 [SIC 3582]
Cistron Biotechnology Inc, p. I-814 [SIC 2836]
Citation Corp, p. II-1197 [SIC 3321]
Citation Homes Inc, p. I-501 [SIC 2439]
Citation Tool Inc, p. II-1563 [SIC 3542]
Citco, p. II-1578 [SIC 3545]
Citel America Inc, p. II-1830 [SIC 3643]
CITGO Asphalt Refining Co, p. I-903 [SIC 2911]
CITGO Petroleum Corp, p. I-903 [SIC 2911]
Citifax Corp, p. II-1713 [SIC 3579]
Citisteel USA Inc, p. II-1172 [SIC 3312]
Citrus Service Inc, p. I-60 [SIC 2037]
Citrus World Inc, p. I-60 [SIC 2037]
City Forge, p. II-1395 [SIC 3462]
City Ice Company Inc, p. I-204 [SIC 2097]
City Machine Tool and Die Co, p. II-1572 [SIC 3544]
City Market Bakery, p. I-101 [SIC 2051]
City Wire Cloth Inc, p. II-1182 [SIC 3315]
Cives Corp, p. II-1344 [SIC 3441]
Cives Steel Co, p. II-1344 [SIC 3441]
CJ Krehbiel Co, p. I-708 [SIC 2732]
CJ Saporito Plating Co, p. II-1420 [SIC 3471]
CJ Vitner Company Inc, p. I-199 [SIC 2096]
CJ Winter Machine Works, p. II-1563 [SIC 3542]
CK Systematics Inc, p. II-1594 [SIC 3548]
CK Worldwide Inc, p. II-1594 [SIC 3548]
CKR Industries Inc, p. I-948 [SIC 3061]
CL Rieckhoff Company Inc, p. II-1377 [SIC 3449]
Claire Manufacturing Company, p. I-825 [SIC 2842]
Clairol Inc, p. I-834 [SIC 2844]
Clalite Concrete Products Inc, p. I-1117 [SIC 3271]
Clamshell Building Inc, p. I-437 [SIC 2394]
Clarcor Inc, p. II-1954 [SIC 3714]
Claremont Company Inc, p. I-1158 [SIC 3296]
Claridge Products&Equipment, p. I-581 [SIC 2531]
Clarion Sintered Metals Inc, p. II-1295 [SIC 3399]
Clark Automotive Products Corp, p. II-1954 [SIC 3714]
Clark Casual Furniture Inc, p. I-567 [SIC 2519]
Clark-Cutler-McDermott Corp, p. I-327 [SIC 2299]
Clark Equipment Co, p. II-1552 [SIC 3537]
Clark Foam, p. I-782 [SIC 2821]
Clark Grave Vault Co, p. II-2191 [SIC 3995]
Clark Manufacturing Inc, p. I-1001 [SIC 3088]
Clark Material Handling Co, p. II-1552 [SIC 3537]
Clark Paper Converting Corp, p. I-674 [SIC 2676]
Clark Refining & Marketing Co, p. I-903 [SIC 2911]
Clark-Schwebel Inc, p. I-236 [SIC 2221]
Clark Specialty Company Inc, p. I-593 [SIC 2542]
Clarke American Checks Inc, p. I-729 [SIC 2759]
Clarke Industries Inc, p. II-1736 [SIC 3589]

Roman numerals I and II indicate the volume in which pages appear. Volume I holds SICs 2011 through 3299; Volume II holds SICs 3312 through 3999.

Roman numerals I and II indicate the volume in which pages appear. Volume I holds SICs 2011 through 3299; Volume II holds SICs 3312 through 3999.

Colorado Serum Co, p. I-814 [SIC 2836]
Colorado Sintered Metals, p. II-1295 [SIC 3399]
Coloray Display Corp, p. II-1883 [SIC 3671]
Colorcon, p. I-897 [SIC 2899]
Colorforms, p. II-2144 [SIC 3944]
Colorhouse Inc, p. I-759 [SIC 2796]
Colorite Plastic Co, p. I-936 [SIC 3052]
Colortran Inc, p. II-1854 [SIC 3648]
Colortricity Inc, p. I-754 [SIC 2791]
Colortronix Corp, p. I-759 [SIC 2796]
Colotone Riverside Press Inc, p. I-718 [SIC 2752]
Colour Graphics Corp, p. I-759 [SIC 2796]
Colour Image, p. I-759 [SIC 2796]
Colson Caster Corp, p. II-1323 [SIC 3429]
Colt Resources Corp, p. I-924 [SIC 2999]
Colt's Manufacturing Company, p. II-1438 [SIC 3484]
Coltec Industries Inc, p. II-1980 [SIC 3728]
Coltene/Whaledent Inc, p. II-2093 [SIC 3843]
Colts Plastics Co, p. I-985 [SIC 3085]
Columbia Aluminum Corp, p. II-1275 [SIC 3365]
Columbia Cascade Co, p. I-501 [SIC 2439]
Columbia Cement Company Inc, p. I-879 [SIC 2891]
Columbia Corrugated Box Co, p. I-643 [SIC 2657]
Columbia Crest Winery, p. I-168 [SIC 2084]
Columbia Forest Products Inc, p. I-491 [SIC 2435]
Columbia Industries Inc, p. II-2150 [SIC 3949]
Columbia Lighting Inc, p. II-1845 [SIC 3646]
Columbia Loose Leaf Corp, p. I-745 [SIC 2782]
Columbia Machine Inc, p. II-1632 [SIC 3559]
Columbia Manufacturing Corp, p. II-1349 [SIC 3442]
Columbia Manufacturing Inc, p. II-2003 [SIC 3751]
Columbia Pacific, p. II-1244 [SIC 3354]
Columbia Paint and Coating Inc, p. I-840 [SIC 2851]
Columbia Panel Manufacturing, p. I-491 [SIC 2435]
Columbia ParCar Corp, p. II-2029 [SIC 3799]
Columbia Scientific Indust Corp, p. II-2051 [SIC 3823]
Columbia Winery, p. I-168 [SIC 2084]
Columbia Woodworking Inc, p. I-486 [SIC 2434]
Columbia Wool Scouring Mills, p. I-327 [SIC 2299]
Columbiana Foundry Co, p. II-1212 [SIC 3325]
Columbus Brick Company Inc, p. I-1081 [SIC 3251]
Columbus Container Inc, p. I-628 [SIC 2653]
Columbus Electric Mfg Co, p. II-2045 [SIC 3822]
Columbus Galvanizing Inc, p. II-1425 [SIC 3479]
Columbus Hydraulic Co, p. II-1747 [SIC 3593]
Columbus Industries Inc, p. II-1764 [SIC 3599]
Columbus Instruments, p. II-2066 [SIC 3826]
Columbus Marble Works Inc, p. I-1140 [SIC 3281]
Columbus McKinnon Corp, pp. Vol II - 1476, 1547 [SICs 3496, 3536]
Columbus Mills Inc, p. I-292 [SIC 2273]
Columbus Roof Trusses Inc, p. I-501 [SIC 2439]
Columbus Seven-Up, p. I-176 [SIC 2086]
Colwell General Inc, p. I-669 [SIC 2675]
Colwell Industries Inc, p. I-745 [SIC 2782]
Colwell Systems, p. I-718 [SIC 2752]
Com-Tal Machine & Engineering, p. II-1764 [SIC 3599]
Comar Inc, p. I-1006 [SIC 3089]
Combe Inc, p. I-834 [SIC 2844]
Combined Fluid Products Co, p. II-1649 [SIC 3563]
Combined Interest Inc, p. I-258 [SIC 2253]
Comdial Corp, p. II-1868 [SIC 3661]
Come Play Products Co, p. II-2144 [SIC 3944]
Comet Confectionery Inc, p. I-127 [SIC 2064]
Comet Industries Inc, p. II-1780 [SIC 3621]
Comet-Montrose Ltd, p. II-2144 [SIC 3944]
Comfort Designs Inc, p. I-548 [SIC 2512]
Comfortex Inc, p. I-558 [SIC 2515]
Cominco Fertilizer, p. I-861 [SIC 2873]
Comlinear Corp, p. II-2061 [SIC 3825]
Commander Aircraft Co, p. II-1970 [SIC 3721]
Commander Packaging Corp, p. I-628 [SIC 2653]
Commerce Clearing House Inc, p. I-713 [SIC 2741]
Commercial & Architectural, p. I-480 [SIC 2431]
Commercial Cam Company Inc, p. II-1578 [SIC 3545]
Commercial Carving Company, p. I-536 [SIC 2499]
Commercial Environmental Syst, p. II-1726 [SIC 3585]
Commercial Forged Products, p. II-1395 [SIC 3462]

Commercial Intertech Corp, p. II-1638 [SIC 3561]
Commercial Metals Co, p. II-1229 [SIC 3341]
Commercial Steel Corp, p. II-1212 [SIC 3325]
Commercial Steel Treating Corp, p. II-1290 [SIC 3398]
Commercial Textiles Inc, p. I-352 [SIC 2326]
Commercial Tool and Die Co, p. II-1413 [SIC 3469]
Commodore Corp, p. I-519 [SIC 2451]
Common Equipment Co, p. II-1552 [SIC 3537]
Commonwealth Aluminum Corp, p. II-1239 [SIC 3353]
Commonwealth Bolt Inc, p. II-1388 [SIC 3452]
Commonwealth Industries, p. II-1290 [SIC 3398]
Commonwealth Scientific Corp, p. II-2061 [SIC 3825]
Commonwealth Toy and Novelty, p. II-2140 [SIC 3942]
CommScope Inc, p. II-1260 [SIC 3357]
CommStar Inc, p. II-1709 [SIC 3578]
Communication Cable Inc, p. II-1260 [SIC 3357]
Communication Coil Inc, p. II-1905 [SIC 3677]
Communications Systems Inc, p. II-1868 [SIC 3661]
Communicolor Inc, p. I-729 [SIC 2759]
Community Playthings, p. II-2144 [SIC 3944]
Como Textile Prints Inc, p. I-279 [SIC 2261]
Compac Corp, p. I-688 [SIC 2679]
Compac Microelectronics Inc, p. II-1699 [SIC 3575]
CompAir Kellogg, p. II-1649 [SIC 3563]
Company Store Inc, p. I-428 [SIC 2392]
Compaq Computer Corp, p. II-1688 [SIC 3571]
Compax Systems Inc, p. II-1822 [SIC 3639]
Component Research Company, p. II-1795 [SIC 3629]
Composing Room Inc, p. I-754 [SIC 2791]
Composing Room of Michigan, p. I-754 [SIC 2791]
Composite Technical Alloys Inc, p. II-1212 [SIC 3325]
Composting Toilet Systems, p. I-523 [SIC 2452]
Comprehensive Industries, p. I-510 [SIC 2448]
Compression Labs Inc, p. II-1873 [SIC 3663]
Compression Services, p. II-1649 [SIC 3563]
Compressor Controls Corp, p. II-2051 [SIC 3823]
Compressor Engineering Corp, p. II-1649 [SIC 3563]
Compressor Systems Inc, p. II-1649 [SIC 3563]
Comptek Research Inc, p. II-1688 [SIC 3571]
Compton Presentation Syst Co, p. I-745 [SIC 2782]
Comptronix Corp, p. II-1887 [SIC 3672]
CompuDyne Corp, p. II-1878 [SIC 3669]
Compumachine Inc, p. II-1563 [SIC 3542]
CompuRegister Corp, p. II-1709 [SIC 3578]
Computer Network Techn Corp, p. II-1704 [SIC 3577]
Computer Power Inc, p. II-1769 [SIC 3612]
Computer Products Inc, p. II-2076 [SIC 3829]
Computer Sales Professional Ltd, p. II-1688 [SIC 3571]
ComputerPREP Inc, p. I-713 [SIC 2741]
Computype Inc, p. I-654 [SIC 2672]
Comstock Creations Inc, p. II-2127 [SIC 3914]
Comstock Michigan Fruit, p. I-45 [SIC 2033]
Comstock-Tivolie, p. I-356 [SIC 2329]
ComStream Corp, p. II-1868 [SIC 3661]
Comtrex Systems Corp, p. II-1699 [SIC 3575]
Comverse Technology Inc, p. II-1688 [SIC 3571]
Con-Cise Lens Co, p. II-2107 [SIC 3851]
Con-Lime Inc, p. I-1132 [SIC 3274]
Con-Lux Coatings Inc, p. I-840 [SIC 2851]
Con-Tech Industries Inc, p. I-1001 [SIC 3088]
ConAgra Consumer, p. I-65 [SIC 2038]
ConAgra Feed Co, p. I-95 [SIC 2048]
ConAgra Flour Milling Co, p. I-70 [SIC 2041]
ConAgra Fresh Meats Co, p. I-2 [SIC 2011]
ConAgra Inc, p. I-95 [SIC 2048]
ConAgra Poultry Co, p. I-12 [SIC 2015]
Conair Corp, p. II-1813 [SIC 3634]
Conair Group Inc, p. II-1632 [SIC 3559]
Conap Inc, p. I-782 [SIC 2821]
Conarc Inc, p. I-1158 [SIC 3296]
Conaway-Winter Inc, p. I-1033 [SIC 3149]
Conax Buffalo Corp, p. II-2076 [SIC 3829]
Concept Seating Inc, p. I-581 [SIC 2531]
Conco Medical Co, p. II-2087 [SIC 3842]
Concord Beverage Co, p. I-176 [SIC 2086]
Concord Camera Corp, p. II-2112 [SIC 3861]

Concord Chemical Co, p. I-819 [SIC 2841]
Concord Fabrics Inc, p. I-231 [SIC 2211]
Concord Instruments Inc, p. II-1850 [SIC 3647]
Concord Litho Company Inc, p. I-718 [SIC 2752]
Concordia Manufacturing, p. I-302 [SIC 2282]
Concordia Publishing House, p. I-703 [SIC 2731]
Concrete Coring Co, p. I-909 [SIC 2951]
Concrete Materials Co, p. I-1127 [SIC 3273]
Concrete Pipe and Products, p. I-1122 [SIC 3272]
Concrete Products Co, p. I-1117 [SIC 3271]
Concrete Supply Co, p. I-1127 [SIC 3273]
Concrete Technology Corp, p. I-1122 [SIC 3272]
Concrete Tie and Anchor Corp, p. I-1122 [SIC 3272]
Concurrent Computer Corp, p. II-1688 [SIC 3571]
Condere Corp, p. I-927 [SIC 3011]
Condor DC Power Supplies Inc, p. II-1830 [SIC 3643]
Condor Systems Inc, p. II-1878 [SIC 3669]
Cone Drive Operations Inc, p. II-1666 [SIC 3566]
Cone Mills Corp, p. I-231 [SIC 2211]
Conestoga Wood Specialties Inc, p. I-480 [SIC 2431]
Congoleum Corp, p. I-1006 [SIC 3089]
Congressional Quarterly Inc, p. I-698 [SIC 2721]
Conklin Company Inc, p. I-865 [SIC 2874]
Conley Frog Switch & Forge Co, p. II-1395 [SIC 3462]
CONMED Corp, p. II-2082 [SIC 3841]
Connaught Laboratories Inc, p. I-814 [SIC 2836]
Connecticut Coining Inc, p. II-2098 [SIC 3844]
Connecticut Container Corp, p. I-628 [SIC 2653]
Connecticut Electric, p. II-1775 [SIC 3613]
Connecticut Fineblanking Corp, p. II-1413 [SIC 3469]
Connecticut Laminating, p. I-973 [SIC 3083]
Connecticut Stamping & Bending, p. II-1334 [SIC 3432]
Connecticut Steel Corp, p. II-1172 [SIC 3312]
Connecting Devices Inc, p. II-1260 [SIC 3357]
Connelly-GPM Inc, p. II-1295 [SIC 3399]
Conner Peripherals Inc, p. II-1694 [SIC 3572]
Connex Pipe Systems Inc, p. II-1486 [SIC 3498]
Connor AGA, p. I-471 [SIC 2426]
Connor Formed Metal Products, p. II-1470 [SIC 3495]
Conoco Inc, p. I-903 [SIC 2911]
Conover Chair Company Inc, p. I-548 [SIC 2512]
Conquest Carpet Mills Inc, p. I-292 [SIC 2273]
Conrad-American Inc, p. II-1509 [SIC 3523]
Conrad Co, p. II-1970 [SIC 3721]
Conrad Industries Inc, p. II-1986 [SIC 3731]
Conroe Creosoting Inc, p. I-528 [SIC 2491]
Consarc Corp, p. II-1671 [SIC 3567]
Consep Inc, p. I-874 [SIC 2879]
Conservatek Industries Inc, p. II-1295 [SIC 3399]
Conso Products Co, p. I-245 [SIC 2241]
Consolidated Biscuit Corp, p. I-107 [SIC 2052]
Consolidated Casting Corp, p. II-1207 [SIC 3324]
Consolidated Coatings Corp, p. I-914 [SIC 2952]
Consolidated Converting Co, p. I-628 [SIC 2653]
Consolidated Counter Inc, p. I-1017 [SIC 3131]
Consolidated Devices Inc, p. II-1312 [SIC 3423]
Consolidated Diesel Co, p. II-1503 [SIC 3519]
Consolidated Distilled Prod, p. I-172 [SIC 2085]
Consolidated Electronic, p. II-1260 [SIC 3357]
Consolidated Fiber, p. I-914 [SIC 2952]
Consolidated Flavor Corp, p. I-181 [SIC 2087]
Consolidated Food, p. I-208 [SIC 2098]
Consolidated Glass, p. I-1071 [SIC 3231]
Consolidated Graphics Inc, p. I-718 [SIC 2752]
Consolidated Industries Corp, p. II-1339 [SIC 3433]
Consolidated Industries Inc, p. II-1400 [SIC 3463]
Consolidated Metal Products Inc, p. II-1377 [SIC 3449]
Consolidated Metco Die Casting, p. II-1275 [SIC 3365]
Consolidated Metco Inc, p. II-1275 [SIC 3365]
Consolidated Papers Inc, pp. Vol. I - 612, 649 [SICs 2621, 2671]
Consolidated Pet Foods Inc, p. I-91 [SIC 2047]
Consolidated Printers Inc, p. I-708 [SIC 2732]
Consolidated Products Corp, p. II-1182 [SIC 3315]
Consolidated Stamp of Chicago, p. II-2164 [SIC 3953]
Consolidated Steel, p. II-1367 [SIC 3446]
Consolidated Systems Inc, p. II-1361 [SIC 3444]

Roman numerals I and II indicate the volume in which pages appear. Volume I holds SICs 2011 through 3299; Volume II holds SICs 3312 through 3999.

2240

CR Daniels Inc, p. I-437 [SIC 2394]
CR Laine Furniture Company, p. I-548 [SIC 2512]
CR/PL LP, p. I-1098 [SIC 3261]
Crackin Good Bakers Inc, p. I-107 [SIC 2052]
Cradle Togs Inc, p. I-390 [SIC 2361]
Craft Die Casting Corp, p. II-1271 [SIC 3364]
Craft House Corp, p. II-2144 [SIC 3944]
Craftex Mills Inc, p. I-231 [SIC 2211]
Craftmation Inc, p. II-1594 [SIC 3548]
Craftsman Press Inc, p. I-735 [SIC 2761]
Craftsmen Machinery Co, p. II-1621 [SIC 3555]
Craftsmen Steel Buildings Co, p. II-1372 [SIC 3448]
Craftstones Inc, p. II-2131 [SIC 3915]
Craig Systems Corp, p. II-1986 [SIC 3731]
Crain Industries Inc, p. I-782 [SIC 2821]
Cramer Coil and Transformer, p. II-1769 [SIC 3612]
CranBarry Inc, p. I-373 [SIC 2339]
Crane Carrier Co, p. II-1948 [SIC 3713]
Crane Carton Co, p. I-643 [SIC 2657]
Crane Co, p. II-1464 [SIC 3494]
Crane Cor Tec, p. I-1066 [SIC 3229]
Crane Electronics Inc, p. II-1910 [SIC 3678]
Crane Manufacturing Company, p. I-373 [SIC 2339]
Crane Mfg & Service Corp, p. II-1547 [SIC 3536]
Crane Plastics Co, p. I-967 [SIC 3082]
Crane Plumbing, p. II-1330 [SIC 3431]
Crane Valve, p. II-1447 [SIC 3491]
Cranesville Block Company Inc, p. I-1117 [SIC 3271]
Cranford Woodcarving Inc, p. I-471 [SIC 2426]
Cranston Print Works Co, p. I-279 [SIC 2261]
Crantex Fabrics, p. I-279 [SIC 2261]
Cratex Manufacturing Company, p. I-1144 [SIC 3291]
Crawford Container Co, p. II-1300 [SIC 3411]
Crawford Doors, p. II-1349 [SIC 3442]
Crawford Furniture Mfg Corp, p. I-542 [SIC 2511]
Cray Research Inc, p. II-1688 [SIC 3571]
Crazy Shirts Inc, p. I-279 [SIC 2261]
CRC-Evans Pipeline Intern, p. II-1632 [SIC 3559]
CRC Industries Inc, p. I-897 [SIC 2899]
Creamland Dairies Inc, p. I-30 [SIC 2024]
Creation Windows Inc, p. I-1066 [SIC 3229]
Creations by Alan Stuart Inc, p. II-2181 [SIC 3991]
Creative Aerosol Corp, p. I-819 [SIC 2841]
Creative Art Activities Inc, p. II-2144 [SIC 3944]
Creative Color Service Corp, p. I-759 [SIC 2796]
Creative Designs International, p. I-536 [SIC 2499]
Creative Expressions Group, p. I-688 [SIC 2679]
Creative Foam Corp, p. I-991 [SIC 3086]
Creative Medical Development, p. I-809 [SIC 2835]
Creative Point Inc, p. I-567 [SIC 2519]
Creative Specialties Inc, p. II-1330 [SIC 3431]
Creative Teaching Press Inc, p. I-708 [SIC 2732]
Creative Technologies Corp, p. II-1813 [SIC 3634]
Credence Systems Corp, p. II-2061 [SIC 3825]
Credo Co, p. II-1323 [SIC 3429]
Creed Co, p. II-1334 [SIC 3432]
Creed Monarch Inc, p. II-1382 [SIC 3451]
Crego Block Co, p. I-1117 [SIC 3271]
Crenlo Inc, p. II-1492 [SIC 3499]
Crescent Brass Mfg Corp, p. II-1285 [SIC 3369]
Crescent Bronze Powder, p. II-2160 [SIC 3952]
Crescent Dental Mfg Co, p. II-2093 [SIC 3843]
Crescent Enterprises Inc, p. I-542 [SIC 2511]
Crescent Iron Works Co, p. II-1377 [SIC 3449]
Crescent Metal Products, p. II-1626 [SIC 3556]
Crescent Metal Products Co, p. II-1486 [SIC 3498]
Crescent Plastics Inc, p. I-973 [SIC 3083]
Crescent Spinning Co, p. I-297 [SIC 2281]
Crescent Woolen Mills Co, p. I-297 [SIC 2281]
Cressona Aluminum Co, p. II-1244 [SIC 3354]
Crest Audio Inc, p. II-1859 [SIC 3651]
Crest Foods Company Inc, p. I-212 [SIC 2099]
Crest/Good Manufacturing, p. II-1334 [SIC 3432]
Crest Manufacturing Co, p. II-2045 [SIC 3822]
Crest Products Inc, p. I-825 [SIC 2842]
Crest Uniform Company Inc, p. I-352 [SIC 2326]
Cresthill Industries Inc, p. II-2177 [SIC 3965]
Crestliner Boats Inc, p. II-1992 [SIC 3732]
Crestwood Cutter Inc, p. II-1308 [SIC 3421]

Cretex Company Inc, p. I-1122 [SIC 3272]
Cri-Tech Inc, p. I-787 [SIC 2822]
Cricket Hosiery Inc, p. I-254 [SIC 2252]
Crimson Industries Inc, p. I-519 [SIC 2451]
Crippen Manufacturing Company, p. II-1626 [SIC 3556]
Criterion Gate & Manufacturing, p. II-1367 [SIC 3446]
Criticare Systems Inc, p. II-2082 [SIC 3841]
CRL Inc, p. II-1438 [SIC 3484]
CRL Industries Inc, p. II-1649 [SIC 3563]
Crocker Technical Papers, p. I-688 [SIC 2679]
Crockett Container Corp, p. I-628 [SIC 2653]
Croda Inks Corp, p. I-889 [SIC 2893]
Croft Metals Inc, p. II-1349 [SIC 3442]
Cromers Inc, p. I-45 [SIC 2033]
Cromwell Leather Company Inc, p. I-1013 [SIC 3111]
Cronkhite Industries Inc, p. II-1961 [SIC 3715]
Cronland Warp Roll Co, p. II-1605 [SIC 3552]
Crosbie Foundry Company Inc, p. II-1280 [SIC 3366]
Crosby Fruit Products Company, p. I-45 [SIC 2033]
Crosby Group Inc, p. II-1531 [SIC 3533]
Crosby-LeBus Manufacturing Co, p. II-1395 [SIC 3462]
Croscill Inc, p. I-424 [SIC 2391]
Crosfield Co, p. I-777 [SIC 2819]
Crosley Canvas, p. I-437 [SIC 2394]
Crosman Corp, p. II-1438 [SIC 3484]
Cross and Peters Co, p. I-199 [SIC 2096]
Cross Creek Apparel Inc, p. I-258 [SIC 2253]
Cross Manufacturing Inc, p. II-1747 [SIC 3593]
Cross Pointe Paper Corp, p. I-612 [SIC 2621]
Crossville Rubber Products Inc, p. I-954 [SIC 3069]
Crotty Corp, p. I-942 [SIC 3053]
Crouse-Hinds Molded Products, p. II-1830 [SIC 3643]
Crowell Corp, p. I-654 [SIC 2672]
Crowley Candy Company Inc, p. I-132 [SIC 2066]
Crowley Company Inc, p. II-1367 [SIC 3446]
Crown Battery Manufacturing, p. II-1920 [SIC 3691]
Crown Castings Inc, p. II-1212 [SIC 3325]
Crown Central Petroleum Corp, p. I-903 [SIC 2911]
Crown Chemical Inc, p. I-819 [SIC 2841]
Crown Clothing Company Inc, p. I-386 [SIC 2353]
Crown Cork and Seal Company, p. II-1300 [SIC 3411]
Crown Crafts Inc, p. I-428 [SIC 2392]
Crown Equipment Corp, p. II-1552 [SIC 3537]
Crown Globe Inc, p. I-258 [SIC 2253]
Crown Industrial Products Co, p. II-1795 [SIC 3629]
Crown Leisure Products Inc, p. I-553 [SIC 2514]
Crown Pacific LP, p. I-471 [SIC 2426]
Crown Pacific Partners LP, p. I-471 [SIC 2426]
Crown Pacific USA Inc, p. I-416 [SIC 2387]
Crown Prince Inc, p. I-283 [SIC 2262]
Crown Products Company Inc, p. II-1361 [SIC 3444]
Crown Roll Leaf Inc, p. II-1482 [SIC 3497]
Crown-Snyder, p. I-997 [SIC 3087]
Crowntuft Manufacturing Corp, p. I-405 [SIC 2384]
CRS Service Inc, p. II-1753 [SIC 3594]
CRT Scientific Corp, p. II-1883 [SIC 3671]
Crucible Specialty Metals, p. II-1172 [SIC 3312]
Crusader Engines, p. II-1503 [SIC 3519]
CrustBuster/Speed King Inc, p. II-1509 [SIC 3523]
Cryopharm Corp, p. I-814 [SIC 2836]
Cryovac, p. I-660 [SIC 2673]
Crystal Brands Inc, p. I-332 [SIC 2311]
Crystal Cabinet Works Inc, p. I-486 [SIC 2434]
Crystal Clear Industries Inc, p. I-1066 [SIC 3229]
Crystal Geyser Water Co, p. I-176 [SIC 2086]
Crystal Inc, p. I-819 [SIC 2841]
Crystal Lake Manufacturing Inc, p. I-428 [SIC 2392]
Crystal Systems Inc, p. I-1167 [SIC 3299]
Crystal-X Corp, p. I-961 [SIC 3081]
Crystals International Inc, p. I-51 [SIC 2034]
CS Crable Sportswear Inc, p. I-279 [SIC 2261]
CSC Industries Inc, p. II-1172 [SIC 3312]
CSC Scientific Company Inc, p. II-2040 [SIC 3821]
CSI Control Systems Intern, p. II-2045 [SIC 3822]
CSI Technologies Inc, p. II-1897 [SIC 3675]
CSR America Inc, p. I-1127 [SIC 3273]
CSS Industries Inc, p. I-688 [SIC 2679]

CST/Auto Weigh Inc, p. II-1759 [SIC 3596]
CST Office Products Inc, p. I-688 [SIC 2679]
CT-Nassau Corp, p. I-245 [SIC 2241]
CTA Insulation Inc, p. I-1158 [SIC 3296]
CTA Manufacturing Inc, p. I-455 [SIC 2399]
CTI-Cryogenics, p. II-1638 [SIC 3561]
CTI Industries Corp, p. I-954 [SIC 3069]
CTS Automotive Products Inc, p. I-948 [SIC 3061]
CTS Corp, pp. Vol II - 1901, 1915 [SICs 3676, 3679]
CTX International Inc, p. II-1704 [SIC 3577]
Cubic Corp, p. II-1873 [SIC 3663]
Cubix Corp, p. II-1878 [SIC 3669]
Cucino Classica Italiana Inc, p. I-21 [SIC 2022]
Cudahy Tanning Company Inc, p. I-1013 [SIC 3111]
Cuddle Knit Inc, p. I-258 [SIC 2253]
Cuddlecoat Inc, p. I-369 [SIC 2337]
Cudner and O'Connor Co, p. I-889 [SIC 2893]
Culinary Foods Inc, p. I-65 [SIC 2038]
Culligan International Co, p. II-1736 [SIC 3589]
Cullman Industries Inc, p. I-258 [SIC 2253]
Cullman Products, p. II-1413 [SIC 3469]
Culp Inc, p. I-231 [SIC 2211]
Culp Woven Velvet, p. I-236 [SIC 2221]
Culpeper Farmers Cooperative, p. I-869 [SIC 2875]
Cultural Exchange Corp, p. II-2140 [SIC 3942]
Cumberland Concrete Corp, p. I-1117 [SIC 3271]
Cumberland Engineering, p. II-1632 [SIC 3559]
Cumberland Lumber, p. I-471 [SIC 2426]
Cumberland Optical Company, p. II-2107 [SIC 3851]
Cumberland Packing Corp, p. I-212 [SIC 2099]
Cumberland-Swan Inc, p. I-834 [SIC 2844]
Cummings Inc, p. II-2186 [SIC 3993]
Cummings-Moore Graphite Co, p. I-1154 [SIC 3295]
Cummins-Allison Corp, p. II-1713 [SIC 3579]
Cummins Engine Company Inc, p. II-1503 [SIC 3519]
Cummins Industrial Center, p. II-1503 [SIC 3519]
Cummins Mid-America Inc, p. II-1600 [SIC 3549]
Cumulus Technology Corp, p. II-1699 [SIC 3575]
Cunningham Brick Company Inc, p. I-1081 [SIC 3251]
Cunningham Manufacturing Co, p. II-1747 [SIC 3593]
Cunningham Pattern, p. II-1568 [SIC 3543]
Cuno Inc, p. II-1682 [SIC 3569]
Cuplex Inc, p. II-1887 [SIC 3672]
Cupples Rubber Co, p. I-927 [SIC 3011]
Curb Records Inc, p. I-745 [SIC 2782]
Currier Manufacturing, p. I-576 [SIC 2522]
Curries Co, p. II-1349 [SIC 3442]
Curt Bullock Builders Inc, p. I-523 [SIC 2452]
Curtain and Drapery Fashions, p. I-424 [SIC 2391]
Curtice Burns Foods Inc, p. I-45 [SIC 2033]
Curtin-Hebert Company Inc, p. II-1605 [SIC 3552]
Curtis 1000 Inc, p. I-729 [SIC 2759]
Curtis Dyna-Fog Ltd, p. I-874 [SIC 2879]
Curtis Instruments Inc, p. II-2061 [SIC 3825]
Curtis Mathes Holding Corp, p. II-1892 [SIC 3674]
Curtis Packaging Corp, p. I-643 [SIC 2657]
Curtis Packing Co, p. I-2 [SIC 2011]
Curtis Products Inc, p. II-1486 [SIC 3498]
Curtis Publishing Co, p. I-698 [SIC 2721]
Curtis Toledo Inc, p. II-1649 [SIC 3563]
Curtis-Young Corp, p. II-2169 [SIC 3955]
Curtiss-Wright Corp, p. II-1747 [SIC 3593]
Curtron Curtains Inc, p. I-424 [SIC 2391]
CurveMaster Inc, p. II-2150 [SIC 3949]
Curwood Inc, p. I-649 [SIC 2671]
Cushing-Malloy Inc, p. I-708 [SIC 2732]
Cushion Cut Inc, p. II-1584 [SIC 3546]
Cushman and Marden Inc, p. I-725 [SIC 2259]
Cushman Inc, p. II-2029 [SIC 3799]
Custom Aircraft Interiors Inc, p. I-581 [SIC 2531]
Custom Alloy Scrap Sales Inc, p. II-1229 [SIC 3341]
Custom Aluminum Products Inc, p. II-1420 [SIC 3471]
Custom Business Forms Inc, p. I-735 [SIC 2761]
Custom Cable Industries Inc, p. II-1878 [SIC 3669]
Custom Camp Vans and Service, p. II-1966 [SIC 3716]
Custom Chemical Corp, p. I-889 [SIC 2893]
Custom Chrome Manufacturing, p. II-2003 [SIC 3751]
Custom Coach Corp, p. II-1948 [SIC 3713]
Custom Coating Inc, p. I-787 [SIC 2822]

Roman numerals I and II indicate the volume in which pages appear. Volume I holds SICs 2011 through 3299; Volume II holds SICs 3312 through 3999.

David Stevens Manufacturing, p. I-365 [SIC 2335]
David Webb Inc, p. II-2122 [SIC 3911]
David Weber Company Inc, p. I-628 [SIC 2653]
David White Inc, p. II-2071 [SIC 3827]
David Witherspoon Inc, p. II-1187 [SIC 3316]
Davidson Instrument Panel, p. I-948 [SIC 3061]
Davidson of Dundee Inc, p. I-127 [SIC 2064]
Davidson-Textron, p. I-936 [SIC 3052]
Davis Clothing Co, p. I-332 [SIC 2311]
Davis Colors, p. I-773 [SIC 2816]
Davis Cookie Company Inc, p. I-107 [SIC 2052]
Davis Industries Inc, p. II-1404 [SIC 3465]
Davis-Lynch Glass Co, p. I-1066 [SIC 3229]
Davis Pipe and Metal Fabricators, p. II-1192 [SIC 3317]
Davis Rubber Co, p. I-927 [SIC 3011]
Davis-Standard Co, p. II-1632 [SIC 3559]
Davis Tool and Engineering Inc, p. II-1404 [SIC 3465]
Davis Water & Waste Industries, p. II-1486 [SIC 3498]
Davis Welding & Mfg Co, p. II-1747 [SIC 3593]
Davis Wood Products Inc, p. I-491 [SIC 2435]
Davisco International Inc, p. I-26 [SIC 2023]
Davis+Geck, p. II-2087 [SIC 3842]
Davol Inc, p. II-2082 [SIC 3841]
Davy-Clicem, p. II-1212 [SIC 3325]
Daw Technologies Inc, p. II-1654 [SIC 3564]
Dawlen Corp, p. II-1382 [SIC 3451]
Dawn Food Products Inc, p. I-83 [SIC 2045]
Dawson AG Service Inc, p. I-869 [SIC 2875]
Dawson Enterprises, p. II-1531 [SIC 3533]
Dawson Heritage Inc, p. I-542 [SIC 2511]
Dawson Home Fashions Inc, p. I-428 [SIC 2392]
Day Companies Inc, p. I-491 [SIC 2435]
Day-Glo Color Corp, p. I-777 [SIC 2819]
Day International Inc, p. II-1605 [SIC 3552]
Day Lumber Corp, p. I-506 [SIC 2441]
Day Plywood Inc, p. I-491 [SIC 2435]
Day Runner Inc, p. I-713 [SIC 2741]
Day-Timers Inc, p. II-2199 [SIC 3999]
Daybrook Holdings Inc, p. I-152 [SIC 2077]
Dayco Products Inc, p. I-936 [SIC 3052]
Daykin Electric Corp, p. II-1905 [SIC 3677]
Daymarc Corp, p. II-2061 [SIC 3825]
Dayon Manufacturing Company, p. II-1470 [SIC 3495]
Dayton Forging, p. II-1395 [SIC 3462]
Dayton Legal Blank Inc, p. I-745 [SIC 2782]
Dayton Progress Corp, p. II-1572 [SIC 3544]
Dayton Reliable Tool & Mfg Co, p. II-1563 [SIC 3542]
Dayton Rogers Mfg Co, p. II-1413 [SIC 3469]
Dayton Showcase Co, p. I-480 [SIC 2431]
Dayton Superior Corp, p. I-1122 [SIC 3272]
Dayton T Brown Inc, p. II-1361 [SIC 3444]
Dayton Tire Co, p. I-927 [SIC 3011]
Dazey Corp, p. II-1813 [SIC 3634]
Dazians Inc, p. I-327 [SIC 2299]
Dazor Manufacturing Corp, p. II-1845 [SIC 3646]
DB Hess Co, p. I-708 [SIC 2732]
DBA Sentry Group, p. II-1492 [SIC 3499]
DBI/Sala, p. II-2150 [SIC 3949]
DC Humphreys Company Inc, p. I-437 [SIC 2394]
DC May Ma-Crepe Corp, p. I-437 [SIC 2394]
DD Bean and Sons Co, p. II-2199 [SIC 3999]
DD Williamson and Company, p. I-181 [SIC 2087]
De Coty Coffee Co, p. I-195 [SIC 2095]
De Lille Oxygen Co, p. I-768 [SIC 2813]
De Moulin Brothers and Co, p. I-420 [SIC 2389]
De-Sta-Co, p. II-1323 [SIC 3429]
De Tomaso Industries Inc, p. II-1943 [SIC 3711]
Dealers Manufacturing Co, p. II-1503 [SIC 3519]
Deals Seafood Company Inc, p. I-190 [SIC 2092]
Dean Co, p. I-491 [SIC 2435]
Dean Foods Co, pp. Vol. I - 26, 35 [SICs 2023, 2026]
Dean Foods Vegetable Co, p. I-60 [SIC 2037]
Dean Markley Strings Inc, p. II-2135 [SIC 3931]
Dean Pickle & Specialty, p. I-40 [SIC 2032]
Dean Sausage Company Inc, p. I-7 [SIC 2013]
Dean Steel Buildings Inc, p. II-1372 [SIC 3448]
Dean/US Range Inc, p. II-1736 [SIC 3589]
Deanna Dee Inc, p. I-369 [SIC 2337]

Dearborn Fabricating, p. II-1541 [SIC 3535]
Dearborn Gage Co, p. II-1578 [SIC 3545]
Dearborn Wire and Cable LP, p. II-1830 [SIC 3643]
Debi Belt Inc, p. I-416 [SIC 2387]
Deborah Mallow, p. I-442 [SIC 2395]
Debra Moises Group, p. I-420 [SIC 2389]
Decatur Casting, p. II-1197 [SIC 3321]
Deccofelt Corp, p. I-310 [SIC 2295]
Decibel Products, p. II-1873 [SIC 3663]
Decimet Sales Inc, p. I-587 [SIC 2541]
Decision Data, p. II-1704 [SIC 3577]
Deck House Inc, p. I-523 [SIC 2452]
Decker Food Co, p. I-2 [SIC 2011]
Decker Manufacturing Corp, p. II-2177 [SIC 3965]
Deckerville Die-Form Co, p. II-1266 [SIC 3363]
Deco Products Co, p. II-1271 [SIC 3364]
Decor Gravure Corp, p. I-496 [SIC 2436]
Decor Home Fashions Inc, p. I-428 [SIC 2392]
Decorating Resources Inc, p. I-729 [SIC 2759]
Decorative Specialties Intern, p. I-654 [SIC 2672]
Decorator Industries Inc, p. I-424 [SIC 2391]
Decouper Industries Inc, p. II-1404 [SIC 3465]
DeCrane Aircraft Holdings Inc, p. II-1980 [SIC 3728]
Decratrend Paints Corp, p. I-840 [SIC 2851]
Decter International Inc, p. I-593 [SIC 2542]
Dee Foundries Inc, p. II-1285 [SIC 3369]
Dee Howard Co, p. II-1970 [SIC 3721]
Dee Inc, p. II-1266 [SIC 3363]
Dee's Manufacturing Co, p. I-390 [SIC 2361]
Dee Zee Inc, p. II-1404 [SIC 3465]
Deena Inc, p. I-378 [SIC 2341]
Deep Ocean Engineering Inc, p. II-1986 [SIC 3731]
Deep Run Packing Company Inc, p. I-91 [SIC 2047]
Deer Creek Pottery, p. I-1086 [SIC 3253]
Deer Park Spring Water Inc, p. I-176 [SIC 2086]
Deere & Co, p. II-1509 [SIC 3523]
Deere and Co, p. II-1509 [SIC 3523]
Deere & Co, p. II-1509 [SIC 3523]
Deere-Hitachi, p. II-1519 [SIC 3531]
Deerfield Plastics Company Inc, p. I-961 [SIC 3081]
Defiance Button Machine, p. II-2177 [SIC 3965]
Defiance Metal Products Co, p. II-1413 [SIC 3469]
Degussa Corp, p. I-894 [SIC 2895]
Deister Machine Company Inc, p. II-1525 [SIC 3532]
Deja Inc, p. I-1025 [SIC 3143]
DeJay Corp, p. II-1938 [SIC 3699]
Dek Tillett Ltd, p. I-279 [SIC 2261]
Del City Wire Company Inc, p. II-1929 [SIC 3694]
Del Electronics Corp, p. II-1769 [SIC 3612]
Del Laboratories Inc, p. I-834 [SIC 2844]
Del Mar Die Casting Company, p. II-1271 [SIC 3364]
Del Monte Foods, p. I-45 [SIC 2033]
Del-Wood Kitchens Inc, p. I-486 [SIC 2434]
Delagra Corp, p. I-60 [SIC 2037]
Delavan Gas Turbine Products, p. II-1980 [SIC 3728]
Delavan Inc, p. II-2076 [SIC 3829]
Delaware Machinery and Tool, p. II-1572 [SIC 3544]
Delaware Punch Co, p. I-181 [SIC 2087]
Delaware Quarries Inc, p. I-1140 [SIC 3281]
Delaware Valley, p. I-506 [SIC 2441]
Delaware Valley Corp, p. I-318 [SIC 2297]
Delco Chassis, p. II-1943 [SIC 3711]
Delco Electronics Corp, p. II-1929 [SIC 3694]
Deleo Clay Tile Inc, p. I-1081 [SIC 3251]
Delevan, p. II-1905 [SIC 3677]
Delex Systems Inc, p. II-1938 [SIC 3699]
Delfield Co, p. II-1626 [SIC 3556]
Delicious/Frookie Company Inc, p. I-107 [SIC 2052]
Delimex, p. I-65 [SIC 2038]
Delisa Pallet Corp, p. I-510 [SIC 2448]
Delker Corporation Inc, p. II-1482 [SIC 3497]
Dellen Wood Products Inc, p. I-536 [SIC 2499]
Delmar Publishers Inc, p. I-703 [SIC 2731]
Delo Screw Products Co, p. II-1382 [SIC 3451]
DeLong Sportswear Inc, pp. Vol. I - 356, 386 [SICs 2329, 2353]
DeLorme Publishing Company, p. I-713 [SIC 2741]
Deloro Stellite Inc, p. II-1594 [SIC 3548]
Delphax Systems, p. II-1621 [SIC 3555]

Delphi Interior & Lighting Syst, p. II-1954 [SIC 3714]
Delroyd Worm Gear, p. II-1666 [SIC 3566]
Delsey Luggage Inc, p. I-1041 [SIC 3161]
Delson Lumber Co, p. I-465 [SIC 2421]
Delsteel Inc, p. II-1584 [SIC 3546]
Delta Beverage Group Inc, p. I-176 [SIC 2086]
Delta Brands Inc, p. II-1600 [SIC 3549]
Delta Brick, p. I-1081 [SIC 3251]
Delta Centrifugal Corp, p. II-1178 [SIC 3313]
Delta Consolidated Industries, p. II-1413 [SIC 3469]
Delta Corp, p. II-1531 [SIC 3533]
Delta Corrugated, p. I-628 [SIC 2653]
Delta Design Inc, p. II-2076 [SIC 3829]
Delta Industries, p. II-1975 [SIC 3724]
Delta Industries Inc, p. I-1127 [SIC 3273]
Delta Lithograph Co, p. I-708 [SIC 2732]
Delta Marine Industries Inc, p. II-1992 [SIC 3732]
Delta Metal Products Company, p. II-1835 [SIC 3644]
Delta Oil Mill, p. I-140 [SIC 2074]
Delta Power Hydraulic Co, p. II-1454 [SIC 3492]
Delta Resins and Refractories, p. I-782 [SIC 2821]
Delta Rubber Co, p. I-948 [SIC 3061]
Delta Scientific Corp, p. II-1372 [SIC 3448]
Delta Tango Inc, p. II-1694 [SIC 3572]
Delta Tanning Corp, p. I-1013 [SIC 3111]
Delta Tooling Co, p. II-1558 [SIC 3541]
Delta-Unibus Corp, p. II-1795 [SIC 3629]
Delta Wire Corp, p. II-1182 [SIC 3315]
Delta Woodside Industries Inc, p. I-231 [SIC 2211]
Deltah Inc, p. II-2173 [SIC 3961]
Deltapaper Corp, p. I-688 [SIC 2679]
Deltec Corp, p. II-1769 [SIC 3612]
Deltech Engineering LP, p. II-1654 [SIC 3564]
Deltic Farm and Timber, p. I-465 [SIC 2421]
Deltona Transformer Corp, p. II-1769 [SIC 3612]
Deltrol Controls, p. II-1790 [SIC 3625]
Deltrol Corp, p. II-1454 [SIC 3492]
Deltrol Fluid Products, p. II-1454 [SIC 3492]
Deltron Inc, p. II-1905 [SIC 3677]
Deltronic Corp, p. II-2071 [SIC 3827]
Deluxe Corp, p. I-729 [SIC 2759]
Deluxe Craft Manufacturing Co, p. I-745 [SIC 2782]
Deluxe Packages, p. I-654 [SIC 2672]
Deluxe Stitcher Company Inc, p. II-1621 [SIC 3555]
Delvest Inc, p. II-1207 [SIC 3324]
Demaco, p. II-1626 [SIC 3556]
Demco Inc, p. I-684 [SIC 2678]
DeMenno-Kerdoon, p. I-903 [SIC 2911]
DeMert and Dougherty Inc, p. I-834 [SIC 2844]
Demmer Corp, p. II-1572 [SIC 3544]
Den-Tal-Ez Inc, p. II-2093 [SIC 3843]
Denar Corp, p. II-2093 [SIC 3843]
Denise Lingerie Corp, p. I-373 [SIC 2339]
Denison Hydraulics Inc, p. II-1753 [SIC 3594]
Denmark Military Equip Co, p. I-455 [SIC 2399]
Dennis Chemical Co, p. I-997 [SIC 3087]
Dennis Inc, p. I-199 [SIC 2096]
Dennis Uniform Mfg Co, p. I-373 [SIC 2339]
Dental Concepts Inc, p. II-2093 [SIC 3843]
Dentsply International Inc, p. II-2093 [SIC 3843]
Denver Bookbinding Company, p. I-750 [SIC 2789]
Denver Lamb Co, p. I-2 [SIC 2011]
Denver Post Corp, p. I-693 [SIC 2711]
Denver Thomas, p. II-1197 [SIC 3321]
DEO Enterprises Inc, p. I-1049 [SIC 3172]
Dep Corp, p. I-834 [SIC 2844]
Department 56 Inc, p. I-1102 [SIC 3262]
Depeche Mode Inc, p. I-365 [SIC 2335]
Dependable Furniture Mfg Co, p. I-571 [SIC 2521]
Deposition Technologies Inc, p. I-840 [SIC 2851]
Der-Tex Corp, p. I-1017 [SIC 3131]
Derby Cap Manufacturing, p. I-386 [SIC 2353]
Derby Cone Company Inc, p. I-107 [SIC 2052]
Deringer Manufacturing Co, p. II-1830 [SIC 3643]
Derlan Inc, p. II-1975 [SIC 3724]
DeRoyal Textiles Inc, p. I-231 [SIC 2211]
Derse Inc, p. II-2186 [SIC 3993]
DESA International Inc, p. II-1584 [SIC 3546]
Desalination Systems Inc, p. I-782 [SIC 2821]

Roman numerals I and II indicate the volume in which pages appear. Volume I holds SICs 2011 through 3299; Volume II holds SICs 3312 through 3999.

Roman numerals I and II indicate the volume in which pages appear. Volume I holds SICs 2011 through 3299; Volume II holds SICs 3312 through 3999.

Domestex USA Ltd, p. I-428 [SIC 2392]
Domine Builders Supply Inc, p. I-1117 [SIC 3271]
Dominion Textile, p. I-231 [SIC 2211]
Dominion Yarn Corp, p. I-297 [SIC 2281]
Domino Sugar Corp, p. I-119 [SIC 2062]
Don Dye Company Inc, p. II-1676 [SIC 3568]
Don-Lin Jewelry Company Inc, p. II-2173 [SIC 3961]
Don Paper Co, p. I-669 [SIC 2675]
Don Post Studios Inc, p. I-420 [SIC 2389]
Don's Manufacturing Company, p. I-390 [SIC 2361]
Don Schreiber and Company Inc, p. I-745 [SIC 2782]
Don Shapiro Industries Inc, p. I-373 [SIC 2339]
Donahue and Associates Intern, p. II-1615 [SIC 3554]
Donaldson Company Inc, p. II-1682 [SIC 3569]
Donco Industries Inc, p. II-1986 [SIC 3731]
Donegal Industries Inc, p. I-373 [SIC 2339]
Donghia Textiles Inc, p. I-236 [SIC 2221]
Doninger Metal Products Corp, p. I-603 [SIC 2599]
Donley Co, p. II-2173 [SIC 3961]
Donnelly Applied Films Corp, p. II-1892 [SIC 3674]
Donnelly Corp, p. I-1071 [SIC 3231]
Donnkenny Apparel Inc, p. I-365 [SIC 2335]
Donnkenny Inc, p. I-361 [SIC 2331]
Donray Co, p. I-991 [SIC 3086]
Donsco Inc, p. II-1197 [SIC 3321]
Door Systems Inc, p. II-1349 [SIC 3442]
Doppelt Industries Inc, p. I-558 [SIC 2515]
Dor-O-Matic Products Corp, p. II-1938 [SIC 3699]
Doran Yarn Mill Inc, p. I-297 [SIC 2281]
Dore Rice Mill, p. I-79 [SIC 2044]
Dorel Hat Company Inc, p. I-386 [SIC 2353]
Dorfile Storage, p. I-587 [SIC 2541]
Dorian America, p. II-1736 [SIC 3589]
Dorian International Inc, p. II-1736 [SIC 3589]
Dorling Kindersley Publishing, p. I-703 [SIC 2731]
Dorma Door Controls Inc, p. II-1323 [SIC 3429]
Dorman Roth Foods Inc, p. I-21 [SIC 2022]
Dorne and Margolin Inc, p. II-1873 [SIC 3663]
Dorner Manufacturing Corp, p. II-1541 [SIC 3535]
Dornier Medical Systems Inc, p. II-2102 [SIC 3845]
Doron Precision System Inc, p. II-1938 [SIC 3699]
Dorr-Oliver Inc, p. II-1682 [SIC 3569]
Dorr Woolen Co, p. I-241 [SIC 2231]
Dorris Lumber and Molding Co, p. I-480 [SIC 2431]
Dorsett Carpet Mills Inc, p. I-292 [SIC 2273]
Dorsey Trailers Inc, p. II-1961 [SIC 3715]
Doskocil Specialty Brands Co, p. I-65 [SIC 2038]
Dossert Corp, p. II-1910 [SIC 3678]
Dot Group, p. I-643 [SIC 2657]
Dot Packaging Printpak Inc, p. I-643 [SIC 2657]
Dotronix Inc, p. II-1883 [SIC 3671]
Doty Lithography, p. I-741 [SIC 2771]
Double D Ranchwear Inc, p. I-373 [SIC 2339]
Double E Company Inc, p. II-1615 [SIC 3554]
Double Envelope Co, p. I-679 [SIC 2677]
Double L Manufacturing Inc, p. II-1509 [SIC 3523]
Double Rainbow, p. I-30 [SIC 2024]
Double Seal Ring Company Inc, p. II-1742 [SIC 3592]
Doughtie's Foods Inc, p. I-7 [SIC 2013]
Douglas and Lomason Co, p. II-1404 [SIC 3465]
Douglas Battery Mfg Co, p. II-1920 [SIC 3691]
Douglas Brothers, p. II-1486 [SIC 3498]
Douglas Company Inc, p. II-2140 [SIC 3942]
Douglas Dynamics Inc, p. II-1519 [SIC 3531]
Douglas Furniture of California, p. I-553 [SIC 2514]
Douglas Machine Corp, p. II-1660 [SIC 3565]
Douglas Press Inc, p. II-2144 [SIC 3944]
Douglas Products, p. II-1818 [SIC 3635]
Dove Audio Inc, p. II-1864 [SIC 3652]
Dover Chemical Corp, p. I-799 [SIC 2833]
Dover Corp, pp. Vol II - 1536, 1649 [SICs 3534, 3563]
Dover Elevator International Inc, p. II-1536 [SIC 3534]
Dover Elevator Systems Inc, p. II-1536 [SIC 3534]
Dover Industries Inc, p. II-1558 [SIC 3541]
Dover Instrument Corp, p. II-1676 [SIC 3568]
Dover Publications Inc, p. I-703 [SIC 2731]
Dow Chemical Co, p. I-777 [SIC 2819]
Dow Corning Corp, p. I-782 [SIC 2821]
Dow Corning STI, p. I-787 [SIC 2822]

Dow Cover Company Inc, p. I-455 [SIC 2399]
Dow Jones and Company Inc, p. I-693 [SIC 2711]
DowBrands LP, p. I-825 [SIC 2842]
Dowcraft Corp, p. II-1372 [SIC 3448]
Down River Casting Co, p. II-1280 [SIC 3366]
Downard Hydraulics Inc, p. II-1525 [SIC 3532]
Downey Manufacturing Inc, p. II-1747 [SIC 3593]
Downing Displays Inc, p. II-2186 [SIC 3993]
Downs Carpet Company Inc, p. I-292 [SIC 2273]
Downs Crane and Hoist, p. II-1547 [SIC 3536]
Dowty Aerospace Aviation Svcs, p. II-1980 [SIC 3728]
Dowty Palmer-Chenard Inc, p. I-948 [SIC 3061]
Doxey Furniture Corp, p. I-542 [SIC 2511]
Doyle Shirt Manufacturing Corp, p. I-352 [SIC 2326]
Dozier Manufacturing Company, p. I-348 [SIC 2325]
DP Packaging Inc, p. II-1434 [SIC 3483]
DPM of Arkansas, p. I-7 [SIC 2013]
DPS Company Inc, p. II-1464 [SIC 3494]
DR Holdings Incorporated, p. II-1688 [SIC 3571]
DR Johnson Lumber Co, p. I-465 [SIC 2421]
DR Kenyon and Son Inc, p. II-1605 [SIC 3552]
Dr Pepper Bottling Holdings Inc, p. I-176 [SIC 2086]
Dr Pepper Bottling of Texas, p. I-176 [SIC 2086]
Dr Pepper/Seven-Up Companies, p. I-181 [SIC 2087]
Drackett Co, p. I-825 [SIC 2842]
Dracon, p. II-2061 [SIC 3825]
Dragon Products Company Inc, p. I-1077 [SIC 3241]
Dragon Systems Inc, p. II-1713 [SIC 3579]
Drainage Products Inc, p. II-1334 [SIC 3432]
Drake Corp, p. II-1318 [SIC 3425]
Drake-Williams Steel Inc, p. II-1344 [SIC 3441]
Dranetz Technologies Inc, p. II-2061 [SIC 3825]
Draper Knitting Company Inc, p. I-267 [SIC 2257]
Draper Shade and Screen, p. II-2112 [SIC 3861]
Draper Valley Farms Inc, p. I-12 [SIC 2015]
Drasin Knitting Mills Inc, p. I-258 [SIC 2253]
Draw-Tite Inc, p. II-2029 [SIC 3799]
Drawform Inc, p. II-1413 [SIC 3469]
Drawn Metal Tube Co, p. II-1234 [SIC 3351]
Draymore Manufacturing Corp, p. I-424 [SIC 2391]
Dreco Holding Co, p. II-1531 [SIC 3533]
Dreis & Krump Mfg Co, p. II-1563 [SIC 3542]
Dreison International Inc, p. II-1764 [SIC 3599]
Dressels Bakeries Inc, p. I-112 [SIC 2053]
Dresser Industries Inc, pp. Vol II - 1503, 1638, 1649 [SICs 3519, 3561, 3563]
Dresser Industries Indust Valve, p. II-1447 [SIC 3491]
Dresser Manufacturing, p. II-1464 [SIC 3494]
Dresser Measurement, p. II-2056 [SIC 3824]
Dresser-Rand Co, p. II-1649 [SIC 3563]
Dresser-Rand Electric Machinery, p. II-1780 [SIC 3621]
Dresser Security, p. II-1531 [SIC 3533]
Drever Co, p. II-1671 [SIC 3567]
Drew Foam Company Inc, p. I-991 [SIC 3086]
Drew Industries Inc, p. II-1349 [SIC 3442]
Drexel Chemical Co, p. I-874 [SIC 2879]
Drexel Heritage Furnishings Inc, p. I-542 [SIC 2511]
Drexel Industries Inc, p. II-1552 [SIC 3537]
Dreyer's Grand Ice Cream Inc, p. I-30 [SIC 2024]
DRG Medical Packaging Inc, p. I-654 [SIC 2672]
Driltech Inc, p. II-1525 [SIC 3532]
Driver-Harris Co, p. II-1255 [SIC 3356]
Drives Inc, p. II-1476 [SIC 3496]
Drug Package Inc, p. I-654 [SIC 2672]
Drug Plastic and Glass Co, p. I-985 [SIC 3085]
Drug Screening Systems Inc, p. I-809 [SIC 2835]
Drulane Co, p. I-455 [SIC 2399]
Drum Workshop Inc, p. II-2135 [SIC 3931]
Drypers Corp, p. I-674 [SIC 2676]
DS America Inc, p. II-1621 [SIC 3555]
DS Brown Company Inc, p. I-948 [SIC 3061]
DS Manufacturing Inc, p. II-1492 [SIC 3499]
DSC Communications Corp, p. II-1868 [SIC 3661]
DSM Copolymer Inc, p. I-787 [SIC 2822]
DT Industries Inc, p. II-1632 [SIC 3559]
Du-Co Ceramics Company Inc, p. I-1109 [SIC 3264]
Du-Mont Co, p. II-1361 [SIC 3444]
Du Pont Tribon Composites Inc, p. II-1644 [SIC 3562]

Du-Wel Products Inc, p. II-1266 [SIC 3363]
Dual-Lite Inc, p. II-1854 [SIC 3648]
Dualite Inc, p. II-2186 [SIC 3993]
Dublin Management, p. I-587 [SIC 2541]
Dubuque Foods Inc, p. I-2 [SIC 2011]
Ducane Corp, p. II-1339 [SIC 3433]
Duchess Bakers' Machinery Co, p. II-1626 [SIC 3556]
Ducktrap River Fish Farm Inc, p. I-186 [SIC 2091]
Ducommun Inc, p. II-1980 [SIC 3728]
Duct-O-Wire Co, p. II-1547 [SIC 3536]
Dudek & Bock Spring Mfg Co, p. II-1470 [SIC 3495]
Duff Maps, p. I-724 [SIC 2754]
Duff-Norton Co, p. II-1682 [SIC 3569]
Duffel Sportswear, p. I-336 [SIC 2321]
Duffin Manufacturing Co, p. II-1382 [SIC 3451]
Dukane Corp, p. II-1878 [SIC 3669]
Duke's Inc, p. II-1753 [SIC 3594]
Duluth Brass and Aluminum Co, p. II-1280 [SIC 3366]
Duluth Spring Co, p. II-1460 [SIC 3493]
Dumore Corp, p. II-1584 [SIC 3546]
Dun-Rite Kitchen Cabinet Corp, p. I-486 [SIC 2434]
Dunbrooke Sportswear Co, p. I-332 [SIC 2311]
Duncan Enterprises, p. I-1113 [SIC 3269]
Dundee Casting Co, p. II-1266 [SIC 3363]
Dundee Mills Inc, p. I-263 [SIC 2254]
Dunham-Bush Inc, p. II-1726 [SIC 3585]
Dunkirk Radiator Corp, p. II-1339 [SIC 3433]
Dunlap Industries Inc, p. II-2177 [SIC 3965]
Dunlop Tire Corp, p. I-927 [SIC 3011]
Dunmore Corp, p. I-961 [SIC 3081]
Dunn Bindery Inc, p. I-750 [SIC 2789]
Dunn Blacktop Company Inc, p. I-909 [SIC 2951]
Dunn Edwards Corp, p. I-840 [SIC 2851]
Dunn Manufacturing Co, p. I-386 [SIC 2353]
Dunsirn Industries Inc, p. I-327 [SIC 2299]
Duo-Fast Corp, p. II-1182 [SIC 3315]
DuPage Machine Products Inc, p. II-1382 [SIC 3451]
Duplex Inc, p. II-1470 [SIC 3495]
Duplex Products Inc, p. I-729 [SIC 2759]
Dupps Co, p. II-1632 [SIC 3559]
Dura Automotive Systems Inc, p. II-1492 [SIC 3499]
Dura-Bilt Products Inc, p. II-1367 [SIC 3446]
Dura-Craft Industries Inc, p. I-486 [SIC 2434]
Dura-Stress Inc, p. I-1122 [SIC 3272]
Dura Supreme Inc, p. I-486 [SIC 2434]
Dura-Ware of America Inc, p. II-1275 [SIC 3365]
Dura-Wood Treating Co, p. I-528 [SIC 2491]
Duracell Inc, p. II-1920 [SIC 3691]
Duracell International Inc, p. II-1920 [SIC 3691]
Duracote Corp, p. I-310 [SIC 2295]
Duracraft Corp, p. II-1813 [SIC 3634]
Duracraft Plastics Inc, p. I-1001 [SIC 3088]
Duraflame Inc, p. I-536 [SIC 2499]
Duralam Inc, p. I-654 [SIC 2672]
Duraloy Technologies Inc, p. II-1212 [SIC 3325]
Duramax Inc, p. I-954 [SIC 3069]
Durametal Corp, p. II-1197 [SIC 3321]
Durametallic Corp, p. I-942 [SIC 3053]
Durand Glass Manufacturing Co, p. I-1071 [SIC 3231]
Durand International, p. I-1071 [SIC 3231]
Durasys Corp, p. II-1699 [SIC 3575]
DuraTech Industries Intern, p. II-1509 [SIC 3523]
Durawear Corp, p. I-1094 [SIC 3259]
Duray Fluorescent Mfg Co, p. II-1840 [SIC 3645]
Duray/JF Duncan Industries, p. I-553 [SIC 2514]
Durbin-Durco Inc, p. II-1212 [SIC 3325]
Durel Corp, p. II-1845 [SIC 3646]
Durex Inc, p. II-1413 [SIC 3469]
Durex Products Inc, p. I-787 [SIC 2822]
Durham Co, p. II-1775 [SIC 3613]
Durham Coca-Cola Bottling Co, p. I-176 [SIC 2086]
DURHAM Products, p. II-1780 [SIC 3621]
Duriron Company Inc, pp. Vol II - 1203, 1454, 1464 [SICs 3322, 3492, 3494]
Durlacher and Company Inc, p. I-373 [SIC 2339]
Duro Art Industries Inc, p. II-2160 [SIC 3952]
Duro Bag Manufacturing Co, p. I-665 [SIC 2674]
Duro Dyne Corp, p. II-1726 [SIC 3585]
Duro Metal Manufacturing, p. I-553 [SIC 2514]

Roman numerals I and II indicate the volume in which pages appear. Volume I holds SICs 2011 through 3299; Volume II holds SICs 3312 through 3999.

Roman numerals I and II indicate the volume in which pages appear. Volume I holds SICs 2011 through 3299; Volume II holds SICs 3312 through 3999.

Edward Mendell Company Inc, p. I-799 [SIC 2833]
Edwards Brothers Inc, p. I-708 [SIC 2732]
Edwards Company Inc, pp. Vol II - 1878, 1938 [SICs 3669, 3699]
Edwards Engineering, p. II-1753 [SIC 3594]
Edwards Heat Treating Inc, p. II-1290 [SIC 3398]
Edwards High Vacuum Intern, p. II-2040 [SIC 3821]
Edwards Lamp and Shade Co, p. II-1840 [SIC 3645]
Edwards Warren Tire, p. I-927 [SIC 3011]
Edwards Wood Products Inc, p. I-510 [SIC 2448]
Eel River Sawmills Inc, p. I-465 [SIC 2421]
EF Data Corp, p. II-1868 [SIC 3661]
EF Johnson Co, p. II-1830 [SIC 3643]
EFCO Corp, p. II-1349 [SIC 3442]
Efco Inc, p. II-1563 [SIC 3542]
Effingham Equity, p. I-95 [SIC 2048]
EFP Corp, p. I-991 [SIC 3086]
EFP South Corp, p. I-991 [SIC 3086]
EG and G Energy Measurements, p. II-2076 [SIC 3829]
EG and G Inc, p. II-2051 [SIC 3823]
EG and G Power Systems, p. II-1769 [SIC 3612]
EG and G Sealol Inc, p. I-942 [SIC 3053]
Egger Steel Co, p. II-1344 [SIC 3441]
Eggers Industries Inc, pp. Vol. I - 480, 491 [SICs 2431, 2435]
Eggo Foods, p. I-55 [SIC 2035]
EH Baare Corp, p. II-1476 [SIC 3496]
EH Kneen Co, p. II-1300 [SIC 3411]
Ehlert Tool Company Inc, p. II-1572 [SIC 3544]
Ehrhorn, p. II-1938 [SIC 3699]
Ehrlich Manufacturing Company, p. I-649 [SIC 2671]
EHV Weidmann Industries Inc, p. I-618 [SIC 2631]
EI du Pont de Nemours, p. I-777 [SIC 2819]
Eickoff Corp, p. II-1525 [SIC 3532]
Eide Industries Inc, p. I-437 [SIC 2394]
Eifel Pattern and Model, p. II-1568 [SIC 3543]
Eight in One Pet Products Inc, p. I-95 [SIC 2048]
EIMCO Coal Machinery Inc, p. II-1525 [SIC 3532]
Eimco Process Equipment Co, p. II-1615 [SIC 3554]
EIS Brake Parts, p. II-1954 [SIC 3714]
Eisenhart Wallcoverings Co, p. I-688 [SIC 2679]
Eitel Presses Inc, p. II-1563 [SIC 3542]
EJ Davis Co, p. I-1158 [SIC 3296]
EJ Houle Inc, p. I-91 [SIC 2047]
EJ Murphy Co, p. II-1943 [SIC 3711]
Ekco Group Inc, p. II-1413 [SIC 3469]
EKCO Products Inc, p. II-1482 [SIC 3497]
Eklund Metal Treating Inc, p. II-1290 [SIC 3398]
Ekstrom Industries Inc, p. II-1795 [SIC 3629]
El Dorado Paper Bag Mfg, p. I-665 [SIC 2674]
El Encanto Inc, p. I-65 [SIC 2038]
EL Gardner Inc, p. I-1127 [SIC 3273]
EL Heacock Company Inc, p. I-1053 [SIC 3199]
El Jay, p. II-1525 [SIC 3532]
El Molino Foods Inc, p. I-40 [SIC 2032]
El Monte Alloys Inc, p. II-1280 [SIC 3366]
El Monte Non-Ferrous Foundry, p. II-1285 [SIC 3369]
EL Mustee and Sons Inc, p. I-1001 [SIC 3088]
EL Smith Air Compressors, p. II-1649 [SIC 3563]
EL Tool and Die Company Inc, p. II-2173 [SIC 3961]
ELA Corp, p. II-2173 [SIC 3961]
Elan Intern Organic Coffees, p. I-195 [SIC 2095]
Elan Technology, p. I-1066 [SIC 3229]
Elano Corp, p. II-1361 [SIC 3444]
Elastic Corporation of America, p. I-245 [SIC 2241]
Elastic Stop Nut, p. II-1388 [SIC 3452]
Elbeco Inc, p. I-352 [SIC 2326]
Elberta Crate and Box Co, p. I-515 [SIC 2449]
Elberton Manufacturing Co, p. I-361 [SIC 2331]
Elco Corp, p. I-897 [SIC 2899]
Elcor Corp, p. I-914 [SIC 2952]
Elder Hosiery Mills Inc, p. I-254 [SIC 2252]
Elder Manufacturing Co, p. I-348 [SIC 2325]
Eldre Corp, p. II-1830 [SIC 3643]
Electri-Cord Manufacturing, p. II-1938 [SIC 3699]
Electri-Flex Co, p. II-1835 [SIC 3644]
Electri-Wire Corp, p. II-1938 [SIC 3699]
Electric and Gas Technology Inc, p. II-1775 [SIC 3613]

Electric M and R Inc, p. II-1826 [SIC 3641]
Electric Mobility Corp, p. II-2087 [SIC 3842]
Electric Steel Castings Co, p. II-1212 [SIC 3325]
Electric Switchboard Company, p. II-1775 [SIC 3613]
Electrical Design and Control, p. II-1775 [SIC 3613]
Electrical Rebuilders Inc, p. II-2029 [SIC 3799]
Electrical Windings Inc, p. II-1905 [SIC 3677]
Electrix Inc, p. II-1826 [SIC 3641]
Electro Abrasives Corp, p. I-1144 [SIC 3291]
Electro Engineering Works, p. II-1905 [SIC 3677]
Electro-Etch Circuits Inc, p. II-1887 [SIC 3672]
Electro Fiberoptics Corp, p. II-1260 [SIC 3357]
Electro-Flex Heat Inc, p. II-1339 [SIC 3433]
Electro-Jet Tool & Mfg, p. II-1975 [SIC 3724]
Electro Machine and Engineering, p. II-1420 [SIC 3471]
Electro Magnetic Products Inc, p. II-1934 [SIC 3695]
Electro-Mechanical Products Inc, p. II-1775 [SIC 3613]
Electro Optical Instruments Inc, p. II-2071 [SIC 3827]
Electro-Optical Systems, p. II-2071 [SIC 3827]
Electro Sales Company Inc, p. II-1780 [SIC 3621]
Electro Scientific Industries Inc, p. II-1915 [SIC 3679]
Electro-Sensors Inc, p. II-2056 [SIC 3824]
Electro Space Fabricator Inc, p. II-1361 [SIC 3444]
Electro Sprayer Systems Inc, p. II-1621 [SIC 3555]
Electro Static Finishing Inc, p. II-1425 [SIC 3479]
Electro-Tec Corp, p. II-1780 [SIC 3621]
ElectroCom Automation Inc, p. II-1713 [SIC 3579]
Electrode Corp, p. II-1915 [SIC 3679]
Electrodyne, p. II-1929 [SIC 3694]
Electrofilm Manufacturing Co, p. II-1795 [SIC 3629]
Electroglas Inc, p. II-2076 [SIC 3829]
Electroid Co, p. II-1790 [SIC 3625]
Electroline Manufacturing Co, p. II-1835 [SIC 3644]
Electromagnetic Sciences Inc, p. II-1915 [SIC 3679]
Electromotive Systems Inc, p. II-1790 [SIC 3625]
Electron Corp, p. II-1197 [SIC 3321]
Electronic Associates Inc, p. II-1699 [SIC 3575]
Electronic Ballast Technology, p. II-1769 [SIC 3612]
Electronic Coil Corp, p. II-1905 [SIC 3677]
Electronic Components Corp, p. II-1769 [SIC 3612]
Electronic Contract Svcs Corp, p. II-1910 [SIC 3678]
Electronic Information Systems, p. II-1868 [SIC 3661]
Electronic Instrumentation, p. II-1887 [SIC 3672]
Electronic Liquid Fillers Inc, p. II-1660 [SIC 3565]
Electronic Plating Service Inc, p. II-1420 [SIC 3471]
Electronic Power Technology Inc, p. II-1795 [SIC 3629]
Electronic Systems International, p. II-1699 [SIC 3575]
Electronic Transformer Corp, p. II-1769 [SIC 3612]
Electronic Voting Systems Inc, p. II-1713 [SIC 3579]
Electroply Inc, p. II-1295 [SIC 3399]
Electrospace Systems Inc, p. II-2035 [SIC 3812]
Electroswitch Corp, p. II-1775 [SIC 3613]
Electrovert USA Corp, p. II-1938 [SIC 3699]
Electruk Battery Co, p. II-1920 [SIC 3691]
Eleja Casuals Corp, p. I-386 [SIC 2353]
Elevator Doors Inc, p. II-1536 [SIC 3534]
Elevator Equipment Corp, p. II-1536 [SIC 3534]
Elevator Systems Inc, p. II-1536 [SIC 3534]
Elf Atochem North America Inc, pp. Vol. I - 777, 874, 918 [SICs 2819, 2879, 2992]
Elgin Business Forms Inc, p. I-735 [SIC 2761]
Elgin-Butler Brick Co, p. I-1081 [SIC 3251]
Elgin E2, p. II-1905 [SIC 3677]
Elgin Sweeper Co, p. II-1943 [SIC 3711]
Eli Lilly and Co, p. I-804 [SIC 2834]
Eli's Chicago Finest Inc, p. I-112 [SIC 2053]
Elias Industries Inc, p. II-1334 [SIC 3432]
Elias Sayour Company Inc, p. I-365 [SIC 2335]
Elika Ltd, p. I-398 [SIC 2371]
Elish and Company Inc, p. I-486 [SIC 2434]
Elite Products Co, p. II-1699 [SIC 3575]
Elitegroup Computer Systems, p. II-1892 [SIC 3674]
Elixir Industries Inc, p. II-1349 [SIC 3442]
Elizabeth Carbide Die Company, p. II-1572 [SIC 3544]
Elizabeth Knits Inc, p. I-258 [SIC 2253]
Elizabeth Lucas Designs, p. I-741 [SIC 2771]

Elizabeth-Meade Hosiery Mills, p. I-254 [SIC 2252]
Eljer Industries Inc, p. II-1334 [SIC 3432]
Eljer Plumbingware, p. I-1098 [SIC 3261]
Elk Corporation of America, p. I-914 [SIC 2952]
Elkay Manufacturing Co, p. II-1330 [SIC 3431]
Elkay Products Inc, p. II-2040 [SIC 3821]
Elkem Metals Co, p. II-1178 [SIC 3313]
Elkhart Bedding Co, p. I-558 [SIC 2515]
Elkhart Brass Manufacturing, p. II-1682 [SIC 3569]
Elkhart Products Corp, p. II-1464 [SIC 3494]
Elkin Valley Apparel Co, p. I-373 [SIC 2339]
Ellanef Manufacturing Corp, p. II-1980 [SIC 3728]
Ellay Inc, p. I-961 [SIC 3081]
Ellehammer Packaging Inc, p. I-961 [SIC 3081]
Eller Manufacturing Company, p. I-1053 [SIC 3199]
Ellicott Machine Corp Intern, p. II-1519 [SIC 3531]
Ellijay Lumber, p. I-528 [SIC 2491]
Ellingson Lumber Co, p. I-465 [SIC 2421]
Elliott Brothers Steel Co, p. II-1187 [SIC 3316]
Elliott Corp, p. I-1037 [SIC 3151]
Elliott Manufactured Homes Inc, p. I-519 [SIC 2451]
Elliott Metal Works Inc, p. II-1605 [SIC 3552]
Elliott Turbomachinery Company, p. II-1498 [SIC 3511]
Elliott Valve and Repair, p. II-1447 [SIC 3491]
Ellis Hosiery Mills Inc, p. I-254 [SIC 2252]
Ellis Pecan Co, p. I-136 [SIC 2068]
Ellisco Inc, p. II-1300 [SIC 3411]
Ellison Bakery Inc, p. I-107 [SIC 2052]
Ellstrom Manufacturing, p. I-491 [SIC 2435]
Ellsworth Cooperative Creamery, p. I-21 [SIC 2022]
Ellwood Group Inc, p. II-1395 [SIC 3462]
Ellwood Texas Forge, p. II-1395 [SIC 3462]
Elm Hill Meats Inc, p. I-7 [SIC 2013]
Elm Packaging Co, p. I-991 [SIC 3086]
Elma Engineering, p. II-1795 [SIC 3629]
Elmer Little and Sons Inc, p. I-1037 [SIC 3151]
Elmhurst Milk and Cream Co, p. I-35 [SIC 2026]
Elmore-Pisgah Inc, p. I-288 [SIC 2269]
Elox Corp, p. II-1764 [SIC 3599]
Elpac Electronics Inc, p. II-1897 [SIC 3675]
Elsag Bailey Process Automation, p. II-2051 [SIC 3823]
Elscint Inc, p. II-2102 [SIC 3845]
Elscott Corp, p. II-1826 [SIC 3641]
Elsevier Science Inc, p. I-698 [SIC 2721]
Elsie Undergarment Corp, p. I-378 [SIC 2341]
Elsinore Ready-Mix Co, p. I-1127 [SIC 3273]
Eltech Research Inc, p. II-1688 [SIC 3571]
Eltech Systems Corp, p. I-897 [SIC 2899]
Eltrex Industries Inc, p. II-1826 [SIC 3641]
Eltron International Inc, p. II-1621 [SIC 3555]
Elwell-Parker Electric Co, p. II-1552 [SIC 3537]
Ely and Walker Co, p. I-336 [SIC 2321]
Elyria Foundry, p. II-1197 [SIC 3321]
Elyria Manufacturing Corp, p. II-1382 [SIC 3451]
EM Chadbourne Inc, p. I-909 [SIC 2951]
EM Cummings Veneers Inc, p. I-491 [SIC 2435]
EM Industries Inc, p. I-773 [SIC 2816]
EM Wiegmann and Company, p. II-1835 [SIC 3644]
EMA Capital Inc, p. II-1826 [SIC 3641]
EMB Giftware Inc, p. I-684 [SIC 2678]
Embassy Embroidery Corp, p. I-271 [SIC 2258]
Embassy Industries Inc, p. II-1339 [SIC 3433]
Embee Inc, p. II-1420 [SIC 3471]
Emby Hosiery Corp, p. I-254 [SIC 2252]
EMC Corp, p. II-1694 [SIC 3572]
EMC Publishing Corp, p. I-703 [SIC 2731]
Emco Industries Inc, pp. Vol. I - 587, Vol. II - 1388 [SICs 2541, 3452]
Emco Specialties Inc, p. II-1349 [SIC 3442]
EMD Associates Inc, p. II-1915 [SIC 3679]
Emdur Metal Products Inc, p. II-2118 [SIC 3873]
Emeco Industries Inc, p. I-576 [SIC 2522]
Emerson & Cuming, p. I-1066 [SIC 3229]
Emerson Electric Co, pp. Vol II - 1780, 2051 [SICs 3621, 3823]
Emerson Electric Co US, p. II-1780 [SIC 3621]
Emerson Leather Inc, p. I-567 [SIC 2519]

Roman numerals I and II indicate the volume in which pages appear. Volume I holds SICs 2011 through 3299; Volume II holds SICs 3312 through 3999.

Emerson Logging Company Inc, p. I-460 [SIC 2411]
Emerson Motor Co, p. II-1780 [SIC 3621]
Emerson Power Transmission, p. II-1753 [SIC 3594]
Emery Group, p. I-855 [SIC 2869]
Emery Winslow Scale Co, p. II-1759 [SIC 3596]
EMF Corp, pp. Vol II - 1713, 1795 [SICs 3579, 3629]
Emge Packing Co, p. I-2 [SIC 2011]
Emglo Products Corp, p. II-1649 [SIC 3563]
Emhart-Hartford, p. II-1632 [SIC 3559]
EMI Co, p. II-1197 [SIC 3321]
Eminence Speaker Corp, p. II-1859 [SIC 3651]
Emory Manufacturing Co, p. II-2003 [SIC 3751]
Empak Inc, p. I-985 [SIC 3085]
Emperor Clock Co, p. II-2118 [SIC 3873]
Empi Inc, p. II-2082 [SIC 3841]
Empire Brass Co, p. II-1334 [SIC 3432]
Empire Carpet Mills Inc, p. I-599 [SIC 2591]
Empire Castings Inc, p. II-1275 [SIC 3365]
Empire Chemical Company Inc, p. I-825 [SIC 2842]
Empire Comfort Systems Inc, p. II-1339 [SIC 3433]
Empire Diamond Corp, p. II-2122 [SIC 3911]
Empire Die Casting Company, p. II-1266 [SIC 3363]
Empire Hard Chrome Inc, p. II-1420 [SIC 3471]
Empire Silver Company Inc, p. II-2127 [SIC 3914]
Empire Steel Company Inc, p. II-1212 [SIC 3325]
Empire Technology Inc, p. II-1584 [SIC 3546]
EMR Photoelectric, p. II-1883 [SIC 3671]
Emsig Manufacturing Corp, p. II-2177 [SIC 3965]
Emson Inc, p. II-1492 [SIC 3499]
Emtex Inc, p. I-310 [SIC 2295]
Emulex Corp, p. II-1694 [SIC 3572]
Emulsicoat Inc, p. I-914 [SIC 2952]
Enamel Products and Plating Co, p. II-1425 [SIC 3479]
Enchanced Imaging Technologies, p. II-2112 [SIC 3861]
Encon Industries Inc, p. II-1813 [SIC 3634]
Encore Computer Corp, p. II-1688 [SIC 3571]
Encore Paper Company Inc, p. I-674 [SIC 2676]
Encore Wire Corp, p. II-1234 [SIC 3351]
Encyclopaedia Britannica, p. I-703 [SIC 2731]
Encyclopaedia Britannica Inc, p. I-703 [SIC 2731]
Endeavor Lumber Co, p. I-471 [SIC 2426]
Endeavor Surgical Products Inc, p. II-2102 [SIC 3845]
Endicott Clay Products Co, p. I-1081 [SIC 3251]
Endicott Forging, p. II-1395 [SIC 3462]
Endicott Johnson Corp, p. I-1025 [SIC 3143]
Endogen Inc, p. I-809 [SIC 2835]
ENDOT Industries Inc, p. I-979 [SIC 3084]
Enercon Engineering Inc, p. II-1790 [SIC 3625]
Energizer Power Systems, p. II-1920 [SIC 3691]
Energy Absorption Systems Inc, p. II-1492 [SIC 3499]
Energy Concepts Inc, p. II-2066 [SIC 3826]
Energy Conversion Devices Inc, p. II-1795 [SIC 3629]
Energy Electric Assembly Inc, p. I-322 [SIC 2298]
Energy Industries Inc, p. II-1649 [SIC 3563]
Enesco Corp, p. II-2199 [SIC 3999]
Enforcer Products Inc, p. I-874 [SIC 2879]
Engelhard Corp, p. I-777 [SIC 2819]
Enger-Kress Co, p. I-1049 [SIC 3172]
Engine Components Inc, p. II-1975 [SIC 3724]
Engineered Air Systems Inc, p. II-1726 [SIC 3585]
Engineered Controls Intern, p. II-1447 [SIC 3491]
Engineered Cooling System Inc, p. II-1654 [SIC 3564]
Engineered Fabrics Corp, p. I-954 [SIC 3069]
Engineered Materials, p. II-1425 [SIC 3479]
Engineered Nonwovens Inc, p. I-318 [SIC 2297]
Engineered Plastics Inc, p. I-997 [SIC 3087]
Engineered Products Inc, p. II-1344 [SIC 3441]
Engineered Support Systems Inc, p. II-1632 [SIC 3559]
Engineered Yarns America Inc, p. I-310 [SIC 2295]
Engineering Group Inc, p. II-2016 [SIC 3769]
Engineering Plastics Inc, p. I-973 [SIC 3083]
Engineering Tube Specialties Inc, p. II-1486 [SIC 3498]
Engineers Tool & Mfg Co, p. II-1271 [SIC 3364]
Engis Corp, p. I-1144 [SIC 3291]
England/Corsair Inc, p. I-548 [SIC 2512]
ENI, p. II-1938 [SIC 3699]
EniChem Elastomers Americas, p. I-855 [SIC 2869]

Ennis Tag and Label Co, p. I-688 [SIC 2679]
Enoch Manufacturing Co, p. II-1382 [SIC 3451]
Enoch Packing Company Inc, p. I-51 [SIC 2034]
Enquirer/Star Group Inc, p. I-698 [SIC 2721]
Enquirer/Star Inc, p. I-698 [SIC 2721]
Ensign-Bickford Industries Inc, p. I-885 [SIC 2892]
Ensign Corp, p. II-1769 [SIC 3612]
Ensolite Inc, p. I-879 [SIC 2891]
Enstrom Helicopter Corp, p. II-1970 [SIC 3721]
Entergy Systems and Service Inc, p. II-1845 [SIC 3646]
Enteron Group Ltd, p. I-759 [SIC 2796]
Enterprise Corrugated, p. I-628 [SIC 2653]
Enterprise Engine Services, p. II-1503 [SIC 3519]
Enterprise Lumber Co, p. I-465 [SIC 2421]
Enterprises International Inc, p. II-1615 [SIC 3554]
Enthone-OMI Inc, p. I-897 [SIC 2899]
Entrepreneur Media Inc, p. I-698 [SIC 2721]
Entron Controls Inc, p. II-1594 [SIC 3548]
Entron Industries LP, p. II-2007 [SIC 3761]
Entwistle Co, p. II-1615 [SIC 3554]
Envel Design Corp, p. II-1845 [SIC 3646]
Envelope Co, p. I-679 [SIC 2677]
Envelope Man Plus, p. II-1679 [SIC 2677]
Envelopes Unlimited Inc, p. I-729 [SIC 2759]
Envirco Corp, p. II-1654 [SIC 3564]
Envirex Inc, p. II-1736 [SIC 3589]
Enviro-Industries Inc, p. II-1372 [SIC 3448]
Enviro Pac Inc, p. I-688 [SIC 2679]
Envirodyne Industries Inc, p. I-961 [SIC 3081]
Environment/One Corp, p. II-1638 [SIC 3561]
Environmental Elements Corp, p. II-1654 [SIC 3564]
Environmental Manufacturing, p. I-528 [SIC 2491]
Environmental Techn Corp, p. I-855 [SIC 2869]
Environmental Technologies, p. II-2076 [SIC 3829]
Environmental Tectonics Corp, p. II-1938 [SIC 3699]
Environments Inc, p. I-587 [SIC 2541]
Enwood Structures Inc, p. I-501 [SIC 2439]
Enzymatic Therapy Inc, p. I-799 [SIC 2833]
EO Wood Company Inc, p. I-991 [SIC 3086]
EP Henry Corp, p. I-1117 [SIC 3271]
EPCO Packaging Products Inc, p. I-639 [SIC 2656]
EPE Corp, p. II-1713 [SIC 3579]
EPE Technologies Inc, p. II-1769 [SIC 3612]
Ephrata Shoe Company Inc, p. I-1033 [SIC 3149]
Epic Eyewear Group Inc, p. II-2107 [SIC 3851]
Epic Metals Corp, p. II-1344 [SIC 3441]
Epitope Inc, p. I-809 [SIC 2835]
Epoxylite Corp, p. I-782 [SIC 2821]
Epp-Tech Oil Co, p. I-918 [SIC 2992]
Eppert Oil Co, p. I-918 [SIC 2992]
Epro Inc, p. I-1086 [SIC 3253]
Epsen Hillmer Graphics Co, p. I-654 [SIC 2672]
Epsen Lithographing Co, p. I-729 [SIC 2759]
Epson America Inc, p. II-1688 [SIC 3571]
Epworth Manufacturing, p. II-1525 [SIC 3532]
Equipment Innovators Inc, p. II-1948 [SIC 3713]
Equipto, p. I-593 [SIC 2542]
Equity Enterprises Inc, p. II-2112 [SIC 3861]
Equity Marketing Inc, p. II-2144 [SIC 3944]
Equity Resource Group, p. II-1404 [SIC 3465]
ER Advanced Ceramics Inc, p. I-1163 [SIC 3297]
ER Wagner Manufacturing Co, p. II-1413 [SIC 3469]
Erath Veneer Corporation, p. I-491 [SIC 2435]
Erbtec Engineering Inc, p. II-1938 [SIC 3699]
ERC Co, p. II-2093 [SIC 3843]
ERC Industries Inc, p. II-1531 [SIC 3533]
Erdle Perforating Co, p. II-1413 [SIC 3469]
Erez Fashions Inc, p. I-1045 [SIC 3171]
ERI, p. II-1395 [SIC 3462]
Eric Javits Hats, p. I-386 [SIC 2353]
Erickson/Beamon Associates, p. II-2122 [SIC 3911]
Erickson Displays Inc, p. I-593 [SIC 2542]
Erickson Metals Corp, p. II-1250 [SIC 3355]
Erickson Wood Products, p. I-515 [SIC 2449]
Erico Inc, p. II-1830 [SIC 3643]
Erie Bronze and Aluminum Co, p. II-1280 [SIC 3366]
Erie Engineered Products Inc, p. II-1304 [SIC 3412]
Erie Foods International Inc, p. I-26 [SIC 2023]
Erie Forge and Steel Inc, p. II-1395 [SIC 3462]

Erie Plating Co, p. II-1420 [SIC 3471]
Erie Press Systems Inc, p. II-1753 [SIC 3594]
Erie Scientific Co, p. I-1057 [SIC 3211]
Eriez Magnetics, p. II-1682 [SIC 3569]
Erincraft Manufacturing Co, p. II-1813 [SIC 3634]
Eritech Inc, p. II-1425 [SIC 3479]
Erlanger Tubular Corp, p. II-1290 [SIC 3398]
ERLY Industries Inc, p. I-79 [SIC 2044]
Ermanco Inc, p. II-1541 [SIC 3535]
Ernest and Julio Gallo Winery, p. I-168 [SIC 2084]
Ernest Maier Inc, p. I-1117 [SIC 3271]
Ernie Ball Inc, p. II-2135 [SIC 3931]
Ero/Goodrich Forest Products, p. I-496 [SIC 2436]
Ertel Manufacturing Corp, p. II-1742 [SIC 3592]
Ertl Company Inc, p. II-2144 [SIC 3944]
Ervin Industries Inc, p. II-1212 [SIC 3325]
Erving Industries Inc, p. I-612 [SIC 2621]
ES Products Inc, p. II-1182 [SIC 3315]
ESA Inc, p. II-2066 [SIC 3826]
ESAB Automation Inc, p. II-1600 [SIC 3549]
ESAB Welding & Cutting Prod, p. II-1594 [SIC 3548]
ESC Electronics Corp, p. II-1905 [SIC 3677]
Escalade Inc, p. II-2150 [SIC 3949]
Escalon Packers Inc, p. I-45 [SIC 2033]
ESCO Company LP, p. I-855 [SIC 2869]
ESCO Electronics Corp, p. II-1915 [SIC 3679]
Escod Industries, p. II-1915 [SIC 3679]
ESI Meats Inc, p. I-7 [SIC 2013]
Eskimo Pie Corp, p. I-30 [SIC 2024]
Eslon Thermoplastics, p. I-973 [SIC 3083]
ESP Lock Products Inc, p. II-1323 [SIC 3429]
Espey Mfg & Electronics Corp, p. II-1905 [SIC 3677]
Esposito Jewelry Inc, p. II-2122 [SIC 3911]
Esprit de Corp, p. I-361 [SIC 2331]
Esprit Systems Inc, p. II-1883 [SIC 3671]
Esquire Novelty Corp, p. II-2144 [SIC 3944]
ESSEF Corp, p. I-1006 [SIC 3089]
Esselte Pendaflex Corp, p. I-669 [SIC 2675]
Essex Machine Works Inc, p. II-1420 [SIC 3471]
Essex Specialty Products Inc, p. I-879 [SIC 2891]
Essex Structural Steel Company, p. II-1372 [SIC 3448]
Essick Air Products Inc, p. II-1654 [SIC 3564]
Essilor of America Inc, p. I-1066 [SIC 3229]
Esskay Inc, p. I-2 [SIC 2011]
ESSROC Materials Inc, p. I-1077 [SIC 3241]
Esstar Holdings Inc, p. II-1349 [SIC 3442]
Esstar Inc, p. II-1349 [SIC 3442]
EST Co, p. II-1400 [SIC 3463]
Esterline Technologies Corp, p. II-1558 [SIC 3541]
Estherville Foods Inc, p. I-12 [SIC 2015]
Estill Manufacturing Co, p. I-332 [SIC 2311]
Estwing Manufacturing Co, p. II-1312 [SIC 3423]
ET Lippert Saw Co, p. II-1318 [SIC 3425]
Etec Systems Inc, p. II-1621 [SIC 3555]
Ethan Allen Inc, pp. Vol. I - 542, 548 [SICs 2511, 2512]
Ethan Allen Interiors Inc, p. I-542 [SIC 2511]
Ethel Maid, p. I-420 [SIC 2389]
Etherington Industries, p. II-1182 [SIC 3315]
Ethicon Inc, p. II-2087 [SIC 3842]
Ethyl Corp, p. I-777 [SIC 2819]
Ethylene Corp, p. I-973 [SIC 3083]
ETI Explosives Techn Intern, p. I-885 [SIC 2892]
Etienne Aigner Inc, p. I-1013 [SIC 3111]
ETM Corp, p. II-2093 [SIC 3843]
Etonic Inc, p. I-1033 [SIC 3149]
ETREMA Products Inc, p. II-1747 [SIC 3593]
Etta Packaging Inc, p. I-643 [SIC 2657]
Ettaco Inc, p. II-1600 [SIC 3549]
Euclid Chemical Co, p. I-897 [SIC 2899]
Euclid Garment Mfg Co, p. I-352 [SIC 2326]
Euclid Universal Corp, p. II-1666 [SIC 3566]
Eudora Garment Corp, p. I-373 [SIC 2339]
Eugene Chemical, p. I-152 [SIC 2077]
Eugene Ernst Products Co, p. II-2056 [SIC 3824]
Eugene Sand and Gravel Inc, p. I-1154 [SIC 3295]
Eureka Brick and Tile Company, p. I-1081 [SIC 3251]
Eureka Co, p. II-1818 [SIC 3635]
Eureka Manufacturing Co, p. I-536 [SIC 2499]
Europe Craft Imports Inc, p. I-348 [SIC 2325]

Roman numerals I and II indicate the volume in which pages appear. Volume I holds SICs 2011 through 3299; Volume II holds SICs 3312 through 3999.

2249

Company Index

European Gas Turbines Inc, p. II-1498 [SIC 3511]
Eurotech Cabinets/Salientte, p. I-486 [SIC 2434]
Eurotherm Controls Inc, p. II-2051 [SIC 3823]
Eurotherm International Inc, p. II-2051 [SIC 3823]
Eutectic Engineering Co, p. II-1285 [SIC 3369]
Evans Adhesive Corp, p. I-879 [SIC 2891]
Evans & Sutherland, p. II-1688 [SIC 3571]
Evans Bakery Inc, p. I-112 [SIC 2053]
Evans Food Products Co, p. I-107 [SIC 2052]
Evans Industries Inc, pp. Vol. I - 973, Vol. II - 1304
  [SICs 3083, 3412]
Evans Machinery Inc, p. II-1610 [SIC 3553]
Evansville Brewing Co, p. I-160 [SIC 2082]
Evansville Veneer and Lumber, p. I-491 [SIC 2435]
Evapco Inc, p. II-1726 [SIC 3585]
Evcon Industries Inc, p. II-1726 [SIC 3585]
Evelands Inc, p. II-2020 [SIC 3792]
Evenflo Products Company Inc, p. I-985 [SIC 3085]
Everbrite Inc, p. II-2186 [SIC 3993]
Eveready Battery Company Inc, p. II-1920 [SIC 3691]
Everest and Jennings Intern Ltd, p. II-2087 [SIC 3842]
Everett Charles Technologies, p. II-2061 [SIC 3825]
Everett Pad and Paper Company, p. I-684 [SIC 2678]
Everett Smith Investment Ltd, p. II-1344 [SIC 3441]
Everex Systems Inc, p. II-1688 [SIC 3571]
Evergood Sausage Co, p. I-7 [SIC 2013]
Evergreen International Inc, p. II-1514 [SIC 3524]
Evergreen Mobile Co, p. I-523 [SIC 2452]
Evergreen Packaging Equipment, p. II-1660 [SIC 3565]
Evergreen Solutions Inc, p. I-654 [SIC 2672]
Evergreen Weigh Inc, p. II-1759 [SIC 3596]
Eversharp Pen Co, p. II-2156 [SIC 3951]
Everson Electric Co, p. II-1780 [SIC 3621]
Evode Tanner Industries, p. I-879 [SIC 2891]
EW Bowman Inc, p. II-1671 [SIC 3567]
EW Knauss and Son Inc, p. I-7 [SIC 2013]
EW Scripps Co, p. I-693 [SIC 2711]
EWC Inc, p. II-1905 [SIC 3677]
EWI Inc, p. II-1404 [SIC 3465]
Ewing Athletics USA Inc, p. I-932 [SIC 3021]
Ex-Tech Plastics Inc, p. I-961 [SIC 3081]
Exact Equipment Corp, pp. Vol II - 1660, 1759 [SICs
  3565, 3596]
Exactacut Steel Company Inc, p. II-1187 [SIC 3316]
Exar Corp, p. II-1892 [SIC 3674]
Excal Inc, p. II-1280 [SIC 3366]
Excalibur Automobile Corp, p. II-1943 [SIC 3711]
Excalibur Extrusions Inc, p. I-979 [SIC 3084]
Excel Corp, p. I-2 [SIC 2011]
Excel Inc, pp. Vol II - 1552, 1887 [SICs 3537, 3672]
Excel Industries Inc, p. II-1954 [SIC 3714]
Excel International Group Inc, p. II-2003 [SIC 3751]
Excel Recycling, p. II-1764 [SIC 3599]
Excel Spring and Stamping Inc, p. II-1470 [SIC 3495]
Excel Trailer Company Inc, p. II-2020 [SIC 3792]
Excell Manufacturing Co, p. II-2131 [SIC 3915]
Excello Specialty Co, p. I-654 [SIC 2672]
Excellon Automation Co, p. II-1578 [SIC 3545]
Excelsior Inc, pp. Vol. I - 373, 942 [SICs 2339, 3053]
Excelsior Marking Products, p. II-2164 [SIC 3953]
Excelsior Mfg & Supply, p. II-1361 [SIC 3444]
Executive Furniture Inc, p. I-571 [SIC 2521]
EXECUTONE Info Systems, p. II-1868 [SIC 3661]
Exemplar Manufacturing Co, p. II-1388 [SIC 3452]
Exhibit Crafts Inc, p. II-1372 [SIC 3448]
Exhibit Systems Inc, p. I-587 [SIC 2541]
Exhibitgroup Inc, p. II-2186 [SIC 3993]
Exide Corp, p. II-1920 [SIC 3691]
Exide Electronics Corp, p. II-1795 [SIC 3629]
Exide Electronics Group Inc, p. II-1795 [SIC 3629]
Eximco Manufacturing Co, p. II-1826 [SIC 3641]
Exmark Corp, p. I-961 [SIC 3081]
Exmet Corp, p. II-1187 [SIC 3316]
Exotherm Corp, p. II-1295 [SIC 3399]
Exotic Materials Inc, p. II-2071 [SIC 3827]
Exotic Metals Forming Co, p. II-1975 [SIC 3724]
ExOxEmis Inc, p. I-814 [SIC 2836]
Expo Inc, p. I-365 [SIC 2335]
Expo Industries Inc, p. I-1167 [SIC 3299]

Exposaic Industries, p. I-1122 [SIC 3272]
Exposaic Industries Incorporated, p. I-1122 [SIC 3272]
Express Container Corp, p. I-628 [SIC 2653]
Extek Inc, p. II-1563 [SIC 3542]
Extreme Sportswear, p. I-446 [SIC 2396]
Extrude Hone Corp, p. II-1558 [SIC 3541]
Extrusion Dies Inc, p. II-1572 [SIC 3544]
Extrusion Painting Inc, p. II-1244 [SIC 3354]
Extrusion Technology Inc, p. II-1244 [SIC 3354]
Extrusions Inc, p. II-1244 [SIC 3354]
Exxel Container Inc, p. II-1649 [SIC 3563]
Exxon Chemical Co, p. I-855 [SIC 2869]
Exxon Corp, p. I-903 [SIC 2911]
Exxtra Corp, p. II-1621 [SIC 3555]
Eye Kraft Optical Inc, p. II-2107 [SIC 3851]
Eyecatcher Screen Printing, p. I-279 [SIC 2261]
Eyelematic Manufacturing, p. II-1413 [SIC 3469]
EZ Loader Boat Trailers Inc, p. II-2029 [SIC 3799]
EZ Steer Inc, p. II-2029 [SIC 3799]

F and A Cheese Corp, p. I-21 [SIC 2022]
F and A Dairy Products Inc, p. I-21 [SIC 2022]
F and B Manufacturing Co, p. II-1413 [SIC 3469]
F and F Laboratories Inc, p. I-127 [SIC 2064]
F and M Hat Company Inc, p. I-386 [SIC 2353]
F and S Carton Co, p. I-643 [SIC 2657]
F and W Publications Inc, p. I-698 [SIC 2721]
F Korbel and Bros Inc, p. I-168 [SIC 2084]
F Schumacher and Co, p. I-236 [SIC 2221]
Fab-Knit Manufacturing, p. I-356 [SIC 2329]
Fabco-Air Inc, p. II-1747 [SIC 3593]
Fabcon Inc, p. I-1122 [SIC 3272]
Fabcor Inc, p. I-1001 [SIC 3088]
Faber Enterprises Inc, p. II-1454 [SIC 3492]
Fabil Manufacturing Corp, p. I-344 [SIC 2323]
Fable Toy Company Inc, p. II-2140 [SIC 3942]
Fablok Mills Inc, p. I-271 [SIC 2258]
Fabri Cote Corp, p. I-310 [SIC 2295]
Fabri Quilt Inc, p. I-442 [SIC 2395]
Fabri-Tech Inc, p. I-455 [SIC 2399]
Fabric Chemical Corp, p. I-819 [SIC 2841]
Fabricated Wood Products Inc, p. I-501 [SIC 2439]
Fabrico Manufacturing Corp, p. I-961 [SIC 3081]
Fabricut Manufacturing Inc, p. I-231 [SIC 2211]
Fabriko Inc, p. I-455 [SIC 2399]
Fabrionics Inc, p. I-310 [SIC 2295]
Fabsco Inc, p. II-1355 [SIC 3443]
Fabwel Inc, p. II-1344 [SIC 3441]
Facelifters Home Systems Inc, p. I-486 [SIC 2434]
Facemate Corp, p. I-279 [SIC 2261]
Facemate PL-GF Inc, p. I-279 [SIC 2261]
Facets by Spectrum, p. II-2173 [SIC 3961]
Factory Direct Disks Inc, p. II-1934 [SIC 3695]
FAFCO Inc, p. II-1339 [SIC 3433]
FAI Inc, p. II-1568 [SIC 3543]
Fair Haven Industries Inc, p. I-446 [SIC 2396]
Fair Publishing House Inc, p. II-2169 [SIC 3955]
Fairbanks Inc, p. II-1759 [SIC 3596]
Fairbanks Morse Engine, p. II-1503 [SIC 3519]
Fairbanks Morse Pump Corp, p. II-1638 [SIC 3561]
Fairchild Aircraft Inc, p. II-1970 [SIC 3721]
Fairchild Corp, p. II-1980 [SIC 3728]
Fairchild Fastener Group, p. II-1980 [SIC 3728]
Fairchild Industrial Products Co, p. II-2051 [SIC 3823]
Fairchild International, p. II-1525 [SIC 3532]
Fairchild of California, p. I-548 [SIC 2512]
Fairfield Chair Company Inc, p. I-548 [SIC 2512]
Fairfield Industries Inc, p. II-2076 [SIC 3829]
Fairfield Manufacturing Co, p. II-1666 [SIC 3566]
Fairfield Processing Corp, p. I-327 [SIC 2299]
Fairlane Inc, p. I-271 [SIC 2258]
Fairmont Corp, p. I-327 [SIC 2299]
Fairmont Homes Inc, p. I-519 [SIC 2451]
Fairmont Minerals Ltd, p. I-1154 [SIC 3295]
Fairmont Snacks Group Inc, p. I-136 [SIC 2068]
Fairmont-Zarda Dairy, p. I-35 [SIC 2026]
Fairprene Industrial Products, p. I-310 [SIC 2295]
Fairview Block and Supply Corp, p. I-1117 [SIC 3271]
Fairview Dairy Inc, p. I-30 [SIC 2024]

Fairview Sintered Metals Inc, p. II-1295 [SIC 3399]
Fairway Spring Company Inc, p. II-1470 [SIC 3495]
Fairwinds Gourmet Coffee, p. I-195 [SIC 2095]
Faith Dairy Inc, p. I-30 [SIC 2024]
Falco Data Products Inc, p. II-1699 [SIC 3575]
Falcon Building Products Inc, p. II-1339 [SIC 3433]
Falcon Business Forms Inc, p. I-735 [SIC 2761]
Falcon Candy Co, p. I-127 [SIC 2064]
Falcon Foundry Co, p. II-1271 [SIC 3364]
Falcon Industries Inc, p. I-409 [SIC 2385]
Falcon Jet Corp, p. II-1970 [SIC 3721]
Falcon Safety Products Inc, p. II-1878 [SIC 3669]
Falcon Shoe Manufacturing Co, p. I-1025 [SIC 3143]
Falcon Systems Inc, p. II-1694 [SIC 3572]
Falk Corp, p. II-1666 [SIC 3566]
Fall River Foundry Co, p. II-1285 [SIC 3369]
Fall River Group Inc, p. II-1271 [SIC 3364]
Fall River Tool and Die, p. II-1271 [SIC 3364]
Family Movie Centers, p. I-587 [SIC 2541]
Famous-Fraternity, p. I-356 [SIC 2329]
Famous Hospitality Inc, p. I-735 [SIC 2761]
Famous Lubricants Inc, p. I-918 [SIC 2992]
Fancy Feet Inc, p. I-1033 [SIC 3149]
Fankhauser Inc, p. I-1117 [SIC 3271]
Fanning-Schuett of New Jersey, p. II-1589 [SIC 3547]
Fansteel Inc, p. II-1980 [SIC 3728]
Fansteel Wellman Dynamics, p. II-1275 [SIC 3365]
Fantastic Graphics Inc, p. II-2160 [SIC 3952]
Fantasy-BlankeBaer, p. I-181 [SIC 2087]
Fantasy Diamond Corp, p. II-2122 [SIC 3911]
FANUC Robotics Corp, p. II-1682 [SIC 3569]
FANUC Robotics North America, p. II-1682 [SIC
  3569]
Faraday Inc, p. II-1878 [SIC 3669]
Farah Inc, p. I-348 [SIC 2325]
Fargo Assembly of Indiana Co, p. II-1929 [SIC 3694]
Fargo Assembly of PA Inc, p. II-1600 [SIC 3549]
Faria Corp, p. II-2076 [SIC 3829]
Faribault Woolen Mill Co, p. I-241 [SIC 2231]
Farinon, p. II-1873 [SIC 3663]
Faris Brothers of California Inc, p. I-378 [SIC 2341]
Farley Foods USA, p. I-127 [SIC 2064]
Farm Fresh Catfish Co, p. I-190 [SIC 2092]
Farm Journal Inc, p. I-698 [SIC 2721]
Farm Progress Co, p. I-698 [SIC 2721]
Farm Service Elevator Co, p. I-95 [SIC 2048]
Farm Services Inc, p. I-869 [SIC 2875]
Farmdale Creamery Inc, p. I-21 [SIC 2022]
Farmers Coop Elevator Co, p. I-95 [SIC 2048]
Farmers Cooperative, p. I-869 [SIC 2875]
Farmers Cooperative Association, p. I-869 [SIC 2875]
Farmers Cooperative Co, p. I-95 [SIC 2048]
Farmers Cooperative Oil Co, p. I-869 [SIC 2875]
Farmers Elevator Co, p. I-95 [SIC 2048]
Farmers Pride Inc, p. I-12 [SIC 2015]
Farmers Union Coop Oil Assoc, p. I-869 [SIC 2875]
Farmers Union Oil Co, p. I-869 [SIC 2875]
Farmington Displays Inc, p. I-587 [SIC 2541]
Farmland Foods Inc, p. I-2 [SIC 2011]
Farmland Industries Inc, p. I-903 [SIC 2911]
Farnam Companies Inc, p. I-804 [SIC 2834]
Farnam Sealing Systems, p. I-942 [SIC 3053]
Faro USA Corp, p. II-2093 [SIC 3843]
Farr Co, p. II-1654 [SIC 3564]
Fasco Consumer Products, p. II-1654 [SIC 3564]
Fasco Motors Group, p. II-1654 [SIC 3564]
Fashion Bed Group, p. I-553 [SIC 2514]
Fashion Cabinet Manufacturing, p. I-486 [SIC 2434]
Fashion Engravers Inc, p. I-759 [SIC 2796]
Fashion Enterprises Inc, p. I-369 [SIC 2337]
Fashion Inc, p. II-1361 [SIC 3444]
Fashion Point Accessories Inc, p. I-344 [SIC 2323]
Fashion Ribbon Company Inc, p. I-446 [SIC 2396]
Fashion World Career Apparel, p. I-332 [SIC 2311]
Fasson Specialty, p. I-654 [SIC 2672]
Fast Clothing Inc, p. I-336 [SIC 2321]
Fastener Industries Inc, p. II-1388 [SIC 3452]
FasTest Inc, p. II-2045 [SIC 3822]
Fat City Cycles Inc, p. II-2003 [SIC 3751]

Roman numerals I and II indicate the volume in which pages appear. Volume I holds SICs 2011 through 3299; Volume II holds SICs 3312 through 3999.

2250

Company Index

Roman numerals I and II indicate the volume in which pages appear. Volume I holds SICs 2011 through 3299; Volume II holds SICs 3312 through 3999.

Fleetwood Enterprises Inc, p. II-1966 [SIC 3716]
Fleetwood Folding Trailers Inc, p. II-2020 [SIC 3792]
Fleetwood Homes of Florida Inc, p. I-519 [SIC 2451]
Fleetwood Motor Homes, p. II-1966 [SIC 3716]
Fleetwood Shirt Corp, p. I-336 [SIC 2321]
Fleetwood Snacks Inc, p. I-107 [SIC 2052]
Fleetwood Travel Trailers, p. II-2020 [SIC 3792]
Fleischmann's Yeast, p. I-212 [SIC 2099]
Fleming Packaging Corp, p. I-718 [SIC 2752]
Fleming-Potter Co, p. I-718 [SIC 2752]
Flender Corp, p. II-1666 [SIC 3566]
Fletcher Granite Company Inc, p. I-1140 [SIC 3281]
Fletcher Industries Inc, p. II-1605 [SIC 3552]
Fletcher Paper Co, p. I-612 [SIC 2621]
Fletcher Terry Co, p. II-1312 [SIC 3423]
Flex Metal Components Co, p. II-1470 [SIC 3495]
Flex-O-Lite Inc, p. II-1845 [SIC 3646]
Flex Products Inc, p. I-649 [SIC 2671]
Flex Technologies Inc, p. II-1726 [SIC 3585]
Flexcon and Systems Inc, p. I-455 [SIC 2399]
Flexcon Company Inc, p. I-961 [SIC 3081]
Flexcraft Inc, p. I-759 [SIC 2796]
Flexfab Inc, p. I-936 [SIC 3052]
Flexfirm Products Inc, p. I-310 [SIC 2295]
Flexi-Mat Corp, p. I-558 [SIC 2515]
Flexible Material Handling, p. II-1476 [SIC 3496]
Flexible Materials Inc, p. I-491 [SIC 2435]
Flexlon Fabrics Inc, p. I-275 [SIC 2259]
Flexo Transparent Inc, p. I-649 [SIC 2671]
Flexseal Intern Packaging Corp, p. I-688 [SIC 2679]
Flexsteel Industries Inc, p. I-548 [SIC 2512]
Flexwrap Corp, p. I-961 [SIC 3081]
Flight Equip & Eng Corp, p. I-581 [SIC 2531]
Flight Insulation, p. I-1158 [SIC 3296]
Flight Suits Ltd, p. I-332 [SIC 2311]
Flightline Electronics Inc, p. II-1873 [SIC 3663]
FlightSafety International Inc, p. II-1938 [SIC 3699]
Flink Company Inc, p. II-1514 [SIC 3524]
Flint and Walling Industries Inc, p. II-1638 [SIC 3561]
Flint Asphalt and Paving Co, p. I-909 [SIC 2951]
Flint Ink Corp, p. I-889 [SIC 2893]
Flint Manufacturing Co, p. II-1404 [SIC 3465]
Flint River Mills Inc, p. I-95 [SIC 2048]
Flittie, Marshall Concrete Prod, p. I-1117 [SIC 3271]
Flo-Bend Inc, p. II-1486 [SIC 3498]
Flo-Pac Corp, p. II-2181 [SIC 3991]
Flo-Tork Inc, p. II-1747 [SIC 3593]
Flohr Metal Fabricators Inc, p. II-1626 [SIC 3556]
Flora Springs Wine Co, p. I-168 [SIC 2084]
Florasynth Inc, p. I-855 [SIC 2869]
Florence Eiseman Inc, p. I-390 [SIC 2361]
Florida Coca-Cola Bottling Co, p. I-176 [SIC 2086]
Florida Drum Company Inc, p. II-1304 [SIC 3412]
Florida Drum Delta Co, p. II-1304 [SIC 3412]
Florida Engineered Constr, p. I-1122 [SIC 3272]
Florida Extruders International, p. II-1244 [SIC 3354]
Florida Favorite Fertilizers Inc, p. I-869 [SIC 2875]
Florida Forest Products Ltd, p. I-501 [SIC 2439]
Florida Furniture Industries Inc, p. I-542 [SIC 2511]
Florida Mining, p. I-1122 [SIC 3272]
Florida Pneumatic Mfg Corp, p. II-1753 [SIC 3594]
Florida Production Engineering, p. II-1404 [SIC 3465]
Florida Rock Industries Inc, p. I-1127 [SIC 3273]
Florida Steel Corp, pp. Vol II - 1172, 1182 [SICs 3312, 3315]
Florida Tile Industries Inc, p. I-1086 [SIC 3253]
Florida Wire and Cable Inc, pp. Vol II - 1182, 1476 [SICs 3315, 3496]
Floridin Co, p. I-1154 [SIC 3295]
Florin Box and Lumber Co, p. I-506 [SIC 2441]
Florin Tallow Co, p. I-152 [SIC 2077]
Florsheim Shoe Co, p. I-1025 [SIC 3143]
Flos Inc, p. II-1840 [SIC 3645]
Floturn Inc, p. II-1764 [SIC 3599]
Flour City Brush Co, p. II-2181 [SIC 3991]
Flow-Eze Co, p. I-446 [SIC 2396]
Flow International Corp, p. II-1558 [SIC 3541]
Flow Products Inc, p. II-1753 [SIC 3594]
Floway Pumps Inc, p. II-1638 [SIC 3561]

Flower City Builders Supply, p. I-1117 [SIC 3271]
Flowers Baking of Jacksonville, p. I-101 [SIC 2051]
Flowers Baking of Lynchburg, p. I-101 [SIC 2051]
Flowers Baking of Thomasville, p. I-101 [SIC 2051]
Flowers Baking of Tyler, p. I-101 [SIC 2051]
Flowers Distributors Co, p. I-101 [SIC 2051]
Flowers Indust Specialty, p. I-112 [SIC 2053]
Flowers Industries Inc, p. I-101 [SIC 2051]
Flowline, p. II-1464 [SIC 3494]
Floyd Manufacturing, p. II-2020 [SIC 3792]
FLS Holdings Inc, p. II-1172 [SIC 3312]
Flue-Cured Tobacco, p. I-228 [SIC 2141]
Fluid Power Systems, p. II-1454 [SIC 3492]
Fluid Regulators Corp, p. II-1753 [SIC 3594]
Fluid Systems Corp, p. II-1736 [SIC 3589]
Fluidex, p. II-1454 [SIC 3492]
Fluidmaster Inc, p. II-1464 [SIC 3494]
Fluidyne Ansonia, p. II-1382 [SIC 3451]
Fluke Corp, p. II-2061 [SIC 3825]
Fluoro Chemicals, p. I-777 [SIC 2819]
Fluoroware Inc, p. I-1006 [SIC 3089]
Flushing Shirt Manufacturing Co, p. I-336 [SIC 2321]
Flute Inc, p. II-1840 [SIC 3645]
Flyer Printing Company Inc, p. I-713 [SIC 2741]
Flynn Enterprises Inc, p. I-348 [SIC 2325]
Flynt Fabrics and Finishing Inc, p. I-258 [SIC 2253]
Flyte Tyme Productions Inc, p. II-1864 [SIC 3652]
FM Browns Sons Inc, p. I-95 [SIC 2048]
FM Brush Company Inc, p. II-2160 [SIC 3952]
FM Howell and Company Inc, p. I-623 [SIC 2652]
FM Manufacturing Inc, p. II-1736 [SIC 3589]
FMC Corp, pp. Vol. I - 777, Vol. II - 1443, 1447, 1541, 1626 [SICs 2819, 3489, 3491, 3535, 3556]
FMC Corp Food and Machinery, p. II-1626 [SIC 3556]
FMC Corp Steel Products, p. II-1395 [SIC 3462]
FMP/Rauma Co, p. II-1615 [SIC 3554]
FN Burt Company Inc, p. I-623 [SIC 2652]
FN Manufacturing Inc, p. II-1438 [SIC 3484]
FNT Industries Inc, p. I-322 [SIC 2298]
Foam Pack Inc, p. I-991 [SIC 3086]
Foam Plastics of New England, p. I-991 [SIC 3086]
Foam Rubber Products Inc, p. I-991 [SIC 3086]
Foamade Industries, p. I-991 [SIC 3086]
Foamex International Inc, p. I-782 [SIC 2821]
Foamex LP, pp. Vol. I - 782, 954 [SICs 2821, 3069]
Foamseal Inc, p. I-787 [SIC 2822]
Focus Enhancements Inc, p. II-1694 [SIC 3572]
Foell Packing Co, p. I-40 [SIC 2032]
Foerster Enterprises Inc, p. I-750 [SIC 2789]
Foilmark Inc, p. II-2164 [SIC 3953]
Fojtasek Companies Inc, p. II-1349 [SIC 3442]
Fold-Pak Corp, p. I-643 [SIC 2657]
Foldcraft Co, p. I-581 [SIC 2531]
Foley-Belsaw Co, p. II-1682 [SIC 3569]
Foley Material Handling, p. II-1519 [SIC 3531]
Foley Pattern Company Inc, p. II-1275 [SIC 3365]
Folger Adam Co, p. II-1323 [SIC 3429]
Fonar Corp, p. II-2102 [SIC 3845]
Fonda Group Inc, p. I-639 [SIC 2656]
Fontaine Modification Co, p. II-1943 [SIC 3711]
Fontana Steel Co, p. II-1344 [SIC 3441]
Food Automation, p. II-1626 [SIC 3556]
Food City USA Inc, p. I-208 [SIC 2098]
Food Ingredients, p. I-814 [SIC 2836]
Food Oils Corp, p. I-156 [SIC 2079]
Food Traditions, p. I-55 [SIC 2035]
Foodbrands America Inc, p. I-7 [SIC 2013]
Foods C'est Bon Ltd, p. I-65 [SIC 2038]
FoodScience Corp, p. I-799 [SIC 2833]
Foote and Davies Lincoln, p. I-708 [SIC 2732]
Foote-Jones/Illinois Gear, p. II-1666 [SIC 3566]
Fora Inc, p. II-1688 [SIC 3571]
Forbes Products Corp, p. I-745 [SIC 2782]
Force Control Industries Inc, p. II-1676 [SIC 3568]
Ford Motor Co, pp. Vol. I - 1071, Vol. II - 1943 [SICs 3231, 3711]
Ford New Holland Americas, p. II-1509 [SIC 3523]
Fordick Corp, p. I-536 [SIC 2499]
Forecaster of Boston Inc, p. I-409 [SIC 2385]

Foredom Electric Co, p. II-1584 [SIC 3546]
Foreign Accents, p. I-292 [SIC 2273]
Foreign and Domestic Woods, p. I-480 [SIC 2431]
Foreign Embroidery Inc, p. I-446 [SIC 2396]
Foremost Dairies-Hawaii, p. I-35 [SIC 2026]
Foremost Manufacturing, p. II-1854 [SIC 3648]
Forenta LP, p. II-1722 [SIC 3582]
Forest City Technologies Inc, p. I-942 [SIC 3053]
Forest Laboratories Inc, p. I-804 [SIC 2834]
Forester Boats, p. II-1992 [SIC 3732]
Forestex Co, p. I-491 [SIC 2435]
Foretravel Inc, p. II-1966 [SIC 3716]
Forever Living Products Intern, p. I-834 [SIC 2844]
Forged Products Inc, p. II-1395 [SIC 3462]
Forged Vessel Connections Inc, p. II-1395 [SIC 3462]
Form A Feed Inc, p. I-95 [SIC 2048]
Forma Scientific Inc, p. II-2040 [SIC 3821]
Formation Inc, p. II-1694 [SIC 3572]
Formetal Inc, p. II-1377 [SIC 3449]
Formflex, p. I-745 [SIC 2782]
Formica Corp US, p. I-973 [SIC 3083]
Formico Food Co, p. I-7 [SIC 2013]
Formosa Plastics Corp USA, p. I-764 [SIC 2812]
Forms Manufacturing Inc, p. I-735 [SIC 2761]
Formsprag-Warren, p. II-1676 [SIC 3568]
Fornaca Family Bakery Inc, p. I-101 [SIC 2051]
Forney Corp, p. II-2051 [SIC 3823]
Forney Industries Inc, p. II-1594 [SIC 3548]
Forsch Corp, p. I-1066 [SIC 3229]
Forstmann and Company Inc, p. I-241 [SIC 2231]
Fort Biscuit Co, p. I-107 [SIC 2052]
Fort Dearborn Lithograph Co, p. I-718 [SIC 2752]
Fort Howard Corp, p. I-674 [SIC 2676]
Fort Howard Steel Inc, p. II-1187 [SIC 3316]
Fort Lock Corp, p. II-1323 [SIC 3429]
Fort Orange Paper Co, p. I-643 [SIC 2657]
Fort Payne Dekalb Hosiery Mills, p. I-254 [SIC 2252]
Fort Recovery Industries Inc, p. II-1334 [SIC 3432]
Fort Smith Table & Furniture, p. I-542 [SIC 2511]
Fort Wayne Wire Die Inc, p. II-1572 [SIC 3544]
Fort Worth Star-Telegram Inc, p. I-693 [SIC 2711]
Fortifiber Corp, p. I-654 [SIC 2672]
Fortress Inc, p. I-576 [SIC 2522]
Fortunately Yours, p. I-107 [SIC 2052]
Fortune Fabrics Inc, p. I-236 [SIC 2221]
Fortune Plastics Inc, p. I-660 [SIC 2673]
Forward Technology Industries, p. II-1632 [SIC 3559]
Fosroc Inc, p. I-879 [SIC 2891]
Fossil Inc, p. II-2118 [SIC 3873]
Fosston Cooperative Association, p. I-869 [SIC 2875]
Fosta-Tek Optics Inc, p. II-2107 [SIC 3851]
Foster Forbes Glass, p. I-1062 [SIC 3221]
Foster Inc, p. II-2122 [SIC 3911]
Foster Industries Inc, p. I-405 [SIC 2384]
Foster Needle Company Inc, p. II-2177 [SIC 3965]
Foster Poultry Farms Inc, p. I-12 [SIC 2015]
Foster Transformer Co, p. II-1769 [SIC 3612]
Foster Valve Corp, p. II-1531 [SIC 3533]
Fotec Inc, p. II-2071 [SIC 3827]
Foulds Inc, p. I-208 [SIC 2098]
Foundry and Steel Inc, p. II-1764 [SIC 3599]
Foundry Inc, p. II-1275 [SIC 3365]
Foundry Service Co, p. II-1197 [SIC 3321]
Fountain Inc, p. II-1290 [SIC 3398]
Fountain Plating Company Inc, p. II-1420 [SIC 3471]
Fountain Powerboat Industries, p. II-1992 [SIC 3732]
Four B's Sales Co, p. II-2140 [SIC 3942]
Four County Agricultural Svcs, p. I-869 [SIC 2875]
Four Paws Products Ltd, p. I-95 [SIC 2048]
Four Seasons Garment Co, p. I-356 [SIC 2329]
Four Winds International Corp, p. II-1966 [SIC 3716]
Four Winns Inc, p. II-1992 [SIC 3732]
Fourth Street Rock Crusher, p. I-1154 [SIC 3295]
Fowler Products Co, p. II-1660 [SIC 3565]
Fowler's Milling Co, p. I-70 [SIC 2041]
Fownes Brothers and Company, p. I-1037 [SIC 3151]
Fox Point Sportswear Inc, p. I-356 [SIC 2329]
Fox Pool Corp, p. II-2150 [SIC 3949]
Fox River Mills Inc, p. I-254 [SIC 2252]

Roman numerals I and II indicate the volume in which pages appear. Volume I holds SICs 2011 through 3299; Volume II holds SICs 3312 through 3999.

2252

Roman numerals I and II indicate the volume in which pages appear. Volume I holds SICs 2011 through 3299; Volume II holds SICs 3312 through 3999.

Company Index

Gannon Manufacturing, p. II-1552 [SIC 3537]
Gano Welding Supplies Inc, p. I-768 [SIC 2813]
Gans Ink and Supply Company, p. I-889 [SIC 2893]
Gans Tire Company Inc, p. I-927 [SIC 3011]
Garan Inc, p. I-356 [SIC 2329]
Garber Co, p. I-718 [SIC 2752]
Garden Jewelry Mfg Corp, p. II-2122 [SIC 3911]
Garden State Paper Company, p. I-612 [SIC 2621]
Garden State Tanning, p. I-1013 [SIC 3111]
Garden Way Inc, p. II-1514 [SIC 3524]
Gardiner Metal Co, p. II-1255 [SIC 3356]
Gardinier Associates Inc, p. I-420 [SIC 2389]
Gardner Asphalt Corp, p. I-914 [SIC 2952]
Gardner Baking, p. I-101 [SIC 2051]
Gardner Candies Inc, p. I-127 [SIC 2064]
Gardner Cryogenics, p. II-1355 [SIC 3443]
Gardner Denver Machinery Inc, p. II-1649 [SIC 3563]
Gardner Machinery Corp, p. II-1547 [SIC 3536]
Gare Inc, p. I-1113 [SIC 3269]
Garelick Farms Inc, p. I-35 [SIC 2026]
Garfield Industries, p. I-1144 [SIC 3291]
Garland Commercial Industries, p. II-1736 [SIC 3589]
Garland Company Inc, p. I-1150 [SIC 3292]
Garland Floor Co, p. I-825 [SIC 2842]
Garland Industries Inc, p. II-2156 [SIC 3951]
Garlock Mechanical Packing, p. I-942 [SIC 3053]
Garlock Printing, p. I-688 [SIC 2679]
Garment Corporation, p. I-352 [SIC 2326]
Garner Glass Co, p. I-1066 [SIC 3229]
Garr Tool, p. II-1578 [SIC 3545]
Garrett Metal Detectors, p. II-1878 [SIC 3669]
Garrity Industries, p. II-1854 [SIC 3648]
Garry Packing Inc, p. I-51 [SIC 2034]
Garsite TSR Inc, p. II-1552 [SIC 3537]
Garvey Corp, p. II-1541 [SIC 3535]
Gary Concrete Products Inc, p. I-1122 [SIC 3272]
Gary's Leather Creations, p. I-1053 [SIC 3199]
Gary Steel Products Corp, p. II-1361 [SIC 3444]
Gary-Williams Energy Corp, p. I-903 [SIC 2911]
Gas Arc Supply Inc, p. I-768 [SIC 2813]
Gas Drying Inc, p. II-1649 [SIC 3563]
Gas Tech Inc, p. II-2076 [SIC 3829]
Gaska-Tape Inc, p. I-991 [SIC 3086]
Gasket Engineering Co, p. I-942 [SIC 3053]
GaSonics International Corp, p. II-1632 [SIC 3559]
Gaspro, p. I-768 [SIC 2813]
Gasser Chair Co, p. I-571 [SIC 2521]
Gast Manufacturing Corp, p. II-1638 [SIC 3561]
Gaston Copper Recycling Corp, p. II-1234 [SIC 3351]
Gaston Cty Dyeing Machine, p. II-1605 [SIC 3552]
Gates Albert Inc, p. II-1382 [SIC 3451]
Gates Corp, p. I-936 [SIC 3052]
Gates Export Corp, p. I-936 [SIC 3052]
Gates Flag and Banner Co, p. I-455 [SIC 2399]
Gates Formed-Fibre Products, p. I-236 [SIC 2221]
Gates Mills Inc, p. I-1037 [SIC 3151]
Gates Rubber Co, p. I-936 [SIC 3052]
Gateway Amsafe Inc, p. II-2107 [SIC 3851]
Gateway Construction Company, p. II-1344 [SIC 3441]
Gateway Press Inc, p. I-729 [SIC 2759]
Gatewood Products Inc, p. I-510 [SIC 2448]
Gator Asphalt Co, p. I-909 [SIC 2951]
Gator Industries Inc, p. I-1033 [SIC 3149]
Gatorade, p. I-176 [SIC 2086]
Gaudette Leather Goods Inc, p. I-1053 [SIC 3199]
Gauntlet Inc, p. II-2173 [SIC 3961]
Gauss Corp, p. II-1929 [SIC 3694]
Gaw-O'Hara Envelope Co, p. I-679 [SIC 2677]
Gayle Manufacturing Co, p. II-1344 [SIC 3441]
Gayle's Chocolates Inc, p. I-132 [SIC 2066]
Gaylord Bag Partnership, p. I-665 [SIC 2674]
Gaylord Container Corp, p. I-628 [SIC 2653]
GB Products International Corp, p. I-759 [SIC 2796]
GBC Film Products, p. I-973 [SIC 3083]
GBC Scientific Equipment Inc, p. II-2066 [SIC 3826]
GC Broach Co, p. II-1671 [SIC 3567]
GC Quality Lubricants Inc, p. I-918 [SIC 2992]
GC Technologies Inc, p. I-1066 [SIC 3229]
GCH Systems Inc, p. II-1699 [SIC 3575]

GD Searle and Co, p. I-804 [SIC 2834]
GDE Systems Inc, p. II-2035 [SIC 3812]
GE Aircraft Engines, p. II-1975 [SIC 3724]
GE Appliances, p. II-1800 [SIC 3631]
GE Control Products, p. II-1790 [SIC 3625]
GE Environmental Systems, p. II-1654 [SIC 3564]
GE Medical Systems Group, p. II-2102 [SIC 3845]
GE Motors, p. II-1780 [SIC 3621]
GE Transportation Systems, p. II-1998 [SIC 3743]
GEA Power Cooling System Inc, p. II-1498 [SIC 3511]
GEA Rainey Corp, p. II-1355 [SIC 3443]
Geac/Fasfax, p. II-1709 [SIC 3578]
Gear Motions Inc, p. II-1666 [SIC 3566]
Gear Research Inc, p. II-1666 [SIC 3566]
Gear Works, Seattle Inc, p. II-1666 [SIC 3566]
Geartronics Industries Inc, p. II-1666 [SIC 3566]
GEC-Marconi, p. II-1915 [SIC 3679]
GEC-Marconi Aerospace Inc, p. II-1980 [SIC 3728]
GEC-Marconi Avionics Inc, p. II-2035 [SIC 3812]
Geckostufs Inc, p. II-2140 [SIC 3942]
Gehl's Guernsey Farms Inc, p. I-26 [SIC 2023]
Gehring Textiles Inc, p. I-271 [SIC 2258]
Geiger and Peters Inc, p. II-1344 [SIC 3441]
Gelman Sciences Inc, p. II-1682 [SIC 3569]
Gelok International, p. I-674 [SIC 2676]
Gem Dandy Inc, p. I-416 [SIC 2387]
Gem East Corp, p. II-2122 [SIC 3911]
Gem Electric Manufacturing, p. II-1854 [SIC 3648]
Gem Products Inc, p. II-1726 [SIC 3585]
Gem Southeast Inc, p. I-567 [SIC 2519]
GEM Textile Company Inc, p. II-1605 [SIC 3552]
Gem Urethane Corp, p. I-310 [SIC 2295]
Gemco-Ware Inc, p. I-1066 [SIC 3229]
Gemini Shirtmakers Inc, p. I-390 [SIC 2361]
Gemini Sound Products Corp, p. II-1859 [SIC 3651]
Gemsco Inc, p. I-455 [SIC 2399]
Gemtron Corp, p. I-1057 [SIC 3211]
Genal Strap Company Inc, p. I-1049 [SIC 3172]
Genco Corp, p. I-1037 [SIC 3151]
Gencor Industries Inc, p. II-1519 [SIC 3531]
GenCorp Automotive, p. I-1006 [SIC 3089]
GenCorp Inc, p. II-2011 [SIC 3764]
GenCorp Polymer Products, p. I-782 [SIC 2821]
Genencor International Inc, p. I-809 [SIC 2835]
Genentech Inc, p. I-814 [SIC 2836]
General American Door Co, p. II-1349 [SIC 3442]
General Asphalt Company Inc, p. I-909 [SIC 2951]
General Automation Inc, p. II-1382 [SIC 3451]
General Automotive Corp, p. II-1948 [SIC 3713]
General Automotive Mfg Co, p. II-1413 [SIC 3469]
General Automotive Specialty, p. II-1915 [SIC 3679]
General Bag Corp, p. I-669 [SIC 2675]
General Bearing Corp, p. II-1212 [SIC 3325]
General Bindery Company Inc, p. I-750 [SIC 2789]
General Binding Corp, pp. Vol. I - 745, Vol. II - 1713 [SICs 2782, 3579]
General Broach & Engineering, p. II-1558 [SIC 3541]
General Business Envelope, p. I-679 [SIC 2677]
General Business Forms Inc, p. I-735 [SIC 2761]
General Casting Co, p. II-1197 [SIC 3321]
General Chemical Corp, p. I-764 [SIC 2812]
General Cigar Co, p. I-221 [SIC 2121]
General Clay Co, p. II-2186 [SIC 3993]
General Crushed Stone Co, p. I-1140 [SIC 3281]
General DataComm Industries, p. II-1868 [SIC 3661]
General Devices Company Inc, p. II-1660 [SIC 3565]
General Die Casting Co, p. II-1271 [SIC 3364]
General Dynamics Corp, pp. Vol II - 1780, 1970, 1986, 2025 [SICs 3621, 3721, 3731, 3795]
General Dynamics Services Co, p. II-2025 [SIC 3795]
General Electric Co, pp. Vol II - 1498, 1769, 1780, 1826, 1897 [SICs 3511, 3612, 3621, 3641, 3675]
General Electro-Mechanical, p. II-1563 [SIC 3542]
General Electrodynamics Corp, p. II-1759 [SIC 3596]
General Engineering Co, p. II-1747 [SIC 3593]
General Engines Company Inc, p. II-1961 [SIC 3715]
General Equip Manufacturers, p. II-2040 [SIC 3821]
General Equipment & Mfg, p. II-1775 [SIC 3613]
General Extrusions Inc, p. II-1244 [SIC 3354]

General Fiber Optics Inc, p. I-1066 [SIC 3229]
General Films Inc, p. I-961 [SIC 3081]
General Filter Co, p. II-1736 [SIC 3589]
General Fire Extinguisher Corp, p. II-2199 [SIC 3999]
General Foam Plastics Corp, p. I-954 [SIC 3069]
General Formulations, p. I-654 [SIC 2672]
General Hoist Corp, p. II-1547 [SIC 3536]
General Housewares Corp, p. II-1275 [SIC 3365]
General Industries Co, p. II-1780 [SIC 3621]
General Instrument Corp, pp. Vol II - 1892, 1915 [SICs 3674, 3679]
General Kinematics Corp, p. II-1541 [SIC 3535]
General Latex & Chemical Corp, p. I-897 [SIC 2899]
General Loose Leaf Bindery Co, p. I-745 [SIC 2782]
General Machine Corp, p. II-1339 [SIC 3433]
General Machine Products, p. II-1312 [SIC 3423]
General Magnaplate Corp, p. II-1425 [SIC 3479]
General Magnetic Co, p. I-1109 [SIC 3264]
General Marble Co, p. I-542 [SIC 2511]
General Marine Industries LP, p. II-1992 [SIC 3732]
General Mechatronics Corp, p. II-1970 [SIC 3721]
General Media International Ltd, p. I-698 [SIC 2721]
General Metal Heat Treating Inc, p. II-1290 [SIC 3398]
General Metal Products, p. II-1413 [SIC 3469]
General Metalcraft Inc, p. I-576 [SIC 2522]
General Metalworks Corp, p. II-1854 [SIC 3648]
General Mills Inc, p. I-75 [SIC 2043]
General Monitors Inc, p. II-2076 [SIC 3829]
General Motors Corp, p. II-1943 [SIC 3711]
General Office Manufacturing, p. I-587 [SIC 2541]
General Packaging Corp, p. I-991 [SIC 3086]
General Parametrics Corp, p. II-1699 [SIC 3575]
General Partitions Mfg Corp, p. I-593 [SIC 2542]
General Pencil Co, p. II-2160 [SIC 3952]
General Plasma Inc, p. II-1290 [SIC 3398]
General Plastics Corp, p. I-593 [SIC 2542]
General Plastics Mfg Co, p. I-991 [SIC 3086]
General Plating Inc, p. II-1420 [SIC 3471]
General Plug & Mfg Co, p. II-1212 [SIC 3325]
General Porcelain Mfg Co, p. I-1113 [SIC 3269]
General Processor Inc, p. I-228 [SIC 2141]
General Products Company Inc, p. II-1349 [SIC 3442]
General Railway Signal Corp, p. II-1873 [SIC 3663]
General Refractories Co, p. I-1090 [SIC 3255]
General Ribbon Corp, p. II-2169 [SIC 3955]
General Safety Equipment Inc, p. II-1943 [SIC 3711]
General Scientific Corp, p. II-2071 [SIC 3827]
General Shale Products Corp, p. I-1090 [SIC 3255]
General Ship Corp, p. II-1986 [SIC 3731]
General Shoe Lace Company Inc, p. I-245 [SIC 2241]
General Signal Corp, p. II-1638 [SIC 3561]
General Smelting and Refining, p. II-1229 [SIC 3341]
General Spice Inc, p. I-212 [SIC 2099]
General Sportswear Company, p. I-394 [SIC 2369]
General Spring Inc, p. II-1470 [SIC 3495]
General Switch Co, p. II-1775 [SIC 3613]
General Time Corp, p. II-2118 [SIC 3873]
General Tire Inc, p. I-927 [SIC 3011]
General Tools Manufacturing, p. II-1312 [SIC 3423]
General Trailer Company Inc, p. II-1961 [SIC 3715]
General Tube Co, p. II-1192 [SIC 3317]
General Valve Company Inc, p. II-1447 [SIC 3491]
General Wire Products Inc, p. II-1260 [SIC 3357]
Genesco Inc, p. I-1025 [SIC 3143]
Genesee Brewing Company Inc, p. I-160 [SIC 2082]
Genesee Corp, p. I-160 [SIC 2082]
Genesee Leroy Stone Corp, p. I-1140 [SIC 3281]
Genesis Clothing Consultants, p. I-258 [SIC 2253]
Genesis Systems Group, p. II-1594 [SIC 3548]
Genetics Institute Inc, p. I-814 [SIC 2836]
Genetronics Inc, p. II-2066 [SIC 3826]
Geneva Steel, p. II-1172 [SIC 3312]
GENICOM Corp, p. II-1704 [SIC 3577]
Genie Co, p. II-1780 [SIC 3621]
Genie Industries, p. II-1547 [SIC 3536]
Genlyte Group Inc, pp. Vol II - 1840, 1845 [SICs 3645, 3646]
Genmar Industries Inc, p. II-1992 [SIC 3732]

Roman numerals I and II indicate the volume in which pages appear. Volume I holds SICs 2011 through 3299; Volume II holds SICs 3312 through 3999.

2254

Company Index

Roman numerals I and II indicate the volume in which pages appear. Volume I holds SICs 2011 through 3299; Volume II holds SICs 3312 through 3999.

Golden Guernsey Dairy Coop, p. I-35 [SIC 2026]
Golden Harvest Foods Inc, p. I-40 [SIC 2032]
Golden Harvest Products Inc, p. I-127 [SIC 2064]
Golden Manufacturing Co, p. I-356 [SIC 2329]
Golden Needles, p. I-401 [SIC 2381]
Golden Poultry Company Inc, p. I-12 [SIC 2015]
Golden Ribbon Corp, p. II-2169 [SIC 3955]
Golden Ruling and Binding Co, p. I-750 [SIC 2789]
Golden State Foods Corp, p. I-101 [SIC 2051]
Golden Stream Quality Foods, p. I-127 [SIC 2064]
Golden Technologies Company, p. I-95 [SIC 2048]
Golden Valley Microwave Foods, p. I-212 [SIC 2099]
Golden Walnut Specialty Foods, p. I-107 [SIC 2052]
Goldens Foundry & Machine, p. II-1197 [SIC 3321]
Goldman Arts Inc, p. I-567 [SIC 2519]
Golten's NY Corp, p. II-1986 [SIC 3731]
Gomaco Corp, p. II-1519 [SIC 3531]
Gomoljak Block, p. I-1117 [SIC 3271]
Gonnella Baking Co, p. I-101 [SIC 2051]
Gooch Foods Inc, p. I-208 [SIC 2098]
Good-All Electric Inc, p. II-1795 [SIC 3629]
Good Humor Ice Cream, p. I-30 [SIC 2024]
Good Lad Co, p. I-394 [SIC 2369]
Good Life Feed Additives, p. I-95 [SIC 2048]
Good Tables Inc, p. I-542 [SIC 2511]
Goodhart Sons Inc, p. II-1355 [SIC 3443]
Goodman Equipment Corp, p. II-1525 [SIC 3532]
Goodman Jewelers Inc, p. II-2122 [SIC 3911]
Goodman Knitting Company Inc, p. I-356 [SIC 2329]
Goodmark Foods Inc, p. I-2 [SIC 2011]
Goodmark Inc, p. I-7 [SIC 2013]
Goodyear Tire, p. I-936 [SIC 3052]
Goodyear Tire and Rubber Co, p. I-927 [SIC 3011]
Gooseneck Trailer Mfg, p. II-1948 [SIC 3713]
Gopher Motor Rebuilding, p. II-1503 [SIC 3519]
Gopher Pattern Works Inc, p. II-1568 [SIC 3543]
Gorant Candies Inc, p. I-127 [SIC 2064]
Gordon Brush Manufacturing, p. II-2181 [SIC 3991]
Gordon Bryans County Farm, p. I-2 [SIC 2011]
Gordon Paper Company Inc, p. I-684 [SIC 2678]
Gordon Publications Inc, p. I-698 [SIC 2721]
Gordon Rubber and Packing, p. I-948 [SIC 3061]
Goria Enterprises, p. I-1122 [SIC 3272]
Gorilla Rack, p. I-593 [SIC 2542]
Gorman-Rupp Co, p. II-1638 [SIC 3561]
Gorman-Rupp Industries, p. II-1638 [SIC 3561]
Gorton's Seafood, p. I-190 [SIC 2092]
Goshen Dairy Inc, p. I-35 [SIC 2026]
Goshen Rubber Company Inc, p. I-942 [SIC 3053]
Goss Inc, p. II-1594 [SIC 3548]
Gotcha International LP, p. I-356 [SIC 2329]
Gotham Apparel Corp, p. I-258 [SIC 2253]
Gotham Ink and Color Company, p. I-889 [SIC 2893]
Gougler Industries Inc, p. II-1271 [SIC 3364]
Gould and Goodrich Leather Inc, p. I-1053 [SIC 3199]
Gould Electronics Inc, p. II-1482 [SIC 3497]
Gould Inc, p. II-1830 [SIC 3643]
Gould-Mersereau Company Inc, p. I-599 [SIC 2591]
Goulds Pumps Inc, p. II-1638 [SIC 3561]
Goulds Pumps Vertical Products, p. II-1638 [SIC 3561]
Gourmet Concepts International, p. I-112 [SIC 2053]
Gourmet Goodies Inc, p. I-107 [SIC 2052]
Gourmet's Fresh Pasta Inc, p. I-208 [SIC 2098]
Governair Corp, p. II-1726 [SIC 3585]
Government Micro Resources, p. II-1688 [SIC 3571]
Gow-Mac Instrument Co, p. II-2066 [SIC 3826]
Gowanda Electronics Corp, p. II-1905 [SIC 3677]
Gowans-Knight Company Inc, p. II-1943 [SIC 3711]
Gower Corp, p. I-593 [SIC 2542]
GP Putnam's Sons, p. I-703 [SIC 2731]
Grace Engineering Corp, p. I-759 [SIC 2796]
Grace-Lee Products Inc, p. I-819 [SIC 2841]
Graceland Fruit Cooperative, p. I-51 [SIC 2034]
Graco Fertilizer Co, p. I-869 [SIC 2875]
Graco Inc, p. II-1732 [SIC 3586]
Graco-LTI, p. II-1632 [SIC 3559]
Graco Robotics Inc, p. II-1649 [SIC 3563]
Gradall Co, p. II-1552 [SIC 3537]
Graduate Supply House Inc, p. I-420 [SIC 2389]

Grady Brothers Inc, p. I-909 [SIC 2951]
Grady Garment, p. I-373 [SIC 2339]
Grady White Boats Inc, p. II-1992 [SIC 3732]
Graetz Manufacturing Inc, p. II-1676 [SIC 3568]
Graham Architectural, p. II-1349 [SIC 3442]
Graham Corp, p. II-1649 [SIC 3563]
Graham-Field Health Products, p. II-2082 [SIC 3841]
Graham-Field Inc, p. II-2087 [SIC 3842]
Graham Magnetics Inc, p. II-1934 [SIC 3695]
Graham Manufacturing, p. II-1649 [SIC 3563]
Graham Steel Corp, p. II-1344 [SIC 3441]
Graham-White Mfg Co, p. II-1998 [SIC 3743]
Grain Millers of Iowa Inc, p. I-75 [SIC 2043]
Grain Processing Corp, p. I-172 [SIC 2085]
Grain Systems Inc, p. II-1509 [SIC 3523]
Granco-Clark Inc, p. II-1671 [SIC 3567]
Grand Blanc Cement Products, p. I-1117 [SIC 3271]
Grand Haven Brass Foundry, p. II-1280 [SIC 3366]
Grand Haven Stamped Prod Co, p. II-1404 [SIC 3465]
Grand Isle Shipyard Inc, p. II-1344 [SIC 3441]
Grand Metropolitan Inc, p. I-101 [SIC 2051]
Grand Rapids Alloys Inc, p. II-1229 [SIC 3341]
Grand Rapids Die Casting, p. II-1285 [SIC 3369]
Grand Rapids Gravel Co, p. I-1127 [SIC 3273]
Grand Rapids Spring & Wire, p. II-1470 [SIC 3495]
Grand Transformers Inc, p. II-1769 [SIC 3612]
Grandma Brown's Beans Inc, p. I-40 [SIC 2032]
Grandoe Corp, p. II-2150 [SIC 3949]
Grandview Products Co, p. I-486 [SIC 2434]
Grange Cooperative, p. I-95 [SIC 2048]
Granit Bronz CSG Inc, p. I-1140 [SIC 3281]
Granite Communications Inc, p. II-1699 [SIC 3575]
Granite Knitwear Inc, p. I-258 [SIC 2253]
Granite State Packing Company, p. I-2 [SIC 2011]
Graniteville Co, p. I-231 [SIC 2211]
Granitize Products Inc, p. I-825 [SIC 2842]
Grannas Brothers, p. I-909 [SIC 2951]
Granse and Associates Inc, p. I-1140 [SIC 3281]
Grant Assembly Technologies, p. II-1563 [SIC 3542]
Grant Chemical, p. I-855 [SIC 2869]
Grant Gear Inc, p. II-1666 [SIC 3566]
Grant's Dairy Inc, p. I-30 [SIC 2024]
Grant TFW Inc, p. I-948 [SIC 3061]
Grape Links Wine Productions, p. I-168 [SIC 2084]
Graphic Color Plate Inc, p. I-759 [SIC 2796]
Graphic Color Systems Inc, p. I-759 [SIC 2796]
Graphic Controls Corp, p. II-2112 [SIC 3861]
Graphic Converting Inc, p. I-669 [SIC 2675]
Graphic Enterprises of Ohio Inc, p. II-2112 [SIC 3861]
Graphic Finishers Inc, p. I-750 [SIC 2789]
Graphic Forms and Labels Inc, p. I-735 [SIC 2761]
Graphic Industries Inc, p. I-729 [SIC 2759]
Graphic Innovators Inc, p. I-708 [SIC 2732]
Graphic Looseleaf Products, p. I-745 [SIC 2782]
Graphic Media Corp, p. I-724 [SIC 2754]
Graphic Packaging Corp, p. I-660 [SIC 2673]
Graphic Prints Inc, p. I-279 [SIC 2261]
Graphic Techn Inc, p. I-718 [SIC 2752]
Graphics Atlanta Inc, p. I-759 [SIC 2796]
Graphics LX Corp, p. II-1621 [SIC 3555]
Graphics Technology Company, p. II-1699 [SIC 3575]
Graphite Metallizing Corp, p. II-1786 [SIC 3624]
Graphite Sales Inc, p. II-1786 [SIC 3624]
Graphite Systems Inc, p. II-1786 [SIC 3624]
GraphOn Corp, p. II-1699 [SIC 3575]
Graseby Controls Inc, p. II-1676 [SIC 3568]
Grass Roots Publishing Co, p. I-698 [SIC 2721]
Grass Valley Group Inc, p. II-1873 [SIC 3663]
Grasshopper Co, p. II-1514 [SIC 3524]
Grassland Dairy Products Inc, p. I-17 [SIC 2021]
Graver Co, p. II-1736 [SIC 3589]
Graver Tank and Manufacturing, p. II-1355 [SIC 3443]
Gray and Prior Machine Co, p. II-1676 [SIC 3568]
Gray Envelope Co, p. I-679 [SIC 2677]
Gray Goods, p. I-231 [SIC 2211]
Gray Syracuse Inc, p. II-1207 [SIC 3324]
Grayline Housewares, p. II-1182 [SIC 3315]
Graymills Corp, p. II-1638 [SIC 3561]
Grays Harbor Veneer, p. I-491 [SIC 2435]

Grayson Controls, p. II-2045 [SIC 3822]
Great, p. I-542 [SIC 2511]
Great Amer Wirebound Box Co, p. I-515 [SIC 2449]
Great American Knitting Mills, p. I-250 [SIC 2251]
Great American Management, p. II-1476 [SIC 3496]
Great American Packaging Inc, p. I-660 [SIC 2673]
Great Cover Up, p. I-420 [SIC 2389]
Great Dane Holdings Inc, p. II-1961 [SIC 3715]
Great Eastern Industries Inc, p. I-660 [SIC 2673]
Great Lakes Business Forms Inc, p. I-735 [SIC 2761]
Great Lakes Carbon Corp, p. I-924 [SIC 2999]
Great Lakes Castings Corp, p. II-1197 [SIC 3321]
Great Lakes Chemical Corp, p. I-777 [SIC 2819]
Great Lakes Instruments Inc, p. II-2051 [SIC 3823]
Great Lakes Paper Co, p. I-310 [SIC 2295]
Great Lakes Tissue Co, p. I-674 [SIC 2676]
Great Neck Saw Manufacturing, p. II-1318 [SIC 3425]
Great Northern Mfg Corp, p. II-1182 [SIC 3315]
Great Northern Paper Inc, p. I-612 [SIC 2621]
Great Pacific Iron Works, p. I-356 [SIC 2329]
Great Plains Manufacturing Inc, p. II-1509 [SIC 3523]
Great Southern Wood Preserving, p. I-528 [SIC 2491]
Great Valley Products Inc, p. II-1694 [SIC 3572]
Great Western Airgas Inc, pp. Vol. I - 768, Vol. II - 1594 [SICs 2813, 3548]
Great Western Carpet Cushion, p. I-991 [SIC 3086]
Great Western Directories Inc, p. I-713 [SIC 2741]
Great Western Malting Co, p. I-165 [SIC 2083]
Great Western Publishing Inc, p. I-729 [SIC 2759]
Great Western Tortilla Co, p. I-199 [SIC 2096]
Greater New York Box Co, p. I-628 [SIC 2653]
Greater Omaha Packing Co, p. I-2 [SIC 2011]
Grede Foundries Inc, p. II-1197 [SIC 3321]
Grede Foundry Inc, p. II-1197 [SIC 3321]
Grede Perm Cast Inc, p. II-1197 [SIC 3321]
Grede-Vassar Foundry, p. II-1197 [SIC 3321]
Green Apparel Inc, p. I-340 [SIC 2322]
Green Ball Bearing Co, p. II-1644 [SIC 3562]
Green Bay Dressed Beef Inc, p. I-2 [SIC 2011]
Green Bay Packaging Inc, pp. Vol. I - 618, Vol. II - 1615 [SICs 2631, 3554]
Green Garden Inc, p. II-1334 [SIC 3432]
Green Industries Corp, p. II-1420 [SIC 3471]
Green Manufacturing Company, p. II-1920 [SIC 3691]
Green Manufacturing Inc, p. II-1747 [SIC 3593]
Green Mountain Products Inc, p. II-1318 [SIC 3425]
Green Spring Dairy Inc, p. I-35 [SIC 2026]
Green Technologies Inc, p. II-1830 [SIC 3643]
Green Tree Packing Co, p. I-7 [SIC 2013]
Greenbrier Companies Inc, p. II-1998 [SIC 3743]
Greendale Bicycle Co, p. II-2003 [SIC 3751]
Greene Line Mfg Corp, p. II-1660 [SIC 3565]
Greene, Tweed and Co, p. I-942 [SIC 3053]
Greenfield Industries Inc, p. II-1578 [SIC 3545]
Greenleaf Corp, p. II-1578 [SIC 3545]
Greensburg Concrete Block Co, p. I-1117 [SIC 3271]
Greensteel, p. I-581 [SIC 2531]
GreenStone Industries Inc, p. I-688 [SIC 2679]
Greenville Machinery Corp, p. II-1605 [SIC 3552]
Greenville Metals Inc, p. II-1229 [SIC 3341]
Greenwich Air Services Inc, p. II-1975 [SIC 3724]
Greenwich Company Ltd, p. II-1975 [SIC 3724]
Greenwich Mills, p. I-195 [SIC 2095]
Greenwich Workshop Inc, p. I-713 [SIC 2741]
Greenwood Fixture, p. I-587 [SIC 2541]
Greenwood Mills Inc, p. I-231 [SIC 2211]
Greenwood Mop and Broom Inc, p. I-428 [SIC 2392]
Greer Industries Inc, p. I-1187 [SIC 3316]
Greer Limestone Co, p. I-909 [SIC 2951]
Gregg Industries Inc, p. II-1197 [SIC 3321]
Gregg's Foods, p. I-156 [SIC 2079]
Gregory Galvanizing, p. II-1425 [SIC 3479]
Gregory Group Inc, p. II-1552 [SIC 3537]
Gregson Furniture Industries, p. I-571 [SIC 2521]
Greif Bros Corp, p. I-634 [SIC 2655]
Grenecker Wolf and Vine Inc, p. II-2199 [SIC 3999]
Greystone Concrete Products, p. I-1117 [SIC 3271]
Greystone Inc, p. II-1382 [SIC 3451]
Grieve Corp, p. II-1671 [SIC 3567]

Roman numerals I and II indicate the volume in which pages appear. Volume I holds SICs 2011 through 3299; Volume II holds SICs 3312 through 3999.

2256

Griffin Corp, p. I-874 [SIC 2879]

Griffin Entities Inc, p. I-119 [SIC 2062]

Griffin Envelope Inc, p. I-679 [SIC 2677]

Griffin Food Co, p. I-119 [SIC 2062]

Griffin Industries Inc, p. I-152 [SIC 2077]

Griffin Lamp Co, p. II-1850 [SIC 3647]

Griffin Pipe Products Co, p. II-1197 [SIC 3321]

Griffin Printing & Lithograph, p. I-708 [SIC 2732]

Griffin Technology Inc, p. II-1938 [SIC 3699]

Griffin Wheel, p. II-1395 [SIC 3462]

Griffith Laboratories Inc, p. I-212 [SIC 2099]

Griffiths Corp, p. II-1413 [SIC 3469]

Grimes Aerospace Co, p. II-1850 [SIC 3647]

Grimmer-Schmidt Corp, p. II-1649 [SIC 3563]

Grindmaster Corp, p. II-1626 [SIC 3556]

Grinnell Corp, p. II-1682 [SIC 3569]

Grinnell Systems Company Inc, p. II-1878 [SIC 3669]

Gripco Fastener, p. II-1388 [SIC 3452]

Grist Mill Co, pp. Vol. I - 75, 95, 127 [SICs 2043, 2048, 2064]

Grob Inc, p. II-1563 [SIC 3542]

Grobet File Company of America, p. II-1312 [SIC 3423]

Grocers Baking Co, p. I-101 [SIC 2051]

Groendyk Manufacturing, p. I-787 [SIC 2822]

Groovfold Inc, p. I-532 [SIC 2493]

Grosfillex Inc, p. I-567 [SIC 2519]

Gross-Given Manufacturing, p. II-1718 [SIC 3581]

Gross-Medick-Barrows Inc, p. I-669 [SIC 2675]

Grossman Steel, p. II-1367 [SIC 3446]

Groth Corp, p. II-1464 [SIC 3494]

Grotnes Metalforming Systems, p. II-1563 [SIC 3542]

Ground Control, p. I-446 [SIC 2396]

Group Dekko International Inc, p. II-1830 [SIC 3643]

Group Technologies Corp, p. II-1892 [SIC 3674]

Grove North America, p. II-1519 [SIC 3531]

Grove Valve and Regulator Co, p. II-1464 [SIC 3494]

Grover Industries Inc, p. I-297 [SIC 2281]

Grow Group Inc, pp. Vol. I - 840, 879 [SICs 2851, 2891]

Growers Container Cooperative, p. I-515 [SIC 2449]

Growers Fertilizer Corp, p. I-861 [SIC 2873]

Growing Healthy Inc, p. I-40 [SIC 2032]

Gruen Marketing Corp, p. II-2118 [SIC 3873]

Grumman Allied, p. II-1948 [SIC 3713]

Grumman Olson, p. II-1948 [SIC 3713]

GS Electric, p. II-1780 [SIC 3621]

GS Fibers, p. I-327 [SIC 2299]

GS Metals Corp, p. II-1367 [SIC 3446]

GS Roofing Products Co, p. I-914 [SIC 2952]

GS Technologies Inc, p. I-1144 [SIC 3291]

GST Steel Co, p. II-1476 [SIC 3496]

GT Industries of Texas, p. I-936 [SIC 3052]

GT Sales and Manufacturing Inc, p. I-936 [SIC 3052]

GTE Directories Corp, p. I-713 [SIC 2741]

GTE Government Systems Corp, p. II-1873 [SIC 3663]

GTE Metal Erectors Inc, p. II-1377 [SIC 3449]

GTI Corp, p. II-1234 [SIC 3351]

Guarantee Specialties Inc, p. II-1764 [SIC 3599]

Guard-Line Inc, p. I-1037 [SIC 3151]

Guardian Chemical Co, p. I-825 [SIC 2842]

Guardian Electric Mfg Co, p. II-1790 [SIC 3625]

Guardian Industries Corp, p. I-1071 [SIC 3231]

Guardian Metal Sales Inc, p. II-1225 [SIC 3339]

Guardian Products Company Inc, p. II-1234 [SIC 3351]

Guardsman Products Inc, pp. Vol. I - 310, 840 [SICs 2295, 2851]

Gudebrod Inc, p. I-245 [SIC 2241]

Guerdon Homes Inc, p. I-523 [SIC 2452]

Guerlain Inc, p. I-834 [SIC 2844]

Guernsey Dell Inc, p. I-181 [SIC 2087]

Guest Company Inc, p. II-1938 [SIC 3699]

Guest Supply Inc, p. I-834 [SIC 2844]

Guhring Inc, p. II-1578 [SIC 3545]

Guida-Seibert Dairy Co, p. I-26 [SIC 2023]

Guidant Corp, p. II-2102 [SIC 3845]

Guilford Corp, p. I-587 [SIC 2541]

Guilford Mills, p. I-267 [SIC 2257]

Guilford Mills Inc, p. I-271 [SIC 2258]

Guilford Mills Inc Automotive, p. I-455 [SIC 2399]

Guilford of Maine Inc, p. I-241 [SIC 2231]

Guiltless Gourmet Inc, p. I-199 [SIC 2096]

Guittard Chocolate Corp, p. I-132 [SIC 2066]

Gulbransen Inc, p. II-2135 [SIC 3931]

Gulf Aluminum Products Inc, p. II-1367 [SIC 3446]

Gulf Asphalt Corp, p. I-909 [SIC 2951]

Gulf Chemical, p. II-1229 [SIC 3341]

Gulf Coast Fertilizer Inc, p. I-865 [SIC 2874]

Gulf Coast Pre-Stress, p. I-1122 [SIC 3272]

Gulf Coast Pre-Stress Inc, p. I-1122 [SIC 3272]

Gulf Forge Co, p. II-1395 [SIC 3462]

Gulf Marine Repair Corp, p. II-1986 [SIC 3731]

Gulf Met Holdings Corp, p. II-1229 [SIC 3341]

Gulf Reduction, p. II-1285 [SIC 3369]

Gulf State Abrasive Mfg, p. I-1144 [SIC 3291]

Gulf State Airgas Inc, p. I-768 [SIC 2813]

Gulf States Asphalt Company, p. I-914 [SIC 2952]

Gulf States Canners Inc, p. I-176 [SIC 2086]

Gulf States Manufacturers Inc, p. II-1372 [SIC 3448]

Gulf States Paper Corp, p. I-618 [SIC 2631]

Gulf States Steel Incorporated, p. II-1172 [SIC 3312]

Gulf States Tube, p. II-1192 [SIC 3317]

Gull Laboratories Inc, p. I-809 [SIC 2835]

Gunderson Inc, p. II-1998 [SIC 3743]

Gundlach-Bundschu Winery, p. I-168 [SIC 2084]

Gundle Environmental Systems, p. I-961 [SIC 3081]

Gunlocke Co, p. I-571 [SIC 2521]

Gunnebo Fastening Corp, p. II-1395 [SIC 3462]

Gunnison Brothers Inc, p. I-1013 [SIC 3111]

Gunter and Cooke, p. II-1605 [SIC 3552]

Gunton Corp, p. I-480 [SIC 2431]

Gusher Pumps Inc, p. II-1638 [SIC 3561]

Gussco Manufacturing Inc, p. I-669 [SIC 2675]

Gustafsons Dairy Inc, p. I-35 [SIC 2026]

Guth Lighting, p. II-1845 [SIC 3646]

Guthrie Corp, p. I-869 [SIC 2875]

Guthrie Trailer Sales, p. II-1961 [SIC 3715]

Gutmann Leather Company Inc, p. I-1013 [SIC 3111]

Guy Gannett Publishing Co, p. I-693 [SIC 2711]

Guy's Foods Inc, p. I-199 [SIC 2096]

Guyco Corp, p. II-1854 [SIC 3648]

Guzzler Manufacturing Inc, p. II-1943 [SIC 3711]

GVC Technologies Inc, p. II-1868 [SIC 3661]

GVK America Inc, p. I-532 [SIC 2493]

GW Lisk Company Inc, p. II-1915 [SIC 3679]

Gwaltney of Smithfield Ltd, p. I-7 [SIC 2013]

GWI Engineering Inc, p. II-1600 [SIC 3549]

Gyp-Crete Corp, p. I-1136 [SIC 3275]

H and H Machine Tool of Iowa, p. II-1578 [SIC 3545]

H and H Manufacturing, p. II-1498 [SIC 3511]

H and H Products Inc, p. II-1382 [SIC 3451]

H and H Tooling, p. II-1312 [SIC 3423]

H & H Tube & Mfg Co, p. II-1234 [SIC 3351]

H and L Enterprises, p. I-688 [SIC 2679]

H and L Products Inc, p. I-446 [SIC 2396]

H and L Tool Company Inc, p. II-1382 [SIC 3451]

H and M Food Systems Co, p. I-65 [SIC 2038]

H and R 1871 Inc, p. II-1438 [SIC 3484]

H and R Electronics Co, p. II-1718 [SIC 3581]

H and S Bakery Inc, p. I-101 [SIC 2051]

H and W Shoe Supplies Co, p. I-283 [SIC 2262]

H Cross Co, p. II-1255 [SIC 3356]

H Kramer and Co, p. II-1229 [SIC 3341]

H Maimin Company Inc, p. II-1605 [SIC 3552]

H Meyer Dairy Co, p. I-35 [SIC 2026]

H Miller Spring & Mfg Co, p. II-1460 [SIC 3493]

H Nagel and Son Co, p. I-70 [SIC 2041]

H-P Products Inc, p. II-1486 [SIC 3498]

H-R Industries Inc, p. II-1887 [SIC 3672]

H Warshow and Sons Inc, pp. Vol. I - 236, 275 [SICs 2221, 2259]

HA Davidson Box Co, p. I-515 [SIC 2449]

Haag Laboratories Inc, p. I-819 [SIC 2841]

Haagen-Dazs Company Inc, p. I-30 [SIC 2024]

Haarmann and Reimer Corp, p. I-855 [SIC 2869]

Haartz Corp, p. I-241 [SIC 2231]

Haas Tailoring Co, p. I-332 [SIC 2311]

HAB Industries Inc, p. I-241 [SIC 2231]

HABCO Steel Service Inc, p. II-1486 [SIC 3498]

Habitec Security, p. II-1878 [SIC 3669]

Hach Co, p. II-2066 [SIC 3826]

Hackett Brass Foundry, p. II-1280 [SIC 3366]

Hackney, p. II-1355 [SIC 3443]

Hackney and Sons Inc, p. II-1948 [SIC 3713]

Hackney Brothers Inc, p. II-1948 [SIC 3713]

Hadco Corp, p. II-1887 [SIC 3672]

Haden Schweitzer Corp, p. II-1632 [SIC 3559]

Hadley Cos, p. I-713 [SIC 2741]

Hadley Gear Manufacturing Co, p. II-1666 [SIC 3566]

Haeger Industries Inc, p. I-1113 [SIC 3269]

Haeger Potteries, p. I-1113 [SIC 3269]

Haeger Potteries of Macomb, p. I-1113 [SIC 3269]

Haemonetics Corp, p. II-2082 [SIC 3841]

Hagale Industries Inc, p. I-348 [SIC 2325]

Hagale Manufacturing Co, p. I-348 [SIC 2325]

Haggar Apparel Co, p. I-348 [SIC 2325]

Haggar Corp, p. I-348 [SIC 2325]

Haggerty Enterprises Inc, p. II-1854 [SIC 3648]

Hagie Manufacturing Co, p. II-1509 [SIC 3523]

Hago Manufacturing Company, p. II-1334 [SIC 3432]

Hahn Machinery Inc, p. II-1610 [SIC 3553]

Hako Minuteman Inc, p. II-1736 [SIC 3589]

Hal Leonard Corp, p. I-713 [SIC 2741]

Hal Reed Co, p. I-1066 [SIC 3229]

Halaco Engineering Co, p. II-1229 [SIC 3341]

Hale Products Inc, p. II-1638 [SIC 3561]

Hale's Ales Ltd, p. I-160 [SIC 2082]

Halethorpe Extrusions Inc, p. II-1244 [SIC 3354]

Halex Co, p. II-1271 [SIC 3364]

Haley Brothers Inc, p. I-480 [SIC 2431]

Halifax Paper Board Company, p. I-618 [SIC 2631]

Halifax Paving Inc, p. I-909 [SIC 2951]

Halkey Roberts Corp, p. II-1447 [SIC 3491]

Hall Chemical Co, p. I-855 [SIC 2869]

Hall China Co, p. I-1102 [SIC 3262]

Hall Industries Inc, p. II-1382 [SIC 3451]

Hall Laboratories Inc, p. I-804 [SIC 2834]

Hallagan Manufacturing, p. I-548 [SIC 2512]

Hallcrest Inc, p. II-2076 [SIC 3829]

Hallmark Industries Inc, p. I-593 [SIC 2542]

Hallmark Marketing Inc, p. I-587 [SIC 2541]

Hallmark Precious Metals Inc, p. II-1229 [SIC 3341]

Halltown Paperboard Co, p. I-618 [SIC 2631]

Halmar Robicon Group, p. II-1790 [SIC 3625]

Halo Optical Corp, p. II-2107 [SIC 3851]

Halsted Corp, p. I-433 [SIC 2393]

Halves/Coppersource, p. II-1280 [SIC 3366]

Hamakua Sugar Company Inc, p. I-116 [SIC 2061]

Hamburg Broom Works Inc, p. II-2181 [SIC 3991]

Hamden Metal Service Co, p. II-1182 [SIC 3315]

Hamilton Beach/Proctor-Silex, p. II-1813 [SIC 3634]

Hamilton Die Cast Inc, p. II-1266 [SIC 3363]

Hamilton Foundry & Machine, p. II-1197 [SIC 3321]

Hamilton Glass Products Inc, p. I-1057 [SIC 3211]

Hamilton Materials Inc, p. I-1136 [SIC 3275]

Hamilton Precision Metals Inc, p. II-1255 [SIC 3356]

Hamilton Sorter Company Inc, p. I-576 [SIC 2522]

Hamilton Standard Electr Syst, p. II-2045 [SIC 3822]

Hamilton-Stevens Group Inc, p. II-1621 [SIC 3555]

Hamlin Inc, p. II-1790 [SIC 3625]

Hammary Furniture Company, p. I-542 [SIC 2511]

HammerBlow Corp, p. II-2029 [SIC 3799]

Hammerman Brothers Inc, p. II-2122 [SIC 3911]

Hammill Manufacturing Co, p. II-1572 [SIC 3544]

Hammond and Irving Inc, p. II-1395 [SIC 3462]

Hammond Inc, p. I-713 [SIC 2741]

Hammond Lumber Co, p. I-465 [SIC 2421]

Hampco Apparel Inc, p. I-405 [SIC 2384]

Hampden Papers Inc, p. II-1482 [SIC 3497]

Hampshire Hosiery Inc, p. I-250 [SIC 2251]

Hampton Industries Inc, p. I-336 [SIC 2321]

Hamrick Mills Inc, p. I-231 [SIC 2211]

Hancock Manufacturing, p. II-1492 [SIC 3499]

Hancock Wood Engineering Inc, p. II-2164 [SIC 3953]

Hancor Co, p. I-1001 [SIC 3088]

Roman numerals I and II indicate the volume in which pages appear. Volume I holds SICs 2011 through 3299; Volume II holds SICs 3312 through 3999.

2257

Company Index

Handcraft Company Inc, p. I-250 [SIC 2251]
Handcraft Tile Inc, p. I-1086 [SIC 3253]
Handi Craft Co, p. II-2144 [SIC 3944]
Handicaps Inc, p. II-1536 [SIC 3534]
Handy & Harman, p. I-936 [SIC 3052]
Handy and Harman, p. II-1225 [SIC 3339]
Handy & Harman, p. II-1420 [SIC 3471]
Handy Button Machine Co, p. II-1323 [SIC 3429]
Handy Wacks Corp, p. I-654 [SIC 2672]
Hanel Lumber Company Inc, p. I-465 [SIC 2421]
Hanford Brick Company Inc, p. I-1081 [SIC 3251]
Hangsterfers Laboratories Inc, p. I-918 [SIC 2992]
Hankins Lumber Co, p. I-465 [SIC 2421]
Hankison, p. II-1682 [SIC 3569]
Hankscraft Motors, p. II-1813 [SIC 3634]
Hanna Car Wash International, p. II-1736 [SIC 3589]
Hanna Corp, p. II-1747 [SIC 3593]
Hannaco Knives and Saws, p. II-1312 [SIC 3423]
Hannibal Carbide Tool Inc, p. II-1578 [SIC 3545]
Hannibal Industries, p. II-1187 [SIC 3316]
Hannon Hydraulics Inc, p. II-1747 [SIC 3593]
Hanover Wire Cloth, p. II-1476 [SIC 3496]
Hanovia Inc, p. II-2102 [SIC 3845]
Hans C Egloff Inc, p. I-603 [SIC 2599]
Hans Sumpf Company Inc, p. I-1094 [SIC 3259]
Hansel 'N Gretel Brand Inc, p. I-7 [SIC 2013]
Hansen Beverage Co, p. I-176 [SIC 2086]
Hansen Corp, p. II-1780 [SIC 3621]
Hansen Coupling, p. II-1464 [SIC 3494]
Hansen Natural Corp, p. I-176 [SIC 2086]
Hansford Manufacturing Corp, p. II-1632 [SIC 3559]
Hansman Industries Inc, p. II-1492 [SIC 3499]
Hanson General Products, p. II-1514 [SIC 3524]
Hanson Graphics of Memphis, p. I-759 [SIC 2796]
Happ Controls Inc, p. II-1790 [SIC 3625]
Happiness Express Inc, p. II-2140 [SIC 3942]
Happy Refrigerated Services, p. I-204 [SIC 2097]
Happy's Potato Chip Co, p. I-199 [SIC 2096]
Happyknit Inc, p. I-258 [SIC 2253]
Harbison-Fischer Manufacturing, p. II-1531 [SIC 3533]
Harbison-Walker, p. I-1090 [SIC 3255]
Harbor Electronics Inc, p. II-1260 [SIC 3357]
Harbor Industries Inc, p. I-587 [SIC 2541]
Harbor Pallet Co, p. I-510 [SIC 2448]
Harbour Group Ltd, p. II-1578 [SIC 3545]
Harco Graphic Products Inc, p. II-1605 [SIC 3552]
Harco Industries Inc, p. II-1709 [SIC 3578]
Harco Laboratories Inc, p. II-1498 [SIC 3511]
Harcourt Brace and Co, p. I-703 [SIC 2731]
Harcros Chemicals Inc, pp. Vol. I - 830, 879 [SICs 2843, 2891]
HARD Manufacturing Company, p. I-603 [SIC 2599]
Harden Furniture Company Inc, p. I-542 [SIC 2511]
Harden Industries Inc, p. II-1334 [SIC 3432]
Hardie-Tynes Manufacturing Co, p. II-1649 [SIC 3563]
Harding Co, p. II-1547 [SIC 3536]
Hardinge Inc, p. II-1558 [SIC 3541]
Hardrives Inc, p. I-909 [SIC 2951]
Hardware Products Co, p. II-1470 [SIC 3495]
Hardway Concrete Co, p. I-1127 [SIC 3273]
Hardwick Clothes Inc, p. I-332 [SIC 2311]
Hardwicke Chemical Co, p. I-855 [SIC 2869]
Hardwood Dimensions Inc, p. I-471 [SIC 2426]
Hardy Instruments Inc, p. II-1795 [SIC 3629]
Hargro Associates, p. I-665 [SIC 2674]
Hargro Packaging Corp, p. I-649 [SIC 2671]
Harkham Industries Inc, p. I-365 [SIC 2335]
Harlan Corp, p. II-1552 [SIC 3537]
Harlequin Designs Inc, p. I-258 [SIC 2253]
Harley-Davidson Inc, p. II-2003 [SIC 3751]
Harlo Products Corp, p. II-1552 [SIC 3537]
Harloc Inc, p. II-1323 [SIC 3429]
Harmal Industries Inc, p. I-416 [SIC 2387]
Harman Automotive Inc, p. II-1954 [SIC 3714]
Harman International Industries, p. II-1859 [SIC 3651]
Harmeson Manufacturing, p. II-2186 [SIC 3993]
Harmon Industries Inc, p. II-1878 [SIC 3669]
Harmon Publishing Co, p. I-698 [SIC 2721]
Harmon Wood Company Inc, p. I-460 [SIC 2411]

Harmonic Lightwaves Inc, p. I-1066 [SIC 3229]
Harmony Blue Granite Co, p. I-1140 [SIC 3281]
Harnischfeger Corp, p. II-1525 [SIC 3532]
Harnischfeger Industries Inc, p. II-1615 [SIC 3554]
Harold Johnson Optical Labs, p. II-2071 [SIC 3827]
Harper Bros Inc, p. I-909 [SIC 2951]
Harper Corp, p. II-2181 [SIC 3991]
Harper Electric Furnace, p. II-1671 [SIC 3567]
Harper Foundry&Machine Co, p. II-1197 [SIC 3321]
Harper Furniture Co, p. I-542 [SIC 2511]
Harper Industries Inc, p. I-336 [SIC 2321]
Harper Leather Goods Mfg Co, p. I-1049 [SIC 3172]
Harper Rubber Products, p. I-948 [SIC 3061]
Harper-Wyman Co, p. II-2056 [SIC 3824]
HarperCollins Publishers, p. I-703 [SIC 2731]
Harriet and Henderson Yarns, p. I-297 [SIC 2281]
Harrington Tools Inc, p. II-1312 [SIC 3423]
Harris Corp, pp. Vol II - 1868, 1892, 1915 [SICs 3661, 3674, 3679]
Harris Hardware Sales Corp, p. II-1323 [SIC 3429]
Harris Lamps Co, p. II-1840 [SIC 3645]
Harris Mallow and Contempra, p. II-2118 [SIC 3873]
Harris Marcus Group, p. II-1840 [SIC 3645]
Harris Potteries, p. I-1086 [SIC 3253]
Harris Ranch Beef Co, p. I-2 [SIC 2011]
Harris-Tarkett Inc, p. I-471 [SIC 2426]
Harris Thomas Drop Forge Co, p. II-1395 [SIC 3462]
Harrisburg/Woolley, p. II-1531 [SIC 3533]
Harrison Baking Co, p. I-101 [SIC 2051]
Harrison Steel Casting Co, p. II-1212 [SIC 3325]
Harrisons and Crosfield, p. I-777 [SIC 2819]
Harriss & Covington, p. I-250 [SIC 2251]
Harrow Corp, p. II-1361 [SIC 3444]
Harry J Bosworth Co, p. II-2093 [SIC 3843]
Harry London's Candies Inc, p. I-132 [SIC 2066]
Harry Miller Company Inc, p. I-437 [SIC 2394]
Harry N Abrams Inc, p. I-703 [SIC 2731]
Harsco Corp, pp. Vol II - 1339, 1367, 1447 [SICs 3433, 3446, 3491]
Harsh International Inc, p. II-1547 [SIC 3536]
Hart and Cooley Inc, p. II-1361 [SIC 3444]
Hart Bindery Co, p. I-750 [SIC 2789]
Hart Brewing Co, p. I-160 [SIC 2082]
Hart Graphics Inc, p. I-703 [SIC 2731]
Hart Metals Inc, p. II-1295 [SIC 3399]
Hart Tie and Lumber Co, p. I-486 [SIC 2434]
Harte-Hanks Communications, p. I-693 [SIC 2711]
Hartel Corp, p. II-2045 [SIC 3822]
Harter Corp, p. I-576 [SIC 2522]
Hartford Ball Co, p. II-1644 [SIC 3562]
Hartford Bearing Co, p. II-1644 [SIC 3562]
Hartford Brewery Ltd, p. I-160 [SIC 2082]
Hartford Prospect Industries Inc, p. I-1066 [SIC 3229]
Harting Electronik Inc, p. II-1910 [SIC 3678]
Hartley Co, p. II-2156 [SIC 3951]
Hartman Electrical Mfg, p. II-1775 [SIC 3613]
Hartman-Fabco Inc, p. II-1187 [SIC 3316]
Hartman Products Inc, p. II-1813 [SIC 3634]
Hartman USA Inc, p. I-567 [SIC 2519]
Hartmann Luggage Co, p. I-1041 [SIC 3161]
Hartmarx Corp, p. I-332 [SIC 2311]
Hartness International Inc, p. II-1626 [SIC 3556]
Hartstone Inc, p. I-1113 [SIC 3269]
Hartung Agalite Glass Co, p. I-1057 [SIC 3211]
Hartz and Co, p. I-332 [SIC 2311]
Hartz Mountain Corp, p. I-91 [SIC 2047]
Hartzell Fan Inc, p. II-1654 [SIC 3564]
Hartzell Manufacturing Inc, p. II-1266 [SIC 3363]
Hartzell Propeller Inc, p. II-1980 [SIC 3728]
Harvard Folding Box Company, p. I-643 [SIC 2657]
Harvard Industries Inc, p. I-1071 [SIC 3231]
Harvard Interiors Manufacturing, p. I-576 [SIC 2522]
Harvard Sports Inc, p. II-2150 [SIC 3949]
Harve Benard Ltd, p. I-332 [SIC 2311]
Harvest Land Cooperative, p. I-95 [SIC 2048]
Harvey Universal Inc, p. I-819 [SIC 2841]
Harwood Companies Inc, p. I-340 [SIC 2322]
Hasbro Inc, p. II-2144 [SIC 3944]
Hasco Electric Corp, p. II-1845 [SIC 3646]

Hasco Industries Inc, p. II-1460 [SIC 3493]
Haskel International Inc, p. II-1638 [SIC 3561]
Haskell-Dawes Inc, p. II-1605 [SIC 3552]
Haskell of Pittsburgh Inc, p. I-576 [SIC 2522]
Hassell and Hughes Lumber, p. I-471 [SIC 2426]
Hastings Coop Creamery Co, p. I-35 [SIC 2026]
Hastings Fiber Glass Products, p. I-967 [SIC 3082]
Hastings Lighting Company Inc, p. II-1845 [SIC 3646]
Hasty Plywood Co, p. I-491 [SIC 2435]
Hat Brands Inc, p. I-386 [SIC 2353]
Hatch and Kirk Inc, p. II-1503 [SIC 3519]
Hatco Corp, p. II-1736 [SIC 3589]
Hater Industries Inc, p. II-1275 [SIC 3365]
Hatfield Quality Meats Inc, p. I-7 [SIC 2013]
Hatteras Hammocks, p. I-553 [SIC 2514]
Hatteras Yachts, p. II-1986 [SIC 3731]
Hauck Manufacturing Co, p. II-1339 [SIC 3433]
Haulpak, p. II-1519 [SIC 3531]
Hauppauge Computer Works, p. II-1887 [SIC 3672]
Hauser Chemical Research Inc, p. I-855 [SIC 2869]
Hausman Corp, p. II-1344 [SIC 3441]
Hausted Inc, p. I-603 [SIC 2599]
Havatampa Inc, p. I-221 [SIC 2121]
Haven Homes Inc, p. I-523 [SIC 2452]
Haven Steel Products Inc, p. II-1460 [SIC 3493]
Havens Steel Co, p. II-1344 [SIC 3441]
Haverhill Paperboard Corp, p. I-618 [SIC 2631]
Haviland Enterprises Inc, p. II-1736 [SIC 3589]
Hawaii Baking Company Inc, p. I-101 [SIC 2051]
Hawaiian Candies and Nuts Ltd, p. I-127 [SIC 2064]
Hawaiian Flour Mills Inc, p. I-70 [SIC 2041]
Hawker Energy Products Inc, p. II-1920 [SIC 3691]
Hawkeye Steel Products Inc, p. II-1509 [SIC 3523]
Hawkins Chemical Inc, p. I-777 [SIC 2819]
Hawkins Indust Resource Corp, p. I-787 [SIC 2822]
Haworth Inc, p. I-571 [SIC 2521]
Hawthorne Metal Products Co, p. II-1361 [SIC 3444]
Hay and Forage Industries, p. II-1509 [SIC 3523]
Haydon Corp, p. II-1339 [SIC 3433]
Haydon Switch and Instrument, p. II-1830 [SIC 3643]
Hayes Manufacturing Company, p. I-548 [SIC 2512]
Hayes Manufacturing Group Inc, p. I-639 [SIC 2656]
Hayes Microcomputer Products, p. II-1704 [SIC 3577]
Hayes Wheels International Inc, p. II-1954 [SIC 3714]
Haynes International Inc, p. II-1212 [SIC 3325]
Haynes Manufacturing Co, p. I-918 [SIC 2992]
Haysite Reinforced Plastics, p. I-973 [SIC 3083]
Hayssen, p. II-1660 [SIC 3565]
Hayworth Roll and Panel Co, p. I-496 [SIC 2436]
Haz-Stor Co, p. II-1372 [SIC 3448]
Hazama, p. I-496 [SIC 2436]
Hazelden Educational Materials, p. I-703 [SIC 2731]
Hazeltine Corp, p. II-1980 [SIC 3728]
Hazelwood Farms Bakeries Inc, p. I-112 [SIC 2053]
Hazen Paper Co, p. I-654 [SIC 2672]
Hazleton Pumps Inc, p. II-1638 [SIC 3561]
HB Fuller Automotive Products, p. I-879 [SIC 2891]
HB Fuller Co, pp. Vol. I - 825, 879 [SICs 2842, 2891]
HB Fuller Co Specialty Group, p. I-840 [SIC 2851]
HB Hunter Co, p. I-127 [SIC 2064]
HB Ives Co, p. II-1323 [SIC 3429]
HB Larkin Corp, p. II-1486 [SIC 3498]
HB Sherman Manufacturing, p. I-936 [SIC 3052]
HB Williamson Co, p. I-536 [SIC 2499]
HBC Barge Inc, p. II-1986 [SIC 3731]
HBD Industries Inc, p. I-936 [SIC 3052]
HC Duke and Son Inc, p. II-1626 [SIC 3556]
HC Miller Co, p. I-745 [SIC 2782]
HC Osvold Co, p. I-587 [SIC 2541]
HC Starck Inc, p. II-1225 [SIC 3339]
HCC Inc, p. II-1509 [SIC 3523]
HD Campbell Mfg Co, p. I-865 [SIC 2874]
HD Hudson Manufacturing Co, p. II-1509 [SIC 3523]
HDK Industries Inc, p. I-318 [SIC 2297]
Head Sportswear International, p. I-356 [SIC 2329]
Header Products Inc, p. II-1563 [SIC 3542]
Health-Chem Corp, p. I-310 [SIC 2295]
Health-Mor Inc, p. II-1818 [SIC 3635]
Health o meter Products Inc, p. II-1759 [SIC 3596]

Roman numerals I and II indicate the volume in which pages appear. Volume I holds SICs 2011 through 3299; Volume II holds SICs 3312 through 3999.

Healthdyne Technologies Inc, p. II-2082 [SIC 3841]
HealthTex Inc, p. I-348 [SIC 2325]
Healthy Planet Products Inc, p. I-741 [SIC 2771]
Healthy Roman Foods Inc, p. I-60 [SIC 2037]
Healthy Times Inc, p. I-75 [SIC 2043]
Hearst Corp, p. I-693 [SIC 2711]
Heart Interface Corp, p. II-1795 [SIC 3629]
Heat and Control Inc, p. II-1626 [SIC 3556]
Heat Controller Inc, p. II-1726 [SIC 3585]
Heat-N-Glo Fireplaces, p. II-1339 [SIC 3433]
Heat Treat Corporation, p. II-1290 [SIC 3398]
Heat Treating of Minnesota, p. II-1290 [SIC 3398]
Heatbath Corp, p. I-897 [SIC 2899]
Heatcraft Inc, pp. Vol II - 1355, 1726 [SICs 3443, 3585]
Heaters Engineering Inc, p. II-1830 [SIC 3643]
Heath and Co, p. II-2186 [SIC 3993]
Heath Ceramics Inc, p. I-1113 [SIC 3269]
Heath Manufacturing Co, p. I-536 [SIC 2499]
Heath Northwest Inc, p. II-2186 [SIC 3993]
Heath Printers Inc, p. I-724 [SIC 2754]
Heath Tecna Aerospace Co, p. II-1980 [SIC 3728]
Heatilator Inc, p. I-1122 [SIC 3272]
Heatron Inc, p. II-1671 [SIC 3567]
Heaven Hill Distilleries Inc, p. I-172 [SIC 2085]
Heckler Mfg, p. I-378 [SIC 2341]
Heckman Bindery Inc, p. I-750 [SIC 2789]
HECO Pacific Manufacturing, p. II-1547 [SIC 3536]
Hedaya Brothers, p. I-231 [SIC 2211]
Hedland Flow Meters, p. II-2056 [SIC 3824]
Heel Rite Co, p. I-1017 [SIC 3131]
Heick Die Casting Corp, p. II-1266 [SIC 3363]
HEICO Aerospace Corp, p. II-1975 [SIC 3724]
Heico Corp, p. II-1975 [SIC 3724]
Heidelberg Harris Inc, p. II-1621 [SIC 3555]
Heidi's Joy Ltd, p. I-378 [SIC 2341]
Heidtman Steel Products Inc, p. II-1187 [SIC 3316]
Heil Co, p. II-1948 [SIC 3713]
Heinemann's Bakeries Inc, p. I-101 [SIC 2051]
Heinmann Products, p. II-1775 [SIC 3613]
Heinn Trend Corp, p. I-745 [SIC 2782]
Heinrich Envelope Corp, p. I-679 [SIC 2677]
Heisler Industries Inc, p. II-1660 [SIC 3565]
Heisler's Cloverleaf Dairy Inc, p. I-30 [SIC 2024]
Hekimian Laboratories Inc, p. II-2061 [SIC 3825]
Hekman Furniture Co, p. I-542 [SIC 2511]
Helac Corp, p. II-1747 [SIC 3593]
Helbros Watches Inc, p. II-2118 [SIC 3873]
Helen of Troy Corp, p. II-1813 [SIC 3634]
Helen's Tropical Exotics Inc, p. I-55 [SIC 2035]
Helena Cotton Oil Company Inc, p. I-140 [SIC 2074]
Helene Curtis Industries Inc, p. I-834 [SIC 2844]
Helga Inc, p. I-365 [SIC 2335]
Heli Coil, p. II-1388 [SIC 3452]
Helical Products Company Inc, p. II-1470 [SIC 3495]
Helicoflex Co, p. II-2164 [SIC 3953]
Helicomb International Inc, p. II-1970 [SIC 3721]
Helikon Furniture Company Inc, p. I-581 [SIC 2531]
Helix Technology Corp, p. II-1649 [SIC 3563]
Hella North America Inc, p. II-1790 [SIC 3625]
Hellam Hosiery Company Inc, p. I-250 [SIC 2251]
Heller Seasonings, p. I-212 [SIC 2099]
Hellyer Steel Parts Co, p. II-1584 [SIC 3546]
Helmac Products Corp, p. II-2181 [SIC 3991]
Helman Holding Inc, p. II-1800 [SIC 3631]
Helmont Mills Inc, p. I-271 [SIC 2258]
Helsel Inc, p. II-1295 [SIC 3399]
Heluva Good Cheese Inc, p. I-21 [SIC 2022]
Helwig Carbon Products Inc, p. II-1786 [SIC 3624]
Hemagen Diagnostics Inc, p. I-809 [SIC 2835]
Hemco Inc, p. I-288 [SIC 2269]
Hemmerich Industries Inc, p. I-283 [SIC 2262]
Hemphill Spring Co, p. II-1470 [SIC 3495]
Hendershot Tool Co, p. II-1531 [SIC 3533]
Henderson Brick, p. I-1081 [SIC 3251]
Henderson Camp Products Inc, p. I-455 [SIC 2399]
Henderson Manufacturing, p. II-1948 [SIC 3713]
Hendrick Manufacturing Co, p. II-1413 [SIC 3469]
Hendrickson Stamping, p. II-1413 [SIC 3469]
Hendrix Batting Company Inc, p. I-327 [SIC 2299]

Hendrix Manufacturing Company, p. II-1519 [SIC 3531]
Hendry Telephone Products, p. II-1775 [SIC 3613]
Henges Associates Inc, p. II-1372 [SIC 3448]
Henkel Adhesives Corp, p. I-879 [SIC 2891]
Henkel Corp, p. I-855 [SIC 2869]
Henkel Corp Textile Chemicals, p. I-897 [SIC 2899]
Hennegan Co, p. I-718 [SIC 2752]
Henningsen Foods Inc, p. I-12 [SIC 2015]
Henny Penny Corp, p. II-1626 [SIC 3556]
Henredon Furniture Industries, p. I-542 [SIC 2511]
Henri's Food Products Company, p. I-55 [SIC 2035]
Henry A Jacobs and Company, p. I-245 [SIC 2241]
Henry Brick Company Inc, p. I-1081 [SIC 3251]
Henry County Plywood Corp, p. I-491 [SIC 2435]
Henry Heide Inc, p. I-127 [SIC 2064]
Henry Mali Company Inc, p. I-241 [SIC 2231]
Henry Segal Co, p. I-332 [SIC 2311]
Henry Valve Co, p. II-1464 [SIC 3494]
Henry Vogt Machine Co, p. II-1464 [SIC 3494]
Henry Wurst Inc, p. I-718 [SIC 2752]
Hepa Corp, p. II-1654 [SIC 3564]
Her Majesty Industries Inc, p. I-373 [SIC 2339]
Herbert Melarky Roofing Co, p. I-914 [SIC 2952]
Herbert Ritts Inc, p. I-567 [SIC 2519]
Herco Technology Corp, p. II-1887 [SIC 3672]
Hercules Cement Co, p. I-1077 [SIC 3241]
Hercules Inc, p. I-782 [SIC 2821]
Hercules Machine Tool, p. II-1572 [SIC 3544]
Hercules Sheet Metal Inc, p. II-1361 [SIC 3444]
Hercules Welding Products, p. II-1594 [SIC 3548]
Herculite Products Inc, p. I-310 [SIC 2295]
Herff Jones, p. I-420 [SIC 2389]
Herff Jones Inc, p. II-2122 [SIC 3911]
Heritage Bag Co, p. I-660 [SIC 2673]
Heritage Display, p. II-2199 [SIC 3999]
Heritage Home Inc, p. I-1102 [SIC 3262]
Heritage Springfield Inc, p. I-745 [SIC 2782]
Herker Screw Products Inc, p. II-1382 [SIC 3451]
Herley-Vega Systems, p. II-1938 [SIC 3699]
Herman Goelitz Candy Company, p. I-127 [SIC 2064]
Herman Miller Inc, p. I-571 [SIC 2521]
Herman Schwabe Inc, p. II-1615 [SIC 3554]
Hermann Companies Inc, p. I-782 [SIC 2821]
Hermann Oak Leather Company, p. I-1013 [SIC 3111]
Hermes Automotive Mfg Corp, p. II-1395 [SIC 3462]
Hermetic Coil Inc, p. II-1905 [SIC 3677]
Hermetic Seal Corp, p. II-1910 [SIC 3678]
Herr and Sacco Inc, p. II-1344 [SIC 3441]
Herr Foods Inc, p. I-199 [SIC 2096]
Herr Manufacturing Co, p. I-558 [SIC 2515]
Herr Manufacturing Company, p. II-1584 [SIC 3546]
Herr's Potato Chips, p. I-199 [SIC 2096]
Herrin Brothers Coal and Ice Co, p. I-204 [SIC 2097]
Herschel Corp, p. II-1509 [SIC 3523]
Hershey Foods Corp, p. I-132 [SIC 2066]
Hertel Cutting Technologies Inc, p. II-1572 [SIC 3544]
Herzman and Company Inc, p. I-369 [SIC 2337]
Hess and Clark Inc, p. I-95 [SIC 2048]
Hess Collection Winery, p. I-168 [SIC 2084]
Hess Engineering Inc, p. II-1572 [SIC 3544]
Hess Industries Inc, p. II-1572 [SIC 3544]
Hessco Industries Inc, p. I-1001 [SIC 3088]
Hesse Corp, p. II-1948 [SIC 3713]
Hettich America LP, p. II-1323 [SIC 3429]
Heublein Inc, p. I-168 [SIC 2084]
Hewitt-Robins Corp, p. II-1541 [SIC 3535]
Hewitt Soap Company Inc, p. I-819 [SIC 2841]
Hewlett-Packard Co, p. II-1688 [SIC 3571]
Hexacomb Corp, p. I-688 [SIC 2679]
Hexacon Electric Co, p. II-1594 [SIC 3548]
Hexcel Corp, p. II-1413 [SIC 3469]
Hey Ice Cream, p. I-30 [SIC 2024]
HF Coors China Co, p. I-1102 [SIC 3262]
HF Henderson Industries, p. II-1887 [SIC 3672]
HF Livermore Corp, p. II-1605 [SIC 3552]
HFI Inc, p. I-446 [SIC 2396]
HGP Industries Inc, p. I-1057 [SIC 3211]
HH Brown Shoe Company Inc, p. I-1025 [SIC 3143]

HH Fessler Knitting Company, p. I-267 [SIC 2257]
HH Hiatt Furniture Mfg, p. I-548 [SIC 2512]
Hi-Country Foods Corp, p. I-60 [SIC 2037]
Hi Lo Manufacturing Co, p. II-1666 [SIC 3566]
Hi-Lo Trailer Company Inc, p. II-2020 [SIC 3792]
Hi-Mill Manufacturing Co, p. I-1234 [SIC 3351]
Hi-Ram Inc, p. II-1790 [SIC 3625]
Hi-Shear Corp, p. II-1382 [SIC 3451]
Hi-Shear Industries Inc, p. II-1323 [SIC 3429]
Hi-Shear Technology Corp, p. II-1443 [SIC 3489]
Hi-Tech Engineering Inc, p. II-1610 [SIC 3553]
Hi-Tech Hose Inc, p. I-936 [SIC 3052]
Hi-Tech Seating Products Inc, p. I-581 [SIC 2531]
HI TecMetal Group Inc, p. II-1425 [SIC 3479]
Hi-Temp Inc, p. II-1671 [SIC 3567]
Hi-Temp Insulation Inc, p. I-1158 [SIC 3296]
Hickey-Freeman Company Inc, p. I-332 [SIC 2311]
Hickok Inc, p. II-2061 [SIC 3825]
Hickory Brands Inc, p. I-245 [SIC 2241]
Hickory Business Furniture, p. I-571 [SIC 2521]
Hickory Chair Co, p. I-542 [SIC 2511]
Hickory Craft Furniture Inc, p. I-548 [SIC 2512]
Hickory Dyeing and Winding, p. I-302 [SIC 2282]
Hickory Fry Furniture Company, p. I-548 [SIC 2512]
Hickory Hill Furniture Corp, p. I-548 [SIC 2512]
Hickory Printing Group, p. I-729 [SIC 2759]
Hickory Specialties Inc, p. I-846 [SIC 2861]
Hickory Springs Manufacturing, p. I-991 [SIC 3086]
Hickory Springs Mfg Co, p. II-1470 [SIC 3495]
Hickory White Co, p. I-548 [SIC 2512]
Higbee Gaskets & Sealing Prod, p. I-942 [SIC 3053]
Higgins Brick Co, p. I-1117 [SIC 3271]
High End Systems, p. II-1845 [SIC 3646]
High Frequency Technology, p. II-1594 [SIC 3548]
High Plains Corp, p. I-855 [SIC 2869]
High Point Chemical Corp, p. I-830 [SIC 2843]
High Point Furniture Industries, p. I-571 [SIC 2521]
High's Ice Cream Inc, p. I-30 [SIC 2024]
High Street Associates Inc, p. I-777 [SIC 2819]
High Vacuum Apparatus Mfg, p. II-1447 [SIC 3491]
High Voltage Engineering Corp, p. II-2051 [SIC 3823]
Higher Octave Music Inc, p. II-1864 [SIC 3652]
Highfield Manufacturing Co, p. II-1447 [SIC 3491]
Highland House Inc, p. I-548 [SIC 2512]
Highland Industries Inc, p. I-236 [SIC 2221]
Highland Manufacturing Co, p. II-1413 [SIC 3469]
Highland Plating Company Inc, p. II-1420 [SIC 3471]
Highland Supply Corp, p. I-654 [SIC 2672]
Highland Tank & Mfg Co, p. II-1355 [SIC 3443]
Highland Yarn Mills, p. I-297 [SIC 2281]
Highlights for Children Inc, p. I-698 [SIC 2721]
Highline Products Corp, p. II-1854 [SIC 3648]
Highway Equipment Co, p. II-1509 [SIC 3523]
Highway Machine Company Inc, p. II-1676 [SIC 3568]
Highway Materials Inc, p. I-909 [SIC 2951]
Highway Safety Corp, p. II-1361 [SIC 3444]
Hiland Dairy Co, p. I-35 [SIC 2026]
Hilb and Company Inc, p. I-394 [SIC 2369]
Hilec Inc, p. II-1835 [SIC 3644]
Hill Brothers Chemical Co, p. I-897 [SIC 2899]
Hill Manufacturing Co, p. II-1961 [SIC 3715]
Hill Rom Company Inc, p. I-581 [SIC 2531]
Hill Wood Products Inc, p. I-536 [SIC 2499]
Hillenbrand Industries Inc, p. I-1041 [SIC 3161]
Hiller Industries Inc, p. I-750 [SIC 2789]
Hilliard Corp, p. II-1676 [SIC 3568]
Hillsboro Glass Co, p. I-1062 [SIC 3221]
Hillsdale Industries Inc, p. I-1057 [SIC 3211]
Hillside Capital Inc, p. I-1006 [SIC 3089]
Hillside Industries Inc, p. I-1006 [SIC 3089]
Hilltop Slate Inc, p. I-1140 [SIC 3281]
Hillwood Manufacturing Co, p. II-2177 [SIC 3965]
Hillyard Industry Inc, p. I-825 [SIC 2842]
Hilman Rollers Inc, p. II-1589 [SIC 3547]
Hilmar Cheese Company Inc, p. I-21 [SIC 2022]
Hilti Inc, p. II-1584 [SIC 3546]
Hilton Active Apparel, p. I-361 [SIC 2331]
Hilton Davis Co, p. I-855 [SIC 2869]
Himolene Inc, p. I-660 [SIC 2673]

Roman numerals I and II indicate the volume in which pages appear. Volume I holds SICs 2011 through 3299; Volume II holds SICs 3312 through 3999.

2259

Himont Inc, p. I-782 [SIC 2821]
Hinchcliff Lumber Co, p. I-510 [SIC 2448]
Hinchen Brothers Shingle, p. I-476 [SIC 2429]
Hinckley Co, p. II-1992 [SIC 3732]
Hind Inc, p. I-356 [SIC 2329]
Hinderliter Heat Treating Inc, p. II-1290 [SIC 3398]
Hindostone Products Inc, p. I-1140 [SIC 3281]
Hiniker Co, p. II-1509 [SIC 3523]
Hinkley Lighting Inc, p. II-1840 [SIC 3645]
Hino and Malee Boutique, p. I-361 [SIC 2331]
Hintzsche Feed and Grain Inc, p. I-95 [SIC 2048]
Hippographics Co, p. I-729 [SIC 2759]
Hipwell Manufacturing Co, p. II-1854 [SIC 3648]
Hirata Corporation of America, p. II-1682 [SIC 3569]
Hirsch International Corp, p. II-1605 [SIC 3552]
Hirschfeld Steel Company Inc, p. II-1344 [SIC 3441]
Hirshfield's Inc, p. I-840 [SIC 2851]
Hirzel Canning Co, p. I-45 [SIC 2033]
Hisonic Inc, p. II-1905 [SIC 3677]
Hitachi Construction Machinery, p. II-1519 [SIC 3531]
Hitachi Magnetics Corp, p. II-1492 [SIC 3499]
Hitchcock Industries Inc, p. II-1975 [SIC 3724]
Hitchiner Manufacturing Co, p. II-1212 [SIC 3325]
Hitech Holding Inc, p. I-885 [SIC 2892]
HITECH Inc, p. I-885 [SIC 2892]
Hitech Instruments, p. II-2061 [SIC 3825]
Hitox Corporation of America, p. I-773 [SIC 2816]
Hitran Corp, p. II-1769 [SIC 3612]
HJ Baker and Brother Inc, p. I-869 [SIC 2875]
HJ Heinz Co, p. I-45 [SIC 2033]
HKS Marketing, p. I-40 [SIC 2032]
HL Blachford Inc, p. I-991 [SIC 3086]
HL Bouton Company Inc, p. II-2107 [SIC 3851]
HL Miller and Son Inc, p. I-365 [SIC 2335]
HLC Industries Inc, p. II-1486 [SIC 3498]
HM Quackenbush Inc, p. II-1420 [SIC 3471]
HM Smyth Co, p. I-718 [SIC 2752]
HME Inc, p. II-1943 [SIC 3711]
HMG Digital Technologies Corp, p. II-1864 [SIC 3652]
HMG Worldwide, p. II-2186 [SIC 3993]
HMK Enterprises Inc, p. I-576 [SIC 2522]
HNU Systems Inc, p. II-2076 [SIC 3829]
HO Trerice Co, p. II-2051 [SIC 3823]
Hobart Lasers, p. II-1938 [SIC 3699]
Hobbs Bonded Fibers, p. I-327 [SIC 2299]
Hobbs Corp, p. II-2061 [SIC 3825]
Hobbs Industries, p. I-795 [SIC 2824]
Hobbs International Inc, p. II-1948 [SIC 3713]
Hobe Cie Ltd, p. II-2173 [SIC 3961]
Hobie Cat Co, p. II-1992 [SIC 3732]
Hobson and Motzer Inc, p. II-1413 [SIC 3469]
Hobson Brothers Inc, p. II-1910 [SIC 3678]
Hodgdon Powder Company Inc, p. I-885 [SIC 2892]
Hodge Foundry Division, p. II-1197 [SIC 3321]
Hoechst Celanese Corp, pp. Vol. I - 782, 795 [SICs 2821, 2824]
Hoechst CeramTec, p. I-1167 [SIC 3299]
Hoechst Diafoil Co, p. I-961 [SIC 3081]
Hoeck Metal Fabricators Inc, p. II-1536 [SIC 3534]
Hoeganaes Corp, p. II-1295 [SIC 3399]
Hoegh Industries Inc, p. II-2191 [SIC 3995]
Hoerbiger Corporation, p. II-1454 [SIC 3492]
Hoertig Iron Works Inc, p. II-1377 [SIC 3449]
Hofer Valve and Supply Inc, p. II-1447 [SIC 3491]
Hoffco Inc, p. II-1514 [SIC 3524]
Hoffers Inc, p. I-1057 [SIC 3211]
Hoffman Air & Filtration Syst, p. II-1654 [SIC 3564]
Hoffman Brothers Packing, p. I-2 [SIC 2011]
Hoffman Engineering Co, p. II-1835 [SIC 3644]
Hoffman Mills Inc, p. I-231 [SIC 2211]
Hofley Manufacturing Co, p. II-1404 [SIC 3465]
Hogan Manufacturing Inc, p. II-1536 [SIC 3534]
Hoge Lumber Co, p. I-465 [SIC 2421]
Hogue Cellars Ltd, p. I-168 [SIC 2084]
Hoke Inc, p. II-1454 [SIC 3492]
Hol-Mac Corp, p. II-1747 [SIC 3593]
Hol N One Donut, p. I-83 [SIC 2045]
Holaday Circuits Inc, p. II-1887 [SIC 3672]

Holan Manufacturing Inc, p. II-1948 [SIC 3713]
Holcroft, p. II-1671 [SIC 3567]
Hold Company Inc, p. II-1864 [SIC 3652]
Holden Business Forms Co, p. I-735 [SIC 2761]
Hole-in-None Hosiery, p. I-254 [SIC 2252]
Holiday Rambler LLC, p. II-2020 [SIC 3792]
Holiday Togs Inc, p. I-361 [SIC 2331]
Holland American Wafer Co, p. I-107 [SIC 2052]
Holland Company Inc, p. I-1140 [SIC 3281]
Holland Dairies Inc, p. I-35 [SIC 2026]
Holland Manufacturing, p. I-654 [SIC 2672]
Holland Wire Products Inc, p. II-1470 [SIC 3495]
Hollander Industries, p. II-1266 [SIC 3363]
Hollander-Stapo Industries Inc, p. II-2199 [SIC 3999]
Holley Automotive, p. II-1954 [SIC 3714]
Hollingsworth and Vose Co, p. I-688 [SIC 2679]
Hollingsworth Saco Lowell Corp, p. II-1605 [SIC 3552]
Hollister Inc, p. II-2082 [SIC 3841]
Holliston Mills Inc, p. I-231 [SIC 2211]
Holloway Sportswear Inc, p. I-356 [SIC 2329]
Holly Bra of California, p. I-373 [SIC 2339]
Holly Corp, p. I-903 [SIC 2911]
Holly Decorations Inc, p. I-1071 [SIC 3231]
Holly Hill Fruit Products Inc, p. I-60 [SIC 2037]
Holly Sugar Corp, p. I-123 [SIC 2063]
Hollyvogue Ties Corp, p. I-344 [SIC 2323]
Hollywood Alloy Casting Co, p. II-1275 [SIC 3365]
Hollywood Needlecraft Inc, p. I-390 [SIC 2361]
Hollywood Records, p. II-1864 [SIC 3652]
Holman Boiler Works Inc, p. II-1355 [SIC 3443]
Holmes and Company Inc, p. I-471 [SIC 2426]
Holmes Cheese Co, p. I-21 [SIC 2022]
Holmes Garage Door Company, p. I-480 [SIC 2431]
Holmes Typography, p. I-754 [SIC 2791]
Hologic Inc, p. II-2098 [SIC 3844]
Holometrix Inc, p. II-2066 [SIC 3826]
Holophane Corp, p. II-1845 [SIC 3646]
Holoubek Inc, p. I-279 [SIC 2261]
Holson Burnes Group Inc, p. I-745 [SIC 2782]
Holston Defense Corp, p. I-885 [SIC 2892]
Holston Gases Inc, p. I-768 [SIC 2813]
Holsum, p. I-101 [SIC 2051]
Holsum Foods, p. I-156 [SIC 2079]
Holt Hosiery Mills Inc, p. I-250 [SIC 2251]
Holt Manufacturing Company, p. I-288 [SIC 2269]
Holum and Sons Company Inc, p. I-745 [SIC 2782]
Holyoke Card and Paper Co, p. I-688 [SIC 2679]
Holyoke Machine Co, p. II-1615 [SIC 3554]
Holz Rubber Company Inc, p. I-948 [SIC 3061]
Holzma-US, p. II-1610 [SIC 3553]
Homac MFG Co, p. II-1910 [SIC 3678]
HOMARK Company Inc, p. I-519 [SIC 2451]
Home Care Industries Inc, p. I-660 [SIC 2673]
Home Fashions Del Mar, p. I-599 [SIC 2591]
Home Fashions Inc, p. I-599 [SIC 2591]
Home Innovations, p. I-428 [SIC 2392]
Home of the Hebert Candies Inc, p. I-127 [SIC 2064]
Home Orders Inc, p. I-212 [SIC 2099]
Home-Style Industries Inc, p. I-548 [SIC 2512]
Home Theater Products Intern, p. II-1859 [SIC 3651]
HomeCrest Corp, p. I-486 [SIC 2434]
Homecrest Industries Inc, p. I-553 [SIC 2514]
Homedics Inc, p. II-1813 [SIC 3634]
Homer Laughlin China Co, p. I-1102 [SIC 3262]
Homera Homes of Minnesota, p. I-523 [SIC 2452]
Homes by Keystone Inc, p. I-523 [SIC 2452]
Homes of Merit Inc, p. I-519 [SIC 2451]
Homeshield Fabricated, p. II-1239 [SIC 3353]
Homestead Valve, p. II-1447 [SIC 3491]
Homette, p. I-519 [SIC 2451]
Homogeneous Metals Inc, p. II-1295 [SIC 3399]
HON Industries Inc, p. I-571 [SIC 2521]
Honda Power Equipment Mfg, p. II-1514 [SIC 3524]
Honeymead Products Co, p. I-144 [SIC 2075]
Honeywell Envir Air Control, p. II-1654 [SIC 3564]
Honeywell Inc, pp. Vol II - 2035, 2045 [SICs 3812, 3822]
Honor Gard, p. II-1718 [SIC 3581]

Hood and Company Inc, p. II-1255 [SIC 3356]
Hood Enterprises Inc, p. II-1992 [SIC 3732]
Hood Furniture Mfg Co, p. I-542 [SIC 2511]
Hood Sailmakers Inc, p. I-437 [SIC 2394]
Hoogovens Aluminium Corp, p. II-1244 [SIC 3354]
Hooper Industries Inc, p. I-245 [SIC 2241]
Hoosier Gasket Corp, p. I-942 [SIC 3053]
Hoosier Magnetics Group, p. II-1295 [SIC 3399]
Hoover Co, p. II-1818 [SIC 3635]
Hoover Color Corp, p. I-773 [SIC 2816]
Hoover Group Inc, pp. Vol II - 1323, 1476 [SICs 3429, 3496]
Hoover Inc, p. I-1127 [SIC 3273]
Hoover Industries Inc, p. I-437 [SIC 2394]
Hoover North America, p. II-1736 [SIC 3589]
Hoover Precision Products Inc, p. II-1644 [SIC 3562]
Hop-Cap Inc, p. II-2020 [SIC 3792]
Hope Haven Inc, p. II-1382 [SIC 3451]
Hopeman Brothers Inc, p. II-1986 [SIC 3731]
Hopes Architectural Products, p. II-1349 [SIC 3442]
Hopkes Logging Company Inc, p. I-460 [SIC 2411]
Hopkinsville Milling Company, p. I-70 [SIC 2041]
Hopple Plastics Inc, p. I-991 [SIC 3086]
Hoppmann Corp, p. II-1541 [SIC 3535]
Hops Extract Corporation, p. I-160 [SIC 2082]
Horace Small Apparel Co, p. I-332 [SIC 2311]
Horiba Instruments Inc, p. II-2066 [SIC 3826]
Horizon Enterprises Inc, p. II-1382 [SIC 3451]
Horizon Music Inc, p. II-1859 [SIC 3651]
Horizon Technology Group, p. II-1382 [SIC 3451]
Horizons Inc, p. II-1420 [SIC 3471]
Hormel Foods Corp, p. I-2 [SIC 2011]
Horn Badge Co, p. II-2177 [SIC 3965]
Horn Packaging Corp, p. I-991 [SIC 3086]
Hornell Brewing Company Inc, p. I-176 [SIC 2086]
Horner Flooring Company Inc, p. I-471 [SIC 2426]
Horning Wire Corp, p. II-1260 [SIC 3357]
Horowitz-Rae, p. I-708 [SIC 2732]
Horsburgh and Scott Co, p. II-1666 [SIC 3566]
Horstmann Mix and Cream Co, p. I-30 [SIC 2024]
Hortie-Van Innovation, p. I-455 [SIC 2399]
Horton Automatics, p. II-1349 [SIC 3442]
Horton Homes Inc, p. I-519 [SIC 2451]
Horton Manufacturing Co, p. II-1676 [SIC 3568]
Horween Leather Co, p. I-1013 [SIC 3111]
Hoskins Manufacturing Co, p. II-1929 [SIC 3694]
Hosokawa Bepex Corp, p. II-1626 [SIC 3556]
Hosokawa Micron Powder Syst, p. II-1682 [SIC 3569]
Hospital Specialty Co, p. I-674 [SIC 2676]
Hospitality Systems Inc, p. II-1709 [SIC 3578]
Hosposable Products Inc, p. I-271 [SIC 2258]
Host Apparel Inc, p. I-340 [SIC 2322]
Hostess Cake, p. I-101 [SIC 2051]
Hot-Mix Asphalt Inc, p. I-909 [SIC 2951]
Hot Shot Products Company Inc, p. II-1795 [SIC 3629]
Hot Tools, p. II-1584 [SIC 3546]
Hotpack Corp, p. II-2040 [SIC 3821]
Hotwatt Inc, p. II-1671 [SIC 3567]
Hotz Manufacturing Company, p. I-471 [SIC 2426]
Houbigant Inc, p. I-834 [SIC 2844]
Hougen Manufacturing Inc, p. II-1578 [SIC 3545]
Houghton International Inc, p. I-918 [SIC 2992]
Houghton Mifflin Co, p. I-703 [SIC 2731]
Houma Fabricators Inc, p. II-1986 [SIC 3731]
House Autry Mills Inc, p. I-70 [SIC 2041]
House-O-Lite Corp, p. II-1845 [SIC 3646]
House of Bianchi Inc, p. I-365 [SIC 2335]
House of Hansen, p. I-420 [SIC 2389]
House of Perfection Inc, p. I-373 [SIC 2339]
House of Raeford Farms Inc, p. I-12 [SIC 2015]
House of Ronnie Inc, p. I-390 [SIC 2361]
House of Trophies, p. II-2127 [SIC 3914]
House of Tsang Ltd, p. I-55 [SIC 2035]
Houston Chronicle Pub Co, p. I-693 [SIC 2711]
Houston Engineers Inc, p. II-1531 [SIC 3533]
Houston Post Co, p. I-693 [SIC 2711]
Houston Wiper, p. I-327 [SIC 2299]
Howard B Wolf Inc, p. I-365 [SIC 2335]
Howard Brothers Mfg, p. I-245 [SIC 2241]

Roman numerals I and II indicate the volume in which pages appear. Volume I holds SICs 2011 through 3299; Volume II holds SICs 3312 through 3999.

Howard Eldon Ltd, p. II-2173 [SIC 3961]
Howard Engineering Co, p. II-2177 [SIC 3965]
Howard Furniture Inc, p. I-548 [SIC 2512]
Howard H Sweet and Son Inc, p. II-2122 [SIC 3911]
Howard Industries Inc, pp. Vol. I - 1071, Vol. II - 1654, 1769 [SICs 3231, 3564, 3612]
Howard Miller Clock Co, p. I-542 [SIC 2511]
Howard Plating Industry Inc, p. II-1420 [SIC 3471]
Howard Price Turf Equipment, p. II-1509 [SIC 3523]
Howard Publications Inc, p. I-693 [SIC 2711]
Howatt Company Inc, p. I-510 [SIC 2448]
Howden Airdynamics, p. II-1649 [SIC 3563]
Howden Compressors Inc, p. II-1649 [SIC 3563]
Howden Fan Co, p. II-1654 [SIC 3564]
Howden Food Equipment Inc, p. II-1626 [SIC 3556]
Howe Furniture Corp, p. I-571 [SIC 2521]
Howell Corp, p. II-1795 [SIC 3629]
Howell Industries Inc, p. II-1404 [SIC 3465]
Howell Manufacturing Co, p. I-297 [SIC 2281]
Howell Metal Co, p. II-1234 [SIC 3351]
Howmedica Inc, p. II-2087 [SIC 3842]
Howmet Corp, p. II-1498 [SIC 3511]
Howmet Tempcraft Inc, p. II-1572 [SIC 3544]
Howmet Whitehall Casting, p. II-1207 [SIC 3324]
Howse Implement Company Inc, p. II-1509 [SIC 3523]
Hoyle Products, p. I-729 [SIC 2759]
Hoyt Corp, p. II-1722 [SIC 3582]
HP Hood Inc, p. I-35 [SIC 2026]
HPH Apparel Manufacturing Co, p. I-348 [SIC 2325]
HPM Corp, p. II-1632 [SIC 3559]
HR Kaminsky and Sons Inc, p. I-348 [SIC 2325]
HR Krueger Machine Tool Inc, p. II-1558 [SIC 3541]
HR Terryberry Co, p. II-2122 [SIC 3911]
HR Textron Inc, p. II-2035 [SIC 3812]
HRR Enterprises, p. I-156 [SIC 2079]
HSI Scales Inc, p. II-1759 [SIC 3596]
Hu-Friedy Manufacturing, p. II-2093 [SIC 3843]
Hub City Inc, p. II-1666 [SIC 3566]
Hub Folding Box Company Inc, p. I-643 [SIC 2657]
Hub Pen Company Inc, p. II-2156 [SIC 3951]
Hub Plastics Inc, p. I-985 [SIC 3085]
Hubbard Company Inc, p. I-332 [SIC 2311]
Hubbard Milling Co, p. I-95 [SIC 2048]
Hubbell Inc, p. II-1830 [SIC 3643]
Hubbell Industrial Controls Inc, p. II-1790 [SIC 3625]
Hubbell Lighting Inc, p. II-1845 [SIC 3646]
Huck Fixture Co, p. I-587 [SIC 2541]
Huck International Inc, p. II-1388 [SIC 3452]
Hudson Cloth, p. I-302 [SIC 2282]
Hudson Foods Inc, p. I-12 [SIC 2015]
Hudson ICS, p. I-536 [SIC 2499]
Hudson Industries Inc, p. I-55 [SIC 2035]
Hudson Lock Inc, p. II-1323 [SIC 3429]
Hudson Neckware Company Inc, p. I-344 [SIC 2323]
Hudson RCI Inc, p. II-2082 [SIC 3841]
Hudson Standard Corp, p. II-1813 [SIC 3634]
Hudson Universal Ltd, p. II-2107 [SIC 3851]
Huey Co, p. II-2112 [SIC 3861]
Hufcor Inc, p. I-587 [SIC 2541]
Huffy Bicycle, p. II-2003 [SIC 3751]
Huffy Corp, pp. Vol II - 2003, 2150 [SICs 3751, 3949]
Huge Company Inc, p. I-874 [SIC 2879]
Hughes Aircraft Co, pp. Vol II - 1594, 1883, 2035 [SICs 3548, 3671, 3812]
Hughes Brothers Inc, p. II-1425 [SIC 3479]
Hughes Corp, p. II-1775 [SIC 3613]
Hughes Danbury Optical Systems, p. II-2071 [SIC 3827]
Hughes Enterprises Inc, p. II-2199 [SIC 3999]
Hughes Identification Devices, p. II-2199 [SIC 3999]
Hughes-JVC Technology Corp, p. II-2112 [SIC 3861]
Hughes LAN Systems Inc, p. II-1868 [SIC 3661]
Hughes Network Systems Inc, p. II-1868 [SIC 3661]
Hughes Products Co, p. II-2199 [SIC 3999]
Hughes Resources Inc, p. I-465 [SIC 2421]
Hughes Training Inc, p. II-1938 [SIC 3699]
Hugo Bosca Company Inc, p. I-1049 [SIC 3172]
Hulman and Co, p. I-212 [SIC 2099]
Huls America Inc, p. I-782 [SIC 2821]

Human Designed Systems Inc, p. II-1699 [SIC 3575]
Humane Manufacturing Co, p. II-1486 [SIC 3498]
Humboldt Manufacturing Co, p. II-2066 [SIC 3826]
Humbug Mountain Research, p. II-1795 [SIC 3629]
Hummer Sportswear Inc, p. I-356 [SIC 2329]
Humphrey Products Co, p. II-1464 [SIC 3494]
Humphrey's Inc, p. I-1041 [SIC 3161]
Hunt and Behrens Inc, p. I-95 [SIC 2048]
Hunt Manufacturing Co, p. II-2156 [SIC 3951]
Hunt Refining Co, p. I-903 [SIC 2911]
Hunt Valve Company Inc, p. II-1447 [SIC 3491]
Hunt-Wesson Inc, p. I-45 [SIC 2033]
Huntco Inc, p. II-1172 [SIC 3312]
Huntco Steel Inc, p. II-1172 [SIC 3312]
Hunter Associates Laboratory, p. II-2076 [SIC 3829]
Hunter Container Corp, p. II-1300 [SIC 3411]
Hunter Douglas Fabrication, p. I-599 [SIC 2591]
Hunter Douglas Inc, p. I-599 [SIC 2591]
Hunter Engineering Co, p. II-1632 [SIC 3559]
Hunter Engineering Company, p. II-1589 [SIC 3547]
Hunter Fan Co, p. II-1813 [SIC 3634]
Hunter Graphics Inc, p. I-754 [SIC 2791]
Hunter Marine Corp, p. II-1992 [SIC 3732]
Hunter Publishing Co, p. I-708 [SIC 2732]
Hunter Woodworks Inc, p. I-510 [SIC 2448]
Hunterdon Concrete Co, p. I-1127 [SIC 3273]
Huntingdon Throwing Mills Inc, p. I-302 [SIC 2282]
Huntingdon Yarn Mill Inc, p. I-327 [SIC 2299]
Huntington Electric Inc, p. II-1901 [SIC 3676]
Huntington Laboratories Inc, p. I-825 [SIC 2842]
Huntington Mechanical Labs, p. II-1682 [SIC 3569]
Huntington Park Rubber, p. II-2164 [SIC 3953]
Huntsman Film Products Corp, p. I-961 [SIC 3081]
Huntway Partners LP, p. I-903 [SIC 2911]
Hurco Companies Inc, p. II-1558 [SIC 3541]
Hurco Manufacturing Company, p. II-1558 [SIC 3541]
Hurd Lock & Manufacturing Co, p. II-1323 [SIC 3429]
Hurn Shingle Company Inc, p. I-476 [SIC 2429]
Huron Casting Inc, p. II-1212 [SIC 3325]
Huron Inc, p. II-1382 [SIC 3451]
Hurst Boiler and Welding, p. II-1339 [SIC 3433]
Hurst Graphics Inc, p. I-889 [SIC 2893]
Hurst Labeling Systems Inc, p. I-654 [SIC 2672]
Hurst Manufacturing, p. II-1780 [SIC 3621]
HUSCO International Inc, p. II-1454 [SIC 3492]
Husky Injection Molding, p. II-1764 [SIC 3599]
Husman Snack Foods Co, p. I-199 [SIC 2096]
Husman Snackfoods, p. I-199 [SIC 2096]
Hussey Copper Ltd, p. II-1234 [SIC 3351]
Hussey Seating Company Inc, p. I-581 [SIC 2531]
Hussmann Corp, p. II-1726 [SIC 3585]
Hutchins Manufacturing Co, p. II-1584 [SIC 3546]
Hutchinson/Mayrath, p. II-1312 [SIC 3423]
Hutchinson Technology Inc, p. II-1704 [SIC 3577]
Huttig Sash and Door Co, p. I-480 [SIC 2431]
Hutton Industries Inc, p. I-428 [SIC 2392]
Huxley Envelope Corp, p. I-679 [SIC 2677]
HV Component Associates Inc, p. II-1795 [SIC 3629]
HWH Corp, p. I-523 [SIC 2452]
Hy-Cal Engineering, p. II-2076 [SIC 3829]
Hy-Form Products Inc, p. II-1404 [SIC 3465]
HY-JO Mfg Imports Corp, p. I-536 [SIC 2499]
Hy-Ko Products Co, p. I-446 [SIC 2396]
Hy-Matic Manufacturing Inc, p. II-1447 [SIC 3491]
Hy-Production Inc, p. II-1382 [SIC 3451]
Hybritech Inc, p. I-804 [SIC 2834]
Hycalog, p. II-1531 [SIC 3533]
Hycorr Machine Corp, p. II-1615 [SIC 3554]
Hyde Athletic Industries Inc, p. I-1033 [SIC 3149]
Hyde Manufacturing Co, p. II-1308 [SIC 3421]
Hydra-Electric Co, p. II-1830 [SIC 3643]
Hydra-Mac Inc, p. II-1552 [SIC 3537]
HydraForce Inc, p. II-1454 [SIC 3492]
HydraMechanica Corp, p. II-1676 [SIC 3568]
Hydranamics, p. II-1747 [SIC 3593]
Hydranautics Inc, p. II-1736 [SIC 3589]
Hydraulic Accessories Co, p. II-1454 [SIC 3492]
Hydraulic & Pneumatic, p. II-1621 [SIC 3555]
Hydraulic Drives Inc, p. II-1747 [SIC 3593]

Hydraulic Press Brick Co, p. I-1154 [SIC 3295]
Hydraulic Tubes and Fittings Inc, p. II-1486 [SIC 3498]
Hydril Co, p. II-1531 [SIC 3533]
Hydrite Chemical Co, p. I-799 [SIC 2833]
Hydro-Aire, p. II-1980 [SIC 3728]
Hydro Aluminum-Adrian, p. II-1239 [SIC 3353]
Hydro Carbide, p. II-1572 [SIC 3544]
Hydro Conduit Corp, p. I-1122 [SIC 3272]
Hydro-Craft Inc, p. II-1377 [SIC 3449]
Hydro Honing Laboratories Inc, p. II-1290 [SIC 3398]
Hydro-Mill Co, p. II-1980 [SIC 3728]
Hydro Tube Corp, p. II-1486 [SIC 3498]
Hydroacoustics Inc, p. II-1443 [SIC 3489]
Hydrolabs Inc, p. I-897 [SIC 2899]
HydroTech Labs LP, p. I-834 [SIC 2844]
Hygenic Corp, p. II-2093 [SIC 3843]
Hygrade Metal Moulding, p. II-1250 [SIC 3355]
Hygrade Printing Corp, p. I-735 [SIC 2761]
Hyland, p. I-814 [SIC 2836]
Hynes Industries Inc, p. II-1187 [SIC 3316]
Hynite Corp, p. I-861 [SIC 2873]
Hyosung, p. I-318 [SIC 2297]
Hyper Alloys Inc, p. II-1255 [SIC 3356]
Hypercom Inc, p. II-1868 [SIC 3661]
Hyperion Inc, p. II-2040 [SIC 3821]
Hypertherm Inc, p. II-1558 [SIC 3541]
Hyphen Inc, p. II-1621 [SIC 3555]
Hypnotic Hats Ltd, p. I-386 [SIC 2353]
Hyponex Corp, p. I-861 [SIC 2873]
Hypress Technologies Inc, p. II-1747 [SIC 3593]
Hypro Corp, p. II-1638 [SIC 3561]
Hypro Inc, p. II-1764 [SIC 3599]
Hyster Co, p. II-1552 [SIC 3537]
Hytek Finishes Co, p. II-1420 [SIC 3471]
Hytrol Conveyor Company Inc, p. II-1541 [SIC 3535]
Hytronics Corp, p. II-1905 [SIC 3677]

I Appel Corp, p. I-405 [SIC 2384]
I-C Manufacturing Co, p. I-378 [SIC 2341]
I Ginsberg and Sons Inc, p. I-587 [SIC 2541]
I Johns Company Inc, p. I-442 [SIC 2395]
I-K-I Manufacturing Company, p. II-2199 [SIC 3999]
I Kassoy Inc, p. II-2131 [SIC 3915]
I/N Kote, p. II-1425 [SIC 3479]
I Rokeach and Sons Inc, p. I-212 [SIC 2099]
I Schumann and Co, p. II-1225 [SIC 3339]
I Shalom and Company Inc, p. I-420 [SIC 2389]
IBC Corp, p. II-1835 [SIC 3644]
Iberia Sugar Cooperative Inc, p. I-116 [SIC 2061]
IBG Corp, p. I-874 [SIC 2879]
IBP Inc, p. I-2 [SIC 2011]
IBR Corp Harbortown, p. I-603 [SIC 2599]
IC Isaacs and Co Newton Co, p. I-348 [SIC 2325]
ICA Inc, p. I-1158 [SIC 3296]
Icare Industries Inc, p. II-2107 [SIC 3851]
ICC Industries Inc, p. I-855 [SIC 2869]
Ice-O-Matic, p. II-1726 [SIC 3585]
ICEE-USA Corp, p. II-1626 [SIC 3556]
Iceland Seafood Corp, p. I-190 [SIC 2092]
ICI Advanced Ceramics Inc, p. I-1113 [SIC 3269]
ICI Composites Inc, p. I-782 [SIC 2821]
ICI Explosives USA Inc, p. I-885 [SIC 2892]
ICI Films Group, p. I-961 [SIC 3081]
ICI Security Systems, p. II-1878 [SIC 3669]
Icicle Seafoods Inc, p. I-186 [SIC 2091]
ICN Biomedicals Inc, p. II-2102 [SIC 3845]
ICN Pharmaceuticals Inc, p. I-804 [SIC 2834]
Icore International Inc, p. II-1835 [SIC 3644]
ICT Manufacturing Inc, p. II-1747 [SIC 3593]
IDAB Inc, p. II-1541 [SIC 3535]
Idaho Candy Co, p. I-127 [SIC 2064]
Idaho Cedar Sales Inc, p. I-536 [SIC 2499]
Idaho Pacific Corp, p. I-51 [SIC 2034]
Idaho Pulp and Paperboard, p. I-608 [SIC 2611]
Idaho Timber Corp, p. I-465 [SIC 2421]
IDB WorldCom Services Inc, p. II-1868 [SIC 3661]
IDEA Inc, p. II-2051 [SIC 3823]
Ideal Baking Company Inc, p. I-101 [SIC 2051]

Roman numerals I and II indicate the volume in which pages appear. Volume I holds SICs 2011 through 3299; Volume II holds SICs 3312 through 3999.

Ideal Box Co, p. I-628 [SIC 2653]
Ideal Door Co, p. I-480 [SIC 2431]
Ideal Electric Co, p. II-1780 [SIC 3621]
Ideal Fastener Corp, p. II-2177 [SIC 3965]
Ideal Frame Company Inc, p. I-471 [SIC 2426]
Ideal Tape Company Inc, p. I-649 [SIC 2671]
Ideal Textile Company Inc, p. I-279 [SIC 2261]
Ideas in Motion Inc, p. II-1541 [SIC 3535]
Identix Inc, p. II-1938 [SIC 3699]
IDEX Corp, p. II-1638 [SIC 3561]
IDEXX Corp, p. II-1704 [SIC 3577]
IDEXX Laboratories Inc, p. I-809 [SIC 2835]
IDL Inc, p. II-2186 [SIC 3993]
IEC Electronics Corp, p. II-1887 [SIC 3672]
IEH Corp, p. II-1910 [SIC 3678]
IFC Disposable Inc, p. I-674 [SIC 2676]
IFE, p. II-1980 [SIC 3728]
IFR Systems Inc, p. II-2061 [SIC 3825]
IGEN Inc, p. I-809 [SIC 2835]
Igloo Holdings Inc, p. I-1006 [SIC 3089]
Igloo Products Corp, p. I-1006 [SIC 3089]
IGM Robotic Systems Inc, p. II-1594 [SIC 3548]
IGS Inc, p. I-1066 [SIC 3229]
IGT-North America, p. II-2199 [SIC 3999]
IGYS Inc, p. II-1878 [SIC 3669]
IH Marshall Ltd, p. I-365 [SIC 2335]
Ihling Bros Everard Co, p. I-352 [SIC 2326]
II-VI Inc, p. II-2071 [SIC 3827]
IICON Corp, p. II-1699 [SIC 3575]
IKG Industries, p. II-1344 [SIC 3441]
IKS American Corp, p. II-1644 [SIC 3562]
ILC Dover Inc, p. II-2087 [SIC 3842]
ILC Industries Inc, p. II-1892 [SIC 3674]
ILC Technology Inc, p. II-2071 [SIC 3827]
Ilco Unican Corp, p. II-1323 [SIC 3429]
Ilex, p. II-2071 [SIC 3827]
ILFORD Photo Corp, p. II-2112 [SIC 3861]
Ilie Wacs Inc, p. I-369 [SIC 2337]
Illinois Capacitor Inc, p. II-1897 [SIC 3675]
Illinois Cement Co, p. I-1077 [SIC 3241]
Illinois Cereal Mills Inc, p. I-70 [SIC 2041]
Illinois Concrete Company Inc, p. I-1117 [SIC 3271]
Illinois Oil Products Inc, p. I-918 [SIC 2992]
Illinois Tool Works Inc, p. II-1388 [SIC 3452]
Illmo RX Services Inc, p. II-2071 [SIC 3827]
Ilsco, p. II-1830 [SIC 3643]
IMA Fashions Phillippe, p. I-1045 [SIC 3171]
Image Industries Inc, p. I-292 [SIC 2273]
Image Laboratories Inc, p. I-834 [SIC 2844]
Image Yarn, p. I-297 [SIC 2281]
Imaging Products International, p. II-2066 [SIC 3826]
Imatron Inc, p. II-2102 [SIC 3845]
IMC Fertilizer Group Inc, p. I-777 [SIC 2819]
Imco Inc, p. I-948 [SIC 3061]
IMCOA, p. I-1158 [SIC 3296]
IMED Corp, p. II-2082 [SIC 3841]
IMI Cash Valve Inc, p. II-1464 [SIC 3494]
IMI Cornelius Inc, p. II-1718 [SIC 3581]
Immucor Inc, p. I-809 [SIC 2835]
Immunex Corp, p. I-814 [SIC 2836]
Imo Industries Inc, p. II-2076 [SIC 3829]
IMP Inc, p. II-1892 [SIC 3674]
Impact Industries Inc, p. II-1266 [SIC 3363]
Impact Label Corp, p. II-1425 [SIC 3479]
Impact Plastics Inc, p. I-961 [SIC 3081]
Impco, p. II-1615 [SIC 3554]
IMPCO Technologies Inc, p. II-1742 [SIC 3592]
Imperial Adhesives Inc, p. I-879 [SIC 2891]
Imperial Bedding Co, p. I-558 [SIC 2515]
Imperial Bondware Inc, p. I-639 [SIC 2656]
Imperial Defense Systems Inc, p. II-2011 [SIC 3764]
Imperial Die Casting, p. II-1266 [SIC 3363]
Imperial Eastman, p. I-936 [SIC 3052]
Imperial Electric Co, p. II-1780 [SIC 3621]
Imperial Fabrics, p. I-424 [SIC 2391]
Imperial Graphics Inc, p. I-735 [SIC 2761]
Imperial Headwear Inc, p. I-386 [SIC 2353]
Imperial Holly Corp, p. I-119 [SIC 2062]
Imperial Industries Inc, p. I-1167 [SIC 3299]

Imperial of Morristown Inc, p. I-548 [SIC 2512]
Imperial Printing Co, p. I-698 [SIC 2721]
Imperial Spring Company Inc, p. II-1460 [SIC 3493]
Imperial Stamp and Engraving, p. II-1621 [SIC 3555]
Imperial Wallcoverings, p. I-688 [SIC 2679]
Imperial Woodworks Inc, p. I-581 [SIC 2531]
Implant Innovations Inc, p. II-2093 [SIC 3843]
Impulse Designs Inc, p. I-718 [SIC 2752]
IMS Products Inc, p. II-2003 [SIC 3751]
IMT, p. II-1578 [SIC 3545]
IMT Accessories Inc, p. I-1045 [SIC 3171]
In Focus Systems Inc, p. II-1915 [SIC 3679]
In Private Inc, p. I-336 [SIC 2321]
In-Sink-Erator, p. II-1822 [SIC 3639]
In Step Promotions Inc, p. I-932 [SIC 3021]
In Vitro International, p. I-809 [SIC 2835]
In Vitro Scientific Products Inc, p. I-1066 [SIC 3229]
INA Bearing Company Inc, p. II-1644 [SIC 3562]
INBRAND Corp, p. I-674 [SIC 2676]
Inclinator Company of America, p. II-1536 [SIC 3534]
Inco United States Inc, p. II-1212 [SIC 3325]
Incom Inc, p. II-1883 [SIC 3671]
INCSTAR Corp, p. II-2082 [SIC 3841]
Indak Manufacturing Corp, p. II-2045 [SIC 3822]
Indepane IG Inc, p. I-1057 [SIC 3211]
Independence Tube Corp, p. II-1192 [SIC 3317]
Independent Can Co, p. II-1300 [SIC 3411]
Independent Cement Corp, p. I-1077 [SIC 3241]
Independent Concrete Pipe Corp, p. I-1122 [SIC 3272]
Independent Container Inc, p. I-628 [SIC 2653]
Independent Forge Company Inc, p. II-1400 [SIC 3463]
Independent Stave Company Inc, p. I-476 [SIC 2429]
Independent Steel Castings, p. II-1207 [SIC 3324]
Index Print Inc, p. I-669 [SIC 2675]
India Ink, p. I-288 [SIC 2269]
Indian Country Inc, p. I-532 [SIC 2493]
Indian Head, p. I-491 [SIC 2435]
Indian Industries Inc, p. II-2150 [SIC 3949]
Indian Refining Ltd, p. I-903 [SIC 2911]
Indian Ribbon Inc, p. I-688 [SIC 2679]
Indian Ridge Canning Company, p. I-190 [SIC 2092]
Indian River Foods Inc, p. I-60 [SIC 2037]
Indiana Carbon Company Inc, p. II-2169 [SIC 3955]
Indiana Desk Company Inc, p. I-571 [SIC 2521]
Indiana Furniture Industries Inc, p. I-571 [SIC 2521]
Indiana General Motor Products, p. II-1780 [SIC 3621]
Indiana Harness and Saddlery, p. I-1053 [SIC 3199]
Indiana Knitwear Corp, p. I-336 [SIC 2321]
Indiana Rolling Mill Baling Corp, p. II-1229 [SIC 3341]
Indiana Steel and Wire Co, p. II-1182 [SIC 3315]
Indiana Tube Corp, p. II-1192 [SIC 3317]
Indianapolis Bakery, p. I-101 [SIC 2051]
Indianapolis Newspapers Inc, p. I-693 [SIC 2711]
Indium Corporation of America, p. II-1229 [SIC 3341]
Indmar Products Inc, p. II-1503 [SIC 3519]
Indramat, p. II-1780 [SIC 3621]
INDRESCO Inc, p. II-1525 [SIC 3532]
Indspec Chemical Corp, p. I-850 [SIC 2865]
Inductametals Corp, p. II-1304 [SIC 3412]
Inductoheat Inc, p. II-1671 [SIC 3567]
Inductotherm Corp, p. II-1671 [SIC 3567]
Inductotherm Industries Inc, p. II-1671 [SIC 3567]
Inductoweld Tube Corp, p. II-1486 [SIC 3498]
Industrial Acoustics Company, p. II-1492 [SIC 3499]
Industrial Airsystems Inc, p. II-1671 [SIC 3567]
Industrial Alloys Inc, p. II-1182 [SIC 3315]
Industrial and Wholesale Lumber, p. I-515 [SIC 2449]
Industrial Battery Engineering, p. II-1925 [SIC 3692]
Industrial Boiler Company Inc, p. II-1355 [SIC 3443]
Industrial Brush Company Inc, p. II-2181 [SIC 3991]
Industrial Brush Corp, p. II-2181 [SIC 3991]
Industrial Castings, p. II-1271 [SIC 3364]
Industrial Clutch Corp, p. II-1676 [SIC 3568]
Industrial Coatings Group Inc, p. I-961 [SIC 3081]
Industrial Coils Inc, p. II-1905 [SIC 3677]
Industrial Color Inc, p. I-850 [SIC 2865]
Industrial Computer Source, p. II-1688 [SIC 3571]

Industrial Crane and Equipment, p. II-1547 [SIC 3536]
Industrial Custom Prod LLC, p. I-942 [SIC 3053]
Industrial Data Entry, p. II-1699 [SIC 3575]
Industrial Devices Corp, p. II-1795 [SIC 3629]
Industrial Dielectrics Inc, p. II-1835 [SIC 3644]
Industrial Dynamics Ltd, p. II-2051 [SIC 3823]
Industrial Eng & Equipment Co, p. II-1339 [SIC 3433]
Industrial Fabrics Corp, p. I-245 [SIC 2241]
Industrial Fasteners Corp, p. II-1388 [SIC 3452]
Industrial Gasket and Shim, p. I-942 [SIC 3053]
Industrial Gasket Inc, p. I-942 [SIC 3053]
Industrial General Corp, p. II-1780 [SIC 3621]
Industrial Heat Treating, p. II-1290 [SIC 3398]
Industrial Ladder Co, p. I-536 [SIC 2499]
Industrial Lead & Plastics Constr, p. II-1229 [SIC 3341]
Industrial Louvers Inc, p. II-1244 [SIC 3354]
Industrial Metal Products Corp, p. II-1558 [SIC 3541]
Industrial Metalworking Services, p. II-1584 [SIC 3546]
Industrial Minerals Co, p. I-1090 [SIC 3255]
Industrial Parts Depot Inc, p. II-1503 [SIC 3519]
Industrial Piping Inc, p. II-1486 [SIC 3498]
Industrial Powder Coatings Inc, p. II-1295 [SIC 3399]
Industrial Sales Co, p. II-1676 [SIC 3568]
Industrial Saws Inc, p. II-1610 [SIC 3553]
Industrial Scientific Corp, p. II-2066 [SIC 3826]
Industrial Seaming Co, p. I-378 [SIC 2341]
Industrial Services Technologies, p. II-1355 [SIC 3443]
Industrial Steel Treating Co, p. II-1290 [SIC 3398]
Industrial Structures Inc, p. II-1367 [SIC 3446]
Industrial Technologies Inc, p. II-2071 [SIC 3827]
Industrial Tectonics Bearings, p. II-1644 [SIC 3562]
Industrial Tectonics Inc, p. II-1644 [SIC 3562]
Industrial Terminal Systems Inc, p. I-897 [SIC 2899]
Industrial Wire Products Corp, p. II-1182 [SIC 3315]
Industrial Woodworking Machine, p. II-1610 [SIC 3553]
Indy Lighting Inc, p. II-1845 [SIC 3646]
Inertia Dynamics Inc, p. II-1790 [SIC 3625]
Inex Vision Systems, p. II-2076 [SIC 3829]
Infilco Degremont Inc, p. II-1736 [SIC 3589]
Infinity Systems Inc, p. II-1859 [SIC 3651]
Infinity USA Inc, p. I-759 [SIC 2796]
Information Conservation Inc, p. I-750 [SIC 2789]
Information Display Technology, p. I-581 [SIC 2531]
Information Packaging Corp, p. I-679 [SIC 2677]
Information Storage Devices Inc, p. II-1694 [SIC 3572]
Information Transmission, p. II-1873 [SIC 3663]
Informer Computer Systems Inc, p. II-1699 [SIC 3575]
InfoWorld Publishing Corp, p. I-698 [SIC 2721]
Infusion Technology Inc, p. II-2082 [SIC 3841]
Ingalls Shipbuilding Inc, p. II-1986 [SIC 3731]
Ingersoll Cutting Tool Co, p. II-1632 [SIC 3559]
Ingersoll-Dresser Pump Co, p. II-1638 [SIC 3561]
Ingersoll Equipment Co, p. II-1514 [SIC 3524]
Ingersoll Products, p. II-1509 [SIC 3523]
Ingersoll-Rand Co, pp. Vol II - 1519, 1547, 1600, 1644, 1649 [SICs 3531, 3536, 3549, 3562, 3563]
Ingleside Plantation Inc, p. I-168 [SIC 2084]
Ingleside Plantation Winery, p. I-168 [SIC 2084]
Ingomar Packing Co, p. I-45 [SIC 2033]
Ingraham Time Products, p. II-2118 [SIC 3873]
Ingram Enterprises Inc, p. I-1127 [SIC 3273]
Initial Trends Inc, p. I-231 [SIC 2211]
Ink Co, p. I-889 [SIC 2893]
Inland Buildings, p. II-1372 [SIC 3448]
Inland Container Corp, p. I-628 [SIC 2653]
Inland Mills Co, p. I-70 [SIC 2041]
Inland Products Inc, p. I-152 [SIC 2077]
Inland Refractories Co, p. I-1163 [SIC 3297]
Inland Sintered Metals, p. II-1295 [SIC 3399]
Inland Steel Co, p. II-1172 [SIC 3312]
Inland Steel Industries Inc, p. II-1172 [SIC 3312]
Inman Mills Inc, p. I-231 [SIC 2211]
Inncom International Inc, p. II-2045 [SIC 3822]
InnerAsia Trading Company Inc, p. I-292 [SIC 2273]
InnoServ Technologies Inc, p. II-2082 [SIC 3841]
Innovative Data Technology, p. II-1934 [SIC 3695]
Innovative Folding Carton Co, p. I-643 [SIC 2657]

Roman numerals I and II indicate the volume in which pages appear. Volume I holds SICs 2011 through 3299; Volume II holds SICs 3312 through 3999.

2262

Company Index

Roman numerals I and II indicate the volume in which pages appear. Volume I holds SICs 2011 through 3299; Volume II holds SICs 3312 through 3999.

ITI MOVATS Inc, p. II-2076 [SIC 3829]
Ito Ham USA Inc, p. I-2 [SIC 2011]
Itron Inc, p. II-1878 [SIC 3669]
ITT Aerospace, p. II-1878 [SIC 3669]
ITT Aerospace Controls, p. II-1830 [SIC 3643]
ITT Automotive, p. II-1266 [SIC 3363]
ITT Automotive Inc, p. II-1954 [SIC 3714]
ITT Bell and Gossett, p. II-1638 [SIC 3561]
ITT Cannon, p. II-1910 [SIC 3678]
ITT Communications, p. I-713 [SIC 2741]
ITT Conoflow, p. II-1454 [SIC 3492]
ITT Corp, pp. Vol II - 1883, 1954 [SICs 3671, 3714]
ITT Defense and Electronics Inc, p. II-1915 [SIC 3679]
ITT Engineered Valves, p. II-1464 [SIC 3494]
ITT Fluid Technology Corp, p. II-1753 [SIC 3594]
ITT General Controls, p. II-1790 [SIC 3625]
ITT Gilfillan, p. II-2035 [SIC 3812]
ITT Pomona Electronics, p. II-2061 [SIC 3825]
ITT Sealectro, p. II-1910 [SIC 3678]
ITW Devcon, p. I-879 [SIC 2891]
ITW Fasterners, p. II-1388 [SIC 3452]
ITW Fastex, p. II-1413 [SIC 3469]
ITW Heartland Components, p. II-1584 [SIC 3546]
ITW Linx, p. II-1769 [SIC 3612]
ITW Paktron, p. II-1897 [SIC 3675]
ITW Ransburg Electrostatic Syst, p. II-1649 [SIC 3563]
ITW Shakeproof, p. II-1388 [SIC 3452]
ITW Southern Gage, p. II-1578 [SIC 3545]
ITW Waterbury Buckle Co, p. II-2177 [SIC 3965]
ITW Woodworth, p. II-1578 [SIC 3545]
IVAX Corp, p. I-855 [SIC 2869]
Ivex Packaging Corp, pp. Vol I - 618, 688 [SICs 2631, 2679]
Ivy Hill Corp, p. I-718 [SIC 2752]
IW Industries Inc, p. II-1382 [SIC 3451]
IXL Manufacturing Company, p. I-536 [SIC 2499]

J Allan Steel Co, p. II-1344 [SIC 3441]
J and B Sausage Company Inc, p. I-7 [SIC 2013]
J and C Dyeing Inc, p. I-288 [SIC 2269]
J and H Clasgens Company Inc, p. I-297 [SIC 2281]
J and J Enterprise of Louisiana, p. I-510 [SIC 2448]
J and J Flock Products Inc, p. I-283 [SIC 2262]
J and J Industries Inc, p. I-292 [SIC 2273]
J and J Mill and Lumber Co, p. I-1122 [SIC 3272]
J and J Pallet, p. I-510 [SIC 2448]
J and J Register, p. II-2045 [SIC 3822]
J and J Snack Foods Corp, p. I-107 [SIC 2052]
J and J South Central, p. I-628 [SIC 2653]
J and J Spring Company Inc, p. II-1460 [SIC 3493]
J and K Button Company Inc, p. II-2177 [SIC 3965]
J and K Sales Company Inc, p. II-2173 [SIC 3961]
J and L Metrology Company Inc, p. II-1759 [SIC 3596]
J and L Specialty Steel Inc, p. II-1187 [SIC 3316]
J and M Dyers Inc, p. I-288 [SIC 2269]
J and M Industries Inc, p. I-433 [SIC 2393]
J and R Film Company Inc, p. II-1859 [SIC 3651]
J and W Typesetting Co, p. I-754 [SIC 2791]
J-B Tool and Machine Inc, p. II-1404 [SIC 3465]
J Jenkins Sons Company Inc, p. II-2122 [SIC 3911]
J Josephson Inc, p. I-688 [SIC 2679]
J Kinderman and Sons Inc, p. II-2199 [SIC 3999]
J Lamb Inc, p. I-428 [SIC 2392]
J Lee Milligan Inc, p. I-909 [SIC 2951]
J-M Manufacturing Company, p. I-1006 [SIC 3089]
J Manheimer Inc, p. I-855 [SIC 2869]
J Marie Martin Co, p. I-428 [SIC 2392]
J Press Inc, p. I-420 [SIC 2389]
J Ray McDermott Inc, p. II-1986 [SIC 3731]
J Robert Scott Textiles Inc, p. I-236 [SIC 2221]
J Schoeneman Inc, p. I-332 [SIC 2311]
J-Star Industries, p. II-1509 [SIC 3523]
J-Von LP, p. I-973 [SIC 3083]
Jabel Inc, p. II-2122 [SIC 3911]
Jabil Circuit Inc, p. II-1887 [SIC 3672]
Jack Daniel's Distillery, p. I-172 [SIC 2085]
Jack Frost Inc, p. I-12 [SIC 2015]
Jack-Post Corp, p. I-553 [SIC 2514]
Jack Stone Company Inc, p. II-2186 [SIC 3993]

Jackson Furniture of Danville, p. I-548 [SIC 2512]
Jackson Lumber Harvester, p. II-1610 [SIC 3553]
Jackson Manufacturing Co, p. I-548 [SIC 2512]
Jackson-Mitchell Inc, p. I-26 [SIC 2023]
Jackson Ready Mix Inc, p. I-1127 [SIC 3273]
Jackson Tube Service Inc, p. II-1192 [SIC 3317]
Jackson Typesetting Co, p. I-754 [SIC 2791]
Jackson Wheeler Metals, p. II-1255 [SIC 3356]
JacksonLea, p. I-1144 [SIC 3291]
Jaco Manufacturing Co, p. II-1572 [SIC 3544]
Jacob Leinenkugel Brewing, p. I-160 [SIC 2082]
Jacob Siegel Company Inc, p. I-332 [SIC 2311]
Jacobs Electronics, p. II-1929 [SIC 3694]
Jacobs Management Corp, p. I-212 [SIC 2099]
Jacobsen Manufacturing Inc, p. I-519 [SIC 2451]
Jacquard Fabrics Inc, p. I-241 [SIC 2231]
Jacques Bobbe and Associates, p. I-160 [SIC 2082]
Jacques Moret Inc, p. I-258 [SIC 2253]
Jacuzzi Brothers, p. II-1638 [SIC 3561]
JAD Corporation of America Inc, p. I-660 [SIC 2673]
Jadair Inc, p. II-1525 [SIC 3532]
Jade Corp, p. II-1572 [SIC 3544]
Jaeckel Inc, p. II-2135 [SIC 3931]
Jahns Quality Pistons, p. II-1742 [SIC 3592]
Jalate Ltd, p. I-390 [SIC 2361]
Jameco Metal Products Co, p. I-416 [SIC 2387]
James A Murphy and Son Inc, p. II-2131 [SIC 3915]
James Austin Co, p. I-819 [SIC 2841]
James Burn/American Inc, p. II-1476 [SIC 3496]
James Hardie Building Products, p. I-1122 [SIC 3272]
James Hardie Irrigation Inc, p. II-1509 [SIC 3523]
James Industries Inc, p. II-2144 [SIC 3944]
James Jones Co, p. II-1447 [SIC 3491]
James L Taylor Mfg Co, p. II-1610 [SIC 3553]
James Machine Works Inc, p. II-1764 [SIC 3599]
James Page Brewing Co, p. I-160 [SIC 2082]
James River Corporation, pp. Vol. I - 674, 991 [SICs 2676, 3086]
James River Iron Co, p. II-1367 [SIC 3446]
James River Limestone Co, p. I-1127 [SIC 3273]
James River Paper Company Inc, p. I-612 [SIC 2621]
James Skinner Baking Co, p. I-112 [SIC 2053]
James Spring and Wire Co, p. II-1470 [SIC 3495]
James Thompson and Company, p. I-231 [SIC 2211]
Jameslee Corp, p. I-1041 [SIC 3161]
Jameson Home Products Inc, p. II-2045 [SIC 3822]
Jamestown Laminating Co, p. I-587 [SIC 2541]
Jamestown Metal Products Inc, p. II-2040 [SIC 3821]
Jamison Bedding Inc, p. I-558 [SIC 2515]
Jamison Door Co, p. II-1349 [SIC 3442]
Jan Bell Marketing Inc, p. II-2122 [SIC 3911]
Jan Michaels Inc, p. I-420 [SIC 2389]
Jana's Classics Inc, p. I-112 [SIC 2053]
Janco Designs Inc, p. I-536 [SIC 2499]
Jandy Industries Inc, p. II-2150 [SIC 3949]
Janel Inc, p. II-1312 [SIC 3423]
Janelle Products Inc, p. II-1840 [SIC 3645]
Janesville Products, p. I-791 [SIC 2823]
Janex International Inc, p. II-2144 [SIC 3944]
Jansko Inc, p. I-571 [SIC 2521]
Janson Industries, p. I-424 [SIC 2391]
Jansport Inc, p. I-356 [SIC 2329]
January and Wood Company Inc, p. I-322 [SIC 2298]
Japenamelac Corp, p. I-446 [SIC 2396]
Japler Aquisition Co, p. II-1795 [SIC 3629]
Japs-Olson Co, p. I-718 [SIC 2752]
Jarett Industries Inc, p. II-1498 [SIC 3511]
Jarp Industries Inc, p. II-1747 [SIC 3593]
Jarratt Studios, p. I-741 [SIC 2771]
Jarrett Rifles Inc, p. II-1438 [SIC 3484]
Jarvis Airfoil Inc, p. II-1975 [SIC 3724]
Jarvis East, p. II-1323 [SIC 3429]
Jarvis Pemco, p. II-1492 [SIC 3499]
Jarvis Products Corp, p. II-1626 [SIC 3556]
Jasco Fabrics Inc, p. I-267 [SIC 2257]
Jasco Products Company Inc, p. II-1859 [SIC 3651]
Jasco Tools Inc, p. II-1572 [SIC 3544]
Jason Inc, p. II-1498 [SIC 3511]
Jason Pharmaceuticals Inc, p. I-65 [SIC 2038]

Jasper Desk Co, p. I-571 [SIC 2521]
Jasper Farmers Elevator, p. I-95 [SIC 2048]
Jasper Glove Co, p. I-401 [SIC 2381]
Jasper Rubber Products Inc, p. I-954 [SIC 3069]
Jasper Seating Inc, p. I-571 [SIC 2521]
Jasper Veneers Inc, p. I-491 [SIC 2435]
Jasper Wyman and Son Inc, p. I-45 [SIC 2033]
Jaton Corp, p. II-1704 [SIC 3577]
Jaunty Company Inc, p. I-292 [SIC 2273]
Jaunty Textile, p. I-231 [SIC 2211]
Jaxson Roll Forming Inc, p. II-1212 [SIC 3325]
Jay-El Products Inc, p. II-1830 [SIC 3643]
Jay Franco and Sons Inc, p. I-428 [SIC 2392]
Jay Garment Co, p. I-348 [SIC 2325]
Jay-K Independent Lumber, p. I-480 [SIC 2431]
Jayar Manufacturing Co, p. I-973 [SIC 3083]
Jayark Corp, p. I-428 [SIC 2392]
Jayco Inc, p. II-1966 [SIC 3716]
Jaycraft Corp, p. II-1498 [SIC 3511]
Jays Foods LLC, p. I-199 [SIC 2096]
JB Battle Uniform Co, p. I-352 [SIC 2326]
JB Group Inc, p. I-310 [SIC 2295]
JB Kunz Co, p. I-745 [SIC 2782]
JB Martin Company Inc, p. I-236 [SIC 2221]
JBF Scientific Company Inc, p. II-1986 [SIC 3731]
JBG Inc, p. I-224 [SIC 2131]
JBJ Fabrics Inc, p. I-279 [SIC 2261]
JBL Consumer Products, p. II-1859 [SIC 3651]
JBL Inc, p. II-1859 [SIC 3651]
JC Carter Company Inc, p. II-1980 [SIC 3728]
JC Decker Company Inc, p. I-1053 [SIC 3199]
JC Goss Co, p. I-437 [SIC 2394]
JC Potter Sausage Co, p. I-7 [SIC 2013]
JC Renfroe and Sons Inc, p. II-1547 [SIC 3536]
JcAIR Inc, p. II-2061 [SIC 3825]
JCM Industries Inc, p. II-1552 [SIC 3537]
JD Gould Company Inc, p. II-1454 [SIC 3492]
JD Wilkins Co, p. II-1367 [SIC 3446]
JDI Group Inc, p. I-542 [SIC 2511]
JE Baker Co, p. I-1163 [SIC 3297]
JE Hanger Inc, p. II-2087 [SIC 3842]
JE Siebel Son's, p. I-181 [SIC 2087]
JE Simon Co, p. I-1127 [SIC 3273]
Jean Philippe Fragrances Inc, p. I-834 [SIC 2844]
Jeannette Shade and Novelty, p. I-1066 [SIC 3229]
Jebco Inc, p. I-576 [SIC 2522]
Jeffboat, p. II-1986 [SIC 3731]
Jefferson Ice Company Inc, p. I-204 [SIC 2097]
Jefferson Mills Inc, p. I-302 [SIC 2282]
Jefferson Smurfit Corp, pp. Vol. I - 618, 729 [SICs 2631, 2759]
Jefferson Wood Working, p. I-471 [SIC 2426]
Jeffrey Chain Corp, p. II-1676 [SIC 3568]
Jel Sert Co, p. I-181 [SIC 2087]
Jeld-Wen Inc, p. I-480 [SIC 2431]
Jeld-Wen Inc Young Door Co, p. I-480 [SIC 2431]
Jem Sportswear Inc, p. I-336 [SIC 2321]
Jemison Investment Company, p. I-465 [SIC 2421]
Jen Cel Lite Corp, p. I-455 [SIC 2399]
Jeneric-Pentron Inc, p. II-2093 [SIC 3843]
Jenkins Metal Corp, p. II-1605 [SIC 3552]
Jenn-Air Co, p. II-1800 [SIC 3631]
Jennifer Dale Inc, p. I-378 [SIC 2341]
Jennifer Dawn Inc, p. I-373 [SIC 2339]
Jennison Enterprises Inc, p. II-1280 [SIC 3366]
Jennmar Corp, p. II-1388 [SIC 3452]
Jensen Cabinet Inc, p. I-587 [SIC 2541]
Jensen Corp, p. II-1722 [SIC 3582]
Jensen Industries Inc, p. II-1800 [SIC 3631]
Jensen International Inc, p. II-1638 [SIC 3561]
Jensen Manufacturing Co, p. I-136 [SIC 2068]
Jensen Mixers International Inc, p. II-1531 [SIC 3533]
Jensen's Inc, p. II-2160 [SIC 3952]
Jeppesen Sanderson Inc, p. I-713 [SIC 2741]
Jerebar Corp, p. II-1275 [SIC 3365]
Jerell Inc, p. I-365 [SIC 2335]
Jergens Inc, p. II-1572 [SIC 3544]
Jerome Cheese Co, p. I-21 [SIC 2022]
Jerome Foods Inc, p. I-12 [SIC 2015]

Roman numerals I and II indicate the volume in which pages appear. Volume I holds SICs 2011 through 3299; Volume II holds SICs 3312 through 3999.

Roman numerals I and II indicate the volume in which pages appear. Volume I holds SICs 2011 through 3299; Volume II holds SICs 3312 through 3999.

Just Toys Inc, p. II-2140 [SIC 3942]
Justice Record Co, p. II-1864 [SIC 3652]
Justin Boot Co, p. I-1025 [SIC 3143]
Justin Industries Inc, p. I-1081 [SIC 3251]
Justin's Ice Cream Company Inc, p. I-30 [SIC 2024]
JusTins Ltd, p. II-1300 [SIC 3411]
JW Allen and Co, p. I-83 [SIC 2045]
JW Fergusson and Sons, p. I-724 [SIC 2754]
JW Peters and Son Inc, p. I-1122 [SIC 3272]
JW Rex Co, p. II-1290 [SIC 3398]
JW Speaker Corp, p. II-1850 [SIC 3647]
JWI Inc, p. II-1682 [SIC 3569]
JY Taylor Manufacturing Co, p. II-2056 [SIC 3824]

K and B Brothers Inc, p. II-2140 [SIC 3942]
K and B Manufacturing Inc, p. II-2144 [SIC 3944]
K & C Organic Soil Prod Corp, p. I-861 [SIC 2873]
K and F Manufacturing Co, p. II-1621 [SIC 3555]
K and H Finishing Inc, p. II-1425 [SIC 3479]
K and K Screw Products Inc, p. II-1382 [SIC 3451]
K and L Feed Mill Corp, p. I-95 [SIC 2048]
K and M Electronics Inc, p. II-1901 [SIC 3676]
K and M Manufacturing Co, p. I-361 [SIC 2331]
K and P Manufacturing, p. II-2029 [SIC 3799]
K and R Sportswear Inc, p. I-373 [SIC 2339]
K and W Products, p. I-918 [SIC 2992]
K-Bin Inc, p. I-782 [SIC 2821]
K-Byte Manufacturing, p. II-1887 [SIC 3672]
K-C Aviation Inc, p. II-1970 [SIC 3721]
K-C Products Company Inc, p. I-428 [SIC 2392]
K-D Manitou Inc, p. II-1552 [SIC 3537]
K-III Communications Corp, p. I-703 [SIC 2731]
K-Lath, p. II-1476 [SIC 3496]
K-Line Industries Inc, p. II-1312 [SIC 3423]
K/P Corp, p. I-718 [SIC 2752]
K-Ply Inc, p. I-496 [SIC 2436]
K-Products Inc, p. I-356 [SIC 2329]
K-Swiss Inc, p. I-1033 [SIC 3149]
K-Tech Manufacturing Inc, p. II-1388 [SIC 3452]
K-Tel International, p. II-1864 [SIC 3652]
K-Tron International Inc, p. II-2051 [SIC 3823]
K-Tron North America Corp, p. II-2051 [SIC 3823]
K2 Corp, p. II-2150 [SIC 3949]
Kahr Bearing, p. II-1644 [SIC 3562]
Kaibab Industries Inc, p. I-465 [SIC 2421]
Kaiser Aerotech, p. II-2011 [SIC 3764]
Kaiser Aluminum Corp, p. II-1239 [SIC 3353]
Kaiser Cement Corp, p. I-1077 [SIC 3241]
Kaiser Ceramic Composites, p. I-1167 [SIC 3299]
Kaiser Electronics, p. II-1980 [SIC 3728]
Kaiser Electroprecision Inc, p. II-1980 [SIC 3728]
Kaiser Marquardt, p. II-2011 [SIC 3764]
Kaiser Sand and Gravel Co, p. I-909 [SIC 2951]
Kal Kan Foods Inc, p. I-91 [SIC 2047]
Kalama Chemical Inc, p. I-181 [SIC 2087]
Kalamazoo Container, p. I-628 [SIC 2653]
Kalamazoo Industries Inc, p. I-1144 [SIC 3291]
Kalas Manufacturing Inc, p. II-1260 [SIC 3357]
Kalex Chemical Products Inc, p. I-649 [SIC 2671]
Kalibre, p. II-2122 [SIC 3911]
Kalikow Brothers Inc, p. I-348 [SIC 2325]
Kalil Bottling Co, p. I-176 [SIC 2086]
Kalmar AC Inc, p. II-1552 [SIC 3537]
Kalmbach Publishing Co, p. I-698 [SIC 2721]
Kalmus and Associates, p. II-1887 [SIC 3672]
Kalsec Inc, p. I-212 [SIC 2099]
Kama Corp, p. I-961 [SIC 3081]
Kamaka Hawaii Inc, p. II-2135 [SIC 3931]
Kaman Aerospace Corp, p. II-1980 [SIC 3728]
Kaman Corp, p. II-1970 [SIC 3721]
Kaman Music Corp, p. II-2135 [SIC 3931]
Kamatics Corp, p. II-1644 [SIC 3562]
Kan Build Inc, p. I-523 [SIC 2452]
Kane Hardware, p. I-460 [SIC 2411]
Kane Industries Inc, p. I-332 [SIC 2311]
Kangaroo Technologies Corp, p. II-1699 [SIC 3575]
Kangaroos Inc, p. I-1053 [SIC 3199]
Kansas Brick and Tile Company, p. I-1081 [SIC 3251]
Kansas City Drumworks, p. II-2135 [SIC 3931]

Kansas City Star Co, p. I-693 [SIC 2711]
Kansas Oxygen Inc, p. I-768 [SIC 2813]
Kanthal Corp, p. II-1182 [SIC 3315]
Kantor Brothers Neckwear, p. I-344 [SIC 2323]
Kanzaki Specialty Papers Inc, p. I-654 [SIC 2672]
Kao Corporation of America Inc, p. I-777 [SIC 2819]
Kapak Corp, p. I-1049 [SIC 3172]
Kaplan Building Systems Inc, p. II-1372 [SIC 3448]
Kappler USA Inc, p. II-2087 [SIC 3842]
Kar Nut Products Co, p. I-136 [SIC 2068]
Karbra Co, p. II-2122 [SIC 3911]
Kard Corp, p. II-1563 [SIC 3542]
Kardex Systems Inc, p. I-593 [SIC 2542]
Kards for Kids Inc, p. I-741 [SIC 2771]
Karl Schmidt Unisia Inc, p. II-1742 [SIC 3592]
Karlshamns USA Inc, p. I-148 [SIC 2076]
Karman Kitchens Inc, p. I-486 [SIC 2434]
Kasbar National Industries Inc, p. I-327 [SIC 2299]
Kasco Corp, p. II-1318 [SIC 3425]
Kaser Corp, p. I-1154 [SIC 3295]
Kaspar Wire Works, p. II-1476 [SIC 3496]
Kass and Co, p. I-373 [SIC 2339]
Kasson and Keller Inc, p. I-1071 [SIC 3231]
Kaswer Custom Inc, p. II-1430 [SIC 3482]
Kater-Crafts Bookbinders Inc, p. I-750 [SIC 2789]
Kato Engineering, p. II-1780 [SIC 3621]
Katty Industries Inc, p. II-1742 [SIC 3592]
Katzenberg Brothers Inc, p. I-356 [SIC 2329]
Kaufman Knitting Company Inc, p. I-258 [SIC 2253]
Kaukauna Cheese Wisconsin LP, p. I-21 [SIC 2022]
Kavlico Corp, p. II-1915 [SIC 3679]
Kawada Industries USA Inc, p. II-1372 [SIC 3448]
Kawai America Corp, p. II-2135 [SIC 3931]
Kawai America Manufacturing, p. II-2135 [SIC 3931]
Kawasaki Motors, p. II-2003 [SIC 3751]
Kawasaki Robotics, p. II-1682 [SIC 3569]
Kay Dee Feed Company Inc, p. I-95 [SIC 2048]
Kay Home Products, p. I-515 [SIC 2449]
Kay Home Products Inc, pp. Vol. I - 553, Vol. II - 1492 [SICs 2514, 3499]
Kay Lynn Sportswear Inc, p. I-369 [SIC 2337]
Kay Pneumatics Inc, p. II-1454 [SIC 3492]
Kaydon Corp, pp. Vol II - 1644, 1682 [SICs 3562, 3569]
Kaydon Ring and Seal Inc, p. II-1742 [SIC 3592]
Kaye-Smith Business Graphics, p. I-735 [SIC 2761]
Kayem Foods Inc, p. I-7 [SIC 2013]
Kayex Corp, p. II-1671 [SIC 3567]
Kaynar Technologies Inc, p. II-1388 [SIC 3452]
Kayo of California Inc, p. I-373 [SIC 2339]
Kaytee Products Inc, p. I-95 [SIC 2048]
KB Electronics Inc, p. II-1790 [SIC 3625]
KB Lighting Manufacturing, p. II-1840 [SIC 3645]
KB Socks Inc, p. I-250 [SIC 2251]
KBA-Motter Corp, p. II-1621 [SIC 3555]
KBD Inc, p. II-1840 [SIC 3645]
KBI Ltd, p. II-1503 [SIC 3519]
KC Abrasive Company Inc, p. I-1144 [SIC 3291]
KC Photo Engraving Co, p. I-759 [SIC 2796]
KCI Communications Inc, p. I-713 [SIC 2741]
KCI Inc, p. II-1649 [SIC 3563]
KCL Corp, p. I-660 [SIC 2673]
KCS Industries Inc, p. II-2186 [SIC 3993]
KCS International Inc, p. II-1992 [SIC 3732]
KD Kanopy Inc, p. I-437 [SIC 2394]
KD Lamp Co, p. II-1850 [SIC 3647]
KDI Precision Products Inc, p. II-2061 [SIC 3825]
Keane Monroe Corp, p. I-1071 [SIC 3231]
Kearfott Guidance, p. II-2035 [SIC 3812]
Kearney K-P-F Electric, p. II-1775 [SIC 3613]
Kearse Manufacturing Company, p. I-491 [SIC 2435]
Keco Industries Inc, p. II-1726 [SIC 3585]
Kedem Food Products Co, p. I-40 [SIC 2032]
Kedie Image Systems, p. I-759 [SIC 2796]
Kedman Co, p. II-1312 [SIC 3423]
Keeler Hardware Group, p. II-1323 [SIC 3429]
Keene, p. I-510 [SIC 2448]
Keepers International Inc, p. I-254 [SIC 2252]
Keiper Recaro Seating Inc, p. I-581 [SIC 2531]
Keith Clark, p. I-729 [SIC 2759]

Keith Smith Company Inc, p. I-95 [SIC 2048]
Keithley Instruments Inc, p. II-2061 [SIC 3825]
Kelch Corp, p. II-1275 [SIC 3365]
Kelco Division, p. I-212 [SIC 2099]
Kelco Industries Inc, p. II-1382 [SIC 3451]
Keller Aluminum Products, p. II-1349 [SIC 3442]
Keller Extrusions of Texas, p. II-1244 [SIC 3354]
Keller Group Inc, p. II-1395 [SIC 3462]
Keller Holding Inc, p. II-1349 [SIC 3442]
Keller Industries Inc, p. II-1349 [SIC 3442]
Keller Laboratories Inc, p. II-2093 [SIC 3843]
Keller Manufacturing Co, p. I-542 [SIC 2511]
Keller Products Inc, p. I-491 [SIC 2435]
Keller Ticket Co, p. I-654 [SIC 2672]
Kelley Company Inc, p. II-1536 [SIC 3534]
Kellogg Brush Manufacturing Co, p. II-2181 [SIC 3991]
Kellogg Co, p. I-75 [SIC 2043]
Kellogg Industries, p. II-2087 [SIC 3842]
Kellogg Supply Co, p. I-861 [SIC 2873]
Kellogg USA Inc, p. I-75 [SIC 2043]
Kellwood Co, p. I-365 [SIC 2335]
Kelly Foods Inc, p. I-40 [SIC 2032]
Kelly Micro Systems Inc, p. II-1892 [SIC 3674]
Kelly-Moore Paint Company Inc, p. I-840 [SIC 2851]
Kelly-Springfield Tire Co, p. I-927 [SIC 3011]
Kelty Pack Inc, p. I-455 [SIC 2399]
Kemco Systems Inc, p. II-1822 [SIC 3639]
KEMET Corp, p. II-1897 [SIC 3675]
Kemet Electronics Corp, p. II-1897 [SIC 3675]
Kemin Industries Inc, p. I-95 [SIC 2048]
Kemlon Prod & Development, p. II-1910 [SIC 3678]
Kemp Industries Inc, p. II-1747 [SIC 3593]
Kemper Enterprises Inc, p. II-1312 [SIC 3423]
Kemper Industries Inc, p. I-576 [SIC 2522]
Kempsmith Machine Co, p. II-1615 [SIC 3554]
Ken Specialties Inc, p. II-1312 [SIC 3423]
Kenall Manufacturing Co, p. II-1845 [SIC 3646]
Kenda Knits Inc, p. I-267 [SIC 2257]
Kendale Industries Inc, p. II-1388 [SIC 3452]
Kendall Co, p. II-2087 [SIC 3842]
Kendall Healthcare Products Co, p. II-2082 [SIC 3841]
Kendall International Inc, p. II-2087 [SIC 3842]
Kendrick Carpets Inc, p. I-292 [SIC 2273]
Kendrick Company Inc, p. I-245 [SIC 2241]
KENETECH Corp, p. II-1498 [SIC 3511]
Kenfair Manufacturing Co, p. I-599 [SIC 2591]
Kenkor Molding, p. I-991 [SIC 3086]
Kenmar Comfort Company Inc, p. I-442 [SIC 2395]
Kennametal Inc, p. II-1558 [SIC 3541]
Kennebunk Weavers Inc, p. I-428 [SIC 2392]
Kennebunkport Brewing Co, p. I-160 [SIC 2082]
Kennedy Co, p. II-1694 [SIC 3572]
Kennedy Die Castings Inc, p. II-1266 [SIC 3363]
Kennedy Endeavors Inc, p. I-199 [SIC 2096]
Kennedy Manufacturing Co, p. II-1413 [SIC 3469]
Kennedy Sawmills Inc, p. I-528 [SIC 2491]
Kennedy Value, p. II-1464 [SIC 3494]
Kennedy Wood Products Inc, p. I-476 [SIC 2429]
Kenner Products, p. II-2144 [SIC 3944]
Kenneth Cole Productions Inc, p. I-1029 [SIC 3144]
Kenneth Fox Supply Co, p. I-433 [SIC 2393]
Kenneth Gordon New Orleans, p. I-336 [SIC 2321]
Kenney Drapery Associates Inc, p. I-424 [SIC 2391]
Kenney Manufacturing Co, p. I-599 [SIC 2591]
Kenney Steel Treating Co, p. II-1290 [SIC 3398]
Kenosha Beef International Ltd, p. I-2 [SIC 2011]
Kenro Corp, p. II-2112 [SIC 3861]
Kensington Lamp Co, p. II-1840 [SIC 3645]
Kent-Bragaline Inc, p. I-283 [SIC 2262]
Kent Co, p. II-1736 [SIC 3589]
Kent Corp, p. I-593 [SIC 2542]
Kent Electronics Corp, p. II-1915 [SIC 3679]
Kent Feeds Inc, p. I-95 [SIC 2048]
Kent Manufacturing Co, p. I-297 [SIC 2281]
Kent-Moore Tool Group, p. II-1632 [SIC 3559]
Kent Sportswear Inc, p. I-356 [SIC 2329]
KENTECH Windpower Inc, p. II-1498 [SIC 3511]
Kenton Industries Inc, p. II-1840 [SIC 3645]

Roman numerals I and II indicate the volume in which pages appear. Volume I holds SICs 2011 through 3299; Volume II holds SICs 3312 through 3999.

Roman numerals I and II indicate the volume in which pages appear. Volume I holds SICs 2011 through 3299; Volume II holds SICs 3312 through 3999.

Kolene Corp, p. II-1671 [SIC 3567]
Koller Group, p. II-1476 [SIC 3496]
Kollmann Monumental Works, p. I-1140 [SIC 3281]
Kollmorgen Corp, pp. Vol II - 1780, 2071, 2112 [SICs 3621, 3827, 3861]
Kollsman, p. II-2035 [SIC 3812]
Kolmar Laboratories Inc, p. I-834 [SIC 2844]
Kolpak, p. II-1726 [SIC 3585]
Komag Inc, p. II-1934 [SIC 3695]
Komar Industries Inc, p. II-1764 [SIC 3599]
Komatsu Dresser Co, p. II-1519 [SIC 3531]
KOMET of AMERICA Inc, p. II-1578 [SIC 3545]
Komline-Sanderson Engineering, p. II-1682 [SIC 3569]
Komo Machine Inc, p. II-1558 [SIC 3541]
Komtek Inc, p. II-1395 [SIC 3462]
Kona Corp, p. II-1632 [SIC 3559]
Kona Kai Farms, p. I-195 [SIC 2095]
Konami, p. II-2144 [SIC 3944]
Kone Holdings Inc, p. II-1790 [SIC 3625]
KONE-Landel Inc, p. II-1547 [SIC 3536]
Koneta Rubber-LRV, p. I-954 [SIC 3069]
Konrad Corp, p. II-1589 [SIC 3547]
Konrad Marine Inc, p. II-1503 [SIC 3519]
Kontur Kontact Lens Co, p. II-2107 [SIC 3851]
Konz Wood Products Inc, p. I-510 [SIC 2448]
Kooima Manufacturing, p. II-1492 [SIC 3499]
Kooltronic Inc, p. II-1654 [SIC 3564]
Koolvent Aluminum Products, p. II-1349 [SIC 3442]
Koos Inc, p. I-897 [SIC 2899]
Kopp Glass Inc, p. I-1066 [SIC 3229]
Koppel Steel Corp, p. II-1172 [SIC 3312]
Koppers Industries Inc, p. I-528 [SIC 2491]
Koppy Corp, p. II-1563 [SIC 3562]
Koral Industries Inc, p. I-1001 [SIC 3088]
Korber Hats Inc, p. I-386 [SIC 2353]
Koret Inc, pp. Vol. I - 361, 1045 [SICs 2331, 3171]
Koret of California Inc, p. I-361 [SIC 2331]
Korex Co, p. I-819 [SIC 2841]
Koring Brothers Inc, p. I-322 [SIC 2298]
Korn Industries Inc, p. I-542 [SIC 2511]
Korry Electronics Co, p. II-1980 [SIC 3728]
Kortick Manufacturing Co, p. II-1835 [SIC 3644]
Koryn Rolstad/Bannerworks Inc, p. I-455 [SIC 2399]
Kosakura and Associates Inc, p. I-587 [SIC 2541]
Koss Classics Ltd, p. II-1864 [SIC 3652]
Koss Corp, p. II-1859 [SIC 3651]
Kosto Food Products Co, p. I-181 [SIC 2087]
Kotecki Monuments Inc, p. I-1140 [SIC 3281]
Kottler Industries Inc, p. I-536 [SIC 2499]
Kowa Printing Corp, p. I-735 [SIC 2761]
KP Iron Foundry Inc, p. II-1275 [SIC 3365]
KPT Inc, p. I-1086 [SIC 3253]
KR Edwards Leaf Tobacco, p. I-228 [SIC 2141]
KR Industries Inc, p. I-581 [SIC 2531]
Kraft, p. I-608 [SIC 2611]
Kraft Food Ingredients Corp, p. I-212 [SIC 2099]
Kraft Hardware Inc, p. II-1334 [SIC 3432]
Kraftile Co, p. I-1086 [SIC 3253]
Kraftube Inc, p. II-1192 [SIC 3317]
Kraftware Corp, p. I-1066 [SIC 3229]
Kramer Co, p. II-1726 [SIC 3585]
Krames Communications Inc, p. I-703 [SIC 2731]
Kranco Crane Services Inc, p. II-1547 [SIC 3536]
Krause Milling Co, p. I-70 [SIC 2041]
Kreber Graphic Inc, p. I-759 [SIC 2796]
Kremco Inc, p. II-1531 [SIC 3533]
Kremlin Inc, p. II-1649 [SIC 3563]
Krengel Manufacturing Company, p. II-2164 [SIC 3953]
Kreonite Inc, p. II-2112 [SIC 3861]
Kress Corp, p. II-1519 [SIC 3531]
Kreuter Manufacturing Company, p. II-2045 [SIC 3822]
Krispy Kreme Doughnut Corp, p. I-101 [SIC 2051]
Krist Gudnason Inc, p. I-365 [SIC 2335]
Krogh Pump Co, p. II-1753 [SIC 3594]
Krohn Industries Inc, p. II-2093 [SIC 3843]
Krone Casting Corp, p. II-1266 [SIC 3363]
Krones Inc, p. II-1660 [SIC 3565]

Kronos, p. I-773 [SIC 2816]
Kronos Inc, p. II-1713 [SIC 3579]
Kropp Forge, p. II-1395 [SIC 3462]
Kroy Inc, p. II-1713 [SIC 3579]
Krueger International Inc, p. I-576 [SIC 2522]
Krueger Sheet Metal Co, p. I-1094 [SIC 3259]
Kruger and Sons Inc, p. I-55 [SIC 2035]
Krupp Gerlach Co, p. II-1395 [SIC 3462]
Krupp Robins Inc, p. II-1541 [SIC 3535]
KS Enterprises Inc, p. I-501 [SIC 2439]
KSC Industries Inc, p. II-1905 [SIC 3677]
KSG Industries Inc, p. II-1742 [SIC 3592]
KSM Seafood Corp, p. I-190 [SIC 2092]
KTR Corp, p. II-1676 [SIC 3568]
Kue-Ken Corp, p. II-1525 [SIC 3532]
Kuehne Chemical Co, p. I-777 [SIC 2819]
Kufner Textile Corp, p. I-420 [SIC 2389]
Kugel Komponents Inc, p. II-1943 [SIC 3711]
Kugler Oil Co, p. I-869 [SIC 2875]
Kuhlman Corp, pp. Vol. I - 1127, Vol. II - 1769 [SICs 3273, 3612]
Kuka Welding, p. II-1541 [SIC 3535]
Kukla Press Inc, p. I-718 [SIC 2752]
Kulicke and Soffa Industries Inc, p. II-1632 [SIC 3559]
Kunde Enterprises Inc, p. I-168 [SIC 2084]
Kunkle Industries Inc, pp. Vol II - 1464, 1492 [SICs 3494, 3499]
Kunzler and Company Inc, p. I-7 [SIC 2013]
Kurdziel Industries Inc, p. II-1197 [SIC 3321]
Kurfees Coatings Inc, p. I-840 [SIC 2851]
Kurt Manufacturing Co, p. II-1492 [SIC 3499]
Kurtz Bros Inc, p. I-684 [SIC 2678]
Kurtztown Foundry, p. II-1212 [SIC 3325]
Kurz-Hastings Inc, p. II-1482 [SIC 3497]
Kush Industries, p. II-1709 [SIC 3578]
Kusters Corp, p. II-1605 [SIC 3552]
Kut-Kwick Corp, p. II-1514 [SIC 3524]
Kutters Cheese Factory, p. I-21 [SIC 2022]
KUX Manufacturing Co, p. II-2186 [SIC 3993]
KVAL Inc, p. II-1610 [SIC 3553]
KVP Systems Inc, p. I-936 [SIC 3052]
KW Thompson Tool Co, p. II-1438 [SIC 3484]
Kwik-File Inc, p. I-576 [SIC 2522]
Kwik Wall Co, p. I-587 [SIC 2541]
Kwik-Way Industries Inc, p. II-1558 [SIC 3541]
Kwik-Way Manufacturing Co, p. II-1578 [SIC 3545]
KWS Manufacturing Company, p. II-1541 [SIC 3535]
KYB Corporation of America, p. II-1323 [SIC 3429]
KYD Inc, p. I-45 [SIC 2033]
Kysor Industrial Corp, p. II-1726 [SIC 3585]

L-7 Inc, p. I-369 [SIC 2337]
L and B Products West Corp, p. I-603 [SIC 2599]
L and D Manufacturing Inc, p. I-378 [SIC 2341]
L and E Packaging Inc, p. I-688 [SIC 2679]
L & H Wood Manufacturing Co, p. I-510 [SIC 2448]
L and J Engineering Inc, p. II-2056 [SIC 3824]
L and J Technologies Inc, p. II-1464 [SIC 3494]
L and JG Stickley Inc, p. I-542 [SIC 2511]
L & L Fittings Mfg Co, p. II-1464 [SIC 3494]
L and L Manufacturing Co, p. I-356 [SIC 2329]
L and L Products Inc, p. I-879 [SIC 2891]
L and LK Inc, p. II-1255 [SIC 3356]
L and M Castings, p. II-2122 [SIC 3911]
L & P Aluminum Smelting, p. II-1229 [SIC 3341]
L and P Machine Inc, p. II-1438 [SIC 3484]
L&R Furniture Manufacturing, p. I-567 [SIC 2519]
L and R Manufacturing Co, p. II-1938 [SIC 3699]
L and S Machine Company Inc, p. II-1975 [SIC 3724]
L and S Products Inc, p. I-593 [SIC 2542]
L and W Engineering Company, p. II-1404 [SIC 3465]
L Farber Company Inc, p. I-1017 [SIC 3131]
L Foppiano Wine Co, p. I-168 [SIC 2084]
L'Garde Inc, p. II-2007 [SIC 3761]
L Gordon Packaging Inc, p. I-623 [SIC 2652]
L Hardy Co, p. II-1308 [SIC 3421]
L Karp and Sons Inc, p. I-101 [SIC 2051]
L Ray Packing Co, p. I-186 [SIC 2091]
L-S Plate and Wire Corp, p. II-1225 [SIC 3339]

L Thorn Company Inc, p. I-1117 [SIC 3271]
La Barge Inc, p. I-542 [SIC 2511]
La Canasta of Minnesota Inc, p. I-212 [SIC 2099]
La Choy Food Products Co, p. I-45 [SIC 2033]
La Cie Ltd, p. II-1694 [SIC 3572]
La-Co Industries Inc, p. I-879 [SIC 2891]
LA Darling Co, p. I-593 [SIC 2542]
La-Del Manufacturing Company, p. I-369 [SIC 2337]
LA Gear Inc, p. I-1033 [SIC 3149]
La Gloria Oil and Gas Co, p. I-903 [SIC 2911]
La Habra Products Inc, p. I-1167 [SIC 3299]
La-Man Corp, p. II-1753 [SIC 3594]
La Marche Manufacturing Co, p. II-1795 [SIC 3629]
La Mode Sportswear Group, p. I-336 [SIC 2321]
La Reina Inc, p. I-212 [SIC 2099]
La Roche Chemicals Inc, p. I-777 [SIC 2819]
LA Rockler Fur Co, p. I-398 [SIC 2371]
La Tempesta Bakery Confections, p. I-107 [SIC 2052]
La Victoria Foods Inc, p. I-45 [SIC 2033]
La-Z-Boy Chair Co, pp. Vol. I - 548, 571 [SICs 2512, 2521]
La-Z-Boy South, p. I-548 [SIC 2512]
Lab Glass, p. I-1071 [SIC 3231]
LaBarge Electronics, p. II-1323 [SIC 3429]
LaBarge Inc, p. II-2035 [SIC 3812]
LaBarge Pipe and Steel Co, p. II-1486 [SIC 3498]
Labconco Corp, p. II-2040 [SIC 3821]
Label-Aire Inc, p. II-1660 [SIC 3565]
Label America Inc, p. I-724 [SIC 2754]
Label Art Inc, p. I-654 [SIC 2672]
Label Systems/OSI Corp, p. I-879 [SIC 2891]
Labeljet, p. II-1660 [SIC 3565]
Labelon Corp, p. I-688 [SIC 2679]
LaBounty Manufacturing Inc, p. II-1519 [SIC 3531]
Labs International Inc, p. II-2118 [SIC 3873]
Lacey Milling Co, p. I-70 [SIC 2041]
Lacey Rug Mills Inc, p. I-292 [SIC 2273]
Lachina Draperies Co, p. I-424 [SIC 2391]
Lackawanna Leather Company, p. I-1013 [SIC 3111]
Lacks Enterprises Inc, pp. Vol II - 1404, 1420 [SICs 3465, 3471]
Lacks Industries Inc, p. II-1954 [SIC 3714]
Laclede Steel Co, p. II-1172 [SIC 3312]
LaCrosse Footwear Inc, p. I-932 [SIC 3021]
Lactona Corp, p. II-2181 [SIC 3991]
Lacy Foundries Inc, p. II-1568 [SIC 3543]
LADD Furniture Inc, p. I-542 [SIC 2511]
Ladish Company Inc, p. II-1395 [SIC 3462]
Ladshaw Explosives Inc, p. I-885 [SIC 2892]
Lady Carol Dresses Inc, p. I-365 [SIC 2335]
Lady Ester Lingerie Corp, p. I-378 [SIC 2341]
Lady Hope Dress Company Inc, p. I-365 [SIC 2335]
Lady Linda Covers Inc, p. I-424 [SIC 2391]
Lady Madison Inc, p. I-292 [SIC 2273]
Ladyfair Mills Inc, p. I-378 [SIC 2341]
Lafarge Corp, p. I-1077 [SIC 3241]
Lafayette Manufacturing Co, p. I-465 [SIC 2421]
Lafitte Frozen Foods Corp, p. I-190 [SIC 2092]
Laher Spring Corp, p. II-1460 [SIC 3493]
Laidlaw Corp, p. II-1476 [SIC 3496]
Laird and Co, p. I-172 [SIC 2085]
Laird Plastics Inc, p. I-967 [SIC 3082]
Lake Erie Design Inc, p. II-1568 [SIC 3543]
Lake Erie Screw Corp, p. II-1388 [SIC 3452]
Lake Odessa Group, p. II-1404 [SIC 3465]
Lake Shore Litho Inc, p. I-759 [SIC 2796]
Lake Superior Paper Industries, p. I-612 [SIC 2621]
Lake Union Drydock Co, p. II-1986 [SIC 3731]
Lake Zurich Film, p. I-961 [SIC 3081]
Lakefront Brewery Inc, p. I-160 [SIC 2082]
Lakeside Dairy, p. I-30 [SIC 2024]
Lakeside Foods Inc, p. I-45 [SIC 2033]
Lakeside Machine Inc, p. II-1764 [SIC 3599]
Lakeside Metals Inc, p. II-1250 [SIC 3355]
Laketronics Inc, p. II-1929 [SIC 3694]
Lakeview Brewery LP, p. I-160 [SIC 2082]
Lakeville Laminating Co, p. I-310 [SIC 2295]
Lakewood Eng & Mfg Co, p. II-1654 [SIC 3564]
Lakso Co, p. II-1660 [SIC 3565]

Roman numerals I and II indicate the volume in which pages appear. Volume I holds SICs 2011 through 3299; Volume II holds SICs 3312 through 3999.

2268

Roman numerals I and II indicate the volume in which pages appear. Volume I holds SICs 2011 through 3299; Volume II holds SICs 3312 through 3999.

Roman numerals I and II indicate the volume in which pages appear. Volume I holds SICs 2011 through 3299; Volume II holds SICs 3312 through 3999.

Roman numerals I and II indicate the volume in which pages appear. Volume I holds SICs 2011 through 3299; Volume II holds SICs 3312 through 3999.

2271

Lunt Manufacturing Co, p. II-1271 [SIC 3364]
Luseaux Laboratories Inc, p. I-819 [SIC 2841]
Luster-on Products Inc, p. I-773 [SIC 2816]
Luster Products Co, p. I-834 [SIC 2844]
Lustro Plastics Co, p. I-961 [SIC 3081]
Lutco Bearings, p. II-1644 [SIC 3562]
Luther L Smith and Son Fish, p. I-190 [SIC 2092]
Luthi Machine and Engineering, p. II-1660 [SIC 3565]
Lutron Electronics Company Inc, p. II-1775 [SIC 3613]
Luwa Bahnson Inc, p. II-1632 [SIC 3559]
Luxar Corp, p. II-2102 [SIC 3845]
Luxo Corp, p. II-1845 [SIC 3646]
Luxor, p. I-576 [SIC 2522]
LV Myles Inc, p. I-378 [SIC 2341]
LVD Corp, p. II-1563 [SIC 3542]
LW Packard and Company Inc, p. I-241 [SIC 2231]
LXE Inc, p. II-1873 [SIC 3663]
Ly-Line Products Inc, p. I-973 [SIC 3083]
Lyall Assemblies Inc, p. II-1830 [SIC 3643]
LYCON Inc, p. I-1127 [SIC 3273]
Lydall Inc, pp. Vol. I - 532, 688 [SICs 2493, 2679]
Lydon-Bricher Manufacturing, p. I-428 [SIC 2392]
Lyman Lumber Co, p. I-480 [SIC 2431]
Lyman Products Corp, p. II-1430 [SIC 3482]
Lyn-Flex West Inc, p. I-1017 [SIC 3131]
Lyn-Tron Inc, p. I-1835 [SIC 3644]
Lynch Machinery-Miller Hydro, p. II-1632 [SIC 3559]
Lynk Corp, p. II-1699 [SIC 3575]
Lynn Electronics Corp, p. II-1182 [SIC 3315]
Lynn Ladder and Scaffolding, p. I-536 [SIC 2499]
Lynn Plastics Corp, p. I-997 [SIC 3087]
Lynn Products Company Inc, p. I-1154 [SIC 3295]
Lynwood Battery Mfg Co, p. II-1920 [SIC 3691]
Lyon and Healy Harps Inc, p. II-2135 [SIC 3931]
Lyon Food Products Inc, p. I-186 [SIC 2091]
Lyon Metal Products Inc, p. I-593 [SIC 2542]
Lyondell-CITGO Refining Ltd, p. I-903 [SIC 2911]
Lyondell Petrochemical Co, p. I-903 [SIC 2911]
Lyons Diecasting Co, p. II-1271 [SIC 3364]
Lyons-Magnus Inc, p. I-132 [SIC 2066]
Lytton Inc, p. II-1887 [SIC 3672]

M/A-COM Inc, pp. Vol II - 1873, 1892 [SICs 3663, 3674]
M/A-COM Omni Spectra Inc, p. II-1915 [SIC 3679]
M and B Headwear Company, p. I-386 [SIC 2353]
M and D Flexographic Printers, p. I-991 [SIC 3086]
M and I Electric Industries Inc, p. II-1775 [SIC 3613]
M and L Products Inc, p. I-553 [SIC 2514]
M and M Designs Inc, p. I-446 [SIC 2396]
M and N Cigar Manufacturing, p. I-221 [SIC 2121]
M and R Industries Inc, p. II-1192 [SIC 3317]
M and R Marking Systems, p. II-2164 [SIC 3953]
M and R Printing Equipment Inc, p. II-1621 [SIC 3555]
M and S Drapery Workroom, p. I-424 [SIC 2391]
M & W Electric Mfg Co, p. II-1275 [SIC 3365]
M and W Pump Corp, p. II-1753 [SIC 3594]
M and W Sportswear Co, p. I-348 [SIC 2325]
M-C Power Corp, p. II-1795 [SIC 3629]
M Chasen and Son Inc, p. I-327 [SIC 2299]
M-E-C Co, p. II-1626 [SIC 3556]
M Grossman and Sons Inc, p. I-386 [SIC 2353]
M Joseph Machine Company, p. II-1822 [SIC 3639]
M Kimerling and Sons Inc, p. II-1221 [SIC 3334]
M Lyon and Co, p. I-1013 [SIC 3111]
M Pressner and Company Inc, p. II-2144 [SIC 3944]
M Putterman and Company Inc, p. I-437 [SIC 2394]
M Rubin and Sons Inc, p. I-336 [SIC 2321]
M Shapiro and Company Inc, p. I-369 [SIC 2337]
M Stephens Manufacturing Inc, p. II-1835 [SIC 3644]
M Swift and Sons Inc, p. II-1482 [SIC 3497]
M W Windows, p. I-480 [SIC 2431]
M4 Data Inc, p. II-1694 [SIC 3572]
MA Bell Co, p. II-1280 [SIC 3366]
MA Bruder and Sons Inc, p. I-840 [SIC 2851]
MA Ford Manufacturing, p. II-1578 [SIC 3545]
MA Gedney Co, p. I-55 [SIC 2035]

MA Hanna Co, p. I-782 [SIC 2821]
MA Harrison Manufacturing, p. II-1275 [SIC 3365]
MA Industries Inc, p. I-782 [SIC 2821]
MA Patout and Son Ltd, p. I-116 [SIC 2061]
Maas-Rowe Carillons Inc, p. II-2135 [SIC 3931]
Maasdam Pow'R-Pull Inc, p. II-1547 [SIC 3536]
Maass Manufacturing Inc, p. II-1486 [SIC 3498]
MAB Paints, p. I-840 [SIC 2851]
Mac Corp, p. II-1764 [SIC 3599]
MAC Equipment Inc, p. II-1541 [SIC 3535]
Mac Products Inc, p. II-1830 [SIC 3643]
Mac Tools Inc, p. II-1312 [SIC 3423]
Macalloy Corp, p. II-1178 [SIC 3313]
MacAndrews and Forbes Corp, p. I-181 [SIC 2087]
MacBeath Hardwood Company, p. I-491 [SIC 2435]
MacBee Engineering Corp, p. II-1600 [SIC 3549]
MacDermid Inc, p. I-897 [SIC 2899]
Mace Security, p. II-2199 [SIC 3999]
Mace Security International Inc, p. I-855 [SIC 2869]
MacGillis and Gibbs Co, p. I-528 [SIC 2491]
Machinery Products Co, p. II-1854 [SIC 3648]
Machining Enterprises Inc, p. II-1285 [SIC 3369]
Mack-Chicago Corp, p. I-628 [SIC 2653]
Mack Trucks Inc, p. II-1503 [SIC 3519]
Mackay Communications Inc, p. II-1873 [SIC 3663]
Mackay Envelope Corp, p. I-679 [SIC 2677]
MacKenzie Laboratories Inc, p. II-1864 [SIC 3652]
MacKenzie Machine and Design, p. II-1454 [SIC 3492]
Mackie Designs Inc, p. II-1859 [SIC 3651]
Mackintosh of New England, p. I-369 [SIC 2337]
MacKissic Inc, p. II-1514 [SIC 3524]
Macklanburg-Duncan Co, p. II-1244 [SIC 3354]
MacLean-Fogg Co, p. II-2177 [SIC 3965]
MacMillan Bloedel, p. I-628 [SIC 2653]
MacMillan Bloedel Inc, p. I-618 [SIC 2631]
MacNaughton Lithography Corp, p. I-718 [SIC 2752]
Maco Corp, p. II-1275 [SIC 3365]
Macristy Industries Inc, p. I-973 [SIC 3083]
Macromedia Inc, p. I-693 [SIC 2711]
Macrotech Polyseal Inc, p. I-942 [SIC 3053]
MacSema Inc, p. II-1709 [SIC 3578]
MacSteel, p. II-1172 [SIC 3312]
MacWhyte Co, p. II-1476 [SIC 3496]
Mad Bomber Co, p. I-386 [SIC 2353]
Maddak Inc, p. II-2199 [SIC 3999]
Made Rite Foods Inc, p. I-212 [SIC 2099]
Madelaine Chocolate Novelties, p. I-132 [SIC 2066]
Madge Networks Inc, p. II-1704 [SIC 3577]
Madico Inc, p. I-961 [SIC 3081]
Madison Chemical Company Inc, p. I-819 [SIC 2841]
Madison Dairy Produce, p. I-17 [SIC 2021]
Madison Industries Inc, p. I-428 [SIC 2392]
Madison-Kipp Corp, p. II-1266 [SIC 3363]
Madix Inc, p. I-593 [SIC 2542]
Madson Line, p. I-1041 [SIC 3161]
Maeve Carr Designs Inc, p. II-2173 [SIC 3961]
Mag Instrument Inc, p. II-1854 [SIC 3648]
Mag-Tek Inc, p. II-1699 [SIC 3575]
Magee Co, p. I-536 [SIC 2499]
Magee Industrial Enterprises Inc, p. I-292 [SIC 2273]
Magic Chef Co, p. II-1800 [SIC 3631]
Magic Novelty Company Inc, p. II-2131 [SIC 3915]
Magic Valley Foods Inc, p. I-60 [SIC 2037]
Magid Glove, p. I-401 [SIC 2381]
Magie Brother Oil Co, p. I-918 [SIC 2992]
Magliano Pants Co, p. I-348 [SIC 2325]
Magna Design Inc, p. I-571 [SIC 2521]
Magna Industrial Tools, p. II-1312 [SIC 3423]
Magnacast Corp, p. II-1280 [SIC 3366]
Magnat Rolls Inc, p. II-1589 [SIC 3547]
Magnavox Electronics Syst Co, pp. Vol II - 1915, 2035 [SICs 3679, 3812]
Magneco/Metrel Inc, p. I-1086 [SIC 3253]
Magnecraft Electric Co, p. II-1790 [SIC 3625]
Magnesium Alloy Products Co, p. II-1225 [SIC 3339]
Magnesium Aluminum Corp, p. II-1271 [SIC 3364]
Magnesium Casting Co, p. II-1266 [SIC 3363]
Magnesium Elektron Inc, p. I-777 [SIC 2819]
MAGNET Inc, p. II-1492 [SIC 3499]

MAGNETCARDS, p. I-1109 [SIC 3264]
MagneTek, p. II-1905 [SIC 3677]
Magnetek Century Inc, p. II-1780 [SIC 3621]
MagneTek Electric Inc, p. II-1769 [SIC 3612]
MagneTek Inc, p. II-1769 [SIC 3612]
MagneTek Incc, p. II-1780 [SIC 3621]
Magnetek National, p. II-1780 [SIC 3621]
Magnetek Ohio, p. II-1769 [SIC 3612]
Magnetic Coils Inc, p. II-1905 [SIC 3677]
Magnetic Data Inc, p. II-1688 [SIC 3571]
Magnetic Metals Corp, p. II-1563 [SIC 3542]
Magnetic Products Corp, p. II-1934 [SIC 3695]
Magnetrol International Inc, p. II-2051 [SIC 3823]
Magnolia, p. I-523 [SIC 2452]
Magnolia Brush Manufacturing, p. II-2181 [SIC 3991]
Magnolia Coca-Cola Bottling Co, p. I-176 [SIC 2086]
Magnolia Hosiery Mill Inc, p. I-378 [SIC 2341]
Magnolia Metal Corp, p. II-1400 [SIC 3463]
Magnum Corp, p. II-2045 [SIC 3822]
Magnum Importers Inc, p. I-1049 [SIC 3172]
Magnum Marine Corp, p. II-1992 [SIC 3732]
Magnum Resources Inc, p. II-1552 [SIC 3537]
Magparts, p. II-1285 [SIC 3369]
Mahle Inc, p. II-1742 [SIC 3592]
Maida Development Co, p. II-1897 [SIC 3675]
Mail-Well Envelope Co, p. I-679 [SIC 2677]
Main Iron Works Inc, p. II-1986 [SIC 3731]
Main Street Muffins Inc, p. I-112 [SIC 2053]
Maine Biotechnology Services, p. I-814 [SIC 2836]
Maine Brand Manufacturing Inc, p. I-401 [SIC 2381]
Maitlen and Bensen Inc, p. II-1594 [SIC 3548]
Majco Building Specialties LP, p. II-1339 [SIC 3433]
Majestic Athletic Wear Ltd, p. I-356 [SIC 2329]
Majestic Distilling Co, p. I-172 [SIC 2085]
Major Partitions Inc, p. I-593 [SIC 2542]
Major Products Company Inc, p. I-51 [SIC 2034]
Major Tool and Machine Inc, p. II-1572 [SIC 3544]
Makins Hats Ltd, p. I-386 [SIC 2353]
Maklihon Group Corp, p. I-361 [SIC 2331]
Mako Marine Inc, p. II-1992 [SIC 3732]
Malbon Co, p. I-373 [SIC 2339]
MALCO Inc, p. I-446 [SIC 2396]
Malco Interconnector Syst, p. II-1910 [SIC 3678]
Malco Modes Inc, p. I-369 [SIC 2337]
Malco Products Inc, p. II-1312 [SIC 3423]
Malden International Designs, p. I-536 [SIC 2499]
Malden Novelty Company Inc, p. I-428 [SIC 2392]
Malina Inc, p. I-394 [SIC 2369]
Mallery Lumber Corp, p. I-510 [SIC 2448]
Mallet and Co, p. I-212 [SIC 2099]
Malley and Co, p. I-394 [SIC 2369]
Mallinckrodt Anesthesiology, p. II-2082 [SIC 3841]
Mallinckrodt Group Inc, p. II-2102 [SIC 3845]
Mallinckrodt Medical Inc, p. II-2082 [SIC 3841]
Mallinckrodt Spec Chemicals Co, p. I-804 [SIC 2834]
Mallinckrodt Veterinary Inc, p. I-804 [SIC 2834]
Mallory and Kraft Ltd, p. I-378 [SIC 2341]
Mallory Controls, p. II-1790 [SIC 3625]
Mallory Inc, p. II-1929 [SIC 3694]
Malloy Lithographing Inc, p. I-708 [SIC 2732]
Malnove Inc, p. I-643 [SIC 2657]
Malnove Incorporated of Florida, p. I-643 [SIC 2657]
Maloney-Crawford Inc, p. II-1531 [SIC 3533]
Mamco Corp, p. II-1780 [SIC 3621]
Mamco Manufacturing Inc, p. II-1980 [SIC 3728]
Mameco International Inc, p. I-879 [SIC 2891]
Mammoth Inc, p. II-1726 [SIC 3585]
Mamo Howell Inc, p. I-369 [SIC 2337]
Man Gill Chemical Co, p. I-825 [SIC 2842]
MAN Roland Inc, pp. Vol II - 1615, 1621 [SICs 3554, 3555]
Manchester Knitted Fashion Inc, p. I-258 [SIC 2253]
Manchester Plastics, p. I-1006 [SIC 3089]
Manchester Tool Co, p. II-1572 [SIC 3544]
Manchester Wood Inc, p. I-536 [SIC 2499]
Mancillas International Ltd, p. I-332 [SIC 2311]
Manco Inc, p. I-654 [SIC 2672]
Mancuso Cheese Co, p. I-21 [SIC 2022]
Mandel Metals Inc, p. II-1225 [SIC 3339]

Roman numerals I and II indicate the volume in which pages appear. Volume I holds SICs 2011 through 3299; Volume II holds SICs 3312 through 3999.

2272

Roman numerals I and II indicate the volume in which pages appear. Volume I holds SICs 2011 through 3299; Volume II holds SICs 3312 through 3999.

Master Products Manufacturing, p. II-1713 [SIC 3579]
Master Shield Building Prod LP, p. I-961 [SIC 3081]
Master Typographers Inc, p. I-754 [SIC 2791]
Master Woodcraft Inc, p. I-536 [SIC 2499]
Masterack, p. I-603 [SIC 2599]
MasterBrand Industries Inc, p. II-1334 [SIC 3432]
Mastercraft Furniture, p. I-548 [SIC 2512]
Mastercraft Industries Inc, p. I-486 [SIC 2434]
Mastercrafters Corp, p. II-1992 [SIC 3732]
Masterlooms Inc, p. I-292 [SIC 2273]
Masters Gallery Foods Inc, p. I-21 [SIC 2022]
Matador Processors Inc, p. I-65 [SIC 2038]
Matanzas Creek Winery, p. I-168 [SIC 2084]
Mate Punch and Die Co, p. II-1572 [SIC 3544]
MATEC Corp, p. II-1425 [SIC 3479]
Mateer-Burt Company Inc, p. II-1660 [SIC 3565]
Material Handling Services Inc, p. II-1552 [SIC 3537]
Material Sciences Corp, p. II-1425 [SIC 3479]
Material Service Corp, p. I-1127 [SIC 3273]
Material Supply Inc, p. I-1122 [SIC 3272]
Materials Research Corp, p. II-1621 [SIC 3555]
Materials Transportation Co, p. II-1552 [SIC 3537]
Matheson Gas Products, p. I-768 [SIC 2813]
Mathews Co, p. II-1509 [SIC 3523]
Mathews Conveyor, p. II-1541 [SIC 3535]
Mathis-Akins Concrete Block, p. I-1117 [SIC 3271]
Matot Inc, p. II-1536 [SIC 3534]
Matrix Unlimited Inc, p. I-759 [SIC 2796]
Matsi Inc, p. II-1920 [SIC 3691]
Matt Brewing Company Inc, p. I-160 [SIC 2082]
Matt-Son Inc, p. II-1736 [SIC 3589]
Mattco Manufacturing Inc, p. II-1531 [SIC 3533]
Mattel Inc, p. II-2144 [SIC 3944]
Matthew Bender and Company, p. I-703 [SIC 2731]
Matthews International Corp, p. I-759 [SIC 2796]
Mattison Machine Works, p. II-1558 [SIC 3541]
Mattison Woodworking Co, p. II-1610 [SIC 3553]
Mattituck Aviation Corp, p. II-1970 [SIC 3721]
Mattson Spray Equipment Inc, p. II-1649 [SIC 3563]
Maugansville Elevator & Lumber, p. I-501 [SIC 2439]
Maugus Manufacturing Inc, p. II-2181 [SIC 3991]
Maui Land and Pineapple, p. I-45 [SIC 2033]
Maui Pineapple Company Ltd, p. I-45 [SIC 2033]
Maule Air Inc, p. II-1970 [SIC 3721]
Maurice Lenell Cooky Co, p. I-107 [SIC 2052]
Maurice Silvera Inc, p. I-394 [SIC 2369]
Maury Office Systems Inc, p. I-576 [SIC 2522]
Mautner Company Inc, p. I-623 [SIC 2652]
Mautz Paint Co, p. I-840 [SIC 2851]
Maverick Industries Inc, p. II-1938 [SIC 3699]
Maverick Tube Corp, p. II-1192 [SIC 3317]
Max Daetwyler Corp, p. II-1621 [SIC 3555]
Max Duraffourg Gem Company, p. II-2131 [SIC 3915]
Max Kahn Curtain Corp, p. I-424 [SIC 2391]
Max Leather Group Inc, p. I-416 [SIC 2387]
Max Machinery Inc, p. II-2056 [SIC 3824]
Max Marx Color Co, p. I-850 [SIC 2865]
Maxcor Manufacturing Co, p. II-1492 [SIC 3499]
Maxim Integrated Products Inc, p. II-1892 [SIC 3674]
Maxitrol Co, p. II-2051 [SIC 3823]
Maxon America Inc, p. II-1873 [SIC 3663]
Maxon Corp, p. II-1339 [SIC 3433]
Maxon Systems Inc, p. II-1873 [SIC 3663]
MaxTech Corp, p. II-1868 [SIC 3661]
Maxtor Corp, p. II-1694 [SIC 3572]
Maxwell Laboratories Inc, p. II-1795 [SIC 3629]
Maxwell Shoe Company Inc, p. I-1029 [SIC 3144]
Maxx Inc, p. I-1045 [SIC 3171]
MAXXAM Group Inc, p. I-460 [SIC 2411]
MAXXAM Inc, p. II-1244 [SIC 3354]
Maxxim Medical Inc, p. II-2082 [SIC 3841]
May Apparel Group Inc, p. I-390 [SIC 2361]
May Foundry and Machine, p. II-1212 [SIC 3325]
Mayco Oil, p. I-918 [SIC 2992]
Mayer Brothers Apple Products, p. I-45 [SIC 2033]
Mayer China, p. I-1102 [SIC 3262]
Mayer Textile Machine Corp, p. II-1605 [SIC 3552]
Mayes Brothers Tool Mfg, p. II-1312 [SIC 3423]
Mayes Manufacturing Co, p. I-369 [SIC 2337]

Mayfair Industries, p. I-258 [SIC 2253]
Mayfair Packing Company Inc, p. I-51 [SIC 2034]
Mayfield Dairy Farms Inc, p. I-35 [SIC 2026]
Mayfield Manufacturing, p. I-428 [SIC 2392]
Mayfran International, p. II-1541 [SIC 3535]
Mayline Company Inc, p. I-576 [SIC 2522]
Maynard H Moore Jr Inc, p. I-1017 [SIC 3131]
Maynard Manufacturing Inc, p. II-1388 [SIC 3452]
Maynard Steel Casting Co, p. II-1212 [SIC 3325]
Maysteel Corp, p. II-1492 [SIC 3499]
Maytag Clarence, p. II-1800 [SIC 3631]
Maytag Corp, p. II-1809 [SIC 3633]
Maytown Shoe Manufacturing, p. I-1033 [SIC 3149]
Mayville Engineering Company, p. II-1492 [SIC 3499]
Maywood Inc, p. I-480 [SIC 2431]
Mazer Corp, p. I-750 [SIC 2789]
Mazzella Wire Rope & Sling Co, p. II-1447 [SIC 3491]
MBL USA Corp, p. I-936 [SIC 3052]
MBM Corp, p. II-1713 [SIC 3579]
MBO America, p. II-1615 [SIC 3554]
MBP Neckwear Inc, p. I-344 [SIC 2323]
MBX Packaging Specialists, p. I-515 [SIC 2449]
MC Canfield Sons, p. II-1229 [SIC 3341]
MC Davis Company Inc, p. II-1905 [SIC 3677]
MC Decorating Co, p. I-1071 [SIC 3231]
MC Gill Corp, p. II-1980 [SIC 3728]
MC Snack Inc, p. I-51 [SIC 2034]
MCA Sign Co, p. II-2186 [SIC 3993]
McArthur Dairy Inc, p. I-35 [SIC 2026]
McBee Manufacturing Co, p. I-373 [SIC 2339]
MCC Inc, p. I-1127 [SIC 3273]
McCain Citrus Inc, p. I-60 [SIC 2037]
McCain Ellio's Foods Inc, p. I-65 [SIC 2038]
McCain Foods Inc, p. I-65 [SIC 2038]
McCain Manufacturing Corp, p. II-1621 [SIC 3555]
McCain USA Inc, p. I-65 [SIC 2038]
McCanna Inc, p. II-1747 [SIC 3593]
McCarthy Cabinet Co, p. I-486 [SIC 2434]
McCarthy's Portable Structures, p. I-519 [SIC 2451]
McCauley Accessory, p. II-1980 [SIC 3728]
McClatchy Newspapers Inc, p. I-693 [SIC 2711]
McCleary Industries Inc, p. I-199 [SIC 2096]
McCleery-Cumming Co, p. I-724 [SIC 2754]
McClendon's Dairy Products Inc, p. I-21 [SIC 2022]
McClinton-Anchor, p. I-909 [SIC 2951]
McCollister and Co, p. I-918 [SIC 2992]
McConnell Cabinets Inc, p. I-486 [SIC 2434]
McCormick and Company Inc, p. I-212 [SIC 2099]
McCormick Distilling Company, p. I-172 [SIC 2085]
McCormick International, p. I-212 [SIC 2099]
McCormick/Schilling, p. I-212 [SIC 2099]
McCowat-Mercer Press Inc, p. I-618 [SIC 2631]
McCoy-Ellison Inc, p. II-1605 [SIC 3552]
McCoy Manufacturing Inc, p. I-348 [SIC 2325]
McCracken Concrete Pipe Mach, p. II-1632 [SIC 3559]
McCulloch Corp, p. II-1584 [SIC 3546]
McCurdy Circuits Inc, p. II-1887 [SIC 3672]
McDATA Corp, p. II-1835 [SIC 3644]
McDermott International Inc, p. II-1780 [SIC 3621]
McDonald Welding and Machine, p. II-1594 [SIC 3548]
McDonnell Douglas, p. II-1873 [SIC 3663]
McDonnell Douglas Corp, p. II-1970 [SIC 3721]
McDonough Manufacturing Co, p. II-1610 [SIC 3553]
McDougal Littell Inc, p. I-703 [SIC 2731]
McDowell & Craig Mfg Co, p. I-576 [SIC 2522]
McEnroe Inc, p. II-1840 [SIC 3645]
McFarland Cascade, p. I-528 [SIC 2491]
McGean-Rohco Inc, p. I-777 [SIC 2819]
McGehee Industries Inc, p. I-352 [SIC 2326]
McGill Electric Switch Prod, p. II-1830 [SIC 3643]
McGill Inc, p. II-1713 [SIC 3579]
McGill-Jensen Inc, p. I-718 [SIC 2752]
McGill Manufacturing Company, p. II-1644 [SIC 3562]
McGinley Mills Inc, p. I-236 [SIC 2221]
McGraw-Hill Inc, pp. Vol. I - 698, 703 [SICs 2721, 2731]
McGraw Manufacturing Inc, p. II-1280 [SIC 3366]
McGregor Agri Corporation Inc, p. I-437 [SIC 2394]

McGregor Architectural Iron, p. II-1367 [SIC 3446]
McGregor Printing Corp, p. I-735 [SIC 2761]
McGuire Furniture Co, p. I-567 [SIC 2519]
McGuire-Nicholas Company Inc, p. I-1053 [SIC 3199]
McInnes Steel Co, p. II-1395 [SIC 3462]
McIntosh Box and Pallet Co, p. I-515 [SIC 2449]
McIntosh Farm Service Co, p. I-70 [SIC 2041]
McIntyre Tile Company Inc, p. I-1086 [SIC 3253]
McKee Button Company Inc, p. II-2177 [SIC 3965]
McKee Door Inc, p. II-1349 [SIC 3442]
McKee Foods Corp, p. I-101 [SIC 2051]
McKees Rocks Forging Co, p. II-1519 [SIC 3531]
McKesson Water Products Co, p. I-176 [SIC 2086]
McKinstry Inc, p. II-1835 [SIC 3644]
McKissick Products Inc, p. II-1531 [SIC 3533]
McKnight Plywood Inc, p. I-491 [SIC 2435]
MCL Inc, p. II-1873 [SIC 3663]
McLanahan Corp, p. II-1525 [SIC 3532]
McLane Manufacturing Inc, p. II-1514 [SIC 3524]
McLaughlin Body Company Inc, p. II-1948 [SIC 3713]
McLaughlin Gormley King Co, p. I-777 [SIC 2819]
McLaughlin International Inc, p. I-310 [SIC 2295]
McLellan Equipment Inc, p. II-1525 [SIC 3532]
MCM Enterprises Inc, p. II-1182 [SIC 3315]
McMillan Fiberglass Stocks Inc, p. I-236 [SIC 2221]
McMinnville Manufacturing Co, p. I-471 [SIC 2426]
McNally Industries Inc, p. II-1764 [SIC 3599]
McNally Wellman, p. II-1541 [SIC 3535]
McNaughton and Gunn Inc, p. I-708 [SIC 2732]
McNish Corp, p. II-1736 [SIC 3589]
MCO/EASTECH Inc, p. II-2056 [SIC 3824]
MCP Industries Inc, p. I-948 [SIC 3061]
McPhilben, p. II-1845 [SIC 3646]
McPhillips Manufacturing, p. I-480 [SIC 2431]
McQuay International, p. II-1726 [SIC 3585]
McRae Footwear, p. I-1025 [SIC 3143]
McRae Industries Inc, p. I-1025 [SIC 3143]
McRoskey Airflex Mattress, p. I-558 [SIC 2515]
MCS Industries Inc, p. I-536 [SIC 2499]
MCT Custom Truck Bodies Inc, p. II-1948 [SIC 3713]
McWhorter Technologies Inc, p. I-782 [SIC 2821]
McWilliams Forge Company Inc, p. II-1395 [SIC 3462]
MD-Both Industries, p. II-1295 [SIC 3399]
MD Knowlton Co, p. II-1536 [SIC 3534]
MD Stetson Co, p. I-825 [SIC 2842]
MDC Industries Inc, p. I-1144 [SIC 3291]
MDC Vacuum Products Corp, p. II-1492 [SIC 3499]
MDH Industries Inc, p. II-2098 [SIC 3844]
ME Cunningham International, p. II-2164 [SIC 3953]
ME Heuck Company Inc, p. II-1312 [SIC 3423]
Mead Coated Board Inc, p. I-618 [SIC 2631]
Mead Corp, p. I-612 [SIC 2621]
Mead Fine Paper, p. I-612 [SIC 2621]
Mead Fluid Dynamics Inc, p. II-1454 [SIC 3492]
Meade Instruments Corp, p. II-2071 [SIC 3827]
Meadow Brook Dairy Corp, p. I-35 [SIC 2026]
Meadow Gold Dairies of Hawaii, p. I-35 [SIC 2026]
Meadox Medicals Inc, p. II-2087 [SIC 3842]
Meadville Forging Co, p. II-1395 [SIC 3462]
Means Industries Inc, p. II-1404 [SIC 3465]
Mearl Corp, p. I-773 [SIC 2816]
Measurements Group Inc, p. II-2076 [SIC 3829]
Measurex Corp, p. II-1790 [SIC 3625]
Mebane Hosiery Inc, p. I-254 [SIC 2252]
Mebane Packaging Group, p. I-643 [SIC 2657]
Meca Sportswear Inc, p. I-455 [SIC 2399]
Mechanic's Time Savers Inc, p. II-1584 [SIC 3546]
Mechanical Equipment Company, p. II-1632 [SIC 3559]
Mechanical Mirror Works Inc, p. I-1071 [SIC 3231]
Mechanical Products Inc, p. II-1775 [SIC 3613]
Mechanical Technology Inc, p. II-2076 [SIC 3829]
Meco Inc, p. II-1975 [SIC 3724]
Med Service Inc, p. I-581 [SIC 2531]
Med-Tek Inc, p. II-1290 [SIC 3398]
MEDA Inc, p. I-809 [SIC 2835]
Medalist Apparel Inc, p. I-263 [SIC 2254]
Medalist Industries Inc, p. II-1388 [SIC 3452]
Medallion Kitchens of Minnesota, p. I-486 [SIC 2434]

Roman numerals I and II indicate the volume in which pages appear. Volume I holds SICs 2011 through 3299; Volume II holds SICs 3312 through 3999.

Roman numerals I and II indicate the volume in which pages appear. Volume I holds SICs 2011 through 3299; Volume II holds SICs 3312 through 3999.

Roman numerals I and II indicate the volume in which pages appear. Volume I holds SICs 2011 through 3299; Volume II holds SICs 3312 through 3999.

Mills Manufacturing Corp, p. I-455 [SIC 2399]
Mills Resistor Co, p. II-1901 [SIC 3676]
Millstein Industries, p. II-1840 [SIC 3645]
Millstone Coffee Inc, p. I-195 [SIC 2095]
Milnot Co, p. I-26 [SIC 2023]
Milor Corp, p. II-2181 [SIC 3991]
Milprint Inc, p. I-649 [SIC 2671]
Milton Bradley Co, p. II-2144 [SIC 3944]
Milton Can Company Inc, p. II-1300 [SIC 3411]
Milton Roy Co, pp. Vol II - 1638, 2056 [SICs 3561, 3824]
Miltope Group Inc, p. II-1704 [SIC 3577]
Milwaukee Bearing & Machining, p. II-1676 [SIC 3568]
Milwaukee Cylinder, p. II-1747 [SIC 3593]
Milwaukee Dustless Brush Co, p. II-2181 [SIC 3991]
Milwaukee Electric Tool Corp, p. II-1584 [SIC 3546]
Milwaukee Envelope Inc, p. I-679 [SIC 2677]
Milwaukee Gear Company Inc, p. II-1666 [SIC 3566]
Milwaukee Malleable, p. II-1203 [SIC 3322]
Milwaukee Marble and Granite, p. I-1140 [SIC 3281]
Milwaukee Resistor Corp, p. II-1901 [SIC 3676]
Milwaukee Sign Co, p. II-2186 [SIC 3993]
Milwaukee Tool & Equip Co, p. II-1312 [SIC 3423]
Milwaukee Valve Company Inc, p. II-1454 [SIC 3492]
Milwaukee Wire Products, p. II-1476 [SIC 3496]
Mims and Thomas Inc, p. I-480 [SIC 2431]
Minalex Corp, p. II-1244 [SIC 3354]
Minarik Corp, p. II-1790 [SIC 3625]
Minco Products Inc, p. II-2076 [SIC 3829]
Mine Safety Appliances Co, p. II-2087 [SIC 3842]
Miner Enterprises Inc, p. II-1998 [SIC 3743]
Minerallac Electric Co, p. II-1835 [SIC 3644]
Minerals Technologies Inc, p. I-777 [SIC 2819]
Minerva Cheese Factory Inc, p. I-21 [SIC 2022]
Minette-Bates Inc, p. I-428 [SIC 2392]
Mini World Inc, p. I-390 [SIC 2361]
Minimatic Implant Technology, p. II-2093 [SIC 3843]
MiniMed Technologies, p. II-2082 [SIC 3841]
Mining Services Intern Corp, p. I-885 [SIC 2892]
Minisink Rubber Company Inc, p. I-948 [SIC 3061]
Minn-Dak Farmers Cooperative, p. I-123 [SIC 2063]
Minn-Dak Growers Ltd, p. I-55 [SIC 2035]
Minnesota Automation, p. II-1660 [SIC 3565]
Minnesota Brewing Co, p. I-160 [SIC 2082]
Minnesota Corn Processors Inc, p. I-87 [SIC 2046]
Minnesota Knitting Mills Inc, p. I-258 [SIC 2253]
Minnesota Malting Co, p. I-165 [SIC 2083]
Minnesota Mining & Mfg, pp. Vol I - 649, 654, 1154, Vol. II -1910, 2087 [SICs 2671, 2672, 3295, 3678, 3842]
Minnesota Rubber Co, p. I-954 [SIC 3069]
Minnesota Valley Engineering, p. II-1355 [SIC 3443]
Minntech Corp, p. II-2087 [SIC 3842]
Minolta Corp, p. II-2112 [SIC 3861]
Minowitz Manufacturing Co, p. II-1929 [SIC 3694]
Minstar Inc, p. II-1992 [SIC 3732]
Minster Machine Co, p. II-1563 [SIC 3542]
MINTEQ International Inc, p. I-1163 [SIC 3297]
Minuteman International Inc, p. II-1514 [SIC 3524]
Minuteman Laboratories Inc, p. II-2071 [SIC 3827]
Miracle Steel Corp, p. II-1372 [SIC 3448]
Mirada Research & Mfg Co, p. II-1447 [SIC 3491]
Mirax Chemical Products Corp, p. II-1304 [SIC 3412]
Mirro-Foley Co, p. II-1413 [SIC 3469]
Misomex North America Inc, p. I-759 [SIC 2796]
Misonix Inc, p. II-2040 [SIC 3821]
Miss Elliette Inc, p. I-365 [SIC 2335]
Mission Clay Products Corp, p. I-1094 [SIC 3259]
Mission Foods Corp, p. I-199 [SIC 2096]
Mission Kleensweep Inc, p. I-825 [SIC 2842]
Mission Stucco Company Inc, p. I-1167 [SIC 3299]
Mississippi Blending Co, p. I-87 [SIC 2046]
Mississippi Chemical Corp, p. I-861 [SIC 2873]
Mississippi Materials Co, p. I-1127 [SIC 3273]
Missoula White Pine Sash, p. I-480 [SIC 2431]
Mister Coats Inc, p. I-332 [SIC 2311]
Mister Cookie Face Inc, p. I-30 [SIC 2024]
MIT International Inc, p. I-428 [SIC 2392]

Mitchel and Scott Machine, p. II-1382 [SIC 3451]
Mitchel Manufacturing Company, p. I-424 [SIC 2391]
Mitchell Industrial Tire Co, p. I-927 [SIC 3011]
Mitchell Manufacturing Co, p. I-603 [SIC 2599]
Mitchell Rubber Products Inc, p. I-954 [SIC 3069]
Mitchellace Inc, p. I-322 [SIC 2298]
Mitco Inc, p. I-897 [SIC 2899]
Mitek Industries Inc, p. II-1610 [SIC 3553]
Mitel Inc, p. II-1868 [SIC 3661]
Mitsubishi Caterpillar Forklift, p. II-1552 [SIC 3537]
Mitsubishi Chemical America, p. II-1892 [SIC 3674]
Mitsubishi Consumer, p. II-1859 [SIC 3651]
Mitsubishi Electric Power Prod, p. II-1775 [SIC 3613]
Mitsubishi Electronics America, p. II-1859 [SIC 3651]
Mitsubishi Elevator Co, p. II-1536 [SIC 3534]
Mitsumi Electronics Corp, p. II-1915 [SIC 3679]
Mitternight Boiler Works Inc, p. II-1355 [SIC 3443]
Mity-Lite Inc, p. I-576 [SIC 2522]
Mixon Inc, p. II-1920 [SIC 3691]
MJ Enterprises Inc, p. II-2173 [SIC 3961]
MJ Wood Products Inc, p. I-471 [SIC 2426]
MK Diamond Products Inc, p. I-1086 [SIC 3253]
MK Morse Co, p. II-1318 [SIC 3425]
MK Products Inc, p. II-1594 [SIC 3548]
MK Rail Corp, p. II-1998 [SIC 3743]
MKM Machine Tool Company, p. II-1382 [SIC 3451]
MKS Instruments Inc, p. II-2076 [SIC 3829]
ML Rongo Inc, p. II-1308 [SIC 3421]
MLP Seating Corp, p. I-576 [SIC 2522]
MLX Corp, p. II-1492 [SIC 3499]
MM Systems Corp, p. II-1367 [SIC 3446]
MMC Systems, p. II-1722 [SIC 3582]
MMG Corp, p. I-344 [SIC 2323]
MMI Products Inc, p. II-1476 [SIC 3496]
MMO Music Group Inc, p. II-1864 [SIC 3652]
MNP Corp, p. II-1388 [SIC 3452]
Mobil Chemical Co, p. I-660 [SIC 2673]
Mobil Corp, p. I-903 [SIC 2911]
Mobil Mining and Minerals Co, p. I-865 [SIC 2874]
Mobil Oil Corp, p. I-903 [SIC 2911]
Mobil Oil Corp US, p. I-903 [SIC 2911]
Mobilcraft Wood Products, p. I-486 [SIC 2434]
Mobile Aerospace Engineering, p. II-1970 [SIC 3721]
Mobile Hydraulics, p. II-1454 [SIC 3492]
Mobile Paint Manufacturing Co, p. I-840 [SIC 2851]
Mobile Pulley & Machine Works, p. II-1519 [SIC 3531]
Mobile Structures Inc, p. I-523 [SIC 2452]
Mobile Tool International Inc, p. II-1552 [SIC 3537]
MoCaro Industries Inc, p. I-267 [SIC 2257]
Moco Thermal Industries Inc, p. II-1671 [SIC 3567]
MOD-PAC Corp, p. I-643 [SIC 2657]
Mod-U-Kraf Homes Inc, p. I-523 [SIC 2452]
Modar Inc, p. I-587 [SIC 2541]
Mode Corp, p. I-571 [SIC 2521]
Model Pattern Co, p. II-1568 [SIC 3543]
Model Stone and Ready Mix Co, p. I-1127 [SIC 3273]
Modern Arts Package Inc, p. I-665 [SIC 2674]
Modern Contract Furniture Co, p. I-581 [SIC 2531]
Modern Curriculum Press, p. I-703 [SIC 2731]
Modern Drop Forge Co, p. II-1395 [SIC 3462]
Modern Faucet Mfg Co, p. II-1334 [SIC 3432]
Modern Gloves Inc, p. I-1037 [SIC 3151]
Modern Home and Equipment, p. I-523 [SIC 2452]
Modern Inc, p. II-1961 [SIC 3715]
Modern Industries Inc, p. II-1764 [SIC 3599]
Modern Manufacturing Inc, p. II-1764 [SIC 3599]
Modern Marketing Aids Inc, p. I-750 [SIC 2789]
Modern Metal Products, p. II-2098 [SIC 3844]
Modern Milltex Corp, p. I-991 [SIC 3086]
Modern of Marshfield Inc, p. I-548 [SIC 2512]
Modern Packaging Corp, p. I-623 [SIC 2652]
Modern Plating Inc, p. II-1420 [SIC 3471]
Modern Plating Corp, p. II-1420 [SIC 3471]
Modern Products Inc, p. I-83 [SIC 2045]
Modern Prototype Co, p. II-1404 [SIC 3465]
Modern Quilters Inc, p. I-442 [SIC 2395]
Modern Ribbon and Carbon Inc, p. II-2169 [SIC 3955]
Modern Steel Treating Co, p. II-1290 [SIC 3398]

Modern Welding Company Inc, p. II-1344 [SIC 3441]
Modern Window Shade Co, p. I-599 [SIC 2591]
Modern Woodcrafts Inc, p. I-587 [SIC 2541]
Modesto Tallow Company Inc, p. I-95 [SIC 2048]
Modicon Inc, p. II-2051 [SIC 3823]
Modine Manufacturing Co, p. II-1954 [SIC 3714]
Modineer Co, p. II-1572 [SIC 3544]
Modoc Lumber Co, p. I-465 [SIC 2421]
Modular and Plastic Products, p. I-1071 [SIC 3231]
Modular Casework Systems Inc, p. I-587 [SIC 2541]
Modular Devices Inc, p. II-1910 [SIC 3678]
Modular Panel Company Inc, p. II-1372 [SIC 3448]
Modular Systems Inc, pp. Vol. I - 587, Vol. II - 1372 [SICs 2541, 3448]
Moebius Printing Company Inc, p. I-718 [SIC 2752]
Moeller Manufacturing Company, p. II-1312 [SIC 3423]
Moeller Products Company Inc, p. II-1486 [SIC 3498]
Moen Inc, p. II-1334 [SIC 3432]
MOEX Corp, p. II-1961 [SIC 3715]
Moffatt Products Inc, p. II-1845 [SIC 3646]
Mohasco Corp, p. I-548 [SIC 2512]
Mohawk Carpet Corp, p. I-292 [SIC 2273]
Mohawk Fabric Company Inc, p. I-271 [SIC 2258]
Mohawk Finishing Products Inc, p. I-840 [SIC 2851]
Mohawk Industries Inc, pp. Vol. I - 292, Vol. II - 1563 [SICs 2273, 3542]
Mohawk Metal Products, p. II-1355 [SIC 3443]
Mohawk Paper Mills Inc, p. I-612 [SIC 2621]
Mohawk Plastics Inc, p. I-660 [SIC 2673]
Mohawk Resources Ltd, p. II-1552 [SIC 3537]
Mohawk Western Plastics Inc, p. I-961 [SIC 3081]
Mohl Fur Company Inc, p. I-398 [SIC 2371]
Mold Ex Rubber Co, p. I-954 [SIC 3069]
Mold-In Graphics Systems Inc, p. II-2112 [SIC 3861]
Moldcast Lighting, p. II-1845 [SIC 3646]
Molded Acoustical Products, p. I-1158 [SIC 3296]
Mole-Richardson Co, p. II-2112 [SIC 3861]
Molecu Wire Corp, p. II-1182 [SIC 3315]
Molecular Devices Corp, p. II-2066 [SIC 3826]
Molecular Dynamics Inc, p. II-2071 [SIC 3827]
Molecular Genetic Resources, p. I-814 [SIC 2836]
Molex-ETC Inc, p. II-1910 [SIC 3678]
Molex Inc, p. II-1910 [SIC 3678]
Moline Bearing Co, p. II-1644 [SIC 3562]
Moline Forge Inc, p. II-1395 [SIC 3462]
Moline Paint Manufacturing Co, p. I-840 [SIC 2851]
Moller International Inc, p. II-1970 [SIC 3721]
Molon Motor and Coil Inc, p. II-1780 [SIC 3621]
Moltrup Steel Products Company, p. II-1187 [SIC 3316]
Mom's Best Cookies Inc, p. I-107 [SIC 2052]
Momar Inc, p. I-825 [SIC 2842]
Mona Industries Inc, p. I-777 [SIC 2819]
Monaco Coach Corp, p. II-1966 [SIC 3716]
Monadnock Co, p. II-2177 [SIC 3965]
Monadnock Paper Mills Inc, p. I-612 [SIC 2621]
Monarch Company Inc, p. I-181 [SIC 2087]
Monarch Hosiery Mills Inc, p. I-254 [SIC 2252]
Monarch Hydraulics Inc, p. II-1747 [SIC 3593]
Monarch Industrial Tire Corp, p. I-927 [SIC 3011]
Monarch Knit and Sportswear, p. I-361 [SIC 2331]
Monarch Machine Tool Co, p. II-1558 [SIC 3541]
Monarch Marking Systems Inc, p. II-1704 [SIC 3577]
Monarch Paint Co, p. I-840 [SIC 2851]
Monarch Rubber Company Inc, p. I-954 [SIC 3069]
Monarch Systems, p. I-576 [SIC 2522]
Monarch Tile Inc, p. I-1086 [SIC 3253]
Monarch Wear Inc, p. II-1425 [SIC 3479]
Monfort Inc, p. I-2 [SIC 2011]
Monfort Pork Inc, p. I-2 [SIC 2011]
Mongold Lumber Enterprises, p. I-465 [SIC 2421]
Monier Roof Tile Inc, p. I-1122 [SIC 3272]
Monitor Aerospace Corp, p. II-1980 [SIC 3728]
Monitor Labs Inc, p. II-2076 [SIC 3829]
Monitor Sugar Co, p. I-123 [SIC 2063]
Monk-Austin Inc, p. I-228 [SIC 2141]
Monofrax Refractories, p. I-1163 [SIC 3297]
Monogram Models Inc, p. II-2144 [SIC 3944]

Roman numerals I and II indicate the volume in which pages appear. Volume I holds SICs 2011 through 3299; Volume II holds SICs 3312 through 3999.

2277

MonoLith, p. I-754 [SIC 2791]
Monon Corp, p. II-1961 [SIC 3715]
Monona Tube and Welding Inc, p. II-1486 [SIC 3498]
Monotype Typography Inc, p. I-754 [SIC 2791]
Monroc Inc, p. I-1127 [SIC 3273]
Monroe Brick and Tile, p. I-1081 [SIC 3251]
Monroe Co, p. I-581 [SIC 2531]
Monroe Fluid Technology Inc, p. I-918 [SIC 2992]
Monroe Systems for Business, p. II-1709 [SIC 3578]
Monroe Truck Equipment Inc, p. II-1948 [SIC 3713]
Monsanto Co, pp. Vol. I - 782, 874, 897 [SICs 2821, 2879, 2899]
Monsey Products Co, p. I-914 [SIC 2952]
Monster Motorsports, p. II-1943 [SIC 3711]
Montague Machine Company, p. II-1615 [SIC 3554]
Montana Refining Co, p. I-903 [SIC 2911]
Montana Silversmiths Inc, p. II-2177 [SIC 3965]
Montek, p. II-2035 [SIC 3812]
Montello Heel Manufacturing, p. I-1017 [SIC 3131]
Monterey Inc, p. I-245 [SIC 2241]
Montgomery Elevator Co, p. II-1536 [SIC 3534]
Montgomery Hosiery Mill Inc, p. I-254 [SIC 2252]
Montgomery Truss and Panel, p. I-501 [SIC 2439]
Montgomery Wire Corp, p. II-1260 [SIC 3357]
Monticello Spring Corp, p. II-1470 [SIC 3495]
Montpelier Glove Company Inc, p. I-1037 [SIC 3151]
Montrose/CDT, p. II-1476 [SIC 3496]
Monumental Sales Inc, p. I-1140 [SIC 3281]
Moody Dunbar Inc, p. I-45 [SIC 2033]
Moody Optical, p. II-2107 [SIC 3851]
Moody's Investors Service, p. I-698 [SIC 2721]
Moog Automotive Inc, p. II-1954 [SIC 3714]
Moog Inc, p. II-2035 [SIC 3812]
Moog Inc Engine Controls, p. II-1975 [SIC 3724]
Mooney Aircraft Corp, p. II-1970 [SIC 3721]
Mooney Holding Corp, p. II-1970 [SIC 3721]
Moorco International Inc, p. II-2056 [SIC 3824]
Moore American Graphics Inc, p. I-745 [SIC 2782]
Moore and Munger Marketing, p. I-903 [SIC 2911]
Moore Co, p. I-245 [SIC 2241]
Moore Data Management, p. I-718 [SIC 2752]
Moore Industries-International, p. II-2051 [SIC 3823]
Moore-O-Matic Inc, p. II-1938 [SIC 3699]
Moore Products Co, p. II-2051 [SIC 3823]
Moore's Quality Snack Foods, p. I-199 [SIC 2096]
Moore Tool Co, p. II-1558 [SIC 3541]
Mooresville Ice Cream Company, p. I-30 [SIC 2024]
Moorman Manufacturing Co, p. I-95 [SIC 2048]
Moose Products, p. II-1878 [SIC 3669]
Moot's Mountain Bikes Inc, p. II-2003 [SIC 3751]
Moretz Hosiery Mills Inc, p. I-250 [SIC 2251]
Morey Fish Co, p. I-186 [SIC 2091]
Morflex Inc, p. I-855 [SIC 2869]
Morgan, p. II-1992 [SIC 3732]
Morgan Adhesives Co, p. II-1409 [SIC 3466]
Morgan Brothers Bag Company, p. I-433 [SIC 2393]
Morgan Buildings and Spas, p. I-523 [SIC 2452]
Morgan Grain and Feed Co, p. I-95 [SIC 2048]
Morgan Lumber Sales Company, p. I-510 [SIC 2448]
Morgan Products Ltd, p. I-480 [SIC 2431]
Morgan Shirt Corp, p. I-336 [SIC 2321]
Morganite Inc, p. II-1780 [SIC 3621]
Morgen Manufacturing Co, p. II-1632 [SIC 3559]
Mori Lee Associates, p. I-365 [SIC 2335]
Moriarty Hat and Sweater Co, p. I-386 [SIC 2353]
Moritz Embroidery Works Inc, p. I-451 [SIC 2397]
Morland Valve Co, p. II-1454 [SIC 3492]
Morley Candy Makers Company, p. I-127 [SIC 2064]
Morning Glory Products, p. I-455 [SIC 2399]
Morning Sun Inc, p. I-442 [SIC 2395]
Morningstar Inc, p. I-698 [SIC 2721]
Moroso Performance Products, p. II-1395 [SIC 3462]
Morpac Industries Inc, p. II-1775 [SIC 3613]
Morrill Motors Inc, p. II-1780 [SIC 3621]
Morris Bean and Co, p. II-1377 [SIC 3449]
Morris Communications Corp, p. I-693 [SIC 2711]
Morris Communications Inc, p. I-693 [SIC 2711]
Morris Newspaper Corp, p. I-693 [SIC 2711]
Morrisette Paper Co, p. I-669 [SIC 2675]

Morrison Berkshire Inc, p. II-1605 [SIC 3552]
Morrison Group, p. I-889 [SIC 2893]
Morrison Knudsen Corp, p. II-1998 [SIC 3743]
Morrison Milling Co, p. I-70 [SIC 2041]
Morrison Molded Fiber, p. I-1006 [SIC 3089]
Morrison Textile Machinery Co, p. II-1605 [SIC 3552]
Morse Industrial, p. II-1676 [SIC 3568]
Morse Manufacturing Inc, p. II-1948 [SIC 3713]
Mortara Instrument Inc, p. II-2102 [SIC 3845]
Morton Automotive Coating Inc, p. I-840 [SIC 2851]
Morton Electronic Materials, p. I-897 [SIC 2899]
Morton International, p. II-1954 [SIC 3714]
Morton International Inc, p. I-879 [SIC 2891]
Morton Manufacturing Co, p. II-1367 [SIC 3446]
Morton Metalcraft Co, p. II-1361 [SIC 3444]
Morton Salt, p. I-897 [SIC 2899]
Moser Corp, p. I-571 [SIC 2521]
Mosher Company Inc, p. I-1144 [SIC 3291]
Mosier Industries Inc, p. II-1747 [SIC 3593]
Mosinee Paper Corp, p. I-654 [SIC 2672]
Mosler Inc, p. II-1878 [SIC 3669]
Moss Inc, p. I-437 [SIC 2394]
Moss Supply Co, p. I-480 [SIC 2431]
Mosstype Corp, p. II-1621 [SIC 3555]
Mosstype Corporation of Illinois, p. II-1621 [SIC 3555]
Motch Corp, p. II-1558 [SIC 3541]
Motek Eng & Manufacturing, p. II-1764 [SIC 3599]
Mother Come Home Inc, p. I-741 [SIC 2771]
Mother Murphys Laboratories, p. I-181 [SIC 2087]
Motif Inc, p. II-1699 [SIC 3575]
Motion Analysis Corp, p. II-2102 [SIC 3845]
Motion Analysis Systems, p. II-2112 [SIC 3861]
Motor Castings Co, p. II-1197 [SIC 3321]
Motor Coils Manufacturing Co, p. II-1780 [SIC 3621]
Motor Wheel Corp, p. II-1954 [SIC 3714]
Motorcar Parts and Accessories, p. II-1929 [SIC 3694]
Motorola AIEG, p. II-1929 [SIC 3694]
Motorola Inc, p. II-1873 [SIC 3663]
Motown Record Company LP, p. II-1864 [SIC 3652]
Mott Metallurgical Corp, p. II-1492 [SIC 3499]
Mott's of Mississippi, p. I-12 [SIC 2015]
Moultrie, p. II-1244 [SIC 3354]
Moultrie Manufacturing Co, p. II-1367 [SIC 3446]
Mount Hope Machine Co, p. II-1605 [SIC 3552]
Mount Joy Wire Corp, p. II-1182 [SIC 3315]
Mount Olive Pickle Company, p. I-55 [SIC 2035]
Mount Pulaski Products Inc, p. I-1144 [SIC 3291]
Mount Rose Ravioli, p. I-65 [SIC 2038]
Mount Vernon Mills Inc, p. I-231 [SIC 2211]
Mount Vernon Press Fabrics, p. I-327 [SIC 2299]
Mountain Equipment Inc, p. I-433 [SIC 2393]
Mountain Lumber Co, p. I-471 [SIC 2426]
Mountain Safety Research Inc, p. II-1339 [SIC 3433]
Mountain States Bindery, p. I-750 [SIC 2789]
Mountain View Fabricating, p. II-1718 [SIC 3581]
MountainGate Data Systems Inc, p. II-1694 [SIC 3572]
Mountaire Feeds Inc, p. I-95 [SIC 2048]
Mousefeathers Inc, p. I-390 [SIC 2361]
Movie Star Inc, p. I-405 [SIC 2384]
Moxness Products Inc, p. I-942 [SIC 3053]
Moxvil Manufacturing Company, p. I-258 [SIC 2253]
Moyco Industries Inc, p. I-1144 [SIC 3291]
Mozzarella Co, p. I-21 [SIC 2022]
MPB Corp, p. II-1644 [SIC 3562]
MPD Inc, p. II-1873 [SIC 3663]
MPI Inc, p. I-782 [SIC 2821]
MPI International Inc, p. II-1404 [SIC 3465]
MPI Label Systems, p. I-654 [SIC 2672]
MPL Technologies, p. II-2093 [SIC 3843]
MPM, p. II-1694 [SIC 3572]
MPO Videotronics Inc, p. II-1859 [SIC 3651]
Mr Carmen Inc, p. I-378 [SIC 2341]
MR Christmas Inc, p. II-1938 [SIC 3699]
Mr Remo of California Inc, p. I-373 [SIC 2339]
Mr T's Apparel Inc, p. I-352 [SIC 2326]
MRC Bearings, p. II-1644 [SIC 3562]
MRC Inc, p. II-1420 [SIC 3471]
Mrs Alison's Cookie Company, p. I-107 [SIC 2052]
Mrs Baird's Bakeries Inc, p. I-101 [SIC 2051]

Mrs Barry's Kona Cookies, p. I-107 [SIC 2052]
Mrs Smith's Inc, p. I-112 [SIC 2053]
MS Chambers and Sons Inc, p. I-759 [SIC 2796]
MS Co, p. II-2131 [SIC 3915]
MS Willett Inc, p. II-1572 [SIC 3544]
MSD Inc, p. II-1790 [SIC 3625]
MSI Corp, p. II-1413 [SIC 3469]
MSM Industries Inc, p. II-1361 [SIC 3444]
Mt Clemens Mineral Prod Co, p. I-799 [SIC 2833]
Mt Hope Finishing Co, p. I-231 [SIC 2211]
Mt Valley Farms & Lumber, p. I-510 [SIC 2448]
MTD Technologies Inc, p. II-1572 [SIC 3544]
MTE Corp, p. II-1769 [SIC 3612]
MTI Inc, p. II-1694 [SIC 3572]
MTI Technology Corp, p. II-1694 [SIC 3572]
MTS Systems Corp, p. II-2076 [SIC 3829]
MTU Detroit Diesel, p. II-1503 [SIC 3519]
MTX, p. II-1859 [SIC 3651]
MU Industries Inc, p. I-386 [SIC 2353]
Mueller Brass Co, p. II-1234 [SIC 3351]
Mueller Industries Inc, p. II-1234 [SIC 3351]
Mueller Streamline, p. II-1234 [SIC 3351]
Muench-Kreuzer Candle Co, p. II-2199 [SIC 3999]
Mulay Plastics, p. I-1006 [SIC 3089]
Mule Skins, p. I-336 [SIC 2321]
Mulherin Architectural, p. I-1140 [SIC 3281]
Mulholland Harper Co, p. II-2186 [SIC 3993]
Muller-Ray Corp, p. II-1600 [SIC 3549]
Mullins Food Products, p. I-45 [SIC 2033]
Multi-Arc Inc, p. II-1425 [SIC 3479]
Multi-Cast Corp, p. II-1275 [SIC 3365]
Multi-Color Corp, p. I-724 [SIC 2754]
Multi-Electric Manufacturing Inc, p. II-1854 [SIC 3648]
Multi-Line Cans Inc, p. I-634 [SIC 2655]
Multi-Local Media Info, p. I-713 [SIC 2741]
Multi-Pak Corp, p. II-1822 [SIC 3639]
Multi-Plex Inc, p. II-1404 [SIC 3465]
Multi Products International, p. II-1905 [SIC 3677]
Multi-Tech Systems Inc, p. II-1694 [SIC 3572]
Multiform Desiccants Inc, p. I-1154 [SIC 3295]
Multimedia Inc, p. I-693 [SIC 2711]
Multiples, p. I-258 [SIC 2253]
Multiplex Company Inc, p. II-1764 [SIC 3599]
Multiplex Display Fixtures, p. I-587 [SIC 2541]
Multipress, p. II-1563 [SIC 3542]
Multiquip Inc, p. II-1584 [SIC 3546]
Multiscore Inc, p. II-1610 [SIC 3553]
Multitex Corporation, p. I-297 [SIC 2281]
Multitex Corporation of America, p. I-292 [SIC 2273]
Munchkin Bottling Inc, p. I-1062 [SIC 3221]
Muncy Building Enterprises LP, p. I-523 [SIC 2452]
Mundet Inc, p. I-612 [SIC 2621]
Munro and Co, p. I-1025 [SIC 3143]
Munroe Inc, p. II-1192 [SIC 3317]
Munsingwear Inc, p. I-340 [SIC 2322]
Munters Corp, p. II-1682 [SIC 3569]
Mupac Corp, p. II-1795 [SIC 3629]
Muralo Company Inc, p. I-840 [SIC 2851]
Murata/Muratec, p. II-1868 [SIC 3661]
Murata Wiedemann Inc, p. II-1563 [SIC 3542]
Murmur Corp, p. II-1430 [SIC 3482]
Murphy Body Co, p. II-1948 [SIC 3713]
Murphy Co, p. I-496 [SIC 2436]
Murphy-Miller Co, p. I-571 [SIC 2521]
Murray Cabinet and Fixtures, p. I-486 [SIC 2434]
Murray Corp, p. II-1323 [SIC 3429]
Murray Ohio Manufacturing Co, p. II-1514 [SIC 3524]
Muscle Bound Bindery, p. I-750 [SIC 2789]
Museum Reproductions Inc, p. II-2173 [SIC 3961]
Musgrave Pencil Company Inc, p. II-2160 [SIC 3952]
Mushroom Cooperative Co, p. I-45 [SIC 2033]
Music Maestro Inc, p. II-2135 [SIC 3931]
Muskin Leisure Products Inc, p. II-2150 [SIC 3949]
Mustang Manufacturing, p. II-1519 [SIC 3531]
Mutual Engraving Company Inc, p. I-724 [SIC 2754]
Mutual Industries Inc, p. I-446 [SIC 2396]
Mutual Materials Co, p. I-1081 [SIC 3251]
Mutual Tool and Die Inc, p. II-1572 [SIC 3544]

Roman numerals I and II indicate the volume in which pages appear. Volume I holds SICs 2011 through 3299; Volume II holds SICs 3312 through 3999.

Roman numerals I and II indicate the volume in which pages appear. Volume I holds SICs 2011 through 3299; Volume II holds SICs 3312 through 3999.

Roman numerals I and II indicate the volume in which pages appear. Volume I holds SICs 2011 through 3299; Volume II holds SICs 3312 through 3999.

NLC Inc, p. II-1594 [SIC 3548]
NMB, p. II-1704 [SIC 3577]
NMB Technologies Inc, p. II-1704 [SIC 3577]
NN Ball and Roller Co, p. II-1295 [SIC 3399]
No Moon Co, p. I-420 [SIC 2389]
No-Sag Foam, p. I-991 [SIC 3086]
Noahs Potato Chip Company Inc, p. I-199 [SIC 2096]
Nobelpharma USA Inc, p. II-2093 [SIC 3843]
Nobleworks Inc, p. I-741 [SIC 2771]
Nocona Boot Co, p. I-1025 [SIC 3143]
Noel Corp, p. I-176 [SIC 2086]
Noel Joanna Inc, p. I-236 [SIC 2221]
Noevir Inc, p. I-834 [SIC 2844]
Nog Inc, p. I-132 [SIC 2066]
Nokomis Mill, p. I-70 [SIC 2041]
Nolan Co, p. II-1998 [SIC 3743]
Noll Manufacturing Co, p. II-1361 [SIC 3444]
Noll Printing Company Inc, p. I-718 [SIC 2752]
NoMac Energy Systems Inc, p. II-1498 [SIC 3511]
Nomadics Tipi Makers, p. I-437 [SIC 2394]
Non-Ferrous Metals Inc, p. II-1285 [SIC 3369]
Non-Fluid Oil Corp, p. I-918 [SIC 2992]
Nonferrous Products Inc, p. II-1400 [SIC 3463]
Nonpareil Corp, p. I-51 [SIC 2034]
Noodles By Leonardo Inc, p. I-208 [SIC 2098]
Noon Hour Food Products, p. I-186 [SIC 2091]
Nooter Corp, p. II-1172 [SIC 3312]
NOR-AM Chemical Co, p. I-874 [SIC 2879]
Nor-Cal Beverage Company Inc, p. I-176 [SIC 2086]
Nor-Cote Inc, p. II-1290 [SIC 3398]
Nor-Cote International Inc, p. I-889 [SIC 2893]
Norac Company Inc, p. I-782 [SIC 2821]
NORAN Instruments Inc, p. II-2066 [SIC 3826]
Norand Corp, p. II-1688 [SIC 3571]
Noranda Aluminum Inc, p. II-1221 [SIC 3334]
Norca Corp, p. II-1334 [SIC 3432]
Norco, p. I-536 [SIC 2499]
Norco Windows Inc, p. I-480 [SIC 2431]
Norcold, p. II-1805 [SIC 3632]
Norcraft Companies Inc, p. I-486 [SIC 2434]
Norcross Footwear Inc, p. I-1025 [SIC 3143]
Nord Gear Corp, p. II-1666 [SIC 3566]
Nord Photo Engineering Inc, p. II-2112 [SIC 3861]
NORDAM Group, p. II-1975 [SIC 3724]
Nordberg Inc, p. II-1525 [SIC 3532]
Nordex Inc, p. II-1666 [SIC 3566]
Nordic Group, p. I-1006 [SIC 3089]
Nordic Packaging Inc, p. I-643 [SIC 2657]
NordicTrack Inc, p. II-2150 [SIC 3949]
Nordigo Peripherals, p. II-1694 [SIC 3572]
Nordson Corp, p. II-1682 [SIC 3569]
Nordyne Inc, p. II-1726 [SIC 3585]
Norelco Consumer Products, p. II-1813 [SIC 3634]
Norfab Inc, p. II-1229 [SIC 3341]
Norfield Industries, p. II-1610 [SIC 3553]
Norfolk Conveyor, p. II-1541 [SIC 3535]
Norforge and Machine Inc, p. II-1395 [SIC 3462]
Norland Corp, p. II-2098 [SIC 3844]
Norman Enterprises Inc, p. II-1854 [SIC 3648]
Norman, Fox and Co, p. I-819 [SIC 2841]
Norman Perry Co, p. II-1840 [SIC 3645]
Norman W Paschall Company, p. I-327 [SIC 2299]
Normandy Industries, p. I-979 [SIC 3084]
Norment Industries WSA Inc, p. II-1878 [SIC 3669]
Norpac Foods Inc, p. I-60 [SIC 2037]
Norpak Corp, p. I-649 [SIC 2671]
Norris Cylinder Co, p. II-1747 [SIC 3593]
Norris Knitting Company Inc, p. I-258 [SIC 2253]
Norris Plumbing Fixtures, p. I-1098 [SIC 3261]
Norriseal, p. II-1531 [SIC 3533]
Norse Furniture Co, p. I-581 [SIC 2531]
Nortek Inc, p. II-1822 [SIC 3639]
North Amer Power Tools, p. II-1584 [SIC 3546]
North Amer Spring & Stamping, p. II-1470 [SIC 3495]
North Amer Textile Machinery, p. II-1605 [SIC 3552]
North America Packaging, p. II-1300 [SIC 3411]
North American Advanced, p. I-1167 [SIC 3299]
North American Biologicals Inc, p. I-814 [SIC 2836]
North American Capacitor Co, p. II-1897 [SIC 3675]

North American Chemical Co, p. I-1017 [SIC 3131]
North American Clutch Corp, p. II-1676 [SIC 3568]
North American Container Corp, p. I-628 [SIC 2653]
North American Drager Inc, p. II-2082 [SIC 3841]
North American Enclosures Inc, p. I-536 [SIC 2499]
North American Housing Corp, p. I-523 [SIC 2452]
North American Industries Inc, p. II-1547 [SIC 3536]
North American Lighting Inc, p. II-1850 [SIC 3647]
North American Mfg Co, p. II-1339 [SIC 3433]
North American Oxide Inc, p. II-1225 [SIC 3339]
North American Plastics Corp, p. I-660 [SIC 2673]
North American Printing Ink, p. I-889 [SIC 2893]
North American Products Corp, p. II-1558 [SIC 3541]
North American Rayon Corp, p. I-791 [SIC 2823]
North American Refractories Co, p. I-1090 [SIC 3255]
North American Transformer, p. II-1769 [SIC 3612]
North American Watch Corp, p. II-2118 [SIC 3873]
North American Wood Products, p. I-491 [SIC 2435]
North and Judd Inc, p. II-2177 [SIC 3965]
North Arkansas Poultry Co, p. I-12 [SIC 2015]
North Carolina Finishing Co, p. I-279 [SIC 2261]
North Carolina Granite Corp, p. I-1140 [SIC 3281]
North Carolina SRT Inc, p. II-2066 [SIC 3826]
North Coast Brewing Company, p. I-160 [SIC 2082]
North Florida Concrete Inc, p. I-1117 [SIC 3271]
North Florida Shipyards Inc, p. II-1986 [SIC 3731]
North Hickory Furniture Co, p. I-548 [SIC 2512]
North Metal and Chemical Co, p. II-1178 [SIC 3313]
North Pacific Paper Corp, p. I-612 [SIC 2621]
North Pacific Processors Inc, p. I-190 [SIC 2092]
North Plains Textiles Inc, p. I-302 [SIC 2282]
North River Apparel Inc, p. I-420 [SIC 2389]
North Sails Group Inc, p. I-437 [SIC 2394]
North Santiam Plywood Inc, p. I-496 [SIC 2436]
North Side Packing Co, p. I-7 [SIC 2013]
North Star Company Inc, p. II-1377 [SIC 3449]
North Star Concrete Co, p. I-1122 [SIC 3272]
North Star Distributors Inc, p. I-603 [SIC 2599]
North Star Glove Co, p. I-1037 [SIC 3151]
North Star Steel Co, p. II-1172 [SIC 3312]
North Starlighting Inc, p. II-1854 [SIC 3648]
North State Garment Company, p. I-373 [SIC 2339]
North State Pyrophyllite, p. I-1090 [SIC 3255]
Northcoast Oil Inc, p. I-918 [SIC 2992]
Northcutt Woodworks LP, p. I-480 [SIC 2431]
Northeast Graphics Inc, p. I-729 [SIC 2759]
Northeast Steel & Machine Prod, p. II-1182 [SIC 3315]
Northeastern Culvert, p. II-1334 [SIC 3432]
Northeastern Envelope, p. I-679 [SIC 2677]
Northeastern Envelope Co, p. I-679 [SIC 2677]
Northeastern Log Homes, p. I-523 [SIC 2452]
Northern Can Systems Inc, p. II-1300 [SIC 3411]
Northern Cap Manufacturing Co, p. I-386 [SIC 2353]
Northern Cross Industries Inc, p. I-599 [SIC 2591]
Northern Hardwoods, p. I-471 [SIC 2426]
Northern Michigan Veneers Inc, p. I-491 [SIC 2435]
Northern Pride Inc, p. I-12 [SIC 2015]
Northern Products Log Homes, p. I-523 [SIC 2452]
Northern Telecom Inc, p. II-1868 [SIC 3661]
Northfield Foundry, p. II-1610 [SIC 3553]
Northgate Computer Systems, p. II-1688 [SIC 3571]
Northlake Engineering Inc, p. II-1905 [SIC 3677]
Northland Aluminum Products, p. II-1275 [SIC 3365]
Northland Corp, p. II-1805 [SIC 3632]
Northland Shoe Corp, p. I-1029 [SIC 3144]
Northrop Corp, pp. Vol II - 1970, 2035 [SICs 3721, 3812]
Northrop Grumman, p. II-1915 [SIC 3679]
Northrop Grumman Corp, p. II-1970 [SIC 3721]
Northrop Grumman Intern, p. II-1970 [SIC 3721]
Northstar Computer Forms Inc, p. I-735 [SIC 2761]
Northwest Aluminum Products, p. II-1349 [SIC 3442]
Northwest Automatic Products, p. II-1382 [SIC 3451]
Northwest Bodies Inc, p. II-1948 [SIC 3713]
Northwest Cooperage Company, p. II-1304 [SIC 3412]
Northwest Design Products Inc, p. I-536 [SIC 2499]
Northwest Futon Co, p. I-558 [SIC 2515]
Northwest Grating Products Inc, p. II-1367 [SIC 3446]
Northwest Hardwoods, p. I-465 [SIC 2421]

Northwest Paper, p. I-612 [SIC 2621]
Northwest Pea and Bean, p. I-51 [SIC 2034]
Northwest Pipe and Casing Co, p. II-1192 [SIC 3317]
Northwest Publications Inc, p. I-693 [SIC 2711]
Northwest Spec Baking, p. I-83 [SIC 2045]
Northwest Swiss-Matic Inc, p. II-1382 [SIC 3451]
Northwest Windows, p. II-1349 [SIC 3442]
Northwestern Casket Co, p. II-2191 [SIC 3995]
Northwestern Colorgraphics Inc, p. I-759 [SIC 2796]
Northwestern Extract Co, p. I-165 [SIC 2083]
Northwestern Golf Co, p. II-2150 [SIC 3949]
Northwestern Industries Inc, p. I-1071 [SIC 3231]
Northwestern Motor Co, p. II-1552 [SIC 3537]
Northwestern Steel and Wire Co, p. II-1172 [SIC 3312]
Northwestern Tile & Marble Co, p. I-1140 [SIC 3281]
Northwood Asphalt Products, p. I-909 [SIC 2951]
Northwoods Log Homes Inc, p. I-523 [SIC 2452]
Norton-Alcoa Proppants, p. II-1531 [SIC 3533]
Norton Battery Mfg Co, p. II-1920 [SIC 3691]
Norton Chemical, p. I-897 [SIC 2899]
Norton Co, p. I-1144 [SIC 3291]
Norton Construction Products, p. II-1578 [SIC 3545]
Norton McNaughton Inc, p. I-369 [SIC 2337]
Norton McNaughton of Squire, p. I-369 [SIC 2337]
Norton Performance, p. I-1006 [SIC 3089]
Norwalk Company Inc, p. II-1649 [SIC 3563]
Norwalk Powdered Metals Inc, p. II-1295 [SIC 3399]
Norwell Manufacturing Inc, p. II-1854 [SIC 3648]
Norwood Coated Products, p. I-879 [SIC 2891]
Norwood Marking Systems, p. II-2164 [SIC 3953]
Norwood Sash, p. I-501 [SIC 2439]
Nosco Inc, p. I-718 [SIC 2752]
Noteworthy Co, p. I-660 [SIC 2673]
Nouveau International Inc, p. II-1718 [SIC 3581]
Nova Biomedical Corp, p. II-2040 [SIC 3821]
Nova Materials Inc, p. I-1127 [SIC 3273]
Nova Office Furniture Inc, p. I-571 [SIC 2521]
Nova Precision Casting Corp, p. II-1207 [SIC 3324]
Novacor Chemical Inc, p. I-782 [SIC 2821]
Novamax Technologies Inc, p. I-918 [SIC 2992]
Novar Controls Corp, p. II-2045 [SIC 3822]
Novatron Corp, p. II-1938 [SIC 3699]
Novellus Systems Inc, p. II-1632 [SIC 3559]
Novelty Cord and Tassel, p. I-327 [SIC 2299]
Novelty Textile Mills Inc, p. I-275 [SIC 2259]
Novi American Inc, p. I-1001 [SIC 3088]
Novikoff Inc, p. I-571 [SIC 2521]
Novo Card Publishers Inc, p. I-741 [SIC 2771]
Novo Products Inc, p. II-1845 [SIC 3646]
Novus International Inc, p. I-95 [SIC 2048]
Now Products Inc, p. I-567 [SIC 2519]
NRP Inc, p. I-713 [SIC 2741]
NS Group Inc, p. II-1187 [SIC 3316]
NSA International Inc, p. II-1654 [SIC 3564]
NSC Corp, p. I-437 [SIC 2394]
NSK Corp, p. II-1644 [SIC 3562]
NT Jenkins Manufacturing, p. I-480 [SIC 2431]
NTF Inc, p. II-1486 [SIC 3498]
NTT Inc, p. II-1830 [SIC 3643]
Nu-Air Manufacturing Co, p. II-1349 [SIC 3442]
Nu-Art, p. I-729 [SIC 2759]
Nu-Art Lighting & Mfg, p. II-1845 [SIC 3646]
Nu Container Corp, p. I-634 [SIC 2655]
Nu-kote Holding Inc, p. II-2169 [SIC 3955]
Nu Life Fertilizers, p. I-861 [SIC 2873]
Nu-Look Fashions Inc, p. I-332 [SIC 2311]
Nu-Method Pest Control Prod, p. I-874 [SIC 2879]
Nu Quaker Dyeing Inc, p. I-279 [SIC 2261]
Nu-Wa Industries Inc, p. II-2020 [SIC 3792]
Nu-Way Speaker Products Inc, p. II-1859 [SIC 3651]
Nu-West Industries Inc, p. I-865 [SIC 2874]
NuArc Company Inc, p. II-2112 [SIC 3861]
Nuclear Cooling Inc, p. II-1726 [SIC 3585]
Nuclear Metals Inc, p. II-1295 [SIC 3399]
Nucor Bearing Products Inc, p. II-1644 [SIC 3562]
Nucor Corp, p. II-1172 [SIC 3312]
Nucor-Yamato Steel Co, p. II-1212 [SIC 3325]
Nucraft Furniture Co, p. I-571 [SIC 2521]
Numa Tool Co, p. II-1531 [SIC 3533]

Roman numerals I and II indicate the volume in which pages appear. Volume I holds SICs 2011 through 3299; Volume II holds SICs 3312 through 3999.

2281

NUMAR Corp, p. II-1531 [SIC 3533]
Number Nine Visual, p. II-1704 [SIC 3577]
Numberall Stamp and Tool, p. II-1713 [SIC 3579]
Numerx Corp, p. II-2061 [SIC 3825]
Nupla Corp, p. II-1312 [SIC 3423]
NuSil Technology, p. II-1225 [SIC 3339]
Nutek Inc, p. II-1929 [SIC 3694]
NuTone Inc, p. II-1813 [SIC 3634]
NutraMax Products Inc, p. I-834 [SIC 2844]
NutraSweet Co, p. I-855 [SIC 2869]
Nutri-Fruit Inc, p. I-60 [SIC 2037]
NutriBasics LP, p. I-95 [SIC 2048]
Nutrilite Products Inc, p. I-804 [SIC 2834]
Nutritional Life Support Syst Co, p. I-799 [SIC 2833]
Nuway-Microflake Partnership, p. I-224 [SIC 2131]
Nuwoods Inc, p. I-532 [SIC 2493]
NVF Co, pp. Vol. I - 782, 1006 [SICs 2821, 3089]
NWL Transformers Inc, p. II-1769 [SIC 3612]
Nycor Inc, p. II-1726 [SIC 3585]
Nygaard Logging Company Inc, p. I-460 [SIC 2411]
Nylint Corp, p. II-2144 [SIC 3944]
Nylon Net Co, p. I-322 [SIC 2298]
Nyman Marine Corp, p. II-1547 [SIC 3536]
Nypro Inc, p. I-1006 [SIC 3089]
NYSCO Products Inc, p. I-634 [SIC 2655]
Nystrom Co, p. I-713 [SIC 2741]

O Ames Co, p. II-1312 [SIC 3423]
O and E Machine Corp, p. II-1578 [SIC 3545]
O & G Spring & Wire, p. II-1470 [SIC 3495]
O and H Manufacturing, p. I-263 [SIC 2254]
O and K American Corp, p. II-1182 [SIC 3315]
O and W Heat Treat Inc, p. II-1290 [SIC 3398]
O-AT-KA Milk Products Coop, p. I-26 [SIC 2023]
O'Brien Consolidated Industries, p. II-1377 [SIC 3449]
O'Brien Gear Company Inc, p. II-1666 [SIC 3566]
O'Brien Powder Products Inc, p. I-840 [SIC 2851]
O'Bryan Brothers Inc, p. I-378 [SIC 2341]
O-Cedar/Vining, p. II-2181 [SIC 3991]
O'Connell Machinery Co, p. II-1563 [SIC 3542]
O'Day Equipment Inc, p. II-1732 [SIC 3586]
O'Dell Industries Inc, p. I-1071 [SIC 3231]
O'Dell Williams Inc, p. I-310 [SIC 2295]
O'Gara, Hess, p. II-1943 [SIC 3711]
O'Grady Containers Inc, p. I-688 [SIC 2679]
O'Hara Metal Products Co, p. II-1470 [SIC 3495]
O'Keeffe's Inc, p. I-1057 [SIC 3211]
O'Malley Wood Products Inc, p. I-515 [SIC 2449]
O'Neill, p. I-327 [SIC 2299]
O Seal, p. I-942 [SIC 3053]
O'Sullivan Corp, p. I-1006 [SIC 3089]
O'Sullivan Industries Holdings, p. I-571 [SIC 2521]
O'Sullivan Industries Inc, p. I-571 [SIC 2521]
Oak Brook Equities Inc, p. I-548 [SIC 2512]
Oak Canyon Inc, p. I-542 [SIC 2511]
Oak Creek Homes Inc, p. I-519 [SIC 2451]
Oak Hill Sportswear Corp, p. I-361 [SIC 2331]
Oak Industries Inc, p. II-1915 [SIC 3679]
Oak Land Furniture Mfg, p. I-548 [SIC 2512]
Oak-Mitsui Inc, p. II-1482 [SIC 3497]
Oak Paper Products Company, p. I-628 [SIC 2653]
Oak Products Inc, p. II-1563 [SIC 3542]
Oak Technology Inc, p. II-1892 [SIC 3674]
Oakdale Cotton Mills, p. I-297 [SIC 2281]
Oakdale Knitting Mills Inc, p. I-258 [SIC 2253]
Oakhurst Dairy, p. I-35 [SIC 2026]
Oakland Corp, p. II-1388 [SIC 3452]
Oakland Metal Treating, p. II-1290 [SIC 3398]
Oakland National Engraving Co, p. I-759 [SIC 2796]
Oakley Industries Inc, p. II-1192 [SIC 3317]
Oakloom Clothes Inc, p. I-332 [SIC 2311]
Oakwood Homes Corp, p. I-519 [SIC 2451]
Oates Flag Company Inc, p. I-455 [SIC 2399]
Oatmeal Studios Inc, p. I-741 [SIC 2771]
OB Macaroni Co, p. I-208 [SIC 2098]
OB Systems and Mining Inc, p. II-1769 [SIC 3612]
OBBCO Consolidated Industries, p. II-1400 [SIC 3463]
Oberdorfer Industries Inc, p. II-1275 [SIC 3365]

Oberg Industries Inc, p. II-1764 [SIC 3599]
Oberheim, p. II-2135 [SIC 3931]
Oberlin Farms Dairy Inc, p. I-35 [SIC 2026]
Obron Atlantic Corp, p. II-1295 [SIC 3399]
OCA MicroCoatings, p. II-2071 [SIC 3827]
Occidental Chemical Corp, pp. Vol. I - 764, 825 [SICs 2812, 2842]
Occidental Petroleum Corp, p. I-764 [SIC 2812]
Ocean Beauty Seafoods Inc, p. I-186 [SIC 2091]
Ocean Bio-Chem Inc, p. I-825 [SIC 2842]
Ocean Foods of Astoria Inc, p. I-190 [SIC 2092]
Ocean of America Inc, p. II-2144 [SIC 3944]
Ocean Spray Cranberries Inc, p. I-45 [SIC 2033]
Ocean Yachts Inc, p. II-1992 [SIC 3732]
Oceantrawl Inc, p. I-190 [SIC 2092]
Ocello Inc, p. I-258 [SIC 2253]
Ochoco Lumber Co, p. I-465 [SIC 2421]
Ochs Brick Co, p. I-1081 [SIC 3251]
Ockerlund Industries Inc, p. I-506 [SIC 2441]
Ockerlund Wood Products, p. I-506 [SIC 2441]
Oconomowoc Canning Co, p. I-45 [SIC 2033]
Octel Communications Corp, p. II-1868 [SIC 3661]
October Company Inc, p. I-576 [SIC 2522]
OddzOn Products Inc, p. I-954 [SIC 3069]
Odetics Inc, p. II-1694 [SIC 3572]
ODL Inc, p. I-1057 [SIC 3211]
Odom's Tenn Pride Sausage Co, p. I-7 [SIC 2013]
Odwalla Inc, p. I-45 [SIC 2033]
OE Clark Paper Box Co, p. I-623 [SIC 2652]
OEA Aerospace Inc, p. I-885 [SIC 2892]
OEA Inc, p. II-1980 [SIC 3728]
OEC Medical Systems Inc, p. II-2098 [SIC 3844]
Oeser Co, p. I-528 [SIC 2491]
OF Mossberg and Sons Inc, p. II-1438 [SIC 3484]
OFC Corp, p. II-2071 [SIC 3827]
Office Electronics Inc, p. I-735 [SIC 2761]
Office Systems Inc, p. II-1713 [SIC 3579]
Officially for Kids Inc, p. I-390 [SIC 2361]
Offset Paperback Manufacturers, p. I-708 [SIC 2732]
Ogden Engineering Corp, p. II-1594 [SIC 3548]
Ogden Manufacturing Co, p. II-1671 [SIC 3567]
Ogden Projects Inc, p. II-1671 [SIC 3567]
Ogemaw Forge Co, p. II-1395 [SIC 3462]
Ogihara America Corp, p. II-1413 [SIC 3469]
Ogontz Corp, p. II-1447 [SIC 3491]
Oh Boy Corp, p. I-65 [SIC 2038]
OH Kruse Grain, p. I-95 [SIC 2048]
Ohaus Corp, p. II-1759 [SIC 3596]
Ohi Automotive, p. II-1404 [SIC 3465]
Ohio Aluminum Industries Inc, p. II-1275 [SIC 3365]
Ohio Art Co, p. II-2144 [SIC 3944]
Ohio Bag Corp, p. I-1045 [SIC 3171]
Ohio Brass Co, p. II-1830 [SIC 3643]
Ohio Brush Co, p. II-2181 [SIC 3991]
Ohio Carbon Co, p. II-1786 [SIC 3624]
Ohio Decorative Products Inc, p. II-1285 [SIC 3369]
Ohio Electric Motors Inc, p. II-1780 [SIC 3621]
Ohio Electronic Engravers Inc, p. I-759 [SIC 2796]
Ohio Health Care Products Inc, p. II-2093 [SIC 3843]
Ohio Jacobson Co, p. II-1388 [SIC 3452]
Ohio Knitting Mills Inc, p. I-258 [SIC 2253]
Ohio Magnetics Inc, p. I-1109 [SIC 3264]
Ohio Moulding Corp, p. II-1377 [SIC 3449]
Ohio Nut and Bolt Co, p. II-1388 [SIC 3452]
Ohio Packing Co, p. I-2 [SIC 2011]
Ohio Rod Products Inc, p. II-2177 [SIC 3965]
Ohio Screw Products Inc, p. II-1382 [SIC 3451]
Ohio Sealants Inc, p. I-879 [SIC 2891]
Ohio Steel Industries Inc, p. II-1344 [SIC 3441]
Ohlinger Industries Inc, p. II-1600 [SIC 3549]
Ohmstede Inc, p. II-1355 [SIC 3443]
Ohmtek Inc, p. II-1901 [SIC 3676]
Oil Center Research Inc, p. I-918 [SIC 2992]
Oil Chem Inc, p. I-918 [SIC 2992]
Oil-Dri Corp, p. I-825 [SIC 2842]
Oil-Dri Corporation of America, p. I-1154 [SIC 3295]
Oil Dynamics Inc, p. II-1531 [SIC 3533]
Oildyne, p. II-1747 [SIC 3593]
Oiles America Corp, p. II-1644 [SIC 3562]

Oilgear Co, p. II-1753 [SIC 3594]
OK-1 Manufacturing Co, p. II-2087 [SIC 3842]
OK Industries Inc, pp. Vol. I - 12, Vol. II - 1312 [SICs 2015, 3423]
Okaloosa Asphalt Inc, p. I-909 [SIC 2951]
Okidata Group, p. II-1704 [SIC 3577]
Okla Homer Smith, p. I-542 [SIC 2511]
Oklahoma Graphics, p. I-729 [SIC 2759]
Oklahoma Leather Products Inc, p. I-1053 [SIC 3199]
Oklahoma Publishing Co, p. I-693 [SIC 2711]
Oklahoma Waste&Wiping Rag, p. I-327 [SIC 2299]
Okleelanta Corp, p. I-116 [SIC 2061]
Okonite Company Inc, p. II-1260 [SIC 3357]
Okuma Machinery Inc, p. II-1558 [SIC 3541]
Olathe Boot Co, p. I-1025 [SIC 3143]
Old Colony Box Company Inc, p. I-623 [SIC 2652]
Old Colony Envelope Co, p. I-679 [SIC 2677]
Old Dominion Box Company Inc, p. I-623 [SIC 2652]
Old Dominion Peanut Inc, p. I-127 [SIC 2064]
Old Dutch Foods Inc, p. I-212 [SIC 2099]
Old Dutch Mustard Company, p. I-45 [SIC 2033]
Old Fashion Kitchen Inc, p. I-65 [SIC 2038]
Old Fashioned Foods Inc, p. I-21 [SIC 2022]
Old Forge Lamp and Shade Inc, p. II-1840 [SIC 3645]
Old Mansion Foods Inc, p. I-195 [SIC 2095]
Old Quaker Paint Co, p. I-840 [SIC 2851]
Oldcastle Precast East Inc, p. I-1122 [SIC 3272]
Olde Country Reproductions Inc, p. II-2127 [SIC 3914]
Olde Towne Tavern, p. I-160 [SIC 2082]
Oldenburg Group Inc, p. II-1525 [SIC 3532]
Olds Products Company Inc, p. I-55 [SIC 2035]
Oles Envelope Corp, p. I-679 [SIC 2677]
Olicom USA Inc, p. II-1704 [SIC 3577]
Olin Aerospace Co, p. II-2011 [SIC 3764]
Olin Brass, Indianapolis, p. II-1234 [SIC 3351]
Olin Corp, p. I-764 [SIC 2812]
Olin Corp Brass Group, p. II-1280 [SIC 3366]
Olive Can Co, p. II-1300 [SIC 3411]
Oliver and Williams Elevator Inc, p. II-1536 [SIC 3534]
Oliver Products Co, p. II-1626 [SIC 3556]
Oliver Rubber Co, p. I-954 [SIC 3069]
Olivieri Reeds, p. II-2135 [SIC 3931]
Olmec Toys Inc, p. II-2144 [SIC 3944]
Olson Metal Products Co, p. II-1413 [SIC 3469]
Olsonite Corp, p. I-536 [SIC 2499]
Olsun Electrics Corp, p. II-1905 [SIC 3677]
Olympia Cheese Co, p. I-21 [SIC 2022]
Olympia Foundry&Fabrication, p. II-1275 [SIC 3365]
Olympia Lighting Inc, p. II-1840 [SIC 3645]
Olympic Controls Inc, p. II-2056 [SIC 3824]
Olympic Fish Co, p. I-190 [SIC 2092]
Olympic Foods Inc, p. I-45 [SIC 2033]
Olympic Home Care Prod Co, p. I-840 [SIC 2851]
Olympic Structures Inc, p. I-501 [SIC 2439]
Olympic Wire and Cable, p. II-1260 [SIC 3357]
Olympus America Inc, pp. Vol II - 2051, 2071 [SICs 3823, 3827]
Olympus Flag and Banner Inc, p. I-455 [SIC 2399]
OM Group Inc, p. I-777 [SIC 2819]
OM Scott and Sons Co, p. I-861 [SIC 2873]
Omaha Steel Castings Co, p. II-1212 [SIC 3325]
Omak Wood Products Inc, p. I-465 [SIC 2421]
Omanhene Cocoa Bean Co, p. I-132 [SIC 2066]
OMC Milwaukee, p. II-1503 [SIC 3519]
OMC SysteMatched, p. II-1503 [SIC 3519]
Omega Acquisition Corp, p. II-2112 [SIC 3861]
Omega Carpet Mills Inc, p. I-292 [SIC 2273]
Omega Engineering Inc, p. II-2045 [SIC 3822]
Omega Optical Co, p. II-2107 [SIC 3851]
OMEGA Pultrusions Inc, p. I-973 [SIC 3083]
Omega Rug Works, p. I-292 [SIC 2273]
Omega Sunspaces Inc, p. II-1372 [SIC 3448]
Omega Wire Inc, p. II-1476 [SIC 3496]
Omhaline Hydraulic Co, p. II-1747 [SIC 3593]
OMI Georgia Inc, p. I-297 [SIC 2281]
Omni International Inc, p. I-571 [SIC 2521]
Omni International Trading Inc, p. I-45 [SIC 2033]

Roman numerals I and II indicate the volume in which pages appear. Volume I holds SICs 2011 through 3299; Volume II holds SICs 3312 through 3999.

2282

Company Index

Pacific Fabric Reels Inc, p. II-1605 [SIC 3552]
Pacific Fibre and Rope Company, p. II-1786 [SIC 3624]
Pacific Fisherman Inc, p. II-1986 [SIC 3731]
Pacific Fixture Company Inc, p. I-587 [SIC 2541]
Pacific Foods Inc, p. I-212 [SIC 2099]
Pacific Forest Industries Inc, p. I-669 [SIC 2675]
Pacific Grain Products Inc, p. I-79 [SIC 2044]
Pacific Great Lakes Corp, p. I-1037 [SIC 3151]
Pacific Grinding Wheel Company, p. I-1144 [SIC 3291]
Pacific Handy Cutter Inc, p. II-1308 [SIC 3421]
Pacific Hardwoods, p. I-471 [SIC 2426]
Pacific/Hoe, Saw and Knife Co, p. II-1610 [SIC 3553]
Pacific Home Products Inc, p. II-1334 [SIC 3432]
Pacific International, p. I-1122 [SIC 3272]
Pacific International Rice Mills, p. I-79 [SIC 2044]
Pacific Lumber Co, p. I-460 [SIC 2411]
Pacific Micro Data Inc, p. II-1694 [SIC 3572]
Pacific Miniatures Inc, p. II-2144 [SIC 3944]
Pacific Modern Homes Inc, p. I-523 [SIC 2452]
Pacific Pipe Co, p. II-1464 [SIC 3494]
Pacific Piston Ring Company Inc, p. II-1742 [SIC 3592]
Pacific Precision Metals Inc, p. II-1413 [SIC 3469]
Pacific Press and Shear Inc, p. II-1563 [SIC 3542]
Pacific Resistor Co, p. II-1901 [SIC 3676]
Pacific Rim Diesel Inc, p. II-1503 [SIC 3519]
Pacific Scientific Co, pp. Vol. I - 885, Vol. II - 1498, 1780 [SICs 2892, 3511, 3621]
Pacific Seacraft Corp, p. II-1992 [SIC 3732]
Pacific Ship Repair & Fabrication, p. II-1986 [SIC 3731]
Pacific Snax Corp, p. I-199 [SIC 2096]
Pacific Softwoods Co, p. I-496 [SIC 2436]
Pacific Steel Casting Co, p. II-1212 [SIC 3325]
Pacific Trail Inc, p. I-356 [SIC 2329]
Pacific Tube Co, p. II-1187 [SIC 3316]
Pacific Utility Body Co, p. II-1948 [SIC 3713]
Pacific Western Extruded Plastics, p. I-979 [SIC 3084]
Pacific Western Resin Co, p. I-782 [SIC 2821]
Pack-Rite Inc, p. I-506 [SIC 2441]
Package Industries Inc, p. II-1372 [SIC 3448]
Package Machinery Co, p. II-1660 [SIC 3565]
Package Pavement Company Inc, p. I-1122 [SIC 3272]
Packaging Corporation, p. I-618 [SIC 2631]
Packaging Enterprises Inc, p. I-991 [SIC 3086]
Packaging Industries Inc, p. I-649 [SIC 2671]
Packaging Resources Inc, p. I-1006 [SIC 3089]
Packaging Systems International, p. II-1660 [SIC 3565]
Packaging Un-Limited Inc, p. I-628 [SIC 2653]
Packard Bell Electronics Inc, p. II-1688 [SIC 3571]
Packard Electric, p. II-1954 [SIC 3714]
Packard Industries Inc, p. I-593 [SIC 2542]
Packerland Packing Co, p. I-2 [SIC 2011]
Packing Material Co, p. I-510 [SIC 2448]
Paco Pharmaceutical Services Inc, p. I-804 [SIC 2834]
Paco Pumps Inc, p. II-1638 [SIC 3561]
Paco Winders Manufacturing Inc, p. II-1615 [SIC 3554]
Pacon Corp, p. I-688 [SIC 2679]
PacOrd Inc, p. II-1986 [SIC 3731]
Pacquet Oneida Inc, p. I-660 [SIC 2673]
Padco Inc, p. II-2181 [SIC 3991]
Paddy-Lee Fashions Co, p. I-373 [SIC 2339]
Page and Hill Forest Products, p. I-528 [SIC 2491]
Page Belting Co, p. I-314 [SIC 2296]
Page Foam Cushion Products, p. I-428 [SIC 2392]
Page Packaging Corp, p. I-628 [SIC 2653]
Pageland Manufacturing Inc, p. I-361 [SIC 2331]
PAGES Inc, p. I-703 [SIC 2731]
PairGain Technologies Inc, p. II-1868 [SIC 3661]
PAJ America Inc, p. II-1813 [SIC 3634]
Pajco Products Inc, p. I-310 [SIC 2295]
Pak-Mor Manufacturing Co, p. II-1948 [SIC 3713]
Pak-Sak Industries Inc, p. I-961 [SIC 3081]
Pal Graphics Inc, p. I-754 [SIC 2791]
Palermo's Villa Inc, p. I-65 [SIC 2038]
Pall Aeropower Corp, p. II-1975 [SIC 3724]

Pall Corp, p. II-1682 [SIC 3569]
Pall Trinity Micro, p. II-1682 [SIC 3569]
Pallet Masters Inc, p. I-510 [SIC 2448]
Palmer Asphalt Co, p. I-914 [SIC 2952]
Palmer Candy Co, p. I-127 [SIC 2064]
Palmer Manufacturing Company, p. II-1975 [SIC 3724]
Palmer Tube Mills Inc, p. II-1192 [SIC 3317]
Palmetto Baking Co, p. I-101 [SIC 2051]
Palmetto Box Company Inc, p. I-623 [SIC 2652]
Palmetto Brick Co, p. I-1081 [SIC 3251]
Palmetto Spinning Corp, p. I-297 [SIC 2281]
Palnut Co, p. II-1388 [SIC 3452]
Palomar Medical Technologies, p. II-2082 [SIC 3841]
Palos Verdes Building Corp, p. II-1920 [SIC 3691]
Paltier, p. I-593 [SIC 2542]
Pamarco Inc, pp. Vol II - 1621, 2164 [SICs 3555, 3953]
Pamrod Products, p. I-1158 [SIC 3296]
Pan Abode Cedar Homes Inc, p. I-523 [SIC 2452]
Pan American Metal Products, p. II-1382 [SIC 3451]
Pan American Tanning Corp, p. I-1013 [SIC 3111]
Panache, p. I-548 [SIC 2512]
Panametrics Inc, p. II-2051 [SIC 3823]
Panavision, p. II-2112 [SIC 3861]
Panda Motors Corp, p. II-1943 [SIC 3711]
Panduit Corp, p. II-1830 [SIC 3643]
Panel Concepts Inc, p. I-576 [SIC 2522]
Panel Prints Inc, p. I-718 [SIC 2752]
Panel Processing Inc, p. I-536 [SIC 2499]
Panel Processing of Texas Inc, p. I-532 [SIC 2493]
Panel Techn Building Systems, p. I-491 [SIC 2435]
Panelfold Inc, p. I-480 [SIC 2431]
Panex Corp, p. I-563 [SIC 2517]
Pangborn Corp, p. II-1632 [SIC 3559]
Pangburn Candy Corp, p. I-127 [SIC 2064]
Panhandle Foods Inc, p. I-65 [SIC 2038]
Panhandle Industrial Company, p. II-1649 [SIC 3563]
Panhandle Slim, p. I-352 [SIC 2326]
Panlmatic Co, p. II-1775 [SIC 3613]
Panola Pepper Corp, p. I-55 [SIC 2035]
Panoramic Corp, p. II-2098 [SIC 3844]
Pants Plus Inc, p. I-369 [SIC 2337]
PanVera Corp, p. I-814 [SIC 2836]
Paola Yarns Inc, p. I-297 [SIC 2281]
Paoli Inc, p. I-571 [SIC 2521]
Papac Logging Inc, p. I-460 [SIC 2411]
Paper, Calmenson and Co, p. II-1519 [SIC 3531]
Paper Coating Co, p. I-654 [SIC 2672]
Paper Converting Machine Co, p. II-1615 [SIC 3554]
Paper Magic Group Inc, p. I-741 [SIC 2771]
Paper-Pak Products Inc, p. I-674 [SIC 2676]
Paper Systems Inc, p. I-688 [SIC 2679]
Paperades Inc, p. I-674 [SIC 2676]
Papercon Inc, p. I-649 [SIC 2671]
Paperfold/Graphic Finishers Inc, p. I-750 [SIC 2789]
Par-Tee Company Inc, p. II-1732 [SIC 3586]
Par-Way Group, p. II-1626 [SIC 3556]
Para-Chem Southern Inc, p. I-782 [SIC 2821]
Para-Flite Inc, p. I-455 [SIC 2399]
Para Laboratories Inc, p. I-834 [SIC 2844]
Para Systems Inc, p. II-1795 [SIC 3629]
ParaBody Inc, p. II-2150 [SIC 3949]
Parade Packaging Materials, p. I-991 [SIC 3086]
Paradise Products Corp, p. I-55 [SIC 2035]
Paraflex Industries Inc, p. II-1854 [SIC 3648]
Paragon Die and Engineering Co, p. II-1572 [SIC 3544]
Paragon Electric Company Inc, p. II-1790 [SIC 3625]
Paragon Packaging Inc, p. I-623 [SIC 2652]
Paragon Pattern & Mfg Co, p. II-1568 [SIC 3543]
Paragon Spring Co, p. II-1470 [SIC 3495]
Paramount Citrus Association, p. I-60 [SIC 2037]
Paramount Fabricating Inc, p. II-1404 [SIC 3465]
Paramount Fitness Corp, p. II-2150 [SIC 3949]
Paramount Foods Inc, p. I-55 [SIC 2035]
Paramount Industrial Companies, p. I-558 [SIC 2515]
Paramount Industries Inc, p. II-1845 [SIC 3646]
Paramount Perlite Co, p. I-1154 [SIC 3295]
Paramount Plywood Products, p. I-491 [SIC 2435]
Paramount Sales Company Inc, p. II-2173 [SIC 3961]

Parana Supplies Corp, p. II-2169 [SIC 3955]
Parasitix Corp, p. I-814 [SIC 2836]
Parco Inc, p. I-942 [SIC 3053]
Parent Metal Products, p. I-553 [SIC 2514]
Parenti and Raffaelli Ltd, p. I-480 [SIC 2431]
Paris Accessories Inc, p. I-420 [SIC 2389]
Paris Blues Inc, p. I-378 [SIC 2341]
Paris Business Forms Inc, p. I-735 [SIC 2761]
Paris Ceramics Inc, p. I-1086 [SIC 3253]
Paris Food Corp, p. I-60 [SIC 2037]
Paris Lace Inc, p. I-271 [SIC 2258]
Paris Manufacturing Co, p. I-348 [SIC 2325]
Parish Light Vehicle, p. II-1954 [SIC 3714]
Park Electrochemical Corp, p. II-1915 [SIC 3679]
Park Lane Neckwear Inc, p. I-344 [SIC 2323]
Park Manufacturing Company, p. I-361 [SIC 2331]
Park-Ohio Industries Inc, pp. Vol. I - 1006, Vol. II - 1986 [SICs 3089, 3731]
Park Place Corp, p. I-558 [SIC 2515]
Park Scientific Instruments, p. II-2071 [SIC 3827]
Parke Industries Inc, p. II-1845 [SIC 3646]
Parkell Products Inc, p. II-2093 [SIC 3843]
Parker Bertea Aerospace Group, p. II-1975 [SIC 3724]
Parker Brothers, p. II-2144 [SIC 3944]
Parker Compumotor, p. II-1682 [SIC 3569]
Parker Hannifin Corp, pp. Vol. I - 954, Vol. II - 1447, 1454, 1464, 1682, 1726, 1747 [SICs 3069, 3491, 3492, 3494, 3569, 3585, 3593]
Parker Inc, p. II-1961 [SIC 3715]
Parker McCrory Mfg Co, p. II-1938 [SIC 3699]
Parker Pen USA, p. II-2156 [SIC 3951]
Parker School Uniforms, p. I-332 [SIC 2311]
Parker Seal Group, p. I-942 [SIC 3053]
Parker Station Inc, p. I-168 [SIC 2084]
Parker Sweeper Co, p. II-1514 [SIC 3524]
Parker Tobacco Company Inc, p. I-228 [SIC 2141]
Parker+Amchem, p. I-897 [SIC 2899]
Parkhurst Manufacturing, p. II-1948 [SIC 3713]
Parking Products Inc, p. II-1713 [SIC 3579]
Parkinson Machinery, p. II-1605 [SIC 3552]
Parkline Inc, p. II-1372 [SIC 3448]
Parks and Woolson Machine Co, p. II-1605 [SIC 3552]
Parks Sausage Co, p. I-2 [SIC 2011]
Parkson Corp, p. II-1736 [SIC 3589]
Parkview Metal Products Inc, p. II-1413 [SIC 3469]
Parlux Fragrances Inc, p. I-834 [SIC 2844]
Parmatech Corp, p. II-1388 [SIC 3452]
Parmelee Industries Inc, p. II-2107 [SIC 3851]
Parsons and Whittemore Inc, p. I-608 [SIC 2611]
Parsons Footwear, p. I-932 [SIC 3021]
Partech Inc, p. II-1531 [SIC 3533]
Partek Insulations Inc, p. I-1158 [SIC 3296]
Parthenon Metal Works, p. II-1192 [SIC 3317]
Particle Measuring Systems Inc, p. II-2076 [SIC 3829]
Partlow Corp, p. II-2045 [SIC 3822]
Parts and Systems Company Inc, p. II-1605 [SIC 3552]
Party Professionals Inc, p. I-420 [SIC 2389]
Party Time Manufacturing Co, p. I-688 [SIC 2679]
Pasco Scientific Corp, p. II-2040 [SIC 3821]
Pascoe Building Systems Inc, p. II-1372 [SIC 3448]
Pass and Seymour/Legrand, p. II-1830 [SIC 3643]
Pasta USA Inc, p. I-208 [SIC 2098]
Pastorelli Food Products Inc, p. I-40 [SIC 2032]
Patcraft Commercial, p. I-292 [SIC 2273]
Pate Foods Corp, p. I-199 [SIC 2096]
Patent Construction Systems, p. II-1367 [SIC 3446]
Paterson Pacific Parchment Co, p. I-674 [SIC 2676]
Pathe Technologies Inc, p. II-1470 [SIC 3495]
PathFinder Operations, p. II-1552 [SIC 3537]
Pathway Bellows Inc, p. II-1344 [SIC 3441]
Patricia Knitting Inc, p. I-236 [SIC 2221]
Patrician Furniture Co, p. I-571 [SIC 2521]
Patrick Cudahy Inc, p. I-2 [SIC 2011]
Patrick Metals, p. II-1244 [SIC 3354]
Patriot Printing Ink Co, p. I-889 [SIC 2893]
Patriot Sensors and Controls, p. II-1790 [SIC 3625]
PATS Inc, p. II-1975 [SIC 3724]
Patsy Aiken Designs Inc, p. I-390 [SIC 2361]
Patterson Frozen Foods Inc, p. I-60 [SIC 2037]

Roman numerals I and II indicate the volume in which pages appear. Volume I holds SICs 2011 through 3299; Volume II holds SICs 3312 through 3999.

2284

Company Index

Roman numerals I and II indicate the volume in which pages appear. Volume I holds SICs 2011 through 3299; Volume II holds SICs 3312 through 3999.

2285

Perry Tritech Inc, p. II-2029 [SIC 3799]
Perrygraf, p. II-2199 [SIC 3999]
PerSeptive Biosystems Inc, p. II-2102 [SIC 3845]
Personal Marketing Co, p. I-713 [SIC 2741]
Persons-Majestic Mfg Co, p. II-2003 [SIC 3751]
Perstorp Components Inc, p. I-991 [SIC 3086]
Perstorp Inc, p. I-782 [SIC 2821]
Perstorp Polyols Inc, p. I-855 [SIC 2869]
Pertron Controls, p. II-1594 [SIC 3548]
PerTronix Inc, p. II-1929 [SIC 3694]
Pet Cards Inc, p. I-741 [SIC 2771]
Pet Inc, p. I-26 [SIC 2023]
Pet Life Foods Inc, p. I-91 [SIC 2047]
Pet Products Plus Inc, p. I-91 [SIC 2047]
Petco Inc, p. I-322 [SIC 2298]
Pete's Brewing Co, p. I-160 [SIC 2082]
Peter Baker and Son Co, p. I-909 [SIC 2951]
Peter Sachs, p. I-292 [SIC 2273]
Petersen Aluminum Corp, p. II-1239 [SIC 3353]
Peterson American Corp, p. II-1470 [SIC 3495]
Peterson Builders Inc, p. II-1992 [SIC 3732]
Peterson Farms, p. I-12 [SIC 2015]
Peterson Manufacturing Co, p. II-1850 [SIC 3647]
Peterson Nut Co, p. I-136 [SIC 2068]
Peterson Pattern Works Inc, p. II-1568 [SIC 3543]
Peterson's Guides Inc, p. I-703 [SIC 2731]
Petoskey Manufacturing Co, p. II-1271 [SIC 3364]
Petoskey Plastics Inc, p. I-961 [SIC 3081]
Petricca Industries Inc, p. I-1122 [SIC 3272]
Petrillo Brothers Inc, p. I-1090 [SIC 3255]
Petro-Lube Inc, p. I-918 [SIC 2992]
Petrolite Corp, pp. Vol I - 855, 897 [SICs 2869, 2899]
Petrx Inc, p. I-91 [SIC 2047]
Pettibone Michigan, p. II-1552 [SIC 3537]
Petty Printing Company Inc, p. I-729 [SIC 2759]
Pevco Systems International Inc, p. II-1541 [SIC 3535]
Pez Candy Inc, p. I-127 [SIC 2064]
Pez Manufacturing Corp, p. I-127 [SIC 2064]
PF Labs Inc, p. I-804 [SIC 2834]
PFA Inc, p. II-1747 [SIC 3593]
Pfanstiehl Laboratories Inc, p. I-799 [SIC 2833]
Pfaudler Companies Inc, p. I-1066 [SIC 3229]
Pfister and Vogel Leather Co, p. I-1013 [SIC 3111]
Pfister Chemicals Inc, p. I-850 [SIC 2865]
Pfizer Animal Health Group, p. I-799 [SIC 2833]
Pfizer Consumer Health Care, p. I-804 [SIC 2834]
Pfizer Hospital Products Group, p. II-2087 [SIC 3842]
Pfizer Inc, pp. Vol I - 212, 804 [SICs 2099, 2834]
Pfizer Pharmaceutical, p. I-804 [SIC 2834]
PG Publishing Co, p. I-693 [SIC 2711]
PH Hydraulics and Automation, p. II-1563 [SIC 3542]
Phar-Shar Manufacturing, p. I-332 [SIC 2311]
Pharmaceutical Formulations Inc, p. I-804 [SIC 2834]
Pharmachem Corp, p. I-799 [SIC 2833]
Pharmacia Biotech Inc, p. II-2082 [SIC 3841]
Pharmacia P-L Biochemicals Inc, p. I-855 [SIC 2869]
Pharmacy Ellegant Paperbox, p. I-623 [SIC 2652]
Pharmasol Corp, p. II-2199 [SIC 3999]
Pharmavite Corp, p. I-799 [SIC 2833]
PharmTech Ltd, p. I-95 [SIC 2048]
Pharr Yarns Inc, p. I-297 [SIC 2281]
Pharr Yarns of Georgia, p. I-297 [SIC 2281]
Phase X Systems Inc, p. II-1699 [SIC 3575]
Phases Inc, p. I-378 [SIC 2341]
PHB Die Castings Inc, p. II-1266 [SIC 3363]
PHB Inc, p. II-1275 [SIC 3365]
PHB Machining, p. II-1578 [SIC 3545]
PHD Inc, p. II-1747 [SIC 3593]
Phelps Cement Products Inc, p. I-1117 [SIC 3271]
Phelps Dodge Corp, p. II-1234 [SIC 3351]
Phelps Dodge Magnet Wire Co, p. II-1260 [SIC 3357]
Phelps Dodge Refining Corp, p. II-1217 [SIC 3331]
Phelps-Tointon Inc, p. I-1122 [SIC 3272]
Phelps Tool and Die Company, p. II-1572 [SIC 3544]
Phibro-Tech Inc, p. I-777 [SIC 2819]
Phil Coca-Cola, p. I-176 [SIC 2086]
Philadelphia Gear Corp, p. II-1666 [SIC 3566]
Philadelphia Insulated Wire Co, p. II-1260 [SIC 3357]
Philadelphia Macaroni Co, p. I-208 [SIC 2098]

Philadelphia Mixers Corp, p. II-1519 [SIC 3531]
Philadelphia Newspapers Inc, p. I-693 [SIC 2711]
Philadelphia Tramrail Co, p. II-1547 [SIC 3536]
Philatron International, p. II-1260 [SIC 3357]
Philbrick Booth Spencer Inc, p. II-1212 [SIC 3325]
Philip A Stitt Agencies, p. I-437 [SIC 2394]
Philip Klein Neckwear Inc, p. I-344 [SIC 2323]
Philip Morris Companies Inc, p. I-218 [SIC 2111]
Philipp Brothers Chemicals Inc, p. I-874 [SIC 2879]
Philips Broadband Network Inc, p. II-1873 [SIC 3663]
Philips Components, p. II-1915 [SIC 3679]
Philips Consumer Electronics Co, p. II-1859 [SIC 3651]
Philips Display Components Co, p. II-1883 [SIC 3671]
Philips Electric, p. II-1859 [SIC 3651]
Philips Laser Magnetic Storage, p. II-1694 [SIC 3572]
Philips Lighting Co, p. II-1826 [SIC 3641]
Philips Products, p. II-1349 [SIC 3442]
Philips Technologies, pp. Vol II - 1780, 1929 [SICs 3621, 3694]
Philips Ultrasound, p. II-2102 [SIC 3845]
Phillips Brothers Inc, p. I-708 [SIC 2732]
Phillips Chemical Co, p. I-782 [SIC 2821]
Phillips Components, p. II-1915 [SIC 3679]
Phillips Construction Co, p. II-1377 [SIC 3449]
Phillips Corp, p. II-1764 [SIC 3599]
Phillips Driscopipe Inc, p. I-979 [SIC 3084]
Phillips Foundry Inc, p. II-1275 [SIC 3365]
Phillips Industries Inc, p. I-236 [SIC 2221]
Phillips Manufacturing Company, p. I-501 [SIC 2439]
Phillips Pattern and Castings, p. II-1285 [SIC 3369]
Phillips-Van Heusen Corp, p. I-336 [SIC 2321]
Philomath Forest Products Co, p. I-496 [SIC 2436]
Philway Products Inc, p. II-1887 [SIC 3672]
Phoenix Cement Co, p. I-1077 [SIC 3241]
Phoenix Contact Inc, p. II-1830 [SIC 3643]
Phoenix Controls Corp, p. II-2045 [SIC 3822]
Phoenix Designs Inc, p. I-571 [SIC 2521]
Phoenix Down Corp, p. I-428 [SIC 2392]
Phoenix Fabricators and Erectors, p. II-1355 [SIC 3443]
Phoenix Forging Company Inc, p. II-1395 [SIC 3462]
Phoenix Gold International Inc, p. II-1859 [SIC 3651]
Phoenix Housewares Corp, p. II-1813 [SIC 3634]
Phoenix Industries of Huntsville, p. I-437 [SIC 2394]
Phoenix Manufacturing Company, p. I-318 [SIC 2297]
Phoenix Manufacturing Inc, p. II-1948 [SIC 3713]
Phoenix Marine Enterprises Inc, p. II-1992 [SIC 3732]
Phoenix Newspapers Inc, p. I-693 [SIC 2711]
Phoenix Oil Co, p. I-918 [SIC 2992]
Phoenix Redi-Mix Company Inc, p. I-1127 [SIC 3273]
Phoenix Refrigeration Systems, p. II-1726 [SIC 3585]
Phoenix Wire Cloth Inc, p. II-1182 [SIC 3315]
Phoenix Zinc, p. I-861 [SIC 2873]
Phoenixware Ltd, p. I-428 [SIC 2392]
Photo Control Corp, p. II-2112 [SIC 3861]
Photo Sciences Inc, p. I-759 [SIC 2796]
Photo-Sonics Inc, p. II-2112 [SIC 3861]
Photocircuits Corp, p. II-1887 [SIC 3672]
Photometrics Ltd, p. II-2112 [SIC 3861]
Photon Kinetics Inc, p. II-2076 [SIC 3829]
Photon Technology International, p. II-2071 [SIC 3827]
Photonics Systems Inc, p. II-1699 [SIC 3575]
Phototype Color Graphics Inc, p. I-759 [SIC 2796]
Photronics Inc, p. I-1066 [SIC 3229]
Physio-Control Corp, p. II-2102 [SIC 3845]
Piacere International Inc, p. I-195 [SIC 2095]
Piad Precision Casting, p. I-1280 [SIC 3366]
PianoDisc, p. II-2135 [SIC 3931]
Piccolino USA Inc, p. I-356 [SIC 2329]
Pick Fisheries Inc, p. I-190 [SIC 2092]
Picker International Inc, p. II-2102 [SIC 3845]
Picker International Inc Dunlee, p. II-2098 [SIC 3844]
Pickett Hosiery Mills Inc, p. I-250 [SIC 2251]
Pico Products Inc, p. II-1873 [SIC 3663]
Picoma Industries Inc, p. II-1486 [SIC 3498]
PictureTel Corp, p. II-1878 [SIC 3669]
Picut Manufacturing Company, p. II-1377 [SIC 3449]

Piedmont Home Textile Corp, p. I-428 [SIC 2392]
Piedmont Manufacturing Co, p. II-1447 [SIC 3491]
Piedmont Poultry Processing Inc, p. I-12 [SIC 2015]
Pierce Co, p. II-1638 [SIC 3561]
Pierce Pacific Manufacturing, p. II-1519 [SIC 3531]
Pieri Creations Inc, p. II-1840 [SIC 3645]
Pierre Frozen Foods Inc, p. I-7 [SIC 2013]
Pies Inc, p. I-112 [SIC 2053]
Piggie Park Enterprises Inc, p. I-55 [SIC 2035]
Pike Industries Inc, p. I-909 [SIC 2951]
Pikeville Manufacturing Co, p. I-394 [SIC 2369]
Piknik Products Company Inc, p. I-55 [SIC 2035]
Pilgrim Glass Corp, p. I-1066 [SIC 3229]
Pilgrim House Rug Co, p. I-292 [SIC 2273]
Pilgrim's Pride Corp, p. I-12 [SIC 2015]
Pillar Corp, p. II-1671 [SIC 3567]
Pillar Industries, p. II-1671 [SIC 3567]
Pillowtex Corp, pp. Vol. I - 231, 428 [SICs 2211, 2392]
Pillsbury Bakeries, p. I-112 [SIC 2053]
Pillsbury Co, p. I-70 [SIC 2041]
Pilot Chemical of California, p. I-819 [SIC 2841]
Pilot Hosiery Mills Inc, p. I-254 [SIC 2252]
Pilot Industries Inc, p. II-1172 [SIC 3312]
PILZ America Inc, p. II-1864 [SIC 3652]
Pima Valve Inc, p. II-1454 [SIC 3492]
Pinahs Company Inc, p. I-107 [SIC 2052]
Pincus Brothers Inc, p. I-332 [SIC 2311]
Pine Hall Brick Company Inc, p. I-1081 [SIC 3251]
Pine Mountain Corp, p. I-536 [SIC 2499]
Pine Place Sportswear, p. I-361 [SIC 2331]
Pine Point Wood Products Inc, p. I-506 [SIC 2441]
Pine Ridge Winery, p. I-168 [SIC 2084]
Pine State Knitwear Co, p. I-258 [SIC 2253]
Pine Valley Meats Inc, p. I-2 [SIC 2011]
Pines Manufacturing Inc, p. II-1486 [SIC 3498]
Pines Trailer LP Kewanee, p. II-1961 [SIC 3715]
Pinkham Lumber, p. I-465 [SIC 2421]
Pinnacle Automation Inc, p. II-1541 [SIC 3535]
Pinnacle Micro Inc, p. II-1694 [SIC 3572]
Pinnacle Systems Inc, p. II-2112 [SIC 3861]
Pinole Point Steel Co, p. II-1172 [SIC 3312]
Pioneer Aerospace Corp, p. I-455 [SIC 2399]
Pioneer Asphalt Corp, p. I-894 [SIC 2895]
Pioneer Astro Industries Inc, p. II-1355 [SIC 3443]
Pioneer Concrete of America Inc, p. I-1127 [SIC 3273]
Pioneer Concrete of Texas Inc, p. I-1127 [SIC 3273]
Pioneer Container Corp, p. I-665 [SIC 2674]
Pioneer-Eclipse Corp, p. II-1736 [SIC 3589]
Pioneer Metal Finishing, p. II-1420 [SIC 3471]
Pioneer New Media Technologies, p. II-1859 [SIC 3651]
Pioneer Paper Corp, p. I-649 [SIC 2671]
Pioneer Pipe Inc, p. II-1486 [SIC 3498]
Pioneer Southern Inc, p. I-536 [SIC 2499]
Pioneer Steel Co, p. II-1377 [SIC 3449]
Pipe Fabricating and Supply Co, p. II-1486 [SIC 3498]
Piper Aircraft Corp, p. II-1970 [SIC 3721]
Piper Impact Inc, p. II-1400 [SIC 3463]
Piper Products Inc, p. I-593 [SIC 2542]
Piper Sonoma, p. I-168 [SIC 2084]
Piping Companies Inc, p. II-1486 [SIC 3498]
Piqua Engineering Inc, p. II-1764 [SIC 3599]
Piqua Minerals Inc, p. I-1140 [SIC 3281]
Pirelli Armstrong Tire Corp, p. I-927 [SIC 3011]
Pirelli Cable Corp, p. II-1260 [SIC 3357]
Pisciotta Inc, p. I-199 [SIC 2096]
Pisgah Yarn and Dyeing, p. I-288 [SIC 2269]
Pitaria Products Co, p. I-101 [SIC 2051]
Pitco Frialator Inc, p. II-1736 [SIC 3589]
Pitman-Dreitzer, p. I-1066 [SIC 3229]
Pitney Bowes Inc, p. II-1713 [SIC 3579]
Pitt Plastic Inc, p. I-660 [SIC 2673]
Pittcon Industries, p. I-1136 [SIC 3275]
Pittman, p. II-1780 [SIC 3621]
Pittsburg Knitting Mills, p. I-254 [SIC 2252]
Pittsburgh Brewing Co, p. I-160 [SIC 2082]
Pittsburgh Corning Corp, p. I-1066 [SIC 3229]
Pittsburgh Flatroll Co, p. II-1255 [SIC 3356]
Pittsburgh International, p. II-1192 [SIC 3317]

Roman numerals I and II indicate the volume in which pages appear. Volume I holds SICs 2011 through 3299; Volume II holds SICs 3312 through 3999.

2286

Roman numerals I and II indicate the volume in which pages appear. Volume I holds SICs 2011 through 3299; Volume II holds SICs 3312 through 3999.

Precision Anodizing and Plating, p. II-1420 [SIC 3471]
Precision Cable Mfg Corp, p. II-1260 [SIC 3357]
Precision Cast Products Inc, p. II-1285 [SIC 3369]
Precision Castparts Corp, p. II-1207 [SIC 3324]
Precision Coil Spring Co, p. II-1470 [SIC 3495]
Precision Diamond Tool Co, p. I-1144 [SIC 3291]
Precision Electronic Glass Inc, p. I-1066 [SIC 3229]
Precision Enterprises Inc, p. II-1275 [SIC 3365]
Precision Extrusions Inc, p. II-1244 [SIC 3354]
Precision Filaments Inc, p. II-1826 [SIC 3641]
Precision Form Inc, p. II-1388 [SIC 3452]
Precision Gasket Co, p. I-942 [SIC 3053]
Precision General Inc, p. II-1447 [SIC 3491]
Precision Heat Treatment, p. II-1290 [SIC 3398]
Precision Husky Corp, p. II-1610 [SIC 3553]
Precision Inc, p. II-1905 [SIC 3677]
Precision Industrial, p. II-1666 [SIC 3566]
Precision Industries Inc, p. II-1187 [SIC 3316]
Precision-Kidd Steel Co, p. II-1187 [SIC 3316]
Precision Laminates Corp, p. I-973 [SIC 3083]
Precision Machine & Engineering, p. II-2016 [SIC 3769]
Precision Metal Products Inc, p. II-1764 [SIC 3599]
Precision Multiple Controls Inc, p. II-1790 [SIC 3625]
Precision Optics Corporation Inc, p. II-2071 [SIC 3827]
Precision Optics Inc, p. II-2107 [SIC 3851]
Precision Pallets Co, p. I-510 [SIC 2448]
Precision Paper Tube Co, p. I-634 [SIC 2655]
Precision Plating Company Inc, p. II-1420 [SIC 3471]
Precision Printing and Packaging, p. I-724 [SIC 2754]
Precision Products Inc, p. II-1514 [SIC 3524]
Precision Rings Inc, p. II-1742 [SIC 3592]
Precision Rubber Plate Company, p. II-1621 [SIC 3555]
Precision Scientific Inc, p. II-2040 [SIC 3821]
Precision Screen Machines Inc, p. II-1605 [SIC 3552]
Precision Shooting Equip Co, p. II-2150 [SIC 3949]
Precision Spring & Stamping Co, p. II-1470 [SIC 3495]
Precision Stamping, p. II-1413 [SIC 3469]
Precision Standard Inc, p. II-2035 [SIC 3812]
Precision Truss Systems Inc, p. I-501 [SIC 2439]
Precision Tube Company Inc, p. II-1234 [SIC 3351]
Precision Twist Drill Co, p. II-1558 [SIC 3541]
Precision Valve Corp, p. II-1447 [SIC 3491]
Precision Wire Products Inc, p. II-1476 [SIC 3496]
Precision Wood Products Inc, p. I-510 [SIC 2448]
Preco Corp, p. I-608 [SIC 2611]
Preco Turbine Services Inc, p. II-1498 [SIC 3511]
Precoat Metals-Chicago, p. II-1425 [SIC 3479]
Predco Inc, p. II-1541 [SIC 3535]
Preferred Foundations, p. I-382 [SIC 2342]
Preferred Meal Systems Inc, p. I-65 [SIC 2038]
Preferred Products Inc, p. I-212 [SIC 2099]
Preferred Utilities Mfg Corp, p. II-1339 [SIC 3433]
Premarc Corp, p. I-1122 [SIC 3272]
Premark International Inc, p. I-1006 [SIC 3089]
Premco Forge Inc, p. II-1400 [SIC 3463]
Premdor Corp, p. I-480 [SIC 2431]
Premier Aluminum Inc, p. II-1275 [SIC 3365]
Premier Coatings Inc, p. II-1425 [SIC 3479]
Premier Furniture Inc, p. I-571 [SIC 2521]
Premier Gear & Machine Works, p. II-1610 [SIC 3553]
Premier Industries Inc, pp. Vol. I - 688, 991 [SICs 2679, 3086]
Premier Manufacturing Corp, p. II-1476 [SIC 3496]
Premier Plastics Co, p. I-997 [SIC 3087]
Premier Sleep Products Inc, p. I-558 [SIC 2515]
Premier Tool and Die Cast Corp, p. II-1266 [SIC 3363]
Premiere Candy Co, p. I-127 [SIC 2064]
Premiere Manufacturing Corp, p. I-254 [SIC 2252]
Premiere Polymers, p. I-997 [SIC 3087]
Premium Allied Tool Inc, p. II-1413 [SIC 3469]
Premium Beverage Packers Inc, p. I-176 [SIC 2086]
Premix Acrocrete Inc, p. I-1167 [SIC 3299]
Premix Inc, p. I-1006 [SIC 3089]
Prent Corp, p. I-967 [SIC 3082]
Prentice Hall Computer Pub, p. I-703 [SIC 2731]

Prentiss Inc, p. I-874 [SIC 2879]
Prepared Products Company Inc, p. I-212 [SIC 2099]
PrePress Solutions Inc, p. II-1621 [SIC 3555]
Pres Glas Corp, p. I-1158 [SIC 3296]
Pres-On Abrasives, p. I-1144 [SIC 3291]
Prescription Optical Supply Inc, p. II-2107 [SIC 3851]
Presmet Corp, p. II-1295 [SIC 3399]
Presray Corp, p. I-942 [SIC 3053]
Presrite Corp, p. II-1395 [SIC 3462]
Press of Ohio, p. I-698 [SIC 2721]
Press-Seal Gasket Corp, p. I-942 [SIC 3053]
Presscut Industries Inc, p. I-942 [SIC 3053]
Pressed Steel Tank Company Inc, p. II-1355 [SIC 3443]
Pressman-Gutman Company Inc, p. I-258 [SIC 2253]
Pressman Toy Corp, p. II-2144 [SIC 3944]
Presstek Inc, p. II-1621 [SIC 3555]
Pressure Castings Inc, p. II-1275 [SIC 3365]
Pressure Cool Co, p. II-2045 [SIC 3822]
Pressware International Inc, p. I-688 [SIC 2679]
Prestige Products Inc, p. I-1854 [SIC 3648]
Prestige Stamping Inc, p. II-1388 [SIC 3452]
Prestigeline Inc, p. II-1840 [SIC 3645]
Prestini Musical Instrument, p. II-2135 [SIC 3931]
Presto Casting Co, p. II-1285 [SIC 3369]
Presto Food Products Inc, p. I-26 [SIC 2023]
Prestolite Electric Inc, p. II-1929 [SIC 3694]
Prestolite Wire Corp, pp. Vol II - 1260, 1929 [SICs 3357, 3694]
Pretty Bird International Inc, p. I-91 [SIC 2047]
Pretty Made Coat Company Inc, p. I-369 [SIC 2337]
Pretty Products Inc, p. I-954 [SIC 3069]
Pretzels Inc, p. I-107 [SIC 2052]
Priamo Designs Ltd, p. I-263 [SIC 2254]
Price Brothers Co, p. I-1122 [SIC 3272]
Price Candy Company Inc, p. I-127 [SIC 2064]
Price Rubber Corp, p. I-936 [SIC 3052]
Pride-Made Products Inc, p. I-639 [SIC 2656]
Pride Manufacturing Co, p. I-536 [SIC 2499]
Pride Refining LP, p. I-903 [SIC 2911]
Pridgeon and Clay Inc, p. II-1413 [SIC 3469]
Priester Pecan Company Inc, p. I-136 [SIC 2068]
Prima Die Casting Inc, p. II-1266 [SIC 3363]
Prima Royale Enterprises Ltd, p. I-1029 [SIC 3144]
Primark Tool Group, p. II-1318 [SIC 3425]
Primax Electronics, p. II-1704 [SIC 3577]
Prime Alloy Castings Inc, p. II-1285 [SIC 3369]
Prime Corp, p. II-1998 [SIC 3743]
Prime Leather Finishes Co, p. I-825 [SIC 2842]
Prime Tanning Company Inc, p. I-1013 [SIC 3111]
Prime Technology Inc, p. II-1901 [SIC 3676]
Prime Wood Inc, p. I-486 [SIC 2434]
Primeway Tool and Engineering, p. II-1764 [SIC 3599]
Primex Plastics Corp, p. I-973 [SIC 3083]
Primore Inc, p. II-1464 [SIC 3494]
Prince Castle Inc, p. II-1736 [SIC 3589]
Prince Corp, p. I-446 [SIC 2396]
Prince Gardner Inc, p. I-1049 [SIC 3172]
Prince Manufacturing Co, p. I-95 [SIC 2048]
Prince Manufacturing Corp, p. II-1747 [SIC 3593]
Prince Michel Vineyards, p. I-168 [SIC 2084]
Prince Potato Chip Co, p. I-199 [SIC 2096]
Prince St Technologies Ltd, p. I-292 [SIC 2273]
Princess Belt and Novelty Inc, p. I-416 [SIC 2387]
Princeton Gamma-Tech Inc, p. II-2076 [SIC 3829]
Princeton Inc, p. I-292 [SIC 2273]
Princeton Upholstery Co, p. I-571 [SIC 2521]
Principle Business Enterprises, p. I-674 [SIC 2676]
Print Northwest Company LP, p. I-718 [SIC 2752]
Printed Circuit Corp, p. II-1887 [SIC 3672]
Printed Fabrics Corp, p. I-231 [SIC 2211]
Printers Bindery Inc, p. I-750 [SIC 2789]
Printing Developments Inc, p. I-759 [SIC 2796]
Printing Holdings LP, p. I-698 [SIC 2721]
Printing House Inc, p. I-718 [SIC 2752]
Printing Prep Inc, p. I-754 [SIC 2791]
Printpack Inc, p. I-649 [SIC 2671]
Priscilla of Boston Inc, p. I-365 [SIC 2335]

Prism Group Inc, p. I-713 [SIC 2741]
Pristine Foods Inc, p. I-190 [SIC 2092]
Prive Inc, p. I-361 [SIC 2331]
Pro Battery Inc, p. II-1920 [SIC 3691]
Pro-Edge Ltd, p. I-95 [SIC 2048]
Pro-Line Corp, p. I-834 [SIC 2844]
Pro-Log Corp, p. II-2051 [SIC 3823]
Pro-Serve Inc, p. I-874 [SIC 2879]
Procedyne Corp, p. II-1671 [SIC 3567]
Process Combustion Corp, p. II-1671 [SIC 3567]
Process Displays Inc, p. II-2186 [SIC 3993]
Process Engineering Corp, p. I-310 [SIC 2295]
Processall Inc, p. II-1649 [SIC 3563]
Processed Plastic Co, p. II-2144 [SIC 3944]
Procom Technology Inc, p. II-1694 [SIC 3572]
Procon Products, p. II-1638 [SIC 3561]
Procter & Gamble, p. I-804 [SIC 2834]
Procter and Gamble Co, p. I-819 [SIC 2841]
Proctor Products Inc, p. I-1017 [SIC 3131]
Producers Cooperative Feed Mill, p. I-95 [SIC 2048]
Producers Cooperative Oil Mill, p. I-140 [SIC 2074]
Producers Rice Mill Inc, p. I-79 [SIC 2044]
Producers Supply Cooperative, p. I-869 [SIC 2875]
Productigear Company Inc, p. II-1666 [SIC 3566]
Production Aluminum Co, p. II-1266 [SIC 3363]
Production Anodizing Corp, p. II-1420 [SIC 3471]
Production Equipment Company, p. II-1547 [SIC 3536]
Production Equipment Service, p. II-1531 [SIC 3533]
Production Experts Inc, p. II-1632 [SIC 3559]
Production Industries Inc, p. I-603 [SIC 2599]
Production Management Cos, p. II-1986 [SIC 3731]
Production Management Indust, p. II-1986 [SIC 3731]
Production Pattern Shop Inc, p. II-1568 [SIC 3543]
Production Saw and Machine Co, p. II-1747 [SIC 3593]
Production Stamping Inc, p. II-1404 [SIC 3465]
Production Typographers Inc, p. I-754 [SIC 2791]
Producto Machine Co, p. II-1572 [SIC 3544]
Products Engineering Corp, p. II-1312 [SIC 3423]
Products Enterprises Inc, p. I-181 [SIC 2087]
Products Unlimited Corp, p. II-1905 [SIC 3677]
Proel Systems USA Inc, p. II-2056 [SIC 3824]
Proferas Pizza Bakery Inc, p. I-65 [SIC 2038]
Professional Binding Co, p. I-750 [SIC 2789]
Professional Medical Products, p. II-2087 [SIC 3842]
Proffitt Manufacturing Co, p. I-292 [SIC 2273]
Profile Plastics Inc, p. I-961 [SIC 3081]
Profile Records Inc, p. II-1864 [SIC 3652]
Programmed Composites Inc, p. II-2016 [SIC 3769]
Progress Casting Group Inc, p. II-1266 [SIC 3363]
Progress Castings, p. II-1275 [SIC 3365]
Progress Graphics Inc, p. I-759 [SIC 2796]
Progress Lighting, p. II-1840 [SIC 3645]
Progress Pattern Corp, p. II-1568 [SIC 3543]
Progress Printing Co, p. I-729 [SIC 2759]
Progress Wire Products Inc, p. II-1709 [SIC 3578]
Progressive Crane Inc, p. II-1547 [SIC 3536]
Progressive Dynamics Inc, p. II-1850 [SIC 3647]
Progressive Info Technologies, p. I-754 [SIC 2791]
Progressive International Corp, p. I-603 [SIC 2599]
Progressive Plastics Inc, p. I-985 [SIC 3085]
Progressive Steel Treating Inc, p. II-1290 [SIC 3398]
Progressive Tool & Indust Co, p. I-1594 [SIC 3548]
ProGroup Inc, p. II-2150 [SIC 3949]
Projects Inc, p. II-1975 [SIC 3724]
Proler International Corp, p. II-1229 [SIC 3341]
Proma Inc, p. II-2093 [SIC 3843]
Promega Corp, p. I-809 [SIC 2835]
Propellex Corp, p. I-885 [SIC 2892]
Propipe Corp, p. II-1486 [SIC 3498]
Proscotech Industries Inc, p. I-1167 [SIC 3299]
Prospect Foundry Inc, p. II-1197 [SIC 3321]
Prospect Industries Inc, p. II-1300 [SIC 3411]
Prosser-Enpo Industries, p. II-1525 [SIC 3532]
ProtectAide Inc, p. I-420 [SIC 2389]
Protection Controls Inc, p. II-1775 [SIC 3613]
Protectoseal Co, p. II-1300 [SIC 3411]
Protein Techn Intern Holdings, p. I-144 [SIC 2075]

Roman numerals I and II indicate the volume in which pages appear. Volume I holds SICs 2011 through 3299; Volume II holds SICs 3312 through 3999.

2288

Rainfair Inc

Roman numerals I and II indicate the volume in which pages appear. Volume I holds SICs 2011 through 3299; Volume II holds SICs 3312 through 3999.

Roman numerals I and II indicate the volume in which pages appear. Volume I holds SICs 2011 through 3299; Volume II holds SICs 3312 through 3999.

RELA Inc, p. II-2102 [SIC 3845]
Reliability Inc, p. II-2061 [SIC 3825]
Reliable Automatic Sprinkler, p. II-1682 [SIC 3569]
Reliable Knitting Works Inc, p. I-258 [SIC 2253]
Reliable Power Products, p. I-1066 [SIC 3229]
Reliable Products, p. II-1244 [SIC 3354]
Reliance COMM-TEC, p. II-1260 [SIC 3357]
Reliance Electric Co, p. II-1780 [SIC 3621]
Reliance Lamp Company Inc, p. II-1840 [SIC 3645]
Reliance Trailer Manufacturing, p. II-1961 [SIC 3715]
Reliv' International Inc, p. I-212 [SIC 2099]
Relton Corp, p. II-1318 [SIC 3425]
Remanco International Inc, p. II-1709 [SIC 3578]
Remco Hydraulics Inc, p. II-1747 [SIC 3593]
REMEC Inc, p. II-1915 [SIC 3679]
Remee Products Corp, p. II-1260 [SIC 3357]
Remfo, p. II-1260 [SIC 3357]
Remington Apparel Company, p. I-344 [SIC 2323]
Remington Group, p. I-795 [SIC 2824]
Remington Lamp Co, p. II-1840 [SIC 3645]
Remmele Engineering Inc, p. II-1764 [SIC 3599]
Remo Inc, p. II-2135 [SIC 3931]
Remstar International Inc, p. I-593 [SIC 2542]
Renaissance Eyewear Inc, p. II-2107 [SIC 3851]
Renaissance Greeting Cards Inc, p. I-741 [SIC 2771]
Renaissance Inc, p. I-258 [SIC 2253]
Renard Machine, p. II-1615 [SIC 3554]
Renco Encoders Inc, p. II-2071 [SIC 3827]
Renco Machine Co, p. II-1660 [SIC 3565]
Renfro Corp, p. I-254 [SIC 2252]
Rennoc Corp, p. II-2199 [SIC 3999]
Renold Inc, p. II-1676 [SIC 3568]
Renosol Corp, p. I-991 [SIC 3086]
Rensen Products Inc, p. II-2177 [SIC 3965]
Renton Coil Spring Co, p. II-1470 [SIC 3495]
Renton Folding, p. I-643 [SIC 2657]
Replogle Globes Inc, p. II-2199 [SIC 3999]
Republic Aluminum Inc, p. II-1349 [SIC 3442]
Republic Container Co, p. II-1304 [SIC 3412]
Republic Converting Company, p. I-279 [SIC 2261]
Republic Die and Tool Co, p. II-1572 [SIC 3544]
Republic Engineered Steels Inc, p. II-1212 [SIC 3325]
Republic Gypsum Co, p. I-1136 [SIC 3275]
Republic Industries Inc, p. II-1938 [SIC 3699]
Republic Machinery Company, p. II-1558 [SIC 3541]
Republic Storage Systems, p. I-593 [SIC 2542]
Republic Tool & Mfg Co, p. II-1312 [SIC 3423]
Request Foods Inc, p. I-65 [SIC 2038]
Resco Products Inc, p. I-1163 [SIC 3297]
Research-Cottrell Companies, p. II-2045 [SIC 3822]
Research Medical Inc, p. II 2087 [SIC 3842]
Research Nets Inc, p. I-322 [SIC 2298]
Research Products Company Inc, p. I-874 [SIC 2879]
Research Products Corp, p. II-1813 [SIC 3634]
Resinoid Engineering Corp, p. I-973 [SIC 3083]
Resistance Technology Inc, p. II-2087 [SIC 3842]
Resistoflex Co, p. I-979 [SIC 3084]
ReSound Corp, p. II-2087 [SIC 3842]
Resources Conservation Co, p. II-1632 [SIC 3559]
Respironics Inc, p. II-2082 [SIC 3841]
Restwell Mattress Co, p. I-558 [SIC 2515]
Results Media, p. I-713 [SIC 2741]
Retech Inc, p. II-1671 [SIC 3567]
Reuter Organ Co, p. II-2135 [SIC 3931]
Reuter-Stokes Inc, p. II-2076 [SIC 3829]
Revco/Limberg, p. II-1726 [SIC 3585]
Revcor Inc, p. II-1654 [SIC 3564]
Revelation Brassiere Company, p. I-382 [SIC 2342]
Revell-Monogram Inc, p. II-2144 [SIC 3944]
Revere Copper Products Inc, p. II-1234 [SIC 3351]
Revere Ware Corp, p. II-1413 [SIC 3469]
Revlon Implement Corp, p. II-1308 [SIC 3421]
Revo Inc, p. II-2107 [SIC 3851]
Revtech Inc, p. II-1382 [SIC 3451]
Rex Cut Products Inc, p. I-1144 [SIC 3291]
Rex Forge, p. II-1395 [SIC 3462]
Rex-Hide Inc, p. I-954 [SIC 3069]
Rex Lumber Co, p. I-465 [SIC 2421]
Rex Packaging Inc, p. I-643 [SIC 2657]

Rex-Rosenlew International Inc, p. I-660 [SIC 2673]
Rex Roto Corp, p. I-1158 [SIC 3296]
REXA Corp, p. II-1747 [SIC 3593]
Rexair Inc, p. II-1818 [SIC 3635]
Rexair Manufacturing Inc, p. II-1818 [SIC 3635]
Rexene Corp, p. I-782 [SIC 2821]
Rexford Paper Company, p. I-654 [SIC 2672]
Rexhall Industries Inc, p. II-1966 [SIC 3716]
Rexham Inc, p. I-991 [SIC 3086]
Rexham Packaging, p. I-643 [SIC 2657]
Rexnord Corp, p. II-1954 [SIC 3714]
Rexnord Corp Link-Belt, p. II-1666 [SIC 3566]
REXON Inc, p. II-1694 [SIC 3572]
Rexroth Corp, pp. Vol II - 1454, 1747 [SICs 3492, 3593]
Rextrude Co, p. I-967 [SIC 3082]
Rexworks Inc, p. II-1519 [SIC 3531]
Reynolds Machine & Tool Corp, p. II-1578 [SIC 3545]
Reynolds Metals Co, pp. Vol II - 1221, 1300 [SICs 3334, 3411]
Rezex Corp, p. I-288 [SIC 2269]
Reznor, p. II-1726 [SIC 3585]
RF Communications, p. II-1873 [SIC 3663]
RF Power Components Inc, p. II-1901 [SIC 3676]
RF Power Products Inc, p. II-1905 [SIC 3677]
RF Simmons Company Inc, p. II-2173 [SIC 3961]
RFD Publications Inc, p. I-698 [SIC 2721]
RFI Corp, p. II-1905 [SIC 3677]
RFS Corp, p. I-773 [SIC 2816]
RG Barry Corp, p. I-1021 [SIC 3142]
RG Ray Corp, p. II-1323 [SIC 3429]
RGT Enterprises Inc, p. II-2127 [SIC 3914]
RH Bauman and Company Inc, p. I-181 [SIC 2087]
RH Little Co, p. II-1644 [SIC 3562]
RHC/Spacemaster Corp, p. I-593 [SIC 2542]
Rhe Tech Inc, p. I-782 [SIC 2821]
Rheem Air Conditioning, p. II-1726 [SIC 3585]
Rheem Manufacturing Co, p. II-1726 [SIC 3585]
Rheem Ruud, p. II-1822 [SIC 3639]
Rheodyne Inc, p. II-2040 [SIC 3821]
Rheometrics Inc, p. II-2066 [SIC 3826]
Rheon USA, p. II-1626 [SIC 3556]
Rheox Inc, p. I-855 [SIC 2869]
Rhino Foods Inc, p. I-112 [SIC 2053]
RHM Fluid Power Inc, p. II-1747 [SIC 3593]
Rhodes American, p. I-1144 [SIC 3291]
Rhodes International Inc, p. I-101 [SIC 2051]
Rhone-Poulenc Ag Co, p. I-874 [SIC 2879]
Rhone-Poulenc Rorer Inc, p. I-804 [SIC 2834]
Rhopac Fabricators Inc, p. I-942 [SIC 3053]
Rhyne Lumber Company Inc, p. I-471 [SIC 2426]
Rhyne Mills Inc, p. I-297 [SIC 2281]
RI Lampus Co, p. I-1117 [SIC 3271]
Ribbon Technology Corp, p. II-1482 [SIC 3497]
Ribi ImmunoChem Research Inc, p. I-799 [SIC 2833]
Riblet Products Corp, p. II-1344 [SIC 3441]
Riblet Tramway Co, p. II-2029 [SIC 3799]
Rice Barton Corp, p. II-1615 [SIC 3554]
Rice Chadwick Rubber Co, p. I-954 [SIC 3069]
Rice Engineering Corp, p. I-1158 [SIC 3296]
Rice Oil Company Inc, p. I-204 [SIC 2097]
Rice Packaging Inc, p. I-623 [SIC 2652]
Riceland Foods Inc, p. I-79 [SIC 2044]
Rich Lumber Co, p. I-471 [SIC 2426]
Rich Maid Kabinetry Inc, p. I-486 [SIC 2434]
Rich Products Corp, pp. Vol. I - 26, 83, 112 [SICs 2023, 2045, 2053]
Rich Tool and Die Co, p. II-1503 [SIC 3519]
Richard D Irwin Inc, p. I-703 [SIC 2731]
Richard F Kline Inc, p. I-909 [SIC 2951]
Richard Leather Company Inc, p. I-1013 [SIC 3111]
Richard N Bird and Company, p. II-2131 [SIC 3915]
Richards & Malloy Mfg, p. I-519 [SIC 2451]
Richards Brick Co, p. I-1081 [SIC 3251]
Richards Brush Co, p. II-2181 [SIC 3991]
Richards Corp, p. II-1280 [SIC 3366]
Richards Graphic Commun, p. I-754 [SIC 2791]
Richards Industries Inc, p. II-1464 [SIC 3494]
Richards Manufacturing Ltd, p. II-1910 [SIC 3678]

Richards-Wilcox Mfg Co, p. II-1541 [SIC 3535]
Richardson Electronics Ltd, p. II-1883 [SIC 3671]
Richardson Foods Corp, p. I-181 [SIC 2087]
Richardson Industries Inc, p. I-501 [SIC 2439]
Richardson Seating Corp, p. I-581 [SIC 2531]
Richfield Iron Works Inc, p. II-1476 [SIC 3496]
Richheimer Coffee Co, p. I-195 [SIC 2095]
Richland Glass Company Inc, p. I-1066 [SIC 3229]
Richland Ltd, p. II-1503 [SIC 3519]
Richmark International Inc, p. I-424 [SIC 2391]
Richmond Baking Company Inc, p. I-107 [SIC 2052]
Richmond Gravure, p. I-643 [SIC 2657]
Richmond Industries Inc, p. II-1280 [SIC 3366]
Richmond Newspapers Inc, p. I-693 [SIC 2711]
Richmond Screw Anchor Co, p. I-1122 [SIC 3272]
Richter Manufacturing Corp, p. I-688 [SIC 2679]
Richter Precision Inc, p. II-1425 [SIC 3479]
Richter's Bakery Inc, p. I-101 [SIC 2051]
Richtex Corp, p. I-1081 [SIC 3251]
Rickard Circular Folding Co, p. I-750 [SIC 2789]
Rico Industries Inc, p. I-1049 [SIC 3172]
Rico International, p. II-2135 [SIC 3931]
Rico Sportswear Inc, p. I-446 [SIC 2396]
RICOH Corp, p. II-1868 [SIC 3661]
Ricon Corp, p. II-1536 [SIC 3534]
Ricwil Piping Systems LP, p. II-1486 [SIC 3498]
Riddell Inc, p. II-2150 [SIC 3949]
Riddell Sports Inc, p. II-2150 [SIC 3949]
Ridg-U-Rak Inc, p. I-593 [SIC 2542]
Ridge Tool Co, p. II-1312 [SIC 3423]
Ridgeview Inc, p. I-258 [SIC 2253]
Ridgeway Clock Co, p. II-2118 [SIC 3873]
Ridgways Inc, p. I-718 [SIC 2752]
Riedell Shoes Inc, p. II-2150 [SIC 3949]
Rigid Hitch Inc, p. II-2029 [SIC 3799]
Rigid Products, p. II-1312 [SIC 3423]
Rigidized Metals Corp, p. II-1187 [SIC 3316]
Riker Products Inc, p. II-2029 [SIC 3799]
Riley and Geehr Inc, p. I-660 [SIC 2673]
Riley and Scott Inc, p. II-1943 [SIC 3711]
Riley Consolidated Inc, p. II-1355 [SIC 3443]
Riley Creek Lumber Co, p. I-465 [SIC 2421]
Rima Manufacturing Co, p. II-1382 [SIC 3451]
Ring Can Corp, p. I-985 [SIC 3085]
Ring-O-Valve Inc, p. II-1447 [SIC 3491]
Ring Specialty Co, p. II-2122 [SIC 3911]
Ringer Corp, p. I-861 [SIC 2873]
Ringier America Inc, pp. Vol. I - 708, 724 [SICs 2732, 2754]
Rinker Materials Corp, p. I-1127 [SIC 3273]
Rinn Corp, p. II-2093 [SIC 3843]
Rio Grande Val Sugar, p. I-116 [SIC 2061]
Ripon Foods Inc, p. I-107 [SIC 2052]
Rippel Architectural Metals Inc, p. II-1367 [SIC 3446]
Rising Paper, p. I-612 [SIC 2621]
Ristance Corp, p. II-1600 [SIC 3549]
Rita's Sportswear Inc, p. I-361 [SIC 2331]
Rite-Off Inc, p. I-825 [SIC 2842]
Rite-On Industries Inc, p. II-1764 [SIC 3599]
Rite Sole Corp, p. I-1017 [SIC 3131]
Rite-Style Optical Co, p. II-2107 [SIC 3851]
Ritepoint Inc, p. II-2156 [SIC 3951]
Rival Co, p. II-1813 [SIC 3634]
Rival Manufacturing Co, p. II-1813 [SIC 3634]
Rivco, p. I-352 [SIC 2326]
River Cement Co, p. I-1077 [SIC 3241]
River City Sound Productions, p. II-1864 [SIC 3652]
River Oaks Furniture Inc, p. I-548 [SIC 2512]
River Recycling Industries, p. II-1229 [SIC 3341]
Rivera Manufacturing Inc, p. I-336 [SIC 2321]
Rivera Prodion Tooling Group, p. II-1572 [SIC 3544]
Riverdale Chemical Co, p. I-874 [SIC 2879]
Riverdale Decorative Products, p. I-428 [SIC 2392]
Riverhead Circuits, p. II-1887 [SIC 3672]
Riverside Book and Bible House, p. I-703 [SIC 2731]
Riverside Brick and Supply Co, p. I-1081 [SIC 3251]
Riverside Cement Co, p. I-1077 [SIC 3241]
Riverside Furniture Corp, p. I-542 [SIC 2511]
Riverside Group, p. I-674 [SIC 2676]

Roman numerals I and II indicate the volume in which pages appear. Volume I holds SICs 2011 through 3299; Volume II holds SICs 3312 through 3999.

Riverside Ice Company Inc, p. I-204 [SIC 2097]
Riverside Manufacturing Co, p. I-352 [SIC 2326]
Riverside Millwork Company Inc, p. I-480 [SIC 2431]
Riverside Paper Corp, p. I-654 [SIC 2672]
Riverside Products, p. II-1395 [SIC 3462]
Riverside Publishing Co, p. I-713 [SIC 2741]
Riverside Refractories Inc, p. I-1090 [SIC 3255]
Riverside Seafoods Inc, p. I-190 [SIC 2092]
Riverton Corp, p. I-1077 [SIC 3241]
Riverview Sportswear Inc, p. I-373 [SIC 2339]
Riverwood Intern Corp, p. I-618 [SIC 2631]
Riverwood International Georgia, p. I-618 [SIC 2631]
Riviana Foods Inc, p. I-79 [SIC 2044]
Riviera Convertibles Inc, p. I-558 [SIC 2515]
RJ Manufacturing Co, p. II-2173 [SIC 3961]
RJ Reynolds Tobacco Co, p. I-218 [SIC 2111]
RJ Taylor Corp, p. I-576 [SIC 2522]
RJ Tower Corp, p. II-1404 [SIC 3465]
RJM Manufacturing Inc, p. I-654 [SIC 2672]
RJP Electronics Inc, p. II-1709 [SIC 3578]
RJR Nabisco Holdings Corp, p. I-218 [SIC 2111]
RJR Nabisco Inc, p. I-218 [SIC 2111]
RL Drake Co, p. II-1873 [SIC 3663]
RL Ziegler Company Inc, p. I-2 [SIC 2011]
RLC Electronics Inc, p. II-1915 [SIC 3679]
RM Crow Co, p. I-433 [SIC 2393]
RM Engineered Products Inc, p. I-954 [SIC 3069]
RM Kerner Co, p. II-1764 [SIC 3599]
RM Wieland Company Inc, p. I-548 [SIC 2512]
RMA/KOLKO Corp, p. I-669 [SIC 2675]
RMC Lone Star, p. I-1127 [SIC 3273]
RMI Titanium Co, p. II-1255 [SIC 3356]
RMS Co, p. II-1764 [SIC 3599]
Ro-An Industries Corp, p. II-1660 [SIC 3565]
Ro-Lab American Rubber, p. I-948 [SIC 3061]
Road Rescue Inc, p. II-1943 [SIC 3711]
Road Systems Inc, p. II-1961 [SIC 3715]
Roadmaster Corp, p. II-2003 [SIC 3751]
Roadmaster Industries Inc, p. II-2003 [SIC 3751]
Roanoke Coca-Cola Bottling Inc, p. I-176 [SIC 2086]
Roanoke Electric Steel Corp, p. II-1172 [SIC 3312]
Roaring Spring Blank Book Co, p. I-684 [SIC 2678]
Rob-Ran Corp, p. I-361 [SIC 2331]
Robbins and Myers Inc, p. II-1638 [SIC 3561]
Robbins Co, p. II-2122 [SIC 3911]
Robbins Inc, p. I-471 [SIC 2426]
Robbins Manufacturing Co, p. I-528 [SIC 2491]
Roberds-Johnson Industries Inc, p. II-1764 [SIC 3599]
Robert A Main and Sons Inc, p. II-1605 [SIC 3552]
Robert Bosch Corp, p. II-1954 [SIC 3714]
Robert Bosch Fluid Power Corp, p. II-1753 [SIC 3594]
Robert Busse and Company Inc, p. I-688 [SIC 2679]
Robert E Derecktor Inc, p. II-1992 [SIC 3732]
Robert F Lewis Inc, p. I-1053 [SIC 3199]
Robert Lee Morris Inc, p. I-1013 [SIC 3111]
Robert Manufacturing Co, p. II-1464 [SIC 3494]
Robert Manufacturing Company, p. I-1041 [SIC 3161]
Robert Mitchell Company Inc, p. II-1464 [SIC 3494]
Robert Mondavi Corp, p. I-168 [SIC 2084]
Robert S Fisher and Co, p. II-2122 [SIC 3911]
Robert Simmons Inc, p. II-2160 [SIC 3952]
Robert Talbott Inc, p. I-344 [SIC 2323]
Robert Wooler Company Inc, p. II-1290 [SIC 3398]
Robertet Flavors Inc, p. I-181 [SIC 2087]
Robertet Inc, p. I-181 [SIC 2087]
Roberts Automatic Products Inc, p. II-1382 [SIC 3451]
Roberts Dairy Co, p. I-35 [SIC 2026]
Roberts-Gordon Inc, p. II-1339 [SIC 3433]
Roberts Metal Mfg Co, p. II-1600 [SIC 3549]
Roberts Oxygen Company Inc, p. I-768 [SIC 2813]
Roberts Pharmaceutical Corp, p. I-804 [SIC 2834]
Roberts Systems Inc, p. II-1660 [SIC 3565]
Robertshaw Controls Co, p. II-2045 [SIC 3822]
Robertson-Ceco Corp, p. II-1349 [SIC 3442]
Robertson Factories Inc, p. I-424 [SIC 2391]
Robertson Furniture Company, p. I-603 [SIC 2599]
Robertson Shake Mill Inc, p. I-476 [SIC 2429]
Robertson Transformer, p. II-1769 [SIC 3612]
Robespierre Inc, p. I-361 [SIC 2331]

Robin International, p. I-394 [SIC 2369]
Robin Rose America Inc, p. I-30 [SIC 2024]
Robinson Brick Co, p. I-1081 [SIC 3251]
Robinson Engineering Corp, p. II-1600 [SIC 3549]
Robinson Furniture Mfg, p. I-542 [SIC 2511]
Robinson-Halpern Products, p. II-1905 [SIC 3677]
Robinson Helicopter Co, p. II-1970 [SIC 3721]
Robinson Hosiery Mill Inc, p. I-254 [SIC 2252]
Robinson Industries Inc, p. I-991 [SIC 3086]
Robinson Manufacturing, p. I-356 [SIC 2329]
Robinson Manufacturing Co, pp. Vol. I - 241, 297
 [SICs 2231, 2281]
Robinson Nugent Inc, p. II-1830 [SIC 3643]
Robinson-Ransbottom, p. I-1113 [SIC 3269]
Robison-Anton Textile Co, p. I-306 [SIC 2284]
Robot Aided Mfg Center, p. II-1764 [SIC 3599]
Robot Research Inc, p. II-1878 [SIC 3669]
Robotic Vision Systems Inc, p. II-1682 [SIC 3569]
Robotics Inc, p. II-1732 [SIC 3586]
Robotron Corp, p. II-1795 [SIC 3629]
Robroy Industries Inc, p. I-1006 [SIC 3089]
Robroy Industries-Texas Inc, p. II-1835 [SIC 3644]
Rocco Further Processing Inc, p. I-2 [SIC 2011]
Rocco Turkeys Inc, p. I-12 [SIC 2015]
Roche Diagnostic Systems Inc, p. II-2066 [SIC 3826]
Rochester Bronze and Aluminum, p. II-1280 [SIC 3366]
Rochester Coca-Cola, p. I-176 [SIC 2086]
Rochester Empire, p. I-759 [SIC 2796]
Rochester Gauges Incorporated, p. II-2051 [SIC 3823]
Rochester Instrument Systems, p. II-2061 [SIC 3825]
Rochester Manufacturing Co, p. II-1503 [SIC 3519]
Rochester Midland Corp, p. I-825 [SIC 2842]
Rochester Optical Mfg Co, p. II-2107 [SIC 3851]
Rochester Sand and Gravel Inc, p. I-909 [SIC 2951]
Rochester Shoe Tree Company, p. I-536 [SIC 2499]
Rock-Tenn Co, p. I-643 [SIC 2657]
Rock Valley Oil and Chemical, p. I-918 [SIC 2992]
Rock Wool Manufacturing Co, p. I-1158 [SIC 3296]
Rockdale Sash and Trim, p. I-480 [SIC 2431]
Rocketdyne, p. II-2011 [SIC 3764]
Rockford Corp, p. II-1859 [SIC 3651]
Rockford International, p. I-942 [SIC 3053]
Rockford Powertrain Inc, p. II-1676 [SIC 3568]
Rockford Products Corp, p. II-1388 [SIC 3452]
Rockford Spring Co, p. II-1470 [SIC 3495]
Rockland Industries Inc, p. I-231 [SIC 2211]
Rockline Industries Inc, p. I-688 [SIC 2679]
Rockmount Ranch Wear Mfg Co, p. I-356 [SIC 2329]
Rocknel Fastener Inc, p. II-2177 [SIC 3965]
Rockwell Automotive, p. II-1954 [SIC 3714]
Rockwell Graphic Systems, p. II-1621 [SIC 3555]
Rockwell International Corp, pp. Vol II - 1868, 1887,
 1954 [SICs 3661, 3672, 3714]
Rockwell Lime Company Inc, p. I-1132 [SIC 3274]
Rockwood, p. II-1943 [SIC 3711]
Rockwood Swendeman Corp, p. II-1447 [SIC 3491]
Rocky Mount Cord Co, p. I-322 [SIC 2298]
Rocky Mount Mills, p. I-297 [SIC 2281]
Rocky Mountain Chocolate, p. I-132 [SIC 2066]
Rocky Mountain Food Factory, p. I-40 [SIC 2032]
Rocky Mountain Log Homes, p. I-523 [SIC 2452]
Rocky's Brewing Co, p. I-160 [SIC 2082]
Rocky Shoes and Boots Inc, p. I-1025 [SIC 3143]
Rococo, p. I-390 [SIC 2361]
Roctronics Lighting, p. II-1840 [SIC 3645]
Rod's Food Products Inc, p. I-55 [SIC 2035]
Rodan, p. II-1901 [SIC 3676]
Rodel Inc, p. I-782 [SIC 2821]
Rodgers and McDonald Graphics, p. I-713 [SIC 2741]
Rodgers Instrument Corp, p. II-2135 [SIC 3931]
Rodman Industries, p. I-532 [SIC 2493]
Rodney Hunt Co, p. II-1344 [SIC 3441]
Rodney Strong Vineyards, p. I-168 [SIC 2084]
Rodriguez Festive Foods Inc, p. I-65 [SIC 2038]
Roesch Inc, p. II-1425 [SIC 3479]
Roffe Inc, p. I-356 [SIC 2329]
Roger Cleveland Golf Co, p. II-2150 [SIC 3949]
Rogers Brothers Corp, p. II-1961 [SIC 3715]

Rogers Corp, p. I-973 [SIC 3083]
Rogers Corp Molding Material, p. I-782 [SIC 2821]
Rogers Foam Corp, p. I-991 [SIC 3086]
Rogers Foods Inc, p. I-51 [SIC 2034]
Rogers Foods Inc Chili Products, p. I-212 [SIC 2099]
Rogers Loose Leaf Company Inc, p. I-745 [SIC 2782]
Rogers Machinery Company Inc, p. II-1649 [SIC 3563]
Rogersol Inc, p. I-889 [SIC 2893]
Rogerson Aircraft Corp, p. II-1980 [SIC 3728]
Rohm and Haas Co, p. I-782 [SIC 2821]
Rohm and Haas Texas Inc, p. I-855 [SIC 2869]
Rohm Tech Inc, p. I-879 [SIC 2891]
Rohr Inc, pp. Vol II - 1975, 1980 [SICs 3724, 3728]
Rohr Lingerie Inc, p. I-378 [SIC 2341]
Rolex Co, p. II-1460 [SIC 3493]
Roll-A-Way Conveyor Co, p. II-1552 [SIC 3537]
Roll Coater Inc, p. II-1425 [SIC 3479]
Roll Forming Corp, p. II-1250 [SIC 3355]
Rolled Wire Products Co, p. II-1187 [SIC 3316]
Roller Bearing of America, p. II-1644 [SIC 3562]
Rollex Corp, p. II-1361 [SIC 3444]
Rollform Inc, p. II-1367 [SIC 3446]
Rollform of Jamestown Inc, p. II-1377 [SIC 3449]
Rolligon Corp, p. II-2029 [SIC 3799]
Rollin J Lobaugh Inc, p. II-1382 [SIC 3451]
Rollpak Corp, p. I-660 [SIC 2673]
Rolls-Royce Inc, p. II-1975 [SIC 3724]
Rollyson Aluminum Products, p. II-1349 [SIC 3442]
Rolo Manufacturing Company, p. II-2173 [SIC 3961]
Roma Color Inc, p. I-773 [SIC 2816]
Romac Industries Inc, p. II-1447 [SIC 3491]
Romac Metals Inc, p. II-1192 [SIC 3317]
Roman Adhesives Inc, p. I-879 [SIC 2891]
Roman Empire, p. I-471 [SIC 2426]
Roman Knit Inc, p. I-271 [SIC 2258]
Roman Meal Co, p. I-83 [SIC 2045]
Roman Meal Milling Co, p. I-70 [SIC 2041]
Roman Research Inc, p. II-2122 [SIC 3911]
Romany Ceramics Inc, p. I-1086 [SIC 3253]
Rome Cable Corp, p. II-1260 [SIC 3357]
Rome Fastener Corp, p. II-2177 [SIC 3965]
Rome Specialty Company Inc, p. I-322 [SIC 2298]
Rome Strip Steel Company Inc, p. II-1187 [SIC 3316]
Romero Foods Inc, p. I-199 [SIC 2096]
Ron Ink Company Inc, p. I-889 [SIC 2893]
Ronan Engineering Co, p. II-2051 [SIC 3823]
Rondy and Company Inc, p. I-954 [SIC 3069]
Ronnie Manufacturing Inc, p. I-373 [SIC 2339]
RonPak Inc, p. I-665 [SIC 2674]
Ronsil Rubber, p. I-954 [SIC 3069]
Roodhouse Envelope Co, p. I-679 [SIC 2677]
Root Corp, p. II-1476 [SIC 3496]
Ropak Corp, p. I-991 [SIC 3086]
Roper Industries Inc, p. II-1638 [SIC 3561]
Roper Pump Co, p. II-1638 [SIC 3561]
Roper Whitney of Rockford Inc, p. II-1563 [SIC 3542]
Ropkey Graphics Inc, p. I-759 [SIC 2796]
Roppe Corp, p. I-954 [SIC 3069]
Rosalco Inc, p. I-428 [SIC 2392]
Rosboro Lumber Co, p. I-465 [SIC 2421]
Rosbro Sportswear Company Inc, p. I-394 [SIC 2369]
Rosco Laboratories Inc, p. II-2112 [SIC 3861]
Roscoe Moss Co, p. II-1192 [SIC 3317]
Roscommon Manufacturing Co, p. II-1676 [SIC 3568]
Rose Art Industries Inc, p. II-2144 [SIC 3944]
Rose City Awning Co, p. I-437 [SIC 2394]
Rose City Paper Box Inc, p. I-643 [SIC 2657]
Rose Hill Company Inc, p. I-542 [SIC 2511]
Rose Hill Linen Co, p. I-231 [SIC 2211]
Rose Industries Inc, p. II-1519 [SIC 3531]
Rose Packing Company Inc, p. I-2 [SIC 2011]
Rose Printing Company Inc, p. I-708 [SIC 2732]
Roseboro Manufacturing Co, p. I-336 [SIC 2321]
Rosebud Manufacturing Co, p. I-486 [SIC 2434]
Roseburrough Tool Inc, p. II-1584 [SIC 3546]
Rosecraft Inc, p. II-2173 [SIC 3961]
Roselin Manufacturing Co, p. I-245 [SIC 2241]
Roselon Industries Inc, p. I-302 [SIC 2282]
Rosemar Silver Company Inc, p. II-2127 [SIC 3914]

Roman numerals I and II indicate the volume in which pages appear. Volume I holds SICs 2011 through 3299; Volume II holds SICs 3312 through 3999.

2292

Roman numerals I and II indicate the volume in which pages appear. Volume I holds SICs 2011 through 3299; Volume II holds SICs 3312 through 3999.

Salz Leathers Inc, p. I-1013 [SIC 3111]
Sam Kane Beef Processors Inc, p. I-2 [SIC 2011]
Sam Moore Furniture Industries, p. I-548 [SIC 2512]
Sama Plastics Corp, p. I-593 [SIC 2542]
Sames Electrostatic Inc, p. II-1649 [SIC 3563]
Samplemasters Inc, p. I-750 [SIC 2789]
Sams Manufacturing Company, p. I-581 [SIC 2531]
Samsill Corp, p. I-745 [SIC 2782]
Samson Ocean Systems Inc, p. I-322 [SIC 2298]
Samson Technologies Corp, p. II-1859 [SIC 3651]
Samsonite Corp, p. I-1041 [SIC 3161]
Samsonite Furniture Co, p. I-553 [SIC 2514]
Samsons Manufacturing Corp, p. I-336 [SIC 2321]
Samtec Inc, p. II-1830 [SIC 3643]
Samuel Cabot Inc, p. I-840 [SIC 2851]
Samuel Ehrman Company Inc, p. I-451 [SIC 2397]
Samuel Platzer Company Inc, p. II-2122 [SIC 3911]
Samuel Strapping Systems, p. II-1492 [SIC 3499]
Samuel-Whittar Inc, p. II-1187 [SIC 3316]
Samuels Glass Co, p. I-1071 [SIC 3231]
San Antonio Developers Inc, p. I-914 [SIC 2952]
San Brushardi Inc, p. I-420 [SIC 2389]
San Carlin Manufacturing, p. I-365 [SIC 2335]
San Diego Paper Box Co, p. I-643 [SIC 2657]
San Diego Wood Preserving Co, p. I-528 [SIC 2491]
San Francisco French Bread Co, p. I-101 [SIC 2051]
San Francisco Newspaper Agency, p. I-693 [SIC 2711]
San Giorgio Macaroni, p. I-208 [SIC 2098]
San-J International Inc, p. I-55 [SIC 2035]
San Joaquin Blocklite Inc, p. I-1117 [SIC 3271]
San Joaquin Packaging Corp, p. I-643 [SIC 2657]
San Joaquin Refining Company, p. I-903 [SIC 2911]
San Joaquin Valley, p. I-35 [SIC 2026]
San Jose Mercury News Inc, p. I-693 [SIC 2711]
San Martin Bridals, p. I-365 [SIC 2335]
Sanborn Manufacturing Co, p. II-1649 [SIC 3563]
Sanborn Wire Products Inc, p. II-1470 [SIC 3495]
Sancor Inc, p. I-365 [SIC 2335]
Sand and Siman Inc, p. I-401 [SIC 2381]
Sand Mountain Industries Inc, p. I-378 [SIC 2341]
Sandberg Furniture Mfg, p. I-542 [SIC 2511]
Sanden International, p. II-1726 [SIC 3585]
Sanders, p. II-1915 [SIC 3679]
Sanders Company Inc, p. II-1676 [SIC 3568]
Sanders Lead Company Inc, p. II-1225 [SIC 3339]
Sanders Manufacturing Co, p. II-2186 [SIC 3993]
Sanders Manufacturing Company, p. I-491 [SIC 2435]
Sanderson Farms Inc, p. I-12 [SIC 2015]
Sanderson-MacLeod Inc, p. II-2181 [SIC 3991]
Sandess Manufacturing Company, p. I-332 [SIC 2311]
Sandler Brothers Inc, p. I-674 [SIC 2676]
Sandler Sanitary Wiping Cloth, p. I-275 [SIC 2259]
SandMold Systems Inc, p. II-1541 [SIC 3535]
Sandoz Chemicals Corp, p. I-850 [SIC 2865]
Sandoz Corp, p. I-804 [SIC 2834]
Sandoz Crop Protection Corp, p. I-874 [SIC 2879]
Sandoz Nutrition Corp, p. I-799 [SIC 2833]
Sandpoint Design Inc, p. II-2003 [SIC 3751]
Sands, Taylor and Wood Co, p. I-70 [SIC 2041]
Sandusky Cabinets Inc, p. I-576 [SIC 2522]
Sandusky International Inc, p. II-1615 [SIC 3554]
Sandvik Inc, p. II-1558 [SIC 3541]
Sandvik Milford Corp, p. II-1318 [SIC 3425]
Sandvik Rhenium Alloys Inc, p. II-1255 [SIC 3356]
Sandvik Rock Tools Inc, p. II-1525 [SIC 3532]
Sandvik Special Metals Corp, p. II-1255 [SIC 3356]
Sandvik Steel Co, p. II-1182 [SIC 3315]
Sandy-Alexander Inc, p. I-718 [SIC 2752]
Sanford Chemical Company Inc, p. I-819 [SIC 2841]
Sanford Corp, p. II-2156 [SIC 3951]
Sanford Finishing Co, p. I-288 [SIC 2269]
Sani-Co Products, p. I-819 [SIC 2841]
Sani-Dairy, p. I-35 [SIC 2026]
Sani-Top Inc, p. I-973 [SIC 3083]
SaniServ, p. II-1626 [SIC 3556]
Sanitary-Dash Manufacturing, p. II-1334 [SIC 3432]
Sanitek Product Inc, p. I-819 [SIC 2841]
Sanitized Inc, p. I-819 [SIC 2841]
Sankyo Seiki America Inc, p. II-1694 [SIC 3572]

Sanmina Corp, p. II-1887 [SIC 3672]
Sanofi Bio-Industries, p. I-51 [SIC 2034]
Sanofi Bio-Industries Inc, pp. Vol. I - 181, 212 [SICs 2087, 2099]
Sanofi Inc, p. I-804 [SIC 2834]
Sanolite Corp, p. I-819 [SIC 2841]
Santa Barbara Engineering Inc, p. II-1934 [SIC 3695]
Santa Clara Folding, p. I-628 [SIC 2653]
Santa Cruz Industries Inc, p. I-587 [SIC 2541]
Santa Fe Rubber Products Inc, p. I-948 [SIC 3061]
Santa Fe Systems Co, p. II-1649 [SIC 3563]
Santa's Best, p. II-2199 [SIC 3999]
Santana Cycles Inc, p. II-2003 [SIC 3751]
Santana Ltd, p. I-336 [SIC 2321]
Santee Print Works, p. I-279 [SIC 2261]
Santens of America Inc, p. I-231 [SIC 2211]
Santomo Partners, p. I-45 [SIC 2033]
Sanyo E and E Corp, p. II-1726 [SIC 3585]
Sanyo Fisher, p. II-1859 [SIC 3651]
Sanyo/Icon, p. II-1694 [SIC 3572]
Sapona Manufacturing Company, p. I-302 [SIC 2282]
Sara Lee Bakery North America, p. I-65 [SIC 2038]
Sara Lee Corp, p. I-7 [SIC 2013]
Sargent and Greenleaf Inc, p. II-1323 [SIC 3429]
Sargent Controls and Aerospace, p. II-1980 [SIC 3728]
Sargent-Fletcher Co, p. II-1980 [SIC 3728]
Sargent Manufacturing Co, p. II-1323 [SIC 3429]
Sargent-Sowell Co, p. II-2186 [SIC 3993]
Sargento Food Services Corp, p. I-21 [SIC 2022]
Sarns Inc, p. II-2082 [SIC 3841]
Sartron Inc, p. II-1905 [SIC 3677]
SAS Industries Inc, p. II-2169 [SIC 3955]
Sasgen and Derrick Co, p. II-1547 [SIC 3536]
Sasib Bakery North America, p. II-1541 [SIC 3535]
Sate-Lite Manufacturing, p. II-1850 [SIC 3647]
Satellite Transmission Systems, p. II-1915 [SIC 3679]
Sato America Inc, p. II-1704 [SIC 3577]
Sattex Corp, p. I-1144 [SIC 3291]
Saturn Corp, p. II-1943 [SIC 3711]
Saturn Industries Inc, p. II-1594 [SIC 3548]
Sau-Sea Foods Inc, p. I-190 [SIC 2092]
Sauder Manufacturing Company, p. I-581 [SIC 2531]
Sauer-Sundstrand, p. II-1753 [SIC 3594]
Sauk Rapids Dairy Inc, p. I-30 [SIC 2024]
Saul Brothers and Company Inc, p. I-373 [SIC 2339]
Saulsbury Fire Equipment Corp, p. II-1682 [SIC 3569]
Saunders Brothers Inc, p. I-536 [SIC 2499]
Sauquoit Industries Inc, p. I-310 [SIC 2295]
Savaage, p. I-1041 [SIC 3161]
Savage Arms Inc, p. II-1438 [SIC 3484]
Savair Inc, p. II-1594 [SIC 3548]
Savanna Pallet Inc, p. I-510 [SIC 2448]
Savannah Foods and Industries, p. I-119 [SIC 2062]
Savannah Luggage Works Inc, p. I-1041 [SIC 3161]
Savant Instruments Inc, p. II-2040 [SIC 3821]
Savel Corp, p. II-1447 [SIC 3491]
Savoie Industries Inc, p. I-116 [SIC 2061]
Savoy Brass, p. II-1334 [SIC 3432]
Savoy Manufacturing, p. I-587 [SIC 2541]
Saw Pipes USA Inc, p. II-1192 [SIC 3317]
Sawhill Tubular, p. II-1192 [SIC 3317]
Saxonburg Ceramics Inc, p. I-1167 [SIC 3299]
Sayers Communications Group, p. I-708 [SIC 2732]
Saylor Beall Manufacturing Co, p. II-1649 [SIC 3563]
Sazerac Company Inc, p. I-172 [SIC 2085]
SB Foot Tanning Co, p. I-1013 [SIC 3111]
Sbicca of California, p. I-1029 [SIC 3144]
SBS Engineering Inc, p. II-1938 [SIC 3699]
SBS Enterprises Inc, p. II-2087 [SIC 3842]
SC Johnson and Son Inc, p. I-825 [SIC 2842]
Scale-Tronix Inc, p. II-1759 [SIC 3596]
Scaled Composites Inc, p. II-1970 [SIC 3721]
Scan-Pac Manufacturing Inc, p. I-1167 [SIC 3299]
Scandia Packaging Machinery Co, p. II-1660 [SIC 3565]
Scandura Inc, p. I-936 [SIC 3052]
Scarlett Nite, p. I-365 [SIC 2335]
SCCA Enterprises Inc, p. II-1943 [SIC 3711]
Schad Industries Inc, p. I-156 [SIC 2079]

Schaefer Equipment Inc, p. II-1395 [SIC 3462]
Schaefer Manufacturing Inc, p. II-1395 [SIC 3462]
Schaefer-Ross Corp, p. II-1615 [SIC 3554]
Schaeff Inc, p. II-1552 [SIC 3537]
Schaeffer Manufacturing Co, p. I-918 [SIC 2992]
Schafer Gear Works Inc, p. II-1666 [SIC 3566]
Schaff Piano Supply Co, p. II-2135 [SIC 3931]
Schaffner Manufacturing, p. I-825 [SIC 2842]
Schantz Organ Co, p. II-2135 [SIC 3931]
Schatz Bearing Corp, p. II-1644 [SIC 3562]
Schawk Inc, p. I-759 [SIC 2796]
SCHEBLER Co, p. II-1541 [SIC 3535]
Scheffer Inc, p. II-1621 [SIC 3555]
Schenck Turner Inc, p. II-1558 [SIC 3541]
Scher Fabrics Inc, p. I-236 [SIC 2221]
Schering-Plough Corp, p. I-804 [SIC 2834]
Scheu Manufacturing Co, p. II-1339 [SIC 3433]
Schick Safety Razor, p. II-1308 [SIC 3421]
Schiffenhaus Industries Inc, p. I-623 [SIC 2652]
Schiffenhaus Packaging Corp, p. I-628 [SIC 2653]
Schilberg Integrated Metals, p. II-1229 [SIC 3341]
Schildberg Construction, p. I-1132 [SIC 3274]
Schlegel Corp, p. I-942 [SIC 3053]
Schlegel Tennessee Inc, p. I-954 [SIC 3069]
Schlueter Company Inc, p. II-1626 [SIC 3556]
Schlumberger, p. II-1732 [SIC 3586]
Schlumberger Indust Water, p. II-2056 [SIC 3824]
Schlumberger Industries Inc, p. II-2056 [SIC 3824]
Schlumberger Technologies Inc, p. II-1883 [SIC 3671]
Schmidt Baking Company Inc, p. I-101 [SIC 2051]
Schmidt Cabinet Co, p. I-486 [SIC 2434]
Schnadig Corp, p. I-548 [SIC 2512]
Schnee-Morehead Inc, p. I-879 [SIC 2891]
Schneeberger Inc, p. II-1644 [SIC 3562]
Schneider, p. II-2082 [SIC 3841]
Schneller Inc, p. II-1980 [SIC 3728]
Schnitzer Steel Industries Inc, p. II-1172 [SIC 3312]
Schnuck's Midstate Dairy, p. I-35 [SIC 2026]
Schoeller Technical Papers Inc, p. II-2112 [SIC 3861]
Scholastic Corp, p. I-698 [SIC 2721]
Scholastic Inc, p. I-698 [SIC 2721]
Scholle Corp, p. I-660 [SIC 2673]
Schonberg Research Corp, p. II-2098 [SIC 3844]
School Apparel Inc, p. I-369 [SIC 2337]
School House Candy Co, p. I-127 [SIC 2064]
School Stationers Corp, p. I-684 [SIC 2678]
Schott Brothers Inc, p. I-413 [SIC 2386]
Schott Corp, pp. Vol. I - 1066, 1071, Vol. II - 1905 [SICs 3229, 3231, 3677]
Schott Fiber Optics Inc, p. II-2071 [SIC 3827]
Schott Glass Technologies Inc, p. I-1066 [SIC 3229]
Schott Process Systems Inc, p. I-1066 [SIC 3229]
Schrafft's Ice Cream Ltd, p. I-30 [SIC 2024]
Schramm Inc, p. II-1531 [SIC 3533]
Schreier Malting Co, p. I-165 [SIC 2083]
Schrock-WCI Cabinet Group, p. I-486 [SIC 2434]
Schroeder Industries, p. II-1682 [SIC 3569]
Schubert Industries Inc, p. I-558 [SIC 2515]
Schuessler Knitting Mills Inc, p. I-258 [SIC 2253]
Schuler Industries Inc, p. II-1764 [SIC 3599]
Schuller International Inc, p. I-1158 [SIC 3296]
Schulmerich Carillons Inc, p. II-2135 [SIC 3931]
Schult Homes Corp, p. I-519 [SIC 2451]
Schumacher Architectural, p. I-1098 [SIC 3261]
Schumacher Electric Corp, p. II-1795 [SIC 3629]
Schumacher Elevator Company, p. II-1536 [SIC 3534]
Schumag-Kieserling Machinery, p. II-1600 [SIC 3549]
Schundler Co, p. I-1154 [SIC 3295]
Schurman Machine, p. II-1610 [SIC 3553]
Schutte and Koerting, p. II-1339 [SIC 3433]
Schuylkill Haven, p. I-267 [SIC 2257]
Schuylkill Holdings Inc, p. II-1172 [SIC 3312]
Schuylkill Metals Corp, p. II-1229 [SIC 3341]
Schuylkill Products Inc, p. I-1122 [SIC 3272]
Schwaab Inc, p. II-2164 [SIC 3953]
Schwartz Electro-Optics Inc, p. II-2102 [SIC 3845]
Schwartz Industries Inc, p. II-1943 [SIC 3711]
Schwarz Paper Co, p. I-729 [SIC 2759]
Schweiger Industries Inc, p. I-548 [SIC 2512]

Roman numerals I and II indicate the volume in which pages appear. Volume I holds SICs 2011 through 3299; Volume II holds SICs 3312 through 3999.

Roman numerals I and II indicate the volume in which pages appear. Volume I holds SICs 2011 through 3299; Volume II holds SICs 3312 through 3999.

Sewer Rodding Equipment Co, p. II-1736 [SIC 3589]
Sexton Metalcraft Inc, p. I-536 [SIC 2499]
Seyfert Foods Inc, p. I-199 [SIC 2096]
Seymour Housewares Corp, p. II-1492 [SIC 3499]
Seymour Manufacturing, p. II-1312 [SIC 3423]
SFI Inc, p. I-669 [SIC 2675]
SFS Corp, p. I-603 [SIC 2599]
SGS Tool Co, p. II-1578 [SIC 3545]
SH Clausin and Company Inc, p. II-2122 [SIC 3911]
SH Leggitt Co, p. II-2051 [SIC 3823]
Shachihata Incorporated USA, p. II-2164 [SIC 3953]
Shade/Allied Inc, p. I-735 [SIC 2761]
Shade Foods Inc, p. I-132 [SIC 2066]
Shade Pasta Inc, p. I-208 [SIC 2098]
Shady Character Ltd, p. I-378 [SIC 2341]
Shaer Shoe Corp, p. I-1029 [SIC 3144]
Shafer Commercial Seating Inc, p. I-603 [SIC 2599]
Shafer Valve Co, p. II-1464 [SIC 3494]
Shaffer, p. II-1531 [SIC 3533]
Shakertown 1992 Inc, p. I-476 [SIC 2429]
Shakespeare Fishing Tackle, p. II-2150 [SIC 3949]
ShaLor Designs Inc, p. I-369 [SIC 2337]
Shamrock Cabinet & Fixture, p. I-486 [SIC 2434]
Shandon Lipshaw, p. II-2102 [SIC 3845]
Shane Group Inc, p. II-2150 [SIC 3949]
Shane Hunter Inc, p. I-361 [SIC 2331]
Shane Steel Processing Inc, p. II-1187 [SIC 3316]
Shanklin Corp, p. II-1660 [SIC 3565]
Shannon Group, p. II-1726 [SIC 3585]
Shape Corp, p. II-1187 [SIC 3316]
Shape Electronics Inc, p. II-1769 [SIC 3612]
Shape Inc, p. I-1006 [SIC 3089]
Shaped Wire Inc, p. II-1250 [SIC 3355]
Sharif Designs Ltd, p. I-1049 [SIC 3172]
Sharilove Fashions Inc, p. I-361 [SIC 2331]
Sharon Concepts Inc, p. I-1071 [SIC 3231]
Sharp Electronics Corp, p. II-1800 [SIC 3631]
Sharp International Corp, p. II-2118 [SIC 3873]
Sharpe Manufacturing Co, p. II-1649 [SIC 3563]
Shartle, p. II-1615 [SIC 3554]
Shasta Beverages Inc, p. I-176 [SIC 2086]
Shasta Industries Inc, p. II-2020 [SIC 3792]
Shaver-Howard Co, p. I-548 [SIC 2512]
Shaw Group Inc, p. II-1486 [SIC 3498]
Shaw Industries Inc, p. I-292 [SIC 2273]
Shaw Lumber Co, p. I-480 [SIC 2431]
Shaw Manufacturing Company, p. I-548 [SIC 2512]
Shaw Resource Services Inc, p. II-1531 [SIC 3533]
Shawnee Plastics Inc, p. II-1572 [SIC 3544]
Shea Communications Co, p. I-718 [SIC 2752]
Sheaffer Inc, p. II-2156 [SIC 3951]
Sheet Metal Manufacturing, p. II-1361 [SIC 3444]
Sheetmetal Inc, p. II-1361 [SIC 3444]
Sheffer Corp, p. II-1747 [SIC 3593]
Sheffield Furniture Corp, p. I-548 [SIC 2512]
Sheffield Lumber and Pallet Inc, p. I-510 [SIC 2448]
Sheffield Plastics Inc, p. I-961 [SIC 3081]
Shelby Die Casting Co, p. II-1572 [SIC 3544]
Shelby Dyeing and Finishing Inc, p. I-267 [SIC 2257]
Shelby Elastics Inc, p. I-245 [SIC 2241]
Shelby Steel Fabricators Inc, p. II-1344 [SIC 3441]
Shelby Williams Industries Inc, p. I-571 [SIC 2521]
Sheldahl Inc, p. II-1915 [SIC 3679]
Sheldon Machine, p. II-1558 [SIC 3541]
Sheldons' Inc, p. II-2150 [SIC 3949]
Shelf Stable Foods Inc, p. I-40 [SIC 2032]
Shell Oil Co, p. I-903 [SIC 2911]
Shelly and Sands Inc, p. I-909 [SIC 2951]
Shelter Components Corp, p. I-292 [SIC 2273]
Shelter-Kit Inc, p. I-523 [SIC 2452]
Shelter Systems of NJ Inc, p. I-501 [SIC 2439]
Sheltex Manufacturing Company, p. I-245 [SIC 2241]
Shenago China, p. I-1102 [SIC 3262]
Shenandoah Mills Inc, p. I-70 [SIC 2041]
Shenango Steel Buildings Inc, p. II-1372 [SIC 3448]
Shepard Clothing Company Inc, p. I-332 [SIC 2311]
Shepard Niles Inc, p. II-1547 [SIC 3536]
Shepard Poorman, p. I-718 [SIC 2752]
Shepard's/McGraw-Hill Inc, p. I-703 [SIC 2731]

Shepherd Chemical Co, p. I-777 [SIC 2819]
Shepherd Electric Company Inc, p. II-1769 [SIC 3612]
Shepherd Products US Inc, p. II-1323 [SIC 3429]
Sher Woven Label Co, p. I-245 [SIC 2241]
Sheridan Corp, p. I-1140 [SIC 3281]
Sheridan Group, p. I-724 [SIC 2754]
Sherman and Reilly Inc, p. II-1835 [SIC 3644]
Sherman Concrete Pipe, p. I-1122 [SIC 3272]
Sherman International, p. II-1344 [SIC 3441]
Sherman Lumber Co, p. I-471 [SIC 2426]
Sherman Prestressed Concrete, p. I-1122 [SIC 3272]
Sherman Wire, p. II-1182 [SIC 3315]
Sherman Wire of Caldwell Inc, p. II-1182 [SIC 3315]
Sherri Cup Inc, p. I-639 [SIC 2656]
Sherrill Furniture Co, p. I-548 [SIC 2512]
Sherrod Vans Inc, p. II-1948 [SIC 3713]
Sherry Manufacturing Company, p. I-336 [SIC 2321]
Sherwin-Williams Co, p. I-840 [SIC 2851]
Sherwood Industries Inc, pp. Vol. I - 639, Vol. II - 1671 [SICs 2656, 3567]
Sherwood Kimtron Corp, p. II-1699 [SIC 3575]
Sherwood Tool Inc, p. II-1615 [SIC 3554]
Shick Tube Veyor Corp, p. II-1541 [SIC 3535]
Shields Bag and Printing Co, p. I-782 [SIC 2821]
Shiely Masonry Products, p. I-1117 [SIC 3271]
Shiloh Industries Inc, p. II-1413 [SIC 3469]
Shimadzu Scientific Instruments, p. II-2066 [SIC 3826]
Shindaiwa Inc, p. II-1514 [SIC 3524]
Shippers Paper Products Co, p. I-618 [SIC 2631]
Shiseido Cosmetics, p. I-834 [SIC 2844]
Shivvers Inc, p. II-1509 [SIC 3523]
Shockey Brothers Inc, p. I-1122 [SIC 3272]
Shogren Hosiery Mfg Co, p. I-250 [SIC 2251]
Shook Builder Supply Co, p. I-501 [SIC 2439]
Shop Vac Corp, p. II-1736 [SIC 3589]
Shopsmith Inc, p. II-1584 [SIC 3546]
Shorebreak, p. I-348 [SIC 2325]
Shorewood Packaging Co, p. I-643 [SIC 2657]
Shorewood Packaging Corp, p. I-688 [SIC 2679]
Short Order Fortune, p. I-107 [SIC 2052]
Showbest Fixture Corp, p. I-587 [SIC 2541]
Showeray Corp, p. I-428 [SIC 2392]
Shred Pax Systems Inc, p. II-1578 [SIC 3545]
Shryock Brothers Inc, p. I-618 [SIC 2631]
Shuffle Master Inc, p. II-1713 [SIC 3579]
Shugart Corp, p. II-1915 [SIC 3679]
Shuqualak Lumber Company Inc, p. I-465 [SIC 2421]
Shur-Lok Corp, p. II-1388 [SIC 3452]
Shurclose Seal Rubber & Plastic, p. I-787 [SIC 2822]
Shure Manufacturing Corp, p. I-603 [SIC 2599]
Shurflo, p. II-1638 [SIC 3561]
Shurtape, p. I-654 [SIC 2672]
Shuster-Mettler Corp, p. II-1600 [SIC 3549]
Shuttleworth Inc, p. II-1541 [SIC 3535]
SI Handling Systems Inc, p. II-1541 [SIC 3535]
SI Jacobson Manufacturing Co, p. I-961 [SIC 3081]
Siberian Salmon Egg Co, p. I-186 [SIC 2091]
SIBIA Inc, p. I-814 [SIC 2836]
SIBV/MS Holdings Inc, p. I-618 [SIC 2631]
Sibyl Shepard Inc, p. I-288 [SIC 2269]
Sico Inc, p. I-553 [SIC 2514]
Sid E Parker Boiler Mfg, p. II-1339 [SIC 3433]
Sid Greenberg Inc, p. I-378 [SIC 2341]
Sid Harvey Industries Inc, p. II-1339 [SIC 3433]
Sid Jones Dental Studios, p. II-2093 [SIC 3843]
Sid Richardson, p. I-894 [SIC 2895]
Sideffects of California Inc, p. I-373 [SIC 2339]
Sidmak Laboratories Inc, p. I-804 [SIC 2834]
Sidran Inc, p. I-332 [SIC 2311]
Siebe North, p. I-954 [SIC 3069]
Siecor Corp, p. II-1260 [SIC 3357]
Siegel-Robert Inc, p. II-1420 [SIC 3471]
Siemens Components Inc, p. II-1915 [SIC 3679]
Siemens Corp, p. II-1892 [SIC 3674]
Siemens Duewag Corp, p. II-1998 [SIC 3743]
Siemens Energy and Automation, pp. Vol II - 1769, 1780 [SICs 3612, 3621]
Siemens Industrial Automation, p. II-2098 [SIC 3844]
Siemens Medical Systems, pp. Vol II - 2098, 2102

[SICs 3844, 3845]
Siemens Medical Systems Inc, pp. Vol II - 2082, 2102 [SICs 3841, 3845]
Siemens Nixdorf Printing Syst, p. II-1704 [SIC 3577]
Siemens Pacesetter Systems Inc, p. II-2102 [SIC 3845]
Siemens Power Corp, p. I-777 [SIC 2819]
Siemens Solar Industries Inc, p. II-1795 [SIC 3629]
Siemens Stromberg-Carlson, p. II-1868 [SIC 3661]
Siemer Milling Co, p. I-70 [SIC 2041]
Sierra Ag Chemical, p. I-861 [SIC 2873]
Sierra Alloys Company Inc, p. II-1225 [SIC 3339]
Sierra Lumber Manufacturers, p. I-480 [SIC 2431]
Sierra Nevada Brewing Co, p. I-160 [SIC 2082]
Sierra Pacific Apparel Company, p. I-348 [SIC 2325]
Sierra Pacific Industries, p. I-465 [SIC 2421]
Sierra Scientific, p. II-2098 [SIC 3844]
Sierra Semiconductor Inc, p. II-1892 [SIC 3674]
Sierracin Corp, p. I-942 [SIC 3053]
Sierracin/Sylmar Corp, p. I-1071 [SIC 3231]
SIFCO Custom Machining, p. II-1975 [SIC 3724]
SIFCO Industries Inc, p. II-1395 [SIC 3462]
SIFCO Selective Plating, p. II-1420 [SIC 3471]
Sifton Motorcycle Products Inc, p. II-2003 [SIC 3751]
Sig Manufacturing Company Inc, p. II-2144 [SIC 3944]
Sigma, p. II-1275 [SIC 3365]
Sigma-Aldrich Corp, p. I-799 [SIC 2833]
Sigma Circuits Inc, p. II-1887 [SIC 3672]
Sigma Designs Imaging Systems, p. II-1934 [SIC 3695]
Sigmaform, p. I-973 [SIC 3083]
Sigman Meat Company Inc, p. I-7 [SIC 2013]
Sigmapower Inc, p. II-1905 [SIC 3677]
SigmaTron International Inc, p. II-1887 [SIC 3672]
Sigmund Cohn Corp, p. II-1255 [SIC 3356]
Signal Apparel Company Inc, p. I-356 [SIC 2329]
Signal Control Co, p. II-1790 [SIC 3625]
Signal Mountain Cement Co, p. I-1077 [SIC 3241]
Signal Technology Corp, p. II-2061 [SIC 3825]
Signal Transformer Company, p. II-1905 [SIC 3677]
Signature Foods Corp, p. I-55 [SIC 2035]
Signature Lighting Corp, p. II-1840 [SIC 3645]
Signet Systems Inc, p. II-1726 [SIC 3585]
Signicast Corp, p. II-1207 [SIC 3324]
Signmark, p. II-2186 [SIC 3993]
Sihi Pumps Inc, p. II-1649 [SIC 3563]
Sikama International Inc, p. II-1594 [SIC 3548]
Sikorsky Aircraft, p. II-1970 [SIC 3721]
Silberline Manufacturing, p. I-773 [SIC 2816]
Silbrico Corp, p. I-1154 [SIC 3295]
Silent Knight Security Systems, p. II-1878 [SIC 3669]
Silgan Holdings Inc, p. I-1006 [SIC 3089]
Silgan Plastics Corp, p. I-1006 [SIC 3089]
Silicon Graphics Inc, p. II-1688 [SIC 3571]
Silicon Valley Group Inc, p. II-1632 [SIC 3559]
Siliconix Inc, p. II-1892 [SIC 3674]
Silk Screen Technology Inc, p. I-446 [SIC 2396]
Silkcraft of Oregon, p. II-1605 [SIC 3552]
Siltron Illumination Inc, p. II-1854 [SIC 3648]
Silvanus Products Inc, p. I-684 [SIC 2678]
Silver Cloud Manufacturing Co, p. II-1775 [SIC 3613]
Silver Engineering Works Inc, p. II-1626 [SIC 3556]
Silver Furniture Co, p. I-542 [SIC 2511]
Silver King, p. II-1736 [SIC 3589]
Silver Lake Cookie Co, p. I-107 [SIC 2052]
Silver Line Building Products, p. II-1349 [SIC 3442]
Silver Oak Wine Cellars, p. I-168 [SIC 2084]
Silver Springs Citrus Cooperative, p. I-60 [SIC 2037]
Silver Star Meats Inc, p. I-7 [SIC 2013]
Silver Star RV, p. II-2020 [SIC 3792]
Silverado Vineyards, p. I-168 [SIC 2084]
Silverlight Corp, p. II-1854 [SIC 3648]
Silverwood Products Inc, p. I-1071 [SIC 3231]
Silvestri Studio Inc, p. II-2199 [SIC 3999]
Silvi Concrete Products Inc, p. I-1127 [SIC 3273]
Sima Products Corp, p. II-1859 [SIC 3651]
Simco Company Inc, p. II-1795 [SIC 3629]
Simco Leather Co, p. I-1053 [SIC 3199]
Simco Leather Corp, p. I-1013 [SIC 3111]
Simco Manufacturing Jewelers, p. II-2122 [SIC 3911]
Simer Pump Co, p. II-1638 [SIC 3561]

Roman numerals I and II indicate the volume in which pages appear. Volume I holds SICs 2011 through 3299; Volume II holds SICs 3312 through 3999.

2296

Roman numerals I and II indicate the volume in which pages appear. Volume I holds SICs 2011 through 3299; Volume II holds SICs 3312 through 3999.

Somerset Refinery Inc, p. I-903 [SIC 2911]
Somerset Technologies Inc, p. II-1654 [SIC 3564]
Somerset Welding and Steel, p. II-1948 [SIC 3713]
Sommer Awning Co, p. I-437 [SIC 2394]
Sommer Metalcraft Corp, p. II-1476 [SIC 3496]
Sona and Hollen Foods Inc, p. I-55 [SIC 2035]
Sonex International Corp, p. II-1813 [SIC 3634]
Sonfarrel Inc, p. II-1764 [SIC 3599]
Sonic Solutions, p. II-2061 [SIC 3825]
Sonicraft Inc, p. II-1878 [SIC 3669]
Sonics and Materials Inc, p. II-1594 [SIC 3548]
Sonobond Ultrasonics Inc, p. II-1563 [SIC 3542]
Sonoco Products Co, pp. Vol. I - 618, 628, 634 [SICs 2631, 2653, 2655]
Sonoma Pacific Co, p. I-510 [SIC 2448]
Sonstegard Foods Inc, p. I-12 [SIC 2015]
Sony Electronics Inc, pp. Vol II - 1704, 2102 [SICs 3577, 3845]
Sony Music Entertainment Inc, p. II-1864 [SIC 3652]
Sony Trans Com Inc, p. II-1873 [SIC 3663]
Sony Tree Music Publishing Inc, p. I-713 [SIC 2741]
SOR Inc, p. II-2051 [SIC 3823]
Sorbent Products Company Inc, p. I-327 [SIC 2299]
Sore Saddle Cyclery Inc, p. II-2003 [SIC 3751]
Sorg Paper Co, p. I-612 [SIC 2621]
Sorrento Cheese Company Inc, p. I-21 [SIC 2022]
Sossner Tap and Tool Corp, p. II-1578 [SIC 3545]
Soule' Steam Feed Works, p. II-1610 [SIC 3553]
Sound Casket Co, p. II-2191 [SIC 3995]
Sound Enhancements Inc, p. II-2135 [SIC 3931]
Soundcoat Company Inc, p. I-991 [SIC 3086]
SoundTech Inc, p. I-567 [SIC 2519]
Source North West Inc, p. I-599 [SIC 2591]
South Bay Cable, p. II-1260 [SIC 3357]
South Bend Lathe Corp, p. II-1558 [SIC 3541]
South Carolina Elastic Co, p. I-245 [SIC 2241]
South Carolina Steel Corp, p. II-1344 [SIC 3441]
South Central Co-op, p. I-869 [SIC 2875]
South Georgia Pecan Co, p. I-136 [SIC 2068]
South Haven Coil Inc, p. II-1905 [SIC 3677]
South Holland Metal Finishing, p. II-1420 [SIC 3471]
South Monroe Sportswear Inc, p. I-348 [SIC 2325]
South Point Ethanol, p. I-855 [SIC 2869]
South Texas Can Company Inc, p. II-1300 [SIC 3411]
South-Western Publishing Co, p. I-703 [SIC 2731]
Southampton Textile Co, p. I-288 [SIC 2269]
Southbend, p. II-1736 [SIC 3589]
Southdown Inc, p. I-1127 [SIC 3273]
Southeast Atlantic Corp, p. I-176 [SIC 2086]
Southeast Paper, p. I-612 [SIC 2621]
Southeast Paper Mfg Co, p. I-612 [SIC 2621]
Southeastern Container Inc, p. I-985 [SIC 3085]
Southeastern Metals Mfg Co, p. II-1349 [SIC 3442]
Southeastern Mills Inc, p. I-70 [SIC 2041]
Southern Aluminum Castings, p. II-1275 [SIC 3365]
Southern Audio Services Inc, p. II-1859 [SIC 3651]
Southern Belle Dairy Co, p. I-35 [SIC 2026]
Southern Binders, p. I-750 [SIC 2789]
Southern Brick and Tile, p. I-1081 [SIC 3251]
Southern Brick Co, p. I-1081 [SIC 3251]
Southern California Machinery, p. II-1610 [SIC 3553]
Southern Cellulose Products Inc, p. I-608 [SIC 2611]
Southern Coatings Inc, p. I-840 [SIC 2851]
Southern Coil Processing Inc, p. II-1212 [SIC 3325]
Southern Cross, p. I-480 [SIC 2431]
Southern Devices Inc, p. II-1830 [SIC 3643]
Southern Die Casting & Eng, p. II-1266 [SIC 3363]
Southern Ductile Casting Co, p. II-1197 [SIC 3321]
Southern Energy Homes Inc, p. I-519 [SIC 2451]
Southern Equipment Co, p. II-1736 [SIC 3589]
Southern Film Extruders Inc, p. I-961 [SIC 3081]
Southern Frozen Foods, p. I-60 [SIC 2037]
Southern Glove Manufacturing, p. I-401 [SIC 2381]
Southern Iowa Mfg Co, p. II-1531 [SIC 3533]
Southern Litho Plate Inc, p. II-1621 [SIC 3555]
Southern Mattress Co, p. I-558 [SIC 2515]
Southern Mercerizing Co, p. I-297 [SIC 2281]
Southern Metal Industries Inc, p. I-593 [SIC 2542]
Southern Mfg Chemists, p. I-819 [SIC 2841]

Southern Mills Inc, p. I-327 [SIC 2299]
Southern Ohio Fabricators Inc, p. II-1344 [SIC 3441]
Southern Optical Company Inc, p. II-2107 [SIC 3851]
Southern Pallet and Crate, p. I-510 [SIC 2448]
Southern Plastics of Louisiana, p. I-979 [SIC 3084]
Southern Precision Corp, p. II-1568 [SIC 3543]
Southern Progress Corp, p. I-698 [SIC 2721]
Southern Quilters, p. I-428 [SIC 2392]
Southern Ready-Mix Inc, p. I-1127 [SIC 3273]
Southern Saw Service Inc, p. II-1318 [SIC 3425]
Southern Shell Fish Company, p. I-186 [SIC 2091]
Southern Soya Corp, p. I-144 [SIC 2075]
Southern Specialty Printing Inc, p. I-735 [SIC 2761]
Southern Spring and Stamping, p. II-1460 [SIC 3493]
Southern States Asphalt Co, p. I-909 [SIC 2951]
Southern States Inc, p. II-1775 [SIC 3613]
Southern Structures Inc, p. II-1372 [SIC 3448]
Southern Systems Inc, p. II-1541 [SIC 3535]
Southern Tea Co, p. I-212 [SIC 2099]
Southern Tier Hide and Tallow, p. I-1013 [SIC 3111]
Southern Tool Company Inc, p. II-1660 [SIC 3565]
Southern Tool Inc, p. II-1207 [SIC 3324]
Southern Webbing Mills Inc, p. I-245 [SIC 2241]
Southern Wipers Inc, p. I-267 [SIC 2257]
Southland Foods Inc, p. I-12 [SIC 2015]
Southland Industries Inc, p. I-942 [SIC 3053]
Southland Manufacturing, p. I-348 [SIC 2325]
Southland Mower Corp, p. II-1514 [SIC 3524]
Southside Manufacturing Corp, p. I-581 [SIC 2531]
Southtec Inc, p. II-2177 [SIC 3965]
Southwall Technologies Inc, p. I-961 [SIC 3081]
Southwest Canners Inc, p. I-176 [SIC 2086]
Southwest Canners of Texas Inc, p. I-176 [SIC 2086]
Southwest Lubricants, p. I-918 [SIC 2992]
Southwest Oilfield Products Inc, p. II-1531 [SIC 3533]
Southwest Plastic Binding Co, p. I-750 [SIC 2789]
Southwest Rebar, p. II-1377 [SIC 3449]
Southwest Steel Company Inc, p. II-1187 [SIC 3316]
Southwest Textiles Inc, p. I-297 [SIC 2281]
Southwestern Bell Publications, p. I-713 [SIC 2741]
Southwestern Glass Company, p. I-1066 [SIC 3229]
Southwestern/Great American, p. I-703 [SIC 2731]
Southwestern Petroleum Corp, p. I-840 [SIC 2851]
Southwestern Pipe Inc, p. II-1192 [SIC 3317]
Southwire Co, pp. Vol II - 1229, 1260 [SICs 3341, 3357]
Southwire Specialty Products, p. II-1234 [SIC 3351]
Southwood Furniture Corp, p. I-548 [SIC 2512]
Southworth Co, p. I-684 [SIC 2678]
Sovereign Yachts Corp, p. II-1992 [SIC 3732]
SP/Sheffer International Inc, p. II-1578 [SIC 3545]
SP Systems Armor, p. II-1395 [SIC 3462]
Space, p. II-1970 [SIC 3721]
Space Age Laminating & Bindery, p. I-750 [SIC 2789]
Space Systems/Loral Inc, p. II-1873 [SIC 3663]
Space Vector Corp, p. II-2011 [SIC 3764]
SpaceLabs Medical Inc, p. II-2102 [SIC 3845]
Spalding and Evenflo Companies, p. II-2150 [SIC 3949]
Spalding Knitting Mills, p. I-254 [SIC 2252]
Spalding Sports Worldwide, p. II-2150 [SIC 3949]
Span America Medical Systems, p. II-2087 [SIC 3842]
Spanco Industries Inc, p. I-245 [SIC 2241]
Spancrete Midwest Co, p. I-1122 [SIC 3272]
Spancrete Northeast Inc, p. I-1122 [SIC 3272]
Spangler Candy Co, p. I-127 [SIC 2064]
Spangler's Flour Mills Inc, p. I-70 [SIC 2041]
Spangler Valve Co, p. II-2056 [SIC 3824]
Spanjer Brothers Inc, p. II-2186 [SIC 3993]
Spantek, p. II-1377 [SIC 3449]
Sparkletts Drinking Water Corp, p. I-176 [SIC 2086]
Sparkomatic Corp, p. II-1859 [SIC 3651]
Sparks Belting Co, p. II-1676 [SIC 3568]
Sparrer Sausage Company Inc, p. I-7 [SIC 2013]
Sparta Brush Company Inc, p. II-2181 [SIC 3991]
Sparta Foods Inc, p. I-40 [SIC 2032]
Sparta Spoke Factory, p. I-471 [SIC 2426]
Spartan Aluminum Products Inc, p. II-1266 [SIC 3363]
Spartan Chemical Co, p. I-825 [SIC 2842]

Spartan Mills Inc, p. I-231 [SIC 2211]
Spartan Motors Inc, p. II-1943 [SIC 3711]
Spartan Products Inc, p. II-2029 [SIC 3799]
Spartan Saw Works, p. II-1318 [SIC 3425]
Spartanburg Steel Products Inc, p. II-1413 [SIC 3469]
Spartech Compounding, p. I-782 [SIC 2821]
Spartech Corp, p. I-973 [SIC 3083]
Sparton Corp, pp. Vol II - 1915, 2035 [SICs 3679, 3812]
Spartus Corp, p. II-1840 [SIC 3645]
Spaulding Composites Company, p. I-973 [SIC 3083]
Spaulding Lumber Co, p. I-506 [SIC 2441]
Spax Inc, p. II-1192 [SIC 3317]
SPD Magnet Wire Co, p. II-1234 [SIC 3351]
SPD Technologies, p. II-1775 [SIC 3613]
Speaco Foods Inc, p. I-45 [SIC 2033]
Speakman Co, p. II-1334 [SIC 3432]
Spear Inc, p. I-718 [SIC 2752]
Spearhead Industries Inc, p. II-2144 [SIC 3944]
Spec-Built Systems Inc, p. I-603 [SIC 2599]
Special Devices Inc, p. II-1434 [SIC 3483]
Special Metals Corp, p. II-1225 [SIC 3339]
Special Mine Services Inc, p. II-1910 [SIC 3678]
Specialized Bicycle Components, p. II-2003 [SIC 3751]
Specialized Printed Forms Inc, p. I-735 [SIC 2761]
Specialty Bakers Inc, p. I-107 [SIC 2052]
Specialty Brands, p. I-212 [SIC 2099]
Specialty Cable Corp, p. II-1260 [SIC 3357]
Specialty Carpets, p. I-292 [SIC 2273]
Specialty Chemical Resources, p. I-768 [SIC 2813]
Specialty Connector Co, p. II-1830 [SIC 3643]
Specialty Container Corp, p. I-660 [SIC 2673]
Specialty Equipment Companies, p. II-1626 [SIC 3556]
Specialty Lighting Inc, p. II-1840 [SIC 3645]
Specialty Machine and Supply, p. II-1531 [SIC 3533]
Specialty Manufacturing Co, p. II-1464 [SIC 3494]
Specialty Paperboard Inc, p. I-618 [SIC 2631]
Specialty Plastics Inc, p. I-1066 [SIC 3229]
Specialty Records Corp, p. II-1864 [SIC 3652]
Specialty Rice Marketing Inc, p. I-79 [SIC 2044]
Specialty Silicone Products Inc, p. I-787 [SIC 2822]
Specialty Steel Treating Inc, p. II-1290 [SIC 3398]
Specialty Tires of America Inc, p. I-927 [SIC 3011]
Specification Rubber Products, p. I-948 [SIC 3061]
Specified Components Inc, p. II-1221 [SIC 3334]
Spectator Sports Services Inc, p. I-446 [SIC 2396]
Spectra-Mat Inc, p. II-1883 [SIC 3671]
Spectra-Physics Lasers Inc, p. II-2102 [SIC 3845]
Spectra-Physics Scanning Systems, p. II-2051 [SIC 3823]
SpecTran Corp, p. I-1066 [SIC 3229]
Spectranetics Corp, p. II-2102 [SIC 3845]
Spectravest Inc, p. I-373 [SIC 2339]
Spectrex Inc, p. II-2071 [SIC 3827]
Spectrian Corp, p. II-1873 [SIC 3663]
Spectro Oils of America, p. I-918 [SIC 2992]
Spectronics Corp, p. II-1854 [SIC 3648]
Spectrowax Corp, p. I-825 [SIC 2842]
Spectrulite Consortium Inc, p. II-1250 [SIC 3355]
Spectrum Astro Inc, p. II-2007 [SIC 3761]
Spectrum Construction Group, p. I-909 [SIC 2951]
Spectrum Control Inc, p. II-1897 [SIC 3675]
Spectrum Engineering Inc, p. II-1694 [SIC 3572]
Spectrum Fabrics, p. I-279 [SIC 2261]
Spectrum Glass Co, p. I-1057 [SIC 3211]
Spectrum Industries Inc, p. I-576 [SIC 2522]
Spectrum Metals, p. II-1486 [SIC 3498]
Spectrum Systems Inc, p. II-2066 [SIC 3826]
SPEDETOOL Manufacturing, p. I-1144 [SIC 3291]
Speed Queen Co, p. II-1809 [SIC 3633]
Speed Selector Inc, p. II-1666 [SIC 3566]
Speedring Inc, p. II-1600 [SIC 3549]
Speizman Industries Inc, p. II-1605 [SIC 3552]
Spence Engineering Company, p. II-1454 [SIC 3492]
Spencer Boat Company Inc, p. II-1992 [SIC 3732]
Spencer Products Inc, p. II-1476 [SIC 3496]
Spencer Turbine Co, p. II-1654 [SIC 3564]
Spencer Wright Industries Inc, p. II-1605 [SIC 3552]
Spenco Medical Corp, p. II-2087 [SIC 3842]

Roman numerals I and II indicate the volume in which pages appear. Volume I holds SICs 2011 through 3299; Volume II holds SICs 3312 through 3999.

2298

Sperry Marine Inc, p. II-2035 [SIC 3812]
Sperry Rubber and Plastics, p. I-948 [SIC 3061]
Spex Chemical and Sample Prep, p. II-2066 [SIC 3826]
Spex Group Inc, p. II-2066 [SIC 3826]
Spherical Roller Bearing, p. II-1644 [SIC 3562]
Spicer Driveshaft, p. II-1954 [SIC 3714]
Spicer Heavy Axle, p. II-1954 [SIC 3714]
Spider Staging Corp, p. II-1367 [SIC 3446]
Spiegel Neckwear Company Inc, p. I-344 [SIC 2323]
Spiller Spring Co, p. I-558 [SIC 2515]
Spinco Metal Products Inc, p. II-1234 [SIC 3351]
Spindletop Draperies Inc, p. I-236 [SIC 2221]
Spinnerin Inc, p. I-283 [SIC 2262]
Spinweld Inc, p. II-1594 [SIC 3548]
Spiral Binding Company Inc, p. II-1713 [SIC 3579]
Spirit of America Corp, p. II-2056 [SIC 3824]
Spirol Intern Holding Corp, p. II-1600 [SIC 3549]
Spirol International Corp, p. II-1600 [SIC 3549]
Splendor Form International Inc, p. I-382 [SIC 2342]
SPN Inc, p. I-365 [SIC 2335]
Spokane Industries Inc, p. II-1212 [SIC 3325]
Spokane Steel Foundry, p. II-1212 [SIC 3325]
Sponge-Cushion Inc, p. I-954 [SIC 3069]
Spontex Inc, p. I-791 [SIC 2823]
Sport Court Inc, p. II-2195 [SIC 3996]
Sport-Craft Inc, p. II-1992 [SIC 3732]
Sport Supply Group Inc, p. II-2150 [SIC 3949]
Sportif USA Inc, p. I-356 [SIC 2329]
Sports Apparel Corp, p. I-356 [SIC 2329]
Sports Belle Inc, p. I-356 [SIC 2329]
Sportsprint Inc, p. I-446 [SIC 2396]
Sportswear Unlimited Inc, p. I-361 [SIC 2331]
Spot-Bilt Inc, p. I-1033 [SIC 3149]
Spot International Inc, p. I-336 [SIC 2321]
Spot Weld Inc, p. II-1594 [SIC 3548]
Spotlight Company Inc, p. I-378 [SIC 2341]
Spray Cotton Mills, p. I-297 [SIC 2281]
Spray-N-Grow Inc, p. I-861 [SIC 2873]
Spraying Systems Co, p. II-1649 [SIC 3563]
Spraylat Corp, p. II-1425 [SIC 3479]
Sprecher Brewing Company Inc, p. I-160 [SIC 2082]
Spreckels Industries Inc, p. I-123 [SIC 2063]
Spreckels Limestone, p. I-1132 [SIC 3274]
Spreckels Sugar Company Inc, p. I-123 [SIC 2063]
Spreuer and Son Inc, p. II-2029 [SIC 3799]
Sprigg Lane Investment Corp, p. II-2199 [SIC 3999]
Spring Air Bedding Inc, p. I-558 [SIC 2515]
Spring Air Co, p. I-558 [SIC 2515]
Spring Air Mattress of California, p. I-558 [SIC 2515]
Spring Air Mattress of Colorado, p. I-558 [SIC 2515]
Spring City Electrical Mfg Co, p. II-1854 [SIC 3648]
Spring Crest Company Inc, p. I-599 [SIC 2591]
Spring Dynamics Inc, p. II-1460 [SIC 3493]
Spring Engineers Inc, p. II-1470 [SIC 3495]
Spring Glen Fresh Foods Inc, p. I-55 [SIC 2035]
Spring Replacement Company, p. II-1460 [SIC 3493]
Spring Wood Products Inc, p. I-510 [SIC 2448]
Springfield Aluminum Co, p. II-1275 [SIC 3365]
Springfield Forest Products LP, p. I-496 [SIC 2436]
Springfield Inc, p. II-1438 [SIC 3484]
Springfield Tablet Mfg Co, p. I-679 [SIC 2677]
Springhouse Corp, p. I-698 [SIC 2721]
Springs Industries Inc, p. I-231 [SIC 2211]
Springs Window Fashions, p. I-599 [SIC 2591]
Springwall Mattress, p. I-558 [SIC 2515]
Sprint Publishing and Advertising, p. I-713 [SIC 2741]
Spruce Pine Mica Company Inc, p. I-1167 [SIC 3299]
Sprunger Corp, p. II-1610 [SIC 3553]
SPS Technologies Inc, p. II-1388 [SIC 3452]
Spudnik Equipment Co, p. II-1509 [SIC 3523]
Spuhl-Anderson Machine Co, p. II-1605 [SIC 3552]
SPX Corp, p. II-1584 [SIC 3546]
SQP Inc, p. I-674 [SIC 2676]
Square Shooter Candy Co, p. I-127 [SIC 2064]
SR Smith Inc, p. II-2150 [SIC 3949]
SRDS Inc, p. I-713 [SIC 2741]
SS/F Inc, p. II-2173 [SIC 3961]
SS Pierce Co, p. I-168 [SIC 2084]
SS Steele and Company Inc, p. I-501 [SIC 2439]

SS White Technologies Inc, p. I-322 [SIC 2298]
SSAC Inc, p. II-1790 [SIC 3625]
SSE Technologies Inc, p. II-1873 [SIC 3663]
SSE Telecom Inc, p. II-1873 [SIC 3663]
SSI Technologies Inc, p. II-1295 [SIC 3399]
SSP Fittings Corp, p. II-1486 [SIC 3498]
St Albans Cooperative Creamery, p. I-17 [SIC 2021]
St Anthony Publishing, p. I-703 [SIC 2731]
St Clair Foods Inc, p. I-212 [SIC 2099]
St Clair Pakwell, p. I-688 [SIC 2679]
St George Packing Company Inc, p. I-190 [SIC 2092]
St Henry Tile Inc, p. I-1086 [SIC 3253]
St James Sugar Cooperative Inc, p. I-116 [SIC 2061]
St Joe Container Co, p. I-628 [SIC 2653]
St Joe Forest Products Co, p. I-618 [SIC 2631]
St Joe Paper Co, p. I-618 [SIC 2631]
St John Knits Inc, p. I-258 [SIC 2253]
St Jude Medical Inc, p. II-2087 [SIC 3842]
St Julian Wine Co, p. I-168 [SIC 2084]
ST Laminating Corp, p. I-973 [SIC 3083]
St Louis Bearing Company Inc, p. II-1676 [SIC 3568]
St Louis Cold Drawn Inc, p. II-1187 [SIC 3316]
St Louis Embroidery, p. I-442 [SIC 2395]
St Louis Gear Company Inc, p. II-1666 [SIC 3566]
St Louis Lithographing, p. I-718 [SIC 2752]
St Louis Steel Casting Inc, p. II-1212 [SIC 3325]
St Paul Brass Foundry Co, p. II-1280 [SIC 3366]
St Paul Corrugating Co, p. II-1187 [SIC 3316]
St Paul Metalcraft Inc, p. II-1572 [SIC 3544]
ST Products Inc, p. I-745 [SIC 2782]
ST Research Corp, p. II-1878 [SIC 3669]
St Vincent DePaul, p. I-515 [SIC 2449]
Sta-Rite Ginnie Lou Inc, p. II-2177 [SIC 3965]
Sta-Rite Indust Water Syst, p. II-1638 [SIC 3561]
Sta-Rite Industries Inc, p. II-1753 [SIC 3594]
STAAR Surgical Co, p. II-2107 [SIC 3851]
Stack On Products Co, p. II-1413 [SIC 3469]
Stackpole Corp, p. II-1786 [SIC 3624]
Stackpole Magnet, p. I-1109 [SIC 3264]
Staco Energy Products Co, p. II-1769 [SIC 3612]
Stacoswitch Inc, p. II-1775 [SIC 3613]
Stadco Corp, p. II-1578 [SIC 3545]
Stafast Products Inc, p. II-1388 [SIC 3452]
Stafford County Flour Mills Co, p. I-70 [SIC 2041]
Stafford Higgins Industries Inc, p. I-373 [SIC 2339]
Stafford (USA) Ltd, p. I-332 [SIC 2311]
Stage II Apparel Corp, p. I-336 [SIC 2321]
Stagg Foods Inc, p. I-40 [SIC 2032]
Stahl/Scott Fetzer Company Inc, p. II-1948 [SIC 3713]
Stahl Specialty Co, p. II-1266 [SIC 3363]
Stahl USA Inc, p. I-825 [SIC 2842]
Stainless Foundry & Engineering, p. II-1207 [SIC 3324]
Stainless Inc, pp. Vol. I - 553, Vol. II - 1372 [SICs 2514, 3448]
Staley Elevator Company Inc, p. II-1536 [SIC 3534]
Stallion Technologies Inc, p. II-1887 [SIC 3672]
Stamco, p. II-1600 [SIC 3549]
Stamco Industries Inc, p. II-1413 [SIC 3469]
Stamler Corp, p. II-1525 [SIC 3532]
Stamper Black Hills Gold Jewelry, p. II-2122 [SIC 3911]
Stamptastics, p. II-2164 [SIC 3953]
Stanbury Uniforms Inc, p. I-420 [SIC 2389]
Stanco Corp, p. I-567 [SIC 2519]
Stanco Metal Products Inc, p. II-1492 [SIC 3499]
Standard and Poor's Corp, p. I-698 [SIC 2721]
Standard Cabinet Works Inc, p. I-587 [SIC 2541]
Standard Candy Company Inc, p. I-127 [SIC 2064]
Standard Cigar Company Inc, p. I-221 [SIC 2121]
Standard Concrete Products, p. I-1127 [SIC 3273]
Standard Container of Edgar, p. I-567 [SIC 2519]
Standard Cos, p. I-674 [SIC 2676]
Standard Fittings Co, p. II-1192 [SIC 3317]
Standard Foods Inc, p. I-212 [SIC 2099]
Standard Furniture Mfg Co, p. I-542 [SIC 2511]
Standard Fusee Corp, p. I-897 [SIC 2899]
Standard Glass and Screen Co, p. I-1057 [SIC 3211]
Standard Havens, p. II-1519 [SIC 3531]

Standard Homeopathic Company, p. I-814 [SIC 2836]
Standard Industries Inc, pp. Vol. I - 991, Vol. II - 1920 [SICs 3086, 3691]
Standard Insert Company Inc, p. II-1382 [SIC 3451]
Standard Iron and Wire Works, p. II-1492 [SIC 3499]
Standard Iron Inc, p. II-1367 [SIC 3446]
Standard-Keil Hardware Mfg Co, p. II-1323 [SIC 3429]
Standard-Knapp Inc, p. II-1660 [SIC 3565]
Standard Locknut Inc, p. II-1644 [SIC 3562]
Standard Machine & Mfg Co, p. II-1447 [SIC 3491]
Standard Machine and Tool Co, p. II-1558 [SIC 3541]
Standard Mattress Co, p. I-558 [SIC 2515]
Standard Microsystems Corp, p. II-1688 [SIC 3571]
Standard Motor Products Inc, p. II-1954 [SIC 3714]
Standard Nipple Works Inc, p. II-1486 [SIC 3498]
Standard Paper Box Machine Co, p. II-1615 [SIC 3554]
Standard Plywood Inc, p. I-491 [SIC 2435]
Standard Products Co, p. II-1954 [SIC 3714]
Standard Publishing Co, p. I-698 [SIC 2721]
Standard Refrigeration Co, p. II-1726 [SIC 3585]
Standard Register Co, p. I-735 [SIC 2761]
Standard Rendering Co, p. I-152 [SIC 2077]
Standard Safety Equipment Co, p. I-455 [SIC 2399]
Standard Steel Specialty Co, p. II-1367 [SIC 3446]
Standard Tallow Corp, p. I-152 [SIC 2077]
Standco Industries Inc, pp. Vol. I - 942, 1150 [SICs 3053, 3292]
Standex International Corp, p. II-1626 [SIC 3556]
Stanford Research Systems Inc, p. II-2066 [SIC 3826]
Stanford Telecommunications, p. II-1915 [SIC 3679]
Stangenes Industries Inc, p. II-1769 [SIC 3612]
Stanislaus Food Products Co, p. I-45 [SIC 2033]
Stanley Air Tools, p. II-1584 [SIC 3546]
Stanley Blacker Inc, p. I-332 [SIC 2311]
Stanley-Bostitch Inc, p. II-1312 [SIC 3423]
Stanley Creations Inc, p. II-2122 [SIC 3911]
Stanley Door Systems Inc, p. II-1349 [SIC 3442]
Stanley Furniture, p. I-542 [SIC 2511]
Stanley Furniture Company Inc, p. I-542 [SIC 2511]
Stanley G Flagg Co, p. II-1486 [SIC 3498]
Stanley Hardware, p. II-1323 [SIC 3429]
Stanley-Proto Industrial Tools, p. II-1312 [SIC 3423]
Stanley Spring & Stamping Corp, p. II-1460 [SIC 3493]
Stanley-Vidmar Inc, p. I-593 [SIC 2542]
Stanley W Ferguson Inc, p. I-195 [SIC 2095]
Stanley Works, p. II-1312 [SIC 3423]
Stanly Fixtures Company Inc, p. I-587 [SIC 2541]
Stanspec, p. II-1547 [SIC 3536]
Stant Corp, p. II-1954 [SIC 3714]
Stanwood Corp, p. I-332 [SIC 2311]
Stanwood Drapery Company Inc, p. I-424 [SIC 2391]
Stanwood Mills Inc, p. I-245 [SIC 2241]
Staodyn Inc, p. II-2102 [SIC 3845]
Stapling Machines Co, p. II-1660 [SIC 3565]
Star Binding and Trimming Corp, p. I-245 [SIC 2241]
Star Brass Foundry, p. II-1212 [SIC 3325]
Star Building Systems, p. II-1372 [SIC 3448]
Star Childrens Dress Co, p. I-390 [SIC 2361]
Star Corrugated Box Company, p. I-628 [SIC 2653]
Star Cutter Co, p. II-1558 [SIC 3541]
Star Dynamics Inc, p. II-1747 [SIC 3593]
Star Enterprise, p. I-903 [SIC 2911]
Star Gate Technologies Inc, p. II-1887 [SIC 3672]
Star Glove Company Inc, p. I-401 [SIC 2381]
Star Hydraulics Inc, p. II-1753 [SIC 3594]
Star-Kist Foods Inc, p. I-186 [SIC 2091]
Star Knitwear Inc, p. I-336 [SIC 2321]
Star Linear Systems Co, p. II-1676 [SIC 3568]
Star Milling Co, p. I-91 [SIC 2047]
Star of Phoenix Aircraft Corp, p. II-1970 [SIC 3721]
Star of West Mill Inc, p. I-70 [SIC 2041]
Star Packaging Corp, p. I-660 [SIC 2673]
Star Paper Tube Inc, p. I-634 [SIC 2655]
Star Porcelain Co, p. I-1109 [SIC 3264]
Star Synthetic Mfg Corp, p. I-322 [SIC 2298]
Star-Telegram Inc, p. I-693 [SIC 2711]
Star Tex Corp, p. I-649 [SIC 2671]
Star Tool and Engineering, p. II-1666 [SIC 3566]

Roman numerals I and II indicate the volume in which pages appear. Volume I holds SICs 2011 through 3299; Volume II holds SICs 3312 through 3999.

2299

Star Transport Trailers, p. II-1961 [SIC 3715]
Star Tubular Products Co, p. II-1486 [SIC 3498]
Star Valley Cheese Inc, p. I-21 [SIC 2022]
Starboard Industries Inc, p. II-1404 [SIC 3465]
Starbuck Creamery Co, p. I-869 [SIC 2875]
Starensier Inc, p. I-310 [SIC 2295]
Starfire Lumber Co, p. I-460 [SIC 2411]
Starflo Corp, p. II-1742 [SIC 3592]
Stark Ceramics Inc, p. I-1081 [SIC 3251]
Stark Manufacturing Inc, p. II-1486 [SIC 3498]
Starke Uniform Mfg Co, p. I-352 [SIC 2326]
Starkey Labs Inc, p. II-2087 [SIC 3842]
Starmark Inc, p. I-486 [SIC 2434]
Starr National Mfg Corp, p. I-825 [SIC 2842]
Start Master, p. II-1503 [SIC 3519]
Startex Mills, p. I-231 [SIC 2211]
State Chemical Mfg Co, p. I-819 [SIC 2841]
State Coat Front Co, p. I-446 [SIC 2396]
State Fair Foods Inc, p. I-65 [SIC 2038]
State Heat Treat Inc, p. II-1290 [SIC 3398]
State Line Snacks Corp, p. I-199 [SIC 2096]
State of the Art Inc, p. II-1901 [SIC 3676]
State Printing Co, p. I-708 [SIC 2732]
State Wide Aluminum Inc, p. II-1349 [SIC 3442]
States Electric Mfg Co, p. II-1775 [SIC 3613]
States Industries Inc, p. I-491 [SIC 2435]
Statesville Brick Company Inc, p. I-1081 [SIC 3251]
Stationers Engraving Inc, p. I-724 [SIC 2754]
Statler Industries Inc, p. I-674 [SIC 2676]
Statler Tissue Corp, p. I-674 [SIC 2676]
Stature Electric Inc, p. II-1780 [SIC 3621]
Stauffer Cheese Inc, p. I-21 [SIC 2022]
Stauffer Edition Binding, p. I-750 [SIC 2789]
Staver Foundry Company Inc, p. II-1203 [SIC 3322]
STB Systems Inc, p. II-1704 [SIC 3577]
Ste Chapelle Inc, p. I-168 [SIC 2084]
Ste Genevieve Manufacturing, p. I-471 [SIC 2426]
Steadley Co, p. II-1470 [SIC 3495]
Stearns Manufacturing Co, p. I-954 [SIC 3069]
Stearns Prod Development Corp, p. II-1584 [SIC 3546]
Stearnswood Inc, p. I-506 [SIC 2441]
Stebco Products Corp, p. I-1053 [SIC 3199]
Steccone Products Company Inc, p. I-825 [SIC 2842]
Steck-Vaughn Co, p. I-703 [SIC 2731]
Steck-Vaughn Publishing Corp, p. I-703 [SIC 2731]
Steco Inc, p. II-1961 [SIC 3715]
Steel Ceilings Inc, p. II-1367 [SIC 3446]
Steel City Corp, p. II-1413 [SIC 3469]
Steel Craft Fluorescent Inc, p. II-1845 [SIC 3646]
Steel Heddle Manufacturing Co, p. II-1605 [SIC 3552]
Steel Inc, p. II-1344 [SIC 3441]
Steel Industries Inc, p. II-1395 [SIC 3462]
Steel King Industries Inc, p. II-1682 [SIC 3569]
Steel of West Virginia Inc, p. II-1178 [SIC 3313]
Steel Parts Corp, p. II-1404 [SIC 3465]
Steel Sportswear Inc, p. I-369 [SIC 2337]
Steel Technologies Inc, p. II-1492 [SIC 3499]
Steel Treaters Inc, p. II-1290 [SIC 3398]
Steelcase Inc, p. I-576 [SIC 2522]
Steelco Inc, p. II-2029 [SIC 3799]
Steelcraft Corp, p. II-1654 [SIC 3564]
Steelcraft Manufacturing Co, p. II-1349 [SIC 3442]
Steele Canvas Basket Corp, p. I-437 [SIC 2394]
Steelfab Inc, p. II-1344 [SIC 3441]
Steelox Systems Inc, p. II-1372 [SIC 3448]
Steelstran Industries, p. I-245 [SIC 2241]
Steeltech Manufacturing Inc, p. II-1355 [SIC 3443]
Steelweld Equipment Company, p. II-1948 [SIC 3713]
Steelworks Inc, p. I-576 [SIC 2522]
Steffen Bookbinders Inc, p. I-750 [SIC 2789]
Stegner Food Products Co, p. I-40 [SIC 2032]
Stein Inc, p. II-1626 [SIC 3556]
Stein Seal Company Inc, p. II-1644 [SIC 3562]
Steiner Company Inc, p. I-593 [SIC 2542]
Steinfeld's Products Co, p. I-55 [SIC 2035]
Stella D'oro Biscuit Company, p. I-107 [SIC 2052]
Stella D'oro Biscuit of California, p. I-107 [SIC 2052]
Stella Foods Inc, p. I-21 [SIC 2022]
Stemaco Products Inc, p. II-2107 [SIC 3851]

Stemco Inc, p. I-942 [SIC 3053]
StemWood Corp, p. I-491 [SIC 2435]
Stenograph Corp, p. II-1713 [SIC 3579]
STEP Co, p. II-2150 [SIC 3949]
Stepan Co, p. I-830 [SIC 2843]
Stephen Gould Paper Company, p. I-628 [SIC 2653]
Stephen Lawrence Co, p. I-688 [SIC 2679]
Stephenson and Lawyer Inc, p. I-991 [SIC 3086]
Stephenson Enterprises Inc, p. I-348 [SIC 2325]
Sterile Concepts Holdings Inc, p. II-2082 [SIC 3841]
Sterile Concepts Inc, p. II-2082 [SIC 3841]
Sterile Products Corp, p. I-231 [SIC 2211]
Sterile Recoveries Inc, p. I-420 [SIC 2389]
Sterilite Corp, p. I-1006 [SIC 3089]
STERIS Corp, p. II-2087 [SIC 3842]
Steris Laboratories Inc, p. I-804 [SIC 2834]
Sterling Chemicals Inc, p. I-782 [SIC 2821]
Sterling China Company Inc, p. I-1102 [SIC 3262]
Sterling Die Operation, p. II-1578 [SIC 3545]
Sterling Electric Inc, p. II-1780 [SIC 3621]
Sterling Hydraulics Inc, p. II-1454 [SIC 3492]
Sterling Inc, p. II-2045 [SIC 3822]
Sterling Instrument, p. II-1666 [SIC 3566]
Sterling Lumber and Supply, p. I-510 [SIC 2448]
Sterling Plumbing Group Inc, p. II-1334 [SIC 3432]
Sterling Radiator, p. II-1339 [SIC 3433]
Sterling Scale Company Inc, p. II-1759 [SIC 3596]
Sterling Spring Corp, p. II-1470 [SIC 3495]
Sterling Steel Foundry Inc, p. II-1212 [SIC 3325]
Sterling Vision Shoppes Inc, p. II-2107 [SIC 3851]
Sterlings Name Tape Co, p. I-288 [SIC 2269]
Stern and Stern Industries Inc, p. I-236 [SIC 2221]
Stern Metals Inc, p. II-1225 [SIC 3339]
Stern's Miracle-Gro Products Inc, p. I-861 [SIC 2873]
Sterner Lighting Systems Inc, p. II-1840 [SIC 3645]
Steuben, p. I-1066 [SIC 3229]
Steuben Foods Inc, p. I-35 [SIC 2026]
Stevcoknit Fabrics Co, p. I-267 [SIC 2257]
Steven Fabrics Co, p. I-424 [SIC 2391]
Steven Krauss Menswear Cies, p. I-344 [SIC 2323]
Steven Madden Ltd, p. I-1029 [SIC 3144]
Steven Manufacturing Co, p. II-2144 [SIC 3944]
Stevens Graphics Inc, p. I-713 [SIC 2741]
Stevens Industries Inc, p. I-587 [SIC 2541]
Steves and Sons Inc, p. I-480 [SIC 2431]
Stevison Ham Co, p. I-7 [SIC 2013]
Steward Machine Company Inc, p. II-1764 [SIC 3599]
Stewart and Stevenson Services, p. II-1498 [SIC 3511]
Stewart Boot Company Inc, p. I-1025 [SIC 3143]
Stewart-Decatur Security Systems, p. II-1878 [SIC 3669]
Stewart Filmscreen Corp, p. II-2112 [SIC 3861]
Stewart Manufacturing Co, p. II-1813 [SIC 3634]
Stewart R Browne Mfg Co, p. II-1854 [SIC 3648]
Stewart's Forest Products, p. I-515 [SIC 2449]
Stewart Sutherland Inc, p. I-660 [SIC 2673]
Stewart Warner Instrument, p. II-2045 [SIC 3822]
Stewarts Private Blend Foods, p. I-195 [SIC 2095]
Stewert and Stevenson, p. II-1649 [SIC 3563]
Stidham Horse Trailers Inc, p. II-2029 [SIC 3799]
Stiefel Laboratories Inc, p. I-804 [SIC 2834]
Stiffel Co, p. II-1840 [SIC 3645]
Stihl Inc, p. II-1584 [SIC 3546]
Stillwater Milling Co, p. I-95 [SIC 2048]
Stillwater Sales Inc, p. I-231 [SIC 2211]
Stimpert Enterprises Inc, p. I-523 [SIC 2452]
Stimson Lane Wine, p. I-168 [SIC 2084]
Stimson Lumber Co, p. I-465 [SIC 2421]
Stinson Seafood Co, p. I-186 [SIC 2091]
STK Electronics Inc, p. II-1897 [SIC 3675]
Stock Drive Products, p. II-1666 [SIC 3566]
Stock Equipment Co, p. II-1509 [SIC 3523]
Stocker and Yale Inc, p. II-1769 [SIC 3612]
Stockham Valves and Fittings, p. II-1447 [SIC 3491]
Stockpot Soups Inc, p. I-51 [SIC 2034]
Stockwell Rubber Company Inc, p. I-942 [SIC 3053]
Stoddard-Hamilton Aircraft Inc, p. II-1970 [SIC 3721]
Stoelting Inc, p. II-1626 [SIC 3556]
Stokes Canning Co, p. I-40 [SIC 2032]

Stone Construction Equipment, p. II-1519 [SIC 3531]
Stone Container Corp, pp. Vol. I - 618, 665 [SICs 2631, 2674]
Stone Creek Brick Co, p. I-1081 [SIC 3251]
Stone Creek Yarn Mill Company, p. I-297 [SIC 2281]
Stone Mfg Co Menswear, p. I-340 [SIC 2322]
Stone Southwest, p. I-496 [SIC 2436]
Stonelight Tile Inc, p. I-1086 [SIC 3253]
StoneMark Inc, p. I-1033 [SIC 3149]
Stoneridge Inc, p. II-1954 [SIC 3714]
Stoneway Concrete Inc, p. I-1127 [SIC 3273]
Stonhard Inc, p. II-2199 [SIC 3999]
Stonington Corp, p. I-634 [SIC 2655]
Stonyfield Farm Inc, p. I-30 [SIC 2024]
Stop-Shock Inc, p. I-874 [SIC 2879]
Stora Newton Falls Inc, p. I-654 [SIC 2672]
Storage Dimensions Inc, p. II-1694 [SIC 3572]
Storage Technology Corp, p. II-1694 [SIC 3572]
Storck Baking Company Inc, p. I-101 [SIC 2051]
Storck USA LP, p. I-132 [SIC 2066]
Store Kraft Manufacturing Co, p. I-587 [SIC 2541]
Stored Energy Systems, p. II-1795 [SIC 3629]
Stores Automated Systems Inc, p. II-1709 [SIC 3578]
Storm Products Company Inc, p. II-1260 [SIC 3357]
Storopack Inc, p. I-991 [SIC 3086]
Story and Clark Piano Co, p. II-2135 [SIC 3931]
Stoudt's Brewing Inc, p. I-160 [SIC 2082]
Stouffer Corp, p. I-65 [SIC 2038]
Stouffer Foods Corp, p. I-65 [SIC 2038]
Stoughton Composites Inc, p. II-2199 [SIC 3999]
Stoughton Trailers Inc, p. II-1961 [SIC 3715]
Stout Industries Inc, p. II-2186 [SIC 3993]
Stow Davis, p. I-480 [SIC 2431]
Stow Manufacturing Co, p. II-1519 [SIC 3531]
Straits Steel and Wire Co, p. II-1476 [SIC 3496]
Strand Lighting Inc, pp. Vol II - 1775, 1845 [SICs 3613, 3646]
Strandflex, p. I-322 [SIC 2298]
Strata Design Inc, p. I-587 [SIC 2541]
StrataCom Inc, p. II-1868 [SIC 3661]
Stratford Homes LP, p. I-523 [SIC 2452]
Strathmore Paper Co, p. I-612 [SIC 2621]
Strato Medical Corp, p. II-2102 [SIC 3845]
Stratoflex Aerospace, p. II-1464 [SIC 3494]
Stratton Hats Inc, p. I-386 [SIC 2353]
Stratus Computer Inc, p. II-1704 [SIC 3577]
Stratus Specialty Vehicles Inc, p. II-1943 [SIC 3711]
Strauser Manufacturing Inc, p. I-536 [SIC 2499]
Strauss Industries Inc, p. II-1300 [SIC 3411]
Strauss Jewelry Manufacturing, p. II-2122 [SIC 3911]
Streamline Plastics Company Inc, p. I-973 [SIC 3083]
Streat Garment Co, p. I-394 [SIC 2369]
Streator Brick Systems Inc, p. I-1081 [SIC 3251]
Street Life Inc, p. I-361 [SIC 2331]
Streimer Sheet Metal Works Inc, p. II-1361 [SIC 3444]
Strescon Industries Inc, p. I-1122 [SIC 3272]
Stresscon Corp, p. I-1122 [SIC 3272]
Strick Corp, p. II-1961 [SIC 3715]
Stride Rite Corp, p. I-1033 [SIC 3149]
Striker Industries Inc, p. I-914 [SIC 2952]
Strilich Technologies Inc, p. II-1558 [SIC 3541]
Strip Steel Inc, p. II-1187 [SIC 3316]
Stroehmann Bakeries Inc, p. I-101 [SIC 2051]
Stroh Brewery Co, p. I-160 [SIC 2082]
Stroh Die Casting Company Inc, p. II-1271 [SIC 3364]
Stroh's Ice Cream Co, p. I-30 [SIC 2024]
Strohm Manufacturing Inc, p. I-386 [SIC 2353]
Strombecker Corp, p. II-2144 [SIC 3944]
Strongheart Products Co, p. I-91 [SIC 2047]
Stronglite Inc, p. I-603 [SIC 2599]
Stroupe Mirror Company Inc, p. I-1071 [SIC 3231]
Strouse, Adler Co, p. I-382 [SIC 2342]
Structural Display Inc, p. II-2186 [SIC 3993]
Structural Instrumentation Inc, p. II-1759 [SIC 3596]
Structural Instrumentation Mfg, p. II-1759 [SIC 3596]
Structural Laminates Co, p. II-1295 [SIC 3399]
Structural Steel Services Inc, p. II-1344 [SIC 3441]
Structural Test Products, p. II-2076 [SIC 3829]
Structural Wood Systems, p. I-501 [SIC 2439]

Roman numerals I and II indicate the volume in which pages appear. Volume I holds SICs 2011 through 3299; Volume II holds SICs 3312 through 3999.

Roman numerals I and II indicate the volume in which pages appear. Volume I holds SICs 2011 through 3299; Volume II holds SICs 3312 through 3999.

SWIBCO Inc, p. I-1071 [SIC 3231]
Swid Powell Designs Inc, p. I-1113 [SIC 3269]
Swift Adhesives, p. I-879 [SIC 2891]
Swift-Cor Tool Engineering Co, p. II-1572 [SIC 3544]
Swift Glass Company Inc, p. I-1071 [SIC 3231]
Swift Textile Metalizing Corp, p. I-310 [SIC 2295]
Swift Textiles Inc, p. I-231 [SIC 2211]
Swindell-Dressler Intern Co, p. II-1671 [SIC 3567]
Swing-A-Way Manufacturing, p. II-1312 [SIC 3423]
Swingster Marketing, p. I-356 [SIC 2329]
Swire Pacific Holdings Inc, p. I-176 [SIC 2086]
Swiss Maid Inc, p. I-442 [SIC 2395]
Swiss Valley Farms Co, p. I-35 [SIC 2026]
Swisstronics Inc, p. II-1382 [SIC 3451]
Switching Power Inc, p. II-1775 [SIC 3613]
Switlik Parachute Co, p. I-455 [SIC 2399]
Swivelier Company Inc, p. II-1840 [SIC 3645]
Swords Veneer and Lumber Co, p. I-491 [SIC 2435]
Sybron Chemicals Inc, p. I-799 [SIC 2833]
Sybron International Corp, p. II-2093 [SIC 3843]
SyDOS, p. II-1694 [SIC 3572]
Sylray Inc, p. I-378 [SIC 2341]
Sylvan Sales Inc, p. I-532 [SIC 2493]
Sylvest Farms Inc, p. I-12 [SIC 2015]
Symbiosis Corp, p. II-2082 [SIC 3841]
Symbol Technologies Inc, p. II-1709 [SIC 3578]
Syme Inc, p. II-1300 [SIC 3411]
Symetrics Inc, p. II-1975 [SIC 3724]
Symmetricom Inc, p. II-1868 [SIC 3661]
Symmons Industries Inc, p. II-1334 [SIC 3432]
Symons Corp, pp. Vol. I - 897, Vol. II - 1361 [SICs 2899, 3444]
Symons Frozen Foods Inc, p. I-60 [SIC 2037]
Syn-Tech Systems Inc, p. II-1443 [SIC 3489]
Synalloy Corp, p. II-1172 [SIC 3312]
Syncom Technologies Inc, p. II-1934 [SIC 3695]
Syncro Corp, p. II-1929 [SIC 3694]
Syncro Machine Co, p. II-1600 [SIC 3549]
Syndicate Store Fixtures Inc, p. I-593 [SIC 2542]
SynTechnics Inc, p. I-236 [SIC 2221]
Syntex Agribusiness Inc, p. I-804 [SIC 2834]
Syntex Corp, p. I-804 [SIC 2834]
Synthes USA, p. II-2087 [SIC 3842]
Synthetic Genetics Inc, p. I-814 [SIC 2836]
Synthetic Industries Inc, p. I-236 [SIC 2221]
Synthetic Products Co, p. I-855 [SIC 2869]
Synthetic Thread Company Inc, p. I-306 [SIC 2284]
Synthetic Vision Systems, p. II-2071 [SIC 3827]
Syntro Corp, p. I-814 [SIC 2836]
Syntronic Instruments Inc, p. II-1883 [SIC 3671]
SyQuest Technology Inc, p. II-1694 [SIC 3572]
Syracuse China Corp, p. I-1102 [SIC 3262]
Syracuse Rubber Products, p. I-954 [SIC 3069]
Syracuse Safety-Lites Inc, p. II-1854 [SIC 3648]
Syratech Corp, p. II-2127 [SIC 3914]
Syro Steel Co, p. II-1361 [SIC 3444]
Syroco Inc, p. I-567 [SIC 2519]
Syscon International Inc, p. II-2045 [SIC 3822]
Syscon-RKC, p. II-2045 [SIC 3822]
Systec Inc, p. II-2040 [SIC 3821]
SYSTECH Corp, p. II-1878 [SIC 3669]
System and Methods Inc, p. I-735 [SIC 2761]
Systron Donner Inertial, p. II-2035 [SIC 3812]

T and B Leather Fashions Inc, p. I-413 [SIC 2386]
T and D Metal Products Inc, p. II-1395 [SIC 3462]
T and H Machine Inc, p. II-1589 [SIC 3547]
T and N Industries Inc, p. II-1954 [SIC 3714]
T and R Electric Supply, p. II-1769 [SIC 3612]
T and R Engraving Inc, p. I-759 [SIC 2796]
T and S Brass and Bronze Works, p. II-1334 [SIC 3432]
T and W Forge Inc, p. II-1395 [SIC 3462]
T-Chem Products Inc, p. I-819 [SIC 2841]
T-L Irrigation Co, p. II-1509 [SIC 3523]
T Marzetti Co, p. I-55 [SIC 2035]
T-PAC, p. I-1122 [SIC 3272]
T Sardelli and Sons Inc, p. II-2122 [SIC 3911]
T Sendzimir Inc, p. II-1589 [SIC 3547]

T Williams and Son Inc, p. I-437 [SIC 2394]
TA Manufacturing Co, p. II-2016 [SIC 3769]
TA Sullivan and Son, p. II-1140 [SIC 3281]
Tab McGraw-Hill Inc, p. I-703 [SIC 2731]
TABC Inc, p. II-1948 [SIC 3713]
Taber Metals LP, p. II-1244 [SIC 3354]
Tabernash Brewing Co, p. I-160 [SIC 2082]
TACC International Corp, p. I-879 [SIC 2891]
Taco Inc, p. II-1638 [SIC 3561]
Tacoma Lime, p. I-1132 [SIC 3274]
Tacoma Rubber Stamp, p. I-844 [SIC 3953]
Tacoma Truss Systems Inc, p. I-501 [SIC 2439]
Taconite Eng & Mfg Co, p. II-1525 [SIC 3532]
TAFA Inc, p. II-1425 [SIC 3479]
Tafco Equipment Co, p. II-1948 [SIC 3713]
Tagsons Papers Inc, p. I-612 [SIC 2621]
Tailored Baby Inc, p. I-428 [SIC 2392]
Takara Belmont USA Inc, p. II-2093 [SIC 3843]
Takara Sake USA, p. I-168 [SIC 2084]
Takata Inc, p. I-455 [SIC 2399]
Takka, p. II-1813 [SIC 3634]
Talbert Manufacturing Inc, p. II-1961 [SIC 3715]
Talbot Holdings Ltd, p. II-1558 [SIC 3541]
Talladega Castings, p. II-1212 [SIC 3325]
Talladega Foundry and Machine, p. II-1197 [SIC 3321]
Talley Defense Systems Inc, p. I-777 [SIC 2819]
Talley Industries Inc, p. II-2035 [SIC 3812]
Tallgrass Technologies Corp, p. II-1694 [SIC 3572]
Talmo Inc, p. I-460 [SIC 2411]
Talsol Corp, p. I-825 [SIC 2842]
Tam Ceramics Inc, p. I-1167 [SIC 3299]
Tam Industries Inc, p. I-378 [SIC 2341]
Tamaqua Cable Products Corp, p. II-1260 [SIC 3357]
Tambrands Inc, p. I-674 [SIC 2676]
Tamco Industries Inc, p. II-2025 [SIC 3795]
Tamko Asphalt Products Inc, p. I-914 [SIC 2952]
Tampa Brass & Aluminum Corp, p. II-1280 [SIC 3366]
Tampa Soap and Chemical, p. I-152 [SIC 2077]
Tampella Power Corp, p. II-1339 [SIC 3433]
Tamrock/EJC USA Inc, p. II-1525 [SIC 3532]
Tamron Industries Inc, p. II-2112 [SIC 3861]
Tandem Computers Inc, p. II-1688 [SIC 3571]
Tandem Fabrics, p. I-318 [SIC 2297]
Tandem Products Inc, p. II-1425 [SIC 3479]
Tandy Brands Accessories Inc, p. I-416 [SIC 2387]
Tandy Electronics, p. II-1260 [SIC 3357]
Tandy Leather Co, p. I-1049 [SIC 3172]
Tandy Wire and Cable, p. II-1260 [SIC 3357]
Tanel Corp, p. I-1033 [SIC 3149]
Taney Corp, p. I-480 [SIC 2431]
Tankinetics Inc, p. I-1158 [SIC 3296]
Tannenbaum and Sons Inc, p. I-1041 [SIC 3161]
Tanner Companies Inc, p. I-365 [SIC 2335]
Tanner Industries, p. I-897 [SIC 2899]
Tannewitz Inc, p. II-1610 [SIC 3553]
TANO Corp, p. II-1503 [SIC 3519]
Tapco International Inc, p. II-1464 [SIC 3494]
Tapco Plumbing, p. II-1334 [SIC 3432]
Tape Inc, p. I-654 [SIC 2672]
TAPEMARK Company Inc, p. I-688 [SIC 2679]
Tapistron International Inc, p. II-1605 [SIC 3552]
Taptite Products, p. II-1388 [SIC 3452]
Tara-Lee Sportswear Company, p. I-361 [SIC 2331]
Taracorp Evans Inc, p. II-1225 [SIC 3339]
Taracorp Industries Inc, p. II-1486 [SIC 3498]
Targ-It-Tronics Inc, p. II-1887 [SIC 3672]
Target Rock Corp, p. II-1454 [SIC 3492]
Target Therapeutics Inc, p. II-2082 [SIC 3841]
TargeTech Inc, p. I-814 [SIC 2836]
Tarmac America Inc, p. II-1122 [SIC 3272]
Tarpenning-LaFollette Company, p. II-1377 [SIC 3449]
Tastee Apple Inc, p. I-51 [SIC 2034]
Tastemaker, p. I-181 [SIC 2087]
Tasty Baking Co, p. I-101 [SIC 2051]
Tate and Lyle Inc, p. I-119 [SIC 2062]
Taubensee Steel and Wire Co, p. II-1182 [SIC 3315]
Taunton Press Inc, p. I-698 [SIC 2721]
Tavco Inc, p. II-1454 [SIC 3492]

Tax Forms Printing, p. I-724 [SIC 2754]
Taylor Chair Co, p. I-571 [SIC 2521]
Taylor Cheese Corp, p. I-21 [SIC 2022]
Taylor Clay Products Company, p. I-1081 [SIC 3251]
Taylor Clothing Inc, p. I-332 [SIC 2311]
Taylor Co, p. II-1626 [SIC 3556]
Taylor Corp, p. I-718 [SIC 2752]
Taylor Desk Co, p. I-571 [SIC 2521]
Taylor-Dunn Manufacturing Co, p. II-1552 [SIC 3537]
Taylor Forge Engineered Syst, p. II-1355 [SIC 3443]
Taylor Guitars, p. II-2135 [SIC 3931]
Taylor Industries Inc, p. I-1098 [SIC 3261]
Taylor Lumber and Treating Inc, p. I-465 [SIC 2421]
Taylor Machine Works Inc, p. II-1552 [SIC 3537]
Taylor-Pohlman Inc, p. II-1266 [SIC 3363]
Taylor Publishing Co, p. I-713 [SIC 2741]
Taylor-Ramsey Corp, p. I-465 [SIC 2421]
Taylor-Ramsey Dimensions, p. I-581 [SIC 2531]
Taylor Ready Mix Inc, p. I-909 [SIC 2951]
Taylor Togs Inc, p. I-348 [SIC 2325]
Taylor-Wharton Cryogenics, p. II-1300 [SIC 3411]
Taylor-Winfield Corp, p. II-1594 [SIC 3548]
TB Wood's Sons Co, p. II-1676 [SIC 3568]
TBG Inc, p. II-1694 [SIC 3572]
TBG Industries Inc, p. II-1954 [SIC 3714]
TC/American Monorail Inc, p. II-1547 [SIC 3536]
TC Industries Inc, p. II-1290 [SIC 3398]
TCF Aeorovent Co, p. II-1654 [SIC 3564]
TCH Industries Inc, p. II-1271 [SIC 3364]
TCI Aluminum, p. II-1285 [SIC 3369]
TCI International Inc, p. II-1873 [SIC 3663]
TCOM LP, p. II-2035 [SIC 3812]
TCR Corp, p. II-1388 [SIC 3452]
TD Industrial Coverings Inc, p. I-437 [SIC 2394]
TD Williamson Inc, p. II-1531 [SIC 3533]
TDG Aerospace Inc, p. I-603 [SIC 2599]
TDK Ferrites Corp, p. I-1109 [SIC 3264]
TDK Magnetic Tape Corp, p. II-1934 [SIC 3695]
TDM Inc, p. I-713 [SIC 2741]
TE Brown Inc, p. I-515 [SIC 2449]
TEAC America Inc, p. II-1694 [SIC 3572]
Tec-Cast Inc, p. II-1275 [SIC 3365]
TEC Inc, p. I-879 [SIC 2891]
Tech-Etch Inc, p. II-1492 [SIC 3499]
Tech Group Inc, p. I-1006 [SIC 3089]
Tech Industries Inc, p. I-991 [SIC 3086]
Tech-Interactive Inc, p. I-754 [SIC 2791]
Tech/Ops Sevcon Inc, p. II-1790 [SIC 3625]
Tech Spray Inc, p. I-897 [SIC 2899]
Tech-Sym Corp, p. II-1915 [SIC 3679]
Techalloy Company Inc, pp. Vol II - 1182, 1594 [SICs 3315, 3548]
Techdyne Inc, p. II-1830 [SIC 3643]
TechKnits Inc, p. I-373 [SIC 2339]
Techna Type Inc, p. I-754 [SIC 2791]
Techne Corp, p. II-2082 [SIC 3841]
Techne Inc, p. II-2066 [SIC 3826]
Technetics Corp, p. II-1295 [SIC 3399]
Techni-Braze Inc, p. II-1290 [SIC 3398]
Techni-Cast Corp, p. II-1275 [SIC 3365]
Technic Equipment Corp, p. I-787 [SIC 2822]
Technic Inc, p. II-1420 [SIC 3471]
Technical Chemical Co, p. I-897 [SIC 2899]
Technical Chemicals & Products, p. I-809 [SIC 2835]
Technical Devices Co, p. II-1938 [SIC 3699]
Technical Instrument Co, p. II-2061 [SIC 3825]
Technical Materials Inc, p. II-1234 [SIC 3351]
Technical Metals Co, p. II-1187 [SIC 3316]
Technical Oil Products Inc, p. I-156 [SIC 2079]
Technical Ordnance Inc, p. I-885 [SIC 2892]
Technical Prod & Precision Mfg, p. II-1454 [SIC 3492]
Technicorp Inc, p. II-2061 [SIC 3825]
Technipower Inc, p. II-1795 [SIC 3629]
Technitrol Inc, p. II-1915 [SIC 3679]
Techno Components Inc, p. II-1901 [SIC 3676]
Technocell, p. II-1920 [SIC 3691]
Technology Research Corp, p. II-2061 [SIC 3825]
TechnoTrim Inc, p. I-455 [SIC 2399]
Techsonic Industries Inc, p. II-2035 [SIC 3812]

Roman numerals I and II indicate the volume in which pages appear. Volume I holds SICs 2011 through 3299; Volume II holds SICs 3312 through 3999.

2302

Roman numerals I and II indicate the volume in which pages appear. Volume I holds SICs 2011 through 3299; Volume II holds SICs 3312 through 3999.

Company Index

Thiele Industries Inc, p. II-1948 [SIC 3713]
Thiesing Veneer Company Inc, p. I-491 [SIC 2435]
Thilmany, p. I-612 [SIC 2621]
Thinking Machines Corp, p. II-1688 [SIC 3571]
Thiokol Corp, pp. Vol II - 1434, 2011 [SICs 3483, 3764]
This Week Publications Inc, p. I-713 [SIC 2741]
Tho-Ro Products Inc, p. I-245 [SIC 2241]
Thomas A Schutz Co, p. II-2186 [SIC 3993]
Thomas and Betts Corp, p. II-1910 [SIC 3678]
Thomas and Skinner Inc, p. II-1492 [SIC 3499]
Thomas Built Buses Inc, p. II-1948 [SIC 3713]
Thomas C Wilson Inc, p. II-1584 [SIC 3546]
Thomas Concrete Inc, p. I-1127 [SIC 3273]
Thomas Conveyor Company Inc, p. II-1541 [SIC 3535]
Thomas Creative Apparel Inc, p. I-420 [SIC 2389]
Thomas Die and Stamping Inc, p. II-1404 [SIC 3465]
Thomas Electronics Inc, p. II-1883 [SIC 3671]
Thomas Engineering Inc, p. II-1682 [SIC 3569]
Thomas Industries Inc, p. II-1649 [SIC 3563]
Thomas J Lipton Co, p. I-55 [SIC 2035]
Thomas L Green and Company, p. II-1626 [SIC 3556]
Thomas Nelson Inc, p. I-703 [SIC 2731]
Thomas Publishing Co, p. I-713 [SIC 2741]
Thomas Steel Strip Corp, p. II-1187 [SIC 3316]
Thomas Tape Co, p. I-654 [SIC 2672]
Thomas Taylor and Sons Inc, p. I-245 [SIC 2241]
Thomaston Mills Inc, p. I-231 [SIC 2211]
Thompson and Formby, p. I-840 [SIC 2851]
Thompson Casting Co, p. II-1275 [SIC 3365]
Thompson Creek Metals Co, p. II-1178 [SIC 3313]
Thompson Maple Products Inc, p. I-536 [SIC 2499]
Thompson-McCully Co, p. I-909 [SIC 2951]
Thompson Pipe and Steel Co, p. II-1486 [SIC 3498]
Thompson's Pet Pasta Products, p. I-91 [SIC 2047]
Thompson Steel Company Inc, p. II-1187 [SIC 3316]
Thomson Co, p. I-348 [SIC 2325]
Thomson Components, p. II-1883 [SIC 3671]
Thomson Consumer Electronics, p. II-1859 [SIC 3651]
Thomson Corporation, p. II-1938 [SIC 3699]
Thomson National Press Co, p. II-1615 [SIC 3554]
Thomson Oak Flooring Co, p. I-471 [SIC 2426]
Thomson Packing Company Inc, p. I-2 [SIC 2011]
Thomson Precision Ball, p. II-1644 [SIC 3562]
Thomson Saginaw Ball Screw, p. II-1747 [SIC 3593]
Thomson-Shore Inc, p. I-708 [SIC 2732]
Thomson Training & Simulation, p. II-1938 [SIC 3699]
Thona Corp, p. I-948 [SIC 3061]
Thonet Industries Inc, p. I-571 [SIC 2521]
Thor Industries Inc, p. II-2020 [SIC 3792]
Thor-Lo Inc, p. I-254 [SIC 2252]
Thore Inc, p. II-1775 [SIC 3613]
Thorn Apple Valley Inc, p. I-2 [SIC 2011]
Thorn Automated Systems Inc, p. II-1878 [SIC 3669]
Thornton Winery, p. I-168 [SIC 2084]
Thoro-Packaging, p. I-643 [SIC 2657]
Thorpe Technologies Inc, p. II-1671 [SIC 3567]
Threaded Rod Company Inc, p. II-1563 [SIC 3542]
Three-D Investment Inc, p. II-1430 [SIC 3482]
Three-Five Systems Inc, p. II-1892 [SIC 3674]
Three Rivers Aluminum Co, p. II-1349 [SIC 3442]
Three Sixty Services Inc, p. II-1938 [SIC 3699]
Three Weavers Inc, p. I-428 [SIC 2392]
Thrifty Oil Co, p. I-903 [SIC 2911]
THT Inc, p. I-688 [SIC 2679]
Thunder Pallet Inc, p. I-510 [SIC 2448]
Thunderbird Products, p. II-1992 [SIC 3732]
Thunderline Corp, p. I-948 [SIC 3061]
Thurman Manufacturing Co, p. II-1759 [SIC 3596]
Thurman Scale Co, p. II-1759 [SIC 3596]
Thurmont Shoe Co, p. I-1025 [SIC 3143]
Thurner Heat Treating Corp, p. II-1290 [SIC 3398]
Thurston Sails Inc, p. I-437 [SIC 2394]
TI-Brook Inc, p. II-1961 [SIC 3715]
TI Holdings Inc, p. II-1868 [SIC 3661]
Tiara Corp, p. II-2122 [SIC 3911]
Tibor Machine Products, p. II-1764 [SIC 3599]
TIC United Corp, p. II-1509 [SIC 3523]
TIDI Products Inc, p. I-654 [SIC 2672]

TIE communications Inc, p. II-1868 [SIC 3661]
Tie Down Engineering Inc, p. II-1395 [SIC 3462]
Tiff and Griff Designs Ltd, p. I-365 [SIC 2335]
Tiffany Furniture Industries Inc, p. I-558 [SIC 2515]
Tiffany Industries Inc, p. I-571 [SIC 2521]
Tiffen Manufacturing Corp, p. II-2112 [SIC 3861]
Tighe Industries Inc, p. I-420 [SIC 2389]
TII Industries Inc, p. II-1769 [SIC 3612]
Tilcon Maine Inc, p. I-909 [SIC 2951]
Tile Helper Inc, p. I-1086 [SIC 3253]
TileCera Inc, p. I-1086 [SIC 3253]
TiLine Inc, p. II-1285 [SIC 3369]
Tillamook Cty Creamery, p. I-35 [SIC 2026]
Tillamook Lumber Co, p. I-465 [SIC 2421]
Tillotson Healthcare Corp, p. I-954 [SIC 3069]
Tilton Truss Manufacturers Inc, p. I-501 [SIC 2439]
Timber Crest Farms, p. I-51 [SIC 2034]
Timber Products Co, p. I-496 [SIC 2436]
Timber Products Co Medford, p. I-491 [SIC 2435]
Timberland Co, p. I-1025 [SIC 3143]
Timberland Homes Inc, p. I-523 [SIC 2452]
Timco Manufacturing Company, p. II-1404 [SIC 3465]
Time Inc, p. I-698 [SIC 2721]
Time Products Inc, p. I-819 [SIC 2841]
Time To Market Associates Inc, p. II-2164 [SIC 3953]
TimeMed Labeling Systems Inc, p. I-654 [SIC 2672]
Times Fiber Communications, p. II-1260 [SIC 3357]
Times Mirror, p. I-708 [SIC 2732]
Times Mirror Co, p. I-693 [SIC 2711]
Times Publishing Co, p. I-693 [SIC 2711]
Timesavers Inc, p. II-1610 [SIC 3553]
Timeter Instrument Corp, p. II-2102 [SIC 3845]
Timex Corp, p. II-2118 [SIC 3873]
Timken Co, p. II-1644 [SIC 3562]
Tingley Rubber Corp, p. I-932 [SIC 3021]
Tingstol Co, p. II-1887 [SIC 3672]
Tinsley Laboratories Inc, p. II-2071 [SIC 3827]
Tionesta Sand and Gravel Inc, p. I-1154 [SIC 3295]
Tip Top Poultry Inc, p. I-12 [SIC 2015]
Tippecanoe Laboratories, p. I-804 [SIC 2834]
Tippins Inc, p. II-1589 [SIC 3547]
Tipton Box Company Inc, p. I-515 [SIC 2449]
Tiro Industries Inc, p. I-834 [SIC 2844]
Tishcon Corp, p. I-799 [SIC 2833]
Tishken Products Co, p. II-1563 [SIC 3542]
Titan Corp, p. II-1688 [SIC 3571]
Titan Specialties Inc, p. II-1531 [SIC 3533]
Titanium Metals Corp, p. II-1255 [SIC 3356]
Titmus Optical Inc, p. II-2107 [SIC 3851]
Tivolie Fashions Inc, p. I-413 [SIC 2386]
TIW Corp, p. II-1447 [SIC 3491]
TJ Cope Inc, p. II-1835 [SIC 3644]
TJ International, p. I-501 [SIC 2439]
TJ International Inc, p. I-501 [SIC 2439]
TJFC Holding Co, p. I-336 [SIC 2321]
TKC Inc, p. II-2107 [SIC 3851]
TKM Specialty Fasteners Inc, p. II-1182 [SIC 3315]
TL Systems Corp, p. II-1660 [SIC 3565]
TLK Industries Inc, p. I-327 [SIC 2299]
TM Analytic Inc, p. II-2056 [SIC 3824]
TM Athletics Corp, p. I-258 [SIC 2253]
TM Digital Solutions Inc, p. II-1704 [SIC 3577]
TM Morris Manufacturing Co, p. II-1929 [SIC 3694]
TMC Group Inc, p. I-745 [SIC 2782]
TMI Industries, p. II-1764 [SIC 3599]
Tnemec Company Inc, p. I-840 [SIC 2851]
TNI Inc, p. II-1905 [SIC 3677]
TO Dey Service Corp, p. I-1025 [SIC 3143]
Toastmaster Inc, p. II-1813 [SIC 3634]
Tobacco Processors Inc, p. I-228 [SIC 2141]
Tobias Guitars Inc, p. II-2135 [SIC 3931]
Tobin-Hamilton Company Inc, p. I-1033 [SIC 3149]
Tocco Inc, p. II-1671 [SIC 3567]
Toccoa Casket Co, p. II-2191 [SIC 3995]
Todd Seafoods Inc, p. I-190 [SIC 2092]
Todd Shipyards Corp, p. II-1986 [SIC 3731]
Todd Uniform Inc, p. I-352 [SIC 2326]
Toddco General Inc, p. II-1594 [SIC 3548]
Todhunter International Inc, p. I-172 [SIC 2085]

Tokheim Corp, p. II-1732 [SIC 3586]
Tokico, p. II-1954 [SIC 3714]
Toko America Inc, p. II-1873 [SIC 3663]
Tol-O-Matic Inc, p. II-1747 [SIC 3593]
Tolaram Fibers Inc, p. I-322 [SIC 2298]
Tolas Health Care, p. I-649 [SIC 2671]
Toledo Metals Furniture, p. I-576 [SIC 2522]
Toledo Technologies, p. II-1404 [SIC 3465]
Toll-Gate Garment Co, p. I-336 [SIC 2321]
Tolleson Lumber Company Inc, p. I-528 [SIC 2491]
Tollycraft Corp, p. II-1986 [SIC 3731]
Toluca Garment Co, p. I-332 [SIC 2311]
Tom McClain Co, p. I-30 [SIC 2024]
Tom Smith Industries Inc, p. I-782 [SIC 2821]
Tom Sturgis Pretzels Inc, p. I-107 [SIC 2052]
Tom Thumb Glove Co, p. I-401 [SIC 2381]
Tom Togs Inc, p. I-373 [SIC 2339]
Tomar Electronics Inc, p. II-1826 [SIC 3641]
Tomco Auto Products Inc, p. II-1742 [SIC 3592]
Tomen-Ein Inc, p. I-310 [SIC 2295]
Tomkins Corp, p. II-1438 [SIC 3484]
Tomkins Industries Inc, pp. Vol I - 480, 979, Vol. II - 1349 [SICs 2431, 3084, 3442]
Tomlinson Industries, p. II-1626 [SIC 3556]
Tommy Armour Golf Co, p. II-2150 [SIC 3949]
Tommy's Foods Inc, p. I-65 [SIC 2038]
Tomoegawa, p. II-2112 [SIC 3861]
Ton-Tex Corp, p. I-936 [SIC 3052]
Tone Brothers Inc, p. I-212 [SIC 2099]
Tony Downs Foods Co, p. I-12 [SIC 2015]
Tony Lama Co, p. I-1025 [SIC 3143]
Tony Lambert Design Group Inc, p. I-258 [SIC 2253]
Tool Die Engineering Inc, p. II-1266 [SIC 3363]
Tool Products, p. II-1266 [SIC 3363]
Tooling Systems, p. II-1578 [SIC 3545]
Tools For Bending Inc, p. II-1563 [SIC 3542]
Tootsie Roll Industries Inc, p. I-127 [SIC 2064]
Top Catch Inc, p. I-190 [SIC 2092]
Top Die Casting Co, p. II-1266 [SIC 3363]
Top Flight Inc, p. I-684 [SIC 2678]
Top Flight Paper Products, p. I-684 [SIC 2678]
Top Manufacturing Company, p. I-973 [SIC 3083]
TOP Tobacco LP, p. I-224 [SIC 2131]
Topcon America Corp, p. II-2076 [SIC 3829]
Topflite, p. I-927 [SIC 3011]
TopoMetrix Corp, p. II-2066 [SIC 3826]
Toppan West Inc, p. II-1887 [SIC 3672]
Topps Company Inc, p. I-729 [SIC 2759]
Topps Manufacturing Co, p. I-352 [SIC 2326]
TOPS Business Forms, p. I-735 [SIC 2761]
Tops Manufacturing Company, p. II-1813 [SIC 3634]
Topstone Industries Inc, p. II-2144 [SIC 3944]
Toray Industries, p. I-318 [SIC 2297]
Torco Inc, p. II-1382 [SIC 3451]
Toro Co, pp. Vol II - 1514, 1840 [SICs 3524, 3645]
Toroid Corp, p. II-1759 [SIC 3596]
Torpedo Wire and Strip Co, p. II-1476 [SIC 3496]
Torque Engineering Corp, p. II-1503 [SIC 3519]
Torrey Investments Inc, p. II-2061 [SIC 3825]
Torrington Co, p. II-1644 [SIC 3562]
Torstenson Glass Co, p. I-1057 [SIC 3211]
Tosca Ltd, p. I-506 [SIC 2441]
Tosco Corp, p. I-903 [SIC 2911]
Tosco Refining Co, p. I-903 [SIC 2911]
Toshiba America Info Systems, pp. Vol II - 1688, 1713 [SICs 3571, 3579]
Toshiba International Corp, pp. Vol II - 1498, 1780 [SICs 3511, 3621]
Tosoh SMD Inc, p. II-1892 [SIC 3674]
Total Containment Inc, p. I-979 [SIC 3084]
Total Energy Services Inc, p. II-1536 [SIC 3534]
Total Performance Inc, p. II-1943 [SIC 3711]
Total Petroleum, p. I-903 [SIC 2911]
Tote Cart Co, p. II-1476 [SIC 3496]
Touch Books Inc, p. I-708 [SIC 2732]
Touchstone Inc, p. II-1998 [SIC 3743]
Tough Traveler Inc, p. I-433 [SIC 2393]
Tower Asphalt Inc, p. I-909 [SIC 2951]
Tower Automotive Inc, p. II-1404 [SIC 3465]

Roman numerals I and II indicate the volume in which pages appear. Volume I holds SICs 2011 through 3299; Volume II holds SICs 3312 through 3999.

Tower Extrusions Ltd, p. II-1244 [SIC 3354]
Tower Media Inc, p. I-693 [SIC 2711]
Tower Mills Inc, p. I-250 [SIC 2251]
Towle Manufacturing Co, p. II-2127 [SIC 3914]
Town and Country Cedar Homes, p. I-523 [SIC 2452]
Town and Country Corp, p. II-2122 [SIC 3911]
Town Food Service Equip Co, p. II-1800 [SIC 3631]
Town Talk Cap Mfg Co, p. I-386 [SIC 2353]
Townsends Inc, p. I-12 [SIC 2015]
Toy Biz Inc, p. II-2144 [SIC 3944]
Toyo Seat USA Corp, p. II-1492 [SIC 3499]
Toyoda Machinery USA Inc, p. II-1558 [SIC 3541]
Toyoshima Indiana Inc, p. II-1460 [SIC 3493]
Toyota Motor Mfg USA, p. II-1943 [SIC 3711]
TPA Inc, p. I-903 [SIC 2911]
TPD California Trailers Inc, p. II-2020 [SIC 3792]
TPH Graphics Inc, p. I-754 [SIC 2791]
TPI Corp, p. II-1813 [SIC 3634]
TR Metals Corp, p. I-1154 [SIC 3295]
TR Miller Mill Company Inc, p. I-465 [SIC 2421]
Trachte Building Systems Inc, p. II-1372 [SIC 3448]
Tracker Marine Corp, p. II-1992 [SIC 3732]
Trackmobile Inc, p. II-1998 [SIC 3743]
Tracor Inc, p. II-2035 [SIC 3812]
Tracy-Luckey Company Inc, p. I-212 [SIC 2099]
Trade Press Typography, p. I-754 [SIC 2791]
Trade Service Corp, p. I-703 [SIC 2731]
Trader Publishing Co, p. I-698 [SIC 2721]
Tradewinds Outdoor, p. I-553 [SIC 2514]
Trafalgar Ltd, p. I-416 [SIC 2387]
Trail-Rite Inc, p. II-2029 [SIC 3799]
Trailmaster Tanks Inc, p. II-1961 [SIC 3715]
Trailmobile Inc, p. II-1961 [SIC 3715]
TRAK International Inc, p. II-1519 [SIC 3531]
Trans-Apparel Group, p. I-348 [SIC 2325]
Trans-Lux Corp, p. II-2186 [SIC 3993]
Trans-Matic Manufacturing Co, p. II-1413 [SIC 3469]
Trans-Rim Enterprises, p. I-608 [SIC 2611]
Trans World Marketing, p. II-2186 [SIC 3993]
Trans World Textile Corp, p. I-390 [SIC 2361]
Transco Products Co, p. II-1864 [SIC 3652]
Transcolor Corp, p. I-288 [SIC 2269]
Transcolor East Inc, p. II-1605 [SIC 3552]
Transcolor West Inc, p. I-446 [SIC 2396]
Transcraft Corp, p. II-1961 [SIC 3715]
TransDigm Corp, p. II-1644 [SIC 3562]
Transelco, p. I-777 [SIC 2819]
Transfer Print Foils Inc, p. II-1482 [SIC 3497]
Transformer Manufacturers Inc, p. II-1769 [SIC 3612]
Transhumance Corp, p. I-2 [SIC 2011]
Transilwrap Company Inc, p. I-961 [SIC 3081]
Transitional Technology Inc, p. II-1694 [SIC 3572]
Transkrit Corp, p. I-735 [SIC 2761]
TransLogic Corp, p. II-1541 [SIC 3535]
Transmet Corp, p. II-1229 [SIC 3341]
Transpo Electronics Inc, p. II-1929 [SIC 3694]
Transportation Mfg Corp, p. II-1943 [SIC 3711]
TransTechnology Corp, p. II-1552 [SIC 3537]
TransTechnology Syst & Services, p. II-1699 [SIC 3575]
Transwall Corp, p. I-576 [SIC 2522]
TransWestern Publishing LP, p. I-713 [SIC 2741]
Tranter Inc, p. II-1355 [SIC 3443]
Tranter Industries Inc, p. I-510 [SIC 2448]
Tranzonic Cos, p. I-674 [SIC 2676]
Trask River Lumber Co, p. I-496 [SIC 2436]
Traub Container Corp, p. I-628 [SIC 2653]
Travel Leather Company Inc, p. I-1013 [SIC 3111]
Travis Meats Inc, p. I-2 [SIC 2011]
Treage Ltd, p. I-365 [SIC 2335]
Treasure Chest Advertising, p. I-718 [SIC 2752]
Treasure Craft Co, p. I-1113 [SIC 3269]
Treat Ice Cream Co, p. I-30 [SIC 2024]
Trece Corp, p. I-874 [SIC 2879]
Tredegar Aluminum, p. II-1229 [SIC 3341]
Tree Saver Inc, p. I-433 [SIC 2393]
Tree Top Inc, p. I-60 [SIC 2037]
Treen Box and Pallet Corp, p. I-510 [SIC 2448]
Trefethen Vineyards Winery Inc, p. I-168 [SIC 2084]

Trek Bicycle Corp, p. II-2003 [SIC 3751]
Tremont Corp, p. II-1255 [SIC 3356]
Tremont Nail, p. II-1182 [SIC 3315]
Trencor Jetco Inc, p. II-1519 [SIC 3531]
Trend Offset Printing Services, p. I-718 [SIC 2752]
Trendar Corp, p. II-1709 [SIC 3578]
Trendex Inc, p. I-745 [SIC 2782]
Trendlines Inc, p. I-486 [SIC 2434]
Trends of Hawaii, p. I-424 [SIC 2391]
Trendway Corp, p. I-571 [SIC 2521]
Trent Tube, p. II-1192 [SIC 3317]
Trenwyth Industries Inc, p. I-1117 [SIC 3271]
Tri-Bio Laboratories Inc, p. I-814 [SIC 2836]
Tri-Glas Inc, p. II-2020 [SIC 3792]
Tri/Mark Corp, p. II-1323 [SIC 3429]
Tri-Mark Metal Corp, p. II-1361 [SIC 3444]
Tri-Pack, p. I-628 [SIC 2653]
Tri-Pak Machinery Inc, p. II-1660 [SIC 3565]
Tri-Star Electronics International, p. II-1830 [SIC 3643]
Tri-Sum Potato Chip Company, p. I-199 [SIC 2096]
Tri-Tec Engineering Inc, p. II-1260 [SIC 3357]
Tri Tech Laboratories Inc, p. I-834 [SIC 2844]
Tri Valley Growers, p. I-45 [SIC 2033]
TriAm Inc, p. II-1699 [SIC 3575]
Triana Industries Inc, p. II-1563 [SIC 3542]
Triangle Brick Co, p. I-1081 [SIC 3251]
Triangle Dies and Supplies Inc, p. I-850 [SIC 2865]
Triangle Lingerie Corp, p. I-378 [SIC 2341]
Triangle Pacific Corp, p. I-471 [SIC 2426]
Triangle Tool Corp, p. II-1572 [SIC 3544]
Triangle Wire and Cable Inc, p. II-1260 [SIC 3357]
Triarc Companies Inc, p. I-231 [SIC 2211]
Triax-Davis Inc, p. II-1998 [SIC 3743]
Tribol Inc, p. I-918 [SIC 2992]
Tribotech, p. II-1795 [SIC 3629]
Tribune Co, p. I-693 [SIC 2711]
Tribune Publishing Co, p. I-693 [SIC 2711]
Tricord Systems Inc, p. II-1704 [SIC 3577]
Trident Inc, p. II-1621 [SIC 3555]
Trident Microsystems Inc, p. II-1704 [SIC 3577]
Trident Seafoods Corp, p. I-186 [SIC 2091]
Trig Inc, p. II-2045 [SIC 3822]
Trilectron Industries Inc, p. II-1975 [SIC 3724]
Trilogy Communications Inc, p. II-1260 [SIC 3357]
Trilogy Plastics Inc, p. I-787 [SIC 2822]
Trim-Line Foundations, p. I-382 [SIC 2342]
Trim Trends Inc, p. II-1404 [SIC 3465]
TriMas Corp, p. II-2029 [SIC 3799]
Trimble Navigation Ltd, p. II-2035 [SIC 3812]
Trimco Mfg & Eng, p. I-1041 [SIC 3161]
Trimedyne Inc, p. II-2102 [SIC 3845]
Trimfit Inc, p. I-250 [SIC 2251]
Trimfoot Co, p. I-1033 [SIC 3149]
Trimtex Company Inc, p. I-245 [SIC 2241]
Trina Inc, p. I-1045 [SIC 3171]
Trinacria Specialty Mfg, p. II-1470 [SIC 3495]
Trine Products Co, p. II-1769 [SIC 3612]
Triner Scale and Manufacturing, p. II-1759 [SIC 3596]
Trinetics Inc, p. II-1905 [SIC 3677]
Trinitas Corp, p. I-70 [SIC 2041]
Trinity Furniture Inc, p. I-496 [SIC 2436]
Trinity Indust Moss Point Marine, p. II-1986 [SIC 3731]
Trinity Industries Inc, pp. Vol II - 1344, 1998 [SICs 3441, 3743]
Trinity Marine Group, p. II-1986 [SIC 3731]
Trinity Products Inc, p. II-2144 [SIC 3944]
TRINOVA Corp, p. II-1753 [SIC 3594]
Trintex Corp, p. I-927 [SIC 3011]
Trio Aviation Inc, p. II-1929 [SIC 3694]
Trio Dyeing and Finishing, p. I-283 [SIC 2262]
Trio Manufacturing Co, p. I-297 [SIC 2281]
Trio-Tech International, p. II-2061 [SIC 3825]
Trion Inc, p. II-1654 [SIC 3564]
Triple A Tube Inc, p. II-1192 [SIC 3317]
Triple F Inc, p. I-95 [SIC 2048]
Tripp Lite, p. II-1795 [SIC 3629]
Triumph Industries, p. II-1425 [SIC 3479]

Triumph of California Inc, p. I-336 [SIC 2321]
Triumph Twist Drill Co, p. II-1578 [SIC 3545]
Trojan Corp, p. I-885 [SIC 2892]
Trojan Inc, p. II-1826 [SIC 3641]
Trojan Tube Company Inc, p. II-1234 [SIC 3351]
Trompeter Electronics Inc, p. II-1910 [SIC 3678]
Tronex Chemical Corp, p. I-819 [SIC 2841]
Trooper Inc, p. I-332 [SIC 2311]
Trophy Radiology Inc, p. II-2098 [SIC 3844]
Tropical Paper Box Co, p. I-643 [SIC 2657]
Tropical Preserving Co, p. I-45 [SIC 2033]
Tropitone Furniture Company, p. I-567 [SIC 2519]
Trouvailles Inc, p. I-603 [SIC 2599]
Troxel Co, p. II-1344 [SIC 3441]
Troxel Co Cycling and Fitness, p. II-2150 [SIC 3949]
Troy, p. II-1621 [SIC 3555]
Troy Corp, p. I-874 [SIC 2879]
Troy Mattress Company Inc, p. I-558 [SIC 2515]
Troy Mills Inc, p. I-327 [SIC 2299]
Troy Wood Products Inc, p. I-510 [SIC 2448]
Troydon Hosiery Mills Inc, p. I-254 [SIC 2252]
Troyer Potato Products Inc, p. I-199 [SIC 2096]
Tru Die Cast Corp, p. II-1271 [SIC 3364]
Tru-Form Tool & Mfg Indust, p. II-1470 [SIC 3495]
Tru-Link Fence Co, p. II-1367 [SIC 3446]
Tru-Stone Corp, p. I-1140 [SIC 3281]
Truax Engineering Inc, p. II-2016 [SIC 3769]
Truck Cab Manufacturers Inc, p. II-1948 [SIC 3713]
Truck Components Inc, p. II-1954 [SIC 3714]
Truck Equipment Service, p. II-1961 [SIC 3715]
Truck-Lite Company Inc, p. II-1850 [SIC 3647]
Truco Inc, p. I-914 [SIC 2952]
True Fitness Technology Inc, p. II-2150 [SIC 3949]
True Temper Hardware Inc, p. II-1514 [SIC 3524]
Truing Systems Inc, p. I-1144 [SIC 3291]
Truitt Brothers Inc, p. I-45 [SIC 2033]
Truly Yours Inc, p. I-369 [SIC 2337]
Trumpf Inc, p. II-1563 [SIC 3542]
Trusco Tank Inc, p. II-1355 [SIC 3443]
Truss-Com Co, p. I-501 [SIC 2439]
Truss Manufacturing Company, p. I-501 [SIC 2439]
Truss-Span Corp, p. I-501 [SIC 2439]
Truss-T Structures Inc, p. II-1372 [SIC 3448]
Trussway Inc, p. I-501 [SIC 2439]
Truth Hardware Corp, p. II-1323 [SIC 3429]
TRW Inc, pp. Vol II - 1868, 1954 [SICs 3661, 3714]
TRW Steering & Suspension Syst, p. II-1954 [SIC 3714]
TRW Technar Inc, p. II-2061 [SIC 3825]
TRW Vehicle Safety Systems Inc, p. I-455 [SIC 2399]
TS Trim Industries Inc, p. I-446 [SIC 2396]
TSC, p. II-1938 [SIC 3699]
Tseng Labs Inc, p. II-1892 [SIC 3674]
TSI Center for Diagnostic Prod, p. I-809 [SIC 2835]
TSI Inc, p. II-2051 [SIC 3823]
TST/Impreso Inc, p. I-612 [SIC 2621]
TST Inc, p. II-1221 [SIC 3334]
TTX Co, p. II-1998 [SIC 3743]
Tubbs Rope Works Inc, p. I-322 [SIC 2298]
Tube-Alloy Corp, p. II-1212 [SIC 3325]
Tube Forgings of America Inc, p. II-1464 [SIC 3494]
Tubed Products Inc, p. I-967 [SIC 3082]
Tubetech Inc, p. II-1192 [SIC 3317]
Tubetronics, p. II-1600 [SIC 3549]
Tuboscope Vetco Intern Corp, p. II-1531 [SIC 3533]
Tubular Specialties Mfg, p. II-1377 [SIC 3449]
Tubular Textile Machinery, p. II-1605 [SIC 3552]
Tufco Technologies Inc, p. I-612 [SIC 2621]
Tuffaloy Products Inc, p. II-1594 [SIC 3548]
Tuftco Corp, p. II-1605 [SIC 3552]
Tuftco Finishing System Inc, p. II-1605 [SIC 3552]
Tuftex Carpet Mills Inc, p. I-292 [SIC 2273]
Tug Manufacturing Corp, p. II-1552 [SIC 3537]
Tug River Armature, p. II-1525 [SIC 3532]
Tule River Cooperative Dryer, p. I-51 [SIC 2034]
Tulip Corp, p. II-1563 [SIC 3542]
Tulkoff Products Company Inc, p. I-55 [SIC 2035]
Tulsa Fittings Inc, p. II-1207 [SIC 3324]
Tulsa Tube Bending Co, p. II-1486 [SIC 3498]

Roman numerals I and II indicate the volume in which pages appear. Volume I holds SICs 2011 through 3299; Volume II holds SICs 3312 through 3999.

2305

Tultex Corp, p. I-258 [SIC 2253]
Tumac Industries Inc, p. II-1509 [SIC 3523]
Tumi Luggage Inc, p. I-1041 [SIC 3161]
Tumwater Lumber Co, p. I-496 [SIC 2436]
Tuohy Furniture Corp, p. I-581 [SIC 2531]
Tupperware, p. I-1006 [SIC 3089]
Tur-bo Jet Products Company, p. II-1905 [SIC 3677]
Turbo Power&Marine Systems, p. II-1498 [SIC 3511]
Turbomeca Engine Corp, p. II-1975 [SIC 3724]
Turbonetics Energy Inc, p. II-1498 [SIC 3511]
Turck Inc, p. II-1790 [SIC 3625]
Turco Products Inc, p. I-897 [SIC 2899]
Turnbull Cone Baking Co, p. I-107 [SIC 2052]
Turner & Seymour Mfg Co, p. II-1476 [SIC 3496]
Turner Bellows Inc, p. II-2112 [SIC 3861]
Turner-Cooper Hand Tools, p. II-1312 [SIC 3423]
Turner Dairies Inc, p. I-35 [SIC 2026]
Turner, Day, p. I-536 [SIC 2499]
Turner Electric Corp, p. II-1775 [SIC 3613]
Turtle Island Herbs Inc, p. I-799 [SIC 2833]
Turtle Wax Inc, p. I-825 [SIC 2842]
Tuscaloosa Steel Corp, p. II-1187 [SIC 3316]
Tuscan Dairy Farms Inc, p. I-35 [SIC 2026]
Tuscarora Inc, pp. Vol. I - 782, 991 [SICs 2821, 3086]
Tuscarora Yarns Inc, p. I-297 [SIC 2281]
Tusco Display Co, p. II-2186 [SIC 3993]
Tutco Inc, p. II-1813 [SIC 3634]
Tuthill Corp, p. II-1638 [SIC 3561]
TV/COM International, p. II-1873 [SIC 3663]
TV Host Inc, p. I-713 [SIC 2741]
Tweave Inc, p. I-236 [SIC 2221]
Twentieth Century Machine Co, p. II-1644 [SIC 3562]
Twin City Die Castings Co, p. II-1266 [SIC 3363]
Twin City Fan and Blower Co, p. II-1654 [SIC 3564]
Twin City Foods Inc, p. I-60 [SIC 2037]
Twin City Knitting Co, p. I-254 [SIC 2252]
Twin City Wire-MFI Inc, p. II-1476 [SIC 3496]
Twin Disc Inc, p. II-1666 [SIC 3566]
Twinhead Corp, p. II-1688 [SIC 3571]
Twinpoint Inc, p. II-1901 [SIC 3676]
Twist Inc, p. II-1470 [SIC 3495]
TWP Inc, p. II-1182 [SIC 3315]
TWS Industries Inc, p. I-825 [SIC 2842]
TXT Inc, p. II-1464 [SIC 3494]
Tyca Corp, p. I-279 [SIC 2261]
Tyco International Ltd, p. II-1682 [SIC 3569]
Tyco Playtime Inc, p. II-2144 [SIC 3944]
Tyco Toys Inc, p. II-2144 [SIC 3944]
Tygart Moulding Corp, p. I-536 [SIC 2499]
TYK Refractories Co, p. I-1163 [SIC 3297]
TyKel Inc, p. II-1295 [SIC 3399]
Tylan General Inc, p. II-2051 [SIC 3823]
Tyler Building Systems Co, p. II-1372 [SIC 3448]
Tyler Corp, p. II-1197 [SIC 3321]
Tyler LP, p. II-1509 [SIC 3523]
Tyler Machinery Company Inc, p. II-1610 [SIC 3553]
Tyler Pipe Industry Inc, p. II-1486 [SIC 3498]
Tyler Refrigeration Corp, p. II-1726 [SIC 3585]
TyLink Corp, p. II-1878 [SIC 3669]
Tylok International Inc, p. II-1454 [SIC 3492]
Typesetting Inc, p. I-754 [SIC 2791]
Typography Plus Inc, p. I-754 [SIC 2791]
Tyson Foods Inc, p. I-12 [SIC 2015]
Tyson's of Rogers, p. I-12 [SIC 2015]
Tysons in Nashville, p. I-12 [SIC 2015]

U S WEST Marketing, p. I-713 [SIC 2741]
UBC Inc, p. I-501 [SIC 2439]
Uber Glove Co, p. I-1013 [SIC 3111]
UC Milk Company Inc, p. I-35 [SIC 2026]
UCAR Carbon Company Inc, p. I-942 [SIC 3053]
UCO Inc, p. II-1244 [SIC 3354]
UEC Equipment Co, p. II-1519 [SIC 3531]
UEC Industries Inc, p. II-1519 [SIC 3531]
UFP Technologies Inc, p. I-991 [SIC 3086]
Ugg Holdings Inc, p. I-1029 [SIC 3144]
UGIMAG Inc, p. II-1492 [SIC 3499]
UIP Engineered Products, p. II-1764 [SIC 3599]
Ujena Inc, p. I-373 [SIC 2339]

UKI Yarns Inc, p. I-297 [SIC 2281]
Ullenberg Corp, p. II-2122 [SIC 3911]
Ultimate Precision Inc, p. II-1764 [SIC 3599]
Ultimate Products Inc, p. I-258 [SIC 2253]
Ultimate Support Systems Inc, p. II-2135 [SIC 3931]
Ultimate Technology Corp, p. II-1699 [SIC 3575]
Ultra Clean Techn Syst & Service, p. II-1883 [SIC 3671]
Ultra Light Arms Inc, p. II-1438 [SIC 3484]
Ultrak Inc, p. II-1878 [SIC 3669]
Ultralife Batteries Inc, p. II-1925 [SIC 3692]
Ultramar Corp, p. I-903 [SIC 2911]
Ultramar Inc, p. I-903 [SIC 2911]
Ultramet, p. II-1229 [SIC 3341]
Ultrasonic Seal Co, p. II-1594 [SIC 3548]
UltraStor Corp, p. II-1694 [SIC 3572]
Ultronix Inc, p. II-1901 [SIC 3676]
UMBRA USA Inc, p. I-603 [SIC 2599]
Umpqua Dairy Products Co, p. I-30 [SIC 2024]
Unaflex Inc, p. I-936 [SIC 3052]
Unaka Company Inc, p. II-1800 [SIC 3631]
Unarco Industries Inc, pp. Vol II - 1344, 1492 [SICs 3441, 3499]
UNC All Fab Inc, p. II-1361 [SIC 3444]
UNC Helicopter Inc, p. II-1970 [SIC 3721]
UNC Johnson Technology Inc, p. II-1764 [SIC 3599]
Uncle Dave's Kitchens Inc, p. I-40 [SIC 2032]
Unelko Corp, p. I-897 [SIC 2899]
Ungar, p. II-1312 [SIC 3423]
Uni-Seal Valve Co, p. II-1454 [SIC 3492]
Unibev, p. I-168 [SIC 2084]
UniCargo Group International, p. I-442 [SIC 2395]
Unicell Body Co, p. II-1948 [SIC 3713]
Unichema North America, p. I-855 [SIC 2869]
Unico Inc, pp. Vol. I - 991, Vol. II - 1790 [SICs 3086, 3625]
Unicon Concrete Inc, p. I-1127 [SIC 3273]
Unicorn Concrete Inc, p. I-1127 [SIC 3273]
Unicut Corp, p. II-1318 [SIC 3425]
Unifi Inc, p. I-302 [SIC 2282]
Unifi Spun Yarn Inc, p. I-302 [SIC 2282]
Uniflex Inc, p. I-660 [SIC 2673]
Uniflow Corp, p. II-1395 [SIC 3462]
Unifoil Corp, p. II-1425 [SIC 3479]
UniGrace Inc, p. II-1992 [SIC 3732]
Unilens Corporation USA, p. II-2107 [SIC 3851]
Unimac Company Inc, p. II-1722 [SIC 3582]
Unimast Inc, p. II-1377 [SIC 3449]
Unimetrics Corp, p. II-2066 [SIC 3826]
Union Butterfield Corp, p. II-1558 [SIC 3541]
Union Camp Corp, p. I-612 [SIC 2621]
Union Camp Corp Kansas City, p. I-628 [SIC 2653]
Union Camp Corp Lakeland, p. I-628 [SIC 2653]
Union Carbide Corp, pp. Vol. I - 782, 855 [SICs 2821, 2869]
Union City Body Company Inc, p. II-1948 [SIC 3713]
Union Ice Co, p. I-204 [SIC 2097]
Union Knitting Mills Inc, p. I-263 [SIC 2254]
Union Metal Corp, p. II-1344 [SIC 3441]
Union Oil Mill Inc, p. I-140 [SIC 2074]
Union Spring & Mfg Corp, p. II-1460 [SIC 3493]
Union Standard Equipment Co, p. II-1660 [SIC 3565]
Union Texas Petroleum, p. I-855 [SIC 2869]
Union-Tribune Publishing Co, p. I-693 [SIC 2711]
Union Underwear, p. I-340 [SIC 2322]
Union Wire Rope, p. II-1476 [SIC 3496]
UnionTools Inc, p. II-1312 [SIC 3423]
Uniphase Corp, p. II-2051 [SIC 3823]
Unique Bathing Supplies Inc, p. I-1001 [SIC 3088]
Unique Fabricating Inc, p. I-942 [SIC 3053]
Unique Functional Products, p. II-2029 [SIC 3799]
Uniroyal Chemical Company Inc, p. I-897 [SIC 2899]
Uniroyal Engineered Products, p. I-310 [SIC 2295]
Uniroyal Technology Corp, p. I-787 [SIC 2822]
Unisea Inc, p. I-190 [SIC 2092]
Uniseal Inc, p. I-879 [SIC 2891]
Unisen Inc, p. II-2150 [SIC 3949]
Unistrut International Corp, p. II-1255 [SIC 3356]
Unisys, p. II-1868 [SIC 3661]

Unisys Corp, p. II-1688 [SIC 3571]
Unit Instruments Inc, p. II-2051 [SIC 3823]
Unitech, p. II-1726 [SIC 3585]
United Air Specialists Inc, p. II-1654 [SIC 3564]
United Aluminum Corp, p. II-1239 [SIC 3353]
United Bags Inc, p. I-433 [SIC 2393]
United Bakery Equipment, pp. Vol II - 1626, 1660 [SICs 3556, 3565]
United Barcode Industries Inc, p. II-1713 [SIC 3579]
United Belt of California, p. I-416 [SIC 2387]
United Bindery Service Inc, p. I-684 [SIC 2678]
United Biomedical Inc, p. I-814 [SIC 2836]
United Brands International, p. I-390 [SIC 2361]
UNITED BRASS Manufacturers, p. II-1400 [SIC 3463]
United Brick and Tile Co, p. I-1081 [SIC 3251]
United Catalysts Inc, p. I-777 [SIC 2819]
United Chair Company Inc, p. I-576 [SIC 2522]
United Coatings Inc, p. I-840 [SIC 2851]
United Communications Group, p. I-713 [SIC 2741]
United Conveyor Corp, p. II-1541 [SIC 3535]
United Dairy Farmers Inc, p. I-35 [SIC 2026]
United Dominion, p. II-1361 [SIC 3444]
United Dominion Industries, p. II-1671 [SIC 3567]
United Dominion Industries Inc, p. II-1372 [SIC 3448]
United Elastic Corp, p. I-245 [SIC 2241]
United Electric Co, p. II-1244 [SIC 3354]
United Electric Controls Co, p. II-2076 [SIC 3829]
United Embroidery Inc, p. I-442 [SIC 2395]
United Feeds Inc, p. I-95 [SIC 2048]
United Fixtures Co, p. II-1323 [SIC 3429]
United Gasket Corp, p. II-1312 [SIC 3423]
United Graphics Inc, p. I-708 [SIC 2732]
United Grinding Technologies, p. II-1584 [SIC 3546]
United Hospital Supply Corp, p. II-2040 [SIC 3821]
United Industrial Corp, p. II-2035 [SIC 3812]
United Industries Inc, pp. Vol II - 1192, 1568 [SICs 3317, 3543]
United Laboratories Inc, p. I-897 [SIC 2899]
United Lighting and Ceiling Co, p. II-1854 [SIC 3648]
United McGill Corp, p. II-1361 [SIC 3444]
United Merchants, p. I-258 [SIC 2253]
United Metal Fabricators Inc, p. II-2040 [SIC 3821]
United Musical Instruments, p. II-2135 [SIC 3931]
United Oil Co, p. I-918 [SIC 2992]
United Pioneer Corp, p. I-352 [SIC 2326]
United Plating Inc, p. II-1420 [SIC 3471]
United Premix Concrete, p. I-1127 [SIC 3273]
United Rail Car Manufacturing, p. II-1998 [SIC 3743]
United Refining&Smelting Co, p. II-1225 [SIC 3339]
United Refining Co, p. I-903 [SIC 2911]
United Sciences Inc, p. II-2066 [SIC 3826]
United Screw and Bolt Corp, pp. Vol. I - 1006, Vol. II -1413 [SICs 3089, 3469]
United Seal and Rubber Co, p. I-942 [SIC 3053]
United Shellfish Inc, p. I-190 [SIC 2092]
United Solar Systems Corp, p. II-1339 [SIC 3433]
United Speaker Systems, p. II-1859 [SIC 3651]
United Spinners Corp, p. I-297 [SIC 2281]
United States Aluminum, p. II-1244 [SIC 3354]
United States Aluminum Corp, p. II-1349 [SIC 3442]
United States Alumoweld, p. II-1250 [SIC 3355]
United States Bakery Inc, p. I-101 [SIC 2051]
United States Banknote Corp, p. I-729 [SIC 2759]
United States Can Co, p. II-1300 [SIC 3411]
United States Ceramic Tile Co, p. I-1086 [SIC 3253]
United States Energy Corp, p. II-1929 [SIC 3694]
United States Filter Corp, p. II-1682 [SIC 3569]
United States Gauge, p. II-2051 [SIC 3823]
United States Gypsum Co, p. I-1136 [SIC 3275]
United States Leather Holdings, p. I-1013 [SIC 3111]
United States Lime and Minerals, p. I-1140 [SIC 3281]
United States Pipe & Foundry, p. II-1197 [SIC 3321]
United States Playing Card Co, p. I-724 [SIC 2754]
United States Shoe Corp, p. I-1017 [SIC 3131]
United States Sugar Corp, p. I-116 [SIC 2061]
United States Surgical Corp, p. II-2082 [SIC 3841]
United States Tobacco Co, p. I-224 [SIC 2131]
United Supply Company Inc, p. II-1753 [SIC 3594]

Roman numerals I and II indicate the volume in which pages appear. Volume I holds SICs 2011 through 3299; Volume II holds SICs 3312 through 3999.

Roman numerals I and II indicate the volume in which pages appear. Volume I holds SICs 2011 through 3299; Volume II holds SICs 3312 through 3999.

Company Index

Roman numerals I and II indicate the volume in which pages appear. Volume I holds SICs 2011 through 3299; Volume II holds SICs 3312 through 3999.

Roman numerals I and II indicate the volume in which pages appear. Volume I holds SICs 2011 through 3299; Volume II holds SICs 3312 through 3999.

Company Index

Roman numerals I and II indicate the volume in which pages appear. Volume I holds SICs 2011 through 3299; Volume II holds SICs 3312 through 3999.

Westlake Monomers Corp, p. I-782 [SIC 2821]
Westland Oil Co, p. I-903 [SIC 2911]
Westland Packaging Inc, p. I-665 [SIC 2674]
Westlectric Castings Inc, p. II-1212 [SIC 3325]
Westmont Industries Inc, p. II-1547 [SIC 3536]
Westmoor Manufacturing, p. I-332 [SIC 2311]
Weston Engraving Company Inc, p. I-759 [SIC 2796]
Weston Wear Inc, p. I-361 [SIC 2331]
WestPoint Stevens Inc, p. I-231 [SIC 2211]
Westvaco Corp, pp. Vol I - 612, 618, 643 [SICs 2621, 2631, 2657]
Wetsel-Oviatt Lumber Co, p. I-465 [SIC 2421]
Wex-Tex of Ashford Inc, p. I-340 [SIC 2322]
WEYCO Group Inc, p. I-1025 [SIC 3143]
Weyerhaeuser Co, pp. Vol I - 608, 628 [SICs 2611, 2653]
Weyerhaeuser Paper Co, pp. Vol I - 618, 628 [SICs 2631, 2653]
WFI Corp, p. I-55 [SIC 2035]
WFI Industries Ltd, p. II-1339 [SIC 3433]
WG Block Co, p. I-1122 [SIC 3272]
WH Bagshaw Co, p. II-1605 [SIC 3552]
WH Brady Co, p. I-879 [SIC 2891]
WH Maze Co, p. II-1182 [SIC 3315]
WH Milroy and Company Inc, p. II-1525 [SIC 3532]
WH Porter Inc, p. I-1372 [SIC 3448]
WH Salisbury and Co, p. I-936 [SIC 3052]
Wheatland Tube Co, p. II-1192 [SIC 3317]
Wheatland Waters Inc, p. I-897 [SIC 2899]
Wheatly TXT Corp, p. II-1638 [SIC 3561]
Wheaton Dumont Coop Elevator, p. I-869 [SIC 2875]
Wheaton Plastic Products, p. I-1006 [SIC 3089]
Wheaton Science Products Inc, p. I-1062 [SIC 3221]
Wheeled Coach Industries Inc, p. II-1943 [SIC 3711]
Wheeler Manufacturing Corp, p. II-1676 [SIC 3568]
Wheeling-Pittsburgh Corp, p. II-1187 [SIC 3316]
Wheeling-Pittsburgh Steel Corp, p. II-1187 [SIC 3316]
Whessoe Varec Inc, p. II-1578 [SIC 3545]
Whink Products Co, p. I-825 [SIC 2842]
Whip-Mix Corp, p. II-2093 [SIC 3843]
Whirlpool Corp, pp. Vol II - 1800, 1805, 1809 [SICs 3631, 3632, 3633]
Whirlwind Steel Building Inc, p. II-1372 [SIC 3448]
Whisper Knits Inc, p. I-336 [SIC 2321]
Whistler Corp, p. II-2061 [SIC 3825]
Whitacre-Greer Fireproofing Co, p. I-1090 [SIC 3255]
Whitcraft Corp, p. II-1975 [SIC 3724]
White Aluminum Products Inc, p. II-1361 [SIC 3444]
White Cap Inc, p. I-1006 [SIC 3089]
White Cloud, p. I-195 [SIC 2095]
White Coffee Corp, p. I-195 [SIC 2095]
White Directory Publishers Inc, p. I-713 [SIC 2741]
White Industries LLC, p. II-1726 [SIC 3585]
White Lightning Products Corp, p. I-879 [SIC 2891]
White-Rodgers, p. II-2045 [SIC 3822]
White-Rodgers Inc, p. II-1726 [SIC 3585]
White Way Sign, p. II-2186 [SIC 3993]
White Way Sign Co, p. II-2186 [SIC 3993]
Whitecraft Rattan Inc, p. I-567 [SIC 2519]
Whiteford Packing Co, p. I-60 [SIC 2037]
Whitehall Corp, p. II-2035 [SIC 3812]
Whitehall Furniture Inc, p. I-571 [SIC 2521]
Whitehall Industries, p. II-1244 [SIC 3354]
Whitestone Products Inc, p. I-674 [SIC 2676]
Whiteville Apparel Co, p. I-332 [SIC 2311]
Whitford Corp, p. II-1425 [SIC 3479]
Whitin-Roberts Co, p. II-1605 [SIC 3552]
Whiting Corp, p. II-1547 [SIC 3536]
Whiting Manufacturing Company, p. I-428 [SIC 2392]
Whitinsville Spinning Rings, p. II-1605 [SIC 3552]
Whitley Co, p. I-708 [SIC 2732]
Whitley Products Inc, p. II-1486 [SIC 3498]
Whitman Corp, p. I-176 [SIC 2086]
Whitney Blake of Vermont, p. II-1938 [SIC 3699]
Whitney Yarn, p. I-302 [SIC 2282]
Whittaker Controls Inc, p. II-1980 [SIC 3728]
Whittaker Corp, pp. Vol II - 1925, 2035 [SICs 3692, 3812]
Whittier Wood Products Inc, p. I-542 [SIC 2511]

WHK Inc, p. I-361 [SIC 2331]
Wholesome and Hearty Foods, p. I-65 [SIC 2038]
Wholly Cow Foods, p. I-75 [SIC 2043]
WHX Corp, p. II-1187 [SIC 3316]
Whyco Chromium Company Inc, p. II-1420 [SIC 3471]
Wichita Coca-Cola Bottling Co, p. I-176 [SIC 2086]
Wick Building Systems Inc, p. I-519 [SIC 2451]
Wicks Organ Company Inc, p. II-2135 [SIC 3931]
WICO Corp, p. II-1718 [SIC 3581]
Widder Brothers Inc, p. I-236 [SIC 2221]
Wide-Lite, p. II-1845 [SIC 3646]
Widmer Brewing Co, p. I-160 [SIC 2082]
Widmer's Wine Cellars Inc, p. I-168 [SIC 2084]
Wiegand Industrial, p. II-1671 [SIC 3567]
Wiegardt Brothers Inc, p. I-190 [SIC 2092]
Wieland Inc, p. II-1910 [SIC 3678]
Wiener Laces Inc, p. I-442 [SIC 2395]
Wiggins Connectors, p. II-1323 [SIC 3429]
Wiggins Lift Company Inc, p. II-1552 [SIC 3537]
Wigwam Mills Inc, p. I-254 [SIC 2252]
WIKA Instrument Corp, p. II-2051 [SIC 3823]
Wikoff Color Corp, p. I-889 [SIC 2893]
Wil-Rich Manufacturing, p. II-1509 [SIC 3523]
Wilbanks International Inc, p. I-1167 [SIC 3299]
Wilbert Inc, p. I-1122 [SIC 3272]
Wilbur Chocolate Company Inc, p. I-132 [SIC 2066]
Wilcar Products Inc, p. I-310 [SIC 2295]
Wilcox Electric Inc, p. II-2035 [SIC 3812]
Wild Goose Brewery, p. I-160 [SIC 2082]
Wilde Tool Company Inc, p. II-1584 [SIC 3546]
Wilden Pump & Engineering Co, p. II-1638 [SIC 3561]
Wildwood Lamps Co, p. II-1840 [SIC 3645]
Wilen Manufacturing Company, p. I-825 [SIC 2842]
Wilkata Packaging Corp, p. I-643 [SIC 2657]
Wilkens-Anderson Co, p. II-2040 [SIC 3821]
Wilkerson Corp, p. II-1454 [SIC 3492]
Wilkins, p. II-1447 [SIC 3491]
Wilkins Industries Inc, p. I-373 [SIC 2339]
Wilkins-Rogers Inc, p. I-70 [SIC 2041]
Will-Burt Co, p. II-1492 [SIC 3499]
Willamette Electric Products Co, p. II-1929 [SIC 3694]
Willamette Graystone Inc, p. I-1117 [SIC 3271]
Willamette Industries Inc, p. I-618 [SIC 2631]
Willamette Valley Co, p. I-840 [SIC 2851]
Willard Grain and Feed Inc, p. I-861 [SIC 2873]
Willard Marine Inc, p. II-1992 [SIC 3732]
Willert Home Products Inc, p. I-825 [SIC 2842]
William Barnet and Son Inc, pp. Vol. I - 327, 795 [SICs 2299, 2824]
William Exline Inc, p. I-745 [SIC 2782]
William H Harvey Co, p. I-825 [SIC 2842]
William Haynes, p. II-2135 [SIC 3931]
William J Dixon Company Inc, p. I-288 [SIC 2269]
William L Bonnell Company Inc, p. II-1244 [SIC 3354]
William Morrow and Company, p. I-703 [SIC 2731]
William Pearson Inc, p. I-361 [SIC 2331]
William Powell Co, p. II-1464 [SIC 3494]
William Rand Inc, p. II-2173 [SIC 3961]
William Schneider Inc, p. II-2122 [SIC 3911]
William Steinen Mfg Co, p. II-1334 [SIC 3432]
Williamhouse-Regency Inc, p. I-729 [SIC 2759]
Williamhouse Sales Corp, p. I-679 [SIC 2677]
Williams Brothers Co, p. I-70 [SIC 2041]
Williams Cheese Company Inc, p. I-21 [SIC 2022]
Williams Enterprise of Georgia, p. II-1355 [SIC 3443]
Williams Forest Products Corp, p. I-460 [SIC 2411]
Williams Furnace Co, p. II-1339 [SIC 3433]
Williams Inc, p. II-1845 [SIC 3646]
Williams International Corp, p. II-1975 [SIC 3724]
Williams Optical Laboratory Inc, p. II-2107 [SIC 3851]
Williams Patent Crusher, p. II-1525 [SIC 3532]
Williams Printing Co, p. I-729 [SIC 2759]
Williams, White and Co, p. II-1563 [SIC 3542]
Williamsburg Millwork, p. I-510 [SIC 2448]
Williamsburg Soap & Candle Co, p. I-819 [SIC 2841]
Williamshouse, p. I-679 [SIC 2677]
Williamson-Dickie Mfg Co, p. I-348 [SIC 2325]
Williamsport Wirerope Works, p. II-1476 [SIC 3496]

Willis Hosiery Mills Inc, p. I-250 [SIC 2251]
Willitts Design International Inc, p. I-1113 [SIC 3269]
Willmar Cookie and Nut, p. I-107 [SIC 2052]
Willoughby Industries Inc, p. I-603 [SIC 2599]
Willow Manufacturing Inc, p. II-1795 [SIC 3629]
Willson Safety Products, p. II-2107 [SIC 3851]
Wilmad Glass, p. I-1071 [SIC 3231]
Wilmer Service Line, p. I-735 [SIC 2761]
Wilmington Finishing Co, p. I-267 [SIC 2257]
Wilroy Inc, p. I-365 [SIC 2335]
Wilshire Corp, p. II-1718 [SIC 3581]
Wilshire Designs Inc, p. II-2107 [SIC 3851]
Wilson Concrete Company Inc, p. I-1122 [SIC 3272]
Wilson Engraving Company Inc, p. I-759 [SIC 2796]
Wilson Foods Corp, p. I-2 [SIC 2011]
Wilson Greatbatch Ltd, p. II-1925 [SIC 3692]
Wilson Pallet Co, p. I-510 [SIC 2448]
Wilson Sporting Goods Co, pp. Vol II - 1207, 2150 [SICs 3324, 3949]
Wilson Steel and Wire Co, p. II-1182 [SIC 3315]
Wilson Trailer Co, p. II-1961 [SIC 3715]
Wilton Corp, p. II-1312 [SIC 3423]
Wiltron Co, p. II-1873 [SIC 3663]
Wimer Logging Company Inc, p. I-460 [SIC 2411]
WIN Laboratories Ltd, p. II-1688 [SIC 3571]
Winamac Coil Spring Inc, p. II-1470 [SIC 3495]
Winandy Greenhouse Company, p. II-1372 [SIC 3448]
Winchester-Auburn Mills Inc, p. I-322 [SIC 2298]
Winco Inc, p. II-1780 [SIC 3621]
WinCraft Inc, p. I-729 [SIC 2759]
Wind Mill Woodworking Inc, p. I-587 [SIC 2541]
Windle Industries Inc, p. I-327 [SIC 2299]
Windmere Corp, p. II-2199 [SIC 3999]
Windowmaster Products, p. II-1349 [SIC 3442]
Windsor Door, p. II-1349 [SIC 3442]
Windsor Industries Inc, p. II-1736 [SIC 3589]
Windsor Manufacturing Co, p. II-1975 [SIC 3724]
Wine Alliance, p. I-168 [SIC 2084]
Wine World Estates, p. I-168 [SIC 2084]
Winegard Co, p. II-1878 [SIC 3669]
Winfield Brooks Company Inc, p. I-819 [SIC 2841]
Winfield Cover Company Inc, p. I-413 [SIC 2386]
Wingra Stone Co, p. I-1127 [SIC 3273]
Winnebago Industries Inc, p. II-1966 [SIC 3716]
Winning Moves Inc, p. I-356 [SIC 2329]
Winning Ways Inc, p. I-356 [SIC 2329]
Winnsboro Plywood Company, p. I-491 [SIC 2435]
Winona Knitting Co, p. I-258 [SIC 2253]
Winona Knitting Mills Inc, p. I-258 [SIC 2253]
Winona Monument Company, p. I-1140 [SIC 3281]
Winona Van-Norman, p. II-1600 [SIC 3549]
Winrock Enterprises Inc, p. I-979 [SIC 3084]
WinsLoew Furniture Inc, p. I-542 [SIC 2511]
Winsted Precision Ball Co, p. II-1644 [SIC 3562]
Winston Manufacturing Corp, p. II-1531 [SIC 3533]
Winters Industries, p. II-1275 [SIC 3365]
Wintor Swan Associates Inc, p. I-718 [SIC 2752]
Winzeler Stamping Co, p. II-1413 [SIC 3469]
Wippette International Inc, p. I-409 [SIC 2385]
Wippette Kids, p. I-409 [SIC 2385]
Wippette Rainthings Inc, p. I-409 [SIC 2385]
Wire Products Co, p. II-1182 [SIC 3315]
Wire Rope Corporation, p. I-322 [SIC 2298]
Wiremold Co, p. II-1830 [SIC 3643]
Wirtz Manufacturing Company, p. II-1572 [SIC 3544]
Wis-Pak Inc, p. I-176 [SIC 2086]
Wiscassett Mills Co, p. I-297 [SIC 2281]
Wisconsin Aluminum Foundry, p. II-1275 [SIC 3365]
Wisconsin Automated Machinery, pp. Vol II - 1610, 1621 [SICs 3553, 3555]
Wisconsin Box Co, p. I-515 [SIC 2449]
Wisconsin Cheeseman Inc, p. I-21 [SIC 2022]
Wisconsin Homes Inc, p. I-523 [SIC 2452]
Wisconsin Hydraulics Inc, p. II-1753 [SIC 3594]
Wisconsin Label Corp, p. I-729 [SIC 2759]
Wisconsin Machine Tool Corp, p. II-1578 [SIC 3545]
Wisconsin Ordnance Works Ltd, p. II-1676 [SIC 3568]
Wisconsin Pharmacal Company, p. I-954 [SIC 3069]
Wisconsin Tissue Mills Inc, p. I-674 [SIC 2676]

Roman numerals I and II indicate the volume in which pages appear. Volume I holds SICs 2011 through 3299; Volume II holds SICs 3312 through 3999.

2311

Company Index

Roman numerals I and II indicate the volume in which pages appear. Volume I holds SICs 2011 through 3299; Volume II holds SICs 3312 through 3999.

Roman numerals I and II indicate the volume in which pages appear. Volume I holds SICs 2011 through 3299; Volume II holds SICs 3312 through 3999.

OCCUPATION INDEX

This index lists those occupations in manufacturing that account for 1 percent or more of employment. This limitation excludes many occupations employed in the sector in small numbers—physicians and nurses, for instance. After the name of each occupation, a value in parentheses shows the number of 3-digit manufacturing industry groups in which the occupation occurs. One or more page numbers follow; these are marked with Roman numerals I or II to indicate the volume. After the page numbers, the 3-digit SICs are shown inside brackets. *Please note:* page and SIC references are sorted so that they point to industry groups with descending employment (the first page reference is to the largest employing industry group). Only the top ten industry groups are referenced.

Roman numerals I and II indicate the volume in which pages appear. Volume I holds SICs 2011 through 3299; Volume II holds SICs 3312 through 3999.

Roman numerals I and II indicate the volume in which pages appear. Volume I holds SICs 2011 through 3299; Volume II holds SICs 3312 through 3999.

2317